Organic Chemistry

Organic Chemistry

Theory, Reactivity and Mechanisms in Modern Synthesis

With a Foreword by Robert H. Grubbs

Pierre Vogel
Kendall N. Houk

WILEY-VCH

Authors

Prof. Pierre Vogel
EPFL
SB-DO
Avenue F.-A. Forel 2
1015 Lausanne
Switzerland

Prof. Kendall N. Houk
Dept. of Chemistry and Biochemistry
University of California
Los Angeles, CA 90095–1569
United States

Cover: The cover features a computed transition state structure with frontier molecular orbitals for the Diels-Alder reaction of SO2 and butadiene, catalyzed by another SO2 (J. Am. Chem. Soc. 1998, 120, 13276–13277). Pierre Vogel established the mechanism of this reaction and applied it to the total synthesis of natural product (-)-dolabriferol (Angew. Chem. Int. Ed. 2010, 49, 8525–8527), the structure of which shown in the green hexagon, originally from dolabrifera dolabrifera the sea slug (also shown in its vivid UCLA colors). A potential energy diagram in the red hexagon and blackboard writings in the background (courtesy P. Vogel) are key concepts discussed extensively in this book to describe mechanism and reactivity.

Library of Congress Card No.:
applied for

British Library Cataloguing-in-Publication Data
A catalogue record for this book is available from the British Library.

Bibliographic information published by the Deutsche Nationalbibliothek
The Deutsche Nationalbibliothek lists this publication in the Deutsche Nationalbibliografie; detailed bibliographic data are available on the Internet at <http://dnb.d-nb.de>.

Print ISBN: 978-3-527-34532-8
ePDF ISBN: 978-3-527-81925-6
ePub ISBN: 978-3-527-81927-0

Cover Design Fang Liu, DesignOne, Nanjing, China 210095
Typesetting SPi Global, Chennai, India
Printing and Binding Himmer GmbH Druckerei, Augsburg, Germany

Printed on acid-free paper

10 9 8 7 6 5 4 3 2 1

Contents

Preface

Scientists interested in molecular sciences with basic knowledge in chemistry might retain this book as their second textbook in organic chemistry. This book is also a reference manual for chemists and chemical engineers who invent new reactions and design new procedures for the conversion of simple chemicals into high value-added materials. All answers to the problems in this book, and references to the original literature relevant to the problems, are contained in our companion Workbook of the same name as this book. We plan to produce another book describing reaction intermediate and their reactions, as well as solvation and weak molecular interactions.

Chemistry is an empirical science but is increasingly influenced by understanding and prediction. Before starting a new experiment in the laboratory, a chemist would like to know the following:

(1) Is the reaction possible thermodynamically?
(2) How long is it going to take?
(3) What will be the properties of the reaction products?

This book introduces and documents models that enable chemists to answer these questions and to understand the reasons behind the answers. The methods will be illustrated with a large number of reactions that have a wide practical value in synthesis and biology. Reactions involving organic, organometallic, and biochemically important reactants and catalysts will be presented. We teach the tools that can be used to understand Nature and to control and create new chemistry to achieve a better world. Given specific combinations of solvent, concentration, temperature, pressure, the presence or absence of catalysts and inhibitors, light, or other types of radiation, a given system of reactants will be converted into a mixture of products. Rates of product formation or attainment of equilibria define chemical reactivity. Living systems are made of ensembles of molecules that are connected through ensembles of chemical reactions. We like to think of most chemists, biochemists, molecular biologists, material physicists, and all those who study molecular phenomena as molecular scientists. They (and we) try to understand Nature and to imitate its efficiency and diversity. Molecular scientists, especially chemists, are not passive observers. Chemists can even surpass Nature, by inventing new molecular entities – chemicals! – and new reactions that have not been observed yet in our Universe, at least on our planet! Through chemical knowledge, combined with serendipity, molecular scientists are creating a new world, consisting of useful chemicals such as pharmaceuticals, crop protection agents, food protective agents, perfumes, aromas, optical and electronic materials, fabric for clothes and other applications, construction materials for energy-saving houses and vehicles, and coatings and paints. The new world of nanoscience is molecular and supramolecular science. Chemists – many of whom are really molecular engineers – strive to obtain targeted compounds by chemical or biochemical synthesis as rapidly as possible and by the most economic routes possible. Nowadays, chemists invent procedures that are environmental friendly and contribute significantly to a more sustainable development, with more respect for the limited resources of our Earth.

Chemical structures, stability, and reactivity are governed by thermodynamics (Chapters 1 and 2) and kinetics (Chapter 3). Thermodynamics dictates how atoms assemble into stable molecules and how molecules assemble into supramolecular systems. Kinetics quantitates the rates at which molecules are transformed into other molecules or assemblies of molecules under specific conditions. Our preface gives a brief history of chemistry and shows how heat exchange is fundamental to produce and modify chemicals. All chemical changes are accompanied by absorption (endothermic reactions: $\Delta_r H^T > 0$) or release (exothermic reactions: $\Delta_r H^T < 0$) of heat. The heat of any reaction can be measured by calorimetry. It is the variation of enthalpy ($H = E + PV$) during the time between when the reactants are mixed and when the equilibrium with the products is reached, for a reaction at constant temperature and pressure. The first reaction used by man was fire, the combustion of dry grass or wood in the air to produce carbon dioxide + water (fumes), heat, and light next to ashes that are inorganic carbonates, hydroxides, and oxides. Any chemical or biochemical reaction equilibrates reactants (also called substrates and reagents, or starting materials) with products (and coproducts). At temperature T, the reaction equilibrium is characterized by an equilibrium constant, K^T, which depends on the nature of reactants and products and on the reaction conditions (temperature, pressure, concentration, and solvent). For instance, if equilibrium $\mathbf{A}+\mathbf{B} \rightleftarrows \mathbf{P}+\mathbf{Q}$ (one molecule of reactant **A** and one molecule of reactant **B** equilibrate with one molecule of product **P** and one molecule of coproduct **Q**) can be considered as an ideal solution, $K^T = [\mathbf{P}][\mathbf{Q}]/[\mathbf{A}]][\mathbf{B}]$; [**P**], [**Q**], [**A**], and [**B**] are the concentration of products **P**, coproduct **Q**, and of reactants **A** and **B**, respectively. Under constant pressure and temperature, the Gibbs free energy of the reaction $\Delta_r G^T = -RT \ln K^T = \Delta_r H^T - T\Delta_r S^T$, with $\Delta_r H^T$ = heat of the reaction and $\Delta_r S^T$ = entropy variation of the reaction (or reaction entropy). Those reactions that convert reactants into products with a good conversion have $K^T > 1$ and correspond to $\Delta_r G^T < 0$. They are said to be exergonic. For endergonic reactions with $K^T < 1$, $\Delta_r G^T > 0$, products can be obtained with good conversion if they can be separated selectively from the reactants (e.g. precipitation of one product from an homogenous solution and evaporation of one product or coproduct from the solid or liquid reaction mixture) they are equilibrating with (equilibrium shift). As a general rule, condensations that convert small molecules into larger molecules (the number of molecules diminishes from reactants to products) have negative reaction entropies ($\Delta_r S^T < 0$) and fragmentations that convert large molecules into smaller molecules (the number of molecules increases from reactants to products) have positive entropies ($\Delta_r S^T > 0$). The heat absorbed or released in a reaction, $\Delta_r H^T$, represents a powerful tool to understand chemical transformations at the molecular level (molecular chemistry). This textbook shows how thermochemical data such as standard (1 atm, 25 °C) heats of formation ($\Delta_f H°$), standard entropies ($S°$), homolytic bond dissociation enthalpies ($DH°(R^{\cdot}/X^{\cdot})$), gas-phase heterolytic bond dissociation enthalpies ($DH°(R^+/X^-)$), gas-phase acidities ($\Delta_f G°(A-H \rightleftarrows A^- + H^+)$ and proton affinities (PA $= DH°(A^-/H^+)$), ionization energies (EIs), electron affinities (−EAs), and solution acidity constants (K_a, pK_a) from the literature (tables of data collected before references to Chapter 1, p. 53–91) and online data banks can be used to understand molecular properties and reaction equilibria, including equilibria involving charged species (Chapters 1 and 2). We give simple techniques ("back of the envelope methods") that allow one to estimate thermochemical data of reactants, products, and reactive intermediates for which these data have not been measured. This permits one to evaluate the equilibrium constants of any organic reactions for systems that can be considered as ideal gases or ideal solutions, which is the case for a large number of organic and organometallic reactions run in the laboratory. Equilibria between two phases find multiple applications in preparative chemistry (e.g. solution/solid: crystallization) and analytical chemistry (e.g. solid of liquid stationary phase/mobile liquid or gaseous phase: chromatography). They are exploited in the resolution of racemates into enantiomers and in thermodynamically controlled deracemizations. Isotopic substitution affects equilibria and gives important information about bonding in molecules.

A chemical or biochemical reaction is characterized by its rate of reaction and its rate of law (Chapter 3). Both depend on the nature of the reactants, the reaction mechanism, and the reaction conditions (temperature, pressure, concentration, solvent, and presence of catalyst(s) and inhibitor(s)). For instance, for the irreversible reaction (with a large K^T value) $\mathbf{A}+\mathbf{B} \rightarrow \mathbf{P}+\mathbf{Q}$, the disappearance of reactant **A** may follow the second-order rate law $d[\mathbf{A}]/dt = -k[\mathbf{A}][\mathbf{B}]$ with k being the rate constant. Chemical kinetics (the measure of reaction rate constant k as a function of temperature) allows one to evaluate activation parameters using the empirical Arrhenius relationship: $k = A\, e^{-E_a/RT}$. This gives the empirical activation parameters E_a = activation energy and A = frequency factor. Eyring considers the transition state of a reaction to be an activated complex in a quasi-equilibrium with the reactants (equilibrium $\mathbf{A}+\mathbf{B} \rightleftarrows [\mathbf{A} \cdot \mathbf{B}]^{\ddagger}$). Thermodynamics applied to this equilibrium defines the Eyring activation parameters $\Delta^{\ddagger}H$ = activation enthalpy ($\Delta^{\ddagger}H = E_a - RT$), $\Delta^{\ddagger}S$ = activation entropy, $\Delta^{\ddagger}G = \Delta^{\ddagger}H - T\Delta^{\ddagger}S$ = free energy of activation and permits the delineation of mechanistic limits (nature of the transition state of the rate-determining step) at the molecular level. Conversely, if the reaction mechanism is known, the activation parameters can be estimated and can be used to predict under which conditions (pressure, concentration, and temperature) the reaction will occur and how long it will take for a given conversion. For systems in solution, rates can be enhanced or reduced by applying high pressures. This provides activation volumes ($\Delta^{\ddagger}V$) that are important information about reaction mechanisms. Rates of reaction also depend on chirality, a phenomenon exploited in asymmetric synthesis (the preparation of enantiomerically enriched or pure compounds) that is extremely important in modern medicinal chemistry and material sciences. The most important tools of modern asymmetric synthesis will be presented (Section 3.6) and illustrated throughout the book. The question of how chirality appeared on Earth will be addressed (e.g. asymmetric autocatalysis). Isotopic substitution can also affect the rate of a given reaction. Kinetic isotopic effects are powerful tools to study reaction mechanisms.

Quantum mechanical calculations have become routine molecular models for chemists, biochemists, and biologists. They are the basis of simpler molecular orbital theories (Hückel method, Coulson and Longuet–Higgins approach, and the perturbation of molecular orbital (PMO) theory) that help to describe molecular properties and their reactions and to establish bridges between molecular organic, organometallic, and inorganic chemistry (Chapter 4 and Section 7.6). Notions such as conjugation, hyperconjugation, Hückel and Heilbronner aromaticity, and antiaromaticity find a solid basis in quantum mechanical calculations. Modern computational methods have proven to be a robust way to establish mechanisms; continuing increases in computer power and the accuracy of methods make computations an increasingly valuable way to establish the favored mechanisms of reactions.

Mechanistically, reactions can be classified into one-step and multistep reactions. Pericyclic reactions (electrocyclic ring closures and openings, cycloadditions and cycloreversions, cheletropic additions and eliminations, sigmatropic rearrangements, dyotropic rearrangements, and ene-reactions) for long were considered as "no-mechanism reactions." They have played a key role in our understanding of reaction mechanisms (concerted vs. nonconcerted mechanism, importance of diradical and zwitterion intermediates, and the diradicaloid theory for transition states) and chemical reactivity in general (Chapter 5). These reactions are extremely useful synthetic tools, including in asymmetric synthesis.

Without sunlight, green plants do not grow. The color of natural or painted objects fades away when they are exposed to the sun. Light can induce chemical and biochemical reactions. The concepts that enable us to understand the interaction of light with organic compounds and how light can make them to react in ways different from under heating are presented in Chapter 6. Interpretation of the UV–visible spectra of organic molecules has played a major role in structural analytical chemistry and in the design of dying agents. Phenomena such as fluorescence and phosphorescence, chemiluminescence, and bioluminescence teach us about the nature of the electronically excited state of molecules (singlet vs. triplet states) and their unimolecular and bimolecular reactions. The photochemistry of functional compounds (isomerization, bond cleavage, cycloadditions, photooxidations, photocatalysis, etc.) represents a powerful tool of preparative chemistry. The photoreactions in which light initiates chain processes, or induces electronic transfers, are extremely useful. Photoinduced electron transfer is fundamental to dye-sensitized solar cells.

Humans have survived eating animals, plants, and parts of plants. Animals also survive consuming other animals or plants. Photosynthesis ($n\mathrm{CO_2} + n\mathrm{H_2O} \rightarrow \mathrm{C}_n\mathrm{H}_{2n}\mathrm{O_2}$ [carbohydrates]) in plants has for long produced more biomass than necessary for all living species on Earth. Geological phenomena have permitted the storage of large parts of past biomass underground in the form of coal, tars, petroleum, and natural gas (fossil fuels). When human beings started to control fire (c. 1.6×10^6 years ago), they found that heat can be used to convert biomass and minerals into valuable materials. This is obvious with the development of pottery and metallurgy, which represent the first chemical industries. Then, biomass fermentative processes (wine and beer) and wood distillation have become the next chemical industries. The Industrial Revolution, which began in late 1700s in the UK, has led to mass production and, consequently, to a new consumer society. The processes applied have produced a lot of unwanted secondary products (waste) and are consuming larger and larger amounts of energy, mostly burning fossil fuels. This cannot be continued without affecting irreversibly our environment (emission of CO_2, nitrogen oxides, methane, nanoparticles, etc.) and our quality of life. It is urgent to develop cleaner processes that do not reject any waste and require much less energy. Today, chemists invent new procedures that contribute to a more sustainable economy ("green chemistry"). The new procedures rely upon new reactions that are atomic economically (no coproducts, no secondary products, and no solvent) and require no heating or no cooling. Most chemists create new compounds by combining reagents in C-heteroatom or C—C bond forming reactions. For 150 years, this required polar starting materials (organometallic reagents and halogenated compounds) that can combine in substitution and addition reactions. Quite often, these reactions produce coproducts and side products that cannot be recycled in an economical manner. Organometallic reagents and halogenated starting materials require several synthetic steps for their obtainment from available resources. For instance, the very much applied Friedel–Crafts acylation Ar–H (aromatic hydrocarbon) + $RCOCl + AlCl_3 \rightarrow ArCOR + HAlCl_4$ first requires the conversion $RCOOH + SOCl_2 \rightarrow RCOCl + SO_2 + HCl$. The process produces HCl and SO_2 washed with alkaline water-producing large amounts of waste. Another example is the classical preparation of secondary alcohols from alcohols and aldehydes using Grignard reagents, e.g. $R–Br + Mg \rightarrow RMgBr$; then $R'CHO + RMgBr \rightarrow RCH(OMgBr)R'$, then $RCH(OMgBr)R' + H_2O \rightarrow RCH(OH)R' + Mg(OH)Br$ (waste). In general, bromides are not readily available; they can be made according to $ROH + BBr_3 \rightarrow R–Br + B(OH)Br_2$ (waste). Mg and other reactive metals such as Li, Na, and K require a lot of energy for their preparation. Direct hydrocarbation of unsaturated compounds is much more atomic economically. Examples are the aldol reaction ($RCHO + R'CH_2COR'' \rightleftarrows RCH(OH)–CH(R')–COR''$) and many newer reactions presented in this book such as $RCH_2OH + CH_2 = CHR' \rightarrow RCH(OH)–CH(Me)R'$. The latter reaction can generate four possible stereoisomers (two diastereomers as racemic mixtures) as two new stereogenic centers are created. If the reaction should not be regioselective, one further isomeric product can form. In this latter case, $RCH_2OH + CH_2 = CHR' \rightarrow RCH(OH)–CH_2CH_2R'$ (racemate). We shall see that suitable catalysts are available that make it possible to form only one major product enantiomerically enriched, if not enantiomerically pure. Emphasis today is to use readily available starting materials extracted from renewable resources such as the biomass and chemicals derived from it. For that, chemists invent new catalysts that are either heterogeneous (do not dissolve in the reactants and solvent) of homogeneous (dissolve in the reactants and solvent) and perform better and better. Chapters 7 and 8 are devoted to catalytic reactions with examples applied in the bulk chemical industry and many others applied in fine chemistry, including in the asymmetric synthesis of compounds of biological interest. These chapters give the concepts to understand how homogeneous catalysts work at the molecular level. They should help the reader to invent further catalysts and new reactions that are high yielded, chemoselective (e.g. hetero-Diels–Alder reaction vs. (4+1)-cheletropic addition of SO_2 to 1,3-dienes) site-selective (selective between similar functions of multifunctional reactants), regioselective (e.g. Markovnikov or anti-Markovnikov orientation), diastereoselective (e.g. *erythro* or *threo* through *anti* or *syn* addition), and enantioselective (e.g. π-face selective), requiring no heating or cooling and that are completely atomic economical.

History, enthalpy, and entropy in the transformation of matter

As mentioned above, fire is the oldest reaction used by man (most of the material presented in this section can be found in the Internet: Wikipedia, the free encyclopedia, www.wikipedia.org). The earliest reactions induced by heat has been the smelting of lead and tin (6500 BC). A common lead ore is galena (PbS). When heated in the air, lead sulfite is obtained (equilibrium: $2PbS + 3O_2 \rightleftarrows 2PbSO_3$). Oxygen of air burns lead sulfide in a exothermic reaction that condenses five molecules into two, a process disfavored entropically, but it occurs because of the exothermicity ($\Delta_r H^T < 0$) of the reaction, which pays for the entropy cost ($-T\Delta_r S^T > 0$). Upon heating, lead sulfite decomposes into solid lead oxide and volatile sulfur dioxide (equilibrium $PbSO_3 \rightleftarrows PbO + SO_2$). Although $PbSO_3$ is a stable compound at room temperature, heating induces its fragmentation into two smaller molecules. At high temperature, the reaction is favored entropically and also by the "Le Châtelier principle" (SO_2 flies away from the reaction mixture). This reaction is like limestone calcining: $CaCO_3 \rightarrow CaO + CO_2$. Incomplete combustion of charcoal produces carbon monoxide, CO, which reduces lead oxide into metallic lead and CO_2 according to equilibrium $PbO + CO \rightleftarrows Pb + CO_2$. The variation of entropy for this reaction is small as it does not change the number of molecules between reactants and products. Metallic lead forms because the C—O bonds in CO_2 are stronger than the Pb—O bond in solid lead oxide. This is demonstrated by the heats of combustion $\Delta_r H^{298\,K}(CO + ½O_2 \rightleftarrows CO_2$, gas phase) = $-67.6\,kcal\,mol^{-1}$ and $\Delta_r H^{298\,K}$(Pb(solid) + $½O_2 \rightleftarrows$ PbO(solid) = $-52.4\,kcal\,mol^{-1}$. Overall, the reduction of lead oxide by CO is exothermic by $\Delta_r H^{298\,K}$(PbO(solid) + CO(gas) $\rightleftarrows$ Pb(solid) + CO_2 (gas)) = $-15.2\,kcal\,mol^{-1}$ (NIST WebBook of Chemistry, National Institute of Standards and Technology, http://webbook.nist.gov/chemistry/). The Bronze age started with the discovery that a better metallic material, the alloy bronze, can be obtained by smelting tin (e.g. cassiterite: SnO_2) and copper (e.g. malachite: $[Cu_2CO_3(OH)_2]$, chalcocite: CuS, chalcopyrite: $CuFeS_2$) ores together with carboneous materials such as charcoal (c. 3500 BC). Iron Age (c. 1500 BC) started with the discovery of smelting of iron oxide with charcoal. Overall, $\Delta_r H^{298\,K}(2Fe_2O_3$(solid) +

$3C(solid) \rightleftarrows 4Fe(solid)+3CO_2(gas)) = 112.6\,kcal\,mol^{-1}$, which is highly endothermic, but profits of the positive entropy of reaction and of the le Châtelier principle (formation of CO_2 that flies away) at high temperature. Concomitant burning of charcoal compensates for the overall endothermicity. The process implies several reactions: First $4C + 2O_2 \rightarrow 4CO$, then three successive reductions with CO: $3Fe_2O_3 + CO \rightarrow 2Fe_3O_4 + CO_2$; $Fe_3O_4 + CO \rightarrow 3FeO + CO_2$; $FeO + CO \rightarrow Fe + CO_2$. The overall process $Fe_2O_3 + 4C + 2O_2 \rightleftarrows 2Fe + 3CO_2 + CO$ is exothermic by c. $-110\,kcal\,mol^{-1}$.

Fermentative processes (biochemical transformations catalyzed by a microorganism; e.g. $C_6H_{12}O_6$ (D-glucose in water) $\rightarrow 2CH_3CH_2OH$ (ethanol in water) $+ 2CO_2$, $\Delta_rH^{298\,K} = -17.8\,kcal\,mol^{-1}$) such as beer and wine making have been known for at least 8000 years. Acetic acid (CH_3COOH, IUPAC name: ethanoic acid) in the form of sour wine has also been known for the same time. The process of distillation permits the isolation of pure organic chemicals such as ethanol and acetic acid as described for the first time by the Alexandrians (500 BC). One of the earliest organic chemistry reaction (2800 BC) is the formation of soap (e.g. sodium stearate: $Me(CH_2)_{16}COONa$, sodium palmitate: $Me(CH_2)_{14}COONa$) obtained by reacting olive oil or palm oil with ashes (NaOH, Na_2CO_3). The reaction ($RCOOCH_2$–CH(OCOR)–CH_2OCOR (triglyceride) + 3NaOH $\rightleftarrows$ 3RCOONa (soap) + $HOCH_2$–CH(OH)–CH_2OH (glycerin) + heat) occurs already at room temperature. Soap manufacturers have observed very early that heating the reaction mixture would accelerate the process. The rate of the reaction increases with temperature. It also depends on the type of ashes used for saponification. Some are more active (contain more NaOH) than others. Aged ashes are less reactive because they contain more hydrogenocarbonates and carbonates. This results from the slow absorption of CO_2 present in the air, which reacts with oxides, hydroxides (e.g. $NaOH + CO_2 \rightleftarrows NaHCO_3$). At low temperature, condensation is favored thermodynamically, whereas the reverse reaction, fragmentation (decarboxylation), is favored upon heating.

Charcoal required by the early metallurgy was produced by partial combustion of wood. With time, various techniques of wood pyrolysis (also called destructive distillation) have been developed, which have led to the production and isolation of several chemicals such as methanol, turpentine (volatiles), and tar (nonvolatiles). Turpentine (from pine tree), used as paint thinner, was prepared first by the Persians (3000 BC). It is mentioned in European literature in the thirteenth century. It is mostly composed of (−)-*α*-pinene (European pine), (+)-*α*-pinene (North American pine), *β*-pinene, (+)-3-carene, and lesser amounts of (−)-camphene, dipentene (racemic limonene = (±)-limonene = 1 : 1 mixture of (+)-limonene and (−)-limonene), and *α*-terpinolene (Figure 1). Except for *α*-terpinolene, these monoterpenes are chiral compounds that can be obtained with high enantiomeric purity. These odorous compounds are found in several plants (essential oils). Nowadays, they are used as starting materials in the perfume industry and in the asymmetric synthesis of drugs (part of the chiral pool).

IUPAC names: (−)-*α*-pinene = (−)-(1*S*,5*S*)-2,6,6-trimethylbicyclo[3.1.1]hept-2-ene; (−)-*β*-pinene: (−)-(1*S*,5*S*)-6,6-dimethyl-2-methylidenebicyclo[3.1.1] heptane; (+)-3-carene: (+)-(1*S*,6*S*)-3,7,7-trimethylbicyclo[4.1.0]hept-3-ene; (−)-camphene: (−)-(1*S*,4*R*)-2,2-dimethyl-3-methylidenebicyclo[2.2.1]heptane (as these rigid bicyclic hydrocarbons with two stereogenic centers cannot have diastereomers, they exist as two enantiomers only, the second stereomarker can be dropped); (+)-limonene (orange odor): (+)-(*R*)-1-methyl-4-(prop-2-en-1-yl)cyclohexene; (−)-limonene (lemon odor): (−)-(*S*)-1-methyl-4-(prop-2-en-1-yl)cyclohexene.

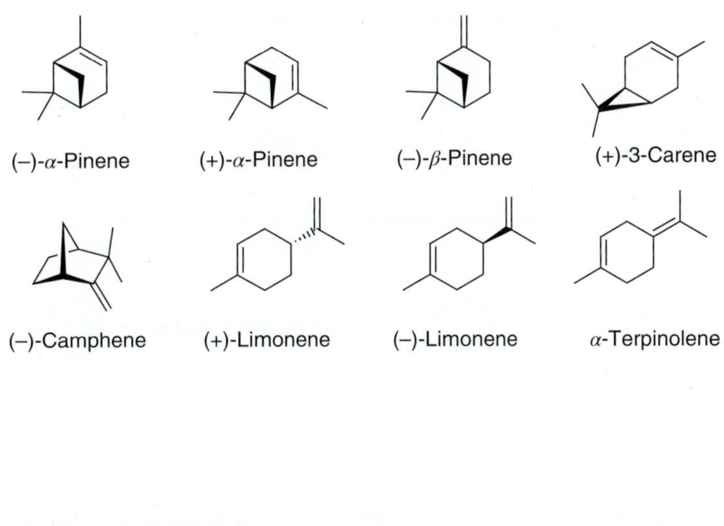

Figure 1 Examples of monoterpenes obtained from the pyrolysis of pine trees.

In 1800, about 500 organic compounds were known. Around 1850 pyrolysis (carbonization or destructive distillation) of hard coal produced many new substances, and this launched the chemical industry of organic compounds. When the first edition of Beilstein's Handbook of Organic Chemistry appeared in 1882, already 20 000 organic compounds were cited. Isolation of compounds from plants and animals also contributed to this number. In 1912, about 150 000 organic substances were known. Today, over 50 million chemicals have been registered. Pyrolysis of coal produces coke (70%), NH_3/H_2O(10%), coal gas (town gas: mostly H_2 and CH_4; contains lesser amounts of CO, ethane, ethylene, benzene, toluene, and cyclopentadiene) and coal tar as the main source of starting materials (benzene, toluene, phenols, anilines, pyridine, naphthalene, anthracene, phenanthrene, chrysene, carbazole, quinoline, and pyrrole) for the manufacture of soap, fats, dyes, plastics, perfumes, drugs, pesticides, explosives, etc. Industrial organic chemistry started with the manufacture of mauveine (a purple dying agent called also Perkin mauve, aniline purple, and Tyrian purple) suitable as a dye of silk and other textiles. In 1856, William Henry Perkin oxidized aniline using $K_2Cr_2O_7$ in H_2SO_4, whose toluidine impurities reacted with the aniline and yielded the dye (Figure 2) [1–4].

Invented in 1888, the industrial production of calcium carbide combines lime and coke in an electric arc furnace at 2000 °C. This highly endothermic reaction ($\Delta_r H^{298\,K}$[CaO(solid) + C(graphite) $\rightleftarrows$ CaC_2(solid) + CO(gas)] = 112 kcal mol^{-1} [5] is possible because of the formation of CO that is eliminated while it forms (equilibrium shift). Calcium carbide reacts with water to give acetylene ($CaC_2 + H_2O \rightarrow CaCO_3 + H{-}C{\equiv}C{-}H$), and with nitrogen to give calcium cyanamide ($CaC_2 + N_2 \rightarrow C + Ca^{++}/^{-}N{=}C{=}N^{-}$), a fertilizer ($CaCN_2 + 3H_2O \rightarrow 2NH_3 + CaCO_3$). Acetylene is an important compound used in welding (metal industry) and in the production of several chemicals such as acetaldehyde (MeCHO), acetic acid (MeCOOH), ethanol ($MeCH_2OH$), vinyl acetate ($H_2C{=}CH{-}OCOMe$), vinyl chloride ($CH_2{=}CH{-}Cl$), acrylic acid and esters ($CH_2{=}CHCOOR$, R = H, alkyl), acrylonitrile ($CH_2{=}CH{-}C{\equiv}N$), and chloroprene ($CH_2{=}CH{-}C(Cl){=}CH_2$). Except in China where the calcium carbide procedure enjoys a renaissance, acetylene is produced today utilizing natural gas or petroleum as sources.

Isoprene (2-methylbutadiene) was first obtained by the distillation of natural rubber [6, 7]. In 1835, Liebig observed that the most volatile fraction of rubber produces a high boiling (230 °C) oil by means of concentrated sulfuric acid [8]. In 1879, Bouchardat

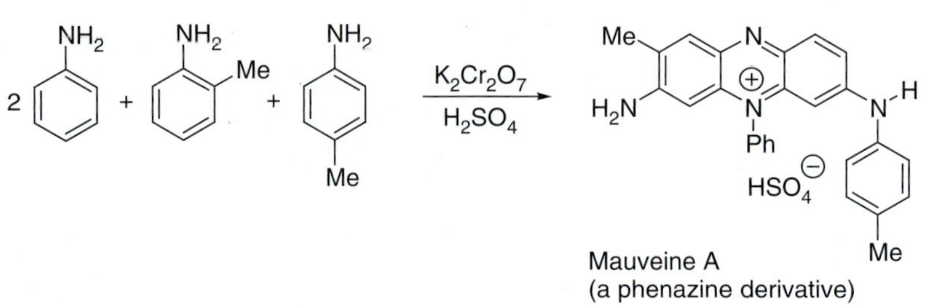

Figure 2 The birth of the industry of organic chemistry is based on coal tar.

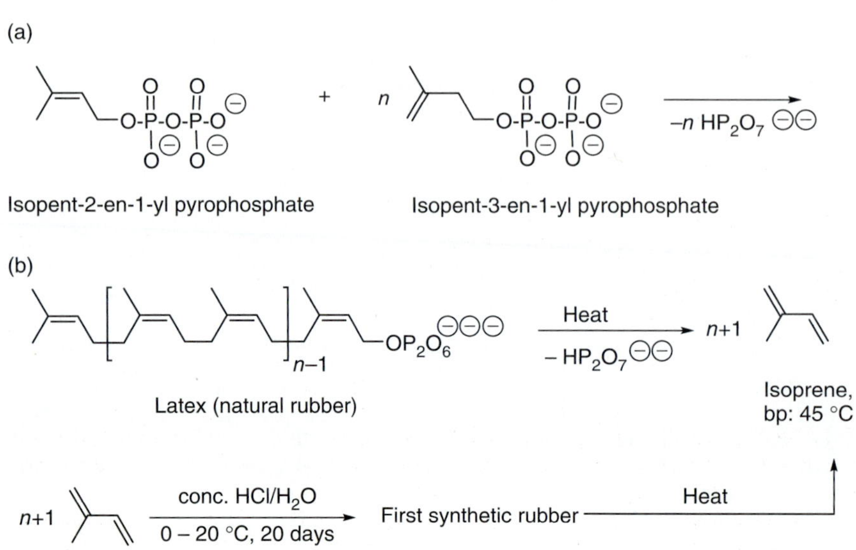

Figure 3 (a) Biosynthesis of natural rubber latex and its pyrolysis into isoprene and (b) protic-acid-induced polymerization of isoprene that produces synthetic rubber.

reported the polymerization of isoprene to an elastic product that again gave isoprene on distillation (Figure 3) [9]. This discovery opened the field of polymer chemistry that our civilization could not exist without today [10]. Thus, heat breaks C—C bonds in a large organic molecule (rubber is a long polymer with a molecular mass of 10^5–10^6) and produces smaller molecules; in this case, isoprene. In the presence of a suitable catalyst, the polymer can be formed again at a lower temperature. Isoprene is protonated by the protic acid equilibrating with 2-methylbut-3-en-2-yl cation intermediate that adds to another molecule of isoprene, giving an another carbocation intermediate that continues the polymerization process (an example of cationic polymerization).

Nowadays, petroleum is by far the most important raw material for producing chemicals. Although most of it is utilized for the manufacture of gasoline, diesel fuel, jet fuel, heating oil, and power plant fuel, 10% of it is used to produce chemicals. In refineries, petroleum is first rectified to give various fractions having different boiling temperatures. These fractions are then upgraded to fuels, mostly applying catalytic processes. Steam cracking of hydrocarbons at c. 850 °C without catalyst produces mostly ethylene (CH_2=CH_2), propylene (Me—CH=CH_2), and by-products such hydrogen as (H_2), methane (CH_4), C_4 hydrocarbons (butane: Me—CH_2—CH_2—Me), isobutane: Me_2CHMe, (*E*)- and (*Z*)-but-2-ene: (*E*)-and (*Z*)-Me—CH=CH—Me, but-1-ene: Me—CH_2—CH=CH_2), and the "BTX-aromatics" (benzene: C_6H_6=Ph–H, toluene: PhMe, *ortho-*, *meta-*, and *para*-xylene: $C_6H_4Me_2$).

Reaction energy hypersurfaces

At the macroscopic level, any equilibrium at constant temperature T and pressure P is characterized by a Gibbs energy or free energy diagram (Figure 4) and an enthalpy diagram (Figure 5). Measurement of the equilibrium constant K^T gives $\Delta_r G^T$ and calorimetry provides $\Delta_r H^T$. Kinetics (measurement of the rate constant at different temperature, Section 3.2) gives the activation parameters $\Delta^{\ddagger}G^T$ and $\Delta^{\ddagger}H^T$ for the transition states with the highest free energy and enthalpy, respectively. For the

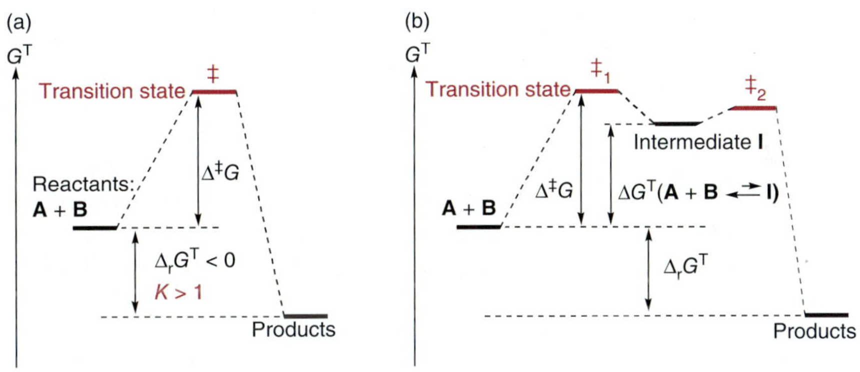

Figure 4 Free energy diagram (a) for an equilibrium that does not involve a reactive intermediate and (b) for an equilibrium that involves a single reactive intermediate. In this case, the rate-determining step (the slowest step) is associated with transition state $\ddagger_1$, the highest in free energy.

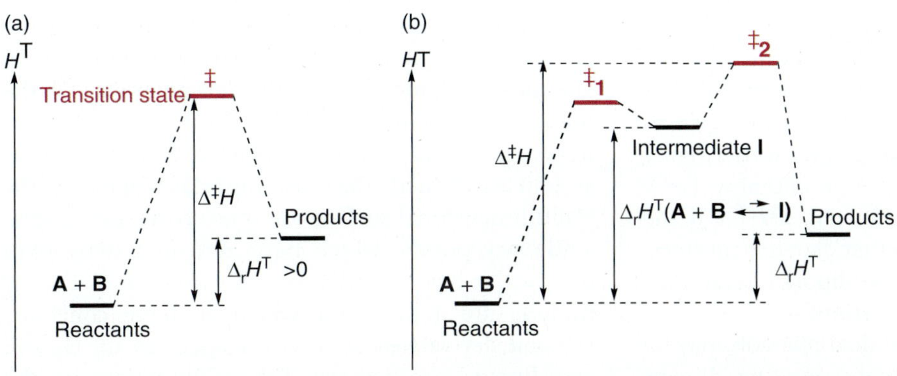

Figure 5 Enthalpy diagram for an equilibrium (a) that does not involve a reactive intermediate and (b) that involves a single reactive intermediate. In the case chosen, $\ddagger_2$ is higher in enthalpy than $\ddagger_1$, which corresponds to the transition state of the rate-determining step in the free energy diagram of Figure 4b. This is possible because of $\Delta^{\ddagger}G = \Delta^{\ddagger}H - T\Delta^{\ddagger}S$. Both reactions chosen here are endothermic ($\Delta_r H^T > 0$) and have a positive entropy variations ($\Delta_r S^T > 0$) making $\Delta_r G^T = \Delta_r H^T - T\Delta_r S^T < 0$. The reaction illustrated in Figures 4b and 5b has a more negative activation entropy ($\Delta^{\ddagger}S < 0$) for the slowest step involving transition state $\ddagger_1$ than for $\ddagger_2$.

other transition states and the intermediates that are involved in the reaction, their thermochemical data can be estimated by quantum mechanical calculations or by applying various theories on chemical activation. For neutral reactive intermediates such as radicals and diradicals, their standard heats of formation can be estimated readily from gas-phase homolytic bond dissociation enthalpies ($DH°(R^{\cdot}/X^{\cdot})$). Therefore, $\Delta_r H°$(reactants ⇄ intermediate) can be obtained through a simple thermochemical calculation. To a first approximation entropy variations, $\Delta_r S°$(reactants ⇄ intermediate) is estimated readily by considering the change of number of species between the intermediate and the reactants, by considering their molecular masses and whether rotations about single bonds are lost or gained between the intermediate and the reactants. For reactions generating ion pairs such as acid/base equilibria, $\Delta_r G°$(reactants ⇄ intermediate) = $1.36\cdot(\Delta pK_a)$. For other heterolyses in solution, the gas-phase heterolytic bond dissociation enthalpies (e.g. ($DH°(R^+/X^-)$) can often be used applying well-defined corrections for reactions in solutions. In several instances, substituent effects on the relative stability of charged intermediates in the gas phase correlate with the substituent effects on the same species in solution. When a reaction has a relatively high barrier and a slowly varying entropy (e.g. an isomerization has a relatively small positive or negative $\Delta^{\ddagger}S$ value, a fragmentation has a slightly positive $\Delta^{\ddagger}S$ value; however, a reaction following an associative mechanism has a highly negative $\Delta^{\ddagger}S$ value) in the region of the transition state, its energy and geometry correspond closely to those of the reactive intermediate it is connected with. This is the Hammond postulate. In the case of Figure 5b, the reactive intermediate resembles in the geometry and enthalpy transition state $\ddagger_1$ that separates it from the reactants. It also resembles transition state $\ddagger_2$ that separates it from the products. This postulate is in fact a theorem demonstrated by the Bell–Evans–Polanyi theory and reflected in the Dimroth principle for one-step reactions: $\Delta^{\ddagger}H^T = \alpha\Delta_r H^T + \beta$ (with α varying between 0 and 1). The higher the exothermicity of a reaction, the lower its activation enthalpy. For a thermoneutral equilibrium ($\Delta_r H^T = 0$), $\Delta^{\ddagger}H^T = \beta$, the intrinsic barrier of the reaction that depends on steric factors, electronic factors (dipole/dipole interactions and electron exchange), and solvation.

At the molecular level, a chemical reaction may be represented in $N+1$ dimensional space. One dimension represents the potential energy, E, of the system, whereas the other N dimensions are the coordinates that describe the geometries of the chemical species undergoing change. For a reaction involving a single,

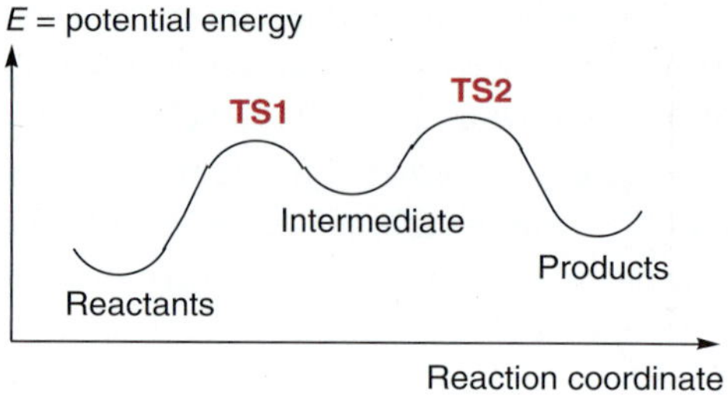

Figure 6 A one-dimensional slice through a reaction energy hypersurface (potential energy vs. reaction coordinate diagram) corresponds to the enthalpy diagram of Figure 5b.

nonlinear molecule, it takes $N = 3n - 6$ (coordinates where n = number of atoms in the molecule) to fully describe the molecule and the reaction. For example, each atom can be defined in space by an X, Y, and Z coordinate, giving $3n$ total coordinates. Only $3n - 6$ are needed to define the internal structure of a molecule, three more give the position of the molecule in space with respect to some reference, while three more tell how the molecule is oriented in space. The potential energy $E = f$(coordinates) will have minima, maxima, and saddle points as shown in Figure 6 for the two-step reaction illustrated in Figures 4b and 5b.

The minima correspond to reactants, products, or reactive intermediates (**I**), whereas the saddle points are transition structures TS1 and TS2 that are associated with the transition states $\ddagger_1$ and $\ddagger_2$ of the reaction, respectively.

Such a one-dimensional slice is just a glimpse of the whole story, as a full description of a molecule actually involves all $3n - 6$ internal coordinates. Energy *versus* reaction coordinate diagrams in Figure 6 show energy as a function of one coordinate change only. Quantum mechanical calculations incorporating the Born–Oppenheimer approximation (the motion of the nuclei can be separated from the motion of the electrons) can be applied to determine the potential energies E of molecules with any geometry of the nuclei. When a large number of these calculations are done, a potential energy hypersurface for vibrationless system is obtained. The most important regions of the multidimensional surface are those corresponding to stationary points, which have zero first derivatives of E with respect to the $3n - 6$ coordinates. Energy minima are a point for which all force constants (second derivatives of E with respect to the $3n - 6$ coordinates) are positive. The saddle points are the transition structures (Figure 7); they have one, and only one, negative second derivative, the remaining $3n - 7$ second derivatives are positive. The negative second derivation of E corresponds to a force constant

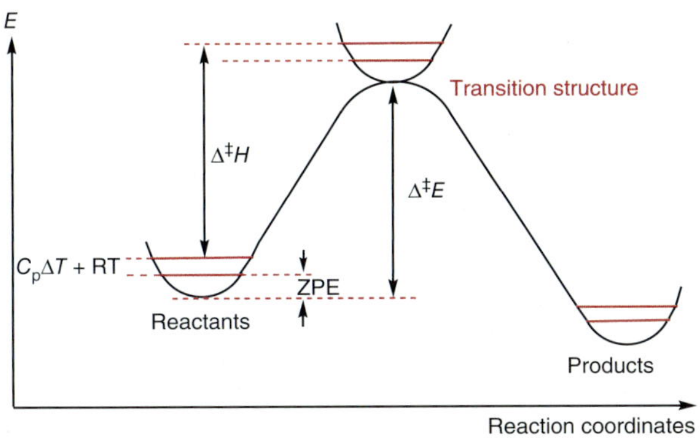

Figure 7 Relationship between $\Delta^{\ddagger}H$ (macroscopic activation parameter) and calculated $\Delta^{\ddagger}E$ (microscopic level) at $T > 0$ K along a reaction pathway. ZPE, zero-point energy and C_p, calorific capacity at constant pressure p.

for the motion along the reaction coordinate, which is referred to an "imaginary vibrational frequency" as the vibrational frequency is proportional to the square root of the force constant [11, 12].

When a reaction has a low barrier or rapidly varying entropy in the region of the potential energy maximum, the transition state may have a geometry different from that of the calculated transition structure. Furthermore, a transition state might be associated with more than one transition structure.

In 1931, about 40 years after Arrhenius's empirical observation, Eyring and Polanyi developed the first potential energy hypersurface for the degenerate reaction of hydrogen atom ($H^{\cdot}$) with dihydrogen (H_2) [13, 14]. Then, Hirschfelder, Eyring, and Topley performed the first trajectory calculation with femtosecond steps in 1936 [15]. These theoretical developments constituted the birth of reaction dynamics, and chemists began to think in terms of motions of atoms and molecules (dynamics) on potential energy surfaces. In 1973, Wang and Karplus [16] were the first to carry out a trajectory calculation of this type for a simple organic reaction: $CH_2 + H_2 \rightarrow CH_4$. Such calculations have become more commonplace, but only in the last decade have organic chemists begun to recognize how dynamics may alter the static picture of a reaction given by the potential surface [17].

The Arrhenius *A* frequency factor is typically 10^{13} Hz (per second) for a unimolecular reaction, a typical value of the frequency of a molecular vibration. In the mid-1930s, experimental temporal resolution of only seconds to milliseconds was possible in chemistry by means of the stopped-flow technique. Norrish and Porter [18] introduced in 1949 the flash photolysis technique reaching millisecond timescale. By exposing a solution to a heat, pressure, or electrical shock (the so-called temperature-jump method, etc.), Eigen achieved microsecond (10^{-6} seconds) temporal resolution [19]. The advent of the pulsed nanosecond (10^{-9} seconds) laser in the mid-1960s [20, 21], and soon after of the picosecond (10^{-12} seconds) laser [22, 23], brought a million times improvement in temporal resolution of chemical elementary processes. However, even on the short picosecond timescale, molecular states already reside in eigenstates (the static limit), and only the change of population of that state with time is observable, not the change of geometry of the molecules. The advent of femtosecond (10^{-15} seconds) laser technology of Shank [24–26] finally opened the possibility to probe molecular motion and chemical reactions in real time [27]. Transition states as well as reactive intermediates can now be visualized as demonstrated by Zewail for a large number of chemical and biological processes [28–32]. Attosecond temporal resolution is now possible and even permits the observation of electron dynamics [33, 34].

Can we see reactions in real time?

To take what amounts to a movie of a simple chemical reaction, Zewail and coworkers used two beams of femtosecond pulses and a mass spectrometer. A first pulse of light, called the pump pulse, strikes the molecule and energizes it. If the photon energy is sufficient, it induces a chemical reaction that can break the molecule apart into molecular fragments. In order to follow the birth and order of appearance of these fragments, a second pulse traveling just a few femtoseconds behind the first hits the fragments and ionizes them. The nature of fragments can be followed by mass spectrometry. The second pulse, called the probe pulse, can be timed precisely at different intervals to reveal how long it takes for various chemical species to appear and in what order they do so. The experiment that gave birth to femtochemistry in 1987 involved the dissociation of cyanogen iodide (ICN), in which the appearance of a free CN fragment was found to occur in about 200 fs [35]. Figure 8 is a colorful popular presentation of the way Zewail's technique works, from the Nobel Prize lecture cover of the journal [36].

Laser irradiation of 1,2-diiodo-1,1,2,2-tetrafluoroethane generates a molecule of tetrafluoroethylene and two iodine radicals. The first C—I bond cleavage takes about 200 fs, whereas the second follows on a timescale 100 times longer [37]. This

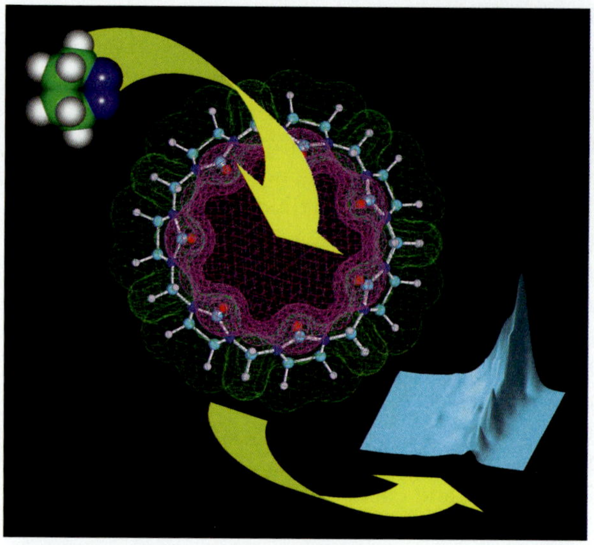

Figure 8 How Zewail's technique obtains a "movie" of a reaction. Source: Drawing supplied by Werner M. Nau, International University, Bremen, Germany.

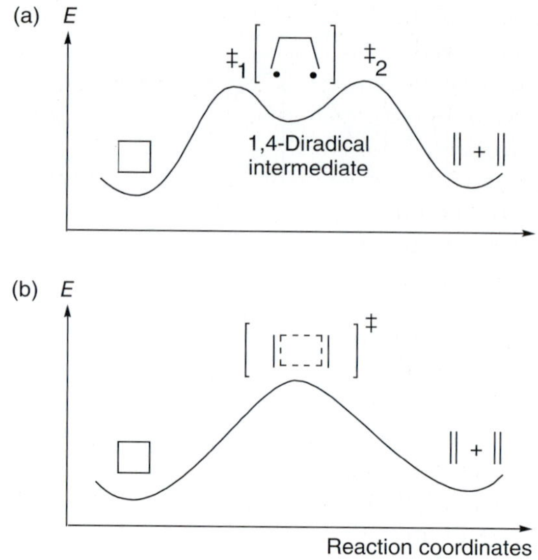

Figure 9 Two possible mechanisms for the thermal (2 + 2) cycloreversion of cyclobutane to ethylene. (a) a stepwise intermediate involving a diradical intermediate and two transition states; (b) a concerted mechanism involving one transition state.

demonstrates that the photoinduced fragmentation is a two-step process with the formation of a 1,1,2,2-tetrafluoro-2-iodoethyl radical intermediate. This conclusion may, or may not, apply to a reaction in solution induced by heating but provides a fast snapshot of the radical process in the gas phase [38].

Concerted or nonconcerted?

For the past 70 years, the concept of diradicals as intermediates of reactions has been considered as the archetype of chemical transformations in many classes of thermally activated, as well as photochemical, reactions, including the broad class of pericyclic reactions (Chapter 5). In one classical example, the ring opening of cyclobutane and its fragmentation into two molecules of ethylene ((2+2)-cycloreversion) may proceed directly through a transition state at the saddle point of an activation barrier (Figure 9a) or through a two-step, nonconcerted mechanism involving first the cleavage of one of the σ(C—C) bonds to yield a tetramethylene diradical (buta-1,4-diyl diradical) intermediate (Figure 9b). A reactive intermediate is expected to be longer lived than a transition state, such that the dynamics of its nuclear motion (vibration and rotation), unlike a concerted motion (translation), determines the outcome of the reaction.

By combining femtosecond spectroscopy with time-of-flight mass spectrometry and molecular beams, and by generating the diradical from an alternate source, Zewail and coworkers established the existence of this 1,4-diradical (Figure 9a) as a distinct molecular species [39]. Femtochemistry has been applied to the condensed phase to pinpoint the details of solvation dynamics and to biomolecules [40, 41]. It provides insight into the function of biological systems. The ability to visualize motion in a protein enables one to study the relationship between nuclear motion and biological functions. As an example, it is known that hydrogen bonds bind the double-stranded DNA helix and determine the complementarity of pairing. With ultrafast laser spectroscopy, Zewail and coworkers have identified different timescales of the structural relaxation and cooling of the tautomers [42–44]. These studies have demonstrated that we can now watch reactions occur in ideal systems, and they give us the hope that one day we will obtain a detailed molecular picture of the nuclear dynamics that govern the fundamentals of chemical reactivity in biological systems. Femtochemistry has been applied to the study of reactions at metal surfaces [45–47].

Structures of species on the reaction hypersurface

In an ultrafast laser experiment, the data collected do not give the direct structure of the species under study, as fluorescence or mass spectra have to be translated into structures. Actually, the only species well characterized on a reaction hypersurface are the starting materials (or reactants) and the final products that are long-lived and thus can be analyzed by X-ray crystallography and neutron diffraction for crystalline compounds or by electron diffraction for gaseous substances. In some cases, reactive intermediates can be "frozen out" by some special techniques and thus analyzed as any other substances. The geometry of a transition state cannot be analyzed by these means as it is too short-lived (less than the time necessary to a molecular vibration, by definition; 300 fs for the conversion of (*Z*)-stilbene into its (*E*)-isomer). Transition structures must be inferred from theories and models, and by the interpretation of spectroscopic fingerprints in the case of ultrafast laser spectroscopy.

At 25 °C, simple molecules or atoms of a gas travel with a speed of 10^4–10^5 cm s^{-1} (i.e. 10^{12}–10^{13} Å s^{-1}). The lifetime of an activated complex (or transition state) [48] resulting from the collision of molecule **AB** with **C** to generate products, **A** + **BC**, can be estimated as follows. The distance traveled by the ensemble **AB** + **C** undergoing through the activated complex [48] to give product **A** + **BC** amounts to about 1–10 Å or 0.1–1 nm (10^{-9} m), which, at 10^{12}–10^{13} Å s^{-1}, requires about 10^{-13}–10^{-11} seconds (100 fs to 10 ps) [49]. The timescale for movements of valence electrons is 10^{-17} seconds (10 as), and for molecular rearrangements by movements of nuclei, 10^{-13} seconds (100 fs) [34, 50].

Through a combination of light and electron probes, it is possible to record single-molecule dynamics with simultaneous sub-angstrom spatial and femtosecond temporal resolution. Single-molecule femtochemistry is becoming possible through a melding of laser spectroscopy and electron microscopy techniques [51, 52]. The computational study of organic reaction dynamics is becoming increasingly common, and a time-resolved understanding of the timing of bond formation has enriched our views of the details of organic reaction mechanisms [53, 54].

Lausanne, June 1st, 2019 — *Pierre Vogel*
Los Angeles, June 1st, 2019 — *Kendall N. Houk*

References

1 Perkin, W.H. On mauvein and allied colouring matters. *Journal of the Chemical Society, Transactions* 1879: 717–732.

2 Travis, A.S. (1990). Perkins Mauve – ancestor of the organic-chemical industry. *Technology and Culture* 31 (1): 51–82.

3 Huebner, K. (2006). History – 150 years of mauveine. *Chemie in unserer Zeit* 40 (4): 274–275.

4 Sousa, M.M., Melo, M.J., Parola, A. et al. (2008). A study in mauve: unveiling Perkin's dye in historic samples. *Chemistry - A European Journal* 14 (28): 8507–8513.

5 Geiseler, G. and Buchner, W. (1966). Uber Die Bildungsenthalpie des Calciumcarbids. *Zeitschrift für Anorganische und Allgemeine Chemie* 343 (5–6): 286–293.

6 Whitby, G.S. and Katz, M. (1933). Synthetic rubber. *Industrial and Engineering Chemistry* 25: 1204–1211.

7 Whitby, G.S. and Katz, M. (1933). Synthetic rubber (concluded). *Industrial and Engineering Chemistry* 25: 1338–1348.

8 von Liebig, J. (1835). Bemerkung über die methoden der Darstellung und Reinigung flüchtiger, durch trockene destillation organischer materien erhaltene producte. *Annalen der Pharmacie* 16 (1): 61–62.

9 Bouchardat, G. (1879). Action des hydracides sur l'isoprène; reproduction du caoutchouc. *Comptes rendus hebdomadaires des séances de l'Académie des Sciences Paris* 89: 1117–1120.

10 Long, J.C. (2001). The history of rubber – a survey of sources about the history of rubber. *Rubber Chemistry and Technology* 74 (3): 493–508.

11 Hehre, W.J., Radom, L., Schleyer, P.v.R., and Pople, J.A. (1986). *Ab initio Molecular Orbital Theory*. New York: Wiley.

12 Simons, J. (1991). An experimental chemists guide to *ab initio* quantum-chemistry. *Journal of Physical Chemistry* 95 (3): 1017–1029.

13 Eyring, H. and Polanyi, M. (1931). Concerning simple gas reactions. *Zeitschrift fuer Physikalische Chemie* B12 (4): 279–311.

14 Polanyi, M. (1932). *Atomic Reactions*. London: Williams and Norgate.

15 Hirschfelder, J., Eyring, H., and Topley, B. (1936). Reactions involving hydrogen molecules and atoms. *Journal of Chemical Physics* 4 (3): 170–177.

16 Wang, I.S.Y. and Karplus, M. (1973). Dynamics of organic reactions. *Journal of the American Chemical Society* 95 (24): 8160–8164.

17 Carpenter, B.K. (1998). Dynamic behavior of organic reactive intermediates. *Angewandte Chemie International Edition* 37 (24): 3340–3350.

18 Norrish, R.G.W. and Porter, G. (1949). Chemical reactions produced by very high light intensities. *Nature* 164 (4172): 658–658.

19 Eigen, M. (1954). Methods for investigation of ionic reactions in aqueous solutions with half-times as short as 10^{-9} s – application to neutralization and hydrolysis reactions. *Discussions of the Faraday Society* 17: 194–205.

20 Hellwarth, R.W. (1961). *Advances in Quantum Electronics*. New York: Columbia University Press.

21 McClung, F.J. and Hellwarth, R.W. (1962). Giant optical pulsations from ruby. *Journal of Applied Physics* 33 (3): 828–829.

22 Hargrove, L.E., Fork, R.L., and Pollack, M.A. (1964). Locking of He–Ne laser modes induced by synchronous intracavity modulation (diffraction by phonons in crystals e). *Applied Physics Letters* 5 (1): 4–5.

23 McDuff, O.P. and Harris, S.E. (1967). Nonlinear theory of internally loss-modulated laser. *IEEE Journal of Quantum Electronics* QE 3 (3): 101–111.

24 Shank, C.V. (1988). *Ultrashort Laser Pulses*. Berlin: Springer-Verlag.

25 Shank, C.V. (1986). Investigation of ultrafast phenomena in the femtosecond time domain. *Science* 233 (4770): 1276–1280.

26 Zewail, A.H. (2000). Femtochemistry: atomic-scale dynamics of the chemical bond. *Journal of Physical Chemistry A* 104 (24): 5660–5694.

27 Leone, S.R., McCurdy, C., Burgdoerfer, J. et al. (2014). What will it take to observe processes in 'real time'? *Nature Photonics* 8 (3): 162–166.

28 Zewail, A.H. and Bernstein, R.B. (1988). Special report – real-time laser femtochemistry – viewing the transition from reagents to products. *Chemical & Engineering News* 66 (45): 24–43.

29 Khundkar, L.R. and Zewail, A.H. (1990). Ultrafast molecular reaction dynamics in real-time – progress over a decade. *Annual Review of Physical Chemistry* 41: 15–60.

30 Zewail, A.H. (1993). Femtochemistry. *Journal of Physical Chemistry* 97 (48): 12427–12446.

31 Zewail, A.H. (1990). The birth of molecules. *Scientific American* 263 (6): 76–82.

32 Zewail, A.H. (2000). Femtochemistry. Past, present, and future. *Pure and Applied Chemistry* 72 (12): 2219–2231.

33 Kling, M.F. and Vrakking, M.J. (2008). Attosecond electron dynamics. *Annual Review of Physical Chemistry* 59: 463–492.

34 Sansone, G., Kelkensberg, F., Perez-Torres, J. et al. (2010). Electron localization following attosecond molecular photoionization. *Nature* 465 (7299): 763–766.

35 Dantus, M., Rosker, M.J., and Zewail, A.H. (1988). Femtosecond real-time probing of reactions. 2. The dissociation reaction of ICN. *Journal of Chemical Physics* 89 (10): 6128–6140.

36 Zewail, A.H. (2000). Femtochemistry: atomic-scale dynamics of the chemical bond using ultrafast lasers – (Nobel lecture). *Angewandte Chemie International Edition* 39 (15): 2586–2631.

37 Khundkar, L.R. and Zewail, A.H. (1990). Picosecond photofragment spectroscopy. 4. Dynamics of consecutive bond breakage in the reaction $C_2F_4I_2 \rightarrow C_2F_4 + 2I$. *Journal of Chemical Physics* 92 (1): 231–242.

38 Carley, R.E., Heesel, E., and Fielding, H.H. (2005). Femtosecond lasers in gas phase chemistry. *Chemical Society Reviews* 34 (11): 949–969.

39 Pedersen, S., Herek, J.L., and Zewail, A.H. (1994). The validity of the diradical hypothesis – direct femtosecond studies of the transition-state structures. *Science* 266 (5189): 1359–1364.

40 Brixner, T. and Gerber, G. (2003). Quantum control of gas-phase and liquid-phase femtochemistry. *ChemPhysChem* 4 (5): 418–438.

41 Brixner, T., Damrauer, N.H., Kiefer, B., and Gerber, G. (2003). Liquid-phase adaptive femtosecond quantum control: removing intrinsic intensity dependencies. *Journal of Chemical Physics* 118 (8): 3692–3701.

42 Douhal, A., Kim, S.K., and Zewail, A.H. (1995). Femtosecond molecular-dynamics of tautomerization in model base-pairs. *Nature* 378 (6554): 260–263.

43 Fiebig, T., Chachisvilis, M., Manger, M. et al. (1999). Femtosecond dynamics of double proton transfer in a model DNA base pair: 7-azaindole dimers in the condensed phase. *Journal of Physical Chemistry A* 103 (37): 7419–7431.

44 Wan, C.Z., Fiebig, T., Schiemann, O. et al. (2000). Femtosecond direct observation of charge transfer between bases in DNA. *Proceedings of the National Academy of Sciences of the United States of America* 97 (26): 14052–14055.

45 Frischkorn, C. and Wolf, M. (2006). Femtochemistry at metal surfaces: nonadiabatic reaction dynamics. *Chemical Reviews* 106 (10): 4207–4233.

46 Mehlhorn, M., Gawronski, H., Nedelmann, L. et al. (2007). An instrument to investigate femtochemistry on metal surfaces in real space. *Review of Scientific Instruments* 78 (3): 033905.

47 Petek, H. (2012). Photoexcitation of adsorbates on metal surfaces: one-step or three-step. *Journal of Chemical Physics* 137 (9): 091704.

48 Tommos, C. and Babcock, G.T. (2000). Proton and hydrogen currents in photosynthetic water oxidation. *Biochimica et Biophysica Acta-Bioenergetics* 1458 (1): 199–219.

49 Polanyi, J.C. and Zewail, A.H. (1995). Direct observation of the transition-state. *Accounts of Chemical Research* 28 (3): 119–132.

50 Bucksbaum, P.H. (2007). The future of attosecond spectroscopy. *Science* 317 (5839): 766–769.

51 Lee, J., Perdue, S.M., Perez, A.R., and Apkarian, V.A. (2014). Vibronic motion with joint angstrom-femtosecond resolution observed through fano progressions recorded within one molecule. *ACS Nano* 8 (1): 54–63.

52 Petek, H. (2014). Single-molecule femtochemistry: molecular imaging at the space-time limit. *ACS Nano* 8 (1): 5–13.

53 Yang, Z. and Houk, K.N. (2018). The Dynamics of Chemical Reactions: Atomistic Visualizations of Organic Reactions, and Homage to van't Hoff. *Chem. Eur. J.* 24: 3916–3924.

54 Yang, Z., Jamieson, C.S., Xue, X.-S., Garcia-Borras, M., Benton, T., Dong, X., Liu, F. and Houk, K.N. (2019). Mechanisms and Dynamics of Reactions Involving Entropic Intermediates *Trends in Chemistry*. 1: 22–34.

Foreword

The determination of natural product structure and the discovery of new reactions defined early organic chemistry, followed by the synthesis of preparing known molecules and creating new molecules. Physical organic chemistry came later [1] in the beginning of 1920s and 1930s with the book of Hammett, "Physical Organic Chemistry," published in 1940 [2]. Physical organic chemistry was defined as the "Application of quantitative mathematical methods to Organic Chemistry." This was the union of organic chemistry – the discovery of molecules in Nature, then transformation into molecules that never before existed – with physical chemistry – the determination of structures with spectroscopy, measurements of rates of reaction, and theoretical descriptions of chemistry. Physical Organic Chemistry has become the foundation of organic chemistry.

Chemists, biochemists, physical chemists, and chemical engineers invent procedures to transform matter either empirically, by trial and error, by intuition, by serendipity, or by applying theoretical models. This new book by Pierre Vogel of the EPFL (Swiss Federal Institute of Technology in Lausanne, Switzerland) and Kendall Houk of the University of California, Los Angeles, is the twentyfirst century paradigm of the field of organic chemistry, combining the extraordinary power of thermodynamics, thermochemical data banks, kinetics, quantum mechanics, and spectroscopy to understand and control the diversity of chemical reactivity and the modern synthetic methods in a novel fashion. Studies on the mechanism of reaction of organic molecules in solution dominated physical organic chemistry at its beginning, but contemporary synthetic methods use the whole periodic table, photochemistry, and reactions in the vapor phase, solution, and in solid state and enzymes to create new chemistry to apply to the problems of both commercial and intellectual interest.

Since Hammett's treatise [2] there have been many mechanistic books, such as those due to Ingold in the 1950s [3], and Hine [4] and Gould (the book that inspired me) in the 1960s [5]. Lowry and Richardson dominated the field in the 1970s and 1980s [6], and Anslyn and Dougherty (2005) have dominated mechanistic and physical organic chemistry teaching in the past decade [7]. More general books such as March (1968ff) [8], now Smith [9], and Carey and Sundberg (1977) covered the synthesis and mechanisms [10]. Other books by Isaacs [11], Carroll [12], and Maskill [13] were more directed at physical organic chemistry. In 2017, an Encyclopedia of Physical Organic Chemistry has been published [14]. Now, Vogel and Houk unite the challenging diversity of modern synthetic methodology, including asymmetric synthesis and catalysis, with modern theories to present a new text that will also serve as a useful resource for the chemical and biochemical communities. The Vogel–Houk book is a textbook and a reference manual at the same time; it provides a new way to think about the chemical reactivity and a powerful toolbox to inventors of new reactions and new procedures.

Caltech, Pasadena, CA *Robert H. Grubbs*

References

1 Mayr, H. (2016). Physical organic chemistry–development and perspectives. *Isr. J. Chem.* 56: 30–37.
2 Hammett, L.P. (1940). *Physical Organic Chemistry*, 1–404. New York, NY: MacGraw-Hill Co.
3 Ingold, C.K. (1953). *Structure and Mechanisms in Organic Chemistry*, 1–826. Ithaca, NY: Cornell University Press.
4 Hine, J. (1962). *Physical Organic Chemistry (Advanced Chemistry)*, 2e, 1–552. New York, NY: McGraw-Hill.
5 Gould, E.S. (1959). *Mechanism and Structure in Organic Chemistry*, 1–790. New York, NY: Henry Holt & Co.
6 Lowry, T.H. and Richardson, S.K. (1976). *Mechanism and Theory in Organic Chemistry*, 1–748.

New York, NY: Harper & Row; International 2nd revised edition, 1987, pp. 1–1090.

7 Anslyn, E.V. and Dougherty, D.A. (2006). *Modern Physical Organic Chemistry*, 1–1095. Sausalito, CA: University Science Books.

8 March, J. *Advanced Organic Chemistry: Reactions, Mechanisms, and Structure,* 4th edition. Wiley., New York, NY, 1992, pp. 1–1495.

9 Smith, M.B. (2013). *March's Advanced Organic Chemistry: Reactions, Mechanisms and Structure*, 7e, 1–2047. Hoboken, NJ: Wiley.

10 (a) Carey, F.A. and Sundberg, R.J. (2000). *Advanced Organic Chemistry, Part A: Structure and Mechanisms*, 4e, 1–823. New York, NY: Springer Science & Business Media. (b) Sundberg, R.J. and Carey, F.A. (2001). *Advanced Organic Chemistry, Part B: Reactions and Synthesis*, 4e, 1–958. New York, NY: Kluwer Academic/Plenum Publishers.

11 (a) Isaacs, N.S. (1987). *Physical Organic Chemistry*, 1–828. New York, NY: Wiley. (b) Isaacs, N.S. (1995). *Physical Organic Chemistry*, 2e, 1–877. New York, NY: Wiley.

12 (a) Carroll, F.A. (1997). *Perspectives on Structure and Mechanism in Organic Chemistry*, 1–919. Pacific Grove, CA: Brooks & Cole. (b) Carroll, F.A. (2014). *Perspectives on Structure and Mechanism in Organic Chemistry*, 2e, 1–972. New York, NY: Wiley.

13 (a) Maskill, H. (1986). *The Physical Basis of Organic Chemistry*, 1–490. Oxford University Press. (b) Aldabbagh, F., Atherton, J.H., Bentley, W. et al. (2006). *The Investigation of Organic Reactions and their Mechanisms* (ed. H. Maskill), 1–370. Oxford: Blackwell Publishing.

14 Wang, Z. (ed.) (associate eds. U. Wille and E. Juaristi) (2017). *Encyclopedia of Physical Organic Chemistry*, 6 Volume set, 1–4464. New York, NY: Wiley.

1

Equilibria and thermochemistry

1.1 Introduction

This chapter introduces the quantitative treatment of the energetics of molecules and equilibria and describes how to interpret these quantities. It presents tables of thermochemical data, including standard heats of formation and standard entropies (Tables 1.A.1–1.A.4), Pauling electronegativities (Table 1.A.5), bond lengths (Table 1.A.6), bond dissociation energies (BDEs) or standard homolytic bond dissociation enthalpies (Tables 1.A.7–1.A.11, 1.A.13, 1.A.14), gas-phase heterolytic bond dissociation enthalpies (Tables 1.A.13–1.A.16), gas-phase proton affinities (Tables 1.A.13, 1.A.15, 1.A.18), gas-phase hydride affinities (Tables 1.A.14 and 1.A.16), ionization enthalpies (Tables 1.A.13, 1.A.20, 1.A.21), electron affinities (Tables 1.A.13, 1.A.20, 1.A.22), gas-phase acidities (Table 1.A.17), and substituent effects on the relative stabilities of reactive intermediates in the gas phase such as radicals (Tables 1.A.9 and 1.A.12), carbenium ions (Table 1.A.14) and anions (Tables 1.A.19), and solution acidities (Tables 1.A.23 and 1.A.24) for selected species.

Thermochemistry is "the study of heat produced or required by a chemical reaction" [1]. Thermochemistry is closely associated with calorimetry, an experimental technique that can be used to measure the thermodynamics of chemical reactions. First developed by Black, Lavoisier, and Laplace in the eighteenth century, and further by Berthelot and Thomsen in the nineteenth century [2], the golden years of calorimetry began in the 1930s; Rossini [3] at the National Bureau of Standards determined the thermodynamic quantities for a number of organic compounds. The thermochemical studies of organometallic compounds were pioneered by Skinner and coworkers [4, 5]. Calorimetry has been the main source of thermodynamic quantities, such as the standard enthalpies of selected reactions ($\Delta_r H°$), and, for pure compounds, standard enthalpies of combustion ($\Delta_c H°$), standard enthalpies of hydrogenation ($\Delta_h H°$), standard enthalpies of vaporization ($\Delta_{vap} H°$), standard enthalpies of sublimation ($\Delta_{sub} H°$), standard enthalpies of solubilization ($\Delta_{sol} H°$), standard enthalpies of formation ($\Delta_f H°$), standard entropies ($S°$), and heat capacities ($C_p°$) [6, 7].

1.2 Equilibrium-free enthalpy: reaction-free energy or Gibbs energy

The Le Châtelier principle states "On modifying pressure or temperature of a stable equilibrium, the latter is modified until cancelation of the effects imposed by the external changes; concentrations of reactants and products are modified such as to oppose the effects of the external changes." In other words, an equilibrium (reaction (1.1)) between **A**, **B**, etc., and **P**, **Q**, etc., as reactants and products, respectively, can be written as:

$$\alpha\mathbf{A} + \beta\mathbf{B} + \cdots \underset{}{\overset{K}{\rightleftarrows}} \pi\mathbf{P} + \theta\mathbf{Q} + \cdots \tag{1.1}$$

Interestingly, a few months before Le Châtelier, Van't Hoff had announced the same principle [8–10]. At equilibrium, the free energies G^T of the reactants and products are equal. At constant temperature (T) and pressure (p), and for reactants and products in their standard states (that is, 1 M in solution or 1 atm in the gas phase), the second law of thermodynamics gives Eq. (1.2), from which the change in Gibbs energy, $\Delta_r G^T$, between the moment reactants **A**, **B**, … are mixed and the moment equilibrium (1.1) is reached can be determined. $\Delta_r G^T$ is called the Gibbs energy of reaction (free enthalpy or just free energy of reaction).

$$\Delta_r G^T = -RT \ln K \tag{1.2}$$

where R is the gas constant (1.987 cal K^{-1} mol^{-1} = 1.987 eu (entropy units) ≅ 8.314 472 J K^{-1} mol^{-1}), and T is the temperature in K (Kelvin) and

$$K = \frac{a_P^\pi a_Q^\theta \cdots}{a_A^\alpha a_B^\beta \cdots} \tag{1.3}$$

Organic Chemistry: Theory, Reactivity and Mechanisms in Modern Synthesis, First Edition. Pierre Vogel and Kendall N. Houk.

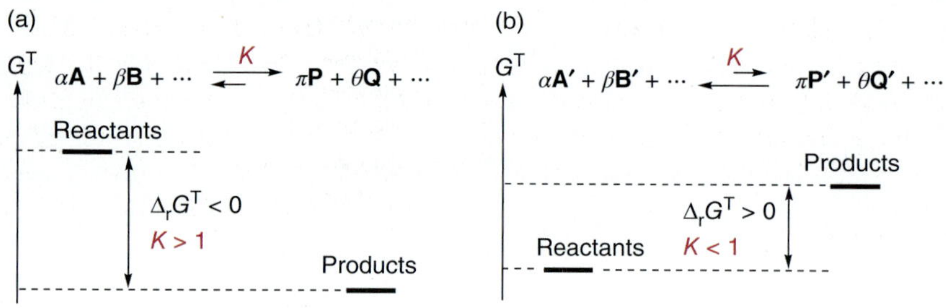

Figure 1.1 Free enthalpy diagrams: variation of Gibbs energy for (a) an exergonic reaction ($K > 1$) and (b) for an endergonic reaction ($K < 1$) (reactants: **A**, **B**, …; products: **P**, **Q**, …).

Here, a_P, a_Q, … and a_A, a_B, … are the activities (or relative activities) of products **P**, **Q**,… and reactants **A**, **B**,..., respectively, at equilibrium, and α, β,… π, θ are the stoichiometric factors of equilibrium (1.1) in solution.

Concentrations are generally used in place of activities; this is equivalent to assuming that the activity coefficients, γ, (e.g. $a_A = \gamma_A[\mathbf{A}]$, $a_B = \gamma_B[\mathbf{B}]$, $a_P = \gamma_P[\mathbf{P}]$, and $a_Q = \gamma_Q[\mathbf{Q}]$) are equal to unity.

If $\Delta_r G^T < 0$, the reaction is exergonic: $K > 1$ (e.g. Figure 1.1a)
If $\Delta_r G^T > 0$, the reaction is endergonic: $K < 1$ (e.g. Figure 1.1b).

The terms exergonic and endergonic are related to the more familiar ones exothermic and endothermic that refer to enthalpies (see below).

For a reaction in the gas phase,

$$K = \frac{p_P^\pi p_Q^\theta \cdots}{p_A^\alpha p_B^\beta \cdots} \tag{1.4}$$

where p_P, p_Q, p_A, and p_B are the partial pressures of **P**, **Q**, … and **A**, **B**, … respectively.

If equilibrium (1.1) is considered to be an ideal solution, then

$$K = \frac{[\mathbf{P}]^\pi[\mathbf{Q}]^\theta \ldots}{[\mathbf{A}]^\alpha[\mathbf{B}]^\beta \ldots} \tag{1.5}$$

where [**P**], [**Q**], … are the concentrations of the products and [**A**], [**B**], … are the concentrations of the reactants. A large number of organic reactions can be treated as ideal solutions, as long as dilute solutions are used under conditions of temperature and pressure that do not differ too greatly from: 298.15 K and 1 atm.

The Gibbs free energy of reaction is directly related to the relative amounts of two or more than two species at equilibrium: at temperature, T. This ratio can be determined from Eq. (1.2),

$$\ln K = -\Delta_r G^T/RT, \quad \text{or}$$
$$K = \exp(-\Delta_r G^T/RT) \tag{1.6}$$

As proposed first by Guldberg and Waage in 1879 [11], the equilibrium constant, K, is a ratio of rate constants (Chapter 3) k_{forward} (k_1) and k_{reverse} (k_{-1}), where k_{forward} is for the forward reaction (pure reactants equilibrating with products) and k_{reverse} is for the reverse reaction (pure products equilibrating with reactants), at the same temperature T: $K = k_{\text{forward}}/k_{\text{reverse}}$.

We shall show later that a free energy difference can be used to compare not only the forward and reverse reaction rate constants but also any two reaction rate constants k_1 and k_2:

$$\Delta\Delta_r G^T = -RT\ln(k_1/k_2) \tag{1.7}$$

As equilibria are usually discussed as existing at room temperature (25 °C, 298.15 K), it is useful to plot in R, T and to convert ln K to log K to obtain the following relationship (1.8):

$$\Delta_r G^\circ = -RT\ln K = (-1.987\ \text{eu})$$
$$\times\ 298.15 \cdot 2.303 \cdot \log K = -1.36 \cdot \log K, \quad \text{or}:$$
$$\Delta_r G^\circ = -1.36\ \log K \tag{1.8}$$

In Eq. (1.8), $\Delta_r G^\circ$ is in kcal/mol (IUPAC: kcal mol^{-1}), and the measurement is at room temperature (° for 298.15 K). This is a very useful relationship! Rounding off a bit, this expression shows that a 1.4 kcal mol^{-1} (5.9 kJ mol^{-1}) free energy difference results in a factor of 10 in equilibrium constant at 25 °C. Another way to say this is that $K = 10$ corresponds to a 1.4 kcal mol^{-1} difference in free energy, whereas $\Delta_r G^\circ = -2.8$ kcal mol^{-1} corresponds to $K = 100$ at 25 °C, and so forth.

1.3 Heat of reaction and variation of the entropy of reaction (reaction entropy)

Free energy provides a way to quantify experimental equilibria. Gibbs free energy at temperature T_x is written as $\Delta_r G^T$ (or $\Delta_r G(T_x)$). It can be separated into two other thermodynamic quantities: $\Delta_r H^T$ (or $\Delta_r H(T_x)$), the change in enthalpy or heat of reaction at temperature T_x, and $\Delta_r S^T$ (or $\Delta_r S(T_x)$), the change in entropy or reaction entropy. The heat of reaction is related to the internal energy U ($H = U + RT$) or heat content of a system. The reaction entropy is the variation of entropy between the beginning (when reactants **A**, **B**,... are mixed) and the end of the reaction (when equilibrium (1.1) is reached) at temperature T_x. It gives a quantitative measure of "disorder." The thermodynamic definitions of these quantities are given in the following section. Under constant pressure p, the Gibbs–Helmholtz equation (1.9) provides for equilibrium the relationship between $\Delta_r G^T$ and temperature T,

$$\mathrm{d}(\Delta_r G^T/T)/\mathrm{d}T = \mathrm{d}(\Delta_r H^T/T - \Delta_r S^T)/\mathrm{d}T = -\Delta_r H^T/T^2 \tag{1.9}$$

The heat of reaction, $\Delta_r H^T$, is the heat produced (exothermic) or absorbed (endothermic) between the beginning of the reaction (time t_0, the moment of mixing the reactants) and the end of the reaction (time t_∞, when the equilibrium **reactants ⇄ products** is reached, see Figure 1.2). The reaction entropy, $\Delta_r S^T$, expresses the change of order, or disorder, between products and reactants. This thermodynamic quantity will be discussed further in Section 1.4.

The Van't Hoff equation provides the relationship between the equilibrium constant, K, or rate constant ratio, $K = k\text{forward}/k\text{reverse}$, and the heat of reaction:

$$\ln K = -\Delta_r H^T/RT + \text{constant} \tag{1.10}$$

The slope of the plot of ln K vs. $1/T$ provides the value of $-\Delta_r H^T/R$. By measuring the equilibrium constant of a given equilibrium at two different temperatures, the average heat of reaction $\overline{\Delta_r H}$ can be determined roughly using Eq. (1.11):

$$\log\frac{(K)_2}{(K)_1} = -\frac{\overline{\Delta_r H}}{2.303\,R}\left(\frac{l}{T_2} - \frac{l}{T_1}\right) \tag{1.11}$$

This method is one of the most widely used methods to determine thermochemical parameters of reactions evolving to equilibria. It is not absolutely rigorous because it assumes a constant heat of reaction for the whole temperature range of investigation. However, in reality, the heat content of a substance changes with temperature, and this variation of heat content with temperature is given by Kirchhoff law (1.12):

$$H(T_2) - H(T_1) = \int_{T_1}^{T_2} C_p\,\mathrm{d}T \tag{1.12}$$

where $C_p = \left(\frac{\delta H}{\delta T}\right)_p$ is the molar heat capacity at a constant pressure.

The second law of thermodynamics states that the difference in free energy, $\Delta_r G^T$, between the initial state of a reaction and the final state at equilibrium depends on these two states only; it does not depend on the path followed to reach the equilibrium. Consequently,

$$\Delta_r H(T_2) - \Delta_r H(T_1) = \int_{T_1}^{T_2} \Delta_r C_p\,\mathrm{d}T \tag{1.13}$$

where $\Delta_r C_p = \Sigma C_p(\text{products}) - \Sigma C_p(\text{reactants})$

Equation (1.13) is a way to calculate $\Delta_r H(T_2)$ at temperature T_2, as long as $\Delta_r H(T_1)$ is known at temperature T_1, and C_p is known for all reactants and products ($\Delta_r C_p$). Often, it is assumed that $\Delta_r C_p$ has a constant value, leading to the simple approximation (1.14):

$$\Delta_r H(T_2) - \Delta_r H(T_1) = \Delta_r C_p\,(T_2 - T_1) \tag{1.14}$$

The standard Gibbs free energies for equilibrium (1.1) at T_1 and T_2 are given by Eqs. (1.15) and (1.16), respectively.

$$\Delta_r G(T_1) = \Delta_r H(T_1) - T_1 \Delta_r S(T_1) \tag{1.15}$$

$$\Delta_r G(T_2) = \Delta_r H(T_2) - T_2 \Delta_r S(T_2) \tag{1.16}$$

For small temperature differences $T_1 - T_2$, the entropies of reaction $\Delta_r S(T_1)$ and $\Delta_r S(T_2)$ can be

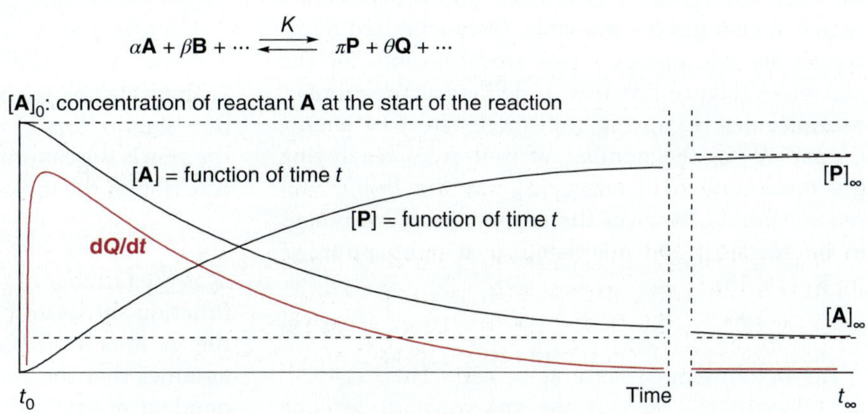

Figure 1.2 Reaction kinetics showing the disappearance of one reactant **A** (rate law d[**A**]/dt) and the appearance of one product **P** (rate law d[**P**]/dt) from the beginning of the reaction (time: t_0) to the end of the reaction (time: t_∞). The red curve is the heat flow (heat produced by time unit: dQ/dt) for an exothermic reaction ($\Delta_r H^T < 0$). $[A]_\infty$, $[M]_\infty$ are the concentrations of reactant **A** and product **P**, respectively, at equilibrium. The integral $\int_0^\infty Q\,\mathrm{d}t = \Delta_r H^T$ = heat of reaction at constant temperature T.

assumed to be identical. Consequently, a measurement of K_1 at T_1 and K_2 at T_2 allows one to estimate the average heat of the reaction $\overline{\Delta_r H}$.

The standard entropies of reaction (in cal K^{-1} mol^{-1} = eu = entropy units) at 298.15 K and under 1 atm (pure compounds that can be considered as ideal gases) can be calculated from Eq. (1.17), applying the third law of thermodynamics:

$$\Delta_r S^o = \Sigma S^o(\text{products}) - \Sigma S^o(\text{reactants}) \quad (1.17)$$

The standard entropy values S° are tabulated for a large number of gaseous compounds in the NIST Webbook of Chemistry (http://webbook.nist.gov) (Table 1.A.2). Alternatively, if the products and reactants are ideal gases (ideal gas law: $pV = NRT$; p = pressure, V = volume, N = number of moles, R = ideal gas constant, and T = temperature in K), the entropies can be calculated from statistical thermodynamics.

1.4 Statistical thermodynamics

Statistical thermodynamics establishes a relationship between the microscopic world of quantum mechanics and the macroscopic world that we readily observe [12, 13]. Thermodynamics has its origin in steam engines, and much of the language used to describe these engines persists to this day and is used to describe chemical processes and chemical themselves. We are able to derive thermodynamic properties of any compound from the structures of molecules. The thermodynamic parameters (internal energy U, enthalpy H ($H = U + pV$), entropy S, and free energy G) of an ensemble of molecules can be determined from spectroscopic data or quantum mechanical treatments of the molecules. The total energy of one molecule is the sum of the nuclear (E_{nucl}), electronic (E_{elec}), vibrational (E_{vib}), rotational (E_{rot}), and translational energies (E_{trans}). All these energies are quantized and only discrete values of energies are available. Only a limited number of discrete energy levels are accessible for the molecules (Figure 1.3). If N_i is defined as the number of molecules occupying the microstate i of energy E_i, and N_o is the number of molecules occupying the microstate o of energy $E_0 = 0$, the Boltzmann relationship (1.18) gives the proportion of molecules in microstate i and microstate o at temperature T (in K) [14, 15]:

$$N_i/N_o = e^{-E_i/k_b T} \quad (1.18)$$

The Boltzmann constant $k_b = 3.30 \times 10^{-24}$ cal K^{-1}, or 1.38×10^{-23} J K^{-1}, is the gas constant for one

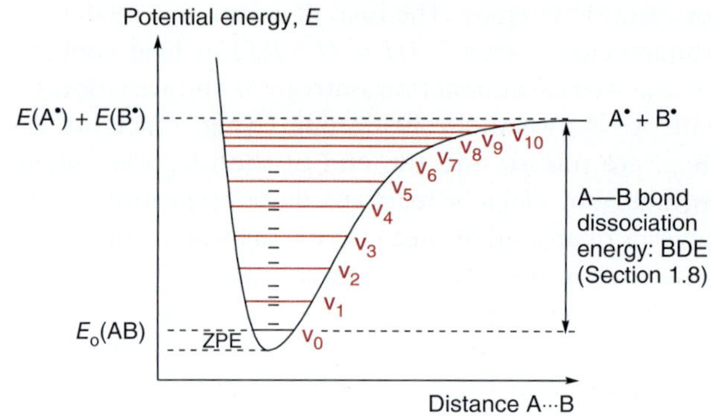

Figure 1.3 Representation of the Morse potential for a diatomic molecule A—B in its electronic ground state. The red full horizontal lines represent the vibrational energy levels (as given by infrared spectroscopy, or calculated by quantum mechanics; the energy difference between the vibrational levels $\Delta E = h\nu$ decreases on increasing E (nonharmonic oscillator). The black horizontal lines represent the rotational levels (as given by microwave spectroscopy or by quantum mechanical calculations, the energy difference between the rotational levels increases on increasing energy E). The translational levels are not shown; they are separated by very small energy differences. E_o(AB) = energy of molecule A–B at 0 K; ZPE = zero-point energy (or quantum vacuum zero-point energy) = $h\nu/2$ with ν = the vibrational frequency of oscillator A—B and h = Planck's constant; $E(A^\bullet)$ and $E(B^\bullet)$ energies of atoms $A^\bullet$ and $B^\bullet$. Similar Morse potentials can be represented for doubly bonded diatomic molecules A=B and triply bonded diatomic molecules A≡B.

molecule, i.e. $k_b = R/L$, where L = the Avogadro constant (also named Avogadro's number and also noted as N_A), the number of molecules in 1 mol = 6.02×10^{23} mol^{-1}. If there are several energy levels of the same energy, the proportion N_i/N_o becomes:

$$N_i/N_o = (g_i/g_o)e^{-E_i/k_b T} \quad (1.19)$$

where g_i and g_o are the statistical factors enumerating the number of identical microstates available, for energy levels E_i and E_o, respectively. If N is the total number of molecules of the system under investigation, then:

$$N = \Sigma N_i = \frac{N_o}{g_o}\Sigma g_i e^{-E_i/k_b T} \quad (1.20)$$

The factor N_o/g_o has been taken out of the summation sign as it is a constant. The quantity defined by the sum is the canonical partition function or partition function of the molecule:

$$Z = \Sigma g_i e^{-E_i/k_b T} \quad (1.21)$$

Z is the number of molecules in all microstates i: this function shows how the molecules "partition" among the various available microstates of energy E_i. One assumes that the energies of the molecules are independent of each other, which is the case for an ideal gas. From the partition function Z, the thermodynamic parameters U, H, S, and G of the macroscopic system can be calculated. For most chemical systems, $U°$, the lowest internal energy, is the sum of electronic (E_{el}) and nuclear energies (E_{nucl}) at $T = 0$ K for all the molecules of the system. Generally, there are very large differences between the energies of different nuclear and electronic quantum states, so that the accessible energy levels E_i of microstates i for a molecule correspond to quantized translation (E_{trans}), rotation (E_{rot}), and vibration (E_{vib}) energies, all for a single electronic state of energy.

To determine the internal energy change $\Delta U^T = \Sigma N_i E_i$ from 0 K to some finite temperature, T, the partition function can be used to obtain Eq. (1.22).

$$\Delta U^T = \left(\frac{N_o}{g_o}\right) \Sigma g_i E_i e^{-E_i/k_b T} \tag{1.22}$$

Differentiation of the partition function (Eq. (1.21)) with respect to temperature, at a constant volume, followed by rearrangement of the resulting expression yields Eq. (1.23) for one mole of ideal gas:

$$\Delta U^T = RT^2 \left(\frac{\delta \ln Z}{\delta T}\right)_V \tag{1.23}$$

The derivative of this Eq. (1.23) with respect to T, at a constant volume V, is the heat capacity of an ideal gas:

$$C_v = \left(\frac{\delta U}{\delta T}\right)_V = \frac{\delta}{\delta T}\left[RT^2\left(\frac{\delta \ln Z}{\delta T}\right)\right]_V \tag{1.24}$$

The entropy varies with temperature according to Eq. (1.25):

$$S^T - S^{\circ} = \Delta S = \int_0^T \frac{C_v}{T}\,\mathrm{d}T \tag{1.25}$$

Therefore:

$$S^T - S^{\circ} = RT\left(\frac{\delta \ln Z}{\delta T}\right)_V + R\ln Z - R\ln Z_o \tag{1.26}$$

At $T = 0$ K, all N molecules occupy microstates of energy level E_o. The partition function $Z_o = g_o$. For an ideal gas, $S° = k_b \cdot \ln[(g_o)^N/N!] = R \cdot \ln(g_o) - k_b \cdot \ln(N!)$ (applying Boltzmann–Planck equation: $S^T = k_b \cdot \ln \Omega$, with Ω the number of microstates available; for N distinguishable molecules, Ω would be $(g_o)^N$, but as the molecules in a gas are not distinguishable, this probability must be divided by N!. At a higher temperature, the entropy S^T of one mole of an ideal gas is

$$S^T = RT\left(\frac{\delta \ln Z}{\delta T}\right)_V + R\ln Z - k_b \ln N! \tag{1.27}$$

(Note the entropy $S°$ of a perfectly ordered crystal at 0 K is 0 eu, which is defined below.)

The internal energy ΔU^T can be calculated from relationship (1.23), the C_V from Eq. (1.24), and the entropy S^T from Eq. (1.27). Quantum mechanical calculations give estimates of the partition functions of isolated molecules in the gas phase; the accuracy can be very high when state-of-the-art quantum mechanical methods are used. The relationships between computed properties of an ideal gas molecule and the partition function are described below.

1.4.1 Contributions from translation energy levels

For translational energy levels, the partition function is given by:

$$Z_{trans} = \frac{(2\pi m k_b T)^{3/2}}{Lh^3} V \tag{1.28}$$

where m = mass of the molecule and h = Planck's constant (=6.626 068 96 × 10^{-34} J s). Combining Eqs. (1.23) and (1.24) gives:

$$\Delta U^T{}_{trans} = 1.5RT \text{ and } C_v = 1.5R$$

The translational entropy at temperature T ($S = S^T$ here below) becomes (using the Sterling approximation for large numbers: $\ln N! = N \cdot \ln N - N$):

$$S_{trans} = R \cdot \left\{\frac{(2\pi m k_b T)^{3/2}}{Lh^3} V + 5/2\right\} \tag{1.29}$$

where L is the Avogadro constant.

Using the mass of one molecule $m = M_r$(molecular mass)/L, volume $V = RT/p$ (ideal gas), and the values given for the constants h, R, L at pressure $p = 1$ atm, and using molecular mass in g units:

$$S_{trans} = 2.98 \cdot \ln M_r(\text{g}) + 4.97 \cdot \ln T - 2.31 \text{ eu} \tag{1.30}$$

or, converting to base 10 logs:

$$S_{trans} = 6.86 \cdot \log M_r(\text{g}) + 11.44 \cdot \log T - 2.31 \text{ eu} \tag{1.31}$$

S_{trans} is the entropy of a gas made of monoatomics (e.g. He, Ne, and Ar). Monoatomics have neither rotational energy levels nor vibrational levels, so that the calculation of entropy requires only the mass and temperature.

1.4.2 Contributions from rotational energy levels

A diatomic molecule can be assumed to be a rigid molecule that does not change its interatomic distance (bond length) with its frequency of rotation.

The partition function for the rotational energy levels in this rigid rotor is given by:

$$Z_{rot} = \frac{8\pi^2 I k_b T}{\sigma h^2} \quad (1.32)$$

where I = the inertia moment of the molecule. The moment of inertia $I = m_i r_i^2$, where m_i = mass of the atom i at distance r_i from the rotation axis. The symbol σ = symmetry number of the molecule, that is $\sigma = 1$, for diatomic molecules made of two different atoms or isotopes, or $\sigma = 2$ for symmetrical molecules made of two identical atoms or isotopes.

Combining Eqs. (1.27) and (1.32), the rotational entropy at temperature T for a rigid diatomic molecule becomes:

$$S_{rot} = 1.987 \cdot (\ln I + \ln T - \ln\sigma + 89.4)\ \text{eu} \quad (1.33)$$

or, in $\log_{10}$ units,

$$S_{rot} = 4.576 \cdot (\log I + \log T - \log\sigma + 32.82)\ \text{eu}$$

Values of I can be determined by rotational spectroscopy or by quantum mechanical calculations.

For a nonlinear polyatomic molecule, the partition function for its rotational energy levels is more complicated, as there are three moments of inertia.

$$Z_{rot} = \frac{(8\pi^2 k_b T)^{3/2}}{\sigma h^3}(\pi ABC)^{1/2} \quad (1.34)$$

The rotational entropy at temperature T becomes:

$$S_{rot} = 1.987 \cdot (0.5 \cdot \ln ABC + 1.5 \cdot \ln T - \ln\sigma + 134.68)\ \text{eu} \quad (1.35)$$

or, converting in $\log_{10}$ units,

$$S_{rot} = 2.288 \cdot \log ABC + 6.864 \cdot \log T - 4.576 \cdot \log\sigma + 267.74\ \text{eu}$$

A, B, and C are the three moments of inertia of the molecule in cgs units, and σ is the symmetry number. σ is the number of times the molecule is superposed upon itself rotating about each rotation axis of symmetry (e.g. $\sigma = 3\times2 = 6$ for cyclohexane in a chair conformation, $\sigma = 3$ for $CHCl_3$, $\sigma = 2$ for CH_2Cl_2, $\sigma = 6\times2\times2 = 24$ for benzene, and $\sigma = 2$ for toluene). According to Eq. (1.35), the entropy is reduced as the symmetry of the molecule increases. If two chemical systems with the same heat of reaction can evolve toward two different types of products, the lower symmetry products will be preferred, as $\Delta_r S$ (higher symmetry) < $\Delta_r S$ (lower symmetry). Nature dislikes symmetry, at least where entropy is concerned.

1.4.3 Contributions from vibrational energy levels

For a real diatomic molecule, vibrations are also present and make a contribution to entropy. For an idealized diatomic system vibrating as a perfectly elastic harmonic oscillator, the partition function for the vibrational energy levels is:

$$Z_{vib} = (1 - e^{-x})^{-1}, \quad \text{with} \quad x = hc\omega/k_b T = h\nu/k_b T \quad (1.36)$$

where $x = hc\omega/k_b T = 1.439{\cdot}\omega/T$, with c = light velocity in a vacuum and ω (in cm^{-1} units) is the vibrational frequency of the molecule determined by infrared (IR) absorption spectroscopy or by quantum mechanics calculations. Alternatively, the equation is written in terms of the frequency, ν, of vibration in units of s^{-1}. Combining Eqs. (1.27) and (1.36), the vibrational entropy of a harmonic diatomic molecule is:

$$S_{vib} = 1.987 \cdot x/(e^x - 1) - 4.576 \cdot \log(1 - e^{-x})\ \text{eu} \quad (1.37)$$

For small and rigid molecules of molecular mass < 500, the relative importance of the partition functions is $Z_{trans} > Z_{rot} > Z_{vib}$ because the energy differences between the translational levels are much smaller than those between rotational levels and because the energy differences between rotational levels are smaller than those between vibrational levels. At any given temperature T, more excited translational and rotational states are occupied than higher energy vibrational states. For small and rigid molecules of molecular mass < 500, Hooke's law is the spring equation $F = -kx$. It relates the force F exerted by a spring to the distance x it is stretched by a spring constant k. The negative sign indicates that F is a "restoring force" as it tends to restore the system to equilibrium. The potential energy (PE) stored in the spring is given by $\text{PE} = 0.5kx^2$. If a mass m is attached to the end of the spring, the system might be seen as a harmonic oscillator that vibrates with an angular frequency $\omega = \sqrt{k}/\sqrt{m}$, or with a natural frequency $\nu = \omega/2\pi$. The solution to the Schrödinger equation for such system gives the eigenvalues $E_i = (i + 1/2){\cdot}h\nu$, where $h\nu$ is the energy difference between two vibrational levels, and ν is the frequency of the vibration. The larger the spring constant k, the "stiffer the spring," the larger the vibrational frequency and the greater the energy difference between two vibrational levels. Molecules that can be deformed easily have small force constants for vibrational deformation. When the spring constant k is small, the energy difference between the corresponding vibrational is relatively small, and this mode of deformation can contribute significantly to the partition function Z_{vib}, and to the entropy of the molecule.

The entropy of an ideal gas can be measured "macroscopically" from the relationship:

$$\Delta S = S_2 - S_1 = C_v \int_{T_1}^{T_2} \frac{dT}{T} + R \int_{V_1}^{V_2} \frac{dV}{V}$$

$$= C_v \ln \frac{T_2}{T_1} + R \ln \frac{V_2}{V_1}$$

$$\Delta S = \int_{T_1}^{T_2} C_p \frac{dT}{T} = \int_{T_1}^{T_2} C_p \, d(\ln T) \qquad (1.38)$$

1.4.4 Entropy of reaction depends above all on the change of the number of molecules between products and reactants

For reactions occurring in the gas phase or in ideal solutions and for rigid reactants equilibrating with rigid products (Z_{rot} and Z_{vib} contributions to the entropy are roughly identical for products and reactants), $\Delta_r S^T \cong 0$ when the number of molecules does not change between products and reactants. When this number decreases as in addition reactions, $\Delta_r S^T \ll 0$. In the case of fragmentations, $\Delta_r S^T \gg 0$ (Section 2.6). For instance, the isomerization of (*Z*)-but-2-ene into (*E*)-but-2-ene, a reaction that does not change the number of molecules between the product and the reactant, and using experimental standard entropies for these compounds (Table 1.A.2), one finds $\Delta_r S° = -1.2 \pm 2$ eu at 298 K. As the reactant and the product maintain the same type of σ(C—H), σ(C—C), and π(C=C) bonds and the same number of symmetry ($\sigma = 2$, C_2 axis of symmetry, see Eq. (1.34)), the partition functions Z_{rot} and Z_{vib} are expected to be nearly the same for both the reactant and the product.

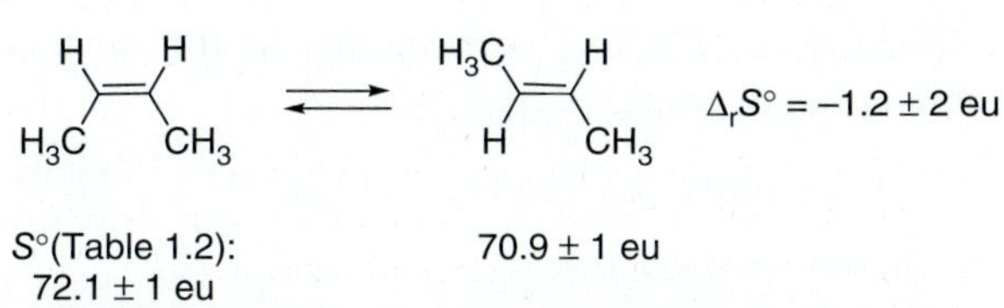

In the case of Diels–Alder reaction that condenses a diene with an alkene (dienophile) into a cyclohexene derivative (Section 5.3.8), a negative entropy of reaction is expected. In the case of prototype reaction, involving conversion butadiene with ethylene into cyclohexene, experimental standard entropies (Table 1.A.2) permit to calculate $\Delta_r S° = -44.8 \pm 3$ eu for this reaction. If one considers only the contributions from the translation degrees of freedom (Z_{trans}), Eq. (1.31) gives $\Delta_r S°_{trans} = -34.67$ eu. This confirms that Z_{rot} and Z_{vib} contributions to the entropy (c. −10 eu) of this condensation are less important than the Z_{trans} contribution (c. −35 eu).

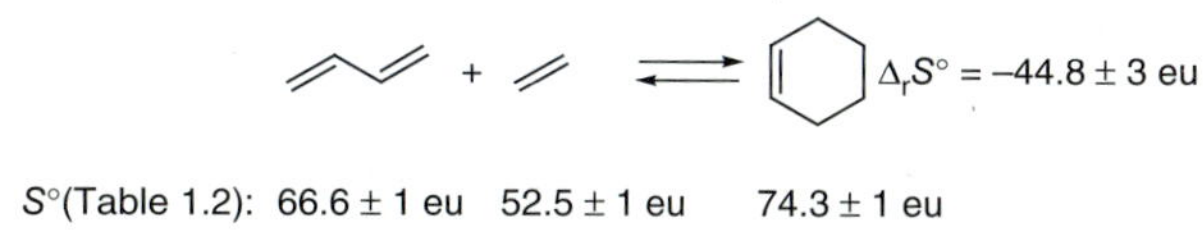

1.4.5 Additions are favored thermodynamically on cooling, fragmentations on heating

As condensations have negative $\Delta_r S^T$ values, the $-T\Delta_r S^T$ term in Eq. (1.15) ($\Delta_r G^T = \Delta_r H^T - T\Delta_r S^T$) is positive. For exergonic reactions ($\Delta_r G^T < 0, K > 1$), their $\Delta_r H^T$ must be smaller than $T\Delta_r S^T$. Exothermicity is "the glue" that permits the reactants to remain attached in the product, as long as the temperature in not too high. On lowering the reaction temperature, additions have higher equilibrium constants, K, because the $-T\Delta_r S^T$ term becomes less positive. Fragmentations feature a positive $\Delta_r S^T$, yielding a negative $-T\Delta_r S^T$ term favored thermodynamically on heating, and for reactions in the gas phase, on lowering the pressure (Le Châtelier's principle, for examples of reactions of preparative interest, see Section 2.11).

Most addition reactions are exothermic ($\Delta_r H^T < 0$); thus, care must be taken when running them in the laboratory or in a factory. Reactants should never be mixed at once because of the risk of explosion. The danger is real if the heat generated by the reaction cannot be extracted efficiently. Safe practice is to add slowly one of the reactants into the stirred mixture of the other reactants + catalyst (if any). The addition must be stopped if the temperature increases. A simple way to avoid overheating is to carry out the reaction in a boiling solvent under reflux, adapting the addition rate of the reactant with the rate of boiling. Unsaturated compounds such as alkenes, alkynes, dienes, etc., can undergo polymerizations under storage. Reactions involving transformation of a π(C=C) bond into a σ(C—C) bond are typically exothermic by −20 to −24 kcal mol^{-1} (see reaction (1.48)). Polymerization of unsaturated compounds is induced by initiators such as oxy and peroxy radicals resulting from exposure to air (Section 6.9.1). In order to avoid "accidental" polymerization (that

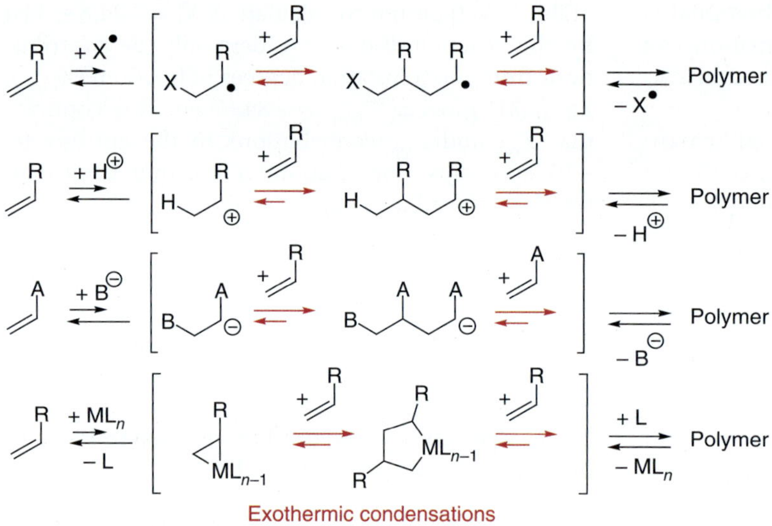

Scheme 1.1 Possible mechanisms for the polymerization of alkenes.

can lead to sudden explosion), one "stabilizes" the unsaturated compounds by radical scavenging agents or one keeps them below room temperature under inert atmosphere (vacuum, Ar, and N_2). Polymerization (Scheme 1.1) can also be induced by protic or Lewis acids, by bases, or by metallic complexes (Section 7.7) or by thermal self-initiation via the formation of 1,4-diradical ↔ zwitterion intermediates (Section 5.5). Storage and shipping of unsaturated compounds such as acetylene (HC≡CH), propyne (CH_3C≡CH), butadiene (CH_2=CH—CH=CH_2), styrene (PhCH=CH_2), acrolein (CH_2=CH—CHO), acrylonitrile (CH_2=CH—CN), acrylic esters (CH_2=CH—COOR), methacrylates (CH_2=CMe—COOR), methyl vinyl ketone (CH_2=CH—COMe), etc., all important industrial chemicals, are risky operations. In this textbook, we teach how one can evaluate the heat of any organic reactions and predict their rates under given conditions.

Problem 1.1 A hydrocarbon, RH, can be reversibly isomerized into two isomeric compounds $\mathbf{P_1}$ and $\mathbf{P_2}$ with the same heat of reaction. Both have C_1 symmetry. $\mathbf{P_1}$ is a rigid compound and $\mathbf{P_2}$ is a flexible one adopting several conformations of similar enthalpies. Which product will be preferred at equilibrium?

Problem 1.2 Define the symmetry numbers, σ, of methane, ethane, propane, cyclopropane, cyclobutane, cyclohexanone, ferrocene, bicyclo[2.2.1]hepta-2,5-diene (norbornadiene), 1,4-difluorobenzene, *meso*-tartaric acid, and (*R,R*)-tartaric acid (see Figure 1.24 for structure of the two latter compounds).

Problem 1.3 What is the Gibbs energy of the racemization of an enantiomerically pure α-amino acid at 25 °C?

1.5 Standard heats of formation

The standard heat of formation, $\Delta_f H°$, of a pure compound is the change in enthalpy for the conversion of the elements into the chosen compound in the standard state, i.e. 1 mol, at 298.15 K, under 1 atm. By convention, the standard heats of formation of the pure elements are set equal to zero. Thus, $\Delta_f H°$(graphite, solid) = 0, $\Delta_f H°(Cl_2$, gas) = 0, $\Delta_f H°(H_2$, gas) = 0, $\Delta_f H°(O_2$, gas) = 0, etc.

The standard heat of formation of H_2O corresponds to the heat of combustion of H_2:

$$H_2(\text{gas}) + \tfrac{1}{2}O_2(\text{gas}) \rightarrow H_2O(\text{liquid}) \qquad (1.39)$$

For this reaction, the standard heat of reaction can be computed from the standard heats of formation:

$$\begin{aligned}\Delta_r H^\circ(1.39) &= \Delta_f H^\circ(H_2O,\text{liquid})\\ &\quad -\Delta_f H^\circ(H_2) - \tfrac{1}{2}\Delta_f H^\circ(O_2)\\ &= -68.3\ \text{kcal mol}^{-1}\end{aligned}$$

Similarly, $\Delta_f H°(CO_2)$ corresponds to the heat of combustion of graphite, $\Delta_c H°$(C):

$$C(\text{graphite}) + O_2(\text{gas}) \rightarrow CO_2(\text{gas}) \qquad (1.40)$$

$$\begin{aligned}\Delta_r H^\circ(1.40) &= \Delta_f H^\circ(CO_2) - \Delta_f H^\circ(\text{graphite})\\ &\quad -\Delta_f H^\circ(O_2)\\ &= \Delta_f H^\circ(CO_2) = \Delta_c H^\circ(C)\\ &= -94.05\ \text{kcal mol}^{-1}\end{aligned}$$

At 298.15 K and under 1 atm, water and carbon dioxide are more stable than the elements from which they are composed. By contrast, HI in the gas phase has a positive heat of formation, $\Delta_f H°$(HI, gas)) = 6.2 kcal mol^{-1}, so this compound is unstable thermodynamically ($\Delta_r G°(1.41) \cong \Delta_r H°(1.41) > 0$ as $\Delta_r S° \sim 0$, two molecules in the reactants and two molecules in the products). This compound does not decompose instantaneously, as the activation barrier ($\Delta^{\ddagger}G$, see Section 3.3) for its decomposition is relatively high. HI is a metastable compound in the gas phase, whereas in water, HI ionizes to give stable ion pair H_3O^+/I^- that is strongly solvated.

$$H_2(\text{gas}) + I_2(\text{solid}) \rightleftarrows 2HI(\text{gas}) \tag{1.41}$$

$$2 \cdot \Delta_f H^{\circ}(\text{HI}) = 12.4 \text{ kcal mol}^{-1} \quad (25\,^{\circ}\text{C, 1 atm})$$

The heat for reaction (1.41) is 12.4 kcal mol^{-1}, so the heat of formation of HI is half of that, or 6.2 kcal mol^{-1}.

The heats of formation of most organic and organometallic compounds cannot be measured directly by calorimetry, which measures $\Delta_r H^T$, or by measuring the equilibrium constants K of the formation reactions at different temperatures (Van't Hoff plot). It is also very rare that the rate constant for the conversion of the elements into the pure substance of interest, or that of the reverse reaction, the decomposition of the substance into its pure elements, can be measured directly. Instead, thermodynamic cycles (Born–Haber cycles) are used to determine the heats of formation (see Figure 1.4 and Eq. (1.42)) for the determination of the standard heats of formation of the hydrocarbons C_nH_m. The heat of combustion of n moles of graphite to produce n moles of CO_2 plus the heat of combustion of $m/2$ moles of H_2 to produce $m/2$ moles of water can be compared to the heat of combustion of hydrocarbon C_nH_m to give the same amount of CO_2 and H_2O (Figure 1.4).

$$\Delta_f H^{\circ}(C_nH_m) = n \cdot \Delta_c H^{\circ}(C) + m/2 \cdot \Delta_c H^{\circ}(H_2) - \Delta_c H^{\circ}(C_nH_m) \tag{1.42}$$

In some cases, reactions other than combustions can be used in Born–Haber cycles. Calorimetry can be applied, for instance, to hydrogenations of unsaturated compounds or to catalyzed isomerizations. A major difficulty encountered in calorimetry is the formation of secondary products (isomers, polymers, and products of fragmentation) in addition to the desired products of a given reaction under investigation. If the reaction is not perfectly clean (when it competes with other reactions), deviations of the measured heats from the quantity of evaluation become large. This problem is less serious when applying the Van't Hoff method, i.e. measuring equilibrium constants at various temperatures. Despite this, very accurate heats of formation are now available for a

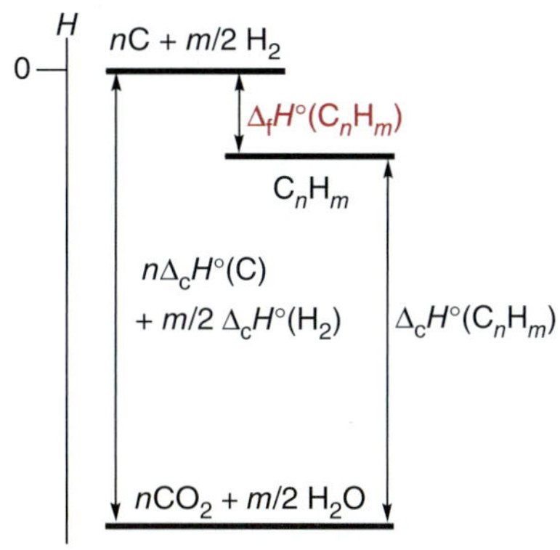

Figure 1.4 A thermodynamic cycle from which the heat of formation of a hydrocarbon can be determined by combustion calorimetry.

large number of organic and organometallic compounds. With high-pressure mass spectrometry (MS) and ion cyclotron resonance, the thermochemistry of ionized species as well as of transient neutral species such as radicals, diradicals, and carbenes is now possible (Sections 1.10–1.12). Today, accurate heats of formation for almost any kinds of chemical species of relatively small molecular weight ($M_r < 500$) can be reliably determined. To estimate the standard heat of reaction, $\Delta_r H°$, of a given reaction (1.1) from standard heats of formation, Hess's law (Eq. (1.43)) can be used (reactants **A**, **B**, …; products **P**, **Q**, …):

$$\Delta_r H^{\circ}(1.1) = \pi \Delta_f H^{\circ}(\mathbf{P}) + \theta \Delta_f H^{\circ}(\mathbf{Q}) + \cdots - \alpha \Delta_f H^{\circ}(\mathbf{A}) - \beta \Delta_f H^{\circ}(\mathbf{B}) - \cdots \tag{1.43}$$

If a value for a given compound is not available from the NIST Webbook of Chemistry or from another source, the Benson's group additivity method proposed in 1958 [16, 17] (Section 2.2) can be used instead to estimate these quantities [18]. Other additivity methods for the calculations of thermochemical parameters have been proposed such as Laidler's bond enthalpy method presented in 1956 [19]. The use of high-level accuracy quantum mechanical methods has also become increasingly important, as well.

1.6 What do standard heats of formation tell us about chemical bonding and ground-state properties of organic compounds?

Table 1.A.1 gives a compilation of the standard heats of formation for a selected number of inorganic compounds in the gas phase under standard conditions. Tables 1.A.2–1.A.4 give standard heats of formation and standard entropies of selected organic compounds in the gas phase (for more values, see [6, 7, 20–24]). These values can be used for calculating the heats and entropies of reactions or equilibria in ideal solutions. This simplification leads to satisfactory predictions for the thermodynamic parameters of a large number of organic reactions involving nonpolar reagents in nonpolar solvents. Some illustrative examples of the use of these methods are given in the next chapters.

Problem 1.4 What products do you expect to be formed combining $HO^{\bullet}$ with organic compounds? What happens to NO in the air and to SO_2 in the air?

Problem 1.5 Propose a reaction for diimide (diazene: HN=NH) + cyclohexene and calculate its heat of reaction.

1.6.1 Effect of electronegativity on bond strength

The reactions of hydrogen (dihydrogen: H_2) with fluorine (F_2), chlorine (Cl_2), bromine (Br_2), and iodine (I_2) generate the corresponding hydrogen halides HF, HCl, HBr, and HI (called hydrohalic acids when dissolved in water). Although HI has a positive gas-phase standard heat of formation, the other hydrogen halides have negative standard heats of formation in the gas phase (equilibria (1.44)). As already mentioned above, HI in the gas phase is a metastable compound with respect to its decomposition into its elements, whereas HF, HCl, and HBr are stable with respect to their decomposition into their respective elements. On heating in the gas phase, HI will equilibrate with H_2 and I_2, whereas the same type of decomposition will not occur with the other hydrogen halides. The entropies of reaction, $\Delta_r S^T$, are estimated to be small for all of these reactions, as the number of molecules does not change between products and reactants.

$$\tfrac{1}{2}\,H_2 + \tfrac{1}{2}\,X_2 \underset{}{\overset{\text{Gas}}{\rightleftharpoons}} HX \qquad (1.44)$$

X =	F	Cl	Br	I
$\Delta_r H°(1.44) = \Delta_f H°(HX)$:	−65.1	−22.1	−8.6	6.2 kcal mol^{-1}

The relatively large variations in the $\Delta_r H°$ (1.44) values are the result of the difference in electronegativity between atoms H and X (Table 1.A.5). HF is more stable than HI because it combines two different atoms with the highest possible (Pauling) electronegativity difference and gives the shortest and strongest bond as a result [25]. Similar observations can be made about the standard heats of formation of the other derivatives of hydrogen of Table 1.A.1. For instance, water is more stable than H_2S, and NH_3 is more stable than PH_3, for the same reasons.

Fluorination, chlorination, and bromination of (1.45) of propane ($CH_3CH_2CH_3$) into the corresponding *n*-propyl halides ($CH_3CH_2CH_2$—X: *n*-Pr—X) are all exothermic. However, direct iodination of propane is endothermic. In fact, the fluorination reaction is an explosive transformation because of the very high exothermicity of −108.4 kcal mol^{-1} (as a comparison, the standard heat of combustion of hydrogen [H_2] amounts to −57.8 kcal mol^{-1} only). These results illustrate the role of electronegativity on bond strength and therefore on the stabilities of organic compounds. In both examples described above, the polarities (electronegativity difference) of the bonds (C—F or H—F) formed in the products are much larger than the polarities of the bonds (F—F or C—H) cleaved in the reactants.

$$\text{Pr–H}_2 + X_2 \overset{\text{Gas}}{\rightleftharpoons} n\text{-Pr–X} + HX \qquad (1.45)$$

X =	F	Cl	Br	I
$\Delta_r H°(1.45)$:	−108.4	−28.6	−4.8	21.7 kcal mol^{-1}

Similar observations are made for the direct monohalogenations (1.46) of benzene (C_6H_6 = Ph—H)

$$\text{Ph–H} + X_2 \overset{\text{Gas}}{\rightleftharpoons} \text{Ph–X} + HX \qquad (1.46)$$

X =	F	Cl	Br	I
$\Delta_r H°(1.46)$:	−112.5	−29.4	−3.1	26.0 kcal mol^{-1}

The effect of the electronegativity differences between the C—X bonds and the H—X bonds also explains the different standard heats of acid-catalyzed additions (1.47) to propene, giving isopropyl derivatives *i*-Pr—X:

$$\text{Propene} + HX \overset{\text{Gas}}{\rightleftharpoons} i\text{-Pr–X} \qquad (1.47)$$

X =	F	Cl	Br	I	CN
$\Delta_r H°(1.47)$:	−10.0	−17.4	−20.1	−20.7	−31.8 kcal mol^{-1}

The exothermicity of the additions (1.48) of water, ammonia, and hydrogen sulfide to propene to give isopropanol (propan-2-ol), isopropylamine (2-aminopropane), and isopropylmercaptan (propane-2-thiol), respectively, is the highest for Y = SH and the lowest for Y = OH because the "preference" for hydrogen (more electropositive than carbon) to be bonded to an oxygen atom (more electronegative) is greater than hydrogen's "preference" to be bonded to sulfur. The exothermicities of hydrocyanation (reaction (1.47), X = CN), of hydrocarbation (reaction (1.48), Y = CH_2—CH_3: hydroethylation; Y = CH=CH_2: hydrovinylation; Y = Ph: hydrophenylation; Y = C≡CH: hydroethynylation; Y = CHO: hydroformylation), and of hydrogenation (Y = H) of alkenes (more precisely, dihydrogenation, as the reaction involves the addition of two hydrogen atoms) are higher than for the heteropolar additions (reaction (1.47), for X = F, Cl, Br, I, and reaction (1.48) for Y = OH, NH_2, SH):

$$\text{Propene} + \text{HY} \xrightleftharpoons{\text{Gas}} i\text{-Pr–Y} \qquad (1.48)$$

Y =	OH	NH_2	SH	CH_2CH_3	CH=CH_3
$\Delta_rH°$ (1.48):	−12.2	−13.8	−18.1	−21.5	−23.8

Y =	PH	C≡CH	CHO	H
$\Delta_rH°$ (1.48):	−23.5	−26.7	−28.7	−29.8 kcal mol^{-1}

Under thermodynamic control, substitutions of alkyl halides (1.49) and of alkenyl halides (1.50) by other halides generally favor the formation of HF:

$$\text{Pr–F} + \text{HBr} \xrightleftharpoons{\text{Gas}} \text{Pr–Br} + \text{HF} \qquad (1.49)$$

$\Delta_rH°$(1.49): −9.0 kcal mol^{-1}

$$\text{vinyl–F} + \text{HBr} \xrightleftharpoons{\text{Gas}} \text{vinyl–Br} + \text{HF} \qquad (1.50)$$

$\Delta_rH°$(1.50): −4.4 kcal mol^{-1}

Problem 1.6 Among the amino acids serine and cysteine, which of these give stable adducts with cyclohex-2-enone at 37 °C when they are part of a protein?

1.6.2 Effects of electronegativity and of hyperconjugation

In contrast with equilibria (1.49) and (1.50) that favor the formation of HF, equilibria (1.51) that exchange the fluoride of acetyl fluoride by chloride, bromide, or iodide with the corresponding hydrogen halide HX disfavor the formation of HF, meaning that fluorine prefers to be bonded to an acyl carbon rather than a hydrogen atom. In contrast, equilibria (1.52) of acetyl chloride with the corresponding bromide and iodide are nearly thermoneutral.

$$\text{AcF} + \text{HX} \xrightleftharpoons{\text{Gas}} \text{AcX} + \text{HF} \qquad (1.51)$$

X =	Cl	Br	I
$\Delta_rH°$(1.51):	4.6	4.0	4.0 kcal mol^{-1}

$$\text{AcCl} + \text{HX} \xrightleftharpoons{\text{Gas}} \text{AcX} + \text{HCl} \qquad (1.52)$$

X =	Br	I
$\Delta_rH°$(1.52):	−0.6	−0.6 kcal mol^{-1}

Why does fluorine prefer the right side of the equilibrium shown in equilibrium (1.51)? Donation of nonbonding electrons of the oxygen atom of the carbonyl group stabilizes the polar form of the acetyl–halide bond. This hyperconjugation effect (n(C=O:)/σ interaction) involves the interaction of the nonbonding, or lone pair, orbitals n(CO:) of the carbonyl group and the antibonding, empty orbital σ*(C—F) of the C—F bond (molecular orbital theory, Sections 4.5.15 and 4.8.1). This interaction is not possible in alkyl, alkenyl, and hydrogen halides, which do not possess lone pair electrons. Of all acyl halides, this hyperconjugative interaction is strongest in acyl fluorides where the difference in electronegativity between carbon and fluorine is larger than in other acyl halides. Thus, because of the large electronegativity difference between F and C, σ*(C—F) is the best sigma acceptor of all C—X bonds. Furthermore, the conjugation n(X:) → π*(C=O) (donation from the nonbonded electron pairs of X: to the carbonyl double bond), which stabilizes the reactant, is the weakest for X = F and the strongest for amino groups (Figure 1.6). The infrared carbonyl stretching frequencies of acyl derivatives ($\nu_{C=O}$) increase with the C=O bond strength as shown in Figure 1.5.

Quantum mechanical calculations give an indication of the differences between halogen atoms attached to alkyl and acyl groups. Quantum calculations predict a C—F bond length of 1.383 Å for methyl fluoride and a C—Cl bond length of 1.804 Å for methyl chloride [26] (experimentally, these are 1.385 ± 0.004 and 1.66 ± 0.05 Å [27–30], respectively, Å = 10^{-10} m). In the cases of formyl fluoride and formyl chloride, the C—F and C—Cl bond lengths are calculated to be 1.345 and 1.797 Å, respectively [31, 32]. These represent lengthening of the bond lengths of 0.04 and 0.01 Å, respectively. The carbonyl bond length is predicted to be shorter in formyl fluoride (1.186 Å) than in formyl chloride (1.200 Å), consistently with the interpretation given above (Figure 1.5) [33].

Problem 1.7 Explain the difference in C=O bond stretching frequencies between ethyl (*Z*)-3-fluorocinnamate (1736 cm^{-1}) and ethyl cinnamate ((*E*)-PhCH=CHCOOEt: 1715 cm^{-1}) [34].

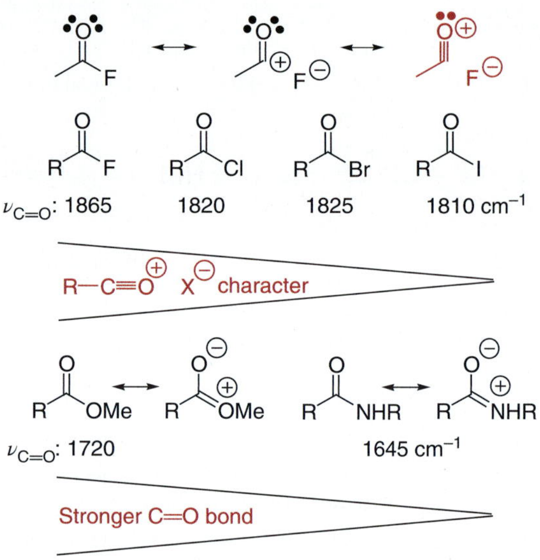

Figure 1.5 Hyperconjugation in acetyl halides (donation from the carbonyl group n(CO:) nonbonded electron-pairs to the σ(C—X) bond) competes with the n(X:)/π(C=O) conjugation. This competition also exists in carboxylic esters and carboxamides.

1.6.3 π-Conjugation and hyperconjugation in carboxylic functions

Esterification equilibrium (1.53) and amidification equilibrium (1.54) are exothermic. In contrast, the formation of ethyl thioacetate from ethanethiol and acetic acid (equilibrium (1.55)) is endothermic by 2.1 kcal mol^{-1}. Anhydride formation (equilibrium (1.56)) is even more endothermic (c. 12 kcal mol^{-1}).

$$\text{AcOH} + \text{ROH} \xrightleftharpoons{\text{Gas}} \text{AcOR} + \text{H}_2\text{O} \quad (1.53)$$

R =	Me	*i*-Pr
$\Delta_r H°$(1.53):	−5.0	−5.3 kcal mol^{-1}

$$\text{AcOH} + \text{RR'NH} \xrightleftharpoons{\text{Gas}} \text{AcNR'R} + \text{H}_2\text{O} \quad (1.54)$$

RR′NH =	Me_2NH	$PhNH_2$
$\Delta_r H°$(1.54):	−4.4	−6.0 kcal mol^{-1}

$$\text{AcOH} + \text{EtSH} \xrightleftharpoons{\text{Gas}} \text{AcSEt} + \text{H}_2\text{O} \quad (1.55)$$

$\Delta_r H°$(1.55): 2.1 kcal mol^{-1}

$$2\ \text{AcOH} \xrightleftharpoons{\text{Gas}} \text{Ac}_2\text{O} + \text{H}_2\text{O} \quad (1.56)$$

$\Delta_r H°$(1.56): 12.2 kcal mol^{-1}

These data can be explained by invoking both electronegativity differences between the atom pairs that are exchanged in these reactions and by differential conjugation effects involving the nonbonding electron pair of the nucleophile (O of esters, N of amides, S of thioesters, and O of the carboxylic anhydride) and the carbonyl groups depicted in Figure 1.6. In a classical view, n/π conjugation is proposed to involve some electron transfer from the nucleophilic center Y: to the electrophilic carbonyl group, noted by n(Y:) → π^*(CO) or n(Y:)/π^*(CO) (Section 4.5.15). The charge and geometry analysis by Wiberg and coworkers (Section 2.7.6) show that the carbonyl C=O bond length and oxygen charges are about the same in an aldehyde and an amide, whereas an aldehyde can be represented by resonance structures **A** and **B** and an amide has an additional limiting structure **E**, which represents interactions between donor Y: and the carbonyl group.

The relative importance of resonance structure **E** depends on the ionization energy (IE(Y:) = $\Delta_f H°(\text{Y}^{\bullet+})$ − $\Delta_f H°$(Y:), Section 1.8) of the nucleophilic center Y: and the overlap of n(Y:) orbitals with the empty 2p orbital at the carbon center (theory of perturbation molecular orbitals, PMO theory, Section 4.4.2). The ionization energy of Y: is another expression of the electronegativity of center Y: (Table 1.A.5). The less Y: is electronegative, the lower its ionization energy, and the easier it can release electrons to the neighboring carbonyl group. In terms of molecular orbital theory (Section 4.5.15), this is expressed by the energy difference between the LUMO (lowest unoccupied molecular orbital) of the carbonyl group and the HOMO (highest occupied molecular orbital) of center Y:. In acetic anhydride (Ac_2O), the Y: center is an oxygen atom stabilized by the acyl group of the carboxylate moiety; the HOMO of AcO moiety is lying lower than that of the alkoxy group in the corresponding ester. The n(alkoxy) → π^*(C=O) interaction is more stabilizing than the n(acyloxy) → π^*(C=O) interaction, rendering esters more stable than the corresponding carboxylic anhydrides, as shown by the standard heats of equilibria (1.53) and (1.56).

Since the electronegativity decreases from oxygen to nitrogen, and then from nitrogen to sulfur (Table 1.A.5), this factor would cause the n(Y:) → π^*(C=O) stabilizing interaction to increase from esters to amides and then from amides to thioesters. The thermochemical data given for equilibria (1.53)–(1.55) are inconsistent with this hypothesis. The increased stabilization of esters and amides (equilibria (1.53) and (1.54)) can be attributed, in part, to the energy necessary to planarize the amine group that maximizes the n(Y:)/π_{CO} overlap. The

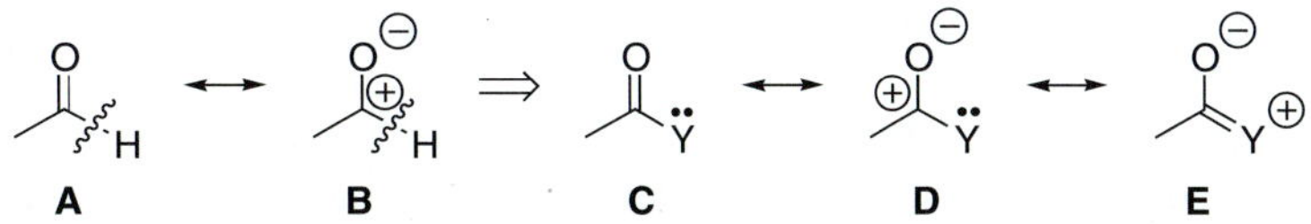

Figure 1.6 Classical limiting structures of aldehydes (**A**, **B**), esters, and amides (**C**, **D**, **E**).

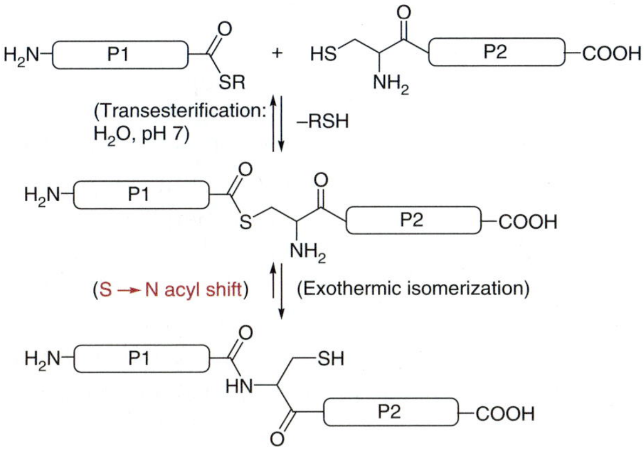

Scheme 1.2 Native chemical ligation: a tool for chemical protein synthesis.

lower stabilization of thioesters compared with esters and amides arises from the poorer overlap and mixing of the high-lying 3p sulfur orbital with the 2p orbital of the vicinal carbon center (see the shape of the 3p(S) orbital and compare it with that of a 2p(O) orbital: the 3p(S) orbital occupies a much larger space than the 2p(O) orbital; as a consequence, the C—S bond is longer than the C—O bond [Table 1.A.6]). The lower stabilities of thioesters compared with amides have been exploited in "native chemical ligations," transformations used, for example, to construct large peptides from two or more unprotected peptides (Scheme 1.2) [35–37].

The relative importance of n(Y:)/π conjugation as a function of the heteroatom will be discussed again in Section 2.7.6 when comparing the heats of hydrogenation of enol ethers, and enamines with the heats of hydrogenation of alkenes, and also in Section 2.7.8 when comparing the stabilizations by the aromaticity of furan, pyrrole (azole), thiophene, and phosphole (for molecular orbital theory applications, see Section 4.6).

Problem 1.8 Estimate the standard heat of esterification of methanol with acetic acid. Estimate the variation of entropy of this reaction at 298.15 K and calculate the equilibrium constant at the same temperature and under 1 atm in tetrahydrofuran (THF) solution. Is the equilibrium constant the same under the same conditions for the esterification of anthracene-2-carboxylic acid with 2-hydroxynaphthacene?

Problem 1.9 The Newman–Kwart rearrangement is a valuable synthetic technique for converting phenols to thiophenols via their *O*- and *S*-thiocarbamates [38–40]. Explain why the *S*-thiocarbamates are more stable than their isomeric *O*-thionocarbamates.

1.6.4 Degree of chain branching and Markovnikov's rule

The stability of alkanes increases with the their degree of chain branching [41, 42]. Electron correlation is largely responsible for this observation. Branched alkanes have greater number of attractive 1,3-alkyl/alkyl group interactions; there are three such stabilizing 1,3-"protobranching" dispositions in isobutane (2-methylpropane), but only two in *n*-butane. Neopentane (2,2-dimethylpropane) has six protobranches, but *n*-pentane has only three [43]. In the cases of functional systems such as alcohols, amines, thiols, and alkyl halides, secondary derivatives are more stable than their primary isomers. The same trend is found for the isomerization equilibria (1.58) and (1.59): tertiary systems are more stable than their secondary isomers.

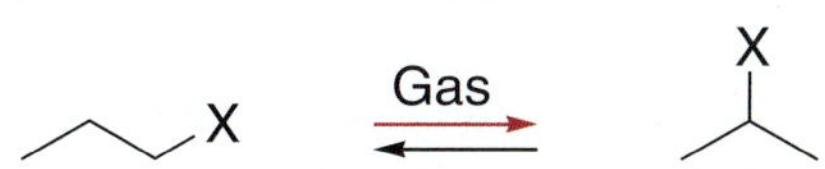

(1.57)

X = Me Et *n*-Pr OH SH NH_2 F Cl Br I

$\Delta_r H°$ (1.57): −2.0 −1.6 −1.8 −4.2 −4.0 −3.2 −1.8 −3.1 −3.0 −2.4 kcal mol^{-1}

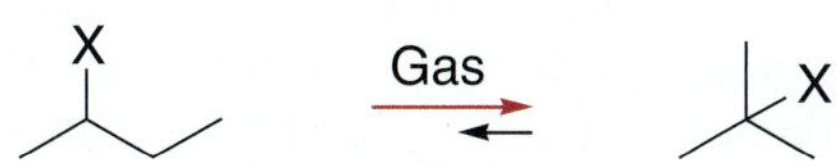

(1.58)

X = Me OH SH NH_2 Cl Br I

$\Delta_r H°$(1.58): −3.5 −4.8 −3.0 −3.8 −5.0 −5.0 −2.2 kcal mol^{-1}

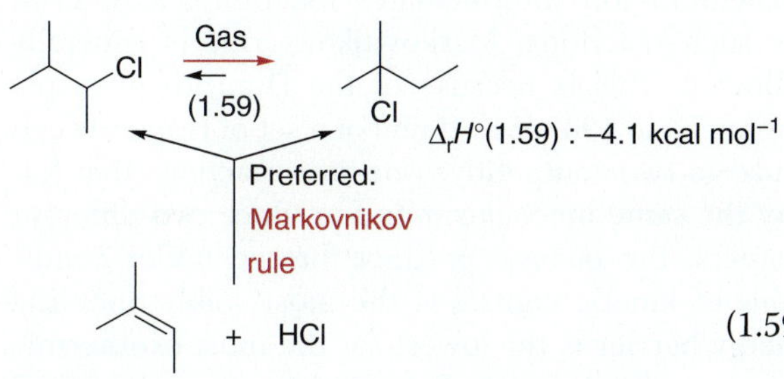

(1.59)

In solution, the additions of hydrogen halides HX (or hydrohalic acids, HX in water) to alkenes give in preference secondary and tertiary alkyl halides instead of the isomeric primary and secondary isomers, respectively. This is Markovnikov's rule, which is often explained in terms of a kinetic control (product selectivity given by the ratio of rate constants of product formation [parallel reactions, Section 3.2.5], no equilibration of products with reactants) rather than in terms of thermodynamic control (the product selectivity is governed by their relative stability) [44–47]. The same rule applies to the additions of water, alcohols, and carboxylic acids to alkenes. The versatility of Markovnikov's rule can be attributed to the large stability difference between primary, secondary, and tertiary carbenium ion intermediates (c. -15 kcal mol^{-1} for acyclic alkyl cations in strongly ionizing media; for the gas phase, Table 1.A.14 gives $\Delta_r H°(n\text{-Pr}^+ \rightarrow i\text{-Pr}^+) = -20$ kcal mol^{-1}, $\Delta_r H°(n\text{-Bu}^+ \rightarrow i\text{-Bu}^+) = -17$ kcal mol^{-1}, and $\Delta_r H°$ $(i\text{-Bu}^+ \rightarrow t\text{-Bu}^+) = -16$ kcal mol^{-1}) [48–50] that are generally considered to be formed in the rate-determining steps of these reactions (protonation of the alkenes). The formulation of Markovnikov's rule as a kinetic effect is not always valid. Additions to alkenes are exothermic but have negative entropies (condensations) that can cause the reactions to be reversible (with $\Delta_r G^T = \pm 1$ kcal mol^{-1}). For example, addition of water to unstrained alkenes are exothermic by c. -12 kcal mol^{-1}. A value very similar to the entropy cost of the addition. For instance, $\Delta_r H°$ (2-methylpropene + $H_2O \rightleftarrows t$-butanol) $\cong -12.6$ kcal mol^{-1} and $\Delta_r S°$(2-methylpropene + $H_2O \rightleftarrows t$-butanol) $\cong -37$ eu.: at 25 °C, the entropy cost $-T\Delta_r S°$ amounts to $-298(-37$ cal mol^{-1} K$^{-1}) \cong 11.0$ kcal mol^{-1}. Only additions that give rise to highly stable carbenium ion intermediates such as tertiary alkyl cations, cyclopropylmethyl cations, allylic, and benzylic cations proceed via a "cationic mechanism." Additions of HX to 1,2-dialkylethenes, instead, avoid the generation of secondary carbenium intermediates and follow other mechanisms that do not involve carbenium ion intermediates. The reverse reactions, eliminations, may also follow concerted mechanisms avoiding carbenium ion intermediates (Section 3.9.3). Even for such reactions, Markovnikov's rule is generally followed. This is because of the Dimroth principle enounced in 1933 [51]. If one or a set of reactants can undergo two competitive one-step reactions that follow the same mechanism and produce two different isomers, the favored product formed under conditions of kinetic control is the most stable one. The energy barrier is the lowest for the most exothermic reaction (Bell–Evans–Polanyi theory established in 1936–1938 for radical exchange reactions such as $\text{R–X} + \text{Y}^\bullet \rightarrow \text{R}^\bullet + \text{X–Y}$ and proton transfers $\Delta^{\ddagger}H = \alpha\Delta_r H + \beta$) [52, 53]. In 1928 already, Brønsted had found a linear relationship between the rate of proton transfer of an acid and its acidity constant [54].

Problem 1.10 A mixture of 1 mmol cyclohex-2-enone, 1 mmol of thiophenol, and 5 mg of Et_3N is kept at 25 °C in 1 ml of CH_2Cl_2. After 30 minutes at 25 °C, the ^{1}H-NMR spectra of the reaction mixture shows that the corresponding 1,4-adduct is formed almost completely. Attempted purification of the adduct by column chromatography on silica gel gives, however, only a low yield of adduct (10–20%) and recovered cyclohex-2-enone (80%) and thiophenol (80%). Why?

1.7 Standard heats of typical organic reactions

Alkanes are reference compounds (basis set) for organic chemists (Chapter 2). Their didehydrogenation (elimination of H_2) generates alkenes, and their tetradehydrogenation produces alkynes, allenes, 1,3-dienes or n,n+2-dienes (if the two double bonds of 1,3-dienes are coplanar, they are said conjugated dienes), 1,4-dienes or n,n+3-dienes (are often said homoconjugated dienes), and $n,n+\omega$-dienes ($\omega > 3$, are usually said nonconjugated dienes). Hydrogenation (addition of H_2) converts unsaturated hydrocarbons into alkanes. Formally, cycloalkanes can be hydrogenated into ring-opened alkanes (Section 1.7.1). These reactions are reference reactions and the corresponding standard heats of hydrogenation ($\Delta_h H°$) are part of a thermochemical basis set. The same can be said for compounds containing heteroatoms. For instance, the hydrogenation of aldehydes and ketones convert them into alcohols that are related to alkanes through C—H oxidations (Section 1.7.2). Thus, the standard heats of these reactions have become reference thermochemical data.

1.7.1 Standard heats of hydrogenation and hydrocarbation

The standard heats of hydrogenation (addition of H—H across a double bond), $\Delta_h H°$, of ethylene, acyclic terminal alkenes, and unstrained (E)-1,2-dialkylalkenes are about -32, -30, and -28 kcal mol^{-1} (Table 1.A.2), respectively, consistently with c. 2 kcal mol^{-1} stabilization of an alkene by each alkyl substituent. The standard heats of hydrogenation of acetylene, terminal alkynes, and dialkylethynes to give the corresponding alkenes amount to about -42, -40, and -37 kcal mol^{-1}, respectively. They indicate that alkyl substitution of alkenes increases relative stabilities of these

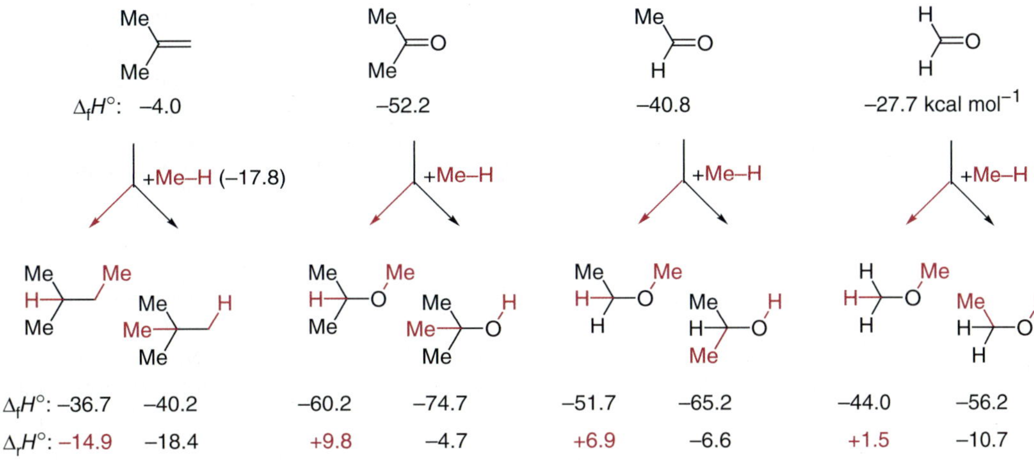

Figure 1.7 Heats of methanation of 2-methylpropene (isobutylene), acetone, acetaldehyde, and formaldehyde (examples of hydrocarbations).

compounds, just as it does for alkanes. Hydrocarbations (addition of R—H across a double bond, Figure 1.7) of alkenes have exothermicities of c. −20 kcal mol^{-1}, whereas hydrocarbations of alkynes to give alkenes have exothermicities of c. −30 kcal mol^{-1}. A π-bond of an alkene is c. 10 kcal mol^{-1} stronger than one π-bond of an alkyne. Both hydrogenations and hydrocarbations of aldehydes and ketones are much less exothermic than those of alkenes (Figure 1.7). As for alkenes and alkynes, alkyl groups stabilize the carbonyl group, but the effects are much larger. Additions to formaldehyde are easier than those to the larger aldehydes, for thermodynamic and steric reasons. Ketones are expected to undergo additions less readily than aldehydes as their hydrocarbations are less exothermic than those of the corresponding aldehydes.

The methanation of 2-methypropene (isobutylene) that gives 2,2-dimethylpropane (neopentane) is more exothermic (−18.4 kcal mol^{-1}) than the methanation that yields the less branched product, 2-methylbutane (−14.9 kcal mol^{-1}). The methanation of acetone, acetaldehyde, and formaldehyde is much less exothermic. The formation of isopropyl methyl ether, ethyl methyl ether, and dimethyl ether, respectively, is endothermic, whereas the formation of the corresponding alcohols is moderately exothermic. Thus, the conversion of a C=C doubly bonded system into a C—C singly bonded system is generally more exothermic than the conversion of a C=O doubly bonded system (aldehyde and ketone) into a C—O singly bonded system (alcohol and ether). For instance, the Diels–Alder reaction of butadiene with ethylene (Section 5.3.8) has a standard exothermicity of −39.9 kcal mol^{-1} in the gas phase (Table 1.A.2), whereas the hetero-Diels–Alder reaction (Section 5.3.15) of butadiene with formaldehyde equilibrating with 1,4-dihydro-2*H*-pyran is exothermic by −23.8 kcal mol^{-1} only ($\Delta_f H°$(1,4-dihydro-2*H*-pyran) = −53.3 kcal mol^{-1} estimated from $\Delta_f H°$(tetrahydropyran, Table 1.A.4) + 28 kcal mol^{-1} (dehydrogenation of cyclohexane into cyclohexene, Table 1.A.2)).

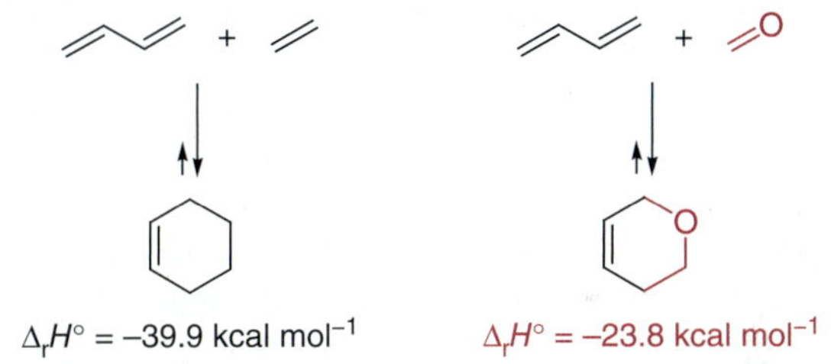

Hydrocarbation of propene with formaldehyde (an example of hydroformylation) giving 2-methylpropanal is more exothermic ($\Delta_r H°$ = −28.7 kcal mol^{-1}) than hydrocarbation of propene with ethylene (an example of hydrovinylation) giving 2-methylbut-1-ene ($\Delta_r H°$ = −23.8 kcal mol^{-1}) because alkyl substitution stabilizes carbonyl groups to a greater extent than it stabilizes alkenes (Figure 1.8) (the partial positive charge on the carbon center of the C=O function is significantly larger than that on the olefinic carbon atoms).

1.7.2 Standard heats of C–H oxidations

The standard heats of oxidation of alkanes into corresponding alcohols vary from −30 kcal mol^{-1} for the conversion of methane into methanol to about −36 kcal mol^{-1} for the oxidation of other alkanes into primary alcohols. The enthalpy change is c. −40 kcal mol^{-1} for the formation of secondary alcohols and c. −43 kcal mol^{-1} for the oxidation of branched alkanes into the corresponding tertiary alcohols. The oxidation of an alcohol into the corresponding vicinal diol (*n*,*n*+1-diol) has nearly the same exothermicity as the oxidation of corresponding

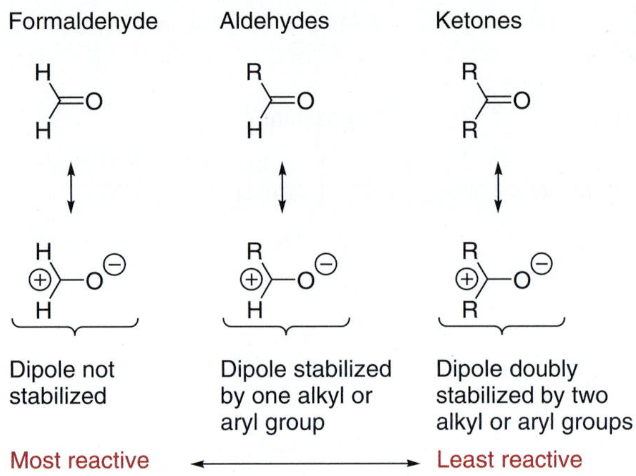

Figure 1.8 Interpretation of the difference in heats of hydrogenations (and hydrocarbations) of formaldehyde, aldehydes, and ketones.

alkane (see below data for the oxidations of propane). Thus, the interaction of two hydroxy groups on vicinal carbon centers does not introduce any significant stabilization or destabilization (gas phase). This suggests that conformers of *n*,*n*+1-diols (and also *n*,*n*+2-diols: compare the heats of oxidation of propane → *n*-PrOH and of *n*-PrOH → $HOCH_2CH_2CH_2OH$) that would permit stabilizing intramolecular hydrogen bonding between the two hydroxy groups (chelation: C—O—H···O(H)—C) are not favored, or if such interactions do exist, they are compensated by gauche effects (Section 2.5.1), by torsional strain (Section 2.6), or/and by electrostatic dipole/dipole repulsions between the two electronegative oxygen atoms (Section 2.7.9). Note that the heat of oxidation (oxygen atom insertion between a C—C bond) of *n*-propyl methyl ether (*n*-Pr-OMe) into 1,2-dimethoxyethane ($MeOCH_2CH_2OMe$) is not very much less exothermic (-25 ± 0.4 kcal mol^{-1}) than the oxidation of *n*-butane into *n*-PrOMe (-26.9 ± 0.3 kcal mol^{-1}). The electrostatic repulsion revealed in 1,2-dialkoxyethane is not larger than 1.9 ± 0.7 kcal mol^{-1}. The oxidation of benzene into phenol and of phenol into benzene-1,2-diol have nearly the same exothermicity of -42.7 ± 0.6 kcal mol^{-1} (almost the same exothermicity than for the oxidation of a branched alkane into a tertiary alcohol). If a repulsive electrostatic effect between the two oxygen atoms does destabilize benzene-1,2-diol, then it must be counteracted by stabilizing intramolecular hydrogen bonding between the two hydroxy groups (chelation) or by another effect. Oxidation of phenol into benzene-1,4-diol is exothermic by -43.2 ± 0.6 kcal mol^{-1}, nearly the same exothermicity as for the oxidation of benzene into phenol. There is no possibility for intramolecular hydrogen bridging in benzene-1,4-diol and electrostatic repulsion must be smaller than it is for benzene-1,2-diol. The finding that benzene-1,3-diol is more stable than benzene-1,2-diol and benzene-1,4-diol by c. 2 kcal mol^{-1} suggests that other factors may contribute to the relative stability of these compounds. Consistent with what has been described above, the oxidation of benzene-1,3-diol into benzene-1,3,5-triol is slightly more exothermic than its oxidation into benzene-1,2,3-triol and benzene-1,2,4-triol. In contrast to vicinal or *n*,*n*+1-dioxy-substitution, geminal or *n*,*n*-dioxy-substitution leads to a significant stabilization effect (enthalpic anomeric effect, Section 2.7.9).

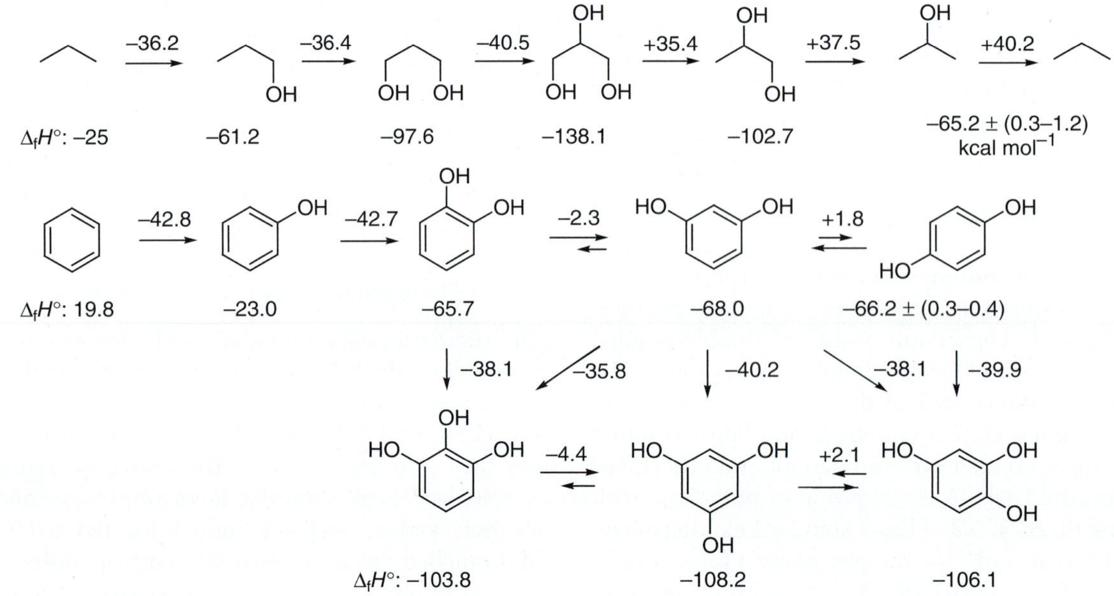

1.7.3 Relative stabilities of alkyl-substituted ethylenes

Table 1.A.2 shows that simple acyclic (E)-alkenes are in general more stable than their (Z)-isomers. If the substituents of the ethene moiety are bulky, this stability difference increases because of increased steric interactions between these substituents in the (Z)-isomers, as demonstrated with equilibria (1.60).

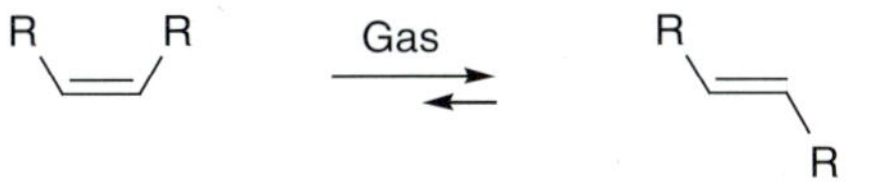

(1.60)

R =	Me	Et	i-Pr	t-Bu	Ph (gas)	Ph (solid)	COOH (solid)
$\Delta_r H°$ (1.60):	−0.75	−0.8	−2.0	−9.6	−3.9	−11.0	−5.1 kcal mol^{-1}

The thermochemical data reported below for C_6H_{12} alkenes and C_6H_{14} alkanes show that dialkylethylenes are more stable than monoalkylethylene isomers. Similarly, trialkylethylenes are more stable than dialkylethylene isomers, and finally, tetraalkylethylenes are more stable than trialkylethylene isomers.

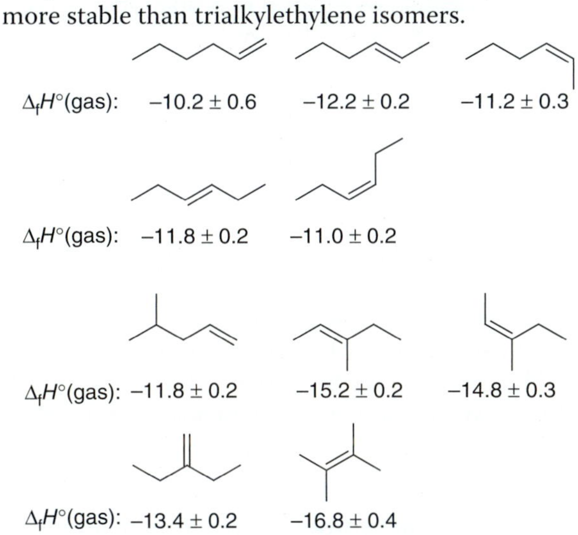

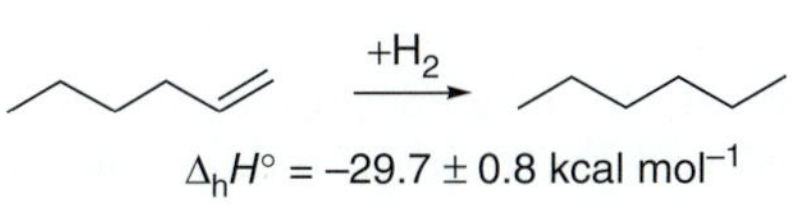

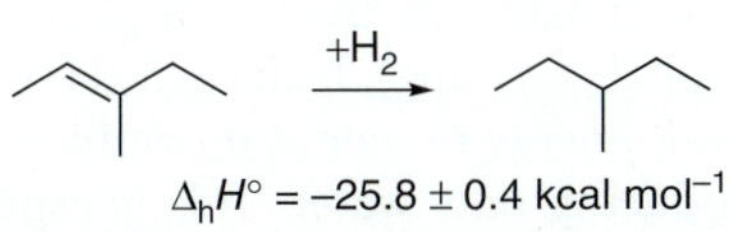

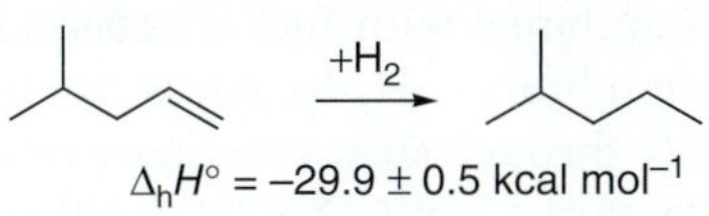

(1.61)

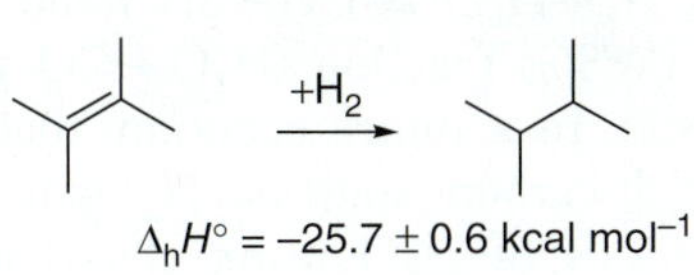

(1.62)

As for alkanes, "branched" alkenes are more stable than linear isomers, as evidenced by the fact that hydrogenation of monoalkylethylenes (e.g. $\Delta_h H°(1.61) \cong -30$ kcal mol^{-1}) are more exothermic than the hydrogenations of tri- and tetraalkylethylenes (e.g. $\Delta_h H°(1.62) \cong -26$ kcal mol^{-1}).

1.7.4 Effect of fluoro substituents on hydrocarbon stabilities

Using data from NIST Chemistry Webbook, the following heats of reaction are estimated for the Diels–Alder addition of butadiene to ethylene, propene, chloroethylene, and fluoroethylene.

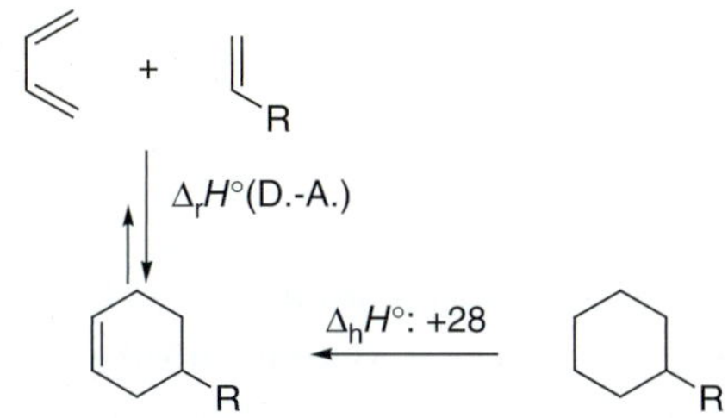

R = H $\Delta_r H°$(D.-A.) = −39.9 ± 1 kcal mol^{-1}

R = Me $\Delta_r H°$(D.-A.) = −42.9 ± 2 kcal mol^{-1}

R = Cl $\Delta_r H°$(D.-A.) = −44.8 ± 2 kcal mol^{-1}

R = F $\Delta_r H°$(D.-A.) = −46.1 ± 2 kcal mol^{-1}

These estimates use $\Delta_f H°$(butadiene) = 26.0 kcal mol^{-1}, $\Delta_f H°$(ethylene) = 12.5 kcal mol^{-1}, $\Delta_f H°$(cyclohexene) = −1.0 kcal mol^{-1}, $\Delta_f H°$(propene) = 4.9 kcal mol^{-1}, $\Delta_f H°$(4-methylcyclohexene) = −12.0 kcal mol^{-1} (estimated from $\Delta_f H°$(methylcyclohexane) = −40.0 and 28 kcal mol^{-1} for the heat of didehydrogenation), $\Delta_f H°$ (chloroethylene) = 7.0 kcal mol^{-1}, $\Delta_f H°$(4-chlorocyclohexene) = −11.8 kcal mol^{-1} (estimated from $\Delta_f H°$(chlorocyclohexane) = −39.8 kcal mol^{-1} and assuming 28 kcal mol^{-1} for the heat of didehydrogenation into 4-chlorocyclohexene), $\Delta_f H°$(fluoroethylene) = −32.4 kcal mol^{-1}, and $\Delta_f H°$(4-fluorocyclohexene) = −52.5 kcal mol^{-1} (estimated from $\Delta_f H°$(fluorocyclohexane) = −80.5 kcal mol^{-1} and assuming 28 kcal mol^{-1} for the heat of its didehydrogenation). The data show that substitution of a sp^3-hybridized carbon center is more stabilizing than substitution of a sp^2-hybridized carbon center. The small increase of exothermicity for the Diels–Alder reactions of fluoroethylene and chloroethylene with 1,3-butadiene compared to that of propene and 1,3-butadiene is attributed to the electronegative nature of Cl and F substituents, which confers a dipolar character to their C—Cl and C—F bonds, respectively, which is usually denoted by the limiting structures:

Vinyl cations are less stable than secondary alkyl cations as given by comparison of the following hydride affinities (reaction (1.90)): $DH°(CH_2{=}CH^+/H^-)$ = 291 kcal mol^{-1}, $DH°(i\text{-}Pr^+/H^-)$ = 251 kcal mol^{-1}, and $DH°(c\text{-}C_5H_9{}^+/H^-)$ = 249.8 kcal mol^{-1} (Table 1.A.14). Thus, the polar C—X bond is expected to be more important for secondary (and tertiary) alkyl systems than for alkenyl systems. Furthermore, the differential substitution effect between alkene and alkane giving by the above $\Delta_r H°$ (Diels–Alder reaction) values is expected to be larger for fluoro than for the other substituents, as indicated by the heats of the isodesmic reactions shown below. Cyclohexane is stabilized to a greater extent than methane by substitution of a hydrogen atom by a methyl, chloro, or fluoro group. For the reasons invoked (relative stability of methyl vs. cyclohexyl cation and difference in electronegativity between C and the substituent), the effect is the largest for fluoro substitution [55].

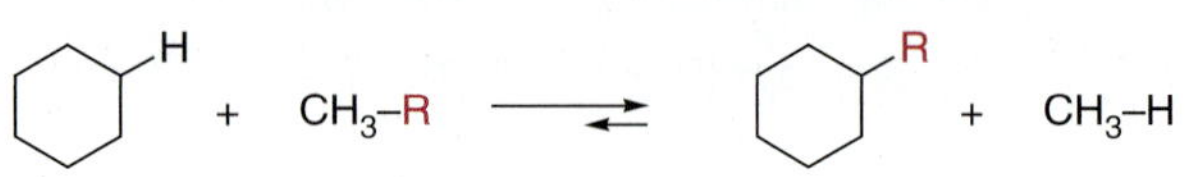

R = Me $\Delta_r H°$ = −8.1 ± 2 kcal mol^{-1}
R = Cl $\Delta_r H°$ = −7.8 ± 2 kcal mol^{-1}
R = F $\Delta_r H°$ = −12.6 ± 2 kcal mol^{-1}

The following equilibria reveal important differences for the substituent effects of chloro, fluoro, and methoxy groups on sp^3 vs. sp^2-hybridized carbon centers. Geminal dichlorinated alkanes and alkenes are equally stabilized by the substituents. Conversely, fluorine "prefers to reside" at sp^3-hybridized carbon centers. In the case of geminal dimethoxy disubstitution, the ketene acetal ($(MeO)_2C{=}CH_2$) is stabilized by n(O:)/$\pi_{C=C}$ conjugation that is not present in acetaldehyde dimethyl acetal ($(MeO)_2CH{-}CH_3$). The analogous n(F:)/$\pi_{C=C}$ and n(Cl:)/$\pi_{C=C}$ interactions are relatively weak interactions (Section 2.7.5).

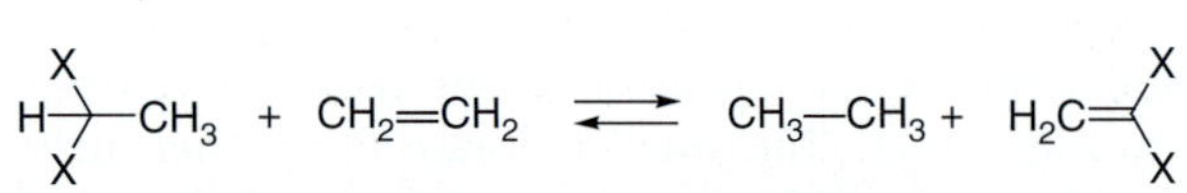

X = Cl $\Delta_r H°$ = −0.4 ± 1.5 kcal mol^{-1}
X = F $\Delta_r H°$ = 3.9 ± 4.0 kcal mol^{-1}
X = MeO $\Delta_r H°$ = −6.7 ± 1.5 kcal mol^{-1}

The next two equilibria reveal once more that chloride substituent stabilizes sp^3- and sp^2-hybridized carbon centers equally. This is not the case for fluoro substitution as 1,1,2-trifluoroethane is highly preferred with respect to 1,1,2-trifluoroethylene. The reason for that is that the three C—F bonds in 1,1,2-trifluoroethane are stabilized by three hyperconjugative interactions (σ(C—H) → σ^*(C—F) donation) involving two antiperiplanar H—C(2)/C(1)—F bond pairs and one antiperiplanar H—C(1)/C(2)—F bond pair (Section 4.8.1), whereas 1,1,2-trifluorethylene has only one C—F bond that can be stabilized through this mechanism.

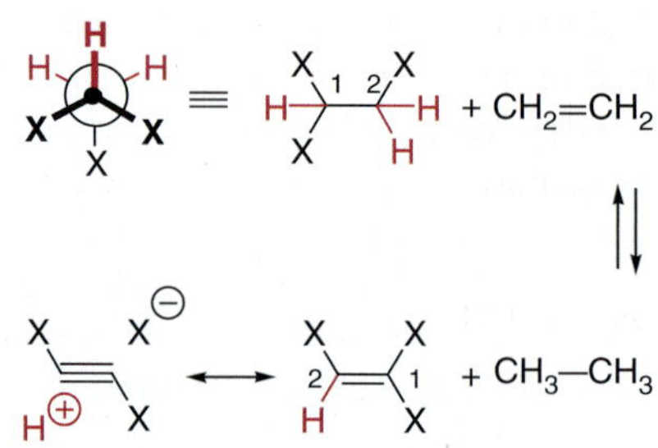

X = Cl $\Delta_r H°$ = −2.5 ± 1.5 kcal mol^{-1}
X = F $\Delta_r H°$ = 19.2 ± 5.0 kcal mol^{-1}

Problem 1.11 On heating, 1,1-dideuteriohexa-1,5-diene equilibrates with 3,3-dideuteriohexa-1,5-diene (reversible Cope rearrangement: section "Cope Rearrangements"). Similarly 1,1-difluorohexa-1,5-diene equilibrates with 3,3-difluorohexa-1,5-diene. Which of these two last isomeric compounds is most stable? [56]

1.7.5 Storage of hydrogen in the form of formic acid

Because of the limited sources of oil, natural gas, and coal and the need to reduce the concentration of carbon dioxide in the atmosphere (greenhouse effect, global warming), hydrogen (H_2) is considered as a "clean" energy carrier as it combines with oxygen (O_2) producing only water. This is especially attractive when combined with fuel cells based on proton exchange membranes. Hydrogen is produced industrially mainly through steam methane reforming at c. 800 °C ($CH_4 + H_2O \rightleftarrows CO + 3H_2$) and by the water shift gas (WSG) reaction ($H_2O + CO \rightleftarrows H_2 + CO_2$, Section 8.2). In a future economy that cannot rely upon fossil carbon sources, H_2 will be obtained by electrolysis of H_2O using electricity generated by water or wind turbines, for instance, or by photovoltaics. Sooner or later direct sunlight-driven photochemical water splitting into H_2 and O_2 will become economical [57]. To use it, H_2 must be stored and transported in the form of compressed or/and absorbed gas. This remains problematic in terms of cost and safety. Conversion of H_2 in a nontoxic liquid that can be handled at atmospheric pressure and useful temperatures (−50 to 50 °C) is highly desirable. The liquid should be used as fuel directly without producing toxic products and coproducts (nitrogen oxides and CO). It should decompose cleanly back into H_2 at the site where it is needed. In the presence of suitable catalysts (Section 7.8.8), cheap and abundant carbon dioxide (CO_2) can be hydrogenated into formic acid (HCO_2H) [58, 59]. The NIST Chemistry Webbook gives the following data for the hydrogenation of carbon dioxide under 1 atm and at 25 °C:

	H_2(gas) +	CO_2(gas)	$\rightleftarrows$ HCO_2H (liquid)
$\Delta_f H°$:	0.0	−94.05 ± 0.04	−101.7 ± 0.1 kcal mol^{-1}
$S°$:	31.2	47.2	31.5 eu

One calculates the standard heat of reaction $\Delta_r H°(H_2 + CO_2 \rightleftarrows HCO_2H) = -7.65 \pm 0.14$ kcal mol^{-1} and the standard variation of entropy of reaction $\Delta_r S° = -46.9$ eu, which is much more negative than for other condensations of two gaseous compounds of similar molecular weight. This arises from the smaller entropy of a compound in its liquid phase than in the vapor phase ($S°(HCO_2H$, gas) = 59.4 eu). Because of this negative entropy of reaction (condensation of two gases into a liquid), the hydrogenation of carbon dioxide is endergonic at 25 °C by $\Delta_r G°(H_2(gas) + CO_2(gas) \rightleftarrows HCO_2H(liquid)) = -7.65 - 298(-0.0469) = 6.33$ kcal mol^{-1} (equilibrium constant $K^{298\,K} = 10^{-4.65}$ at 25 °C). Thus, for this reaction to occur at 25 °C, high pressure must be applied (Le Châtelier principle, mass law effect). If the reaction is carried out in a solvent giving an ideal solution, one can use $S°(HCO_2H$, gas), which leads to $\Delta_r S° = -19$ eu. Using $\Delta_f H°(HCO_2H$, gas)=−90.5 kcal mol^{-1}, one calculates $\Delta_r H°(H_2(gas)+CO_2(gas) \rightleftarrows HCO_2H(gas)) = 3.55 \pm 0.14$ kcal mol^{-1} and $\Delta_r G°(H_2(gas) + CO_2(gas) \rightleftarrows HCO_2H(gas)) = 3.55 - 298(-0.019) = 9.2$ kcal mol^{-1}: the reaction becomes even more endergonic (see Section 7.8.8). An option is to convert formic acid into a stable derivative such as a salt with an inexpensive base. The base **B** must have a $pK_a(\mathbf{BH^+}) = -\log([\mathbf{B}][H^+]/[\mathbf{BH^+}])$ for its conjugate acid $\mathbf{BH^+}$:

$pK_a(\mathbf{BH^+}) > pK_a(HCO_2H) - \log\ K(H_2 + CO_2 \rightleftarrows HCO_2H)$

Using data given in Table 1.A.23 (pK_a(HCOOH) = 3.75) for diluted water solution at 25 °C, and above estimated $\Delta_r G°(H_2(gas) + CO_2(gas) \rightleftarrows HCO_2H(liquid)) = 6.33$ kcal mol^{-1} = 1.36 log([HCOOH]/[H_2][CO_2]) = 1.36 (−4.65), one predicts $pK_a(\mathbf{BH^+}) > 3.75 + 4.65 = 8.3$.

Thus, bases like NH_3 ($pK_a(NH_4^+$, aq.) = 9.2) or Na_2CO_3 ($pK_a(HCO_3^-$, aq.) = 10.3) are suitable, which is verified experimentally. The catalytic decomposition of HCO_2H/HCO_2NH_4 mixtures into $H_2 + CO_2$ can be carried out at 25 °C in the presence of all kinds of transition metal complexes [60], or by electrocatalytic oxidation at platinum electrodes [61]. The direct photoinduced hydrogenation of CO_2 into HCO_2H is an elegant way to generate a liquid fuel using solar energy. It is expected to become a reality in the future. The method couples CO_2 reduction and light-driven water splitting [62]. Formic acid can be decomposed into water and carbon monoxide. Thermochemical data predicts this reaction to be exergonic ($\Delta_r G°(HCO_2H(gas) \rightleftarrows H_2O(gas) + CO(gas)) = -2.65$ kcal mol^{-1}) at 25 °C, and under 1 atm. Thus, the challenge is to find catalysts that selectively decompose formic acid into $H_2 + CO_2$, without concurrent decomposition into $H_2O + CO$. Carbon monoxide is toxic and quite often it combines with transition metal catalysts (Section 7.7) inhibiting the desired reaction.

Problem 1.12 Is the hydrogenation of carbon monoxide into formaldehyde a feasible reaction provided that a suitable catalyst is available to catalyze this reaction?

Problem 1.13 Are the hydrocarbations of carbon monoxide by alkanes, alkenes, or alkynes (or the C—H carbonylations of alkanes, alkenes, and alkynes) possible reactions provided suitable catalysts are available?

Problem 1.14 Are the hydrocarbations of carbon dioxide possible for alkanes, alkenes, and alkynes provided suitable catalysts are available to catalyze them?

Problem 1.15 Can one fix CO_2 with epoxides? What products are expected (catalysts: mixed Mg/Al oxides and dimethylformamide [DMF] as a solvent) to be formed at 100 °C?

Problem 1.16 Which of the two products **B** and **C** is the favored product of cyclization of **A** catalyzed by $Bu_4N^+F^-$ (base catalyst) in DMF.

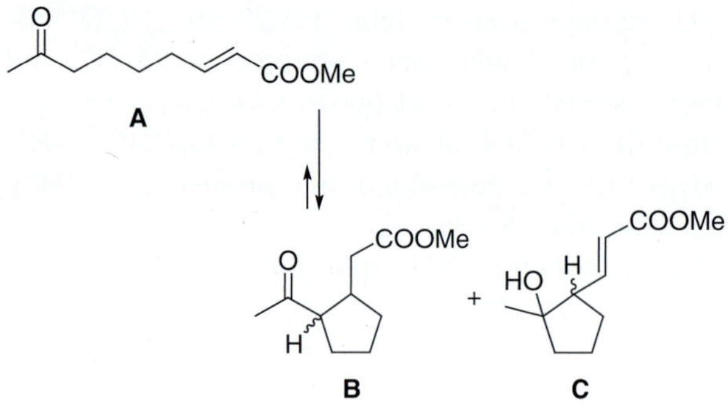

Problem 1.17 Are the Diels–Alder reactions ((4+2)-cycloadditions) of benzene with ethylene, acetylene, and allene thermodynamically possible at room temperature? Are the intramolecular versions possible? [63–65]

1.8 Ionization energies and electron affinities

The ionization energy IE(**A**) of an atom **A** is the energy required to remove an electron from it in the gas phase generating a cation $\mathbf{A}^+$ in its ground state (reaction (1.63)).

$$\mathbf{A}(\text{ground state}) \rightarrow \mathbf{A}^+(\text{ground state}) + e^- \quad (1.63)$$

$$\text{IE}(\mathbf{A}) = \Delta_r H^\circ(1.63) = \Delta_f H^\circ(\mathbf{A}^+) - \Delta_f H^\circ(\mathbf{A})$$

Experimentally, the ionization energy of an atom can be determined by mass spectrometry (MS) or by photoelectron spectroscopy (PES) [66, 67]. IE(**A**)s can be determined by mass spectroscopy in the following way (Figure 1.9): the gaseous atoms are bombarded by electrons accelerated by potential V (in volts). Once their energy, measured in eV, reaches the value of IE(**A**), ions are formed (appearance potential or appearance energy AP($\mathbf{A}^+$)) and can be counted by the ion detector [21–23, 43, 68].

PES uses (Figure 1.10) a source of monochromatic light (e.g. that emitted by excited helium atoms: $h\nu = 21.21$ eV) that strikes atoms **A**. They expel electrons with kinetic energy KIE = $h\nu$ − IE(**A**) (reaction (1.64)). Vacuum ultraviolet laser light can also be used for the ionization of organic molecules [69].

$$\begin{aligned} \mathbf{A} + h\nu &\rightarrow \mathbf{A}^+(\Psi_0) + e^- \quad \text{with KIE}_0 \\ \mathbf{A} + h\nu &\rightarrow \mathbf{A}^+(\Psi_1^*) + e^- \quad \text{with KIE}_1 \\ \mathbf{A} + h\nu &\rightarrow \mathbf{A}^+(\Psi_2^*) + e^- \quad \text{with KIE}_2 \end{aligned} \quad (1.64)$$

Several ionization energies can be measured for atoms **A**. The lowest values $\text{IE}_0(\mathbf{A})$ correspond to

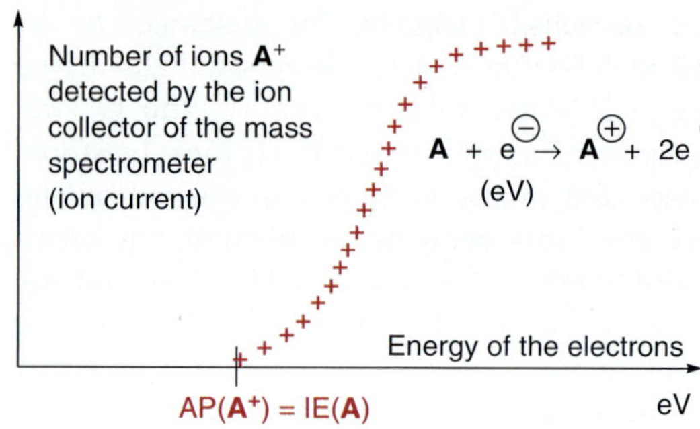

Figure 1.9 Measurement of the appearance potential AP($\mathbf{A}^+$) for the ionization of atoms **A** by accelerated electrons. For atoms AP($\mathbf{A}^+$) = IE(**A**).

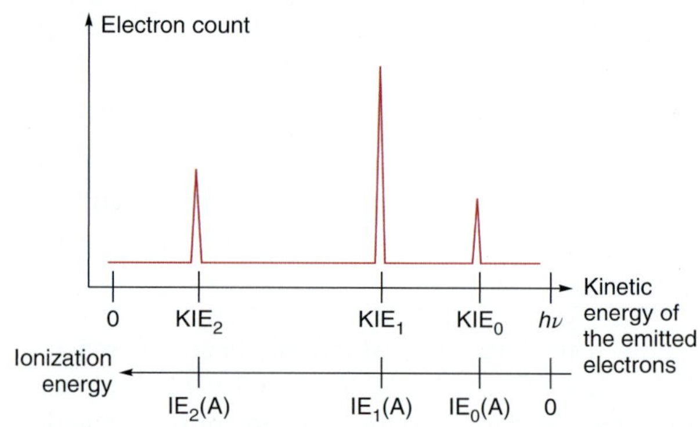

Figure 1.10 Photoelectron spectrum of atom **A** showing three ionizations giving cations $\mathbf{A}^+$ in its ground state Ψ_0 and in its electronically excited states and, corresponding to electrons emitted with kinetic energies $\text{KIE}_0 = h\nu - \text{IE}_0(\mathbf{A})$, $\text{KIE}_1 = h\nu - \text{IE}_1(\mathbf{A})$, and $\text{KIE}_2 = h\nu - \text{IE}_2(\mathbf{A})$; $h\nu$ being the energy of the monochromatic light source.

the ionization of **A** into cation $\mathbf{A}^+$ in its ground state Ψ_0. The higher values are associated with the formation of electronically excited states of cation $\mathbf{A}^+$. The photoelectron spectrum of atoms is made of lines, whereas those of molecules are made of bands (Franck–Condon contours) whose width and shape can vary quite significantly. Ionization of a molecule by electron impact or by photoionization is governed by the Franck–Condon principle, which states that the most probable ionization transition will be that in which the geometry and the momentum of the ion are the same as those of the molecule in its ground state (Figure 1.11). The time required for the ionization is much shorter (10^{-16} seconds) than the time necessary for a vibration (10^{-13} to 10^{-12} seconds).

When the equilibrium geometries of an ion and its corresponding neutral species are nearly identical, the onset of ionization will be a sharp step function leading to the ion vibrational ground state (Figure 1.11a). However, when the equilibrium geometry of the ion

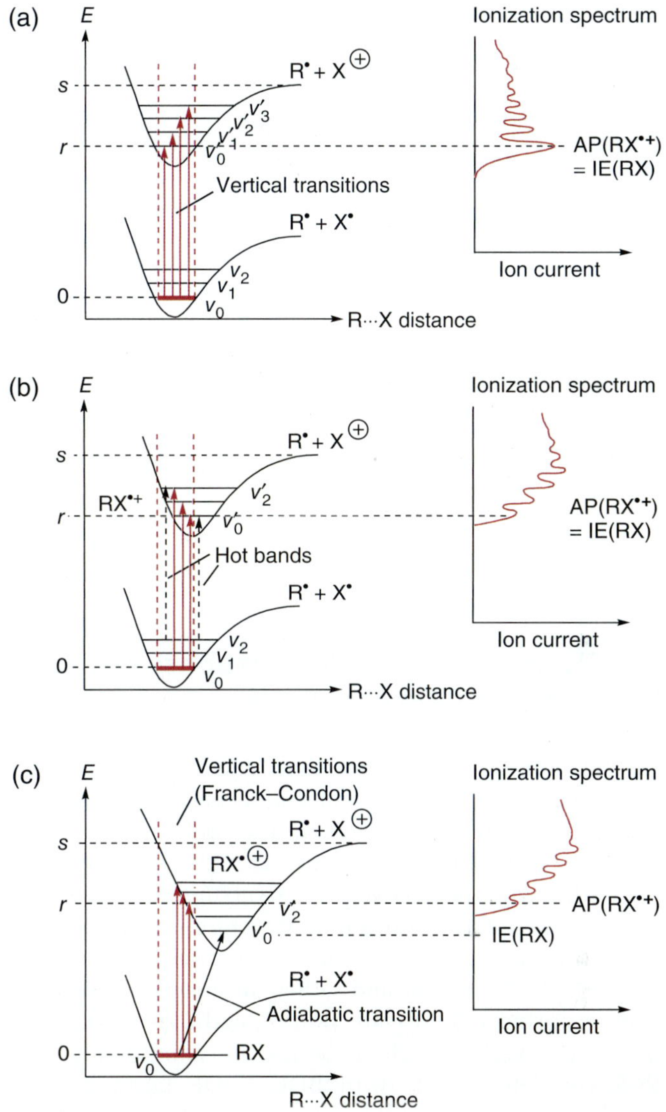

Figure 1.11 Examples of ionization of molecules RX; r = minimal value for the appearance potential of ion $RX^{\bullet+}$ ($AP(RX^{\bullet+})$); s = minimal value for the appearance potential of ion X^+ ($AP(X^+)$), resulting from the fragmentation of $RX^{\bullet+}$. (a) The ground states of RX and $RX^{\bullet+}$ have the same geometry. The vertical transition of highest probability corresponds to the adiabatic transition. (b) The geometries of RX and $RX^{\bullet+}$ are not the same but similar: the transition is still probable, but not of the highest intensity. (c) The geometries of RX and $RX^{\bullet+}$ are different; the adiabatic transition is of lower energy than the first observable transition.

involves a significant change in one or more bond lengths/angles from that of the neutral species, the transition to the lowest vibrational level of the ion is no longer the most intense, and the maximum transition probability (the vertical ionization energy) will favor population of a higher vibrational level of the ion (Figure 1.11b); if the geometry change is significant, then the transition to the lowest vibrational level of the ion may not be observed (Figure 1.11c). For such cases, adiabatic ionization energies can be obtained by determining the equilibrium constant for charge transfer to another molecule (or atom) of known ionization energy (equilibrium (1.65)) by high-pressure mass spectrometry [70], flow tube [71–73], or ion cyclotron resonance (ICR) spectrometry [74, 75].

$$RX + A^+ \rightleftarrows RX^{\bullet+} + A^{\bullet} \tag{1.65}$$

The electron affinity (EA) of an atom or a molecule is equal to the enthalpy difference between the heat of formation of a neutral species and the heat of formation of the negative ion of the same structure (Eq. (1.66)), that is the energy change upon addition of an electron. Several methods are available to measure electron affinities of isolated molecules [76]. Electron transmission spectroscopy, charge transfer reactions in a mass spectrometer, collisional ionization with fast alkali–metal beams, plasma and optogalvanic spectroscopy, and collisional activations have been used. The most effective methods to measure electron affinities of solid substance rely upon the photoelectric effect. A typical experiment involves A target anion, R^-, which is bombarded with a light beam of frequency ν; either the photodestruction of R^- or the appearance of the scattered electrons, e^- are monitored [21–23, 43, 68, 77]. For experimental ionization energies (IE) and electron affinities (−EA), see Tables 1.A.13, 1.A.20–1.A.22.

$$\Delta_r H^\circ(\mathrm{M} + \mathrm{e}^- \rightarrow \mathrm{M}^-) = -\mathrm{EA}(\mathrm{M}) = \Delta_f H^\circ(\mathrm{M}^-) - \Delta_f H^\circ(\mathrm{M}) \quad (1.66)$$

$$\Delta_r H^\circ(\mathrm{R}^- + h\nu \rightarrow \mathrm{R}^\bullet + \mathrm{e}^-) = -(-\mathrm{EA}(\mathrm{R}^\bullet)) = \Delta_f H^\circ(\mathrm{R}^\bullet) - \Delta_f H^\circ(\mathrm{R}^-) \quad (1.67)$$

Problem 1.18 Interpret the relative ionization energies of $HO^\bullet$, $HS^\bullet$, and $HSe^\bullet$. Interpret the relative ionization energies of halide radicals. What makes the trends observed?

Problem 1.19 Why is the hydride anion more stable than hydrogen radical in the gas phase?

Problem 1.20 Interpret the differences in electron affinities between alkyl, alkenyl, and alkynyl radicals.

1.9 Homolytic bond dissociations; heats of formation of radicals

The relationship between reactivity and structure of molecules, or fragments of molecules, constitutes the basis of the molecular sciences. The knowledge of reliable specific BDEs, i.e. particularly the variation in bond strength with changes in structure, provides quantitative information about structure–reactivity relationships. Various correlations between bond lengths (Table 1.A.6) and other properties have been reported [25, 78–80]. Bond lengths (r, in Å) of typical carbon–carbon bonds in compounds R–R correlate linearly with BDEs (BDE taken as $\Delta_r H^\circ(\mathrm{R{-}R} \rightleftarrows 2\mathrm{R}^\bullet)$, kcal mol^{-1}) in the full range of single, double, and triple bonds with equation $r = 1.748 - 0.002\,371\cdot(\mathrm{BDE})$ for which a correlation coefficient of 0.999 998 4 is obtained using a data set of 41 compounds with C—C bonds ranging from 1.20 to 1.71 Å [81]. Examples are given below:

	r (Å)	BDE (±2 kcal mol^{-1})
$\mathrm{HC{\equiv}CH} \rightleftarrows 2{:}\mathrm{CH}^\bullet$	1.203	229.9
$\mathrm{H_2C{=}CH_2} \rightleftarrows 2{:}\mathrm{CH_2}$	1.339	172.2
$\mathrm{HC{\equiv}C{-}C{\equiv}CH} \rightleftarrows 2\mathrm{HC{\equiv}C}^\bullet$	1.384	155.0
$\mathrm{Ph{-}Ph} \rightleftarrows 2\mathrm{Ph}^\bullet$	1.48	118.0
$\mathrm{CH_2{=}CH{-}CH{=}CH_2} \rightleftarrows 2\mathrm{CH_2{=}CH}^\bullet$	1.467	116.0
$\mathrm{Me{-}Me} \rightleftarrows 2\mathrm{Me}^\bullet$	1.535	89.7
t-Bu—t-Bu $\rightleftarrows$ 2t-Bu$^\bullet$	1.572	68.8
$\mathrm{PhCH_2{-}CH_2Ph} \rightleftarrows 2\mathrm{Bn}^\bullet$	1.58	66.6
$\mathrm{Et_3C{-}CEt_3} \rightleftarrows 2\mathrm{Et_3C}^\bullet$	1.635	51.0
$\mathrm{Ph_3C{-}CPh_3} \rightleftarrows 2\mathrm{Ph_3C}^\bullet$	1.72	16.6

1.9.1 Measurement of bond dissociation energies

The BDEs of diatomic molecule A–B have been defined in Figure 1.3 (Morse potential). We consider now the standard homolytic bond dissociation enthalpies $DH^\circ(\mathrm{R}^\bullet/\mathrm{X}^\bullet)$ and the enthalpy change involved in breaking 1 mol of compound R–X under 1 atm. and at 25 °C into two fragments $\mathrm{R}^\bullet$ and $\mathrm{X}^\bullet$ (Tables 1.A.7–1.A.11, 1.A.13, 1.A.14). In practice, $DH^\circ(\mathrm{R}^\bullet/\mathrm{X}^\bullet)$ is taken as BDE of R–X [82–84]. Thus, for equilibrium (1.68):

$$\mathrm{R-X} \rightleftarrows \mathrm{R}^\bullet + \mathrm{X}^\bullet$$

$$\text{the BDE} \cong DH^\circ(\mathrm{R}^\bullet/\mathrm{X}^\bullet) = \Delta_f H^\circ(\mathrm{R}^\bullet) + \Delta_f H^\circ(\mathrm{X}^\bullet) - \Delta_f H^\circ(\mathrm{RX}) \quad (1.68)$$

where $\Delta_f H^\circ(\mathrm{R}^\bullet)$ and $\Delta_f H^\circ(\mathrm{X}^\bullet)$ are the standard heats of formation of radicals $\mathrm{R}^\bullet$ and $\mathrm{X}^\bullet$, respectively [85]. To measure $DH^\circ(\mathrm{R}^\bullet/\mathrm{X}^\bullet)$, the equilibrium constants $K(1.68)$ must be measured at different temperatures (Van't Hoff experiments). An example is given by the low-pressure pyrolysis of hexa-1,5-diene that equilibrates with 2 equiv. of allyl radical (Eq. (1.69)) between 625 and 900 K [86].

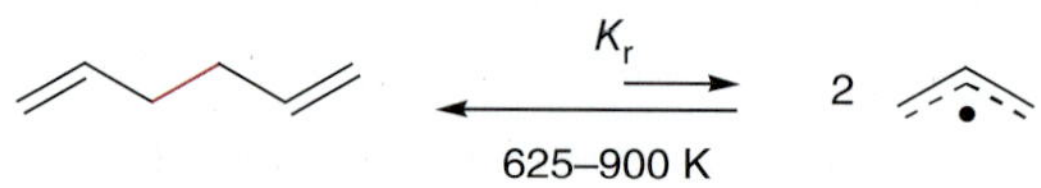

(1.69)

$DH^\circ(\mathrm{allyl}^\bullet/\mathrm{allyl}^\bullet) = \Delta_r H^\circ(1.69) = 2\cdot\Delta_f H^\circ(\mathrm{allyl}^\bullet) - \Delta_f H^\circ$ (hexa-1,5-diene) $= 56.1$ kcal mol^{-1};

$\Delta_r S^\circ(1.69) = 34.6 \pm 10.6$ eu

$\Delta_f H^\circ(\mathrm{allyl}^\bullet) = \frac{1}{2}\{\Delta_r H^\circ(1.69) - \Delta_f H^\circ(\text{hexa-1,5-diene})\} = 39.1 \pm 1.6$ kcal mol^{-1}

In this method, both Le Châtelier's principle (low pressure) and translational entropy makes the fragmentation favored at high temperature (positive reaction entropy, the $-T\Delta_r S^T$ term is negative and thus tends to cancel the positive heat of reaction $\Delta_r H^T$). Standard homolytic bond dissociation enthalpy (or bond energy) $DH^\circ(\mathrm{allyl}^\bullet/\mathrm{allyl}^\bullet) = \Delta_r H^\circ(1.69)$. It measures the "glue" that keeps the two allyl radicals together in hexa-1,5-diene, whereas the $-T\Delta S$ term is a probability factor that favors two molecules (or molecular fragments), more so at higher temperatures (Section 2.6). Most organic and organometallic compounds decompose on heating. Therefore, low-pressure pyrolysis cannot be used to measure equilibrium constants $K(1.68)$ for such fragmentations, except for systems generating relatively stable product radicals that can equilibrate with the precursor at relatively low temperature. One such example is given above with equilibrium

(1.69). A second example is given with equilibrium (1.70) for which the hexasubstituted ethane $(CF_3S)_3C{-}C(SCF_3)_3$ is split into 2 equiv. of $(CF_3S)_3C^{\bullet}$ radical, the concentration of which is measured by electron spin resonance (ESR) for benzene solutions [87]. One obtains $\Delta_r H(1.70) = 13.7$ kcal mol^{-1} and $\Delta_r S(1.70) = 3.6$ eu, with $K(1.70) = 7.5 \times 10^{-10}$ M at 20 °C. The relatively small value found for the entropy of reaction suggests that a specific solvation effect influences the equilibrium (1.70). Because of the higher polarizability of radicals $(CF_3S)_3C^{\bullet}$ compared with its precursor $(CF_3S)_3C{-}C(SCF_3)_3$, benzene molecules interact more strongly with the radicals than with the reactant. This interaction immobilizes molecules of benzene around the $(CF_3S)_3C^{\bullet}$ solute, so that the increase of the number of species (increase of translation entropy) arising from the fragmentation is compensated by loss of translation and rotation degrees of freedom for solvent molecules.

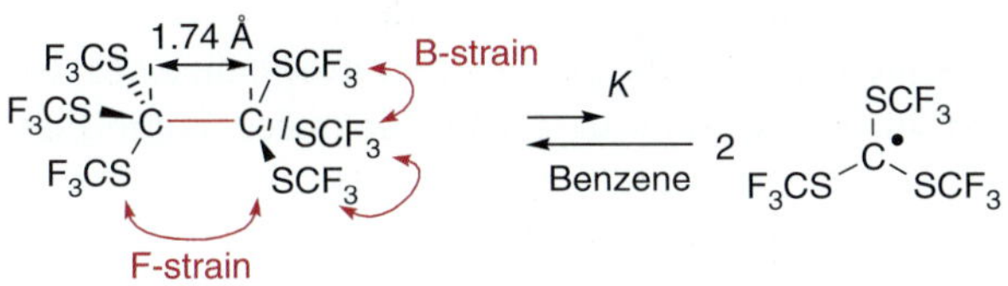

(1.70)

If equilibrium (1.70) was not affected by specific solvation effects, and assuming its variation of entropy to be associated exclusively by changes in the degree of translational levels, one calculates, with $S_{trans} = 6.86{\cdot}\log M_r(g) + 11.44{\cdot}\log\ T - 2.31$ eu, $\Delta_r S°(1.70)_{trans} = 31$ eu, which is significantly larger than the experimental value (3.6 eu). The crystal structure of $(CF_3S)_3C{-}C(SCF_3)_3$ demonstrates that this compound possesses an unusually long C—C single bond length of 1.74 Å (see above, Section 1.8). Furthermore, one observes that the three C—S bonds about each carbon center of the ethane moiety are almost coplanar (strongly deviating from the classical tetrahedral structure). This is due to the bulk of the CF_3S groups, which repel each other. In the $(CF_3S)_3C$ moiety, there is "Back strain" (or B-strain) among the CF_3S groups that makes the S—C—S bond angle larger than 109°, and "Front strain" (or F-strain) between the two $(CF_3S)_3C$ moieties (see Section 2.5.1). These steric, repulsive interactions destabilize $(CF_3S)_3C{-}C(SCF_3)_3$ with respect to the two radicals $^{\bullet}C(SCF_3)_3$ and reduce the bond energy of the hexasubstituted ethane.

Activation parameters (Section 3.3) for reaction (1.70) are $\Delta^{\ddagger}H = 21$ kcal mol^{-1} and $\Delta^{\ddagger}S = 7.5$ eu (Figure 1.12). Apparently, there is less order in the transition state of the homolytical process than in the

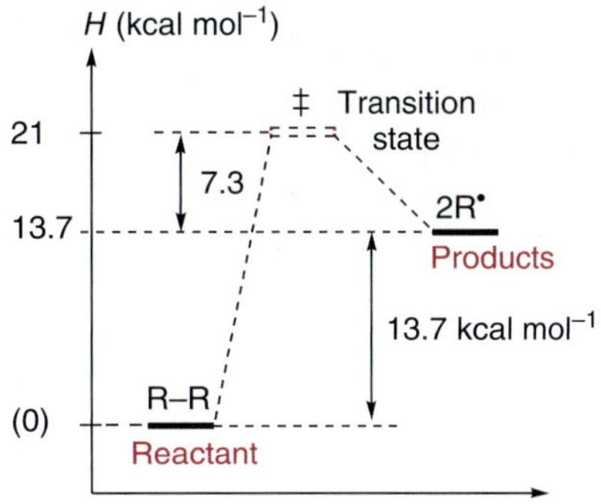

Figure 1.12 Enthalpy diagram for homolysis (1.70).

products of the reaction. This is consistent with the intervention of a specific solvation effect that implies fewer molecules of benzene around the transition state than around the radicals. In the case of the homolysis of ethane into two methyl radicals in the gas phase (equilibrium (1.71)), $\Delta^{\ddagger}S(1.71) = 17.2$ eu, which corresponds to a fraction (about 50%) of the reaction entropy, estimated to be $\Delta_r S°(1.71) \approx 32$ eu.

$$CH_3 - CH_3 \rightleftarrows 2CH_3^{\bullet} \quad (1.71)$$

$$\Delta_r H^{\circ}(1.71) = \Delta H^{\circ}(Me^{\bullet}/Me^{\bullet}) = 2 \cdot \Delta_f H^{\circ}(Me^{\bullet}) - \Delta_f H^{\circ}(\text{ethane}) = 89.7\ \text{kcal mol}^{-1}$$

$$t\text{-Bu} - CH_3 \rightleftarrows t\text{-Bu}^{\bullet} + CH_3^{\bullet} \quad (1.72)$$

$$= DH^{\circ}(t\text{-Bu}^{\bullet}/Me^{\bullet}) = \Delta_f H^{\circ}(t\text{-Bu}^{\bullet}) + \Delta_f H^{\circ}(Me^{\bullet}) - \Delta_f H^{\circ}(\text{neopentane}) = 81.8\ \text{kcal mol}^{-1}$$

$$t\text{-Bu} - t\text{-Bu} \rightleftarrows 2t\text{-Bu}^{\bullet}\Delta_r H^{\circ} \quad (1.73)$$

$$= DH^{\circ}(t\text{-Bu}^{\bullet}/t\text{-Bu}^{\bullet}) = 2 \cdot \Delta_f H^{\circ}(t\text{-Bu}^{\bullet}) - \Delta_f H^{\circ}(Me_3C - CMe_3) = 68.8\ \text{kcal mol}^{-1}$$

Figure 1.12 indicates that the recombination of two radicals $R^{\bullet} = (CF_3S)_3C^{\bullet}$ to form $R{-}R = (CF_3S)_3C{-}C(SCF_3)_3$ has an activation enthalpy of 7.3 kcal mol^{-1}. This renders the recombination of the two radicals much slower than diffusion limited ($k_D \approx 10^{10}$ M^{-1}s^{-1}). The data demonstrate that the activation enthalpy $\Delta^{\ddagger}H$ of a homolytic dissociation cannot be considered as equivalent to the bond dissociation enthalpy (energy), $DH°(R^{\bullet}/X^{\bullet})$. The recombination of two radicals that are polyatomic species requires a change in geometry and a change in solvation (for reactions in solution) as a bond is formed. Most homolytic bond dissociation enthalpies have been measured by indirect methods, for instance by the method of radical buffers that can be applied to homolyses in the gas phase and in solution [88–90]. The equilibrium constant of equilibrium (1.74) is measured directly as a function of temperature, providing $\Delta_r H^T$ (1.74). One can write equilibrium (1.74′) for which

$\Delta_r H°(1.74')$=$DH°(R^•/X^•)$–$DH°(X^•/Y^•) = \Delta_r H°(1.74)$. If $DH°(X^•/Y^•)$ is known, $DH°(R^•/X^•)$ is determined by the relationship (1.75). In practice, and for reactions in the gas phase, $Y^• = Cl^•, Br^•$ or $I^•$. Concentrations of reactants and products are determined by mass spectrometry [88–90].

$$Y^• + R - X \rightleftarrows R^• + X - Y \tag{1.74}$$

$$Y^• + X^• + R - X \rightleftarrows R^• + X - Y + X^• \tag{1.74'}$$

$$DH^°(R^•/X^•) = \Delta_r H^°(1.74) + DH^°(X^•/Y^•) \tag{1.75}$$

For reactions in solution, the concentration of the radicals can be determined by ESR. An example is given with the measurements of the heats of formation of alkyl radicals in isooctane with that of the methyl radical serving as standard ($\Delta_f H°(Me^•) = 34.8$ kcal mol^{-1}). Typically, an isooctane solution of 0.5 M di-*tert*-butyl hyponitrite, 0.5 M triphenylarsine or triphenylboron, and 0.1–2.0 M of the two alkyl iodides is heated in the ESR spectrometer cavity and the radical concentration measured. The reaction of phenyl radical ($Ph^•$) produced by reaction (1.77) of the *tert*-butoxy radical (t-BuO$^•$) with $AsPh_3$ is irreversible. The phenyl radical reacts rapidly ($k > 10^5$ M^{-1}s^{-1}) with alkyl iodides RI and R′I generating radicals the corresponding $R^•$ and $R'^•$ and stable PhI. This reaction (1.78) is also essentially irreversible. Radicals $R^•$ and $R'^•$ equilibrate with their iodides RI and R′I, respectively (equilibria (1.79)). Thus, the relative concentration of radicals $R^•$ and $R'^•$ is given by $K(1.79)$.

$$t\text{-Bu} - O - N = N - O - t\text{-Bu} \rightarrow 2t\text{-BuO}^• + N_2 \tag{1.76}$$

$$t\text{-BuO}^• + Ph_3As \rightarrow t\text{-BuO} - AsPh_2 + Ph^• \tag{1.77}$$

$$Ph^• + R - I/R' - I \rightarrow Ph - I + R^•/R'^• \tag{1.78}$$

$$R^• + R' - I \rightleftarrows R - I + R'^• \tag{1.79}$$

Tributylstannyl radical produced by the photolysis of hexabutyldistannane (reaction (1.80)) can also be used as a radical initiator. Irreversible reaction (1.81) generates radicals $R^•$ and $R'^•$ that are equilibrated, as above, with their iodides [91].

$$Bu_3Sn - SnBu_3 \xrightarrow{h\nu} 2\ Bu_3Sn^• \tag{1.80}$$

$$Bu_3Sn^• + R - I/R' - I \rightarrow Bu_3Sn - I + R^•/R'^• \tag{1.81}$$

Photoacoustic calorimetry (PAC) [92] and time-resolved photoacoustic calorimetry (TR-PAC) [93] are increasingly being used to determine bond dissociation enthalpies in solution [94]. The PAC technique involves the measurement of a volume change that occurs when a laser pulse strikes a solution containing the reactants and initiates a chemical reaction. This sudden volume change generates an acoustic wave that can be recorded by a sensitive microphone such as ultrasonic transducer. The resulting photoacoustic signal can be analyzed in terms of rate and equilibrium constants of the reaction under investigation. Tables 1.A.7 and 1.A.8 give standard homolytic bond dissociation enthalpies for a collection of σ-bonded, organic compounds, as well as the enthalpies of formation of the radicals formed in these homolyses. These data constitute the basis of quantitative physical organic chemistry. Bordwell has proposed a simple method of estimating the BDEs $DH°(A^•/H^•)$ in weak acids A–H. The method uses empirical equation (1.82), an equation based on a thermodynamic cycle (Figure 1.13).

$$DH^°(A^•/H^•) = 1.36 \cdot pK_a + E_{ox}(A^-) + 73.3\ \text{kcal mol}^{-1} \tag{1.82}$$

The pK_a values in the DMSO of the acids AH (Table 1.A.24), which are accurate to ±0.2 pK_a unit, are multiplied by 1.36 (see Eq. (1.8)) to convert pK_a units to kcal mol^{-1} (at 298 K, $\Delta_r G°(AH \rightleftarrows A^- + H^+) = 1.36 \cdot pK_a$ with $pK_a = -\log([A^-][H^+]/[AH])$, and the oxidation potentials of their conjugate bases, $E_{ox}(A^-)$ in DMSO, in kcal mol^{-1}. The method has been applied to estimate $DH°(A^•/H^•)$ of acidic C—H, N—H, O—H, and S—H bonds [95–98].

Problem 1.21 Why is there no linear relationship for the homolytic bond dissociation enthalpies $DH°(Me^•/Me^•) = 89$ kcal mol^{-1}, $DH°(t\text{-Bu}^•/Me^•) = 81.8$ kcal mol^{-1}, and $DH°(t\text{-Bu}^•/t\text{-Bu}^•) = 68.8$ kcal mol^{-1}?

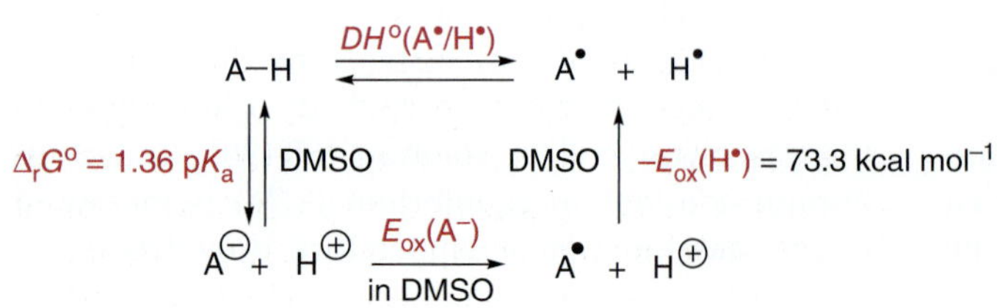

Figure 1.13 Thermodynamic cycle for the estimation of $DH°(A^•/H^•)$ of weak acids in DMSO (E_{ox} = IE in DMSO).

Problem 1.22 Why is the homolytic dissociation enthalpy of HF higher than that of HCl?

Problem 1.23 Why are oxygen-centered radicals more reactive than analogous carbon-centered radicals?

1.9.2 Substituent effects on the relative stabilities of radicals

Substituent effects are a fundamental aspect of physical organic chemistry. As we shall see (Chapter 2), the additivity rules for thermodynamic properties of molecules allow one to "transport" a property of a molecular fragment from one molecule to another one. The substituent effects on the relative stabilities of primary alkyl radicals can be defined by Eq. (1.83) [99].

$$E_S = DH^\circ(CH_3^\bullet/X^\bullet) - DH^\circ(R-CH_2^\bullet/X^\bullet) \quad (1.83)$$

For X = H, the values of E_S given in Table 1.A.9 are obtained.

All substituents R stabilize carbon-centered radicals, except CF_3 that destabilizes a primary alkyl radical by about 3 kcal mol^{-1} (this arises presumably from the strong hyperconjugative stabilization of the trifluoroalkane CF_3CH_3, Sections 1.7.4 and 4.8.5). For monosubstituted methanes with one substituent R, the radical stabilization enthalpy RSE(S) = E_S is given by Eq. (1.83). If X is larger than H, or more polar than H, there can be stabilization (see enthalpic anomeric effect, hydrogen bridging, Section 2.7.9) or destabilization (cyclic strain, front-strain, see reaction (1.70) and Section 2.6.8) effects in the precursors that are not present in the radicals R—$CH_2^\bullet$ and $X^\bullet$. Accordingly, polysubstituted methane derivatives may show RSE that are not additive with the E_S values defined by Eq. (1.83). For secondary and tertiary alkyl radicals, C—H bond homolytic dissociation enthalpies in DMSO and RSE are given in Table 1.A.10. These values were derived by measuring the heat of dissociation of $\Delta_r H$(1.84) (van't Hoff plot, ESR in solutions, no effect of solvent polarity). They include corrections for the "Back-strain" and "Front-strain" (Section 2.5.1) in the precursors that are not present in RR′R″CH and in radicals RR′R″C$^\bullet$, as well as corrections for differential electrostatic interactions between substituents R, R′, and R″ (Rüchardt's values [100–103]).

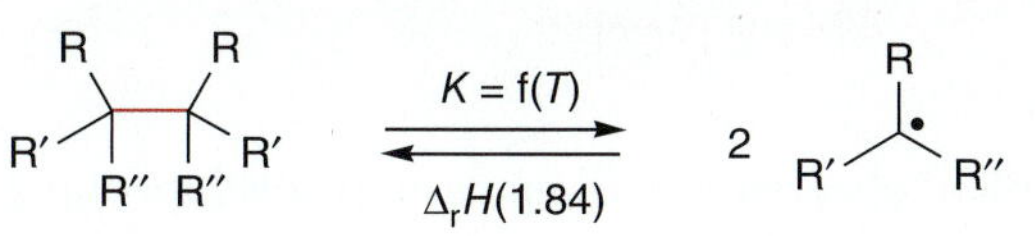

(1.84)

Problem 1.24 Give an interpretation for the better stabilizing effect of 4-amino than of 4-CN and 4-nitro group in 4-substituted phenoxyl radicals.

Problem 1.25 Compare the N—H homolytic bond dissociation enthalpies of amines (Table 1.A.11) and explain the nonadditivity of phenyl substitution on the stability of nitrogen-centered radical.

Problem 1.26 Calculate the standard gas-phase heat of reaction of the following additions:

$$R^\bullet + CO \rightleftarrows RCO^\bullet \quad \text{for } R = Ph, Me, t\text{-Bu}$$

$$Et^\bullet + MeCOMe \rightleftarrows EtC(Me)_2C{-}O^\bullet$$

$$Et^\bullet + \text{acetone} \rightleftarrows EtO{-}C^\bullet Me_2$$

$$Et^\bullet + \text{cyclopentene} \rightleftarrows \text{2-ethylcyclopentyl}^\bullet$$

Problem 1.27 What is the preferred regioisomeric adduct of the following equilibria?

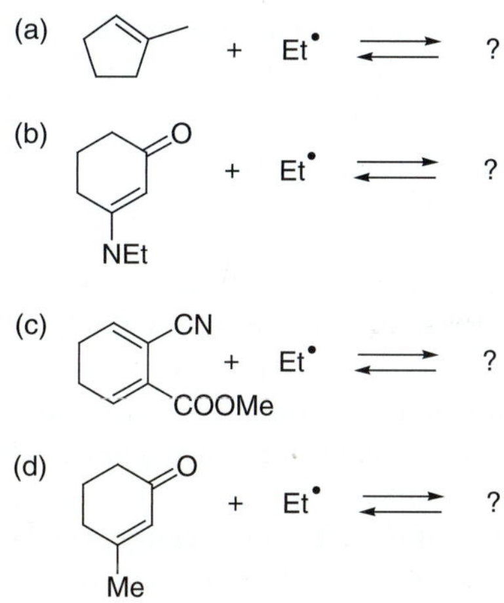

1.9.3 π-Conjugation in benzyl, allyl, and propargyl radicals

Table 1.A.11 gives the homolytic O—H, N—H, and C—H bond dissociation enthalpies as obtained by the Bordwell's method in DMSO (Eq. (1.82)). A simple estimate of the phenyl substituent effect on the stability of primary alkyl radicals is given by the difference $DH^\circ(n\text{-}Pr^\bullet/H^\bullet) - DH^\circ(PhCH_2^\bullet/H^\bullet) = 101.1 - 89.0 \cong$ 12 kcal mol^{-1}. For primary nitrogen-centered radicals, the same phenyl substituent effect is found by comparing $DH^\circ(MeNH^\bullet/H^\bullet)$ = 100 kcal mol^{-1} and $DH^\circ(PhNH^\bullet/H^\bullet)$ = 88 kcal mol^{-1}. In the case of oxygen-centered radical, the phenyl substituent stabilization is significantly larger; it amounts to 18 kcal mol^{-1} by comparing $DH^\circ(EtO^\bullet/H^\bullet)$=104.5 kcal

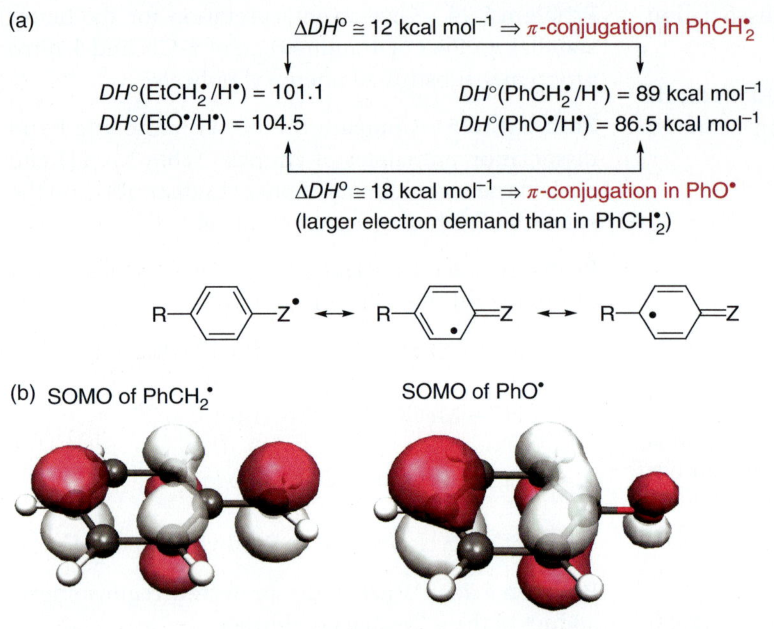

Figure 1.14 Interpretation of the relative stability of benzyl and phenyloxy radicals by (a) the valence bond theory, (b) the quantum calculations; SOMO's of $PhCH_2^•$ and $PhO^•$ showing electron delocalization to the ortho and para positions of the phenyl substituent (UHF/6-31G* calculations).

mol^{-1} and $DH°(PhO^•/H^•)$ = 86.5 kcal mol^{-1} (gas phase). This can be attributed to the greater electronegativity of oxygen atom than those of nitrogen and carbon atoms. The higher the electronegativity of the radical center, the higher its electron demand; phenyl donates electron density and stabilizes the radical. The stabilization effect introduced by phenyl substitution of a radical can be interpreted in terms of valence bond or resonance theory that implies delocalization of the unshared electron into the benzene ring as shown in Figure 1.14a. This interpretation is supported by quantum mechanical calculations. Delocalization of the spin by π-conjugation is indicated by the computed highest energy singly occupied molecular orbitals (SOMOs, Section 4.5.2) calculated for these species (Figure 1.14b).

Substitution of the phenyl group in $PhCH_2^•$ and $PhO^•$ at the ortho or para positions will influence the relative stabilities of these radicals, as those are the sites of significant odd electron density. In agreement with these predictions, phenol and 3-methoxyphenol (Table 1.A.11) have similar OH homolysis energies, whereas *para*-methoxy substitution provides an extra stabilization of c. 6 kcal mol^{-1}. Because of the radical delocalization into the phenyl ring, substitution of benzyl radical by a phenyl group to make a benzhydryl (diphenylmethyl) radical introduces a stabilization of c. 7.5 kcal mol^{-1}, which is less than the 12 kcal mol^{-1} observed for the exchange of an ethyl substituent for a phenyl group in primary alkyl radical (Figure 1.15), demonstrating the nonadditivity of phenyl substituent effects on the stability of alkyl radicals. Comparison of $DH°(Ph_2CH^•/H^•)$ = 82 kcal mol^{-1} to $DH°(Ph_3C^•/H^•)$ = 81 kcal mol^{-1} (Bordwell's values in DMSO, Table 1.A.11) provides another example. The absence of stabilization of the benzhydryl radical by a third phenyl substitution is due both to the diminishing effect of delocalization and because the three geminal phenyl rings cannot all be coplanar for optimal radical delocalization. Gauche interactions between the *ortho* hydrogen atoms force the trityl (triphenylmethyl) radical to adopt a propeller shape (Figure 1.15a).

$$DH°(PhCH_2^•/H^•) = 89\ \text{kcal mol}^{-1}\ \text{(gas phase)}$$

$$DH°(Ph_2CH^•/H^•) = 81.4\ \text{kcal mol}^{-1}\ \text{(82 in DMSO)}$$

$$DH°(Ph_3C^•/H^•) = 81\ \text{kcal mol}^{-1}\ \text{(DMSO)}$$

Comparison of the homolytic C—H bond dissociation enthalpies $DH°(CH_3CH_2CH_2^•/H^•)$ and $DH°(CH_2{=}CHCH_2^•/H^•)$ (Table 1.A.7) shows a difference of c. 15 kcal mol^{-1} that can be attributed to the delocalization of the allyl radical.

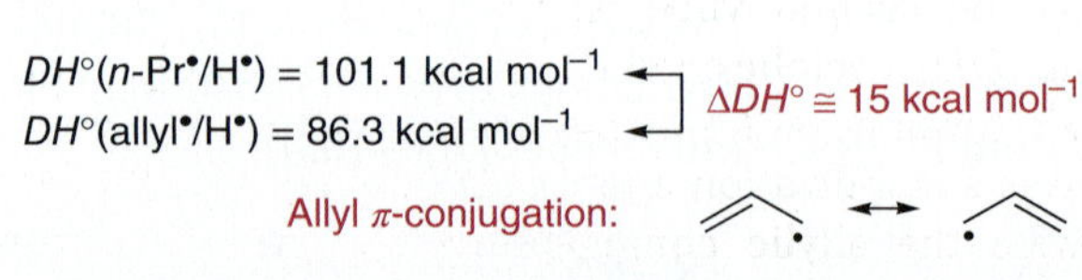

A similar observation is made with cyclopentane and cyclopentene.

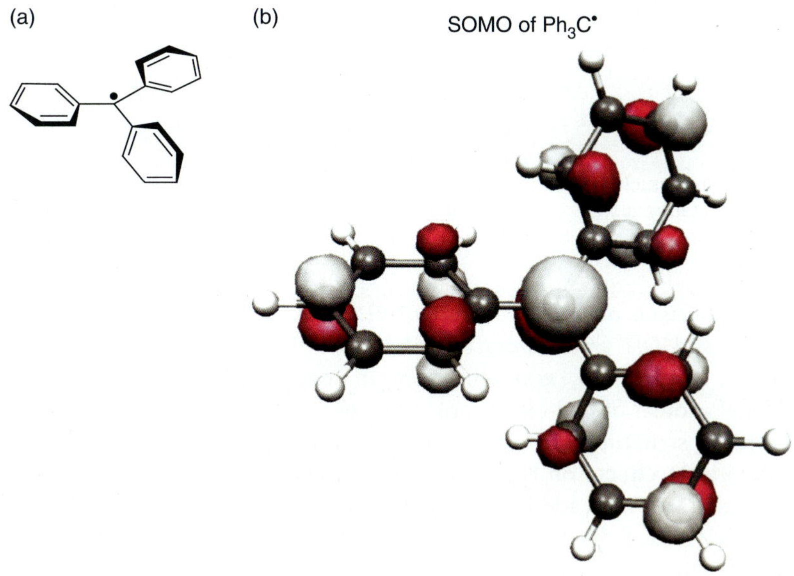

Figure 1.15 (a) Structure of trityl radical obtained by UHF calculations, (b) representation of its SOMO (singly occupied molecular orbital, Section 4.5).

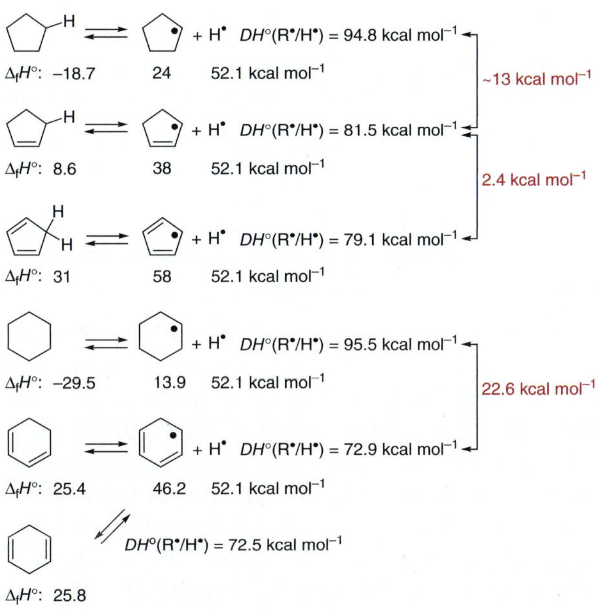

Introduction of a second double bond in cyclopentene generates cyclopentadiene for which the C—H homolytic bond dissociation enthalpy $DH°(c\text{-}C_5H_5^{\bullet}/H^{\bullet})$ = 79.1 kcal mol^{-1}, only 2.4 kcal mol^{-1} lower than $DH°$(cyclopent-2-enyl$^•$/H$^•$) = 81.5 kcal mol^{-1}. In this case, there is a massive nonadditivity effect for the second vinyl group conjugation. Comparing $DH°$(cyclohexyl$^•$/H$^•$) = 95.5 kcal mol^{-1} with $DH°$(cyclohexa-2,4-dienyl$^•$/H$^•$) = 72.9 kcal mol^{-1} gives a stabilization energy of 22.6 kcal mol^{-1}, about twice the allylic conjugation effect observed with cyclopentane and cyclopentene (c. 13 kcal mol^{-1}). It is thus evident that cyclopentadienyl radical suffers from some destabilization effect compared with its homolog, the cyclohexa-2,4-dienyl radical ($DH°$(cyclopentyl$^•$/H$^•$)−$DH°$(cyclopentadienyl$^•$/H$^•$) = 15.7 kcal mol^{-1} vs. $DH°$(cyclohexyl$^•$/H$^•$) − $DH°$(cyclohexa-2,4-dien-1-yl$^•$/H$^•$) = 22.6 kcal mol^{-1}). Simple applications of resonance theory imply that the stability of π-conjugated cation, anion, or radical depends on the number of equivalent limiting structures one can write for these species. However, a larger number of limiting structures does not always lead to a more stable species for cyclic systems, a result of the fact that not all resonance structures can mix with each other. Hückel's rule and aromaticity/antiaromaticity of cyclic conjugated systems are the result of such factors (Section 4.5). Not like cyclopent-2-enyl, cyclohex-2-enyl, and cyclohexadienyl radicals, cyclopentadienyl radical does not have any C—H bonds that hyperconjugate with the π-system (Section 4.8).

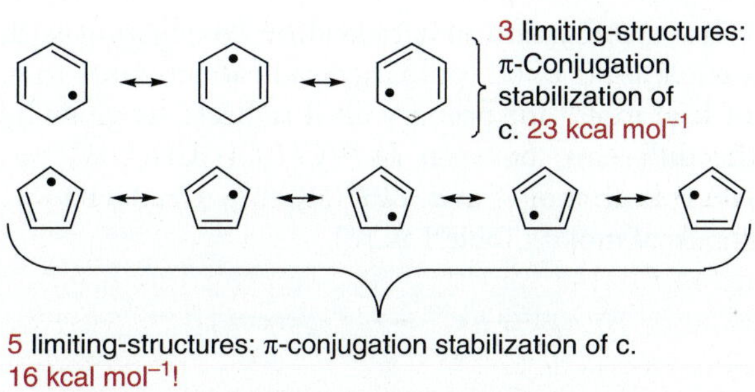

A similar observation is made for the C—H homolytic bond dissociation enthalpies of cyclopropane vs. cyclopropene and propane vs. propene. In this case, neither allyl radical nor cyclopropenyl radical have C—H bonds that hyperconjugate with the π-systems.

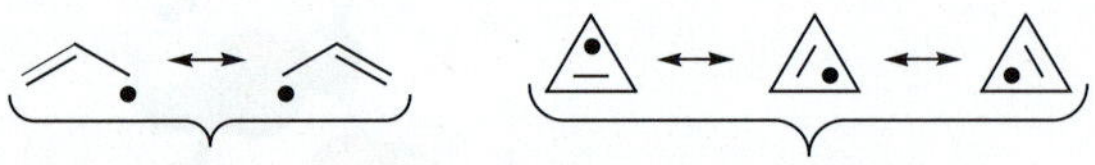

2 limiting-structures: allylic stabilization ~13.6 kcal mol^{-1}

3 limiting-structures: cyclopropenyl stabilization ~14 kcal mol^{-1}

Comparison of $DH°$(cycloheptyl$^•$/H$^•$) = 95.5 kcal mol^{-1} and $DH°$(cycloheptatrienyl$^•$/H$^•$) = 67.4 kcal mol^{-1} suggests a π-stabilization in cyclopentatrienyl radical of c. 28 kcal mol^{-1} for which 7 equiv. limiting structures can be written. Although the substituent effects on the relative stabilities of alkyl radicals are not additive, further conjugation of an allyl radical by vinyl substitution does increase its stability. β-Carotene has antiradical properties, and anticancer activity, because it can equilibrate with diradicals arising from the rotation about its C(13)—C(14) and C(15)—C(16) double bonds, at 37 °C already. This hypothesis is confirmed by the activation parameters ($\Delta^{\ddagger}H$, $\Delta^{\ddagger}S$, Section 3.3) measured for the (*E*) ⇄ (*Z*) isomerizations of the polyolefins of Table 1.A.12. These data allow the evaluation of stabilization due to π-conjugation in allyl, penta-2,4-dien-1-yl, hepta-2,4,6-trien-1-yl, and nona-2,4,6,8-octatetraen-1-yl radical. They amount to c. 13.5, 17, 19, and 21 kcal mol^{-1}, respectively. In this case, one finds that the larger the number of limiting structures of the conjugated radical, the greater is its relative stability [27].

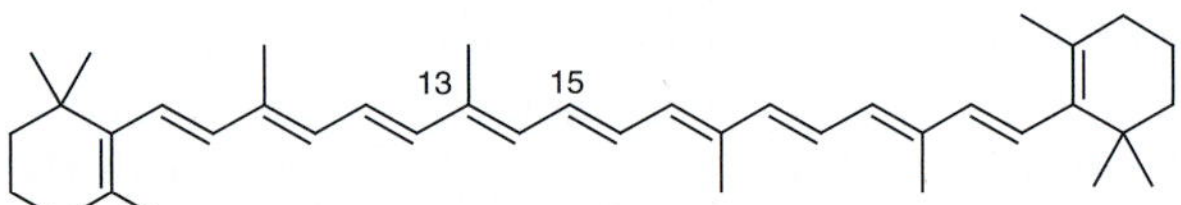

β-Carotene

The stabilization of a radical by conjugation with a triple C≡C bond (propargyl radical) amounts to c. 14 kcal mol^{-1} for primary alkyl radicals, as given by the difference between $DH°(CH_3CH_2CH_2{}^•/H^•)$ = 100.1 kcal mol^{-1} and $DH°(HC{\equiv}C{-}CH_2{}^•/H^•)$ = 86.5 kcal mol^{-1} (Table 1.A.13).

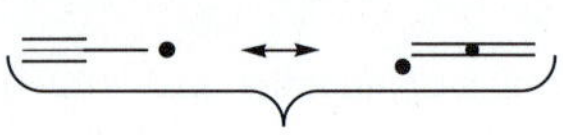

c. 14 kcal mol^{-1} of stabilization

Problem 1.28 If you had to propose good radical scavenging agents, which compounds listed in Table 1.A.11 would you choose?

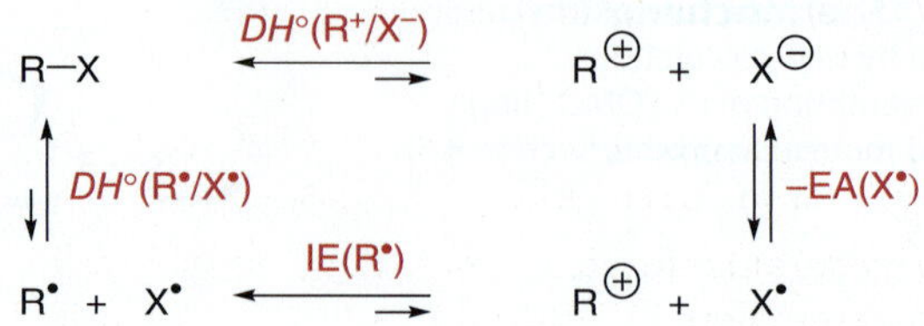

Figure 1.16 Thermodynamic cycle for the determination of heterolytic bond dissociation enthalpy: $DH°(R^+/X^-) = DH°(R^•/X^•) + IE(R^•) + (-EA(X^•))$.

1.10 Heterolytic bond dissociation enthalpies

The standard heterolytic bond dissociation enthalpy $DH°(R^+/X^-)$ for a compound, R—X, is given by the heat of the equilibrium reaction (1.85) in the gas phase, at 25 °C and under 1 atm. (Tables 1.A.13–1.A.16).

$$R - X \rightleftarrows R^+ + X^- \tag{1.85}$$

$$DH^{\circ}(R^+/X^-) = \Delta_f H^{\circ}(R^+) + \Delta_f H^{\circ}(X^-) - \Delta_f H^{\circ}(RX)$$

1.10.1 Measurement of gas-phase heterolytic bond dissociation enthalpies

The direct measurement of equilibrium constant K(1.85) in the gas phase is not possible, as pyrolysis of R–X in the gas phase will not give cations and anions, but rather radicals. In order to evaluate $DH°(R^+/X^-)$, a Born–Haber thermodynamic cycle must be applied as shown in Figure 1.16. The homolytic bond dissociation enthalpy $DH°(R^•/X^•)$, the ionization enthalpy IE(R$^•$) of radical R$^•$, and the electron affinity −EA(X$^•$) of radical X$^•$ can be used to estimate $DH°(R^+/X^-)$ [104–108].

Problem 1.29 Compare the gas-phase heterolytic BDEs ROH ⇄ RO$^-$ + H$^+$ for methanol, ethanol, isopropanol, and *tert*-butanol. Is the trend the same in solution?

Problem 1.30 Diazotization of primary alkyl amines with $NaNO_2/HCl/H_2O$ at 0 °C leads to mixtures of alcohols, chlorides, and alkenes with the evolution of N_2. The same reaction with aniline and other aromatic primary amines generates at 0 °C persistent diazonium salts that decompose with N_2 evolution on heating above 60 °C. Why is there this difference in behavior between the diazonium salts resulting from alkyl and aryl primary amines?

1.10.2 Thermochemistry of ions in the gas phase

High-pressure mass spectrometry (MS) and ICR techniques [109–121] allow one to measure equilibrium constants for ion/molecule reactions such as proton transfers (1.86) [113b, 122, 123], hydride transfers (1.87) [124, 125], and halide transfers (1.88) [126]. Van't Hoff plots provide the heats and entropies of these equilibria in the gas phase with high accuracy [127, 128].

$$A + AH^+ \rightleftarrows A + BH^+ \quad (1.86)$$

$$R-H + R'^+ \rightleftarrows R^+ + R'-H \quad (1.87)$$

$$R-X + R'^+ \rightleftarrows R^+ + R'-X \quad (1.88)$$

The gas-phase proton affinity PA(**A**) of a compound **A** is defined as

$$PA(\mathbf{A}) = \Delta_r H^\circ(1.89) = \Delta_f H^\circ(\mathbf{A}) + \Delta_f H^\circ(H^+) - \Delta_f H^\circ(\mathbf{AH^+})$$

$$\mathbf{AH^+} \rightleftarrows \mathbf{A} + H^+ \quad (1.89)$$

The hydride affinities HA(R^+) of carbenium ions R^+ are defined as

$$HA(R^+) = \Delta_r H^\circ(1.90) = \Delta_f H^\circ(R^+) + \Delta_f H^\circ(H^-) - \Delta_f H^\circ(R-H)$$

$$R-H \rightleftarrows R^+ + H^- \quad (1.90)$$

and halide affinities of carbenium ions R^+ are defined as

$$\Delta_r H^\circ(1.91) = \Delta_f H^\circ(R^+) + \Delta_f H^\circ(X^-) - \Delta_f H^\circ(RX)$$

$$R-X \rightleftarrows R^+ + X^- \quad (X^- = \text{halide anion}) \quad (1.91)$$

A collection of proton affinities is given in Table 1.A.15 for amines and other Lewis bases and for alkenes and benzene derivatives. Comparison of these data allows one to define substituent effects on the relative stabilities of ammonium ions, alkoxonium ions, sulfonium ions, phosphonium ions, and carbenium ions. We note that the stabilization effect of methyl substituents on cations with an octet of electrons (ammonium, oxonium, sulfonium, and phosphonium ions) is weaker than for carbenium ions with a sextet of electrons. For carbenium ions in the gas phase, the ethyl substituent stabilizing effect is larger than the methyl substituent stabilization effect (e.g. $\Delta_r H^\circ(Me_3C^+ + Me_2(Et)C{-}H \rightarrow Me_3C{-}H + Me_2(Et)C^+) = -2.8$ kcal mol^{-1}). In solution, steric hindrance to solvatation compensates for this difference and generally makes the larger cation or anion less stable than the smaller ones (Baker–Nathan effect). The larger the acyclic alkyl cation, the more stable it is because $2p^{(+)}/\sigma(C{-}C)$ hyperconjugation is more stabilizing than $2p^{(+)}/\sigma(C{-}H)$. However, hyperconjugation (π/σ interaction) is not the unique cause of alkyl substituent effects. Under the influence of charge (positive or negative), σ-bonds are also polarized. This creates induced dipoles that contribute to the stabilization of the cation or anion. The electrostatic field model for substituent effects considers two contributions: the substituent dipole/charge stabilization $V_c = -(q\mu|\cos\theta|)/\varepsilon r^2$ or destabilization $V_c = (q\mu|\cos\theta|)/\varepsilon r^2$ and the stabilization due to the substituent polarizability (induced dipole, hyperconjugation, conjugation) $V_I = -(q^2\alpha)/2\varepsilon r^4$, where q is the charge, μ is the permanent dipole of the substituent, θ is the angle it makes with the lines of the electrical field created by the charge, ε is the dielectric constant of the medium, α is the polarizability of the substituent, and r is the distance separating it from the charged center. Substituent effects are not strictly additive. The larger an ionized species, the more delocalized is the charge, and the weaker is the substituent effect. Substituent effects depend on the electronic demand of the ions as shown by equilibria (1.92)–(1.94).

$$H_2C{=}CH_2 + i\text{-}Pr^+ \rightleftarrows Me{-}CH{=}CH_2 + Et^+ \quad (1.92)$$

$$\Delta_r H^\circ(1.92) = PA(\text{propene}) - PA(\text{ethylene}) = 19.4\ \text{kcal mol}^{-1}$$

$$PhCH{=}CH_2 + Ph(Me)_2C^+ \rightleftarrows Ph(Me)C{=}CH_2 + Ph(Me)CH^+$$

$$\Delta_r H^\circ(1.93) = PA(\alpha\text{-methylstyrene}) - PA(\text{styrene}) = 5.9\ \text{kcal mol}^{-1} \quad (1.93)$$

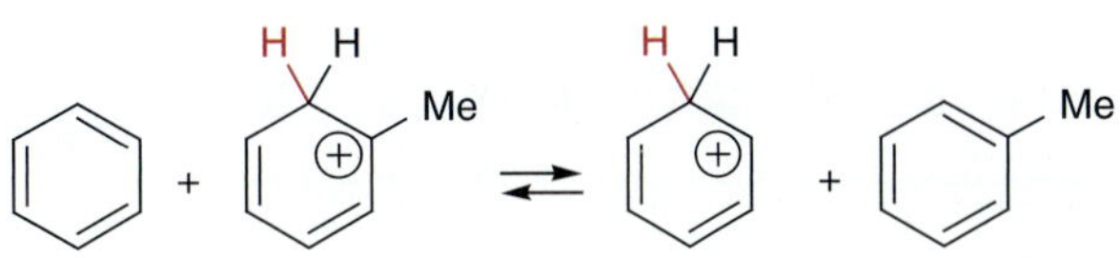

(1.94)

$$\Delta_r H^\circ(1.94) = PA(\text{toluene}) - PA(\text{benzene}) = 6.5\ \text{kcal mol}^{-1} = \Delta_f H^\circ(C_6H_7^+) + \Delta_f H^\circ(\text{toluene}) - \Delta_f H^\circ(PhH) - \Delta_f H^\circ(MeC_6H_6^+)$$

Comparison of proton affinities of H_2O (170.3), H_2S (173.9), H_2Se (174.8), AsH_3 (180.9), PH_3 (187.3), and NH_3 (202.3 kcal mol^{-1}) shows that the proton affinity of these bases diminishes with the increasing electronegativity of the heteroatoms for the series

O, S, and Se but increases with electronegativity for series N, P, and As! The fact that HC≡N (174.5) and H_2C=O (174.6 kcal mol^{-1}) have similar proton affinities suggests that the triply bonded nitrogen atom has similar electronegativity as the doubly bonded oxygen center. Substitution of an alkyl group by a fluoro substituent destabilizes a cation because of the permanent dipole created by the C—X bond. The same trend is observed for chloro substitution of alkyl groups. However, these substituents (F and Cl) can also stabilize small carbenium ions (e.g. $\Delta_r H°(Me^+ + FCH_2{-}H \rightarrow Me{-}H + F{-}CH_2{}^+) = -22.6$ kcal mol^{-1}, $\Delta_r H°(FCH_2{}^+ + F_2CH{-}H \rightarrow FCH_2{-}H + F_2CH^+) = -5.7$ kcal mol^{-1}, but $\Delta_r H°(F_2CH^+ + F_3C{-}H \leftarrow F_2CH{-}H + F_3C^+) = 15.1$ kcal mol^{-1}). Thus, when the electron demand is high (localized positive charge), n(X:)/$2p^{(+)}$ conjugation intervenes and stabilizes the cation. The proton affinity of methanol (182.2) is 7.6 kcal mol^{-1} higher than that of formaldehyde (174.6). Thus, oxygen centers in alcohols appear to be more basic than doubly bonded oxygen centers in similar environment (alkyl and hydrogen substituents). Similarly, one finds PA($MeNH_2$) about 9 kcal mol^{-1} higher than PA (CH_2=NH), itself 36.8 kcal mol^{-1} higher than PA (HC≡N).

Carbenium ions react in the gas phase with alkanes [117, 118, 127, 129] and silanes [127, 128, 130–135] via bimolecular processes involving hydride transfers (1.87) (Tables 1.A.15 and 1.A.16). These reactions occur with rate constants from 2×10^6 to 3×10^{10} $dm^3\,mol^{-1}\,s^{-1}$, which correspond to high collisional efficiencies of 10^{-4} to 1. Groups other than hydride can be transferred in collisions of carbenium ions with neutral molecules, such as halide (1.88), methide, and ethide anions (equilibria (1.95)).

$$R^+ + R'{-}X \rightleftarrows R{-}X + R'^+ \qquad X = Me, Et \tag{1.95}$$

Problem 1.31 Explain the difference in heterolytic bond dissociation enthalpies $DH°$(cyclopent-2-enyl$^+$/H^-), $DH°$(cyclopentadienyl$^+$/H^-), and $DH°$ (tropylium$^+$/H^-).

Problem 1.32 Ethyl cation and methylsilicenium ion are both primary cations. Explain the data of Table 1.A.16 in particular $\Delta_r H°(Et^+ + MeSiH_3 \rightleftarrows MeSiH_2{}^+ + EtH) = -20$ kcal mol^{-1}. Why is this value not closer to zero?

1.10.3 Gas-phase acidities

The Gibbs free energies of equilibria $\Delta_r G°(1.96)$ in the gas phase at 298 K give the gas-phase acidities of compounds X—H [136].

$$X - H(gas) \rightleftarrows H^+(gas) + X^-(gas) \tag{1.96}$$

Selected gas-phase acidities $\Delta_r G(A^-/H^+)$ are given in Table 1.A.17 and gas-phase proton affinities $\Delta_r H(R^-/H^+)$ are collected in Table 1.A.18. Acidities in water and in DMSO (Me_2SO) are given as pK_a values in Tables 1.A.23 and 1.A.24, respectively. As shown with Eq. (1.8), $\Delta_r G°(1.96)$ in kcal mol^{-1} = $-1.36 \cdot \log K_a$, where K_a is the acidity constant and $pK_a = -\log K_a$. Using kcal mol^{-1} for the Gibbs energies $pK_a(1.36) = \Delta_r G°(1.96)/1.36$; they are measured by evaluating the equilibrium constants $K(1.97)$ for the proton exchange reactions (1.97) in the gas phase by high-pressure mass spectrometry or by Fourier transform ICR mass spectrometry [114–116].

$$X - H + A^- \rightleftarrows X:^- + A - H \tag{1.97}$$

$$NF_3 + e^- \rightarrow F^- + NF_2 \tag{1.98}$$

For instance, fluoride anion, F^-, can be produced through electron capture by NF_3 (reaction (1.98)). F^- reacts with acids, XH, generating the conjugate base X^- and HF. The concentrations of XH, F^-, X^-, and HF are measured once equilibrium is reached in the reaction chamber of the spectrometer [137–141]. De Puy et al. have used an alternative method involving a flowing afterglow-selected ion flow tube [142–148]. In one example, an alkyltrimethylsilane reacts with hydroxide anion to form a siliconate anion **A** that expels an alkyl (k_1) or a methyl anion (k_2), irreversibly. Either of these anions then reacts irreversibly with the silanol **B** and **C** to form alkanes RH, CH_4, and the corresponding trialkylsiloxide anions **D** and **E**, respectively (Figure 1.17). The relative amount of the two siloxide anions **D** and **E**, reflects the ease of formation of the two ion–dipole complexes **C** and **B**, which in turn is determined by the relative ease of formation of the two carbanions R^- and Me^- after a statistical correction is made because of the presence of three methyl groups and only one alkyl group. The enthalpy difference between ion dipole complexes **B** and **C** depends on the sum of the differences in the R—Si and Me—Si bond strengths as well as the electron affinities of $R^\bullet$ and $Me^\bullet$ as $\Delta\Delta H(\mathbf{B} \rightleftarrows \mathbf{C}) = [DH°(R^\bullet/Si^\bullet) - DH°(Me^\bullet/Si^\bullet)] + [(-EA(R^\bullet)) - (-EA(Me^\bullet))]$ [142, 148].

The existence of a correlation between these two processes implies a linear correlation between $DH°(R^\bullet/H^\bullet)$ and $DH°(R^\bullet/Si^\bullet)$. This is verified if

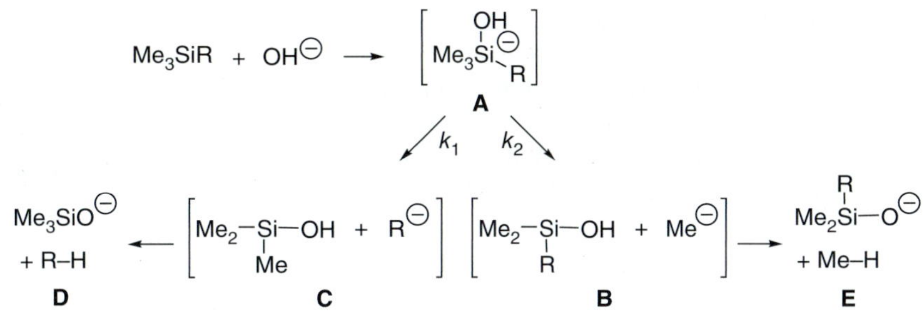

Figure 1.17 A kinetic method for the determination of relative gas-phase acidities of hydrocarbons R–H. The ratio [**D**]/[**E**] is correlated by the relative basicities of the alkyl and methyl anions (calibration with $\Delta_r H°(CH_4 \rightleftarrows Me^- + H^+) = 416.6\ \text{kcal mol}^{-1}$, $\Delta_r H°(PhH \rightleftarrows Ph^- + H^+) = 401.7\ \text{kcal mol}^{-1}$, see Table 1.A.13).

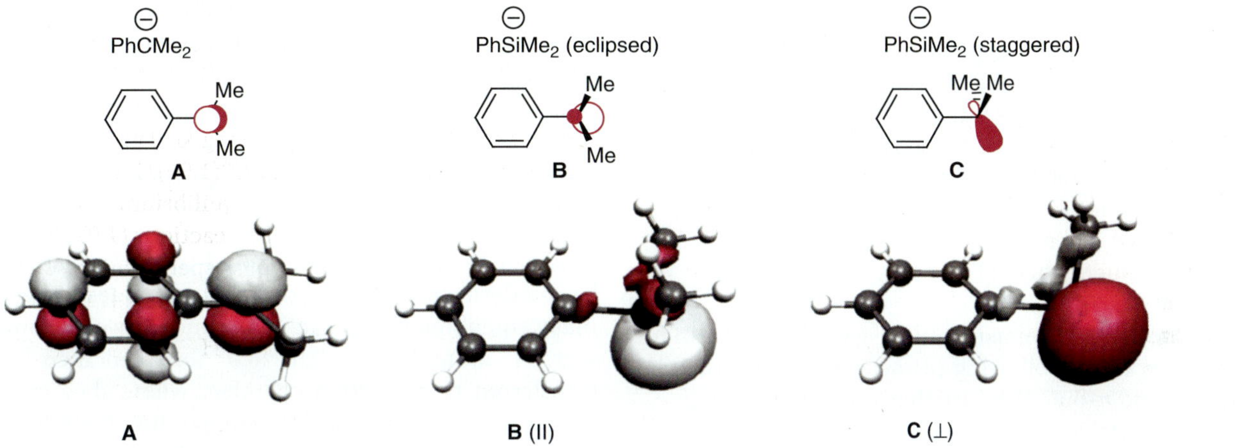

Figure 1.18 Calculated dimethylphenylmethyl anion (**A**) and phenyldimethylsilyl anion (**B**) and (**C**). The carbanion $PhMe_2C^-$ adopts a planar structure for the carbanion, whereas the silicon atom is pyramidal in the silyl anion. The RHF/6-31G* HOMO of each is shown. $PhSiMe_2^-$ has very little preference in terms of the rotation around the Ph—Si bond. The structure (**B**, eclipsed) in which the lone pair of the Si center is conjugated with π-electrons of the phenyl ring is only $0.16\ \text{kcal mol}^{-1}$ more stable than the conformer (**C**, staggered) in which the Si lone pair resides in the π-plane (perpendicular to the π-orbitals of the phenyl ring).

Eq. (1.99) holds (experiments for alkane acidities led to $\beta = 0.221$).

$$\ln[k_1/k_2] = -\beta[\Delta_r H^\circ(\text{RH} \rightleftarrows \text{R}^- + \text{H}^+) - \Delta_r H^\circ(\text{Me}-\text{H} \rightleftarrows \text{Me}^- + \text{H}^+)] \quad (1.99)$$

Squires and coworkers have evaluated gas-phase acidities of hydrocarbons by measuring the appearance potentials or AP(R^-) of ions R^- resulting from the decomposition of carboxylate anion $RCOO^-$ by collision with a flux of helium (reaction (1.100)) according to Eq. (1.101) [149, 150].

$$\text{RCOO}^- \xrightarrow[\text{(collision)}]{\text{He}} \text{R}^- + \text{CO}_2 \quad (1.100)$$

$$DH^\circ(\text{R}^-/\text{H}^+) = \Delta_r H^\circ(1.100) + DH^\circ(\text{RCOO}^-/\text{H}^+) + \Delta_f H^\circ(\text{RCOOH}) - \Delta_f H^\circ(\text{RH}) - \Delta_f H^\circ(\text{CO}_2) \quad (1.101)$$

Van't Hoff plots of equilibria (1.97) allow the determination of proton affinities defined as $\Delta_r H°$ (R—H $\rightleftarrows$ $R^- + H^+$) = $DH°(R^-/H^+)$ (Table 1.A.18). Gas-phase acidities provide intrinsic substituent effects on the relative stabilities of negatively charges species (Table 1.A.19). Here, we compare the acidities of alkanes and silanes (Eq. (1.101)). In the gas phase, silyl anions are much more stable than carbanions of similar size; for instance: [21–23, 43, 68]

$$\Delta_r G^\circ(\text{Me}-\text{H} \rightleftarrows \text{Me}^- + \text{H}^+) = 408.5\ \text{kcal mol}^{-1}$$
$$\Delta_r G^\circ(\text{H}_3\text{Si}-\text{H} \rightleftarrows \text{H}_3\text{Si}^- + \text{H}^+) = 363.8\ \text{kcal mol}^{-1}$$
$$\Delta_r G^\circ(\text{MeSiH}_3 \rightleftarrows \text{MeSiH}_2^- + \text{H}^+) = 369.6\ \text{kcal mol}^{-1}$$
$$\Delta_r G^\circ(\text{Me}_2\text{SiH}_2 \rightleftarrows \text{Me}_2\text{SiH}^- + \text{H}^+) = 373.2\ \text{kcal mol}^{-1}$$
$$\Delta_r G^\circ(\text{Me}_3\text{SiH} \rightleftarrows \text{Me}_3\text{Si}^- + \text{H}^+) = 377\ \text{kcal mol}^{-1}$$
$$\Delta_r G^\circ(\text{PhSiH}_3 \rightleftarrows \text{PhSiH}_2^- + \text{H}^+) = 361.0\ \text{kcal mol}^{-1}$$
$$\Delta_r G^\circ(\text{PhSiMe}_2\text{H} \rightleftarrows \text{PhSiMe}_2^- + \text{H}^+) = 366.5\ \text{kcal mol}^{-1}$$

This is related to the higher electron affinities of silyl than alkyl radicals, e.g. $-\text{EA}(\text{SiH}_3^\bullet) = 32.4\ \text{kcal mol}^{-1}$, $-\text{EA}(\text{Me}^\bullet) = 1.8\ \text{kcal mol}^{-1}$. Methyl substitution of methane increases its acidity, except for the first methyl substitution (see Table 1.A.18). In contrast, the gas-phase acidity of silane is found to decrease by approximately 3–5 kcal mol^{-1} with each successive methyl substitution [151]. The substitution of a phenyl group has essentially no effect on the acidity of silanes, suggesting that the π-delocalization does not stabilize silyl anions [152]. This is in contrast to observed π-effects in carbanions, e.g. $\Delta_r G°(PhCH_3 \rightleftarrows PhCH_2^- + H^+) = 373.7$ kcal mol^{-1} vs. $\Delta_r G°(CH_4 \rightleftarrows Me^- + H^+) = 408.5$ kcal mol^{-1} [153].

The striking difference in substituent effects on the relative stability of carbanions and silyl anions has been explored by quantum mechanical calculations: silyl anions are pyramidal, and the C—Si bond in $PhMe_2Si^-$ anion is not shorter than the C—Si bond in silane $PhMe_2SiH$. The situation is quite different in carbanions that adopt planar structures, in which anion can donate its electrons into a π-substituent through resonance as illustrated in Figure 1.18. Data reported in Table 1.A.19 show similar substituent effects on the relative stability of alcoholates, amides, and carbanions. The data also show that the more delocalized (the more stabilized) an anion, the weaker the substituent effect (Table 1.1).

Problem 1.33 Compare the gas-phase hydride affinities of the following primary alkyl cations: Et^+, n-Pr^+, and n-Bu^+. Why are they not the same?

1.11 Electron transfer equilibria

Heats and entropies of ionization of organic compounds can be determined by Van't Hoff plots (measurements of equilibrium constants K as function of T) of equilibria (1.102).

$$R^1 - H + [R^2 - H]^{\bullet +} \rightleftarrows [R^1 - H]^{\bullet +} + R^2 - H \quad (1.102)$$

For many radical cations of type $RH^{\bullet +}$, heat of formation follow an additivity rule [154].

1.12 Heats of formation of neutral, transient compounds

We have seen already that modern techniques of mass spectrometry, ion cyclotron resonance, and laser spectroscopy allow one to measure thermodynamic parameters of many charged molecules in the gas phase. We now show that the same techniques can be applied to measure the gas-phase thermodynamic data of transient neutral species such as carbenes, diradicals, and unstable organic compounds.

1.12.1 Measurements of the heats of formation of carbenes

Ions can be generated and stored in the gas phase by modern mass spectrometry techniques. Thermally equilibrated ions can be produced by collisions with He atoms. These "thermal" ions can then be transferred to a reaction chamber, where they may react with all kinds of gaseous compounds. Collisions with jets of Ar atoms of specific kinetic energies can be used to determine the energetic threshold for inducing reactions such as the formation of carbenes: CX_2 from anions CX_3^- according to reaction (1.103)

$$CX_3^- + \text{Ar(kinetic energy)} \rightarrow :CX_2 + X^- + \text{Ar} \quad (1.103)$$

With that method, Squires and coworkers have determined the heats of formation of the following carbenes [155–157]:

Table 1.1 Nonlinearity of substituent effects on the relative stability of carbanions and amide anions.[a)]

Substituent	XCH_3	X_2CH_2	Diff.	X_3CH	Diff.	Substituent	XNH_2	X_2NH	Diff.
X = H	(0.0)	(0.0)		(0.0)		X = H	(0.0)	(0.0)	
F	≤18.0	27.0	*≤9*	39.3	*12.3*	F		32.8	
C_6H_5	34.8	50.3	*15.5*	55.7	*5.4*	$(CH_3)_3Si$	25.1	43.2	*18.1*
CF_3	38.0	64.6	*26.6*	81.7	*17.1*	C_6H_5	37.0	52.3	*15.3*
CN	44.5	80.2	*35.7*	115.0	*34.8*	CH_3CO	41.0	56.3	*15.3*
CH_3CO	46.6	71.8	*25.2*	79.6	*7.8*	C_6F_5	54.8	79.7	*24.9*
C_6F_5	53.8			90.9		CF_3CO	59.4	88.6	*29.2*
CF_3CO	66.4	97.8	*31.4*	107.6	*9.8*	4-C_5F_4N	63.3	90.0	*26.7*
CF_3SO_2	68.7	106.7	*38.0*	117.0	*10.3*	CF_3SO_2	74.8	104.3	*29.5*
$C_4F_9SO_2$		119.8				$C_4F_9SO_2$	81.0	112.0	*31.0*

Blue italics are used to indicate differential substituent effects, to emphasize the non-linearity of the substituent effects on the stability of carbon-centered and nitrogen-centered anions.

a) Taken from [153].

	$:CH_2$	$:CHCl$	$:CHF$	$:CClF$	$:CCl_2$	$:CF_2$	$:CHCH{=}CH_2$	$:CH{-}Ph$
$\Delta_f H°$:	92.9	78.0	34.2	7.4	55.0	−44.0	93.3	108.2 kcal mol^{-1}
	±0.6	±2.0	±3.0	±3.2	±2.0	±2.0	±2.6	±3.5

Similarly, by Ar atom bombardment of carbene radical anion $:CXCl^{\bullet -}$ that produces Cl^- and carbynes $:CX^\bullet$, the following heats of formation have been obtained: $\Delta_f H°(:CH^\bullet) = 142.2 \pm 3.2$, $\Delta_f H°(:CF^\bullet) = 60.6 \pm 3.4$, and $\Delta_f H°(:CCl^\bullet) = 105.9 \pm 3.1$ kcal mol^{-1} [158].

1.12.2 Measurements of the heats of formation of diradicals

In a similar manner, the heats of formation of diradicals have been obtained. For instance, *o*-, *m*-, and *p*-benzyne were generated from the corresponding *o*-, *m*-, and *p*-chlorophenyl anions, which were generated in turn through either deprotonation or desilylation (Figure 1.19). The heats of formation of the neutral benzynes have been determined using dissociation induced by collision with argon atoms (collision-induced dissociation: CID) [159]. Electron affinities and singlet/triplet energy gap for *o*-, *m*-, and *p*-benzyne have also been determined [160]. *o*-, *m*-, and *p*-benzyne can be generated from 1,2-, 1,3-, and 1,4-dibromobenzene, respectively, using a molecular beam. Using femtosecond mass spectrometry, Zewail and coworkers established that the time required for the two successive C—Br bond cleavages is less than 100 fs. Based on this, these didehydrobenzenes have a lifetime of at least 400 ps [161].

In a similar manner, the heat of formation of trimethylenemethane, a 1,3-diradical, was determined to be 90 ± 5 kcal mol^{-1} [162]. The unstable α-lactone (oxooxirane) can be generated in the gas phase by Ar atom bombardment of chloroacetate anion, and its heat of formation has also been estimated (1.104) [163].

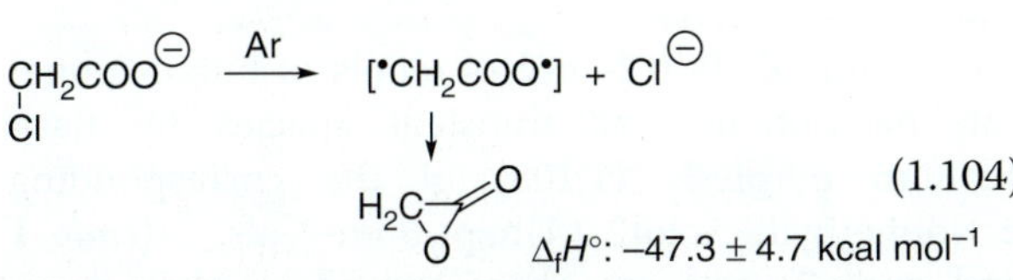

(1.104)

Another technique to produce unstable neutral species in the gas phase implies the deprotonation of carbocations with bases **B**. The reaction only occurs if the proton affinity of the base equals or surpasses that of the neutral species to be studied. By this method, the heats of formation of *o*-xylylene, *m*-xylylene, and *p*-xylylene have been measured (1.105). Methyl-substituted benzyl cations are generated by ionization of the corresponding bromides under electron impact. The nascent cations are cooled by a helium "bath" and stored in the cavity of a cyclotron resonance mass spectrometer and are then allowed to react with different bases **B** of known proton affinities.

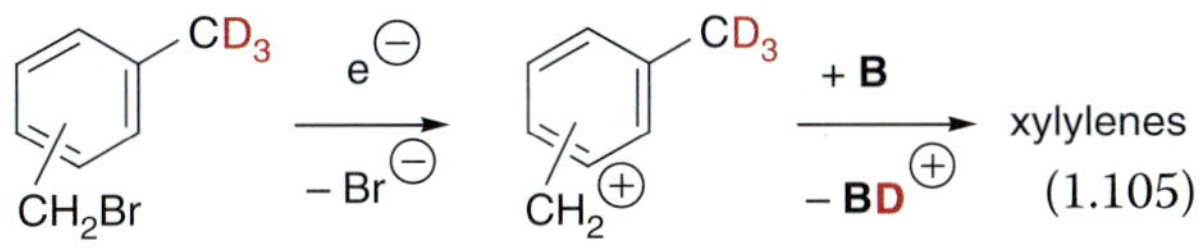

(1.105)

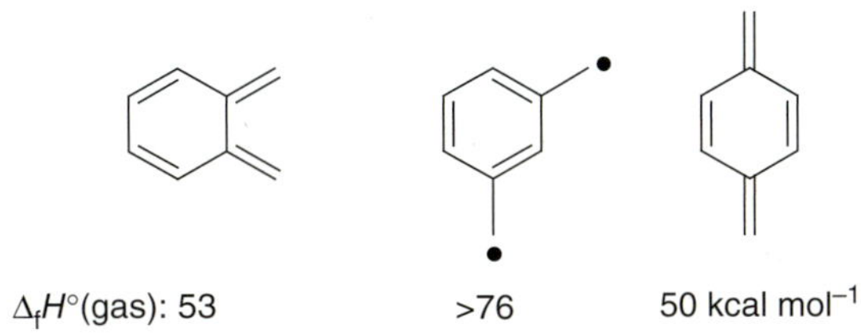

Problem 1.34 Calculate the heat of cyclopropanation of ethylene with CH_2: (methylene).

Problem 1.35 Calculate the heat of the addition of trimethylenemethane to ethylene giving methylenecyclopentane.

Problem 1.36 What is the major product of cyclodimerization of *o*-xylylene?

1.12.3 Keto/enol tautomerism

In 1904, Lapworth first suggested that enol formation is the rate-limiting step in the α-halogenation of ketones. In contrast to the keto tautomer, which reacts with nucleophiles, the enol (and enolate) is the reactive form on which electrophilic additions occur [164].

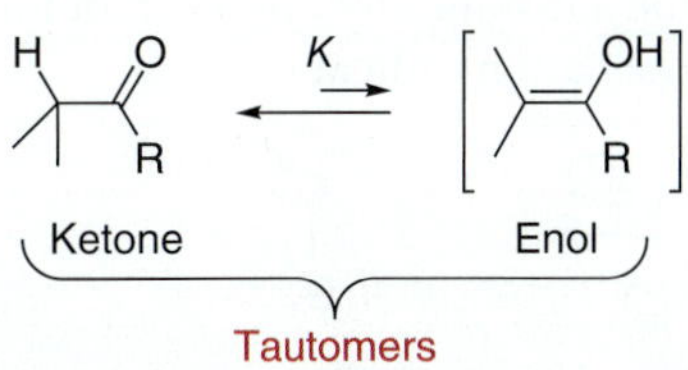

The mass spectrum of 3-methylhexan-2-one shows a $C_4H_8O^{\bullet +}$ ion. It results from the McLafferty fragmentation (1.106), producing propene and a C_4-enol radical cation. By measuring the appearance potential for the formation of this ion, one can evaluate its heat of formation:

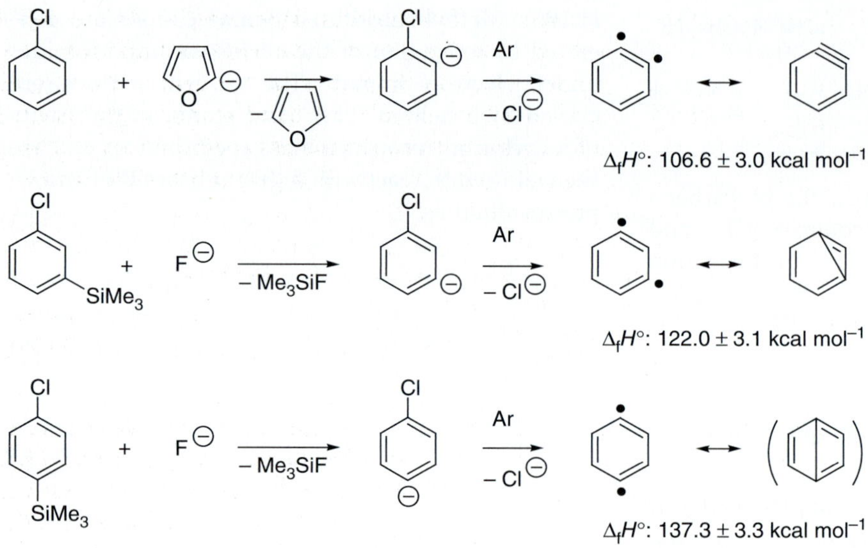

Figure 1.19 Examples of carbanions generated in the gas phase and undergoing collision-induced dissociations to form *o*-, *m*-, and *p*-benzyne. The contribution of the resonance structure in parenthesis is negligible.

$$
\begin{aligned}
&\Delta_f H^\circ(\mathrm{MeC^{(+)}(OH)\text{—}C^\bullet HMe}) \\
&\quad = \Delta_f H^\circ(\text{3-methylhexan-2-one}) \\
&\qquad - \Delta_f H^\circ(\text{propene}) + \mathrm{AP}(\text{enol cation}) \\
&\quad \cong 144\ \mathrm{kcal\ mol^{-1}}
\end{aligned}
$$

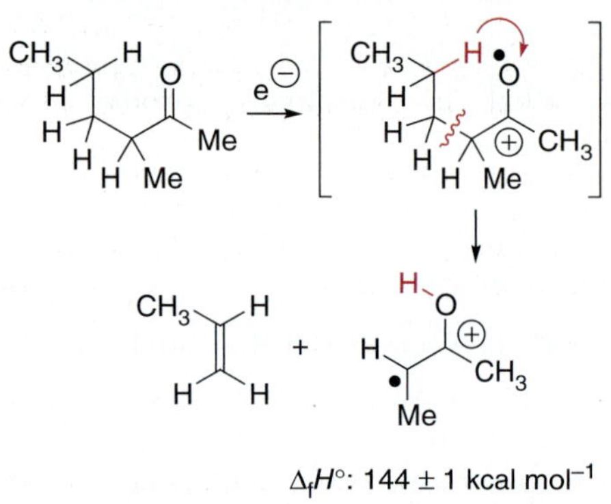

(1.106)

The experimental heat of formation of the resulting keto radical cation is $\Delta_f H°[\mathrm{Et(CO)Me^{\bullet+}}] = 162 \pm 1$ kcal mol^{-1}, a value substantially higher than for the corresponding enol cation radical.

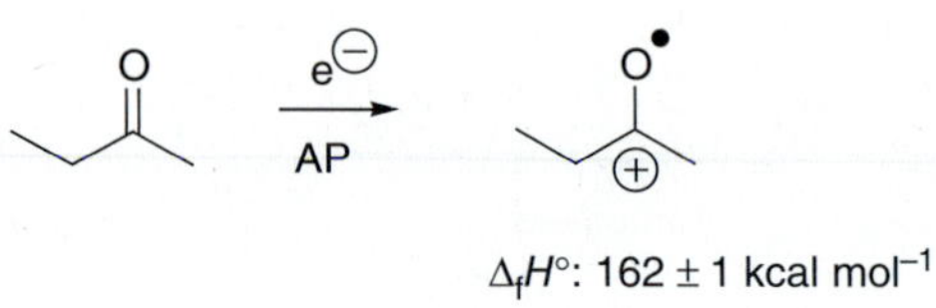

(1.107)

This can be interpreted in terms of the difference in homolytic bond dissociation enthalpies between a O—H and secondary C—H bond (see e.g. $DH°(\mathrm{MeO^\bullet/H^\bullet}) = 104.2$ kcal mol^{-1}, $DH°(i\text{-Pr}^\bullet/\mathrm{H}^\bullet) = 99.4$ kcal mol^{-1}, Table 1.A.7) and by the fact that the oxygen center of the keto radical cation is not able to effectively share its electrons with the carbenium ion, whereas the hydroxycarbenium ion moiety of the enol cation radical can adopt a oxonium-limiting structure (n/π conjugation). Furthermore, the carbenium center of the latter can stabilize the unshared electron of the adjacent radical as shown below through resonance.

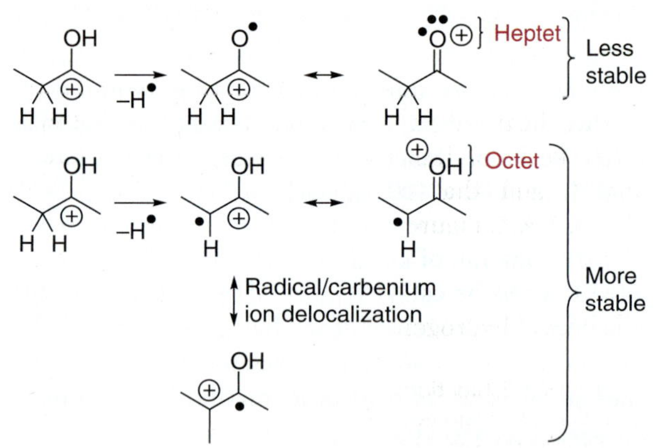

(*E*)- and (*Z*)-But-2-en-2-ol (enols of butan-2-one) can be obtained as transient species by flash vacuum pyrolysis (1.108) of the corresponding 2,3-dimethylbicyclo[2.2.1]hept-5-en-2-ols (*endo*-**1** and *endo*-**2**) and are characterized by their mass spectra. The threshold ionization energies (appearance potentials) for the formation of enol radical cations can be used to evaluate the heats of formation of (*E*)- and (*Z*)-but-2-en-2-ol, using the heat of formation established above by the McLafferty fragmentation appearance potential (1.108). This leads to $\Delta_f H°((Z)\text{-but-2-en-2-ol}) \cong -51.1\ \text{kcal mol}^{-1}$ [165].

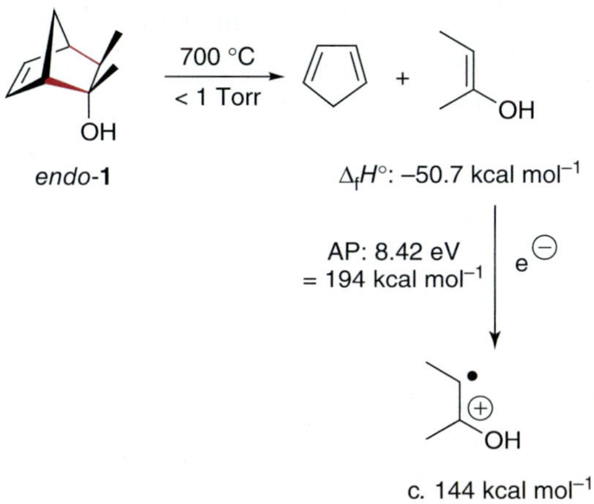

(1.108)

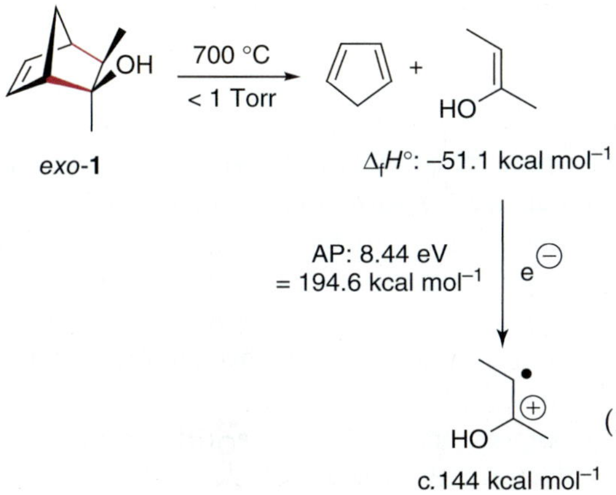

(1.108′)

The heat of hydrogenation of (*Z*)-butene is $-27.3\ \text{kcal mol}^{-1}$, that of vinyl ethyl ether is -26.5 kcal mol^{-1}, and that of pent-1-ene is $-30\ \text{kcal mol}^{-1}$ (Table 1.A.1, Figure 1.19). From the two latter values, the stabilization of vinyl ether due to the n(O)/π conjugation can be estimated to c. 3.5 kcal mol^{-1}. Thus, the heat of hydrogenation of (*E*)-but-2-en-2-ol should be that of (*Z*)-but-2-ene corrected by 3.5 kcal mol^{-1} to account for the n(O:)/π conjugation in this enol, which is assumed to be similar to that of a vinyl ether. The heat of didehydrogenation of butan-2-ol into (*E*)-but-2-en-2-ol is thus estimated to be $27.3 - 3.5 \cong 24\ \text{kcal mol}^{-1}$. Using $-70\ \text{kcal mol}^{-1}$ for the gas-phase heat of formation of butan-2-ol (Table 1.A.4), $\Delta_f H°((E)\text{-but-2-en-2-ol}) = -70 + 24 = -46\ \text{kcal mol}^{-1}$, which is c. 5 kcal mol^{-1} more than the value obtained by the mass spectrometric technique. Thus, one estimates $\Delta_r H°$(butan-2-one ⇄ (*E*)-but-2-en-2-ol) = $-46 - (-57.0) = 11\ \text{kcal mol}^{-1}$. In aqueous solution and at 25 °C, the Gibbs energy for the ketone ⇄ enol equilibrium measured for butan-2-one amounts to $\Delta_r G°$(butan-2-one ⇄ (*E*)-but-2-en-2-ol)= 10.2 kcal mol^{-1} [166], a value similar to that estimated (11 kcal mol^{-1}) in Figure 1.20, assuming $\Delta_r G° \cong \Delta_r H°$ (isomerization, the same number of molecules in reactants and products), and larger than the value $(-51.1 - (-57.0) \cong 4\ \text{kcal mol}^{-1})$ obtained by mass spectrometry (reaction (1.108')). A possible cause for this deviation could be nonadiabatic appearance potentials measured by mass spectrometry.

Applying an ICR mass spectrometric technique, Pollack and Hehre have found that deprotonation of $CD_3C^{(+)}(OH)CD_3$ (protonated form of hexadeuterated acetone) in the gas phase takes away H^+ for weak bases such as THF or $(i\text{-Pr})_2S$. With stronger bases such as amines, D^+ is transferred concurrently to the base, which generates the enol of acetone $CD_2{=}C(OH)CD_3$. $(i\text{-Pr})_2S$ is the strongest base for which D^+ transfer is not observed (reaction (1.109)), and aniline is the weakest base for which D^+ is transferred (reaction (1.111)). Considering the enthalpies measured for reactions (1.109) and (1.110), it was concluded that the relative thermochemical stabilities

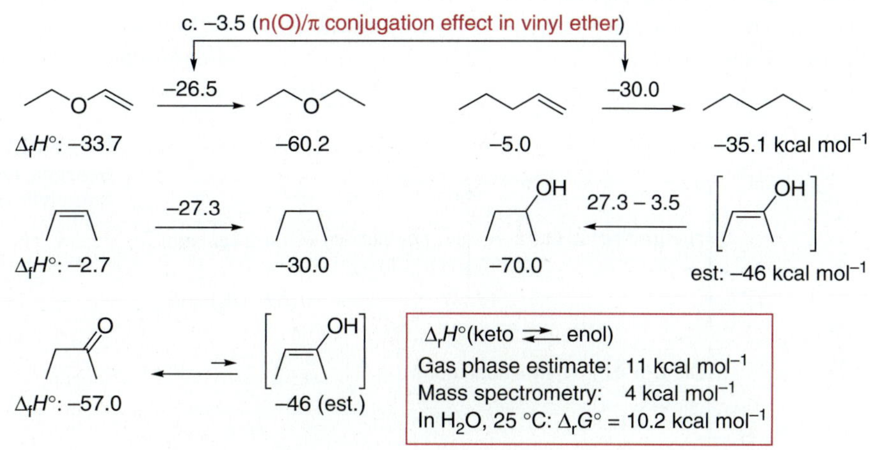

Figure 1.20 Estimation of the gas-phase standard heat of formation of the enol of butan-2-one and comparison with experimental values.

of the keto and enol tautomers of acetone is approximately the same as the difference in Gibbs energies of protonation of aniline and acetone in the gas phase, or 13.9 ± 2 kcal mol^{-1} [167].

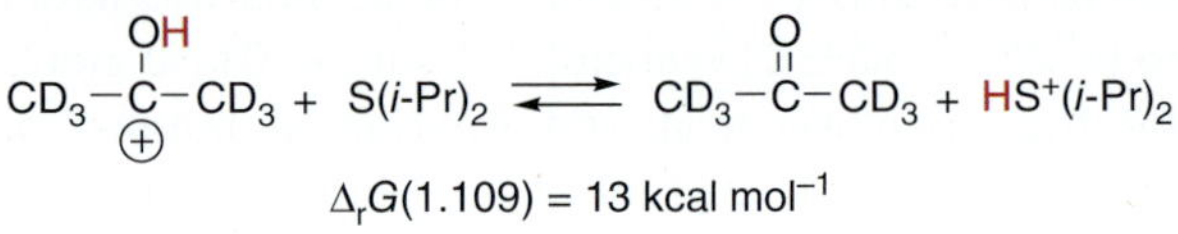

(1.109)

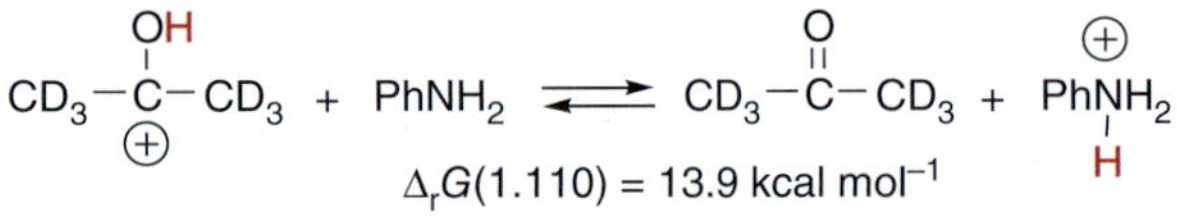

(1.110)

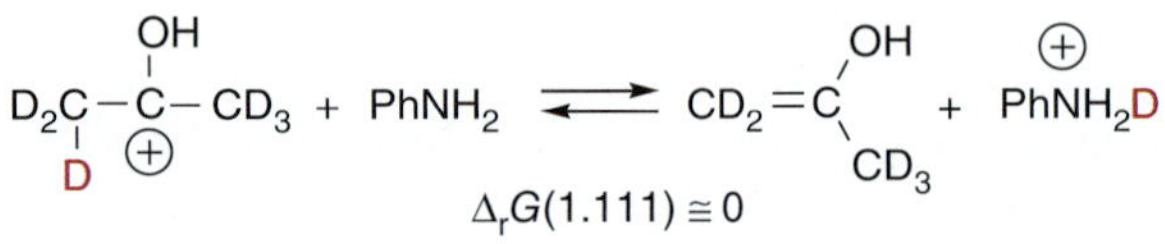

(1.111)

Problem 1.37 Estimate the equilibrium constant at 25 °C for phenol ⇄ cyclohex-2,4-dien-1-one [168, 169].

1.12.4 Heat of formation of highly reactive cyclobutadiene

Applying a similar technique for the reaction of cyclobutenyl cation with amines in the gas phase, PA(cyclobutadiene) = 224 ± 2.7 kcal mol^{-1} and $\Delta_f H°$(cyclobutadiene) = 102.3 ± 4 kcal mol^{-1} has been determined [170]. As cyclobutadiene undergoes quick cyclodimerization at low temperature, its heat of formation cannot be determined experimentally by its heat of hydrogenation into cyclobutene or/and cyclobutane (see antiaromaticity of cyclobutadiene, Section 4.5.6).

1.12.5 Estimate of heats of formation of diradicals

The benzene-1,4-diyl diradical (intermediate in the Bergman cyclization, Section 3.6.5) can be formed as the result of two successive homolytic C—H dissociations of benzene. This process generates the benzene-1,4-diyl diradical and two H$^•$ atoms; it requires twice the bond dissociation enthalpy $DH°(\text{Ph}^•/\text{H}^•)$ = 111.2 kcal mol^{-1}. The two H$^•$ radicals can be combined to form H_2, thus allowing the recovery of 104.2 kcal mol^{-1} = $DH°(\text{H}^•/\text{H}^•)$. As shown in Figure 1.21, this leads to an estimate for the standard heat of formation of benzene-1,4-diyl diradical of 138.3 kcal mol^{-1}, a value very similar to that (137.3 ± 3.3 kcal mol^{-1}) determined experimentally by Squires and coworkers (Figure 1.18).

The experimental standard heats of formation of benzene-1,3-diyl diradical and benzyne (benzene-1,2-diyl diradical) are definitively smaller than that found for benzene-1,4-diyl diradical. As the experimental heat of formation of benzene-1,4-diyl diradical equals that calculated by the thermodynamic cycle of Figure 1.21, the two unshared electrons of this diradical seem to ignore each other. The partial bond formation between the unshared electrons of the benzene-1,3-diyl and benzene-1,2-diyl diradicals are represented by the limiting structures given in Figure 1.19. The triple bond character of benzene-1,2-diyl diradical (benzyne) is only partial. By comparing the heat of hydrogenation of but-2-yne to form (Z)-but-2-ene (−37.5 kcal mol^{-1}) with that of benzyne to give benzene (−86.9 kcal mol^{-1}), it is clear that the bonding between the two electrons of the benzene-1,2-diyl diradical is much lower than for two 2p electrons in an alkyne. This analysis explains the much higher reactivity of benzyne compared to that of an alkyne as electrophile, nucleophile, or dienophile.

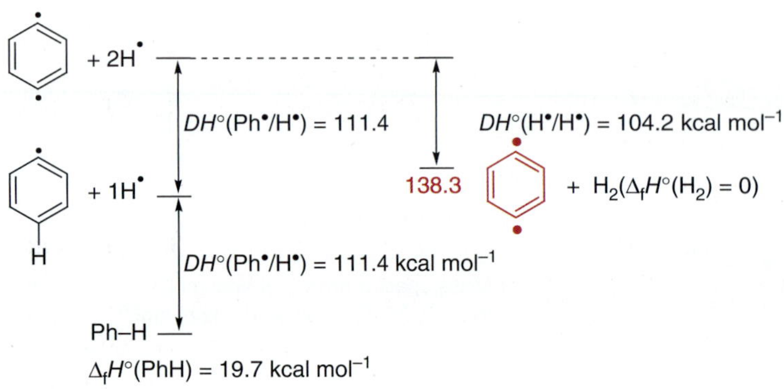

Figure 1.21 Estimate of the heat of formation of benzene-1,4-diyl diradical. See Table 1.A.7 for homolytic bond dissociation enthalpies.

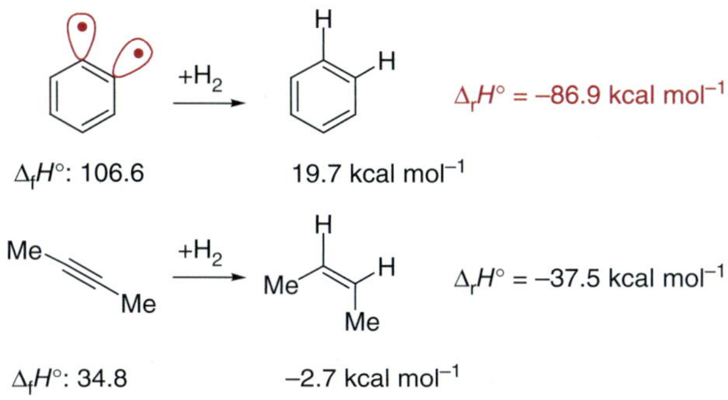

The heat of formation of *m*-xylylene (**2**) can be estimated by considering the heat of formation of *m*-xylene ($\Delta_f H°$(*m*-xylene) = 17.3 ± 0.6 kcal mol^{-1}) and the homolytic bond dissociation enthalpies $DH°(PhCH_2^\bullet/H^\bullet)$ = 89 kcal mol^{-1} and $DH°(H^\bullet/H^\bullet)$ = 104.2 kcal mol^{-1}. This leads to an estimate of $\Delta_f H°$(*m*-xylylene) = $17.3 + 2(89) - 104.2 \cong 91$ kcal mol^{-1}, a value much higher than that determined for *o*-xylylene (**3**; $\Delta_f H°$ = 53 kcal mol^{-1}) and *p*-xylylene (**4**, $\Delta_f H°$ = 50 kcal mol^{-1}). The latter two compounds are Kékulé systems, whereas *m*-xylylene (**2**) is a non-Kékulé π-conjugated system [171].

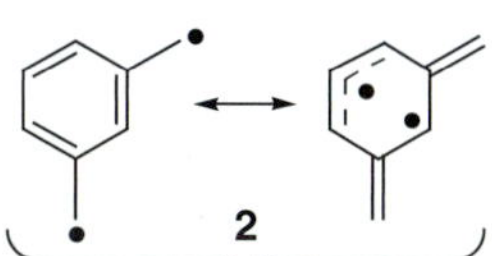

2

Non-Kékulé structure

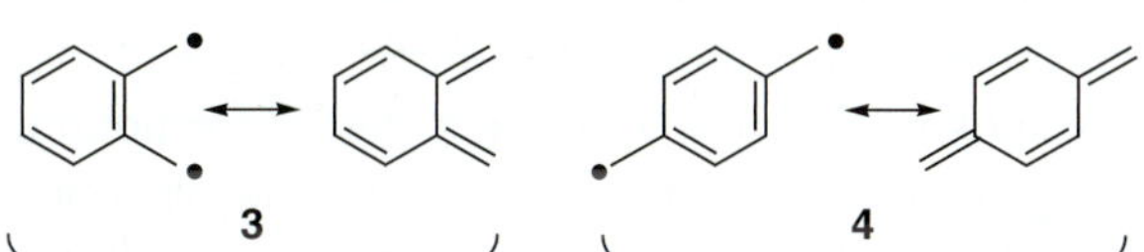

3 Kékulé structure

4 Kékulé structure

Problem 1.38 Can the Diels–Alder addition of methylidenemalonodinitrile with 1-phenylbutadiene undergo though a diradical intermediate? Which one?

Problem 1.39 Why are the yields of Diels–Alder additions usually better when adding a small amount of radical scavenging agent such as 1,4-dihydroxybenzene (parahydroquinone) or 4-methyl-2,6-bis(tertiobutyl)phenol?

Problem 1.40 Estimate the heat of formation of naphthalene-1,4-diyl diradical and compare it with the experimental values proposed by Roth et al. [172].

1.13 Electronegativity and absolute hardness

As shown in Section 1.7, ionization energies (IE) and electron affinities (−EA) can be measured for atoms, for ions, for neutral and charged molecules (Tables 1.A.20–1.A.22), and for stable or transient species. Mulliken defines the absolute electronegativity of a species **A** by $\chi = [\mathrm{IE}(\mathbf{A}) - \mathrm{EA}(\mathbf{A})]/2$, while Pearson [173–179] defines the absolute hardness of a species **A** by $\eta = [\mathrm{IE}(\mathbf{A}) + \mathrm{EA}(\mathbf{A})]/2$, and absolute softness by $\sigma = 1/\eta$. A graphical representation of these values for a few atoms and molecules is given in Figure 1.22.

Absolute softness represents the ease by which a chemical entity $\mathbf{A}^+$ can accept electrons, or species $\mathbf{B}^-$ can lose electrons. When two neutral species **A** and **B** (atoms or molecular species) combine to form compound **A–B**, in addition to the electron exchange that binds **A** to **B** and **B** to **A** (as for $H^\bullet + H^\bullet \rightleftarrows H_2$ or $Br^\bullet + Br^\bullet \rightleftarrows Br_2$), there is an electrostatic contribution to the binding in **A–B** due to the electron flux that goes from the more electropositive species (with a small χ value) to the more electronegative partner (with a large χ value). In the case of the formation of a salt (ionic bond) or complex by combining a Lewis acid **A** with a Lewis base **B**:, the portion of charge transfer from **B**: to **A** is given, to a first approximation, by:

$$\Delta N = \frac{\chi_A - \chi_B}{2(\eta_A + \eta_B)}$$

The smaller the η_A and η_B values, the greater the electron transfer from **B**: to **A**. The quantities η_A and η_B can be considered as a resistance to the flow of electrons from **B**: to **A** driven by the potential difference $\chi_A - \chi_B$. If η_A is small, the softness of **A** $1/\eta_A$ is large. This is the case for soft acids. Similarly, if η_B is small, the softness of **B**: $1/\eta_B$ is large, what is the case for soft bases. Therefore, a strong bond will exist between a soft acid **A** and a soft base **B**: in complex **B:A**. For nonpolar (no permanent dipole) and noncharged Lewis acids **A** and bases **B**:, the bond strength of their complex **B:A** is the largest for pairs **A/B**: having the highest softness (Table 1.A.20).

For a charged acid $\mathbf{A}^+$ and a charged base $\mathbf{B}^-$, their combination into salt $\mathbf{A}^+\mathbf{B}^-$ or neutral species **A–B** will be binding because of the electrostatic interaction (Coulomb's law). The shorter the distance between $\mathbf{A}^+$ and $\mathbf{B}^-$ in **A–B**, and the smaller **A** and **B**, the stronger their binding. For noncharged acid **A** and base **B**:, the ease of generating radical anion $\mathbf{A}^{\bullet-}$ and radical cation $\mathbf{B}^{\bullet+}$, respectively, correlates with the strength of their bonding interaction. This can be expressed by writing

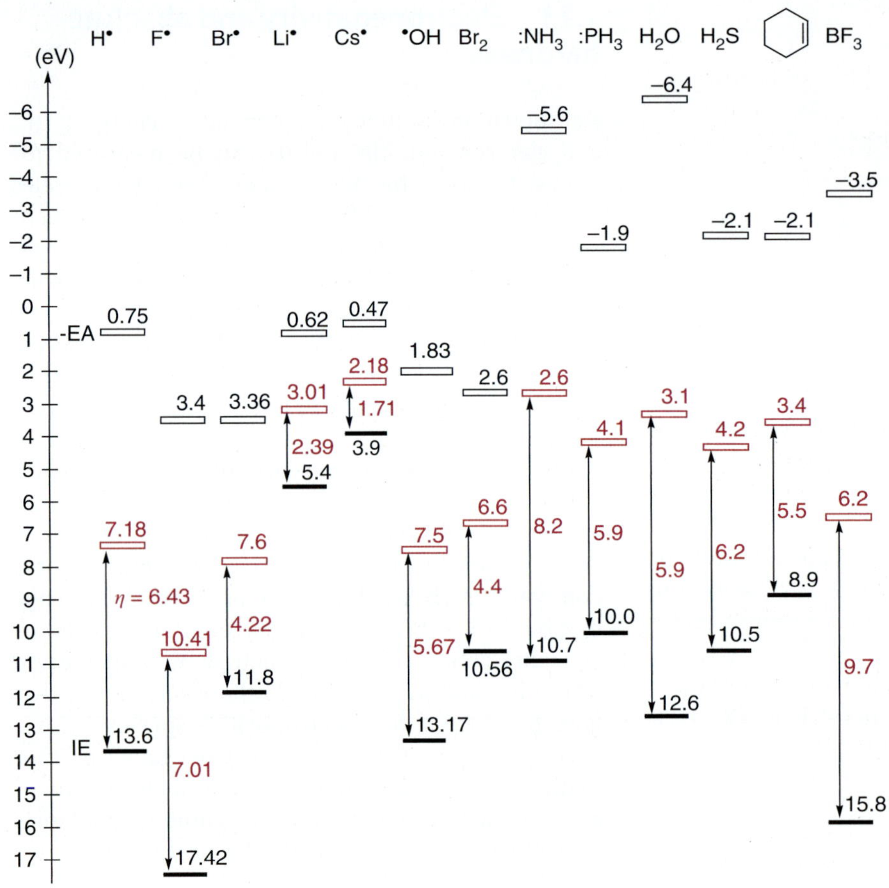

Figure 1.22 Ionization energies (▬), electron affinities (▭), and absolute electronegativities (▭) of a few atoms and molecules. Absolute hardnesses are given in italics (η). (1 eV/molecule ^= 23.06 kcal mol^{-1} = 96.48 kJ mol^{-1}, 1 eV = 1.602 18 × 10^{-19} J).

limiting structures involving the charge transfer from the base **B**: to the acid **A**. The easier this charge transfer, the higher the relative importance of configuration $A^{•-}B^{•+}$ (charge transfer configuration or limiting structure), and thus the stronger their binding.

In the case of a hard and charged acid A^+ interacting with a soft, uncharged base **B**:, a relatively weakly bound complex **B**: → A^+ will be obtained, and the binding between A^+ and **B**: arises mostly from a charge/dipole interaction involving the charged center of A^+ and the permanent and induced dipoles of **B**:. Electron exchange between A^+ and **B**: in complex **B**: → A^+ results essentially from an electron transfer from **B**: to $A^{•+}$, giving the charge transfer configuration $A^• + B^{•+}$. It involves another electrostatic dipole $A^•$/charge of $B^{•+}$ interaction that may not be much larger than that for the ground-state configuration $A^+ + B: \rightleftarrows B{:}A^+$ as represented in Figure 1.23b. This is the reason why Pearson states that the binding between a soft base and a hard acid or between a hard base and a soft acid is weaker than for hard/hard (Figure 1.23a) or soft/soft (Figure 1.23c) pair of acids and bases. This concept (or hard and soft acid base theory or HSAB theory of Pearson) has been applied to the formation of stable inorganic, organometallic, and organic (ionic and covalent) compounds and complexes resulting from the combination of two molecular or atomic fragments (any neutral or charged species). It is a unifying theory for bonding in organic and inorganic chemistry. It has also been applied to evaluate the relative stability of transition states of inorganic, organometallic, and organic reactions [180]. It is a simplification of the earlier theory of Bell–Evans–Polanyi (BEP theory) developed to model transition states of one-step, concerted reactions (transition state (‡) = ground-state configuration of reactants ↔ charge transfer configurations of reactants ↔ ground-state configuration of intermediates ↔ ground-state configuration of products; e.g. $R{-}Y + X^• \rightleftarrows [R{-}Y/X^• \leftrightarrow R{-}Y^{•-}/X^+ \leftrightarrow R{-}Y^{•+}/X^- \leftrightarrow R^•/Y^•/X^• \leftrightarrow R^•/Y^+/X^- \leftrightarrow R^•/Y{-}X]^‡ \rightleftarrows R^• + X{-}Y)$ [51–53]. In the case of regioselectivity of ambident nucleophiles (e.g. O- vs. N-alkylation of nitrite anion, O- vs. C-alkylation of enols and enolates, and O- vs. C-acylation and silylation of enolate anions) and electrophiles (e.g. 1,2- vs. 1,4-addition of nucleophiles to α,β-unsaturated aldehydes, ketones,

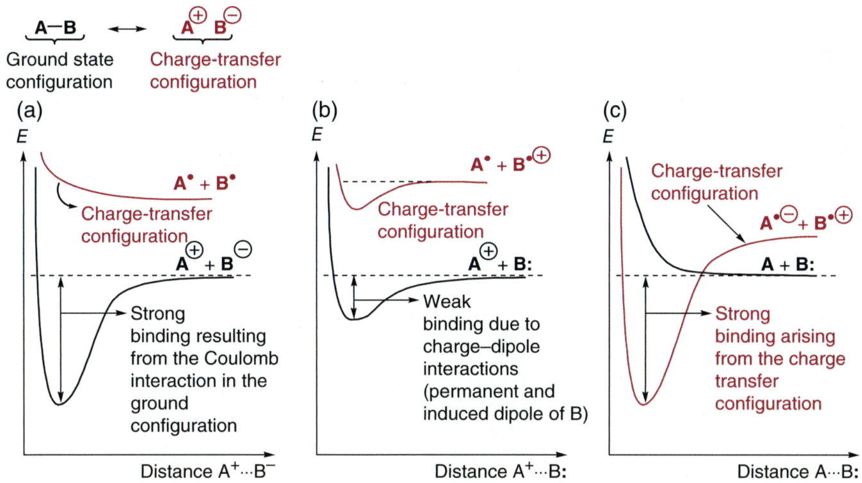

Figure 1.23 Pearson's hard and soft acid base (HSAB) theory. Representation of (a) the strong bonding for the combination of a hard (positively charged) acid $\mathbf{A}^+$ and hard (negatively charged) base $\mathbf{B}^-$, (b) of the weak bonding between a hard acid $\mathbf{A}^+$ and a soft, uncharged base **B:**, (c) of the strong bonding between a soft acid **A** and a soft base **B:**.

carboxylic esters, and carbonitriles), the HSAB theory encounters some difficulties [181]. In the BEP theory ($\Delta^{\ddagger}H = \alpha\Delta_r H + \beta$), the activation enthalpy of a concerted reaction depends on the heat of reaction (Dimroth principle), the ease by which two reacting partners can exchange an electron (given by the sum IE(**A**) + (−EA(**B**)) as in the Pearson theory), steric, geometry, conformational, and solvent effects. For a thermoneutral reaction ($\Delta_r H = 0$), the β term represents its intrinsic barrier, which is made of several contributions such as steric hindrance, geometry distortion of reactants when they reach the transition state, electron exchange between the reactants, solvation, and desolvation effects between reactants and transition state. The α parameter characterizes the position of the transition state along the reaction coordinates ($0 \leq \alpha \leq 1$). The rate constant k of a concerted reactions (Section 3.2) depends on activation-free enthalpy $\Delta^{\ddagger}G$ that depends on two terms ($\Delta^{\ddagger}G = \Delta^{\ddagger}H - T\Delta^{\ddagger}S$) that are $\Delta^{\ddagger}H$= the activation enthalpy and $\Delta^{\ddagger}S$ = the activation entropy of the reaction (Section 3.3). The BEP theory will be applied extensively in this textbook. This can be done when thermochemical data ($\Delta_f H°$, $S°$) of reactants and products are available (in this chapter) or can be estimated (Chapter 2), and when ionization energies (EI) and electron affinities (−EA) of reactants are known or can be evaluated. For reactions forming reactive intermediates in their rate-determining steps (transient species that are much less stable than reactants), the activation-free enthalpy can often be taken as $\Delta^{\ddagger}G \approx \Delta_r G^T$(reactants ⇄ intermediate). Thus, thermochemical data and knowledge of the reactivity of radicals, diradicals, carbenes, cations and carbocations, anions, radical anions, and radical cations, as well as solvent effects, are fundamental to understand the chemical reactivity.

Problem 1.41 Captopril® is one of the early discovered inhibitors of angiotensin converting enzymes and that is an antihypertensive agent. It can be prepared according to the method presented below [182]. Why is thionoacetic acid represented as $CH_3C(S)OH$ rather than $CH_3C(O)SH$? Why is the sulfur moiety of $CH_3C(S)OH$ adding to methacrylic acid and not the oxygen moiety? Why does one add hydroquinone to this reaction mixture? The second step of the procedure is amidification. What is the role of dicyclohexylcarbodiimide (DCCI) added to these reaction mixture? The third step implies the acidic hydrolysis of the t-butyl ester. What is the mechanism of this reaction?

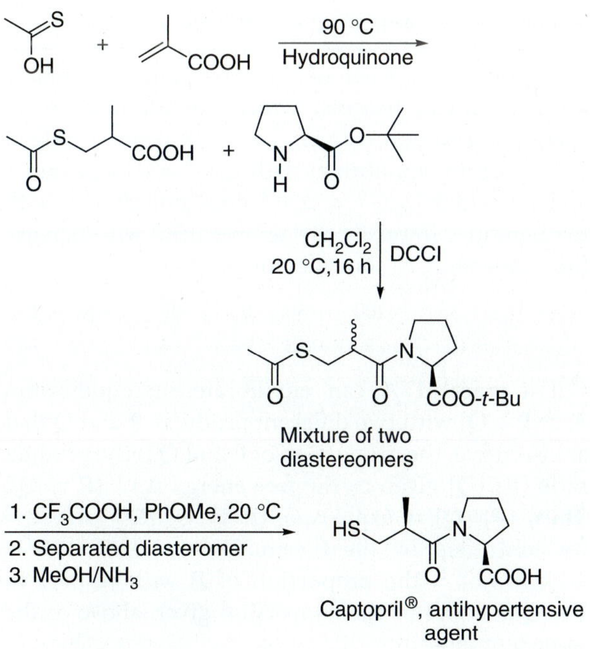

1.14 Chemical conversion and selectivity controlled by thermodynamics

Chemical conversion is the ratio of the amount (in moles) of product $\mathbf{P}_t$ formed to the number of moles of reactant $\mathbf{A}_0$ (at time $t = 0$) used as starting material under given conditions for a given time t of reaction. After infinite time ($t = \infty$), the equilibrium of the reaction is reached. For a simple reaction $\mathbf{A} \rightleftarrows \mathbf{P}$ (in the gas phase or in an ideal solution) that does not give any side products, the equilibrium constant is considered to be $K = [\mathbf{P}_\infty]/[\mathbf{A}_\infty]$, which corresponds to a maximum conversion of $[\mathbf{P}_\infty]/([\mathbf{P}_\infty]+[\mathbf{A}_\infty])$. After isolation and purification of product **P**, the number of moles of **P** recovered reported to the number of moles of **A** engaged in the reaction defines the yield (η) in **P**. $\eta \leq$ maximum conversion. For a reaction in which one reactant **A** equilibrates with one product **P**, the maximum conversion at 25 °C is given by the second law of thermodynamics, Eq. (1.2) (for ideal gas or solution). Examples are given below:

$\Delta_r G°$(A $\rightleftarrows$ P) (kcal mol^{-1})	K = [product P]/[reactant A]	Maximum conversion; maximum possible yield in product P (%)
−1	5.41	84.4
−1.36	10	90.9
−2	29.29	96.7
−3	158.5	99.4
−4	858.0	99.88
−5	4643.8	99.98

In practice, a yield of 96.7% is quite acceptable for a chemical process, which corresponds to an exergonic reaction with $\Delta_r G = -2\,\text{kcal mol}^{-1}$ only at 25 °C! For an equilibrium with $\Delta_r G° = -1\,\text{kcal mol}^{-1}$ and for which $\Delta_r G^T = \Delta_r G°$ does not change with temperature, lowering the temperature will increase the conversion, as shown below.

T (°C)	−100	−50	0	25	100	200
Conversion (%)	94.8	90.5	86.3	84.4	79.4	74.3

If a reactant **A** can equilibrate (in equilibrium $\mathbf{A} \rightleftarrows \mathbf{P}+\mathbf{Q}$) with two different products **P** and **Q** that are isomeric, the proportion of **P** and **Q** is the product ratio [**P**]/[**Q**] given by the free energy $\Delta_r G^T(\mathbf{P} \rightleftarrows \mathbf{Q})$. Thus, $[\mathbf{P}]/[\mathbf{Q}] = \exp[-\Delta_r G^T(\mathbf{P} \rightleftarrows \mathbf{Q})/RT]$ (selectivity controlled by the thermodynamics). If $\Delta_r G^T = -1\,\text{kcal mol}^{-1}$, the proportion of **P** with respect to **P** + **Q** is equal to the conversion given above at the same temperature.

For equilibria with negative or positive reaction entropies, $\Delta_r G^T$ values vary with temperature as $\Delta_r G^T = \Delta_r H^T - T\Delta_r S^T$. Thus, selectivity determined by the thermodynamics (ratio [**P**]/[**Q**]) may decrease, or decrease, on lowering temperature!

1.14.1 Equilibrium shifts (Le Chatelier's principle in action)

In the presence of an acid catalyst, aldehydes and ketones react with alcohols to form the corresponding acetals (products) and water (coproduct) ($R^1COR^2 + R^3OH \rightleftarrows R^1(R^2)C(OR^3) + H_2O$) [183]. With small nonbranched aldehydes and a small primary alcohols, $K > 1$, the equilibrium lies in favor of the acetal at 25 °C (see Problems 2.29 and 2.30). Because the entropy of the reaction ($\Delta_r S^T$) decreases with the size of the reactants (Section 2.9.2), K is smaller than unity with large aldehydes and alcohols. It is also smaller than unity for reactions engaging ketones that are not destabilized by electron-withdrawing α-substituents or by ring strain. The equilibrium can be shifted by the removal of water from the reaction mixture. This can be done by azeotropic distillation, ordinary distillation, or the use of a drying agent such as acidic alumina (Al_2O_3) or molecular sieves (crystalline metal aluminosilicate). Another way to shift equilibrium is to precipitate the product(s) or one of the coproducts as they form. Alternatively, one can shift the equilibrium by selective extraction of the product or of one of the reaction coproducts with a solvent not miscible with the reaction mixture. An example of diastereoselective reaction under thermodynamics controlled is shown in Scheme 1.3. The reversible Diels–Alder reaction of 2,4-dimethylfuran (**5**) and 1-cyanovinyl (1′*S*)-camphanate (**6**) can yield up to eight diastereoisomeric cycloadducts ((−)-**7**, its diastereoisomers **8** + **9** + **10**, and its regioisomers **11** + **12** + **13** + **14**). In the absence of solvent, in the presence of ZnI_2 as a catalyst and under sonication (ultra-sound stirring), all cycloadducts are in equilibrium with the cycloaddents **5** + **6**, but only (−)-**7** crystallizes, thus shifting the equilibrium in favor of this product. (−)-**7** is obtained in 85% yield with a purity of 95% (contaminated with 5% of other diastereioisomers). A simple recrystallization from ethyl acetate and petroleum ether gives 61% of (−)-**7** with a purity > 99.5%. The mother liquor can be evaporated, leaving a residue that can be added to a mixture of **5** + **6** to provide more (−)-**7** [184]. Alkaline hydrolysis of (−)-**7** gives the corresponding cyanohydrines that react with added formaline (H_2CO/H_2O) to provide the enantiometrically

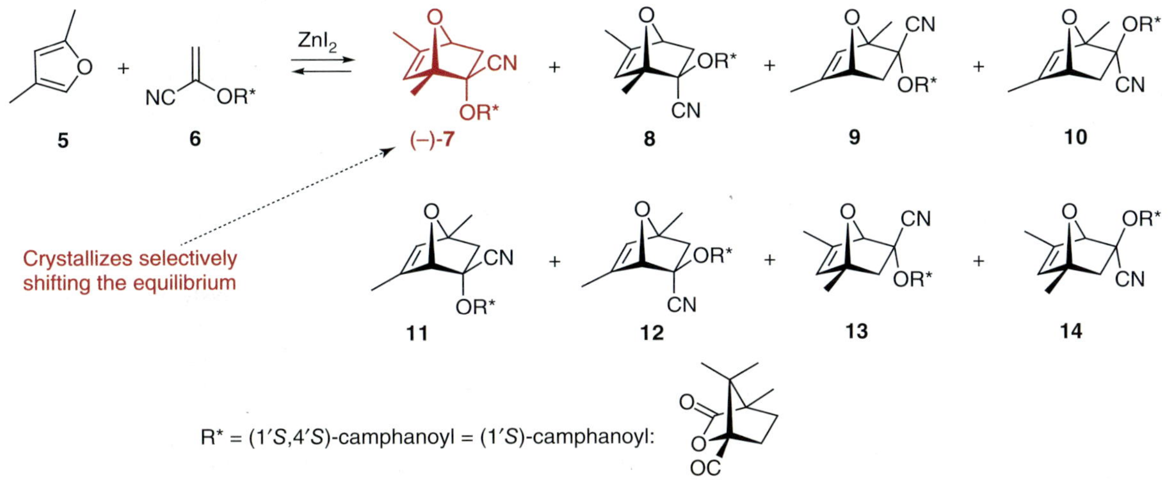

Scheme 1.3 Example of a reversible Diels–Alder reaction. The cycloaddents (reactants) and eight isomeric cycloadducts (products) are initially in equilibrium. One adduct can be selectively crystallized, which shifts the equilibrium in favor of it. Following the CIP priority rules, the diastereoisomeric products are named: (1*S*,2*R*,4*S*)- ((−)-**7**), (1*S*,2*S*,4*S*)- (**8**), (1*R*,2*S*,4*R*)- (**9**), (1*R*,2*R*,4*R*)-2-cyano-1,5-dimethyl-7-oxabicyclo[2.2.1]hept-5-en-2-yl (1′*S*,4′*S*)-camphanate (**10**), and their regioisomers: (1*S*,2*R*,4*S*)- (**11**), (1*S*,2*S*,4*S*)- (**12**), (1*R*,2*S*,4*R*)- (**13**), (1*R*,2*R*,4*R*)-2-cyano-4,6-dimethyl-7-oxabicyclo[2.2.1]hept-5-en-2-yl (1′*S*,4′*S*)-camphanate (**14**). Because of the small, rigid bicyclo[2.2.1]heptane skeletons of the 7-oxanorbornene and camphanoyl systems, the stereomarkers (4*R*),(4*S*), and (4′*S*) can be dropped. Indeed, a (1*R*,4*R*)-bicyclo[2.2.1]heptyl derivative cannot have a (1*R*,4*S*)-diastereoisomer but only a (1*S*,4*S*)-enantiomer.

enriched (1*S*,4*S*)-(−)-1,6-dimethyl-7-oxabicyclo[2.2.1]hept-5-en-2-one (a useful synthetic intermediate in the construction of polypropionates) and recovery of (*S*)-camphanic acid, the chiral auxiliary. A chiral auxiliary is an enantiomerically enriched compound (see Section 3.7.5) incorporated temporarily in a synthesis to control stereoselectivity [185–187]. Figure 1.24 gives examples of chiral compounds with their molecular chirality notations (CIP, Cahn–Ingold–Prelog, priority rule) [188, 189] that will be used throughout this textbook. In these examples, chirality arises from tetrahedric carbon $R^1(R^2)(R^3)CR^4$ that is chiral because it bears four different R^1, R^2, R^3, and R^4 substituents. Compounds with a chiral heteroelement are known. They include tetrahedric silicium compounds [190], pyramidal nitrogen [190, 191], phosphorous [192], arsenic [193, 194] antimony and bismuth organocompounds ($R^1(R^2)(R^3)X$:, X = N, P, As, Sb, Bi and with three different R^1, R^2, and R^3 groups) [195], and organosulfur compounds such as sulfites ($R^1OS(=O)OR^2$), sulfinates (RSO_2R) [196], sulfoxides [197–199], sulfimines ($R^1(R^2)S(=Z)$, Z = O, NR with two different R^1 and R^2 groups) [200], sulfoximines ($R^1(R^2)S(=O)=NR$ with two different R^1 and R^2 groups) [201], sulfur ylides ($R^1(R^2)S=CHR$) [202], and sulfonium salts ($R^1(R^2)(R^3)S:^{(+)}$, with three different R^1, R^2, and R^3 groups) [203]. Organometallic compounds with chiral metal atom are known also [204–207].

Problem 1.42 Estimate the equilibrium constant of the isomerization butane ⇄ isobutane (2-methylpropane) at 300 K, and at 600 K. For reactions and their mechanisms that isomerize alkanes, see [208].

Problem 1.43 Methyl isopropyl ketone (MIPK) is an efficient high-octane (>100) oxygenate gasoline additive, without many of the undesirable effects of the widely used methyl *tert*-butyl ether. One of the methods of preparation of MIPK involves the rearrangement of pivalaldehyde catalyzed by strong acids [209]. Evaluate the equilibrium constant for this rearrangement at 25 °C using gas-phase standard heats of formation for MIPK and *t*-BuCHO.

1.14.2 Importance of chirality in biology and medicine

Chirality is a fundamental symmetry property of three-dimensional objects. A molecule is said chiral if it cannot be superimposed upon its mirror image. Such an object has no symmetry elements of the second kind (a mirror plane, $\sigma \rightarrow S_1$, a center of inversion, $\mathbf{i} \rightarrow S_2$, and a rotation–reflection axis $\rightarrow S_{2n}$). If the object is superposable on its mirror image, the object is described as being achiral [210]. Putting ones' shoes or shaking hands confronts us with macroscopic chirality. Although there is no obvious relationship between macroscopic chirality and chirality at the molecular level, it is accepted

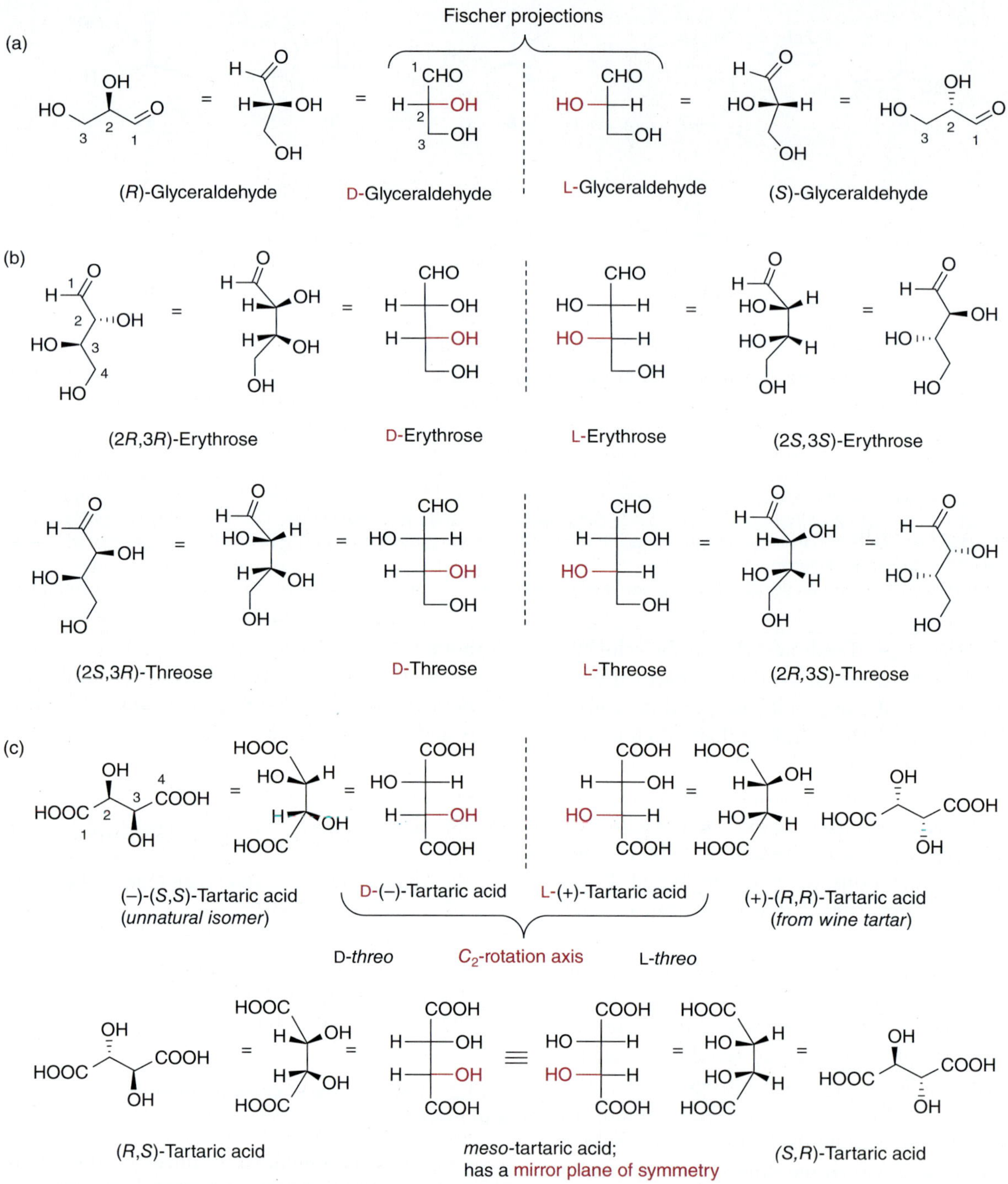

Figure 1.24 Representations of the stereoisomers of (a) glyceraldehyde (2,3-dihydroxypropanal: aldotriose, an example of aldose), (b) of erythrose (2,3,4-trihydroxybutanal: an aldotetrose), and (c) of tartaric acid (2,3-dihydroxybutanedioic acid: an example of aldaric acid). (*R*)- and (*S*)-glyceraldehydes are stereoisomers called enantiomers. (*R*)-glyceraldehyde is D-glyceraldehyde in the Fischer–Rosanoff convention because its hydroxy group at C(2) is drawn right in the Fischer projection. (2*R*,3*R*)-Erythrose is D-erythose because the heavy substituent (OH) of the last stereogenic center (C(3)) is drawn right in the Fischer projection. L-Erythrose is the enantiomer of D-erythrose. (2*S*,3*R*)-Threose is D-threose because the heavy substituent (OH) of the last stereogenic center (C(3)) is drawn right in the Fischer projection. D-Threose and D-erythrose are diastereomers, L-threose, and L-erythrose are also diastereomers; D-threose and L-threose are enantiomers. The prefixes *erythro* and *threo* are often used in place of *anti* and *syn*, respectively, for the nongeminal disubstitution of a chain. (2*R*,3*R*)- and (2*S*,3*S*)-Tartaric acid are chiral and enantiomers. Because 2 and 3 can be exchanged, one drops them: these acids are called (*R*,*R*)- and (*S*,*S*)-tartaric acid, respectively. They have a C_2 axis of rotation as an element of symmetry. They are also called L- and D-tartaric acid, respectively, and, in water, they are dextrorotatory (+) and levorotatory (−), respectively. (*R*,*S*)-Tartaric acid is identical to (*S*,*R*)-tartaric acid; this compound is achiral because it contains a mirror plane of symmetry. It is called *meso*-tartaric acid, which is a diastereoisomer of (*R*,*R*)- and (*S*,*S*)-tartaric acid. L means left in the Fischer projection, not levorotatory. A priori, there is no relationship between L of the Fischer projection and *l* or (−) for levorotatory. The priority Cahn–Ingold–Prelog (CIP) rule defines the (*S*) or (*R*) chirality of a stereogenic center. There is no relationship between (*R*),(*S*) and D,L stereomakers.

that homochirality is one of the fundamental aspects of life on Earth. Parity violation discovered in the weak nuclear force (the fourth type of fundamental forces, next to gravity, electromagnetism, and the strong nuclear force) led to the experimental observation that the β-particles emitted from radioactive nuclei have an intrinsic asymmetry: left-handed (L) electrons are preferentially formed relative to right-handed (R) electrons. The major consequence of this finding is that chirality exists at the level of elemental particles, allowing two enantiomers of a chiral molecule to differ in energy [211]. For a compound like glyceraldehyde (Figure 1.24) with one stereogenic center, the difference in energy between the D- or (*R*)- and L- or (*S*)-enantiomer amounts to no more than 2×10^{-15} cal mol^{-1}, corresponding to an excess of c. 10^6 molecules of the most stable D-(+)-enantiomer (in the form of its hydrate in water) per mol (number of Avogadro $L = 6.022 \times 10^{23}$ mol^{-1}) of racemate in thermodynamic equilibrium at 25 °C. Similarly, D-ribose, the central furanose of nucleic acids, is slightly more stable in water than its L-enantiomer. L-α-Amino acids are slightly more stable than their enantiomers in water [212–214]. Most natural products (amino acids, sugars, terpenes, alkaloids, steroids, etc.) are chiral and are found enantiomerically enriched, if not enantiometrically pure. The building blocks (amino acids, carbohydrates, etc.) of life are chiral. The biopolymers derived from them are also chiral. When a drug interacts with its receptor site, or with an enzyme which is chiral, it is not a surprise that its two enantiomers interact differently and may lead to different biological effects [215, 216]. The tragedy that occurred in the 1960s after racemic thalidomide was administered to pregnant women is a convincing example of the relationship of pharmacological activity to absolute chirality. The (*R*)-enantiomer of thalidomide exhibits desirable analgetic properties; however, the (*S*)-enantiomer does not [217]. Instead, it is a teratogen and induces fetal malformations or death (Figure 1.25). Following this tragedy, the marketing regulations for synthetic drugs have become significantly more severe. The alkaloid (−)-levorphanol is a powerful narcotic analgesic with an activity five to six times stronger than morphine [218]. Its enantiomer (+)-dextrorphan is not an analgesic but is active as a cough suppressant [219]. During the 1960s, (−)-propanolol was introduced as a β-blocker for the treatment of heart disease [220]. Its (+)-enantiomer acts as a contraceptive. The orange aroma extracted from oranges is (*R*)-(+)-limonene, whereas its enantiomer, extracted from lemon, (*S*)-(−)-limonene is responsible for lemon aroma. (*S*)-(+)-Carvone has an odor of caraway, whereas the (*R*)-enantiomer has a spearmint smell [221]. D-Asparagine has a sweet taste, whereas natural L-asparagine is bitter [222]. L-Dopa (L or (*S*)-(3,4-dihydroxyphenyl)alanine) is used in the treatment of Parkinson's disease under the form of oral pills. The active drug is the achiral dopamine that cannot cross the "blood–brain" barrier to reach the required site of action. It forms by decarboxylation of L-Dopa, a reaction catalyzed in the brain by dopamine decarboxylate, an enzyme specific for L-Dopa. Thus, L-Dopa is a prodrug that crosses the "blood–brain" barrier. D-Dopa also crosses it but is not decarboxylated by dopamine decarboxylase. Administration of racemic Dopa would be dangerous because of the build up of D-Dopa [223, 224].

Chirality exists also in minerals. In 1801, Haüy noticed that quartz crystals are hemihedral, i.e. certain facets of one kind of crystals make them to be objects that cannot be superimposed with those that are mirror images [225]. In 1809, Malus observed that quartz crystals induce the polarization of light [226] and, in 1812, Biot and coworker found that a quartz plate cut at right angles to one particular axis rotates the plane of polarized light to an angle proportional to the thickness of the plate. Right and left forms of quartz crystals rotate the plane of polarized light in a different direction. Quartz is the second most abundant mineral in the Earth crust after feldspars ($KAlSi_3O_8$, $NaAlSi_3O_8$, and $CaAl_2Si_2O_8$ make 60% of Earth's crust). In 1815, Biot and coworker noted that solutions of natural organic compounds can rotate the plane of polarized light also and that the optical rotation of the solution depends on the individual molecules [227].

1.14.3 Resolution of racemates into enantiomers

Despite the recent developments in enantioselective synthesis (or asymmetric synthesis, Section 3.7, several examples given in the following chapters) [228, 229] and chromatographic separation methods (high-performance liquid chromatography and gas-phase chromatography) using chiral stationary phases [230–239], physical and chemical resolution of racemates remains the most inexpensive method for producing enantiomerically enriched or pure enantiomers [240, 241]. The three most used methods of racemate resolution are among the oldest that have been developed by Pasteur between 1848 and 1858 [242–244]. The first one, an autocatalytical crystallization process, is the spontaneous resolution of racemates that Pasteur, first, observed in 1848 [245]. The sodium ammonium salt of (racemic) tartaric acid (Figure 1.24) crystallized as homochiral hemihedric crystals (conglomerates) that can be distinguished visually and separated by hand (manual sorting of conglomerate, triage) [246, 247]. The method is not

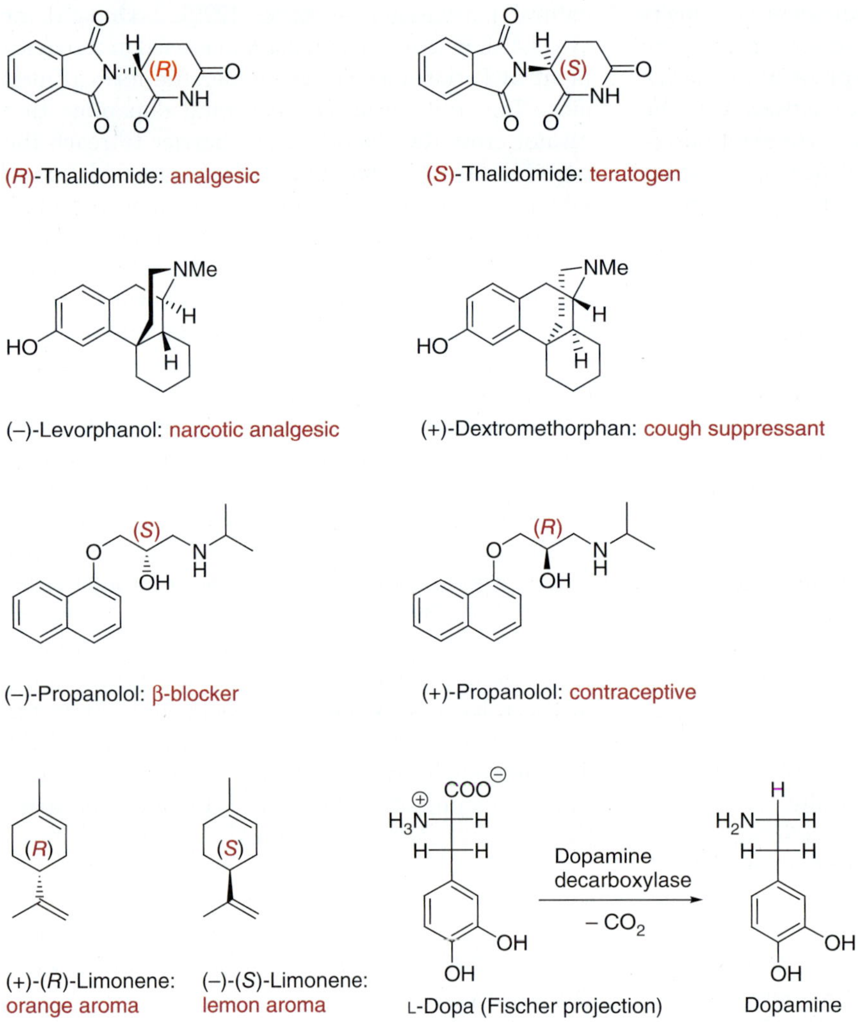

Figure 1.25 Examples of biological response that depends on chirality.

practical for large-scale resolutions but can be used (if chromatographic techniques fail) to obtain small amounts of crystalline enantiomerically pure material that can be used to inoculate (or seed) a saturated solution of the racemate and promote the crystallization of a single enantiomer (entrainment method). After separation of the crystalline material, the solution is enriched in the other enantiomer. The crystallization of the latter can then be induced by an enantiomeric crystalline inoculate. The method (initiated resolution) is applied industrially for the resolution of glutamic acid and threonine [248]. This method requires experimental optimization of a number of parameters (search for solvent, solvent mixture, temperature, temperature program) as many compounds crystallize as racemic compounds rather than as enantioenriched conglomerates [249]. Enantioselective crystallization can also be induced by ultrasound irradiation [250–252]. For example, the crystallization a 5% ee enriched saturated solution of D-threonine ((2*R*,3*S*)-MeCH(OH)-CH(NH_2)-COOH)) gives crystals of D-threonine with 87% ee at the beginning of the crystallization, thus realizing a chiral amplification in crystallization under ultrasound radiation [253]. Kondepudi et al. have induced total chiral discrimination by stirred crystallization of sodium chlorate and other chiral systems [254–256]. Viedma showed that chiral amplification can be achieved by abrasive grinding of saturated solutions of enantiomorphous crystals ($NaClO_4$) to obtain crystals with single handedness [257–259].

In 1999, Mikami and coworkers reported the first example of spontaneous enantioresolution of racemic compound into three-dimensional conglomerate in a fluid liquid-crystalline phase [260]. Enantioselective absorption on crystalline quartz [261] has been argued as a possible mechanism for chiral bias in natural and living systems [262]. Adsorption on achiral surfaces has been used to induce chiral symmetry breaking without the formation of diastereoisomeric pairs [263–268]. For instance, spontaneous separation of chiral phases in oriented monolayers of rigid, chiral amphiphiles on mica has been reported. Atomic force microscopy of the ordered films reveals domains of mirror image two-dimensional structures [269]. Chiral recognition can also occur in systems that are racemic in three dimensions by preferential alignment of groups of particles induced by the surface, generating domains of the separated pure enantiomers at the surface [270, 271]. Ordered supramolecular chiral structures have been observed after deposition of either enantiomerically pure (*P*)- or (*M*)-[7]-helicene (see Section 6.3.5) on Cu(111) surface, thus realizing chirality transfer from a single molecule into self-assembled monolayers [272, 273]. Chiral quartz crystals can promote enantioselective organic reactions through asymmetric autocatalysis (Section 3.7.8) [274].

Most optical resolutions separate diastereoisomers. The method involves converting the racemate (*R*)-**P** + (*S*)-**P** into a mixture (can be 1 : 1 mixture or not) of diastereoisomers (*R*,*R*)-**P**-**A** and (*S*,*R*)-**P**-**A** (can be salts, covalent compounds, complexes) by combining each of its enantiomers with an enantiomerically pure chiral auxiliary (*R*)-**A** (or (S)-**A**). The diastereoisomers are then separated by fractional crystallization, by chromatography, by extraction, or by another technique. Once each diastereoisomeric product has been purified, a suitable reaction is applied to liberate the desired enantiomers (*R*)-**P** and (*S*)-**P**, and, if possible to recover the chiral auxiliary (Scheme 1.4). This is the second method reported by Pasteur in 1853 for the resolution of racemic tartaric acid (1 : 1 mixture of D- and L-tartaric acid) via the formation of salts with natural enantiomerically pure amines [275]. He had established that the cinchona alkaloids, quinine, and cinchonidine, are stereoisomers of quinidine and cinchonine, respectively, and that these compounds are converted into quinicine and cinchonicine, respectively, upon heating under acidic conditions (Scheme 1.4) [276]. Pasteur also determined that the salt of cinchonicine with (−)-tartaric acid crystallizes from a saturated aqueous solution before the salt of (+)-tartaric acid. In contrast, with quinicine, the salt with (+)-tartaric acid crystallizes first. Quinicine is also called quinitoxine; it has been converted into quinine in three steps by Rabe and Kindler in 1918 [277, 278].

In some cases, it is possible to realize the resolution by using less than 1 equiv. of the chiral auxiliary [279–281]. The ideal situation is when conditions are found (solvent, concentration, and temperature) under which one of the two possible diastereoisomeric products, e.g. (*S*,*R*)-**P**-**A** is much less soluble than the other one. In theory, one can end up with an enantioenriched precipitate. This method of separation of crystalline diastereoisomers (that can be recrystallized to a high degree of purity) is often the best method to obtain the final chiral product **P** with high enantiomeric purity. Enantiomeric purity (or enantiopurity) can be defined as the enantiomeric ratio: er = [(*R*)-**P**]/[(*S*)-**P**] or as the enantiomeric excess: ee = ([(*R*)-**P**]-[(*S*)-**P**])/[(*R*)-**P**]+[(*S*)-**P**]).

Instead of fractional crystallization of diastereomers (partitioning between a liquid and a solid phase), one can use enantioselective partitioning between two chiral, nonmiscible liquids (biphasic recognition chiral extraction) [282, 283]. Supercritical fluids have been employed in such resolutions [284]. Enantioselective separation can be achieved on chiral surfaces [285] that are obtained by adsorption of enantiomerically pure chiral molecules, or by cleavage

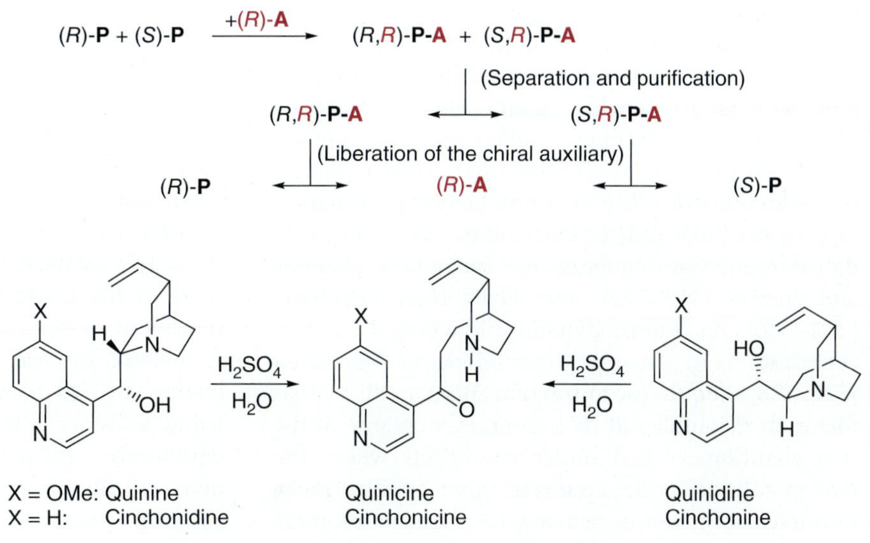

Scheme 1.4 Optical resolution to separate diastereoisomer. Quinicine and cinchonicine, isomers of the alkaloid pairs quinine/quididine, and cinchonidine/cinchonine, respectively, were used by Pasteur to resolve racemic tartaric acid.

of crystals to produce particles with chiral surfaces [268, 286–292].

The third method of resolution of racemates into enantiomers reported by Pasteur in 1858 is a kinetic resolution (Section 3.7.1) of racemic tartaric acid with yeast (from beer). The organism *Penicillium glauca* destroys ammonium (+)-tartrate much more rapidly than its (−)-enantiomer [293]. As recognized by Pasteur, microorganisms and living systems are made of enantiomerically pure molecules that have a very high ability to discriminate between two enantiomers of racemic substrates. Modern asymmetric synthesis has fostered the development of much smaller catalysts (organocatalysts or organometallic complexes) than the yeast enzymes upon which Pasteur relied. The chemical catalysts are usually much more tolerant of changes in temperature, solvent, and pH than enzymes and can induce useful enantioselectivities either through kinetic resolution (Section 3.7.1) [294, 295], through parallel kinetic resolution (Section 3.7.2), or through dynamic kinetic resolution or kinetic deracemization (Section 3.7.3).

1.14.4 Thermodynamically controlled deracemization

The resolution of racemates (e.g. 1 : 1 mixture of (*R*)-**P** + (*S*)-**P**) gives a maximum yield of 50% of a desired enantiomer (e.g. (*R*)-**P**). If the enantiomer (e.g. (*S*)-**P**) cannot be used, or be sold, it must be converted into the desired enantiomer either through a reaction (or a sequence of reactions) that converts it with a good yield and reasonable cost, or by a racemization process. Most useful are processes that do not separate enantiomers but convert the racemate into one or the other enantiomer with a high chemical yield and high enantiomeric excess (deracemization). For compounds that possess one stereogenic center, the deracemization process requires a reversible epimerization that can be coupled with an irreversible enantioselective reaction. This sort of resolution is called dynamic kinetic resolution (Section 3.7.3) [296–304]. Most common procedures involve the enantioselective protonation of enolates [305–314] or carbanions [315–317], oxidation/reduction sequence for secondary alcohols and amines [318–322], and allylic rearrangements [323–326]. In thermodynamically controlled deracemizations or dynamic thermodynamic resolution [327, 328], 1 equiv. (or more) of a chiral auxiliary that forms an insoluble salt or a complex with one of the two enantiomers that under conditions where the two enantiomers are epimerizing reversibly. Three examples are given in Scheme 1.5. The enantiomers of ketone **15** are epimerized rapidly under basic conditions via their achiral enolate. Enantiomer (*S*)-**15** forms an insoluble complex with the TADDOL derivative **16** (derived from L-tartaric or (*R*,*R*)-tartaric acid [329, 330]) and precipitates selectively. Adduct (*R*)-**15** +**16** remains soluble. After washing the precipitate to remove the base used for epimerization, it is placed on a silica gel column, yielding pure (*S*)-**15** and the chiral auxiliary **16** (Scheme 1.5a) [331]. Deracemizations of compounds with chiral quaternary carbon centers are less common [332]. An example is given with deracemization of 5-(4-hydroxyphenyl)-5-phenylhydantoin (HPPH: **17**). In boiling methanol, NaOH and 1 equiv. of brucine, a natural alkaloid, (*S*)-**17** precipitates selectively with brucine. In this case, epimerization (*R*)-**17** ⇄ (*S*)-**17** implies the formation of the achiral quinonic intermediate **18**. The crystalline precipitate is extracted with HCl/H_2O (recovery of brucine). The remaining solid is pure (*S*)-(−)-**17** (59% yield) with ee > 99.5% (Scheme 1.5b) [333]. Enantiomerically pure atropisomers (*R*)- and (*S*)-1,1′-binaphthyl-2,2′-diol ((*R*)-and (*S*)-BINOL)), and their derivatives are extremely useful chiral auxiliaries (Section 3.7.5) and ligands for both stoichiometric and catalytic asymmetric synthesis (Section 3.7.6) due to their axial chirality and molecular flexibility [334, 335]. Enantioselective oxidative coupling of β-naphthol (2-naphthol) or its derivatives catalyzed with chiral metal complexes provides one of the most efficient routes to enantiomerically enriched BINOLs [336]. In 1978, Feringa and Wynberg first reported an enantioselective oxidative coupling of β-naphthol with $Cu(II)(NO_3)_2$ and (*S*)-phenylethylamine producing (*S*)-(−)-BINOL with 63% yield and 8% ee. In 1983, Brussee and Jansen found that in the presence of a large excess of (*S*)-(1-methyl-2-phenylethyl)amine ((+)-amphetamine), (*S*)-BINOL is obtained in 98% yield and 96% ee [337]. The enantioenrichment results form the selective precipitation of the Cu(II)/(*S*)-(+)-amphetamine/(*S*)-(−)-BINOL complex with simultaneous racemization of (*R*)-(+)-BINOL (Scheme 1.5c) [338].

In 1962, Barton and Kirby reported the first synthesis of (−)-galanthamine, an *Amaryllidaceae* alkaloid used in the clinic for more than 40 years for the treatment of several neurological illnesses including Alzheimer disease. Its synthesis involves deracemization of (±)-narwedine (**19**). In boiling EtOH, using Et_3N as a base, the two enantiomers of **19** equilibrate with achiral dienone **20** via intramolecular 1,4-elimination and 1,4-addition. When the solution contains (−)-galanthamine, Barton and

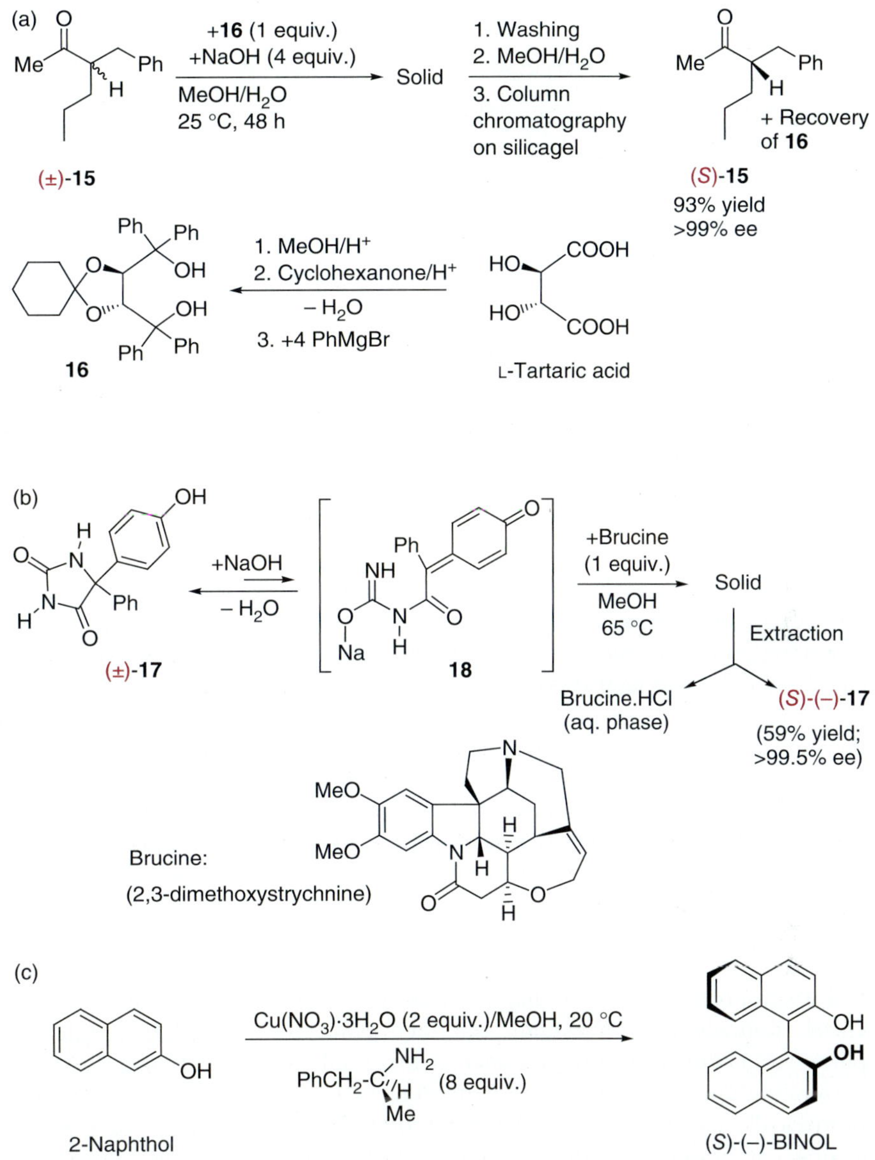

Scheme 1.5 Deracemization of (a) an enolizable ketone by enantioselective formation of a crystalline complex with a enantiomerically pure chiral diol (Seebach's TADDOL derivative), (b) of HPPH via enantioselective crystallization of a salt with brucine, (c) of BINOL via simultaneous racemization of (*R*)-(+)-BINOL during precipitation of a Cu(II)/(+)-amphetamine/(*S*)-(−)-BINOL complex.

Kirby observed that (+)-**19** would precipitate selectively [339]. Alternatively, when a seed (inoculate) of pure (−)-**19** is added to this mixture, selective crystallization of (−)-**19** is induced that forms a conglomerate in 80% yield after cooling and solvent evaporation (Scheme 1.6). The method has been applied in a pilot-scale process for the synthesis of (−)-galanthamine [340]. A similar method (dynamic preferential crystallization) has been developed for the deracemization of N-substituted 3-hydroxy-3-phenylisoindolin-1-ones using DBU (1,8-diazabicyclo[5.4.0]undec-1-ene) as catalyst for the epimerization [341]. In 1941, Havinga reported an example of spontaneous deracemization (in the absence of a chiral auxiliary) through spontaneous crystallization of enantiomerically pure hemihedral crystals of *N*-allyl-*N*-ethyl-*N*-methylanilinium iodide in chloroform. In this case, the quaternary ammonium salt is racemized quickly in $CHCl_3$. However, the method is not practical because both enantiomers can crystallize and performing the resolution with reproducibility is difficult [342]. In 1971, Pincock and Wilson have shown that the crystallization of 1,1′-binaphthyl from its racemic melt is another example of spontaneous deracemization. Right- or left-handed crystallites are formed with equal probability. It can be made stereospecific by addition of low-concentration chiral additives [343–345]. Vlieg and coworkers [346], as well as Blackmond and coworkers [347], have shown that the stirred crystallization of a suspension of nearly racemic amino acid derivative can yield crystals made of a single

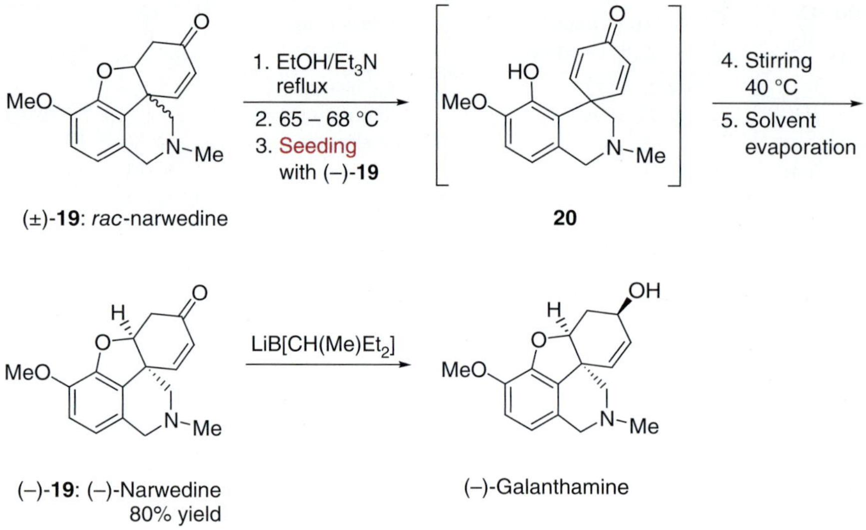

Scheme 1.6 Industrial enantioenrichment of a racemate based on the crystallization of a conglomerate by seeding.

enantiomer so long as the amino acid can be inverted into either enantiomer.

Irradiation of racemic amino acid derivatives with circularly polarized light (CPL) might induce a small ee, which can be amplified by conglomerate formation. A racemate composed of equal amounts of left-and right-handed crystals in contact with the irradiated solution is converted into crystals of single handedness through abrasive grinding when racemization is effected in the solution [348].

1.14.5 Self-disproportionation of enantiomers

Self-disproportionation of enantiomers (SDE) of chiral, nonracemic compounds can take place in phase transitions, in gravitational fields, and in chromatography. It leads to the formation of enantiomerically enriched and depleted fractions under achiral conditions. In many instances, recrystallization of enantiomerically enriched compounds allows one to obtain an enantiomerically pure fraction next to an enantiomerically depleted fraction, if not a fraction containing a pure racemic compound [349]. Because of the fact that racemic and enantiomerically pure crystals have different densities, they can be separated via density gradient ultracentrifugation [350] or suspension precipitation [351]. Sublimation of a crystalline enantiomerically enriched product may provide a sublimed fraction that is enantiomerically enriched and a remaining fraction that is enantiomerically depleted, or vice versa [352]. This sort of enantioenrichment has also been observed in distillations [353]. Very often, SDE is observed during achiral chromatography [354] as reported first by Cundy and Crooks in 1983 [355]. They observed two different peaks for racemic and enantiomerically pure fractions in the chromatogram of enantiomerically enriched nicotine under the conditions of achiral (both the stationary and mobile phase were achiral) high-performance liquid chromatography (HPLC). This phenomenon can be explained simply in the following way. In solution, enantiomers (*R*)-**P** and (*S*)-**P** migrate with the same retention time. They can equilibrate with a homochiral dimeric complexes (*R*,*R*)-**P**-**P** and (*S*,*S*)-**P**-**P**, and with heterochiral dimeric complex (*R*,*S*)-**P**-**P**. The homochiral complexes have the same stability and interact with the stationary phase and mobile phase in the same way. They migrate in the chromatography column with the same retention time, which might be different from that of the heterochiral complex and of the monomeric enantiomers (*R*)-**P** and (*S*)-**P**. In such a situation, all the fractions are racemic if the initial product **P** is racemic (1 : 1 mixture of the two enantiomers). If one considers now an enantiomerically enriched compounds made of two parts of (*R*)-**P** and one part of (*S*)-**P**, the same dimeric complexes will also form. Assuming that the heterochiral dimeric complex (*R*,*S*)-**P**-**P** is more stable than the homochiral complexes (*R*,*R*)-**P**-**P** and (*S*,*S*)-**P**-**P** and the monomeric enantiomers (*R*)-**P** and (*S*)-**P**, there will be an excess of (*R*)-**P** in the solution. If one assumes monomeric (*R*)-**P** to be eluted faster than (*R*,*S*)-**P**-**P**, this gives a first fraction containing enantiomerically pure (*R*)-**P** and then migrates the heterochiral complex giving a fraction containing racemic (±)-**P**. This is an ideal situation. In practice, more complexes can equilibrate with the monomeric and dimeric complexes and their relative stability might be very similar; their relative amounts will depend on concentration (mass law effect). Nevertheless, their retention time may differ enough and lead to enantiomerically enriched fractions of the enantiomer in excess in the initial mixture. Separation by HPLC is not always necessary for SDE. For example, simple preparative flash chromatography on silica gel (EtOAc as eluent) of (*R*)-methyl *p*-tolyl sulfoxide of 86% ee gives a first fraction of 99% ee and a last fraction of 63% ee [356]. Chromatography with achiral phases of nonracemic mixtures of binaphthol furnishes a first fraction with an ee close to 100% and the following fraction has an ee close to 0% (racemic mixture) [357]. The ^{1}H-NMR spectra of enantiopure (*S*)-binaphthol ((*S*)-(−)-1,1′-bi(2-naphthol), or (*S*)-(−)-1,1′-binaphthalene-2,2′-diol: (*S*)-BINOL), and of (±)-binaphthol in $CDCl_3$ show different signals for the OH proton at $\delta_H = 6.30$ and 6.37 ppm, respectively, when taken at the same concentration, which indicates the occurrence of diastereomeric interactions of binaphthol enantiomers in $CDCl_3$. Similar effects are reported with the NMR spectra of optically active and racemic dihydroquinine [358] and with the NMR and IR spectra of leucine dipeptide [359, 360] and organothiophosphorous depsipeptides in solution [361].

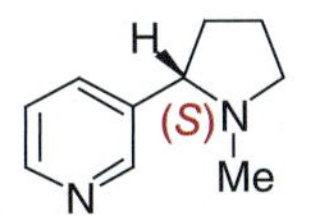

Nicotine

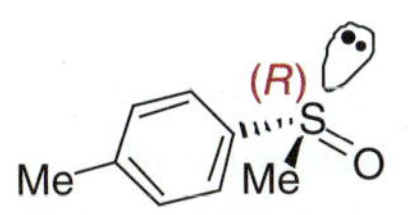

(*R*)-Methyl *p*-tolyl sulfoxide

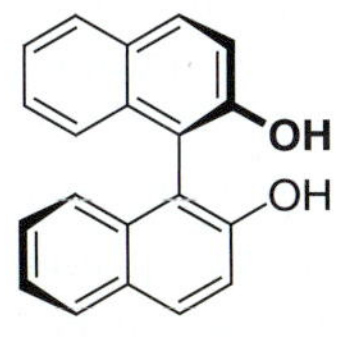

(*R*)-(+)-1,1′-Bi(2-naphthol)

In 2006, Blackmond and coworkers [362, 363], and Hayashi et al. [364], independently found that enantioenriched proline may not be fully soluble in some solvents [365]. Proline is almost insoluble in $CHCl_3$. In the presence of 1% EtOH, proline with low ee (1–10%) separates to give solutions of proline with high ee (97–99%). The enantiomeric enrichment in solution is linked to the different crystal packing in the racemic compound and in the conglomerate crystals, as revealed by powder X-ray diffraction studies. In crystals of (±)-proline, the crystal packing is more compact due to NH···O hydrogen bonds and weak CH···O interactions, whereas the crystal packing in the conglomerate is extended only by NH···O hydrogen bonding interactions. When (±)-proline is crystallized from $CHCl_3$ the crystals are 1 : 1 : 1 D-proline/L-proline/$CHCl_3$. Powder X-ray diffraction studies show a more compact packing in these cocrystals, with extensive hydrogen bonding including an hydrogen bond from the C—H group of $CHCl_3$. The cocrystals are more stable than usual racemate crystals and thus have lower solubility. Enantiomerically pure L- or D-proline are organocatalysts [366, 367] in several reactions in which they form with aldehydes and ketones ene-amine nucleophiles [368] (e.g. Stork enamine reaction [369, 370], aldol reaction (Section 5.7.6) [371], and Michael–Stork addition [372]) or iminium electrophiles (e.g. acid-catalyzed Mannich reaction [373–376], Eschenmoser salts [377], and Pictet–Spengler isoquinoline synthesis [378, 379]). Because of the partial solubility of nonenantiopure proline, the ee of the product formed in a proline-catalyzed aldol reaction might be much higher than that of the amino acid employed as catalyst (see nonlinear effects in asymmetric synthesis, Section 3.7.7) [362, 363].

1.15 Thermodynamic (equilibrium) isotopic effects

Isotopic labeling is an extremely useful tool for the study of reaction mechanisms and equilibria in chemistry and biochemistry and for structural analysis [380, 381]. Isotopic substitution of a molecule causes no change in electronic structure. However, the change in nuclear mass causes a change in vibrational frequency ν and small changes in average bond lengths. Molar volumes of hydrocarbons such as benzene, toluene, cyclohexane, and methylcyclohexane are c. 3% greater than their perdeuterated analogs [382]. This is a manifestation of the bond stretching anharmonicity, which makes the C—H bonds longer than the corresponding C—D bonds [383–385]. Bonds X—H and X—D differ in their zero-point energies (ZPE) as shown in Figure 1.26 [386–390]. Bending modes are also affected by isotopic substitution and contribute to the relative stability and geometry of the molecules [391]. The effect of isotopic substitution on an equilibrium constant is referred to as a thermodynamic or equilibrium isotope effect (EIE) (see kinetic isotope effects, KIE, Section 3.9). For example, self-dissociation constant of H_2O in H_2O ($2H_2O \rightleftarrows H_3O^+ + HO^-$) is larger than that of D_2O in D_2O by a factor of 6.88 at 25 °C [392, 393]. Similarly, dissociation constants K_a of acids A–H in H_2O are generally larger than those of corresponding A–D in D_2O. For acetic acid, one reports $pK_a(AcOH/H_2O) = 4.73$ and $pK_a(AcOD/D_2O) = 5.25$ (dissociation constant is larger for the O—H than for the O—D bonds) [394–396]. The linear relationship (1.112) has been found for Brønsted acids of different types (mineral and carboxylic acids, phenols, thiols, ammonium salts, amino acids, imidazoles, etc.) (138 experimental pairs, correlation coefficient of 0.998) [397].

$$pK_a(A - D/D_2O) = 0.32 + 1.044 \cdot pK_a(A - H/H_2O) \quad (1.112)$$

The dissociation constant ratio $K_a(A\text{–}H + H_2O \rightleftarrows A^- + H_3O^+)/K_a(A\text{–}D + D_2O \rightleftarrows A^- + D_3O^+) = K_H/K_D$ is an example of primary deuterium thermodynamic isotopic effect as the O—H/O—D bond is exchanged in these dissociation processes. This ratio of K_H/K_D is usually larger than unity ($K_H/K_D > 1$) for such primary isotope effects. When this is the case, it is termed a normal equilibrium deuterium isotopic effect. In these equilibria, a strong σ(O—H) bond of H_2O is exchanged with a weaker σ(O—H) bond of H_3O^+. The consequence is that the stretching and bending vibration frequencies of the O—H bond in H_3O^+ are smaller than those of the O—H bond in H_2O. If one considers the Morse potential of a diatomic molecules X—H (Figure 1.26), the zero-point energy ZPE(X–H) (energy of the lowest vibrational level ν_o) is given by Hooke's law (Figure 1.26, see also Figure 3.19). The zero-point energy level ZPE(X–D) is of lower energy for the deuterated X–D molecule. The difference ZPE(X–H) – ZPE(X–D) is given by $\frac{1}{2}(h\nu_{XH} - h\nu_{XD})$, with ν_{XH} and ν_{XD} the stretching vibration frequency of the X–H and X–D molecules, respectively.

The difference ZPE(X–H) – ZPE(X–D) is larger for a strong bond (large force constant, f) than for a weak bond (small force constant, f). Thus, if the equilibrium exchanges a strong σ(X–H,D) bond in the reactant with a weak σ(Y–H,D) bond in the product, the energy difference $\Delta\Delta E(H/D) = \Delta E_H - \Delta E_D < 0$ (Figure 1.27). One can take $\Delta\Delta_r H^T(H/D) = \Delta\Delta E(H/D)$. If the isotope substitution does not affect the entropy of the reaction under study, one can assume $\Delta\Delta_r S^T(H/D) = 0$. This gives $\Delta\Delta_r G^T(H/D) < 0$ and $K_H/K_D > 1$. If one exchanges a weak X—H bond of the reactant for a strong Y—H bond in the product, $K_H/K_D < 1$, which is referred to an inverse equilibrium deuterium isotopic effect.

The difference in the energy of X—H and X—D bond is due to change in nuclear mass, which, in turn, causes a change in vibrational frequency for vibrations involving that nucleus (H or D). A X—H stretch possesses vibrational frequency, $\nu = 1303\sqrt{f/\mu}$ in cm^{-1}, where μ is the reduced mass of m_x and m_y: $\mu = (m_X \cdot m_Y)/(m_X + m_Y)$. For example, for X–H and X–D, with m_X = mass of fragment X, the frequency ratio ν_{XH}/ν_{XD} is given by Eq. (1.113).

$$\frac{\nu_{XH}}{\nu_{XD}} = \frac{\sqrt{(m_X \cdot 2)/(m_X + 2)}}{\sqrt{(m_X \cdot 1)/(m_X + 1)}} \quad (1.113)$$

As $m_D(=2)$ and $m_H(=1)$ are much less than m_X, the frequency ratio ν_{XH}/ν_{XD} becomes almost equal to $\sqrt{2}/\sqrt{1} = \sqrt{2} = 1.414$.

Because the energy E+ZPE(X—D) of an X—D bond is lower than that of E + ZPE(X—H) of an X—H bond, it is harder to break an X—D bond than an X—H bond. This is general for any nuclei and is the origin of the primary equilibrium isotope effects. As a rule (however, see below) the heavier isotope always prefers the more constrained site, the bond the

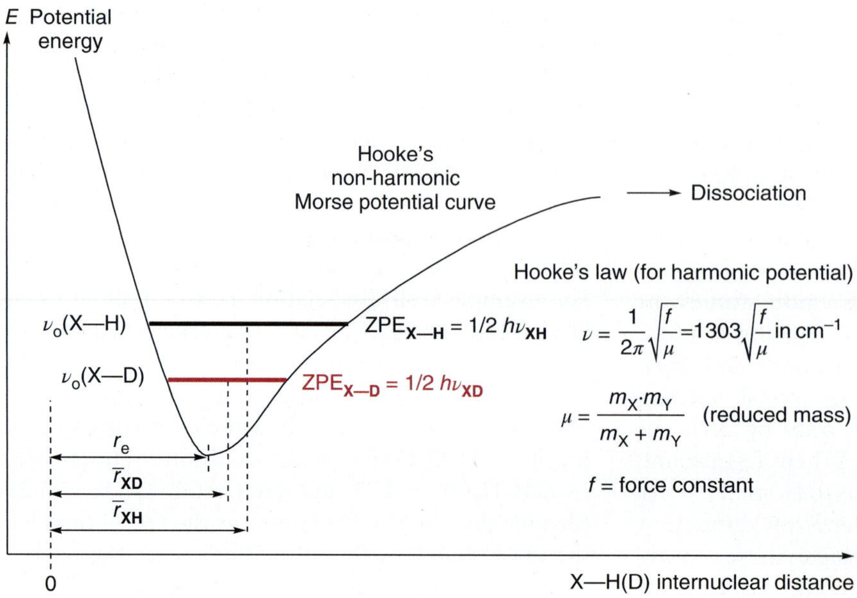

Figure 1.26 Anharmonic Morse potential for molecule X—H. r_e = equilibrium internuclear distance; $\bar{r}_{XH}$ = average X—H bond length; $\bar{r}_{XD}$ = average X—D bond length; and ZPE = zero-point energy.

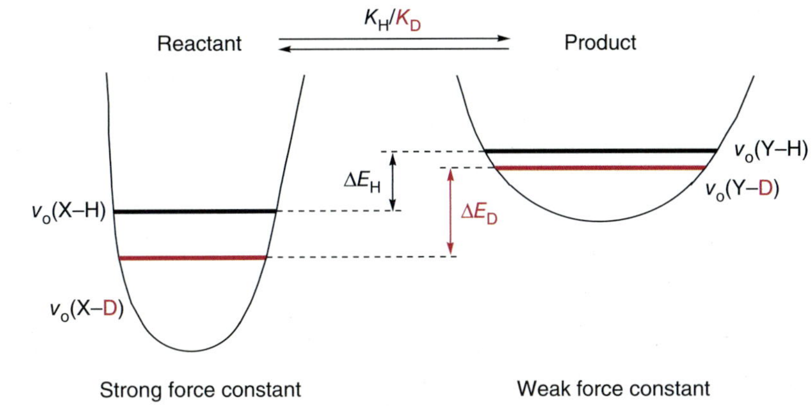

Figure 1.27 Representation of the Morse potential energy curves for a product equilibrating with a reactant in which the Y—H bond in the product is weaker than in the corresponding X—H bond in the reactant. This usually leads to a normal equilibrium isotopic effect $K_H/K_D > 1$; however, see equilibrium (1.115).

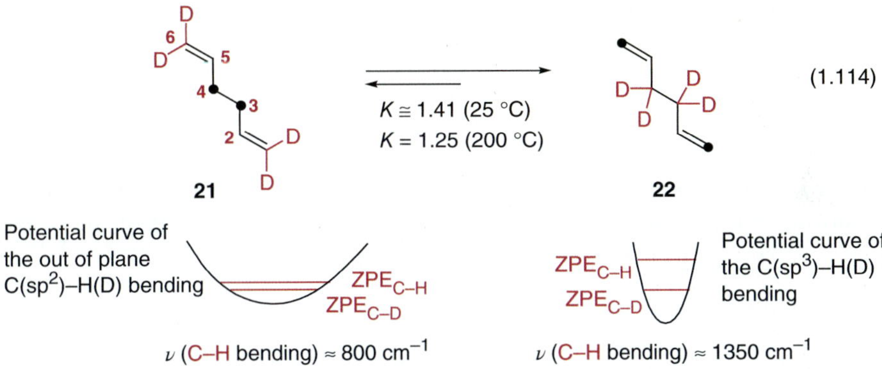

(1.114)

Figure 1.28 Equilibrium deuterium isotope effect on the Cope rearrangement of 1,5-hexadiene. Deuterium "prefers" to be bonded to C(sp³) centers because of the larger difference in ZPE energies (equilibrium isotope effect corresponds to $K(1.114) \cong 1.10$ per deuterium atom at 25 °C).

least easy to deform; the heavier isotope prefers to be located in the highest frequency oscillator. In the case of C—H and C—D bonds, the energy difference due to the difference in ZPE can be estimated from the vibrational frequency, which is about 3000 cm^{-1} for a C—H bond (infrared spectroscopy):

$$\mathrm{ZPE(C{-}H)} - \mathrm{ZPE(C{-}D)} \cong \tfrac{1}{2} h\nu - \frac{1}{2\sqrt{2}} h\nu = \left(\frac{1}{2} - \frac{1}{2\sqrt{2}}\right)(8.6\ \mathrm{kcal\ mol^{-1}}) = 1.3\ \mathrm{kcal\ mol^{-1}} \quad (1.114)$$

This means that the BDE ($DH°(C^•/H^•)$ of a C—H bond is about 1.3 kcal mol^{-1} lower than the BDE of the corresponding C—D bond. Although such a difference in activation energy corresponds to nearly a 10-fold difference in the rate of a reaction at 25 °C, in hydrogen transfers or other processes that break C—H or C—D bonds, the usual primary kinetic isotope effects vary between $k_H/k_D = 2$ and 7 (at room temperature), unless hydrogen atom tunneling occurs (Section 3.9.1). An example of secondary deuterium thermodynamic effect is given (Figure 1.28) for the Cope rearrangement (1.114) (Section 3.4.3 and section "Cope Rearrangements") [398–402].

Oxidative addition of H_2 to transition metal complexes is an important step in the catalytic hydrogenation of unsaturated organic compounds (Sections 7.7.8 and 7.8). An inverse primary equilibrium isotopic effect $K_H/K_D = 0.63$ is measured for equilibrium (1.115) at 60 °C. This is a surprise if one compares the strong σ(H—H) bond with a high stretching frequency with the much weaker σ(W—H) bond stretching frequency that should lead to a normal primary isotopic effect $K_H/K_D > 1$. On measuring EIE(1.115) as a function of temperature, one establishes that the contribution of entropic effects to the isotope effect is small. The addition of D_2 is more exothermic than that of H_2. If one includes the bending vibration modes (Figure 1.29), one finds that the ZPE for $W(PMe_3)_4D_2I_2$ is significantly lower than that of $W(PMe_3)_4H_2I_2$ to the extent that an inverse EIE is observed [403, 404].

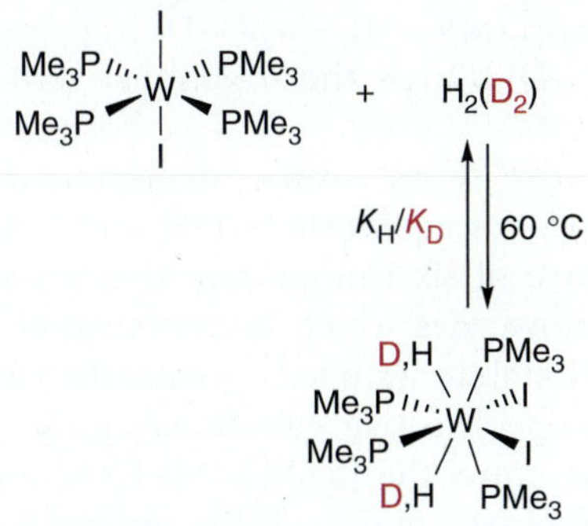

(1.115)

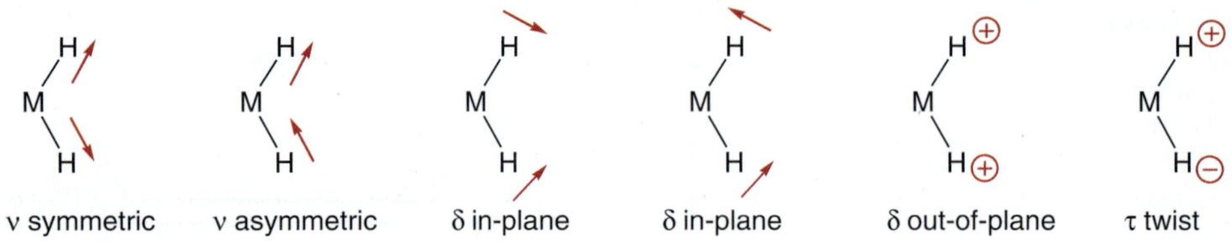

Figure 1.29 Vibrational modes associated with a C_2 symmetric $[MH_2]$ type of dihydride. In addition to the two stretching modes (ν), the $[MH_2]$ fragment has four more low-energy vibration modes (δ and τ). H_2 has only one vibration (stretching) mode.

In the case of the Vaska system (Section 7.7.3), the oxidative addition (1.116) shows a primary EIE = K_H/K_D = 0.41 at 25 °C. It becomes unity near 90 °C and 1.41 at 130 °C [405].

$$H_2(D_2) + Ir(PMe_2Ph)_2(CO)Cl \rightleftarrows Ir(PMe_2Ph)_2(CO)(Cl)H_2(D_2) \quad (1.116)$$

A more complete analysis of isotope effects considers other factors that add to the ZPE term. In fact, EIEs are determined from the molecular translational, rotational, and vibrational partition function ratios, according to the expression (1.117) [406, 407].

$$EIE = K_H/K_D = SYM \cdot MMI \cdot EXC \cdot ZPE \quad (1.117)$$

where SYM = $[\sigma(R—H)/\sigma(R—D)]/[\sigma(P—H)/\sigma(P—D)]$ is the symmetry number that is factored out of the rotational partition function, R—H (nondeuterated) and R—D (deuterated) refer to the reactant, P—H and P—D to the product;

$$\begin{aligned} MMI = \{&[m(P—H)/m(R—H)]^{3/2} \\ &\times [I(P—H)/I(R—H)]^{1/2}\} \\ &\times /\{[m(P—D)/m(R—D)]^{3/2} \\ &\times [I(P—D)/I(R—D)]^{1/2}\} \end{aligned}$$

is the mass moment of inertia term that is factored out of the translational and rotational partition functions with m for molecular masses and I for inertia moments;

EXC = $\{\Pi\{[1-\exp(-u(R—H)_i)]/[1-\exp(-u(R—D)_i)]\}\}/\{\Pi\{[1-\exp(-u(P—H)_i)]/[1-\exp(-u(P—D)_i)]\}\}$ is the excitation term that takes into account vibrationally excited states, with $u(R—H)_i$, $u(R—D)_i$, $u(P—H)_i$, and $u(P—D)_i$ correspond to the respective $h\nu_i/k_bT$, $h\delta_i/k_bT$, and $h\tau_i/k_bT$ values. Finally, ZPE = $\{\exp[\Sigma(u(R—H)_i - u(R—D)_i)/2]\}/\{\exp[\Sigma(u(P—H)_i - u(P—D)_i)/2]\}$ is the traditional zero-point energy term.

The occurrence of an inverse deuterium EIE at low temperature for equilibria (1.115) and (1.116) is the consequence of six isotope-sensitive vibrational modes in the dihydrides, which, in combination, result in the total ZPE stabilization to be greater than that for the single isotope-sensitive vibrational mode in H_2. At high temperature, the [SYM·MMI·EXC] entropy component dominates and the EIE is normal because the entropy of D_2 is greater than that of H_2. If the ZPE contribution reaches unity, the [SYM·MMI·EXC] entropy component will dominate the EIE already at relatively low temperature [407].

As mentioned above, on average, the C—D bond is slightly shorter than the C—H bond. At room temperature, the ^{1}H- and ^{13}C-NMR spectra of 1,1,3,3-tetramethylcyclohexane show single ^{1}H and ^{13}C signals, respectively, for the four methyl groups because of the fast chair/chair interconversion that equilibrates the two axial (Me_{ax}) with the two equatorial methyl groups (Me_{eq}). At −100 °C, the chair/chair interconversion is slowed down ($\Delta^{\ddagger}G \cong 9$ kcal mol^{-1}) and two ^{1}H and ^{13}C signals are observed for Me_{ax} and Me_{eq}. A chemical shift difference $\delta_{eq} - \delta_{ax} = \Delta\delta$ = 9.03 ppm is measured in the ^{13}C-NMR spectrum. If one of the methyl groups is exchange for a trideuterated methyl groups (CD_3), the room temperature spectrum displays two signals for the methyl groups separated by $\delta\delta$ = 0.184 ppm at 17 °C in CS_2. From these data, the equilibrium constant is determined to be $K(1.118) = (\Delta\delta + \delta\delta)/(\Delta\delta - \delta\delta) = 1.042 \pm 0.001$ at 17 °C in CS_2, which shows that the deuterated methyl group is sterically smaller and is better "able" to occupy the more crowded axial position than the nondeuterated methyl group [408, 409] (Tables 1.A.23 and 1.A.24).

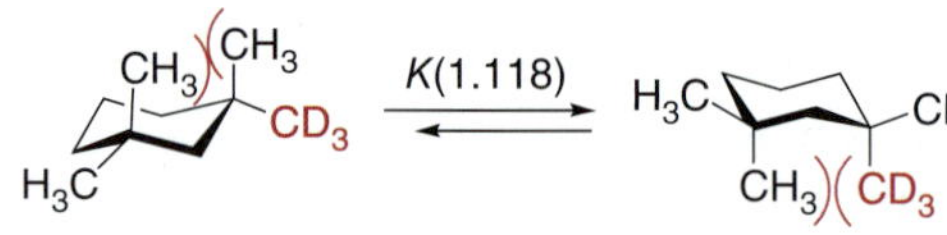

(1.118)

Problem 1.44 Complex **A** equilibrates with complexes **B**, **C**, and **D** via intermediate **I**. At equilibrium, the relative proportions of **A**/**B**/**C**/**D** are 2.7 : 2 : 2 : 1. What are **B**, **C**, **D**, and **I**? Why the relative proportions **A**/**B**/**C**/**D** are not 1 : 2 : 2 : 1? [410]

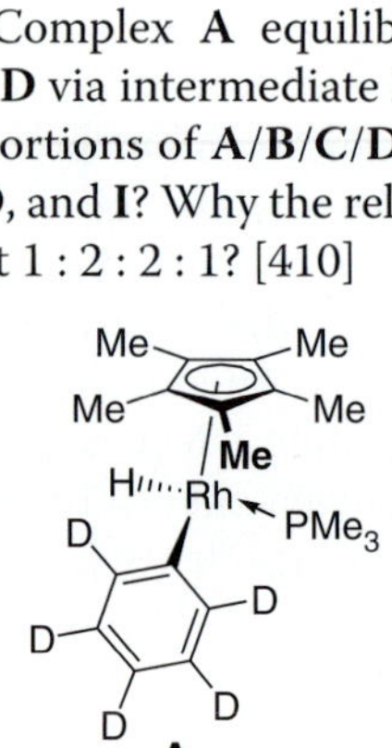

A

1.A Appendix

Table 1.A.1 Standard heats of formation $\Delta_f H°$ (gas phase) in kcal/mol [1 cal = 4.184 J] for selected **inorganic and organometallic compounds**[a,b].

HF	−65.1 ± 0.2	H_2O	−57.8 ± 0.01	NH_3	−11.0	CH_4	−17.89 ± 0.07	BH_3	25.5	BeH	76.7	LiH	33.6
HCl	−22.1 ± 0.03	H_2S	−4.9 ± 0.1	PH_3	1.3	SiH_4	8.2	B_2H_6	9.8	$Mg(OH)_2$	−136.8	NaH	29.7
HBr	−8.7 ± 0.04	H_2Se	7	AsH_3	16	GeH_4	22	BH_2	30.0	$Fe(OH)_2$	−79.0	KH	29.4
HI	6.3 ± 0.03	H_2Te	24	SbH_3	35	SnH_4	39	BF_3	−217.5			LiOH	−56.0
HOCl	−17.8	ICN	54.0	IF	−22.7			BCl_3	−96.3			NaOH	−47.3
H_2O_2	−32.5	HO•	9.3	HO_2•	0.5	HNO	23.8	BBr_3	−48.8			KOH	−56
NO	21.6	NO_2	7.9	N_2O	19.6	HNO_2	−18.3	HNO_3	−32			CsOH	−62.0
H_2S_2	4											EtLi	15.2
S_2O	−13.5	SO_2	−70.9 ± 0.05	SO_3	−94.6	$SOCl_2$	−51	H_2SO_4	−175.7			BuLi	−0.6
CO	−26.4 ± 0.04	OF_2	5.86	NF_3	−31.6	CF_4	−221					Li	38.1
CO_2	−94.0 ± 0.03	SF_2	−70.9	PF_3	−229.1	SiF_4	−386.0	Me_4Si	−68.8			Na	25.7
CS_2	27.95	SCl_2	−4.2	AsF_3	−188	GeF_4	−284.5	B_2O_3	−199.8			K	21.3
OCl_2	21.0	$GeCl_2$	−41			UF_6	−513	O_3(ozone)	34.1			Cs	76.5

(Z)-HN=NH	50.9	H_2N—NH_2	22.8	H_2N—OH	−10	HN_3	70.3	Urea	−58.7	Thiourea	5.5	HCN	32.2
ClF	−12.0	ICl	4.2	LiCl	−47	$(LiCl)_2$	−143	$(LiCl)_3$	−240	NaCl	−43.4	$(NaCl)_2$	−135.3
KCl	−51.3	$(KCl)_2$	−147.6	RbCl	−55	$(RbCl)_2$	−150	CsCl	−57.4	$(CsCl)_2$	−157.7	$MgCl_2$	−93.8
CuCl	21.8	$FeCl_2$	−33.7	$NiCl_2$	−17.7	Cp_2Ni	85.3	Cp_2Os	73.8				
$MeSiF_3$	−294.6	$MeSiCl_3$	−126.4	MeHgCl	−13.2	MeHgBr	−4.5	MeCd	49.9			MeZn	45.5
Me_2Bi	63.3	Me_2Cd	29.4	Me_2Ga	19.3	Me_2Hg	22.6	Me_2In	63.7	Me_2Sb	33.6	Me_2Zn	13.2
Me_3Bi	46.5	Me_3Al	−20.7	Me_3Ga	−11.0	Me_3Ge	22.2	Me_3In	43.9	Me_3Sb	7.7	Me_3Zn	31.6
Me_3Pb	54.4	Me_4Pb	32.5	Me_4Sn	−5.0	$CpTiCl_3$	−124.5	Cp_2TiCl_2	−63.3	Cp_2HfCl_2	−102.4	Cp_2MoCl_2	1.0
Cp_2Mg	32.9	Cp_2Ti	−2.7	Cp_2V	48.6	Cp_2Cr	59.6	Cp_2Mn	66.2	Cp_2Fe	57.9	Cp_2Co	73.8

a) NIST WebBook of Chemistry: http://webbook.nist.gov/chemistry. Standard deviations are not available for all compounds.
b) Me = CH_3; Cp = C_5H_5.

Table 1.A.2 Standard heats of formation $\Delta_f H°$ (gas phase) in kcal/mol [1 cal = 4.184 J] [http://webbook.nist.gov] and standard entropies ($S°$, gas phase, in eu) *in italics* for selected **hydrocarbons.**

Alkanes				Alkenes			
CH_4	Methane	−17.8 ± 0.07	*45.0 ± 0.1*	C_2H_4	Ethene	12.54 ± 0.1	*52.5*
C_2H_6	Ethane	−20.0 ± 0.1	*54.8*	C_3H_6	Propene	4.88 ± 0.27	*63.8*
C_3H_8	Propane	−25.02 ± 0.12	*64.5*	C_4H_8	But-1-ene	0.15 ± 0.19	*73.8*
C_4H_{10}	Butane	−30.04 ± 0.16	*74.1*		(*Z*)-But-2-ene	−1.83 ± 0.30	*72.1*
	2-Me-propane	−32.07 ± 0.15	*70.4*		(*E*)-But-2-ene	−2.58 ± 0.24	*70.9*
C_5H_{12}	Pentane	−35.08 ± 0.14	83.1 ± 0.2		2-Me-propene	−4.29 ± 0.26	*70.2*
	2-Me-butane	−36.7 ± 0.14	*82.1*	C_5H_{10}	Pent-1-ene	−5.0 ± 2.0	*83.3*
	2,2-Me_2-propane	−40.14 ± 0.15	*72.2*		(*Z*)-Pent-2-ene	−7.0 ± 1.0	*81.6*
C_6H_{14}	Hexane	−39.93 ± 0.20	*92.8*		(*E*)-Pent-2-ene	−7.7 ± 0.4	*82.0*
	2-Me-pentane	−41.66 ± 0.25			2-Me-but-1-ene	−8.4 ± 0.2	*81.2*
	3-Me-pentane	−41.02 ± 0.23			3-Me-but-1-ene	−6.1 ± 0.2	
	2,2-Me_2-butane	−44.35 ± 0.23			2-Me-but-2-ene	−9.92 ± 0.21	*80.9*
	2,3-Me_2-butane	−42.49 ± 0.24	*87.4*	C_6H_{12}	Hex-1-ene	−10.2 ± 0.6	*92.7*
C_7H_{16}	Heptane	−44.89 ± 0.19			(*Z*)-Hex-2-ene	−11.2 ± 0.26	
	2-Me-hexane	−46.6 ± 0.30			(*E*)-Hex-2-ene	−12.2 ± 0.24	
	3-Me-hexane	−45.96 ± 0.30			(*Z*)-Hex-3-ene	−11.0 ± 0.2	
	3-Et-pentane	−45.34 ± 0.28			4-Me-pent-1-ene	−11.8 ± 0.16	
	2,2-Me_2-pentane	−49.29 ± 0.32			(*Z*)-3-Me-pent-2-ene	−14.79 ± 0.21	
	2,3-Me_2-pentane	−47.62 ± 0.30			(*E*)-3-Me-pent-2-ene	−15.18 ± 0.21	
	2,4-Me_2-pentane	−48.30 ± 0.23			2,3-Me_2-but-2-ene	−16.8 ± 0.36	
	3,3-Me_2-pentane	−48.17 ± 0.22		C_7H_{14}	Hept-1-ene	−15.1	
	2,2,3-Me_3-butane	−48.96 ± 0.27	*91.6*		(*E*)-Hept-3-ene	−17.4 ± 0.2	*82.2*
C_8H_{18}	Octane	−49.88 ± 0.16	*111.6*		(*Z*)-Hept-3-ene	−16.4 ± 0.1	
	2-Me-heptane	−51.50 ± 0.31			(*Z*)-Hept-2-ene	−16.6 ± 0.2	
	3-Me-heptane	−50.82 ± 0.27			2,3,3-TriMe-but-1-ene	−20.4	
	3-Et-hexane	−50.40 ± 0.28		C_3H_{16}	Oct-1-ene	−19.8	
	2,2-Me_2-hexane	−53.71 ± 0.24			(*E*)-2,5-Me_2hex-3-ene	−28.6	
	2,4-Me_2-hexane	−52.44 ± 0.27			(*Z*)-2,5-Me_2hex-3-ene	−26.6	
	2,5-Me_2-hexane	−53.21 ± 0.26		$C_{10}H_{20}$	Dec-1-ene	−29.5	
	2,2,3-Me_3-pentane	−52.61 ± 0.36			(*E*)-2,2,5,5-Me_4hex-3-ene	−39.9	
	2,2,4-Me_3-pentane	−53.57 ± 0.32			(*E*)-2,2,5,5-Me_4hex-3-ene	−30.3	
	2,2,3,3-Me_4-butane	−54.06 ± 0.36	*93.1*	$C_{16}H_{32}$	Hexadec-1-ene	−59.4	

C_3H_4	Propadiene	45.5	*58.3*	Cyclopropane	12.74 ± 0.14	*56.8*
C_4H_6	Buta-1,2-diene	38.77 ± 0.14	*70.0*	Cyclopropene	66.2 ± 0.6	*58.4*
	Buta-1,3-diene	26.0 ± 0.2	*66.6*	Methylenecyclopropene	48.0 ± 0.4	
C_5H_8	Penta-1,2-diene	33.6 ± 0.16	*79.7*	Cyclobutane	6.6	*63.4*
	(*Z*)-Penta-1,3-diene	19.77 ± 0.22	*76.5*	Cyclobutene	37.5 ± 0.4	*63.0*
	(*E*)-Penta-1,3-diene	18.11 ± 0.16	*76.4*	Cyclobutadiene (Section 4.5.6)	102.3 ± 4	
	2-Me-buta-1,3-diene	18.09 ± 0.24	*75.2*	Methylidenecyclobutane	25.4	*71.4*
	Penta-1,3-diene	25.41 ± 0.31	*79.8*	Cyclopentane	−18.26 ± 0.19	*70*
C_6H_{10}	Hexa-1,5-diene	20.4 ± 0.4		Cyclopentene	8.5	*69.2*
	2,3-Me_2-buta-1,3-diene	10.8		Cyclopenta-1,3-diene	33.2	
C_2H_2	Ethyne	54.19 ± 0.2	*48.0*	Methylidenecyclopentane	2.4	
C_3H_4	Propyne	44.2 ± 0.2	*59.3*	Methylcyclopentane	−25.3 ± 0.2	*81.2*
	Propadiene	45.5		Cyclohexane	−29.5 ± 0.2	*71.3*
C_4H_6	But-1-yne	39.5 ± 0.2	*69.5*	Cyclohexene	−1.03 ± 0.23	*74.2*
	But-2-yne	34.8 ± 0.25	*67.7*	1-Methylcyclopentene	−1.0 ± 0.2	*78.2*
				5-Methylcylopentene	2.0 ± 0.3	
C_5H_6	(*Z*)-Pent-3-en-1-yne	61.7		1,2-Dimethylidenecyclobutane	48.8	
	(*E*)-Pent-3-en-1-yne	61.9		Cyclohexa-1,3-diene	25.4	
C_5H_8	Pent-1-yne	35.5 ± 0.5	*79.4*	Cyclohexa-1,4-diene	24.0	
	Pent-2-yne	30.8 ± 0.5	*79.3*	Methylidenecyclohexane	−6.0	
	3-Me-but-1-yne	32.6 ± 0.5	*76.3*	Cyclohepta-1,3,5-triene	44.6	*75.4*
C_6H_{10}	Hex-1-yne	29.23 ± 0.29		Cyclooctane	−30.14 ± 0.38	*87.7*
	Hex-3-yne	25.19 ± 0.46		(*Z*)-Cyclooctene	−6.5	
C_4H_4	CH_2=CH—C≡CH	70.4		(*Z*,*Z*)-Cycloocta-1,3-diene	20.1	
	HC≡C—C≡CH	111.0		(*E*,*Z*)-Cycloocta-1,3-diene	35.0	
C_6H_8	Me—C≡C—C≡C—Me	90.2		(*Z*,*Z*)-Cycloocta-1,5-diene	24.14 ± 0.33	
Trienes				(*E*,*Z*)-Cycloocta-1,5-diene	37.8	
C_4H_4	Butatriene	83		(*Z*,*Z*,*Z*)-Cycloocta-1,3,6-triene	46.9	
C_6H_8	(*Z*)-Hexa-1,3,5-triene	41		Cyclooctatetraene	71.13 ± 0.31	*78.1*
C_6H_6	*c*-[H_2C=C]$_3$	95.6 ± 2.9		Cyclooctyne	43	
C_7H_{10}	2-Ethylidenepenta-1,4-diene	38.0		3,4-Dimethylidenebicyclo[4.2.0]octa-1,5-diene	85.8	
Tetraene	Octa-1,2,6,7-tetraene	88.1		*ortho*-Xylylene	53	

(*continued*)

Table 1.A.2 (Continued)

Benzene derivatives, see also Tables 2.A.2 and 2.2

PhH	19.8 ± 0.2	*64.3*	1,2-Me_2benzene	4.54 ± 0.26	*84.5*	1,1-DiPh(ethane)	58.8	
PhMe	12.0 ± 0.26	*76.6*	1.3-Me_2benzene	4.12 ± 0.18	*85.6*	1,2-Diphenylethane (dibenzyl)	32.4 ± 0.3	
PhEt	7.12 ± 0.20	*86.2*	1,4-Me_2benzene	4.29 ± 0.24		(*E*)-Stilbene	52.5	
PhPr	1.87 ± 0.20	*95.0*	1,2,3-Me_3benzene	−2.29 ± 0.30	*92.0*	(*Z*)-Stilbene	60.3	
Ph*i*-Pr	0.94 ± 0.26	*92.4*	1,2,4-Me_3benzene	−3.33 ± 0.27	*94.6*	Naphthalene	36.0 ± 2.0	*80.5*
PhBu	−3.06 ± 0.3		1,3,5-Me_3benzene	−3.8	*92.1*	Azulene	73.5	
Ph*t*-Bu	−5.42 ± 0.34		1,2,3,4-Me_4benzene	−8.61 ± 0.34		Anthracene	53.0 ± 4.0	
$PhCH_2$iPr	−5.15 ± 0.34		1,2,3,5-Me_4benzene	−10.33 ± 0.30		Phenanthrene	48.3 ± 0.6	
Ph-*c*-Hex	−3.98 ± 0.35		1,2,4,5-Me_4benzene	−11.30 ± 0.45		Acenaphthene	37.3	
Styrene	35.11 ± 0.24	*82.5*	Me_5benzene	−16.1 ± 0.53		Pyracene	41.7 ± 1.3	
PhC≡CH	73.27 ± 0.41		Me_6benzene	−18.5 ± 0.6		9,10-Dihydrophenanthrene	37.1 ± 0.4	
PhC≡CMe	64.1 ± 0.52		1-*t*-Bu,2-Mebenzene	−8.0 ± 0.4		Benz[*a*]anthracene	70.0	
PhC≡CPh	92.0 ± 0.64		1-*t*-Bu,3-Mebenzene	−13.0 ± 0.4		Benzocyclopropene	89	
Ph—Ph	43.1 ± 0.7	*93.9*	1-*t*-Bu,4-Mebenzene	−13.6 ± 0.4		Benzocyclobutene	47.7 ± 0.2	
$PhCH_2Ph$	39.4 ± 0.53		Triphenylmethane	64.8		Indene	38.5 ± 0.6	
Ph-*c*-Pr	36.02 ± 0.24		1,3,5-Triphbenzene	87.8		Indane	14.6 ± 0.5	
Chrysene	64.2 ± 1.1		Triphenylene	64.6 ± 1.1		Naphthacene	81.9 ± 1.4	

Isomers of benzene

	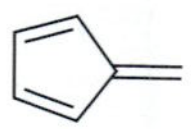	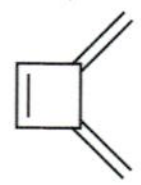				
19.8 ± 0.2	Fulvene 53.6	80.4	[3]Radialene 94.6 ± 2.9	Dewar benzene 87	Benzvalene 87	
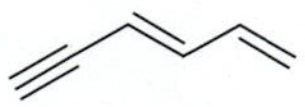			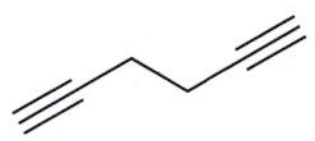	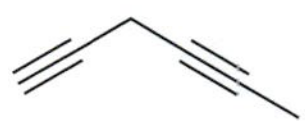	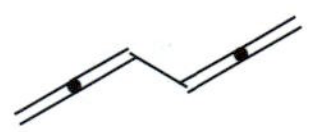	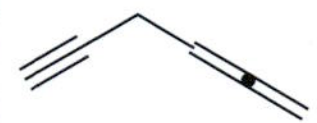
87	84	93.6	99.4	98.1	98	99

Standard deviations are not available for several compounds. $S^\circ(H_2) = 32.1$ eu. See also Tables 2.A.2, 2.A.3, and 2.2 [20].

Table 1.A.3 Standard heats of formation $\Delta_f H°$ (gas phase) in kcal/mol (1 cal = 4.184 J) of **polycyclic hydrocarbons.**[a)]

					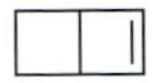		
51.9 ± 0,2	79.6	37.7	49.7	44.23 ± 0.18	61	54.5	9.3 ± 0.8
		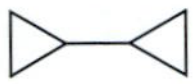					
29.8	15.3	31.0 ± 1.2	40	norbornadiene 57. ± 6.	85	quadricyclane 80.4	0.4 ± 1.0
						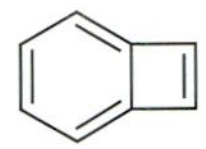	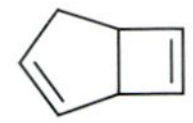
45.7	*trans*-decaline −43.54 ± 0.55	*cis*-decaline −40.45 ± 0.55	norbornene 21. ± 7.	norbornane −13.13 ± 0.25	nortricyclane 19. ± 6.	c. 118	63.2

(continued)

Table 1.A.3 (Continued)

	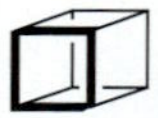						
barrelene 73	cubane 148.7 ± 1.0	96	12	−23	4.9	34	48.1

52	−1.2	97	−1.	11	−2.	8.	13.
	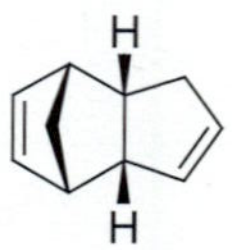					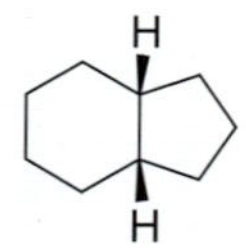	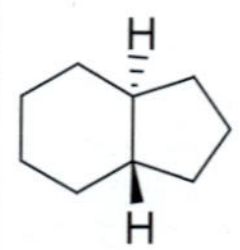
30.5	45	adamantane −32.12 ± 0.56	−14.4 ± 0.9	−0.2	11	−30.4 ± 0.5	−31.5 ± 0.5

a) Taken from [21–23, 43, 68].

Table 1.A.4 Standard gas-phase heats of formation $\Delta_f H°$ in kcal/mol, (1 cal = 4.184 J) and standard entropies $S°$ (in eu, *in italics*) of selected **functional organic compounds.**[a)]

Alcohols			**Diols**			**Polyalkoxyalkanes**	
Methanol (MeOH)	−49.0 ± 3.0	*57.3*	Ethyleneglycol	−94.2 ± 0.67	*74.5*	$(MeO)_2CH_2$	−83.3
Ethanol (EtOH)	−56.0 ± 0.5	*67.6*	Propa-1,2-diol	−102.7 ± 1.0		$(MeO)_2CHMe$	−93.15 ± 0.20
Propan-1-ol (PrOH)	−61.2 ± 0.7	*77.1*	Propa-1,3-diol	−97.6 ± 1.2		1,2-$(MeO)_2$ethane	−81.9 ± 0.2
Propan-2-ol (*i*-PrOH)	−65.2	*74.1*	Buta-1,3-diol	−103.5 ± 0.7		$(MeO)_2CMe_2$	−101.5 ± 0.3
Butan-1-ol (BuOH)	−66.0 ± 1.0	*86.5*	Buta-1,4-diol	−101.8 ± 1.4		$(EtO)_2CH_2$	−98.7 ± 0.2
Butan-2-ol (*i*-BuOH)	−70.0 ± 0.08	*85.8*	Penta-1,2-diol	−111.2		1,1-$(EtO)_2$-ethane	−108.4 ± 0.74
t-Butanol (*t*-BuOH)	−74.72 ± 0.21	*78.0*	Penta-1,5-diol	−105.6		1,2-$(EtO)_2$-ethane	−98.1
Pentan-1-ol	−71.0 ± 1.0	*96.1*	Pinacol	−129.2 ± 2.2		1,3-$(EtO)_2$-propane	−104.3 ± 0.36
Pentan-2-ol	−75.0 ± 0.26		***Ethers***			2,2-$(EtO)_2$-propane	−121.1
Pentan-3-ol	−75.8 ± 0.3		MeOMe	−44.0 ± 0.12	*63.7*	1,1-$(MeO)_2$butane	−101.7 ± 0.4
3-Me-butan-2-ol	−74.9		EtOMe	−51.73 ± 0.16		1,1-$(MeO)_2$ cyclopentane	−95.0 ± 0.4
2-Me-butan-2-ol	−78.7		EtOEt	−60.40 ± 0.47	*81.8*		
Cyclopentanol	−58.1 ± 0.4		*n*-BuOMe	−61.7 ± 0.27			
Benzyl alcohol	−24.0	*84.6*	*t*-BuOMe	−68.1	*85.5*	$(MeO)_3CH$	−127.1 ± 0.77
HC≡COH	9.94		$(i\text{-}Pr)_2O$	−76.3		$(MeO)_3CMe$	−136.4 ± 0.21
CH_2=CHOH	−30.6		$(t\text{-}Bu)_2O$	−86.3 ± 0.2		$(EtO)_3CH$	−150.7
Allyl alcohol	−29.55 ± 0.35	*77.3*	*n*-BuOEt	−61.7			
CH_2=C(OH)Me	−40		*n*-PrOEt	−65.0		$(EtO)_4C$	−206.0 ± 0.5
Me_2C=C(OH)Me	−57.6		EtO-vinyl	−33.5 ± 0.27			
Phenols			$(vinyl)_2O$	−3.03 ± 0.20		2-Methyl-1,3-dioxane	−95.0 ± 0.67
Phenol	−23.0 ± 0.14		CH_2=CHO-*n*-Pr	−38.72 ± 0.53	*58.0*	4-Methyl-1,3-dioxane	−90.1
o-$(OH)_2$benzene	−65.7 ± 0.3		Oxirane	−12.6 ± 0.15	*63.4*	1,3-Dioxaheptane	−80.9 ± 1.1
m-$(OH)_2$benzene	−68.0 ± 0.3		Methyloxirane	−22.6 ± 0.15		1,2-Dioxaheptane	−34
p-$(OH)_2$benzene	−66.2 ± 0.33		Oxetane	−19.3		*cis*-2,4-Me_2-1,3-dioxane	−101.6 ± 1
1,2,3-$(OH)_3$benzene	−103.8 ± 0.26		Tetrahydrofuran	−44.0 ± 0.17		*trans*-4,5-Me_2-1,3-dioxane	−98.2 ± 0.6.
1,2,4-$(OH)_3$benzene	−106.1 ± 0.38		2,3-$(H)_2$furan	−72.25 ± 0.41		1,4,6-Me_3-1,3-dioxane	−106.8 ± 0.8
1,3,5-$(OH)_3$benzene	−108.2 ± 0.26		Tetrahydropyran	−53.5 ± 0.24		$(MeOCH_2CH_2)_2O$	−124.6
1-Naphthol	−7.36 ± 0.38		3,4-$(H)_2$-2*H*-pyran	−29.96 ± 0.22		1,4,7-Trioxocane	−138.9
2-Naphthol	−7.15 ± 0.41	*87.6*	2,3-$(H)_2$benzofuran	−11.1 ± 0.2		1,3,6-Trioxocane	−111.7 ± 0.29
Naphthalene-1,2-diol	−47.92 ± 0.43		Phthalane	−7.19 ± 0.24		Paraformaldehyde	152.1 ± 0.7

(continued)

Table 1.A.4 (Continued)

Hydroperoxides and peroxides				Ketenes		Carbonates	
EtOOH	−50.0	MeOOMe	−30.0 ± 0.1	CH_2=C=O	−20.8	Ethylene carbonate	−120.2 ± 1.0
t-BuOOH	−56.1 ± 1.2	EtOOEt	−46.1	Me_2C=C=O	−32	Methylethylene carbonate	−134.7
c-C_6H_{11}OOH	−52.3	*t*-BuOO*t*-Bu	−81.5	NCCH=C=O	24.0 ± 5.0		
$(MeCOO)_2$	−119.0						

Aldehydes			Ketones			Dicarbonyl compounds	
Formaldehyde	−27.7		Acetone	−52.23 ± 0.14	*70.5*	O=CH—CH=O (glyoxal)	−50.66 ± 0.19
Ethanal	−40.8 ± 0.35	*63.2*	Butan-2-one	−57.02 ± 0.20	*80.8*	MeCOCH=O	−64.8 ± 1.2
Propanal	−45.09 ± 0.18	*73.4*	Pentan-2-one	−61.91 ± 0.26	*90.5*	Buta-2,3-dione	−78.1
Butanal	−50.61 ± 0.22	*82.6*	Pentane-3-one	−60.6 ± 0.2	*86.6*	Pent-2,4-dione	−91.87 ± 0.31
2-Me-propanal	−51.57 ± 0.37		3-Me-butan-2-one	−62.76 ± 0.21	*88.5*	Hexa-2,4-dione	−105.1
Pentanal	−54.6		3,3-DiMe-butan-2-one	−69.47 ± 0.21		3-Methylpenta-2,4-dione	−102.5
O=CH—CH=O	−50.7 ± 0.2		Cyclopropanone	3.8 ± 1.0			
Enals, enones			Cyclobutanone	−21.9 (−24.2)		Cyclohexa-1,4-dione	−79.49 ± 0.29
PhCHO	−8.80 ± 0.72		Cyclopentanone	−47.19 ± 0.30		Cyclohexa-1,3-dione	−80.21 ± 0.38
Crotonaldehyde	−26.22 ± 0.57		Cyclohexanone	−55.23 ± 0.21		*p*-Benzoquinone	−29.4
Methacrolein	−25.4 ± 0.5		MeCOCHMe_2	−62.76 ± 0.21		Me_4-cylobuta-1,3-dione	−73.54 ± 0.38
Furfuraldehyde	−35.4		Hexan-2-one	−66.87 ± 0.26		5,5-Me_2-cyclohexa-1,3-dione	−91.7 ± 0.45
CH_2=CHCOMe	−27.4 ± 2.6	*76.5*	Hexan-3-one	−66.50 ± 0.26			
Acetophenone	−20.71 ± 0.40		MeCOCH_2CHMe_2	−69.60 ± 0.34		***Cyclic anhydrides***	
Benzophenone	11.93 ± 0.71		MeCO-*t*-Bu	−69.47 ± 0.26			
Cyclopent-2-enone	−19		Cycloheptanone	−59.3 ± 0.31		Maleic anhydride	−97.9
Cyclohex-2-enone	−28		Bicyclo[2.2.1]heptan-2-one	−40.8 ± 1.2		Succinic anhydride	−126.2 ± 0.41
Me_2C=CHCOMe	−42.61 ± 0.15		Bicyclo[2.2.1]heptan-7-one	−32.0 ± 0.7		Glutaric anhydride	−127.2 ± 0.43

Carboxylic acids			Esters			Ene-esters	
Formic acid (HCOOH)	−90.5 ± 0.1	*59.4*	HCOOMe	−80.5	*84.6*	Methyl acrylate	−79.6
Acetic acid (MeCOOH)	−103.5 ± 0.6	*67.6*	HCOO-*t*-Bu	−109.2 ± 1.3		Methyl methacrylate	−83.3
Propanoic acid	−108.9 ± 0.48		AcOMe	−98.0		Et (*E*)-but-2-enoate	−89.8 ± 0.6
Butanoic acid	−113.7 ± 0.96	*84.4*	AcOEt	−106.46 ± 0.20	*86.7*	Me (*E*)-but-2-enoate	−81.7 ± 0.5
Pentanoic acid	−117. ± 4.0	*105*	AcO-*i*-Pr	−117.0 ± 0.88		Et (*E*)-pent-2-enoate	−94.2 ± 0.9

3-Methylbutanoic ac.	−120.1 ± 1.6	AcO-*t*-Bu	−123.4 ± 0.31	Et (*Z*)-pent-2-enoate	−94.2 ± 0.7
2-Methylbutanoic acid	−118.4 ± 1.6	Ethyl pentanoate	−121.2 ± 0.4	Et (*E*)-pent-3-enoate	−92.6 ± 0.8
Pivalic acid	−117.4	Ethyl 2-methylbutanoate	−123.35 ± 0.34	Et pent-4-enoate	−92.1 ± 0.6
Hexanoic acid	−122 ± 1.0	Ethyl 3-methylbutanoate	−126.0 ± 0.21	Ethyl acrylate	−79.2
Benzoic acid	−70.3	Et 2,2-Me_2-propanoate	−125.6 ± 0.25	*n*-Pr (*E*)-but-2-enoate	−94.4 ± 0.7
Acrylic acid	−79.0 ± 1.0	$AcOCH_2{=}CH_2$	−73.8 ± 2.4	Et pent-2-ynoate	−59.8 ± 0.6
Methacrylic acid	−87.8 ± 0.57	PhCOOMe	−64.4 ± 1.2	Et pent-3-ynoate	−56.8 ± 0.7
(*E*)-Crotonic acid	−88.0	MeCOOPh	−66.84 ± 0.29	***Diesters***	
				MeOOC—COOMe	−169.5 ± 0.13
		Lactones		$MeOOCCH_2COOMe$	−176.35 ± 0.24
				MeOOCCH(Me)COOMe	−183.7 ± 0.2
		β-Propiolactone	−68.4 ± 0.2	***Anhydrides***	
Dicarboxylic acids		γ-Butyrolactone	−87.0 ± 0.6	Ac_2O	−136.8
Oxalic acid $(COOH)_2$	−173	δ-Valerolactone	−89.9 ± 0.8	Propanoic anhydride	−49.7
Butanedioic acid	−196.7	ε-Caprolactone	−94.7 ± 0.6	PhCOOCOPh	−76.3 ± 1.1

Thiols			**Thioethers**			**Disulfides**		**Sulfones**	
MeSH	−5.5	***70***	Me_2S	−8.9	***68.3***	MeSSMe	0.5	$MeSO_2Me$	−89.2 ± 0.8
EtSH	−11.0	***70.7***	EtSMe	−14.2	***79.6***	EtSSEt	−0.4	$EtSO_2Et$	−102.6 ± 0.62
PrSH	−16.2	***80.4***	EtSEt	−19.9	***87.9***	*t*-BuSS*t*Bu	−47.1	$(Vinyl)_2SO_2$	−37.4 ± 1.2
i-PrSH	−18.2	***77.5***	***Thioic acid***			PhSSPh	58.4	$PhSO_2$vinyl	−30.8 ± 0.7
BuSH	−21.0	***89.4***	MeCOSH	−41.8				$(PhCH_2)_2SO_2$	−37.55 ± 0.76
i-BuSH	−23.2					***Sulfoxides***		(*p*-Toluyl)$_2O_2$	−48.2 ± 0.7
t-BuSH	−26.2	***80.8***				MeSOMe	−36.2	$PhSO_2C{\equiv}CMe$	10.4 ± 1.2
n-DecSH	−50.4		***Thioic acid esters***			EtSOEt	−49.1	$PhSO_2CH_2C{\equiv}CH$	8.7 ± 0.8
PhSH	26.7	***80.5***	AcSEt	−54.5				***Other thio-compounds***	
$PhCH_2SH$	21.9		AcS-*t*-Bu	−67.8		***Disulfone***		Thiophene	27.5 ± 0.3
			AcSAc	−75.8		$PhSO_2SO_2Ph$	−115.0 ± 1.1	2,3-Dihydrothiophene	21.7 ± 0.3
Dithiols								2,5-Dihydrothiophene	20.9 ± 0.3
Ethan-1,2-dithiol		−2.3						2,3-$(H)_2$-thiopheneO_2	−62.6 ± 0.73
Propane-1,3-dithiol		−7.1						2,5-$(H)_2$-thiophenO_2	−61.1 ± 0.40
								Diethyl sulfite	−125.2
								Diethyl sulfate	−180.8

(*continued*)

Table 1.A.4 (Continued)

Primary amines			Other amines		Carboxamides, lactams		Nitro compounds	
$MeNH_2$	−5.6	*57.9*	Me_2NH	−4.7 ± 0.5	Formamide	−44.5	$MeNO_2$	−19.3 ± 0.3
$EtNH_2$	−13.8		Et_2NH	−23.9	Acetamide	−56.96 ± 0.19	$EtNO_2$	−24.5
$PrNH_2$	−16.7 ± 0.2		$(i\text{-Pr})_2NH$	−32.58 ± 0.62	$EtCONH_2$	−61.9 ± 0.16	$PrNO_2$	−29.6
$i\text{-PrNH}_2$	−20.0 ± 0.2		$PhCH_2NHMe$	21.6	Acrylamide	−31.12 ± 0.41	$i\text{-PrNO}_2$	−33.2
$n\text{-BuNH}_2$	−22.7 ± 0.4		Pyrrolidine	−0.82 ± 0.23	$n\text{-PrCONH}_2$	−66.72 ± 0.2	$t\text{-BuNO}_2$	−42.32 ± 0.79
$i\text{-BuNH}_2$	−23.6 ± 0.13		Piperidine	−11.27 ± 0.15	$i\text{-PrCONH}_2$	−67.5 ± 0.2	$PhNO_2$	16.38 ± 0.16
$t\text{-BuNH}_2$	−28.8 ± 0.2		Azepane	−10.8 ± 0.4	$PhCONH_2$	−24.11 ± 0.29	2-Nitroaniline	15.1
2-MePrNH_2	−23.6 ± 0.13		Me_3N	−5.6 ± 0.2	$H_2NCO{-}CONH_2$	−92.52 ± 0.31	$1,3,5\text{-Me}_3C_6H_2NO_2$	−6.4 ± 0.5
$n\text{-Hex-NH}_2$	−46.3		Et_3N	−32.05	MeCONHMe	−59.3 ± 1.3	2-NO_2 phenol	−31.62 ± 0.33
Aniline	20.8 ± 0.21		$(n\text{-Pr})_3N$	−38.4 ± 0.2	Succinimide	−89.75 ± 0.36	3-NO_2 phenol	−26.12 ± 0.26
Benzyl amine	21.0 ± 0.65	*76.4*	$PhNMe_2$	−24.0 ± 0.8	$HCONMe_2$	−46.0	4-NO_2 phenol	−27.41 ± 0.29
$(H_2NCH_2)_2$	−4.07 ± 0.14		$(Me_2N)_2CH_2$	−4.3	$MeCONMe_2$	−54.5		
$c\text{-Pr}{-}NH_2$	18.42 ± 0.16		Pyridine	33.50	$EtCONMe_2$	−59.8	$CH_2(NO_2)_2$	−14.1
$c\text{-Bu}{-}NH_2$	9.8 ± 0.1		Quinoline	47.9	2-Azetidinone	−22.9 ± 0.2	$CH(NO_2)_3$	0.0
$c\text{-Pent-NH}_2$	−13.11 ± 0.22		Isoquinoline	48.9	2-Pyrrolidinone	−47.17 ± 0.73	$C(NO_2)_4$	21.1
$c\text{-HexNH}_2$	−26.3		Quinuclidine	−1.03 ± 0.31	Caprolactam	−57.27	$C_2(NO_2)_6$	42.8

Carbonitriles		Nitrites		Nitroso compounds, isonitriles		Imines	
MeCN	17.6	MeONO	−15.8 ± 0.2	Me—N=O	16.7	$CH_2{=}NH$	16*
EtCN	12.3	EtONO	−25.9	Ph—N=O	48.1 ± 1.0	MeCH=NH	6 ± 2
$CH_2{=}CHCN$	44	*n*-PrONO	−28.4 ± 1.0	Ethyl isonitrile	33.8	1,2-Didehydropyrrolidine	15.1 ± 0.3
n-Pr—CN	7.5	*i*-PrONO	−31.9 ± 1.0				
Me_2CHCN	5.4	*t*-BuONO	−41.0 ± 1.0	***Oximes***		***Azides***	
n-Bu—CN	2.6			MeCH=NOH	−5.4 ± 0.1	$Me{-}N_3$	67
t-Bu—CN	−0.8	***Nitrates***		Cyclohexanone-oxime	−17.9	Cyclopentyl-N_3	52.8
(*E*)-EtCH=CHCN	28.62 ± 0.25	$MeONO_2$	−29.2 ± 0.3			$Ph{-}N_3$	92
(*Z*)-EtCH=CHCN	27.48 ± 0.28	$EtONO_2$	−37.0 ± 0.8	***Hydrazines***		Cyclohexyl-N_3	36.9
PhCN	52.3			$H_2N{-}NH_2$	22.8	***Unsaturated dinitriles***	
NC—CN	73.9			$MeNHNH_2$	22.6	(*E*)-NCCH=CHCN	81.3 ± 0.6
$NC{-}CH_2{-}CN$	63.6 ± 0.24			$Me_2N{-}NH_2$	19.9 ± 0.9	NC—C≡C—CN	127.5
$NC(CH_2)_2CN$	50.1 ± 0.2						

Fluorides		Chlorides		Bromides		Iodides		Acyl halides	
CH_3F	−56.0	Me—Cl	−20.0 ± 0.4	Me—Br	−8.2 ± 0.2	Me—I	3.4 ± 0.4	COF_2	−152.7 ± 1.5
Pr—F	−68.3	Et—Cl	−26.0 ± 2.0	Et—Br	−15.2 ± 1.5	Et—I	−1.7 ± 0.2	$COCl_2$	−52.4
i-Pr—F	−70.1	*n*-Pr—Cl	−31.7 ± 0.2	*n*-Pr—Br	−19.8 ± 0.8	*n*-Pr—I	−7.4	$COBr_2$	−27.1 ± 0.2
c-Hex—F	−80.5 ± 0.3	*i*-Pr—Cl	−34.0 ± 1.0	*i*-Pr—Br	−22.9 ± 0.3	*i*-Pr—I	−9.6	HCOF	−90.0
$PhCH_2F$	−30.2 ± 0.2	*n*-Bu—Cl	−37.0 ± 0.3	*n*-Bu—Br	−25.7 ± 0.4	*t*-Bu—I	−17.2 ± 0.3	AcF	−106.4 ± 0.5
Vinyl—F	−33.4	*s*-Bu—Cl	−39.8 ± 0.2	*s*-Bu—Br	−28.7 ± 0.4	Allyl—I	23.8 ± 0.3	AcCl	−60.1 ± 0.2
HC≡CF	30.0	*t*-Bu—Cl	−43.0	*t*-Bu—Br	−31.6 ± 0.4	$PhCH_2$—I	30.4 ± 0.4	AcBr	−46.8 ± 0.6
Ph—F	−27.7	*c*-Hex—Cl	−39.8 ± 0.5	Vinyl—Br	18.9 ± 0.5	Ph—I	39.4 ± 1.4	AcI	−29.9 ± 0.9
(*E*)-FCH=CHMe	−41.3	Vinyl—Cl	5.3 ± 0.7	HC≡CBr	64.2 ± 1.5	(*E*)-ICH=CHMe	22.3 ± 0.8	$CHCl_2COCl$	−80.3 ±1.5
(*Z*)-FCH=CHMe	−42.1	(*E*)-ClCH=CHMe	−2.8	AllylBr	11.4 ± 1.6	(*Z*)-ICH=CHMe	20.7 ± 0.8	CCl_3COCl	−57.7 ± 2.1
PhCHFPh	−10.2 ± 0.5	(*Z*)-ClCH=CHMe	−3.7	(*E*)-BrCH=CHMe	10.5 ± 1.0	2-MeC_6H_4I	31.7 ± 1.4	PhCOCl	−26.1 ± 1.0
CH_2F_2	−107.7	2-Cl—propene	−5.9	(*Z*)-BrCH=CHMe	9.7 ± 1.0	3-MeC_6H_4I	31.9 ± 1.4	PhCOBr	−11.6 ± 1.5
$MeCHF_2$	−118.8	Allyl—Cl	−1.3	PhBr	25.2	4-MeC_6H_4I	29.1 ± 1.4	PhCOI	2.6 ± 1.0
Me_2CF_2	−129.8 ± 3.0	Benzyl—Cl	4.5 ± 0.8	$PhCH_2Br$	20.0 ± 0.9	CH_2l_2	28.1 ± 1.0		
CH_2=CF_2	−82.2 ± 2.4	Ph—Cl	13.0	$MeCHBr_2$	−9.8	I—CH_2CH_2—I	17.5 ± 0.3	ClCOCOCl	−80.3 ± 1.5
C_2F_2	−263.0	CH_2ClF	−62	1,4-$Br_2C_4H_8$	−20.6	I—$CH_2CH_2CH_2$—I	10.8 ± 0.4		
1,2-F_2benzene	−67.6 ± 0.2	CH_2Cl_2	−22.8 ± 0.6	1,2-$Br_2C_4H_8$	−22.2	$MeCHICH_2$—I	8.5 ± 0.8		
1,3-F_2benzene	−73.9 ± 0.3	$MeCHCl_2$	−31.7 ± 0.8	2,2-$Br_2C_4H_8$	−21.9	(*E*)-CHI=CHI	49.6 ± 0.2		
1.4-F_2benzene	−73.3 ± 0.3	Me_2CCl_2	−42.2 ± 2.	2,3-$Br_2C_4H_8$	−24.6	(*Z*)-CHI=CHI	49.6 ± 0.2		
CHF_3	−166.2	$Cl(CH_2)_2Cl$	−31.5 ± 1.	$CHBr_3$	13.2 ± 0.8	2-IC_6H_4I	60.2 ± 1.4		
CH_3CF_3	−178.9 ± 0.8	CH_2=CCl_2	0.6 ± 0.5	$MeCBr_3$	−1.1	CHI_3	60.0		
CF_2H—CH_2F	−165.2 ± 2.4	(*E*)-$(ClCH)_2$	−0.1 ± 0.5	CBr_4	12.0	CI_4	−64.0		
CF_3CH=CH_2	−146.8 ± 1.6	(*Z*)-$(ClCH)_2$	−0.7 ± 0.5						
CF_3Ph	−138.9	1,2-$Cl_2C_6H_4$	7.9	***Chlorofluorides***		***Iodofluorides***			
CF_2=CHF	−113.3 ± 2.0	1,3-$Cl_2C_6H_4$	6.7	FCH_2Cl	−62.6	$MeCF_2I$	−98.0 ± 2.0		
CF_4	−221.0 ± 6.0	1,4-$Cl_2C_6H_4$	5.9	MeCHFCl	−74.9 ± 0.6	CF_3I	−140.8		
CF_2=CF_2	−157.4 ± 0.8	$CHCl_3$	−24.6 ± 0.6	F_2CHCl	−115.1	CF_3CH_2I	−155.0 ± 1.		
$PhCF_2CF2_ph$	−164.7 ± 0.7	$MeCCl_3$	−34.6 ± 0.5	CF_2Cl_2	−117.5				
1,2,4,5-$F_4C_6H_2$	−154.6 ± 0.8	$ClCH_2CHCl_2$	−35.4 ± 1.	$CFCl_3$	−69.0	***Bromochlorides***			
F_5C_6H	−192.6 ± 0.3	1,2,3-$Cl_3C_3H_5$	−43.8 ±0.5	CF_3Cl	−170. ± 6.	$BrCH_2Cl$	−5. ± 2.		
C_2F_6	−321.2 ± 1.2	$PhCCl_3$	3.0 ± 0.6	CF_2=CFCl	−120.8 ± 1.0	MeCHBrCl	−19.4 ± 0.6		
CF_3CF=CF_2	−275.3	1,2,3-$Cl_3C_6H_3$	2.0 ± 0.4			$BrCCl_3$	10.0 ± 0.3		
CF_2=CFCF=CF_2	−253.4	1,2,4-$Cl_3C_6H_3$	1.1 ± 0.4						

(*continued*)

Table 1.A.4 (Continued)

Fluorides		**Chlorides**		**Bromofluorides**		**Iodochlorides**	
−101.6 ± 0.2	−242.5	1,3,5-$Cl_3C_6H_3$	−0.6 ± 0.4	F_2CHBr	−101.6 ± 0.2		
$CF_3CF_2CF_3$	−426.6 ± 2.1	CCl_4	−25 ± 5.0	$MeCF_2Br$	−113.4 ± 2.0	ICH_2CH_2Cl	−11.4 ± 1.2
F_8-Cyclobutane	−355.7	$Cl_2CHCHCl_2$	−37.5 ± 0.8	CF_3Br	−155.1 ± 0.8		
		Cl_3CCH_2Cl	−36.4 ± 0.6	CF_3CH_2Br	−166.0 ± 0.5		

Heterocyclic compounds with one heteroelement in the ring

Oxirane	−12.6 ± 0.15	Aziridine	30.3	Thiirane	19.7 ± 0.24
Oxetane	−19.25 ± 0.15			Thietane	4.4
Tetrahydrofuran	−44.0 ± 0.2	Pyrrolidine	−0.8 ± 0.25	Tetrahydrothiophene	−8.0 ± 0.3
2,3-Dihydrofuran	−17.3 ± 0.1	1-Pyrroline	15.1 ± 0.3	2,3-Dihydrothiophene	21.7 ± 0.3
2-Furylmethanol	−50.7	2,5-Dihydropyrrole	26		
Furan	−6.6	Azole (pyrrole)	34.2	2,5-Dihydrothiophene	20.9 ± 0.3
Tetrahydropyran	−53.5 ± 0.25	Piperidine	−11.3 ± 0.2	Thiophene	27.8
2,3-Dihydropyran	−27.0 ± 0.25	3,4-Didehydropiperidine	7.1	Tetrahydro-2*H*-thiopyran	−15.1 ± 0.25
Pyridine	33.5 ± 0.4	2-Methylpyridine	24.8 ± 0.3	3-Methylpyridine	24.8 ± 0.2
2,6-Dimethylpyridine	14.0	2,4-Dimethylpyridine	15.3	*N*-Methylpyrrole	24.6
Pyridine N-oxide	21.0 ± 0.60	2-Hydroxypyridine	−19 ± 0.5	3-Hydroxypyridine	−10.4 ± 0.41
		α-Pyridone	−18 ± 0.5	4-Hydroxypyridine	−7.2 ± 1.3
β-Lactone	−68.4	2-Azetidinone	−22.9 ± 0.2		
γ-Lactone	−87.0 ± 0.8	2-Pyrrolidinone	−47.2 ± 0.7		
δ-Lactone	−89.9 ± 0.8				
2-Oxepanone	−94.7 ± 0.6	Caprolactam	−57.3		
1,3-Dioxolane	−72.1 ± 0.5	2,4-Dimethyl-1,3-dioxole	−101.1	1*H*-Pyrazole	42.4 ± 0.2
2-Methyl-1,3-dioxolane	−83.7 ± 0.8			1*H*-Imidazole	30.9 ± 0.1
1,2-Dioxane	−14.0	Pyridazine (1,2-diazine)	66.5 ± 0.3	1*H*-1,2,4-Triazole	46.1 ± 0.2
1,3-Dioxane	−80.9 ± 0.3	Pyrimidine (1,3-diazine)	46.8 ± 0.4	1*H*-Tetrazole	76.6 ± 0.7
1,4-Dioxane	−75.4 ± 0.2	Pyrazine (1,4-diazine)	46.9 ± 0.4	Oxazole	−3.7 ± 0.13
Morpholine	12	1,3,5-Triazine	54.0 ± 0.2	Isoxazole	19.6 ± 0.1
1,3,5-Trioxane	−111.4	Barbituric acid	−132	4-Methylthiazole	26.7 ± 0.2
		2-Me-4,5-dihydrooxazole	−31.2 ± 0.2	Benzothiazole	48.8 ± 0.1

See also Tables 2.A.2 and 2.2. For further values, see: http://webbook.nist.gov.

a) Taken from NIST Chemistry WebBook: http://webbook.nist.gov or from [6d] $S°(H_2O,gas) = 45.1$ eu.

Table 1.A.5 Periodic table of the elements showing atomic masses[a)] **Pauling electronegativities**, ionization enthalpies $\Delta H_r(M - e^- \xrightarrow{gas} M^{\cdot+})$ (in eV), electron affinities $-\Delta H_r(M + e^- \xrightarrow{gas} M^{\cdot-})$ (in eV). (1 eV/atom ^ = 23.06 kcal mol^{-1} = 96.48 kJ mol–; 1 eV = 1.602 19 × 10^{-19} J.

1	H 1 1.0079 **2.20** 13.598eV 0.754eV	atomic mass **Pauling electronegativity** ionization enthalpy electron affinity																He 2 4.0026 24.587 b)
2	Li 3 6.9410 **1.00** 5.392 0.618	Be 4 9.0123 **1.83** 9.322 b)											B 5 10.8100 **2.00** 8.298 0.277	C 6 12.0110 **2.60** 11.260 1.2629	N 7 14.0067 **3.05** 14.534 b)	O 8 15.9994 **3.50** 13.618 1.46112	F 9 18.9984 **4.00** 17.422 3.399	Ne 10 20.1790 21.564 b)
3	Na 11 22.9898 **0.90** 5.139 0.54793	Mg 12 24.305 **1.20** 7.646 b)											Al 13 26.9815 **1.50** 5.986 0.441	Si 14 28.0855 **1.90** 8.151 1.385	P 15 30.9738 **2.15** 10.486 0.7465	S 16 32.0600 **2.60** 10.360 2.07712	Cl 17 35.4530 **3.15** 12.967 3.617	Ar 18 39.9480 15.759 b)
4	K 19 39.0983 **0.80** 4.341 0.50147	Ca 20 40.080 **1.00** 6.113 b)	Sc 21 44.9559 **1.30** 6.540 0.188	Ti 22 47.880 **1.50** 6.820 0.079	V 23 50.9415 **1.60** 6.740 0.525	Cr 24 51.9960 **1.60** 6.766 0.666	Mn 25 54.9380 **1.50** 7.435 b)	Fe 26 55.8470 **1.80** 7.870 0.151	Co 27 58.9332 **1.80** 7.860 0.662	Ni 28 58.6900 **1.80** 7.635 1.156	Cu 29 63.5460 **1.90** 7.726 1.235	Zn 30 65.3800 **1.60** 9.394 b)	Ga 31 69.7200 **1.60** 5.9999 0.3	Ge 32 72.5900 **1.90** 7.899 1.233	As 33 74.9216 **2.00** 9.810 0.81	Se 34 78.9600 **2.45** 9.752 2.0207	Br 35 79.9040 **2.85** 11.814 3.365	Kr 36 83.800 13.999 b)
5	Rb 37 85.4678 **0.80** 4.177 0.48592	Sr 38 87.6200 **1.00** 5.695 b)	Y 39 88.9059 **1.30** 6.380 0.307	Zr 40 91.2200 **1.60** 6.840 0.426	Nb 41 92.9064 **1.60** 6.880 0.893	Mo 42 95.9400 **1.80** 7.099 0.746	Tc 43 98 **1.90** 7.280 0.55	Ru 44 101.0700 **2.20** 7.370 1.05	Rh 45 102.9055 **2.20** 7.470 1.137	Pd 46 106.4200 **2.20** 8.340 0.557	Ag 47 107.8682 **1.90** 7.576 1.302	Cd 48 112.4100 **1.70** 8.993 b)	In 49 114.8200 **1.70** 5.786 0.30	Sn 50 118.6900 **1.80** 7.344 1.112	Sb 51 121.7500 **2.05** 8.641 1.07	Te 52 127.600 **2.30** 9.009 1.9708	I 53 126.9045 **2.65** 10.451 3.0591	Xe 54 131.2900 12.130 b)
6	Cs 55 132.9054 **0.70** 3.894 0.47163	Ba 56 137.3300 **0.90** 5.212	La 57 138.9055 5.577 0.5	Hf 72 178.4900 **1.30** 6.65 0	Ta 73 180.9479 **1.30** 7.890 0.322	W 74 183.8500 **1.70** 7.980 0.815	Re 75 186.2070 **1.90** 7.880 0.15	Os 76 190.2000 **2.20** 8.700 1.10	Ir 77 192.2200 **2.20** 9.100 1.565	Pt 78 195.0800 **2.20** 9.000 2.128	Au 79 196.9865 **2.40** 9.225 2.3086	Hg 80 200.580 **1.90** 10.437 b)	Tl 81 204.3830 **1.80** 6.108 0.2	Pb 82 207.2000 **1.80** 7.416 0.364	Bi 83 208.9804 **1.90** 7.289 0.940	Po 84 209 **2.00** 8.420 1.9	At 85 210 **2.20**	Rn 86 222 10.748 b)
7	Fr 87 233 **0.65**	Ra 88 226.0254 **0.90** 5.279	Ac 89 227.0278 **1.10** 5.170															

Ce 58 140.1200 5.470	Pr 59 140.9077 5.420	Nd 60 144.2400 5.490	Pm 61 145 5.550	Sm 58 150.3600 5.630	Eu 63 151.9600 5.666	Gd 64 157.2500 6.140	Tb 65 158.9254 5.850	Dy 66 162.5000 5.928	Ho 67 164.9304 6.020	Er 68 167.2600 6.10	Tm 69 168.9342 6.184	Yb 70 173.0400 6.254	Lu 71 174.8670 5.426
Th 90 232.0381 **1.30** 6.080	Pa 91 231.0359 **1.50** 5.890	U 92 238.0289 **1.70** 6.050	Np 93 237.0482 **1.30** 6.190	Pu 94 244 **1.30** 6.060									

a) Isotope C: 12.000 000
b) instable anion (gas)

Relation between electronegativity difference (Δ) between atoms A and B and the partial ionic character of bond A-B (Pauling).

Δ	0.1	0.5	1	1.5	2	2.5	3	3.5
%	0.5	6	22	43	63	79	89	94.6

Table 1.A.6 Interatomic distances of organic compounds in Å (1 Å = 100 pm = 10^{-10} m).[a)]

σ_{C-H}	CH_4	XCH_3	$XYCH_2$	XYZCH	$CH_2=CH_2$	C_6H_6
	1.091	1.101 ± 0.003	1.073 ± 0.004	1.07 ± 0.007	1.07 ± 0.01	1.084 ± 0.006
σ_{C-C}	Alkanes	$MeC(Me)=CH_2$	MeCHO	HOOCCOOH	MeC≡CH	$H_2C=CHCN$
	1.541 ± 0.003	1.53 ± 0.01	1.516 ± 0.005	1.49 ± 0.01	1.46 ± 0.003	1.44 ± 0.01
$\pi_{C=C}$	R—CH=CH—R	$RCH=C=CH_2R$	Benzene	$\pi_{C\equiv C}$	CH≡CH	$CH_3C\equiv C-C\equiv CH$
	1.337 ± 0.006	1.309 ± 0.005	1.395 ± 0.003		1.204 ± 0.002	1.206 ± 0.004
σ_{C-N}	$(\equiv C)_4-N^+$	$(\equiv C)_3-N$	CH_3-NO_2	C_6H_5NHAc	$C_6H_5NH_2$	$C_6H_5-NO_2$
	1.479 ± 0.005	1.472 ± 0.005	1.475 ± 0.01	1.426 ± 0.012	1.375 ± 0.025	1.468 ± 0.014
$\pi_{C=N}$	Pyridines	$HCONH_2$	$\pi_{C\equiv N}$	RC≡N	$C_6H_5C\equiv N$	$^{-}S-C\equiv N$
	1.352 ± 0.005	1.322 ± 0.003		1.158 ± 0.002	1.138 ± 0.007	1.155 ± 0.012
σ_{C-O}	Ethers, alcohols	Epoxides	Oxetanes	Carboxylic acids	Esters	Enols
	1.43 ± 0.01	1.47 ± 0.01	1.46 ± 0.015	1.36 ± 0.01	1.45 ± 0.014	1.333 ± 0.017
$\pi_{C=O}$	Aldehydes	Ketones	Esters	RCOCl	R—N=C=O	Phenols/σ_{C-O}
	1.192 ± 0.005	1.210 ± 0.008	1.196 ± 0.009	1.17 ± 0.01	1.240 ± 0.003	1.362 ± 0.015
σ_{C-S}	R—SH	Ph—S—Ph	$\pi_{C=S}$	$S=C(NH_2)$	S=C=O	CH≡CX/$\pi_{C\equiv C}$
	1.81 ± 0.01	1.73 ± 0.01		1.71 ± 0.02	1.558 ± 0.003	1.056 ± 0.003
	CH_3X	CH_2X_2	CH_2CHX	Ph—X	HC≡C—X	HC≡CC≡CH/σ_{C-C}
σ_{C-F}	1.381 ± 0.005	1.334 ± 0.004	1.32 ± 0.1	1.30 ± 0.01		1.373 ± 0.004
σ_{C-Cl}	1.767 ± 0.002	1.767 ± 0.002	1.72 ± 0.01	1.70 ± 0.01	1.635 ± 0.004	CH_3CN/σ_{C-H}
σ_{C-Br}	1.937 ± 0.003	1.937 ± 0.003	1.89 ± 0.01	1.85 ± 0.01	1.79 ± 0.01	1.115 ± 0.004
σ_{C-I}	2.13 ± 0.01	2.13 ± 0.1	2.092 ± 0.005	2.05 ± 0.01	1.99 ± 0.02	
σ_{C-Be}	σ_{C-Hg}	σ_{C-B}	σ_{C-Al}	σ_{C-Si}	σ_{C-Ge}	σ_{C-Sn}
1.93	2.07 ± 0.01	1.56 ± 0.01	2.24 ± 0.04	1.84–1.88	1.95–2.01	2.135–2.20
σ_{C-P}	σ_{C-As}	σ_{C-Sb}	σ_{C-Bi}	σ_{C-S}	σ_{C-Se}	σ_{C-Te}
1.87 ± 0.02	1.98 ± 0.02	2.202 ± 0.016	~2.30	1.55–1.81	1.71–1.98	2.05 ± 0.14

Examples of organic compounds (gas phase)

CH_3CHO		CH_3CONH_2		CH_3CN		$CH_3CN\rightarrow O$		CH_3COCl		CH_3COF	
C=O	1.210	C=O	1.220	C≡N	1.159	N→O	1.217	C—Cl	1.798	C—F	1.362
C—C	1.515	C—N	1.380	C—C	1.468	C≡N	1.169	C=O	1.187	C=O	1.185
C(1)—H	1.128	C—C	1.519	C—H	1.107	C—C	1.442	C—C	1.506	C—C	1.505
C(2)—H	1.107	N—H	1.022		C—H	1.105	C—H	1.105	C—H	1.101	
	C—H	1.124									

$CH_2{=}C{=}CH_2$		$CH_2{=}C{=}C{=}CH_2$		$CH_2{=}CH{-}CH{=}CH_2$		$CH_3{-}CH{=}CH{-}CH_3$					
C—C	1.3084	C(1)=C(2)	1.32	C(1)—C(2)	1.349		(*E*)	(*Z*)			
C—H	1.087	C(2)=C(3)	1.28	C(2)—C(3)	1.467	C(1)—C(2)	1.508	1.506			
		C—H	1.08	C—H	1.108	C(2)—C(3)	1.347	1.346			
				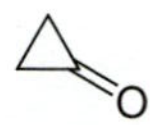							
C=C	1.304	C—C	1.512	C=O	1.191	C—N	1.475	C—O	1.431	C—S	1.815
C—H	1.519	C—H	1.083	C(1)—C(2)	1.475	C—C	1.481	C—C	1.466	C—C	1.484
C(1)—H	1.077			C(2)—C(3)	1.575	N—H	1.016	C—H	1.085	C—H	1.083
C(3)—H	1.112		C—H	1.086	C—H	1.084					
$CH_2{=}CH{-}CHO$		$CH_2{=}CH{-}C{\equiv}N$				$HC{\equiv}C{-}CN$		$H_2N{-}C{\equiv}N$		$NC{-}CN$	
C=O	1.217	C≡N	1.167		C≡N	1.159	C≡N	1.160	C≡N	1.163	
C(1)—C(2)	1.484	C(1)—C(2)	1.438		C≡C	1.205	C—N	1.346	C—C	1.393	
C=C	1.345	C=C	1.343		C—C	1.378	NH	1.00			
C(1)—H	1.13	C(3)—H	1.114		C—H	1.058					
C(3)—H	1.10										
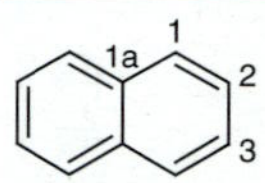		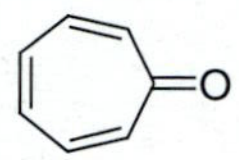		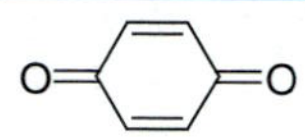							
C(1)—C(2)	1.37	C=O	1.23	C=O	1.225	C(1)—C(4)	1.574	C(1)—C(3)	1.497		
C(2)—C(3)	1.41	C(1)—C(2)	1.45	C=C	1.344	C(1)—C(2)	1.524	C(1)—C(2)	1.498		
C(1)—C(1a)	1.42	C(2)—C(3)	1.36	C—C	1.481	C(2)—C(3)	1.345	C(1)—H	1.071		
		C(3)—C(4)	1.46				C(2)—H	1.093			

a) Taken from [31, 32].

Table 1.A.7 Homolytic bond dissociation enthalpies $DH°(R^•/X^•)$ of selected compounds R—X and heats of formation of radicals $R^•$ and $X^•$, in kcal mol^{-1} [1 cal = 4.184 J][a] ($\Delta H_f^\circ(H^•) = 52.103 \pm 0.001$ kcal mol^{-1} [http://webbook.nist.gov]).

	X	H	Me	Et	F	Cl	Br	I	OH	OMe	NH_2
	$\Delta H_f^o(X^•)$	52.1	34.8	28.0	19.0	29.0	26.7	25.5	9.3	4.1 ± 1	45.5
R	$\Delta H_f^o(R^•)$										
H	52.1 ± 0.001	104.2	104.7	100.1	136.2	103.2	87.8	71.3	119.2	104.2	108.2
Me	34.8 ± 0.2	104.7	89.6	87.6	112.8	83.4	70.6	56.9	92.3	82.5	85.5
Et	28.4 ± 0.5	100.7	87.8	86.0	110	83.8	69.6	55.7	93.5	83.4	84.6
n-Pr	23.5 ± 0.5	101.1	88.8	87.1	111.3	84.5	71.5	56.7	94.3	84.5	85.9
i-Pr	22.0 ± 0.5	99.4	87.1	87.0	111.4	85.6	72.8	57.4	96.8	86.2	87.4
n-Bu	18.0	101.1	87.9	85.9		83.9	70.3	55.5	93.0	83.4	85.1
2-Bu	16.0 ± 0.5	99.1	88.5	86.1		84	72.6	57.5	96.3		87.1
Me_2CHCH_2	17 ± 0.5	100.1	87.5	85.7							
t-Bu	11.0 ± 0.7	95.2	86.0	83.4		83.5	69.7	53.7	95.0	82.5	85.0
1-C_5H_{11}	13	100.2	87.7						81.2		
2-C_5H_{11}	12	99.2	88.5						96.3		
t-Amyl	6.7 ± 0.7	95.3	85.7						94.9		
Neopentyl	8	100.3	87.2								
c-C_3H_5	61.3	106.3	90		110.9	85.4	73.7	58.6	97.5	87.8	91.2
c-C_4H_7	51.2	96.5	86		107.2	81.7	69.9	54.9	93.7	84	87.4
c-C_5H_9	24.3	94.8	84		105.2	79.7	67.9	52.9	91.3	82	85.4
c-C_6H_{11}	13.9	95.5	85		106	82	68.7	51.6	91.3	82.4	86.2
CH_2=CH	71. ± 1.	110.6	93.4	91.4	115.6	87.4	71.2	57.9	102.7	91.1	
CH_2=CHCH_2	40.9 ± 0.7	88.1	73.8	72.0	95	69.3	54.3	40.7	78.3	67.7	73.1
CH≡CCH_2	81. ± 1.0	88.8	77		103.2	76.7	63.9	49.3	86.9	76.9	80.7
c-C_3H_3	105. ± 5.0	91.1 ± 5									
CH_2=CHCH_2CH_2	46	98.2									
CH_2=CH—(Me)CH	31.7	83.9									
CH_2=C(Me)CH_2	29	85.1									
CH_2=CH(Et)CH	25	82.2									
CH_2=CH$(Me)_2$C	19	77.6									
MeCH=CH(Me)CH	22	81.6									

Ph	81. ± 2.0	113.3	101.7	99.8	125.7	95.6	80.5	65.1	111.3	98.9	103.3
$PhCH_2$	49.5 ± 1	89.7	76.6	75.1		74	59.7	49.5	82.3	70.7	74.1
c-C_7H_7	59	67.4									
HCO	10	88	83						109.1	96.1	101.2
MeCO	−2.9 ± 0.7	86	80		119.5	81.6	66.5	49.8	106.5	95.5	99.2
PhCO	26	87	81			80.5	64.4	49.1	105.2	95.1	94.8
FCH_2	−8.0 ± 2.0	103	90		119	83					
$ClCH_2$	31	102.7	92.6		83	83					
$BrCH_2$	42	103.2	91.7				68.7				
$HOCH_2$	−6.2	94.1	84.7								
$MeOCH_2$	−3	93.1	83.8			77					
H_2NCH_2	38.0 ± 2.0	95.6	84.1								
Me_2NCH_2	26	83.8	71.8								
$Me_2(CN)C$	40.3 ± 2.2	86.6±2.4									
NC	104.0 ± 2.0	123.8	120.8	120	114	100	87.7	75.6			
F_2CH	−57	103	98		128						
Cl_2CH	26	101	92		112						
Br_2CH	54	106	98								
CF_3	−112.4	108	104		132	89	72				
CCl_3	19.0	96	88		102	71	56				
CH_3O	4.1 ± 1.0	103.8	82.5	83.4							
PhO	13.0 ± 1.0	86.5	62.4	63.7							
MeS	29.4	87	73								
MeSS	16.0		56.6								
PhS	55.0 ± 2.0	77.2									
H_3Si	46.4	90.3 ± 0.5									

Frequently used values

$DH°(Et^•/H^•) = 100.5 \pm 0.6$
$DH°(H_2C{=}CH^•/H^•) = 110.6 \pm 1.1$
$DH°(HC{\equiv}C^•/H^•) = 111.9 \pm 2.0$
$DH°(MeCO^•/H^•) = 90.0 \pm 1.0$
$DH°(MeO^•/H^•) = 105.2 \pm 4.0$
$DH°(MeS^•/H^•) = 87$
$DH°(Et^•/Et^•) = 86.8 \pm 1.0$
$DH°(MeO^•/MeO^•) = 38 \pm 2$
$DH°(MeS^•/MeS^•) = 64.6$
$DH°(MeS^•/MeSS^•) = 48.4$
$DH°(PhS^•/PhS^•) = 52$
$DH°(HO^•/HO^•) = 51.2$
$DH°(H^•/{}^•OOH) = 87.2$
$DH°(t\text{-}BuO^•/{}^•O{-}t\text{-}Bu) = 42.9$
$DH°(F_3CO^•/{}^•CF_3) = 47$

a) Values taken from http://webbook.nist.gov or from [82, 83].

Table 1.A.8 Homolytic C—H bond dissociation enthalpies $DH°(R^\bullet/H^\bullet)$ of C_1 and C_2 hydrocarbons in kcal mol^{-1} (1 cal = 4.184 J).

	CH_4	$\xrightarrow{-H^\bullet}$	$CH_3^\bullet$	$\xrightarrow{-H^\bullet}$	$:CH_2$	$\xrightarrow{-H^\bullet}$	$:CH^\bullet$	$\xrightarrow{-H^\bullet}$	$:C:$
$\Delta_f H°$(gas)	−17.8		34.8		93		142.4		171.3
$DH°(R^\bullet/H^\bullet)$:		104.7		110.3		101.5		81	
	$CH_3{-}CH_3$	$\xrightarrow{-H^\bullet}$	$CH_3CH_2^\bullet$	$\xrightarrow{-H^\bullet}$	$CH_2{=}CH_2$	$\xrightarrow{-H^\bullet}$	$CH_2{=}CH^\bullet$	$\xrightarrow{-H^\bullet}$	$CH{\equiv}CH$
$\Delta_f H°$(gas)	−20.24		28.4		12.5		71		54.5
$DH°(R^\bullet/H^\bullet)$:		100.7		36.2		110.6		43.2	
	$CH{\equiv}CH$	$\xrightarrow{-H^\bullet}$	$CH{\equiv}C^\bullet$	$\xrightarrow{-H^\bullet}$	C_2	$\rightarrow$	2C		
$\Delta_f H°$(gas)	54.5		114		198.8		342.6		
$DH°(R^\bullet/H^\bullet)$:		111.9		136.9	$DH°(C^\bullet/C^\bullet) = 143.8$				

Table 1.A.9 Substituent S effects (E_S) on the relative stability of primary alkyl radicals as given by Eq. (1.83).[a)]

S	H	CH_3	$CH_3{-}CH_2$	*t*-Bu
$DH°(SCH_2^\bullet/H^\bullet)$	104.7 ± 0.2	100.7 ± 0.2	101.1 ± 0.3	100.3 ± 2.0
E_S	(0)	4.6 ± 1.0	3.6 ± 1.0	4.4 ± 2

S	$CH_2{=}CH$	$CH{\equiv}C$	Ph		CH_3OOC	$N{\equiv}C$	COMe
$DH°(SCH_2^\bullet/H^\bullet)$	86.3 ± 1.5	89.4 ± 2.0	89. ± 1.0	76	95.3	93 ± 2.5	98 ± 1.8
E_S	18.4 ± 1.7	15.3 ± 2.2	16 ± 1.2	29 ± 2	9.7 ± 2	12 ± 2.7	7 ± 2

S	CHO	SMe	F	Cl	Br	I	NH_2	NMe_2	OH	OMe
$DH°(SCH_2^\bullet/H^\bullet)$	94.8 ± 2.0	92. ± 1.0	103. ± 1.0	102.7 ± 2	103.2 ± 2	103 ± 2	95.6 ± 2	84.1 ± 2.2	94.1 ± 2	93.1 ± 2
E_S	10 ± 2.2	12.7 ± 1.2	2 ± 1.2	2 ± 2.2	1.5 ± 2.2	2 ± 2.2	11 ± 2.2	20.5 ± 2.2	10.5 ± 2.2	11.5 ± 2.2

S	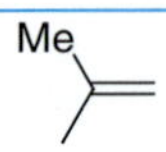	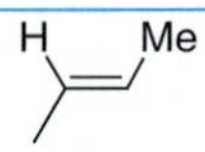	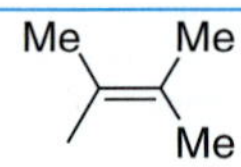	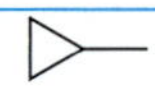	CF_3
DHo$(SCH_2^\bullet/H^\bullet)$	85.6 ± 1.0	85.6 ± 1.5	78.0 ± 1.1	97.4 ± 1.6	108 ± 1.1
E_S	19.4 ± 1.2	19.4 ± 1.7	27 ± 1.2	7.6 ± 1.8	−3 ± 1.3

a) Values taken from Table 1.A.8 and from [88, 89, 411].

Table 1.A.10 Rüchardt's homolytic C—H bond dissociation enthalpies $DH°(RR'R''C^•/H^•)$ in DMSO and radical stabilization enthalpies (RSE) relative to methyl-substituted radicals, including corrections arising from differential R, R′, and R″ group interactions between precursors and radicals (see text), in kcal mol^{-1} [1 cal = 4.184 J].[a)]

Secondary carbon-centered radicals $RR'HC^•$				Tertiary carbon-centered radicals $RR'R''C^•$			
R	R′, R″=H	$DH°(C^•/H^•)$	RSE	R	R′, R″=Me	$DH°(C^•/H^•)$	RSE
Me	Me	98.7	(0)	Me	Me	95.7	(0)
Ph	Me	90.3	−8.4	Ph	Me	87.3	−8.4
t-BuO	Me	96.9	−5.9	*i*-PrO	Me	93.9	−5.9
NH_2	Me	97.1	−3.9	NH_2	Me	94.2	−3.9
COPh	Me	92.9	−6.0	COPh	Me	89.9	−6.0
COOEt	Me	95.6	−2.8	COOEt	Me	92.6	−2.8
CN	Me	94.9	−3.4	CN	Me	91.9	−3.4
COOEt	Ph	88.6	−10.1	COOEt	Ph	85.6	−10.1
COMe	COMe	87.7	−11.4	COMe	COMe	84.7	−11.4
CN	CN	87.6	−2.5	CN	CN	84.6	−2.5
CN	OMe	90.3	−8.1	CN	OH	87.3	−8.1
—CON(Bu)CH_2CON(Bu)—		92.4	−6.3	—CON(Bu)CH_2CON(Bu)—		89.4	−6.3
CN	NH_2	84.9	−13.8	COO—*t*-Bu	NH_2	81.3	−14.8
COOMe	NH_2	84.3	−14.8	Ph	Ph	82.8	−12.9
COPh	NMe_2	78.3	−21.6	9,10-Dihydroanthryl-9-yl		83.5	−15.2
Ph	Ph	85.8	−12.9				
Fluoren-9-yl		81.7	−16.0	9-Methylfluoren-9-yl		79.7	−16.0
Xanth-9-yl		80.6	−18.1	9-Methylxanth-9-yl		77.6	−18.1
CH_2=CH_2	Me	86.1	−12.6	CH_2=CH	Me	83.1	−12.6
Inden-1-yl		83.0	−15.7	Pentamethylcyclopentadienyl		82.0	−13.7
Anthron-10-yl		80.9	−17.8				

a) Taken from [412].

Table 1.A.11 Bordwell's homolytic bond dissociation enthalpies $DH°(X^•/H^•)$ measured in DMSO by Eq. (1.82), in kcal mol^{-1} (1 cal = 4.184 J)[a] (in square parentheses: gas-phase values, Table 1.A.8)

O—H bonds					
MeOH	104.6 [103.8]	4-AcC_6H_4OH	92.8	(TEMPOH) 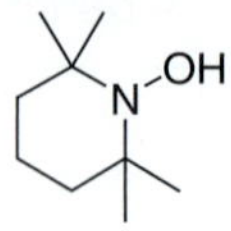	
EtOH	103.0 [104.5]	4-NCC_6H_4OH	94.2		68.7
t-BuOH	105.5 [105]	4-O_2NC_6H_4OH	94.7		
PhOH	90.4 [86.5]	2,4,6-tri(*t*-Bu) C_6H_2OH	82.3		
3-MeOC_6H_4OH	90.8	Et_2NOH	75.9		
4-MeOC_6H_4OH	84.6	$(t\text{-}Bu)_2$NOH	68.2	(*t*-Bu)(*i*-Pr)C=NOH	86.0
4-ClC_6H_4OH	88.7	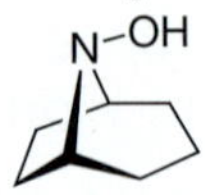	78.0	$(t\text{-}Bu)_2$C=NOH	82.6
4-PhC_6H_4OH	87.6			PhCON(*i*-Pr)OH	81.2
4-H_2NC_6H_4OH	77.3			PhCON(*t*-Bu)OH	79.9
4-(O^-)C_6H_4OH	73.0			*N*-Hydroxyphthalimide	89
(α-tocopherol) 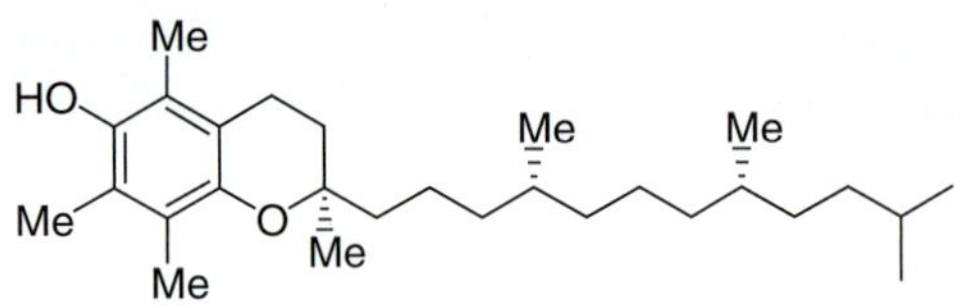			80.9		
N—H-bonds					
NH_2—H	[108]	PhNH_2	92 [88]	Pyrrole	97
MeNH—H	[100]	PhN(Me)H	89 [87.5]	HN_3	94
Me_2N—H	[91.5]	Ph_2NH	87.5		
C—H bonds					
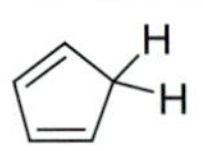	81 [82.8]	PhHCHCN	81.9	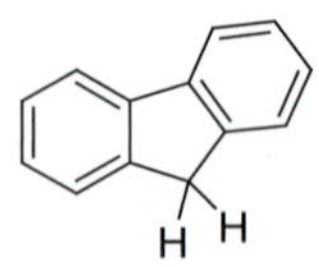	80
		HCH$(CN)_2$	90		
		HCH_2COPh	93		
		Ph_2CH_2	82 [81.4]		
Ph_3CH	81	PhCH_2—H	[89]	PhHCHSO_2Ph	90.2
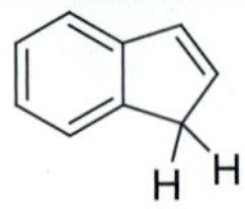	78.9 [83.9]	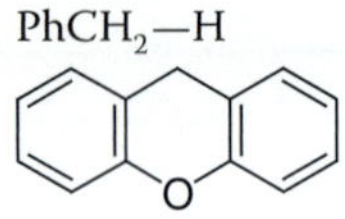	75.5	PhHC$(CN)_2$	77
				HCH_2COOH_3	94
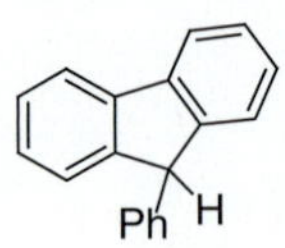	74			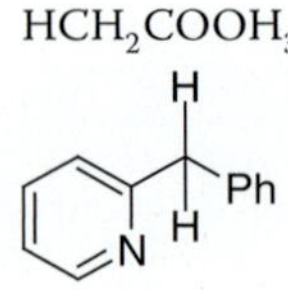	82

a) Taken from [95, 96a, 97, 98].

Table 1.A.12 Activation parameters[a)] for the (E) ⇄ (Z)-isomerization of polyolefins.[b)]

$\Delta^{\ddagger}S$ (eu)	$\Delta^{\ddagger}H$ (kcal mol^{-1})	$\Delta\Delta^{\ddagger}H$	Radical	Number of limiting structures	E_{π} [a)]
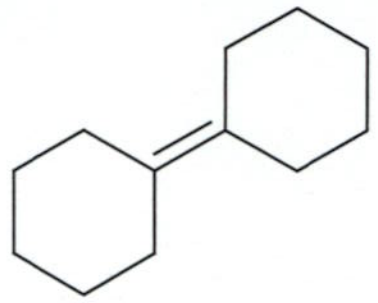	58.1		Model		
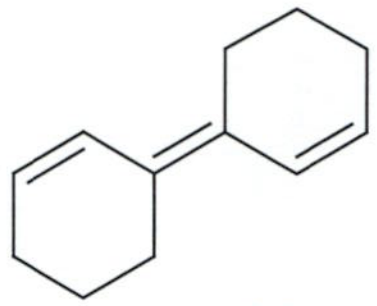					
−2.9	38.9	19.2			
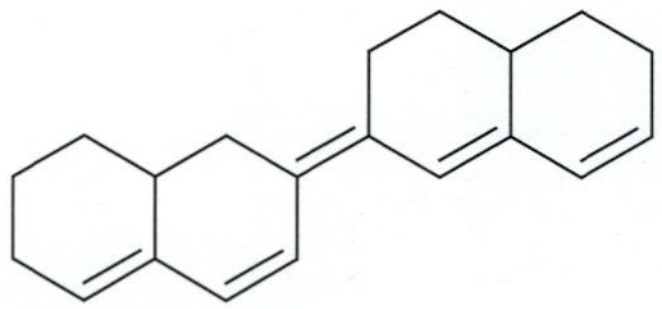				2	~13.5[a)]
−4.4	32.1	26.0			
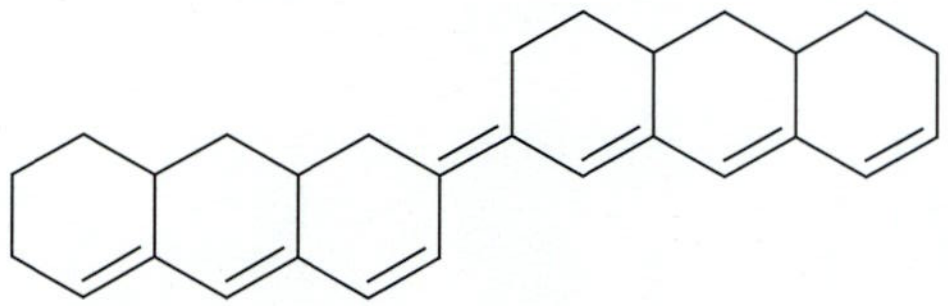				3	~16.9
−4.4	27.5	30.6			
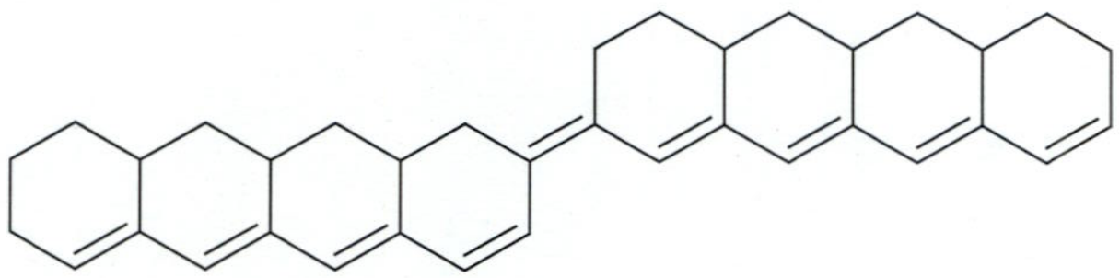				4	~19.2
	24.5	33.6		5	~20.7

a) $E_{\pi} = \frac{1}{2}\Delta\Delta^{\ddagger}H$ corrected by 4 kcal mol^{-1} per alkyl substituent.
b) Taken from [27].

Table 1.A.13 Standard heats of formation of radicals X•, anions X⁻, electron affinities $-\text{EA}(X^\bullet) = \Delta_f H°(X^-) - \Delta_f H°(X^\bullet) = DH°(X^-/H^+) - DH°(X^\bullet/H^\bullet) + EI(H^\bullet)$, ionization energies $\text{IE}(X^\bullet) = \Delta_f H°(X^+) - \Delta_f H°(X^\bullet)$, homolytical bond dissociation enthalpies $DH°(R^\bullet/X^\bullet)$, heterolytical bond dissociation enthalpies $DH°(R^+/X^-)$, and heats of formation of RX, in kcal mol⁻¹.

X•	$\Delta_f H°(X^\bullet)$	$\Delta_f H°(X^-)$	−EA(X•)	IE(X•)	$DH°(X^\bullet/H^\bullet)$	$DH°(X^-/H^+)$	$\Delta_f H°$(X—H)
H•	52.1	34.7	−17.4	313.6	104.2	400.4	0
HO•	9.3	−32.7	−42	299.8	119.2	390.8	57.8
HS•	33.3	−19.4	−52.7	239.1	89.9	351.1	−4.9
HSe•	43	−8	−51	226	88	350	7.0
$H_2N^\bullet$	45.5	27	−17	257	107	404	−11.0
$H_2P^\bullet$	33.3	6.4	−27	226	84.6	371	1.3
$CH_3^\bullet$	34.8	33.2	−1.8	227	104.7	416.6	−17.8
$MeCH_2^\bullet$	28.4	34.2	6.4	190	100.7	420.1	−20.0
$EtCH_2^\bullet$	23.5	24.1	1.9	186.5	101.1	415.6	−25.0
$Me_2CH^\bullet$	22.0	28.9	9.5	170	99.4	419.4	−25.0
$Me_3C^\bullet$	11.0	14.3	5	154.5	95.2	413.1	−32.1
$ClCH_2^\bullet$	31	10.8	−18.4	198	100.9	396	−20.0
$BrCH_2^\bullet$	42	18	−24	182	102	393	−8.2
$ICH_2^\bullet$		24.4				386	3.4
$MeOCH_2^\bullet$	−3	−2.6	1	160	93.0	407	−44.0
$MeSCH_2^\bullet$	35.4	18.4	−17	158.6	96.6	393	−8.9
$Me_2NCH_2^\bullet$	26	35	9	131	83.9	>406	−5.6
$Mc_2PCH_2^\bullet$		1.0 ± 4				391	−24.1
N≡C•	104	17.7	−86.2		123.8	351	32.2
HC≡C•	114	65.5	−69	270	111.9	377	54.2
$H_2C{=}C{=}CH^\bullet$	82	60.5	−20.6	200	87.7	380	45.5
MeC≡C•	124.6	60	−64.5		132	381	44.2
$CH_2{=}CH^\bullet$	71	52.8	−18.5	205	110.6	406	12.5
$N{\equiv}C{-}CH_2^\bullet$	59	25.1	−33.7	230	93	372.8	17.6
$MeCOCH_2^\bullet$	20				98	368.8	−52.2
$HC{\equiv}C{-}CH_2^\bullet$	81	60.5	−21	200	88.8	381	44.2
$H_2C{=}CH{-}CH_2^\bullet$	40.9	29.5	−9.5	187	88.1	391	4.9
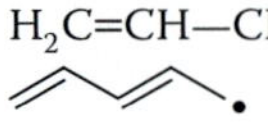	49	28.2	−21	167	76.0	369	18.1
$c\text{-}C_5H_5^\bullet$	58	19.6	−38.5	194	78.5	354	33.2
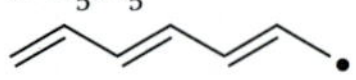			−29.3				31.8
Ph•	81	53.4	−25.3	195	113.3	401.7	19.8

Table 1.A.13 (Continued)

$X^•$	$\Delta_f H°(X^•)$	$\Delta_f H°(X^-)$	$-EA(X^•)$	$IE(X^•)$	$DH°(X^•/H^•)$	$DH°(X^-/H^+)$	$\Delta_f H°(X{-}H)$
$Ph{-}CH_2^•$	49.5	27	−20.7	166	89.7	381	12.0
$c\text{-}C_7H_7^•$	67.6	53	−14.3	144	76	375	44.6
		91				400	57.0
$CH_3O^•$	4.1	−33.2	−37.4	198	104.2	380.5	−49.0
$CH_3CH_2O^•$	−4.1	−44.5	−40.4		104.2	377.4	−56.0
$CH_3(CH_2)_2O^•$	−10.5	−50.7	−40	~212	103.5	376	−66.0
$MeOCH_2CH_2O^•$	−35	−79	−43.8		104	374	−87.5
$(CH_3)_2CHO^•$	−13	−56	−43	~212	104.5	375	−65.2
$t\text{-}BuO^•$	−21	−65	−44		105	374.5	−74.7
$H_2C{=}CHC^•HMe$	31.7			170.3	83.9		−0.15
$H_2C{=}CH{-}CH_2{-}CH_2^•$	46			185	98.2		−0.15
$H_2C{=}C(Me){-}CH_2^•$	29				85.1		−4.3
$H_2C{=}C{-}C^•HEt$	25						−5.0
$H_2C{=}CHC^•Me_2$	19						−6.1
$CH_3S^•$	30	−14	−44	~186	87	357	−5.5
$CH_3CH_2S^•$	24	−21.5	−45.5		87	355	−11.0
$(CH_3)_2CHS^•$	17	−30.5	−47.5		87	353	−16.2
$t\text{-}BuS^•$	11	−39	−48.2		87	352	−26.2
$F^•$	19.0	−59.5	−78.4	401.8	136.2	371	−65.1
$Cl^•$	29.2	−54	−83.4	299.0	103.4	333	−22.1
$Br^•$	26.7	−51	−77.6	272.4	87.5	324	−8.7
$I^•$	25.6	−45	−70.5	241.0	71.4	314	6.3
$N_3^•$	110.7 ± 5	48.5 ± 4.0	−62		92.5 ± 5	344 ± 3.0	70.3
$NO_2^•$	8	−45.2	−53		79 ± 2	340 ± 5.0	−18.3
$NO_3^•$	17	−73.3	−90		101 ± 5	325	−32.0
$SiH_3^•$	48.5	15	−33.4	187.7	92.2	372	8.2
$Li^•$(gas)	38.5	24	−14	124	57.3	356	38.1
$Na^•$(gas)	25.6	13	−13	118			25.7
$K^•$(gas)	21.3	10	−11.5	100	44.4	346	21.3

$IE(H^•) = 313.1$ kcal mol^{-1}, $\Delta_f H°(H^+) = 365.2$ kcal mol^{-1} a) (1 cal = 4.184 J, 1 eV/molecule ≙ 23.0603 kcal mol^{-1}].a,b)

a) See Tables 1.A.7 and 1.A.8. Accuracy, in general, better than ±2 kcal mol^{-1}.

b) Taken from [7].

Table 1.A.14 Thermochemical parameters for selected carbenium ions R^+ and related radicals in the gas phase, in kcal mol^{-1} (1 cal = 4.184 J).

R^+	$DH°(R^+/H^-)$	$\Delta_f H°(R^+)$[a]	$\Delta_f H°(R^•)$[b]	$\Delta_f H°(RH)gas$[c]	$DH°(R^•/H^•)$[b]
H^+	400	365.7	52.1	0.0	104.2
CH^+	327	387	142.4	93.9	100
CH_2^+	334	334	93	34.8	110
CH_3^+	313.4	261.3	34.8	−17.8	104.7
$CH_2{=}CH^+$	291	~269	63.4	12.5	110.6
$CH_3CH_2^+$	270	216	28.0	−20.2	100.7
$HC{\equiv}C{-}CH_2^+$	271	281	82	44.7	88.8
$CH_2{=}CH{-}CH_2^+$	258	~226	39	4.9	88.1
$CH_2{=}C^+{-}CH_3$	266	237	57	4.9	104
$CH_3CH_2CH_2^+$	267	211	24.0	−24.8	101.1
$CH_3{-}C^+H{-}CH_3$	251	190.9	22.3	−24.8	99.4
$CH_3CH_2CH_2CH_2^+$	265	200	18.0	−30.4	101.1
$CH_3CH_2C^+HCH_3$	248	183	17.0	−30.4	99.1
$(CH_3)_2CHCH_2^+$	265	198	16	−32.4	100.1
$(CH_3)_3C^+$	233	167[d]	11.0	−32.4	95.2
$CH_2{=}C(CH_3)CH_2^+$	248	211	29	−4.0	85.1
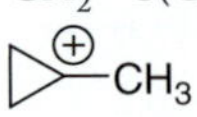	249	222	51	5.7	97
$CH_3C^+HCH_2CH_3CH_3$	241	173		−35.1	
$H_2N{-}CH_2^+$	218	178	38	−5.5	95.6
$NC{-}CH_2^+ \leftrightarrow {}^+N{=}C{=}CH_2$	318	301.8	59	17.6	93
$O{=}CH^+ \leftrightarrow {}^+O{\equiv}CH$	255	195	10	−26	88
$HO{-}CH_2^+ \leftrightarrow H^+O{=}CH_2$	254	172	−6.2	−48	94.1
$CH_3CO^+ \leftrightarrow MeC{\equiv}O^+$	224	151	−6	−39.7	86
$(CH_3O)CH_2^+$	243	163	−3	−44	−93.1
Ph^+	~287	272	79	19.8	113.3
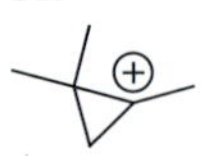	243	195		−15	97
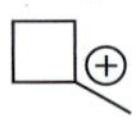	231	194		−40.7	
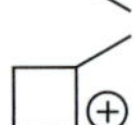	227	185		−9	
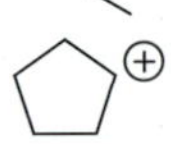	249.8	198.2	24.3	−18.4	94.8
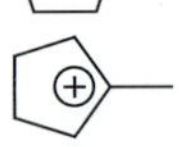	226.5	168		−25.3	
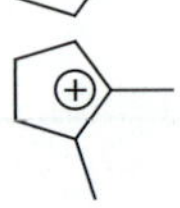	226	160		−32.7	
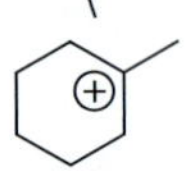	227.5	157		−37	

Table 1.A.14 (Continued)

R^+	$DH°(R^+/H^-)$	$\Delta_fH°(R^+)$[a)]	$\Delta_fH°(R^\bullet)$[b)]	$\Delta_fH°(RH)gas$[c)]	$DH°(R^\bullet/H^\bullet)$[b)]
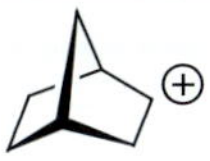	232	187		−12.4	
	225	172		−19.6	
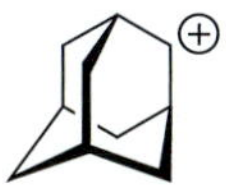	233.8	167.5			96.0
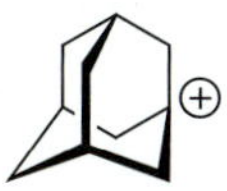	224.3	158			98.5
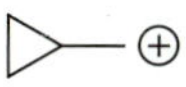	241	214		6	97.4
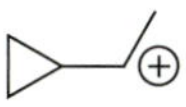	230	197		1	
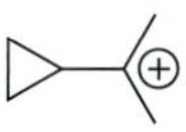	218	179		−6	
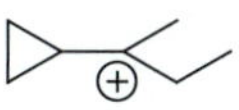	221	182		−6	
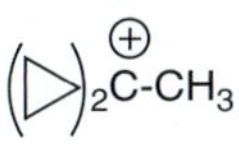	207	196		22	
	225	199		8.6	82
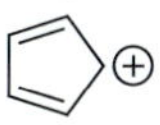	255.6	252		31	81
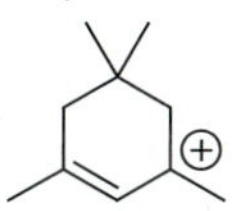	207	149		−25	
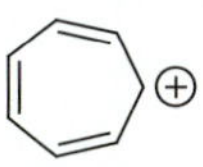	194	203	59	43.7	67.4
$PhCH_2^+$	234	213		12.0	89.7
PhC^+HCH_3	226	200		7.2	
$PhC^+(CH_3)_2$	220	188		1.0	
$Ph_2C^+CH_3$	215	213.6		32	
FCH_2^+	289.6	200.3		−55.9	103.1
F_2CH^+	283.9	142.4		−108.1	101.0
F_3C^+	299	99.3		−166.3	106.2

a) $\Delta_fH°(H^-) = 34.2\ kcal\ mol^{-1}$ is used; other value: $33.23 \pm 0.005\ kcal\ mol^{-1}$, see [413].
b) See Table 1.A.14.
c) Taken from [24].
d) $\Delta_fH°(t\text{-}Bu^+) = 162.1 \pm 0.8\ kcal\ mol^{-1}$ is obtained by photoionization coupled with MS; $\Delta_fH°(t\text{-}Bu^+) = 170\ kcal\ mol^{-1}$ by proton affinity of isobutene.

Table 1.A.15 Proton affinities PA(B) of compounds B, 1 atm, 25 °C, gas phase. Substituent effects on the relative stability of cations BH^+ given by PA(substituted B) – PA(unsubstituted B).

$B + H^+ \rightarrow BH^+$		PA(B)[a)]	Substituent effects
NH_3	$NH_4{}^+$	202.3	(0.0)
CH_3NH_2	$CH_3NH_3{}^+$	211.3	9.0
$(CH_3)_2NH$	$(CH_3)_2NH_2{}^+$	217.9	15.6
$(CH_3)_3N$	$(CH_3)_3NH^+$	222.1	19.8
$CH_3CH_2NH_2$	$CH_3CH_2NH_3{}^+$	214.0	11.7
$(CH_3CH_2)_2NH$	$(CH_3CH_2)_2NH_2{}^+$	224.5	22.2
$(CH_3CH_2)_3N$	$(CH_3CH_2)_3NH^+$	229.0	26.7
$FCH_2CH_2NH_2$	$FCH_2CH_2NH_3{}^+$	210.2	7.9
$F_2CHCH_2NH_2$	$F_2CHCH_2NH_3{}^+$	205.9	3.6
$F_3CCH_2NH_2$	$F_3CCH_2NH_3{}^+$	200.3	−2.0
$F_3C(CH_3)_2N$	$F_3C(CH_3)_2NH^+$	192	−10.3
H_2O	H_3O^+	170.3	(0.0)
CH_3OH	$CH_3OH_2{}^+$	182.2	11.9
$(CH_3)_2O$	$(CH_3)_2O^+H$	191.1	20.8
H_2Se	H_3Se^+	174.7	—
H_2S	H_3S^+	173.8	(0.0)
CH_3SH	$CH_3SH_2{}^+$	185.8	11.9
$(CI\,I_3)_2S$	$(CH_3)_2SH^+$	197.6	23.7
AsH_3	$AsH_4{}^+$	180.9	—
PH_3	$PH_4{}^+$	187.4	(0.0)
CH_3PH_2	$CH_3PH_3{}^+$	201.8	14.4
$(CH_3)_2PH$	$(CH_3)_2PH_2{}^+$	214.0	26.6
$(CH_3)_3P$	$(CH_3)_3PH^+$	223.5	36.1
	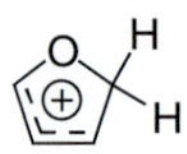	195.5	
CH_4	$CH_5{}^+$	131.6	
CH_3CH_3	$C_2H_7{}^+$	143.6	
$CH_3CH_2CH_3$	$C_3H_9{}^+$	150	
Me_2CHMe	$C_4H_{11}{}^+$	163.3	
Cyclopropane	H-cycloproponium ion	179.8	
$CH_2{=}CH_2$	$CH_3CH_2{}^+$	160.6	(0.0)
$CH_3{-}CH{=}CH_2$	$CH_3{-}C^+H{-}CH_3$	180.4	19.8
$(CH_3)_2C{=}CH_2$	$(CH_3)_3C^+$	193.5	32.9
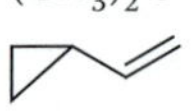	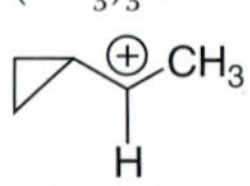	195.2	34.6
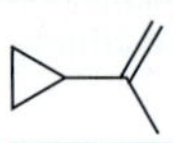	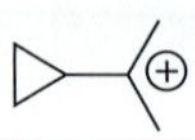	207.1	46.5

Table 1.A.15 (Continued)

$B + H^+ \rightarrow BH^+$	B	$\Delta_f H°$ (B)	BH^+	PA(B)[a]
Ph—CH=CH_2	Ph—C^+H—CH_3	35.1	199.2	38.7
Ph—C(CH_3)=CH_2	Ph—C^+(CH_3)$_2$	28.3	205.2	44.6
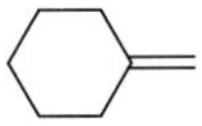	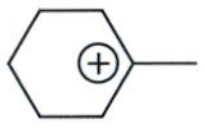	−16.0	196.6	
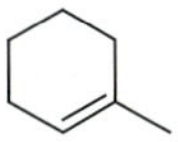	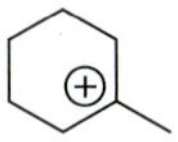	−19.4	200	
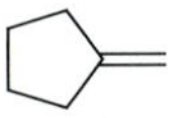	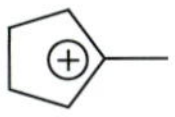	2.4	195.9	
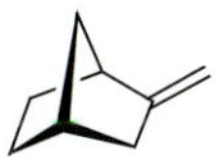	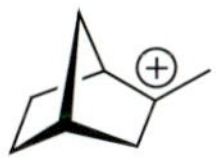	12	207	
	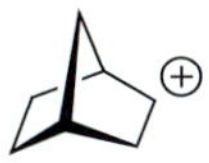	21	201.6	
	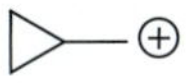	37.5	191	
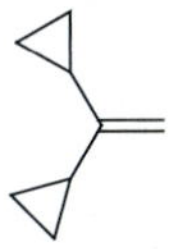	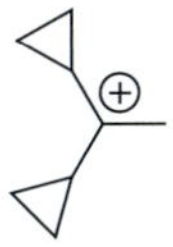	46	217.5	
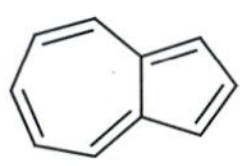	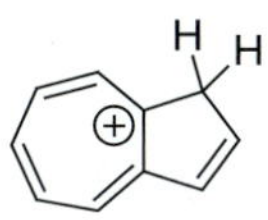	73.5	224.5	
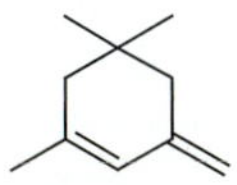	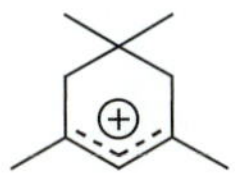	−2	216.1	
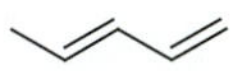		18.1	201.8	(0.0)
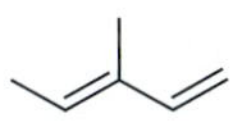	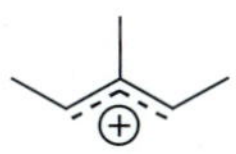	11	205.7	3.9
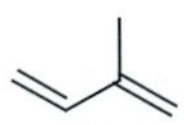	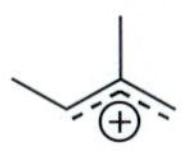	18.1	200.4	
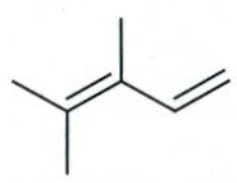	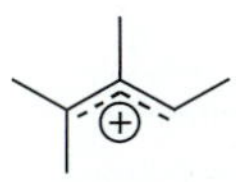	10.8	202.1	

(continued)

Table 1.A.15 (Continued)

$B + H^+ \rightarrow BH^+$	B	$\Delta_f H°$ (B)	BH^+	PA(B)[a]
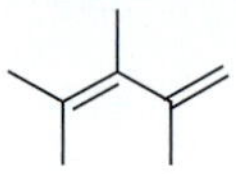	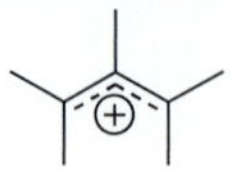	11.0	207.9	
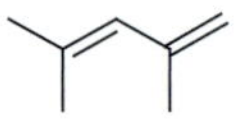		4	213.1	
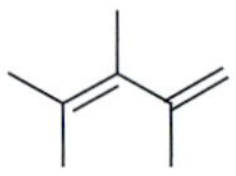	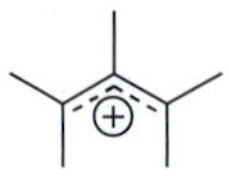	−3	210.6	
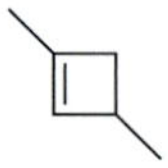		48.3	212	
		33.2	200	
$O{=}CH_2$	$HO{-}CH_2^+ \leftrightarrow HO^+{=}CH_2$		174.6	(0.0)
$O{=}CH(CH_3)$	$HO{-}C^+H(CH_3) \leftrightarrow HO^+{=}CH(CH_3)$		185.0	10.4
$O{=}C(CH_3)_2$	$HO{-}C^+(CH_3)_2$		193.9	19.3
$O{=}C{=}CH_2$	$^+O{\equiv}C{-}CH_3$		195.6	—
HCOOH	$HC^+(OH)_2$		180.1	(0.0)
CH_3COOH	$CH_3C^+(OH)_2$		188.7	8.6
CF_3COOH	$CF_3C^+(OH)_2$		173.3	−6.8
$HCON(CH_3)_2$	$H{-}C^+(OH)[N(CH_3)_2]$		209.8	(0.0)
$CH_3CON(CH_3)_2$	$CH_3C^+(OH)[N(CH_3)_2]$		213.8	4.0
$CH_2{=}NH$	$CH_2{=}N^+H_2$		202.3	—
$H_2C{=}NCH_3$	$H_2C{=}NHCH_3^+$		211.4	−7.6
$CH_3CH{=}NH$	$CH_3CH{=}NH_2^+$		211.5	−7.7
Pyrrolidine	$(CH_2)4{=}NH_2^+$		226.6	
Piperidine	$(CH_2)5{=}NH_2^+$		228	
1-Methylpyrrolidine	$(CH_2)4{=}N(CH_3)H^+$		230.8	
$(CH_3)_2C{=}NCH_2CH_3$	$(CH_3)_2C{=}NC(Et)H^+$		233.3	
$(CH_3)_2S{=}O$	$(CH_3)_2S{=}OH^+$		211.4	

Table 1.A.15 (Continued)

$B + H^+ \rightarrow BH^+$		PA(B)[a]	Substituent effects
Tetrahydrofuran	$(CH_2)4{=}OH^+$	196.5	
HCN	$H{-}C^+{=}NH \leftrightarrow HC{\equiv}N^+H$	174.5	(0.0)
CH_3CN	$CH_3{-}C{\equiv}N^+H$	187	12.5
$ClHC_2CN$	$ClCH_2{-}C{\equiv}N^+H$	180.9	6.4
CCl_3CN	$CCl_3{-}C{\equiv}N^+H$	177.5	3.0
$CH_2(CN)_2$	$NCCH_2{-}C{\equiv}N^+H$	177.7	3.2
Ph—H	$C_6H_7{}^+ \equiv$ 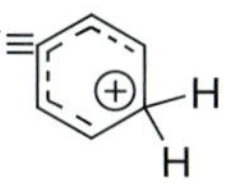	185.0	(0.0)
Ph—CH_3	$C_6H_6CH_3{}^+$	191.5	6.5
PhF	$C_6H_6F^+$	184.1	−0.9
PhCl	$C_6H_6Cl^+$	184.0	−1.0
He	HeH^+	42.7	
Ne	NeH^+	48.1	
Ar	ArH^+	88.6	
Xe	XeH^+	118.6	
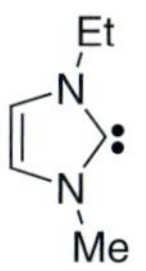	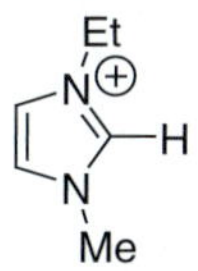	251.3 ± 4	
$:CF_2$	$CHF_2{}^+$	172 ± 2	
$:CCl_2$	$CHCl_2{}^+$	208.3 ± 2	
$:SiH_2$	$SiH_3{}^+$	201 ± 3	

Reference is $PA(NH_3) = 202.3 \pm 2\ \mathrm{kcal\ mol^{-1}}$.[a] $\Delta_f H°(H^+) = 365.7\ \mathrm{kcal\ mol^{-1}}$ (1 cal = 4.184 J).[a]
a) Taken from [109, 110a, 111, 112, 414–418].

Table 1.A.16 Heats of reactions equilibrating carbenium ions and silicenium ions with corresponding hydrocarbons and silanes, in kcal mol^{-1} (1 cal = 4.184 J).[a]

							$\Delta_r H°$
Et^+	+	SiH_4	$\rightleftharpoons$	$SiH_3{}^+$	+	Et–H	– 8 kcal/mol
Et^+	+	CH_3SiH_3	$\rightleftharpoons$	$CH_3SiH_2{}^+$	+	Et–H	– 20 kcal/mol
Et^+	+	CH_3SiH_3	$\rightleftharpoons$	$SiH_3{}^+$	+	$CH_3CH_2CH_3$	– 1 kcal/mol
Et^+	+	$(CH_3)_4Si$	$\rightleftharpoons$	$(CH_3)_3Si^+$	+	$CH_3CH_2CH_3$	– 46 kcal/mol

a) Taken from [109, 110a, 111, 112, 414].

Table 1.A.17 Gas-phase acidities $\Delta G(\mathrm{A{-}H} \rightleftarrows \mathrm{A^-} + \mathrm{H^+})$ as measured by ion cyclotron resonance mass spectrometry, usually with partial pressure of the equilibrating compounds ($\mathrm{A_1H} + \mathrm{A_2^-} \rightleftarrows \mathrm{A_1^-} + \mathrm{A_2H}$) in the 10^{-8} to 10^{-7} Torr range, in kcal mol^{-1} (1 cal = 4.184 J).[a)]

Acid	$\Delta G°$ (298 K)	Acid	$\Delta G°$ (298 K)	Acid	$\Delta G°$ (298 K)
CH_4	408.5	$C_6F_5NH_2$	341.3	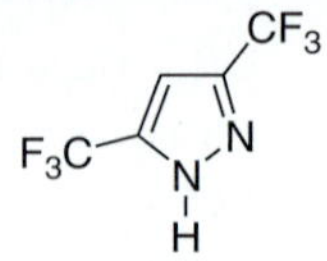	317.3
NH_3	396.1	CH_3CO_2H	341.1	C_6F_5COOH	316.6
$Si(CH_3)_4$	390.7	p-$CF_3SO_2C_6H_4CH_3$	340.7	$(C_6F_5)_2NH$	316.5
H_2O	384.1	CF_3OH	340.7	CF_3COOH	316.1
$PhCH_3$	373.7	$(CH_3CO)_2NH$	339.8	p-$CF_3SO_2C_6H_4OH$	315.7
$(CH_3)_3SiNH_2$	371.0	$CF_3SO_2CH_3$	339.8	p-$MeC_6H_4CH(CN)_2$	315.7
CH_3CN	364.0	p-$(NCCO)C_6H_4CH_3$	339.1	$C_4F_9SO_2NH_2$	315.1
F_2NH	363.3	$CH_3SO_2NH_2$	338.8	CH_3SO_3H	315.0
CH_3COCH_3	361.9	HCO_2H	338.4	$(C_6F_5)_2CHCN$	312.4
p-$CF_3C_6H_4CH_3$	359.8	CH_3COCN	337.7	$(CF_3CO)_2CH_2$	310.3
$PhNH_2$	359.1	$(CH_3CO)_2CH_2$	336.7	HI	309.2
CH_3CHO	359.0	CF_3CONH_2	336.7	2,4,6-$(NO_2)_3C_6H_2CH_3$	309.0
Ph_2CH_2	358.2	p-$NO_2C_6H_4NH_2$	336.2	2,4-$(NO_2)C_6H_3OH$	308.6
$CH_3SO_2CH_3$	358.2	$PhSO_2NH_2$	333.2	$(CF_3CO)_2NH$	307.5
$(CH_3)_3SiOH$	356.0	$PhCO_2H$	333.0	$(FSO_2)_2CH_2$	307.3
CH_3CONH_2	355.1	4-$NH_2C_5F_4N$	332.7	$PhCH(SO_2F)_2$	307.0
$C_6F_5CH_3$	354.7	p-$CF_3SO_2C_6H_4NH_2$	331.3	m-$CF_3C_6H_4CH(CN)_2$	307.0
$PhCOCH_3$	354.5	p-$CF_3C_6H_4OH$	330.1	$(CF_3CO)_2CHCF_3$	305.0
p-$(NC)C_6H_4CH_3$	353.6	$(CH_3CO)_3CH$	328.9	$C_6F_5CH(CN)_2$	303.6
$HCONH_2$	353.0	$(C_6F_5)_2CHPh$	328.4	H_2SO_4	302.2
$((CH_3)_3Si)_2NH$	352.9	$CH_2(CN)_2$	328.3	FSO_3H	299.8
Ph_3CH	352.8	$C_6F_5CH_2CN$	327.6	$PhCH(CN)_2$	314.3
p-$HCOC_6H_4CH_3$	352.6	$(CF_3)_3CH$	326.6	CF_3SO_2NHPh	313.5
$(CF_3)_3CNH_2$	350.1	p-$(NC)C_6F_4NH_2$	326.2	CF_3CSOH	312.5
CH_3NO_2	349.7	p-$HCOC_6H_4OH$	326.1	m-$NO_2{-}C_6H_4CH(CN)_2$	303.0
o-$NO_2C_6H_4CH_3$	348.6	p-$(NC)C_6H_4OH$	325.3	Picric acid	302.8
$PhCONH_2$	347.0	$(CF_3)_2NH$	324.3	$(CF_3SO_2)_2CH_2$	301.5
p-$CF_3C_6H_4NH_2$	346.0	$(CF_3)_3COH$	324.0	$(CF_3SO_2)_2CHPh$	301.3
p-$(CF_3)_3CC_6H_4NH_2$	345.6	CHF_2COOH	323.8	CF_3SO_3H	299.5
p-$NO_2C_6H_4CH_3$	345.3	3,5-$(CF_3)_2C_6H_3OH$	322.9	$CF_3CONHSO_2CF_3$	298.2
$PhCH_2CN$	344.1	CF_3COCH_2COOMe	322.0	$(CF_3SO_2)_2NH$	291.8
H_2NCN	344.1	$CF_3SO_2NH_2$	321.3	2,4,6-$(CF_3SO_2)_3C_6H_2OH$	291.8
$(CF_3)_2CH_2$	343.9	p-$NO_2{-}C_6H_4OH$	320.9	$CF_3SO_2NHSO_2C_2F_2$	290.3
Ph_2NH	343.8	C_6F_5OH	320.8	$(C_2F_5SO_2)NH$	289.6
p-$HCOC_6H_4NH_2$	342.3	$CHCl_2COOH$	320.8	$(CF_3SO_2)_3CH$	289.0
PhOH	342.2	HBr	318.3	$(C_4F_9SO_2)_2CH_2$	288.7
CF_3COCH_3	342.1	HNO_3	317.8	$(C_4F_9SO_2)_2NH$	284.1
p-$(NC)C_6H_4NH_2$	341.5	$(C_6F_5)_3CH$	317.6		

a) Taken from [415].

Table 1.A.18 Heat of heterolytic dissociations $DH°(R^-/H^+)$ (=proton affinities of R^-) in kcal mol^{-1} (1 cal = 4.184 J).[a)]

R—H	$DH°(R^-/H^+)$[a)]	R—H	$DH°(R^-/H^+)$[a)]	R—H	$DH°(R^-/H^+)$[a)]
CH_3—H	416.6	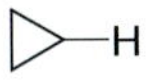	411.5	$PhCH_2CH_2$—H	406 ± 4.6
CH_3CH_2—H	420.1	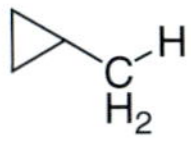	410.5	$MeCOCH_2CH_2$—H	401 ± 4
Me_2CH—H	419.4	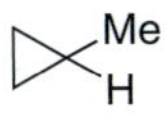	409.2	CH_2=CH—CH_2—H	391 ± 2.5
c-C_4H_7—H	417.4	Me_3CCH_2—H	408.9	MeC(O)—H	387 ± 8
c-C_5H_9—H	416.1	CH_2=CH—H	407.5	F_2CH—H	389 ± 3.5
Me(Et)CH—H	415.7	CH_2=C(Me)—H	405.8	$PhCH_2$—H	377 ± 3.5
$CH_3CH_2CH_2$—H	415.6	Ph—H	401.7	F_3C—H	376 ± 4.5
t-BuH	413.1		411 ± 3.5[b)]	$NCCH_2$—H	369 ± 4.5
$(Me)_2CHCH_2$—H	412.9	FCH_2—H	409 ± 4	HC≡CCH_2—H	383 ± 3[c)]
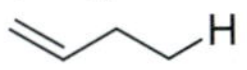	412.0	$CF_3CH_2CH_2$—H	406 ± 3.5	MeC≡CCH_2—H	381 ± 1

a) Taken from [137].
b) Taken from [142a, 143].
c) Taken from [415].

Table 1.A.19 Substituent effects on the stability of anions in the gas phase given as gas-phase acidity differences $\Delta G°(HOH \rightleftarrows HO^- + H^+) - \Delta G°(XOH \rightleftarrows XO^- + H^+)$, $\Delta G°(H_2NH \rightleftarrows H_2N^- + H^+) - \Delta G°(XNH_2 \rightleftarrows XH_3N^- + H^+)$, and $\Delta G°(H_3CH \rightleftarrows CH_3^- + H^+) - \Delta G°(XCH_3 \rightleftarrows XCH_2^- + H^+)$ in kcal mol^{-1} (1 cal = 4.184 J).[a)]

Substituent, X	XOH	XNH_2	XCH_3	Substituent, X	XOH	XNH_2	XCH_3
H	(0.0)	(0.0)	(0.0)	p-CNC_6H_4	58.8	54.6	54.9
$(CH_3)_3Si$	28.1	25.1	17.8	$(CF_3)_3C$	60.1	46.0	
C_6H_5	41.8	37.0	34.8	p-$NO_2C_6H_4$	63.2	59.9	63.2
CH_3CO	43.0	41.0	46.6	NO_2	66.3		58.8
CF_3	43.4		38.1	CF_3CO	67.8	59.4	66.4
HCO	45.7	43.1	49.5	p-$CF_3SO_2C_6H_4$	68.4	64.8	67.8
C_6H_5CO	51.1	49.1	54.0	CH_3SO_2	69.1	57.3	50.3
p-$CF_3C_6H_4$	54.0	50.1	48.7	CF_3SO_2	84.6	74.8	68.7
p-$(CF_3)_3CC_6H_4$		50.2		p-$COCNC_6H_4$			69.4
CN		52.0	44.5	COCN			70.8
C_6F_5	63.3	54.8	53.8	$C_4F_9SO_2$		81.0	
4-C_5F_4N	72.8	63.3		p-$CF_3C_6H_4S(O)(=NSO_2CF_3)$		82.7	
p-$HCOC_6H_4$	58.0	53.8	55.9				

a) Taken from [153].

Table 1.A.20 Examples of ionization energies (IE), electron affinities (−EA) (in eV) of cations, and neutral compounds[a] (1 eV/molecule = 23.0603 kcal mol^{-1}).

	EI	−EA	χ	η		EI	−EA	χ	η		EI	−EA	χ	η		EI	−EA	χ	η
Li^+	75.64	5.39	40.52	35.12	Ag^+	21.49	7.58	14.53	6.96	H_2O	12.6	−6.4	3.1	9.5	BF_3	15.81	−3.5	6.2	9.7
N^+	47.29	5.14	26.21	21.08	Au^+	20.5	9.23	14.9	5.6	H_2S	10.5	−2.1	4.2	6.2	SO_3	12.7	1.7	7.2	5.5
K^+	31.63	4.34	17.99	13.64	Cl^+	23.81	12.97	18.89	5.42	H_3N	10.7	−5.6	2.6	8.2	SO_2	12.3	1.1	6.7	5.6
Rb^+	27.28	4.18	15.77	11.55	Br^+	21.8	11.81	16.8	5.0	H_3P	10.0	−1.9	4.1	6.0	N_2	15.6	−2.2	6.7	8.9
Cs^+	25.1	3.89	14.5	10.6	I^+	19.13	10.45	14.79	4.34	C_2H_2	11.4	−2.6	4.4	7.0	O_2	12.2	0.4	6.3	5.9
Cu^+	20.29	7.74	14.01	6.28											CO	14.0	−1.8	6.1	7.9

a) See Table 1.A.14 for more data.

Table 1.A.21 Examples of ionization energies (IE) of neutral organic compounds, in eV/molecule = 23.0306 kcal mol^{-1} (NIST Chemistry WebBook) (standard deviation do not surpass ±0.1 eV).

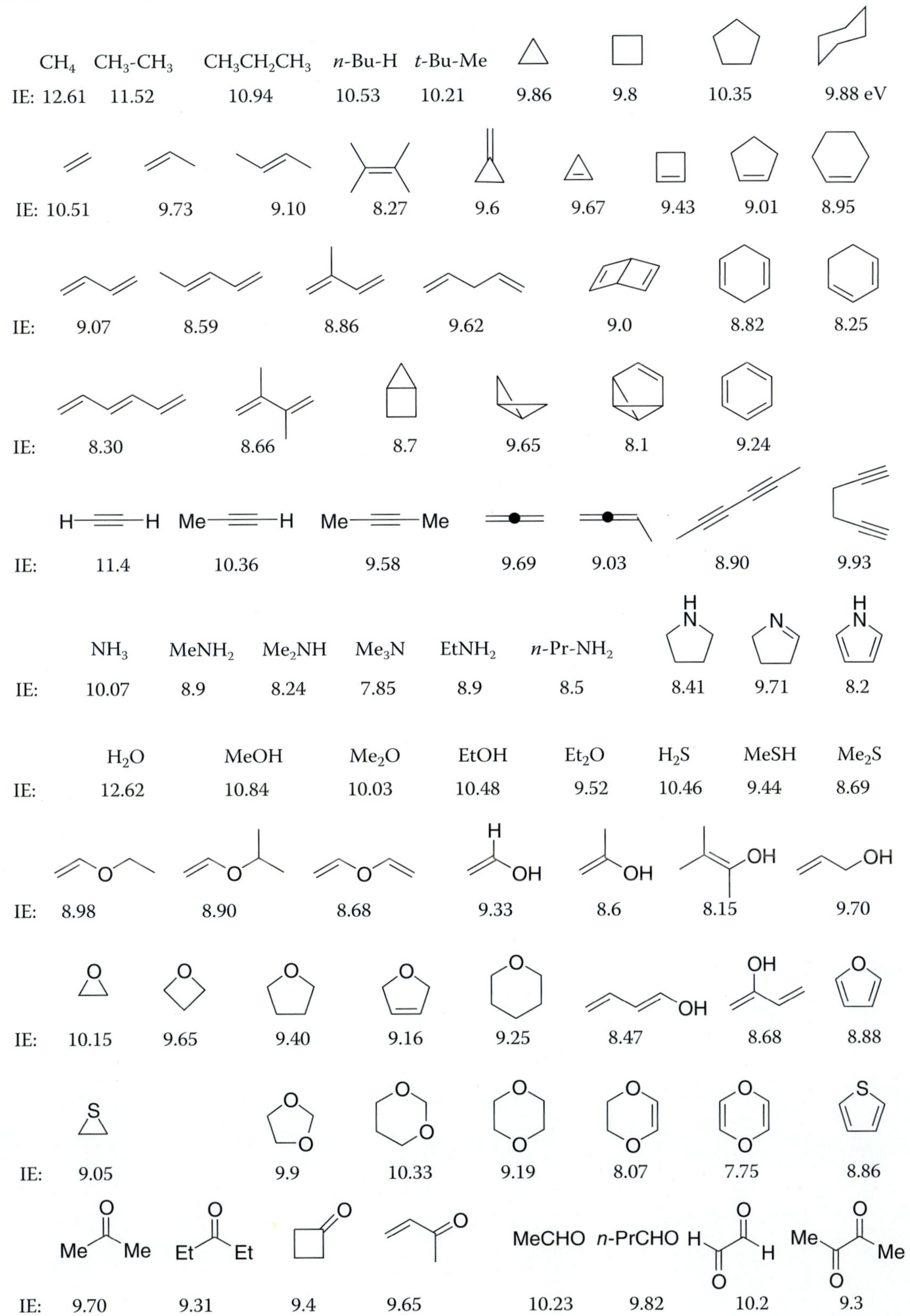

Table 1.A.22 Electron affinities –EA(**A**) = $\Delta_r H°$ (A + e ⇄ **A**$^-$) of selected organic compounds in kcal mol^{-1} (1 eV/molecule = 23.0306 kcal mol^{-1}) (1 cal = 4.184 J).

MeCl[a]	CH_2Cl_2	$HCCl_3$	$CHFCl_2$	CF_2Cl_2	$CH_2{=}CH_2$	CH=CHMe	$Me_2C{=}CH_2$	(*Z*)-MeCH=CHMe
+79.5	+28.3	+8.0	+22.1	+22.6	+41.0	+45.8	+50.4	+51.1

(*E*)-MeCH=CHMe[a]	FCH=CH_2	ClCH=CH_2	(*E*)-FCH=CHF	(*E*)-ClCH=CHCl	$F_2C{=}CH_2$	$Cl_2C{=}CH_2$
+45.4	+44.0	+29.5	+42.4	+18.4	+55.0	+17.5

HC≡CH[a]	MeC≡CH	$CH_2{=}C{=}CH_2$	Butadiene	[b] (*E*)-NC-CH=CH-CN	$MeNO_2$	CS_2	SO_2
+59.8	+64.5	+43.7	+12.4	−28.6	−11.1	−11.7	−25.2

PhH[c]	C_6F_6[b]	Naphthalene[c]	Anthracene[b]	Azulene[b]	$H_2C{=}O$[a]	MeCH=O[a]	EtCH=O[d]
+25.8	−12.0	+4.4	−13.8	−15.2	+19.8	+27.2	+37.3

PhCHO[e]	(*E*)-PhCH=CHCHO[e]	Naphthalene-2-CHO[e]	Acetone[d]	Cyclobutanone	Cyclopentanone
−9.9	−18.9	−14.3	−40.5	−41.5	−36.8

CF_3COMe[d]	MeCOCOMe	$Ph_2C{=}O$	benzoquinone[a]	naphthoquinone[a]	3,4-$(Cl)_2$naphtoquinone[a]
−60.5	−15.9	−14.3	−44.1	−41.1	−50.7

X-C_6H_4CN[g]	X = H	X = *m*-CF_3	*m*-CHO	*m*-NO_2	*p*-CF_3	*p*-CN	*p*-CHO	*p*-NO_2
	−5.5	−10.0	−17.7	−29.3	−13.1	−20.3	−22.6	−33.6

X-$C_6H_4NO_2$[g]	X = H	X = *m*-CF_3	*m*-CHO	*m*-NO_2	*p*-CF_3	*p*-CN	*p*-CHO	*p*-NO_2
		−8.8	−8.9	−14.1	−11.0	−15.9	−14.9	−21.5

Pyridine[h]	Styrene	PhCOOMe	Cyclooctatetraene	Maleic anhydride	Butadienehexacarbonitrile
+14.3	+5.8	−4.6	−13.8	−33.2	−74.6

a) Taken from [419].
b) Taken from [77a].
c) Taken from [420].
d) Taken from [421].
e) Taken from [422].
f) Taken from [423].
g) Taken from [424].
h) Taken from [425].

Table 1.A.23 Selected pK_a data in water (or extrapolated for water in the case of strong and weak acids) at 25 °C.[a)]

(a) Inorganic acids

AgOH	3.96	H_2O	15.6	NH_4^+	9.2	HTe^-	11.0
$Al(OH)_3$	11.2	H_3O^+	−1.7	HN_3	4.7	H_2Te	2.6
$B(OH)_3$	9.23	HS^-	12.9	HNO_2	3.3	$HTeO_3^-$	8.0
HF	3.0	H_2S	7.0	HNO_3	−1.3	H_2TeO_3	2.7
HCl	−7.0	HSCN	0.8	$H_2PO_2^-$	2.2	HCO_3^-	10.3
HBr	−9.0	HSO_3^-	6.9	H_3PO_2	2.0	H_2CO_3	6.4
HI	−10.0	H_2SO_3	1.8	HPO_4^-	12.3	HCN	9.4
HClO	7.5	HSO_4^-	1.9	$H_2PO_4^-$	7.2	HOCN	3.9
$HClO_2$	2.0	H_2SO_4	−3.0	H_3PO_4	2.1	H_2NNH_2	8.2
$HClO_3$	−1.0	HSe^-	11.0	$H_2P_2O_7^-$	6.6	$H_2NNH_3^+$	−0.9
$HClO_4$	−10.0	H_2Se	3.9	$H_3P_2O_7^-$	2.4	$HONH_2$	6.0
H_2O_2	11.8	$HC_2O_4^-$	4.2	$H_4P_2O_7$	1.5	$NH_2CONH_3^+$	0.2

(b) Organic acids

Carboxylic acids ($RCOOH \rightleftarrows RCOO^- + H^+$)

HCOOH	3.75	FCH_2COOH	2.66	PhCOOH	4.20
MeCOOH	4.76	$ClCH_2COOH$	2.86	$2\text{-}(NO_2)C_6H_4COOH$	2.17
$MeCOCH_2COOH$	3.6	$BrCH_2COOH$	2.86	$3\text{-}(NO_2)C_6H_4COOH$	3.45
O_2NCH_2COOH	1.7	ICH_2COOH	3.12	$4\text{-}(NO_2)C_6H_4COOH$	3.44
$^+H_3NCH_2COOH$	2.3	$F_2CHCOOH$	1.24	$2\text{-}(MeO)C_6H_4COOH$	4.09
$MeSO_2CH_2COOH$	2.4	$Cl_2CHCOOH$	1.29	$3\text{-}(MeO)C_6H_4COOH$	4.09
$MeOCH_2COOH$	3.53	F_3CCOH	0.23	$4\text{-}(MeO)C_6H_4COOH$	4.47

Dicarboxylic acids[b)] $R'(COOH)_2 \xrightleftharpoons{pK_{a_1}} R'(COOH)COO^- + H^+ \xrightleftharpoons{pK_{a_2}} R'(COO^-)_2 + 2H^+$

Oxalic acid	1.23;4.19	D-Tartaric acid	3.03;4.45	Phthalic acid	2.98;5.28
Malonic acid	2.83;5.69	Glutaric acid	4.34;5.42	Fumaric acid	3.02;4.38

Sulfinic, sulfonic acids (O—H acids)

$PhSO_2H$	2.0	$PhSO_3H$	−6.5	$MeSO_3H$	−2.0	CF_3SO_3H	~−13.0

Alcohols (O—H acids)

$MeC{=}O^{(+)}{-}H$	−7.0	$MeCOOH_2^+$	−6.0	$EtOH_2^+$	−2.0	MeOH	15.5
EtOH	16.0	$HO(CH_2)_2OH$	14.8	*i*-PrOH	18.8	*t*-BuOH	19.9
$CH_2{=}CH{-}CH_2OH$	15.5	$CH{\equiv}C{-}CH_2OH$	13.5	$MeOCH_2CH_2OH$	14.8	CHF_2CH_2OH	12.7
CF_3CH_2OH	12.4	CCl_3CH_2OH	12.2	$(CF_3)_2CHOH$	9.3		

Phenols (O—H acids)

PhOH	9.9	$2\text{-}(NO_2)C_6H_4OH$	7.2	$3\text{-}(NO_2)C_6H_4OH$	8.3
$4\text{-}(NO_2)C_6H_4OH$	7.2	$2,4\text{-}(NO_2)_2C_6H_3OH$	4.1	$2,4,6\text{-}(NO_2)_3C_6H_2OH$	0.3
$2\text{-}(HO)C_6H_4OH$	9.5	$3\text{-}(HO)C_6H_4OH$	9.4	$4\text{-}(HO)C_6H_4OH$	9.9

Enols (C—H acids)

Tropolone	6.7	Squaric acid	1.5;3.1	Ascorbic acid	4.2;11.6

(continued)

Table 1.A.23 (Continued)

Oximes (O—H acids)					
MeCONHO—H	9.4	Ph_2C=NO—H	11.3	Me_2C=NO—H	12.2

Ammonium salts (N—H acids)					
Guanidinium ($(H_2N)_2C = NH_2^+$)	13.6	Piperidinium	11.0	c-C_6H_{11}-NH_3^+	10.7
n-Bu-NH_3^+	10.6	*N*-Methylpyrolidinium	10.7	Me_3NH^+	9.8
Imidazolium	7.0	Pyridinium	5.2	$PhNH_3^+$	4.6
Quinolinium	4.5	Pyridazinium	2.1	Pyrimidium	1.1
Pyrazinium	0.4	Azolium	0.0	$RCONH_3^+$	−10.0

Amides (N—H acids)					
$RCONH_2$	17.0	Succinimide	9.6	Phthalimide	8.3

Amines (N—H acids) ($R_2NH \rightleftarrows R_2N^- + H^+$)					
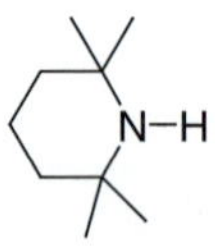	37.0	(i-Pr)$_2$N—H	35.7	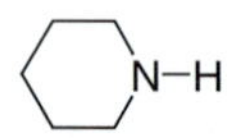 piperidine	33.0
Ph—NH—H	27.0	$(Me_3Si)_2$N—H	26.0	4-(NO_2)C_6H_4NH—H	19.0
Pyrrole	15.0	Imidazole	14.5		

Thiols (S—H acids)					
n-PrS—H	10.7	$PhCH_2S$—H	9.4	PhS—H	7.8

Carbonyl compounds (C—H acids)					
H—CH_2—COOEt	24.5	H—CH_2COMe	20.0	Cyclohexanone	16.7
H—CH$(COOEt)_2$	13.3	H—CH$(COMe)_2$	9.0	H—CH(COMe)COOEt	11.0
H—C$(COMe)_3$	5.8	H—C$(COMe)_3$	5.8	H—CH(COMe)CHO	6.9

Nitro compounds (C—H acids)					
H—CH_2NO_2	10.3	H—CH(Me)NO_2	8.6	H—CH$(NO_2)_2$	3.6

Carbonitriles (C—H acids)					
H—CH_2CN	25.0	H—CH$(CN)_2$	11.0	H—C$(CN)_3$	−5.0

Hydrocarbons (C—H acids)							
H—Me	~56	H—CH=CH_2	~44	H—CH_2CH=CH_2	~43	H—Ph	~43
H—CH_2Ph	~41	H—CH$(Ph)_2$	~32	H—C$(Ph)_3$	30.6	H—C≡CPh	28.6
H—C≡C—H	25.0	Indene	22.6	Fluorene	22.6	Cyclopentadiene	15.0

a) Acidity constant of acid HA in water: $K_a = @ [H_3O^+][A^-]/[HA]$. Basicity constant for A^-: K_b @ $[HO^-][HA]/[A^-]$. Ion product constant for water: $K_w = [H_3O^+][HO^-] = 10^{-14}$ at 25 °C. $K_w = K_a \times K_b$; $pK_a = -\log K_a$; $pK_b = -\log K_b \times pK_a + pK_b = 14$; pK_a's taken from [Williams, R.; Jencks, W. P.; Westheimer, F. H. http://www.webqc.org/pkaconstants.phd/2011.

b) pK_a of two successive deprotonation equilibria.

Table 1.A.24 Selected pK_a data in DMSO (Bordwell's table).[a)]

(a) Inorganic acids

HBr	0.9	H_2O	31.4	HN_3	7.9	$H_2N—CN$	16.9
HCl	1.8	H_3N	41.0	NH_4^+	10.5		
HF	15.0	HCN	12.9	HONO	7.5		

(b) Organic acids

Carboxylic acids (O—H acids)		Alcohols (O—H acids)		Phenols (O—H acids)		Thiol (S—H acids)	
AcOH	12.6	H_2O	31.4	PhOH	18.0	PhS—H	10.3
BzOH	11.1	MeOH	29.0	4-$MeOC_6H_4OH$	19.1	*t*-BuS—H	17.9
4-NO_2—C_6H_4COOH	9.1	EtOH	29.8	4-$NO_2C_6H_4OH$	10.8	*n*-BuS—H	17.0
		i-PrOH	30.3	α-Naphthol	16.2	$PhCH_2S$—H	15.3
		t-BuOH	32.2	β-Naphthol	17.2	PhCOS—H	5.2
		CF_3CH_2OH	23.5	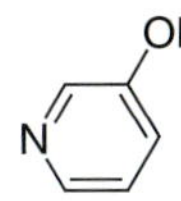	15.7	PhSe—H	7.1
		$(CF_3)_2CHOH$	17.9				
		$(CF_3)_3COH$	10.7				

Ammoniums salts (N—H acids)		Amines (N—H acids)		Amides, carbamates (N—H acids)		Various (N—H acids)	
NH_4^+	10.5	NH_3	~41.0	$HCONH_2$	23.5	$MeSO_2NH$—H	17.5
$BuNH_3^+$	11.1	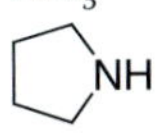	~44.0	$MeCONH_2$	25.5	CF_3SO_2NH—H	9.7
$PhNH_3^+$	3.8	$PhNH_2$	30.6	MeCONHPh	21.5	$PhSO_2NH$—H	16.1
Et_3NH^+	9.0	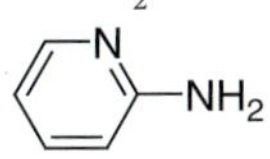	27.7	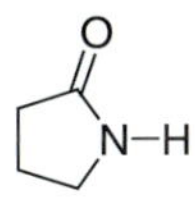	24.2	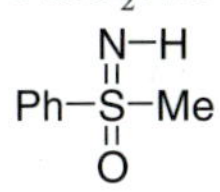	24.3
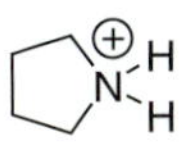	11.1	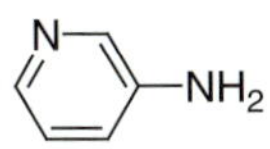	28.5	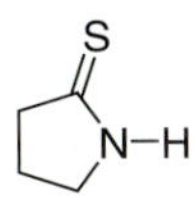	13.3	$PhSO_2NHNH$—H	17.1
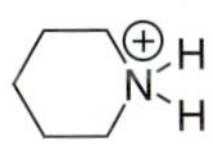	10.9	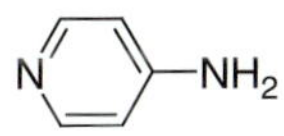	26.5	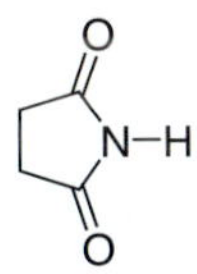	14.7	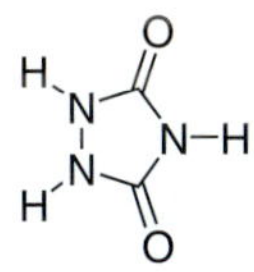	13.1
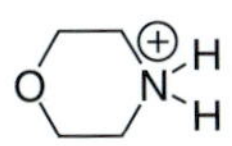	9.2	N≡C—NH_2	17.0	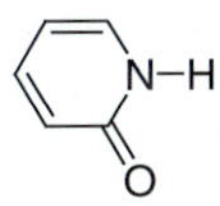	17.0	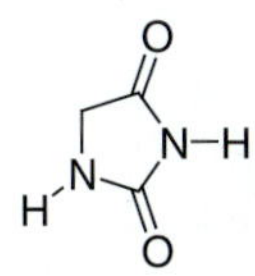	15.1
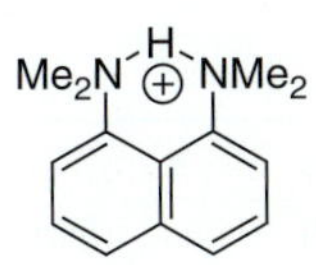	7.5	Ph_2NH	25.0	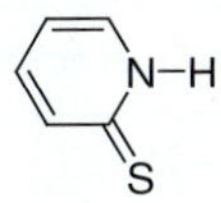	13.3	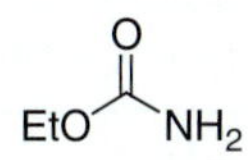	24.2
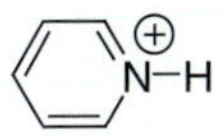	3.4			EtOCO—NH_2	24.2	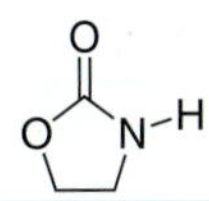	20.8

(continued)

Table 1.A.24 (Continued)

Ureas (N—H acids)		Amidines (N—H acids)		Guadinines (N—H acids)	
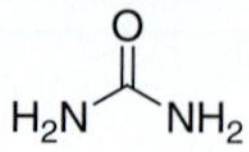	26.9	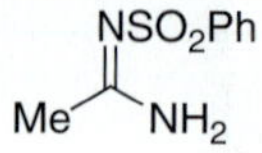	17.3	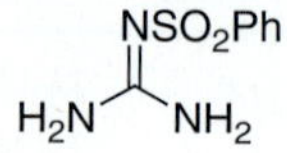	19.4
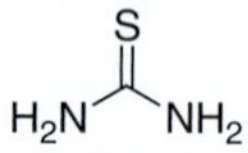	21.0				

N-heterocycles (N—H acids)							
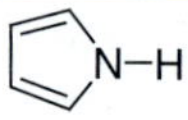	23.0	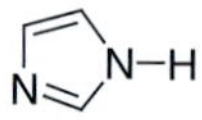	18.6	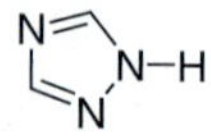	14.8	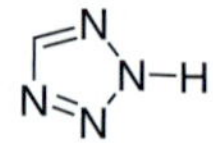	8.2
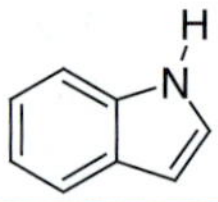	21.0	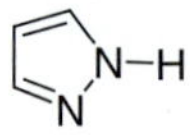	19.8	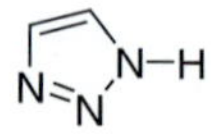	13.9		

Carbonyl compounds (C—H acids)							
EtOOC—CH_2—H	29.5	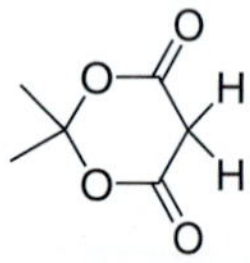	7.3	MeCOMe	26.5	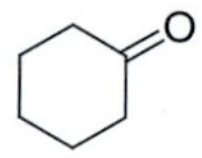	26.4
t-BuOOC—CH_2—H	30.3	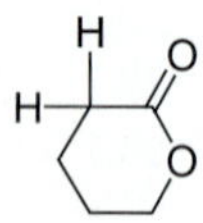	25.2	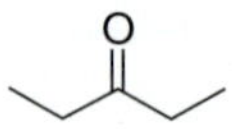	27.1	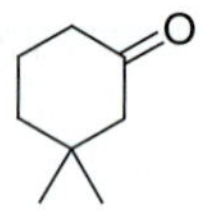	24.8
EtOOC—CH(CN)—H	13.1	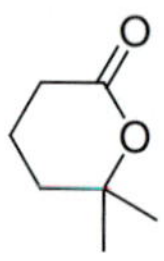	24.5	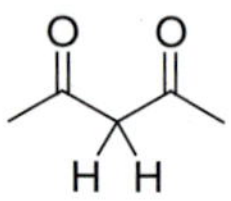	13.3	$(CH_3CO)_3C$—H	8.6
EtOOC—CH(NO_2)—H	9.1	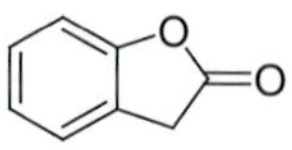	13.5	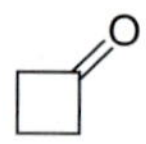	26.2	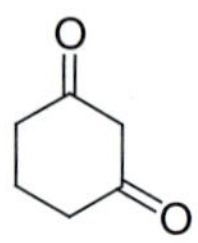	10.3
EtOOC—CH(SO_2CF_3)—H	6.4	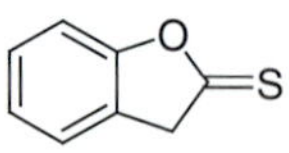	10.7	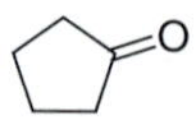	25.8		
PhOOC—CH(Ph)—H	18.7	PhSOC—CH(Ph)—H	16.9	EtOOC—CH(MeCO)—H	14.2	$(EtOOC)_2CH$—H	16.4
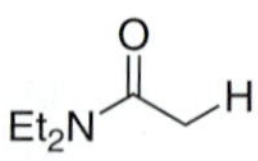	~35.0	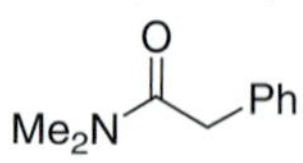	26.6	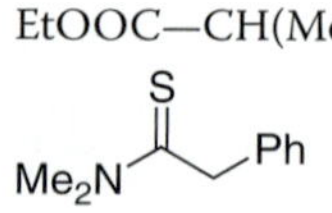	21.3	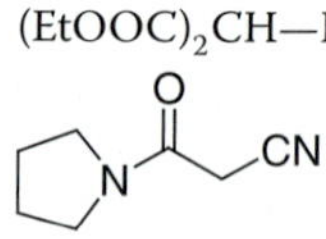	17.2
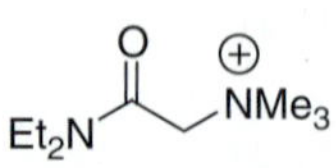	24.9						

Phosphonium salt (*C—H acids*)		Phosphines (*C—H acids*)		Phosphine oxides (*C—H acids*)	
Ph_3P^+—CH_2—H	22.4	$(Ph_2P)_2CH$—H	29.9	$(Ph)_2POCH(SPh)$—H	24.9
Ph_3P^+—CH(Ph)—H	17.4	$(Ph_2P)(PhSO_2)CH$—H	20.2	$(Ph)_2POCH(CN)$—H	16.9
Ph_3P^+—CH(SPh)—H	14.9			$(Ph)_2PSCH(CN)$—H	16.3

Table 1.A.24 (Continued)

Ph_3P^+—CH(COOEt)—H	8.5	*Phosphonates*			
Ph_3P^+—CH(COMe)—H	7.1	$(EtO)_2PO$—CH(Ph)—H	27.6		
Ph_3P^+—CH(CHO)—H	6.1	$(EtO)_2PO$—CH(CN)—H	16.4		
Ph_3P^+—CH_2—CH=CH_2	18.5	$(MeO)_2PO$—CH(Cl)—H	26.3		
Organosulfur compounds (C—H acids)					
$MeSCH_2$—H	~45	$PhSOCH_2$—H	33.0	$PhSO_2CH_2$—H	29.0
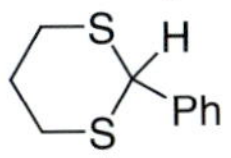	30.7	PhSOCH(Ph)—H	27.2	$PhSO_2CH(SMe)$—H	23.4
$Me_2\overset{+}{S}(O)CH_2$—H	18.2	$(PhSO)_2CH$—H	18.2	$MeSO_2CH_2$—H	31.1
$Ph(Me)S^+$—CH(Ph)—H	16.3	$PhSOC(Ph)_2$—H	24.6	$(MeSO_2)_2CH$—H	15.0
$PhOSO_2CH_2$—H	25.2			$(PhSO_2)_2C(Me)$—H	14.3
$FSO_2CH(Ph)$—H	16.9			$CF_3SO_2CH_2$—H	18.8
PhN(Me)—SO_2—CH_2Ph	24.1			$(CF_3SO_2)_2CH$—H	2

Hydrocarbons (C—H acids)							
CH_4	~56.0	$PhCH_2$—H	~43	$(CH_2{=}CH)$—CH_2—H	~44.0	Indene	20.1
Cyclopentane	~59.0	Ph_2CH—H	32.3	$(CH_2{=}CH)_2CH$—H	32.3	Fluorene	22.6
Cyclopentadiene	18.0	Ph_3C—H	30.6	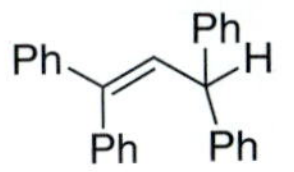	25.8	PhC≡C—H	28.8

Heterocycles (C—H acids)							
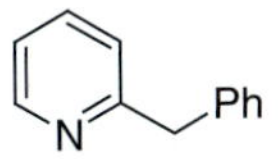	28.2	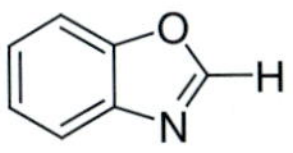	24.4	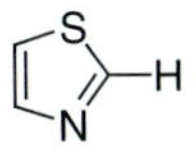	29.4	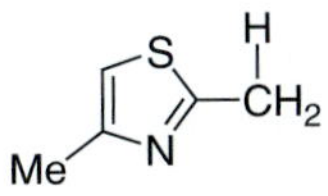	30.3
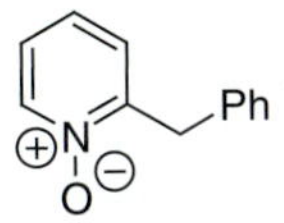	25.2	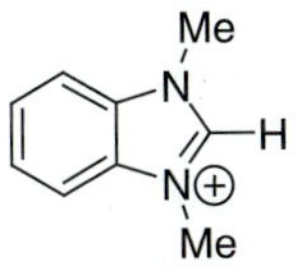	18.6	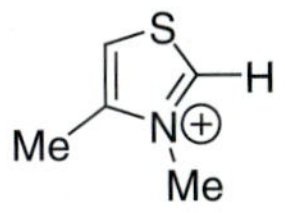	16.5	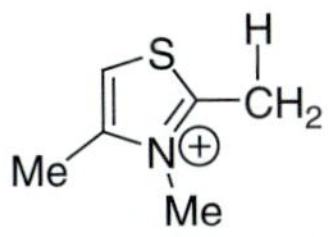	13.9
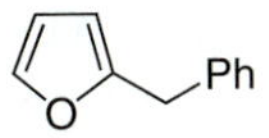	30.2	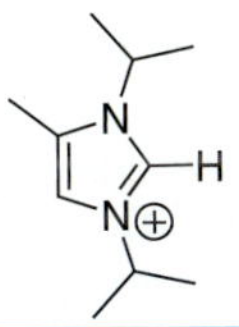	24.0	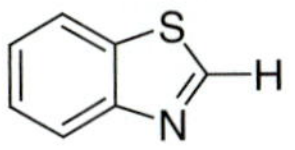	27.0	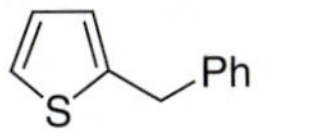	30.0

Carbonitriles (C—H acids)		*Nitroalcanes (C—H acids)*	
NC—CH_2—H	31.3	O_2NCH_2-H	17.2
NC—CH(SMe)—H	24.3	$O_2NCH(Me)$—H	16.9
NC—CH(Ph)—H	21.9	$O_2NCH(Ph)$—H	12.2
NC—CH(SO_2Ph)—H	12.0	$O_2NCH(SPh)$—H	11.8
$(NC)_2CH$—H	11.1	$O_2NCH(CH{=}CH)$—H	11.3
		$O_2NCH(SO_2Ph)$—H	7.1

a) Taken from Bordwell pK_a table (acidity in DMSO), by Hans J. Reich, University of Wisconsin, USA: http://www.chem.wis.edu/areas/reich/pKa/pKatable.

References

1 Atkins, P.W. and de Paula, J. (2010). Thermochemistry. In: *Physical Chemistry*, 9e, Chapter 2.7–2.9, 65–73. Oxford and New York: Oxford University Press.

2 Laidler, K.J. (1995). Thermodynamics. In: *The World of Physical Chemistry*, Chapter 4. Oxford: Oxford University Press.

3 Rossini, F.D. (1976). 50 years of thermodynamics and thermochemistry. *J. Chem. Thermodyn.* 8 (9): 805–834.

4 Skinner, H.A. (1978). Thermochemistry of organometallic compounds. *J. Chem. Thermodyn.* 10 (4): 309–320.

5 Pilcher, G. and Professor, H.A. (1996). Skinner (30 January 1916–14 May 1996) - obituary. *J. Chem. Thermodyn.* 28 (11): 1195–1196.

6 (a) Pedley, J.B., Naylor, R.D., and Kirby, S.P. (1986). *Thermochemical Data of Organic Compounds*, 2e, 1–792. London and New York: Chapman & Hall. (b) Cox, J.D., Wagman, D.D., and Medvedev, V.A. (1989). *CODATA Key Values for Thermodynamics*, 1–271. New York: Hemisphere Publishing Corp. (c) Gurvich, L.V., Veyts, I.V., and Alcock, C.B. (1989). *Thermodynamic Properties of Individual Substances*, 4e, vol. 1. New York: Hemisphere Publishing Corp. (d) Pedley, J.B. (1994). *Thermochemical Data and Structures of Organic Compounds*, TRC Data Series, vol. 1, 1–807. College Station, TX: Thermodynamics Research Center. (e) Gurvich, L.V., Veyts, I.V., and Alcock, C.B. (1994). *Thermodynamic Properties of Individual Substances*, 4e, 1–991. Boca Raton, FL: CRC Press.

7 Linstrom, P.J. and Mallard, W.G. (eds.) (2009). *NIST Chemistry WebBook*, NIST Standard Reference Database. Gaithersburg, MD: National Institute of Standards and Technology.

8 (a) Le Chatelier, H.L. (1884). Sur un énoncé général des lois des équilibres chimiques. *Comptes rendus hebdomadaires des Séances de l Académie des Sciences* 99: 786–789. (b) Hillert, M. (1995). Le Chatelier's principle – restated and illustrated with phase diagrams. *J. Phase Equilib.* 16 (5): 403–410.

9 Van't Hoff, J.H. (1884). *Etudes de dynamique chimique*, 1–214. Amsterdam, The Netherlands: Frederik Muller & Co.

10 Laidler, K.J. (1995). Lessons from the history of chemistry. *Acc. Chem. Res.* 28 (4): 187–192.

11 Guldberg, C.M. and Waage, P. (1879). Concerning chemical affinity. *Erdmann's Journal für praktische Chemie* 127: 69–114.

12 Schrödinger, E. (1946). *Statistical Thermodynamics*, New Publication: 1989, 1–95. New York: Dover Publications, Inc.

13 Perrot, P. (1998). *A to Z of Thermodynamics*, 1–336. Oxford, New York, and Tokyo: Oxford University Press.

14 Carter, A.H. (2001). *Classical and Statistical Thermodynamics*, 1–432. Upper Saddle River, NJ: Prentice-Hall Inc. (A Pearson Education Co.).

15 Cercignani, C.L. (2006). *Boltzmann - The Man Who Trusted Atoms*, 1–348. Oxford: Oxford University Press.

16 (a) Benson, S.W. and Buss, J.H. (1958). Additivity rules for the estimation of molecular properties - thermodynamic properties. *J. Chem. Phys.* 29 (3): 546–572. (b) Cohen, N. and Benson, S.W. (1993). Estimation of heats of formation of organic compounds by additivity methods. *Chem. Rev.* 93: 2419–2438.

17 Cohen, N. (1996). Revised group additivity values for enthalpies of formation (at 298 K) of carbon–hydrogen and carbon–hydrogen–oxygen compounds. *J. Phys. Chem. Ref. Data* 25 (6): 1411–1481.

18 (a) Zavitsas, A.A., Rogers, D.W., and Matsunaga, N. (2010). Heats of formation of organic compounds by simple calculations. *J. Org. Chem.* 75: 2502–2515. (b) Hukkerikar, A.S., Meier, R.J., Sin, G., and Gani, R. (2013). A method to estimate the enthalpy of formation of organic compounds with chemical accuracy. *Fluid Phase Equilib.* 348: 23–32.

19 Laidler, K.J. (1956). A system of molecular thermochemistry for organic gases and liquids. *Can. J. Chem.* 34: 626–648.

20 Benson, S.W., Cruickshank, F.R., Golden, D.M. et al. (1969). Additivity rules for estimation of thermochemical properties. *Chem. Rev.* 69 (3): 279–324.

21 Lias, S.G., Bartmess, J.E., Liebman, J.F. et al. (1988). Gas-phase ion and neutral thermochemistry. *J. Phys. Chem. Ref. Data* 17 (Suppl. 1): 861.

22 Rienstra-Kiracofe, J.C., Tschumper, G.S., Schaefer, H.F.I. et al. (2002). Atomic and molecular electron affinities: photoelectron experiments and theoretical computations. *Chem. Rev.* 102 (1): 231–282.

23 Sablier, M. and Fujii, T. (2002). Mass spectrometry of free radicals. *Chem. Rev.* 102 (9): 2855–2924.

24 Cox, J.D. and Pilcher, G. (1970). *Thermochemistry of Organic and Organometallic Compounds*, 1–643. London and New York: Academic Press.

25 Pauling, L. (1960). *The Nature of the Chemical Bond and the Structure of Molecules and Crystals*, 3e, 1–643. Ithaca, NY: Cornell University Press.

26 Bettens, R.P.A. (2003). Bound state potential energy surface construction: Ab initio zero-point energies and vibrationally averaged rotational constants. *J. Am. Chem. Soc.* 125 (2): 584–587.

27 von Doering, W. and Sarma, K. (1992). Stabilization energy of polyenyl radicals-all-trans-nonatetraenyl radical by thermal rearrangement of a semirigid (4-1-2) heptaene - model for thermal lability of β-carotene. *J. Am. Chem. Soc.* 114 (15): 6037–6043.

28 Duncan, J.L. (1970). Equilibrium geometries of methyl halides. *J. Mol. Struct.* 6 (6): 447–456.

29 Allen, P.W. and Sutton, L.E. (1951). The molecular structures of the acetyl halides. *Trans. Faraday Soc.* 47 (3): 236–254.

30 Sutherland, G.B.B.M. (1937). Carbon–halogen distance in the methyl halides. *Nature (London)* 140: 239–240.

31 Allen, F.H., Kennard, O., Watson, D.G. et al. (1987). Tables of bond lengths determined by X-ray and neutron-diffraction. 1. Bond lengths in organic compounds. *J. Chem. Soc., Perkin Trans. 2* 12: S1–S19.

32 Tsuchiya, S. (1974). Molecular structures of acetyl fluoride and acetyl iodide. *J. Mol. Struct.* 22: 77–95.

33 O'Hagan, D. (2008). Understanding organofluorine chemistry. An introduction to the C—F bond. *Chem. Soc. Rev.* 37: 308–319.

34 Peng, S., Qing, F.L., Li, Y.Q., and Hu, C.M. (2000). Palladium(0)/copper(I)-cocatalyzed cross-coupling of the zinc reagent of ethyl 3-bromo-3,3-difluoropropionate with aryl (alkenyl) halides: an efficient stereoselective synthesis of β-fluoro-α,β-unsaturated esters. *J. Org. Chem.* 65 (3): 694–700.

35 Wieland, T., Bokelmann, E., Bauer, L. et al. (1953). Über Peptidsynthesen. 8. Bildung von S-haltigen Peptiden durch intramolekulare Wanderung von Aminoacylresten. *Justus Liebigs Ann. Chem.* 583 (2): 129–149.

36 Dawson, P.E., Muir, T.W., Clarklewis, I., and Kent, S.B.H. (1994). Synthesis of proteins by native chemical ligation. *Science* 266 (5186): 776–779.

37 Lahiri, S., Brehs, M., Olschewski, D., and Becker, C.F. (2011). Total chemical synthesis of an integral membrane enzyme: diacylglycerol kinase from *Escherichia coli*. *Angew. Chem. Int. Ed.* 50 (17): 3988–3992.

38 Kwart, H. and Evans, E.R. (1966). Vapor phase rearrangement of thioncarbonates and thioncarbamates. *J. Org. Chem.* 31 (2): 410–413.

39 Newman, M.S. and Karnes, H.A. (1966). Conversion of phenols to thiophenols via dialkylthiocarbamates. *J. Org. Chem.* 31 (12): 3980–3984.

40 Lloyd-Jones, G.C., Moseley, J.D., and Renny, J.S. (2008). Mechanism and application of the Newman-Kwart O→S rearrangement of O-aryl thiocarbamates. *Synthesis* (5): 661–689.

41 Rossini, F.D. (1934). Heats of combustion and of formation of the normal paraffin hydrocarbons in the gaseous state, and the energies of their atomic linkages. *J. Res. Nat. Bur. Stand.* 13 (1): 21–35.

42 Rossini, F.D. (1940). Heats of formation of gaseous hydrocarbons. *Chem. Rev.* 27 (1): 1–16.

43 Wodrich, M.D., Wannere, C.S., Mo, Y. et al. (2007). The concept of protobranching and its many paradigm shifting implications for energy evaluations. *Chem. Eur. J.* 13 (27): 7731–7744.

44 Kerber, R.C. (2007). Markovnikov's rule. *J. Chem. Educ.* 84 (7): 1109–1109.

45 Ilich, P.P. (2007). Markovnikov's rule - replies. *J. Chem. Educ.* 84 (7): 1109–1109.

46 Hughes, P. (2006). Was Markovnikov's rule an inspired guess? *J. Chem. Educ.* 83 (8): 1152–1154.

47 Aizman, A., Contreras, R., Galvan, M. et al. (2002). The Markovnikov regioselectivity rule in the light of site activation models. *Chem. A Eur. J.* 106 (34): 7844–7849.

48 Abboud, J.L.M., Alkorta, I., Davalos, J.Z. et al. (2002). Thermodynamic stabilities of carbocations. *Adv. Phys. Org. Chem.* 37: 57–135.

49 Aue, D.H. (2011). Carbocations. *Wiley Interdiscip. Rev. Comput. Mol. Sci.* 1 (4): 487–508.

50 McClelland, R.A. (2012). Carbocations. *Org. React. Mech.* 257–273.

51 Dimroth, O. (1933). Relationships between affinity and reactions speed. *Angew. Chem.* 46: 571–576.

52 Evans, M.G. and Polanyi, M. (1936). Further considerations on the thermodynamics of chemical equilibria and reaction rates. *Trans. Faraday Soc.* 32 (2): 1333–1359.

53 Bell, R.P. (1938). The kinetics of proton transfer reactions. *Trans. Faraday Soc.* 34 (1): 0229–0236.

54 Brønsted, J.N. (1928). Acid and basic catalysis. *Chem. Rev.* 5 (3): 231–338.

55 Pieniazek, S.N. and Houk, K.N. (2006). The origin of the halogen effect on reactivity and reversibility of Diels-Alder cycloadditions involving furan. *Angew. Chem. Int. Ed.* 45 (9): 1442–1445.

56 Dolbier, W.R., Alty, A.C., and Phanstiel, O. (1987). Kinetic and stereochemical effect of a fluorine substituent on the Cope and the homodienyl 1,5-hydrogen shift rearrangements. *J. Am. Chem. Soc.* 109 (10): 3046–3050.

57 (a) Walter, M.G., Warren, E.L., McKone, J.R. et al. (2010). Solar water splitting cells. *Chem. Rev.* 110 (11): 6446–6473. (b) Shaner, M.R., Atwater, H.A., Lewis, N.S., and McFarland, E.W. A comparative technoeconomic analysis of renewable hydrogen production using solar energy. *Energy Environ. Sci.* 2016 (7): 9, 2353–2371. (c) Preuster, P., Alexander, A., and Wasserscheid, P. (2017). Hydrogen storage technologies for future energy systems. *Annu. Rev. Chem. Biomol. Eng.*, Prausnitz, J.M. Ed. 8: 445–471.

58 (a) Joo, F. (2008). Breakthroughs in hydrogen storage-formic acid as a sustainable storage material for hydrogen. *ChemSusChem* 1 (10): 805–808. (b) Mellman, D., Sponholz, P., Junge, H., and Beller, M. (2016). Formic acid as a hydrogen storage material - development of homogeneous catalysts for selective hydrogen release. *Chem. Soc. Rev.* 45 (14): 3954–3988.

59 (a) Federsel, C., Jackstell, R., and Beller, M. (2010). State-of-the-art catalysts for hydrogenation of carbon dioxide. *Angew. Chem. Int. Ed.* 49 (36): 6254–6257. (b) Bertini, F., Glatz, M., Gorgas, N. et al. (2017). Carbon dioxide hydrogenation catalysed by well-defined Mn(I) PNP pincer hydride complexes. *Chem. Sci.* 8: 5024–5029. (c) Kanega, R., Onishi, N., Szalda, D.J. et al. (2017). CO_2 hydrogenation catalysts with deprotonated picolinamide ligands. *ACS Catal.* 7 (10): 6426–6429. (d) Song, Q.-W., Zhou, Z.-H., and He, L.-N. (2017). Efficient, selective and sustainable catalysis of carbon dioxide. *Green Chem.* 19 (16): 3707–3728.

60 Boddien, A., Gaertner, F., Jackstell, R. et al. (2010). ortho-Metalation of Fe(0)-tribenzylphosphine complexes: homogeneous catalysts for the generation of hydrogen from formic acid. *Angew. Chem. Int. Ed.* 49 (47): 8993–8996.

61 Aldous, L. and Compton, R.G. (2010). Clean, efficient electrolysis of formic acid via formation of eutectic, ionic mixtures with ammonium formate. *Energy Environ. Sci.* 3 (10): 1587–1592.

62 Richardson, R.D., Holland, E.J., and Carpenter, B.K. (2011). A renewable amine for photochemical reduction of CO_2. *Nat. Chem.* 3 (4): 301–303.

63 Himbert, G. and Henn, L. (1982). Intramolecular Diels-Alder reaction of allenecarboxanilides. *Angew. Chem. Int. Ed. Eng.* 21 (8): 620–620.

64 Henn, L., Himbert, G., Diehl, K., and Kaftory, M. (1986). Cycloadditions. 6. Intramolecular Diels-Alder reaction of allenecarboxanilides - variation of the residue geminal to the carboxamide group. *Chem. Ber. Recl.* 119 (6): 1953–1963.

65 (a) Diehl, K., Himbert, G., and Henn, L. (1986). Cycloadditions. 7. Intramolecular Diels-Alder reactions of allenecarboxanilides - variation of the substituents in the para-position of the aniline nucleus. *Chem. Ber. Recl.* 119 (8): 2430–2443. (b) Schmidt, Y., Lam, J.K., Pham, H.V. et al. (2013). Studies on the Himbert intramolecular arene/allene Diels-Alder cycloaddition. Mechanistic studies and expansion of scope to all-carbon tethers. *J. Am. Chem. Soc.* 135 (19): 7339–7348.

66 Bock, H. and Mollere, P.D. (1974). Photoelectron spectra - experimental approach to teaching molecular orbital models. *J. Chem. Educ.* 51 (8): 506–514.

67 Hüfner, S. (2003). *Photoelectron Spectroscopy, Principles and Applications*, 1–634. Berlin, Heidelberg and New York: Springer-Verlag.

68 National Institute of Standards and Technology (2000). *Chemistry WebBook*, NT Standard Reference Data-base Number 69. Gaithersburg, MD. http://webbook.nist.gov.

69 Shi, Y.J. and Lipson, R.H. (2005). An overview of organic molecule soft ionization using vacuum ultraviolet laser radiation. *Can. J. Chem.* 83 (11): 1891–1902.

70 (a) Likholyot, A., Hovey, J.K., and Seward, T.M. (2005). A new high-pressure mass spectrometer with improved reaction temperature measurement. *Int. J. Mass Spectrom.* 243 (1): 49–62. (b) Hanley, L. and Zimmermann, R. (2009). Light and molecular ions. The emergence of vacuum UV single-photon ionization in MS. *Anal. Chem.* 81 (11): 4174–4182.

71 (a) Schultz, R.H., Crellin, K.C., and Armentrout, P.B. (1991). Sequential bond-energies of Fe(CO)X+ (X = 1-5) - systematic effects on collision-induced dissociation measurements. *J. Am. Chem. Soc.* 113 (23): 8590–8601. (b) Sunderlin, L.S., Wand, D., and Squires, R.R. (1992). Metal carbonyl bond strengths in $Fe(CO)_n$- and $Ni(CO)_n$. *J. Am. Chem. Soc.* 114: 2788–2796.

72 Howle, C.R., Mayhew, C.A., and Tuckett, R.P. (2005). Selected ion flow tube study of the reactions between gas phase cations and $CHCl_2F$,

$CHClF_2$, and CH_2ClF. *Chem. A Eur. J.* 109 (16): 3626–3636.

73 Mikhailov, V.A., Parkes, M.A., Simpson, M.J. et al. (2008). Selected ion flow tube study of the ion-molecule reactions of monochloroethene, trichloroethene, and tetrachloroethene. *Chem. A Eur. J.* 112 (38): 9012–9022.

74 McIver, R.T. (1978). Pulsed ion-cyclotron resonance mass spectrometer for studying ion-molecule reactions. *Rev. Sci. Instrum.* 49 (1): 111–118.

75 Leito, I., Koppel, I.A., Burk, P. et al. (2010). Gas-phase basicities around and below water revisited. *Chem. A Eur. J.* 114 (39): 10694–10699.

76 Simpson, M.J. and Tuckett, R.P. (2011). Vacuum-UV negative photo-ion spectroscopy of gas phase polyatomic molecules. *Int. Rev. Phys. Chem.* 30 (2): 197–273.

77 (a) Kebarle, P. and Chowdhury, S. (1987). Electron affinities and electron transfer reactions. *Chem. Rev.* 87 (3): 513–534. Femtosecond real-time probing of reactions. 21 (b)Cheng, P.Y., Zhong, D., and Zewail, A.H. (1996). Direct observation of transition-state dynamics and structure in charge-transfer reactions. *J. Chem. Phys.* 105: 6216–6248.

78 Pauling, L. (1939). *The Nature of the Chemical Bond and the Structure of Molecules and Crystals: An Introduction to Modern Structural Chemistry*, 1–430. Ithaca, NY: Cornell University Press.

79 Grimme, S. (1996). Theoretical bond and strain energies of molecules derived from properties of the charge density at bond critical points. *J. Am. Chem. Soc.* 118 (6): 1529–1534.

80 Exner, K. and Schleyer, P.V. (2001). Theoretical bond energies: a critical evaluation. *Chem. A Eur. J.* 105 (13): 3407–3416.

81 Zavitsas, A.A. (2003). The relation between bond lengths and dissociation energies of carbon–carbon bonds. *Chem. A Eur. J.* 107 (6): 897–898.

82 Berkowitz, J., Ellison, G.B., and Gutman, D. (1994). 3 Methods to measure R—H bond-energies. *J. Phys. Chem.* 98 (11): 2744–2765.

83 Blanksby, S.J. and Ellison, G.B. (2003). Bond dissociation energies of organic molecules. *Acc. Chem. Res.* 36 (4): 255–263.

84 Holmes, J.L., Aubry, C., and Wang, X. (2007). Assessing thermochemical data. *Int. J. Mass Spectrom.* 267 (1–3): 263–267.

85 Ruscic, B., Boggs, J.E., Burcat, A. et al. (2005). IUPAC critical evaluation of thermochemical properties of selected radicals. Part I. *J. Phys. Chem. Ref. Data* 34 (2): 573–656.

86 Rossi, M., King, K.D., and Golden, D.M. (1979). Equilibrium constant and rate constant for allyl radical recombination in the gas-phase. *J. Am. Chem. Soc.* 101 (5): 1223–1230.

87 Chupka, W.A. and Russell, M.R. (1968). Photoionization study of ion-molecule reactions in mixtures of hydrogen and rare gases. *J. Chem. Phys.* 49 (12): 5426–5437.

88 McMillen, D.F. and Golden, D.M. (1982). Hydrocarbon bond dissociation energies. *Annu. Rev. Phys. Chem.* 33: 493–532.

89 Hiatt, R. and Benson, S.W. (1972). New method for measuring ratios of rate constants for radical recombination. *J. Am. Chem. Soc.* 94 (1): 25–29.

90 (a) Tsang, W. (1996). Heats of formation of organic free radicals and kinetic methods. In: *Energetics of Organic Free Radicals*, Part of The Structure Energetics and Reactivity in Chemistry (SEARCH) Series, vol. 4 (ed. J.A.M. Simões, A. Greenberg and J.F. Liebman), 32–58. Dordrecht, The Netherlands: Springer. (b) Traeger, J.C. and Kompe, B.M. (1996). Thermochemical data for free radicals from studies of ions. In: *Energetics of Organic Free Radicals*, Part of The Structure Energetics and Reactivity in Chemistry (SEARCH) Series, vol. 4 (ed. J.A.M. Simões, A. Greenberg and J.F. Liebman), 59–109. Dordrecht, The Netherlands: Springer. (c) Holmes, J.L. and Aubry, C. (2014). Methods for critically assessing old and for estimating new organic gas-phase neutral and ion thermochemical data. A user's guide. *Int. Rev. Phys. Chem.* 33 (2): 209–228.

91 Castelhano, A.L., Marriott, P.R., and Griller, D. (1981). Heats of formation of tert-butyl and ethyl radicals. *J. Am. Chem. Soc.* 103 (14): 4262–4263.

92 Laarhoven, L.J.J., Mulder, P., and Wayner, D.D.M. (1999). Determination of bond dissociation enthalpies in solution by photoacoustic calorimetry. *Acc. Chem. Res.* 32 (4): 342–349.

93 Peters, K.S. (1994). Time-resolved photoacoustic calorimetry - from carbenes to proteins. *Angew. Chem. Int. Ed. Eng.* 33 (3): 294–302.

94 Muralha, V.S.F., dos Santos, R.M.B., and Simões, J.A.M. (2004). Energetics of alkylbenzyl radicals: a time-resolved photoacoustic calorimetry study. *Chem. A Eur. J.* 108 (6): 936–942.

95 Bordwell, F.G., Cheng, J.P., Ji, G.Z. et al. (1991). Bond-dissociation energies in DMSO related to the gas phase. *J. Am. Chem. Soc.* 113 (26): 9790–9795.

96 (a) Bordwell, F.G. and Cheng, J.P. (1991). Substituent effects on the stabilities of phenoxyl radicals and the acidities of phenoxyl radical cations. *J. Am. Chem. Soc.* 113 (5): 1736–1743.

(b) Wu, Y.D. and Lai, D.K.W. (1996). A density functional study of substituent effects on the O—H and O—CH_3 bond dissociation energies in phenol and anisole. *J. Org. Chem.* 61 (22): 7904–7910.

97 Bordwell, F.G. and Liu, W.Z. (1996). Solvent effects on homolytic bond dissociation energies of hydroxylic acids. *J. Am. Chem. Soc.* 118 (44): 10819–10823.

98 Ingold, K.U. and Wright, J.S. (2000). Understanding trends in C—H, N—H, and O—H bond dissociation enthalpies. *J. Chem. Educ.* 77 (8): 1062–1064.

99 (a) Wodrich, M.D., McKee, W., and Schleyer, P.v.R. (2011). On the advantages of hydrocarbon radical stabilization energies based on R—H bond dissociation energies. *J. Org. Chem.* 76 (8): 2439–2447. (b) Hudzik, J.M., Bozzelli, J.W., and Simmie, J.M. (2014). Thermochemistry of C_7H_{16} to $C_{10}H_{22}$ alkane isomers: primary, secondary, and tertiary C—H bond dissociation energies and effects of branching. *Chem. A Eur. J.* 118 (40): 9364–9379.

100 Luo, Y.-R. (2007). *Comprehensive Handbook of Chemical Bond Energies*, 1–1688. Boca Raton, FL: CRC Press.

101 Rüchardt, C. and Beckhaus, H.D. (1986). Steric and electronic substituent effects on the Carbon–Carbon bond. In: *Synthetic Organic Chemistry*, Topics in Current Chemistry, vol. 130 (ed. F.L. Boschke), 1–22. Berlin, Heidelberg: Springer-Verlag.

102 Welle, F.M., Beckhaus, H.D., and Rüchardt, C. (1997). Thermochemical stability of α-amino-α-carbonylmethyl radicals and their resonance as measured by ESR. *J. Org. Chem.* 62 (3): 552–558.

103 Brocks, J.J., Beckhaus, H.D., Beckwith, A.L.J., and Rüchardt, C. (1998). Estimation of bond dissociation energies and radical stabilization energies by ESR spectroscopy. *J. Org. Chem.* 63 (6): 1935–1943.

104 Traeger, J.C. and McLoughlin, R.G. (1981). Absolute heats of formation for gas phase cations. *J. Am. Chem. Soc.* 103 (13): 3647–3652.

105 Luo, Y.-R. and Holmes, J.L. (1994). Homolytic and heterolytic X—C bond energies. 1. Homolytic bond energies in common unsaturated organic compounds. *J. Phys. Chem.* 98 (1): 303–312.

106 Armentrout, P.B. and Simons, J. (1992). Understanding heterolytic bond cleavage. *J. Am. Chem. Soc.* 114 (22): 8627–8633.

107 Tolbert, M.A. and Beauchamp, J.L. (1986). Homolytic and heterolytic bond dissociation energies of the 2nd-row Group-8, Group-9, and Group-10 diatomic transition-metal hydrides - correlation with electronic structure. *J. Phys. Chem.* 90 (21): 5015–5022.

108 Luo, Y.-R. and Pacey, P.D. (1991). Heterolytic energies of X-alkyl bonds. *J. Phys. Chem.* 95 (23): 9470–9476.

109 McMahon, T.B. (2000). Thermochemical ladders: scaling the ramparts of gaseous ion energetics. *Int. J. Mass Spectrom.* 200 (1-3): 187–199.

110 (a) Szulejko, J.E. and McMahon, T.B. (1993). Progress toward an absolute gas phase proton affinity scale. *J. Am. Chem. Soc.* 115 (17): 7839–7848. (b) Hunter, E.P.L. and Lias, S.:.G. (1998). Evaluated gas phase basicities and proton affinities of molecules. An up-date. *J. Phys. Chem. Ref. Data* 27 (3): 413–656. (c) Moser, A., Range, K., and York, D.M. (2010). Accurate proton affinity and gas phase basicity values for molecules important in bio-catalysis. *J. Phys. Chem. B* 114 (43): 13911–13921.

111 Houriet, R., Schwarz, H., and Zummack, W. (1980). Proton and hydrogen affinity of furan and the site of protonation in the gas-phase. *Angew. Chem. Int. Ed. Eng.* 19 (11): 905–906.

112 Yamdagni, R. and Kebarle, P. (1976). Gas-phase basicities and proton affinities of compounds between water and ammonia and substituted benzenes from a continuous ladder of proton-transfer equilibria measurements. *J. Am. Chem. Soc.* 98 (6): 1320–1324.

113 (a) Meot-Ner, M. and Sieck, L.W. (1991). Proton affinity ladders from variable-temperature equilibrium measurements. 1. A evaluation of the upper proton affinity range. *J. Am. Chem. Soc.* 113 (12): 4448–4460. (b) Meot-Ner, M. (2003). The proton affinity scale, and effects of ion structure and solvation. *Int. J. Mass Spectrom.* 227 (3): 525–554.

114 Laude, D.A. Jr., Johlman, C.L., Brown, R.S. et al. (1986). Fourier-transform mass spectrometry - recent instrumental developments and applications. *Mass Spectrom. Rev.* 5 (2): 107–166.

115 Russell, D.H. (1986). An evaluation of Fourier-transform mass spectrometry for high mass applications. *Mass Spectrom. Rev.* 5 (2): 167–189.

116 Baykut, G. and Eyler, J.R. (1986). Fourier-transform ion cyclotron resonance mass-spectrometry. *TrAC Trends Anal. Chem.* 5 (2): 44–49.

117 Lias, S.G., Shold, D.M., and Ausloos, P. (1980). Proton-transfer reactions involving alkyl ions and alkenes - rate constants, isomerization processes,

and the derivation of thermochemical data. *J. Am. Chem. Soc.* 102 (8): 2540–2548.

118 Aue, D.H., Davidson, W.R., and Bowers, M.T. (1976). Heats of formation of C3H5+ ions - allyl, vinyl, and cyclopropyl cations in gas-phase proton-transfer reactions. *J. Am. Chem. Soc.* 98 (21): 6700–6702.

119 Kebarle, P. (2000). Gas phase ion thermochemistry based on ion equilibria from the ionosphere to the reactive centers of enzymes. *Int. J. Mass Spectrom.* 200 (1–3): 313–330.

120 Ervin, K.M. (2001). Experimental techniques in gas phase ion thermochemistry. *Chem. Rev.* 101 (2): 391–444.

121 Holmes, J.L. and Aubry, C. (2009). Neutral and ion thermochemistry: its present status and significance. *Mass Spectrom. Rev.* 28 (4): 694–700.

122 Wolf, J.F., Staley, R.H., Koppel, I. et al. (1977). Gas-phase basicities and relative proton affinities of compounds between water and ammonia from pulsed ion–cyclotron resonance thermal equilibria measurements. *J. Am. Chem. Soc.* 99 (16): 5417–5429.

123 Bouchoux, G. (2007). Gas-phase basicities of polyfunctional molecules. Part 1: Theory and methods. *Mass Spectrom. Rev.* 26 (6): 775–835.

124 Lossing, F.P. and Holmes, J.L. (1984). Stabilization energy and ion size in carbocations in the gas-phase. *J. Am. Chem. Soc.* 106 (23): 6917–6920.

125 Mormann, M. and Kuck, D. (2003). Hydride abstraction from 1,3,5-cycloheptatriene by gaseous carbenium ions, as studied by Fourier transform ion cyclotron resonance kinetics and deuterium labeling. *J. Phys. Org. Chem.* 16 (10): 746–752.

126 Müller, P., Milin, D., Feng, W.G.Q. et al. (1992). The stability of bridgehead and rigid tertiary carbenium ions in the gas phase and their rate of formation in solvolysis reactions. *J. Am. Chem. Soc.* 114 (15): 6169–6172.

127 Cooks, R.G., Patrick, J.S., Kotiaho, T., and Mcluckey, S.A. (1994). Thermochemical determinations by the kinetic method. *Mass Spectrom. Rev.* 13 (4): 287–339.

128 Cooks, R.G., Koskinen, J.T., and Thomas, P.D. (1999). Special feature: commentary - the kinetic method of making thermochemical determinations. *J. Mass Spectrom.* 34 (2): 85–92.

129 Boronat, M., Viruela, P., and Corma, A. (1998). A theoretical study of the mechanism of the hydride transfer reaction between alkanes and alkenes catalyzed by an acidic zeolite. *Chem. A Eur. J.* 102 (48): 9863–9868.

130 Meot-Ner, M. and Field, F.H. (1978). Correlation between exothermicity, rate, and negative temperature-dependence in slow ion-molecule reactions. *J. Am. Chem. Soc.* 100 (5): 1356–1359.

131 Meot-Ner, M., Solomon, J.J., and Field, F.H. (1976). Stability of some C7 tertiary alkyl carbonium ions. *J. Am. Chem. Soc.* 98 (4): 1025–1026.

132 Solomon, J.J. and Field, F.H. (1975). Reversible reactions of gaseous ions. 9. Stability of C_4-C_7 tertiary alkyl carbonium ions. *J. Am. Chem. Soc.* 97 (10): 2625–2628.

133 Goren, A. and Munson, B. (1976). Thermochemistry of alkyl ions. *J. Phys. Chem.* 80 (26): 2848–2854.

134 Williamson, A.D. and Beauchamp, J.L. (1975). Ion-molecule reactions of NO+ with organic molecules by ion–cyclotron resonance spectroscopy. *J. Am. Chem. Soc.* 97 (20): 5714–5718.

135 Holznagel, C.M., Bakhtiar, R., and Jacobson, D.B. (1991). Elucidating the structures of isomeric silylenium ions ($SiC_3H_9^+$, $SiC_4H_{11}^+$, $SiC_5H_{13}^+$) by using specific ion molecule reactions. *J. Am. Soc. Mass Spectrom.* 2 (4): 278–291.

136 Ervin, K.M. and Deturo, V.F. (2002). Anchoring the gas phase acidity scale. *Chem. A Eur. J.* 106 (42): 9947–9956.

137 Kebarle, P. (1977). Ion thermochemistry and solvation from gas phase ion equilibria. *Annu. Rev. Phys. Chem.* 28: 445–476.

138 Goodloe, G.W. and Lampe, F.W. (1979). Reactions of methyl cation with ethylsilanes. *J. Am. Chem. Soc.* 101 (19): 5649–5653.

139 Staley, R.H., Wieting, R.D., and Beauchamp, J.L. (1977). Carbenium ion stabilities in gas phase and solution - ion–cyclotron resonance study of bromide transfer reactions involving alkali ions, alkyl carbenium ions, acyl cations, and cyclic halonium ions. *J. Am. Chem. Soc.* 99 (18): 5964–5972.

140 Cumming, J.B. and Kebarle, P. (1978). Summary of gas phase acidity measurements involving acids A—H - entropy changes in proton transfer reactions involving negative-ions - bond dissociation energies DH(A—H) and electron affinities EA(A). *Can. J. Chem.* 56 (1): 1–9.

141 (a) Harrison, A.G. (1992). *Chemical Ionization Mass Spectrometry*, 2e, 1–224. Baco Raton, FL: CRC Press. (b) O'Hair, R.A.J. (2016). Chemical ionization mass spectrometry: 50 years on. *J. Am. Soc. Mass Spectrom.* 27 (11): 1787–1788.

142 (a) DePuy, C.H. and Bierbaum, V.M. (1981). Gas-phase reactions of organic anions as studied by the flowing afterglow technique. *Acc. Chem. Res.* 14 (5): 146–153. (b) DePuy, C.H., Grabowski,

J.J., and Bierbaum, V.M. (1982). Chemical reactions of anions in the gas phase. *Science* 218 (4576): 955–960. (c)DePuy, C.H. (2002). Understanding organic gas phase anion molecule reactions. *J. Org. Chem.* 67 (8): 2393–2401.

143 DePuy, C.H., Gronert, S., Barlow, S.E. et al. (1989). The gas phase acidities of the alkanes. *J. Am. Chem. Soc.* 111 (6): 1968–1973.

144 Bernasconi, C.F., Stronach, M.W., DePuy, C.H., and Gronert, S. (1990). Reaction of anions with activated olefins in the gas phase - a flowing afterglow selected ion flow tube study. *J. Am. Chem. Soc.* 112 (25): 9044–9052.

145 Davico, G.E., Bierbaum, V.M., DePuy, C.H. et al. (1995). The C—H bond energy of benzene. *J. Am. Chem. Soc.* 117 (9): 2590–2599.

146 Robinson, M.S., Polak, M.L., Bierbaum, V.M. et al. (1995). Experimental studies of allene, methylacetylene, and the propargyl radical - bond-dissociation energies, gas phase acidities, and ion-molecule chemistry. *J. Am. Chem. Soc.* 117 (25): 6766–6778.

147 Ellison, G.B., Davico, G.E., Bierbaum, V.M., and DePuy, C.H. (1996). Thermochemistry of the benzyl and allyl radicals and ions. *Int. J. Mass Spectrom. Ion Process.* 156 (1–2): 109–131.

148 Lee, H.S., DePuy, C.H., and Bierbaum, V.M. (1996). Reactivity and thermochemistry of quadricyclane in the gas phase. *J. Am. Chem. Soc.* 118 (21): 5068–5073.

149 Squires, R.R. (1992). Gas phase carbanion chemistry. *Acc. Chem. Res.* 25 (10): 461–467.

150 Graul, S.T. and Squires, R.R. (1990). Gas phase acidities derived from threshold energies for activated reactions. *J. Am. Chem. Soc.* 112 (7): 2517–2529.

151 Brinkman, E.A., Salomon, K., Tumas, W., and Brauman, J.I. (1995). Electron affinities and gas phase acidities of organogermanium and organotin compounds. *J. Am. Chem. Soc.* 117 (17): 4905–4910.

152 Wetzel, D.M., Salomon, K.E., Berger, S., and Brauman, J.I. (1989). Gas phase acidities of organosilanes and electron affinities of organosilyl radicals. *J. Am. Chem. Soc.* 111 (11): 3835–3841.

153 Irie, M., Kikukawa, K., Shimizu, N., and Mishima, M. (2001). Gas-phase acidities of aryldimethylsilanes. *J. Chem. Soc., Perkin Trans. 2* (6): 923–928.

154 Holmes, J.L., Fingas, M., and Lossing, F.P. (1981). Towards a general scheme for estimating the heats of formation of organic ions in the gas phase. 1. Odd-electron cations. *Can. J. Chem.* 59 (1): 80–93.

155 Paulino, J.A. and Squires, R.R. (1991). Carbene thermochemistry from collision-induced dissociation threshold energy measurements - the heats of formation of X^1A_1 CF_2 and X^1A_1 CCl_2. *J. Am. Chem. Soc.* 113 (15): 5573–5580.

156 Poutsma, J.C., Paulino, J.A., and Squires, R.R. (1997). Absolute heats of formation of CHCl, CHF, and CClF. A gas phase experimental and G2 theoretical study. *Chem. A Eur. J.* 101 (29): 5327–5336.

157 Poutsma, J.C., Nash, J.J., Paulino, J.A., and Squires, R.R. (1997). Absolute heats of formation of phenylcarbene and vinylcarbene. *J. Am. Chem. Soc.* 119 (20): 4686–4697.

158 Jesinger, R.A. and Squires, R.R. (1999). Carbyne thermochemistry from energy-resolved collision-induced dissociation. The heats of formation of CH, CF, and CCl. *Int. J. Mass Spectrom.* 185: 745–757.

159 Wenthold, P.G., Hu, J., and Squires, R.R. (1996). *o*-, *m*-, and *p*-benzyne negative ions in the gas phase: synthesis, authentication, and thermochemistry. *J. Am. Chem. Soc.* 118 (47): 11865–11871.

160 Wenthold, P.G., Squires, R.R., and Lineberger, W.C. (1998). Ultraviolet photoelectron spectroscopy of the *o*-, *m*-, and *p*-benzyne negative ions. Electron affinities and singlet-triplet splittings for *o*-, *m*-, and *p*-benzyne. *J. Am. Chem. Soc.* 120 (21): 5279–5290.

161 Diau, E.W.G., Casanova, J., Roberts, J.D., and Zewail, A.H. (2000). Femtosecond observation of benzyne intermediates in a molecular beam: Bergman rearrangement in the isolated molecule. *Proc. Natl. Acad. Sci. U. S. A.* 97 (4): 1376–1379.

162 Clifford, E.P., Wenthold, P.G., Lineberger, W.C. Jr. et al. (1998). Properties of tetramethyleneethane (TME) as revealed by ion chemistry and ion photoelectron spectroscopy. *J. Chem. Soc., Perkin Trans. 2* (5): 1015–1022.

163 Schröder, D., Goldberg, N., Zummack, W. et al. (1997). Generation of α-acetolactone and the acetoxyl diradical $^{\bullet}CH_2COO^{\bullet}$ in the gas phase. *Int. J. Mass Spectrom.* 165/166: 71–82.

164 Toullec, J. (1982). Enolization of simple carbonyl compounds and related reactions. *Adv. Phys. Org. Chem.* 18: 1–77.

165 Turecek, F., Brabec, L., and Korvola, J. (1988). Unstable enols in the gas phase. Preparation, ionization energies, and heats of formation of (*E*)-2-buten-2-ol and (*Z*)-2-buten-2-ol, 2-methyl-1-propen-1-ol, and 3-methyl-2-buten-2-ol. *J. Am. Chem. Soc.* 110 (24): 7984–7990.

166 Keeffe, J.R., Kresge, A.J., and Schepp, N.P. (1990). Keto-enol equilibrium constants of simple monofunctional aldehydes and ketones in aqueous solution. *J. Am. Chem. Soc.* 112 (12): 4862–4868.
167 Pollack, S.K. and Hehre, W.J. (1977). Enol of acetone. *J. Am. Chem. Soc.* 99 (14): 4845–4846.
168 Capponi, M., Gut, I., and Wirz, J. (1986). The phenol reversible 2,4-cyclohexadienone equilibrium in aqueous solution. *Angew. Chem. Int. Ed. Eng.* 25 (4): 344–345.
169 Zhu, L. and Bozzelli, J.W. (2003). Kinetics and thermochemistry for the gas phase keto-enol tautomerism of phenol = 2,4-cyclohexadienone. *Chem. A Eur. J.* 107 (19): 3696–3703.
170 Fattahi, A., Lis, L., Tian, Z.X., and Kass, S.R. (2006). The heat of formation of cyclobutadiene. *Angew. Chem. Int. Ed.* 45 (30): 4984–4988.
171 Zhang, D.Y. and Borden, W.T. (2002). Do deviations from bond enthalpy additivity define the thermodynamic stabilities of diradicals? *J. Org. Chem.* 67 (12): 3989–3995.
172 Roth, W.R., Wasser, T., Zimmermann, H. et al. (1996). 1,4-didehydronaphthalin. *Liebigs Ann.* (10): 1691–1695.
173 Pearson, R.G. (1987). Recent advances in the concept of hard and soft acids and bases. *J. Chem. Educ.* 64 (7): 561–567.
174 Pearson, R.G. (1989). Absolute electronegativity and hardness - applications to organic chemistry. *J. Org. Chem.* 54 (6): 1423–1430.
175 Pearson, R.G. (1991). Absolute electronegativity and hardness. *Chem. Br.* 27 (5): 444–447.
176 Reed, J.L. (1997). Electronegativity: chemical hardness. 1. *Chem. A Eur. J.* 101 (40): 7396–7400.
177 Pearson, R.G. (1988). Absolute electronegativity and hardness - application to inorganic chemistry. *Inorg. Chem.* 27 (4): 734–740.
178 Pearson, R.G. (1988). Chemical hardness and bond dissociation energies. *J. Am. Chem. Soc.* 110 (23): 7684–7690.
179 Pearson, R.G. (1995). The HSAB principle - more quantitative aspects. *Inorg. Chim. Acta* 240 (1-2): 93–98.
180 Pearson, R.G. (2005). Chemical hardness and density functional theory. *J. Chem. Sci.* 117 (5): 369–377.
181 Mayr, H., Breugst, M., and Ofial, A.R. (2011). Farewell to the HSAB treatment of ambident reactivity. *Angew. Chem. Int. Ed.* 50 (29): 6470–6505.
182 Ondetti, M.A., Rubin, B., and Cushman, D.W. (1977). Design of specific inhibitors of angiotensin-converting enzyme: new class of orally active antihypertensive agents. *Science* 196 (4288): 441–444.
183 Meskens, F.A.J. (1981). Methods for the preparation of acetals from alcohols or oxiranes and carbonyl compounds. *Synthesis* (7): 501–522.
184 Sevin, A.F. and Vogel, P. (1994). A new stereoselective and convergent approach to the synthesis of long-chain polypropionate fragments. *J. Org. Chem.* 59 (20): 5920–5926.
185 Ensley, H.E., Parnell, C.A., and Corey, E.J. (1978). Convenient synthesis of a highly efficient and recyclable chiral director for asymmetric induction. *J. Org. Chem.* 43 (8): 1610–1612.
186 Gnas, Y. and Glorius, F. (2006). Chiral auxiliaries - principles and recent applications. *Synthesis* (12): 1899–1930.
187 Evans, D.A., Helmchen, G., and Rüping, M. (2007). Chiral auxiliaries in asymmetric synthesis. In: *Asymmetric Synthesis - The Essentials*, 2e (ed. M. Christmann and S. Bräse), 3–9. Weinheim: Wiley-VCH.
188 Cahn, R.S., Ingold, C., and Prelog, V. (1966). Specification of molecular chirality. *Angew. Chem. Int. Ed. Eng.* 5 (4): 385–415.
189 Prelog, V. and Helmchen, G. (1982). Basic principles of the CIP-system and proposals for a revision. *Angew. Chem. Int. Ed. Eng.* 21 (8): 567–583.
190 Weickgenannt, A. and Oestreich, M. (2012). The Renaissance of silicon-stereogenic silanes: a personal account. In: *Asymmetric Synthesis II: More Methods and Applications* (ed. M. Christmann and S. Bräse), 35–42. Weinheim: Wiley-VCH.
191 Prelog, V. and Wieland, P. (1944). Concerning the fission of the Tröger base in optical antipodes, an article on the stereochemistry of triple value nitrogen. *Helv. Chim. Acta* 27: 1127–1134.
192 Pietrusiewicz, K.M. and Zablocka, M. (1994). Preparation of scalemic P-chiral phosphines and their derivatives. *Chem. Rev.* 94 (5): 1375–1411.
193 Horner, L. and Fuchs, H. (1962). Optisch Aktive tertiäre Arsine aus optisch aktiven Quartären Arsoniumsalzen. *Tetrahedron Lett.* 3 (5): 203–204.
194 Yambushev, F.D. and Savin, V.I. (1979). Stereochemistry of organoarsenic compounds. *Usp. Khim.* 48 (6): 1093–1118.
195 Yamamoto, Y., Chen, X., and Akiba, K. (1992). Synthesis and crystal structure of intramolecularly coordinated organobismuth compounds and edge inversion at trivalent bismuth. *J. Am. Chem. Soc.* 114 (20): 7906–7907.
196 Drabowicz, J., Kielbasinski, P., Krasowska, D., and Kikolojczyk, M. (2009). Asymmetric synthesis of optically active sulfinic acid derivatives. In:

Organosulfur Chemistry in Asymmetric Synthesis (ed. T. Toru and C. Bolm), 31–54. Weinheim: Wiley-VCH.
197 Carreno, M.C. (1995). Applications of sulfoxides to asymmetric synthesis of biologically active compounds. *Chem. Rev.* 95 (6): 1717–1760.
198 Pellissier, H. (2006). Use of chiral sulfoxides in asymmetric synthesis. *Tetrahedron* 62 (24): 5559–5601.
199 Kagan, H.B. (2009). Chiral sulfoxides. In: *Organosulfur Chemistry in Asymmetric Synthesis* (ed. T. Toru and C. Bolm), 1–29. Weinheim: Wiley-VCH.
200 Senanayake, C.H., Han, X., and Kirshnamurphy, D. (2009). Synthesis and use of chiral sulfinamides. In: *Organosulfur Chemistry in Asymmetric Synthesis* (ed. T. Toru and C. Bolm), 233–264. Weinheim: Wiley-VCH.
201 Worch, C., Mayer, A.C., and Bolm, C. (2009). Synthesis and use of chiral sulfoximines. In: *Organosulfur Chemistry in Asymmetric Synthesis* (ed. T. Toru and C. Bolm), 209–232. Weinheim: Wiley-VCH.
202 Brière, J.-F. and Metzner, P. (2009). Synthesis and use of chiral sulfur ylides. In: *Organosulfur Chemistry in Asymmetric Synthesis* (ed. T. Toru and C. Bolm), 179–208. Weinheim: Wiley-VCH.
203 Bentley, R. (2005). Role of sulfur chirality in the chemical processes of biology. *Chem. Soc. Rev.* 34 (7): 609–624.
204 Knight, P.D. and Scott, P. (2003). Predetermination of chirality at octahedral centres with tetradentate ligands: prospects for enantioselective catalysis. *Coord. Chem. Rev.* 242 (1–2): 125–143.
205 Bauer, E.B. (2012). Chiral-at-metal complexes and their catalytic applications in organic synthesis. *Chem. Soc. Rev.* 41 (8): 3153–3167.
206 Crassous, J. (2009). Chiral transfer in coordination complexes: towards molecular materials. *Chem. Soc. Rev.* 38 (3): 830–845.
207 Crassous, J. (2012). Transfer of chirality from ligands to metal centers: recent examples. *J. Chem. Soc. Chem. Commun.* 48 (78): 9684–9692.
208 Ono, Y. (2003). A survey of the mechanism in catalytic isomerization of alkanes. *Catal. Today* 81 (1): 3–16.
209 Olah, G.A., Mathew, T., Marinez, E.R. et al. (2001). Acid-catalyzed isomerization of pivalaldehyde to methyl isopropyl ketone via a reactive protosolvated carboxonium ion intermediate. *J. Am. Chem. Soc.* 123 (47): 11556–11561.
210 Moss, G.P. (1996). Basic terminology of stereochemistry. *Pure Appl. Chem.* 68 (12): 2193–2222.
211 Mason, S.F. and Tranter, G.E. (1984). The parity-violating energy difference between enantiomeric molecules. *Mol. Phys.* 53 (5): 1091–1111.
212 Ulbricht, T.L.V. (1981). Reflections on the origin of optical asymmetry on earth. *Orig. Life Evol. Biosph.* 11 (1–2): 55–70.
213 Mason, S.F. (1989). The development of concepts of chiral discrimination. *Chirality* 1 (3): 183–191.
214 Feringa, B.L. and van Delden, R.A. (1999). Absolute asymmetric synthesis: the origin, control, and amplification of chirality. *Angew. Chem. Int. Ed.* 38 (23): 3419–3438.
215 Nguyen, L.A., He, H., and Chuong, P.-H. (2006). Chiral drugs. An overview. *Int. J. Biomed. Sci. (Monterey Park, CA, USA)* 2 (2): 85–100.
216 Davies, N.M. and Teng, X.W. (2003). Importance of chirality in drug therapy and pharmacy practice. Implication for psychiatry. *Adv. Pharm.* 1 (3): 242–252.
217 Teo, S.K., Colburn, W.A., Tracewell, W.G. et al. (2004). Clinical pharmacokinetics of thalidomide. *Clin. Pharmacokinet.* 43 (5): 311–327.
218 Wiffer, P.J., Wee, B., Derry, S. et al. (2017). Opioids for cancer pain – an overview of Cochran Reviews. *Cochrane Database Syst. Rev.* 7: 1–22. (Art. No.: CD0112592).
219 Wong, B.Y., Coulter, D.A., Choi, D.W., and Prince, D.A. (1988). Dextrorphan and dextromethorphan, common antitussives, are antiepileptic and antagonize N-methyl-d-aspartate in brain-slices. *Neurosci. Lett.* 85 (2): 261–266.
220 Black, J.W., Crowther, A.F., Shanks, R.G. et al. (1964). A new adrenergic beta receptor antagonist. *Lancet* 1 (7342): 1080–1081.
221 Leitereg, T.J., Guadagni, D.G., Harris, J. et al. (1971). Evidence for difference between odours of optical isomers (+)-carvone and (-)-carvone. *Nature* 230 (5294): 455–456.
222 Gal, J. (2012). The discovery of stereoselectivity at biological receptors: Arnaldo Piutti and the taste of the asparagine enantiomers - history and analysis on the 125th anniversary. *Chirality* 24 (12): 959–976.
223 Hauser, R.A. (2009). Levodopa: past, present, and future. *Eur. Neurol.* 62 (1): 1–8.
224 Müller, T., van Laar, T., Cornblath, D.R. et al. (2013). Peripheral neuropathy in Parkinson's disease: levodopa exposure and implications for duodenal delivery. *Parkinsonism Relat. Disord.* 19 (5): 501–507.
225 Haüy, R.J. (1801). *Traité de Minéralogie*, vol. 5. Paris: Chez Louis.

226 Kahr, B. and Claborn, K. (2008). The lives of malus and his bicentennial law. *ChemPhysChem* 9 (1): 43–58.

227 Crosland, M.P. and Biot, J.-B. (2008). *Dictionary of Scientific Biography*, vol. 2, 133–140. New York: Charkes Scribner's Sons.

228 Lin, G.-Q., Li, Y.-M., and Chan, A.S.C. (2001). *Principles and Applications of Asymmetric Synthesis*, 1–507. New York: Wiley.

229 Gawley, R.E. and Aubé, J. (2012). *Principles of Asymmetric Synthesis*, 2e, 1–555. Amsterdam, The Netherlands: Elsevier Ltd.

230 Pirkle, W.H. and Pochapsky, T.C. (1989). Considerations of chiral recognition relevant to the liquid-chromatographic separation of enantiomers. *Chem. Rev.* 89 (2): 347–362.

231 Francotte, E. and Junkerbuchheit, A. (1992). Preparative chromatographic separation of enantiomers. *J. Chromatogr. B Biomed. Appl.* 576 (1): 1–45.

232 Gubitz, G. and Schmid, M.G. (2001). Chiral separation by chromatographic and electromigration techniques. A review. *Biopharm. Drug Dispos.* 22 (7–8): 291–336.

233 Schurig, V. (2002). Chiral separations using gas chromatography. *TrAC Trends Anal. Chem.* 21 (9–10): 647–661.

234 Turiel, E. and Martin-Esteban, A. (2004). Molecularly imprinted polymers: towards highly selective stationary phases in liquid chromatography and capillary electrophoresis. *Anal. Bioanal. Chem.* 378 (8): 1876–1886.

235 Bléhaut, J., Franco, P., Zhang, T. et al. (2012). Industrial applications of chiral chromatography. In: *Comprehensive Chirality*, vol. 9. (ed. E.M. Carreira and H. Yamamoto), 400–456. Amsterdam, The Netherlands: Elsevier.

236 Jac, P. and Scriba, G.K.E. (2013). Recent advances in electrodriven enantioseparations. *J. Sep. Sci.* 36 (1): 52–74.

237 Scriba, G.K.E. (2013). Chiral recognition in separation science: an overview. In: *Chiral Separations: Methods and Protocols*, Methods in Molecular Biology, 2e, vol. 970 (ed. G.K.A. Scriba), 1–27. New York: Humana Press.

238 Schurig, V. and Kreidler, D. (2013). Gas chromatographic enantioseparation of unfunctionalized chiral hydrocarbons: an overview. In: *Chiral Separations: Methods and Protocols*, Methods in Molecular Biology, 2e, vol. 970 (ed. G.K.A. Scriba), 45–67. New York: Humana Press.

239 Spivak, D.A. (2013). Enantioseparation by high-performance liquid chromatography using molecularly imprinted polymers. In: *Chiral Separations: Methods and Protocols*, Methods in Molecular Biology, 2e, vol. 970 (ed. G.K.A. Scriba), 209–220. New York: Humana Press.

240 Sheldon, R.A. (1996). Chirotechnology: designing economic chiral syntheses. *J. Chem. Technol. Biotechnol.* 67 (1): 1–14.

241 Kellogg, R.M. and Leeman, M. (2012). *Crystallization As a Tool in Industrial Applications of Asymmetric Synthesis*, vol. 9 (ed. E.M. Carreira and H. Yamamoto), 367–399. New York: Academic Press.

242 Crosby, J. (1991). Synthesis of optically active compounds - a large-scale perspective. *Tetrahedron* 47 (27): 4789–4846.

243 Collet, A. (1998). Resolution of racemates: did you say "classical"? *Angew. Chem. Int. Ed. Eng.* 37 (23): 3239–3241.

244 Faigl, F., Fogassy, E., Nogradi, M. et al. (2008). Strategies in optical resolution: a practical guide. *Tetrahedron Asymmetry* 19 (5): 519–536.

245 Pasteur, L. (1848). Mémoire sur a relation qui peut exister entre la forme crystalline et la composition chimique, et sur la cause de la polarisation rotatoire. *C. r. h. Séances Acad. Sci.* 26: 535–538.

246 Jacques, J., Collet, A., and Wilen, S.H. (1994). Resolution by direct crystallization. In: *Enantiomers, Racemates, and Resolution*, 217–250. Malabar, FL: Krieger Publishing Company.

247 Hager, O., Llamas-Saiz, A.L., Foces-Foces, C. et al. (1999). Complexes between 1,1'-binaphthyl-2,2'-dicarboxylic acid and pyrazoles: a case of manual sorting of conglomerate crystals (triage). *Helv. Chim. Acta* 82 (12): 2213–2230.

248 Profir, V.M. and Matsuoka, M. (2000). Processes and phenomena of purity decrease during the optical resolution of DL-threonine by preferential crystallization. *Colloids Surf. A Physicochem. Eng. Asp.* 164 (2–3): 315–324.

249 Lorenz, H., Polenske, D., and Seidel-Morgenstern, A. (2006). Application of preferential crystallization to resolve racemic compounds in a hybrid process. *Chirality* 18 (10): 828–840.

250 Avalos, M., Babiano, R., Cintas, P. et al. (1998). Absolute asymmetric synthesis under physical fields: facts and fictions. *Chem. Rev.* 98 (7): 2391–2404.

251 Cintas, P. (2008). On cavitation and chirality: a further assessment. *Cryst. Growth Des.* 8 (8): 2626–2627.

252 Song, Y., Chen, W., and Chen, X. (2008). Ultrasonic field induced chiral symmetry breaking of

$NaClO_3$ crystallization. *Cryst. Growth Des.* 8 (5): 1448–1450.
253 Medina, D.D., Gedanken, A., and Mastai, Y. (2011). Chiral amplification in crystallization under ultrasound radiation. *Chem. Eur. J.* 17 (40): 11139–11142.
254 Kondepudi, D.K., Kaufman, R.J., and Singh, N. (1990). Chiral symmetry-breaking in sodium chlorate crystallization. *Science* 250 (4983): 975–976.
255 Kondepudi, D.K., Laudadio, J., and Asakura, K. (1999). Chiral symmetry breaking in stirred crystallization of 1,1 '-binaphthyl melt. *J. Am. Chem. Soc.* 121 (7): 1448–1451.
256 Kondepudi, D.K. and Asakura, K. (2001). Chiral autocatalysis, spontaneous symmetry breaking, and stochastic behavior. *Acc. Chem. Res.* 34 (12): 946–954.
257 Viedma, C. (2001). Enantiomeric crystallization from DL-aspartic and DL-glutamic acids: implications for biomolecular chirality in the origin of life. *Orig. Life Evol. Biosph.* 31 (6): 501–509.
258 Viedma, C. (2005). Chiral symmetry breaking during crystallization: complete chiral purity induced by nonlinear autocatalysis and recycling. *Phys. Rev. Lett.* 94 (6): 65505–64508.
259 Viedma, C. (2007). Selective chiral symmetry breaking during crystallization: parity violation or cryptochiral environment in control? *Cryst. Growth Des.* 7 (3): 553–556.
260 Takanishi, Y., Takezoe, H., Suzuki, Y. et al. (1999). Spontaneous enantiomeric resolution in a fluid smectic phase of a racemate. *Angew. Chem. Int. Ed.* 38 (16): 2354–2356.
261 Kavasmaneck, P.R. and Bonner, W.A. (1977). Adsorption of amino acid derivatives by D-quartz and L-quartz. *J. Am. Chem. Soc.* 99 (1): 44–50.
262 Mason, S.F. (1987). Universal dissymmetry and the origin of biomolecular chirality. *Biosystems* 20 (1): 27–35.
263 Viswanathan, R., Zasadzinski, J.A., and Schwartz, D.K. (1994). Spontaneous chiral-symmetry breaking by achiral molecules in a Langmuir-Blodgett film. *Nature* 368 (6470): 440–443.
264 De Feyter, S., Grim, P.C.M., Rucker, M. et al. (1998). Expression of chirality by achiral coadsorbed molecules in chiral monolayers observed by STM. *Angew. Chem. Int. Ed. Eng.* 37 (9): 1223–1226.
265 Bohringer, M., Morgenstern, K., Schneider, W.D., and Berndt, R. (1999). Separation of a racemic mixture of two-dimensional molecular clusters by scanning tunneling microscopy. *Angew. Chem. Int. Ed.* 38 (6): 821–823.
266 Ohtani, B., Shintani, A., and Uosaki, K. (1999). Two-dimensional chirality: self-assembled monolayer of an atropisomeric compound covalently bound to a gold surface. *J. Am. Chem. Soc.* 121 (27): 6515–6516.
267 Paci, I., Szleifer, I., and Ratner, M.A. (2007). Chiral separation: mechanism modeling in two-dimensional systems. *J. Am. Chem. Soc.* 129 (12): 3545–3555.
268 Ernst, K.H. (2010). Amplification of chirality at solid surfaces. *Orig. Life Evol. Biosph.* 40 (1): 41–50.
269 Eckhardt, C.J., Peachey, N.M., Swanson, D.R. et al. (1993). Separation of chiral phases in monolayer crystals of racemic amphiphiles. *Nature* 362 (6421): 614–616.
270 Stevens, F., Dyer, D.J., and Walba, D.M. (1996). Direct observation of enantiomorphous monolayer crystals from enantiomers by scanning tunneling microscopy. *Angew. Chem. Int. Ed. Eng.* 35 (8): 900–901.
271 Kuhnle, A., Linderoth, T.R., and Besenbacher, F. (2003). Self-assembly of monodispersed, chiral nanoclusters of cysteine on the Au(110)-(1 × 2) surface. *J. Am. Chem. Soc.* 125 (48): 14680–14681.
272 Fasel, R., Parschau, M., and Ernst, K.H. (2003). Chirality transfer from single molecules into self-assembled monolayers. *Angew. Chem. Int. Ed.* 42 (42): 5178–5181.
273 Fasel, R., Parschau, M., and Ernst, K.H. (2006). Amplification of chirality in two-dimensional enantiomorphous lattices. *Nature* 439 (7075): 449–452.
274 Soai, K., Osanai, S., Kadowaki, K. et al. (1999). *d*- and *l*-quartz-promoted highly enantioselective synthesis of a chiral organic compound. *J. Am. Chem. Soc.* 121 (48): 11235–11236.
275 Pasteur, L. (1853). Transformation des acides tartriques en acide tartrique racémique. Découverte de l'acide tartrique inactif. Nouvelle méthode de séparation de l'acide racémique en acides tartriques droit et gauche. *Comptes rendus hebdomadiares des Séances de l'Académie des Sciences* 37: 162–166.
276 Pasteur, L. (1853). Recherches sur les alcaloïdes des quinquinas. *Comptes rendus hebdomadiares des Séances de l'Académie des Sciences* 37: 110–114.
277 Rabe, P. and Kindler, K. (1918). Concerning the partial synthesis of chinine - information on china-alkaloids XIX. *Ber. Dtsch. Chem. Ges.* 51: 466–467.
278 Smith, A.C. and Williams, R.M. (2008). Rabe rest in peace: confirmation of the Rabe-Kindler

conversion of *d*-quino-toxine into quinine: experimental affirmation of the Woodward-Doering formal total synthesis of quinine. *Angew. Chem. Int. Ed.* 47 (9): 1736–1740.

279 Pope, W.J. and Peachey, S.J. (1899). The application of powerful optically active acids to the resolution of externally compensated basic substances. Resolution of tetrahydroquinaldine. *J. Chem. Soc.* 75: 1066–1093.

280 Eliel, E.L., Wilen, S.H., and Mander, L.N. (1994). *Stereochemistry of Organic Compounds*, 1–1190. New York: Wiley.

281 Cervinka, O. (1995). Resolution of racemates into enantiomers. In: *Enantioselective Reactions in Enantioselective Chemistry*, 5–16. London: Ellis Horwood.

282 Schuur, B., Verkuijl, B.J.V., Minnaard, A.J. et al. (2011). Chiral separation by enantioselective liquid–liquid extraction. *Org. Biomol. Chem.* 9 (1): 36–51.

283 Sunsandee, N., Rashatasakhon, P., Ramakul, P. et al. (2014). Selective enantioseparation of racemic amlodipine by biphasic recognition chiral separation system. *Sep. Sci. Technol.* 49 (9): 1357–1365.

284 West, C. (2014). Enantioselective separations with supercritical fluids - review. *Curr. Anal. Chem.* 10 (1): 99–120.

285 Barlow, S.M. and Raval, R. (2003). Complex organic molecules at metal surfaces: bonding, organisation and chirality. *Surf. Sci. Rep.* 50 (6–8): 201–341.

286 Karagunis, G. and Coumoulos, G. (1938). A new method of resolving a racemic compound. *Nature* 142: 162–163.

287 Bonner, W.A., Kavasman, P.R., Martin, F.S., and Flores, J.J. (1974). Asymmetric adsorption of alanine by quartz. *Science* 186 (4159): 143–144.

288 Bonner, W.A. and Kavasmaneck, P.R. (1976). Asymmetric adsorption of DL-alanine hydrochloride by quartz. *J. Org. Chem.* 41 (12): 2225–2226.

289 Hazen, R.M., Filley, T.R., and Goodfriend, G.A. (2001). Selective adsorption of L- and D-amino acids on calcite: implications for biochemical homochirality. *Proc. Natl. Acad. Sci. U. S. A.* 98 (10): 5487–5490.

290 Ortega-Lorenzo, M., Baddeley, C.J., Muryn, C., and Raval, R. (2000). Extended surface chirality from supramolecular assemblies of adsorbed chiral molecules. *Nature* 404 (6776): 376–379.

291 Behzadi, B., Romer, S., Fasel, R., and Ernst, K.H. (2004). Chiral recognition in surface explosion. *J. Am. Chem. Soc.* 126 (30): 9176–9177.

292 Yun, Y. and Gellman, A.J. (2013). Enantioselective separation on naturally chiral metal surfaces: d,l-aspartic acid on Cu(3,1,17)$^{R\&S}$ surfaces. *Angew. Chem. Int. Ed.* 52 (12): 3394–3397.

293 Pasteur, L. (1858). Mémoire sur la fermentation de l'acide tartrique. *C. r. h. Séances Acad. Sci.* 46: 615–618.

294 Keith, J.M., Larrow, J.F., and Jacobsen, E.N. (2001). Practical considerations in kinetic resolution reactions. *Adv. Synth. Catal.* 343 (1): 5–26.

295 Pellissier, H. (2011). Catalytic non-enzymatic kinetic resolution. *Adv. Synth. Catal.* 353 (10): 1613–1666.

296 Noyori, R., Tokunaga, M., and Kitamura, M. (1995). Stereoselective organic synthesis via dynamic kinetic resolution. *Bull. Chem. Soc. Jpn.* 68 (1): 36–55.

297 Ward, R.S. (1995). Dynamic kinetic resolution. *Tetrahedron Asymmetry* 6 (7): 1475–1490.

298 Huerta, F.F., Minidis, A.B.E., and Backvall, J.E. (2001). Racemisation in asymmetric synthesis. Dynamic kinetic resolution and related processes in enzyme and metal catalysis. *Chem. Soc. Rev.* 30 (6): 321–331.

299 Pellissier, H. (2003). Dynamic kinetic resolution. *Tetrahedron* 59 (42): 8291–8327.

300 Andraos, J. (2003). Quantification and optimization of dynamic kinetic resolution. *Chem. A Eur. J.* 107 (13): 2374–2387.

301 Pellissier, H. (2008). Recent developments in dynamic kinetic resolution. *Tetrahedron* 64 (8): 1563–1601.

302 Pellissier, H. (2011). Recent developments in dynamic kinetic resolution. *Tetrahedron* 67 (21): 3769–3802.

303 Pellissier, H. (2011). Organocatalyzed dynamic kinetic resolution. *Adv. Synth. Catal.* 353 (5): 659–676.

304 Steinreiber, J., Faber, K., and Griengl, H. (2008). De-racemization of enantiomers versus de-epimerization of diastereomers - classification of dynamic kinetic asymmetric transformations (DYKAT). *Chem. Eur. J.* 14 (27): 8060–8072.

305 Duhamel, L. and Plaquevent, J.C. (1977). Deracemization by enantioselective protonation. *Tetrahedron Lett.* (26): 2285–2288.

306 Gerlach, U. and Hunig, S. (1987). Enantioselective protonation of carbanions with chiral proton sources. *Angew. Chem. Int. Ed. Eng.* 26 (12): 1283–1285.

307 Vedejs, E., Lee, N., and Sakata, S.T. (1994). Deracemization via highly enantioselective enolate

protonation using a chiral aniline as the acid. *J. Am. Chem. Soc.* 116 (5): 2175–2176.

308 Fehr, C. (1991). Enantioselective protonation of enolates in natural product synthesis. *Chimia* 45 (9): 253–261.

309 Fehr, C. (1996). Enantioselective protonation of enolates and enols. *Angew. Chem. Int. Ed. Eng.* 35 (22): 2567–2587.

310 Camps, P. and Gimenez, S. (1996). Deracemization of α-substituted arylacetic acids. *Tetrahedron Asymmetry* 7 (4): 1227–1234.

311 Vedejs, E., Kruger, A.W., Lee, N. et al. (2000). Enantioselective enolate protonation with chiral anilines: scope, structural requirements, and mechanistic implications. *J. Am. Chem. Soc.* 122 (19): 4602–4607.

312 Solladié-Cavallo, A., Sedy, O., Salisova, M., and Schmitt, M. (2002). Stereodifferentiation in a chiral 1,4-oxazin-2-one derived from 2-hydroxy-2-methyl-1-tetralone - a reagent for deracemization of amino acids. *Eur. J. Org. Chem.* (17): 3042–3049.

313 Kato, D., Mitsuda, S., and Ohta, H. (2002). Microbial deracemization of α-substituted carboxylic acids. *Org. Lett.* 4 (3): 371–373.

314 Clericuzio, M., Degani, I., Dughera, S., and Fochi, R. (2003). Deracemization of thiol esters of α-arylpropionic acids. *Tetrahedron Asymmetry* 14 (1): 119–125.

315 Schlosser, M. and Limat, D. (1995). Sparteine-mediated α-lithiation of *N*-Boc-*N*-methylbenzylamine - rapid racemization and subsequent deracemization. *J. Am. Chem. Soc.* 117 (49): 12342–12343.

316 Prat, L., Mojovic, L., Levacher, V. et al. (1998). Deracemization of diarylmethanes via lateral lithiation–protonation sequences by means of sparteine. *Tetrahedron Asymmetry* 9 (14): 2509–2516.

317 Burton, A.J., Graham, J.P., and Simpkins, N.S. (2000). Enantioselective protonation of organolithiums having the tetrahydroisoquinoline skeleton. *Synlett* (11): 1640–1642.

318 Goswami, A., Mirfakhrae, K.D., and Patel, R.N. (1999). Deracemization of racemic 1,2-diol by biocatalytic stereoinversion. *Tetrahedron Asymmetry* 10 (21): 4239–4244.

319 Persson, B.A., Larsson, A.L.E., Le Ray, M., and Bäckvall, J.E. (1999). Ruthenium- and enzyme-catalyzed dynamic kinetic resolution of secondary alcohols. *J. Am. Chem. Soc.* 121 (8): 1645–1650.

320 Dijksman, A., Elzinga, J.M., Li, Y.X. et al. (2002). Efficient Ru-catalyzed racemization of secondary alcohols: application to dynamic kinetic resolution. *Tetrahedron Asymmetry* 13 (8): 879–884.

321 Turner, N.J. (2004). Enzyme catalysed deracemisation and dynamic kinetic resolution reactions. *Curr. Opin. Chem. Biol.* 8 (2): 114–119.

322 Lackner, A.D., Samant, A.V., and Toste, F.D. (2013). Single-operation deracemization of 3H-indolines and tetrahydroquinolines enabled by phase separation. *J. Am. Chem. Soc.* 135 (38): 14090–14093.

323 Trost, B.M., Krueger, A.C., Bunt, R.C., and Zambrano, J. (1996). On the question of asymmetric induction with acyclic allylic substrates. An asymmetric synthesis of (+)-polyoxamic acid. *J. Am. Chem. Soc.* 118 (27): 6520–6521.

324 Trost, B.M., Tsui, H.C., and Toste, F.D. (2000). Deracemization of Baylis-Hillman adducts. *J. Am. Chem. Soc.* 122 (14): 3534–3535.

325 Lussem, B.J. and Gais, H.J. (2003). Palladium-catalyzed deracemization of allylic carbonates in water with formation of allylic alcohols: hydrogen carbonate ion as nucleophile in the Pd-catalyzed allylic substitution and kinetic resolution. *J. Am. Chem. Soc.* 125 (20): 6066–6067.

326 Martin-Matute, B. and Backvall, J.E. (2007). Dynamic kinetic resolution catalyzed by enzymes and metals. *Curr. Opin. Chem. Biol.* 11 (2): 226–232.

327 Beak, P., Anderson, D.R., Curtis, M.D. et al. (2000). Dynamic thermodynamic resolution: control of enantioselectivity through diastereomeric equilibration. *Acc. Chem. Res.* 33 (10): 715–727.

328 Lee, W.K., Park, Y.S., and Beak, P. (2009). Dynamic thermodynamic resolution: advantage by separation of equilibration and resolution. *Acc. Chem. Res.* 42 (2): 224–234.

329 Seebach, D., Beck, A.K., and Heckel, A. (2001). TADDOLs, their derivatives, and TADDOL analogues: versatile chiral auxiliaries. *Angew. Chem. Int. Ed.* 40 (1): 92–138.

330 Kaku, H., Takaoka, S., and Tsunoda, T. (2002). The nature of the thermodynamically controlled deracemization of 2-benzylcyclohexanone using (*R,R*)-(-)-*trans*-2,3-bis(hydroxydiphenylmethyl)-1,4-dioxaspiro[5.4]decane: a crystallographic result of inclusion complex. *Tetrahedron* 58 (17): 3401–3407.

331 Kaku, H., Imai, T., Kondo, R. et al. (2013). A method to prepare optically active acyclic α-benzyl ketones by thermodynamically controlled deracemization. *Eur. J. Org. Chem.* 2013 (36): 8208–8213.

332 Trost, B.M., Tang, W.P., and Schulte, J.L. (2000). Asymmetric synthesis of quaternary centers. Total synthesis of (-)-malyngolide. *Org. Lett.* 2 (25): 4013–4015.

333 Riedner, J. and Vogel, P. (2004). Deracemization of 5-(4-hydroxyphenyl)-5-phenylhydantoin (HPPH): practical synthesis of (-)-(*S*)-HPPH. *Tetrahedron Asymmetry* 15 (17): 2657–2660.

334 Noyori, R. (1994). *Asymmetric Catalysis in Organic Chemistry*, 1–378. Chichester: Wiley.

335 Brunel, J.M. (2005). BINOL: a versatile chiral reagent. *Chem. Rev.* 105 (3): 857–897.

336 Wang, H. (2010). Recent advances in asymmetric oxidative coupling of 2-naphthol and its derivatives. *Chirality* 22 (9): 827–837.

337 Brussee, J. and Jansen, A.C.A. (1983). A highly stereoselective synthesis of (-)-(*S*)-[1,1'-binaphthalene]-2,2'-diol. *Tetrahedron Lett.* 24 (31): 3261–3262.

338 Brussee, J., Groenendijk, J.L.G., Tekoppele, J.M., and Jansen, A.C.A. (1985). On the mechanism of the formation of (-)-(*S*)-(1,1'-binaphthalene)-2,2'-diol via Cu(II)amine complexes. *Tetrahedron* 41 (16): 3313–3319.

339 Barton, D.H.R. and Kirby, G.W. (1962). Phenol oxidation and biosynthesis. 5. Synthesis of galanthamine. *J. Chem. Soc.* 806–817.

340 Küenburg, B., Czollner, L., Fröhlich, J., and Jordis, U. (1999). Development of a pilot scale process for the anti-Alzheimer drug (-)-galanthamine using large-scale phenolic oxidative coupling and crystallisation-induced chiral conversion. *Org. Process. Res. Dev.* 3 (6): 425–431.

341 Yagishita, F., Ishikawa, H., Onuki, T. et al. (2012). Total spontaneous resolution by deracemization of isoindolinones. *Angew. Chem. Int. Ed.* 51 (52): 13023–13025.

342 Havinga, E. (1954). Spontaneous formation of optically active substances. *Biochim. Biophys. Acta* 13 (2): 171–174.

343 Pincock, R.E. and Wilson, K.R. (1971). Solid state resolution of racemic 1,1'-binaphthyl. *J. Am. Chem. Soc.* 93 (5): 1291–1292.

344 Pincock, R.E., Perkins, R.R., Ma, A.S., and Wilson, K.R. (1971). Probability distribution of enantiomorphous forms in spontaneous generation of optically active substances. *Science* 174 (4013): 1018–1020.

345 Amabilino, D.B. and Kellogg, R.M. (2011). Spontaneous deracemization. *Isr. J. Chem.* 51 (10): 1034–1040.

346 Noorduin, W.L., Izumi, T., Millemaggi, A. et al. (2008). Emergence of a single solid chiral state from a nearly racemic amino acid derivative. *J. Am. Chem. Soc.* 130 (4): 1158–1159.

347 Viedma, C., Ortiz, J.E., de Torres, T. et al. (2008). Evolution of solid phase homochirality for a proteinogenic amino acid. *J. Am. Chem. Soc.* 130 (46): 15274–15275.

348 Noorduin, W.L., Bode, A.A., van der Meijden, M. et al. (2009). Complete chiral symmetry breaking of an amino acid derivative directed by circularly polarized light. *Nat. Chem.* 1 (9): 729–732.

349 Faigl, F., Fogassy, E., Nogradi, M. et al. (2010). Separation of non-racemic mixtures of enantiomers: an essential part of optical resolution. *Org. Biomol. Chem.* 8 (5): 947–959.

350 Mastai, Y., Voelkel, A., and Coelfen, H. (2008). Separation of racemate from excess enantiomer of chiral nonracemic compounds via density gradient ultracentrifugation. *J. Am. Chem. Soc.* 130 (8): 2426–2427.

351 Kozma, D., Kassai, C., and Fogassy, E. (1995). Enantiomeric enrichment by the use of density differences between racemic compounds and optically active enantiomers. *Tetrahedron Lett.* 36 (18): 3245–3246.

352 Ueki, H., Yasumoto, M., and Soloshonok, V.A. (2010). Rational application of self-disproportionation of enantiomers via sublimation-a novel methodological dimension for enantiomeric purifications. *Tetrahedron Asymmetry* 21 (11–12): 1396–1400.

353 Katagiri, T., Takahashi, S., Tsuboi, A. et al. (2010). Discrimination of enantiomeric excess of optically active trifluorolactate by distillation: evidence for a multi-center hydrogen bonding network in the liquid state. *J. Fluor. Chem.* 131 (4): 517–520.

354 Soloshonok, V.A., Roussel, C., Kitagawa, O., and Sorochinsky, A.E. Self-disproportionation of enantiomers via achiral chromatography: a warning and an extra dimension in optical purifications. *Chem. Soc. Rev.* 2012 (11): 41, 4180–4188.

355 Cundy, K.C. and Crooks, P.A. (1983). Unexpected phenomenon in the high-performance liquid-chromatographic analysis of racemic ^{14}C-labeled nicotine - separation of enantiomers in a totally achiral system. *J. Chromatogr.* 281: 17–33.

356 Diter, P., Taudien, S., Samuel, O., and Kagan, H.B. (1994). Enantiomeric enrichment of sulfoxides by preparative flash chromatography on an achiral phase. *J. Org. Chem.* 59 (2): 370–373.

357 Nicoud, R.M., Jaubert, J.N., Rupprecht, I., and Kinkel, J. (1996). Enantiomeric enrichment

of non-racemic mixtures of binaphthol with non-chiral packings. *Chirality* 8 (3): 234–243.

358 Williams, T., Pitcher, R.G., Bommer, P. et al. (1969). Diastereomeric solute-solute interactions of enantiomers in achiral solvents. Nonequivalence of nuclear magnetic resonance spectra of racemic and optically active dihydroquinine. *J. Am. Chem. Soc.* 91 (7): 1871–1872.

359 Cung, M.T., Marraud, M., and Neel, J. (1976). Experimental study of self-aggregation of dipeptide model molecules. 2. Stereoselective aggregation of enantiomer molecules. *Biopolymers* 15 (10): 2081–2095.

360 Cung, M.T., Marraud, M., and Neel, J. (1977). Experimental study on aggregation of model dipeptide molecules. 3. Influence of stereoselective dimerization on proton magnetic-resonance spectra. *Biopolymers* 16 (4): 715–729.

361 Kabachnik, M.I., Mastryukova, T.A., Fedin, E.I. et al. (1976). NMR-Study of optical isomers in solution. *Tetrahedron* 32 (14): 1719–1728.

362 Klussmann, M., Iwamura, H., Mathew, S.P. et al. (2006). Thermodynamic control of asymmetric amplification in amino acid catalysis. *Nature* 441 (7093): 621–623.

363 Klussmann, M., White, A.J., Armstrong, A., and Blackmond, D.G. (2006). Rationalization and prediction of solution enantiomeric excess in ternary phase systems. *Angew. Chem. Int. Ed.* 45 (47): 7985–7989.

364 Hayashi, Y., Matsuzawa, M., Yamaguchi, J. et al. (2006). Large nonlinear effect observed in the enantiomeric excess of proline in solution and that in the solid state. *Angew. Chem. Int. Ed.* 45 (28): 4593–4597.

365 Kellogg, R.M. (2007). The crystallization behavior of proline and its role in asymmetric organocatalysis. *Angew. Chem. Int. Ed.* 46 (4): 494–497.

366 List, B. (2006). The ying and yang of asymmetric aminocatalysis. *J. Chem. Soc. Chem. Commun.* (8): 819–824.

367 Melchiorre, P., Marigo, M., Carlone, A., and Bartoli, G. (2008). Asymmetric aminocatalysis - gold rush in organic chemistry. *Angew. Chem. Int. Ed.* 47 (33): 6138–6171.

368 Cook, A.G. (ed.) (1988). *Enamines, Synthesis, Structures, and Reactions*, 2e, 1–720. New York: Marcel Dekker.

369 Whitesell, J.K. and Whitesell, M.A. (1983). Alkylation of ketones and aldehydes via their nitrogen derivatives. *Synthesis* (7): 517–536.

370 Hickmott, P.W. (1984). Recent advances in the chemistry of conjugated enamines. *Tetrahedron* 40 (16): 2989–3051.

371 Bahmanyar, S. and Houk, K.N. (2001). Transition states of amine-catalyzed aldol reactions involving enamine intermediates: theoretical studies of mechanism, reactivity, and stereoselectivity. *J. Am. Chem. Soc.* 123 (45): 11273–11283.

372 Silvestri, M.A., Bromfield, D.C., and Lepore, S.D. (2005). Michael-Stork addition of cyclopentyl enamine to allenyl ketones and esters. *J. Org. Chem.* 70 (20): 8239–8241.

373 Benkovic, S.J., Benkovic, P.A., and Comfort, D.R. (1969). Kinetic detection of imminium cation in formaldehyde-amine condensations in neutral aqueous solution. *J. Am. Chem. Soc.* 91 (7): 1860–1861.

374 Liras, S., Lynch, C.L., Fryer, A.M. et al. (2001). Applications of vinylogous Mannich reactions. Total syntheses of the *Ergot* alkaloids rugulovasines A and B and setoclavine. *J. Am. Chem. Soc.* 123 (25): 5918–5924.

375 Tanaka, Y., Hasui, T., and Suginome, M. (2008). Diarylborinic acid derivatives as a catalytic iminium ion generator in the Mannich-type reaction using secondary amines, aldehydes, and ketene silyl acetals. *Synlett* (8): 1239–1242.

376 Mengozzi, L., Gualandi, A., and Cozzi, P. (2014). A highly enantioselective acyl-Mannich reaction of isoquinolines with aldehydes promoted by proline derivatives: an approach to 13-alkyl-tetrahydroprotoberberine alkaloids. *Chem. Sci.* 5 (10): 3915–3921.

377 Schreiber, J., Maag, H., Hashimoto, N., and Eschenmoser, A. (1971). Synthetic methods. 2. Dimethyl(methylene)ammonium iodide. *Angew. Chem. Int. Ed. Eng.* 10 (5): 330–330.

378 Pictet, A. and Spengler, T. (1911). On the formation of isochinolin derivatives through the development of methylial on phenyl ether, phenylalanin and tyrosin. *Ber. Dtsch. Chem. Ges.* 44: 2030–2036.

379 Cox, E.D. and Cook, J.M. (1995). The Pictet-Spengler condensation - a new direction for an old reaction. *Chem. Rev.* 95 (6): 1797–1842.

380 (a) Wolfsberg, M. (1972). Theoretical evaluation of experimentally observed isotope-effects. *Acc. Chem. Res.* 5 (7): 225–233. (b) Scheiner, S. (2000). Calculation of isotope effects from first principles. *Biochim. Biophys. Acta Bioenerg.* 1458 (1): 28–42. (c) Bigeleisen, J. (2006). Theoretical basis of isotope effects from an autobiographical perspective. In: *Isotope Effects in Chemistry and Biology*, Chapter 1 (ed. A. Kohen and H. Limbach), 1–40. Boca Raton, FL: CRC Press, Taylor & Francis Group.

381 Hanson, J.R. (2011). *The Organic Chemistry of Isotopic Labelling*, 1–200. London: Royal Society of Chemistry.

382 Bartell, L.S. and Roskos, R.R. (1966). Isotope effects on molar volume and surface tension - simple theoretical model and experimental data for hydrocarbons. *J. Chem. Phys.* 44 (2): 457–463.

383 Bartell, L.S. (1960). Secondary isotope effects and mass-sensitive amplitudes of vibration. *Tetrahedron Lett.* (6): 13–16.

384 Bartell, L.S. (1961). Role of non-bonded repulsions in secondary isotope effects. 1. α and β-substitution effects. *J. Am. Chem. Soc.* 83 (17): 3567–3571.

385 Bartell, L.S. (1965). Secondary isotope effects on amplitudes of molecular vibrations. *J. Chem. Phys.* 42 (5): 1681–1987.

386 Soddy, F. (1913). Radioactivity. *Annu. Rep. Prog. Chem.* 10: 262–288.

387 Urey, H.C. (1947). The thermodynamic properties of isotopic substances. *J. Chem. Soc.* 562–581.

388 Bigeleisen, J. and Mayer, M.G. (1947). Calculation of equilibrium constants for isotopic exchange reactions. *J. Chem. Phys.* 15 (5): 261–267.

389 Bigeleisen, J. (1949). Isotope effect in the decarboxylation of labelled malonic acids. *J. Chem. Phys.* 17 (4): 425–426.

390 Melander, L. On the mechanism of electrophilic aromatic substitution - an investigation by means of the effect of isotopic mass on reaction velocity. *Arkiv for Kemi* 1950 (3): 2, 213–292.

391 Wolfsberg, M. (1969). Isotope effects. *Annu. Rev. Phys. Chem.* 20: 449–478.

392 Halpern, O. (1935). On the dissociation constants of acids in light and heavy water. *J. Chem. Phys.* 3 (8): 456–457.

393 Kingerley, R.W. and Lamer, V.K. (1941). Exchange and transfer equilibria of acids, bases, and salts in deuterium-protium oxide mixtures. The ion product constant of deuterium oxide. *J. Am. Chem. Soc.* 63: 3256–3262.

394 Lewis, G.N. and Schutz, P.W. (1934). The ionization of some weak electrolytes in heavy water. *J. Am. Chem. Soc.* 56 (7): 1913–1915.

395 Rule, C.K. and La Mer, V.K. (1938). Dissociation constants of deutero acids by E.M.F measurements. *J. Am. Chem. Soc.* 60: 1974–1981.

396 Wiberg, K.B. (1955). The deuterium isotope effect. *Chem. Rev.* 55 (4): 713–743.

397 Delgado, R., Dasilva, J.J.R.F., Amorim, M.T.S. et al. (1991). Dissociation constants of Brønsted acids in D_2O and H_2O - studies on polyaza and polyoxa-polyaza macrocycles and a general correlation. *Anal. Chim. Acta* 245 (2): 271–282.

398 Humski, K., Malojcic, R., Borcic, S., and Sunko, D.E. (1970). Thermodynamic and kinetic secondary isotope effects in Cope rearrangement. *J. Am. Chem. Soc.* 92 (22): 6534–6538.

399 Gajewski, J.J. and Conrad, N.D. (1979). Variable transition-state structure in 3,3-sigmatropic shifts from α-secondary deuterium isotope effects. *J. Am. Chem. Soc.* 101 (22): 6693–6704.

400 Gajewski, J.J. and Conrad, N.D. (1979). Aliphatic Claisen rearrangement transition state structure from secondary α-deuterium isotope effects. *J. Am. Chem. Soc.* 101 (10): 2747–2748.

401 Houk, K.N., Gustafson, S.M., and Black, K.A. (1992). Theoretical secondary kinetic isotope effects and the interpretation of transition-state geometries. 1. The Cope rearrangement. *J. Am. Chem. Soc.* 114 (22): 8565–8572.

402 Quast, H., Seefelder, M., Becker, C. et al. (1999). Extension of Saunders' isotopic perturbation method as probe for the structures in solution of 2,4,6,8-substituted barbaralanes - NMR-spectroscopic evidence for the coexistence of localised and delocalised states. *Eur. J. Org. Chem.* 11: 2763–2779.

403 Rabinovich, D. and Parkin, G. (1993). A mechanistic study of the oxidative addition of H_2 to $W(PMe_3)_4I_2$ - observation of an inverse equilibrium isotope effect. *J. Am. Chem. Soc.* 115 (1): 353–354.

404 Hascall, T., Rabinovich, D., Murphy, V.J. et al. (1999). Mechanistic and theoretical analysis of the oxidative addition of H_2 to six-coordinate molybdenum and tungsten complexes $M(PMe_3)_4X_2$ (M = Mo, W; X = F, Cl, Br, I): an inverse equilibrium isotope effect and an unprecedented halide dependence. *J. Am. Chem. Soc.* 121 (49): 11402–11417.

405 Janak, K.E. and Parkin, G. (2003). Experimental evidence for a temperature dependent transition between normal and inverse equilibrium isotope effects for oxidative addition of H_2 to $Ir(PMe_2Ph)_2(CO)Cl$. *J. Am. Chem. Soc.* 125 (43): 13219–13224.

406 Melander, L. and Saunders, W.H. Jr., (1980). *Reactions Rates of Isotopic Molecules*, 1–331. New York: Wiley.

407 Parkin, G. (2009). Temperature-dependent transitions between normal and inverse isotope effects pertaining to the interaction of H—H and C—H bonds with transition metal centers. *Acc. Chem. Res.* 42 (2): 315–325.

408 Anet, F.A.L., Basus, V.J., Hewett, A.P.W., and Saunders, M. (1980). Isotopic perturbation of

degenerate conformational equilibria. *J. Am. Chem. Soc.* 102 (11): 3945–3946.

409 Saunders, M., Wolfsberg, M., Anet, F.A.L., and Kronja, O. (2007). A steric deuterium isotope effect in 1,1,3,3-tetramethylcyclohexane. *J. Am. Chem. Soc.* 129 (33): 10276–10281.

410 Jones, W.D. and Feher, F.J. (1986). Isotope effects in arene C—H bond activation by $[(C_5Me_5)Rh(PMe_3)]$. *J. Am. Chem. Soc.* 108 (16): 4814–4819.

411 Tsang, W. (1996). *Energetics of Organic Free Radicals*. London: Black Academic and Professional.

412 Brocks, J.J., Beckhaus, H.D., Beckwith, A.L.J., and Ruechardt, C. (1998). Estimation of bond dissociation energies and radical stabilization energies by ESR spectroscopy. *J. Org. Chem.* 63 (6): 1935–1943.

413 Chase, M.W., Davies, C.A., Downey, J.R. et al. (1985). JANAF thermochemical tables - 3Rd edition .1. Al-Co. *J. Phys. Chem. Ref. Data* 14: 1–926.

414 Meot-Ner, M. and Sieck, L.W. (1991). Proton affinity ladders from variable-temperature equilibrium measurements .1. A reevaluation of the upper proton affinity range. *J. Am. Chem. Soc.* 113 (12): 4448–4460.

415 (a) Koppel, I.A., Taft, R.W. Jr., Anvia, F. et al. (1994). The gas-phase acidities of very strong neutral Bronsted acids. *J. Am. Chem. Soc.* 116 (7): 3047–3057. (b) Koppel, I.A., Koppel, J., Pihl, V. et al. (2000). Comparison of Bronsted acidities of neutral CH acids in gas phase and dimethyl sulfoxide. *J. Chem. Soc., Perkin Trans.* 2 (6): 1125–1133.

416 (a) Chen, H., Justes, D.R., and Cooks, R.G. (2005). Proton affinities of *N*-heterocyclic carbene super bases. *Org. Lett.* 7 (18): 3949–3952. (b) Vogt, J. and Beauchamp, J.L. (1975). Reactions of CHF2 + with n-donor bases by ion-cyclotron resonance spectroscopy - proton affinity of difluorocarbene. *J. Am. Chem. Soc.* 97 (23): 6682–6685.

417 Ausloos, P. and Lias, S.G. (1978). Proton affinity of dichlorocarbene. *J. Am. Chem. Soc.* 100 (14): 4594–4595.

418 Shin, S.K. and Beauchamp, J.L. (1986). Proton affinity and heat of formation of silylene. *J. Phys. Chem.* 90 (8): 1507–1509.

419 Heinrich, N., Koch, W., and Frenking, G. (1986). On the use of Koopmans theorem to estimate negative electron affinities. *Chem. Phys. Lett.* 124 (1): 20–25.

420 Burrow, P.D., Michejda, J.A., and Jordan, K.D. (1987). Electron transmission study of the temporary negative-ion states of selected benzenoid and conjugated aromatic-hydrocarbons. *J. Chem. Phys.* 86 (1): 9–24.

421 Brinkman, E.A., Berger, S., Marks, J., and Brauman, J.I. (1993). Molecular rotation and the observation of dipole-bound states of anions. *J. Chem. Phys.* 99 (10): 7586–7594.

422 Wentworth, W.E. and Chen, E. (1967). Experimental determination of electron affinity of several aromatic aldehydes and ketones. *J. Phys. Chem.* 71 (6): 1929–1931.

423 Paul, G. and Kebarle, P. (1989). Electron affinities of cyclic unsaturated dicarbonyls - maleic anhydrides, maleimides, and cyclopentenedione. *J. Am. Chem. Soc.* 111 (2): 464–470.

424 Mishima, M., Huh, C., Nakamura, H. et al. (1993). Electron affinities of benzaldehydes - substituent effects on stabilities of aromatic radical-anions. *Tetrahedron Lett.* 34 (26): 4223–4226.

425 Chen, E.C.M. and Wentworth, W.E. (1981). Correlation and prediction of electron capture response from molecular parameters. *J. Chromatogr.* 217 (6): 151–166.

426 Trouillas, P.D.M.F., Gierschner, J., Linares, M. et al. (2015). Optical properties of wine pigments: theoretical guidelines with new methodological perspectives. *Tetrahedron* 71: 3079–3088.

427 Williams, R., Jencks, W.P., and Westheimer, F.H. (2011). Personal communication.

428 Gao, Y., DeYonker, N.J., Garrett, E.C. III et al. (2009). Enthalpy of formation of the cyclohexadienyl radical and the C—H bond enthalpy of 1,4-cyclohexadiene: an experimental and computational re-evaluation. *J. Phys. Chem. A* 113 (25): 6955–6963.

429 Chen, H., Justes, D.R., and Cooks, R.G. (2005). Proton affinities of *N*-heterocyclic carbene super bases. *Org. Lett.* 7 (18): 3949–3952.

430 Vogt, J. and Beauchamp, J.L. (1975). Reactions of CHF_2^+ with n-donor bases by ion-cyclotron resonance spectroscopy - proton affinity of difluorocarbene. *J. Am. Chem. Soc.* 97 (23): 6682–6685.

431 Ausloos, P. and Lias, S.G. (1978). Proton affinity of dichlorocarbene. *J. Am. Chem. Soc.* 100 (14): 4594–4595.

432 Shin, S.K. and Beauchamp, J.L. (1986). Proton affinity and heat of formation of silylene. *J. Phys. Chem.* 90 (8): 1507–1509.

2

Additivity rules for thermodynamic parameters and deviations

2.1 Introduction

There are very useful relationships between the structure of simple organic compounds and their standard heats of formation $\Delta_f H°$(gas) and standard entropies $S°$(gas). This chapter presents the classical group additivity method of Benson–Buss for the calculation of these thermochemical parameters [1–5]. This method leads to accurate estimates and should be used to test the consistency of computed values obtained by more sophisticated methods such as those of mechanical mechanics and quantum mechanics [6]. Accurate estimates are obtained if suitable corrections to the additivity rules are considered. They include corrections for gauche interactions and steric effects ("Back-strain," "Front-strain," and bond eclipsing), ring strain, allylic-1,2 and allylic-1,3 strain, π/π conjugation, cyclic conjugation (aromaticity and homoaromaticity), n/π conjugative effects, hyperconjugation effects, enthalpic and conformational anomeric effects (AEs), dipole/dipole effects, and repulsive electronic interactions (antiaromaticity, barrelene effect). This method can be applied to estimate the heat of almost any organic reactions, including those involving transient intermediates such as radicals, diradicals, carbenes, and ion pairs. Thermochemical calculations represent a powerful method to evaluate mechanistic limits, as shown in the following chapters. The estimation of Gibbs free energy of a given equilibrium (e.g. (1.1)) in the gas phase (25 °C, 1 atm) or in an ideal solution (25 °C) is trivial if the standard heats of formation, $\Delta_f H°$(gas), and standard entropies, $S°$(gas), of reactants (**A**, **B**, …) and products (**P**, **Q**, …) are all known. The standard Gibbs free energy (1.15) of equilibrium (1.1) can be written as Eq. (2.1):

$$\alpha\mathbf{A} + \beta\mathbf{B} + \ldots \rightleftarrows \pi\mathbf{P} + \theta\mathbf{Q} + \ldots \tag{1.1}$$

$$\Delta_r G°(1.1) = \Delta_r H°(1.1) - T\Delta_r S°(1.1) = -RT \ln K^{298\mathrm{K}} \tag{2.1}$$

with

$$\Delta_r H°(1.1) = [\pi\Delta_f H°(\mathbf{P}) + \theta\Delta_f H°(\mathbf{Q}) + \ldots] - [\alpha\Delta_f H°(\mathbf{A}) + \beta\Delta_f H°(\mathbf{B}) + \ldots] \tag{2.2}$$

and

$$\Delta_r S°(1.1) = [\pi S°(\mathbf{P}) + \theta S°(\mathbf{Q}) + \ldots] - [\alpha S°(\mathbf{A}) + \beta S°(\mathbf{B}) + \ldots] \tag{2.3}$$

For a large number of reactions, the entropy of reaction (or variation of entropy of reaction), $\Delta_r S°$, can be evaluated just by considering the change in the number of molecules going from the reactants to the products. A good approximation to $\Delta_r S°$ is given by the difference in translation entropies between the products and the reactants, as calculated with $S_{trans} = 6.86 \log(\mathrm{M(g)}) + 11.44 \log T - 2.31$ for the entropy of monoatomic, ideal gas (Eq. (1.31), Section 1.4.1). As demonstrated by statistical thermodynamics (Section 1.4), $\Delta_r S°$ is negative for condensations, positive for fragmentations, and close to zero for isomerizations (Section 1.4.4). Exceptions to this latter rule will be presented below. They include change in symmetry and mobility between the products and the reactants. Thus, $\Delta_r S° \cong S_{trans}(\text{products}) - S_{trans}(\text{reactants})$ + corrections for changes in flexibility and symmetry. We shall see that entropy can be used as a synthetic tool (Section 2.11). If the standard heats of formation, $\Delta_f H°$, of some of the products and reactants are not available, they can be estimated by quantum mechanical calculations (expensive and sometimes inaccurate for large organic and organometallic compounds), by molecular mechanics (much cheaper, sometimes reliable estimates are obtained if the standard compounds of reference are quite similar to those for which $\Delta_f H°$ is calculated), or, more simply, by applying additivity rule (2.4) [7]. The standard heat of formation of a given compound is the sum of group equivalents associated with the

Organic Chemistry: Theory, Reactivity and Mechanisms in Modern Synthesis, First Edition. Pierre Vogel and Kendall N. Houk.

constituent molecular groups and any corrections that will be defined in Sections 2.5–2.8.

$$\Delta_f H^\circ(\text{compound}) = \sum(\Delta_f H \text{ increments associate with molecular groups (Table 2.1)} + \text{corrections (Sections 2.5–2.8)} \quad (2.4)$$

Problem 2.1 Evaluate the entropy variations of the following reactions at 25 °C.

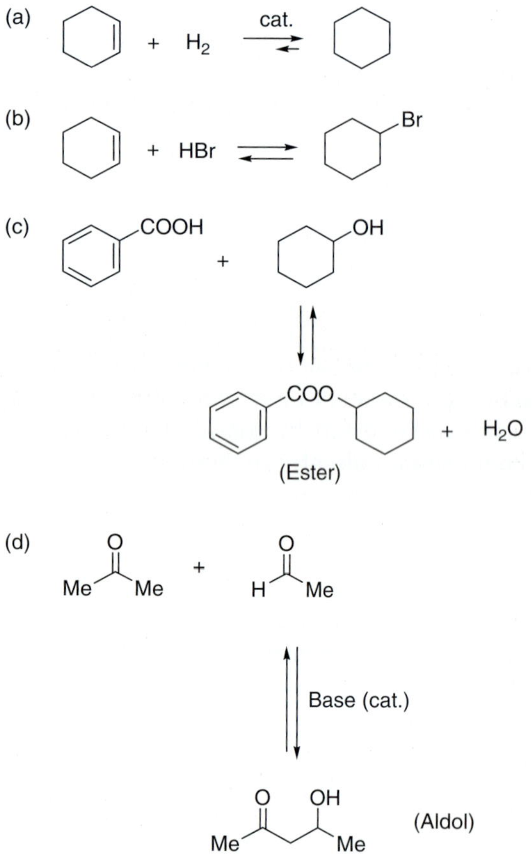

Problem 2.2 An enzyme **E** of molecular mass 120 000 activates a substrate **A** of molecular mass 200 making a complex **E·A** of molecular mass 120 200. Estimate the entropy variation of equilibrium **E** + **A** ⇄ **E·A**, assuming that the flexibilities of **E** and **A** are the same as in the complex **E·A**. If the same enzyme releases one, two, three, or four molecules of water during the formation of complex **E·A**, what are the entropy variations of these equilibria?

2.2 Molecular groups

The additivity of molecular properties has been recognized for a long time. As early as 1946, Prosen et al. found for hydrocarbons in the gas phase that the standard heats of formation can be defined by relationship (2.5) [8]:

$$\Delta_f H^\circ(\text{Y} - (\text{CH}_2)_m - \text{H, gas}) = A + m\,B + C \quad (2.5)$$

where A is a constant typical of molecular group Y, B is a constant typical of linear hydrocarbons, and C is a constant. It appears, therefore, that the increment of standard heat of formation associated with the sp^3-hybridized methylene group ($-CH_2-$) may depend on its neighboring groups, that is, on the nature of the atoms and the molecular fragments to which it is attached. This has led to the definition of molecular fragments (groups) given below and to the group additivity methods of Laidler [9, 10] and of Benson–Buss [1, 4].

[C—(C)(H)$_3$]	Methyl group attached to a tetragonal carbon atom (sp^3-hybridized C, saturated hydrocarbon)
[C—(C)$_2$(H)$_2$]	Methylene group inserted between two tetragonal carbon atoms (sp^3-C)
[C—(C)$_3$(H)]	Methine group inserted between three tetragonal carbon atoms
[C$_d$—(C$_d$)(H)$_2$]	Methylidene group, sp^2-hybridized carbon atom bearing two hydrogen atoms and attached to a trigonal carbon atom of an alkene (d = double bond of alkenes)
[C$_d$—(C$_d$)(C)(H)]	sp^2-Hydridized carbon atom of an alkene connected to an sp^2-carbon center, an sp^3-carbon center, and bearing a H-atom;
[C—(C)$_4$]	Quaternary sp^3-carbon center of doubly branched hydrocarbons, sp^3-carbon atom attached to four tetragonal, sp^3-hybridized carbon atoms
[C$_b$—(C$_b$)$_2$(H)]	Methine group of a benzene derivative bearing one hydrogen atom and inserted between two benzenic carbon centers (b = benzene; the increment associated with this group contains the contribution of benzene aromaticity, see Section 2.7.4)
[C$_t$—(C$_t$)(H)]	Methine group of a terminal alkyne, sp-hybridized carbon atom bearing one hydrogen atom attached to an sp-hybridized carbon center through a triple bond
[C$_d$—(C$_d$)(C)$_2$]	sp^2-Hybridized carbon of an alkene attached to an sp^2-hydribized carbon atom through a double bond and to two sp^3-hybridized carbon atoms of two alkyl groups
[C$_t$—(C$_b$)]	sp^1-Hybridized carbon atom of an alkynyl group attached to a benzene derivative;

[CO—(C)$_2$]	Carbonyl group of a ketone
[CO—(C)(H)]	Carbonyl group of an aldehyde
[O—(C)(H)]	Hydroxyl group of a saturated alcohol; oxygen atom bearing a hydrogen atom attached to a tetragonal carbon atom
[O—(C)$_2$]	Oxygen atom of a saturated ether, attached to two tetragonal carbon atoms
[O—(O)H]	Hydroxyl group of a hydroperoxide
[O—(O)(C)]	Oxygen atom of a peroxide attached to an oxygen atom and an sp^3-hybridized carbon atom
[O—(CO)$_2$]	Oxygen atom of a carboxylic anhydride
[C$_d$—(C$_d$)(Br)(H)]	Bromomethylidene group, sp^2-hybridized atom of an alkene bearing a hydrogen and a bromine atom, attached at an sp^2-hybridized carbon atom of an alkene through a double bond
[C—(N)(H)$_3$]	Methyl group of a methylamine, sp^3-hybridized carbon atom bearing three hydrogen atoms attached to the nitrogen center of an amine
[C—(N$_A$)(C)$_2$H]	sp^3-Hybridized carbon atom bearing a hydrogen atom and attached to two sp^3-hybridized carbon atom and the nitrogen center of an azo compound
[N—(C$_b$)$_2$(H)]	NH group of pyrrole
[CO—(O)(C)]	Carbonyl group of an ester attached to an sp^3-hybridized carbon center
[CO—(N)(C)]	Carbonyl group of an amide attached to an sp^3-hybridized carbon center
[N—(CO)(C)(H)]	Nitrogen atom of a *N*-alkyl amide; nitrogen atom bearing one hydrogen atom attached to a carbonyl group and an sp^3-hybridized carbon center
[N$_I$—(C)]	Nitrogen atom of a *N*-alkylated imine
[N$_I$—(C$_b$)]	Nitrogen atom of pyridine and derivatives
[S—(C)(H)]	S—H group of an alkanethiol; sulfur atom bearing a hydrogen atom attached to a tetragonal, sp^3-hybridized carbon atom
[SO—(C)$_2$]	Sulfoxide of a saturated sulfoxide; SO group attached to two sp^3-hybridized carbon atoms
[C—(SO$_2$)(H)$_3$]	Methyl group of a methyl sulfone
[C—(Sn)(H)$_3$]	Methyl group of a methylstannane
[Sn—(C$_3$)(Br)]	Tin atom of a trialkylstannyl bromide; tin atom bearing a bromine atom and attached to three sp^3-hybridized carbon atoms

More groups are given in Table 2.A.1 with their increments proposed by Benson et al. for the calculation of the standard heat of formation, the standard entropy, and the molecular heat capacity. Sometimes, it is simpler and more accurate to derive the heat of formation of a given compound by considering it as a substituted derivative of a compound for which the standard heat of formation is available. For instance, the standard heat of formation of *n*-propylbenzene in the gas phase can be derived from that of toluene. It is only necessary to substitute the one hydrogen atom of methyl group of toluene by an ethyl group. This is done on adding to the heat of formation of toluene an increment associated with this substitution and noted [C—(C$_b$)(C)(H)$_2$] – [C—(C$_b$)(H)$_3$]. Statistically, the increment associated with this substitution amounts to 4.9 kcal mol^{-1} (increments of other group exchanges are also given in Table 2.A.1). Then, the increments associated with the ethyl group are added as shown below:

$$\begin{aligned}\Delta_f H^\circ(n - \mathrm{Pr} - \mathrm{Ph, gas}) &= \Delta_f H^\circ(\mathrm{Me} - \mathrm{Ph, gas}) + [\mathrm{C} - (\mathrm{C_b})(\mathrm{C})(\mathrm{H})_2] \\ &\quad - [\mathrm{C} - (\mathrm{C_b})(\mathrm{H})_3] + [\mathrm{C} - (\mathrm{C})_2(\mathrm{H})_2] \\ &\quad + [\mathrm{C} - (\mathrm{C})(\mathrm{H})_3] \\ &= 12.0 + 4.9 – 4.92 – 10.12 = 1.86\ \mathrm{kcal\ mol^{-1}}.\end{aligned}$$

This result is in perfect agreement (within experimental error) with the experimental standard heat of formation of *n*-propylbenzene (1.89 kcal mol^{-1}).

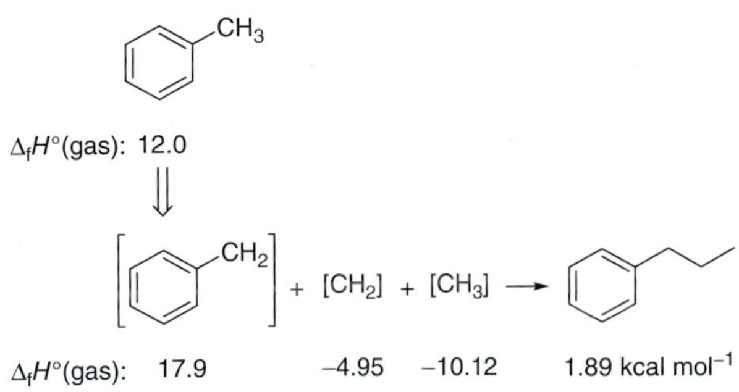

2.3 Determination of the standard group equivalents (group equivalents)

Comparison of the measured standard heats of formation of *n*-alkanes (see below) shows that the insertion of one methylene (CH$_2$) group in ethane, propane, *n*-butane, etc., leads to very similar decreases in heats of formation. Using the data available in 1969 (see below, they differ only slightly from those of Table 1.A.2, recent NIST values), Benson et al. proposed a value of –4.92 kcal mol^{-1} for the increment associated with group [C—(C)$_2$(H)$_2$] as obtained for a large number of hydrocarbons (–5.0 kcal mol^{-1} is commonly used for this increment) [2]. The insertion of a methylene group in the C—H bond of methane, to generate ethane, is associated with a value of −2.35 kcal mol^{-1} instead of −4.92 kcal mol^{-1}, showing the importance of the environment of the methylene group: it is flanked by one hydrogen and one sp^3-hybridized carbon in ethane and by two sp^3-hybridized carbons in propane and the other larger *n*-alkanes.

	$\Delta_f H°$(gas)	[C–(C)(H)(H$_2$)]
CH_4	−17.89	
CH_3–CH_2–H	−20.24	−2.35
CH_3–CH_2–CH_3	−24.82	−4.58
CH_3–CH_2–CH_2–CH_3	−30.36	−5.54
CH_3–CH_2–CH_2–CH_2–CH_3	−35.10	−4.74
CH_3–CH_2–CH_2–CH_2–CH_2–CH_3	−39.92	−4.82

Statistically: −4.92 kcal mol^{-1} for [C–(C)$_2$(H)$_2$]

$\Delta_f H°(CH_3–CH_3)$ = 2 group equivalents of [C–(C)(H)$_3$]

Statistically: −10.12 kcal mol^{-1} for [C–(C)(H)$_3$]

The standard heat of formation of ethane gives a first value for the group equivalent associated with the methyl group ([C—(C)(H)$_3$]) of *n*-alkanes of −10.12 kcal mol^{-1} = ½$\Delta_f H°$(ethane). The same value is found for a large number of alkanes (−10.0 kcal mol^{-1} is commonly used for this increment). The same group equivalent can be used for methyl groups on alkenes, alkynes, and benzenes, as the equivalents for the sp^2- or sp-hybridized carbon centers are derived appropriately (Table 2.A.1).

Comparing the standard heats of formation of linear and branched hydrocarbons, the group equivalents associated with methine group [C—(C)$_3$(H)] of monobranched alkanes, and of carbon atom [C—(C)$_4$] of geminal doubly branched hydrocarbons, can be determined.

n-Pentane

$\Delta_f H°$(gas): −35.1

2 [C–(C)(H)$_3$]:	−20.24
3 [C–(C)$_2$(H)$_2$]:	−14.76
First estimate:	
Total:	−36.00

2-Methylbutane

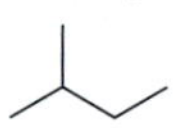

−36.7

3 [C–(C)(H)$_3$]:	−30.36
1 [C–(C)$_2$(H)$_2$]:	−4.92
1 [C–(C)$_3$(H)]:	−1.57
$\Delta_f H°$(gas)exp:	−36.85

2,2-Dimethylpropane

−40.14 kcal mol^{-1}

4 [C–(C)(H)$_3$]:	−40.48
1 [C–(C)$_4$]:	0.34
$\Delta_f H°$(gas)exp:	−40.14 kcal mol^{-1}

A statistical analysis led Benson et al. propose the group equivalents (see Table 2.A.1) of −1.75 kcal mol^{-1} for [C—(C)$_3$(H)] and −0.06 kcal mol^{-1} for [C—(C)$_4$] (−2.0 and 0.0 kcal mol^{-1} are commonly used for these two increments) [2].

Comparison of the standard heats of formation of alkenes (gas phase) allows the statistical determination of the increments associated with the groups found in these compounds. A first estimate of the increment associated with a methylidene group [C$_d$—(C$_d$)(H)$_2$] is given by ½$\Delta_f H°(CH_2{=}CH_2)$ = (12.54)/2 = 6.27 kcal mol^{-1}. With $\Delta_f H°(CH_3{—}CH{=}CH_2)$ = 4.88 kcal mol^{-1} and taking 6.27 kcal mol^{-1} for the methylidene group equivalent and −10.12 kcal mol^{-1} for the group equivalent associated with the methyl group of alkanes, a first value of 8.73 kcal mol^{-1} is obtained for the group equivalent associated with [C$_d$—(C$_d$)(C)(H)] of propene: ($\Delta_f H°(CH_3{—}CH{=}CH_2)$ = sum of increments associated with [C—(C)(H)$_3$] + [C$_d$—(C$_d$)(C)(H)] + [C$_d$—(C$_d$)(H)$_2$] = −10.12 + 8.73 + 6.27 = 4.88 kcal mol^{-1}.

Statistically, the increments associated with groups [C—(C$_d$)(H)$_3$], [C$_d$—(C$_d$)(H)$_2$] and [C$_d$—(C$_d$)(C)(H)] of alkenes have been found to be −10.12, 6.22, and 8.65 kcal mol^{-1}, respectively. With these group equivalents, $\Delta_f H°$((*E*)-but-2-ene, est.) = −10.12 + 8.65 + 8.65−10.12 = −2.54 kcal mol^{-1} is estimated. This estimate compares well with the experimental value $\Delta_f H°$((*E*)-but-2-ene, exp.) = −2.58 ± 0.24 kcal mol^{-1}. For (*Z*)-but-2-enc, an experimental heat of formation of −1.83 ± 0.24 kcal mol^{-1} is found, showing that steric repulsions between the two methyl groups destabilize the (*Z*)-alkene with respect to its (*E*)-isomer by 0.75 ± 0.54 kcal mol^{-1} (Table 1.A.2, see Section 1.7.3).

In the gas phase, the heat of isomerization of 2-methylbut-1-ene into 2-methylbut-2-ene amounts to −1.51 ± 0.41 kcal mol^{-1} (Table 1.A.2). Wiberg and Hao reported for the same isomerization in aqueous CF_3COOH a value of −1.82 ± 0.06 kcal mol^{-1}, very similar to the gas-phase value [11]. This encourages us to use the gas-phase standard heats of formation of hydrocarbons to estimate the heat of reactions of these hydrocarbons in polar and ionizing solutions such as in H_2O/CF_3COOH.

$\Delta_r H°$(soln.) = −1.82 ± 0.06 kcal mol^{-1}

$\Delta_r H°$ (gas) = −1.52 ± 0.41 kcal mol^{-1}

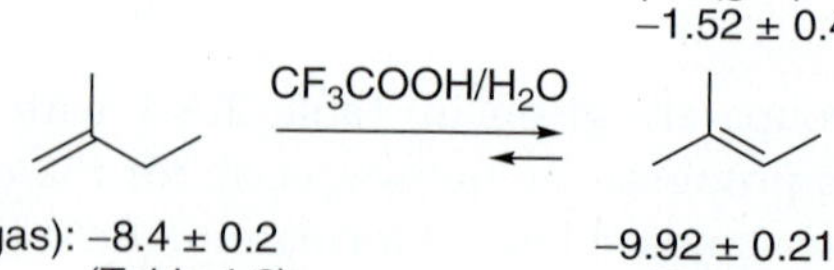

$\Delta_f H°$(gas): −8.4 ± 0.2 (Table 1.2)

−9.92 ± 0.21 kcal mol^{-1} (Table 1.2)

Applying the group equivalents given in Table 2.A.1, one estimates the heats of formation of *n*-butanol, *sec*-butanol (butan-2-ol), and *tert*-butanol (2-methylpropan-2-ol) to be $\Delta_f H°(n\text{-BuOH, est.}) = -65.81$, $\Delta_f H°(s\text{-BuOH, est.}) = -70.04$, and $\Delta_f H°(t\text{-BuOH, est.}) = -74.87\,\text{kcal mol}^{-1}$. These values are quite similar to the experimental values $\Delta_f H°(n\text{-BuOH, exp.}) = -66.0 \pm 1.0$, $\Delta_f H°(s\text{-BuOH, exp.}) = -70.0 \pm 0.08$, and $\Delta_f H°(t\text{-BuOH, exp.}) = -74.72 \pm 0.21\,\text{kcal mol}^{-1}$ (Table 1.A.4), respectively. When an experimental value for the heat of formation deviates by more than $2\,\text{kcal mol}^{-1}$ from the value estimated by the additivity rule, it is worth repeating the measurements!

n-Butanol		Butan-2-ol		2-Methylpropan-2-ol	
[C—(C)(H)$_3$]	−10.12	2[C—(C)(H)$_3$]	−20.24	3[C—(C)(H)$_3$]	−30.36
2[C—(C)$_2$(H)$_2$]	−9.84	[C—(C)$_2$(H)$_2$]	−4.92	[C—(O)(C)$_3$]	−6.66
[C—(O)(C)(H)$_2$]	−8.00	[C—(O)(C)$_2$(H)]	−7.03	[O—(C)(H)]	−37.85
[O—(C)(H)]	−37.85	[O—(C)(H)]	−37.85		
Estimate	−65.81		−70.04		−74.87 kcal mol^{-1}

For certain classes of compounds, their experimental standard heats of formation and entropies deviate significantly (>2 kcal mol^{-1} for $\Delta_f H°$, >5 eu for $S°$) from the values calculated by simply adding the increments associated with the constituent fragments (Table 2.A.1). Quite often, these deviations are associated with specific structural features, and thus, they become increments themselves, or corrections that can be added to the sum of the increments (Eqs. (2.4) and (2.6)). These deviations arise from several factors, as discussed in the next chapters. They include gauche interactions and other steric interactions between groups that are close in space ("Front-strain," "Back-strain" allylic strain, and bond eclipsing), geometrical deformations (ring strain), electrostatic interactions (repulsive or attractive dipole/dipole interaction and enthalpic and conformational anomeric effects), conjugation due to π/π and n/π overlap, cyclic conjugation (aromaticity and antiaromaticity), and σ/π interaction (hyperconjugation).

Problem 2.3 Using data of Table 1.A.4, determine the group equivalents for the molecular group of thiols.

2.4 Determination of standard entropy increments

Comparing the experimental values of the standard entropies of alkanes, the entropy increments given below can be determined [2]:

$$[\text{C}-(\text{C})(\text{H})_3]:\ 30.41\ \text{eu};\quad [\text{C}-(\text{C})_2(\text{H})_2]:\ 9.42\ \text{eu};$$
$$[\text{C}-(\text{C})_3(\text{H})]:\ -12.07\ \text{eu};\quad [\text{C}-(\text{C})_4]:\ -35.10\ \text{eu}$$

Table 1.A.2 gives the standard entropy increments for other groups. The standard entropies of organic compounds can be estimated from the additivity rule (2.6).

$$\begin{aligned} &S°(\text{pure compound, gas}) \\ &\quad = \sum(\text{entropy increments associated with molecular groups, (Table 2.1)} \\ &\quad + \text{corrections (see below)} \end{aligned} \tag{2.6}$$

A comparison of the experimental standard entropies of *n*-octane and 2,2,3,3-tetramethylbutane shows that the entropy of a hydrocarbon decreases with the degree of chain branching. This is due to the local C_3 symmetry of the methyl groups and to other changes in symmetry between the two compounds. Internal rotation of a methyl group generates three identical molecules (not different rotamers) during one full rotation around the σ(C,C) bond. This leads to a decrease of entropy by $-R\ln3$ (Eq. (1.33)). Local axial symmetry, as well as molecular axial symmetry, diminishes the number of rotamers available. Isomerization (2.7) of *n*-octane into 2,2,3,3-tetramethylbutane is associated with $\Delta_r S° = -18.5$ eu. This arises from the greater symmetry of 2,2,3,3-tetramethylbutane compared with *n*-octane. Indeed, there are four more methyl groups, two *tert*-butyl groups and a C_3 axis of rotation in 2,2,3,3-tetramethylbutane not present in *n*-octane. The *tert*-butyl groups as well as the methyl groups have local C_3 axis that reduces the entropy by $-R\ln_3$ each. Therefore, a first estimate is $\Delta_r S°(2.7)_{est} = 4(-R\ln3) + 2(-R\ln3) - R\ln3 = -15.3$ eu. This value is less negative than the experimental value of −18.5 eu. The difference is associated with the reduced moment of inertia in the more compact 2,2,3,3-tetramethylbutane compared with *n*-octane (entropy contribution x), as well as some correlated rotation of the Me_3C groups [12], i.e. a "gear effect" in the quadruply branched hydrocarbon (entropy contribution y). Because of gauche interactions (Section 2.5.1) between the methyl groups of the two *tert*-butyl groups opposing each other, the rotation of their methyl group is not completely independent:

these groups rotate more or less synchronously. The preferred linear zigzag conformation of n-octane (and of other n-alkanes) will be discussed below.

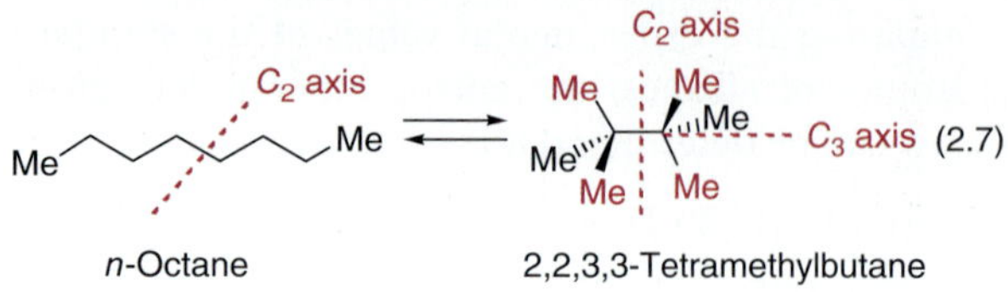

$\Delta_f H°$(gas): −49.88 ± 0.16 | −54.06 ± 0.36 kcal mol^{-1}

$S°$(gas): 111.6 eu | 93.1 eu

$\Delta_r H°(2.7) = -4.2 \pm 0.52$ kcal mol^{-1}

$\Delta_r S°(2.7) = -18.5$ eu

in n-octane:
Two methyl groups, one C_2 axis, larger inertia moment

projection of 2,2,3,3-tetramethylbutane in a plane perpendicular to the C_3 axis:

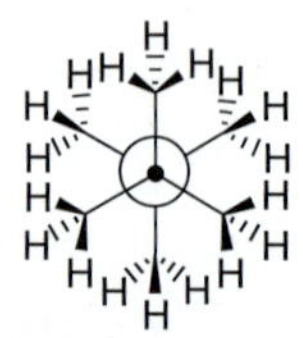

Four more methyl groups:	4(−Rln3)
One more C_3 axis:	(−Rln3)
Two *tert*-butyl groups	2(−Rln3)
Smaller inertia moment	x
Gear effect	y

$\Delta_r S°(2.7) = -15.26 + x + y$ eu

For alkenes, one also finds that their standard entropies decrease with the degree of chain branching, as illustrated below.

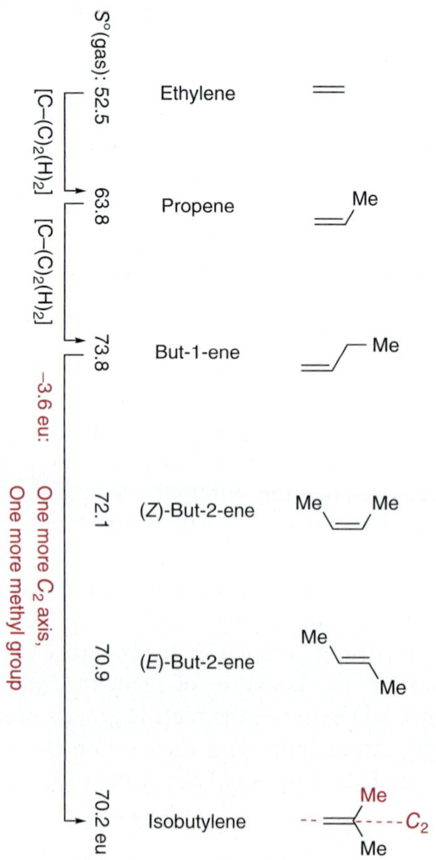

Problem 2.4 Calculate the equilibrium constant of isomerization (2.7) at 25 °C. At which temperature should this isomerization be carried out in order to obtain a yield of 80% in 2,2,3,3-tetramethylbutane by isomerization of n-octane?

Problem 2.5 n-Heptane is the hydrocarbon used in engine fuels with the highest knocking effect, while 2,2,4-trimethylpentane has the smallest knocking effect. A gasoline with 95% octane means that its knocking effect is the same as a 5 : 95 mixture of n-heptane and 2,2,4-trimethylpentane. Isomerize n-octane into 2,2,4-trimethylpentane in order to obtain a 5 : 95 mixture of octane and 2,2,4-trimethylpentane. At which temperature does the isomerization have to be carried out under 1 atm?

Hint: Use the $\Delta_f H°$(gas) values given in Table 1.A.2 and estimate the standard entropy of that isomerization.

2.5 Steric effects

2.5.1 Gauche interactions: the preferred conformations of alkyl chains

The n-alkanes adopt zigzag conformations in order to avoid repulsive gauche interactions, as illustrated in Figure 2.1 for n-butane, n-pentane, and n-hexane, using Newman projections along all nonterminal C—C bonds [13]. Long-chain substituted alkanes bearing a polar function at one terminus such as COONa (soap), NR_3Cl (ammonium detergents), SO_3Na (sulfonate detergents), as well as alcohols and polyols tend to form micelles in an aqueous medium. The same phenomena are responsible for the formation of cytoplasmic membranes of living cells (glycerophospholipids).

The same analysis of the Newman projections of 2-methylbutane, 2,2-trimethylbutane, 2,2,3-trimethylbutane, and 2,2,3,3-tetramethylbutane shows that all these alkanes cannot avoid gauche interactions between vicinal methyl groups in their most stable staggered conformations. One, two, four, and six gauche interactions are present in 2-methylbutane, 2,2-dimethylbutane, 2,2,3-trimethylbutane, and 2,2,3,3-tetramethylbutane, respectively.

In these molecules, the heats of formation are 0.33, 0.71, 3.45, and 6.66 kcal mol^{-1}, respectively, higher than expected if no correction is made for Me/Me interactions. However, there is no linear relationship between these energetic variations and the number of gauche interactions. The strain due to gauche interactions increases with their accumulation along bond C(2)—C(3) of the butane system. This is explained by the fact that the “ideal” 60° dihedral angle present in ethane and in 2,2,3,3-trimethylbutane for their stable

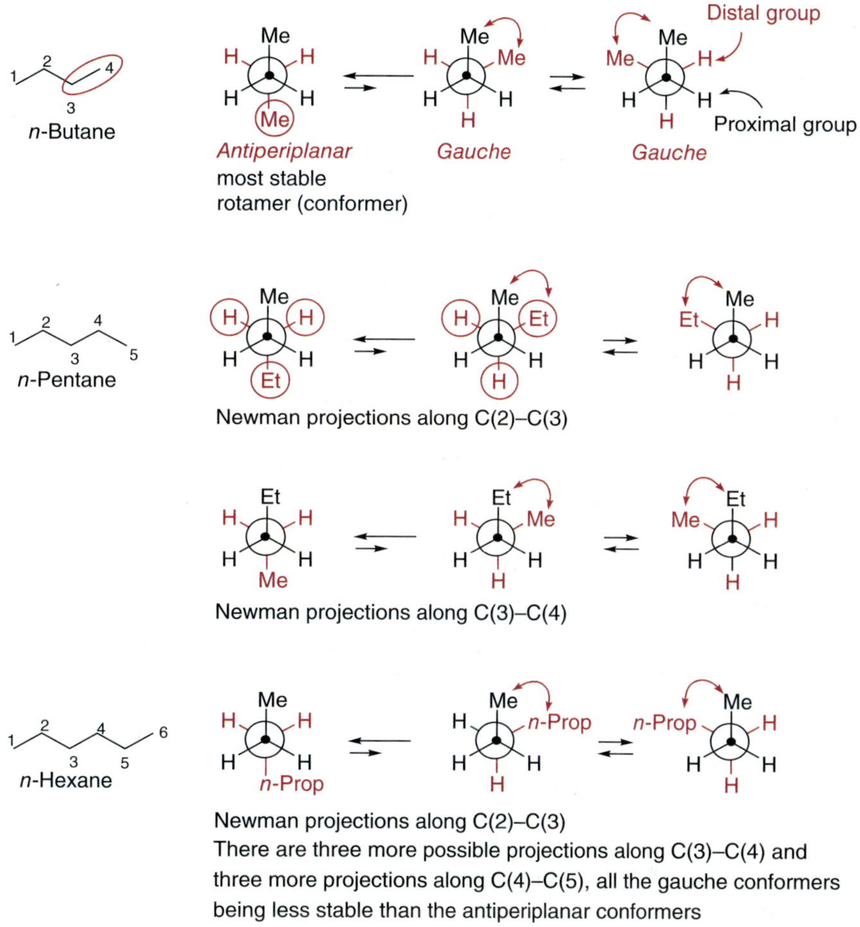

Figure 2.1 Newman projections for *n*-alkanes showing the staggered conformations. Antiperiplanar conformers are devoid of repulsive gauche interactions.

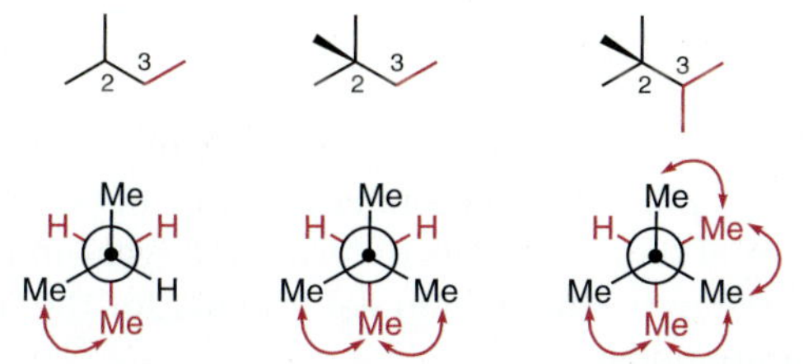

optimal torsional angle does not have to be 60 degrees

Σ group equivalents:	−37.03	−45.06	−52.41
$\Delta_f H°$(gas):	−36.7 ± 0.14	−44.35 ± 0.23	−48.96 ± 0.27
Deviations:	0.33	0.71	3.45

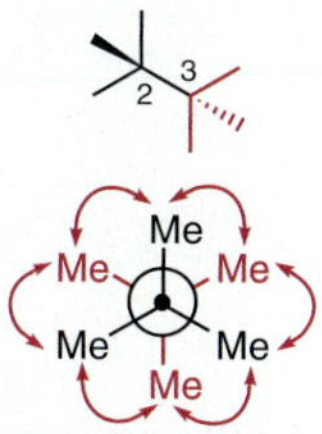

six gauche interactions, optimal torsional angle: 60 degrees,

−60.72 kcal mol^{-1}
−54.06 ± 0.36 kcal mol^{-1}

Deviation: 6.66 kcal mol^{-1}

staggered conformers does not occur in the branched hydrocarbons. Because of the shallow energy potential for rotation about C—C bonds in alkanes, repulsion between the two gauche, vicinal methyl groups in 2-methylbutane increases the dihedral angle between Me–C(3) and the gauche Me–C(2) groups to a value larger than 60°. This type of torsional deformation is not available to the other branched hydrocarbons. In 2,2-dimethylbutane, the molecule can open the bond angle between the two geminal methyl groups at C(2), but torsional deformation cannot occur. Similarly, in 2,2,3-trimethylbutane and 2,2,3,3-tetramethylbutane, only bond angle deformations and bond elongations can alleviate repulsive interactions between the vicinal methyl groups. The X-ray crystal data for the hexasubstituted ethane derivatives **1** [14] and **2** show these deformations [15]. The σC(2,3) bond of **1** is remarkably elongated to 1.639 Å instead of 1.54 Å as found for standard alkanes (Table 1.A.6). For **2**, an extra short σC(1,1′) bond length of 1.445 Å is measured. Similarly, with the tetrahedranyltetrahedrane derivative **3**, an extra short single C(1)—C(1′) bond of 1.436 Å is measured [16]. This is explained by the fact that the repulsion between the substituents at C(1) and C(1′) of **2** and **3** is dramatically reduced as two and three, respectively, of their bond angles (cyclopropanes) are about 60°, instead of 110° for standard alkanes.

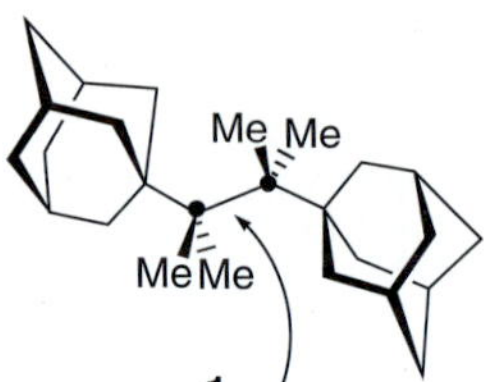

1

1.639 Å: extra-long bond due to front-strain

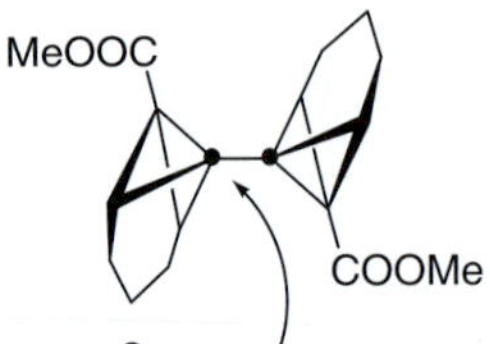

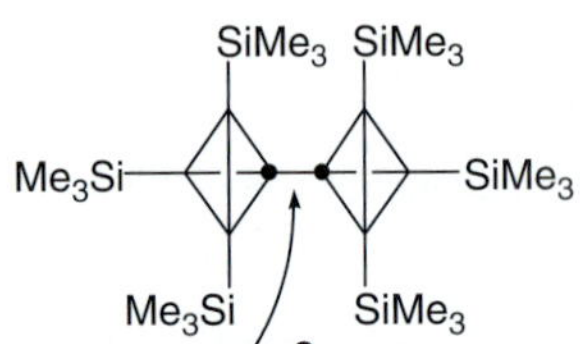

2 **3**

1.445 Å: extra-short bond: reduced repulsion between vicinal groups 1.436 Å

These examples show that it is difficult to define a single increment for any gauche interaction that exists in an organic compound. The number of gauche interactions is not sufficient to define quantitatively the stability of alkanes; for example, for 2,3-dimethylbutane, the *trans* (with two gauche interactions) and *gauche* rotamers (with three gauche interactions) have nearly the same stability [17].

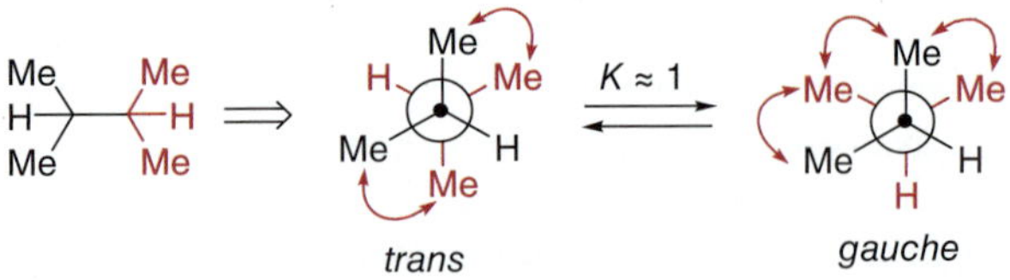

Nevertheless, when comparing normal alkanes such as 2-methyl- and 3-methylpentane, the difference in their experimental standard heats of formation (0.6 kcal mol^{-1}) corresponds to the fact that the former compound has only one gauche interaction, whereas the latter has two. Applying this analysis to a large number of monobranched alkanes, one obtains an increment of 0.7 kcal mol^{-1} per gauche interaction.

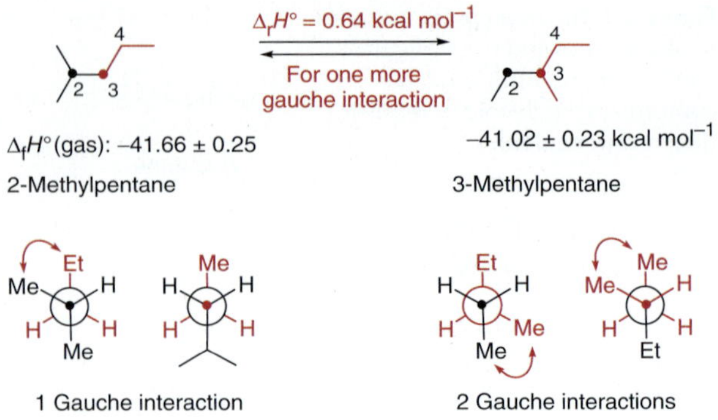

Comparison of the experimental standard heats of formation of alkanes **4**, **5**, and **6** with the estimated values that consider a correction of 0.7 kcal mol^{-1} for each gauche interaction leads to deviations of 1.63, 1.22, and 5.24 kcal mol^{-1}, respectively. They are larger than the experimental standard deviations given for the heat of formation of these hydrocarbons. These observations demonstrate that on top of the gauche repulsive interaction defined above, there are further steric repulsions; these are assigned to "Front-strain" (F-strain). "F-strain" is important between the two *tert*-butyl groups of 2,2,4,4-tetramethylpentane (**6**).

	4	**5**	**6**
Σ group equivalents + corrections for gauche interactions:	−55.2	−53.83	−62.95 kcal mol^{-1}
$\Delta_f H°$(gas, WebBook):	−53.57 ± 0.32	−52.61 ± 0.36	−57.71 ± 0.35
Deviations ("Front-strain")	1.63	1.22	5.24 kcal mol^{-1}

Without corrections for the 1,4-gauche interactions, the following deviations are found between experimental standard heats of formation (gas) and the sum of the molecular group equivalents for the following compounds:

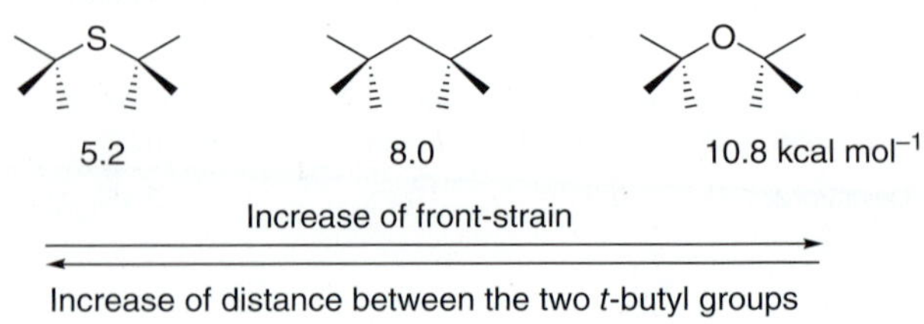

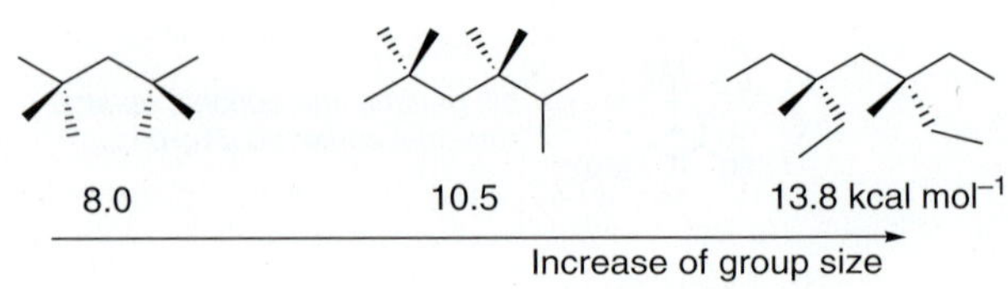

The increase of front-strain between di(t-butyl) sulfide and di(t-butyl) ether is explained by considering the difference in bond lengths for sulfides (C—S: 1.73 Å) and ethers (C—O: 1.43 Å, Table 1.A.6).

2.5.2 (*E*)- vs. (*Z*)-alkenes and *ortho*-substitution in benzene derivatives

As already seen in Section 1.7.3, (E)-alkenes are more stable than their (Z)-isomers by about 0.8 kcal mol^{-1} for 1,2-di(n-alkyl)ethene derivatives. For larger substituents, the difference increases [18]. Sometimes, especially in solution, (Z)-alkenes can be as stable as or more stable than their (E)-isomers. For instance:

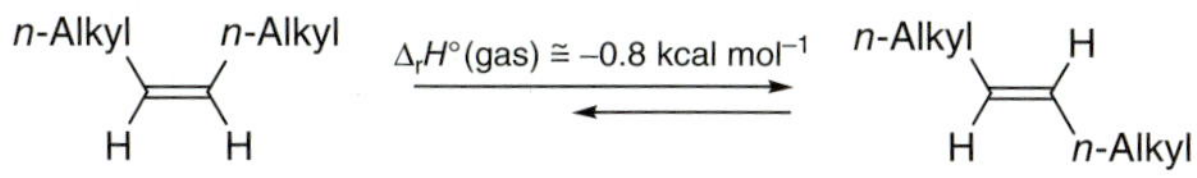

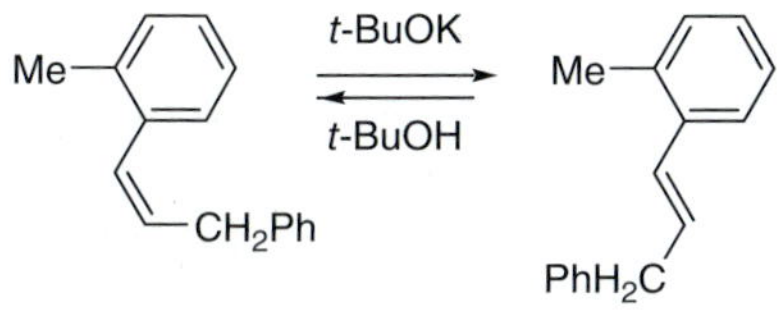

$\Delta_rH = 0.3 \pm 1.7$ kcal mol^{-1}

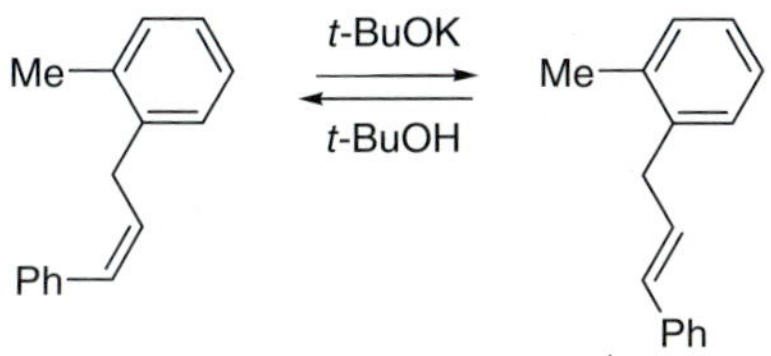

$\Delta_rH = 1.3 \pm 1.5$ kcal mol^{-1}

ortho-disubstituted benzenes (and other related aromatic heterocyclic compounds) are usually less stable than the *meta*- and *para*-disubstituted isomers for steric reasons (eclipsing of the *ortho*-substituents). The destabilization effect varies between 1 kcal mol^{-1} for n-alkyl substituents to several kilocalories per mole for more bulky groups (Table 2.A.2). When polar substituents are present, electrostatic repulsions or/and attractions (Sections 2.7.9 and 2.8) can add to the steric, repulsive eclipsing effect. An extreme case of repulsive *ortho*-disubstitution is realized in dimethyl 3,4,5,6-tetrakis(*tert*-butyl)phthalate (7). This compound is converted at 120 °C to the Dewar benzene derivative **8** in 90% yield. Although the parent Dewar benzene is 57.3 kcal mol^{-1} less stable than benzene (Table 1.A.2), the relief of back-strain and front-strain when going from **7** to **8** makes this thermal isomerization possible [19]. Quantum mechanical calculations predicted that tetrahedrane (tricyclo[1.1.0.0^{2,4}]butane) is 26 kcal mol^{-1} less stable than cyclobutadiene [20]. Tetrakis(trimethylsilyl)cyclobutadiene (**9**) is destabilized by F-strain to a greater extent than tetrakis(trimethylsilyl)tetrahedrane (**10**). In the presence of $B(C_6F_5)$ as the catalyst (acting as the electron acceptor, catalysis through single electron transfer, Section 7.5.8), **9** is isomerized into **10** at 20 °C through a formal intramolecular $[\pi^2a + \pi^2a]$-cycloaddition [21].

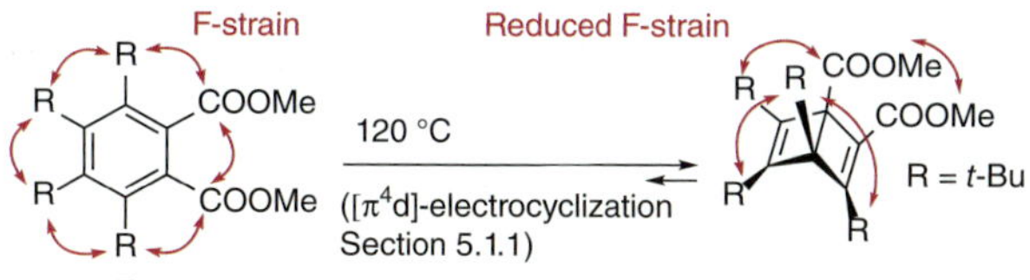

7 **8**

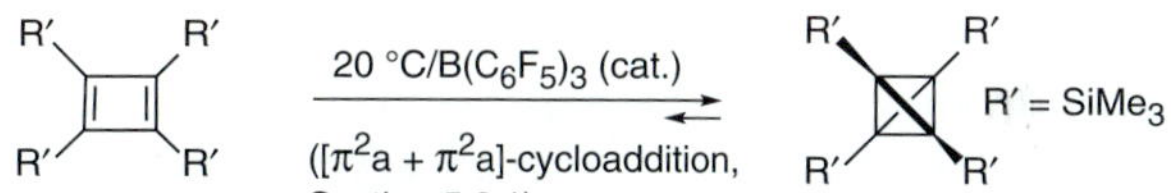

9 **10**

2.6 Ring strain and conformational flexibility of cyclic compounds

The heats of formation of the fullerenes C_{60} (buckminsterfullerene) and C_{70} have been measured by combustion calorimetry. Compared with graphite (planar sheets), C_{60} and C_{70} are strained by c. 10 kcal mol^{-1} and c. 9 kcal mol^{-1} per carbon atom, respectively [22–25].

	C_{60}	C_{70}
$\Delta_fH°$(gas):	597.7 ± 4.1	631.5 ± 3 kcal mol^{-1}
$\Delta_fH°$(solid):	556.2 ± 4.1	610.6 ± 3 kcal mol^{-1}

The ring strain, [S], of a cycloalkane, C_nH_{2n}, is defined by Eq. (2.8).

$$[S] = \Delta_fH°(\text{cycloalkanes}, C_nH_{2n}, \text{gas}) - n(-4.92), \quad \text{in kcal/mol} \tag{2.8}$$

For other monocyclic and polycyclic systems, the ring strain energy is given by the difference between the experimental heat of formation $\Delta_f H°$(gas) and the sum of enthalpy increments of the component groups. Table 2.A.3 gives the ring strains, [*S*], for some cycloalkanes, cycloalkenes, and polycyclic hydrocarbons and their derivatives [26]. Ring strain arises from geometrical deformations, such as bond angle deformation (that is, deviation from the standard C—C/C—C bond angle of 110°), C—C and C—H bond eclipsing, and transannular repulsions between methylene groups. Bond angle deformation is often referred to as "Baeyer strain", whereas eclipsing strain is also called "Pitzer strain." 1,3-Transannular interactions are also known as "Dunitz–Schomaker strain" and 1,4 and the longer transannular interactions are also called "Prelog strain". [27–32] Analysis of the ring strains of cycloalkanes possessing three to eight carbon atoms as a function these factors permits to predict their most stable conformations and whether they are rigid (adopt a single conformation) or flexible compounds (adopt several conformations that can be interconverted with low energy barriers).

2.6.1 Cyclopropane and cyclobutane have nearly the same strain energy

Cyclopropane ([*S*] = 27.5 kcal mol^{-1}) and cyclobutane ([*S*] = 26.5 kcal mol^{-1}) are the most strained of all monocyclic cycloalkanes. Cyclopentane ([*S*] = 6.3 kcal mol^{-1}) and medium-sized cycloalkanes ($n = 7$–14, [*S*] = 4–12 kcal mol^{-1}), on the other hand, are less strained. Large rings ($n = 20$) are not strained as they resemble long-chain *n*-alkanes. The cyclopropane molecule is planar (Schoenflies notation: D_{3h}, that are, the symmetry point group is characterized by the presence of a threefold axis of rotation (C_3) and three twofold axis of rotation (C_2) that is perpendicular to the C_3 axis and a mirror plane σ of symmetry perpendicular to the C_3 axis) [33]. It has all of its six vicinal C—H bonds eclipsed with one another, while the three C—C bonds are arranged at the small bond angles of 60° instead of 110°. As shown below, because the energy barrier for rotation about the C—C bond of ethane, which contains three pairs of vicinal C—H bonds, is about 3 kcal mol^{-1}, it is convenient to assign a penalty of 1 kcal mol^{-1} per pair of C—H/C—H eclipsing. Thus, from the six pairs of eclipsing C—H bonds in cyclopropane, one estimates that the C—H/C—H eclipsing amounts to 6 kcal mol^{-1} of the total cyclopropane strain (Pitzer strain or torsional strain) [34]. The remaining ring strain of cyclopropane (27.5 − 6.0 = 21.5 kcal mol^{-1}) is associated with the total bond angle deformation, which amounts to 3(110° − 60°) = 150° of Bayer strain or angle strain.

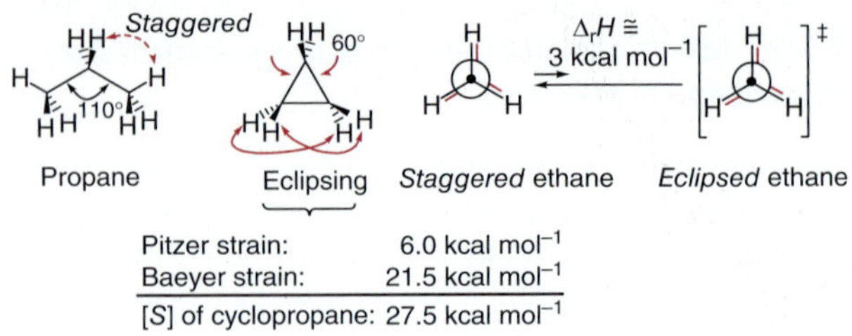

The conformational analysis of butane shows that the *eclipsed syn*-periplanar conformer transition structure for rotation about the central C—C bond is c. 4.4 kcal mol^{-1} above the most stable *staggered anti*-periplanar conformer [35]. Eclipsing of the σ(C(1),C(2))/σ(C(3),C(4)) bond in *syn*-periplanar butane leads to a strain of 2.4 kcal mol^{-1} if one considers that eclipsing of the two pairs of C—H bonds of the $H_2C(2)$ and $H_2C(3)$ groups contributes by 2 kcal mol^{-1}, as in ethane. When the cyclobutane ring is planar (point group D_{4h}: one C_4 axis of rotation, four twofold (C_2) axes of rotation perpendicular to the C_4 axis, and a mirror plane perpendicular to the C_4 axis), there are eight pairs of eclipsing C—H/C—H bonds (8 kcal mol^{-1} of Pitzer strain) and two pairs of eclipsing σ(C,C)/σ(C,C) bonds (4.8 kcal mol^{-1} of strain). Cyclobutane equilibrates between two nonplanar conformers having a puckering angle of about 30° (point group C_{2v}: one C_2 rotation axis and two mirror planes σ containing the axis). This reduces the repulsions due to the C—H and C—C bond eclipsing to a fraction of 8 + 4.8 kcal mol^{-1}, or approximately 7 kcal mol^{-1}. Baeyer strain in cyclobutane corresponds to a total angle deformation of about 4(110° − 90°) = 80°. Assuming a linear relationship for the energy potential with angle deformation and considering that in cyclopropane the Baeyer strain amounts to 21.5 kcal mol^{-1} for 150° (0.15 kcal mol^{-1} per degree), the contribution of Baeyer strain in cyclobutane is about 0.15(80) = 12 kcal mol^{-1}. Pitzer and Baeyer strain in cyclobutane amounts to 7 + 12 = 19 kcal mol^{-1}. The difference with the total ring strain of cyclobutane 26.5 − 19 = 7.5 kcal mol^{-1} is attributed to 1,3-transannular repulsions between the methylene moieties (Dunitz–Schomaker strain) [29, 36]. The latter arises from the fact that the 1,3-distance in cyclobutane is shorter than in *n*-alkanes because of the smaller bond angle (approximately 90° instead of 110°).

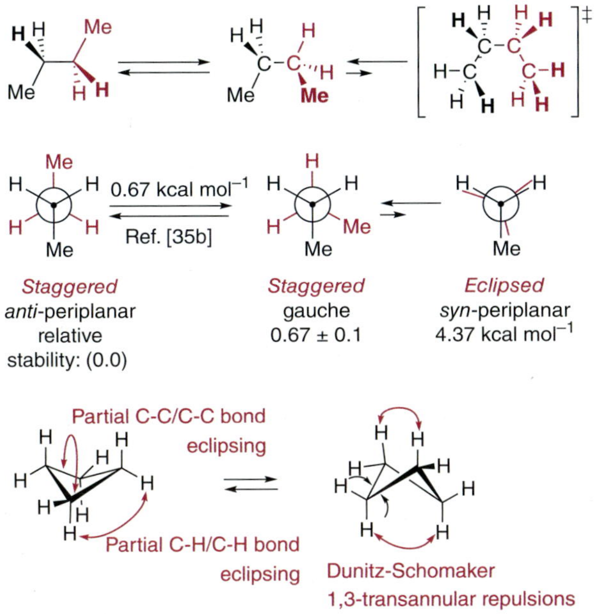

Pitzer strain:	~7.0 kcal mol^{-1}
Baeyer strain:	~12.0 kcal mol^{-1}
Dunitz-Schomaker:	~7.5 kcal mol^{-1}
[S] of cyclobutane:	26.5 kcal mol^{-1}

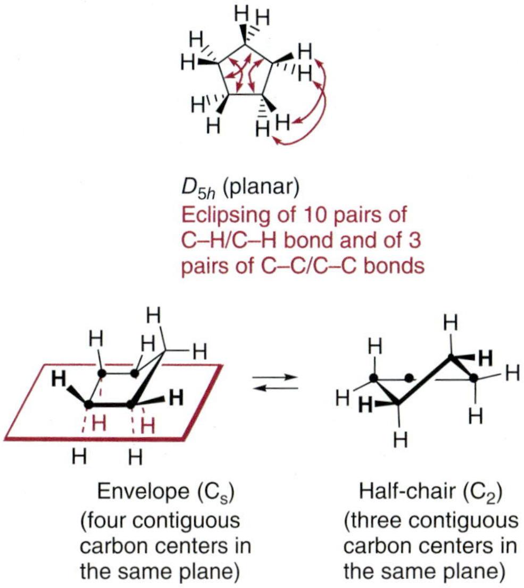

2.6.2 Cyclopentane is a flexible cycloalkane

If cyclopentane were planar (point group D_{5h}), the ring strain would be mostly due to the eclipsing of 10 pairs of C—H/C—H bonds and three 3 pairs of C—C/C—C bonds. This would amount to a strain of more than 10 kcal mol^{-1}. However, the actual ring strain of cyclopentane is only 6.3 kcal mol^{-1}. The planar conformation of cyclopentane is avoided because all of its methylene groups undulate up and down. A cyclopentane molecule adopts 5 degenerate (that is, same enthalpy) envelopes of C_s symmetry (that is, containing one mirror plane of symmetry) and 10 degenerate half-chair (twisted) forms with three coplanar carbon atoms and C_2 symmetry (that is, containing only one C_2 rotation axis). In the C_2 conformers, the destabilization due to C—H/C—H eclipsing is reduced significantly to make the total ring strain reach 6.3 kcal mol^{-1}.

Cyclopentane is a conformationally mobile molecule because the interconversion of the envelopes and the half-chairs does not occur through any planar transition structures, and the barriers of interconversion are low. In other words, the pseudo-rotation about the σ(C—C) bonds of cyclopentane is facile (Figure 2.2). This is in contrast to the conformational dynamics of cyclohexane (Figure 2.3).

2.6.3 Conformational analysis of cyclohexane

Cyclohexane is unstrained. The chair conformation (four carbon centers C(1),C(2),C(4), and C(5) in the same plane, point group D_{3d}: one C_3 axis of rotation, and three C_2 axes of rotation perpendicular to the C_3 axis, and three mirror planes containing the C_3 axis) has C—C/C—C bond angles of 111.5° (112.5° in n-alkanes, 109.5° for the tetrahedral angle [van't Hoff], and bond angle in methane) [37–39]. The dihedral angles are nearly 60°, corresponding to those in the staggered conformations of n-alkanes. There are six gauche interactions, but these do not introduce any strain, as the hydrogen atoms in gauche-butane are replaced by bonded carbon atoms in cyclohexane. The hydrogen atoms (or any substituents) are designated equatorial or axial if they lie parallel with the equatorial plane or nearly perpendicularly to it, respectively. At room temperature, the ^{1}H-NMR spectrum of cyclohexane shows a single signal. However, at −100 °C, two separate signals are seen for the axial and equatorial hydrogens ($\Delta\delta_H = 0.46$ ppm). The set of equatorial hydrogens (H_e) and the set of axial hydrogens (H_a) are interchanged through the chair ⇄ chair interconversion (Figure 2.3). From the line-shape analysis of the ^{1}H-NMR spectrum of cyclohexane at various temperatures between −100 and 0 °C, one calculates the activation parameters for the chair ⇄ chair interconversion as $\Delta^{\ddagger}H = 10.6$ kcal mol^{-1} and $\Delta^{\ddagger}S = 0$ eu. At 800 °C, cyclohexane contains about c. 25% of a twist-boat conformer (point group D_2: has a C_2 axis of rotation and an additional C_2 axis perpendicular to it). Passing a jet of cyclohexane into a frozen matrix at −253 °C traps this c. 75 : 25 equilibrium mixture (vibrational spectrum of

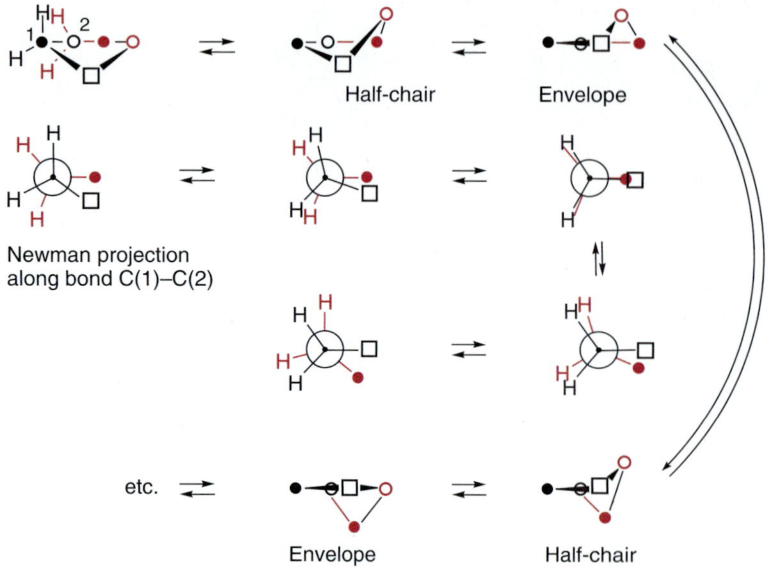

Figure 2.2 Representation of the pseudo-rotation in cyclopentane. The envelope conformers equilibrate with the nearly isoenthalpic half-chair conformers; the bond angles vary little during the interconversion, and C–H/C–H eclipsing are minimized throughout.

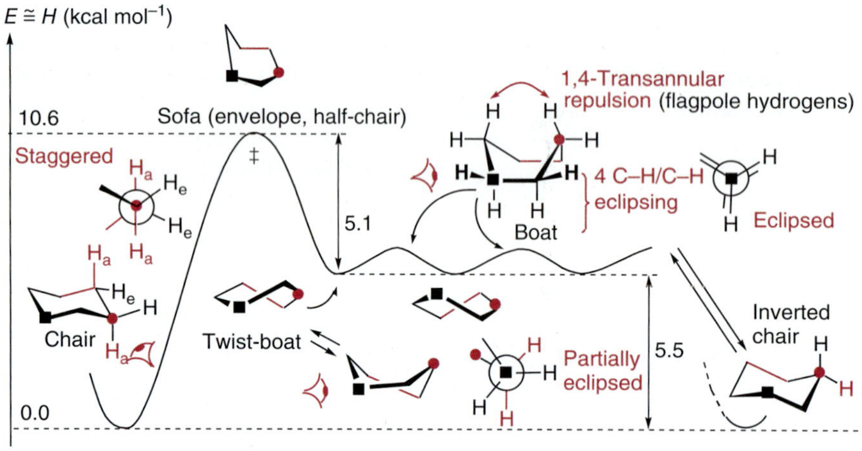

Figure 2.3 Potential energy hypersurface for the interconversion of chair (D_{3d}) ⇄ twist-boat (D_2) ⇄ boat (C_{2v}) ⇄ chair (D_{3d}) conformers of cyclohexane.

twist-boat cyclohexane has been recorded [40]) and the kinetics of the isomerization of the twist-boat conformer to the more stable chair conformer can be followed. Anet and coworkers determined a free energy of activation $\Delta^{\ddagger}G = 5.3\,\text{kcal mol}^{-1}$ ($\Delta^{\ddagger}H \cong 5.1\,\text{kcal mol}^{-1}$) and an enthalpy difference $\Delta_r H$(twist boat ⇄ chair) = 5.5 kcal mol^{-1} [41]. For the hypothetical planar structure of cyclohexane, the ring strain is estimated to be about 21 kcal mol^{-1} because of the eclipsing of 12 pairs of C—H/C—H bonds (Pitzer strain energy of 12 kcal mol^{-1}) and bond angle deformation of 6(120–110) = 60° corresponding to 0.15(60) = 9 kcal mol^{-1} (Baeyer strain energy). It is, thus, clear that planar cyclohexane cannot be the transition structure of the chair ⇄ chair interconversion. An alternative structure with less Pitzer and Baeyer strain such as the sofa (or envelope or half-chair) conformer places five carbon atoms in the same plane and can be isomerized into the twist-boat conformers. Pitzer strain of less than 8 kcal mol^{-1} for the sofa (or envelope) can be estimated. The Baeyer strain is less than that in planar cyclohexane in both the sofa and the envelope. The twist-boat conformers pseudo-rotate *via* boat conformer transition structures. Although Baeyer strain is the same in boat and chair conformers, four pairs of C–H/C–H eclipsing lead to a Pitzer strain of 4 kcal mol^{-1} in the boat. The 1,4-transannular interactions in the boat conformer make the boat less stable by 1–2 kcal mol^{-1} than the twist boat (Figure 2.3).

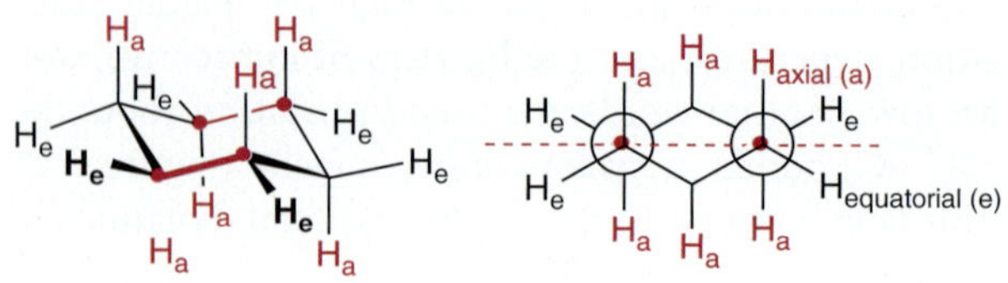

Thus, at a minimum, the interconversion between the two-chair conformations involves the following sequence: chair ⇄ sofa (or half-chair or envelope) ⇄ twist boat ⇄ sofa ⇄ chair. The relative stabilities of the various conformers of the cyclohexane ring follow the order: chair > twist boat > boat > sofa. However, this order might not hold for substituted cyclohexanes and saturated six-membered heterocycles because of steric and electrostatic interactions between the substituents. Although the chair-to-chair interconversion does not occur through any boat conformations, the boats are often included to describe this interconversion because their energies are considerably lower than those of the sofa conformations. In reality, there are multiple pathways by which cyclohexane in its twist-boat conformations can equilibrate with its chair conformations. The deformation of the chair to the sofa (envelope: point group C_s has a mirror plane of symmetry) transforms a perfectly staggered conformer into an eclipsed one; there is a severe torsional barrier for this process, thus making the chair conformer relatively rigid. On the other hand, equilibrium between the twist boat and the boat conformers requires low torsional barriers, resulting in high mobilities between these conformers [42, 43].

Monosubstituted cyclohexanes exist as equilibria (2.9). The free enthalpy $-\Delta_r G°(2.9)$ defines the A values of substituted cyclohexanes [44–47]. Because of gauche interactions, conformers with equatorial substituents are usually preferred (Table 2.1) [48–50]. In 1950, Barton proposed the correct structure of the steroid nucleus, which contains three annulated cyclohexane units all in the chair conformation [51].

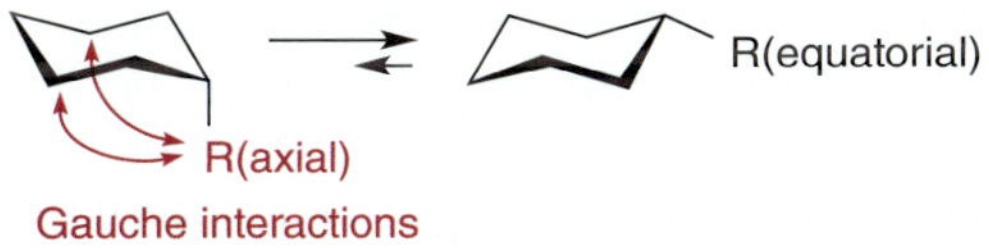

(2.9)

Steroid nucleus: except for the methyl groups all alkyl substituents of each cyclohexane are equatorial

In polysubstituted, six-membered rings, the preference for equatorial substitution is not always observed. For instance, all-*trans*-1,2,3,4,5,6-hexa(isopropyl)cyclohexane prefers a conformation in which all of the isopropyl groups are axial. This is because repulsive steric interactions and torsional effects between the substituents were the ring to adopt the alternative chair conformer in which the isopropyl groups are all equatorial [52, 53].

Table 2.1 Gibbs energy differences between equatorial (more stable) and axial monosubstituted cyclohexanes (*A*-values), in kcal mol^{-1} (1 cal = 4.184 J).

R	$\Delta_r G$	R	$\Delta_r G$
F	0.35	*i*-Pr	2.2
Cl	0.53	*t*-Bu	4.9
Br	0.53	Ph	2.9
I	0.48	C≡CH	0.41
OH	0.90	C≡N	0.24
OMe	0.63	$CH{=}CH_2$	1.49
OAc	0.71	COMe	1.02
SMe	1.0	COOMe	1.31
NO_2	1.05	SO_2Me	2.50
Me	1.6	OSO_2Me	0.56
CF_3	2.5	NH_2	1.41
Et	1.8	MgBr	0.46

2.6.4 Conformational analysis of cyclohexanones

Oxo-substitution in cyclohexane generates cyclohexanone. The stability difference between the boat and the chair is narrowed to only 2.5–3.0 kcal mol^{-1}. In cyclohexanone, either the 1,4-transannular repulsions or the C—H/C—H eclipsing interactions in the boat conformers are reduced [54].

Cyclohexanone:

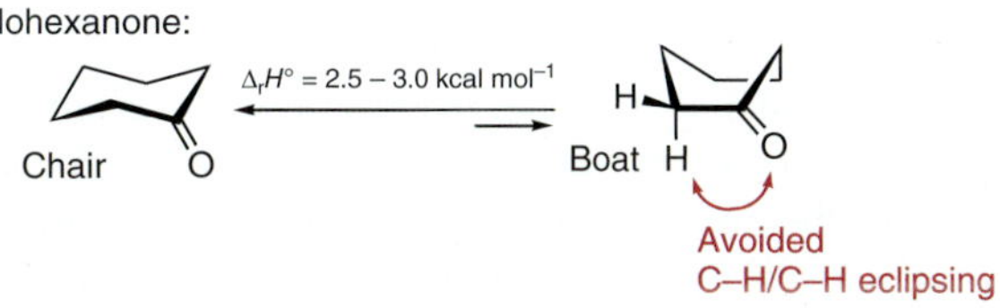

In 4-methylcyclohexanone, one estimates the following energy difference between the two-chair conformers and the twist-boat conformer [55, 56]:

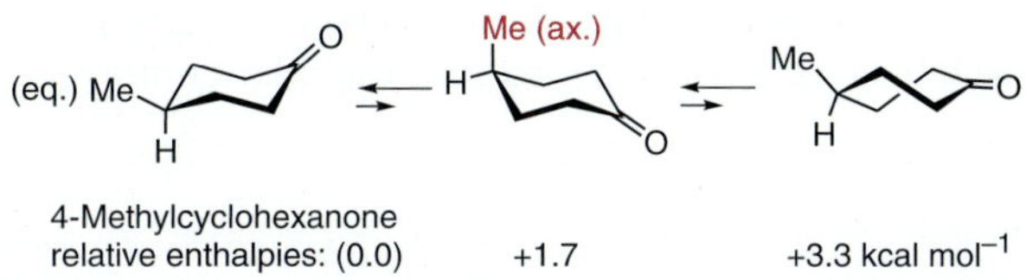

The same stability differences are expected for 3-methylcyclohexanone. Interestingly, cyclohexane-1,4-dione prefers the boat conformation in solution and in the solid state [57–60]. For 2-halocyclohexanones as well as cyclohexanones substituted by other polar substituents, the positions of the conformational equilibria depend heavily on solvent [46, 61, 62]. Quantum mechanical calculations predict that the axial conformer of 2-fluorocyclohexanone is more stable than the equatorial conformer by 0.45 kcal in the gas phase. In CCl_4, acetone, MeCN, or DMSO (dimethylsulfoxide), however, the equatorial conformer is preferred. For 2-chloro-, 2-bromo-, and 2-iodocyclohexanones, the axial conformers are also more stable than the equatorial conformers in the gas phase (by 1.05, 1.50, and 1.90 kcal mol^{-1}, respectively). For 2-iodocyclohexanone, the axial conformer prevails in both nonpolar (CCl_4) and polar solvents (MeCN and DMSO). Quantum mechanical calculations suggest that the interaction between 2-halogeno substituents and the carbonyl oxygen in the equatorial conformer is strongly attractive for F, much less attractive for Cl, nil for Br, and repulsive for I [63, 64].

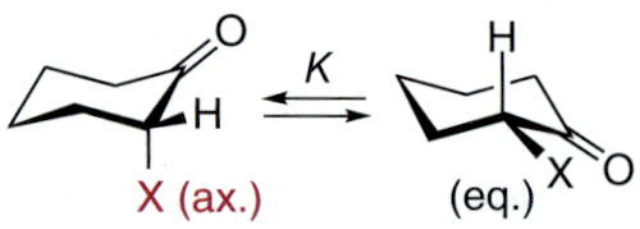

2.6.5 Conformational analysis of cyclohexene

In the gas phase, cyclohexene prefers a half-chair conformation, as proposed by Barton et al. in 1954 [65, 66]. A barrier $\Delta^{\ddagger}H$ = 5.3 kcal mol^{-1} has been evaluated by ^{1}H-NMR for the half-chair ⇄ half-chair inversion [67]. Quantum mechanical calculations estimate the potential energy profile shown in Figure 2.4 for the ring inversion of cyclohexene [68]. In contrast to cyclohexane (Section 2.6.3), for which the sofa and the boat conformers are 10.6 and 5.1 kcal mol^{-1}, respectively, above the chair conformer, the corresponding sofa and boat conformers of cyclohexene have nearly the same stability and are both 6.5 kcal mol^{-1} above the half-chair conformer.

The ring strain of cyclohexene (1.3 kcal mol^{-1}, Table 2.A.3) arises from the partial eclipsing of vicinal C—H bonds between C(1)—H and pseudo-equatorial C(6)—H and between C(2)—H and pseudo-equatorial C(3)—H bonds. When the half-chair conformer of cyclohexene is transformed into the boat conformer, two pairs of vicinal C—H bonds are eclipsed between the $H_2C(4)$ and $H_2C(5)$ moieties, thus contributing to c. 2 kcal mol^{-1} to the ring strain. This is less than the 5.5 kcal mol^{-1} of stability difference between the chair and boat conformers of cyclohexane. The difference $\Delta_f H°$(cyclohexene boat) – eclipsing interactions = 6.5–2 = 4.5 kcal mol^{-1} must be attributed to repulsive 1,4-interactions between the $H_2C(3)$ and $H_2C(6)$ groups and of bond angle deformations (Baeyer strain) arising from the flattening of the boat conformer.

2.6.6 Medium-sized cycloalkanes

The conformational behavior of saturated rings with more than six centers is more complicated than the common rings just described [69]. The potential energy hypersurfaces of these rings have been delineated by molecular mechanics calculations. The representative conformers and their calculated relative enthalpies are given in Figure 2.5. As a general rule, the most stable conformers minimize C—H/C—H eclipsing (Pitzer strain) and 1,*n*-transannular repulsions (Prelog strain). Cycloalkanes with an odd number of carbon atoms (7, 9, 11, 13, and 15) are more mobile and adopt a larger number of conformers at equilibrium than the cycloalkanes with an even number of carbon atoms [80–83]. Cycloheptane prefers a twist-chair conformation with a C_2 axis of rotation. This conformation is flexible, and it equilibrates with a chair conformation, which is only 1.4 kcal mol^{-1} higher in enthalpy. Cyclooctane adopts a boat-chair conformation that equilibrates with a chair–chair (also known as "crown") conformation, which is 2 kcal mol^{-1} higher in enthalpy. The energy barrier for interconversion between the boat-chair and the chair–chair is about 11 kcal mol^{-1}. The boat-chair conformation is not flexible, having a significant barrier to conformational change. Cyclononane

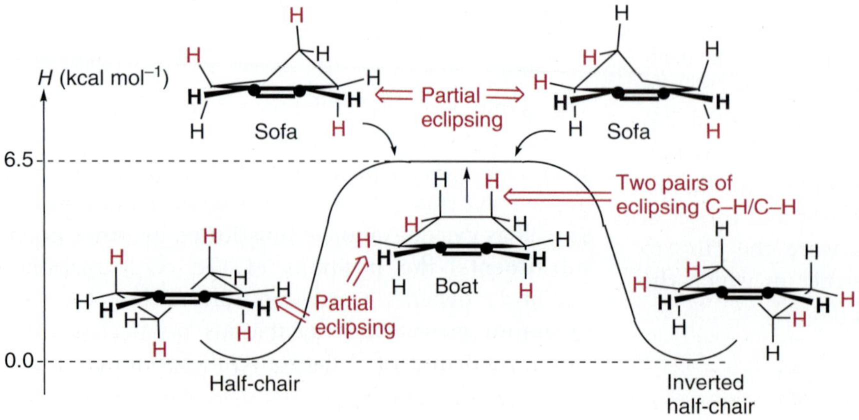

Figure 2.4 Potential energy hypersurface (quantum mechanical calculations) for the interconversion half-chair (C_2) ⇄ half-chair (C_2) conformer of cyclohexene. Axial and pseudo-axial hydrogen atoms (in red) in one half-chair conformer become equatorial and pseudo-equatorial, respectively, in the inverted half-chair conformer (see Section 2.6.10).

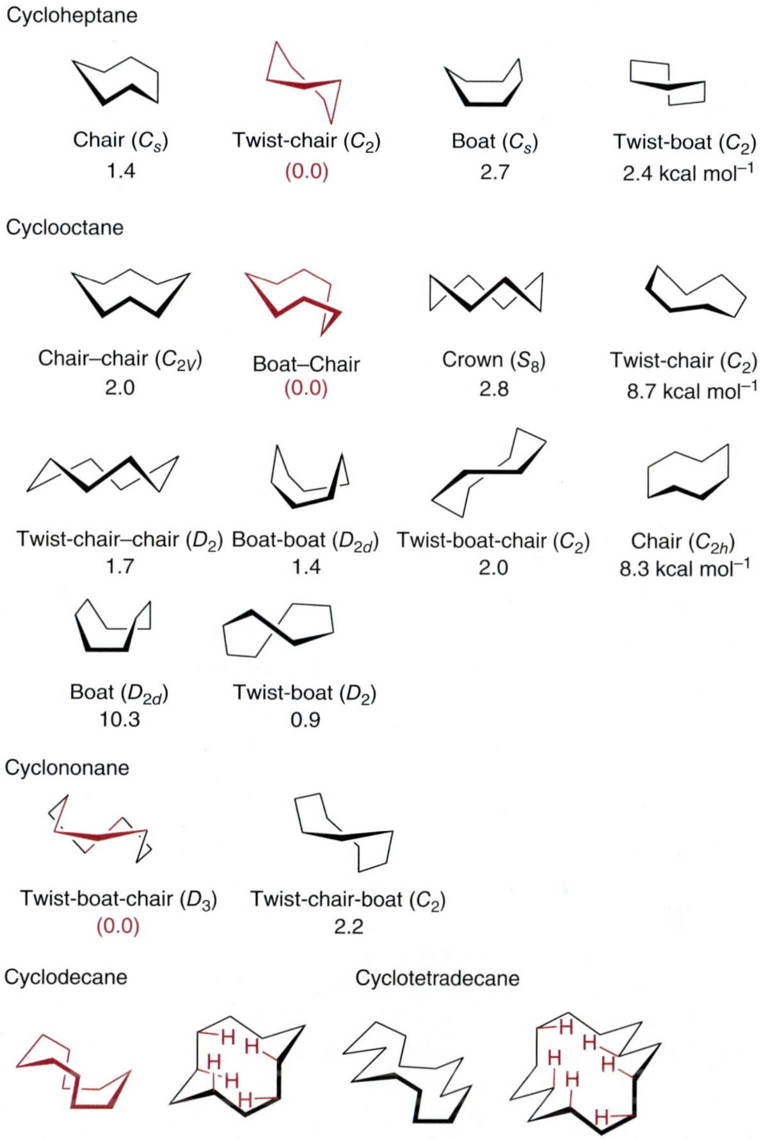

Figure 2.5 Most stable conformations of cycloheptane [70], cyclooctane [71], cyclononane [70–75], cyclodecane [76], and cyclotetradecane [77–79].

is quite mobile. Its most stable conformation is the twist-boat chair, and it equilibrates with the twist-chair-boat conformation with an energy barrier of about 6 kcal mol^{-1}. Cyclodecane and cyclotetradecane adopt lowest energy conformations that can be superimposed on the diamond lattice [84–88]. However, the odd-membered cycloalkanes cannot be superimposed on a diamond lattice. Consequently, they have more conformations of similar energies and consequently are mobile. For example, cycloheptadecane has 262 conformations within 3 kcal mol^{-1} of the lowest energy conformer; according to MM2 calculations and even at room temperature, the amount of global minimum contributing to the equilibrium mixture of conformers is very small [72, 73, 89, 90].

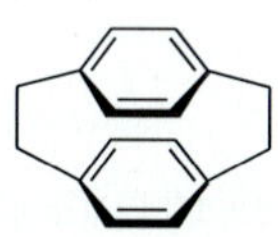

A
[2.2]Paracyclophane
$\Delta_f H°$(gas): 58.8 ± 0.8

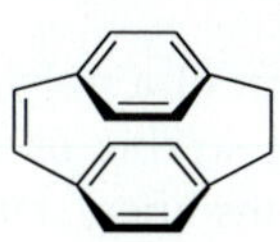

B
[2.2]Paracyclophane-1-ene
86.8 ± 1.0

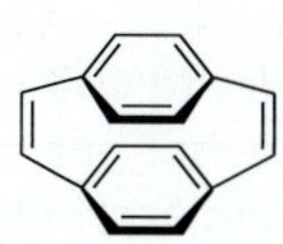

C
[2.2]Paracyclophane-1,9-diene
117.6 ± 1.2 kcal mol^{-1}

Problem 2.6 Evaluate the strain energies of [2.2] paracyclophane (**A**), [2.2]paracyclophane-1-ene (**B**), and [2.2]paracyclophane-1,9-diene (**C**) [91].

Problem 2.7 Hexoses (aldoses with six carbon centers, sugars) in water exist as equilibria of pyranoses and furanoses. Which hexose should show the highest pyranose content? [92]

Problem 2.8 Estimate the standard heats of formation of 1,2-, 1,3-, and 1,4-dimethylcyclohexane isomers. Compare with the experimental values given below [93, 94].

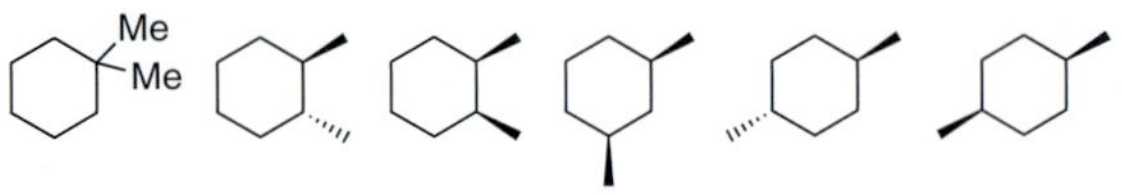

−43.2   −42.2   −41.1   −44.1   −44.1   −42.2 kcal mol^{-1}

2.6.7 Conformations and ring strain in polycycloalkanes

The decahydronaphthalenes (*cis*- and *trans*-decalin) have two cyclohexane rings fused together with two carbon atoms common to each ring. By simply adding the group equivalents given in Table 2.A.1 for these compounds, $\Delta_f H°(C_{10}H_{18}, \text{gas}) = -42.86\,\text{kcal mol}^{-1}$; using the simplified group equivalents ($[C-(C)_2(H)_2] = -5.0\,\text{kcal mol}^{-1}$, $[C-(C)_3(H)] = -1.75\,\text{kcal mol}^{-1}$) and neglecting the stereochemistry, a value of $\Delta_f H°(C_{10}H_{18}, \text{gas}) = -43.5\,\text{kcal mol}^{-1}$ is calculated. This value agrees well with the experimental heat of formation of *trans*-decalin ($-43.54 \pm 0.55\,\text{kcal mol}^{-1}$). Thus, *trans*-decalin is unstrained. *trans*-Decalin is 3 kcal mol^{-1} more stable than *cis*-decalin because all of the alkyl groups in *trans*-decalin occupy equatorial positions, whereas two of the alkyl rings are in axial positions in *cis*-decalin. Gauche interactions between the axial substituents and the cyclohexane moieties are significantly larger than the gauche interactions between equatorial substituents (Table 2.1). This phenomenon is significantly reduced with *cis*-($\Delta_f H° = -30.41 \pm 0.47\,\text{kcal mol}^{-1}$) and *trans*-octahydro-1*H*-indene ($\Delta_f H° = -31.45 \pm 0.50\,\text{kcal mol}^{-1}$) as they have nearly the same standard heats of formation in the gas phase.

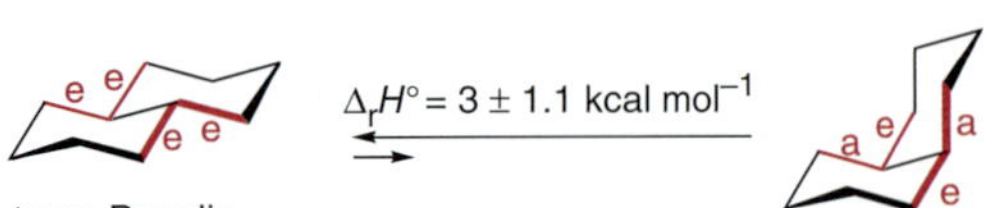

trans-Decalin
$\Delta_f H°$(gas): −43.54 ± 0.55 kcal mol^{-1}

cis-Decalin
−40.45 ± 0.55 kcal mol^{-1}

In order to estimate the ring strain of a polycyclic hydrocarbon, one can add the ring strains of all of the component small rings. For instance, cubane contains six cyclobutane rings; thus, $\Delta_f H°(\text{cubane, est.}) = 8(-1.75) + 6(26.5) = 145\,\text{kcal mol}^{-1}$ (compare gas-phase experimental value: $148.7 \pm 1.0\,\text{kcal mol}^{-1}$, Table 1.A.3). For semi-bullvalene, an estimated value of $\Delta_f H°(\text{semi-bullvalene, est.}) = 68.7\,\text{kcal mol}^{-1}$ is obtained (compare gas-phase experimental value; 73.7 kcal mol^{-1}). For quadricyclane, the heat of formation can be estimated as $\Delta_f H°(\text{quadricyclane, est.}) = 78.5\,\text{kcal mol}^{-1}$ (compare gas-phase experimental value: 80.4 kcal mol^{-1}, Table 1.A.3) [95]. In these cases, the rule slightly underestimates the total ring strain. In contrast, for [1.1.1.1]pagodane and dodecahedrane [96], the rule overestimates the total ring strain because it overcounts the number of C—H/C—H eclipsing bonds, the main source of strain in cyclopentane. These comparisons show the limits of the rule for complicated polycyclic systems. In practice, however, it is fairly accurate for bicyclic and tricyclic systems.

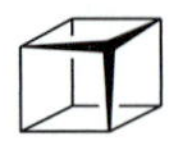

Cubane		Semi-bullvalene	
Σ increments:	−14		27.6
6 cyclobutanes:	−159	1 cyclopropane:	27.5
		2 cyclopentenes:	13.6
$\Delta_f H°_{est}$:	145		68.7
$\Delta_f H°(\text{gas})_{exp}$:	148.7 kcal mol^{-1}		73.7

Quadricyclane	
Σ increments:	−15.4
2 cyclopropanes:	55.0
1 cyclobutane:	26.5
2 cyclopentanes:	12.4
$\Delta_f H°_{est}$:	78.5
$\Delta_f H°(\text{gas})_{exp}$:	80.4 kcal mol^{-1}

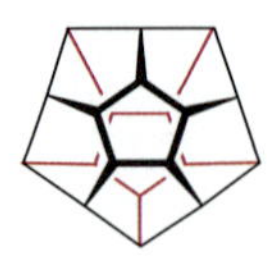

[1.1.1]Pagodane ($C_{20}H_{20}$)		Dodecahedrane ($C_{20}H_{20}$)	
$\Delta_f H°$(gas):	47.0	$\Delta_f H°$(gas):	22.4 ± 1 kcal mol^{-1}
Σ increments:	41	Σ increments:	35
1 cyclobutane:	26.5		
12 cyclopentanes:	74.4	12 cyclopentanes:	74.4
$\Delta_f H°(C_{20}H_{20})_{est}$:	59.9	$\Delta_f H°(C_{20}H_{20})_{est}$:	39.4 kcal mol^{-1}

Problem 2.9 In 3β-cholestanol, the A/B, B/C, and C/D ring junctions are all *trans*. The two methyl groups at the ring junctions are in axial positions. Estimate the heat of formation of this natural compound.

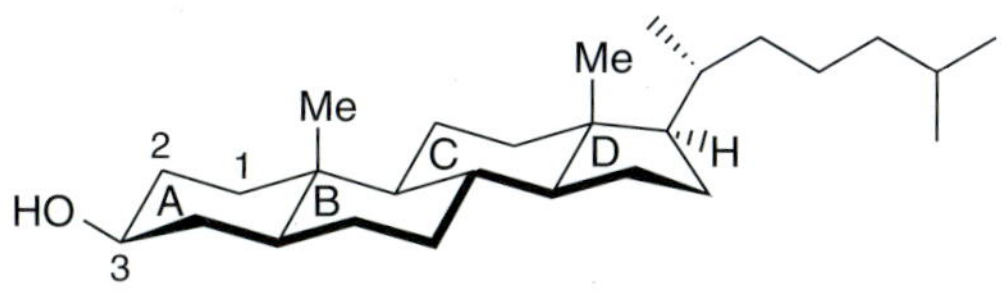

3β-Cholestanol

2.6.8 Ring strain in cycloalkenes

The heats of hydrogenation of cyclopropene, cyclobutene, cyclopentene, and cyclohexene to give cyclopropane, cyclobutane, cyclopentane, and cyclohexane, respectively, are given below. It is clear from these data that cyclopropene is much more strained than cyclopropane [95]. This is mostly due to bond angle deformation (Baeyer strain) imposed on the alkene sp^2 carbons. Cyclobutene is more strained than cyclobutane by only 3–4 kcal mol^{-1}. Cyclopentene is slightly less strained than cyclopentane because some of the Pitzer strain is removed by introduction of an alkene moiety. However, for the six-membered rings, cyclohexene suffers 1.3 kcal mol^{-1} of strain that is absent in cyclohexane. This difference manifests itself when comparing the heat of isomerization of methylidenecyclohexane into 1-methylcyclohexene ($\Delta_r H = -1.94$ kcal mol^{-1}) with that of the isomerization of methylidenecyclopentane into 1-methylcyclopentene ($\Delta_r H = -3.65$ kcal mol^{-1}) [11]. In the case of the isomerization of methylidenecyclobutane into 1-methylcyclobutene [11, 97], the reaction is less exothermic as the Baeyer strain increases with the number of sp^2-carbon centers incorporated into the small cycloalkane. Baeyer angular strain is much higher in cyclopropene than in methylidenecyclopropane as demonstrated by theendothermicity of 10.2 ± 0.8 kcal mol^{-1} measured for

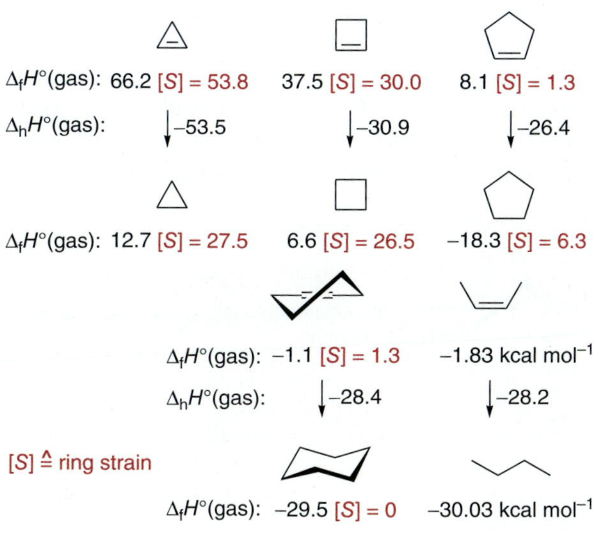

the isomerization of methylidenecyclopropane into 1-methylcyclopropene.

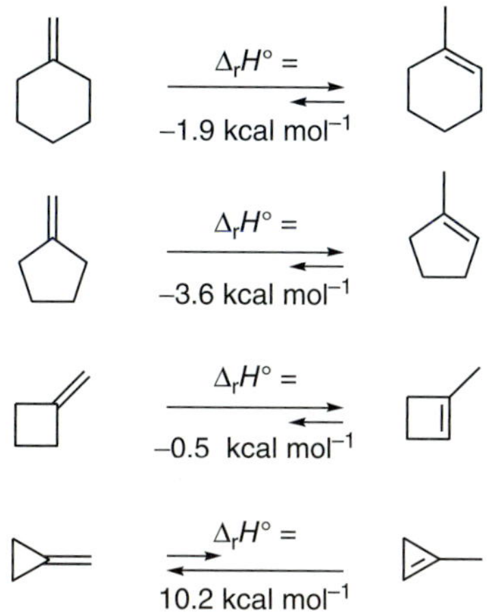

The hydrogenation of (*E*)-cyclooctene is more exothermic than that of (*Z*)-cyclooctene, showing that (*E*)-cyclooctene is less stable than the (*Z*)-isomer. This is because transannular repulsions are more severe in (*E*)-cyclooctene than in (*Z*)-cyclooctene [98].

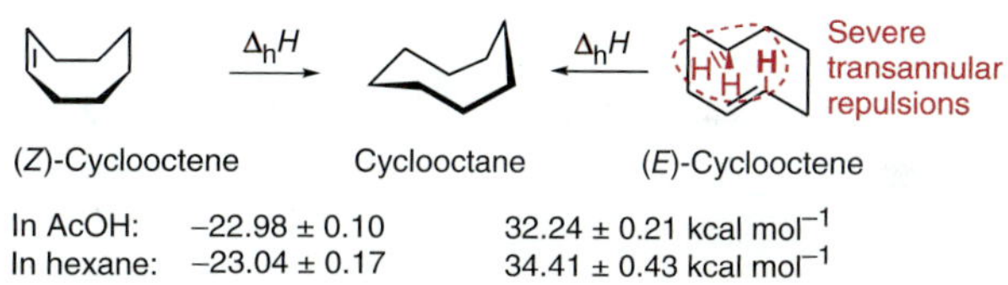

Problem 2.10 On heating cyclohexanol on acidic alumina in the gas phase below 300 °C, rapid dehydration occurs to give cyclohexene. Above 350 °C, 3.5% of methylcyclopentenes (mostly 1-methylcyclopentene, some 4-methylcyclopentene) are also formed. At 450 °C, up to 19.5% of methylcyclopentenes are present. Explain these results (see, e.g. Refs. [99, 100]).

2.6.9 Bredt's rule and "*anti*-Bredt" alkenes

In 1924, Bredt et al. proposed a rule stating that bridgehead olefins are unstable because they include a *trans*-double bond ((*E*)-alkene) in a relatively small ring [101, 102]. The first violation of this rule was presented in 1950 by Prelog who obtained the bridgehead enone **12** through intramolecular aldol/crotonalization of diketone **11** [103]. The "*anti*-Bredt" alkene **12** is a bridged (*E*)-cyclodecene derivative. In 1967 and independently, the groups of Wiseman [104] and Marshall and Faubl [105] obtained bicyclo[3.3.1]non-1-ene (**13**), which can be seen as a bridged analog of (*E*)-cyclooctene for which a ring strain of 23 kcal mol^{-1} is estimated. The ring strain of the bicyclic **13** is not much higher than that evaluated for (*E*)-cyclooctene itself (16.7 kcal mol^{-1}, Table 2.A.2). In 11-bromo-9-*endo*-chloro-7-ethoxybicyclo[5.3.1]undec-1(11)-ene (**14**), X-ray diffraction studies demonstrate that part of the Baeyer strain expected for (*E*)-cycloalkene is relieved by torsion about the C=C double bond [106–109].

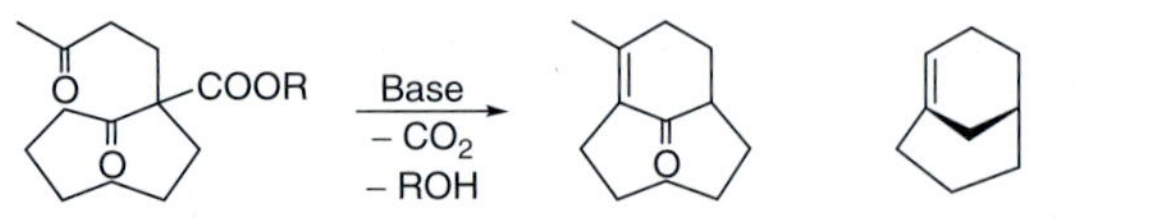

11 **12** **13** [*S*] = 23 kcal mol^{-1}

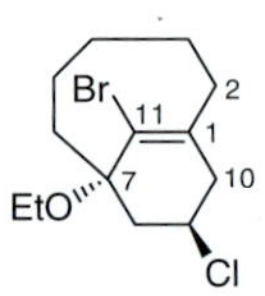

14

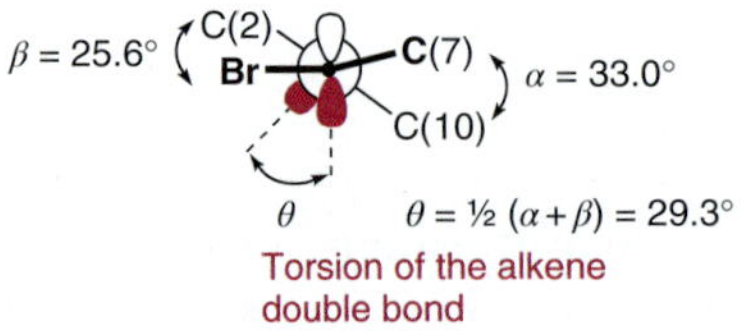

Problem 2.11 What are the products of the following reactions for $n = 3, 7, 9$?

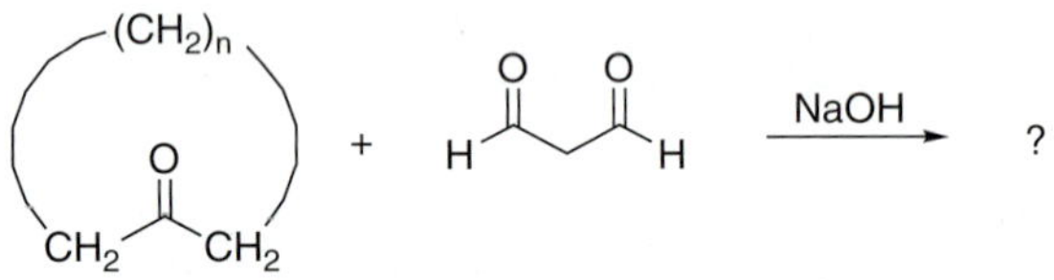

Problem 2.12 What is the product of the following hydrolysis? [108]

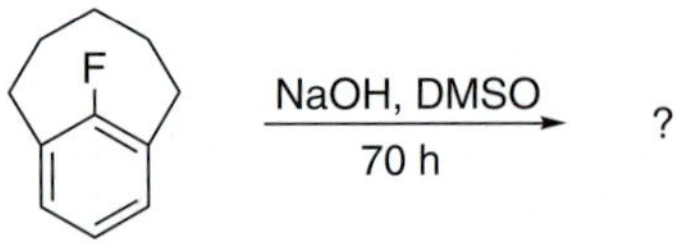

2.6.10 Allylic 1,3- and 1,2-strain: the model of banana bonds

The concept of allylic 1,3-strain ($A^{1,3}$-strain) was first proposed by Johnson in 1968 [110]. It has proven to be one of the most powerful tools to determine the conformations of acyclic systems [111, 112]. This involves the 1,3-interaction of an allylic group with a *cis*-vinylic group. Quantum mechanical calculations on 3-methylbut-1-ene show that both conformers **15a** and **15b** are populated to a considerable extent [113]. However, a different situation is found for (*Z*)-4-methylpent-2-ene (and other related systems) [114]. Conformer **16a** is strongly favored over **16b** and **16c**. Conformer **16b** is destabilized substantially by allylic 1,3-strain.

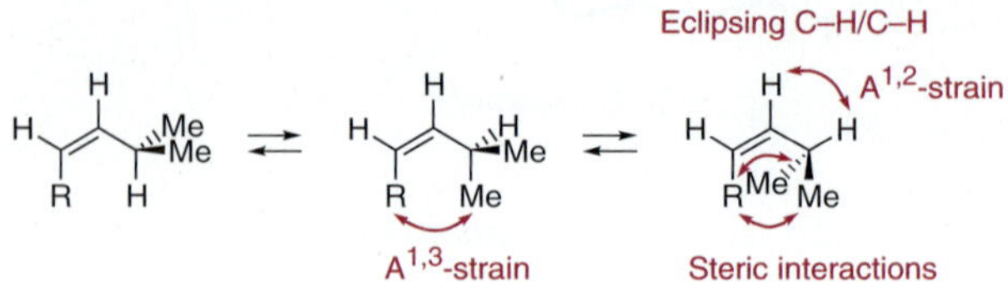

R = H:	**15a**	**15b**	**15c**
ΔH_{rel}:	(0.0)	0.73	2.48 kcal mol^{-1}
R = Me	**16a**	**16b**	**16c**
ΔH_{rel}:	(0.0)	4.86	3.44 kcal mol^{-1}

The most stable rotamers of the two alkenes, which arise from the rotation of the isopropyl group, are **15a, 16a,** in which the isopropyl group's C—H bond lies in the plane of the C=C double bond. According to Hückel [115], an alkene comprises a σ(C—C) and a π(C=C) bond, involving sp^2-hybridized carbon atoms. If sp^3-hybridized carbon atoms were used instead, the two methylene fragments of ethylene would be joined by curved "banana" bond or τ bonds (Figure 2.6). Using the banana bond model to represent alkenes, one finds that the most stable conformers **15a, 16a** correspond to the staggered conformations of *n*-alkanes. Thus, the origin of allylic 1,3-strain is similar to the (steric) gauche effects, or front-strain described for saturated systems, as proposed by Pauling [116] and Slater [117–119]. A modern discussion of the equivalence of σ, π, and banana bond models can be found in Refs. [120–123].

In cycloalkenes, however, conformers of type **15c**, **16c** might be populated. It appears, therefore, that substitution of centers C(2) or/and C(3) by groups larger than hydrogen atoms will introduce repulsive steric effects. This repulsion is called allylic-1,2-strain ($A^{1,2}$-strain). The allylic groups are involved in repulsions with vicinal vinylic groups. In cyclohexene, which adopts a half-chair conformation in its ground state, two of the C—H bonds at C(4) and C(5) are axial, while the other two C—H bonds are equatorial, analogous to the C—H bonds in cyclohexane. At the allylic

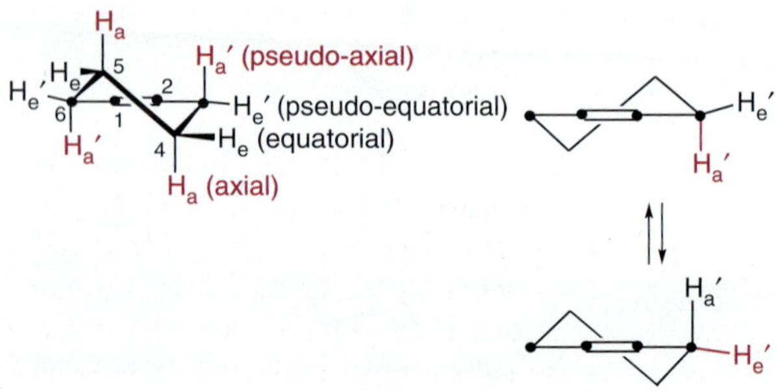

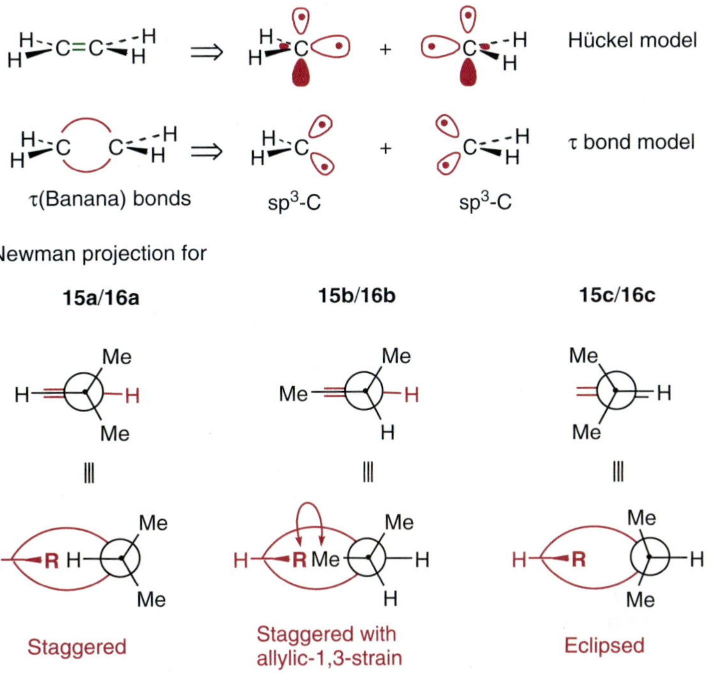

Figure 2.6 τ(Banana) bond model for alkenes; representation of the rotamers of (*Z*)-alkenes substituted with an isopropyl group.

centers C(3) and C(6) of cyclohexene, two C—H bonds lie nearly eclipsed with the C=C bond; these bonds are said to occupy pseudo-equatorial positions, whereas the other two C—H bonds, which are nearly perpendicular to the alkene system, are pseudo-axial. Cyclohexenes that are monosubstituted at C(3), C(4), C(5), or C(6) adopt pseudo-chair conformations in which the substituent resides at a 4-, 5-equatorial or 3-, 6-pseudo-equatorial position. This may not be the case, however, for 1,6-disubstituted derivatives. Here, $A^{1,2}$-strain between the substituents is higher when the six-substituent occupies a pseudo-equatorial position. Examples are given below for 1-phenylcyclohexene derivatives **17** and **18** [124].

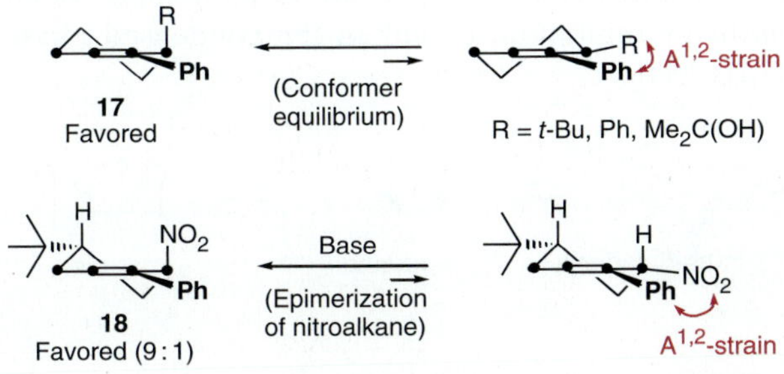

For polysubstituted cyclohexenes, competition between $A^{1,2}$-strain and other steric repulsions between substituents can intervene. In cyclohexenes **19**, repulsions between the methyl groups at C(4) and the six-substituents make the pseudo-chair conformers with pseudo-equatorial six-substituent more stable than the other pseudo-chair conformers.

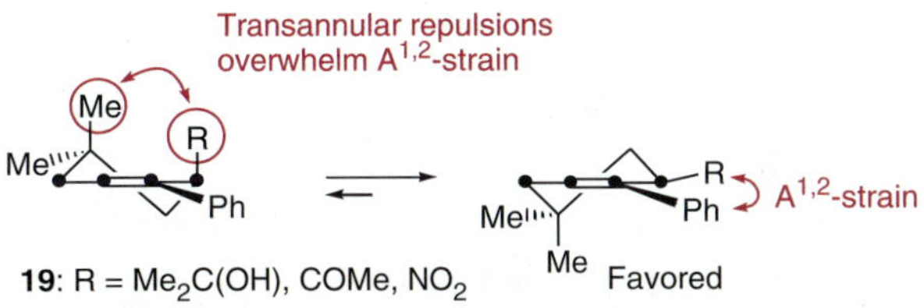

Problem 2.13 What is the most stable conformation of propene? Estimate the energy barrier for the methyl group rotation [125].

Problem 2.14 Predict the favored products (diastereoselectivity) of the ene-reactions (Section 5.7) of 3,4-dimethylpent-3-en-2-ol with singlet oxygen ($^1O_2$), *N*-phenyltriazolinedione (PTAD), and 4-nitrosobenzene (PhN=O) [126].

2.7 π/π-, n/π-, σ/π-, and n/σ-interactions

2.7.1 Conjugated dienes and diynes

Hydrogenation of buta-1,3-diene to form but-1-ene (Table 1.A.2) is less exothermic ($\Delta_h H° = -26.3 \pm 0.36\,\text{kcal mol}^{-1}$) than the hydrogenation of but-1-ene to form butane ($\Delta_h H° = -30.18 \pm 0.35\,\text{kcal mol}^{-1}$). This comparison suggests that the conjugation of two alkene moieties stabilizes them by about 3 kcal mol^{-1}. A similar comparison derives from the hydrogenations of (*E*)-penta-1,3-diene and cyclohexa-1,3-diene vs. their monounsaturated counterparts.

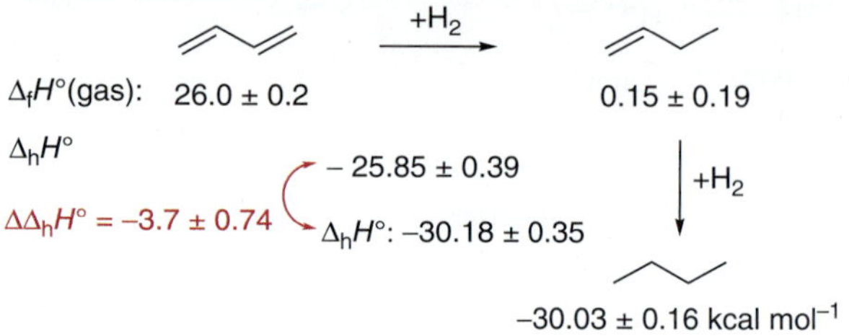

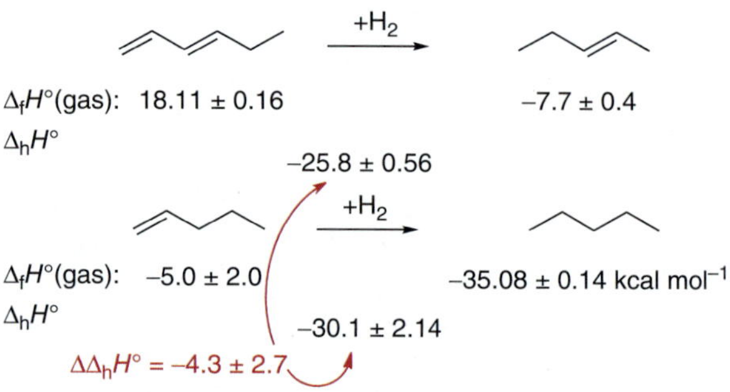

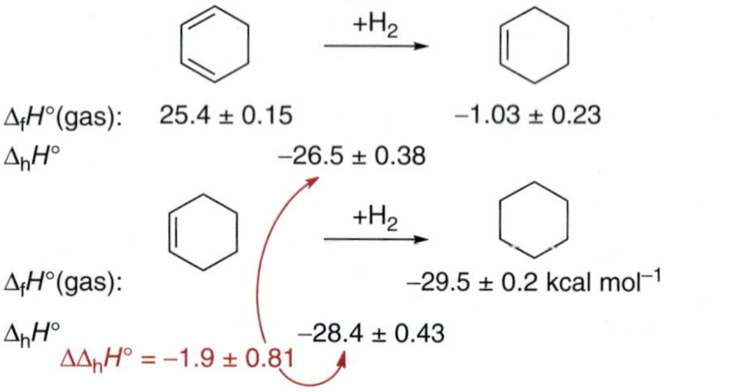

This method of assessing conjugative stabilization was used first by Kistiakowsky et al. in the 1930s [127]. In fact, it underestimates the conjugative stabilization. This is because of a hyperconjugative stabilization of the alkene, which amounts to 2.7 kcal mol^{-1}, as measured by the enthalpy change of the following reaction:

$$\Delta_r H°(CH_2{=}CH_2 + CH_3CH_2CH_2CH_3 \rightleftarrows CH_2{=}CHCH_2CH_3 + CH_3CH_3) = -2.7\ \text{kcal/mol}$$

When corrected for hyperconjugative stabilization, the stabilization arising from attaching two vinyl groups is 6.5 kcal mol^{-1} [128].

$$\Delta_r H°(CH_2{=}CH_2 + CH_3CH_2CH{=}CH_2 \rightleftarrows CH_2{=}CH-CH{=}CH_2 + CH_3CH_3) = -6.5\ \text{kcal/mol}$$

The standard heats of isomerization $\Delta_r H°$(CH_2=CH—CH_2—CH=CH_2 ⇄ (*E*)-CH_3—CH=CH—CH=CH_2) = −7.3 ± 0.3 kcal mol^{-1} and $\Delta_r H°$(CH_2=CH—CH_2—Ph ⇄ (*E*)-CH_3—CH=CH—Ph) = −5.6 ± 0.6 kcal mol^{-1} have been reported [129]. Comparison of the heats of hydrogenation of conjugated diynes with unconjugated isomers led Rogers et al. to conclude that in buta-1,3-diyne, the stabilization afforded by conjugation is zero [130]. This is surprising when one considers the remarkably short (1.38 Å) carbon–carbon single bond of buta-1,3-diyne. Comparison of the heats of hydrogenation of buta-1,4-diyne into but-1-yne (−71.5 kcal mol^{-1}) and of but-1-yne into butane (−69.5 kcal mol^{-1}) evaluates not only conjugation effects but also other structural and electronic differences between the conjugated molecule and its hydrogenated products. Specifically, but-1-yne, the reference compound for buta-1,3-diyne, is stabilized significantly by hyperconjugation, which is not present in but-1,3-diyne. The hyperconjugative stabilization in but-1-yne compared with ethyne is given by the heat (−4.7 kcal mol^{-1}) of the following isodesmic equilibrium:

$$\Delta_r H°(CH\equiv CH + CH_3CH_2CH_2CH_3 \rightleftarrows HC\equiv CCH_2CH_3 + CH_3CH_3) = -4.7\ \text{kcal/mol}$$

Isomerization of hexa-1,3-diyne into hexa-1,4-diyne (3.7 kcal mol^{-1}, Table 1.A.2) gives a first conjugative stabilization of 3.7 kcal mol^{-1}. However, this value is not hyperconjugation-balanced as the product (hexa-1,4-diyne) is stabilized more than the reactant (hexa-1,3-diyne). When the hyperconjugation stabilization of each species is taken into account, the change in energy for the isomerization of hexa-1,3-diyne into hexa-1,4-diyne is 9.2 kcal mol^{-1} [128].

The most stable conformation of buta-1,3-diene in the gas phase is the *s-trans*-conformer. The second stable conformer is the *s-gauche*. The *s-cis*-conformer, in which π/π conjugation would be optimal, is not an enthalpy minimum, except in the condensed phase [131–134].

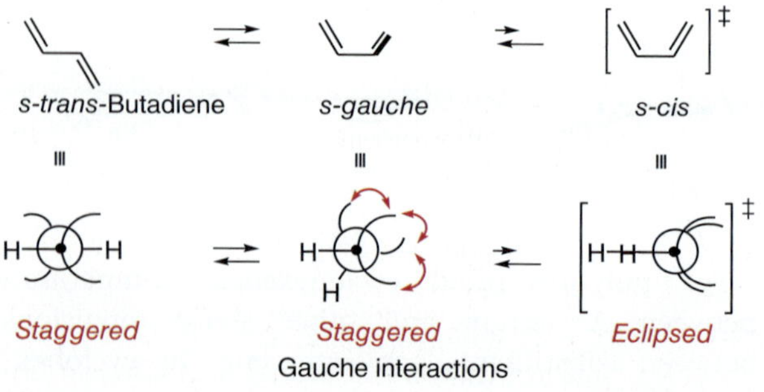

$\Delta^{\ddagger}H$ (*s-trans* ⇄ *s-cis*-butadiene) ≅ 3.9 ± 0.2 kcal mol^{-1}

The model of "banana" bonds for alkenes perfectly rationalizes these observations. There are two possible "staggered" conformations, one allowing hypothetical π/π conjugation (*s-trans*-conformer) and the second introducing more gauche interactions and reducing the π/π conjugation (*s-gauche*). The *s-cis*-conformer is disfavored because it corresponds to an "eclipsed" conformer, although it allows for a complete π/π conjugation [135].

2.7.2 Atropisomerism in 1,3-dienes and diaryl compounds

Because of allylic strain, highly substituted 1,3-dienes adopt nonplanar conformations that are chiral [136]. One of the two conformers is a right-handed helix described as *P* (plus) or Δ and the enantiomeric one is a left-handed helix described as *M* (minus) or Λ [137]. A free enthalpy of activation $\Delta^{\ddagger}G = 17.6\,\text{kcal mol}^{-1}$ has been evaluated for the rotation about the central single bond in dimethyl (*E*,*E*)-2,3,4,5-tetramethyl-2,4-hexadienedioate (Eq. (2.10)) [138].

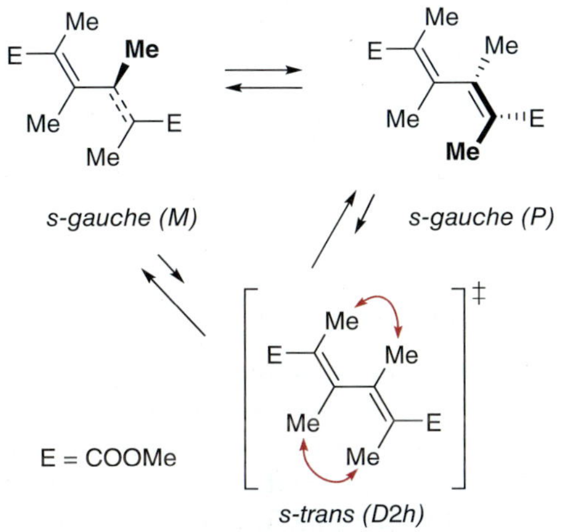

(2.10)

When the energy barrier for rotation about a single bond becomes large, atropisomers can be separated as enantiomers at room temperature. This phenomenon is called atropisomerism (derived from the Greek for "not turning"). Atropisomeric dienes have been isolated enantiomerically enriched and employed as chiral catalysts [139]. Atropoisomerism is manifested typically in *ortho*-substituted biphenyls (and, more generally, biaryls) where steric congestion between the substituents restricts rotation about the sp^2–sp^2 carbon–carbon single bond (e.g. equilibria (2.11)) [140, 141]. The majority of tetra-*ortho*-substituted biphenyls possess a barrier to rotation about the single sp^2—sp^2 bond that is sufficiently high to prevent the interconversion (or racemization) of atropisomers at room temperature and above. Consequently, the different atropisomers can be resolved [142]. Atropisomeric stability is, however, reduced when two or more of the *ortho*-substituents are small. The half-lives of racemization of the substituted biphenyls **20** are given below [137]:

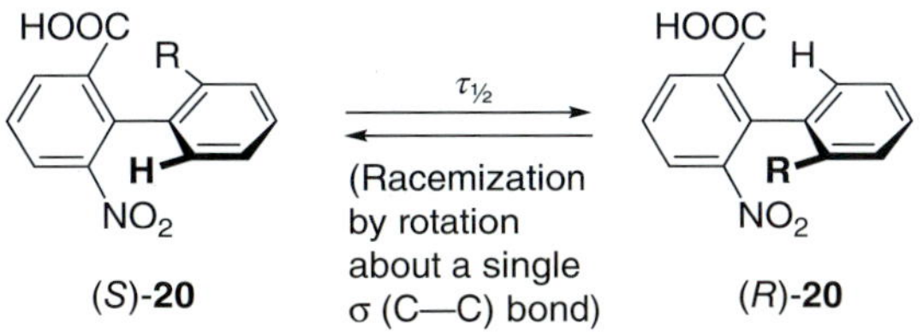

(2.11)

		$\tau_{1/2}$
20	R = Me	$\tau_{1/2}$ = 179 min at 118 °C
	R = NO_2	125 min at 118 °C
	R = COOH	91 min at 118 °C
	R = OMe	9.4 min at 25 °C

Cahn-Ingold-Prelog rule for the assignment of absolute configuration:

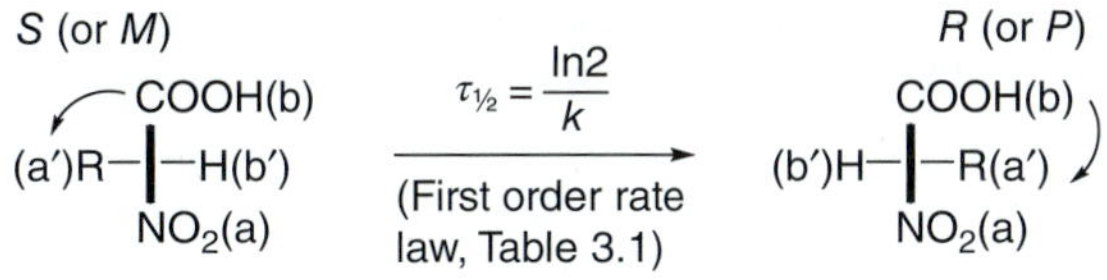

Regardless of the substituent pattern, in solution, the aromatic rings in the biphenyls in their lowest energy conformations are usually neither coplanar nor perpendicular. The steric bulk of the *ortho*-substituents tends to enforce noncoplanarity, but the stabilization of the biphenyl π-system is the greatest when the aryl rings are coplanar. These opposing effects mean that the inter-ring torsional angles are usually between 42° and 90°. In the crystal, however, the parent biphenyl molecule is planar [143]. Atropisomerism is not restricted to 1,3-dienes, biaryls, and biheteroaryls [144] but also arises in sterically impeded substituted styrenes and some aromatic amides and anilides [145–147]. Atropisomerism about single bonds between an sp^2 carbon and an sp^3 carbon is also known [148]. Less commonly, rotations about some sterically highly hindered sp^3–sp^3 single bonds are restricted, permitting the resolution of atropisomers. Atropoisomerism at C—N bonds [149–153], C—O bonds [154–157], and C—S bonds has been reported as well [158]. The biaryl subunit is found in a wide variety of natural products, including alkaloids, coumarins, flavonoids, lignans, polyketides, tannins, terpenes, and peptides (see, for example, the vancomycin class of glycopeptide antibiotics) [159]. Compounds incorporating biaryls also find applications as chiral reagents, chiral ligands for metal complex catalysts

(Sections 3.7, 5.2.6, 5.2.14, 5.2.16–17, 5.6.2, 5.6.5, 7.4.6, 7.5.1–4, 7.5.7–8, 7.7.5–6, 7.8.2, 8.3.2–4, 8.3.6, 8.3.11, 8.3.13–17, 8.4.5. 8.5.2, 8.6.2–4, 8.7.4, 8.7.6–7), chiral phases in chromatography, and chiral liquid crystals [160–164].

2.7.3 α,β-Unsaturated carbonyl compounds

For conjugated enals (Table 1.A.4), one finds conjugative stabilization energies up to 4 kcal mol^{-1} by measuring the heats of hydrogenation and ignoring hyperconjugative stabilization of the hydrogenation products. Comparison of the heat of hydrogenation of methyl vinyl ketone (−29.6 kcal mol^{-1}) with that of a terminal alkene (c. −30 kcal mol^{-1}) suggests, however, that there is no stabilization in methyl vinyl ketone due to π/π-conjugation. According to perturbation of molecular orbital (PMO) theory (Section 4.5.15), a conjugative stabilization in conjugated enals and enones should exist and is associated with a possible electron transfer from the alkene to the carbonyl moiety when the two π-systems overlap (are not orthogonal). For propenal in the gas phase, the *s-cis*-conformer is 1.7 ± 0.04 kcal mol^{-1} less stable than the *s-trans*-conformer [165]. Quantum mechanical calculations estimate $\Delta_r H°$(CH_2=CH—CH—CHO ⇄ (*E*)-CH_3—CH=CH—CHO) = −7.5 kcal mol^{-1} and $\Delta_r H°$(CH_2=CH—CH—COMe ⇄ (*E*)-CH_3—CH=CH—COMe) = −5.3 kcal mol^{-1}.

Interestingly, the thermodynamic preferences for α,β-unsaturated isomer over the β,γ-unsaturated isomer of sulfoxides, sulfones, sulfonamides, and sulfonic esters have been found by computations to be smaller (less than 1 kcal mol^{-1}) [129].

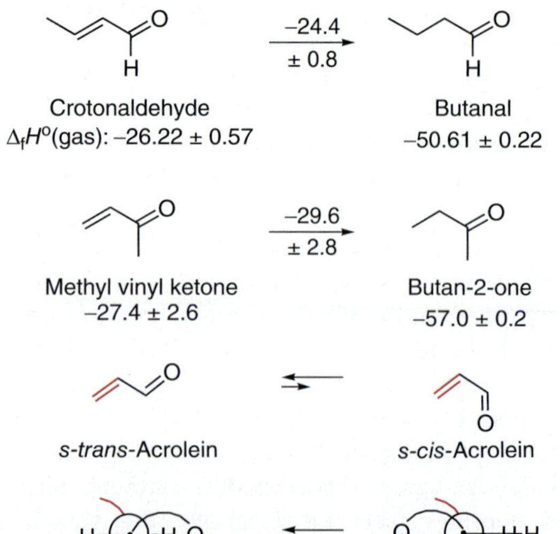

The heats of hydrogenation of acrylic acid, ethyl acrylate, and acrylamide are similar (c. −30 kcal mol^{-1}) to those of terminal alkenes. This suggests that these α,β-unsaturated carboxylic derivatives are not stabilized by $\pi_{C=O}/\pi_{C=C}$ conjugation more than the corresponding saturated carboxylic derivatives stabilized by σ(α-CH_2)/$\pi_{C=O}$ hyperconjugation.

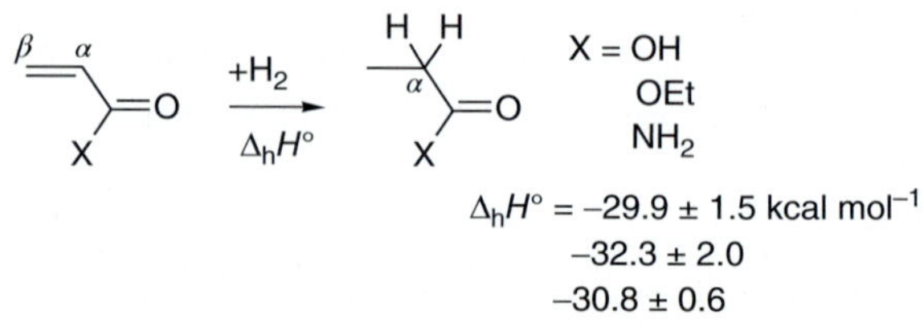

Problem 2.15 What is the most stable conformation of 2,3-dimethylbuta-1,3-diene?

Problem 2.16 What experiments do you propose to carry out in order to prove that a given 2-substituted-3-methylbutadiene prefers a *s-gauche* (nonplanar) conformation? [138]

Problem 2.17 Estimate the energy of π-conjugation of *p*-benzoquinone.

Problem 2.18 Do you expect an increase or a decrease in the stability difference between the *s-cis* and the *s-trans*-conformers of acryloyl chloride and acryloyl fluoride compared with propenal? [166, 167]

2.7.4 Stabilization by aromaticity

The history of aromaticity, a fundamental chemical concept, [168–171] began with the isolation of benzene by Faraday in 1825. Many definitions or criteria for characterizing aromaticity have been considered subsequently. Benzene was characterized by its "aromatic" smell. In 1865, Kekulé proposed a first structure for benzene. Erlenmeyer, in 1866, recognized that substitution of benzene is more facile than addition. In 1910, Pascal observed that "aromatic" compounds have exalted diamagnetic susceptibilities, which was attributed, by Armit and Robinson in 1925, to the sextet of electrons. A milestone in the understanding of aromaticity is the Hückel theory, published in 1931. Hückel theory predicts that planar monocyclic conjugated polyenes ([*N*]-annulenes) with $N = 4n + 2\pi$ electrons (where n is any integer) are stabilized by "aromaticity" (Section 4.5.10). In 1937, London associates the enhanced magnetic susceptibility of benzene with the π-electron current (London diamagnetism). In 1956, Pople proposed a quantum interpretation of the ring current effects on NMR chemical shifts. Dauben, in the late 1960s, revisited the diamagnetic susceptibility exaltation. In 1970, Flygare proposed magnetic susceptibility anisotropy as criteria for aromaticity. Garratt and coworkers defined molecules with an induced diamagnetic ring current as diatropic and those with a paramagnetic ring current as paratropic [172, 173]. In 1996, Schleyer et al. proposed that diamagnetic susceptibility exaltation is the only measurable property uniquely associated with aromaticity and defined aromaticity in terms of this exaltation. He and others contend that the various indices of aromaticity are well correlated. Schleyer et al. introduced nucleus-independent chemical shifts (NICSs) [174], the negative of the absolute magnetic shielding, as a means of evaluating the aromaticity and antiaromaticity (see e.g. cyclobutadiene, Sections 1.11.4 and 4.5.6) of a π-system [175]. In this chapter, we concentrate on the thermodynamic criteria to define aromaticity [176, 177]. We refer to "aromatic stabilization energies" (ASEs), which is defined analogously to resonance stabilization energies of conjugated π-systems [178–180].

Using the group equivalents associated with groups $[C_b—(C_b)_2(H)]$, $[C_b—(C_b)_2(C)]$, and $[C_b—(C_b)_2(X)]$ (Table 2.A.1), the conjugative stabilization, or ASE, of benzene derivatives is considered. If benzene were simply a cyclohexa-1,3,5-triene devoid of any π/π conjugation stabilization, a deviation of -36.2 ± 1.8 kcal mol^{-1} (or -36 kcal mol^{-1}) is found between the gas-phase heat of hydrogenation (Table 1.A.2) of actual benzene (-49.3 ± 0.4 kcal mol^{-1}) and that of the hypothetical, nonaromatic cyclohexa-1,3,5-triene (three times the hydrogenation heat of cyclohexene: $3(-28.5 \pm 0.45) = -85.5 \pm 1.35$ kcal mol^{-1} (Table 2.2)). As will be discussed in Section 4.5.10 and 4.6, the deviation of -36 kcal mol^{-1} does not accurately estimate the π-conjugation energy

Table 2.2 Standard heats of formation of benzene and derivatives (gas phase) and their heats of hydrogenation in kcal mol^{-1} (1 cal = 4.184 J).[a)]

		$+H_2$ →		$+H_2$ →		$+H_2$ →		$-3H_2$ →	PhH
$\Delta_f H°$	19.8 ± 0.2	5.8 ± 0.34	25.0 ± 0.14	−26.0 ± 0.37	−1.03 ± 0.23	−28.5 ± 0.43	−29.5 ± 0.2	49.3 ± 0.4	
	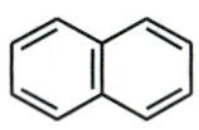	$+2H_2$ →	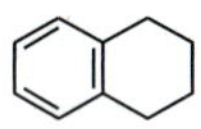	$+3H_2$ →	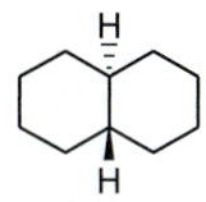				
$\Delta_f H°$	36.0 ± 2.0	−29.8 ± 2.3	7.2 ± 0.3	−50.7 ± 0.9	−43.54 ± 0.55				
	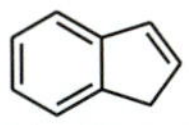	$+H_2$ →	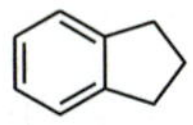			$+H_2$ →			
$\Delta_f H°$	38.5 ± 0.6	−23.9 ± 1.1	14.6 ± 0.5		8.5 ± 0.2	−26.8 ± 0.4	−18.26 ± 0.19		
	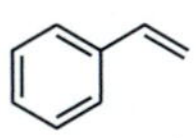	$+H_2$ →	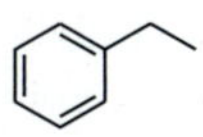	$+3H_2$ →	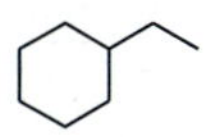				
$\Delta_f H°$	35.11 ± 0.24	−28.0 ± 0.44	7.12 ± 0.20	−48.4 ± 0.6	−41.25 ± 0.4				
	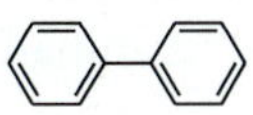	$+3H_2$ →	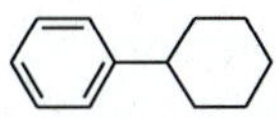	$+3H_2$ →	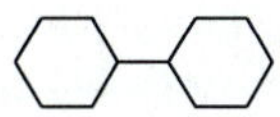				
	1,1′-Biphenyl		Cyclohexylbenzene		1,1′-bicyclohexyl				
$\Delta_f H°$	43.1 ± 0.7	−47.1 ± 1.1	−3.38 ± 0.35	−48.1 ± 1.1	−52.19 ± 0.74				
	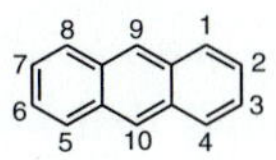	$+H_2$ →	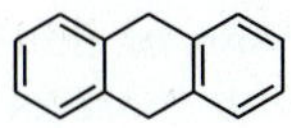		Ph—≡—Ph	$+2H_2$ →	$PhCH_2CH_2Ph$		
$\Delta_f H°$	55.2	c. −17	38.2		92.0 ± 0.64	−59.6 ± 1.0	32.4 ± 0.3		
	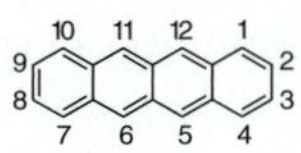	$+H_2$ →	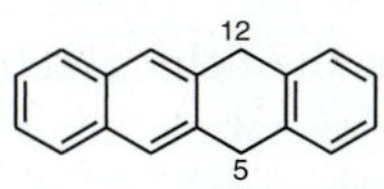		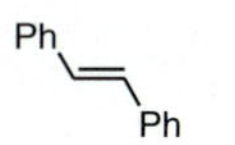	$+H_2$ →	$PhCH_2CH_2Ph$		
	Tetracene		5,12-Dihydrotetracene		(E)-Stilbene	→	Dibenzyl		
$\Delta_f H°$	81.9 ± 1.4	−28.2 ± 2.0	53.7		52.5	c. −19.3	32.4 ± 0.3		

a) See: http://webbook.nist.gov.

(or aromaticity) of benzene because the method ignores changes in ring strain between cyclohexane, cyclohexene, and benzene, as well as the σ(CH)/π(C=C) hyperconjugative interactions in cyclohexene.

The heat of hydrogenation of naphthalene into *trans*-decalin amounts to -79.5 ± 2.5 kcal mol^{-1} to be compared with $5(-28.5 \pm 0.45) = -142.5 \pm 2.25$ kcal mol^{-1} for five times the heat of hydrogenation of cyclohexene. The difference of -63.0 ± 4.8 kcal mol^{-1} is less than twice the aromaticity of benzene, i.e. $2(-36.2 \pm 1.8) = -72.4 \pm 3.6$ kcal mol^{-1}. Dihydrogenation of benzene into cyclohexa-1,3-diene is endothermic by 5.8 ± 0.34 kcal mol^{-1}, whereas the dihydrogenation of naphthalene into 1,2-dihydronaphthalene must be thermoneutral as the heat of double hydrogenation of naphthalene into benzocyclohexene is -29.8 ± 2.3 kcal mol^{-1}, a value similar to the heat of hydrogenation of cyclohexene into cyclohexane (Table 2.2). Dihydrogenation of anthracene into 9,10-dihydroanthracene is endothermic by -17 kcal mol^{-1}. Similarly, the dihydrogenation of tetracene (or naphthacene) into 5,12-dihydrotetracene is not less exothermic (-28.2 ± 2.0 kcal mol^{-1}) than the hydrogenation of cyclohexene. These data demonstrate that the gain in stability due to the aromaticity is not additive with the number of annulated benzene rings. Furthermore, they suggest that tetracene and anthracene can be engaged in Diels–Alder reactions, hydrogenation, and other reactions that convert the middle benzene rings to a cyclohexa-1,4-diene moiety more readily than the conversion of either naphthalene or benzene to their (benzo)cyclohexa-1,4-diene or (benzo)cyclo-1,3-diene derivatives. It has been suggested that, as the number of annulated benzene rings increases (one generates the poly(acene)s), the resulting structures increasingly resemble two polyacetylenes connected by relatively long σ-bonds [181, 182]. Hexacene, which is unstable in solution, undergoes dimerization and oxidation with air to give endoperoxides [183].

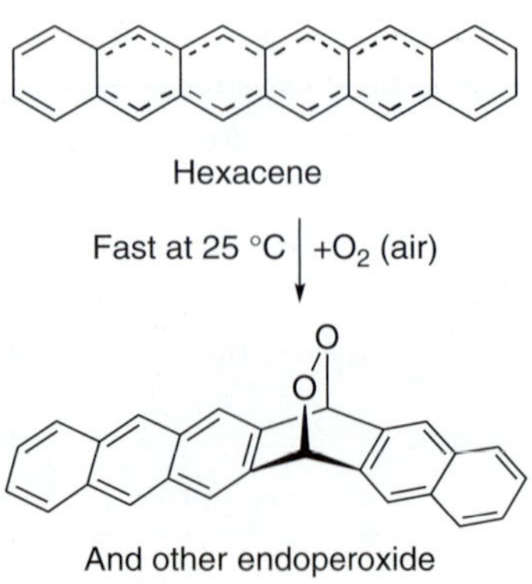

The hydrogenation of indene into indane is less exothermic (-23.9 ± 1.1 kcal mol^{-1}) than that of cyclopentene into cyclopentane (-26.8 ± 0.4 kcal mol^{-1}), suggesting that conjugation of an alkene double bond with a phenyl ring introduces a π/π conjugation stabilization of 2–3 kcal mol^{-1}. A similar conclusion is reached on comparing the heats of hydrogenation of styrene (-28.0 ± 0.44 kcal mol^{-1}) with that of monoalkylethylene (c. -30 kcal mol^{-1}). The π/π conjugation stabilization energies as well as the stabilization by aromaticity are well predicted by the Hückel theory (Section 4.5) and confirmed by higher level quantum mechanical calculations. These stabilization effects can be explained by the PMO theory (Sections 4.5.5 and 4.5.10). According to the Hückel theory, [*N*]annulenes, with $N = 4n + 2$ conjugated 2p electrons and n = 0, 1, 2, 3, and so on, are stabilized by aromaticity [168]. This should be the case also for 1,6-methano[10]annulenes and 1,5-methano[10]annulene. According to Roth, the ASEs of these bridged [10]annulenes amount to 17.2 and 6.5 kcal mol^{-1} (Section 4.5.13), only, to be compared with that of azulene (16.1 kcal mol^{-1}) and that of naphthalene (63 kcal mol^{-1}) [184, 185].

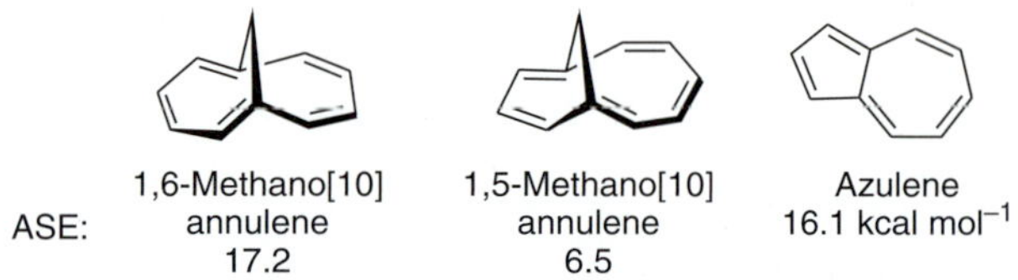

Problem 2.19 Can one obtain a product of addition of DCl to naphthalene? At which temperature would you expect a Gibbs energy for that equilibrium to be zero? Can one observe a product of addition of DCl to anthracene? If so, at which temperature would you propose to run the reaction using 1 M concentration of the reactants? Compare with the bromination of naphthalene and of anthracene.

2.7.5 Stabilization by n(Z:)/π conjugation

Comparison of the heats of hydrogenation of vinyl ethers with those of allyl ethers shows that n(O:)/π conjugation stabilizes the vinyl ethers by about 3 kcal mol^{-1} (Table 2.3). The stabilization may arise from the delocalization of the nonbonding electrons between the ether oxygen and the π-electrons of the alkene [186]. $\Delta_r H°(CH_2{=}CH{-}CH_2{-}OMe \rightleftarrows (E)\text{-}CH_3{-}CH{=}CH{-}OMe) = -4.8$ kcal mol^{-1} has been reported [187]. The more electron-rich an aryl propenyl ether, the greater its relative stability with respect to the corresponding isomeric aryl allyl ether [188].

Alcohols can be protected as allyl (prop-2-enyl) ethers, which are stable under basic and acidic conditions [189]. Deprotections of these ethers can be affected by their isomerization to the corresponding prop-1-enyl ethers, which are then hydrolyzed under weakly acidic conditions [190]. Isomerization can be induced by a relatively strong base such as t-BuOK (deprotonation, protonation), by a transition metal

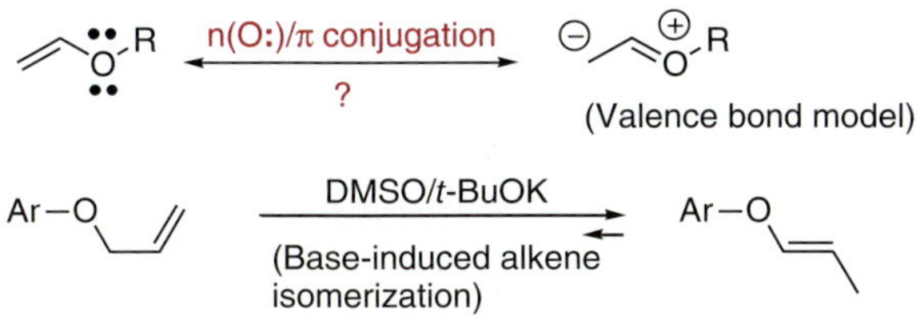

Ar = 4-MeOC$_6$H$_4$: $\Delta_rG = -4.20 \pm 0.05$ kcal mol^{-1}; $\Delta_rH = -4.2$ kcal mol^{-1}; $\Delta_rS = 0.0$ eu

Ar = 2,5-F$_2$C$_6$H$_3$: $\Delta_rG = -3.08 \pm 0.03$ kcal mol^{-1}; $\Delta_rH = -3.35$ kcal mol^{-1}; $\Delta_rS = 1.3$ eu

complex (formation of allyl(hydride)metal intermediate) or, for methallyl and other alkyl substituted allyl ethers, by reversible hydrogen atom transfer to $PhSO_2^{\bullet}$ engendered by homolysis of PhSO$_2$SO$_2$Ph [191]. The thermochemical data for enamines are scarce, and quantitative estimates of the n(N:)/π conjugation effect in these compounds are not available. Nevertheless, it has been reported that 2,5-dihydroazoles are isomerized into their 2,3-dihydro isomers (equilibria (2.12)) in the presence of palladium catalysts, thus suggesting that the latter, in which n(N:)/π conjugation is possible, are more stable than the former compounds [192]. For the isomerization of allyl methyl thioether into methyl (E)-prop-2-en-1-yl thioether, $\Delta_rH°$(CH$_2$=CH—CH$_2$—SMe ⇄ (E)-CH$_3$—CH=CH—SMe) < −2.7 kcal mol^{-1} has been proposed [129].

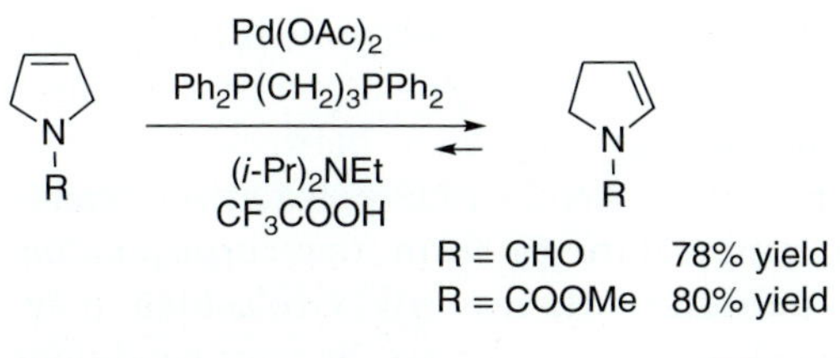

(2.12)

2.7.6 π/π-Conjugation and σ/π-hyperconjugation in esters, thioesters, and amides

The conjugation of an ether with a carbonyl function as in esters (Section 1.6.3) leads to a stabilization of about −25 kcal mol^{-1}, as given by comparison of the heats of formation of ethyl acetate ($\Delta_fH°$(gas) = −106.5 ± 0.2 kcal mol^{-1}) with that of methoxypropanone (estimated: −80 kcal mol^{-1}). Traditionally, this stabilization is attributed to the n(O:)/π_{CO} conjugation that involves the transfer of electrons from the ether moiety to the carbonyl group. However, the valence bond model implies σ_{C-O} bond shortening and $\pi_{C=O}$ bond lengthening, which is not observed (Table 1.A.5)!

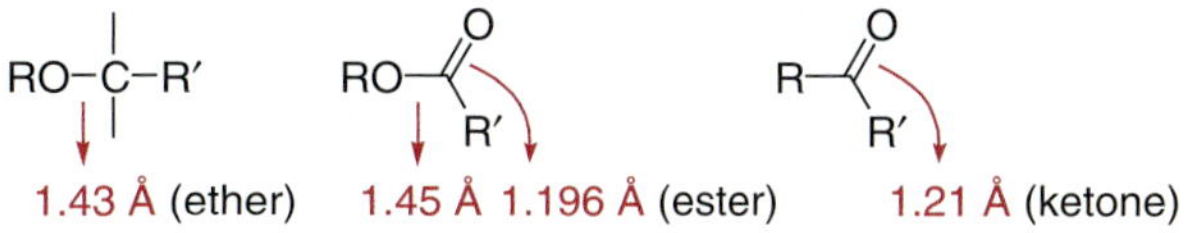

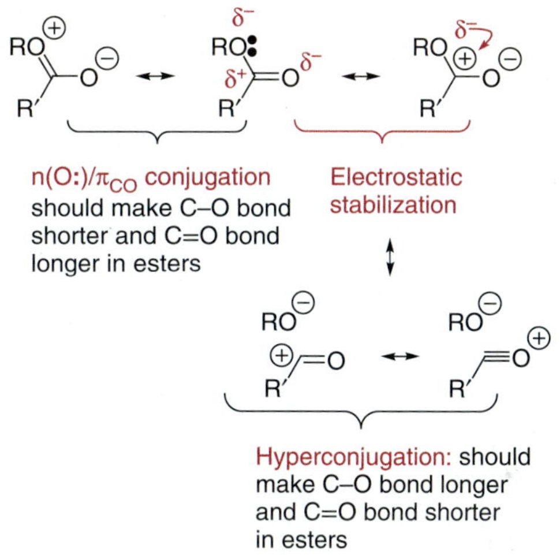

Esters adopt a planar conformation, and this is consistent with the partial double bond between the carbonyl and ether oxygen centers. The planar conformations of esters (and thioesters) are also predicted by the "banana" bond model (Figure 2.6). The latter model also predicts that the (Z)-conformers are more stable than their (E)-conformers (equilibria (2.13)) [46, 193–196]. This is more frequently rationalized as a consequence of electrostatic effects [197]. In the (Z)-conformer, the positively charged end of the Z–R dipole is closer to the negatively charged end of the stronger C=O dipole. This leads to a stronger electrostatic stabilization in the (Z)-conformer.

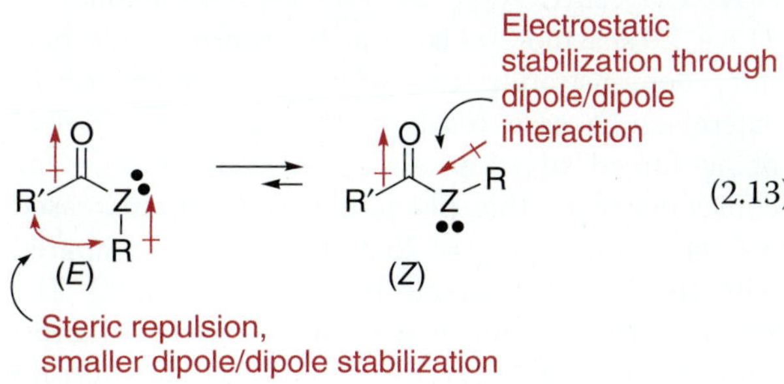

(2.13)

The larger the acyl group, R′, of an ester, the more favored the (*Z*)-conformer. For formyl esters (R′ = H), the larger the alkyl group, R, the less favored the (*Z*)-conformer (H being smaller than O). The difference in stability between the (*Z*)- and the (*E*)-conformers is smaller in thioesters than in esters. Small- and medium-sized lactones are forced to adopt (*E*)-conformations. Thus, the nucleophilic additions to lactones are expected to be more exothermic than the corresponding additions to acyclic esters.

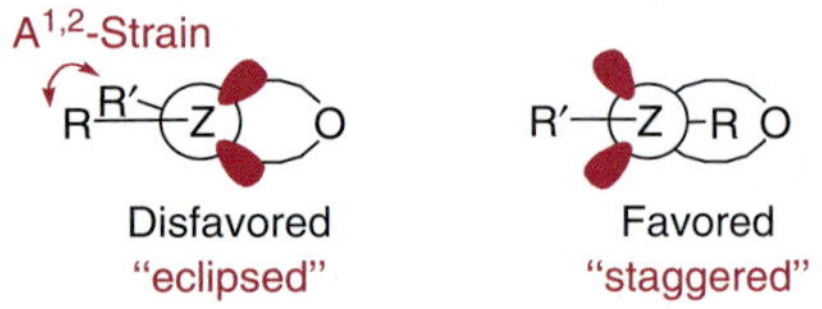

The hydrolyses of ethyl esters in 60% aqueous ethanol containing 0.8 M NaOH are slightly endothermic, with heats of reaction between 0.5 and 1.0 kcal mol^{-1} [198]. For instance, the heat of hydrolysis of ethyl propyonate amounts to 0.745 ± 0.045 kcal mol^{-1}. In the gas phase, this reaction is also endothermic by 4.4 ± 0.7 kcal mol^{-1}. By contrast, the hydrolyses of γ-butyrolactone, δ-valerolactone, and η-octanolactone are more strongly exothermic, with heats of reaction of −12.3, −14.0, and -18.2 ± 0.07 kcal mol^{-1} (H_2O/EtOH, NaOH), respectively. The dehydrogenation of *n*-hexane into cyclohexane (exchange of two methyl groups for two methylene units, note that ring strain is absent in cyclohexane) is endothermic by 10 ± 0.4 kcal mol^{-1} in the gas phase. The heat of didehydrogenation $\Delta_r H°$(hexan-3-one ⇄ H_2 + cyclohexanone) = 11.3 ± 0.4 kcal mol^{-1} shows a ring strain of c. 1 kcal mol^{-1} in cyclohexanone. The standard heats of didehydrogenations $\Delta_r H°$(Et-O-*n*-propyl ⇄ H_2 + tetrahydro-2*H*-pyran) = 11.5 ± 0.6 kcal mol^{-1} and $\Delta_r H°$((Z)-hex-3-ene ⇄ H_2 + cyclohexene) = 10 ± 0.5 kcal mol^{-1}. For the four didehydrogenations considered above, either no or very small ring strain is present in the six-membered rings formed. On the contrary, didehydrogenation $\Delta_r H°$(CH_3CH_2COOEt ⇄ H_2 + δ-valerolactone) = 21.6 ± 1.0 kcal mol^{-1} is much more endothermic, demonstrating the relative instability of δ-valerolactone as a result of the carboxylic moiety being forced to adopt the (*E*)-conformation. One consequence of this phenomenon is an increased α-C-H acidity (pK_a = 25.2 in DMSO) compared with that of ethyl acetate (29.5, Table 1.A.24). The effect of ring strain on α-C–H acidity is insignificant in the case of cycloalkanones as shown with pK_a(pentan-3-one) = 27.1 and pK_a(cyclohexanone) = 26.4 (DMSO, Table 1.A.24). The gas-phase acidities of lactones are also higher than those of the corresponding acyclic esters. Because of the negative charge delocalization in the conjugate base of ethyl acetate and of δ-valerolactone, the electrostatic interaction between the O–R dipole and the negative charge of the enolate moiety is not affected significantly by the conformation. Thus, the increase of acidity of δ-valerolactone compared with that of ethyl acetate is mostly due to a smaller dipole/dipole stabilization (see equilibrium (2.13)) in δ-valerolactone ((*E*)-carboxylic ester) than in ethyl acetate ((*Z*)-carboxylic ester).

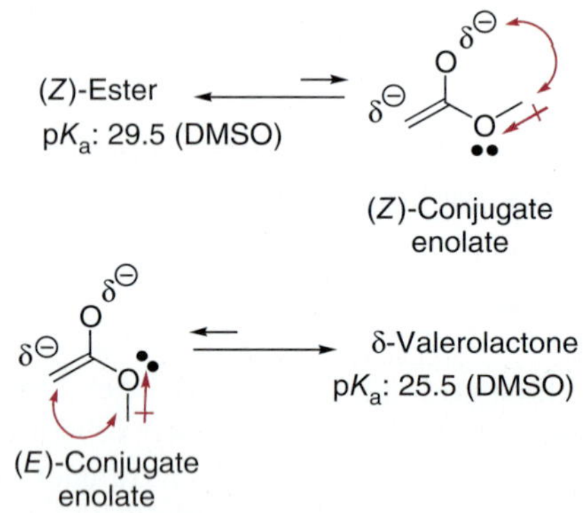

In the case of didehydrogenation of ethyl acetate into γ-butyrolactone, the endothermicity amounts to 19.5 ± 1 kcal mol^{-1}, which has to be compared with $\Delta_r H°$(penta-3-one ⇄ H_2 + cyclopentanone) = 12.4 ± 0.5 kcal mol^{-1}, $\Delta_r H°$((Z)-pent-2-ene ⇄ H_2 + cyclopentene) = 15.5 ± 2.0 kcal mol^{-1} and $\Delta_r H°$(2-ethylbut-1-ene ⇄ H_2 + methylidenecyclopentane) = 15.8 ± 0.5 kcal mol^{-1}. Taking the pair pent-3-one/cyclopentanone as the reference for the didehydrogenation of EtOAc into γ-butyrolactone, one realizes that the ring strain increase in γ-butyrolactone compared with that in cyclopentanone ($19.5 - 12.4 \pm 1.3 = 7.1 \pm 1.3$ kcal mol^{-1}) is less than the ring strain increase calculated for δ-valerolactone ($21.6 - 11.3 \pm 1.0 = 10.3 \pm 1.0$ kcal mol^{-1}). One of the consequences of this difference in ring strain between these two lactones is that δ-valerolactone polymerizes readily at room temperature to form the corresponding aliphatic polyester, whereas γ-butyrolactone polymerizes much more slowly under the same conditions [199].

The theory of resonance assigns a stabilization energy of 20 kcal mol^{-1} in amides to the juxtaposition of an amine and a carbonyl function (as given by the comparison of the heats of formation of **21** and **22**) [186]. The same theory explains the short N—C(O)

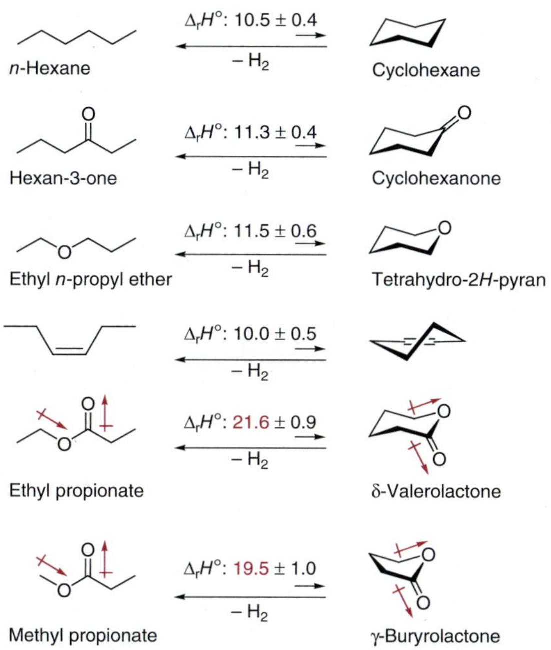

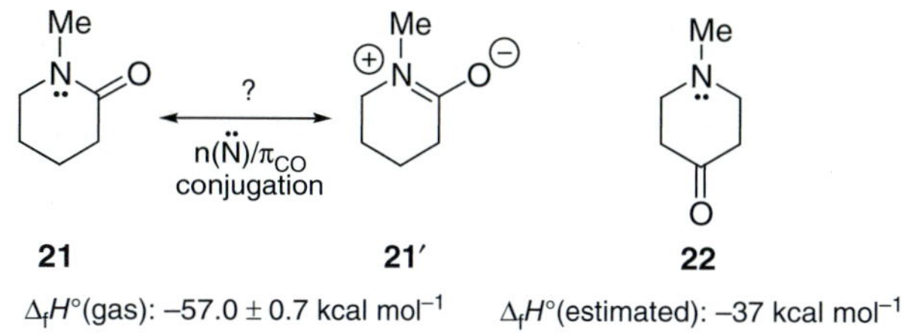

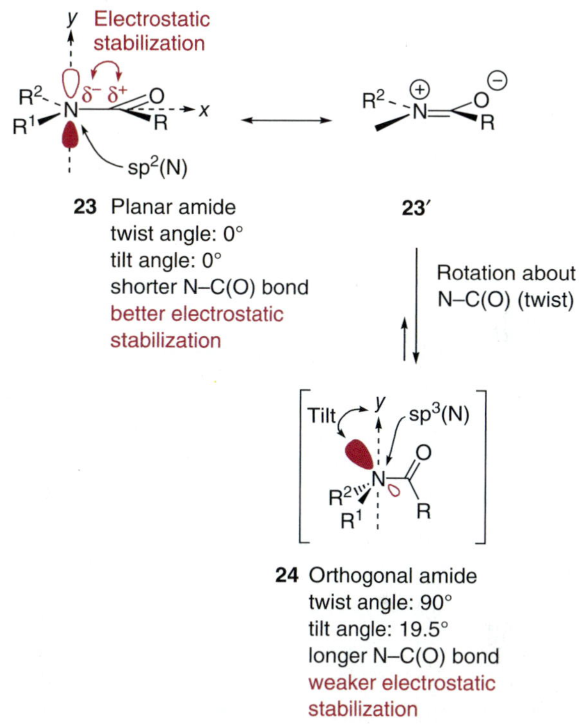

bond length [200], the planar geometry, the high N—C(O) rotational barrier [201, 202], the infrared C=O stretching frequencies, and the kinetic stability of amides toward nucleophilic attack (specifically, hydrolysis) [203]. Wiberg and Laidig on the one hand [197] and Bader and Nguyen-Dang on the other hand [204] calculated the N—C(O) bond to lengthen when going from the planar conformation **23** to the orthogonal conformation **24**; however, the C=O bond length remains nearly unchanged.

These authors proposed a modified elaboration of the resonance theory, which can be summarized as described earlier for esters. Rotation in the planar amide shortens the N—C(O) bond because this lowers the potential energy associated with the attractive interaction between the electron-rich nitrogen center and the electron-poor carbon atom of the carbonyl group. Structural studies are consistent with this theory. For instance, when going from the nearly planar amide **25** to the bicyclic lactam **26**, the N—C(O) bond length increases by 0.07 Å while the C=O bond remains nearly unchanged. In the same time, N-atom pyramidalization is observed (as indicated by the tilt angle) while deviation from planarity is realized with the rotation about the N—C(O) bond (indicated by the twist angle) [205]. For bicyclic lactam **27**, in which the nonbonding electrons at the nitrogen atom (n(N:)) are orthogonal with the π-system of the carbonyl group (limiting structure **27′** cannot be realized), the amide moiety is stabilized by hyperconjugation **27** ↔ **27″** as the quantum mechanical calculations predict an extra-long σ(N(1)—C(2)) bond and short π(C=O) bond [206]. The most twisted amide is probably 3,5,7-trimethyl-1-azadamantan-2-one (**28**) for which the N—C and C=O bond lengths are substantially longer and shorter, respectively, than expected for a tertiary δ-lactam (1.352, 1.233 Å). Amide **28** behaves like an amino-ketone. The C=O group of the latter absorbs at 1732 cm^{-1} and shows a ^{13}C-chemical shift δ_C = 200 ppm. Unstrained lactams and amides do not undergo Wittig olefination; they do not form acetal with propane-1,3-diol, contrary to **28** that gives alkene **29B** and acetal **29B** readily [207]. In sulfonylamides, the nitrogen atom is pyramidal [208]. The group equivalents for esters (e.g. [O—(CO)(C)] and [CO—(O)(C)]), amides (e.g. [CO—(N)(H)], [CO_2—(N)(C)], [N—(CO)$(H)_2$], [N—(CO)(C)(H)], and [N—(CO)$(C)_2$]), and imides (e.g. [N—$(CO)_2$(H)] and [N—$(CO)_2$(C)]) given in Table 2.A.1 include the n(O:)/π_{CO} and n(N:)/π_{CO} juxtaposition (conjugation) effects.

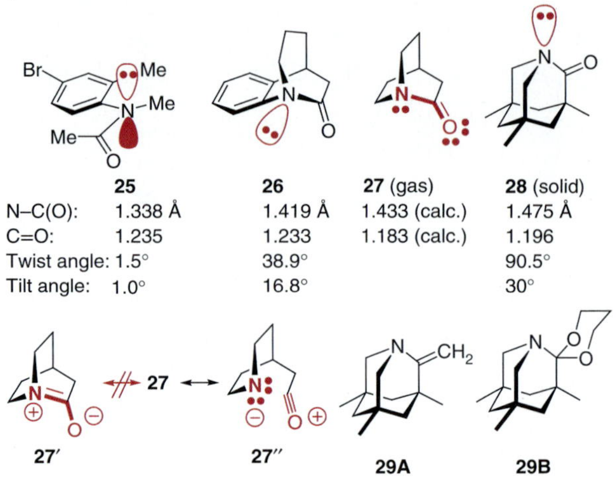

Problem 2.20 What is the most stable conformation of alkyl vinyl ethers? [209, 210]

Problem 2.21 Predict whether (*E*)- or (*Z*)-planar conformers are preferred for planar amides [193].

2.7.7 Oximes are more stable than imines toward hydrolysis

Aldehydes and primary amines equilibrate with the corresponding imines and water. The reaction is nearly thermoneutral as shown with equilibrium (2.14) for which $\Delta_r H°(2.14) = 0 \pm 2.5\,\text{kcal mol}^{-1}$ is found in the gas phase.

$$CH_3CHO + NH_3 \rightleftarrows CH_3CH{=}NH + H_2O \quad \text{(thermoneutral)} \qquad (2.14)$$

$$CH_3CHO + NH_2OH \rightleftarrows CH_3CH{=}NOH + H_2O \quad \text{(exothermic)} \qquad (2.15)$$

The formation of oximes from aldehydes and hydroxylamine are exothermic by about 10 kcal mol^{-1}. For the gas-phase equilibrium (2.15), a value of $\Delta_r H°(2.15) = -11.3 \pm 0.8\,\text{kcal mol}^{-1}$ is found. The exothermicity arises from the higher stability of oximes because of n(O:)/π(N=C) conjugation.

2.7.8 Aromatic stabilization energies of heterocyclic compounds

The comparison of the standard heats of hydrogenation (Table 1.A.3) of furan ($\Delta_f H°$(tetrahydrofuran) − $\Delta_f H°$(furan) = −44.0 − (−6.6) = −37.4 kcal mol^{-1}), azole or pyrrole ($\Delta_f H°$(pyrrolidine) − $\Delta_f H°$(azole) = −0.8 − (34.2) = −35.0 kcal mol^{-1}), and thiophene ($\Delta_f H°$(tetrahydrothiophene) − $\Delta_f H°$(thiophene) = −8.0 − (27.8) = −35.8 kcal mol^{-1}) to give tetrahydrofuran, pyrrolidine, and thiolane, respectively, with the standard heat of hydrogenation (Table 1.A.2) of cyclopentadiene into cyclopentane ($\Delta_f H°$(cyclopentane) − $\Delta_f H°$(cyclopentadiene) = −18.3 − (33.2) = −51.5 kcal mol^{-1}) suggests that furan and azole are stabilized by 14.1 and 16.5 kcal mol^{-1} by π-conjugation, respectively, whereas thiophene appears to be destabilized by −51.5 − (−60.2) = 8.7 kcal mol^{-1}. The origin of stabilization energies in furan and azole is explained by Hückel theory (Section 4.5) and PMO theory (Section 4.6.15, Figure 4.26), as well as high-level quantum mechanical calculations.

The heat of hydrogenation of pyridine ($\Delta_f H°$(piperidine) − $\Delta_f H°$(pyridine) = −11.3 − (33.5) = −44.8 ± 0.6 kcal mol^{-1}; Table 1.A.3) is 4.5 ± 1.0 kcal mol^{-1} less than that of benzene (($\Delta_f H°$(cyclohexane) − $\Delta_f H°$(benzene) = −29.5 − 19.8 = −49.3 kcal mol^{-1} ± 0.4 kcal mol^{-1}; Tables 1.A.2 and 2.2). This does not mean that pyridine is more stabilized by cyclic π-conjugation than benzene. Pyridine implies the cyclic conjugation of an imine, a π(C=N) double bond, with two π(C=C) double bonds. If one wishes to evaluate the ASE of pyridine, one should compare the standard heats of hydrogenation shown below. As the heat of formation of 1,2-didehydropiperidine is not known, we estimate it to be $\Delta_f H°$(1,2-didehydropiperidine) = $\Delta_f H°$(piperidine) − $\Delta_h H°$(1,2-didehydropyrrolidine) = −11.3 − (−16) = 4.7 kcal mol^{-1} (Table 1.A.3). The comparison of the standard heats of hydrogenation $\Delta_h H°$(pyridine → 1, 2-didehydropiperidine) = −28.8 kcal mol^{-1} and $\Delta_h H°$(benzene → cyclohexene) = −18.8 kcal mol^{-1} suggests that the ASE in pyridine is about 10 kcal mol^{-1} less than that in benzene [211]. Note that the heat of hydrogenation of an imine (c. −16 kcal mol^{-1}) is, like that of a ketone (Section 1.7.1), much less than the heat of hydrogenation of an alkene (c. −28 kcal mol^{-1}). According to the NICS index method, pyridine and all azines containing any number of nitrogen atoms should have the same ASE as benzene [212].

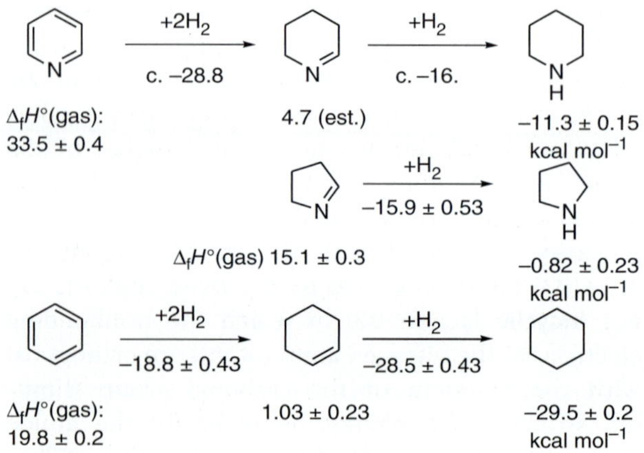

Scheme 2.1 Diels–Alder vs. Michael additions of aromatic heterocyclic compounds.

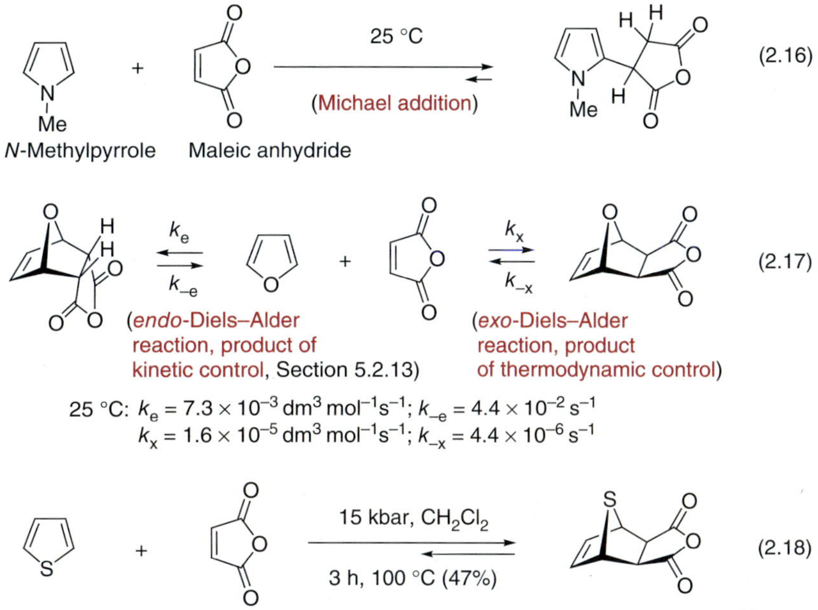

In 1931, Diels and Alder noticed that *N*-methylpyrrole (*N*-methylazole) did not give product of (4+2)-cycloaddition with maleic anhydride (Diels–Alder adducts) but products of Michael addition (Scheme 2.1) [213, 214]. Because of its ASE of about 17 kcal mol^{-1}, *N*-methylpyrrole (like benzene and naphthalene) prefers to undergo substitutions (e.g. Michael addition (2.16)) rather than cycloadditions, in order to recover the aromaticity in the product of reaction. Unsubstituted (1*H*-pyrrole) rarely gives products of cycloaddition. An exception is reaction (2.19) involving the extremely reactive dienophile hexafluoro-Dewar benzene [215]. Partial release of ring strain during the cycloaddition of hexafluoro-Dewar benzene (cyclobutene → cyclobutane) and differential fluoro substituent effects between sp^2- and sp^3-hybridized carbon centers (Section 1.7.4) facilitate this cycloaddition thermodynamically. Furan reacts with maleic anhydride in a Diels–Alder reaction at room temperature (equilibria (2.17)). Thiophene has low reactivity because the long distance between C(2) and C(5) retards the Diels–Alder reactions (Section 5.2.12). Nevertheless, under forcing conditions (high pressure, 100 °C), the Diels–Alder reaction (2.18) of thiophene is observed (Schemes 2.1 and 2.2).

α-Pyridones are much more stable than γ-pyridones. This has been interpreted in terms of the necessity to create a zwitterionic character to realize the pyridinium-type aromaticity by $n(\mathrm{N:})/\pi_{\mathrm{CO}}$ conjugation. The positive and negative charges are less separated for α-pyridone than for γ-pyridone. In terms of the theory of Wiberg and Laidig [197] and Bader and Nguyen-Dang [204] for the amide moiety (Section 2.7.6), N/C=O stabilization is much larger in α-pyridone than in γ-pyridone because of the difference in distances between N and CO.

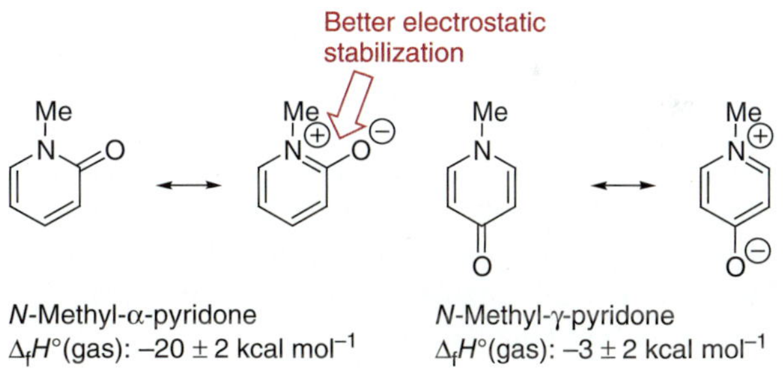

The parent phosphole is poorly stabilized by aromaticity and is the least aromatic of the simple five-membered heterocyclic compounds [216–219]. 1-Alkyl and 1-arylphospholes show the same ASE as parent phosphole, whereas 1-cyanophosphole is slightly less stabilized by aromaticity. 1-Alkoxy- and 1-halogenophospholes lose most of the cyclic delocalization. Pyramidality at phosphorus does not explain these variations. Variations in conjugative ($n(\mathrm{P:})/\pi$) and hyperconjugative effects are responsible for the change in cyclic delocalization. It has been observed that the weaker is the ASE of the P-substituted phosphole, the higher is its Diels–Alder reactivity. There are no experimentally derived thermochemical data for the parent phosphole, a compound first prepared by Mathey and coworkers [220]. Applying quantum mechanical calculations,

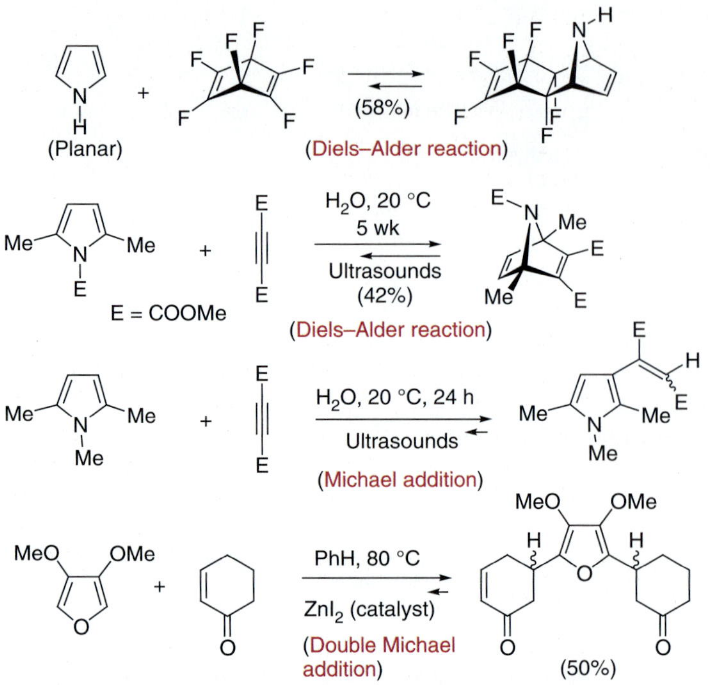

(2.19)

(2.20)

(2.21)

(2.22)

Scheme 2.2 1H-pyrrole can undergo Diels–Alder cycloaddition only with extremely reactive dienophiles or when its aromaticity is reduced by N-substitution with electron-withdrawing groups. Otherwise, it undergoes Michael additions. This can also occur with other dienes.

the ASE of phosphole is estimated to amount to only $-13\,\text{kcal mol}^{-1}$ [216–220]. For several *P*-alkylated derivatives, Mislow and coworkers measured energy barriers of c. $16\,\text{kcal mol}^{-1}$ for phosphorus inversion (pyramidal ⇄ planar ⇄ pyramidal interconversion) [221]. This is about $20\,\text{kcal mol}^{-1}$ lower than the barrier of phosphorus inversion in the corresponding phospholanes. It has been proposed that this difference represents the gain in stability of planar phosphole because of its aromaticity. Thus, phospholes do not resemble azoles as the former are pyramidal and the latter planar (as amides).

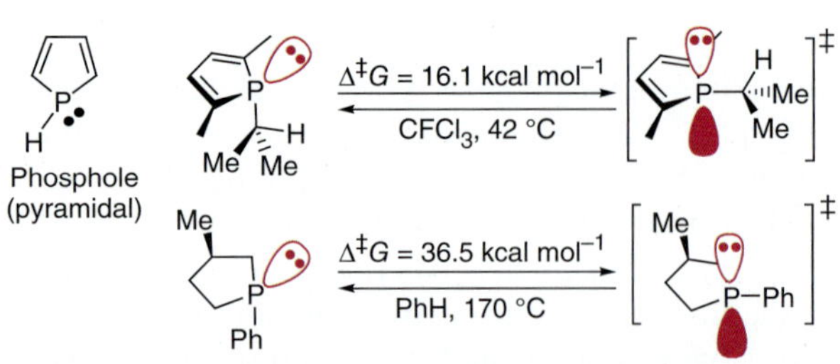

Phospholes are little reactive in Diels–Alder reactions, not only because of their aromaticity but also because the distance between C(2) and C(5) is much longer than in furans and azoles (Section 5.2.12) [222]. Only extremely reactive dienophiles will react with phospholes in a Diels–Alder reaction, as shown with 3,4-dimethyl-1-phenylphosphole and *N*-phenylmaleimide at 40 °C (Scheme 2.3). By oxidation, sulfurization, quaternization, or complexation of

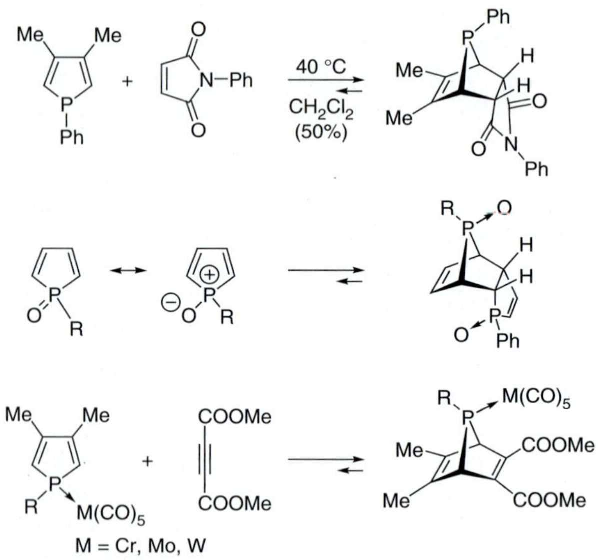

Scheme 2.3 Quaternization of the phosphine moiety of phospholes generates more reactive derivatives.

the phosphine moiety, the phospholes become much more reactive as they lose their aromaticity arising from the $n(\text{P:})/\pi$ conjugation. Examples are given in Scheme 2.3.

Problem 2.22 Evaluate the Gibbs energy of the Diels–Alder reactions of 1*H*-pyrrole, furan, and cyclopentadiene with maleic anhydride giving the corresponding *exo* adducts at 300 K. Compare these values with those evaluated for the corresponding Michael additions.

Problem 2.23 From the heats of formation of dibenzofuran (19.9 kcal mol^{-1}), diphenyl ether (10.6 kcal mol^{-1}), dibenzocyclopentadiene (42.2 kcal mol^{-1}), and diphenylmethane (36.8 kcal mol^{-1}), evaluate the extra amount of aromaticity introduced by the furan ring of dibenzofuran. Compare your answers with the aromaticity of anthracene.

2.7.9 Geminal disubstitution: enthalpic anomeric effects

Common sense would predict that geminal (i,i)-disubstitution by strongly electron-withdrawing groups should generate compounds less stable than their vicinal (i,i + 1)- or (i,i + n)-disubstituted isomers ($n > 1$). This is the case when comparing the heats of formation of malonodinitrile with that of succinodinitrile. Going from the former to the latter compound, a decrease of −5 kcal mol^{-1} would be expected for the standard heat of formation because of the insertion of a methylene fragment. Thus, if there were no repulsive interaction between the carbonitrile groups of malonodinitrile, the estimated heat of formation of succinodinitrile would be 63.5 – 5 = 58.5 kcal mol^{-1}, instead of the measured value of 50.1 kcal mol^{-1}. This demonstrates that malonodinitrile suffers about 8 kcal mol^{-1} of electrostatic destabilization. Another interpretation would be to invoke more stabilizing C–H/CN hyperconjugative stabilization in succinodinitrile than in malonodinitrile.

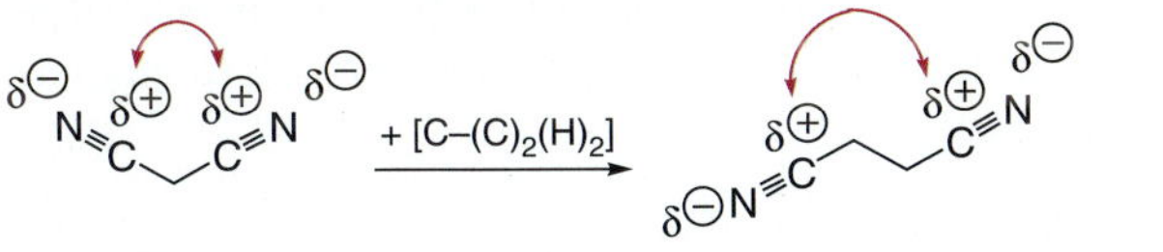

Malodinitrile $\Delta_f H°$(gas): 63.5 kcal mol^{-1} — Succinodinitrile $\Delta_f H°$(gas): 50.1 kcal mol^{-1}

Using standard heats of formation (Tables 1.A.2–1.A.4), one obtains the heats of reaction (2.23) for various substituents X given below. With the nonpolar methyl group (X = Me), the exothermicity of −2.5 kcal mol^{-1} expresses the difference in [C—(C)$_2$(H)$_2$] and [C—(H)(C)(H)$_2$] group equivalent in propane and methane. Compared with equilibrium (2.23, X = Me), geminal disubstitution by X = CN in malonodinitrile (equilibrium (2.23), X = CN) introduces a destabilization effect of 12–13 kcal mol^{-1}. With X = COMe (penta-2,4-dione), a stabilization is found instead. It arises from the formation of a stabilized enol (tautomerization), an equilibrium that is not favorable in malonodinitrile. With X = Cl, Br, a destabilization of c. 3 kcal mol^{-1} is observed for the geminal disubstitution. This destabilization amounts to c. 5 kcal mol^{-1} for X = NO_2. In contrast, important stabilization effects of −9 to −12.5 kcal mol^{-1} are found for X = F, OH, O-alkyl, and Me_2N. A stabilization of c. −2 kcal mol^{-1} is also found for the chlorofluoro geminal disubstitution. These stabilizing geminal disubstitution effects are called enthalpic anomeric effects [223–225]. An example of geminal dioxy disubstitution effect is manifested by the difference $\Delta_f H°$(1,4-dioxane, gas) – $\Delta_f H°$(1,3-dioxane, gas) = −75.4 ± 0.2 – (−80.9 ± 0.3) = −5.5 ± 0.5 kcal mol^{-1}. The origin of these effects is essentially electrostatic, analogous to those for esters and amides (Section 2.7.6). Electron delocalization through orbital interactions (hyperconjugation) cannot be excluded, although, according to recent quantum mechanical calculations, are now considered to be a minor contribution [226].

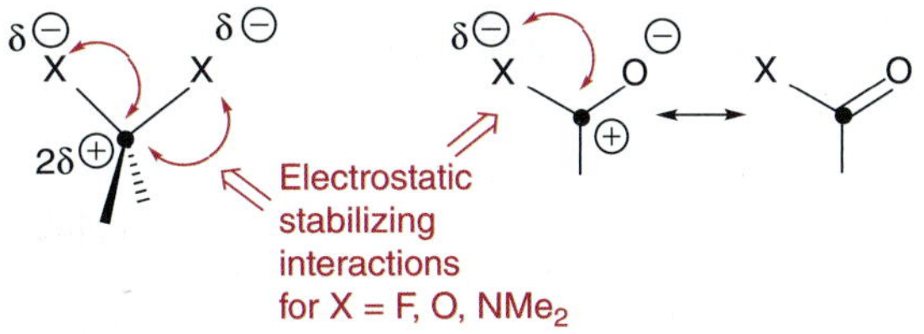

$\Delta_r H°(CH_3F + CH_3Cl \rightleftharpoons CH_4 + CH_2FCl) = -4.5$ kcal mol^{-1}

$\Delta_r H°(CH_3F + CH_3Br \rightleftharpoons CH_4 + CH_2FBr) = 5.3$ kcal mol^{-1}

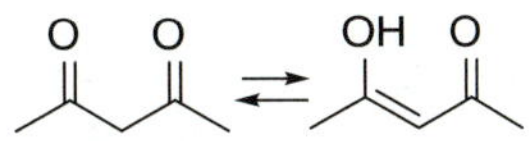

$$2CH_3X \xrightleftharpoons{\text{Gas}} CH_4 + CH_2X_2 \quad (2.23)$$

X:	$\Delta_r H°$(2.23)	X:	$\Delta_r H°$
F	−12.3	Cl	−0.2 kcal mol^{-1}
OH	−15.0	Br	−0.7
OMe	−13.0	CH_3	−2.5
OEt	−13.6	CN	10.3
NMe_2	−11.0	COMe	−5.5
NO_2	2.4		

Methyl substitution of n-alkanes C_mH_{2m+2} on other positions than the terminal carbon centers generates mono-branched higher homologs $C_{m+1}H_{2m+4}$ that have standard heats of formation equal to that of the lower homologs −7 kcal mol^{-1} (see differences in heats of formation of reactions (2.25) for X = Me, Et). This is due to the exchange of a methylene group for a methine group and addition of a methyl substituent, as noted by the substitution equivalent [C—(C)(H)$_3$] + [C—(C)$_3$(H)] – [C—(C)$_2$(H)$_2$].

Comparing the heats of formation of CH_3CH_2OMe and $(CH_3)_2CHOMe$, one finds a more negative value of −8.5 kcal mol^{-1} for this α-methyl substituent effect. This effect is enhanced to −11.6 and −14.0 kcal mol^{-1} when comparing the heats of formation of 1,3-dioxolane with that of its 2-methyl-substituted homolog and of 1,3-dioxane with that of 2-methyl-1,3-dioxane, respectively. Interestingly, the α-methyl substituent effect is less negative (−9.6 kcal mol^{-1}) when substituting 1,3-dioxane at position C(4). Thus, α-methyl substitution of an ether and of an acetal introduces stabilizations that are larger than for alkanes, alkyl chlorides and bromides, and malonodinitrile. The effect is larger for acetals than for ethers as the partial positive charge at their α-carbon centers is larger in the former (α-carbon with one polar C—O bond) than in the latter compounds (α-carbon with two polar C—O bonds).

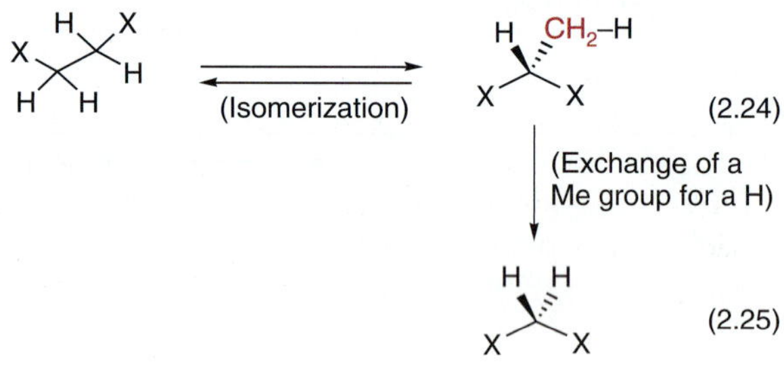

X =	Et	Me	Cl	CN	OMe	OEt
$\Delta_r H°$(2.24):	−1.1 ± 0.4	−2.0 ± 0.3	0.0 ± 1.2	ca. 6.5	−11.2 ± 0.4	−10.4 ± 1.0 kcal mol^{-1}
$\Delta_r H°$(2.25):	6	7	8.7	7	10	10 kcal mol^{-1}

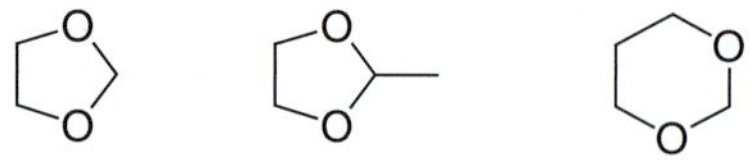

$\Delta_f H°$(gas): −72.1 ± 0.5 −83.7 ± 0.8 −80.9 ± 0.3

$\Delta_f H°$(gas): −90.1 −95.0 ± 0.7 kcal mol^{-1}

If one considers the two oxy or two fluoro substituents to realize two dipoles (C–O, C–F), electrostatic theory predicts that there are families of geometries for the two dipoles that lead to a stabilization and other families of geometries that lead to a destabilization. This phenomenon can be modeled by relationship (2.26) for an ensemble of two positive charges q$^+$ and two negative charges q$^-$. The attraction between the negative and positive ends of two different dipoles overcomes the repulsion arising from the juxtaposition of the two positive ends to each other and the two negative ends to each other [227].

$$E_{electro} = \left[\underbrace{\frac{1}{r_{--}} + \frac{1}{r_{++}}}_{\text{Repulsions}} - \underbrace{\left(\Sigma\frac{1}{r_{+-}}\right)}_{\text{Attractions}}\right] \times \frac{|q^+||q^-|}{4\pi\varepsilon} \qquad (2.26)$$

where
- r_{--}: distance separating negative extremities of the dipoles
- r_{++}: distance separating positive extremities of the dipoles
- r_{+-}: distances separating the positive and negative extremities of the dipoles
- ε: dielectric constant (ε = 1, vacuum)

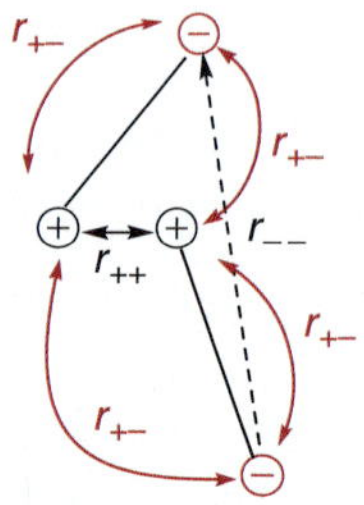

The balance between repulsive and attractive electrostatic effects depends on charges q$^+$ and q$^-$ (electronegativity differences of the C—X bonds) and on the nature of the dipoles (type of bonds) and their orientation (Section 2.7.10). In the case of acetals, one needs to consider an ensemble of four dipoles, as shown below:

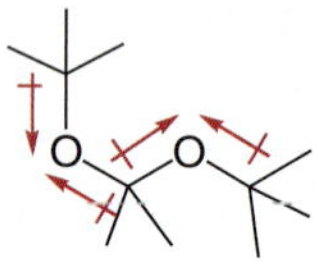

In the presence of a Lewis acid catalyst, glycals undergo allylic rearrangements into 2,3-unsubstituted glycosyl derivatives that are stabilized by enthalpic anomeric effects. This is the Ferrier rearrangement [228, 229].

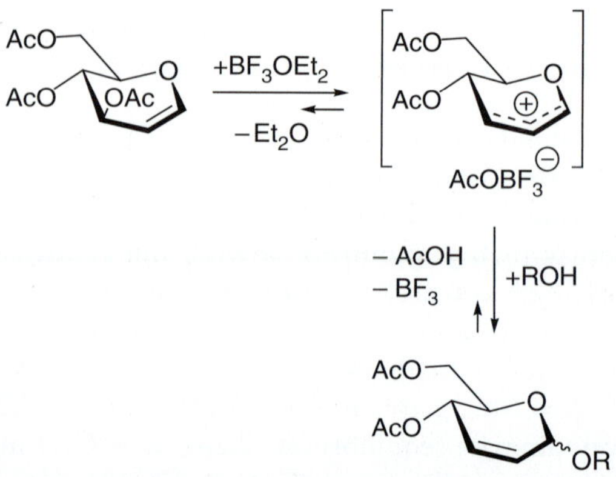

Problem 2.24 On treating 2,3,4,6-tetra-O-acetyl-α-D-glucopyranosyl bromide (**A**) with Bu_3SnH and AIBN ($Me_2C(CN)N{=}N(CN)Me_2$) in benzene at 80 °C,

2,3,4,6-tetra-O-acetyl-1,5-anhydro-D-glucitol (**B**) is formed concurrently with 1,3,4,6-tetra-O-acetyl-2-deoxy-α-D-arabinohexose (**C**). The proportion of these two products of reduction depends on the concentration of Bu_3SnH. With low concentration of Bu_3SnH, **C** is the major product. Explain [230, 231].

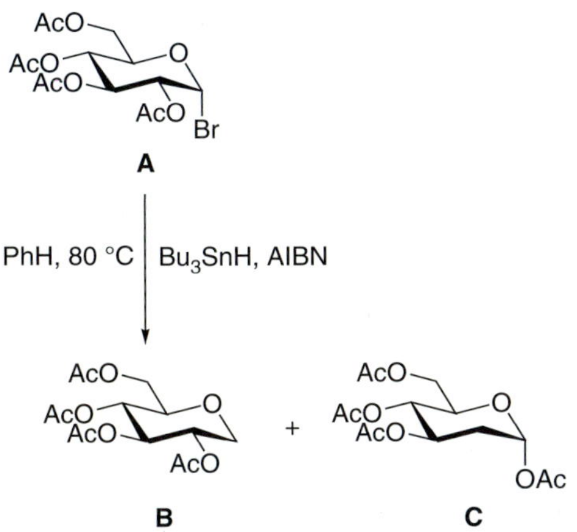

2.7.10 Conformational anomeric effect

The conformational anomeric effects constitute contrasteric effects observed in acetals, making the *gauche/gauche* conformers more stable, in spite of their higher steric encumbrance, than the *anti/gauche* and *anti/anti* conformers. As a general rule, the latter enthalpy differences are much smaller than the enthalpic anomeric effects (*gem*-dioxy substitution effects) defined above (Section 2.7.9). Conformational anomeric effects were first evidenced by Jungins in 1905 but were rediscovered by Edward in 1955 [232] and by Lemieux and coworkers in 1958 [233]. They observed that the α-anomers of alkyl D-glucopyranosides are more stable than the β-anomers (Scheme 2.4) [234, 235].

Following these observations, it was established that the conformational anomeric effect is not restricted to carbohydrates but applies to fragments of type R–X–A–Y, where X stands for an atom bearing a lone pair of electrons, A represents an atom of intermediate electronegativity (typically C, P, and S), and Y is an atom more electronegative than A (typically N, O, and the halogens) [236–238]. These systems often display a preference for the substituents on the anomeric carbon to adopt a synclinal (*gauche*) position rather than an antiplanar (*anti*) position (Figure 2.7).

One recognizes the conformational *endo*-anomeric effect for pyranosides with polar X substituents at C(1) (contrasteric electronic stabilization effect: Figure 2.8a) and conformational *exo*-anomeric effect for glycosides (acetals) in which the alkyl group R of the exocyclic moiety is synclinal (Figure 2.8b,c).

In many pyranoses (aldoses that exist mostly as six-membered cyclic hemiacetals; $C_mH_{2m}O_m$, with m most commonly 5, 6, and 7), the proportion of the anomer with an axial OH group is higher than the ratio expected based on the consideration of only steric factors (Section 2.6.3: equilibria (2.9)). Thus, although the predicted ratio in aqueous solution at 25 °C based on the AE value for axial vs. equatorial cyclohexanol is 11/89 (see below), the observed ratio is 34/64 for α-D-glucopyranose/β-D-glucopyranose [239, 240]. Figure 2.9a gives the mechanisms of the isomerization of various forms of D-glucose, and Figure 2.9b presents the other D-aldohexoses in their most stable form in aqueous solution (nomenclature, see Figure 2.9a: D means C(5)-OH is placed right in the Fischer projection of a hexose; D means C(4)-OH is placed right in the Fisher projection of a pentose (D does not mean dextrogyre), *α* means the hydroxy groups C(1)–OH and C(5)–OH are drawn in the same

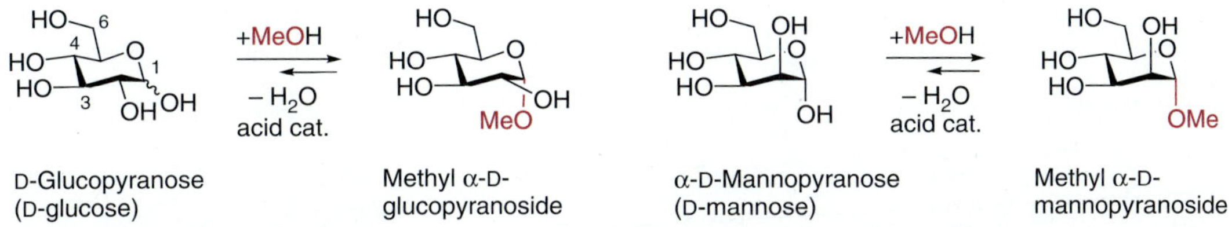

Scheme 2.4 Fischer's acid-catalyzed glycosylation of D-glucose and D-mannose.

Figure 2.7 Generalized conformational anomeric effect in systems of type R–X–A–Y.

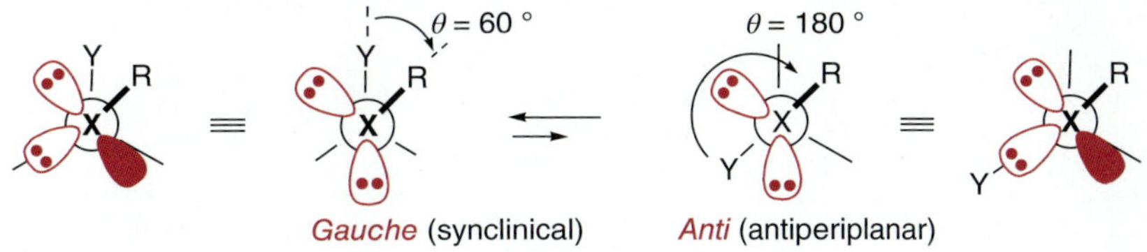

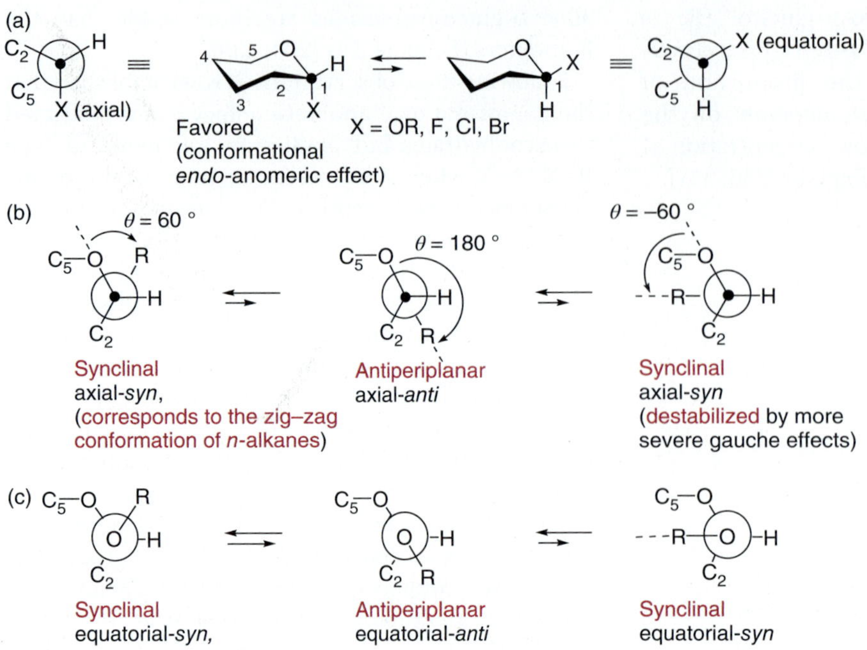

Figure 2.8 (a) Conformational *endo*-anomeric effect; (b) *exo*-anomeric effect in axial *O*-pyranosides (X = O-alkyl, O–R); and (c) conformational *exo*-anomeric effect in equatorial *O*-pyranosides (X = O-alkyl, O–R).

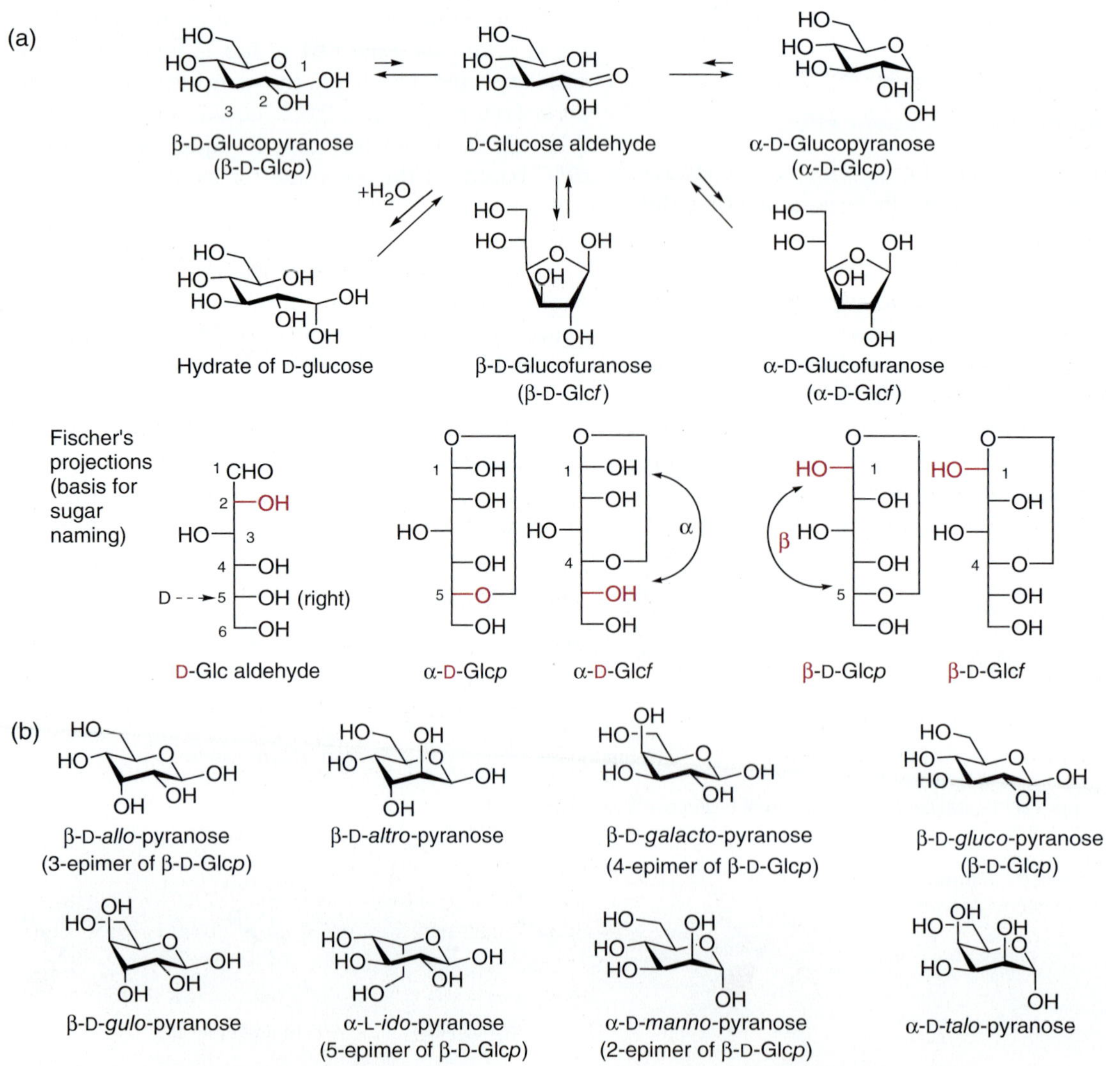

Figure 2.9 (a) In water, D-glucose exists as an equilibrium among two pyranose and two furanose forms, as well as an open aldehyde form and its hydrate. Similar equilibria exist with all aldoses having 5, 6, 7, etc., carbons. (b) Representation of the most stable forms of all D-aldohexoses in water (Table 2.3).

Table 2.3 Percentage of cyclic α-pyranose (α-p), β-pyranose (β-p), α-furanose (α-f), β-furanose (β-f), and acyclic forms of D-aldohexoses (aldehyde, reducing sugar) in water at 30 °C.

Monosaccharide	α-p	β-p	α-f	β-f	Hydrate	Aldehyde
D-*allo*	14.6	77.1	3.0	5.3	0.0063	0.0032
D-*altro*	26.8	41.1	8.6	13.4	0.079	0.014
D-*galacto*	31.2	62.7	2.3	3.7	0.046	0.006
D-*gluco*	37.6	62.0	0.1	0.28	0.0059	0.004
D-*gulo*	12.2	83.7	0.94	3.04	0.077	0.005
D-*ido*	33.7	37.3	12.14	16.12	0.7	0.094
D-*manno*	66.2	32.9	0.64	0.24	0.022	0.0044
D-*talo*	42.1	28.7	17.9	11.1	0.052	0.029

For fructose (a ketose, nonreducing sugar) see Section 3.2.8, Scheme 3.1.

side of the Fischer projection of the aldohexose; and α does not mean axial). *Ab initio* molecular dynamics simulations applying the Carr–Parrinello method [241] have revealed different solvation behaviors about the anomeric center (hemiacetal at C(1)) of D-glucose in water. The analysis of hydrogen bonds around the anomeric oxygen shows distinct solvation behaviors for the two anomers of D-glucopyranose. The β anomer allows water molecules to flow in a disorderly manner, whereas the α anomer tends to bind the water molecules more tightly [242, 243]. Using ^{13}C-NMR, Serianni and coworkers have measured the ratios of the different forms of aldohexoses ($C_6H_{12}O_6$) in Table 2.3 [92].

As proposed by Edward [232], the higher stability of alkyl α-D-glucopyranosides with respect to their β-anomers (a contrasteric effect) arises from more favorable electrostatic interactions in the former. For 2-methoxytetrahydro-2*H*-pyran (IUPAC nomenclature of heterocyclic compounds, number 1 for the heteroatom part of the ring), the methoxy-axial conformer (**30a**) is favored to a greater extent in nonpolar solvents than in polar solvents [233]. The equilibrium constant K(2.27) varies with solvent as follows: 17/83 in CCl_4 (ε = 2.2), 18/82 in PhH (ε = 2.3), 28/72 in acetone (ε = 20.7), 32/68 in MeCN (ε = 37.5), and 48/52 in water (ε = 78.5). The decrease in conformational anomeric effect as solvent polarity increases has been reproduced by quantum mechanical calculations [244].

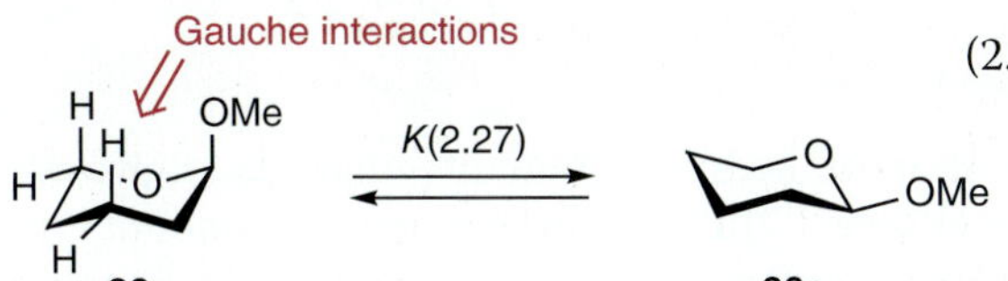

30a **30e** (2.27)

One defines the conformational AE of two-substituted tetrahydro-2*H*-pyran as the difference in free energies of equilibria (2.28) ($\Delta_r G°$ = A values, Section 2.6.3) and Eq. (2.29) [245] as already seen above. AEs depend on solvent [246, 247] and the nature of the two-substituent [248].

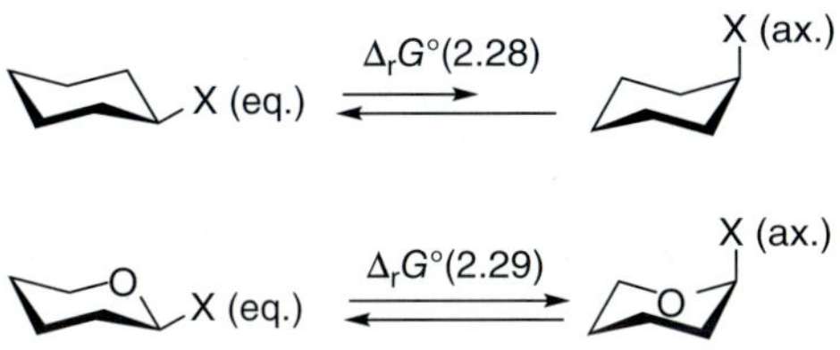

(2.28)

(2.29)

Anomeric effect AE = $\Delta_r G°$(2.28) − $\Delta_r G°$(2.29)

X:	AE:	X:	AE:
Cl	2.4	OH	0.8
Br	2.3	NHMe	0.4
OMe	1.7	($\Delta\Delta_r H°$ = 0.0)	
($\Delta\Delta_r H°$ = 0.74)		COOMe	−0.1
OEt	1.6	CN	0.17
SMe	1.5	Alkyl	c. 0.0

Cyclohexanecarbonitrile prefers a chair conformation with the CN group equatorial. This is also the case for tetrahydro-2*H*-pyran-2-carbonitrile but not for piperidine-2-carbonitrile [249]. In the latter compound, the entropy variation is negative, which implies a stronger solvation for the conformer with the axial CN group than for the other conformer with the equatorial CN group.

In the chair conformers of thiane-2-carbonitrile and selane-2-carbonitrile, the cyano group also prefers the axial position [250]. This has been attributed to a through-space interaction between the lone pair of the heteroatom (N:, S:, Se:) and the lowest lying π^*-orbital (lowest unoccupied molecular orbital [LUMO], Section 4.5.5), which is stronger when the CN group is axial than when it is equatorial. In CCl_4 and MeCN, 2-alkoxythianes adopt chair conformations in which the MeO, *t*-BuO, or OH group prefers the axial position by c. 80%. In 2-alkylthiothianes, axial preferences of 33–50% have been observed for *n*-PrS, *n*-BuS, and PhS substituents [251, 252]. Thiane-1-oxide adopts chair conformations in which the S=O bond resides in equatorial and axial position almost equally (equilibrium (2.31), $Y = Z = CH_2$) [253]. In contrast, in 1,2-oxathiane-2-oxide, 1,2-azathiane-2-oxide, 1,2-dithiane-2-oxide, and 1,6-dioxa-2-thiane-2-oxide, the S=O bond prefers the axial position by 3–5 kcal mol^{-1} [254, 255]. Quantum mechanical calculations suggest that the axial preference for the S=O bond arises from a weaker dipole moment in axial than in equatorial conformations [256].

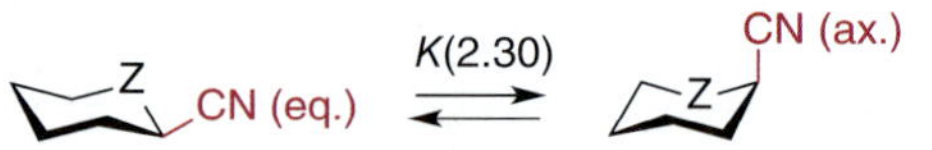

(2.30)

Z:	$\Delta_r H°$ (2.30):	$\Delta_r S°$ (2.30):
Z = CH_2	0.182 ± 0.007 kcal mol^{-1}	0.0 ± 0.04 eu
O	0.36 ± 0.007	0.84 ± 0.04
NH	−2.22 ± 0.56	−5.65 ± 2.67

Preferred for Y = CH_2, Z = O, NH, S and Y = Z = O

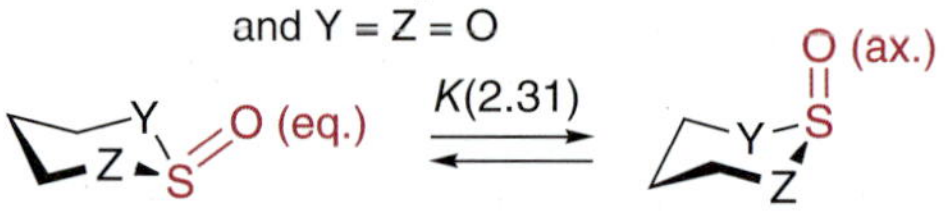

(2.31)

2,6-Dihydro-1,2-oxathiin-2-oxide (sultine **31**: Diels–Alder cycloadduct of sulfur dioxide and butadiene, Section 5.2.16) adopts a pseudo-chair or half-chair conformation with axial S=O bond [257]. In agreement with quantum mechanical calculations, 6-fluorosultines of type **32c** and **32t** (resulting from the reaction of SO_2 with 1-fluoro-1,3-dienes) prefer sofa conformations with axial S=O bond. This demonstrates that the highly polar fluoro substituent introduces dipole/dipole interactions that force the six-membered ring to adopt unusual conformations. In cyclohexene, the sofa conformation is associated with the transition state of the half-chair ⇄ half-chair interconversion (Section 2.6.5). Thus, the sofa transition structure of cyclohexene becomes the ground-state structure in 6-fluorosultines [258].

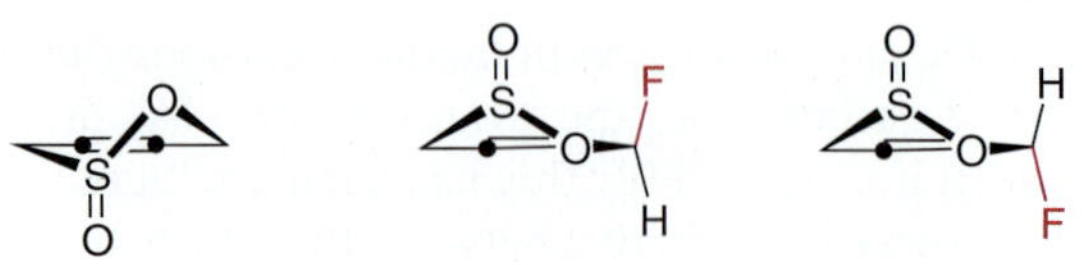

31 (Pseudo-chair) **32c** (Sofa) **32t** (Sofa)

2.8 Other deviations to additivity rules

According to Dewar planar, [4*N*]annulenes such as cyclobutadiene ($n = 1$) are destabilized by antiaromaticity, whereas [4*N*+2]annulenes are stabilized by aromaticity [259–262]. This is the case for $N = 0$, 1, 2, and less so for $N > 3$ (Section 4.5). The overlap between the π-orbitals of the three ethylene units of bicyclo[2.2.2]octa-2,5,7-triene, or barrelene, is responsible of the c. 8 kcal mol^{-1} of electronic destabilization in this compound (Section 4.7.6: the "barrelene effect"). Cyclopropane/π interaction makes bicyclo[2.1.0]pent-2-ene resemble cyclobutadiene, as it is destabilized by c. 10 kcal mol^{-1}. This is evidenced by the heat of hydrogenation of bicyclo[2.1.0]pent-2-ene into bicyclo[2.1.0]pentane, which amounts to −41.9 kcal mol^{-1} instead of $\Delta_h H°$(cyclobutene → cyclobutane) = −30.9 kcal mol^{-1} (Section 2.6.8). The σ/π type of electronic repulsion (destabilization through hyperconjugation) found in bicyclo[2.1.0]pent-2-ene is not observed with bicyclo[2.2.0]hex-2-ene as its hydrogenation into bicyclo[2.2.0]hexane has an exothermicity of c. −31.2 kcal mol^{-1}, nearly the same as $\Delta_h H°$(bicyclo[3.2.0]hept-6-ene → bicyclo[3.2.0]heptane) = −32.7 kcal mol^{-1}. Destabilizing hyperconjugative interactions are also evidenced in bicyclo[2.2.1]hept-2-ene (norbornene) and bicyclo[2.2.1]hepta-2,5-diene (norbornadiene) (Section 4.8.3).

Cyclobutadiene: antiaromaticity (Section 4.4.6) Barrelene: π/π-repulsion (Section 4.6.6) Bicyclo[2.1.0]pent-2-ene: σ/π-repulsions

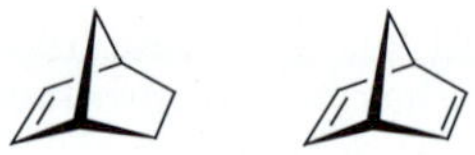

Norbornene: σ/π-repulsions (Section 4.7.3) Norbornadiene: σ/π-repulsions (Section 4.7.3)

The heat of hydrogenation of ethanedial (glyoxal) into ethane-1,2-diol (ethylene glycol) amounts to −43.5 ± 0.9 kcal mol^{-1}, about 10 kcal mol^{-1} more than twice the heat of hydrogenation of ethanol

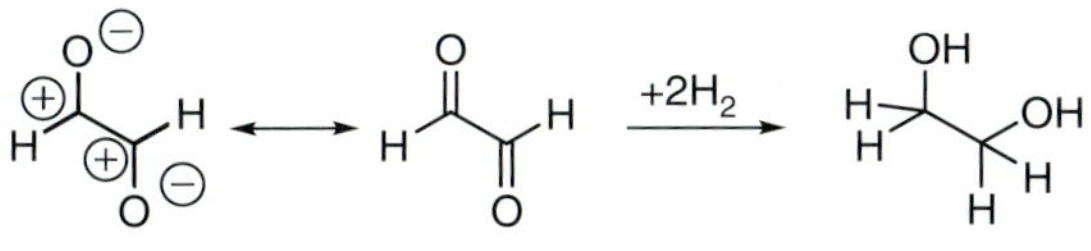

$\Delta_f H°$(gas): -50.7 ± 0.2 kcal mol^{-1} $\quad -94.2 \pm 0.67$

$\Delta_r H°$: $-43.5 \pm 0.9 < 2\,(-15.2)$

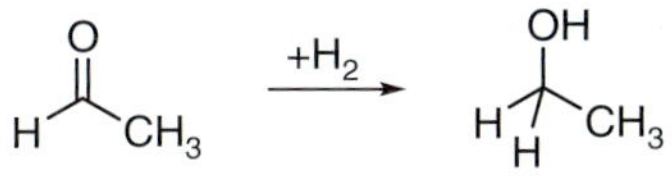

$\Delta_f H°$(gas): -40.8 ± 0.35 $\quad -56.0 \pm 0.5$

$\Delta_r H°$: -15.2 ± 0.9

$$CH_3–CH_3 \xrightarrow{+O} CH_3CH_2OH$$

$\Delta_f H°$(gas): -20.0 ± 0.1 $\quad -56.0 \pm 0.5$

$\Delta_r H°$: -36.0 ± 0.6

$$CH_3CH_2OH \xrightarrow{+O} HOCH_2CH_2OH$$

$\Delta_f H°$(gas): -56.0 ± 0.5 $\quad -94.2 \pm 0.67$

$\Delta_r H°$: -38.2 ± 1.2

$$CH_4 \xrightarrow{+O} CH_2OH$$

$\Delta_f H°$(gas): -17.8 ± 0.07 $\quad -49.0 \pm 3.0$

$\Delta_r H°$: -31.2 ± 3.1

$$CH_3OH \xrightarrow{+O} CH_2(OH)_2$$

$\Delta_f H°$(gas): -49.0 ± 3.0 $\quad -93.5$

$\Delta_r H°$: $-44.5 \pm 3.5 < -31.2$ kcal mol^{-1} (enthalpic anomeric effect: Section 2.7.9)

into ethanol ($2(-15.2 \pm 0.9$ kcal mol^{-1})). Interestingly, the heat of oxidation of ethane into ethanol (-36.0 ± 0.6 kcal mol^{-1}) and that of ethanol into ethylene glycol (-38.2 ± 1.2 kcal mol^{-1}) are nearly the same; this shows that vicinal OH substitution does not introduce any stabilization or destabilization effect. Alternatively, one could argue that vicinal dioxy-substitution induces a destabilizing electrostatic interaction between the two hydroxy groups that is compensated by the bridging hydrogen bond between them. In Section 1.7.2, we have shown that, if indeed operative, these effects amount to less than 2 kcal mol^{-1}. Thus, the high exothermicity of the reduction of the carbonyl moieties of glyoxal is due to an electrostatic destabilization effect arising from the α-juxtaposition of the two carbonyl functions. Note that glyoxal is also destabilized with respect to two aldehydes because of lack of α-CH_2 groups that can stabilize the carbonyl moieties through C–H/CO hyperconjugation.

Indane-1,2,3-trione contains three carbonyl groups in a planar geometry [263]. This compound is highly electrophilic and readily adds 1 mol of water to yield ninhydrin. Penta-2,3,4-trione adopts a nonplanar geometry in the solid state [264], the pairs of carbonyl groups making dihedral angles of c. 120°. Diphenylpropanetrione is also a very reactive electrophile toward many nucleophiles, including water (2.32). It can be used as a drying agent [46, 265, 266].

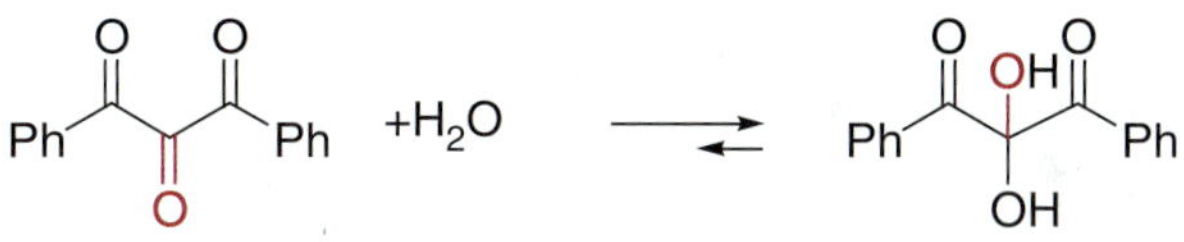

(2.32)

Ninhydrin is used to identify primary α-amino acids as shown in Scheme 2.5. It is one of the techniques used in the forensic sciences to visualize finger prints.

Problem 2.25 Bicyclo[2.1.0]pent-2-ene reacts with cyclopentadiene at 25 °C giving a Diels–Alder

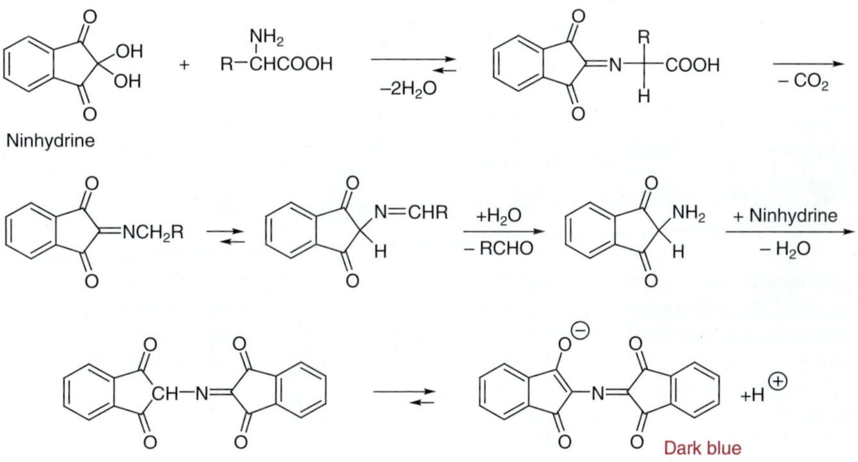

Scheme 2.5 Ninhydrin test for primary α-amino acids.

cycloadduct 3000 times as fast as bicyclo[2.2.0]hex-2-ene. Why? [267]

Problem 2.26 Using heat group increments (Table 2.A.1), estimate the standard heats of formation of CH_3COCHO and of $CH_3COCOCH_3$ and compare them with the experimental data in Table 1.A.3. Explain the deviations. Evaluate the heat of formations of penta-2,4-dione and compare it with the experimental value given in Table 1.A.4. Explain the deviation.

Problem 2.27 Which of the aldehydes below has the largest equilibrium constant for an aldol reaction with an enolizable ketone? Aldehydes: acetaldehyde, formaldehyde, benzaldehyde, and ethyl glyoxylate (CHOCOOEt).

2.9 Major role of translational entropy on equilibria

As already seen (Section 1.3), translational entropy makes reaction entropies negative for condensations, positive for fragmentations, and nearly nil for isomerizations. As a general rule, condensations are favored on lowering temperature, whereas fragmentations are favored upon heating. At elevated temperatures, molecules tend to decompose into smaller molecular fragments and finally into atoms. The bond energies that maintain atoms together in molecules, the positive $\Delta_r H$ term of fragmentation, are overcome by the negative $-T\Delta_r S$ term at high temperature. Molecules are objects in a box and the binding energy between molecular fragments and atoms is the glue that keeps these objects from falling apart. Increasing the temperature is analogous to shaking the box increasingly vigorously: the objects are fragmentized into smaller objects and finally into atoms as illustrated in Figure 2.10.

2.9.1 Aldol and crotonalization reactions

The aldol reaction is probably one of the most important synthetic reactions for the formation of C—C bonds, both for preparative chemistry (Sections 5.7.6 and 7.6.1) and biosynthesis [268, 269]. This reaction is important in the chemical and pharmaceutical industries [270, 271]. The reaction combines two carbonyl compounds to form new β-hydroxy carbonyl compounds. When an enolizable aldehyde ($R^1CH(X)CHO$) is combined with an aldehyde (R^2CHO), the β-hydroxyaldehyde ($R^2CH(OH)C(R^1)(X)CHO$) obtained is an aldol. This C—C bond forming reaction usually requires an acid or a base as the catalyst. The enol tautomer of an enolizable aldehyde or ketone (Section 1.11.3) undergoes nucleophilic addition to the carbonyl function of aldehydes, or ketones. As expected for any condensation reactions, aldol reactions have negative entropies of reaction, and their equilibrium constants increase on lowering the temperature.

Table 2.4 gives the equilibrium constants for various aldol reactions in water at 25 °C. The largest equilibrium constants are observed for formaldehyde condensing with acetaldehyde ($\log K = 7.78$) and acetone ($\log K = 6.67$). Acetaldehyde reacting with itself (Wurtz reaction described in 1872 [272, 273]) and with acetone give the corresponding aldols with $\log K = 2.60$ and 1.59, respectively. Formaldehyde is more reactive than acetaldehyde because the C=O double bond in formaldehyde is less stable than in acetaldehyde, as shown by comparing their heats of hydrogenation ($\Delta_r H°(CH_2{=}O + H_2 \rightarrow CH_3OH) = -22.3\,\text{kcal mol}^{-1}$; $\Delta_r H°(CH_3{-}CHO + H_2 \rightarrow CH_3CH_2OH) = -16.5\,\text{kcal mol}^{-1}$, Section 1.7.1). The heat of hydrogenation of acetone ($-13.5\,\text{kcal mol}^{-1}$) is even less negative than that of ethanol because two alkyl groups stabilize the dipole of the C=O moiety (Figure 1.8). This explains why acetone and other ketones are bad electrophilic partners in the aldol

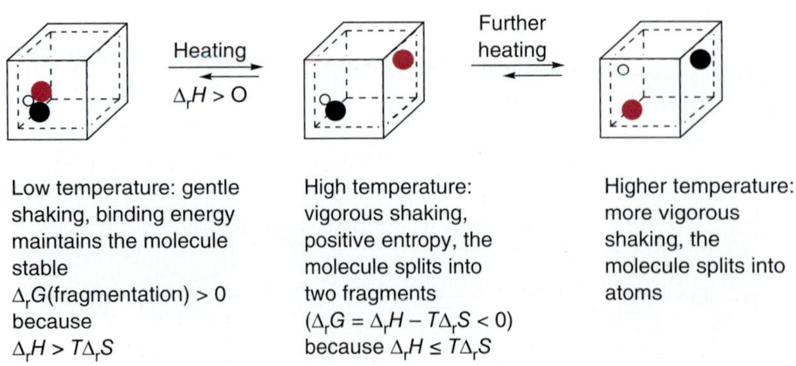

Figure 2.10 Representation of the effect of temperature on the stability of molecules.

Table 2.4 Equilibrium constant (log *K*) for aldol reactions in water at 25 °C.[a)]

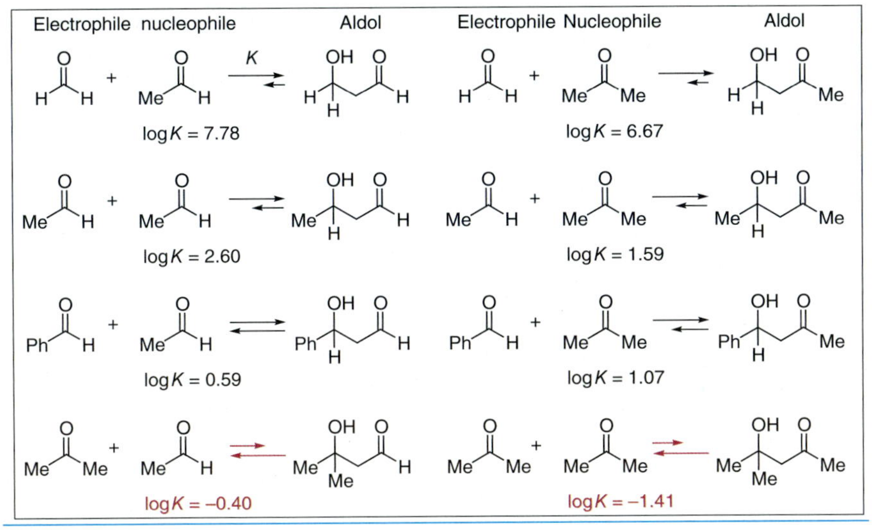

a) Taken from Ref. [280].

reaction, as illustrated by the $\log K = -0.40$ and -1.41 measured for the two last equilibria shown in Table 2.4. There is also a correlation between the heats of hydrogenation and the Gibbs energies of hydration of the aldehydes and the ketones, as shown below.

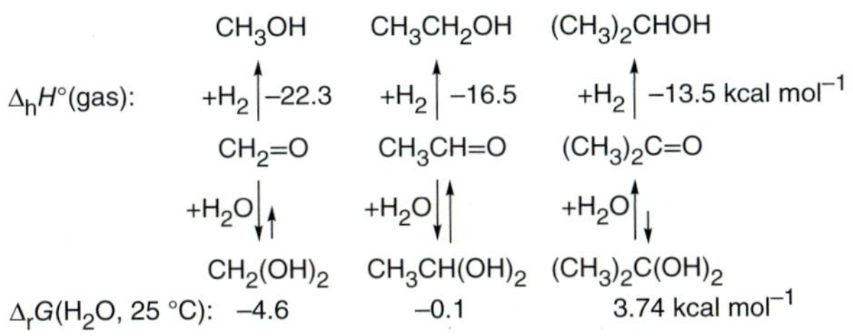

As predicted for the translational entropy changes of condensations, the larger the reactants of an aldol reaction, the smaller is their equilibrium constant at a given temperature. Because it has a positive reaction entropy, water elimination from aldols to form the α,β-unsaturated carbonyl compound will be favored at high temperature. In the case of 3-hydroxybutanal, the aldol resulting from the self-condensation of acetaldehyde, an equilibrium constant K (20 °C) = 3.2 M has been measured in aqueous solution for the elimination (2.33) giving crotonaldehyde: this corresponds to $\Delta_r G(2.33, H_2O) = -0.7\ \text{kcal mol}^{-1}$.

$$CH_3\text{–}CH(OH)CH_2CHO \underset{}{\overset{K}{\rightleftharpoons}} H_2O + CH_3CH{=}CHCHO \tag{2.33}$$

3-hydroxybutanal → crotonaldehyde

	3-hydroxybutanal	H_2O	crotonaldehyde
$\Delta_f H°(\text{gas})_{est}$:	−89.5	−57.8	−24.0 kcal mol^{-1}
$S°(\text{gas})_{est}$:	94.3	46	74.3 eu

$\Delta_r G°(\text{gas})_{est}$: −0.048 kcal mol^{-1}; $\Delta G_r(2.33, H_2O) = -0.7$ kcal mol^{-1}

The standard entropy of 3-hydroxybutanal can be derived from that of butanal ($S° = 82.6$ eu, Table 1.A.4) by adding the entropy increment associated with its oxidation into 3-hydroxybutanal, which is taken as the entropy difference between butan-2-ol ($S° = 85.8$ eu) and butane ($S° = 74.1$ eu), i.e. 11.7 eu. This leads to an estimated $S°(\text{3-hydroxybutanal}) = 82.6 + 11.7 = 94.3$ eu. In this calculation, one assumes that the possible bridging hydrogen bonds between the hydroxy and carbaldehyde moiety of 3-hydroxybutanal are not important. That is, it does not stabilize this compound at 25 °C, nor block rotations about σ(C(1)—C(2)) and σ(C(2)—C(3)) bonds. The standard entropy of (*E*)-but-2-enal can be derived from that of butanal ($S° = 82.6$ eu, Table 1.A.4) by correcting it for the didehydrogenation of butanal and by considering the blocking of the rotation about σ(C(1),C(2)) bond because of the π(C=C)/π(C=O) conjugation (Section 2.7.3). The difference $S°((E)\text{-butene}) - S°(\text{butane}) = 70.86 - 74.12$ eu $= -3.26$ eu will be applied, as well as a entropy correction of −5 eu (Section 2.10) for the loss of free rotation about one σ-bond. This leads to an estimated $S°((E)\text{-but-2-enal}) = 82.6\text{–}3.26 - 5 \cong 74.3$ eu. Thus, one calculates $\Delta_r S°(2.33, \text{gas}) = 26$ eu and $\Delta_r G°(2.33, \text{gas}) = 7.7\text{–}298{\cdot}(0.026) = -0.048\ \text{kcal mol}^{-1}$, a value very similar to that determined experimentally for dehydration in H_2O at 25 °C!

The self-aldol reaction of acetone is not favored at 25 °C (Table 2.4) because the $-T\Delta_r S°$ term overcompensates the heat of reaction $\Delta_r H°$ which is smaller for additions to ketones than for additions to aldehydes. In the case of hexane-2,5-dione and heptane-2,6-dione, their intramolecular aldol reactions occur with equilibrium constants $K \gg 1$ at 25 °C. This is due to the fact that the cyclization entropies of these reactions (Section 2.10) are smaller than the condensation entropies of intermolecular aldol reactions [274–279].

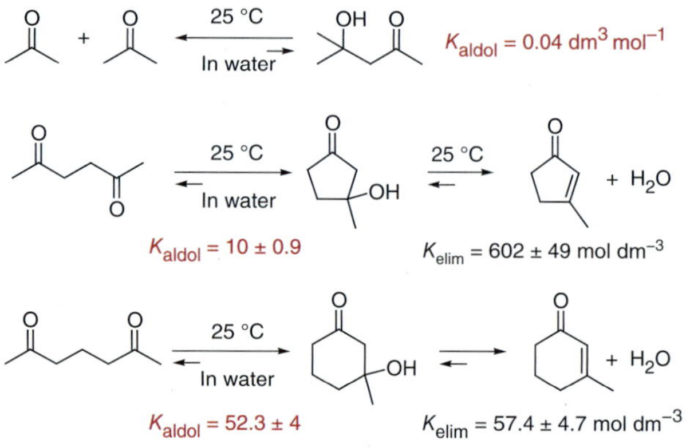

Problem 2.28 Highly branched hydrocarbons are usually obtained by a process known as cracking, in which higher boiling alkanes are decomposed into isobutene, but-1-ene and (*E*)-, and (*Z*)-but-2-ene. This mixture is then reacted over an acidic catalyst producing high proportion to triply branched octanes. What are these products? Estimate the heats and entropies of these reactions. Are they reversible below 100 °C? Explain your answer. Propose mechanisms for these reactions using H_2SO_4 as the catalyst.

Problem 2.29 Calculate the heat of the reaction that equilibrates 3-methylcyclohexanone and methanol with the corresponding hemiacetal knowing that $K = 0.4\,dm^3\,mol^{-1}$ at 25 °C and compare it with the values measured for the following equilibrium [280]:

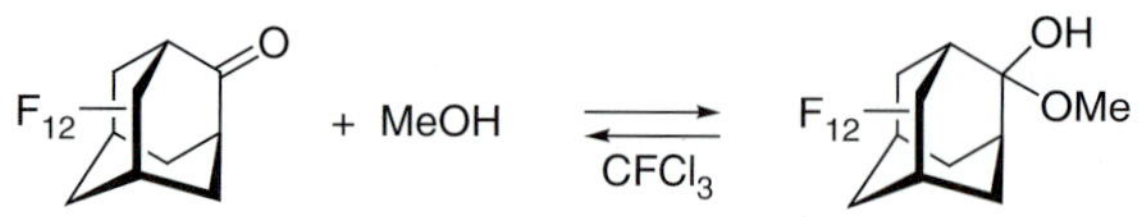

$\Delta_r H° = -8.8$ kcal mol^{-1}
$\Delta_r S° = -26$ eu

Problem 2.30 The hydration of acetone in aqueous solution at 25 °C has a Gibbs energy of 3.74 kcal mol^{-1}, whereas that of bicyclo[2.2.1]heptan-7-one has a $\Delta_r G°$ value of −0.42 kcal mol^{-1} ($\Delta_r H° = -7.34$ kcal mol^{-1}, $\Delta_r S° = -23.2$ eu) [281, 282]. Explain why the hydration of bicyclic ketone is more complete than that of acetone.

Problem 2.31 ZnI_2-catalyzed reaction of furan (excess) with 1-cyanovinyl acetate gives a mixture of 2-cyano-7-oxabicyclo[2.2.1]hept-5-en-2-yl acetates (**A**). Upon alkaline hydrolysis, the corresponding cyanohydrines **B** are obtained. What can one add to the reaction mixture to generate 7-oxabicyclo [2.2.1]hept-5-en-2-one (**P**)?

Problem 2.32 The binding of many carbohydrates and lectins by enzymes has negative enthalpy values, where entropy plays a relatively minor role [283–285]. The thermodynamic parameters for the binding of nojirimycin (**N**) and isofagomine (**I**), two competitive inhibitors of almond β-glucosidase, are given below. Interpret these data [286].

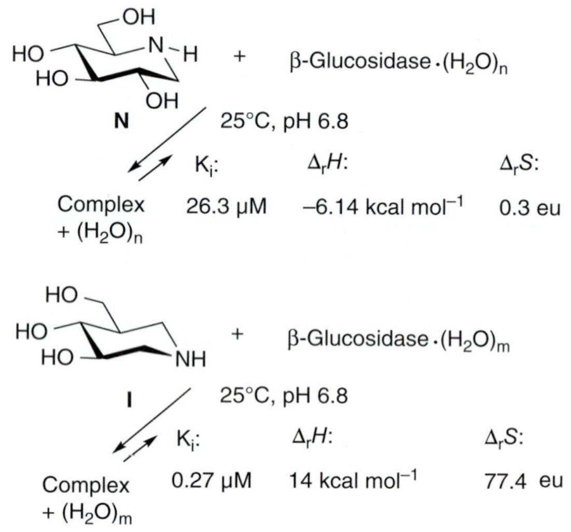

2.9.2 Aging of wines

It is well known that white and especially red wines get better on aging in a dark cellar at 10–12 °C. As all chemical reactions are accelerated on heating (Arrhenius law: rate constant $k = A\exp(-E_a/RT)$; Section 3.3), why not heat wines to bring them to maturity more quickly? The answer is no! At higher temperature, the composition of the substances evolving during aging of wines will not be the same for a thermodynamic reason. The improvements of organoleptic properties as well as the color change observed during aging of wines are associated with the condensation of tannins (polyphenols) with themselves (white wines) and with anthocyanins (dyestuff of red wines). Although benzylation of benzene by benzyl alcohol giving $PhCH_2Ph + H_2O$ is exothermic by −15.6 kcal mol^{-1} (NIST WebBook of Chemistry) for an equilibrium in the gas phase (variation of translation entropy amounts to c. −3 eu), the condensations of tannins are less exothermic and are closer to reversibility for steric reasons and because of the high concentration of H_2O in wine. As we shall see below, reversibility is reached on heating. Catechin, epicatechin (flavan-3-ols), leucoanthocyanidins (flavan-3,4-diols), and their products of condensation such as epicatechin (4β→8)-catechin (procyanidin B-1) and epicatechin (4β→6)-catechin (procyanidin B-7) are examples of tannins. In red wines, the colored components can be complex mixtures that mainly include delphinin, malvidin, petunidin, cyanidin, and peonidin (Figure 2.11) [287–291]. The phenolic substances in wine (and other drinks and food containing them) are recognized to have beneficial effects in human health [292]. On aging, these aromatic compounds undergo electrophilic benzylations shown below (equilibria (2.34)) [293], and if stored in a oak barrel, are cocondensed with substances extracted from the oak wood [294].

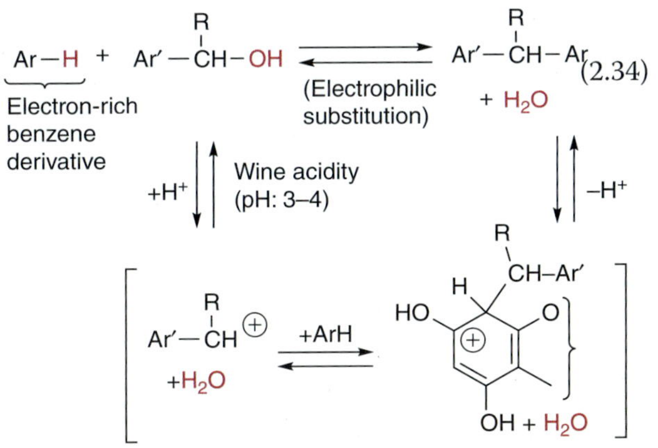

(2.34)

The average molecular mass of the tannins of a young wine (<1 year) varies between 500 and 700. For maturated wines stored at 10–12 °C, tannins can have molecular masses of 2000–5000. This condensation process transforms astringent polyphenols into fewer molecules that are less astringent. In the same time, low molecular weight tannins that are nearly transparent become yellow with the increase of their molecular weight (proximity of electron-rich aromatic rings). Young white wines are slightly yellow, whereas a "Sauternes" of 20–30 years is deep yellow or brownish. A young red wine is brighter than an older one because the "heavy" yellow tannins contribute to the color (tile color) and also because the anthocyanins can react with the polyphenols through electrophilic substitution [295, 296]. These reactions destroy the flavylium chromophores as shown in Figure 2.12.

This explains why old Port wines are brownish to yellow, whereas young Port wines are dark red or cyan. Pouring old red wine into an open jar before serving allows it to mix with air, improving its organoleptic properties. This aeration causes a change in color. The oxidation of these yellowish adducts leads to flavylium cations, which are dark red. Additionally, the color of red wines depends on the acidity and the presence of other nucleophiles, such as bisulfite (arising from the addition of SO_2 to sanitize the grapes), as given by equilibria (2.35) [297, 298].

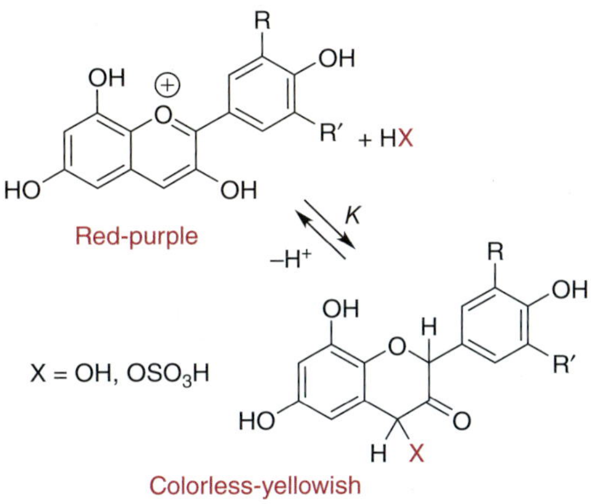

(2.35)

Equilibria between reactants **A**, **B** and products **P**, **Q** of similar mobility have their entropies given by differences in translational entropies. One obtains the relationship in Eq. (2.36).

$$\Delta_r S(\mathbf{A}+\mathbf{B} \rightleftarrows \mathbf{P}+\mathbf{Q}) \cong \Delta_r S_{trans} = 1.5\ R\ln(M_P M_Q/M_A M_B) \tag{2.36}$$

where M_P, M_Q, and M_A and M_B are the molecular mass of products **P**, **Q** and reactants **A**, **B** (Section 1.4, Eq. (1.30)). In the case of equilibrium (2.34) in which a leucoanthocyanidin condenses with (+)-catechin to give a product of benzylation and an equivalent of water (products and reactants with same mobility), $\Delta_r S_{trans}(2.34) = 1.5\ R\ln(578.538 \times 18.016/306.277 \times 290.277) = -6.4$ eu is calculated. Although the number of molecules does not change between products and reactants, a negative change in translation entropy of reaction is expected. This is always the case for the condensation of two compounds of medium size giving two products, one being a large molecule and the second a small molecule like water. For the formation of a "trimeric" polyphenol, one calculates $\Delta_r S_{trans} = -7.2$ eu. For an equilibrium in which a leucoanthocyanidin would benzylate a "nonameric" polyphenol to equilibrate

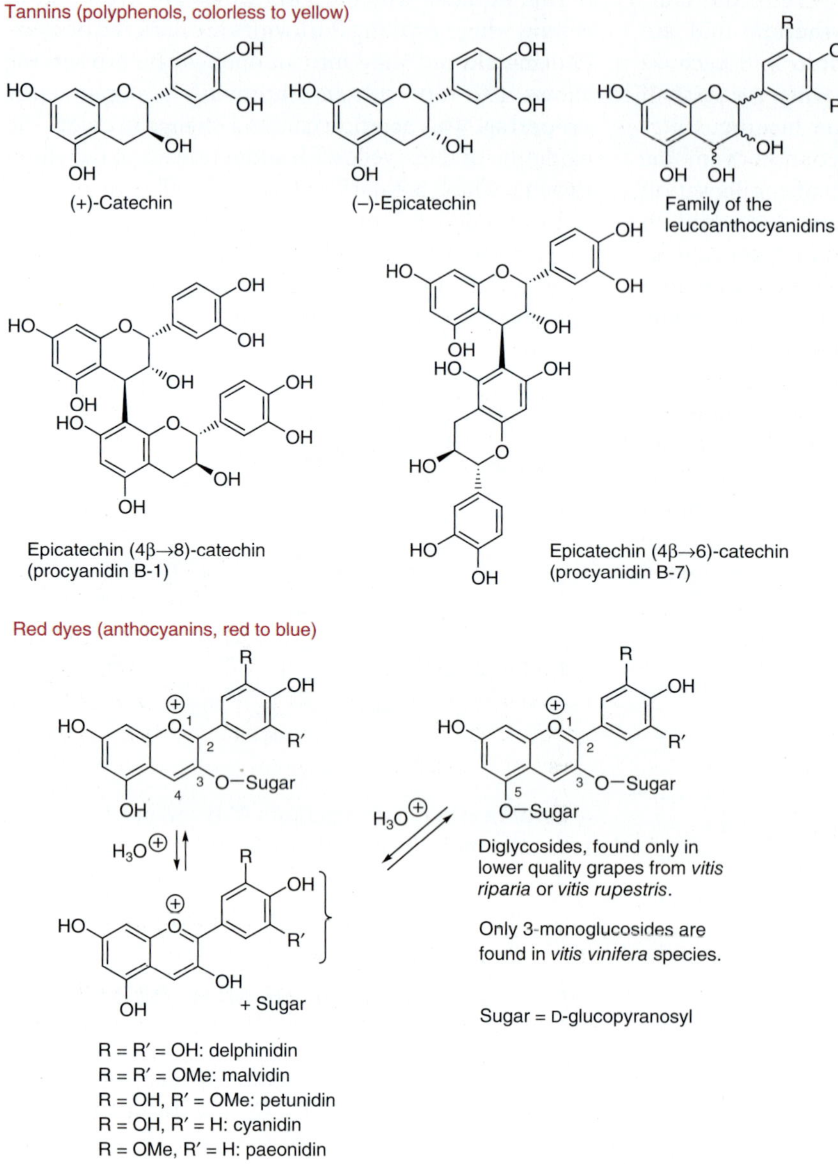

Figure 2.11 Examples of tannins and dyes found in wines.

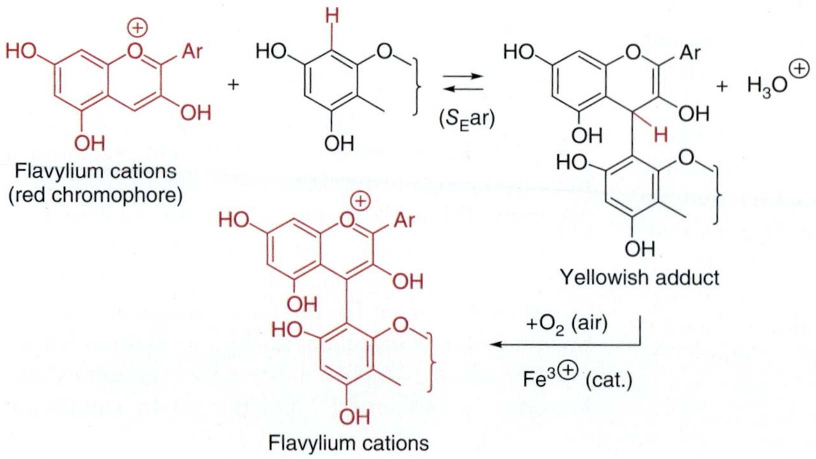

Figure 2.12 Loss of the red color of red wines on aging; recovery of red color on aerating old red wines. The latter process involves the radical oxidation of labile allylic dibenzylic C—H bonds.

with a "decameric" polyphenol and a molecule of H_2O, one calculates $\Delta_r S_{trans} = -8.1$ eu. Thus, the greater the degree of condensation of the polyphenols, the more negative are the entropies of the reactions forming them. Thus, the lower the temperature of storing of a wine, the higher the degree of condensation the polyphenols can attain. On the contrary, on heating, high molecular weight tannins will have no chance to be formed, thus altering the intrinsic organoleptic properties of the wine compared with a wine stored in an adequate cellar. Storing a wine below 10 °C slows down the process of condensation of its polyphenols, thus retarding its aging. A wine kept in a freezer will not improve with time for a kinetic reason, not for a thermodynamic reason.

Problem 2.33 Why is the oxidation of the yellow adduct of Figure 2.12 an easy process?

2.10 Entropy of cyclization: loss of degrees of free rotation

Selected cyclization reactions are presented in Table 2.5 with their thermochemical parameters calculated from the data in Table 1.A.2. All of these isomerizations have negative reaction entropies because they block a number of free rotations about σ(C—C) bonds due to the reaction. For the isomerization (equilibrium (2.37)) of hex-1-ene ($S^\circ = 92.7$ eu) into cyclohexane ($S^\circ = 71.3$ eu), four single σ(C—C) bonds cannot rotate any more. The terminal methyl group in hex-1-ene reduces the entropy by $-R\ln 3 = -2.18$ eu because of local C_3 symmetry. The symmetry of cyclohexane ($\sigma = 3 \times 2 = 6$) reduces the entropy in this compound by $-R\ln 6 = -3.6$ eu.

Thus, the four free rotations about σ(C—C) bonds in hex-1-ene contribute to the reaction entropy by $-21.4 - 2.18 + 3.6 \cong -20$ eu. On average, an entropy change of −5 eu occurs per bond for which the rotation is blocked. Benson has proposed a value of −4.7 ± 0.3 eu for the blockage of rotation about each σ(C—C) bond [299].

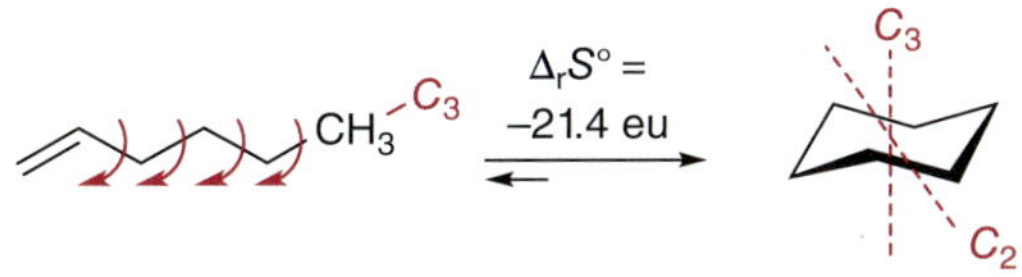

(2.37)

S°(gas): 92.7 eu S°(gas): 71.3

Local symmetry: − Rln3

$\Delta_r S^\circ$(4 blocked rotations): $-21.4 + R\ln 3 - R\ln 6 = -20$ eu

2.11 Entropy as a synthetic tool

Cracking heavy hydrocarbons gives lighter alkanes and alkenes. Below 500 °C and under 0.3–2 atm. in the presence of clays or mixtures of Al_2O_3 and SiO_2, mostly C_4-hydrocarbons are formed; they are used in the synthesis of high-octane gasoline. Above 1000 °C, CH_4, H_2, and graphite are the major products.

2.11.1 Pyrolysis of esters

One of the oldest methods of the preparation of alkenes is the low-pressure pyrolysis of alkyl acetates (0.01–1 Torr, 300–600 °C) [300, 301]. The prototype reaction (2.38) that transforms ethyl acetate into ethylene (ethene) and acetic acid (ethanoic acid) has a heat of reaction $\Delta_r H^\circ(2.38) = 15.5$ kcal mol^{-1} and a reaction entropy $\Delta_r S^\circ(2.38) = 29.9$ eu at 25 °C

Table 2.5 Heats (in kcal mol^{-1}) and entropies (in cal K^{-1} mol^{-1}) for cyclizations of hydrocarbons (1 cal = 4.184 J).[a)]

Reaction	$\Delta_r H^\circ$	$\Delta_r S^\circ$	Reaction	$\Delta_r H^\circ$	$\Delta_r S^\circ$
	7.8	−7		11.3	−3.6
	7.0	−10.4		−10.4	−7.2
	−13.1	−13.2		−19.6	−21.4
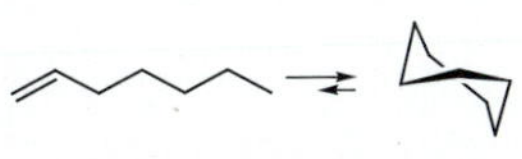	−13.4	−20.3	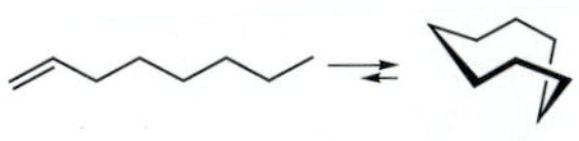	−10.3	−23.8

a) See: http://webbook.nist.gov.

(1 atm). Using these data and assuming the pyrolysis to occur at 1 atm, one estimates a temperature $T = \Delta_r H°/\Delta_r S° = 15\,500/29.9 = 518$ K (245 °C) for an equilibrium constant $K(2.38) = [\text{AcOH}] \times [\text{CH}_2{=}\text{CH}_2]/[\text{EtOAc}] = 1\ \text{mol dm}^{-3}$ ($\Delta_r G° = 0$). The reaction follows a concerted mechanism (E_i elimination that can be seen as a retro-dioxa-ene reaction, Section 5.6.2).

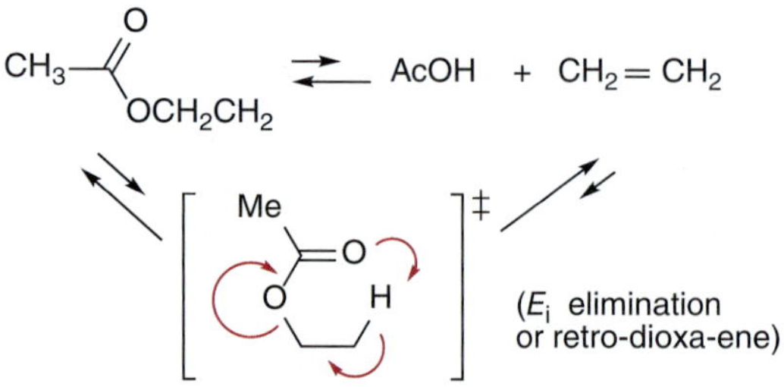

(2.38)

$\Delta_r H°(2.38) = 15.5$ kcal mol^{-1}; $\Delta_r S°(2.38) = 29.9$ eu

On lowering the pressure, fragmentation is favored in accord with the Le Châtelier principle. The technique of "flash high-vacuum pyrolysis" allows the synthesis of a wide variety of high-energy alkenes and polyenes as illustrated in Figure 2.13 [302–318].

Problem 2.34 Propose a mechanism for the elimination of AcOH from alkyl acetate involving diradical intermediates. Which diradicals?

2.11.2 Method of Chugaev

Much easier than the pyrolysis of alkyl acetates is the Chugaev (Tschugaeff) reaction (2.39) that fragmentizes xanthates at 100–250 °C to form the corresponding alkenes, methanethiol, and O=C=S. These fragmentations are highly favored by entropy as three molecules are formed from one [319–322].

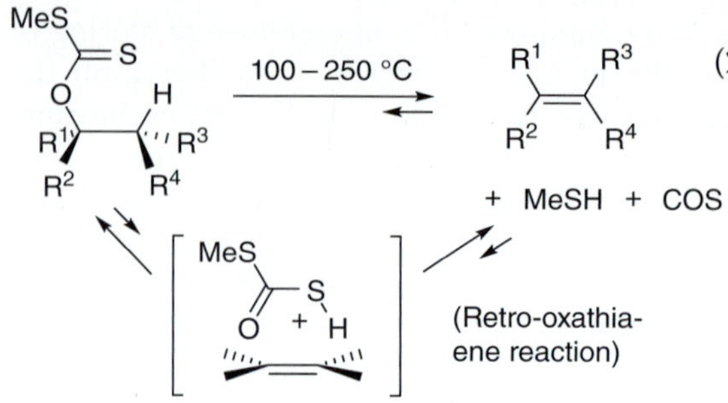

(2.39)

2.11.3 Eschenmoser–Tanabe fragmentation

For thermodynamic reasons (translation entropy), fragmentations in which four compounds are formed by reaction of a single compound are highly favored on heating, as illustrated by the Eschenmoser–Tanabe fragmentation (2.40) [323–326].

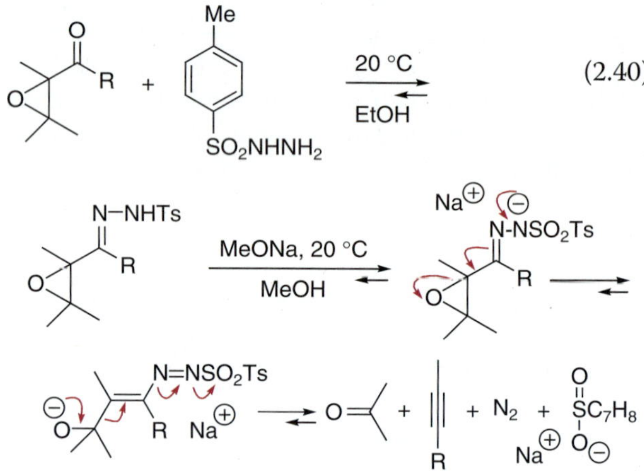

(2.40)

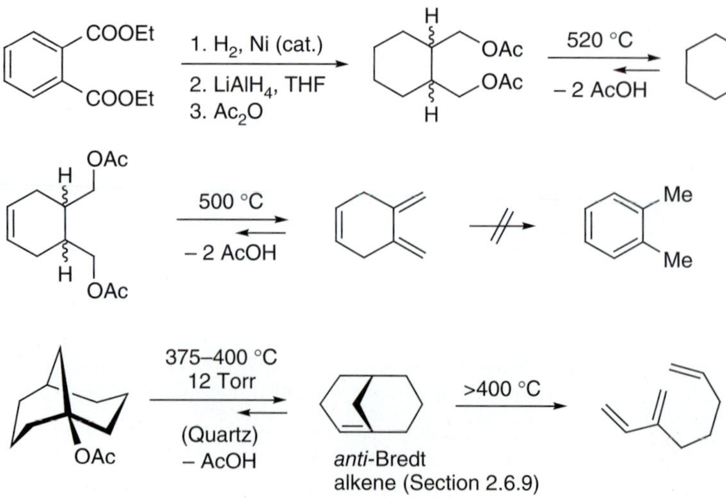

Figure 2.13 Examples of alkene synthesis by flash vacuum pyrolysis of alkyl acetates.

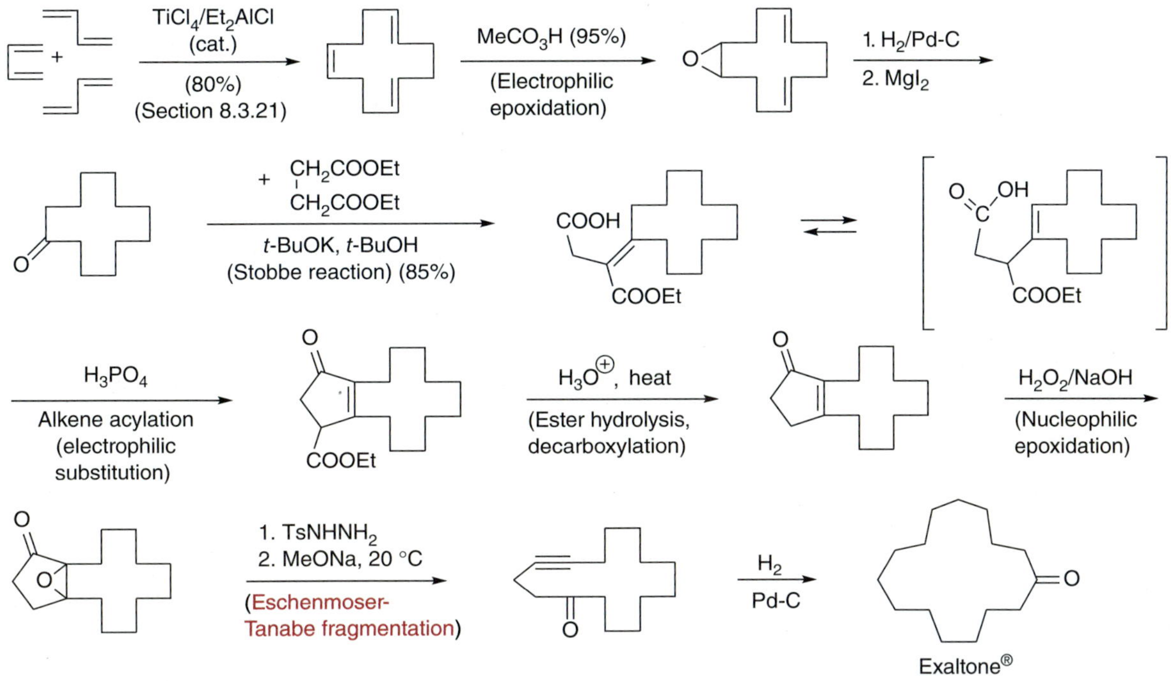

Scheme 2.6 Synthesis of Exaltone® starting from buta-1,3-diene.

This reaction has been applied to the synthesis of Exaltone®, an important perfume (Scheme 2.6). Exaltone (cyclopentadecanone or normuscone) has musk-like odor. Muscone is (−)-(3*R*)-3-methylcyclopentadecanone; it is the major constituent of natural musk. An earlier synthesis of exaltone was reported by Ruzicka et al. in 1926 through the Ruzicka large ring synthesis ($HOOC(CH_2)_nCOOH \rightleftarrows (CH_2)_n = CO + CO_2 + H_2O$, heating in the presence of ThO_2) [327].

2.11.4 Eschenmoser fragmentation

Hydrazones derived from the condensation of 1-aminoaziridines and α,β-epoxyketones are fragmented into four compounds, namely, alkene, dinitrogen, alkyne, and acetone (reaction (2.41)) [328].

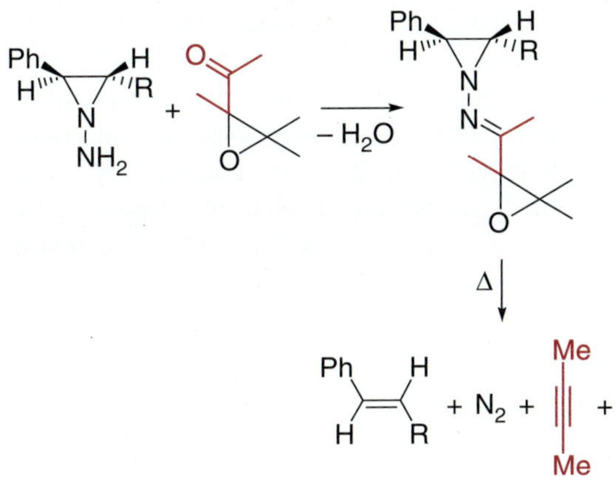

(2.41)

This method has been applied to convert α-ionone into α-citral, an important product of the perfume industry, as shown below.

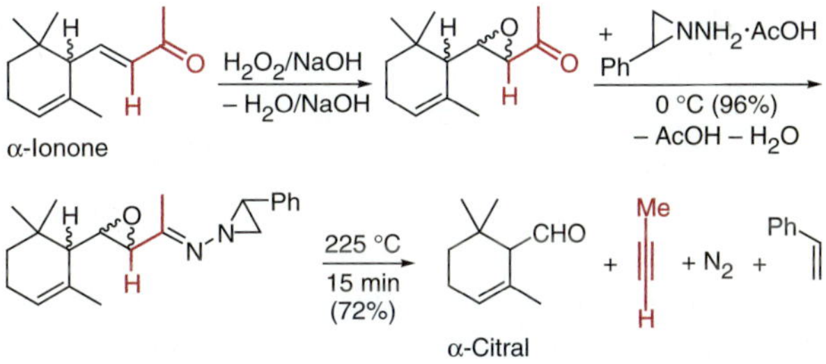

2.11.5 Thermal 1,4-eliminations

Dehydration of allylic alcohols by heating in the presence of KOH generates the corresponding conjugated dienes as illustrated below [329].

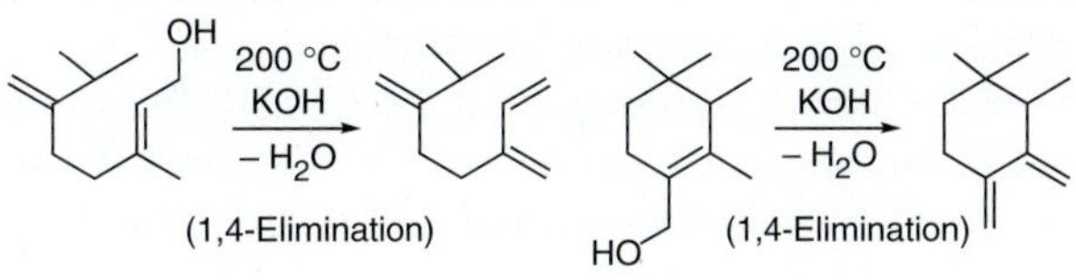

Benzocyclobutenes equilibrate with the corresponding *ortho*-xylylenes, which are highly reactive dienes of practical interest. They can be obtained by the elimination of HCl from the corresponding

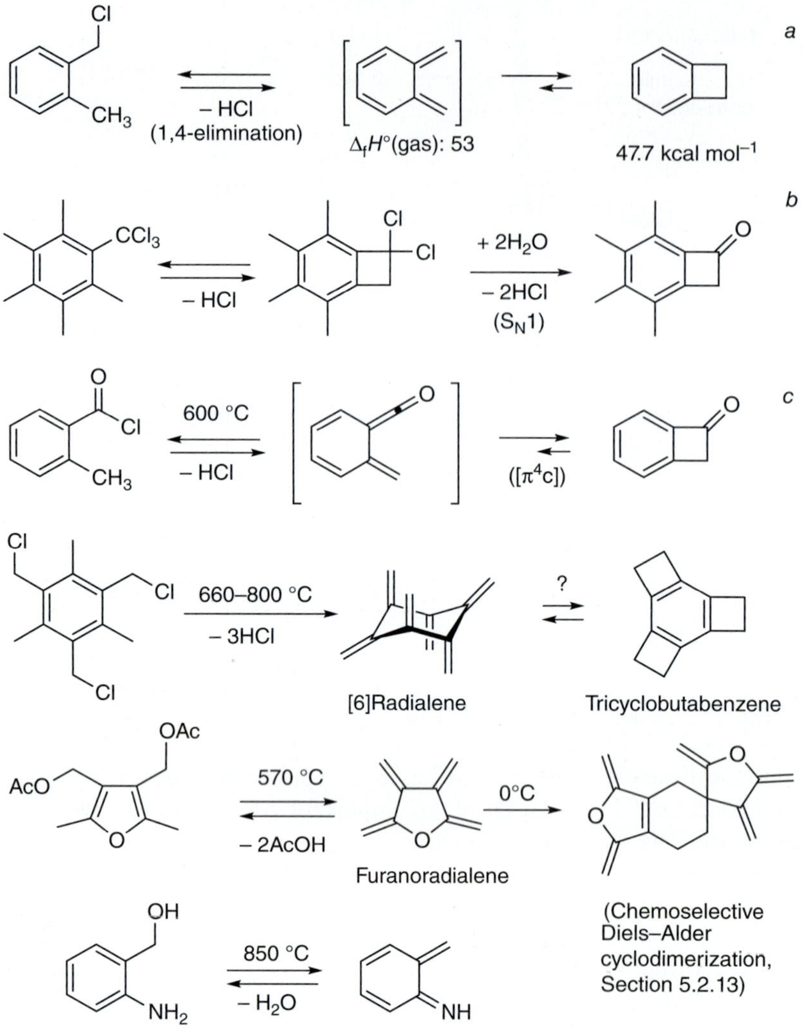

Figure 2.14 Thermal 1,4-eliminations generating unstable polyenes and their further reactions. [a]Refs. [330, 331], [b]Refs. [332–334], and [c]Ref. [335].

ortho-methylbenzyl chloride, as illustrated in Figure 2.14 [336–343]. Translation entropy and flash vacuum pyrolysis allows one to construct high-energy molecules from stable, aromatic compounds. This technique has been used to generate benzocyclobutenone by pyrolysis of *o*-methylbenzoyl chloride [335]. The reaction involves the intermediacy of α-methylideneketene. In 1978, Schiess and Heitzmann [344] and Boekelheide and coworkers have derived [6]radialene (hexamethylidenecyclohexane) from 1,3,5-tris(chloromethyl)-2,4,6-trimethylbenzene [345]. [6]Radialene is stable in dilute solutions at low temperature only. On concentrating or/and heating, it forms a polymeric material. It does not seem to equilibrate with tricyclobutabenzene (see [π^4c]-electrocyclic reactions, Section 5.1). 1,4-Elimination of acetic acid is also a mean to generate highly reactive 1,3-dienes as shown in Figure 2.14 with the pyrolysis of (2,5-dimethylfur-3,4-diyl)dimethyl diacetate to provide furanoradialene. In the condensed state at 0 °C, the latter tetraene undergoes a fast cyclodimerization giving a single Diels–Alder adduct with high chemoselectivity [346]. Ripoll and coworkers have obtained 6-methylidenecyclohexa-2,4-dien-1-imine by flash pyrolysis of 2-aminobenzyl alcohol (Figure 2.14) [347, 348]. A very elegant application of 1,4-elimination

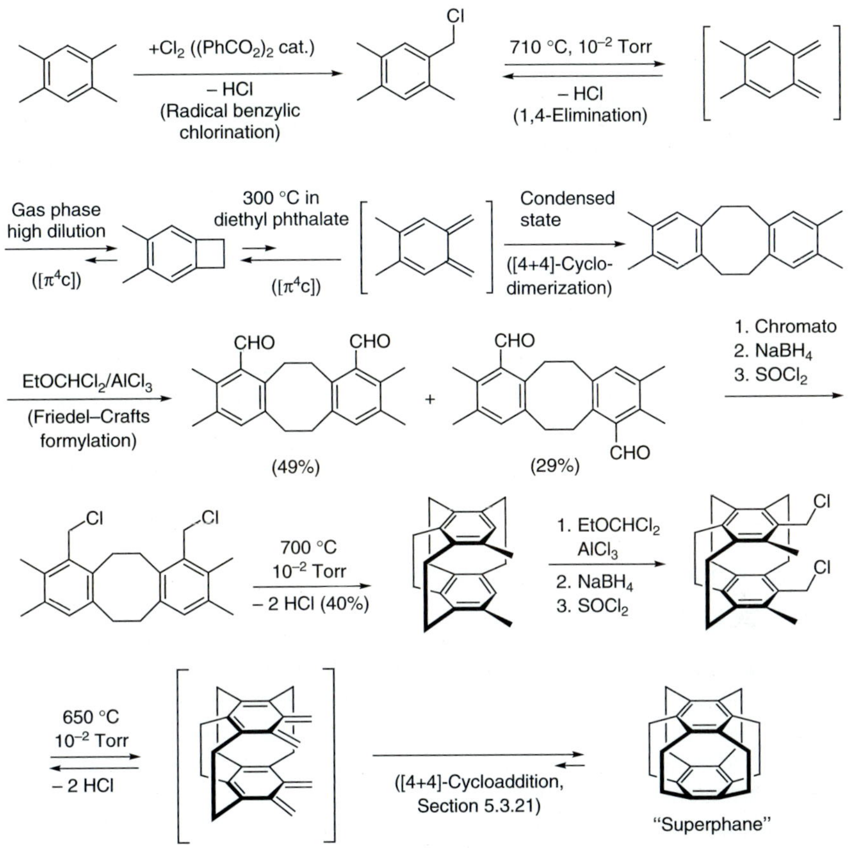

Figure 2.15 Synthesis of "superphane"([2.2.2.2.2.2](1,2,3,4,5,6)-cyclophane) according to Boekelheide.

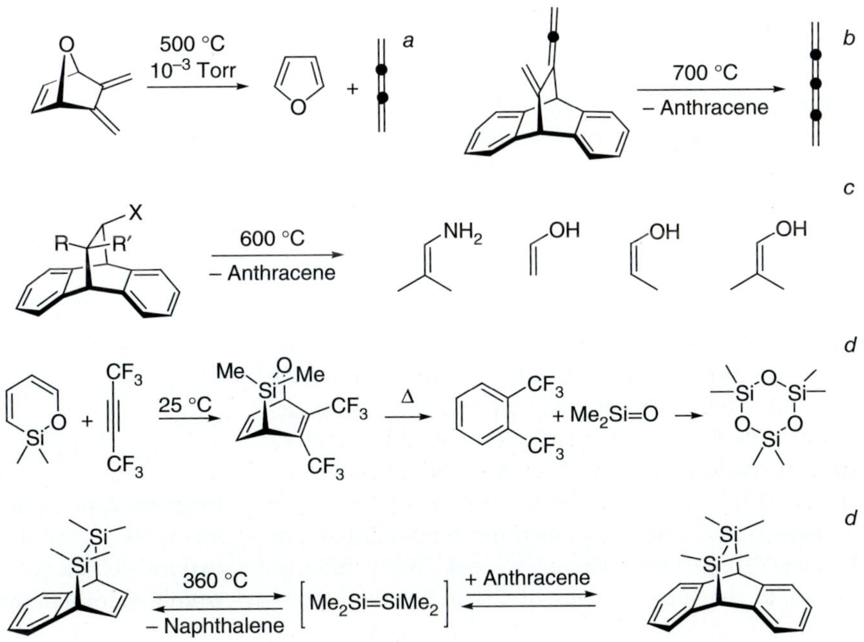

Figure 2.16 Synthesis of unstable compounds by retro-Diels–Alder cycloreversions. [a]Refs. [349–353], [b]Refs. [347, 348], [c]Refs. [358, 359], and [d]Refs. [360].

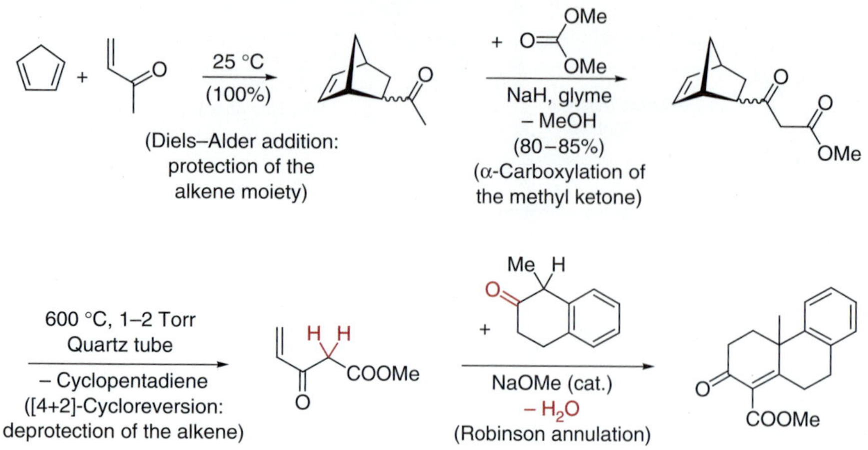

Scheme 2.7 Application of the Diels–Alder addition/cycloreversion for the protection/deprotection of alkenes.

induced by flash pyrolysis is the synthesis of "superphane" reported by Boekelheide and coworkers (Figure 2.15) [349–353].

Problem 2.35 Estimate the heat of isomerization of [6]radialene into tricyclobutabenzene assuming a heat of isomerization of butadiene into cyclobutene of 11.3 kcal mol^{-1} and an ASE of −37 kcal mol^{-1} for tricyclobutabenzene. Can [6]radialene be isomerized into dodeca-1,5,9-triyne? [344, 354]

Problem 2.36 What is the product of flash high-vacuum pyrolysis of 2,3,6,7-tetrakis(chloromethyl)-1,4,5,8-tetramethylnaphthalene? [355, 356]

Problem 2.37 What is the major product of pyrolysis of 2,3-dimethylidene-7-oxabicyclo[2.2.1]heptane under high vacuum. What are the favored products if the pyrolysis tube contains acidic chips? [357]

2.11.6 Retro-Diels–Alder reactions

Retro-Diels–Alder reactions have been used extensively for the construction of unstable compounds such as buta-1,2,3-triene, penta-1,2,3,4-tetraene, 1-aminoalkenes, enols, and other highly reactive species such as $Me_2Si = O$ and $Me_2Si = SiMe_2$ (Figure 2.16).

Diels–Alder reactions (Sections 5.3.9–5.3.16) can be used to protect alkene moieties. After modification of the other groups of the initial dienophile moiety, heating liberates the modified alkene by cycloreversion (retro-Diels–Alder reaction). An example is given for the synthesis of methyl acryloylacetate [361], an important synthetic intermediate for Robinson annulations (Scheme 2.7) [362].

Problem 2.38 Pyrolysis of (2-methylfur-3-yl)methyl benzoate at 640 °C under 0.0001 Torr gives a diene that dimerizes in the condensed state at 0–20 °C. What is this diene? What is its dimer? [363, 364]

Problem 2.39 (*E*)-1-Methoxybutadiene and methyl vinyl ketone give a mixture of Diels–Alder adducts above 40 °C. The adducts eliminate MeOH on further heating. What conditions of concentration and temperature to apply in order to avoid this elimination?

Problem 2.40 Methyl isopropyl ketone (MIPK) is an efficient high-octane (>100) oxygenate gasoline additive without many of the undesirable effects of the widely used methyl *tert*-butyl ether. One method of preparing MIPK involves the rearrangement of pivalaldehyde catalyzed by strong acids [365]. Evaluate the equilibrium constant for this rearrangement at 25 °C using the gas-phase standard heats of formation for MIPK and *t*-BuCHO.

Problem 2.41 Using gas-phase standard heats of formation, what base should be used to generate oxirane (ethylene oxide) by 1,2-elimination of HCl from 2-chloroethanol?

2.A Appendix

Table 2.A.1 Benson's group equivalents for standard heats of formation ($\Delta_f H°$ in kcal mol^{-1}), for *standard entropies* ($S°$, in eu = cal mol^{-1} K^{-1}) and standard heat capacities (C_p^o, in eu = cal mol^{-1} K^{-1}) of gaseous compounds (25 °C, 1 atm) (1 cal = 4.184 J) [2, 4].

	$\Delta_f H°$	$S°$	C_p^o		$\Delta_f H°$	$S°$	C_p^o
Hydrocarbons							
[C—(C)(H)$_3$]	−10.12 [2]	30.4	6.20	[C$_t$—(C$_b$)]	24.5		
[C—(C$_b$)(H)$_3$],[C—(C$_t$)(H)$_3$], [C—(C$_d$)(H)$_3$]	−10.12 −10.12			[C$_b$—(C$_b$)$_2$(H)]	3.30	11.5	3.25
[C—(C)$_2$(H)$_2$]	−4.92	9.42	5.45	[C$_b$—(C$_b$)$_2$(C)]	5.62		
[C—(C$_d$)(C)(H)$_2$]	−4.92			[C—(C$_b$)(C)(H)$_2$]	−5.18		
[C—(C$_t$)(C)(H)$_2$]	−4.92			[C—(C$_b$)(C)$_2$(H)]	−1.02		
[C—(C)$_3$(H)]	−1.75	−12	4.47	[C—(C$_b$)(C)$_3$]	2.88		
[C—(C$_d$)(C)$_2$(H)]	−1.75			[C$_d$—(C$_d$)(C)(H)] + [C—(C$_d$)(H)$_3$]	−1.20	38.5	10.0
[C—(C$_t$)(C)$_2$(H)]	−1.7			[C$_d$—(C$_d$)(C)$_2$] + 2[C—(C$_d$)(H)$_3$]	−9.85	48.7	16.1
[C—(C)$_4$]	−0.06	35.1	4.35	[C—(C$_d$)(C)$_2$] − [C—(C$_d$)(H)$_3$]	5.2	−21	−0.8
[C—(C$_t$)(C)$_3$]	0.6			[C—(C$_d$)(C)$_2$(H)] − [C—(C$_d$)(H)$_3$]	8.2	−43	1.0
[C—(C$_t$)(C)$_3$], [C—(C$_d$)(C)$_3$]	−0.06			[C—(C$_d$)(C)$_3$] − [C—(C$_d$)(H)$_3$]	10.9	−67	−3.5
[C—(C$_t$)$_2$(C)$_2$]	2.0			[C$_t$—(C$_t$)(C)] − [C—(C$_t$)(H)$_3$]	17.5	36.7	9.3
[C$_d$—(C$_d$)(H)$_2$]	6.22	27.6	5.2	[C—(C$_t$)(C)(H)$_2$] − [C—(C$_t$)(H)$_3$]	5.2	−20	−1.1
[C$_d$—(C$_d$)(C)(H)]	8.65			[C—(C$_t$)(C)$_2$(H)] − [C—(C$_t$)(H)$_3$]	8.4	−42	−2.0
[C$_d$—(C$_d$)(C)$_2$]	9.75			[C$_b$—(C$_b$)$_2$(C)] + [C—(C$_b$)(H)$_3$]	−4.20	22.0	9.25
[C$_t$—(C$_t$)(H)]	27.7	24.7	5.3	[C—(C$_b$)(C)(H)$_2$] − [C—C$_b$)(H)$_3$]	4.9	−20	−0.8
[C$_t$—(C$_t$)(C)]	27.41			[C—(C$_b$)(C)$_2$(H)] − [C—(C$_b$)(H)$_3$]	9.0	−42	−0.8
[C$_t$—(C$_d$)]	28.5			[C—(C$_b$)$_2$(C$_d$)] + [C$_d$—(C$_b$)(C$_d$)(H)]	12.5	−1.5	7.7
Alcohols, phenols, ethers, and peroxides							
[O—(C)(H)]	−37.85			[O—(C$_b$)(C)]	−22.4		
[O—(C)$_2$]	−23.80			[O—(C$_b$)$_2$]	−18.7		
[O—(O)(H)]	−16.25	28.5	5.2	[C$_b$—(O)(C$_b$)$_2$]	−1.19		
[O—(O)(C)]	−5.13			[O—(C)(H)] + [C—(O)(H)$_3$]	−48.1	59.5	10.5
[C—(O)(C)(H)$_2$]	−8.00			[O—(C)$_2$] + 2[C—(O)(H)$_3$]	−45.3	69.4	15.7
[C—(O)(C)$_2$(H)]	−7.03			[C—(O)(C)(H)$_2$] − [C—(O)(H)$_3$]	2.0	−20	−1.1
[C—(O)(C)$_3$]	−6.66			[C—(O)(C)$_2$(H)] − [C—(O)(H)$_3$]	3.0	−44	−1.2
[O—(C$_b$)(H)]	−37.85						
Aldehydes and ketones							
[CO—(C)(H)]	−29.1			[CO—(C$_b$)(H)]	−30.92		
[CO—(C)$_2$]	−31.66			[CO—(C$_b$)(C)[	−32.71		
[CO—(C)(H)] + [C—(CO)(H)$_3$]	−39.7	65.3	13.1	[CO—C$_b$)$_2$]	−29.64		
[CO—(C)$_2$] + 2[C—(CO)(H)$_3$]	−51.9	76.2	17.9				

Table 2.A.1 (Continued)

	$\Delta_f H°$	$S°$	$C_p°$		$\Delta_f H°$	$S°$	$C_p°$
Esters, carboxylic acids, and anhydrides							
[CO—(O)(C)]	−33.4	14.8	6.0	[O—(CO)(H)]	−60.3	24.5	3.8
[CO—(O)(C_d)]	−33.5			[O—(CO)(C_d)]	−41.3		
[CO—(O)(C_b)]	−46.0			[O—(CO)(O)]	−19.0		
[CO—(O)(H)]	−29.5	34.9	7.0	[O—$(CO)_2$]	−50.9		
[CO—(C_d)(H)]	−31.7			[C_d—(CO)(O)]	9.4		
[O—(CO)(C)]	−41.3	8.4					
Organohalides							
[C—(F)(C)$(H)_2$]	−52.56			[C—$(Cl)_3$(C)]	−23.8	50.1	15.8
[C—(F)$(C)_2$(H)]	−47.92			[C_d—(C_d)(Cl)(H)]	2.7	35.3	7.9
[C—(F)$(C)_3$]	−41.73			[C—(Cl)$(C_b)_2$]	−4.29		
[C—$(F)_2$$(C)_2$]	−97.64			[C—(Br)(C)$(H)_2$]	−5.08	40.5	9.1
[C—$(F)_2$(C)(H)]	−105.2			[C—(Br)$(C)_2$(H)]	−3.26	20.3	
[C—$(F)_3$(C)]	−160.6	42.0	12.5	[C—(Br)$(C)_3$]	−1.50	−4.2	9.3
[C—(F)$(C_b)_2$]	−44.26			[C—(Br)$(C_b)_2$]	8.70		
[C—(Cl)(C)$(H)_2$]	−15.98	37.8	8.8	[C_d—(C_d)(Br)(H)]	12.4	38.1	8.1
[C—(Cl)$(C)_2$(H)]	−13.36	17.7	8.5	[C—(I)(C)$(H)_2$]	8.02		
[C—(Cl)$(C)_3$]	−12.50	−7.6	8.7	[C—(I)$(C)_2$(H)]	10.84		
[C—$(Cl)_2$$(C)_2$]	−19.97			[C—(I)$(C)_3$]	12.5		
[C—$(Cl)_2$(C)(H)]	−20.68	44.2	12.2	[C—(I)$(C_b)_2$]	22.6		
Amines (N), imines (N_I), pyridines , (N_I—(C_b)), azo compounds, nitriles, and nitro compounds							
[C—(N)$(H)_3$]	−10.08	30.4	6.19	[N—$(C_b)_2$(H)]	16.3		
[C—(N)(C)$(H)_2$]	−6.6			[C_b—(N)]	−0.5	−9.7	3.95
[C—(N)$(C)_2$(H)]	−5.2			[N_A—(N)]	23.0		
[C—(N)$(C)_3$]	−3.2			[CO—(N)(H)]	−29.6	34.9	7.03
[C—(N_A)(C)$(H)_2$]	−5.5 [2]	(9.8)		[CO—(N)(C)]	−32.8	16.2	5.37
[C—(N_A)$(C)_2$(H)]	(3.3)	−11.7		[N—(CO)$(H)_2$]	−14.9	24.7	4.07
[C—(N_A)$(C)_3$]	(−1.9)	−34.7		[N—(CO)(C)(H)]	−4.4		
[N—(C)$(H)_2$]	4.8	29.7	5.72	[N—(CO)$(C)_2$]			
[N—$(C)_2$(H)][	15.4	8.94	4.20	[N—(CO)(C_b)(H)]	0.4		
[N—$(C)_3$]	24.4	13.4	3.48	[N—$(CO)_2$(H)]	−18.5		
[N—(N)$(H)_2$]	11.4	29.1	6.10	[N—$(CO)_2$(C)]	−5.9		
[N—(N)(C)(H)]	20.9	9.61	4.82	[N—$(CO)_2$(C_b)]	−0.5		
[N—(N)$(C)_2$]	29.2	−14		[C—(CN)(C)$(H)_2$]	22.5	40.2	11.0
[N—(N)(C_b)(H)]	22.1			[C—(CN)$(C)_2$(H)]	25.8	19.8	11.0
[N_I—(C)]	21.3			[C—(CN)$(C)_2$]		−2.8	
[N_I—(C_b)]	16.7			[C—$(CN)_2$$(C)_2$]	28.40		
[N_A—(H)]	25.1	26.8	4.38	[C_b—(CN)]	35.8	20.5	9.8
[N_A—(C)]	(32.5)	(8.0)		[C_t—(CN)]	63.8	35.4	10.3
[N—(C_b)$(H)_2$]	4.8	29.7	5.72	[C—(NO_2)(C)$(H)_2$]	−15.1		
[N—(C_b)(C)(H)]	14.9			[C—(NO_2)$(C)_2$(H)]	−15.8		
[N—(C_b)$(C)_2$]	26.2			[C—$(NO_2)_2$(C)(H)]	−14.9		

(Continued)

Table 2.A.1 (Continued)

	$\Delta_f H°$	$S°$	$C_p°$		$\Delta_f H°$	$S°$	$C_p°$
Thiols, sulfides, disulfides, sulfoxides, sulfones, thioesters, thiocyanates, thioureas, sulfinylamides, and sulfonylamides							
[C—(H)$_3$(S)]	−10.08	30.4	6.19	[C—(SO$_2$)(H)$_3$]	−10.08	30.4	6.19
[C—(C)(H)$_2$(S)]	−5.65	9.88	5.38	[C—(C)(SO$_2$)(H)$_2$]	−7.68		
[C—(C)$_2$(H)(S)]	−2.64	−11	4.85	[C—(C)$_2$(SO$_2$)(H)]	−2.62		
[C—(C)$_3$(S)]	−0.55	−34	4.57	[C—(C)$_2$(SO$_2$)]	−0.61		
[C—(C$_b$)(H)$_2$(S)]	−4.73			[C—(C$_d$)(SO$_2$)(H)$_2$]	−7.14		
[C—(C$_d$)(H)$_2$(S)]	−6.45			[C—(C$_b$)(SO$_2$)(H)$_2$]	−5.54		
[C$_b$—(S)]	−1.8	10.2	3.90	[C$_b$—(SO$_2$)]	2.3		
[C$_d$—(H)(S)]	8.56	8.0	4.16	[C$_d$—(H)(SO$_2$)]	12.53		
[C$_d$—(C)(S)]	10.93	−12	3.50	[C$_d$—(C)(SO$_2$)]	14.47		
[S—(C)(H)]	4.62	32.7	5.86	[SO$_2$—(C)$_2$]	−69.74	20.9	11.5
[S—(C$_b$)(H)]	11.96	12.6	5.12	[SO$_2$—(C)(C$_b$)]	−72.29		
[S—(C)$_2$]	11.51	13.1	4.99	[SO$_2$—(C$_b$)$_2$]	−68.58		
[S—(C)(C$_d$)]	9.97			[SO$_2$—(C$_d$)(C$_b$)]	−68.58		
[S—(C$_d$)$_2$]	−4.54	16.5	4.79	[SO$_2$—(C$_d$)$_2$]	−73.58		
[S—(C$_b$)(C)]	19.16			[SO$_2$—(SO$_2$)(C$_b$)]	−76.25		
[S—(C$_b$)$_2$]	25.90			[CO—(S)(C)]	−31.56	15.4	5.59
[S—(S)(C)]	7.05	12.4	5.23	[S—(H)(CO)]	−1.41	31.2	7.63
[S—(S)(C$_b$)]	14.5			[S—(C)(CN)]	37.18	41.1	9.51
[S—(S)$_2$]	3.04	13.4	4.66	[CS—(N)$_2$]	−31.56	15.4	5.59
[C—(SO)(H)$_3$]	−10.08	30.4	6.19	[N—(CS)(H)$_2$]	12.78	29.2	6.07
[C—(C)(SO)(H)$_2$]	−7.72			[N—(S)(C)$_2$]	29.9		
[C—(C)$_2$(SO)]	−3.05			[SO—(N)$_2$]	−31.56		
[C—(C$_d$)(SO)(H)$_2$]	−7.35			[N—(SO)(C)$_2$]	16.0		
[C$_b$—(SO)]	2.3			[SO$_2$—(N)$_2$]	−31.56		
[SO—(C)$_2$]	−14.41	18.1	8.88	[N—(SO$_2$)(C)$_2$]	−20.4		
[SO—(C$_b$)]	−12.0						

Organo-P	$\Delta_f H°$		$\Delta_f H°$		$\Delta_f H°$		$\Delta_f H°$
[C—(P)(H)$_3$]	−10.08	[P—(C)$_3$]	7.04	[O—(H)(PO)]	−65.0	[O—(P:N)(C)]	−40.7
[C—(P)(C)(H)$_2$]	−2.47	[PO—(C)(O)$_2$]	−99.5	[O—(PO)$_2$]	−54.5	[N—(P)(C)$_2$]	32.2
[C—(PO)(H)$_3$]	−10.08	[PO—(O)$_3$]	−104.6	[P—(C)(Cl)$_2$]	−50.1	[N—(PO)(C)$_2$]	17.8
[C—(PO)(C)(H)$_2$]	−3.4	[PO—(O)$_2$F]	−167.7	[P—(C$_b$)$_3$]	28.3	[P:N—(C)$_3$(C)]	0.50
[C—(P:N)(H)$_3$]	−10.08	[PO—(C$_b$)$_3$]	−52.9	[P—(O)$_3$]	−66.8	[P:N—(C$_b$)$_3$(C)]	−25.7
[C—(N:P)(C)(H)$_2$]	19.4	[PO—(N)$_3$]	−104.6	[P—(N)$_3$]	−66.8	[P:N—(N:P)(C)$_2$(P:N)]	−15.5
[C$_b$—(P)]	−1.8	[O—(C)(P)]	−23.5	[PO—(C)$_3$]	−72.8	[P:N—(N:P)(C$_b$)$_2$(P:N)]	−22.9
[C$_b$—(PO)]	2.3	[O—(H)(P)]	−58.7	[PO—(C)(Cl)$_2$]	−123.0	[P:N—(N:P)(Cl)$_2$(P:N)]	−58.2
[C$_b$—(P:N)]	2.3	[O—(C)(PO)]	−40.7	[PO—(C)(O)(Cl)]	−112.6	[P:N—(N:P)(O)$_2$(P:N)]	−43.4

Organo-B	$\Delta_f H°$		$\Delta_f H°$		$\Delta_f H°$		$\Delta_f H°$
[C—(B)(H)$_3$]	−10.08	[B—(O)$_2$(Cl)]	−19.7	[B—(C)(F)$_2$]	−187.9	[O—(B)(H)]	−115.
[C—(B)(C)(H)$_2$]	−2.22	[B—(O)$_2$(Cl)$_2$]	−61.2	[B—(C)$_2$(Cl)]	−42.7	[O—(B)(C)]	−69.3
[C—(B)(C)$_2$(H)]	1.1	[B—(O)$_2$(H)]	19.9	[B—(C)$_2$(Br)]	−26.9	[N—(B)(C)$_2$]	−9.93
[C—(BO$_3$)(H)$_3$]	−10.08	[B—(N)$_3$]	24.4	[B—(C)$_2$(I)]	−8.9	[B—(S)$_3$]	24.4
[C—(BO$_3$)(C)(H)$_2$]	−2.2	[B—(N)$_2$(Cl)]	−23.8	[B—(C)$_2$(O)]	29.3	[S—(B)(C)]	−14.5
[C$_d$—(B)(H)]	15.6	[B—(N)(Cl)$_2$]	−67.9	[B—(C$_d$)(F)$_2$]	−192.9	[S—(B)(C$_b$)]	−7.8
[B—(C)$_3$]	0.9	[BO$_3$—(C)$_3$]	−208.7	[B—(O)$_3$]	24.4		

Table 2.A.2 Standard heats of formation ($\Delta_f H°$, gas) of polysubstituted benzenes and pyridines in kcal mol^{-1} (1 cal = 4.148 J).

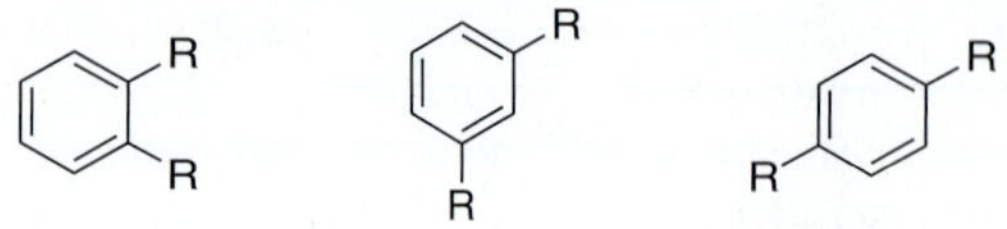

	ortho	meta	para	
$\Delta_f H°$	4.3	4.1	4.3 ± 0.2	R = Me
	−4	−5	−5 ± 0.2	R = Et
	7.9	6.7	5.9	R = Cl
	−70.2	−73.9	−73.3 ± 0.3	R = F
	21	14	14 ± 0.7	R = NO_2
	−65	−65.6	−63 ± 0.5	R = OH

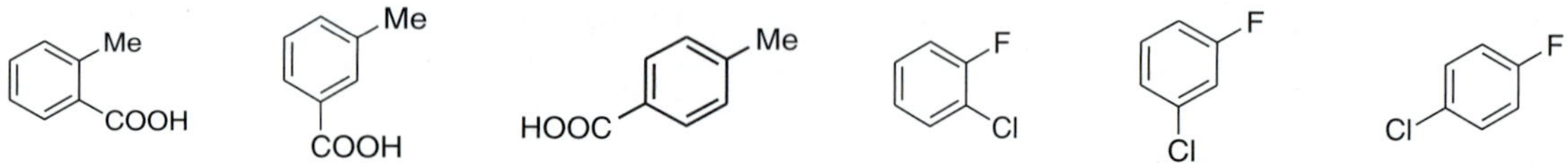

$\Delta_f H°$	−76.5	−79.0	−79.0 ± 0.2	−31	−33	−34

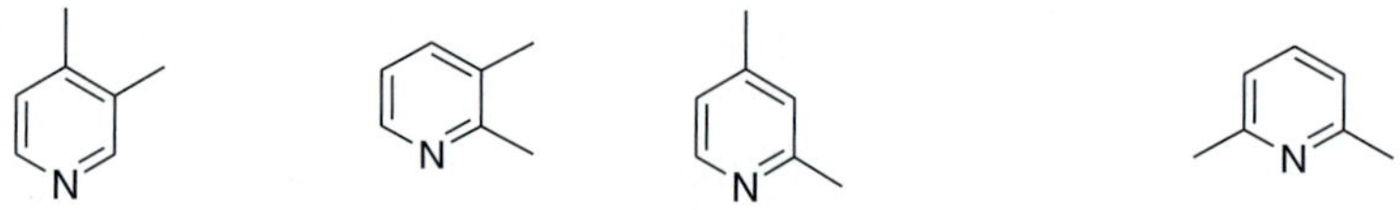

$\Delta_f H°$	16.7	16.3	15.3	14.0	15.9

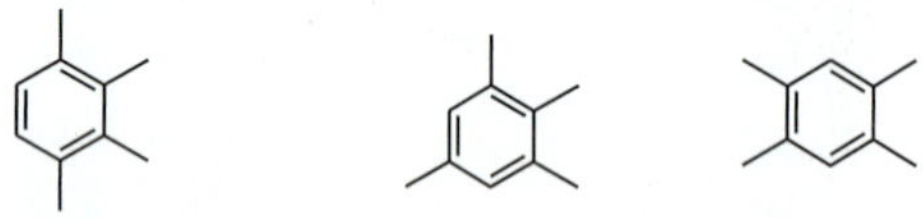

$\Delta_f H$	−8	−10	−11

Table 2.A.3 Ring stain energies, kcal mol^{-1} (1 cal = 4.184 J).

$CH_2{=}CH_2$

[S] = 22.3	27.5	26.5	6.3	0.0	6.2	9.5

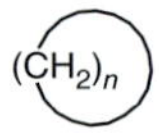

$n = 9$	10	11	12	13	14	15
12.5	12.3	11.2	4	5	11.8	1.8

27.2	25.4	5.6	27.1	5.9	19.9	19.7	2.0

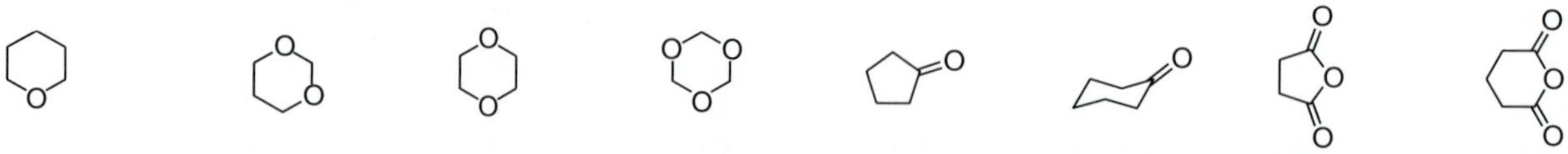

2.2	3.5	5.4	3.4	6.0	3.4	1.1	1.4

65.2	64.0	162.7	55.3	32.3	28.4	29.2	26.6

62	23	6	53.8	41.7	30.0	28.8	6.0

5.6	6.3	1.3	1.9	4.7	6.7	27.7	7.4

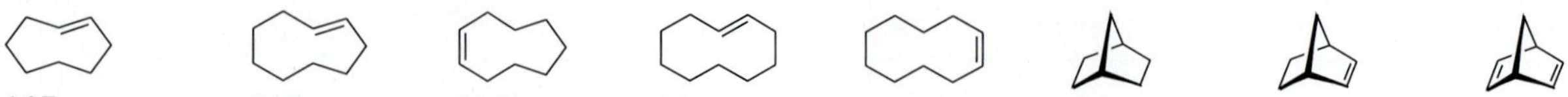

16.7	14.4	11.5	12	11.6	16.2	22.5	32.5

a) Using values from Tables 1.A.2–1.A.4.

References

1 Benson, S.W. and Buss, J.H. (1958). Additivity rules for the estimation of molecular properties - thermodynamic properties. *J. Chem. Phys.* 29 (3): 546–572.

2 Benson, S.W., Cruickshank, F.R., Golden, D.M. et al. (1969). Additivity rules for estimation of thermochemical properties. *Chem. Rev.* 69 (3): 279–324.

3 Cohen, N. and Benson, S.W. (1993). Estimation of heats of formation of organic compounds by additivity methods. *Chem. Rev.* 93 (7): 2419–2438.

4 Cohen, N. (1996). Revised group additivity values for enthalpies of formation (at 298 K) of carbon–hydrogen and carbon–hydrogen–oxygen compounds. *J. Phys. Chem. Ref. Data* 25 (6): 1411–1481.

5 Wodrich, M.D. and Schleyer, P.V.R. (2006). New additivity schemes for hydrocarbon energies. *Org. Lett.* 8 (10): 2135–2138.

6 Fishtik, I. and Datta, R. (2003). Group additivity vs *ab initio*. *Chem. A Eur. J.* 107 (34): 6698–6707.

7 Vitale, D.E. (1986). Relative stabilities of organic compounds using Benson additivity rules. *J. Chem. Educ.* 63 (4): 304–306.

8 Prosen, E.J., Johnson, W.H., and Rossini, F.D. (1946). Heats of formation and combustion of the normal alkylcyclopentanes and cyclohexanes and the increment per CH_2 group for several homologous series of hydrocarbons. *J. Res. Natl. Bur. Stand.* 37 (1): 51–56.

9 Laidler, K.J. (1956). A system of molecular thermochemistry for organic gases and liquids. *Can. J. Chem.* 34 (5): 626–648.

10 Leal, J.P. (2006). Additive methods for prediction of thermochemical properties. The Laidler method revisited. 1. Hydrocarbons. *J. Phys. Chem. Ref. Data* 35 (1): 55–76.

11 Wiberg, K.B. and Hao, S. (1991). Enthalpies of hydration of alkenes. 4. Formation of acyclic tert-alcohols. *J. Org. Chem.* 56 (17): 5108–5110.

12 (a) Dix, D.T., Fraenkel, G., Karnes, H.A., and Newman, M.S. (1966). Hindered rotation in 1,2-dineopentyltetramethylbenzene. *Tetrahedron Lett.* 517–521. (b) Roussel, C., Chanon, M., and Metzger, J. (1971). Gear effect – definition and clear cut NMR evidence. *Tetrahedron Lett.* 1861–1864. (c) Hounshell, W.D., Iroff, L.D., Wroczynski, R.J., and Mislow, K. (1978). Correlated rotation of tert-butyl groups in tri-tert-butylsilane. *J. Am. Chem. Soc.* 100: 5212–5213. (d) Casarini, D., Foresti, E., Lunazzi, L., and Mazzanti, A. (1999). Correlated rotation of the Me_3Si groups in tris(trimethylsilyl)methanes. *Chem. Eur. J.* 5 (12): 3501–3508.

13 (a) Newman, M.S. (1955). A notation for the study of certain stereochemical problems. *J. Chem. Educ.* 33: 344–347. (b) Thomas, L.L., Christakis, T.J., and Jorgensen, W.L. (2006). Conformation of alkanes in the gas phase and pure liquids. *J. Phys. Chem. B* 110: 21198–21204. (c) Ferguson, A.L., Debenedetti, P.G., and Panagiotopoulos, A.Z. (2009). Solubility and molecular conformations of *n*-alkane chains in water. *J. Phys. Chem. B* 113: 6405–6414. (d) Balabin, R.M. (2011). Enthalpy difference between conformations of normal alkanes: effects of basis set and chain length on intramolecular basis set superposition error. *Mol. Phys.* 109 (6): 943–953.

14 (a) Flamm-termeer, M.A., Beckhaus, H.D., Peters, K. et al. (1985). Thermolabile hydrocarbons. 27. 2,3-Di-1-adamantyl-2,3-dimethylbutane – long bonds and low thermal stability. *Chem. Ber. Recl.* 118 (12): 4665–4673. (b) Fokin, A.A., Chernish, L.V., Gunchenko, P.A. et al. (2012). Stable alkanes containing very long carbon–carbon bonds. *J. Am. Chem. Soc.* 134 (33): 13641–13650.

15 Ermer, O., Bell, P., Schäfer, J., and Szeimies, G. (1989). Shortened C—C bonds in coupled bicyclo[1.1.0]butanes. *Angew. Chem. Int. Ed. Eng.* 28 (4): 473–476.

16 Huntley, D.R., Markopoulos, G., Donovan, P.M. et al. (2005). Squeezing C—C bonds. *Angew. Chem. Int. Ed.* 44 (46): 7549–7553.

17 (a) Verma, A.L., Murphy, W.F., and Bernstein, H.J. (1974). Rotational isomerism. 11. Raman-spectra of butane, 2-methylbutane, and 2,3-dimethylbutane. *J. Chem. Phys.* 60 (4): 1540–1544. (b) Churchill, G.B., Dombrowski, J.P., Maa, L. et al. (2010). Microwave spectroscopy and conformations of 2-methylbutane and 2,3-dimethylbutane. *J. Mol. Struct.* 978: 11–13.

18 Matos, M.A.R., Miranda, M.S., Morais, V.M.F., and Liebman, J.F. (2003). Thermochemistry of (E)- and (Z)-disubstituted alkene species: a combined experimental and theoretical investigation of isomeric dimethyl fumarate and dimethyl maleate. *Org. Biomol. Chem.* 1 (16): 2930–2934.

19 Maier, G. and Schneider, K.A. (1980). Reversal of stability in the Benzene/Dewar-Benzene system. *Angew. Chem. Int. Ed. Eng.* 19 (12): 1022–1023.

20 (a) Jursic, B.S. (2000). Theoretical study of structural properties, infrared spectra, and energetic

properties of C_4H_4 isomers. *J. Mol. Struct. THEOCHEM* 507: 185–192. (b) Cremer, D., Kraka, E., Joo, H. et al. (2006). Exploration of the potential energy surface of C_4H_4 for rearrangement and decomposition reactions of vinylacetylene: a computational study. Part I. *Phys. Chem. Chem. Phys.* 8: 5304–5316.

21 Nakamoto, M., Inagaki, Y., Ochiai, T. et al. (2011). Cyclobutadiene to tetrahedrane: valence isomerization induced by one-electron oxidation. *Heteroat. Chem.* 22 (3–4): 412–416.

22 (a) Hirsch, A. and Brettreich, M. (2006). *Fullerenes: Chemistry and Reactions*, 1–440. Weinheim: Wiley-VCH. (b) Fowler, P.V. and Manolopoulos, D.E. (2007). *An Atlas of Fullerenes* (Dower Book on Chemistry), 1–392. New York: Dover Publications. (c) Acquah, S.F.A., Penkova, A.V., Markeov, D.A. et al. (2017). Review—the beautiful molecule: 30 years of C_{60} and its derivatives. *ECS J. Solid State Sci. Technol.* 6 (6): M3155–M3162.

23 (a) Diederich, F., Isaacs, L., and Philp, D. (1994). Syntheses, structures, and properties of methanofullerenes. *Chem. Soc. Rev.* 23 (4): 243–255. (b) Prato, M. (1997). [60]Fullerene chemistry for materials science applications. *J. Mater. Chem.* 7 (7): 1097–1109.

24 Beckhaus, H.D., Verevkin, S., Rüchardt, C. et al. (1994). C_{70} Is more stable than C_{60} – experimental determination of the heat of formation of C_{70}. *Angew. Chem. Int. Ed. Eng.* 33 (9): 996–998.

25 Kiyobayashi, T., Nagano, Y., Sakiyama, M. et al. (1995). Formation enthalpy of corannulene – microbomb combustion calorimetry. *J. Am. Chem. Soc.* 117 (11): 3270–3271.

26 (a) Schleyer, P.v.R., Williams, J.E., and Blanchard, K.R. (1970). Evaluation of strain in hydrocarbons – strain in adamantane and its origin. *J. Am. Chem. Soc.* 92 (8): 2377–2386. (b) Bach, R.D. and Dmitrenko, O. (2004). Strain energy of small ring hydrocarbons. Influence of C—H bond dissociation energies. *J. Am. Chem. Soc.* 126 (13): 4444–4452.

27 Pitzer, K.S. (1940). Chemical equilibria, free energies, and heat contents for gaseous hydrocarbons. *Chem. Rev.* 27 (1): 39–57.

28 Pitzer, K.S. (1944). The molecular structure and thermodynamics of propane – the vibration frequencies, barrier to internal rotation, entropy, and heat capacity. *J. Chem. Phys.* 12 (7): 310–314.

29 Dunitz, J.D. and Schomaker, V. (1952). The molecular structure of cyclobutane. *J. Chem. Phys.* 20 (11): 1703–1707.

30 a) Bauld, N.L., Cessac, J., and Holloway, R.L. (1977). 1,3(Nonbonded) carbon–carbon interactions – common cause of ring strain, puckering, and inward methylene rocking in cyclobutane and of vertical nonclassical stabilization, pyramidalization, puckering, and outward methylene rocking in cyclobutyl cation. *J. Am. Chem. Soc.* 99 (25): 8140–8144. (b) Kummli, D.S., Frey, H.M., and Leutwyler, S. (2007). High-accuracy structure of cyclobutane by femtosecond rotational Raman four-wave mixing. *J. Phys. Chem.* 111: 11936–11942. (c) Stein, A., Lehmann, C.W., and Luger, P. (1992). Crystal structure of cyclobutane at 117 K. *J. Am. Chem. Soc.* 114 (20): 7684–7687.

31 (a) Cremer, D. and Gauss, J. (1986). Theoretical determination of molecular structure and conformation. 20. Reevaluation of the strain energies of cyclopropane and cyclobutane – C—C and C—H bond energies, 1,3-interactions, and σ-aromaticity. *J. Am. Chem. Soc.* 108 (24): 7467–7477. (b) Barić, D. and Maksić, Z.B. (2005). On the origin of Baeyer strain in molecules – an *ab initio* and DFT analysis. *Theor. Chem. Accounts* 114: 222–228.

32 Wiberg, K.B. (1986). The concept of strain in organic chemistry. *Angew. Chem. Int. Ed. Eng.* 25 (4): 312–322.

33 Cotton, F.A. (1990). *Chemical Applications of Group Theory*, 3e, 1–461. New York: Wiley.

34 Mo, Y. (2011). Rotational barriers in alkanes. *Wiley Interdiscip. Rev. Comput. Mol. Sci.* 1 (2): 164–171.

35 (a) Allinger, N.L., Grev, R.S., Yates, B.F., and Schaefer, H.F.I. (1990). The syn rotational barrier in butane. *J. Am. Chem. Soc.* 112 (1): 114–118. (b) Herrebout, W. A., van der Veken, B. J., Wang, A., Durig, J. R. (1995). Enthalpy difference between conformers of *n*-butane and the potential function governing conformational interchange. *J. Phys. Chem.* 99 (2): 578–585; see also Balabin, R. M. (2009). Enthalpy difference between conformations of normal alkanes: Raman spectroscopy. Study of *n*-pentane and *n*-butane. *J. Phys. Chem. A.* 113 (6): 1012–1019. (c) Murko, M. A., Castejon, H., Wiberg, K. B. (1996). Carbon-carbon rotational barriers in butane, 1-butene, and 1,3-butadiene. *J. Phys. Chem.* 100 (40): 16162–16168.

36 Eyring, H. (1932). Steric hindrance and collision diameters. *J. Am. Chem. Soc.* 54: 3191–3203.

37 Barton, D.H.R. and Cookson, R.C. (1956). The principles of conformational analysis. *Q. Rev.* 10 (1): 44–82.

38 (a) Nelson, D.J. and Brammer, C.N. (2011). Toward consistent terminology for cyclohexane conformers in introductory organic chemistry. *J. Chem. Educ.* 88 (3): 292–294. (b) Eliel, E.L. (1965). Conformational analysis in mobile cyclohexane systems. *Angew. Chem. Int. Ed. Eng.* 4 (9): 761–774. (c) Franklin, N.C. and Feltkamp, H. (1965). Conformational analysis of cyclohexane derivatives by Nuclear Magnetic Resonance spectroscopy. *Angew. Chem. Int. Ed. Eng.* 4 (9): 774–783.

39 Barton, D.H.R. (1970). Principles of conformational analysis. *Science* 169 (3945): 539–544.

40 Offenbach, J.L., Fredin, L., and Strauss, H.L. (1981). Vibrational spectra of twist-boat cyclohexane. *J. Am. Chem. Soc.* 103 (5): 1001–1006.

41 Squillacote, M., Sheridan, R.S., Chapman, O.L., and Anet, F.A.L. (1975). Spectroscopic detection of twist-boat conformation of cyclohexane – direct measurement of free-energy difference between chair and twist-boat. *J. Am. Chem. Soc.* 97 (11): 3244–3246.

42 Dixon, D.A. and Komornicki, A. (1990). Ab initio conformational analysis of cyclohexane. *J. Phys. Chem.* 94 (14): 5630–5636.

43 Kakhiani, K., Lourderaj, U., Hu, W. et al. (2009). Cyclohexane isomerization. Unimolecular dynamics of the twist-boat intermediate. *Chem. A Eur. J.* 113 (16): 4570–4580.

44 Eliel, E.L., Allinger, N.L., Angyal, S.J., and Morrison, G.A. (1965). *Conformational Analysis.* New York: Interscience Publishers.

45 Allinger, N.L., Miller, M.A., Vancatle, F.A., and Hirsch, J.A. (1967). Conformational analysis .57. Calculation of conformational structures of hydrocarbons by Westheimer-Hendrickson-Wiberg method. *J. Am. Chem. Soc.* 89 (17): 4345–4357.

46 Eliel, E.L., Wilen, S.H., and Mander, L.N. (1994). *Stereochemistry of Organic Compounds*, 1–1267. New York: Wiley.

47 Abraham, R.J. and Ribeiro, D.S. (2001). Conformational analysis. Part 36. A variable temperature ^{13}C NMR study of conformational equilibria in methyl substituted cycloalkanes. *J. Chem. Soc., Perkin Trans. 2* (3): 302–307.

48 Barton, D.H.R., Hassel, O., Pitzer, K.S., and Prelog, V. (1954). Nomenclature of cyclohexane bonds. *Science* 119 (3079): 49–49.

49 Barton, D.H.R. (1953). Conformational analysis of substituted cyclohexanes. *Chem. Ind.* 27: 664–666.

50 Papulov, Y.G., Vinogradov, M.G., Lyamtseva, O.I. et al. (2005). Classification of conformations of substituted cyclohexanes. *Russ. J. Phys. Chem.* 79 (11): 1803–1806.

51 Barton, D.H.R. (1950). The conformation of the steroid nucleus. *Experientia* 6 (8): 316–320.

52 Goren, Z. and Biali, S.E. (1990). All-trans-1,2,3,4,5,6-hexaisopropylcyclohexane - an all-axial hexaalkyl cyclohexane. *J. Am. Chem. Soc.* 112 (2): 893–894.

53 Golan, O., Goren, Z., and Biali, S.E. (1990). Axial equatorial stability reversal in all-trans-polyalkylcyclohexanes. *J. Am. Chem. Soc.* 112 (25): 9300–9307.

54 (a) Allinger, N.L. (1959). Conformational analysis. 3. Applications to some medium ring compounds. *J. Am. Chem. Soc.* 81 (21): 5727–5733. (b) Allinger, N.L. and Blatter, H.M. (1961). Conformational analysis. 16. The energy of the boat form of cyclohexanone. Direct measurements of 2-alkyl ketone effects. *J. Am. Chem. Soc.* 83 (4): 994–995.

55 Allinger, N.L., Blatter, H.M., Freiberg, L.A., and Karkowsk, F.M. (1966). Conformational analysis. 51. Conformations of cyclohexanone rings in simple molecules. *J. Am. Chem. Soc.* 88 (13): 2999–3011.

56 Taskinen, E. (2010). Conformations and enthalpies of formation of methylenecyclohexane, 1,4-dimethylenecyclohexane, cyclohexanone, 4-methylenecyclohexanone, 1,4-cyclohexanedione, and their mono- and dioxa derivatives by G3(MP2)//B3LYP calculations. *J. Phys. Org. Chem.* 23 (2): 105–114.

57 Groth, P. and Hassel, O. (1963). Crystal and molecular structure of cyclohexane-1,4-dione. *Proc. Chem. Soc. London* 218.

58 Mossel, A., Romers, C., and Havinga, E. (1963). Investigations into the conformation of non-aromatic ring compounds XII. 1. Cyclohexanedione 1,4 and related compounds. *Tetrahedron Lett.* (19): 1247–1249.

59 Bhatt, M.V., Srinivas, G., and Neelakan, P. (1965). On conformation of cyclohexane-1,4-dione and iIts derivatives. I. Infrared and Raman spectra of cyclohexane 1,4-dione and Infrared spectrum of its octadeutero analogue. *Tetrahedron* 21 (2): 291–297.

60 Gryff-Keller, A. and Szczecinski, P. (2010). Structures of cyclohexane-1,3-dione and 4-hydroxycyclohexane-1,3-dione in solutions – an NMR and GIAO-DFT computational study. *J. Mol. Struct.* 967 (1–3): 94–98.

61 Allinger, J. and Allinger, N.L. (1958). The conformers of 2-bromocyclohexanone. *Tetrahedron* 2 (1–2): 64–74.

62 Freitas, M.P., Tormena, C.F., Garcia, J.C. et al. (2003). NMR, solvation and theoretical investigations of conformational isomerism in 2-X-cyclohexanones (X = NMe_2, OMe, SMe and SeMe). *J. Phys. Org. Chem.* 16 (11): 833–838.
63 Yoshinaga, F., Tormena, C.F., Freitas, M.P. et al. (2002). Conformational analysis of 2-halocyclohexanones: an NMR, theoretical and solvation study. *J. Chem. Soc., Perkin Trans. 2* (9): 1494–1498.
64 Azarakhshi, F., Nori-Shargh, D., Attar, H. et al. (2011). Conformational behaviours of 2-substituted cyclohexanones: a complete basis set, hybrid-DFT study and NBO interpretation. *Mol. Simul.* 37 (14): 1207–1220.
65 Barton, D.H.R., Cookson, R.C., Klyne, W., and Shoppee, C.W. (1954). The conformation of cyclohexene. *Chem. Ind.* (1): 21–22.
66 Naumov, V.A., Dashevsk, V.G., and Zaripov, N.M. (1970). Determination of molecular cyclohexene structure. *Zh. Strukt. Khim.* 11 (5): 793–800.
67 Anet, F.A.L. and Haq, M.Z. (1965). Ring inversion in cyclohexene. *J. Am. Chem. Soc.* 87 (14): 3147–3150.
68 Anet, F.A.L., Freedberg, D.I., Storer, J.W., and Houk, K.N. (1992). On the potential energy surface for ring inversion in cyclohexene and related molecules. *J. Am. Chem. Soc.* 114 (27): 10969–10971.
69 Verdal, N., Wilke, J.J., and Hudson, B.S. (2006). Conformational distinguishability of medium cycloalkanes in crystals via inelastic neutron scattering. *J. Phys. Chem. A* 110 (8): 2639–2646.
70 Bocian, D.F. and Strauss, H.L. (1977). Vibrational spectra, conformations, and potential functions of cycloheptane and related oxepanes. *J. Am. Chem. Soc.* 99 (9): 2866–2876.
71 Pawar, D.M., Moody, E.M., and Noe, E.A. (1999). Conformational study of oxacyclooctan-2-one by dynamic NMR spectroscopy and computational methods. *J. Org. Chem.* 64 (13): 4586–4589.
72 Weinberg, N. and Wolfe, S. (1994). A Comprehensive approach to the conformational-analysis of cyclic compounds. *J. Am. Chem. Soc.* 116 (22): 9860–9868.
73 Saunders, M., Houk, K.N., Wu, Y.D. et al. (1990). Conformations of cycloheptadecane – a comparison of methods for conformational searching. *J. Am. Chem. Soc.* 112 (4): 1419–1427.
74 Dorofeeva, O.V., Mastryukov, V.S., Allinger, N.L., and Almenningen, A. (1990). Molecular-structure and conformations of cyclononane as studied by electron diffraction and molecular mechanics. *J. Phys. Chem.* 94 (21): 8044–8048.
75 Anet, F.A.L. and Wagner, J.J. (1971). Conformation of cyclononane – evidence from 251-MHz ^1H nuclear magnetic resonance and 63-MHz ^{13}C Fourier transform nuclear magnetic resonance. *J. Am. Chem. Soc.* 93 (20): 5266–5268.
76 Pawar, D.M., Smith, S.V., Mark, H.L. et al. (1998). Conformational study of cyclodecane and substituted cyclodecanes by dynamic NMR spectroscopy and computational methods. *J. Am. Chem. Soc.* 120 (41): 10715–10720.
77 Anet, F.A.L., Cheng, A.K., and Wagner, J.J. (1972). Determination of conformational energy barriers in medium-ring and large-ring cycloalkanes ^{1}by H and ^{13}C nuclear magnetic-resonance. *J. Am. Chem. Soc.* 94 (26): 9250–9252.
78 Shannon, V.L., Strauss, H.L., Snyder, R.G. et al. (1989). Conformation of the cycloalkanes $C_{14}H_{28}$, $C_{16}H_{32}$, and $C_{22}H_{44}$ in the liquid and high-temperature crystalline phases by vibrational spectroscopy. *J. Am. Chem. Soc.* 111 (6): 1947–1958.
79 Chickos, J.S., Hesse, D.G., Panshin, S.Y. et al. (1992). The strain energy of cyclotetradecane is small. *J. Org. Chem.* 57 (6): 1897–1899.
80 Dale, J. (1973). Exploratory calculations of medium and large rings. 1. Conformational minima of cycloalkanes. *Acta Chem. Scand.* 27 (4): 1115–1129.
81 Dale, J. (1973). Exploratory calculations of medium and large rings. 2. Conformational interconversions in cycloalkanes. *Acta Chem. Scand.* 27 (4): 1130–1148.
82 Dale, J. (1973). Exploratory calculations of medium and large rings. 3. Mono-gem-dimethyl and bis(gem-dimethyl) cycloalkanes. *Acta Chem. Scand.* 27 (4): 1149–1158.
83 Anet, F.A.L., St.Jacques, M., Henrichs, P.M. et al. (1974). Conformational analysis of medium ring ketones. *Tetrahedron* 30 (12): 1629–1637.
84 Dale, J. (1963). Macrocyclic compounds .3. Conformations of cycloalkanes and other flexible macrocycles. *J. Chem. Soc.* 93–111.
85 Dale, J. (1966). Conformational aspects of many-membered rings. *Angew. Chem. Int. Ed. Eng.* 5 (12): 1000–1021.
86 Keller, T.H., Neeland, E.G., Rettig, S. et al. (1988). Conformational-analysis of 14-membered macrolides using X-ray crystallography and molecular mechanics calculations. *J. Am. Chem. Soc.* 110 (23): 7858–7868.
87 Ounsworth, J.P. and Weiler, L. (1987). Use of polar maps in conformational analysis. *J. Chem. Educ.* 64 (7): 568–572.

88 Still, W.C. and Galynker, I. (1981). Chemical consequences of conformation in macrocyclic-compounds – an effective approach to remote asymmetric induction. *Tetrahedron* 37 (23): 3981–3996.

89 Ngo, J.T. and Karplus, M. (1997). Pseudosystematic conformational search. Application to cycloheptadecane. *J. Am. Chem. Soc.* 119 (24): 5657–5667.

90 Wang, C.S. (1997). Efficient algorithm for conformational search of macrocyclic molecules. *J. Comput. Chem.* 18 (2): 277–289.

91 de Meijere, A., Kozhushkov, S.I., Rauch, K. et al. (2003). Heats of formation of [2.2] paracyclophane-1-ene and [2.2]paracyclophane-1, 9-diene – an experimental study. *J. Am. Chem. Soc.* 125 (49): 15110–15113.

92 Zhu, Y.P., Zajicek, J., and Serianni, A.S. (2001). Acyclic forms of [1-^{13}C]aldohexoses in aqueous solution: quantitation by ^{13}C NMR and deuterium isotope effects on tautomeric equilibria. *J. Org. Chem.* 66 (19): 6244–6251.

93 Cox, J.D. and Pilcher, G. (1970). *Thermochemistry of Organic and Organometallic Compounds*, 1–643. London and New York: Academic Press.

94 Fliszar, S., Vauthier, E.C., and Chretien, S. (2004). On simple, accurate calculations of atomic charges, bond properties and molecular energies. IV. Butane gauche interactions and vibrational energies of alkylcyclohexanes and related polycyclic hydrocarbons. *J. Mol. Struct. THEOCHEM* 682 (1–3): 153–158.

95 Liebman, J.F. and Greenberg, A. (1989). Survey of the heats of formation of 3-membered-ring species. *Chem. Rev.* 89 (5): 1225–1246.

96 Beckhaus, H.D., Rüchardt, C., Lagerwall, D.R. et al. (1994). Experimental enthalpies of formation and strain energies for the caged $C_{20}H_{20}$ pagodane and dodecahedrane frameworks. *J. Am. Chem. Soc.* 116 (26): 11775–11778.

97 Herling, J., Shabtai, J., and Gil-Av, E. (1965). Relative stability of cyclic olefins. I. Equilibrium isomerization of monocyclic olefins containing 4-to 6-membered rings. *J. Am. Chem. Soc.* 87 (18): 4107–4110.

98 Rogers, D.W., von Voithenberg, H., and Allinger, N.L. (1978). Heats of hydrogenation of *cis* and trans isomers of cyclooctene. *J. Org. Chem.* 43 (2): 360–361.

99 Pines, H. and Haag, W.O. (1961). Alumina – catalyst and support. 11. Alumina catalyzed dehydration of alcohols. *J. Am. Chem. Soc.* 83 (13): 2847–2852.

100 Tsitsishvili, G.V., Dolidze, A.V., and Balardzhishvili, G.I. (1986). Synthesis of methylcyclopentenes by dehydration of cyclohexanol on zeolite catalysts. *USSR Izv. Akad. Nauk Gruzs. SSR, Ser. Khim.* 12 (4): 305–307.

101 Bredt, J., Thouet, H., and Schmitz, J. (1924). Steric hindrance in the bridge ring (Bredt's rule) and the meso-trans-position in condensed ring systems of the hexamethylenes. *Justus Liebigs Ann. Chem.* 437: 1–13.

102 Bredt, J. (1927). Is the stability of ketopinic acid a special case of Bredt's rule? The course of the ketone decomposition of β-ketonic acids. *Annales Academiæ Scientiarum Fennicæ* 29A (2): 20.

103 Prelog, V. (1950). Prelog – newer developments of newer developments of the chemistry of many-membered ring compounds. *J. Chem. Soc.* 420–428.

104 Wiseman, J.R. (1967). Bredt's rule. Bicyclo[3.3.1]non-1-ene. *J. Am. Chem. Soc.* 89 (23): 5966–5968.

105 Marshall, J.A. and Faubl, H. (1967). Bicyclo[3.3.1]non-1-ene. *J. Am. Chem. Soc.* 89 (23): 5965–5966.

106 Wijsman, G.W., de Wolf, W.H., Bickelhaupt, F. et al. (1992). Synthesis and X-ray crystal-structure of 11-bromo-endo-9-chloro-7-ethoxybicyclo[5.3.1]undec-1(11)-ene – a stable, highly strained anti-Bredt olefin. *J. Am. Chem. Soc.* 114 (23): 9191–9192.

107 Wijsman, G.W., Iglesias, G.A., Beekman, M.C. et al. (2001). X-ray crystal structures and conformational analysis of bicyclo[5.3.1]undec-1(11)-enes: twisting versus pyramidalization in anti-Bredt olefins. *J. Org. Chem.* 66 (4): 1216–1227.

108 Wijsman, G.W., Boesveld, W.M., Beekman, M.C. et al. (2002). Unusual reactions of halo[5]metacyclophanes. *Eur. J. Org. Chem.* (4): 614–629.

109 Warner, P.M. (1989). Strained bridgehead double-bonds. *Chem. Rev.* 89 (5): 1067–1093.

110 Johnson, F. (1968). Allylic strain in 6-membered rings. *Chem. Rev.* 68 (4): 375–413.

111 Hoffmann, R.W. (1989). Allylic 1,3-strain as a controlling factor in stereoselective transformations. *Chem. Rev.* 89 (8): 1841–1860.

112 Hoffmann, R.W. (1992). Flexible molecules with defined shape – conformational design. *Angew. Chem. Int. Ed. Eng.* 31 (9): 1124–1134.

113 Broeker, J.L., Hoffmann, R.W., and Houk, K.N. (1991). Conformational analysis of chiral alkenes and oxonium ions: *ab initio* molecular orbital

calculations and an improved MM_2 force field. *J. Am. Chem. Soc.* 113 (13): 5006–5017.

114 Tietze, L.F. and Schulz, G. (1996). Ab initio molecular orbital calculations on allylic 1,3-strain of electron-donor- and electron-acceptor-substituted alkenes. *Liebigs Ann.* (10): 1575–1579.

115 Hückel, E. (1930). Quantum theory of double linkings. *Z. Phys.* 60: 423–456.

116 Pauling, L. (1931). The nature of the chemical bond. Application of results obtained from the quantum mechanics and from a theory of paramagnetic susceptibility to the structure of molecules. *J. Am. Chem. Soc.* 53: 1367–1400.

117 Slater, J.C. (1931). Directed valence in polyatomic molecules. *Phys. Rev.* 37 (5): 481–489.

118 Wiberg, K.B. (1996). Bent bonds in organic compounds. *Acc. Chem. Res.* 29 (5): 229–234.

119 Bader, R.F.W. (1985). Atoms in molecules. *Acc. Chem. Res.* 18 (1): 9–15.

120 Dobado, J.A., Martínez-García, H., Molina, J.M., and Sundberg, M.R. (2000). Chemical bonding in hypervalent molecules revised. 3. Application of the atoms in molecules theory to $Y_3X—CH_2$ (X = N, P, or As ; Y = H or F) and $H_2X—GH(2)$ (X = O, S, or Se) ylides. *J. Am. Chem. Soc.* 122 (6): 1144–1149.

121 Gilheany, D.G. (1994). No d-orbitals but Walsh diagrams and maybe banana bonds – chemical bonding in phosphines, phosphine oxides, and phosphonium ylides. *Chem. Rev.* 94 (5): 1339–1374.

122 Purcell, K.F. and Zapata, J. (1978). Banana and non-banana bonds of carbonyl group. *J. Am. Chem. Soc.* 100 (8): 2314–2319.

123 Kaldor, U. (1967). Localized orbitals for NH_3, C_2H_4, and C_2H_2. *J. Chem. Phys.* 46 (5): 1981–1987.

124 Garbisch, E.W. Jr., (1965). Cycloalk-2-enones and $\alpha,\beta,\alpha',\beta'$-cycloalkadienones. A. synthesis. B. On direction of bromination of 2-substituted cycloalkanones and their ketals. *J. Org. Chem.* 30 (7): 2109–2120.

125 Squillacote, M.E., Semple, T.C., and Mui, P.W. (1985). The geometries of the *s-cis*-conformers of some acyclic 1,3-dienes – planar or twisted. *J. Am. Chem. Soc.* 107 (24): 6842–6846.

126 Adam, W., Bottke, N., Krebs, O. et al. (2002). Ene reaction of singlet oxygen, triazolinedione, and nitrosoarene with chiral deuterium-labeled allylic alcohols: the interdependence of diastereoselectivity and regioselectivity discloses mechanistic insights into the hydroxy-group directivity. *J. Am. Chem. Soc.* 124 (48): 14403–14409.

127 Kistiakowsky, G.B., Ruhoff, J.R., Smith, H.A., and Vaughan, W.E. (1936). Heats of organic reactions. IV. Hydrogenation of some dienes and of benzene. *J. Am. Chem. Soc.* 58: 146–153.

128 Jarowski, P.D., Wodrich, M.D., Wannere, C.S. et al. (2004). How large is the conjugative stabilization of diynes? *J. Am. Chem. Soc.* 126 (46): 15036–15037.

129 Lee, P.S., Du, W., Boger, D.L., and Jorgensen, W.L. (2004). Energetic preferences for α,β versus β,γ-unsaturation. *J. Org. Chem.* 69 (16): 5448–5453.

130 Rogers, D.W., Matsunaga, N., McLafferty, F.J. et al. (2004). On the lack of conjugation stabilization in polyynes (polyacetylenes). *J. Org. Chem.* 69 (21): 7143–7147.

131 Kofranek, M., Karpfen, A., and Lischka, H. (1992). Gauche- versus *s-cis*-butadiene revisited – a molecular-dynamics simulation of the Ar matrix effect. *Chem. Phys. Lett.* 189 (3): 281–286.

132 Fisher, J.J. and Michl, J. (1987). *S-cis* vs gauche-1,3-butadiene – evidence for planarity from polarized matrix-isolation IR spectroscopy. *J. Am. Chem. Soc.* 109 (4): 1056–1059.

133 Arnold, B.R., Balaji, V., and Michl, J. (1990). Structure of matrix-isolated *s-cis*-1,3-butadiene from polarized IR-spectra – a closer look. *J. Am. Chem. Soc.* 112 (5): 1808–1812.

134 Squillacote, M.E., Sheridan, R.S., Chapman, O.L., and Anet, F.A.L. (1979). Planar *s-cis*-1,3-butadiene. *J. Am. Chem. Soc.* 101 (13): 3657–3659.

135 Breulet, J., Lee, T.J., and Schaefer, H.F.I. (1984). Comparison between the *s-cis* and gauche conformers of 1,3-butadiene. *J. Am. Chem. Soc.* 106 (21): 6250–6253.

136 Becher, G. and Mannschreck, A. (1981). Chiral butadienes. 9. Preparative enrichment of enantiomeric 2,3,4,5-tetrabromo-2,4-hexadienes by liquid chromatography. *Chem. Ber.* 114 (6): 2365–2368.

137 Moss, G.P. (1996). Basic terminology of stereochemistry. *Pure Appl. Chem.* 68 (12): 2193–2222.

138 Becher, G. and Mannschreck, A. (1983). Chiral butadienes. 10. Barriers to enantiomerization of 1,2,3,4-tetramethyl-1,3-butadienes. *Chem. Ber.* 116 (1): 264–272.

139 Ogasawara, M., Kotani, S., Nakajima, H. et al. (2013). Atropisomeric chiral dienes in asymmetric catalysis: C_2-symmetric (Z,Z)-2,3-bis[1-(diphenylphosphinyl)ethylidene]tetralin as a highly active Lewis base organocatalyst. *Angew. Chem. Int. Ed.* 52 (51): 13798–13802.

140 Adams, R. and Yuan, H.C. (1933). The stereochemistry of diphenyls and analogous compounds. *Chem. Rev.* 12 (2): 261–338.

141 Haberhauer, G., Tepper, C., Wölper, C., and Bläser, D. (2013). Tropos, nevertheless conformationally stable biphenyl derivatives. *Eur. J. Org. Chem.* (12): 2325–2333.

142 Wallace, T.W. (2006). Biaryl synthesis with control of axial chirality. *Org. Biomol. Chem.* 4 (17): 3197–3210.

143 Brock, C.P. and Minton, R.P. (1989). Systematic effects of crystal-packing forces – biphenyl fragments with H-atoms in all 4 *ortho* positions. *J. Am. Chem. Soc.* 111 (13): 4586–4593.

144 Alkorta, I., Elguero, J., Roussel, C. et al. (2012). Atropisomerism and axial chirality in heteroaromatic compounds. *Adv. Heterocycl. Chem.* 105: 1–188.

145 Curran, D.P., Geib, S., and DeMello, N. (1999). Rotational features of carbon–nitrogen bonds in *N*-aryl maleimides. Atroposelective reactions of *o-tert*-butylphenylmaleimides. *Tetrahedron* 55 (18): 5681–5704.

146 Clayden, J. and Lai, L.W. (1999). Enantioselective synthesis of atropisomeric amides by dynamic resolution: thermodynamic control with a proline-derived diamine resolving agent. *Angew. Chem. Int. Ed.* 38 (17): 2556–2558.

147 Godfrey, C.R.A., Simpkins, N.S., and Walker, M.D. (2000). High stereochemical fidelity in reactions of an atropisomeric *N*-acyliminium intermediate. *Synlett* (3): 388–390.

148 Murrison, S., Glowacki, D., Einzinger, C. et al. (2009). Remarkably slow rotation about a single bond between an sp^3-hybridised carbon atom and an aromatic ring without *ortho* substituents. *Chem. Eur. J.* 15 (9): 2185–2189.

149 Ototake, N., Morimoto, Y., Mokuya, A. et al. (2010). Catalytic enantioselective synthesis of atropisomeric indoles with an N—C chiral axis. *Chem. Eur. J.* 16 (23): 6752–6755.

150 Tietze, L.F., Schuster, H.J., von Hof, J. et al. (2010). Atropisomerism of aromatic carbamates. *Chem. Eur. J.* 16 (42): 12678–12682.

151 Takahashi, H., Wakamatsu, S., Tabata, H. et al. (2011). Atropisomerism observed in indometacin derivatives. *Org. Lett.* 13 (4): 760–763.

152 Mino, T., Yamada, H., Komatsu, S. et al. (2011). Atropisomerism at C—N bonds of acyclic amines: synthesis and application to Pd-catalyzed asymmetric allylic alkylations. *Eur. J. Org. Chem.* (24): 4540–4542.

153 Tabata, H., Kayama, S., Takahashi, Y. et al. (2014). A complete gear system in *N*-benzoyl-carbazole derivatives. *Org. Lett.* 16 (5): 1514–1517.

154 Fuji, K., Oka, T., Kawabata, T., and Kinoshita, T. (1998). The first synthesis of an optically active molecular bevel gear with only two cogs on each wheel. *Tetrahedron Lett.* 39 (11): 1373–1376.

155 Betson, M.S., Clayden, J., Worrall, C.R., and Peace, S. (2006). Three groups good, four groups bad? Atropisomerism in *ortho*-substituted diaryl ethers. *Angew. Chem. Int. Ed.* 45 (35): 5803–5807.

156 Clayden, J., Worrall, C.P., Moran, W.J., and Helliwell, M. (2008). Enantioselective synthesis of an atropisomeric diaryl ether. *Angew. Chem. Int. Ed.* 47 (17): 3234–3237.

157 Koizumi, Y., Suzuki, S., Takeda, K. et al. (2008). Synthesis and characteristic stereostructure of a biphenanthryl ether. *Tetrahedron Asymmetry* 19 (12): 1407–1410.

158 Clayden, J., Senior, J., and Helliwell, M. (2009). Atropisomerism at C—S bonds: asymmetric synthesis of diaryl sulfones by dynamic resolution under thermodynamic control. *Angew. Chem. Int. Ed.* 48 (34): 6270–6273.

159 Bringmann, G., Mortimer, A.J.P., Keller, P.A. et al. (2005). Atroposelective synthesis of axially chiral biaryl compounds. *Angew. Chem. Int. Ed.* 44 (34): 5384–5427.

160 Bringmann, G., Walter, R., and Weirich, R. (1990). The directed synthesis of biaryl compounds - modern concepts and strategies. *Angew. Chem. Int. Ed. Eng.* 29 (9): 977–991.

161 Stanforth, S.P. (1998). Catalytic cross-coupling reactions in biaryl synthesis. *Tetrahedron* 54 (3–4): 263–303.

162 Pu, L. (1998). 1,1′-binaphthyl dimers, oligomers, and polymers: molecular recognition, asymmetric catalysis, and new materials. *Chem. Rev.* 98 (7): 2405–2494.

163 Lloyd-Williams, P. and Giralt, E. (2001). Atropisomerism, biphenyls and the Suzuki coupling: peptide antibiotics. *Chem. Soc. Rev.* 30 (3): 145–157.

164 Berthod, M., Mignani, G., Woodward, G., and Lemaire, M. (2005). Modified BINAP: the how and the why. *Chem. Rev.* 105 (5): 1801–1836.

165 Blom, C.E., Grassi, G., and Bauder, A. (1984). Molecular structure of *s-cis*-acrolein and *s-trans*-acrolein determined by microwave spectroscopy. *J. Am. Chem. Soc.* 106 (24): 7427–7431.

166 Hagen, K. and Hedberg, K. (1984). Conformational analysis. 8. Propenoyl chloride – an electron-diffraction investigation of the molecular structure, composition, syn/anti energy and

entropy differences, and potential hindering internal rotation. *J. Am. Chem. Soc.* 106 (21): 6150–6155.

167 Hagen, K. and Hedberg, K. (1989). Conformational analysis. 12. 2-Chloropropenoyl chloride – an electron diffraction investigation of the molecular structure, conformational composition, anti/syn energy and entropy differences, and potential hindering internal-rotation. *J. Am. Chem. Soc.* 111 (18): 6905–6909.

168 Schleyer, P.v.R. and Jiao, H.J. (1996). What is aromaticity? *Pure Appl. Chem.* 68 (2): 209–218.

169 Schleyer, P.V.R. (2001). Introduction: aromaticity. *Chem. Rev.* 101 (5): 1115–1117.

170 Feixas, F., Matito, E., Poater, J., and Sola, M. (2008). On the performance of some aromaticity indices: a critical assessment using a test set. *J. Comput. Chem.* 29 (10): 1543–1554.

171 D'Auria, M. (2013). A new proposal for the estimation of the aromatic character. *Lett. Org. Chem.* 10 (4): 277–282.

172 Sondheimer, F., Calder, I.C., Elix, J.A. et al. (1967). Annulenes and related compounds From Special Publication . *Chem. Soc.* 21: 75–107.

173 Sondheimer, F. (1972). Annulenes. *Acc. Chem. Res.* 5 (3): 81–91.

174 Schleyer, P.v.R., Maerker, C., Dransfeld, A. et al. (1996). Nucleus-independent chemical shifts: a simple and efficient aromaticity probe. *J. Am. Chem. Soc.* 118 (26): 6317–6318.

175 Gomes, J.A.N.F. and Mallion, R.B. (2001). Aromaticity and ring currents. *Chem. Rev.* 101 (5): 1349–1384.

176 Slayden, S.W. and Liebman, J.F. (2001). The energetics of aromatic hydrocarbons: an experimental thermochemical perspective. *Chem. Rev.* 101 (5): 1541–1566.

177 Schleyer, P.V. and Puhlhofer, F. (2002). Recommendations for the evaluation of aromatic stabilization energies. *Org. Lett.* 4 (17): 2873–2876.

178 Krygowski, T.M. and Cyranski, M.K. (2001). Structural aspects of aromaticity. *Chem. Rev.* 101 (5): 1385–1419.

179 Cyranski, M.K., Krygowski, T.M., Katritzky, A.R., and Schleyer, P.v.R. (2002). To what extent can aromaticity be defined uniquely? *J. Org. Chem.* 67 (4): 1333–1338.

180 Ciesielski, A., Krygowski, T.M., Cyranski, M.K., and Balaban, A.T. (2011). Defining rules of aromaticity: a unified approach to the Hückel, Clar and Randic concepts. *Phys. Chem. Chem. Phys.* 13 (9): 3737–3747.

181 Nendel, M., Houk, K.N., Tolbert, L.M. et al. (1998). Bond alternation and aromatic character in cyclic polyenes: assessment of theoretical methods for computing the structures and energies of bismethano[14]annulenes. *J. Phys. Chem. A* 102 (36): 7191–7198.

182 Choi, C.H. and Kertesz, M. (1998). New interpretation of the valence tautomerism of 1,6-methano[10]annulenes and its application to fullerene derivatives. *J. Phys. Chem. A* 102 (19): 3429–3437.

183 Mondal, R., Adhikari, R.M., Shah, B.K., and Neckers, D.C. (2007). Revisiting the stability of hexacenes. *Org. Lett.* 9 (13): 2505–2508.

184 Roth, W.R. (1983). Heats of hydrogenation. *Nachrichten aus Chemie Technik und Laboratorium* 31 (12): 964–966.

185 George, P., Glusker, J.P., and Bock, C.W. (1991). Homodesmotic reaction energies show little or no aromatic stabilization in 1,6-methano[10]annulene. *J. Mol. Struct. THEOCHEM* 81 (1–2): 193–196.

186 Pauling, L. (1960). *The Nature of the Chemical Bond and the Structure of Molecules and Crystals*, 3e, 1–664. Ithaca, NY: Cornwell University Press.

187 Taskinen, E. (1993). Relative thermodynamic stabilities of isomeric alkyl allyl and alkyl (Z)-propenyl ethers. *Tetrahedron* 49 (48): 11389–11394.

188 Taskinen, E. (2001). Thermodynamic, spectroscopic, and density functional theory studies of ally aryl and prop-1-enyl aryl ethers. Part 1. Thermodynamic data of isomerization. *J. Chem. Soc., Perkin Trans. 2* (9): 1824–1834.

189 Kocienski, P.J. (2005). *Protecting Groups*, 3e, 1–679. Stuttgart and New York: Georg Thieme Verlag.

190 Guibe, F. (1997). Allylic protecting groups and their use in a complete environment. 1. Allylic protection of alcohols. *Tetrahedron* 53 (40): 13509–13556.

191 Markovic, D. and Vogel, P. (2004). Allyl, methallyl, prenyl, and methylprenyl ethers as protected alcohols: their selective cleavage with diphenyldisulfone under neutral conditions. *Org. Lett.* 6 (16): 2693–2696.

192 Sonesson, C. and Hallberg, A. (1995). Preparation of *N*-formyl-2,3-dihydropyrrole and *N*-carbomethoxy-2,3-dihydropyrrole by Pd-catalyzed isomerization of the corresponding *N*-acyl-2,5-dihydropyrrole. *Tetrahedron Lett.* 36 (25): 4505–4506.

193 Pawar, D.M., Khalil, A.A., Hooks, D.R. et al. (1998). E and Z conformations of esters, thiol

esters, and amides. *J. Am. Chem. Soc.* 120 (9): 2108–2112.

194 Huisgen, R. and Ott, H. (1959). Die Konfiguration der Carbonestergruppe und die Sondereigenschaften der Lactone. *Tetrahedron* 6 (3): 253–267.

195 Exner, O. (1977). Dipole moments, configurations and conformations of molecules containing X=Y groups. In: *Supplement A, The Chemistry of Double-Bonded Functional Groups*, Part 1, Chapter 1 (ed. S. Patai), 1–92. London: Wiley.

196 (a) Liebman, J.F. and Greenberg, A. (1974). Origin of rotational barriers in amides and esters. *Biophys. Chem.* 1 (3): 222–226. (b) Fischer, G. (2000). Chemical aspects of peptide bond isomerisation. *Chem. Soc. Rev.* 29 (2): 119–127. (c) Demaison, J., Császar, A.G., Kleiner, I., and Møllendal, H. (2007). Equilibrium vs ground-state planarity of the CONH linkage. *J. Phys. Chem. A* 111 (13): 2571–2586.

197 Wiberg, K.B. and Laidig, K.E. (1987). Barriers to rotation adjacent to double-bonds. 3. The C—O barrier in formic acid, methyl formate, acetic acid, and methyl acetate – the origin of ester and amide resonance. *J. Am. Chem. Soc.* 109 (20): 5935–5943.

198 Wiberg, K.B. and Waldron, R.F. (1991). Lactones. 2. Enthalpies of hydrolysis, reduction, and formation of the C_4–C_{13} monocyclic lactones - strain energies and conformations. *J. Am. Chem. Soc.* 113 (20): 7697–7705.

199 Houk, K.N., Jabbari, A., Hall, H., and Aleman, C. (2008). Why δ-valerolactone polymerizes and γ-butyrolactone does not. *J. Org. Chem.* 73 (7): 2674–2678.

200 Chakrabarti, P. and Dunitz, J.D. (1982). Structural characteristics of the carboxylic amide group. *Helv. Chim. Acta* 65 (5): 1555–1562.

201 Stewart, W.E. and Siddall, T.H. (1970). Nuclear magnetic resonance studies of amides. *Chem. Rev.* 70 (5): 517–551.

202 Toriumi, Y., Kasuya, A., and Itai, A. (1990). Crystallographic studies on retinoidal active and retinoidal inactive aromatic anilides. *J. Org. Chem.* 55 (1): 259–263.

203 Deslongchamps, P. (1983). *Stereoelectronic Effects in Organic Chemistry*, 1–375. Elmsford, NY: Pergamon Press, Inc., Maxwell House.

204 Bader, R.F.W. and Nguyen-Dang, T.T. (1981). Quantum theory of atoms in molecules – Dalton revisited. *Adv. Quantum Chem.* 14: 63–124.

205 Bennet, A.J., Wang, Q.P., Slebocka-Tilk, H. et al. (1990). Relationship between amidic distortion and ease of hydrolysis in base –if amidic resonance does not exist, then what accounts for the accelerated hydrolysis of distorted amides. *J. Am. Chem. Soc.* 112 (17): 6383–6385.

206 Morgan, J. and Greenberg, A. (2014). Novel bridgehead bicyclic lactams: (a) molecules predicted to have O-protonated and N-protonated tautomers of comparable stability; (b) hyperstable lactams and their O-protonated tautomers. *J. Chem. Thermodyn.* 73: 206–212.

207 Kirby, A.J., Komarov, I.V., Wothers, P.D., and Feeder, N. (1998). The most twisted amide: structure and reactions. *Angew. Chem. Int. Ed. Eng.* 37 (6): 785–786.

208 Ohwada, T., Okamoto, I., Shudo, K., and Yamaguchi, K. (1998). Intrinsic pyramidal nitrogen of N-sulfonylamides. *Tetrahedron Lett.* 39 (43): 7877–7880.

209 Afonin, A.V., Khi'lko, M.Y., Komel'kova, V.I. et al. (1991). Conformation analysis of substituted alkylvinyl ethers according to ^{1}H and ^{13}C NMR data. *Zh. Org. Khim.* 27 (1): 161–170.

210 Taskinen, E. (1980). Thermodynamics of vinyl ethers. 24. Relative stabilities and structures of some isomeric derivatives of phenyl vinyl ether. *Acta Chem. Scand. Ser. B* 34 (3): 203–208.

211 Mandado, M., Otero, N., and Mosquera, R.A. (2006). Local aromaticity study of heterocycles using *n*-center delocalization indices: the role of aromaticity on the relative stability of position isomers. *Tetrahedron* 62 (52): 12204–12210.

212 Wang, Y., Wu, J.I.-C., Li, Q., and Schleyer, P.v.R. (2010). Aromaticity and relative stabilities of azines. *Org. Lett.* 12 (21): 4824–4827.

213 Diels, O. and Alder, K. (1931). Syntheses in the hydroaromatic range. *Justus Liebigs Ann. Chem.* 490: 267–276.

214 Diels, O. and Alder, K. (1931). Syntheses in the hydro-aromatic sequence; ('dien-syntheses' with pyrrol and its homologes). *Justus Liebigs Ann. Chem.* 486: 211–225.

215 Barlow, M.G., Haszeldi, R.N., and Hubbard, R. (1969). Diels–Alder reactions of para-bonded valence isomer of hexafluorobenzene and of benzvalene isomer of hexakis(trifluoromethyl)benzene – pyrrole as a diene. *J. Chem. Soc. Chem. Commun.* (6): 301–302.

216 Schleyer, P.v.R., Jiao, H.J., Goldfuss, B., and Freeman, P.K. (1995). Aromaticity and antiaromaticity in 5-membered C_4H_4X ring systems – classical and magnetic concepts may not be orthogonal. *Angew. Chem. Int. Ed. Eng.* 34 (3): 337–340.

217 Mattmann, E., Mathey, F., Sevin, A., and Frison, G. (2002). De-aromatizing phosphole. *J. Org. Chem.* 67 (4): 1208–1213.

218 Nyulaszi, L. (2001). Aromaticity of phosphorus heterocycles. *Chem. Rev.* 101 (5): 1229–1246.

219 Mathey, F. (1988). The organic chemistry of phospholes. *Chem. Rev.* 88 (2): 429–453.

220 Charrier, C., Bonnard, H., de Lauzon, G., and Mathey, F. (1983). Proton (1,5)-shifts in P-unsubstituted 1*H*-phospholes – synthesis and chemistry of 2*H*-phosphole dimers. *J. Am. Chem. Soc.* 105 (23): 6871–6877.

221 Egan, W., Tang, R., Zon, G., and Mislow, K. (1971). Barriers to pyramidal inversion at phosphorus in phospholes, phosphindoles, and dibenzophospholes. *J. Am. Chem. Soc.* 93 (23): 6205–6216.

222 Sustmann, R., Böhm, M., and Sauer, J. (1979). Significance of diene-1,4-distance on the reactivity in Diels–Alder reactions. *Chem. Ber.* 112 (3): 883–889.

223 Beckhaus, H.D., Dogan, B., Haedrich, J. et al. (1990). Geminal substituent effects. 2. Dependence of anomeric stabilization on structure in acetals. *Angew. Chem. Int. Ed. Eng.* 29 (3): 320–321.

224 Hati, S. and Datta, D. (1992). Anomeric effect and hardness. *J. Org. Chem.* 57 (22): 6056–6057.

225 Juaristi, E. and Cuevas, G. (1999). Enthalpic anomeric effect in 2-Y-1,3-dithianes (Y = SC_6H_5, $CO_2CH_2CH_3$, and COC_6H_5). Experimental and theoretical evaluation. Solvent effects. *Tetrahedron* 55 (2): 359–372.

226 (a) Mo, Y. (2010). Computational evidence that hyperconjugative interactions are not responsible for the anomeric effect. *Nat. Chem.* 2 (8): 666–671. (b) Bauerfeldt, G.F., Cardozo, T.M., Pereira, M.S., and da Silva, C.O. (2013). The anomeric effect: the dominance of exchange effects in closed-shell systems. *Org. Biomol. Chem.* 11 (2): 299–308.

227 Streitwieser, A. Jr., and Swanson, J.T. (1983). Carbon acidity. 64. 1st and 2nd Acidity constants for some indenyl and fluorenyl hydrocarbons – coulombic effects in ion triplets. *J. Am. Chem. Soc.* 105 (8): 2502–2503.

228 Ferrier, R.J. (1979). Unsaturated carbohydrates. 21. Carbocyclic ring closure of a hex-5-enopyranoside derivative. *J. Chem. Soc., Perkin Trans. 1* (6): 1455–1458.

229 Ferrier, R.J. and Hoberg, J.O. (2003). Synthesis and reactions of unsaturated sugars. *Adv. Carbohydr. Chem. Biochem.* 58: 55–119.

230 Giese, B., Groninger, K.S., Witzel, T. et al. (1987). Synthesis of 2-deoxy sugars. *Angew. Chem. Int. Ed. Eng.* 26 (3): 233–234.

231 Giese, B., Gilges, S., Groninger, K.S. et al. (1988). Synthesis of 2-deoxy sugars. *Liebigs Ann. Chem.* (6): 615–617.

232 Edward, J.T. (1955). Stability of glycosides to acid hydrolysis – a conformational analysis. *Chem. Ind.* (36): 1102–1104.

233 Lemieux, R.U., Kullnig, R.K., Bernstein, H.J., and Schneider, W.G. (1958). Configurational effects on the proton magnetic resonance spectra of 6-membered ring compounds. *J. Am. Chem. Soc.* 80: 6098–6105; see also: Praly, J.-P. and Lemieux, R.U. (1987). Influence of solvent on the magnitude of the anomeric effect. *Can. J. Chem.* 65: 213–223.

234 Schmidt, O.T. (1963). *Methods in Carbohydrate Chemistry*, vol. 2 (ed. R.L. Whistler and M.L. Wolfrom), 318. New York: Academic Press.

235 Gerber-Lemaire, S. and Vogel, P. (2009). Anomeric effects in pyranosides and related acetals. In: *Carbohydrate Chemistry*, vol. 35, 13–32. Wiley.

236 Juaristi, E., Bandala, Y., and Katritzky, A.R. (2012). Anomeric effect in saturated heterocyclic ring systems. 105: 189–222.

237 Juaristi, E. and Cuevas, G. (1992). Recent studies of the anomeric effect. *Tetrahedron* 48 (24): 5019–5087.

238 Juaristi, E., Tapia, J., and Mendez, R. (1986). Study of the anomeric effect in 2-substituted 1,3-dithianes. *Tetrahedron* 42 (5): 1253–1264.

239 Durette, P.L. and Horton, D. (1971). Conformational analysis of sugars and their derivatives. *Adv. Carbohydr. Chem.* 26: 49–125.

240 Horton, D., Durette, P.L., and Wander, J.D. (1993). Conformational properties of sugars and their derivatives. *Ann. N. Y. Acad. Sci.* 222: 884–914.

241 Hassanali, A.A., Cuny, J., Verdolino, V., and Parrinello, M. (2014). Aqueous solutions: state of the art in *ab initio* molecular dynamics. *Philos. Trans. R. Soc. A Math. Phys. Eng. Sci.* 372 (2011).

242 Molteni, C. and Parrinello, M. (1998). Glucose in aqueous solution by first principles molecular dynamics. *J. Am. Chem. Soc.* 120 (9): 2168–2171.

243 Momany, F.A., Appell, M., Willett, J.L., and Bosma, W.B. (2005). B3LYP/6-311++G** geometry-optimization study of pentahydrates of α- and β-D-glucopyranose. *Carbohydr. Res.* 340 (9): 1638–1655.

244 Tvaroska, I. and Kozar, T. (1983). Theoretical studies on the conformation of saccharides. 5. Hydration of the acetal segment in glycosides. *Int. J. Quantum Chem.* 23 (2): 765–778.

245 Kirby, A.J. (1987). Stereoelectronic effects on bond length and reactivity. *Pure Appl. Chem.* 59 (12): 1605–1612.

246 Booth, H. and Readshaw, S.A. (1990). Experimental studies of the anomeric effect. 4. Conformational equilibria due to ring inversion in tetrahydropyrans substituted at position-2 by the groups ethoxy, 2′-fluoroethoxy, 2′,2′-difluoroethoxy, and 2′, 2′, 2′-trifluoroethoxy. *Tetrahedron* 46 (6): 2097–2110.

247 Booth, H., Dixon, J.M., and Readshaw, S.A. (1992). Experimental studies of the anomeric effect. 5. The influence of some solvents on the conformational equilibria in 2-methoxy-tetrahydropyran and 2-(2′, 2′, 2′-trifluoroethoxy)-tetrahydropyran. *Tetrahedron* 48 (29): 6151–6160.

248 Perrin, C.L. (1995). Reverse anomeric effect – fact or fiction. *Tetrahedron* 51 (44): 11901–11935.

249 Booth, H., Dixon, J.M., and Khedhair, K.A. (1992). Experimental studies of the anomeric effect. 6. Ring inversion equilibria in cyclohexane, tetrahydropyran and piperidine rings substituted by a carbomethoxy or a cyano group. *Tetrahedron* 48 (29): 6161–6174.

250 Benn, M.H., Huang, Y.Y., Johannsen, F. et al. (2010). Concerning the conformational preferences of the 2-cyano derivatives of oxane, thiane, and selenane. *Can. J. Chem.* 88 (8): 831–838.

251 Zefirov, N.S., Blagoveshchenskii, V.S., Kazimirchik, I.V., and Yakovleva, O.P. (1971). Stereochemistry. 9. Conformation of 2-alkoxythiotetrahydrothiopyrans and 2-alkylthiotetrahydrothiopyrans. *Zh. Org. Khim.* 7 (3): 594–598.

252 Trapp, M.L., Watts, J.K., Weinberg, N., and Pinto, B.M. (2006). Component analysis of the X—C—Y anomeric effect (X = O, S ; Y = F, OMe, NHMe) by DFT molecular orbital calculations and natural bond orbital analysis. *Can. J. Chem.* 84 (4): 692–701.

253 Allinger, N.L., Hirsch, J.A., Miller, M.A., and Tyminski, I.J. (1969). Conformational analysis. 65. Calculation by Westheimer method of structures and energies of a variety of organic molecules containing nitrogen, oxygen and halogen. *J. Am. Chem. Soc.* 91 (2): 337–343.

254 Harpp, D.N. and Gleason, J.G. (1971). Conformational analysis of sulfur containing heterocycles – dipolar effect. *J. Org. Chem.* 36 (9): 1314–1316.

255 Buchanan, G.W., Sharma, N.K., Reinachhirtzbach, F.D., and Durst, T. (1977). Conformational analysis of substituted 1,2-oxathiane 2-oxides by ^{13}C and ^{1}H nuclear magnetic-resonance spectroscopy. *Can. J. Chem.* 55 (1): 44–49.

256 Zeinahnezhad, A., Nori-Shargh, D., Abbasi-Bakhtiari, Z., and Boggs, J.E. (2010). *Ab initio* and NBO analysis of the conformational properties of 1,2-oxathiane mono-S-oxide, 1,2-dithiane mono-S-oxide and 1,2-thiaselenane mono-S-oxide. *J. Mol. Struct. THEOCHEM* 947 (1–3): 52–57.

257 Markovic, D., Roversi, E., Scopelliti, R. et al. (2003). The hetero-Diels–Alder addition of sulfur dioxide: the pseudo-chair conformation of a 4,5-dialkylsultine. *Chem. Eur. J.* 9 (20): 4911–4915.

258 Roversi, E., Scopelliti, R., Solari, E. et al. (2002). The hetero-Diels–Alder addition of sulfur dioxide to 1-fluorobuta-1,3-dienes: the sofa conformations preferred by 6-fluorosultines (6-fluoro-3,6-dihydro-1,2-oxathiin-2-oxides) are stabilized by enthalpic and conformational anomeric effects. *Chem. Eur. J.* 8 (6): 1336–1355.

259 Dewar, M.J.S. (1969). *The Molecular Orbital Theory of Organic Chemistry*, McGraw-Hillr Series in Advanced Chemistry. New York: McGraw-Hill.

260 Baird, N.C. (1971). Dewar resonance energy. *J. Chem. Educ.* 48 (8): 509–514.

261 Haddon, R.C., Haddon, V.R., and Jackman, L.M. (1971). *Nuclear Magnetic Resonance Spectroscopy of Annulenes*, Topics in Current Chemistry, vol. 16, 103–220. Heidelberg: Springer-Verlag.

262 Haddon, R.C. (1979). Unified theory of resonance energies, ring currents, and aromatic character in the $(4N+2)\pi$-electron annulenes. *J. Am. Chem. Soc.* 101 (7): 1722–1728.

263 Bolton, W. (1965). Crystal structure of triketoindane (anhydrous ninhydrin). A structure showing close C=O⋯C interactions. *Acta Crystallogr.* 18: 5–10.

264 Bischof, P., Gleiter, R., and Hofmann, P. (1975). Photoelectron spectra of vicinal tricarbonyls. *Helv. Chim. Acta* 58 (7): 2130–2135.

265 Rubin, M.B. (1975). Chemistry of vicinal polyketones. *Chem. Rev.* 75 (2): 177–202.

266 Beck, E., Hofmann, P., and Sieber, A. (1981). The electron structure of vicinal triketones – bicyclo[3.1.0]hexane-2,3,4-trione. *Tetrahedron Lett.* 22 (47): 4683–4686.

267 Roth, W.R., Klärner, F.G., and Lennartz, H.W. (1980). Heats of hydrogenation. 2. Heats of hydrogenation of bicyclo[2.1.0]pent-2-ene, an antiaromatic system. *Chem. Ber.* 113 (5): 1818–1829.

268 Marhwald, R. (ed.) (2004). *Modern Aldol Reactions*, 1e, vol. 1 and 2, 1–679. Weinheim: Wiley-VCH.

269 (a) Scheffler, U. and Mahrwald, R. (2011). Asymmetric organocatalyzed direct aldol additions of enolizable aldehydes. *Synlett* (12): 1660–1667. (b) Mahrwald, R. (2012). Selected diastereoselective reactions: aldol type additions. In: *Comprehensive Chirality*, vol. 2 (ed. E.M. Carreira and H. Yamamoto), 370–397. Amsterdam: Elsevier. (c) Perrin, C.L. and Chang, K.-L. (2016). The complete mechanism of an aldol condensation. *J. Org. Chem.* 81: 5631–5635.

270 Green, M.M. and Wittcoff, H.A. (2003). *Organic Chemistry Principles and Industrial Practice*, 1–341. Weinheim: Wiley-VCH.

271 Mestres, R. (2004). A green look at the aldol reaction. *Green Chem.* 6: 583–560.

272 Wurtz, C.A. (1872). Sur un aldéhyde-alcohol. *Comptes rendus hebdomadaires des Séances de l'Académie des Sciences* Ref. 292b: *Molecules* 22(2), article number 292.

273 Wurtz, C.A. (1872). Über einen Aldehyd-Alkohol. *J. Prakt. Chem.* 5: 457–464.

274 Guthrie, J.P. (1974). Aldol condensation of acetaldehyde – equilibrium constant for reaction and rate constant for hydroxide-catalyzed retroaldol reaction. *Can. J. Chem.* 52 (11): 2037–2040.

275 Guthrie, J.P. (1978). Equilibrium constants for a series of simple aldol condensations, and linear free energy relations with other carbonyl addition reactions. *Can. J. Chem.* 56 (7): 962–973.

276 Guthrie, J.P. and Wang, X.P. (1991). The aldol condensation of acetophenone with acetone. *Can. J. Chem.* 69 (2): 339–344.

277 Guthrie, J.P. (1991). Rate equilibrium correlations for the aldol condensation – an analysis in terms of Marcus theory. *J. Am. Chem. Soc.* 113 (19): 7249–7255.

278 Arnett, E.M., Fisher, F.J., Nichols, M.A., and Ribeiro, A.A. (1989). Thermochemistry of a structurally defined aldol reaction. *J. Am. Chem. Soc.* 111 (2): 748–749.

279 Richard, J.P. and Nagorski, R.W. (1999). Mechanistic imperatives for catalysis of aldol addition reactions: partitioning of the enolate intermediate between reaction with Brønsted acids and the carbonyl group. *J. Am. Chem. Soc.* 121 (20): 4763–4770.

280 Adcock, J.L. and Luo, H. (1992). Synthesis and nucleophilic and photochemical reactions of F-adamantanone. *J. Org. Chem.* 57 (15): 4297–4300.

281 Wiberg, K.B., Morgan, K.M., and Maltz, H. (1994). Thermochemistry of carbonyl reactions. 6. A study of hydration equilibria. *J. Am. Chem. Soc.* 116 (24): 11067–11077.

282 Guthrie, J.P. and Guo, J.N. (1996). Intramolecular aldol condensations: rate and equilibrium constants. *J. Am. Chem. Soc.* 118 (46): 11472–11487.

283 Spohr, U., Hindsgaul, O., and Lemieux, R.U. (1985). Molecular recognition. 2. The binding of the Lewis-B and Y-human blood-group determinants by the lectin-IV of *Griffonia simplicifolia*. *Can. J. Chem.* 63 (10): 2644–2652.

284 Hindsgaul, O., Khare, D.P., Bach, M., and Lemieux, R.U. (1985). Molecular recognition. 3. The binding of the H-type-2 human-blood group determinant by the lectin-I of *Ulex europaeus*. *Can. J. Chem.* 63 (10): 2653–2658.

285 Lemieux, R.U., Du, M.H., Spohr, U. et al. (1994). Molecular recognition. 13. The binding of the H-type-2 human blood-group determinant by a winged bean (*Psophocarpus tetragonolobus*) acidic lectin. *Can. J. Chem.* 72 (1): 158–163.

286 Bülow, G., Plesner, I.W., and Bols, M. (2000). A large difference in the thermodynamics of binding of isofagomine and 1-deosoxynojirimycin to β-glucosidase. *J. Am. Chem. Soc.* 122 (35): 8567–8568.

287 Bergmann, W.R., Barkley, M.D., Hemingway, R.W., and Matticc, W.L. (1987). Heterogeneous fluorescence decay of (4-6)-linked and (4-8)-linked dimers of (+)-Catechin and (−)-epicatechin a result of rotational isomerism. *J. Am. Chem. Soc.* 109 (22): 6614–6619.

288 Thompson, R.S., Jacques, D., Haslam, E., and Tanner, R.J.N. (1972). Plant proanthocyanidins. 1. Introduction – isolation, structure, and distribution in nature of plant procyanidins. *J. Chem. Soc., Perkin Trans. 1* (11): 1387–1399.

289 Jacques, D., Opie, C.T., Porter, L.J., and Haslam, E. (1977). Plant proanthocyanidins. 4. Biosynthesis of procyanidins and observations on metabolism of cyanidin in plants. *J. Chem. Soc., Perkin Trans. 1* (14): 1637–1643.

290 Botha, J.J., Viviers, P.M., Young, D.A. et al. (1982). Synthesis of condensed tannins. 5. The 1st angular [4,6-4,8]-triflavanoids and their natural counterparts. *J. Chem. Soc., Perkin Trans. 1* (2): 527–533.

291 (a) Hemingway, R.W., Karchesy, J.J., McGraw, G.W., and Wielesek, R.A. (1983). Heterogeneity of interflavanoid bond location in loblolly pine bark procyanidins. *Phytochemistry* 22 (1): 275–281. (b) McRae, J.M. and Kennedy, J.A. (2011). Wine

and grape tannin interactions with salivary proteins and their impact on astringency: a review of current research. *Molecules* 16 (3): 2348–2364.

292 (a) Crozier, A., Jaganath, I.B., and Clifford, M.N. (2009). Dietary phenolics: chemistry, bioavailability and effects on health. *Nat. Prod. Rep.* 26 (8): 1001–1043. (b) Fernandes, I., Pérez-Gregorio, R., Soares, S. et al. (2017). Wine flavonoids in health and disease prevention. *Molecules* 22 (2): 292. (c) Zuniga-López, M.C., Laurie, V.F., Barriga-González, G. et al. (2017). Chemical and biological properties of phenolics in wine: analytical determinations and health benefits. *Curr. Org. Chem.* 21 (4): 357–367.

293 McRae, J.M., Dambergs, R.G., Kassara, S. et al. (2012). Phenolic compositions of 50 and 30 year sequences of Australian red wines: the impact of wine age. *J. Agric. Food Chem.* 60 (40): 10093–10102.

294 (a) de Beer, D., Joubert, E., Marais, J. et al. (2008). Characterisation of Pinotage wine during maturation on different oak products. *S. Afr. J. Enol. Vitic.* 29 (1): 39–49. (b) Watrelot, A.A., Badet-Murat, M.-L., and Waterhouse, A.L. (2018). Oak barrel tannin and toasting temperature: effects on red wine condensed tannin chemistry. *LWT Food Sci. Technol.* 91: 330–338.

295 (a) Brouillard, R., Chassaing, S., and Fougerousse, A. (2003). Why are grape/fresh wine Anthocyanins so simple and why is it that red wine color lasts so long? *Phytochemistry* 64 (7): 1179–1186. (b) Salas, E., Fulcrand, H., Meudec, E., and Cheynier, V. (2003). Reactions of anthocyanins and tannins in model solutions. *J. Agric. Food Chem.* 51 (27): 7951–7961.

296 Oliveira, J., Mateus, N., and de Freitas, V. (2017). Wine-inspired chemistry: anthocyanin transformations for a portfolio of natural colors. *Synlett* 28 (8): 898–906.

297 (a) Ribéreau-Gayon, P. (1974). The chemistry of red wine color. In: *Chemistry of Winemaking*, Chapter 3, Advances in Chemistry Series, vol. 137 (ed. A.D. Webb), 50–87. Washington, DC. (b) Tao, J.-X., Dykes, S.I., and Kilmartin, P.A. (2007). Effect of SO_2 concentration on polyphenol development during red wine micro-oxygenation. *J. Agric. Food Chem.* 55 (15): 6104–6109.

298 Brouillard, R. (1983). The in-vivo expression of anthocyanin color in plants. *Phytochemistry* 22 (6): 1311–1323.

299 Benson, S.W. (1968). *Thermochemical Kinetics Methods for the Estimation of Thermochemical Data and Rate Parameters*, 1–223. New York: Wiley.

300 Karpf, M. (1986). Organic synthesis at high temperatures – gas-phase flow thermolysis. *Angew. Chem. Int. Ed. Eng.* 25 (5): 414–430.

301 DePuy, C.H. and King, R.W. (1960). Pyrolytic *cis*-eliminations. *Chem. Rev.* 60 (5): 431–457.

302 Becker, K.B. and Pfluger, R.W. (1979). Facile synthesis of strained bridgehead olefins by a gas phase elimination reaction. *Tetrahedron Lett.* (39): 3713–3716.

303 Bailey, W.J. and Rosenberg, J. (1955). Cyclic dienes. 9. Synthesis of 1,2-dimethylene-4-cyclohexene. *J. Am. Chem. Soc.* 77 (1): 73–75.

304 Bailey, W.J. and Miller, S.S. (1963). Cyclic dienes. 28. 3,4-dimethylenetetrahydrofuran. *J. Org. Chem.* 28 (3): 802–805.

305 Bailey, W.J., Liao, C.W., and Coleman, G.H. (1955). Cyclic dienes. 10. 2,3-dimethylenedecalin. *J. Am. Chem. Soc.* 77 (4): 990–991.

306 Bailey, W.J. and Hudson, R.L. (1956). Cyclic dienes.15. 3-Methyl-1,2-dimethylenecyclohexane. *J. Am. Chem. Soc.* 78 (3): 670–672.

307 Bailey, W.J. and Quigley, S.T. (1959). Cyclic dienes. 24. 1,2-dimethylenedecalin. *J. Am. Chem. Soc.* 81 (21): 5598–5601.

308 Bailey, W.J., Rosenberg, J., and Young, L.J. (1955). Cyclic dienes. 13. Substituted 1,2-dimethylene-4-cyclohexenes. *J. Am. Chem. Soc.* 77 (5): 1163–1166.

309 Bailey, W.J., Hudson, R.L., and Liao, C.W. (1958). Cyclic dienes. 22. Substituted 1,2-dimethylene-4-cyclohexenes. *J. Am. Chem. Soc.* 80 (16): 4358–4360.

310 Bailey, W.J. and Lawson, W.B. (1957). Cyclic dienes.19. 2,3-dimethylenebicyclo[2.2.2]octane. *J. Am. Chem. Soc.* 79 (6): 1444–1447.

311 Alder, K. and Mölls, H.H. (1956). Über den sterischen Verlauf der dien-synthese von cycloheptadien-(1.3) mit Maleinsaure-Anhydrid – Ein Bietrag zur Kenntnis bicyclischer Systeme mit Brucken. *Chem. Ber.* 89 (8): 1960–1971.

312 Bailey, W.J. and Lawson, W.B. (1955). Cyclic dienes.14. 2,3-dimethylenebicyclo [2.2.1]heptane. *J. Am. Chem. Soc.* 77 (6): 1606–1608.

313 Bowe, M.A.P., Miller, R.G.J., Rose, J.B., and Wood, D.G.M. (1960). Preparation and polymerisation of some bicyclic dienes. *J. Chem. Soc.* 1541–1547.

314 Bailey, W.J., Fetter, E.J., and Economy, J. (1962). Cyclic dienes. 27. 1,2,4,5-tetramethylenecyclohexane. *J. Org. Chem.* 27 (10): 3479–3482.

315 Bailey, W.J. and Nielsen, N.A. (1962). Pyrolysis of esters. 23. 2,3-divinyl-1,3-butadiene. *J. Org. Chem.* 27 (9): 3088–3091.

316 Hurd, C.D. and Blunck, F.H. (1938). The pyrolysis of esters. *J. Am. Chem. Soc.* 60: 2419–2425.

317 Barton, D.H.R. and Rosenfelder, W.J. (1951). The stereochemistry of Steroids. 4. The concept of equatorial and polar bonds. *J. Chem. Soc.* 1048–1054.

318 Barton, D.H.R. (1949). *cis*-Elimination in thermal decompositions. *J. Chem. Soc.* 2174–2178.

319 Tschugaeff, L. (1899). New method for preparing unsaturated hydrocarbons. *Ber. Dtsch. Chem. Ges.* 32: 3332–3335.

320 Nace, H.R. (1962). The preparation of olefins by the pyrolysis of xanthates. The Chugaev reaction. *Org. React.* 12: 57–100.

321 Gerlach, H. (1972). Synthesis of tricyclo[4.4.1.113,8]dodecane-4,9-dione and tricyclo[4.4.1.113,8]dodeca-4,9-diene. *Helv. Chim. Acta* 55 (8): 2962–2964.

322 De Groot, A., Evenhuis, B., and Wynberg, H. (1968). Syntheses and properties of sterically hindered butadienes. A modification of Chugaev reaction. *J. Org. Chem.* 33 (6): 2214–2217.

323 Eschenmoser, A., Felix, D., and Ohloff, G. (1967). Eine neuartige Fragmentierung Cyclischer α,β-ungesattigter Carbonylsysteme – synthese von Exalton und rac-Muscon aus Cyclododecanon. *Helv. Chim. Acta* 50 (2): 708–713.

324 Tanabe, M., Crowe, D.F., Dehn, R.L., and Detre, G. (1967). Synthesis of secosteroid acetylenic ketones. *Tetrahedron Lett.* (38): 3739–3743.

325 Reese, C.B. and Sanders, H.P. (1981). Use of mesitylene-2-sulfonyl hydrazide in the synthesis of medium-ring cycloalkynones by the Eschenmoser fragmentation reaction. *Synthesis* (4): 276–278.

326 Nakajima, R., Ogino, T., Yokoshima, S., and Fukuyama, T. (2010). Total synthesis of (−)-mersicarpine. *J. Am. Chem. Soc.* 132 (4): 1236–1239.

327 Ruzicka, L., Stoll, M., and Schinz, H. (1926). On the understanding of carbon rings II. Synthesis of carbocyclic ketones with from ten up to eighteen rings. *Helv. Chim. Acta* 9: 250–264.

328 Felix, D., Müller, R.K., Horn, U. et al. (1972). Methods F Synthesis. 4. α,β-epoxyketone-alkynone fragmentation. 2. Pyrolytic decay of hydrazones from α,β-epoxyketones and N-amino-aziridines. *Helv. Chim. Acta* 55 (4): 1276–1319.

329 Ohloff, G. (1957). Zur basenkatalysierten Dehydratisierung. 1. Eine neue Methode zur Darstellung Aliphatischer 1,3-Diene der Terpenreihe. *Chem. Ber.* 90 (8): 1554–1559.

330 Hart, H. and Fish, R.W. (1960). A novel synthesis of benzocyclobutene derivatives. *J. Am. Chem. Soc.* 82 (3): 749–750.

331 Hart, H., Hartlage, J.A., Fish, R.W., and Rafos, R.R. (1966). Synthesis of benzocyclobutenes from trichloromethylbenzenes. *J. Org. Chem.* 31 (7): 2244–2250.

332 Cava, M.P. and Napier, D.R. (1956). Benzocyclobutene and benzocyclobutadiene dimer. *J. Am. Chem. Soc.* 78 (2): 500–500.

333 Sanders, A. and Giering, W.P. (1973). New synthesis of benzocyclobutene. *J. Org. Chem.* 38 (17): 3055–3055.

334 Kambe, N., Tsukamoto, T., Miyoshi, N. et al. (1986). Ortho-quinodimethane generation from α,α′-dihalo-ortho-xylenes by use of sodium benzenetellurolate. *Bull. Chem. Soc. Jpn.* 59 (10): 3013–3018.

335 Schiess, P. and Heitzmann, M. (1977). Formation of benzocyclobutenone by gas phase pyrolysis. *Angew. Chem. Int. Ed. Eng.* 16 (7): 469–470.

336 Thummel, R.P. (1980). Benzocyclobutene and related compounds. *Acc. Chem. Res.* 13 (3): 70–76.

337 Laduranty, J., Le Page, L., and Lepage, Y. (1980). Capture of para-dimethoxyorthoquinodimethane – synthesis of quinones and an daunomycinone intermediate. *Can. J. Chem.* 58 (12): 1161–1167.

338 Oppolzer, W. (1978). Intramolecular cycloaddition reactions of ortho-quinodimethanes in organic synthesis. *Synthesis* (11): 793–802.

339 Mann, J. and Piper, S.E. (1982). The preparation of lignans via intermolecular cyclo-additions with quinodimethanes. *J. Chem. Soc. Chem. Commun.* (8): 430–432.

340 Shishido, K., Hiroya, K., Komatsu, H. et al. (1986). The total synthesis of (+/−)-geneserine. *J. Chem. Soc. Chem. Commun.* (12): 904–905.

341 Azadi-Ardakani, M. and Wallace, T.W. (1983). 3,6-Dimethoxybenzocyclobutenone – a versatile quinone precursor. *Tetrahedron Lett.* 24 (17): 1829–1832.

342 Gleicher, G.J., Newkirk, D.D., and Arnold, J.C. (1973). Calculations on quinonoid compounds. 1. Quinododimethides. *J. Am. Chem. Soc.* 95 (8): 2526–2531.

343 Aihara, J. (1976). New definition of Dewar-type resonance energies. *J. Am. Chem. Soc.* 98 (10): 2750–2758.

344 Schiess, P. and Heitzmann, M. (1978). Hexakis(methylidene)cyclohexane ([6]radialene) –

chemical and spectral properties. *Helv. Chim. Acta* 61 (2): 844–847.

345 Harruff, L.G., Brown, M., and Böckelheide, V. (1978). Hexaradialene – precursors and structure. *J. Am. Chem. Soc.* 100 (9): 2893–2894.

346 Trahanovsky, W.S. and Cassady, T.J. (1984). Preparation of furanoradialene by the flash vacuum pyrolysis of diesters of 3,4-bis(hydroxymethyl)-2,5-dimethylfuran. *J. Am. Chem. Soc.* 106 (26): 8197–8201.

347 Letulle, M., Guenot, P., and Ripoll, J.L. (1991). The syntheses of 6-methylene-2,4-cyclohexadien-1-imine and related ortho-quinonoids by FVT of 1-hetero-1,2,3,4-tetrahydronaphthalenes. *Tetrahedron Lett.* 32 (18): 2013–2016.

348 Sander, W. and Morawietz, J. (1993). Matrix isolation of a reactive o-quinoid compound – 6-methylene-2,4-cyclohexadiene-1-imine. *Tetrahedron Lett.* 34 (12): 1913–1916.

349 (a) Sekine, Y., Brown, M., and Böckelheide, V. (1979). [2.2.2.2.2.2](1,2,3,4,5,6)cyclophane – superphane. *J. Am. Chem. Soc.* 101 (11): 3126–3127. (b) Böckelheide, V. (1980). [2N]Cyclophanes – paracyclophane to superphane. *Acc. Chem. Res.* 13 (3): 65–70.

350 Neuschwander, B. and Böckelheide, V. (1980). 2 Syntheses of [2.2.2](1,2,3)cyclophane. *Isr. J. Chem.* 20 (3–4): 288–290.

351 Eaton, B.E., Laganis, E.D., and Böckelheide, V. (1981). Gas phase pyrolytic formation and dimerization of benzocyclobutenes – synthesis of [2-4](1,2,4,5)cyclophane. *Proc. Natl. Acad. Sci. U. S. A.* 78 (11): 6564–6566.

352 Kang, H.C. and Böckelheide, V. (1984). Study of the syntheses and properties of 5,12-diaza[24](1,2,4,5)cyclophane and 5,15-diaza[24](1,2,4,5)cyclophane. *J. Am. Chem. Soc.* 106 (9): 2672–2680.

353 Gleiter, R. and Roers, R. (2004). Superphanes. In: *Modern Cyclophane Chemistry* (ed. R. Gleiter and H. Hopf), 105–129. Weinheim: Wiley-VCH.

354 Hopfner, T., Jones, P.G., Ahrens, B. et al. (2003). Alicyclic chemistry – 8 – [6]Radialenes revisited. *Eur. J. Org. Chem.* (14): 2596–2611.

355 Hart, H., Jeffares, M., Teuerstein, A., and Ward, D.L. (1978). Evidence for naphtharadialene (1,2,3,4,5,6,7,8-octakis(methylene)-δ-9-octalin). *J. Am. Chem. Soc.* 100 (25): 8012–8013.

356 Hart, H., Teuerstein, A., Jeffares, M. et al. (1980). Naphtharadialene, related polyenes and naphthocyclobutenes. *J. Org. Chem.* 45 (19): 3731–3735.

357 Vogel, P. and Hardy, M. (1974). Pyrolysis of 2,3-dimethylidene-7-oxabicyclo[2.2.1]heptane. *Helv. Chim. Acta* 57 (1): 196–199.

358 Ripoll, J.L., Rouessac, A., and Rouessac, F. (1978). Recent applications of retro-Diels–Alder reaction in organic synthesis. *Tetrahedron* 34 (1): 19–40.

359 Ripoll, J.L. (1979). Retrodiene reactions. 7. General approach to thermodynamically unstable enols by flash thermolysis of 11-hydroxy-9,10-ethano-9,10-dihydroanthracene. *New J. Chem.* 3 (3): 195–198.

360 Hussmann, G., Wulff, W.D., and Barton, T.J. (1983). Direct thermal and photochemical generation of silanones. *J. Am. Chem. Soc.* 105 (5): 1263–1269.

361 Stork, G. and Guthikon, R.N. (1972). General synthesis of esters of acryloyl acetic acid and their homologs. *Tetrahedron Lett.* (27): 2755–2758.

362 Wenkert, E., Afonso, A., Bredenberg, J.B. et al. (1964). Synthesis of some resin acids. *J. Am. Chem. Soc.* 86 (10): 2038–2043.

363 Trahanovsky, W.S., Cassady, T.J., and Woods, T.L. (1981). Preparation of 2,3-dimethylene-2,3-dihydrofurans by the flash vacuum pyrolysis of substituted furylmethyl esters. *J. Am. Chem. Soc.* 103 (22): 6691–6695.

364 Chou, C.H. and Trahanovsky, W.S. (1986). The (4+4)-dimerization of 2,3-dimethylene-2,3-dihydrofuran – secondary deuterium kinetic isotope effect evidence for a 2-step mechanism. *J. Am. Chem. Soc.* 108 (14): 4138–4144.

365 Olah, G.A., Mathew, T., Marinez, E.R. et al. (2001). Acid-catalyzed isomerization of pivalaldehyde to methyl isopropyl ketone via a reactive protosolvated carboxonium ion intermediate. *J. Am. Chem. Soc.* 123 (47): 11556–11561.

3

Rates of chemical reactions

3.1 Introduction

Chemical kinetics describe and quantify the rates of chemical reactions [1–6]. These rates depend on a variety of factors such as temperature, pressure, concentration of the reactants and products, the presence or absence of catalysts and inhibitors, the nature of the surface of the reaction vessel, and its volume/surface ratio. A standard kinetics measurement involves the variation of the concentration or pressure of reactants and products as a function of time at a constant temperature (isothermal kinetics). The reaction progress can also be followed by calorimetry that measures the heat flow during the reaction. Analysis of nonisothermal kinetics is also possible [7–9]. In 1846, Dubrunfaut observed that the optical rotation angle varied with time when a cold aqueous solution of D-glucose was freshly prepared [10]. The reaction (called mutarotation or anomerization of aldoses) results from the balanced reaction involving α- and β-anomers (α-D-glucopyranose ⇄ β-D-glucopyranose, see Section 2.7.10, Figure 2.9, Table 2.3) [11]. In 1864, Guldberg and Waage pioneered the development of chemical kinetics by formulating the law of mass action, which states that the reaction rate is proportional to the quantity of the reacting substances [12]. For equilibrium (1.1), the variations of the concentration of reactants and products during a time interval, dt, are given by the degree of reaction progress $d\delta$ (3.1).

$$\alpha\mathbf{A} + \beta\mathbf{B} + \cdots \overset{K}{\rightleftharpoons} \pi\mathbf{P} + \theta\mathbf{Q} + \cdots \tag{1.1}$$

$$d\delta = -\frac{d[\mathbf{A}]}{\alpha} = -\frac{d[\mathbf{B}]}{\beta} = \cdots = +\frac{d[\mathbf{P}]}{\pi} = +\frac{d[\mathbf{Q}]}{\theta} \tag{3.1}$$

The rate of a reaction that transforms reactants **A**, **B**, … into products **P**, **Q**, … is defined by Eq. (3.2).

$$\underbrace{-\frac{1}{\alpha}\frac{d[\mathbf{A}]}{dt} = -\frac{1}{\beta}\frac{d[\mathbf{B}]}{dt} =}_{\text{rates of disappearance of reactants}} \cdots = \underbrace{+\frac{1}{\pi}\frac{d[\mathbf{P}]}{dt} = \frac{1}{\theta}\frac{d[\mathbf{Q}]}{dt}}_{\text{rates of appearance of products}} = \cdots \tag{3.2}$$

In the case of reaction **A** → **P**, the rate of concentration change of [**A**] and [**P**] during time interval t is proportional to the slope the line tangent to the concentration vs. t curve (Figure 3.1). The rate of concentration change depends on time, except for reactions of zero order (see below), and approaches zero as the reaction approaches equilibrium **A** ⇄ **P**. Chemical reactions can have one step or consist of a sequence of single-step processes called elementary processes, elementary reactions, or elementary steps. Elementary processes typically involve the collisions between two reactant molecules in gas or solution phase. In solution, the collision can take place between one molecule of reactant and one molecule of solvent (bimolecular step). At low pressure, gas-phase reactions involve collisions between reactant molecules and the wall of the reaction vessel (Section 3.2.2). Only under high pressure may a termolecular step (simultaneous collision of three molecules) occur.

3.2 Differential and integrated rate laws

For the uncatalyzed reactions that convert reactants **A**, **B**, … into products **P**, **Q**, …, the rate of disappearance of reactant **A** may depend on the concentrations of **A**, **B**, … and **P**, **Q**, … Equation (3.3) is called the

Organic Chemistry: Theory, Reactivity and Mechanisms in Modern Synthesis, First Edition. Pierre Vogel and Kendall N. Houk.

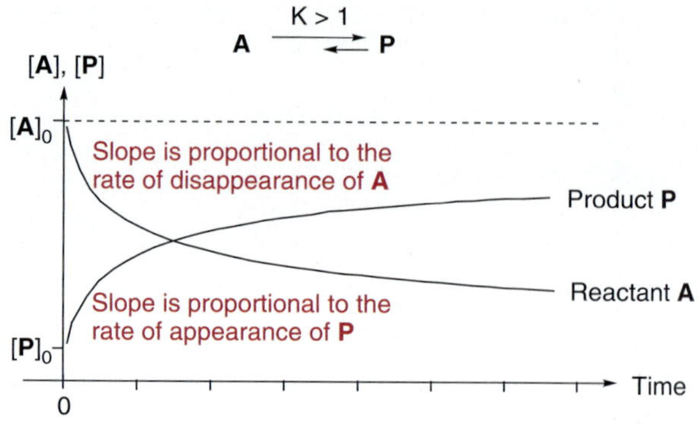

Figure 3.1 Example of a kinetic measurement of reaction-converting reactant **A** into product **P**, $[\mathbf{A}]_0$ = initial concentration of **A**, $[\mathbf{P}]_0$ = initial concentration of **P**.

rate law of the forward reaction in which k_1 is the rate constant (not the Boltzmann constant k_b!). The rate constant for the reverse reaction is defined as k_{-1}, where **P** and **Q** are the reactants and **A** and **B** are the products. As demonstrated in Section 3.2.4, $K(1.1) = k_1/k_{-1}$.

$$-\frac{1}{\alpha}\frac{d[\mathbf{A}]}{dt} = k_1[\mathbf{A}]^a[\mathbf{B}]^b \cdots [\mathbf{P}]^p[\mathbf{Q}]^q \cdots \qquad (3.3)$$

The rate law may be much more complicated than Eq. (3.3). For instance, for the cheletropic addition of sulfur dioxide to (*E*)-1-methoxybutadiene (**A**) to give sulfolene (**P**) (reaction (3.4)), the rate law (3.5) has been derived from the measured concentration of **A** as a function of time and concentration of reactants: three different contributions (with rate constants k_1, k_2, and k_{-1}) are needed to account for the results [13].

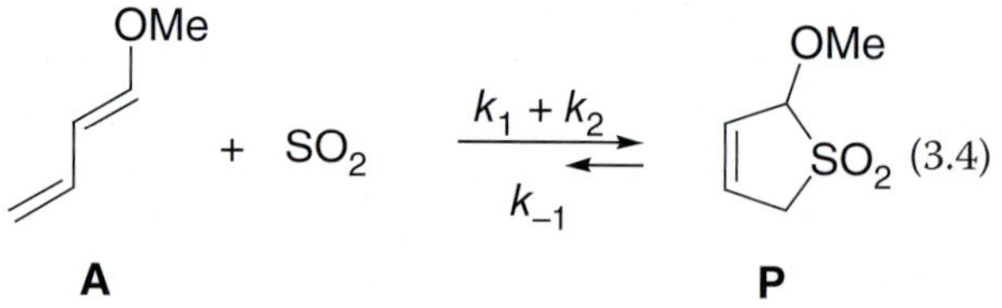

$$-\frac{d[\mathbf{A}]}{dt} = k_1[\mathbf{A}][SO_2]^3 + k_2[\mathbf{A}][SO_2]^2 - k_{-1}[\mathbf{P}][SO_2] \qquad (3.5)$$

3.2.1 Order of reactions

In the case of rate law (3.3), the overall order of the reaction is defined as the sum of the exponents $\rho = a + b + \cdots + p + q + \cdots$. The order of the reaction with respect to a particular species is defined as the power ($a, b, \ldots p, q, \ldots$) to which its concentration is raised in the rate law. The rate law gives clues to the reaction mechanism by eliminating many possibilities from consideration. Please note that the overall order of a reaction cannot be obtained simply from its stoichiometry. For instance, the burning of hydrogen with bromine to produce 2 equiv. of hydrogen bromide (reaction (3.6)) is not second order because the reaction consumes one molecule of H_2 and one molecule of Br_2. In fact, the rate law of the forward reaction is determined experimentally according to Eq. (3.7). It depends on the concentration of H_2 and on the square root of the concentration of Br_2, as shown in Eq. (3.7).

$$H_2 + Br_2 \rightleftarrows 2\,HBr \qquad (3.6)$$

$$\frac{d[HBr]}{2dt} = k_1[H_2][Br_2]^{1/2} \qquad (3.7)$$

One method to determine the order of a given reaction is the method of initial slopes. If $[Br_2]$ is constant and the initial rate of formation of HBr is measured, the production of HBr will depend only on the initial concentration of hydrogen $[H_2]_0$. By doubling $[H_2]_0$, the rate of formation of HBr would double. However, fixing $[H_2]$ and monitoring the initial rate of formation of HBr results in a rate increase of $\sqrt{2}$-fold increase when the concentration of Br_2 is doubled, instead of a doubling of the rate.

A common technique to determine the reaction order relies on computer line-fitting. The computer simulates a number of different kinetic expressions and fits these to experimental data. An example concentration vs. time plots for reactions of zero order (Figure 3.2), first order (Figure 3.3), and second order (Figure 3.4) are presented below.

Other kinetics laws are given in Table 3.1.

A very important consequence of the order of the reaction is that the time required to reach 99% of conversion increases with the order. For

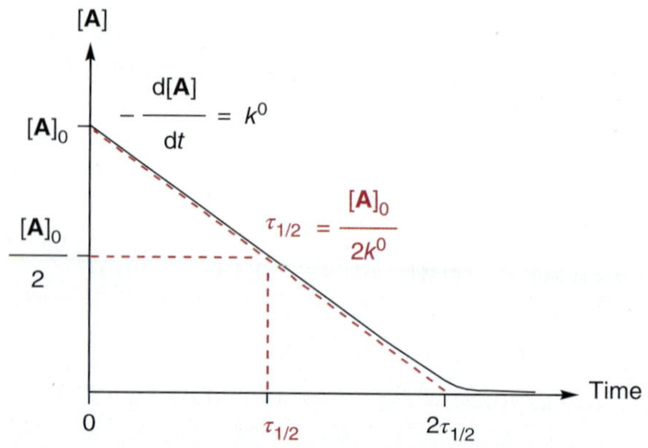

Figure 3.2 Kinetics of an irreversible reaction (reactant **A** is converted into 100% product **P**) following a zero-order rate law with respect to [**A**], at least until 80% conversion.

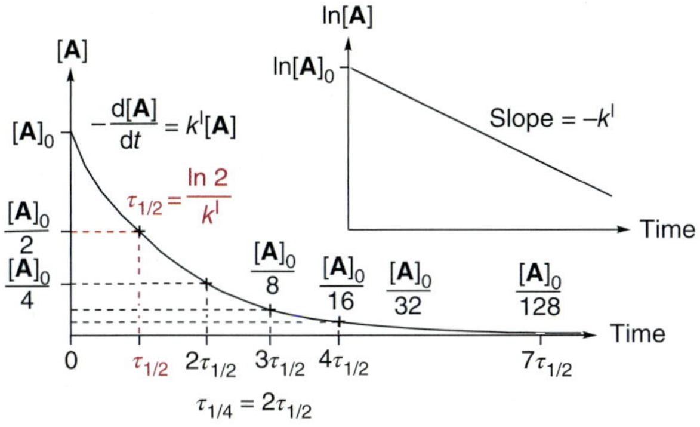

Figure 3.3 Kinetics of an irreversible reaction following a first-order rate law with respect to [**A**].

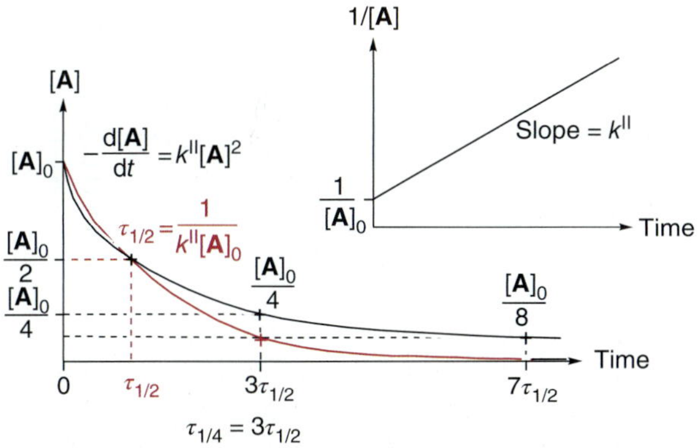

Figure 3.4 Kinetics of an irreversible reaction with a second-order rate law with respect to the concentration of reactant **A**. The kinetics for a reaction following a first-order rate law with the same half-life and initial concentration of **A** is given with the red line.

reactions with the same half-life (time to reach 50% conversion), a zero-order reaction is complete after two half-lives. First-order reactions require about 7 half-lives, whereas a second-order reaction requires about 100 half-lives to achieve 99% completion (Table 3.1). The order of the reaction plays a key role for chemical engineers developing processes in industry. Oftentimes, the desired reaction is second order, whereas the undesired decomposition or side product formation follows a lower order reaction. One solution to this problem is to convert a second-order process, $\mathbf{A}+\mathbf{B}\rightarrow\mathbf{P}$, with a rate law $d[\mathbf{A}]/dt = -k^{II}[\mathbf{A}][\mathbf{B}]$ into a pseudo-first rate law reaction (3.8) by using a large excess of one of the two reactants, e.g.:

$$-\,d[\mathbf{A}]/dt = -k^{II}[\mathbf{B}][\mathbf{A}] \cong k'[\mathbf{A}] \quad \text{with}$$
$$k' = k^{II}[\mathbf{B}] \cong \text{ constant if } [\mathbf{B}] \gg [\mathbf{A}] \qquad (3.8)$$

For industrial purposes, the excess reactant should be the cheapest or one that can be recovered readily.

Problem 3.1 Estimate the time for 99% conversion of $\mathbf{A}\rightarrow\mathbf{P}$, an irreversible reaction following a first-order rate law with $k = 10^{-3}\ \mathrm{s^{-1}}$. What is the half-life of this reaction?

Problem 3.2 Estimate the time for 99% conversion of reaction $\mathbf{A}+\mathbf{B}\rightarrow\mathbf{P}$ following a first-order rate law in [**A**] and a first-order rate law in [**B**]. A 20-fold excess of **B** is used and $k = 10^{-3}\ \mathrm{dm^3\ mol^{-1}\ s^{-1}}$. Estimate the half-life of this reaction.

3.2.2 Molecularity and reaction mechanisms

As shown for the $Br_2 + H_2$ reaction (Eq. (3.6)), many rate laws do not parallel the overall stoichiometry. The macroscopic rate law is the consequence of a mechanism consisting of one or more elementary steps, each one of which describes a process that takes place on the microscopic level. If the reaction implies only one elementary step (microscopic level), the order of the reaction rate law (macroscopic level) is often defined as the molecularity of the reaction. Thus, it is commonly said that a unimolecular reaction follows the first-order rate law, a bimolecular reaction follows the second-order rate law, and a termolecular reaction follows the third-order rate law. Reactions occur when molecules collide with one another or of one molecule of the reactants with the wall of the reaction vessel. Unimolecular reactions involve the reaction of a single molecule that has been activated by collisions with the wall of the vessel (gas-phase reactions) or with solvent molecules (reactions in solutions): The Lindemann–Hinshelwood mechanism is the simplest theory of unimolecular reaction rates [14–17]. Molecule **A** collides with the vessel wall (**M**), the activation step, followed by a rate-determining unimolecular step.

$$\mathbf{A} + \mathbf{M} \underset{k_{-1}}{\overset{k_1}{\rightleftharpoons}} \mathbf{A}^* + \mathbf{M}$$

$$\mathbf{A}^* \xrightarrow{k_2} \mathbf{P}$$

Applying the steady-state approximation (Section 3.2.6) to [**A***], one obtains

$$[\mathbf{A}^*] = \frac{k_1[\mathbf{A}][\mathbf{M}]}{k_{-1}[\mathbf{M}] + k_2}$$

Table 3.1 Rate laws for simple isothermal, irreversible reactions (with $K > 100$).

Order ρ Stoichiometry	Differential equations	Integrated equations	Units	$\tau_{1/2}$	$\tau_{1/4}$	Time for 99% conversion
0	$-\frac{dA}{dt} = k^0 = \frac{dP}{dt}$	$A_0 - A = P = k^0 t$ (for $P_0 = 0$)	$M\ s^{-1}$	$\frac{A_0}{2k^0}$	$\frac{3}{2}\tau_{1/2}$	$<2\ \tau_{1/2}$
X + A → P			$mol\ dm^{-3}s^{-1}$			
1	$-\frac{dA}{dt} = k^{I}A$	$A = A_0 e^{-kt}$		$\frac{\ln 2}{k^{I}}$	$2\tau_{1/2}$	c. $7\ \tau_{1/2}$
A → P		$\log A = \log A_0 \frac{k^{I}t}{2.303}$	s^{-1}			
	$-\frac{dA}{dt} = k^{I}[A_0 - P]$	$P = A_0(1 - e^{-kt})$				
2	$-\frac{dA}{dt} = k^{II}A^2$	$\frac{1}{A} - \frac{1}{A_0} = k^{II}t$	$M^{-1}\ s$	$\frac{1}{k^{II}A_0}$	$3\tau_{1/2}$	c. $99\ \tau_{1/2}$
2A → P			$dm^3\ mol^{-1}\ s^{-1}$			
	$\frac{dP}{dt} = k^{II}\frac{1}{2}[A_0 - 2P]^2$	$\frac{1}{A_0 - 2P} - \frac{1}{A_0} = k^{II}t$		For A		
A + B → P (A ≠ B)	$-\frac{dA}{dt} = k^{II}A[B_0 - A_0 + A]$	$\log\frac{A + B_0 - A_0}{A} = \frac{B_0 - A_0}{2.303}k^{II}t + \log\frac{B_0}{A_0}$		$\frac{2.303}{k^{II}(A_0 - B_0)}\log\frac{B_0}{2B_0 - A_0}$		
	$\frac{dP}{dt} = k^{II}[A_0 - P][B_0 - P]$	$\log\frac{B_0 - P}{A_0 - P} = \frac{B_0 - A_0}{2.303}k^{II}t + \log\frac{B_0}{A_0}$				
3	$-\frac{dA}{dt} = k^{III}A^3$	$\frac{1}{A^2} - \frac{1}{A_0^2} = 2k^{III}t$	$M^{-2}\ s^{-1}$	$\frac{3}{2}\frac{1}{k^{III}A_0^2}$	$5\tau_{1/2}$	c. $350\,000\ \tau_{1/2}$
3A → P			$dm^6\ mol^{-2}s^{-1}$			
	$\frac{dP}{dt} = \frac{k^{III}}{3}(A_0 - 3P)^3$	$\frac{1}{(A_0 - 3P)^2} - \frac{1}{A^2} = 2k^{III}t$				
2A + B → P (A ≠ 2B)	$-\frac{dA}{dt} = k^{III}A^2\left(B_0\frac{A_0}{2} + \frac{A}{2}\right)$	$\frac{2}{2B_0 - A_0}\left(\frac{1}{A} - \frac{1}{A_0}\right) + \frac{2(2.303)}{(2B_0 - A_0)^2}\log\frac{B_0 A}{A_0 B} = k^{III}t$				
A + B + C → P (A ≠ B ≠ C)	$-\frac{dA}{dt} = k^{III}A(B_0 - A_0 + A)(C_0 - A_0 + A)$	$\frac{2.303}{(A_0 - B_0)(B_0 - C_0)(C_0 - A_0)}\log\left(\frac{A}{A_0}\right)^{(B_0 - C_0)}\cdot\left(\frac{B}{B_0}\right)^{(C_0 - A_0)}\left(\frac{C}{C_0}\right)^{(A_0 - B_0)} = k^{III}t$				
ρ	$-\frac{dA}{dt} = k^{\rho}A^{\rho}$	$1/A^{(\rho-1)} - 1/A_0^{(\rho-1)} = (\rho - 1)k^{\rho}t$	$\frac{(dm^3)^{(\rho-1)}}{mol^{(\rho-1)}\ s}$	$\frac{2^{(\rho-1)} - 1}{k(\rho - 1)A_0^{(\rho-1)}}$	$[1 + 2(\rho - 1)]\tau_{1/2}$	
ρA → P ($\rho \neq 1$)	$\frac{dP}{dt} = \frac{k^{\rho}(A_0 - \rho P)^{\rho}}{\rho}$	$1/(A_0 - \rho P)^{(\rho-1)} - 1/A_0^{(\rho-1)} = (\rho - 1)k^{\rho}t$				

Rate constant k, $\tau_{1/2}$ = half-life = time for 50% of conversion, $\tau_{1/4}$ = time for 75% of conversion. Reactants are **A**, **B**, and **C**, product is **P**, with their respective concentrations (reactions in solution) or partial pressure (reactions in the gas phase) [**A**] = A, [**B**] = B, [**C**] = C, and [**P**] = P. At time $t = 0$, concentrations or partial pressures are A_0, B_0, C_0, and P_0, respectively.

For a zeroth order reaction in [**A**] ($d[\mathbf{A}]/dt = -k^0$) with $k^0 = 0.001\ mol\ dm^3\ s^{-1} = 0.001\ M\ s^{-1}$, time for 50% conversion $\tau_{1/2} = 693$ s if the initial concentration of reactant **A** is $[\mathbf{A}]_0 = A_0 = 2\ln 2 = 1.386\ mol\ dm^{-3} = 1.386$ M.

For a first-order reaction in [**A**] ($d[\mathbf{A}]/dt = -k^{I}[\mathbf{A}]$) with $k^{I} = 0.001\ s^{-1}$ and $\tau_{1/2} = 693$ s = 11 min and 33 s.

For the dimerization of reactant **A** that is second order in [**A**] ($d[\mathbf{A}]/dt = -k^{II}[\mathbf{A}]^2$) with $k^{II} = 0.001\ dm^3\ mol^{-1}\ s^{-1} = 0.001\ M^{-1}\ s^{-1}$ $\tau_{1/2} = 693$ s if $1/[\mathbf{A}]_0 = 1/A_0 = \ln 2\ M^{-1} = 0.693\ M^{-1}$.

The overall rate of formation of product **P** is thus

$$\frac{d[\mathbf{P}]}{dt} = k_2[\mathbf{A}^*] = \frac{k_2k_1[\mathbf{A}][\mathbf{M}]}{k_{-1}[\mathbf{M}] + k_2} = k_{eff}[\mathbf{A}]$$

where $k_{eff} = \dfrac{k_2k_1[\mathbf{M}]}{k_{-1}[\mathbf{M}] + k_2}$ is an effective first-order rate constant

$$\mathbf{A} \rightarrow \text{products} \tag{3.9}$$

The number of product molecules produced per time unit will be proportional to the number of reactants, **A**, $d[\text{products}]/dt = -d[\mathbf{A}]/dt = k[\mathbf{A}]$. Unimolecular reactions are first order ($\rho = 1$). Bimolecular reactions involve two molecules in the rate-determining step:

$$\mathbf{A} + \mathbf{B} \rightarrow \text{products} \tag{3.10}$$

The number of products formed per time unit (typically seconds) will be directly proportional to the number of collisions between **A** and **B**. The rate of this bimolecular reaction will be proportional to the product of the concentrations of **A** and **B**: $d[\text{products}]/dt = k[\mathbf{A}][\mathbf{B}]$. Bimolecular reactions are second order ($\rho = 2$).

Termolecular reactions are encountered much less frequently than unimolecular or bimolecular reactions but are important at high pressures or in the condensed phase. A termolecular reaction is one that requires the combination of three species, usually in two consecutive collisions.

$$\mathbf{A} + \mathbf{B} + \mathbf{C} \rightarrow \text{products} \tag{3.11}$$

The number of termolecular collisions will be proportional to $[\mathbf{A}][\mathbf{B}][\mathbf{C}]$, so that the termolecular reactions are third order ($\rho = 3$). The termolecular process might start with a "sticky" bimolecular collision one in which the collision pair, **A** and **B**, is held together by weak intermolecular forces for a finite period. If the pair is struck during this period by the third reactant **C**, then products are formed; if not, then the reactants **A** and **B** separate. Whether one regards the overall process as two consecutive bimolecular collisions or as a termolecular one depends, in part, on the length of time that **A** and **B** spend together. If the time is longer than 1–100 stretching vibrations (10^{-13} to 10^{-11} seconds), the mechanism is considered to be termolecular. Catalytic reactions can be three component. For instance, reactants **A** and **B** equilibrate first with a complex **A—B*** (**A** + **B** ⇄ **A—B***), which interacts then with the catalyst **C** (**A—B*** + **C** ⇄ **A—B—C***). The rate-determining step might be the formation of products (**A—B*** + catalyst → products + catalyst) [18]. Reactions of higher molecularity are not observed because of very small probability of simultaneous interaction between four or more molecules.

3.2.3 Examples of zero order reactions

The thermal decomposition of *ortho*-benzenediazonium carboxylate (**1**) into 1,2-didehydrobenzene (benzyne) follows a zero-order rate law. The rate of reaction is constant, regardless of the concentration of **1**! The reactant is insoluble in most organic solvents because of its zwitterionic (highly polar) nature. The rate of reaction can be followed by measuring the volume of the gaseous products formed ($N_2 + CO_2$). The rate laws are zero order in $CHCl_3$, THF (tetrahydrofuran), and benzene, with rate constants $k^0 = 54 \times 10^{-7}$, 6.8×10^{-7}, and 1.7×10^{-7} M s^{-1}, respectively, at 41 °C. The highly reactive benzyne reacts with furan via a Diels reaction (Sections 5.3.9–5.3.16), which yields 7-oxabenzobicyclo[2.2.1]hepta-2,5-diene (**2**) (reaction (3.12)). In these solvents, the rates of the reactions are limited by the solubility of zwitterion **1**. This is verified by running reaction (3.12) in 2-methoxyethanol. **1** dissolves completely in this solvent at 41 °C, and the appearance of $CO_2 + N_2$ follows a first-order rate law with $k^1 = 2 \times 10^{-3}$ s^{-1}.

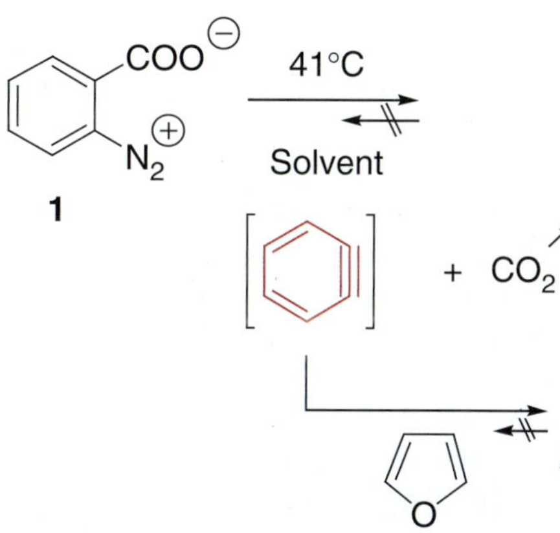

(3.12)

Simple photochemical reactions (Chapter 6) of type (3.13) are zero order if the flux of absorbed light J is constant (does not vary with time). Reactions involve excited state **A***, for which the concentration is relatively small and constant.

$$\mathbf{A} \underset{k'}{\overset{h\nu(J)}{\rightleftharpoons}} [\mathbf{A}^*] \xrightarrow{k} \text{Products} \tag{3.13}$$

Quantum yield Ø is defined as the number of molecules of photoproduct produced divided by the number of photons absorbed (Section 6.3). The integrated form for the rate law becomes

$$[\mathbf{A}] - [\mathbf{A}]_o = J\text{Ø}t = k^0t \tag{3.14}$$

[A] is the concentration of A at time t and $[A]_0$ is the initial concentration of A at time $t = 0$.

Catalyzed reactions (Chapters 7 and 8) may also follow zero-order rate laws, as the catalyst produces a constant concentration of activated reactant A* or reactive intermediate I. This occurs when the number of active sites of the solid catalyst is very small compared with the number of reactants A at any time during the reaction, and when the rate of formation of activated A* is not diffusion limited (Sections 3.2.8 and 3.2.10). This may also occur in the case of homogeneous catalytic processes in which the catalyst is used in a very low relative concentration.

3.2.4 Reversible reactions

Consider first the opposing first-order elementary reactions, $A \rightleftarrows P$, with the rate constant k_1 for the forward reaction and k_{-1} for the reverse reaction. The differential rate law is

$$-d[A]/dt = d[P]/dt = k_1[A] - k_{-1}[P] \quad (3.15)$$

If $[A]_0$ and $[P]_0$ are nonzero, then $[A(t)] = [A]_0 - x$ and $[P(t)] = [P]_0 + x$, where x is the concentration of A that has been converted into P at any time. In 1924, Tolman discovered the principle of microscopic reversibility: the forward and reverse of a particular reaction pathway are related as mirror images [6, 19]. It is a consequence of the time reversal symmetry of classical or quantum mechanics. A macroscopic manifestation of this principle is called the principle of detailed balance: in a system at equilibrium, any process and its reverse proceed at the same rate. At equilibrium, $k_1[A]_e = k_{-1}[P]_e$. At equilibrium, the concentrations of the species are constant: $-d[A]/dt = d[P]/dt = 0$ when $[A] = [A]_e$ and $[P] = [P]_e$. As the equilibrium constant for the reaction is $K_e = [P]_e/[A]_e = k_1/k_{-1}$, the principle of detailed balance also shows that the equilibrium constant for opposing first-order reactions is equal to the ratio of the forward and reverse rate constants. Rate law (3.15) can be rewritten as Eq. (3.16) in terms of the single variable x, which can be considered as the reaction progress:

$$dx/dt = k_1([A]_0 - x) - k_{-1}([P]_0 + x) \quad (3.16)$$

The equilibrium concentrations are $[A]_e = [A]_0 - x_e$ and $[P]_e = [P]_0 + x_e$. Substituting these into Eq. (3.16) gives

$$dx/dt = k_1([A]_e + x_e - x) - k_{-1}([P]_e - x_e + x), \text{ or} \quad (3.17)$$

$$dx/dt = d[P]/dt = (k_1 + k_{-1})(x_e - x) \quad (3.18)$$

Since $k_1[A]_e = k_{-1}[P]_e$, then

$$\frac{dx}{x_e - x} = (k_1 + k_{-1})dt \quad (3.19)$$

Integrating both sides leads to

$$-\int_0^x d\ln(x_e - x) = (k_1 + k_{-1})\int_0^t dt \text{ or} \quad (3.20)$$

$$-\ln\frac{x_e - x}{x_e} = (k_1 + k_{-1})t \quad (3.21)$$

This leads to $(x_e - x)/x_e = \exp[-(k_1 + k_{-1})t]$ or

$$x_e - x = x_e \exp[-(k_1 + k_{-1})t] \quad (3.22)$$

Since $x_e = [A]_0 - [A]_e$ and $x = [A]_0 - [A]$, one obtains $[A]_0 - [A]_e - [A]_0 + [A] = x_e \exp[-(k_1 + k_{-1})t]$ or

$$[A] = [A]_e + x_e \exp[-(k_1 + k_{-1})t] \quad (3.23)$$

At the beginning of the reaction, [A] is $[A]_e + x_e = [A]_0$ and exponentially decreases to $[A]_e$. Similarly, [P] starts at $[P]_e - x_e = [P]_0$ and increases to $[P]_e$, exponentially with the same rate constant (Figure 3.5).

When the forward rate constant is much larger than the reverse, i.e. $k_1 \gg k_{-1}$, the completed reaction produces P and minor traces of A. Consequently, $[A]_e = 0$ and $x_e = [A]_0$, so that Eq. (3.23) reduces to $[A] = [A]_0 \exp(-k_1 t)$ (Figure 3.3).

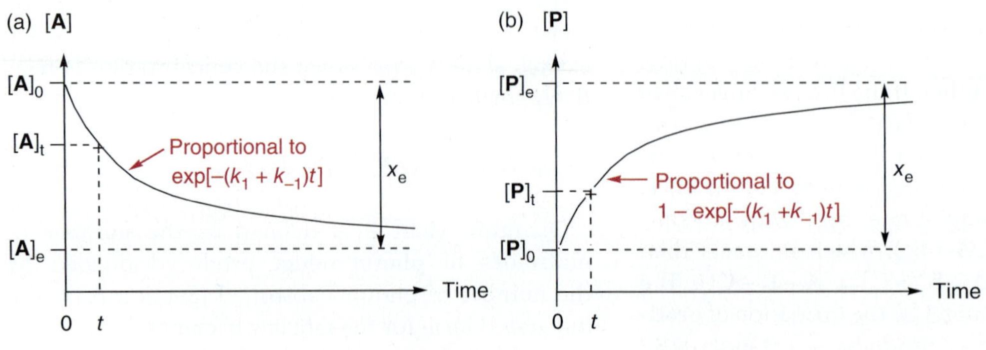

Figure 3.5 Concentrations of A and P as a function of time for opposing reactions $A \rightarrow P$ and $P \rightarrow A$.

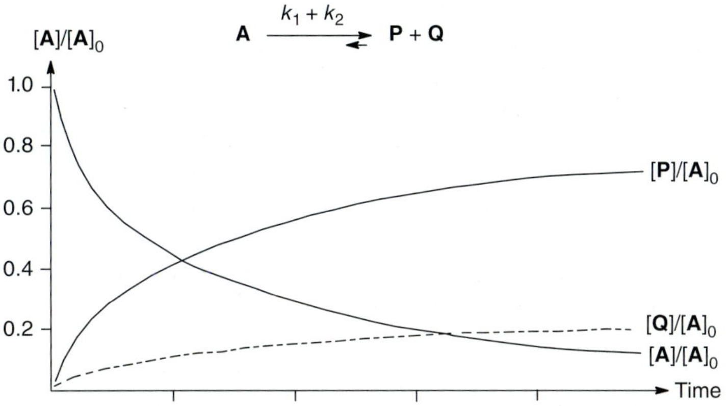

Figure 3.6 Relative concentration of **P** and **Q** as a function of time for the parallel reactions according to Eq. (3.24). Concentration of [**Q**] has been arbitrarily set as lower than that of [**P**], i.e. $k_1 > k_2$.

3.2.5 Parallel reactions

One chemical species can often react following more than one pathway. For simplicity, the parallel reactions are modeled here as competing, irreversible first-order processes:

$$\mathbf{P} \xleftarrow{k_1} \mathbf{A} \xrightarrow{k_2} \mathbf{Q} \tag{3.24}$$

The differential equations describing the parallel reactions are

$$-\mathrm{d}[\mathbf{A}]/\mathrm{d}t = k_1[\mathbf{A}] + k_2[\mathbf{A}] = (k_1 + k_2)[\mathbf{A}] \tag{3.25}$$

$$\mathrm{d}[\mathbf{P}]/\mathrm{d}t = k_1[\mathbf{A}] \tag{3.26}$$

$$\mathrm{d}[\mathbf{Q}]/\mathrm{d}t = k_2[\mathbf{A}] \tag{3.27}$$

The solution to the first of these equations, for a starting concentration of $[\mathbf{A}]_0$, is

$$[\mathbf{A}] = [\mathbf{A}]_0 \exp[-(k_1 + k_2)t] \tag{3.28}$$

Substitution of this equation for [**A**] into the second differential equation (3.26) leads to

$$\mathrm{d}[\mathbf{P}]/\mathrm{d}t = k_1[\mathbf{A}]_0 \exp[-(k_1 + k_2)t]$$

and

$$\int_{[\mathbf{P}]=0}^{[\mathbf{P}(t)]} \mathrm{d}[\mathbf{P}] = k_1[\mathbf{A}]_0 \int_0^t \exp[-(k_1 + k_2)t]\, \mathrm{d}t$$

therefore,

$$[\mathbf{P}] = \frac{k_1}{k_1 + k_2}[\mathbf{A}]_0 \{1 - \exp[-(k_1 + k_2)t]\} \tag{3.29}$$

A similar equation can be derived for [**Q**]:

$$[\mathbf{Q}] = \frac{k_2}{k_1 + k_2}[\mathbf{A}]_0 \{1 - \exp[-(k_1 + k_2)t]\} \tag{3.30}$$

There are two important points to note in comparing Eqs. (3.29) and (3.30). First, both [**P**] and [**Q**] increase exponentially with a rate constant $k_1 + k_2$, as shown in Figure 3.6. The ratio of products, sometimes called the branching ratio, is $[\mathbf{P}]/[\mathbf{Q}] = k_1/k_2$, which is constant throughout the reaction.

3.2.6 Consecutive reactions and steady-state approximation

Many reactions involve several steps and one or more intermediates. An example is the sequence of two irreversible first-order elementary reactions:

$$\mathbf{A} \xrightarrow{k_1} \mathbf{B} \xrightarrow{k_2} \mathbf{P} \tag{3.31}$$

for which the differential equations are

$$-\mathrm{d}[\mathbf{A}]/\mathrm{d}t = k_1[\mathbf{A}] \tag{3.32}$$

$$\mathrm{d}[\mathbf{B}]/\mathrm{d}t = k_1[\mathbf{A}] - k_2[\mathbf{B}] \tag{3.33}$$

$$\mathrm{d}[\mathbf{P}]/\mathrm{d}t = k_2[\mathbf{B}] \tag{3.34}$$

Assuming that the initial concentration of **A** is $[\mathbf{A}]_0$ and that the initial concentrations of **B** and **P** are $[\mathbf{B}]_0 = [\mathbf{P}]_0 = 0$, integration of (3.32) gives (3.35)

$$[\mathbf{A}] = [\mathbf{A}]_0 \exp(-k_1 t) \tag{3.35}$$

Substitution of this solution into Eq. (3.32) yields the differential equation:

$$\mathrm{d}[\mathbf{B}]/\mathrm{d}t = k_1[\mathbf{A}]_0 \exp(-k_1 t) - k_2[\mathbf{B}] \tag{3.36}$$

The solution to this equation, verified by direct differentiation, is

$$[\mathbf{B}] = \frac{k_1}{k_2 - k_1}[\mathbf{A}]_0 [\exp(-k_1 t) - \exp(-k_2 t)] \tag{3.37}$$

Finally, since by mass balance $[\mathbf{A}]_0 = [\mathbf{A}] + [\mathbf{B}] + [\mathbf{P}]$, $[\mathbf{P}] = [\mathbf{A}]_0 - [\mathbf{B}] - [\mathbf{A}]$

$$[\mathbf{P}] = [\mathbf{A}]_0 \left\{1 - \frac{k_1}{k_2 - k_1}[\exp(-k_1 t) - \exp(-k_2 t)] - \exp(-k_1 t)\right\} \tag{3.38}$$

B is a reactive intermediate in a process transforming **A** into **P**, then $k_1 \ll k_2$. After an initial transient rise, called the induction period, the concentration of **B** can be expressed as $[\mathbf{B}] \approx (k_1/k_2)[\mathbf{A}]_0 \exp(-k_1 t) = (k_1/k_2)[\mathbf{A}]$. Rearrangement of this last expression yields, $k_2[\mathbf{B}] \approx k_1[\mathbf{A}]$, or, after inserting this approximation into Eq. (3.33), $d[\mathbf{B}]/dt \approx 0$ is obtained. Because k_1/k_2 is small, the concentration of **B** is always much less than [**A**] and [**P**] when the reaction has started.

The steady-state approximation describes this set of reaction kinetics. After an initial induction period, the concentration of any intermediate species (i.e. **B**) in a consecutive reaction can be calculated by setting its time derivative equal to zero. Thus, if $d[\mathbf{B}]/dt = 0$ in Eq. (3.33), then $[\mathbf{B}] = (k_1/k_2)[\mathbf{A}]$. As the concentration of **A** as a function of time is the solution to Eq. (3.32), then $[\mathbf{B}] = (k_1/k_2)[\mathbf{A}]_0 \exp(-k_1 t)$. Equation (3.38) gives [**P**] as $[\mathbf{A}]_0 - [\mathbf{A}] - [\mathbf{B}]$ or [**P**] can be found by inserting the solution for [**B**] into Eq. (3.34) and then integrating.

In the case of $k_2 = 10k_1$, the time-dependent concentrations of [**A**], [**B**], and [**P**] are given in Figure 3.7.

The rate of formation of **P** is controlled by the first step $\mathbf{A} \rightarrow \mathbf{B}$, when $k_1 \ll k_2$. The slowest step is defined as the rate-determining step of the overall process converting **A** into **P**.

3.2.7 Consecutive reactions: maximum yield of the intermediate product

For the two consecutive irreversible reactions (3.31) for which $k_1 \gg k_2$, Eq. (3.37) becomes

$$[\mathbf{B}] = [\mathbf{A}]_0[\exp(-k_2 t) - \exp(-k_1 t)] \tag{3.39}$$

The second term in the square brackets of (3.37) rapidly approaches zero while the first term is still near unity because $k_1 \gg k_2$. Consequently, the concentration of the intermediate **B** rapidly approaches a value nearly equal to $[\mathbf{A}]_0$, whereas [**B**] decays slowly throughout the reaction according to $[\mathbf{B}] = [\mathbf{A}]_0 \exp(-k_2 t)$.

The exact solution for $k_1 = 10k_2$ is shown in Figure 3.8. The physical situation that the equations represent can be described with an analogy involving three stacked buckets. The top bucket has a large hole and leaks into the second, which itself leaks through a much smaller hole into a third. Water placed in the first bucket would thus flow rapidly into the second

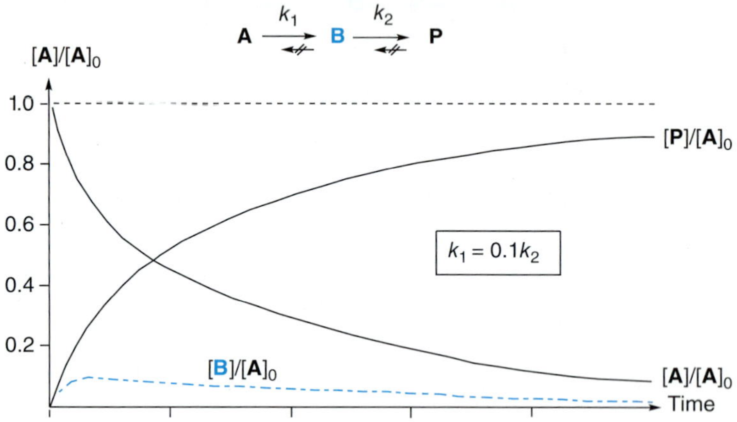

Figure 3.7 Relative concentrations in consecutive reactions with $k_2 = 10k_1$, initial concentrations $[\mathbf{A}] = [\mathbf{A}]_0$, $[\mathbf{B}]_0 = [\mathbf{P}]_0 = 0$, first-order irreversible reactions with rate constants k_1 ($\mathbf{A} \rightarrow \mathbf{B}$) and k_2 ($\mathbf{B} \rightarrow \mathbf{P}$).

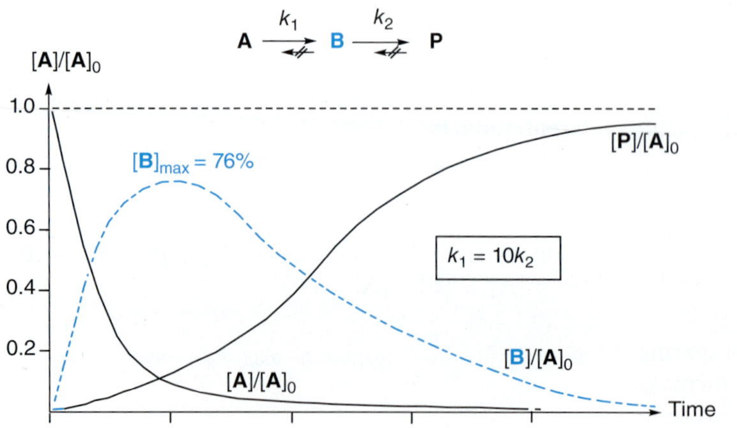

Figure 3.8 Calculated relative concentration of **A**, **B**, and **P** of two consecutive irreversible first-order (or pseudo-first-order) elementary reactions k_1 ($\mathbf{A} \rightarrow \mathbf{B}$) and k_2 ($\mathbf{B} \rightarrow \mathbf{P}$). If $k_1 = 10k_2$, the maximum yield for **B** would be 76%!

bucket, from which it would then leak slowly into the third.

To obtain an optimum yield of product **B**, the reaction must be stopped before reactant **A** is completely consumed. For instance, if $k_1 = 10k_2$, the maximum yield in **B** can only be 76%. Considering other consecutive irreversible reactions $\mathbf{A} \rightarrow \mathbf{B} \rightarrow \mathbf{P}$ with $k_1 = 2k_2$ results in a 49% maximum yield of **B**. If $k_1 = k_2$, the maximum yield in **B** is lower still, just 36%!

3.2.8 Homogeneous catalysis: Michaelis–Menten kinetics

A yeast enzyme called invertase (or β-fructofuranosidase, sucrose, or saccharase: EC 3.2.1.26) catalyzes the hydrolysis of sucrose (cane or beet sugar) into a mixture of D-fructose and D-glucose called inverted sugar syrup (Thompson's inversion process) [20]. For industrial use, invertase is usually derived from yeast. Optimum temperature at which the rate of reaction is at its greatest is 60 °C and an optimum pH of 4.5 [21]. Bees use invertase to make honey from nectar. D-Glucose and D-fructose (Scheme 3.1) sugars exist as an equilibrium mixture of various hemiacetals in water (for glucose, see Section 2.7.10, Figure 2.9, Table 2.3).

In 1902, Henri proposed a general theory for the sucrase-catalyzed hydrolysis of sucrose in which he recognized that the enzyme equilibrates with a complex containing the substrate [22]. In 1914, Michaelis and Menten proposed the following mechanism for the reaction shown in Scheme 3.1 and derived the rate law. These kinetics are general to any process utilizing an enzyme (**E**), substrate (**S**), resulting in an intermediate (**X**) to form product (**P**) [23]. First, enzyme **E** (invertase) combines with the reactant **S** (sucrose) to form **X**. The enzyme then converts **X** into product, **P**, which is then released, returning the enzyme to its original state (reaction (3.40)).

$$\mathbf{S} + \mathbf{E} \underset{k_{-1}}{\overset{k_1}{\rightleftarrows}} \mathbf{X} \xrightarrow{k_2} \mathbf{P} + \mathbf{E} \qquad (3.40)$$

Using the steady-state approximation (Section 3.2.6) for the formation of intermediate **X**, the differential equation is obtained:

$$\mathrm{d}[\mathbf{X}]/\mathrm{d}t = k_1[\mathbf{E}][\mathbf{S}] - k_{-1}[\mathbf{X}] - k_2[\mathbf{X}] = 0 \qquad (3.41)$$

If $[\mathbf{E}]_0$ is the original concentration of the enzyme (catalyst), then the sum of free enzyme, $[\mathbf{E}]$, and intermediate $[\mathbf{X}]$ is always, $[\mathbf{E}]_0 = [\mathbf{E}] + [\mathbf{X}]$. Similarly, if $[\mathbf{S}]_0$ is the initial concentration of reactant **S**, then by mass balance, $[\mathbf{S}]_0 = [\mathbf{S}] + [\mathbf{X}] + [\mathbf{P}]$. Substitution of these into (3.41) gives the differential equation (3.42).

$$\mathrm{d}[\mathbf{X}]/\mathrm{d}t = k_1([\mathbf{E}]_0 - [\mathbf{X}])([\mathbf{S}]_0 - [\mathbf{X}] - [\mathbf{P}]) - k_{-1}[\mathbf{X}] - k_2[\mathbf{X}] = 0 \qquad (3.42)$$

For enzyme catalysis, the initial concentration of the catalyst is generally much smaller than the initial concentration of the reactant, $[\mathbf{E}]_0 \ll [\mathbf{S}]_0$. Since $[\mathbf{X}]$ can never be larger than $[\mathbf{E}]_0$, $[\mathbf{X}] \ll [\mathbf{S}]_0$, so that $[\mathbf{S}]_0 - [\mathbf{X}] - [\mathbf{P}] \cong [\mathbf{S}]_0 - [\mathbf{P}]$. At $t \rightarrow 0$, $[\mathbf{P}] \rightarrow 0$; thus, Eq. (3.42) can be simplified into $k_1[\mathbf{E}]_0([\mathbf{S}]_0 - [\mathbf{X}]) - k_{-1}[\mathbf{X}] - k_2[\mathbf{X}] = k_1[\mathbf{E}]_0[\mathbf{S}]_0 - k_1[\mathbf{E}]_0[\mathbf{X}] - k_{-1}[\mathbf{X}] - k_2[\mathbf{X}] = k_1[\mathbf{E}]_0[\mathbf{S}]_0 - [\mathbf{X}](k_1[\mathbf{E}]_0 + k_{-1} + k_2) = 0$. Finally, Eq. (3.43) is obtained.

$$[\mathbf{X}] = k_1[\mathbf{E}]_0[\mathbf{S}]_0/(k_1[\mathbf{E}]_0 + k_{-1} + k_2) \qquad (3.43)$$

Defining $v_0 = \mathrm{d}[\mathbf{P}]/\mathrm{d}t$, the initial velocity of formation of **P**, one obtains at $t \rightarrow 0$:

$$v_0 = (\mathrm{d}[\mathbf{P}]/\mathrm{d}t)_{t\rightarrow 0} = k_2[\mathbf{X}]$$

$$= k_1k_2[\mathbf{E}]_0 \Big/ \left[k_1 + \frac{k_{-1} + k_2}{[\mathbf{S}]_0}\right]$$

$$= k_2[\mathbf{E}]_0 \Big/ \left[1 + \frac{k_{-1} + k_2}{k_1[\mathbf{S}]_0}\right]$$

$$\text{or} \quad (\mathrm{d}[\mathbf{P}]/\mathrm{d}t)_{t\rightarrow 0} = V_{\mathrm{max}} \Big/ \left[1 + \frac{K_{\mathrm{m}}}{[\mathbf{S}]_0}\right]$$

where $V_{\mathrm{max}} = k_2[\mathbf{E}]_0 = k_{\mathrm{cat}}[\mathbf{E}]_0$ and $K_{\mathrm{m}} = (k_{-1} + k_2)/k_1$.

Scheme 3.1 Invertase-catalyzed hydrolysis of sucrose giving D-fructose and D-glucose.

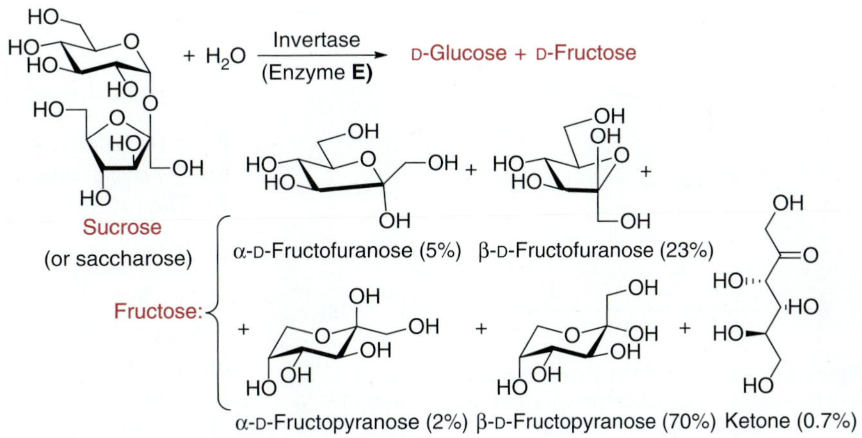

This is commonly written as

$$\text{Initial rate} = v_0 = d[\mathbf{P}]/dt = k_{cat}[\mathbf{E}]_0[\mathbf{S}]/([\mathbf{S}] + K_m)$$
$$= V_{max}[\mathbf{S}]/([\mathbf{S}] + K_m) \quad (3.44)$$

Equation (3.44) is known as the Michaelis–Menten equation. It is based on the steady-state assumption plus the use of $[\mathbf{E}]_0$, rather than $[\mathbf{E}]$, as the latter is small and often nonmeasurable. The initial rate of the catalyzed reaction is proportional to the initial concentration of enzyme ($[\mathbf{E}]_0$) and to the rate constant $k_2 = k_{cat}$, known as the turnover number of the catalyst, defined as the number of molecules of product per molecule of enzyme that can be created per time unit for a saturated enzyme. Typical values are 10^2–10^3 s^{-1}, but values as large as 10^6 s^{-1} have been observed. At very high initial substrate concentrations, the initial rate approaches $v_0 = V_{max} = k_{cat}[\mathbf{E}]_0$ or $d([\mathbf{P}]/[\mathbf{E}]_0)/dt = k_{cat}$. At low substrate concentration $[\mathbf{S}] \ll K_m$, then $v = k_{cat}/(K_m[\mathbf{E}]_0[\mathbf{S}])$. V_{max} represents the maximum rate achieved by the system, at maximum (saturating) substrate concentrations. The Michaelis constant K_m is the substrate concentration at which the reaction rate is half of V_{max}. K_m is roughly the dissociation constant of the enzyme/substrate complex, k_{-1}/k_1, provided that k_2 is relatively small. It is expressed in M, and common values of K_m are from mM to μM [24]. A plot of Eq. (3.44) is shown in Figure 3.9. Initially, the rate increases linearly with $[\mathbf{S}]_0$ and then approaches V_{max}. Although Figure 3.9 visually indicates how initial rate changes, it cannot be utilized to obtain the reaction rate constant. However, this can be achieved by taking reciprocal of both sides of Eq. (3.44) to obtain the Lineweaver–Burk form

$$\frac{1}{v_0} = \frac{1}{V_{max}} + \frac{K_m}{V_{max}[\mathbf{S}]_0} \quad (3.45)$$

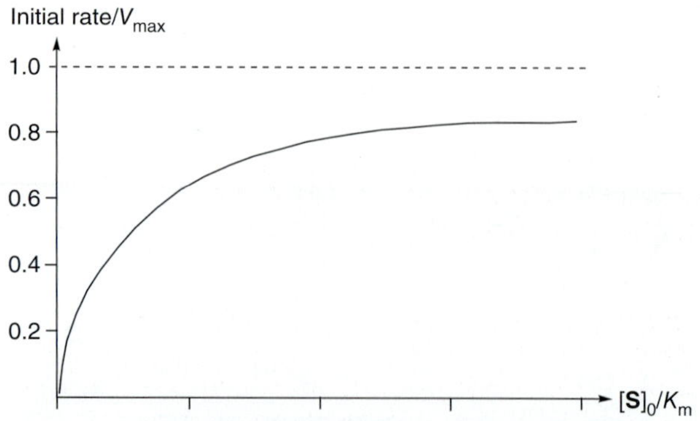

Figure 3.9 A plot of the initial rate, in units of V_{max}, as a function of the initial substrate concentration, in units of K_m, for the Michaelis–Menten mechanism.

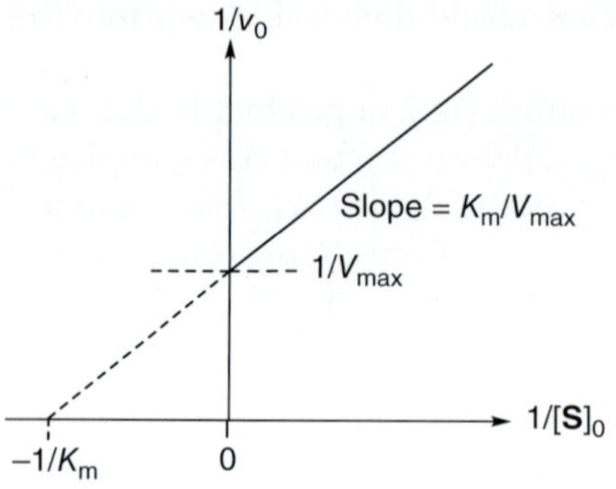

Figure 3.10 Lineweaver–Burk plot for an enzyme-catalyzed reaction obeying the Michaelis–Menten mechanism.

This form shows that a plot of the reciprocal of the initial rate as a function of $[\mathbf{S}]_0^{-1}$ should yield a straight line with intercept V_{max}^{-1} and slope K_m/V_{max}, as shown in Figure 3.10.

3.2.9 Competitive vs. noncompetitive inhibition

Inhibitors retard catalysis by interacting with the catalyst in a way that reduces its catalytic activity. Many important drugs inhibit enzyme activity (biological catalysts). Competitive inhibition occurs when the inhibitor competes with the substrate (reactant) for binding at the active site of the enzyme (or homogenous catalyst in general). The mechanism can be represented by

$$\mathbf{E} + \mathbf{S} \underset{k_{-1}}{\overset{k_1}{\rightleftarrows}} \mathbf{X} \overset{k_2}{\rightarrow} \mathbf{P} + \mathbf{E} \quad (3.40)$$

$$\mathbf{E} + \mathbf{I} \underset{k_{-i}}{\overset{k_i}{\rightleftarrows}} \mathbf{EI} \quad (3.46)$$

It is assumed that reaction (3.46) is always at equilibrium and that complex **EI** cannot catalyze the reaction. Application of the steady-state approximation yields

$$v_0 = \frac{V_{max}}{1 + \frac{K_m}{[\mathbf{S}]_0}\left[1 + \frac{[\mathbf{I}]}{K_I}\right]} \quad (3.47)$$

$K_I = [\mathbf{E}][\mathbf{I}]/[\mathbf{EI}] = k_{-i}/k_i$ is the equilibrium constant for the dissociation of the enzyme/inhibitor complex **EI**. The Lineweaver–Burk equation then becomes

$$\frac{1}{v_0} = \frac{1}{V_{max}} + \left[1 + \frac{[\mathbf{I}]}{K_I}\right]\frac{K_m}{V_{max}[\mathbf{S}]_0} \quad (3.48)$$

A plot of the inverse of the initial rate as a function of the inverse of the initial substrate concentration gives the same intercept as in the absence of inhibitor, but a different slope, as shown in

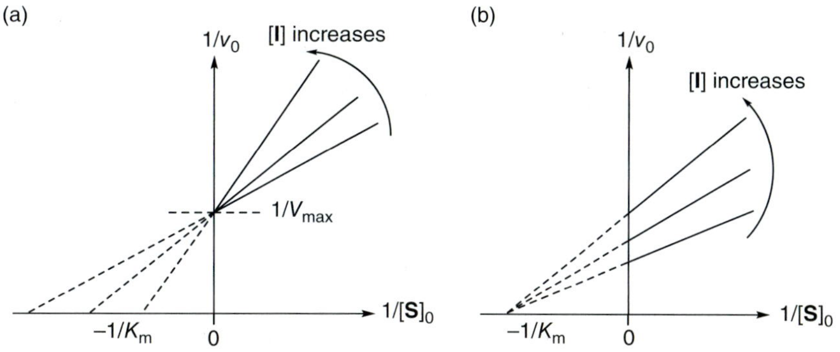

Figure 3.11 (a) Competitive inhibition and (b) noncompetitive inhibition on the initial rate of catalyzed reaction following the Michaelis–Menten mechanism ([I] = concentration of the inhibitor).

Figure 3.11a. An example of a competitive inhibitor is malonic acid, $CH_2(COOH)_2$, which resembles succinic acid (HOOC—CH_2—CH_2—COOH) sufficiently to bind to succinic dehydrogenase and inhibit it from converting succinic acid to fumaric acid ((*E*)-HOOC—CH=CH—COOH).

Noncompetitive inhibition occurs when the inhibitor does not bind to the active site of the enzyme but still inhibits product formation:

$$\mathbf{E} + \mathbf{S} \underset{k_{-1}}{\overset{k_1}{\rightleftarrows}} \mathbf{X} \overset{k_2}{\rightarrow} \mathbf{P} + \mathbf{E} \tag{3.40}$$

$$\mathbf{E} + \mathbf{I} \rightleftarrows \mathbf{EI} \tag{3.46}$$

$$\mathbf{X} + \mathbf{I} \rightleftarrows \mathbf{XI} \tag{3.49}$$

If the last two equilibria (3.46) and (3.49) are assumed to have same equilibrium constant: K_I = $[\mathbf{E}][\mathbf{I}]/[\mathbf{EI}] = [\mathbf{X}][\mathbf{I}]/[\mathbf{XI}]$. The initial rate of reaction is

$$v_o = \frac{V_{max}[\mathbf{S}]_0}{\{[\mathbf{S}]_0 + K_m\}\left[1 + \frac{[\mathbf{I}]}{K_I}\right]} \tag{3.50}$$

and the Lineweaver–Burk equation is

$$\frac{1}{v_0} = \left[\frac{1}{V_{max}} + \frac{K_m}{V_{max}[\mathbf{S}]_0}\right]\left[1 + \frac{[\mathbf{I}]}{K_I}\right] \tag{3.51}$$

Note that the presence of the inhibitor affects both the slope and intercept of the Lineweaver–Burk plot for noncompetitive inhibition, as shown in Figure 3.11b [5]. The smaller the K_I value, the more strongly bound is the inhibitor. A "picomolar drug" binds tightly to its receptor ($K_I = 10^{-12}$ M), whereas a "micromolar drug" is a much weaker binder ($K_I = 10^{-6}$ M).

3.2.10 Heterogeneous catalysis: reactions at surfaces

Heterogeneous catalysts, such as solids and metals, exhibit different kinetics. In many cases, diffusion of reactant **A** onto the solid catalyst surface and desorption of the product **P** off the surface are faster than the rate-limiting step. A subsequent unimolecular elementary step in a reaction that occurs at the surface has a first-order rate law in the surface coverage of the reactant:

$$\mathrm{d}[\mathbf{P}]/\mathrm{d}t = k[\theta] = kK[\mathbf{A}]/(1 + K[\mathbf{A}]) \tag{3.52}$$

where **P** is the product of the surface reaction and $kK[\mathbf{A}]/(1 + K[\mathbf{A}])$ follows the Langmuir model, i.e. the absorption and desorption processes can be represented as an equilibrium:

$$\mathbf{A} + \text{surface} \underset{k_d}{\overset{k_a}{\rightleftarrows}} \mathbf{A}_a \text{ (reactant absorbed)}$$

$[\mathbf{A}]$ represents the concentration of reactant **A** and θ is the fraction of the surface sites covered by **A**, then the rate of absorption is proportional to $k_a[\mathbf{A}](1 - \theta)$ and the rate of desorption is proportional to $k_d[\theta]$. Here k_a and k_d are the rate constants for the two processes. At equilibrium, we have

$$k_a[\mathbf{A}](1 - \theta) = k_d\theta \text{ or } \frac{\theta}{(1 - \theta)} = \frac{k_a}{k_d}[\mathbf{A}]$$

Denoting $K = k_a/k_d$ and solving for the fraction of occupied sites of the catalyst, we obtain what is known as the Langmuir adsorption isotherm (3.53):

$$\theta = K[\mathbf{A}]/(1 + K[\mathbf{A}]) \tag{3.53}$$

The order of the overall reaction $\mathbf{A} \rightarrow \mathbf{P}$ catalyzed by heterogeneous catalyst depends on the conditions. For low values of $K[\mathbf{A}]$, the reaction rate increases linearly with $[\mathbf{A}]$ because absorption is the rate-limiting step; the reaction is thus first order in $[\mathbf{A}]$. When $K[\mathbf{A}]$ is large, the surface is saturated with **A** and the absorption and desorption processes are not limited by the diffusion, the rate of the overall reaction $\mathbf{A} \rightarrow \mathbf{P}$ becomes zero order in $[\mathbf{A}]$ and depends only on the rate constant k. This situation is analogous to the enzyme or homogeneous catalyst-catalyzed reactions following the Michaelis–Menten mechanism (Section 3.2.8). For most bimolecular surface reactions,

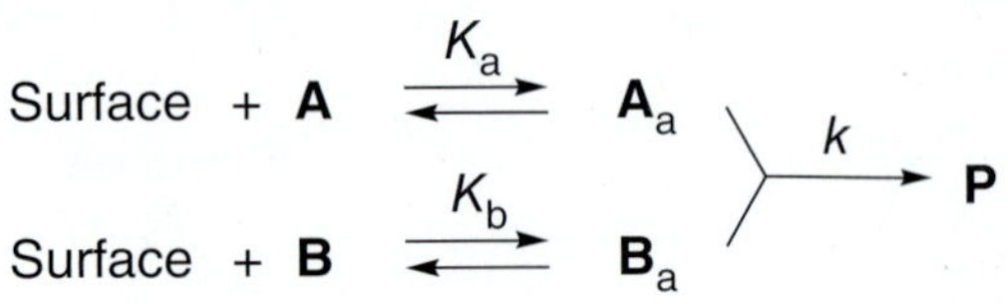

(3.54)

the reaction rate is proportional to the product of the surface coverage of the two reactants $d[\mathbf{P}]/dt = k\theta_A\theta_B = kK_aK_b[\mathbf{A}][\mathbf{B}]/(1+K_a[\mathbf{A}]+K_b[\mathbf{B}])$ [2, 5].

3.3 Activation parameters

The kinetic theory of gases gives an estimate of the collision frequency between two gaseous molecules [25]. Although this frequency varies somewhat with the size ("cross-section") of the molecules, it is of the order of $10^{29}\ s^{-1}$ at 25 °C and 1 atm. If every collision were followed by a reaction, the rate would be c. $10^{10}\ M^{-1}\ s^{-1}$, which corresponds to the rate limited by diffusion in an ideal gas. Most organic reactions are much slower, as only a small fraction of the collisions leads to reaction. The kinetic energy of the colliding molecules must be sufficient to overcome repulsive forces between two species. The higher the temperature, the higher the kinetic energy of the molecules and thus the higher the rate of reaction [2, 3]. In 1889, Arrhenius observed that for most simple reactions, k follows the empirical relationship (3.55).

$$k = A\exp(-E_a/RT) \tag{3.55}$$

where A is a pre-exponential factor called the frequency factor, independent from temperature T (in kelvin), and E_a is the Arrhenius activation energy [26, 27]. As a rule of thumb, the rate constant of simple reactions doubles for every 10 K increase in temperature. From Eq. (3.55), this can be shown to be strictly true for $E_a \sim 20$ kcal mol^{-1}. At an infinite temperature, or if the activation energy is zero, the maximum rate constant will be $k = A$.

In Eyring's theory (1935) [28, 29], an equilibrium (3.56) is assumed between reactants and the [activated complex]‡.

(3.56)

Using the normal thermodynamic relationship between free energy and equilibrium (3.56), relationship (3.57) is obtained:

$$\Delta^\ddagger G^T = -RT\ln K^\ddagger \tag{3.57}$$

where $K^\ddagger = k^\ddagger/k^{-\ddagger}$ and $\Delta^\ddagger G^T$ (often written $\Delta^\ddagger G$ *or* $\Delta G^\ddagger$) is the free energy of activation of reaction (3.56).

The activation enthalpy, $\Delta^\ddagger H^T$ (often written $\Delta^\ddagger H$ *or* $\Delta H^\ddagger$), and the activation entropy, $\Delta^\ddagger S^T$ (often written $\Delta^\ddagger S$ *or* $\Delta S^\ddagger$), are defined as $\Delta^\ddagger G^T = \Delta^\ddagger H^T - T\Delta^\ddagger S^T$ or $\Delta^\ddagger G = \Delta^\ddagger H - T\Delta^\ddagger S$.

Relationship (3.57), or its integral form (3.58), demonstrates Eyring's assumption that the conversion of reactant to transition state, ‡, can be taken as an equilibrium.

$$K^\ddagger = \exp(-\Delta^\ddagger G/RT) \tag{3.58}$$

The kinetic theory of ideal gases gives for the rate constant, k_c, for the collision of two molecules:

$$k_c = RT/Lh \tag{3.59}$$

R is the ideal gas constant, T is the temperature (kelvin), L is the Avogadro number (also noted N_A), and h is the Planck constant. If the rate constant for the conversion of the activated complex into products is taken as a fraction of $k^\ddagger$, say $\kappa k^\ddagger$, it becomes the rate of collision, κkc, times the fraction of molecules with energy to traverse the transition state, $K^\ddagger$. The rate constant k of the overall reaction becomes $k = \kappa k^\ddagger = \kappa k_c K^\ddagger$. This gives

$$\begin{aligned} k &= \kappa(RT/Lh)\exp(-\Delta^\ddagger G/RT)\\ &= (\kappa RT/Lh)\exp(-\Delta^\ddagger H/RT)\exp(\Delta^\ddagger S/R) \end{aligned} \tag{3.60}$$

or

$$k = (\kappa k_b T/h)\exp(-\Delta^\ddagger H/RT)\exp(\Delta^\ddagger S/R) \tag{3.60'}$$

considering the Boltzmann constant $k_b = R/L$.

Kappa, κ, the so-called "transmission coefficient," represents the fraction of transition states that continue to products. If one chooses it to be equal to unity, Eq. (3.60) can be written under the more useful form:

$$\ln(k/T) = -\Delta^\ddagger H/RT + \ln(k_b/h) + \Delta^\ddagger S/R \tag{3.60''}$$

Differentiating equation (3.60) with respect to temperature T, Eq. (3.61) is obtained:

$$\begin{aligned} dk/dT &= \kappa\frac{R}{Lh}\exp(\Delta^\ddagger S/R)\exp(-\Delta^\ddagger H/RT)\\ &[1-\Delta^\ddagger H/RT] = \frac{k}{T}\left[\frac{RT+\Delta^\ddagger H}{RT}\right] \end{aligned} \tag{3.61}$$

Correspondingly, differentiating the Arrhenius equation (3.55) with respect to temperature gives Eq. (3.62):

$$dk/dT = A\exp(-E_a/RT)\cdot E_a/RT^2 = \frac{k}{T}\cdot\frac{E_a}{RT} \tag{3.62}$$

Comparison of Eqs. (3.61) and (3.62) establishes

$$E_a = \Delta^\ddagger H + RT \text{ and } \Delta^\ddagger H = E_a - RT \tag{3.63}$$

Substituting $\Delta^{\ddagger}H$ of Eq. (3.60) by $E_a - RT$, the rate constant k becomes

$$k = \kappa \frac{RT}{Lh} \exp(-E_a/RT) \exp(RT/RT) \exp(\Delta^{\ddagger}S/R) \tag{3.64}$$

Comparing Eqs. (3.64) and (3.57), one obtains

$$A = RT/Lh \exp(\Delta^{\ddagger}S/R) = 2.085 \times 10^{10} T_m \times \exp(\Delta^{\ddagger}S/R) \text{ in s}^{-1} \tag{3.65}$$

$$\text{and } \Delta^{\ddagger}S = R(\ln A - \ln T_m - 23.76) \text{ in eu} \tag{3.66}$$

$$(\text{eu} = \text{entropy units} = \text{cal K}^{-1}\text{mol}^{-1})$$

where T_m is the average temperature at which the rate constants have been measured. This can be combined with the expression for ln k from (3.60) to give

$$\Delta^{\ddagger}G = RT(23.76 + \ln T - \ln k) = RT \ln k + RT(\ln T + 23.76) \text{ in kcal mol}^{-1} \tag{3.67}$$

Examples of $\Delta^{\ddagger}G$ values as a function of temperature are given in Table 3.2.

Generally, first-order, irreversible reactions with rate law $d[\mathbf{A}]/dt = -k^{I}[\mathbf{A}]$ and $k^{I} = 0.001\ s^{-1}$ ($\tau_{1/2}$ = 693 seconds, see Section 3.2.1), and $\Delta^{\ddagger}G$ = 21.5 kcal mol^{-1} at 25 °C, will be practically done when it has reached 99% conversion of reactant **A** into products. This corresponds to $7\tau_{1/2} = 7 \times 693 = 4851$ seconds (Table 3.1), or c. 1 hour and 20 minutes. A zeroth order reaction with an initial concentration of $[\mathbf{A}]_0 = 2\ln 2$ M and with $\Delta^{\ddagger}G$ = 21.5 kcal mol^{-1} at 25 °C will complete in $2\tau_{1/2} = 2 \times 693$ seconds = 1386 seconds, or c. 22 minutes, whereas a second-order dimerization with $1/[\mathbf{A}]_0$ = ln 2 M^{-1} will be done in $99\tau_{1/2} = 99 \times 693$ seconds = 68 607 seconds (Table 3.1), or c. 19 hours. This illustrates how preparative chemists and chemical engineers prefer reactions with zeroth and first-order rate laws for economical considerations.

$\Delta^{\ddagger}G$ values are a function of temperature, and for k^{I} = 0.0001, 0.001, and 0.01 s^{-1}, reactions with first-order rate law of type $d[\mathbf{A}]/dt = -k^{I}[\mathbf{A}]$ are summarized in Figure 3.12.

Table 3.2 Eyring activation-free energies ($\Delta^{\ddagger}G$ in kcal mol^{-1}) as a function of temperature (in °C) for zeroth order, first-order, and second-order reactions with k^0 = 0.001 M s^{-1}, k^{I} = 0.001 s^{-1}, and k^{II} = 0.001 M^{-1} s^{-1}, respectively.

T (°C)	−180	−150	−120	−90	−78	−60	−30	0	**25**	50	75	100	125
$\Delta^{\ddagger}G^T$	6.5	8.7	10.8	13.0	13.9	15.3	17.5	19.7	**21.5**	23.4	24.3	27.1	29.0
T (°C)	150	175	200	225	250	300	350	400	450	500	550	600	
$\Delta^{\ddagger}G^T$	30.9	32.7	34.6	36.5	28.4	42.2	45.9	49.7	53.5	57.3	61.1	64.9	

1 cal = 4.184 J.

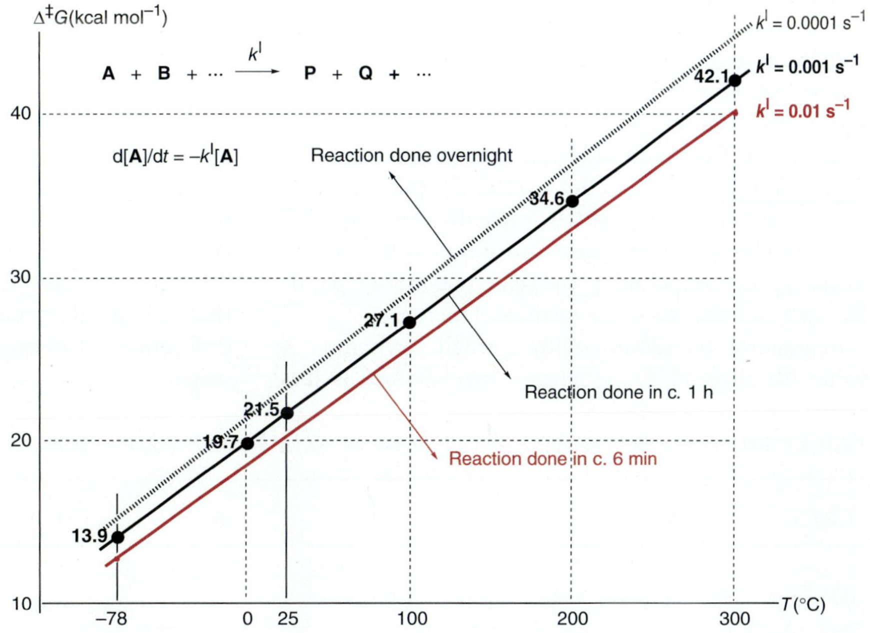

Figure 3.12 Eyring free energy of activation $\Delta^{\ddagger}G = RT(23.76 + \ln T - \ln k)$ in kcal mol^{-1} as a function of rate constant k^{I} and temperature T (1 cal = 4.184 J).

The activation parameters $\Delta^{\ddagger}G$ and $\Delta^{\ddagger}H$ can be used to construct free energy and enthalpy diagrams (macroscopic model) for reactions (e.g. Figures 4, 5, 1.12, and below). The activation entropy, $\Delta^{\ddagger}S$, can reveal a great deal about the mechanism of the reaction. $\Delta^{\ddagger}S$ represents the change in disorder between the reactants and the transition state. It expresses a probability factor in terms of energetic quantities. Associative processes in which two or more molecules are associated into an activated complex or transition state can be characterized by a negative $\Delta^{\ddagger}S$, which suggests the unlikely event of bringing two or more molecules together. However, dissociative processes in which a strong bond is broken in the transition state, or a reaction that increases the number of molecules of products relative to reactants, can have a positive value of $\Delta^{\ddagger}S$. Several examples of reactions with their activation parameters will be discussed in Section 3.4. Table 3.3 gives the relationship between log A (Arrhenius) and $\Delta^{\ddagger}S$ (Eyring) parameters at two different temperatures.

3.3.1 Temperature effect on the selectivity of two parallel reactions

For reaction (3.24) (Section 3.2.5), the branching ratio, k_1/k_2, and the product ratio [**P**]/[**Q**] for two parallel first-order (or pseudo-first-order) reactions depends on temperature as given by Eq. (3.68):

$$k_1/k_2 = [\mathbf{P}]/[\mathbf{Q}] = \exp(-\Delta\Delta^{\ddagger}G^T/RT) \quad (3.68)$$

where $\Delta\Delta^{\ddagger}G^T = \Delta^{\ddagger}G(\mathbf{A}\rightarrow\mathbf{P}) - \Delta^{\ddagger}G(\mathbf{A}\rightarrow\mathbf{Q})$ at temperature T.

If $\Delta\Delta^{\ddagger}G = \Delta\Delta^{\ddagger}H - T\Delta\Delta^{\ddagger}S$ does not vary with T, the branching ratio, and thus the product selectivity, will increase on lowering the temperature of reaction, unless $\Delta\Delta^{\ddagger}G = 0$. Usually, $\Delta\Delta^{\ddagger}H$ does not depends significantly on temperature.

If the $\Delta\Delta^{\ddagger}S$ term differs from zero (different mechanisms for the two parallel reactions), the $-T\Delta\Delta^{\ddagger}S$ term might either be positive or negative and thus will vary with temperature. Therefore, it is possible that on lowering the temperature, the selectivity of two parallel reactions will decrease instead of increase!

In general, a reaction with a small Arrhenius A value (negative $\Delta^{\ddagger}S$, associative mechanism) is less retarded on lowering the temperature than a reaction with large A value ($\Delta^{\ddagger}S \geq 0$, dissociative mechanism). As a rule of thumb, heating will favor dissociative processes, whereas cooling will favor associative processes.

Problem 3.3 Consider the two parallel irreversible reactions **A** → **P** (rate constant k_1, $\Delta^{\ddagger}H_1$ = 23 kcal mol^{-1}, $\Delta^{\ddagger}S_1 = -1$ cal K^{-1}mol^{-1}) and **A** → **Q** (rate constant k_2, $\Delta^{\ddagger}H_2$ = 17.4 kcal mol^{-1}, $\Delta^{\ddagger}S_2 = -20$ cal K^{-1} mol^{-1}) following the first-order rate laws. Estimate the temperature T_1 at which 99% conversion of **A** will be realized in two hours. What will be the product ratio [**P**]/[**Q**] at this temperature T_1. What will this product ratio be at temperature $T_i = T_1 - 50$ K, $T_1 - 30$ K, $T_1 + 30$ K, and $T_1 + 50$ K. What are the Arrhenius pre-exponential factors A_1 and A_2 at temperature T_1 for these two reactions?

3.3.2 The Curtin–Hammett principle

Consider a reactant that exists as equilibrium between two isomers (tautomers, conformers, rotamers, etc.) **A** ⇄ **B** with the equilibrium constant K_{eq} = [**B**]/[**A**] = k_1/k_{-1} ($\Delta_r G^T$(**A** ⇄ **B**)). **A** and **B** become **P** and **Q** irreversibly (**A** → **P** and **B** → **Q**) with rate constants k_P and k_Q, respectively. The product ratio [**P**]/[**Q**] may not depend only on the isomeric ratio [**A**]/[**B**] [30–32]. If k_1 and k_{-1} are much larger than k_P and k_Q as represented in Figure 3.13, and if the two reactions follow the first-order rate laws d[**P**]/dt = k_P[**A**] and d[**Q**]/dt = k_Q[**B**], one can write d[**P**]/d[**Q**] = k_P[**A**]/k_Q[**B**] = $(k_P/k_Q)/K_{eq}$. As the last term can be considered as a constant, integration from time 0 to time ∞ gives the product ratio = [**P**]/[**Q**] = $(k_P/k_Q)/K_{eq}$ = [exp($-\Delta^{\ddagger}G$(**A** → **P**)/RT)/exp($-\Delta^{\ddagger}G$(**B** → **Q**)/RT)]/exp($-\Delta G^T$(**A** ⇄ **B**)/RT)

$$\text{or} \quad [\mathbf{P}]/[\mathbf{Q}] = \exp(-\mathrm{DD}^{\ddagger}G/RT) \quad (3.69)$$

with DD$^{\ddagger}G$ = D$^{\ddagger}G$(**A** ® **P**) – D$_r G^T$(**A** ⇄ **B**) – D$^{\ddagger}G$ (**B** ® **Q**).

In this case, the Curtin–Hammett principle states that the product composition is controlled by the difference in free energies of the respective transition states.

Table 3.3 Relationship between log A and $\Delta^{\ddagger}S$ (in eu) values ($\Delta^{\ddagger}S$ = 4.576 (log A – log T – 10.319)).

T: 25 °C	log A	5	7	9	11	13	15	17
	$\Delta^{\ddagger}S^{298\,K}$	−35.7	−26.5	−17.4	−8.2	0.9	10.1	19.3
T: 325 °C	$\Delta^{\ddagger}S^{698\,K}$	−37.0	−27.9	−18.7	−9.6	−0.4	8.7	17.9

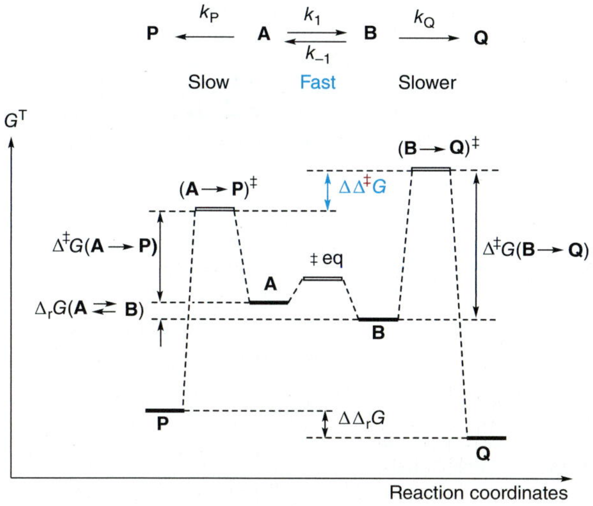

Figure 3.13 Free energy diagram representing the Curtin–Hammett principle for two competitive reactions involving a fast equilibrium between two reactants **A** and **B**. **P** is the major product that arises from the minor reactant **A** ($\Delta^{\ddagger}G$(**A** → **P**) − $\Delta_r G^T$(**A** ⇄ **B**) < $\Delta^{\ddagger}G$(**B** → **Q**); $\Delta\Delta^{\ddagger}G < 0$).

A typical example is the alkylation of tropane (reaction (3.70)) for which the least stable conformer reacts faster than the energetically preferred conformer [33].

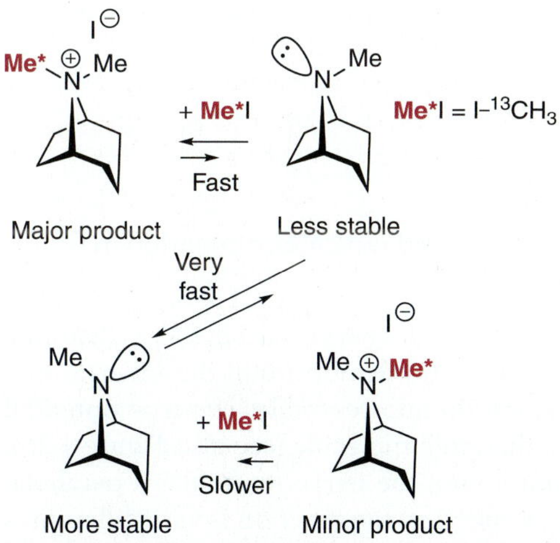

(3.70)

The Curtin–Hammett principle explains the product composition [**P**]/[**Q**] when it does not reflect the relative stability of reactants **A** and **B** as in the case described in Figure 3.14a. If reactions **A** → **P** and **B** → **Q** are not highly exergonic, and time is allowed for the reverse reactions **P** → **A** and **Q** → **B** to compete with the equilibration **A** ⇄ **B**, the product ratio [**P**]/[**Q**] is determined by the relative stability of product **P** and **Q** (Figure 3.14b) [30–32].

Winstein and Holness have developed an equation (Winstein–Holness equation (3.71)) that considers an empirical reaction rate constant k_{WH} defined by $d[\mathbf{P}]/dt + d[\mathbf{Q}]/dt = k_P[\mathbf{A}] + k_Q[\mathbf{B}] = k_{WH}([\mathbf{A}] + [\mathbf{B}])$. Solving for $k_{WH} = k_P[\mathbf{A}]/([\mathbf{A}]+[\mathbf{B}]) + k_Q[\mathbf{B}]/([\mathbf{A}]+[\mathbf{B}])$ at time t, if one considers $K_{eq} = [\mathbf{B}]/[\mathbf{A}]$, one obtains

$$k_{WH} = (k_P + 1/K_{eq}k_Q)/(1/K_{eq} + 1) \quad (3.71)$$

$$\text{and } 1/K_{eq} = (k_P - k_{WH})/(k_{WH} - k_Q) \quad (3.72)$$

With Eq. (3.72), $K_{eq} = [\mathbf{B}]/[\mathbf{A}]$ can be estimated by measuring k_{WH} if k_P and k_Q can be taken from model reactions. A classic example is the oxidation of cyclohexanol into cyclohexanone by chromic acid (Scheme 3.2) [34]. The oxidations of *cis*-4-*t*-butylcyclohexanol (k'_P) represents cyclohexanol oxidation from its minor conformation with an axial hydroxy group (k_P), and the oxidation of *trans*-4-*t*-butylcyclohexanol (k'_Q) models the oxidation of cyclohexanol in its major conformation with an equatorial hydroxy group (k_Q). If one assumes $k_P = k'_P$ and $k_Q = k'_Q$, one finds $1/K_{eq} = 7.5 = [\mathbf{A}]/[\mathbf{B}]$ at 25 °C in AcOH. This leads to $\Delta_r G°(\mathbf{A} \rightleftarrows \mathbf{B}) = -1.2\ \text{kcal mol}^{-1}$ where **A** is the minor axial cyclohexanol and **B** is the most stable cyclohexanol conformer with an equatorial hydroxy group (Figure 3.13). This gives an A value $= -\Delta_r G°(\mathbf{A} \rightleftarrows \mathbf{B}) = 1.2\ \text{kcal mol}^{-1}$, which can be compared with the A values of 0.9 measured by NMR in CCl_4 (Table 2.1) [35].

If the product ratios $[\mathbf{P}]/[\mathbf{Q}] = P$ and $K_{eq} = [\mathbf{B}]/[\mathbf{A}]$ are known, and the **A** ⇄ **B** equilibrium is fast relative to their reactions (Curtin–Hammett conditions: $k_1, k_{-1} \gg k_P, k_Q$), the rate constants k_P and k_Q can be evaluated by measuring k_{WH} and using Eqs. (3.73) and (3.74).

$$k_P = k_{WH}\left(\frac{1/K_{eq} + 1}{P + 1}\right) \quad (3.73)$$

$$k_Q = k_{WH}\left(\frac{1/K_{eq} + 1}{1/K_{eq}}\right)\left(\frac{P}{P + 1}\right) \quad (3.74)$$

Defining $1/K_{eq} = K = [\mathbf{A}]/[\mathbf{B}]$, these equations simplify into

$$k_P = k_{WH}\left(\frac{K + 1}{P + 1}\right) \quad (3.73')$$

$$k_Q = k_{WH}\left(\frac{K + 1}{K}\right)\left(\frac{P}{P + 1}\right) \quad (3.74')$$

Problem 3.4 The *N*-methyl-2-phenylpyrrolidine shows $K = [\mathbf{A}]/[\mathbf{B}] = 17.0$. One measures $k_{WH} = 3 \times 10^{-3}\ \text{M}^{-1}\ \text{s}^{-1}$ for its quaternization with $^{13}CH_3I$ and a product ratio $[\mathbf{M}]/[\mathbf{N}] = 1.72$ is found. Calculate the rate constants $k(\mathbf{A} \rightarrow \mathbf{M})$ and $k(\mathbf{B} \rightarrow \mathbf{N})$ [36].

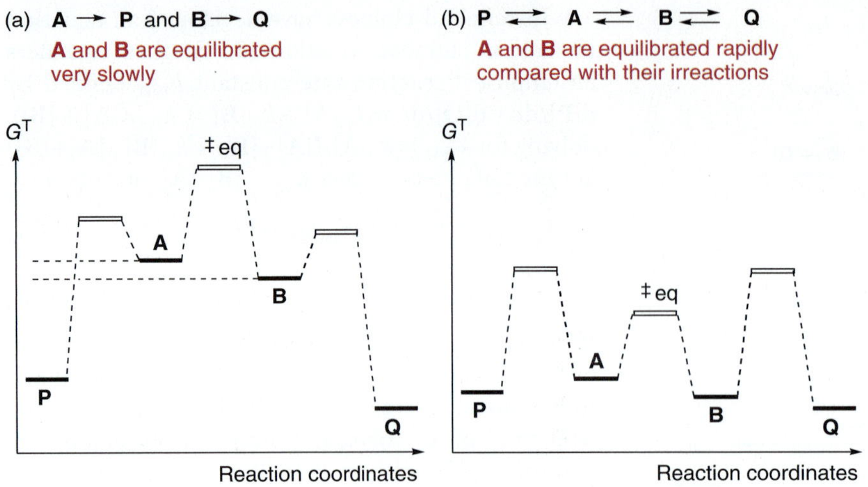

Figure 3.14 Free energy diagram (a) for two competing irreversible reactions **A** → **P** and **B** → **Q** in which reactants **A** and **B** have a slow equilibrium relative to the product formation step. The product ratio reflects initial proportion of reactants (e.g. flash photolysis, slow addition of an acid that induces quick reactions of both reactants: *kinetic quench*). (b) For two competitive reversible reactions of type **P** ⇄ **A** ⇄ **B** ⇄ **Q**; the product composition [**P**]/[**Q**] = $\exp(-\Delta_r G^T(\mathbf{P} \rightleftarrows \mathbf{Q})/RT)$ (*thermodynamic control*).

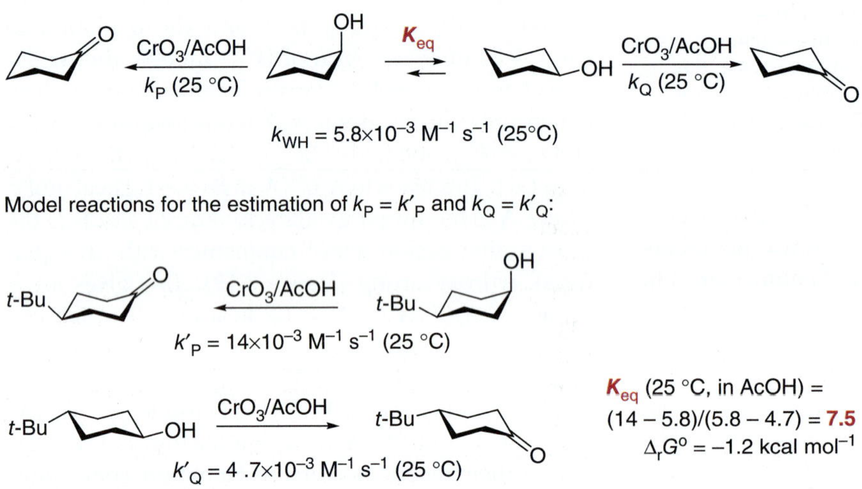

Scheme 3.2 Application of the Winstein–Holness kinetic method of conformational analysis to cyclohexanol.

3.4 Relationship between activation entropy and the reaction mechanism

Entropy of activation, $\Delta^{\ddagger}S^T$, measures the change in order or disorder between the reactants and transition states. Negative activation entropy is expected for associative mechanisms (condensation, ring-closing reaction, and formation of polar transition states that are more solvated than the reactants), nearly zero for simple isomerization that do not change the number of molecules between the transition state and reactants or do not liberate or block rotations in the transition state, and positive for a dissociative mechanism in which the transition state has much weaker bonds than reactants, or is less polar than reactants (desolvation of reactants in the transition state).

3.4.1 Homolysis and radical combination in the gas phase

The Morse potential energy curve for a diatomic molecule (Figure 3.15) demonstrates the anharmonicity of the vibration and resembles the true potential energy of a diatomic molecule (see also Figure 1.26). Near the minimum, the curve resembles a parabola, but at higher energies, the function is very different.

At high vibrational excitations, the vibrational motion of the atoms allows the molecule to reach other regions of the potential energy curve leading to dissociation. The motion becomes anharmonic: the restoring force is not proportional to the square of the change in bond length. Because the actual potential curve is less confining than a parabola, the energy difference between the vibrational levels decreases as the dissociation energy is approached. According to statistical thermodynamics (Section 1.4.3, Eq. (1.36)), the entropy of a diatomic molecule approaching the

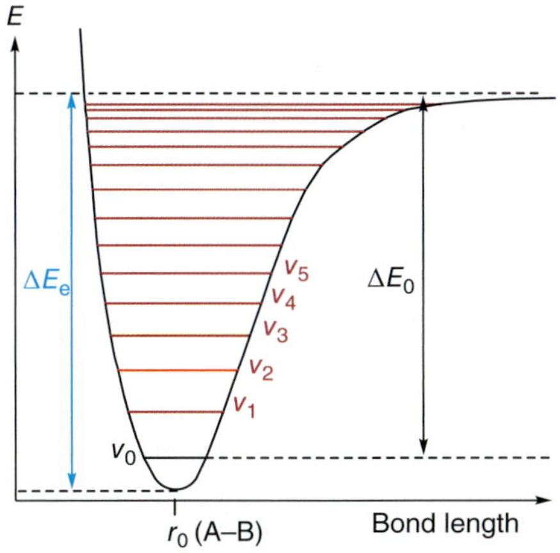

Figure 3.15 The Morse potential energy curve reproduces the general shape of a molecular potential energy curve. The corresponding Schrödinger equation can be solved, and the values of the energies are obtained. The number of bound levels is finite. ΔE_0 = dissociation energy, ΔE_e = minimum energy of the potential energy curve.

dissociation energy level increases because there are so many nearly degenerate vibrational levels that can be populated.

Table 3.4 shows that the elementary homolysis reactions have positive activation entropies in the gas phase for dissociative mechanisms. The energy barrier for the recombination of two monoatomic radicals should be zero, so that the activation free energy should be controlled by the entropy of condensation. Such associations have negative activation entropies, as for all other bimolecular elementary reactions (associative mechanism), as illustrated with the following examples [37, 38]:

$$2\,\text{Me}^\bullet \rightarrow \text{Me}-\text{Me} \quad E_a = 0\,\text{kcal mol}^{-1},$$
$$\log A = 10.5, \Delta^{\ddagger}S \cong -12\,\text{eu}$$
$$2t\text{-BuO}^\bullet \rightarrow t\text{-BuO—O—}t\text{-Bu} \quad E_a = 0\,\text{kcal mol}^{-1},$$
$$\log A = 8.8, \Delta^{\ddagger}S \cong -19\,\text{eu}$$

When two large polyatomic radicals combine to form a stable compound, the activation energy of the reaction might be positive because of the necessity to deform one or both radicals before allowing them to combine in the transition state. This phenomenon was discussed earlier (Section 1.9.1, Figure 1.12).

3.4.2 Isomerizations in the gas phase

Thermolysis of cyclopropane produces propene (reaction (3.75)) [39–41]. The overall process involves an elementary reaction of C—C bond breakage, forming a trimethylene diradical intermediate in the rate-determining step. Then, a (1,2)-hydrogen shift (Section 5.5.2) generates the product. Consistent with other homolysis (Table 3.4), reaction (3.75) has positive activation entropy.

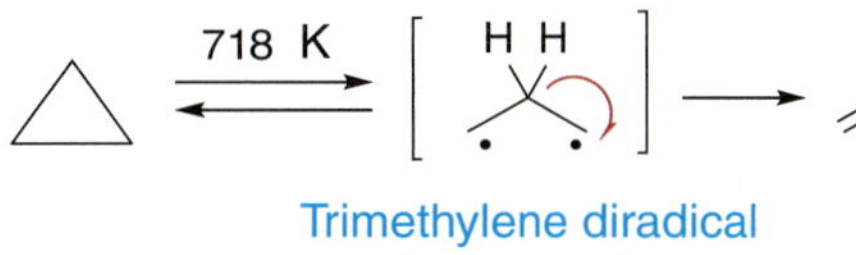

(3.75)

E_a = 62.5 kcal mol^{-1}, log A = 15.1,
$\Delta^{\ddagger}S$ = 8.8 cal K^{-1} mol^{-1} (eu) at 718 K

Most thermal (Z) ⇄ (E) alkene isomerizations (e.g. reactions (3.76)) in the gas phase (in the absence of radical catalyst or surface catalyst) have activation entropies near zero. The bond length of the twisting bond does not vary significantly in the term of bond strength (double bond ⇄ single bond) between reactant and the transition structure: one does not gain more vibration levels as the energy separation between them is not reduced significantly; the transition states of reactions (3.76) are not much more flexible than reactants.

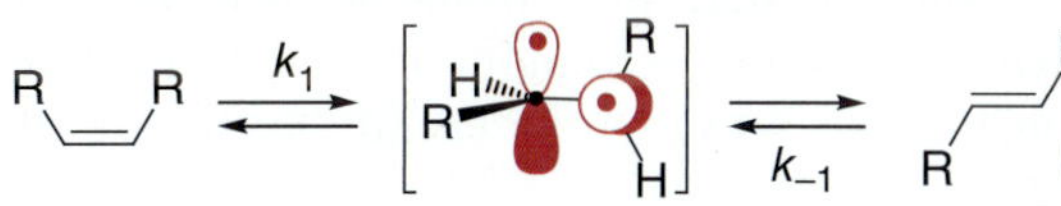

(3.76)

R = D E_a = 65 kcal mol^{-1} log A = 13: $\Delta^{\ddagger}S$ = 0 eu
R = Me E_a = 62 kcal mol^{-1} log A = 13.6
R = F E_a = 60.7 kcal mol^{-1} log A = 13.4
R = Ph E_a = 42.8 kcal mol^{-1} log A = 12.8

Table 3.4 Activation parameters for homolysis in the gas phase (E_a, $\Delta^{\ddagger}H$ in kcal mol^{-1}, $\Delta^{\ddagger}S$ in cal K^{-1} mol^{-1} or eu).

			T_m (K)	E_a	log A	$\Delta^{\ddagger}H$	$\Delta^{\ddagger}S$
CH_3—H	→	$Me^\bullet + H^\bullet$	1900	103	14.88	101	7
CH_3—CH_3	→	2 $Me^\bullet$	800	91.7	17.45	90	20
Et—Et	→	2 $Et^\bullet$	400	82	17.4	81.2	20.5
CF_3—CF_3	→	2 $CF_3^\bullet$	350	88	17.4	87.3	21
$(MeO)_2$	→	2 $MeO^\bullet$	410	36.1	15.4	35.3	10
$(AcO)_2$	→	2 $AcO^\bullet$	330	31	14.9	30.4	9.5

Isomerization (3.77) of cyclopropylmethyl radical into the homoallyl radical occurs with log $A \cong 13$, $\Delta^{\ddagger}S \cong 0$. This is also observed for reaction (3.78) and the isomerization (3.79) of cyclobutylmethyl radical into pent-4-en-1-yl radical.

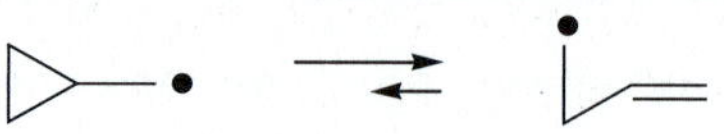

(3.77)

E_a = 7.05 kcal mol^{-1}, log A = 13.15

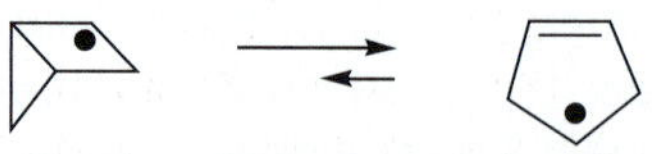

(3.78)

E_a = 5.2 kcal mol^{-1}, log A = 13.05

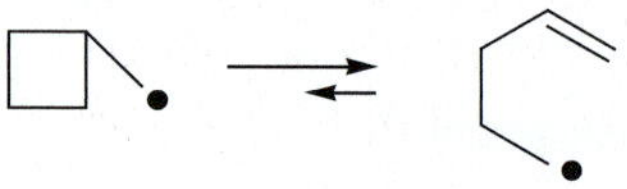

(3.79)

E_a = 11.9 ± 2 kcal mol^{-1}, log A = 13.1 ± 1.4

This indicates that σ(C—C) does not break appreciably in the transition states; instead, the intermediate alkyl radicals interact strongly with the π bond generated in the reaction.

The *exo*-trig cyclization of hept-6-en-2-yl radical into (2-methylcyclopentyl)methyl radical (reaction (3.80)) can be followed by electron spin resonance (ESR) spectroscopy in benzene at 40 °C. The reaction has an activation entropy, $\Delta^{\ddagger}S = 15$ eu. Notice how free rotation about four σ(C—C) bonds is lost in the transition state due to alkene formation. A negative entropy of cyclization of pent-1-ene into cyclopentane of −13.1 eu is found (Section 2.10, Table 2.4) [42].

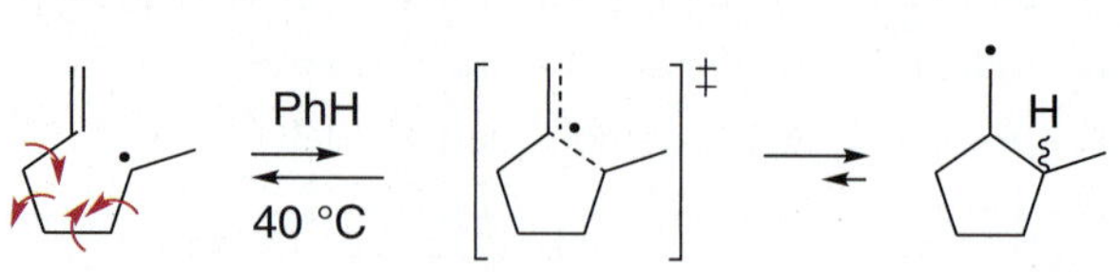

$\Delta^{\ddagger}S = -15$ eu (*exo*-Trig-cyclization) (3.80)

The (1,5)-sigmatropic shift (Section 5.5.6) responsible for the isomerization of (*Z*)-hexa-1,3-diene to (*Z*,*Z*)-hexa-2,4-diene (reaction (3.81)) has an activation entropy $\Delta^{\ddagger}S(3.81) = -12$ eu, typical of a concerted mechanism. This implies a cyclic transition structure for which the rotations about two σ(C—C) bonds are blocked [43].

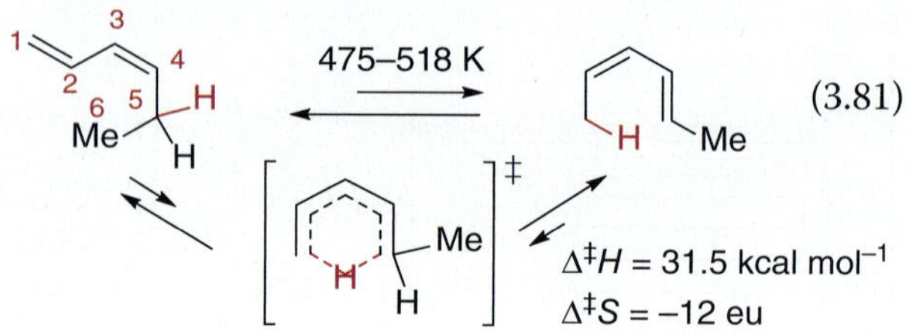

(3.81)

Problem 3.5 The racemization of 1,3-dimethylallene (−)-**A** has activation parameters $\Delta^{\ddagger}H = 45.0$ kcal mol^{-1} and $\Delta^{\ddagger}S = 0.5$ eu (at 280 °C) [44]. Two mechanisms can be proposed for that isomerization. Hypothesis I: rotation about one double C=C bond via **B**. Hypothesis II: equilibration with cyclopropylidene intermediates **C** and **D**. Applying thermochemical calculations, which of these two hypotheses is more consistent with the experimental data?

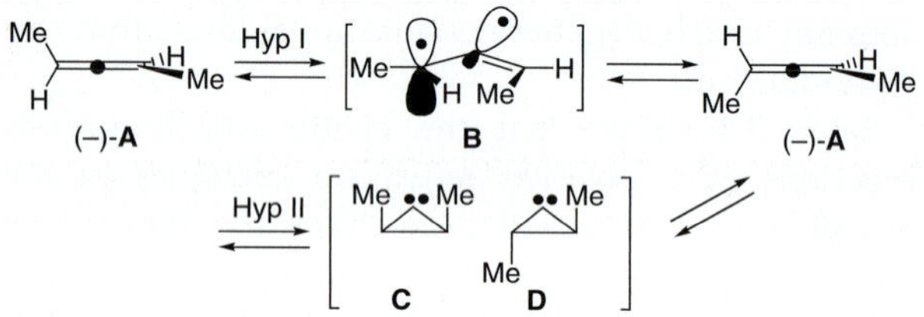

Problem 3.6 Devise a synthetic procedure to prepare (−)-**A** and (+)-**A** (problem 3.5). Compare with references [45] and [46].

Problem 3.7 Give possible mechanisms for the following isomerizations and estimate their activation entropies.

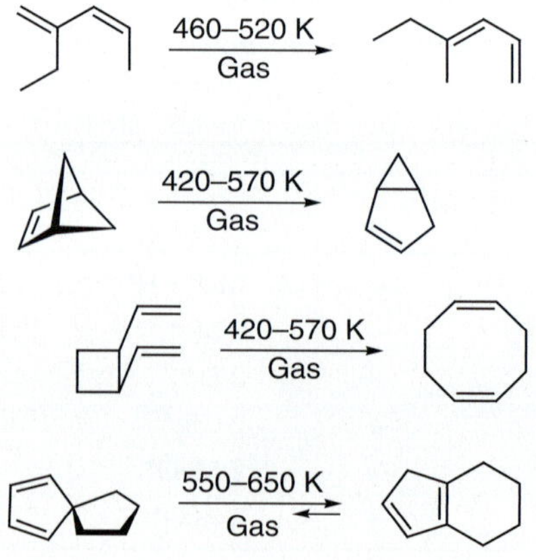

3.4.3 Example of homolysis assisted by bond formation: the Cope rearrangement

For the prototype of the Cope rearrangement (3.82) (an example of (3,3)-sigmatropic rearrangement, Section 5.5.9), $\Delta^{\ddagger}H = 33.5 \pm 0.5$ kcal mol^{-1} and $\Delta^{\ddagger}S = -13.8 \pm 1$ eu between 480 and 530 K [47–49]. As shown below, these data are not consistent with a dissociative mechanism that would generate two allyl radicals by the breakage of the σ(C(3),C(4)) bond (kinetic analysis). The fact that 1,1,4,4-tetradeuterohexa-1,5-diene (**5**) is not observed as product excludes by itself such dissociative mechanism (product analysis).

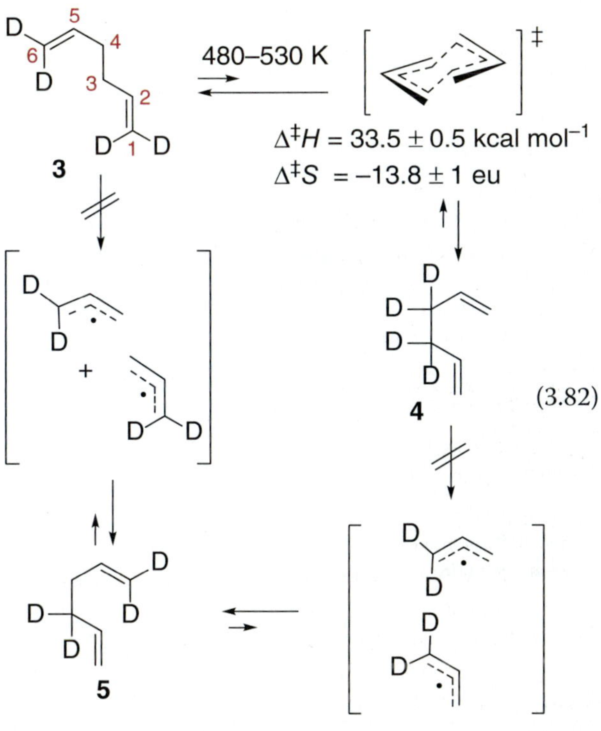

(3.82)

The comparison of $\Delta^{\ddagger}H$(3.82) with the homolytic bond dissociation enthalpy $DH°(CH_2{=}CHCH_2^{\bullet}/CH_2{=}CHCH_2^{\bullet})$ = $\Delta_r H°$(3.82) = $\Delta_r H°$(1.69) = 56.1 kcal mol^{-1} (Section 1.9.1) also excludes the dissociative mechanism (Figure 3.16). An associative mechanism must be deduced because of the negative activation entropy. The mechanism of Grob that implies the formation of a chair-like transition state for the Cope reaction is a possible transition state based on the negative activation entropy (and the stereoselectivity of the reaction). Indeed, cyclization of the reactant blocks the rotation of three σ(C—C) bonds (diminution of entropy by c. −15 eu, Section 2.10). The heat of formation of cyclohexa-1,4-diyl diradical is 57.3 kcal mol^{-1} according to $\Delta_f H°$(cyclohexane) + 2 $DH°$(cyclohexyl$^{\bullet}$/H$^{\bullet}$) − $DH°$(H$^{\bullet}$/H$^{\bullet}$) = −29.5 + 2(95.5) − 104.2 kcal mol^{-1} (Tables 1.A.2 and 1.A.7, Section 1.12.5). Compared with the heat of formation of hexa-1,5-diene (20.1 kcal mol^{-1}, Table 1.A.2), one estimates a minimum activation enthalpy for the Grob mechanism $\Delta^{\ddagger}H \geq 57.3 - 20.1 = 37.2$ kcal mol^{-1}, which surpasses by 37.2 − 33.5 = 3.7 kcal mol^{-1} the experimental activation enthalpy measured for the Cope rearrangement (3.82). Uncertainty on this estimate can be evaluated to 0.5 (experimental $\Delta^{\ddagger}H$) + 3 (uncertainty of 1 kcal mol^{-1} for the homolytic dissociation enthalpies used) = 3.5 kcal mol^{-1}. It therefore appears that the Grob mechanism may compete with another associative mechanism in which the breakage of bond σ(C(3),C(4)) of hexa-1,5-diene occurs in a concerted manner with the formation of the σ bond between C(1) and C(6). The latter mechanism corresponds to a concerted (3,3)-sigmatropic shift (Section 5.5.9) in which the bond breakage is assisted by the bond-forming process.

Cope rearrangements of substituted 1,5-hexadienes may either follow one-step, concerted mechanisms or follow two-step mechanisms. The latter can be associative Grob-type mechanisms in which the cyclohexane-1,4-diyl diradical intermediates are stabilized by the substituents at C(2) and C(5) or be dissociative mechanisms where the two allyl radical intermediates that formed are stabilized by the substituents at C(1), C(3), C(4), and C(6) (section 5.5.9.6) [50–57]. The timing of bond formation of Cope rearrangements has been studied by molecular dynamics simulations, and the simple examples are dynamically concerted, with bond-making and breaking both occurring with in 60 fs. [57b].

Problem 3.8 What is the probable mechanism of the Claisen rearrangement of allyl vinyl ether into pent-4-enal for which $\Delta^{\ddagger}H = 25.7 \pm 0.3$ kcal mol^{-1}, $\Delta^{\ddagger}S = -14.1 \pm 0.7$ eu have been determined between 118 and 153 °C in deuterated benzene (see section "Claisen Rearrangement and Its Variants (5.5.9.2) [58]. Estimate the heat of this reaction.

3.4.4 Example of homolysis assisted by bond-breaking and bond-forming processes: retro-carbonyl-ene reaction

Gas-phase pyrolysis of pent-1-ene at 550–650 °C gives a mixture of $\leq C_6$ alkanes and alkenes resulting from the formation of radicals (C—H and C—C homolysis) [59–61]. In contrast, the gas-phase pyrolysis of allyl ethyl ether (**6**) at 285–365 °C produces propene and acetaldehyde as unique products of reaction with activation parameters $\Delta^{\ddagger}H$ = 42.3

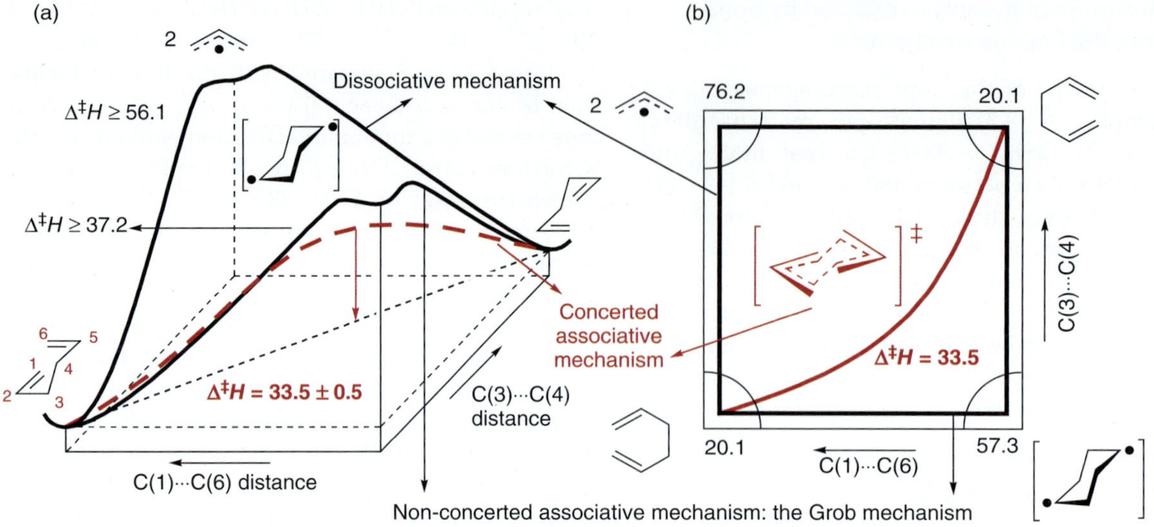

Figure 3.16 (a) Representation of the enthalpy hypersurface of the Cope rearrangement following a dissociative mechanism giving two allyl radicals, an associative mechanism generating the cyclohexa-1,4-diyl diradical (Grob mechanism) and (in black) a concerted (associative) mechanism. (b) More O'Ferrall–Jencks type diagram representing the same mechanism limits with the elevations given by the heats of formation of reactant and intermediates.

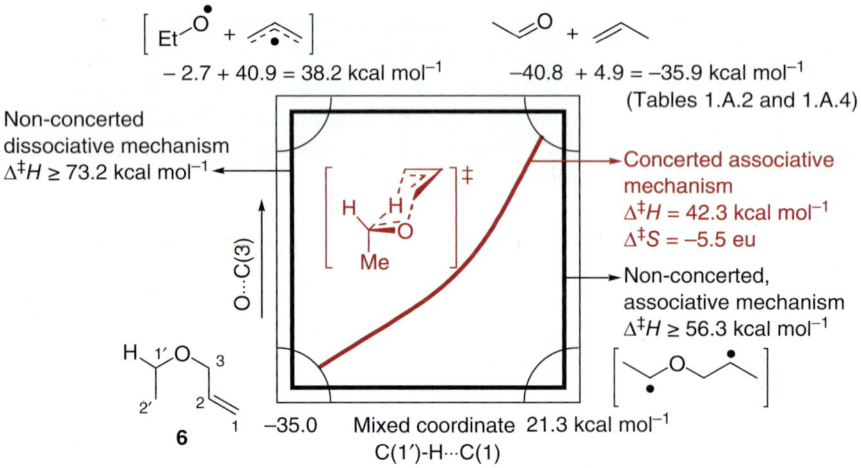

Figure 3.17 More O'Ferrall–Jencks type diagram showing two limiting nonconcerted mechanisms and a concerted (in black) mechanism for the retro-carbonyl-ene reaction allyl ethyl ether → ethanol + propene.

$\pm$ 0.8 kcal mol^{-1} and $\Delta^{\ddagger}S$ = -5.5 ± 2.0 eu [62]. This is an example of retro-carbonyl-ene reaction that follows a monomolecular, one-step mechanism (see Section 5.7.2 for the reverse reaction, the carbonyl-ene reaction). Although the retro-ene reaction has positive entropy of reaction (fragmentation), its activation entropy is negative. This rules out the hypothesis of a dissociative mechanism such as the homolysis generating the allyl and ethoxy radicals (Figure 3.17). The latter process would require an activation enthalpy $\Delta^{\ddagger}H$(homolysis) > 73.2 kcal mol^{-1}, much higher than the measured barrier ($\Delta^{\ddagger}H$). The following thermochemical parameters are used: $\Delta_f H°$(Et–O–allyl) = $\Delta_f H°$(Et–O–n-Pr) + 30 kcal mol^{-1} = $-65.0 + 30$ = -35.0 kcal mol^{-1} (Table 1.A.4); $\Delta_f H°$(Et-O$^{\bullet}$) = -5 kcal mol^{-1} [63, 64]; $\Delta_f H°$(allyl$^{\bullet}$) = 40.9 kcal mol^{-1} (Table 1.A.7); and $\Delta_f H°$(MeCH$^{\bullet}$–O–CH$_2$CH$^{\bullet}$–CH$_3$) = $\Delta_f H°$(CH$_3$CH$_2$–O–CH$_2$CH$_2$CH$_3$) + $DH°$(MeOCH$_2{}^{\bullet}$/H$^{\bullet}$) − 2 + $DH°$(i-Pr$^{\bullet}$/H$^{\bullet}$) − $DH°$(H$^{\bullet}$/H$^{\bullet}$) (Table 1.A.7) = $-65.0 + 93.1 - 2 + 99.4 - 104.2 = 21.3$ kcal mol^{-1}.

An associative mechanism involving the transfer of a hydrogen atom to generate the MeCH$^{\bullet}$–O–CH$_2$–CH$^{\bullet}$–Me 1,4-diradical is consistent with $\Delta^{\ddagger}S < 0$ but not with $\Delta^{\ddagger}H = 42.3$ kcal mol^{-1} as one estimates $\Delta_f H°$(MeCH$^{\bullet}$–O–CH$_2$CH$^{\bullet}$–CH$_3$) = 21.3 kcal mol^{-1}, thus requiring a minimum activation enthalpy $\Delta^{\ddagger}H$(1,4-diradical) $\geq 21.3 - (-35.0) \cong 56.3$ kcal mol^{-1}. The conclusion is that the retro-carbonyl-ene reaction follows a concerted mechanism in which the σ(C—O) bond breakage is assisted by the hydrogen atom transfer that implies itself a C—H

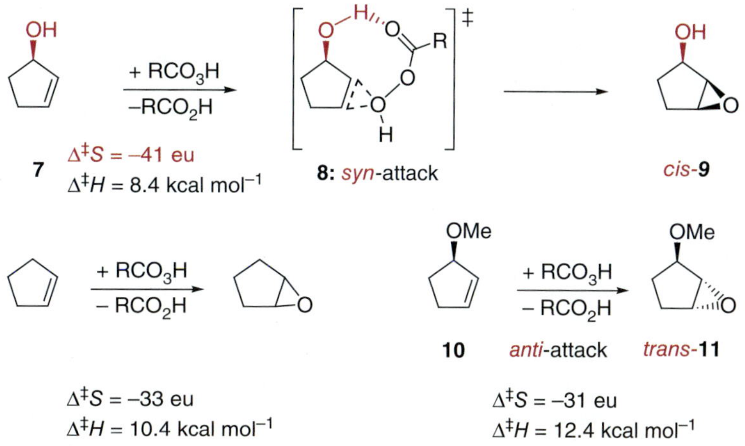

Scheme 3.3 Stereochemical control by the allylic alcohol moiety in its epoxidation.

bond-breaking and a C—H bond-forming process (Figure 3.17) [65, 66]. Pyrolysis of hepta-1,6-diene at 355–470 °C produces propene and butadiene as major products with $\Delta^{\ddagger}H = 45 \pm 0.8$ kcal mol^{-1} and $\Delta^{\ddagger}S = -11.3 \pm 2.0$ eu, also consistently with a concerted retro–ene reaction [67]. The gas-phase pyrolysis of 3-methylpent-4-yn-2-ol at 520–700 °C produces buta-1,2-diene (methylallene) and acetaldehyde also in a retro–ene process [68]. The gas-phase pyrolysis of but-3-enoic acid into propene + CO_2 has $\Delta^{\ddagger}H = 37$ kcal mol^{-1} and $\Delta^{\ddagger}S = -10.3$ eu at 658 K, also consistently with a concerted retro–carbonyl–ene reaction [69, 70].

Cycloadditions (Section 5.3) often follow concerted one-step mechanisms in which two bonds are formed simultaneously. The transition states of the forward and reverse (cycloreversion) reactions are the same if both reactions occur under the same conditions (principle of microscopic reversibility [19]). For instance, at 175 °C, the Diels–Alder reaction (Section 5.3.9) butadiene + ethylene → cyclohexene (that competes with the cyclodimerization of butadiene, Section 5.3.4) follows a concerted mechanism, whereas the reverse reaction cyclohexene → butadiene + ethylene, which must be carried out at a much higher temperature (Section 2.11.6: fragmentation on heating), might follow a nonconcerted mechanism with the formation of hex-5-en-1-yl diradical as a reactive intermediate. The diradical has three freely rotating single C—C bonds and a higher entropy than a transition structure resembling cyclohexene. Molecular dynamics studies of the reaction of butadiene and ethylene at high temperature show that the stepwise reaction begins to compete with the concerted process at high temperatures. [57c]

Problem 3.9 What are the major products of gas-phase pyrolysis allyl methyl amine [66], of diallyl ether [65]?

3.4.5 Can a reaction be assisted by neighboring groups?

Reactions involving neighboring group directing effects have usually more negative activation entropies than those that do not. Scheme 3.3 shows an example: the epoxidation of cyclopent-2-enyl alcohol (**7**) with a carboxylic peracid. More ordered transition structures typically occur for reactions involving a neighboring group. The reagent is typically guided by coordination; the hydroxyl group of the allylic alcohol makes a hydrogen bond with the carbonyl group of the peracid as shown with transition state **8** [71]. This hypothesis is confirmed by the *syn*-stereoselectivity of the epoxidation of the allylic alcohol, which gives selectively the *cis*-epoxy-alcohol **9**, whereas the corresponding methyl ether **10** gives preferentially the *anti*-epoxide **11** under the same conditions.

3.5 Competition between cyclization and intermolecular condensation

Before the structure elucidation of civetone by Ruzicka in 1926 [72], it was believed that all macrocycles would be unstable because of their angle strain (Baeyer strain, Section 2.6). In 1933, Ziegler and coworkers found that alkane-1,ω-dinitriles can be cyclized into α-iminocycloalkanecarbonitriles on treatment with Et_2NLi [73]. Yields were poor for the preparation of 9-, 10-, 11-, 12-, and 13-membered cycles but better for rings containing more than 15 members [74]. Hydrolysis of the α-iminocarbonitriles provides the corresponding cycloalkanones as exemplified by the synthesis of Exaltone® (perfume of Naef and Ruzicka; Scheme 3.4).

In 1934, Stoll and coworkers measured the rate constants for the lactonization (intramolecular

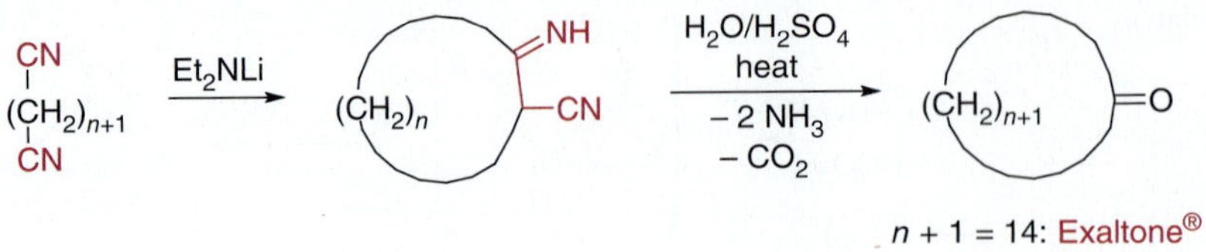

Scheme 3.4 Ziegler's synthesis of cyclopentadecanone and other macrocycloalkanones.

esterification) of ω-hydroxycarboxylic acids to give the corresponding lactones, esters, and polyesters, under high dilution conditions [75–77]. It was found that the rate constant of lactonization was the highest for the formation of γ-lactone (five-membered ring lactone) and smaller for the formation of other lactones (reaction (3.83)).

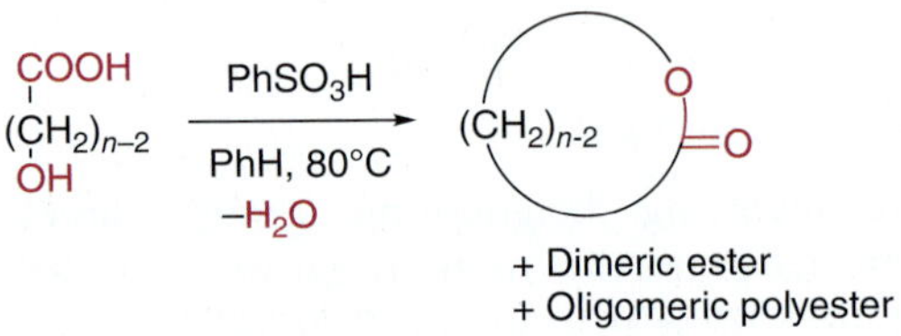

(3.83)

	$n = 5$	7	9	11
k_{rel} (lactonization):	>0.5	2.8×10^{-3}	8×10^{-6} Slowest	2.8×10^{-5}
	13	15	19	24
	2.6×10^{-4}	2.3×10^{-3}	4.9×10^{-3}	3.4×10^{-3}

The second-order rate constants of S_N2 displacements (3.84) are almost the same in dimethyl sulfoxide (DMSO) at 50 °C for alkanoates and primary alkyl bromides of different sizes. In the case of intramolecular displacement reactions (intramolecular nucleophilic substitutions: S_Ni) (3.85), their first-order rate constants strongly depend on the ring size of the lactone formed. Their activation enthalpies and activation entropies depend on the ring size [78].

$$RCO_2^{\ominus} + Br{-}CH_2R' \xrightarrow[\text{DMSO, 50 °C } (S_N2)]{k^{II}} RCOOCH_2R' + Br^{\ominus} \quad (3.84)$$

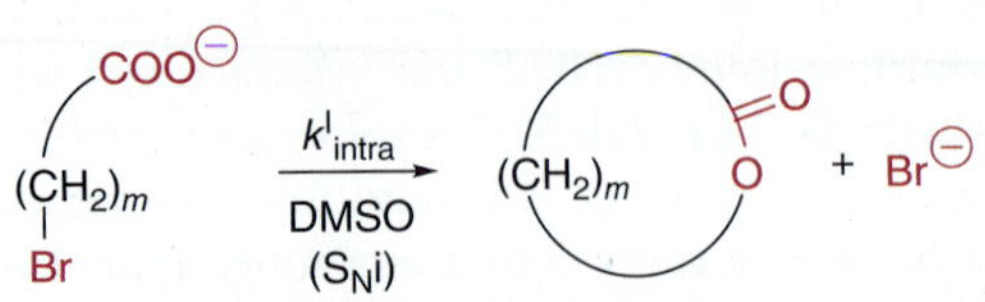

(3.85)

Lactonizations (3.85) have higher activation enthalpies than intermolecular esterification reaction (3.84). The rate decreases with the increase of the ring size. For large rings ($n > 24$), $\Delta^{\ddagger}H(3.85) \cong \Delta^{\ddagger}H(3.84)$.

For the formation of α-lactone ($m = 1$), β-lactone ($m = 2$), and medium-size lactones ($m = 6$–18), $\Delta^{\ddagger}H(3.85)$ roughly parallels the ring strain of the lactone formed (Table 2.A.3). The activation entropies for the formation of α-, β-, and γ-lactones are zero or slightly negative, as expected for monomolecular processes that do not have to block several free rotations about σ(C—C) bonds. As also expected, the larger the ring size of the lactone formed, the more negative the activation entropy. Now, a number of bond rotations have to be blocked as the ring size increases. For large ring size ($m > 18$), the activation entropy of lactonization becomes as negative as that of the bimolecular condensation. To avoid the competitive intermolecular displacements that have lower activation enthalpies than the intramolecular reactions, high dilution conditions are used. The slow addition of a dilute solution of the ω-bromoalkanoic acid to an excess of base (to form the ω-bromoalkanoate salt) in DMSO (Me_2SO) maintained at a temperature at which the displacement reaction is rapid insures optimal conditions for a good yield of macrolactonization.

3.5.1 Thorpe–Ingold effect

The Thorpe–Ingold effect (also called *gem*-dimethyl effect or angle compression effect) was first reported by Beesley, Thorpe, and Ingold in 1915 as part of a study of the cyclization reactions generating spiro-compounds. Geminal substitution of two hydrogen atoms by two methyl groups (or larger substituents) increases the bond angle between them (Scheme 3.5). As a result, the bond angle between the other two substituents decreases as shown with **12** and **13**. By moving them closer together, reactions between them are accelerated. Thus, it is a kinetic effect [79, 80]. Geminal disubstitution of an alkane chain limits the number of conformers that the alkane can adopt. The two chains departing from the quaternary center in **13** are in gauche conformations more often than in the nonsubstituted system **12** [81–83]. This explanation has been criticized [84]. The Thorpe–Ingold effect controls peptide conformation containing α-C-tetrasubstituted amino acids [85]. It accelerates the cyclopolymerization of

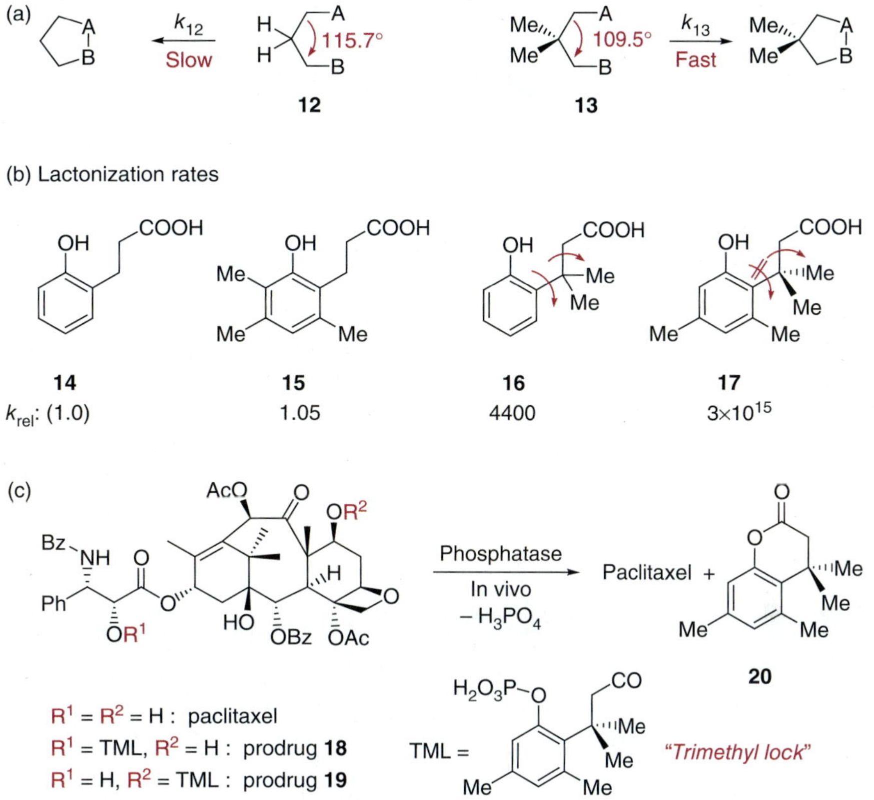

Scheme 3.5 The Thorpe–Ingold effect: (a) geminal disubstitution changes the geometry of the chain; (b) relative rates of lactonization of *ortho*-hydroxydihydrocinnamic acid derivatives; and (c) example of the use of "trimethyl lock" to form more water soluble derivatives (prodrugs) than anticancer paclitaxel.

4,4-disubstituted octa-1,7-diynes [86]. Lactonization of *ortho*-hydroxydihydrocinnamic acid (**14**) is accelerated by a factor of 4440 upon *β*,β-dimethyl substitution (compare **16** and **14**, Scheme 3.5). The propionic acid side chains in **14**–**16** can rotate freely. This is not the case anymore with derivative **17** because of the *ortho*-methyl substituent that intercalates between the benzylic *gem*-dimethyl group, realizing a "trimethyl lock" that forces the carboxylic function to approach more often the phenol moiety [87, 88]. Derivatives of **17** are used for molecular release in chemistry, biology, and pharmacology [89, 90]. For instance, esters of paclitaxel and **17** are more water soluble than paclitaxel (Taxol®), an important anticancer drug. Hydrolysis of the dihydrogenophosphate moieties of **18** and **19** is catalyzed in vivo by alkaline phosphatase. This releases the phenol group that immediately generates lactone **20** liberating paclitaxel and H_3PO_4 (Scheme 3.5c) [91].

In the case of the lactonization of 5-bromo-3,3-dimethylpentanoate, its rate constant at 50 °C is 38.5 times as large as that of the lactonization of 5-bromopentanoate under the same conditions [82, 92–94]. This effect is important for the formation of six-membered rings, but not always for others as shown with reaction (3.86) [95].

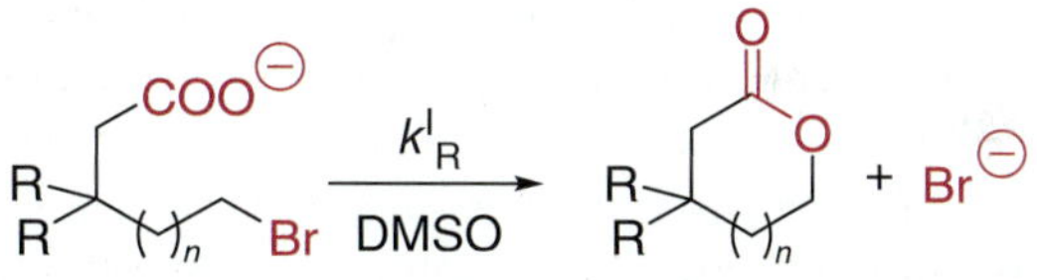

(3.86)

$$\frac{k^{I}_{R} = \mathrm{Me}}{k^{I}_{R} = \mathrm{H}} = \begin{array}{ccccc} n=1 & 2 & 3 & 4 & 9 \\ 38.5 & 6.6 & 1.1 & 0.6 & 1.2 \end{array}$$

Acylfuran **21** does not undergo the expected intramolecular Diels–Alder addition on heating. In contrast, its synthetic precursor **22** (Scheme 3.6), which bears gem-cyano and trimethylsilyloxy substituents on the side chain, undergoes a reversible Diels–Alder addition giving two diastereomeric cycloadducts **23** and **24**. This is explained by invoking the intervention of a Thorpe–Ingold effect: the *gem*-disubstitution limits the number of possible conformations of the alkenyl side chain in the ground state, thus making the entropy of the intramolecular Diels–Alder addition less negative and the activation energy smaller. Another factor that can intervene is the fact that the bond angle about the carbonyl moiety in **21** is wider than that about the corresponding carbon center of **22** bearing the CN and Me_3SiO substituents, thus making the approach of the alkene

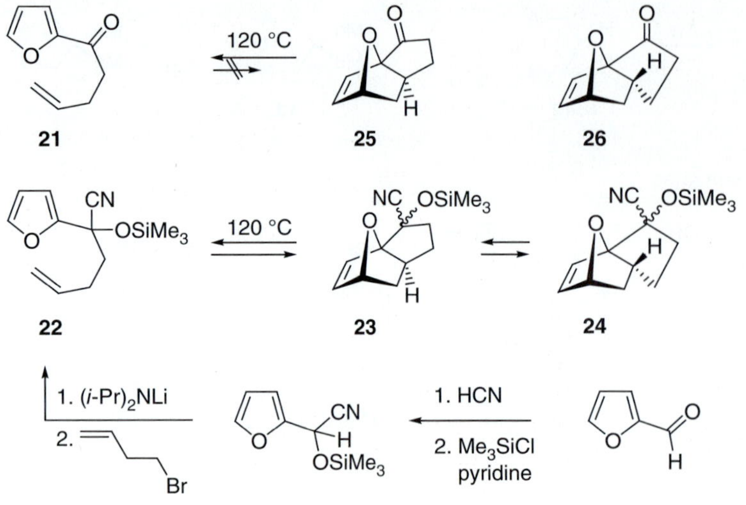

Scheme 3.6 Substituent effect on the equilibria resulting from intramolecular Diels–Alder additions of furan derivatives.

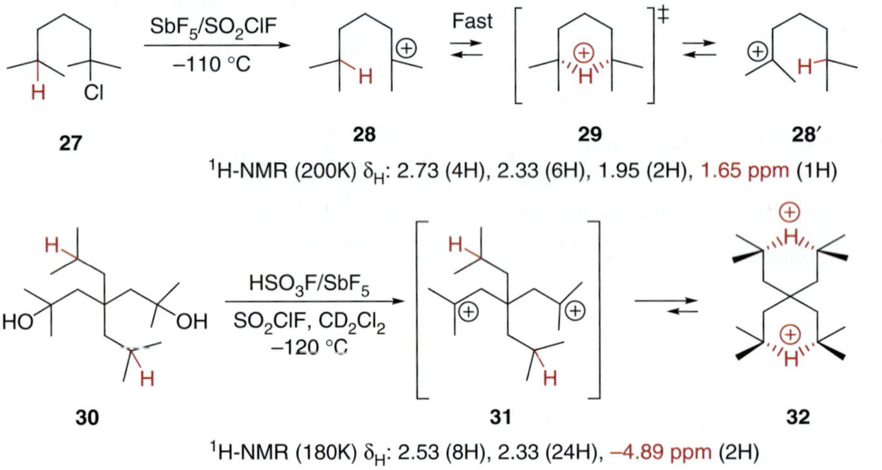

Scheme 3.7 The Thorpe–Ingold effect converts the transition structure of a 1,5-hydride shift into the ground state in the bis-μ-hydrido species **32**.

moiety (dienophile) to the furan unit (diene) easier for **22** than for **21** [96–101]. Essentially, any factor that destabilizes the transition-state geometry without affecting the ground state will retard the reaction, whereas destabilizing the reactant accelerates the reaction.

The ^{1}H-NMR spectrum of 2,6-dimethyloct-2-yl cation (**28**) obtained by ionization of the corresponding chloride **27** in a mixture of SbF_5 and SO_2ClF at −110 °C (super-ionizing medium) is characteristic of a tertiary carbenium ion undergoing a very fast, degenerate 1,5-hydride shift (equilibrium **28** ⇄ **28′**), with a transition state that is the bridged μ-hydrido species **29** (Scheme 3.7). Ionization of diol **30** in HSO_3F/SbF_5 (magic acid)/SO_2ClF/CD_2Cl_2 at −120 °C gives a dication, the ^{1}H-NMR of which is not consistent with a classical dicarbenium ion **31**, but with a doubly bridged di-μ-hydrido dication **32** (the high field proton at $\delta_\pi = -4.89$ ppm is typical for μ-hydrido species, see e.g. the ^{1}H-NMR spectrum of B_2H_6). The geminal 4,4-disubstitution of each 2,6-dimethylhept-2-yl moieties of dication **31** favors its cyclization and makes the μ-hydrido structure **32** more stable than the open, classical tertiary carbenium ions [102].

Problem 3.10 The ^{1}H-NMR spectrum of the dication obtained by dissolving 4-(2-methylpropyl)-2,6-dimethylocta-2,6-diol in magic acid and SO_2ClF at −120 °C shows the following signals at δ_H: 3.20 (6H), 2.88 (18H), 2.88 (1H), and −0.78 ppm (1H). Propose a structure for this dication.

Problem 3.11 On treating nona-1,8-dien-5-one (**A**) with the molybdenum complex **C** (catalyst for metathesis of alkenes, Section 8.6.2), a polymer is obtained together with ethylene. Under similar conditions, 4,4,6,6-tetramethylnona-1,8-dien-5-one generates 2,2,7,7-tetramethylcyclohept-4-en-1-one (**B**) in 95% yield [103]. Explain the difference in behavior between **A** and **B**.

C:

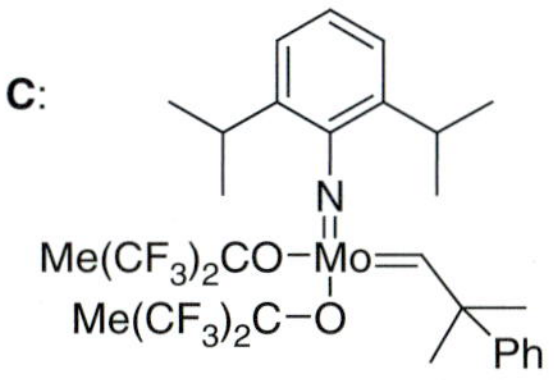

Problem 3.12 Explain why enone **21** (Scheme 3.6) refuses to equilibrate with **25** and **26**, whereas **D** equilibrates with **E** at 80 °C in benzene.

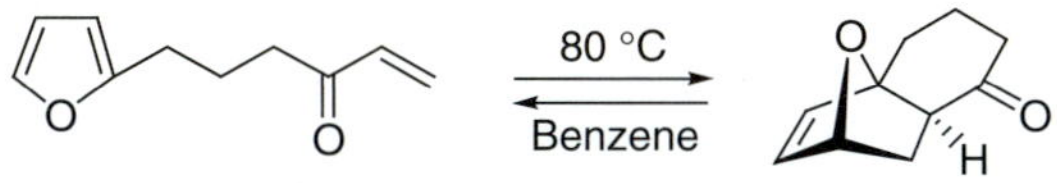

D **E**

3.6 Effect of pressure: activation volume

In 1887, Planck defined the reaction volume $\Delta_r V$ (in $cm^3\ mol^{-1}$) as [104]

$$\left(\frac{\partial \ln K}{\partial P}\right)_T = \frac{-\Delta_r V}{RT} \tag{3.87}$$

at constant temperature. The reaction volume varies with pressure P (1 Pa = 1 N m^{-1}; 1 MPa = 10 bar, 1 GPa = 10 kbar). The variation of the rate constant of a given reaction as a function of pressure and at constant temperature is given by van't Hoff with the Eq. (3.88) [105]. Measuring the rate constants k as a function of pressure allows one to determine the activation volume, $\Delta^\ddagger V$, parameter of the reaction.

$$\Delta^\ddagger V = -RT\left(\frac{\mathrm{d}\ln k}{\mathrm{d}P}\right)_T \tag{3.88}$$

3.6.1 Relationship between activation volume and the mechanism of reaction

Depending on the reaction mechanism, the activation volume, $\Delta^\ddagger V$, of a given reaction might be positive (dissociative mechanism), zero (isomerization), or negative (associative mechanism) [106–111]. In 1892, Röntgen observed that the HCl-catalyzed hydrolysis of sucrose slows down with increasing pressure to P = 50 MPa, compared with ambient pressure (P = 0.1 MPa) [112]. Soon after, Rothmund observed that some acid-catalyzed ester hydrolyzes are accelerated on increasing the pressure [105]. For some reactions, one observes a linear relationship for ln k and pressure (P); for other reactions, this is not the case [113]. The activation volume, $\Delta^\ddagger V$, gives information concerning the reaction mechanism [114, 115]. For instance, the thermal intermolecular carbonyl–ene reaction (3.89) (Section 5.7.2) has an activation volume $\Delta^\ddagger V = -8.7\ cm^3\ mol^{-1}$ at 25 °C, which is somewhat more negative than the variation of reaction volume, $\Delta_r V = -5.7\ cm^3\ mol^{-1}$ [116]. This can be taken as a proof for a concerted mechanism in which the hydrogen transfer from 3-methylbut-1-ene to dimethyl oxomalonate occurs in concert with the formation of a new σ(C—C) bond. For the intermolecular ene-reaction of cyclohexene with 4-phenyl1,2.4-triazoline-3,5-dione, for which $\Delta^\ddagger H = -12.7$ kcal mol^{-1}, $\Delta^\ddagger S = -25.3$ eu, $\Delta^\ddagger V = -29.1\ cm^3\ mol^{-1}$, $\Delta_r H° = -37.3$ kcal mol^{-1}, and $\Delta_r V = -25.0\ cm^3\ mol^{-1}$ have been measured at 25 °C in toluene, a concerted mechanism analogous to that retained for reaction (3.89) has been proposed [117].

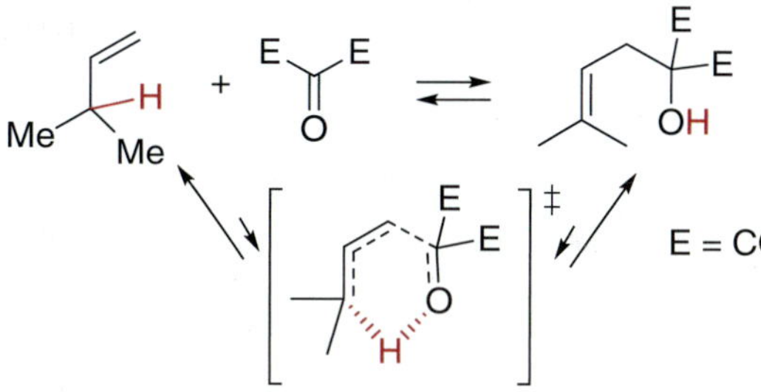

(3.89)

$\Delta V^\ddagger = -8.7\ cm^3\ mol^{-1}$ at 25 °C,
$\Delta_r V = -5.7\ cm^3\ mol^{-1}$

For Diels–Alder additions (Section 5.3.9), activation volumes of −25 to −45 $cm^3\ mol^{-1}$ are observed, in agreement with the hypothesis that the transition states of these cycloadditions resemble more the cycloadducts than the reactants; that is, the reaction volumes are similar to the activation volumes [118, 119]. For intramolecular Diels–Alder additions, their activation volumes are also negative and can reach −45 $cm^3\ mol^{-1}$ [120, 121]. For intermolecular Diels–Alder addition, solvent effects on the activation volume have been observed [122–125].

The activation volume of the degenerate Wagner–Meerwein rearrangement (Section 5.5.1) of 1,2-dimethoxybicyclo[2.2.1]hept-2-yl salts **33** in superionizing media is slightly positive $\Delta^\ddagger S(3.90) = 8\ cm^3\ mol^{-1}$ [126]. This observation has been interpreted in terms of solvation effects: there is greater positive charge delocalization in the transition state **34** of the reaction than in the carbenium ions **33**.

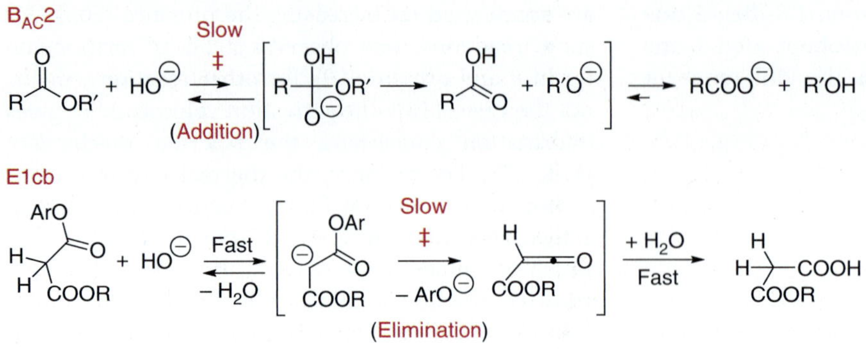

Scheme 3.8 Two limiting mechanisms for the base-catalyzed hydrolysis of esters.

This makes the solvent interaction with the transition structure weaker than with the reactants, leading to overall larger volume in the transition state than in the reactants when the solvation shell is included.

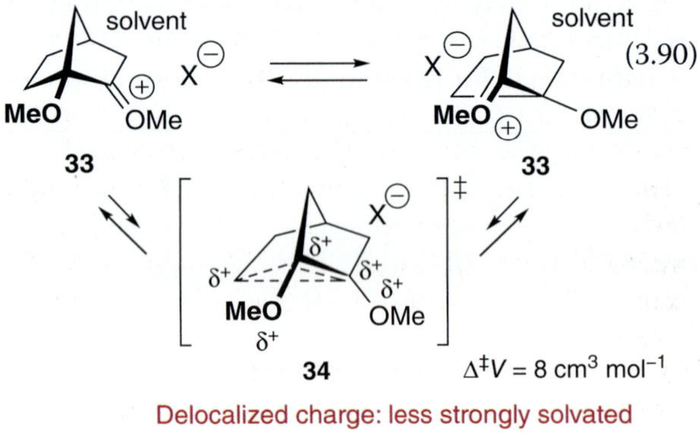

Problem 3.13 Ethylenetetracarbonitrile (tetracyanoethylene) adds to 1,1-dimethylbutadiene competitively in the (4+2) and (2+2) mode giving 3,3-dimethylcyclohex-4-ene-1,1,2,2-tetracarbonitrile and 3-(2-methylprop-1-enyl) cyclobutane-1,1,2,2-tetracarbonitrile, respectively. In toluene, the activation volume of the (2+2)-cycloaddition is more negative than that of the (4+2)-cycloaddition. In butyronitrile, reverse results are found. Explain [127].

3.6.2 Detection of change of mechanism

The activation volume is an indicator of the position of the transition state along the reaction coordinates. It permits the detection of subtle mechanistic details such as alteration of mechanism with variations of substrates [128–132]. Base-catalyzed hydrolysis of esters usually follows the $B_{AC}2$ mechanism (Ingold classification, Table 7.1, Section 7.2.1). Ester that has an acidic proton α to the carbonyl group or in a vinylogous position and a good leaving group (e.g. 4-nitrophenolate) may undergo hydrolysis by elimination to an intermediate ketene (E1cb mechanism), which rapidly adds water (Scheme 3.8).

A distinction between these two mechanisms can be made by measuring the activation volume. For instance, at pH 10.3 and 29.8 °C, hydrolysis of 2,4-dinitrophenyl 4-methoxybenzoate (**35**) has a negative activation volume $\Delta^{\ddagger}V = -19$ cm^3 mol^{-1} (rate-determining step implies bond formation between the base and the ester) and that of 2,4-dinitrophenyl 4-hydroxybenzoate (**36**) under the same conditions has a positive activation volume, $\Delta^{\ddagger}V$ = 16 cm^3 mol^{-1} (dissociative process in the rate-determining step) [133].

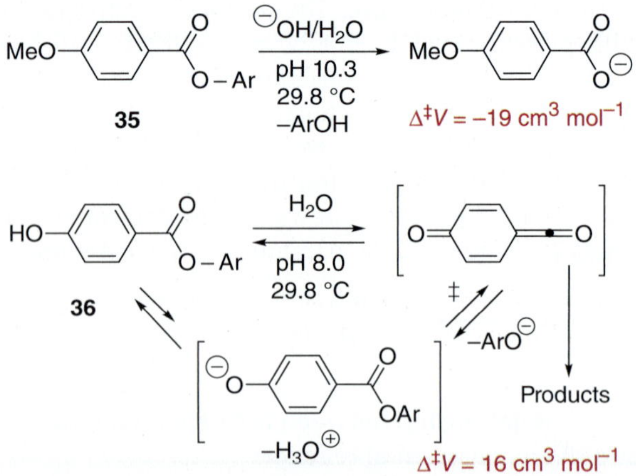

Problem 3.14 Explain the change of $\Delta^{\ddagger}V$ values from 16, 7.1 to −17.9 cm^3 mol^{-1} for the hydrolysis of 2,4-dinitrophenyl 4-hydroxybenzoate at pH 8.0, 10.1, and 12.5 (at 29 °C) [133].

Problem 3.15 Evaluate whether the Cope rearrangement of *rac*-(*E*,*E*)- and *meso*-(*E*,*E*)-1,3,4,6-tetraphenylhexa-1,5-diene can occur by a dissociative mechanism with the formation of 1,3-diphenylallyl radicals, considering the following activation parameters for their interconversion: $\Delta^{\ddagger}H = 30.7 \pm 0.2$ kcal mol^{-1}, $\Delta^{\ddagger}S = 2.1 \pm 0.4$ eu, and $\Delta^{\ddagger}V = 13.5 \pm 0.1$ cm^3 mol^{-1} between 77 and 115 °C. Resolution of *rac*-(*E*,*E*)-1,3,4,6-tetraphenylhexa-1,5-diene can be done by chiral chromatography. This allowed one to observe a fast racemization at 40–65 °C with $\Delta^{\ddagger}H = 21.3 \pm 0.1$ kcal mol^{-1}, $\Delta^{\ddagger}S = -13.2 \pm 0.3$ eu, and $\Delta^{\ddagger}V = -7.4 \pm 0.4$ cm^3 mol^{-1}. What is the probable mechanism of this racemization [134]?

3.6.3 Synthetic applications of high pressure

Reactions under high pressure occur everywhere in the Universe, for instance, in the deep sea, inside the Earth, and on other planets. Pressure is a critical parameter in coal and oil formation. Ultrahigh pressure (>100 kbar) has led to the preparation of synthetic diamonds and rubies [135]. In organic chemistry, high pressure refers to 1–20 kbar and uses batch reactors of 0.1–20 cm^3 [136–140]. With the development of microreactor synthesis under medium pressure (20–1000 bar), the technique becomes accessible to preparative laboratories [139]. Dauben and coworkers have developed a two-step synthesis of cantharidin [141, 142], the active principle of the aphrodisiac "spanish fly" (Scheme 3.9). The first step involves the Diels–Alder addition of furan to 2,5-dihydrothiophene-3,4-dicarboxylic anhydride (**37**), which leads to a 1 : 4 mixture of cycloadducts **38** and **39**. The reaction requires a high pressure of 7 kbar at 20 °C. After desulfurization and alkene hydrogenation with H_2 and Raney nickel as a catalyst, a mixture of cantharidin (**40**) and *epi*-cantharidin (**41**) is obtained from which pure **40** is isolated in 51% yield after selective recrystallization from EtOAc. Dauben also found that furan adds to **37** at 80 °C under 1 atm in the presence of the Grieco's catalyst (5 M $LiClO_4$ in ether) [143–146].

Several syntheses based on high-pressure Diels–Alder reactions have been reported [147–161]. After five hours at 25 °C and under 5 kbar, a 1 : 1 mixture of 1,1-bis(3,5-dimethylfur-2-yl)ethane (**42**; made in four steps from acetone) and diethyl (*E*,*E*)-4-oxohepta-1,7-dioate (**43**) gives a single adduct **44** isolated in 95% yield. This *meso* compound contains nine stereogenic centers. Desymmetrization (Section 3.7.6) is realized by hydroboration of one of the two alkene moieties of **44** with monoisopinocampheylborane ((+)-IpcBH_2) [162–164] in THF at −25 °C. Work-up with $NaBO_3{\cdot}(H_2O)_4$ provides the enantiomerically enriched alcohol (+)-**45** in 59% yield and 78% ee. In only two steps, an achiral compound (**42**) has been converted into an enantiomerically enriched compound ((+)-**45**) containing 11 stereogenic centers. The latter compound has been converted into complicated polypropionate derivatives (Scheme 3.10) [165].

Chemoselectivity can be affected by pressure. An example is given with the thermolysis of (*Z*)-nona-1,3,8-triene (**46**) that gives a 2.2 : 1 mixture of (*E*,*Z*)-1,5,7-nonatriene (**47**) via a concerted (1,5)-hydrogen shifts and *cis*-bicyclo[4.3.0]non-2-ene (**48**) via an intramolecular Diels–Alder addition at 150 °C in *n*-pentane solution and at 1 bar (Scheme 3.11). The intramolecular Diels–Alder addition is accelerated by pressure: at 7.7 kbar and 150 °C, it becomes the favored reaction by a factor of 4.2/1. The Diels–Alder cycloaddition has a more negative activation volume than the (1,5)-sigmatropic shift ($\Delta\Delta^{\ddagger}V \cong -10$ cm^3 mol^{-1}) [121].

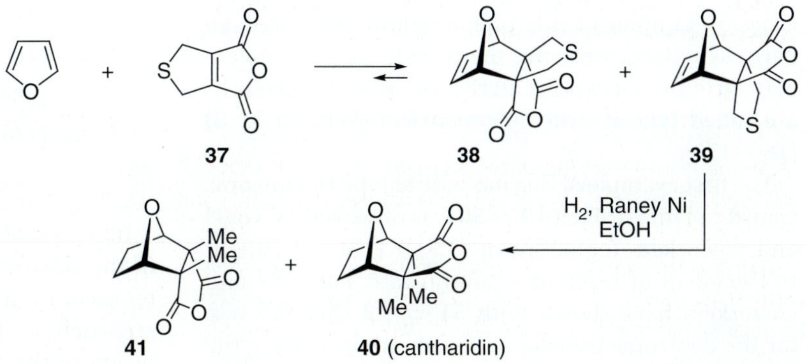

Scheme 3.9 Dauben's synthesis of cantharidin and *epi*-cantharidin

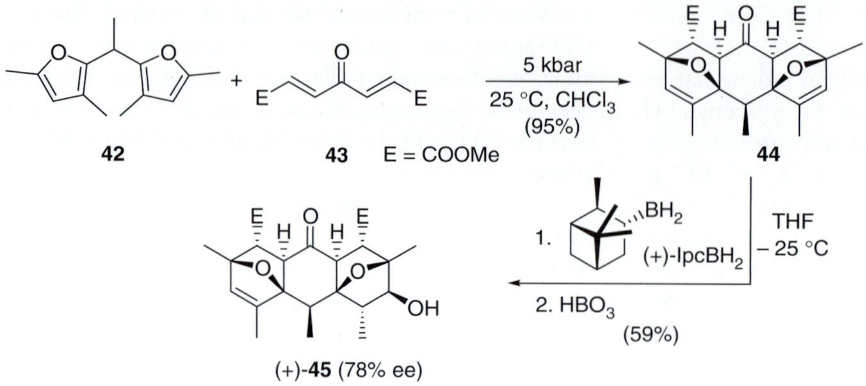

Scheme 3.10 Highly stereoselective double Diels–Alder reaction facilitated by pressure and desymmetrization of the adduct by hydroboration with an enantiomerically enriched alkylborane (desymmetrization by chirality, Section 3.7.4).

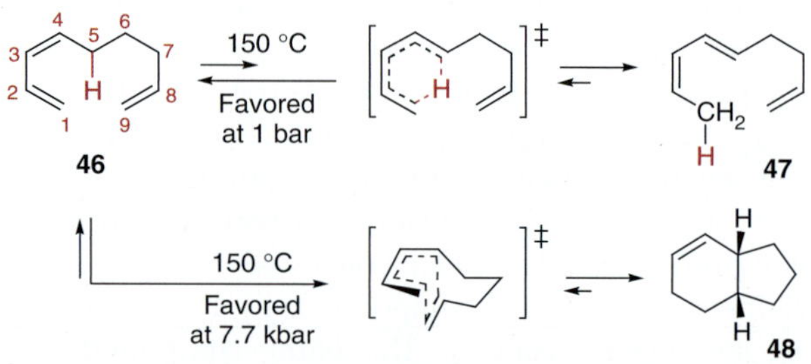

Scheme 3.11 Competition between (1,5)-hydrogen shift and intramolecular Diels–Alder addition (example of chemoselectivity). The latter is favored at high pressure only.

3.6.4 Rate enhancement by compression of reactants along the reaction coordinates

There are a number of reactions that can be accelerated by incorporation into structures or environments where the reactants are forced into proximity. This is one of the possible mechanisms of the Thorpe–Ingold effect (Section 3.5.1). In such cases, the reaction is often said to be accelerated by stress or strain. Effectively, the ground-state free energy is raised by such interactions, relative to the transition state. The activation free energy is lowered and the rate is increased. Classic examples of this phenomenon are discussed here. Reactions involving the transfer of two molecular groups intramolecularly, or intermolecularly, are called type II dyotropic transfers (Section 5.6.3) [166–170].

The first example of thermoneutral type II dyotropic transfer of dihydrogen **49** ⇄ **50** was observed by Vogel and coworkers (equilibrium (3.91)) [171]. Pertinent to the effect of reactant strain on rate, Paquette and coworkers have shown with **51** ⇄ **52** that the rate for the dyotropic transfer increases on decreasing the distance between the ethano and ethylene moieties (reaction (3.92)) [172, 173]. This demonstrates that the compression of the two reactants along the main reaction coordinates reduces the activation enthalpy as they are placed closer to the transition structure in their ground state. Distance provides a handy metric, although it is the energetic consequences of forcing reactant groups into proximity that accelerates the reaction.

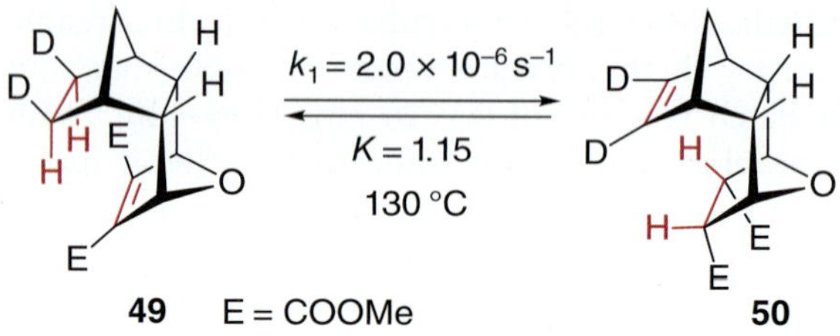

(3.91)

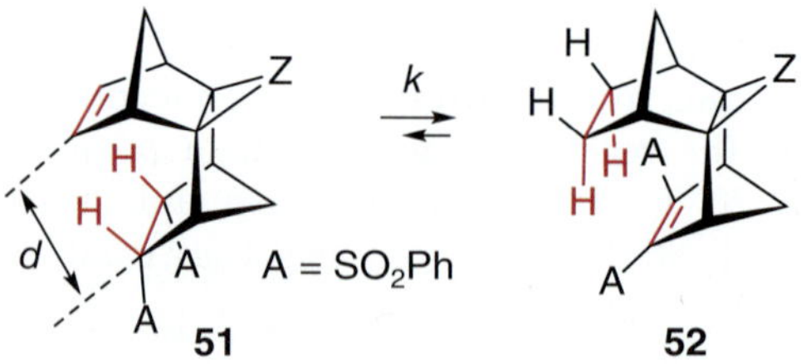

(3.92)

Z:	O	H H	NPh	CH_2	CHMe
d:	2.54	2.41	2.40	2.32	2.28 Å
k_{rel}:	(1.0)	24	440	4 100	48 000

In enzymology, the catalysis of a reaction by strain of the substrate by the catalyst is one factor involved to account for catalysis [174–177]. The distortions are such as to weaken the existing bonds and to promote the formation of new ones. In general, the enzyme-active site has evolved to fit the transition state better than the substrate in its ground state [178]. This accounts for the lowering of the activation enthalpy of the process **ES** ⇄ **EP** in which the substrate **S** is converted into product **P** (Section 3.2.8). An experimental argument in favor of this theory comes from the finding that molecules that resemble the transition structure of the process **S** → **P**, but cannot form the product (transition structure mimics), often bind very strongly to the enzyme **E** and inhibit it effectively. Many transition-state mimics, also called transition-state analogs, make useful drugs.

Another example of reaction rate enhancement by reactant compression is given by the comparison of the rates of the proton transfer for the intramolecular process (3.93) that is very fast, with that of the intermolecular reaction (3.94), which is very slow. In the solid state, the distance between C—H and :N is 2.34 Å for **53**, which is significantly shorter than the sum of the van der Waals radii (2.75 Å) of these two functions [179–181]. Reaction (3.93) can be taken as a model of the activation by enzymes that use proton transfer during the catalytic process.

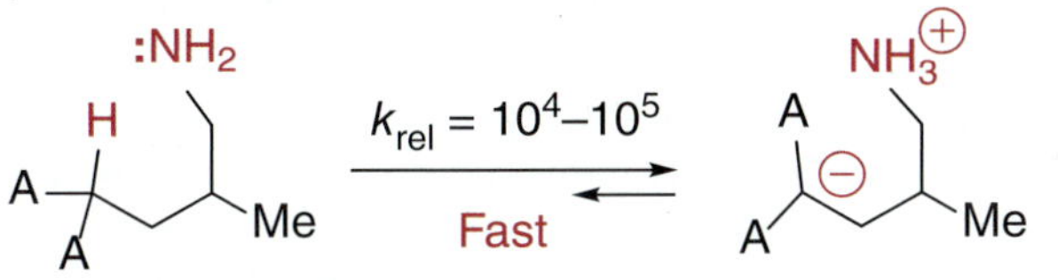

53

(3.93)

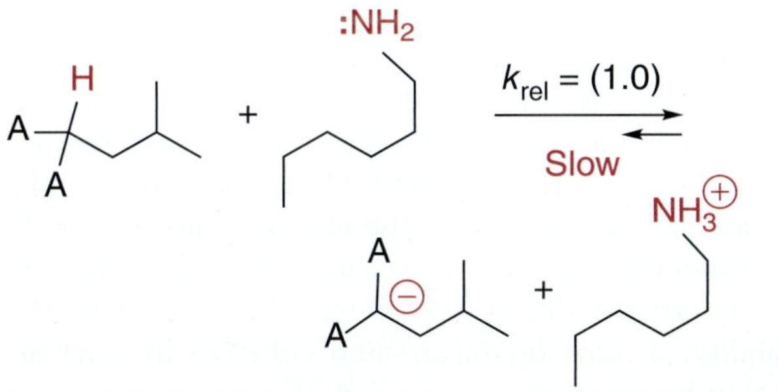

(3.94)

It is not the distance itself that determines reactivity but the fact that the compressed geometry of the molecule is now unfavorable for the reactant and favorable for the transition state; this reduces the free energy of activation and accelerates the reaction. Lively debates over the origins of such effects have been published; Menger favors the emphasis on the distance between reactants [180, 181] and Houk emphasizes the free energy change upon compression [182]. Dynamic contributions must also be considered to understand enzyme catalysis [177].

Problem 3.16 What occurs to the polycyclic system **49** or **50** upon heating above 160 °C? Propose a synthesis for **49** (see Section 5.6.3).

3.6.5 Structural effects on the rate of the Bergman rearrangement

Another example of rate enhancement by functional group compression is the Bergman rearrangement of enediynes into benzene-1,4-diyl diradicals (Section 1.12.2) [183, 184]. For the parent compound, 1,6-dideuterohex-3-ene-1,5-diyne $\Delta^{\ddagger}G(3.95) \cong 32\ \text{kcal mol}^{-1}$ at 200 °C($\Delta^{\ddagger}H(3.95) \cong 28.2\ \text{kcal mol}^{-1}$, $\Delta^{\ddagger}S \cong -10.2$ eu) has been determined ($\tau_{1/2}$ = 30 seconds at 200 °C). In the case of cyclodec-3-ene-1,5-diyne (**54** (n = 2)), the Bergman rearrangement occurs rapidly at 25 °C, whereas the homologs **54** (n = 3,4) are stable at this temperature. As the distance between the two alkyne moieties that have to form a σ(C—C) bond decreases, the activation enthalpy of the cyclization decreases (reaction (3.96)). This reaction is essential for the mechanism of action of antitumor drugs such as esperamycin A_1. A simple analog of these drugs is enone **55**, which is a stable compound at 25 °C, whereas its product of intramolecular conjugate addition **56** undergoes fast Bergman rearrangement at 25 °C. Converting the sp^2-hybridized carbon bridgehead center of **55** into a sp^3-hybridized carbon center as in **56** makes the 1,6-distance of the enediyne moiety shorter in **56** than in **55**, therefore, the rate enhancement for the Bergman rearrangement of **56** compared with that of **55** [185–188].

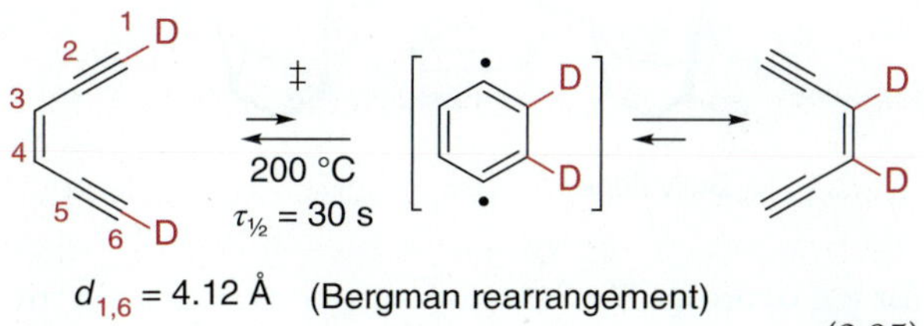

(3.95)

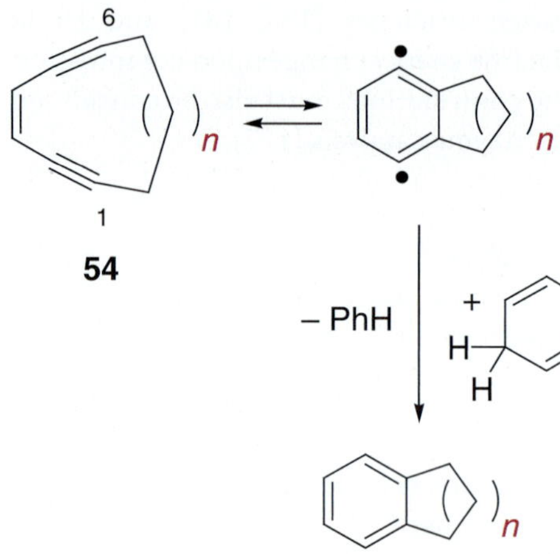

(3.96)

$n = 2$ $d_{1,6} = 3.25$ Å: fast reaction at 25 °C

$n = 3$ $d_{1,6} = 3.61$ Å: stable at 25 °C

$n = 7$ $d_{1,6} = 4.33$ Å: stable at 25 °C

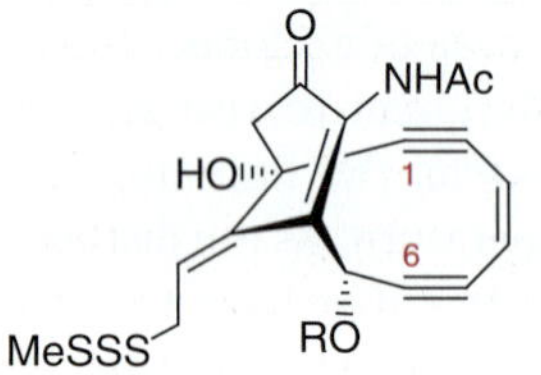

55: $d_{1,6} = 3.35$ Å
Stable at 25 °C

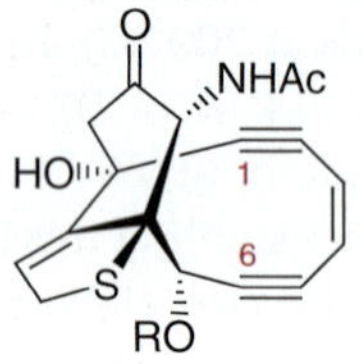

56: $d_{1,6} = 3.16$ Å
Fast cyclization at 25 °C

Problem 3.17 The degenerate Cope rearrangement of bullvalene that exchanges all the proton signals in its ^{1}H-NMR spectrum and all the ^{13}C signals in its ^{13}C-NMR spectrum at room temperature has activation parameters $\Delta^{\ddagger}H = 12.6 \pm 0.05$ kcal mol^{-1} and $\Delta^{\ddagger}S = 0.8 \pm 0.2$ eu (233–413 K). Why is this isomerization so much easier than the Cope rearrangement (3.82) (Section 3.4.3) or that of 1,4-dimethylidenecyclohexane for which $\Delta^{\ddagger}H = 39.0 \pm 1$ kcal mol^{-1}, $\Delta^{\ddagger}S = -13.8 \pm 2$ eu have been determined at 600 K [48]?

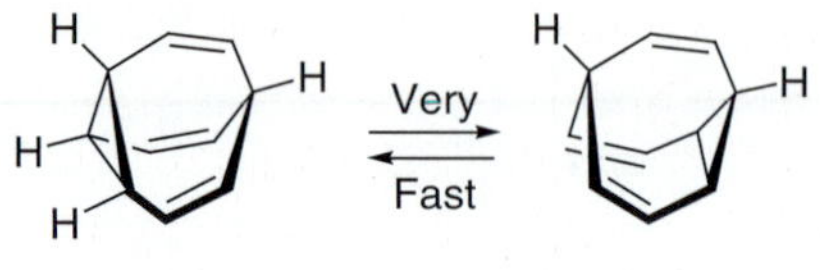

Bullvalene

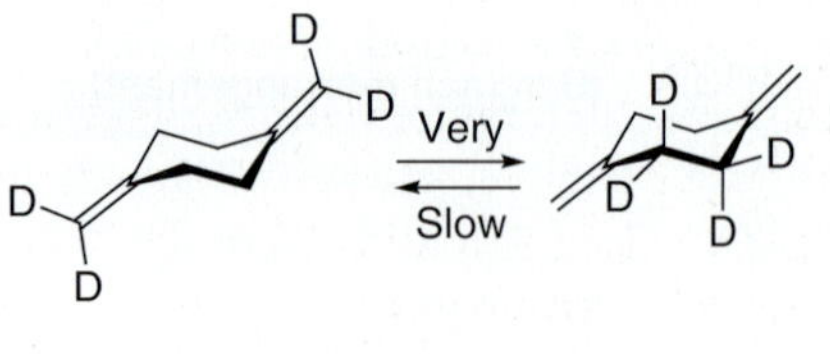

3.7 Asymmetric organic synthesis

Because living systems are made of enantiomerically pure molecules and biopolymers, chiral therapeutic agents should not be administered as racemic mixtures but as enantiomerically pure compounds (Section 1.14.2) [189, 190]. The need of enantiomerically pure chiral compounds also exists in material sciences because polymers made out of enantiomerically pure monomers might have specific mechanical, physical, and chemical properties that polymers made from racemic monomers have not [191–198]. The preparation of enantiomerically enriched compounds follows two general methods: the resolution of racemic mixtures [190] and asymmetric synthesis [199–202]. The term asymmetric synthesis was first coined by Emil Fischer [203] referring to the Kiliani–Fischer aldose chain elongation method [204–206] that produces diastereoisomeric cyanhydrines (HCN addition to carbaldehydes) in a ratio different from 1/1 [207]. Morrison and Mosher give a wider definition: asymmetric synthesis is a reaction or a sequence of reactions in which an achiral moiety in an ensemble of substrate molecules is converted by a reactant into a chiral unit in such a manner that the stereoisomeric products are produced in unequal amounts. Reactants include reagents and catalysts [208]. In Section 1.14, we presented a few methods based on thermodynamic control for the obtainment of enantiomerically enriched compounds. We now describe a selection of methods that generate enantiomerically enriched compounds under kinetic control [209, 210].

3.7.1 Kinetic resolution

In 1858, Pasteur reported the first kinetic resolution in which racemic ammonium tartrate was reacted with *Penicillium glaucum* [211]. The enantiomerically pure enzymes present in the microorganism catalyze the metabolism of (*R*,*R*)-tartrate selectively, leaving an (*S*,*S*)-tartrate almost unchanged (Section 1.14.3). The ability of microorganisms and enzymes to discriminate between enantiomers of racemic substrates is one of the best documented chapters of biochemical methodology in asymmetric synthesis [212–219]. Most kinetic resolutions of industrial importance are realized through the use of hydrolases such as lipases from *Candida cylindracea*, from *Candida antarctica* B, from *Pseudomonas* sp. (= *aeruginosa*), from *Pseudomonas fluorescens* (= *cepacia*), from *Mucor miehei*, from *Humicola lanuginosa*, from *Aspergillus niger*, from *Geotrichum candidum*, from *Rhizopus delemar*, and from porcine pancreas. Other enzymes that catalyze enantioselective hydrolyzes are esterases from pig liver and horse liver; acetyl esterase from orange *flavedo*; proteases such as α-chrymotrypsin, papain, and subtilisin A from *Bacillus licheniformis*, thermolysin from *Bacillus thermoproteolyticus*; protease from *Aspergillus oryzae*; amino acylase from porcine kidney and from *Aspergillus* sp.; penicillin acylase from *Escherichia coli*; D- and L-hydantoinases (dihydropyrimidinases) from *Agrobacterium radiobacter* and *Brevibacillus brevis*; and nitrilase and nitrile hydratase from *Brevibacterium* or from *Rhodococcus* pp. These enzymes are produced in bulk amount (detergent and food industry) and have the advantage of not requiring expensive coenzymes [220]. They can be active toward several substrates and may tolerate organic solvents [221]. Several enzymes have been immobilized on solid supports and can be used as heterogeneous catalysts [222–226]. An example of industrial application is the resolution of (±)-2-aminobutan-1-ol ((±)-**57**) used in the preparation of ethambutol, an antibiotic for the treatment of tuberculosis [227]. Several methods have been developed [228]. One of the simplest applies the enantioselective hydrolysis of racemic *N*-phenylacetyl derivative (±)-**58** with penicillin G acylase immobilized on Eupergit C. Penicillin G acylase is used mainly for the production of (+)-6-aminopenicillanic acid [229]; it is commercially available in large quantities. Product (*S*)-(+)-**57** is obtained in 99% ee (for conversion rates up to 40%) and the enzyme is recovered by simple filtration [230].

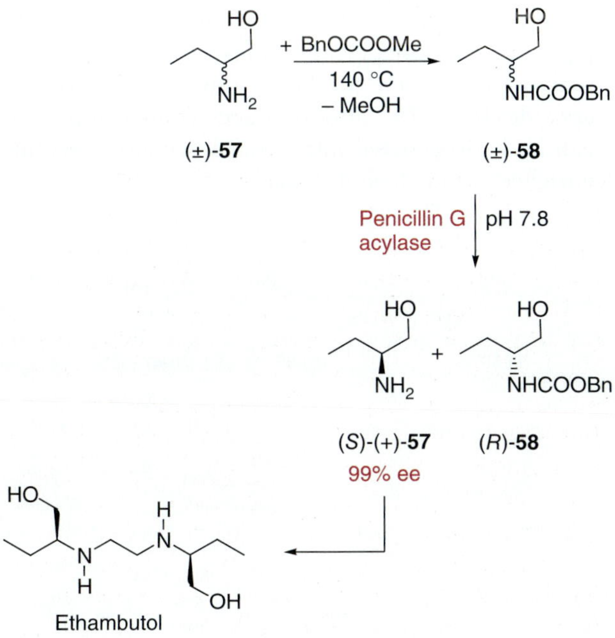

Epoxides constitute valuable synthetic intermediates [231–242]. The yeast *Rhodotorula glutinis* is a microorganism-effective catalyst for the enantioselective hydrolysis of various epoxides [243–246]. For instance, (*S*)-**59** is obtained in a 48% yield and 99% ee. Diol (*R*)-**60** that remains is isolated with an ee of 79% [247].

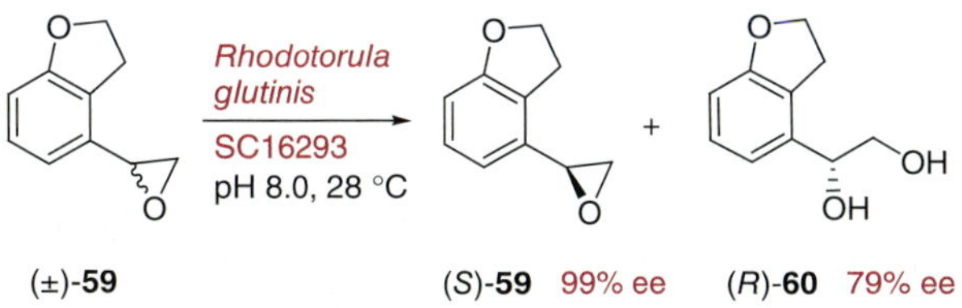

One of the problems with many enzymes and microorganisms is that they react with a limited number of substrates. For some reactions, enantioselectivity may be low. In principle, enzymatic activity can be improved through exchange of specific amino acid in the enzyme by using site-directed mutagenesis [248]. Alternatively, improved enantioselectivity can be obtained by creating new enzymes by "test tube directed evolution." [249–251] One starts with a wild enzyme that has a low selectivity and creates a library of mutants from which a more enantioselective variant is identified, and one repeats the process as many times as necessary by choosing in each case an improved mutant for the next round of mutagenesis. In doing so, random mutagenesis is not performed on the enzyme itself but on the gene (DNA segment) of the microorganism that encodes for the enzyme of interest [252–261]. Kinetic resolution by synthetic means [262] was first reported by Marckwald and McKenzie in 1899 in the esterification of (±)-mandelic acid with optically active (−)-menthol in which (+)-mandelic acid reacts faster than its enantiomer [207]. Kinetic resolution of a 1 : 1 mixture of two enantiomeric compounds (*R*)-**A** and (*S*)-**A** with a single enantiomerically pure reagent, or an enantiomerically pure catalyst, or in a chiral medium giving products **P** and **Q** leads to a yield of less than 50% of the slow reacting enantiomer (e.g. (*S*)-**A**) with a high ee if the rate constant ratio $k_R/k_S = s$ (selectivity factor) is larger than 200 [263–266] (Scheme 3.12).

Products **P** and **Q** can be diastereomers, enantiomers, or identical depending on the type of substrate (±)-**A** and reaction conditions. For instance, the esterification of a racemic alcohol with an enantiomerically pure chiral acylation agent gives two diastereoisomeric esters, whereas the esterification of a racemic secondary alcohol with an achiral acylation agent catalyzed by an enantiomerically pure base gives

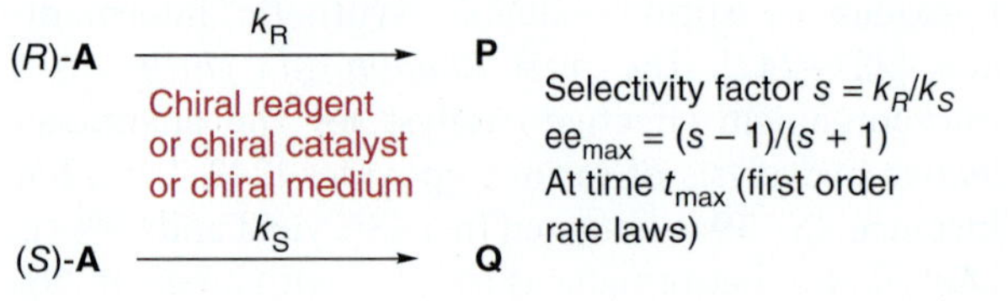

Scheme 3.12 Kinetic resolution with an enantiomerically pure chiral reagent, catalyst, or medium.

two enantiomeric esters. Alternatively, the oxidation (or didehydrogenation) of a racemic secondary alcohol with an enantiomerically pure oxidant or catalyst give the same achiral ketone from both enantiomeric alcohols. At time $t = 0$, $[(S)\text{-}\mathbf{A}] - [(R)\text{-}\mathbf{A}] = 0$. At $t = \infty$, there is no ee for products **P** and **Q** should they be enantiomeric products. There is an optimal time of reaction t_{max} for which $[(S)\text{-}\mathbf{A}] - [(R)\text{-}\mathbf{A}]$ reaches a maximum value $\{[(S)\text{-}\mathbf{A}] - [(R)\text{-}\mathbf{A}]\}_{max}$. If one stops the reaction at time t_{max}, the ee of unreacted (*S*)-**A** is $ee_{max} = \{[(S)\text{-}\mathbf{A}] - [(R)\text{-}\mathbf{A}]\}/\{[(S)\text{-}\mathbf{A}] + [(R)\text{-}\mathbf{A}]\} = (s - 1)/(s + 1)$. For longer reaction time, the yield in (*S*)-**A** diminishes. If **P** and **Q** are enantiomers, their ee′ is the highest at the beginning of the reaction: $ee'_0 = \{[\mathbf{P}] - [\mathbf{Q}]\}/\{[\mathbf{P}] + [\mathbf{Q}]\} = (s - 1)/(s + 1)$. With a selectivity factor s = 10 (first-order rate laws), 99% ee in (*S*)-**A** is possible with approximately 70% conversion and 30% yield. The method of kinetic resolution is useful if an enantiomerically pure catalyst is the source of asymmetry and if it can be used in small amounts. Easy separation of unreacted substrate, products of reaction, and catalyst is also a prerequisite for making the method useful. Furthermore, products **P** and **Q** should not become wastes; they should be valued, for instance, by their easy conversion into the starting substrate [267]. Examples are given below. Catalytic hydrogenation (4 atm) of racemic 4-hydroxycyclopent-2-en-1-one ((±)-**61**) with 0.1% enantiomerically pure catalyst (*S*)-**62** (Noyori's (*S*)-(BINAP)Ru(OAc)$_2$), Section 7.8.6) proceeds with $k_S/k_R = s = 11 : 1$. At 68% conversion, a maximum yield of slow reacting (*R*)-**61** with 98% ee is obtained, together with (*R*)-enriched product of hydrogenation **63**. Silylation of (*R*)-**61** gives (*R*)-**64** that can be purified by crystallization. It is an important enantiomerically enriched synthetic intermediate (a chiron) [268] in the Noyori's synthesis of prostaglandins (e.g.: (−)-PGE$_2$) [269–271]. Prostaglandins are a group of hormone-like lipids that are mediators and have a variety of physiological effects such as regulating the contraction and relaxation of smooth muscle tissues [272]. Compound **63** is not a waste as it is converted into valuable cyclopent-2-en-1-one upon treatment with a base such 1,8-diazabicyclo[5.4.0]undec-7-ene (DBU) [273].

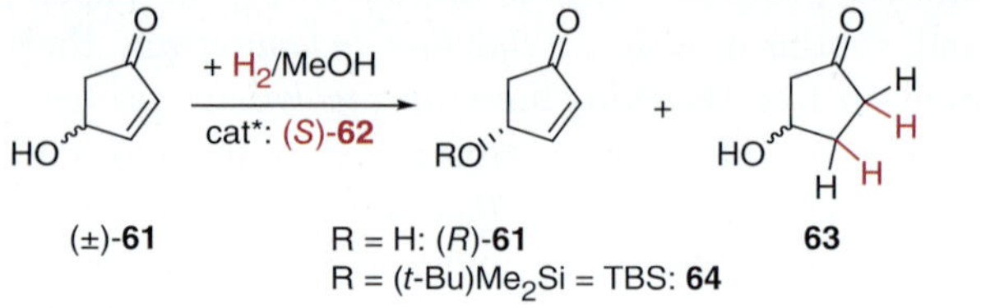

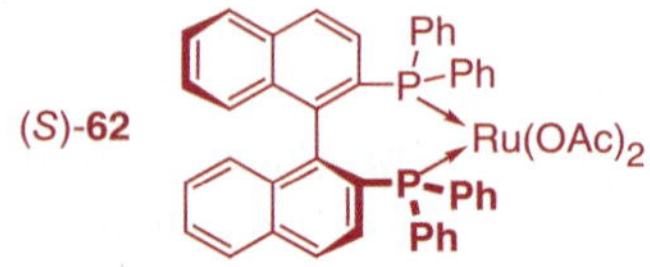

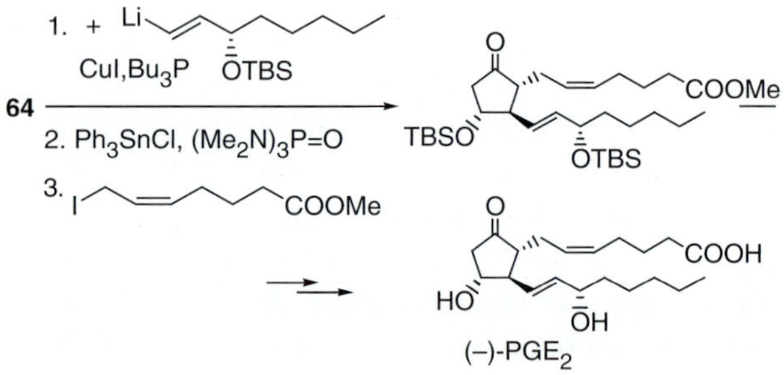

Jacobsen and coworkers have developed the enantiomerically pure cobalt(III) catalyst (*R*,*R*)-**66** (Jacobsen's (salen)Co-OAc catalyst). It catalyzes the enantioselective hydrolysis of terminal epoxides with a selectivity factor approaching 400. For example, racemic propylene oxide ((±)-**65**) adds water in the presence of only 0.2–0.8 mol% of (*R*,*R*)-**66** producing epoxide (*S*)-**65** and (*R*)-propane-1,2-diol ((*R*)-**67**) with high yield and excellent enantiomeric excesses [274]. Similar kinetic resolutions have been realized with RCOOH and Me$_3$SiN$_3$ additions to terminal epoxides [275]. The latter produces azidoalcohols that can be hydrogenated into valuable enantiomerically enriched aminoalcohols [276].

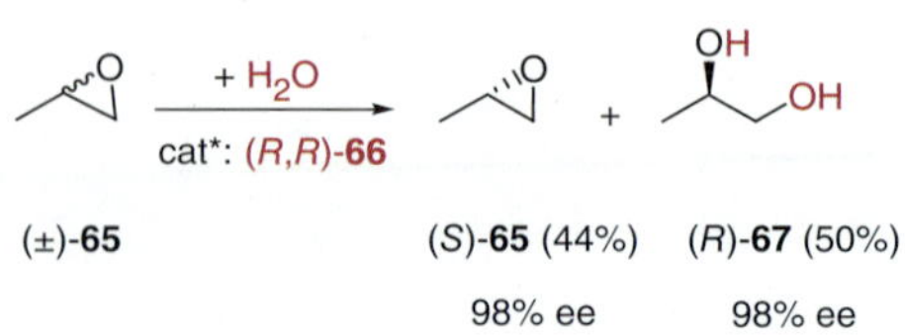

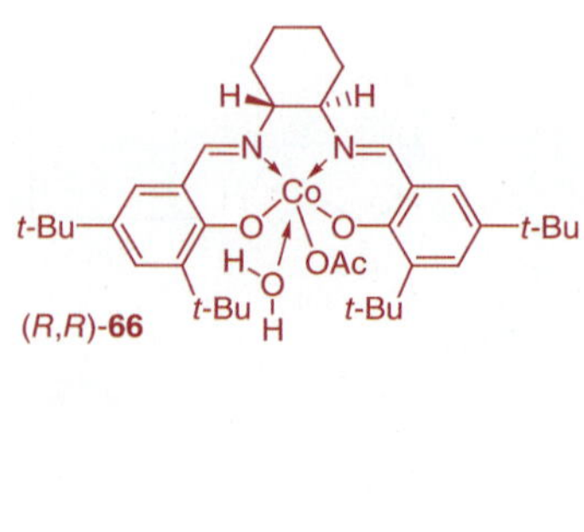

4-Dimethylaminopyridine (DMAP) is a well-known catalyst for acyl transfer reactions including esterification of alcohols by carboxylic anhydrides and acyl chlorides (Section 7.2.4). Fu and coworkers have developed enantiomerically pure DMAP analogs capable to catalyze the enantioselective esterification of secondary alcohols [277], of arylalkylcarbinols [278], of propargylic alcohols [279] and of allylic alcohols with high enantioselectivities [280]. For instance, 0.5% planar–chiral iron complex (−)-**69** (Fu's catalyst) permits the preparation of diol (*S*,*S*)-**68** and diacetate (*R*,*R*)-**70** with good yields and high ee values by enantioselective acetylation of racemic diol (±)-**68**. Kinetic resolution of thiols through acylation with cyclic dicarboxylic anhydrides catalyzed by thiourea-modified cinchona alkaloid catalysts has also been reported [281].

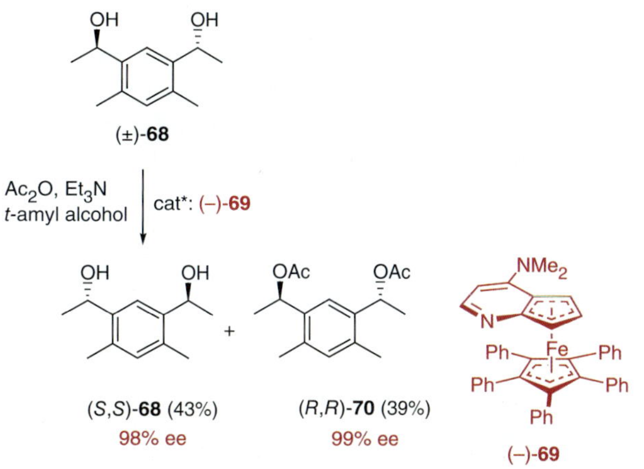

An example of kinetic resolution of carboxylic acids (±)-**70** through catalyzed enantioselective esterification (reaction (3.97)) is presented in Scheme 3.13. It has been applied to the preparation of (*S*)-ibuprofen

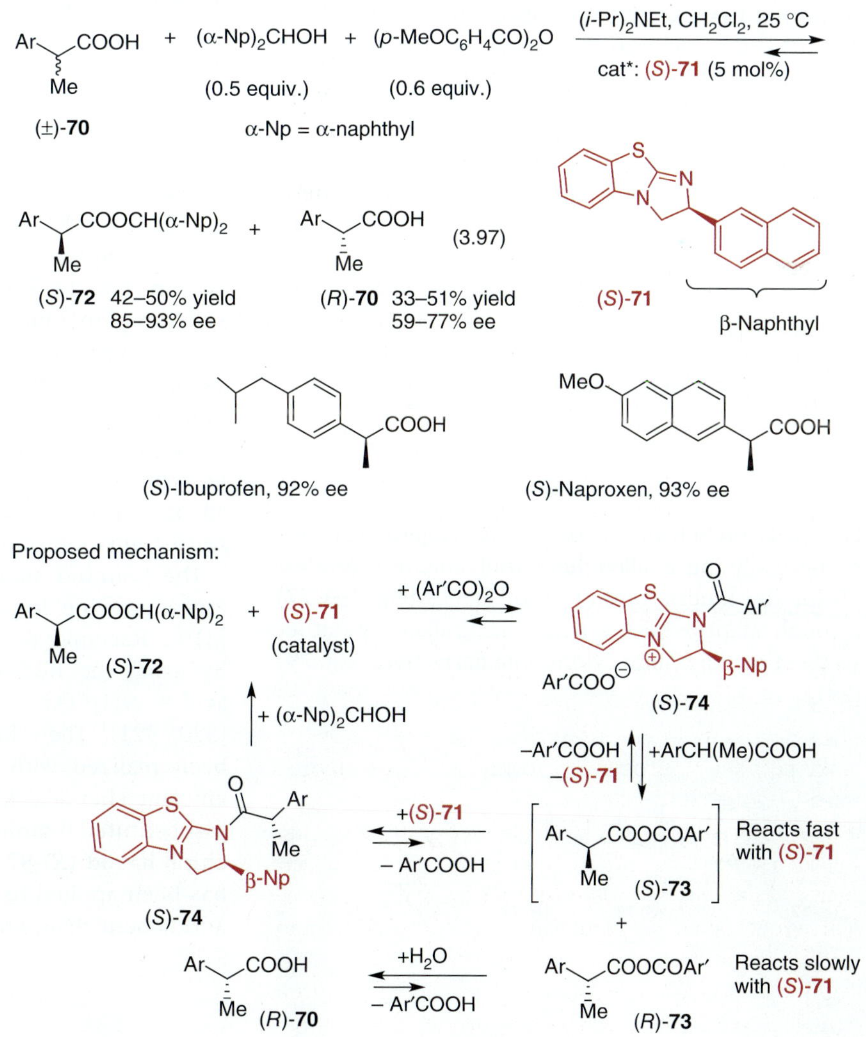

Scheme 3.13 Kinetic resolution of α-arylalkanoic acids with an achiral alcohol via enantioselective esterification using an enantiomerically pure acyl transfer catalyst.

((*S*)-2-[4-(2-methylpropyl)phenyl]propanoic acid) and (*S*)-naproxen ((*S*)-6-methoxy-α-methyl-2-napthaceneacetic acid), two important nonsteroidal anti-inflammatory drugs. The method realizes an acyl transfer to a hindered secondary alcohol (di(α-naphthyl)methanol) via mixed anhydrides (*S*)-**73** and (*R*)-**73** resulting from the combination of (±)-**70** with 4-methoxybenzoic anhydride. The enantiomeric mixed anhydrides react with the enantiomerically pure chiral catalyst (*S*)-**71** (a benzotetramisole-type catalyst) [282–284] giving preferentially (*S*)-**74** and unreacted (*R*)-**73**. After aqueous work-up, carboxylic acid (*R*)-**70** is isolated next to ester (*S*)-**72**. Catalytic hydrogenolysis of (*S*)-**72** provides the corresponding acids (*S*)-**70** [285]. A silylation-based kinetic resolution for α-hydroxylactones and lactams employing isothiourea catalyst of type **71** and Ph_3SiCl has been reported [286].

Enantioselective didehydrogenation (Section 7.8.6) of benzylic and allylic alcohols (reaction (3.98)) has permitted their kinetic resolution. The method uses the ruthenium catalysts **75** of Noyori [287, 288] or **76** proposed by Uemura and Hidai [289].

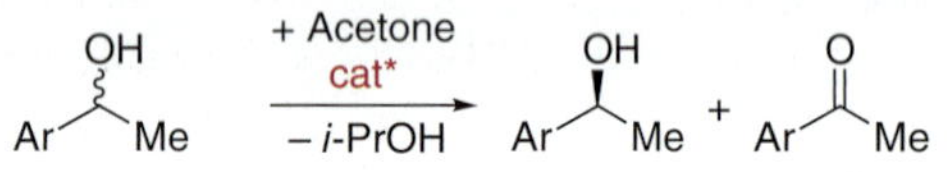

(3.98)

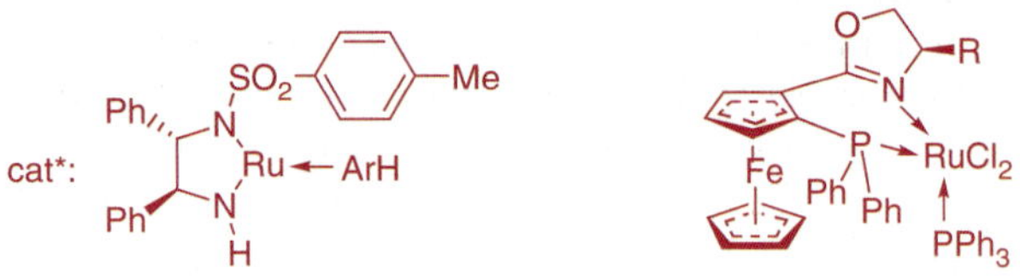

75 ArH = *p*-cymene, mesitylene

76 R = *i*-Pr, Ph

Hoveyda and Schrock [290] have developed the enantiomerically pure alkylidene molybdenum catalyst **78** for the kinetic resolution of dienes (e.g. (±)-**77**) through enantioselective ring-closing alkene metathesis (Section 8.6.2) into cylohexene derivatives (e.g. **79**) [291].

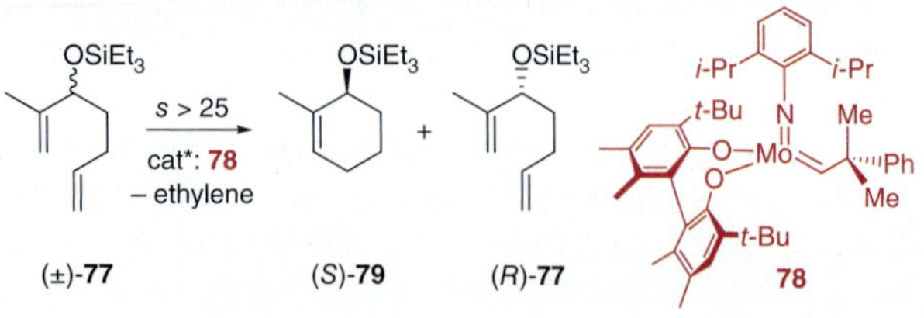

The silver-catalyzed isomerization of α-allenic alcohols into 2,4-dihydrofurans [292] can be enantioselective in the presence of a chiral silver phosphate catalyst as illustrated with the kinetic resolution of (±)-**80** applying catalyst (*S*)-**81** [293]. The reaction is an intramolecular addition of an alcohol onto an alkene, which implies the intermediacy of an alkene–silver cation complex intermediate.

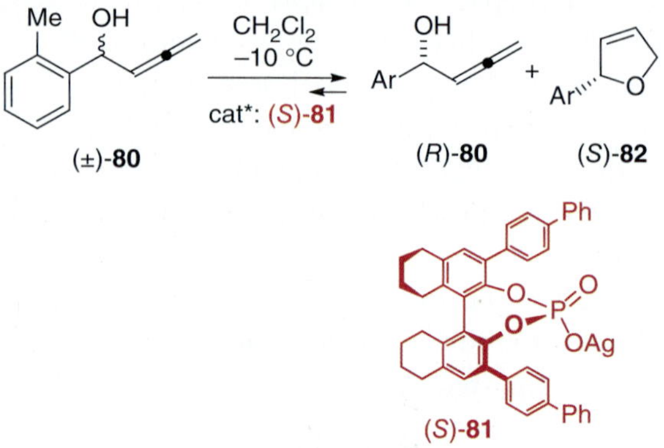

Asymmetric epoxidation of alkenes to generate enantiomerically enriched epoxides is one of the most important tools for asymmetric synthesis [294–302]. Among the most successful enantioselective oxidants are chiral dioxiranes derived from chiral ketones [303–308] and oxaziridium salts obtained by oxidation of iminium salts typically with oxone ($K_2SO_5 + 0.5\ KHSO_4 + 0.5\ K_2SO_4$) [309–313], with tetraphenylphosphonium monoperoxybisulfate [314] or with H_2O_2 [315] in the presence of PhSeSePh [316]. An example of kinetic alkene resolution through enantioselective epoxidation is given with the reaction of chromene (±)-**83** with tetraphenylphosphonium monoperoxysulfate in the presence of nonracemic chiral iminium salts **84** [317]. Oxygen atom transfer from the oxaziridinium salt intermediate to the alkene is probably a concerted process that adopts a spiro-transition state [318].

The β-amino acids and their derivatives are taking a growing importance in medicinal chemistry [319]. Racemic β-amino acids are obtained readily using the Rodionov reaction (RCHO + malonic acid + $NH_4OAc \rightarrow$ β-amino acid + CO_2 + AcOH) [320, 321]. Their kinetic enzymatic resolution has been realized with success [322]. Alternatively, an efficient chemical kinetic resolution of unprotected β-substituted β-amino acids (±)-**86** using a recyclable chiral ligand (*R*)-**87** has been reported. The method has been applied to the asymmetric synthesis of the antidiabetic drug sitagliptin [323a].

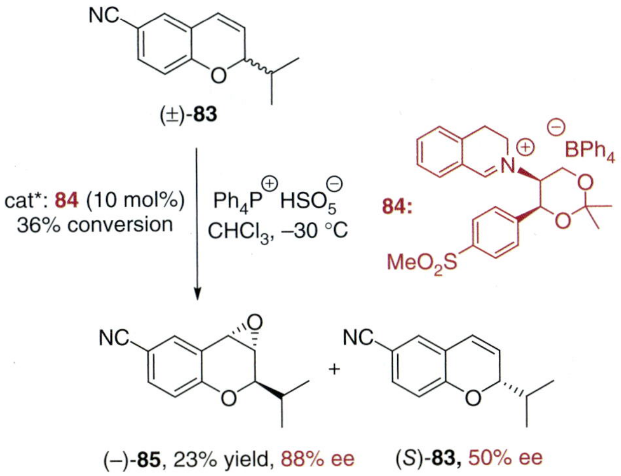

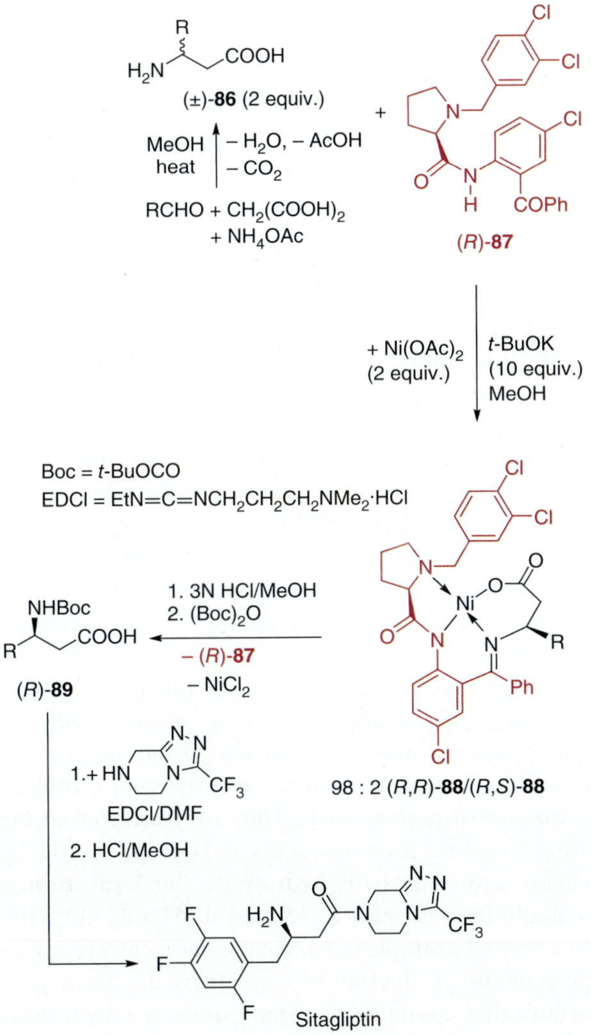

Kinetic resolution of 2-subtituted 2,3-dihydro-4-pyridones **90** by palladium-catalyzed asymmetric allylic alkylation has been proposed. As shown with reaction (3.99) [323b].

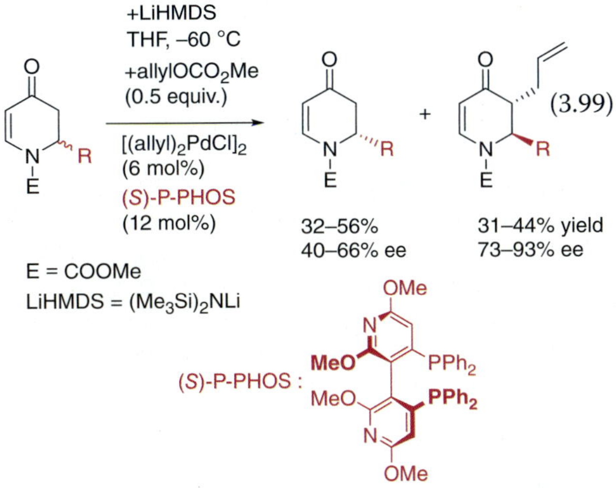

(3.99)

3.7.2 Parallel kinetic resolution

Kinetic resolution of a racemic compound (±)-**A** leads to a maximum yield of 50% of one enantiomer of product **P**, say (+)-**P**, with ee > 99%, provided the selectivity factor $s > 200$. It leaves (−)-**A** unreacted, but as its relative concentration increases in the reaction mixture, its competitive reaction gives (−)-**P** that contaminates the desired product (+)-**P** and thus decreases its enantiomeric purity. In a parallel kinetic resolution, the slow reacting enantiomer (−)-**A** is removed from the reaction mixture through an enantioselective reaction giving another product **Q**, which is not the enantiomer of (+)-**P** and is readily separated from it after completion of the reaction. The ideal situation is realized when both parallel reactions are highly enantioselective and have the same rate. The reactions can be stereodivergent, regiodivergent, or structurally divergent [324–326]. An elegant application of this strategy was reported by Vedejs and Chen in 1997 (Scheme 3.14) with a chiral DMAP acyl transfer reaction. Activation of the quasi-enantiomeric pyridines (*R*)-**91** and (*S*)-**92** with the hindered chloroformate **93** and (+)-fenchyl chloroformate (**94**) gives the acyl transfer agents **95** and **96**, respectively. The fenchyl group in **96** is irrelevant to the selectivity. On mixing the racemic 1-(α-napthyl)ethanol with a 1 : 1 mixture of **95** and **96**

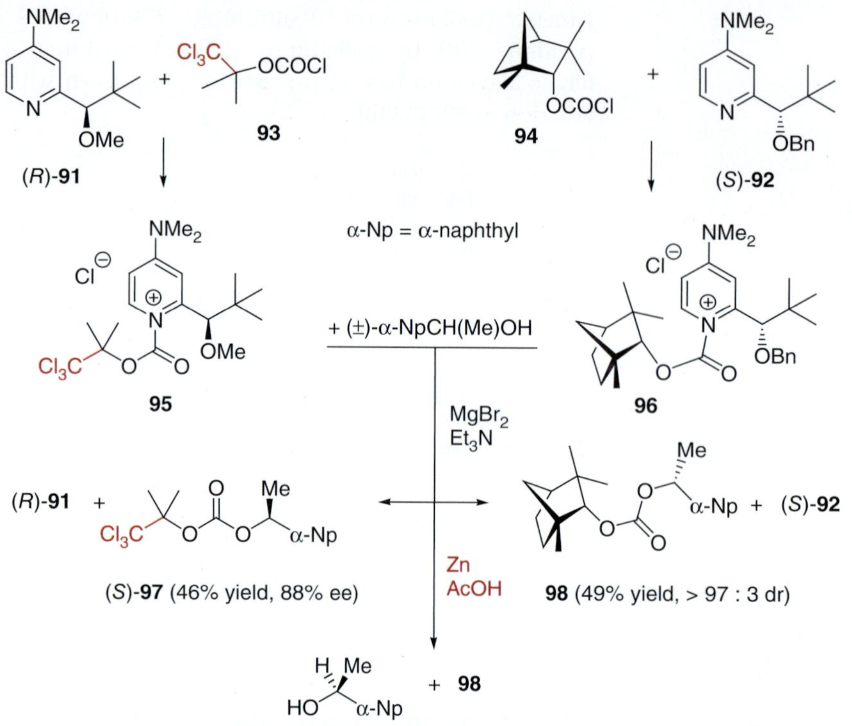

Scheme 3.14 Parallel kinetic resolution of a secondary alcohol by acyl transfer using quasi-enantiomeric DMAP derivatives.

and an excess of $MgBr_2$ and Et_3N leads to the formation of mixed carbonates (*S*)-**97** and **98**, respectively. Treatment of this mixture with Zn/AcOH removes the trichlorotertiobutyl group chemoselectively to give (*S*)-1-(α-naphthyl)ethanol and leaving **98** unreacted. These two products are separated readily. The quasi-enantiomeric chiral DMAP equivalents (*R*)-**91** and (*S*)-**92** liberated during the acyl transfers are fully recyclable [327].

The reaction of a racemic compound with an enantiomerically pure reagent or with an enantiomerically pure catalyst giving two different diastereoisomeric products is another way to realize parallel kinetic resolution [328–330].

3.7.3 Dynamic kinetic resolution: kinetic deracemization

Deracemization is the conversion of a racemic compound into one of its enantiomer. We have seen that this can be carried out under thermodynamic control (Section 1.14.4). We present here examples of deracemization realized under kinetic control [331–335]. Dynamic kinetic resolution occurs when the racemic reactant (±)-**A** is able to epimerize while it reacts enantioselectively to give an enantiomerically enriched product, e.g. (+)-**P** which, in theory, can form quantitatively, not only in less than 50% yield as in kinetic and parallel kinetic resolutions. The kinetic scheme is that of the Curtin–Hammett principle (Section 3.3.2) [32]. The enantiomeric ratio [(+)-**P**]/[(−)-**P**] is given by the rate constant ratio k_+/k_- if racemization of (−)-**A** is faster than reaction (−)-**A** → (−)-**P**. Note that **A** may contain more than one element of dissymmetry.

Natural and unnatural D- or L-α-amino acids can be prepared via the racemic hydantoins (±)-**99** (5-substituted imidazolidine-2,4-dione or glycolylurea) obtained by the Bucherer–Bergs reaction [336–338] that condenses aldehydes, KCN, and ammonium carbonate (Scheme 3.15a). Hydantoinases catalyze the enantioselective hydrolytic ring opening of hydantoins to form the corresponding *N*-carbamoyl-α-amino acids. Both D- and L-hydantoinases are available, depending on the microbial source of the enzyme. The D- (D-**100**) or L-carbamoyl derivatives (L-**100**) are converted to the corresponding α-amino acids D-**101** and L-**101** by chemical or enzymatic hydrolysis. The 5-substituted hydantoins are readily racemized at pH > 8 [339, 340]. In a second example of enzymatic deracemization, we present the production of L-lysine by the Toray procedure that combines the enantioselective hydrolysis of D,L-α-amino ε-caprolactam to L-lysine catalyzed by a yeast (*Cryptococcus laurentii*) with a bacterial racemase-catalyzed epimerization of the remaining D-α-amino caprolactam (*Achromobacter obae*).

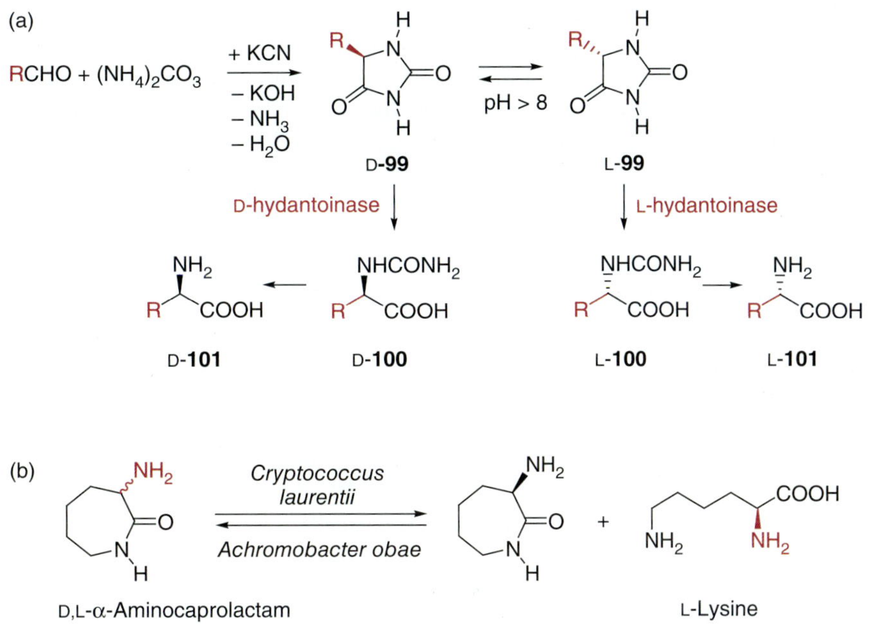

Scheme 3.15 Example of production of α-amino acids via enzymatic deracemization.

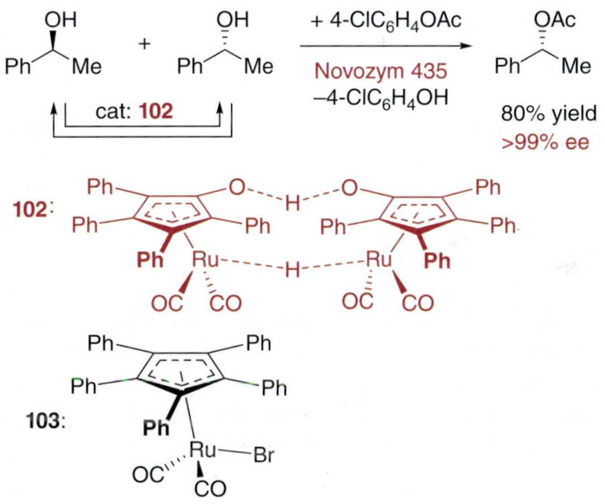

The combined operation (Scheme 3.15b) of both enzymatic activities, in the form of microorganism whole cells, results in nearly quantitative yield of L-lysine [341]. Racemic natural and non-natural *N*-Boc(*t*-BuOOC)-amino acid thioesters undergo enzyme-catalyzed ammoniolysis and aminolysis with concomitant base-catalyzed racemization of the unreacted enantiomer with the formation of the corresponding amides in good yields and ee's up to 98% [342].

Polymers are submitted to mechanical forces generated by ultrasounds [343–346] and can be used to surmount thermally inaccessible racemization barriers. This phenomenon has been used to prepare enantiomerically pure (*S*)-BINOL ((*S*)-1,1′-bi-2-naphthol) by cholesterol esterase-catalyzed hydrolysis of a polyacrylic derived diester of racemic BINOL at 6–9 °C [347]. Enzymes and transition metal complexes can work in tandem in dynamic kinetic resolution [348, 349]. For instance, (±)-1-phenylethanol has been converted into (*R*)-1-phenylethyl acetate in 80% yield and high enantiomeric purity using an immobilized lipase from *C. antarctica* (Novozym 435) as catalysts for the enantioselective acyl transfer from 4-chlorophenol acetate. The diruthenium complex **102** (2 mol%) catalyzes the racemization of the unreacted alcohol through an oxidation/reduction mechanism (analogous to Oppenauer oxidation by acetone and Meerwein–Ponndorf–Verley reduction by aluminum isopropoxide) [350–352].

Processes that combine enzymatic and chemically catalyzed reactions have been called chemoenzymatic dynamic kinetic resolutions [353, 354]. Secondary alcohols have been resolved by Bäckvall and coworkers with yields up to 99% and >99% ee utilizing *C. antarctica* lipase B together with Bäckvall's ruthenium catalyst **103** [355, 356]. Dynamic kinetic resolution of acyclic allylic acetates using lipase to catalyze the enantioselective acetate hydrolyzes and allylpalladium complexes to catalyze the racemization of unreacted allyl acetates have been reported [357]. Alternatively, enantioselective acetylation with 4-chlorophenyl acetate of racemic allylic alcohols can be catalyzed by a lipase and the racemization of the unreacted allyl alcohol can be catalyzed by an achiral ruthenium complex [358]. Nonenzymatic dynamic kinetic resolution of secondary alcohols via enantioselective acylation has been realized by Fu and coworkers using their chiral DMAP analog (+)-**69** and Ru(Ph_5C_5)$(CO)_2$Cl (chloride analog of **103**) as catalysts and carbonate MeOCOCOO-*i*-Pr as acylating agent [359].

The Noyori enantioselective hydrogenation of ketones is an early example of chemical dynamic kinetic resolution (Scheme 3.16a) [360–362]. In the presence of (*R*)-Ru(BINAP)$(OAc)_2$ catalyst, hydrogenation of racemic methyl 2-oxocyclopentanecarboxylate ((±)-**104**) produces the β-hydroxycarboxylic ester (1*R*,2*R*)-**105** with high diastereoisomeric and enantiomeric purity. Genêt and coworkers have developed the ligands (*R*)- and (*S*)-SYNPHOS, BINAL analogs that form ruthenium complexes, and excellent catalysts for enantioselective hydrogenations. Hydrogenation of racemic α-amino-β-ketoesters in the presence of (*R*)-Ru(SYNPHOS)Br_2 catalyst produces corresponding *syn*-1,2-aminoalcohols (2*S*,3*R*)-**107** in good yield and high ee's (Scheme 3.16b) [363]. Diastereoisomeric *anti*-1,2-aminoalcohols can also be prepared in a similar way [364, 365]. Catalytic systems permitting enantioselective hydrogenation of compounds of type (±)-**106** in water have been proposed [366].

Asymmetric synthesis of *anti*-β-amino-α-hydroxy esters via dynamic kinetic resolution of β-amino-α-keto esters has also been realized using a ruthenium catalyst and HCOOH/Et_3N as a hydrogen source [367]. Deracemization of 2-aryl-3H-indolines (±)-**108** through their one-pot oxidation by the oxopiperidinium tetrafluoroborate **109** into the corresponding 1,2-didehydro derivatives **110** and enantioselective reduction of the latter with Hantzsch ester **110** using a chiral phosphoric acid catalyst such as **111** have been developed by Toste and coworkers. The method can also be applied to the deracemization of 2-aryltetrahydroquinolines [368].

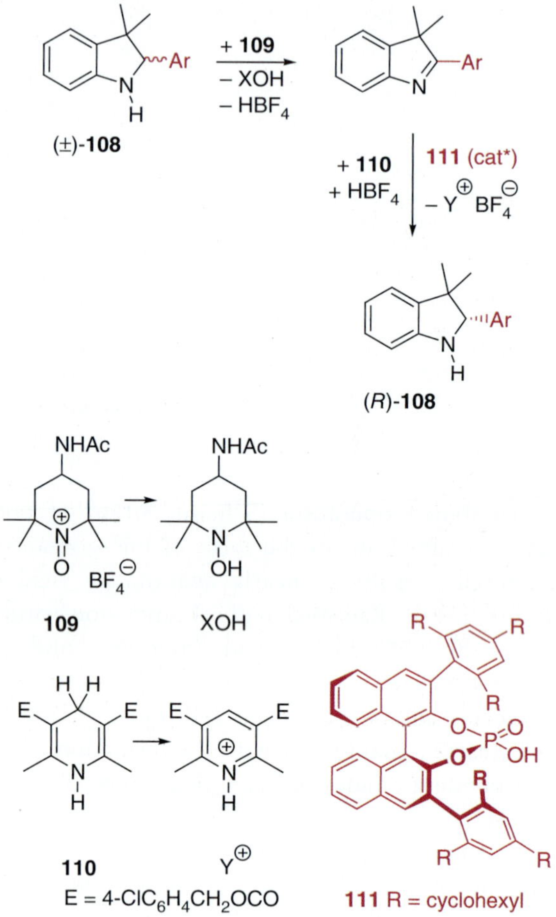

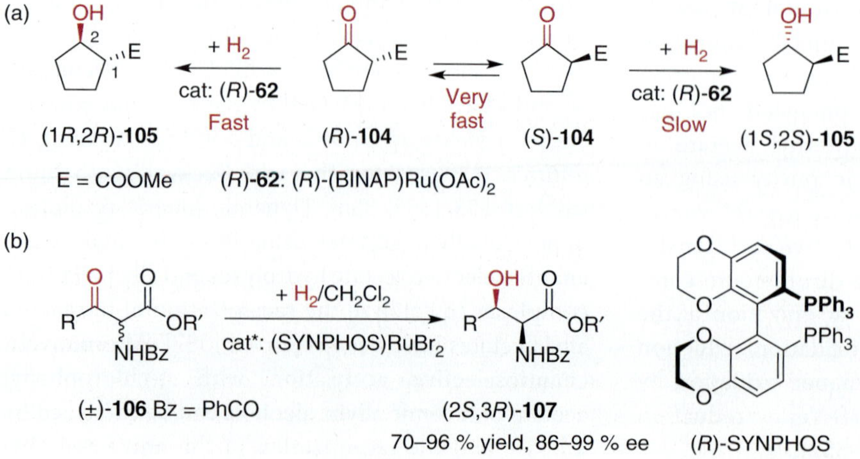

Scheme 3.16 Examples of chemical dynamic kinetic resolutions through enantioselective catalytic hydrogenation of β-ketoesters that are readily epimerized.

Problem 3.18 Devise a synthetic procedure to prepare (−)- and (+)-penta-2,3-diene (see Problem 3.5). Compare with references [45] and [46].

Problem 3.19 Explain the enantioselective conversion of (±)-**A** into **P** [369, 370].

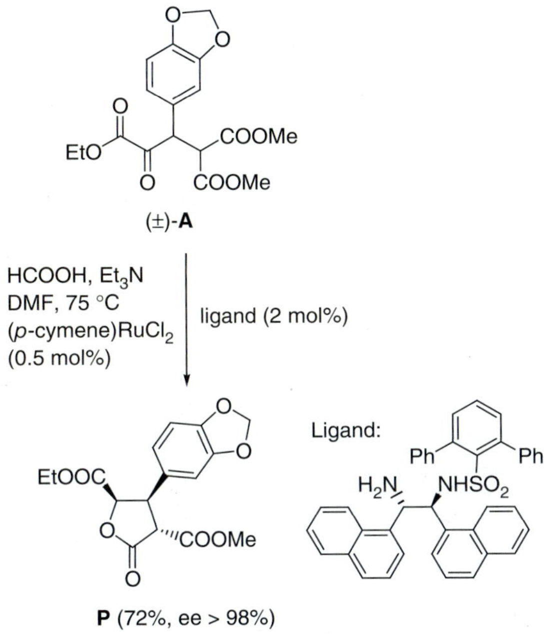

Problem 3.20 Give a possible mechanism for the following dynamic kinetic asymmetric transformation [371].

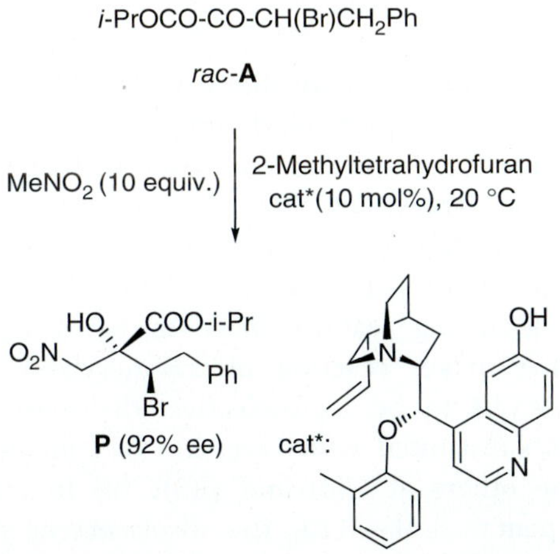

3.7.4 Synthesis starting from enantiomerically pure natural compounds

Nature provides us with many enantiomerical compounds that can be used as starting materials in asymmetric synthesis. This is an important part of the chiral pool that is composed of alkaloids (e.g. cinchonidine, cinchonine, (+)-ephedrine, (−)-nicotine, quinidine, quinine, (+)-pseudoephedrine, (−)-pseudoephedrine, and (−)-sparteine) of amino acids (e.g. L-alanine, L-arginine, D-asparagine, L-asparagine, L-aspartic acid, L-cysteine, L-glutamic acid, L-isoleucine, L-leucine, L-lysine, L-methionine, L-ornithine, L-phenylalanine, D-phenylglycine, L-proline, L-pyroglutamic acid, L-serine, L-tryptophan, L-tyrosine, and L-valine) [372], carbohydrates (e.g. D-arabinose, L-arabinose, L-ascorbic acid, chiro-inositols, diacetone-D-glucose, D-fructose, D-galactonic acid γ-lactone, D-galactose, D-gluconic acid δ-lactone, L-gluconic acid δ-lactone, D-glucosamine (2-amino-2-deoxy-D-glucose), D-glucose, D-mannitol, D-mannose, D-quinic acid, D-ribolactone, D-ribose, D-sorbitol, L-sorbose, and D-xylose) [373–376], hydroxy acids (e.g. L-lactic acid, D-lactic acid, (*S*)-malic acid, poly-(*R*)-3-hydroxybutyrate, (−)-shikimic acid, L-tartaric acid, D-tartaric acid, D-threonine, and L-threonine), and terpenes (e.g. (−)-borneol, (+)-camphene, (+)-camphor, (+)-camphoric acid, (+)-3-carene, (−)-carvone, (+)-citronellal, (+)-fenchone, (+)-isomenthol, (+)-limonene, (−)-limonene, (−)-menthol, (+)-menthol, (−)-menthone, (−)-α-phellandrene, (−)-α-pinene, (−)-β-pinene, and (*R*)-(+)-pulegone). The other part is constituted of enantiomerically enriched or enantiopure non-natural compounds that can be obtained nowadays through efficient enantioselective catalytic processes (as illustrated in the next chapters). The most efficient syntheses based on the chiral pool are those that require a limited number of synthetic steps, or better require the purification of a limited number of synthetic intermediates (see, e.g. the industrial synthesis of vitamin C, Scheme 3.27, Section 3.8). This is realized when selective protection of polyfunctional starting material and of intermediates can be avoided and when functional modifications are highly chemo-, regio- (site-), and stereoselective. As an illustration of the use of the chiral pool, we present (Scheme 3.17) the synthesis of Oseltamivir phosphate (Tamiflu) developed by Roche's scientists Karpf and Trussardi [377], one of the many asymmetric syntheses reported for this compound [378, 379]. It is a prodrug of the active neuraminidase inhibitor **117** that was developed by Kim and coworkers at Gilead Sciences [380] and launched on the market in 1999. It is used as an orally active drug for the treatment and prophylaxis of both type A and type B human influenza. It is also active against the H5N1 Avian Flu virus. (−)-Shikimic acid is available in large amounts by extraction of Chinese star anise and by a fermentation process using a genetically engineered *E. coli* strain [381]. The ethyl ester of (−)-shikimic acid is converted into

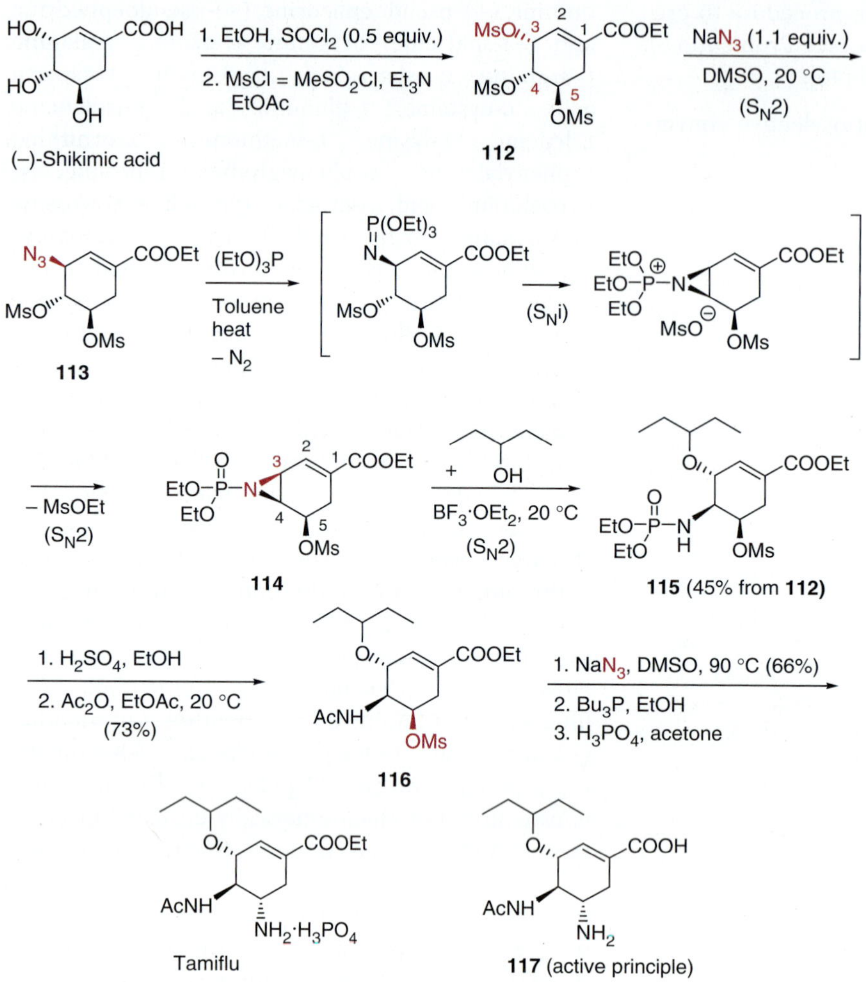

Scheme 3.17 Synthesis of Oseltamivir phosphate (Tamiflu) starting from (−)-shikimic acid.

the tris(mesylate) **112**. The allylic mesylate is displaced chemo-, regio-, and stereoselectively with the azide anion (S_N2 substitution, no allylic rearrangement, the other secondary mesylate substitutions are slower because they are not activated by the allylic C=C double bond) into azide **113**. Staudinger phosphite reduction [382] of azide **113** with triethyl phosphite gives the corresponding primary phosphorimidate that undergoes in toluene under a reflux 1,3-elimination of ethyl mesylate ($MeSO_3Et$), providing aziridine **114**. Chemo-, regio-, and stereoselective displacement of the allylic C—N bond of **114** with pentan-3-ol under activation with BF_3 etherate furnishes **115** in 45% overall yield based on (−)-shikimic acid, without the need to isolate and purify any of the synthetic intermediates! Acid-catalyzed hydrolysis of the phosphorimidate **115** liberates the corresponding primary amine that is subsequently *N*-acetylated with acetic anhydride giving **116**. Nucleophilic displacement (S_N2) of the last mesylic ester by NaN_3, followed by Staudinger reduction of the azide intermediate and treatment with phosphoric acid yields Tamiflu. This synthesis is particularly efficient as it requires only three work-ups and purifications. It illustrates important aspects of S_N2 displacements of type $M^+Y:^- + R—X \rightarrow R—Y + M^+X:^-$ ($MY = NaN_3$) and of type $Nu: + R—X \rightarrow R—Nu^+ + X:^-$ ($Nu: = R'OH$).

The following features have to be recognized: (i) What makes reaction of tris(mesylate) **112** + $NaN_3 \rightarrow$ **113** to be regioselective: the mesylate at C(3) is substituted with inversion of configuration not the others at C(4) and C(5)? (ii) In reaction **114** + pentan-3-ol → **115**, the displacement of the aziridine nucleofugal group is preferred to the departure of the mesylate and is regioselective with preferred attack at C(3); why? (iii) Why reaction **116** + NaN_3 displaces preferentially the mesytate at C(6) and does not give any product of conjugate addition to the acrylic moiety?

Steric hindrance to the nucleophilic attack at C(3) in **112** is probably weaker than at C(4) and C(5). There is also an electronic contribution that renders attack at C(3) the preferred one. As other concerted, one-step reactions S_N2 nucleophilic displacements usually follow the Bell–Evans–Polanyi theory for which $\Delta^{\ddagger}H = \alpha\Delta_r H + \beta$. The activation enthalpy depends on steric factors, on the exothermicity of the reaction and on the ease with which reactants can exchange an electron. The α term depends on $\Delta_r H$: $\alpha \rightarrow 1$ for endothermic reaction (late transition state) and $\alpha \rightarrow 0$ for exothermic reaction (early transition state). The β term represents the contributions from steric factors that repel the reactants, differential solvation effects between transition state and reactants, and stabilizing contributions arising from the ease with which the reactants can exchange an electron, as given by *IE* (nucleophile) + (−*EA* (electrophile)). The transition state of reaction $Y:^- + R—X \rightarrow Y—R + X:^-$ can be represented by the combination of configurations $[Y:^-R^\bullet X^\bullet \leftrightarrow Y{:}\ R^+X^- \leftrightarrow Y^\bullet\ R^\bullet X:^- \leftrightarrow Y^\bullet R:^-X^\bullet]^{\ddagger}$. Thus, any structural feature that stabilizes the tight ion-pair $[R^+X^-]$, the radical $[R^\bullet]$, and/or carbanion $[R^-]$ will accelerate the S_N2 reaction. In the case of reaction **112** + NaN_3 → **113**, the allylic mesylate displacement reaction (attack at C(3)) leads to stabilized charge transfer configurations of the transition state. This is not the case for the displacements of two other mesylate moieties of **112** at C(4) and C(5). The same electronic factor (allylic conjugation) intervenes to render the nucleophilic opening of the aziridine moiety of **114** by pentan-3-ol regioselective. The chemoselective displacement of the azirine phosphoramide arises from the ring strain release (makes the reaction more exothermic), which overwhelms the poorer nucleofugacity of the phosphoramido group compared with that of mesyloxy group ($(RO)_2P(=NR')—O^-$ anion is somewhat less stable than $MeSO_3^-$ anion). In addition, HN_3 + cyclohexene → cyclohexyl azide is highly exothermic (−32.4 kcal mol^{-1}) in the gas phase, as estimated from the standard heats of formation $\Delta_f H°(HN_3)$ = 70.3 kcal mol^{-1} (Table 1.A.1), $\Delta_f H°$(cyclohexene) = 1.0 kcal mol^{-1} (Table 1.A.2), and $\Delta_f H°(c\text{-}C_6H_{11}—N_3)$ = 36.9 kcal mol^{-1} (Table 1.A.4). Conjugate addition of NaN_3 to **116** would generate a sodium enolate of an ester. In DMSO, $pK_a(HN_3) = 7.9$ and $pK_a(CH_3COOEt) = 29.5$ (Table 1.A.24). Thus, the 1,4-addition of the azide (NaN_3) equilibrates with a carbanion (sodium enolate) that is $1.36 \times \Delta pK_a = 1.36 \times 21.6 \cong 29$ kcal mol^{-1} less stable than the azide anion. This leads to $\Delta_r H°(NaN_3 + \mathbf{116}$ → sodium azidoenolate$) \cong -32.4 + 29 \cong -4$ kcal mol^{-1}, a reaction not exothermic enough at 90 °C to compensate for the 10–12 kcal mol^{-1} of its entropy cost (addition is a condensation with $\Delta_r S° < -30$ eu, whereas the substitution with $\Delta_r S° \approx 0$ is not a condensation). As a result, the S_N2 displacement of the cyclohexyl mesylate is the favored reaction.

Asymmetric synthesis can be realized by applying enantiomerically pure reagents. An example has been given in Scheme 3.10 with the asymmetric oxidative hydroboration of a dialkene using isopinocampheylborane (also named (−)-tetra-3-pinanyldiborane, (−)-$Pn_4B_2H_2$, or (−)-tetraisopinocampheylborane), a reagent derived from inexpensive (+)-α-pinene [383]. With reaction (3.100), we illustrate the possibility to apply a chiral environment to induce asymmetry. The chiral reagent is a mixed aggregate that combines the achiral lithium trimethylsilylacetylide (initial reagent) and a chiral lithium alcoholate (external additive) [384, 385]. Another example of asymmetric induction by the environment (external chiral ligand) [386, 387] will be given with the synthesis of efavirenz, an important anti-AIDS drug (see Section 3.7.8).

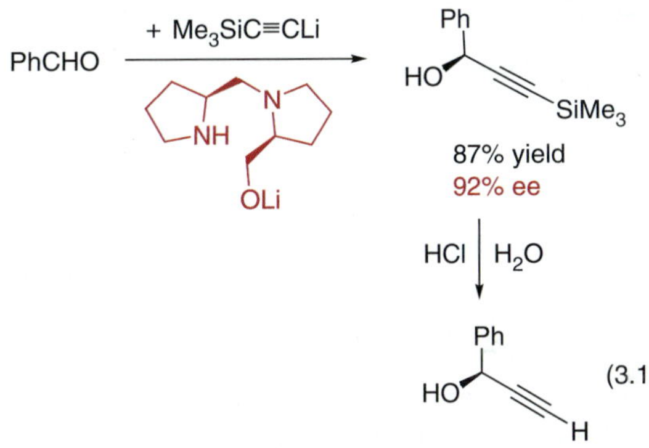

(3.100)

3.7.5 Use of recoverable chiral auxiliaries

An asymmetric synthesis starting from a natural product may prove expensive because of the number of isolation and purification steps it requires and because of the cost of the reagents. An alternative approach is to attach an inexpensive enantiopure chiral auxiliary to an inexpensive achiral substrate. This generates an enantiopure synthetic intermediate, the reaction of which can be stereoselective. After the removal of the chiral auxiliary, an enantiomerically enriched intermediate or final product is obtained [388–394]. The method is most useful when the chiral auxiliary is recovered without racemization or isomerization at an early stage of a multistep synthesis. Chiral auxiliaries are also used to generate enantiomerically pure octahedral metal complexes [395]. A classical example of the application of chiral auxiliary to the asymmetric synthesis of prostaglandins has been reported by Corey and coworkers. It employs a Lewis-acid-catalyzed

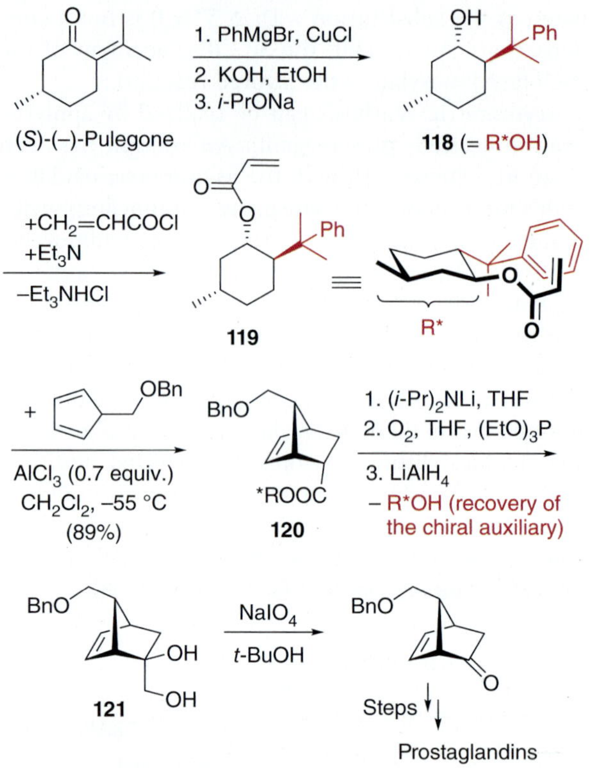

Diels–Alder addition (Sections 5.3.8 and 7.6.6) of 5-(benzyloxymethyl)cyclopentadiene to the enantiomerically pure 8-phenylmenthyl acrylate (**119**) [396, 397]. Alcohol **118**, the chiral auxiliary, derives from the non-natural (*S*)-(−)-pulegone that is obtained from (*S*)-(−)-citronellol. It is recovered after conversion of the Diels–Alder adduct **120** into diol mixture **121**. A large number of other enantiomerically pure alcohols have been proposed as chiral auxiliaries [398–403].

Enders and coworkers have developed the (*S*)- and (*R*)-proline derivatives **SAMP** ((*S*)-1-amino-2-methoxymethylpyrrolidine) and **RAMP** ((*R*)-1-amino-2-methoxymethylpyrrolidine), which can be used for the asymmetric α-alkylation of aldehydes and ketones (Scheme 3.18) [404–406]. Most of the problems in carbonyl chemistry, such as aldol self-condensation, *α,α′*-dialkylation, control of the diastereoselectivity, side reactions of products, and lack of reactivity of the corresponding enolates (mostly for thermodynamic reasons: reversibility) are solved by the use of the hydrazones derived from dialkylhydrazines such as **SAMP** and **RAMP** [407]. The corresponding chiral hydrazones of ketones deprotonate and react with the electrophile on the least hindered center [408–410].

Evans' oxazolidinones (e.g. **122**R, see Table 5.7, Section 5.7.7, for more examples) obtained by the reaction of enantiomerically pure 1,2-amino alcohols with diethyl carbonate have proven extremely effective for controlling reactions of attached acyl fragments such as α-alkylation (Scheme 3.19a), α-acylation, α-amination, α-azidation, α-bromination, α-hydroxylation, aldol reaction (Section 5.7.7) [411], Diels–Alder reaction (Sections 5.3.8–5.3.15), and conjugate addition. Natural α-amino acids are reduced (e.g. $LiAlH_4$, $BH_3{\cdot}Me_2S/BF_3{\cdot}Et_2O$, and $LiBH_4/Me_3SiCl$) into enantiomerically pure 1,2-amino alcohols. The latter can also be obtained from enantiomerically pure epoxides for which several asymmetric syntheses have been proposed. Davies' 5,5-disubstituted SuperQat **123**R [412, 413] and Gibson's variants **124** [414] also appear to be useful auxiliaries because of their better recyclability and crystallinity (importance to obtain diastereomerically pure products). Chiral oxazolidinethiones and thiazolinethiones have also found interesting applications [415]. The samarium-diiodide-mediated asymmetric Reformatsky reaction (Scheme 3.19b) of enantiomerically pure bromoacetamide **125** produces β-hydroxy carboximides **126** in high yield and diastereoselectivities. Treatment of **126** with aqueous LiOH at 0 °C liberates the corresponding β-hydroxy acids and permits recovery of the chiral auxiliaries [416]. Equally accessible Oppolzer's camphorsultam derivatives (Section 5.7.7) are useful chiral auxiliaries [417–422]. They generate products of α-alkylation, α-acylation, α-amination, and aldol reaction with high crystallinity [423].

Chiral auxiliaries can be used to desymmetrize *meso*-compounds as illustrated below. In the first example, the key asymmetric step of the formal total asymmetric synthesis of rifamycin reported by Harada and coworkers (Scheme 3.20) [424, 425], the silyl enol ether of (−)-menthone reacts with (achiral) *meso*-tetrol **128** giving a 45 : 1 mixture of diastereoisomeric monoacetals **129** and **129′** (61% yield) together with the corresponding bis-acetal (12%) and unreacted **127** (10%). This is an example of kinetically controlled desymmetrization by chirality. First, total synthesis of rifamycin S has been reported by Kishi and coworkers [426–428]. Rifamycin S belongs to the ansamycin antibiotics. It is particularly effective against tuberculosis and leprosy [429–431].

In the second example that presents a synthesis of (*S*)-(−)-3-methyl-δ-valerolactone (Scheme 3.21), achiral diacid **130** is reacted with 2 equiv. of 1,3-thiazoline-2-thione **131** derived from L-cysteine. The chiral (contains a 1,5-*syn* and a 1,5-*anti* structural element) diamide **132** so obtained undergoes diastereoselective transamidification (steric factor) with piperidine at −30 °C giving **133** as the major product. Both carboxylic moieties of **133** are differentiated chemically. Sodium *p*-bromothiophenolate reacts preferentially with the 1,3-thiazolidine-2-thione carboxamide, giving a thioester that is reduced

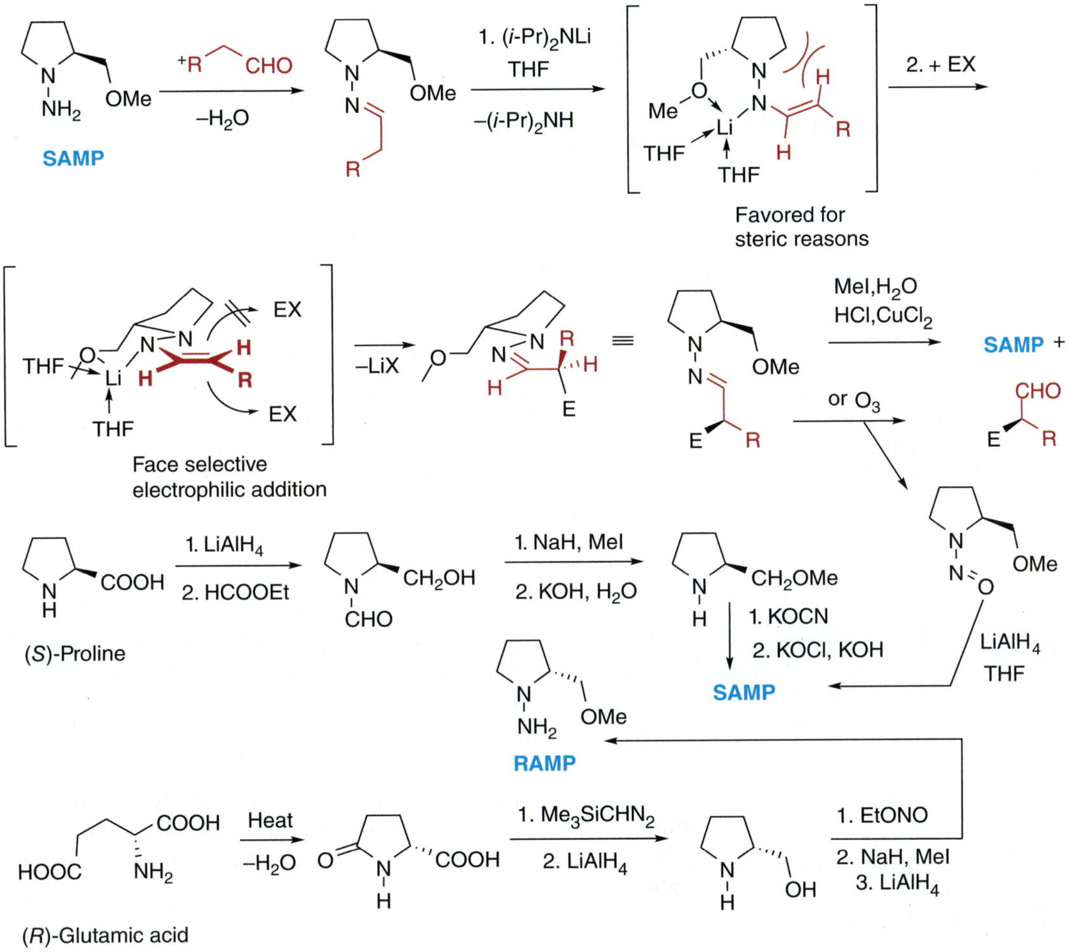

Scheme 3.18 Asymmetric α-alkylation of aldehydes via their homochiral hydrazones.

Scheme 3.19 (a) Face-selective reactions of enolates derived from chiral *N*-acyl oxazolidinones; and (b) asymmetric Reformatsky-type reactions.

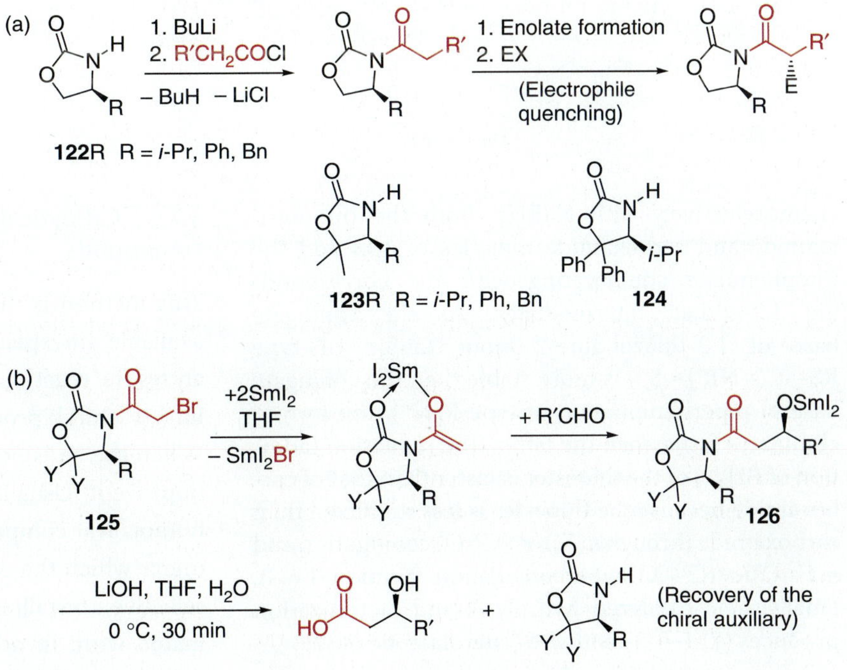

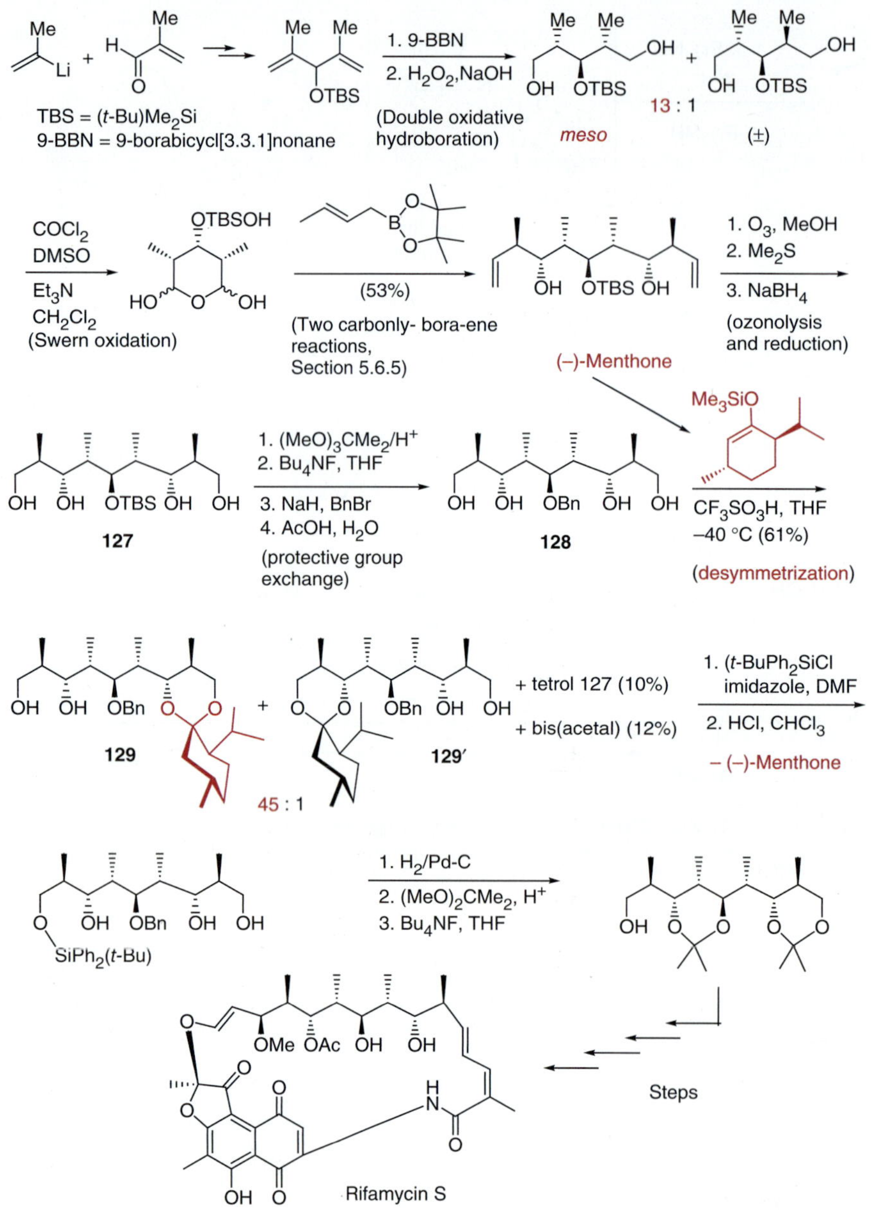

Scheme 3.20 Harada's formal total synthesis of rifamycin S.

chemoselectively with $NaBH_4$. Both the thionothioamide and carboxamide moieties of **133** add the thiophenolate equilibrating with the corresponding 1,2-adduct intermediates. As the conjugate base of 1,3-thiazolidine-2-thione (anion of type RS—C(=NR′)—S$^-$) is more stable than the conjugate base of piperidine (anion of type R_2N^-), the former eliminates faster than the latter. The reduction (addition of BH_4^-) of the thioester is faster than that of carboxamide because the thioester is less stabilized than carboxamide through n(X:)/π*(C=O) conjugation and n(C=O)/σ*(C—X) hyperconjugation (Section 1.6.3). Finally, acid-catalyzed hydrolysis and lactonization produces (S)-(−)-3-methyl-δ-valerolactone [432].

3.7.6 Catalytic desymmetrization of achiral compounds

This method is the most interesting one if a readily available (inexpensive) enantiopure compound catalyzes the reaction of an inexpensive achiral substrate into a chiral product with high enantioselectivity. When the ee is not high enough, crystallization may improve it. Usually, racemates crystallize better than homochiral compounds. Conditions might be found under which the ee of a product can be improved by selective crystallization of the racemate it is contaminated with. In other cases, it may happen that the

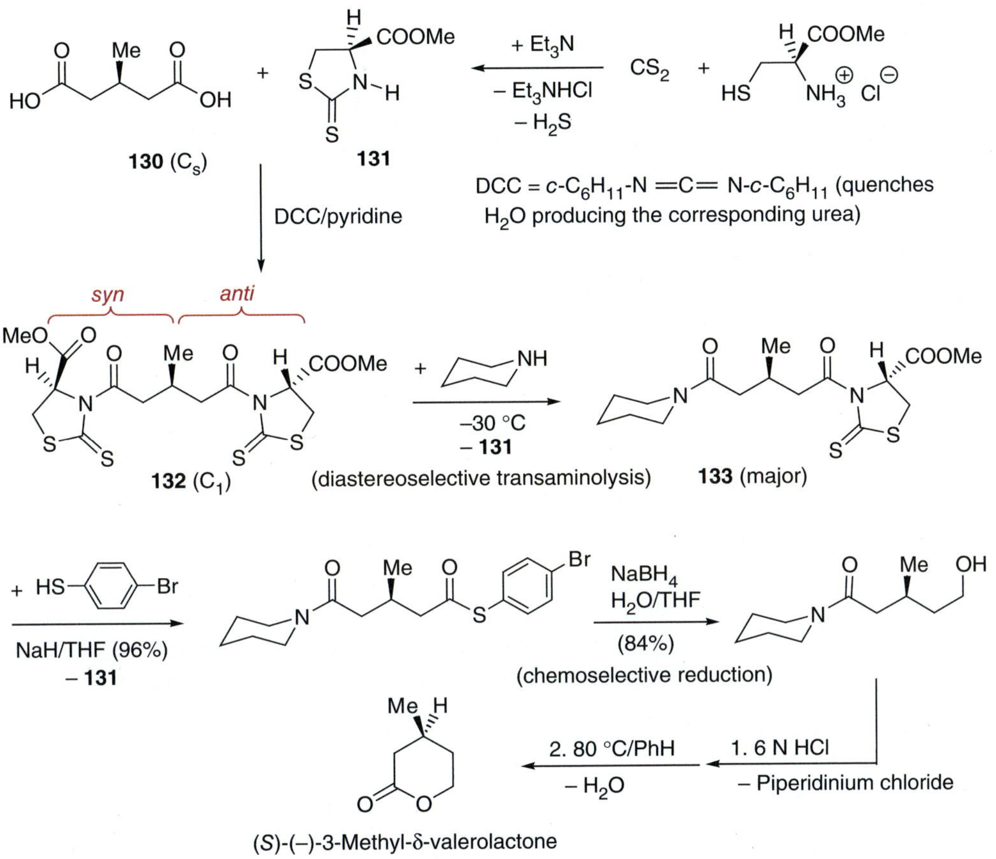

Scheme 3.21 Example of desymmetrization of an achiral dicarboxylic acid via the formation of a diamide with a chiral amine.

major enantiomer crystallizes with enantioenrichment. However, this is not always the case, which may render the other methods presented above (resolution of racemates via formation of diastereomers, Section 1.14.3; conglomerate crystallization by seeding, thermodynamically controlled deracemization, Section 1.14.4; kinetic resolution, Section 3.7.1; kinetic deracemization, Section 3.7.3; use of enantiomerically pure natural products, Section 3.7.4, and use of chiral auxiliaries, Section 3.7.5) competitive with catalytic enantioselective synthesis. With crystallization (and recrystallization) of diastereomers, the chances to obtain a pure compound are much greater. After cleavage of the chiral auxiliary, an enantiomerically pure compound is obtained if racemization does not compete with the process that liberates the desired compound from its chiral auxiliary. With catalytic enantioselective reactions, the enantiomerically pure catalyst employed in a small amount leads to a large quantity of enantiomerically enriched product (chiral amplification). Usually, the catalyst is prepared from a chiral auxiliary, which is itself derived from a natural product or by resolution of a racemic precursor. In some cases, the catalyst does not have to be enantiomerically pure because of positive nonlinear effects (NLEs) [433] (Section 3.7.7). There are also examples of reactions that benefits from asymmetric autocatalysis [434], an enantioselective synthesis in which the chiral product acts as enantioselective catalyst for its own production (Section 3.7.8). After Pasteur discoveries, several attempts were made to generate optically active products from achiral substrates. Until 1970, most of them relied on fermentation in the presence of microorganisms (bio-catalysis). Today, a large number of enantiomerically pure compounds used in the fine chemical and pharmaceutical industry can be produced applying enzyme-catalyzed processes [236]. An early example of an enzyme-catalyzed enantioselective synthesis reported in 1908 by Rosenthaler is the addition of HCN to benzaldehyde to give optically active mandelonitrile (PhCH(OH)CN) catalyzed by emulsin, an enzyme isolated from almonds [435]. The same year, Hayashi obtained optically active mandelic acid (PhCH(OH)COOH, 95% ee) by rearrangement of phenylglyoxal hydrate (PhCOCHO·H_2O) in the presence of *Bacillus proteus* [436]. The enantioselective monoacylation of *meso*-polyol **127** (Scheme 3.20) by vinyl acetate in the presence of porcine pancreas lipase gives (+)-**134** (75% yield and >98% ee) [437]. This is another example of kinetic desymmetrization by chirality.

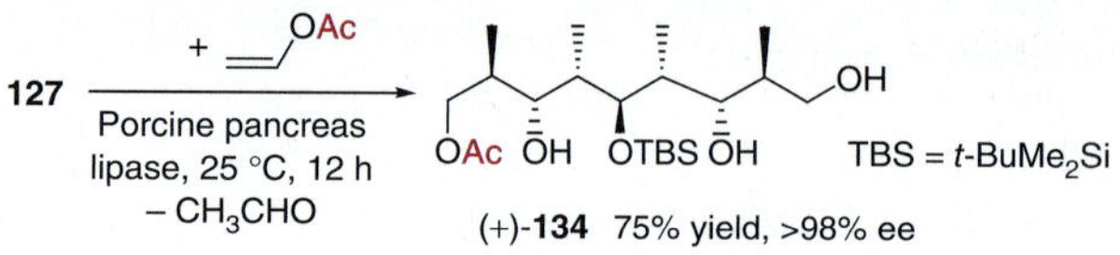

Baker's yeast is the most common agent used for the enantioselective reduction of achiral ketones into secondary alcohols [438–443]. A rule has been developed by Prelog that predicts that the hydride transfer from alcohol dehydrogenase of *Curvularia falcata* occurs on the *Re* face of the ketone [212]. Reductions, when performed by an enzymatic system, usually require a cofactor. That is why organisms are used instead of purified enzymes.

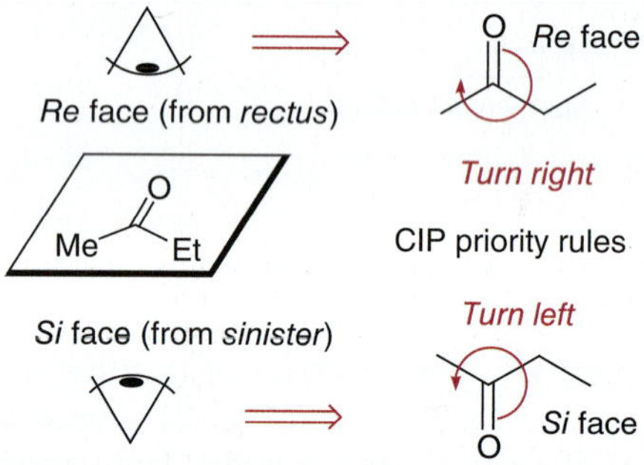

In some cases, microorganisms are available that allow for complementary, enantioselective production of either one of the two possible enantiomeric alcohols as illustrated here [444]. Immobilized microorganisms can be used, making the purification procedure simpler [445–447].

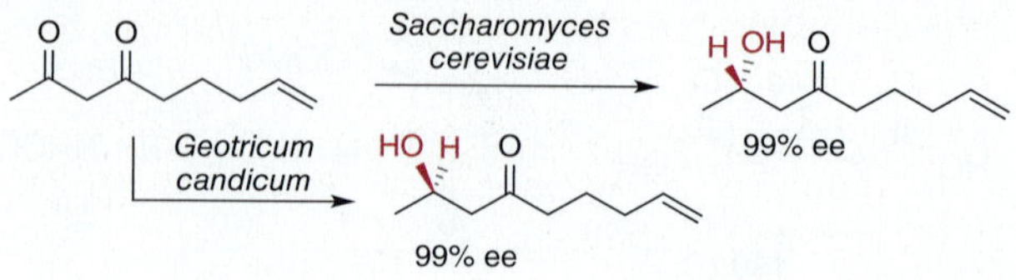

The enantioselective reduction of the C=C double bond in α,β-unsaturated aldehydes, ketones, and esters is possible with NADH-dependent enoate reductases from microorganisms such as *Clostridium*, *Proteus* sp., and Bakers' yeast (NADH: nicotinamide adenine dinucleotide hydride, the reduced formed of NAD^+). NADH-dependent reductive amination of α-keto acids produces α-amino acids [212]. Microbial oxidation of benzene derivatives produces enantiomerically pure cyclohexa-3,5-diene-1,2-diol derivatives [448, 449]. Toluene dioxygenase, the active enzyme in these oxidations, has been expressed in recombinant strains of *E. coli* JM109 (pDTG601) for improved efficiency [450, 451]. Enantiomerically pure halocyclohexadienediols have been converted into a large variety of products of biological interest such as aminocyclitols, conduritols, rare carbohydrates, and alkaloids [452–459]. Microbial oxidation of inactivated methylene groups has been catalyzed by the fungus *Beauveria bassiana* [460] and by a *Bacillus megaterium* strain isolated from topsoil [461, 462]. Nature condenses a carbonyl carbon center via Umpolung to another carbonyl unit using thiamine pyrophosphate cofactor (acyloin condensation, Section 7.5.1) with yeast pyruvate decarboxylase. With benzaldehyde as a cosubstrate and using yeast whole cells in the presence of glucose or pyruvic acid, this constitutes the key step in the industrial process for the manufacture of (−)-ephedrine (Scheme 3.22) [212].

Diastereo- and enantioselective aldol reactions of 1,3-dihydroxyacetone monophosphate (DHAP) with aldehydes **135** to give ketones **136** are catalyzed by fructose diphosphate aldolase (e.g. from rabbit

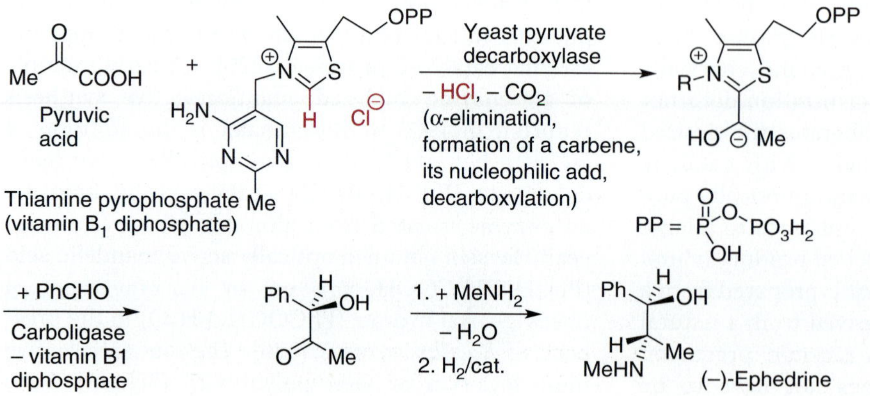

Scheme 3.22 Industrial synthesis of (−)-ephedrine based on a microbial mediated acyloin condensation.

muscle aldolase, RAMA). Wong and coworkers have converted compounds **136** into a large variety of rare carbohydrates, glycomimetics, and heterocyclic compounds of biological interest [463–468] and key synthetic intermediates [469].

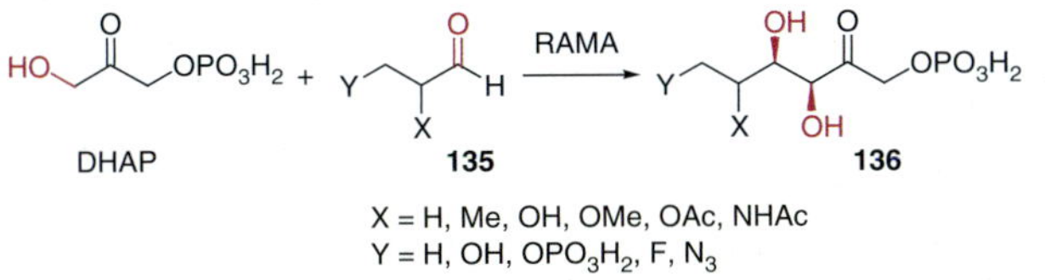

During the 1980s and 1990s, initial protein engineering technologies, typically structure based, extended the substrate range of enzymes to allow the synthesis of synthetic intermediates important for the fine chemical and pharmaceutical industry. Nowadays, molecular biology modifies biocatalysts via an in vitro version of Darwinian evolution commonly called directed evolution. The objectives are to improve enzyme stability (up to 80 °C), substrate specificity, enantioselectivity, tolerance to organic solvents, and ease of their recovery (e.g. through immobilization). Industrial-scale biocatalysis focuses primarily on hydrolases, ketoreductases (KREDs), aldolases, and cofactor regeneration [470, 471]. "Improved" aldolases simplifies the synthesis of the side chain of atorvastatin (Lipitor) [472], an important cholesterol-lowering drug. Among the various enzymatic approaches reported for the preparation of the key intermediate **137** (Scheme 3.23), one of them applies a double aldolase-catalyzed reaction condensing 2 equiv. of acetaldehyde with 3-azidopropanal.

Several examples of enantioselective reactions catalyzed by organic molecules or organometallic complexes will be given along the next chapters. Examples of catalytical enantioselective aldol reactions will be

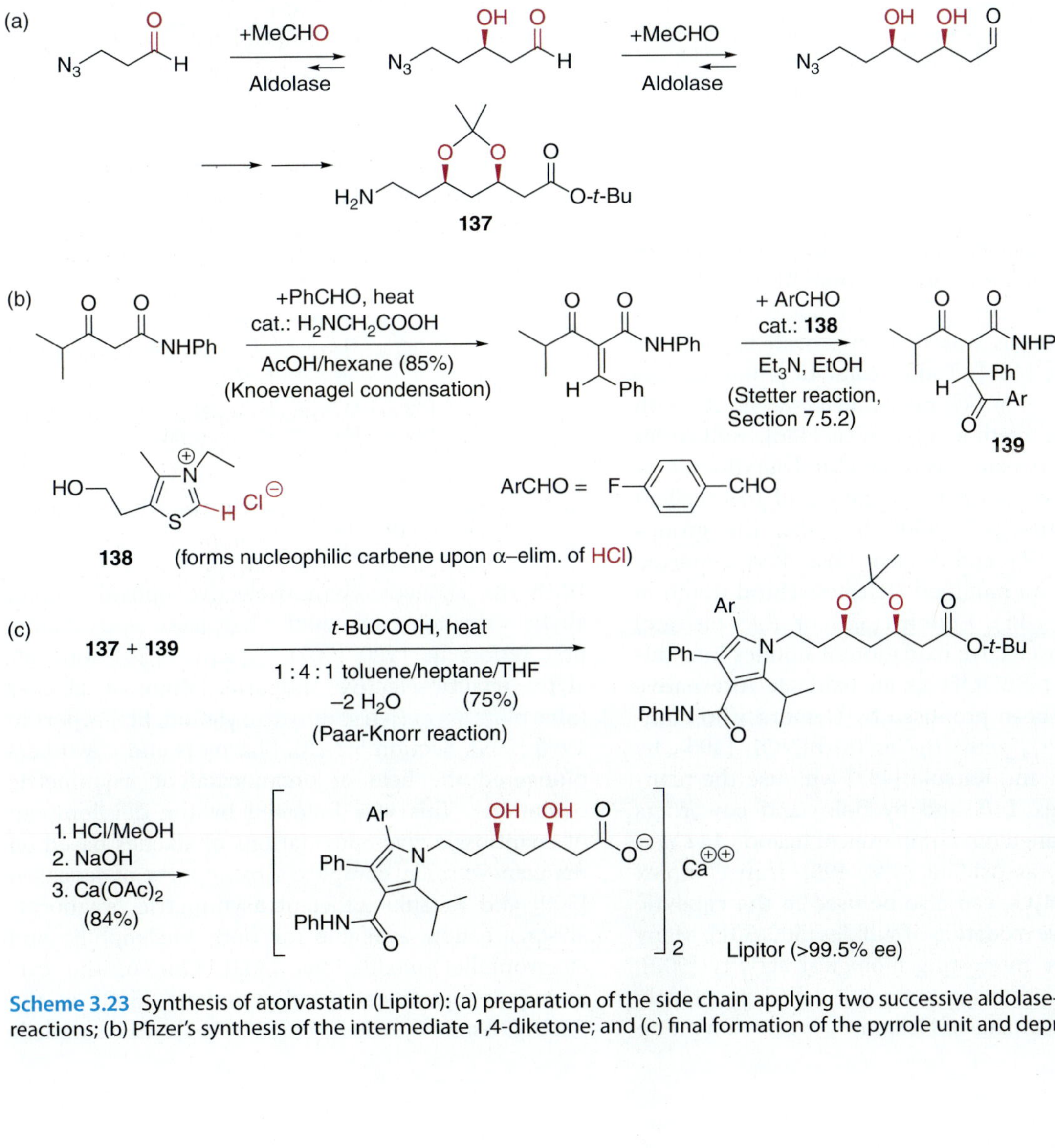

Scheme 3.23 Synthesis of atorvastatin (Lipitor): (a) preparation of the side chain applying two successive aldolase-catalyzed reactions; (b) Pfizer's synthesis of the intermediate 1,4-diketone; and (c) final formation of the pyrrole unit and deprotection.

described in Section 5.7.6. In 1909, Dakin recognized that certain α-amino acids catalyze the Knoevenagel reaction ($RCHO + CH_2(COOH)_2 \rightleftarrows RCH{=}CH(COOH) + H_2O$, like aldol reaction followed by water elimination) in water [473]. In 1971, Hajos and Parrish at Hoffmann-La Roche [474–476], on the one hand, and Eder, Sauer, and Wiechert at Schering [477], on the other hand, found that catalytical amount of readily available L-proline catalyzes the asymmetric intramolecular aldol reactions (3.101). In 2000, List, Lerner, and Barbas reported the first examples of direct, asymmetric intermolecular aldol reactions of acetone with various aldehydes also catalyzed by L-proline (Section 5.7.6) [478]. In the meantime, other chiral α-amino acids, small peptides, chiral amines and diamines, amides, sulfonamides and their combinations, as well as chiral metal complexes and Lewis acids have been found to catalyze the direct enantioselective aldol reaction [479–482].

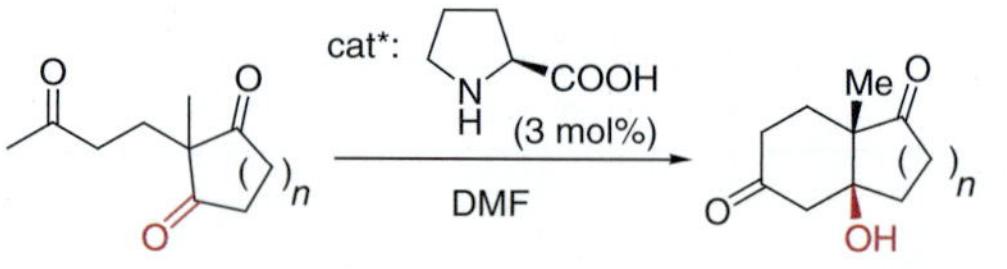

(3.101)

$n = 1$: 100% yield, 93% ee
$n = 2$: 52% yield, 74% ee

In 1962, Andersen reported the synthesis of enantiopure sulfoxides by reaction of menthyl *p*-toluenesulfinate with a Grignard nucleophile [483, 484]. The synthesis was then improved by Solladié and coworkers [485]. Today, catalytic enantioselective oxidation of prochiral sulfides ($R^1{-}S{-}R^2$ with $R^1 \neq R^2$) is the method of choice [486]. Sulfoxides have been used extensively as chiral auxiliaries or chiral substrates as their reactions lead to excellent stereoselectivities [487–490]. In 1984, the groups of Modena [491] and Kagan [492–494] reported independently on modified Sharpless chiral titanium reagents ($Ti(O\text{-}i\text{-}Pr)_4$ + diethyl (*S,S*)- or (*R,R*)-tartrate) for the enantioselective oxidation of sulfides into sulfoxides using *t*-BuOOH as an oxidant. Alternative catalysts have been proposed by Uemura who combines $Ti(O\text{-}i\text{-}Pr)_4$ with (*R*)- or (*S*)-BINOL [495], by Jacobsen [496] and Katsuki [497] who use the manganese complex **140**, and by Bolm and coworkers who employ vanadium complexes of ligands **141** and aqueous H_2O_2 as oxidant [498, 499]. Iron complex of **141b** and H_2O_2 can also be used in the catalytic enantioselective oxidation of sulfides [500, 501]. Many sulfoxides have interesting biological activity [502]. Esomeprazole (its magnesium salt called perprazole is sold under the name of Nexium™) is a proton pump inhibitor used as an antiulcer agent [503]. The last key step of its synthesis applies Kagan's method using **142** and cumyl hydroperoxide as oxidant (inexpensive, from the air oxidation of cumene, synthesis of phenol, and acetone, Hock rearrangement) [504, 505] giving esomeprazole (**143**) with 94% ee. Crystallization of its sodium salt **144** improves the enantiopurity (>99.5% ee) [506].

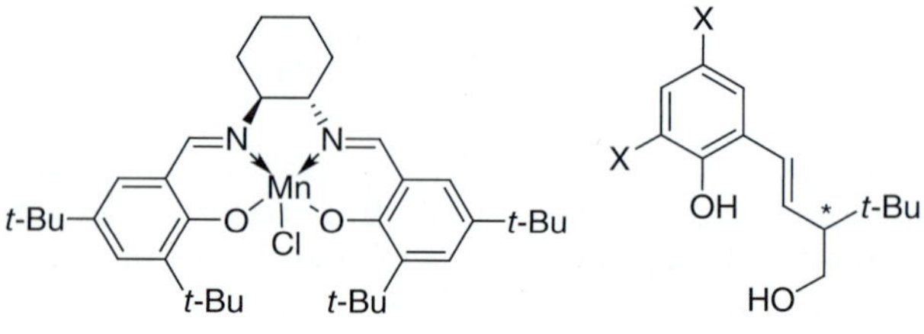

140

141a: X = *t*-Bu
141b: X = I

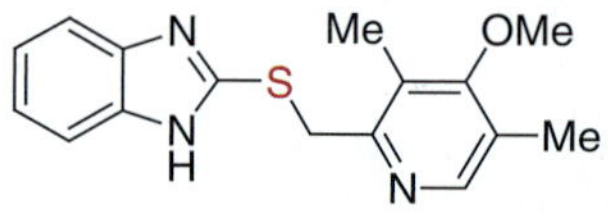

142

1. $Ti(O\text{-}i\text{-}Pr)_4$
(*S,S*)-D-(−)-DET
H_2O, PhMe, 54 °C

2. $(i\text{-}Pr)_2NEt$, 30 °C,
then CHP (1 equiv.)
(92%, >94% ee)

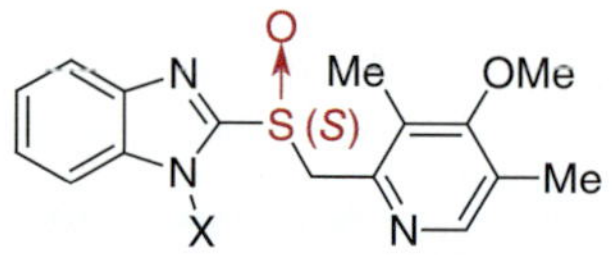

143 X = H (esomeprazole)
144 X = Na (>99.5% ee, cryst.)

DET = diethyl tartrate
CHP = cumyl hydroperoxide

With the catalytic enantioselective epoxidation of allylic alcohols (Katzuki–Sharpless epoxidation, first report in 1980 [507], Scheme 3.24) and catalytic enantioselective dihydroxylation of alkenes (Sharpless asymmetric dihydroxylation, first report in 1988 [508], Section 5.3.18), Sharpless and coworkers pioneered the field of organocatalytic asymmetric oxidations. This was followed by the development of enantioselective epoxidations of alkenes based on manganese/salen complexes by reports of Jacobsen [509] and Katsuki. Efficient asymmetric organocatalysts are now available for both nucleophilic and electrophilic epoxidations [302]. One of the earliest polypeptide-catalyzed reactions [510] is the Juliá–Colonná epoxidation [511–514]. An example is

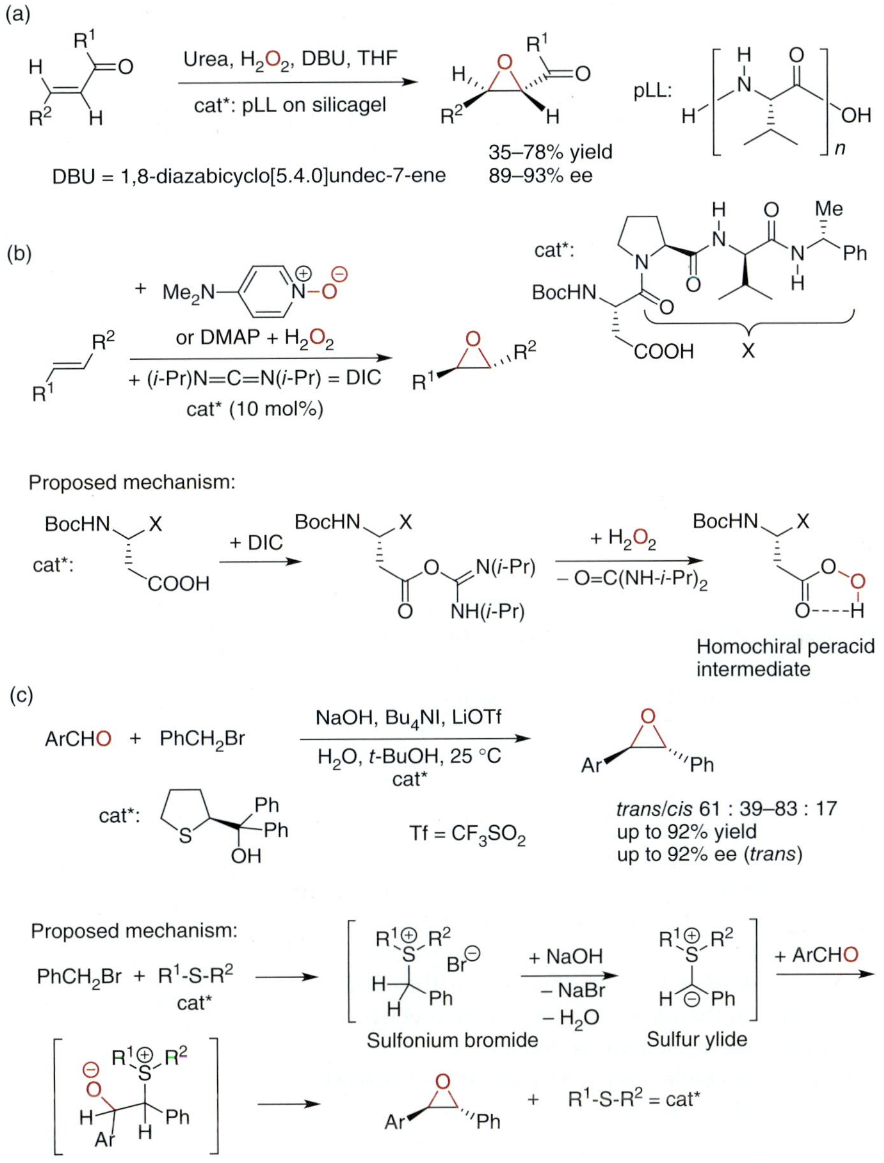

Scheme 3.24 Examples of enantioselective epoxidation catalyzed by organocatalysts: (a) Juliá–Colunná epoxidation (nucleophilic epoxidation), (b) peptide-derived peracid epoxidation (electrophilic epoxidation), and (c) catalyzed asymmetric version of the Corey–Chaykovsky reaction (sulfur ylide reaction with carbonyl compounds).

given in Scheme 3.24a that uses poly-L-aspartic acid (pLL) immobilized on silica gel, which renders the recovery of the catalyst simple. The peptide makes a complex with the prochiral ketone by hydrogen bonding (N—H···O=C) and renders one of the two faces of the alkene moiety more available than the other to the hydroxyperoxide anion (nucleophilic conjugate addition) [515, 516]. In Scheme 3.24b, an example of electrophilic epoxidation of alkenes is shown. It uses a tripeptide catalyst that generates a homochiral peracid intermediate by reaction with the oxidant [517, 518]. Several other peptides can be used as catalysts in the regioselective and enantioselective epoxidation of polyenes [519]. In the Corey–Chaykovsky reaction, sulfur ylides react with aldehydes forming epoxides [520], not alkenes as phosphor ylides do (Wittig olefination, Section 5.3.7). This reaction can be catalytic in the sulfide [521, 522]. Using homochiral sulfide catalysts, enantioselective epoxidations have been reported [523–530]. An example is shown in Scheme 3.24c [529].

The Baeyer–Villiger oxidation [531, 532] of acyclic and cyclic ketones produces esters and lactones, respectively [533, 534]. Biocatalysts [535–540] and metal complexes [541–545] have been used for the enantioselective formation of lactones. An example of desymmetrization of *meso*-cyclohexanones is given with reaction (3.102) [546].

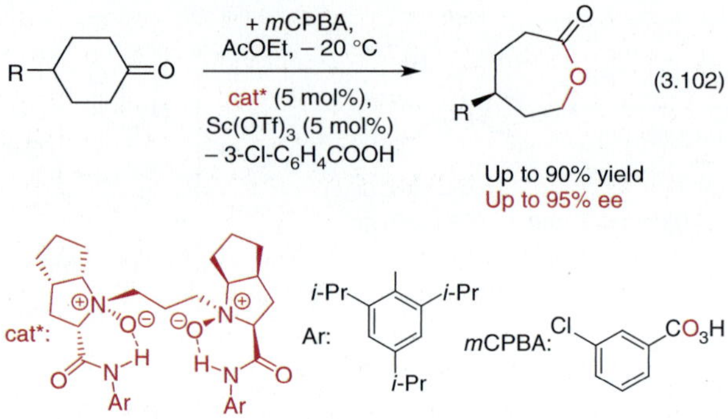

(3.102)

3.7.7 Nonlinear effects in asymmetric synthesis

If a chiral reagent or catalyst is not enantiopure (enantiomeric excess $ee_{aux} < 100\%$, one speaks of a scalemic mixture, a nonracemic mixture at a ratio other than 1 : 1), the enantiopurity of the product $ee_{prod} = ee_{max} \cdot ee_{aux}$, where ee_{max} is the enantiomeric excess obtained for the enantioselective reaction employing an enantiopure reagent or catalyst. This is not always the case; if the reactive species imply autoassociation of the chiral reagent or formation of multiligand catalysts made of chiral ligands, NLEs can be observed (Figure 3.18). They can be positive ($ee_{prod} > ee_{max} \cdot ee_{aux}$: (+)-NLE) or negative ($ee_{prod} < ee_{max} \cdot ee_{aux}$: (−)-NLE) and depend on the conversion; concentration of substrate, reactants, and catalysts; solvent; and temperature. S-shape curves $ee_{prod} = \text{function}(ee_{aux})$ can be observed [433]. Kinetic studies as well as NMR studies of the reaction mixtures, in addition to quantification of the NLE, can contribute to the understanding of the mechanism of these asymmetric reactions [547].

In a first example reported by Kagan and coworkers in 1986 [548], the Katsuki–Sharpless epoxidation (Scheme 3.25) [507, 549, 550] of geraniol with 1 equiv. of a 1 : 2 : 1 mixture of $Ti(O\text{-}i\text{-}Pr)_4$/diethyl (*S*,*S*)-(+)-tartrate/*t*-BuOOH (reaction (3.103)) showed a moderate (+)-NLE. The reaction is chemoselective for allylic alcohols and converts geraniol into (−)-**145** exclusively. Mechanistic studies suggested an intermediate of type **146** comprising two chiral ligands (L = dialkyl tartrate). When the ligand is a nonracemic mixture of both enantiomeric ligands (*R*)-L and (*S*)-L, three diastereoisomeric intermediates of type **146** can form: the two homochiral species $M[(R)\text{-L}]_2$ and $M[(S)\text{-L}]_2$ and the heterochiral or *meso* structure M[(*R*)-L·(*S*)-L)]. The rate of reaction of the homochiral intermediates (rate constant k) may or may not be the same as that of the *meso* intermediate (rate constant k_m).

If the heterochiral or *meso* intermediate is less reactive, (+)-NLE will be observed. This will also be

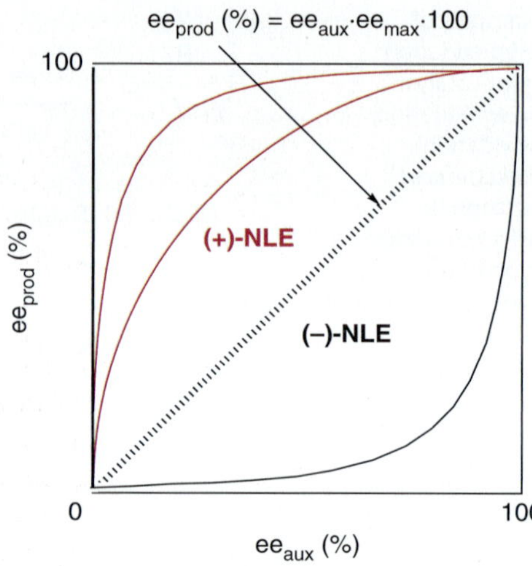

Figure 3.18 Representation of linear and nonlinear effects in enantioselective reactions.

the case if the *meso* intermediate should precipitate. As seen in Section 1.14.5, the racemic catalyst might be less soluble than the enantiomerically pure catalyst, which leads to enantioenrichment of the catalyst in the solution and thus to a (+)-NLE as reported for the aldol reaction (3.105) of acetone with 2-chlorobenzaldehyde catalyzed by serine [551, 552], and the α-aminoxylation (3.106) of propanal with nitrosobenzene catalyzed by proline [553]. Depending on the reaction conditions and the nature of the catalyst, PhN=O reacts with aldehydes and ketones giving either products of α-aminoxylation (CH–ONHPh) [554] or products of α-hydroxyamination (CH–N(OH)Ph) [555].

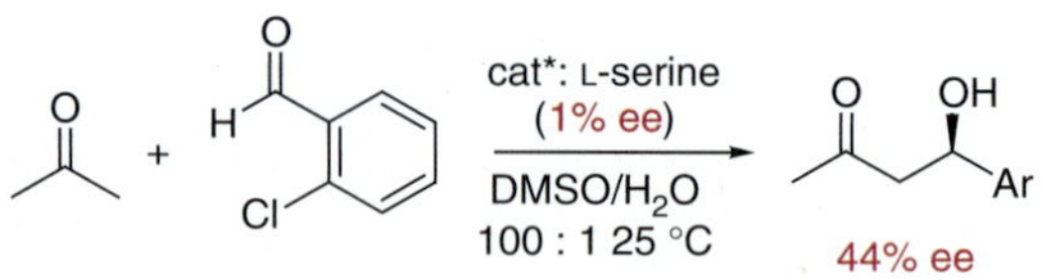

(3.105)

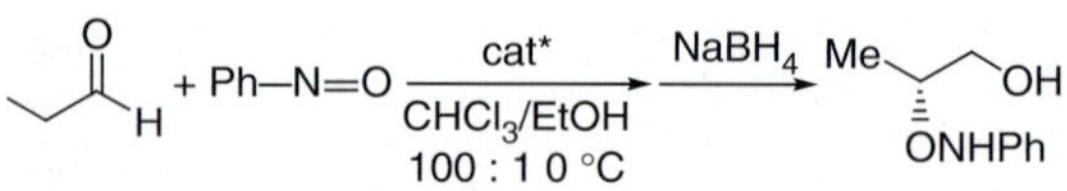

(3.106)

cat*: solid L-proline with 10% ee leads to product with 19% ee. After filtration of the catalyst, product with 96% ee is isolated!

If the *meso* intermediate should be more reactive than the homochiral intermediates, (−)-NLE will be observed. Taking $g = k_m/k$, the relative reactivity of the *meso* and homochiral intermediate, and β, their relative concentration (β = concentration of M[(*R*)-L·(*S*)-L)]/{concentration of $M[(R)\text{-L}]_2$ + concentration of $M[(S)\text{-L}]_2$}), a simple kinetic treatment gives Eq. (3.104) where the enantiomeric excess of the product and ee_{aux} = enantiomeric excess of

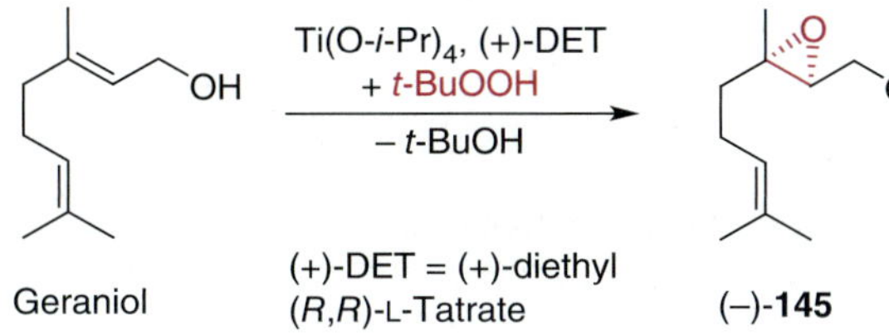

(3.103)

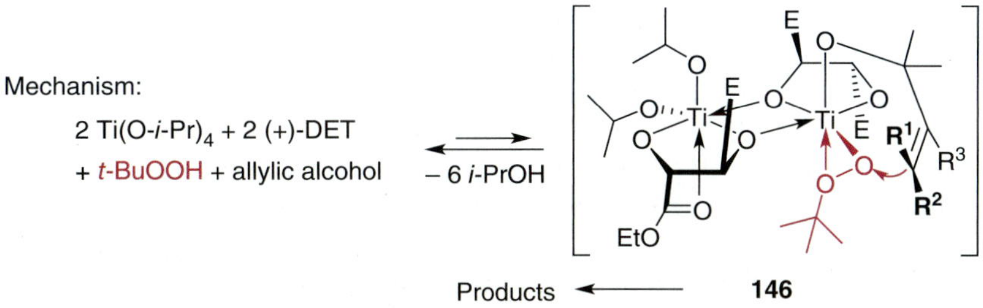

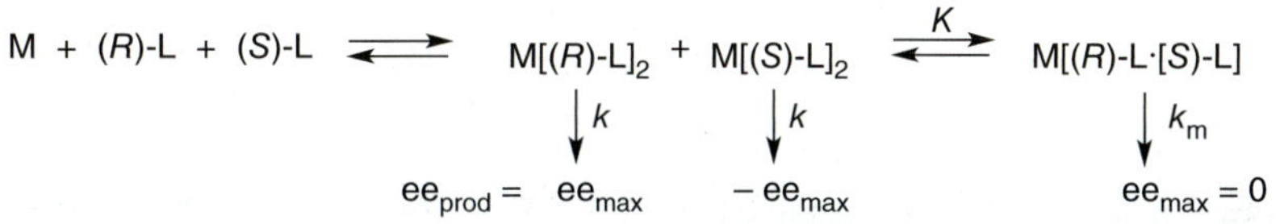

$$ee_{prod} = (ee_{max} \cdot ee_{aux})(1 + \beta)/(1 + g\beta) \quad (3.104)$$

Scheme 3.25 The Katsuki–Sharpless epoxidation of geraniol shows a (+)-NLE due to the formation of a reactive intermediate comprising two chiral ligands of type ML_2 that is more reactive when both ligands are homochiral than heterochiral.

the ligands in the case of reaction (3.103). In this model, called the ML_2 model (the two ligands are equilibrated into the set of ML_2 intermediates, no other reactive intermediates), a (+)-NLE corresponds to $g < 1$, a linear effect corresponds to $g = 1$, and (−)-NLE corresponds to $g > 1$. The strength of the NLE will be higher when diastereomeric intermediates ML_2 are formed irreversibly than when they are formed reversibly.

Inanaga and coworkers have reported a strong (+)-NLE (ligand with 40% ee leads to product with 99% ee) in the enantioselective epoxidation of enones (e.g. reaction (3.107)) using a chiral lanthanum complex generated in situ from lanthanum triisopropoxide, (*R*)-BINOL, triarylphospine oxide, and cumyl hydroperoxide as the oxidant. The (+)-NLE was explained by the formation of thermodynamically stable heterochiral aggregates, with the homochiral binuclear μ-complex **147** being the probable catalytically active species. The enone displaces a triphenylphospine oxide ligand of **147** giving **148**. The latter undergoes intramolecular oxygen transfer on one of the prochiral face of the enone [556].

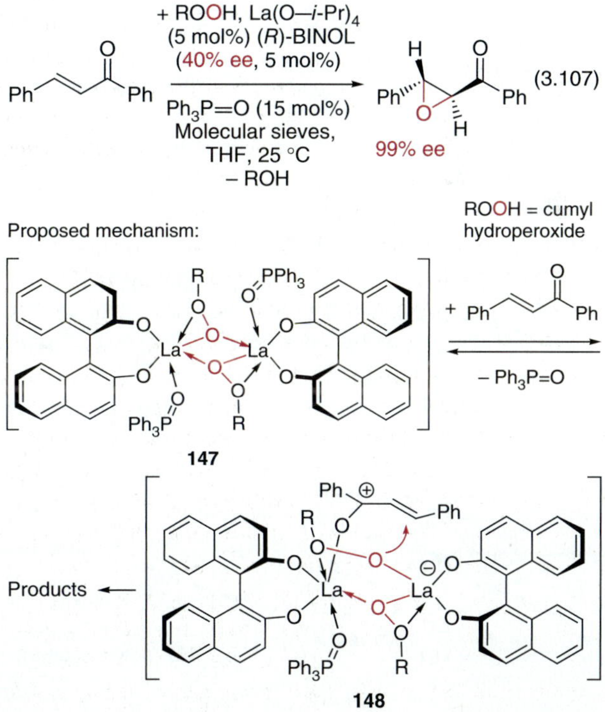

A priori, there are many more possibilities of aggregation in reactive intermediates than that retained in the ML_2 model. Intermediates of type ML_2 can equilibrate with LM, M_2L_2, M_2L_3, M_2L_4, etc. species that can be homochiral combinations of the ligands, or not, and have different reactivities. The reservoir model proposed by Kagan and coworkers describes the case when several intermediates are generated during the catalyst formation, one being the catalytically active species, e.g. ML [557]. The other aggregates are not reactive, or/and precipitate. Under these conditions, the enantiomer of L in excess will be found in ML, which leads to a (+)-NLE. The Noyori asymmetric hydrogenation of ketones employs chiral ruthenium complexes as catalysts (Section 7.8.6) [558–561]. The reaction is highly enantioselective and is applied in the industry to produce several drugs and other valuable chemicals [200]. A strong (+)-NLE is observed in the asymmetric catalytic hydrogenation (3.108) of ethyl acetylacetate catalyzed by [(BINAP)RuBr$_2$] prepared in situ from scalemic (*S*)-BINAP. The extent of the NLE depends on the conversion, with a lower asymmetric induction for extended reaction times. The homo- (less stable) and heterochiral (more stable) dinuclear complexes **149** or **150** precipitate while the trinuclear complex **151** is soluble in the supernatant (^{31}P-NMR studies) [562–564]. During hydrogenation, the homochiral dinuclear complex (pre-catalyst) generates the catalytically active species **152** (of type ML, contains a single BINAP ligand). On increasing the temperature, more heterodimeric complex is soluble, and this leads to lower (+)-NLE [565].

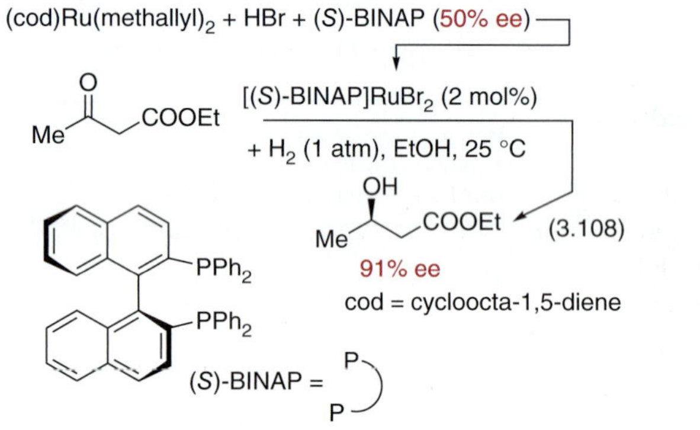

(3.108)

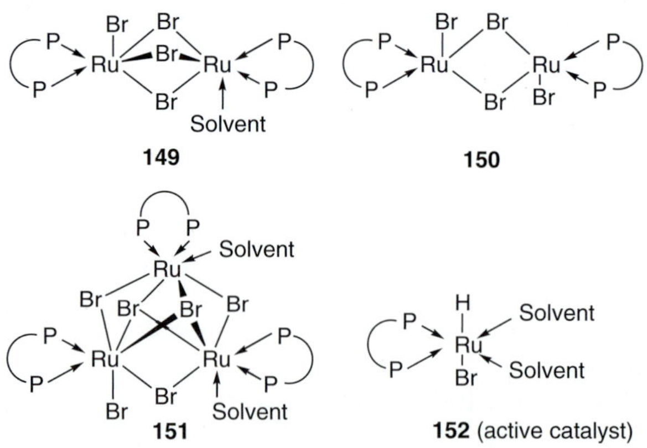

3.7.8 Asymmetric autocatalysis

Several theories have been proposed to explain the origin of molecular homochirality in living species and how it occurred in prebiotic era [566–579]. Frank has envisaged an asymmetric autocatalytic model, where one of the enantiomer catalyzes its own production and at the same time inhibits the formation of its opposite enantiomer [580]. In this model, a reaction that is not 100% enantioselective can provide a very high asymmetric amplification in an autocatalytic process [581]. It is similar to the reservoir model of Kagan and coworkers (Section 3.7.7) as one possible explanation for NLE. A spontaneous asymmetric synthesis arising from a minor imbalance in the enantiomeric ratio because of classical statistical fluctuation can be envisioned with this model [582]. An autocatalytic process without (+)-NLE cannot propagate asymmetric amplification during the course of a reaction, unless the autocatalytic process is 100% enantioselective [583]. Soai and coworkers have found a spectacular example of an asymmetric autocatalytic process (3.109) in which the addition of an isopropyl group to aldehyde **153** is catalyzed by the product (*S*)-**154** [584–586]. Applying catalyst (*S*)-**154** with only 0.000 05% ee, and after several runs using the product so obtained, (*S*)-**154** with >99.6% ee was produced [587, 588].

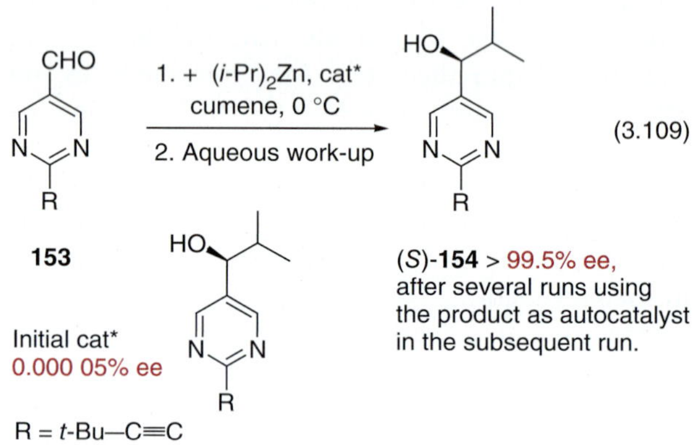

(3.109)

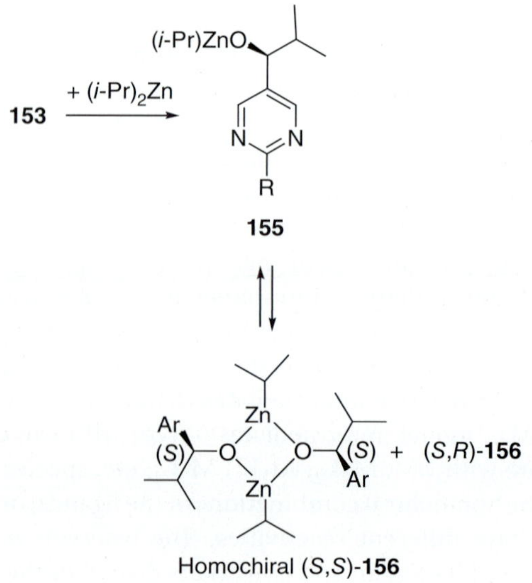

In 1999, Soai and coworkers reported the first example of asymmetric synthesis with a high ee using a chiral inorganic crystal as the catalyst. When 2-(*t*-butylethynyl)pyrimidine-5-carbaldehyde (**153**) is treated with (i-Pr)$_2$Zn in the presence of *d*-quartz powder in toluene, (*S*)-(2-*t*-butylethynyl)-5-(2-methylprop-1-yl)pyrimidine ((*S*)-**154**) with 89% ee is isolated. In the presence of *l*-quartz, the opposite enantiomer is produced [589]. The autocatalytic reaction rate law is first order in aldehyde **153** and (i-Pr)$_2$Zn and second order in alkylzinc alkoxide **155**. This suggests that the active catalyst is the homochiral dimer **156**. The overall reaction profile (conversion versus time) is S-shaped, which is an indication of an autocatalytic reaction [590]. Kinetic studies using microcalorimetry on the reaction of 2-methylpyrimidine-5-carbaldehyde + (i-Pr)$_2$Zn and NMR studies on reaction (3.108) suggested that both the homochiral and heterochiral dimeric zinc alkoxide intermediates have similar stabilities [591], the homochiral dimers ((*S*,*S*)-**156**) being more reactive than the corresponding heterochiral dimer ((*R*,*S*)-**156**) [592]. It was also found that dimers of type **156** can give rise to oligomers or dissociate into monomers, which renders the rate law quite complicated [593, 594]. Carreira and coworkers have developed an economical synthesis of efavirenz (Sustiva, Stocrin) [595], a key drug for the treatment of HIV, that applies asymmetric autocatalysis. The process uses substoichiometric quantities of homochiral ligand **159** ((1*R*,2*S*)-*N*-pyrrolidine norephedrine), of product (*S*)-**158**, of Et$_2$Zn (reaction (3.110)) [596], whereas the Merck process employs 1.5 equiv. of **159**, 1.2 equiv. of Et$_2$Zn, and 1.2 equiv. of magnesium chloride cyclopropylacetylide [597, 598].

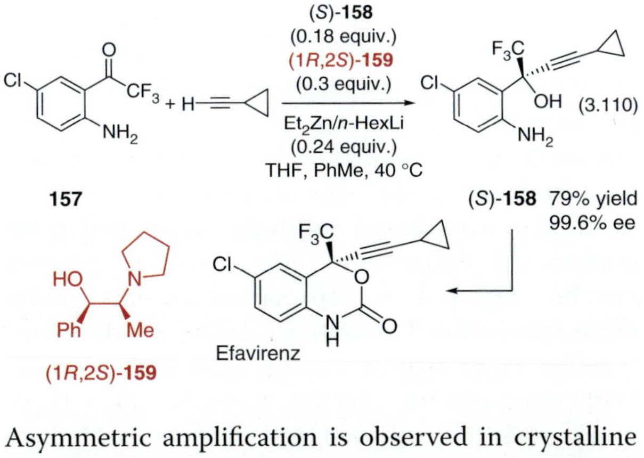

(3.110)

Asymmetric amplification is observed in crystalline (*S*)-**160** when it is vigorously stirred in toluene (slurry) at room temperature in the presence of racemic urea **162**. For instance, in the presence of 30 mol% of **162**, initial (*S*)-**160** with 17.5% ee is isomerized into crystals (*S*)-**160** with ee reaching 100%. Racemization occurs in the solution phase via a reversible Mannich-type reaction forming acetone and hydrazone **161**. It is assumed that the major enantiomer (*S*)-**160** is removed from the solution faster than the minor enantiomer owing to nonlinear crystal growth driven by the nonequilibrium distribution of crystal sizes (kinetically controlled crystal growth) [599, 600]. Autoamplification of molecular chirality through the induction of supramolecular chirality has been reported [601].

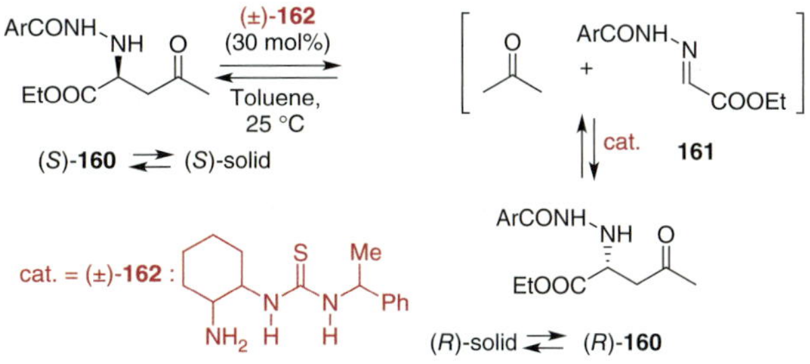

3.8 Chemo- and site-selective reactions

Synthetic chemists are often confronted with the chemo-, site-, or regioselective manipulation of polyfunctional compounds such a carbohydrates, polyketide antibiotics, peptides, or polyenes. When steric factors and electronic factors do not differentiate the reactivity of the various functions of the substrate, the chemists are forced to apply multistep processes that selectively protect the functions that should not be affected by the reaction conditions. Once the desired chemical transformation on the selected function is achieved, conditions for deprotection of the other functions must be found that liberate the desired product in good yield. This is usually complicated, costly, and time consuming. In several cases, microorganisms and enzymes are available that catalyze selectively the planned reaction on the desired function without protection of the other functions that should stay intact. This is illustrated with the industrial synthesis of L-ascorbic acid (vitamin C), which was first isolated in 1928 and subsequently identified as the antiscorbutic factor [602–605]. Vitamin C is used in the feed, food industry as a food preservative, and nutritional supplement [606, 607], making use of its antioxidant properties [608]. In 1933, Haworth and coworkers reported the first syntheses of D- and L-ascorbic acid and derivatives [609]. A year later, Reichstein and Grüssner presented a simpler synthesis of L-ascorbic acid (Scheme 3.26a) [610] that has become an industrial

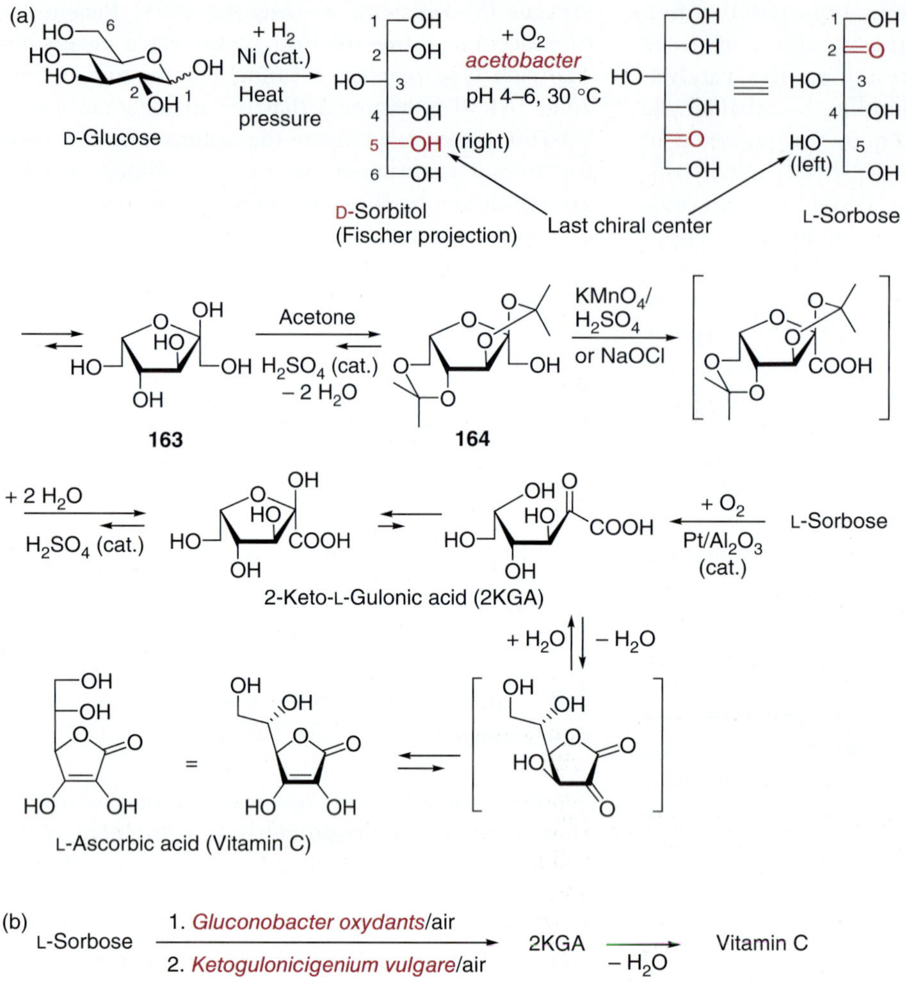

Scheme 3.26 Industrial production of vitamin C: (a) Reichstein–Grüssner synthesis and (b) two successive fermentative oxidations without protection/deprotection steps.

process (c. 60% yield from D-glucose) exploited first by Hoffmann–La Roche. Synthetic L-ascorbic acid is the first synthetic vitamin sold under several brand names, e.g. Redoxon (Roche) and Cebion (Merck) [611]. Catalytic hydrogenation of D-glucose gives D-sorbitol, hexol containing two primary and four secondary alcoholic moieties. Without selective protection, the secondary alcohol at C(5) is oxidized selectively by fermentation with *Acetobacter aceti* subsp. *xylinum* into L-sorbose, a process discovered in 1896 by Bertrand [612, 613]. L-Sorbose equilibrates with the furanose **163**, a pentol that reacts selectively with acetone under acidic conditions furnishing the diacetonide **164** containing an unprotected primary alcohol group. The latter is oxidized into the corresponding carboxylic acid with $KMnO_4$ or NaClO. A more economical route is the air oxidation on a platinum-based solid catalyst [614–616]. As the reaction is carried out under aqueous acidic conditions, the acetals are hydrolyzed liberating 2-keto-L-gulonic acid (2KGA) that eliminates 1 equiv. of water forming L-ascorbic acid. This industrial process has been used until the 1990s. Novel processes have eliminated the chemical protection–oxidation–deprotection steps **163** → 2KGA and rely upon two fermentative oxidations applying genetically modified bacteria (Scheme 3.26b) [617–619]. The method reduces the number of synthetic steps, it is more economical, and produces less coproducts.

Man-designed and man-made catalysts can render the search for chemo-, site-, and regioselective reactions of polyfunctional substrates easier and more economical. Furthermore, the chemical catalysts can be employed in nonaqueous solvents, under more concentrated conditions and in a larger temperature range than microorganisms and enzymes. With molecular weights much smaller than those of enzymes, the man-designed catalysts can imitate enzymes in their reactivity and selectivity. An example is given with the selective 4-O esterification of alkyl D-glucopyranoside that will be presented in Section 7.2.4 [620].

3.9 Kinetic isotope effects and reaction mechanisms

In the early 1930s, H_2 and D_2 were predicted to react at different rates because of the difference in zero-point energy (ZPE) [621, 622]. Kinetic isotope effects (KIEs) refer to the change in the rate of a chemical reaction upon substitution of an atom in one of the reactants with one of its isotopes [623–626]. Formally, it is defined as the ratio $\mathrm{KIE} = k_{\mathrm{light}}/k_{\mathrm{heavy}}$, where k_{light} is the rate constant of the reaction for the reactant with the light isotope, and k_{heavy} is the rate constant for the reaction for the reactant with the heavy isotope (for instance, ^{1}H, ^{2}H = D (deuterium); ^{3}H = T (tritium); ^{12}C, ^{13}C, ^{14}C; ^{14}N, and ^{15}N; ^{16}O, ^{17}O, and ^{18}O; ^{18}F and ^{19}F; ^{35}Cl, ^{37}Cl, etc.). KIEs represent the most powerful tools to investigate reaction mechanisms, including biochemical processes [627–641]. They give information about which bonds are broken or modified in a transition state of a reaction, and in some cases, about the properties of the transition states through which these bonds are modified [642, 643]. Kinetic isotope effects are more pronounced when the relative mass change is greatest, as the effect is related to vibrational frequencies of the affected bonds (Section 1.15). Substitution of a hydrogen atom by a deuterium atom represents a 100% increase in mass, whereas in replacing ^{12}C with ^{13}C, the mass increases by only 8%. Two types of KIEs are recognized: (i) primary KIEs in which the bond that contains the light or heavy isotope is broken in the transition state of the reaction under study and (ii) secondary KIEs for reactions where the bond that contains the light or heavy isotope is not broken. Primary kinetic isotope effects are usually larger than secondary kinetic isotope effects as illustrated with the deprotonation of ethylbenzene reported by Streitwieser and Van Sickle [644]. A primary kinetic deuterium isotope effect $k_H/k_D \cong 12$ for deprotonation (3.111) is found, whereas a much smaller k_H/k_{D3} value of 1.10 (1.03–1.04 per deuterium atom) is measured for reaction (3.112).

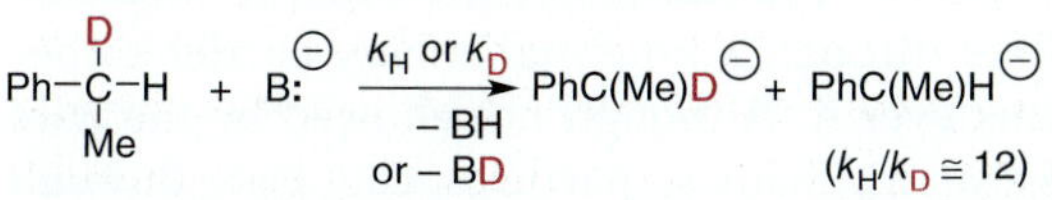
(3.111)

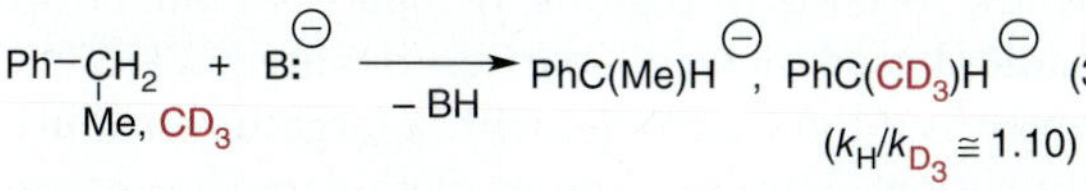
(3.112)

3.9.1 Primary kinetic isotope effects: the case of hydrogen transfers

In the case of C—H and C—D bonds, the energy difference due to the difference in ZPE (Eq. (3.113)) can be estimated from the vibrational frequency (Eq. (3.113)), which corresponds to 3000 cm^{-1} = 8.6 kcal mol^{-1} for a C—H bond (Section 1.15).

$$\mathrm{ZPE}_{\mathrm{C-H}} - \mathrm{ZPE}_{\mathrm{C-D}} \cong \tfrac{1}{2}h\nu - \frac{1}{2\sqrt{2}}h\nu \quad (3.113)$$

One obtains $\mathrm{ZPE}_{\mathrm{C-H}} - \mathrm{ZPE}_{\mathrm{C-D}} = \left(\frac{1}{2} - \frac{1}{2\sqrt{2}}\right)$ (8.6 kcal mol^{-1}) = 1.3 kcal mol^{-1}.

This means that the bond dissociation energy ($DH°(\mathrm{C}^\bullet/\mathrm{H}^\bullet)$) of a C—H bond is about 1.3 kcal mol^{-1} easier than the breaking of the corresponding C—D bond. At room temperature, this difference in activation energy corresponds to nearly a 10-fold difference in the rate of a reaction, in hydrogen transfers or other processes that break C—H or C—D bonds. An earlier example is the radical bromination with *N*-bromosuccinimide (NBS) of the methyl group of toluene reported by Wiberg and Slaugh in 1958 (reaction (3.114), in CCl_4 at 77 °C) for which the proportion of monodeuterated and nondeuterated benzyl bromide formed corresponds to $k_H/k_D = 4.86 \pm 0.03$. For the bromination with Br_2 (CCl_4, 77 °C), a similar KIE is found: $k_H/k_D = 4.59 \pm 0.03$, which is consistent with a mechanism in which the rate-determining step is the hydrogen transfer to bromine radical (Ph—CH_2—H + $Br^\bullet$ → $PhCH_2^\bullet$ + HBr). Interestingly, for the chlorination at 110 °C in toluene solution with *N*-chlorosuccinimide (NCS) and Cl_2, $k_H/k_D = 1.59 \pm 0.05$ and 1.47 ± 0.02, respectively, are reported [645]. One can conclude that the C—H bond breakage occurs to a greater extent in the transition state of the bromination than for the chlorination. One speaks of an earlier transition state for the chlorination than for the bromination. This is predicted by the Bell–Evans–Polanyi theory on concerted reactions ($\Delta^{\ddagger}H = \alpha\Delta_r H + \beta$). The transition state is earlier for the more exothermic hydrogen transfer (in the gas phase: $\Delta_r H°(\mathrm{Bn{-}H} + \mathrm{Cl}^\bullet \rightarrow \mathrm{Bn}^\bullet + \mathrm{HCl}) = -13.6 \pm 1.3$ kcal mol^{-1}; $\Delta_r H°(\mathrm{Bn{-}H} + \mathrm{Br}^\bullet \rightarrow \mathrm{Bn}^\bullet + \mathrm{HBr}) = 2.1 \pm 1.3$ kcal mol^{-1}). The transition state of the radical chlorination resembles the reactants (C—H, C—D bonds nearly unchanged), whereas the transition state of the radical bromination resembles the products (C—H, C—D bonds partially broken).

$$\mathrm{PhCH_2D} + \mathrm{NBS} \rightarrow \mathrm{PhCHDBr} + \mathrm{PhCH_2Br} + \text{succinimide} \quad (3.114)$$

A classical example of primary kinetic isotope effect reported by Westheimer and Nicolaides is the oxidation of secondary alcohols by chromic acid (Figure 3.19) [646–648]. For instance, the oxidation of 2-propanol gives acetone (reaction (3.115), rate constant k_H is six times faster than the oxidation of

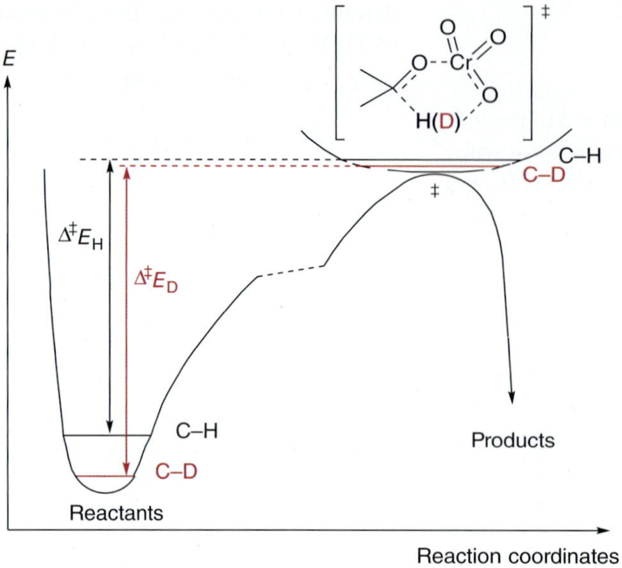

Figure 3.19 Hydrogen is transferred in the rate-determining step of the oxidation of *i*-propanol by chromic acid (Westheimer's mechanism).

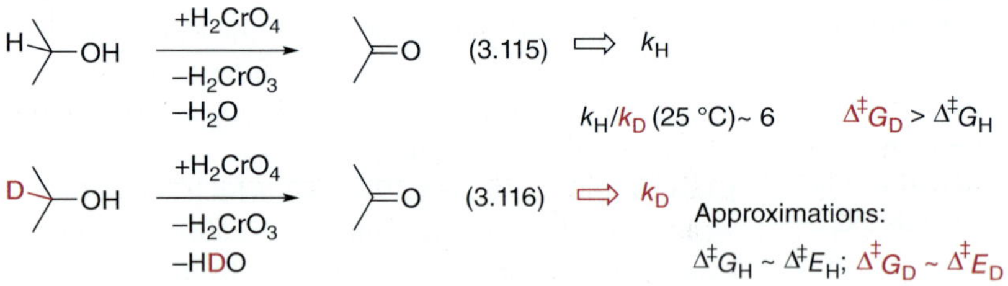

2-deuterio-2-propanol (reaction (3.116), rate constant k_D) at 25 °C.

The relatively large rate constant ratio k_H/k_D (25 °C) ≈ 6 for chromic acid oxidation of isopropanol into acetone is interpreted in terms of a rate-determining step in which H(D) is removed from C(2) of the isopropyl group of the hemiester of chromic acid and isopropanol (Figure 3.19). If one considers the breaking of the C—H bond, its bond strength diminishes between the reactant and the transition state. Thus, the ZPE of the C—H(D) bond decreases from reactant to transition state, and more so for the C—H bond than for the C—D bond. Swain and coworkers measured the KIEs $k_H/k_D = 5.02$ and $k_T/k_H = 10.2$ for the racemization of D-(+)-α-phenylisocaprophenone (8.4 M in dioxane, 1 : 1 AcOH/AcONa, NaCl, 97.8 °C) for which the rate-determining step is a proton transfer to the medium to form the corresponding enolate intermediate ($PhC(O^-){=}C(Ph){-}CH_2{-}CHMe_2$). Rate constant k_H was evaluated by polarimetry and k_D and k_T by measuring the rate of α-D/H and α-T/H exchange with the nonlabeled medium using racemic labeled ketones [649]. Assuming that the KIEs are uniquely due to changes in ZPE, $k_H/k_T = (k_H/k_D)^{1.442}$. This is called the Swain–Schaad relationship, which is followed by the above KIEs. This equation holds as long as the temperature is sufficiently low to have the great majority of the X—H, X—D, and X—T bonds in their lowest vibrational state.

3.9.2 Tunneling effects

Not only the stretching force constant changes during the C—H bond elongation but also the bending forces change. The C—D bond being shorter than the C—H bond, a steric factor (see Section 3.9.4), may also intervene. For light atoms, quantum mechanical tunneling may also make a large contribution to the kinetic isotope effect. Tunneling occurs when a molecule penetrates through a potential energy barrier rather than runs over it. Although not allowed by the laws of classical mechanics, particles can pass through classically forbidden regions of space in quantum mechanics based on wave–particle duality [650–652]. Hydrogen is a light particle, with a large uncertainty in its position. A measurement of this uncertainty is the deBroglie wavelength, $\lambda = h/(2mE)^{1/2}$, in which h is the Planck constant, m is the mass of the particle, and E is the energy. For E = 5 kcal mol^{-1}, the deBroglie wavelength is calculated to be 0.6, 0.5, and 0.4 Å for H^+, D^+, and T^+, respectively. As hydrogen is transferred over similar distances (<1 Å), this positional uncertainty is significant and implicates considerable nonclassical properties for hydrogen (H^+, $H^\bullet$, and H^-) transfers. Primary KIE $k_H/k_D > 10$ at 25 °C can be observed when tunneling effects are operating [653]. As tunneling effects occur with lightest isotopes, the Swain–Schaad relationship beaks down, with $k_H/k_T > (k_H/k_D)^{1.442}$. An example of primary KIE affected by tunneling effect is given with the intermolecular proton transfer reaction (3.117) between 4-nitrophenylnitromethane and tetramethylguadinidine [654]. In the less polar solvent toluene, the KIE is exceptionally large. The rate constant k_H measured at different temperatures shows a positive deviation from linear Arrhenius plot ($\ln k_H = \ln A - E_a/RT$), which is strongly indicative of a tunneling effect for proton transfer as discussed by Bell during the 1930s [624, 655, 656]. Usually, for KIE not affected by tunneling effect, the Arrhenius intercept $A_H/A_D \approx$ unity, whereas when tunneling effect operates for the transfer of the lighter isotope, $A_H/A_D < 1$. In more polar solvents such as CH_2Cl_2 and MeCN, the KIEs are closer to the range predicted by semiclassical theory (without tunneling effect). A possible interpretation for this solvent dependence on KIEs is to invoke solvent–solute interactions that affect the height of the reaction barrier and that motions of solvent molecules are coupled with the motion of the proton in the more polar solvents but not in the less polar ones; reorganization of solvent molecules accompanies the proton transfer in the more polar solvents, but only electron polarization in the less polar. Tunneling effect is large in toluene where the proton is the only atom that moves in the rate-determining step, but much smaller in more polar solvents where solvent motion is coupled to that of the proton. A pictorial representation of tunneling effect is given in Figure 3.20 [657]. For the hydroxide, anion-catalyzed H^+/D^+ transfer from 3,5-dimethylpentan-3-one k_H/k_D (25 °C, H_2O) = 6.09 is measured. This primary KIE is normal and the Arrhenius plot is linear, suggesting that no tunneling effect operates in this case [658].

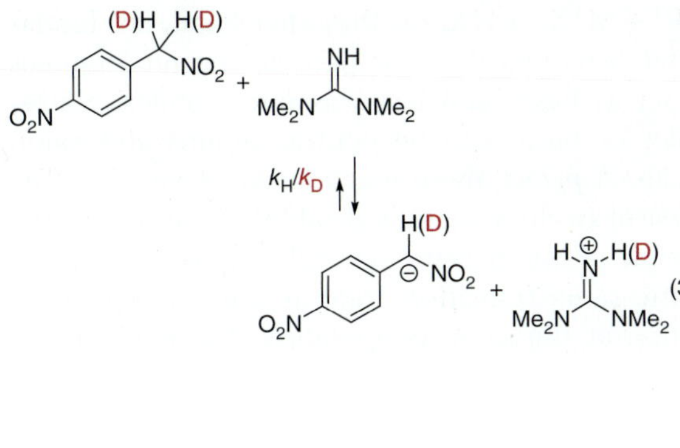

(3.117)

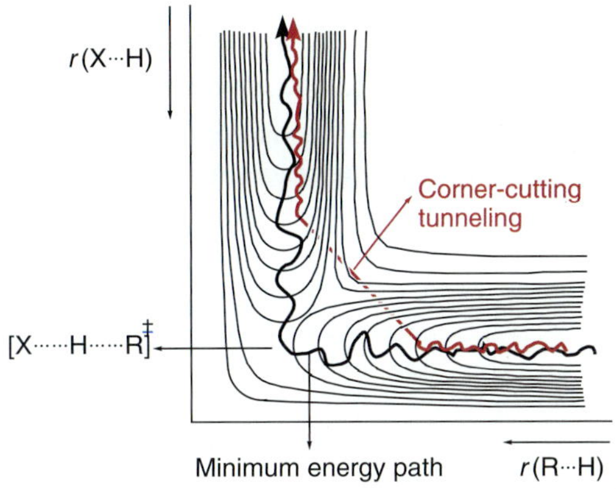

Figure 3.20 Representation of a "corner-cutting" tunneling effect (red trajectory).

In toluene, 25 °C: k_H/k_D (3.117) $= 45 \pm 2$, which corresponds to $\Delta E_a = 4.3 \pm 0.3$ kcal mol^{-1} and $\log(A_H/A_D) = 0.66 \pm 0.1$. The reaction is exergonic:

with $K_H = 174 \pm 6$, $\Delta_r H°(H) = -17.9 \pm 1.0$ kcal mol^{-1} and $\Delta_r S°(H) = -33 \pm 6$ eu and
with $K_D = 144 \pm 10$, $\Delta_r H°(D) = -9.4 \pm 1.5$ kcal mol^{-1} and $\Delta_r S°(D) = -22 \pm 5$ eu.

In CH_2Cl_2, 25 °C: k_H/k_D (3.117) $= 11.4 \pm 0.2$,

with $K_H = 6540 \pm 400$, $\Delta_r H°(H) = -8.7 \pm 1.0$ kcal mol^{-1} and $\Delta_r S°(H) = -12 \pm 4$ eu and
with $K_D = 4915 \pm 800$, $\Delta_r H°(D) = -9.1 \pm 0.6$ kcal mol^{-1} and $\Delta_r S°(D) = -14 \pm 3$ eu.

In MeCN, 25 °C: k_H/k_D(3.117) $= 11.8 \pm 0.2$.

The Bigeleisen treatment of KIEs (Eq. (3.118)) is based on Eyring's absolute rate theory in which the reactants are in equilibrium with the transition state ($\mathbf{A} + \mathbf{B} \rightleftarrows \ddagger$) [659, 660]. The transition state has the properties of a stable molecule except that one vibrational degree of freedom has become imaginary and that its energy is converted into motion along the reaction coordinate. The κ terms are the transmission coefficients and the Q and $Q^\ddagger$ terms are the complete partition functions for reactants **A** and **B** and the transition state, respectively.

$$k_L/k_H = (\kappa_L/\kappa_H)(Q^\ddagger{}_L/Q^\ddagger{}_H)(Q_{AH}Q_{BH}/Q_{AL}Q_{BL}) \tag{3.118}$$

A simplified version of this KIE equation is given with Eq. (3.119) [637, 650, 661].

$$k_L/k_H = (\kappa_L/\kappa_H)_T(v^\ddagger{}_L/v^\ddagger{}_H) \times[1 + \Sigma G(u_i)\Delta u_i - \Sigma G(u_i{}^\ddagger)\Delta u_i{}^\ddagger] \tag{3.119}$$

where $\Sigma G(u_i) = [1/2 - 1/u_i + 1/(\exp(u_i) - 1)]$, $\Sigma G(u_i{}^\ddagger) = [1/2 - 1/u_i{}^\ddagger + 1/(\exp(u_i{}^\ddagger) - 1)]$, $\Delta u_i = hc/k_b T(\Delta v_i)$, and

$\Delta u_i^{\ddagger} = hc/k_bT(\Delta v_i^{\ddagger})$. The terms h, c, k_b, and T are Planck's constant, the speed of light, Boltzmann's constant, and the absolute temperature, respectively. The Δv_i and $\Delta v_i^{\ddagger}$ terms represent the isotope effect on the vibrational frequencies of the reactant and transition state, respectively. In the absence of tunneling effect, $KIE_T = (\kappa_L/\kappa_H)_T = 1$. For many reactions, the imaginary frequency ratio or temperature-independent factor TIF $= (v^{\ddagger}_L/v^{\ddagger}_H)$ is close to unity and much smaller than the temperature-dependent factor TDF $= [1 + \Sigma G(u_i)\Delta u_i - \Sigma G(u_i^{\ddagger})\Delta u_i^{\ddagger}]$.

Biocatalysts such as enzymes speed up reactions by many orders of magnitude using fundamental physical processes such as stabilization of the transition state, destabilization of the reactants, and/or bringing them closer to the transition state by compression (see Section 3.6.4), and/or by barrier compression [662, 663]. Hydrogen tunneling also contributes to enzyme-catalyzed reactions [664–669]. An early example reported by Klinman and coworkers in 1989 concerns the hydride transfer reactions (3.120) between benzyl alcohol and the cofactor NAD+ (nicotinamide adenine diphosphate). The reaction is catalyzed by yeast alcohol dehydrogenase (YADH) and shows large k_H/k_D values [670].

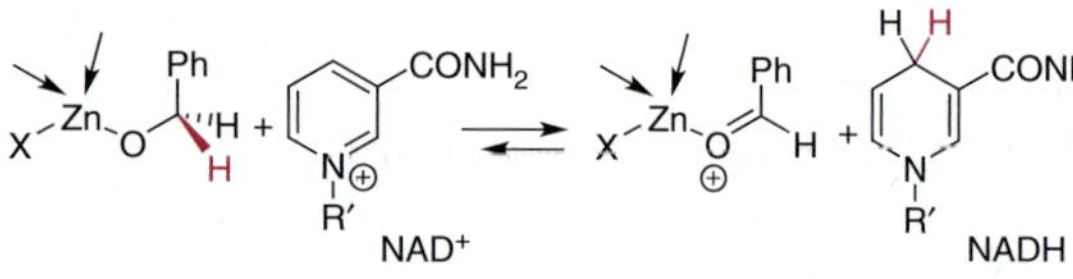

(3.120)

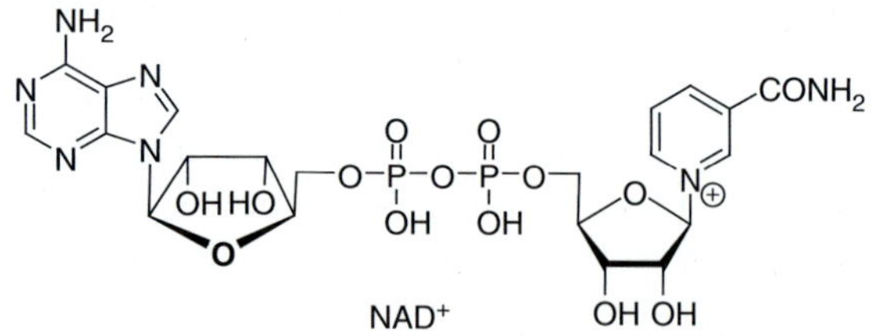

The same year, Grant and Klinman reported a first example of enzyme-catalyzed proton abstraction that is affected by quantum mechanical effects [671]. At 25 °C, $k_H/k_T = 35.2 \pm 0.8$ and $k_H/k_D = 3.07 \pm 0.07$ were found for the oxidation of benzyl amine into benzaldehyde + NH_3 catalyzed by bovine serum amine oxidase. The rate-determining step of this reaction is a proton transfer. Temperature dependence (0–45 °C) of the KIEs gave (Arrhenius intercepts) $A_H/A_T = 0.12 \pm 0.04$ and $A_D/A_T = 0.51 \pm 0.10$, values significantly smaller than unity, which confirm the occurrence of tunneling effects for both H^+ and D^+ transfers.

Problem 3.21 Acetophenone reacts with Br_2 in alkaline solutions with the rate law d[PhCOMe]/dt = $-k$[PhCOMe][HO^-]. The reaction shows important KIE k_D/k_H and k_T/k_H as shown with the following Arrhenius activation parameters for $PhCOCH_2T$, $E_a(T) = 15.35$ kcal mol^{-1}, $A(T) = 10^{8.58}$ M^{-1} s^{-1}; for $PhCOCD_3$, $E_a(D) = 14.6$ kcal mol^{-1}, $A(D) = 10^{9.24}$ M^{-1} s^{-1}; and for $PhCOCH_3$, $E_a(H) = 12.85$ kcal mol^{-1}, $A(H) = 10^{8.82}$ M^{-1} s^{-1}. What are the products of this reaction and what is the rate-determining step [672]?

Problem 3.22 In MeCN, 9-phenylxanthium tetrafluoroborate (**A**) adds isopropanol giving **B**. Using *i*-PrOD, EIE $K_H/K_D = 2.67$ is measured at 62 °C. Explain this EIE. A slower reaction converts **A** into **P** + **Q**. At 62 °C, a KIE $k_H/k_D = 4.4$ is measured using *i*-PrOH and $Me_2C(D)$-OH. Give a possible mechanism for the reduction of **A** by isopropanol [673].

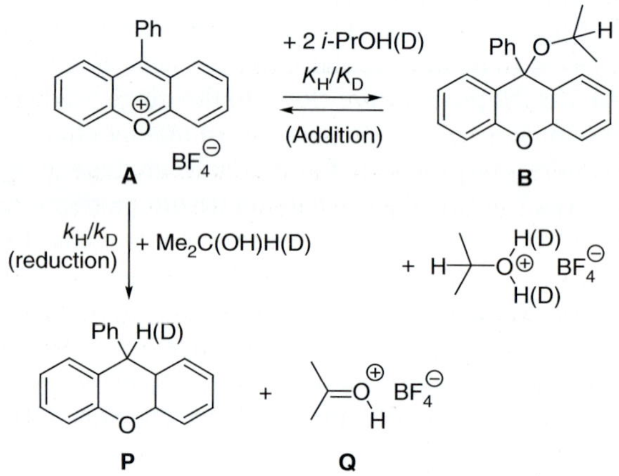

3.9.3 Nucleophilic substitution and elimination reactions

Bimolecular nucleophilic substitutions R—X (substrate = electrophile) + B: (nucleophile) ⇄ R—B$^+$/X$^-$ and R—X + M$^+$Y$^-$ ⇄ R—Y + M$^+$X$^-$, on the one hand, and base-induced eliminations R^4(R^3)CH—C(R^1)(R^2)—X + B: (base) ⇄ R^4(R^3)C=C(R^2)R^1 + BH$^+$/X$^-$ and R^4(R^3)CH—C(R^1)(R^2)—X + M$^+$Y$^-$ ⇄ R^4(R^3)C=C(R^2)R^1 + M$^+$X$^-$ + HX, on the other hand, are fundamental reactions that compete as the nucleophiles can act as bases and bases as nucleophiles. Nucleophiles or bases can be neutral, anions and anion radicals. *A priori*, these reactions are reversible. The reverse of an elimination is an addition. Because of the principle of microscopic reversibility, an elimination has the same transition state as the corresponding addition at the same temperature. For nucleophilic substitutions, they are summarized in Figure 3.21, and for eliminations in Figure 3.22.

Substitutions, eliminations, and additions all follow associative mechanisms if one considers solvent interactions. The order of the rate laws of these reactions permits a first classification. Displacement reactions with rate laws depending only on the concentration of the substrate [RX] follow S_N1 mechanisms. Eliminations with rate laws depending only on the substrate concentration $[R^4(R^3)CH—C(R^1)(R^2)—X]$ follow E1 mechanisms. These mechanisms imply the generation of carbocationic intermediates (tight, solvent-separated, or free ion pairs) in their rate-determining steps. Follows the fast quenching of the carbocationic intermediates by the nucleophile or medium in S_N1 substitutions, or attack of a α-, β-, or γ-proton by the medium in α-, β-, or γ-eliminations following E1 mechanisms. Mechanisms E2 imply transition states in which the α-C—X and β-H—C bonds are broken to the same extent, whereas for mechanism E2H the C—H bond is broken to a larger extent than the C—X bond and the proton is transferred to a large extent to the base. The transition state of the E2C mechanism is similar to that of a S_N2 reaction, the β-C—H bond is nearly intact, and the base makes a partial bond with α-C while the C—X bond is partially broken. Mechanism E1cb is a two-step elimination, with the conjugate base of the substrate that undergoes loss of the nucleofugal group (X^-). The transition state may precede or be after the carbanionic intermediate. The actual mechanism can be found in between the mechanism limits represented in Figures 3.21 and 3.22 (mechanism spectra). The transition state of a concerted process (S_N2, E2, E2H, and E2C) can be found anywhere between the reactant and the product, and this for all the reaction coordinates (see, however, the Bell–Evans–Polanyi theory, $\Delta^{\ddagger}H = \alpha\Delta_r H + \beta$, which tell us where to find the transition state as a function of the exothermicity of the one-step, concerted reaction). Primary and secondary KIEs can help us in distinguishing between possible mechanisms and, in some cases, can tell us about the position of the transition state (early or reactant-like transition state, late or product-like transition state) [627]. Stereochemistry offers us other criteria to establish mechanisms such as inversion for S_N2 and the requirement of *antiperiplanar* conformation for the β-C—H and α-C—X bonds of the substrate $R^4(R^3)CH—C(R^1)(R^2)—X$ for E2 and E2H mechanisms (*anti*elimination). The departing anion X^- is the nucleofugal group or leaving group. The departing proton H^+ in eliminations is the electrofugal group.

Primary KIEs of α-carbon center are nearly the same for S_N1 and S_N2 substitutions [674, 675], but secondary KIEs of α- and β-hydrogen atoms are not the same and permit mechanism distinction [629, 675]. Because of hyperconjugative interactions in carbenium ion intermediates that weakens the β-C—H bonds, $k_{\beta\text{-H}}/k_{\beta\text{-D}}$ depends on the relative amount of positive charge developed in the transition state of the nucleophilic substitution [676]. This is illustrated with the following hydrolyzes [677]. The hydrolysis of primary alkyl bromides follows pure S_N2 mechanism and the hydrolysis of tertiary alkyl halides follows S_N1 mechanisms. For secondary alkyl halides, S_N2 and S_N1 mechanisms may compete.

$$CH_3(D_3)CH_2 - Br + H_2O \rightarrow CH_3(D_3)CH_2 - OH$$
$$+HBr \quad k_H/k_{\beta\text{-}3D} = 1.03(80^\circ C)$$
$$[CH_3(D_3)]_2CH - Br + H_2O \rightarrow i\text{-}Pr - OH$$
$$+HBr \quad k_H/k_{\beta\text{-}6D} = 1.33(60^\circ C)$$
$$[CH_3(D_3)]_3C - Cl + H_2O \rightarrow t\text{-}Bu - OH$$
$$+HCl \quad k_H/k_{\beta\text{-}9D} = 2.57(2^\circ C)$$

In the case of the Menschutkin reaction, a typical S_N2 reaction that condenses alkyl halides with tertiary amines to generate the corresponding tetraalkylammonium salts, secondary α-deuterium isotope effects $k_H/k_D \cong 0.87$–0.91 (0.95–0.97 per D atom) are found for the reactions of CH_3I vs. CD_3I with Et_3N, Pr_3N, Bu_3N, pyridine, 2-picoline, and 2,6-lutidine [678]. With larger electrophiles such as EtBr, EtI, *n*-Pr–Br, *n*-Pr–I, and *i*-Pr–Br, their reactions with pyridine have smaller α-deuterium isotope effects ($k_H/k_D = 0.98$–1.0 per D atom) and no significant kinetic β-deuterium isotope effects ($k_H/k_D \approx 1.0$) [679].

For the S_N2 substitution (3.121) of ethyl chloride by cyanide anion in DMSO at 30 °C ($\Delta^{\ddagger}G$(25 °C) = 22.6 ± 0.1, $\Delta^{\ddagger}H$ = 18.7 ± 0.1 kcal mol^{-1}, $\Delta^{\ddagger}S$ = −13.2 ± 0.1 eu), the KIEs given here have been reported [680, 681].

$$CH_3 - CH_2 - Cl + Bu_4N^+/CN^- \rightarrow CH_3 - CH_2 - CN + Bu_4N^+/Cl^- \quad (3.121)$$

Primary KIEs	**Secondary KIEs**
$k_{\alpha\text{-}12C}/k_{\alpha\text{-}13C} = 1.0711 \pm 0.0078$	$k_{\alpha\text{-}H2}/k_{\alpha\text{-}D2} = 0.990 \pm 0.004$
$k_{\alpha\text{-}11C}/k_{\alpha\text{-}14C} = 1.21 \pm 0.02$	$k_{\beta\text{-}H3}/k_{\beta\text{-}D3} = 1.014 \pm 0.003$
$k_{35Cl}/k_{37Cl} = 1.0070 \pm 0.00026$	
$k_{12C(CN)}/k_{13C(CN)} = 1.0009 \pm 0.0007$	$k_{\beta\text{-}14N}/k_{\beta\text{-}14N} = 1.0002 \pm 0.0006$

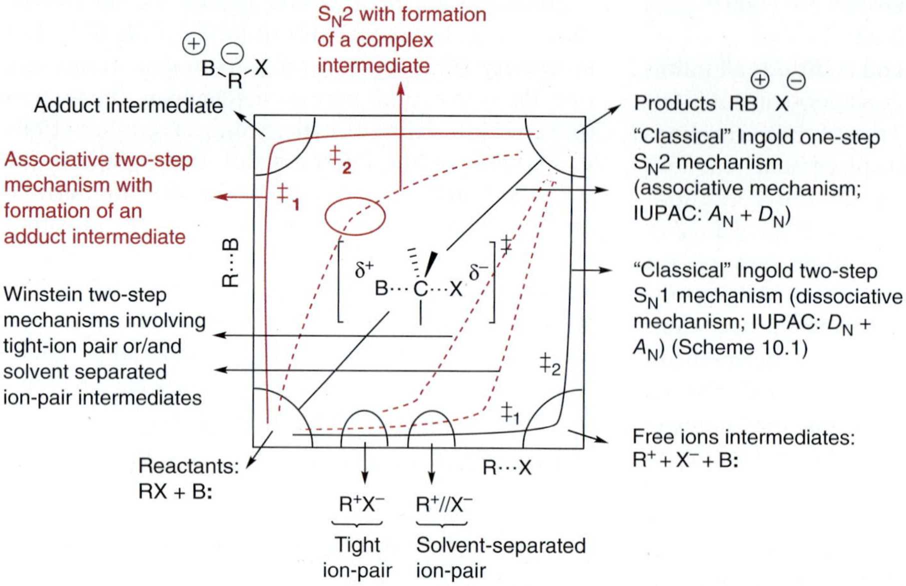

Figure 3.21 More O'Ferrall–Jencks diagram showing an ensemble of possible mechanisms for the displacement reactions RX + B: ⇄ RB^+X^-; the potential energy increases perpendicular to the plane defined by the two main reaction coordinates, i.e. distance separating R· · ·X and R· · ·B. RX is the substrate or electrophile, B: the base or nucleophile, X^-, the nucleofugal group or leaving group. If B: = H—Y:, the product of substitution might be R—Y + HX instead of R—$Y^{(+)}$—H + X^-. If B: = M^+Y^-, the products are R—Y + M^+X^-. By definition, S_N1 mechanisms obey rate laws of type $d[RX]/dt = -k_s{}^{I}[RX]$, and S_N2 mechanisms obey rate laws of type $d[RX]/dt = -k_s{}^{II}[RX][B:]$.

Corrections and improvements

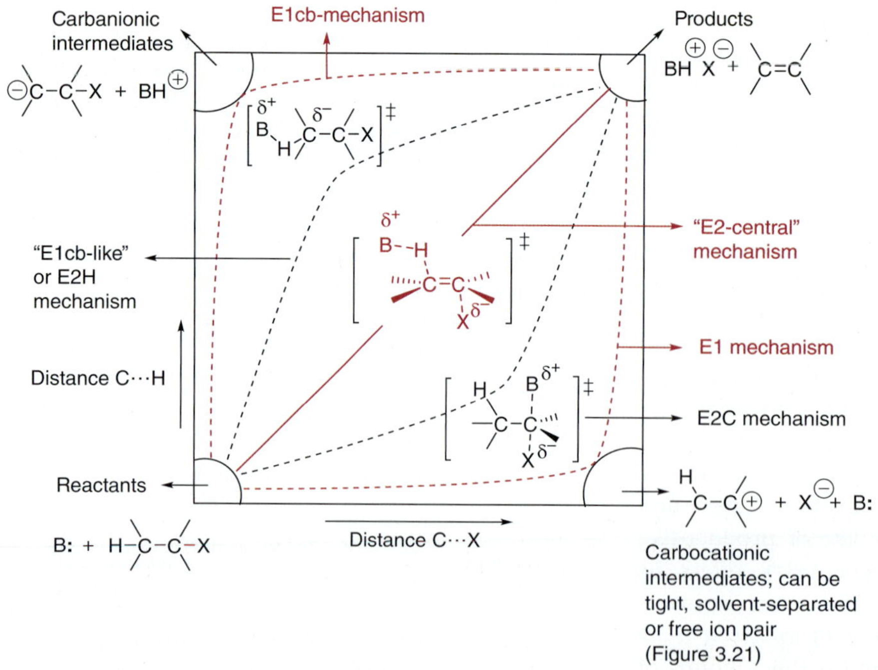

Figure 3.22 More O'Ferrall–Jencks diagram representing the energy surface associated with the E1cb-E2-E1 mechanistic spectrum for an elimination reaction B: + H—C—C—X → BH^++C=C + X^-. Depending on the type of substrate or electrophile (H—C—C—X), base (B:), solvent, and temperature, other intermediates may exist. By definition, mechanism E1 obeys rate laws of type $d[RX]/dt = -k_e{}^{I}[RX]$; E2C, E2, and E2H mechanisms obey rate laws of type $d[RX]/dt = -k_e{}^{II}[RX][B:]$, and mechanisms E1cb obey rate laws of type $d[RX]/dt = -k_b[RX]$.

Quantum mechanical calculations on MeCl reacting with a wide variety of anionic, neutral, and radical anion nucleophiles predicted that the C—Cl bond orders varied widely from c. 0.32 to 0.78, a range from reactant-like (early) to product-like (late) transition states. Unfortunately, this is not evidenced by the primary KIE $k_{35\text{Cl}}/k_{37\text{Cl}}$ that fall in a very small range (1.0056–1.0091) [682, 683]. A similar conclusion has been reached for primary α-carbon KIE on the S_N2 reaction [684]. The primary hydrogen–deuterium kinetic isotopic effect for the S_N2 displacement of *para*-substituted benzyl chlorides **165** by $NaBH_4/NaBD_4$ in DMSO at 30 °C is small (average: 1.246 ± 0.010) and insensitive to a change in the substitution at the α-carbon. A relatively large, constant secondary α-deuterium kinetic isotopic effect of 1.089 ± 0.002, and a large chlorine leaving group kinetic isotope effect $k_{35\text{Cl}}/k_{37\text{Cl}} = 1.0076 \pm 0.0003$ are found. These data suggest that the transition states **166** for these reactions are unsymmetrical with short H—C_α and long B· · ·H and C_α· · ·Cl bonds [685]:

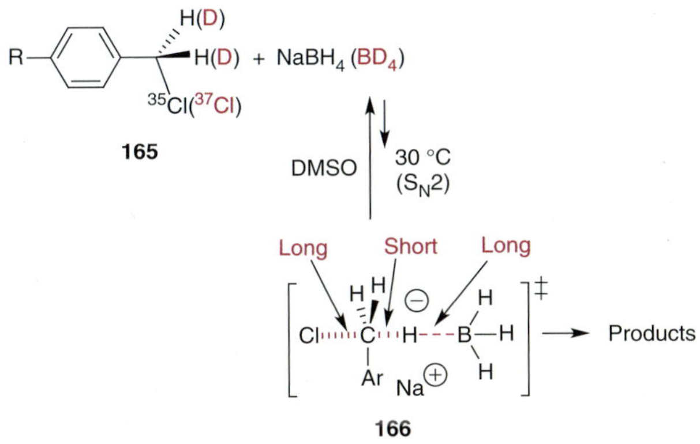

Lee and coworkers have proposed that displacement reactions of benzyl sulfonates **167** and **169** by anilines may follow two different mechanisms: the classical S_N2 mechanism that implies backside attack and formation of a three-center trigonal bipyramidal transition state **168**, and frontside attack that corresponds to an S_Ni process of an intermediate complex **170** that forms (hydrogen bridging) between the sulfonate and aniline reactants and evolves to a six-center cyclic transition state **171**. The latter mechanism competes with the S_N2 process. This was suggested by the finding that $k_{\alpha\text{-H}}/k_{\alpha\text{-D}} < 1$ for deuterated aniline on primary benzyl sulfonates and $k_{\alpha\text{-H}}/k_{\alpha\text{-D}} > 1$ for secondary 1-phenylethyl benzenesulfonates. In transition state **168**, the N—H(D) bonds are not broken, whereas in transition state **171**, one N—H(D) bond is partially broken. Thus, the proton transfer from the aniline is coupled with the displacement reaction at the benzylic center. Note that if a second molecule of aniline should be hydrogen bridging, the aniline moiety (Ph—NH—H· · ·:NH_2—Ph) that attacks the sulfonates, $k_{\alpha\text{-H}}/k_{\alpha\text{-D}} > 1$ would also be found [686]. As the rate laws are strictly first order in sulfonates and anilines, this latter hypothesis is not retained [687–689].

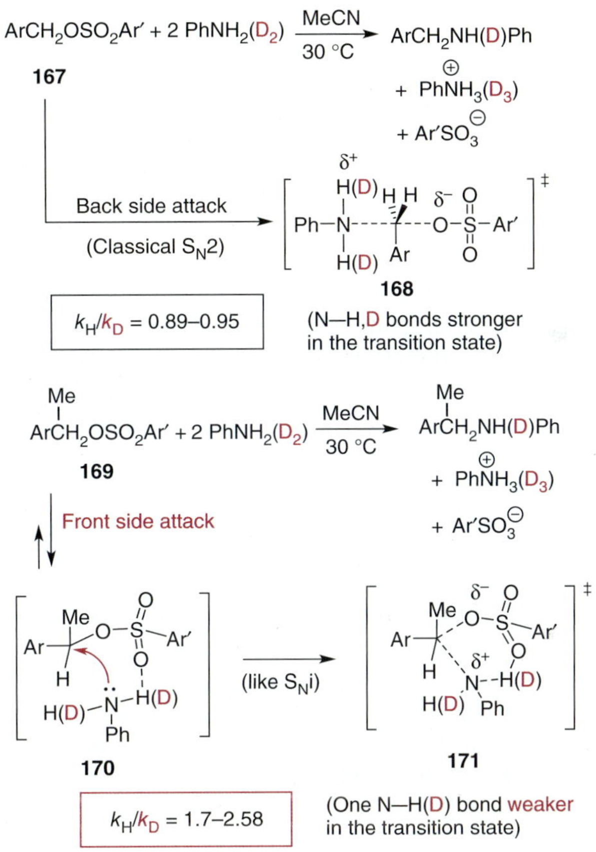

A similar competition between backside and frontside attacks by anilines has been evidenced applying the above KIE criterion for the nucleophilic substitution reactions of phosphoryl (e.g. $X_2P(=O)(OEt)—Cl$) [690–694] and thiophosphoryl (e.g. $X_2P(=S)(OEt)Cl$) systems [695–697].

Fluorine KIE [698, 699] has been measured for nucleophilic aromatic substitution (S_NAr) of 2,4-dinitrofluorobenzene (**172**) with piperidine and is found to depend on solvent, with $k_{18\text{F}}/k_{19\text{F}} = 0.9982 \pm 0.0004$ in MeCN and 1.0262 ± 0.0007 in THF at 22 °C [700]. The current reaction mechanism (Scheme 3.27) implies the formation of adduct **173**, which can equilibrate with its conjugate base **174** (Meisenheimer complex). Then follows the elimination of the nucleofugal group, in this case, the fluoride anion, a reaction that is acid catalyzed and implies transition state **175** (addition–elimination mechanism). In the polar solvent MeCN, the addition of the nucleophile to the aromatic substrate to generate **173** (rate constant k_1) is the rate-determining step, whereas in THF, the departure of the leaving group (formation of **176**) is the rate-determining step. The very small inverse k_{18F}/k_{19F} in MeCN might be attributed to a very small secondary KIE for the rate-limiting formation of intermediate **173**. Going from reactant to the transition state of the formation of adduct **173**, the hybridization of the carbon atom of the C—F moiety changes from sp^2 to sp^3. This is associated with a small increase of the force field for ^{18}F as one normal mode of out-of-plane bending vibration is transformed to a valence angle bending vibration possessing a somewhat higher frequency. In MeCN, k_{18F}/k_{19F} does not vary with the concentration of the nucleophile (piperidine). Being a better hydrogen donor than THF (makes stronger hydrogen bridging with the nucleofugal group ($F^-\cdots H—CH_2CN$) than THF), MeCN is a general acid catalyst that accelerates the conversion of **174** into transition state **175** and then into the product **176**. Overall, the elimination steps are more rapid than the addition step (k_1). The reaction does not need the protonated nucleophile as acid catalyst. Thus, in MeCN, KIE = $k_1(18F)/k_1(19F)$ = 0.9982 ± 0.004. In THF, departure of the nucleofugal group is the slowest process. The apparent rate constant k_{app} of the overall S_NAr represented in Scheme 3.27 is given by expression (3.122) based on the steady-state assumption for the intermediates **173** and **174**, where K_3 is the equilibrium constant for deprotonation of **173** to **174** and k_4 is the rate constant for the general-acid-catalyzed expulsion of the leaving group from **174**. If $k_{-1} \gg k_2 + k_3[\mathbf{B}]$, the observed KIE = $[K_1(18F)/K_1(19F)][k_4(18F)/k_4(19F)][(K_3(18F)/K_3(19F)]$. The first term $[K_1(18F)/K_1(19F)]$ is an EIE that must be inverse and small for the argument discussed above in relation with KIE = $k_1(18F)/k_1(19F)$. The third term $[(K_3(18F)/K_3(19F)]$ is the EIE for the proton transfer reaction **174** + **B** ⇄ **176** + $\mathbf{BH}^+/F^-$, which is expected to be close to unity since the force fields for the ^{18}F and ^{19}F atoms are hardly changed between reactant and product. Thus, the observed KIE in THF amounts approximately to $[k_4(18F)/k_4(19F)]$.

$$k_{app} = (k_1k_2 + k_1k_2K_3[\mathbf{B}])(k_{-1} + k_2 + k_4K_3[\mathbf{B}]) \tag{3.122}$$

KIE for the following eliminations depend on the nature of the nucleofugal group. They show that the extent of β-C—H bond breaking in the transition state for the elimination from the bromide is larger (E2H mechanism) than for the elimination from the ammonium salts (E2 or E2C mechanism) [701, 702]. For large $k_{\beta\text{-H}}/k_{\beta\text{-D}}$ values, tunneling effect might operate [703].

$$PhCH_2(D_2) - CH_2 - X + EtONa/EtOH \rightarrow PhCH(D) = CH_2 + EtOH(D) + NaX$$

$$X = Br(50°C) \quad k_{\beta-2H}/k_{\beta-2D} = 6.79 \pm 0.19$$
$$X = TsO(30°C) \quad 5.66 \pm 0.57$$
$$X = S^{(+)}Me_2/Br^-(30°C) \quad 5.07 \pm 0.22$$
$$X = N^{(+)}Me_3/Br^-(50°C) \quad 2.98 \pm 0.08$$

In general, $k_{\beta\text{-H}}/k_{\beta\text{-D}} \approx 6$–7 for an elimination following a E2H mechanism and 1.1–2.5 for an elimination

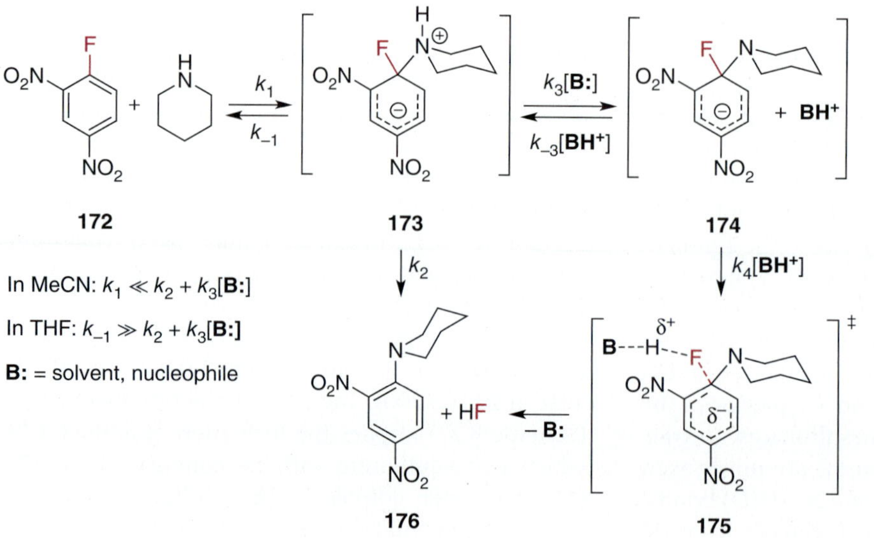

Scheme 3.27 Fluorine kinetic isotope effect on the nucleophilic aromatic substitution reaction (S_NAr) of 2,4-dinitrofluorobenzene by piperidine demonstrates that the rate-determining step is the formation of the Meisenheimer complex in MeCN, and the acid-catalyzed departure of the leaving group (F^-) from it in THF.

following a E2C mechanism [704–706]. In the case of the elimination of *p*-toluenesulfonic acid from tosylate **177** induced by EtONa in EtOH $k_{\beta\text{-H}}/k_{\beta\text{-D}} = 2.6$ for the formation of the major product, the Saytzeff alkene 2-methylbut-2-ene, and c. 6 for the formation of the minor product, the Hofmann olefin 3-methylbut-1-ene [707].

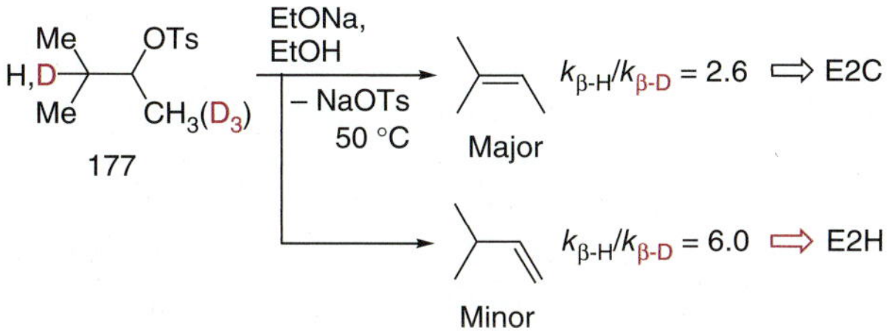

Problem 3.23 Base-induced elimination of HBr from bromocyclohexane gives cyclohexene. KIEs $k_{\alpha\text{-H}}/k_{\alpha\text{-D}}$ = 1.13–1.15 and $k_{\beta\text{-H}}/k_{\beta\text{-D}}$ = 1.15–1.25 have been measured for this elimination. Propose a mechanism for this reaction [708–710].

Problem 3.24 Eliminations of AcOH(D) from 3-(2-acetoxyprop-2-yl)-(1*H*)-indene and from its 1-deuterated analog, promoted by tertiary amines, show $k_{\beta\text{-H}}/k_{\beta\text{-D}}$ = 0.9–2.0 in solvents ranging from *t*-BuOH to MeOH/H_2O. Using MeONa as base in MeOH, $k_{\beta\text{-H}}/k_{\beta\text{-D}} = 7.6$. Propose mechanisms for these reactions [711].

Problem 3.25 For the aminolysis of methyl ethyl (**A**) and di(isopropyl) chlorothiophosphate (**B**) with 3-chloroaniline, $k_H/k_D = 0.659 \pm 0.006$ and 1.21 ± 0.002, respectively, have been measured. Propose mechanisms for these reactions [712].

3.9.4 Steric effect on kinetic isotope effects

KIE isotope effects of reactions that do not involve bond breaking of bond formation have been reported. They are attributed to different vibrational amplitudes of analogs substituted by isotopes. An early example reported by Breslow and coworkers in 1964 is the KIE of the racemization (3.123) of 9,10-dihydro-4,5-dimethylphenanthrene [713]. The deuterium KIE in the racemization of (−)-1,1′-binaphthyl-2.2′-d_2 amounts to k_H/k_D = 0.83–0.87 between 20 and 65 °C. Eyring plots gives $\Delta^{\ddagger}H_H - \Delta^{\ddagger}H_D = 0.27 \pm 0.14$ kcal mol^{-1} and $\Delta^{\ddagger}S_H - \Delta^{\ddagger}S_D = 0.54 \pm 0.43$ eu [714]. These KIE are often said to manifest a differential steric effects in the transition state of the reaction associated with the fact that the C—H bonds are slightly longer than the corresponding C—D bonds (see Section 1.15).

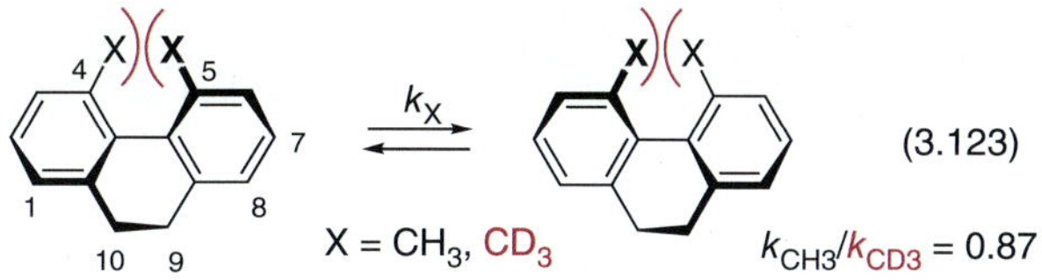

(3.123)

Problem 3.26 Explain the following secondary KIE [715, 716].

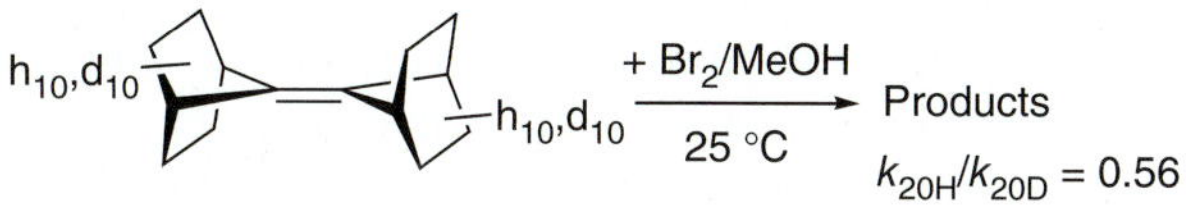

3.9.5 Simultaneous determination of multiple small kinetic isotope effects at natural abundance

As any reaction proceeds, the starting materials are enriched in isotopically slower reacting components. The proportion of a minor isotopic component in recovered material compared to the original reactants (R/R_o) is related to the fractional conversion of reactants (f) and the kinetic isotope effect KIE (relative rate for major/minor isotopic components by Eq. (3.124)) [650]. As a reaction approaches completion ($f \to 1$), R/R_o approaches ∞, and KIEs become greatly magnified in the observable R/R_o. For example, a KIE of 1.05 results in ~25% enrichment of a slower reacting isotopic component at 99% conversion. This ordinary effect of kinetic fractionation can improve the precision of KIE determinations [717–719], but the degree of improvement possible depends critically on the magnitude of the KIE, the precision of the analytical technique ($\Delta R/R_o$), and the uncertainly in f (Δf).

$$R/R_o = (1-f)^{(1/\mathrm{KIE})-1} \tag{3.124}$$

$$\mathrm{KIE}_{\mathrm{calcd}} = \frac{\ln(1-f)}{\ln[(1-f)R/R_o]} \tag{3.125}$$

$$\Delta\mathrm{KIE}_f = \frac{\partial KIE}{\partial f}\Delta f = \frac{-\ln(R/R_o)}{(1-f)\ln^2[(1-f)R/R_o]}\Delta f \tag{3.126}$$

$$\Delta\mathrm{KIE}_R = \frac{\partial KIE}{\partial(R/R_o)}\Delta(R/R_o) = \frac{-\ln(1-f)}{(R/R_o)\ln^2[(1-f)R/R_o]}\Delta(R/R_o) \tag{3.127}$$

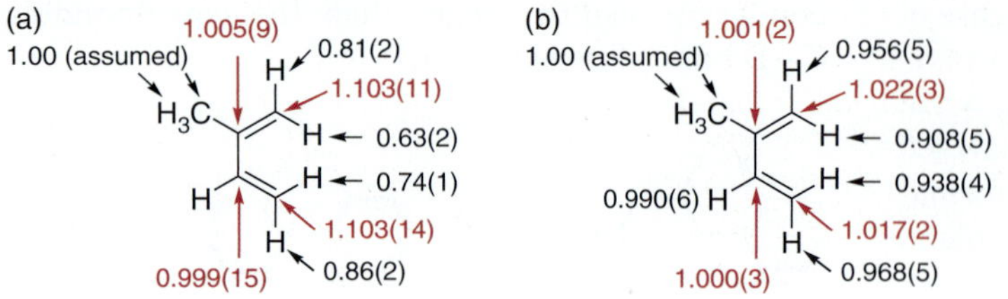

Figure 3.23 (a) ^2H and ^{13}C isotopic composition of isoprene recovered from a Diels–Alder reaction taken to 98.9% completion, relative to starting isoprene, with standard deviations in parenthesis ($n = 11$ (3 samples) for ^{13}C data; $n = 5$ (2 samples) for ^2H data). (b) ^2H and ^{13}C KIEs (k_H/k_D and $k_{12}C/k_{13}C$) calculated from the results in (a) and Eq. (3.125).

The uncertainties in calculated KIEs (Eq. (3.125)) due to $\Delta R/R_o$ (ΔKIE_R) and Δf (ΔKIE_f) are shown in Eqs. (3.126) and (3.127), respectively. Most commonly, either the KIE is large or the measurement of R/R_o is highly precise ($\Delta R/R_o$ small). In these cases, the uncertainty in the KIE is dominated by ΔKIE_f (not counting any systematic error), and no advantage is gained at high conversion. However, for the relatively low precision of NMR integrations, the uncertainty in the KIE is dominated by ΔKIE_R, which decreases greatly as f increases, and ΔKIE_f is nearly negligible. For example, with a KIE of 1.02 and a $1.0\% \pm 0.1$ measurement of unreacted starting material ($1 - f = 0.01 \pm 0.001$), ΔKIE_f is only 0.0004, and a 1.5% uncertainly in NMR integrations results in an uncertainty ($\approx \Delta KIE_R$) of only 0.003 in the KIE.

For the Diels–Alder addition of isoprene and maleic anhydride (3.128), the relative proportion of ^2H(D) and ^{13}C at the various positions of isoprene compared to the original starting material can be determined by deuterium and carbon-13 NMR using the methyl group as an "internal standard" on the assumption that its isotopic composition does not change. As the reaction proceeds, the relative proportion of ^{13}C at C(1) and C(4) of isoprene increases, and the proportion of deuterium in these positions decreases. Isoprene recovered by distillation from a reaction taken to 98.9(1)% completion was analyzed, and the results are shown in Figure 3.23a. The ^2H and ^{13}C KIEs (k_H/k_D and $k_{^{12}C}/k_{^{13}C}$) could be calculated from Eq. (3.125) and are shown in Figure 3.23b [720–722]. The above method has been applied by Singleton and coworkers to determine the mechanisms of several reactions such as the addition of Bu_2CuLi to cyclohexanone [723–725], the Sharpless asymmetric dihydroxylation [726], the dichlorocarbene cycloaddition to alkenes [727], the bromination of 1-pentene [728], the Et_2AlCl-catalyzed Diels–Alder addition of isoprene to methyl vinyl ketone [729], the cycloaddition of trimethylenemethanepalladium to unsaturated esters [730], the Baeyer–Villiger oxidation of cyclohexanone [731], the Claisen rearrangement [732], ene-reactions [733, 734], the allylic oxidation of alkenes with SeO_2 [735], the (2+2)-cycloaddition of ketenes with aldehydes [736], the electron-transfer-catalyzed Diels–Alder additions [737], the rhodium-catalyzed cyclopropanations [738], and the palladium-catalyzed allylic alkylation [739].

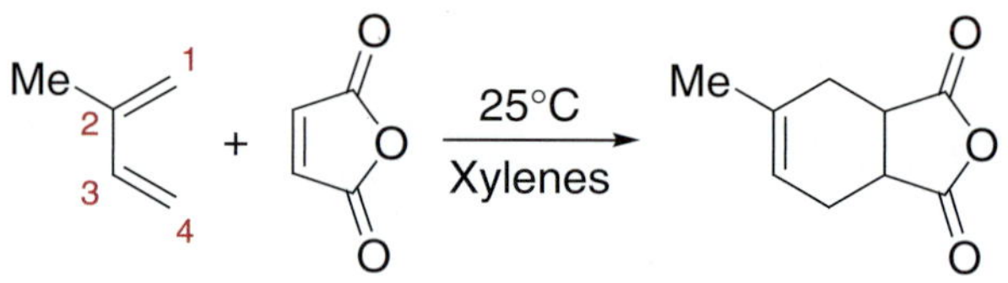

(3.128)

References

1 Atkins, P.W. and de Paula, J. (2014). *Atkins' Physical Chemistry: Thermodynamics, Structure and Change*, 10e, 1–1040. Oxford: Oxford University Press.

2 Laidler, K.J. (1997). *Chemical Kinetics*, 3e, 1–490. Upper Saddle River, NJ: Prentice Hall Inc.

3 Laidler, K.J. (1995). *The World of Physical Chemistry*, 1–488. Oxford: Oxford University Press.

4 Pilling, M.J. and Seakins, P.W. (1996). *Reaction Kinetics* (Oxford Science Publications), 2e, 1–320. Oxford: Oxford University Press.

5 Houston, P.L. (2001). *Chemical Kinetics and Reaction Dynamics*, 1–330. New York, NY: McGraw Hill Higher Education.

6 Tolman, R.C. (1924). Duration of molecules in upper quantum states. *Phys. Rev.* 23 (6): 693–709.

7 Brown, M.E. and Phillpotts, C.A.R. (1978). Nonisothermal kinetics. *J. Chem. Educ.* 55 (9): 556–560.

8 Maeder, M., Molloy, K.J., and Schumacher, M.M. (1997). Analysis of non-isothermal kinetic measurements. *Anal. Chim. Acta* 337 (1): 73–81.

9 Arhangel'skii, I.V., Dunaev, A.V., Makarenko, I.V. et al. (2013). Non-isothermal kinetic methods. In: *Max Planck Research Library for the History and Development of Knowledge*, Textbook 1 (ed. J. Renn, R. Schlögl and B.F. Schutz),

1–65. Berlin: Edition Open Access, Epubli www.edition-open-access.de.

10 Dubrunfaut, A.P. (1846). Note sur quelques phénoménes rotatoires et sur quelques propriétés des sucres. *Comptes rendus hebdomadaires des Séances de l'Académie des Sciences* 23: 38–44.

11 Hudson, C.S. (1910). A review of discoveries on the mutarotation of the sugars. *J. Am. Chem. Soc.* 32: 889–894.

12 Waage, P. and Guldberg, C.M. (1864). Studies concerning affinity. *Forhandlinger: Videnskabs-Selskabet i Christiana* 35.

13 Monnat, F., Vogel, P., Rayon, V.M., and Sordo, J.A. (2002). Ab initio and experimental studies on the hetero-Diels–Alder and cheletropic additions of sulfur dioxide to (*E*)-1-methoxybutadiene: a mechanism involving three molecules of SO_2. *J. Org. Chem.* 67 (6): 1882–1889.

14 Lindemann, F.A., Arrhenius, S., Langmuir, I. et al. (1922). Discussion on "the radiation theory of chemical action". *Trans. Faraday Soc.* 17: 598–606.

15 Hinshelwood, C.N. and Thompson, H.W. (1926). An apparently unimolecular reaction – the homogeneous decomposition of gaseous propionic aldehyde. *Proc. R. Soc. Lond. A Math. Phys. Eng. Sci.* 113 (763): 221–229.

16 Rice, O.K. and Ramsperger, H.C. (1927). Theories of unimolecular gas reactions at low pressures. *J. Am. Chem. Soc.* 49: 1617–1629.

17 Kassel, L.S. (1928). Studies in homogeneous gas reactions I. *J. Phys. Chem.* 32 (2): 225–242.

18 Temkin, O.N. (2012). *Homogeneous Catalysis with Metal Complexes: Kinetic Aspects and Mechanisms*, 1–830. New York, NY: Wiley.

19 Tolman, R.C. (2010). *The Principles of Statistical Mechanics* (first published in1938), 1–704. Mineola, NY: Dover Publications Inc.

20 O'Sullivan, C. and Tompson, F.W. (1890). Invertase: a contribution to the history of an enzyme or unorganized ferment. *J. Chem. Soc. Trans.* 57: 834–931.

21 Schiwerk, H., Clarke, M., and Pollach, G. (2007). *Sugar*, Ullmann's Encyclopedia of Industrial Chemistry, vol. 34, 558–628. Weinheim: Wiley.

22 Henri, V. (1902). Théorie générale de l'action de quelques diastases. *Comptes rendus hebdomadaires des Séances de l'Académie des Sciences* 135: 916–919.

23 Michaelis, L. and Menten, M.L. (1913). Die Kinetik der Invertinwirkung. *Biochem. Z.* 49: 333–369.

24 Houk, K.N., Leach, A.G., Kim, S.P., and Zhang, X.Y. (2003). Binding affinities of host-guest, protein-ligand, and protein-transition-state complexes. *Angew. Chem. Int. Ed.* 42 (40): 4872–4897.

25 Chapman, S. and Cowling, T.G. (1970). *The Mathematical Theory of Non-uniform Gas: An Account of the Kinetic Theory of Viscosity, Thermal Conduction and Diffusion in Gases*, 3e, 1–423. Cambridge: Cambridge University Press.

26 Arrhenius, S.A. (1889). Über die Dissociationwärme und den Einfluss der Temperatur auf den Dissociationsgrad der Electrolyte. *Z. Phys. Chem.* 4: 96–116.

27 Arrhenius, S.A. (1889). Über die Reaktiongeschwindigkeit bei der Inversion von Rohrzucker durch Säuren. *Z. Phys. Chem.* 4: 226–248.

28 Eyring, H. (1935). The activated complex and the absolute rate of chemical reactions. *Chem. Rev.* 17 (1): 65–77.

29 Eyring, H. (1935). The activated complex in chemical reactions. *J. Chem. Phys.* 3 (2): 107–115.

30 Seeman, J.I. (1983). Effect of conformational change on reactivity in organic chemistry. Evaluations, applications, and extensions of Curtin–Hammett Winstein–Holness kinetics. *Chem. Rev.* 83 (2): 83–134.

31 Seeman, J.I. (1986). The Curtin–Hammett principle and the Winstein-Holness equation: new definition and recent extensions to classical concepts. *Can. J. Chem. (Rev. Can. Chim.)* 63 (1): 42–48.

32 Andraos, J. (2003). Quantification and optimization of dynamic kinetic resolution. *J. Phys. Chem. A* 107 (13): 2374–2387.

33 Jones, A.J., Beeman, C.P., Hasan, M.U. et al. (1976). Morpholines: stereochemistry and preferred steric course of quaternization. *Can. J. Chem.* 54 (1): 126–135.

34 Winstein, S. and Holness, N.J. (1955). Neighboring carbon and hydrogen. 19. Tert-butylcyclohexyl derivatives. Quantitative conformational analysis. *J. Am. Chem. Soc.* 77 (21): 5562–5578.

35 (a) Eliel, E.L. (1965). Conformational analysis in mobile cyclohexane systems. *Angew. Chem. Int. Ed. Eng.* 4 (9): 761–774. (b) Eliel, E.L., Willen, S.H., and Mander, L.N. (1994). *Stereochemistry of Organic Compounds*, Chapter 11.4, 688–699. New York, NY: Wiley.

36 Seeman, J.I., Secor, H.V., Hartung, H., and Galzerano, R. (1980). Steric effects in conformationally mobile systems. The iodomethylation of 1-methyl-2-arylpyrrolidines related to Nicotine. *J. Am. Chem. Soc.* 102 (26): 7741–7747.

37 Bowry, V.W., Lusztyk, J., and Ingold, K.U. (1991). Calibration of a new horologery of

fast radical clocks. Ring opening rates for ring-alkyl-substituted and α-alkyl substituted cyclopropylcarbinyl radicals and for the bicyclo[2.1.0]pent-2-yl radical. *J. Am. Chem. Soc.* 113 (15): 5687–5698.

38 Nagahara, K., Ryu, I., Kambe, N. et al. (1995). Rate constants for addition of a primary alkyl radical to carbon monoxide. *J. Org. Chem.* 60 (23): 7384–7385.

39 Chambers, T.S. and Kistiakowsky, G.B. (1934). Kinetics of the thermal isomerization of cyclopropane. *J. Am. Chem. Soc.* 56: 399–405.

40 Rabinovitch, B.S., Schlag, E.W., and Wiberg, K.B. (1958). Geometrical and structural unimolecular isomerization of sym-cyclopropane-D2. *J. Chem. Phys.* (3): 28, 504–505.

41 Schlag, E.W. and Rabinovitch, B.S. (1960). Kinetics of the thermal unimolecular isomerization reactions of cyclopropane-D2. *J. Am. Chem. Soc.* 82 (23): 5996–6000.

42 Maeda, Y. and Ingold, K.U. (1979). Kinetic applications of electron paramagnetic resonance spectroscopy. 34. Rate constants for spin trapping. 2. Secondary alkyl radicals. *J. Am. Chem. Soc.* 101 (17): 4975–4981.

43 Spangler, C.W. (1976). Thermal [1,*j*]-sigmatropic rearrangements. *Chem. Rev.* 76 (2): 187–217.

44 Roth, W.R., Ruf, G., and Ford, P.W. (1974). Rotational barrier in 1,2-dienes – resonance energy of allylic radical. *Chem. Ber. Recl.* 107 (1): 48–52.

45 Waters, W.L., Linn, W.S., and Caserio, M.C. (1968). Stereochemistry of addition reactions of allenes. I. Methoxymercuration and halogenation of 1,3-dimethylallene. *J. Am. Chem. Soc.* 90 (24): 6741–6749.

46 Borden, W.T. and Corey, E.J. (1969). A Synthesis of optically active 1,3-di-t-butylallene from propargylic derivatives by hydride displacement. *Tetrahedron Lett.* (5): 313–316.

47 Doering, W.v.E., Toscano, V.G., and Beasley, G.H. (1971). Kinetics of Cope rearrangement of 1,1-dideuteriohexa-1,5-diene. *Tetrahedron* 27 (22): 5299–5306.

48 Gajewski, J.J., Hoffman, L.K., and Shih, C.N. (1974). Hydrocarbon thermal degenerate rearrangements. 6. Boat Cope transition state in self-interconversion of 1,4-dimethylenecyclohexane. *J. Am. Chem. Soc.* 96 (11): 3705–3706.

49 McGuire, M.J. and Piecuch, P. (2005). Balancing dynamic and nondynamic correlation for diradical and aromatic transition states: a renormalized coupled-cluster study of the Cope rearrangement of 1,5-hexadiene. *J. Am. Chem. Soc.* 127 (8): 2608–2614.

50 Dewar, M.J.S. and Wade, L.E. (1977). Study of mechanism of Cope rearrangement. *J. Am. Chem. Soc.* 99 (13): 4417–4424.

51 Gajewski, J.J. and Conrad, N.D. (1978). Mechanism of Cope rearrangement. *J. Am. Chem. Soc.* 100 (19): 6268–6269.

52 Gajewski, J.J. and Conrad, N.D. (1979). Variable transition state structure in (3,3)-sigmatropic shifts from α-secondary deuterium isotope effects. *J. Am. Chem. Soc.* 101 (22): 6693–6704.

53 Kaufmann, D. and de Meijere, A. (1984). On the influence of substituents on the Cope rearrangement of 1,5-hexadienes – cyclopropyl-substituted and methyl substituted model systems. *Chem. Ber.* 117 (3): 1128–1138.

54 Dewar, M.J.S. and Jie, C.X. (1987). Mechanism of the Cope rearrangement. *J. Am. Chem. Soc.* 109 (20): 5893–5899.

55 Wessel, T.E. and Berson, J.A. (1994). Stabilization of a putative cyclohexane-1,4-diyl intermediate elicits an antarafacial Cope rearrangement via a stepwise mechanism. Pyrolysis of (*R*,*E*)-5-methyl-1,2,6-octatriene to 4-methyl-3-methylene-1,5-heptadiene. *J. Am. Chem. Soc.* 116 (2): 495–505.

56 Staroverov, V.N. and Davidson, E.R. (2000). Transition regions in the Cope rearrangement of 1,5-hexadiene and its cyano derivatives. *J. Am. Chem. Soc.* 122 (30): 7377–7385.

57 (a) Black, K.A., Wilsey, S., and Houk, K.N. (2003). Dissociative and associative mechanisms of Cope rearrangements of fluorinated 1,5-hexadienes and 2,2′-bis-methylenecyclopentanes. *J. Am. Chem. Soc.* 125 (22): 6715–6724. (b) Mackey, J.L., Yang, Z., and Houk, K.N. (2017). Dynamically Concerted and Stepwise Trajectories of the Cope Rearrangement of 1,5-Hexadiene. *Chem. Phys. Lett.*, 683: 253–257. (c) Black, K., Liu, P., Xu, L., Doubleday, C., and Houk, K.N. (2012). Dynamics, Transition States, and Timing of Bond Formation in Diels-Alder Reactions. *Proc. Nat Acad. Sci. USA*, 109: 12860–12865.

58 Gajewski, J.J., Gee, K.R., and Jurayj, J. (1990). Energetic and rate effects of the trifluoromethyl group at C-2 and C-4 on the aliphatic Claisen rearrangement. *J. Org. Chem.* 55 (6): 1813–1822.

59 Kopinke, F.D., Bach, G., Ondruschka, B., and Zimmermann, G. (1983). On the pyrolysis of 1-^{14}C-pent-1-ene. *J. Prakt. Chem.* 325 (2): 283–292.

60 Kopinke, F.D., Bach, G., Ondruschka, B. et al. (1983). On the pyrolysis of pent-1-ene-3,3-d_2. *J. Prakt. Chem.* 325 (3): 375–381.

61 Kopinke, F.D., Bach, G., Ondruschka, B., and Zimmermann, G. (1983). On the pyrolysis of pent-1-ene-4,4,5,5,5-d_5. *J. Prakt. Chem.* 325 (5): 699–707.

62 Egger, K.W. and Vitins, P. (1974). Thermochemical kinetics of retroene reactions of molecules with general structure (allyl)Xyh in gas phase. 9. Thermal unimolecular decomposition of ethyallylether in gas phase. *In. J. Chem. Kinet.* 6 (3): 429–435.

63 Tumanov, V.E. and Denisov, E.T. (2004). Estimation of enthalpies of alkoxy radical formation and bond strengths in alcohols and ether. *Kinet. Catal.* 45 (5): 621–627.

64 Matus, M.H., Nguyen, M.T., and Dixon, D.A. (2007). Theoretical prediction of the heats of formation of $C_2H_5O^\bullet$ radicals derived from ethanol and of the kinetics of β-C-C scission in the ethoxy radical. *J. Phys. Chem. A* 111 (1): 113–126.

65 Vitins, P. and Egger, K.W. (1974). Thermochemical kinetics of retro-ene reactions of molecules with general structure (allyl)Xyh in gas phase. 10. Unimolecular thermal-decomposition of diallyl ether. *J. Chem. Soc., Perkin Trans. 2* (11): 1292–1293.

66 Vitins, P. and Egger, K.W. (1974). Thermochemical kinetics of retro-ene reactions of molecules with general structure (allyl)Xyh in gas phase. 9. Unimolecular thermal-decomposition of allylmethylamine. *J. Chem. Soc., Perkin Trans. 2* (11): 1289–1291.

67 Egger, K.W. and Vitins, P. (1974). Thermochemical kinetics of retro reaction ene reactions of molecules with general structure (allyl)Xyh in gas phase. 6. Concerted unimolecular decomposition of hepta-1,6-diene. *J. Am. Chem. Soc.* 96 (9): 2714–2719.

68 Hopf, H. and Kirsch, R. (1985). Thermal rearrangements. 16. Novel retro-ene reactions of β-hydroxyacetylenes. *Angew. Chem. Int. Ed. Eng.* 24 (9): 783–784.

69 Bigley, D.B. and Clarke, M.J. (1982). Studies in decarboxylation. 14. The gas phase decarboxylation of but-3-enoic acid and the intermediacy of isocrotonic (cis-but-2-enoic) acid in its isomerization to crotonic (trans-but-2-enoic) acid. *J. Chem. Soc., Perkin Trans. 2* (1): 1–6.

70 Wang, Y.L. and Poirier, R.A. (1994). Ab-initio study on the thermal decarboxylation of but-3-enoic acid and its derivatives. *Can. J. Chem. (Rev. Can. Chim.)* 72 (5): 1338–1346.

71 Hoveyda, A.H., Evans, D.A., and Fu, G.C. (1993). Substrate-directable chemical-reactions. *Chem. Rev.* 93 (4): 1307–1370.

72 Ruzicka, L. (1926). Information on carbon ring VIII. Further articles on the constitution of muscone. *Helv. Chim. Acta* 9: 1008–1017.

73 Ziegler, K., Eberle, H., and Ohlinger, H. (1933). Polynomial ring systems I. The preparative, productive synthesis of polymethyl ketones with more than 6 members of the ring. *Liebigs Ann. Chem.* 504: 94–130.

74 Ziegler, K. and Hechelhammer, W. (1937). The vaious member ring systems VI. The formation tendency of polymethyl ketones with more than 20 carbon atoms. *Liebigs Ann. Chem.* 528: 114–142.

75 Stoll, M., Rouvé, A., and Stoll-Comte, G. (1934). Multiple membered heterocyclic combinations V. On knowledge of ring ends in the formation of high membered lactone. *Helv. Chim. Acta* 17: 1289–1308.

76 Stoll, M. and Rouvé, A. (1935). Polynomial heterocyclic compounds VIII. Information on high membered mono and poly lactone rings. *Helv. Chim. Acta* 18: 1087–1125.

77 Stoll, M. and Rouvé, A. (1947). Synthèse de produits macrocycliques a odeur musquée. 3. Sur les acyloïnes cycliques. *Helv. Chim. Acta* 30 (6): 1822–1836.

78 Galli, C. and Mandolini, L. (1977). Ring closure reactions. 6. Intermolecular models for intramolecular reactions. Reactions between long-chain alkanoate ions and alkyl bromides as intermolecular analogs of lactonization of ω-bromoalkanoate ions. *J. Chem. Soc., Perkin Trans. 2* (4): 443–445.

79 Beesley, R.M., Ingold, C.K., and Thorpe, J.F. (1915). The formation and stability of spiro-compounds. Part I. Spiro-compounds from cyclohexane. *J. Chem. Soc.* 107: 1080–1106.

80 Kirby, A.J. and Lloyd, G.J. (1976). Structure and efficiency in intramolecular and enzymic catalysis: intramolecular general base catalysis. Hydrolysis of monoaryl malonates. *J. Chem. Soc., Perkin Trans. 2* (14): 1753–1761.

81 Allinger, N.L. and Zalkow, V. (1960). Conformational analysis. 9. The gem-dimethyl effect. *J. Org. Chem.* 25 (5): 701–704.

82 Jung, M.E. and Gervay, J. (1991). Gem-dialkyl effect in the intramolecular Diels–Alder reaction of 2-furfuryl methyl fumarates. The reactive rotamer effect, enthalpic basis for acceleration, and evidence for a polar transition state. *J. Am. Chem. Soc.* 113 (1): 224–232.

83 Jager, J., Graafland, T., Schenk, H. et al. (1984). Intramolecular catalyzed sulfonamide hydrolysis. 7. The Thorpe–Ingold effect in the intramolecular carboxyl catalyzed hydrolysis of sulfonamides. *J. Am. Chem. Soc.* 106 (1): 139–143.

84 Parrill, A.L. and Dolata, D.P. (1994). Evidence against the reactive rotamer explanation of the gem-dialkyl effect. *Tetrahedron Lett.* 35 (40): 7319–7322.

85 Toniolo, C., Crisma, M., Formaggio, F., and Peggion, C. (2001). Control of peptide conformation by the Thorpe–Ingold effect (C^{α}-tetrasubstitution). *Biopolymers* 60 (6): 396–419.

86 Park, H., Lee, H.K., and Choi, T.L. (2013). Faster cyclopolymerisation of 4,4-disubstituted 1,7-octadiynes through an enhanced Thorpe–Ingold effect. *Polym. Chem.* 4 (17): 4676–4681.

87 Milstien, S. and Cohen, L.A. (1970). Rate acceleration by stereopopulation control: models for enzyme action. *Proc. Natl. Acad. Sci. U. S. A.* 67 (3): 1143–1147.

88 Milstein, S. and Cohen, L.A. (1972). Stereopopulation control. 1. Rate enhancement in lactonizations of ortho-hydroxyhydrocinnamic acids. *J. Am. Chem. Soc.* 94 (26): 9158–9165.

89 Nicolaou, M.G., Yuan, C.S., and Borchardt, R.T. (1996). Phosphate prodrugs for amines utilizing a fast intramolecular hydroxy amide lactonization. *J. Org. Chem.* 61 (24): 8636–8641.

90 Levine, M.N. and Raines, R.T. (2012). Trimethyl lock: a trigger for molecular release in chemistry, biology, and pharmacology. *Chem. Sci.* 3 (8): 2412–2420.

91 Ueda, Y., Mikkilineni, A.B., Knipe, J.O. et al. (1993). Novel water-soluble phosphate prodrugs of Taxol® possessing in vivo antitumor activity. *Bioorg. Med. Chem. Lett.* 3 (8): 1761–1766.

92 Ingold, C.K. (1921). The conditions underlying the formation of unsaturated and of cyclic compounds from halogenated open-chain derivatives Part I. Products derived from α-halogenated glutaric acids. *J. Chem. Soc. Trans.* 119: 305–329.

93 Eliel, E.L. (1962). *Stereochemistry of Carbon Compounds*, 1–486. New York, NY: McGraw-Hill.

94 Kirby, A.J. (1980). Effective molarities for intramolecular reactions. *Adv. Phys. Org. Chem.* 17: 183–278.

95 Galli, C., Giovannelli, G., Illuminati, G., and Mandolini, L. (1979). Ring-closure reactions. 12. Gem-dimethyl effect in some medium and large rings. *J. Org. Chem.* 44 (8): 1258–1261.

96 Fischer, K. and Hünig, S. (1987). Trimethylsilyl cyanide – a reagent for Umpolung. XIII. Furfural, a convenient precursor for intramolecular Diels–Alder reactions via Umpolung with trimethylsilyl cyanide. *J. Org. Chem.* 52 (4): 564–569.

97 Sternbach, D.D. and Rossana, D.M. (1982). Intramolecular Diels–Alder reactions of the furan diene: substitutent and solvent effects. *Tetrahedron Lett.* 23 (3): 303–306.

98 Sternbach, D.D., Rossana, D.M., and Onan, K.D. (1984). Stereochemical control in the intramolecular Diels–Alder reaction with furan. *J. Org. Chem.* 49 (18): 3427–3428.

99 Parker, K.A. and Adamchuk, M.R. (1978). Intramolecular Diels–Alder reactions of furan diene. *Tetrahedron Lett.* 19 (19): 1689–1692.

100 Jung, M.E. and Street, L.J. (1984). Synthetic studies on the Avermectins: substituent effects on intramolecular Diels–Alder reactions of N-furfurylacrylamides and further reactions of the cycloadducts. *J. Am. Chem. Soc.* 106 (26): 8327–8329.

101 Klein, L.L. (1985). Structural effects on intramolecular furan cycloadditions. *J. Org. Chem.* 50 (10): 1770–1773.

102 Sun, F. and Sorensen, T.S. (1993). Mono-μ-hydrido-bridged and di-μ-hydrido-bridged carbodications in acyclic systems. *J. Am. Chem. Soc.* 115 (1): 77–81.

103 Forbes, M.D.E., Patton, J.T., Myers, T.L. et al. (1992). Solvent-free cyclization of linear dienes using olefin metathesis and the Thorpe Ingold effect. *J. Am. Chem. Soc.* 114 (27): 10978–10980.

104 Planck, M. (1887). Über das Princip der Vermehrung der Entropie. *Ann. Phys. Chem.* 268 (11): 462–503.

105 Rothmund, V. (1896). The influence of pressure on reaction velocities. *Z. Phys. Chem.* 20: 168–179.

106 (a) Diedrich, M.K., Hochstrate, D., Klärner, F.G., and Zimny, B. (1994). The pressure effect in pericyclic rearrangements: the Cope rearrangement, electrocyclization, and the intramolecular Diels–Alder reaction. *Angew. Chem. Int. Ed. Eng.* 33 (10): 1079–1081. (b) Chen, B., Hoffmann, R., and Cammi, R. (2017). The effect of pressure on organic reactions in fluids – a new theoretical perspective. *Angew. Chem. Int. Ed.* 56 (37): 11126–11142.

107 van Eldik, R. (1992). Pressure as a mechanistic indicator in substitution reactions of metal complexes. *Pure Appl. Chem.* 64 (10): 1439–1448.

108 (a) Lawrance, G.A. and Stranks, D.R. (1979). Role of activation volume in the elucidation of reaction mechanisms in octahedral coordination complexes. *Acc. Chem. Res.* 12: 403–409. (b) Eldik, v. (2006). The interpretation and mechanistic significance of activation volumes for organometallic reactions. *Adv. Phys. Org. Chem.* 41: 1–78.

109 Drljaca, A., Hubbard, C.D., van Eldik, R. et al. (1998). Activation and reaction volumes in solution. 3. *Chem. Rev.* 98 (6): 2167–2289.

110 van Eldik, R. and Klärner, F.-G. (eds.) (2008). *High Pressure Chemistry: Synthetic, Mechanistic, and Supercritical Applications*, 1–474. Weinheim: Wiley-VCH.

111 Davies, M.B. (2003). Mechanisms of reactions in solution. *Annu. Rep. Prog. Chem. Sect. A: Inorg. Chem.* 99: 505–543.

112 Röntgen, W.C. (1892). Kurze Mitteilung von Versuchen über den Einfluss des Druckes auf einige physikalische Erscheinungen. *Ann. Phys.* 281 (1): 98–107.

113 Van Eldik, R., Asano, T., and Le Noble, W.J. (1989). Activation and reaction volumes in solution. 2. *Chem. Rev.* 89 (3): 549–688.

114 Wiebe, H., Spooner, J., Boon, N. et al. (2012). Calculation of molecular volumes and volumes of activation using molecular dynamics simulations. *J. Phys. Chem. C* 116 (3): 2240–2245.

115 Spooner, J., Wiebe, H., Boon, N. et al. (2012). Molecular dynamics calculation of molecular volumes and volumes of activation. *Phys. Chem. Chem. Phys.* 14 (7): 2264–2277.

116 Jenner, G. and Papadopoulos, M. (1982). Concertedness in thermal intermolecular ene reactions of dimethyl mesoxalate and substituted alkenes. *J. Org. Chem.* 47 (22): 4201–4204.

117 Kiselev, V.D., Kornilov, D.A., Kashaeva, H.A. et al. (2014). 4-Phenyl-1,2,4-triazoline-3,5-dione in the ene reactions with cyclohexene, 1-hexene and 2, 3-dimethyl-2-butene. The heat of reaction and the influence of temperature and pressure on the reaction rate. *J. Phys. Org. Chem.* 27 (5): 401–406.

118 Vor der Brück, D., Bühler, R., Heuck, C.C. et al. (1970). High-pressure reactions, supplement to organic syntheses. *Chem. Ztg, Chem. Apparatur* 94 (6): 183–189.

119 McCabe, J.R. and Eckert, C.A. (1974). Role of High pressure kinetics in studies of transition states of Diels–Alder reactions. *Acc. Chem. Res.* 7 (8): 251–257.

120 Isaacs, N.S. and Van der Beeke, P. (1982). Reactions at high pressure. 10. The volume profile of an intramolecular Diels–Alder reaction. *Tetrahedron Lett.* 23 (20): 2147–2148.

121 Diedrich, M.K. and Klärner, F.G. (1998). The effect of pressure on cyclizations. The ring-size dependent reaction volumes of the cyclization of 1-alkenes to cycloalkanes. Experimental measurement of activation and reaction volumes of the intramolecular Diels–Alder reaction of 1,3,8-nonatriene and 1,3,9-decatriene. Temperature dependence of activation and reaction volumes. *J. Am. Chem. Soc.* 120 (25): 6212–6218.

122 Kiselev, V.D., Konovalov, A.I., Asano, T. et al. (2001). Solvent effect on the volume of activation and volume of the Diels–Alder reaction. *J. Phys. Org. Chem.* 14 (9): 636–643.

123 Kiselev, V.D., Kashaeva, E.A., Potapova, L.N. et al. (2005). Volume parameters of some Diels–Alder reactions involving C=C, C=S, and N=N bonds. *Russ. Chem. Bull.* 54 (9): 2034–2041.

124 Kiselev, V., Kornilov, D., Kashaeva, E. et al. (2014). Effect of pressure on the rate of the Diels–Alder reaction of diethyl azodicarboxylate with 9,10-dimethylanthracene. *Russ. J. Org. Chem.* 50 (4): 489–493.

125 Kiselev, V.D., Bolotov, A.V., Shakirova, I.I. et al. (2012). Anomalies in the change of volumetric parameters of the Diels–Alder reaction in solution. *J. Solut. Chem.* 41 (3): 525–535.

126 Le Noble, W.J., Merbach, A.E., and Schulman, E.M. (1981). Volume profile of the degenerate equilibration of a non-classical 7-norbornadienyl cation. *J. Org. Chem.* 46 (16): 3352–3353.

127 Klärner, F.G. (1989). Chemistry under high pressure – effects of high pressure on reactions in organic chemistry. *Chemie in unserer Zeit* 23 (2): 53–63.

128 Jenner, G. (1997). High pressure and selectivity in organic reactions. *Tetrahedron* 53 (8): 2669–2695.

129 Jenner, G. (2002). High-pressure mechanistic delineation based on activation volumes. *J. Phys. Org. Chem.* 15 (1): 1–13.

130 Schmidt, R. (1998). Interpretation of reaction and activation volumes in solution. *J. Phys. Chem. A* 102 (45): 9082–9086.

131 Klärner, F.G. and Wurche, F. (2000). The effect of pressure on organic reactions. *Journal für praktische Chemie: Practical Applications and Applied Chemistry* 342 (7): 609–636.

132 Tietze, L.F. and Steck, P.L. (2002). High pressure in organic synthesis: influence on selectivity. In: *High Pressure Chemistry: Synthetic, Mechanistic, and Supercritical Applications*, Chapter 8 (ed. R. van Eldik and F.-G. Klärner). Weinheim: Wiley-VCH.

133 Isaacs, N.S. and Najem, T.S. (1986). Activation volumes for ester hydrolysis via elimination addition. *Can. J. Chem. (Rev. Can. Chim.)* 64 (6): 1140–1144.

134 von Doering, W., Birladeanu, L., Sarma, K. et al. (2000). A non-Cope among the Cope rearrangements of 1,3,4,6-tetraphenylhexa-1,5-dienes. *J. Am. Chem. Soc.* 122 (2): 193–203.

135 Hamann, S.D. (1957). *Physico-Chemical Effects of Pressure*. London: Butterworths Scientific Publications, Ltd.

136 Matsumoto, K. (1981). Synthesis of organic compounds under high pressure. 3. High pressure Michael addition catalyzed by fluoride ions. *Angew. Chem. Int. Ed. Eng.* 20 (9): 770–771.

137 Matsumoto, K., Sera, A., and Uchida, T. (1985). Organic synthesis under high pressure. 1. *Synthesis* (1): 1–26.

138 Matsumoto, K. and Sera, A. (1985). Organic synthesis under high pressure. 2. *Synthesis* (11): 999–1027.

139 Jenner, G. (2005). Pressure activation in organic synthesis. *Adv. Org. Synth.* 1: 25–80.

140 Thomassigny, C., Sineriz, F., Greck, C., and Dumas, F. (2012). Ultra-high pressure as a green and promising tool in glycochemistry. *Mini-Rev. Org. Chem.* 9 (3): 228–242.

141 Dauben, W.G., Kessel, C.R., and Takemura, K.H. (1980). Simple, efficient total synthesis of Cantharidin via a high pressure Diels–Alder reaction. *J. Am. Chem. Soc.* 102 (22): 6893–6894.

142 Dauben, W.G., Gerdes, J.M., and Smith, D.B. (1985). Organic reactions at high pressure. The preparative scale synthesis of Cantharidin. *J. Org. Chem.* 50 (14): 2576–2578.

143 Grieco, P.A., Nunes, J.J., and Gaul, M.D. (1990). Dramatic rate accelerations of Diels–Alder reactions in 5 M lithium perchlorate–diethyl ether: the Cantharidin problem reexamined. *J. Am. Chem. Soc.* 112 (11): 4595–4596.

144 Grieco, P.A. (1991). Organic chemistry in unconventional solvents. *Aldrichimica Acta* 24 (3): 59–66.

145 Waldmann, H. (1991). $LiClO_4$ in ether – an unusual solvent. *Angew. Chem. Int. Ed. Eng.* 30 (10): 1306–1308.

146 Handy, S.T., Grieco, P.A., Mineur, C., and Ghosez, L. (1995). Lithium trifluoromethanesulfonimide in acetone or diethyl ether as a safe alternative to lithium perchlorate in diethyl ether for effecting Diels–Alder reactions. Unexpected influence of the counterion on exo/endo selectivity. *Synlett* (5): 565–567.

147 Heiner, T., Michalski, S., Gerke, K. et al. (1995). High pressure mediated intramolecular Diels–Alder reactions of furans with unactivated methylenecyclopropane terminators. *Synlett* (4): 355–357.

148 Beusker, P.H., Aben, R.W.M., Seerden, J.P.G. et al. (1998). Exploration of high pressure cycloadducts of furans and citraconic anhydride as precursors for CD-ring fragments of paclitaxel and its analogues. *Eur. J. Org. Chem.* (11): 2483–2492.

149 Tietze, L.F., Henrich, M., Niklaus, A., and Buback, M. (1999). The effect of high pressure on the diastereoselectivity of intermolecular all-carbon Diels–Alder reactions. *Chem. Eur. J.* 5 (1): 297–304.

150 Al-Badri, H., Collignon, N., Maddaluno, J., and Masson, S. (2000). Efficient synthesis of new phosphono-substituted dihydrothiopyrans via hetero Diels–Alder reaction, under thermal and high pressure conditions. *Tetrahedron* 56 (24): 3909–3919.

151 Biolatto, B., Kneeteman, M., Paredes, E., and Mancini, P.M.E. (2001). Reactions of 1-tosyl-3-substituted indoles with conjugated dienes under thermal and/or high-pressure conditions. *J. Org. Chem.* 66 (11): 3906–3912.

152 Rao, H.S.P., Murali, R., Taticchi, A., and Scheeren, H.W. (2001). High-pressure-promoted Diels–Alder reactions between cycloalkenones and 3-methylsulfanylfuran and 3-phenylsulfanylfuran. *Eur. J. Org. Chem.* (15): 2869–2876.

153 Malinowska, M., Salanski, P., Caille, J.C., and Jurczak, J. (2002). The high-pressure [4+2] cycloaddition of activated 1,3-dienes to non-activated heterodienophiles derived from glycolaldehyde. *Synthesis* (18): 2707–2710.

154 Minuti, L., Taticchi, A., Marrocchi, A. et al. (2004). High pressure Diels–Alder reactions of (+)-nopadiene with cycloalkenones. *Tetrahedron: Asymmetry* 15 (7): 1187–1192.

155 Ballerini, E., Minuti, L., Piermatti, O., and Pizzo, F. (2009). High pressure Diels–Alder approach to hydroxy substituted 6a-cyano-tetrahydro-6H-benzo[c]chromen-6-ones: a route to δ(6)-cis-cannabidiol. *J. Org. Chem.* 74 (11): 4311–4317.

156 Minuti, L. and Ballerini, E. (2011). High pressure access to the δ(9)-cis- and δ(9)-trans-tetrahydrocannabinols family. *J. Org. Chem.* 76 (13): 5392–5403.

157 Minuti, L., Temperini, A., and Ballerini, E. (2012). High pressure promoted Diels–Alder approach to biaryls: application to the synthesis

of the cannabinols family. *J. Org. Chem.* 77 (18): 7923–7931.

158 Mori, K., Yamauchi, T., Maddaluno, J. et al. (2011). Organocatalytic asymmetric hetero-Diels–Alder reaction of oxindoles under high pressure. *Synlett* (14): 2080–2084.

159 Krayushkin, M., Sedishev, I., Yarovenko, V. et al. (2008). Synthesis of pyridines from 1,2,4-triazines under high pressure. *Russ. J. Org. Chem.* 44 (3): 407–411.

160 Rogachev, V.O. and Metz, P. (2007). Stereoselective preparation of γ- and δ-sultams by thermal and high-pressure intramolecular Diels–Alder reaction of vinylsulfonamides. *Arkivoc* 167–190.

161 Kosior, M., Kwiatkowski, P., Asztemborska, M., and Jurczak, J. (2005). Stereochemistry of the Diels–Alder reaction at high pressure: diastereo- and enantioselective [4+2]-cycloaddition of buta-1,3-diene to glyoxylic acid derivatives catalysed by (salen)chromium(III) complexes. *Tetrahedron: Asymmetry* 16 (17): 2897–2900.

162 Brown, H.C., Schwier, J.R., and Singaram, B. (1978). Simple synthesis of monoisopinocampheylborane of high optical purity. *J. Org. Chem.* 43 (22): 4395–4397.

163 Brown, H.C. and Singaram, B. (1984). Improved procedures for the synthesis of diisopinocampheylborane of high optical purity. *J. Org. Chem.* 49 (5): 945–947.

164 Brown, H.C. and Singaram, B. (1984). Hydroboration. 68. Chiral synthesis via organoboranes. 1. A simple procedure to achieve products of essentially 100-percent optical purity in hydroboration of alkenes with monoisopinocampheylborane. Synthesis of boronic esters and derived products of very high enantiomeric purities. *J. Am. Chem. Soc.* 106 (6): 1797–1800.

165 Marchionni, C. and Vogel, P. (2001). Asymmetric synthesis of complicated bicyclic and tricyclic polypropanoates via the double Diels–Alder addition of 2,2′-ethyiidenebis[3,5-dimethylfuran]. *Helv. Chim. Acta* 84 (2): 431–472.

166 (a) Woodward, R.B. and Hoffmann, R. (1968). Conservation of orbital symmetry. *Acc. Chem. Res.* 1 (1): 17–22. (b) Meriu, H., Ross, J., and Whitesides, G.M. (1979). A basis for orbital symmetry rules. *Angew. Chem. Int. Ed. Eng.* 18: 377–392. (c) Yadav, V.K. (2016). The conservation of orbital symmetry (Woodward–Hoffmann Rules). In: *Steric and Stereoelectronic Effects in Organic Chemistry*, 127–146. Singapore: Springer.

167 Reetz, M.T. (1973). Primary and secondary orbital effects in dyotropic rearrangements. *Tetrahedron* 29 (14): 2189–2194.

168 Grimme, W., Pohl, K., Wortmann, J., and Frowein, D. (1996). Dyotropic hydrogen transfer in dihydrosesqui- and in dihydro-syn-sesterbicyclo[2.2.2]octenes. *Liebigs Ann. Chem.* 11: 1905–1916.

169 Mackenzie, K., Astin, K.B., Gravett, E.C. et al. (1998). Intramolecular thermal $(4\sigma + 2\pi)$ dyotropy: primary H-2 kinetic isotope effects. Experimental limiting barrier parameters for quantum tunnelling in 2H transfer processes and mechanistically significant substituent effects. *J. Phys. Org. Chem.* 11 (12): 879–886.

170 Mackenzie, K. (2008). Intramolecular $(4\sigma + 2\pi)$ thermal dyotropy of pre-stressed molecules. The correlation of reactivity with experimental crystallographic and calculated structural features, substituent and kinetic isotope effects. *Recent Res. Dev. Org. Chem.* 4: 295–326.

171 Hagenbuch, J.P., Stampfli, B., and Vogel, P. (1981). Thermoneutral dyotropic transfer of hydrogen in a hydrocarbon. *J. Am. Chem. Soc.* 103 (13): 3934–3935.

172 Paquette, L.A., Kesselmayer, M.A., and Rogers, R.D. (1990). Quantitation of proximity effects on rate. A case study involving dyotropic hydrogen migration within syn-sesquinorbornene disulfones carrying central substituents having different spatial demands. *J. Am. Chem. Soc.* 112 (1): 284–291.

173 Mackenzie, K., Howard, J.A.K., Mason, S. et al. (1993). On intramolecular dyotropy: structural effects on reaction rates, crystal structure–molecular mechanics correlations and primary deuterium kinetic isotope effects. (Parameters for intramolecular recognition). *J. Chem. Soc., Perkin Trans. 2* (6): 1211–1228.

174 Jencks, W.P. (1969). *Catalysis in Chemistry and Enzymology*, 1–644. New York, NY: McGraw Hill Book Co., Inc.; Mineola, NY: Dover Publications Inc., 1987.

175 Jencks, W.P. (1994). Reaction mechanisms, catalysis, and movement. *Protein Sci.* 3 (12): 2459–2464.

176 Jencks, W.P. (1997). From chemistry to biochemistry to catalysis to movement. *Annu. Rev. Biochem.* 66: 1–18.

177 (a) García-Viloca, M., Gao, J., Karplus, M., and Truhlar, D.G. (2004, 2004). How enzymes work: analysis by modern rate theory and computer simulations. *Science* 303 (5655): 186–195. (b) Olsson, M.H.M., Parson, W.W., and Warshel, A. (2006). Dynamical contributions to enzyme catalysis: critical tests of a popular hypothesis. *Chem. Rev.* 106 (5): 1737–1756. (c) Foigel, A.G.

(2011). Is the enzyme a powerful reactant of the biochemical reaction? *Mol. Cell. Biochem.* 352: 87–89.

178 (a) Page, M.I. and Jencks, W.P. (1971). Entropic contributions to rate accelerations in enzymic and intramolecular reactions and the chelate effect. *Proc. Natl. Acad. Sci. U. S. A.* 68: 1678–1683. (b) Bruice, T.C. and Lightstone, F.C. (1999). Ground state and transition state contributions to the rates of intramolecular and enzymatic reactions. *Acc. Chem. Res.* 32 (2): 127–136.

179 Menger, F.M. and Gabrielson, K. (1992). An unusually rapid intramolecular general-base catalysis. Relevance to enzymology. *J. Am. Chem. Soc.* 114 (9): 3574–3575.

180 Menger, F.M. (1985). On the source of intramolecular and enzymatic reactivity. *Acc. Chem. Res.* 18 (5): 128–134.

181 Houk, K.N., Tucker, J.A., and Dorigo, A.E. (1990). Quantitative modeling of proximity effects on organic reactivity. *Acc. Chem. Res.* 23 (4): 107–113.

182 Nicolaou, K.C., Zuccarello, G., Riemer, C. et al. (1992). Design, synthesis, and study of simple monocyclic conjugated enediynes. The 10-membered ring enediyne moiety of the enediyne anticancer antibiotics. *J. Am. Chem. Soc.* 114 (19): 7360–7371.

183 Jones, R.R. and Bergman, R.G. (1972). *p*-Benzyne. Generation as an intermediate in a thermal isomerization reaction and trapping evidence for 1,4-benzenediyl structure. *J. Am. Chem. Soc.* 94 (2): 660–661.

184 Bergman, R.G. (1973). Reactive 1,4-dehydroaromatics. *Acc. Chem. Res.* 6 (1): 25–31.

185 Nicolaou, K.C. and Dai, W.M. (1991). Chemistry and biology of the enediyne anticancer antibiotics. *Angew. Chem. Int. Ed. Eng.* 30 (11): 1387–1416.

186 Lee, M.D., Ellestad, G.A., and Borders, D.B. (1991). Calicheamicins: discovery, structure, chemistry, and interaction with DNA. *Acc. Chem. Res.* 24 (8): 235–243.

187 Yu, L., Golik, J., Harrison, R., and Dedon, P. (1994). The deoxyfucose-anthranilate of Esperamicin A1 confers intercalative DNA-binding and causes a switch in the chemistry of bistranded DNA lesions. *J. Am. Chem. Soc.* 116 (21): 9733–9738.

188 Giese, B., Beyrich-Graf, X., Erdmann, P. et al. (1995). The chemistry of single-stranded 4′-DNA radicals – influence of the radical precursor on anaerobic and aerobic strand cleavage. *Chem. Biol.* 2 (6): 367–375.

189 (a) Lin, G.-Q., Li, Y.-M., and Chan, A.S.C. (2001). *Principles and Applications of Asymmetric Synthesis*, 1–536. New York, NY: Wiley. (b) Gawley, R.E. and Aubé, J. (2012). *Principles of Asymmetric Synthesis*, 2e, 1–555. Amsterdam: Elsevier Ltd.

190 Lorenz, H. and Seidel-Morgenstern, A. (2014). Processes to separate enantiomers. *Angew. Chem. Int. Ed.* 53 (5): 1218–1250.

191 Luisi, P.L. and Bonsigno, O. (1972). Conformational equilibria in isotactic copolymers of enantiomeric α-olefins. *J. Chem. Phys.* 56 (9): 4298–4302.

192 Totland, K.M., Boyd, T.J., Lavoie, G.G. et al. (1996). Ring opening metathesis polymerization with binaphtholate or biphenolate complexes of molybdenum. *Macromolecules* 29 (19): 6114–6125.

193 Uno, T., Habaue, S., and Okamoto, Y. (1998). Stereospecific anionic polymerization of chiral benzyl α-[(1-phenylethoxy)methyl]acrylate. *Chirality* 10 (7): 711–716.

194 Duxbury, C.J., Hilker, I., de Wildeman, S.M., and Heise, A. (2007). Enzyme-responsive materials: chirality to program polymer reactivity. *Angew. Chem. Int. Ed.* 46 (44): 8452–8454.

195 Ejfler, J., Krauzy-Dziedzic, K., Szafert, S. et al. (2009). Novel chiral and achiral benzoxazine monomers and their thermal polymerization. *Macromolecules* 42 (12): 4008–4015.

196 Gómez, R.V. and Varela, O. (2009). Synthesis of polyhydroxy [n]-polyurethanes derived from a carbohydrate precursor. *Macromolecules* 42 (21): 8112–8117.

197 Kramer, J.W., Treitler, D.S., Dunn, E.W. et al. (2009). Polymerization of enantiopure monomers using syndiospecific catalysts: a new approach to sequence control in polymer synthesis. *J. Am. Chem. Soc.* 131 (44): 16042–16044.

198 Bleschke, C., Schmidt, J., Kundu, D.S. et al. (2011). A Chiral microporous polymer network as asymmetric heterogeneous organocatalyst. *Adv. Synth. Catal.* 353 (17): 3101–3106.

199 Knowles, W.S. (2002). Asymmetric hydrogenations (Nobel lecture). *Angew. Chem. Int. Ed.* 41 (12): 1999–2007.

200 Noyori, R. (2002). Asymmetric catalysis: science and opportunities (Nobel lecture). *Angew. Chem. Int. Ed.* 41 (12): 2008–2022.

201 Sharpless, K.B. (2002). Searching for new reactivity (Nobel lecture). *Angew. Chem. Int. Ed.* 41 (12): 2024–2032.

202 Hugl, H. (2008). Asymmetric synthesis in industry. In: *Asymmetric Synthesis: The Essentials* (ed. M. Christmann and S. Bräse), 305–310. Weinheim: Wiley-VCH.

203 Fischer, E. (1894). Synthesen in der Zucker Gruppe. II. *Ber. Dtsch. Chem. Ges.* 27: 3189–3232.

204 (a) Kiliani, H. (1885). Über das Cyanhydrin der Lävulose. *Ber. Dtsch. Chem. Ges.* 18: 3066–3072. (b) Fischer, E. (1889). Reduction von Säure der Zuckergruppe. *Ber. Dtsch. Chem. Ges.* 22: 2204–2205.

205 Kuhn, R. and Grassner, H. (1958). Mechanismus der katalytischen Hydrierung von Hydroxynitrilen. α-D-Fructoheptose aus D-Fructosecyanhydrin. *Liebigs Ann. Chem.* 612 (1–3): 55–64.

206 Marckwald, W. (1904). Über asymmetrische Synthese. *Ber. Dtsch. Chem. Ges.* 37: 1368–1370.

207 Kagan, H.B. and Gopalaiah, K. (2011). Early history of asymmetric synthesis: who are the scientists who set up the basic principles and the first experiments. *New J. Chem.* 35: 1933–1937.

208 (a) Morrison, J.D. and Mosher, H.S. (1971). *Asymmetric Organic Reactions*. Englewood Cliffs, NJ: Prentice Hall. (b) Wynberg, H. (1986). Asymmetric catalysis by alkaloids. In: *Topics in Stereochemistry*, vol. 16 (ed. E.L. Eliel, S.H. Wilen and N.L. Allinger), 87–129. New York, NY: Wiley.

209 Rachwalski, M., Vermue, N., and Rutjes, F.P. (2013). Recent advances in enzymatic and chemical deracemisation of racemic compounds. *Chem. Soc. Rev.* 42 (24): 9268–9282.

210 Rouf, A. and Taneja, S. (2014). Synthesis of single enantiomer bioactive molecules: a brief overview. *Chirality* 26 (2): 63–78.

211 Pasteur, L. (1858). Mémoire sur la fermentation de l'acide tartrique. *Comptes rendus hebdomadaires des Séances de l'Académie des Sciences* 46: 615–618.

212 Azerad, R. (1995). Application of biocatalysts in organic synthesis. *Bull. Soc. Chim. Fr.* 132 (1): 17–51.

213 Patel, R.N. (2001). Enzymatic synthesis of chiral intermediates for drug development. *Adv. Synth. Catal.* 343 (6-7): 527–546.

214 Contesini, F.J., Lopes, D.B., Macedo, G.A. et al. (2010). Aspergillus sp lipase: potential biocatalyst for industrial use. *J. Mol. Catal. B Enzym.* (Elsevier) 67 (3–4): 163–171.

215 Patel, R.N. (2011). Biocatalysis: synthesis of key intermediates for development of pharmaceuticals. *ACS Catal.* 1 (9): 1056–1074.

216 Kuriata-Adamusiak, R., Strub, D., and Lochynski, S. (2012). Application of microorganisms towards synthesis of chiral terpenoid derivatives. *Appl. Microbiol. Biotechnol.* 95 (6): 1427–1436.

217 Ghislieri, D., Green, A.P., Pontini, M. et al. (2013). Engineering an enantioselective amine oxidase for the synthesis of pharmaceutical building blocks and alkaloid natural products. *J. Am. Chem. Soc.* 135 (29): 10863–10869.

218 Hall, M., Kroutil, W., and Faber, K. (2012). The evolving role of biocatalysis in asymmetric synthesis. In: *Asymmetric Synthesis II: More Methods and Applications* (ed. M. Christmann and S. Bräse), 221–232. Weinheim: Wiley-VCH.

219 Reetz, M.T. (2013). Biocatalysis in organic chemistry and biotechnology: past, present, and future. *J. Am. Chem. Soc.* 135 (34): 12480–12496.

220 Clouthier, C.M. and Pelletier, J.N. (2012). Expanding the organic toolbox: a guide to integrating biocatalysis in synthesis. *Chem. Soc. Rev.* 41 (4): 1585–1605.

221 Klibanov, A.M. (1990). Asymmetric transformations catalyzed by enzymes in organic-solvents. *Acc. Chem. Res.* 23 (4): 114–120.

222 García-Galan, C., Berenguer-Murcia, Á., Fernández-Lafuente, R., and Rodrigues, R.C. (2011). Potential of different enzyme immobilization strategies to improve enzyme performance. *Adv. Synth. Catal.* 353 (16): 2885–2904.

223 Ansari, S.A. and Husain, Q. (2012). Potential applications of enzymes immobilized on/in nano materials: a review. *Biotechnol. Adv.* 30 (3): 512–523.

224 Hwang, E.T. and Gu, M.B. (2013). Enzyme stabilization by nano/microsized hybrid materials. *Eng. Life Sci.* 13 (1): 49–61.

225 Sheldon, R.A. and van Pelt, S. (2013). Enzyme immobilisation in biocatalysis: why, what and how. *Chem. Soc. Rev.* 42 (15): 6223–6235.

226 Magner, E. (2013). Immobilisation of enzymes on mesoporous silicate materials. *Chem. Soc. Rev.* 42 (15): 6213–6222.

227 Wilkinson, R., Thomas, J.P., Shepherd, R.G., and Baughn, C. (1961). Stereospecificity in a new type of synthetic antituberculous agent. *J. Am. Chem. Soc.* 83 (9): 2212–2213.

228 Fernandez, S., Brieva, R., Rebolledo, F., and Gotor, V. (1992). Lipase-catalyzed enantioselective acylation of N-protected or unprotected 2-aminoalkan-1-ols. *J. Chem. Soc., Perkin Trans. 1* (21): 2885–2889.

229 Batchelor, F.R., Doyle, F.P., Nayler, J.H.C., and Rolinson, G.N. (1959). Synthesis of penicillin: 6-aminopenicillanic acid in penicillin fermentations. *Nature* 183 (4656): 257–258.

230 Fadnavis, N.W., Sharfuddin, M., and Vadivel, S.K. (1999). Resolution of racemic 2-amino-1-butanol with immobilised penicillin G acylase. *Tetrahedron: Asymmetry* 10 (23): 4495–4500.

231 Alcon, M., Poch, M., Moyano, A. et al. (1997). Enantioselective synthesis of (*S*)-Vigabatrin®. *Tetrahedron: Asymmetry* 8 (17): 2967–2974.

232 Aggarwal, V.K., Barrell, J.K., Worrall, J.M., and Alexander, R. (1998). Highly diastereoselective epoxidation of ketene dithioacetal dioxides: a new approach to the asymmetric synthesis of α-amino amides. *J. Org. Chem.* 63 (21): 7128–7129.

233 Medina, E., Moyano, A., Pericas, M.A., and Riera, A. (2000). Enantioselective syntheses of conformationally rigid, highly lipophilic mesityl-substituted amino acids. *Helv. Chim. Acta* 83 (5): 972–988.

234 Benedetti, F., Berti, F., and Norbedo, S. (2002). Epoxyalcohol route to hydroxyethylene dipeptide isosteres. Stereodivergent synthesis of the diamino alcohol core of ritonavir and its C-2 epimer. *J. Org. Chem.* 67 (24): 8635–8643.

235 Taylor, S.K., Ivanovic, M., Simons, L.J., and Davis, M.M. (2003). An efficient asymmetric synthesis of key intermediates in the synthesis of aphanorphine and eptazocine. *Tetrahedron: Asymmetry* 14 (6): 743–747.

236 Breuer, M., Ditrich, K., Habicher, T. et al. (2004). Industrial methods for the production of optically active intermediates. *Angew. Chem. Int. Ed.* 43 (7): 788–824.

237 Taber, D.F., Pan, Y.C., and Zhao, X. (2004). A flexible enantioselective synthesis of the isofurans. *J. Org. Chem.* 69 (21): 7234–7240.

238 Murruzzu, C. and Riera, A. (2007). Enantioselective synthesis of hydroxylated pyrrolidines via Sharpless epoxidation and olefin metathesis. *Tetrahedron: Asymmetry* 18 (1): 149–154.

239 Dias, L.C. and Ferreira, M.A. (2012). Total synthesis of (−)-goniotrionin. *J. Org. Chem.* 77 (8): 4046–4062.

240 Lee, J.C., Francis, S., Dutta, D. et al. (2012). Synthesis and evaluation of eight- and four-membered iminosugar analogues as inhibitors of testicular ceramide-specific glucosyltransferase, testicular β-glucosidase 2, and other glycosidases. *J. Org. Chem.* 77 (7): 3082–3098.

241 Wang, X.G., Wang, A.E., Hao, Y. et al. (2013). Modular enantioselective synthesis of 8-aza-prostaglandin E-1. *J. Org. Chem.* 78 (18): 9488–9493.

242 Chen, W., Lin, L., Cai, Y. et al. (2014). Catalytic asymmetric (3+2)-cycloaddition of aromatic aldehydes with oxiranes by C–C bond cleavage of epoxides: highly efficient synthesis of chiral 1,3-dioxolanes. *J. Chem. Soc. Chem. Commun.* 50 (17): 2161–2163.

243 Weijers, C.A.G.M. (1997). Enantioselective hydrolysis of aryl, alicyclic and aliphatic epoxides by Rhodotorula glutinis. *Tetrahedron: Asymmetry* 8 (4): 639–647.

244 Botes, A.L., Weijers, C.A.G.M., Botes, P.J., and van Dyk, M.S. (1999). Enantioselectivities of yeast epoxide hydrolases for 1,2-epoxides. *Tetrahedron: Asymmetry* 10 (17): 3327–3336.

245 Weijers, C.A., Koenst, P., Franssen, M.C., and Sudhoelter, E.J. (2007). Stereochemical preference of yeast epoxide hydrolase for the O-axial C3 epimers of 1-oxaspiro[2.5] octanes. *Org. Biomol. Chem.* 5 (19): 3106–3114.

246 Lin, H., Liu, J.Y., Wang, H.B. et al. (2011). Biocatalysis as an alternative for the production of chiral epoxides: a comparative review. *J. Mol. Catal. B Enzym.* 72 (3-4): 77–89.

247 Goswami, A., Totleben, M.J., Singh, A.K., and Patel, R.N. (1999). Stereospecific enzymatic hydrolysis of racemic epoxide: a process for making chiral epoxide. *Tetrahedron: Asymmetry* 10 (16): 3167–3175.

248 Desantis, G. and Jones, J.B. (1999). Chemical modification of enzymes for enhanced functionality. *Curr. Opin. Biotechnol.* 10 (4): 324–330.

249 Kaur, J. and Sharma, R. (2006). Directed evolution: an approach to engineer enzymes. *Crit. Rev. Biotechnol.* 26 (3): 165–199.

250 Seelig, B. and Szostak, J.W. (2007). Selection and evolution of enzymes from a partially randomized non-catalytic scaffold. *Nature* 448 (7155): 828–U13.

251 Kumar, A. and Singh, S. (2013). Directed evolution: tailoring biocatalysts for industrial applications. *Crit. Rev. Biotechnol.* 33 (4): 365–378.

252 Reetz, M.T. and Jaeger, K.E. (2000). Enantioselective enzymes for organic synthesis created by directed evolution. *Chem. Eur. J.* 6 (3): 407–412.

253 Reetz, M.T. (2001). Combinatorial and evolution-based methods in the creation of enantioselective catalysts. *Angew. Chem. Int. Ed.* 40 (2): 284–310.

254 Reetz, M.T. (2006). Directed evolution of enantioselective enzymes as catalysts for organic synthesis. *Adv. Catal.* 49: 1–69.

255 Reetz, M.T. (2007). Evolution in the test-tube as a means to create selective biocatalysts. *Chimia* 61 (3): 100–103.

256 Reetz, M.T., Bocola, M., Wang, L.W. et al. (2009). Directed evolution of an enantioselective epoxide

hydrolase: Uncovering the source of enantioselectivity at each evolutionary stage. *J. Am. Chem. Soc.* 131 (21): 7334–7343.
257 Bougioukou, D.J., Kille, S., Taglieber, A., and Reetz, M.T. (2009). Directed evolution of an enantioselective enoate-reductase: testing the utility of iterative saturation mutagenesis. *Adv. Synth. Catal.* 351 (18): 3287–3305.
258 Reetz, M.T. and Wu, S. (2009). Laboratory evolution of robust and enantioselective Baeyer–Villiger monooxygenases for asymmetric catalysis. *J. Am. Chem. Soc.* 131 (42): 15424–15432.
259 Reetz, M.T. (2009). Directed evolution of enantioselective enzymes: an unconventional approach to asymmetric catalysis in organic chemistry. *J. Org. Chem.* 74 (16): 5767–5778.
260 Reetz, M.T. (2011). Laboratory evolution of stereoselective enzymes: a prolific source of catalysts for asymmetric reactions. *Angew. Chem. Int. Ed.* 50 (1): 138–174.
261 Reetz, M.T. (2012). Laboratory evolution of stereoselective enzymes as a means to expand the toolbox of organic chemists. *Tetrahedron* 68 (37): 7530–7548.
262 Vedejs, E. and Jure, M. (2005). Efficiency in nonenzymatic kinetic resolution. *Angew. Chem. Int. Ed.* 44 (26): 3974–4001.
263 Brown, J.M. (1987). Directed homogeneous hydrogenation. *Angew. Chem. Int. Ed. Eng.* 26 (3): 190–203.
264 Fiaud, J.C. and Kagan, H.B. (1988). Kinetic resolution. In: *Topics in Stereochemistry* (ed. E.L. Eliel and S.H. Wilen), 249–340. New York, NY: Wiley.
265 Robinson, D.E.J.E. and Bull, S.D. (2003). Kinetic resolution strategies using non-enzymatic catalysts. *Tetrahedron: Asymmetry* 14 (11): 1407–1446.
266 (a) Pellissier, H. (2011). Catalytic non-enzymatic kinetic resolution. *Adv. Synth. Catal.* 353 (10): 1613–1666. (b) Zhang, R.K., Rommey, D.K., Kan, S.B., and Arnold, F.H. (2018). Directed evolution of artificial metalloenzymes: bridging synthetic chemistry and biology. In: Artificial Metalloenzymes and MetalloDNAzymes in Catalysis. *From Design to Applications*, Chapter 5 (ed. M. Dieguez, J.-E. Bäckvall, and O. Pamies), 37–170. Wiley-VCH.
267 Keith, J.M., Larrow, J.F., and Jacobsen, E.N. (2001). Practical considerations in kinetic resolution reactions. *Adv. Synth. Catal.* 343 (1): 5–26.
268 Hanessian, S., Franco, J., and Larouche, B. (1990). The psychobiological basis of heuristic synthesis planning man, machine and the chiron approach. *Pure Appl. Chem.* 62 (10): 1887–1910.
269 Suzuki, M., Yanagisawa, A., and Noyori, R. (1988). Prostaglandin synthesis. 16. The 3-component coupling synthesis of prostaglandins. *J. Am. Chem. Soc.* 110 (14): 4718–4726.
270 Noyori, R. and Suzuki, M. (1993). Organic synthesis of Prostaglandins: advancing biology. *Science* 259 (5091): 44–45.
271 Noyori, R. (1989). Prostaglandins made simple. *Chem. Br.* 25: 883.
272 Buchmann, B. and Rehwinkel, H. (2007). Prostaglandins. In: *Ullmann's Encyclopedia of Industrial Chemistry*, vol. 30, 341–364. Weinhein: Wiley-VCH.
273 Kitamura, M., Kasahara, I., Manabe, K. et al. (1988). Kinetic resolution of racemic allylic alcohols by (BINAP)Ru(II)-catalyzed hydrogenation. *J. Org. Chem.* 53 (3): 708–710.
274 Tokunaga, M., Larrow, J.F., Kakiuchi, F., and Jacobsen, E.N. (1997). Asymmetric catalysis with water: efficient kinetic resolution of terminal epoxides by means of catalytic hydrolysis. *Science* 277 (5328): 936–938.
275 Jacobsen, E.N., Kakiuchi, F., Konsler, R.G. et al. (1997). Enantioselective catalytic ring opening of epoxides with carboxylic acids. *Tetrahedron Lett.* 38 (5): 773–776.
276 Larrow, J.F., Schaus, S.E., and Jacobsen, E.N. (1996). Kinetic resolution of terminal epoxides via highly regioselective and enantioselective ring opening with TMSN3. An efficient, catalytic route to 1,2-amino alcohols. *J. Am. Chem. Soc.* 118 (31): 7420–7421.
277 Ruble, J.C., Latham, H.A., and Fu, G.C. (1997). Effective kinetic resolution of secondary alcohols with a planar-chiral analogue of 4-(dimethylamino)pyridine. Use of the $Fe(C_5Ph_5)$ group in asymmetric catalysis. *J. Am. Chem. Soc.* 119 (6): 1492–1493.
278 Ruble, J.C., Tweddell, J., and Fu, G.C. (1998). Kinetic resolution of arylalkylcarbinols catalyzed by a planar-chiral derivative of DMAP: a new benchmark for nonenzymatic acylation. *J. Org. Chem.* 63 (9): 2794–2795.
279 Tao, B.T., Ruble, J.C., Hoic, D.A., and Fu, G.C. (1999). Nonenzymatic kinetic resolution of propargylic alcohols by a planar-chiral DMAP derivative: crystallographic characterization of the acylated catalyst. *J. Am. Chem. Soc.* 121 (21): 5091–5092.
280 Bellemin-Laponnaz, S., Tweddell, J., Ruble, J.C. et al. (2000). The kinetic resolution of allylic

alcohols by a non-enzymatic acylation catalyst; application to natural product synthesis. *J. Chem. Soc. Chem. Commun.* (12): 1009–1010.

281 Peschiulli, A., Procuranti, B., O' Connor, C.J., and Connon, S.J. (2010). Synergistic organocatalysis in the kinetic resolution of secondary thiols with concomitant desymmetrization of an anhydride. *Nat. Chem.* 2 (5): 380–384.

282 Birman, V.B. and Li, X.M. (2006). Benzotetramisole: a remarkably enantioselective acyl transfer catalyst. *Org. Lett.* 8 (7): 1351–1354.

283 Birman, V.B. and Guo, L. (2006). Kinetic resolution of propargylic alcohols catalyzed by benzotetramisole. *Org. Lett.* 8 (21): 4859–4861.

284 Zhang, Y. and Birman, V.B. (2009). Effects of methyl substituents on the activity and enantioselectivity of homobenzotetramisole-based catalysts in the kinetic resolution of alcohols. *Adv. Synth. Catal.* 351 (14-15): 2525–2529.

285 Shiina, I., Nakata, K., Ono, K. et al. (2010). Kinetic resolution of racemic α-arylalkanoic acids with achiral alcohols via the asymmetric esterification using carboxylic anhydrides and acyl-transfer catalysts. *J. Am. Chem. Soc.* 132 (33): 11629–11641.

286 Clark, R.W., Deaton, T., Zhang, Y. et al. (2013). Silylation-based kinetic resolution of α-hydroxy lactones and lactams. *Org. Lett.* 15 (24): 6132–6135.

287 Hashiguchi, S., Fujii, A., Haack, K.J. et al. (1997). Kinetic resolution of racemic secondary alcohols by Ru(II)-catalyzed hydrogen transfer. *Angew. Chem. Int. Ed. Eng.* 36 (3): 288–290.

288 Noyori, R. and Hashiguchi, S. (1997). Asymmetric transfer hydrogenation catalyzed by chiral ruthenium complexes. *Acc. Chem. Res.* 30 (2): 97–102.

289 Nishibayashi, Y., Takei, I., Uemura, S., and Hidai, M. (1999). Extremely high enantioselective redox reaction of ketones and alcohols catalyzed by $RuCl_2(PPh_3)$(oxazolinylferrocenylphosphine). *Organometallics* 18 (12): 2291–2293.

290 Hoveyda, A.H. and Schrock, R.R. (2001). Catalytic asymmetric olefin metathesis. *Chem. Eur. J.* 7 (5): 945–950.

291 Malcolmson, S.J., Meek, S.J., Sattely, E.S. et al. (2008). Highly efficient Mo-based catalysts for enantioselective alkene metathesis. *Nature* 456 (7224): 933–937.

292 Leandri, G., Monti, H., and Bertrand, M. (1974). Reactivity in methylenecyclopropane series. 2. Study of electrocyclic opening of 2-bromo-1,1-dialkyl-3-methylenecyclopropanes. *Tetrahedron* 30 (2): 289–293.

293 Wang, Y., Zheng, K., and Hong, R. (2012). Chiral silver phosphate-catalyzed cycloisomeric kinetic resolution of α-allenic alcohols. *J. Am. Chem. Soc.* 134 (9): 4096–4099.

294 Sharpless, K.B., Woodard, S.S., and Finn, M.G. (1983). On the mechanism of Ti-tartrate catalyzed asymmetric epoxidation. *Pure Appl. Chem.* 55 (11): 1823–1836.

295 Sharpless, K.B. (1986). The discovery of the asymmetric epoxidation. *Chem. Ber.* 22 (1): 38–44.

296 Katsuki, T. (1995). Catalytic asymmetric oxidations using optically-active (salen)Mn(III) complexes as catalysts. *Coord. Chem. Rev.* 140: 189–214.

297 Palucki, M., Finney, N.S., Pospisil, P.J. et al. (1998). The mechanistic basis for electronic effects on enantioselectivity in the (salen)Mn(III)-catalyzed epoxidation reaction. *J. Am. Chem. Soc.* 120 (5): 948–954.

298 Jacobsen, E.N., Zhang, W., Muci, A.R. et al. (1991). Highly enantioselective epoxidation catalysts derived from 1,2-diaminocyclohexane. *J. Am. Chem. Soc.* 113 (18): 7063–7064.

299 Dai, W., Shang, S., Chen, B. et al. (2014). Asymmetric epoxidation of olefins with hydrogen peroxide by an in situ-formed Mn complex. *J. Org. Chem.* 79 (14): 6688–6694.

300 Gelalcha, F.G. (2014). Biomimetic Fe-catalyzed asymmetric epoxidations: fundamental concepts, challenges and opportunities. *Adv. Synth. Catal.* 356 (2–3): 261–299.

301 Zhu, Y., Wang, Q., Cornwall, R.G., and Shi, Y. (2014). Organocatalytic asymmetric epoxidation and aziridination of olefins and their synthetic applications. *Chem. Rev.* 114 (16): 8199–8256.

302 Davis, R.L., Stiller, J., Naicker, T. et al. (2014). Asymmetric organocatalytic epoxidations: reactions, scope, mechanisms, and applications. *Angew. Chem. Int. Ed.* 53 (29): 7406–7426.

303 Yang, D., Yip, Y.C., Tang, M.W. et al. (1996). A C-2 symmetric chiral ketone for catalytic asymmetric epoxidation of unfunctionalized olefins. *J. Am. Chem. Soc.* 118 (2): 491–492.

304 Yang, D., Yip, Y.C., Tang, M.W. et al. (1998). Novel cyclic ketones for catalytic oxidation reactions. *J. Org. Chem.* 63 (26): 9888–9894.

305 Denmark, S.E. and Matsuhashi, H. (2002). Chiral fluoro ketones for catalytic asymmetric epoxidation of alkenes with oxone. *J. Org. Chem.* 67 (10): 3479–3486.

306 Armstrong, A., Ahmed, G., Dominguez-Fernandez, B. et al. (2002). Enantioselective epoxidation of alkenes catalyzed by 2-fluoro-N-carbethoxytropinone and related

tropinone derivatives. *J. Org. Chem.* 67 (24): 8610–8617.

307 Tu, Y., Wang, Z.X., and Shi, Y. (1996). An efficient asymmetric epoxidation method for trans-olefins mediated by a fructose-derived ketone. *J. Am. Chem. Soc.* 118 (40): 9806–9807.

308 Wu, X.Y., She, X.G., and Shi, Y.A. (2002). Highly enantioselective epoxidation of α,β-unsaturated esters by chiral dioxirane. *J. Am. Chem. Soc.* 124 (30): 8792–8793.

309 Milliet, P., Picot, A., and Lusinchi, X. (1976). Formation of quaternary oxaziridinium salt via oxazirane methylation – demonstration of oxidizing properties. *Tetrahedron Lett.* (19): 1573–1576.

310 Picot, A., Milliet, P., and Lusinchi, X. (1976). Reaction of para-nitroperbenzoic acid and hydrogen-peroxide with steroid heterocyclic immonium salt and corresponding enamine. *Tetrahedron Lett.* (19): 1577–1580.

311 Hanquet, G., Lusinchi, X., and Milliet, P. (1988). Oxygen-transfer from an oxaziridinium salt to an ethylenic double-bond. *Tetrahedron Lett.* 29 (32): 3941–3944.

312 Bohe, L., Lusinchi, M., and Lusinchi, X. (1999). Oxygen atom transfer from a chiral oxaziridinium salt. Asymmetric epoxidation of unfunctionalized olefins. *Tetrahedron* 55 (1): 141–154.

313 Biscoe, M.R. and Breslow, R. (2005). Oxaziridinium salts as hydrophobic epoxidation reagents: remarkable hydrophobically-directed selectivity in olefin epoxidation. *J. Am. Chem. Soc.* 127 (31): 10812–10813.

314 Page, P.C., Buckley, B.R., Barros, D. et al. (2007). Asymmetric organocatalysis of epoxidation by iminium salts under non-aqueous conditions. *Tetrahedron* 63 (25): 5386–5393.

315 Page, P.C., Parker, P., Rassias, G.A. et al. (2008). Iminium salt-catalysed asymmetric epoxidation using hydrogen peroxide as stoichiometric oxidant. *Adv. Synth. Catal.* 350 (11-12): 1867–1874.

316 Buckley, B.R., Elliott, C.E., Chan, Y. et al. (2013). Activation of hydrogen peroxide by diphenyl diselenide for highly enantioselective oxaziridinium salt mediated catalytic asymmetric epoxidation. *Synlett* 24 (17): 2266–2270.

317 Page, P.C., Appleby, L.F., Chan, Y. et al. (2013). Kinetic resolution in asymmetric epoxidation using iminium salt catalysis. *J. Org. Chem.* 78 (16): 8074–8082.

318 Washington, I. and Houk, K.N. (2000). Transition states and origins of stereoselectivity of epoxidations by oxaziridinium salts. *J. Am. Chem. Soc.* 122 (12): 2948–2949.

319 Seebach, D. and Gardiner, J. (2008). β-Peptidic peptidomimetics. *Acc. Chem. Res.* 41 (10): 1366–1375.

320 Rodionow, W.M. and Postovskaja, E.A. (1929). The mechanism of formation of β-aryl-β-amino fatty acids by the condensation of aromatic aldehydes with malonic acid and its derivatives. *J. Am. Chem. Soc.* 51: 841–847.

321 Tan, C.Y.K. and Weaver, D.F. (2002). A one-pot synthesis of 3-amino-3-arylpropionic acids. *Tetrahedron* 58 (37): 7449–7461.

322 Forro, E. and Fueloep, F. (2012). Recent lipase-catalyzed hydrolytic approaches to pharmacologically important β-and γ-amino acids. *Curr. Med. Chem.* 19 (36): 6178–6187.

323 (a) Zhou, S., Wang, J., Chen, X. et al. (2014). Chemical kinetic resolution of unprotected β-substituted β-amino acids using recyclable chiral ligands. *Angew. Chem. Int. Ed.* 53 (30): 7883–7886. (b) Lei, B.-L., Zhang, Q.-S., Yu, W.-H. et al. (2014). Kinetic resolution of 2-substituted 2,3-dihydro-4-pyridones by palladium-catalyzed asymmetric allylic alkylation: catalytic asymmetric total synthesis of indolizidine (−)-209I. *Org. Lett.* 16 (7): 1944–1947.

324 Eames, J. (2000). Parallel kinetic resolutions. *Angew. Chem. Int. Ed.* 39 (5): 885–888.

325 Dehli, J.R. and Gotor, V. (2002). Parallel kinetic resolution of racemic mixtures: a new strategy for the preparation of enantiopure compounds? *Chem. Soc. Rev.* 31 (6): 365–370.

326 Liao, L.A., Zhang, F., Dmitrenko, O. et al. (2004). A reactivity/affinity switch for parallel kinetic resolution: alpha-amino acid quasienantiomers and the resolution of cyclopropene carboxylic acids. *J. Am. Chem. Soc.* 126 (14): 4490–4491.

327 Vedejs, E. and Chen, X.H. (1997). Parallel kinetic resolution. *J. Am. Chem. Soc.* 119 (10): 2584–2585.

328 Kurosu, M. and Kishi, Y. (1998). A novel example for optical resolution of racemic ketones originating from batrachotoxin synthesis. *J. Org. Chem.* 63 (18): 6100–6101.

329 Pedersen, T.M., Jensen, J.F., Humble, R.E. et al. (2000). Parallel kinetic resolution of racemic aldehydes by use of asymmetric Horner-Wadsworth–Emmons reactions. *Org. Lett.* 2 (4): 535–538.

330 Tanaka, K. and Fu, G.C. (2003). Parallel kinetic resolution of 4-alkynals catalyzed by Rh(I)/Tol-BINAP: synthesis of enantioenriched cyclobutanones and cyclopentenones. *J. Am. Chem. Soc.* 125 (27): 8078–8079.

331 Noyori, R., Tokunaga, M., and Kitamura, M. (1995). Stereoselective organic synthesis via dynamic kinetic resolution. *Bull. Chem. Soc. Jpn.* 68 (1): 36–55.

332 Genêt, J.P. (2003). Asymmetric catalytic hydrogenation. Design of new Ru catalysts and chiral ligands: from laboratory to industrial applications. *Acc. Chem. Res.* 36 (12): 908–918.

333 Pellissier, H. (2003). Dynamic kinetic resolution. *Tetrahedron* 59 (42): 8291–8327.

334 Pellissier, H. (2008). Recent developments in dynamic kinetic resolution. *Tetrahedron* 64 (8): 1563–1601.

335 Pellissier, H. (2011). Recent developments in dynamic kinetic resolution. *Tetrahedron* 67 (21): 3769–3802.

336 Bucherer, H.T. and Fischbeck, H. (1934). Hexahydrodiphenylamine and its derivatives. *J. Prakt. Chem.* (Leipzig) 140 (3/4): 69–89.

337 Bucherer, H.T. and Steiner, W. (1934). On reactions of α-oxy-nitriles and α-amino-nitriles – synthesis of hydantoins. *J. Prakt. Chem.* (Leipzig) 140 (10/12): 291–316.

338 Ware, E. (1950). The chemistry of the hydantoins. *Chem. Rev.* 46 (3): 403–470.

339 Ebbers, E.J., Ariaans, G.J.A., Houbiers, J.P.M. et al. (1997). Controlled racemization of optically active organic compounds: prospects for asymmetric transformation. *Tetrahedron* 53 (28): 9417–9476.

340 Taniguchi, T., Kanada, R.M., and Ogasawara, K. (1997). Lipase-mediated kinetic resolution of tricyclic acyloins, endo-3-hydroxytricyclo [4.2.1.0(2,5)]non-7-en-4-one and endo-3-hydroxytricyclo[4.2.2.0(2,5)]dec-7-en-4-one. *Tetrahedron: Asymmetry* 8 (16): 2773–2780.

341 El Gihani, M.T. and Williams, J.M.J. (1999). Dynamic kinetic resolution. *Curr. Opin. Chem. Biol.* 3 (1): 11–15.

342 Tessaro, D., Cerioli, L., Servi, S. et al. (2011). L-Amino acid amides via dynamic kinetic resolution. *Adv. Synth. Catal.* 353 (13): 2333–2338.

343 Hickenboth, C.R., Moore, J.S., White, S.R. et al. (2007). Biasing reaction pathways with mechanical force. *Nature* 446 (7134): 423–427.

344 Caruso, M.M., Davis, D.A., Shen, Q. et al. (2009). Mechanically-induced chemical changes in polymeric materials. *Chem. Rev.* 109 (11): 5755–5798.

345 Lenhardt, J.M., Ong, M.T., Choe, R. et al. (2010). Trapping a diradical transition state by mechanochemical polymer extension. *Science* 329 (5995): 1057–1060.

346 Black, A.L., Lenhardt, J.M., and Craig, S.L. (2011). From molecular mechanochemistry to stress-responsive materials. *J. Mater. Chem.* 21 (6): 1655–1663.

347 Wiggins, K.M. and Bielawski, C.W. (2012). A Mechanochemical approach to deracemization. *Angew. Chem. Int. Ed.* 51 (7): 1640–1643.

348 Allen, J.V. and Williams, J.M.J. (1996). Dynamic kinetic resolution with enzyme and Pd combinations. *Tetrahedron Lett.* 37 (11): 1859–1862.

349 Stürmer, R. (1997). Enzymes and transition metal complexes in tandem – a new concept for dynamic kinetic resolution. *Angew. Chem. Int. Ed. Eng.* 36 (11): 1173–1174.

350 Persson, B.A., Larsson, A.L.E., Le Ray, M., and Backvall, J.E. (1999). Ruthenium- and enzyme-catalyzed dynamic kinetic resolution of secondary alcohols. *J. Am. Chem. Soc.* 121 (8): 1645–1650.

351 Persson, B.A., Huerta, F.F., and Bäckvall, J.E. (1999). Dynamic kinetic resolution of secondary diols via coupled Ru and enzyme catalysis. *J. Org. Chem.* 64 (14): 5237–5240.

352 Warner, M.C. and Bäckvall, J.E. (2013). Mechanistic aspects on (cyclopentadienyl)Ru complexes in catalytic racemization of alcohols. *Acc. Chem. Res.* 46 (11): 2545–2555.

353 Fransson, A.B.L., Boren, L., Pamies, O., and Bäckvall, J.E. (2005). Kinetic resolution and chemoenzymatic dynamic kinetic resolution of functionalized γ-hydroxy amides. *J. Org. Chem.* 70 (7): 2582–2587.

354 Pamies, O. and Bäckvall, I.-E. (2005). Chemoenzymatic dynamic kinetic resolution. *Trends Biotechnol.* 22: 130–135.

355 Martin-Matute, B., Edin, M., Bogar, K. et al. (2005). Combined Ru(II) and lipase catalysis for efficient dynamic kinetic resolution of secondary alcohols. Insight into the racemization mechanism. *J. Am. Chem. Soc.* 127 (24): 8817–8825.

356 Warner, M.C., Shevchenko, G.A., Jouda, S. et al. (2013). Dynamic kinetic resolution of homoallylic alcohols: application to the synthesis of enantiomerically pure 5,6-dihydropyran-2-ones and δ-lactones. *Chem. Eur. J.* 19 (41): 13859–13864.

357 Choi, Y.K., Suh, J.H., Lee, D. et al. (1999). Dynamic kinetic resolution of acyclic allylic acetates using lipase and Pd. *J. Org. Chem.* 64 (22): 8423–8424.

358 Lee, D., Huh, E.A., Kim, M.J. et al. (2000). Dynamic kinetic resolution of allylic alcohols mediated by Ru- and lipase-based catalysts. *Org. Lett.* 2 (15): 2377–2379.

359 Lee, S.Y., Murphy, J.M., Ukai, A., and Fu, G.C. (2012). Nonenzymatic dynamic kinetic resolution

of secondary alcohols via enantioselective acylation: Synthetic and mechanistic studies. *J. Am. Chem. Soc.* 134 (36): 15149–15153.

360 Noyori, R., Ikeda, T., Ohkuma, T. et al. (1989). Stereoselective hydrogenation via dynamic kinetic resolution. *J. Am. Chem. Soc.* 111 (25): 9134–9135.

361 Kitamura, M., Tokunaga, M., and Noyori, R. (1993). Mathematical treatment of kinetic resolution of chirally labile substrates. *Tetrahedron* 49 (9): 1853–1860.

362 Kitamura, M., Tokunaga, M., and Noyori, R. (1993). Quantitative expression of dynamic kinetic resolution of chirally labile enantiomers: stereoselective hydrogenation of 2-substituted 3-oxo carboxylic esters catalyzed by (BINAP)Ru(II) complexes. *J. Am. Chem. Soc.* 115 (1): 144–152.

363 de Paule, S.D., Jeulin, S., Ratovelomanana-Vidal, V. et al. (2003). Synthesis and molecular modeling studies of SYNPHOS®, a new, efficient diphosphane ligand for Ru-catalyzed asymmetric hydrogenation. *Eur. J. Org. Chem.* (10): 1931–1941.

364 Mordant, C., Reymond, S., Ratovelomanana-Vidal, V., and Genêt, J.P. (2004). Synthesis of both syn and anti diastereoisomers of Boc-dolaproine from (*S*)-proline through DKR using Ru-catalyzed hydrogenation: a dramatic role of N-protecting groups. *Tetrahedron* 60 (43): 9715–9723.

365 Mordant, C., Dunkelmann, P., Ratovelomanana-Vidal, V., and Genêt, J.P. A versatile route to syn- and anti-α-amino β-hydroxy esters from β-keto esters by dynamic kinetic resolution with Ru-SYNPHOS® catalyst. *Eur. J. Org. Chem.* 200 (14): 3017–3026.

366 Seashore-Ludlow, B., Saint-Dizier, F., and Somfai, P. (2012). Asymmetric transfer hydrogenation coupled with dynamic kinetic resolution in water: synthesis of anti-β-hydroxy-α-amino acid derivatives. *Org. Lett.* 14 (24): 6334–6337.

367 Goodman, C.G., Do, D.T., and Johnson, J.S. (2013). Asymmetric synthesis of anti-β-amino-α-hydroxy esters via dynamic kinetic resolution of β-amino-α-keto esters. *Org. Lett.* 15 (10): 2446–2449.

368 Lackner, A.D., Samant, A.V., and Toste, F.D. (2013). Single operation deracemization of 3H-indolines and tetrahydroquinolines enabled by phase separation. *J. Am. Chem. Soc.* 135 (38): 14090–14093.

369 Steward, K.M., Gentry, E.C., and Johnson, J.S. (2012). Dynamic kinetic resolution of α-keto esters via asymmetric transfer hydrogenation. *J. Am. Chem. Soc.* 134 (17): 7329–7332.

370 Canivet, J. and Süss-Fink, G. (2007). Water-soluble (arene)Ru catalysts containing sulfonated diamine ligands for asymmetric transfer hydrogenation of α-aryl ketones and imines in aqueous solution. *Green Chem.* 9 (4): 391–397.

371 Corbett, M.T. and Johnson, J.S. (2014). Dynamic kinetic asymmetric transformations of β-stereogenic α-ketoesters by direct aldolization. *Angew. Chem. Int. Ed.* 53 (1): 255–259.

372 Vicario, J.L., Badia, D., Carrillo, L. et al. (2005). α-Amino acids, β-amino alcohols and related compounds as chiral auxiliaries, ligands and catalysts in the asymmetric aldol reaction. *Curr. Org. Chem.* 9 (3): 219–235.

373 Boysen, M.M.K. (2007). Carbohydrates as synthetic tools in organic chemistry. *Chemistry* (Weinheim an der Bergstrasse, Germany) 13 (31): 8648–8659.

374 Singleton, J. and Hoberg, J. (2009). Chiro-inositols in organic synthesis. *Mini-Rev. Org. Chem.* 6 (1): 1–8.

375 Benessere, V., Del Litto, R., De Roma, A., and Ruffo, F. (2010). Carbohydrates as building blocks of privileged ligands. *Coord. Chem. Rev.* 254 (5-6): 390–401.

376 Lehnert, T., Oezueduru, G., Grugel, H. et al. (2011). More than just sweet – sugar-derived stereodifferentiating agents for asymmetric synthesis. *Synthesis* (17): 2685–2708.

377 Karpf, M. and Trussardi, R. (2009). Efficient access to Oseltamivir phosphate (Tamiflu) via the *O*-trimesylate of shikimic acid ethyl ester. *Angew. Chem. Int. Ed.* 48 (31): 5760–5762.

378 Werner, L., Machara, A., Sullivan, B. et al. (2011). Several generations of chemoenzymatic synthesis of Oseltamivir (Tamiflu): evolution of strategy, quest for a process-quality synthesis, and evaluation of efficiency metrics. *J. Org. Chem.* 76 (24): 10050–10067.

379 Kim, H.-K. and Park, K.-J.J. (2012). A new efficient synthesis of Oseltamivir phosphate (Tamiflu) from (−)-shikimic acid. *Tetrahedron Lett.* 53 (13): 1561–1563.

380 Kim, C.U., Lew, W., Williams, M.A. et al. (1998). Structure-activity relationship studies of novel carbocyclic influenza neuraminidase inhibitors. *J. Med. Chem.* 41 (14): 2451–2460.

381 Chandran, S.S., Yi, J., Draths, K.M. et al. (2003). Phosphoenolpyruvate availability and the biosynthesis of shikimic acid. *Biotechnol. Prog.* 19 (3): 808–814.

382 Serwa, R., Wilkening, I., Del Signore, G. et al. (2009). Chemoselective Staudinger-phosphite reaction of azides for the phosphorylation of proteins. *Angew. Chem. Int. Ed.* 48 (44): 8234–8239.
383 Pelter, A., Ryder, D.J., Sheppard, J.H. et al. (1979). Direct production of monoisopinocampheylborane by the redistribution of diisopinocampheylborane with borane. *Tetrahedron Lett.* (49): 4777–4780.
384 Mukaiyama, T., Suzuki, K., Soai, K., and Sato, T. (1979). Enantioselective addition of acetylene to aldehyde. Preparation of optically-active alkynyl alcohols. *Chem. Lett.* (5): 447–448.
385 Mukaiyama, T., Soai, K., Sato, T. et al. (1979). Enantioface-differentiating (asymmetric) addition of alkyllithium and dialkylmagnesium to aldehydes by using (2S,2′S)-2-hydroxymethyl-1-[(1-alkylpyrrolidin-2-yl)-methyl]pyrrolidines as chiral ligands. *J. Am. Chem. Soc.* 101 (6): 1455–1460.
386 Tomioka, K. (1990). Asymmetric synthesis utilizing external chiral ligands. *Synthesis* (7): 541–549.
387 Noyori, R. and Kitamura, M. (1991). Enantioselective addition of organometallic reagents to carbonyl compounds: chirality transfer, multiplication, and amplification. *Angew. Chem. Int. Ed. Eng.* 30 (1): 49–69.
388 Knauer, S., Kranke, B., Krause, L., and Kunz, H. (2004). Amino sugars and glycosylamines as tools in stereoselective synthesis. *Curr. Org. Chem.* 8 (18): 1739–1761.
389 Gnas, Y. and Glorius, F. (2006). Chiral auxiliaries – principles and recent applications. *Synthesis* (12): 1899–1930.
390 Evans, D.A., Helmchen, G., and Rüping, M. (2007). Chiral auxiliaries in asymmetric synthesis. In: *Asymmetric Synthesis – The Essentials*, 2e (ed. M. Christmann and S. Bräse), 3–9. Weinheim: Wiley-VCH.
391 Nicolaou, K.C. (2008). *Classics in Total Synthesis*, 5e, 485–508. Weinheim: Wiley-VCH.
392 Vega-Pérez, J.M., Periñan, I., Vega, M., and Iglesias-Guerra, F. (2008). Stereoselective cyclopropanation of unsaturated acetals, using carbohydrates with D-gluco, L-rhamno and D-xylo configurations as chiral auxiliaries. *Tetrahedron: Asymmetry* 19 (14): 1720–1729.
393 Davies, S.G., Fletcher, A.M., and Thomson, J.E. (2013). Direct asymmetric syntheses of chiral aldehydes and ketones via N-acyl chiral auxiliary derivatives including chiral Weinreb amide equivalents. *J. Chem. Soc. Chem. Commun.* 49 (77): 8586–8598.
394 Matteson, D.S. (2013). Boronic esters in asymmetric synthesis. *J. Org. Chem.* 78 (20): 10009–10023.
395 Meggers, E. (2010). Chiral auxiliaries as emerging tools for the asymmetric synthesis of octahedral metal complexes. *Chem. Eur. J.* 16 (3): 752–758.
396 Corey, E.J. and Ensley, H.E. (1975). Preparation of an optically-active prostaglandin intermediate via asymmetric induction. *J. Am. Chem. Soc.* 97 (23): 6908–6909.
397 Ensley, H.E., Parnell, C.A., and Corey, E.J. (1978). Convenient synthesis of a highly efficient and recyclable chiral director for asymmetric induction. *J. Org. Chem.* 43 (8): 1610–1612.
398 Whitesell, J.K., Chen, H.H., and Lawrence, R.M. (1985). Trans-2-phenylcyclohexanol – a powerful and readily available chiral auxiliary. *J. Org. Chem.* 50 (23): 4663–4664.
399 Comins, D.L. and Salvador, J.M. (1993). Efficient synthesis and resolution of trans-2-(1-aryl-1-methylethyl)cyclohexanols: practical alternatives to 8-phenylmenthol. *J. Org. Chem.* 58 (17): 4656–4661.
400 García, C. and Martín, V.S. (2006). Asymmetric addition to ketones: enantioselective formation of tertiary alcohols. *Curr. Org. Chem.* 10 (14): 1849–1889.
401 de Parrodi, C.A. and Juaristi, E. (2006). Chiral 1,2-amino alcohols and 1,2-diamines derived from cyclohexene oxide: recent applications in asymmetric synthesis. *Synlett* (17): 2699–2715.
402 Moyano, A. and Ríos, R. (2009). En route to new chiral ferrocene derivatives: dead ends, detours, and avenues. *Synlett* (12): 1863–1886.
403 Ni, Y. and Xu, J.H. (2012). Biocatalytic ketone reduction: a green and efficient access to enantiopure alcohols. *Biotechnol. Adv.* 30 (6): 1279–1288.
404 Enders, D. and Eichenauer, H. (1977). Enantioselective alkylation of aldehydes via metalated chiral hydrazones. *Tetrahedron Lett.* (2): 191–194.
405 Enders, D. and Eichenauer, H. (1976). Asymmetric synthesis of α-substituted ketones by metalation and alkylation of chiral hydrazones. *Angew. Chem. Int. Ed. Eng.* 15 (9): 549–551.
406 Enders, D. and Eichenauer, H. (1979). Asymmetric synthesis via metalated chiral hydrazones – enantioselective alkylation of cyclic ketones and aldehydes. *Chem. Ber.* 112 (8): 2933–2960.
407 Corey, E.J. and Enders, D. (1976). Applications of N,N-dimethylhydrazones to synthesis. Use in

efficient, positionally and stereochemically selective C–C bond formation; oxidative hydrolysis to carbonyl compounds. *Tetrahedron Lett.* (1): 3–6.

408 Enders, D., Eichenauer, H., Baus, U. et al. (1984). Asymmetric syntheses via metalated chiral hydrazones. Overall enantioselective α-alkylation of acyclic ketones. *Tetrahedron* 40 (8): 1345–1359.

409 Enders, D., Wortmann, L., and Peters, R. (2000). Recovery of carbonyl compounds from N,N-dialkylhydrazones. *Acc. Chem. Res.* 33 (3): 157–169.

410 Lazny, R. and Nodzewska, A. (2010). N,N-Dialkylhydrazones in organic synthesis. From simple N,N-dimethylhydrazones to supported chiral auxiliaries. *Chem. Rev.* 110 (3): 1386–1434.

411 Heravi, M.M. and Zadsirjan, V. (2013). Oxazolidinones as chiral auxiliaries in asymmetric aldol reactions applied to total synthesis. *Tetrahedron: Asymmetry* 24 (19): 1149–1188.

412 Bull, S.D., Davies, S.G., Jones, S. et al. (1998). A practical procedure for the multigram synthesis of the SuperQuat chiral auxiliaries. *Synlett* (5): 519–521.

413 Davies, S.G. and Sanganee, H.J. (1995). 4-Substituted-5,5-dimethyl oxazolidin-2-ones as effective chiral auxiliaries for enolate alkylations and Michael additions. *Tetrahedron: Asymmetry* 6 (3): 671–674.

414 Gibson, C.L., Gillon, K., and Cook, S. (1998). A study of 4-substituted 5,5-diaryl oxazolidin-2-ones as efficacious chiral auxiliaries. *Tetrahedron Lett.* 39 (37): 6733–6736.

415 Velázquez, F. and Olivo, H.F. (2002). The application of chiral oxazolidinethiones and thiazolidinethiones in asymmetric synthesis. *Curr. Org. Chem.* 6 (4): 303–340.

416 Fukuzawa, S., Matsuzawa, H., and Yoshimitsu, S. (2000). Asymmetric Sm-Reformatsky reaction of chiral α-bromoacetyl-2-oxazolidinones with aldehydes. *J. Org. Chem.* 65 (6): 1702–1706.

417 Oppolzer, W. and Barras, J.P. (1987). Asymmetric dihydroxylations of β-substituted N-(α,β-enoyl)bornane-10,2-sultams. *Helv. Chim. Acta* 70 (7): 1666–1675.

418 Oppolzer, W., Poli, G., Kingma, A.J. et al. (1987). Asymmetric induction at C(β) and C(α) of N-enoylsultams by organomagnesium 1,4-addition enolate trapping. *Helv. Chim. Acta* 70 (8): 2201–2214.

419 Oppolzer, W. (1987). Camphor derivatives as chiral auxiliaries in asymmetric synthesis. *Tetrahedron* 43 (9): 1969–2004.

420 Oppolzer, W., Blagg, J., Rodriguez, I., and Walther, E. (1990). Bornanesultam-directed asymmetric synthesis of crystalline, enantiomerically pure syn aldols. *J. Am. Chem. Soc.* 112 (7): 2767–2772.

421 Toyota, S. (1999). Camphorsultam – an excellent chiral reagent for enantiomer resolution and determination of absolute stereochemistry. *Enantiomer* 4 (1): 25–32.

422 Miyabe, H., Ushiro, C., Ueda, M. et al. (2000). Asymmetric synthesis of α-amino acids based on carbon radical addition to glyoxylic oxime ether. *J. Org. Chem.* 65 (1): 176–185.

423 Heravi, M.M. and Zadsirjan, V. (2014). Recent advances in the application of the Oppolzer camphorsultam as a chiral auxiliary. *Tetrahedron: Asymmetry* 25: 1061–1090.

424 Harada, T., Kagamihara, Y., Tanaka, S. et al. (1992). A highly convergent asymmetric synthesis of the C(19)–C(27) segment of rifamycin S: an application of enantiodifferentiating acetalization with Menthone. *J. Org. Chem.* 57 (6): 1637–1639.

425 Harada, T. and Oku, A. (1994). Enantiodifferentiating transformation of prochiral polyols by using menthone as chiral template. *Synlett* (2): 95–104.

426 Iio, H., Nagaoka, H., and Kishi, Y. (1980). Total synthesis of rifamycins. 2. Total synthesis of racemic rifamycin S. *J. Am. Chem. Soc.* 102 (27): 7965–7967.

427 Nagaoka, H., Rutsch, W., Schmid, G. et al. (1980). Total synthesis of rifamycins. 1. Stereocontrolled synthesis of the aliphatic building block. *J. Am. Chem. Soc.* 102 (27): 7962–7965.

428 Nagaoka, H. and Kishi, Y. (1981). Further synthetic studies on rifamycin S. *Tetrahedron* 37 (23): 3873–3888.

429 Pozniak, A.L., Miller, R., and Ormerod, L.P. (1999). The treatment of tuberculosis in HIV-infected persons. *AIDS* 13 (4): 435–445.

430 Floss, H.G. and Yu, T.W. (2005). Rifamycin-mode of action, resistance, and biosynthesis. *Chem. Rev.* 105 (2): 621–632.

431 Gumbo, T., Louie, A., Deziel, M.R. et al. (2007). Concentration-dependent Mycobacterium tuberculosis killing and prevention of resistance by Rifampin. *Antimicrob. Agents Chemother.* 51 (11): 3781–3788.

432 Nagao, Y., Ikeda, T., Yagi, M. et al. (1982). A new design for chiral induction: a highly regioselective differentiation between 2 identical groups in an acyclic compound having a prochiral center. *J. Am. Chem. Soc.* 104 (7): 2079–2081.

433 Satyanarayana, T., Abraham, S., and Kagan, H.B. (2009). Nonlinear effects in asymmetric catalysis. *Angew. Chem. Int. Ed.* 48 (3): 456–494.

434 Soai, K., Shibata, T., and Sato, I. (2000). Enantioselective automultiplication of chiral molecules by asymmetric autocatalysis. *Acc. Chem. Res.* 33 (6): 382–390.

435 Rosenthaler, L. (1908). Enzyme effected asymmetrical synthesis. *Biochem. Z.* 14: 238–253.

436 Hayashi, S. (1908). Experimentelle Untersuchungen über die sterische Verhältnisse bei die Dismutation von Phenylglyoxal-hydrat durch verschiedene Bakterien (B. proteus, B. fluorescens, B. pyocyaneum, B. prodigiosum und B. coli). *Biochem. Z.* 46: 7–15.

437 Chênevert, R. and Rose, Y.S. (2000). Enzymatic desymmetrization of a meso polyol corresponding to the C(19)–C(27) segment of rifamycin S. *J. Org. Chem.* 65 (6): 1707–1709.

438 Pereira, R.D. (1998). The use of baker's yeast in the generation of asymmetric centers to produce chiral drugs and others compounds. *Crit. Rev. Biotechnol.* 18 (1): 25–83.

439 Rodriguez, S., Kayser, M.M., and Stewart, J.D. (2001). Highly stereoselective reagents for β-keto ester reductions by genetic engineering of baker's yeast. *J. Am. Chem. Soc.* 123 (8): 1547–1555.

440 Wolberg, M., Hummel, W., and Müller, M. (2001). Biocatalytic reduction of β,δ-diketo esters: a highly stereoselective approach to all four stereoisomers of a chlorinated β,δ-dihydroxy hexanoate. *Chem. Eur. J.* 7 (21): 4562–4571.

441 Moore, J.C., Pollard, D.J., Kosjek, B., and Devine, P.N. (2007). Advances in the enzymatic reduction of ketones. *Acc. Chem. Res.* 40 (12): 1412–1419.

442 Acetti, D., Brenna, E., Fuganti, C. et al. (2010). Baker's yeast reduction of β-hydroxy ketones. *Eur. J. Org. Chem.* (1): 142–151.

443 Yildiz, T., Canta, N., and Yusufoglu, A. (2014). Synthesis of new chiral keto alcohols by baker's yeast. *Tetrahedron: Asymmetry* 25 (4): 340–347.

444 Dauphin, G., Fauve, A., and Veschambre, H. (1989). Use of biological systems for the preparation of chiral molecules. 6. Preparation of stereoisomeric 2,4-diols: synthesis and conformational study of bicyclo derivatives, isomeric components of a pheromone of *Trypodendron lineatum*. *J. Org. Chem.* 54 (9): 2238–2242.

445 Nakamura, K. and Ohno, A. (1991). Stereochemical control in microbial reduction. *J. Synth. Org. Chem. Jpn* 49 (2): 110–117.

446 Ramos, A.d.S., Ribeiro, J.B., Vazquez, L. et al. (2009). Immobilized microorganisms in the reduction of ethyl benzoylacetate. *Tetrahedron Lett.* 50 (52): 7362–7364.

447 de Oliveira Lopes, R., Ribeiro, J.B., de Miranda, A.S. et al. (2014). Continuous flow whole cell bioreduction of fluorinated acetophenone. *Tetrahedron* 70 (20): 3239–3242.

448 Gibson, D.T., Koch, J.R., Schuld, C.L., and Kallio, R.E. (1968). Oxidative degradation of aromatic hydrocarbons by microorganisms. 2. Metabolism of halogenated aromatic hydrocarbons. *Biochemistry* 7 (11): 3795–3802.

449 Gibson, D.T., Hensley, M., Yoshioka, H., and Mabry, T.J. (1970). Formation of (+)-cis-2,3-dihydroxy-1-methylcyclohexa-4,6-diene from toluene by *Pseudomonas putida*. *Biochemistry* 9 (7): 1626–1630.

450 Zylstra, G.J. and Gibson, D.T. (1989). Toluene degradation by Pseudomonas putida F1: nucleotide sequence of the TodclC2Bade genes and their expression in *Escherichia coli*. *J. Biol. Chem.* 264 (25): 14940–14946.

451 Froese, J., Endoma-Arias, M.A., and Hudlicky, T. (2014). Processing of o-halobenzoates by toluene dioxygenase. The role of the alkoxy functionality in the regioselectivity of the enzymatic dihydroxylation reaction. *Org. Process. Res. Dev.* 18 (6): 801–809.

452 Hudlicky, T. (1996). Design constraints in practical syntheses of complex molecules: current status, case studies with carbohydrates and alkaloids, and future perspectives. *Chem. Rev.* 96 (1): 3–30.

453 Hudlicky, T., Entwistle, D.A., Pitzer, K.K., and Thorpe, A.J. (1996). Modern methods of monosaccharide synthesis from non-carbohydrate sources. *Chem. Rev.* 96 (3): 1195–1220.

454 Akgun, H. and Hudlicky, T. (1999). Total syntheses of ent-conduramine A and ent-7-deoxypancratistatin. *Tetrahedron Lett.* 40 (16): 3081–3084.

455 Desjardins, M., Lallemand, M.C., Freeman, S. et al. (1999). Synthesis and biological evaluation of conduritol and conduramine analogs. *J. Chem. Soc., Perkin Trans. 1* (5): 621–628.

456 Hudlicky, T., Gonzalez, D., and Gibson, D.T. (1999). Enzymatic dihydroxylation of aromatics in enantioselective synthesis: Expanding asymmetric methodology. *Aldrichim. Acta* 32 (2): 35–62.

457 Oppong, K.A., Hudlicky, T., Yan, F.Y. et al. (1999). Chemoenzymatic enantiodivergent synthesis of 1,2-dideoxy-2-amino-1-fluoro-allo-inositol. *Tetrahedron* 55 (10): 2875–2880.

458 Duchek, J., Adams, D.R., and Hudlicky, T. (2011). Chemoenzymatic synthesis of inositols, conduritols, and cyclitol analogues. *Chem. Rev.* 111 (7): 4223–4258.

459 Varghese, V. and Hudlicky, T. (2014). Short chemoenzymatic total synthesis of ent-Hydromorphone: an oxidative dearomatization/intramolecular [4+2]-cycloaddition/amination sequence. *Angew. Chem. Int. Ed.* 53 (17): 4355–4358.

460 Hemenway, M.S. and Olivo, H.F. (1999). Syntheses of new phosphorus-containing azabicycloalkanes and their microbial hydroxylation using *Beauveria bassiana*. *J. Org. Chem.* 64 (17): 6312–6318.

461 Adam, W., Lukacs, Z., Saha-Moller, C.R. et al. (2000). Microbial asymmetric C–H oxidations of simple hydrocarbons: a novel monooxygenase activity of the topsoil microorganism *Bacillus megaterium*. *Eur. J. Org. Chem.* (16): 2923–2926.

462 Adam, W., Lukacs, Z., Harmsen, D. et al. (2000). Biocatalytic asymmetric hydroxylation of hydrocarbons with the topsoil-microorganism *Bacillus megaterium*. *J. Org. Chem.* 65 (3): 878–882.

463 Wong, C.H., Halcomb, R.L., Ichikawa, Y., and Kajimoto, T. (1995). Enzymes in organic-synthesis: application to the problems of carbohydrate-recognition. 2. *Angew. Chem. Int. Ed. Eng.* 34 (5): 521–546.

464 Wong, C.H., Halcomb, R.L., Ichikawa, Y., and Kajimoto, T. (1995). Enzymes in organic synthesis. Applications to the problems of carbohydrate recognition. *Angew. Chem. Int. Ed.* 34: 412–432.

465 Liu, J.J. and Wong, C.H. (2002). Aldolase-catalyzed asymmetric synthesis of novel pyranose synthons as a new entry to heterocycles and Epothilones. *Angew. Chem. Int. Ed.* 41 (8): 1404–1407.

466 Whalen, L.J. and Wong, C.H. (2006). Enzymes in organic synthesis: aldolase-mediated synthesis of iminocyclitols and novel heterocycles. *Aldrichim. Acta* 39 (3): 63–71.

467 Dean, S.M., Greenberg, W.A., and Wong, C.H. (2007). Recent advances in aldolase-catalyzed asymmetric synthesis. *Adv. Synth. Catal.* 349 (8-9): 1308–1320.

468 Sugiyama, M., Greenberg, W., and Wong, C.H. (2008). Recent advances in aldolase-catalyzed synthesis of unnatural sugars and iminocyclitols. *J. Synth. Org. Chem. Jpn* 66 (6): 605–615.

469 Machajewski, T.D. and Wong, C.H. (1999). Chemoenzymatic synthesis of key Epothilone fragments. *Synthesis* 1469–1472.

470 Panke, S. and Wubbolts, M. (2005). Advances in biocatalytic synthesis of pharmaceutical intermediates. *Curr. Opin. Chem. Biol.* 9 (2): 188–194.

471 Bornscheuer, U., Huisman, G., Kazlauskas, R. et al. (2012). Engineering the third wave of biocatalysis. *Nature* 485 (7397): 185–194.

472 Greenberg, W.A., Varvak, A., Hanson, S.R. et al. (2004). Development of an efficient, scalable, aldolase-catalyzed process for enantioselective synthesis of statin intermediates. *Proc. Natl. Acad. Sci. U. S. A.* 101 (16): 5788–5793.

473 Dakin, H.D. (1909). The catalytic action of amino-acids, peptones and proteins in effecting certain syntheses. *J. Biol. Chem.* 7 (1): 49–55.

474 Hajos, Z.G. and Parrish, D.R. (1971). Asymmetric synthesis of polycyclic organic compounds. German Patent DE 2102623.

475 Hajos, Z.G. and Parrish, D.R. (1974). Asymmetric synthesis of bicyclic intermediates of natural product chemistry. *J. Org. Chem.* 39 (12): 1615–1621.

476 Hajos, Z.G. and Parrish, D.R. (1974). Synthesis and conversion of 2-methyl-2-(3-oxobutyl)-1,3-cyclopentanedione to isomeric racemic ketols of [3.2.1]bicyclooctane of perhydroindan series. *J. Org. Chem.* 39 (12): 1612–1615.

477 Eder, U., Sauer, G., and Weichert, R. (1971). Total synthesis of optically active steroids. 6. New type of asymmetric cyclization to optically active steroid CD partial structures. *Angew. Chem. Int. Ed. Eng.* 10 (7): 496–497.

478 List, B., Lerner, R.A., and Barbas, C.F.I. (2000). Proline-catalyzed direct asymmetric aldol reactions. *J. Am. Chem. Soc.* 122 (10): 2395–2396.

479 Trost, B.M. and Brindle, C.S. (2010). The direct catalytic asymmetric aldol reaction. *Chem. Soc. Rev.* 39 (5): 1600–1632.

480 Kano, T. and Maruoka, K. (2013). Unique properties of chiral biaryl-based secondary amine catalysts for asymmetric enamine catalysis. *Chem. Sci.* 4 (3): 907–915.

481 Mlynarski, J. and Bas, S. (2014). Catalytic asymmetric aldol reactions in aqueous media – a 5 year update. *Chem. Soc. Rev.* 43 (2): 577–587.

482 Mlynarski, J. and Bas, S. (2014). Catalytic asymmetric aldol reactions in aqueous media – a 5 year update. *Chem. Soc. Rev.* 43 (17): 6470–6470.

483 Andersen, K.K. (1962). Synthesis of (+)-ethyl para-tolyl sulfoxide from (−)-menthyl (−)-para-toluenesulfinate. *Tetrahedron Lett.* 3: 93–95.

484 Andersen, K.K., Foley, J.W., Gaffield, W. et al. (1964). Optically active sulfoxides. The synthesis and rotatory dispersion of some diaryl sulfoxides. *J. Am. Chem. Soc.* 86 (24): 5637–5646.

485 Solladié, G., Hutt, J., and Girardin, A. (1987). Improved preparation of optically active methyl para-tolyl sulfoxide. *Synthesis* (2): 173–173.

486 Legros, J., Dehli, J.R., and Bolm, C. (2005). Applications of catalytic asymmetric sulfide oxidations to the syntheses of biologically active sulfoxides. *Adv. Synth. Catal.* 347 (1): 19–31.

487 Carreno, M.C. (1995). Applications of sulfoxides to asymmetric synthesis of biologically-active compounds. *Chem. Rev.* 95 (6): 1717–1760.

488 Pellissier, H. (2006). Use of chiral sulfoxides in asymmetric synthesis. *Tetrahedron* 62 (24): 5559–5601.

489 Ferber, B. and Kagan, H.B. (2007). Metallocene sulfoxides as precursors of metallocenes with planar chirality. *Adv. Synth. Catal.* 349 (4–5): 493–507.

490 Carreño, M.C., Hernandez-Torres, G., Ribagorda, M., and Urbano, A. (2009). Enantiopure sulfoxides: recent applications in asymmetric synthesis. *J. Chem. Soc. Chem. Commun.* (41): 6129–6144.

491 Difuria, F., Modena, G., and Seraglia, R. (1984). Synthesis of chiral sulfoxides by metal-catalyzed oxidation with tert-butyl hydroperoxide. *Synthesis* (4): 325–326.

492 Pitchen, P. and Kagan, H.B. (1984). An efficient asymmetric oxidation of sulfides to sulfoxides. *Tetrahedron Lett.* 25 (10): 1049–1052.

493 Pitchen, P., Dunach, E., Deshmukh, M.N., and Kagan, H.B. (1984). An efficient asymmetric oxidation of sulfides to sulfoxides. *J. Am. Chem. Soc.* 106 (26): 8188–8193.

494 Brunel, J.M. and Kagan, H.B. (1996). Catalytic asymmetric oxidation of sulfides with high enantioselectivities. *Synlett* (4): 404–406.

495 Komatsu, N., Hashizume, M., Sugita, T., and Uemura, S. (1993). Catalytic asymmetric oxidation of sulfides to sulfoxides with tert-butyl hydroperoxide using binaphthol as a chiral auxiliary. *J. Org. Chem.* 58 (17): 4529–4533.

496 Palucki, M., Hanson, P., and Jacobsen, E.N. (1992). Asymmetric oxidation of sulfides with H_2O_2 catalyzed by (salen)Mn(III) complexes. *Tetrahedron Lett.* 33 (47): 7111–7114.

497 Kokubo, C. and Katsuki, T. (1996). Highly enantioselective catalytic oxidation of alkyl aryl sulfides using Mn-salen catalyst. *Tetrahedron* 52 (44): 13895–13900.

498 Bolm, C. and Bienewald, F. (1995). Asymmetric sulfide oxidation with V catalysts and H_2O_2. *Angew. Chem. Int. Ed.* 34 (23-24): 2640–2642.

499 Bolm, C. and Bienewald, F. (1998). Asymmetric oxidation of dithioacetals and dithioketals catalyzed by a chiral V complex. *Synlett* (12): 1327–1328.

500 Legros, J. and Bolm, C. (2003). Iron-catalyzed asymmetric sulfide oxidation with aqueous hydrogen peroxide. *Angew. Chem. Int. Ed.* 42 (44): 5487–5489.

501 Korte, A., Legros, J., and Bolm, C. (2004). Asymmetric synthesis of Sulindac by Fe-catalyzed sulfoxidation. *Synlett* (13): 2397–2399.

502 Fernandez, I. and Khiar, N. (2003). Recent developments in the synthesis and utilization of chiral sulfoxides. *Chem. Rev.* 103 (9): 3651–3705.

503 Shin, J.M., Cho, Y.M., and Sachs, G. (2004). Chemistry of covalent inhibition of the gastric (H+, K+)-ATPase by proton pump inhibitors. *J. Am. Chem. Soc.* 126 (25): 7800–7811.

504 Hock, H. and Lang, S. (1944). Autoxidation of carbon hydrogen. IX Announcement: concerning benzol derivatives. *Ber. Dtsch. Chem. Ges.* 77 (3/4): 257–264.

505 Weber, M., Weber, M., and Kleine-Boymann, M. (2004). Phenol. In: *Ullmann's Encyclopedia of Industrial Chemistry*, 503–519. Weinheim: Wiley-VCH.

506 Cotton, H., Elebring, T., Larsson, M. et al. (2000). Asymmetric synthesis of Esomeprazole. *Tetrahedron: Asymmetry* 11 (18): 3819–3825.

507 Katsuki, T. and Sharpless, K.B. (1980). The first practical method for asymmetric epoxidation. *J. Am. Chem. Soc.* 102 (18): 5974–5976.

508 Jacobsen, E.N., Marko, I., Mungall, W.S. et al. (1988). Asymmetric dihydroxylation via ligand-accelerated catalysis. *J. Am. Chem. Soc.* 110 (6): 1968–1970.

509 Zhang, W., Loebach, J.L., Wilson, S.R., and Jacobsen, E.N. (1990). Enantioselective epoxidation of unfunctionalized olefins catalyzed by (salen)Mn complexes. *J. Am. Chem. Soc.* 112 (7): 2801–2803.

510 Wennemers, H. (2011). Asymmetric catalysis with peptides. *J. Chem. Soc. Chem. Commun.* 47 (44): 12036–12041.

511 Julia, S., Masana, J., and Vega, J.C. (1980). Synthetic enzymes – highly stereoselective epoxidation of chalcone in a triphasic toluene-water-poly[(*S*)-alanine] system. *Angew. Chem. Int. Ed. Eng.* 19 (11): 929–931.

512 Banfi, S., Colonna, S., Molinari, H. et al. (1984). Asymmetric epoxidation of electron-poor olefins. 5. Influence on stereoselectivity of the structure of poly-α-aminoacids used as catalysts. *Tetrahedron* 40 (24): 5207–5211.

513 Colonna, S., Perdicchia, D., and Di Mauro, E. (2009). Enantioselective reactions catalyzed by

synthetic enzymes. A model for chemical evolution. *Tetrahedron: Asymmetry* 20 (15): 1709–1714.

514 Weiss, K.M. and Tsogoeva, S.B. (2011). Enantioselective epoxidation of electron-deficient olefins: an organocatalytic approach. *Chem. Rec.* 11 (1): 18–39.

515 Carde, L., Davies, H., Geller, T.P., and Roberts, S.M. (1999). PaaSicats: powerful catalysts for asymmetric epoxidation of enones. Novel syntheses of α-arylpropanoic acids including (S)-Fenoprofen. *Tetrahedron Lett.* 40 (29): 5421–5424.

516 Geller, T. and Roberts, S.M. (1999). A new procedure for the Juliá–Colonna stereoselective epoxidation reaction under non-aqueous conditions: the development of a catalyst comprising polyamino acid on silica (PaaSiCat). *J. Chem. Soc., Perkin Trans. 1* (11): 1397–1398.

517 Peris, G., Jakobsche, C.E., and Miller, S.J. (2007). Aspartate-catalyzed asymmetric epoxidation reactions. *J. Am. Chem. Soc.* 129 (28): 8710–8711.

518 Jakobsche, C.E., Peris, G., and Miller, S.J. (2008). Functional analysis of an aspartate-based epoxidation catalyst with amide-to-alkene peptidomimetic catalyst analogues. *Angew. Chem. Int. Ed.* 47 (35): 6707–6711.

519 Lichtor, P.A. and Miller, S.J. (2012). Combinatorial evolution of site- and enantioselective catalysts for polyene epoxidation. *Nat. Chem.* 4 (12): 990–995.

520 Corey, E.J. and Chaykovs, M. (1965). Dimethyloxosulfonium methylide and dimethylsulfonium methylide. Formation and application to organic synthesis. *J. Am. Chem. Soc.* 87 (6): 1353–1364.

521 Aggarwal, V.K. and Winn, C.L. (2004). Catalytic, asymmetric sulfur ylide-mediated epoxidation of carbonyl compounds: scope, selectivity, and applications in synthesis. *Acc. Chem. Res.* 37 (8): 611–620.

522 McGarrigle, E.M., Myers, E.L., Illa, O. et al. (2007). Chalcogenides as organocatalysts. *Chem. Rev.* 107 (12): 5841–5883.

523 Furukawa, N., Sugihara, Y., and Fujihara, H. (1989). Camphoryl sulfide as a chiral auxiliary and a mediator for one-step synthesis of optically-active 1,2-diaryloxiranes. *J. Org. Chem.* 54 (17): 4222–4224.

524 Li, A.H., Dai, L.X., Hou, X.L. et al. (1996). Preparation of enantiomerically enriched (2R,3R)- or (2S,3S)-trans-2,3-diaryloxiranes via camphor-derived sulfonium ylides. *J. Org. Chem.* 61 (2): 489–493.

525 Zanardi, J., Leriverend, C., Aubert, D. et al. (2001). A catalytic cycle for the asymmetric synthesis of epoxides using sulfur ylides. *J. Org. Chem.* 66 (16): 5620–5623.

526 Davoust, M., Briere, J.F., Jaffres, P.A., and Metzner, P. (2005). Design of sulfides with a locked conformation as promoters of catalytic and asymmetric sulfonium ylide epoxidation. *J. Org. Chem.* 70 (10): 4166–4169.

527 Minière, S., Reboul, V., Metzner, P. et al. (2004). Catalytic ferrocenyl sulfides for the asymmetric transformation of aldehydes into epoxides. *Tetrahedron: Asymmetry* 15 (20): 3275–3280.

528 Hansch, M., Illa, O., McGarrigle, E.M., and Aggarwal, V.K. (2008). Synthesis and application of easily recyclable thiomorpholines for use in sulfur ylide mediated asymmetric epoxidation of aldehydes. *Chem. Asian J.* 3 (8–9): 1657–1663.

529 Wu, H.Y., Chang, C.W., and Chein, R.J. (2013). Enantioselective synthesis of (thiolan-2-yl)diphenylmethanol and its application in asymmetric, catalytic sulfur ylide-mediated epoxidation. *J. Org. Chem.* 78 (11): 5788–5793.

530 Bellenie, B.R. and Goodman, J.M. (2004). Sulfonium ylide epoxidation reactions: methylene transfer. *J. Chem. Soc. Chem. Commun.* (9): 1076–1077.

531 Baeyer, A. and Villiger, V. (1899). Einwirkung des Caro'schen Reagens auf Ketone. *Ber. Dtsch. Chem. Ges.* 32: 3625–3633.

532 Levinger, S. (2011). Baeyer–Villiger reactions. In: *Science of Synthesis, Stereoselective Synthesis 3* (ed. P.A. Evans), 747–757. Stuttgart: Georg Thieme Verlag KG.

533 Renz, M. and Meunier, B. (1999). 100 years of Baeyer–Villiger oxidations. *Eur. J. Org. Chem.* (4): 737–750.

534 ten Brink, G.J., Arends, I.W.C.E., and Sheldon, R.A. (2004). The Baeyer–Villiger reaction: new developments toward greener procedures. *Chem. Rev.* 104 (9): 4105–4123.

535 Shipston, N.F., Lenn, M.J., and Knowles, C.J. (1992). Enantioselective whole cell and isolated enzyme catalyzed Bayer-Villiger oxidation of bicyclo[3.2.0]hept-2-en-6-one. *J. Microbiol. Methods* 15 (1): 41–52.

536 Secci, F., Frongia, A., and Piras, P.P. (2013). Stereocontrolled synthesis and functionalization of cyclobutanes and cyclobutanones. *Molecules* 18 (12): 15541–15572.

537 Gagnon, R., Grogan, G., Levitt, M.S. et al. (1994). Biological Baeyer–Villiger oxidation of some monocyclic and bicyclic ketones using monooxygenases from *Acinetobacter calcoaceticus* NCIMB-9871 and *Pseudomonas putida*

NCIMB-10007. *J. Chem. Soc., Perkin Trans. 1* (18): 2537–2543.

538 Gagnon, R., Grogan, G., Roberts, S.M. et al. (1995). Enzymatic Baeyer–Villiger oxidations of some bicyclo[2.2.1]heptan-2-ones using monooxygenases from *Pseudomonas putida* NCIMB-10007: enantioselective preparation of a precursor of azadirachtin. *J. Chem. Soc., Perkin Trans. 1* (12): 1505–1511.

539 Swizdor, A., Panek, A., and Milecka-Tronina, N. (2014). Microbial Baeyer–Villiger oxidation of 5-α-steroids using *Beauveria bassiana*. A stereochemical requirement for the 11-α-hydroxylation and the lactonization pathway. *Steroids* 82: 44–52.

540 Alphand, V., Mazzini, C., Lebreton, J., and Furstoss, R. (1998). A new microorganism for highly stereospecific Baeyer–Villiger oxidation of prochiral cyclobutanones. *J. Mol. Catal. B Enzym.* (Elsevier) 5 (1-4): 219–221.

541 Strukul, G. (1998). Transition metal catalysis in the Baeyer–Villiger oxidation of ketones. *Angew. Chem. Int. Ed.* 37 (9): 1199–1209.

542 Hollmann, F., Taglieber, A., Schulz, F., and Reetz, M.T. (2007). A light-driven stereoselective biocatalytic oxidation. *Angew. Chem. Int. Ed.* 46 (16): 2903–2906.

543 Bolm, C., Frison, J.C., Zhang, Y., and Wulff, W.D. (2004). Vaulted biaryls: efficient ligands for the Al-catalyzed asymmetric Baeyer–Villiger reaction. *Synlett* (9): 1619–1621.

544 Xu, S., Wang, Z., Zhang, X. et al. (2008). Chiral Brønsted acid catalyzed asymmetric Baeyer–Villiger reaction of 3-substituted cyclobutanones by using aqueous H_2O_2. *Angew. Chem. Int. Ed.* 47 (15): 2840–2843.

545 Xu, S., Wang, Z., Li, Y. et al. (2010). Mechanistic investigation of chiral phosphoric acid catalyzed asymmetric Baeyer-Villiger reaction of 3-substituted cyclobutanones with H_2O_2 as the oxidant. *Chem. Eur. J.* 16 (10): 3021–3035.

546 Zhou, L., Liu, X., Ji, J. et al. (2012). Enantioselective Baeyer–Villiger oxidation: desymmetrization of meso cyclic ketones and kinetic resolution of racemic 2-arylcyclohexanones. *J. Am. Chem. Soc.* 134 (41): 17023–17026.

547 Blackmond, D.G. (2000). Kinetic aspects of nonlinear effects in asymmetric catalysis. *Acc. Chem. Res.* 33 (6): 402–411.

548 Puchot, C., Samuel, O., Dunach, E. et al. (1986). Nonlinear effects in asymmetric synthesis. Examples in asymmetric oxidations and aldolization reactions. *J. Am. Chem. Soc.* 108 (9): 2353–2357.

549 Pfenninger, A. (1986). Asymmetric epoxidation of allylic alcohols: the Sharpless epoxidation. *Synthesis* (2): 89–116.

550 Gao, Y., Hanson, R.M., Klunder, J.M. et al. (1987). Catalytic asymmetric epoxidation and kinetic resolution: modified procedures including in situ derivatization. *J. Am. Chem. Soc.* 109 (19): 5765–5780.

551 Klussmann, M., Iwamura, H., Mathew, S.P. et al. (2006). Thermodynamic control of asymmetric amplification in amino acid catalysis. *Nature* 441 (7093): 621–623.

552 Klussmann, M., White, A.J., Armstrong, A., and Blackmond, D.G. (2006). Rationalization and prediction of solution enantiomeric excess in ternary phase systems. *Angew. Chem. Int. Ed.* 45 (47): 7985–7989.

553 Hayashi, Y., Matsuzawa, M., Yamaguchi, J. et al. (2006). Large nonlinear effect observed in the enantiomeric excess of proline in solution and that in the solid state. *Angew. Chem. Int. Ed.* 45 (28): 4593–4597.

554 Merino, P. and Tejero, T. (2004). Organocatalyzed asymmetric α-aminoxylation of aldehydes and ketones: an efficient access to enantiomerically pure α-hydroxycarbonyl compounds, diols, and even amino alcohols. *Angew. Chem. Int. Ed.* 43 (23): 2995–2997.

555 Qin, L., Li, L., Yi, L. et al. (2011). Direct asymmetric N-specific reaction of nitrosobenzene with aldehydes catalyzed by a chiral primary amine-based organocatalyst. *Chirality* 23 (7): 527–533.

556 Daikai, K., Hayano, T., Kino, R. et al. (2003). Asymmetric catalysis with self-organized chiral lanthanum complexes: practical and highly enantioselective epoxidation of α,β-unsaturated ketones. *Chirality* 15 (1): 83–88.

557 Guillaneux, D., Zhao, S.H., Samuel, O. et al. (1994). Nonlinear effects in asymmetric catalysis. *J. Am. Chem. Soc.* 116 (21): 9430–9439.

558 Noyori, R., Ohkuma, T., Kitamura, M. et al. (1987). Asymmetric hydrogenation of β-keto carboxylic esters. A practical, purely chemical access to β-hydroxy esters in high enantiomeric purity. *J. Am. Chem. Soc.* 109 (19): 5856–5858.

559 Mashima, K., Kusano, K.H., Sato, N. et al. (1994). Cationic Binap-Ru(II) halide complexes: highly efficient catalysts for stereoselective asymmetric hydrogenation of α-functionalized and β-functionalized ketones. *J. Org. Chem.* 59 (11): 3064–3076.

560 Ohkuma, T., Ooka, H., Yamakawa, M. et al. (1996). Stereoselective hydrogenation of simple

ketones catalyzed by Ru(II) complexes. *J. Org. Chem.* 61 (15): 4872–4873.

561 Noyori, R. and Ohkuma, T. (2001). Asymmetric catalysis by architectural and functional molecular engineering: practical chemo- and stereoselective hydrogenation of ketones. *Angew. Chem. Int. Ed.* 40 (1): 40–73.

562 James, B.R., Pacheco, A., Rettig, S.J. et al. (1987). Activation of dihydrogen by Ru(II)-chelating phosphine complexes, and activation of dioxygen by Ru(II) porphyrin complexes: an update. *J. Mol. Catal.* (Elsevier) 41 (1–2): 147–161.

563 Genêt, J.P., Pinel, C., Ratovelomananavidal, V. et al. (1994). Novel, general synthesis of the chiral catalysts diphosphine-Ru(II) diallyl complexes and a new practical in situ preparation of chiral Ru(II) catalysts. *Tetrahedron Asymmetry* 5 (4): 665–674.

564 Genêt, J.P., Pinel, C., Ratovelomananavidal, V. et al. (1994). Enantioselective hydrogenation reactions with a full set of preformed and prepared in situ chiral diphosphine-Ru(II) catalysts. *Tetrahedron Asymmetry* 5 (4): 675–690.

565 Girard, C., Genêt, J.P., and Bulliard, M. (1999). Non-linear effects in Ru-catalysed asymmetric hydrogenation with atropisomeric diphosphanes. *Eur. J. Org. Chem.* (11): 2937–2942.

566 Calvin, M. (1969). *Chemical Evolution*, 1–278. Oxford: Oxford University Press.

567 Kondepudi, D.K., Kaufman, R.J., and Singh, N. (1990). Chiral symmetry-breaking in Na-chlorate crystallization. *Science* 250 (4983): 975–976.

568 Bolm, C., Bienewald, F., and Seger, A. (1996). Asymmetric autocatalysis with amplification of chirality. *Angew. Chem. Int. Ed. Eng.* 35 (15): 1657–1659.

569 Avalos, M., Babiano, R., Cintas, P. et al. (2000). Chiral autocatalysis: where stereochemistry meets the origin of life. *J. Chem. Soc. Chem. Commun.* (11): 887–892.

570 Feringa, B.L. and van Delden, R.A. (1999). Absolute asymmetric synthesis: the origin, control, and amplification of chirality. *Angew. Chem. Int. Ed.* 38 (23): 3419–3438.

571 Buschmann, H., Thede, R., and Heller, D. (2000). New developments in the origins of the homochirality of biologically relevant molecules. *Angew. Chem. Int. Ed.* 39 (22): 4033–4036.

572 Kondepudi, D.K. and Asakura, K. (2001). Chiral autocatalysis, spontaneous symmetry breaking, and stochastic behavior. *Acc. Chem. Res.* 34 (12): 946–954.

573 Mislow, K. (2003). Absolute asymmetric synthesis: a commentary. *Collect. Czechoslov. Chem. Commun.* 68 (5): 849–864.

574 Bissette, A.J. and Fletcher, S.P. (2013). Mechanisms of autocatalysis. *Angew. Chem. Int. Ed.* 52 (49): 12800–12826.

575 Saito, Y. and Hyuga, H. (2013). Colloquium: homochirality: symmetry breaking in systems driven far from equilibrium. *Rev. Mod. Phys.* 85 (2): 603–621.

576 Hein, J.E., Gherase, D., and Blackmond, D.G. (2013). Chemical and physical models for the emergence of biological homochirality. In: *Biochirality: Origins, Evolution and Molecular Recognition*, Topics in Current Chemistry, vol. 333 (ed. P. Cintas), 83–108. Berlin, Heidelberg: Springer-Verlag.

577 Hein, J.E. and Blackmond, D.G. (2012). On the origin of single chirality of amino acids and sugars in biogenesis. *Acc. Chem. Res.* 45 (12): 2045–2054.

578 Avalos, M., Babiano, R., Cintas, P. et al. (2010). Homochirality and chemical evolution: new vistas and reflections on recent models. *Tetrahedron: Asymmetry* 21 (9-10): 1030–1040.

579 Plasson, R., Kondepudi, D.K., Bersini, H. et al. (2007). Emergence of homochirality in far-from-equilibrium systems: mechanisms and role in prebiotic chemistry. *Chirality* 19 (8): 589–600.

580 Frank, F.C. (1953). On spontaneous asymmetric synthesis. *Biochim. Biophys. Acta* 11 (4): 459–463.

581 Tsogoeva, S.B. (2010). When chiral product and catalyst are the same: discovery of asymmetric organoautocatalysis. *J. Chem. Soc. Chem. Commun.* 46 (41): 7662–7669.

582 Soai, K. and Shibata, T. (1999). Asymmetric autocatalysis and biomolecular chirality. In: *Advances in Biochirality* (ed. G. Pályi, L. Caglioti and C. Zucchi), 125–136. Elsevier.

583 Girard, C. and Kagan, H.B. (1998). Nonlinear effects in asymmetric synthesis and stereoselective reactions: ten years of investigation. *Angew. Chem. Int. Ed.* 37 (21): 2923–2959.

584 Soai, K., Shibata, T., Morioka, H., and Choji, K. (1995). Asymmetric autocatalysis and amplification of enantiomeric excess of a chiral molecule. *Nature* 378 (6559): 767–768.

585 Shibata, T., Yonekubo, S., and Soai, K. (1999). Practically perfect asymmetric autocatalysis with (2-alkynyl-5-pyrimidyl)alkanols. *Angew. Chem. Int. Ed.* 38 (5): 659–661.

586 Gehring, T., Busch, M., Schlageter, M., and Weingand, D. (2010). A concise summary of experimental facts about the Soai reaction. *Chirality* 22 (1E): E173–E182.

587 Sato, I., Urabe, H., Ishiguro, S. et al. (2003). Amplification of chirality from extremely low to greater than 99.5% ee by asymmetric autocatalysis. *Angew. Chem. Int. Ed.* 42 (3): 315–317.

588 Soai, K., Kawasaki, T., and Matsumoto, A. (2014). The origins of homochirality examined by using asymmetric autocatalysis. *Chem. Rec.* 14 (1): 70–83.

589 Soai, K., Osanai, S., Kadowaki, K. et al. (1999). d- and l-quartz-promoted highly enantioselective synthesis of a chiral organic compound. *J. Am. Chem. Soc.* 121 (48): 11235–11236.

590 Sato, I., Omiya, D., Tsukiyama, K. et al. (2001). Evidence of asymmetric autocatalysis in the enantioselective addition of diisopropylzinc to pyrimidine-5-carbaldehyde using chiral pyrimidyl alkanol. *Tetrahedron: Asymmetry* 12 (14): 1965–1969.

591 Blackmond, D.G., McMillan, C.R., Ramdeehul, S. et al. (2001). Origins of asymmetric amplification in autocatalytic alkylzinc additions. *J. Am. Chem. Soc.* 123 (41): 10103–10104.

592 Gridnev, I.D., Serafimov, J.M., and Brown, J.M. (2004). Solution structure and reagent binding of the zinc alkoxide catalyst in the Soai asymmetric autocatalytic reaction. *Angew. Chem. Int. Ed.* 43 (37): 4884–4887.

593 Gridnev, I.D. (2006). Chiral symmetry breaking in chiral crystallization and Soai autocatalytic reaction. *Chem. Lett.* 35 (2): 148–153.

594 Gehring, T., Quaranta, M., Odell, B. et al. (2012). Observation of a transient intermediate in Soai's asymmetric autocatalysis: insights from 1H NMR turnover in real time. *Angew. Chem. Int. Ed.* 51 (38): 9539–9542.

595 Young, S.D., Britcher, S.F., Tran, L.O. et al. (1995). L-743,726 (Dmp-266) – a novel, highly potent nonnucleoside inhibitor of the human-immunodeficiency-virus type-1 reverse-transcriptase. *Antimicrob. Agents Chemother.* 39 (12): 2602–2605.

596 Chinkov, N., Warm, A., and Carreira, E.M. (2011). Asymmetric autocatalysis enables an improved synthesis of Efavirenz. *Angew. Chem. Int. Ed.* 50 (13): 2957–2961.

597 Tan, L., Chen, C.Y., Tillyer, R.D. et al. (1999). A novel, highly enantioselective ketone alkynylation reaction mediated by chiral zinc aminoalkoxides. *Angew. Chem. Int. Ed.* 38 (5): 711–713.

598 Chen, C.Y. and Tan, L.S. (1999). Chemical process evolution of Efavirenz, a potent non-nucleosidal HIV reverse transcriptase inhibitor. *Enantiomer* 4 (6): 599–608.

599 Tsogoeva, S.B., Wei, S., Freund, M., and Mauksch, M. (2009). Generation of highly enantioenriched crystalline products in reversible asymmetric reactions with racemic or achiral catalysts. *Angew. Chem. Int. Ed.* 48 (3): 590–594.

600 Wei, S., Mauksch, M., and Tsogoeva, S.B. (2009). Autocatalytic enantiomerisation at the crystal surface in deracemisation of scalemic conglomerates. *Chem. Eur. J.* 15 (39): 10255–10262.

601 van Dijken, D.J., Beierle, J.M., Stuart, M.C. et al. (2014). Autoamplification of molecular chirality through the induction of supramolecular chirality. *Angew. Chem. Int. Ed.* 53 (20): 5073–5077.

602 Svirbely, J.L. and Szent-Gyogyi, A. (1932). The chemical nature of vitamin C. *Biochem. J.* 26: 865–870.

603 Waugh, W.A. and King, C.G. (1932). Isolation and identification of vitamin C. *J. Biol. Chem.* 97 (1): 325–331.

604 King, C.G. and Waugh, W.A. (1932). The chemical nature of vitamin C. *Science* 75: 357–358.

605 Herbert, R.W., Hirst, E.L., Percival, E.G.V. et al. (1933). The constitution of ascorbic acid. *J. Chem. Soc.* 1270–1290.

606 Michels, A. and Frei, B. (2012). Vitamin C. In: *Biochemical, Physiological, an Molecular Aspects of Human Nutrition*, 3e (ed. M.A. Caudill and M. Rogers), 627–654. Philadelphia, PA: Saunders.

607 Lykkesfeldt, J., Michels, A.J., and Frei, B. (2014). Vitamin C. *Adv. Nutr.* 5 (1): 16–18.

608 Padayatty, S.J., Katz, A., Wang, Y.H. et al. (2003). Vitamin C as an antioxidant: evaluation of its role in disease prevention. *J. Am. Coll. Nutr.* 22 (1): 18–35.

609 Ault, R.G., Baird, D.K., Carrington, H.C. et al. (1933). Synthesis of d- and of l-ascorbic acid and of analogous substances. *J. Chem. Soc.* 1419–1423.

610 Reichstein, T. and Grüssner, A. (1934). A productive synthesis of l-arcorbic acid (C-vitamin). *Helv. Chim. Acta* 17: 311–328.

611 Pappenberger, G. and Hohmann, H.-P. (2014). Industrial production of l-ascorbic acid (vitamin C) and d-isoascorbic acid. *Adv. Biochem. Eng. Biotechnol.* 143: 143–188.

612 Bertrand, G. (1896). Preparation biochimique du sorbose. *Comptes rendus hebdomadaires des Séances de l'Académie des Sciences* 122: 900–903.

613 Gillis, M., Kersters, K., Gossele, F. et al. (1983). Rediscovery of Bertrand's sorbose bacterium (*Acetobacter aceti* subsp. *xylinum*): proposal to designate NCIB-11664 in place of NCIB-4112 (ATCC-23767) As the type strain of *Acetobacter*

aceti subsp *xylinum* – request for an opinion. *Int. J. Syst. Bacteriol.* 33 (1): 122–124.

614 Heyns, K. (1947). Oxydative Umwandlungen an Kohlenhydraten. 2. Katalytische Oxydation von 1-Sorbose zu 2-Keto-1-gulonsäure. *Liebigs Ann. Chem.* 558 (3): 177–187.

615 Heyns, K. and Paulsen, H. (1957). Selective catalytic oxidations by means of noble metal catalysts. *Angew. Chem.* 69 (18–19): 600–608.

616 Brönnimann, C., Bodnar, Z., Hug, P. et al. (1994). Direct oxidation of L-sorbose to 2-keto-L-gulonic acid with molecular oxygen on Pt-based and Pd-based catalysts. *J. Catal.* (Elsevier) 150 (1): 199–211.

617 Hancock, R.D. and Viola, R. (2002). Biotechnological approaches for L-ascorbic acid production. *Trends Biotechnol.* (Elsevier) 20 (7): 299–305.

618 Sauer, M., Branduardi, P., Valli, M., and Porro, D. (2004). Production of L-ascorbic acid by metabolically engineered *Saccharomyces cerevisiae* and *Zygosaccharomyces bailii*. *Appl. Environ. Microbiol.* 70 (10): 6086–6091.

619 Bremus, C., Herrmann, U., Bringer-Meyer, S., and Sahm, H. (2006). The use of microorganisms in L-ascorbic acid production. *J. Biotechnol.* 124 (1): 196–205.

620 Kawabata, T., Muramatsu, W., Nishio, T. et al. (2007). A catalytic one-step process for the chemo- and regioselective acylation of monosaccharides. *J. Am. Chem. Soc.* 129 (42): 12890–12895.

621 Cremer, E. and Polanyi, M. (1932). A report of the "tunnel theory" of heterogeneous catalysis on example of the hydration of styrol. *Zeitschrift für Physikalische Chemie, Abteilung B, Chemie der Elementarprozesse, Aufbau der Materie* 19 (6): 443–450.

622 Eyring, H. and Sherman, A. (1933). Theoretical considerations concerning the separation of isotopes. *J. Chem. Phys.* 1 (6): 345–349.

623 Wiberg, K.B. (1955). The deuterium isotope effect. *Chem. Rev.* 55 (4): 713–743.

624 Bell, R.P. (1974). Recent advances in study of kinetic hydrogen isotope effects. *Chem. Soc. Rev.* 3 (4): 513–544.

625 Giagou, T. and Meyer, M.P. (2010). Kinetic isotope effects in asymmetric reactions. *Chem. Eur. J.* 16 (35): 10616–10628.

626 Klinman, J.P. (2010). A new model for the origin of kinetic hydrogen isotope effects. *J. Phys. Org. Chem.* 23 (7): 606–612.

627 Lee, I. (1995). Secondary kinetic isotope effects involving deuterated nucleophiles. *Chem. Soc. Rev.* 24 (3): 223–229.

628 Saunders, W.H. Jr., (1986). *Techniques of Chemistry*, vol. 6, 565–611.

629 Matsson, O. and Westaway, K.C. (1998). Secondary deuterium kinetic isotope effects and transition state structure. *Adv. Phys. Org. Chem.* (31): 143–248.

630 Berti, P.J. (1999). Determining transition states from kinetic isotope effects. *Methods Enzymol.* 308: 355–397.

631 Scheiner, S. (2000). Calculation of isotope effects from first principles. *Biochim. Biophys. Acta* 1458 (1): 28–42.

632 Leach, A.G. and Houk, K.N. (2002). Diels–Alder and ene reactions of singlet oxygen, nitroso compounds and triazolinediones: transition states and mechanisms from contemporary theory. *J. Chem. Soc. Chem. Commun.* (12): 1243–1255.

633 Cleland, W.W. (2005). The use of isotope effects to determine enzyme mechanisms. *Arch. Biochem. Biophys.* 433 (1): 2–12.

634 Sicinska, D., Rostkowski, M., and Paneth, P. (2005). Chlorine isotope effects on chemical reactions. *Curr. Org. Chem.* 9 (1): 75–88.

635 Wright, S.K., Declue, M.S., Mandal, A. et al. (2005). Isotope effects on the enzymatic and nonenzymatic reactions of chorismate. *J. Am. Chem. Soc.* 127 (37): 12957–12964.

636 Schramm, V.L. (2007). Enzymatic transition state theory and transition state analogue design. *J. Biol. Chem.* 282 (39): 28297–28300.

637 Westaway, K.C. (2007). Determining transition state structure using kinetic isotope effects. *J. Label. Compd. Radiopharm.* 50 (11-12): 989–1005.

638 Gómez-Gallego, M. and Sierra, M.A. (2011). Kinetic isotope effects in the study of organometallic reaction mechanisms. *Chem. Rev.* 111 (8): 4857–4963.

639 Fisher, H.F. (2013). The application of transient-state kinetic isotope effects to the resolution of mechanisms of enzyme-catalyzed reactions. *Molecules* 18 (7): 8230–8242.

640 Roston, D., Islam, Z., and Kohen, A. (2013). Isotope effects as probes for enzyme catalyzed hydrogen-transfer reactions. *Molecules* 18 (5): 5543–5567.

641 Gu, H. and Zhang, S. (2013). Advances in kinetic isotope effect measurement techniques for enzyme mechanism study. *Molecules* 18 (8): 9278–9292.

642 Westheimer, F.H. (1961). The magnitude of the primary kinetic isotope effect for compounds of hydrogen and deuterium. *Chem. Rev.* 61 (3): 265–273.

643 Kwart, H. (1982). Temperature-dependence of the primary kinetic hydrogen isotope effect as a mechanistic criterion. *Acc. Chem. Res.* 15 (12): 401–408.

644 Streitwieser, A. Jr., and van Sickle, D.E. (1962). Acidity of hydrocarbons. 4. Secondary deuterium isotope effects in exchange reactions of toluene and ethylbenzene with lithium cyclohexylamide. *J. Am. Chem. Soc.* 84 (2): 254–258.

645 Wiberg, K.B. and Slaugh, L.H. (1958). The deuterium isotope effect in the side chain halogenation of toluene. *J. Am. Chem. Soc.* 80 (12): 3033–3039.

646 Westheimer, F.H. and Nicolaides, N. (1949). The kinetics of the oxidation of 2-deuteropropanol-2 by chromic acid. *J. Am. Chem. Soc.* 71 (1): 25–28.

647 Harding, C.E., Mitchell, C.W., and Devenyi, J. (2000). Kinetic isotope effect in the chromic acid oxidation of secondary alcohols. *J. Chem. Educ.* 77 (8): 1042–1044.

648 Iskenderian-Epps, W.S., Soltis, C., and O'Leary, D.J. (2013). Visual isotope effects: demonstrating the primary kinetic isotope effect in the Cr(VI) oxidation of 2-propanol-d(8) and methanol-d(4). *J. Chem. Educ.* 90 (8): 1044–1047.

649 Swain, C.G., Stivers, E.C., Reuwer, J.F., and Schaad, L.J. (1958). Use of hydrogen isotope effects to identify the attacking nucleophile in the enolization of ketones catalyzed by acetic acid. *J. Am. Chem. Soc.* 80 (21): 5885–5893.

650 Melander, L. and Saunders, W.H. Jr., (1980). *Reactions Rates of Isotopic Molecules*, 1–391. New York, NY: Wiley.

651 Razavy, M. (2014). *Quantum Theory of Tunneling*, 2e, 1–792. Singapore: Word Scientific Publishing Co.

652 Carpenter, B.K. (2010). Kinetic isotope effects. Unearthing the unconventional. *Nat. Chem.* 2 (2): 80–82.

653 Bell, R.P. (1980). *The Tunnel Effect in Chemistry*, 1–222. New York, NY: Chapman & Hall.

654 Caldin, E.F. and Mateo, S. (1975). Kinetic isotope effects and tunnelling in proton-transfer reaction between 4-nitrophenylnitromethane and tetramethylguanidine in various aprotic solvents. *J. Chem. Soc., Faraday Trans. 1 F* 71: 1876–1904.

655 Bell, R.P. (1933). The application of quantum mechanics to chemical kinetics. *Proc. R. Soc. Lond. A Math. Phys. Eng. Sci.* 139 (838): 466–474.

656 Bell, R.P. (1935). Quantum mechanical effects in reactions involving hydrogen. *Proc. R. Soc. Lond. A Math. Phys. Sci.* 148 (A864): 0241–0250.

657 More O'Ferrall, R.A. (2010). A pictorial representation of zero-point energy and tunnelling contributions to primary hydrogen isotope effects. *J. Phys. Org. Chem.* 23: 572–579.

658 Lynch, R.A., Lin, Y.T., Smucker, L.D. et al. (1972). Anomalous kinetic hydrogen isotope effects on rate of ionization of some dialkyl substituted ketones. *J. Am. Chem. Soc.* 94 (24): 8351–8356.

659 Bigeleisen, J. (1949). The relative reaction velocities of isotopic molecules. *J. Chem. Phys.* 17 (8): 675–678.

660 Bigeleisen, J. and Wolfsberg, M. (1958). Theoretical and experimental aspects of isotope effects chemical kinetics. *Adv. Chem. Phys.* 1: 15–76.

661 Matsson, O., Persson, J., Axelsson, B.S. et al. (1996). Using incoming group C-11/C-14 kinetic isotope effects to model the transition states for the S_N2 reactions between para-substituted benzyl chlorides and labeled cyanide ion. *J. Am. Chem. Soc.* 118 (27): 6350–6354.

662 Hay, S., Johannissen, L.O., Sutcliffe, M.J., and Scrutton, N.S. (2010). Barrier compression and its contribution to both classical and quantum mechanical aspects of enzyme catalysis. *Biophys. J.* 98 (1): 121–128.

663 Pudney, C.R., Johannissen, L.O., Sutcliffe, M.J. et al. (2010). Direct analysis of donor acceptor distance and relationship to isotope effects and the force constant for barrier compression in enzymatic H-tunneling reactions. *J. Am. Chem. Soc.* 132 (32): 11329–11335.

664 Kohen, A. and Klinman, J.P. (1999). Hydrogen tunneling in biology. *Chem. Biol.* 6 (7): R191–R198.

665 Kohen, A., Cannio, R., Bartolucci, S., and Klinman, J.P. (1999). Enzyme dynamics and hydrogen tunnelling in a thermophilic alcohol dehydrogenase. *Nature* 399 (6735): 496–499.

666 Knapp, M.J. and Klinman, J.P. (2002). Environmentally coupled hydrogen tunneling. Linking catalysis to dynamics. *Eur. J. Biochem.* 269 (13): 3113–3121.

667 Knapp, M.J., Rickert, K., and Klinman, J.P. (2002). Temperature-dependent isotope effects in soybean lipoxygenase-1: correlating hydrogen tunneling with protein dynamics. *J. Am. Chem. Soc.* 124 (15): 3865–3874.

668 Hammann, B., Razzaghi, M., Kashefolgheta, S., and Lu, Y. (2012). Imbalanced tunneling ready states in alcohol dehydrogenase model reactions: rehybridization lags behind H-tunneling. *J. Chem. Soc. Chem. Commun.* 48 (92): 11337–11339.

669 Roston, D., Islam, Z., and Kohen, A. (2014). Kinetic isotope effects as a probe of hydrogen transfers to and from common enzymatic cofactors. *Arch. Biochem. Biophys.* 544: 96–104.

670 Cha, Y., Murray, C.J., and Klinman, J.P. (1989). Hydrogen tunneling in enzyme reactions. *Science* 243 (4896): 1325–1330.

671 Grant, K.L. and Klinman, J.P. (1989). Evidence that both protium and deuterium undergo significant tunneling in the reaction catalyzed by bovine serum amine oxidase. *Biochemistry* 28 (16): 6597–6605.

672 Jones, J.R. (1969). Kinetic isotope effects. 5. Bromination of acetophenone-D3 and inter-relationship between deuterium and tritium isotope effects. *Trans. Faraday Soc.* 65 (560P): 2138–2143.

673 Lu, Y., Qu, F.R., Moore, B. et al. (2008). Hydride reduction of NAD(+) analogues by isopropyl alcohol: Kinetics, deuterium isotope effects and mechanism. *J. Org. Chem.* 73 (13): 4763–4770.

674 Bender, M.L. and Hoeg, D.F. (1957). Carbon-14 kinetic isotope effects in nucleophilic substitution reactions. *J. Am. Chem. Soc.* 79 (21): 5649–5654.

675 Leffek, K.T., Llewellyn, J.A., and Robertson, R.E. (1960). Some deuterium kinetic isotope effects. 2. α-Deuterium effects in the water solvolysis of some alkyl compounds. *Can. J. Chem. (Rev. Can. Chim.)* 38 (9): 1505–1510.

676 Shiner, V.J. (1953). Solvolysis rates of some deuterated tertiary amyl chlorides. *J. Am. Chem. Soc.* 75 (12): 2925–2929.

677 Leffek, K.T., Llewellyn, J.A., and Robertson, R.E. (1960). Some deuterium kinetic isotope effects. 4. β-Deuterium effects in water solvolysis of ethyl, isopropyl, and tert-butyl compounds. *Can. J. Chem. (Rev. Can. Chim.)* 38 (11): 2171–2177.

678 Leffek, K.T. and MacLean, J.W. (1965). Secondary kinetic isotope effects in bimolecular nucleophilic substitutions. 2. α-Deuterium effects in Menschutkin reactions of methyl iodide. *Can. J. Chem. (Rev. Can. Chim.)* 43 (1): 40–46.

679 Leffek, K.T. and Matheson, A.F. (1972). Secondary kinetic isotope effects in bimolecular nucleophilic substitutions. 6. Effect of α- and β-deuteration of alkyl halides in their Menschutkin reactions with pyridine in nitrobenzene. *Can. J. Chem. (Rev. Can. Chim.)* 50 (7): 986–991.

680 Fang, Y.R., Gao, Y., Ryberg, P. et al. (2003). Experimental and theoretical multiple kinetic isotope effects for an S_N2 reaction. An attempt to determine transition-state structure and the ability of theoretical methods to predict experimental kinetic isotope effects. *Chem. Eur. J.* 9 (12): 2696–2709.

681 Fang, Y.R., MacMillar, S., Eriksson, J. et al. (2006). The effect of solvent on the structure of the transition state for the S_N2 reaction between cyanide ion and ethyl chloride in DMSO and THF probed with six different kinetic isotope effects. *J. Org. Chem.* 71 (13): 4742–4747.

682 Dybala-Defratyka, A., Rostkowski, M., Matsson, O. et al. (2004). A new interpretation of chlorine leaving group kinetic isotope effects; a theoretical approach. *J. Org. Chem.* 69 (15): 4900–4905.

683 Westaway, K.C., Fang, Y.R., MacMillar, S. et al. (2007). A new insight into using chlorine leaving group and nucleophile carbon kinetic isotope effects to determine substituent effects on the structure of S_N2 transition states. *J. Phys. Chem. A* 111 (33): 8110–8120.

684 Matsson, O., Dybala-Defratyka, A., Rostkowski, M. et al. (2005). A theoretical investigation of alpha-carbon kinetic isotope effects and their relationship to the transition-state structure of S_N2 reactions. *J. Org. Chem.* 70 (10): 4022–4027.

685 Koerner, T., Fang, Y.R., and Westaway, K.C. (2000). Using incoming nucleophile primary hydrogen-deuterium kinetic isotope effects to model the S_N2 transition state. *J. Am. Chem. Soc.* 122 (30): 7342–7350.

686 Oh, H.K., Kim, S.K., Cho, I.H. et al. (2000). Kinetics and mechanism of the aminolysis of aryl phenyldithioacetates in acetonitrile. *J. Chem. Soc., Perkin Trans. 2* (11): 2306–2310.

687 Lee, I., Kim, H.Y., and Kang, H.K. (1987). Cross interaction constants as a mechanistic criterion for the intermolecular S_Ni mechanism. *J. Chem. Soc. Chem. Commun.* (16): 1216–1217.

688 Lee, I., Kim, H.Y., Kang, H.K., and Lee, H.W. (1988). Nucleophilic substitution reactions of 1-phenylethyl benzenesulfonates with anilines in methanol acetonitrile. *J. Org. Chem.* 53 (12): 2678–2683.

689 Lee, I., Koh, H.J., Lee, B.S. et al. (1990). Kinetic isotope effects in the nucleophilic substitution reactions of benzyl benzenesulfonates and 1-phenylethyl benzenesulfonates with deuterated aniline nucleophiles. *Bull. Kor. Chem. Soc.* 11 (5): 435–438.

690 Dey, N.K., Kim, C.K., and Lee, H.W. (2011). Kinetics and mechanism of the anilinolyses of aryl dimethyl, methyl phenyl and diphenyl phosphinates. *Org. Biomol. Chem.* 9 (3): 717–724.

691 Dey, N.K., Hoque, M., Kim, C.K. et al. (2009). Kinetics and mechanism of the aminolysis of dimethyl and methyl phenyl phosphinic chlorides with anilines. *J. Phys. Org. Chem.* 22 (5): 425–430.

692 Barai, H.R. and Lee, H.W. (2011). Kinetics and mechanism of the anilinolysis of ethylene phosphorochloridate in acetonitrile. *Bull. Kor. Chem. Soc.* 32 (12): 4185–4190.

693 Ul Hoque, M. and Lee, H.W. (2012). Kinetics and mechanism of the anilinolysis of dipropyl chlorophosphate in acetonitrile. *Bull. Kor. Chem. Soc.* 33 (6): 1879–1884.

694 Barai, H.R., Adhikary, K.K., and Lee, H.W. (2012). Anilinolysis of dimethyl isothiocyanophosphate

in acetonitrile. *Bull. Kor. Chem. Soc.* 33 (8): 2769–2772.

695 Hoque, M., Dey, S., Guha, A.K. et al. (2007). Kinetics and mechanism of the aminolysis of aryl phenyl chlorothiophosphates with anilines. *J. Org. Chem.* 72 (15): 5493–5499.

696 Hoque, M.E.U. and Lee, H.W. (2012). Kinetics and mechanism of the aminolysis of O-ethyl phenyl phosphonochloridothioate in acetonitrile. *Bull. Kor. Chem. Soc.* 33 (8): 2707–2710.

697 Barai, H.R. and Lee, H.W. (2014). Kinetics and mechanism of anilinolyses of aryl methyl and aryl propyl chlorothiophosphates in acetonitrile. *Bull. Kor. Chem. Soc.* 35 (9): 2797–2802.

698 Matsson, O., Persson, J., Axelsson, B.S., and Långström, B. (1993). Fluorine kinetic isotope effects. *J. Am. Chem. Soc.* 115 (12): 5288–5289.

699 Persson, J. and Matsson, O. (1998). Use of fluorine kinetic isotope effects in the study of steric effects in nucleophilic aromatic substitution reactions. *J. Org. Chem.* 63 (25): 9348–9350.

700 Persson, J., Axelsson, S., and Matsson, O. (1996). Solvent dependent leaving group fluorine kinetic isotope effect in a nucleophilic aromatic substitution reaction. *J. Am. Chem. Soc.* 118 (1): 20–23.

701 Saunders, W.H. and Edison, D.H. (1960). Mechanisms of elimination reactions. 4. Deuterium isotope effects in E2 reactions of some 2-phenylethyl derivatives. *J. Am. Chem. Soc.* 82 (1): 138–142.

702 Yoshida, T., Yano, Y., and Oae, S. (1971). Mechanism of elimination reactions – E2 reactions of substituted α-phenylethyl bromides. *Tetrahedron* 27 (22): 5343–5351.

703 Miller, D.J. and Saunders, W.H. (1981). Mechanisms of elimination reactions. 35. Deuterium kinetic isotope effects and tunneling in the reaction of {[2-para-(trifluoromethyl) phenyl]ethyl}trimethylammonium ion with hydroxide ion in mixtures of water and dimethylsulfoxide. *J. Org. Chem.* 46 (21): 4247–4252.

704 Biale, G., Parker, A.J., Smith, S.G. et al. (1970). E2C-mechanism in elimination reactions. Absence of an extreme form of merged mechanism for elimination and substitution. A comparison of Saytzeff vs Hofmann tendencies and 3F anti-vs-syn eliminations. *J. Am. Chem. Soc.* 92 (1): 115–122.

705 Biale, G., Cook, D., Lloyd, D.J. et al. (1971). E2C mechanism in elimination reactions. 2. Substituent effects on rates of elimination from acyclic systems. *J. Am. Chem. Soc.* 93 (19): 4735–4749.

706 Beltrame, P., Biale, G., Parker, A.J. et al. (1972). E2C mechanism in elimination reactions. 5. Elimination from 5-membered and 6-membered alicyclics. *J. Am. Chem. Soc.* 94 (7): 2240–2255.

707 Biale, G., Winstein, S., Stevens, I.D.R. et al. (1972). E2C mechanism in elimination reactions. 4. Primary hydrogen isotope effects. *J. Am. Chem. Soc.* 94 (7): 2235–2239.

708 Cook, D., Hutchins, R.E., Macleod, J.K., and Parker, A.J. (1974). E2C mechanism in elimination reactions. 6. Primary hydrogen isotope effects on rates of E2 reactions of alicyclics. *J. Org. Chem.* 39 (4): 534–539.

709 Cook, D., Hutchinson, R.E., and Parker, A.J. (1974). E2C mechanism in elimination reactions. 7. Secondary kinetic hydrogen isotope effects in E2 reactions of alicyclics. *J. Org. Chem.* 39 (20): 3029–3038.

710 Cook, D. (1976). E2C mechanism of elimination reactions. 9. Secondary deuterium-isotope effects on rates of bimolecular reactions in alicyclic systems. *J. Org. Chem.* 41 (12): 2173–2179.

711 Koch, H.F., Dahlberg, D.B., Mcentee, M.F., and Klecha, C.J. (1976). Use of kinetic isotope-effects in mechanism studies. Anomalous Arrhenius parameters in study of elimination-reactions. *J. Am. Chem. Soc.* 98 (4): 1060–1061.

712 Barai, H.R., Hoque, M.E.U., and Lee, H.W. (2013). Kinetics and mechanism of anilinolyses of ethyl methyl, ethyl, propyl and diisopropyl chlorothiophosphates in acetonitrile. *Bull. Kor. Chem. Soc.* 34 (12): 3811 3816.

713 Mislow, K., Wahl, G.H., Gordon, A.J., and Graeve, R. (1964). Conformational kinetic isotope effects in racemization of 9,10-dihydro-4,5-dimethylphenanthrene. *J. Am. Chem. Soc.* 86 (9): 1733–1741.

714 Carter, R.E. and Dahlgren, L. (1969). Steric isotope effects. 3. Deuterium isotope effect in racemization of (−)-1,1′-binaphthyl-2,2′-D2. *Acta Chem. Scand.* 23 (2): 504–508.

715 Nagorski, R.W., Slebockatilk, H., and Brown, R.S. (1994). Observation of an unusually large inverse secondary deuterium kinetic isotope effect for the reaction of a sterically congested olefin with Br_2. *J. Am. Chem. Soc.* 116 (1): 419–420.

716 Slebockatilk, H., Motallebi, S., Nagorski, R.W. et al. (1995). Electrophilic bromination of 7-norbornylidene-7′-norbornane – the observation of an unusually large inverse deuterium kinetic isotope effect. *J. Am. Chem. Soc.* 117 (34): 8769–8776.

717 Raaen, V.R., Ropp, G.A., and Raaen, H.P. (1968). *Carbon-14*, 1–240. New York, NY: McGraw-Hill Book Co.

718 Ropp, G.A., Danby, C.J., and Dominey, D.A. (1957). Studies involving isotopically labelled formic acid and its derivatives. 2. Relation of the

absolute reaction rates to the magnitudes of the magnitudes of the isotope effects in the oxidation of formic-C13 acid by halogen atoms. *J. Am. Chem. Soc.* 79 (18): 4944–4948.

719 Jones, W.M. (1951). The relative rates of reaction of hydrogen and tritium hydride with chlorine. *J. Chem. Phys.* 19 (1): 78–85.

720 Singleton, D.A. and Thomas, A.A. (1995). High precision simultaneous determination of multiple small kinetic isotope effects at natural abundance. *J. Am. Chem. Soc.* 117 (36): 9357–9358.

721 Lucero, M.J. and Houk, K.N. (1996). High precision simultaneous determination of multiple small kinetic isotope effects at natural abundance. *Chemt. Org. Chem.* 9: 72–74.

722 Beno, B.R., Houk, K.N., and Singleton, D.A. (1996). Synchronous or asynchronous? An "experimental" transition state from a direct comparison of experimental and theoretical kinetic isotope effects for a Diels–Alder reaction. *J. Am. Chem. Soc.* 118 (41): 9984–9985.

723 Frantz, D.E., Singleton, D.A., and Snyder, J.P. (1997). C-13 kinetic isotope effects for the addition of lithium dibutylcuprate to cyclohexenone. Reductive elimination is rate-determining. *J. Am. Chem. Soc.* 119 (14): 3383–3384.

724 Frantz, D.E. and Singleton, D.A. (2000). Isotope effects and the mechanism of chlorotrimethylsilane-mediated addition of cuprates to enones. *J. Am. Chem. Soc.* 122 (14): 3288–3295.

725 Singleton, D.A. and Hang, C. (2000). C-13 and H-2 kinetic isotope effects and the mechanism of Lewis acid-catalyzed ene reactions of formaldehyde. *J. Org. Chem.* 65 (3): 895–899.

726 DelMonte, A.J., Haller, J., Houk, K.N. et al. (1997). Experimental and theoretical kinetic isotope effects for asymmetric dihydroxylation. Evidence supporting a rate-limiting "(3+2)" cycloaddition. *J. Am. Chem. Soc.* 119 (41): 9907–9908.

727 Keating, A.E., Merrigan, S.R., Singleton, D.A., and Houk, K.N. (1999). Experimental proof of the non-least-motion cycloadditions of dichlorocarbene to alkenes: kinetic isotope effects and quantum mechanical transition states. *J. Am. Chem. Soc.* 121 (16): 3933–3938.

728 Merrigan, S.R. and Singleton, D.A. (1999). C-13 and H-2 kinetic isotope effects and the mechanism of bromination of 1-pentene under synthetic conditions. *Org. Lett.* 1 (2): 327–329.

729 Singleton, D.A., Merrigan, S.R., Beno, B.R., and Houk, K.N. (1999). Isotope effects for Lewis acid catalyzed Diels–Alder reactions. The experimental transition state. *Tetrahedron Lett.* 40 (32): 5817–5821.

730 Singleton, D.A. and Schulmeier, B.E. (1999). Evidence for a concerted mechanism in a palladium trimethylenemethane cycloaddition. *J. Am. Chem. Soc.* 121 (40): 9313–9317.

731 Singleton, D.A. and Szymanski, M.J. (1999). Simultaneous determination of intermolecular and intramolecular C-13 and H-2 kinetic isotope effects at natural abundance. *J. Am. Chem. Soc.* 121 (40): 9455–9456.

732 Meyer, M.P., DelMonte, A.J., and Singleton, D.A. (1999). Reinvestigation of the isotope effects for the Claisen and aromatic Claisen rearrangements: the nature of the Claisen transition states. *J. Am. Chem. Soc.* 121 (47): 10865–10874.

733 Singleton, D.A. and Hang, C. (1999). Isotope effects and the mechanism of triazolinedione ene reactions. Aziridinium imides are innocent bystanders. *J. Am. Chem. Soc.* 121 (50): 11885–11893.

734 Singleton, D.A. and Hang, C. (1999). Isotope effects and the experimental transition state for a prototypical thermal ene reaction. *Tetrahedron Lett.* 40 (51): 8939–8943.

735 Singleton, D.A. and Hang, C. (2000). Isotope effects and the mechanism of allylic hydroxylation of alkenes with selenium dioxide. *J. Org. Chem.* 65 (22): 7554–7560.

736 Singleton, D.A., Wang, Y.C., Yang, H.W., and Romo, D. (2002). Mechanism and origin of stereoselectivity in Lewis acid catalyzed (2+2)-cycloadditions of ketenes with aldehydes. *Angew. Chem. Int. Ed.* 41 (9): 1572–1575.

737 Saettel, N.J., Wiest, O., Singleton, D.A., and Meyer, M.P. (2002). Isotope effects and the mechanism of an electron-transfer-catalyzed Diels–Alder reaction. *J. Am. Chem. Soc.* 124 (38): 11552–11559.

738 Nowlan, D.T.I., Gregg, T.M., Davies, H.M.L., and Singleton, D.A. (2003). Isotope effects and the nature of selectivity in rhodium-catalyzed cyclopropanations. *J. Am. Chem. Soc.* 125 (51): 15902–15911.

739 Singleton, D.A. and Christian, C.F. (2005). Isotope effects and the mechanism of Pd-catalyzed allylic alkylation. *Tetrahedron Lett.* 46 (10): 1631–1634.

4

Molecular orbital theories

4.1 Introduction

Molecular orbital (MO) theories and various approximations have been extremely useful to understand the electronic structures of complicated molecules, bonding in molecules and molecular assemblies, substituent effects on the relative stabilities of neutral or charged species, deviations from thermochemical additivity rules (such as π-conjugation, aromaticity, and antiaromaticity), bonding in π- and σ-complexes, hyperconjugation, hypervalency, and the relative stability of transition structures of concerted reactions. Simple molecular orbital theory has demonstrated the similarities in bonding between organic, organometallic, and inorganic compounds.

4.2 Background of quantum chemistry

Molecular orbital theories are based on quantum chemistry, the application of quantum mechanics and quantum field theory to chemistry [1, 2]. The experimental observations that eventually led to quantum mechanics began in 1828 with the discovery of cathode rays by Faraday (light arc beginning at the negative electrode and ending at the anode [positive electrode] when current passes through rarefied air). In 1859, Kirchhoff stated the blackbody radiation problem, and in 1877, Boltzmann suggested that energy states of physical system (atoms and molecules) could be discrete. In 1900, Planck proposed the quantum hypothesis that energy-irradiating atomic systems can be divided into numbers of discrete energy elements ε such that each of these energy elements is proportional to the frequency ν with which they radiate energy, as defined by $\varepsilon = h\nu$, where h is a numerical value (Planck's constant). In 1905, Einstein explained the photoelectric effect discovered in 1839 by Becquerel while studying the effect of light on electrolytic cells and by Hertz in 1887 when he observed that electrons are emitted from matter when hit by light. Based on Planck's quantum hypothesis, Einstein stated that light itself consists of individual quantum particles, the latter being called photons since 1926. In parallel with developments leading to quantum mechanics, the structural theory of organic chemistry was developed. In 1852, Frankland proposed that atoms of a given element have finite capacity to combine with atoms of other elements, i.e. a definite valence. In 1858, Kekulé and Couper, independently, introduced the concept of valence bonds between atoms, including bonds between two carbon atoms, and proposed the tetravalency of carbon atom. In 1861, Butlerow stated that the properties of a compound are determined by its molecular structure. In 1874, Van't Hoff recognized that the optical activity of carbon compounds, discovered by Pasteur in 1848, can be explained by the postulate that the four-valence bonds of carbon atom are directed in space to the corners of a tetrahedron. With the discovery of the electron, the electronic theory of the chemical bond became possible. In 1916, Lewis stated that the chemical bond between two carbon atoms in alkanes or between one carbon atom and a hydrogen atom consists of a pair of electrons held jointly by the two atoms connected by a single bond. Lewis also suggested that atoms tend to reach the electronic configuration of a noble gas through the sharing of electrons with other atoms or through electron transfer. Although this theory predated quantum mechanics, chemists nevertheless find it a useful model for the understanding of bonding in organic molecules. Dirac recognized at the time when all of chemistry could be understood quantitatively by the equations of quantum mechanics but could not predict the advent of powerful computers of today. In 1911, Rutherford established that atoms consisted of a diffuse cloud of electrons surrounding a small positively charged nucleus and proposed the planetary model atom. In 1913, Bohr proposed that electrons travel in certain orbits at discrete distances from the nucleus, with specific energies. The electrons can lose

Organic Chemistry: Theory, Reactivity and Mechanisms in Modern Synthesis, First Edition. Pierre Vogel and Kendall N. Houk.

or gain a fixed energy amount by jumping from one allowed orbit to another allowed orbit, absorbing or emitting electromagnetic radiation with a determined frequency v. This is the basis of atomic spectroscopy: atoms heated in a flame emit (Na emits yellow light, Sr emits red light, etc.) or absorb (flame atomic absorption spectroscopy established by Bunsen and Kirchhoff during the 1870s) [3].

The foundation of quantum chemistry is the wave model in which the positively charged nucleus of an atom is surrounded by a cloud of electrons moving in orbitals, and their positions are represented by probability distributions rather that discrete points. This model predicts the pattern of similar elements as found in the Mendeleev periodic table of elements. It models properties of electrons and of any other fundamental particles [4]. The mathematical basis of quantum chemistry was made by Dirac and Schrödinger in 1926. Heisenberg in 1932 and Dirac and Schrödinger in 1933 received Nobel Prizes for quantum mechanics. There is perhaps no more famous quote about quantum mechanics than Dirac's in 1929 (three years after his PhD): "The fundamental laws necessary for the mathematical treatment of a large part of physics and the whole of chemistry are thus completely known, and the difficulty lies only in the fact that application of these laws leads to equations that are too complex to be solved [5].

4.3 Schrödinger equation

The starting point of any theoretical description of the energy and electronic properties of a chemical system (stable or unstable species) with a given geometry is the Schrödinger equation:

$$\mathbf{H}\Psi(\mathbf{r}) = E\Psi(\mathbf{r}) \tag{4.1}$$

where $\Psi(\mathbf{r})$ is a stationary, electronic wave function, E is the electronic energy, and $\mathbf{H}$ is the Hamiltonian operator (*symbol* $\hat{H}$ also used) for a system assumed to be in a fixed nuclear configuration. $\hat{H}$ is given by Eq. (4.2):

$$\mathbf{H} = \sum_i \left(\frac{h^2}{2m_e} \nabla_i^2 \right) - \sum_i \sum_m \frac{Z_m e^2}{r_{im}} + \sum_{i<j} \sum \frac{e^2}{r_{ij}} \tag{4.2}$$

where the successive terms represent operators ($\nabla_i^2 = (\partial^2/\partial x_i^2) + (\partial^2/\partial y_i^2) + (\partial^2/\partial z_i^2)$) for the kinetic energy of the electrons (i) of mass m_e, the nuclear–electronic attraction (Z_m is the atomic number of nucleus m and r_{im} is the distance between electron i and nucleus m), and the repulsion between electrons (r_{ij} is the distance between electrons i and j). The wave function $\Psi(\mathbf{r})$, which is understood to depend on the position vectors ($\mathbf{r}$) of all particles, describes the state of the system. $\Psi(\mathbf{r})$ is the probability amplitude, whereas $\Psi(\mathbf{r}){\cdot}\Psi(\mathbf{r})\mathrm{d}\tau_1 \cdots \mathrm{d}\tau_i \cdots \mathrm{d}\tau_N$ ($\mathrm{d}\tau_i$ is the volume element of electron i) represents the probability distribution function for the N particles. For bound states, it is common to normalize the probability to unity.

$$\int \Psi(\mathbf{r}) \cdot \Psi(\mathbf{r}) \ldots \mathrm{d}\tau \ldots = <\Psi(\mathbf{r})^2> = 1 \quad \text{all space} \tag{4.3}$$

In 1927, Burrau presented a solution to the wave Eq. (4.1) for hydrogen molecule ion, H_2^+, which describes the nature of the one-electron bond in this two atomic, one electron species [6]. The same year, Heitler and London [7] proposed an approximate solution to Eq. (4.1) for H_2 and were able to explain for the first time the covalent bond. Their approximate solution requires the two electrons of H_2 to be separated, each being close to one of the two nuclei (protons). Simultaneously, Condon [8] presented another solution to the wave function (4.1) for H_2 that permits the electrons to be distributed between the two nuclei independently of one another, each occupying a wave function similar to Burrau's function for H_2^+. Condon's solution is the prototype of the molecular orbital treatment that will be used in this textbook and which is the basis of Pauling's work on the chemical bond [9, 10]. Pauling introduced in 1932 the concept of electronegativity and the theory of hybridization of atomic orbitals (AOs) [11].

For chemical systems with more than two electrons, the solution to Eq. (4.1) is mathematically impossible unless one makes assumptions. One of the earliest methods has been proposed in 1930 by Hückel for π-systems (Section 4.5) [12–15] and applied by several authors to solve all kind of chemical problems [16–18]. As computers have become more and more powerful, solutions of the Schrödinger equation for complex chemical systems are now possible; although exact analytical solutions are not possible, "chemical accuracy," which is solutions within the experimental error of measurements, often taken as 1 kcal mol^{-1}, is usually possible. One very much applied calculation method to relatively large molecules including organometallic systems is based on Kohn's density functional theory (DFT) of electronic structure [19]. In common calculations, the total energy is estimated by approximate solutions of the time-dependent Schrödinger Eq. (4.1), usually with no relativistic terms and by making use of the Born–Oppenheimer approximation, which separates electronic and atomic motions. Since the 1960s, semiempirical MO calculations (e.g. Hückel, extended Hückel theory, Pariser–Parr–Pople [PPP] treatments, complete neglect of differential overlap [CNDO], Dewar's modified intermediate neglect of differential overlap, version 3 [MINDO/3], modified neglect of differential overlap [MNDO], Austin model 1 [AM1], and now Stewart's parametric model numbers 3 and 6 [PM3 and PM6]) have become routine [20]. These semiempirical methods set various integrals to quantities that are not computed but are obtained from empirical experimental data. The first *ab initio* Hartree–Fock (HF) calculations on diatomic molecules were carried out in 1956 using Slater orbitals as the basis set [21]. *Ab initio* (from the beginning), solve approximations to the Schrödinger equation without recourse to experimental data. However, various approximations are made in order to make the computations feasible. *Ab initio* does not imply that the solution is an exact one because approximations are invariably made. The simplest type of *ab initio* calculation is the HF scheme, commonly known as molecular orbital theory. HF theory makes the assumption that the repulsion between an electron and all other electrons can be treated as the interaction of one electron with the average position of the others. This assumption leads to an error, which is known as electron correlation. Electron correlation is the difference between the HF energy using a very large basis set (mathematical representation of atomic orbitals) and the actual energy. It arises for the fact that electrons do correlate their motions to avoid each other, an energy-reducing effect that is missed when electron–electron repulsion is calculated based on the average position of electrons. Most approximate quantum mechanical theories of electronic structure are independent electron theories. Independent electron theories start off by assuming a solution Ψ of the Schrödinger Eq. (4.1) with precisely the form it would have if $\mathbf{H}$ were a sum of one-electron terms without any instantaneous correlation of the electron positions or electron motions. The full wave function, Ψ, is the product of these one-electron functions. Each of these one-electron functions is a molecular orbital. Mulliken defined an AO as an eigenfunction of one-electron Schrödinger equation for an atom. A MO is defined as an eigenfunction of one-electron Schrödinger equation, which is based on the attractions of two or more nuclei plus the average repulsions of the other electrons. An orbital (AO, MO) is a mathematical function in an ordinary three-dimensional space. When an electron is occupying an orbital, the shape of the orbital gives the fraction of time the electron spends in different regions of space around the nucleus, or nuclei [22, 23]. One of the most powerful methods of quantum mechanics is called the variation theorem and states the following: "Given any approximate wave function satisfying the boundary conditions of the problem, the expectation value of the energy calculated from this function will always be higher than the true energy of the ground state." This principle allows us to guess several functions called trial functions and calculate the expectation value of the energy for each function that can be obtained from the original guesses. One way to be more systematic is to start with a trial function containing several arbitrary parameters. The expectation value of the energy, E_i, is then calculated and is minimized with respect to the arbitrary parameters. In this way, a large number of "guesses" can be made with a single function. The resulting wave function is then the best available for the particular parametric form chosen. The usual way to carry out such a calculation is to make use of a linear variation function, and this is the basis for the "linear combination of basis functions" φ_k (Eq. (4.4)). The basis functions can be taken as atomic orbitals, or now commonly, several functions for each orbital plus other functions that do not necessarily resemble atomic orbitals; this approach is the linear combination of atomic orbitals or LCAO method introduced by Lennard-Jones in 1929 [24].

$$\Psi_i = \sum_k c_{ik}\varphi_k \tag{4.4}$$

As molecules are made up of atoms, it is fairly reasonable to assume that the electron distribution in a molecule can be approximately represented as a sum of atomic electron distributions. This is the physical basis for the molecular orbital or LCAO–MO method of solving quantum mechanical problems. In this method, one chooses as a trial function for a molecular orbital Ψ_i, a linear variation function (4.4) where the coefficients c_{ik} are the parameters that will be obtained so as to minimize the energy. For each set of c_{ik}, one obtains for a given molecule a value E_i. The expectation value of the energy, E_i, is given by the variation principle (4.5) (E_0 is the true ground-state energy). In order to obtain the best possible solution Ψ_i, one chooses the parameters c_{ik} in such a way that they give the smallest possible value for E_i. The minimal value of E_i will be obtained if the conditions (4.6) are satisfied (Figure 4.1).

$$E_i = \frac{\langle \Psi_i|\mathbf{H}|\Psi_i\rangle}{\langle \Psi_i/\Psi_i\rangle} \geq E_0 \tag{4.5}$$

$$\frac{\partial E_i}{\partial c_{il}} = \cdots = \frac{\partial E_i}{\partial c_{ik}} = \cdots = \frac{\partial E_i}{\partial c_{in}} = 0 \tag{4.6}$$

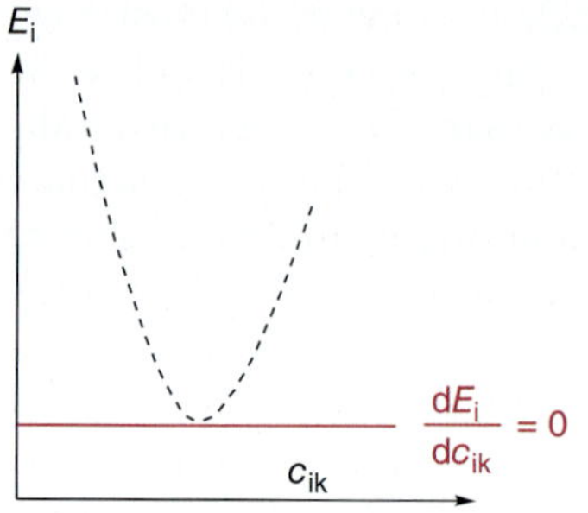

Figure 4.1 Searching for a "good" set of parameters c_{ik} corresponds to finding the minimal value of E_i, for which $\frac{\partial E_i}{\partial c_{ik}} = 0$.

By combining Eqs. (4.4) and (4.5), one obtains

$$E_i = \frac{\sum_k \sum_l c_{ik} c_{il} \langle \varphi_k | \mathbf{H} | \varphi_l \rangle}{\sum_k \sum_l c_{ik} c_{il} \langle \varphi_k / \varphi_l \rangle} \tag{4.7}$$

with $k = 1, 2, \ldots, n$ and $l = 1, 2, \ldots, n$.

The energy values E_i have the form (4.8) for which Eq. (4.9) can be written.

$$E_i = f_1 / f_2 \tag{4.8}$$

$$\frac{\partial E_i}{\partial c_{ik}} = \frac{1}{f_2}\frac{\partial f_1}{\partial c_{ik}} - \frac{f_1}{(f_2)^2}\frac{\partial f_2}{\partial c_{ik}} = 0 \tag{4.9}$$

By multiplying (4.9) with f_2, for all coefficients c_{ik}, one obtains equations

$$\frac{\partial f_1}{\partial c_{ik}} - E_i \frac{\partial f_2}{\partial c_{ik}} = 0 \tag{4.10}$$

We define the integrals $H_{kl} = \langle \varphi_k \mathbf{H} \varphi_l \rangle$ and $S_{kl} = \langle \varphi_k / \varphi_l \rangle$, using a notation proposed by Dirac. The brackets represent integrals over all space of the quantity inside the brackets. In semiempirical theory, these integrals can be determined by parametrization with experimental data (e.g. ionization energies, electronic transition energies, and heats of formation), but in *ab initio* theory, they are computed analytically or numerically without any parameters based on the experiment. We can now write the secular equations (4.11) with unknown coefficients c_{il}, whose nontrivial solutions can be obtained for values of E_i satisfying the condition of secular determinant (4.12):

$$\sum_l c_{il}(H_{kl} - E_i S_{kl}) = 0 \tag{4.11}$$

$$\begin{vmatrix} H_{11} - ES_{11} & H_{12} - ES_{12} & \cdots & H_{1n} - ES_{1n} \\ H_{21} - ES_{21} & H_{22} - ES_{22} & \cdots & H_{2n} - ES_{2n} \\ \vdots & & & \\ H_{n1} - ES_{n1} & H_{n2} - ES_{n2} & \cdots & H_{nn} - ES_{nn} \end{vmatrix} = 0 \tag{4.12}$$

By solving the secular determinant (4.12), one obtains n values E_i for the energy as a function of the integrals H_{kl} and S_{kl}. As the basis functions φ_k and φ_l are real, $H_{kl} = H_{lk}$ and $S_{kl} = S_{lk}$; the secular determinant (4.12) is thus a symmetrical determinant. Values for the coefficients c_{il} are then obtained by using Eqs. (4.11) and the normalization conditions (4.13).

$$\langle \Psi_i^2 \rangle = \left\langle \left(\sum_k c_{ik} \varphi_k \right)^2 \right\rangle = 1 \tag{4.13}$$

4.4 Coulson and Longuet-Higgins approach

The total energy of a given chemical system is the sum of nuclear, electronic, vibrational, rotational, and translational energies. In chemistry, we are mostly interested in the changes of these energies upon alteration of geometry, or when changing one molecule to another. Thus, instead of trying to approximate absolute properties of molecules, it is more reasonable to evaluate the differences between molecules of a given series of compounds. One should also recall that chemical reactions are in fact the result of only a small perturbation when compared with the total energy of the reacting molecules. A chemical system A–B (a stable molecule or a transition structure) can be considered as the result of the interactions between the two molecular subfragments or atoms A and B. If the properties of A and B are known, those of A–B would also be known if one could estimate the result of the mutual perturbations of A and B juxtaposed in the geometry of A–B. A perturbs B, and B perturbs A. We consider a diatomic molecule A–B made up of the two subfragments A˙ and B˙, which are hypothetical atoms with one electron only. Furthermore, we assume that the atomic orbital φ_a of energy H_{aa} is a good description of the properties of A˙ and the atomic orbital φ_b of energy H_{bb} is a good description of B˙. The interactions between A˙ and B˙ will thus be described by trial wave functions that are linear combinations of φ_a and φ_b. The interaction will depend on the overlap S_{ab} between orbitals φ_a and φ_b and on the exchange energy H_{ab} involving the interactions between these two orbitals. Overlap S_{ab} depends on the geometry of A–B (distance between A and B) and H_{ab} will be a function of S_{ab} and other factors [25, 26]. The secular determinant for the above system A–B can be written as

$$\begin{vmatrix} H_{aa} - ES_{aa} & H_{ab} - ES_{ab} \\ H_{ba} - ES_{ba} & H_{bb} - ES_{bb} \end{vmatrix} = 0 \tag{4.14}$$

For reasons of simplicity, we shall write $H_a = H_{aa}$, $H_b = H_{bb}$, $S_{ab} = S_{ba} = S$, $S_{aa} = S_{bb} = 1$ (normalization condition (4.13)), and $H_{ba} = H_{ab}$. Thus, the determinant (4.14) becomes

$$\begin{vmatrix} H_a - E & H_{ab} - ES \\ H_{ab} - ES & H_b - E \end{vmatrix} = 0 \quad (4.15)$$

or

$$(H_a - E)(H_b - E) - (H_{ab} - ES)^2 = 0 \quad (4.16)$$

or finally

$$E^2(1 - S^2) + E(2SH_{ab} - (H_a + H_b)) + H_{ab} - H_{ab}^2 = 0 \quad (4.17)$$

The roots of which are given below:

$$E_1 = \frac{-SH_{ab}}{1 - S^2} + \frac{H_a + H_b}{2(1 - S^2)} - D \quad (4.18)$$

$$E_2 = \frac{-SH_{ab}}{1 - S^2} + \frac{H_a + H_b}{2(1 - S^2)} + D \quad (4.19)$$

with $D = \sqrt{\left\{ \begin{array}{l} [2SH_{ab} - (H_a + H_b)]^2 \\ -4(1 - S^2) \cdot (H_a H_b - H_{ab}^2) \end{array} \right\} / [2(1 - S^2)]}$.

4.4.1 Hydrogen molecule

H_2 arises from the combination of two atoms of hydrogen, and H$^\cdot$, the Morse potential energy of H_2 shown in Figure 4.2, represents the energy of H_2 as a function of the distance separating the two nuclei (protons) [27, 28]. When the H/H distance is long, the energy corresponds to that of two separated hydrogen atoms. The "zero" of energy is generally taken to be that of nuclei plus electrons at infinity. The energy of $H^\cdot$ is given by the negative of the ionization energy of $H^\cdot$, i.e. $-EI(\mathrm{H}^\cdot) = \Delta_f H°(\mathrm{H}^\cdot) - \Delta_f H°(\mathrm{H}^+) = -13.7$ eV = -313.6 kcal mol^{-1} (Table 1.A.13, quantum physics uses Hartree as the atomic unit for energy; 1 Ha = 1 au = 27.21 eV = 4.36×10^{-18} J per atom = 627 kcal mol^{-1} [29]). In its most stable form, the bond distance H$\cdots$H in H_2 is 0.74 Å. For shorter H$\cdots$H distances, internuclear repulsion destabilizes H_2.

For molecular H_2, parameters $H_a = H_b = \alpha_H$ (determinant (4.15)) correspond to $-IE(\mathrm{H}^\cdot)$. $-IE(\mathrm{H}^\cdot)$ is the energy gained by a system made of a proton and an electron when both come from infinity to generate H$^\cdot$ atom. Thus, α_H is a Coulombic interaction called Coulomb integral (negative as one gains the electrostatic energy of the electron gravitating around a proton). Mulliken proposes the H_{ab} term of determinant (4.15) to be proportional to the overlap between the two atomic 1s(H) orbitals of the two hydrogen atoms [30]. Thus, $H_{ab} = kS$, the resonance integral, a negative terms as it corresponds to the energy gain of an electron circulating around two nuclei. S is positive, and k is negative. Determinant (4.15) can now be written as Eq. (4.20):

$$\begin{vmatrix} \alpha_H - E & kS - SE \\ kS - ES & \alpha_H - E \end{vmatrix} = 0 \quad (4.20)$$

$(\alpha_H - E)^2 - (kS - ES)^2 = 0$ for which one finds two roots: Eqs. (4.20) and (4.21) for E:

$$E_1 = \frac{\alpha_H + kS}{1 + S} \quad (4.21)$$

$$E_2 = \frac{\alpha_H - kS}{1 - S} \quad (4.22)$$

These correspond to the eigenfunctions or MOs that are the combinations of two 1s(H) atomic orbitals

$$\Psi_1 = c_{11}(1s(H)) + c_{12}(1s(H))$$
$$\Psi_2 = c_{21}(1s(H)) + c_{22}(1s(H))$$

with $c_{11} = c_{22} = \sqrt{1/2 + 2S}$ and $c_{21} = -c_{22} = \sqrt{1/2 + 2S}$.

This allows one to construct the MO diagram of Figure 4.3. Ψ_1 is the bonding MO, noted $\sigma_{H,H}$. Ψ_2 is

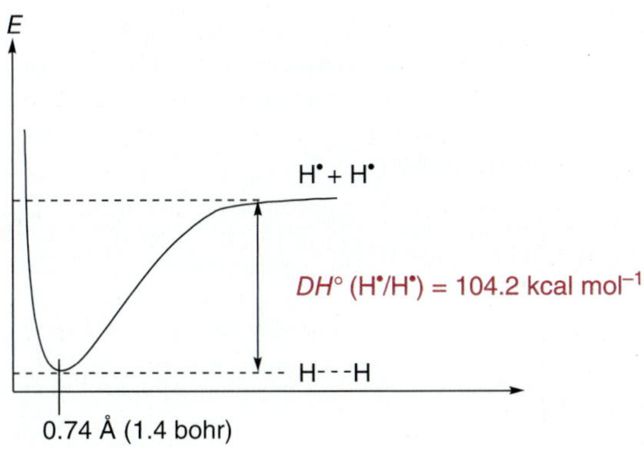

Figure 4.2 Potential energy profile of H_2 as a function of the separation between two H nuclei (Bohr radius: the smallest radius of 1s(H) orbital in the Bohr model of H$^\cdot$ atom 1 bohr = 0.053 nm = 0.53 Å).

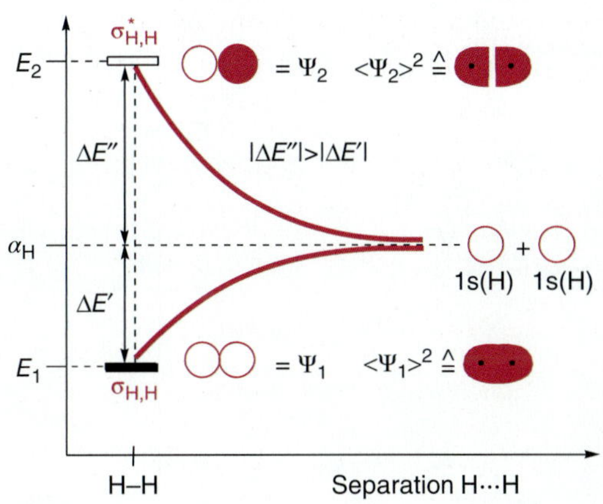

Figure 4.3 MO diagram for the combination of two H$^\cdot$ atoms into H_2 molecule.

the antibonding MO, noted $\boldsymbol{\sigma}^*_{\mathbf{H,H}}$. In $\boldsymbol{\sigma}_{\mathbf{H,H}}$, the contribution of the two 1s(H) atomic orbitals is in phase (white and white); in $\sigma^*_{\mathrm{H,H}}$, they are out of phase (white and red). It is important to realize that the energy difference (degree of bonding) $\Delta E' = |\alpha_{\mathrm{H}} - E_1|$ is smaller than the energy difference (degree of antibonding) $\Delta E'' = |E_2 - \alpha_{\mathrm{H}}|$. This arises from denominator $1 + S$ in Eq. (4.21), which is larger than denominator $1 - S$ in Eq. (4.22).

4.4.2 Hydrogenoid molecules: The PMO theory

Several techniques of semiempirical quantum chemistry (Hückel [12–18], CNDO [31], etc.) make the simplification, $S_{ij} = 0$ for $i \neq j$, the neglect of overlap, or of differential overlap, $\phi_i \phi_j \, d\tau = 0$. In this case, the energy difference $|H_\mathrm{a} - E_1| = |H_\mathrm{b} - E_2|$ (see Figure 4.4a). When $S \neq 0$ and $H_\mathrm{a} < H_\mathrm{b}$, Eqs. (4.40) and (4.41) show that the perturbation of orbital φ_a by φ_b leads to a new orbital $\Psi_\mathrm{a} = c_{\mathrm{aa}}\varphi_\mathrm{a} + c_{\mathrm{ab}}\varphi_\mathrm{b}$ of energy E_1, which is lower than that of φ_a by the value $|H_\mathrm{a} - E_1|$. At the same time, orbital φ_b becomes a new orbital $\Psi_\mathrm{b} = c_{\mathrm{ba}}\varphi_\mathrm{a} + c_{\mathrm{bb}}\varphi_\mathrm{b}$ of energy E_2, which is higher in energy than φ_b by the value $|H_\mathrm{b} - E_2|$. The energy difference $|H_\mathrm{b} - E_2|$ is larger than $|H_\mathrm{b} - E_1|$ because of the fact that $S \neq 0$ (see Figure 4.4b).

The coefficients c_{aa}, c_{ab}, c_{ba}, and c_{bb} of the resulting orbitals Ψ_a and Ψ_b are obtained by solving the secular Eqs. (4.23) and (4.24) and the normalization Eqs. (4.25) and (4.26).

$$c_{\mathrm{aa}}(H_\mathrm{a} - E_1) + c_{\mathrm{ab}}(H_{\mathrm{ab}} - E_1 S) = 0 \tag{4.23}$$

$$c_{\mathrm{ba}}(H_{\mathrm{ab}} - E_2 S) + c_{\mathrm{bb}}(H_\mathrm{b} - E_2) = 0 \tag{4.24}$$

$$\Psi_\mathrm{a}^2 = c_{\mathrm{aa}}^2 + c_{\mathrm{ab}}^2 + 2c_{\mathrm{aa}}c_{\mathrm{ab}}\, S = 1 \tag{4.25}$$

$$\Psi_\mathrm{b}^2 = c_{\mathrm{ba}}^2 + c_{\mathrm{bb}}^2 + 2\, c_{\mathrm{ba}}c_{\mathrm{bb}}\, S = 1, \quad \text{with } S = \varphi_\mathrm{a} \cdot \varphi \tag{4.26}$$

When overlap is neglected ($S = 0$; $H_\mathrm{a} - E_1 = E_2 - H_\mathrm{b}$), then

$$c_{\mathrm{aa}}(H_\mathrm{a} - E_1) = -c_{\mathrm{ab}}H_{\mathrm{ab}} \text{ or } c_{\mathrm{aa}} \cong -c_{\mathrm{ab}}H_{\mathrm{ab}}/(H_\mathrm{a} - E_1)$$

$$c_{\mathrm{bb}}(H_\mathrm{b} - E_2) = -c_{\mathrm{ba}}H_{\mathrm{ab}} \text{ or } c_{\mathrm{bb}} \cong c_{\mathrm{ba}}H_{\mathrm{ab}}/(E_2 - H_\mathrm{b})$$

$$c_{\mathrm{aa}}^2 + c_{\mathrm{ab}}^2 = 1$$

$$c_{\mathrm{ba}}^2 + c_{\mathrm{bb}}^2 = 1$$

If φ_a is a 1s(H) orbital of a hydrogen atom (AO) and φ_b a sp^3 hybridized orbital (AO) of the unshared electron of an alkyl radical, MOs Ψ_a and Ψ_b will have the approximate shapes represented below and can be defined as a bonding σ_{CH} orbital and an antibonding σ^*_{CH} orbital. The coefficients c_{aa} and c_{ab} have the same sign because H_{ab} is a negative energy value (when an electron circulates about two nuclei, it is stabilized). In contrast, c_{bb} and c_{ba} have opposite signs for the same reason.

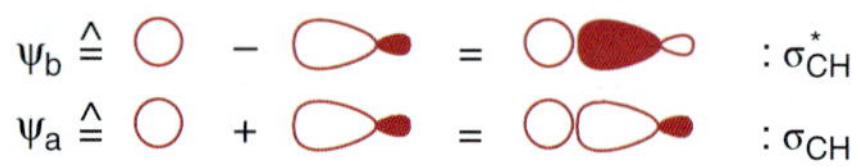

If φ_a and φ_b of subfragments A˙ and B˙ are each populated with one electron, the energy of bond AB is calculated to be proportional to $H_\mathrm{a} - E_1 + H_\mathrm{b} - E_1$.

Because of the Pauli exclusion principle [32] proposed by Wolfgang Pauli in 1925, two identical fermions (particles with half-integer spin) may not occupy the same quantum state simultaneously; consequently, the overlap of occupied orbitals leads to Pauli repulsion or the so-called closed shell repulsion (4.27).

$$\Delta E(\text{four electrons}) = 2(E_1 + E_2) - 2(H_\mathrm{a} + H_\mathrm{b}) \tag{4.27}$$

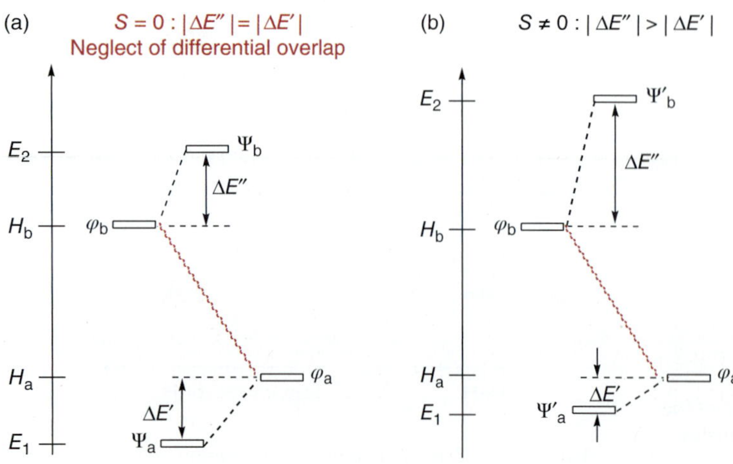

Figure 4.4 The interactions between two nondegenerate orbitals φ_a and φ_b that overlap and lead to the interaction energy H_{ab}. (a) results for E_1 and E_2 when $S = 0$ and (b) when $S \neq 0$ in the calculations.

By considering the values calculated for E_1 and E_2 (Eqs. (4.18) and (4.19)), one obtains

$$\Delta E\,(4\,e^-) = \frac{S(-2\,H_{ab} + S(H_a + H_b))}{1 - S^2} \quad (4.28)$$

$\Delta E(4e^-)$ corresponds to the repulsion energy between the two electron pairs on A: and B:, which occupy φ_a and φ_b, respectively. In the case of the zero overlap approximation (ZOA), $\Delta E(4e^-)$ is calculated to be zero [31, 33].

If φ_a is doubly occupied and φ_b is empty, the change in energy associated with the interaction of A: and B in molecule A: → B will be given by $2\Delta E' = 2(E_1 - H_a)$ (Figure 4.1). It corresponds to a stabilization energy. In general, interaction of an occupied orbital of a molecule or molecular fragment with an empty orbital of another molecule or molecular fragment leads to a stabilization of the system. From Eq. (4.16), one can write for $\Delta E'$:

$$\Delta E' = H_a - E_1 = (H_{ab} - E_1 S)^2/(H_b - E_1) \quad (4.29)$$

If the perturbation of φ_a by φ_b is not too strong, i.e. if $H_a \approx E_1$ and $H_b \approx E_2$, Eq. (4.29) can be approximated by (this is the so-called second-order perturbation equation):

$$\Delta E' = H_a - E_1 \cong \frac{(H_{ab} - H_a S)^2}{H_b - H_a} \quad (4.30)$$

Similarly, for the destabilization energy $\Delta E'' = E_2 - H_b$ when going from orbital φ_b to ψ_b, one obtains

$$\Delta E'' = E_2 - H_b \cong \frac{(H_{ab} - H_b S)^2}{H_a - H_b} \quad (4.31)$$

Equations (4.30) and (4.31) demonstrate that the result of the interaction between fragment orbitals φ_a and φ_b leads to a system of orbitals ψ_a and ψ_b whose energy differences $\Delta E'$ and $\Delta E''$ depend on the square of the interaction energy (resonance integral) H_{ab} (itself dependent on the overlap S between φ_a and φ_b) and the inverse of the energy difference $H_b - H_a$ between orbitals φ_a and φ_b. Eqs. (4.30) and (4.31) are expressions of the second order in the theory of perturbation of molecular orbitals (PMOs) [34–36]. If atomic orbitals φ_a and φ_b are replaced by molecular orbitals ψ_μ and ψ_ν of polyatomic M and N, the PMO theory gives

$$\delta\varepsilon \cong \frac{[H_{\mu\nu}]^2}{E_\mu - E_\nu} \quad (4.32)$$

for the change in energy, $\delta\varepsilon$, associated with the interaction of these orbitals. $H_{\mu\nu}$ is the interaction between MO ψ_μ and ψ_ν of energy E_μ and E_ν, respectively. Within the ZOA and considering the MOs $\psi_\mu = \sum_k c_{\mu k}\varphi_k$ and $\psi_\nu = \sum_l c_{\nu l}\varphi_l$, Eq. (4.32) can be written as

$$\delta\varepsilon \cong \frac{\sum_k \sum_l (c_{\mu k} c_{\nu l}\,\gamma_{kl})^2}{E_\mu - E_\nu} \quad (4.33)$$

γ_{kl} is the interaction energy between the atomic orbitals φ_k and φ_l of MOs ψ_μ and ψ_ν, which overlap in the chemical system M···N. The theory is valid when M and N are radicals, closed-shell molecules, charged molecules, long-lived or short-lived species, transition structures, etc. Although qualitative, the PMO theory and Eqs. (4.32) and (4.33) can be used to rationalize a great deal of organic [37, 38] and organometallic chemistry [39], as illustrated in the forthcoming chapters.

4.5 Hückel method

In this theory, the physicist Erich Hückel made simplifications [12–15] that permit an easy resolution of the nonrelativistic Schrödinger equation (4.1) exclusively for planar π-systems for which the underlying σ-skeletons are considered to be part of the nuclei [19, 40]. Although the method was developed in the 1930s in order to make it possible to compute approximate MOs without the aid of computers, nowadays, the Hückel molecular orbitals (HMOs) of even gigantic π-systems can be computed in fractions of seconds with programs such as simplified Hückel molecular orbital theory (SHMO, which can be used online or downloaded to your laptop). In the Hückel approximation, the electrons in σ-MOs do not interact with those located in the π-systems because they are in MOs that are orthogonal to each other (their product $<\sigma> \cdot <\pi> = 0$ because they differ in symmetry). Thus, in HMO theory, methyl cation, radical, and anion are described by the same AO 2p(C) or 2p. This orbital is vacant in Me^+, singly occupied in $Me^{\bullet}$, and doubly occupied in Me^-.

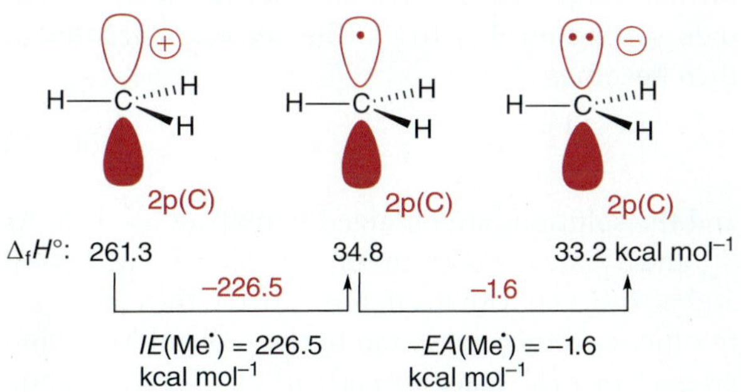

The Coulomb integral associated (H_{aa} or H_{11}) with this 2p(C) orbital is noted as α_C and corresponds to $-EI(\text{Me}^{\cdot})$ (see Tables 1.A.13 and 1.A.14). Obviously, the energy gained (Coulomb integral) on combining an electron and methyl cation to form methyl radical ($-226.5\ \text{kcal mol}^{-1}$) is much higher than that gained by adding a second electron to generate methyl anion ($-1.6\ \text{kcal mol}^{-1}$). This demonstrates that the two electrons in the 2p(C) AO of methyl anion suffer from a significant interelectronic repulsion, thus $\alpha_C(\text{Me}^-) \gg \alpha_C(\text{Me}^{\cdot})$. This will have to be considered when discussing the thermochemical and electronic properties of carbanionic π-systems. In HMO theory, where electron repulsion is neglected, the energy to put an electron into a 2p(C) orbital is α_C regardless of the occupation of the orbital p, obviously a rather drastic assumption! In spite of its inaccuracy, this assumption allows us to do a great many computations that give good qualitative pictures of π-systems.

4.5.1 π-Molecular orbitals of ethylene

A second approximation in HMO theory is to consider the resonance integral between two contiguous sp²-hybridized carbon centers $H_{12} = H_{21} = \beta_{cc}$ and the same in all systems that do not contain heteroatoms, and zero for two noncontiguous sp²-hybridized centers. Furthermore, the differential overlap $S_{i,j} (i \neq j)$ is set to zero (called neglect of overlap). For ethylene, the secular determinant (4.12) becomes

$$\begin{vmatrix} H_{11} - ES_{11} & H_{12} - ES_{12} \\ H_{21} - ES_{21} & H_{22} - ES_{22} \end{vmatrix} = \begin{vmatrix} \alpha_c - E & \beta_{cc} \\ \beta_{cc} & \alpha_c - E \end{vmatrix} = 0 \tag{4.34}$$

We simplify $\alpha_c = \alpha$ and $\beta_{cc} = \beta$, thus Eq. (4.34) becomes Eq. (4.34′).

$$\begin{vmatrix} \alpha - E & \beta \\ \beta & \alpha - E \end{vmatrix} = 0 \tag{4.34′}$$

In some texts, and with the online program SHMO mentioned earlier, the secular determinant is even further simplified by dividing each term by β, and then setting $\alpha - E/\beta$ to X. The secular determinant then becomes

$$\begin{vmatrix} X & 1 \\ 1 & X \end{vmatrix} = 0 \tag{4.34″}$$

and the solutions are obtained in units of $\alpha - E/\beta$. As S_{11} and S_{22} are the overlaps of AOs 2p(C(1)) with itself and of 2p(C(2)) with itself, respectively, they are equal to unity as they correspond to the probability to find an electron in all space around carbon C(1) and carbon C(2), respectively. One finds two roots to Eq. (4.34′) that are

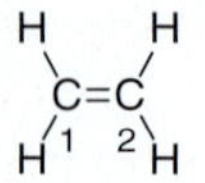

$$\begin{cases} E_1 = \alpha + \beta \\ E_2 = \alpha - \beta \end{cases}$$

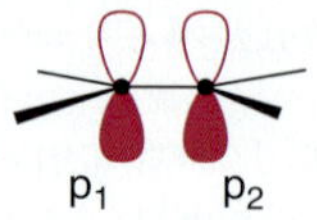

The secular equations for this system become

$$\begin{cases} c_{11}(H_{11} - ES_{11}) + c_{12}(H_{12} - ES_{12}) \\ \quad = c_{11}(\alpha - E_1) + c_{12}(\beta) = c_{11}(-\beta) + c_{12}(\beta) = 0 \\ c_{21}(H_{21} - ES_{21}) + c_{22}(H_{22} - ES_{22}) \\ \quad = c_{21}(\beta) + c_{22}(\alpha - E_2) = c_{21}(\beta) + c_{22}(\beta) = 0 \end{cases}$$

If one considers the normalization conditions and p_1 and p_2 the 2p(C) orbital residing or carbon C(1) and C(2), respectively.

$$\begin{cases} <\pi_2>^2 =< c_{11}p_1 + c_{12}p_2>^2 = 1 \\ <\pi_2>^2 =< c_{21}p_1 + c_{22}p_2>^2 = 1 \end{cases}$$

Assuming terms $2c_{11} \cdot c_{12}(p_1 \cdot p_2) = 2c_{21} \cdot c_{22}(p_1 \cdot p_2) = 0$ (ZOA), one obtains

$$\begin{cases} c_{11}^2 + c_{12}^2 = 1 \\ c_{21}^2 + c_{22}^2 = 1 \end{cases}$$

which gives coefficients $c_{11} = c_{12} = c_{21} = 0.707\ (1/\sqrt{2})$ and $c_{22} = -0.707$.

Ethylene is described (Figure 4.5b) by the π-MOs π and π^*. π is the bonding orbital, the highest occupied molecular orbital (HOMO), and π^* is the antibonding orbital, the lowest unoccupied molecular orbital (LUMO).

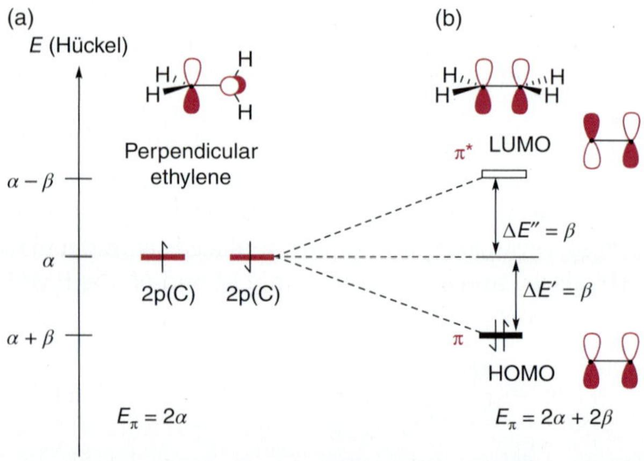

Figure 4.5 MO diagram for (a) perpendicular ethylene (transition structure of the (Z) ⇄ (E)-alkene isomerization) and for (b) parallel ethylene (and alkyl substituted alkenes).

The energy gained on combining two AOs p_1 and p_2 of ethylene is the π-energy of the system $E_\pi = 2$(energy of MO π) − 2(energy of 2p(C)) $= 2(\alpha + \beta) - 2\alpha = 2\beta$. In the transition structures of the monomolecular reactions that isomerize (*Z*)-alkenes into their (*E*)-stereoisomers, the two 2p(C) AOs are perpendicular. Thus, their $E_\pi = 2\alpha$ (Figure 4.5a). The experimental energy barrier for the isomerization of (*Z*)-1,2-dideuteroethylene into (*E*)-1,2-dideuteroethylene amounts to about 65 kcal mol^{-1} [41] β can thus be parameterized to c. −32 kcal mol^{-1}. For twisted ethylene- and alkyl-substituted derivatives, $H_{12} - ES_{12} = H_{21} - ES_{21}$ terms vary between 0 and β for torsional angle varying from 90 to 0 °C. Twisting an alkene decreases the $|\Delta E'|$ and $|\Delta E''|$ values.

4.5.2 Allyl cation, radical, and anion

The allyl systems, cation, radical, and anion, result from the combination of three 2p(C) orbitals p_1, p_2, and p_3 residing on carbon centers C(1), C(2), and C(3), respectively. These three AOs (basis orbitals) will generate three MOs π_1, π_2, and π_3 upon solving the Schrödinger Eq. (4.1) that has the form:

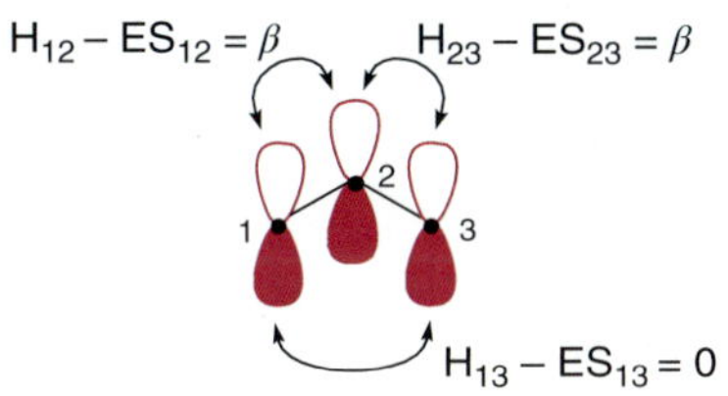

$$\begin{cases} \pi_1 = c_{11}p_1 + c_{12}p_2 + c_{13}p_3 \\ \pi_2 = c_{21}p_1 + c_{22}p_2 + c_{23}p_3 \\ \pi_3 = c_{31}p_1 + c_{32}p_2 + c_{33}p_3 \end{cases} \quad (4.35)$$

The secular determinant for these systems is

$$\begin{vmatrix} \alpha - E & \beta & 0 \\ \beta & \alpha - E & \beta \\ 0 & \beta & \alpha - E \end{vmatrix} = 0 \quad (4.36)$$

In the Hückel method, the off-diagonal term of the secular determinant $H_{ij} - ES_{ij}$ for noncontiguous centers is set equal to zero and, as for ethylene and all conjugated π-systems, the $H_{ij} - ES_{ij}$ terms for two contiguous carbon centers are all taken as $\beta_{cc} = \beta$.

This leads to three roots for the energy (see the MO diagram of Figure 4.6a):

$$E_1 = \alpha + \sqrt{2}\beta = \alpha + 1.41\beta$$
$$E_2 = \alpha$$
$$E_3 = \alpha - 1.41\beta$$

The three secular equations become

$$\begin{Bmatrix} c_{11}(\alpha - E_1) + c_{12}(\beta) + c_{13}(0) = 0 \\ c_{21}(\beta) + c_{22}(\alpha - E_2) + c_{23}(\beta) = 0 \\ c_{31}(0) + c_{32}(\beta) + c_{33}(\alpha - E_3) = 0 \end{Bmatrix} \quad (4.37)$$

Exploiting the C_{2V} symmetry of parallel allyl systems, one can write

$$\begin{Bmatrix} c_{11} = c_{13} \\ c_{21} = c_{23} \\ c_{31} = c_{33} \end{Bmatrix} \quad (4.38)$$

The normalization conditions give three further equations; assuming ZOA ($p_ip_j = 0$ if $i \neq j$):

$$\begin{Bmatrix} c_{11}^2 + c_{12}^2 + c_{13}^2 = 1 \\ c_{21}^2 + c_{22}^2 + c_{23}^2 = 1 \\ c_{31}^2 + c_{32}^2 + c_{33}^2 = 1 \end{Bmatrix} \quad (4.39)$$

The nine equations permit us to solve the nine unknowns that all the coefficients for the three MOs of the allyl systems. This gives

$$\begin{cases} \pi_1 = 0.5\,p_1 + 0.707p_2 + 0.5p_3 \\ \pi_2 = 0.707p_1 - 0.707p_3 \\ \pi_3 = 0.5p_1 - 0.707p_2 + 0.5p_3 \end{cases}$$

The E_π difference between allyl⊥ and allyl// systems gives an estimate of the π-conjugation energy gained on combining an ethylene unit and a 2p(C) moiety in a parallel manner. If one assumes $\beta = -32$ kcal mol^{-1} (as estimated from the rotation barrier in ethylene), the Hückel method predicts the π-resonance energies in allyl cation, radical, and anion to be the same (0.82β) and to amount to −26 kcal mol^{-1} (Figure 4.6b). To a first approximation, the π-resonance energy (RE) of allyl cation can be defined as the difference between the heterolytic dissociation enthalpies of propane and propene; these heterolytic dissociations give the hydride anion and either *n*-propyl or allyl cations (hydride affinities), respectively. Table 1.A.14 gives the data that are reproduced below.

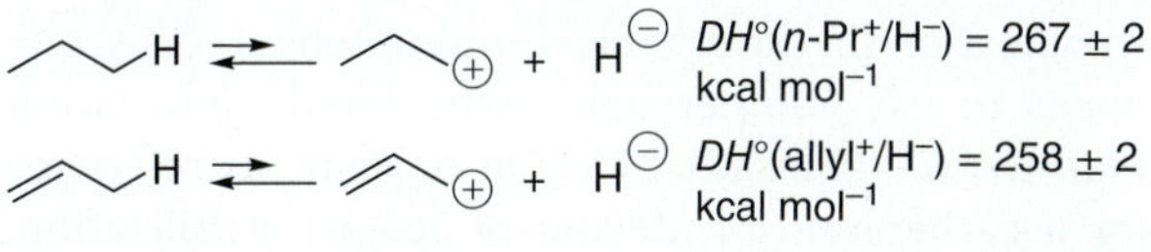

The difference of 9 ± 4 kcal mol^{-1} is much smaller than that predicted by the Hückel method (−26 kcal mol^{-1}). The reason for this is that *n*-propyl cation is not a good model for perpendicular allyl cation.

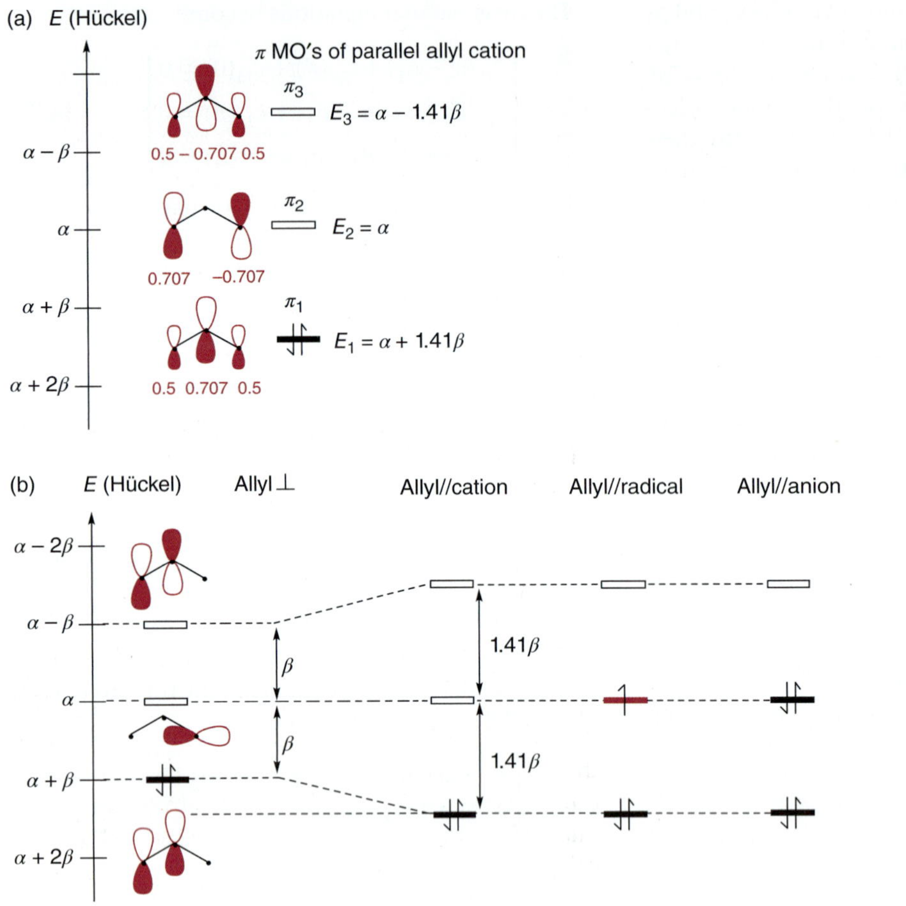

	Allyl//cation	Allyl//radical	Allyl//anion
E_π (allyl//):	$2\alpha + 2.82\beta$	$3\alpha + 2.82\beta$	$4\alpha + 2.82\beta$
for perpendicular allyl systems: E_π (allyl ⊥):	$2\alpha + 2\beta$	$3\alpha + 2\beta$	$4\alpha + 2\beta$
energy of π-resonance in allyl//:	0.82β	0.82β	0.82β

Figure 4.6 (a) MO diagram presenting the Hückel π-MOs of allyl systems in which the three 2p(C) AOs are parallel (allyl//). (b) Perpendicular allyl systems (allyl⊥) taken as the combination of an ethylene moiety with a 2p(C) system, the AOs of the latter being orthogonal to the 2p(C) AO′ of the ethylene moiety. The E_π(allyl⊥) is thus the sum of E_π(ethylene) + E_π(2p(C) system) as these two systems do not perturb each other ($1.41 \cong \sqrt{2}$; $0.707 \cong 1/\sqrt{2}$).

n-Propyl cation is stabilized by C—C—C bond angle deformation that improves hyperconjugative stabilization of the primary carbenium ion (enhanced σ_{CC}/2p(C$^+$) interaction) on the one hand and by interaction of one of the σ_{CH} bonds at C(3) with the empty 2p(C$^+$) AO (formation of an intramolecular σ-complex between the carbenium ion center and this C—H bond) on the other hand. The *n*-propyl cation has a geometry quite different from that of propane [42]. It is said to benefit from nonvertical stabilization (in the Franck–Condon sense). Thus, one needs to use another heterolytic bond dissociation than the C—H bond at C(1) of propane as reference for a carbenium ion devoid of special stabilization. Data given below (Table 1.A.14) for cyclopentane and cyclopentene gives -25 ± 4 kcal mol^{-1} for the stabilization gained by the secondary carbenium ion because of π-conjugation. This is in line with the Hückel prediction.

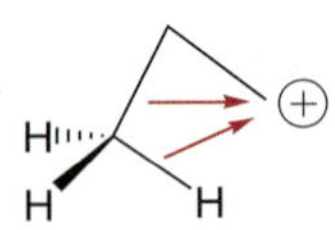

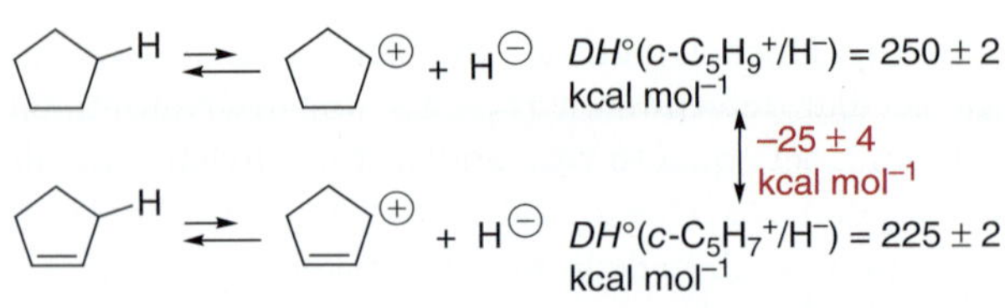

To a first approximation, the stabilization of allyl radical due to π-conjugation can be given by the difference between the homolytic bond dissociation enthalpies of the σ_{CH} bond at C(1) of propane and of the σ_{CH} bond at C(3) of propene. Table 1.A.7 gives the data that are repeated below:

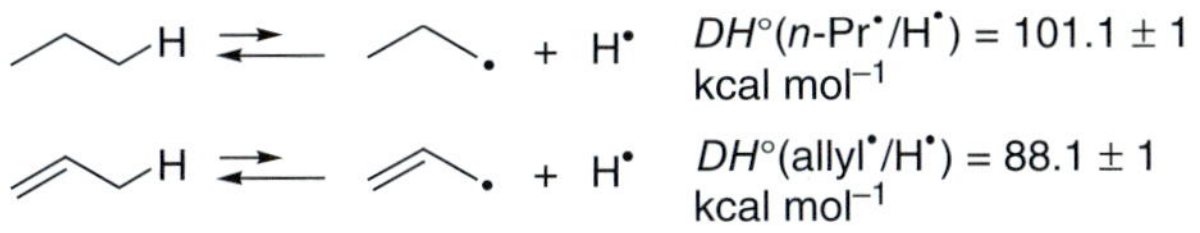

These data allow a first estimate for the π conjugative stabilization in the allyl radical of -13 ± 2 kcal mol^{-1}, a value also much smaller than that predicted (-26 kcal mol^{-1}) by the Hückel method (Figure 4.6B). This discrepancy arises, in part, from the ZOA applied by the Hückel method. If $S_{12} \neq 0$, one expects $|\Delta E''| > |\Delta E'|$ (Figure 4.7A). Thus, when combined with an ethylene π-system, a carbon-centered radical has a 2p(C) orbital singly occupied SOMO(R$^•$) (SOMO, singly occupied molecular orbital) of energy α that lies closer to π (the HOMO) than π^* (the LUMO of the vinyl substituent of the allyl radical). According to the expression (4.33) obtained for the perturbation of two orbitals of different energies, the prediction is that SOMO(R$^•$) is more perturbed by the HOMO(vinyl) than by LUMO(vinyl) as these perturbations depend on the inverse of $|\Delta E'|$ and $|\Delta E''|$ [$\Delta E' = \alpha - E$(HOMO(vinyl)); $\Delta E'' = \alpha - E$(LUMO(vinyl))]. When allyl⊥ radical becomes allyl// radical, SOMO(R$^•$) interacts with HOMO(vinyl), generating two orbitals $\psi_1' =$ in-phase combination of π and 2p(C) and $\psi_2' =$ out-of-phase combination of π and 2p(C). The change in energy associated with these perturbations are $\delta\varepsilon_1$ (stabilization of the HOMO(vinyl)) and $\delta\varepsilon_2$ (destabilization of the radical SOMO(R•)). During the formation of the allyl// radical, the 2p(C) orbital also interacts with π^*, the LUMO(vinyl). The result of this interaction can be seen as to arise from the overlap of ψ_2' with LUMO(vinyl). This generates two new MOs, ψ_2 being an in-phase combination of ψ_2' and π^* that is stabilized with respect to ψ_2' by the energy value $\delta\varepsilon_3$ and ψ_3 being an out-of-phase combination of ψ_2' and π^* that is destabilized by $\delta\varepsilon_4$ with respect to π^*, LUMO(vinyl). Finally, π^* also interacts slightly with ψ_1' to generate ψ_3 and ψ_1. ψ_1 is stabilized by the energy value $\delta\varepsilon_5$ with respect to ψ_1'. In the same time, ψ_3 is destabilized by $|\delta\varepsilon_4| + |\delta\varepsilon_6|$ with respect to LUMO(vinyl). In this treatment, one finds that the two electrons residing in HOMO(vinyl) are stabilized by $|\delta\varepsilon_1| + |\delta\varepsilon_5|$ on forming the allyl// radical. In the same time, the electron of the methyl radical is destabilized by $\delta\varepsilon_2$ but stabilized by $\delta\varepsilon_3$. As SOMO(R$^•$) is much closer in energy to HOMO(vinyl) than to LUMO(vinyl), one expects $|\delta\varepsilon_2| > |\delta\varepsilon_3|$. Thus, contrary to the Hückel method that calculates SOMO(Me$^•$) and SOMO(allyl$^•$) to have the same energy α, the π-allyl conjugation (delocalization) destabilizes the radical electron by $|\delta\varepsilon_2| - |\delta\varepsilon_3|$. The stabilization of 0.82β due to π-conjugation and calculated by the Hückel method must be corrected by the destabilization $|\delta\varepsilon_2| - |\delta\varepsilon_3|$. This explains why the experimental π-conjugation (-13 ± 2 kcal mol^{-1}) is only one half the value calculated (-26 kcal mol^{-1}) by the Hückel method.

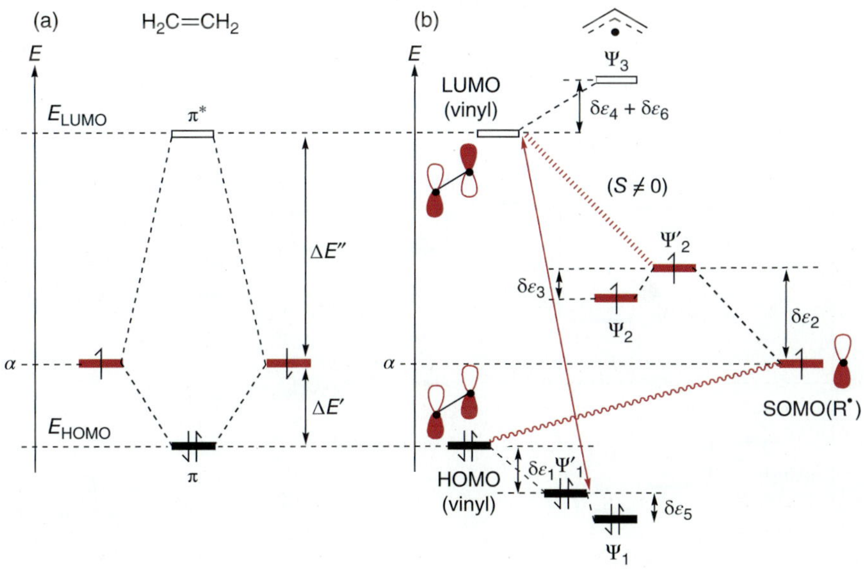

Figure 4.7 (a) PMO diagram representing the π-orbitals of ethylene as a combination of two 2p orbitals of adjacent sp^2 carbon atoms, including $S_{1,2} \neq 0$; (b) PMO diagram, representing the interactions between the π-orbitals of the vinyl substituent in the allyl// radical with the unshared electron of a methyl radical localized in the SOMO(R$^•$) (SOMO, singly occupied molecular orbital).

The PMO diagram represented in Figure 4.7b for the allyl radical is also valid for the allyl cation. The energy of an empty orbital, in this case, the $2p(C^+)$ orbital of the methyl cation, corresponds to the energy that an electron would have when occupying that orbital. Thus, the π-RE in the allyl cation would correspond to $2(|\delta\varepsilon_1 + \delta\varepsilon_5|)$, which is evaluated experimentally to be $-26 \pm 4\,\text{kcal mol}^{-1}$. Accordingly, as the π-resonance stabilization energy in the allyl radical is about $-13 \pm 2\,\text{kcal mol}^{-1}$, the energy difference $|\delta\varepsilon_2| - |\delta\varepsilon_3|$ in Figure 4.7b must be the order of $-13\,\text{kcal mol}^{-1}$.

The simple HMO diagram predicts that the resonance energies of allyl cation, radical, and anion are all the same, as the difference between change from ethylene plus methyl cation, radical, or anion is just the stabilization of the two electrons in the lowest π-orbital. Experimentally, the following gas-phase proton affinities have been reported [43a]: $DH°(\text{allyl}^-/\text{H}^+) = 391.1 \pm 0.3\,\text{kcal mol}^{-1}$ and $DH°(n\text{-propyl}^-/\text{H}^+) = 416\,\text{kcal mol}^{-1}$. This give an estimate of $-25\,\text{kcal mol}^{-1}$ for the π-allylic stabilization energy in allyl anion, a value similar to the π-allylic stabilization energy in allyl cation. We can consider the interaction of the methyl carbanion and an ethylene π-system with the PMO diagram shown in Figure 4.8. Here, the methyl carbanion π-orbital is high in energy, and there is significant stabilization arising from its interaction with the ethylene π^* orbital.

High accuracy calculations on stabilization energies of the allyl cation, radical, and anion, show how the effect of electron repulsion, neglected in Hückel theory, influences resonance stabilization and the rotational barriers about the C_1–C_2 bond in the cation,

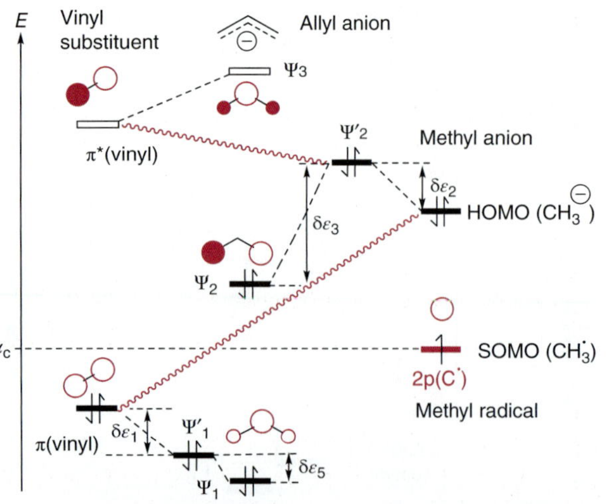

Figure 4.8 PMO diagram representing the π-MOs of the allyl carbanion (Ψ_1, Ψ_2, and Ψ_3) as a result of the interaction between the π-MOs of ethylene and the HOMO of methyl carbanion.

radical, and anion. The respective rotational barriers are 33, 15, and 21 kcal/mol. The result for allyl radical indicates the case where there are not large changes in electron repulsion in the planar p system and the transition state for rotation. In the cation, the 2p electrons are spread over 3 atoms, but in the transition state for rotation, these electrons are localized to two carbons, substantially increasing the electron repulsion. In the anion, the 4 electrons are distributed over 3 atoms in the planar minimum and two electrons become more localized on one carbon in the transition state [43b].

4.5.3 Shape of allyl π-molecular orbitals

The Hückel method applied to an allyl system gives the three MOs π_1, π_2, and π_3 shown in Figure 4.6. In π_1, there is an electron concentration around center C(2) and an electron depletion around this center in π_2. If π_3 were occupied, the electron density would concentrate on C(2) in a antibonding way with the electron densities around centers C(1) and C(3). The general shapes of π_1, π_2, and π_3 imitate the amplitude of vibration of a vibrating string, π_1 with the maximum of amplitude in the middle of the string (tone), one node in the middle of the string for π_2 (first harmonic), and with two nodes in π_3 (second harmonic; Figure 4.9). The π_1, π_2, and π_3 orbitals of allyl cation can be derived, for instance, by considering the interaction of a $2p(C^+)$ of methyl cation and the two π-MOs of an ethylene unit (π, π^*). The LUMO(Me^+)/π ethylene interaction generates the two combinations Ψ'_1 and Ψ'_2 (Figure 4.9). Ψ'_1 and Ψ'_2 are interacting with π^*(ethylene) generating π_1, π_2, and π_1. Upon going from π(ethylene) to π_1, the bonding character (sum of coefficient products $c_{11}{\cdot}c_{12} + c_{12}{\cdot}c_{13}$) is increased. In the case of π_2, which has in allyl cation the energy α, the same Hückel energy as $2p(C^+)$, this orbital is nonbonding ($c_{21}{\cdot}c_{22} + c_{22}{\cdot}c_{23} = 0$).

4.5.4 Cyclopropenyl systems

The Hückel method gives for the cyclopropenyl systems three MOs formed from the three π-MOs: The secular determinant becomes

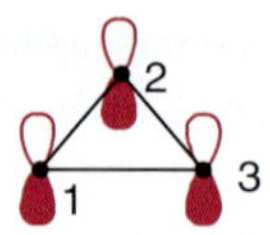

$$\begin{vmatrix} \alpha - E & \beta & \beta \\ \beta & \alpha - E & \beta \\ \beta & \beta & \alpha - E \end{vmatrix} = 0 \Rightarrow \begin{vmatrix} E_1 = \alpha + 2\beta \\ E_2 = \alpha - \beta \\ E_3 = \alpha - \beta \end{vmatrix}$$

Terms $H_{13} - ES_{13} = H_{31} - ES_{31} = \beta$ as C(1) and C(3) are contiguous. This was not the case in the allyl system.

Figure 4.9 The π_1, π_2^*, and π_3^*-MOs of allyl cation can be derived from the interaction of π, π^* (ethylene) with 2p(C⁺) of methyl cation.

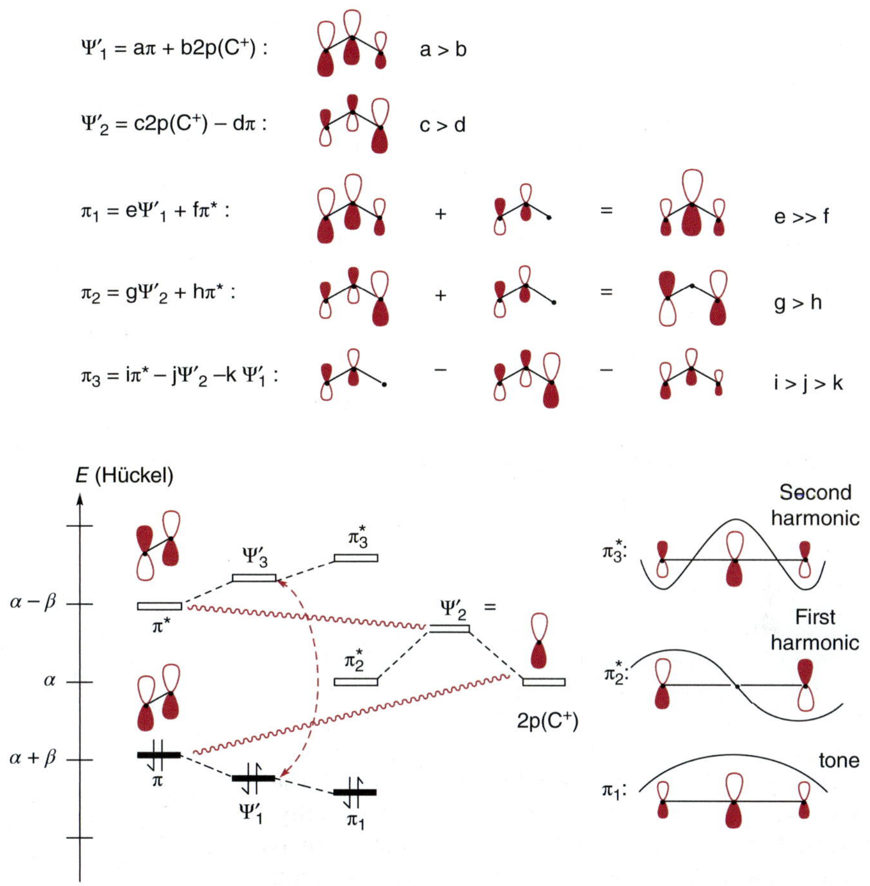

Using the roots E_1, E_2, and E_3 of the above determinant in the three secular equations, using the three normalization conditions $<\pi_1>^2 = <\pi_2>^2 = <\pi_3>^2 = 1$ and the symmetry of the cyclopropenyl system ($c_{11} = c_{13}$; $c_{21} = -c_{23}$; $c_{31} = c_{33}$), one obtains the π-MOs represented in Figure 4.10.

For cyclopropenyl cation, the total π-energy is calculated to be $E_\pi = 2(\alpha + 2\beta)$. Compared with E_π(ethylene) $= 2(\alpha + \beta)$, the cyclopropenyl cation enjoys a special stabilization effect because of cyclic π-conjugation with the carbenium ion center that amounts to E_π(cyclopropenyl cation) − E_π(ethylene) − E_π(methyl cation) $= 2(\alpha + 2\beta) - 2(\alpha + \beta) = 2\beta$. The same analysis for allyl cation gives a vinyl substituent effect on a carbenium ion of 0.82β (Figure 4.4). Hückel proposed that cyclopropenyl cation is stabilized by what we now refer to as aromaticity (cyclic π-system with $4N + 2\pi$ electrons, where $N = 0$). Hydride affinities of cyclopropyl and cyclopropenyl cation confirm the extra stabilization of cyclopropenyl cation by cyclic π-conjugation [44, 45].

The Hückel method predicts $E_\pi(c\text{-}C_3H_3^{\bullet}) = 2(\alpha + 2\beta) + \alpha - \beta = 3\alpha + 3\beta$. This is slightly larger than E_π(allyl radical) $= 3\alpha + 2.82\beta$ (Figure 4.6), which would mean that π-resonance in the cyclic radical

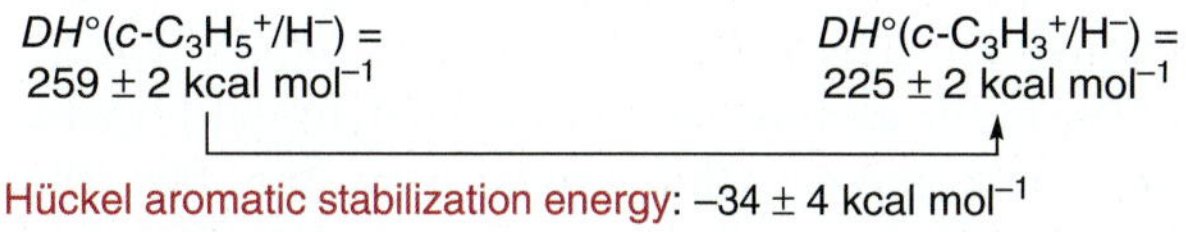

is slightly more stabilizing than in the open-chain radical. Experimental value for $DH°(c\text{-}C_3H_3^{\bullet}/H^{\bullet}) = 104.0 \pm 4\,\text{kcal mol}^{-1}$ has been reported [46], which can be compared with $DH°$(cyclopropyl radical/H$^{\bullet}$) $= 106.3 \pm 1\,\text{kcal mol}^{-1}$. Thus, π-conjugation in cyclopropenyl radical stabilizes the radical by -2 ± 5 kcal mol^{-1}, which has no stabilization or much less than π-conjugation in allyl radical ($-13\,\text{kcal mol}^{-1}$). The ZOA of the Hückel method overestimates $E_\pi(c\text{-}C_3H_3^{\bullet})$ as it did also for E_π(allyl$^{\bullet}$) (Figure 4.7).

For the cyclopropenyl anion, the Hückel method gives $E_\pi(c\text{-}C_3H_3^-) = 2(\alpha + 2\beta) - 2(\alpha - \beta) = 4\alpha + 2\beta$ to be compared with E_π(ethylene) + E_π(methyl anion) $= 2(\alpha + \beta) - 2\alpha = 4\alpha + 2\beta$. Accordingly, cyclopropenyl anion is neither stabilized nor destabilized by cyclic π-conjugation. This is obviously incorrect because of the ZOA and as $\alpha(\text{Me}^-) > \alpha(\text{Me}^{\bullet})$. Breslow has shown that cyclopropenyl anions are among the most basic compounds known in solution [47–49]. This is also the case for cyclopropenyl anions

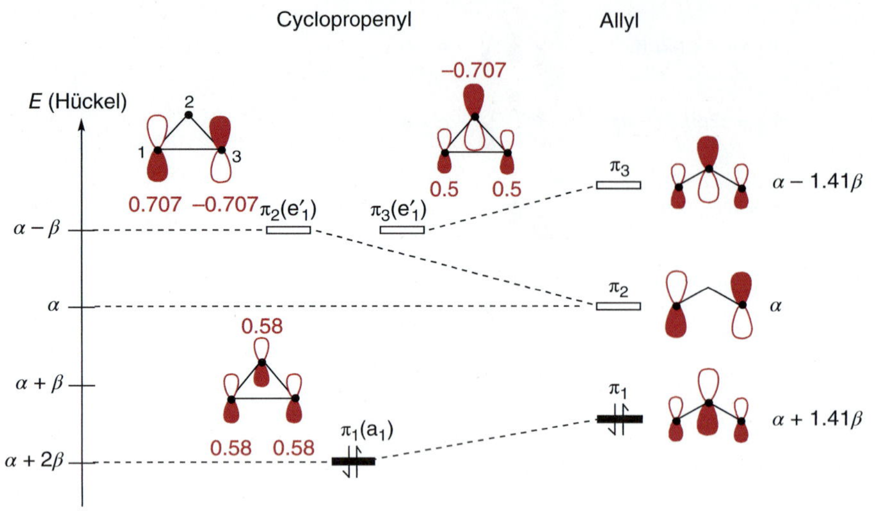

Figure 4.10 Representation of the Hückel π-MOs of cyclopropenyl and allyl cations.

generated in the gas phase (Section 11.3.5) [46]. High-level quantum mechanical calculations predict proton affinity PA(c-$C_3H_3^-$) = 418.9 kcal mol^{-1}, which is higher than PA(Me^-) = 415.2 ± 0.8 kcal mol^{-1} [50] and 5.8 kcal mol^{-1} higher than PA(c-$C_3H_5^-$).

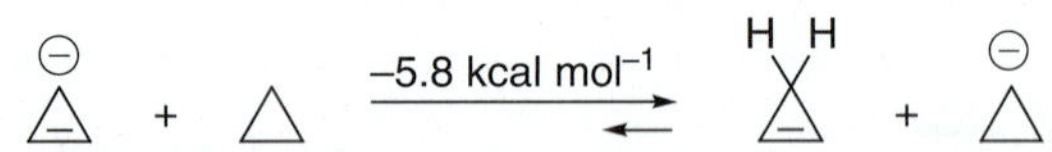

Applying the PMO theory to cyclopropenyl anion and considering $S_{ij} \neq 0$ $(i \neq j)$, one predicts that the interaction of methyl anion with an ethylene unit combining in a C_{2V} or a C_{3V} species (Figure 4.11) must lead to a destabilization because of the cyclic π-conjugation called antiaromatic destabilization energy. This arises from the fact that HOMO(ethylene, π) interacts with HOMO(Me^-), leading to a four-electron repulsive effect, as predicted by Eq. (4.28) and illustrated in Figure 4.4B. Thus, on combining HOMO(ethylene) and MOMO(Me^-), one generates an in-phase combination Ψ_1 stabilized by $|\Delta E'|$ with respect to HOMO(ethylene) and an out-of-phase combination Ψ_2, the latter being destabilized by $|\Delta E''|$ with respect to HOMO(Me^-). As $|\Delta E''| > |\Delta E'|$, this leads to an overall destabilization that cannot be compensated by an eventual LUMO(ethylene)/HOMO(Me^-) interaction as the overlap between these two orbitals is nil in the three-membered ring geometry. Considering the mirror plane of symmetry that cuts cyclopropenyl anion by the corner made of the methyl anion, LUMO(ethylene) is antisymmetrical (a''), whereas both HOMO(ethylene) and HOMO(Me^-) are symmetrical (a') with respect to this element of symmetry.

The π_2 and π_3 orbitals of cyclopropenyl radical and anion are degenerate (they have the same energy), and they appear to distribute electrons unevenly among the three carbon centers of the equilateral three-membered ring (D_{3h} symmetrical, with a C_3

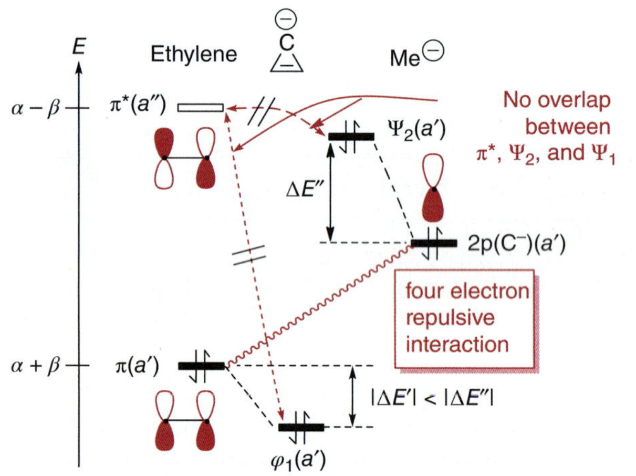

Figure 4.11 PMO diagram representing the cyclopropenyl anion as resulting from the interaction of a methyl carbanion with an ethylene unit and realizing an isocel (D_{2h}) or equilateral (D_{3h}) triangle. Overlap $\pi(a')\cdot 2p(a') \neq 0$ as these two orbitals are both symmetrical with respect to the mirror plane of symmetry of these triangles. Overlap $\pi^*(a'')\cdot 2p(a') = 0$ as these two orbitals do not have same symmetry with respect to the mirror plane of symmetry.

axis of rotation). Electrons in π_2 concentrate on the edge C(1)—C(2) in an antibonding manner, whereas electrons circulating in π_3 concentrate on carbon center C(3) opposite to edge C(1)—C(2). MO π_2 has a mirror plane of symmetry as a nodal surface. In MO π_3, the nodal surface is perpendicular to the mirror plane of the symmetry of π_2. This situation will repeat for SOMOs and HOMOs of symmetrical cyclic compounds that have an odd number of carbon centers (D_{3h}, D_{5h}, D_{7h}, etc.). It turns out that such situations are unstable with respect to what is called the Jahn–Teller distortion. A geometrical distortion to lower C_{2v} symmetry will stabilize the cyclopropenyl radical or anion, by stabilizing the radical or carbanion. The D_{3h} symmetry can only be maintained if both of these degenerate orbitals are equally occupied as in the triplet state of the anion, or in the cation, where both of the orbitals are vacant.

4.5.5 Butadiene

For *s-trans-* and *s-cis-*butadiene, the secular determinant is

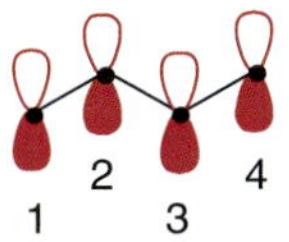

$$\begin{vmatrix} \alpha - E & \beta & 0 & 0 \\ \beta & \alpha - E & \beta & 0 \\ 0 & \beta & \alpha - E & \beta \\ 0 & 0 & \beta & \alpha - E \end{vmatrix} = 0$$

There are four roots: E_1, E_2, E_3, and E_4, and the four MOs, π_1, π_2, π_3^*, and π_4^*, shown in Figure 4.12. E_π(butadiene) $= 2E_1 + 2E_2 = 2(\alpha + 1.62\beta) + 2(\alpha + 0.62\beta) = 4\alpha + 4.48\beta$. For two molecules of ethylene, E_π (2 ethylene) $= 4\alpha + 4\beta$. According to Hückel, butadiene is stabilized by 0.48β due to acyclic π-conjugation.

Using $\beta = -32\,\text{kcal mol}^{-1}$ as parameterized by the energy barrier of rotation about the C=C double bond of ethylene (Section 4.5.1), the π-conjugation of butadiene gives a conjugation stabilization energy of $-15.4\,\text{kcal mol}^{-1}$. This is far too high as we have seen that conjugated 1,3-dienes are stabilized at most by $-7\,\text{kcal mol}^{-1}$ compared with their nonconjugated isomers (Section 2.7.1). Again, it is the neglect of differential overlap (ZOA) that makes the Hückel method overestimate the stabilization because of π-conjugation in 1,3-dienes.

The MO π_1 of butadiene concentrates the electron density on carbon centers C(2) and C(3). The highest amplitude of the stationary wave function π_1 (no nodal plane) is localized in the middle of the molecule (tone frequency of a vibrating string). For MO π_2 that has a nodal surface cutting bond C(2)—C(3) in its middle, the electron density is reduced around C(2) and C(3) that are closed to the nodal surface and augmented around carbon centers C(1) and C(4) that are away from the nodal surface. In π_2, coefficients c_{21} and $-c_{24}$ are larger than coefficients c_{22} and $-c_{23}$. The calculations increase the bonding character of π_1 by making the coefficients c_{12} and c_{13} larger than c_{11} and c_{14} in π_1. Similarly, the antibonding character between C(2) and C(3) in π_2 is reduced by making c_{22} and $-c_{23}$ smaller than c_{21} and $-c_{24}$. The bonding character is giving by the sum of products $c_{11}{\cdot}c_{12} + c_{12}{\cdot}c_{13} + c_{13}{\cdot}c_{14}$ in π_1 and by $c_{21}{\cdot}c_{22} + c_{22}{\cdot}c_{23} + c_{23}{\cdot}c_{24}$ in π_2. π-MOs of butadiene can be derived from those of two ethylene units applying the PMO theory. Using π_a/π_b to refer to the HOMOs of the two ethylene units and π_a^*/π_b^* to designate the LUMOs, the perturbation arising from the degenerate orbitals, π_a and π_b, on the one hand, and π_a^* and π_b^*, on the other hand, are most important and result in the energy changes shown in Figure 4.13. This interaction generates four π-MOs that are combinations $\Psi_1 = 1/\sqrt{2}\pi_a + 1/\sqrt{2}\pi_b$, $\Psi_2 = 1/\sqrt{2}\pi_a - 1/\sqrt{2}\pi_b$, $\Psi_3^* = 1/\sqrt{2}\pi_a^* + 1/\sqrt{2}\pi_b^*$, and $\Psi_4^* = 1/\sqrt{2}\pi_a^* - 1/\sqrt{2}\pi_b^*$. In order to include interactions between empty and filled MOs of the two ethylene units, we need to consider the overlap between Ψ_1, Ψ_2, Ψ_3^*, and Ψ_4^*. This analysis is simplified by the symmetry of *s-trans*-butadiene, which has a C_2 axis of rotation perpendicular to the plane of the molecule and cutting the C(2)—C(3) bond in its middle. Accordingly, $\Psi_1(a)$, $\Psi_2(b)$, $\Psi_3^*(a)$, and $\Psi_4^*(b)$, i.e. Ψ_1 and Ψ_3^* are symmetrical, whereas Ψ_2 and Ψ_4^* are

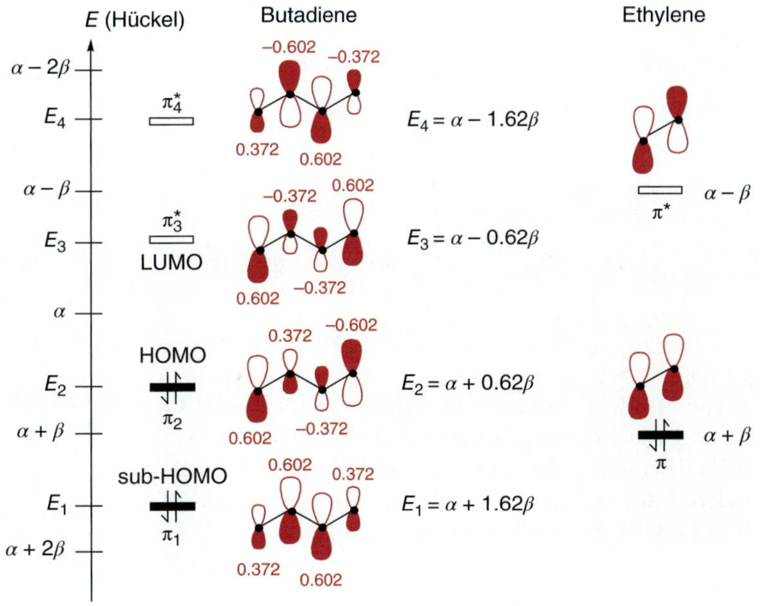

Figure 4.12 Representation of the Hückel MOs of *s-trans*-butadiene. Same MOs and energies are calculated by the Hückel method for *s-cis*-butadiene.

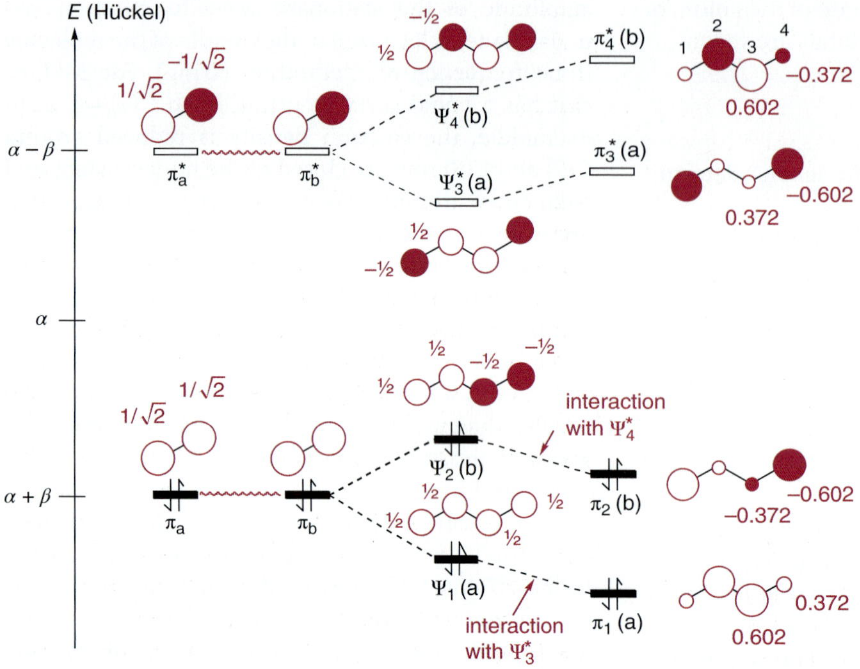

Figure 4.13 PMO diagram showing the π-MOs of *s-trans*-butadiene as derived from the interaction between the π-MOs of two ethylene units. Hückel coefficients are shown (the sum of the square of the coefficients given for each MO must be unity, normalization conditions, ZOA). ((a) ≙ symmetrical with respect to rotation axis C_2 and (b) ≙ antisymmetrical with respect to rotation axis C_2.)

antisymmetrical with respect to the C_2 axis. Rotation of $\Psi_1(a)$ and $\Psi_3^*(a)$ by 180° about this axis reproduces the same MOs, whereas rotation of $\Psi_2(b)$ and $\Psi_4^*(b)$ by 180° about this same axis give $-\Psi_2$ and $-\Psi_4^*$ (the white lobes interchange with the red lobes of the 2p AOs), respectively. Thus, only two interactions need to be considered, i.e. $\Psi_1(a)$ being stabilized by $\Psi_3^*(a)$ and $\Psi_2(b)$ being stabilized by $\Psi_4^*(b)$. This leads to butadiene MO $\pi_1 = a\Psi_1 + b\Psi_3^*$ with $a > b$ and $\pi_3^* = c\Psi_3^* - d\Psi_1$ with $c > d$, and to MO $\pi_2 = a\Psi_2 + b\Psi_4^*$ with $a > b$ and $\pi_4^* = c\Psi_4^* - d\Psi_2$ with $c > d$, as shown graphically below:

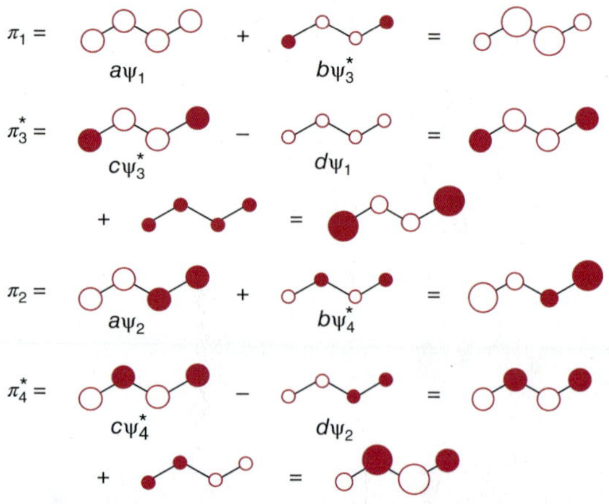

The bonding character in $\Psi_1(b)$ is proportional to $c_{11} \cdot c_{12} + c_{12} \cdot c_{13} + c_{13} \cdot c_{14} = (½)^2 + (½)^2 + (½)^2 = ¾ =$ 0.75. This is smaller than the bonding character calculated for $\pi_1(a)$: $0.372 \times 0.602 + (0.602)^2 + 0.602 \times 0.372 = 0.81$. Similarly, the bonding character in $\Psi_2(b)$ is proportional to $c_{21} \cdot c_{22} + c_{22} \cdot c_{23} + c_{23} \cdot c_{24} = ¼ - ¼ + ¼ = ¼ = 0.25$, which is smaller than that calculated for $\pi_2(b)$: $(0.602)\cdot(0.372) + (0.372)\cdot(-0.372) + (-0.372)\cdot(-0.602) = 0.224 - 0.138 + 0.224 = 0.310$. Thus, stabilization by π-conjugation in *s-trans*-butadiene (π_1 and π_2 being stabilized with respect to Ψ_1 and Ψ_2, respectively) arises from the LUMO/HOMO interactions between the two ethylene units composing this molecule. Applying the Hückel method to *s-cis*-butadiene leads to the same results.

4.5.6 Cyclobutadiene and its electronic destabilization (antiaromaticity)

The Hückel method gives the following secular determinant for cyclobutadiene (**1**):

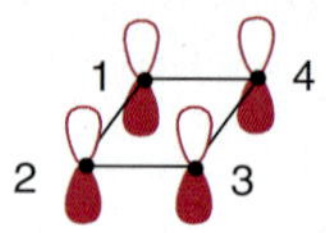

1

$$\begin{vmatrix} \alpha - E & \beta & 0 & \beta \\ \beta & \alpha - E & \beta & 0 \\ 0 & \beta & \alpha - E & \beta \\ \beta & 0 & \beta & \alpha - E \end{vmatrix} = 0$$

for which the four roots E_1, E_2, E_3, and E_4 are found:

$$E_1 = \alpha + 2\beta$$
$$E_2 = E_3 = \alpha$$
$$E_4 = \alpha - 2\beta$$

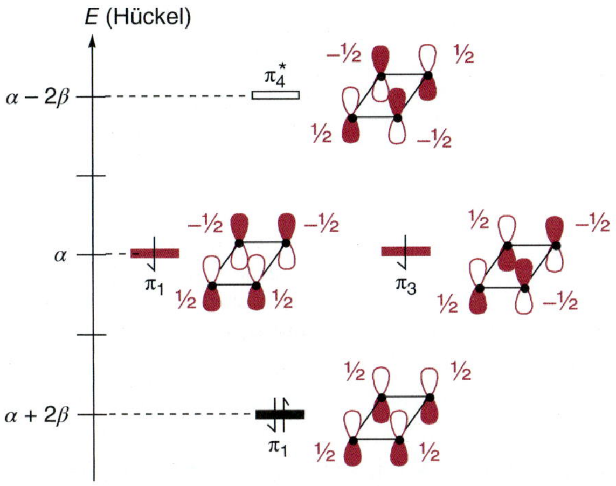

Figure 4.14 Representation of the four π-MOs of cyclobutadiene as calculated by the Hückel method.

Using these energy values in the four secular equations, considering the four normalization conditions $<\pi_1>^2 = <\pi_2>^2 = <\pi_3>^2 = <\pi_4>^2 = 1$ and the symmetry (D_{4h}) of cyclobutadiene, one obtains the MOs shown in Figure 4.14. The total π-energy of cyclobutadiene is $E_\pi(\mathbf{1}) = 2(\alpha+2\beta)+2\alpha = 4\alpha+4\beta$, which is the same E_π energy as for two noninteracting, isolated ethylene units. Accordingly, HMO predicts that cyclic π-conjugation in **1** neither stabilizes, nor destabilizes this compound. In fact, **1** is destabilized by electronic repulsion between the two conjugated ethylene units that compose it because of the D_{4h} symmetry (square **1**) [51–53]. As we shall see (Figure 4.15), this is also the case for rectangular **1** (D_{2h} symmetry).

The four-electron interaction π_a/π_b that is destabilizing (Section 4.4.2, Eq. (4.28)) is not compensated by any π_a^*/π_b and π_b^*/π_a interactions, as the latter leads to zero overlap because of the symmetry of cyclobutadiene. In Figure 4.15, MO overlaps $\Psi_1 \cdot \Psi_3^*, \Psi_1 \cdot \Psi_4^*, \Psi_2 \cdot \Psi_3^*$, and $\Psi_2 \cdot \Psi_4^*$ are all zero because each member of these pairs has different symmetry.

Classical measurement of $\Delta_f H°(\mathbf{1})$ by calorimetry of combustion and/or hydrogenation cannot be carried out, as **1** cyclodimerizes quickly into bicyclo[4.2.0.0^{2,5}]octa-3,7-diene (**3**) [54]. Pettit and coworkers obtained **1** by oxidation of tricarbonyl(cyclobutadiene)iron (**2**) with Ce^{IV} or Fe^{III} salts [55]. Under their conditions, **1** dimerizes into **3**. Nevertheless, in the presence of methyl propionate, **1** can be trapped as a Diels–Alder adduct, the Dewar-benzene derivative **4** (Scheme 4.1) [56].

The cyclodimerization, **1** + **1** → **3**, occurs already at −175 °C as shown by Masamune et al. for **1** generated upon UV irradiation of compound **5** (Scheme 4.2) [57]. Quantum mechanical calculations suggest nonsynchronous (4+2)-cycloadditions for **1** + **1** → **3** with no reaction barrier ($\Delta^{\ddagger}H = 0$) [58] and with passage through diradical species [59].

Photolysis of 2-oxabicyclo[2.2.0]hex-5-en-3-one (**7**) in Ar matrices at low temperature has permitted the direct observation of cyclobutadiene (Scheme 4.3). On melting the matrix, **1** cyclodimerizes quickly [60].

In 1991, Cram et al. obtained **1** in the cavity of a hemicarcerand **8** by photolysis of a complex of 2-pyrone (**6**) and the hemicarcerand **8** at room temperature [61]. As **1** in the hemicarcerand cannot

Figure 4.15 PMO diagram representing rectangular (D_{2h}) and square (D_{4h}) cyclobutadiene π-MOs and relationship to the HOMOs and LUMOs of two ethylene units. As $|\Delta E''| > |\Delta E'|$ and because of the absence of stabilizing LUMO/HOMO interaction between the two ethylene units, cyclobutadiene is destabilized by cyclic π-conjugation. This destabilization is called antiaromatic destabilization energy (a_1 = symmetric with respect to the rotation axis and the "vertical" mirror plane of symmetry, b_2 = antisymmetric with respect to the axis of rotation, antisymmetric with respect to the mirror plane of symmetry).

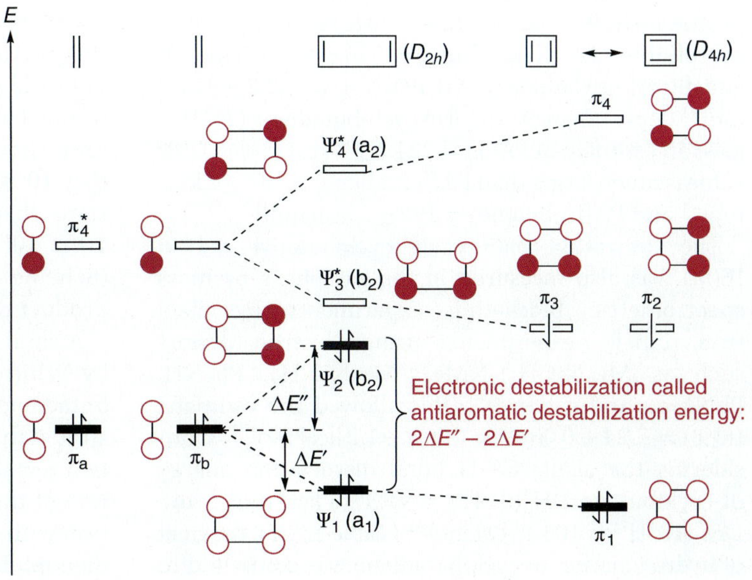

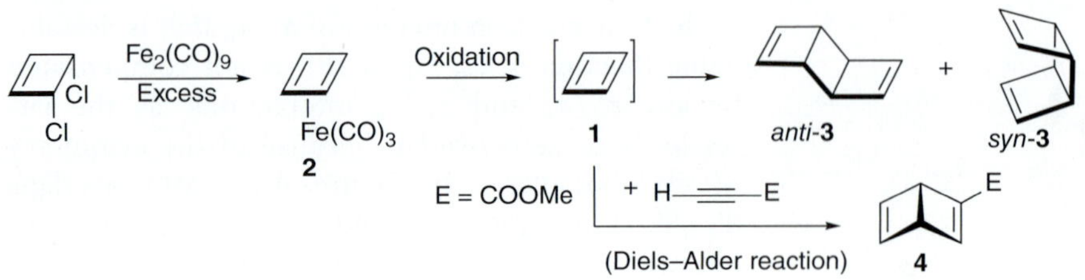

Scheme 4.1 Pettit's generation of cyclobutadiene as a reactive intermediate.

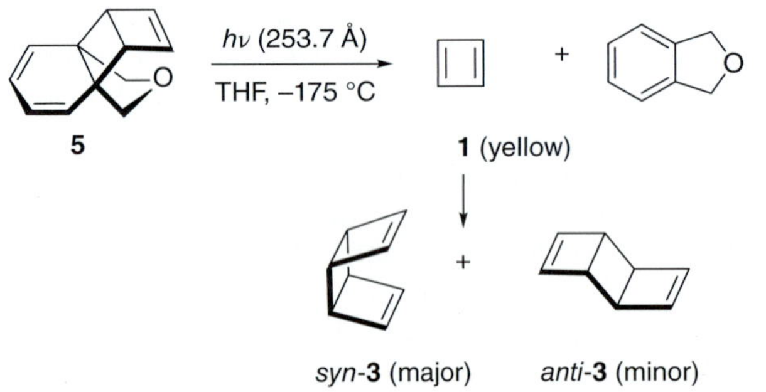

Scheme 4.2 Photochemical generation of cyclobutadiene in THF solution.

collide with another molecule of **1**, its cyclodimerization does not occur (Scheme 4.4) [62].

The gas-phase standard heat of formation of cyclobutadiene has been estimated by a mass spectrometric technique that measures the proton affinity of **1** and ionization energy of cyclobutenyl radical ($\mathbf{9}^{\cdot}$) (Scheme 4.5). The acidity of cyclobutenyl cation $\mathbf{9}^+$ corresponds to the proton affinity of cyclobutadiene, PA(c-C_4H_4). This quantity has been measured in the gas phase by treating $\mathbf{9}^+$ with several bases B: and observing the formation, or not, of the conjugate acids BH^+ [63]. Bases like NH_3 (PA(NH_3) = 204 ± 2 kcal mol^{-1}) and isopropylame PA(i-Pr-NH_2) = 221 ± 2 kcal mol^{-1}) do not abstract a proton form $\mathbf{9}^+$, whereas more basic reagents such as pyrrolidine (PA(pyrrolidine) = 226.6 ± 2 kcal mol^{-1}) and diisopropylamine (PA((i-Pr)$_2$NH) = 232.2 ± 2 kcal mol^{-1} deprotonate $\mathbf{9}^+$ to form cyclobutadiene (**1**). This gave an estimate of PA(**1**) = 224.2 ± 3 kcal mol^{-1}. This value is much larger than PA(butadiene) = 187 ± 2 kcal mol^{-1} and PA(isobutene) = 192 ± 2 kcal mol^{-1}.

The ionization energy of cyclobutenyl radical IE($\mathbf{9}^{\cdot}$) was also measured in the gas phase by mass spectrometric bracketing experiments by electron transfer experiments using electron donors such as $Me_2N(CH_2)_2NMe_2$, Me_2N-NH_2, Ph_2NH, $PhNMe_2$, and cation $\mathbf{9}^+$. This allowed the estimate, IE($\mathbf{9}^{\cdot}$) = 7.24 ± 0.06 eV = 166.8 ± 1.2 kcal mol^{-1}. Considering the allylic C—H bond dissociation energy of cyclobutene $DH°(\mathbf{9}^{\cdot}/H^{\cdot})$ = 91.3 ± 1 kcal mol^{-1} and $DH°(H^{\cdot}/H^{\cdot})$ = 104.2 kcal mol^{-1} (Table 1.A.7), the heat of hydrogenation of cyclobutadiene was computed to be $\Delta_h H°$ = −67.7 ± 4 kcal mol^{-1}. Finally, the standard heat of formation of cyclobutadiene is estimated to be $\Delta_f H°(\mathbf{1})$ = 102.3 ± 4 kcal mol^{-1}. The antiaromatic destabilization energy of cyclobutadiene due to cyclic π-conjugation amounts to 34 ± 4 kcal mol^{-1}. It is given by equilibrium (4.40) if one assumes ring strain to be the same in **1**, cyclobutene, and cyclobutane.

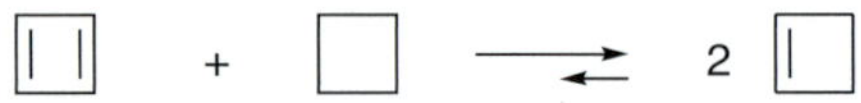

(4.40)

$\Delta_f H°$: 102.3 ± 4   6.6 (Table 1.2)   37.5 (Table 1.2)

$\Delta_r H°$ (4.40) = −34 ± 4 kcal mol^{-1}

4.5.7 Geometries of cyclobutadienes, singlet and triplet states

According to Hund's rule of maximum multiplicity, square (D_{4h}) cyclobutadiene (**1**), which has two degenerate SOMOs π_2 and π_3 ($E_2 = E_3 = \alpha$, Hückel), should have a triplet ground state (π_2 and π_3, occupied each by an electron with parallel spins) and thus be prone to initiate radical polymerization of alkenes. This was not observed in 1969 in Pettit's experiments that generated **1** and substituted derivatives by oxidation of their $Fe(CO)_3$ complexes. For instance, when 1,2-diphenylcyclobutadiene (**10**) was generated by this method, it was quenched by ethylenetetracarbonitrile giving two diastereomers **11** + **12** in a 1 : 7 product ratio. If **10** were a square, a 1 : 1 product ratio should be observed for **11** + **12**. This suggested that **10** adopts a rectangular geometry in its ground state that undergoes valence bond isomerization **10** ⇄ **10′** competitively with its Diels–Alder reaction (Scheme 4.6). With less reactive dienophiles, the product ratio approaches 1 : 1 [64].

A similar experiment was carried out (Scheme 4.7) by Whitman and Carpenter for 1,2-dideuterocyclobutadiene (**13**) [65, 66]. The experiments demonstrated that this compound exists as an equilibrium of two rectangular valence bond isomers **13** ⇄ **13′**, the rate of their equilibration (k_1/k_{-1}) being in competition with the rate of their Diels–Alder reactions with dienophiles (k_2 [dienophiles] and k_2' [dienophiles]).

Scheme 4.3 Argon matrix isolation of cyclobutadiene.

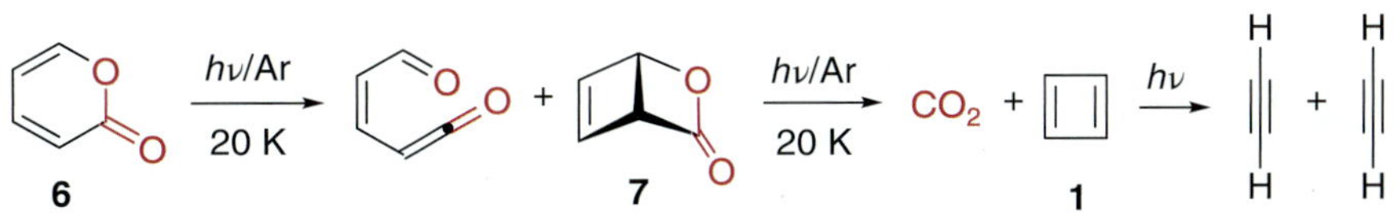

Scheme 4.4 Bottling of a single molecule in a molecular flask: hemicarcerand **8**. Inside **8**, cyclobutadiene (**1**) cannot cyclodimerize at room temperature.

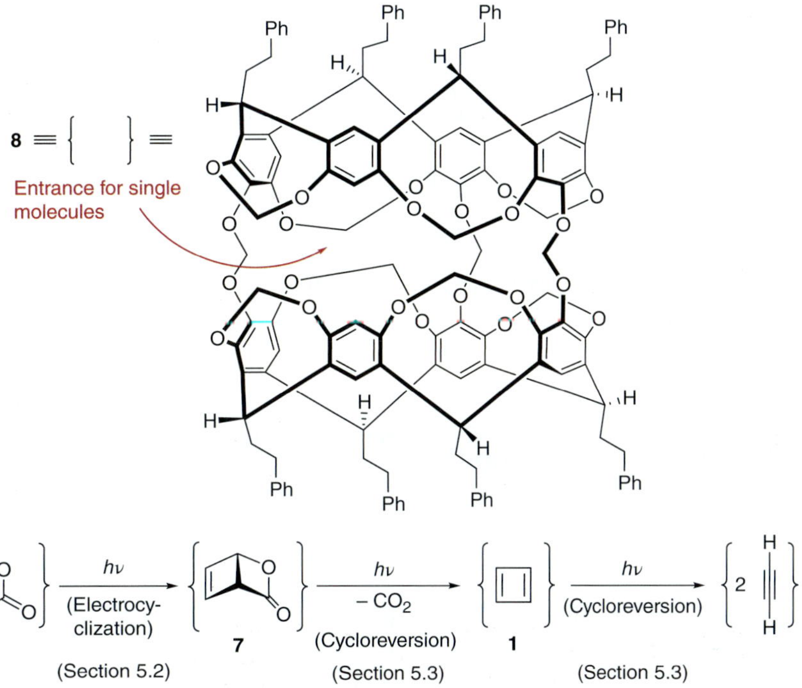

Scheme 4.5 Thermochemical cycle that permits the evaluation of the gas-phase standard heat of formation of cyclobutadiene.

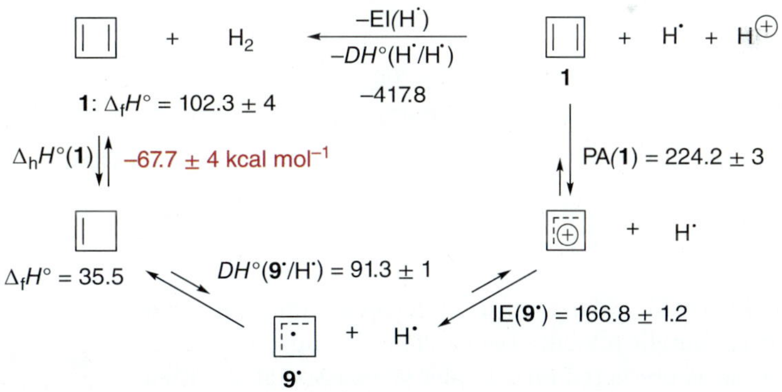

Scheme 4.6 Rectangular geometry of 1,2-diphenylcyclobutadiene.

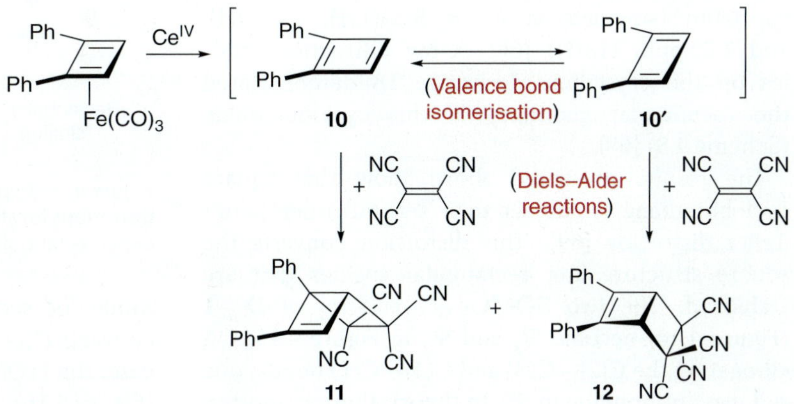

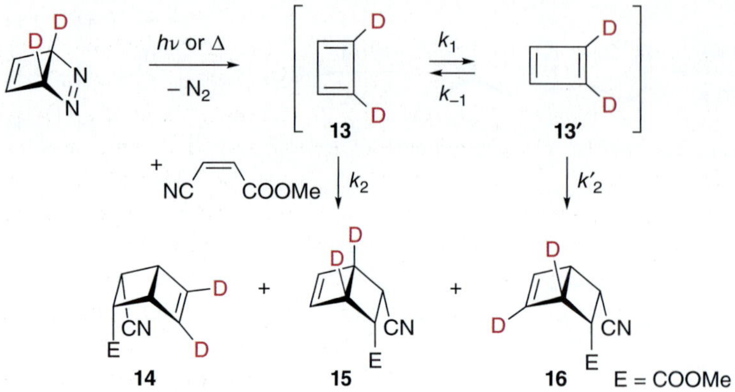

Scheme 4.7 Proof of the rectangular structure of cyclobutadiene in solution. At low concentration of the dienophiles, the proportion of adducts **14**/**15**/**16** reaches 1 : 1 : 2 (statistical ratio, consistently with square **13** or very fast equilibrating **13** ⇄ **13′**). At high concentration of the dienophiles, the product ratio **14**/**15**/**16** is >1 : >1 : <2 (k_2[dienophile] becomes comparable with k_1 and k_{-1}).

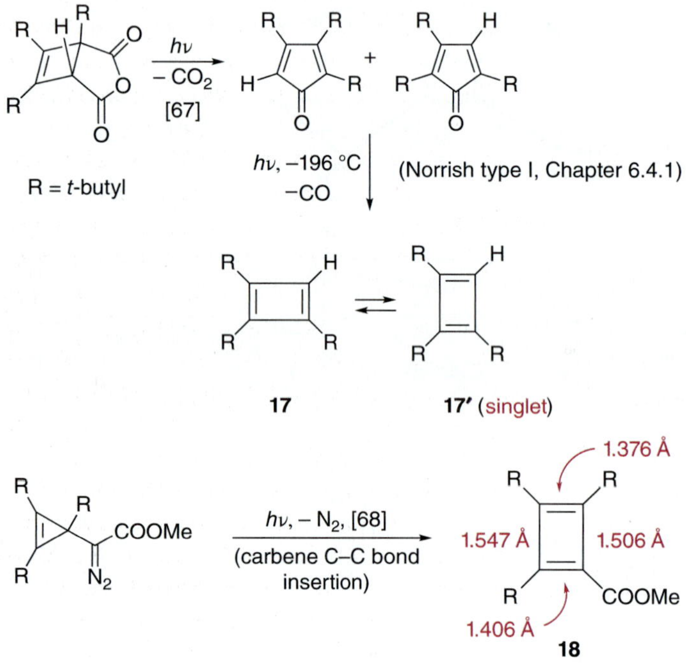

Scheme 4.8 Proof for rectangular geometries of substituted cyclobutadienes and their mode of formation.

In 1973, Maier and Alzérreca reported that tris(t-butyl)cyclobutadiene (**17**) is not paramagnetic as predicted for a triplet state associated with a square geometry and showed a well-defined ^{1}H-NMR spectrum (singulets at δ_H = 5.35(1H), 1.14(9H), and 1.05 ppm (18H)) [67]. X-ray diffraction studies on the crystalline derivative **18** demonstrated the rectangular geometry of this cyclobutadiene (Scheme 4.8) [68].

The results presented above show that square cyclobutadiene is subject to a second-order Jahn–Teller distortion [69]. This distortion converts the square structure into rectangular species that are stabilized. The two SOMOs π_2 and π_3 of D_{4h}-**1** (Figure 4.14) become Ψ_2 and Ψ_3^* in Figure 4.15. On elongating the C(2)—C(3) and C(4)—C(1) bonds, one reduces antibonding in Ψ_2. In theory, there is another mode for second-order Jahn–Teller distortion that converts D_{4h}-**1** into planar rhombic structures. In that case, the HOMO and LUMO of **1** become $\emptyset_2$ and $\emptyset_3^*$ (Figure 4.16).

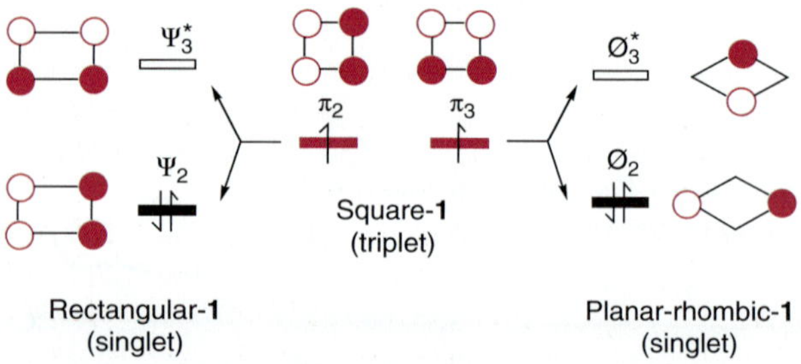

Figure 4.16 Representation of second-order Jahn–Teller distortions for stabilization of square cyclobutadiene into either rectangular structures of planar–rhombic structures.

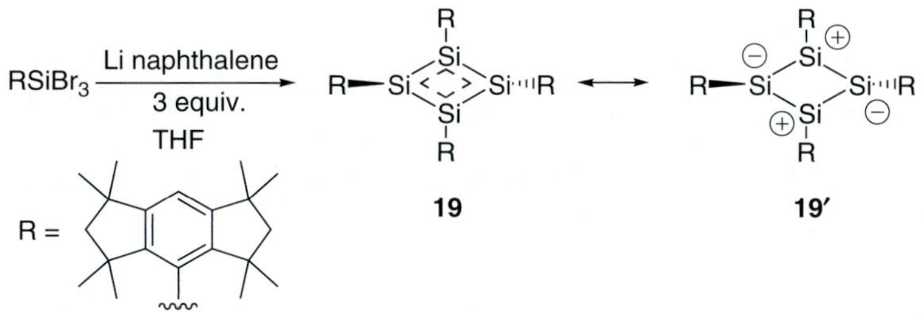

Scheme 4.9 Preparation of a planar rhombic tetrasilacyclobutadiene.

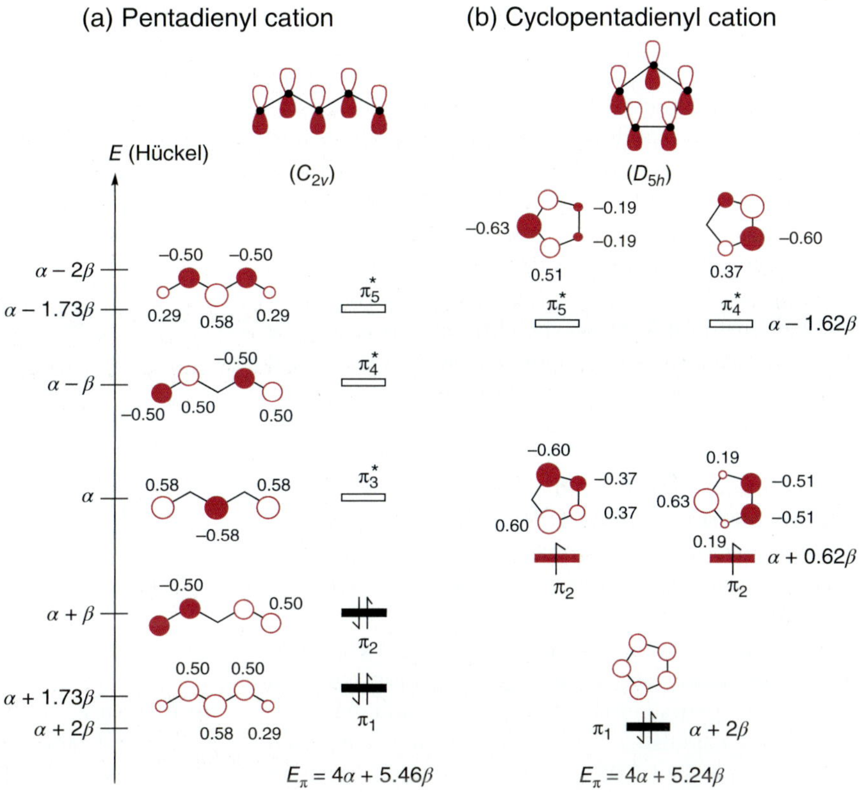

Figure 4.17 Hückel π-MOs for (a) pentadienyl and (b) cyclopentadienyl cation. Hückel calculates same MOs with same energies for corresponding radicals and anions.

This feature has been observed for crystalline tetrasilacyclobutadiene **19** for which the Si_4 ring has a planar rhombic structure with alternating pyramidal and planar configurations for Si atoms (Scheme 4.9). The four Si—Si bond lengths differ by less than 0.03 Å and are intermediate between typical Si—Si single (2.34 Å) and double (2.14 Å) bonds [70].

Problem 4.1 Olah et al. [71, 72] have prepared tetramethylcyclobutenediyl dication as a stable salt in solution by treating 3,4-dichloro-1,2,3,4-tetramethylcyclobutene in SbF_5/SO_2ClF at −75 °C. They also prepared the tetraphenyl-substituted derivative in a similar way [73]. Estimate the Hückel E_π energy of cyclobutenediyl dication. Is it stabilized by aromatic stabilization energy?

4.5.8 Pentadienyl and cyclopentadienyl systems

For pentadienyl and cyclopentadienyl systems, the Hückel method gives the π-MOs shown in Figure 4.17A,B, respectively. For the cations, one calculates $E_\pi(n\text{-}C_5H_7^+) = 4\alpha + 5.46\beta$ and $E_\pi(c\text{-}C_5H_5^+) = 4\alpha + 5.24\beta$. Both cations appear to be stabilized with respect to butadiene + methyl cation, for which one calculates $E_\pi(\text{butadiene}) + E_\pi(\text{Me}^+) = 4\alpha + 4.48\beta$ (Section 4.5.5). As expected, π-conjugation must stabilize the carbenium ion, and this is confirmed by the hydride affinities given below (Table 1.A.14) for cyclopentyl and cyclopentenyl cations. Note that differential hyperconjugation effects (2p(+)/σ(C—H) interactions) also affects the relative stability of these carbenium ions.

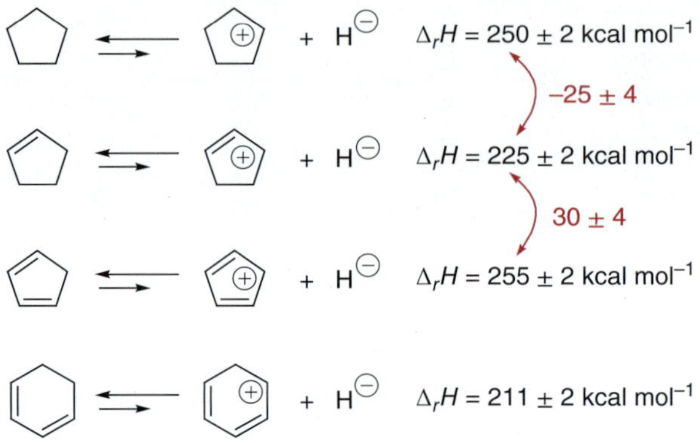

Surprisingly, the hydride affinity of cyclopentadienyl cation is larger than that of cyclopentyl and cyclohexadienyl cation. The latter π-conjugated system models the acyclic pentadienyl cation. Simple valence bond theory argues that the relative stability of a conjugated system (cation, radical, and anion) increases with the number of limiting structures that one can write for it. However, although there are five limiting structures for cyclopentadienyl cation and only two for cyclopentyl and three for cyclohexadienyl cation, cyclopentadienyl cation is the least stable of the four secondary carbenium ions under discussion, as shown by the relatively high hydride affinity measured for it [74]. The Hückel method with no overlap overestimates the stabilization by π-conjugation. In fact, cyclopentadienyl cation is a cyclic π-conjugated system with four electrons; it is antiaromatic, as systems with $4N$ π-electrons are destabilized (see cyclopropenyl anion, Section 4.5.4). For pentadienyl and cyclopentadienyl radical, HMO calculates $E_\pi(n\text{-}C_5H_7^\bullet) = 5\alpha + 5.46\beta$ and $E_\pi(c\text{-}C_5H_5^\bullet) = 5\alpha + 5.86\beta$. Both values are larger than $E_\pi(\text{butadiene}) + E_\pi(\text{methyl radical}) = 4\alpha + 4.48\beta + \alpha = 5\alpha + 4.48\beta$, in agreement with valence bond theory that predicts a stability increase for radicals when conjugated with olefinic and acetylenic substituents. Comparison of the C—H bond dissociation enthalpies for cyclohexane, cyclohexene, and cyclohexa-1,4-diene shows stabilization energies by π-conjugation that amount to $-16.5 \pm 3\,\text{kcal mol}^{-1}$ and $-27.8 \pm 3\,\text{kcal mol}^{-1}$ for cyclohex-2-enyl and cyclohexadienyl radical, respectively [75]. The comparison of the C—H bond dissociation enthalpies for cyclopentane, cyclopentene, and cyclopentadiene shows that π-conjugation stabilization in cyclopentadienyl radical ($-10.5 \pm 3.5\,\text{kcal mol}^{-1}$) is much less than in cyclohexadienyl radical. This is due to ring strain differences between these systems and to a special stabilization of cyclopentadiene by hyperconjugation (Section 4.8.5), which is not present to the same extent in cyclohexa-1,4-diene (Schleyer hyperconjugative aromaticity) [76]. Part of the relative instability of cyclopentadienyl radical arises from cyclic π-destabilization as suggested by the Hückel E_π energies calculated for cyclopentadienyl ($5\alpha + 5.86\beta$) and pentadienyl radical ($5\alpha + 5.46\beta$) [75, 77].

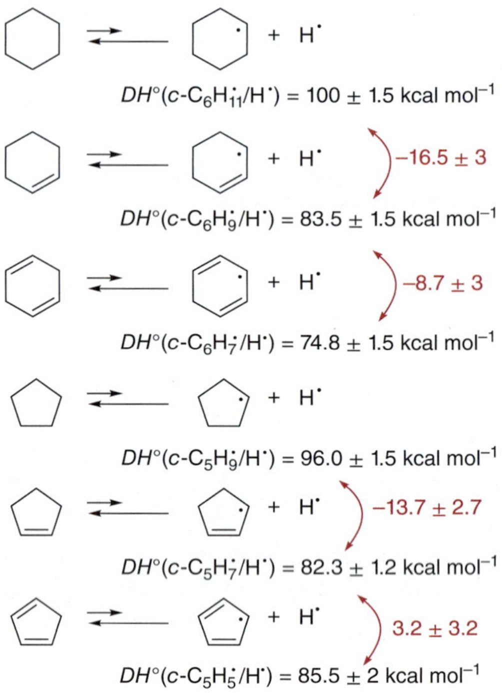

Problem 4.2 Estimate the hydride affinity of cyclohexadienyl cation in the gas phase using proton affinity of benzene ($185\,\text{kcal mol}^{-1}$, Table 1.A.15).

4.5.9 Cyclopentadienyl anion and bishomocyclopentadienyl anions

For cyclopentadienyl anion, the HMO method calculates $E_\pi(c\text{-}C_5H_5^-) = 2(\alpha + 2\beta) + 4(\alpha + 0.62\beta) = 6\alpha + 6.48\beta$, which is significantly more than $E_\pi(\text{butadiene}) + E_\pi(\text{methyl anion}) = 4\alpha + 4.48\beta + 2\alpha = 6\alpha + 4.48\beta$. Thus, cyclic conjugation of methyl anion with a *s-cis*-butadiene unit in cyclopentadienyl anion leads to significant stabilization (aromatic stabilization energy, ASE) because of cyclic π-conjugation. This stabilization is associated with the aromaticity of $c\text{-}C_5H_5^-$, a planar conjugated π-systems with six π-electrons, in accordance with Hückel rule for aromaticity ($4N + 2$, $N = 1$). Cyclopentadienyl anion is isoelectronic with benzene (Section 4.5.10). Table 1.A.24 (Bordwell's acidities in DMSO) gives the following pK_a values in agreement with Hückel's calculations.

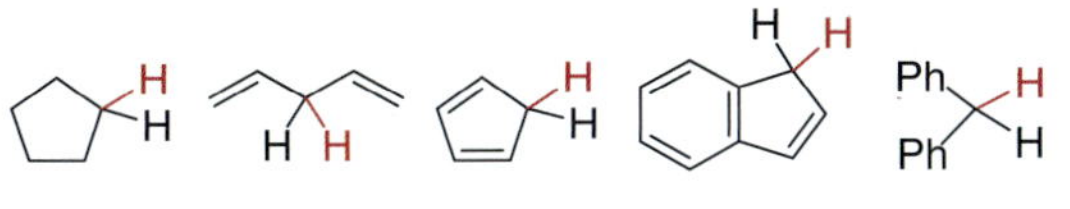

p*K*a: ~59 35 18 20.1 32.2 (DMSO, 25 °C)

The π-system of the cyclopentadienyl anion can be constructed from the allyl anion and an ethylene unit. E_π(allyl anion) + E_π(ethylene) = $4\alpha + 2.82\beta + 2(\alpha + \beta) = 6\alpha + 4.82\beta$, which is significantly less than E_π (c-$C_5H_5^-$) = $6\alpha + 6.48\beta$. The PMO diagram given in Figure 4.18 gives the origin of the ASE of cyclopentadienyl anion and of (1,2:3,4)-bishomocyclopentadienyl anion formed when allyl anion is united with the ethylene unit. For the reason of symmetry (with respect to the mirror plane of symmetry), not all MOs of allyl anion (π_1, π_2, π_3^*) overlap with all π-MOs of the ethylene unit (π,π^*). The sub-HOMO ($\pi_1(a')$) and LUMO ($\pi_3^*(a')$) of allyl anion overlap only with HOMO ($\pi(a')$) of the ethylene unit. The HOMO ($\pi_2(a'')$) of allyl anion interacts only with LUMO ($\pi^*(a'')$) of the ethylene unit. Very importantly, the repulsive four-electron interaction sub-HOMO(allyl anion)/HOMO(ethylene) is overcompensated by two stabilizing LUMO(allyl)/HOMO(ethylene) and LUMO(ethylene)/HOMO(allyl) interactions. The pK_a values measured in cyclohexylamine for bicyclic systems **20**–**24** demonstrate the intervention of a bis-ASE (stabilization due to the bishomoconjugation of an allyl anion with an ethylene moiety) in dienes **21**, **23**, and triene **24**) [78].

			20	**21**
pK_a:	47	16.2	> 40.5	31.4 ± 0.3

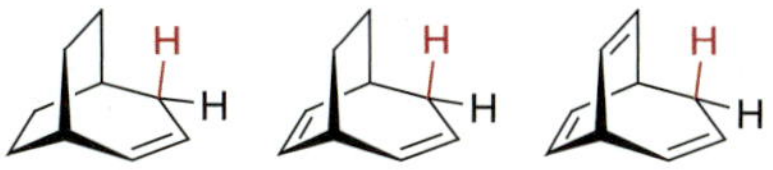

	22	**23**	**24**
pK_a:	> 40.5	34 ± 0.3	32 ± 0.3

The pK_a values given above represent the relative thermodynamic stabilities of the conjugate base generated upon deprotonation of the corresponding hydrocarbon. The bishomoaromatic stabilization energy of bicyclo[3.2.1]octa-3,6-dien-2-yl anion (**26**) is also evidenced by the kinetic acidity of **21** compared with that of **20**. These values are shown below as the relative rate constants for *H*/*D* exchanges in **20** and **21** treated with *t*-BuOK in deuterated DMSO (DMSO-d_6) [79, 80].

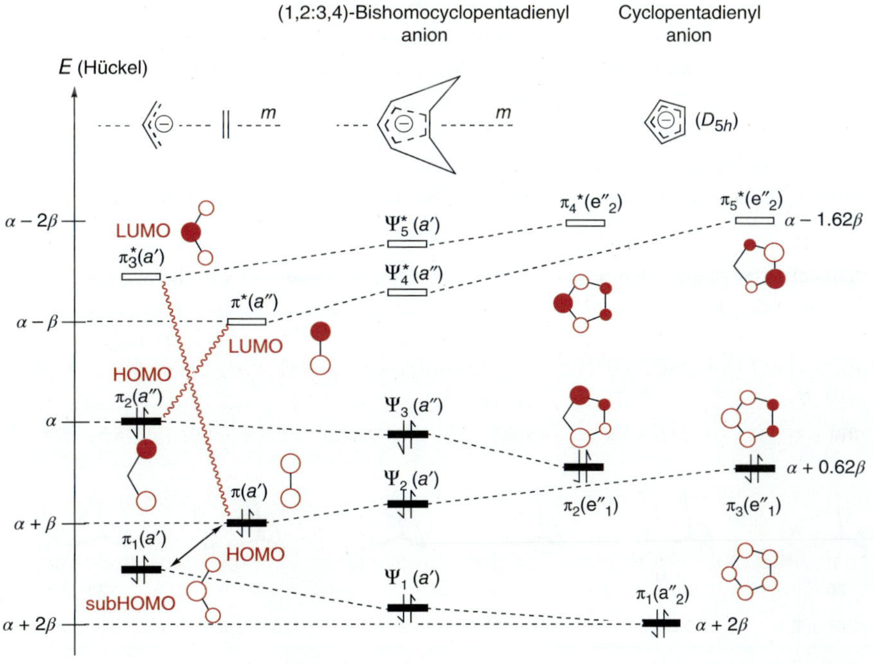

Figure 4.18 PMO diagram representing the origin of the aromatic stabilization energy (cyclic π-conjugation, Hückel rule 4*N* + 2) of cyclopentadienyl anion and bishomocyclopentadienyl anions. LUMO/HOMO interactions between allyl anion and ethylene subunits are responsible of this electronic stabilization (a' ≙ symmetrical, a'' ≙ antisymmetrical with respect to the mirror plane of symmetry *m*).

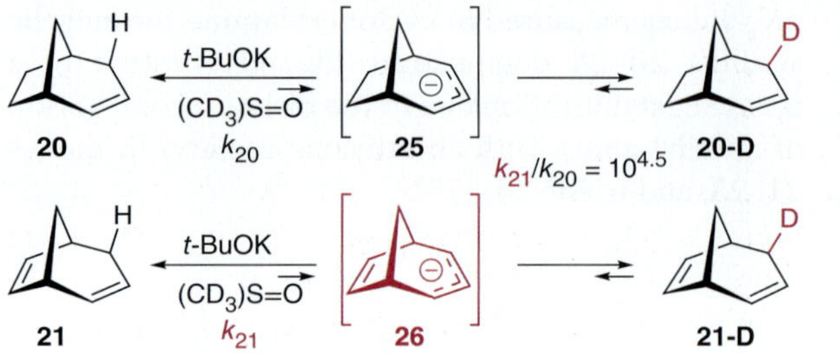

In the presence of a large excess of t-BuOK in DMSO, the bicyclic carbanion **26** is formed as a salt of potassium, the ^{1}H-NMR spectrum of which shows high-field shifted proton signals compared with those of **21**.

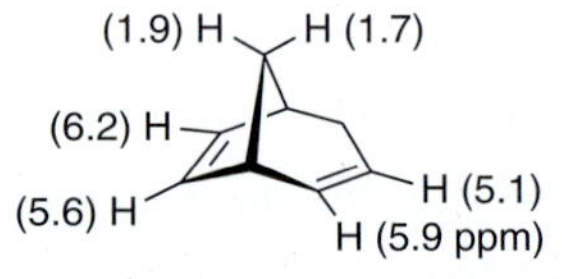

^{1}H-NMR (δ_H) of hydrocarbon **21**

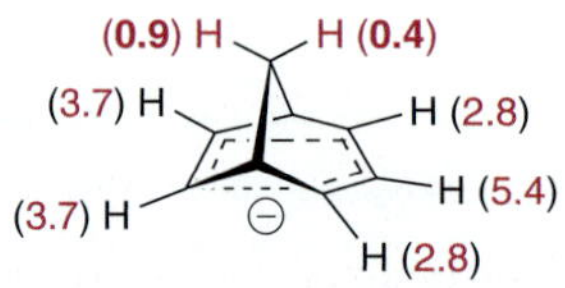

^{1}H-NMR (δ_H) of carbanion **26**

The upfield shift of both $H_2C(8)$ protons suggests an anisotropic effect as found in ansanes due to the ring current of the bishomocyclopentadienyl anion moiety.

Table 4.1 gives gas-phase proton affinities of carbanion ($DH°(R^-/H^+)$) that confirm the bishomoaromatic character of **26**. Bicyclo[3.2.1]octa-2,6-diene (**21**) is more acidic than bicyclo[3.2.1]oct-2-ene (**20**) by c. 10 kcal mol^{-1}. As C(2)-H acidity of norbornadiene is only 3 ± 2.5 kcal mol^{-1} higher than that of norbornene, the higher stability of **26** compared with that of **25** is due to a small extent only to an inductive effect arising by the homoconjugated alkene moiety. This is confirmed by the observation that 5-methylidenebicyclo[2.2.1]hept-2-ene (**28**) and 2-methylidenebicyclo[2.2.1]heptane (**27**) have the same acidity in the gas phase. The concept of bishomoaromaticity was first developed by Winstein for homoallylic carbenium ion intermediates (Section 4.7).

4.5.10 Benzene and its aromatic stabilization energy

The HMO calculation gives E_π(benzene) $= 6\alpha + 8\beta$, which is 2β more than $3E_\pi$(ethylene) $= 6\alpha + 6\beta$. This difference can be taken as the stabilization of benzene by aromaticity (or ASE). Using $\beta = -32$ kcal mol^{-1} (Section 4.5.1), one estimates the ASE of benzene to amount to -64 kcal mol^{-1}, which is much higher than measured by the equilibrium (4.41). Once again, the discrepancy between the ASE estimated by Hückel (-64 kcal mol^{-1}) and $\Delta_r H°(4.41) = -32.7$ kcal mol^{-1} arises in part from the ZOA that neglects four-electron repulsive interactions. This simple analysis ignores differential effects arising from π(C=C)/σ(C—H) hyperconjugation.

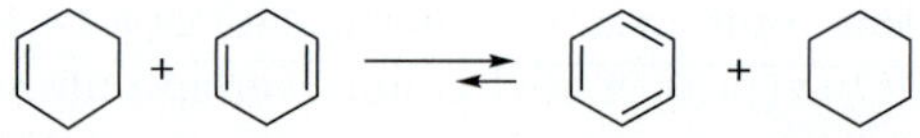 (4.41)

$\Delta_f H°$: −1.1   24.0   19.7   −29.5 kcal mol^{-1}

$\Delta_r H°(4.41) = -32.7$ kcal mol^{-1}

The π-MOs of benzene can be viewed as the result of the interactions between the π-MOs of *s-cis*-butadiene and ethylene. As shown by the PMO diagram of Figure 4.19, the overlap between orbitals that do not have the same symmetry with respect to the mirror plane of symmetry is zero. Thus, the repulsive four-electron interaction π_1(butadiene, sub-HOMO) with π(ethylene, HOMO) is overcompensated by two stabilizing LUMO/HOMO interactions that are π_3^*(butadiene, LUMO) with π_1(ethylene, HOMO) and

Table 4.1 Gas-phase acidities given by $DH°(R^-/H^+)$ in kcal mol^{-1} of selected alkenes.

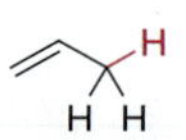	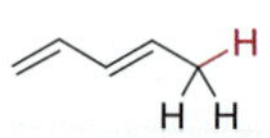	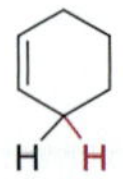	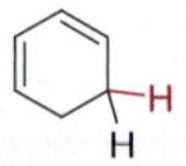	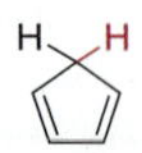	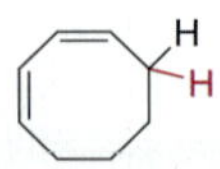
391 ± 1	371.5 ± 1.5	>387	372.2 ± 2.0	356 ± 1.0	375.5 ± 2.0
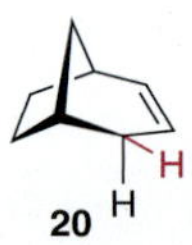 **20**	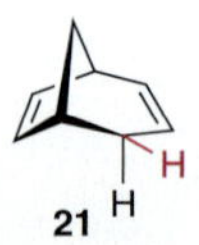 **21**	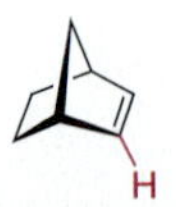	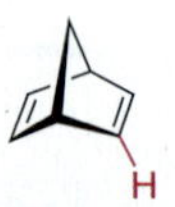	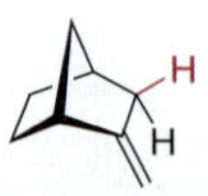 **27**	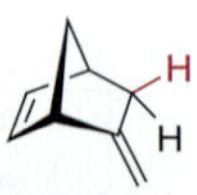 **28**
389 ± 2	379.6 ± 1.0	401 ± 1.0	389 ± 1.5	389 ± 2	389 ± 2

1 cal = 4.184 J.
Source: Taken from [81].

π^*(ethylene, LUMO) with π_2(butadiene, HOMO). Cyclic conjugation in cyclopropenyl cation ($C_3H_3^+$) leads to an electronic stabilization (called aromaticity, Section 4.5.4) that arises from the favorable overlap between a low lying LUMO(C^+) localized on one carbon center with the HOMO of the opposite C=C sub-fragment. Cyclic conjugation in cyclopentadienyl anion ($C_5H_5^-$) also leads to stabilization due to cyclic 6p electron delocalization known as aromaticity (Section 4.5.9). It arises from a favorable LUMO (allyl cation sub-fragment)/HOMO (C=C sub-fragment) overlap. In benzene (C_6H_6), cyclic electronic stabilization arises formally from two such LUMO/HOMO interactions as shown in Figure 4.19. The consequence of these LUMO (sub-fragment)/HOMO (sub-fragment) interactions is to make HOMO's that are lower lying in energy than in related acyclic species allyl cation ($C_3H_5^+$), pentadienyl anion ($C_5H_7^-$), and hexa-1,3,5-diene (C_6H_8), respectively.

Double photoionization of benzene in the gas phase gives the benzene dication ($C_6H_6^{++}$) for which a triplet ground state has been confirmed. This is predicted for a D_{6h}-species in which the two degenerate π-MOs π_2 and π_3 of benzene (Figure 4.19) become singly occupied (Hund's rule of maximum multiplicity). Below 4 K double ionization of helium/benzene complex yields primarily high-energy dications $C_6H_6^{++}$ with a six-membered ring structure. Planar $C_6H_6^{++}$ undergo rearrangement to a more stable pentagonal–pyramidal isomer with a cyclic C_5H_5 base and CH at the apex. By means of isomer-selective heating by a CO_2 laser, infrared predissociation spectra of both the classical and pyramidal dications were obtained [81]. Hogeveen's pyramidal dication $Me_6C_6^{++}$ (with a cyclic Me_5C_5 base and a MeC apex) has been characterized by X-ray diffraction studies for a crystalline salt [81].

4.5.11 3,4-Dimethylidenecyclobutene is not stabilized by π-conjugation

The Hückel method predicts that E_π(3,4-dimethylidenecyclobutene: **29**) $= 6\alpha + 7.2\beta$, which is 1.2β more than E_π(three ethylene units) $= 6\alpha + 6\beta$; therefore,

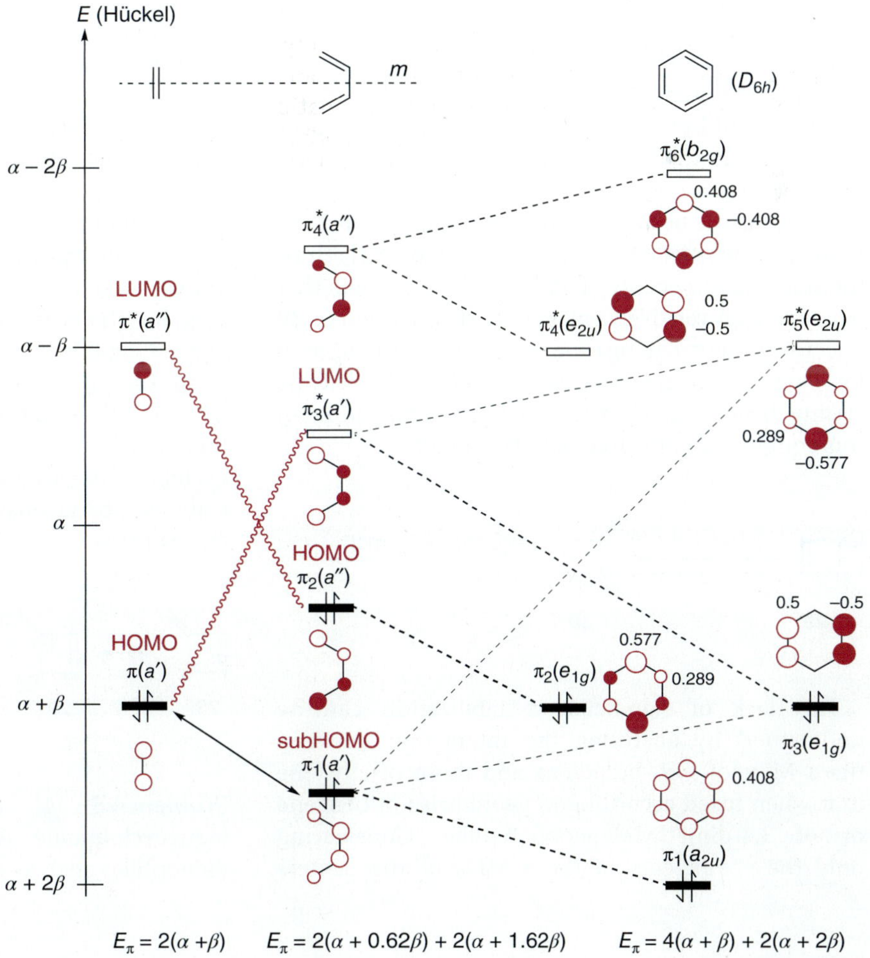

Figure 4.19 PMO diagram representing the π-MOs of benzene as resulting from the interactions of the π-MOs of a *s-cis*-butadiene with those of an ethylene unit sharing a mirror plane of symmetry m (a' ≙ symmetrical, a'' ≙ antisymmetrical with respect to m).

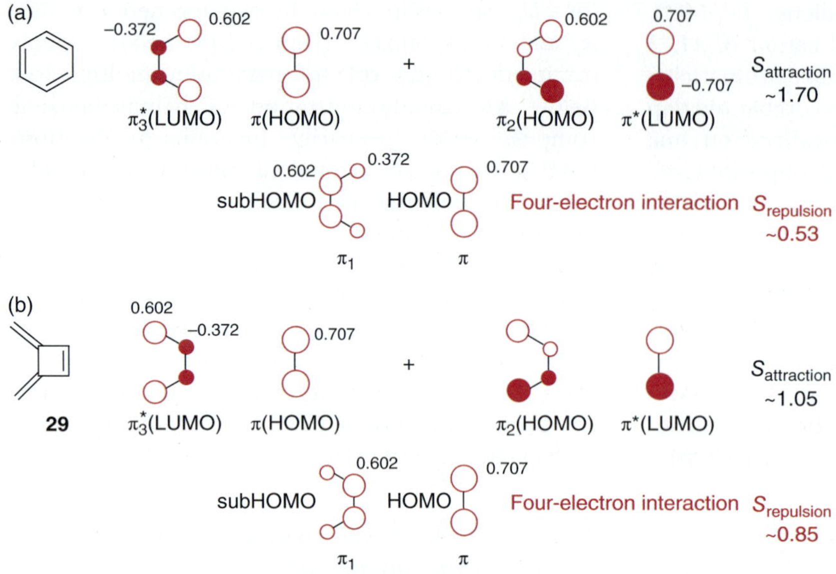

Figure 4.20 Representation of the overlaps between π-MOs of a *s-cis*-butadiene and an ethylene unit generating (a) a benzene molecule and (b) a 3,4-dimethylidenecyclobutene molecule (**29**).

3,4-dimethylidenecyclobutene (**29**) should experience electronic stabilization due to π-conjugation. Compared with E_π(butadiene) + E_π(ethylene) = $6\alpha + 6.48\beta$, **29** is stabilized by c. 0.7β. This is surely less than the stability gained in benzene. For a long time, it was believed that **29** does experience some noncyclic π-conjugative stabilization as predicted by the Hückel method, until Roth et al. measured the heat of hydrogenation of **29** to form 1,2-dimethylidenecyclobutane (**30**) (−31.6 kcal mol^{-1}), which was found to be nearly the same as the heat of hydrogenation of cyclobutene to form cyclobutane (−30.9 kcal mol^{-1}) [82, 83]. Thus, the stabilization energy of **29** is much lower than that of benzene. A relatively small extra stabilization of **29** due to cyclic π-conjugation might be compensated for by a ring strain increase and differential effects arising from π(C=C)/σ(C—H) hyperconjugation in this compound compared with that of **30**.

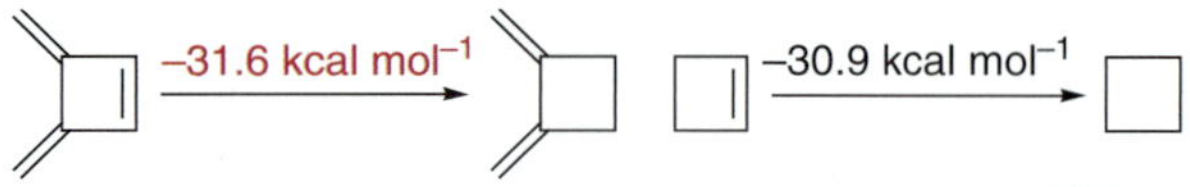

29 **30**

This lack of significant π-stabilization can be understood by analyzing the interaction between the π-MOs of *s-cis*-butadiene and those of ethylene unit when fused according to geometries of benzene or of 3,4-dimethylidenecyclobutene. Considering only the coefficients of the π-MOs of the centers becoming directly connected, one calculates (using Hückel eigenvectors) overlaps S'_k (given by the sum of the products of these eigenvectors) shown in Figure 4.20. The four-electron repulsive interaction sub-HOMO(diene)/HOMO(ethylene) is larger in **29** than in benzene. The two stabilizing LUMO/HOMO interactions are weaker in **29** than in benzene.

The microwave spectrum of 3,4-dimethylidenecyclobutene (**29**) enabled the measurement of the dipole moment of 0.64 D for this compound. It is the results of competitive electron transfer from the cycle to the exocyclic methylidene groups for the π-system and from the methylidene centers to the ring for the σ-framework [84]. The π-contribution can be explained considering that electron transfer from the four-membered ring to the extremities (**29** ↔ **29′** ↔ **29″**) of the butadiene moiety confers an aromatic character to this compound (a cyclobutene-3,4-diyl dication is realized, which is stabilized by aromaticity (4*N* + 2 electrons, *N* = 0)) [85, 86].

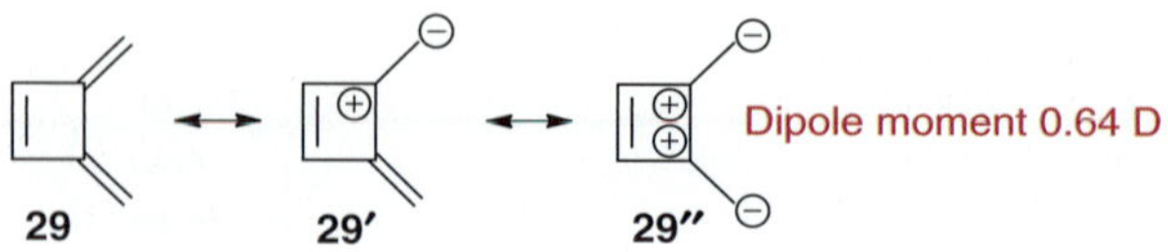

29 **29′** **29″**

Problem 4.3 [4] Radialene (1,2,3,4-tetramethylidenecyclobutane **A**) reacts with highly reactive dienophiles such as ethylenetetracarbonitrile (TCNE:

tetracyanoethylene), 4-phenyl-1,2,4-triazolinedione (PTAD), or diethyl azodicarboxylate (DEAD) giving the corresponding monoadducts. **B**. No bisadducts **C** can be detected under forcing conditions (large excess of dienophiles, prolonged reaction time, and heating) [87]. Why?

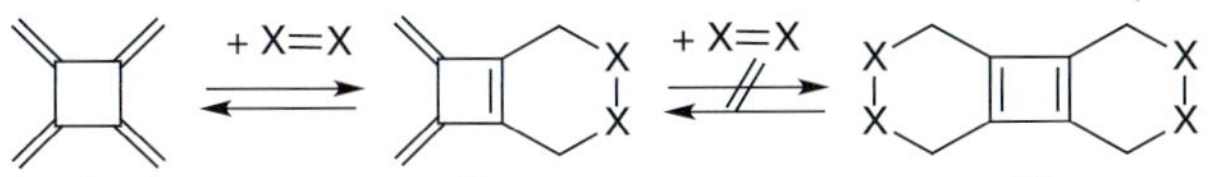

A **B** **C**

4.5.12 Fulvene

The Hückel method calculates for fulvene (**31**) $E_\pi(\mathbf{31}) = 6\alpha + 7.47\beta$, which corresponds to a stabilization due to π-conjugation of 1.47β [88]. This stabilization arise from the interaction of the π-MOs of a *s-cis*-butadiene unit with the π-MOs of an ethylene subunit as shown in Figure 4.21. Contrary to what was found for benzene and 3,4-dimethylidenecyclobutene (**29**), in fulvene (**31**), there is only one LUMO/HOMO stabilizing interaction.

In this interaction, electrons from the HOMO(π) of the exocyclic ethylene moiety are transferred to the LUMO(π_3^*) of the *s-cis*-butadiene subunit. Thus, C(6) becomes positively charged and the other centers of fulvene become negatively charged as indicated by limiting structures **31** ↔ **31′**. This electron transfer confers some aromaticity to the ring as it becomes a cyclopentadienyl anion. As a consequence, fulvene has a dipole moment of 0.42 D [89, 90]. Substitution at C(6) by electron-releasing group stabilizes fulvene. In the case of 6,6-dimethylfulvene (**32**), a dipole moment of 1.48 D is measured [91], much larger than in **31**. This confirms an important contribution of the charge transfer limiting structure **32′**. Interestingly, cyclopentadiene (**33**) also shows a dipole moment of 0.42 Debye, like fulvene (**31**). In this case, hyperconjugation **33** ↔ **33′** (see Section 4.8.5) confers some cyclopentadienyl anion character to this diene, which parallels the enhanced acidity of this hydrocarbon. Contrary to benzene, fulvene (**31**) has low susceptibility exaltations [92], no π-diatropic ring current [93], significant bond length alternation, and high reactivity. Quantum mechanical calculations have suggested that Li·**31** radical complex should be thermodynamically stable [94].

Problem 4.4 Acidity of 6,6-dimethylfulvene (pK_a = 22.7, DMSO) is higher than that of toluene (pK_a ~ 43, DMSO), diphenylmethane (pK_a = 32.3, DMSO), and triphenylmethane (pK_a = 30.6, DMSO, Table 1.A.24). Why?

Problem 4.5 Thiele prepared fulvenes by reaction of cyclopentadiene with aldehydes or ketones under base-catalyzed conditions [95, 96]. They can be obtained in better yields by reaction of imines with cyclopentadiene [97]. Propose a mechanism for the following metathesis of substituted fulvenes catalyzed by *n*-butylamine [98].

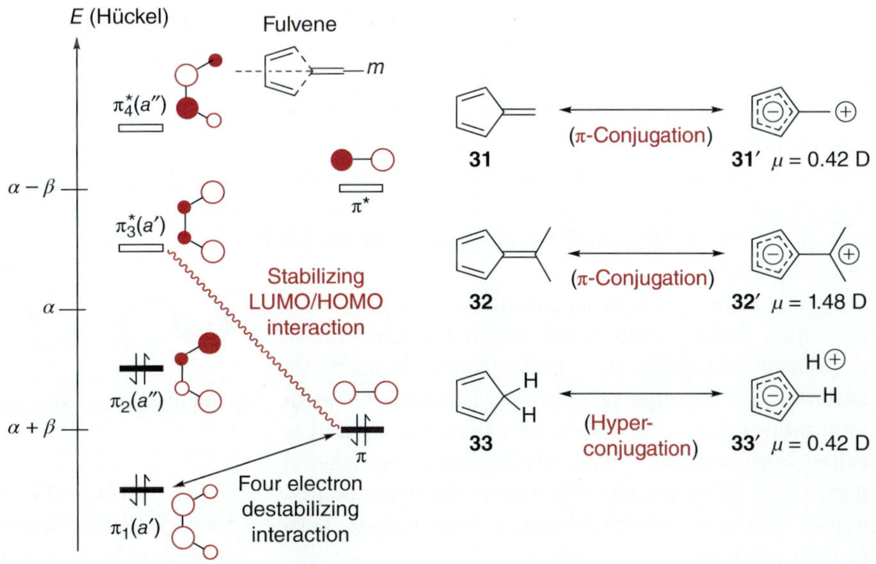

Figure 4.21 PMO diagram representing π-conjugation in fulvene as resulting from the interactions between π-MOs of a *s-cis*-butadiene moiety and an ethylene unit. The repulsive sub-HOMO(butadiene)/HOMO (ethylene) interaction is compensated by only one stabilizing LUMO(butadiene)/HOMO(ethylene) interaction. The LUMO(π^*, ethylene)/HOMO(butadiene) interaction is not possible for the reason of symmetry. π-Electrons are transferred from HOMO(exocyclic ethylene) to LUMO(endocyclic diene) (a' = symmetrical, a'' = antisymmetrical with respect to the mirror place of symmetry m).

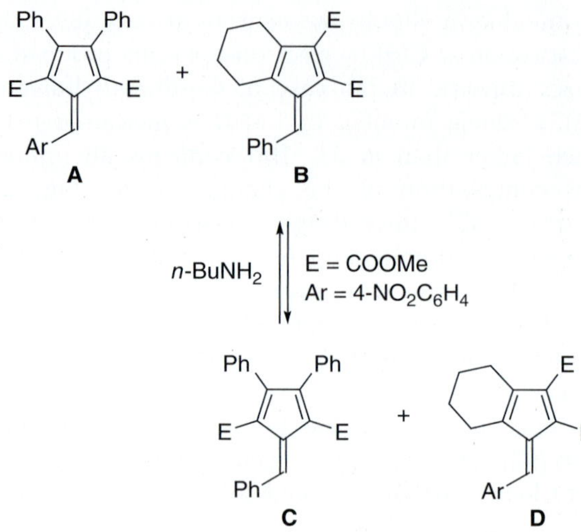

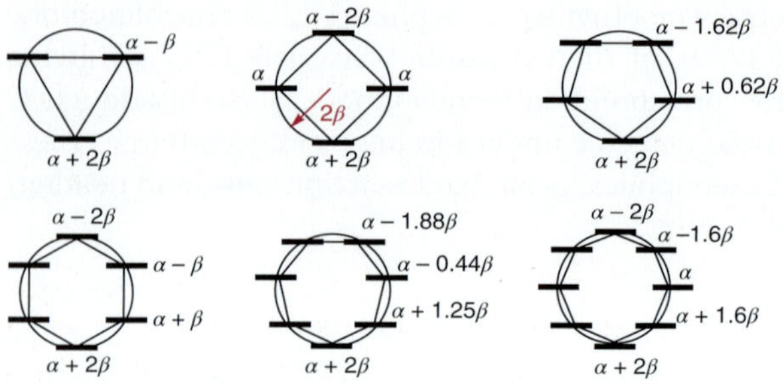

Figure 4.22 Frost circles for [N]annulenes (with $N = 3$–8) showing the Hückel π-MOs energies.

Problem 4.6 Write the Hückel secular determinant for [3]radialene (1,2,3-trimethylidenecyclopropane). Using PMO theory, do you expect this compound to experience significant electronic stabilization energy due to π-conjugation?

4.5.13 [*N*]Annulenes

Hückel rule of aromaticity for [N]annulenes (planar cyclic conjugated π-systems with $4N+2$ electrons, $N = 0, 1, 2, \ldots$) is based on the fact that the filled π-MOs of aromatic rings are well below the energy of nonbonding MOs that makes them "extra" stabilized electronically. [N]Annulenes with $4N$ electrons ($N = 1, 2, \ldots$) have partially filled or completely filled nonbonding MOs and antibonding MOs. They may also have bonding MOs that are unfilled or half-filled. The relative Hückel energies of the π-MOs in [N]annulenes can be determined by drawing Frost circles with radius 2β and inscribing into the circles regular polygon representing the [N]annulenes, with one corner lying on the bottom of the circles. The other corners of the polygon to define the Hückel energies of the π-MOs of the [N]annulene as shown in Figure 4.22.

By filling π-MOs of the [N]annulenes of Figure 4.22, one finds that cyclopropenyl cation (Section 4.5.4), a [3]annulene with two π-electrons, is aromatic and that cyclopropenyl anion, a [3]annulene with four π-electrons, is destabilized electronically and is called "antiaromatic." Similarly, cyclobutene-3,4-diyl dication, a [4]annulene with two π-electrons is aromatic, whereas cyclobutadiene, a [4]annulene with four π-electrons, is "antiaromatic." Cyclopentadienyl cation, a [5]annulene with four π-electrons, is "antiaromatic," whereas cyclopentadienyl anion, a [5]annulene with six π-electrons, is stabilized electronically: it has aromatic electronic stabilization (Section 4.5.9). Benzene is a [6]annulene with six π-electrons that also has aromatic electronic stabilization. Cycloheptatrienyl cation (tropylium ion, Section 4.7.2) is a stable carbocation for which the Hückel method gives E_π(tropylium cation) $= 6\alpha + 9\beta$ and which corresponds to an electronic stabilization due to cyclic π-conjugation of 3β. Cyclooctatetraene is not planar. It adopts a boat or tub D_{2h} conformation to avoid bond angle deformation (Baeyer strain). The planar D_{4h} structure is an intermediate 12 kcal mol^{-1} less stable than the boat or tub D_{2h} conformer on the energy hypersurface of the ring inversion and isomerization of cyclooctatetraene. This will be discussed in Section 4.5.14.

Three isomers of [10]annulene are known. For crystalline **34** (*mono*-(*E*)), five signals are observed in their ^{1}H- and ^{13}C-NMR spectra measured at −160 °C, and only one signal at −40 °C. Masamune suggested that **34** interconverts with the nearly planar structure **35**. A second crystalline [10]annulene **36** was proposed to undergo facile pseudorotation, rendering all CH groups equivalent on the NMR time scale [99–102].

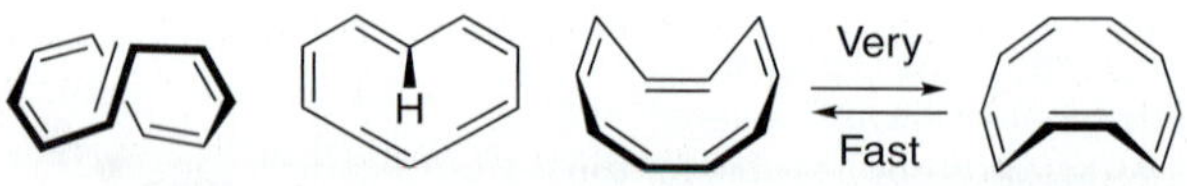

34 (twist) **35** (heart) **36** (boat) **36′** (boat)

At −45 °C, **34** is isomerized irreversibly into *trans*-9,10-dihydronaphthalene (**37**) *via* electrocyclic ring closure. Heart shaped annulene **35** is isomerized at −15 °C into *cis*-9,10-dihydronaphthalene (**38**).

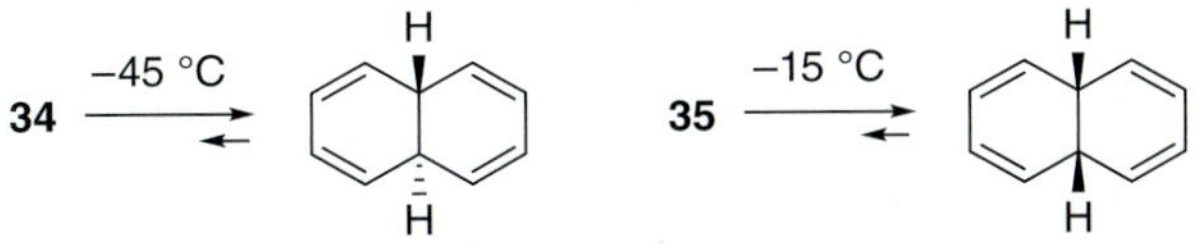

37 **38**

[10]Annulenes **34** and **36** were obtained by irradiating **38** in tetrahydrofuran (THF) at −60 °C (*hν*; λ = 253.7 nm). The UV absorption spectrum of **34** (λ_{max} = 255 nm, ε = 200; 265 nm, and ε = 130) is similar to that of *cis*-cyclonona-1,3,5,7-tetraene and is different from that of 1,6-methano[10]annulene (**39**). This demonstrates the absence or reduced π-conjugation in **34**, which cannot adopt a planar or nearly planar conformation (see Section 6.2.1). Moreover, its ^{1}H-NMR spectrum is consistent with a nonaromatic polyolefinic structure as no diatropism is detected (no ring current like in benzene). The UV absorption spectrum of [10]annulene (**36**: λ_{max} = 257 nm, ε = 29 000; 265 nm, ε = 20 000; 308 nm, ε = 320) demonstrates a better π-conjugation in this system than in **35**. Its NMR spectra show that **36** equilibrates quickly above −50 °C with two boat conformations for which no ring current is detected. Compound **36** is not diatropic (δ_H = 5.86 ppm, single signal at −40 °C; δ_C = 131.9 ppm). 1,6-Methano[10]annulene (**39**) and its heteroanalogs **40** and **41** are stable compounds [103–105]. They are diatropic like benzene and prefer, like benzene, to give products of substitution instead of products of addition with electrophiles. In the case of **39**, bromination at −78 °C generates intermediate adduct **42**, which, at 0 °C, eliminates 1 equiv. of HBr to produce **43**. In **39**, C—C bond lengths are within 1.37 and 1.42 Å, to be compared with benzene (1.39 Å) and butadiene (1.34 Å, 1.48 Å). A stabilization by aromaticity of −17.5 kcal mol^{-1} has been measured for **39** [106, 107]. In the case of 1,5-methano[10]annulene (**44**), a stabilization due to cyclic π-conjugation of only −6.1 kcal mol^{-1} has been evaluated. This is less than that in related azulene (**45**) for which an aromaticity of −12.8 kcal mol^{-1} was estimated. The latter compound can be seen as the annulation of a cyclopentadienyl anion with a tropylium cation. To complete the comparison with other cyclic conjugated systems containing 10 electrons, we note that bicyclo[6.2.0]deca-1,3,5,7,9-pentaene (**46**) does not present any stabilization due to cyclic π-conjugation. Thus, **46** can be seen as the annulation of cyclooctatetraene with cyclobutadiene. Cyclooctatetraene is readily reduced into cyclooctatetraene dianion. If planar, the latter system should be stabilized by aromaticity (4*N* + 2 electrons, *N* = 2). Recent quantum mechanical calculations rise doubt on the existence of such dianion [108].

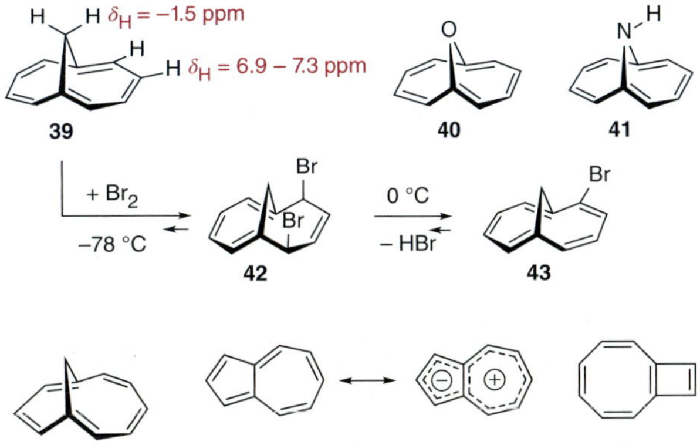

44 **45** (azulene) **46**

[12]Annulene (**47**) is not planar in its ground state but is very mobile as its ^{1}H-NMR shows a single signal at −150 °C [109, 110]. At −170 °C, the ^{1}H-NMR spectrum of **47** shows two types of protons at δ_H = 8.0 ppm (3H pointing inside the ring) and at δ_H = 6.0 ppm (9 H external of the ring. At −50 °C, **47** undergoes an electrocyclic cyclization (Section 5.2.9) giving bicyclo[6.4.0]dodec-2,4,6,9,11-pentaene (**48**). [12]Annulene **47** forms a stable dianion by reduction. A radical anion has been characterized by ESR spectroscopy, and this species is severely distorted from planarity [111].

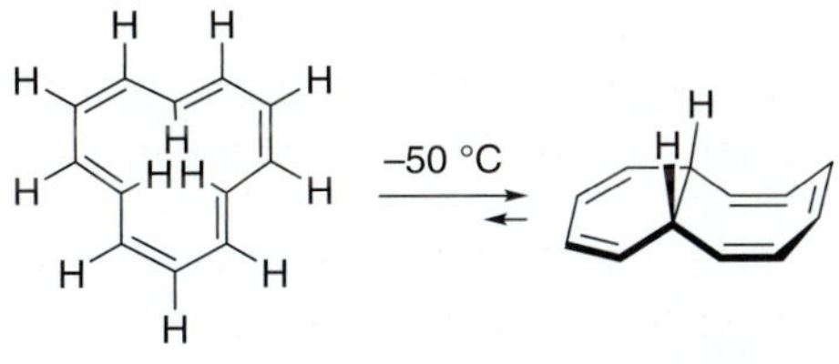

47 **48**

[14]Annulene **49** is nonplanar because of gauche interactions between the H—C groups pointing toward the center of the ring. Carbon–carbon bond lengths are comprised within 1.35 and 1.41 Å. This indicates a cyclic π-conjugation analogous to that of benzene. The diatropic character of **49** is demonstrated by its ^{1}H-NMR spectrum that shows two signals at −126 °C: one for the internal protons (δ_H = −0.61 ppm) and a second for the external protons (δ_H = 7.88 ppm) [112]. Although **49** seems to be aromatic, it is an unstable compound [113] that equilibrates with **50**. At −10 °C, ratio **49**/**50** is c. 92 : 8 [110].

Quantum mechanical calculations suggested that **49** might be converted into tetrahydrophenanthrene **52** by synchronous cycloaddition or independent electrocyclic cyclizations of an intermediate **51** [114].

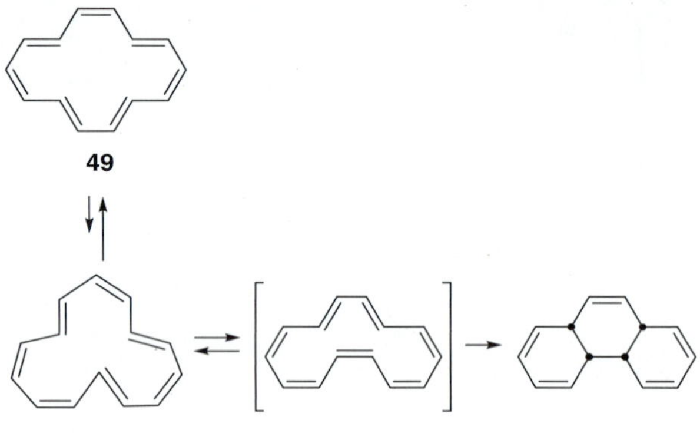

49 **50** **51** **52**

X-ray diffraction studies of 1,8-dimethyl[14] annulene prepared by reaction of Me_2SO_4 with the dilithium salt of octalene provided similar information than with **49**. In solution, there is an equilibrium between isomers **53a**, **53b**, and **53c** [115].

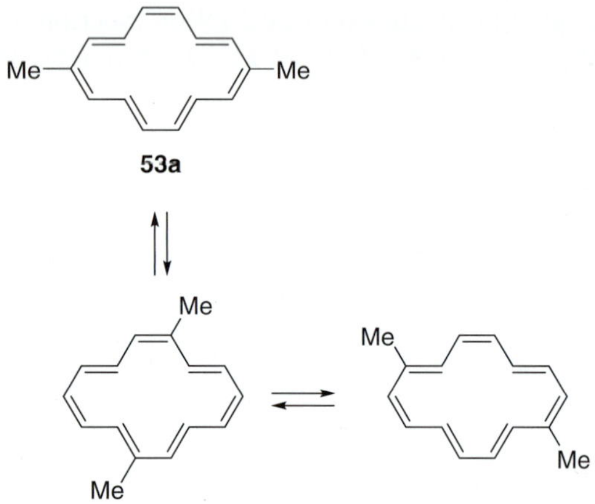

53a **53b** **53c**

Bridged [14]annulenes such as **54–56** are stable compounds for which NMR and X-ray data are consistent with the existence of some aromatic character [116–118].

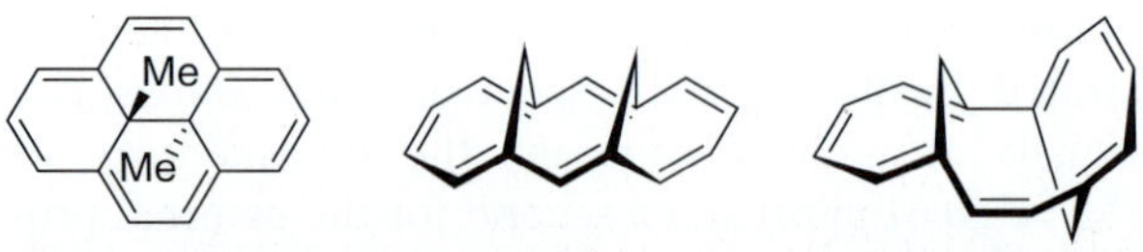

54 **55** **56**

[16]Annulene is prepared by UV irradiation of cyclooctatetraene dimer. It is not planar and is very sensitive to oxygen. It is a very conformationally mobile compound in which the energy minima correspond to **57** ⇄ **58** [119] with a ratio **57**/**58** of 88 : 12 at 163 K [110].

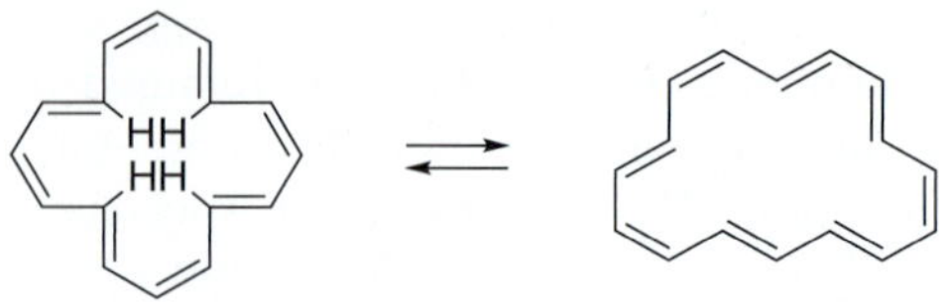

57 **58**

These species equilibrate with intermediate **59** that isomerizes into **60** [120]. At −130 °C, the ^{1}H-NMR spectrum of **57** ⇄ **58** shows two types of protons at $\delta_H = 10.56$ ppm (4H pointing toward the interior of the ring) and at $\delta_H = 5.32$ ppm (12H external of the ring). This is expected for paratropic annulenes with 4 N π-electrons; they induce a paramagnetic ring current. At −100 °C, a single peak is observed in the ^{1}H-NMR spectrum consistently with a fast exchange between the inner and outer protons [121, 122]. Quantum mechanical calculations have located other conformers than **57** and **58** such as twisted **59**, which involves a Möbius twist (Section 4.10) distributed through the all ring. [16]Annulene is readily reduced into its dianion with alkali metals [123]. The latter dianion dimetal salts are diatropic. In the solid state, **58** shows alternating C—C bond lengths for single bonds (1.44–1.47 Å) and double bonds (1.31–1.35 Å) [124].

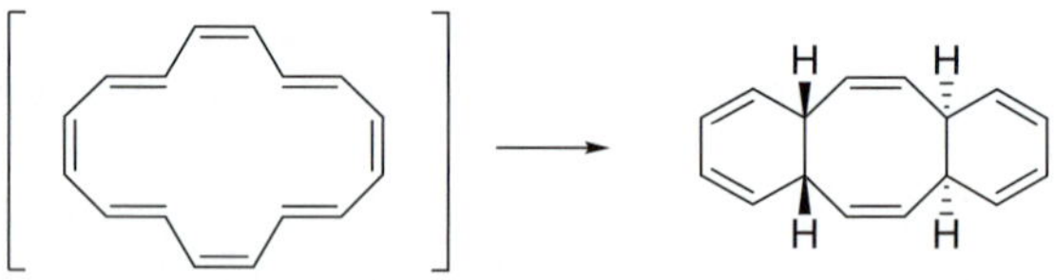

59 **60**

[18]Annulene (**61**) is highly diatropic ($\delta_H = -3.0$; 9.0 ppm) and nearly planar [125]. Its stabilization by cyclic π-conjugation leads to an ASE of c. −37 kcal mol^{-1}, very similar to that of benzene [126]. The Hückel method gives $E_\pi(\mathbf{61}) = 18\alpha + 23.035\beta$, which corresponds to a aromaticity of 5.035β (only 2β for benzene). Single-crystal X-ray radiocrystallography data were interpreted as [18]annulene sharing the D_{6h} symmetry. This might not be correct. Reduction of **61** with alkali metals forms two "antiaromatic" [18]annulene dianions in a 2.3 : 1 ratio, with double-bond localization. The dianions are paratropic [127].

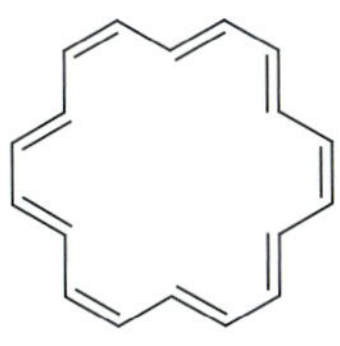

61

Higher [*N*]annulenes up to [30]annulene (except [26] and [28]annulene) have been prepared. [20]Annulene is paratropic [128], [22]annulene is diatropic [129], and [24]annulene is paratropic [130]. Antiaromatic destabilization energy due to cyclic π-conjugation in [24]annulene is estimated to reach c. 9–10 kcal mol^{-1}, in contrast to ASE of −10 kcal mol^{-1} estimated for [14]annulene [131]. [24]Annulene has alternating single and double C—C bonds. As the size of the ring increases, conformational flexibility leads to a loss of π-electron delocalization and the ring current decreases. The ASE is reduced as the ring gets larger, and the energy gap LUMO/HOMO is reduced, which causes nearly degenerate π-levels resulting in pseudo Jahn–Teller distortions. Where to place the transition point between a delocalized π-ring system and a localized system in [4*n*+2]annulenes has provoked a lot of discussions. Quantum mechanical calculations [132] have placed this transition at [30]annulene, a system with localized single and double C—C bonds, but for which the ^{1}H-NMR data suggest this system to be aromatic (diatropic). Thus, bond delocalization is not a prerequisite for diatropicity [133–135]. On treating 1,4,5,8-tetramethylnaphthalene with $SbF_5/Cl_2/SO_2ClF$ at −80 °C, dication **62** is obtained, [136] which is a planar π-conjugated system with eight π-electrons! Dications **63** and **64** are also obtained readily as stable salts in solution [137]. They are planar cyclic π-conjugated systems with 12 and 16 electrons, respectively. *Syn,syn*-1,6:8,17:10,15-trismethano[18]annulene is oxidized readily at −80 °C in $CH_2Cl_2/SO_2ClF/HSO_3F$ generating a solution of a stable dication **65** [138].

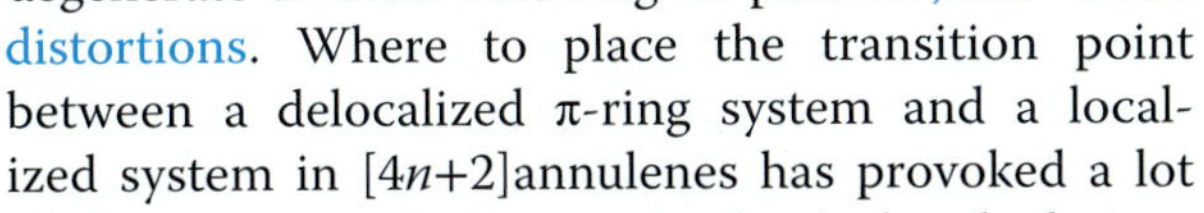

62 63 64 65

4.5.14 Cyclooctatetraene

Cyclooctatetraene (**66**) was first prepared by Willstätter and Waser in 1911 [139]. Unlike benzene, it is highly reactive toward electrophiles just like alkenes. Cyclooctatetraene (**66**) adopts nonplanar tub-shaped D_{2h} geometry (torsional angle between vicinal double bonds is 56°) with alternating single and double C—C bonds (1.48 Å, 1.34 Å) with a bond angle of 126.1° and hence behaves as a nonaromatic polyene rather than as an antiaromatic compound [140]. Standard heat of hydrogenation of D_{2h} cyclooctatetraene (**66**) into cycloocta-1,3,5-triene (**67**) amounts to −27.1 kcal mol^{-1} (Table 1.A.2), which suggests that **66** is neither stabilized nor destabilized by π-conjugation and that ring strain is nearly the same in **66** and **67**. However, this simple analysis ignores the differential effects due to hyperconjugative σ(C—C)/π and σ(C—H)/π interactions, as well as partial π/π conjugation in **66** and **67** [141].

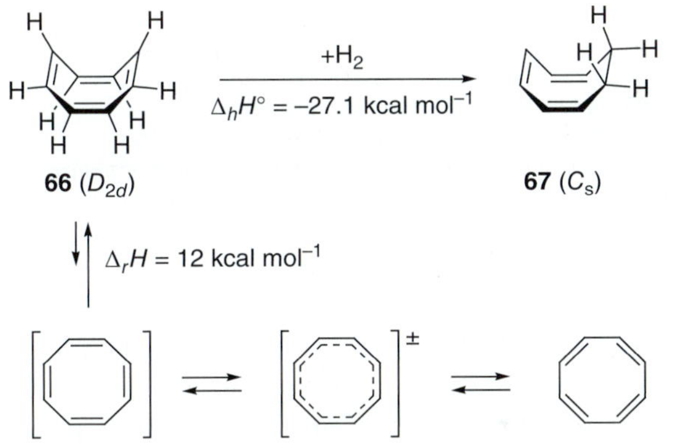

66 (D_{2d}) **67** (C_s) **68** (D_{4h}) **68′** (D_{4h})

Planar D_{4h}-cyclooctatetraene **68** suffers from angular strain (inner bond angle: 135° instead of 120°). The Baeyer strain is associated with a total angle deformation of 120°. If one takes 0.15 kcal mol^{-1} per degree as estimated for cyclopropane (Section 2.6.11), one calculates the ring strain in D_{4h}-cyclooctatetraene to reach 18 kcal mol^{-1}. As we shall see below, D_{4h}-cyclooctatetraene (**68**) is only 12 kcal mol^{-1} above D_{2d}-cyclooctatetraene (**66**). Accordingly, electronic destabilization in the planar D_{4h} geometry (**68**) associated with an hypothetical "antiaromatic destabilization energy" [141] cannot be large [142]. If the π-MOs of D_{4h}-cyclooctatetraene can be represented as the result of the interactions between the π-MOs of two *s-cis*-butadiene subunits, one finds (Figure 4.23) that the two repulsive four-electron interactions between HOMO/HOMO and sub-HOMO/sub-HOMO are probably compensated by four stabilizing two-electron supra-LUMO/HOMO and LUMO/sub-HOMO interactions. There are quantum mechanical calculations supporting this hypothesis [143].

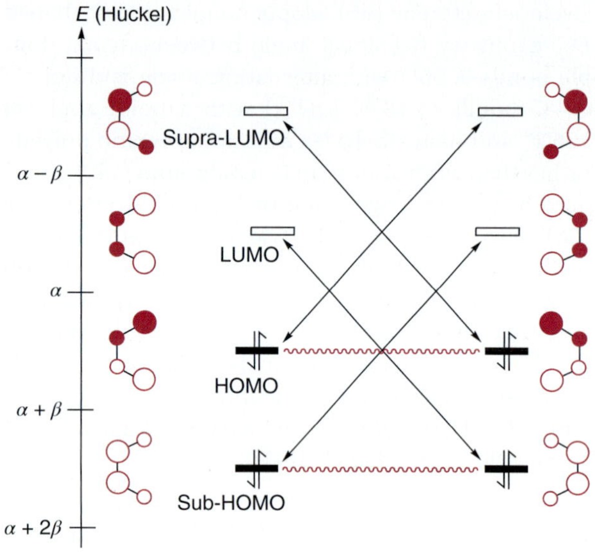

Figure 4.23 PMO diagram showing interactions between π-MOs of two *s-cis*-butadiene subunits in D_{4h}-cyclooctatetraene. The supra-LUMO/HOMO and LUMO/sub-HOMO interactions compensate the destabilizing HOMO/HOMO and sub-HOMO/sub-HOMO interaction.

The Hückel method calculates E_π(**68**) $= 8\alpha + 9.657\beta$, which corresponds to 1.657β of stabilization energy gained by π-conjugation with respect to four non-conjugated ethylene units. With derivative **69**, Anet and Bock have demonstrated that planar D_{4h}-cyclooctatetraene is an intermediate on the energy hypersurface of the tub ⇄ tub interconversion [144]. At 20 °C, the ^{1}H-NMR spectrum of **69** shows four signals for the methyl groups attached to the eight-membered ring. At this temperature, **69** is a slow equilibrium between four diastereomeric compounds **69a**, **69b**, **69c**, and **69d**. Diastereoisomers **69a** and **69b** are favored and represent 85% of the mixture. They are interconverted via planar intermediate **70a** (contains a D_{4h}-cyclooctatetraene moiety), whereas **69c** and **69d** are interconverted via planar intermediate **70b** (Scheme 4.10).

Above 70 °C, the ^{1}H-NMR spectrum of **69** shows that the two methyl signals of **69a** + **69b** coalesce into a single signal without coalescing with the methyl signals of **69c** + **69d**. At this temperature, and on the NMR time scale, the tub/tub interconversion is fast, whereas the valence isomerization that would interconvert **69a** + **69b** with **69c** + **69d** remains relatively slow. By UV irradiation (λ_{irr} = 350 nm) at −50 °C of the above mixture of **69a** + **69b** + **69c** + **69d**, one obtains a mixture in which the proportion [**69a** + **69b**]/[**69c** + **69d**] approaches unity. On letting this mixture warm to −5 °C, the proportion [**69a** + **69b**]/[**69c** + **69d**] varies from 1 : 1 to 85 : 15. This allowed one to evaluate (kinetics) the energy barrier of this phenomenon that implies bond shift (valence isomerization) in intermediates **70a** and **70b**. The process undergoes through an intermediate or transition structure of type **71** that contains a D_{8h}-cyclooctatetraene moiety. This transition structure is only 2–3 kcal mol^{-1} above the "D_{4h} structures" **70a** and **70b**. NMR studies on unsubstituted cyclooctatetraene (**66**) in nematic solvents gave an estimated energy barrier of 10.9 kcal mol^{-1} for the tub/tub + bond shift isomerization (rate: 250 s^{-1} at 0 °C) [145]. Quantum mechanical calculations predicted considerable 8π-antiaromatic paratropicity in planar cyclooctatetraene, for both the D_{4h} and D_{8h} structures [146].

Problem 4.7 Outline the synthesis of cyclooctatetraene presented by Willstätter. Propose another synthesis.

4.5.15 π-Systems with heteroatoms

Aldehydes and ketones are "oxaethylenes"**72** for which the secular determinant, in the Hückel method, writes

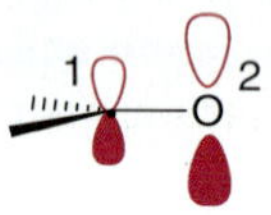

72

$$\begin{vmatrix} H_{11} - ES_{11} & H_{12} - ES_{12} \\ H_{21} - ES_{21} & H_{22} - ES_{22} \end{vmatrix} = \begin{vmatrix} \alpha - E & \beta_{CO} \\ \beta_{CO} & \alpha_O - E \end{vmatrix} = 0$$

The term α_O is the Coulomb integral for an electron in a 2p(O) orbital of the oxygen atom that makes a π-bond with the adjacent carbon center. In this model, the n_+ and n_- nonbonding electron pairs about the oxygen atom are ignored as they lie in the plane of the σ-framework. The resonance integral becomes β_{CO}, Hückel proposes $\alpha_O = \alpha + \beta$ and $\beta_{CO} = \beta$. This permits to write the following secular determinant:

$$\begin{vmatrix} \alpha - E & \beta_{CO} \\ \beta_{CO} & \alpha + \beta - E \end{vmatrix} = 0$$

This give the two roots: $E_1 = \alpha + 1.62\beta$ and $E_2 = \alpha - 0.62\beta$; and the MOs: $\pi_{CO} = 0.525\text{p(C)} + 0.85\text{p(O)}$ and $\pi^*_{CO} = 0.85\text{p(C)} - 0.525\text{p(O)}$ represented in Figure 4.24.

With respect to the π-MOs of alkenes, the π-MOs of aldehydes and ketones are lower lying in energy for both the HOMO(π_{CO}) and LUMO(π^*_{CO}). Importantly, and this is a consequence of the greater electronegativity of oxygen with respect to that of carbon, the carbonyl double bond **72** is polar, with about 50% of

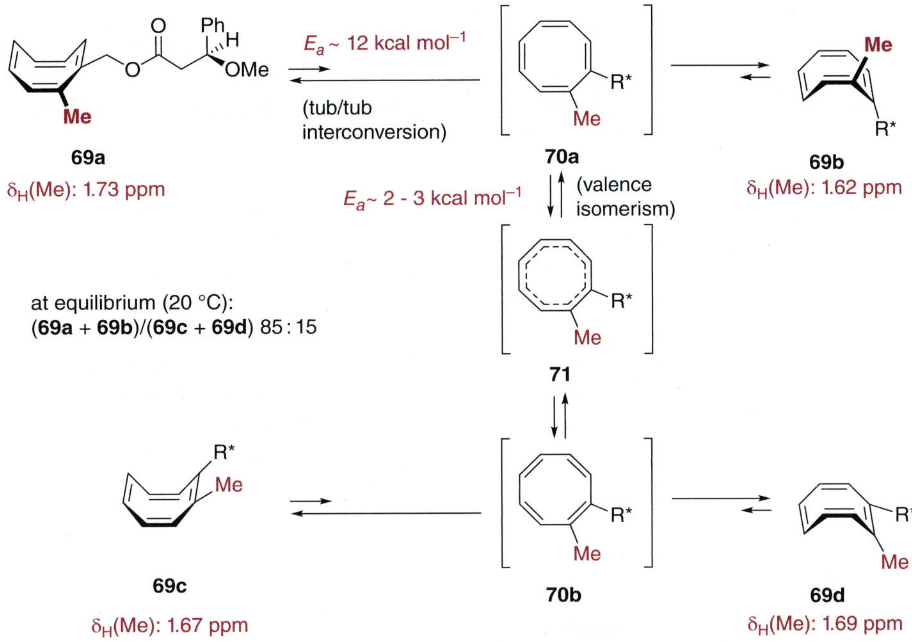

Scheme 4.10 Anet's [144] demonstration of tub ⇄ tub interconversion of cyclooctatetraene and valence bond isomerization in intermediate planar cyclooctatetraene. The stereogenic center of the O-methyl mandelic ester generates diastereomeric pairs **69a** + **69b** and **69c** + **69d**.

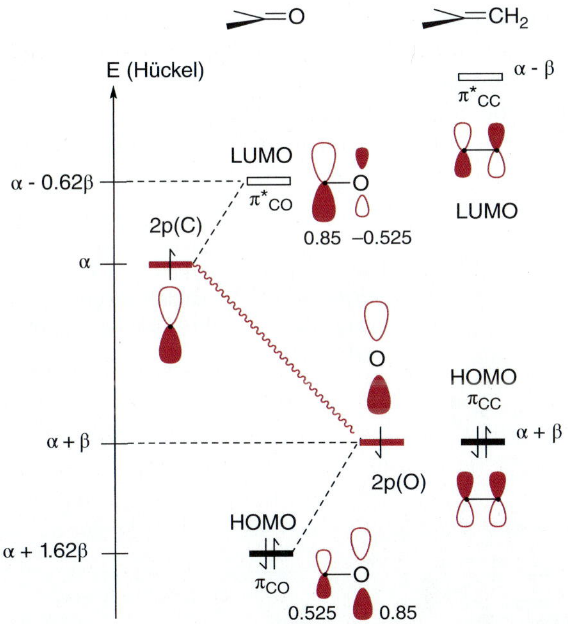

Figure 4.24 Hückel π-MOs of aldehydes and ketones. Comparison with π-MOs of alkenes.

the dipolar limiting structure **72′**. Indeed, in π_{CO}, the coefficient of the oxygen center (0.85) is about twice as large as that at the carbon center (0.525). Thus, the Hückel method reproduces the fundamental properties of aldehydes and ketones in which the oxygen center is basic and the carbon center is a Lewis acid. When an electrophile attacks a carbonyl moiety, it approaches preferentially the oxygen center, whereas a nucleophile H–Nu: prefers the carbon center. In fact, when H–Nu: approaches a carbonyl function, it looks for the best overlap between its HOMO and the LUMO of the carbonyl moiety. Clearly, the Hückel π^*_{CO} shows that the carbon center offers a better HOMO(H–Nu:)/LUMO(carbonyl) interaction as the coefficients at the carbon center is larger than that at the oxygen center (Figure 4.24). One should note, however, that adducts of type H–Nu$^+$–CR$_2$–O$^-$ resulting from the addition of H–Nu: to the carbon center of the carbonyl compound are usually more stable than those of type H–Nu$^+$–O–CR$_2^-$ resulting from the addition of H–Nu: to the oxygen center of the carbonyl compound. Furthermore, the tautomers Nu–CR$_2$–OH are generally more stable than isomeric Nu–O–CR$_2$–H. A priori, the regioselectivity of nucleophilic addition of carbonyl compounds is governed by the relative stability of the regioisomeric adducts that can form. It is fortuitous that the regioselectivity (preference for attack of the nucleophile onto the carbon center rather than the oxygen center) apparently correlates with hypothetical transition structures showing the best HOMO(H–Nu:)/LUMO(carbonyl) overlap.

72 **72′**

Imines have chemical properties intermediate of those of alkenes and carbonyl compounds. Hückel proposes $\alpha_N = \alpha_C + 0.5\beta_{cc} = \alpha + 0.5\beta$ and $\beta_{CN} = \beta_{cc} = \beta$.

Table 4.2 Hückel parameters for heterosubstituted π-system.

Z	—F	—Cl	—Br	—I	—OR	=O	=N	—NR_2	—Me
h_Z	2.84	1.45	1.16	0.78	2.0	1.18	1.47	1.47	0.88
k_{CZ}	0.68	0.57	0.38	0.19	1.31	1.93	1.06	1.30	0.18

When the heteroatom is a substituent of an alkene, diene, etc., it is one of its nonbonding electron pairs that offer a 2p-type orbital for conjugation with the π-system. The parameters collected in Table 4.2 have been proposed. Coulombic integrals are given as $\alpha_Z = \alpha + h_z\beta$ and resonance integrals as $\beta_{zw} = k_{zw}\beta$ (W = C or heteroelement).

For acrolein, the secular determinant becomes

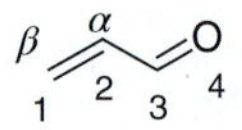

Acrolein

$$\begin{vmatrix} \alpha - E & \beta & 0 & 0 \\ \beta & \alpha - E & \beta & 0 \\ 0 & \beta & \alpha - E & 1.93\beta \\ 0 & 0 & 1.93\beta & \alpha + 1.18\beta - E \end{vmatrix} = 0$$

This gives the four roots for E and the four π-MOs shown in Figure 4.25. These orbitals can be seen as to arise from the interactions between the π-MOs of an ethylene and those of a formaldehyde unit. The destabilizing four-electron interaction HOMO(ethylene)/HOMO($H_2C{=}O$) is overcompensated by the LUMO($H_2C{=}O$)/HOMO(ethylene) interaction and, to a less extent, by the LUMO(ethylene)/HOMO($H_2C{=}O$) interaction. Overall, the four π-electrons of acrolein circulate in molecular orbitals π_1 and π_2 that are lower in energy than π(ethylene) and π_{CO}. This is the origin of the stabilization (−2 to −4 kcal mol^{-1}) found for the conjugation of a carbonyl function with an alkene system of α,β-unsaturated aldehydes, ketones, and esters (Section 2.7.3). The LUMO of acrolein (π_3^*) shows a larger coefficient at C(1) than at C(3), suggesting that a nucleophile can undergo 1,4-addition competitively with 1,2-addition when reacting with α,β-unsaturated aldehydes, ketones, carboxylic esters, and analogs (carboxamides, carbonitrile, nitroalkenes, sulfones, etc.). In fact, this regioselectivity might be controlled thermodynamically (the most stable adduct forms preferentially as the isomeric adducts are equilibrated during their formation) or kinetically (the regioisomeric adducts cannot be equilibrated and their proportion is given by the rate ratio of their formation). The allyl cation

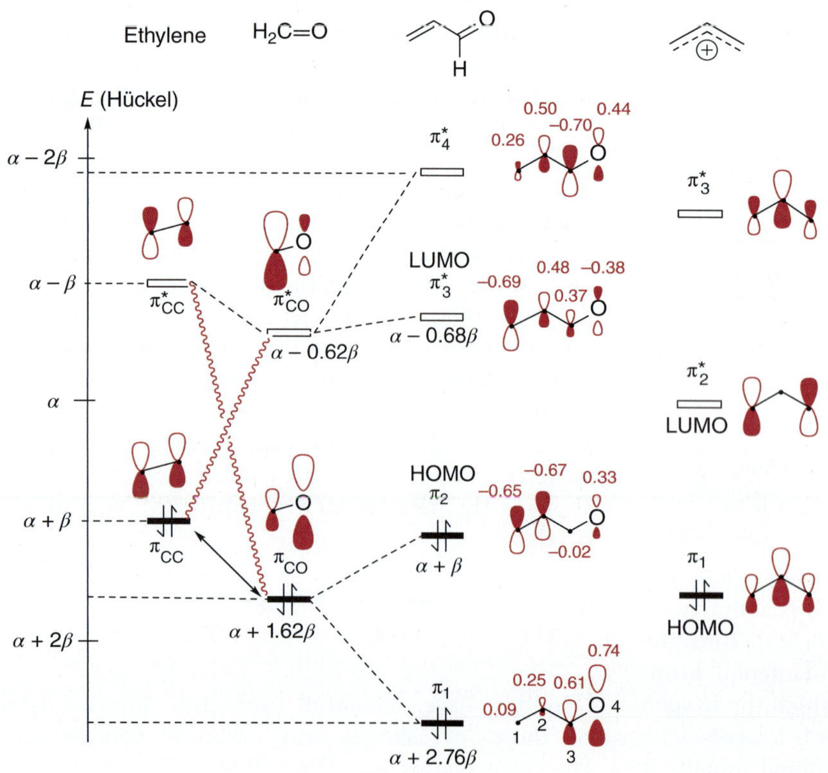

Figure 4.25 PMO diagram representing the π-MOs of acrolein resulting from the interactions between the π-MOs of an ethylene and a formaldehyde unit. Comparison with the Hückel p-MOs of allyl cation.

can be viewed as the model of ethylene substituted by an electron-withdrawing group like a carbonyl function. One consequence of this type of conjugation is to render the β-center of the alkene moiety electrophilic.

Problem 4.8 Write the Hückel secular determinant for methyl vinyl ether. Applying the PMO theory, do you expect this compound to experience electronic stabilization due to $n(\mathrm{O:})/\pi$ conjugation?

Problem 4.9 Write the Hückel determinant for oxirene (oxacyclopropene: $c\text{-}C_2H_2O$). Using the PMO theory, do you predict this compound to experience electronic stabilization due to $n(\mathrm{O:})/\pi$ conjugation or not?

4.6 Aromatic stabilization energy of heterocyclic compounds

In Section 2.7.8, we showed that furan and azole benefit from "aromatic stabilization energies" (ASE) as their hydrogenation into THF and pyrrolidine, respectively, are much less exothermic than that of cyclopentadiene. There has been a lot of debate of which thermochemical cycle should be used to define the ASE in heterocyclic compounds [147–149]. For furan, the difference in standard heats of hydrogenates between furan and cyclopentadiene amounts to $14.2 \pm 0.8\,\mathrm{kcal\,mol^{-1}}$.

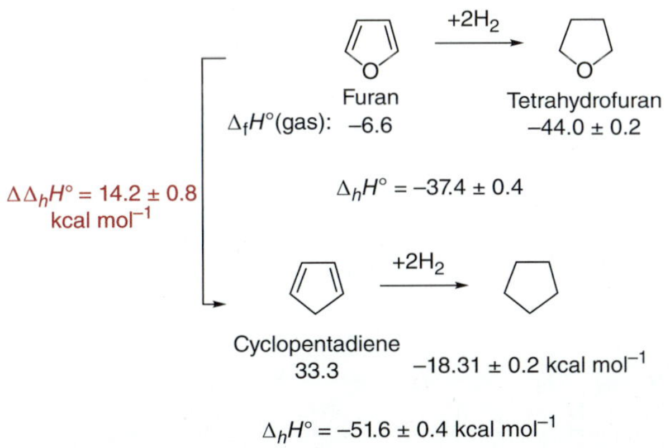

Can this $\Delta\Delta_h H°$ value be considered as the ASE of furan? The answer is no, as this comparison neither considers possible ring strain changes between the four five-membered ring compounds implied nor the "resonance energy" (RE) in cyclopentadiene, a conjugated diene. Ignored is also the possible stabilization of cyclopentadiene through hyperconjugation (Section 4.8.5). Another way to estimate ASE of furan is the standard heat of equilibrium (4.42) for which one obtains $\Delta_r H°(4.42) = -16.0 \pm 0.4\,\mathrm{kcal\,mol^{-1}}$.

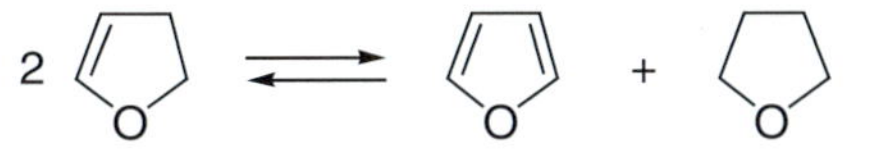 (4.42)

$\Delta_f H°$(gas): −17.3 ± 0.1   −66 ± 0.1   −44.0 ± 0.2

$\Delta_r H°(4.42)$: = −16.0 ± 0.4 kcal mol^{-1}

Assuming that the ring strain is the same in furan, 2,3-dihydrofuran, and THF, equilibrium (4.42) compares the ASE of furan with the RE of two molecules of 2,3-dihydrofuran for which one expects a possible stabilization due to $n(\mathrm{O:})/\pi$ conjugation (Section 2.7.5)! Similar equilibria (4.43) and (4.44) for cyclopentadiene and cyclohexa-1,3-diene reveal the possible RE in these conjugated dienes.

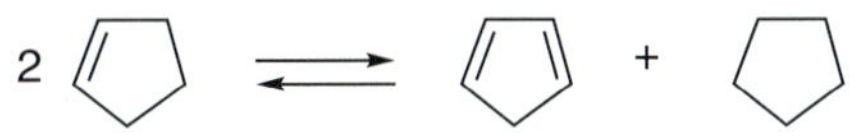 (4.43)

$\Delta_f H°$(gas): 8.5   33.2   −18.3 ± 0.2

$\Delta_r H°(4.43)$: = −2.0 ± 0.6 kcal mol^{-1}

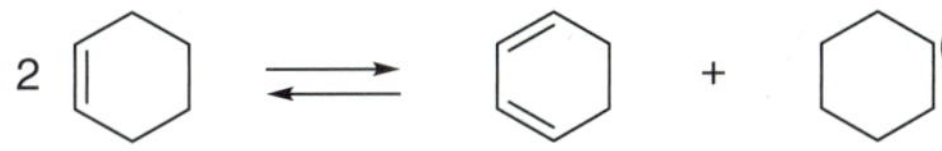 (4.44)

$\Delta_f H°$(gas): −1.0 ± 0.23   25.0 ± 0.15   −29.8 ± 0.2

$\Delta_r H°(4.44)$: = −2.8 ± 0.6 kcal mol^{-1}

Interestingly, equilibrium (4.45) is also endothermic, although cyclohexa-1,4-diene is not a conjugated diene. There must be changes in the ring strain among these compounds implied, or the existence of differential hyperconjugative effect involving the allylic C—H bonds.

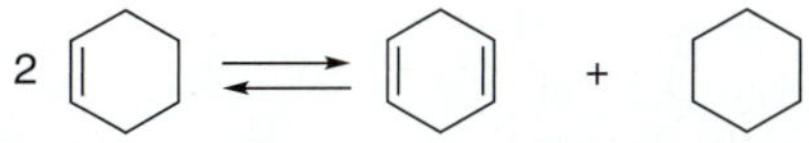

$\Delta_f H°$(gas): −1.0 ± 0.3   24.0 ± 0.7   −29.8 ± 0.2 kcal mol^{-1}

$\Delta_r H°(4.45)$: = −3.8 ± 1.2 kcal mol^{-1}

(4.45)

The above thermochemical analysis demonstrates the difficulties one faces on trying to find a quantitative assessment of ASE of cyclic polyolefins and heterocyclic compounds. Cyranski et al. use the homodesmotic equilibria (4.46) to estimate the ASE [149]. Based on quantum mechanical calculations, the ASE collected in Table 4.3 have been proposed.

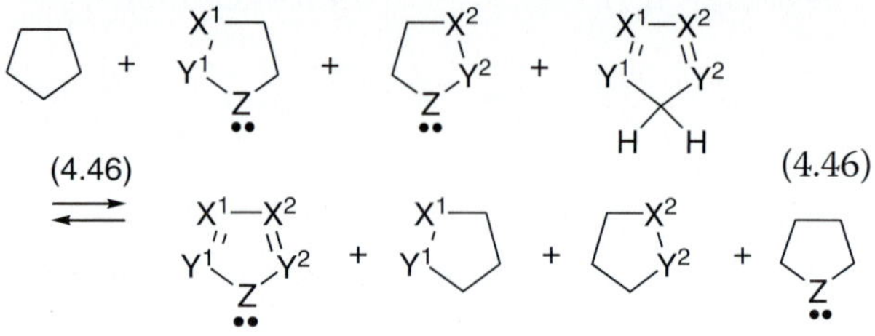

(4.46)

The origin of ASE in furan, azole, and thiophene arises from cyclic n(Z:)/π conjugation. The latter can be seen in the PMO diagram of Figure 4.26.

On approaching the *s-cis*-butadiene unit, the 2p(Z:) nonbonding electron pair of the heteroelement Z: is submitted to a repulsive interaction arising from the overlap sub-HOMO(diene)/HOMO(Z:). This destabilization is overcompensated by the stabilizing LUMO(diene)/HOMO(Z:) interaction. As the latter depends on the energy gap between these orbitals and assuming the same overlap (similar C–Z distances in the hetereocyclic compound) stabilization must be higher in azole than in furan as the energy gap LUMO(diene)/HOMO(N:) is smaller than LUMO(diene)/HOMO(N:) (Section 4.4.2), explaining the larger ASE in azole (−18.0 kcal mol^{-1}) than in furan (−12.3 kcal mol^{-1}, Table 4.3). Even though, sulfur being less electronegative than both nitrogen and oxygen, and thus the LUMO(diene)/HOMO(S:) energy gap is smaller than for azole and furan, the C—S bond being longer than C—O and C—N bonds leads to a smaller overlap between these orbitals. Furthermore, HOMO(S:) is a 3p orbital, not 2p as for O and N; thus, it is more diffuse, which also leads to a smaller overlap with LUMO(diene). This explains why thiophene is not more "aromatic" than furan and azole. Triazoles and tetrazoles are frequent in medicinal chemistry. One reports the following equilibria (tautomerism) [150].

Table 4.3 Calculated ASE (kcal mol^{-1}, 1 cal = 4.184 cal mol^{-1}) for selected heterocyclic and related compounds.

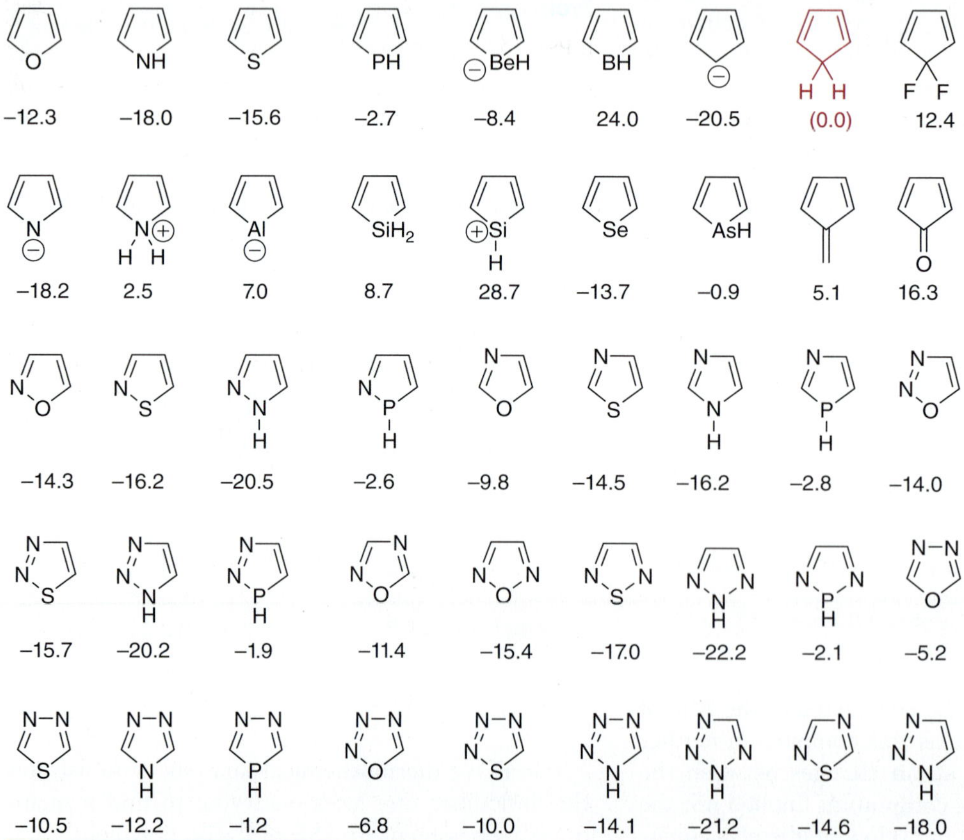

−12.3	−18.0	−15.6	−2.7	−8.4	24.0	−20.5	(0.0)	12.4
−18.2	2.5	7.0	8.7	28.7	−13.7	−0.9	5.1	16.3
−14.3	−16.2	−20.5	−2.6	−9.8	−14.5	−16.2	−2.8	−14.0
−15.7	−20.2	−1.9	−11.4	−15.4	−17.0	−22.2	−2.1	−5.2
−10.5	−12.2	−1.2	−6.8	−10.0	−14.1	−21.2	−14.6	−18.0

ASE < 0, stabilization by aromaticity; ASE > 0, destabilization by antiaromaticity.

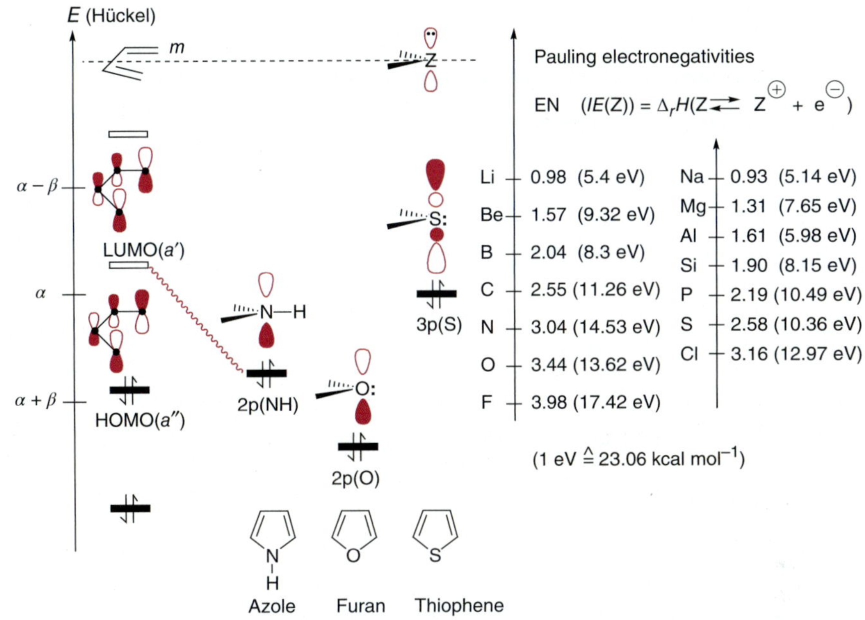

Figure 4.26 PMO diagram representing the electronic stabilization in azole, furan, and thiophene as the result of HOMO(Z:)/LUMO(*s-cis*-butadiene) interactions.

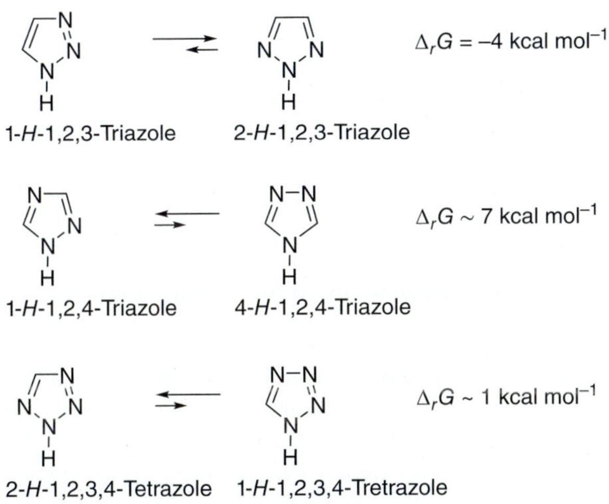

Problem 4.10 What are the products of the following reactions? Z = CH_2, O, NMe, and NSO_2Ph. See Section 2.7.8.

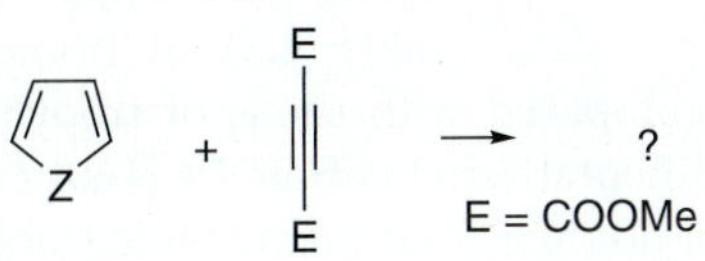

Problem 4.11 What are the products of direct bromination of furan in dimethylformamide [151]?

Problem 4.12 Which of 1-methylazole and pyridine is the best nucleophile to react with MeI?

Problem 4.13 Decide whether pyridine is more or less basic than azole.

Problem 4.14 Why is azole more acidic than piperidine?

Problem 4.15 Why does the acidity of azole increases upon introduction of nitrogen atoms in the ring?

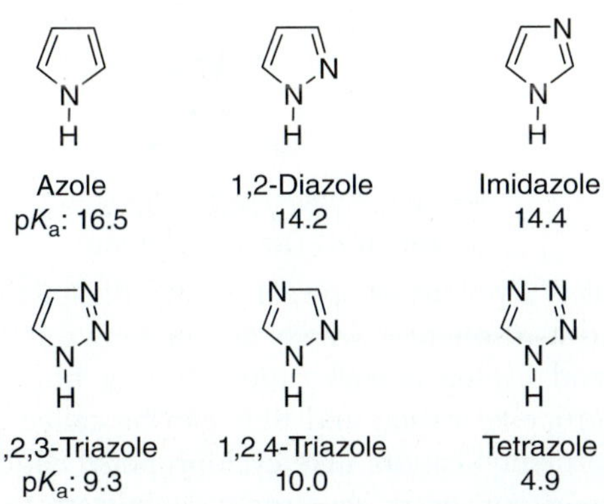

Problem 4.16 Pyridine reacts with electrophiles generating pyridinium salts reversibly. Substitution at the carbon centers is difficult. Clean formation of 3-nitropyridinum can be carried under Bakke's conditions when pyridine reacts with N_2O_5 in SO_2 at 0 °C. Give a mechanism.

4.7 Homoconjugation

4.7.1 Homoaromaticity in cyclobutenyl cation

Trimerization of but-2-yne into hexamethyl-*Dewar*-benzene (hexamethylbicyclo[2.2.0]hexa-2,5-diene) (reaction (4.47)) is catalyzed by Al_2Cl_6. The reaction implies the formation of a cyclobutenyl zwitterion **73** as the intermediate. The latter has been crystallized and analyzed by X-ray radiocrystallography. Interestingly, the distance between carbon center C(1) and C(3) is 1.775 Å, whereas the distance between C(2) and C(4) is 2.202 Å [152]. This demonstrates the existence of a transannular interaction between the 2p AOs at C(1) and C(3), which can be treated by the Hückel method by considering **73** to be a cyclopropenyl cation for which the resonance integral β_{13} is smaller than β, but not nil, as it would be for allyl cation (Figure 4.27).

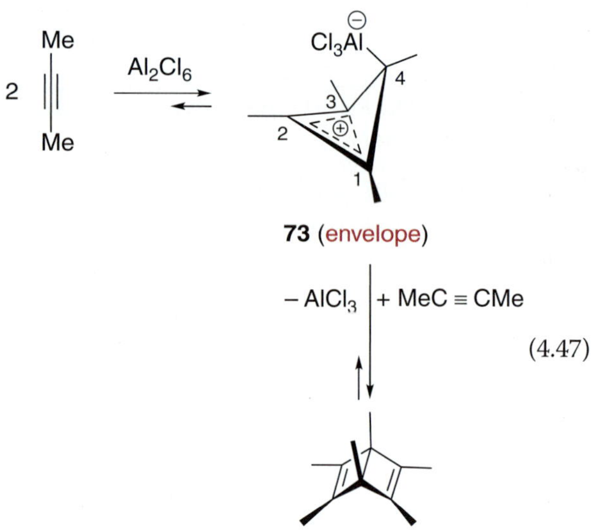

(4.47)

Hexamethylbicyclo[2.2.0]hepta-2,5-diene (hexamethyl-Dewar-benzene)

Thus, **73** possesses special electronic stabilization due to π-resonance, which lies in between that of allyl and cyclopropenyl cation. It is a homolog of cyclopropenyl cation and thus can be called homocyclopropenyl cation. Like cyclopropenyl cation that possesses an extra electronic stabilization called aromatic stabilization energy, homocyclopropenyl cation (**74**) benefits from homoaromatic stabilization energy [153], a concept first introduced by Winstein in 1959 [154]. Quantum mechanical calculations find methylcyclopropenyl cation more stable than **74** [155].

Olah et al. [156] have prepared the cyclobutenyl cation **74** and have determined a ring inversion barrier of $8.4 \pm 0.5\,\text{kcal mol}^{-1}$ termed the "homoaromatization energy" for **74** [157]. NMR data of **74** (Table 4.4) compared with those of cyclopropenyl and cyclopentenyl cation in the $SbF_5/HSO_3F/SO_2ClF$ solution suggested the existence of a diamagnetic ring current in **74** and a contribution of the bicyclic limiting structure **75** in the ground state of this cation. High-level quantum mechanical calculations confirmed the homoaromatic character of **74** and showed nearly equal charges on C(1), C(2), and C(3), the considerable 1,3-bond order, the short C(1)–C(3) distance, and large homoaromatic stabilization energy relative to the allyl cation [158].

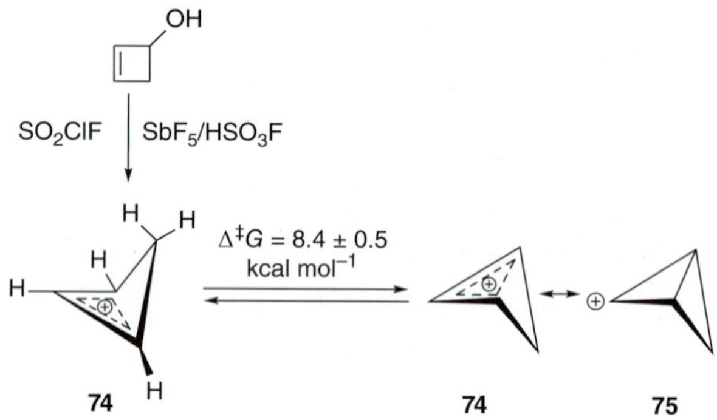

74 **74** **75**

4.7.2 Homoaromaticity in homotropylium cation

Protonation of cyclooctatetraene (**66**) generates a relatively stable homotropylium cation **76** [159]. Using D_2SO_4, monodeuterated **77** is obtained at −10 °C, the NMR spectrum of which slows temperature dependence associated with the ring inversion process equilibrating **77** with **77′** reversibly, and this with a free enthalpy of activation $\Delta^‡G = 22.3\,\text{kcal mol}^{-1}$. This observation demonstrates that the planar transition structure **78** of this process is about $21\,\text{kcal mol}^{-1}$ less stable than the envelop structures **77** and **77′**. Spectroscopic data (Table 4.5) of homotropylium cation **76** compared with those of tropylium cation and acyclic heptatrienyl cation **79** [160] confirm the existence of homoaromatic electronic stabilization in **76** [157, 161]. The diamagnetic current in this "homoaromatic cation" is evidence by the ¹H-chemical

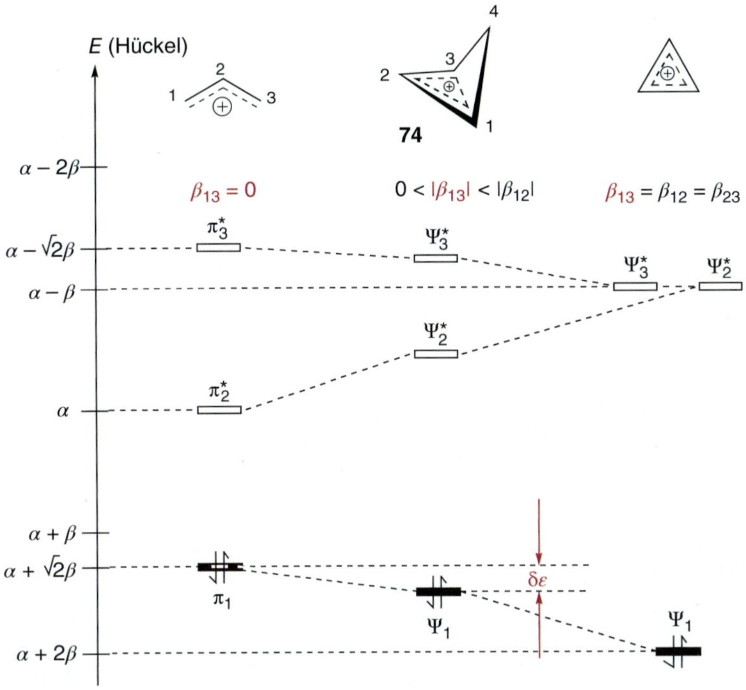

Figure 4.27 Hückel π-MO energies of allyl, cyclobutenyl (**74**), and cyclopropenyl cation. Homoconjugation is cyclobutenyl cation, a homolog of cyclopropenyl cation, leads to "homoaromaticity" because of resonance integral $\beta_{13} > 0$.

Table 4.4 NMR data of cyclopropenyl, cyclobutenyl (**74**), and cyclopentenyl cation as stable salts in $SbF_5/HSO_3/SO_2ClF$ solution (counter-ion; $SbF_5(SO_3F)^-$).

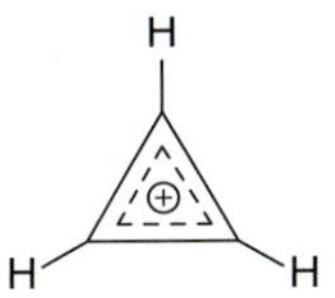	δ_H = 11.5 ppm (highly desheilded due to the positive charge and the diamagnetic ring current of the aromatic cation) δ_H (benzene) = 7.27 ppm δ_C = 175.9 ppm, $^1J_{C,H}$ = 265 Hz
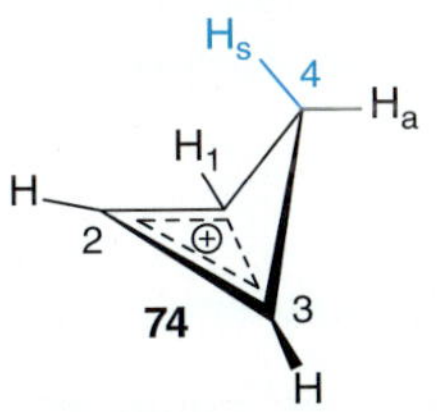 **74**	δ_H = 9.72 ppm (H—C(2)) 7.95 ppm (H—C(1), H—C(3)) 4.53 ppm (H_s—C(4)) (sheilded) 6.62 ppm (H_a—C(4)) (desheilded due to diamagnetic ring current) δ_C = 175.9 ppm, $^1J_{C,H}$ = 236 Hz (C(2)) 130.0 ppm, $^1J_{C,H}$ = 211 Hz (C(1,3)) 54.0 ppm, $^1J_{C,H}$ = 167 Hz (C(4))
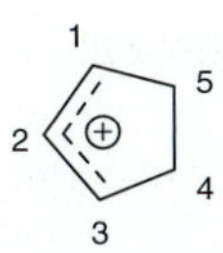	δ_H = 11.26 (H-1, H-3); 8.65 ppm (H-C(2)) δ_C = 234.7 (C(1,3)); 145.7 ppm (C(2))

Table 4.5 Spectral data of homotropylium cation (**76**) compared with those of tropylium cation (aromatic) and acyclic heptatrienyl cation **79** (nonaromatic).

 Tropylium cation cycloheptatrienyl cation	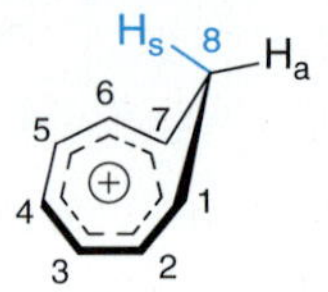 **76**	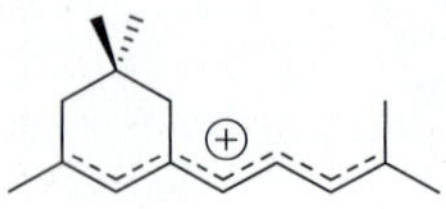 **79**
$\delta_H = 9.0$ ppm	$\delta_H = 8.6$ (H—C(2,3,4,5,6)) 6.6 (H—C(1,7)) 5.2 (H_a—C(8)) −0.4 (H_s—C(8))	
$\lambda_{max} = 217$ ($\log \varepsilon = 4.61$) 274(3.63)	233 ($\log \varepsilon = 4.32$) 313(3.48)	470 nm ($\log \varepsilon = 5.45$)

shift of $H_2C(8)$, the proton *syn* (H_s-C(8)) residing above the ring is strongly shielded [162].

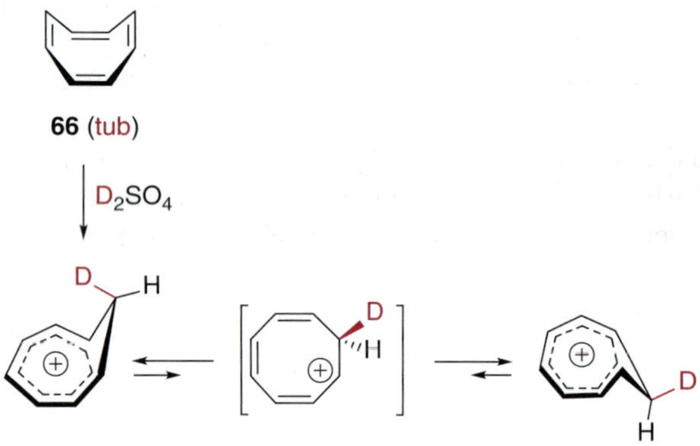

77 (envelope) **78** (planar) **77′** (envelope)

In HCl/SbCl$_5$/CH$_3$NO$_2$, homotropolone **80** generates a salt **81** that precipitates upon addition of benzene. It can be recrystallized from CH_2Cl_2. X-ray radiocrystallography of **81** shows a relatively short distance between C(1) and C(7) of 1.628 Å and delocalized double C=C bonds with bond length C(1)—C(8) and C(7)—C(8) of 1.488 Å. This demonstrates the existence of a strong interaction between centers C(1) and C(7). In the Hückel method, this implies a resonance integral $\beta_{17} > 0$, making **81** to be stabilized by homoaromaticity. The data also confirm contribution of limiting structure **81′** to the ground-state properties of **81** [163].

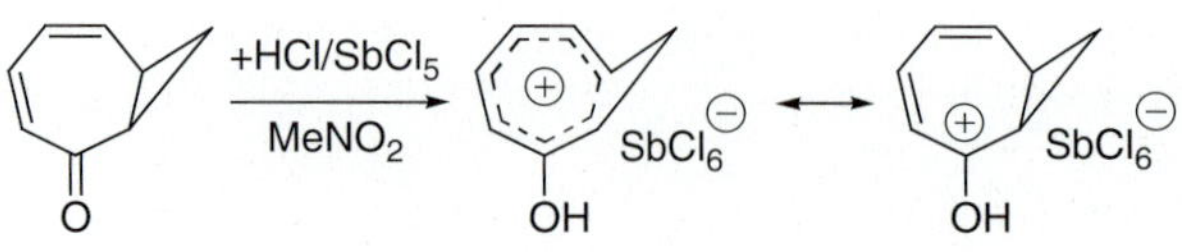

80 **81** **81′**

Homoaromatic carbanions are also known. The prototype would be cyclohexadienyl anion **82**, an intermediate in the Birch reduction of benzene. This anion adopts a planar conformation in its ground state with little or no interaction between the 2p orbitals at C(1) and C(5) [164]. Proton affinity (Table 4.1) of cyclohexadienyl anion (**82**) is slightly smaller than that of cycloocta-2,4-dienyl-1-yl anion. This is consistent with an extra stabilization in **82**.

82 (planar)

Problem 4.17 What type of stable cation in solution is formed upon treatment of 2,7-bis(tertiobutyl) thiepin in HSO_3F/SO_2 [165]?

4.7.3 Homoaromaticity in cycloheptatriene

Cycloheptatriene (**83**) adopts a boat conformation with C_s symmetry in its ground state [166]. Low-temperature ^{1}H-NMR permitted to estimate a free energy of activation of c. 6 kcal mol^{-1} for the boat/boat conformational isomerization through a planar (C_{2V}) transition structure **84** [167].

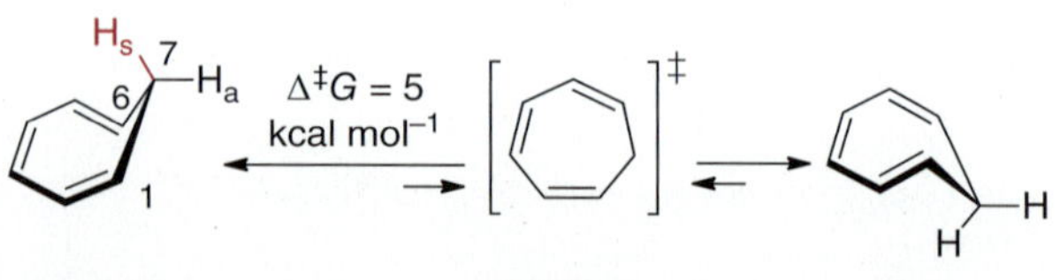

83 (C_S) (boat) **84** (C_{2V}) (planar) **83′** (boat)

Heats of hydrogenation of **83** into cycloheptane ($-70.5\,\mathrm{kcal\,mol^{-1}}$) compared with the heat of hydrogenation of 3 equiv. of cycloheptene ($3(-25.9)$ $\mathrm{kcal\,mol^{-1}}$) suggest a stabilization of $-7.2\,\mathrm{kcal\,mol^{-1}}$ in **83** due to π-conjugation [168]. Existence of a weak interaction between 2p AOs at C(1) and C(6) in **83** responsible of an apparently weak homoaromatic electronic stabilization is indicated by the chemical shift difference of 1.4 ppm for the *syn* (H_s-7) and *anti*-methylene protons (H_a-7) at C(7) in the ^{1}H-NMR spectrum measured at −170 °C. This witnesses the existence of some diatropic ring current in **83**. Furthermore, some magnetic susceptibility exaltation has been found for this compound (c. 59% of that of benzene) [169, 170]. A small bond length alternation is also found in **83** and derivatives [171].

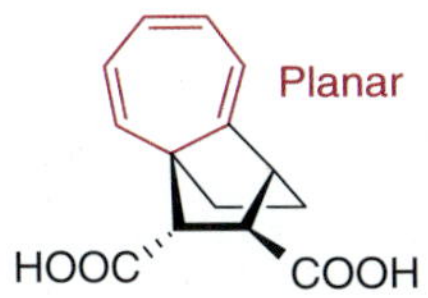

85

Derivative **85** possesses a planar cycloheptatriene moiety for which the UV spectrum and X-ray data confirm that this triene does not share any "homoaromatic features." [172] Quantum mechanical calculations confirm that cycloheptatriene (**83**) is a neutral homoaromatic hydrocarbon. Substitution at C(7) by electron-releasing substituents (e.g. OH, NH_2, Me, and CMe_3) does not significantly affect the homoaromatic characters (geometry, energetic, and magnetic criteria) of cycloheptatrienes. Interestingly, the calculations suggested some antiaromaticity to planar cycloheptatriene **84** attributed to the involvement of the pseudo-2π-electrons of $H_2C(7)$ with the 6π-electrons of the ring to give an 8π-electron system **84′**. A similar hyperconjugative effect is found in cyclopentadiene, which has some "pseudo-aromaticity" (Figure 4.12; Section 4.8.5). Cycloheptatriene equilibrates with the less stable norcaradiene **86** *via* a highly aromatic transition structure, but norcaradiene (**86**) is less aromatic than cycloheptatriene (**83**) [173]. The position of this equilibrium can be modified by annulation and substitution at C(7) (see Figure 5.13)

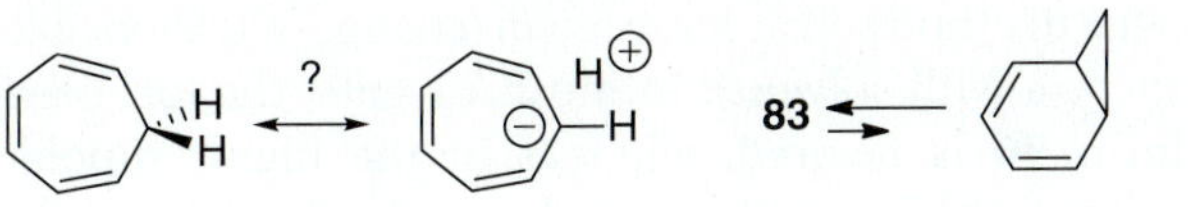

84 (planar) **84′** (antiaromatic) **86** (norcaradiene)

4.7.4 Bishomoaromaticity in bishomotropylium ions

We have seen (Section 4.5.9) that the conjugate bases of bicyclo[3.2.1]octa-2,6-diene (**21**), bicyclo[3.2.2]nona-2,6-diene (**23**), and bicyclo[3.2.2]nona-2,6,8-triene (**24**) involve special electronic stabilization that can be referred to bishomoaromatic stabilization energy as these species can be viewed as bishomocyclopentadienyl anions. The PMO diagram of Figure 4.28 suggests that bishomotropylium ions of type **87** should also have bishomoaromatic stabilization energy resulting from the double homoconjugation between a *s-cis*-butadiene and an allyl cation approaching to each other in geometries maintaining a mirror plane of symmetry (C_s structures). The homoaromatic stabilization energy arises from LUMO(*s-cis*-butadiene)/HOMO(allyl cation) and LUMO(allyl cation)/HOMO(allyl cation) interactions. An example of cation of type **87** is bicyclo[4.3.0]nona-2,4,8-trien-7-yl cation (**89**), also called 1a,3a-dihydro-1*H*-inden-1-yl cation, which is obtained as stable salt in solution by dissolution of tricyclo[4.1.0.0^{4,9}]nona-2,5-dien-7-*exo*-ol (**88**) in HSO_3F/SO_2ClF at −135 °C [174]. Another example of bishomoaromatic cation is bicyclo[4.2.1]deca-2,4,8-trien-7-yl cation (**91**) obtained by protonation of bicyclo[4.2.2]deca-2,4,7-triene (**90**) in HSO_3F/SO_2ClF [175]. ^{1}H-NMR data of **91** confirm the existence of a diatropic ring current.

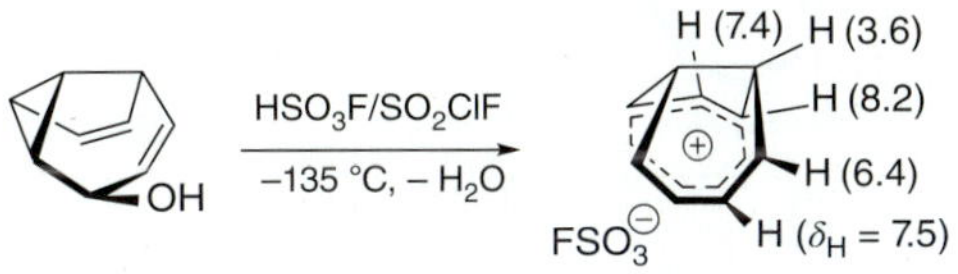

88 **89**

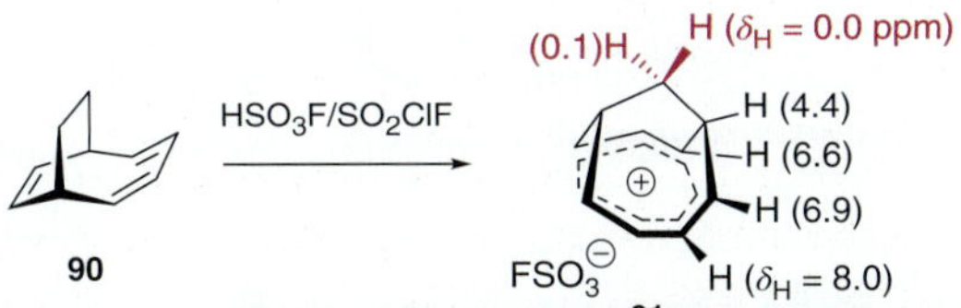

90 **91**

According to Winstein, cations **89** and **91** are examples of 1,4-bishomotropylium ions. An example of 1,3-bishomotropylium ion is given with nona-2,4,7-trien-1-yl cation (**93**) obtained by dissolution of bicyclo[6.1.0]nona-2,4,6-triene (**92**) in HSO_3F/SO_2ClF at −125 °C [176, 177]. The bishomoaromatic character of cation **93** is evidenced by comparing its ^{1}H-NMR data with those of cyclohepta-2,4-dien-1-yl cation, cyclohexene, and benzene. In particular, the chemical

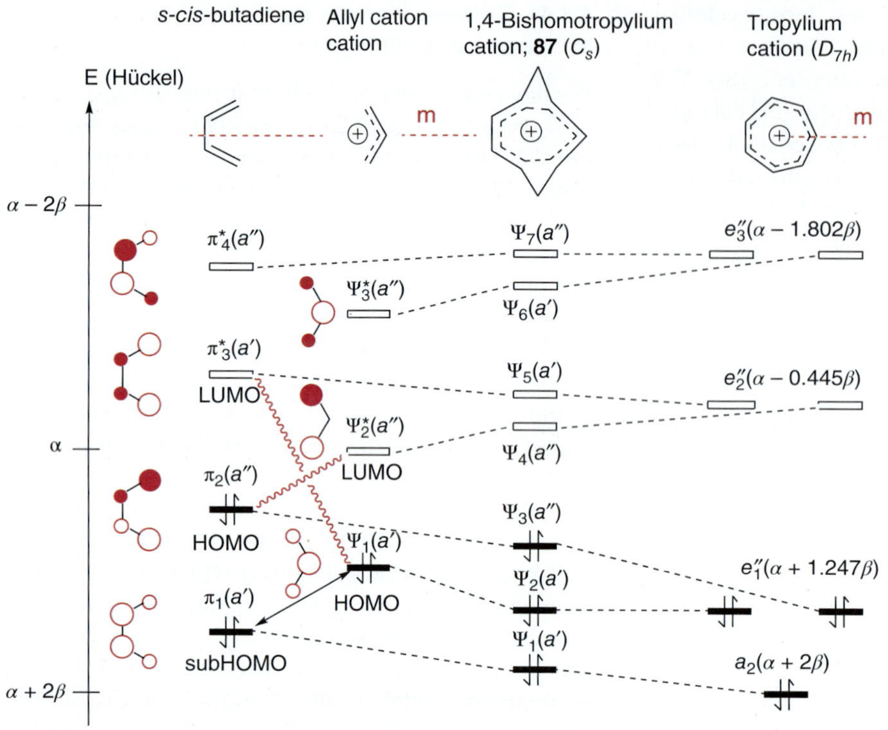

Figure 4.28 PMO diagram showing the π-MOs of bishomotropylium ion **87** as the result of the interaction between π-MOs of *s-cis*-butadiene and allyl cation. π-MOs Ψ_i (**87**) are compared with the Hückel π-MOs of tropylium cation (a' = symmetrical, a'' = antisymmetrical with respect to the mirror plane of symmetry *m*).

shift difference $\Delta\delta_H = 1.9$ ppm for the *syn* H-C(6,9) and *anti*-H-(6,9) protons confirms the existence of a diatropic current in cation **93**.

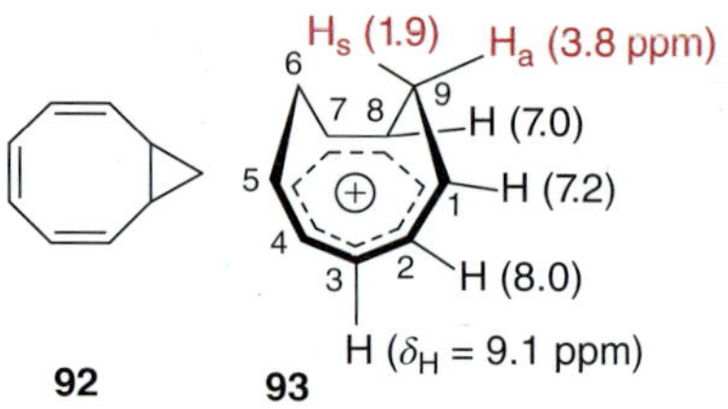

92 **93**

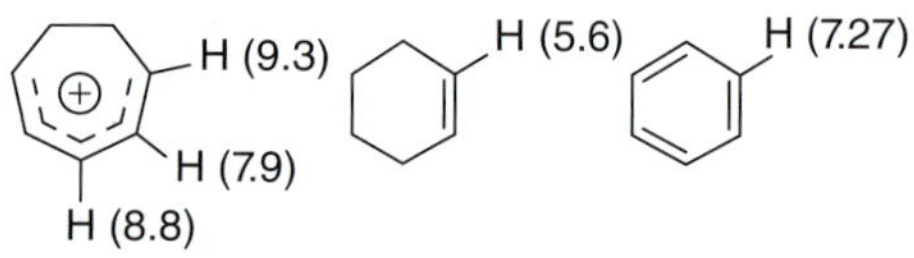

4.7.5 Bishomoaromaticity in neutral semibullvalene derivatives

Parent semibullvalene (**94**) (synthesis: Section 6.3.4) undergoes degenerate Cope rearrangement **94** ⇄ **94′** (sigmatropic shift of order (3,3), Section 5.5.9) through homoaromatic transition structure **95**, with a low activation barrier: $\Delta^{\ddagger}G \simeq 6\ \text{kcal mol}^{-1}$ [178–180].

Structure **95** is a bishomobenzene derivative that can be seen as two allyl radicals approaching to each other in a C_S symmetrical geometry. The PMO theory for this system predicts a substantial stabilization

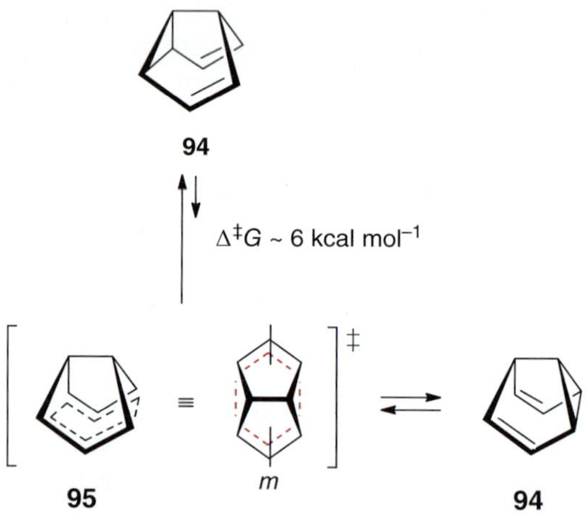

94 **95** **94**

arising from the SOMO(allyl)/SOMO(allyl) overlap and also to a much smaller extent, from the two LUMO(allyl)/HOMO(allyl) interactions (Figure 4.29).

Substitution of semibullvalene affects the energy difference between the localized structure **94** and the delocalized structure **95**. For instance, derivative **96**, which is very closed in energy to **97**, exhibits thermochromic properties (change of color with temperature) and solvatochromism (change of UV–visible spectra with solvent). In most solvents, the localized form **96** is favored, whereas in the highly dipolar solvent *N*,*N′*-dimethylpropyleneurea, the delocalized bishomoaromatic species **97** makes a stable solvate [181]. In contrast, derivative **98** follows the opposite

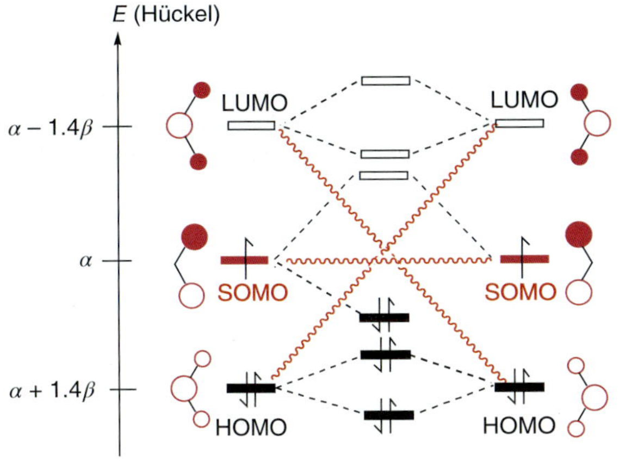

Figure 4.29 PMO diagram representing the π-MOs of bishomobenzene as the result of the mutual perturbations of the π-MOs of two allyl radicals.

trend with the delocalized structure **99** increasing in concentration with decreasing solvent polarity. Infrared spectra for **98** ⇄ **99** proves that in the gas phase, the delocalized form **99** is favored over the localized form **98** [182]. Computational studies suggested that polyaza and polyphosphasemibullvalene might prefer delocalized structures in their ground state [183].

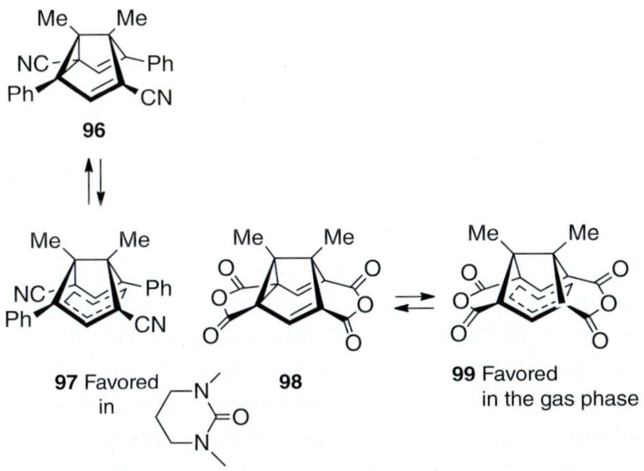

4.7.6 Barrelene effect

In 1957, Hine et al. suggested that bicyclo[2.2.2]octa-2,5,7-triene (**100**) might experience ASE because of three-dimensional π/π overlaps between the three ethylene units, thus realizing a structure resembling that of a barrel, therefore the name "barrelene" giving to this bicyclic triene [184]. The first synthesis of **100** was presented by Zimmerman and Paufler in 1960 [185].

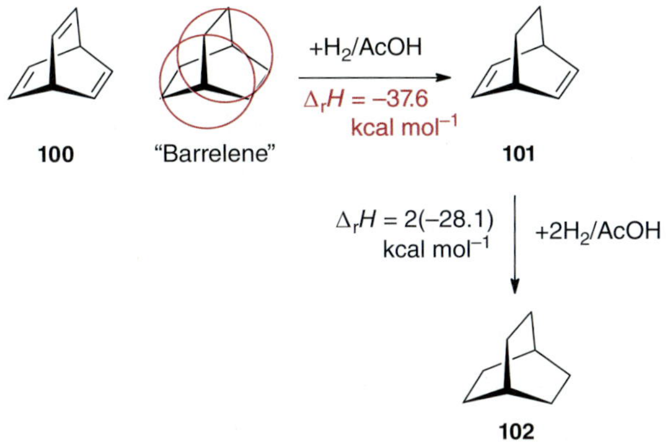

In 1964, Turner reported the heat of hydrogenation of **100** into bicyclo[2.2.2]octane (**102**) in AcOH, which amounts to -93.8 ± 0.3 kcal mol^{-1} [186]. As the heat of hydrogenation of bicyclo[2.2.2]octa-2,5-diene (**101**) into **102** is -56.2 ± 0.1 kcal mol^{-1}, the heat of hydrogenation of **100** into **101** amounts to -37.6 ± 0.4 kcal mol^{-1}, about 10 kcal mol^{-1} more than the hydrogenation of cyclohexene (-28 kcal mol^{-1}) and other (*Z*)-1,2-dialkylethylenes. This demonstrates that barrelene (**100**) is not stabilized but destabilized by the juxtaposition of the three homoconjugated ethylene units contrary to the early prediction of Hine and coworkers. If one considers the π-MOs of barrelene (**100**) to result from the interactions between the π-MOs of a cyclohexa-1,4-diene and of an ethylene subunit approaching to each other in a way to realize barrelene (geometries maintaining the C_{2V} symmetry), the PMO diagram in Figure 4.30 is obtained.

In **100**, there is neither stabilizing LUMO(diene)/HOMO(ethylene) and LUMO(ethylene)/HOMO(diene) interactions nor LUMO/sub-HOMO and supra-LUMO/HOMO interactions compensating for the repulsive HOMO(diene)/HOMO(ethylene) interaction for reason of symmetry. Thus, homoconjugation between the three C=C double bonds of **100** does not lead to any "three-dimensional" electron stabilization but to an electronic destabilization for the reason of symmetry. Essentially, the bis-homocyclobutadiene structures in **100** are anti-aromatic and destabilizing.

Problem 4.18 At 20 °C, [2.2.2]hericene (2,3,5,6,7,8-hexamethylidenebicyclo[2.2.2]octane) adds 2 equiv. of SO_2 giving a bissulfolene. Under forcing conditions (high concentration and long reaction time), no trace of tris-sulfolene can be detected. Propose an explanation knowing that the cheletropic addition of SO_2 to *s-cis*-1,3-dienes gives the corresponding sulfolenes between −10 and 40 °C and that the

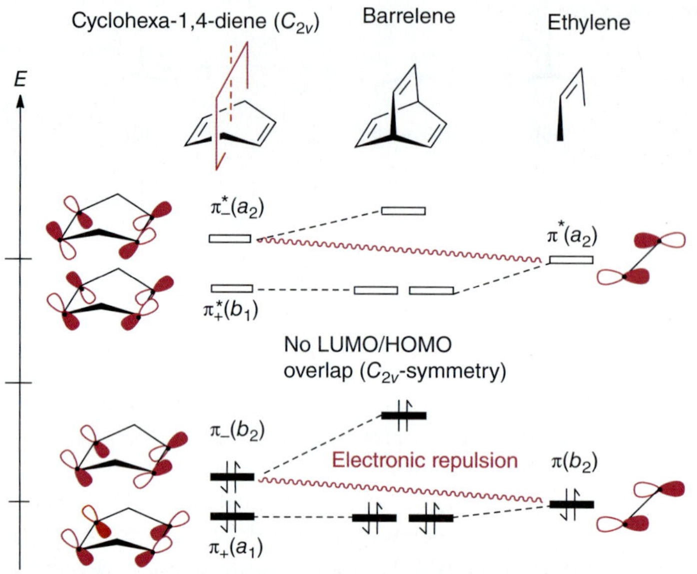

Figure 4.30 PMO diagram representing the interactions between the π-MOs of cyclohexa-1,4-diene and the ethylene subunits of barrelene in geometries maintaining the C_{2v} symmetry (one C_2 axes and a mirror plane containing it: a = symmetrical, b = antisymmetrical with respect to axis C_2; Subscripts: 1, symmetrical and 2, antisymmetrical with respect to the mirror plane of symmetry m chosen to contain C(1) and C(4) of the barrelene system.

sulfolenes undergo cheletropic elimination of SO_2 above 100 °C [187].

4.8 Hyperconjugation

In 1939, Mulliken observed that the absorption maximum in the UV spectra of alkenes and dienes is shifted to longer wavelengths upon alkyl substitution. This bathochromic shift was attributed to a σ(alkyl)/π(alkene, diene) interaction called "hyperconjugation" [188–191] (see Section 6.2.1 for more UV–visible spectra of organic compounds).

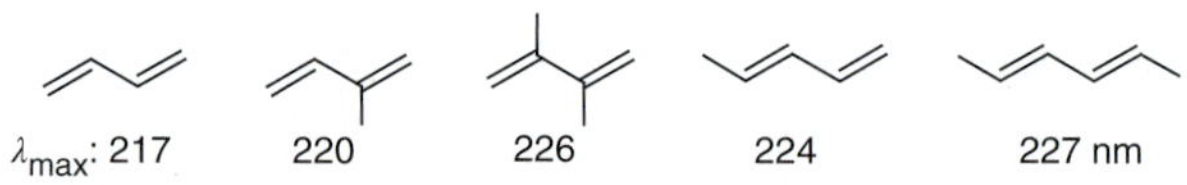

Mulliken also attributed the lower exothermicity of hydrogenation at alkyl-substituted alkenes (−32 kcal mol^{-1} for ethylene, −30 kcal mol^{-1} for 1-alkylethylenes, and c. −28 kcal mol^{-1} for 1,1 and (*E*)-1,2-dialkylethylenes) to hyperconjugation noted (valence bond theory) as [192]

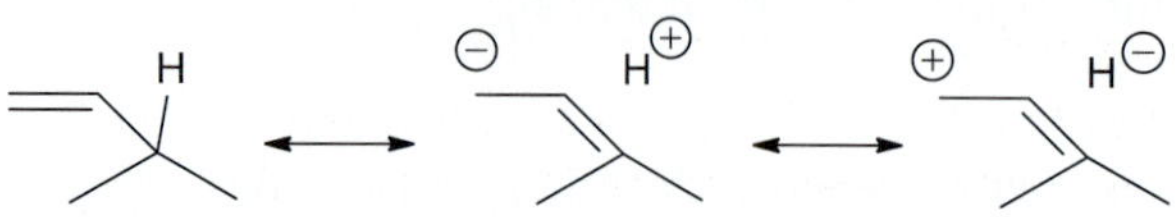

The accelerating effect and orienting effect of alkyl substituents in benzene electrophilic substitutions have been assigned to electron-releasing effects of the alkyl groups. Any substituent effect on a charged species depends on its permanent dipole (stabilizing or destabilizing inductive or field effect) and on its polarizability (ability of the substituent to have its electron density oriented toward the charged center in order to stabilize the species). Hyperconjugation is a molecular theory to interpret polarizability of substituents containing σ-bonds.

4.8.1 Neutral, positive, and negative hyperconjugation

There are different ways to classify hyperconjugative interactions. Historically, common classification is based on their separation into neutral, positive, and negative hyperconjugation (Figure 4.31). For alkenes of type CH_2=CH—CH_2R, its RCH_2 group behaves as an electron-releasing group through donation of electrons from a HOMO(RCH_2) orbital to π^*(ethylene) and as an electron-accepting group *via* retrodonation of electrons from HOMO(ethylene) to a LUMO(RCH_2). In this case, LUMO(RCH_2) and HOMO(RCH_2) have a π character making these group orbitals overlap with the π-molecular orbitals of ethylene (and other unsaturated compounds). In the case of carbenium ions, one speaks of positive hyperconjugation, meaning that the dominant interaction corresponds to electron donation from the HOMO of the RCH_2 substituent to the empty orbital (LUMO) of the cations (HOMO(RCH_2)/LUMO(cation)). By analogy, negative hyperconjugation means that the dominant interaction corresponds to an electron transfer from the anionic center to the RCH_2 substituent (HOMO(anion)/LUMO(RCH_2) interaction) [193].

One possible way to estimate hyperconjugative stabilization of propene and propyne with respect to

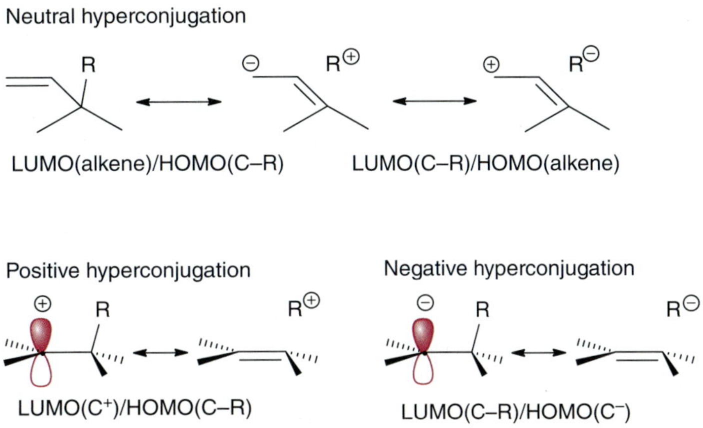

Figure 4.31 Contributing resonance structures for neutral, positive, and negative hyperconjugation.

ethylene and acetylene, respectively, is to measure the heats of the following isodesmic reactions (4.48)–(4.51) (Section 2.7.1).

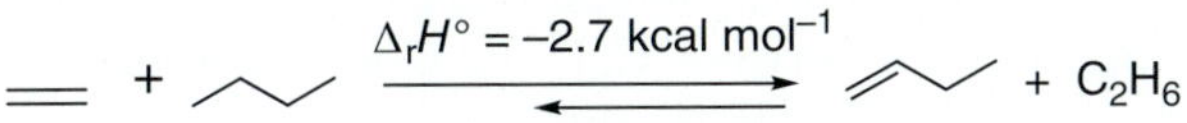

(4.48)

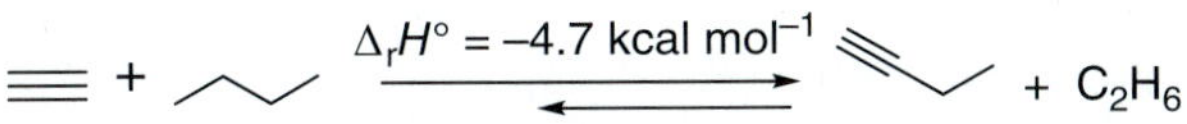

(4.49)

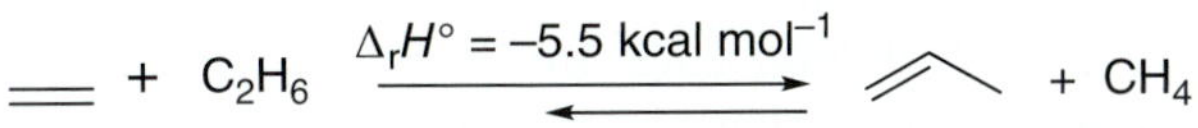

(4.50)

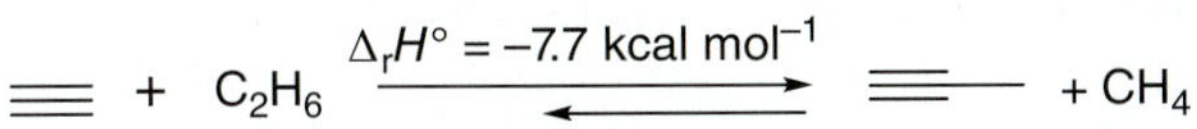

(4.51)

4.8.2 Hyperconjugation in cyclopentadienes

The dipole moment (μ = 0.42 D) of cyclopentadiene was interpreted (Section 4.5.12) as to arise from hyperconjugation between the *s-cis*-butadiene unit and $H_2C(5)$ of this system. This induces cyclopentadiene to possess some aromaticity as it exhibits a slightly enhanced diamagnetic susceptibility anisotropy [194–196]. As seen with equilibrium (4.43) (Section 4.6), ASE in cyclopentadiene amounts to only -2.0 ± 0.6 kcal mol^{-1}, not larger than for cyclohexa-1,3-diene (equilibrium (4.44)) [197]. The ability of a C—E bond to be electron releasing depends on its electrofugacity.

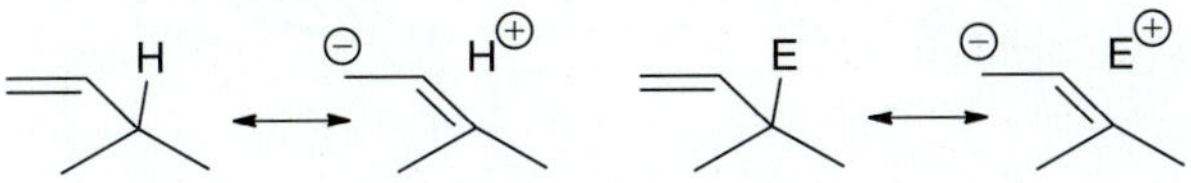

If one replaces the hyperconjugative C—H bond by a C—E bond in which E^+ is a better electrofugal group than H^+, electron transfer from the C—E bond will be larger than from the C—H bond. Thus, 5,5-disubstitution of cyclopentadiene with two silyl, germanyl, or stannyl groups increases the ASE because of enhanced hyperconjugation arising from a greater electron donation to the π-system. This is predicted by quantum mechanical calculations of equilibria (4.52) [76]. They estimate substantial enhanced C—H acidity for silyl and germylcyclopentadienes compared with cyclopentadiene and methylcyclopentadiene [198].

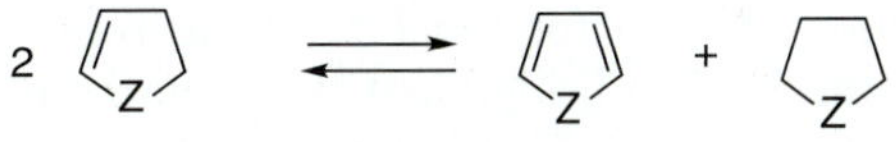

(4.52)

Z =	CH_2	CMe_2	$C(SiH_3)_2$	$C(GeH_3)_2$
$\Delta_r H°$(4.52):	–2.6	–0.8	–9.9	–9.3
Z =	$C(SnH_3)_2$	CCl_2	CF_2	
$\Delta_r H°$(4.52):	–13.8	5.1	9.5 kcal mol^{-1}	

4.8.3 Nonplanarity of bicyclo[2.2.1]hept-2-ene double bond

The heat of hydrogenation of bicyclo[2.2.1]hept-2-ene (norbornene) into bicyclo[2.2.1]heptane (norbornane) amounts to -32.8 ± 1 kcal mol^{-1} and that of norbornadiene into norbornene is -38.7 ± 2 kcal mol^{-1}. Both hydrogenations are more exothermic than those of cyclopentene (-26.8 ± 0.5 kcal mol^{-1}) and cyclohexene (-27.8 ± 0.5 kcal mol^{-1}). This indicates a strain increase of c. 6 kcal mol^{-1} when going from norbornane to norbornene and c. 12 kcal mol^{-1} when going from norbornene to norbornadiene.

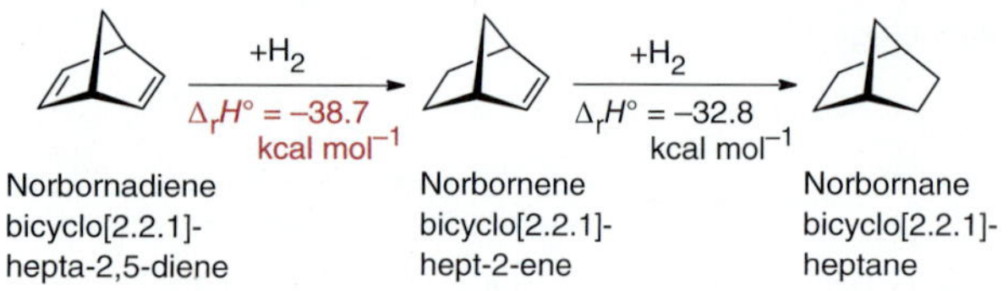

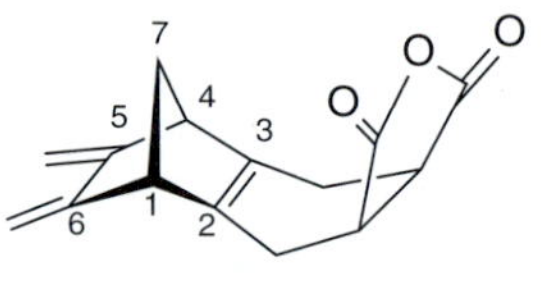

103

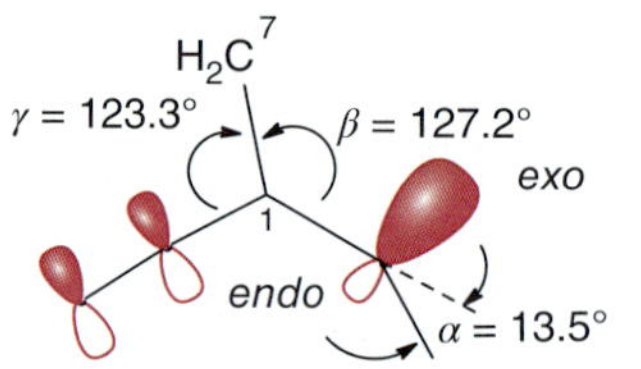

103

Crystalline structure of **103** [199], which contains a norbornene moiety and a *s-cis*-butadiene function, shows that the 2,3-substituents about the endocyclic alkene system are bent toward the *endo* face; the plane average plane R-C(2)-C(3)-R is not in the plane of C(1)-C(2)-C(3)-C(4) as it deviates from it by an angle α = 13.5°. The exocyclic diene moiety, on the contrary, makes a plane that deviates only slightly from the C(1)-C(6)-C(5)-C(4) plane of **103**. Accordingly, the endocyclic alkene moiety at C(2)-C(3) is polarized toward the *exo* face of the bicyclic system. It repels the $H_2C(7)$ bridge as manifested by angle β between the C(1)-C(2)-C(3)-C(4) plane and the C(1)-C(7)-C(4) plane that is larger (127.2°) than angle γ between the C(1)-C(6)-C(5)-C(4) plane and the C(1)-C(7)-C(4) plane (123.3°). Structure of **103** (and of several derivatives) shows that a torsional eclipsing repulsion involves the $H_2C(7)$ bridge and the endocyclic alkene moiety, repulsion that is the origin of the extra ring strain (c. 6 kcal mol^{-1}) of norbornene compared with that of norbornane. This repulsive interaction occurs twice in norbornadienes. The latter suffer from an extra ring strain of c. 12 kcal mol^{-1} compared with that of norbornenes. In *syn*-sesquinorbornene (**104**) and derivatives, X-ray radiocrystallographic data [200, 201] show out-of-plane deformation α of the norbornene double bond that reaches 15°–18° depending on substituents. In contrast, *anti*-sesquinorbornene (**105**) and derivatives usually have nearly planar double bonds. Furthermore, a detailed analysis of the torsional effects and consequences on electronic structure of these molecules has been reported [202a]; **105** is calculated to be c.

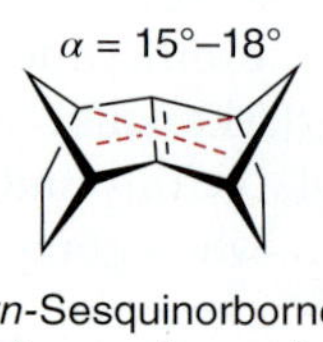

syn-Sesquinorbornene
104: non-planar alkene

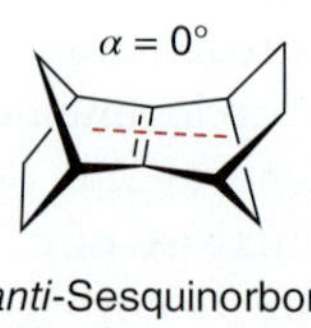

anti-Sesquinorbornene
105: planar alkene

3.5 kcal mol^{-1} less stable than **104** [202b]. In **104**, the nonplanarity of the double bond has a synergic effect on both annulated norbornene moieties.

In **103**, polarization of the π-electron toward the *exo* face of one norbornene unit is not accompanied by the same effect in the second norbornene moiety, therefore its planarity and lower stability. The double-bond polarization toward the *exo* face of norbornenes [203] renders these alkenes highly reactive in electrophilic additions and cycloadditions and this with high *exo* face stereoselectivity (Huisgen's *x*-factor, see Section 5.3.16, Figure 5.35 [204]). NMR has been used to investigate structural features of norbornenes and analogs. Most notably, large chemical shift (downfield shift) is observed upon introduction of an endocyclic double bond in norbornane derivatives for the ^{17}O-, ^{15}N-, ^{29}Si-, and ^{31}P-chemical shifts in corresponding 7-oxa- [205], 7-aza- [206], 7-sila- [207], and 7-phosphanorbornenes [208]. Furthermore, the one-bond C—C scalar coupling constant between C(5) and C(6) in norbornane is increased by c. 10 Hz upon introducing an endocyclic double bond at C(2)—C(3). Hyperconjugation is known to affect ^1J(C,C) constants [209] as well as the hybridization state of the C-atoms participating in the bond [210]. Using a quantum theoretical analysis 202a, 211, it was confirmed that the structural and electronic properties of norbornene and other bicyclo[2.2.1]hept-2-enes are due to hyperconjugative interactions that can be represented by partial retro-Diels–Alder reaction in the ground state of these systems [212].

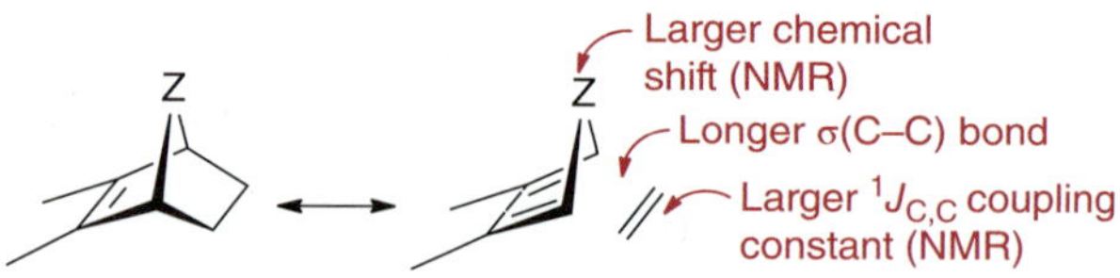

106 **106′**

The nonplanarity of norbornene double bond can also be explained by limiting structures **106** ↔ **106′**. Quantum mechanical calculations on the cycloaddition of ethylene and butadiene show in the transition structure that the H-C(2) and H-C(3) bonds of the butadiene moiety deviate from the C(1)-C(2)-C(3)-C(4) plane and are bending toward the direction of the dienophiles (ethylene) [213].

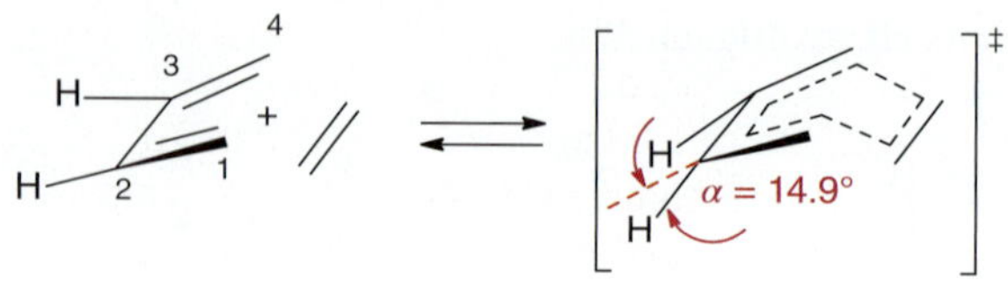

4.8.4 Conformation of unsaturated and saturated systems

Fluoroacetaldehyde adopts two conformations in which the fluoro substituent is either *cis* or *trans* with respect to the carbonyl function. The *cis*-conformer is c. 2 kcal mol^{-1} less stable than the *trans*-conformer because of electrostatic repulsions between the electron pairs about F and O. In both these conformations, an optimal overlap is realized between the $\sigma-(CH_2F)$ and π^*_{CO} group MOs. Any other conformer is less stable than the *trans*-conformer. MO $\sigma-(CH_2F)$ has a π character as it is the out-of-phase combination of two σ_{C-H} bond orbitals (Figure 4.31) [214].

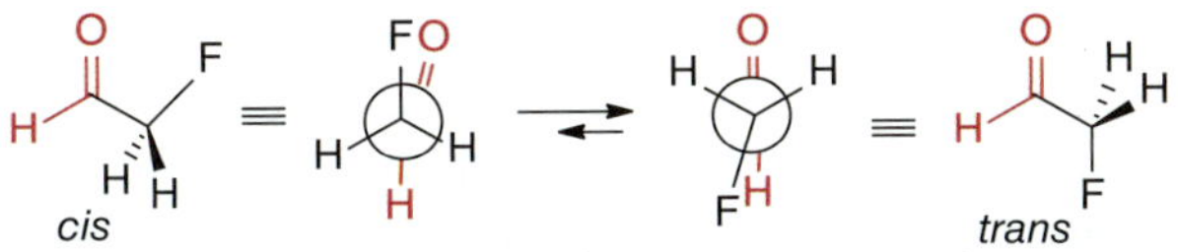

For the FCH_2 group, one can draw the following σ-orbital combinations:

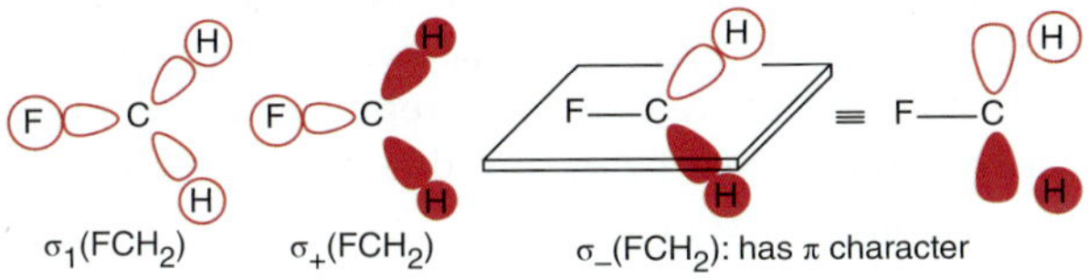

Molecular orbital combinations σ_1 and σ_+ contain contribution from the C–F, which makes these MOs low lying in energy because of the high electronegativity of the fluorine atom. Combination σ_- does not contain any C–F contribution and is antibonding with respect to σ_1. HOMO(FCH_2) is the privileged MO for an interaction with the LUMO of the carbonyl moiety, for which a large coefficient resides on the carbon center (Figure 4.24, Section 4.5.15). In this case, the hyperconjugative interaction that dominates corresponds to an electron donation from the FCH_2 substituent to the carbonyl moiety (Figure 4.32).

(*Z*)-Difluorodiazene is 3.1 ± 0.5 kcal mol^{-1} more stable than (*E*)-difluorodiazene [215]. This "*cis*-effect" is also observed in other mono- and 1,2-dihalogenoalkenes. It can be interpreted as the sum of three different electronic interactions: (i) negative hyperconjugation; i.e. partial delocalization of the lone pair on nitrogen $n(N_1)$ into the $\sigma^*_{N2\text{-}X}$ orbital. This interaction is more stabilizing in the *cis*-form ((*Z*)-isomer) than in the *trans*-form ((*E*)-isomer). Overlap of

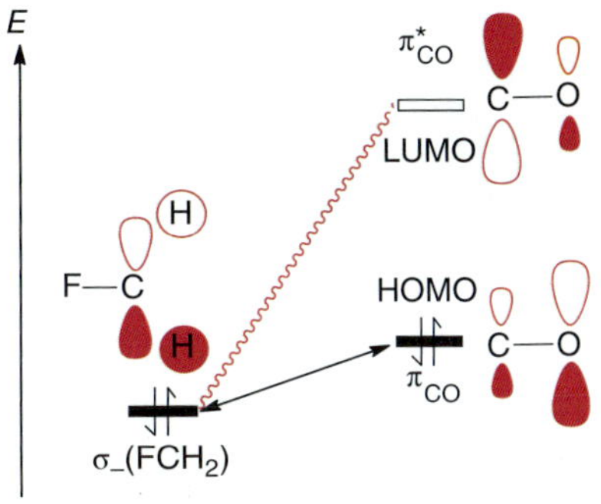

Figure 4.32 PMO diagram showing the stabilizing LUMO(carbonyl)/σ_- interactions in *cis*- and *trans*-fluoroacetaldehyde.

$n(N_1)/\sigma^*_{N2\text{-}X}$ is larger in the *cis*- than in the *trans*-form as shown below:

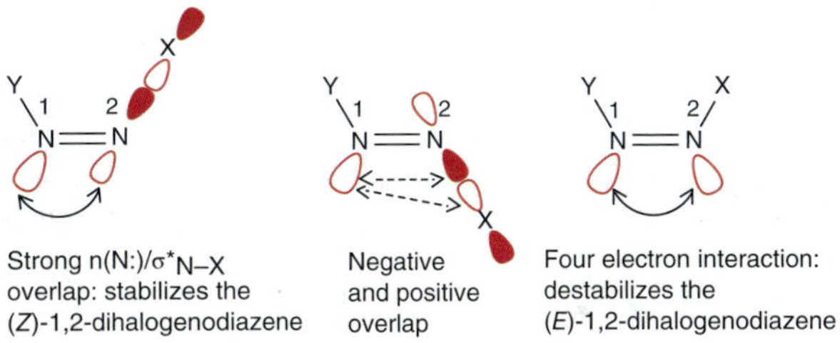

(ii) electrostatic interaction between the two substituents X and Y. This interaction stabilizes the *cis*-form of the unsymmetric diazenes and destabilizes that of the symmetric diazenes; (iii) electrostatic interactions between the two lone pairs of electrons at the two adjacent nitrogen atoms. This interaction ($n(N_1)/n(N_2)$) destabilizes the *cis*-form. Because $\sigma^*_{N\text{-}H}$ orbitals in unsubstituted diazene are much higher lying in energies than σ^*_{N-F} and σ^*_{N-Cl} orbitals, diazene itself prefers the *trans*-form. The effect of the repulsive $n(N_1)/n(N_2)$ interaction dominates the $n(N_1)/\sigma^*_{N_2-H}$ and $n(N_2)/\sigma^*_{N_1-H}$ attractive interactions [216]. In the infrared spectra of alkanes, the C—H frequencies of methyl group have frequencies comprised between 2872 and 2962 cm^{-1}. In methylamine, one measures the three frequencies given below. The weaker the stretching frequency, the weaker the C—H bond. The lowest frequency of 2820 cm^{-1} is assigned to the C—H bond antiperiplanar to the nitrogen nonbonding electron pair. This is the Bohlmann effect [217], which is interpreted as a negative hyperconjugative interaction $n(N:)/\sigma^*_{C-H}$ [218].

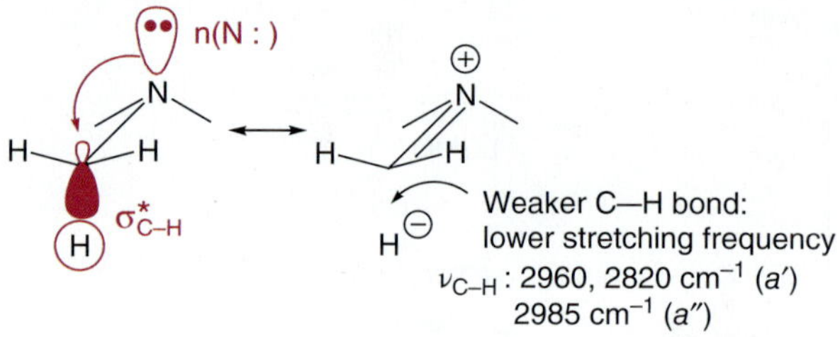

In 1939, Mulliken proposed that the $\sigma_{C-H}/\sigma_{C-H}$ hyperconjugative interactions play an important role in the internal rotation potential of ethane-like molecules [188]. Classically, the greater stability of the staggered conformation of ethane compared with the eclipsed conformation (c. 3 kcal mol^{-1}) is attributed to closed-shell (exchange) repulsions between vicinal σ_{CH} bonds (steric hindrance). This view is still valid today, but several theoretical analyses suggest a 0.5–1.0 kcal mol^{-1} contribution because of hyperconjugation that makes the staggered conformation more stable than the eclipsed conformation (Section 2.6.1) [219].

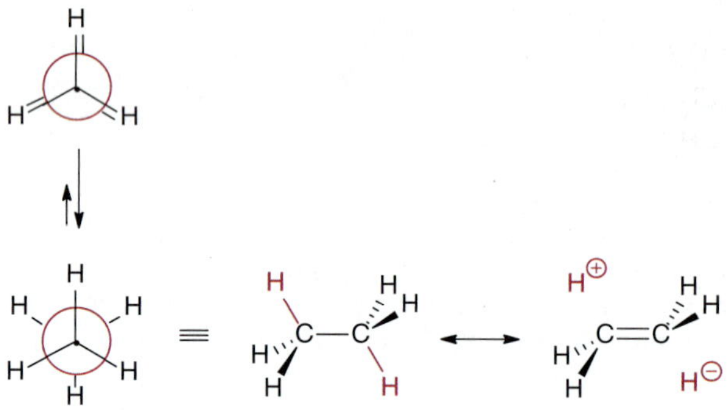

The eclipsed conformation of propene is c. 2 kcal mol^{-1} more stable than the staggered conformation, as predicted by the "banana bond" model (Section 2.6.10). Quantum calculations suggest that hyperconjugation HOMO(CH_3) → σ^*_{C-H} orbital of the antiperiplanar vinyl C—H bond (Figure 4.33) is responsible for the greater stability of the eclipsed conformation [220].

In ^{1}H-NMR spectra of ethane-like compounds, the coupling constant $^3J_{H,H}$ between two vicinal protons depends on the dihedral angle H–C–C–H (Ø) (Karplus rule [221, 222]). It is the largest (10–18 Hz) for *anti*-periplanar pair of protons (Ø = 180°), near zero when the protons are orthogonal (Ø = 90°), and take intermediate values (4–8 Hz) for *syn*-periplanar pair of protons (Ø = 0°).

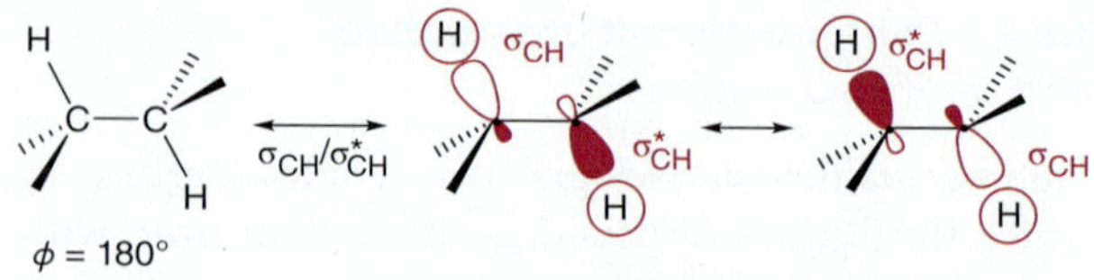

Hyperconjugative interactions of type $\sigma_{CH}/\sigma^*_{CH}$ explain this behavior. Overlap between these MOs is the largest for *anti*-periplanar pair of protons, smaller for *syn*-periplanar pair of protons, and nil when these orbitals are orthogonal [223, 224]. Karplus-type behavior of vicinal NMR ^3J coupling constants are observed for other nuclei than protons [225]. Preference of 1,2-difluoroethane for the gauche conformation over the eclipsed or *anti*-structure (0.8 kcal mol^{-1}) has been attributed to electron donation $\sigma_{C1-H} \rightarrow \sigma^*_{C2-F}$ [226]. This is an example of gauche effect [227]. Other compounds with favored gauche conformations are ethane-1,2-diol, 1,2-dimethoxyethane, 1-fluoro-2-methoxyethane, 1-amino-2-fluoro-ethane, 1-fluoro-2-formyloxyethane, 1-fluoro-2-nitroethane, 1-fluoro-2-formylaminoethane, 1-azido-2-fluoroethane, hydrazine [228], and hydrogen peroxide [229]. For 1,2-dichloro-, 1,2-dibromo-, and 1,2-diiodoethane, the *anti*-conformers are preferred. In these cases, the $\sigma_{C-H} \rightarrow \sigma^*_{C-X}$ stabilizing hyperconjugative interactions cannot override the repulsive interaction between the vicinal halides in the gauche conformation (Coulombic repulsion).

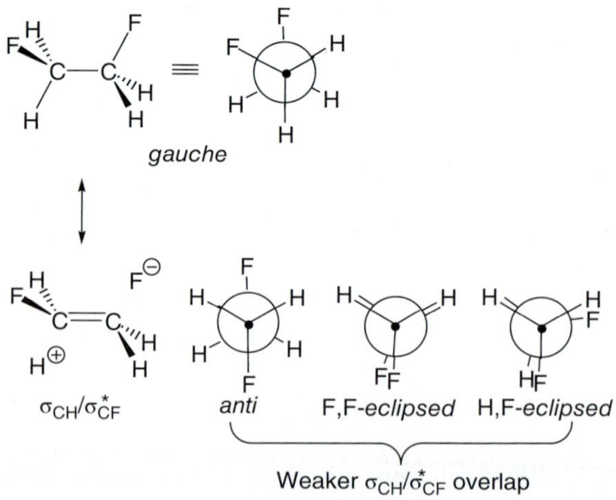

In a similar way, (*Z*)-1,2-difluoro- (0.9 kcal mol^{-1}) and (*Z*)-1,2-dichloroethylene (0.43 kcal mol^{-1}) are more stable than their (*E*)-isomers, whereas (*E*)- and (*Z*)-1,2-dibromoethylene have the same stability.

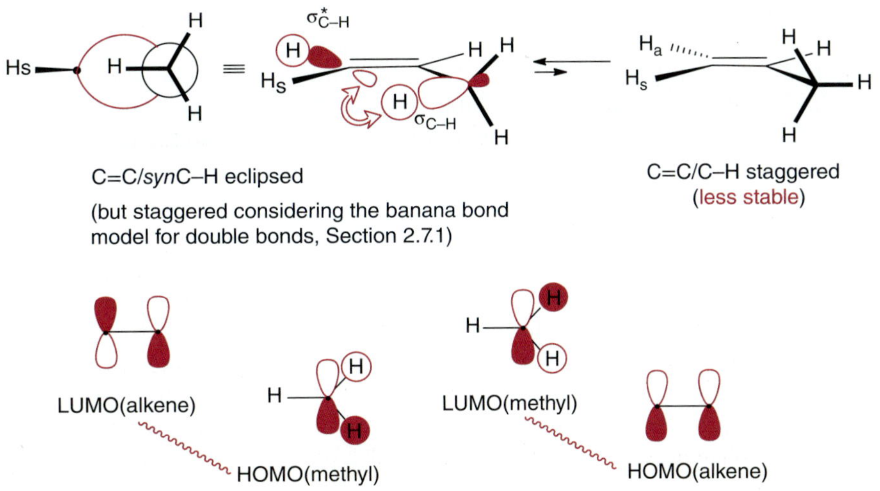

Figure 4.33 The eclipsed conformation of propene is stabilized by hyperconjugation between the methyl and vinyl groups. This implies three types of hyperconjugative interactions.

Problem 4.19 In CH_2=NH, the two C—H bonds have different bond lengths. Give an explanation [230].

Problem 4.20 What are the most stable conformers of methyl fluoroformate [214]?

4.8.5 Hyperconjugation in radicals

At 77 K, most radicals can be trapped as persistent species in solids and be easily detected by electron spin resonance (ESR) spectroscopy. There are two kinds of radicals, the σ-electron radicals (e.g. formyl: HC=O; phenyl: C_6H_5; and vinyl: CH_2=CH) and the more common π-electron radicals (e.g. allyl radicals) [231]. ESR spectra of radicals show hyperfine structure that are due to spin couplings between the electron and nuclei having a spin ≠ 0, such as ^{1}H, ^{13}C, ^{19}F, etc. Very early, the proton hyperfine structures of methyl-substituted semiquinone ions were interpreted as to arise form σ/π interactions [232]. In the case of alkyl radicals, one defines the α and β positions as shown below:

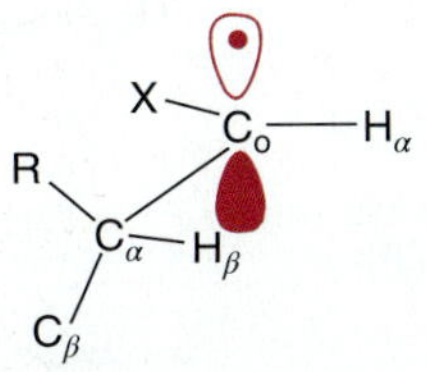

The proton hyperfine structure of π-radicals mainly results from the coupling of protons in α and β positions from an atom bearing a ρ_π spin density. The scalar couplings of α and β protons result from the spin polarization [233, 234] and the hyperconjugation, respectively [235]. The hyperfine coupling constants (hfcc) of these protons are given by

$$a_{\mathrm{H}}^{\alpha} = Q_\alpha \rho_c^\pi \tag{4.53}$$

$$a_{\mathrm{H}}^{\beta} = \rho_c^\pi (Q_1^\beta + Q_2^\beta \cos^2\theta) \tag{4.54}$$

with $Q_\alpha \approx 0.23$ G, $-5 \le Q_1^\beta \le 0$ G, $Q_2^\beta \approx 55$ G. θ is the angle between the $2p_z$ orbital and the H_β-C_α-C_o plane. Table 4.6 gives the ESR data for simple alkyl radicals.

Table 4.6 Hyperfine coupling constants (hfcc) in Gauss (G) of simple alkyl radicals.

Radical	hfcc at C_o		hfcc at methyl substituent	
	$^{13}C_o$	^{1}H: a^αH	$a^\alpha_{^{13}C}$	a^βH
CH_3	38.3	−23.0		
CH_3CH_2	39.1	−22.4	−13.6	26.9
$(CH_3)_2CH$	41.3	−22.1	−13.2	24.7
$(CH_3)_3C$	45.2		−2.4	22.7

Source: Taken from [236].

In the ethyl radical, the carbon atom of the methyl group lies in the average nodal plane of the unpaired electron's $2p_z$ orbital. This carbon acquires spin of the opposite sign to the radical center, i.e. negative spin (spin polarization). The H_3C group's hydrogen atoms acquire positive spin both by hyperconjugation from C_o and by spin polarization from the carbon atom of the methyl group. The three hydrogen atoms in the methyl group produce a characteristic ESR pattern of four lines (quadruplet with relative intensities 1 : 3 : 3 : 1) with hfcc α_H^β = 26.9 G. For some authors, the $\alpha_{\mathrm{H}}^{\beta}$ values multiplied by the number of β-CH bonds correlate with the bond dissociation enthalpies of the corresponding hydrocarbon ($DH°(Me^{\cdot}/H^{\cdot}$ = 104.7 kcal mol^{-1}, $DH°(t\text{-}Bu^{\cdot}/H^{\cdot})$ = 95.2 kcal mol^{-1}, Table 1.A.7) [237]. For other authors, this might not be the case. For instance, Gronert stated that the ESR data do not indicate that hyperconjugation in radicals stabilizes them [238]. He suggests [239] that Dunitz–Schomaker 1,3-repulsions (Section 2.6.1) [240] are released upon planarization of t-butyl radical compared with isobutene and are responsible for the apparent greater stability of t-butyl radical compared with that of methyl radical.

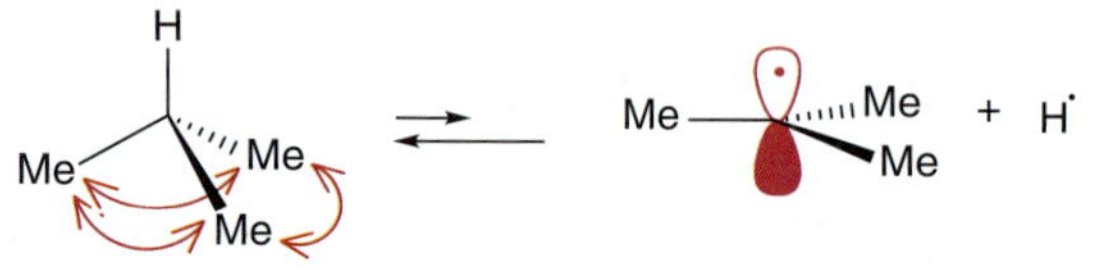

1,3-Repulsive interactions (B-strain)

In 1-adamantyl radical, hfcc of 4.66 and 3.08 G [241] has been measured for γ and δ protons. Quantum mechanical calculations suggest this to be due to a through-bond mechanism in which the σ(C—C) bond that can be aligned with the $2p_z$ orbital at C(1) hyperconjugate with it and transfer the spin at the γ and δ protons [242, 243].

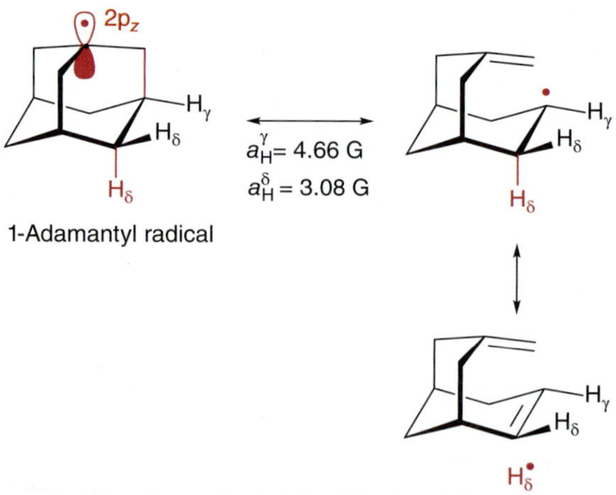

1-Adamantyl radical

summarized in Figure 4.34 [244]. When a substituent is made exclusively of β σ-bonds, the carbenium ion can be stabilized by positive hyperconjugation that implies HOMO(substituent)/2p(C$^+$) interaction. This effect adds or competes with stabilizing or destabilizing inductive effects. In general, a substituent affects the stability of any charged species by two main contributions (electrostatic field model for substituent effects): (i) permanent dipole (μ)/charge interaction ($V_D = q\mu \cos\theta/er^2$) that can be stabilizing of destabilizing depending on the orientation of the dipole with respect to the charged center and the sign of the charge and (ii) by the induced dipole/charge interaction ($V_I = -q^2\alpha/2\varepsilon r^4$) that is always stabilizing whether one deals with a cationic or anionic species. The size of the induced dipole depends on the polarizability α of the substituent and $1/r^2$ (r = distance between the charged center and the substituent). Part of the polarizability α is described by (π,π)-conjugation and (σ,π)-hyperconjugation discussed in this chapter. Note that σ-bond directly attached to a charged center can also be polarized under the influence of a charge.

In the case of cations of type E–CR_2–CR_2^+, the hyperconjugative stabilization effect increases with the electrofugacity of E$^+$. Thus, the cation stabilization increases with the sequence β-C—H < β-C—C < β-C—Si [245] < β-C—Ge < β-C—Sn [246] < β-C—Hg [247, 248]. Unusually stable vinyl cations **107** and **108** have been obtained as stable salts in C_6D_6 solution at 300 K [249]. The two β-silyl substituents are responsible for their high stability arising from hyperconjugation β(C–Si)/2p(C$^+$). Similarly, the secondary carbenium ion **109** containing two CH($SnMe_3$)($SiMe_3$) substituents has been isolated as a stable crystalline salt [250].

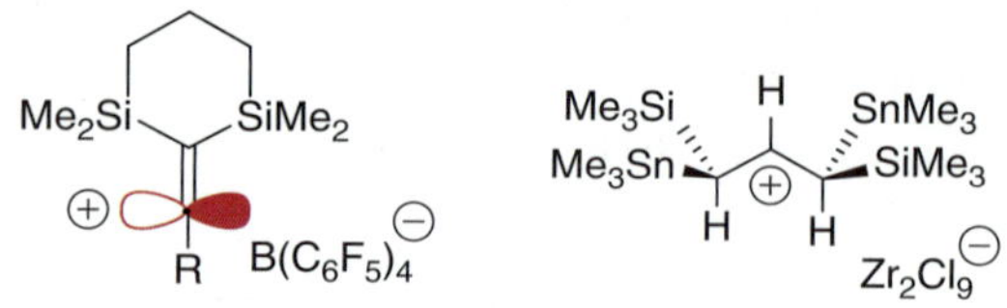

107 R = Me
108 R = Ph

109

4.8.6 Hyperconjugation in carbenium ions

The carbenium ion stability is strongly affected by the hybridization of the carbon sextet (alkyl, alkenyl, and alkynyl cation) and by the nature and number of the substituents, as shown by the gas-phase hydride affinities ($DH°(R^+/H^-$: Table 1.A.14). Various types of substituent effects have been recognized. They are

4.8.7 Hyperconjugation in carbanions

In 1950, Roberts et al. suggested a no-bond resonance structure $CF_3CH_2^-$ ↔ $F^-/CF_2{=}CH_2$ to explain m- and p-trifluoromethyl substituent effects (anionic hyperconjugation) [251]. In terms of the MO theory, an electron donation occurs from the HOMO(anion) to the β–σ^*(C–F) bonds [252].

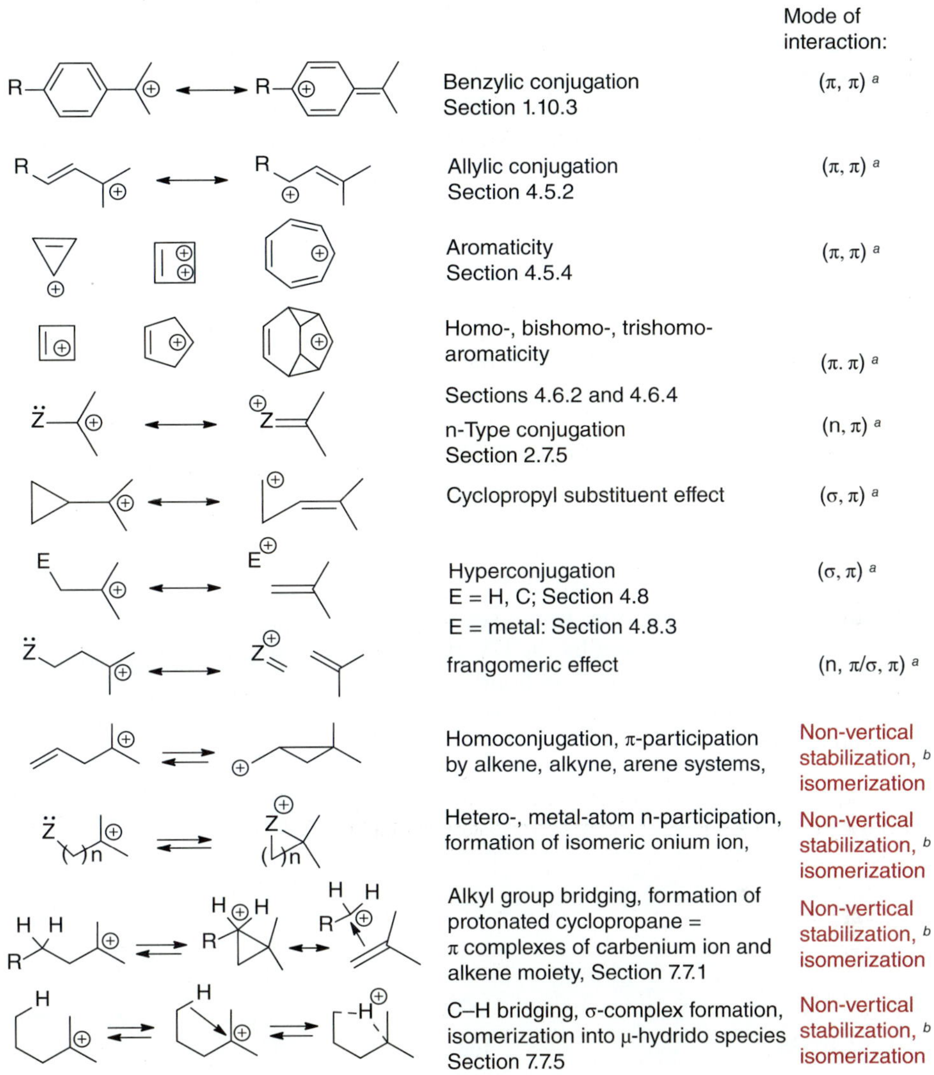

Figure 4.34 Possible modes for substituent effects on the stability of carbenium ions. [a]Relatively small change in the geometry of R^+ compared with those of precursors of type R–X (vertical stabilization). [b]Important change of geometry of R^+ compared with those of precursors R–X (nonvertical stabilization, Section 4.5.2).

Alternative interpretations of the β-C–F stabilization effect are mostly based on inductive and field effects [253–256]. In the gas phase, $(CF_3)_3C—H$ ($\Delta_r G°(RH \rightleftarrows R^- + H^+) = 326.6\,kcal\,mol^{-1}$) is a stronger acid than most C—H acids. It is much stronger acid than ethyne by 89 kcal mol^{-1}, or fluoroform by 42 kcal mol^{-1}.

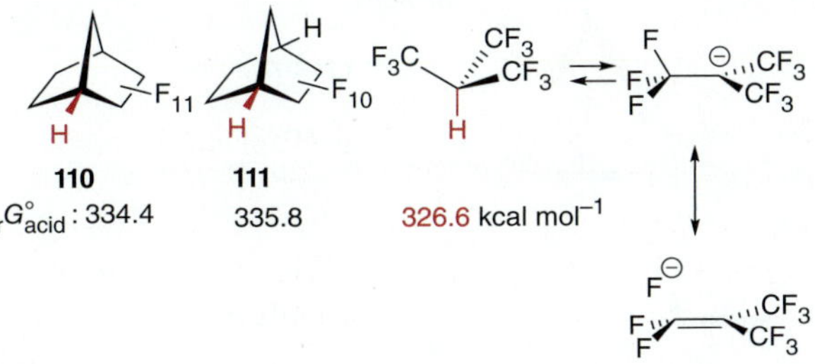

110 $\Delta_r G°_{acid}$: 334.4 **111** 335.8 326.6 kcal mol^{-1}

The greater acidity of open $(CF_3)_3CH$ compared to its bridgehead analogs **110** and **111** is attributed to the inability of the conjugate bases of the rigid bicyclic systems to form planar anionic structures (Bredt's rule, Section 2.6.9) that is a requirement for maximum hyperconjugation 2p(anion) → σ^*_{C-F} . This is taken as an evidence that anionic hyperconjugation does contribute to the relatively high stability of $(CF_3)_3C^-$ anion together with inductive field effects [257, 258].

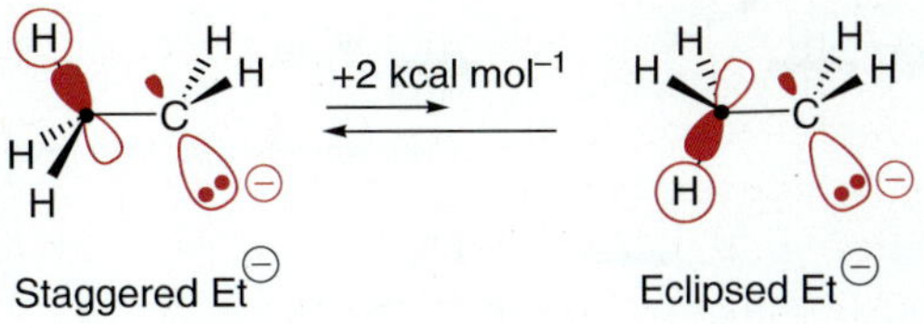

The lifetime of ethyl anion is <25 ms. Quantum mechanical calculations suggested the staggered conformer to be c. 2 kcal mol^{-1} more stable than the eclipsed conformer because of favorable negative hyperconjugation between the lone pair of electron and the *anti*-periplanar β-C—H bond in the staggered conformation [259–261].

4.8.8 Cyclopropyl vs. cyclobutyl substituent effect

Hydride affinities (Table 1.A.14) show that the cyclopropyl substituent stabilizes carbenium ions much better than the isopropyl group and nearly as much as a phenyl group: $DH°(i\text{-Pr-CH}_2^+/\text{H}^-) = 265$, $DH°(c\text{-C}_3\text{H}_5\text{—CH}_2^+/\text{H}^-) = 241$, and $DH°(c\text{-Ph-CH}_2^+/\text{H}^-) = 234\ \text{kcal mol}^{-1}$.

The cyclopropyl group is more like a vinyl group than a cyclobutyl group, although cyclopropane and cyclobutane have almost the same strain energy (Table 2.A.3). This is indicated by the UV absorption data (Section 6.2.1) and the one-bond carbon-13 hydrogen coupling constants in the NMR spectrum, as shown below:

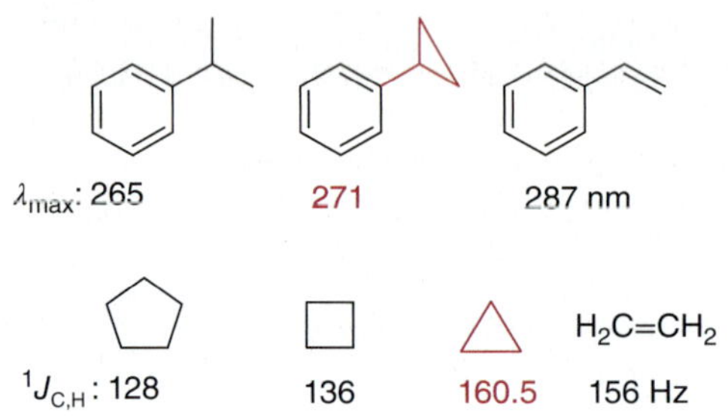

In order to explain the "conjugative ability" of cyclopropane and cyclopropyl group, Walsh [262] and Sugden [263] proposed sp^2 hybridation of the carbon in three-membered rings; this hypothesis is supported by $^1J_{C,H}$ coupling constants [264]. Ignoring the electrons of the C—H bonds of cyclopropane because the σ(C—H) bonds are nearly perpendicular to the σ(C—C) bonds, the orbitals of three-membered ring carbon system can be considered to result from the combination of three sp^2 orbitals pointing toward the center of the cyclopropane ring and three atomic 2p orbitals pointing around the circumference of the ring. A bonding combination of three sp^2 orbitals is the lowest energy (a_1') orbital. There are two degenerate orbitals, Ψ_5^* and Ψ_6^*, with one node and that are strongly antibonding (Figure 4.35a). These orbitals are equivalent to the cyclopropenium π-orbitals in shape and energetic pattern. The three 2p orbitals form three combinations Ψ_2, Ψ_3, and Ψ_4^*. Ψ_2 and Ψ_3 are degenerate bonding orbitals, whereas Ψ_4^* is the all-antibonding combination. Each H_2C fragment brings two electrons to make the three σ(C—C) bonds of cyclopropane; six electrons in total will populate the lowest lying bonding orbitals Ψ_1, Ψ_2, and Ψ_3. The two degenerate HOMOs are thus Ψ_2 and Ψ_3 and the LUMO is Ψ_4^*. Thus, HOMO Ψ_2 concentrates the electron density on centers C(2) and C(3), or in the σC(2,3) bond, making a bent or a "banana" bond (Section 2.6.10). As cyclopropane is not a polar molecule, the other HOMO Ψ_3, which is antisymmetrical with respect to the mirror plane of symmetry, must concentrate the electron density on C(1), which is possible because the 2p orbital at C(1) is antisymmetrical with respect to the mirror plane

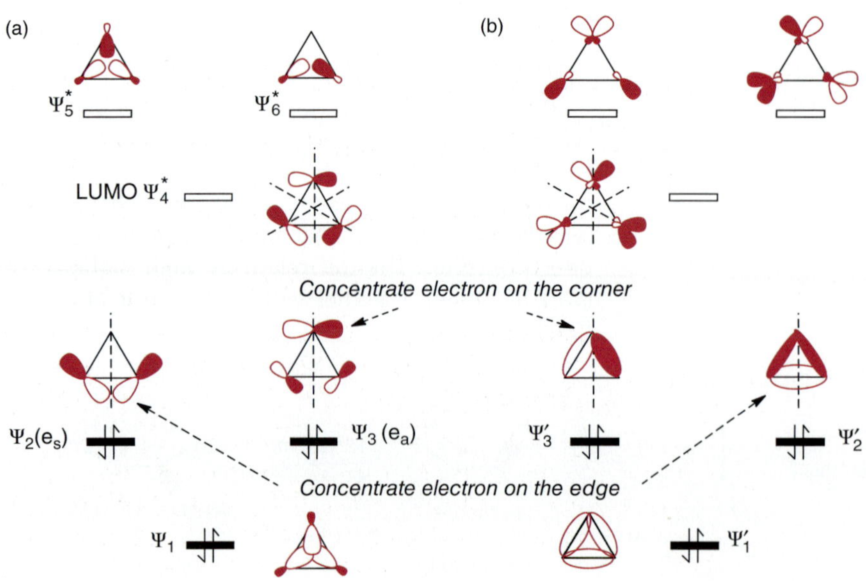

Figure 4.35 (a) The Walsh–Sugden model for the orbitals of the σ(C—C) bonds of cyclopropane as resulting from the combination of three sp^2-hybridized CH_2 groups; (b) the Förster–Coulson–Moffit model of the orbitals of the σ(C—C) framework of cyclopropane as derived from the combination of three σ(C—C) molecular orbitals and three σ^*(C—C) molecular orbitals.

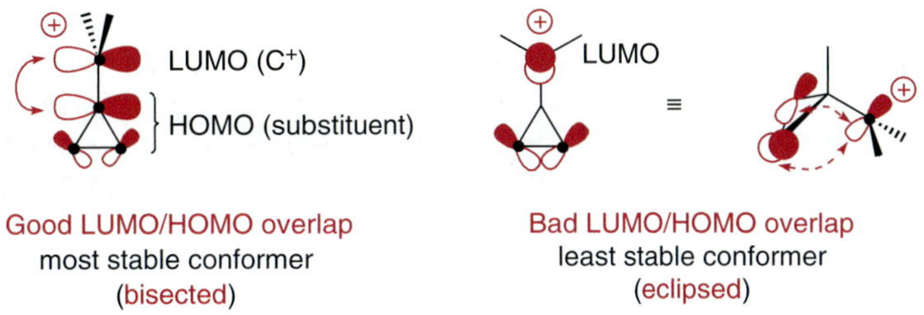

Figure 4.36 Relative stability of bisected and eclipsed conformers of 1-(cyclopropyl)alkyl cations.

of the symmetry. To differentiate Ψ_3 from Ψ_2, we call them $\Psi_2(e_s)$ and $\Psi_3(e_a)$. As they are degenerate, any linear combination of them is acceptable, and this corresponds to a synchronous rotation of the two perpendicular nodal planes that are apparent in $\Psi_2(e_s)$ and $\Psi_3(e_a)$.

When a cyclopropyl group is attached to a carbenium ion, major stabilization will occur when the p orbital of HOMO $\Psi_3(e_A)$ overlaps the LUMO of the carbenium ion; the best rotamer of this cation is the bisected one (Figure 4.36) [265], as it allows an optimal overlap between $\Psi_3(e_a)$ (HOMO) and the empty 2p orbital (LUMO) of the cation [266]. The enhanced electron-releasing ability of the cyclopropyl group arises from its HOMO $\Psi_3(e_A)$ that concentrates the electron density on the center near the carbenium ion and has a relatively high energy.

Förster, Coulson, and Moffit came up with an alternative, but actually equivalent, which is the bent bond model for the bonding in cyclopropane [267]. The sp^2 hybrid and 2p orbital of each methylene group can be combined to form two sp^5 hybrids:

$$\frac{1}{\sqrt{2}}(\Psi(sp_x^2) \pm \Psi(p_y))$$

$$= \frac{1}{\sqrt{2}}\left[\left(\frac{1}{\sqrt{3}}s + \sqrt{\frac{2}{3}}p_x\right) \pm p_y\right]$$

$$= \frac{1}{\sqrt{6}}s + \sqrt{\frac{2}{6}}p_x + \sqrt{\frac{3}{6}}p_y \mathrel{\hat{=}} sp^5(+)$$

and

$$= \frac{1}{\sqrt{6}}s + \sqrt{\frac{2}{6}}p_x - \sqrt{\frac{3}{6}}p_y \mathrel{\hat{=}} sp^5(-)$$

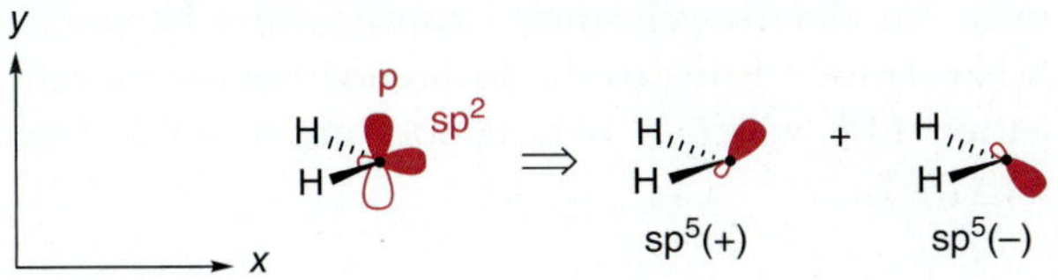

The sp^5 hybrids each contain 1/6s character and 5/6p character. These hybrids can be combined to form three bent bonds involving the overlap of the sp^5 hybrids pointing in the direction of the adjacent carbon center as shown below.

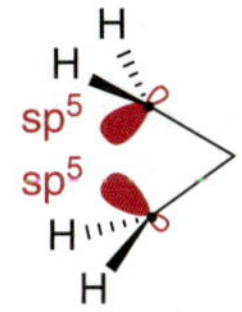

A bent or "banana" bond forms from two sp^5 hybrids. These bent bonds of this type and the corresponding σ^* bent bonds can be combined to form the symmetry-adapted orbitals shown in Figure 4.35B. The relatively small one-bond $^{13}C—^{13}C$ coupling constants (13–15 Hz), which have been observed in the ^{13}C-NMR spectra of cyclopropane derivatives, are in agreement with this concept of sp^5 hybridization of the orbitals used in the forming of the internal bonds [268–270]. The Förster–Coulson–Moffit model [271, 272] also predicts that the electron density in the two degenerate HOMOs is not isotropic, i.e. if one orbital concentrates the electron density in two bonds as in Ψ'_3, the other HOMO concentrates the electron in the opposite direction as in Ψ'_2.

Cyclobutane does not have occupied molecular orbitals that show the electron anisotropy found for the HOMOs of cyclopropane. Considering a square cyclobutane and ignoring the contributions from the σ(C—H) bonds, the occupied orbitals of the σ(C—C) framework of cyclobutane are those shown below. There is never the necessity to concentrate electron unevenly along the four-membered ring; this explains why cyclobutane is not different than another cycloalkyl or than a secondary alkyl group conjugated with a π-system such as carbenium ion [273].

Problem 4.21 Explain the bathochromic shift introduced on the $V \leftarrow N$ transition of cyclohexa-1,3-diene upon substitution with ethane-1,2-diyl groups at C(5) and C(6) [274].

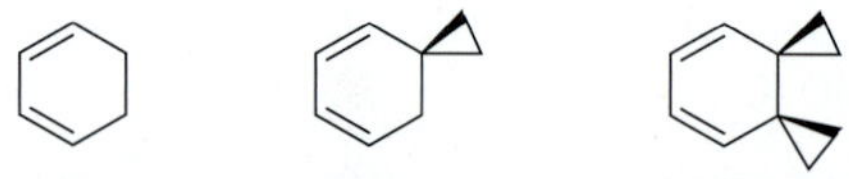

λ_{max}:	256 nm	269 nm	284 nm
ε:	4500	2200	4800

4.9 Heilbronner möbius aromatic [*N*]annulenes

In 1964, Heilbronner recognized that singlet [4*n*] annulenes should have electronic stabilization (Möbius aromatic stabilization energy) in twisted conformations in which the 2p orbitals lie on the surface of a Möbius strip. He pointed out that rings with 20 and more carbon centers might adopt Möbius geometries "without any apparent angle or steric repulsion" [275]. Hückel [*N*]annulenes have at least one mirror plane of symmetry that is not the π-plane of the cyclic hydrocarbon. In contrast (Figure 4.37), the Möbius [*N*]annulenes have a C_2-axis of symmetry instead of a mirror plane of symmetry, as illustrated below [276].

The planar [8]annulene can be formed conceptually from the cyclic π-conjugation of a (*Z,Z*)-hexa-1,3,5-triene with an ethylene moiety; PMO theory predicts that this compound will have little electronic stabilization, as there is no LUMO(triene)/HOMO (ethylene) and LUMO(ethylene)/HOMO(triene) interaction because of the C_s symmetry of this system (Figure 4.38A). If the (*Z,Z*)-hexatriene is twisted by an angle of 180° along its carbon framework (Figure 4.36B), a compound with a C_2-axis of symmetry is formed. The π-conjugation of this twisted compound with an ethylene unit leads to two stabilizing LUMO(triene)/HOMO(ethylene) and LUMO(ethylene)/HOMO(triene) interactions, suggesting that the Möbius [8]annulene is stabilized by Möbius aromaticity.

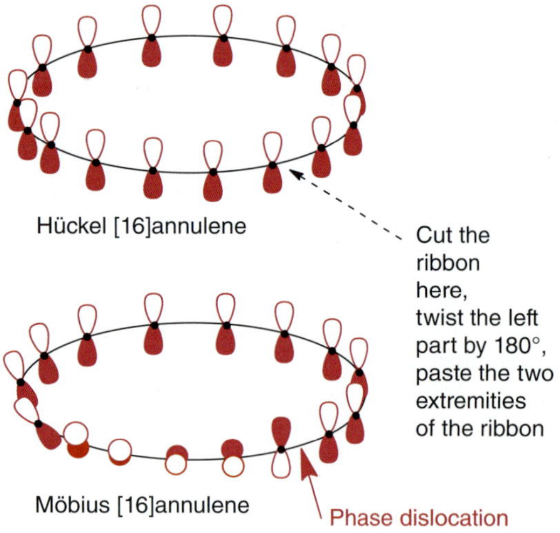

Figure 4.37 Construction of the array of 2p AOs for a Möbius [*N*]annulene from the corresponding Hückel [*N*]annulene. Hückel [*N*]annulenes are characterized by 0, 2, 4, …, phase dislocations (two vicinal 2p AOs have opposite phase), whereas Möbius [*N*]annulenes have 1,3,5, … phase dislocations. The inversion of one 2p AO leads to the increase or decrease by two of the number of phase dislocations.

Quantum mechanical calculations, in agreement with experimental data, predict that the most stable isomer of each [4*n*]annulene, up to $n = 5$, has Hückel topology. Crystal structures for the [4*n*]annulenes with $n = 6$, 8, 14, 16, 18 and all structures identified by NMR exhibit Hückel topology. There are no experimental data available for any neutral, parent Möbius [4*n*]annulene. Schleyer et al. have developed a method that calculates the chemical shift of a hypothetical atom in the center of conjugated ring systems, the so-called nucleus-independent chemical shifts values (NICSs) [277, 278]. These calculations intend to quantify the effect of magnetically induced ring currents in cyclic conjugated molecules. For instance, calculations [279] on some Möbius [12]-, [16]-, and [20]-annulene indicate strongly negative NICS values (NMR upfield shifts). In the case of Möbius [16]annulene, NICS = −14.5 ppm, a value similar to that calculated for benzene (NICS = −11.5 ppm).

The most stable conformation of [9]annulenium cation is a Möbius ring. In 1971, Schleyer and coworkers reported that the *trans*-9-chlorobicyclo[6.1.0] nonatriene labeled with a deuterium atom at C-9 (**112**) gives **115** with complete scrambling of the deuterium atom upon solvolysis. Quantum calculations confirmed that the [9]annulenium cation intermediate (**113**) has a Möbius topology and C_2 symmetry [280]. The barrier for racemization (of the two chiral C_2 symmetrical cation intermediates **113** and *ent*-**113**) is about 1 kcal mol^{-1} only. An electrocyclic ring closure with a barrier of 8.1 kcal mol^{-1} leads to the bishomoaromatic bicyclic cation **114**, which is 16 kcal mol^{-1} more stable than **113** [281].

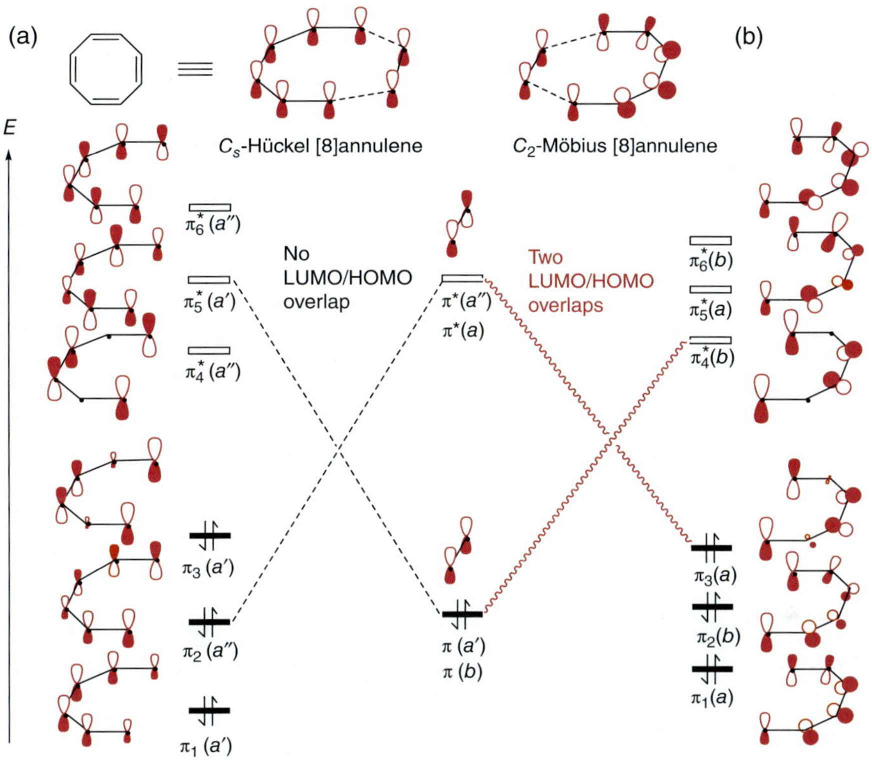

Figure 4.38 PMO diagrams representing the p-electronic interactions between (a) a (*Z,Z*)-hexatriene and an ethylene subunit of a Hückel [8]annulene (planar cyclotetraene) and (b) between a twisted (*Z,Z*)-hexatriene and an ethylene subunit of a Möbius [8]annulene. The Hückel [8]annulene is not aromatic, whereas the Möbius [8]annulene is stabilized by Möbius aromaticity (two LUMO/HOMO interactions).

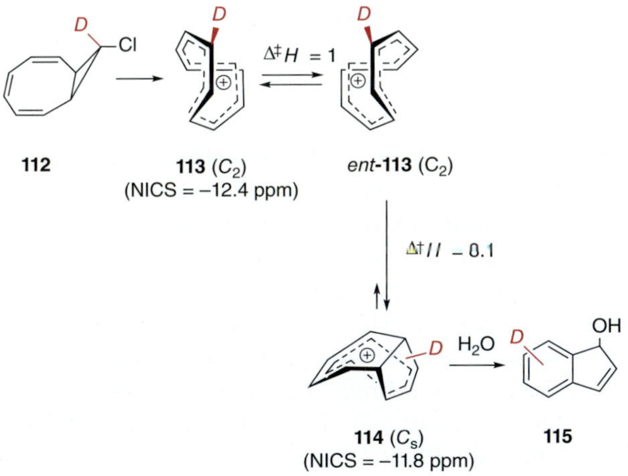

Latos-Grazynski and coworkers reported that porphyrin di-*p*-benzi[28]hexaphyrin(1.1.1.1.1.1) (**116**), an extended porphyrin, can switch between Hückel antiaromatic and Möbius aromatic structures as a function of solvent and temperature [282]. X-ray crystallographic analysis revealed C_2 symmetry with the two nonequivalent phenylene rings, which are aligned in a “edge-to-face” manner. In solution and at low temperature, the Möbius conformer **116** is preferred. At high temperature, the Hückel conformer **117** with parallel phenylene rings becomes dominant [283].

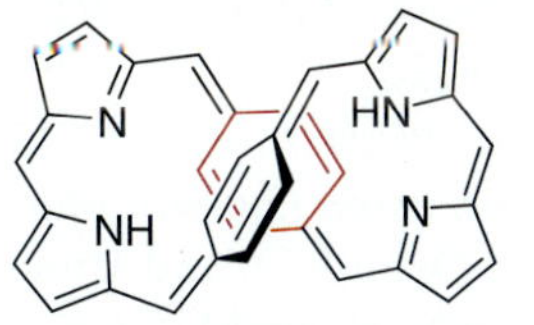

116

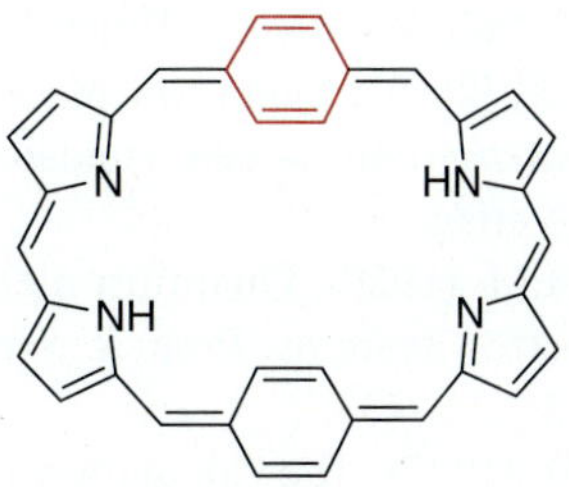

117

4.10 Conclusion

Starting with a qualitative model of the molecule of H_2, a simple molecular orbital theory has been developed to interpret bonding energies, substituent effects on neutral and charged species, and the effect of structure on thermodynamic and electronic properties of simple organic molecules. With modern computers, very sophisticated quantum mechanical calculations can be carried out almost routinely. The complex programs provide quantitative answers, but no insights as to the reasons for these answers, except that good approximations to the Schrödinger equation give those results. For organic chemists, as well as for the specialists in computational chemistry, it is very useful to have a simple model such as the perturbation molecular orbital (PMO) theory to test whether the results are reasonable and to understand qualitatively why they come out that way. For the synthetic chemists, it is useful to have a simple method to evaluate molecular properties just by combining thermochemical data with the PMO theory that requires only a pencil and the back of an envelope. A molecular structure A–B of a stable compound, of a reactive intermediate (radical, diradical ↔ zwitterion, carbocation, anions, and carbene), or a transition structure can be dissected into two molecular or atomic fragments A and B. Then, ionization energies can be taken as negatives of HOMO energies and electron affinities as negatives of LUMO energies and the shapes of the frontier orbitals can be defined by analogy to the MOs discussed in this chapter. The geometry of A–B gives the overlap S between these frontier orbitals, and the PMO theory gives indications of their interactions. If two overlapping orbitals of A and B share four electrons, the interaction is destabilizing. If the two overlapping orbitals share one, two electrons, the interaction is stabilizing. When there are three electrons, the interaction is also stabilizing in most cases. Overlap between two orbitals is always nil when these orbitals are orthogonal with respect to one symmetry element of A–B.

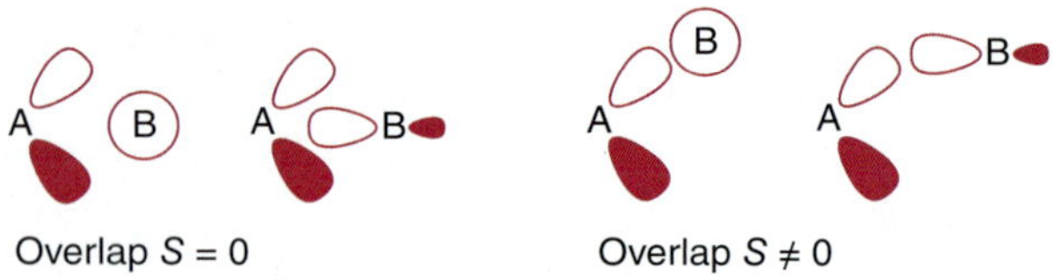

The smaller the energy difference between two overlapping orbitals, the larger the results on the energy and electron distribution of A–B. The PMO theory will be applied extensively in the forthcoming chapters.

References

1 Atkins, P. and Friedman, R. (2005). *Molecular Quantum Mechanics*, 4e, 1–573. Oxford: Oxford University Press; (2011). 5, 1-537. Oxford: Oxford University Press.
2 Atkins, P., de Paula, J., and Friedman, R. (2013). *Physical Chemistry. Quanta, Matter, and Change*, 2e, 1–1008. Oxford: Oxford University Press.
3 Welz, B. and Sperling, M. (1999). *Atomic Absorption Spectrometry*. 3rd Completely Revised Edition, 1–965. Weinheim: Wiley-VCH.
4 Greiner, W. (2001). *Quantum Mechanics: An Introduction*, 4e, 1–486. Heidelberg: Springer-Verlag.
5 Dirac, P.A.M. (1929). Quantum mechanics of many-electron systems. *Proc. R. Soc. London, Ser. A* 123: 714–733.
6 Burrau, O. (1927). The calculation of the energy value of hydrogen molecule ions (H2+) in their normal position. *Naturwissenschaften* 15: 16–17.
7 Heitler, W. and London, F. (1927). Interaction of neutral atoms and homopolar binding according to the quantum mechanics. *Z. Phys.* 44: 455–472.
8 Condon, E.U. (1927). Wave mechanics and the normal state of the hydrogen molecule. *Proc. Natl. Acad. Sci. U. S. A.* 13: 466–470.
9 Pauling, L. (1928). The shared-electron chemical bond. *Proc. Natl. Acad. Sci. U. S. A.* 14: 359–362.
10 Pauling, L. (1960). *The Nature of the Chemical Bond and the Structure of Molecules and Crystals: An Introduction to Modern Structural Chemistry*, 3e, 1–644. Ithaca, NY: Cornell University Press.
11 Pauling, L. (1932). The nature of the chemical bond. III. The transition from one extreme bond type to another. *J. Am. Chem. Soc.* 54: 988–1003.
12 Hückel, E. (1931). Quantum theoretical contributions to the benzene problem. I. The electron configuration of benzene and related compounds. *Z. Phys.* 70: 204–286.
13 Hückel, E. (1931). Quantum theory treatment of the benzene problem. II. Quantum theory of induced polarity. *Z. Phys.* 72: 310–337.
14 Hückel, E. (1932). Quantum theoretical contributions to the problem of aromatic and

non-saturated compounds. III. *Z. Phys.* 76 (628): 648.

15 Hückel, E. (1933). Quantum theoretical contributions to the problem of aromatic and unsaturated compounds. IV. Free radicals in organic chemistry. *Z. Phys.* 83: 632–668.

16 Heilbronner, E. and Bock, H. (1976). *The HMO-Model and its Application, Basis and Manipulation*, vol. 1. London, Weinheim: Wiley-Verlag Chemie.

17 (a) Heilbronner, E. and Bock, H. (1976). *The HMO Model and its Application, Problems with Solutions*, vol. 2. London, Weinheim: Wiley-Verlag Chemie. (b) Heilbronner, E. and Bock, H. (1976). *The HMO Model and its Application, Tables of Hückel Molecular Orbitals*, vol. 3. London, Weinheim: Wiley-Verlag Chemie.

18 (a) Coulson, C.A., O'Leary, B.O., and Mallion, R.B. (1978). *Hückel Theory for Organic Chemists*, 1–182. New York: Academic Press. (b) Kutzelnigg, W. (2007). What I like about Hückel theory. *J. Comput. Chem.* 28 (1): 25–34.

19 (a) Hohenberg, P. and Kohn, W. (1964). Inhomogeneous electron gas. *Phys. Rev.* 136 (3B): B864–B871. (b) Kohn, W. and Sham, L.J. (1965). Self-consistent equations including exchange and correlation effects. *Phys. Rev.* 140 (4A): A1133–A1138. (c) Kohn, W., Becke, A.:.D., and Parr, R.G. (1996). Density functional theory of electronic structure. *J. Phys. Chem.* 100 (31): 12974–12980. (d) Kohn, W. (1999). Nobel lecture: electronic structure of matter - wave functions and density functionals. *Rev. Mod. Phys.* 71 (5): 1253–1266.

20 (a) de Bruijn, S. (1984). Analysis of the inadequacies of some semi-empirical MO methods as theories of structure and reactivity. *Int. J. Quantum Chem.* 25 (2): 367–390. (b) Dewar, M.J.S., Zöbisch, E.G., Healy, E.F., and Stewart, J.J.P. (1985). Development and use of quantum molecular models. 75. Comparative tests of theoretical procedures for studying chemical reactions. *J. Am. Chem. Soc.* 107 (13): 3902–3909. (c) Stewart, J.J.P. (1989). Optimization of parameters for semi-empirical methods I. Method. *J. Comput. Chem.* 10 (2): 209–220. (d) Rocha, G.B., Freire, R.O., Simas, A.M., and Stewart, J.J.P. (2006). *RM1: a re-parameterization of AM1 for H, C, N, O, P, S, F, Cl, Br, and I. J. Comput. Chem.* 27 (10): 1101–1111. (e) Stewart, J.J.P. (2007). *Optimization of parameters for semi-empirical methods V: modification of NDDO approximations and application to 70 elements. J. Mol. Model.* 13 (12): 1173–1213.

21 (a) Slater, J.C. (1930). Atomic shielding constants. *Phys. Rev.* 36: 57–64. (b) Schaefer, H.F.I. (1972). *The Electronic Structure of Atoms and Molecules*, 1–437. Reading, MA: Addison-Wesley Publishing Co. (c) Bouferguene, A., Fares, M., and Hoggan, P.E. (1996). STOP: A slater type orbital package for general molecular electronic structure calculations. *Int. J. Quantum Chem.* 57 (4): 801–810. (d) Guseinov, I.I. (2002). New complete orthonormal sets of exponential-type orbitals and their application to translation of Slater Orbitals. *Int. J. Quantum Chem.* 90 (1): 114–118.

22 (a) Mulliken, R.S. and Ermler, W.C. (1982). *Polyatomic Molecules: Results of ab Initio Calculations*, 1–431. New York: Academic Press. (b) Parr, R.G. (1990). *On the genesis of a theory. Int. J. Quantum Chem.* 37 (4): 327–347. (c) Pople, J.A. (1999). Nobel lecture: quantum chemistry models. *Rev. Mod. Phys.* 71 (5): 1267–1274. (d) Friesner, R.A. (2005). Ab initio quantum chemistry: methodology and applications. *Proc. Natl. Acad. Sci. U. S. A.* 102 (19): 6648–6653. (e) Jensen, F. (2017). *Introduction to Computational Chemistry*, 3e, 1–638. Chichester: *Wiley*.

23 Mulliken, R.S. (1967). Spectroscopy molecular orbitals and chemical bonding, Nobel lecture. *Science* 157 (3784): 13–24.

24 Lennard-Jones, J.E. (1929). The electronic structure of some diatomic molecules. *Trans. Faraday Soc.* 25: 0668–0685.

25 Coulson, C.A. and Longuet-Higgins, H.C. (1948). The electronic structure of conjugated systems - Part-III and Part-IV. 3. Bond orders in unsaturated molecules. *Proc. R. Soc. Lond. A Math. Phys. Sci.* 193 (1035): 447–464.

26 Coulson, C.A. and Longuet-Higgins, H.C. (1947). The electronic structure of conjugated systems. 1. General theory. *Proc. R. Soc. Lond. A Math. Phys. Sci.* 191 (1024): 39–60.

27 Morse, P.M. (1929). Diatomic molecules according to the wave mechanics. II. Vibrational levels. *Phys. Rev.* 34 (1): 57–64.

28 Morse, P.M. (1929). Diatomic molecules according to the wave mechanics I: electronic levels of the hydrogen molecular ion. *Phys. Rev.* 33 (6): 0932–0947.

29 Drake, G.W.F. (ed.) (2006). *Springer Handbook of Atomic, Molecular, and Optical Physics*, 2e. Heidelberg: Springer-Verlag.

30 Mulliken, R.S. (1947). Quantum-mechanical methods and the electronic spectra and structure of molecules. *Chem. Rev.* 41 (2): 201–206.

31 (a) Pople, J.A. (1970). *Approximate Molecular Orbital Theory (Advanced Chemistry)*, 1–214. New York: McGraw-Hill. (b) Bursten, B.E. (1991). Some comments on approximate LCAO molecular orbital theory in organometallic chemistry: getting more by doing less? *Pure Appl. Chem.* 63 (6): 839–844.

32 Pauli, W. (1925). Über den Zusammenhang des Abschusses des Elektronengruppen im Atom mit der Komplexstructur der Spektren. *Z. Phys.* 31: 765–783.

33 Pople, J.A., Santry, D.P., and Segal, G.A. (1965). Approximate self-consistent molecular orbital theory. I. Invariant procedures. *J. Chem. Phys.* 43 (10): S129–S135.

34 Albright, T.A., Burdett, J.K., and Whangbo, M.-H. (2013). *Orbital Interactions in Chemistry*, 2e, 1–834. New York: Wiley.

35 Hoffmann, R. (1988). *Solids and Surfaces: A Chemist's View of Bonding in Extended Structures*, 1–142. New York and Weinheim: VCH Publishers/VCH Verlagsgesellschaft.

36 Seo, D.K., Papoian, G., and Hoffmann, R. (2000). Generalized perturbational molecular orbital (PMO) theory. *Int. J. Quantum Chem.* 77 (1): 408–420.

37 Roberts, J.D. (1961). *Notes on Molecular Orbital Calculations*, pp. 1–156. New York: W.A: Benjamin, Inc.

38 Streitwieser, A.J. (1961). *Molecular Orbital Theory for Organic Chemists*, 1–489. New York: Wiley.

39 Smith, W.B. (1971). Perturbation MO method - quantum mechanics on back of an envelope. *J. Chem. Educ.* 48 (11): 749–752.

40 Hückel, E. (1930). Quantum theory of double linkings. *Z. Phys.* 60: 423–456.

41 Gajewski, J.J. (2004). *Hydrocarbon Thermal Isomerizations*, 2e, 1–442. Amsterdam, The Netherlands: Elsevier Science & Technology Books.

42 Radom, L., Schleyer, P.v.R., Buss, V., and Pople, J.A. (1972). Molecular-orbital theory of electronic structure of organic compounds. 11. Geometries and energies of $C_3H_7^+$ cations. *J. Am. Chem. Soc.* 94 (2): 311–321.

43 (a) Ellison, G.B., Davico, G.E., Bierbaum, V.M., and DePuy, C.H. (1996). Thermochemistry of the benzyl and allyl radicals and ions. *Int. J. Mass Spectrom. Ion Process.* 156 (1–2): 109–131; (b) Li, Z., Bally, T., Houk, K.N., and Borden, W.T. (2016) Variations in rotational barriers of allyl and benzyl cations, anions, and radicals. *J. Org. Chem.* 81: 9576–9584.

44 (a) Lossing, F.P. and Holmes, J.L. (1984). Stabilization energy and ion size in carbocations in the gas phase. *J. Am. Chem. Soc.* 106 (23): 6917–6920. (b) Breslow, R. (2014). Novelaromatic and antiaromatic systems. *Chem. Rec.* 14 (6): 1174–1182.

45 Glukhovtsev, M.N., Laiter, S., and Pross, A. (1996). Thermochemical assessment of the aromatic and antiaromatic characters of the cyclopropenyl cation, cyclopropenyl anion, and cyclopropenyl radical: a high-level computational study. *J. Phys. Chem.* 100 (45): 17801–17806.

46 (a) Kass, S.R. (2013). Cyclopropenyl anion: an energetically nonaromatic ion. *J. Org. Chem.* 78 (14): 1174–1152. (b) Tian, Z., Lis, L., and Kass, S.R. (2013). Carbon–hydrogen bond dissociation energies: the curious case of cyclopropene. *J. Org. Chem.* 78 (24): 12650–12653.

47 Breslow, R., Brown, J., and Gajewski, J.J. (1967). Antiaromaticity of cyclopropenyl anions. *J. Am. Chem. Soc.* 89 (17): 4383–4390.

48 Breslow, R. (1968). Small antiaromatic rings. *Angew. Chem. Int. Ed. Eng.* 7 (8): 565–570.

49 Breslow, R. (1982). Stable 4N-π electron triplet molecules. *Pure Appl. Chem.* 54 (5): 927–938.

50 Berkowitz, J., Ellison, G.B., and Gutman, D. (1994). Three methods to measure R—H bond-energies. *J. Phys. Chem.* 98 (11): 2744–2765.

51 Maier, G. (1974). Cyclobutadiene problem. *Angew. Chem. Int. Ed. Eng.* 13 (7): 425–438.

52 Deniz, A.A., Peters, K.S., and Snyder, G.J. (1999). Experimental determination of the antiaromaticity of cyclobutadiene. *Science* 286 (5442): 1119–1122.

53 (a) Bally, T. (2006). Cyclobutadiene: the antiaromatic paradigm? *Angew. Chem. Int. Ed.* 45 (40): 6616–6619. (b) Fantuzzi, F., Cardozo, T.M., and Nascimento, M.A.C. (2016). The nature of the singlet and triplet states of cyclobutadiene as revealed by quantum interference. *ChemPhysChem* 17 (2): 288–295.

54 Nenitzescu, C.D., Dinulescu, I.G., Avram, M. et al. (1963). Zum dimeren cyclobutadien. *Angew. Chem. Int. Ed. Eng.* 75 (1): 88–88.

55 Watts, L., Fitzpatrick, J.D., and Pettit, R. (1965). Cyclobutadiene. *J. Am. Chem. Soc.* 87 (14): 3253–3254.

56 Schmidt, E.K.G., Brener, L., and Pettit, R. (1970). Cyclobutadiene via dehalogenation of dihalocyclobutenes. *J. Am. Chem. Soc.* 92 (10): 3240–3242.

57 Masamune, S., Ona, H., Suda, M., and Leichter, L.M. (1972). Cyclobutadiene. *J. Chem. Soc. Chem. Commun.* (23): 1268–1269.

58 Houk, K.N., Li, Y., Storer, J. et al. (1994). Concerted and stepwise mechanisms in cycloaddition reactions - potential surfaces and isotope effects. *J. Chem. Soc. Faraday Trans.* 90 (12): 1599–1604.

59 Li, Y. and Houk, K.N. (1996). The dimerization of cyclobutadiene. An ab initio CASSCF theoretical study. *J. Am. Chem. Soc.* 118 (4): 880–885.

60 Lin, C.Y. and Krantz, A. (1972). Matrix preparation of cyclobutadiene. *J. Chem. Soc. Chem. Commun.* (19): 1111–1112.

61 Cram, D.J., Tanner, M.E., and Thomas, R. (1991). The taming of cyclobutadiene. *Angew. Chem. Int. Ed. Eng.* 30 (8): 1024–1027.

62 Dodziuk, H. (2009). Short-lived species stabilized in 'molecular' or 'supramolecular flasks'. In: *Strained Hydrocarbons: Beyond the van't Hoff and Le Bel hypothesis*, Chapter 10 (ed. H. Dodziuk), 449–471. Weinheim: Wiley-VCH.

63 (a) Fattahi, A., Lis, L., Tian, Z.X., and Kass, S.R. (2006). The heat of formation of cyclobutadiene. *Angew. Chem. Int. Ed.* 45 (30): 4984–4988. (b) Hohlneicher, G., Packschies, L., and Weber, J. (2007). On the σ,π-energy separation of the aromatic stabilization energy of cyclobutadiene. *Phys. Chem. Chem. Phys.* 9: 2517–2530.

64 Reeves, P., Devon, T., and Pettit, R. (1969). On possible rectangular nature of cyclobutadiene. *J. Am. Chem. Soc.* 91 (21): 5890–5891.

65 Whitman, D.W. and Carpenter, B.K. (1980). Experimental evidence for non-square cyclobutadiene as a chemically significant intermediate in solution. *J. Am. Chem. Soc.* 102 (12): 4272–4274.

66 Whitman, D.W. and Carpenter, B.K. (1982). Limits on the activation parameters for automerization of cyclobutadiene-1,2-D2. *J. Am. Chem. Soc.* 104 (23): 6473–6474.

67 Maier, G. and Alzérreca, A. (1973). Small rings. 15. Tri-*tert*-butylcyclobutadiene. *Angew. Chem.* 85 (23): 1056–1057.

68 Masamune, S., Nakamura, N., Suda, M., and Ona, H. (1973). Properties of [4]annulene system - induced paramagnetic ring current. *J. Am. Chem. Soc.* 95 (25): 8481–8483.

69 (a) Jahn, H.A. and Teller, E. (1937). Stability of polyatomic molecules in degenerate electronic states. I. Orbital degeneracy. *Proc. R. Soc. London, Ser. A Math. Phys. Eng. Sci.* 161 (A905): 220–235. (b) Liu, Y., Bersuker, I.B., and Boggs, J.E. (2013). Pseudo Jahn–Teller origin of puckering in ${C_4H_4}^{2+}$, ${Si_4H_4}^{2+}$, and ${C_4F_4}^{2+}$ dications. *Chem. Phys.* 417: 26–29.

70 Suzuki, K., Matsuo, T., Hashizume, D. et al. (2011). A Planar rhombic charge-separated tetrasilacyclobutadiene. *Science* 331 (6022): 1306–1309.

71 Olah, G.A., Bollinger, J.M., and White, A.M. (1969). Stable carbonium ions. 89. Tetramethylcyclobutenium dication an aromatic 2 π-electron system. *J. Am. Chem. Soc.* 91 (13): 3667–3669.

72 Olah, G.A. and Staral, J.S. (1976). Novel aromatic systems. 4. Cyclobutadiene dications. *J. Am. Chem. Soc.* 98 (20): 6290–6304.

73 Olah, G.A. and Mateescu, G.D. (1970). Stable carbonium ions. CI. Tetraphenylcyclobutadiene dication. *J. Am. Chem. Soc.* 92 (5): 1430–1432.

74 Rosenberg, M., Ottosson, H., and Kilsa, K. (2010). Proton and hydride affinities in excited states: magnitude reversals in proton and hydride affinities between the lowest singlet and triplet states of annulenyl and benzannulenyl anions and cations. *J. Org. Chem.* 75 (7): 2189–2196.

75 (a) Agapito, F., Nunes, P.A., Costa Cabral, B.J. et al. (2008). Energetic differences between the five- and six-membered ring hydrocarbons: strain energies in the parent and radical molecules. *J. Org. Chem.* 73 (16): 6213–6223. (b) Bertho, F., Rayez, M.-T., and Lesclaux, R. (1999). UV absorption spectrum and self-reaction kinetics of the cyclohexadienyl radical, and stability of a series of cyclohexadienyl-type radicals. *J. Phys. Chem. A* 103 (28): 5501–5509.

76 Nyulaszi, L. and Schleyer, P.v.R. (1999). Hyperconjugative π-aromaticity: how to make cyclopentadiene aromatic. *J. Am. Chem. Soc.* 121 (29): 6872–6875.

77 Wodrich, M.D., McKee, W., and Schleyer, P.v.R. (2011). On the advantages of hydrocarbon radical stabilization energies based on R—H bond dissociation energies. *J. Org. Chem.* 76 (8): 2439–2447.

78 (a) Washburn, W.N. (1983). Bishomoaromatic stabilization of bicyclic anions. *J. Org. Chem.* 48 (23): 4287–4289. (b) Lee, R.E. and Squires, R.R. (1986). Anionic homoaromaticity - a gas phase experimental study. *J. Am. Chem. Soc.* 108 (17): 5078–5086.

79 (a) Brown, J.M. and Occolowi, J.L. (1965). A non-classical carbanion. *J. Chem. Soc. Chem. Commun.* (16): 376–376. (b) Dauben, W.G. and Twieg, R.J. (1974). Facile cleavage of homocuneone involving bishomocyclopentadienyl anion. *Tetrahedron Lett.* 15 (6): 531–534.

80 Trimitsis, G.B. and Tuncay, A. (1975). Effect of stabilizing groups on extent of long-range π-electron participation in carbanions. *J. Am. Chem. Soc.* 97 (24): 7193–7194.

81 (a) Jasik, J., Gerlich, D., and Roithova, J. (2014). Probing isomers of the benzene dication in a low-temperature trap. *J. Am. Chem. Soc.* 136: 2960–2962. (b) Roithova, J., Gray, A., Andris, E. et al. (2016). Helium tagging infrared photo-dissociation spectroscopy of reactive ions. *Acc. Chem. Res.* 49 (2): 223–230. (c) Fantuzzi, F., Oliveira de Sousa, D.W., and Nascimento, M.A.C. (2017). Chemical bonding in the pentagonal-pyramidal benzene dication and analogous isoelectronic hexa-coordinate species. *Comput. Theor. Chem.* 1116: 225–233. (d) Malischewski, M. and Seppelt, K. (2017). Crystal structure determination of the pentagonal-pyramidal hexamethylbenzene dication: $C_6(Me)_6^{++}$. *Angew. Chem. Int. Ed.* 56 (1): 368–370.

82 Roth, W.R. (1983). Heats of hydrogenation. *Nachrichten aus Chemie, Technik und Laboratorium* 31 (12): 964–966.

83 Roth, W.R., Lennartz, H.W., Vogel, E. et al. (1986). Heats of hydrogenation. 6. Resonance energy of fused [4N]annulenes. *Chem. Ber. Recl.* 119 (3): 837–843.

84 (a) Leibovic, C. (1972). Valence shell calculations of structure and spectroscopic properties of 3,4-dimethylenecyclobutene. *J. Mol. Struct.* 12 (3): 343–349. (b) Ligabue, A. and Lazzaretti, P. (2002). Theoretical determination of magnetic properties of planar benzene isomers. *J. Chem. Phys.* 116 (3): 964–973.

85 Brown, R.D., Godfrey, P.D., Hart, B.T. et al. (1983). Bond order/bond length relationships in conjugated hydrocarbons - the microwave-spectrum and structure of 3,4-dimethylenecyclobutene. *Aust. J. Chem.* 36 (4): 639–648.

86 Brown, R.D., Burden, F.R., Jones, A.J., and Kent, J.E. (1967). Rotational constants and dipole moment of 3,4-dimethylenecyclobutene. *J. Chem. Soc. Chem. Commun.* (16): 808–809.

87 Trabert, L. and Hopf, H. (1980). Alicyclic compounds. 3. New method of preparation and reactions of [4]radialene. *Liebigs Ann. Chem.* (11): 1786–1800.

88 (a) Wheland, G.W. (1941). The resonance energies of unsaturated and aromatic molecules. *J. Am. Chem. Soc.* 63: 2025–2027. (b) Hosoya, H. (2017). Genealogy of conjugated acyclic polyenes. *Molecules* 22 (6): 896.

89 Baron, P.A., Burden, F.R., Brown, R.D. et al. (1972). Microwave-spectrum and structure of fulvene. *J. Mol. Spectrosc.* 43 (3): 401–410.

90 Replogle, E.S., Trucks, G.W., and Staley, S.W. (1991). Electron correlation effects on the calculated dipole moments of fulvene and cyclopentadiene. *J. Phys. Chem.* 95 (18): 6908–6912.

91 Wheland, G.W. and Mann, D.E. (1949). The dipole moments of fulvene and azulene. *J. Chem. Phys.* 17 (3): 264–268.

92 Dauben, H.J. Jr.,, Wilson, J.D., and Laity, J.L. (1969). Diamagnetic susceptibility exaltation in hydrocarbons. *J. Am. Chem. Soc.* 91 (8): 1991–1998.

93 Havenith, R.W.A., Fowler, P.W., and Steiner, E. (2002). Substituent effects on induced current densities in penta- and heptafulvenes. *J. Chem. Soc., Perkin Trans. 2* (3): 502–507.

94 Oziminski, W., Krygowski, T., Fowler, P., and Soncini, A. (2010). Aromatization of fulvene by complexation with lithium. *Org. Lett.* 12 (21): 4880–4883.

95 Thiele, J. (1900). Keton reactions via cyclopentadien. *Ber. Dtsch. Chem. Ges.* 33: 666–673.

96 Thiele, J. and Balhorn, H. (1906). The decendants of fulvene. 4. Condensation products of cyclopentadiene. *Liebigs Ann. Chem.* 348 (1/3): 1–15.

97 Bergmann, E.D. (1968). Fulvenes and substituted fulvenes. *Chem. Rev.* 68 (1): 41–84.

98 Bandar, J.S., Coscia, R.W., and Lambert, T.H. (2011). Demonstration of the facile reversibility of fulvene formation. *Tetrahedron* 67 (24): 4364–4370.

99 Masamune, S. and Seidner, R.T. (1969). [10]Annulenes. *J. Chem. Soc. Chem. Commun.* (10): 542–544.

100 Masamune, S., Hojo, K., Bigam, G., and Rabenstein, D.L. (1971). Geometry of [10]annulenes. *J. Am. Chem. Soc.* 93 (19): 4966–4968.

101 Masamune, S. and Darby, N. (1972). [10]Annulenes and other $(CH)_{10}$ hydrocarbons. *Acc. Chem. Res.* 5 (8): 272–281.

102 Castro, C., Karney, W.L., McShane, C.M., and Pemberton, R.P. (2006). [10]Annulene: bond shifting and conformational mechanisms for automerization. *J. Org. Chem.* 71 (8): 3001–3006.

103 Vogel, E. (1982). Recent advances in the chemistry of bridged annulenes. *Pure Appl. Chem.* 54 (5): 1015–1039.

104 Vogel, E. (2011). From small carbocyclic rings to porphyrins: a personal account of 50 years of research. *Angew. Chem. Int. Ed.* 50 (19): 4278–4287.

105 Vogel, E., Klug, W., and Breuer, A. (1974). 1,6-methano[10]annulene. *Org. Synth.* 54: 11–18.

106 Roth, W.R., Adamczak, O., Breuckmann, R. et al. (1991). Resonance energy calculation - the MM2ERW force-field. *Chem. Ber.* 124 (11): 2499–2521.

107 Bremser, W., Hagen, R., Heilbronner, E., and Vogel, E. (1969). Enthalpy of formation of 1,6-bridged[10]annulenes. *Helv. Chim. Acta* 52 (2): 418–431.

108 Dominikowska, J. and Palusiak, M. (2011). Cyclooctatetraene dianion - an artifact? *J. Comput. Chem.* 32 (7): 1441–1448.

109 Oth, J.F.M., Röttele, H., and Schröder, G. (1970). [12]Annulene. *Tetrahedron Lett.* 11 (1): 61–66.

110 Oth, J.F.M. (1971). Conformational mobility and fast bond shift in the annulenes. *Pure Appl. Chem.* 25 (3): 573–622.

111 Stevenson, G.R., Concepcion, R., and Reiter, R.C. (1983). The [12]annulene anion-radical. *J. Org. Chem.* 48 (16): 2777–2778.

112 Chiang, C.C. and Paul, I.C. (1972). Crystal and molecular structure of [14]annulene. *J. Am. Chem. Soc.* 94 (13): 4741–4743.

113 Okoronkwo, T., Nguyen, P.T., Castro, C., and Karney, W.L. (2010). [14]Annulene: *cis*/*trans* isomerization via two-twist and nondegenerate planar bond shifting and Möbius conformational minima. *Org. Lett.* 12 (5): 972–975.

114 Martin-Santamaria, S., Lavan, B., and Rzepa, H.S. (2000). Hückel and Möbius aromaticity and trimerous transition state behaviour in the pericyclic reactions of [10], [14], [16] and [18]annulenes. *J. Chem. Soc., Perkin Trans. 2* (7): 1415–1417.

115 Vogel, E., Engels, H.W., Huber, W. et al. (1982). Octalene-[14]annulene conversion - 1,8-dimethyl[14]annulene. *J. Am. Chem. Soc.* 104 (13): 3729–3731.

116 Vogel, E. (1971). Aromatic and nonaromatic 14π-electron systems. *Pure Appl. Chem.* 28 (2–3): 355–377.

117 Simonetta, M. (1980). Aromatic and non-aromatic bridged annulenes. *Pure Appl. Chem.* 52 (6): 1597–1610.

118 Vogel, E., Schieb, T., Schulz, W.H. et al. (1986). Bridged [14]annulenes with a phenanthrene-perimeter - *anti*-1,6-7,12-bismethano[14]annulene. *Angew. Chem. Int. Ed. Eng.* 25 (8): 723–725.

119 Schröder, G. and Oth, J.F.M. (1966). Zur Kenntnis des [16]Annulens (1). *Tetrahedron Lett.* 7 (34): 4083–4088.

120 Schröder, G., Martin, W., and Oth, J.F.M. (1967). Thermal and photochemical behavior of a [16]annulene. *Angew. Chem. Int. Ed. Eng.* 6 (10): 870–871.

121 Stevenson, C.D. and Kurth, T.L. (1999). Perturbations in aromatic and antiaromatic characters due to deuteration: the case of [16]annulene. *J. Am. Chem. Soc.* 121 (7): 1623–1624.

122 Oth, J.F.M., Schröder, G., Gilles, J.M., and Baumann, H. (1972). Radical-anion and dianion of [16]annulene. *J. Am. Chem. Soc.* 94 (10): 3498–3512.

123 Stevenson, C.D. and Kurth, T.L. (2000). Isotopic perturbations in aromatic character and new closely related conformers found in [16]- and [18]annulene. *J. Am. Chem. Soc.* 122 (4): 722–723.

124 Johnson, S.M., Paul, I.C., and King, G.S.D. (1970). [16]Annulene - crystal and molecular structure. *J. Chem. Soc. B* (4): 643–649.

125 Ermer, O. (2005). Concerning the structure of [18]annulene. *Helv. Chim. Acta* 88 (8): 2262–2266.

126 (a) Fiegl, H., Sundholm, D., Taubert, J., and Klopper, W. (2009). Magnetically induced current densities in aromatic, antiaromatic, homoaromatic, and nonaromatic hydrocarbons. *J. Phys. Chem. A* 113 (30): 8668–8676; (b) Kato, T. (2007). The mechanism of occurrence of diamagnetic intramolecular ring currents in the neutral annulenes. *Chem. Phys. Research J.* (Nova Science Publ. Inc.) 1 (1): 60–96.

127 Oth, J.F.M., Woo, E.P., and Sondheimer, F. (1973). Dianion of [18]annulene. *J. Am. Chem. Soc.* 95 (22): 7337–7345.

128 Metcalf, B.W. and Sondheimer, F. (1971). Unsaturated macrocyclic compounds. 86. [20]Annulene. *J. Am. Chem. Soc.* 93 (24): 6675–6677.

129 McQuilki, R.M., Metcalf, B.W., and Sondheimer, F. (1971). [22]Annulene. *J. Chem. Soc. Chem. Commun.* (7): 338–339.

130 Calder, I.C. and Sondheimer, F. (1966). [24]Annulene - dependence of nuclear magnetic resonance spectrum on temperature. *J. Chem. Soc. Chem. Commun.* (24): 904–905.

131 Oth, J.F.M. and De Julien de Zélicourt, Y. (1999). Conformational mobility and migration of the π-bonds of the [24]annulene. *Helv. Chim. Acta* 82 (3): 435–483.

132 Choi, C.H. and Kertesz, M. (1998). Bond length alternation and aromaticity in large annulenes. *J. Chem. Phys.* 108 (16): 6681–6688.

133 García-Cuesta, I., Sánchez-Marín, J., Bondo-Pedersen, T. et al. (2008). Variation of polarizability in the [4*n*+2]annulene series: from

[22]- to [66]-annulene. *Phys. Chem. Chem. Phys.* 10 (3): 361–365.

134 Wannere, C.S., Sattelmeyer, K.W., Schaefer, H.F.I., and Schleyer, P.v.R. (2004). Aromaticity: the alternating C—C bond length structures of [14]-, [18]-, and [22]annulene. *Angew. Chem. Int. Ed.* 43 (32): 4200–4206.

135 Havenith, R.W.A., Jenneskens, L.W., and Fowler, P.W. (2002). Ring currents that survive bond alternation in constrained 8π and 6π monocycles. *Chem. Phys. Lett.* 367 (3, 4): 468–474.

136 Lammertsma, K., Olah, G.A., Berke, C.M., and Streitwieser, A. Jr., (1979). 1,4,5,8-Tetramethylnaphthalene dication and related radical-cations. *J. Am. Chem. Soc.* 101 (22): 6658–6659.

137 Forsyth, D.A. and Olah, G.A. (1976). Oxidation of polycyclic arenes in SbF_5-SO_2ClF - formation of arene dications and observation of electron exchange with radical-cations based on ^{13}C nuclear magnetic-resonance studies. *J. Am. Chem. Soc.* 98 (14): 4086–4090.

138 Deger, H.M., Müllen, K., and Vogel, E. (1978). *Syn,syn*-1,6-8,17-10,15-trismethano[18]annulene dication - dication of a [4*n*+2]annulene. *Angew. Chem. Int. Ed. Eng.* 17 (12): 957–958.

139 Willstaetter, R. and Waser, E. (1911). On cyclooctatetraenes. (Fifth report on the cyclooctane sequence). *Ber. Dtsch. Chem. Ges.* 44: 3423–3445.

140 (a) Kaufman, H.S., Fankuchen, I., and Mark, H. (1948). Structure of cyclooctatetraene. *Nature* 161 (4083): 165–165. (b) Kummli, D.S., Lobsinger, S., Frey, H.M. et al. (2008). Accurate determination of the structure of cyclooctatetraene by femtosecond rotational coherence spectroscopy and ab initio calculations. *J. Phys. Chem. A* 112 (38): 9134–9143.

141 (a) Havenith, R.W.A., Fowler, P.W., and Jenneskens, L.W. (2006). Persistence of paratropic ring currents in nonplanar, tub-shaped geometries of 1,3,5,7-cyclooctatetraene. *Org. Lett.* 8 (7): 1255–1256. (b) Wu, J.I., Fernández, I., Mo, Y., and Schleyer, P.v.R. (2012). Why cyclooctatetraene is highly stabilized: the importance of "two-way" (double) hyperconjugation. *J. Chem. Theory Comput.* 8 (4): 1280–1287.

142 Andres, J.L., Castano, O., Morreale, A. et al. (1998). Potential energy surface of cyclooctatetraene. *J. Chem. Phys.* 108 (1): 203–207.

143 Karadakov, P.B. (2008). Aromaticity and antiaromaticity in the low-lying electronic states of cyclooctatetraene. *J. Phys. Chem. A* 112 (49): 12707–12713.

144 Anet, F.A.L. and Bock, L.A. (1968). Photochemical interconversion of cyclooctatetraene bond shift isomers. *J. Am. Chem. Soc.* 90 (25): 7130–7132.

145 Luz, Z. and Meiboom, S. (1973). Structure and bond shift kinetics of cyclooctatetraene studied by NMR in nematic solvents. *J. Chem. Phys.* 59 (3): 1077–1091.

146 Ohmae, T., Nishinaga, T., Wu, M., and Iyoda, M. (2010). Cyclic tetrathiophenes planarized by silicon and sulfur bridges bearing antiaromatic cyclooctatetraene core: syntheses, structures, and properties. *J. Am. Chem. Soc.* 132 (3): 1066–1074.

147 Katritzky, A.R., Jug, K., and Oniciu, D.C. (2001). Quantitative measures of aromaticity for mono-, bi-, and tricyclic penta- and hexaatomic heteroaromatic ring systems and their interrelationships. *Chem. Rev.* 101 (5): 1421–1449.

148 Cyranski, M.K., Krygowski, T.M., Katritzky, A.R., and Schleyer, P.v.R. (2002). To what extent can aromaticity be defined uniquely? *J. Org. Chem.* 67 (4): 1333–1338.

149 (a) Cyranski, M.K., Schleyer, P.V., Krygowski, T.M. et al. (2003). Facts and artifacts about aromatic stability estimation. *Tetrahedron* 59 (10): 1657–1665. (b) Alonso, M. and Herradon, B. (2010). A universal scale of aromaticity for π-organic compounds. *J. Comput. Chem.* 31 (5): 917–928.

150 Ramsden, C.A. (2010). The influence of aza-substitution on azole aromaticity. *Tetrahedron* 66 (14): 2695–2699.

151 Keegstra, M.A., Klomp, A.J.A., and Brandsma, L. (1990). Convenient synthetic procedures for 2-bromofuran and 2,5-dibromofuran. *Synth. Commun.* 20 (21): 3371–3374.

152 Krüger, C., Roberts, P.J., Tsay, Y.H., and Koster, J.B. (1974). Molecular structure of a complex of tetramethylcyclobutadiene and aluminum chloride, containing a σ-Al—C bond. *J. Organomet. Chem.* 78 (1): 69–74.

153 (a) Kraka, E. and Cremer, D. (1990). Chemical implication of local features of the electron density distribution. In: *Theoretical Models of Chemical Bonding. The Concept of the Chemical Bond, Part 2* (ed. Z.B. Maksic), 453–542. Heidelberg: Springer-Verlag. (b) Childs, R.F. (1984). The homotropylium ion and homoaromaticity. *Acc. Chem. Res.* 17 (10): 347–352.

154 Winstein, S. (1959). Homo-aromatic structures. *J. Am. Chem. Soc.* 81 (24): 6524–6525.

155 Cunje, A., Rodriquez, C.F., Lien, M.H., and Hopkinson, A.C. (1996). The $C_4H_5^+$ potential energy surface. Structure, relative energies, and

enthalpies of formation of isomers of $C_4H_5^+$. *J. Org. Chem.* 61 (16): 5212–5220.

156 Olah, G.A., Staral, J.S., Spear, R.J., and Liang, G. (1975). Novel aromatic systems. 2. Cyclobutenyl cations and question of their homoaromaticity - preparation and study of homocyclopropenium ion, simplest homoaromatic system. *J. Am. Chem. Soc.* 97 (19): 5489–5497.

157 Haddon, R.C. (1975). Perturbational molecular orbital (PMO) theory of homoaromaticity. *J. Am. Chem. Soc.* 97 (13): 3608–3615.

158 (a) Sieber, S., Schleyer, P.v.R., Otto, A.H. et al. (1993). Structure, stabilization energies and chemical shifts of the cyclobutenyl cation - does it have aromatic homocyclopropenium ion character - an ab-initio study. *J. Phys. Org. Chem.* 6 (8): 445–464. (b) Karadakov, P.B. and Cooper, D.L. (2016). Modern valence-bond description of homoaromaticity. *J. Phys. Chem. A* 120 (43): 8769–8779.

159 Winstein, S., Kreiter, C.G., and Brauman, J.I. (1966). Ring inversion ultraviolet spectrum and electronic structure of monohomotropylium ion. *J. Am. Chem. Soc.* 88 (9): 2047–2048.

160 Deno, N.C., Pittman, C.U., and Turner, J.O. (1965). Carbonium ions. 17. Cyclizations of pentadienyl and heptatrienyl cations. *J. Am. Chem. Soc.* 87 (10): 2153–2157.

161 Williams, R.V. (2001). Homoaromaticity. *Chem. Rev.* 101 (5): 1185–1204.

162 Kleinpeter, E. and Koch, A. (2009). Visualization of homoaromaticity in cations, neutral molecules and anions by spatial magnetic properties (through space NMR shieldings)-an $^1H/^{13}C$ NMR chemical shift study. *Tetrahedron* 65 (27): 5350–5360.

163 Childs, R.F., Varadarajan, A., Lock, C.J.L. et al. (1982). Structure of the 2-hydroxyhomotropylium cation - unequivocal evidence for homoaromatic delocalization. *J. Am. Chem. Soc.* 104 (9): 2452–2456.

164 Barzaghi, M. and Gatti, C. (1987). Homoaromaticity versus Möbius aromaticity. *J. Chim. Phys. Physicochim. Biol.* 84 (6): 783–789.

165 Yamamoto, K., Yamazaki, S., Matsukawa, A., and Murata, I. (1984). Protonation of 2,7-di-*tert*-butylthiepines - the 1st observation of homothiopyrylium ions. *J. Chem. Soc. Chem. Commun.* (9): 604–606.

166 Davis, R.E. and Tulinsky, A. (1962). The structure of thujic acid - 7,7 dimethylcycloheptatriene-3-carboxylic acid. *Tetrahedron Lett.* 3 (19): 839–846.

167 Traetteberg, M. (1964). Molecular structure of 1,3,5-cycloheptatriene in vapor phase as determined by sector electron diffraction method. *J. Am. Chem. Soc.* 86 (20): 4265–4270.

168 Turner, R.B., Meador, W.R., Doering, W.v.E. et al. (1957). Heats of hydrogenation. 3. Hydrogenation of cyclooctatetraene and of some 7-membered non-benzenoid aromatic compounds. *J. Am. Chem. Soc.* 79 (15): 4127–4133.

169 Herges, R. and Geuenich, D. (2001). Delocalization of electrons in molecules. *J. Phys. Chem. A* 105 (13): 3214–3220.

170 Geuenich, D., Hess, K., Kohler, F., and Herges, R. (2005). Anisotropy of the induced current density (ACID), a general method to quantify and visualize electronic delocalization. *Chem. Rev.* 105 (10): 3758–3772.

171 Reed, T.B. and Lipscomb, W.N. (1953). The crystallography of cycloheptatriene. *Acta Crystallogr.* 6 (1): 108–108.

172 Ermer, O., Klärner, F.G., and Wette, M. (1986). Planarization of unsaturated rings - a cycloheptatriene with a planar 7-membered ring. *J. Am. Chem. Soc.* 108 (16): 4908–4911.

173 Chen, Z.F., Jiao, H.J., Wu, J.I. et al. (2008). Homobenzene: homoaromaticity and homoantiaromaticity in cycloheptatrienes. *J. Phys. Chem. A* 112 (42): 10586–10594.

174 Ahlberg, P., Harris, D.L., and Winstein, H.S. (1970). Direct observation of totally degenerate 9-barbaralyl cation and bicyclo[4.3.0]nonatrienyl cation, a 1,4-bishomotropylium ion. *J. Am. Chem. Soc.* 92 (14): 4454–4456.

175 Roberts, M., Hamberger, H., and Winstein, S. (1970). Bicyclo[4.3.1]deca-2,4,7-trienyl cation - a homoconjugative protonation route to a 1,4-bishomotropylium ion. *J. Am. Chem. Soc.* 92 (21): 6346–6348.

176 Ahlberg, P., Seidl, P., Sakai, M. et al. (1972). Bishomotropylium ions under conditions of long and short life. *J. Am. Chem. Soc.* 94 (20): 7063–7073.

177 Paquette, L.A., Broadhurst, M.J., Warner, P. et al. (1973). Protonation of *cis*-bicyclo[6.1.0] nona-2,4,6-trienes in superacidic media - ^{13}C nuclear magnetic-resonance study of monohomotropylium and 1,3-bishomotropylium ions under long life conditions. *J. Am. Chem. Soc.* 95 (10): 3386–3387.

178 Cheng, A.K., Anet, F.A.L., Mioduski, J., and Meinwald, J. (1974). Determination of fluxional barrier in semibullvalene by proton and ^{13}C nuclear magnetic-resonance spectroscopy. *J. Am. Chem. Soc.* 96 (9): 2887–2891.

179 Moskau, D., Aydin, R., Leber, W. et al. (1989). Applications of ^{13}C-NMR spectroscopy. 27. Activation parameters of the Cope rearrangements of semibullvalene, 1,5-dimethylsemibullvalene, and 2,6-dibromo-1,5-dimethylsemibullvalene. *Chem. Ber.* 122 (5): 925–931.

180 Seefelder, M., Heubes, M., Quast, H. et al. (2005). Experimental and theoretical study of stabilization of delocalized forms of semibullvalenes and barbaralanes by dipolar and polarizable solvents. Observation of a delocalized structure that is lower in free energy than the localized form. *J. Org. Chem.* 70 (9): 3437–3449.

181 Seefelder, M. and Quast, H. (1999). Solvent effects on the equilibrium between localized and delocalized states of thermochromic semibullvalenes and barbaralanes. *Angew. Chem. Int. Ed.* 38 (8): 1068–1071.

182 Griffiths, P.G., Pivonka, D.E., and Williams, R.V. (2011). The experimental realization of a neutral homoaromatic carbocycle. *Chem. Eur. J.* 17 (33): 9193–9199.

183 Greve, D.R. (2011). Homoaromaticity in aza- and phosphasemibullvalenes. A computational study. *J. Phys. Org. Chem.* 24 (3): 222–228.

184 Hine, J., Brown, J.A., Zalkow, L.H. et al. (1955). The synthesis of bicyclo[2,2,2]octa-2,5-diene. *J. Am. Chem. Soc.* 77 (3): 594–598.

185 Zimmerman, H.E. and Paufler, R.M. (1960). Bicyclo[2,2,2]octa-2,5,7-triene (barrelene), a unique cyclic 6 electron π-system. *J. Am. Chem. Soc.* 82 (6): 1514–1515.

186 Turner, R.B. (1964). Heat of hydrogenation of bicyclo[2.2.2]octa-2,5,7-triene. *J. Am. Chem. Soc.* 86 (17): 3586–3587.

187 Birbaum, J.L. and Vogel, P. (1988). Tandem cheletropic additions of sulfur dioxide to [2.2.2]hericene - the barrelene effect. *Helv. Chim. Acta* 71 (6): 1474–1482.

188 Mulliken, R.S. (1939). Intensities of electronic transitions in molecular spectra. IV. Cyclic dienes and hyperconjugation. *J. Chem. Phys.* 7 (5): 339–352.

189 Woodward, R.B. (1942). Structure and absorption spectra. III. Normal conjugated dienes. *J. Am. Chem. Soc.* 64: 72–75.

190 Carr, E.P., Pickett, L.W., and Stucklen, H. (1942). The absorption spectra of a series of dienes. *Rev. Mod. Phys.* 14 (2/3): 265–269.

191 Forbes, W.F., Balasubramanian, A., and Shilton, R. (1964). Ultraviolet absorption spectra of some conjugated dienes. *J. Org. Chem.* 29 (12): 3527–3531.

192 Mulliken, R.S., Rieke, C.A., and Brown, W.G. (1941). Hyperconjugation. *J. Am. Chem. Soc.* 63: 41–56.

193 Alabugin, I.V., Gilmore, K.M., and Peterson, P.W. (2011). Hyperconjugation. *WIREs Comput. Mol. Sci.* 1 (1): 109–141.

194 Flygare, W.H. and Benson, R.C. (1970). Molecular Zeeman effect of cyclopentadiene and isoprene and comparison of the magnetic susceptibility anisotropies. *J. Am. Chem. Soc.* 92 (26): 7523–7529.

195 Schmalz, T.G., Norris, C.L., and Flygare, W.H. (1973). Localized magnetic susceptibility anisotropies. *J. Am. Chem. Soc.* 95 (24): 7961–7967.

196 Fleischer, U., Kutzelnigg, W., Lazzeretti, P., and Muhlenkamp, V. (1994). IGLO study of benzene and some of its isomers and related molecules - search for evidence of the ring current model. *J. Am. Chem. Soc.* 116 (12): 5298–5306.

197 Jiao, H.J. and Schleyer, P.v.R. (1994). Electrostatic acceleration of the 1,5-H shifts in cyclopentadiene and in penta-1,3-diene by Li^+ complexation: aromaticity of the transition structures. *J. Chem. Soc. Faraday Trans.* 90 (12): 1559–1567.

198 González-Castrillo, A., Hurtado, M., Mo, O. et al. (2010). The role of hyperconjugative -aromaticity in the enhanced acidity of methyl-, silyl and germylcyclopentadienes. *Mol. Phys.* 108 (19–20): 2467–2476.

199 Pinkerton, A.A., Schwarzenbach, D., Birbaum, J.L. et al. (1984). The non-planarity of the bicyclo[2.2.1]hept-2-ene double bond - crystal-structures of bicyclo[2.2.2]oct-2-ene, bicyclo[2.2.1]hept-2-ene, and bicyclo[2.1.1]Hex-2-ene systems. *Helv. Chim. Acta* 67 (4): 1136–1153.

200 Watson, W.H., Galloy, J., Bartlett, P.D., and Roof, A.A.M. (1981). Double bond deformation in two crystalline derivatives of *syn*-sesquinorbornene (Δ-4A,8A-octahydro-1,4,5,8-dimethanonaphthalene). *J. Am. Chem. Soc.* 103 (8): 2022–2031.

201 Paquette, L.A., Kravetz, T.M., and Hsu, L.Y. (1985). Electronic control of stereoselectivity. 31. π-Facial course of Diels–Alder cycloadditions to 10-isopropylideneisodicyclopentadiene. *J. Am. Chem. Soc.* 107 (23): 6598–6603.

202 (a) Houk, K.N., Rondan, N.G., Brown, F.K. et al. (1983). Electronic origins and consequences of pyramidalization of asymmetric alkenes in ground and triplet excited states. *J. Am. Chem. Soc.* 105: 5980–5988; (b) Holthausen, M.C. and Koch, W. (1993). Double bond geometry in norborene, sesquinorbornenes, and

related compounds - a high-level quantum chemical investigation. *J. Phys. Chem.* 97 (39): 10021–10027.

203 Ermer, O., Bell, P., and Mason, S.A. (1989). Double bond geometry of norbornene - neutron-diffraction measurement of a derivative at 15 K. *Angew. Chem. Int. Ed. Eng.* 28 (9): 1239–1241.

204 (a) Huisgen, R. (1981). Recent contributions to the mechanism of concerted and non-concerted cycloadditions. *Pure Appl. Chem.* 53 (1): 171–187; (b) Rondan, N.G., Paddon-Row, M.N., Caramella, P. et al. (1982). Origin of Huisgen's factor "x": staggering of allylic bonds promotes anomalously rapid exo attack on norbornenes. *J. Am. Chem. Soc.* 104 (18): 4974–4976.

205 Thi, T.T.N., Delseth, C., Kintzinger, J.P. et al. (1980). O-17 Nuclear magnetic-resonance - the effects of remote unsaturation on O-17-chemical shifts in polycyclic ethers. *Tetrahedron* 36 (19): 2793–2797.

206 Belkacemi, D., Davies, J.W., Malpass, J.R. et al. (1992). 15 Nuclear-magnetic-resonance studies of azabicycles - unusual deshielding of nitrogen in the 7-azabicyclo[2.2.1]heptyl ring system. *Tetrahedron* 48 (46): 10161–10176.

207 Sakurai, H., Nakadaira, Y., Koyama, T., and Sakaba, H. (1983). Chemistry of organosilicon compounds. 172. Unusually large deshielding of Si-29 in 7-silanorbornadienes and related compounds. *Chem. Lett.* 12 (2): 213–216.

208 Chesnut, D.B., Quin, L.D., and Moore, K.D. (1993). Characterization of NMR shielding in 7-phosphanorbornenes. *J. Am. Chem. Soc.* 115 (25): 11984–11990.

209 Rastelli, A., Cocchi, M., Schiatti, E. et al. (1990). Model calculations of chemical interactions. 2. Intramolecular interactions and double bond pyramidalization in polycyclic alkenes. *J. Chem. Soc. Faraday Trans.* 86 (5): 783–787.

210 Krivdin, L.B. and Kalabin, G.A. (1989). Structural applications of one-bond carbon-carbon spin-spin coupling constants. *Prog. Nucl. Magn. Reson. Spectrosc.* 21: 293–448.

211 Steinmann, S.N., Vogel, P., Mo, Y., and Corminboeuf, C. (2011). The norbornene mystery revealed. *J. Chem. Soc. Chem. Commun.* 47 (1): 227–229.

212 Jones, G.R., Caldarelli, S., and Vogel, P. (1997). Long-distance effect on one-bond C—C scalar couplings upon introduction of a double bond in bicyclic systems. *Helv. Chim. Acta* 80 (1): 59–64.

213 Brown, F.K. and Houk, K.N. (1984). The STO-3G transition structure of the Diels–Alder reaction. *Tetrahedron Lett.* 25 (41): 4609–4612.

214 (a) Sahnoun, R., Fujimura, Y., Kabuto, K. et al. (2006). Hyperconjugative electron delocalization mechanism controlling the conformational preference of fluoroacetaldehyde and methyl fluoroacetate. *Bull. Chem. Soc. Jpn.* 79 (4): 555–560. (b) Freitag, M.P. and Bühl, M. (2012). Density functional study of the one-bond C—F coupling constant in α-fluorocarbonyl and α-fluorosulfonyl compounds. *J. Fluor. Chem.* 140: 82–87.

215 Armstrong, G.T. and Marantz, S. (1963). Heats of formation of two isomers of difluorodiazine. *J. Chem. Phys.* 38 (1): 169–172.

216 Nordhoff, K. and Anders, E. (1999). MO theoretical investigation of the *cis* effect observed in mono- and dihalodiazenes. *J. Org. Chem.* 64 (20): 7485–7491.

217 Bohlmann, F. (1957). Zur Konfigurationsbestimmung von Chinolizin-Derivaten. *Angew. Chem. Int. Ed. Eng.* 69 (20): 641–642.

218 Chen, K.H., Lii, J.H., Fan, Y., and Allinger, N.L. (2007). Molecular mechanics (MM4) study of amines. *J. Comput. Chem.* 28 (15): 2391–2412.

219 (a) Mo, Y. and Gao, J. (2007). Theoretical analysis of the rotational barrier of ethane. *Acc. Chem. Res.* 40: 113–119; (b) Baranac Stojanovic, M.B. (2015). Theoretical analysis of the rotational barrier in ethane: cause and consequences. *Struct. Chem.* 26 (4): 989–996.

220 Lin, B.L., Xie, Z., Liu, R. et al. (2003). Importance of σ-type hyperconjugation to the eclipsed structure of propylene. *J. Mol. Struct. THEOCHEM* 633 (1): 15–19.

221 Karplus, M. (1959). Contact electron-spin coupling of nuclear magnetic moments. *J. Chem. Phys.* 30 (1): 11–15.

222 Karplus, M. (1963). Vicinal proton coupling in nuclear magnetic resonance. *J. Am. Chem. Soc.* 85 (18): 2870–2871.

223 Wilkens, S.J., Westler, W.M., Markley, J.L., and Weinhold, F. (2001). Natural *J*-coupling analysis: interpretation of scalar *J*-couplings in terms of natural bond orbitals. *J. Am. Chem. Soc.* 123 (48): 12026–12036.

224 Esteban, A.L., Galache, M.P., Mora, F. et al. (2001). Vicinal NMR proton-proton coupling constants. An NBO analysis. *J. Phys. Chem. A* 105 (21): 5298–5303.

225 Provasi, P.F., Gómez, C.A., and Aucar, G.A. (2004). Hyperconjugation: the electronic mechanism that may underlie the Karplus curve of vicinal NMR indirect spin couplings. *J. Phys. Chem. A* 108 (29): 6231–6238.

226 (a) Goodman, L., Gu, H.B., and Pophristic, V. (2005). Gauche effect in 1,2-difluoroethane. Hyperconjugation, bent bonds, steric repulsion. *J. Phys. Chem. A* 109 (6): 1223–1229. (b) Freitas, M.P., Bühl, M., and O'Hagan, D. (2012). 2-Difluoroethane: the angular dependance on $^1J_{CF}$ coupling constants is independent of hyperconjugation. *J. Chem. Soc. Chem. Commun.* 1 (48): 2433–2435. (c) O'Hagan, D. (2012). Organofluorine chemistry: synthesis and conformation of vicinal fluoromethylene motifs. *J. Org. Chem.* 77 (8): 3689–3699. (d) Fox, S.J., Gourdain, S., Coulthurst, A. et al. (2015). A computational study of vicinal fluorination in 2,3-difluorobutane: Implications for conformational control in alkane chains. *Chem. Eur. J.* 21: 1682–1691.

227 (a) Wolfe, S. (1972). Gauche effect - some stereochemical consequences of adjacent electron pairs and polar bonds. *Acc. Chem. Res.* 5 (3): 102–111. (b) Hayashi, N., Ujihara, T., and Ikeda, H. (2017). Effect of orbital interactions between vicinal bonds and between hydroxy groups on the conformational stabilities of 1,2-ethanediol and 2,3-butanediols. *J. Phys. Chem. A* 121 (44): 8484–8494.

228 Werpetinski, K.S. and Cook, M. (1997). A new grid-free density-functional technique: application to the torsional energy surfaces of ethane, hydrazine, and hydrogen peroxide. *J. Chem. Phys.* 106 (17): 7124–7138.

229 Wolfe, S., Rauk, A., Tel, L.M., and Csizmadi, I.G. (1971). Theoretical study of Edward–Lemieux effect (the anomeric effect) - stereochemical requirements of adjacent electron pairs and polar bonds. *J. Chem. Soc. B* (1): 136–145.

230 Li, A.Y. (2007). Chemical origin of blue- and redshifted hydrogen bonds: intramolecular hyperconjugation and its coupling with intermolecular hyperconjugation. *J. Chem. Phys.* 126 (15): 154102/1–154102/11.

231 Wertz, J.E. and Bolton, J.R. (2007). *Electron Spin Resonance, Elementary Theory and Practical Applications*, 2e, 1–664. New York: Wiley.

232 Bersohn, R. (1956). Proton hyperfine interactions in semiquinone ions. *J. Chem. Phys.* 24 (5): 1066–1070.

233 McConnell, H.M. (1956). Electron densities in semiquinones by paramagnetic resonance. *J. Chem. Phys.* 24 (3): 632–632.

234 McConnell, H.M. (1956). Indirect hyperfine interactions in the paramagnetic resonance spectra of aromatic free radicals. *J. Chem. Phys.* 24 (4): 764–766.

235 Chesnut, D.B. (1958). Molecular-orbital treatment of isotropic hyperfine interactions in simple aliphatic radicals. *J. Chem. Phys.* 29 (1): 43–47.

236 Ingold, K.U. and DiLabia, G.A. (2006). Bond strengths: the importance of hyperconjugation. *Org. Lett.* 8: 5923–5925.

237 Wodrich, M.D. and Schleyer, P.v.R. (2006). New additivity schemes for hydrocarbon energies. *Org. Lett.* 8 (10): 2135–2138.

238 Gronert, S. (2007). EPR data do not indicate that hyperconjugation stabilizes alkyl radicals. *Org. Lett.* 9 (11): 2211–2214.

239 Gronert, S. (2006). An alternative interpretation of the C—H bond strengths of alkanes. *J. Org. Chem.* 71 (3): 1209–1219.

240 Dunitz, J.D. and Schomaker, V. (1952). The molecular structure of cyclobutane. *J. Chem. Phys.* 20 (11): 1703–1707.

241 Krusic, P.J., Schleyer, P.v.R., and Rettig, T.A. (1972). Electron-spin resonance spectra and structure of bridgehead adamantyl and bicyclo[2.2.2]octyl free-radicals. *J. Am. Chem. Soc.* 94 (3): 995–997.

242 Hoffmann, R. (1971). Interaction of orbitals through space and through bonds. *Acc. Chem. Res.* 4 (1): 1–9.

243 Sauers, R.R. (2011). A natural bond orbital analysis of hydrocarbon radicals. *Comput. Theor. Chem.* 970 (1–3): 73–78.

244 Lambert, J.B. and Ciro, S.M. (1996). The interaction of π-orbitals with a carbocation over three σ-bonds. *J. Org. Chem.* 61 (6): 1940–1945.

245 Lambert, J.B. (1990). The interaction of silicon with positively charged carbon. *Tetrahedron* 46 (8): 2677–2689.

246 Lambert, J.B., Wang, G.T., and Teramura, D.H. (1988). Interaction of the carbon–germanium or carbon tin bond with positive charge on a β-carbon. *J. Org. Chem.* 53 (23): 5422–5428.

247 Lambert, J.B. and Emblidge, R.W. (1993). Nucleophilic catalysis in deoxymercuration - the β-effect of mercury. *J. Phys. Org. Chem.* 6 (10): 555–560.

248 Fernández, I. and Frenking, G. (2007). Hyperconjugative stabilization in alkyl carbocations: direct estimate of the β-effect of group-14 elements. *J. Phys. Chem. A* 111 (32): 8028–8035.

249 Müller, T., Meyer, R., Lennartz, D., and Siehl, H.U. (2000). Unusually stable vinyl cations. *Angew. Chem. Int. Ed.* 39 (17): 3074–3077.

250 Schormann, M., Garratt, S., Hughes, D.L. et al. (2002). Isolation and structure of $[HC\{CH(SiMe_3)(SnMe_3)\}^2]^+$: a remarkably stable *sec*-alkyl cation. *J. Am. Chem. Soc.* 124 (38): 11266–11267.

251 Roberts, J.D., Webb, R.L., and McElhill, E.A. (1950). The electrical effect of the trifluoromethyl group. *J. Am. Chem. Soc.* 72 (1): 408–411.

252 Schleyer, P.v.R. and Kos, A.J. (1983). The importance of negative (anionic) hyperconjugation. *Tetrahedron* 39 (7): 1141–1150.

253 (a) Holtz, D. (1971). Critical evaluation of the concept of fluorine hyperconjugation. In: *Progress in Physical Organic Chemistry*, vol. 8 (ed. A. Streitwieser Jr., and R.W. Taft), 1–74. New York: Wiley. (b) Apeloig, Y. (1981). Negative fluorine hyperconjugation. A theoretical re-examination. *J. Chem. Soc. Chem. Commun.* 396–398. (c) Farnham, W.B., Smart, B.E., Middleton, W.J. et al. (1985). Crystal and molecular structure of $[(CH_3)_2N]_3S^+CF_3O^-$. Evidence for negative fluorine hyperconjugation. *J. Am. Chem. Soc.* 107: 4565–4567. (d) Wiberg, K.B. and Rablen, P.R. (1993). Origin of the stability of carbon tetrafluoride: negative hyperconjugation re-examined. *J. Am. Chem. Soc.* 115 (2): 614–625. (e) Schneider, W.:.F., Nance, B.I., and Wallington, T.J. (1995). Bond strength trends in halogenated methanols - evidence for negative hyperconjugation. *J. Am. Chem. Soc.* 117 (1): 478–485. (f) Exner, O. and Boehm, S. (2008). Negative hyperconjugation of some fluorine containing groups. *New J. Chem.* 32 (8): 1449–1453.

254 Holtz, D. (1971). Critical examination of fluorine hyperconjugation in aromatic systems. *Chem. Rev.* 71 (1): 139–145.

255 (a) Streitwieser, A., Holtz, D., Ziegler, G.R. et al. (1976). Carbon acidity. 53. Kinetic and equilibrium acidities of fluorinated bicycloalkanes. *J. Am. Chem. Soc.* 98 (17): 5229–5234. (b) Castejon, H.J. and Wiberg, K.B. (1998). Effect of fluorine substitution on the carbon acidity of methane, methyl isocyanide, acetonitrile, acetaldehyde, and nitromethane. *J. Org. Chem.* 63 (12): 3837–3942. (c) Min, Z., Rahman Badal, M.M., Pasikowska, M. et al. (2014). Gas-phase acidity of polyfluorinated hydrocarbons. Effects of fluorine and the perfluoroalkyl group on acidity. *Bull. Chem. Soc. Jpn.* 87 (7): 825–834. (d) Mustanir, Abboud, J.-L.M., Mishima, M. et al. (2015). Gas phase acidities of aryl-substituted fluorinated alkanes. The nature of β-fluorine negative hyperconjugation. *Bull. Chem. Soc. Jpn.* 88 (3): 455–464.

256 (a) Friedman, D.S., Francl, M.M., and Allen, L.C. (1985). Anionic hyperconjugation. *Tetrahedron* 41 (3): 499–506. (b) Oomens, J., Berden, G., and Morton, T.H. (2015). Negative hyperconjugation versus electronegativity: vibrational spectra of free fluorinated alkoxide ions in the gas phase. *ChemPhysChem* 16 (9): 1992–1995.

257 Koppel, I.A., Pihl, V., Koppel, I.A. et al. (1994). Thermodynamic acidity of $(CF_3)_3CH$ and 1*H*-undecafluorobicyclo[2.2.1]heptane - the concept of anionic (fluorine) hyperconjugation. *J. Am. Chem. Soc.* 116 (19): 8654–8657.

258 Dixon, D.A., Fukunaga, T., and Smart, B.E. (1986). Structures and stabilities of fluorinated carbanions - evidence for anionic hyperconjugation. *J. Am. Chem. Soc.* 108 (14): 4027–4031.

259 Hoffmann, R., Hehre, W.J., Salem, L. et al. (1972). Strong conformational consequences of hyperconjugation. *J. Am. Chem. Soc.* 94 (17): 6221–6223.

260 Sauers, R.R. (1999). A computational study of proton and electron affinities. *Tetrahedron* 55 (33): 10013–10026.

261 Goodman, L. and Sauers, R.R. (2005). Ethyl anion preferred conformation. *Int. J. Quantum Chem.* 102 (5): 829–837.

262 Walsh, A.D. (1949). The structures of ethylene oxide, cyclopropane, and related molecules. *Trans. Faraday Soc.* 45 (2): 179–190.

263 Sugden, T.M. (1947). Structure of cyclopropane and ethylene oxide. *Nature* 160 (4063): 367–368.

264 Stothers, J.B. (1972). *Carbon-13 NMR Spectroscopy*, vol. 24, 559. New York: Academic Press; vol. 26, 1–16.

265 Kabakoff, D. and Namanworth, D.S. (1970). Nuclear magnetic double resonance studies of dimethylcyclopropylcarbinyl cation-measurement of rotation barrier. *J. Am. Chem. Soc.* 92: 3234–3235.

266 Olah, G.A., Reddy, P.V., and Prakash, G.K.S. (1995). Long-lived cyclopropylcarbinyl cations. In: *Chemistry of the Cyclopropyl Group*, vol. 2 (ed. S. Patai and Z. Rappoport), 813–859. New York: Wiley.

267 Honegger, E., Heilbronner, E., and Schmelzer, A. (1982). Do Walsh-orbitals exist? *New J. Chem.* 6 (11): 519–526.

268 Stocker, M. (1982). The structural dependence of carbon-13, carbon-13 coupling constants in cyclopropanes. *Org. Magn. Reson.* 20 (3): 175–179.

269 Cremer, D., Kraka, E., Wu, A., and Luettke, W. (2004). Can one assess the π character of a C—C

bond with the help of the NMR spin-spin coupling constants? *ChemPhysChem* 5 (3): 349–366.
270 Wiberg, K.B. (1996). Bent bonds in organic chemistry. *Acc. Chem. Res.* 29: 229–234.
271 (a) Förster, T. (1939). The mutual influence of valences on the carbon atom. *Zeitschrift für physikalische Chemie. Abteilung B, Chemie der Elementarprozesse, Aufbau der Materie* 43 (1): 58–78. (b) Coulson, C.A. and Moffitt, W.E. (1947). Strain in non-tetrahedral carbon atoms. *J. Chem. Phys.* 15 (3): 151–151. (c) Honegger, E., Heilbronner, E., Schmelzer, A., and Wang, J.Q. (1982). A reexamination of the Walsh- and Forster-Coulson-Moffitt- (F.C.M.) orbital concept, and its relevance for the interpretation of PE spectra. *Isr. J. Chem.* 22 (1): 3–10.
272 Karadakov, P.B., Gerratt, J., Cooper, D.L., and Raimondi, M. (1995). Modern valence-bond description of bonding in strained three-membered rings: cyclopropane, aziridine, ethene oxide, phosphirane and thiirane. *J. Mol. Struct. THEOCHEM* 341: 13–24.
273 Prakash Reddy, V., Rasul, G., Prakash, G.K.S., and Olah, G.A. (2007). Structural studies on non-classical cyclobutylmethyl cations by the ab initio method. *J. Org. Chem.* 72: 3076–3080.
274 de Meijere, A. (1979). Bonding properties of cyclopropane and their chemical consequences. *Angew. Chem. Int. Ed. Eng.* 18: 809–826.
275 Heilbronner, E. (1964). Hückel molecular orbitals of Möbius-type conformations of annulenes. *Tetrahedron Lett.* 1923–1928.
276 (a) Ajami, D., Oeckler, O., Simon, A., and Herges, R. (2003). Synthesis of a Möbius aromatic hydrocarbon. *Nature* 426: 819–821. (b) Herges, R. (2006). Topology in chemistry: designing Möbius molecules. *Chem. Rev.* 106 (12): 4820–4842. (c) Alonso, M., Geerlings, P., and de Proft, F. (2012). Viability of Möbius topologies in [26]- and [28]hexaphyrins. *Chem. Eur. J.* 18 (25): 10916–10928.
277 (a) Schleyer, P.v.R., Maerker, C., Dransfeld, A. et al. (1996). Nucleus-independent chemical shifts: a simple and efficient aromaticity probe. *J. Am. Chem. Soc.* 118 (26): 6317–6318. (b) Corminboeuf, C., Heine, T., Seifert, G. et al. (2004). Induced magnetic fields in aromatic [N]annulenes—interpretation of NICS tensor components. *Phys. Chem. Chem. Phys.* 6: 273–276.
278 Chen, Z.F., Wannere, C.S., Corminboeuf, C. et al. (2005). Nucleus-independent chemical shifts (NICS) as an aromaticity criterion. *Chem. Rev.* 105 (10): 3842–3888.
279 Castro, C., Isborn, C.M., Karney, W.L. et al. (2002). Aromaticity with a twist: Möbius [4n]annulenes. *Org. Lett.* 4 (20): 3431–3434.
280 Barborak, J.C., Su, T.M., Schleyer, P.v.R. et al. (1971). thermal bicyclo[6.1.0]nonatrienyl chloride-dihydroindenyl chloride rearrangement. *J. Am. Chem. Soc.* 93 (1): 279–281.
281 Mauksch, M., Gogonea, V., Jiao, H., and Schleyer, P.v.R. (1998). Monocyclic $(CH)_9{}^+$—a Heilbronner-Möbius aromatic system revealed. *Angew. Chem. Int. Ed.* 37 (17): 2395–2397.
282 Stepien, M., Latos-Grazynski, L., Sprutta, N. et al. (2007). Expanded porphyrin with a split personality: a Hückel-Möbius aromaticity switch. *Angew. Chem. Int. Ed.* 46 (41): 7869–7873.
283 (a) Yoon, Z.S., Osuka, A., and Kim, D. (2009). Möbius aromaticity and antiaromaticity in expanded porphyrins. *Nat. Chem.* 1 (2): 113–122. (b) Sung, Y.M., Oh, J., Cha, W.-Y. et al. (2017). Control and switching of aromaticity in various all-aza-expanded porphyrins: spectroscopic and theoretical analyses. *Chem. Rev.* 117 (4): 2257–2312. (c) Tanaka, T. and Atsuhiro, O. (2017). Chemistry of meso-aryl-substituted expanded porphyrins: aromaticity and molecular twist. *Chem. Rev.* 117 (4): 2584–2640.

5

Pericyclic reactions

5.1 Introduction

Organic reactions are generally classified into two main types: (i) reactions involving reactive intermediates such as radicals, diradicals, carbenes, and nitrenes; ion pairs; and radical anions and radical cations; and (ii) reactions that do not involve reactive intermediates and for which the bond-breaking and bond-forming processes are concerted such as S_N2 displacements, E_2 eliminations (Section 3.9.3), and several pericyclic reactions. Catalyzed reactions constitute a subclass of reactions (Chapters 7 and 8) involving reactive intermediates; the catalysts permit the equilibrium of reactants and products with intermediates that have lower free energies than the noncatalyzed reactions (change of mechanism). Concerted and nonconcerted reactions can also be catalyzed. Pericyclic reactions, defined by Woodward and Hoffmann, are concerted and are characterized by cyclic transition states associated with electrons reorganizing in a cyclic array of interacting orbitals. They can be induced thermally (this chapter) and photochemically (Sections 6.3, 6.5, 6.8, and 6.9) or catalyzed by Brønsted or Lewis acids (Section 7.6), transition metal complexes (Sections 8.3 and 8.4), single-electron transfer (Sections 7.5.8 and 7.5.10), or enzymes (See sections 5.5.9.2, chorismate mutase). Some of the most powerful synthetic methods involve pericyclic reactions or stepwise analogs, such as the Diels–Alder (Sections 5.3.9–5.3.15, see also Sections 3.6.3, 1.14.1, 3.7.3, and 3.9.5) and hetero-Diels–Alder reactions (Section 5.3.16), 1,3-dipolar cycloadditions (Section 5.3.17), $[\pi^2+\pi^2]$-cycloadditions (Section 5.3), cyclopropanation (Section 5.4.1), epoxidation, the Claisen and the Cope rearrangements (Section 5.5.9), Wagner–Meerwein rearrangements (Section 5.5.1), the Wittig olefination (Sections 5.3.7 and 5.3.8), the 1,2-Wittig (Section 5.5.3) and 2,3-Wittig rearrangements (Section 5.5.8), electrocyclic ring-closing and ring-opening reactions (Section 5.2), diimide reductions of alkenes and alkynes (Section 5.6.1), sulfoxide eliminations, ester and xanthate eliminations, and the ene and hetero-ene reactions (Section 5.7). Some annulations have been catalyzed by nucleophiles (Sections 7.5.2 and 7.5.7). Before 1965, concerted pericyclic processes had been called "no-mechanism reactions" to imply that there are no detectable intermediates and everything happens in one step. These reactions imply assistance between bond-forming and bond-breaking processes. One-step, concerted pericyclic reactions are usually highly stereoselective and their rates are little affected by solvent, except for reactants bearing polar substituents. Depending on the nature of the reactants (type of substituents, number of substituents, etc.) and reaction conditions, the pericyclic reactions may not follow concerted, one-step mechanisms but can be multistep processes. The Woodward–Hoffmann rules predict the stereoselectivity of concerted pericyclic reactions [1, 2]. As we shall see, perfectly symmetrical transition structures are rarely the case, especially for nonsymmetrical reactants. The activation enthalpy ($\Delta H^{\ddagger}$) of a concerted reaction is the sum of repulsive steric effects and skeletal distortion energies minus stabilization energies resulting from stabilizing orbital interactions in the transition state (electron exchange). This does not only depend on the exact symmetry of the transition structures of the transition states but can be much more favorable (allowed) for one stereochemistry than for another (forbidden). The Woodward–Hoffmann rules and other related theories such as the Longuet-Higgins correlation of electronic configurations between reactants and products (see Sections 5.2.2, 5.2.3, 5.3.1, 5.3.2) rely upon symmetry in the transition structures. We shall see that the theory of aromaticity of transition structures of pericyclic reactions applies to symmetrical and nonsymmetrical transition structures as well. The frontier molecular orbital (FMO) applied to transition structures not only permits the easy prediction of the stereochemistry but also allows one to estimate substituent and additive effects on the rate and stereoselectivity (regioselectivity and

Organic Chemistry: Theory, Reactivity and Mechanisms in Modern Synthesis, First Edition. Pierre Vogel and Kendall N. Houk.

stereoselectivity) of the reactions. Another approach takes advantage of the fact that most pericyclic reactions involve nonsymmetrical transition structures that are diradicaloids (species with very narrow HOMO–LUMO gaps, or in valence bond terminology, diradical species). Thus, without having to calculate molecular orbitals (MOs), just using thermochemical data (homolytical bond dissociation enthalpies: $DH°(R^•/H^•)$ or $DH°(R^•/X^•)$, substituent effect on the relative stability of radicals, cations, and anions) can be used to predict relative rates, regioselectivity, and stereoselectivitiy of pericyclic reactions.

A chemical system that evolves from reactants to products, and vice versa, can be described by a wave function Ψ_r, function of reaction coordinates and of the properties of reactants and products. The energy E_r associated with Ψ_r is the potential energy of the system. Ψ_r can be taken as a linear combination of reactant and product wave functions Ψ_R and Ψ_P. These functions represent the ground-state properties of reactants and products, respectively. Thus,

$$\Psi_r = c_R \Psi_R + c_P \Psi_P \tag{5.1}$$

For the transition structure of the concerted reaction, this wave function becomes

$$\Psi_r^{\ddagger} = c_R^{\ddagger} \Psi_R + c_P^{\ddagger} \Psi_P \tag{5.2}$$

Wave functions of the ground states of reactants (Ψ_R) and products (Ψ_P) might be taken as linear combinations of electronic configurations, which are combinations of MOs of reactants and products, respectively. As we shall see for the electrocyclic reaction that isomerizes cyclobutene into butadiene and vice versa (Section 5.2.1), and the $[\pi^2+\pi^2]$-cycloaddition of two ethylene into cyclobutane and the $[\sigma^2+\sigma^2]$-cycloreversion of cyclobutane into two ethylene units (Section 5.3.1), the total energy of the transition state obtained by applying the variation principle to linear combinations (5.2) might be lower or higher depending on the stereochemistry of the reactions.

5.2 Electrocyclic reactions

Woodward and Hoffmann define electrocyclic transformations as the formation of a single bond between the termini of a conjugated polyene containing N π-electrons and the converse [3]. Examples are presented in the following sections. The reactions can be induced thermally, photochemically (Sections 6.3.3, 6.3.5, and 6.5), or by catalysis (electron transfer [4, 5] or by other mechanisms). Under thermal conditions, the stereoselectivities (conrotatory and disrotatory) shown in Scheme 5.1 are observed [1, 2]. For electrocyclic reactions involving 2, 6, 10 ($N = 4n+2$) electrons, the disrotatory modes (noted π^2d, π^6d, π^{10}d, etc.) are preferred (Hückel type of orbital arrays in the transition structures, that is, a quasi-planar arrangement with a definable top and bottom); for electrocyclic reactions involving 4, 8, 12 ($N = 4n$) electrons, the conrotatory modes (noted π^4c, π^8c, π^{12}c, etc.) are preferred (Möbius type of orbital arrays in the transition structures, that is, a system where the top of the π-system overlaps with the bottom at the site of the partially formed σ bond – thus, the system has only one face, as in the Möbius strip).

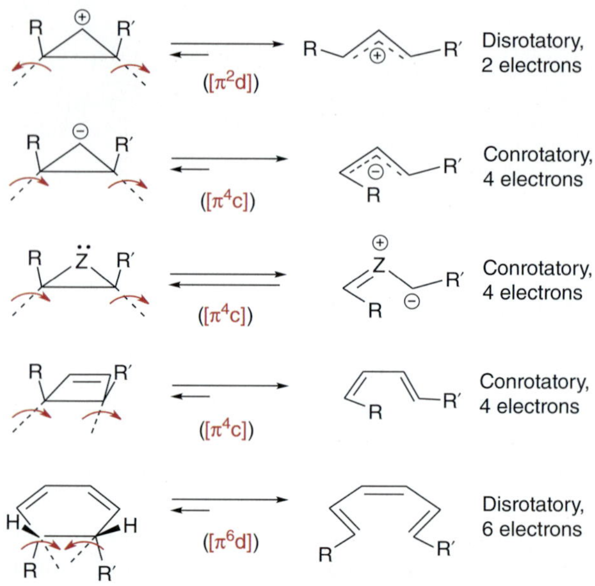

Scheme 5.1 Examples of thermal electrocyclic reactions and their preferred stereoselectivities as predicted by the Woodward–Hoffmann rules.

5.2.1 Stereochemistry of thermal cyclobutene-butadiene isomerization: four-electron electrocyclic reactions

Cyclobutene has been prepared for the first time in 1905 by Willstätter and von Schmaedel [6] by pyrolysis of cyclobutyltrimethylammonium hydroxide (an example of Hofmann elimination) [7]. Butadiene is also formed during this pyrolysis. In 1954, Vogel showed that dimethyl *cis*-cyclobutene-3,4-dicarboxylate is isomerized stereospecifically into dimethyl (*Z*,*E*)-muconate on heating to 120 °C [8], following a four-electron conrotatory mode ([π^4c]) of opening. Criegee and Noll [9] (see also [10]) reported that the pyrolysis at 200 °C of *cis*-1,

2,3,4-tetramethylcyclobutene (**1**) gives (*E*,*Z*)-3,4-dimethylhexa-2,4-diene (**2**) exclusively, whereas *trans*-1,2,3,4-tetramethylcyclobutene (**3**) is isomerized into (*E*,*E*)-3,4-dimethylhexa-2,4-diene (**4**) exclusively. Both reactions imply conrotatory mode of openings. For steric reasons, (Z,Z)-3,4-dimethylhexa-2,4-diene is less stable and does not form. In the case of *cis*-3,4-dimethylcyclobutene, Brauman and Archie [11] established that the conrotatory mode (E_a = 34.3 kcal mol^{-1}, log A = 13.9) [12] of opening that gives (*E*,*Z*)-hexa-2,4-diene has an activation free energy ($\Delta^{\ddagger}G$) that is 10.7 kcal mol^{-1} lower than that of the disrotatory mode ([π^4d]) of opening producing (*E*,*E*)-hexa-2,4-diene.

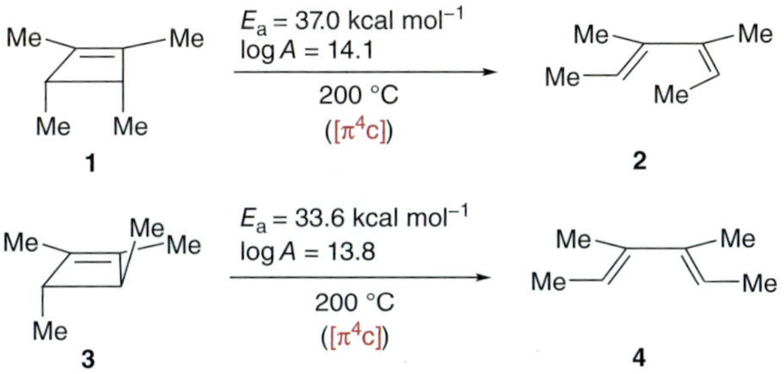

On comparing the rate constant k_d of the thermal isomerization of (*Z*,*Z*)-1,4-dideuteriobutadiene ((*Z*,*Z*)-**5**) into (*E*,*E*)-1,4-dideuteriobutadiene ((*E*,*E*)-**5**) with the rate constant k_s of the isomerization of (*Z*,*Z*)-**5** into (*Z*,*E*)-1,4-dideuteriobutadiene ((*Z*,*E*)-**5**), Brauman and coworkers established that the double rotation (k_d) is 50 times faster than the simple rotation (k_s) about a C=C double bond of butadiene at 910 K (Scheme 5.2) [13]. This allowed them to estimate the difference in activation free energies ($\Delta\Delta^{\ddagger}G$) between these two processes to 7.1 kcal mol^{-1}. Assuming that isomerization of (*Z*,*Z*)-**5** into (*E*,*E*)-**5** involves the conrotatory ring closing into cyclobutene **6** and its conrotatory ring opening, one can estimate that rotation about a C=C double bond of butadiene involves the intermediacy of the allyl/alkyl radical **7**, which is 50.6 kcal mol^{-1} higher in enthalpy than butadiene (Figure 5.1). The activation free energy ($\Delta^{\ddagger}G$) of isomerization (*Z*,*Z*)-**5** into (*E*,*E*)-**5** is estimated to 43.5 kcal mol^{-1}. Considering the stability difference of 11.5 kcal mol^{-1} between cyclobutene and butadiene, the activation free energy of the disrotatory ring opening of cyclobutene **6** amounts to 32 kcal mol^{-1}.

Considering that the activation enthalpy ($\Delta^{\ddagger}H = E_a - RT$) of rotation about the ethylene C=C double bond is 64 kcal mol^{-1} (trans/cis isomerization of 1,2-dideuterioethylene between 450 and 550 °C, E_a = 65 kcal mol^{-1}, log A = 13) [14] and the allylic resonance energy in allyl radical (Section 4.5.2) is 13 kcal mol^{-1}, an energy difference of 51 kcal mol^{-1} (compares 50.6 kcal mol^{-1} above) is estimated between diradical **7** and butadiene; this estimate

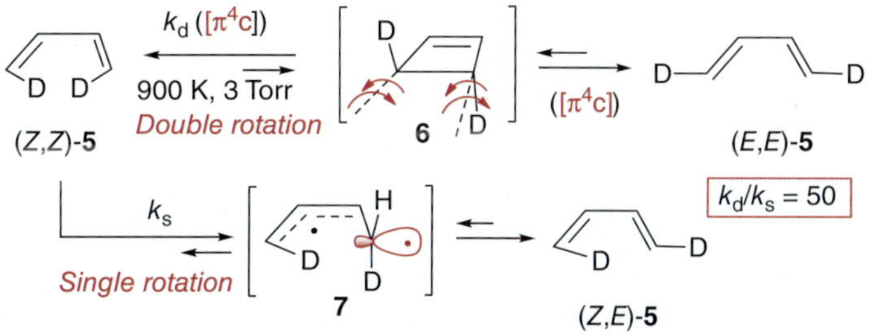

Scheme 5.2 The thermal isomerization of (*Z*,*Z*)-1,4-dideuterobutadiene implies the conrotatory ([π^4c]) ring closing into *trans*-3,4-dideuterocyclobutene intermediate and subsequent conrotatory ring opening giving either (*Z*,*Z*)- or (*E*,*E*)-1,4-dideuterobutadiene (double rotation of C(1) and C(4) of butadiene).

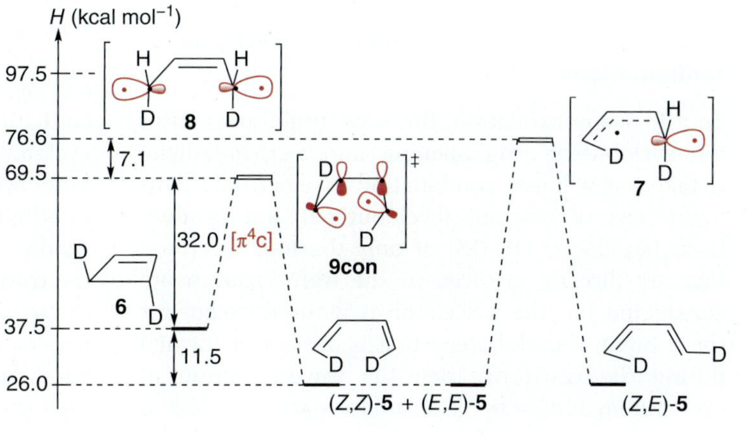

Figure 5.1 Enthalpy diagram of the cyclobutene/*s-cis*-butadiene isomerization. The favored transition state corresponds to a conrotatory mode of electrocyclic ring opening or closing ([π^4c]).

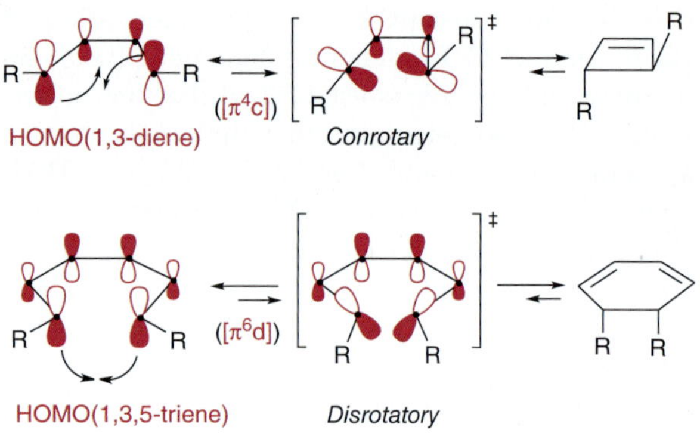

Scheme 5.3 Relationship between the symmetry of the HOMO of π-systems undergoing thermal electrocyclic cyclizations and their stereoselectivity (conrotatory vs. disrotatory).

is consistent with Brauman's hypothesis that rotation about a C=C double of butadiene involves diradical 7 as the intermediate.

Oosterhoff in 1961 (as quoted by Havinga and Schlatmann [15]) and Woodward and Hoffmann in 1965 offered a first, simple theoretical interpretation of the high stereoselectivity of the thermal four-electron cyclobutene/butadiene isomerization (and of other electrocyclic reactions) [3]. They pointed out that the stereochemical course of the [π^4c] electrocyclic ring closure of butadiene into cyclobutene and the [π^6d] ring closure of (*Z*)-hexa-1,3,5-triene into cyclohexa-1,3-diene is determined by the symmetry of the highest occupied molecular orbital (HOMO) of butadiene and hexatriene, respectively (Scheme 5.3). Considering the ring-closing reaction of butadiene into cyclobutene, the 2p-lobes at C(1) and C(4) of its HOMO have opposite phases. Thus, centers C(1) and C(4) combine imposing a conrotatory mode of ring closing into cyclobutene. For the ring closing of (*Z*)-hexa-1,3,5-triene into cyclohexa-1,3-diene, the 2p lobe at C(1) and C(6) of its HOMO has the same phase; thus, they combine imposing a disrotatory mode of cyclization.

5.2.2 Longuet-Higgins correlation of electronic configurations

To a first approximation, the wave function (5.2) of the cyclobutene ring opening into *s-cis*-butadiene is taken as a linear combination of electronic configurations of reactant (cyclobutene) and product (*s-cis*-butadiene) [16–18]. If only the four electrons that are directly involved in the isomerization are considered (i.e. the σ-skeleton is ignored except for the σ-bond of cyclobutene that is broken or formed during the reaction), then the wave function of cyclobutene is $\phi_0 = [\sigma]^2[\pi]^2$ and for *s-cis*-butadiene it is $\varphi_0 = [\pi_1]^2[\pi_2]^2$ (the two lowest lying MOs of reactant and product are occupied by two electrons each). Thus, $\Psi_r^{\ddagger} = c_R^{\ddagger}\Psi_R + c_P^{\ddagger}\Psi_P$ can be approximated by $\Psi_r^{\ddagger} = c_R^{\ddagger}\phi_0 + c_P^{\ddagger}\varphi_0$.

If one applies the variation theorem (Section 4.3) to find coefficients $c_R^{\ddagger}$ and $c_P^{\ddagger}$, one obtains two eigenfunctions associated with two eigenvalues. In the case of two 2p atomic orbitals (AOs) of ethylene, the result of their interaction is nil if the two 2p orbitals are orthogonal (overlap S between the two AOs is nil, $\beta = 0, E_1 = E_2 = \alpha$ in the Hückel method: an electron in 2p at C(1) cannot be stabilized by interaction with the nuclei at C(2); Section 4.5.2), but different from zero, if the two AOs are not orthogonal. In the latter case, one eigenvalue (E_1) is stabilized with respect to the Coulombic integral of the AOs ($E_1 < \alpha$ in the Hückel method, an electron in AO 2p at C(1) is stabilized by interaction with the nuclei at C(2)). One can thus predict for $\Psi_r^{\ddagger} = c_R^{\ddagger}\phi_0 + c_P^{\ddagger}\varphi_0$ that electronic energy might be gained because of the combination of the wave functions of reactant (Ψ_R) and product (Ψ_P) only if the corresponding configurations ϕ_0 and φ_0 are not orthogonal, i.e. if they have the same symmetry. This is the case for the concerted, synchronous (the termini carbon centers of the forming butadiene rotate in phase during the reaction) conrotatory mode ([π^4c]) of isomerization. It is not the case for the disrotatory mode of reaction ([π^4d]). We first examine the symmetry of the electronic configurations of cyclobutene (Figure 5.2) and of *s-cis*-butadiene (Figure 5.3) interconverting through a disrotatory reaction ([π^4d]). A mirror plane of symmetry is maintained during the concerted, synchronous disrotatory electrocyclic reaction (transition structure with C_s-symmetry).

Longuet-Higgins and Abrahamson obtain the correlation diagram of Figure 5.4 for the concerted, synchronous [π^4d] reaction [16]. The lowest energy electronic configurations noted $\phi_0(a')^2(a')^2$ and $\varphi_0(a')^2(a'')^2$ do not have the same symmetry and consequently their overlap is zero: they cannot interact. None of the singly excited electronic configurations ϕ_1, ϕ_2, ϕ_3, and ϕ_4 of cyclobutene and φ_1, φ_2, φ_2, and φ_4 of butadiene share the symmetry of ϕ_0 and φ_0. Interestingly, the doubly excited electronic configurations noted $\phi_5(a')^2(a)^2$ and $\varphi_5(a')^2(a')^2$ of cyclobutene and butadiene, respectively, have the same symmetry as the lowest energy configurations of butadiene (φ_0) and cyclobutene (ϕ_0), respectively. As the energy difference between the doubly excited electronic configurations and the lowest energy electronic configurations is very large, the energy hypersurfaces connecting ϕ_0 with φ_5, on the one hand, and φ_0 with ϕ_5, on the other hand, cross at a high energy level above the ground states of reactant

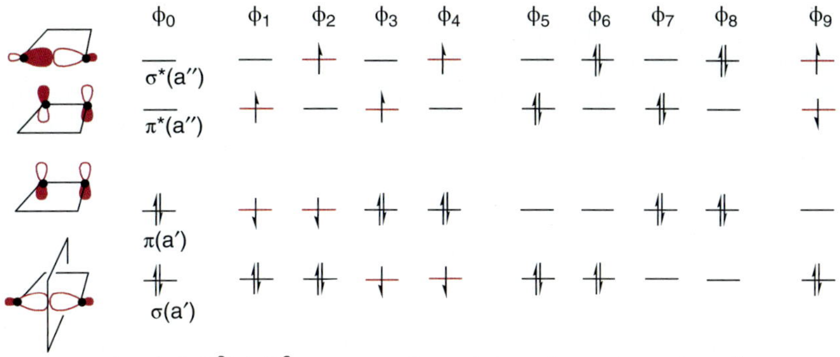

$\phi_0 = [\sigma(a')]^2[\pi(a')]^2$ Ground state electronic configuration (lowest energy)

$\phi_1 = [\sigma(a')]^2[\pi(a')][\pi^*(a'')]$
$\phi_2 = [\sigma(a')]^2[\pi(a')]\sigma^*(a'')]$
$\phi_3 = [\sigma(a')][\pi(a')]^2[\pi^*(a'')]$
$\phi_4 = [\sigma(a')][\pi(a')]^2[\sigma^*(a'')]$
} Singly excited electronic configurations

$\phi_5 = [\sigma(a')]^2[\pi^*(a'')]^2$
$\phi_6 = [\sigma(a')]^2[\sigma^*(a'')]^2$
$\phi_7 = [\pi(a')]^2[\pi^*(a'')]^2$
$\phi_8 = [\pi(a')]^2[\sigma^*(a'')]^2$
$\phi_9 = [\sigma(a')]^2[\pi^*(a'')][\sigma^*(a'')]$
} Doubly excited electronic configurations

Figure 5.2 Representation of the lowest energy electronic configurations of cyclobutene undergoing a synchronous disrotatory electrocyclic reaction ($[\pi^4 d]$) giving butadiene. A mirror plane of symmetry is maintained during the reaction (C_s symmetry: a′, symmetrical and a″, antisymmetrical with respect to the mirror plane of symmetry).

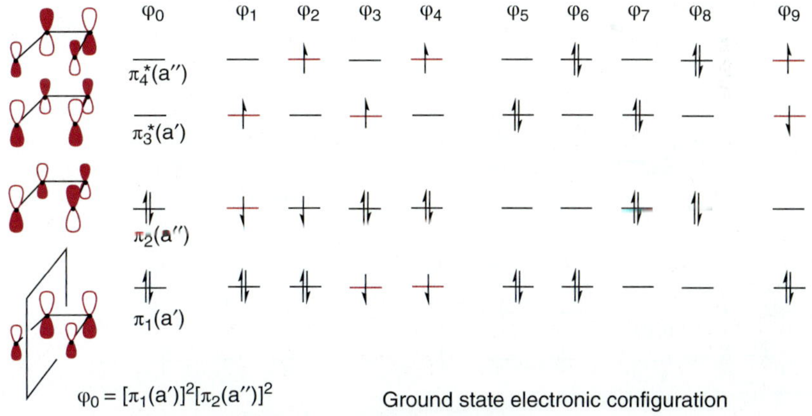

$\varphi_0 = [\pi_1(a')]^2[\pi_2(a'')]^2$ Ground state electronic configuration (lowest energy)

$\varphi_1 = [\pi_1(a')]^2[\pi_2(a'')][\pi_3{}^*(a')]$
$\varphi_2 = [\pi_1(a')]^2[\pi_2(a'')][\pi_4{}^*(a'')]$
$\varphi_3 = [\pi_1(a')][\pi_2(a'')]^2[\pi_3{}^*(a')]$
$\varphi_4 = [\pi_1(a')][\pi_2(a'')]^2[\pi_4{}^*(a'')]$
} Singly excited electronic configurations

$\varphi_5 = [\pi_1(a')]^2[\pi_3{}^*(a')]^2$
$\varphi_6 = [\pi_1(a')]^2[\pi_4{}^*(a'')]^2$
$\varphi_7 = [\pi_2(a'')]^2[\pi_3{}^*(a')]^2$
$\varphi_8 = [\pi_2(a'')]^2[\pi_4{}^*(a'')]^2$
$\varphi_9 = [\pi_1(a')]^2[\pi_3{}^*(a')][\pi_4{}^*(a'')]$
} Doubly excited electronic configurations

Figure 5.3 Representation of the lowest energy electronic configurations of *s-cis*-butadiene resulting from a disrotatory ring opening ($[\pi^4 d]$) of cyclobutene (C_s symmetry).

and product, thus demonstrating the insignificant assistance reactant and product can offer to each other in the thermal [π^4d] reaction. The state function $\Psi_r = c_R\Psi_R + c_P\Psi_P$ can be approximated by $\Psi_r = c_1\phi_0 + c_2\varphi_0 + c_3\phi_5 + c_4\varphi_5$. As ϕ_0 and φ_0 do not overlap and the energy differences between electronic configurations of same symmetry ϕ_0/φ_5 and φ_0/ϕ_5 are very large, the transition state represented by the wave function $\Psi_r^\ddagger$ has an energy that is not very much lower than that of the crossing point of hypersurfaces $\phi_0 - \varphi_5$ and $\varphi_0 - \phi_5$ (Figure 5.4).

In the case of conrotatory ring opening of cyclobutene into butadiene ([π^4c]), and *vice versa*, the lowest energy electronic configurations of cyclobutene (Figure 5.5) and of *s-cis*-butadiene (Figure 5.6) are built with MOs that share a C_2-axis of symmetry. The Longuet-Higgins and Abrahamson correlation diagram reproduced in Figure 5.7 shows that the lowest energy electronic configurations of cyclobutene and butadiene, noted $\phi_0(a)^2(b)^2$ and $\varphi_0(b)^2(a)^2$, have the same symmetry. As they interact, the synchronous [π^4c] reaction is assisted electronically (assistance between σ-bond breaking of cyclobutene and π-bond forming of butadiene). Compared with a reaction that occurs through allyl/alkyl diradical **7**, the conrotatory ring opening of cyclobutene is 7.1 kcal mol^{-1} lower in energy due to stabilizing the electronic energy in the pericyclic

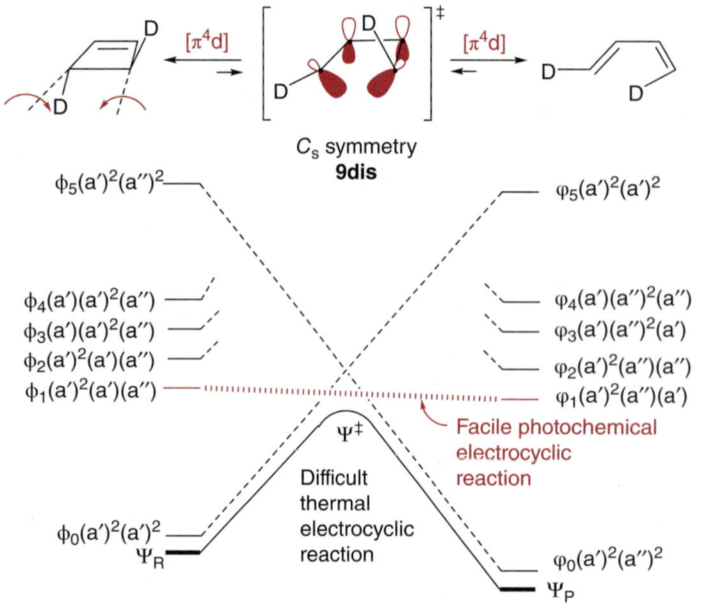

Figure 5.4 Longuet-Higgins and Abrahamson correlation diagram of electronic configurations of cyclobutene and *s-cis*-butadiene undergoing synchronous disrotatory electrocyclic reaction ([π^4d]). The lowest energy configurations of reactant (ϕ_0) and product (φ_0) do not have the same symmetry, they do not interact; as a consequence, the thermal reaction is difficult. It is not stabilized by assistance between σ-bond breaking of cyclobutene and π-bond making in butadiene. In contrast, the first electronic excited configurations ϕ_1 and φ_1 have the same symmetry. Thus, the concerted, synchronous disrotatory electrocyclic reaction that equilibrates cyclobutene with butadiene is said to be facile photochemically. Note that the photochemically induced electrocyclic isomerization of butadiene into cyclobutene, and *vice versa* (Section 6.3.3), might follow a multistep, nonconcerted mechanism, for instance, via a diradical intermediate of type **7** (see Scheme 5.2, Figure 5.1).

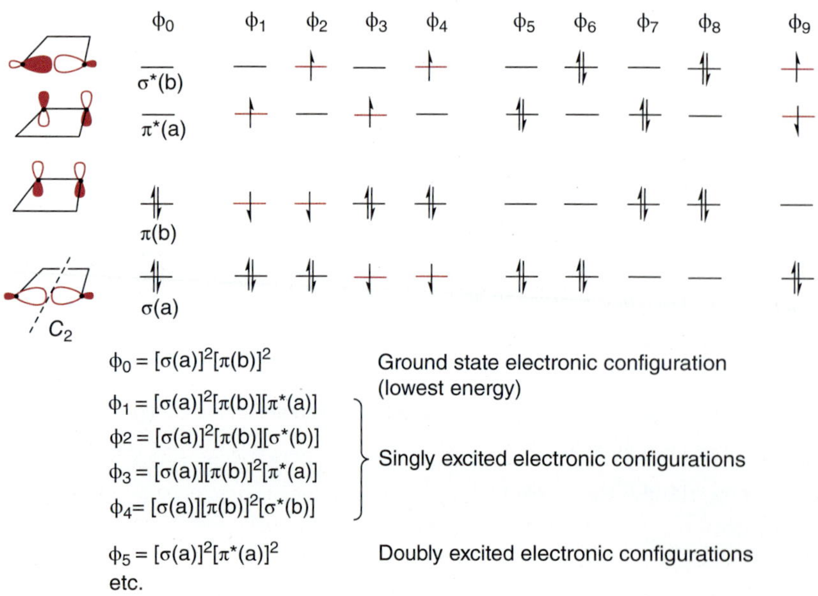

Figure 5.5 Representation of the lowest energy electronic configurations of cyclobutene undergoing conrotatory [π^4c] isomerization with *s-cis*-butadiene. A C_2-axis of symmetry is maintained during this concerted, synchronous ring opening (a, symmetrical and b, antisymmetrical with respect to C_2).

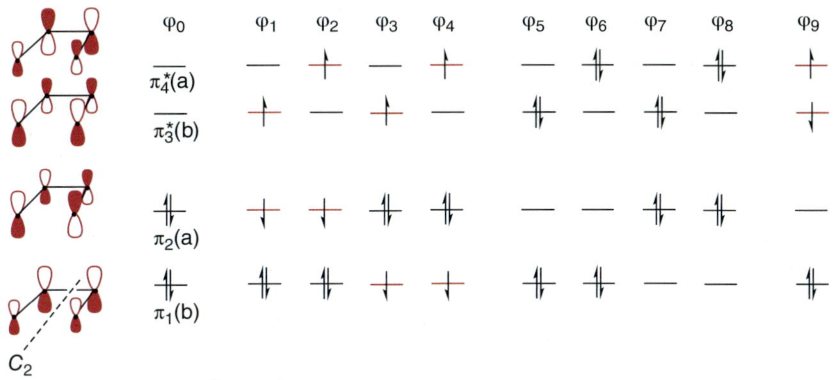

$\varphi_0 = [\pi_1(b)]^2[\pi_2(a)]^2$ Ground state electronic configuration (lowest energy)

$\varphi_1 = [\pi_1(b)]^2[\pi_2(a)][\pi_3^*(b)]$
$\varphi_2 = [\pi_1(b)]^2[\pi_2(a)][\pi_4^*(a)]$
$\varphi_3 = [\pi_1(b)][\pi_2(a)]^2[\pi_3^*(b)]$
$\varphi_4 = [\pi_1(b)][\pi_2(a)]^2[\pi_4^*(a)]$
Singly excited electronic configurations

$\varphi_5 = [\pi_1(b)]^2[\pi_3^*(b)]^2$ Doubly excited electronic configuration

etc.

Figure 5.6 Representation of the lowest energy electronic configurations of *s-cis*-butadiene resulting from the synchronous conrotatory electrocyclic opening ([π^4c]) of cyclobutene (C_2-axis of symmetry).

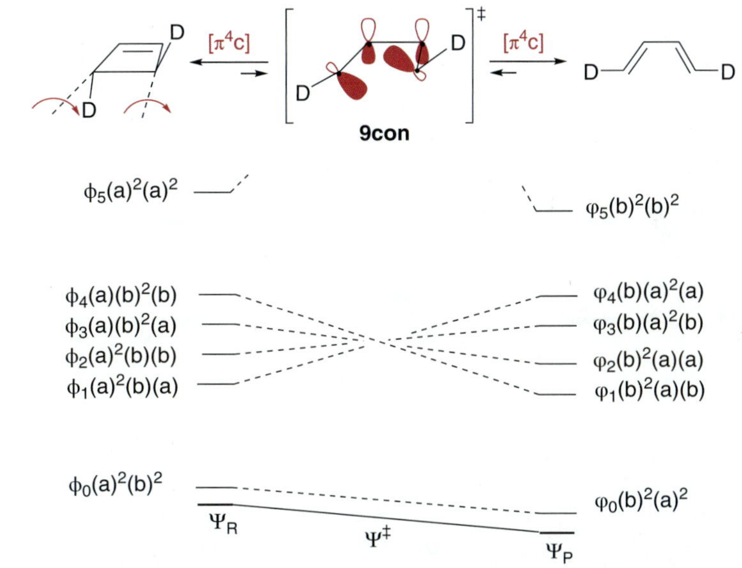

Figure 5.7 Longuet-Higgins and Abrahamson correlation diagram for the electronic configurations of cyclobutene and *s-cis*-butadiene undergoing synchronous conrotatory electrocyclic isomerization ([π^4c]). The reaction is facile thermally because the lowest energy electronic configurations have the same symmetry and thus they can interact during the concerted reaction, providing assistance between the σ-bond breaking and π-bond forming process and *vice versa*.

transition state (Figure 5.1). The activation enthalpy ($\Delta^{\ddagger}H$) of the isomerization of cyclobutene into butadiene amounts to 32 kcal mol^{-1} ($E_a = 32.9$ kcal mol^{-1}, $\log A = 13.4$) [19]. If the rate-determining step of the reaction was the σ-bond breaking of cyclobutene giving but-2-ene-1,4-diyl diradical **8** in which the two radicals at C(1) and C(4) are perpendicular with respect to the π-system at C(2)–C(3), the electronic stabilization energy of the conrotatory electrocyclic reaction can be considered to amount to 7.1 + 13 kcal mol^{-1} (resonance stabilization of allyl radical), thus, to c. 20 kcal mol^{-1}. This energy can be taken as the energy of concert, a term used extensively by the late William von Eggers Doering. It is the energy difference between a nonconcerted process forming diradical intermediate **8** and the synchronous conrotatory ring opening.

5.2.3 Woodward–Hoffmann simplification

Woodward and Hoffmann simplify the Longuet-Higgins analysis and propose to correlate the MOs of reactant and product maintaining the same symmetry in the transition state of the pericyclic reaction [2, 20]. For the thermal conrotatory [π^4c] reaction (C_2-axis of symmetry is maintained during the synchronous reaction), one finds that the two occupied MOs of cyclobutene can be correlated with the two occupied MOs of *s-cis*-butadiene. Thus, the reaction is thermally allowed. In contrast, in the case of the synchronous disrotatory [π^4d] reaction (C_s symmetry maintained during the reaction), only the HOMO-1 of reactant and product have the same symmetry. The HOMO of cyclobutene correlates with the lowest unoccupied molecular orbital (LUMO) of butadiene, and the HOMO of butadiene correlates with the LUMO of cyclobutene (Figure 5.8). This concerted, synchronous disrotatory electrocyclic reaction is said thermally forbidden.

5.2.4 Aromaticity of transition states in cyclobutene/butadiene electrocyclizations

The transition structures **9con** ([π^4c]) and **9dis** ([π^4d]) of the ring opening of cyclobutene can be represented as a twisted diradical located at the termini carbon centers of the forming butadiene interacting with an ethylene moiety. As shown in Figure 5.9a for the [π^4c] reaction, there are two stabilizing interactions involving the HOMO(C_2-diradical)/LUMO(ethylene) and LUMO(ethylene)/HOMO(C_2-diradical). The AO array in transition structure **9con** corresponds to a Möbius strip (Section 4.9) as recognized by Dewar [21, 22] and Zimmerman [23, 24]. It is aromatic

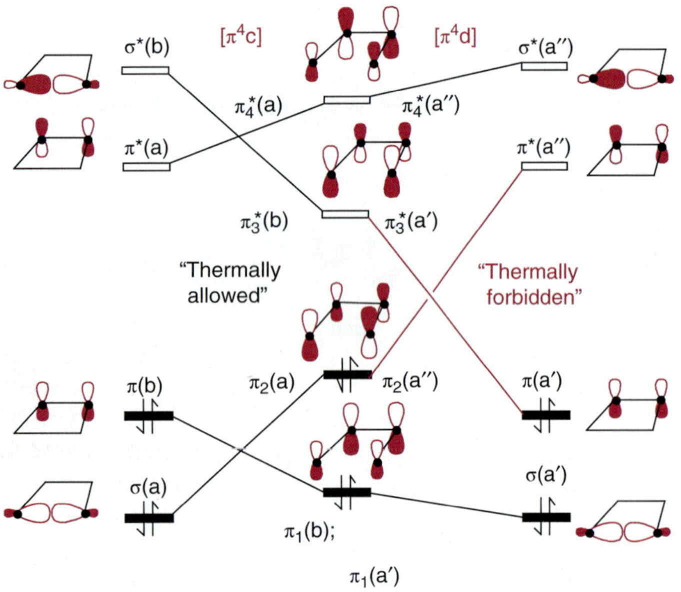

Figure 5.8 Woodward–Hoffmann correlation diagram for the stereoselectivity (conrotatory vs. disrotatory) in the electrocyclic isomerization of cyclobutene and butadiene.

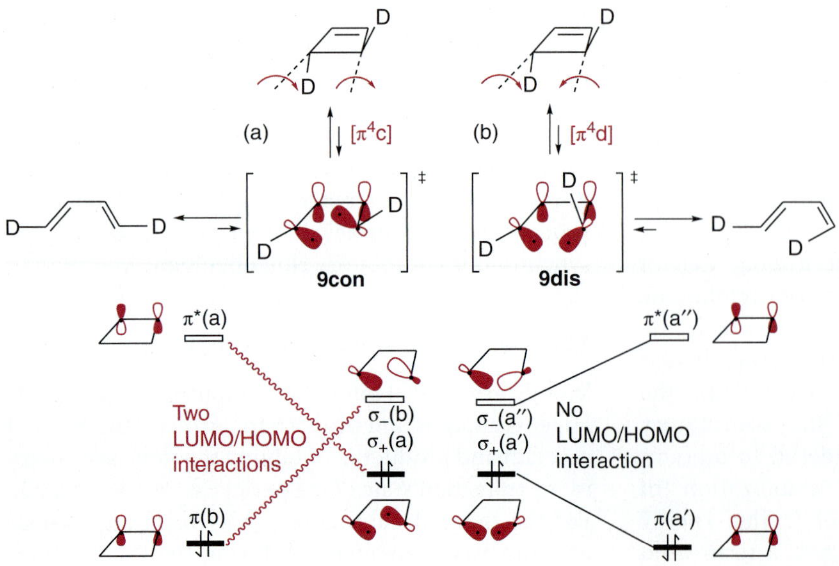

Figure 5.9 (a) FMO diagram demonstrating the Möbius aromaticity of the transition structure of conrotatory electrocyclic isomerization of cyclobutene into butadiene. (b) PMO diagram showing the "*anti*-aromaticity" of the Hückel transition structure of disrotatotary electrocyclic isomerization of cyclobutene into butadiene.

as it counts four electrons, as is demonstrated by the FMO diagram of Figure 5.9a [25]. In the case of the [π^4d] reaction, there is an overlap between HOMO(C_s-diradical) and HOMO(ethylene) that leads to a four-electron destabilization that is not compensated by any LUMO/HOMO interactions (Figure 5.9b). The FMO theory demonstrates that the transition structure **9dis** of the synchronous disrotatory electrocyclic reaction does not have any electronic stabilization. The AO array of transition structure **9dis** is a Hückel strip that is antiaromatic as it contains four electrons, very similar to cyclobutadiene (see Section 4.5.6) [26].

Thermolysis (250–300 °C) of bicyclo[3.2.0]hept-6-ene gives (*Z*,*Z*)-cyclohepta-1,3-diene exclusively with a relatively high activation energy (E_a = 45.4 kcal mol^{-1}, log A = 14.3) [27]. Similarly, thermolysis (235–285 °C) of bicyclo[4.2.0]oct-7-ene gives (*Z*,*Z*)-cycloocta-1,3-diene with E_a = 43.2 kcal mol^{-1} and log A = 14.1 [10]. As these isomerizations have to follow a "forbidden" disrotatory mode of ring opening for structural reasons (the corresponding (*E*,*Z*)-cycloalka-1,3-dienes being unstable cannot form, see Section 2.6.9, Table 2.A.3), they require higher activation energies than for the "allowed" conrotatory ring opening of *cis*-3,4-dimethylcyclobutene (E_a = 34.3 kcal mol^{-1}, log A = 13 [12]). With smaller bicyclic systems like bicyclo[2.2.0]hex-2-ene, bicyclo[2.1.0]pent-2-ene, and bicyclo[2.2.0]hexa-2,5-diene (Dewar benzene, Section 6.5.1), the disrotatory electrocyclic reactions require lower activation energies (Scheme 5.4). One-step, concerted reactions follow the Dimroth principle and the Bell–Evans–Polanyi theory. For a series of analogous reactions, relationship $\Delta^{\ddagger}H = \alpha\Delta_r H + \beta$ exits between the rates and the heats of these reactions. The more exothermic the reaction, the faster it occurs. When the *cis*-bicyclo[*n*.2.0]alk-(n + 3)-ene becomes large enough, the isomeric (*E*,*Z*)-cycloalk-1,3-diene is formed following an expected [π^4c] conrotatory electrocyclic reaction. This is the case with bicyclo[6.2.0]deca-9-ene that isomerizes on heating into (*E*,*Z*)-cyclodeca-1,3-diene [28]. Surprisingly, flash vacuum pyrolysis of *cis*-bicyclo[3.2.0]hept-6-en-2-one (**10**) generates the transient (*Z*,*E*)-cyclohepta-2,4-dien-1-one (**11**) identified by its photoelectron spectrum [29].

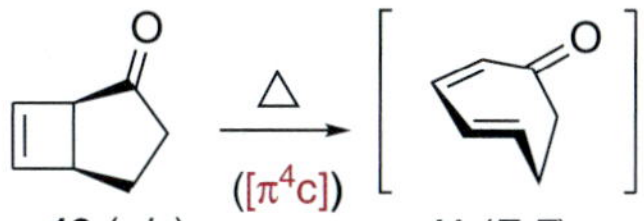

10 (*cis*) **11** (*Z*,*E*)

Problem 5.1 Thermal isomerization of *cis*-bicyclo[3.2.0]hept-6-ene into (*Z*,*Z*)-cyclohepta-1,3-diene has an Arrhenius activation energy E_a = 45.5 kcal mol^{-1}. In contrast, the isomerization of **A** into **B** has a much lower activation energy E_a = 21.3 kcal mol^{-1} [30]. Give an explanation.

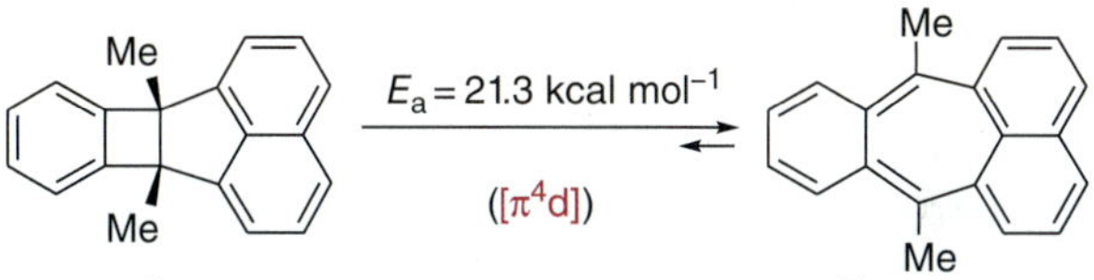

A **B**

5.2.5 Torquoselectivity of cyclobutene electrocyclic reactions

As shown above, the conrotatory isomerization of *trans*-1,2,3,4-tetramethylcyclobutene (**3**) gives exclusively (*E*,*E*)-3,4-dimethylhexa-2,4-diene (**4**), corresponding to an *outward* rotation of the 3,4-dimethyl groups of **3**. (*Z*,*Z*)-3,4-dimethylhexa-2,4-diene resulting from an *inward* rotation of the 3,4-dimethyl group of **3** is not seen. For steric reasons, the latter diene is less stable than **4**. The *outward* rotation of Me–C(3) and Me–C(4) in **3** giving **4** is thus favored kinetically and thermodynamically. In 1969, Frey reported that thermal isomerization of 3-ethyl-3-methylcyclobutene (**12a**) gives preferentially (*Z*)-4-methylhexa-1,3-diene ((*Z*)-**13a**), the larger ethyl group at C(3) having rotated *inward* during the electrocyclic reaction [31]. Similar observations were made with 3-isopropyl-3-methylcyclobutene (**12b**) and other derivatives (Scheme 5.5) [32]. Houk and Rondan discovered through computational studies

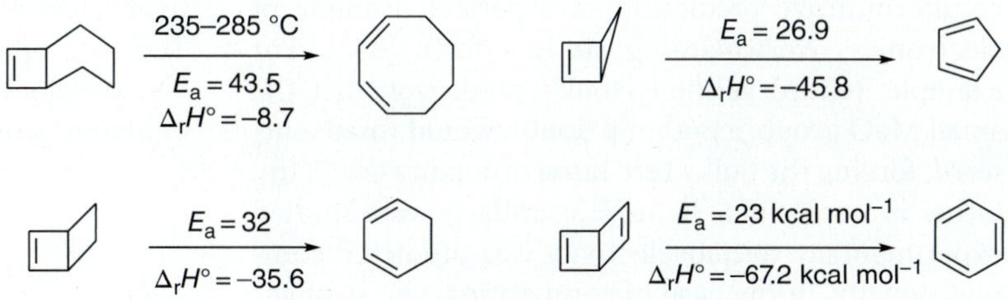

Scheme 5.4 Small bicyclo[*n*.2.0]alkenes containing a cyclobutene moiety are forced to undergo "forbidden" [π^4d] reactions. Their activation energy (E_a) decreases as the exothermicity of the isomerization increases (gas phase, standard conditions) ($\Delta_r H°$, see Table 1.A.3).

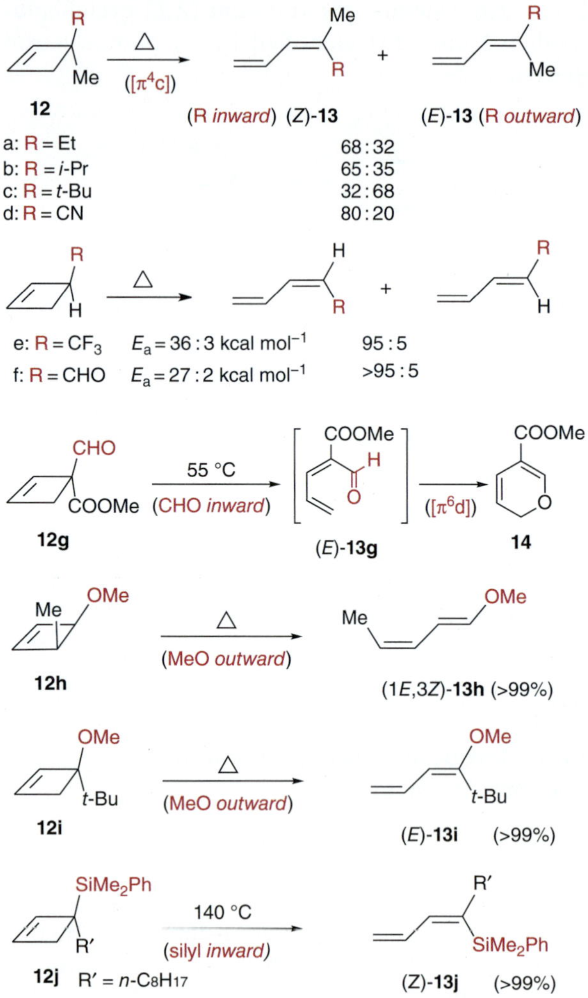

Scheme 5.5 Examples of thermal [π^4c] electrocyclic reactions of substituted cyclobutenes. Electronic factors control the torquoselectivity (*inward* vs. *outward* rotation of the 3-substituent).

the general rule that 3-substituents that are potent electron acceptors such as CHO and NO (and carboxamides, see problem 5.2) prefer the contrasteric *inward* torquoselectivity, whereas 3-substituents that are electron donors such as MeO (**12h**) (or EtO, AcO, Cl) and very bulky groups (e.g. *t*-Bu) prefer the *outward* rotation. The formyl compound, **12f**, was synthesized in the Houk lab after the computational prediction of inward rotation. This successful, and counterintuitive prediction, is a perfect example of electronics overwhelming steric effects [34c]. For example, Jefford verified Houk's prediction that the small MeO group, a potent p donor, would rotate outward, forcing the bulky tert-butyl group inward [34d]. Other examples in Scheme 5.5 similarly were studied experimentally torquoselectivity was predicted computationally. In the case of competition, the stronger donor (or weaker acceptor) always rotates outward. Thus, in the case of **12g** that bears a carbaldehyde and a methoxycarbonyl group at C(3), it is the CHO group that drives the *inward* rotation giving the unstable conjugated dienal (*E*)-**13g** that undergoes easy six-electron electrocyclization (Section 5.2.8) into **14**. The thermal isomerization of **10** into **11** implies the *inward* rotation of the carbonyl moiety that overwhelms the concurrent *inward* rotation of the alkyl group. The methoxy group at C(3) of cyclobutene drives the *outward* rotation (**12h**), even with bulky *t*-butyl group substituent at C(3) as in **12i** [33, 34].

Like π-acceptors, 3-silyl groups exert contrasteric effects leading to preferred *inward* torquoselectivity [35]. The low-lying π^* empty orbitals of π-acceptors (CN, CF_3, CHO, COOR, and $CONR_2$) and σ^* empty orbital of the C—Si bonds of the 3-silyl substituent interact with the HOMO of the C_2-twisted diradical that forms in the transition state (Figure 5.10). These two electron interactions resulting from the LUMO(3-substituent)/HOMO(diradical) overlap are stabilizing. This is more so for the *inward* than for the *outward* rotation of the 3-substituent. If the substituent is a lone pair donor and does not have a low-lying LUMO, the HOMO(3-substituent)/HOMO (C_2-twisted diradical) interaction is repulsive; thus, it avoids the *inward* rotation for which this four-electron interaction is maximized [36]. The contrasteric *inward* electrocyclic reactions represent deviations from the Dimroth principle or the Bell–Evans–Polanyi theory ($\Delta^{\ddagger}H° = \alpha\Delta_r H + \beta$), as the least stable isomers form the fastest.

The activation energies for the cyclobutene ring opening decrease with the relative π-donor ability of the 3-substituent (cyclobutene: E_a = 32.5 kcal mol^{-1}; 3-chlorocyclobutene: E_a = 29.4 kcal mol^{-1}; 3-ethoxy cyclobutene: E_a = 23.5 kcal mol^{-1}). Except for 3-(trifluoromethyl)cyclobutene (E_a = 36.3 kcal mol^{-1}), π-acceptors also accelerate the electrocyclic reaction (cyclobutene-3-carbaldehyde: E_a = 27.2 kcal mol^{-1}), consistently with the FMO model of Figure 5.10 for the interaction between the substituent and the transition structure of the conrotatory ring opening [37].

Benzocyclobutenes are very useful synthetic intermediates that equilibrate on heating with corresponding orthoquinodimethanes by electrocyclic isomerizations. For the unsubstituted system, one estimates an endothermicity of 5.3 kcal mol^{-1} (Section 2.11.5). The electrocyclic reaction is relatively slow and requires a temperature of c. 200 °C. The ring openings occur at lower temperatures for benzocyclobutenes 3-substituted with alkyl (140 °C) or alkoxy groups (110 °C) [38].

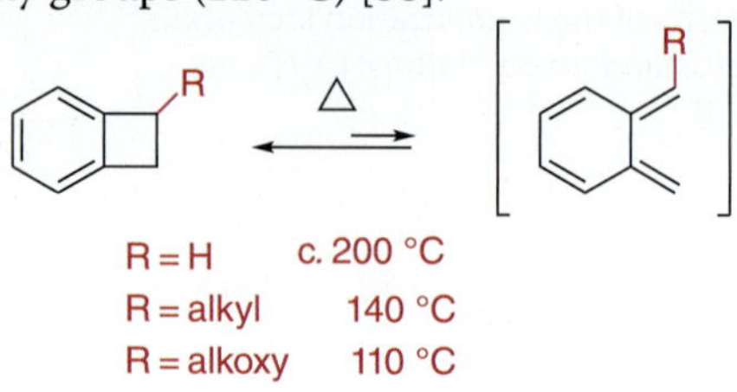

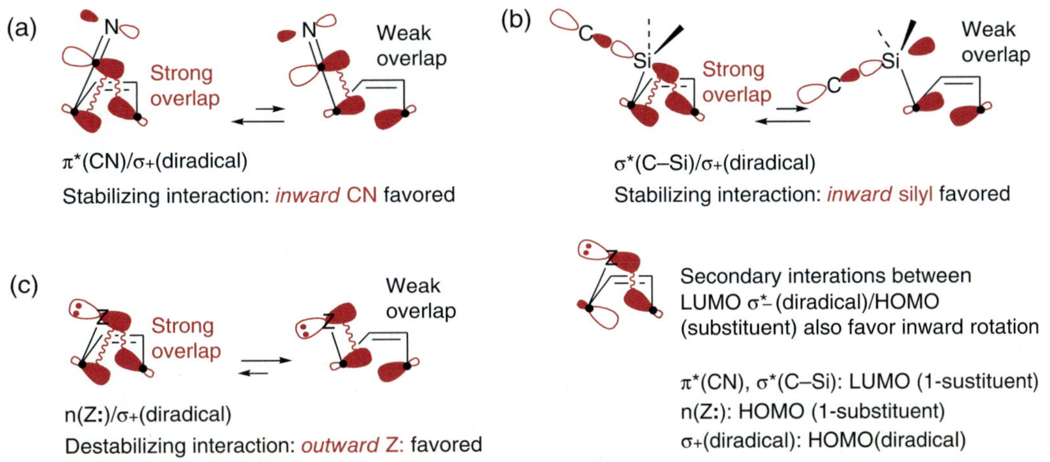

Figure 5.10 Torquoselectivity of [π^4c] reactions of 3-substituted cyclobutenes. FMO interpretation for (a) the favored contrasteric *inward* rotation of 3-substituents that are π-electron acceptors (low lying π^*(C=Z) orbitals), (b) the favored *inward* rotation of 3-silyl substituents (low lying σ^*(Si—C) orbitals), and (c) the favored *outward* rotation of 3-substituents that are electron lone pair donors.

A total synthesis of D-homoestrone has been developed by Kametani et al. that relies upon the intramolecular Diels–Alder reaction of orthoquinodimethane intermediate **16** obtained by thermolysis of benzocyclobutene derivative **15** [39, 40]. In this reaction, *outward* rotation of the bulky 3-alkyl substituent occurs. The transition structure of the subsequent Diels–Alder reaction minimizes strain and gauche interactions leading to **17**, which, upon demethylation, provides (±)-homoestrone (Scheme 5.6), a compound that can be converted into (±)-estrone. D-estrone is an important precursor of 19-norsteroids used as oral contraceptives [41]. An asymmetric total synthesis of estradiol has been realized by the same group applying a similar route using optically active 1-*tert*-butoxy-3-ethenyl-2-[2-(4-methoxybenzocyclobutenyl)-ethyl]-2-methylcyclopentane [42].

Problem 5.2 What are the major products formed upon thermolysis of the benzocyclobutene derivatives **A** and **B**? [43]

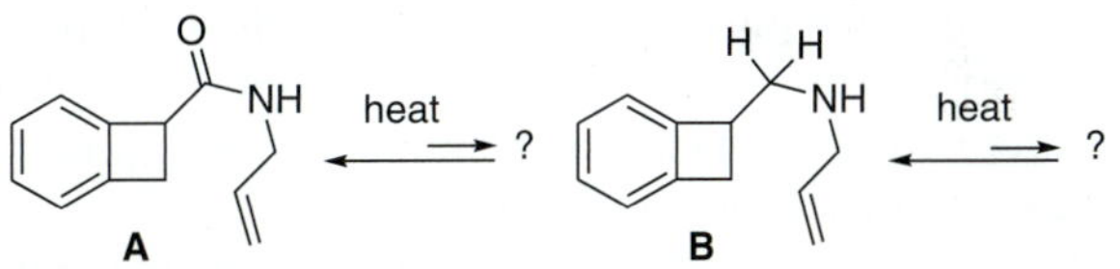

Scheme 5.6 Kametani's synthesis of (±)-homoestrone.

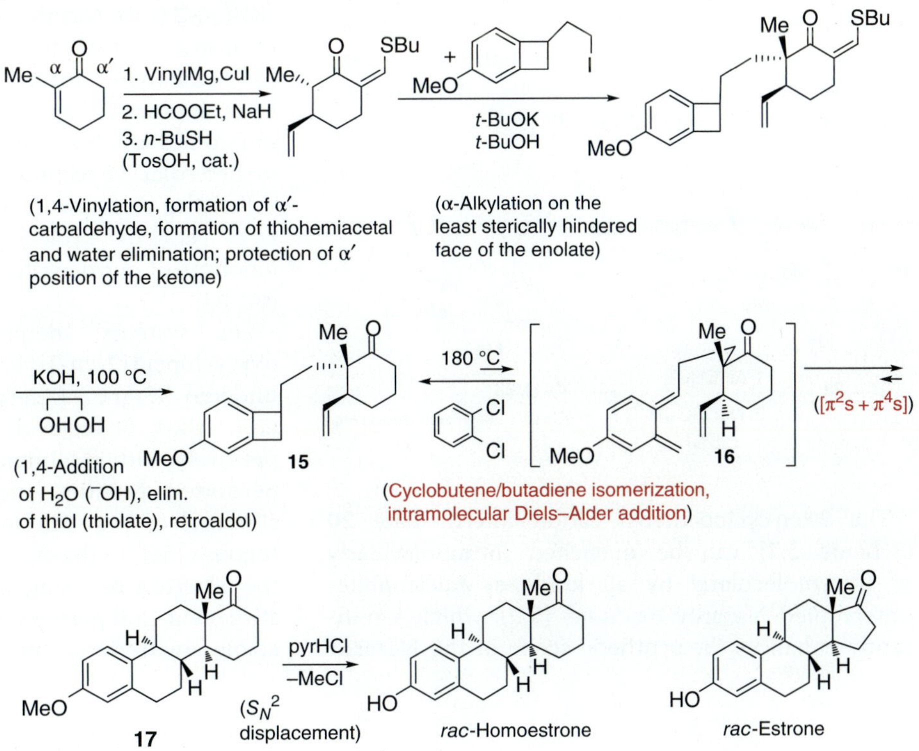

Problem 5.3 On heating to 100 °C, cyclobutene **A** is isomerized into a 85 : 15 mixture of products **P** + **Q** [44]. What are **P** and **Q**?

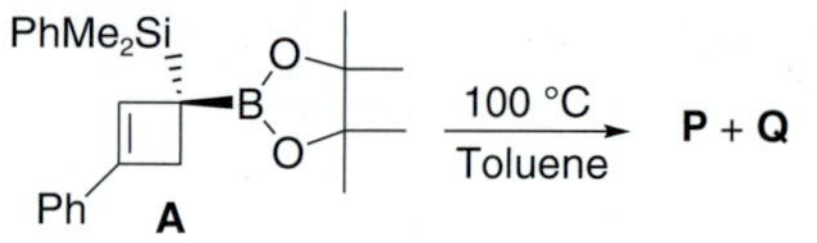

5.2.6 Nazarov cyclizations

Classically, the Nazarov cyclization reaction (often called the Nazavov cyclization) converts divinyl ketones into cyclopent-2-enones using a stoichiometric or super-stoichiometric Lewis (e.g. $TiCl_4$, $BF_3{\cdot}Et_2O$, $Bi(CF_3SO_3)_3$, and $AuCl_3$) or protic acid (e.g. CF_3SO_3H, $MeSO_3H$, and *p*-toluenesulfonic acid: TsOH) as promoter [45–47]. Shoppee proposed the mechanism presented in Scheme 5.7 [48, 49]. Activation of dienones **18** by protic acid generates 3-oxydienyl cation intermediates of type **19** that undergo thermal four-electron conrotatory electrocyclizations. As predicted (Woodward–Hoffmann rules), the 2-oxycyclopent-2-en-1-yl cation intermediates **20** form; the latter lose a proton giving the corresponding cyclopentenones **21**. The reactions may not be regioselective and isomeric cyclopentenones **21′** can form as well. Furthermore, depending on the nature of the substituents of **18**, five-epimers of **21** and **21′** can form concurrently.

Examples of Nazarov cyclizations requiring only a catalytic amount of acid promoters have been reported [50]. For instance, in their total synthesis of (+)-fusicoauritone, a marine natural product, Williams et al. realized the TsOH-catalyzed ring contraction of dienone **22** into **23** [51].

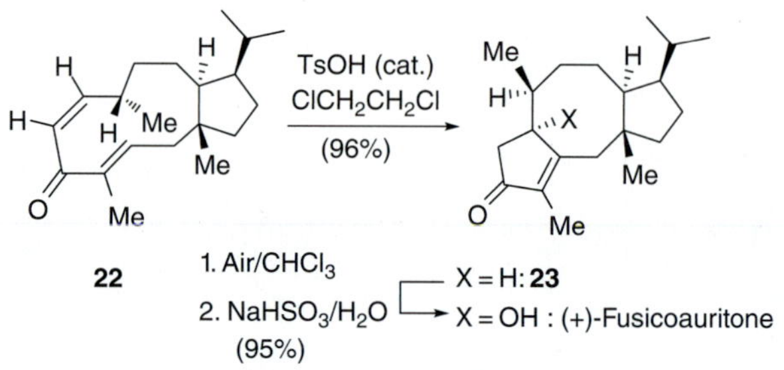

The 2-oxycyclopentenyl cation intermediate **20** (Scheme 5.7) can be quenched intramolecularly or intermolecularly by all kinds of nucleophiles (interrupted Nazarov reactions [52]), which significantly enhances the synthetic scope of the Nazarov electrocyclizations [53]. Furthermore, applying enantiomerically pure Lewis [54, 55], or protic acid promoters [56], enantioselective Nazarov cyclizations are possible (see Scheme 5.12) [57]. If one of the substituents at C(2) or C(4) of **18** is electron releasing, more stable cyclopentenyl cation intermediates **20** form with high regioselectivity. The subsequent reactions (β-elimination, nucleophilic quenching, cycloadditions with alkenes and 1,3-dienes, and Wagner–Meerwein rearrangement) are more selective. Furthermore, weaker acid promoters can be used in stoichiometric amounts or in catalytic amounts. Denmark and Jones applied stoichiometric amounts of $FeCl_3$ to promote the silicon-directed Nazazov cyclizations [58, 59]. For instance, dienone (*E*)-**24** undergoes high yielded and stereoselective conversion into cyclopentenone **26** in high yield. In this case, it is the β-silicon effect (Section 4.8.6) that stabilizes the intermediate cyclopentenyl cation **25**. The latter undergoes β-elimination of Et_3SiCl (Scheme 5.8). The outcome of the Nazarov cyclization depends on the (*E*)- vs. (*Z*)-configuration of the dienone. Under the same conditions as the conversion of (*E*)-**24** into **26**, the reaction of (*Z*)-**24** gives a mixture of **26** and enone **27** in mediocre yield. This is attributed to the torquoselectivity of the conrotatory ring closure that forces the trimethylsilyl group to rotate *inward* (Section 5.2.5) and thus leads to a much less stable cyclopentenyl cation due to gauche interactions [60]. In the case of (*S*)-(−)-**28**, the Nazarov cyclization gives (−)-**30**. The reaction involves transition structure **29** that minimizes steric repulsions and leads to the torquoselectivity observed. In the case of (*R*)-(+)-**28**, its Nazarov cyclization produces (+)-**30** exclusively, which involves transition structure **29′** [59]. Catalytic, formal homo-Nazarov cyclization of cyclopropyl alkenyl ketones interrupted by (hetero)arenes has been reported, which gives access to α-(hetero)aryl cyclohexanones [61].

Frontier and coworkers have developed the polarized Nazarov cyclizations that require catalytic amounts of cupric triflate as promoter (Scheme 5.9) [62, 63].

For systems adequately substituted, the 2-oxycyclopent-2-en-1-yl cation intermediates can undergo Wagner–Meerwein rearrangements ((1,2)-aryl, -alkyl, or -hydride shifts, Section 5.5.1) competitively with β-elimination of a proton. This has permitted Frontier and coworkers to develop a stereoselective synthesis of functionalized cyclopentenones [64]. In the example outlined in Scheme 5.10, the electron-releasing 2,4,6-trimethoxyphenyl substituent at C(5) permits the formation of the relatively stable zwitterionic intermediate **35** arising from

Scheme 5.7 General mechanism of the Nazarov cyclization.

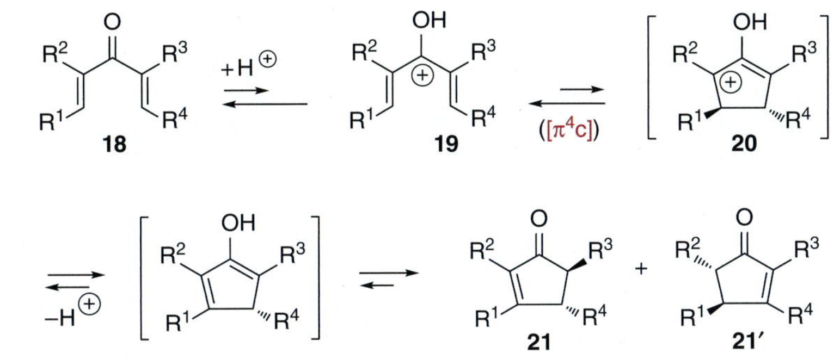

Scheme 5.8 Examples of silicon-directed Nazarov cyclizations.

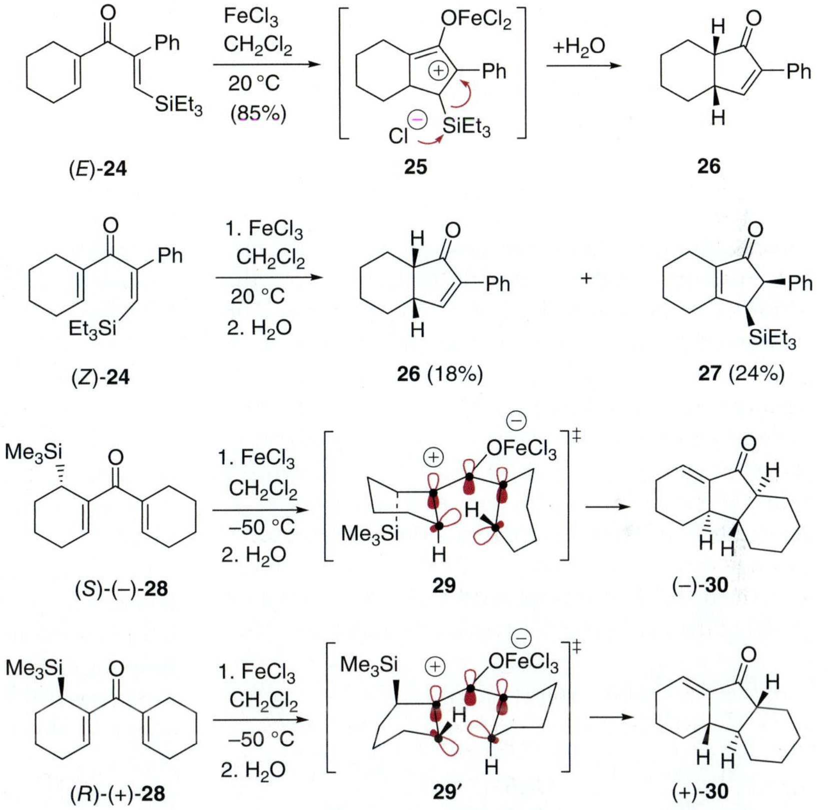

Scheme 5.9 Examples of polarized Nazarov cyclizations.

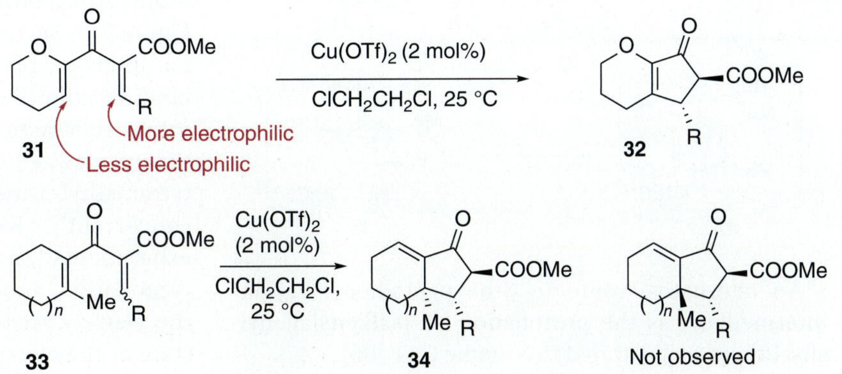

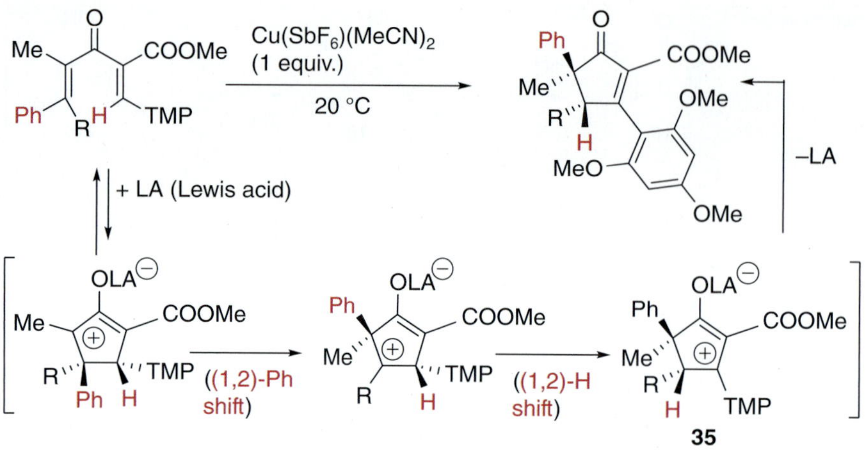

Scheme 5.10 Wagner–Meerwein rearrangements can compete with β-elimination of proton in the Nazarov cyclization (TMP, 2,4,6-trimethoxyphenyl).

the successive (1,2)-shifts of a phenyl and a hydride moiety.

The 3-oxypentadienyl cations can be generated by other routes than through the protonation or coordination of divinylketones to a Lewis acid. For instance, $AgBF_4$-induced ionization of the dichlorocyclopropane derivative **36** generates the 2-chloroallyl cation intermediate **37**, which is in fact a 2-choro-3-silyloxypentadienyl cation. The latter undergoes quick conrotatory ring closure into cyclopentenyl cation intermediate **38**. Desilylation provides cyclopentenone **39** [65]. The cyclopropyl chloride ionization of **36** into allyl ion pair **37** is an example of electrocyclic reaction of cyclopropyl cations. Ionization of the C—Cl bond is concerted with the disrotatory electrocyclic ring opening (noted $[\pi^2 d]$) of the cyclopropyl cationic intermediate.

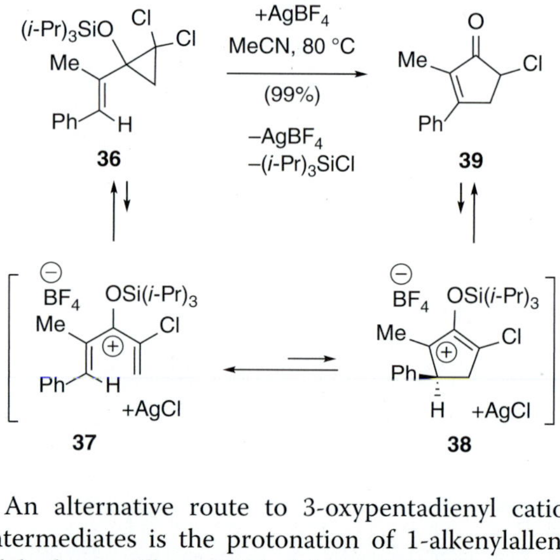

An alternative route to 3-oxypentadienyl cation intermediates is the protonation of 1-alkenylallenyl silyl ethers as illustrated in Scheme 5.11 [66].

In 2007, Rueping et al. have reported the first enantioselective Nazarov cyclization. In the presence of 2 mol% of enantiomerically pure N–H acid **cat1**, dienone **40** is converted into 6 : 1 mixture of cyclopentenones **41** and **42** with ee = 87 and 95%, respectively (Scheme 5.12a) [67]. Tius and coworkers found that the enantiomerically pure thiourea derivative **cat2** (a double N–H acid, Section 7.6) that activates the carbonyl moiety of dienone **43** by double hydrogen bridging (C=O· · ·H—N) induces an asymmetric Nazarov cyclization into optically active cyclopentenone **44** (Scheme 5.12b) [68]. Applying catalysts **cat3** made out of trisoxazoline (TOX) ligand and Lewis acid $Cu(II)[(F_5C_6)_4B]_2$, Tang and coworkers realized higher enantioselectives and higher yields with excellent diastereoisomeric ratio (dr > 99 : 1) (Scheme 5.12c) [69].

In acidic aqueous media, (*E*,*E*)- and (*Z*,*Z*)-3,4,5-trimethylhept-2,5-dien-4-ol are converted into 1,2,3,4,5-pentamethylcyclopentadiene + H_2O. The reaction implies ionization of the alcohol into 2,3,4,5,6-pentamethylhepta-3,5-dien-2-yl cation that undergoes ring closing into 1,2,3,4,5-pentamethycyclopent-2-en-yl cation intermediate and deprotonation of the latter. In the presence of a catalytic amount of self-assembled host of type $[Ga_4L_6]^{2-}$ (L = *N*,*N*-bis(2,3-dihydroxybenzoyl)-1,5-diaminonaphthalene, see Figure 7.6, Section 7.3.4) that has a large hydrophobic interior able to host all sort of molecules with binding affinities of up to $10^5\ M^{-1}$, the rate of the catalyzed reaction can be 2×10^6 times larger than the uncatalyzed reaction. Furthermore, *trans*-1,2,4,5-tetramethyl-3-methylidenecyclopent-1-ene is formed concurrently. Kinetic analysis and ^{18}O exchange experiments showed that encapsulation, alcohol protonation, and loss of H_2O are reversible and that the electrocyclization is irreversible. The transition state of the electrocyclization is more stabilized than the reactant, the protonated alcohol, and the corresponding pentadienyl cation intermediate through encapsulation [70].

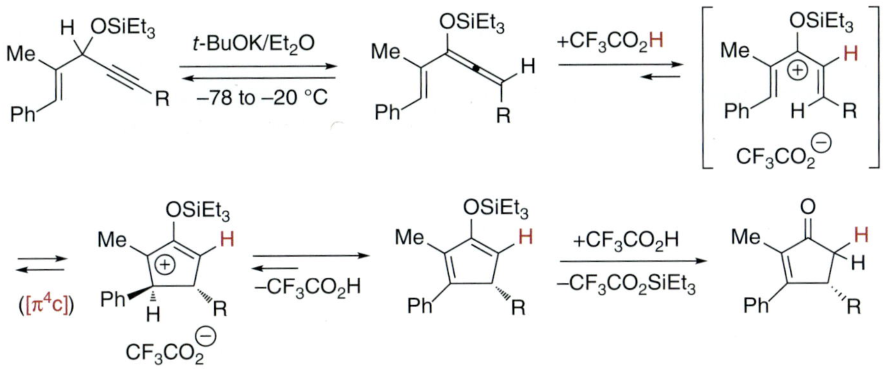

Scheme 5.11 Brønsted-acid-mediated Nazarov cyclization of alkenylallenyl silyl ethers.

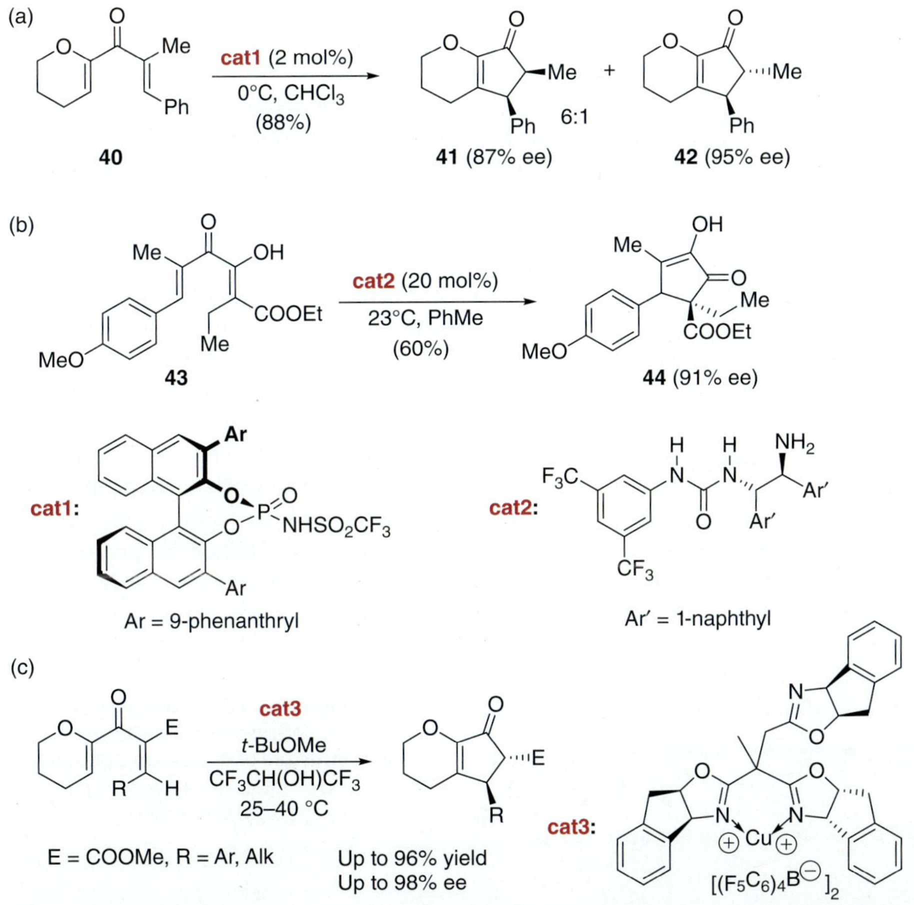

Scheme 5.12 (a, b) Examples of catalyzed asymmetric Nazarov cyclizations applying enantiomerically pure N–H acids to activate electron-rich dialkenylketones; and (c) examples of enantioselective Nazarov cyclizations using a Cu^{++} catalysts.

Problem 5.4 The reaction of phenylacetylene **A** with aldehyde **B** in the presence of SbF_5 and EtOH gives first product **P** that, upon heating in 1,2-dichloroethane, provides the corresponding indanone **Q**. What are **P** and **Q**? [71]

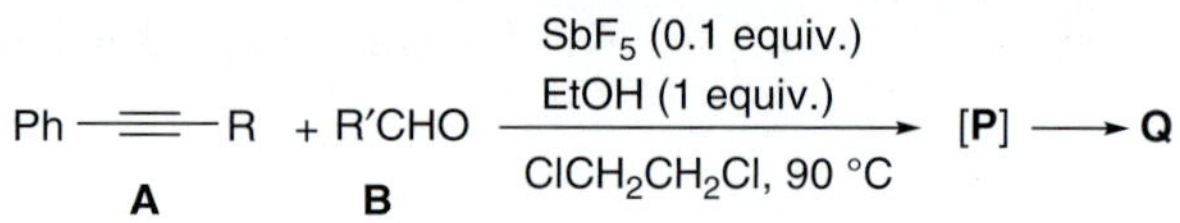

Problem 5.5 What are products **P** + **Q** of the following reaction? [72]

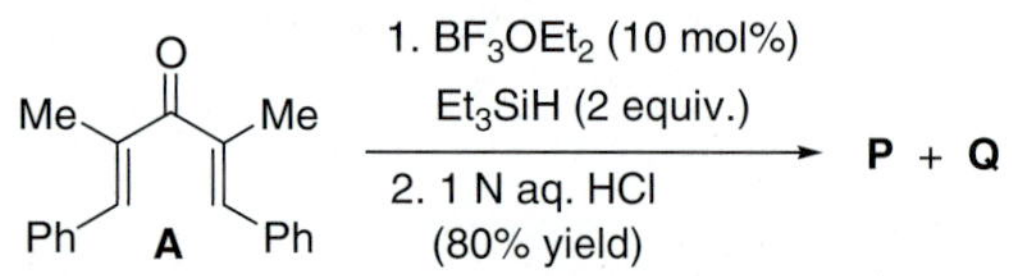

5.2.7 Thermal openings of three-membered ring systems

Being isoelectronic with the cyclobutene/butadiene electrocyclic isomerization, the thermal ring openings of 2,3-disubstituted cyclopropyl carbanions into the corresponding allyl anions should favor the [π^4c] conrotatory mode as predicted by the Woodward–Hoffmann rules (Scheme 5.1). For the C_2-symmetrical transition structure of this reaction, FMO theory shows (Figure 5.11a) that the forming twisted 1,3-diradical interacts with the carbanion at C(2). This interaction is stabilizing because of the overlap between the LUMO(C_2-1,3-diradical)/HOMO (carbanion at C(2)). In the case of a [π^4d] disrotatory ring opening, this stabilizing interaction is not possible because of the C_s symmetry. There is a repulsive HOMO(carbanion)/HOMO(C_s-1,3-diradical) interaction (Figure 5.11b) that makes this electrocyclic reaction difficult: there is no assistance between the breaking of the σ(C(2)—C(3)) bond and the allyl anion π-bond formation. The transition structure for the conrotatory mode is analogous to a four-electron Möbius strip (stabilized by Möbius aromaticity), whereas for the disrotatory mode, its transition structure resembles a four-electron Hückel array of AOs (destabilized by Hückel antiaromaticity).

Boche et al. reported that *trans*-1-cyano-2,3-diphenylcyclopropyllithium (*trans*-**46** resulting from the hydrogen/lithium exchange of aziridine **45**) is stable at 0 °C [73]. At 20 °C, it isomerizes into a mixture of 2-cyano-1,3-diphenylallyllithium composed of c. 92% of (*E*,*Z*)-**47**, 4% of (*E*,*E*)-**47**, and 4% of (*Z*,*Z*)-**47**. The latter are equilibrated rapidly at 20 °C. Similarly, *cis*-**46** is isomerized into the same mixture of **47** at 20 °C (Scheme 5.13). Thus, these experiments alone do not prove that the cyclopropyl carbanions *trans*-**46** and *cis*-**46** have undergone conrotatory ring openings. The electrocyclic reaction of derivative **49**, which has the same type of substituents than *cis*-**46**, can only be disrotatory giving allyl anion **50**. The rate constant of the latter reaction (k_{49} = 1.1×10^{-5} s^{-1}) is much lower (5500 and 740 times, respectively) than the rate constants measured at 20 °C for the electrocyclic ring openings of *trans*-**46** (k_{trans} = 6.1×10^{-2} s^{-1}) and *cis*-**46** (k_{cis} = 7.9×10^{-3} s^{-1}), therefore confirming that the conrotatory mode is preferred, in agreement with the Woodward–Hoffmann rules (Scheme 5.1 and Figure 5.9) and with quantum mechanical calculations [74, 75]. The allyllithium species obtained in these experiments can be quenched by protonation with water (e.g. **50** → **51**) or by alkenes that react with them in (3+2)-cycloadditions (e.g. Scheme 5.14).

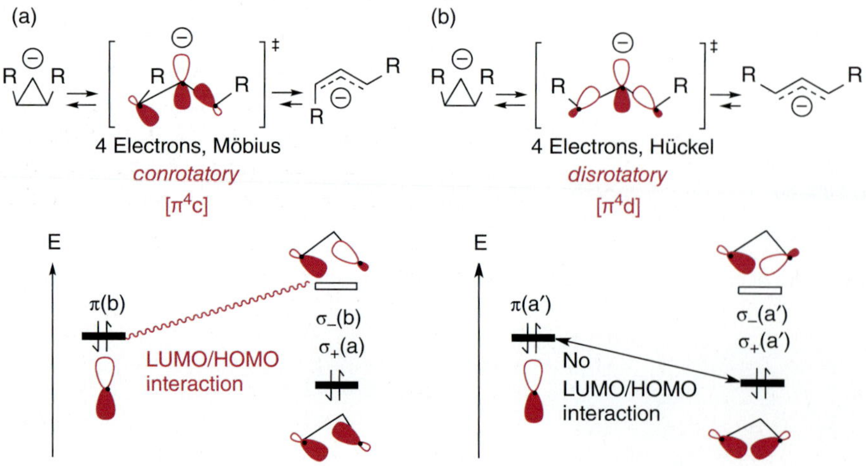

Figure 5.11 FMO diagram representing the transition structure of the (a) conrotatory and (b) disrotatory ring opening of cyclopropyl carbanions.

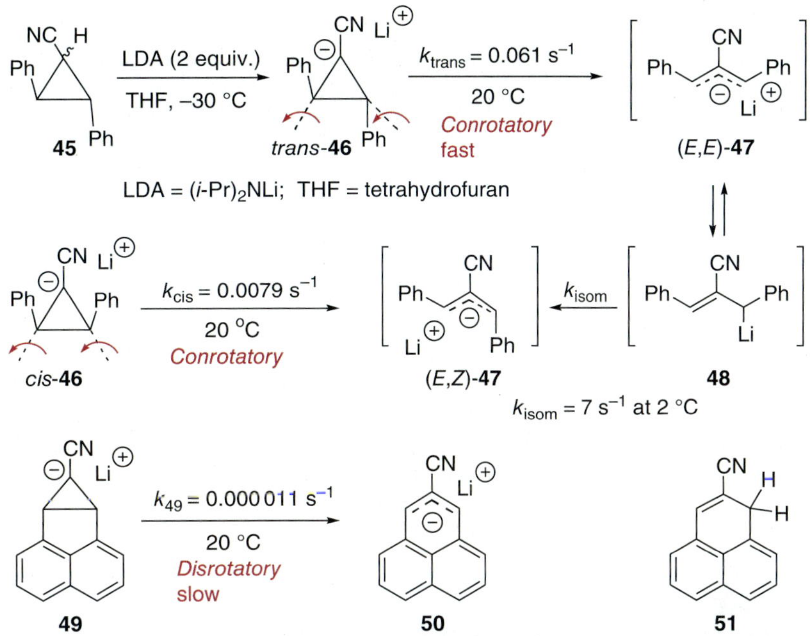

Scheme 5.13 Boche's kinetic measurements confirming that 1-cyano-2,3-diphenylcyclopropyllithium prefer the conrotatory mode of electrocyclic openings into the corresponding 2-cyano-1,3-diphenylallyllithium.

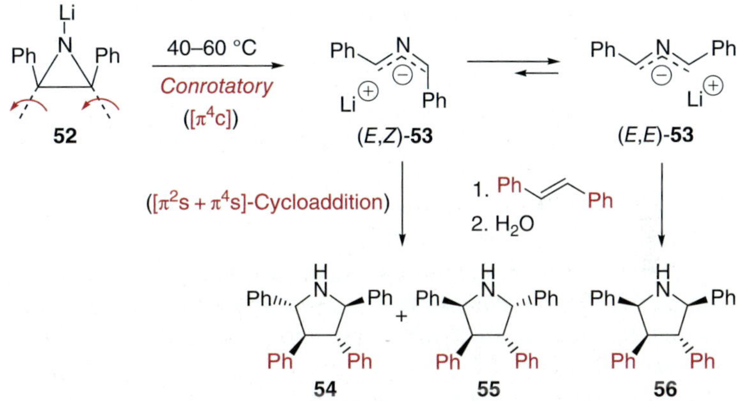

Scheme 5.14 Lithiated aziridines undergo conrotatory ring-opening into 2-azaallyllithium that undergo (3+2)-cycloadditions with (*E*)-stilbene. (*E*,*Z*)-**54**/(*E*,*E*)-**54** isomerization competes with the cycloadditions.

Lithiated *cis*-2,3-diphenylaziridine **52** undergoes conrotatory ring opening (Scheme 5.14) with the formation of (*E*,*Z*)-1,3-diphenyl-2-azaallyllithium ((*E*,*Z*)-**53**). When run in the presence of (*E*)-stilbene, the reaction produces a mixture of pyrrolidinide lithium salts that react with water producing pyrrolidines **54**, **55**, and **56**. The relative amount of **56** diminishes on increasing the concentration of (*E*)-stilbene, thus demonstrating that isomerization of (*E*,*Z*)-**53** into (*E*,*E*)-**53** competes with the $[\pi^4s+\pi^2s]$- (Woodward–Hoffmann electron count) or (3+2)-cycloadditions (number of atomic centers involved in the cycloaddends) [76].

Aziridines **57** and oxiranes **59** possess a lone pair of electrons at the heteroatom and are thus isoelectronic with cyclopropyl anion. The C—C ring openings generate azomethine ylides **58** and carbonyl ylides **60**, respectively (Section 5.3.17, Table 5.5). They are predicted (Woodward–Hoffmann rules) to be conrotatory electrocyclic reactions. The transition structures of these synchronous reactions imply stabilizing HOMO(heteroatom)/LUMO(C_2-1,3-twisted diradical) interactions. This is not the case for disrotatory ring openings.

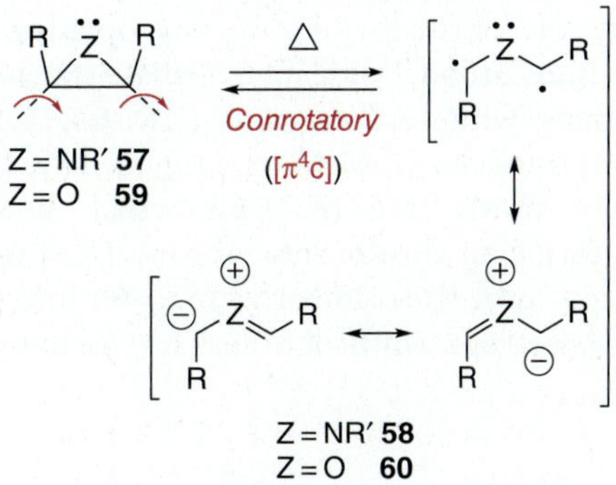

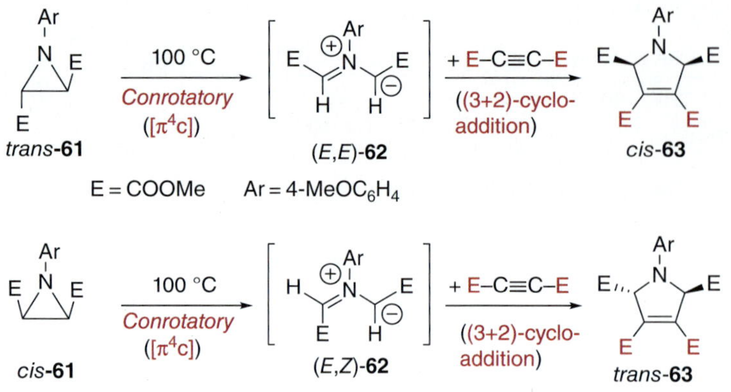

Scheme 5.15 Heating aziridines leads to azomethine ylide intermediates following conrotatory C—C ring opening reactions, as evidenced by the product of (3+2)-cycloadditions.

Electrocyclic ring opening of thiiranes has not been reported yet as their weak C—S bond reacts faster than their strong C—C bond. Quantum mechanical calculations predict the thermal C—C conrotatory electrocyclic opening of thiirane into thione methylide following the conrotatory mode as for cyclopropyl carbanions, aziridines, and oxiranes [77]. The possibility of thione methylides to equilibrate with corresponding thiiranes has been noted [78]. In 1967, Huisgen et al. confirmed the predictions of Woodward–Hoffmann theory with *N*-arylaziridine-2,3-dicarboxylic esters *trans*-**61** and *cis*-**61** that equilibrate with azomethine ylides (*E*,*E*)-**62** and (*E*,*Z*)-**62**, respectively, at 100 °C [79]. This was proven by quenching with dimethyl acetylenedicarboxylate to produce, through (3+2)-cycloadditions (or 1,3-dipolar cycloadditions, Section 5.3.17), the corresponding pyrroline derivatives *cis*-**63** and *trans*-**63**, with high stereoselectivity (Scheme 5.15).

In the gas phase, (+)-2,3-divinyloxirane ((*S*,*S*)-**64**) is racemized (equilibrated with (*R*,*R*)-**64**; $\Delta^{\ddagger}H = 30.1 \pm 0.5$ kcal mol^{-1}, $\Delta^{\ddagger}S = -9 \pm 1$ eu) on heating. At 207 °C, this reaction is 220 times as fast as the isomerization into *meso*-(*R*,*S*)-**64** and faster than rearrangements into (2*S*)-2,3-dihydro-2-vinylfuran (**65**, $\Delta^{\ddagger}H° = 35.1 \pm 0.4$ kcal mol^{-1}, $\Delta^{\ddagger}S = -3 \pm 1$ eu) and 4,5-dihydrooxepine (**66**: $\Delta^{\ddagger}H = 36.8 \pm 0.4$ kcal mol^{-1}, $\Delta^{\ddagger}S = -1 \pm 1$ eu) [80]. The results can be explained in terms of conrotatory C—C ring opening of the oxirane into carbonyl ylide intermediate (*E*,*E*)-**67** that equilibrates with (*S*,*S*)-**64** and (*R*,*R*)-**64**. Carbonyl ylide intermediate (*E*,*E*)-**67** is proposed to isomerize more slowly into (*E*,*Z*)-**67**, which undergoes conrotatory ring closure into oxirane (*R*,*S*)-**64**. Two concurrent processes isomerize (*S*,*S*)-**64** into **65** and **66** via sigmatropic shifts of order (1,3) (Section 5.5.4) [81, 82].

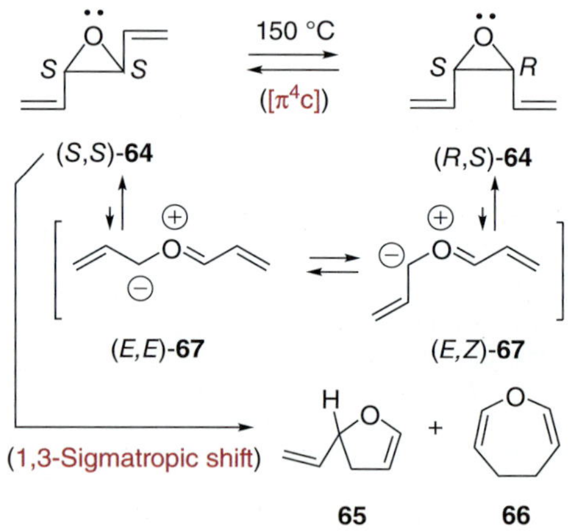

In the cases of electrocyclic ring openings of *trans*- and *cis*-2,3-diphenyloxirane-2,3-dicarbonitrile, nonstereospecific reactions are observed [83] consistently with conrotatory reaction accompanied by concurrent *E*/*Z* isomerization of the carbonyl ylides formed [84].

Woodward and Hoffmann predicted that cyclopropyl cations should undergo disrotatory ring opening (noted [π^2d]) into corresponding allyl cations (Scheme 5.1). For the transition structure of this two-electron electrocyclic reaction, FMO theory (Figure 5.12) shows that it is stabilized by a LUMO(cation)/HOMO(C_s-twisted 1,3-diradical) interaction. This is not the case in the transition structure of a conrotatory ring opening for reasons of C_2-symmetry. The transition structure of the disrotatory mode realizes a Hückel system with two electrons. It is isoelectronic with cyclopropenyl cation and is stabilized by aromatic stabilization. By contrast, the transition structure of the conrotatory mode has a Möbius array with two electrons that is not stabilized by Möbius aromaticity.

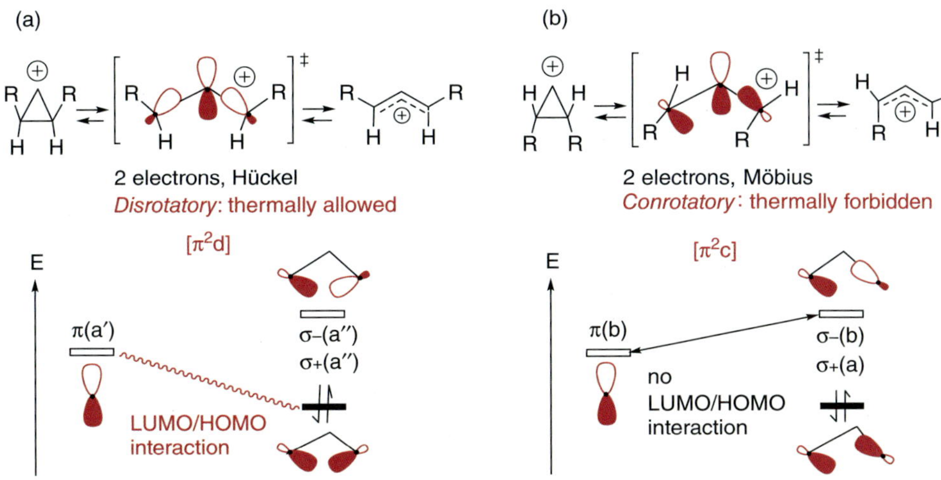

Figure 5.12 FMO diagram representing the transition structure of (a) the disrotatory and (b) of the conrotatory electrocyclic ring opening of cyclopropyl cation into allyl cation.

The Woodward–Hoffmann rules predict that both the conrotatory and disrotatory electrocyclic ring openings of cyclopropyl radical into allyl radical are forbidden as correlation diagrams show that excited electronic arrangements are produced by either pathway [85]. Quantum mechanical calculations predict that both modes of isomerization have similar activation energies [86, 87]. On heating, enantiomerically enriched *trans*-1,2-dideuteriocyclopropane is racemized as fast as it is isomerized into *cis*-1,2-dideuteriocyclopropane and about 20 times more often than it rearranges into dideuteriopropene. The three reactions involve the formation of the trimethylene diradical intermediate, which is c. 58.3 kcal mol^{-1} less stable than cyclopropane. Trimethylene diradical undergoes ring closure without energy barrier.

Problem 5.6 Estimate the heat of isomerization of cyclopropyl carbanion into allyl carbanion in the gas phase.

5.2.8 Six-electron electrocyclic reactions

Thermal electrocyclic ring closure of (3*Z*)-hexa-1,3,5-trienes into cyclohexa-1,3-dienes was first recognized with the rearrangement of vitamin D_2 into precalciferol (through a sigmatropic shift of hydrogen of order (1,7), Section 5.5.7) and subsequent disrotatory ring closure into a mixture of isopyrocalciferol and pyrocalciferol [88].

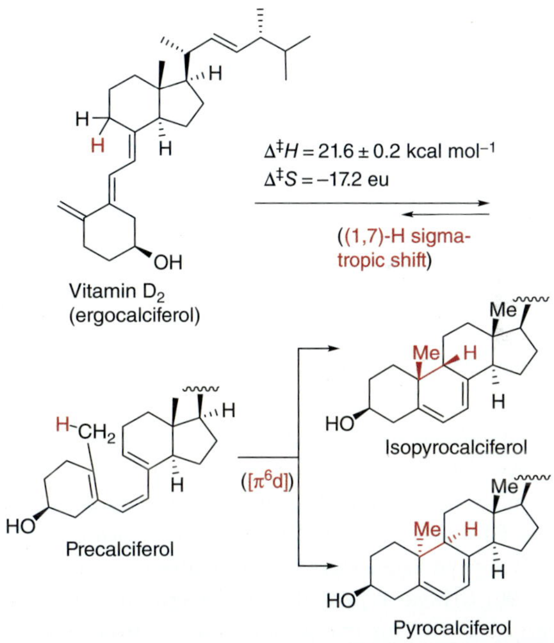

At 130 °C, (*E*,*Z*,*E*)-octa-2,4,6-triene ((*E*,*Z*,*E*)-**68**) undergoes stereospecific [π^6d] disrotatory ring closure into *cis*-5,6-dimethylcyclohexa-1,3-diene (*cis*-**69**) (Scheme 5.16) as predicted by Woodward–Hoffmann rules. The transition structure of this reaction forms a six-electron Hückel system with aromatic stabilization energy. Triene (*E*,*Z*,*Z*)-**68** undergoes disrotatory electrocyclic isomerization into *trans*-**69**. The reaction is slower than isomerization of (*E*,*Z*,*E*)-**68** into *cis*-**69** and the equilibration (*E*,*Z*,*Z*)-**68** into (*Z*,*Z*,*Z*)-**68** that implies two successive (1,7)-hydrogen

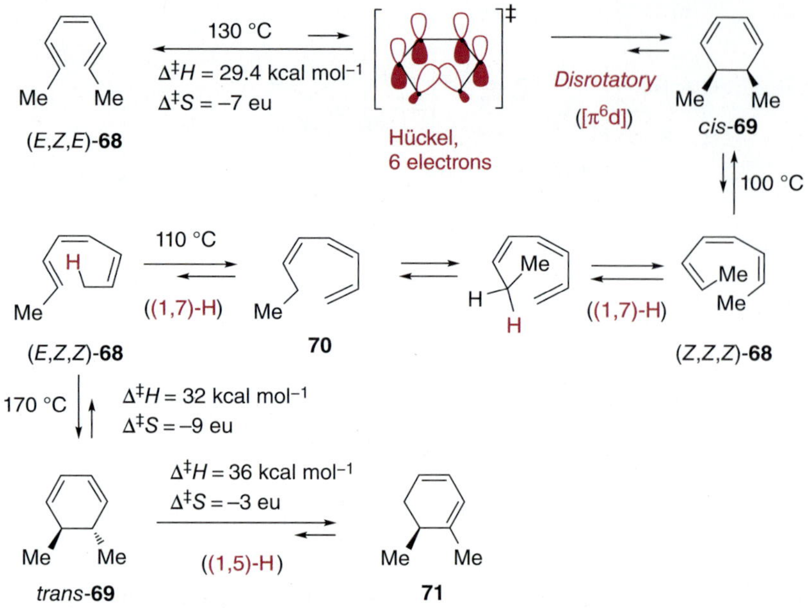

Scheme 5.16 Examples of thermal triene electrocyclizations.

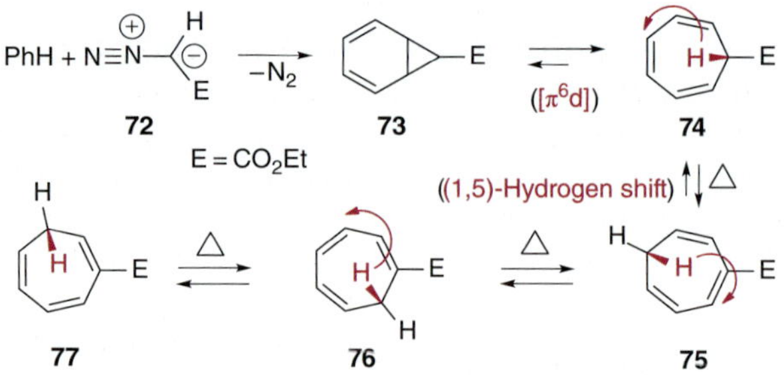

Scheme 5.17 The Büchner reaction.

sigmatropic shifts (Section 5.5.7) via intermediate **70**. At 100 °C, (*Z,Z,Z*)-**68** cyclizes into *cis*-**69**. At 180 °C, cyclohexadienes *trans*-**69** and *cis*-**69** are isomerized into 1,6-dimethylcyclohexa-1,3-diene (**71**) via (1,5)-hydrogen sigmatropic shifts (Section 5.5.6) [89, 90]. These types of rearrangements have also been observed for trienes 1,6-disubstituted by other groups than methyl groups [91].

In 1885, Büchner reported that the thermolysis of ethyl diazoacetate (**72**) in benzene generates norcaradienes derivative **73** (Büchner reaction). Later, it was found that the reaction produces in fact a mixture of cycloheptatrienes **75**, **76**, and **77** resulting from (1,5)-hydrogen sigmatropic shifts (Section 5.5.6) involving cycloheptatriene **74** that equilibrates with **73** [92]. When the Büchner reaction (Scheme 5.17) is catalyzed with $Rh_2(CF_3CO_2)_4$, a quantitative yield of **74** is obtained at 22 °C [93].

Cycloheptatrienes, azepins, and oxepins are in equilibrium with bicyclo[4.1.0]hepta-2,4-dienes (norcaradienes) [94], 7-azabicyclo[4.1.0]hepta-2,4-dienes (benzene imines) [95], and 7-oxabicyclo[4.1.0]hepta-2,4-dienes (benzene oxides) [82], respectively. These equilibria are referred to as valence isomerism or valence tautomerism and are fast reactions at room temperature [96] as they are thermally allowed disrotatory electrocyclic reactions for which the triene termini centers are relatively closed to each other, on the way to the ring-closing process (Section 3.6.4). For unsubstituted systems, the monocyclic heptatrienes **78** are more stable than the corresponding bicyclic dienes **79**, but the planar dienes **79** react faster with dienophiles in Diels–Alder reactions (Section 5.3.9).

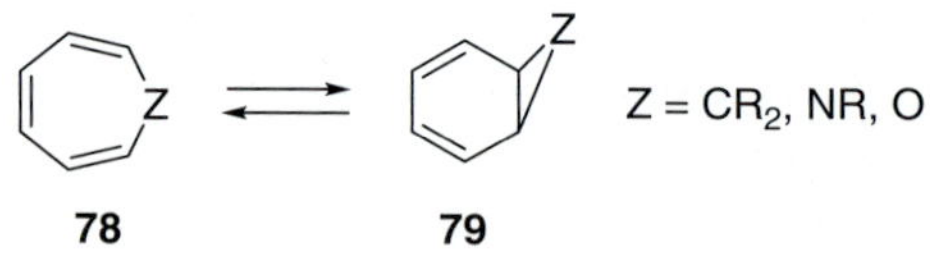

The equilibrium constant between cycloheptatrienes and corresponding bicyclo[4.1.0]hepta-2,4-dienes depends on substitution (Figure 5.13). Annulation of one double bond by a benzo ring fixes either the monocyclic (e.g. benzo[*c*]cycloheptatriene) or the bicyclic system (e.g. benzo[*b*]bicyclo[4.1.0]hepta-2,4-diene). Other type of annulation can also fix one or the other form (e.g. aaptosine, a 5,8-diazabenz[*cd*]azulene extracted from Aaptos *aaptos* sponges found in the Red Sea, a cytotoxic compound [97]). Tropone (cyclohepta-2,4,6-trien-1-one) does not equilibrate with bicyclo[4.1.0]hepta-2,4-dien-7-one as tropone is highly stabilized by aromaticity: its dipolar form is an oxytropylium cation (Section 4.5.13). Norcaradiene is stabilized by π-acceptors at C(7), the introduction of

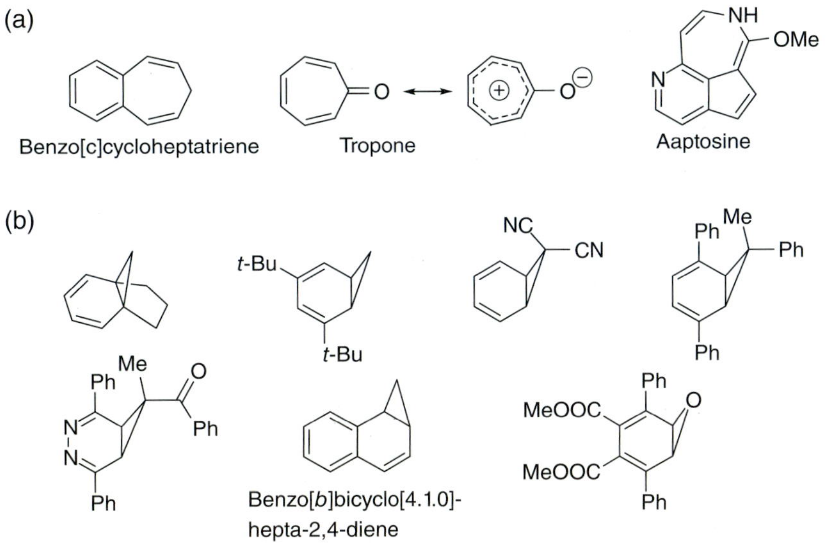

Figure 5.13 Substitution can stabilize either (a) the monocyclic or (b) the bicyclic valence tautomer.

alkyl groups, and extended conjugation with phenyl and/or alkoxycarbonyl substituents [98].

Reactions of C_{60} with diazoalkanes (via 1,3-dipolar cycloadditions, Section 5.3.17) are followed by losses of N_2 and production of two products of types **80** and **81** [99]. The enlarged fullerenes **81** are examples of fulleroids. Fulleroids and corresponding methanofullerenes are known to interconvert in a similar valence tautomerism as cycloheptatriene and norcaradienes [100].

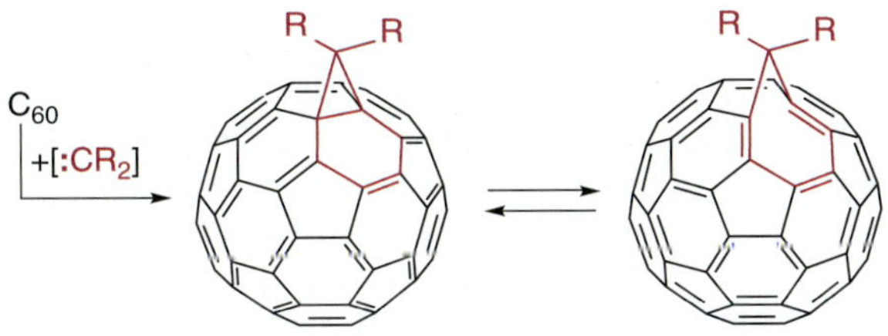

Oxidation of allylic alcohol **82** gives (2*Z*)-hexa-2,4,5-trienal **83** that is immediately isomerized into alkylidene-2*H*-pyran (*E*)-**84** at 20 °C via an apparent six-electron electrocyclization, which is reversible. After 48 hours at 20 °C, the more stable 2*H*-pyran (*Z*)-**84** forms irreversibly. When **83** is reacted with *n*-butyl amine, the corresponding Schiff base **85** is formed, which undergoes isomerization into α-pyridone (*Z*)-**86** [101]. There are two possible mechanisms for these ring-closing reactions. Mechanism (a) is a [π^6d] electrocyclic reaction (Scheme 5.18a). Mechanism (b) does not imply rotation of the termini centers Z(1) and C(6) of the 1-oxa- and 1-azahexa-1,3,5-triene unit but involves in-plane intramolecular nucleophilic addition of the carbaldehyde or imine moiety onto C(6) of the allene function (Scheme 5.18b) [102]. These types of mechanisms have been named pseudopericyclic reactions by Lemal and coworkers as they resemble pericyclic reactions but have no cyclic π overlap [103]. They cannot be said to be allowed or forbidden by orbital symmetry [103–109]. Quantum mechanical calculations on model reactions **87Z** → **88Z** → **89Z** and **87Z** → **90Z** → **89Z** suggested that the [π^6d] pericyclic process is realized for Z = CH_2, but not for Z = O and NH. Thus, cycloisomerization of (2*Z*)-hexa-2,4,5-trienals (considered as 1-oxahepta-1,3,5,6-tetraenes) and their imines (considered as 1-azahepta-1,3,5,6-tetraenes in Scheme 5.18) must be viewed as pseudopericyclic reactions. In their transition structures **90Z**, the C=O and C=NR double bonds have not rotated and maintain complete conjugation with the π-system. However, center C(7) of the allene moiety that becomes negatively charged is rotated in the transition state to ensure some stabilization of the forming carbanionic species by partial π-delocalization.

Treatment of cycloocta-1,4-diene with *n*-butyllithium in tetrahydrofuran (THF) generates salt **91** that undergoes at 35 °C "allowed" disrotatory six-electron electrocyclization (an example of a 1,5-electrocyclization) into bicyclo[3.3.0]oct-3-en-2-yl lithium salt (**92**) [110] (see also the rearrangement of heptatrienyl lithium into cycloheptadienyl lithium [111]).

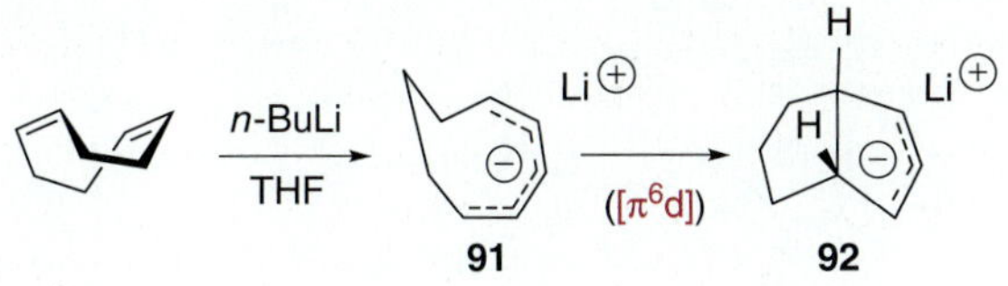

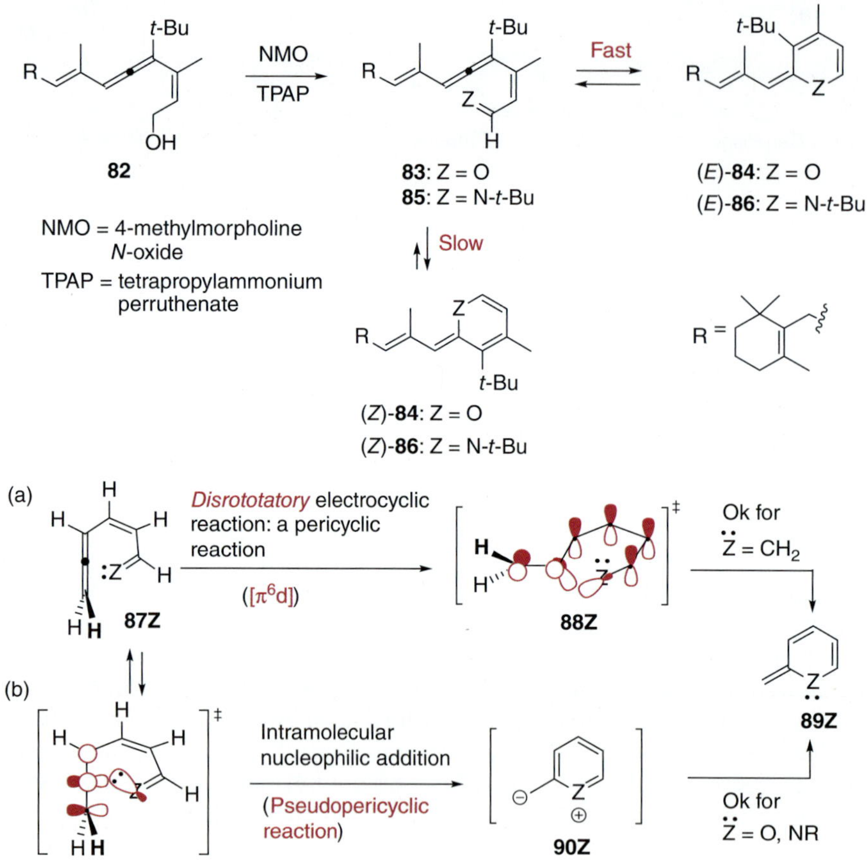

Scheme 5.18 (a) Pericyclic $[\pi^6 d]$ vs. (b) pseudopericyclic electrocyclizations.

The replacement of the $H_2C(1)$ of pentadienyl anion by R_2N or RO and of HC(3) by RN or O generates charge-free resonance structures that undergo 1,5-electrocyclization forming ylides. These might be pseudopericyclic reactions because of the possible intramolecular nucleophilic addition of the NH_2 or RO group, instead of $[\pi^6 d]$ pericyclic reactions that can rearrange by group migration or by deprotonation/protonation into stable five-membered heterocyclic compounds [112], as illustrated herebelow with the hydrazone of (*E*)-pent-3-en-2-one that isomerizes readily into 3,5-dimethyl-4,5-dihydroazole (a 2-pyrazoline) at 20 °C [113].

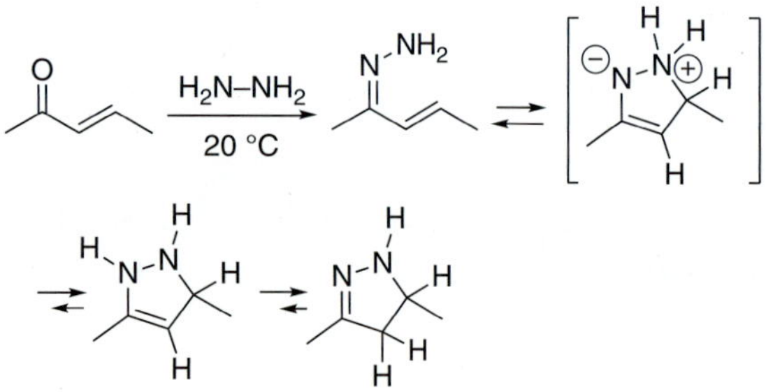

Exchange of HC(2) pentadienyl anion by NR or O produces conjugated 1,3-dipoles (Section 5.3.14) that cyclize into charge-free unsaturated five-membered rings as illustrated with the thermal rearrangement of 2-alkenylaziridines (e.g. *trans*-1-*t*-butyl-2-[(*E*)-prop-1-enyl]-3-phenylaziridine: **93**) into 2-pyrrolines (e.g. *trans*-1-*t*-butyl-4,5-dihydro-4-methyl-5-phenylazole: **94**) [114]. Five-membered sulfur heterocycles are formed by 1,5-electrocyclization of conjugated thiocarbonyl ylides [115].

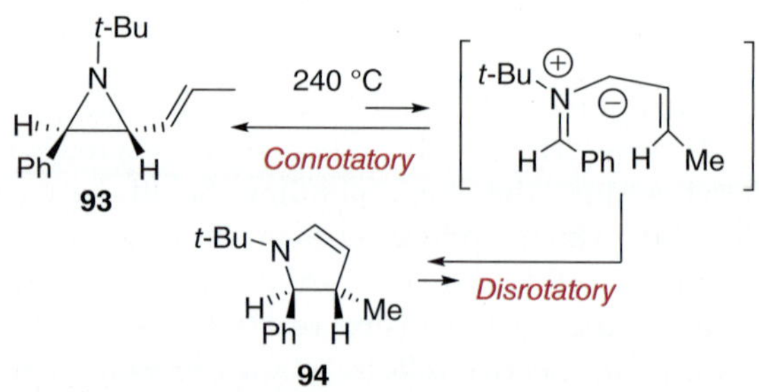

5.2.9 Eight-electron electrocyclic reactions

Upon heating, acyclic (*E*,*Z*,*Z*,*E*)-tetraenes **95** undergo conrotatory 8π-electron cyclization (noted $[\pi^8 c]$)

to corresponding cycloocta-1,3,5-trienes **96** (an example of 1,8-electrocyclization) as predicted by the Woodward–Hoffmann rules [116]. The torquoselectivity is dominated by steric factors, and the substituents at the termini carbon centers prefer *outward* rotation irrespective of their electronic nature [117]. Quantum mechanical calculations confirmed the Möbius aromatic character of the C_2-twisted transition structure [118–121]. In most cases, the [π^8c] electrocyclizations are followed by [π^6d] electrocyclizations to give the corresponding bicyclo[4.2.0]octa-2,4-dienes **97** [116].

This reaction sequence has been applied in the total synthesis of natural products [122, 123] and of fenestrane derivatives [124].

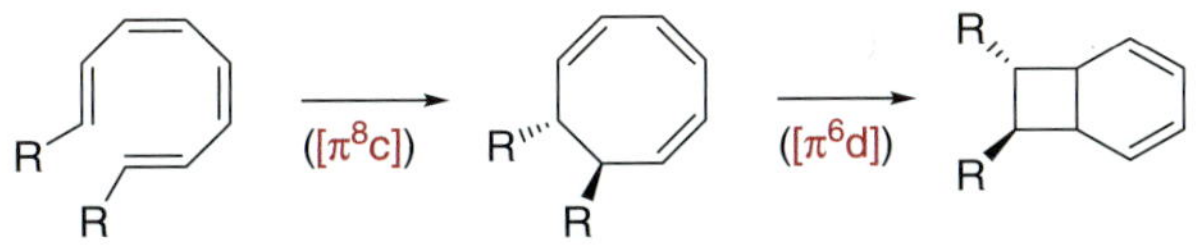

In some cases, tetraenes **95** undergo a first [π^6d] cyclization into corresponding 5-alkenylcyclohexa-1, 3-dienes, followed by intramolecular Diels–Alder reactions to produce tricyclo[3.2.1.0^{2,7}]oct-3-enes [125].

Problem 5.7 Propose a mechanism for the conversion of **A** into **P** [126].

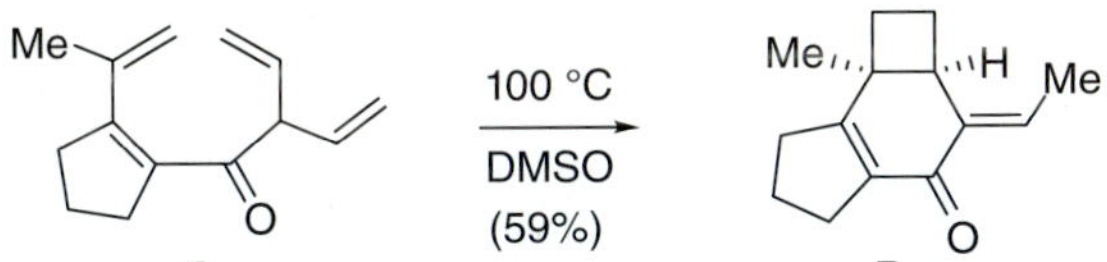

Problem 5.8 Potassium salt **B** is isomerized at −41 °C into **P**. Above 30 °C, **P** is isomerized into **Q** [127]. Propose mechanisms for these isomerizations.

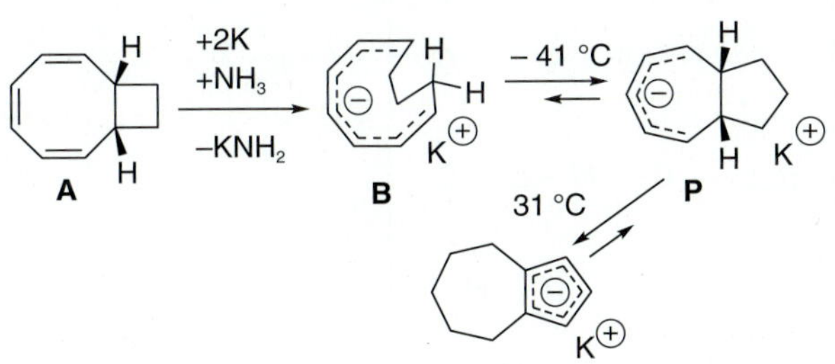

5.3 Cycloadditions and cycloreversions

A cycloaddition combines two or more compounds into a cyclic product (cycloadduct). Most cycloadditions involve unsaturated reactants (cycloaddents) with π double bonds that are used to make σ single bonds between the cycloaddents. There are also cycloadditions in which single σ bond of a cycloaddent is broken to generate other σ bonds in the cycloadducts. One classifies cycloadditions, and their reverse reactions the cycloreversions, by the number of atomic centers involved in the formation of the cyclic products (noted (m+n)-cycloadditions). A second classification considers the number of electrons involved in the pericyclic process (noted [x+y]-cycloadditions, Scheme 5.19).

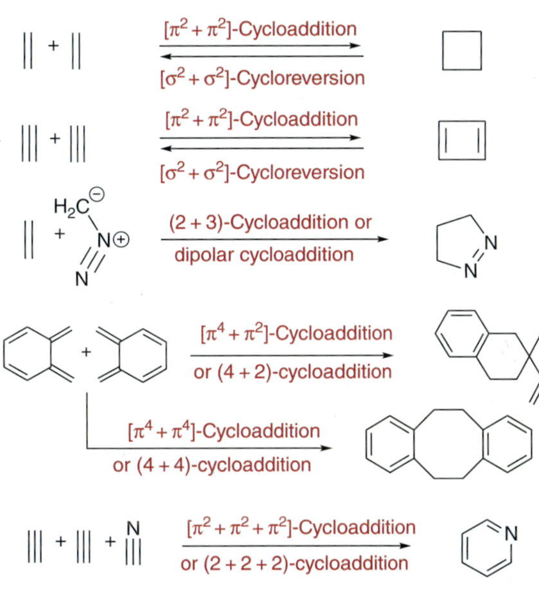

Scheme 5.19 Examples of (m+n)-cycloadditions.

When one cycloaddend has a single atomic center onto which two new σ bonds are formed in the cycloadduct, this type of cycloaddition was named a cheletropic addition by Woodward and Hoffmann. The reverse reaction is a cheletropic elimination (Section 5.4). For instance,

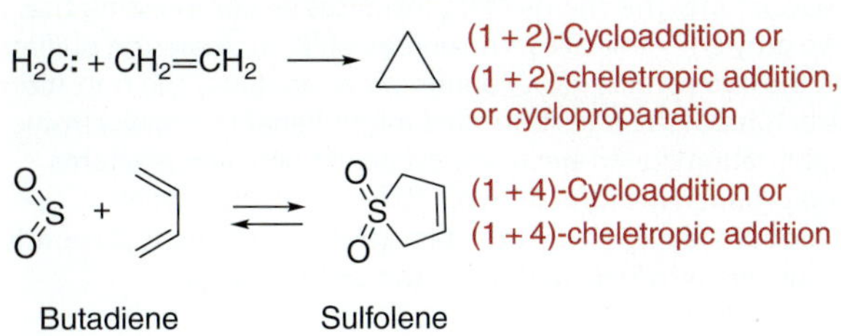

The combination of two carbenes into an alkene is a special case of (1+1)-cheletropic addition.

5.3.1 Stereoselectivity of thermal [$\pi^2+\pi^2$]-cycloadditions: Longuet-Higgins model

In the case of the [$\pi^2+\pi^2$]-cycloaddition of two alkenes giving a cyclobutane derivative, we can simplify the wave functions Ψ_R and Ψ_P by considering only the two π-HOMOs of the two alkenes (π_1 and π_2) and the two σ HOMOs of the two σ(C,C) bonds formed in the cyclobutane (σ_1 and σ_2). Thus, Ψ_R derives from two MOs combinations that are $\pi_+ = (1/\sqrt{2})\pi_1 + (1/\sqrt{2})\pi_2$ and $\pi_- = (1/\sqrt{2})\pi_1 - (1/\sqrt{2})\pi_2$. For the product wave function Ψ_P, one also has two MO combinations $\sigma_+ = (1/\sqrt{2})\sigma_1 + (1/\sqrt{2})\sigma_2$ and $\sigma_- = (1/\sqrt{2})\sigma_1 - (1/\sqrt{2})\sigma_2$ (Figure 5.14).

To a first approximation, the state wave functions can be taken as electronic ground-state configurations (Section 5.2.2), which are defined as combinations of MOs and considering their electron populations. Thus, $\Psi_R \approx [\pi_+]^2[\pi_-]^2$ and $\Psi_P \approx [\sigma_+]^2[\sigma_-]^2$. The wave function $\Psi_r^{\ddagger} = c_R^{\ddagger}\Psi_R + c_P^{\ddagger}\Psi_P$ (see Eq. (5.2)) of the transition structure of a [$\pi^2+\pi^2$]-cycloaddition will be a combination of MOs obtained by solving the Schrödinger equation for a system having four basis functions, the MO combinations π_+, π_-, σ_+, and σ_- for the geometry of the transition structure. Applying the variation theorem (Section 4.3), one finds four eigenfunctions $\Psi_r^{\ddagger}(i) = c_{i1}\pi_+ + c_{i2}\pi_- + c_{i3}\sigma_+ + c_{i4}\sigma_-$ and four eigenvalues E_1, E_2, E_3, and E_4. The energy $E^{\ddagger}$ of the transition structure will be $E^{\ddagger} = 2E_1 + 2E_2$ as four electrons populate the lowest energy eigenfunctions. If one designates by H_{11}, H_{22}, H_{33}, and H_{44} the Coulomb integrals associated with the basis orbitals π_+, π_-, σ_+, and σ_-, respectively, and by $H_{ij} = kS_{ij}$ (Mulliken assumption) the resonance integrals for $i \neq j$, the secular determinant of this transition structure is Eq. (5.3).

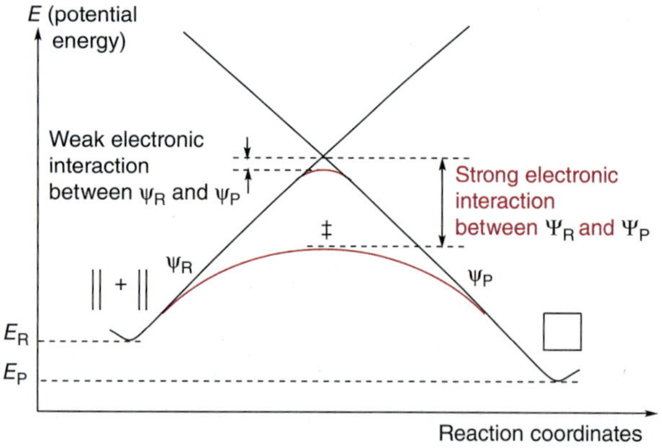

Figure 5.14 Plot of energy of reactants and products as a function of the reaction coordinates of the [$\pi^2+\pi^2$]-cycloaddition of two ethylene units giving cyclobutane. The energy of Ψ_R increases on compressing the two ethylene molecules. The energy of Ψ_P increases on pulling on the two parallel σ(C,C) bonds of cyclobutane. The transition structures in the transition state might benefit from electronic stabilization due to electronic exchange between reactants and product: the formation of the two σ(C,C) bonds of cyclobutane might assist the breaking of the two π(C,C) bonds of the two ethylene moieties in the cycloaddition.

$$\begin{vmatrix} H_{11}-E & H_{12}-ES_{12} & H_{13}-ES_{13} & H_{14}-ES_{14} \\ H_{21}-ES_{21} & H_{22}-E & H_{23}-ES_{23} & H_{24}-ES_{24} \\ H_{31}-ES_{31} & H_{32}-ES_{32} & H_{33}-E & H_{34}-ES_{34} \\ H_{41}-ES_{41} & H_{42}-ES_{42} & H_{43}-ES_{43} & H_{44}-E \end{vmatrix} = 0 \tag{5.3}$$

For the one-step, concerted and synchronous suprafacial/suprafacial cycloaddition of two alkenes (the two σ(C,C) bonds formed or broken in the transition structure have the same length and the same bond order (Figure 5.15a; noted [π^2s+π^2s]), the basis MO combinations are $\pi_+(a_1)$, $\pi_-(b_2)$, $\sigma_+(a_1)$, and $\sigma_-(b_1)$ as the transition structure belongs to the C_{2v} point group of symmetry (one C_2 axis and one mirror plane of symmetry containing it).

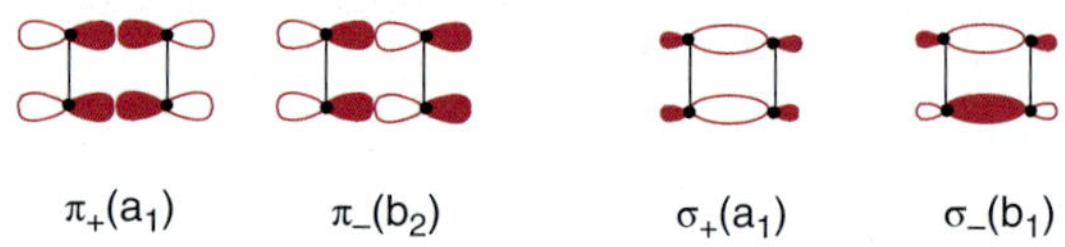

There is an overlap only between orbitals of the same symmetry, i.e. between $\pi_+(a_1)$ and $\sigma_+(a_1)$. Thus, $S_{12} = S_{21} = S_{14} = S_{41} = 0$ and $S_{13} = S_{31} \neq 0$, leading to $H_{13} - ES_{13} = H_{31} - ES_{13} = \beta_{\pi\sigma}$. If one assumes $H_{11} = H_{22} = H_{33} = H_{44} = H$, the secular determinant (5.3) becomes (5.4):

$$\begin{vmatrix} H-E & 0 & \beta_{\pi\sigma} & 0 \\ 0 & H-E & 0 & 0 \\ \beta_{\pi\sigma} & 0 & H-E & 0 \\ 0 & 0 & 0 & H-E \end{vmatrix} = 0 \tag{5.4}$$

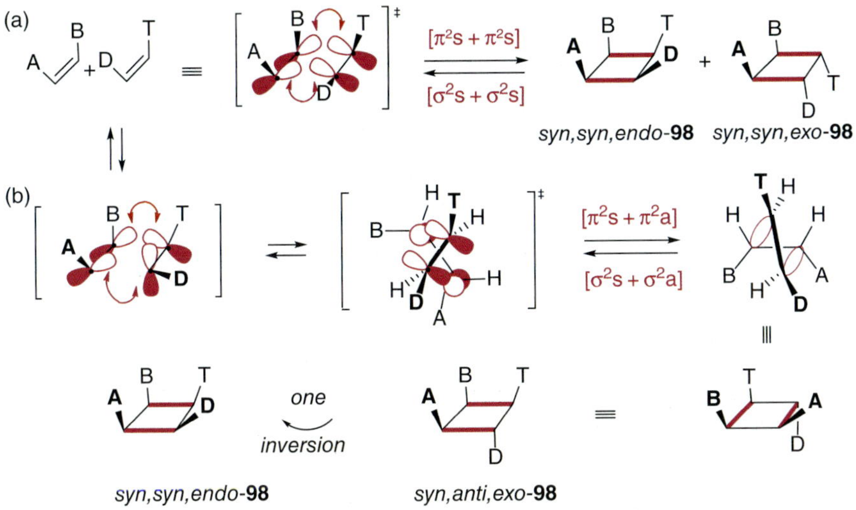

Figure 5.15 Representation of two possible reaction pathways for the $[\pi^2+\pi^2]$-cycloaddition of two ethylene units (a) in a suprafacial/suprafacial approach (noted: $[\pi^2s+\pi^2s]$). (b) In a suprafacial/antarafacial approach (noted $[\pi^2s+\pi^2a]$). Note that cycloadducts *syn,syn,endo*-**98** and *syn,anti,exo*-**98** can be interconverted by epimerization (inversion) of one center.

This gives $(H-E)^4-(\beta_{\pi\sigma})^2(H-E)^2 = 0$ for which the four roots are $E_1 = H+\beta_{\pi\sigma}$, $E_2 = E_3 = H$, and $E_4 = H-\beta_{\pi\sigma}$. As we have four electrons to populate the lowest energy orbitals $\Psi_r^{\ddagger}(i) = c_{i1}\pi_+ + c_{i2}\pi_- + c_{i3}\sigma_+ + c_{i4}\sigma_-$ so-obtained, the total energy $E^{\ddagger} = 2(H+\beta_{\pi\sigma})+2H = 4H+2\beta_{\pi\sigma}$. With respect to the reactants and products, the transition structure experiences electronic stabilization of $2\beta_{\pi\sigma}$ (resonance energy that compensates partially the steric repulsion between reactants and their deformation energy).

For the suprafacial/antarafacial approach (noted $[\pi^2s+\pi^2a]$, Figure 5.15b), the basis MO combinations are $\pi_+(a)$, $\pi_-(b)$, $\sigma_+(a)$, and $\sigma_-(b)$ as the transition structure belongs to the C_2 point group of symmetry.

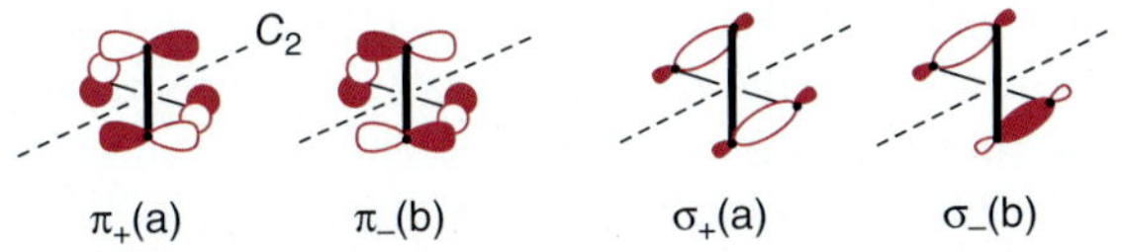

We now have two pairs of orbitals that overlap: $\pi_+(a)/\sigma_+(a)$ and $\pi_-(b)/\sigma_-(b)$. This leads to two resonance integrals $\beta_{\pi\sigma}$ and the secular determinant (5.5) if one makes the same simplifications as above for the $[\pi^2s+\pi^2s]$-cycloaddition.

$$\begin{vmatrix} H-E & 0 & \beta_{\pi\sigma} & 0 \\ 0 & H-E & 0 & \beta_{\pi\sigma} \\ \beta_{\pi\sigma} & 0 & H-E & 0 \\ 0 & \beta_{\pi\sigma} & 0 & H-E \end{vmatrix} = 0 \quad (5.5)$$

This gives $(H-E)^4+(\beta_{\pi\sigma})^4-2(\beta_{\pi\sigma})^2(H-E)^2 = 0$, or $(H-E)^4-2(\beta_{\pi\sigma})^2(H-E)^2+(\beta_{\pi\sigma})^4 = 0$ for which the four roots are $E_1 = E_2 = H+\beta_{\pi\sigma}$ and $E_3 = E_4 = H-\beta_{\pi\sigma}$. The four electrons involved in the transition structure associated with the transition state of the $[\pi^2s+\pi^2a]$-cycloaddition occupy the two lowest energy orbitals with $E_1 = E_2 = H+\beta_{\pi\sigma}$, which gives $E^{\ddagger} = 4H+4\beta_{\pi\sigma}$. Thus, the transition structure of the $[\pi^2s+\pi^2a]$-cycloaddition involves electronic stabilization that amounts to $4\beta_{\pi\sigma}$. This is $2\beta_{\pi\sigma}$ better than for the concerted and synchronous $[\pi^2s+\pi^2s]$-cycloaddition. If one compares with Hückel π-systems, the transition structure of the $[\pi^2s+\pi^2s]$-cycloaddition corresponds to a butadiene for which overlap is possible only between two adjacent 2p AOs, as in perpendicular but-2-en-1,4-diyl diradical. In this case, $E_\pi = 4\alpha+2\beta$. The transition structure of the concerted, synchronous $[\pi^2s+\pi^2a]$-cycloaddition compares with *s-gauche* butadiene in which the two ethylene units do not conjugate and for which $E_\pi = 4\alpha+4\beta$.

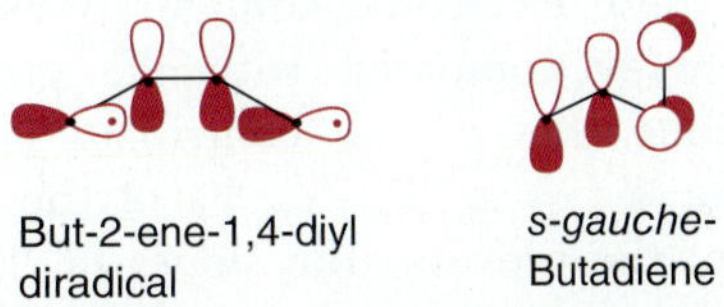

The crossing point of the reactant and product wave functions Ψ_R and Ψ_P in Figure 5.14 corresponds to an energy value that depends on the slopes of these functions. As we shall see (Figure 5.18), the energy of Ψ_R rises sharply when going from two ethylene units to a cyclobutane following the $[\pi^2s+\pi^2s]$

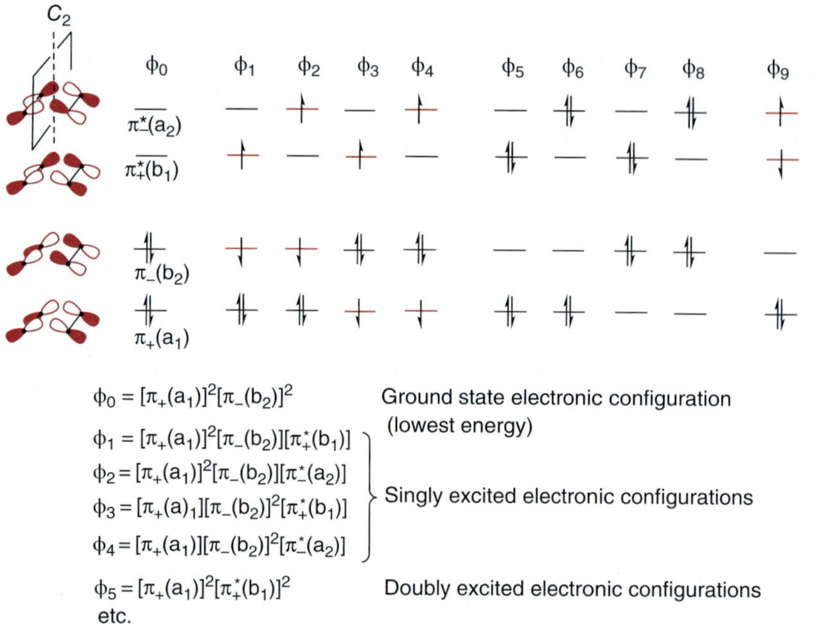

Figure 5.16 Representation of selected electronic configurations of a chemical system constituted of two ethylene units undergoing a concerted, synchronous [π^2s+π^2s]-cycloaddition (C_{2v} point group of symmetry: a, symmetrical; b; antisymmetrical with respect to the C_2 axis; subscript 1, symmetrical; and subscript 2, antisymmetrical with respect to the mirror plane of symmetry that contains the C_2 axis of symmetry).

approach. Similarly, the energy of Ψ_P rises sharply when going from cyclobutane to two ethylene units in a [σ^2s+σ^2s]-cycloreversion. The Longuet-Higgins and Abrahamson diagram of Figure 5.18 correlates the electronic configurations of product and reactants for the [π^2s+π^2s]-cycloaddition (and cycloreversion). They are built (Figure 5.16) as combinations of π_+, π_- (ethylene HOMOs) and π_+^*, π_-^* (ethylene LUMOs). Similarly, one constructs the electronic configurations of cyclobutane considering exclusively the two σ(C,C) bonds that are formed using σ_+, σ_- (cyclobutane HOMOs) and σ_+^*, σ_-^* (cyclobutane LUMOs).

On comparing the electronic configurations ϕ_0, ϕ_1, …, ϕ_i of reactants (two ethylene units) and φ_0, φ_1, …, φ_i of cycloadduct (cyclobutane) undergoing a [π^2s+π^2s]-cycloaddition, one sees that the ground-state configurations ϕ_0 and φ_0 of reactants and products, respectively, do not share the same symmetry (Figure 5.17). These configurations do not correlate by symmetry. They correlate only with the doubly excited configurations φ_5 and ϕ_5, respectively. As these electronic configurations are very high in energy compared with the ground-state configurations, they do not contribute significantly to the state $\Psi_r^\ddagger = c_R{}^\ddagger\Psi_R + c_P{}^\ddagger\Psi_P$ of the transition structure of the process. The slopes of Ψ_R and Ψ_P are steep and thus cross at a very high energy level (Figures 5.14 and 5.18). This thermal cycloaddition is difficult: it has no significant assistance between bond-forming and bond-breaking processes. According to the Woodward–Hoffmann rules, the concerted, synchronous [π^2s+π^2s]-cycloaddition is "forbidden" under thermal conditions (Figure 5.19a). Interestingly, the first excited electronic configurations φ_1 and ϕ_1 correlate by symmetry. This shows that concerted, synchronous [π^2s+π^2s]-cycloadditions are easy photochemically.

For a concerted, synchronous suprafacial/antarafacial [π^2s+π^2a]-cycloaddition, the ground-state electronic configurations $\phi_0(a)^2(b)^2$ of reactants and $\varphi_0(a)^2(b)^2$ of product share the same C_2-symmetry. They are correlated by symmetry and thus the reaction must be much easier than the [π^2s+π^2s]-cycloaddition under thermal conditions [128].

5.3.2 Woodward–Hoffmann rules for cycloadditions

Instead of correlating electronic configurations (first approximations to the states), Woodward and Hoffmann correlate MOs of reactants and products. In the case of the [π^2s+π^2s]-cycloaddition, one obtains the orbital correlation diagram shown in Figure 5.19.

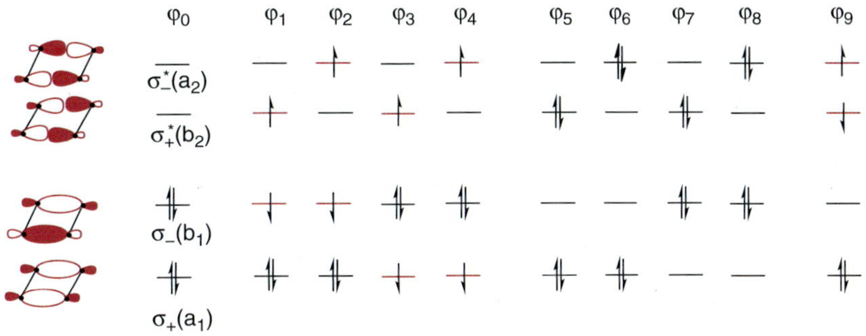

$\varphi_0 = [\sigma_+(a_1)]^2[\sigma_-(b_1)]^2$ Ground state electronic configuration (lowest energy)

$\varphi_1 = [\sigma_+(a_1)]^2[\sigma_-(b_1)][\sigma_+^*(b_2)]$
$\varphi_2 = [\sigma_+(a_1)]^2[\sigma_-(b_1)][\sigma_-^*(a_2)]$
$\varphi_3 = [\sigma_+(a_1)][\sigma_-(b_1)]^2[\sigma_+^*(b_2)]$
$\varphi_4 = [\sigma_+(a_1)][\sigma_-(b_1)]^2[\sigma_-^*(a_2)]$
} Singly excited electronic configurations

$\varphi_5 = [\sigma_+(a_1)]^2[\sigma_+^*(b_2)]^2$ Doubly excited electronic configuration

etc.

Figure 5.17 Representation of selected electronic configurations of a chemical system constituted of two parallel σ(C,C) bonds of a cyclobutane undergoing a concerted, synchronous $[\sigma^2s+\sigma^2s]$-cycloreversion.

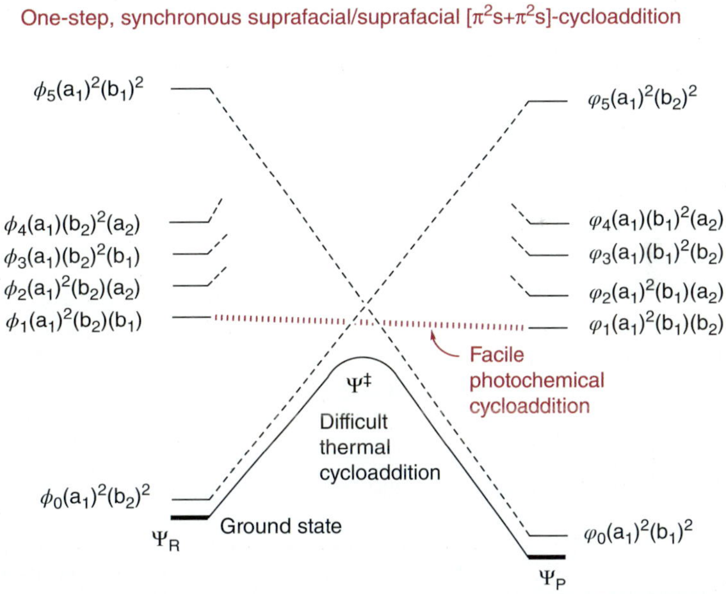

Figure 5.18 Longuet-Higgins and Abrahamson correlation of electronic configurations of reactants (two ethylene units) and product (cyclobutane) of a $[\pi^2s+\pi^2s]$-cycloaddition. Reactant ground state wave function Ψ_R can be approximated by applying the variation theorem in solving the Schrödinger equation for $\Psi_R = c_0\phi_0 + c_5\phi_5$. Similarly, ground-state configuration of product is given by $\Psi_P = c_0'\phi_0 + c_5'\phi_5$. Admixture of monoexcited electronic configurations into Ψ_R and Ψ_P is not possible for reasons of symmetry.

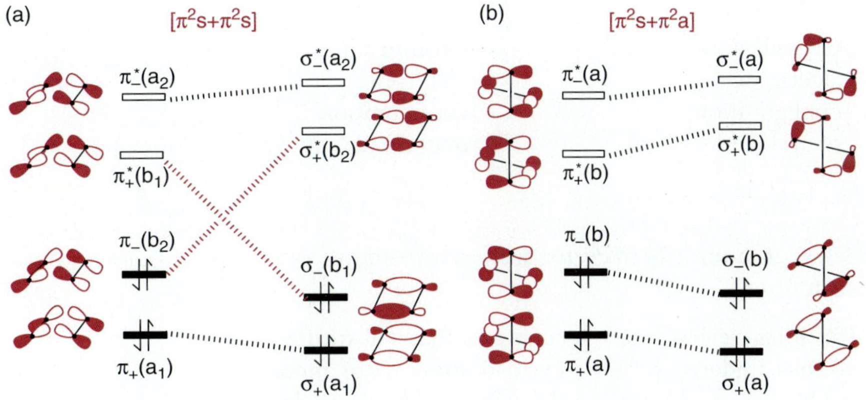

Figure 5.19 Woodward and Hoffmann orbital correlation diagram for (a) a synchronous suprafacial/suprafacial $[\pi^2s+\pi^2s]$-cycloaddition and (b) for a synchronous suprafacial/antarafacial $[\pi^2s+\pi^2a]$-cycloaddition.

One sees that filled $\pi_+(a_1)$ MO combination correlates by symmetry with the filled $\sigma_+(a_1)$ MO combination, and the filled $\pi_-(b_2)$ MO combination correlates by symmetry with empty $\sigma_+^*(b_2)$ MO combination in the case of the $[\pi^2s+\pi^2s]$-cycloaddition (Figure 5.19a). Similarly, filled $\sigma_-(b_1)$ MO combination correlates by symmetry with empty $\pi_+^*(b_1)$ MO combination. In contrast, for the $[\pi^2s+\pi^2a]$-cycloaddition, both filled $\pi_+(b)$ and $\pi_-(a)$ MO combinations of reactants correlate by symmetry with both filled $\sigma_-(b)$ and $\sigma_+(a)$ MO combinations of product, respectively (Figure 5.19b). Woodward and Hoffmann say that concerted and synchronous $[\pi^2s+\pi^2s]$-cycloadditions are "forbidden," whereas $[\pi^2s+\pi^2a]$-cycloadditions are allowed under thermal conditions. The $[\pi^2s+\pi^2s]$-cycloaddition is said to be "allowed photochemically."

Most Diels–Alder reactions follow suprafacial/suprafacial stereochemistry (noted $[\pi^4s+\pi^2s]$) [129]. In general, they are faster than $[\pi^2s+\pi^2s]$-cycloadditions. The MO correlation diagram for concerted and synchronous $[\pi^4s+\pi^2s]$-cycloaddition of butadiene (diene) + ethylene (dienophile) giving cyclohexene (cycloadduct) is shown in Figure 5.20. All occupied reactant MOs correlate by symmetry (mirror plane of symmetry maintained in the transition structure) with occupied product MOs. Woodward and Hoffmann say that this reaction is "allowed" (facile) under thermal conditions. The ground state (or ground electronic configuration approximating it) of reactants (*s-cis*-butadiene + ethylene) also correlate by symmetry with the ground state (or ground electronic configuration approximating it) of cycloadduct (boat conformation of cyclohexene). Thus, one-step, concerted, and synchronous (both new σ bonds are formed to the same extent in the transition structure) Diels–Alder reactions are "allowed" in the ground state:

$$[\pi_1(a')]^2[\pi(a')]^2[\pi_2(a'')]^2 \xleftarrow{\text{same}} \text{symmetry} \xrightarrow[\text{as}]{} [\sigma_+(a')]^2[\sigma_-(a'')]^2[\pi(a')]^2$$

Ground state electronic configuration of cycloaddents		Ground state electronic configuration of cycloadduct

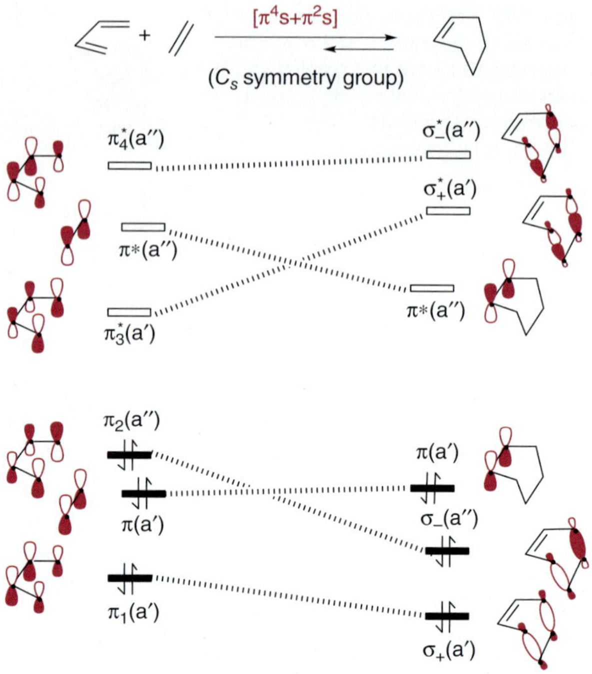

Figure 5.20 Woodward–Hoffmann molecular orbital correlation diagram for the suprafacial/suprafacial Diels–Alder reactions. They are "allowed" under thermal conditions. The transition structure of the $[\pi^4s+\pi^2s]$-cycloaddition shares a plane of symmetry.

5.3.3 Aromaticity of cycloaddition transition structures

For long, it has been recognized that substituted alkenes undergo $[\pi^4+\pi^2]$-cycloadditions (5.6) more readily than $[\pi^2+\pi^2]$-cycloadditions (5.7) under thermal conditions. In 1939, Evans realized that transition structures of Diels–Alder additions ($[\pi^4s+\pi^2s]$-cycloadditions) are isoelectronic with the benzene π-system, whereas transition structures of $[\pi^2s+\pi^2s]$-cycloadditions are isoelectronic with the cyclobutadiene π-system. According to the "Evans' rule," the transition structure of $[\pi^4s+\pi^2s]$-cycloadditions benefit from electronic stabilization like Hückel [6]-annulenes, or benzene, whereas the transition structures of $[\pi^2s+\pi^2s]$-cycloadditions cannot be stabilized electronically in the same way [130].

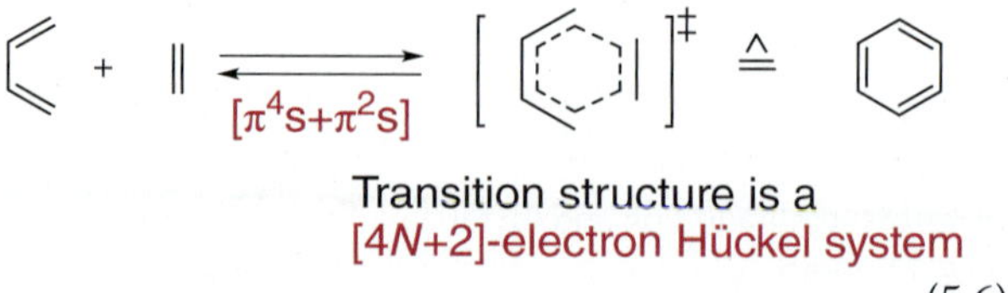

(5.6)

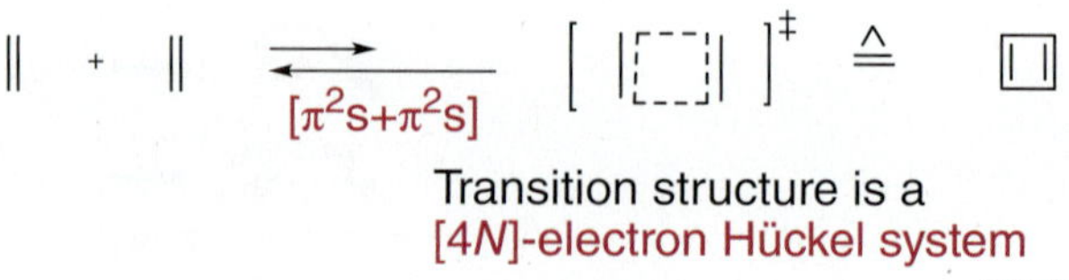

(5.7)

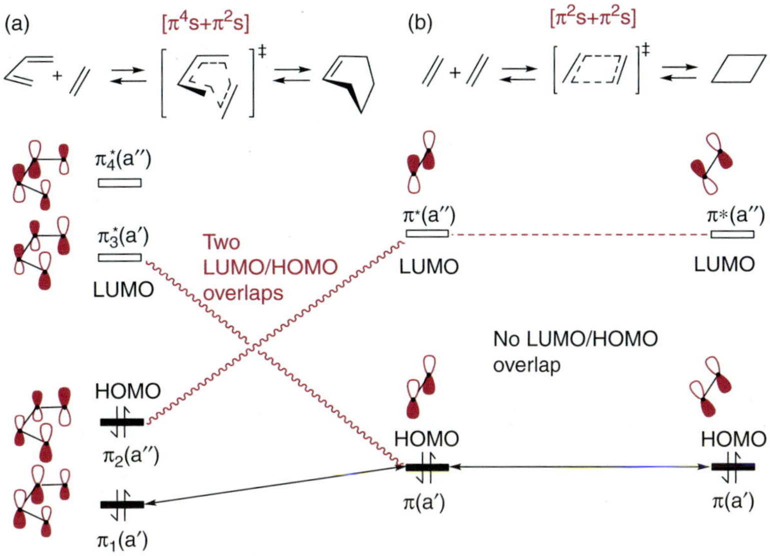

Figure 5.21 PMO diagram showing (a) the origin of the electronic stabilization (aromaticity) of the transition structure of a $[\pi^4s+\pi^2s]$-cycloaddition and (b) the absence of electronic stabilization in the transition structure of a $[\pi^2s+\pi^2s]$-cycloaddition. A mirror plane of symmetry cutting both cycloaddents is present in both transition structures.

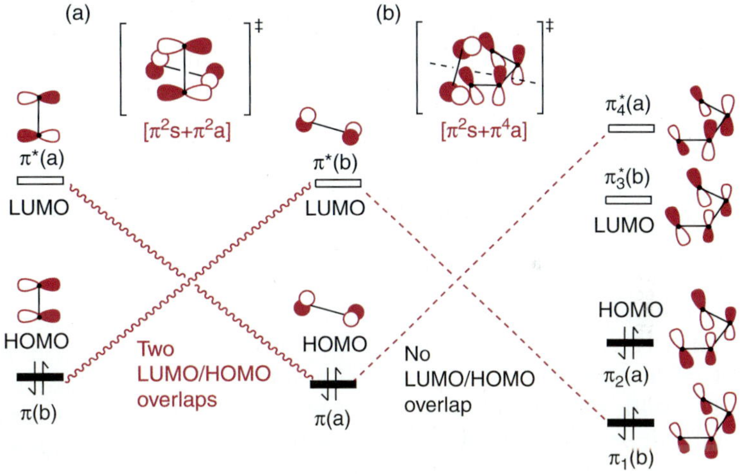

Figure 5.22 PMO diagram for the interactions (a) between π MOs of two alkene units undergoing a suprafacial/antarafacial $[\pi^2s+\pi^2a]$-cycloaddition and (b) between the π MOs of an antarafacial/suprafacial $[\pi^4a+\pi^2s]$-cycloaddition. Both reactions maintain a C_2-axis of symmetry in their transition structures. The $[\pi^2s+\pi^2a]$-cycloaddition benefits from a larger electronic stabilization in its transition structure than the $[\pi^4a+\pi^2s]$-cycloaddition.

The FMO theory supports these hypotheses. In the transition structure of a Diels–Alder addition, the π MOs of the cycloaddends overlap by their termini as shown in Figure 5.21a.

Considering the C_s group of symmetry of such transition structure, one recognizes that the destabilizing four-electron sub-HOMO(diene)/HOMO(dienophile) interaction must be overwhelmed by the two stabilizing two-electron LUMO(dienophile)/HOMO(diene) and LUMO(diene)/HOMO(dienophile) interactions. As for benzene (Section 4.5.10), the two latter interactions are responsible for the electronic stabilization of the transition structure: they diminish the energy barrier of the reaction arising from steric repulsion between reactants and skeleton deformations [22, 131–133]. In the $[\pi^2s+\pi^2s]$-cycloaddition, the transition structure does not benefit from any electronic stabilization as there are no LUMO/HOMO interactions between the π MOs of the cycloaddends (Figure 5.21b). An overlap between both HOMOs leads to an electronic destabilization (Section 4.5.6). In contrast, the transition structure of a concerted, synchronous suprafacial/antarafacial $[\pi^2s+\pi^2a]$-cycloaddition benefits from an electronic stabilization because of two LUMO/HOMO interactions involving the two alkene units (Figure 5.22a).

The transition structure of a hypothetical antarafacial/suprafacial $[\pi^4a+\pi^2s]$-cycloaddition does not share a mirror plane of symmetry but a C_2-axis of symmetry (Figure 5.22b). Contrary to the Diels–Alder reaction (Figure 5.20), there are no LUMO/HOMO interactions between the π MOs of the 1,3-diene and dienophile (alkene and alkyne) of a antarafacial/suprafacial $[\pi^4a+\pi^2s]$-cycloaddition (Scheme 5.20). There is a much weaker electronic stabilization involving LUMO(dienophile)/sub-HOMO

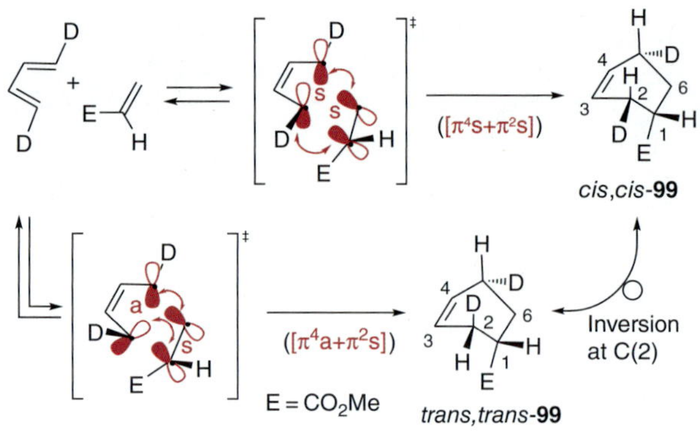

Scheme 5.20 Example of a Diels–Alder reaction ($[\pi^4s+\pi^2s]$-cycloaddition) and of a hypothetical $[\pi^4a+\pi^2s]$-cycloaddition of the same cycloaddents. Note that cycloadducts *cis,cis*-**99** and *trans,trans*-**99** are diastereomers that are interconverted (epimerized) by inversion of center C(2).

(diene) and supra-LUMO(diene)/HOMO(dienophile) interactions. For steric and geometrical reasons, $[\pi^4a+\pi^2s]$- and $[\pi^4s+\pi^2a]$-cycloadditions are much more difficult than $[\pi^4s+\pi^2s]$-cycloadditions as they imply important geometrical distortions of the cycloaddents to realize these transition structures.

As for the $[\pi^4c]$ ring opening of cyclobutene into butadiene, Zimmerman recognized that the transition structures of $[\pi^2s+\pi^2a]$-cycloadditions have Möbius cyclic arrays of atomic orbitals [23, 134]. Accordingly, these transition structures that involve four electrons is stabilized electronically because of Heilbronner–Möbius aromaticity (Section 4.9) [24]. With this observation, the concept of Hückel and Möbius aromaticity of transition structures of cycloadditions (and other pericyclic reactions) became equivalent to the Woodward and Hoffmann rules [26], with the advantage that transition structures do not have to share any element of symmetry to be either stabilized or not by an electronic factor. As a consequence (Figure 5.23), cycloadditions with transition structures that realize Hückel types of atomic orbital arrays are facile ("allowed") if they involve $4N+2$ electrons and difficult ("forbidden") if they involve $4N$ electrons. Similarly, cycloadditions with transition structures realizing Möbius arrays of atomic orbitals are facile ("allowed") if they involve $4N$ electrons and difficult ("forbidden") if they involve $4N+2$ electrons.

5.3.4 Mechanism of thermal $[\pi^2+\pi^2]$-cycloadditions and $[\sigma^2+\sigma^2]$-cycloreversions: 1,4-diradical/zwitterion intermediates or diradicaloid transition structures

Thermochemical data (Table 1.2) for the cyclodimerization (5.8) of ethylene into cyclobutane (gas phase, standard conditions) give an exothermicity of $\Delta_r H° = -18.4$ kcal mol^{-1} and a reaction entropy variation $\Delta_r S° = -41.6$ cal K^{-1} mol^{-1} (eu). Thus, for 1 M concentration in solution or one atm. pressure for gaseous ethylene, the equilibrium constant $K = 1$ ($\Delta_r G° = \Delta_r H° - T\Delta_r S° = 0$) is realized at a relatively low temperature estimated to $\Delta_r H°/\Delta_r S° \approx 18\,400/41.6 \approx 442$ K = 169 °C. Above this temperature, the cyclodimerization is endergonic ($K < 1$). The cycloreversion of cyclobutane into ethylene occurs between 350 and 450 °C with a barrier $\Delta^‡H = 61.1$ kcal mol^{-1}: it involves the formation of tetramethylene diradical intermediate (**100**). The reaction is not concerted; there is no assistance between the σ-bond breaking and π-bond forming processes. The thermally "allowed" concerted $[\pi^2s+\pi^2a]$ cycloreversion is not followed as it implies too severe distortions of the cyclobutane system to reach the twisted transition structures predicted by the Longuet-Higgins theory and the Woodward Hoffmann rules. The thermal $[\pi^2+\pi^2]$-cyclodimerization of ethylene has thus an energy barrier $\Delta^‡H$ corresponding to the formation diradical **100** (principle of microscopic reversibility: the transition states is the same, the transition structures are the same for the forward and reverse reactions under the same conditions) for which one calculates $\Delta_f H°$ (**100**) = $\Delta_f H°$ (cyclobutane) + $DH°(Et^•/Et^•)$ − ring strain of cyclobutane = 6.6 + 86.0 − 26.5 = 66.1 kcal mol^{-1}. Thus, a barrier $\Delta^‡H \geq 66.1 - 2(12.5) \approx 41$ kcal mol ^{1} is estimated for the cyclodimerization of ethylene. The entropy of activation $\Delta^‡S$ of this condensation is that of the reaction $\Delta_r S°$ corrected for the three free rotations about the three single C—C bonds of the 1,4-diradical (three times 5 eu, Section 2.10). One thus estimates $\Delta^‡S = -41.6 + 15 \approx -27$ eu, which leads to $\Delta^‡G = \Delta^‡H - T\Delta^‡S = 41 - 442(-0.027) \approx 53$ kcal mol^{-1}. Using Eq. (3.67), $\ln k = -\Delta^‡G/RT + \ln T + 23.76 = -53\,000/1.987 \cdot 442 + \ln(442) + 23.76 \approx -30.7$, one obtains $k \approx 4.6 \times 10^{-14}$ dm^{3}mol^{-1}s^{-1}, or a half-life (time for 50% conversion, Table 3.1) $\tau_{1/2} = 1/k$ [initial concentration of ethylene = 1] ≈ 6 82 400 years! This reaction is too slow to be observed at 442 K.

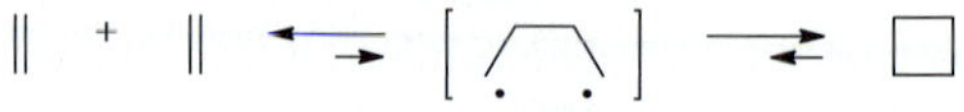

100

$\Delta^‡H$(5.8) = 41 kcal mol^{-1} $\Delta_r H°$(5.8) = −18.4 kcal mol^{-1}
$\Delta^‡S$(5.8) = −28 eu $\Delta_r S°$(5.8) = −41.6 eu

(5.8)

Diradical **100** can be stabilized by substitution at C(1) and/or C(4) (see Table 1.A.9 for the substituent effects on the relative stability of radicals). Thus, the thermal $[\pi^2+\pi^2]$-cycloadditions of alkenes bearing radical stabilizing substituents are predicted to be easier than the cyclodimerization of

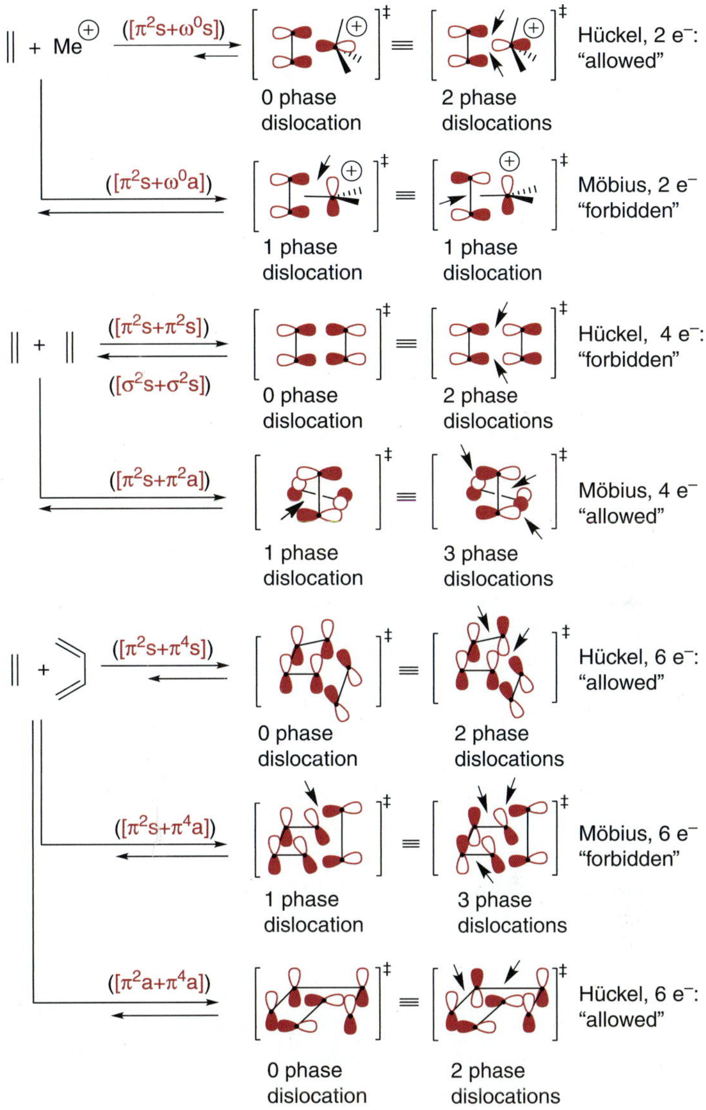

Figure 5.23 Examples of thermal cycloadditions realizing Hückel (0, 2, 4, 6, …, phase dislocations) or Möbius atomic orbitals arrays (1, 3, 5, 7, …, phase dislocations) in their transition structures.

ethylene [135]. For instance, Reed [136] found c. 5% of *trans*-1,2-divinycyclobutane (*trans*-102) (reaction (5.9)) next to the major Diels–Alder (Section 5.3.8) cyclodimer 4-vinylcyclohexene [137] and a small amount of cycloocta-1,5-diene upon heating butadiene in an autoclave at 150 °C for 18 hours in the presence of hydroquinone (radical scavenger: inhibits the polymerization of butadiene) (Scheme 5.21) [138]. In 1956, Vogel [139] showed that cycloocta-1,5-diene arises from a Cope rearrangement (Section 5.5.9.6) of *cis*-1,2-divinylcyclobutane (*cis*-**102**) that forms competitively during the cyclodimerization of butadiene. Both *trans*- and *cis*-1,2-divinylcyclobutane arise from the condensation of butadiene into (*E*,*E*)-octa-1,7-dien-3,6-diyl diradical (**101**). If one considers the vinyl substituent effect on primary alkyl radicals to be −13 kcal mol^{-1} ($DH°(n\text{-prop}^\bullet/H^\bullet) - DH°(\text{allyl}^\bullet/H^\bullet) = 101.1 - 88.1 =$ 13 kcal mol^{-1}, Table 1.A.7) and a difference in relative stability between but-1-yl and but-2-yl radical of −2 kcal mol^{-1} (Table 1.A.7), tetramethylene diradical (**100**) is stabilized by 30 kcal mol^{-1} upon divinyl substitution in **101**. Assuming the same activation entropy than for the equilibrium of ethylene with **100**, the activation free energy of the formation of **101** from butadiene must be $\Delta^{\ddagger}G \geq 53.4 - 30 \geq 23.4$ kcal mol^{-1}. At 150 °C, the rate of formation of this intermediate is estimated (Eq. (3.67)) to $\ln k \leq -23\,400/1.987(423) + \ln(423) + 23.76 = 1.967$, or $k \leq 7.1$ M^{-1} s^{-1}. This highest limit for the rate constant of the formation of diradical **101** at 150 °C is compatible with the mechanism proposed.

In the presence of a radical scavenger such as 2-nitrophenol, styrene and acrylonitrile are cyclodimerized into the corresponding 1,2-disubstituted cyclobutanes with relatively low energy barriers and

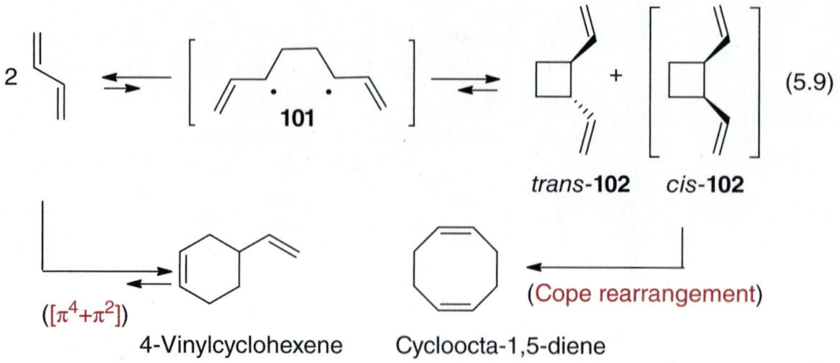

Scheme 5.21 Thermal cyclodimerization of butadiene in the presence of radical scavengers.

consistently with a mechanism involving the rate-determining formation of the corresponding 1,4-disubstituted tetramethylene diradicals [140]. For the cyclodimerization of *cis*-1,2-dideuterioacrylonitrile, a stereorandom process is observed that does not support "even a vestige of concerted [π^2s+π^2a] reaction" [141]. At 0 °C, bicyclo[2.2.0]hex-1(4)-ene (**103**) undergoes quick polymerization and dimerization into **106**. The latter reaction has an enthalpy of activation of only $\Delta^{\ddagger}H = 11.5$ kcal mol^{-1} (second-order rate law reaction with $\Delta^{\ddagger}S = -25$ eu) corresponding to the formation of diradical **104** that cyclizes into **105** and, on its turn, undergoes [σ^2+σ^2]-cycloreversion to **106** [142]. Strain of alkene **103** is relieved upon forming **104** due to pyramidalization of C(1) and C(4) [143]. The perfluorinated analog of **103** does not cyclodimerize because the reaction is endothermic [144].

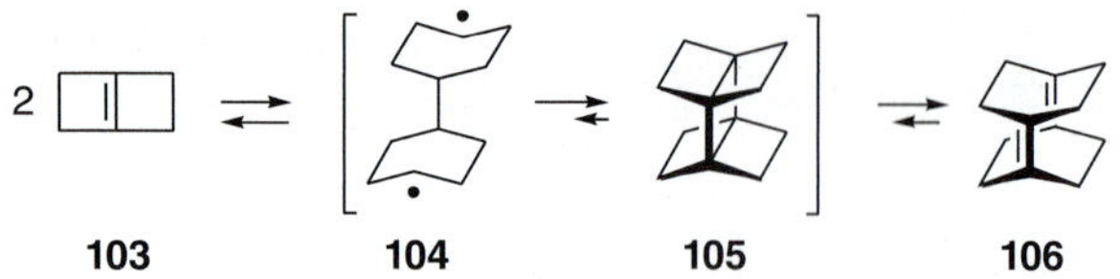

Kinetics for the gas phase cyclodimerization (5.10) of tetrafluoroethylene gave $\Delta^{\ddagger}H = 23.7 \pm 1.0$ kcal ;mol^{-1} and $\Delta^{\ddagger}S = -34.0 \pm 1.8$ eu [145]. Table 1.A.7 shows that difluorination of an alkyl radical stabilizes it by only 1–2 kcal mol^{-1} ($DH°(F_2CH^•/H^•) = 103$ kcal mol^{-1}; $DH°(Me^•/H^•) = 104.7$ kcal mol^{-1}). To a first approximation, 1,1,4,4-tetraflurotetramethylene diradical (**107**) should not be stabilized by more than 4 kcal mol^{-1} with respect to **100**. Compared with the cyclodimerization of ethylene that has an activation enthalpy $\Delta^{\ddagger}H \approx 41$ kcal mol^{-1}, cyclodimerization of $CF_2{=}CF_2$ should be easier but has an activation enthalpy of c. 37 kcal mol^{-1}, not 23.7 kcal mol^{-1} (Scheme 5.22). This is without accounting for the special fluoro substituent effects on the relative stability of alkenes and alkanes (the heat of hydrogenation of $CH_2{=}CF_2$ into $CH_3{-}CHF_2$ amounts to −38.7 kcal mol^{-1}, to be compared with −32.5 kcal mol^{-1} for the standard heat of hydrogenation of ethylene; Section 1.7.1). The cyclodimerization of $CF_2{=}CF_2$ into octafluorocyclobutane is much more exothermic (−40.9 kcal mol^{-1}) than the cyclodimerization of ethylene (−18.4 kcal mol^{-1}). Thus, diradical **107** gains part of this differential stabilization effect of the fluoro substitution that stabilizes sp^3-hybridized carbon centers with respect to sp^2-hybridized carbon centers. In the case of the cyclodimerization (5.11) of $ClCF{=}CF_2$, the reaction is highly regioselective and its enthalpy of activation ($\Delta^{\ddagger}H = 23.7$ kcal mol^{-1}) is the same as for the cyclodimerization of $CF_2{=}CF_2$. These results are interpreted in terms of the formation of diradical **108** in the rate-determining step. Regioisomeric diradical **109** does not form (reaction (5.12)), probably because it contains a $ClCF{-}CF_2$ moiety instead of a $CF_2{-}CF_2$ moiety as in **108** (Scheme 5.22): four fluoro substituents at sp^3-carbon centers are more stabilizing than three fluoro substituents. It must be noted here that a chloro is not better than a fluoro substituent in stabilizing an alkyl radical ($DH°(ClCH_2^•/H^•) = 102.7$ kcal mol^{-1}; $DH°(FCH_2^•/H^•) = 103$ kcal mol^{-1}, Table 1.A.7).

Ethylenetetracarbonitrile (tetracyanoethylene, TCNE) adds to verbenene at 60 °C giving the (2+2)-cycloadduct **111** resulting from the reaction of the least sterically hindered alkene moiety. The reaction rate constant is strongly solvent dependent. It is 700 times faster in 1,2-dichloroethane (polar) than in dibutyl ether (less polar) [146]. A similar solvent effect on the rates of (2+2)-cycloadditions of TCNE to enol ethers has been reported (Scheme 5.23). For instance, the reaction of TCNE and ethyl isobutenyl ether giving cycloadduct **113** the rate constant at 15 °C varies from 0.00 014 M^{-1} s^{-1} in cyclohexane (nonpolar) to 0.629 M^{-1} s^{-1} in acetonitrile (polar) [147].

Reactions of Scheme 5.23 can be interpreted in terms of the formation diradical ↔ zwitterion intermediates or transition structures **110** ↔ **110′** and **112** ↔ **112′**. This mechanism allows one to predict

Scheme 5.22 Examples of thermal $[\pi^2+\pi^2]$-cyclodimerizations of flurorinated ethylenes.

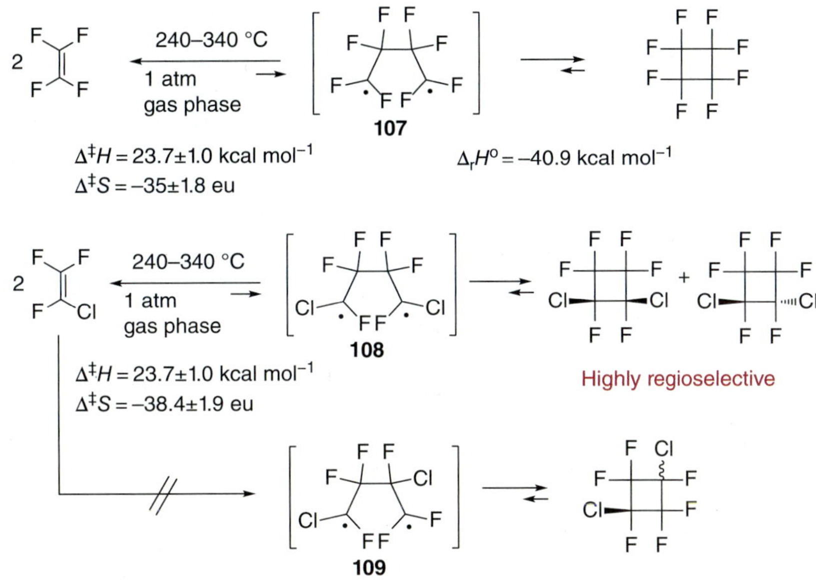

(5.10)

(5.11)

(5.12)

Scheme 5.23 Examples of facile thermal $[\pi^2+\pi^2]$-cycloadditions the rates of which depend strongly on solvent.

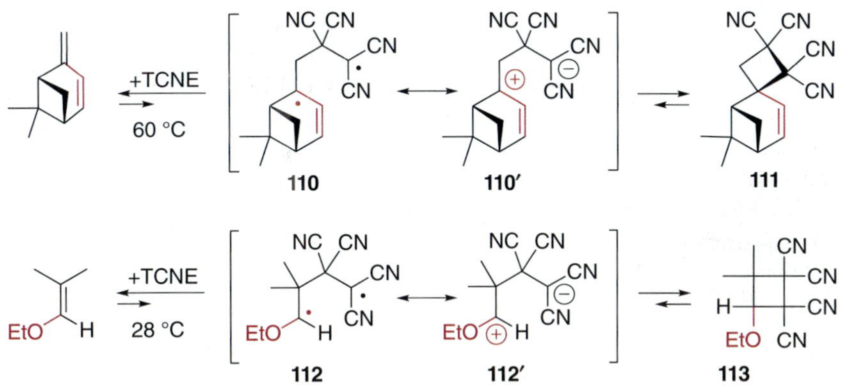

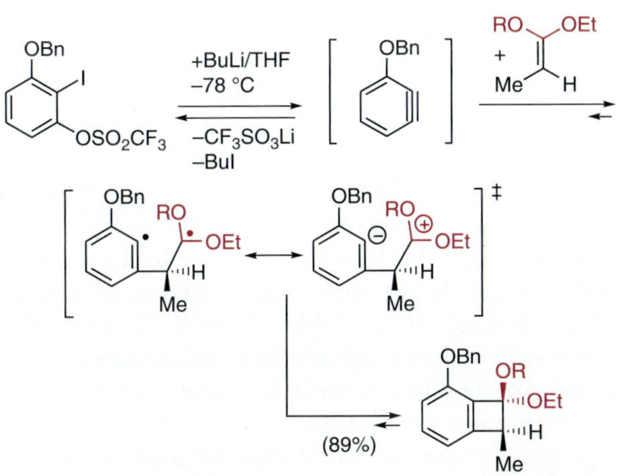

the regioselectivity of the $[\pi^2+\pi^2]$-cycloadditions: the favored regioisomeric cycloadducts obtained under conditions of kinetic control ($\Delta_r G < 0$: cycloreversion does not compete with the cycloaddition) are those arising from the most stable diradical ↔ zwitterion species that can form upon combination of the two cycloaddents. Thus, the substituent effects on the relative stability of radicals, cations, and anions permit the prediction of the regioselectivities of these cycloadditions. An example is given here-below with the regioselective $[\pi^2+\pi^2]$-cycloaddition of 3-alkoxybenzyne to ketene silyl acetals [148].

Formal $[\pi^2+\pi^2]$-cycloadditions of strong electron acceptors such as TCNE [149, 150] and 7,7,8,8-tetracyanoquinodimethane [151] to electron-rich alkynes, followed by electrocyclic openings of the resulting cyclobutene derivatives, generate nonplanar dienes that are push–pull chromophores featuring intense intramolecular charge transfer and high third-order optical nonlinearities [152]. The process has been applied to alkene-1,1-dicarbonitriles as illustrated below [153]. It implies the formation of relatively stable zwitterionic intermediates.

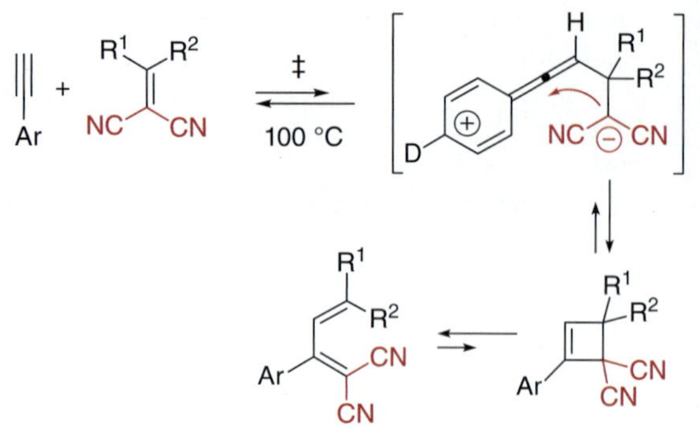

When a cycloaddent has more than one unsaturated function, the chemoselectivity (under conditions of kinetic control: $\Delta_r G < 0$) of the cycloaddition can also be predicted by analyzing the relative stability of the 1,4-diradicals ↔ zwitterions that can form concurrently. Deviations from these predictions might occur; they generally arise from steric factors (e.g. repulsions between substituents, strain difference between isomeric cycloadducts). The $[\pi^2+\pi^2]$-cycloadditions of acrylic esters to electron-rich alkenes are catalyzed by Lewis acids (Section 7.6.8). This is predicted by the diradicaloid model for the transition structures of these reactions. Using enantiomerically pure Lewis acid catalysts, enantioselective cycloadditions have been observed [154–156]. An example is given below, which has permitted an efficient asymmetric synthesis of (*R*)-6-methylbicyclo[4.3.0]non-1-en-3-one, an important synthetic intermediate [157].

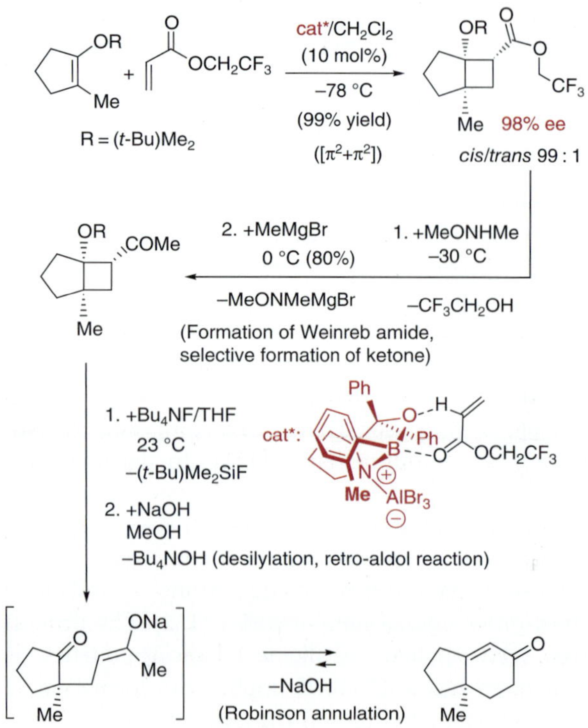

5.3.5 Cycloadditions of allenes

Allene cyclodimerizes above 200 °C via tetramethyleneethane intermediate. Unsymmetrically alkyl substituted allenes give mixtures of all possible regioisomeric cycloadducts [158]. In the case of 3,3-dimethylallenecarbonitrile, the cyclodimerization occurs already at 100 °C because of the radical stabilizing effects of the methyl and cyano groups on intermediate diradical **114**.

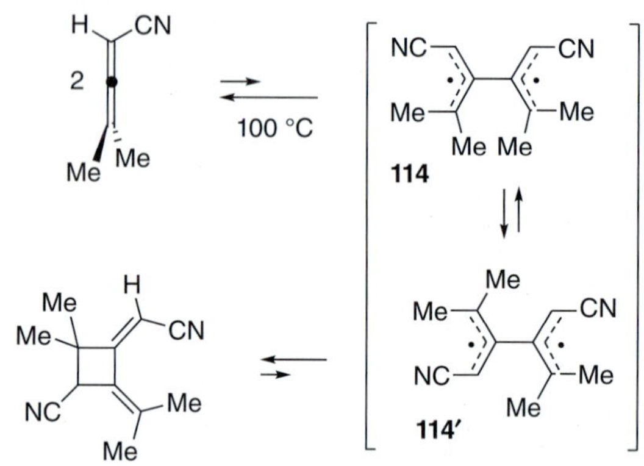

Allenes also add to activated alkenes and alkynes giving the corresponding (2+2)-cycloadducts as mixtures of regioisomers and stereoisomers. The first example of intermolecular (2+2)-cycloaddition of allene to an alkyne was reported by Applequist and Roberts in 1956. They found that allene and phenylacetylene heated in a sealed tube at 130 °C give a 1% yield of 1-phenyl-3-methylidenecyclobutane [159]. When using enantiomerically enriched allenes, partially racemized products are formed because of partial or complete rotation about σ(C—C) in the tetramethyleneethane diradical intermediates [160]. In the case of (*S*)-(+)-1,3-dimethylallene ((*S*)-**115**) reacting with 1,1-diphenylethylene at 160 °C, no cycloadduct is seen but racemization of 1,3-dimethylallene is observed. This is interpreted in terms of reversible formation of diradical **116**, an achiral intermediate responsible for the racemization. It might equilibrate with the (2+2)-cycloadducts **117** and ***ent*-117** in an endergonic reaction at 160 °C (destabilization of the adducts due to front-strain: repulsion between relatively large vicinal substituents of the cyclobutane products, loss of π-conjugation of 1,1,-diphenylethylene) [161].

Intramolecular $[\pi^2+\pi^2]$-cycloadditions of allene-ynes generate fused bicyclic ring structures as illustrated with reaction **118** → **119** induced by microwave irradiation in toluene containing 1-ethyl-3-methylimidazolium hexafluorophosphate (an example of ionic liquid) [162].

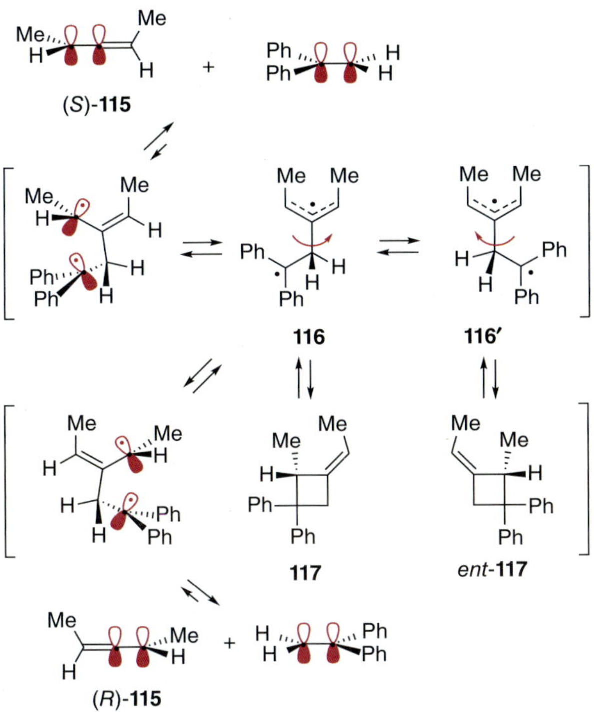

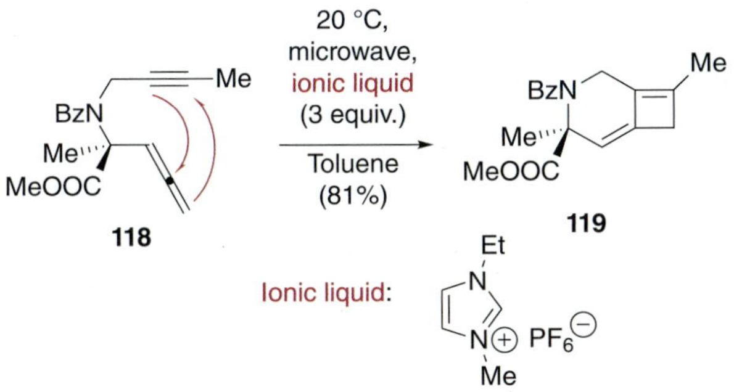

Under similar conditions, allene-yne **120** is converted into a mixture of **121** and **122** (Scheme 5.24), which demonstrates a nonconcerted mechanism involving the intermediacy of diradical **123**. The latter cyclizes more slowly than it is isomerized into diradical **124** through a cyclopropylmethyl/homoallyl radical rearrangement. A hydrogen shift rearranges **124** generating tetraene **125**, which undergoes electrocyclic ring closure into **121** and **122** [163].

Problem 5.9 What is the major product of the reaction of **A** + **B**?

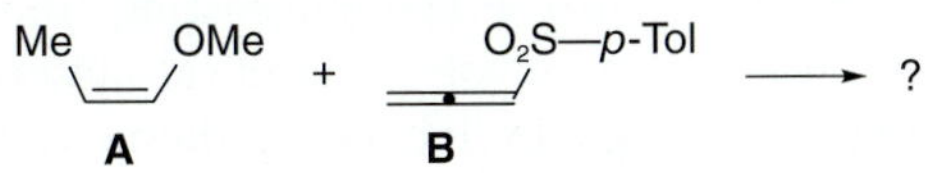

5.3.6 Cycloadditions of ketenes and keteniminium salts

As their discovery by Staudinger [164], ketenes have become very useful synthetic intermediates [165–167]. They are obtained by base-induced elimination of HCl from acyl chlorides possessing an α-hydrogen atom or by the Wolff rearrangement of α-diazoketones [168–171]. In the presence of a base, phenylacetic acid loses 1 equiv. of water to produce phenylketene. Pyrolysis of acetic acid at 700–750 °C produces ketene. Thermal cracking of acetone produces ketene and methane (Schmidlin ketene synthesis [172]). The carbonylation of carbenemetal complexes generates ketenes [173]. As we shall see, the cycloadditions of ketenes to alkenes generate cyclobutenones, the cycloadditions of ketenes to ketones give β-lactones, whereas the cycloadditions with imines (the Staudinger reaction [174]) provide azetidinones = β-lactams. At room temperature, ketenes react with 1,3-dienes giving cyclobutenones in formal (2+2)-cycloadditions, reactions classified as allowed [π^2s+π^2a]-cycloadditions [20]. However, Machiguchi and Yamabe showed in 1996 (low temperature NMR studies) that the initial product of reaction of diphenylketene with cyclopentadiene is 3-(diphenylmethylidene)-2-oxabicyclo[2.2.1]hept-5-ene, which results from a [π^4+π^2]-cycloaddition. The cycloadduct so-obtained then undergoes a Claisen rearrangement (sigmatropic rearrangement of order (3,3), Section 5.5.9.2) into 7,7-diphenylbicyclo[3.2.0]hept-2-en-6-one [175]. Open chain 1,3-dienes such as 2,3-dimethylbutadiene and 1-methoxybutadiene afford initially both the Staudinger ([π^2+π^2]) and the Diels–Alder ([π^4s+π^2s]) cycloadducts. The Staudinger cycloadducts are converted on heating to the Diels–Alder products by retro-Claisen rearrangement [176]. In general, ketenes add to alkenes across their C=C double bond, rather than the C=O bond [177]. However, di(trifluoromethyl)ketene adds to ethyl vinyl ether at −80 °C across its C=O bond giving a colorless oxetane that equilibrates at −20 °C with a deep blue zwitterion **125**, a transient intermediate that generates the corresponding cyclobutanone at 0 °C. Intermediate **125** can be quenched with acetone at −30 °C producing a 1,4-dipolar cycloadduct across the C=O bond of acetone [178]. The two CF_3 substituents play a decisive role in stabilizing the enolate moiety of zwitterion **125**. The rapid formation of the oxetane cycloadduct implies an attractive interaction between the oxygen center of the ketene (partial negative charge) and C(1) of ethyl vinyl ether (partial positive charge) on their way to the transition state of this cycloaddition. As charges develop in the

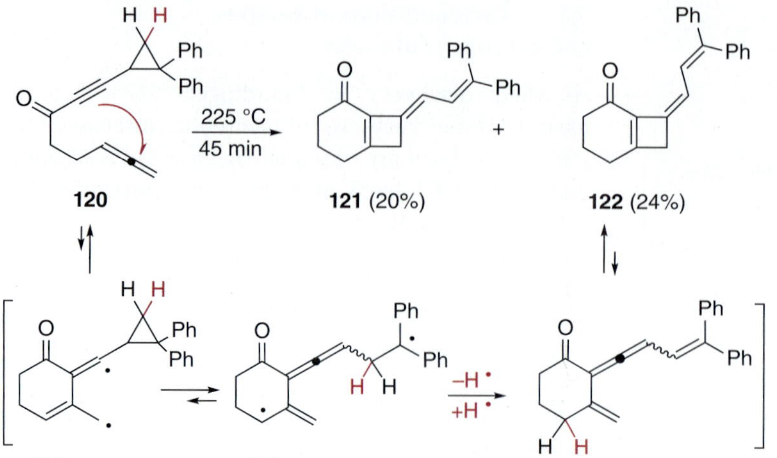

Scheme 5.24 Example of intramolecular [$\pi^2+\pi^2$]-cycloaddition of allene–alkyne. Proof for the formation of 1,4-diradical intermediate by application of a radical clock reaction (relatively fast cyclopropylmethyl radical → homoallyl radical rearrangement, $k_{isom} > 10^{11}\ s^{-1}$).

transition structure of the reaction, The barrier of rotation about the newly formed C—C bond in **125** g, and this permits the ring closing into the oxetane, a process that is much faster than the formation of the cyclobutanone product.

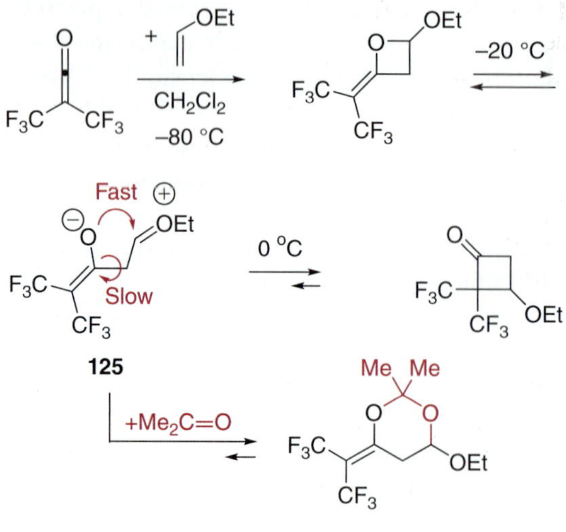

Monosubstituted ketenes added to monosubstituted alkenes are generally cis-stereoselective. Traditionally, this was rationalized in terms of concerted [π^2s+π^2a]-cycloadditions (Figure 5.24a) in which transition structures of type **126** are stabilized by Möbius aromaticity; interactions with the MOs of the carbonyl moiety are ignored [2]. However, quantum mechanical calculations indicate highly asynchronous transition structures in which the central carbon center C(1) of the ketene interacts with both C(1′) and C(2′) carbon centers of the alkene [179, 180]. The interaction of C(2) of the ketene with the two carbon centers of the alkene is much weaker (Figure 5.24b). There is a carbenoid character in the transition structures as the two cycloaddents are almost orthogonal as in **127**: they approach to each other like a carbene approaches an alkene in cyclopropanation, or as a borane approaches the alkene in hydroboration. Transition structure **127** can be seen as a π-complex between the alkene and the carbonyl group of the ketene (donation from $\pi_{C=C}$(alkene) to $\pi^*_{C=O}$ and retrodonation from $\pi_{C=C}$(ketene) to $\pi^*_{C=C}$(alkene)) [181]. The cis-stereoselectivity of the cycloaddition implies that the overlap between these two orbitals is larger when C(2) of ketene approaches C(1′) rather than C(2′) of the alkene. Looking at complex **127**, the 2p orbital at C(2) of the ketene finds the highest coefficient of the LUMO of the π-complex at C(1′) of the alkene. If one considers that the formation of the π-complex **127** involves an electron transfer from the alkene to the carbonyl group (donation $\pi_{C=C}$(alkene) → $\pi^*_{C=O}$ more important than retrodonation $\pi_{C=C}$(ketene) → $\pi^*_{C=C}$(alkene)), it can be represented by a zwitterionic limiting structure of type **127″** in which the positive charge prefers to sit at C(1′) (stabilization by R′) rather than at C(2′). This forces C(2) of the ketene in **127** to combine with C(1′) of the alkene with a rotation that finally realizes the cis-cycloadduct.

Unsubstituted ketene cyclodimerizes into 3-methylidenepropionolactone = diketene. Reaction (5.13) follows a second-order rate law, indicative of a bimolecular process [182]. The rate constant is not affected by triplet oxygen ($^3O_2$, see Section 6.9.1, Figure 6.13), suggesting that the reaction does not involve diradical intermediates [183]. Although cyclobuta-1,3-dione is nearly

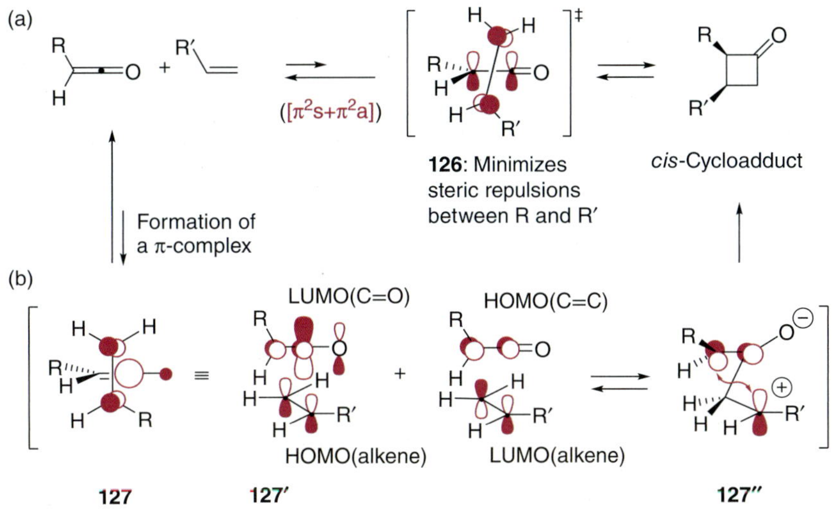

Figure 5.24 Representation of the transition structure of $[\pi^2+\pi^2]$-cycloadditions of monoalkylketenes to 1-alkylalkenes: (a) according to the classical Woodward–Hoffmann rules, and (b) according to quantum mechanical calculations. Both models interpret the cis-stereoselectivity of the cycloadditions and chirality transfer from the alkene to the cycloadducts.

isoenergetic with diketene, it is formed only in trace amounts. However, several substituted ketenes cyclodimerize into cyclobuta-1,3-dione derivatives [184]. Dimethylketene is cyclodimerized into 2,2,4,4-tetramethylcyclobuta-1,3-dione. The reaction also follows second-order kinetics and its rate constant is little affected by the nature of solvent, suggesting that the reaction does not involve the formation of zwitterionic species in the rate-determining step in agreement with a concerted pathway involving an unsymmetrical transition structure **128** [185]. This is supported by quantum mechanical calculations [186] and isotopic secondary deuterium isotope ratios in the formation of diketene measured by natural abundance ^{2}H-NMR spectroscopy. It is found that the ring methylene is enriched in deuterium (k_H/k_D = 0.975 ± 0.026) compared to the methylidene hydrogens and that the latter show unequal deuterium concentrations (ratio: 1.040 ± 0.025) [187, 188].

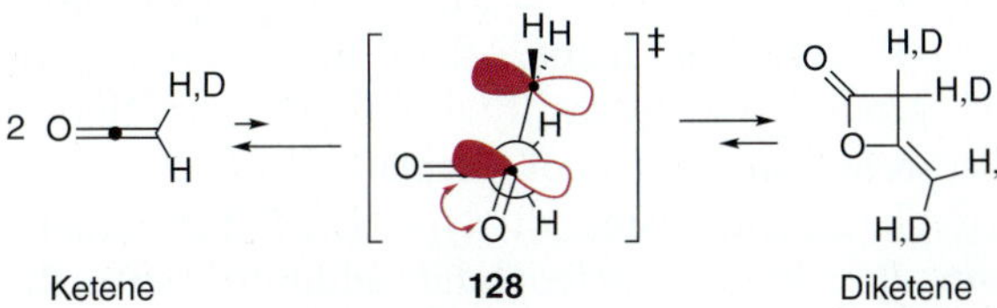

(5.13)

Diphenylketene reacts (20 °C, Et_2O) with *p*-benzoquinone giving propionolactone (a β-lactone) **133** exclusively [166]. The (2+2)-cycloaddition involves the carbonyl moiety of benzoquinone and not the C=C double bond. Diphenylketene reacts with its C=C double bond, not with its C=O double bond. Cyclobutanone **129** is not seen because its formation is endergonic at 25 °C. The standard heat of reaction of ethylene with ketene equilibrating with β-lactone amounts to −13.6 kcal mol^{-1} (Table 1.A.4). If one considers the loss of π-conjugation in diphenylketene when forming **129** and the loss of π-conjugation of one π(C=C) bond of *p*-benzoquinone when forming **129**, this reaction must have an exothermicity of less than 10 kcal mol^{-1} ($\Delta_r H > -10$ kcal mol^{-1}), which is not enough to pay for the entropy cost of this condensation. Taking the standard entropy of cyclodimerization of ethylene into cyclobutane, which amounts to c. −42 eu, the entropy cost at 25 °C is $-T\Delta_r S° = -298(-0.042) = 12.4$ kcal mol^{-1}. The 2-alkylideneoxetane **130** has no chance to form as the exothermicity of this cycloaddition should be < −10 kcal mol^{-1} (for the model equilibrium in the gas phase, one estimates $\Delta_r H°$(ethylene + formaldehyde ⇄ oxetane) = −4.1 kcal mol^{-1}). In the same way, 3-oxooxetane derivative **131** does not form for thermodynamic reasons. Its lactone analog **133**, however, is about 25 kcal mol^{-1} more stable (Section 2.7.6). For the model reaction formaldehyde + ketene ⇄ β-lactone, one estimates $\Delta_r H° = -19.9$ kcal mol^{-1}, which is far enough to pay for the entropy cost of this cycloaddition. Thus, the chemo- and regioselectivity of the cycloaddition of diphenylketene to *p*-benzoquinone is controlled by the thermodynamics. The cycloaddition is rapid at 25 °C because diradical **132** that can form in the rate-determining step of the cycloaddition (alternatively, **132** represents one limiting structure of the transition structure of a concerted reaction) is highly stabilized by its substituents (bisallylic oxyalkyl radical, and diphenylmethyl radical) and by a possible electronic exchange **132** ↔ **132′** ↔ **132″**.

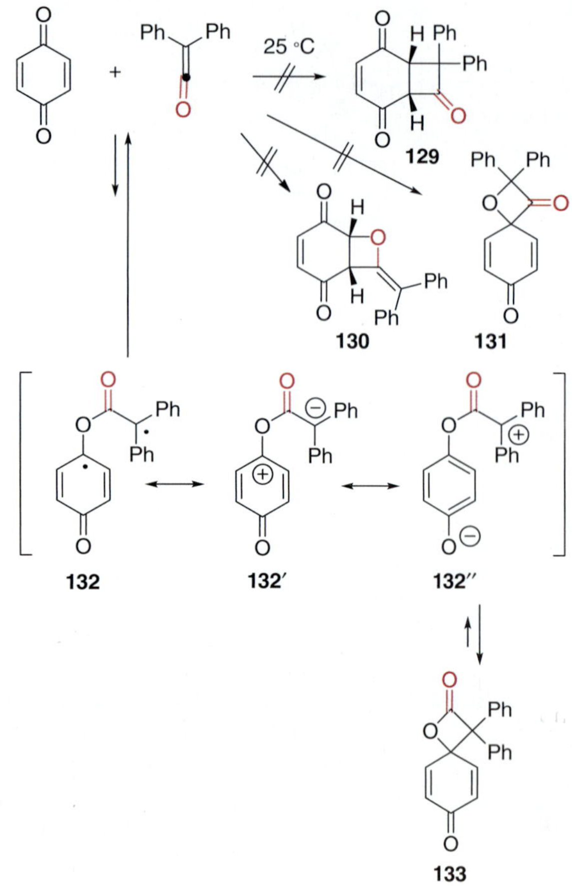

The $[\pi^2+\pi^2]$-cycloaddition of enantiomerically enriched chiral allene (*R*)-**134** to dichloroketene (engendered by reaction of Cl_3CCOCl with Zn/Cu) gives adduct **135** with a chirality transfer better than 98% [189]. The least sterically hindered face of the cyclonona-1,2-diene is attacked by the ketene forming a π-complex **136** that maintains an interaction between C(2′) of the ketene and C(3) of the cycloallene. If a diradical or zwitterionic intermediate **137** should have been formed before ring closure, racemic cycloadduct **135** would have been obtained, which is not the case. Thus, in the transition state **136**, the 2p orbital at C(3) of the allene does not rotate completely (by 90 °) to generate a parallel allyl system, but combines with the 2p orbital at C(2′) of the ketene, as predicted by quantum mechanical calculations for transition structure **127** (Figure 5.24). Alternatively, if an intermediate of type **137** should be formed, it does not allow for free rotation about the C(2)—C(1′) σ bond (Scheme 5.25): it cannot equilibrate with a planar structure and must maintain the original axial chirality of (*R*)-**134**.

An example of enantioselective (2+2)-cycloaddition of dichloroketene to an enantiomerically pure enol ether is given below. This asymmetric alkene + ketene cycloaddition (no (*E*/*Z*) isomerization of the styrene derivative, in agreement with the transition state of type **127**) was applied in the total synthesis of (−)-α-cuprarenone and (+)-β-cuprarenone [190].

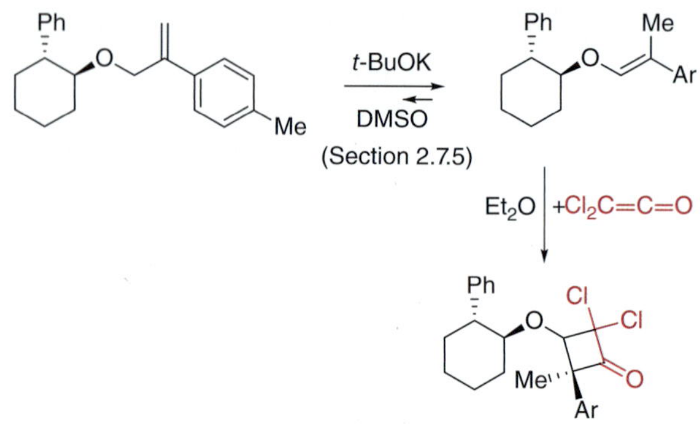

Ketenes and imines (Schiff bases [191]) undergo (2+2)-cycloadditions giving azetidinones (β-lactams) [192]. This is the Staudinger reaction discovered in 1907 [193, 194], which has found a large number of synthetic applications [192, 195–197] including in the synthesis of β-lactam antibiotics (Figure 5.25).

The most frequently accepted mechanism for the ketene–imine cycloadditions is a stepwise reaction with formation of zwitterionic intermediates that are iminium-enolates. They undergo ring closure producing 1,2-*cis*-disubstituted β-lactams concurrently with (*Z*/*E*)-isomerization of the iminium moiety before ring closure into the corresponding 1,2-*trans*-stereomers (Scheme 5.26). For the reaction of ketene **139** (generated by Wolff rearrangement) with arylimines **140**, a linear correlation of log([*cis*-**141**]/[*trans*-**141**]) with the Hammet σ constants (R substituent of **140**) was observed, with a slope ρ = 1.62 and r^2 = 0.98. For electron-rich aryl groups (e.g. with R = MeO), the proportion of *trans*-β-lactams is higher than for electron-poor aryl groups (e.g. with R = CF_3, NO_2). This is interpreted in terms of formation of the zwitterionic intermediates of type (*Z*,*Z*)-**138** arising from the imine nucleophilic addition onto the least sterically hindered face of the ketene carbonyl moiety (*anti* with respect to the largest substituent). When the iminium moiety is stabilized by an electron-releasing substituent, the lifetime of the zwitterion is increased and it has the time to undergo equilibration into stereoisomeric zwitterion (*Z*,*E*)-**138**. Ring closure of (*Z*,*Z*)-**138** gives the corresponding *cis*-1,2-disubstituted β-lactam, whereas ring closure of (*Z*,*E*)-**138** produces *trans*-1,2-disubstituted β-lactams [198].

Scheme 5.25 Intermolecular stereoselective $[\pi^2+\pi^2]$-cycloaddition of a chiral allene to dichloroketene with high degree of chirality transfer from the allene to the cycloadduct.

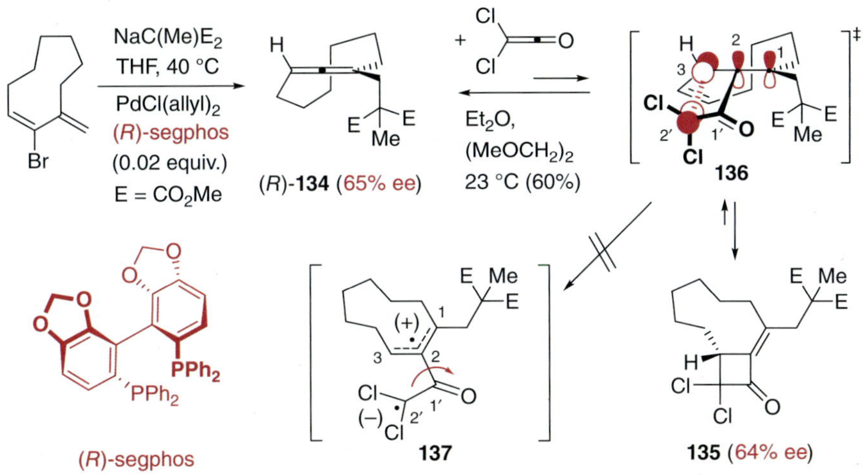

Figure 5.25 Examples of β-lactam antibiotics

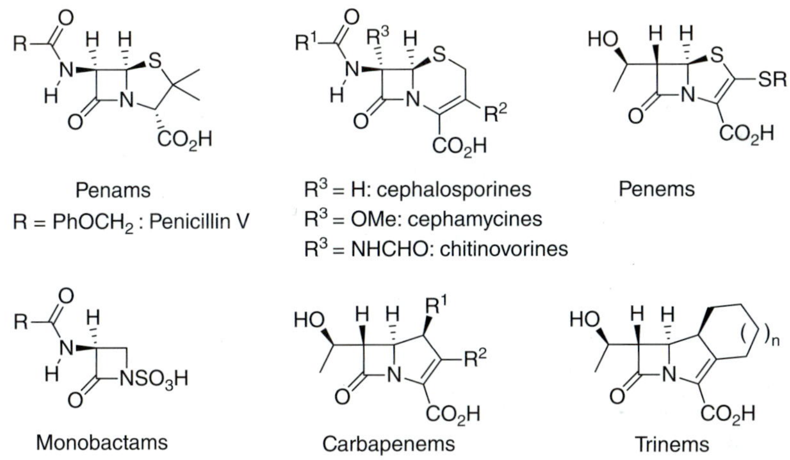

Scheme 5.26 Mechanism of the thermal Staudinger reaction. The ratio of *trans*- vs. *cis*-β-lactam increases with the electron-releasing ability of the R^2 group of the imine.

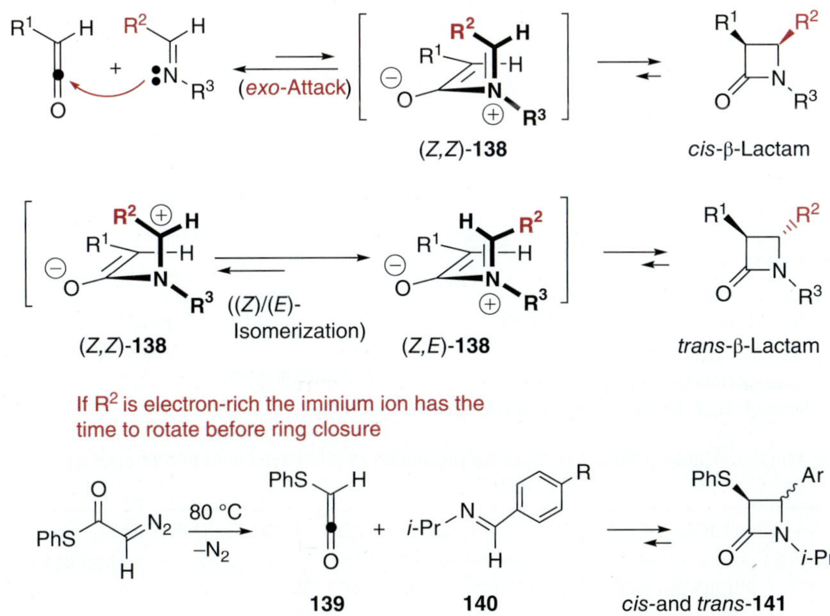

cis/trans Cycloadduct Ratio:	R = NO_2	CF_3	H	MeO
	73:27	42:58	12:88	4:96

In some cases, the zwitterionic intermediates can be quenched by ketenes, sulfur dioxide, or alkenes as shown below [199].

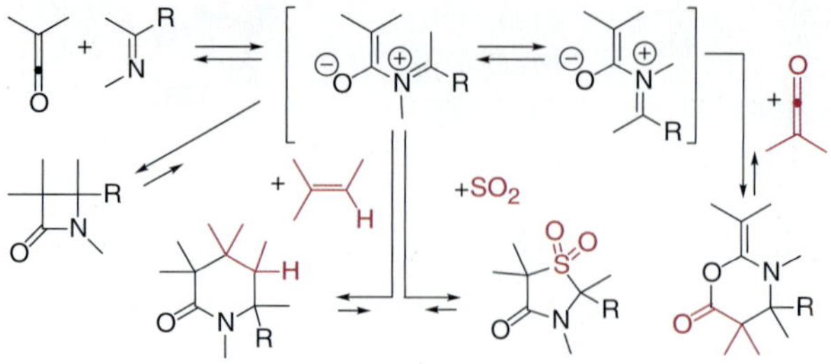

In 1951, Sheehan and Ryan realized the first synthesis of enantiomerically pure β-lactams applying the Staudinger reaction. An example is given in Scheme 5.27 with the synthesis of nocardicin D that starts with (*R*)-4-hydroxyphenylglycine [200].

Nonracemic α-amino-β-lactams can be obtained by the reaction of Evans–Sjögren that engage ketenes and aldimines both derived from enantiomerically pure amino acids [201–203]. This strategy can use resin-bound imines (solid phase synthesis) as illustrated here-below [204].

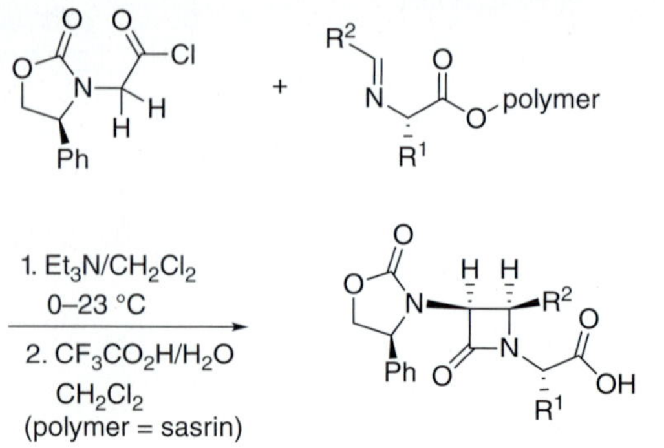

Ketenes react with tertiary amines giving enolates of type **142** that react as nucleophiles onto electrophilic imines, forming corresponding β-lactams.

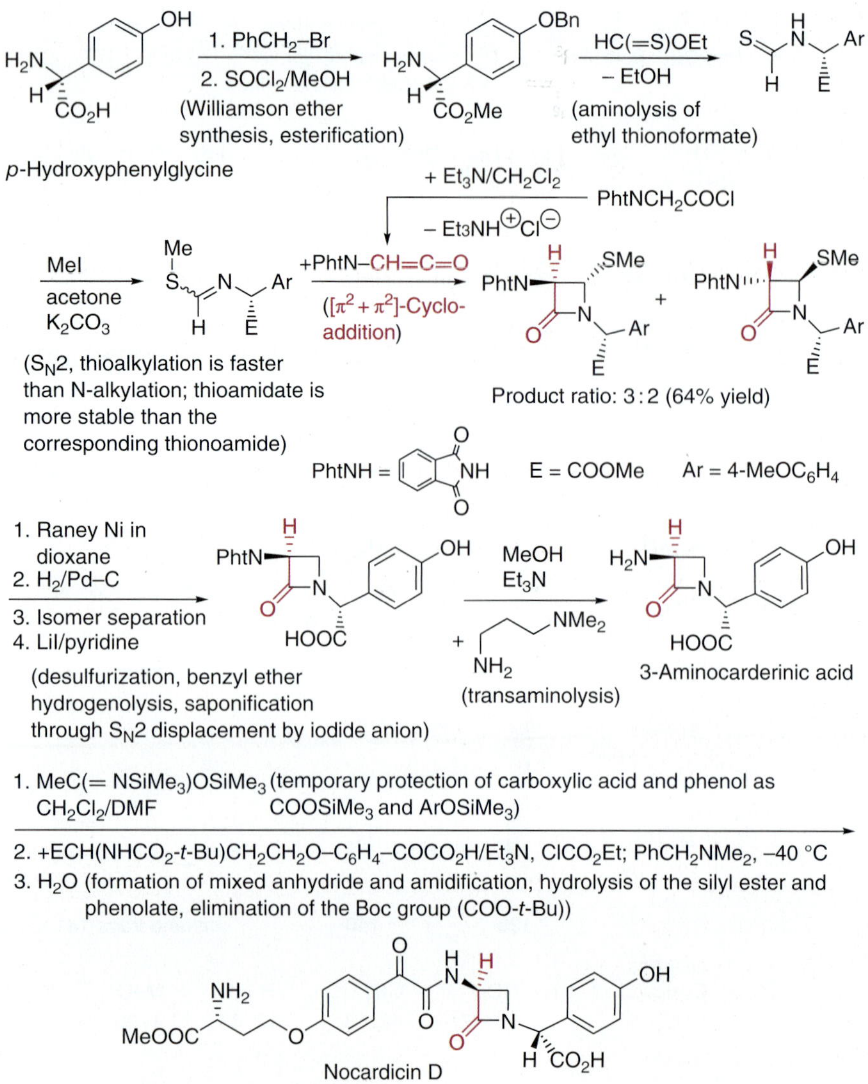

Scheme 5.27 Synthesis of nocardicin D by Sheehan and Ryan.

Scheme 5.28 Asymmetric synthesis of β-lactams via enantiomerically pure ketene–enolate intermediates. Asymmetry is induced by the enantiomerically pure catalyst cat*, which is a stable diaminocarbene (see nucleophilic catalysis, Section 7.7).

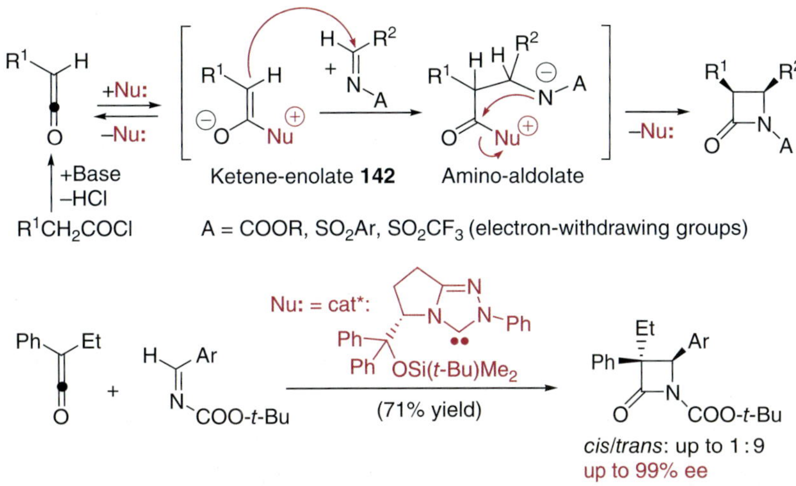

The ketene-enolates **142** can also be formed by reaction of the tertiary amine with acyl chlorides. When an enantiomerically pure tertiary amine is used, it can induce asymmetry in the formation of β-lactam [205]. Instead of a tertiary amine, one can employ an enantiomerically pure chiral *N*-heterocyclic carbene (Sections 7.5.1–7.5.7) as nucleophilic catalyst [206]. An example is given in Scheme 5.28.

In 1972, Ghosez and Marchand disclosed the first preparation of keteniminium salts **145** and their cycloadditions [207]. Treatment of *N*,*N*-dialkylamides **143** with phosgene ($COCl_2$) and Et_3N, or with triflic anhydride (Tf_2O = $(CF_3SO_2)O$) and a tertiary amine, such as collidine or pyridine, generates the corresponding enamines **144** that are ionized into the corresponding iminium salts **145**. Keteniminium salts are more electrophilic than ketenes and do not cyclodimerize or polymerize like many ketenes [208, 209]. Enantiomerically pure derivatives can be obtained readily by *N*-substitution with chiral groups [210]. Their (2+2)-cycloadditions with alkenes are usually highly regioselective and the regioselectivity can be predicted from the shape of the LUMO and HOMO orbitals of the two cycloaddents. Alternatively, transition states similar to **127** (cycloaddition of ketenes to alkenes) can be considered and for which a cationic limiting structure **146** (Scheme 5.29) explains the regioselectivity (the most electron-releasing substituent of the alkene appears at C(3) of the cyclobutyleniminium adduct **147**) [211] Scheme 5.29.

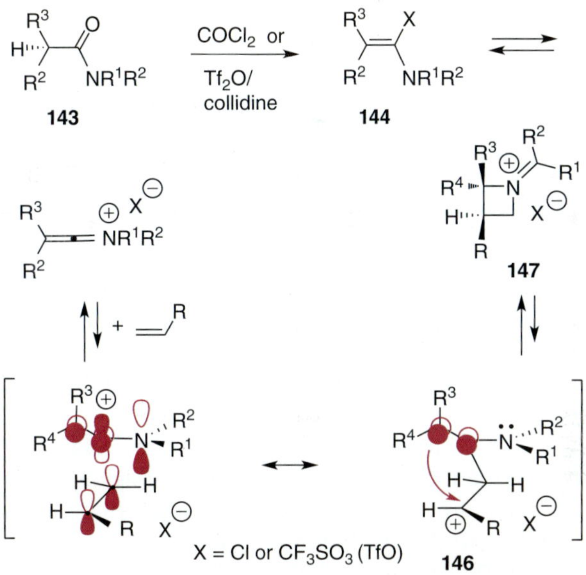

Scheme 5.29 Preparation of keteniminium salts and their (2+2)-cycloaddition with alkenes.

As for the Staudinger reaction, the (2+2)-cycloaddition of keteniminium salts with imines is generally cis-stereoselective and follows a nonconcerted mechanism with the formation of a cationic intermediate **148** that results from the attack of the nitrogen lone pair of the imine onto the C=N double bond of the keteniminium salts. Its ring closure provides the cycloadduct **149**. An asymmetric synthesis of β-lactams has been realized by using enantiomerically pure keteniminium salts derived from L-proline [211–214].

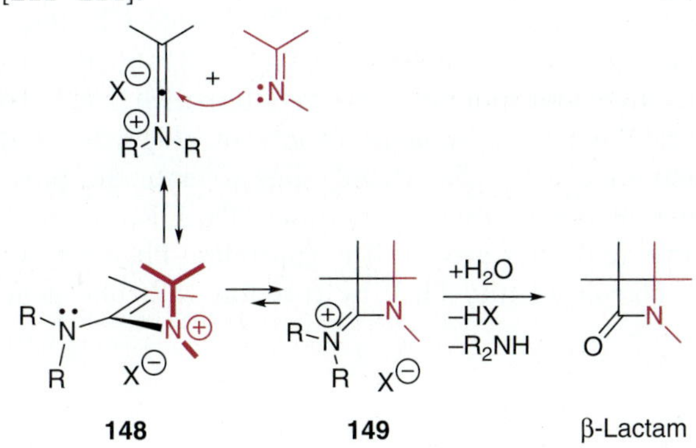

Problem 5.10 What is the product of reaction **A** + **B**? [215]

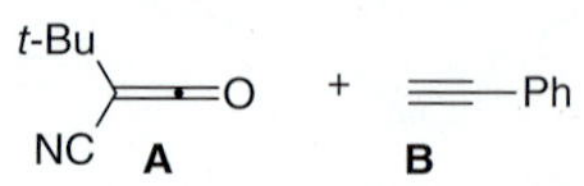

Problem 5.11 Why *trans*- and not *cis*-disubstituted β-lactams are formed in the Staudinger reaction used by Sheehan and Ryan in their synthesis of nocardicin D?

Problem 5.12 What is the product of reaction (three days at 60 °C) of diphenylketene with thiobenzophenone?

Problem 5.13 What is the major (2+2)-cycloadduct of reaction **A** + **B**?

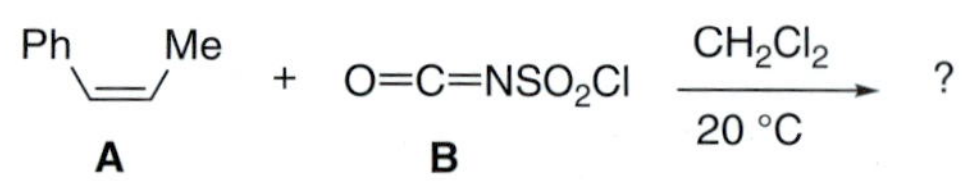

5.3.7 Wittig olefination

The Wittig olefination (or Wittig reaction) discovered in 1954 [216] uses phosphonium ylides (Wittig reagents **150**) to convert aldehydes and ketones into alkenes and phosphine oxide [217]. This reaction and related olefinations like the Horner–Wittig and Horner–Wadsworth–Emmons reactions (Section 5.3.8) are the most useful C—C bond forming transformations [218]. The Wittig reagents **150** are made by deprotonation of phosphonium salts ($R_3P^+CH_2R^1/X^-$) with a strong base such as phenyllithium (PhLi), *n*-butyllithium (BuLi), or $NaNH_2$ (precipitates NaBr [219]) in THF or Et_2O [220]. Salt-free solution of the ylides can be prepared by precipitation of the salt (LiX) formed. The phosphonium salts are made by reaction of a phosphine (usually triphenylphosphine: Ph_3P) with an alkyl halide (R^1CH_2–X). Triphenylphosphine is prepared by the reaction: $3PhCl + 6Na + PCl_3 \rightarrow 6NaCl + Ph_3P$. Alkylation of the simplest ylide $Ph_3P{=}CH_2$ (methylenetriphenylphosphorane) is an alternative route to Wittig reagents. A large variety of aldehydes and ketones can be alkenylated; they tolerate the presence of other functions such as OH, OR, aromatic nitro, and ester groups. The coproduct $Ph_3P{=}O$ can be converted into Ph_3P by reaction with phosgene according to $Ph_3P{=}O + COCl_2 \rightarrow Ph_3PCl_2 + CO_2$. Triphenylphosphine dichloride can be reduced with Al or P into Ph_3P and $AlCl_3$ or PCl_3, respectively [221]. $Ph_3P{=}O$ is reduced into Ph_3P by Ph_2SiH_2, a reagent mild enough that permits its use in the olefination reaction mixture as it does not reduce carbonyl compounds readily [222].

Alkylidenetriphenylphosphoranes are unstabilized ylides: they give (*Z*)-alkenes preferentially (Table 5.1), unless the Schlosser modification is used (Scheme 5.32). If the reaction is done in dimethylformamide (DMF) in the presence of LiI or NaI, the (*Z*)- vs. (*E*)-stereoselectivity is increased.

The reaction of $Ph_3P{=}CHR$ with R′CON(OMe)Me (Weinreb amides) in THF gives the corresponding enamines that are hydrolyzed into ketones RCH_2COR' [223]. Phosphonium ylides containing π-acceptors as in $Ph_3P{=}CHCOOR$ and $Ph_3P{=}CHCN$ are called stabilized Wittig reagents (Table 5.1) and produce (*E*)-alkenes preferentially. Mixtures of (*Z*)- and (*E*)-alkenes are obtained applying semistabilized phosphonium ylides such as $Ph_3P{=}CHAr$ (Ar, aryl groups). These reagents can be generated from tosylhydrazones ArCH=N—NHTs and $P(OMe)_3$ (trimethyl phosphite) and react with aldehydes giving (*E*)-alkenes selectively [224]. There is evidence that the phosphonium ylides **150** can react with the carbonyl compounds in one-step $[\pi^2s+\pi^2a]$-cycloadditions with direct formation of 1,2-oxaphosphetanes **151**. The latter have been isolated in several reactions [225]. Alternatively, nucleophilic additions of the ylides to the carbonyl compounds may generate betaines **152** in a first step, and then, in a second step, they cyclize into the corresponding oxaphosphatanes **151** (Scheme 5.30). The existence of the betaines **152** and their interconversion is still a matter of debate [226–229]. They can be obtained by reaction of phosphines with epoxides and they decompose into alkenes and phosphine oxides.

When phosphorus ylides (engendered by reaction of the corresponding phosphonium bromides with PhLi) are mixed with aldehydes, precipitates may form that are complexes of betaines of type **152** with LiBr. Salt-free betaines can be obtained by adding *t*-BuOK (precipitates KBr) [230]. On heating, the betaines dissolve and decompose into the corresponding alkenes and phosphine oxides. In the absence of LiBr, i.e. when using a base such as $(Me_3Si)_2NK$ in THF to generate the ylide, no precipitate forms. It has been

Table 5.1 Examples of precursors for phosphonium ylides.

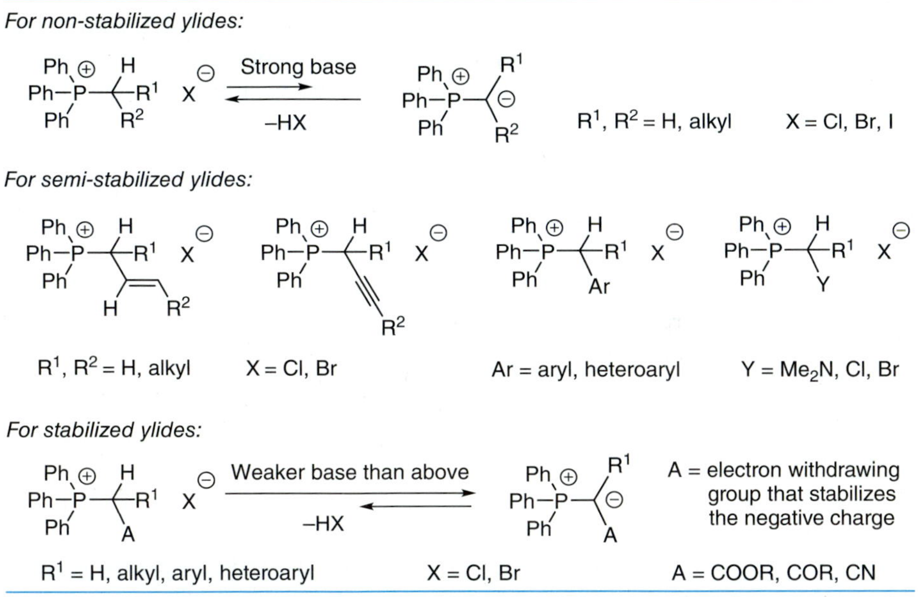

Scheme 5.30 Wittig olefination: preparation of the reagents and possible mechanisms.

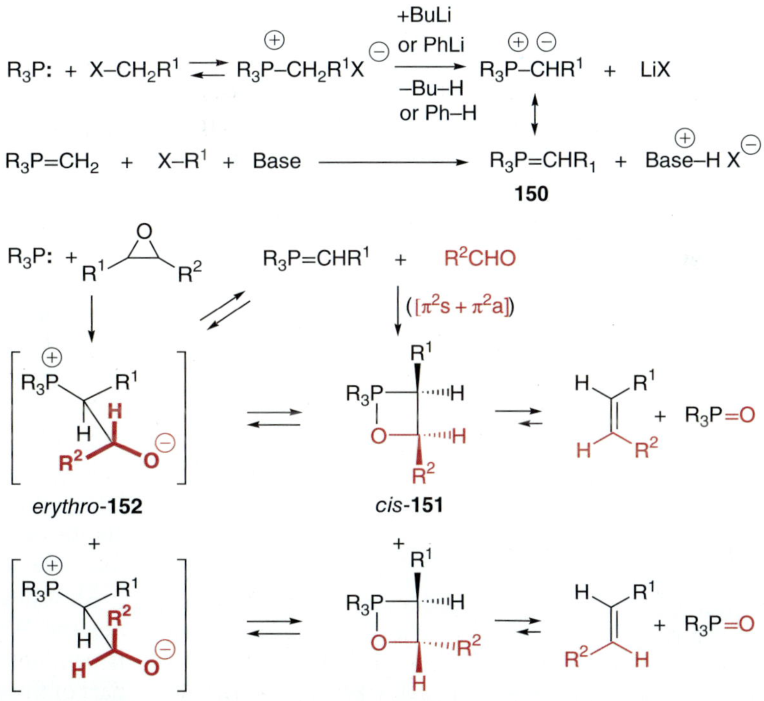

observed that the (*Z*) vs. (*E*)-stereoselectivity of the formed alkenes is better when using the $Ph_3P{=}CHMe$ engendered form $Ph_3(Et)P^+/Br^-$ and NaH in DMF than employing the same ylide generated by a more classical method using BuLi or PhLi in THF. The (*Z*)- vs. (*E*)-stereoselectivity diminishes in the presence of lithium salts [231, 232]. Olefinations of aldehydes with stabilized ylides (Table 5.1) such as $Ph_3P{=}CHCO_2Et$ and $Ph_3P{=}CHCOPh$ have comparable rate constants in polar and nonpolar solvents, which make the formation of betaines unprobable in their rate-determining steps [233, 234]. Olefinations of type $RCHO + Ph_3P{=}CH{-}COR' \rightarrow (E){:}RCH{=}CHCOR' + Ph_3P{=}O$ can be carried out in water [235]. The stabilized ylides $Bu_3P{=}CHCOOR'$ can be formed in water by mixing α-bromoesters with PBu_3 containing a surfactant and using $NaHCO_3$ or Et_3N as a base [236]. Below 0 °C, and under salt-free conditions, nonstabilized ylides such as $Ph_3P{=}CHMe$ react with aldehydes and generate the corresponding 1,2-oxaphosphetanes that have been characterized by their ^{31}P-NMR spectra [237]. In the absence of lithium salts, no trace of the corresponding betaines can be seen. In some cases, the oxaphosphetanes have been crystallized and characterized by X-ray diffraction studies [238, 239]. On heating to 120 °C, crystalline spirophosphorane **153a** is isomerized into **153b** as a consequence of pseudorotation about the phosphorus atom. After prolonged heating to 140 °C, quantitative formation of alkene **154** is observed [240].

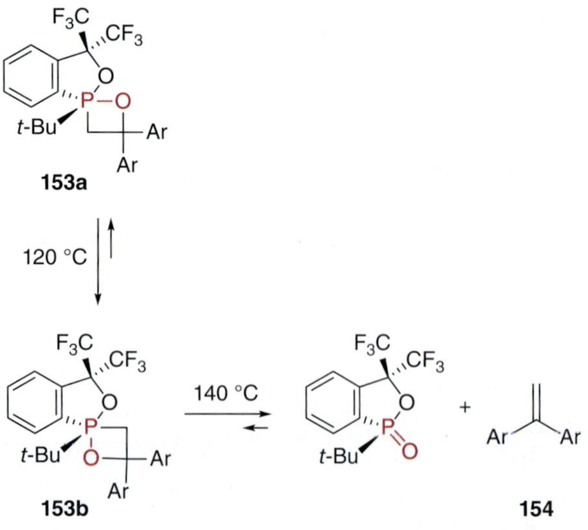

Solvent, substituent, and isotope effects on the rate constants as well as the effect of the nature of reactants on the (*Z*)- vs. (*E*)-stereoselectivity suggest that the salt-free Wittig olefinations of aldehydes are concerted, asynchronous (2+2)-cycloadditions. The transition structures of the reaction can be represented by diradicaloids (diradical ↔ zwitterion) of type **155** ↔ **155′** and **156** ↔ **156′**. As for the corresponding 1,2-oxaphosphetanes that adopt nonplanar conformations for their four-membered rings (X-ray crystalline structures), the diradicaloids may not be planar. The alkene formations are irreversible $[\sigma^2+\sigma^2]$-cycloreversions of the corresponding 1,2-oxaphosphetanes. Under these conditions (kinetic control), the (*Z*)- vs. (*E*)-stereoselectivity is given by the proportion of *cis*- and *trans*-oxaphosphetanes. In the case of reaction $Ph_3P{=}CHR^1 + R^2CHO$ (Scheme 5.31), the cis orientation of the R^1 and R^2 groups is favored for steric reasons because of the phenyl groups of the voluminous Ph_3P moiety (which is twisted [241–243]) that find positions in the transition structures and in the 1,2-oxaphosphetanes that minimize gauche interactions with substituents R^1 and R^2 [244]. This model is supported by the observation that on exchanging the large phenyl groups of the phosphorus ylides for less bulky alkyl groups, the (*Z*)- vs. (*E*)-stereoselectivity is reduced, or even inverted (Scheme 5.31) [227, 245].

When mixture of Wittig ylides and lithium salts (obtained by reaction of triphenylphosphonium halides with PhLi or BuLi) are reacted with aldehydes at −78 °C, mixtures of betaine/lithium salt complexes **157** are formed (Scheme 5.32). Treatment of the latter with PhLi or BuLi in a 1 : 1 mixture of Et_2O/THF at −30 °C generates the organolithium compounds *erythro*-**158** that are quickly isomerized into *threo*-**158**. Their reactions with HCl give corresponding *threo*-**157** that, upon treatment with *t*-BuOK/*t*-BuOH, provide the expected alkenes with high (*E*)- vs. (*Z*)-stereoselectivity [246]. The organolithium compounds **158** can be reacted with aldehydes to generate allylic alcohols [247]. Thioaldehydes and thioketones react with Wittig ylides giving alkenes and phosphine sulfides. Often episulfides are formed concurrently [248].

An industrial synthesis of vitamin A acetate applying the Wittig reaction is outlined in Scheme 5.33. It condenses aldehyde **161** with a stabilized (anionic part of the ylide conjugated with a triene) ylide derived from **160** and obtained by reaction of vinyl-β-ionol under acidic conditions. The corresponding chloride and bromide of this alcohol are unstable. The preparation of vinyl-β-ionol starts with isobutylene, aqueous formaldehyde, and acetone. In the presence of an acid catalyst (200–300 °C, 100 atm), formaldehyde reacts with acetone in an aldol reaction and subsequent water elimination (Section 2.9.1) giving methyl vinyl ketone. Protononation of the latter generates an oxyallyl cation that adds onto isobutylene and

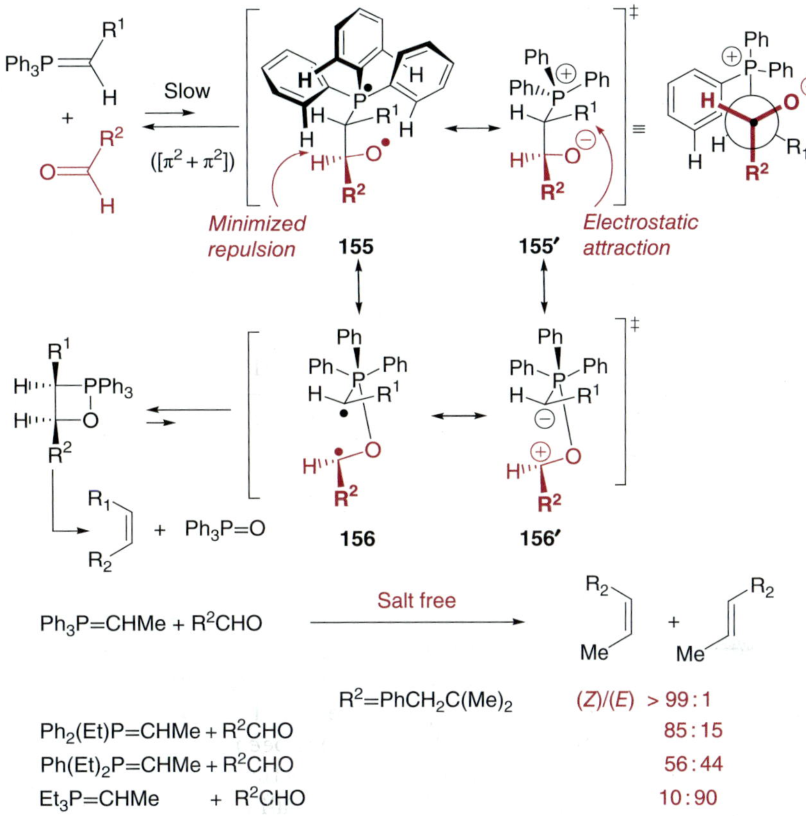

Scheme 5.31 The diradicaloid model of the transition structures of the salt-free Wittig olefinations of aldehydes. Importance of the bulk of phenyl groups of the phosphonium moiety on the (*Z*)- vs. (*E*)-stereoselectivity.

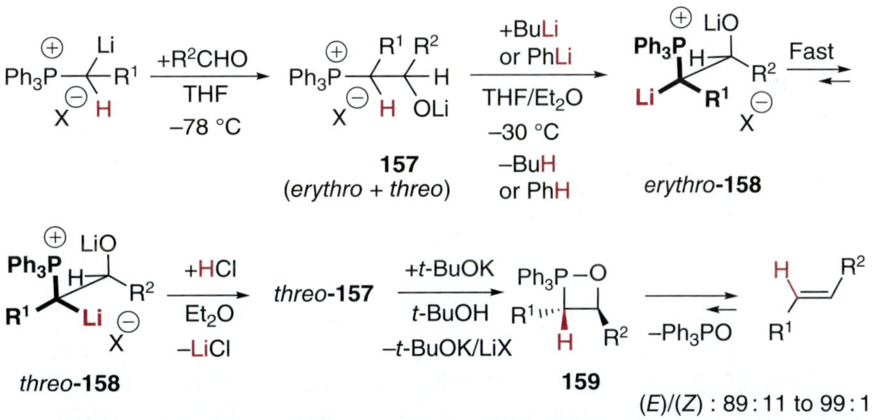

Scheme 5.32 Schlosser modification of the Wittig reaction provides preferentially (*E*)-1,2-dialkylethylenes.

provides 6-methylhept-6-ene-2-one. In the presence of a palladium catalyst, the latter isomerizes into 6-methylhept-5-en-2-one. Nucleophilic addition of acetylene to the ketone moiety gives didehydrolinalool, which undergoes transalcoholysis with ethyl acetoacetate followed by a Carroll reaction (an example of (3,3)-sigmatropic rearrangement, Figure 5.22, Section 5.5.9.2) that produces geranyl acetone. Acid-catalyzed cationic cyclization generates β-ionone. Addition of acetylene and partial hydrogenation provides vinyl-β-ionol. Aldehyde **161** is obtained from acetone via oxidation into methyl glyoxal acetal with MeONO and subsequent addition of acetylene and partial hydrogenolysis. This gives an alcohol that is acetylated with Ac_2O, providing an allyl acetate that undergoes an allylic rearrangement in the presence of a Cu catalyst. Acid-catalyzed hydrolysis of the dimethyl acetal provides **161** [221]. Symmetrical C_{40}-carotenoids are efficiently produced by double Wittig olefination of the corresponding C_{15}-phosphonium salts with C_{10}-dialdehydes [249].

In the Ramizez et al. [250] or Corey–Fuchs [251] olefination of aldehydes and ketones into 1,1-dibromoalkenes ylide, $Ph_3P{=}CBr_2$ (generated from

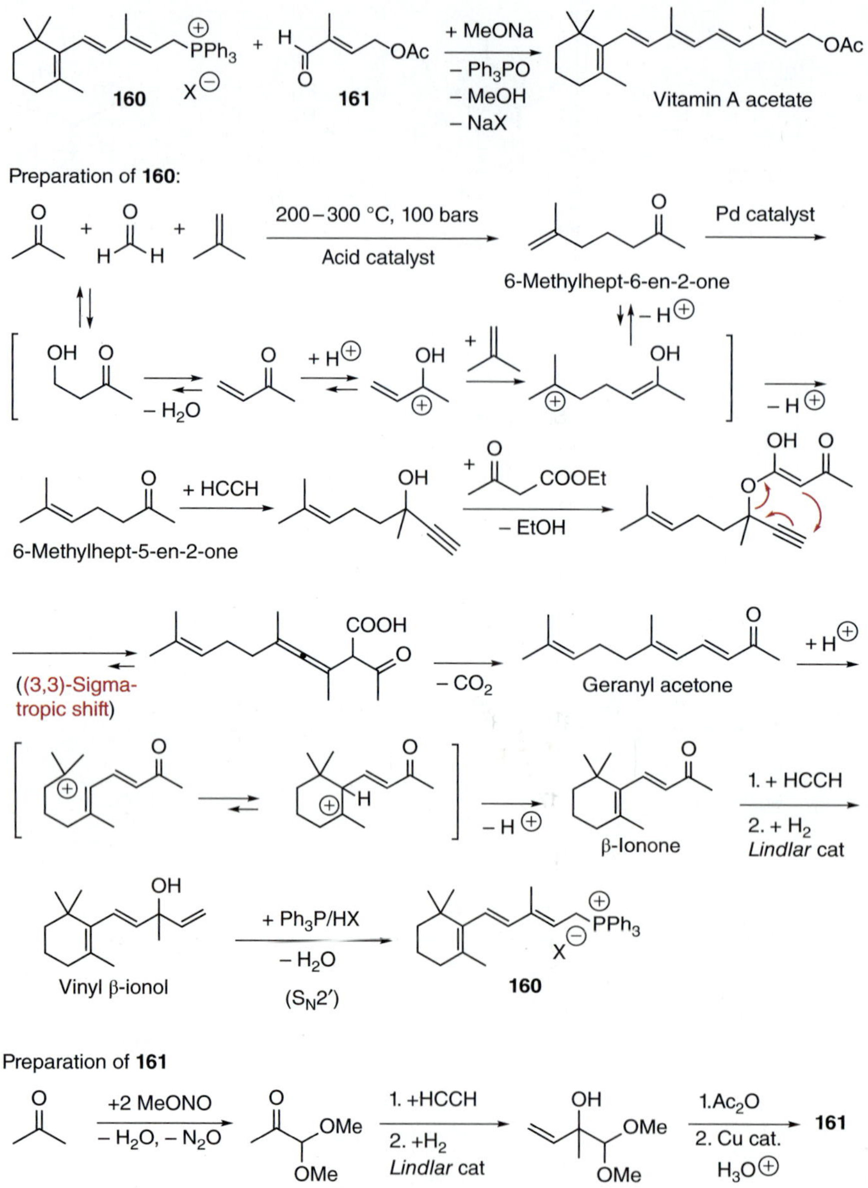

Scheme 5.33 Industrial (Badische Anilin- und Soda-Fabrik [BASF]) synthesis of vitamin A acetate.

the reaction $CBr_4 + 2Ph_3P \rightarrow Ph_3P{=}CBr_2 + Ph_3PBr_2$) reacts with carbonyl compounds. Halogen/metal exchange of the 1,1-dibromoalkenes leads to the formation of alkylidene carbenes or carbenoids that can be quenched by alkenes to form alkylidenecyclopropanes, dimerized into buta-1,2,3-trienes, isomerized by (1,2)-shifts into alkynes (Frisch–Buttenberg–Wiechell rearrangement (Section 5.5.3)) or undergo intramolecular C—H insertion reactions [252].

5.3.8 Olefinations analogous to the Wittig reaction

In 1958, Horner et al. reported a modified Wittig reaction using phosphonate-stabilized carbanions of type **162** [253, 254]. Wadsworth and Emmons further developed the reaction [255]. In contrast to phosphonium ylides, phosphonate-stabilized carbanions are more nucleophilic but less basic. They are alkylated readily and the salt byproducts can be removed easily by simple aqueous extraction. Bases to generate the carbanion salts can be NaH, LiCl + DBU (1,8-diazabicyclo[5.4.0]undec-7-ene) [256], or $MgCl_2 + Et_3N$ [257]. The rate-determining step of the Horner–Wadsworth–Emmons reaction is the nucleophilic addition of the phosphonate carbanion onto the aldehyde or ketone giving alcoholates **163a** and **163b** [258, 259]. Subsequent phosphonate elimination produces alkenes **164**. The (*E*)- vs. (*Z*)-**164** ratio depends on the stereochemical

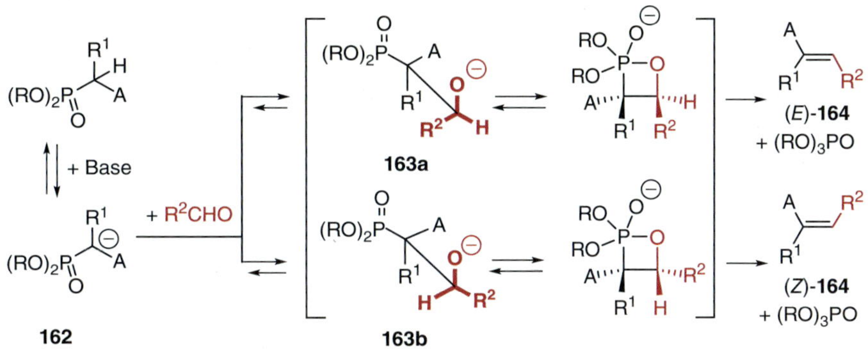

Scheme 5.34 Horner–Wadsworth–Emmons reaction. A, electron-withdrawing group (e.g. COOR).

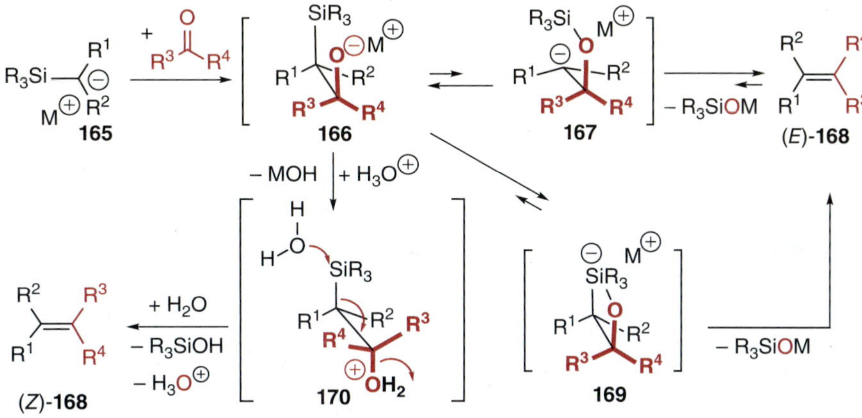

Scheme 5.35 Peterson olefination reaction.

outcome of the initial carbanion addition and on the ability of intermediates **163a** and **163b** to equilibrate (Scheme 5.34). The more this equilibration is complete, the higher the (*E*)- vs. (*Z*)-stereoselectivity. Using trifluoroethyl phosphonates and strong bases such as $(Me_3Si)_2NK$ (KHMDS) and 18-crown-6 in THF at low temperature, Still and Gennari obtain (*Z*)-alkenes with high stereoselectivity [260, 261].

The Peterson olefination is the silicon variation of the Wittig olefination (Scheme 5.35) [262]. α-Silyl stabilized carbanions **165** add to aldehydes and ketones giving β-silylalcoholates **166** that undergo silyl group migration to generate corresponding silyl ethers **167**. Subsequent β-elimination generates alkenes **168**. Alternatively, four-membered ring intermediates **169** with a pentacoordinated silicon atom may form and undergo $[\sigma^2+\sigma^2]$-cycloreversion. When treated under acidic conditions, alcoholates **166** equilibrate with the protonated β-silylalcohols **170** that undergo R_3SiOH elimination into alkenes. Usually, the latter are stereoisomeric with those obtained under alkaline conditions [263, 264].

The Tebbe reagent ((cyclopentadienyl)$_2$TiCH$_2$ClAl(Me)$_2$) treated with a mild base generates the Schrock carbene: (cyclopentadienyl)$_2$Ti=CH$_2$. The latter undergoes (2+2)-cycloadditions with carbonyl compounds and converts them into alkenes [265].

In 1973, Marc Julia and Paris [266] reported the first examples of olefination of carbonyl compounds via 1,2-addition of α-lithiated or α-Grignard sulfones, followed by esterification of the β-hydroxysulfone adducts and subsequent reductive desulfinylation with sodium amalgam in boiling ethanol. Contrary to the Wittig, Horner–Wadsworth–Emmons, and Peterson olefinations that combine the phosphorus and silicon atom with the alcoholate oxygen atom, combination of the sulfur and oxygen atoms does not occur: the β-alkoxysulfone intermediates arising from the addition of the conjugate bases of sulfones to aldehydes are not able to eliminate the corresponding sulfonate salts. In 1977, Lythgoe and Waterhouse found that the β-hydroxysulfones can be esterified as thionobenzoates, the latter undergoing smooth reduction into alkenes on treatment with Bu_3SnH [267]. Thus, the sequence aldehyde + metallated sulfone → alcoholate → ester → β-elimination → α,β-unsaturated sulfone → reductive desulfinylation → generating (*E*)-alkenes is now called the Julia–Lythgoe olefination. Kende showed that 1,2-adducts of aldehydes with 1-methylimidazol-2-yl sulfones are reduced in one step to the corresponding alkenes using SmI_2 [268]. In 1991, Sylvestre Julia and coworkers described a one-pot conversion of carbonyl compounds into alkenes using benzothiazol-2-yl sulfones

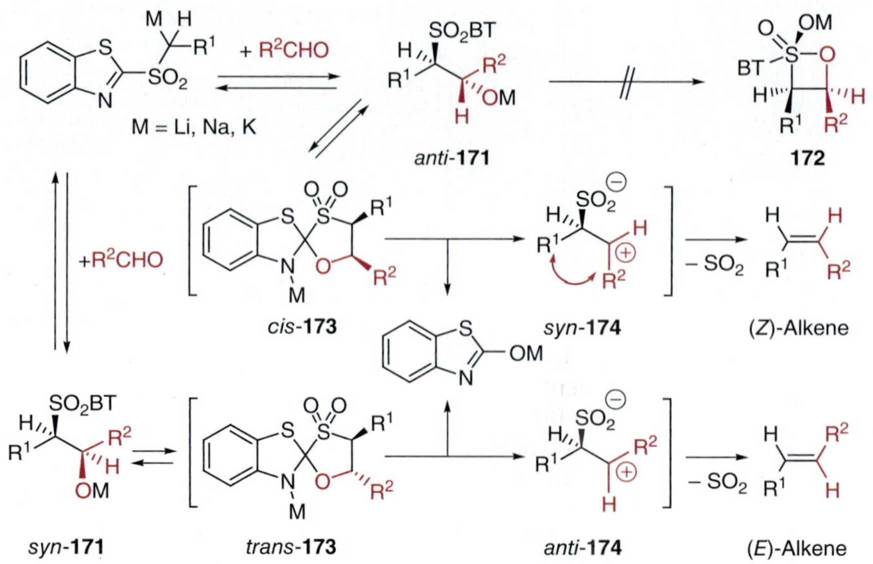

Scheme 5.36 Possible mechanism of the modified Julia olefination using BT sulfones (zwitterionic intermediates).

(BT sulfones) [269]. The process commonly known as the modified Julia reaction implies a Smiles rearrangement [270] of the intermediate lithium alkoxide as shown in Scheme 5.36. Metallated sulfones add to aldehydes giving mixtures of *anti*-**171** and *syn*-**171**. The reversibility of the 1,2-additions has been demonstrated for BT sulfones. The (*E*)-stereoselectivity of the alkene formed increased with electron-releasing group on the benzene ring of starting benzaldehydes. This suggested that the spiro intermediates *cis*-**173** and *trans*-**173** formed during the benzothiazol group migration from the sulfur center to the oxygen center (Smiles rearrangement) undergo elimination of lithium benzothiazol-2-yl oxide with formation of zwitterions *syn*-**174** and *anti*-**174** (benzylic character of the cationic moieties). For steric reasons (gauche interactions), *anti*-**174** is more stable than *syn*-**174**; thus, the former is formed faster than the latter, favoring (*E*)-alkenes.

In 1993, Sylvestre Julia and coworkers introduced the use of pyridin-2-yl sulfones (PYR) and pyrimidin-2-yl sulfones [271, 272]. The modified Julia olefinations was further studied by Kocienski and coworkers who introduced 1-phenyl-1*H*-tetrazol-5-yl sulfones (PT sulfones) in 1998 [273] and 1-*tert*-butyl-1*H*-tetrazol-5-yl sulfones (TBT sulfones) in 2000 [274]. The latter modifications now realize the Julia–Kocienski olefination [274, 275]. Resin-bound BF and PT-sulfones permit the efficient alkene synthesis on solid support [276]. The stereoselectivity of the Julia–Kocienski olefinations depends on the heterocyclic sulfone moiety, on the nature of the carbonyl compound, on solvent polarity, and on the nature of the base used for the deprotonation of the sulfone (role of the counterion, Li^+, K^+, Na^+, MgX^+, etc.) and the presence of additive such as DMF (Me_2NCHO), hexamethylphosphotriamide (HMPA, $(Me_2N)_3P{=}O$), and 1,3-dimethyl-3,4,5,6-tetrahydro-2(1*H*)-pyrimidone (DMPU) or 18-crown-6 ether to coordinate K^+ [277].

Problem 5.14 Propose a one-step conversion of aldehyde **A** into α-santalol, the principal constituent of sandalwood oil [247].

A:

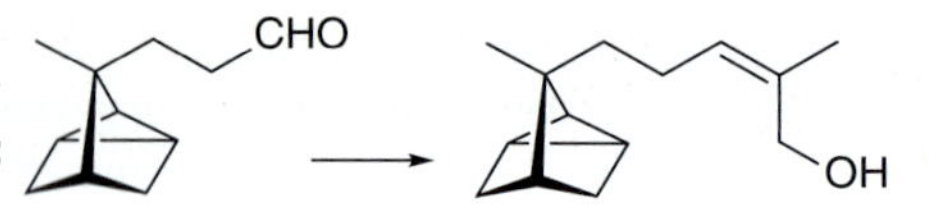

α-Santalol

Problem 5.15 What is the product of the following reaction? [278]

$$(EtO)_2P({=}O)CH_2CO_2R \xrightarrow[\text{2. +R'CHO}]{\text{1. +MeMgBr}} ?$$

Problem 5.16 Give a possible mechanism for the following stereoselective and enantioselective reaction [279].

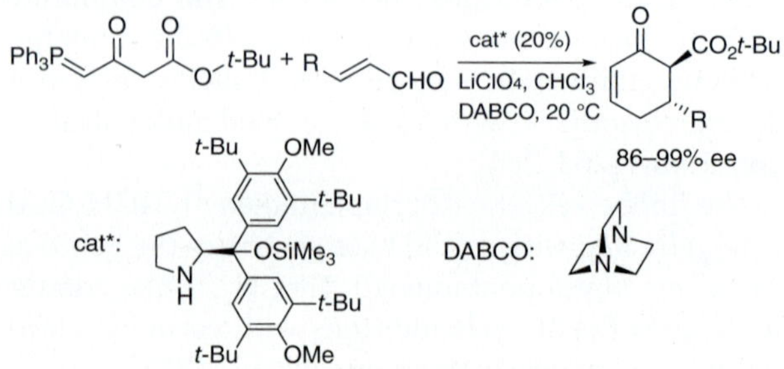

5.3.9 Diels–Alder reaction: concerted and non-concerted mechanisms compete

In 1928, Diels and Alder [129] disclosed their reaction that has become one of the most powerful synthetic tools for the construction of six-membered ring systems [280–284]. They are (4+2)-cycloadditions combining conjugated dienes with alkenes or alkynes (dienophiles), giving cyclohexenes or cyclohexa-1,4-dienes, respectively. Diels and Alder were awarded the Nobel Prize in 1950 "for their discovery and development of the diene synthesis." First applications to the total synthesis of natural products were reported in 1951 by Woodward et al. with their synthesis of racemic cholesterol [285] and cortisone [286] (see also the contribution of Sarett et al. [287]) and by Stork et al. with their synthesis of cantharidin [288]. Since then, the Diels–Alder reaction has been widely used in the synthesis of natural products and analogs [289], including in catalyzed asymmetric synthesis [290–299]. Hetero-Diels–Alder reactions [300, 301] are (4+2)-cycloadditions involving heterodienophiles [302] such as carbonyl compounds (RCHO, RCOR′) [303, 304], thiocarbonyl compounds (RC(R′)=S) [305], selenocarbonyl compounds (RC(R′)=Se) [306], imines (RCH=NR′) [307–309], azo-compounds (R—N=N—R′), nitroso compounds (R—N=O) [310, 311], compounds with C=P double bonds [312], *N*-sulfinyl imines (R—N=S=O, N=S bond as dienophile) [313], sulfonylsulfines (RSO_2(R′S)C=S=O, C=S bond as dienophile) [314], singlet oxygen (^{1}O=O, Section 6.9.3), or sulfur dioxide (SO_2) [315–317]. Heterodienes can be α,β-unsaturated aldehydes [318], α,β-unsaturated imines [319–321], thiones [322], *N*-acylimines [323], 2,3-diazadienes [324–326], nitrosoalkenes (RCH=CH—N=O), nitroalkenes ($RCH=CHNO_2$) [301] and α,α'-dioxothiones (RC(=O)—C=S)—COOR′) [327]. Many of these cycloadditions are catalyzed by Lewis or protic acids (Sections 7.6.6 and 7.6.7), including enantiomerically pure acids that permit asymmetric induction [292, 298, 308, 309].

Diels and Alder were not the first to observe the formation of cyclohexene derivatives from diene and alkene reactions. In 1892, Zincke observed the cyclodimerization of tetrachlorocyclopentadienone, and, in 1910, Lebedev recognized 4-vinylcyclohexene (Scheme 5.21) as the cyclodimer of butadiene [328]. In 1912, Stobbe and Reuss studied the cyclodimerization of cyclopentadiene into dicyclopentadiene [329], both compounds already discovered in 1896 [330]. In solution, the cyclopentadiene cyclodimerization follows a second-order rate law. In pure liquid, cyclopentadiene is polymerized following a first-order rate law, concurrently with its cyclodimerization [331, 332]. For long, dicyclopentadiene was considered to be a (2+2)-cycloadduct [333] rather than (4+2)-cycloadduct **175** [334].

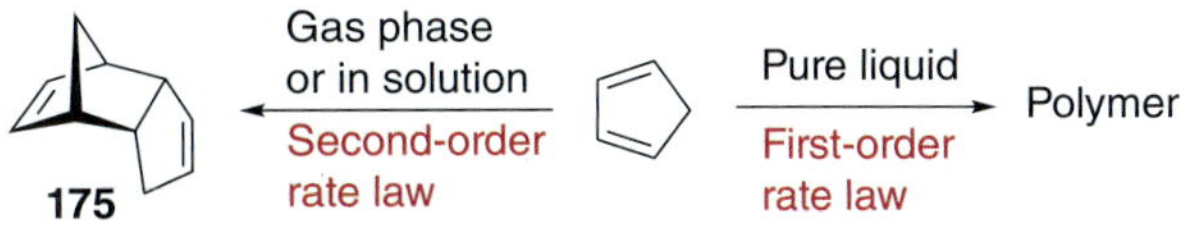

In 1920, Von Euler and Josephson described the double (4+2)-cycloaddition of isoprene to benzoquinone [335] and, in 1925, Diels et al. [336] reported the first hetero-Diels–Alder reaction using diethyl azodicarboxylic diester (**176**) as dienophile:

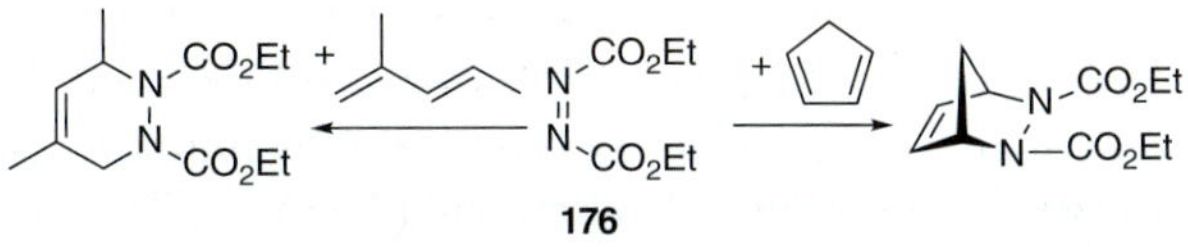

In 1935, Wasserman postulated a concerted mechanism for the Diels–Alder reaction. In 1936, Littmann found that polymers that are formed concurrently during Diels–Alder reactions are copolymers of dienes and dienophiles [337]. This supports the hypothesis of formation of 1,4-diradical intermediates arising from the condensation of dienes and dienophiles. Two types of 1,4-diradicals can form. When an olefinic dienophile combines with a *s-trans*-1,3-diene, a (*E*)-hex-5-ene-1,4-diyl diradical is generated that cannot cyclize into a cyclohexene derivative. In contrast, when the dienophile combines with a *s-cis*-1,3-diene, the corresponding (*Z*)-hex-5-ene-1,4-diyl diradical forms, which can cyclize readily into the cyclohexene product. The latter process (isomerization) can be much faster than polymerization (condensation). In a typical Diels–Alder reaction that combines an electron-poor dienophile (e.g. acrylic ester) with an electron-rich 1,3-diene (e.g. isoprene), the resulting 1,4-diradical is an electron-poor radical at C(1) and an electron-rich radical at C(4) (allyl radical). If the bulk of both radicals (centers C(1), C(4), and C(6)) of the 1,4-diradical are similar, the radical at C(1) adds preferentially to a molecule of 1,3-diene giving an electron-rich (allyl) radical that then adds preferentially to a molecule of the electron-poor dienophile. This leads to an alternating copolymer of the diene and dienophile as observed by Littmann. Depending on substituents, other types of polymers can form [332, 338]. The 1,4-diradical mechanism was adopted by Woodward and Katz [339]

and criticized by Dewar in the same year [340, 341]. For the retro-Diels–Alder reaction of norbornene into cyclopentadiene and ethylene, Zewail and coworkers found in their femtosecond real-time studies that both concerted and nonconcerted trajectories are possible. In both routes, stereochemical retention ($[\sigma^2s+\sigma^2s+\pi^2s]$-cycloreversion) is a consequence of the femtosecond C—C bond dynamical time scale [342]. It is a common practice to add a small amount of a radical scavenger (e.g. hydroquinone or another phenol, amine *N*-oxides, thiols, and tin trialkylhydrides) to the reaction mixture of a Diels–Alder reaction. This leads to higher yields of cycloadducts as one retards the competitive formation of polymers.

The addition of ethylene to butadiene proceeds at 175 °C in the gas phase and competes (12%) with the cyclodimerization (88%) of butadiene (Scheme 5.21). The cycloadducts consist of cyclohexene (99.98%) and vinylcyclobutane (0.02%). The latter product is stable under the conditions of its formation. From the rate constants evaluated by Bartlett and Schueller [343] for the two competitive (4+2)- and (2+2)-cycloadditions, one calculates at 175 °C the free enthalpies of activation given in Scheme 5.37. The Diels–Alder reaction butadiene + ethylene → cyclohexene has an exothermicity of 39.9 kcal mol^{-1} (25 °C, gas phase, 1 atm, Section 1.6.1) and a reaction barrier $\Delta^{\ddagger}H$ = 27.5 kcal mol^{-1} [344]. Thermochemical calculations suggest a concerted mechanism for the formation of cyclohexene, whereas the formation of vinylcyclobutane involves the intermediacy of diradical intermediate **177** ((*E*)-hex-5-ene-1,4-diyl diradical). Tetrafluoroethylene reacts with butadiene giving 1,1,2,2-tetrafluoro-3-vinylcyclobutane exclusively [345]. The reaction involves a 1,4-diradical intermediate [346].

The standard heat of formation of **177** can be derived from the C—C homolysis of cyclohexene: $\Delta_f H°$(**177**) = $\Delta_f H°$(cyclohexene) + $DH°$(CH_2=CH—$CH_2^•$/Et$^•$) = −1.0 (Table 1.2) + 72 (Table 1.7) = 71 kcal mol^{-1}. Alternatively, it can be estimated from the homolytical dissociation enthalpies of C(3)—H and C(6)—H of hex-1-ene: $\Delta_f H°$(**177**) = $\Delta_f H°$(hex-1-ene)+$DH°$(*n*-Bu$^•$/H$^•$)+$DH°$(CH_2=CH—CH(Me)$^•$) − $DH°$(H$^•$/H$^•$) = −10.2 + 101.1 + 84 − 104.2 = 70.7 kcal mol^{-1}. One can thus estimate an activation barrier equal to or larger than $\Delta^{\ddagger}H$(butadiene + ethylene → **177**) = 71 − 26.0 − 12.5 = 33.5 kcal mol^{-1}. The entropy of activation ($\Delta^{\ddagger}S$) of this reaction can be approximated to $S°$(hex-1-ene) − $S°$(butadiene) − $S°$(ethylene) (Table 1.2) + 15 eu (for the liberation of three rotations about three C—C bonds in **177** compared with cyclohexene) = 92.7 − 70 − 52.5 + 15 ≈ −15 eu (calmol^{-1} K^{-1}). Thus, $\Delta^{\ddagger}G$(butadiene + ethylene → **177**) must be equal to or larger than 33.5 − 448 (−0.015) = 40.2 kcal mol^{-1} at 448 K (175 °C). The thermochemical calculations are consistent with a nonconcerted mechanism involving diradical intermediate **177** for the (2+2)-cycloaddition as the measured $\Delta^{\ddagger}G$ (46.5 kcal mol^{-1}) is higher than the minimal value $\Delta^{\ddagger}G$ (40.2 kcal mol^{-1}) estimated for this mechanism. In contrast, the diradical mechanism does not hold for the (4+2)-cycloaddition butadiene + ethylene → cyclohexene as the measured $\Delta^{\ddagger}G$ (38.9 kcal mol^{-1}) is smaller than the minimal value $\Delta^{\ddagger}G$ (40.2 kcal mol^{-1}) estimated for a mechanism involving the intermediacy of (*Z*)-hex-5-ene-1,4-diyl diradical. Quantum mechanical calculations predicted similar stabilities for both (*E*)-(**177**) and (*Z*)-hex-5-ene-1,4-diyl diradical within c. 2 kcal mol^{-1} [347]. This indicates that the Diels–Alder reaction butadiene + ethylene → cyclohexene at 175 °C in the gas phase has an energy of concert of c. 1.3 kcal mol^{-1} (5–8 kcal mol^{-1} by quantum mechanical calculations [348, 349]).

A criterion used frequently for the concertedness of the Diels–Alder reaction is its suprafacial/suprafacial stereoselectivity ($[\pi^4s+\pi^2s]$): this is the "cis rule" formulated by Alder and Stein [350]. In the case of the addition of (*Z*)-1,2-dideuterioethylene to 1,1,4,4-tetradeuteriobutadiene (Scheme 5.38a), the suprafacial mode for ethylene (concerted) is preferred over the formal antarafacial mode (via diradical intermediate?) by a factor of 100 at 185 °C (1800 psi = 122.5 atm) [351]. The addition of ethylene

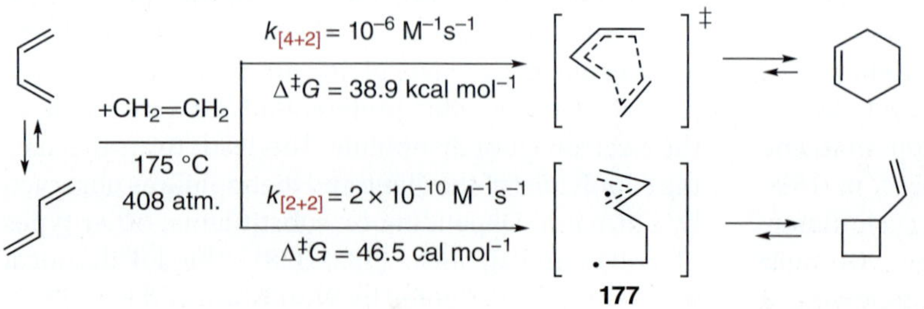

Scheme 5.37 $[\pi^4+\pi^2]$- and $[\pi^2+\pi^2]$-cycloaddition of butadiene to ethylene compete in the gas phase at 175 °C next to the cyclodimerization of butadiene (see Scheme 5.21).

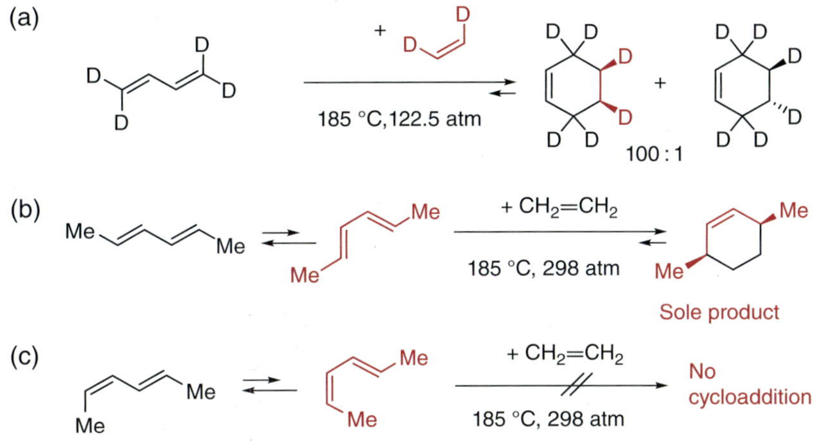

Scheme 5.38 Stereoselectivity of the Diels–Alder reactions of simple dienes with ethylene. (a) The suprafacial mode is preferred for ethylene; (b) the suprafacial mode is preferred for the 1,3-disubstituted 1,3-butadienes; and (c) steric hindrance impedes the formation of the s-cis-conformer of the diene, thus retarding significantly its cycloaddition.

to (*E*,*E*)-hexa-2,4-diene at 185 °C (292.5 atm) is stereospecific and generates *cis*-3,6-dimethylcyclohexene as a unique product of reaction (Scheme 5.38b). Under the same conditions, ethylene (Scheme 5.38c) does not add to (*E*,*Z*)-hexa-2,4-diene [343] because its *s-cis-*diene conformer that is required for the concerted [$\pi^4+\pi^2$]-cycloaddition is unstable for steric reasons. In agreement with the Woodward–Hoffmann rules, the Diels–Alder reactions of many simple 1,3-dienes and ethylenic dienophiles are concerted and prefer the suprafacial/suprafacial mode of cycloaddition. Two-step mechanisms involving diradical intermediates should lead to a loss of stereoselectivity because of the relatively low barrier (3–4 kcal mol^{-1}, see Section 2.6.1) for rotation about the σ(C—C) bonds in (*Z*)-hex-5-ene-1,4-diyl diradical intermediates. Nevertheless, it should be noticed that ring closing of boat-shape (*Z*)-hex-5-ene-1,4-diyl diradical intermediates into corresponding cyclohexenes might have energy barriers significantly smaller than the rotation barriers to be overcome for conformational changes in the diradical intermediates.

The competition between one-step concerted [$\pi^4+\pi^2$]-cycloaddition and the formation of diradical intermediate **101** (Scheme 5.21) is demonstrated for the cyclodimerization of butadiene. Under a pressure of 1 atm and at 120 °C, the Diels–Alder cyclodimerization of (*Z*,*Z*)-1,4-dideuteriobutadiene giving the corresponding 4-vinyl-cyclohexene is 97% stereoselective following the [π^4s+π^2s] mode. Under 8 kbars, the stereoselectivity increases to >99%. Activation volume ($\Delta^‡V$, Section 3.6.1) of the [$\pi^4+\pi^2$]-cycloaddition is 13.3 cm^3 mol^{-1} more negative than that of the [$\pi^2+\pi^2$]-cycloaddition. This demonstrates that the [$\pi^2+\pi^2$]-cycloaddition involves the formation of (*E*,*E*)-octa-2,6-dien-1,8-diyl diradical intermediate **101** (as suggested by thermochemical calculations) that is less compact than the transition structure of the concerted [$\pi^4+\pi^2$]-cycloaddition [352]. Quantum mechanical calculations predicted that the [π^4s+π^2s]-cyclodimerization of butadiene can follow a one-step, concerted mechanism or a two-step process involving (*E*,*Z*)-octa-1,7-diene-3,6-diradical (stereoisomer of diradical **101**) [348].

The cyclodimerization of chloroprene (2-chlorobutadiene) gives mixtures of cycloadducts **178–182**. Products **178** and **179** are the two possible regioisomeric (4+2)-cycloadducts involving the vinyl group of one chloroprene unit as dienophile. The major adduct **180** results from a Diels–Alder reaction in which the chloroalkene moiety of one chloroprene unit plays the role of the dienophile. The latter reaction is highly regioselective contrary to the cycloaddition forming cyclohexenes **178** + **179**. Finally, cycloadducts **181** and **182** result from the concurrent (2+2)-cycloaddition in which only the chloroalkene moiety (CH_2=CCl) of chloroprene is involved in the formation of the cyclobutane products and with high regioselectivity. The pressure dependence of the formation of products of cyclodimerization of chloroprene gives the activation volumes reported below [353].

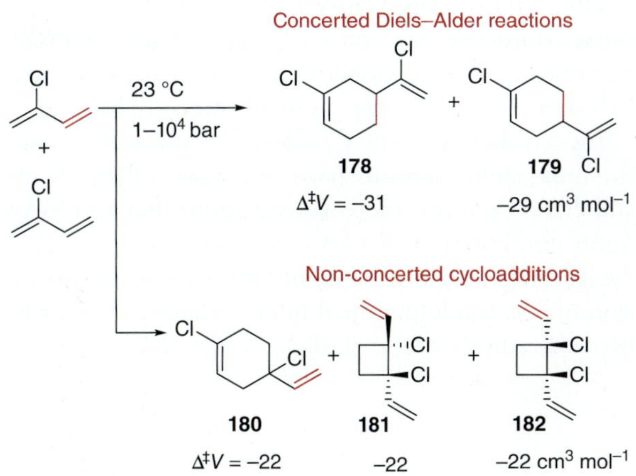

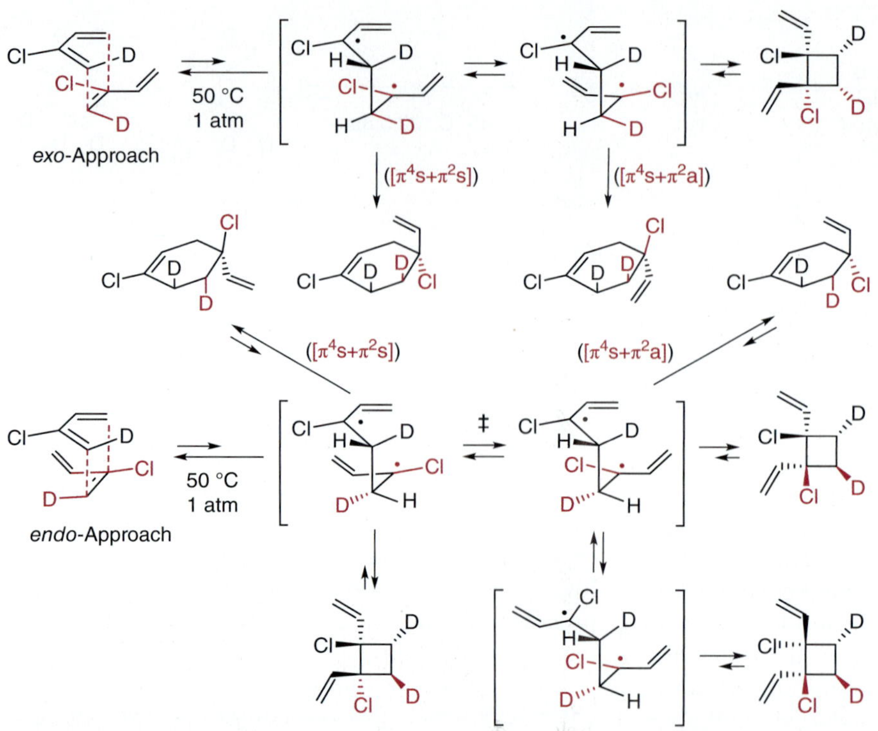

Scheme 5.39 The nonstereoselective cyclodimerization of (*E*)-1-deuteriochloroprene (50 °C, 1 atm): proof for nonconcerted mechanisms involving 3,6-dichloroocta-1,7-dien-3,6-diyl diradicals that have the time to undergo conformational changes before ring closure into 1,4-dichloro-4-vinylcyclohexenes (Diels–Alder reaction) and *trans*-1,2-dichloro-1,2-divinylcyclobutanes ((2+2)-cycloaddition).

The more negative activation volumes observed for the formation of **178** and **179** are consistent with concerted Diels–Alder reactions (tight transition structures). The less negative $\Delta^{\ddagger}V$ values measured for the formation of **180–182** are consistent with a two-step mechanism involving the formation of 3,6-dichloroocta-1,7-diene-3,6-diyl diradical intermediates (less tight transition structures) in their product-determining step. The nonconcerted mechanism is confirmed by the cyclodimerization of (*E*)-1-deuteriochloroprene, which is not stereoselective (gives a mixture of products resulting from formal suprafacial/suprafacial and suprafacial/antarafacial cycloadditions) as shown in Scheme 5.39.

The Diels–Alder cyclodimerization of (*E*)-1-deuteriochloroprene is not stereoselective in a double sense. Both the *exo* and *endo* approaches compete, and both types of cycloadducts resulting from formal [π^4s+π^2s] and [π^4s+π^2a]-cycloaddition are formed.

The 3,6-dichloroocta-1,7-diene-3,6-diyl radicals are the most stable radicals (two secondary allyl radicals stabilized each by chloro substitution) that can form upon combination of two chloroprene units. Other cycloadducts regioisomeric of **180** require the formation of less stable diradical intermediates. Obviously, if the diradicals are unstable, concerted Diels–Alder reactions become competitive as demonstrated with the concurrent formation of adducts **178** + **179**.

Applying the criterion of the difference of activation volumes $\Delta\Delta^{\ddagger}V$ for two competitive cycloadditions of a given pair of cycloaddents, one finds that the (4+2)-cyclodimerization of cyclohexa-1,3-diene follows a concerted mechanism, whereas its (2+2)-cyclodimerization is a stepwise process involving diradical intermediates. This is also the case for the Diels–Alder reaction of butadiene to 1-cyanovinyl acetate giving 1-cyanocyclohex-3-en-1-yl acetate (concerted) that competes with the (2+2)-cycloaddition forming 1-cyano-2-vinylcyclobutyl acetate (stepwise). In contrast, the additions of butadiene to 1,1-dichloro-2,2-difluoroethylene and to dicyclopropylidene, and the reaction of (*E*)-piperilene with 1,2-didehydrobenzene (benzyne), have similar activation volumes for both the (4+2)- and (2+2)-cycloadditions, consistently with nonconcerted mechanisms for these two types of reactions that involve the formation of diradical intermediates. This is also the case for the homo-Diels–Alder reaction of norbornadiene (Section 5.3.17) with didehydrobenzene that competes with the (2+2)-cycloaddition. Similarly, (4+2) and (4+4)-cyclodimerizations

(Section 5.3.21) of *ortho*-quinodimethane and of 3,3,4,4,5,5-hexamethyl-1,2-dimethylidenecyclopentane all involve diradical intermediates [352].

5.3.10 Concerted Diels–Alder reactions with synchronous or asynchronous transition states

Maleic anhydride adds to α-pyrone giving cycloadduct **183**. The latter undergoes a cycloreversion (retro-hetero-Diels–Alder reaction) producing 1,2-dihydrophthalic anhydride (**184**), which then adds to maleic anhydride giving **185**. Carbon and oxygen kinetic primary isotope effects ($k_{^{12}C}/k_{^{13}C} = 1.030$ and $k_{^{16}O}/k_{^{18}O} = 1.014$) have been determined for reaction **183** → **184** + CO_2 and interpreted in terms of a concerted cycloreversion with an asynchronous transition structure **186** in which the C—C bond is nearly broken, whereas the C—O is virtually intact [354].

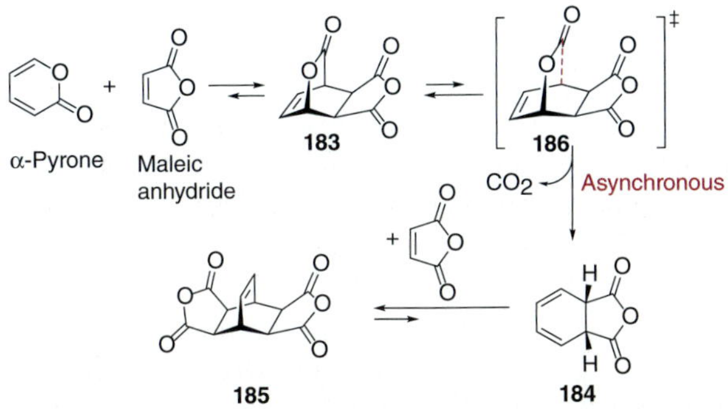

Secondary deuterium isotope effects measured for the retro-Diels–Alder reaction of 9,10-dihydro-9,10-ethanoanthracenes (dibenzobicyclo[2.2.2]octa-2,5-diene) **187d$_0$** (rate constant: k_0), **187d$_2$** (k_2), and **187d$_4$** (k_4) suggested a concerted synchronous mechanism for this cycloreversion. For a concerted synchronous transition structure, $(k_2/k_0)^2 - (k_4/k_0) = x = 0$ is expected (multiplicative effect), whereas for an asynchronous transition structure or a two-step mechanism involving the formation of a 1,4-diradical, x must be different from zero (non-multiplicative effect). As the observed rate constant ratios $k_2/k_0 = 0.924 \pm 0.005$ and $k_4/k_0 = 0.852 \pm 0.007$ are nearly multiplicative, it was concluded that the cycloreversion is concerted with a symmetrical or nearly symmetrical transition structure [355].

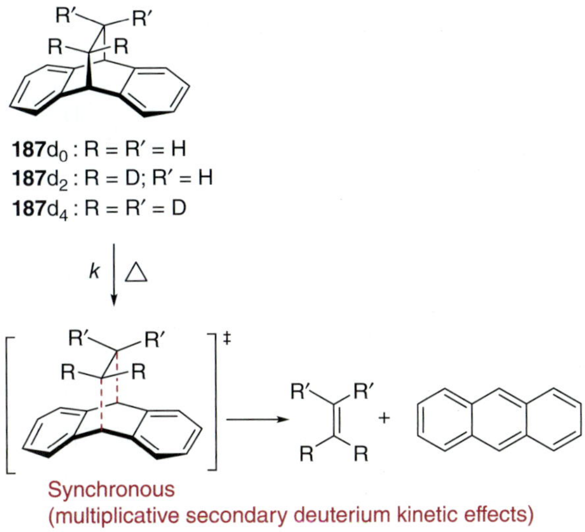

Secondary deuterium isotopic effects in the reaction of fumaronitrile with isoprene (**188**), 1,1-dideuterioisoprene (**189**), and 4,4-dideuterioisoprene (**190**) are nearly the same at 25 °C, thus confirming synchronous concerted mechanism for this Diels–Alder reaction (transition structure **191**), which engages a symmetrical dienophile and a nearly symmetrical diene. The Diels–Alder reaction of isoprene to ethylene-1,1-dicarbonitrile, a nonsymmetrical dienophile, gives the two regioisomeric cycloadducts **194** and **195** in a ratio of 7 : 1 at 25 °C. Both reactions follow concerted asynchronous mechanisms as demonstrated by the secondary deuterium isotope effect given in Scheme 5.40. In the case of the formation of the major product **194**, k_{189}/k_{188} is significantly >1.00, whereas k_{190}/k_{188} equals to unity. This means that the σ(C—C) bond between C(1) of isoprene and C(β) of the dienophile is formed to a large extent in the transition state, whereas the σ(C—C) between C(4) of isoprene and C(α) of the dienophile is almost not formed. The transition state **192** of this reaction can be represented by two limiting structures that are a 1,4-diradical and the corresponding zwitterion (diradicaloid model). Both the diradical and zwitterion-limiting structures are stabilized by the substituents of the diene (2-methyl group) and of the dienophiles (two 1-cyano groups). The alternative transition state **192′** does not form as it cannot benefit from these stabilizing substituent effects. The kinetic deuterium isotope effects measured for this reaction confirm it. The regioselectivity of the reaction comes from the minor transition state

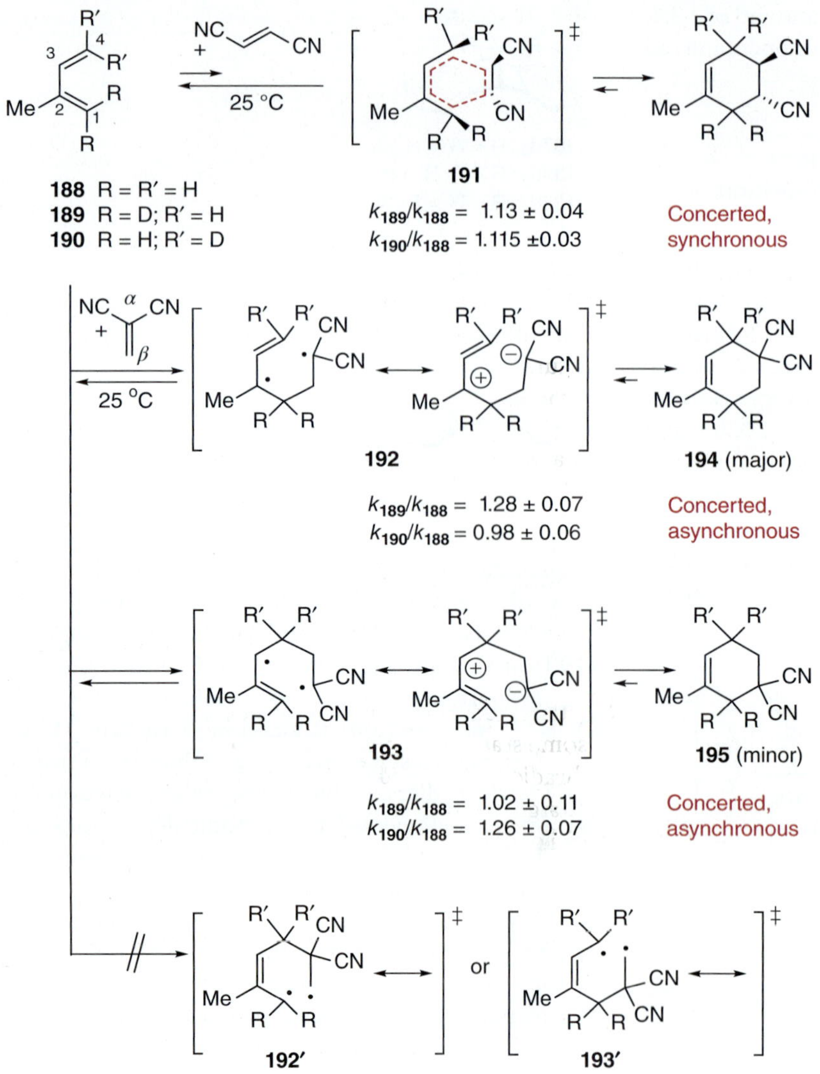

Scheme 5.40 Transition-state structure variations in the Diels–Alder reaction as evidenced by secondary deuterium kinetic isotope effects.

193 builds radical character or positive charge on a 2° carbon where the major transition state builds radical character or positive charge on a 3° carbon, thus explaining the regioselectivity observed. The secondary deuterium isotope effects measured for this reaction confirm that the σ(C—C) bond between C(4) of the diene and C(2′) of the dienophile is formed to a large extent in transition state **193**, whereas the σ(C—C) bond between C(1) of the diene and C(α) of the dienophile is nearly not formed. The alternative transition state **193′** would benefit from the stabilization effect of the methyl group of the diene, but not from the two cyano groups of the dienophiles [356].

Quantum mechanical calculations and experimental secondary kinetic isotope effects confirm that the Diels–Alder reaction of butadiene + ethylene and the cycloreversion of cyclohexene into butadiene + ethylene both follow the same concerted synchronous mechanism [349, 357, 358]. Calculated and experimental kinetic isotope effects are consistent with a moderately asynchronous transition structure for the reaction of isoprene with maleic anhydride [359]. Highly asynchronous transition structures are found for the Diels–Alder reactions of bis(boryl)acetylenes, dialkyl acetylenedicarboxylates, triazolinediones, and dialkyl maleates [360], whereas the (4+2)-cycloaddition of (*E*)-2-phenylnitroethylene to cyclopentadiene follows a concerted, nearly synchronous mechanism [361].

5.3.11 Diradicaloid model for transition states of concerted Diels–Alder reactions

If one considers the two main reaction coordinates that are the C(1)$\cdots$C(β) and C(4)$\cdots$C(α) distances between the diene and dienophile that combine in

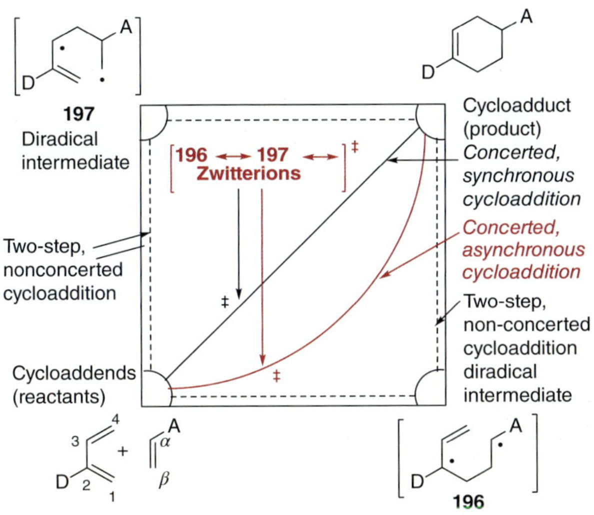

Figure 5.26 More O'Ferrall–Jencks type of diagram representing possible trajectories (family of mechanisms) of $[\pi^4+\pi^2]$-cycloadditions (x-axis: reaction coordinate C(1)$\cdots$C(β) distance; y-axis: reaction coordinate C(4)$\cdots$C(α); z-axis perpendicular to the paper: free energy).

a Diels–Alder reaction, the diagram of Figure 5.26 (More O'Ferrall–Jencks type of diagram) can be constructed for which the x- and y-axes in the plane of the paper are the reaction coordinates defined above and the z-axis vertical to the paper describes the free enthalpy of the cycloaddents evolving to cycloadduct and *vice versa*. There are two limiting trajectories (mechanisms) that correspond to two-step processes involving the formation of 1,4-diradicals **196** and **197**. The diagonal of the diagram is the projection in the x,y-plane of the trajectory for a concerted, synchronous Diels–Alder reaction. In between this trajectory and the diradical mechanisms, there is a lot of room for concerted asynchronous cycloadditions. The transition states of the concerted mechanisms can be represented as a combination of two limiting structures that are diradicals **196** and **197**. Their relative weights depend on their relative stability. Thus, if **196** is more stable than **197**, the transition structure of the concerted cycloaddition will resemble more **196** than **197** and the trajectory will be in the right part of the diagram. Alternatively, if **197** should be more stable than **196**, the transition structure of the cycloaddition will resemble more **197** than **196** and, accordingly, the trajectory of the corresponding reaction will be in the upper-left part of the diagram.

Other limiting structures than diradicals **196** and **197** can contribute to the transition states of concerted (synchronous and asynchronous) Diels–Alder reactions. They are charge transfer configurations **196′** and **197′** as shown in Figure 5.27. Pross and Shaik also consider electronically excited configurations such as triplet state of reactants and products [362]. In analogy with the Bell–Evans–Polanyi theory ($\Delta^{\ddagger}H$(activation enthalpy) $= \alpha\Delta_r H + \beta$ with $\Delta_r H$, heat of reaction; β, intrinsic barrier for thermoneutral reaction, and α varies between 0 and 1) of concerted radical exchanges ($X^\bullet + RY \rightarrow RX + Y^\bullet$) and radical additions ($X^\bullet + RCH{=}CH_2 \rightarrow RCH^\bullet{-}CH_2X$), Diels–Alder reaction transition states result from the interaction of ground-state electronic configuration hypersurface of reactants evolving to products with charge transfer configuration hypersurfaces that involve electron exchange between reactants (Figures 5.27 and 5.28). Projections of a possible reaction trajectory into a plane consisting of a reaction coordinate (x-axis: shortening of C(1)$\cdots$C(β) and C(4)$\cdots$C(α) distances, see Figure 5.26) and enthalpy (y-axis) gives the black curve of Figure 5.27. The ground-state electronic configuration of diene + dienophile sees its enthalpy increasing on compressing them along the reaction coordinate (steric repulsions, molecular deformations, etc.). At some stage, this might lead to the formation of a true diradical intermediate **198**. There are two possible charge transfer configurations **199** and **200** for the cycloaddents. One of them (**199**) corresponds to the electron transfer from the diene to the dienophile: it has an enthalpy higher than that of the reactants by the sum EI(diene) + (−EA(dienophile)). A second charge transfer configuration (**200**) corresponds to an electron transfer from the dienophile to the diene and its enthalpy is EI(dienophile) + (−EA(diene)) above that of reactants. When dienes are electron rich and dienophiles are electron poor, which corresponded to Diels–Alder reactions with normal electronic demand (e.g. furan + maleic anhydride; isoprene + acrylonitrile), charge transfer configurations of type **199** are lower lying in enthalpy than charge transfer configurations of type **200**. In contrast, for Diels–Alder reactions with inverse electronic demand (e.g. enol ethers + acrolein), electron transfer dienophile → diene requires less energy than electron transfer diene → dienophile and renders charge configurations of type **200** more stable than **199**. These reactions have been recognized for the first time in 1949 by Bachmann and Deno [363]. Ionization energies (EI) and electron affinities (−EA) (Section 1.8) are thermodynamic data. They can be associated with HOMO energies (Koopmans' theorem [364]) and LUMO energies (calculated numbers) in the FMO theory [365, 366]. When diene and dienophile have similar ionization energies and electron affinities, electron exchange will also occur

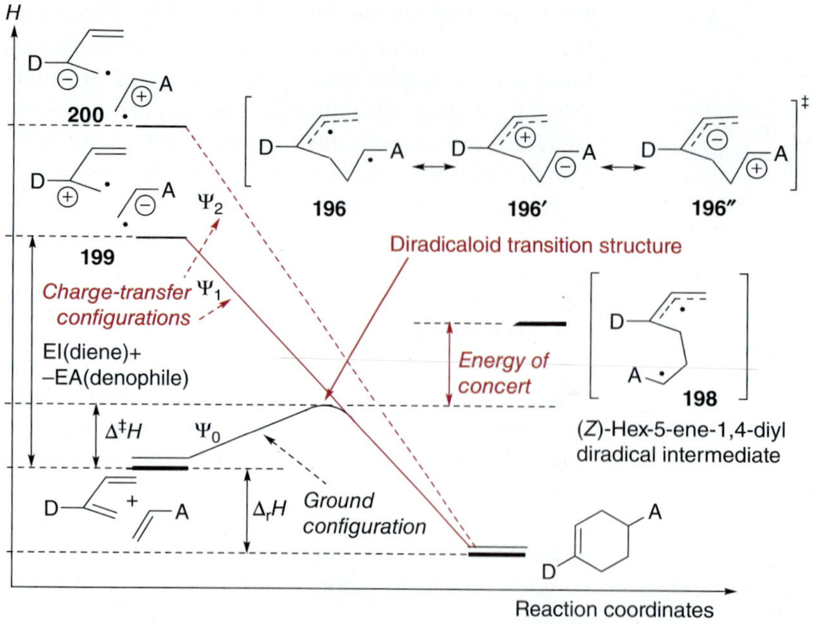

Figure 5.27 The diradicaloid model for transition states of concerted Diels–Alder reactions. The lower the sum EI(diene) + ($-EA$(dienophile)), or EI(dienophile) + ($-EA$(diene)), the faster the cycloaddition, the higher the degree of concert.

and both types of charge transfer configurations **199** and **200** will have comparable enthalpies. The enthalpies of the charge transfer configurations **199** and **200** diminish when the cycloaddents evolve from reactants to product along the reaction path (electrostatic stabilization due to the diminishing distance between a radical cation and a radical anion). This gives the red projections associated with wave functions Ψ_1 and Ψ_2 of the charge transfer configuration trajectories of Figure 5.27. These trajectories may or may not cross the ground-state (black) configuration trajectory associated with wave function Ψ_0 before the formation of true diradical **198**. If overlap is not nil between the wave function pairs Ψ_0/Ψ_1 and Ψ_0/Ψ_2, the energy of the state wave function $\Psi^{\ddagger}$ will be lower than the energy of their crossing point (avoiding crossing, see Longuet-Higgins treatment of $[\pi^4c]$- and $[\pi^4d]$-electrocyclic reactions (Section 5.2.2) and $[\pi^2+\pi^2]$-cycloadditions (Section 5.3.1)). The transition structure of the transition state of the Diels–Alder reaction can thus be represented as a combination of limiting structures that are diradical ↔ zwitterions as shown with **196** ↔ **196′** ↔ **196″** (Figure 5.27). This is the diradicaloid model advocated by Dewar et al. [367–369] and others [356, 370].

The smaller the sum EI(diene) + ($-EA$(dienophile)) and/or EI(dienophile) + ($-EA$(diene)), the more polarizable are the cycloaddents and the lower lying in energy will be the crossing point of wave functions Ψ_0 with Ψ_1 and/or Ψ_2 and, thus, the faster is their cycloaddition. The theory predicts that the Diels–Alder reactions of symmetrically substituted dienes and dienophiles have a better chance to be synchronous than nonsymmetrically substituted cycloaddents. As we shall see, the diradicaloid model allows one to predict the chemo-, regio-, and stereoselectivity of the cycloadditions, as well as the effects of additives (that can coordinate the substituents of cycloaddents) and solvent on reaction rates and regio- and stereoselectivity. Examples of Diels–Alder reactions with normal electron demand are reported in Table 5.2. The highly electrophilic dienophile ethylene–TCNE adds to substituted butadienes. The electron-richest dienes are the most reactive. Interestingly, the acceleration effect observed with (E,E)-1,4-diphenyl- and (E,E)-1,4-dimethoxybutadiene compared with butadiene are not larger than the acceleration effects observed for (E)-1-phenyl- and (E)-1-methoxybutadiene compared with butadiene. If the transition states of these reactions would involve symmetrical transition structures (concerted synchronous mechanisms), a multiplicative effect should be observed. As this is not the case, the data demonstrate that the reactions are asynchronous: they involve nonsymmetrical transition states such as diradicaloids **201** ↔ **201′**, for which the zwitterionic limiting structure **201′** has a relatively important weight (carbanionic moiety stabilized by two cyano groups and carbenium ion moiety stabilized by the diene substituent). Several Diels–Alder reactions of electron-rich dienes with electron-poor dienophiles are not concerted and involve the formation of zwitterion intermediates

Table 5.2 Substi tuent effects on the rate of Diels–Alder reactions with normal electronic demand.

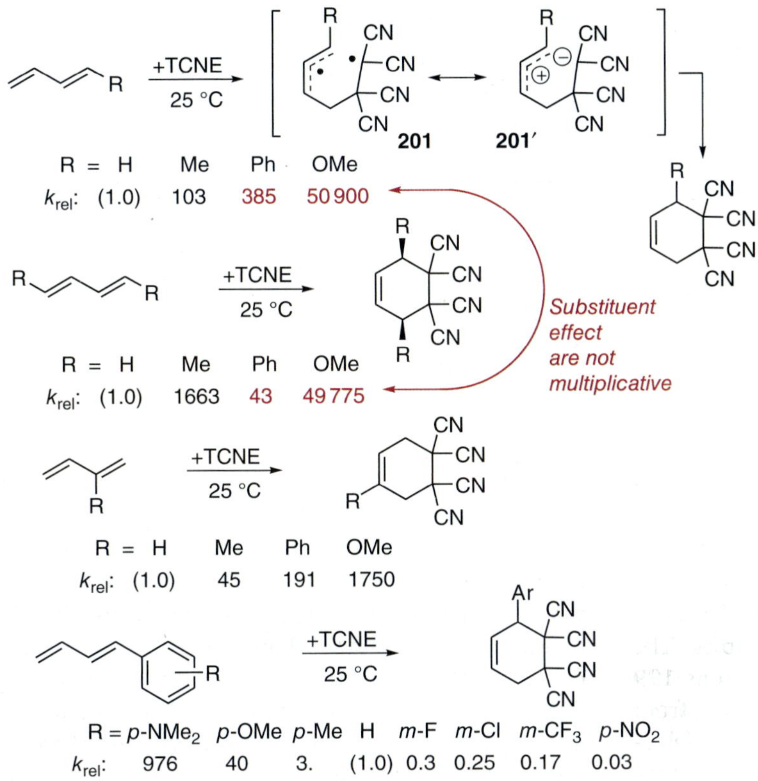

Source: Taken from Ref. [371].

in their rate-determining steps, as exemplified in Scheme 5.41. In the case of the reaction of TCNE with 1,1-dicyclopropylbutadiene, a mixture of (2+2)- and (4+2)-cycloadduct forms at 20 °C in CH_2Cl_2. On heating this mixture to 100 °C, the cyclobutane product is isomerized into the cyclohexene cycloadduct. When run in a more polar solvent such as CH_3CN at 20 °C, the cycloaddents equilibrate with a zwitterion that can be quenched with *p*-toluenethiol before its ring closure into the cycloadduct (Scheme 5.41a) [372]. Dichloroketene adds to electron-rich enones at 5 °C in benzene giving first a zwitterion that cyclizes at 50 °C into the (4+2)-cycloadduct (Scheme 5.41b) [373]. Dienophiles with a β-leaving group X add to electron-rich 1-arylbutadienes giving first zwitterions that can equilibrate with the corresponding ion pairs. The latter induce the formation of diene homopolymers through cationic polymerizations (Scheme 5.41c) [374] (see also the reaction of 4,6-dinitrobenzofuroxan with 1-trimethylsilyloxybuta-1,3-diene [375]).

Diels–Alder reactions with inverse electronic demand such as the cycloadditions of alkenes to dimethyl 1,2,4,5-tetrazine-3,6-dicarboxylate and to 3,6-bis(trifluoromethyl)1,2,4,5-tetrazine [324], and of styrenes to cyclopentadienones, are the fastest for electron-rich dienophiles (see e.g. Table 5.3), and/or for dienes with electron-withdrawing substituents [377]. In the case of the reaction of di(2-pyridyl)-1,2,4,5-tetrazine (**202**) adding to *para*-substituted styrenes, the second-order rate constants are strongly solvent dependent and follow Hammett relationships of type $\log(k_X/k_H) = \rho\sigma$ (σ = substituent (X) constants, ρ = reaction constant). The more negative the ρ-value, the greater the electronic demand in the rate-determining step. The fact that the rate constants are larger in polar solvents like CF_3CH_2OH and water, than in toluene (nonpolar), and the observation that the electronic demand of the transition states is larger in polar and protic solvents than in nonpolar solvents (Table 5.3) are both consistent with asynchronous mechanisms and with the diradicaloid model for their transition states. In polar and protic solvents, the zwitterionic limiting structures are more stabilized relatively than the corresponding diradical limiting structures and cycloaddents. The dipole moments of the transition states are larger than those of the corresponding cycloaddents [376].

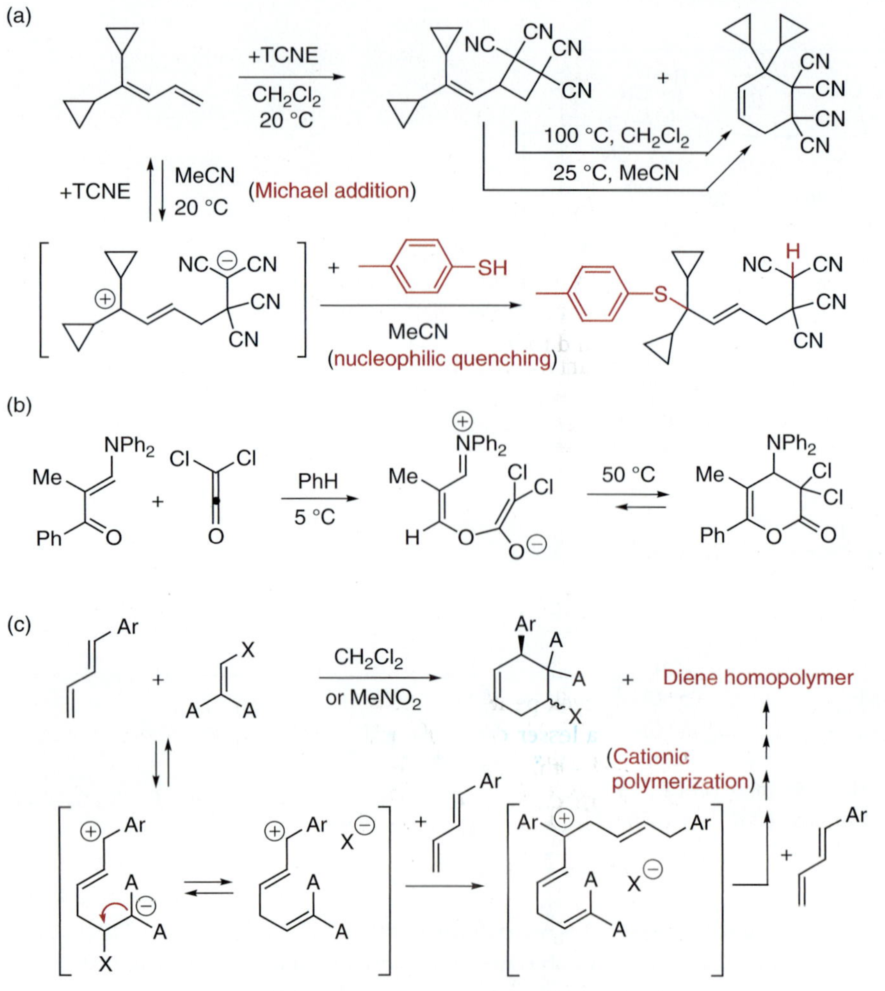

A = CN, COOMe; X = CN, I, Br, Cl, p-$MeC_6H_4SO_3$

Scheme 5.41 Examples of Diels–Alder reactions involving zwitterionic intermediates.

Table 5.3 Substituent and solvent effects on the rate of Diels–Alder reactions with inverse electronic demand.[a]

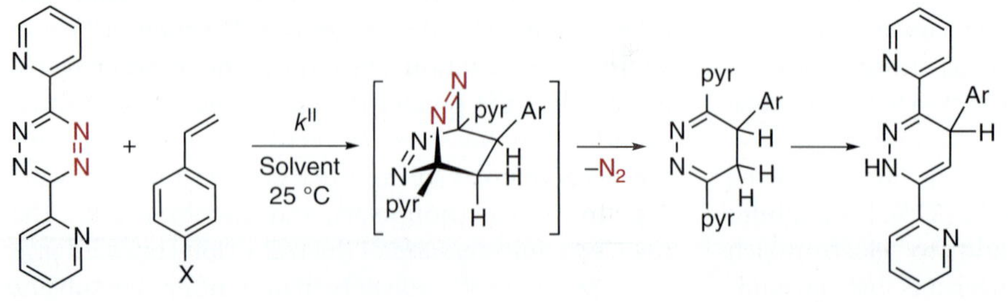

202

In toluene X	OMe	Me	H	Cl	NO_2	ρ
k^{II} (103 $M^{-1}\,s^{-1}$)	5.26	2.85	2.15	1.76	0.83	−0.51
In CF_3CH_2OH	24.5	14.6	7.1	3.40	0.41	−1.64
In 95 : 5 H_2O/t-BuOH	406	218	131	69.1	13.7	−1.32

Source: Taken from Ref. [376].

5.3.12 Structural effects on the Diels–Alder reactivity

The rates of concerted Diels–Alder reactions (and other cycloadditions) depend not only on the ease by which the cycloaddents can exchange electrons. As any other reactions, they are subjected to steric hindrance, geometry, and conformational and flexibility effects (enthalpy required to distort the cycloaddents to reach their geometry in the transition state). At 20 °C, TCNE adds to cyclohexa-1,3-diene and cyclohepta-1,3-diene 2600 and 830 000 times, respectively, as slowly as to cyclopentadiene, even though the ionizing energy of cyclopentadiene is larger (gives an electron less readily) than that of cyclohexa-1,3-diene and that of cyclohepta-1,3-diene (Table 5.4). The reactivity trend cyclopentadiene > cyclohexa-1,3-diene > cyclohepta-1,3-diene is attributed to the C(1)–C(4) distance in these dienes. The shorter the C(1)–C(4) distance, the faster the cycloaddition.

The same explanation has been offered [380] to interpret the lack of reactivity of 1,2-dimethylidenecyclopropane toward TCNE compared with 1,2-dimethylidenecyclobutane and 1,2-dimethylidenecyclopentane that add to TCNE with second-order rate constants $k = 0.04$ and $1.4\ M^{-1}\ s^{-1}$, respectively, at 20 °C. The Diels–Alder reactivity of these exocyclic 1,3-dienes is not correlated with their ionization energies (*EI*) [381]. Although the distance between the termini carbon centers of the nonplanar, chair-shape 1,2-dimethylidenecyclohexane is larger than in planar 1,2-dimethylenecyclopentane, the former diene reacts with TCNE faster ($k = 114\ M^{-1}\ s^{-1}$) than the latter diene. This is attributed to the shorter C(1)–C(4) distance when the exocyclic diene moiety approaches planarity in 1,2-dimethylidenecyclohexane compared with that of 1,2-dimethylidenecyclopentane.

Table 5.4 Diels–Alder reactivity of cyclic conjugated dienes toward ethylenetetracarbonitrile (TCNE, CH_2Cl_2, 20 °C) depends on the C(1)–C(4) distance.

k^{II} ($M^{-1}\ s^{-1}$)	3570	1.37	0.0043
k_{rel}	1.0	1/2600	1/830 000
IE(gas) (eV)	8.58	8.25	8.31
C(1)–C(4) distance (Å)	2.36	2.82	3.15
C(1)–C(2)–C(3)–C(4) torsion angle (°)	0	18.2	0

Source: Taken from Refs [378] and [379].

Little energy is required for the chair-shape 1,2-dimethylidenecyclohexane to adopt a boat conformation with a planar or nearly planar diene moiety (high conformational flexibility). This contrasts with the lack of Diels–Alder reactivity of nearly orthogonal 1,3-dienes such as 2,3-bis(*t*-butyl)butadiene and 3,3,6,6-tetramethyl-4,5-dimethylidenethiepane [382] that cannot adopt the *s-cis*-1,3-diene conformation. The Diels–Alder reactivity of conjugated cycloalkenones toward cyclopentadiene (reaction (5.14)) span over a million-fold range in rate at 25 °C and follows the trend:

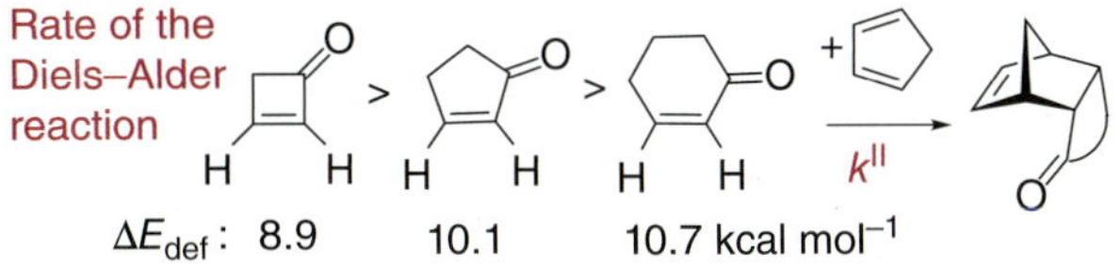

(5.14)

Similar electron affinities are calculated (similar LUMO energies) for all these dienophiles. The higher reactivity of cyclobutenone compared with cyclopent-2-enone and cyclohex-2-enone arises from a lesser distortion energy (ΔE_{def}, see above) required by this dienophile to acquire the geometry it must reach in the transition state of the cycloaddition. The distortion is associated with the bending of C—H bonds out of the plane of the C=C bond to which they are attached, as new C—C bonds are formed. The force constant for bending of the alkene group out of plane is reduced by angle strain in cyclobutenone. This behavior arises from the larger degree of s character in the two olefinic C—H bonds, and the fact that the smaller internal angle in the small rings is more appropriate for the pyramidal transition structure [383]. Transition state distortion enthalpies calculated for polycyclic aromatic hydrocarbons [384], fullerenes, and nanotubes [385] undergoing Diels–Alder reaction with ethylene correlate with the activation enthalpies, even for cases where heats of reaction ($\Delta_r H$) do not correlate with enthalpy activations ($\Delta^{\ddagger}H$) (Figure 5.28).

Benzocyclobutenes equilibrate ([π^4c]-electrocyclic opening, see Section 5.2.5) with the corresponding *ortho*-quinodimethanes. The latter dienes react very quickly in inter- and intramolecular Diels–Alder reactions ((4+2)-cycloadditions (5.15)). In contrast, 3,4-dimethylidenecyclobutene does not add to any dienophiles. This is the consequence of the difference in exothermicity ($\Delta_r H$) for these two cycloadditions. Reaction (5.15) are highly exothermic, whereas reaction (5.16) are much less exothermic. The Dimroth or Bell–Evans–Polanyi principle ($\Delta^{\ddagger}H = \alpha\Delta_r H + \beta$)

states that the activation enthalpy ($\Delta^{\ddagger}H$) of a concerted reaction depends on its exothermicity (Δ_rH) and of an intrinsic barrier β (barrier of thermoneutral reaction). Thus, if two Diels–Alder reactions involve two different dienes and the same dienophile for which the same intrinsic barrier β is realized (the same steric hindrance, the same flexibility, and the same EI and −EA for the two dienes), the reaction with the highest exothermicity is the fastest.

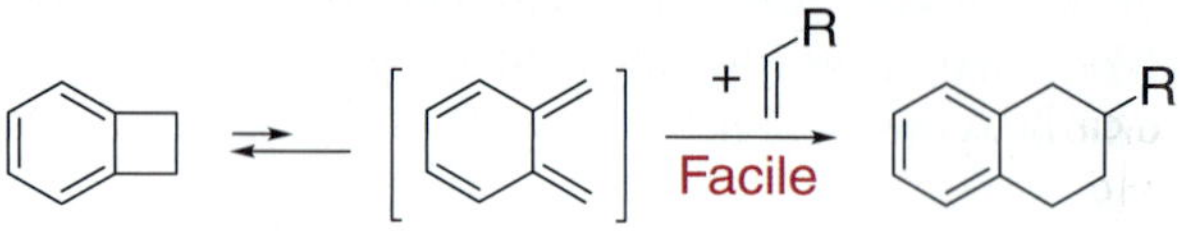

(5.15)

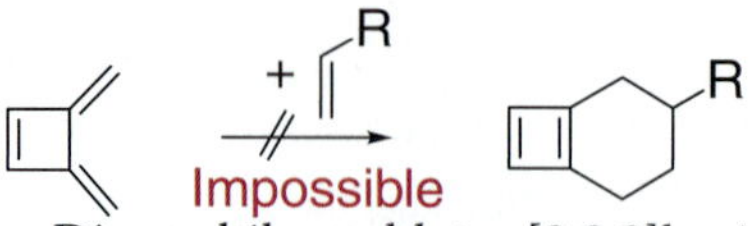

(5.16)

Dienophiles add to [2.2.2]hericene (**203**: hexamethylidenebicyclo[2.2.2]octane) in three successive Diels–Alder reactions with rate constants $k(1)$, $k(2)$, and $k(3)$. For the additions of TCNE in toluene at 25 °C, the kinetic data are given below.

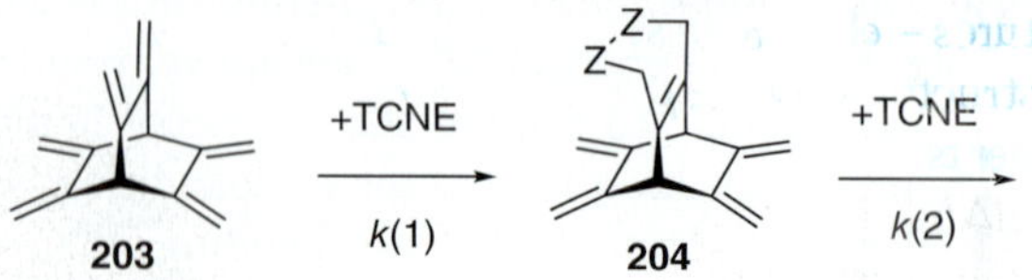

	$k(1)$	$k(2)$
$10^3 \cdot k$(25 °C)($M^{-1}\,s^{-1}$):	136	15.6
$\Delta^{\ddagger}H$ (kcal mol^{-1}):	11.3 ± 0.2	10.9 ± 0.3
$\Delta^{\ddagger}S$ (eu):	−29.5 ± 0.8	−30.1 ± 1.5

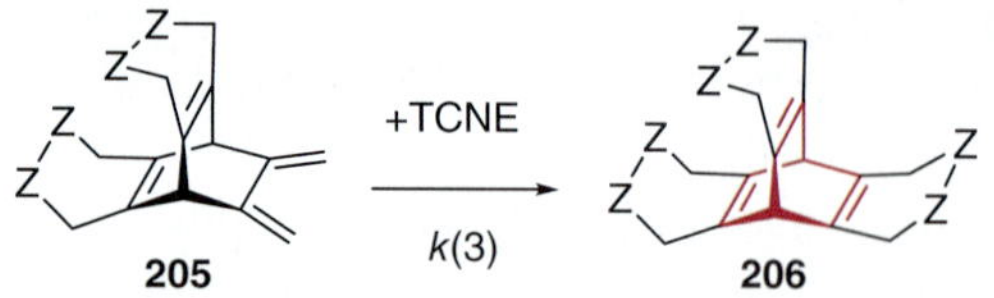

	$k(3)$
$10^3 \cdot k$(25 °C)($M^{-1}\,s^{-1}$):	**0.05**
$\Delta^{\ddagger}H$ (kcal mol^{-1}):	**15.3 ± 1.0**
$\Delta^{\ddagger}S$ (eu):	−28 ± 8

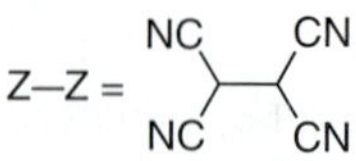

Although the two first cycloadditions have similar rate constants ($k(1) \approx k(2)$), the third addition is significantly slower ($k(3) \ll k(2)$, $k(1)$). As similar activation entropies ($\Delta^{\ddagger}S$) are measured for the three successive reactions, the reactivity difference arises from changes in the activation enthalpy ($\Delta^{\ddagger}H$). The activation enthalpy $\Delta^{\ddagger}H(3)$ is higher than $\Delta^{\ddagger}H(2)$ and $\Delta^{\ddagger}H(1)$ because the third cycloaddition generates a barrelene derivative **206** that is destabilized (see the barrelene effect, Section 4.7.6) with respect to double adduct **205** and monoadduct **204**. Thus, in agreement with the Dimroth principle for one-step, concerted reactions, the lower exothermicity of **205** + TCNE → **206** than for the other cycloadditions is responsible for the smaller rate constant $k(3)$ compared with $k(2)$ and $k(1)$. This interpretation is confirmed by the observation that the two successive Diels–Alder reactions of 2,3,5,6-tetramethylidenebicyclo[2.2.2]octane have similar rates and activation enthalpies, whereas the formation of monoadducts is much faster that the formation of double adducts in the case of the two successive Diels–Alder reactions of 5,6,7,8-tetramethylidenebicyclo[2.2.2]oct-2-ene [386]. Ionization energies (photoelectron spectroscopy in the phase gas) are nearly the same for [2.2.2]hericene (8.38 eV) [387], 5,6,7,8-tetramethylidenebicyclo[2.2.2]oct-2-ene (8.36 eV), 7,8-dimethylidenebicyclo[2.2.2]octa-2,5-diene (8.33 eV) [388], and 2,3,5,6-tetramethylidenebicyclo[2.2.2]octane (8.36 eV) [389]. In the case of reversible (4+1)-cheletropic addition of sulfur dioxide (Section 5.4.4) to [2.2.2]hericene, only the corresponding mono- and double sulfolene are formed. The triple-sulfolene is not formed for thermodynamical reasons [386, 390].

Demonstration of the Dimroth principle or Bell–Evans–Polanyi theory ($\Delta^{\ddagger}H = \alpha\Delta_rH + \beta$) for concerted Diels–Alder reactions is shown in Figure 5.28. One considers the cycloadditions of dienes **D1** and **D2** to dienophile **A** giving the corresponding cycloadducts **D1 − A** and **D2 − A**. Reaction

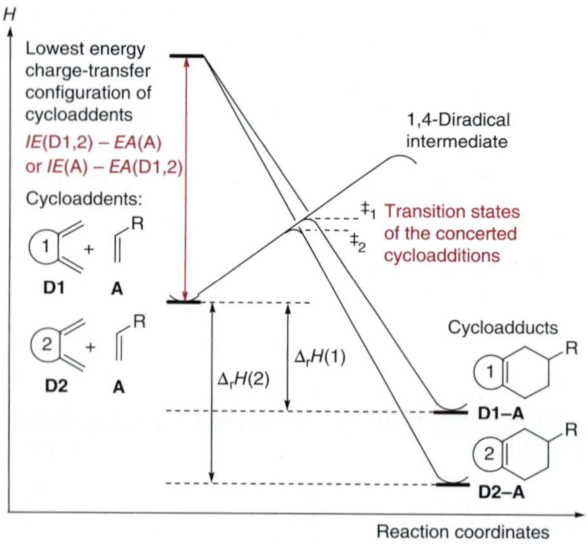

Figure 5.28 Demonstration of the Dimroth or Bell–Evans–Polanyi principle ($\Delta^{\ddagger}H = \alpha\Delta_rH + \beta$) for concerted Diels–Alder additions. The highest the exothermicity for cycloaddents of similar polarizability, the more Δ_rH is negative, the smaller is $\Delta^{\ddagger}H$ and the faster is the reaction.

D1 + **A** → **D1** − **A** is less exothermic ($\Delta_r H(1)$) than reaction **D2** + **A** → **D2** − **A** ($\Delta_r H(2)$). The dienes are chosen to have the same ionization energies (*EI*) (or same HOMO energies, see the Fukui's FMO theory, Section 5.3.11) and electron affinities (−*EA*) (or same LUMO energies); thus, the charge transfer configurations $\mathbf{D1}^{\bullet+} + \mathbf{A}^{\bullet-}$ (or $\mathbf{D1}^{\bullet-} + \mathbf{A}^{\bullet+}$) of cycloaddents **D1** + **A** and $\mathbf{D2}^{\bullet+} + \mathbf{A}^{\bullet-}$ (or $\mathbf{D2}^{\bullet-} + \mathbf{A}^{\bullet+}$) of cycloaddents **D2** + **A** have the same or nearly the same enthalpy. When the cycloaddents approach to each other along the reaction coordinates of the cycloaddition, the energies of the charge transfer configurations decrease more so for the more exothermic than for the less exothermic reaction. Crossing of these hypersurfaces with the ground-state electronic hypersurface occurs earlier for the reaction with the highest exothermicity, in this case for **D2** + **A** → **D2** − **A**. Thus, this cycloaddition is faster than the less exothermic cycloaddition **D1** + **A** → **D1** − **A**.

Problem 5.17 What is the main product of reaction of [4]radialene (tetramethylidenecyclobutane) with 4-phenyl-1,2,4-triazoline-3,5-dione in large excess at 25 °C? [391]

Problem 5.18 Explain why the rate constant ratio (k_1/k_2) for the two successive Diels–Alder reactions of TCNE to 2,3,5,6-tetramethylidenebicyclo[2.2.2] octane is 17, whereas that for the two successive Diels–Alder reactions of TCNE to 2,3,5,6-tetramethylidene-7-oxabicyclo[2.2.1]heptane amounts to 376 at 25 °C [392].

Problem 5.19 Propose a reactivity trend for the following cycloreversions [393].

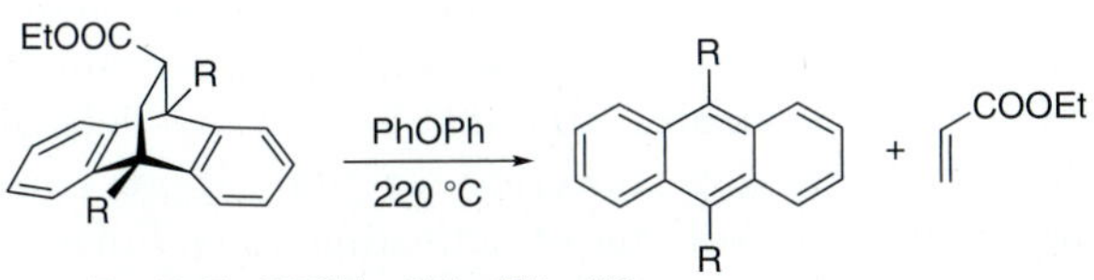

R = H, Et, COOMe, SMe, NH_2, NO_2

Problem 5.20 Propose a reactivity trend for the following Diels–Alder reactions for X = MeO, SO_2Ph_2, and $P(O)Ph_2$ [394].

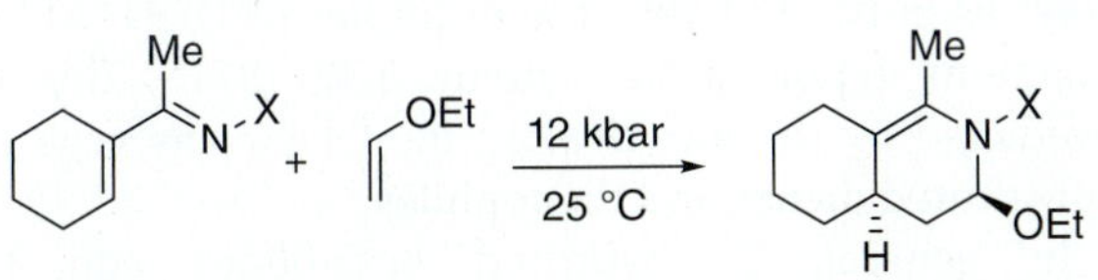

Problem 5.21 Compound **A** is stable under neutral and acidic conditions but it undergoes quick cycloreversion into **P** + **Q** upon treatment with KH in THF at 25 °C. Give an explanation [395].

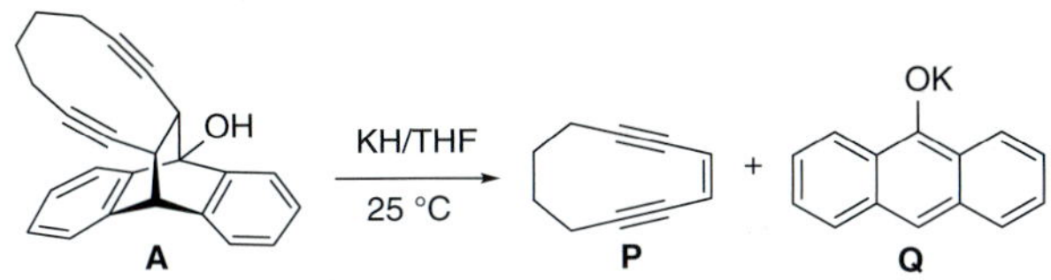

Problem 5.22 Give the reactivity trend of the Diels–Alder reactions of dimethyl 1,2,4,5-tetrazine-3,6-dicarboxylate with styrene, acrylonitrile, vinyl acetate, and ethyl ether [324].

5.3.13 Regioselectivity of Diels–Alder reactions

In the FMO theory developed by Fukui [396–399] (see also the perturbational molecular orbital [PMO] theory. Section 4.4.2), the intrinsic barrier (β term in the Bell–Evans–Polanyi relationship) of a Diels–Alder reaction is the sum of several contributions:

β(cycloaddition) = steric repulsion + energy of deformation of cycloaddents in the transition structures − electronic stabilization (ΔE_{el}) of the transition structures by electron exchange between the cycloaddents.

ΔE_{el} is given by the result (Eq. (5.17)) of the mutual perturbations (two-electron interactions) between FMO of the diene (**D**) + alkene or alkyne (**A**) approaching to each other in the transition state. They depend on the square of their overlaps (*S*) and energy (ε) differences. Thus

$$\Delta E_{el} \approx -[\mathrm{LUMO}(\mathbf{A})\cdot\mathrm{HOMO}(\mathbf{D})]^2/(\varepsilon_{\mathrm{LUMO(A)}}-\varepsilon_{\mathrm{HOMO(D)}}) - [\mathrm{LUMO}(\mathbf{D})\cdot\mathrm{HOMO}(\mathbf{A})]^2/(\varepsilon_{\mathrm{LUMO(D)}}-\varepsilon_{\mathrm{HOMO(A)}}) \quad (5.17)$$

The first term dominates for Diels–Alder reactions with normal electronic demand (electron flow from dienes (**D**) to dienophiles (**A**)) and the second term dominates for Diels–Alder reactions with inverse electron demand (electron flow from dienophiles (**A**) to dienes (**D**)). For nonpolar cycloaddents (e.g. alkyl substituted dienes and dienophiles), both terms contribute to a comparable extent. Thus, the smaller the energy gap $\varepsilon_{\mathrm{LUMO(A)}} - \varepsilon_{\mathrm{HOMO(D)}}$ and/or $\varepsilon_{\mathrm{LUMO(D)}} - \varepsilon_{\mathrm{HOMO(A)}}$, the faster the cycloaddition. This corroborates predictions based on the diradicaloid model (Figure 5.27) for which the rates of concerted cycloadditions increase as the sum *EI*(**D**) + (−*EA*(**A**)) and/or *EI*(**A**) + (−*EA*(**D**)) decreases.

According to the Fukui's FMO theory, regioselectivity (product ratio between regioisomeric cycloadducts

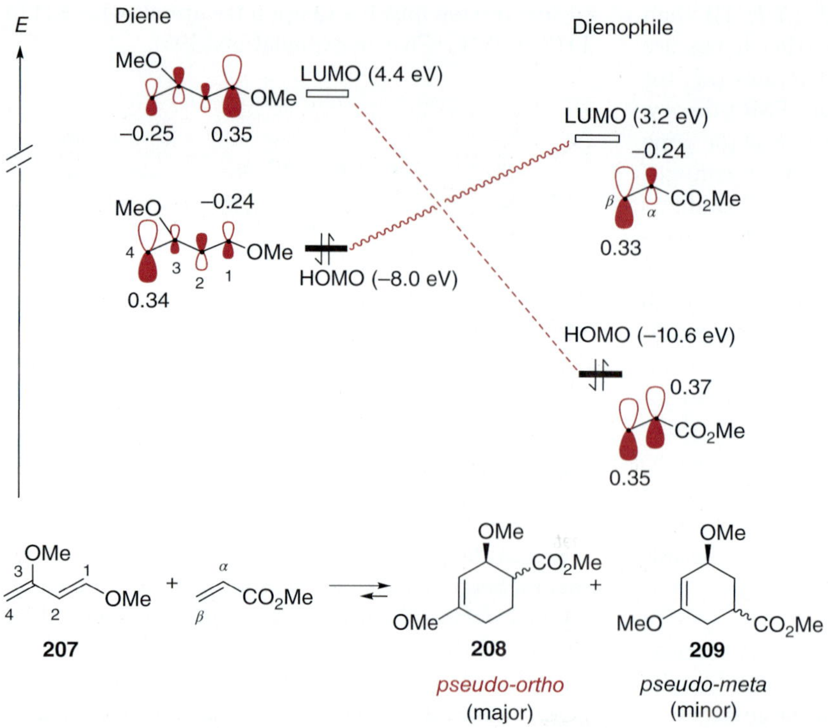

Figure 5.29 Example of application of the FMO theory for the prediction of Diels–Alder regioselectivity (energies (ε) of LUMOs and HOMOs and their atomic coefficients calculated by the HF/6-31G(d) method). Source: Taken from [366].

formed under kinetic control: irreversible reactions), the cycloadditions of nonsymmetrical cycloaddents depend on LUMO(**A**)·HOMO(**D**) overlaps of both possible regioisomeric transition structures for Diels–Alder reactions with normal electronic demand, and on LUMO(**D**)·HOMO(**A**) overlaps for Diels–Alder reactions with inverse electron demand. An example is given in Figure 5.29 with the cycloaddition of (*E*)-1,3-dimethoxybutadiene (**207**) to methyl acrylate, which gives the *pseudo-ortho* cycloadduct **208** as the major product [400, 401]. The overlap between LUMO(methyl acrylate = **A**) and HOMO(**207** = **D**) is larger for the formation of **208** than for the formation of **209**. For the formation of **208**, overlap LUMO(**A**)·HOMO(**D**) ≈ [coefficient at $C(\beta)_{LUMO(A)}$]·[coefficient at $C(4)_{HOMO(D)}$] + [coefficient at $C(\alpha)_{LUMO(A)}$]·[coefficient at $C(1)_{HOMO(D)}$] = 0.33 (0.34) + (−0.33)(−0.34) = 0.1698. For the formation of **209**, overlap LUMO(**A**)·HOMO(**D**) ≈ [coefficient at $C(\beta)_{LUMO(A)}$]·[coefficient at $C(1)_{HOMO(D)}$] + [coefficient at $C(\alpha)_{LUMO(A)}$]·[coefficient at $C(4)_{HOMO(D)}$] = (0.33) (−0.24) + (−0.24)(0.34) = −0.0608 (the negative sign does not count as the square of the overlap intervene in Eq. (5.17)).

The theory has been successful in several cases and has become quite popular even though it necessitates quantum calculations for dienes and dienophiles. When it fails, it has been proposed to consider not only FMOs but also secondary orbital interactions between sub-HOMOs and supra-LUMOs [402–404]. The theory is less successful to predict the regioselectivity of Diels–Alder reactions of polysubstituted cycloaddents [405]. In fact, it is not obvious why the shape of the FMOs calculated for isolated cycloaddents should maintain their "polarity" (their shapes, relative size of their atomic coefficients) once they approach to each other in the transition state [406]. Because most Diels–Alder reactions are exothermic, they have been said to have early transition states, i.e. their transition structures resemble more the cycloaddents than corresponding cycloadducts (Hammond postulate [407, 408], Bell–Evans–Polanyi theory for one-step concerted reaction). If one considers the ratio volume of activation vs. volume of reaction ($\Delta^{\ddagger}V/\Delta_r V$) rather than the exothermicity ($\Delta_r H$) (Sections 5.3.9 and 3.6.1), the transition states of Diels–Alder reactions resemble more products than reactants. In general, 1-substituted butadienes add to 1-substituted dienophiles giving preferentially "*pseudo-ortho*" cycloadducts according to the "ortho rule" as illustrated with the examples of Diels–Alder reactions reported in Scheme 5.42 [371]. This is predicted by the diradicaloid model for any type of substituted dienes and dienophiles.

In general, 2-substituted butadienes add to 1-substituted ethylenes and acetylenes giving preferentially "*pseudo-para*"-cycloadducts according to the "para-rule." This is also predicted by the diradicaloid

Scheme 5.42 The "ortho rule" for the regioselectivity of Diels–Alder reactions of 1-substituted butadienes. Examples are given for (a) Diels–Alder reactions with normal electronic demand (Source: Taken from Ref. [371].) and (b) Diels–Alder reaction with inverse electronic demand. Source: Taken from Ref. [409].

(a)

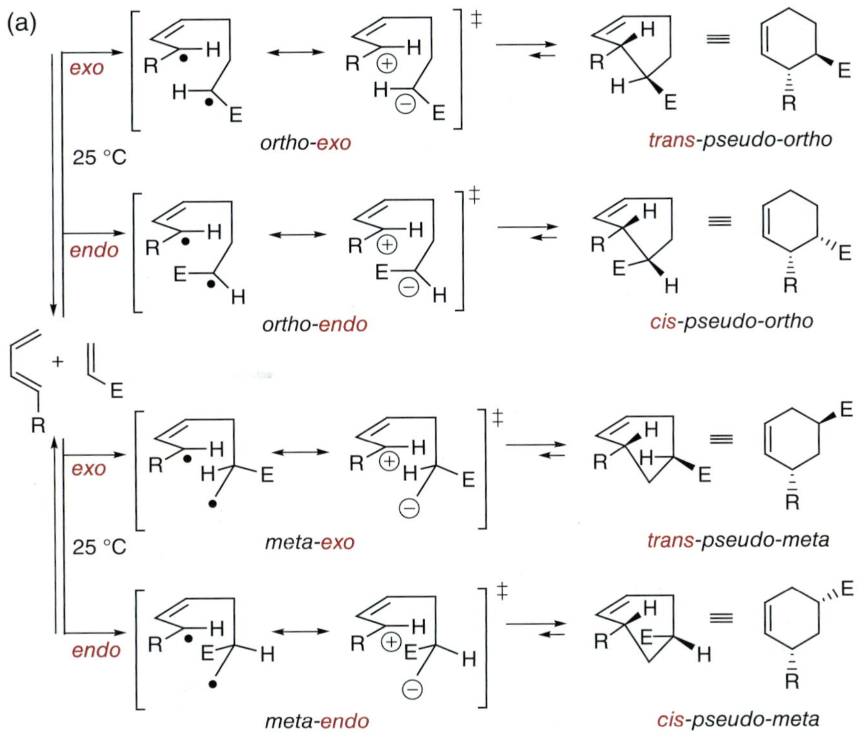

E = COOMe

<table>
<tr><th></th><th>ortho-exo</th><th>ortho-endo</th><th>meta-exo</th><th>meta-endo</th></tr>
<tr><td>R = Me</td><td>44.5%</td><td>42.7%</td><td>6.5%</td><td>6.3%</td></tr>
<tr><td>R = OMe</td><td>24.6%</td><td>75.4%</td><td>—</td><td>—</td></tr>
<tr><td>R = Ph</td><td colspan="2">>98%</td><td colspan="2">< 2%</td></tr>
<tr><td>R = SPh</td><td colspan="2">>98%</td><td colspan="2">< 2%</td></tr>
<tr><td>R = NHCOOEt</td><td colspan="2">98%</td><td colspan="2">2%</td></tr>
<tr><td>R = COOH</td><td colspan="2">83%</td><td colspan="2">17%</td></tr>
</table>

(b)

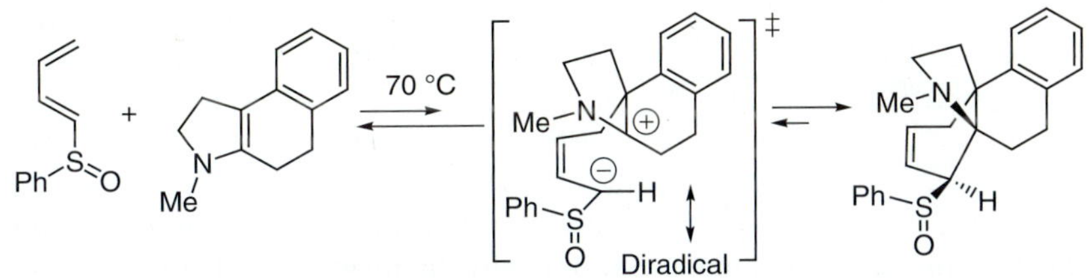

model (Scheme 5.43), which considers that both the diradical and zwitterionic forms of the transition states profit better from the stabilizing substituent effect in the *para* than in the *meta* orientation.

The regioselectivity of the Diels–Alder reaction of a 1,4-disubstituted diene with an electron-poor dienophile is controlled by the substituent of the diene that the best stabilizes a radical and/or a carbocationic center. In the example reported below [411], the carbamoyl group at C(1) is better than the silyloxymethyl substituent. It is also noted that the "para" directing effect of the 2-ethyl substituent of the diene is completely overwhelmed. This Diels–Alder reaction is highly regio- and stereoselective giving a single product in high yield.

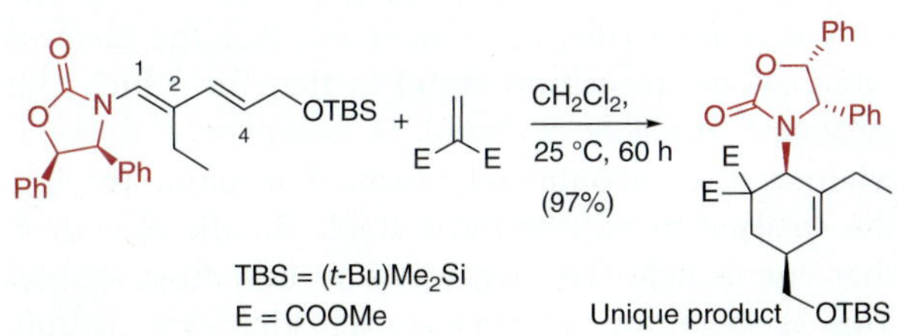

Acrolein reacts with isoprene at 150 °C giving a 59 : 41 mixture of the *pseudo-para* and *pseudo-meta*

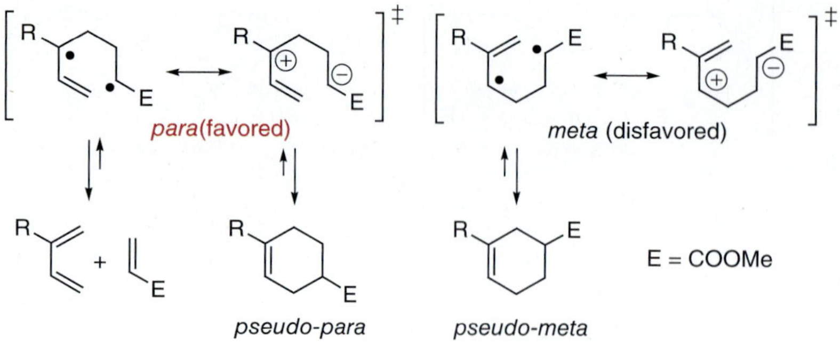

	pseudo-para : pseudo-meta
R = Me	70 : 30
R = Ph	>96 : 4
R = OMe	70 : 30
R = SPh	80 : 20
R = $SiEt_3$	77 : 23

Scheme 5.43 The "para rule" for the regioselectivity of Diels–Alder reactions of 2-substituted butadienes. Source: Taken from Refs. [378, 410].

cycloadducts. In the presence of $SnCl_4{\cdot}(H_2O)_5$, the cycloaddition is much faster (occurs below room temperature) and the regioselectivity is increased to 96 : 4 [412]. The Lewis acid coordinates to the carbonyl group of the dienophile and stabilizes more strongly the zwitterionic limiting structure of the diradicaloid leading to the *pseudo-para* cycloadduct than that leading to the *pseudo-meta* cycloadduct.

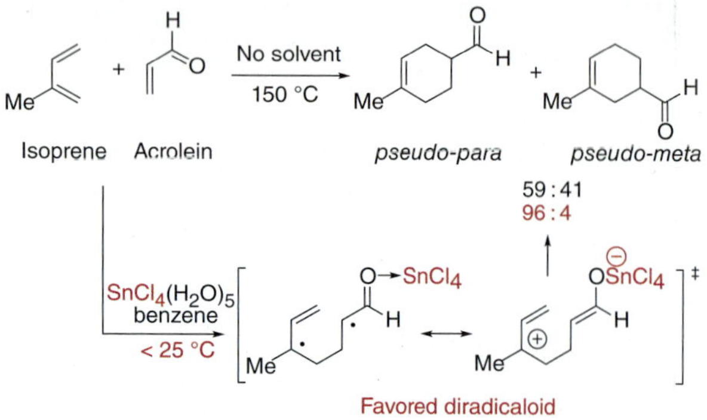

Deuterium and ^{13}C-kinetic isotope effects have been determined for all positions on isoprene in its reactions with methyl vinyl ketone, ethyl acrylate, and acrolein catalyzed by Et_2AlCl. The results support highly asynchronous concerted (4+2)-cycloadditions in agreement with the diradicaloid model [360, 413].

As a general rule, one can retain that the favored diradicaloid (transition state) is that for which the strongest possible σ bond is completely formed, whereas the incompletely formed σ bond implies the weakest bond (the most stable diradical) and/or that for which the zwitterion is the most stable. This is true for all (m+n)-cycloadditions including (2+2)-cycloadditions and hetero-Diels–Alder reactions, with normal or inverse electronic demand.

Thus, thermochemical data such as standard homolytical dissociation enthalpies (Tables 1.A.7, 1.A.13, 1.A.14), gas phase hydride affinities of carbenium ions (Table 1.A.18), and proton affinities of anions (Tables 1.A.13, 1.A.14, 1.A.18), as well as substituent effects on the relative stability of radicals (Table 1.A.9), carbenium ions (Table 1.A.14), and anions (Tables 1.A.13, 1.A.19, and 1.A.20) allow one to estimate the stability difference between the possible diradicaloid transition states for any cycloaddition (see Schemes 5.41 and 5.42), and thus predict the cycloaddition regioselectivity. With acetylenic dienophiles and heterodienophiles such as R—N=O (rapid epimerization about trivalent N in the adduct), only two (regioisomeric) cycloadducts can form. With ethylenic dienophiles (e.g. $R^1(R^2)C{=}CH_2$) and heterodienophiles of types $R^1(R^2)C{=}NR$, $R^1(R^2)C{=}S$, and $R^1(R^2)C{=}O$ ($R^1 \neq R^2$), there are four possible cycloadducts (racemic mixtures) that can form with 1-substituted butadienes. Deviations from the "ortho rule" can be attributed to a thermodynamic control (equilibrium between possible regioisomeric cycloadducts) or to steric factors. In the presence of $AlCl_3$, the Diels–Alder addition of cyclohex-2-enone with isoprene gives the expected "para" cycloadduct **211** in 80% yield. The additive $AlCl_3$ coordinates to the carbonyl group of the dienophile and stabilizes the zwitterionic form **210** of the diradicaloid transition state: this leads to an enhanced rate and higher regioselectivity than for the thermal cycloaddition. Under the same conditions, the Diels–Alder reaction of 2,4,4-trimethylcyclohex-2-enone with isoprene is much less "para"-regioselective because of steric hindrance introduced by the 2-methyl group of the dienophile that retards the σ(C—C) bond formation between C(2) of trimethylcyclohexenone and C(4) of isoprene [414, 415].

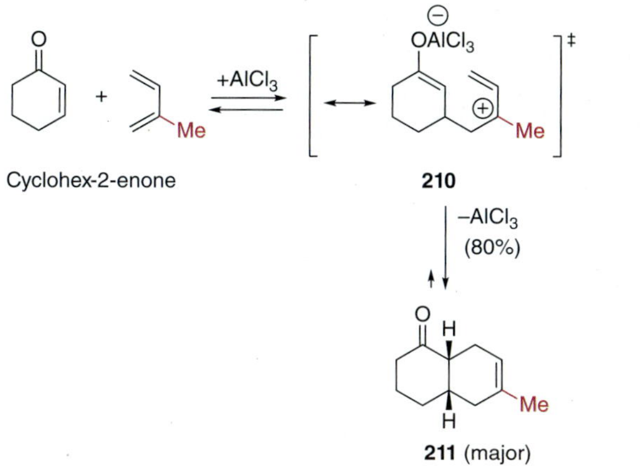

(see Sections 7.6.6 and 7.6.7 for catalyzed Diels–Alder reactions)

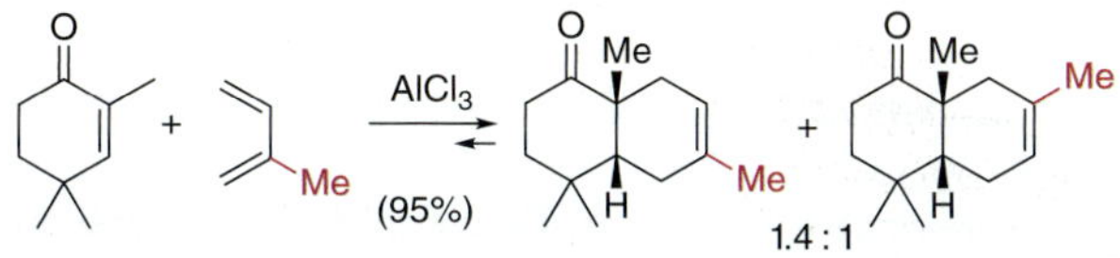

The Diels–Alder reaction of (*E*)-1-trimethylsilylbutadiene with methyl propiolate (propynoate) is little regioselective and gives a 21 : 26 mixture of *pseudo-ortho* and *pseudo-meta* cycloadducts in 47% yield [416, 417]. Steric hindrance can be invoked to explain the formation of the *pseudo-meta* cycloadduct. In fact, one needs to also consider the β-silicon effect that stabilizes radicals [418] and carbenium ions efficiently. This effect contributes to the electronic stability of the *meta*-**212** transition structure as much as the α-silyl substituent in *ortho*-**212** diradicaloid.

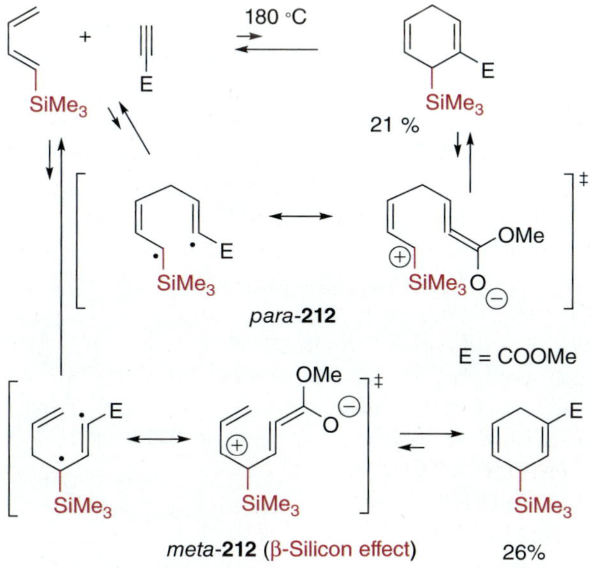

Another example of deviation to the "ortho rule" is the cycloaddition of methyl vinyl ketone to prop-2-en-1-ylidenecyclopropane, which gives the *pseudo-meta* cycloadduct *m*-**214** in 80% yield. Regioisomeric *pseudo-ortho* cycloadduct *o*-**214** is isolated in less than 1% yield [419]. The diradicaloid model explains this results perfectly. The cyclopropyl substituent of the diene cannot be seen as a geminal dialkyl disubstituent. The cyclopropyl group in 1-cyclopropyl-1-alkyl radicals ($DH°(c\text{-}C_3H_5{-}CH_2^{\bullet}/H^{\bullet})$ = 97.4 kcal mol^{-1} (Table 1.A.14)); $DH°(Me_2CH{-}CH_2^{\bullet}/H^{\bullet})$ = 100.1 kcal mol^{-1} (Table 1.A.7) as well as in 1-cyclopropyl-1-alkyl carbenium ions ($DH°(c\text{-}C_3H_5{-}CH_2^{+}/H^{-})$ = 241 kcal mol^{-1}; $DH°(Me_2CH{-}CH_2^{+}/H^{-})$ = 265 kcal mol^{-1} (Table 1.A.14)) is highly stabilizing (Section 4.8.8). This stabilization manifests itself in the *meta* diradicaloid *meta*-**213**, but not in the *ortho* transition structure *ortho*-**213**.

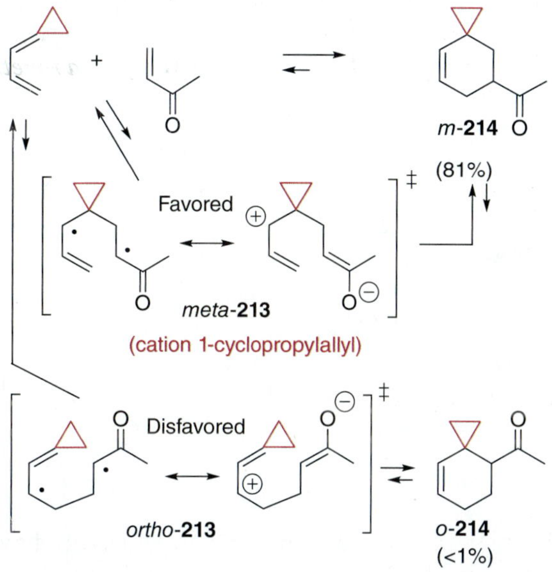

In general, the Diels–Alder reactions of polysubstituted dienes and dienophiles have regioselectivities that are predicted by the diradicaloid model. When the substituents are in conflict with respect to the "ortho rule" (Scheme 5.41) and "para rule" (Scheme 5.42), the substituent that stabilizes the best the diradical and zwitterionic limiting structures (diradicaloid) representing the regioisomeric transition structures controls the regioselectivity. Examples are given here-below.

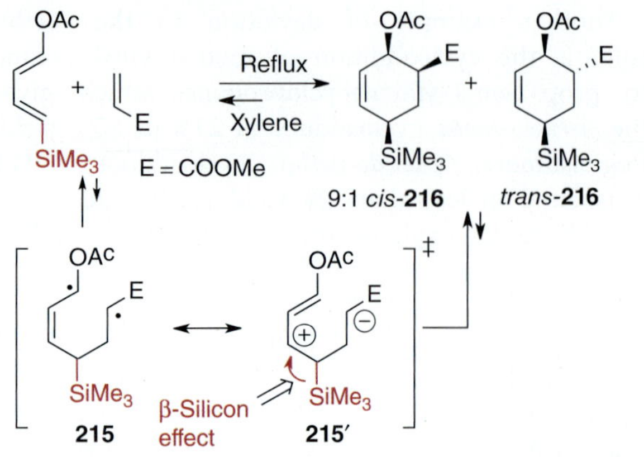

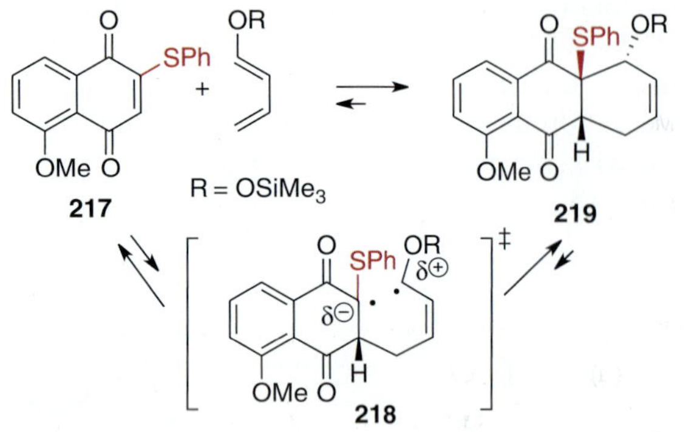

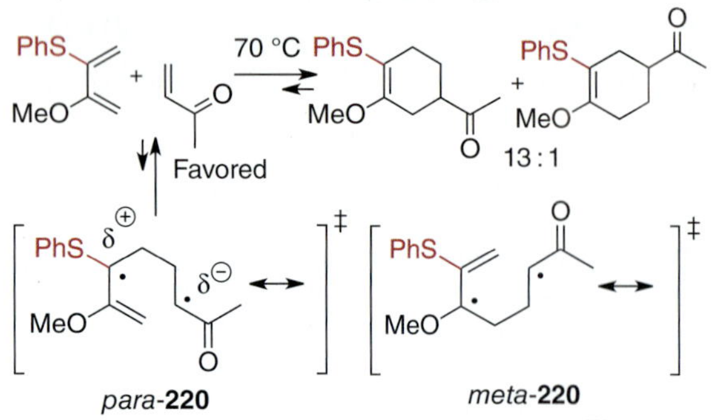

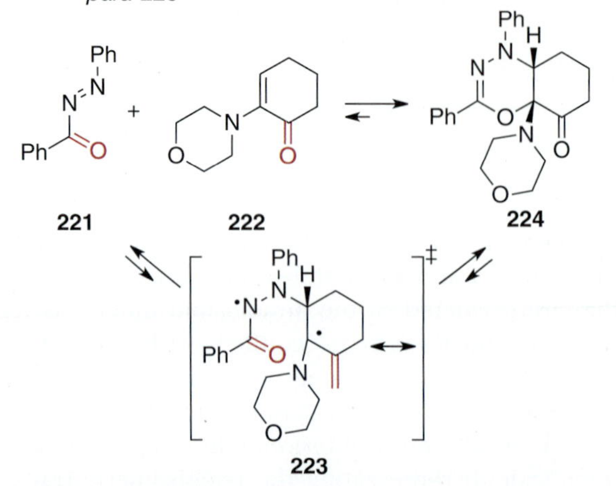

Methyl acrylate adds to (*E,E*)-4-trimethylsilylbutadien-1-acetate giving a 9 : 1 mixture of the *cis*- and *trans-pseudo-ortho* adducts **216** [417, 420]. If one considers the β-silicon effect that stabilizes diradicaloid **215**, both the AcO and Me_3Si substituents lead to the observed regioselectivity. The Diels–Alder reaction of juglone derivative **217** with (*E*)-1-trimethylsilyloxybutadiene generates the major adduct **219** arising from transition state **218** in which the phenylthio substituent plays its stabilizing effect both on the diradical and on the zwitterionic limiting structures of the diradicaloid [421]. Cycloaddition of methyl vinyl ketone to 2-methoxy-3-phenylthiobutadiene gives a 13 : 1 mixture of *pseudo-para-* and *pseudo-meta-*cycloadducts with respect to the phenylthio substituent [422, 423]. The latter stabilizes slightly better a radical than the methoxy group [424, 425]. In the case of the hetero-Diels–Alder reaction **221** + **222** → **224**, the diradical limiting structure **223** of the transition structure experiences strongly stabilizing effects by the two carbonyl moieties and the two amino substituents (Tables 1.A.9 and 1.A.10).

Remote substituents of dienes and/or dienophiles can affect the regioselectivity of their cycloadditions. Juglone adds to 1-methoxycyclohexa-1,3-diene giving **225** as major product [426, 427] (see also [428]). Intramolecular hydrogen bridging (chelation) between the carbonyl group of the benzoquinone moiety and the adjacent hydroxy group renders the latter more electrophilic than the other carbonyl group, thus explaining the regioselectivity observed.

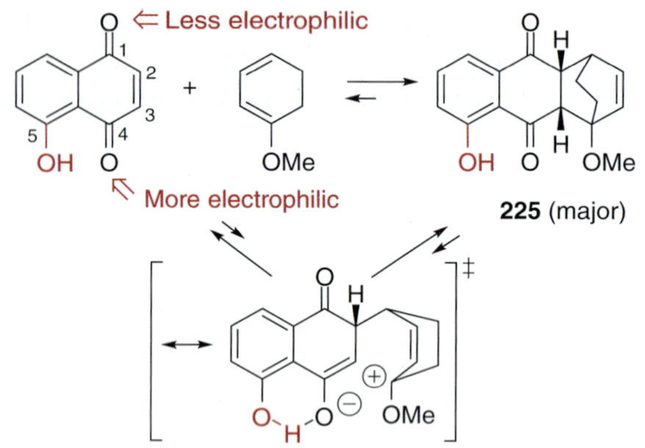

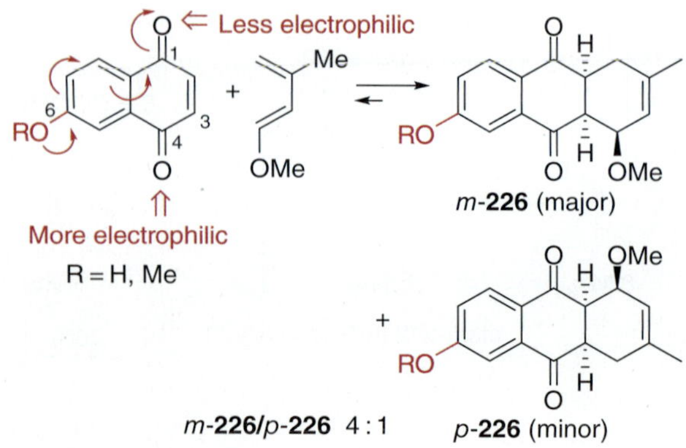

In the case of 6-hydroxy and 6-methoxy naphthoquinone, the 4-carbonyl group is more electrophilic than the 1-carbonyl group because the latter conjugates with the electron-releasing 6-substituent, leading to a 3 : 1 mixture of cycloadducts *m*-**226** and *p*-**226** [429].

A synthesis of daunomycinone (the aglycone of important anticancer agent) has been realized based on the Diels–Alder reaction of methyl vinyl ketone and exocyclic diene **227** [430]. The reaction is promoted by Lewis acid $BF_3{\cdot}Et_2O$ and gives the *pseudo-para*-cycloadduct **228** as major product. The regioselectivity is controlled by the remote alkenyl moiety at C(5) that homoconjugates with the electrodeficient C(3) center in the transition structure **229** ↔ **229′** ↔ **299″** [431].

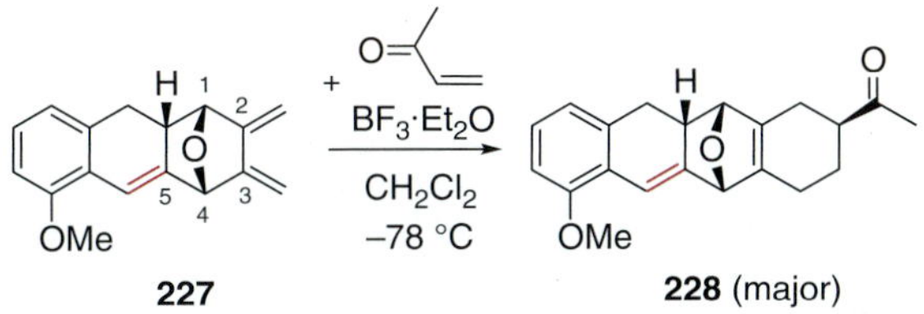

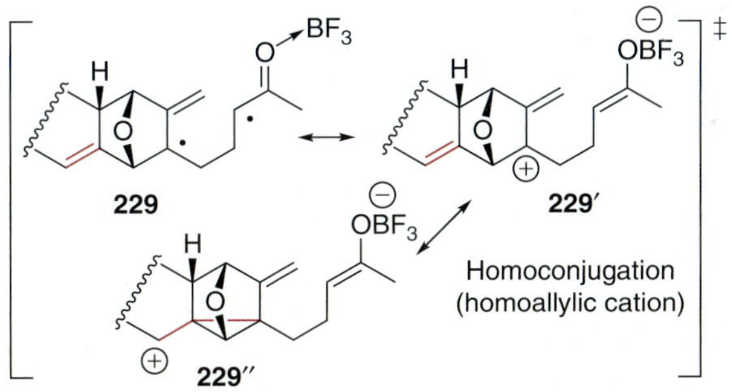

Tetraene **230** (a double diene) adds to the double dienophile **231** with high chemoselectivity (type of diene, type of dienophile) and high regioselectivity giving adduct **233**. Heating **233** leads to an intramolecular Diels–Alder reaction producing the tricyclic adduct **234** [432]. Although the Diels–Alder reactions of acetylenic dienophiles ($\Delta_r H°$(butadiene + acetylene → cyclohexa-1,4-diene, gas phase) = 24.0 − 26.0 − 54.5 = −56.5 kcal mol^{-1}) are about 15 kcal mol^{-1} more exothermic than similar cycloadditions with olefinic dienophiles ($\Delta_r H°$(butadiene + ethylene → cyclohexene, gas phase) = −39.9 kcal mol^{-1}), the enone moiety of **231** (which is not substituted at the β-carbon center and that generates the strongest possible bond in the transition structure **232** of the cycloaddition) is more reactive than the ynone moiety for steric reasons (bulky phenyl substituent of *β′*-carbon center). The regioselectivity of reaction **230** + **231** → **233** is explained by diradicaloid transition structure **232** that involves optimal stabilization for both its diradical and zwitterionic forms.

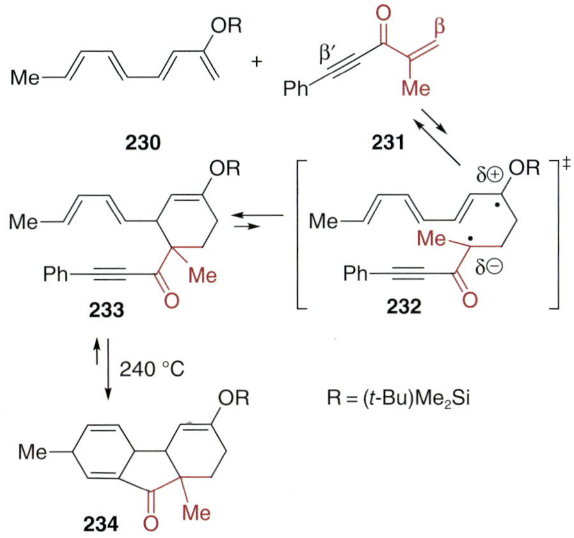

Problem 5.23 Predict the regioselectivity of the following Diels–Alder reactions.

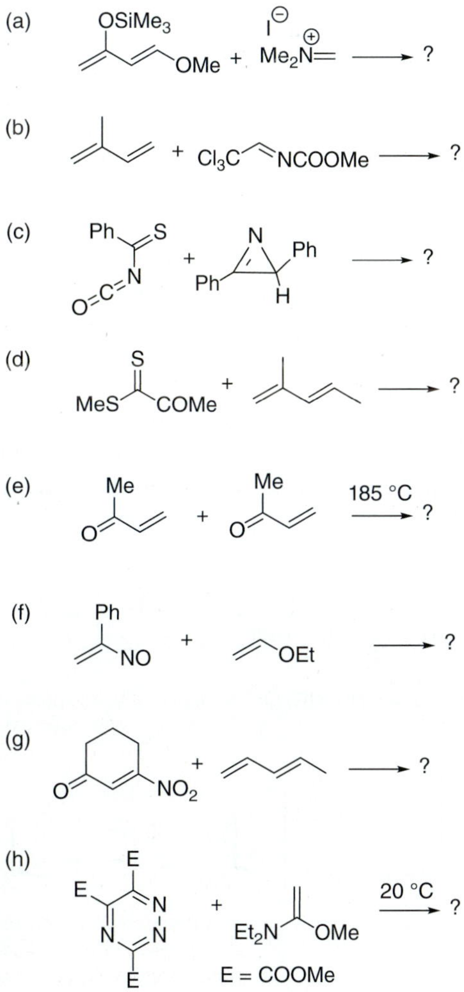

Problem 5.24 What is the major product of the following reaction? [433]

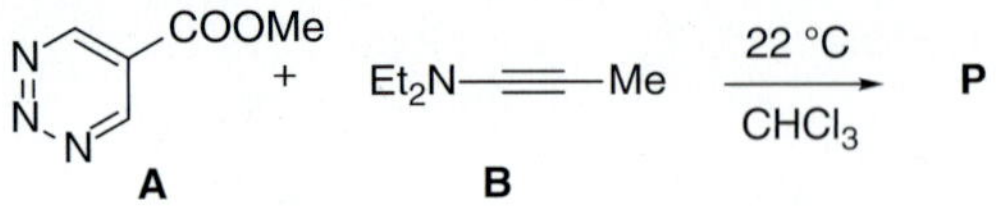

Problem 5.25 Give an explanation for the regioselectivity of the following enantioselective Diels–Alder reaction [434].

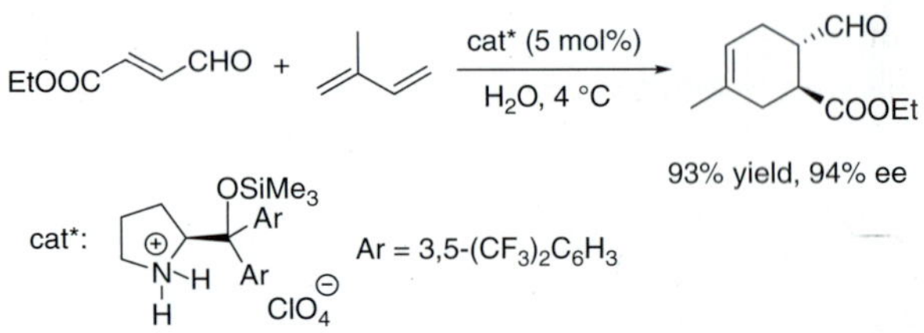

5.3.14 Stereoselectivity of Diels–Alder reactions: the Alder "endo rule"

Alder and Stein demonstrated that cyclopentadiene adds below 100 °C to maleic anhydride, maleic diesters, and acrylic esters with *endo* diastereoselectivity (Alder "endo rule") [435–437]. The *endo*-cycloadducts so-obtained are isomerized into the more stable *exo*-stereomers upon heating (when cycloreversion competes with the Diels–Alder reaction). Similar observations are made with other dienes (e.g. furan + maleic anhydride, reaction (2.17), Section 2.7.8) including (*Z*,*Z*)-1,4-dideuteriobutadiene that adds to maleic anhydride (benzene, 80 °C) with a 85 : 15 *endo*/*exo* stereoselectivity that represents a $\Delta\Delta^{\ddagger}G = 1.2\,\text{kcal mol}^{-1}$ of free energy advantage for the *endo* mode of cycloaddition [438]. *Exo*-adducts may be preferred under conditions of kinetic control for some reactions. The *endo* vs. *exo* stereoselectivity depends on temperature, the nature of dienes and dienophiles, the presence or absence of additives able to coordinate the cycloaddents (e.g. Lewis or protic acid catalysts, Sections 7.6.6 and 7.6.7), and on the medium polarity as illustrated with reaction (5.18) for the reaction of cyclopentadiene with methyl vinyl ketone [439]. Using FMO arguments, Hoffmann and Woodward explain the Alder "endo rule" by invoking the interaction of secondary orbitals as depicted with transition structures **235a** and **235b** [1, 339, 440].

235a

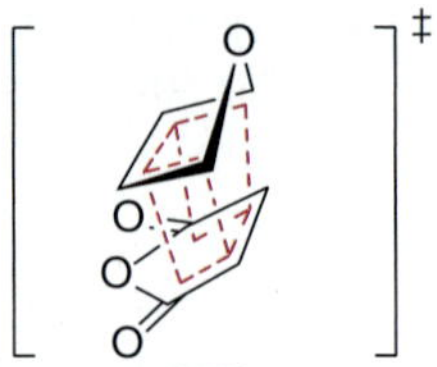

235b

The *endo* stereoselectivity represents a deviation from the Dimroth principle and Bell–Evans–Polaniy theory ($\Delta^{\ddagger}H = \alpha\Delta_r H + \beta$) as the least stable *endo*-cycloadducts form faster than the more stable isomeric *exo*-cycloadducts. This contrasteric effect is predicted by the diradicaloid model. The zwitterion limiting structure of the transition state contains two main dipoles that adopt different orientations depending on the *endo* or *exo* approach of the cycloaddents (Scheme 5.44).

Electrostatic interactions are more stabilizing in the *endo* than in the *exo* transition structure because of the shorter distance between positively and negatively charged centers of the zwitterionic limiting forms of the diradicaloids [441, 442]. The diradicaloid model predicts that the *endo* vs. *exo* stereoselectivity will depend on the polarity of the medium and on additives capable to coordinate the cycloaddents and

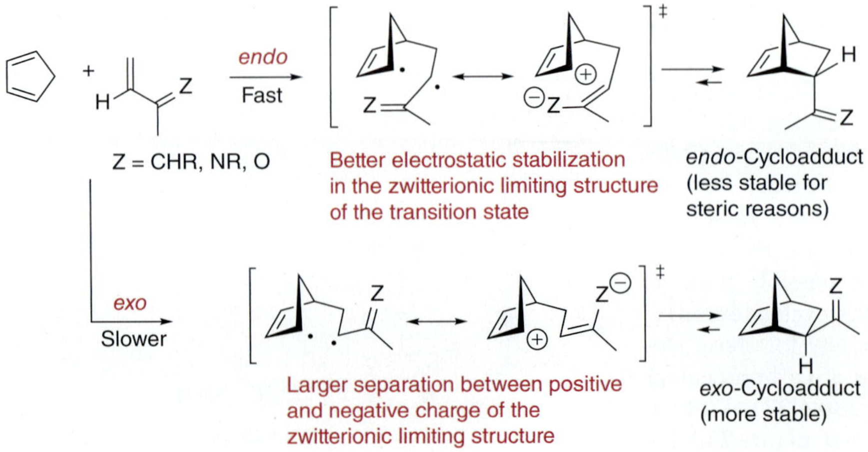

Scheme 5.44 The diradicaloid model applied to the Alder "endo rule."

stabilize the zwitterionic limiting structures. The rate of the reaction and the *endo* vs. *exo* product ratio increase with solvent polarity (reaction (5.18)) [439].

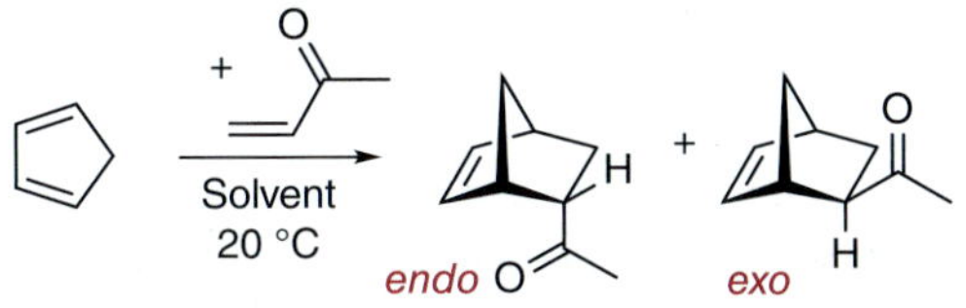

(5.18)

Solvent:	Isooctane	H_2CONH_2	HO OH	H_2O
$10^5 k^{II}$:	6	318	480	4400 $M^{-1}s^{-1}$
[*endo*]/[*exo*]:	8.5	8.9	10.4	25

Methyl acrylate adds to cyclopentadiene at 30 °C in benzene giving a 78 : 22 mixture of *endo*- and *exo*-cycloadduct. When the dienophile is precoordinated to $AlCl_3$, the reaction is much faster and the stereoselectivity is increased to 95 : 5 [443, 444]. In 1975, Corey and coworkers realized a first enantioselective synthesis of prostaglandins based on the $AlCl_3$-catalyzed Diels–Alder reaction (Section 3.7.5) of 5-(benzyloxymethyl)cyclopentadiene with the acrylate ester of 8-phenylmenthol (**236**) (Scheme 5.45) [445]. The reaction gives mostly the *endo* ester **237**. The lithium enolate of **237** reacts with oxygen giving an intermediate peroxide that is reduced by triethyl phosphite into a mixture or *exo*/*endo* α-hydroxyesters

Scheme 5.45 Example of Corey's general synthesis of prostaglandins.

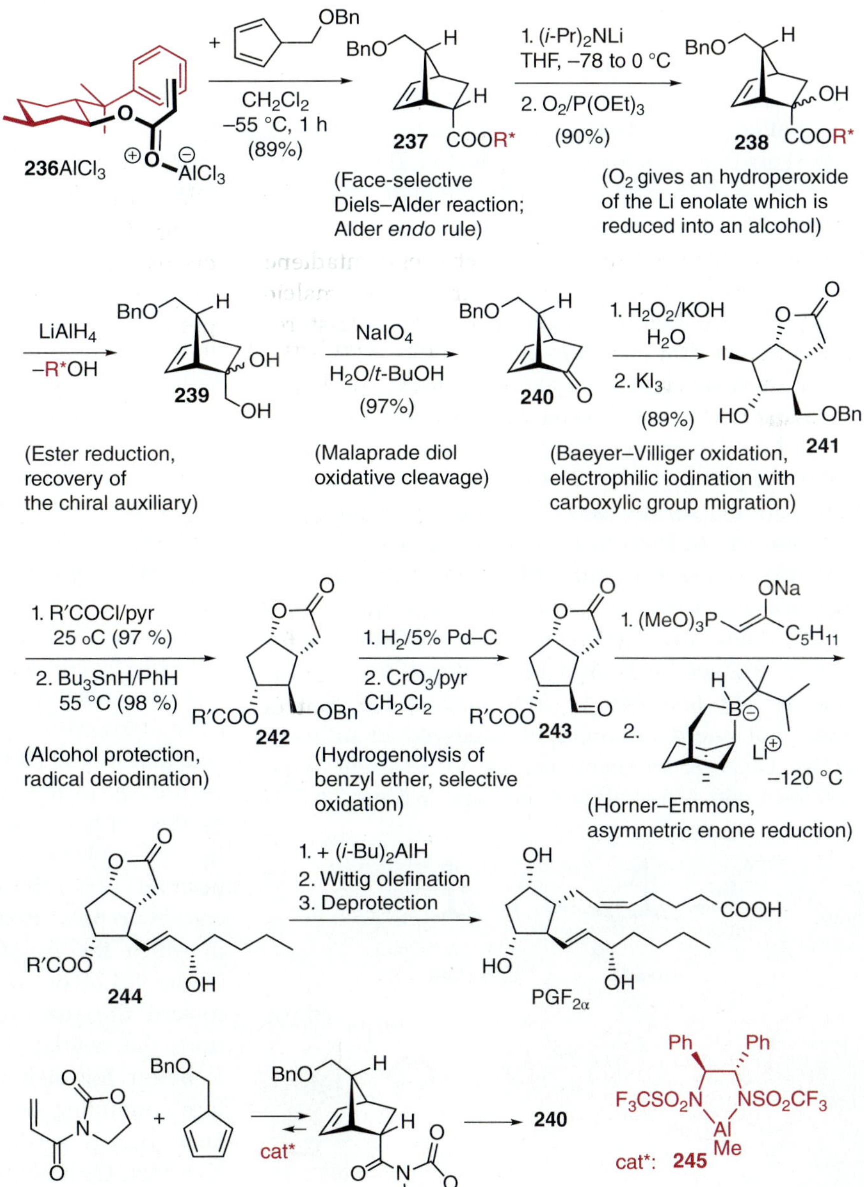

238. Their reduction with $LiAlH_4$ leads to a mixture of diols **239** and (+)-8-phenylmenthol (R*OH: recovery of the chiral auxiliary). Malaprade oxidation of diols **239** affords homoconjugated enone **240**. Its Baeyer–Villiger oxidation generates a 2-oxabicyclo[3.2.1]oct-6-en-3-one derivative that reacts with iodonium ion ($KI + I_2$) with subsequent rearrangement into iodolactone **241**. After esterification of the alcohol and deiodination, **242** is obtained. Debenzylation and subsequent oxidation of the primary alcohol provides aldehyde **243** that undergoes Wittig–Horner olefination into a conjugated enone that is reduced selectively into allylic alcohol **244** (diastereoselectivity 4.5 : 1) [446]. Reduction of lactone **244** with diisobutylaluminum hydride (DIBAH) in THF at low temperature affords the corresponding lactol that can be olefinated by Wittig types of ylides. After deprotection, prostaglandins are obtained, such as $PGF_{2\alpha}$ [447, 448]. A practical enantioselective synthesis of key enone **240** was developed by Corey and coworkers that uses the enantioselective Diels–Alder reaction of 5-(benzyloxymethyl)cyclopentadiene to 3-acryloyl-1,3-oxazolidin-2-one in the presence of 10 mol% of enantiomerically pure diazaaluminolidine **245** [449, 450].

For the cycloaddition (5.19) of cyclopentadiene to methacrolein, the *exo*/*endo* adduct ratio is 86 : 14 under thermal conditions and in CH_2Cl_2. In 1979, Koga and coworkers showed that this ratio increases to 98 : 2 for a reaction carried out in toluene at −78 °C and in the presence of 0.12–0.15 equiv. of menthoxyaluminum dichloride (**247**). Pure *exo*-carbaldehyde *exo*-**246** was isolated in 69% yield and with 72% ee [451, 452] (an early example of a catalyzed asymmetric Diels–Alder reaction). The *exo*/*endo* product ratio is 98 : 2 also when applying catalyst **248** (*exo*-**246** with 97.8% ee) [453], and better than 99 : 1 with catalyst **249** (*exo*-**246** in 84% yield and 97% ee) [454]. For more examples of catalyzed enantioselective Diels–Alder reactions, see for example Refs [297, 449, 450, 455–459] and Sections 7.6.6 and 7.6.7.

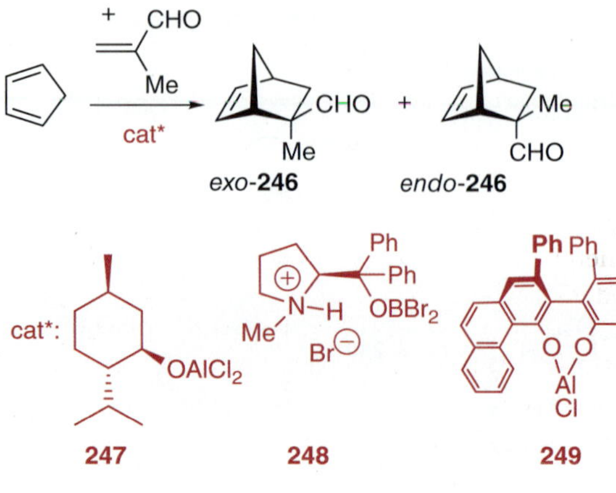

(5.19)

5.3.15 π-Facial selectivity of Diels–Alder reactions

In 1955, Woodward and coworkers reported that the $[\pi^4+\pi^2]$-cycloaddition (5.20) of 5-acetoxycyclopentadiene (CpOAc) to ethylene generates mostly the 7-*anti*-norbornenyl acetate (**250**) [460].

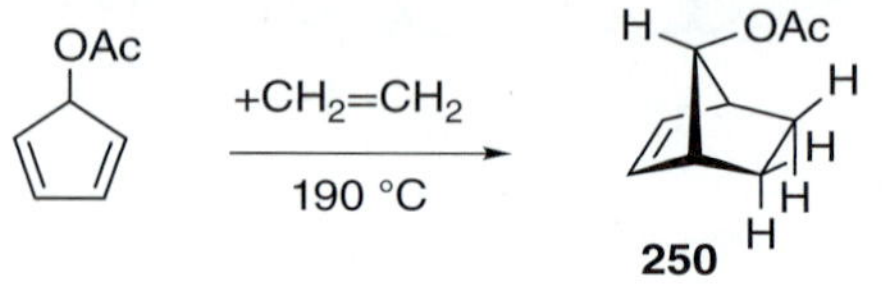

(5.20)

As the acetoxy group in CpOAc is much larger than the hydrogen atom at C(5), the steric factor that should favor the formation of 7-*syn*-norbornenyl acetate is overwhelmed by other factors. Fukui et al. explained the contrasteric face selectivity of the Diels–Alder reactions of 5-acetoxy- and 5-chlorocyclopentadiene in terms of deformation of the HOMO of the diene (dissymmetry with respect to the cyclopentadiene plane). The orbital mixing rule [400] has been proposed to predict the nonequivalent deformation due to asymmetric perturbation of the substituent orbital (Figure 5.30). The occupied FMOs of 5-substituted cyclopentadienes (CpX) are localized π_2(diene), n(X:) of the substituent, and the highest lying σ orbital of the five-membered ring. The latter σ orbital is antisymmetrical with respect to the mirror plane of symmetry of CpX and has higher atomic coefficients at C(1), C(5), C(4) than at C(2), C(3) because of the sp^3-hybridized carbon at C(5). This σ MO combines with n(X:) (have the same a″ symmetry) to give two MO combinations φ_1 and φ_2. On their turn, φ_1 and φ_2 combine with π_2(diene). As the energy gap between φ_1 and π_2(diene) is much larger than the energy gap between φ_2 and π_2(diene), only the latter combination needs to be considered to a first approximation. One can thus generate an in-phase combination that we call sub-HOMO (CpX) and an out-of-phase combination that we call HOMO(CpX) = $e(\pi_2(\text{diene})) - f(\varphi_2)$, with $e > f$. In this MO, the substituent n(X:) orbital carries with it a contribution from the σ-skeleton (loss of π-planar symmetry of cyclopentadiene) and permits σ/π mixing. The resulting HOMO(CpX) can be represented by a distorted π_2(diene) MO in which the 2p AOs at the termini carbon atoms of the 1,3-diene unit have rotated inward, pointing toward the substituent X. Because of this distortion, the overlap HOMO(CpX)/LUMO(dienophile) is better for dienophiles approaching CpX on the face containing the X-substituent than on the other face. This is verified experimentally for CpX with X = OH, OMe, OAc, NH_2, NHAc, F, and Cl. In the cases of CpBr and CpI, the steric repulsive factors

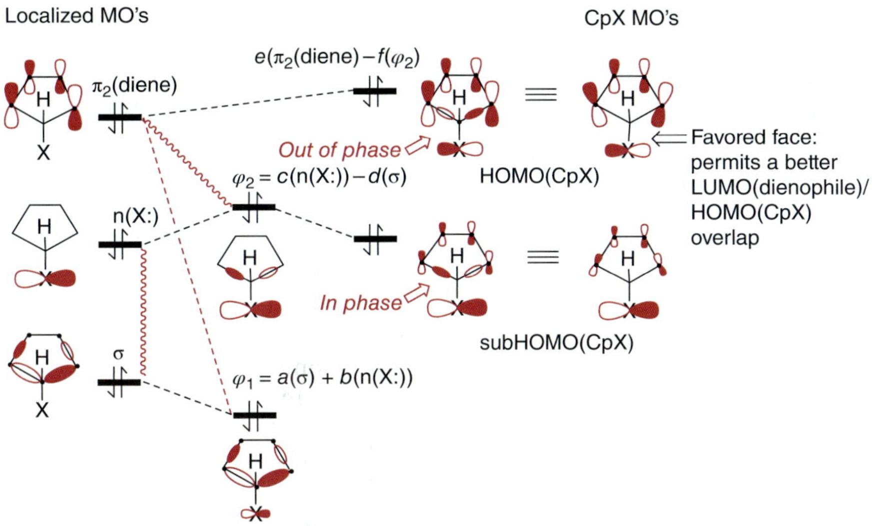

Figure 5.30 Fukui's orbital mixing rule applied to the HOMO of five-substituted cyclopentadienes (CpX) and for which the *n*(X:) orbital localized on the *X*-substituent is lower in energy than π_2(1,3-diene) localized on the diene moiety. The coefficients *a*, *b*, *c*, *d*, *e*, and *f* of the molecular combinations are all positive.

dominate and their Diels–Alder additions occur on the face *anti* with respect to X [461]. If the substituent X has a localized n(X:) orbital higher in energy than π_2(diene), its out-of-phase combination φ_2 with the σ MOs will also combine with π_2(diene), but, this time, it is the in-phase combination $\pi_2(\text{diene}) + \varphi_2$ = sub-HOMO(CpX) that will be important for the Diels–Alder face selectivity. The latter combination will lead to a distortion of the 2p AOs of the diene termini that rotate inward in the face opposite to that of the substituent X. This favors Diels–Alder reactions on the face opposite to that of the X substituent as verified experimentally for the addition of maleic anhydride to 5-methylthio-1,2,3,4,5-pentamethylcyclopentadiene and to other dienes of type Me_5C_5–X with X = SCH_2Ph, SPh, SOMe, and SO_2Me [462]. In the latter cases, the facial selectivity might also be controlled by a steric factor. For disymmetric acyclic and semicyclic (e.g. vinylcyclohexenes) 1,3-dienes, the face selectivity of their Diels–Alder reactions is, in general, controlled by steric factors [463].

In Fukui's theory of orbital mixing [400], one admits that the orbital properties calculated for the cycloaddents are maintained in the transition states of their cycloadditions. As we have seen, the activation volumes of Diels–Alder reactions suggest transition structures that resemble more the cycloadducts than the corresponding cycloaddents. With the diradicaloid model, this fact is not ignored. When the dienophiles attack CpX on the same face as substituent X, transition structure *syn*-**251** is generated in which the C—H bond of the H—C—X moiety hyperconjugates with the electro-deficient center. In the transition structures *anti*-**251** the donor ability (or electrofugacity) of the s(C—X) bond increases in the following sequence X = F < OR < Cl < Me < alkyl < H < SR < SeR < SiR_3. Thus, contrasteric *syn*-facial selectivity (transition structure *syn*-**251**) is preferred for CpX with X = F, OR, Cl, Me, and alkyl, whereas *anti*-facial selectivity (transition structure *anti*-**251**) is favored for CpX with X = SR, SeR, and SiR_3. This is analogous to the hyperconjugation theory developed by Cieplak to interpret facial selectivity of reactions of all π-systems [464].

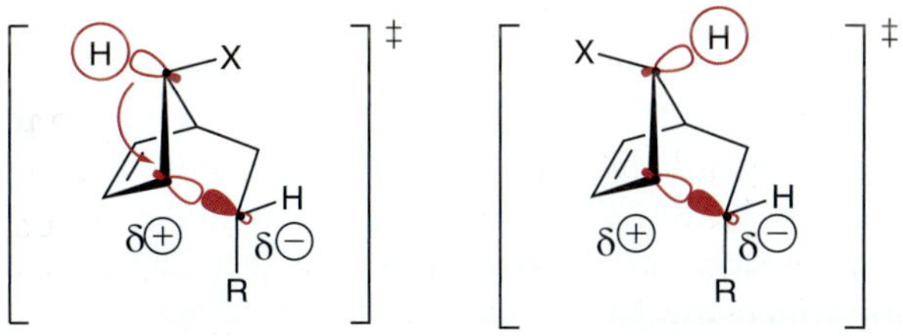

syn-**251** *anti*-**251**

syn-**251** favored for X = F, OR, Cl, NH_2, NHAc, alkyl: the antiperiplanar electrofugal C-H group hyperconjugates better than in *anti*-**251**

The stabilizing effect of the Me_3Si group in *anti*-**251** has been evidenced by comparing the activation parameters of the cycloreversions (5.21) [465]. It should be noted here that 5-trimethylsilacyclopentadiene has a greater hyperconjugative stabilization (π/σ(C—Si)) than cyclopentadiene (π/σ(C—H)) itself (Section 4.8.2) and thus renders the latter cycloreversion less endothermic than the former cycloreversion.

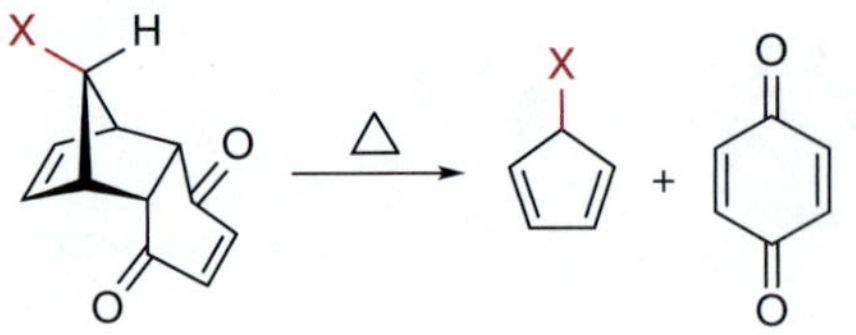

(5.21)

X = H $\Delta^{\ddagger}H$ = 29 ± 1.5 kcal mol^{-1}; $\Delta^{\ddagger}S$ = −3 eu
X = $SiMe_3$ $\Delta^{\ddagger}H$ = 24.8 ± 1 kcal mol^{-1}; $\Delta^{\ddagger}S$ = −5.8 ± 0.5 eu

Isodicyclopentadiene (4,5,6,7-tetrahydro-4,7-methano-2*H*-indene: **252**) adds to maleic anhydride giving a mixture of **253** and **254** arising from the *top* and *bottom* face Diels–Alder reaction of **252**, respectively (reaction (5.22)). The *top* vs. *bottom* facial selectivity varies between 55 : 45 and 35 : 65 depending on the solvent and temperature [466]. The cycloaddition of **252** to methyl acrylate and methyl propynoate are both highly *bottom* face selective [467]. Thus, for small dienophiles, the *bottom* face is preferred and corresponds to a contrasteric preference attributed to a kinetic stereoelectronic control. Gleiter and Paquette interpreted this facial selectivity as the result of secondary orbital interactions between cycloaddents. Based on STO-3G MO quantum mechanical calculations, σ/π-orbital mixing makes the sub-HOMO of **252** twisted in a way that it repels the incoming dienophiles more in the *top* face than in the *bottom* face [468]. On their side, Brown and Houk calculated the transition structure of synchronous cycloaddition of butadiene + ethylene and found that the H—C(2) and H—C(3) bonds of the diene are bent by 14.9° toward the direction of the approaching dienophile in the transition state [469]. Norbornenes do not have planar double bond as their σ(C(2)—R) and σ(C(3)—R) are bent downward, toward the *endo* face of the bicyclic alkenes (Section 4.8.3). When **252** adds a dienophile onto its *bottom* face, a *syn*-sesquinorbornene (e.g. **254**) forms for which the two fused norbornene units have all their alkyl substituents of the endocyclic double bond bending toward their *endo* face. Thus, the transition state of the *bottom* face Diels–Alder reaction of **252** incorporates the required bending of the C(2) and C(3) substituents of the butadiene moiety reacting with a dienophile. This is not the case for the *top* face addition. Indeed, when **252** adds onto its *top* face, an *anti*-sesquinorbornene forms for which one of two fused norbornene units has is alkyl substituents of the endocyclic double bond bent toward the *endo* face and the other toward the *exo* face if the system should not have a nonplanar endocyclic alkene. As a consequence, *syn*-sesquinorbornenes are more stable than corresponding *anti*-sesquinorbornenes. The *bottom* face selectivity follows the Dimroth principle, i.e. the most exothermic cycloaddition is the fastest. This is verified with the Diels–Alder reactions (5.23) of (norborn-2-eno)[*c*]furan (**255**) that are highly *bottom* face selective both under kinetic and thermodynamic control. For instance, **255** adds to maleic anhydride giving exclusively the *syn*-11-oxasesquinorbornene **256**. On heating, **256** is equilibrated with cycloaddents **255** and maleic anhydride and it never forms the isomeric *anti*-11-oxasesquinorbornene **257** [470].

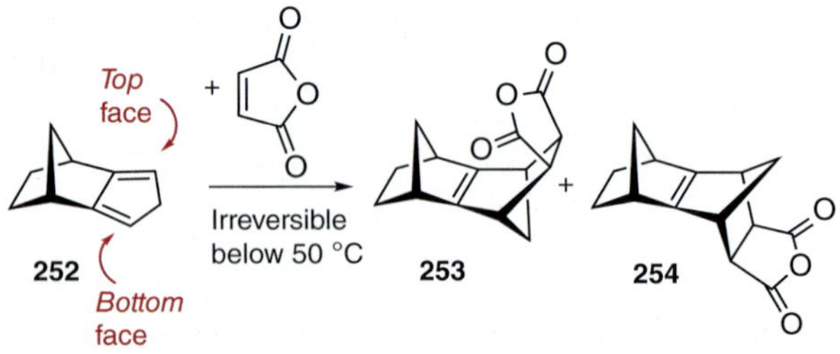

252 253 254

(5.22)

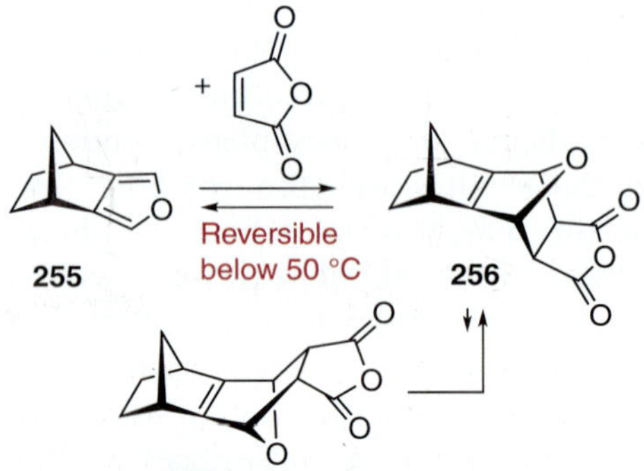

255 256 257 (not seen)

(5.23)

The Diels–Alder reactions of maleic anhydride and benzoquinone to a 4 : 1 mixture of (*E*)- and (*E*)-**258** (monodeuterated 2,3-dimethylidenenorbornane) are *bottom* face selective and do not follow the Alder "endo rule" (Scheme 5.46). With dimethyl acetylenedicarboxylate, the cycloaddition of (*E*)-**258** is also *bottom* face selective (>9 : 1), whereas with the more bulky dienophiles TCNE (ethylenetetracarbonitrile) and *N*-phenyltriazolinedione, the cycloadditions are *top* face selective. All these reactions are irreversible below 100 °C. Interestingly, the (4+1)-cheletropic addition of SO_2 (Section 5.4.4) to (*E*)-**258** is also *bottom* face selective under kinetic control [471].

The stereoelectronic factor responsible of this π-facial selectivity is related to the nonplanarity of norbornene double bond, itself related to hyperconjugative interactions involving the π-skeleton (σ/π mixing [400, 472]) of the bicyclic system and to torsional effects (avoided H—C(1)/R—C(2) and H—C(4)/R—C(3) eclipsing, Section 4.8.3). If one considers the cycloreversion of **259** and the diradicaloid models **260** and **260′** for their transition states (the same transition states as for the corresponding

Scheme 5.46 π-Facial and Alder stereoselectivity of the Diels–Alder reactions of 2,3-dimethylidenebicyclo[2.2.1]heptane with maleic anhydride and *p*-benzoquinone. A contrasteric stereoelectronic factor favors the *bottom* face attack.

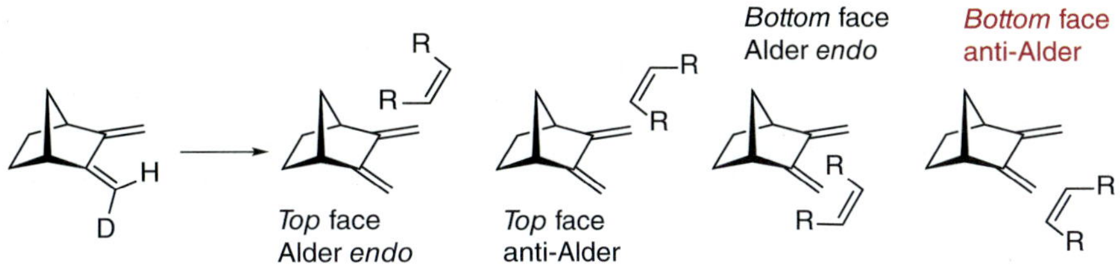

(*E*)-**258**

	Top face Alder *endo*	Top face anti-Alder	Bottom face Alder *endo*	Bottom face anti-Alder
Maleïc anhydride:	17%	13%	14%	56%
p-benzoquinone:	5%	6%	6%	83%

cycloadditions of (*E*)-**258**, principle of microscopic reversibility), one realizes that the π-electrons of the norbornene moiety that are polarized toward the *exo* face of the bicyclic system push the breaking σ(C—C) bonds of a boat cyclohexene moiety annulated to it better when the cyclohexene leans toward the *bottom* face (**260**) than toward the *top* face (**260′**) of the norbornene unit. This relates to the *anti*-selectivity in E_2-eliminations (antiperiplanar alignment of the electrofugal and nucleofugal groups) [473].

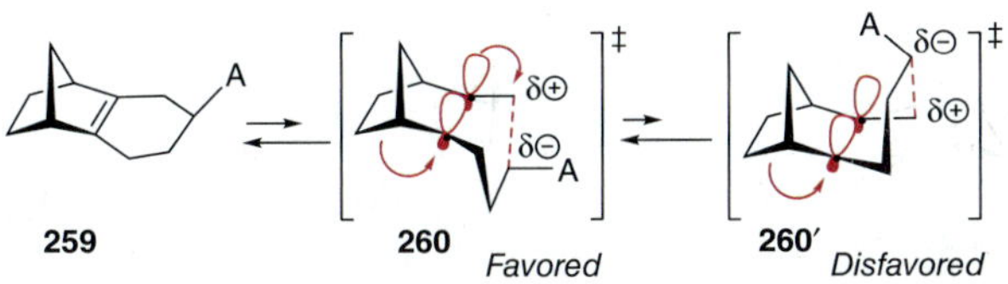

As for disymmetric 1,3-dienes, disymmetric dienophiles might also lead to π-facial selectivity in their cycloadditions for steric, stereoelectronic [474], or electrostatic reasons [475].

Problem 5.26 Thiophene 1-oxides adds to electron-rich and electron-poor dienophiles preferentially onto the face of the diene containing the S=O moiety [476–480]. Give an explanation.

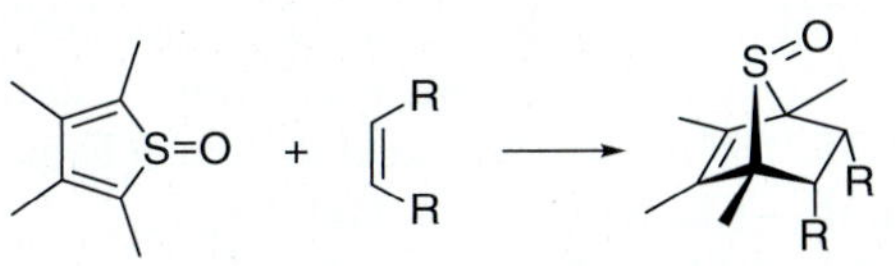

5.3.16 Examples of hetero-Diels–Alder reactions

Aldehydes and ketones add to conjugated dienes in oxa-Diels–Alder reactions to generate 3,4-dihydro-2*H*-pyrans [301, 377]. Dihydro- and tetrahydropyran derivatives are prevalent structural subunits in many bioactive natural products, including carbohydrates, pheromones, iridoids, and polyether antibiotics. As these oxa-Diels–Alder reactions are less exothermic ($\Delta_r H > -23.8$ kcal mol^{-1}; Figure 5.31) than classical Diels–Alder reactions ($\Delta_r H$(butadiene + ethylene → cyclohexene) = −39.9 kcal mol^{-1}, gas phase, Section 1.7.1), the lower driving force ($\Delta_r H$ in the Bell–Evans–Polanyi relationship: $\Delta^{\ddagger}H = \alpha\Delta_r H + \beta$) must be compensated by a lower intrinsic barrier β for one-step concerted reactions. According to the FMO theory (Section 4.4.2) and to the diradicaloid model (Section 5.3.10–11), this can be done by choosing electron-rich 1,3-dienes such as 1,3-dioxy-substituted 1,3-dienes (the Danishefsky's dienes [481, 482]), the 1-alkoxy-1,3-bis(silyloxy)-1,3-butadienes (the Brassard's dienes [483]), the (*E*)-1-dimethylamino-3-silyloxy-1,3-dienes (the Rawal's dienes [484]), or 6-methoxy-1-vinyl-3,4-dihydronaphthalene (the Dane's diene [485]) that permit better stabilizing LUMO(dienophile)/HOMO(diene) interactions and/or better electron transfers from the dienes to the carbonyl dienophiles in the transition states. As shown first by Danishefsky and coworkers in 1982 for the reaction of benzaldehyde with (*E*)-1-methoxy-3-trimethylsilyloxy-1,3-butadiene [303, 304], one can increase the electronic demand (lower the LUMO energy) of the carbonyl compounds by coordination to a Lewis or protic acid [303, 309, 486, 487]. Moreover, if the concerted process remains too difficult, the cycloaddition may follow a two-step mechanism with the formation of a zwitterionic intermediate, the stability of which is strongly affected by the polarity of the medium and/or the presence of Lewis or protic acid. Quantum mechanical calculations on the reaction of butadiene and formaldehyde coordinated to BH_3 predict two transition states for their (4+2)-cycloaddition (BH_3 is placed either *exo* or *endo* relative to the diene) that are both much lower in energy than for the transition structure of the cycloaddition in the absence of BH_3 [488]. Alternative mechanism involving the formation of diradical intermediates can be ruled out by thermochemical criteria. The activation enthalpies for the formation of diradicals of types **261H** or **262H** must be equal or higher than 42.8 kcal mol^{-1}, to be compared with the reaction of butadiene + ethylene ⇄ hex-5-ene-1,4-diyl diradical (**177**) that requires a much lower activation enthalpy

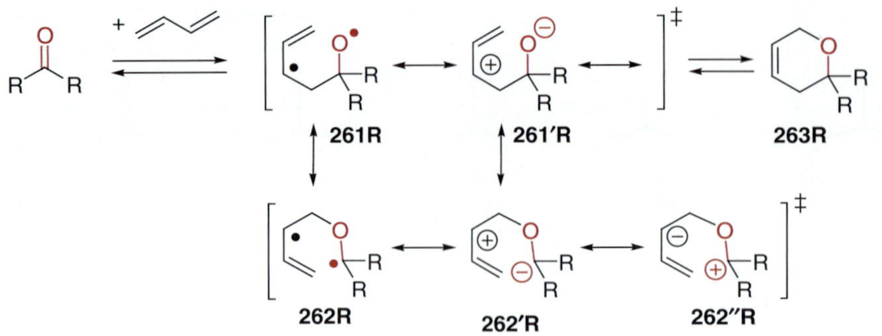

Figure 5.31 Diradicaloid model applied to the hetero-Diels–Alder additions of butadiene to formaldehyde and to acetone. They are much less exothermic than the Diels–Alder reaction of butadiene with ethylene. If diradical intermediates should form, the reactions would be much slower than butadiene + ethylene ⇄ hex-5-ene-1,4-diyl diradical (**177**) → cyclohexene.

of 33.5 kcal mol^{-1} (Figure 5.31). The situation is even worse for the reaction of ketones as an activation enthalpy equal or higher than 48.1 kcal mol^{-1} must be overcome to reach diradical **262Me** by addition of butadiene to acetone. Exothermicity of the cycloadditions can be increased by using the so-called activated aldehydes and ketones: i.e. carbonyl derivatives destabilized electronically by electron-withdrawing α-substituents such as CHO, COOR, COR, CN, and CF_3 (Section 2.8). These substituents might also increase the electron affinity of the dienophiles (Table 1.A.22) and thus enhance their hetero-Diels–Alder reactivity toward electron-rich dienes. In the diradicaloid model, its anionic moiety is conjugated with a carbonyl group, a nitrile group, or stabilized by field effect by the electron-withdrawing substituent such as CF_3. In general, ketones are much less reactive than corresponding aldehydes [489] for steric reasons (increase of the β term in the Bell–Evans–Polanyi relationship) and because of the lower exothermicity of the cycloaddition as shown with thermochemical estimates given here-below:

$\Delta_f H°(H_2C{=}O) = -27.7$ kcal mol^{-1}; $\Delta_f H°(Me_2C{=}O) = -52.2$ kcal mol^{-1} (Table 1.A.4); $\Delta_f H°$(butadiene) = 26.0 kcal mol^{-1} (Table 1.A.2)

$\Delta_f H°$(**261H**) = 42.7; $\Delta_f H°$(**262H**) = 41.1; $\Delta_f H°$(**263H**) = − 25.5 kcal mol^{-1}

$\Delta_f H°$(**261Me**) = 26.6; $\Delta_f H°$(**262Me**) = 19.9; $\Delta_f H°$(**263Me**) = − 41.6 kcal mol^{-1}

$\Delta_r H°(H_2C{=}O$ + butadiene ⇄ **263H**) = − 23.8 kcal mol^{-1}

$\Delta_r H°(Me_2C{=}O$ + butadiene ⇄ **263Me**) = − 15.5 kcal mol^{-1}

If diradical intermediates should be formed in the rate-determining step:

$\Delta^{\ddagger} H(H_2C{=}O$ + butadiene ⇄ **261H**) ≥ 44.4 kcal mol^{-1}
$\Delta^{\ddagger} H(H_2C{=}O$ + butadiene ⇄ **262H**) ≥ 42.8 kcal mol^{-1}
$\Delta^{\ddagger} H(Me_2C{=}O$ + butadiene ⇄ **261Me**) ≥ 52.8 kcal mol^{-1}
$\Delta^{\ddagger} H(Me_2C{=}O$ + butadiene ⇄ **262Me**) ≥ 46.1 kcal mol^{-1}
$\Delta^{\ddagger} H$(ethylene + butadiene ⇄ **177**) ≥ 33.5 kcal mol^{-1}

The standard heats of formation of 2,2-dimethyl-3,4-dihydro-2*H*-pyran (butadiene + acetone → **263Me**) is deduced from that of 3,4-dihydro-2H-pyran (**263H**) and by considering the difference $\Delta_f H°$(*t*-BuOMe) − $\Delta_f H°$(EtOMe) = −67.8 − (−51.7) = −16.1 kcal mol^{-1} (Table 1.A.4): $\Delta_f H°$(**263Me**) = $\Delta_f H°$(**263H**) − 16.1 = −41.6 kcal mol^{-1}. The lower exothermicity of the hetero-Diels–Alder reaction of nonactivated ketones compared to formaldehyde and aldehydes arises from the fact that the π(C=O) bond in ketones is stronger than the π(C=O) bond in aldehydes, as shown by the standard heats of hydrogenation and hydrocarbation of carbonyl compounds (Section 1.7.1). The standard heat of formation of diradicals **261R** is derived from $\Delta_f H°$(**263R**) by considering $DH°$(*n*-Bu•/MeO•) = 83.4 kcal mol^{-1} and the allylic stabilization energy of the carbon-centered radical using $DH°$(CH$_2$=CH—C(Me)CH•/H•) = 83.9 kcal mol^{-1} and $DH°$(2-butyl•/H•) = 99.1 kcal mol^{-1} (Table 1.7). Thus, $\Delta_f H°$(**261H**) = −25.5 + 83.4 − 15.2 = 42.7 kcal mol^{-1}, and $\Delta_f H°$(**261Me**) = −41.6 + 83.9 − 15.2 = 26.6 kcal mol^{-1}. The standard heats of formation of diradicals **262R** are estimated from $\Delta_f H°$(CH$_2$=CHCH$_2$CH$_2$OCHR$_2$, R = H, Me) and $DH°$(CH$_2$=CH—C(Me)H•/H•) = 83.9 kcal mol^{-1} and $DH°$(MeOCH$_2$•/H•) = 93.1 kcal mol^{-1} (Table 1.7). Assuming −5 kcal mol^{-1} of stabilization of tertiary radical MeO(Me)$_2$C• compared with primary radical MeOCH$_2$•, $DH°$(MeO(Me)$_2$C•/H•) = 88.1 kcal mol^{-1}. Thus,

$\Delta_f H°$(**262H**) = $\Delta_f H°$(CH$_2$=CHCH$_2$CH$_2$OMe) + $DH°$(CH$_2$=CH − C(Me)H˙/H˙) + $DH°$(MeOCH$_2$˙/H˙)–$DH°$(H˙/H˙) = −31.7 + 83.9 + 93.1–104.2 = 41.1 kcal mol^{-1}

and

$\Delta_f H°$(**262Me**) = $\Delta_f H°$(CH$_2$=CHCH$_2$CH$_2$OCHMe$_2$) + $DH°$(CH$_2$=CH − C(Me)H˙/H˙) + $DH°$(MeO(Me)$_2$C˙/H˙)–$DH°$(H˙/H˙) = −47.9 + 83.9 + 88.1–104.2 = 19.9 kcal mol^{-1}.

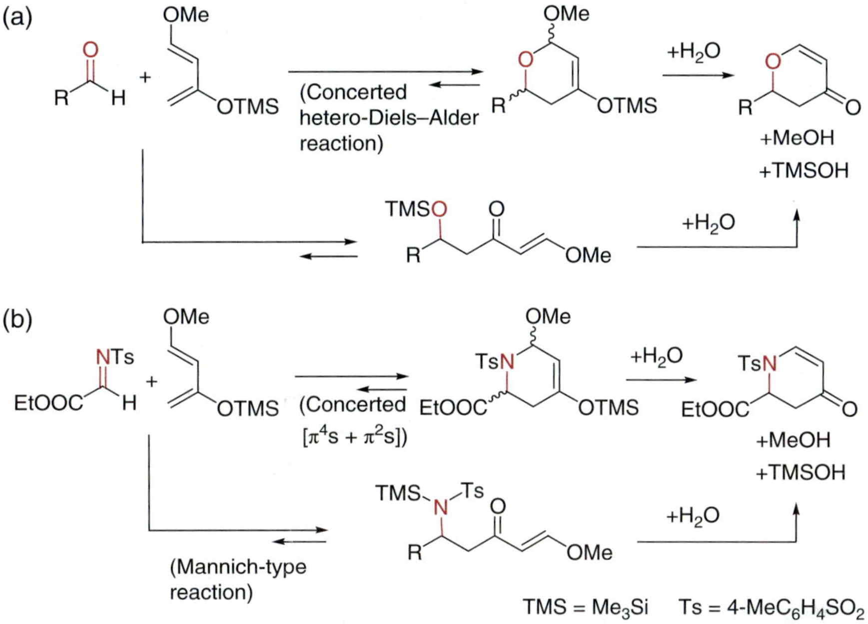

Figure 5.32 Limiting mechanisms for the hetero-Diels–Alder reactions of (a) aldehydes and (b) their imines with a Danishefsky diene.

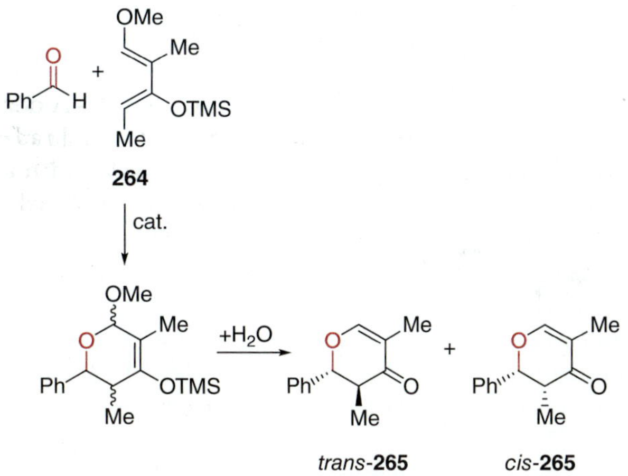

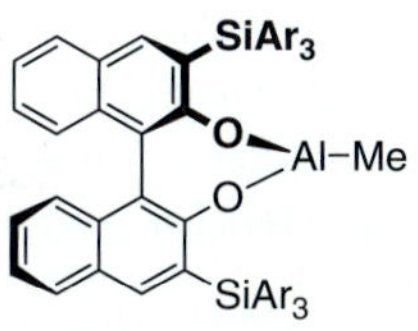

(*R*)-**266** Ar = 3,5-$Me_2C_6H_3$ (3,5-xylyl)

	trans-265 : cis-265
BF_3/CF_3COOH	58% : 23%
$ZnCl_2/CF_3COOH$	<2% : 78%

Two pathways are possible for the reactions of 3-silyloxy-1,3-butadienes with aldehydes catalyzed by Lewis acids that generate, after aqueous work-up, the corresponding 5,6-dihydropyran-4-ones (Figure 5.32a). The first one is a concerted (4+2)-cycloaddition (oxa-Diels–Alder) and the second is a two-step process corresponding to a Mukaiyama aldol condensation. Similarly, the same dienes add to imines either through a concerted (4+2)-cycloaddition (aza-Diels–Alder) or in a two-step process corresponding to a Mannich-type pathway (Figure 5.32b) [307].

In the case of the reaction of benzaldehyde with diene **264**, the proportion of 5,6-dihydropyran-4-ones *trans*-**265** and *cis*-**265** depends on the nature of the Lewis acid catalyst. Danishefsky et al. proposed that with BF_3, the Mukaiyama condensation is followed, whereas with $ZnCl_2$, or lanthanide salts, a concerted (4+2)-cycloaddition is realized [490]. Using (*R*)-**266** as catalyst, Yamamoto and coworkers obtained *cis*-**265** in 90% yield and 97% ee, next to only 3% of *trans*-**265** [491]. For other examples of catalyzed enantioselective hetero-Diels–Alder reaction of aldehydes, see e.g. Refs. [309, 492–499].

The hetero-Diels–Alder reactions of aldehydes and ketones are accelerated by hydrogen bonding. When mixing acetone with $CHCl_3$, heat evolves because of the formation of hydrogen-bridged complex $Me_2C{=}O\cdots H{-}CCl_3$. It is found that the reactions of diene **267** [500, 501] with aldehydes and ketones are faster in $CHCl_3$ and 2-butanol than in aprotic solvents [500, 502]. In the diradicaloid model, the alkoxide moieties of the zwitterionic limiting structures make stronger hydrogen bonds than the carbonyl groups. Using toluene as solvent and the (*R*,*R*)-tartaric acid derived diol (TADDOL) **268**, the corresponding 5,6-dihydro-pyran-4-ones (**269**) were obtained with >92% ee [504, 505]. For more examples of hydrogen bond enantioselective hetero-Diels–Alder reactions, see e.g. Ref. [506]. The chalcogen-Diels–Alder reactions of $H_2C{=}Z$ (Z = S, Se, Te) with 1,3-dienes and $CH_2{=}CH{-}CH{=}Z$ are predicted by quantum mechanical calculations to follow concerted mechanisms rather than two-step pathways involving diradical intermediates [507].

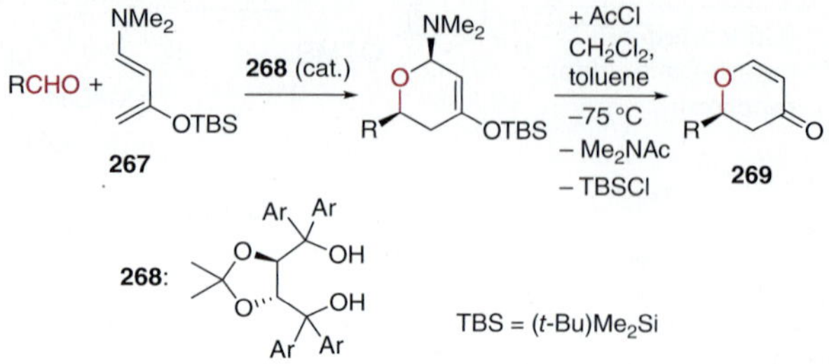

In 1982, Kerwin and Danishefsky reported the first $[\pi^4+\pi^2]$-cycloadditions (5.24) of imines catalyzed by $ZnCl_2$ [508, 509]. The methodology has been applied to the synthesis of alkaloids [510].

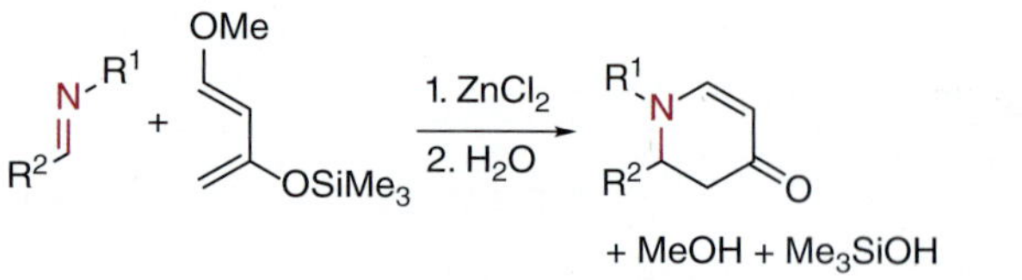

(5.24)

L-Proline catalyzes the enantioselective aza-Diels–Alder reaction of arylimines with conjugated enones as illustrated in Scheme 5.47. L-Proline and the enone equilibrate with the corresponding 2-aminodiene **270** that adds to the imine resulting from the condensation of the arylamine and formaldehyde. The reaction is assumed to involve the formation of the zwitterionic intermediate **271** that cyclizes quickly and irreversibly into the corresponding Diels–Alder cycloadduct [511]. For other examples of catalyzed enantioselective aza-Diels–Alder reactions, see e.g. Refs [308, 512–515].

N-Sulfinyl compounds are good heterodienophiles [516]. Using chiral bis(oxazoline)-copper(II) and zinc(II) triflates in combination with $Me_3SiOSO_2CF_3$ (TMS triflate), high *endo*-stereoselectivities and enantioselectivities have been reported for reactions with (*E*)-piperilene, (*E*,*E*)-hexa-2,4-diene, (*E*,*E*)-1-benzyloxy-hexa-2,4-diene, and cyclohexa-1,3-diene [517]. The inverse electron demand hetero-Diels–Alder reaction of α,β-unsaturated carbonyl compounds (1-oxa-1,3-dienes) with electron-rich alkenes is a very attractive route for the direct formation of 3,4-dihydro-2*H*-pyran derivatives [318]. This reaction follows usually a concerted asynchronous mechanism with retention of the configuration of the dienophile ($[\pi^4s+\pi^2s]$-cycloaddition) and shows high regioselectivity, especially when catalyzed by a Lewis acid (that coordinates to the oxa-diene and stabilizes the zwitterionic limiting structure of the diradicaloid transition state, see e.g. **273**). An example of highly diastereoselective (cycloadduct ratio *cis*-**274**/*trans*-**274** > 95%) and enantioselective reaction (89–98% ee) is given below with the cycloaddition (5.25) of α,β-unsaturated aldehydes **272** to ethyl vinyl ether catalyzed by the (Schiff base)Cr(III) complex (1*S*,2*R*)-**275** [518].

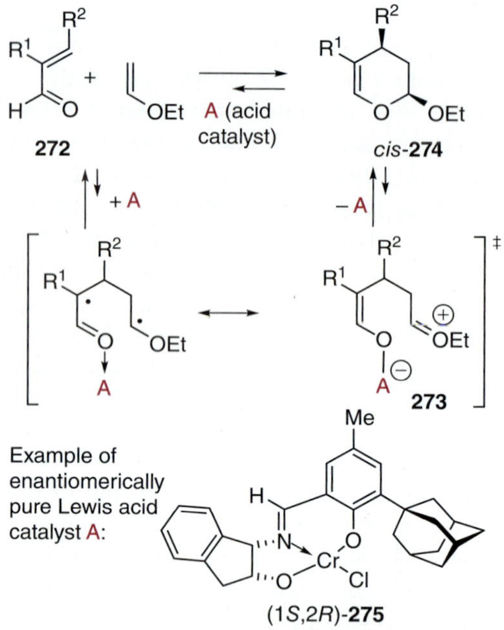

(5.25)

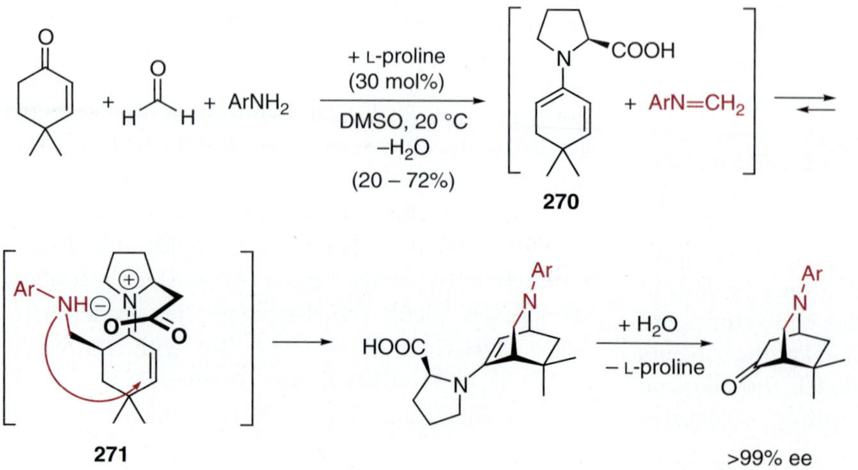

Scheme 5.47 Example of an enantioselective aza-Diels–Alder reaction catalyzed by L-proline.

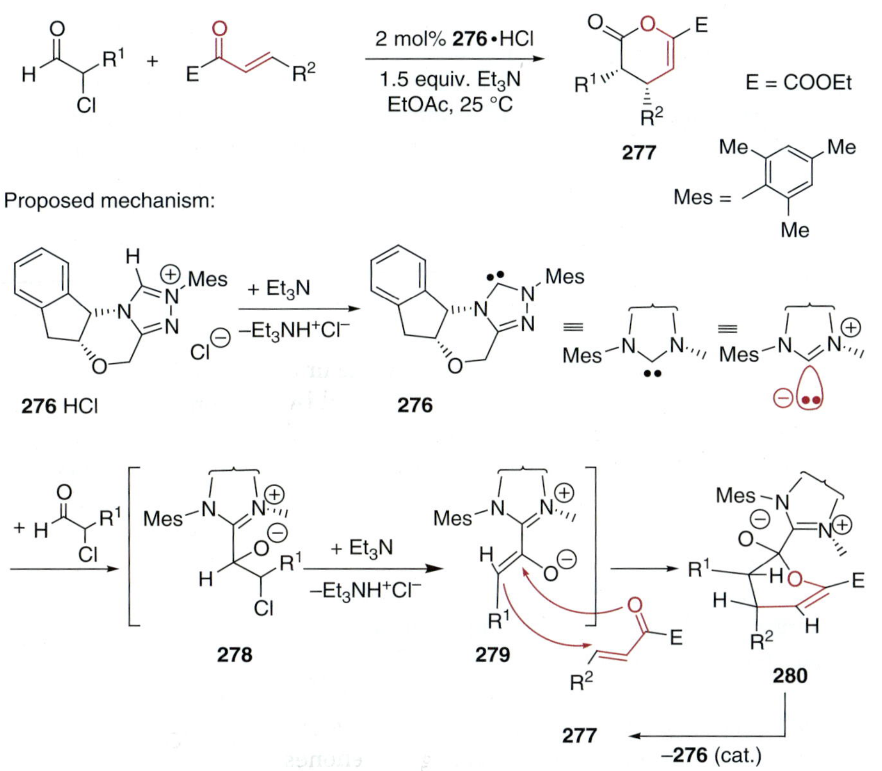

Scheme 5.48 Bode's enantioselective synthesis of 3,4-dihydropyran-2-ones.

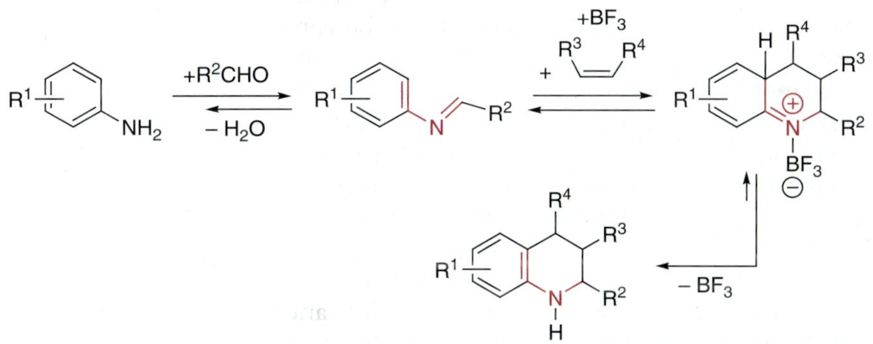

Scheme 5.49 The Povarov reaction.

Enantiomerically pure diaminocarbene **276** catalyzes the enantioselective formation 3,4,6-trisubstituted 3,4-dihydropyran-2-ones **277** (Scheme 5.48). The diamino carbene adds to the aldehyde giving first zwitterion **278** that eliminates 1 equiv. of HCl in the presence of Et_3N providing (*Z*)-enolate **279**, the electron-rich dienophile that adds to the oxadiene on its less sterically hindered face (*anti* with respect to the indenyl moiety of the catalyst), and adopting the Alder *endo* orientation. This generates cycloadduct **280** that eliminates the catalyst (**276**) producing **277** [519]. For other enantioselective cycloadditions of heterodienes, see e.g. Refs. [520–522].

Hetero-Diels–Alder reactions of 1-aza-1,3-dienes with electron-rich alkenes constitute one of the most powerful synthesis of piperidine derivatives [523]. *N*-Sulfonyl 1-aza-1,3-dienes react under thermal conditions with high Alder *endo* diastereoselectivity [320, 394]. Enantioselective versions of this reaction have been reported that use enantiomerically pure Lewis acids [524, 525] or diaminocarbenes [526] as catalysts. One variant of the hetero-Diels–Alder reaction of arylimines is the Povarov reaction that couples aldehydes, anilines, and alkenes under acidic conditions into tetrahydroquinolines (Scheme 5.49) [527]. Generally, the reaction is limited to electron-rich dienophiles or strained alkenes [528].

The nitroso hetero-Diels–Alder reaction provides access to 3,6-dihydro-1,2-oxazines. The latter are useful synthetic intermediates that permit the construction of a large variety of biologically active compounds [310, 311]. The first hetero-Diels–Alder reaction of a nitroso compound was reported by Wichterle [529]. At that time, it was believed that the RN=O moiety adds like SO_2 through a (4+1)-cheletropic reaction. Kresze and Firl found that (*E*)-1-(dimethylamino)butadiene (**281**: electron-rich diene) adds to *p*-chloronitrosobenzene with pseudo-*meta* regioselectivity forming cycloadduct **283** [530]. The regioisomeric diradical limiting structures **282** and **284** are both stabilized by R = Me_2N. Diradical **282** is less stable than diradical **284** because the C—O bond in **284** is stronger than the C—N bond in **282** and the nitrogen center in **284** is arylated (benzylic stabilization), whereas the oxygen center in **282** is not. The corresponding zwitterionic limiting structures **282′** and **284′** are expected to be relatively important because of R = Me_2N. The negative charge of the regioisomeric zwitterionic limiting structures prefers to reside on an oxygen center, as in **282′**, rather than on a nitrogen center, as in **284′**. This might be the reason why the pseudo-*meta* regioselectivity is observed. Quantum mechanical calculations support concerted mechanisms for the nitroso-Diels–Alder reactions that proceed through asynchronous transition structures. Two-step mechanisms via diradical and/or zwitterionic intermediates only compete when the nitroso dienophiles that bear a small nitrogen substituent react with dienes substituted with radical stabilizing groups [531, 532]. It has not been verified whether the regioselectivity of reaction **281** + *p*-chloronitrosobenzene ⇄ **283** is controlled, or not, by thermodynamics: the most stable regioisomer is obtained because of reversible cycloadditions (as for the reactions of sulfur dioxide, Scheme 5.53). The reactions of *p*-chloronitrosobenzene with a large variety of 1-substituted butadienes (R = COOMe, OAc, Ph, *t*-Bu, Me) are also *pseudo-meta* regioselective. The *pseudo-ortho* cycloadducts do not form as they are destabilized by gauche interactions between the aryl and R groups. They form if C(4) of the butadiene system is also substituted [533, 534].

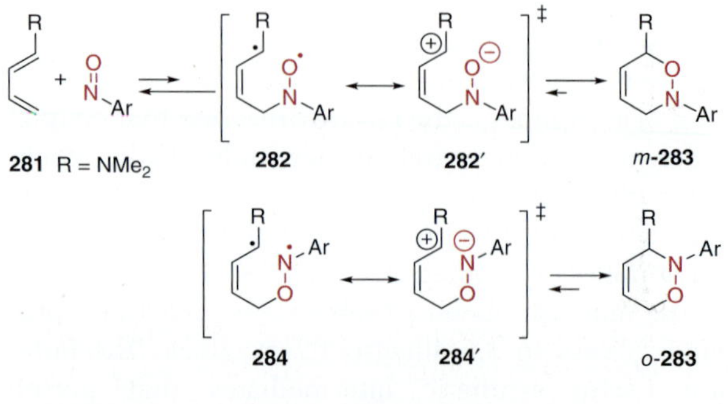

$DH°(PhNH—O^•/H^•) = 77.5$; $DH°(c\text{-}C_5H_{10}N—O^•/H^•) = 77\ kcal\ mol^{-1}$ (Table 1.A.11)

$DH°(CH_2{=}CH—CH_2^•/HO^•) = 78.3$; $DH°(CH_2{=}CH—CH_2^•/H_2N^•) = 73.1\ kcal\ mol^{-1}$ (Table 1.A.7); $DH\ (PhN^•(OH)/H^•) = 77\ kcal\ mol^{-1}$

Vasella and coworkers have prepared important synthetic intermediates of type **286** with high enantiomeric purity by reaction of D-mannose-derived nitroso compound **285** with all kinds of acyclic and cyclic dienes. The example shown in Scheme 5.50 demonstrates the suprafaciality of the hetero-Diels–Alder reaction with respect to the diene [535].

A catalytic asymmetric nitroso-Diels–Alder reaction has been reported by Yamamoto and Yamamoto as illustrated here-below [536]. The enantiomerically pure ligand (*S*)-segphos substitutes two molecules of MeCN in the Cu(I) salt that activates the nitroso dienophile by formation of a Lewis acid–base complex.

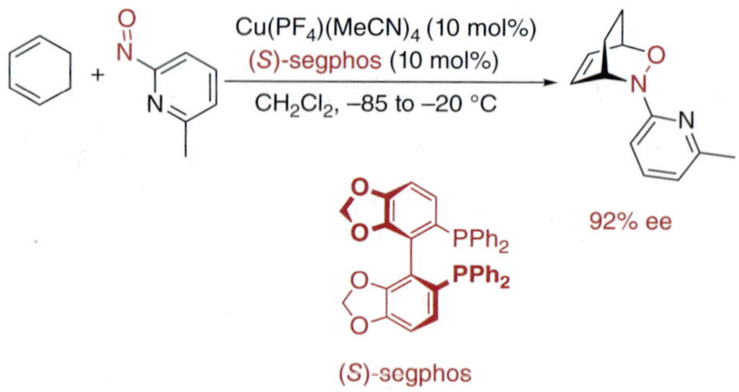

In 1953, Wichterle and Rocek described the first hetero-Diels–Alder reaction of *N*-sulfinylarylamines [537]. The reaction of cyclopentadiene with *N*-sufinylbenzenesulfonamide is reversible. The cycloadduct can be isolated at low temperature but decomposes into its cycloaddents at room temperature (Scheme 5.51) [538]. Under kinetic control (5 °C), the reaction of (*E*)-1-aryl-1,3-butadiene with *N*-sulfinyl-*p*-toluenesulfonamide gives the pseudo-*ortho* cycloadduct **287**, as predicted by the diradicaloid model **288** ↔ **288′**. When run under reflux of benzene (80 °C), the pseudo-*meta* cycloadduct **287′** is obtained. The latter is more stable than **287** because the gauche interactions between the large aryl and sulfonamide groups are avoided. Interestingly, only the N=S double bond of the *N*-sulfinylsulfonamides undergoes the cycloaddition, neither the terminal S=O nor sulphonamide S=O double bonds react. Sulfodimides of types $RSO_2N{=}S{=}NSO_2R$ and $RSO_2N{=}S{=}NAr$ undergo hetero-Diels–Alder reactions also through their S=N double bonds [313].

Scheme 5.50 Vasella's asymmetric hetero-Diels–Alder reaction of D-mannose-derived dienophile.

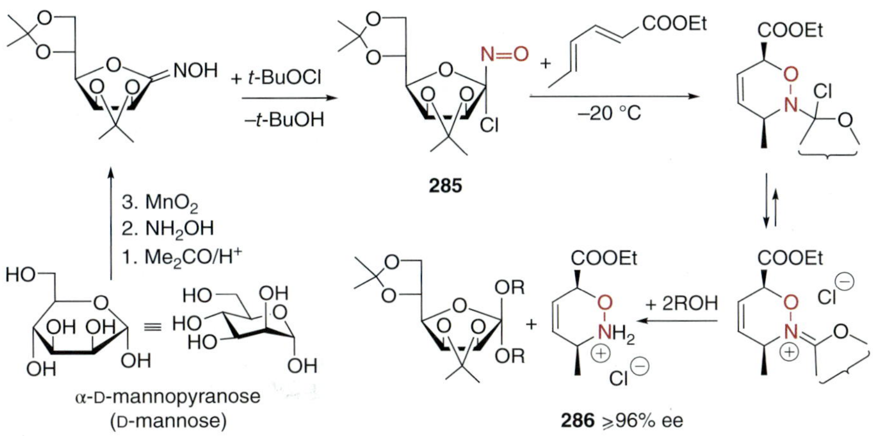

Scheme 5.51 Examples of reversible hetero-Diels–Alder reactions of *N*-sulfinylsulfonamides.

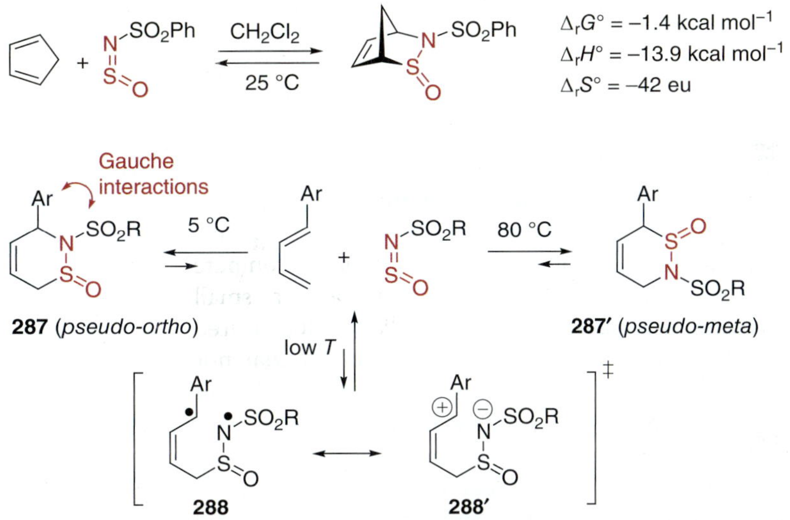

At room temperature, any 1,3-dienes that can adopt the *s-cis*-conformation add to sulfur dioxide (SO_2) giving the corresponding sulfolenes (2,5-dihydrothiophene-1,1,-dioxides) in (4+1)-cheletropic additions (Section 5.4.4), a reaction discovered in 1914 [539]. The hetero-Diels–Alder reaction of SO_2 across one of its S=O double bond was reported first in the 1970s for two extremely reactive dienes, i.e. 1,4,5,6-tetramethyl-2,3-dimethylidenetricyclo[2.1.1.0^{5,6}]hexane [540] and orthoquinodimethane [541]. In the presence of acid catalysts (e.g. CF_3COOH) and at low temperature, alkyl substituted 1,3-dienes [315], and butadiene itself [542] add to SO_2 in the suprafacial hetero-Diels–Alder mode, giving the corresponding sultines (3,6-dihydro-oxathiin-2-oxides). These hetero-Diels–Alder reactions are at least 100 times as fast as the corresponding (4+1)-cheletropic additions and the sultines (e.g. **289**) obtained are about 8–10 kcal mol^{-1} less stable than the corresponding sulfolenes (e.g. **290**) (corresponds to the difference in stability between sulfinic esters and sulfones; Scheme 5.52) [316]. The kinetic chemoselectivity of SO_2 + 1,3-dienes is the hetero-Diels–Alder reaction and the thermodynamic chemoselectivity is the (4+1)-cheletropic addition.

With 1-substituted 1,3-dienes, the Alder *endo* rule is followed leading to *cis*-sultines *cis*-**289** that are isomerized upon warming into the more stable trans-isomer *trans*-**289**. With 1-fluorodienes, SO_2 generates the corresponding sultines exclusively as their isomeric sulfolenes are less stable (see Sections 2.7.9 and 2.7.10). In these cases, kinetic and thermodynamic chemoselectivity is the hetero-Diels–Alder reaction. With electron-rich 1,3-dienes such as (*E*)-1-alkoxy-, 1-trialkylsilyloxy-, 1-aryl-, 1-alkythio-, 1-alkylseleno-, and 1-cyclopropyl-1,3-butadiene, the corresponding sultines are not seen at −100 °C because the dienes are much more solvated (formation of yellow to red complexes) than the corresponding sultines; only the corresponding sulfolenes form. The hetero-Diels–Alder reaction of SO_2 with 1,2-dimethylidenecyclohexane does not require

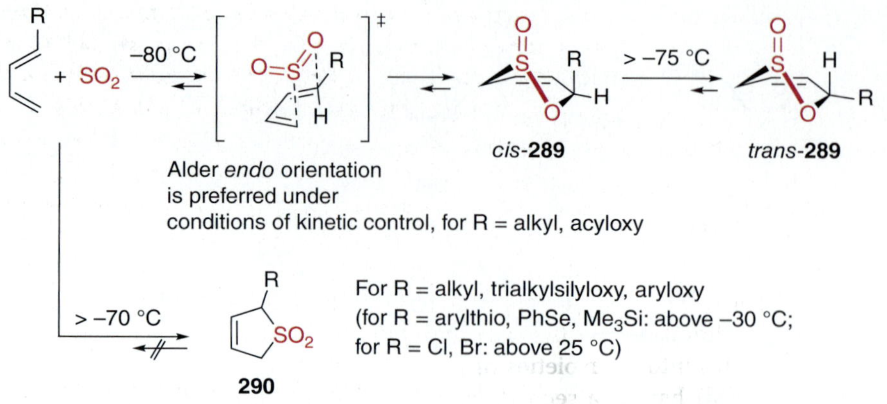

Scheme 5.52 Competition between hetero-Diels–Alder reaction and (4+1)-cheletropic addition of sulfur dioxide.

promotion by an acid. The reaction is promoted by SO_2 itself as the rate law of the cycloaddition is of second order in $[SO_2]$ and first order in the diene [543]. With the dideuterated derivative **291**, the two regioisomeric sultines **292a** and **292b** are obtained with a rate constant ratio $k_a/k_b = 1.11 \pm 0.1$ at −75 °C and 1.0798 ± 0.0003 at −54 °C . Upon staying at −54 °C, **292a** and **292b** are equilibrated in a product ratio [**292a**]/[**292b**] = 0.73 ± 0.04: this is a unique case for which it is demonstrated that the kinetic isotopic effect is inverse of the thermodynamic isotopic effect. In agreement with quantum mechanical calculations, it demonstrates unambiguously that the SO_2-promoted hetero-Diels–Alder reaction of SO_2 follows a concerted asynchronous mechanism with a transition state in which the C—S bond is formed to a greater extent than the C—O bond (Scheme 5.53). This supports the diradicaloid model **293** ↔ **293′** for the transition state of the $[\pi^4+\pi^2]$-cycloaddition. Bending frequencies of the C—H bonds of the methylidene groups in **291** are weaker than those in the sultine **292** (sp^2-C—H bonds in reactants are transformed into sp^3-C—H bonds in the product). As deuterium prefers to be in the C—D bonds that have the strongest force constant for their bending, it prefers diradicaloid **293** ↔ **293′** to diradicaloid **294** ↔ **294′**. The stretching frequency of the C—H bond α of the S=O group in sultine **292** is expected to be lower than that of the C—H of the CH_2—O—S(=O) moiety as the former bond is weaker than the latter. Thus, **292a** is less stable than **292b** because deuterium prefers to substitute the strongest C—H bonds (see Section 1.5 and 3.9). The kinetic deuterium isotopic effect reported for the hetero-Diels–Alder reaction SO_2 + **291** → **292** is not dominated by the thermodynamic deuterium isotopic effects as the Bell–Evans–Polanyi theory would have predicted it (the most stable regioisomeric cycloadduct should form the fastest because of $\Delta^{\ddagger}H = \alpha\Delta_r H + \beta$). This suggests that the diradical

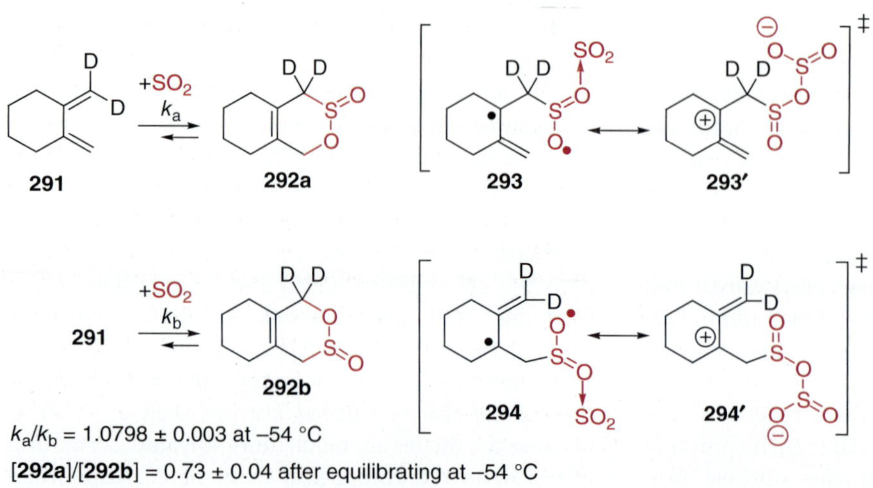

Scheme 5.53 Kinetic vs. thermodynamic deuterium isotopic effect in the hetero-Diels–Alder reaction of SO_2 with 1,2-dimethylidenecyclohexane.

and zwitterionic limiting structures contribute more to the transition structure of the cycloaddition than reactants and product.

Although sultines resulting from the reaction of 1-alkoxy-3-acyloxy-1,3-dienes **295** are not seen in equilibrium with the dienes, they form quickly at −78 °C and are converted into zwitterionic intermediates of type **296** under acidic media. The latter are electrophilic reagents that can be quenched in situ by enoxysilanes or allylsilanes [544] permitting the diastereoselective formation of C—C bonds. The method (Umpolung of electron-rich dienes into oxyallyl cationic intermediates with SO_2 [545]) has been applied to the total synthesis of natural polyketide and polypropionate antibiotics [546–549]. For instance, with the (*Z*)-silyl enol ether of pentan-3-one, the *syn,anti*-stereotriads (polypropionate fragments) **299** are obtained in one step with high diastereoselectivity from the enantiomerically enriched dienes **295** (transfer of chirality between R* and **296**). The reaction cascade involves the formation of **297** which, under protic conditions, liberates the β,γ-unsaturated sulfinic acid intermediates **298** that are not stable but undergo above 0 °C quick stereoselective retro-ene eliminations of SO_2 (Section 5.7.3) producing the *syn,anti*-stereotriads **299** (Scheme 5.54). Using the (*E*)-silyl enol ether of pentan-3-one, the corresponding *anti,anti*-stereotriads are obtained with the same ease. Without deprotection and activation step, stereotriads **299** are ready for cross-aldol reactions with aldehydes R^1CHO on their ethyl ketone moiety. After reduction of the aldols obtained and protection of the corresponding diols into **300**, the enol ester moieties of the latter can be combined directly with a second series of aldehydes R^2CHO in cross-aldol reactions, leading to large collections (libraries) of long-chain polyketides and polypropionates of type **301** [550].

Problem 5.27 Give the major products formed in the following cycloadditions (for reactions a–f) see Ref. [377]; for reaction (g) see Ref. [315]; for reaction (h) see Ref. [551]; for reaction (i) see Ref. [301].

Scheme 5.54 One-pot synthesis of a stereotriad and its two successive cross-aldol reactions: efficient preparation of libraries of long-chain polyketides and polypropionates.

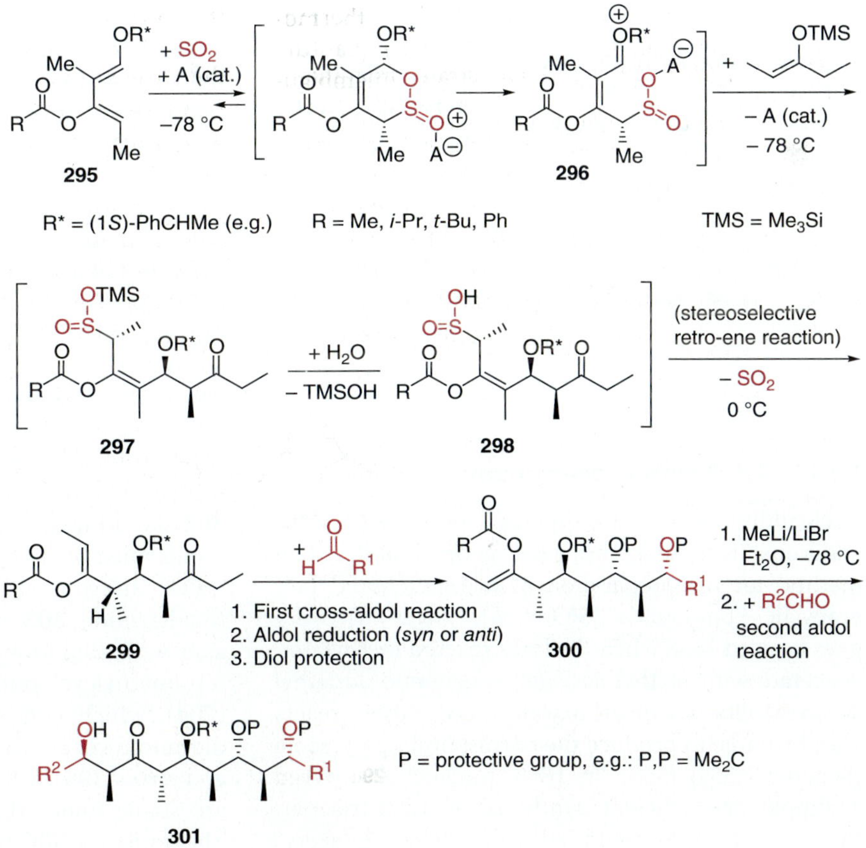

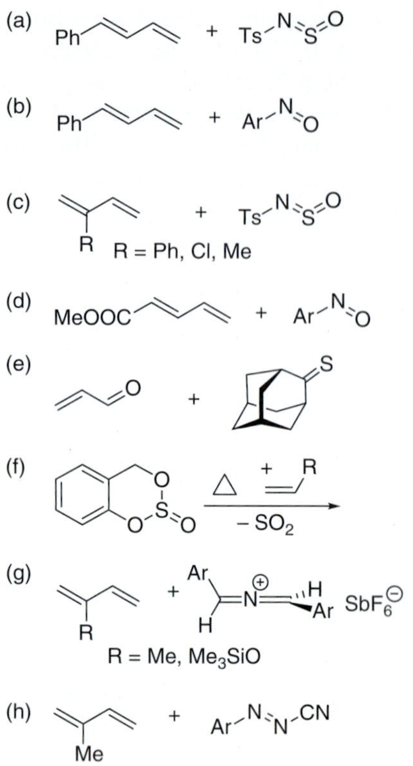

Problem 5.28 In most cases, arylnitroso and acylnitroso dienophiles lead to the same regioselectivity. However, opposite regioselectivities are observed for the following hetero-Diels–Alder reactions [310, 311, 552]. Explain why.

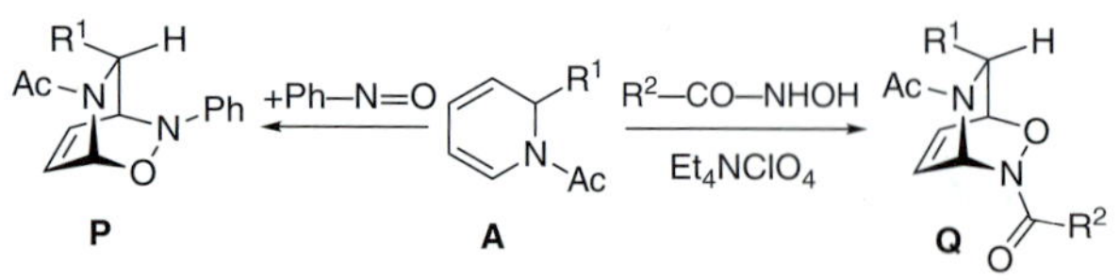

5.3.17 1,3-Dipolar cycloadditions

Cycloadditions of 1,3-dipoles, or (3+2)-cycloadditions, represent the most general and useful method for the preparation of five-membered heterocyclic compounds [553, 554]. Their discovery goes back to 1888 when Büchner reacted methyl diazoacetate with methyl acrylate to generate dimethyl Δ^2-pyrazoline-3,5-dicarboxylate. The latter reacts with bromine to produce the corresponding pyrazole (Scheme 5.55a) [555]. In 1893, Michael discovered a simple and efficient synthesis of 1,2,3-triazoles by reacting azides with alkynes, including acetylene (Scheme 5.55b) [556]. In 1898, Pechmann reported his method of pyrazole synthesis that combines acetylene with diazomethane (Scheme 5.55c) [557]. Five-membered carbocyclic can form by cycloaddition of all carbon two- and three-centered cycloaddents. These reactions are presented in Sections 7.5.9 and 8.3.6.

Huisgen recognized the 1,3-dipolar character of diazo compounds and azides and defined many other 1,3-dipoles (Figure 5.33, Table 5.5) and their 1,3-dipolar cycloadditions to unsaturated systems (alkenes, alkynes, carbonyl compounds, imines, thiocarbonyl compounds, and nitriles) [587]. They are now called the "Huisgen cycloadditions."

Huisgen postulated concerted mechanisms for 1,3-dipolar cycloadditions based on kinetics and stereochemical observations [588]. Early quantum mechanical calculations agreed with this hypothesis [589] for which the two new σ bonds are both formed in single five-membered transition structures. A stereoselectivity better than 99.997% was found for the cycloaddition of diazomethane ($CH_2{=}N_2$) to methyl tiglate (methyl (*E*)-2-methylbut-2-enoate) [590]. Firestone proposed two-step mechanisms with the formation of diradical intermediates [591]. One example of such mechanism was found later by Huisgen et al. (Scheme 5.57) [592, 593]. In the two-stage mechanism, one bond is formed first, generating a short-lived unstable diradical or zwitterion that may cyclize before rotation about the σ bonds, thus retaining the suprafacialityof the (3+2)-cycloaddition making it analogous to the Diels–Alder reaction ([π^4s+π^2s]-cycloaddition) [594]. Quantum mechanical calculations and molecular dynamic investigations have shown that the concerted, one-step mechanism is favored for the reactions of unsubstituted 1,3-dipoles with ethylene or acetylene [595–598]. On the other hand, when the 1,3-dipoles and dipolarophiles are substituted by radical stabilizing groups, the stepwise mechanism is favored [599]. The reaction of 4-nitrobenzonitrile oxide with (*Z*)- and (*E*)-1,2-dideuterioethylene (Scheme 5.56) is stereoselective to better than 98%. If a diradical intermediate **302** should form, the barrier for rotation about its σ bonds must be higher than 2.3 kcal mol^{-1} [600].

On heating (THF, 40 °C), the spiro-1,3,4-thiadiazoline **303** undergoes a dipolar cycloreversion with evolution of N_2 and formation of 2,2,4,4-tetramethylcyclobutan-1-one-3-thione *S*-methylide (**304**), which combines in situ with dimethyl 2,3-dicyanofumarate to give a 48 : 52 mixture of *cis*-**306** and *trans*-**306** in 94% yield. The separated adducts are stable under the conditions of their formation. The ratio *cis*-**306**/*trans*-**306** increases with solvent polarity in accord with a mechanism involving the formation of zwitterionic intermediates **305a** and **305b** (Scheme 5.57) [593]. Quantum mechanical

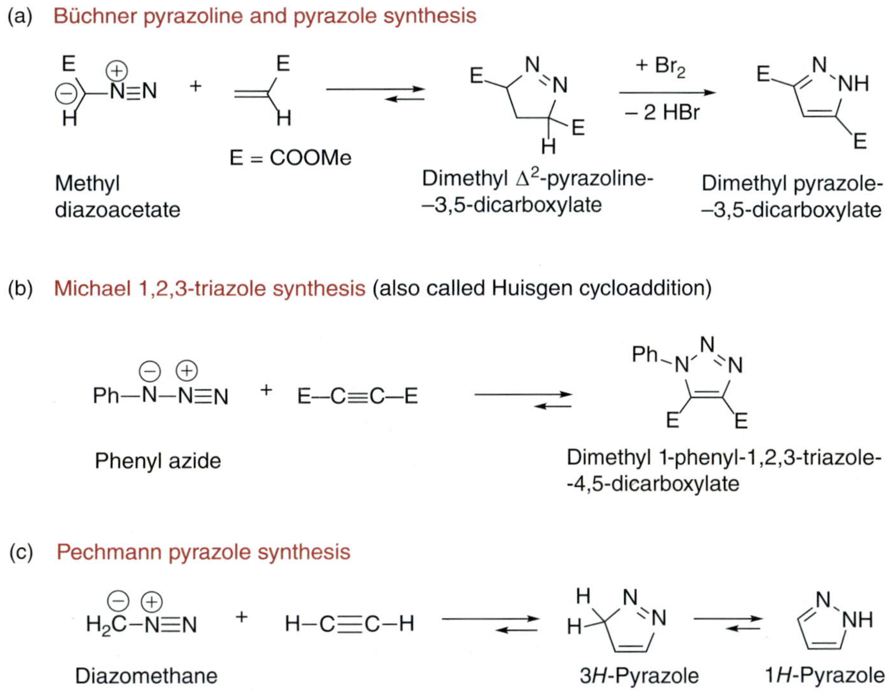

Scheme 5.55 Early examples of 1,3-dipolar cycloaddition reactions.

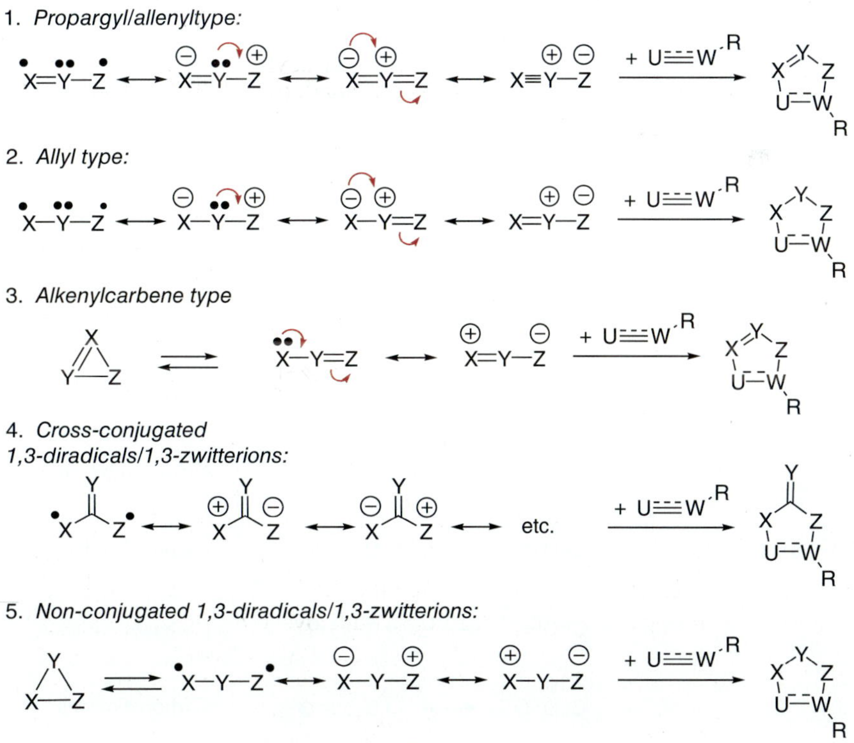

Figure 5.33 Types of 1,3-dipoles and their Huisgen cycloadditions.

calculations agree with this conclusion: the intermediates are zwitterionic and open-shell species referred to as polarized diradicals [599].

Sustmann classifies the 1,3-dipoles according to their type of reactivity (Figure 5.34). Depending on their HOMO and LUMO energies, the 1,3-dipoles react as nucleophiles, electrophiles, or ambiphiles against ethylene, the dipolarophile of reference. In the transition states of Huisgen cycloadditions of nucleophilic dipoles, HOMO(1,3-dipole)/LUMO

Table 5.5 Classification of conjugated 1,3-dipoles containing carbon, nitrogen, oxygen or/and sulfur.[a)]

Propargyl-allenyl types[a)]		
Nitrilium betaïnes		Nitrile ylides
		Nitrile imines
		Nitrile oxides
		Nitrile sulfides[b)]
Diazonium betaïnes	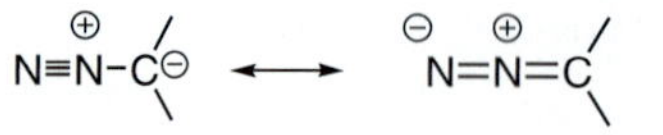	Diazo compounds
		Azides
	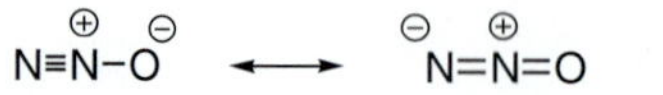	Nitrogen peroxide
		Dinitrogen sulfide[c)]
Allyl types[a)]		
N in the center	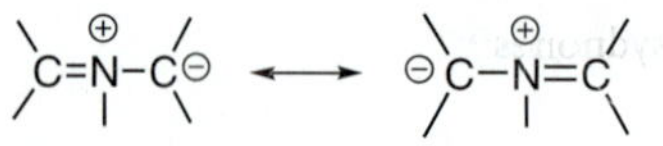	Azomethine ylides
	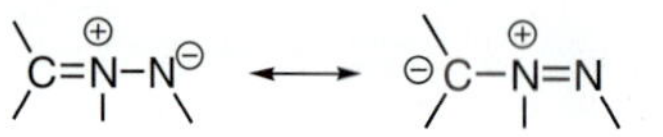	Azomethine imines
		Nitrones
	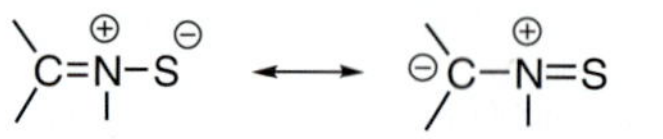	Imine thiones[d)]
	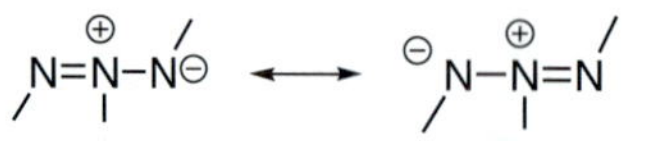	Azimines
	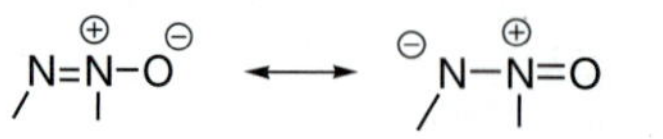	Azoxy compounds
	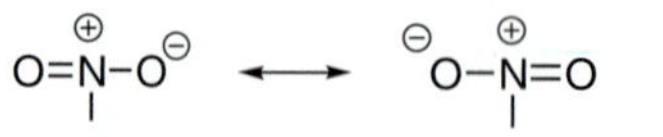	Nitro compounds
O in the center	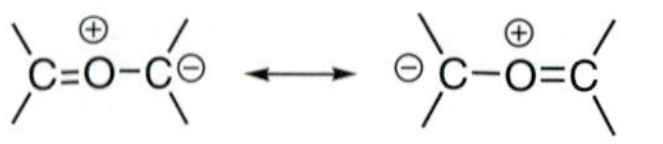	Carbonyl ylides
	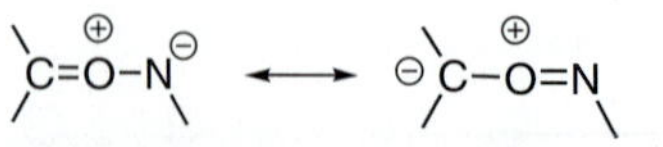	Carbonyl imines
	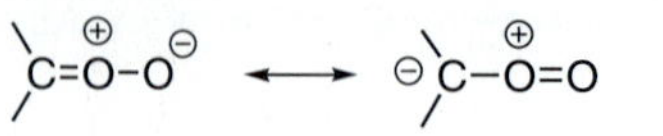	Carbonyl oxides
		Nitrosimines
	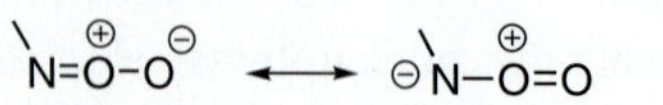	Nitrosoxydes
		Ozone
S in the center	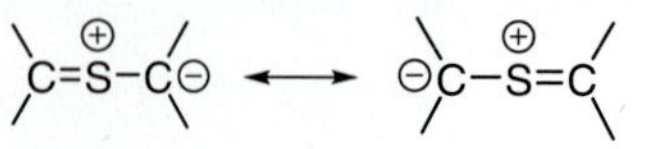	thiocarbonyl ylides[e)]
	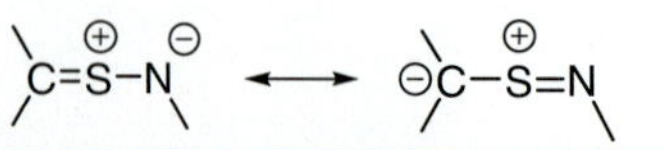	thione *S*-imines[f)]

Table 5.5 (Continued)

Allyl types[a)]		
S in the center		sulfines[f),g)]
		thiocarbonyl *S*-sulfides[h)]
		sulfur diimides[i),j)]
		thionyl imides[i),j)]
		thionitroso *S*-sulfides[k)]
	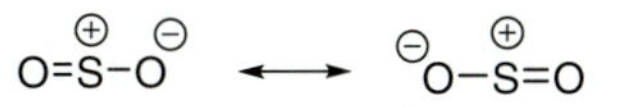	sulfur dioxide[l)]
		disulfur monoxide[m)]
		trithiooxone (thiozone)[n)]
Mesoionic compounds:		
	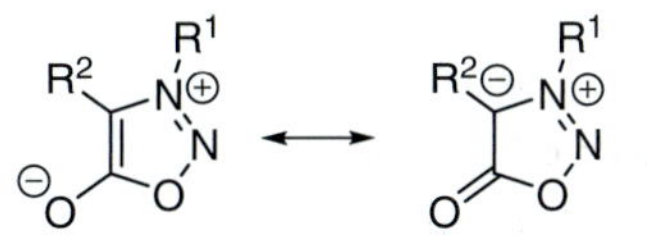	sydnones[a),o)]
	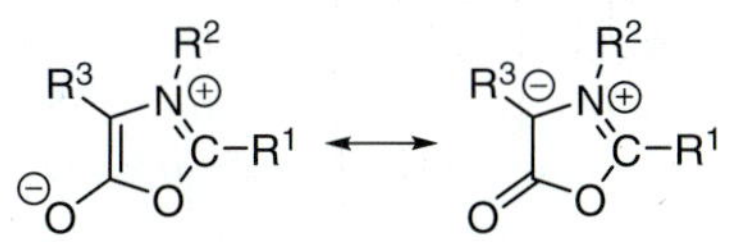	münchnones[a),o)]
	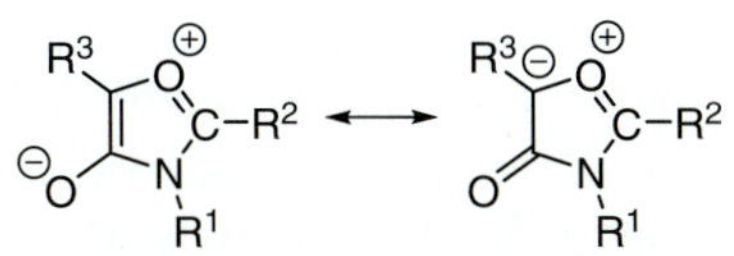	isomünchnones[a),p)]
	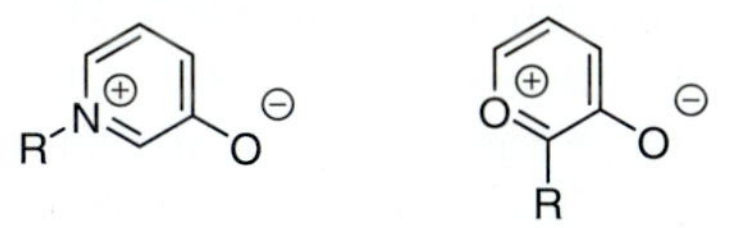	3-oxidopyridiniums & 3-oxidopyrylims[q),r),s)]

a) See e.g. Refs. [558, 559]
b) See e.g. Ref. [560]
c) Unstable compound generated by vacuum pyrolysis of 5-phenyl-1,2,3,4-thia-triazole [561].
d) See e.g. Ref. [562]
e) See e.g. Refs. [563–565]
f) See e.g. Ref. [566–568]
g) See e.g. Ref. [569]
h) See e.g. Refs. [570, 571]
i) See e.g. Ref. [572]
j) See e.g. Refs. [313, 573]
k) See e.g. Ref. [574]
l) Although SO_2 and ozone (O_3) are valence isoelectronics, they show quite different reactivity toward alkenes and alkynes, whereas the 1,3-dipolar cycloadditions of O_3 are highly exothermic, those of SO_2 are endothermic [575].
m) Like SO_2 and RN=S=O, S=S=O acts as a dienophile, not as a 1,3-dipole; furthermore, it disportionates quickly into S_3 and SO_2 [576–578].
n) Like O_3, S_3 reacts as a 1,3-dipole [579].
o) See e.g. Ref. [580]
p) See e.g. Ref. [581]
q) See e.g. Refs. [582, 583]
r) See e.g. Refs. [584–586]
s) These betaines can be seen as four electron 2-oxidopentadienyl cations undergoing (5+*n*)-cycloadditions (Sections 5.3.20 and 5.3.21).

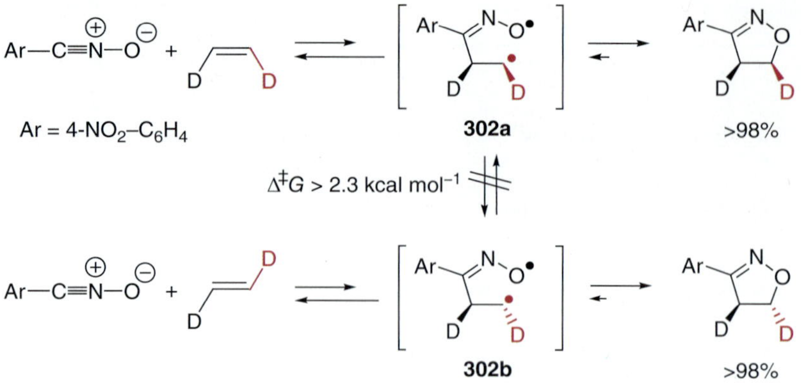

Scheme 5.56 Experimental test for the suprafaciality of the 1,3-dipolar cycloaddition of 4-nitrobenzonitrile oxide to ethylene.

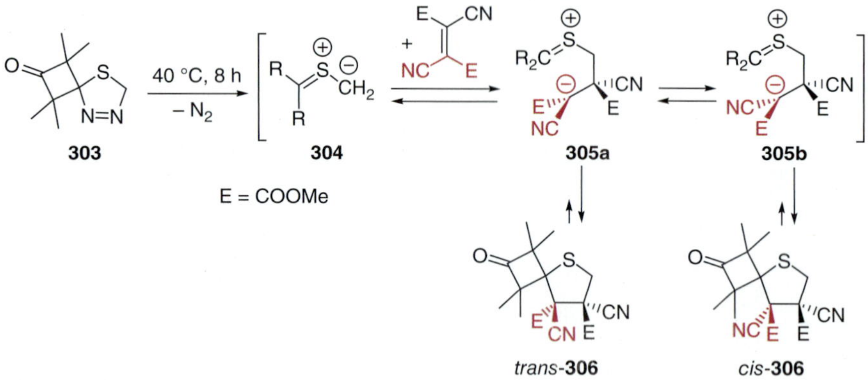

Scheme 5.57 Example of a nonconcerted 1,3-dipolar cycloaddition: loss of the stereoselectivity.

(dipolarophile) interactions are responsible of most of their stabilization energy, whereas for reactions of electrophilic dipoles, the LUMO(1,3-dipole)/HOMO (dipolarophile) interactions dominate. When both types of stabilizing interactions intervene more or less equally, one considers the 1,3-dipoles to be ambiphilic.

1,3-Dipoles with electron-withdrawing substituents react faster with electron-rich dipolarophiles, whereas 1,3-dipoles with electron-releasing substituents react faster with electron-poor dipolarophiles. Depending on the nature of the dipolarophile, phenyl azide reacts either as an electrophile, as an ambiphile, or as a nucleophile (Figure 5.35) [603]. With its relatively high-lying LUMO (Figure 5.34), diazomethane reacts as a nucleophilic 1,3-dipole as shown for reaction (5.26) [604–607].

(5.26)

112 000   44.5   40   0.27 $M^{-1}s^{-1}$ (10^5k^{II})

Quantum mechanical calculations on several 1,3-dipolar cycloadditions to ethylene and acetylene suggest that the 1,3-dipole first distorts so as to reach a reactive state that is independent of the dipolarophile; then it adds to the dipolarophile with a barrier that does not depend on the exothermicity of the cycloaddition and on the FMO interactions [608]. In 1972, Harcourt stressed the importance of the diradical structure of the 1,3-dipoles (see Figure 5.33) and its importance in the barrier of the concerted 1,3-dipolar cycloaddition [609–611]. It has been proposed that 1,3-dipoles are distorted to reach reactive electronic states that have a significant diradical character and that add with little or no barrier to the various dipolarophiles [612]. The FMO interaction energies between the 1,3-dipole and the dipolarophile differentiate reactivity when transition state distortion energies are nearly constant.

Several 1,3-dipolar cycloadditions can be catalyzed. In the example of Scheme 5.58a, a nitrone adds to a vinyl ether giving the corresponding isoxazole **307** with up to 89% ee when applying a suitable enantiomerically pure Lewis acid [613]. The regioselectivity of the cycloaddition is predicted by the diradicaloid model **308** ↔ **308′**. The Lewis acid coordinates the oxide moiety of the zwitterionic limiting structure **308′** and thus stabilizes it, leading to rate and regioselectivity enhancement. Steric factors make the *exo* approach favored and

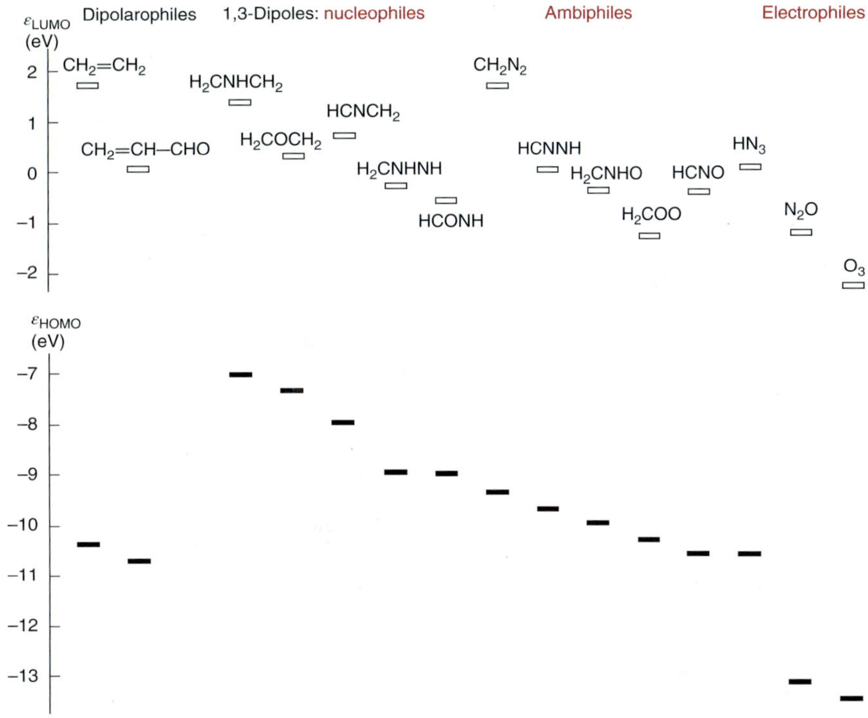

Figure 5.34 FMO energies of common 1,3-dipoles. Source: Taken from Refs. [601, 602].

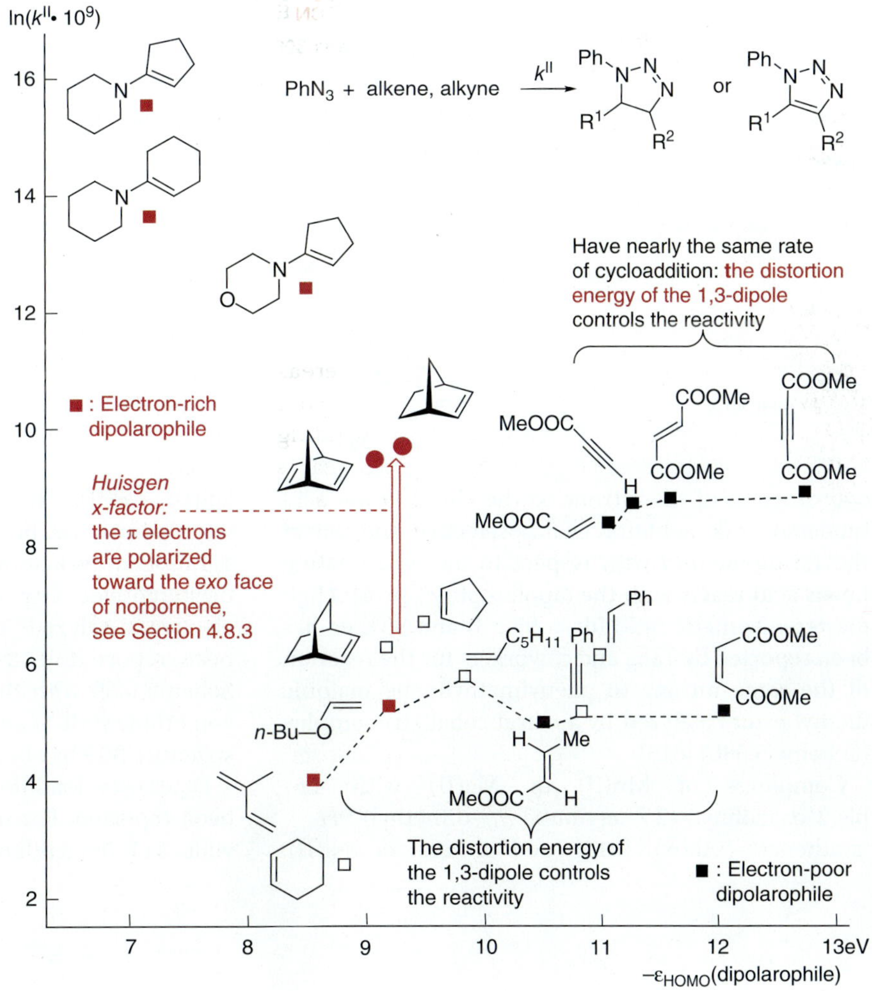

Figure 5.35 1,3-Dipolar cycloadditions of phenyl azide to alkenes and alkynes: correlation of $\ln k^{II}$ vs. HOMO energies of the dipolarophiles.

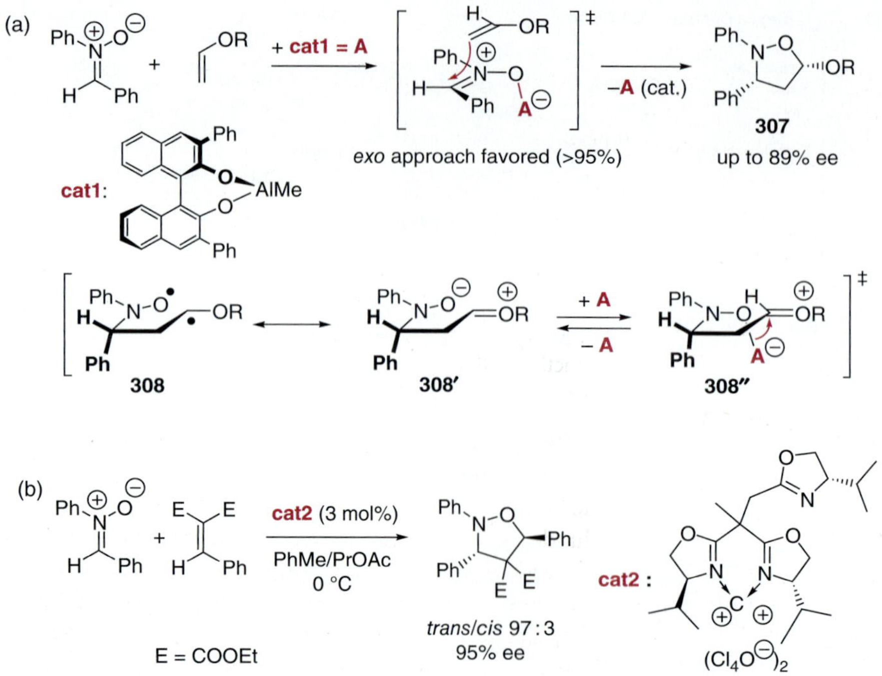

Scheme 5.58 Examples of regioselective and enantioselective catalyzed cycloadditions of a nitrone.

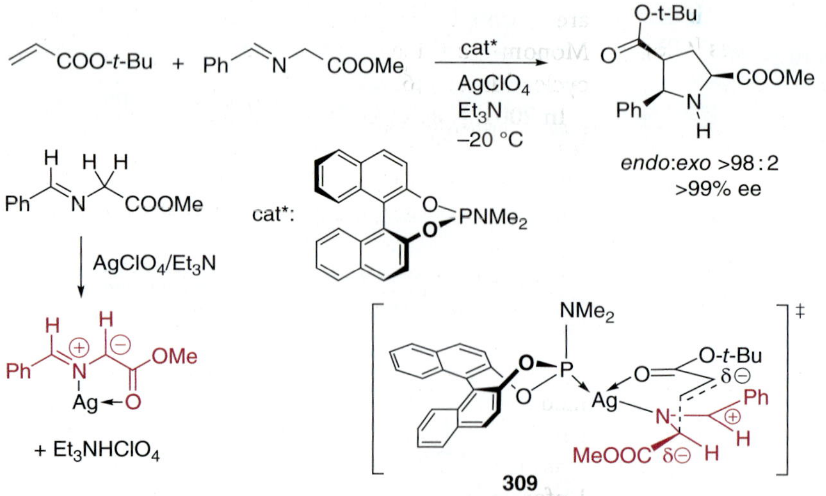

Scheme 5.59 In situ generation of a stabilized azomethine ylide and its enantioselective 1,3-dipolar cycloaddition.

coordination of the nitrone to the chiral Lewis acid renders its cycloaddition enantioselective (the face of the 1,3-dipole *anti* with respect to the coordinating Lewis acid reacts with the dipolarophile) [614]. High diastereoisomeric ratio (dr = 97 : 3) and 95% ee has been reported by Tang and coworkers for the reaction of the same nitrone to phenylmethylidene malonic diethyl ester catalyzed by a chiral cobalt(II) complex (Scheme 5.58b) [615].

Complexes of Mn(II) or Mg(II) with 4,5-bis(2-oxazolinyl)-(2,7-*tert*-butyl-9,9-dimethyl)-9*H*-xanthenes (xabox) [616] and others of Ni(II), Zn(II), Co(II), Ti(IV), Rh(I), Ru(II), and Ir(I) [617] were found to be efficient catalysts for nitrone 1,3-dipolar cycloadditions resulting in high *endo*/*exo* diastereoselectivity and enantioselectivity. Enantioselective catalyzed 1,3-dipolar cycloadditions have been reported [618–620]. An example is given in Scheme 5.59 with the cycloaddition of an argentoazomethine ylide to an acrylic ester for which transition structure **309** has been proposed [621].

Dipolar cycloaddition to carbonyl compounds have been reported. For instance, reaction of azomethine ylide **311** (engendered by thermal isomerization of aziridine **310**, Section 5.2.7) adds to aldehydes and give mixtures of diastereomeric oxazolidines **312**. The acid-catalyzed hydrolysis of the latter provides the corresponding amino-alcohols **313**, the *anti* stereoisomers (*anti*-**313**) being the major products [622].

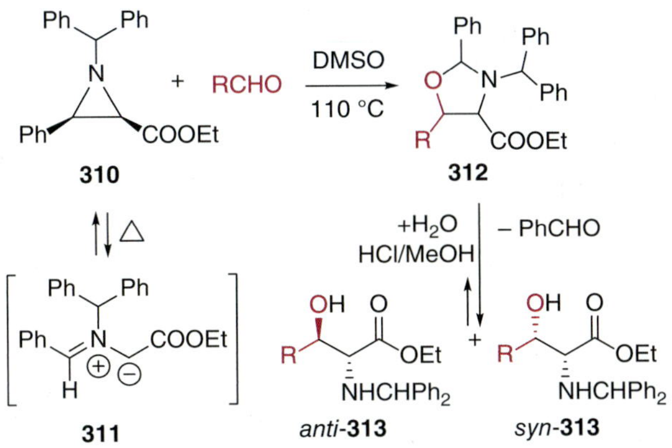

Examples of cycloadditions of diarylnitrilimines **314** (generated in situ by dehydrohalogenation of *N*-phenylbenzhydrazonoyl chlorides) to the imine moieties of pyridines, quinolines, isoquinolines [623], pyrimidines, and pyrazine [624] are given below.

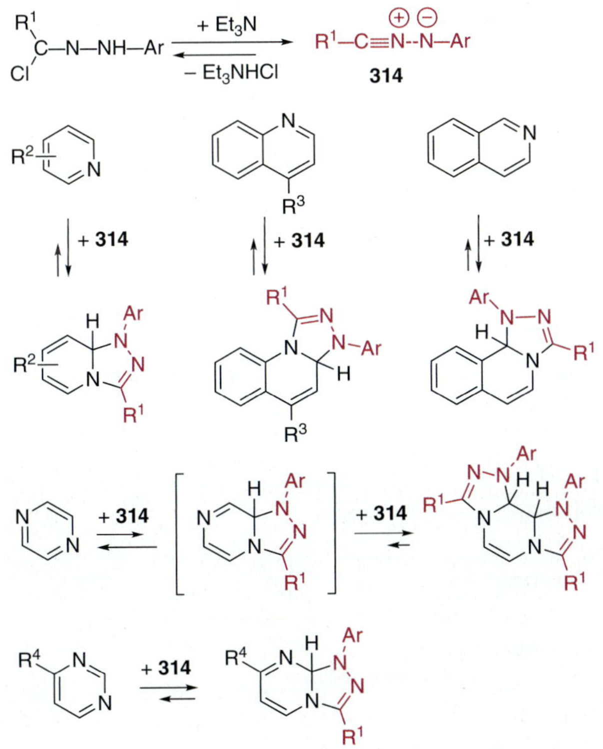

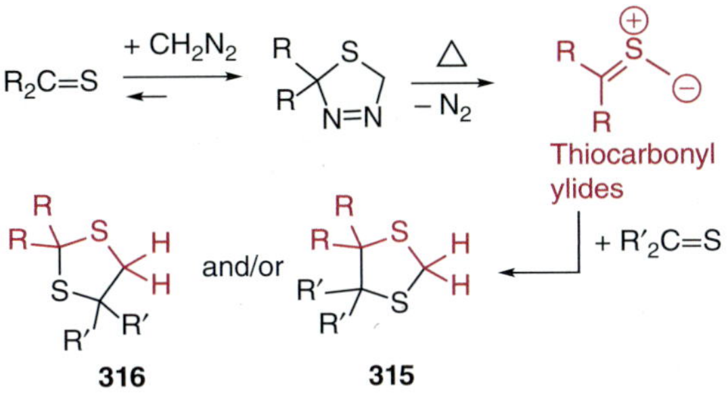

Scheme 5.60 Preparation of thiocarbonyl ylides and their reaction with thiones.

The C=S double bonds of thiones are excellent dipolarophiles [625]. They add to thiocarbonyl ylides with formation of 1,3-dithiolanes **315** and **316**. Sterically hindered cycloalkanethione *S*-methylides and dialkylthioketone *S*-methylides react with alicyclic or aliphatic thiones to give exclusively the 2,2,4,4-tetrasubstituted 1,3-dithiolanes **315** [626], whereas when an aryl group substitutes one of both cycloaddents, the 4,4,5,5-tetrasubstituted isomers **316** are formed preferentially (Scheme 5.60) [627]. Monomeric thioaldehydes also undergo 1,3-dipolar cycloadditions [628].

In 2001, Kolb et al. proposed the concept of "click" reaction [629] for a reaction able to bind quickly two molecular building blocks in a facile, selective, high-yielded reaction, under mild conditions, including water solution, tolerant to a large variety of functional groups, and with little or no side products. Cycloadditions do not generate coproducts (atom economy [630]) and tolerate a large variety of substituents, of functional groups and aqueous media. Furthermore, they are exothermic enough to pay for the entropy cost of condensations, for the mass law effect when using diluted conditions [631]. Unfortunately, a few only are quick enough, unless they can be catalyzed adequately. This is the case with the Cu(I)-catalyzed azide/alkyne cyclocondensation (CuAAC) reported in 2002 independently by Sharpless [632], Meldal [633], and coworkers [634]. Other less efficient catalysts include Ni(II), Pd(II), Pt(II), and Au(I) complexes [635]. Under thermal conditions, the reaction of azides with terminal alkynes produces mixtures of 1,4- and 1,5-disubstituted triazoles, whereas the CuAAC reaction is completely regioselective with the exclusive formation 1,4-disubtituted triazoles (Scheme 5.61). Commonly, the Cu(I) catalyst is made in situ by mixing $CuSO_4$

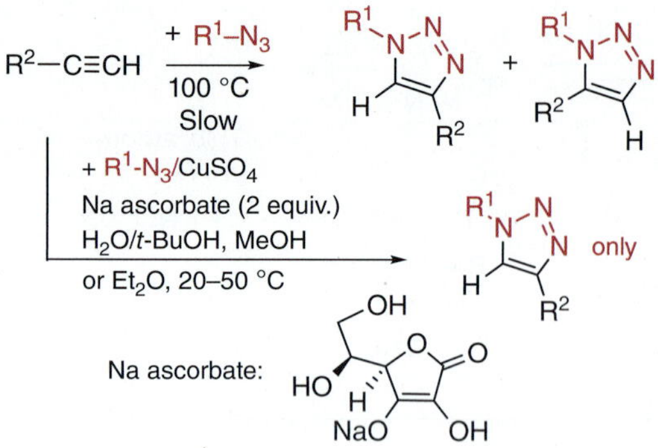

Scheme 5.61 Thermal (Michael, Huisgen) and Cu(I)-catalyzed (Sharpless, Meldal) 1,3-dipolar reaction of azides with terminal alkynes.

with sodium ascorbate (Na salt of vitamin C: sodium (2*R*)-[(1*S*)-1,2-dihydroxyethyl]-4-hydroxy-5-oxo-2*H*-furan-3-olate) in aqueous alcoholic solution.

5.3.18 Sharpless asymmetric dihydroxylation of alkenes

Conversion of alkenes into 1,2-diols by formal addition of two hydroxy groups is a very useful reaction. The most common reagents employed are alkaline $KMnO_4$ [636] and OsO_4 [637]. In 1936, Criegee reported that pyridine catalyzes the alkene dihydoxylation using stoichiometric OsO_4 [638]. As OsO_4 is toxic and expensive, methods have been developed to use it in catalytical amount together with another oxidant such as $KClO_4$ or $NaClO_4$ [639], H_2O_2 (Milas hydroxylation) [640, 641], *N*-methylmorpholine *N*-oxide (Upjohn dihydroxylation) [642, 643], *t*-BuOOH [644, 645], or $K_3Fe(CN)_6/K_2CO_3$ [646–648]. Using chiral tertiary amines instead of pyridine, Hentges and Sharpless reported in 1980 an asymmetric dihydroxylation method (Sharpless Asymmetric Dihydroxylation, SAD) [649]. Amines giving the best results derive from dihydroquine, for instance $(DHQD)_2PHAL$ (dihydroquinidine, DHQD) [650, 651]. Kinetic isotope effects and quantum mechanical calculations support a mechanism in which the rate-determining step is a (3+2)-cycloaddition [652].

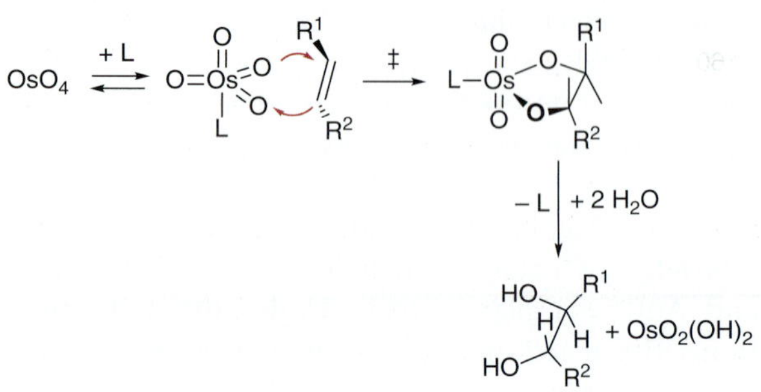

$(DHQD)_2PHAL$:

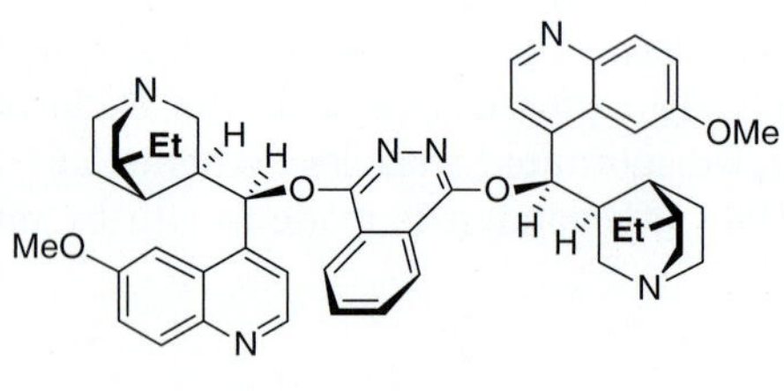

Problem 5.29 What are the products of reaction of sydnones **B** with dimethyl acetylenedicarboxylate? [653]

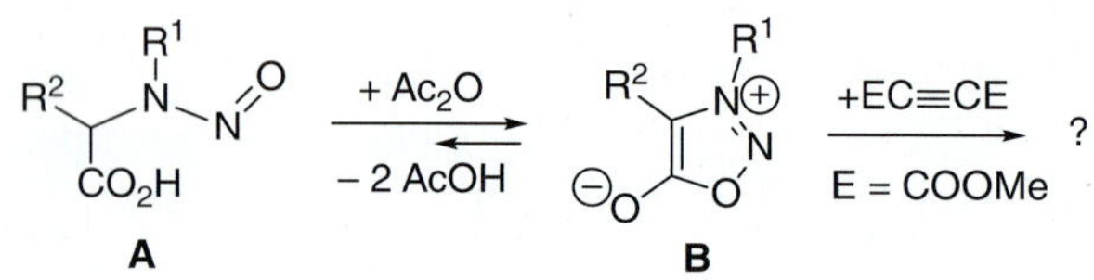

Problem 5.30 What are the products of reaction of münchnones **D** with dimethyl acetylenedicarboxylate? [654]

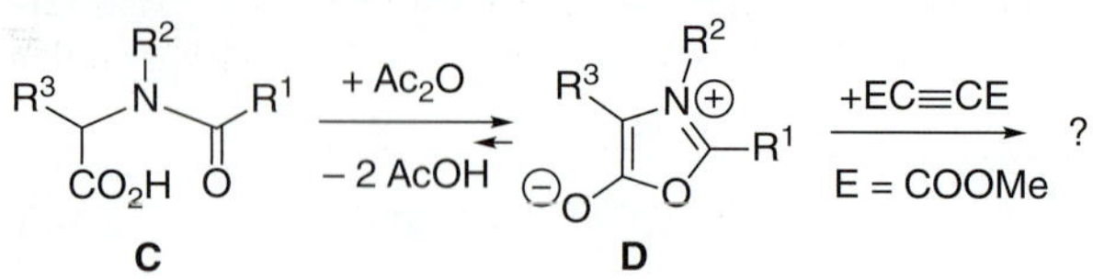

Problem 5.31 What are the main products of the following reactions? [568]

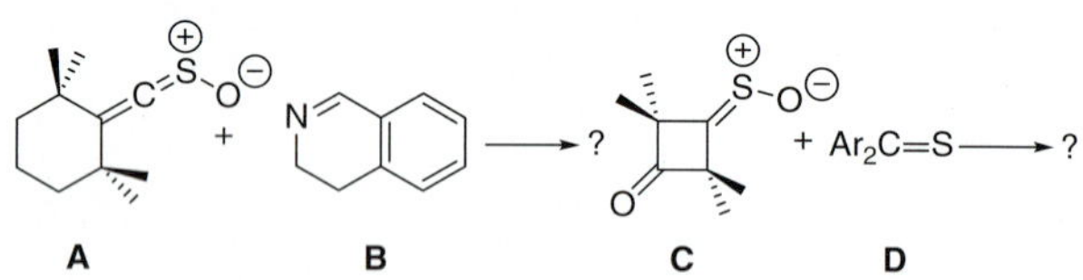

Problem 5.32 What are the major products of the following reaction? [655]

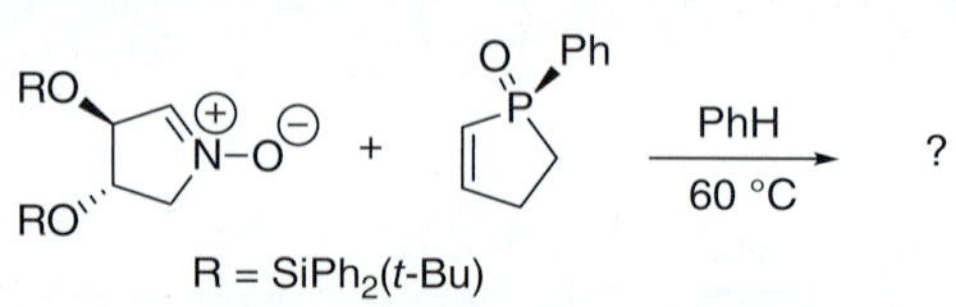

5.3.19 Thermal (2+2+2)-cycloadditions

Alkyne cyclotrimetrization is a very important method to prepare benzene derivatives. It is generally performed in the presence of catalysts such as transition metals [656–658] (Section 8.4.10), Lewis acids [659, 660], amines [661, 662], or disilanes [663, 664]. Although the cyclotrimerization of acetylene into benzene is highly exothermic ($\Delta_r H°$ = −143.8 kcal mol^{-1}), the uncatalyzed thermal reaction is very difficult. In 1866, Berthelot obtained benzene on heating acetylene to 400 °C in a copper tube [665]. Quantum mechanical calculations predicted an activation energy >50 kcal mol^{-1} for the thermal concerted [π^2s+π^2s+π^2s] cycloaddition of acetylene [666, 667]. An exception to all cyclotrimerization of alkynes is the cyclotrimerization of *t*-butylfluroroacetylene reported by Viehe et al. [668, 669]. The reaction is rapid at 0 °C, yielding Dewar-benzene **317**, benzvalene **318**, and an unidentified tetramer (Scheme 5.62). On heating to 100 °C, **317** and **318** are isomerized into benzene derivatives **319** and **320**, respectively. In 1986, Ballester and coworkers found that perchlorophenylacetylene also undergoes thermal cyclotrimerization into perchloro-1,2,3-triphenyl and 1,2,4-triphenylbenzene [670]. On their side, Hopf and Witulski reported that heating of cyanoacetylene to 160 °C generates 1,2,3- and 1,2,4-tricyanobenzene [671]. Quantum mechanical calculations on the thermal cyclotrimerization of fluoro- and chlororacetylene suggested a mechanism involving a rate-determining, stepwise diradical [π^2+π^2]-cycloaddition giving first the corresponding 1,2-dihalogenocyclobutadiene, followed by a stepwise [π^4+π^2]-cycloaddition. In agreement with this mechanism (Scheme 5.62), the thermal reaction (110 °C) of phenylchloroacetylene produces 1,2,3- and 1,2,4-triphenyltrichlorobenzene together with a tetramer, *cis*-1,2,5,6-tetrachloro-3,4,7,8-tetraphenyltricyclo[4.2.0.0^{2,5}]octa-3,7-diene. The proposed intermediate (1,4-dichloro-2,3-diphenylcyclobutadiene) has been trapped by dienophiles such as maleic anhydride and dimethyl acetylenedicarboxylate [672]. One estimates $\Delta_f H°$ (•CH=CH—CH=CH•) = $\Delta_f H°$(CH$_2$=CH—CH=CH$_2$) + 2*DH*°(CH$_2$=CH•/H•) − *DH*°(H•/H•) = 26.0 + 2(110.6) − 104.2 = 143 kcal mol^{-1}. This gives $\Delta_r H°$(2acetylene → •CH=CH—CH=CH•) = 34 kcal mol^{-1}. Considering the entropy of condensation, this makes a highly endergonic reaction. The heat of hydrogenation of fluoroacetylene into fluroethylene amounts to −62.4 kcal mol^{-1} = $\Delta_f H°$(FCH=CH$_2$) − $\Delta_f H°$(FC≡CH) = −32.4 − (30.0) kcal mol^{-1}. In comparison, $\Delta_f H°$(CH$_2$=CH$_2$) − $\Delta_f H°$(HC≡CH) = 12.5 − (54.5) = −42.0 kcal mol^{-1}. Thus, fluoroethylene is stabilized by about −20 kcal mol^{-1} with respect to fluoroacetylene. This effect makes $\Delta_r H°$(2FC≡CH ⇄ •CH=CF—CF=CH•) ≈ 34–40 kcal mol^{-1} ≈ −6 kcal mol^{-1}, an exothermic reaction! This explains the ease

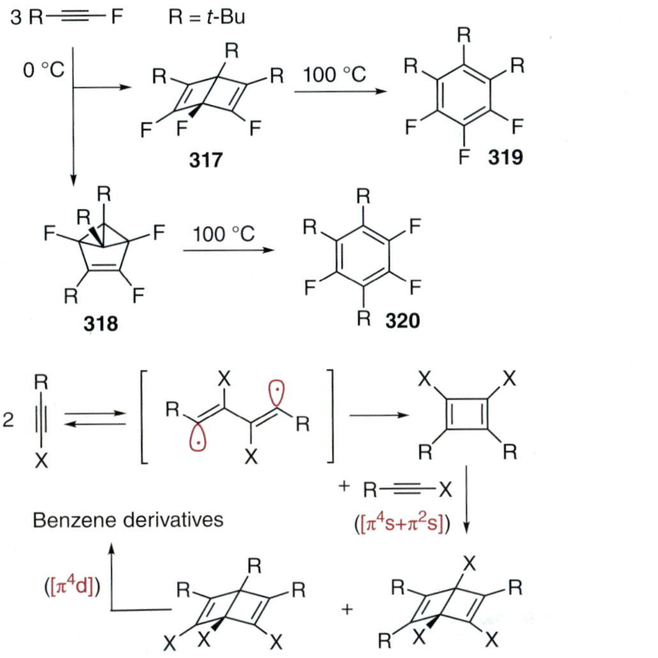

Scheme 5.62 Viehe's thermal cyclotrimerization of *t*-butylfluoroacetylene and possible mechanism.

by which the fluoroalkynes undergo cyclodimerization (Scheme 5.62).

Formal (2+2+2)-cycloadditions of 1,6-diynes with alkenyl and alkynyl dienophiles have been reported as exemplified here-below [673]. The reactions

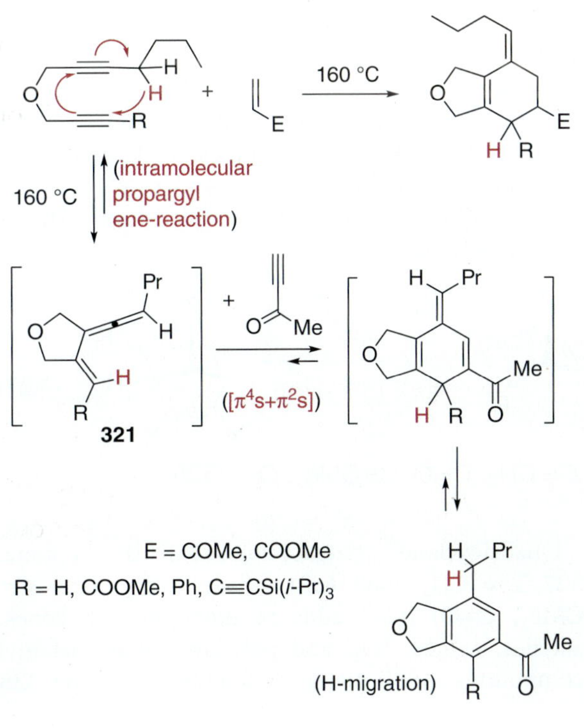

involve first intramolecular propargylic ene-reactions generating the corresponding vinylallene intermediates **321**. The latter intermediates then react with the dienophiles in intermolecular Diels–Alder reactions.

Sakai and Danheiser [674] reported the formation of pyridines via uncatalyzed formal (2+2+2)-cycloaddition. On heating, carbonitriles **322** are isomerized into pyridines of type **323**. The rate-determining steps are the intramolecular propargylic ene-reactions of the two alkyne moieties that generate the corresponding allene–alkene intermediates **324**. The latter undergo intramolecular hetero-Diels–Alder reactions giving tricyclic cycloadducts **325** that are aromatized into **323** through bimolecular hydrogen exchange reactions [675].

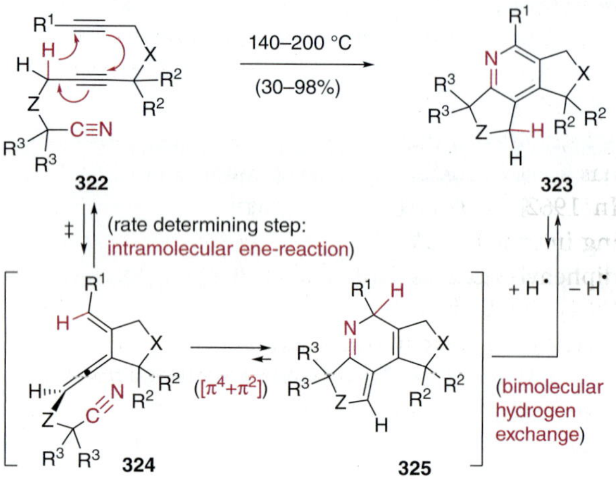

Norbornadienes (bicyclo[2.2.1]hepta-2,5-dienes) add to electron-poor dienophiles [676–678] and strained cycloalkynes dienophiles [679] giving the corresponding tetracyclic $[\pi^{2,1}s+\pi^{2,1}s+\pi^{2,1}s]$-cycloadducts **326**. Reaction (5.27) is called homo-Diels–Alder reactions.

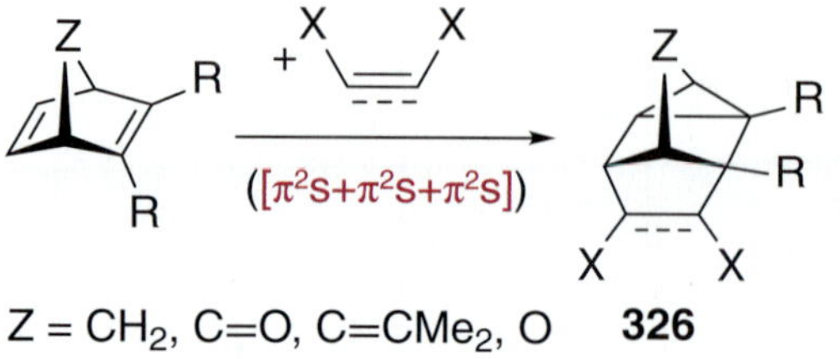

(5.27)

Quadricyclane (tetracyclo[3.2.0.0^{2,7}.0^{4,6}]heptane: **327**, Z = CH_2) [680–683] and derivatives (Z = C=CMe$_2$, C=O) [684] adds to dienophiles (alkenes, alkynes, azines, $^1O_2$, and polyfluorinated carbonyl compounds [683]) giving cycloadducts of type **328** arising from stereoselective $[\pi^2s+\sigma^2s+\sigma^2s]$-cycloadditions (5.28). Sulfur dioxide reacts with quadricyclane giving a mixture of norbornadiene and β-sultine **329** [685]. Quadricyclane also adds to fullerene C_{60} [686]. With diaroyldiazines, the cycloaddition of quadricyclane are faster in MeCN (polar) than in CCl_4 (apolar), suggesting asynchronous processes with charge development in the transition state (diradicaloid transition structure with significant zwitterionic character) [687]. The comparison of activation volume, $\Delta^{\ddagger}V$, and reaction volumes, $\Delta_r V$, of the reactions of quadricyclane with acrylonitrile, methyl acrylate, and diethyl azodicarboxylate confirmed one-step, concerted mechanisms [688]. Monodeuterated quadricyclane-1-*d* and norbornadiene-2-*d* undergoes cycloadditions with electron-poor dienophiles to give products wherein the distribution of deuterium is commensurate with kinetic isotopic effects that confirm asynchronous, concerted mechanisms [689]. Quantum mechanical calculations suggested that the cycloadditions of quadricyclane to ethylene, acetylene, acetylenedicarbonitrile, and dimethyl acetylenedicarboxylate follow stepwise mechanisms with the formation of diradical intermediates [690]. For reaction in the gas phase, the addition of quadricyclane to dimethyl azodicarboxylate generates a diradical intermediate. However, for the reaction on water surface that is much faster [691], a concerted mechanism with a polar transition state **330** (diradicaloid transition structure with a significant zwitterionic character) is proposed [692, 693].

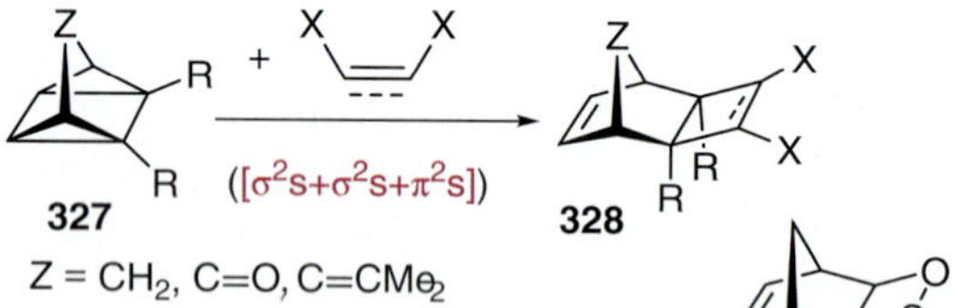

(5.28)

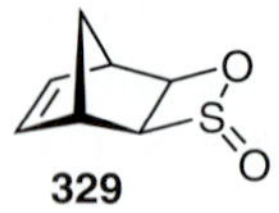

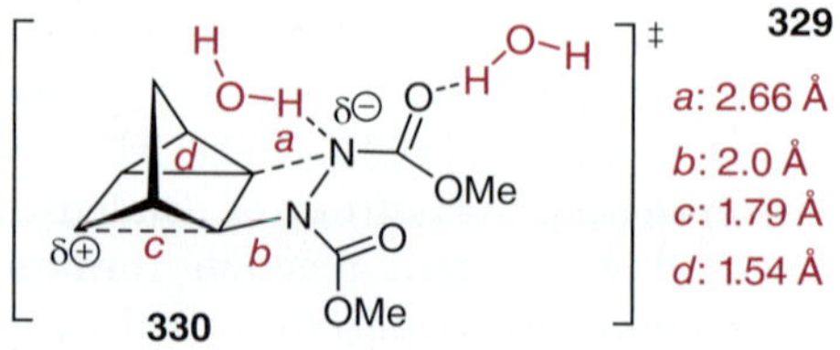

Heating oxaquadricyclanes **331** with alkenyl and alkynyl cycloaddents gives the corresponding cycloadducts **333** that result from the quenching of the carbonyl ylide intermediates **332** [694]. In the absence of dienophiles, electrocyclic opening of intermediates **332** occurs and leads to the formation of corresponding oxepins **334** [695].

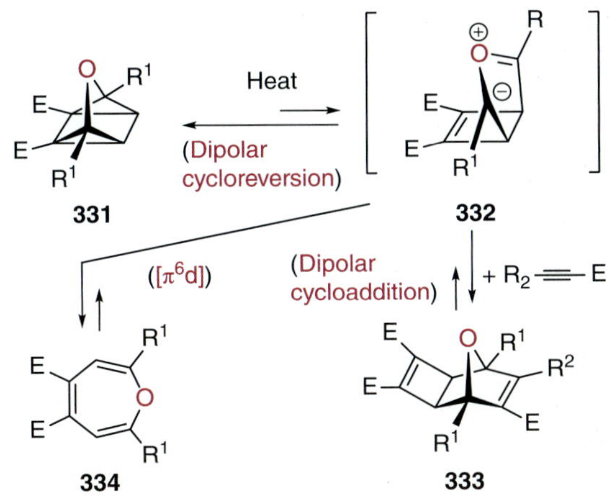

Problem 5.33 What are the products of reaction of cyclooctyne (**C**) with **A**, on the one hand, and with **B**, on the other hand? [679]

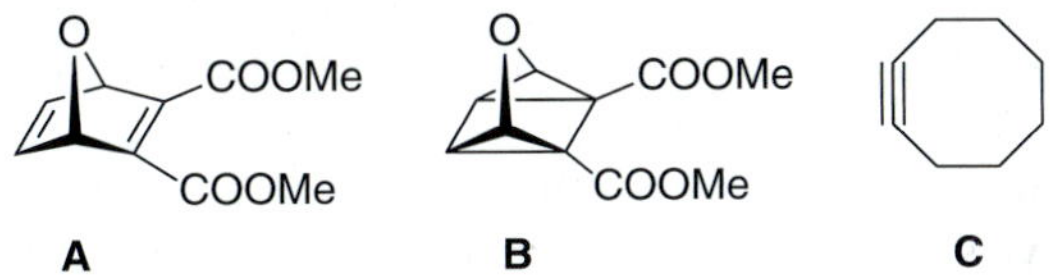

5.3.20 Noncatalyzed (4+3)- and (5+2)-cycloadditions

A large number of catalyzed (4+3)-cycloadditions are known and some of them will be presented in Section 8.4.13. Noncatalyzed (4+3)-cycloadditions involved substituted allyl cations and conjugated dienes and provide a quick access to complex carbocyclic structures containing a seven-membered ring [696–702]. Alternatively, the three-center cycloaddent can be a 1,3-dipole or an ylide, and the diene an heterodiene [703]. Treatment of α-chloro ketones with NaOH, KOH, or with an alcoholate generates the corresponding contracted carboxylic derivatives. This is the Favorskii rearrangement which, in general, involves α-deprotonation of the ketone and subsequent 1,3-elimination of the halide with formation of a substituted cyclopropanone that undergoes hydrolysis or alcoholysis. An example is given in Scheme 5.63a with α-chlorocyclohexanone that is converted into cyclopentanecarboxylic acid. Treatment of α-chlorodibenzyl ketone with 2,6-lutidine (2,6-dimethylpyridine, a base) in MeOH generates α-methoxydibenzyl ketone (**336**). This product is also obtained by treatment of *α,α′*-dibromodibenzyl ketone with NaI (works as a reducing agent) in MeOH. This led Fort to suggest the formation of 1,3-diphenyl-2-oxyallyl cation intermediate (**335**) that has been quenched by MeOH (Scheme 5.63b) [704]. In 1962, he confirmed this hypothesis by quenching intermediate **335** with furan that produced 3,4-diphenyl-8-oxabicyclo[3.2.2]oct-6-en-3-one (**337**) [704]. This is the first example of (4+3)-cycloaddition.

Halogeno-2-oxyallyl cations can be generated by 1,3-elimination of HCl from α-chloro-α-halogenoketones by treatment with a base such as Et_3N or sodium trifluoroethoxide [705, 706]. Alternatively,

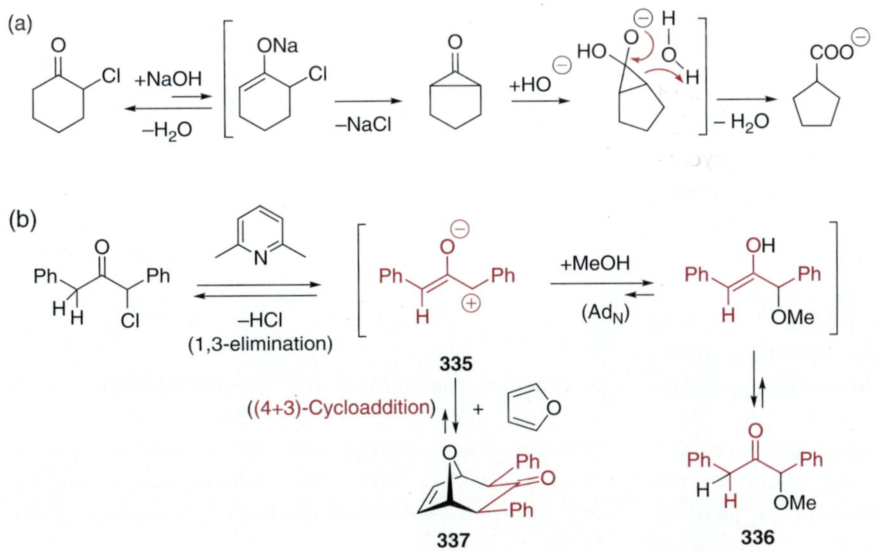

Scheme 5.63 (a) An example of Favorskii rearrangement and (b) an example of 2-oxyallyl cation reacting with furan in a (4+3)-cycloaddition.

halogenooxyallyl cation intermediates can be generated by reduction of polyhalogenoketones. An example is given below for the generation of tetrachloro-2-oxyallyl cation (**339**) that adds to cyclodienes and furans to give the corresponding cycloadducts **340** [707]. Hexachloroacetone reacts with triethyl phosphite to give phosphate **338** (Perkow reaction [708]), which, in the presence of CF_3CH_2ONa, eliminates NaCl and $CF_3CH_2OPO(OEt)_2$ producing oxyallyl cation **339**.

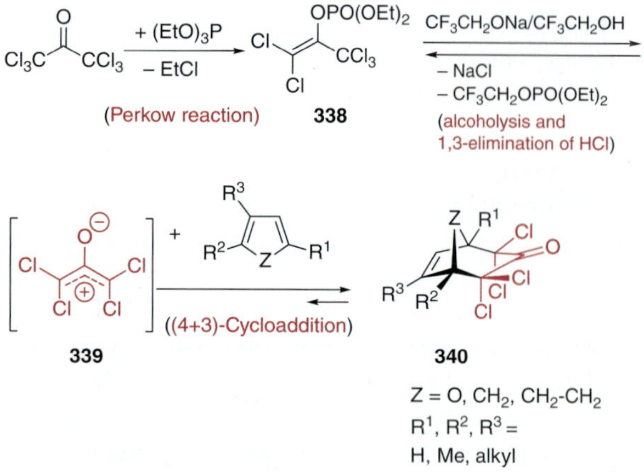

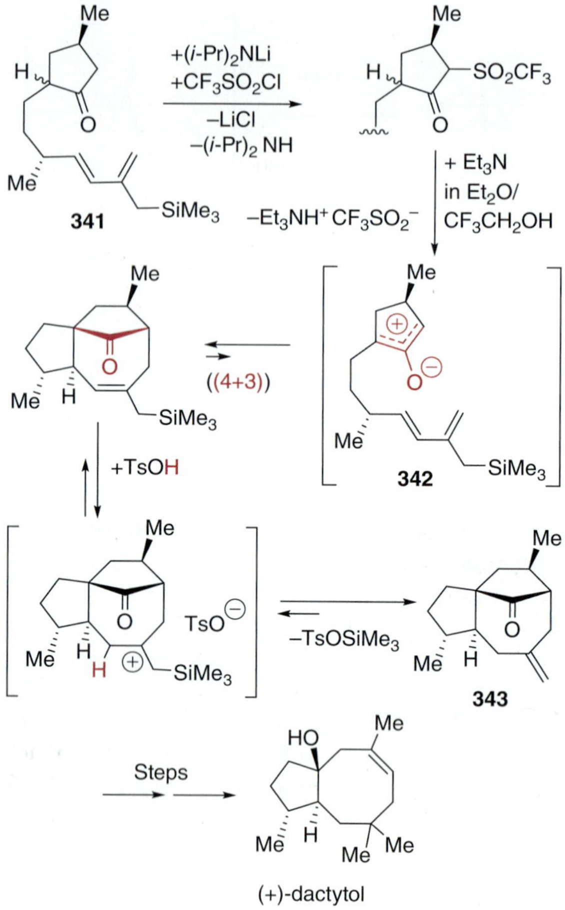

Oxyallyl cation intermediates can be generated by 1,3-eliminations of sulfinate salts from α-sulfonylketones. An example is given below with the conversion of **341** into **343** that involves an intramolecular (4+3)-cycloaddition of intermediate **342**. The tricyclic ketone **343** has been converted into (+)-dactytol [709], a sesquiterpene isolated from the sea hare *Aplysia dactylomela*.

Diethylzinc can be used to generate oxyallyl intermediates from α,α'-dibromoketones. An example is given below. The high diastereoselectivity of the (4+3)-cycloaddition is explained by invoking coordination of the oxygen of the oxyallyl system to the zinc alcoholate as shown with **244** [710].

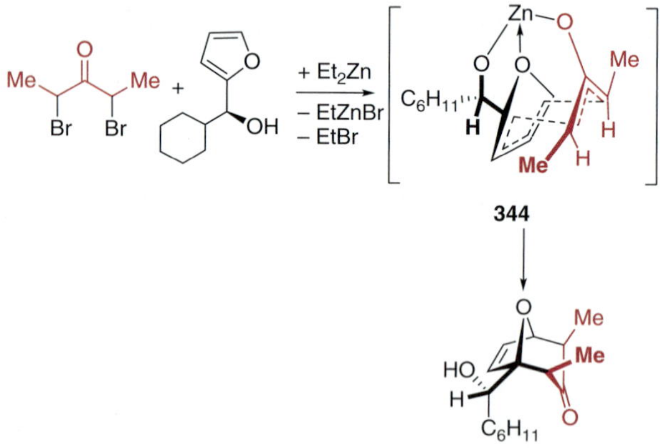

Applying Lewis acid induced ionization of 1,1-dimethoxy-2-trimethylsilyloxyprop-2-ene, Murray and Albizati generated a 1,2-dioxyallyl cation intermediate that adds readily to furan giving *endo*-2-methoxy-8-oxabicyclo[3.2.1]oct-6-en-3-one [711]. Hoffmann and coworkers developed an asymmetric variant of this procedure using mixed acetals **345** derived from chiral α-methyl benzylic alcohols [712]. Quantum mechanical calculations predicted a stepwise mechanism for the cycloaddition with a transition structure of type **346**. Steric repulsion between the α-methyl and allyl groups as well as attractive CH–π interactions between furan and the aryl group in **346** explain the high diastereoselectivity of the reaction [713].

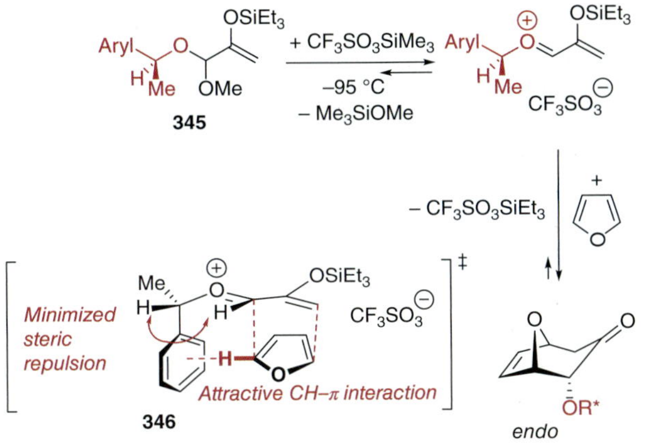

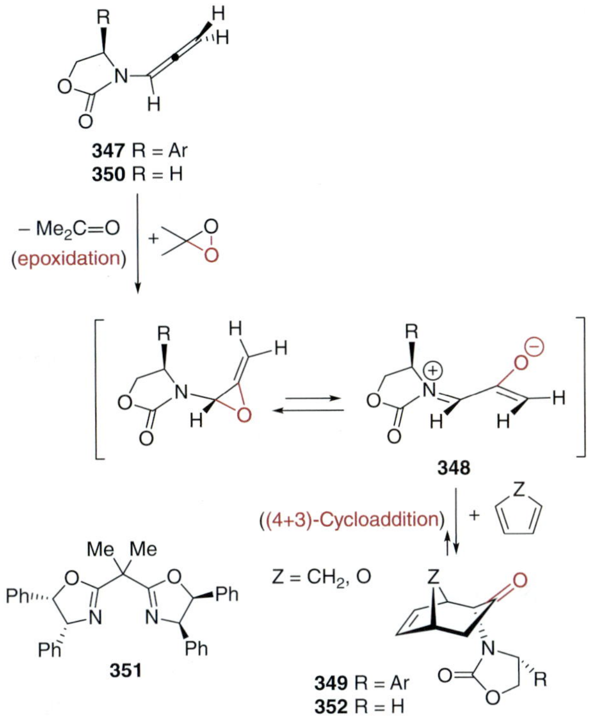

Hsung and coworkers have generated chiral oxazolidinone-stabilized oxyallyl intermediates **348** by epoxidation of allenamides **347**. The oxyallyl intermediates **348** react with cyclopentadiene, furan, and pyrrole in (4+3)-cycloadditions with high diastereoselectivity, which is probably controlled also by CH–π interactions between the cyclodiene and the aryl group of the oxazolidinone group [714]. Huang and Hsung found that oxidation (with MeSOMe) of achiral allenamide **350** in the presence of $Cu(OSO_2CF_3)$ (0.25 equiv.) and ligand **351** (0.25 equiv.) and an excess of furan produces cycloadduct **352** in 90% ee [715].

In 1980, Hendrickson and Farina reported the reaction between acetoxypyranone (**354**) and acrolein that gives 8-oxabicyclo[3.2.1]oct-3-en-2-one (**355**) [716, 717]. The reaction involves the formation of 3-oxidopyrilium betaine (**354**) that reacts as a pentadienyl cation (**354′**) to acrolein. The reaction has been extended to substituted oxidopyrilium betaines and to all kinds (electron-poor, electron-rich) of alkene and alkyne cycloaddents (Scheme 5.64), including to intramolecular reactions [718, 719].

The intermolecular (5+2)-cycloaddition of vinylaziridines (**356**) with electron-poor unsaturated compounds such as hexafluorobut-2-yne and dimethyl acetylenedicarboxylate represents an attractive approach to the synthesis of the azepane framework [720]. With sulfonylisocyanates, seven-membered cyclic ureas **358** are obtained as major products. The proposed reaction mechanism involves the addition of the aziridine onto the electron-poor isocyanate that generates a zwitterionic intermediate of type **357**. The latter undergoes intramolecular S_N2' displacement reaction giving **368** (major) or S_N1 heterolysis into

Scheme 5.64 Preparation of 3-oxidopyrilium betaines and their (5+2)-cycloadditions.

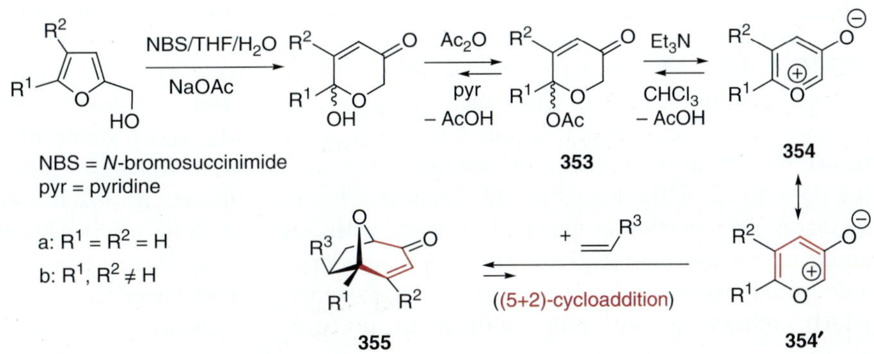

359 and intramolecular quenching furnishing **360** (minor) [721]. Vinyl azetidines undergo similar reactions and produce cycloadducts of formal (6+2)-cycloadditions [722].

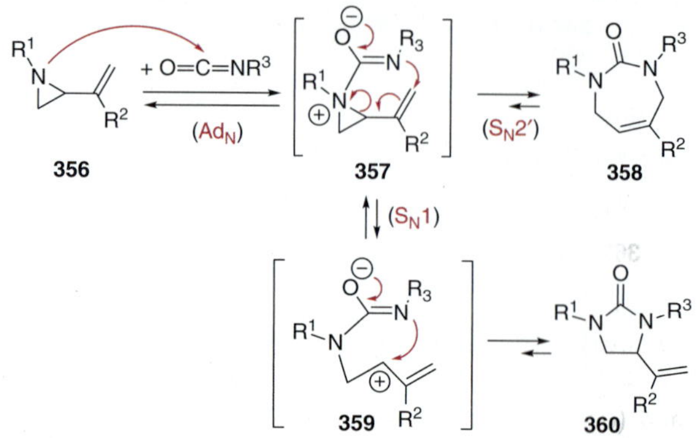

Problem 5.34 What are the possible products of the following reaction? [723]

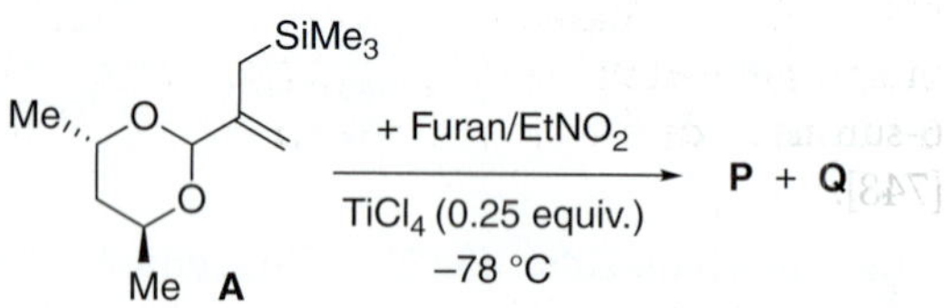

Problem 5.35 (1*S*)-1-Phenylethylamine reacts with 3-chloro-3-methylbutan-2-one, Et_3N (2 equiv.), and $TiCl_4$ (0.5 equiv.) giving an imine **A**. When **A** is reacted with $AgBF_4$ in CH_2Cl_2, a salt **B** forms that reacts with furan providing an iminium salt **P**. The latter is hydrolyzed into a ketone **Q** with 60% ee. What are **A**, **B**, **P**, and **Q**? [724]

5.3.21 Thermal higher order (*m*+*n*)-cycloadditions

Cycloadditions of two cycloaddends contributing with 8, 10, 12, … electrons are generally called higher order cycloadditions [725]. Quite often, they require photochemical activation (Section 6.8), the use of transition metal complexes as catalysts (Section 8.4.15–8.4.21), or the formation of dienamine intermediates by reaction-conjugated enones with an amine as organocatalyst [726]. These reactions provide rapid access to medium-sized ring systems, which, otherwise, are quite difficult to prepare. Unfortunately, and unlike the Diels–Alder reaction that tolerates a large number of substituents and which is highly stereoselective, most higher-order cycloadditions proceed with modest to poor chemical efficiency. The extended π-systems involved in the cycloaddents are prone to participating in multiple, competitive pericyclic processes (lack of chemoselectivity) that result in low chemical yields of cycloaddents [727–729]. For the reaction of cycloheptatriene with cyclopentadiene, quantum mechanical calculations predicted that both the (6+4)- and (4+2)-cycloadditions must compete and that diradical intermediates might form. Furthermore, different types of cycloadducts might be equilibrated through sigmatropic rearrangements [730].

In general, dienes cyclodimerize into (4+2)- rather than (4+4)-cycloadducts. On heating to 150 °C, butadiene cyclodimerizes mostly into 4-vinylcyclohexene, into c. 5% of *trans*-1,2-divinylcyclobutane and a small amount of cycloocta-1,5-diene. The latter cycloadduct does not arise from a direct (4+4)-cycloaddition but from the Cope rearrangement of *cis*-1,2-divinylcyclobutane (Section 5.3.4). Similar observations are done with the cyclodimerization of 2-chlorobutadiene (Section 5.3.8) that generates after seven days at 80 °C c. 14% of 1,6-dichlorocycloocta-1,5-diene [731]. Bulky 3,3,4,4,5,5-hexamethyl-1,2-dimethylidenecyclopentane cyclodimerizes into a mixture of (4+2)- and (4+4)-cycloadducts. The reaction involves the formation of diradical intermediates. The (4+2)-cycloaddition is retarded by steric hindrance introduced by the methyl groups at C(3) and C(5). Thermal (4+4)-cycloadditions are observed with very reactive 1,3-dienes such as orthoquinodimethanes that equilibrate with benzocyclobutenes as illustrated in Figure 2.15 with the synthesis of "superphane." In these cases, and that is shown below, (4+2)-cycloadditions of the dienes leads to products that are less stable (one aromatic ring) than the (4+4)-cycloadducts (two aromatic rings). For instance, **361** cyclodimerizes exclusively into the head-to-head cycloadduct **363** via diradical **362** [732, 733]. Diradical **362** is the most stable diradical that can be obtained on combining two molecules of triene **361**. Each radical moiety in **362** has optimum stabilization because of π- and n(O:)-delocalization (1-oxypentadienyl radical). Any other isomeric diradical contains at least one 1-oxyallyl radical instead of a 1-oxypentadienyl radical.

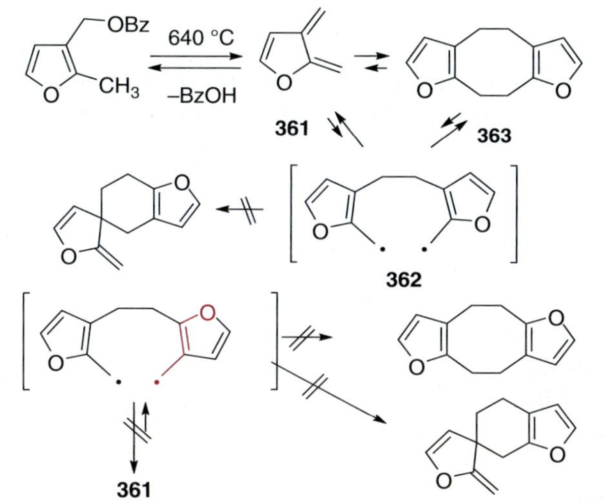

In the absence of alkenes or alkynes, 3-oxidopyrylium undergoes stereocontrolled self-dimerization as illustrated below with **354a**. Reaction (5.29) is an example (5+3)-cycloaddition [716, 734].

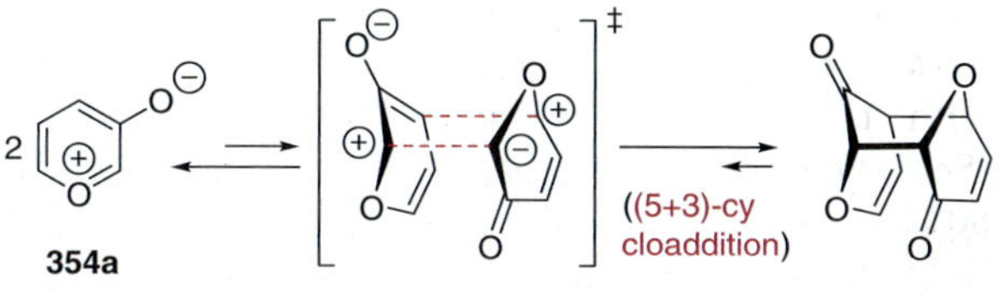

(5.29)

Cycloheptatrienes equilibrate with the corresponding norcaradienes (Section 5.2.8). The latter dienes are usually more rapid than the former trienes in Diels–Alder reactions with alkenes and alkynes, generating (reaction (5.30)) tricyclo[3.2.2.1^{2,4}]nonene derivatives **364** (1,4-distance in butadiene units of cycloheptatrienes is larger than in *s-cis*-butadiene moieties of norcaradienes, see Section 5.3.12). However, with nitrosobenzene, cycloheptatriene undergoes a (6+2)-cycloaddition forming **365** [735]. A similar observation is made with *N*-ethoxycarbonylazepine that gives the corresponding (6+2)-cycloadduct **366** [736].

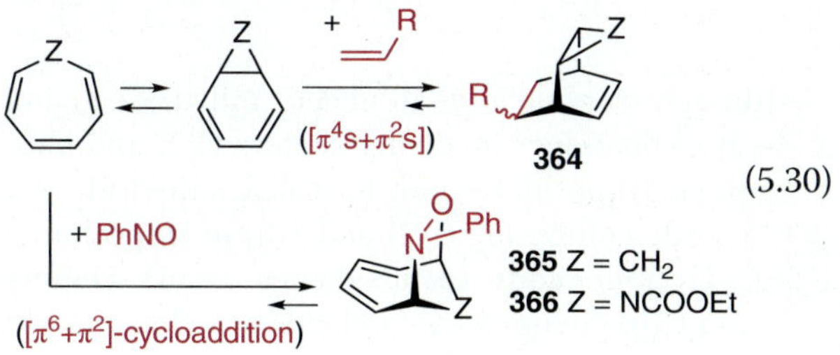

(5.30)

Homofulvenes **367** add to TCNE, maleic anhydride, and dimethyl acetylenedicarboxylate giving the corresponding (6+2)-cycloadducts **368** [737]. The reactions can also be seen as [π^4s+σ^2a+π^2s]-cycloadditions.

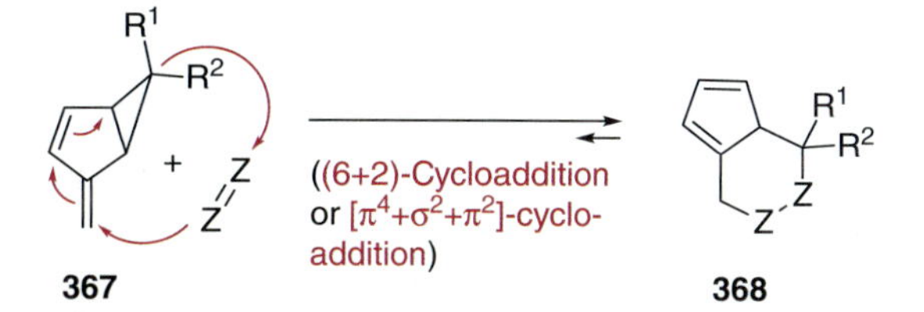

Fulvenes can undergo (2+2) [738], (4+2) [739], (2+4) [740, 741], (4+3) [742], (6+2) [743], (6+3) [744], and (6+4)-cycloadditions [745, 746]. Intramolecular (6+2)-cycloadditions of fulvene-derived enamine **369** generate the corresponding cycloadducts **370** that eliminate readily an equivalent of Et_2NH providing a quick access to linearly fused tricyclopentanoids **371** [747]. In a similar way, intermolecular reactions of in situ generated acetone pyrrolidine enamine with 6-substituted fulvenes give 1,2-dihydropentalenes [748].

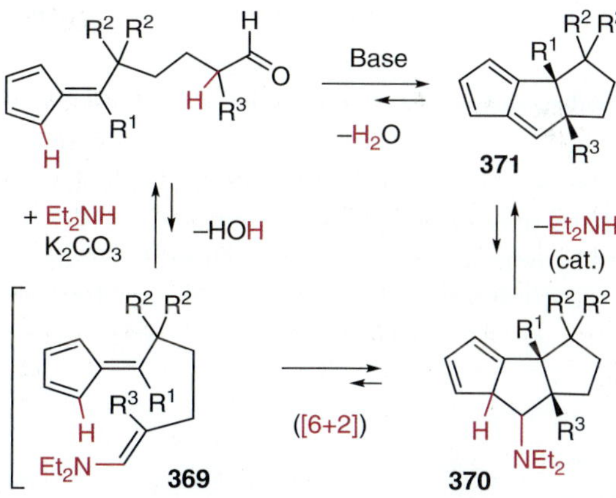

Katritzky's 3-oxidopyridinium betaines may react with conjugated dienes in processes that can be seen as (5+4)-cycloadditions. An example is given below [749]. HCl elimination from the 3-hydroxypyridium chloride **372** generates the 3-oxidopyridimium betaine **373**. It reacts with cyclopentadiene producing a mixture of cycloadducts **374**, **375**, and **376**. Stereoisomeric products **374** and **375** can be seen to result either from (4+3)-cycloadditions if one considers limiting structure **373′** of **373** that is a

3-oxidoallyl cation or from a (5+4)-cycloaddition if one considers limiting structure **373″** being a 2-oxidopentadienyl cation. Cycloadduct **376** results from a (5+2)-cycloaddition of the 2-oxidopentadienyl cation **373″**.

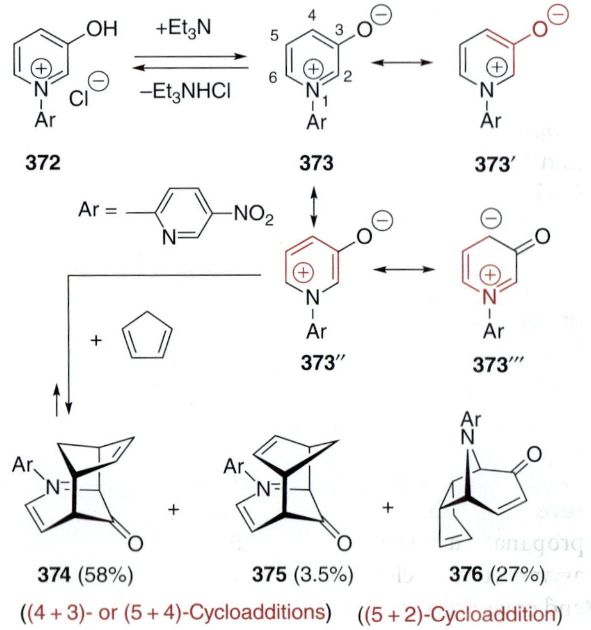

Metal-coordinated trimethylenemethane undergoes (3+2)- (Section 8.4.6), (3+3)- (Section 8.4.11), and (3+4)-cycloadditions (Section 8.4.13). Palladium-stabilized trimethylenemethane derivatives **377** (generated by reaction of $Pd(OAc)_2$, trisopropylphosphite, and 2-[(trimethylsilyl)methyl]allyl carboxylates) undergo (6+3)-cycloadditions (5.31) with tropone [750].

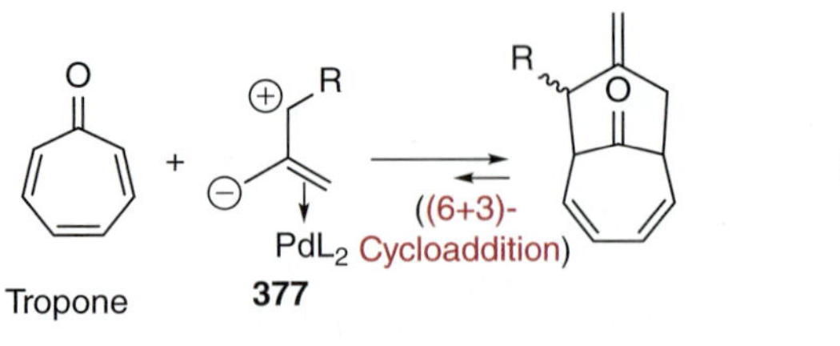 (5.31)

In contrast to the (3+2) (Section 5.3.16) or (4+3)-cycloadditions of *N*-metalated azomethine ylides, lithiated azomethine ylides **378** resulting from the reaction of lithium diisopropylamide (LDA) with *N*-benzylidene glycine ethyl esters react with 6-substituted fulvenes to give products **379** of hetero-(6+3)-cycloadditions (reaction (5.32)) [751].

The first example of (8+2)-cycloaddition was reported by Doering and Wiley [752]: heptafulvene (methylidenecycloheptatriene, obtained by Hofmann elimination of trimethyl 7-cycloaheptatrienylmethylammomnium iodide) is a nonisolable, red-colored, nonbenzoid hydrocarbon that polymerizes quickly upon concentration. Diluted solutions of it react with dimethyl acetylenedicarboxylate. After work-up with Pd on charcoal and air, dimethyl azulene-1,2-dicarboxylate is isolated.

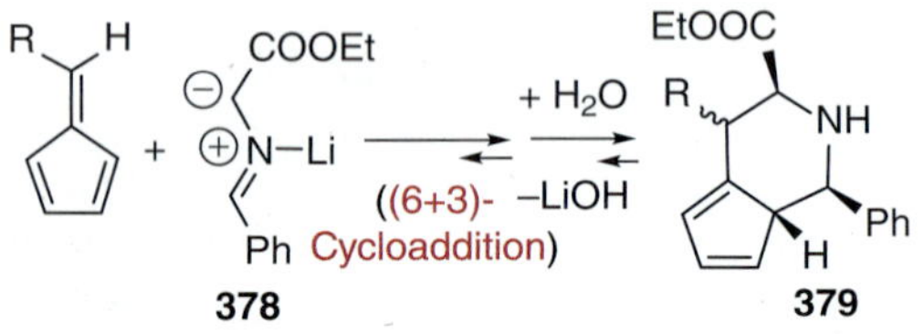 (5.32)

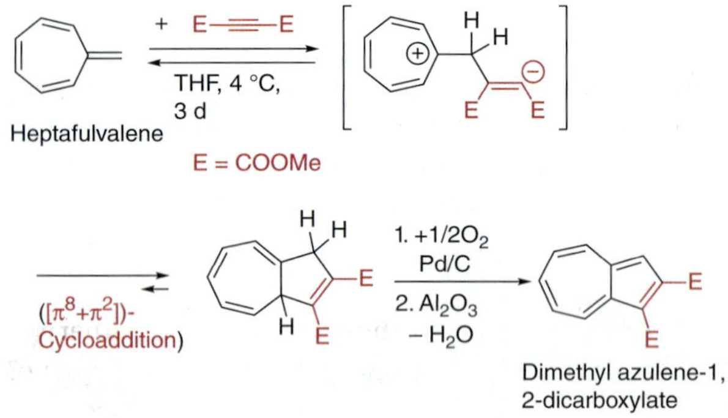

Since then, several examples of (8+2)-cycloadditions of heptafulvene [753] and of other systems have been reported [754]. The reaction of 8,8-dicyano-heptafulvene (7-(dicycanomethylidene) cycloheptatriene) with electron-rich dienes gives (8+2)- or (6+4)-cycloadducts depending on the nature of the diene [755].

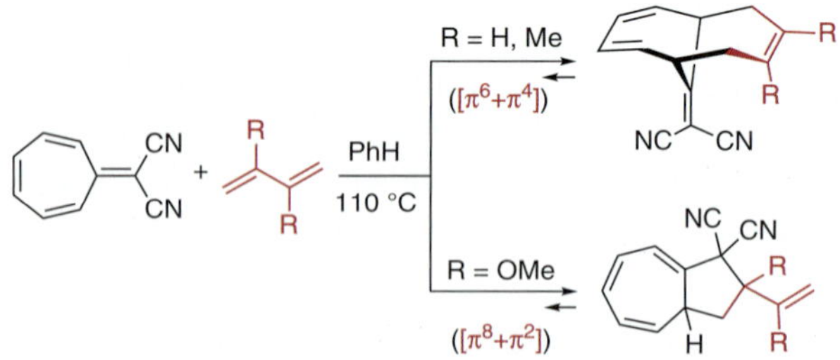

Although tropone adds to maleic anhydride giving a (4+2)-cycloadduct involving centers C(2) and C(5) of tropone, tropothione adds to maleic anhydride in a (8+2) mode combining C(2) and S(8) of tropothione [756]. Tropone adds to 1,3-dienes across centers C(2) and C(6) giving (6+4)-cycloadducts through *exo* transition states [757, 758]. In the following reaction (5.33), the intramolecular (6+4)-cycloaddition is highly stereoselective and generates a useful polycyclic framework [759]. The reaction can be catalyzed by Lewis acids. With Ti(O-*i*-Pr)$_2$Cl$_2$ and *S*-BINOL ((*S*)-(−)-1,1′-bi-2-naphthol = (*S*)-[1,1′-binaphthalene]-2,2′-diol), 40% ee has been observed [760].

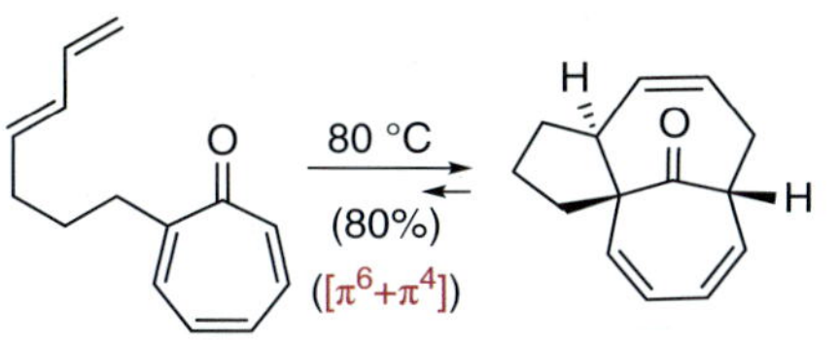

(5.33)

Problem 5.36 Compounds **A** eliminate 1 equiv. of MeOH on treatment with LDA ((*i*-Pr)$_2$NLi) and form the corresponding products **P**. Propose a mechanism [761].

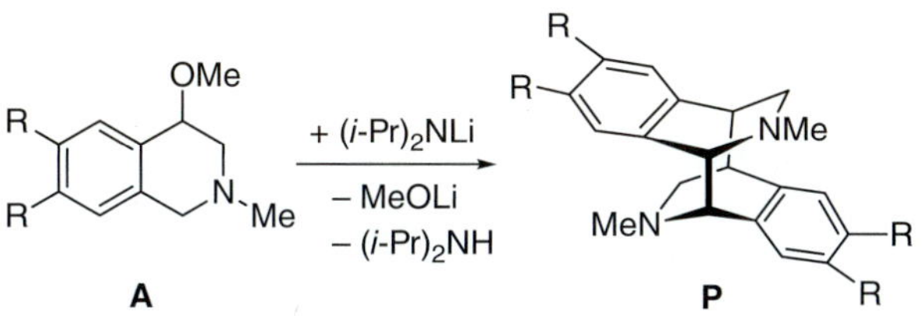

Problem 5.37 What is the major product of thermal reaction (140 °C) of tropone with (*E*)-piperylene? [757, 760]

5.4 Cheletropic reactions

Cheletropic additions are (*n*+1)-cycloadditions across the terminal atoms of a fully π-system (substrate) with formation of two new σ-bonds to a single atom of a reagent. There is formal loss of one π-bond in the substrate and an increase in coordination number of the relevant atom of the reagent. The reverse reactions are called cheletropic eliminations [2, 762]. Examples of reagents are carbenes (R^{1}R^{2}C:), carbon monoxide (:C=O), isocyanides (R—N=C:), nitrenes (RN:), sulfur (S), dinitrogen (N$_2$) [763], nitric oxide (•N=O) [764, 765], nitroso compounds (R—N=O) [766], nitrous oxide (N$_2$O) [767–769], phosphinidenes (RP:) and arsinidenes (RAs:) [770], sulfur dioxide (SO$_2$) [771], and carbene analogs with low-valent group 13 (B(I), Al(I), Ga(I), In(I), and Th(I)) and group 14 elements (silylenes (R^{1}R^{2}Si:), germylenes (R^{1}R^{2}Ge:), stannylenes (R^{1}R^{2}Sn:), and plumbylenes (R^{1}R^{2}Pb:)) [772–774]. More generally, any compound with an unsaturated atomic center (valence shell lacking two electrons: sextet for second- and third-row elements, 14 or 16 electron transition metallic species) can be viewed as a potential reagent for cheletropic reactions (see Section 7.7). Potential substrates are monoalkenes, conjugated dienes, trienes, aromatic π-systems, cumulenes, carbonyl compounds, imines, hetero-1,3-diene (e.g. enals, enones, azadienes), heterocumulenes (e.g. CS$_2$), and alkynes.

5.4.1 Cyclopropanation by (2+1)-cheletropic reaction of carbenes

Cyclopropanations of alkenes with diazo compounds are known since the nineteenth century and have been proposed to involve carbenes adding to alkenes (reaction (5.34)). They are probably the first examples of (2+1)-cheletropic additions. Cyclopropane units are found in many drugs and bioactive natural products [775–777]. Cyclopropanes undergo a wide range of reactions making them quite useful synthetic intermediates [778–781]. Thus, asymmetric cyclopropanation [782, 783] is an important topic of organic chemistry. In1966, Nozaki and coworkers described the first enantioselective cyclopropanation reaction utilizing copper-catalyzed carbene transfer from diazoalkanes to alkenes (see also Section 8.4.2) [784]. Since then, several examples of highly enantioselective cyclopropanation have been reported [785–788]. Catalytic asymmetric cyclopropanations developed by Tang and coworkers are illustrated here-below [788]. Enantioselective cyclopropanations using iodine ylides as carbene precursors (not as dangerous as diazoalkanes) have also been reported [789].

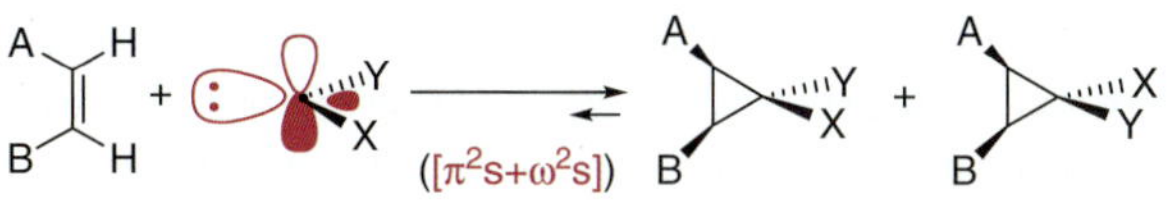

(5.34)

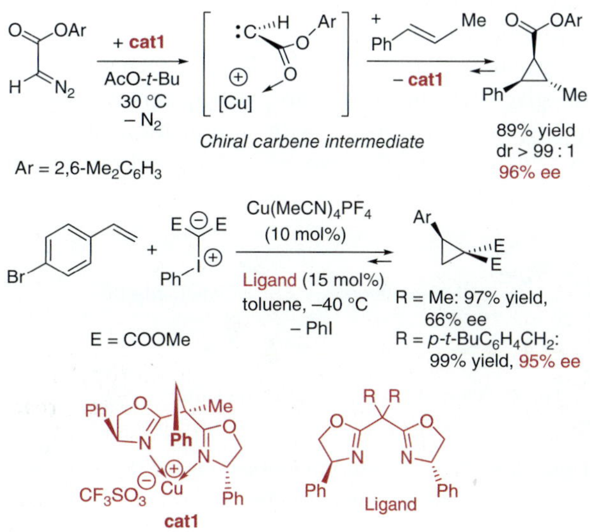

Skell and Garner [790] postulated simultaneous bonding of the carbene carbon to both alkene carbon centers, leading to a "three-center-type" interactions. In 1963, Moore et al. proposed that the bonding in the transition state of the cyclopropanation results from the overlap of the vacant p-orbital of the carbene with the HOMO π-orbital of the alkene [791]. In 1968, Hoffmann showed that the C_{2v} symmetrical cyclic four-electron transition structure is orbital symmetry forbidden [792]. He also pointed out that the main interaction of a singlet carbene such as :CH_2 with an alkene should involve the 2p-orbital (LUMO) of the carbene and the π-HOMO of the alkene, a picture (Figure 5.36) supported by semiempirical calculations [792–795], and later by ab initio quantum mechanical calculations [796–798]. The transition state is very asynchronous, involving more bonding at the left carbon than at the right carbon center (Figure 5.36).

A nonlinear cheletropic model is supported by B3LYP quantum mechanical calculations of the free energy surface for reaction of dichlorocarbene with propene. As shown in Figure 5.37, the free energy maximum occurs at a very unsymmetrical geometry, forming bonding distances differing by 0.6 Å [799]. Computed dynamics trajectories of the reactions of :CF_2 and :CCl_2 with ethylene have nonleast motion approach character. The reaction of :CCl_2 is a dynamically direct and concerted process with average time gap of bond formation of c. 50 fs. The reaction of :CF_2 occurs with biexponential decays of short- and longtime constants. The short component is a direct or concerted process; the long component is dynamically complex, in the sense of a temporally trapped diradical trajectory in a region with no potential energy minimum. The latter trajectory can be seen as a dynamically stepwise process [800].

Kinetic isotopic effects confirm that the cyclopropanation of pent-1-ene by dichlorocarbene, :CCl_2, to give 1,1,-dichloro-2-propylcyclpropane follows a one-step mechanism with a transition structure in which one of the two σ(C—C) bonds is almost

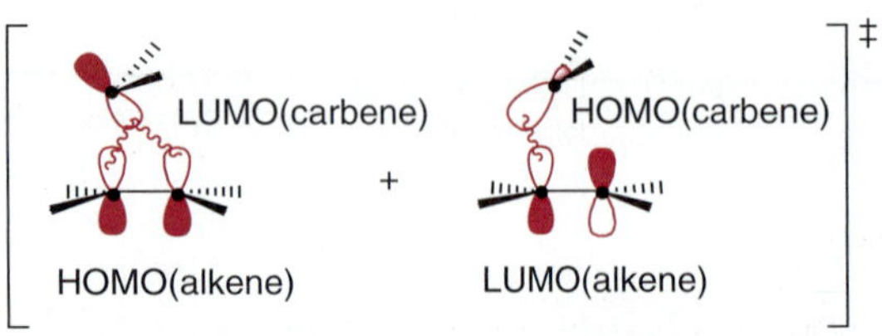

Figure 5.36 The FMO interactions that stabilize the carbene–alkene cycloaddition transition state.

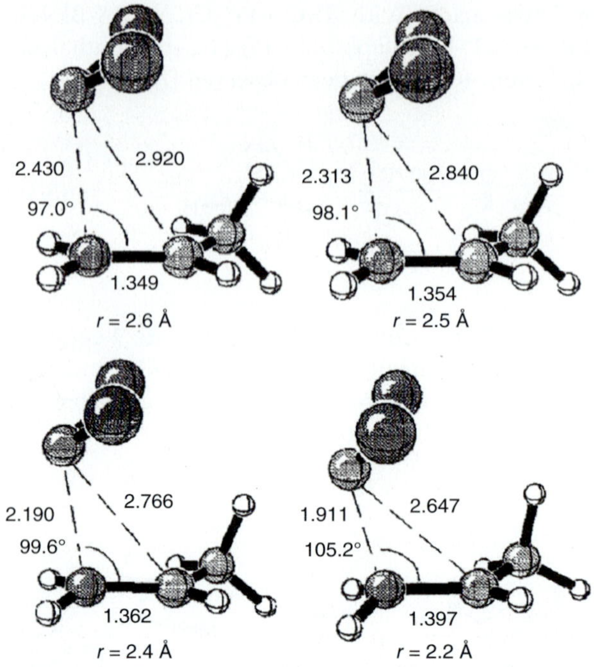

Figure 5.37 B3LYP/3-61G* geometries for :CCl_2 + propene addition at constrained values of *r*. Bond lengths are given in angstrom. Source: Taken from [799].

completely formed while the other is formed to a very small extent [799]. The transition structure of this reaction can also be represented as a 1,3-diradicaloid **380** ↔ **380′**, with a contribution from zwitterionic structure **380′** that is related to an electron transfer from the alkene HOMO to the carbene LUMO.

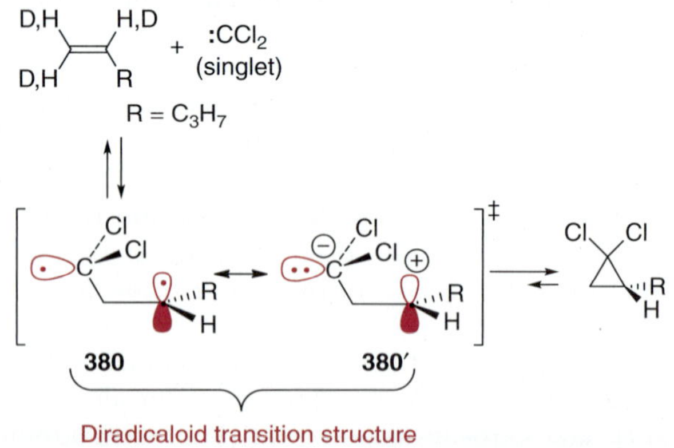

Although the reaction of methylene with an alkene is highly exothermic, the trimethylene that would be formed is a *para*-intermediate, with a very short lifetime. Halocarbenes are unlikely to form diradicals, as their heat of formation is higher than the transition-state enthalpies.

Substitution by electron-releasing groups stabilizes carbenes with respect to their dimerization into corresponding alkenes. For instance, Bertrand's phosphinosilylcarbene **381** [801, 802] and acyclic alkylaminocarbene **382** [803], and Bielawski's diaminocarbene[3]ferrocenophane **383** [804] and *N,N′*-diamidocarbene **384** [805] are isolable free carbenes capable of participating in cyclopropanation and epoxidation reactions.

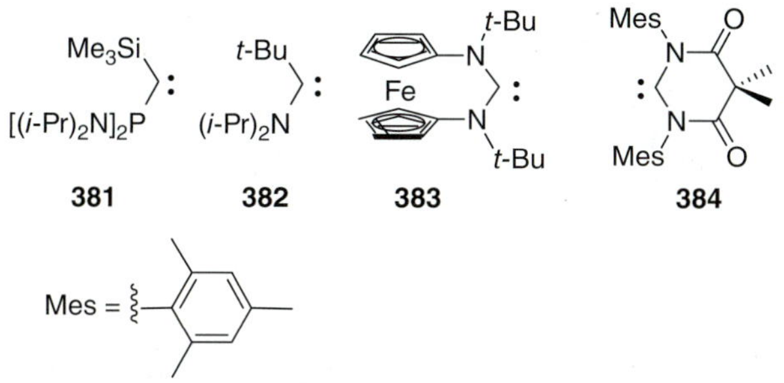

Bertrand's carbene **381** adds to electron-poor alkenes such as dimethyl fumarate with retention of the trans-arrangement between the two ester groups (reaction (5.35)) [802]. Similar results are observed with (*E*)- and (*Z*)-2-deuteriostyrene, supporting concerted suprafacial (2+1)-cheletropic additions. Furthermore, all cyclopropanation reactions of **381** with monosubstituted alkenes are highly diastereoselective, giving exclusively *cis*-cyclopropanes (phosphorous and the substituent are cis). In the case of reaction (5.35), it is the bulky carboxamide group that is cis with respect to the phosphorous substituent. Therefore, electronic rather than steric factors are responsible of the *syn*-stereoselectivity. It has been explained by invoking secondary orbital interactions (Figure 5.38), as for the Alder *endo* rule (Section 5.3.13) of the Diels–Alder reaction

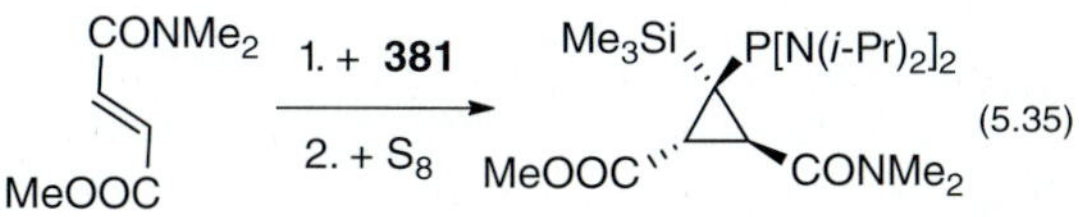

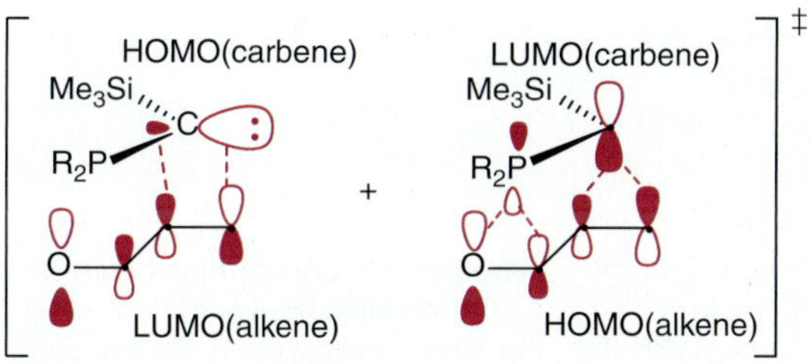

Figure 5.38 *Syn*-stereoselectivity of the cyclopropanation with Bertrand's persistent carbene.

[806]. However, quantum mechanical calculations suggest that electrostatic effects rather than secondary orbital interactions are responsible for the *syn*-stereoselectivity [807].

Diamidocarbene **384** reacts with electron-poor and electron-rich alkenes and aldehydes in (2+1)-cheletropic additions that follow nonconcerted, stepwise mechanisms. With diethyl maleate and fumarate, the reaction occurs at ambient temperature in benzene affording the same product *trans*-**385**. Similarly, (*E*)- and (*Z*)-stilbene react with **384** at 100–120 °C giving exclusively *trans*-**386**. With trisubstituted alkenes such as (*E*)- and (*Z*)-methyl 3-methylbut-2-enoate, no reaction is observed at 100 °C. At 60 °C, a small amount of product of cyclopropanation is observed together with the (*E*)- and (*Z*)-alkene isomerization. This is consistent with a reversible (2+1)-cheletropic reaction that follows a stepwise mechanism. Aldehydes react with carbene **384** giving the corresponding epoxides **387**. With conjugated enones, **384** reacts in formal (4+1)-cheletropic additions providing the corresponding dihydrofurans **388** (Scheme 5.65). The reactions probably proceed via Michael additions of the electron-rich carbene onto the enones giving zwitterionic intermediates that undergo fast ring closures. Alternatively, cyclopropanation or epoxidation occurs first, followed by (1,3)-sigmatropic rearrangements (Section 5.5.4). Reversibility of the (2+1)-cheletropic reactions involving **384** is verified in the following way. Heating **387** (R = Ph) at 80 °C with diethyl fumarate generates a mixture of epoxide **387** (c. 20%) and cyclopropane *trans*-**385** (c. 60%) [805].

Problem 5.38 Dibromocarbene engendered by reaction of bromoform with *t*-BuOK in pentane reacts with diene **D** giving a 9 : 1 mixture of cycloadducts **P** and **Q** in 80% yield. Product **P** is not isomerized into **Q** under the conditions of the reaction. What is the mechanism for the formation of **Q**? [808]

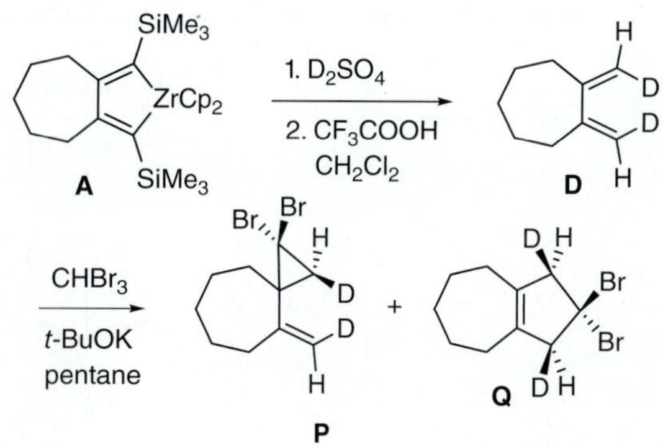

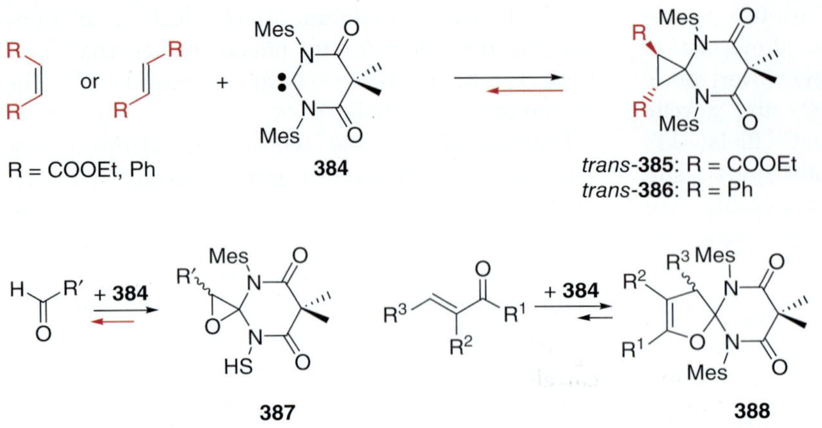

Scheme 5.65 Cheletropic reactions of a stable diamidocarbene. The reactions are reversible and follow multistep mechanisms.

5.4.2 Aziridination by (2+1)-cheletropic addition of nitrenes

The aziridine moiety is found in a number of important naturally occurring compounds (e.g. azinomycins, mitomycins, FR-900482, ficellomycin, miraziridine, maduropeptin, and azicemicins) that possess interesting biological activities. Aziridines are also useful synthetic intermediates. The general methods for their synthesis are summarized in Scheme 5.66. They include nitrene addition to alkenes, carbene and ylide addition to imines, and cyclization of 1,3-aminoalcohols and azidoalcohols. Olefin aziridination is typically accomplished via metal-mediated transfer of nitrene fragments or by conjugate addition of nucleophilic N–X group to an alkene activated by an electron-withdrawing group and subsequent 1,3-HX elimination [809, 810].

The direct aziridination of alkenes with singlet nitrenes generated thermally or photochemically from the corresponding azides are low yielded because most nitrenes have short lifetimes and undergo quick intramolecular rearrangements competitively with their intersystem crossing into more stable triplet nitrenes that dimerize quickly into the corresponding azo compounds, or react with the C—H bonds of the substrate and the solvent (C—H nitrene insertions. R–H+ :N–X → R–N(X)–H). In some cases, solvent may stabilize the singlet nitrene and render its reaction more chemo- (competition between C—H insertions and ((2+1)-cheletropic additions) and stereoselective (formation of mixtures of *cis*- and *trans*-2,3-disubstituted aziridines using either (*Z*) or (*E*)-1,2-disubstituted alkenes as substrates) as illustrated in Scheme 5.67 for the reaction

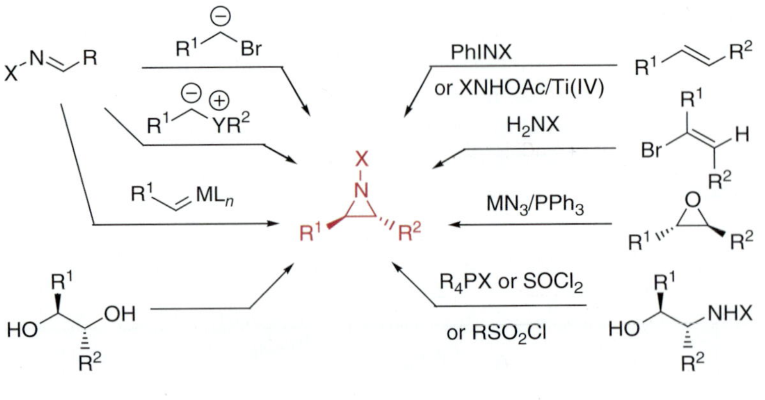

Scheme 5.66 General methods for the synthesis of aziridines.

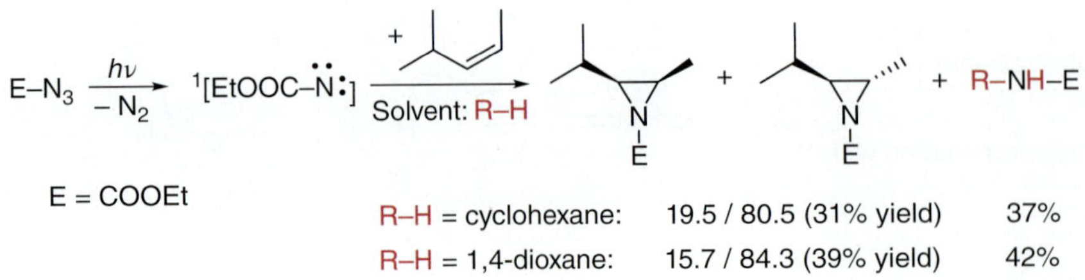

Scheme 5.67 Competition between C—H bond insertion and nonstereospecific (2+1)-cheletropic addition of a nitrene to an alkene.

of ethoxycarbonylnitrene with (*Z*)-4-methylbut-2-ene in cyclohexane and 1,4-dioxane [811].

The first catalyzed synthesis of aziridines by a nitrene transfer reaction was reported by Mansuy et al. [812, 813]. The method uses [*N*-(*p*-toluenesulfonyl)imino]phenyl iodinane [814] (PhI=NTs, preparation: $PhI(OAc)_2 + TsNH_2 \rightarrow 2AcOH + PhI{=}NTs$) as nitrene source and iron porphyrin or manganese porphyrin complexes of type **389** as catalysts (Scheme 5.68) [815]. In 1993, Evans et al. [816], and Jacobsen and coworkers [817] independently reported the copper-catalyzed asymmetric alkene aziridination using PhI=NTs as nitrene source [818, 819]. Ruthenium, cobalt, and iron porphyrin complexes catalyze azididinations using aryl azides [820].

A wide range of metals have been used to catalyze the nitrene transfer from PhI=NTs, including Cu, Ag, Au, Ru, Rh, Mn, Fe, Co, Pd, Ni, In, and Re in combination with suitable coordinating ligands. PhI=NTs has been replaced by *N*-halogenated sulfonylamide salts such as chloramine-T and bromamine-T, and by sulfonamides, carbamates, and sulfamate derivatives in combination with iodonium salts and/or suitable additives [813]. An example of copper-catalyzed olefin aziridination using 5-methyl-2-pyridinesulfonamide (**390**) as nitrene source and $PhI(OAc)_2$ as oxidant is given in Scheme 5.69.

An electrochemical aziridination reaction is shown in Scheme 5.70. Anodic oxidation of *N*-aminophthalimide (**391**) generates the intermediate nitrene that is trapped by AcOH giving the active reagent **392**, which then reacts with the alkene giving the corresponding aziridine **393**. Oxidation of **391** can also be accomplished with $PhI(OAc)_2$.

Problem 5.39 Nitric oxide (NO) is a mediator of many important biochemical pathways [821], including muscle relaxation [822], macrophage activation [823, 824], and neurotransmission [825–827]. Ingold and coworkers have used orthoquinodimethane to quench NO in a (4+1)-cheletropic addition that gives stable paramagnetic nitroxides readily detected by ESR [764]. What is the product of reaction of NO with the product of photoinduced electrocyclic ring

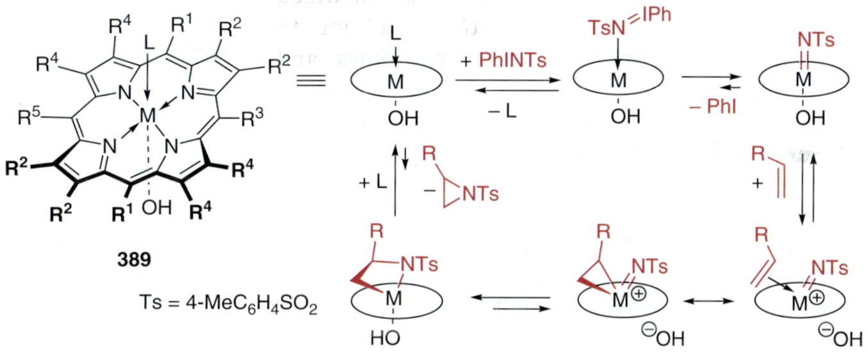

Scheme 5.68 Mansuy's catalyzed nitrene transfer to alkenes (aziridination).

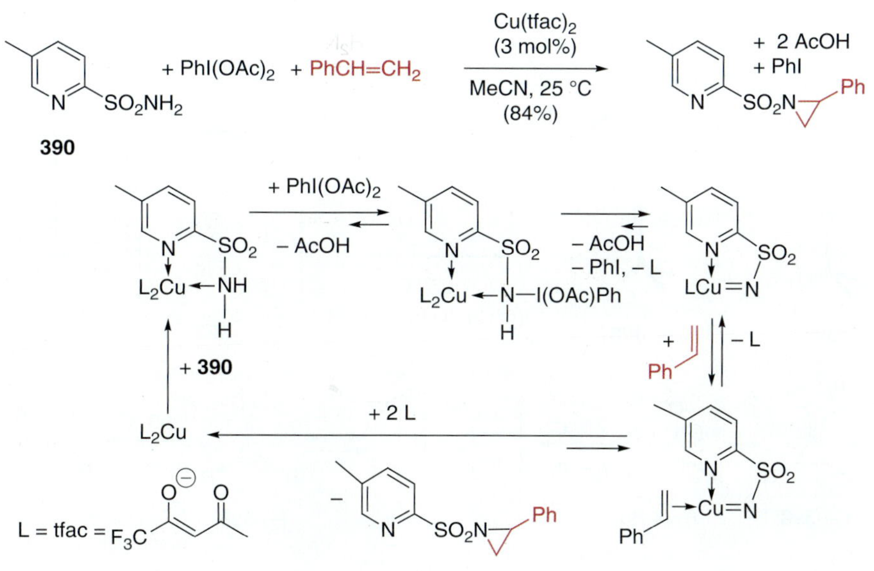

Scheme 5.69 Copper-catalyzed aziridination using a chelating 2-pyridylsulfonylamide moiety.

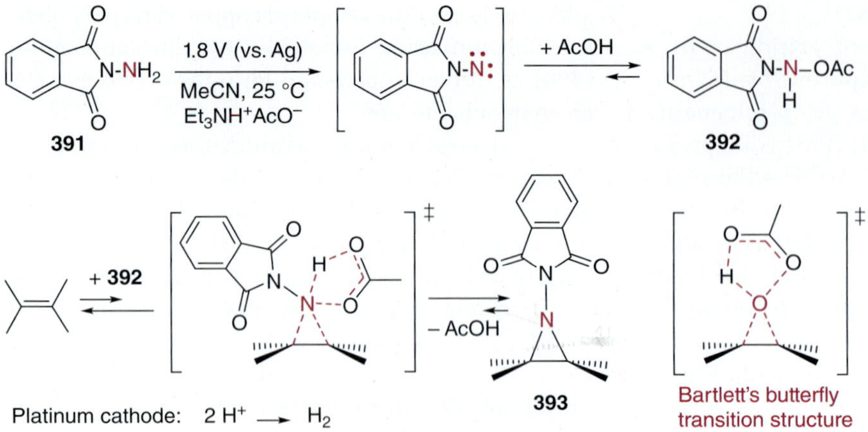

Scheme 5.70 Electrochemical aziridination reaction.

opening of benzocyclobutenedione? (NO can be generated by reaction of HNO_3 with Cu powder) [828].

5.4.3 Decarbonylation of cyclic ketones by cheletropic elimination

Carbon monoxide (:C=O) can be seen as an oxocarbene and thus should be able to undergo (2+1)-cheletropic reactions. Alkenes do not add to CO to generate the corresponding cyclopropanones for thermodynamic reasons. The reaction ethylene + CO ⇄ cyclopropanone is endothermic by $17.7 \pm 1.5\,\text{kcal mol}^{-1}$ and is thus impossible at any temperature (negative entropy of condensation). In solution and at 25 °C, cyclopropanone is polymerized into a polyacetal (Figure 5.39) [829]. In the gas phase, 2,2-dimethylcyclopropanone (**394**) is isomerized into isopropenyl methyl ketone, and in the presence of 2-methylfuran, a 58 : 42 mixture of (4+3)-cycloadducts **396** + **397** is formed [830], consistently with the isomerization of cyclopropanones into 2-oxyallyl zwitterion intermediates (**395**). At 79.6 °C, (+)-*trans*-2,3-bis(*t*-butyl)cyclopropanone ((+)-**398**) is racemized into (±)-**398** with a rate constant of $2.7 \times 10^{-5}\ \text{s}^{-1}$ in isooctane (apolar solvent) and of $33.5 \times 10^{-5}\ \text{s}^{-1}$ in acetonitrile (polar solvent), consistently with the formation of a zwitterionic 2-oxyallyl intermediate (see the disrotatory electrocyclic isomerization of cyclopropyl cation into allyl cation, Section 5.2.7). Heating to 150 °C promotes the suprafacial (2+1)-cheletropic elimination of CO with formation (*E*)-1,2-bis(*t*-butyl)ethylene [831, 832].

The FMO diagram of Figure 5.40 demonstrates for a concerted suprafacial (2+1)-cheletropic elimination of CO from cyclopropanone that its transition state is

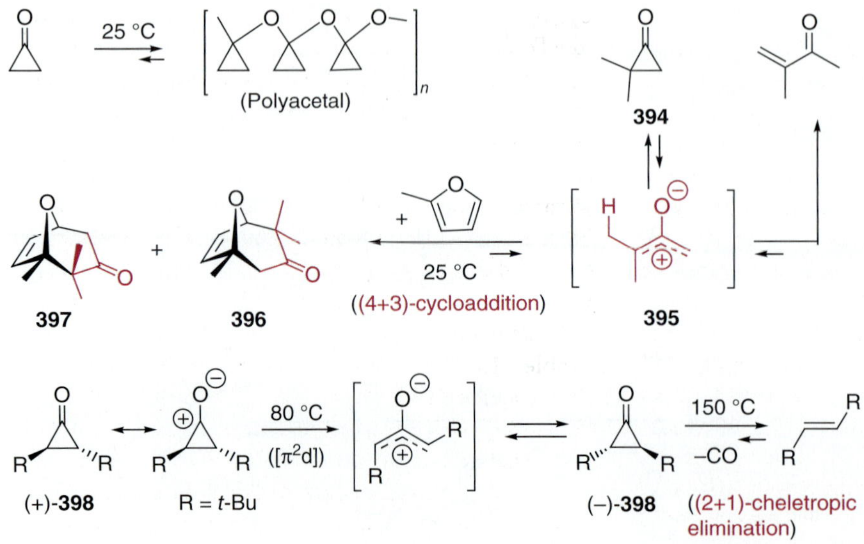

Figure 5.39 Thermal reactions of cyclopropanone and derivatives.

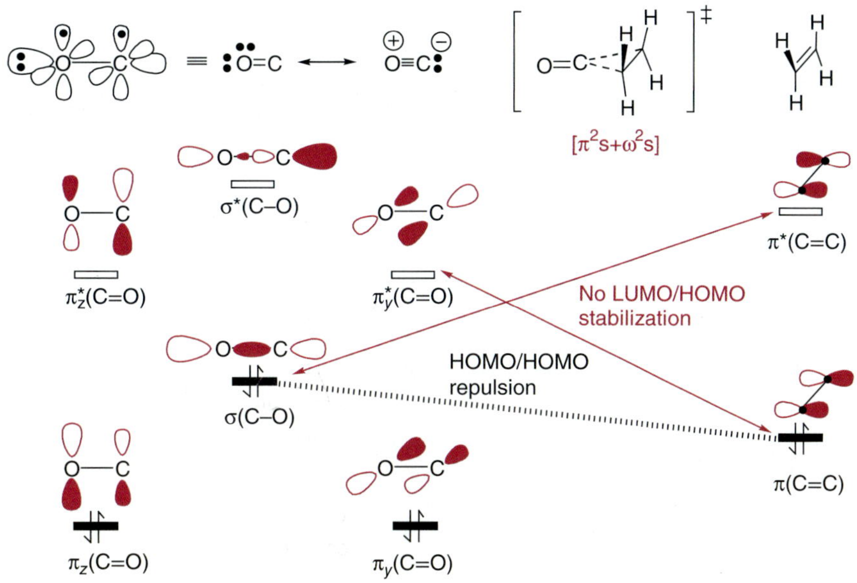

Figure 5.40 FMO diagram representing the FMOs of carbon monoxide and ethylene entering in a suprafacial (2+1)-cheletropic reaction with C_{2v}-symmetrical transition structure. The latter appears to be antiaromatic.

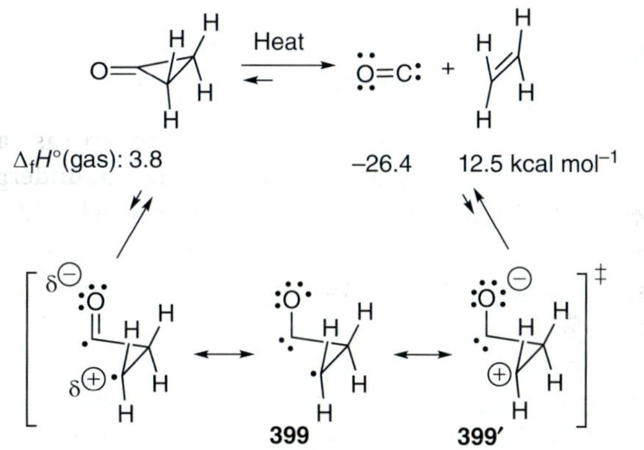

not stabilized by any LUMO(CO)/HOMO(ethylene) and LUMO(ethylene)/HOMO(CO) interaction. Furthermore, it shows a repulsive HOMO(CO)/HOMO(ethylene) interaction. The LUMO(ethylene)/sub-HOMO(CO) is not strong enough to stabilize the C_{2v} transition structure.

MP2/6-31G* quantum mechanical calculations predicted that the departure of CO remains in the molecular plane, but the pathway is not C_{2v} symmetrical: it is very asynchronous with the C=O bond bending over the ethylene moiety. This permits some electron transfer from HOMO(ethylene) to LUMO(C=O) that leads to the development of partial positive charge on C(3) and partial negative charge on the oxygen atom. A stabilizing electrostatic interaction between the oxygen atom and C(3) ensues. It can be interpreted by the diradicaloid model **399** ↔ **399′**. The activation enthalpy for a two-step process involving homolysis of the C(1)—C(2)-bond into diradical intermediate **399** is calculated to be at least $DH°(\mathrm{MeCO^\bullet/Me^\bullet}) = 80\ \mathrm{kcal\ mol^{-1}}$ (Table 1.A.7) corrected for the ring strain release of cyclopropanone ($-45\ \mathrm{kcal\ mol^{-1}}$) and for the stability difference between methyl and ethyl radical ($-4\ \mathrm{kcal\ mol^{-1}}$). The ring strain of cyclopropanone is estimated to $45\ \mathrm{kcal\ mol^{-1}}$ on comparing the heats of hydrogenation: $\Delta_r H°(c\text{-}C_3H_4O + H_2 \rightarrow \mathrm{MeCOMe}) = -56.0\ \mathrm{kcal\ mol^{-1}}$ and $\Delta_r H°(\mathrm{cyclohexane} + H_2 \rightarrow n\text{-hexane}) = -10.4\ \mathrm{kcal\ mol^{-1}}$. This leads to $\Delta_r H°(c\text{-}C_3H_4O \rightarrow$ **399**$) = 80 - 45 - 4 = 31\ \mathrm{kcal\ mol^{-1}}$! The "best" quantum mechanical calculations predict $37\ \mathrm{kcal\ mol^{-1}}$ for the activation enthalpy of the decarbonylation of cyclopropanone [106, 833].

Thermochemical calculations give an endothermicity of $19.2\ \mathrm{kcal\ mol^{-1}}$ for the fragmentation of cyclopent-3-enone into butadiene and carbon monoxide. For a nonconcerted mechanism involving homolysis of C(1)—C(2)-bond into diradical **400**, one estimates a minimal activation enthalpy of $56.8\ \mathrm{kcal\ mol^{-1}}$ ($DH°(\mathrm{MeCO^\bullet/Me^\bullet})$ corrected for the ring strain release of cyclopent-3-enone ($-7.2\ \mathrm{kcal\ mol^{-1}}$, obtained on comparing $\Delta_r H°(c\text{-}C_5H_8O + H_2 \rightarrow \mathrm{EtCOEt}) = -17.6\ \mathrm{kcal\ mol^{-1}}$ and $\Delta_r H°(\mathrm{cyclohexane} + H_2 \rightarrow n\text{-hexane}) = -10.4\ \mathrm{kcal\ mol^{-1}}$) and for the stability difference between methyl and allyl radical ($-16\ \mathrm{kcal\ mol^{-1}} = DH°(\mathrm{allyl^\bullet/H^\bullet}) - DH°(\mathrm{Me^\bullet/H^\bullet})$ (Table 1.A.7)). This is significantly more than the measured barrier $\Delta^{\ddagger}H = 46–51\ \mathrm{kcal\ mol^{-1}}$ [834, 835], which suggests a degree of concentration of $6–11\ \mathrm{kcal\ mol^{-1}}$. Decarbonylation of *cis*-2,5-dimethyl-cyclopent-3-enone occurs exclusively via the disrotatory pathway ([π^4s+ω^2s]-cheletropic elimination)

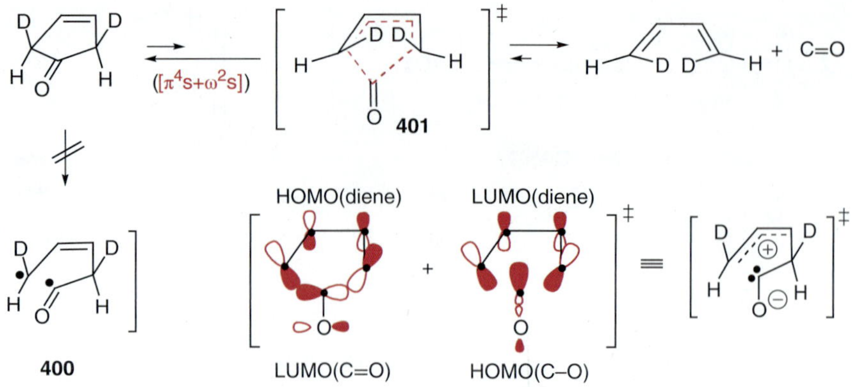

Figure 5.41 Concerted and synchronous suprafacial $[\pi^4s+\omega^2s]$-cheletropic decarbonylation of cyclopent-3-enone. The C_s-symmetrical transition structure is aromatic in character.

consistently with the Woodward–Hoffmann rules. The transition structure **401** of this reaction is C_s-symmetrical (disrotatory, synchronous breaking of the two C(1)—C(2) and C(1)—C(5) bonds), with the carbon atom of the carbonyl group out of the diene plane and the C—O bond bending toward the C(3)=C(4) double bond. It benefits from electronic stabilization because of LUMO(CO)/HOMO(diene) and LUMO(diene)/HOMO(CO) interactions (Figure 5.41). Woodward and Hoffmann recognized a second possible concerted mechanism for the decarbonylation of cyclopent-3-enone with a C_2-transition structure in which the carbon atom moiety of carbon monoxide remains in the plane of cyclopentenone ring, and the diene forms in a conrotatory manner ($[\pi^4a+\omega^2a]$-cheletropic elimination). Quantum mechanical calculations do no find the latter transition structure and estimate to 49 kcal mol^{-1} the activation enthalpy for the disrotatory cheletropic elimination, in agreement with experiments [106].

The thermal fragmentations of furan-2,3-dione (**402**), pyrrole-2,3-dione (**403**), cyclopent-3-ene-1,2-dione (**404**), 3*H*-furan-2-one (**405**), and 3-methylidene-2*H*-furan-2-one (**406**) proceed via planar transition states and are considered to be pseudopericyclic cheletropic reactions [106].

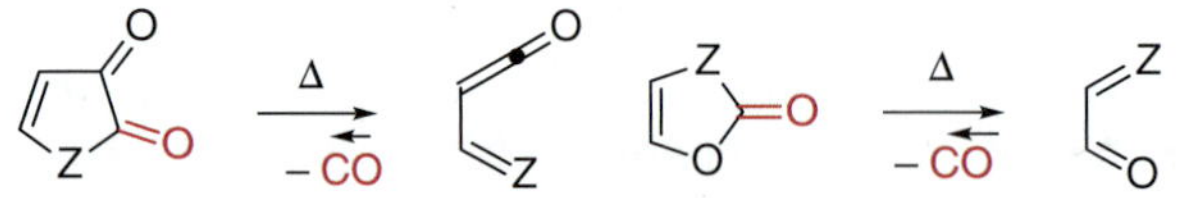

402: Z = O
403: Z = NH
404: Z = CH_2

405: Z = CH_2
406: Z = C=CH_2

Problem 5.40 Decarbonylation of bicyclo[2.2.1]hepta-2,5-dien-7-one in PhH occurs at −60 °C and gives benzene + CO with E_a = 15.2 kcal mol^{-1} [836]. In contrast, decarbonylation of bicyclo[2.2.1]hept-2-en-7-one into 1,3-cyclohexadiene + CO is much slower and has an activation energy E_a = 34.4 kcal mol^{-1} (125 °C) [835, 837]. Both reactions are much faster than the decarbonylation of cyclopent-3-enone. Give an explanation.

Problem 5.41 Zwitterion **A** adds to dimethyl acetylenedicarboxylate giving an unstable cycloadduct **B** that fragmentizes into **P** and phenyl isonitrile. What is **B**? Give a mechanism for the fragmentation. Propose a method to prepare **A** [838].

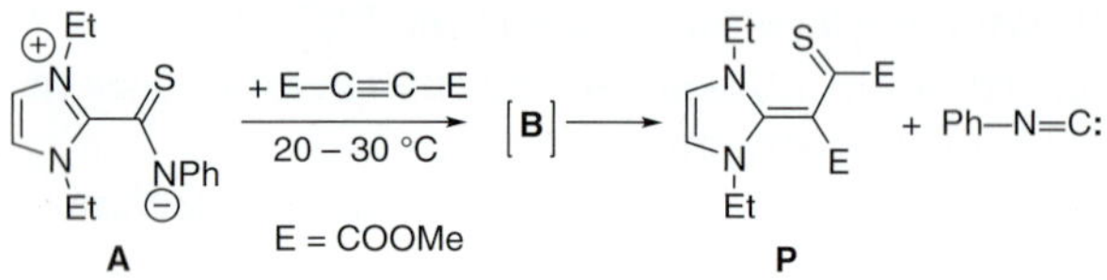

5.4.4 Cheletropic reactions of sulfur dioxide

Conjugated dienes have been known since 1914 to undergo (4+1)-cheletropic additions with sulfur dioxide, generating the corresponding 2,5-dihydrothiophene-1,1-dioxides (sulfolenes) [839]. According to the Woodward–Hoffmann rules [1, 2], the preferred approach of SO_2 to the diene is a linear trajectory in a suprafacial manner. It is a $[\pi^4s+\omega^2s]$-cheletropic addition with C_s-symmetrical transition state that benefits from electronic stabilization because of its aromatic character (Hückel type of AO array with six electrons) as shown in Figure 5.42. At 20 °C, SO_2 adds to (*E*,*E*)-hexa-2,4-diene giving exclusively *cis*-2,5-dimethylsulfolene, whereas (*E*,*Z*)-hexa-2,4-diene gives exclusively *trans*-2,5-dimethylsulfolene. The sulfolenes are stable at ambient temperature but undergo stereospecific suprafacial (4+1)-cheletropic eliminations 100 °C liberating the corresponding dienes (*outward* torquoselectivity for steric reasons) and SO_2 [840]. Thus, an impure 1,4-disubstituted butadiene can be converted to a mixture of sulfolenes. The major one is purified by crystallization. Upon

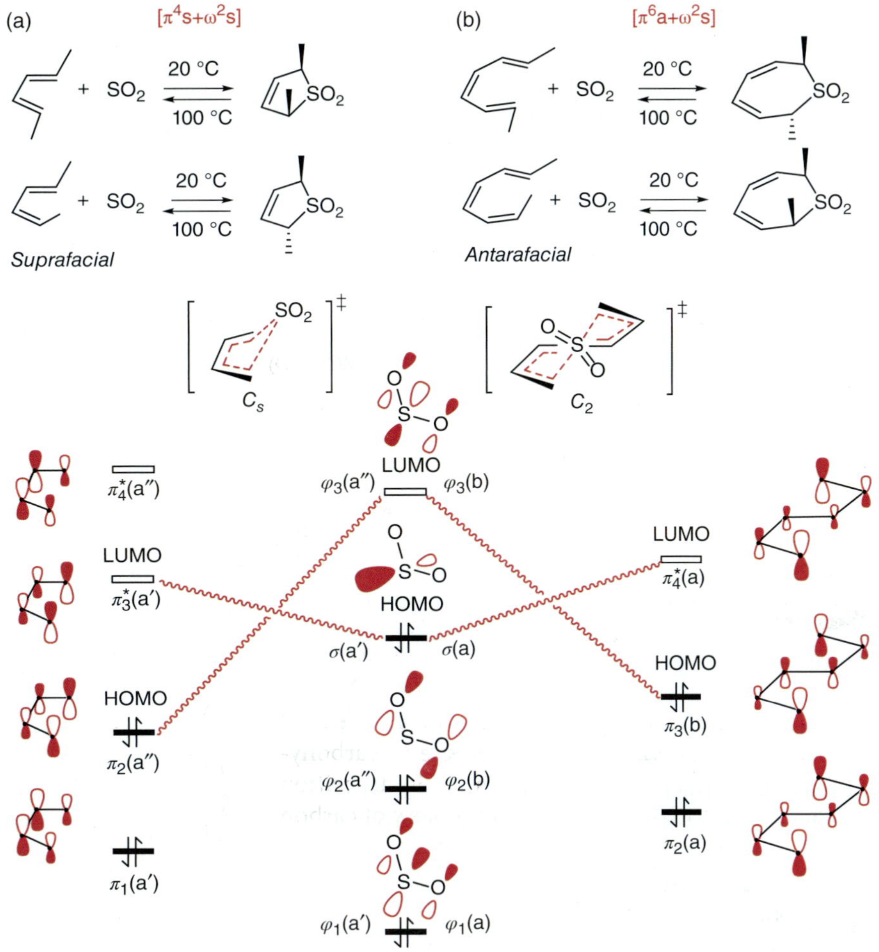

Figure 5.42 Thermally allowed (a) suprafacial $[\pi^4s+\omega^2s]$ and (b) antarafacial $[\pi^6a+\omega^2s]$-cheletropic reactions; FMO diagrams showing the aromatic character of the transition structures of these concerted, reversible reactions.

heating the latter, pure (*E*,*E*) or (*E*,*Z*)-diene can be isolated. At 20 °C, (*E*,*Z*,E)-octa-2,4,6-triene and (*Z*,*Z*,E)-octa-2,4,6-triene react with SO_2 to give *trans*- and *cis*-2,7-dihydro-2,7-dimethylthiepin, respectively. The reactions are antarafacial $[\pi^6a+\omega^2s]$-cheletropic additions that are highly stereoselective. The reverse reactions, antarafacial (6+1)-eliminations of SO_2, occur above 100 °C. These reactions are concerted and involve C_2-symmetrical transition structures that are also stabilized electronically (Möbius orbital arrays with eight electrons) as shown in Figure 5.42b.

The (4+1)-cheletropic addition of SO_2 to 1,2-dimethylidenecyclohexane occurs already at −20 °C and is second order in SO_2 ($[SO_2]$ = 2.6–15.3 M in CD_2Cl_2) showing that SO_2 catalyzes its own cheletropic reaction [841]. In the case of the cheletropic addition of SO_2 to (*E*)-1-methoxybutadiene at −75 °C, the rate law is $d[\text{diene}]/dt = k[\text{diene}][SO_2]^x$ with $x = 2.6 \pm 0.2$. In agreement with quantum mechanical calculations, the reaction follows two parallel mechanisms, one involving two molecules of SO_2 and the other involving three molecules of SO_2 with a transition structure of type **407** [842].

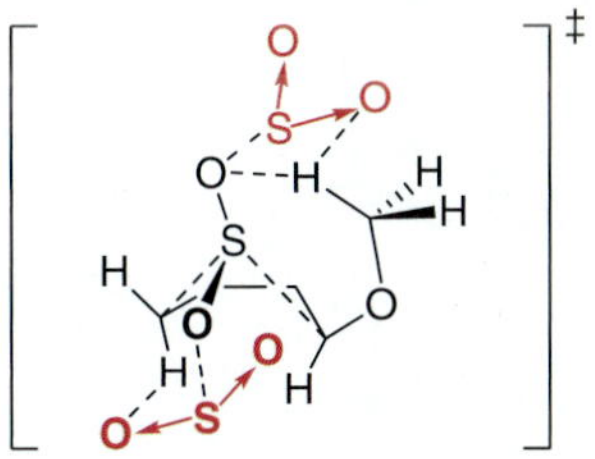

407

An application of (4+1)-cheletropic elimination of SO_2 is presented in Scheme 5.71 with the total synthesis of *rac*-estratrienone. Alkylation of benzosulfolene with tosylate **408** generates **409**. After acetal hydrolysis, heating to 210 °C in di-*n*-butyl phthalate induces SO_2 elimination with formation of the instable *ortho*-quinodimethane **410** (a *ortho*-xylylene) that undergoes intramolecular Diels–Alder reaction with high stereoselectivity providing *rac*-estratrienone [843]. Aza-*ortho*-xylydenes have been generated via thermal extrusion of SO_2 from benzosultams [844].

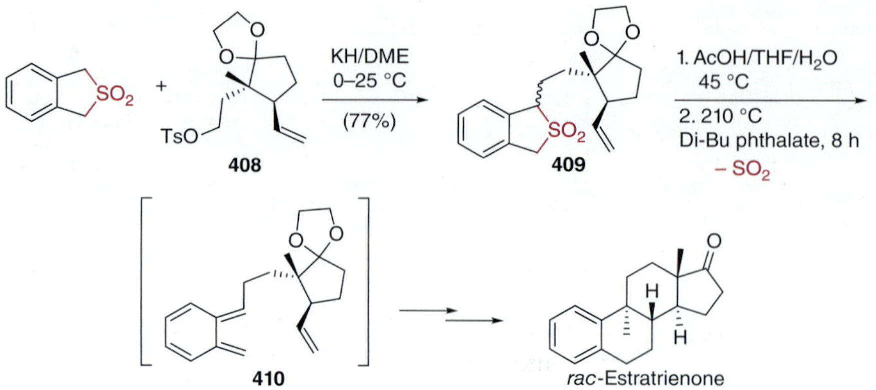

Scheme 5.71 Nicolaou's intramolecular capture via Diels–Alder reaction of an *o*-quinodimethane engendered by SO_2 cheletropic elimination from a benzosulfolene.

At 25 °C, norbornadiene [845] and 3,3-dimethylpenta-1,4-diene react with SO_2 giving the product of $[\pi^2s+\pi^2s+\omega^2s]$-homocheletropic addition **411** and **412**, respectively. The reactions are accompanied by the formation of polysulfone polymers, the proportion of them being reduced in the presence of a radical scavenger such as 2,6-di-*tert*-butyl-*p*-cresol [846]. At −20 °C, SO_2 adds to 7,7-dimethyl[2.2.1]hericene (2,3,5,6-tetrakis (methylidene)-7-isopropylidenebicyclo[2.2.1]heptane: **413**) giving the product of homocheletropic addition **414**. At 20 °C, the latter is equilibrated with **415**, with sulfolane **414** being a little more stable than isomeric sulfolene **415** [847].

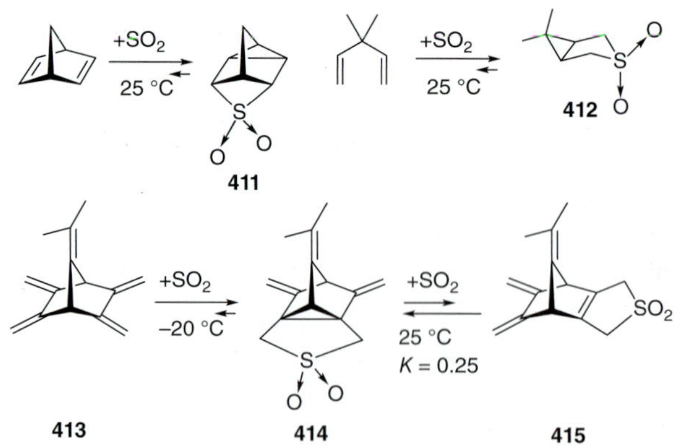

Staudinger and Pfenniger have found that diphenyldiazomethane and SO_2 give tetraphenylthiiran 1,1-dioxide [848]. This method has been used to prepare the parent heterocycle [849] and 2,3-dialkyl- or 2,3-diaryl derivatives [850, 851]. Alternatively, unsymmetrically substituted thiirane-1,1-dioxides can be obtained by cyclopropanation of sulfenes ($R_2C{=}SO_2$) with diazomethane ($CH_2{=}N_2$) [852]. At 25 °C, the *trans*- and *cis*-1,2-diphenylthiirane 1,1-dioxides are fragmented into SO_2 and (*E*)- and (*Z*)-stilbene, respectively. The reactions are slightly faster in polar than in apolar solvents. As for the suprafacial cyclopropanations of alkenes that are forbidden according to the Woodward–Hoffmann rules, concerted and synchronous thermal suprafacial (2+1)-cheletropic eliminations of thiirane 1,1-dioxides are also forbidden (C_s-symmetrical transition structure implies repulsion between HOMO(alkene) and HOMO(SO_2)). Nevertheless, the reactions are fast at ambient temperature and generate diradical intermediates of type **416** for which rotation about the σ(C—C) bond is slow compared with the fragmentation into SO_2 and the corresponding alkenes. With SO_2, alkenes generate polysulfones that are alternating polymers of SO_2 and of the alkenes. With 1,1-dialkylalkenes, the polymerization occurs at ambient temperature. On heating, the polysulfones equilibrate with diradicals and decompose liberating SO_2 [853, 854].

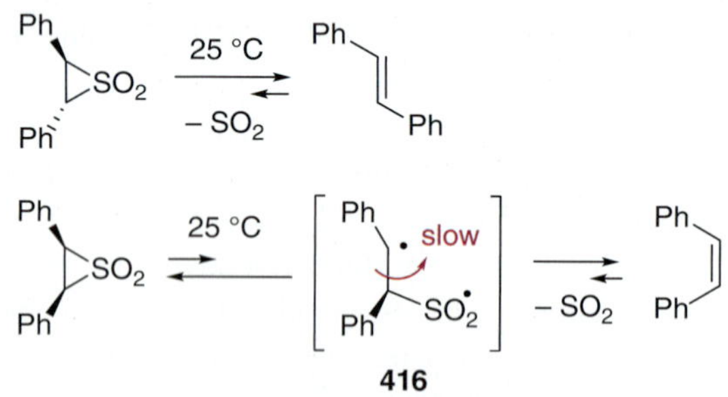

Thiirane 1,1-dioxides are formed as intermediates in the Ramberg–Bäcklund reaction (Scheme 5.72) that convert α-halogenosulfones into alkenes in the presence of strong bases [855].

An application of the Ramberg–Bäcklund reaction is given in Scheme 5.73 with the synthesis of C-isotrehalose, the C-linked analog of β,β-trehalose (β-D-glucopyranosyl-(1→1′)-β-D-glucopyranose). The benzyl-protected exoglycal **417** undergoes face selectivity (steric hindrance by the α-2-benzyloxy substituent), giving an alcohol that is converted into the corresponding iodide **418**. The thioglycoside **419**

Scheme 5.72 Bordwell's mechanism of the Ramberg–Bäcklund reaction

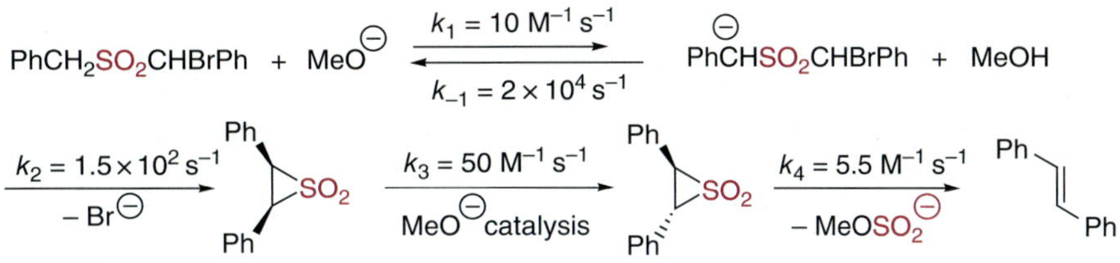

Scheme 5.73 Taylor's synthesis of a C-linked disaccharide.

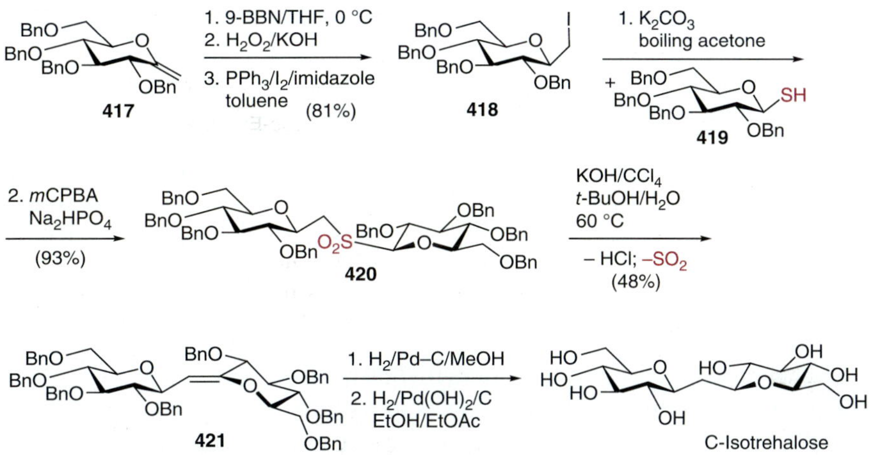

displaces iodide **418** giving a thioether that is oxidized into sulfone **420** with *meta*-chloroperbenzoic acid (*m*CPBA). The Ramberg–Bäcklund reaction is done under Meyers' conditions ($KOH/CCl_4/t\text{-}BuOH/H_2O$) [856] and affords a 91 : 9 mixture of (*Z*)- and (*E*)-enol **421**. Catalytic alkene hydrogenation and subsequent debenzylation furnishes C-isotrehalose [857]. Halogenation and subsequent 1,3-elimination can be carried out under Chan's conditions ($KOH/Al_2O_3/CF_2Br_2/CH_2Cl_2$, 40 °C) [858].

Problem 5.42 What is the major product of the following reaction? [859]

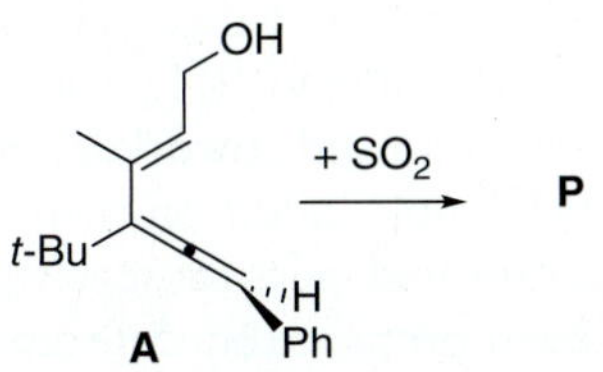

Problem 5.43 What is the major product of the following reaction? [860]

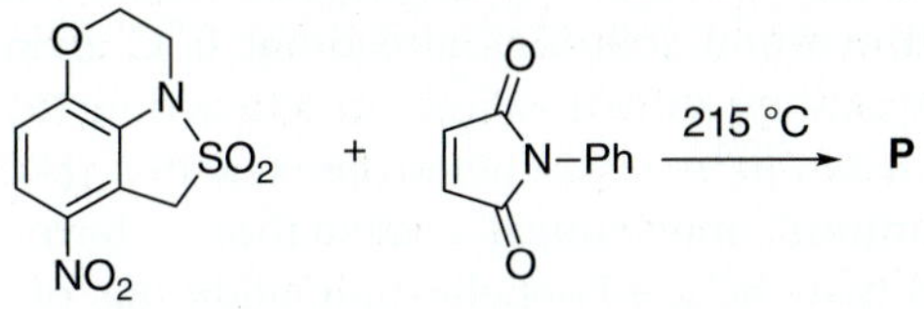

Problem 5.44 What is the main product **P** of the following reaction [861].

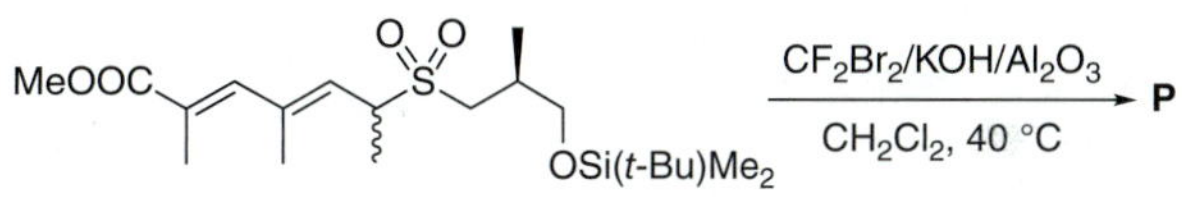

Problem 5.45 Betaine **A** adds to dimethyl acetylenedicarboxylate giving a bicyclic adduct **B** which, upon oxidation with *meta*-chloroperbenzoic acid, fragmentizes into a cycloheptatrienone derivative **P** and nitrosobenzene. What are **B** and **P**? Explain these reactions [766].

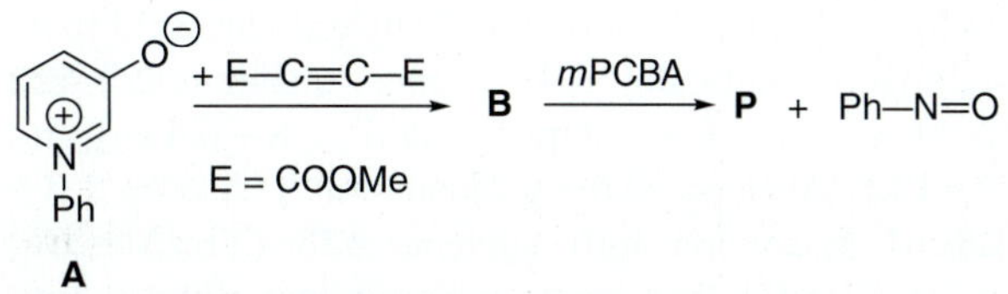

5.4.5 Cheletropic reactions of heavier congeners of carbenes and nitrenes

Divalent silicon species analogous to carbenes are called silylenes. They are important reactive intermediates in organosilicon chemistry [862, 863]. Silylenes undergo cheletropic additions with π-systems such as alkenes, dienes, alkynes, and σ-insertion (formal

[σ^2s+ω^0s]-cycloadditions: (2+1)-cheletropic additions with σ-systems) into C—H (intramolecular), Si—H, Si—O, Si–Hal, Si—N, Si—Si (strained), N—H, O—H, and H—H bond.

Classical methods for the generation of silylene intermediates is photolysis of cyclic [864] and linear oligosilanes, pyrolysis of 7-silanorbornadiene derivatives [865, 866], and thermolysis of silacyclopropanes [867]. For instance, heating hexamethylsilacyclopropane with bis(trimethylsilyl)acetylene to 65 °C provides a mixture of tetramethylethylene and 1, 1-dimethyl-2,3-bis(trimethylsilyl)-1-silacyclopropene [868].

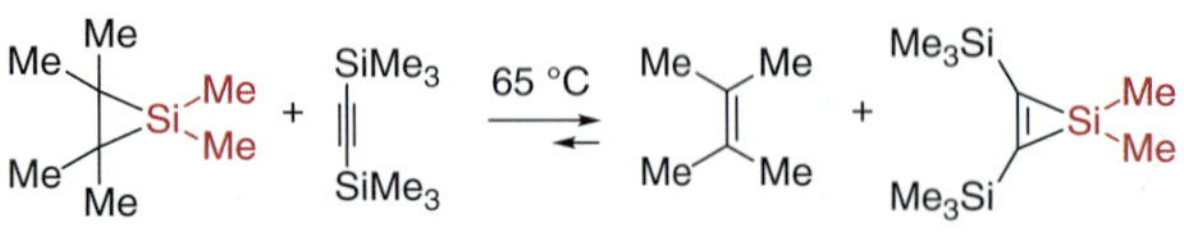

Flash pyrolysis of bis[(methoxy)dimethyl]disilane in the presence of dimethylacetylene yields dimethoxydimethylsilane and tetramethylsilacyclopropene [869].

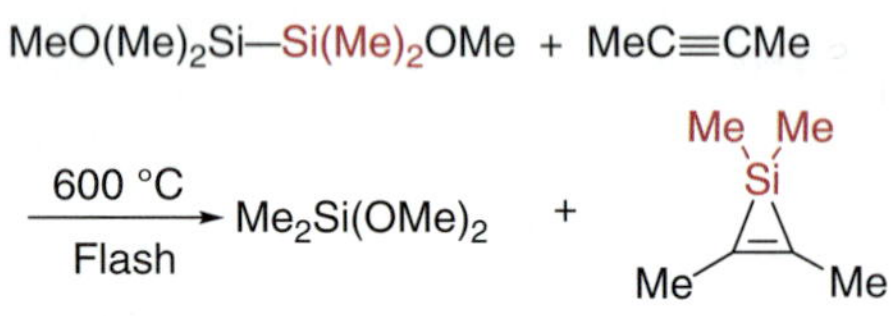

In the absence of efficient trapping agents, silylenes R_2Si: dimerize into disilenes, $R_2Si{=}SiR_2$. On heating, disilenes dissociate into silylenes more readily than corresponding alkenes dissociate into carbenes ($\Delta_r H°(H_2C{=}CH_2 \rightleftarrows 2\ {}^3[H_2C:]) = 165\,\text{kcal mol}^{-1}$; $\Delta_r H°(H_2Si{=}SiH_2 \rightleftarrows 2\ {}^1[H_2Si:]) = 74–82\,\text{kcal mol}^{-1}$ [870]). Disilenes with large substituents are destabilized by Front-strain with respect to their silylenes. As for carbenes, the "crowded" silylenes are said to be stabilized kinetically. For instance, disilenes (*Z*)-**422** ($\Delta^{\ddagger}H = 25.5 \pm 0.4\,\text{kcal mol}^{-1}$, $\Delta^{\ddagger}S = 8 \pm 0.4$ eu) and (*E*)-**422** ($\Delta^{\ddagger}H = 25.6 \pm 0.7\,\text{kcal mol}^{-1}$, $\Delta^{\ddagger}S = 1.3 \pm 1.8$ eu) dissociate into silylene **423** (Tbt(Mes)Si:) at 70 °C [871] that reacts with alkene, alkyne, benzene, carbonitrile derivatives, and carbon disulfide in (2+1)-cheletropic additions (Scheme 5.74). With isonitriles, silylene **423** forms complexes that can be seen as products of (1+1)-cheletropic additions, as for the dimerization of silylenes into the corresponding disilenes [773].

The reactions of silylenes with conjugated dienes give the corresponding 3-silonenes (silacyclopent-3-enes) resulting from initial (4+1)-cheletropic addition of one of C=C double bond generating the corresponding 2-vinylsiliranes and subsequent rearrangement into the 3-silolenes [872]. Complex **424** of silylene **423** and 2,4,6-tris(tertiobutyl)phenylisonitrile reacts with isoprene at 20 °C giving the isolable vinylsilirane **425**. This compound is also isolated from the reaction of disilene **422** with isoprene at 70 °C, which indicates that both reactions proceed via the silylene intermediate **423**. Thus, complex **424** is a masked silylene. On heating to 160 °C, **425** is isomerized into silolene **426**, the product of formal (4+1)-cheletropic addition of **423** to isoprene. The reaction of complex **424** with 2,3-dimethylbutadiene at 25 °C gives a mixture of vinylsilirane **427** ((2+1)-cycloadduct) and silonene **428** ((4+1)-cycloadduct). As heating this mixture to 70 °C does not isomerize **427** into **428**, in this case, the (4+1)-cheletropic addition competes with the (2+1)-cheletropic addition. On heating to 100 °C, **427** isomerizes into **428**.

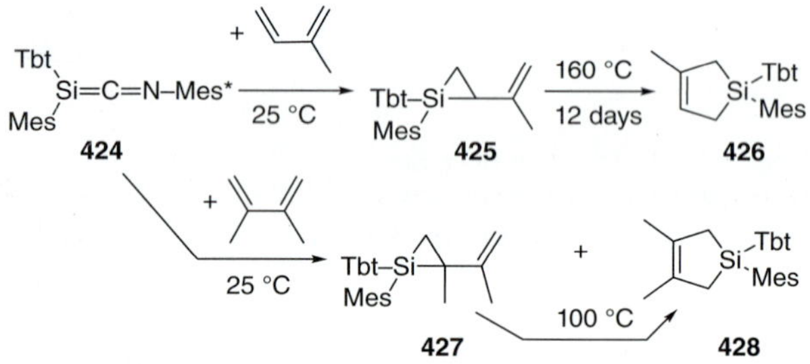

Persistent silylenes (Scheme 5.75) are known since 1986 with the synthesis of decamethylsilicocene **429** by Jutzi et al. [873]. In 1994, West and coworkers [874] reported the first *N*-heterocyclic silylene **430**, which is stabilized by crowding (kinetic stabilization with respect to its dimerization) and n(N:) → 3p(Si) donation [772]. The first alkyl disubstituted silylene **431** was prepared by Kira et al. [875, 876]. The phosphine-stabilized silicon(II) hydride **432** has been prepared by Baceiredo and coworkers [877]. It undergoes (2+1)-cheletropic addition with alkenes and alkynes [878]. The first persistent acyclic dithiosilylene **433** has been prepared by debromination of the corresponding dibromide with Jones Mg(I)–Mg(I) complex [879]. The stable boryl(amino)-substituted silylene **434** has been obtained by redox reaction of the corresponding tribromo(amino)silane with a lithium boryl reagent. Silylene **434** reacts with H_2 in hydrocarbon solutions already at 0 °C giving the corresponding dihydrosilane, in a reaction that can be seen as a [σ^2s+ω^0s]-cheletropic reaction [880].

Quantum mechanical calculations have suggested that the (2+1)-cheletropic additions of stable

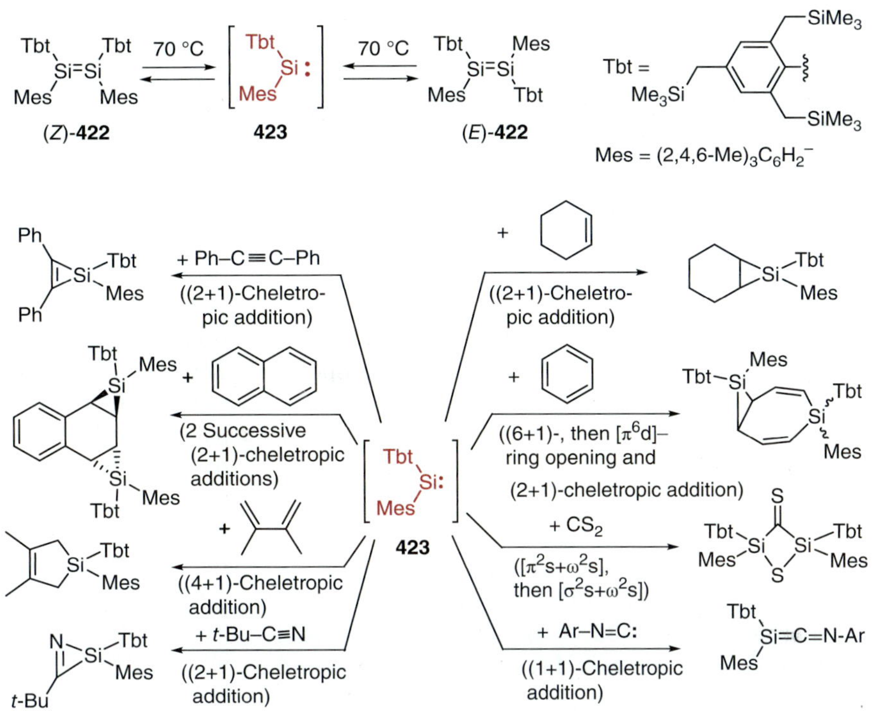

Scheme 5.74 Thermal formation of a transient diarylsilylene at low temperature and its reactions with π-systems.

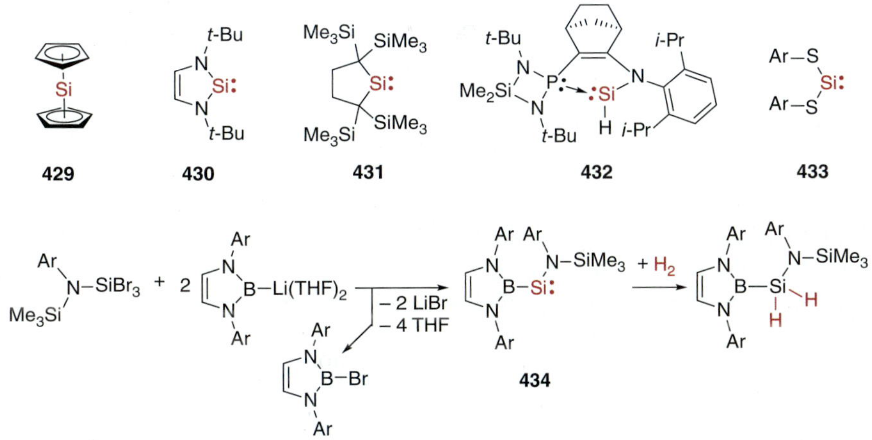

Scheme 5.75 Examples of stable (persistent) silylenes and of a H—H insertion by a bora(amino)silylene.

N-heterocyclic silylenes with ethylene and formaldehyde proceed through concerted mechanisms to form the corresponding siliranes and oxasiliranes, respectively [881]. Calculations also suggested that dimethylsilylene (Me_2Si:) and dimethylgermylene (Me_2Ge:) add to butadiene to form the corresponding vinylmetalliranes, the products of concerted (2+1)-cheletropic addition, and the corresponding metallacyclopent-3-enes, the products of concerted (4+1)-cheletropic addition. The rearrangements vinyl(Si,Ge)irane → metalla(Si,Ge)cyclopen-3-ene occur through (2+1)-cheletropic eliminations and subsequent (4+1)-cheletropic additions. The calculations predicted that dimethylstannylene (Me_2Sn:) adds to butadiene giving both the products of (2+1)- and (4+1)-cycloaddition, but the vinylstannirane is unstable and equilibrates quickly with butadiene and Me_2Sn: [882].

Me_2Ge: has been generated by thermolysis (70–150 °C) of **435**. It reacts with (*E*,*E*)-1,4-diphenylbutadiene giving the product of suprafacial (4+1)-cheletropic addition **436** exclusively [883]. In contrast, free dimethylstannylene, Me_2Sn:, does not react with alkenes or dienes because of its too rapid polymerization into $(Me_2Sn)_n$. This is not the case, although, with $SnCl_2$, $SnBr_2$, SnI_2, and $[(Me_3Si)_2CH]_2Sn$: that undergo $[\pi^4s+\omega^2s]$-cheletropic additions with 1,3-dienes [884].

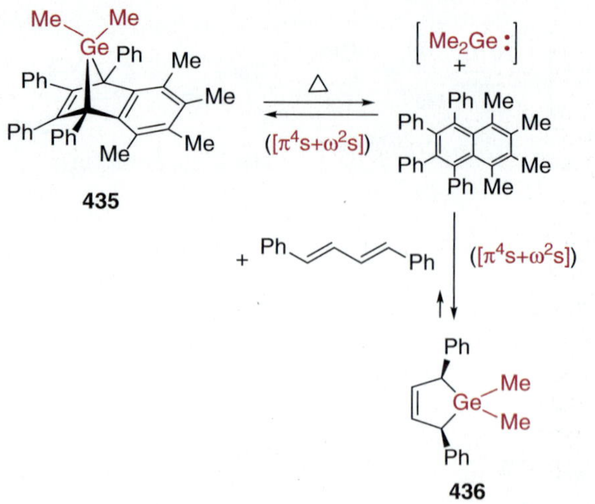

The germylenes **438** have been generated by flash vacuum pyrolysis of the corresponding germacyclopent-3-enes **437** [885]. The first isolable diaminogermylenes **439**, diaminostannylenes **440**, and diaminoplumbylenes **441** have been prepared by Harris and Lappert in 1974 by reaction of $GeCl_2$, $SnCl_2$, and $PbCl_2$, respectively, with $LiNR_2{\cdot}OEt_2$ [886].

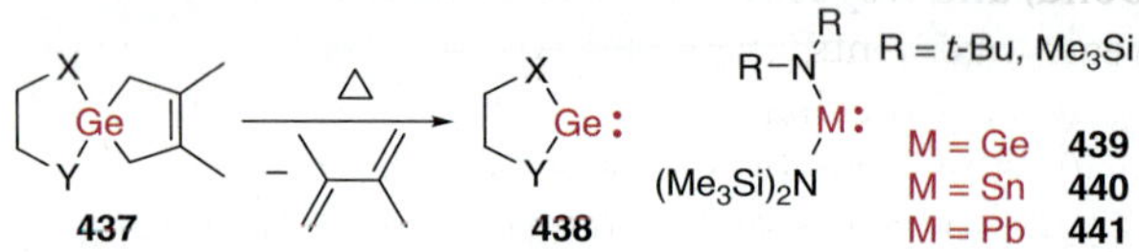

In 1976, the same group has shown that $[(Me_3Si)_2CH]_2M$: (M = Ge, Sn, Pb) are monomeric in solution, but dimeric in the solid state [887]. For equilibrium $[(Me_3Si)_2CH]_2Sn{=}Sn[CH(Me_3Si)_2]_2 \rightleftarrows 2\ [(Me_3Si)_2CH]_2Sn$: $\Delta_rH° = 11.2 \pm 0.3\ \text{kcal mol}^{-1}$ and $\Delta_rS° = 28 \pm 1.3$ eu have been measured [773].

Phosphinidenes , R–P, are the phosphorous analogs of nitrenes. They are extremely reactive intermediates that have been detected only in the gas phase by mass spectrometry and in glassy and cryogenic matrices by electron spin resonance (ESR), IR, and UV spectroscopy [888, 889]. Photolysis of diazide Mes*P$(N_3)_2$ gives phosphaindane **442** that results from an intramolecular C—H insertion by the intermediate phosphinidene Mes*P. The same product **442** is obtained by photolysis of Mes*P=C=O and Mes*P=PMes* [890]. UV irradiation of mesitylphosphinacyclopropane **443** in the presence of hex-3-yne generates ethylene and the product of (2+1)-cheletropic addition **444** (Scheme 5.76). The reaction involves the formation of mesitylphosphinidene. Photolysis of 3-methyl-1-phenylphosphinacyclopent-3-ene (**445**) in the presence of 2,3-dimethylbutadiene generates products **446** and **447**. Phenylphosphinidene is formed as intermediate. It reacts with the diene in a (4+1)-cheletropic addition. Competetively, its dimerization gives Ph—P=P—Ph that undergoes a hetero-Diels–Alder reaction with the diene giving cycloadduct **447** [770].

Terminal transition-metal-complexed phosphinidenes, $L_nM{=}P{—}R$, are the phosphorous analogs of Fischer-type transition metal complex (transition metal complexes of carbenes, Section 7.7). They are valuable synthons [891, 892]. In 1982, Mathey and coworkers [893, 894] reported a first transient electrophilic species $[(OC)_5W{=}P{—}Ph]$, and, in 1987, Lappert and coworkers prepared the first isolable nucleophilic phosphinidene complex $[Cp_2W{=}PMes^*]$ (Cp,

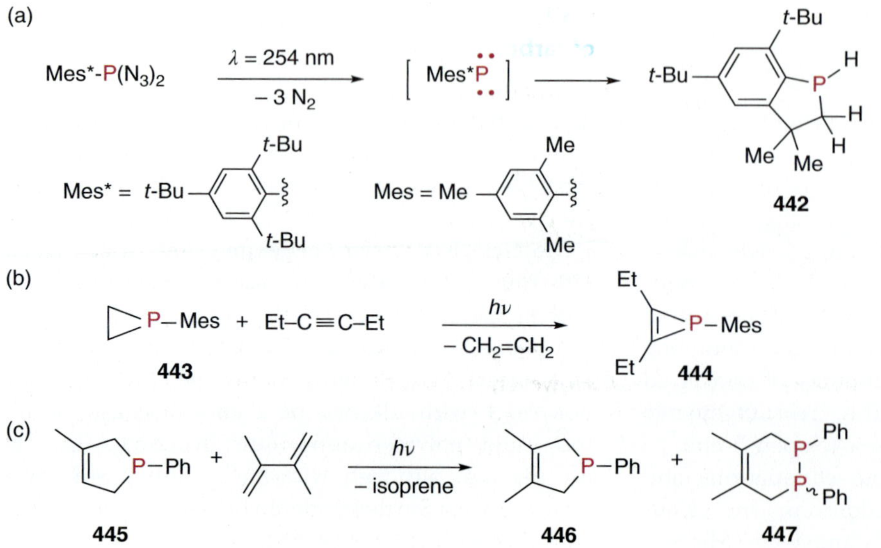

Scheme 5.76 Examples of reactions involving phosphinidene intermediates that can undergo (a) C—H insertions, (b) (2+1)-cheletropic additions, and (c) (4+1)-cheletropic additions.

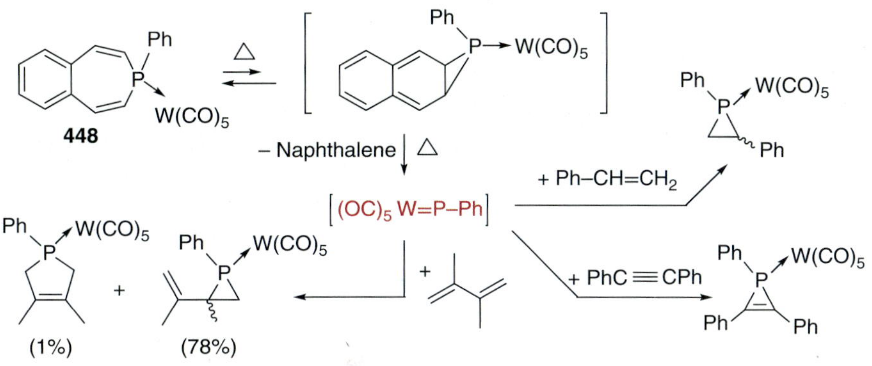

Scheme 5.77 Example of generation of a phosphinidene complex and its cheletropic reactions.

cyclopentadienyl; Mes*, 2,4,6-tris(tertiobutyl)phenyl) [895, 896]. Lammertsma and coworkers [897] have generated $[(OC)_5W{=}P{-}Ph]$ by heating the benzophosphepine complex **448** at 75–80 °C. It undergoes (2+1)-cheletropic additions with alkenes and alkynes, and with 2,3-dimethylbutadiene (Scheme 5.77). It does not react with arenes such as toluene, but with azulenes, 1,6-methano[10]annulene and strained benzenes [770].

Problem 5.46 What are the products of reaction of $[(Me_3Si)_2CH]_2Sn$: with conjugated enals, enones, and 1,2-diketones? [898]

Problem 5.47 What are the products **P** and **Q** of the following reactions? [884]

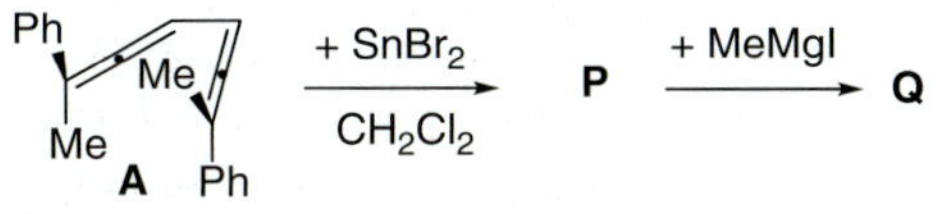

Problem 5.48 What is the product of reaction of $[(Me_3Si)_2CH]_2Sn$: with $Fe_2(CO)_9$? With MeI? [773]

Problem 5.49 On heating the germacyclopropene **A**, a blue solution forms. The blue color fades on adding of 2,3-dimethylbutadiene to this solution. Explain.

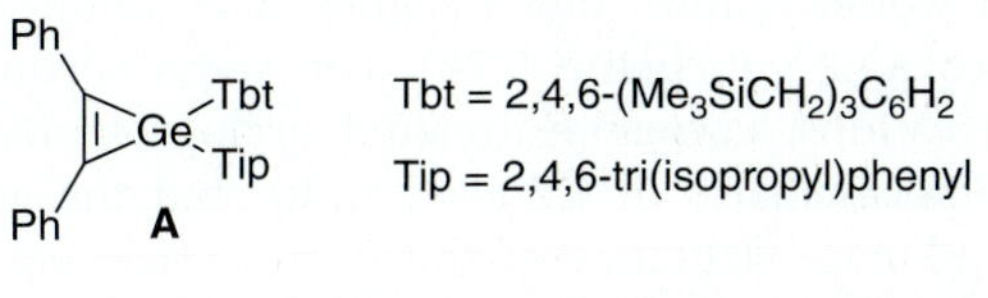

Tbt = 2,4,6-$(Me_3SiCH_2)_3C_6H_2$

Tip = 2,4,6-tri(isopropyl)phenyl

Problem 5.50 In toluene, $[(OC)_5W{=}P{-}Me]$ reacts with 1,6-methano[10]annulene giving a $W(CO)_5$ complex of 3,8-methano-1-methylphosphacycloundecapentaene. Give a mechanism for this reaction [899].

5.5 Thermal sigmatropic rearrangements

A sigmatropic rearrangement is a molecular transformation involving the breaking of a σ-bond and the formation of a new σ-bond between two atomic centers not linked initially. Generally, there is a relocation of π-bonds in the molecule concerned, but the total of number of σ- and π-bonds does not change between reactant and product. The name sigmatropic derives from the Greek *sigma* for single bond, and *tropos* meaning turn. Thermal sigmatropic rearrangements may or may not follow one-step concerted mechanisms. As for other pericyclic reactions, they can be catalyzed by Brønsted or Lewis acids, or by transition metal complexes. The most well-known sigmatropic rearrangements are the Wagner–Meerweein rearrangement of carbenium ions, the Claisen rearrangement and its numerous variants, the Cope rearrangement and its variants, and the Fischer indole synthesis.

5.5.1 (1,2)-Sigmatropic rearrangement of carbenium ions

One of the first sigmatropic rearrangement is the pinacol rearrangement (5.36) discovered by Fittig [900]. The reaction transforms pinacol into pinacolone (the correct structure of which has been established by Butlerow [901]) under acidic conditions. When using ^{18}O-labeled H_2O, ^{18}O-labeled pinacol forms concurrently with pinacolone, demonstrating that reversible heterolysis into a carbocationic intermediate **449** precedes the methyl group migration [902]. However, quantum mechanical calculations have suggested that concerted alkyl migration and C—O heterolysis can compete with the two-step mechanism in which ionization of the C—O bond giving the carbenium precedes the alkyl group migration. This is indeed the case when secondary–secondary 1,2-diols are converted into siladioxolane intermediates and treated with strong and bulky Lewis acid catalysts such as $(C_6F_5)_3B$ [903].

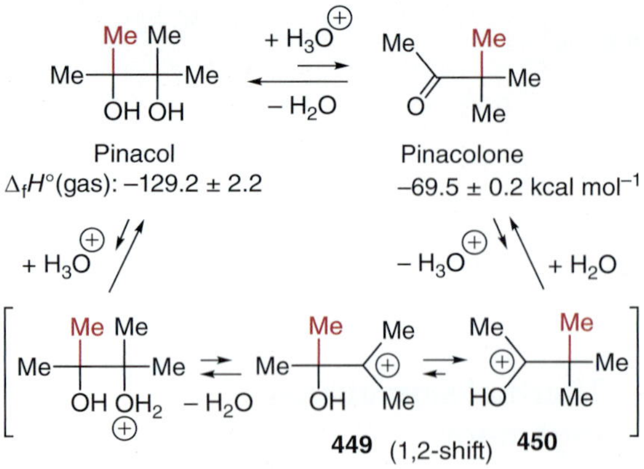

$\Delta_rH°$(5.36, gas) = 1.92 ± 2.4 kcal mol^{-1}
$\Delta_rS°$(5.36, gas) = 35 eu (estimate)
$\Delta_fH°$(H_2O, gas) = −57.8 kcal mol^{-1}

(5.36)

In 1932, Whitemore proposed the theory of "1,2-shifts" that includes the pinacol/pinacolone rearrangement, the Beckmann rearrangement, the Hofmann degradation of amides into amines, and a number of other similar rearrangements [904]. These rearrangements are intramolecular in nature, which has been proven by the nonobservation of "crossed products" when carrying out the reaction of nonlabeled together with labeled reactants. The migrating group does not jump from one molecule to another molecule. In addition, retention of configuration of the migrating center and inversion of the center which it moves to has been demonstrated in some cases employing optically active (enantiomerically enriched), chiral materials. For instance, (4*R*)-2,3,4-trimethylhexane-2,3-diol (**451**) is rearranged into (+)-3,3,4-trimethylhexan-2-one (**453**), indicating that the chiral migrating (*S*)-*s*-butyl group undergoes 1,2-migration with complete retention of configuration consistently with an intramolecular process involving the formation of intermediate or transition structure **452** [905].

The key step of the pinacol/pinacolone rearrangement is the Wagner–Meerwein rearrangement of carbenium ion intermediates. Hoffmann and Woodward classify these rearrangements as (1,2)-sigmatropic rearrangements noted [π^2s+ω^0s] [20]. Contrary to common use, we do not use [1,2]-sigmatropic shift but (1,2)-sigmatropic shift to account for the atomic centers concerned, not the electrons involved in the process. The migrating group is concerned by one atomic center and it migrates over a molecular fragment composed of two contiguous atomic centers.

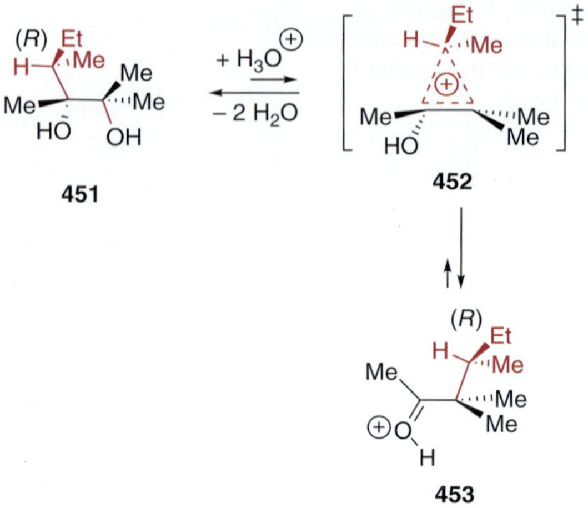

In the original Fittig pinacol rearrangement, a tertiary carbenium ion **449** is formed, which is isomerized into a more stable tertiary hydroxycarbenium ion **450**, the conjugate acid of pinacolone (Table 1.A.14 gives for the gas phase $DH°(MeCH_2^+/H^-) = 270$ kcal mol^{-1} and $DH°(HOCH_2^+/H^-) = 254$ kcal mol^{-1}). The activation enthalpy ($\Delta^‡H$) for the methyl group migration **449** → **450** is very low. Carbocations can be prepared as stable salts in SO_2ClF solution with $Sb_2F_{11}^-$ or $Sb_2F_{10}Cl^-$ counter-ions. Degenerate Wagner–Meerwein rearrangements of acyclic alkyl cations are very fast, have very low activation enthalpies. For instance, the degenerate isomerization (5.37) that equilibrates carbenium ions **454** and **454′** $\Delta^‡H = 3.5 \pm 0.1$ kcal mol^{-1} has been measured at −138 °C [906]. Thermochemical data can help us to limit the number of possible mechanisms for any reactions. As we shall see, the most plausible mechanism for **454** ⇄ **454′** is the formation of corner-protonated cyclopropane **455**, which is an intermediate able to exchange all the hydrogen atoms of the five methyl groups [907]. The very low activation enthalpy measured for **454** ⇄ **455** ⇄ **454′** confirms that the Me—C(3) bond breaking process is assisted by the Me—C(2) bond forming process. We are in the presence of an intramolecular, associative mechanism. As we shall see, any nonconcerted dissociative mechanism would require much higher $\Delta^‡H$ values than 3.5 kcal mol^{-1} (Scheme 5.78). For some secondary alkyl cations, corner-protonated cyclopropanes are not intermediates of 1,2-alkyl shifts, but the actual ground-state structure of the cations (see, e.g. the 2-norbornyl cation). The gas phase dissociation of **454** into methyl cation and 2,3-dimethylbut-2-ene (**456**) would require $\Delta^‡H \approx 100$ kcal mol^{-1} [908]. Another dissociative mechanism involving homolysis of **454** into methyl radical and radical cation **457** (2,3-dimethylbuta-2,3-diyl radical-cation) would

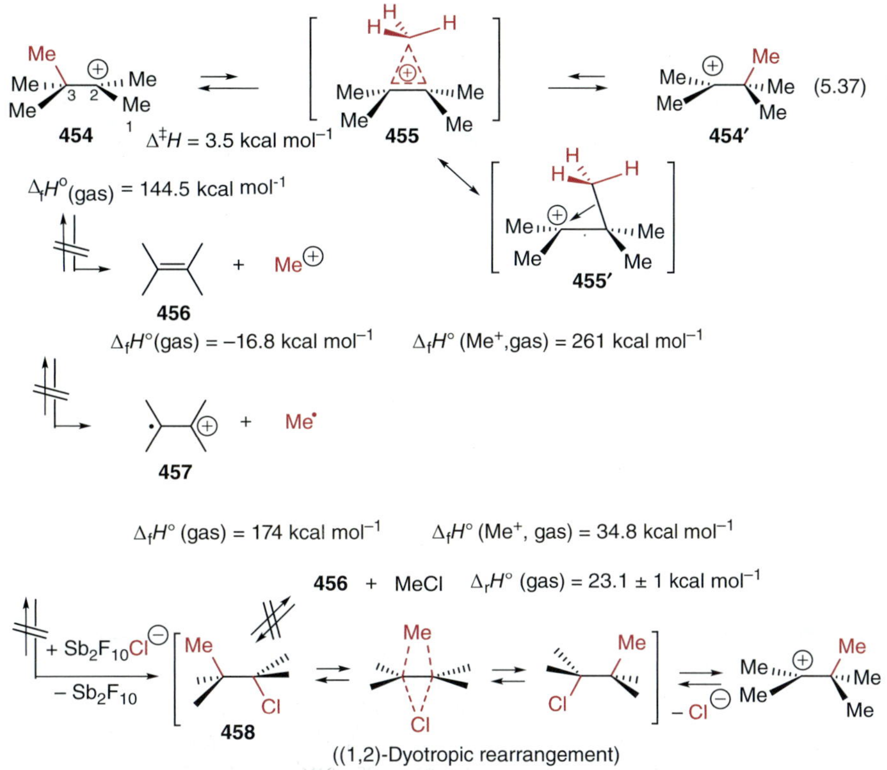

Scheme 5.78 Degenerate Wagner–Meerwein ((1,2)-sigmatropic) rearrangement of 2,3,3-trimethylbut-2-yl cation. Thermochemical data taken from the NIST Chemistry WebBook. $\Delta_f H°(\mathbf{458}) = -59.9\,\text{kcal mol}^{-1}$ is estimated from (t-BuCl) $= -43.0\,\text{kcal mol}^{-1}$ and by incrementing this values by $-16.9\,\text{kcal mol}^{-1} = \Delta_f H°(\text{Me}_3\text{C—CHMe}_2) - \Delta_f H°(\text{Me}_3\text{CH})$; $\Delta_f H°(\text{MeCl}) = -20.0\,\text{kcal mol}^{-1}$

require $\Delta^{\ddagger}H \approx 64\,\text{kcal mol}^{-1}$. In solution, **454** might equilibrate with the corresponding chloride **458** through reaction with the counterion $[\text{Sb}_2\text{F}_{10}\text{Cl}]^-$. If the methyl group migration **454** ⇄ **454′** should involve the reversible fragmentation of **458** into alkene **456** + MeCl, $\Delta^{\ddagger}H \approx 23\,\text{kcal mol}^{-1}$ is required. Entropy of fragmentation (Sections 1.3.1 and 2.9) that amounts to $\Delta_r S \approx 35$ eu will make $\Delta_r G$(**458** ⇄ **456** + MeCl) = $23.1 - 135\,\text{K}(0.035\,\text{kcal mol}^{-1}\,\text{K}^{-1}) \approx 18\,\text{kcal mol}^{-1}$, still much higher than the energy barrier observed for reaction (5.37). The latter mechanism would correspond to mechanism **454** ⇄ **456** + Me^+ ⇄ **454′** assisted by the counterion. The three dissociative mechanisms considered above would exchange a methyl group between two cations (intermolecular process). This is not observed. Furthermore, the measured activation enthalpy $\Delta^{\ddagger}H(5.37)$ does not vary with the concentration of the carbenium salt. In theory, chloride **458** could undergo a 1,2-dyotropic rearrangement (Section 5.6.1) in which the migration of the chloro and methyl groups would occur simultaneously, in concert. In the case of neopentyl chloride, such a process would require an energy barrier above $60\,\text{kcal mol}^{-1}$.

Intermediate **455** can be represented as a π-complex **455′** (Figure 5.43) of a methyl cation with 2,3-dimethylbut-2-ene (**456**). Its relatively high stability arises from a strongly stabilizing HOMO(alkene)/LUMO(Me^+) interaction (Dewar π-complex theory, Section 7.7.1) [909]. When going from carbenium ion **454** to π-complex **455** (an example of corner-protonated cyclopropane), the bond angle Me—C(3)—C(2) varies from 110° to approximately 60°. The strain increase associated with this deformation is compensated by the increasing stabilization energy gained because of enhanced overlap between the σ(Me—C(3))-filled orbital and the vacant $2\text{p}(\text{C}^+)$ orbital of **454**. An optimum is reached in the C_s-symmetrical intermediate **455**. For other carbenium ions, this might not be the case and partially bridged ions might be preferred. The nonsymmetrical species that form when **454** is converted to the corner-protonated cyclopropane **455** can be seen as a succession of intramolecular σ-complexes resulting from the interaction of the σ(C—C) migrating methyl group and the carbenium ion center (stabilization arising from the HOMO(C—C)/LUMO(p^+) interaction).

During the (1,2)-alkyl shift of the methyl group **454** → **455**, the σ(Me—C(3)) bond is never broken. The bending of this bond is accompanied by the formation of the σ(Me—C(2)) bond of **454′**. Along this process, the configuration of the methyl group, as shown using the chiral deuterotritiomethyl group

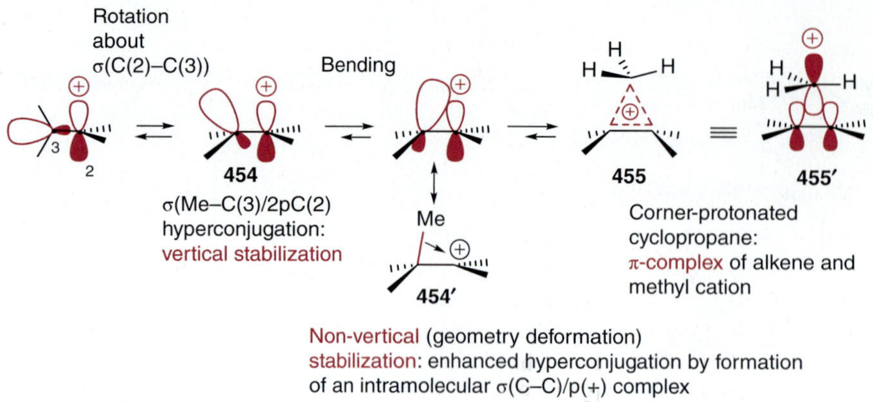

Figure 5.43 Representation of the stabilization gained on moving Me—C(3) toward C(2) in carbenium ion **454** and the nature of bonding in the corner-protonated cyclopropane **455** (Dewar complex of Me^+ with alkene **456**), which compensates for the bond angle deformation.

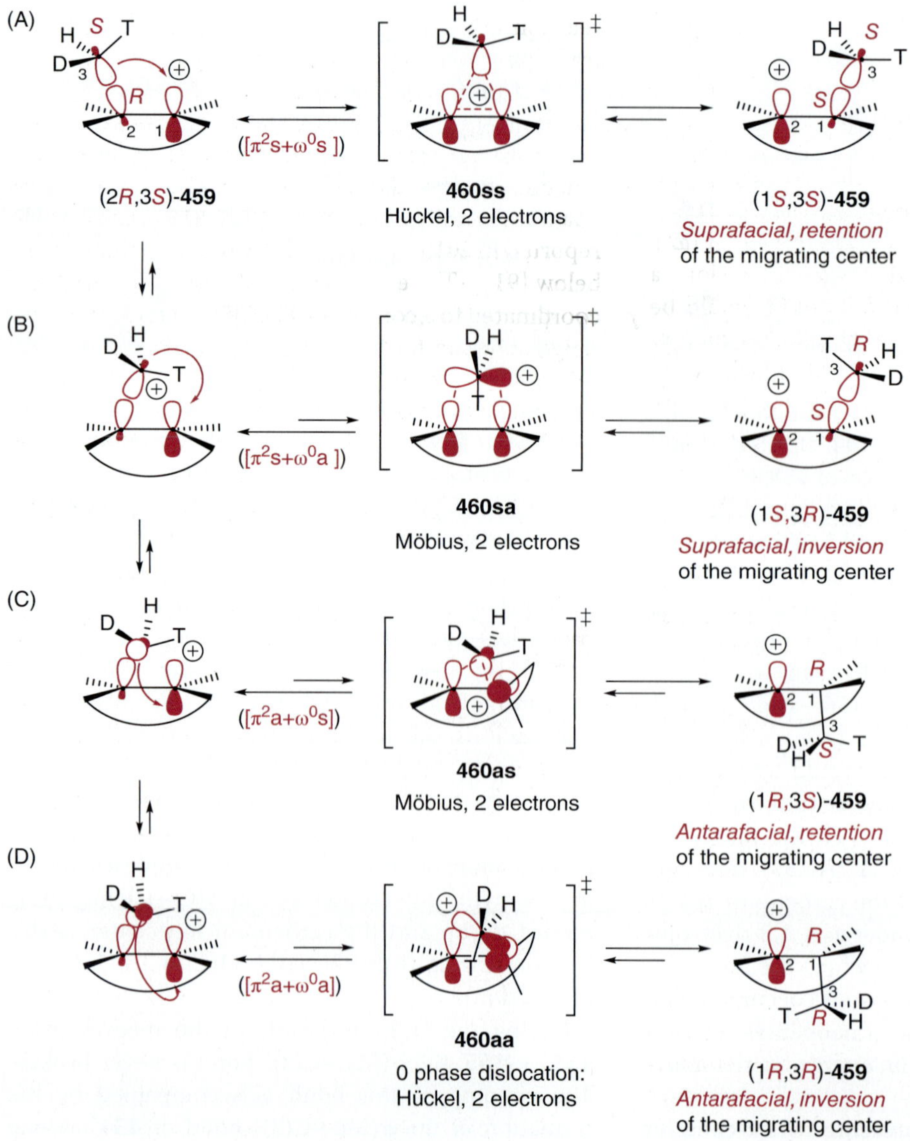

Figure 5.44 Representation of four pathways for the (1,2)-sigmatropic rearrangement of a 1,2,2-trimethylcycloalkyl cation in which the migrating group R is a chiral deuterotritiomethyl group. Only pathway (A) is observed (suprafacial with retention of the configuration of the migrating center). The other pathways (B) and (C) are "forbidden" for stereoelectronic reasons. Pathway (D) is "allowed" but difficult as it requires severe skeleton deformations.

((*S*)-C(T)(D)H [910], Figure 5.44) is maintained while it circulates from C(3) to C(2). Thus, cation (2*R*,3*S*)-**459** is isomerized into diastereoisomeric cation (1*S*,3*S*)-**459** (pathway A) in Figure 5.44. This pathway (noted [π^2s+ω^0s]) involves a transition structure with a pyramidal methyl cation contributing with zero electron that retains its initial configuration (s for retention = suprafacial) and that interacts in a suprafacial manner with a π-system contributing with two electrons (s for suprafacial). At the limit, this reaction pathway could be noted [ω^2s+π^0s] and would represent a transition structure made of methide anion and a π-system doubly positively charged and contributing with zero electrons. Woodward and Hoffmann call the Wagner–Meerwein rearrangement (2*R*,3*S*)-**459** ⇄ (1*S*,3*S*)-**459** a suprafacial sigmatropic shift of order (1,2) with retention of the migrating center.

In theory, one can envision three more reaction pathways for the Wagner–Meerwein rearrangements. For acyclic and cyclic carbenium ions, the suprafacial (1,2)-shift could occur with inversion of configuration of the migrating center, as shown in pathway (B) of Figure 5.44, and which can be noted [π^2s+ω^0a] (ω^0a means an alkyl cation migrating with inversion, a for antarafacial). In this case, (2*R*,3*S*)-**459** would be isomerized into (1*S*,3*R*)-**459**. The transition structure of this process is a π-complex **460as** in which no LUMO(2p(Me^+))/HOMO(alkene) is possible. Thus, transition structure **460sa** is much less stable than **460ss**. Reaction (2*R*,3*S*)-**459** → **460sa** → (1*S*,3*R*)-**459** is said "forbidden" for stereoelectronic reasons. It is also difficult because of the strain resulting from the C—H moieties of the migrating group that are forced to point toward the alkene fragment. In a third pathway (C) that would isomerize (2*R*,3*S*)-**459** into (1*R*,3*S*)-**459** via transition structure **460as**, the migrating methyl group crosses the π-plane of the π-system and maintains it original (*S*)-configuration. This rearrangement noted [π^2a+ω^0s] is an antarafacial sigmatropic (1,2)-shift with retention at the migrating center. It is difficult for stereoelectronic reasons as transition structure **460as** implies zero LUMO(Me^+)/HOMO(alkene) overlap and because of the high strain increase arising from skeleton distortions. A fourth pathway (D) would equilibrate (2*R*,3*S*)-**459** with (1*R*,3*R*)-**459**. This is an antarafacial (1,2)-sigmatropic shift with inversion of configuration of the migrating center, noted [π^2a+ω^0a]. It is also very difficult because of the skeleton deformations it requires. In theory, the latter rearrangement is not "forbidden" for stereoelectronic reasons as its transition structure **460aa** permits a nonzero LUMO(Me^+)/HOMO(alkene) overlap.

In the case of the (1,2)-migration of a hydrogen atom, only the suprafacial [π^2s+ω^0s] path is possible. Related to the pinacol rearrangement, the intramolecular Cannizzaro reaction of α-ketoaldehydes to α-hydroxy carboxylic acids (ArCO—CHO ⇄ ArCH(OH)CHHOH) can be seen as a (1,2)-sigmatropic rearrangement with a hydrogen migrating group (often said a (1,2)-hydride migration: in the intermolecular Cannizzaro aldehyde disproportionation (2ArCHO + NaOH ⇄ ArCH(OH)ONa + ArCHO ⇄ ArCOONa + $ArCH_2OH$) a hydride transfer is implied). Classically, the Cannizzaro reaction [911, 912] requires strong bases and high temperature. Alternatively, $Et_3N + MgBr_2{\cdot}Et_2O$ [913] or organobases [914, 915] can be used as catalysts and the reaction works under much softer reaction conditions. In the presence of enantiomerically pure β-thioamines, enantioselective (ee up to 7.7%) intramolecular Cannizzaro reaction have been observed by Franzen [916]. Better enantioselectivity has been reported for intramolecular Cannizzaro reaction catalyzed by Lewis acids coordinated to enantiomerically pure ligands [917, 918]. An example reported in 2013 by Tang and coworkers is shown here below [919]. The enantiopure catalysts is a Cu(II) salt coordinated to a congested TOX [55] ligand. The rearrangement can be seen as a concerted (1,2)-hydride shift with a transition state that can be represented as a π-complex between an alkene and a proton.

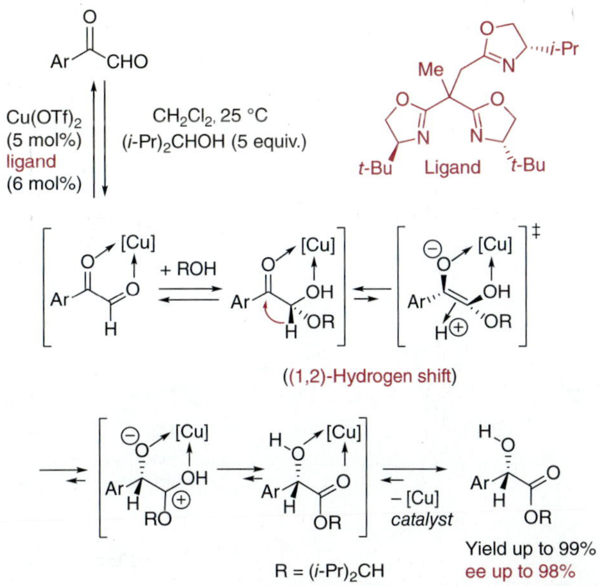

Problem 5.51 The (1,2)-migrations of hydrogen in acyclic carbenium ions are very fast. Give an explanation.

5.5.2 (1,2)-Sigmatropic rearrangements of radicals

The first (1,2)-sigmatropic rearrangement in radicals has been reported in 1911 by Wieland with the conversion of bis(triphenylmethyl)peroxide into 1,2-diphenyloxytetraphenylethane (Scheme 5.79) [920]. The reaction involves homolysis of the peroxide with formation of triphenylmethyloxy radical that undergoes (1,2)-shift of a phenyl group, probably via a cyclohexadienyl radical intermediate (not detected by ESR). This generates diphenylphenyloxymethyl radical that dimerizes into the product of reaction.

Rearrangements of carbon-centered radicals via (1,2)-shift of hydrogen or alkyl group are difficult reactions, much more so than the rearrangement of corresponding carbenium ions. They might occur at high temperature, most probably through non-concerted processes involving β-elimination of $H^\bullet$ or $R^\bullet$ radicals and formation of alkenes, followed by addition (most reactions in solution occur within the solvent cage). Berson and coworkers found that the decomposition of 2-azabornane (**461**) in Ph_2O and *n*-decane (solvent) above 250 °C gives a mixture (Scheme 5.80) containing 2,3,3-trimethylnorbornane (isocamphane) [921]. This product results from a formal Wagner–Meerwein rearrangement ((1,2)-shift of CH_2(6) in radical intermediate **462** to give **464**; subsequent hydrogen atom abstraction from the solvent or from **461** and the products formed gives isocamphane).

A rate constant $k = 5 \times 10^5\ s^{-1}$ at 278 °C (560 K) was measured for the rearrangement **462** → **464**. Assuming a monomolecular process for which the entropy of activation is nil (Arrhenius pre-exponential factor $A = 10^{13}\ s^{-1}$), one obtains $E_a = 18.7$ kcal mol^{-1} ($k = A\ \exp(-E_a/RT)$; $E_a = RT(\ln A - \ln k)$) and $\Delta^{\ddagger}H = 17.6$ kcal mol^{-1} ($\Delta^{\ddagger}H = E_a - RT$) (Section 3.3). This is consistent with mechanism **462** → **463** → **464** (fragmentation/addition) as demonstrated here-below.

A model reaction for **462** → **463** is fragmentation (5.38) of 3,3-dimethylpent-2-yl radical into ethyl radical and 2-methylbut-2-ene. The standard heat of formation $\Delta_f H°(EtC(Me)_2{-}C^\bullet(Me)H)$ is estimated from $\Delta_f H°(Et{-}C(Me)_2{-}Et)$ and $DH°(EtC(Me)_2C^{(\bullet)}(Me)H/H^\bullet)$, the latter value being considered equal to $DH°(2\text{-}C_5H_{11}{}^\bullet/H^\bullet) = 99.2$ kcal mol^{-1}. This gives $\Delta_r H°(EtC(Me)_2C^{(\bullet)}(Me)H \rightleftarrows \text{2-methylbut-2-ene} + Et^\bullet) = 19.6 \pm 1.7$ kcal mol^{-1}. In the case of **462** ⇄ **463**, ring strain release between the bicyclic radical (ring strain of norbornane: 16.2 kcal mol^{-1}) and the monocyclic radical (ring strain of cyclopentene: 5.6 kcal mol^{-1}, Table 2.3) leads to $\Delta_r H$(**462** ⇄ **463**) = 19.6 − 10 = 9.6 kcal mol^{-1}, a value significantly smaller than $\Delta^{\ddagger}H = 17.6$ kcal mol^{-1} measured for this process.

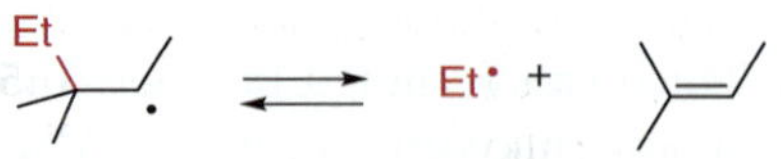

(5.38)

$\Delta_f H°$: −1.1 ± 1   28.4 ± 0.5   −9.9 ± 0.2 kcal mol^{-1}

$\Delta_r H°$ (5.38) = 19.6 ± 1.7 kcal mol^{-1}

The FMO theory (Section 4.4.2) predicts that a concerted suprafacial (1,2)-methyl sigmatropic rearrangement with retention of the migrating center (noted [π^2s+ω^1s], Figure 5.45a) in an alkyl radical involves a transition structure of type **465ss** that is not stabilized as much as a π-complex between an alkene

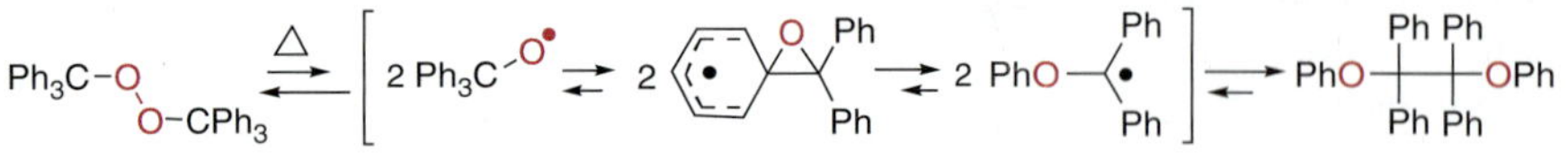

Scheme 5.79 The Wieland's (1,2)-rearrangement of triphenyloxy radical.

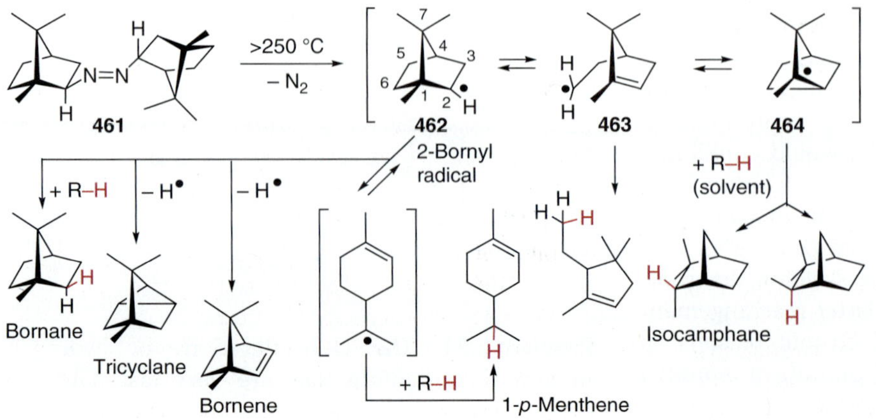

Scheme 5.80 Reactions of 2-bornyl radical in *n*-decane generated above 250 °C.

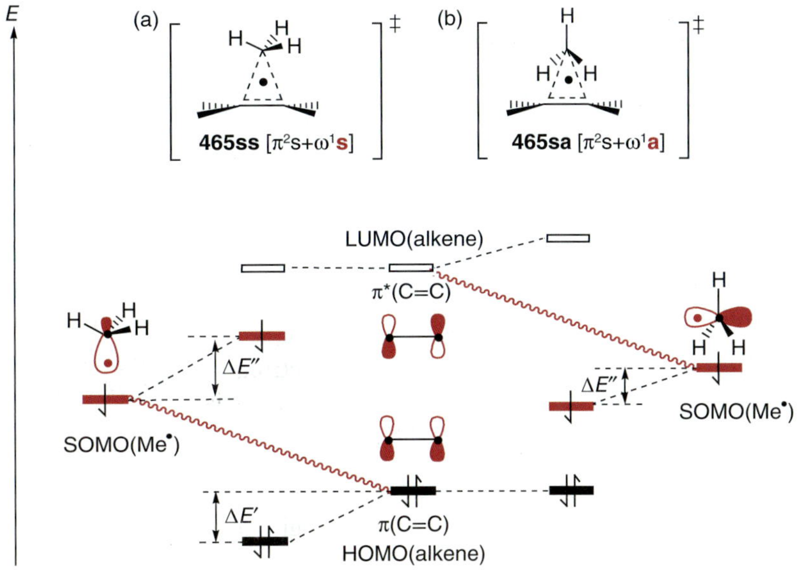

Figure 5.45 FMO diagram representing the transition structure **465** of one-step, concerted (1,2)-methyl shift in alkyl radicals following (a) a suprafacial rearrangement with retention of the migrating center and (b) a suprafacial rearrangement with inversion of configuration of the migrating center. The same PMO diagrams can be used to discuss the transition structures of (1,2)-alkyl shifts in alkyl carbanions (Section 5.5.3).

and a carbenium ion (a corner-protonated cyclopropane), intermediate in the Wagner–Meerwein rearrangement of the corresponding alkyl cation (see e.g. **455**, **460ss**). This arises from the fact that **465ss**, which is a complex of an alkyl radical with an alkene, implies a three-electron/two-orbital interaction rather than a two-electron/two-orbital interaction. Overlap between the radical singly occupied molecular orbital (SOMO) (a 2p or s^xp^y AO) and the HOMO(alkene) leads to MOs φ_1 and φ_2. Although φ_1 is stabilized by $\Delta E'$ with respect to HOMO(alkene), φ_2 is destabilized by $\Delta E''$ with respect to SOMO(Me). As $|\Delta E''| > |\Delta E'|$ (see Section 4.4.2), the stabilization of **465ss** with respect to Me• + alkene, which amounts to $2|\Delta E'| - |\Delta E''|$, is relatively small. The concerted (1,2)-methyl shift implies bond angle deformations responsible for the strain increase between transition structure **465ss** and the starting alkyl radical. As the electronic stabilization of **465ss** is too weak, the (1,2)-methyl shift follows a nonconcerted, elimination/addition mechanism.

For the hypothetical suprafacial (1,2)-shift with inversion of the migrating center (noted $[\pi^2s+\omega^1a]$, Figure 5.45b), one can foresee some stabilization of transition structure **465sa** due to the SOMO(Me)/LUMO(alkene) interaction. As the energy gap between these orbitals is relatively large, and because of the strain of the three-membered ring that forms, stereoselective radical rearrangement of this type has not been reported yet. Note that the strain is particularly important in **465sa** in which at least one of the C—H bond of the migrating group points toward the alkene moiety. Note also that π-complexes of type **465ss** and **465sa** force the three carbon centers of the three-membered ring to share more than eight electrons each. This is not the case in the Wagner–Meerwein rearrangement.

As a general rule, (1,2)-sigmatropic shifts (5.39) of hydrogen and of groups made out of atoms of the second row of the periodic table are difficult [922]. They are much easier for migrating groups made out of atoms of the third row of the periodic table [923]. If one considers the (1,2)-shift (5.40) of chlorine in β-chloroalkyl radicals, the transition structures for one-step, concerted suprafacial rearrangements are of type **466** (Figure 5.46). They imply weakly stabilizing SOMO(Cl•)/HOMO(alkene) interaction and a stabilizing LUMO(alkene)/3p(Cl•) interaction.

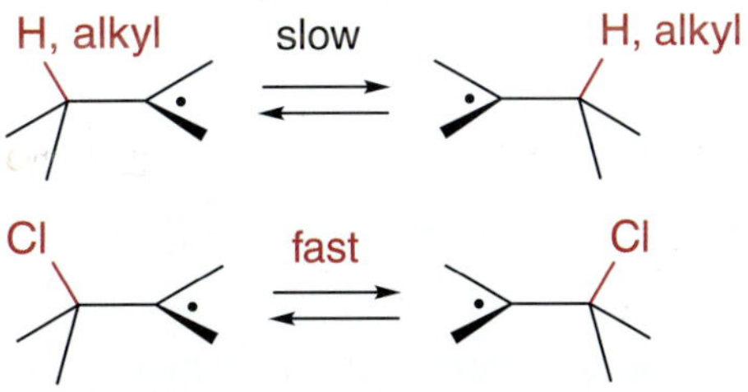

(5.39)

(5.40)

Compared with complex **465ss** ((1,2)-methyl shift of methyl group, Figure 5.45a), **466** has extra-stabilization because of a $3p_x$(Cl•)/LUMO(alkene) interaction. The higher polarizability of Cl• compared with H• (and alkyl•) is responsible for the higher stability of **466** as expressed with diradicaloid model **466** ↔ **466′** ↔ **466″**. The greater relative stabilization of **466** with respect to **465ss** can also be interpreted in terms of the possible electron exchange between Cl• and alkene, on the one hand, and between Me• and alkene, on the other hand. If one choses (*E*)-but-2-ene

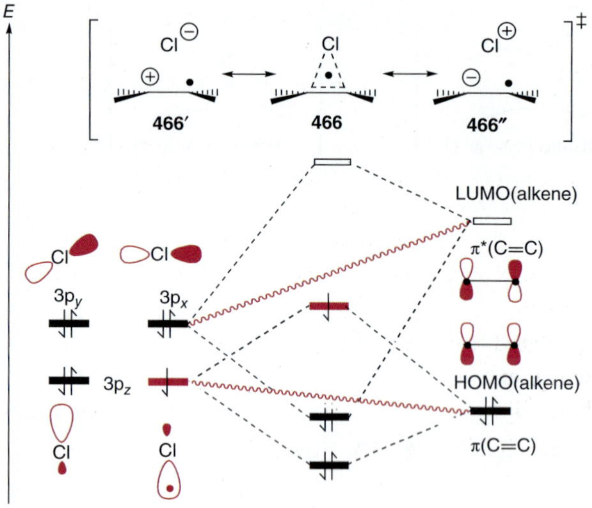

Figure 5.46 PMO diagram representing the π-complex **466** of an alkene with a chlorine radical as the transition structure of a suprafacial (1,2)-chlorine shift in β-chloroalkyl radical.

as alkene, the charge transfer configuration **467′** and **467″** can be envisioned for complexes **467**. Their relative stability is given by EI(alkene) + (−EA(X•)) and EI(X•) + (−EA(alkene)), respectively. One sees that charge transfer configuration of type **467″** is less favorable than that of type **467′** for both methyl and chlorine radicals. In other words, it is easier to take an electron from the alkene and place it onto the migrating X group than to take an electron from X and to attach it to the alkene. Because of the higher electronegativity of chlorine atom with respect to carbon (Table 1.A.5), electron transfer represented by **467** ↔ **467′** is easier for π-complexes of chlorine than for π-complexes of methyl radical [924].

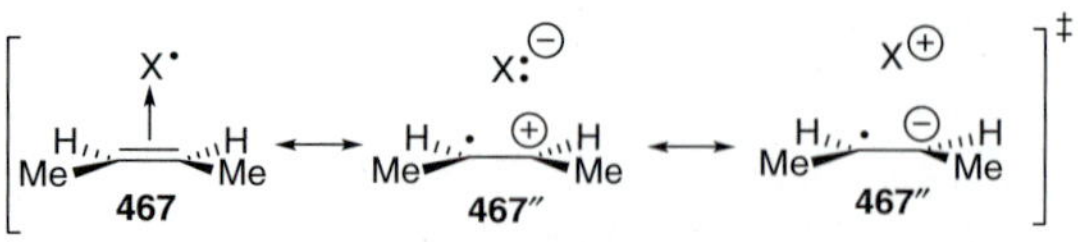

	IE(alkene) + (−EA(X•))	IE(X•) + (−EA(alkene))
X = Me	210.5 − 1.6 = 208.9	227 + 48.4 = 275.4 kcal mol^{-1}
X = Cl	210.5 − 83.4 = 127.1	299 + 48.4 = 347.4 kcal mol^{-1}

If the relative stability of π-complexes **467** would depend only on the relative weight of the charge transfer configurations that they can attain, **467′** would contribute more to the stability of **467** for X = Cl than for X = Me [925]. The (1,2)-sigmatropic rearrangements (5.41) and (5.42) of alkyl radical involving the migration of an alkenyl or aryl groups are facile reactions as they equilibrate quickly with cyclopropylalkyl and cyclohexadienyl radicals, respectively [926]. The latter are transition structures or intermediates of the rearrangements.

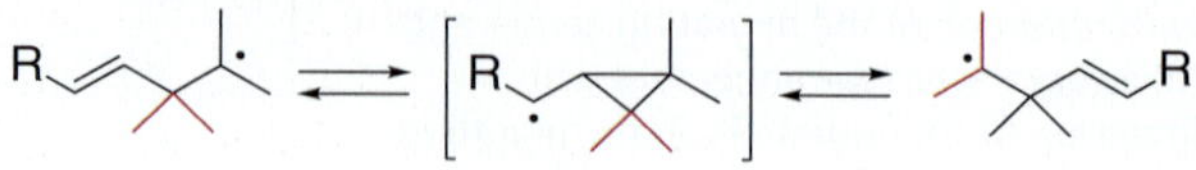

(5.41)

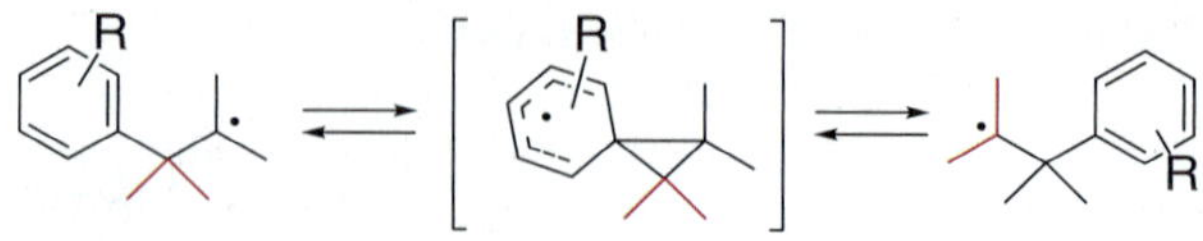

(5.42)

The heat of isomerization of homoallyl radical into cyclopropylmethyl radical amounts to 5.0 ± 2.0 kcal mol^{-1}. For the isomerization of 2-phenylethyl radical into 1,1-ethylidenecyclohexadienyl radical, quantum calculations predict a heat of reaction of c. 10 kcal mol^{-1} [927]. By ESR experiments over the temperature range 283–307 K, Maillard and Ingold [928] determined $\log A = 11.7 \pm 1$ and $E_a = 13.6 \pm 1$ kcal mol^{-1} for the (1,2)-phenyl group migration in the neophyl rearrangement (5.43) ($E_a = 11.8$ kcal mol^{-1} found by Franz et al. [929]).

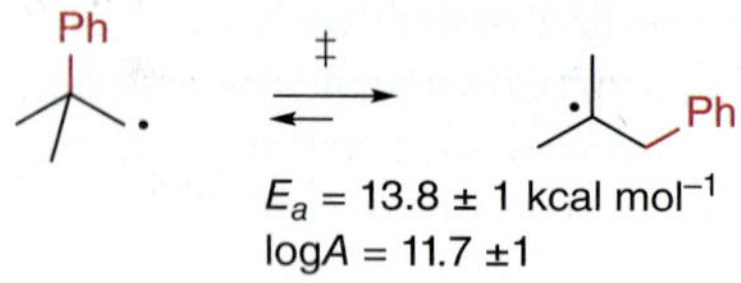

(5.43)

The Surzur–Tanner rearrangement (5.44) of β-(acyloxy)alkyl radicals [930, 931] involves either (1,2)- (noted [ω^1s+π^2s]) or (2,3)-acyloxy shifts (noted [π^3s+π^2s]) for acyclic systems [932–934]. In the case of lactone-derived radicals (configurationally fixed (*E*)-esters), the ring contractions (5.45) imply only (1,2)-acycloxy shifts with rate constants of c. 10^6 s^{-1} (n = 1,2,3) at 80 °C. For the seven-membered lactone (n = 2), $\log A = 11.8 \pm 0.5$ and $E_a = 9.0 \pm 0.8$ kcal mol^{-1} were determined [935].

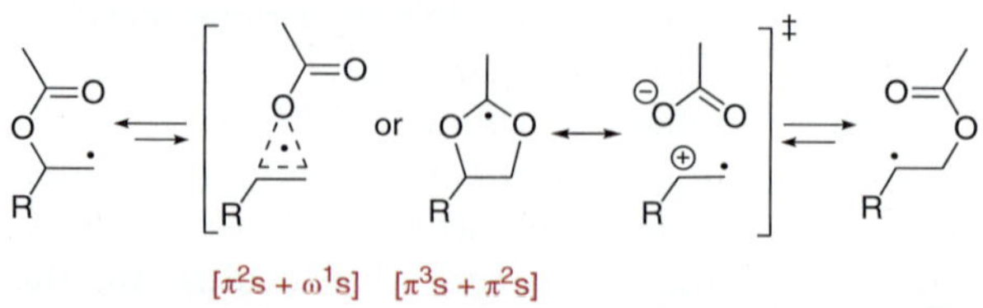

(5.44)

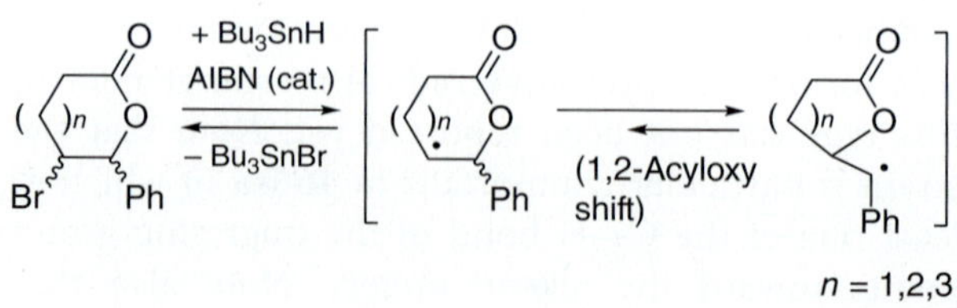

(5.45)

5.5.3 (1,2)-Sigmatropic rearrangements of organoalkali compounds

The FMO theory (Figure 5.45) predicts that the concerted, intramolecular suprafacial (1,2)-alkyl shift in an alkyl anion must be difficult when the alkyl group migrates with retention of configuration (rearrangement noted [π^2s+ω^2s]). An electronic stabilizing interaction LUMO(alkene)/HOMO(anion) is predicted for the (1,2)-alkyl shift with inversion of configuration of the migrating group (rearrangement noted [π^2s+ω^2a]). This rearrangement is prohibited due to strain (gauche effect between C—H bonds of the migrating group and the alkene moiety). For 2,3,3-trimetylbut-2-yl anion (**468**), one estimates its fragmentation (5.46) into 2,3-dimethylbut-2-ene (**456**) and methyl anion to be endothermic by $17 \pm 4\,\text{kcal mol}^{-1}$ in the gas phase. Considering the positive entropy variation of this reaction, it might become possible on heating. Thus, (1,2)-alkyl sigmatropic shifts in carbanions might occur at elevated temperature following a two-step mechanism of elimination and re-addition.

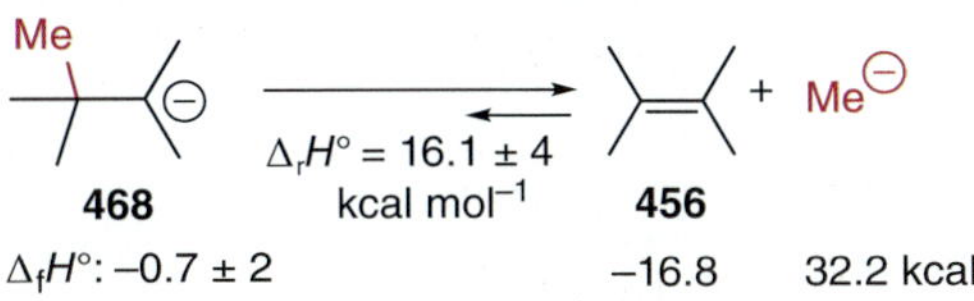

(5.46)

Reaction of chloride **469** with lithium in THF at −75 °C gives alkyllithium **470**. After three hours at 0 °C, 9-methylfluorenyllithium (**471**) forms (65% ± 9%) together with 3% of 9-neopentylfluorenyllithium (**472**). The later compound arises from a formal (1,2)-shift of the *t*-butyl group, which involves either a σ(C—C) bond cleavage with formation of *t*-butyl anion or a *t*-butyl radical (Scheme 5.81), followed by readdition to the resulting 9-methylidenefluorene (**474**) or 9-methylidenefluorene radical anion (**475**), respectively, to give **472**. When the reaction is carried out in the presence of BnLi, benzyl incorporation into the products is not detected, in contrast with a related case of (1,2)-benzyl group migration for which external added isopropyllithium or radioactive benzyllithium has been incorporated in the final products [936]. Thus, intermediates **473** or **474** of reaction **470** → **471** are not “free” but exist in a solvent cage before collapsing. Compound **471** results from a proton transfer from the *t*-Bu group of **475** to the fluorenylmethyl anion giving isobutylene and 9-methyfluorenyl anion [937].

As a rule, (1,2)-alkyl and hydrogen shifts in all carbon alkylalkali compounds are rare and difficult reactions [938, 939]. However, the (1,2)-shits of aryl, alkenyl, or alkynyl groups are facile [940, 941] and involve the formation of relatively stable bridged species **476–478**, respectively, resulting from the intramolecular addition of the carbanion onto the unsaturated migrating moiety. Allyl group migrations occur by intramolecular (2,3)-sigmatropic shifts with allylic inversion, either in a concerted manner, or by intermediacy of organometal **479** [942, 943].

Scheme 5.81 Reactions of 9-*t*-butyl-9-fluorenyllithium in THF at 0 °C.

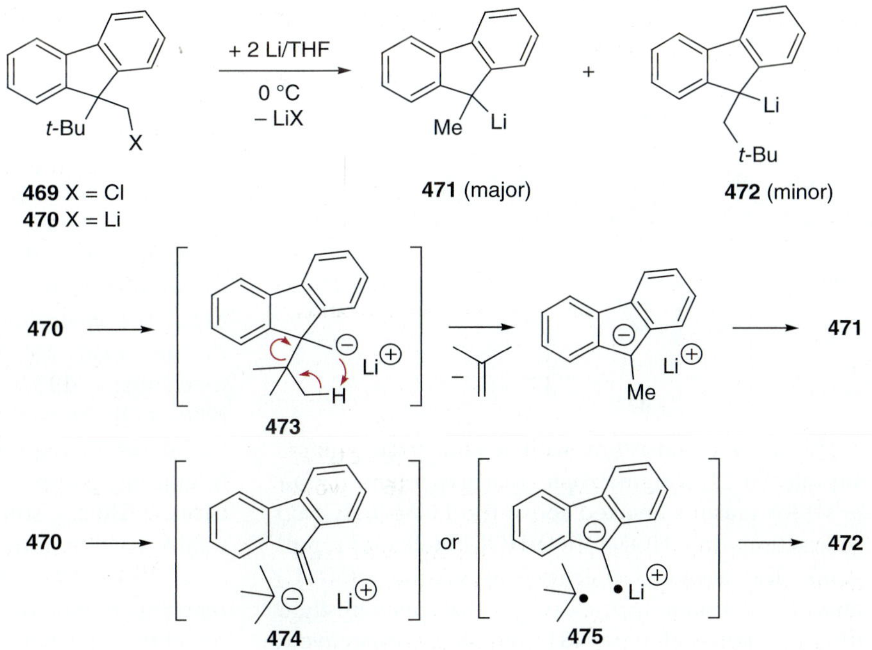

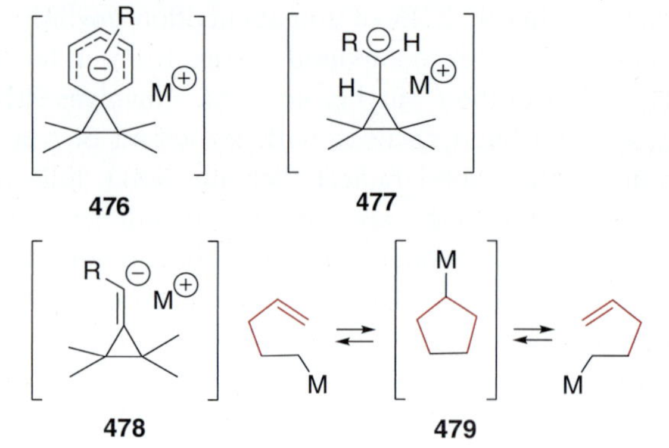

(Intramolecular additions and subsequent eliminations)

(1,2)-Alkyl shifts in carbanions substituted by heteroelements are facile. For instance, the Stevens rearrangements of quaternary ammonium (5.47) [944], sulfonium salts (5.48) [945], and oxonium salts are easy at room temperature [946]. The rearrangements start with the formation of ylides **480** and **483**, respectively, after deprotonation with a base. The alkyl group (R) migrations that follow occur with retention of configuration.

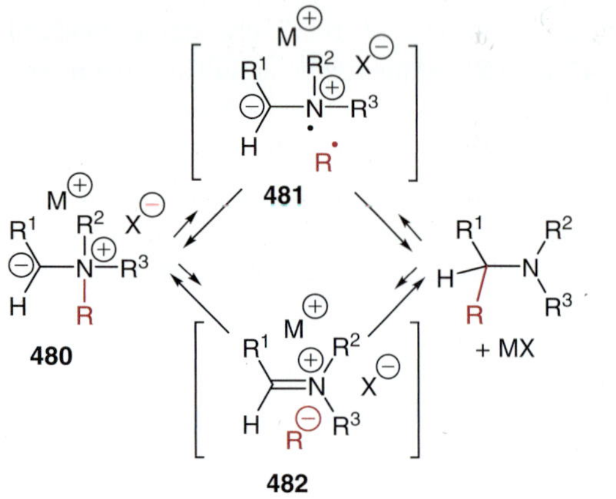

(5.47)

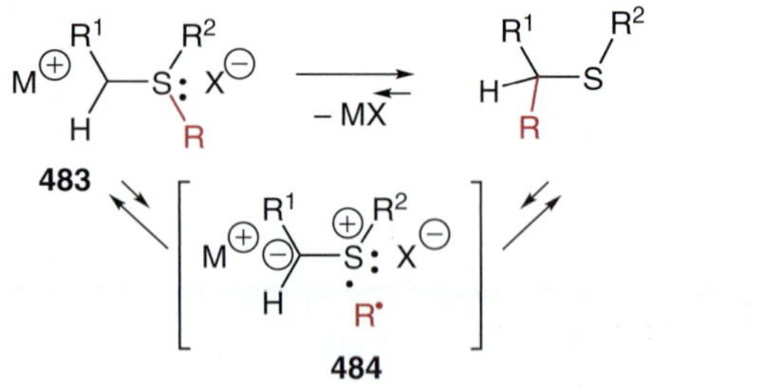

(5.48)

This is not consistent with a concerted anionic suprafacial (1,2)-sigmatropic rearrangement ([π^2s+ω^2s]) for reasons invoked above (no LUMO/HOMO interaction, only HOMO(R^-)/HOMO(alkene) repulsion). The accepted mechanisms involve σ(N—C) and σ(S—C) homolytic cleavages that generate tight diradical pairs of type **481** and **484**, respectively. The retention of configuration of the sp^3-hybridized carbon center of the migrating group is explained by invoking a solvent cage that prohibits the migrating radical $R^\bullet$ to undergo inversion. Another possibility is the β-elimination with formation of an ion pair of type **482**, for instance. Hydrogen/lithium exchange of dibenzylthio ether with BuLi at −78 °C produces **485** that is quenched with MeI giving **486** exclusively. If **485** is left at higher temperature, the subsequent quenching with MeI generates the four products **489–492** (Scheme 5.82). Radical pair **487** leads to the product of Stevens rearrangement **489**, and radical diffusion outside of the solvent cage produces **490** and **491**. Product **492** results from a concurrent Sommelet–Hauser rearrangement, which might involve a concerted (2,3)-sigmatropic rearrangement (see Section 5.5.8) and that implies a phenyl ring (Scheme 5.82) [947, 948]. When LiBr is added to the reaction mixture, the Sommelet–Hauser rearrangement is suppressed. LiBr (common ion effect) suppresses the formation of free ions and, thus, the tight and solvent-separated ion pairs react in the Stevens mode. The Sommelet–Hauser rearrangement implies free ion-pair (carbanion + Li^+) [949].

The Wittig rearrangement (5.49) is a carbanion rearrangement in solution that converts 1-alkoxyalkyllithium **493** into the corresponding lithium alkoholate **496** [950–953]. It may involve intermediate **494** that arises from β-elimination of the migrating alkyl or aryl group R, or **495**, that results from the homolysis of the C—O bond [954, 955].

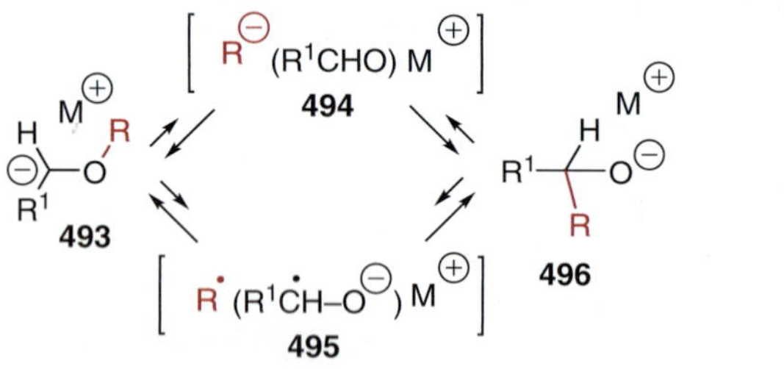

(5.49)

Support for these mechanisms is given with the observation of aldehydes R^1CHO as byproducts. The migratory aptitude of R increases with the intrinsic stability of radical $R^\bullet$. This makes mechanism **493** → **495** → **496** the preferred one [956–959]. However, it might not always be followed as it requires epimerization of the chiral migrating group R. This is not observed in all cases. Thus, some 1,2-Wittig rearrangements might have a ionic character and follow mechanism **493** → **494** → **496** [960]. Gas phase 1,2-Wittig rearrangements occur through anion elimination/readdition mechanisms [961]. The relative migratory

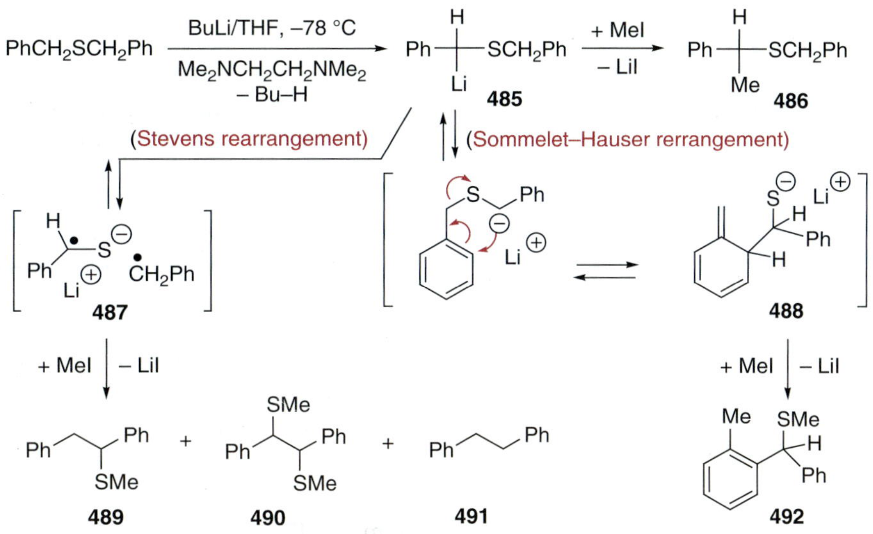

Scheme 5.82 Proof of the intermediacy of radical pairs in the Stevens rearrangement **485** → **489**, which may compete with the Sommelet–Hauser rearrangement **485** → **492**. The Sommelet–Hauser rearrangement requires free carbanion + Li^+ intermediates. The Stevens rearrangement does not!

aptitude of the alkyl anions is t-$Bu^- > i$-$Pr^- > Et^- \gg Me^-$ [962].

In 1894, Fritsch [963], Buttenberg [964], and Wiechell [965] reported conversion (5.50) of 2,2-diphenylethenyl bromide into 1,2-diphenylacetylene.

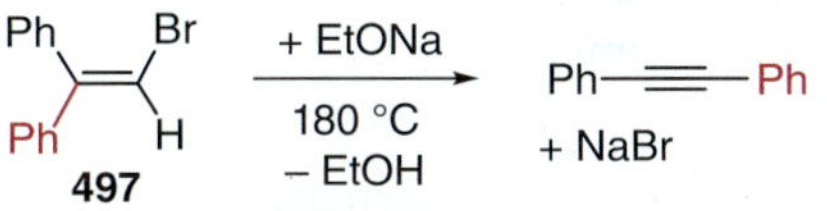

(5.50)

The reaction starts with the hydrogen/sodium exchange **497** → **498** (Scheme 5.83). α-Elimination of NaBr from **498** generates the alkylidene carbene **499** that undergoes a facile (1,2)-shift of a phenyl group, producing diphenylacetylene. Like carbenium ions, carbenes are electron-poor species (sextet of electrons). This mechanism avoids the "forbidden" (1,2)-shift in the carbanion stage **498** → **500**. Studies with bromoalkene **502** suggested that conversion **498** → diphenylacetylene involves a concerted (1,2)-dyotropic shift **498** → **501** → **500′** → diphenylacetylene. In this case, the (1,2)-migration of the phenyl group is assisted by the (1,2)-migration of the bromide anion. This mechanism is supported by the nonobservation of products of quenching of the carbene intermediate.

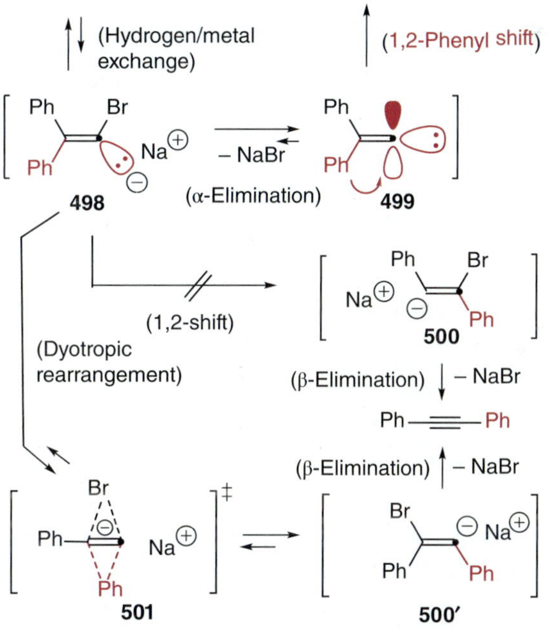

Scheme 5.83 Possible mechanisms for the conversion of 2,2-diphenylethenyl bromide (**497**) into diphenylacetylene.

Treatment of ^{13}C-labeled bromoalkene **502** with t-BuOK at 36 °C produces a major compound **505** resulting from the deprotonation of **502** into carbanion **503**. The latter undergoes double (1,2)-migration of the (methoxymethyl)alkyl group and the bromide into **504**, the protonation of which gives **505**. Interestingly, the (1,2)-alkyl shift occurs with retention of configuration. An alternative mechanism involving α-elimination of HBr with formation of a alkylidenecarbene intermediate is not retained as no product of carbene quenching could be detected [966].

The dyotropic rearrangement **503** → **504** might be seen as a Wagner–Meerwein rearrangement (Section 5.6.1) of **503′**, which has a carbene character (electrophilic) at the C—Br center, and a nucleophilic intramolecular attack of the carbanion onto the bromide generating **504′**, which corresponds to the carbene–carbanion limiting structure of alkenyl carbanion **504**.

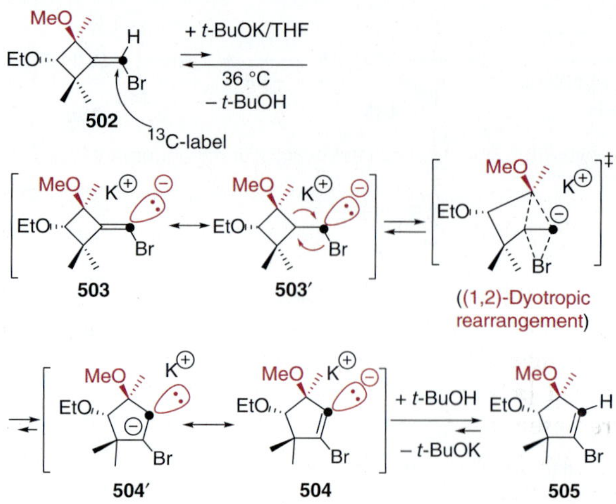

Related to the (1,2)-alkyl shift in carbanions, one finds the intermolecular (1,2)-anionic migration of a silyl group from a carbon center to an oxygen atom, or the (1,2)-Brook rearrangement (5.51) [967, 968]. The reverse process, the intramolecular migration of a silyl group from oxygen to carbon, first reported by Speier [969], is called the retro-Brook reaction. These rearrangements proceed via the intermediacy of pentacoordinated silicon species [970–974].

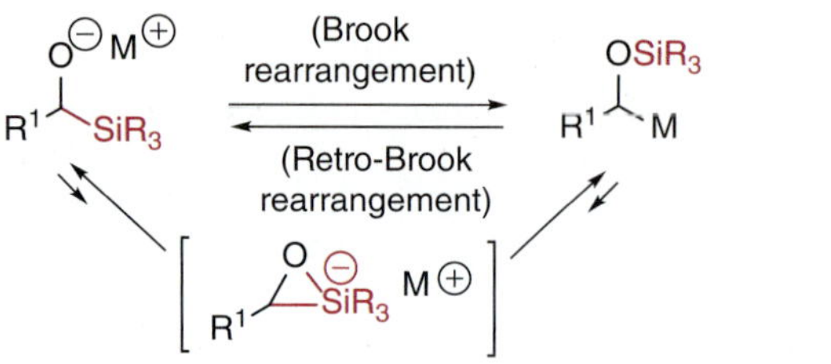

(5.51)

Treatment of enantiomerically enriched alcohol **506** with Na/K in Et_2O and subsequent acidic aqueous work-up produces **507** (Scheme 5.84). This demonstrates that the silyl group migrates with retention of configuration at the silicon center and that the C-metal protolysis also occurs with retention of configuration at the carbon center [975].

Problem 5.52 Estimate the heat of the fragmentation of 2,3,3-trimethylbut-2-yl radical into methyl radical and 2,3-dimethylbut-2-ene (gas phase, 25 °C, 1 atm).

Problem 5.53 What would be the activation enthalpy of isomerization **462** → **463** at 278 °C ($k = 5 \times 10^5\ s^{-1}$) if an entropy of activation $\Delta^{\ddagger}S = 10$ eu would be considered for the two freely rotating σ bonds of **463** (late transition state with no interaction between the radical and alkene moiety)?

Problem 5.54 Using $\Delta_fH°$(PhEt) = 7.1 kcal mol^{-1}, $\Delta_fH°$(1,1-dimethylcyclopropane) = −2.0 kcal mol^{-1}, $\Delta_hH°$(cyclohexa-1,4-diene → cyclohexane) = −53.7 kcal mol^{-1} and $DH°$(cyclohexadienyl$^\bullet$/H$^\bullet$) = 76.9 kcal mol^{-1} (Table 1.13), estimate the standard heat of isomerization 2-phenylethyl radical (**A**$^\bullet$) into 6,6-ethylenecyclohexa-2,4-dien-1-yl radical (**B**$^\bullet$).

5.5.4 (1,3)-Sigmatropic rearrangements

According to the Woodward–Hoffmann rules, the (1,3)-sigmatropic shift of an allylic hydrogen atom in an alkene is "forbidden" (difficult) for the intramolecular one-step suprafacial mode ([π^3s+ω^1s] = [π^2s+ω^2s] = [π^4s+ω^0s]), but allowed (thermally possible) for the antarafacial mode ([π^3a+ω^1s]). If one considers the transition structure of the suprafacial [π^3s+ω^1s] rearrangement as the combination of an hydrogen radical with an allyl radical, the

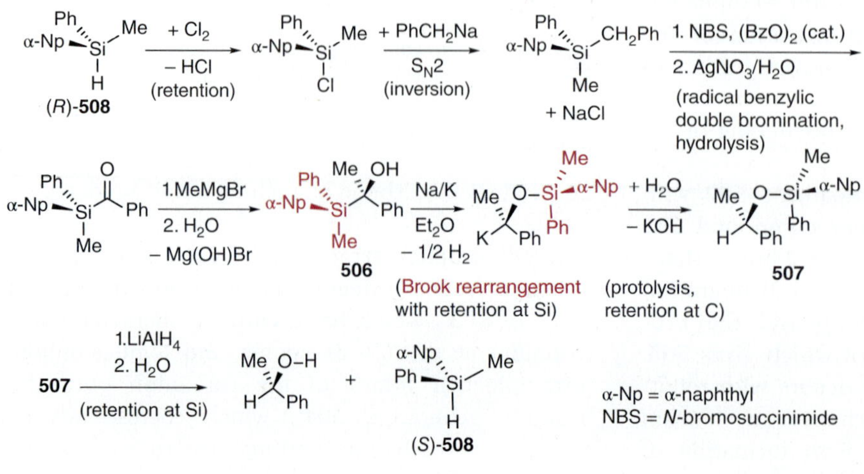

Scheme 5.84 Preparation of an enantiomerically enriched silyl carbinol (**506**) and its conversion into silyl ether (**507**) via a base-induced (1,2)-Brook rearrangement. The configuration of ether **507** was determined by its $LiAlH_4$-reduction into (*S*)-1-phenylethanol and silane (*S*)-**508**.

FMO theory retains the absence of stabilizing interaction between SOMO(H•) and SOMO(allyl•) because their overlap is nil for reasons of symmetry. A weak stabilization might result from the SOMO(H•)/π_1(allyl•) overlap (Figure 5.47a). In contrast, for the antarafacial [π^3a+ω^1s] rearrangement, the SOMO(H•)/SOMO(allyl•) overlap is not nil and leads to a stabilization of the H• + allyl• complex compared with two separated radicals (Figure 5.47b). This means that the rearrangement would benefit from assistance between the C—H bond breaking and C—H bond forming processes. In reality, concerted, one-step (1,3)-shifts of hydrogen are not observed because they require too severe skeleton deformations. Should they occur, they involve complete homolysis of the allylic C—H bond and radical recombination. An exception is given with the (1,3)-hydrogen shift that converts 1-*t*-butyl-3-methylallene (**509**) into 1-*t*-butylbuta-1,3-diene (**511**). This intramolecular process has activation parameters $\Delta^{\ddagger}H = 27.9 \pm 1.0$ kcal mol^{-1} and $\Delta^{\ddagger}S = 1.4 \pm 1.5$ eu (150–196 °C). The transition state **510** has been proposed for this concerted one-step reaction [976]. It involves six electrons and the twisting of the double bond π(C(1)—C(2)). This is an example of pseudopericyclic reaction.

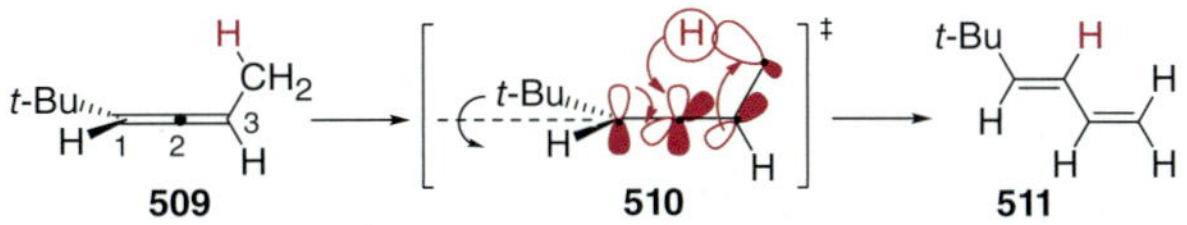

509 **510** **511**

With the FMO diagram of Figure 5.48a, one demonstrates that the suprafacial (1,3)-alkyl shift is "forbidden" if the migrating center maintains its configuration (retention, rearrangement noted [π^3s+ω^1s]). In the case of a suprafacial (1,3)-alkyl shift with inversion of the migrating center (rearrangement noted [π^3s+ω^1a]), the FMO diagram (Figure 5.48b) representing its transition state shows that the concerted process is "allowed" as SOMO(alkyl•) and SOMO(allyl•) overlap.

On heating, vinylcyclopropane isomerizes into cyclopentene [977, 978] and into the three isomeric pentadienes [979, 980]. It is now recognized that the Woodward–Hoffmann rules are not relevant to these reactions. The vinylcyclopropanes equilibrate with conformationally flexible diradical intermediates that, in turn, react further through (1,3)-carbon shift to give cyclopentenes, through (1,2)-hydrogen shift to give pentadienes, or they recombine to produce isomeric vinylcyclopropanes [981]. Similarly, vinylcyclobutanes are isomerized via diradical intermediates (e.g.:

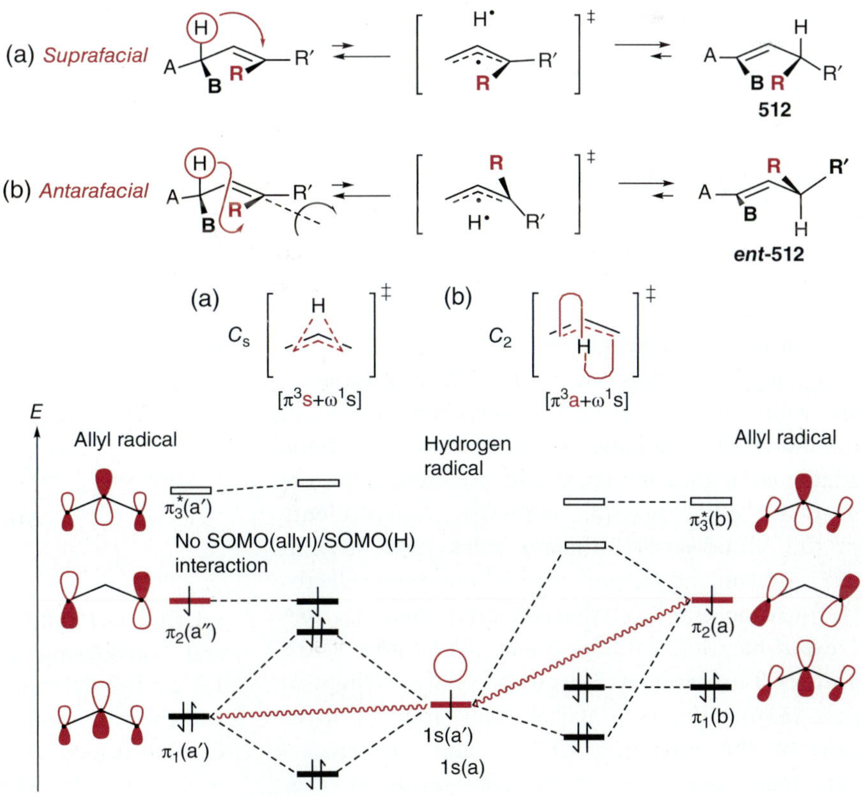

Figure 5.47 MO diagrams representing the transition state of the (1,3)-hydrogen shift in (a) the suprafacial mode ([π^3s+ω^1s]) and (b) in the antarafacial mode ([π^3a+ω^1s]). The suprafacial mode maintains a mirror plane of symmetry, whereas the antarafacial mode has formally a C_2-axis of symmetry in their respective transition states. If R ≠ R′, product **512** is the enantiomer of product *ent*-**512** (a′, symmetrical a″, antisymmetrical with respect to the mirror plane; a, symmetrical, b, antisymmetrical with respect to the C_2-axis). Note that hydrogen migrates with retention only.

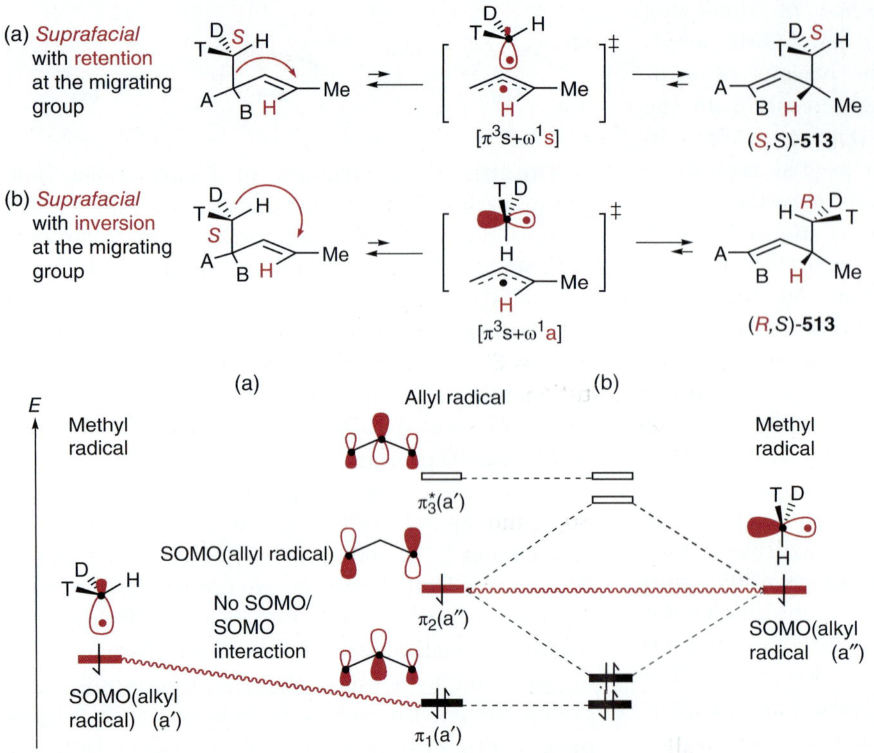

Figure 5.48 FMO diagram representing the transition structure of a suprafacial (1,3)-alkyl shift ((*S*)-deuteriotritiomethyl group) with (a) retention of configuration of the migrating center and (b) with inversion of the migrating center. Note that products (*S*,*S*)-**513** and (*R*,*S*)-**513** are diastereoisomers.

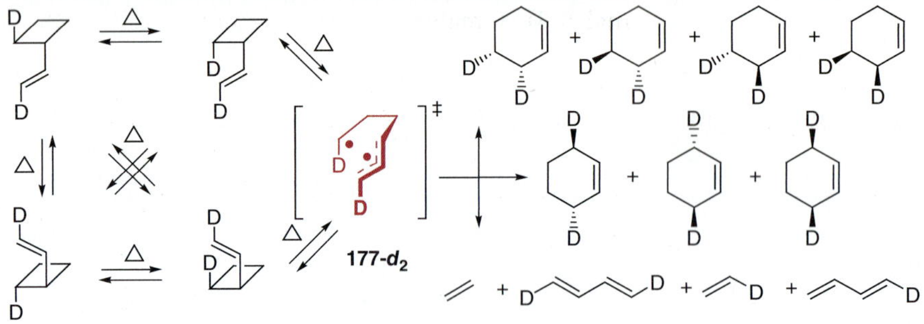

Scheme 5.85 Thermal stereomutation of (1*R*,2*R*)-1-((*E*)-*d*-vinyl)-2-*d*-cyclobutane, its rearrangement into 3,4-d_2-cyclohexenes through nonstereoselective (1,3)-sigmatropic carbon shift, and competitive fragmentation into ethylene + butadiene.

177-d$_2$) into all possible stereoisomeric cyclohexenes on heating (Scheme 5.85) [982–985].

When vinylcyclopropane and vinylcyclobutane are substituted by groups larger than deuterium, the intermediate diradicals may find conformational adjustments that influence the preference for one over the other reaction pathways, independently of the Woodward–Hoffmann rules [986–989]. In 1967, Berson and Nelson reported the stereoselective (1,3)-carbon shift (5.52) that converts 6-*endo*-acetoxy-7-*exo*-*d*-bicyclo[3.2.0]hept-2-ene (**514**) into 2-*exo*-acetoxy-3-*exo*-acetoxy-7-*exo*-*d*-bicyclo[2.2.1]hept-5-ene (**516**). This suprafacial shift implies inversion of the migrating CHD center in apparent agreement with the Woodward–Hoffmann rules [990].

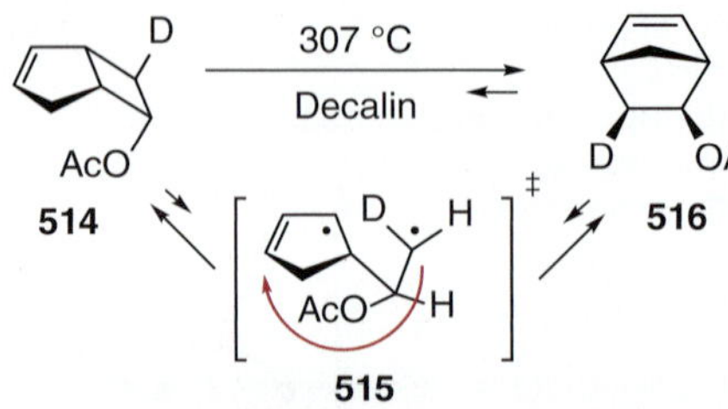

(5.52)

Using $DH°(\text{allyl}^\bullet/\text{Et}^\bullet) = 72\,\text{kcal mol}^{-1}$ (Table 1.7) and considering the fact that diradical **515** is a 1,3-dialkylallyl radical (stabilization of c. 2 kcal mol^{-1}) and the ring strain release of 26.5 kcal mol^{-1} for the C—C breaking of the cyclobutane unit of **514**, one estimates $\Delta_r H°(\mathbf{514} \rightleftarrows \mathbf{515}) = 72 - 2 - 26.5 = 43.5$ kcal mol^{-1}. If one considers the measured rate constant $k(5.52) = 0.0001\,\mathrm{s}^{-1}$ (half-life of approximately two hours) and a pre-exponential factor $A = 10^{13}$ for this reaction of order one, one calculates $E_a = \mathrm{RT}(\ln A - \ln k) = 1.987\ 580(29.9 + 9.2) = 45\,000\ \mathrm{cal\,mol^{-1}} = 45\,\mathrm{kcal\,mol^{-1}}$ at 307 °C. The activation enthalpy of this reaction amounts to $\Delta^{\ddagger}H = E_a - \mathrm{RT} = 45 - 1.1 = 43.9\,\mathrm{kcal\,mol^{-1}}$. This is nearly the same value as $\Delta_r H°$(**514** ⇄ **515**) = 43.5 kcal mol^{-1} evaluated above. Thus, diradical **515** might well be an intermediate of rearrangement (5.52), indicating that there is little or no assistance between the bond-breaking and bond-forming processes involved in this rearrangement. The stereoselectivity observed (inversion of the migrating center) does not prove the intervention of an electronic stabilization as predicted by the Woodward–Hoffmann rules and the FMO theory (Figure 5.48b). For the thermal rearrangement of bicyclo[3.2.0]hept-2-ene into norbornene, quantum mechanical calculations predicted a concerted (1,3)-sigmatropic rearrangement with a diradical transition structure proceeding through a broad, flat energy hypersurface that leads to the predominant $[\pi^3\mathrm{s}+\omega^1\mathrm{a}]$-pathway [991, 992]. This process obeys to the least motion principle (avoids rotation of the alkyl radical).

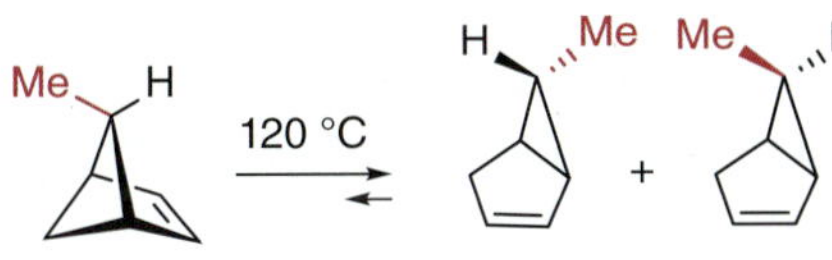

(5.53)

517 **518a** **518s**

[**518a**]/[**518s**] = 200

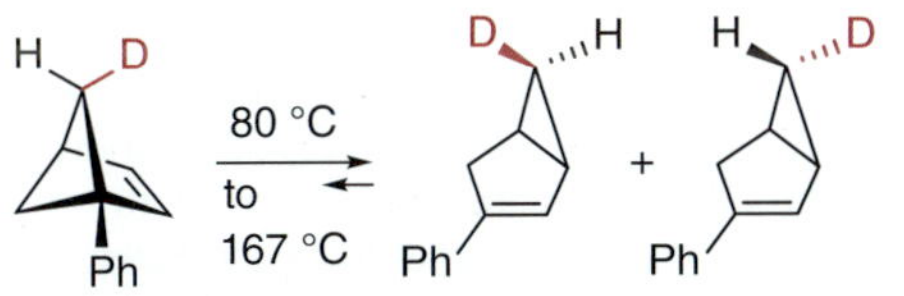

(5.54)

519 **520a** **520s**

[**520a**]/[**520s**] = 10

Thermal (1,3)-sigmatropic rearrangements (5.53) and (5.54) of bicyclo[2.1.1]hex-2-ene derivatives **517** [993] and **519** [994] into bicyclo[3.1.0]hex-2-enes **518** and **520**, respectively, are not stereospecific, but the suprafacial mode with inversion of the migrating alkyl group is preferred. According to Carpenter, the preference for inversion ($[\pi^3\mathrm{s}+\omega^1\mathrm{a}]$) has nothing to do with the Woodward–Hoffmann rules, but is a result of the dynamics of the bond cleavage in the reactants [995]. (1,3)-Sigmatropic rearrangements of optically active 6-methylidenebicyclo[3.1.0]hex-2-enyl derivatives give the corresponding 2-methylidenebicyclo[3.1.0]hex-3-en-6-yl derivatives that are completely racemized. Racemization of the starting material competes with the rearrangement, in agreement with the formation of achiral diradical intermediates (reaction (5.55)). The NIST Chemistry WebBook gives $\Delta_f H°$(**521**) = 60.0 kcal mol^{-1}, $\Delta_f H°$(**523**) = 37.8 kcal mol^{-1}, and $\Delta_f H°$(3-methylcyclopentene) = 2.3 kcal mol^{-1}. One calculates $\Delta_f H°$(diradical **522**) = $\Delta_f H°$(3-methylcyclopentene) + $DH°$(MeCH=CH—C$^{(\bullet)}$(Me)H/H$^\bullet$) + $DH°$(Et$^\bullet$/H$^\bullet$) − $DH°$(H$^\bullet$/H$^\bullet$) = 2.3 + 100.7 + 81.6 − 104.2 = 80.4 kcal mol^{-1}. Therefore, $\Delta_r H°$(**521** ⇄ **522**) = 80.4 − 60.0 = 20.4 kcal mol^{-1}. Methyl substitution as in **517** reduces this enthalpy difference by c. 2 kcal mol^{-1}. Thus, diradicals of type **522** ([(cyclopent-3-en-2-yl)-1-yl]methyl diradical) are viable intermediates in rearrangements occurring at 80 °C and above [996].

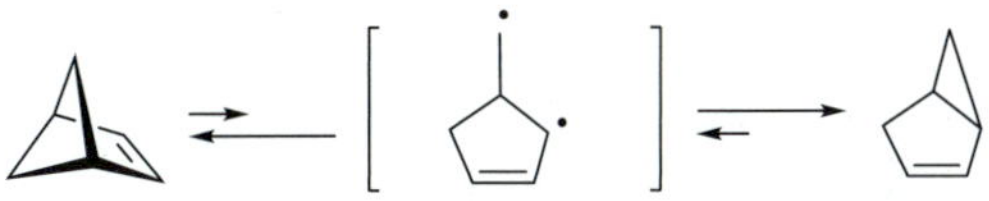

(5.55)

521 **522** **523**

Problem 5.55 Explain the substituent effects on the activation energies measured for the following thermal (1,3)-sigmatropic rearrangements [997–1000].

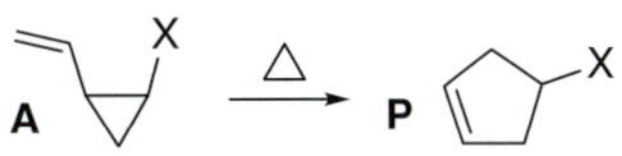

X = Me, E_a = 48.7; X = Ph, E_a = 44.7 kcal mol^{-1}
X = OMe, 38.7; X = NMe$_2$, 31.2

5.5.5 (1,4)-Sigmatropic rearrangements

Protonation of triene **524** with FSO_3H gives the stable salt of carbocation **525**. Its irradiation at −78 °C induces its clean isomerization into the corresponding salt of carbocation **526**. The ^{1}H-NMR spectrum of **526** depends on temperature because of its degenerate rearrangement that implies the walk of the dimethylcyclopropyl group around the cyclopentenyl ring. The process is a suprafacial (1,4)-sigmatropic rearrangement with inversion of the migrating C(6) center. This is demonstrated by the fact that the 6-*exo*- and 6-*endo*-methyl groups are not interchanged while all the methyl substituents of the five-membered ring are interchanged on the NMR time scale. At −89 °C, $\Delta^{\ddagger}G = 9\,\mathrm{kcal\,mol^{-1}}$ is measured for this degenerate rearrangement. At −10 °C, carbocation **526** is isomerized irreversibly into **525** with $\Delta^{\ddagger}G = 19.8\,\mathrm{kcal\,mol^{-1}}$.

This disrotatory rearrangement is a "forbidden" electrocyclic ring opening (Section 5.2.6) corresponding to the inverse of a Nazarov cyclizaton [1001, 1002].

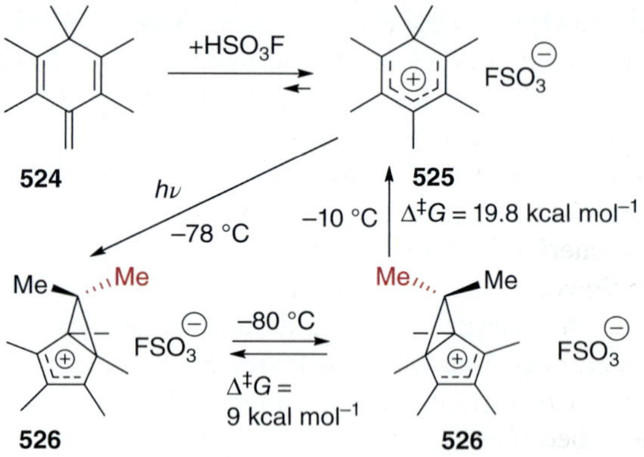

The 2,6-*endo*-dideuterobicyclo[3.1.0]hex-3-en-2-yl cation (**527**) also undergoes a fivefold degenerate (1,4)-sigmatropic rearrangement at −90 °C with $\Delta^\ddagger G = 15 \pm 1\,\text{kcal mol}^{-1}$. During this process, the *endo*-6-deutero group stays *endo* while **527** equilibrates with **527a** ⇄ **527b** ⇄ **527c** ⇄ **527d**: thus, the migration of C(6) occurs with inversion of configuration ([π^3s+ω^1a]-rearrangement). At −20 °C, **527** is isomerized irreversibly into dideuterocyclohexadienyl cation (**528**) with $\Delta^\ddagger G = 19.8\,\text{kcal mol}^{-1}$ [1003].

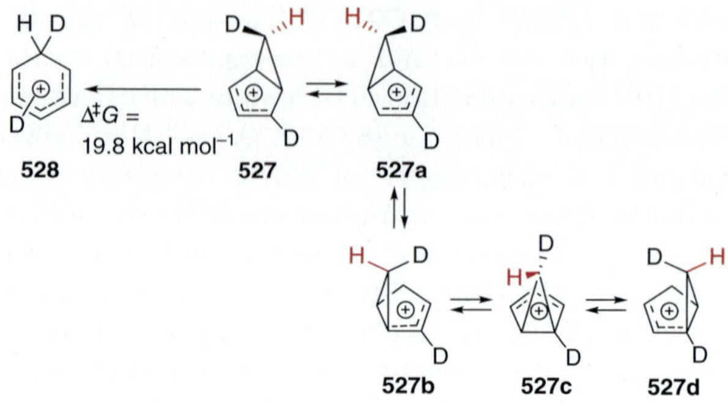

The stereoselectivity of the degenerate (1,4)-sigmatropic rearrangement of **527** is predicted by the FMO theory (Figure 5.49a). The transition structure of this process can be represented by **529a** in which the butadiene moiety interacts with the 2p empty orbital of the alkyl cation of the migrating group. In this structure, overlap is possible between the LUMO(cation) and HOMO(diene moiety) and thus leads to an electronic stabilization. In contrast, transition structure **529s** of the suprafacial (1,4)-sigmatropic rearrangement with retention of configuration of the migrating center C(6) does not enjoy from any LUMO(cation)/HOMO(diene) stabilization (Figure 5.49b).

Thermolysis of 2-deuterobicyclo[3.1.0]hex-2-ene (**530**) in the presence of $(t\text{-BuO})_2$ in PhCl at 130 °C

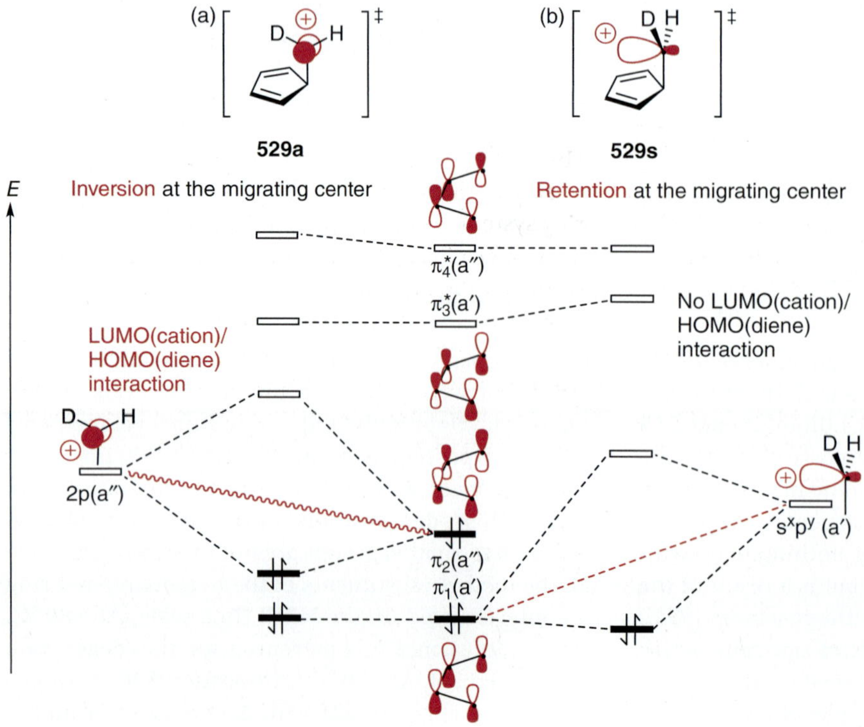

Figure 5.49 FMO diagram representing the transition structure of a (1,4)-sigmatropic rearrangement in bicyclo[3.1.0]hex-3-en-2-yl cation with (a) inversion of configuration and (b) retention of configuration at the migrating C(6) center. Both transition states **529a** and **529s** share a mirror plane of symmetry.

affords monodeuterobenzene via the electrocyclic ring opening involving the inter-ring bond in radical **531** to give cyclohexadienyl radical **533** that loses an hydrogen atom by reaction with the medium. In an adamantane matrix, radical **531** is rearranged into radical **531a** at −50 °C with $\Delta^{\ddagger}G = 14.5\,\text{kcal mol}^{-1}$. In the matrix, termination reactions are retarded; thus, controlled reactions can be observed. By this method, the degenerate (1,4)-sigmatropic rearrangement of **531** has been proven to have $\Delta^{\ddagger}G < 14.5\,\text{kcal mol}^{-1}$ at −50 °C [1004]. The latter process involves the intermediacy of cyclopentadienylmethyl radical **532** [1005].

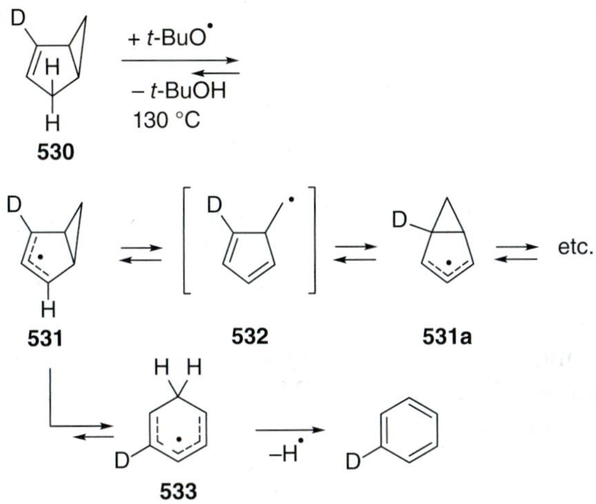

Problem 5.56 Estimate the stability difference between radicals **531** and **532**.

5.5.6 (1,5)-Sigmatropic rearrangements

The most familiar (1,5)-sigmatropic rearrangement is the suprafacial migration of hydrogen around the cyclopentadiene ring [1006] that rapidly interconverts 5-alkylcyclopentadienes with their 1-alkyl and 2-alkyl isomers as illustrated with the isomerization (5.55) of 5-methylcyclopentadiene into 1-methylcyclopentadiene (Scheme 5.86) [1007]. The activation parameters of this reaction are consistent with an intramolecular concerted mechanism (assistance between the C—H breaking C—H bond forming processes; $DH°(c\text{-}C_5H_5^{\bullet}/H^{\bullet}) = 81\,\text{kcal mol}^{-1}$, Table 1.14), with a transition structure **534** representing a Hückel system with six electrons (stabilized by aromaticity). In the cases of 5-trimethylsilyl-, 5-trimethylgermanium-, and trimethyltincyclopenta-1,3-diene, the (1,5)-migrations of the Me_3Si, Me_3Ge, and Me_3Sn groups are faster than that of hydrogen. The processes are fast enough to make these compounds fluxional on the NMR time scale [1008]. This is also the case for cyclopentadienyldimethylborane, which shows a single peak in its ^{1}H- and ^{13}C-NMR spectra for the five-membered ring at −78 °C ($k > 10\,\text{s}^{-1}$). Above −15 °C, the latter compound is isomerized irreversibly into 1- and 2-$Me_2B—C_5H_5$ via slower (1,5)-hydrogen shifts [1009, 1010].

Thermal hydrogen migration are common in acyclic conjugated dienes [1011, 1012]. The possible (1,5)-hydrogen shift (5.56) in (*Z*)-penta-1,3-diene was described first by Wolinsky et al. [1013].

$$(Z) - CH_2{=}CH{—}CH{=}CH{—}CD_3$$
$$\rightleftarrows (Z) - CH_2D{—}CH{=}CH{—}CH{=}CD_2 \qquad (5.56)$$

Using deuterium-labeled (*Z*)-penta-1,3-diene, Roth and König found $\Delta^{\ddagger}H(5.56) = 35\,\text{kcal mol}^{-1}$ and $\Delta^{\ddagger}S(5.56) = -7.1$ eu (gas phase, 185–205 °C) for this (1,5)-migration of hydrogen and a primary kinetic isotopic effect $k_H/k_D = 12.2$ (25 °C) [1014, 1015]. Quantum mechanical calculations confirmed a preferred concerted suprafacial (noted $[\pi^4s+\omega^2s] = [\pi^5s+\omega^1s]$) vs. an antarafacial (noted $[\pi^4a+\omega^2s] = [\pi^5a+\omega^1s]$) migration with a C_s-symmetrical transition structure analogous to **534** [1016]. (1,5)-Hydrogen shifts in heterodienic systems have been reported [1017–1021]. For instance, azoalkenes are isomerized easily into α,β-unsaturated hydrazones [1022]. There are also examples of analogous migrations of alkyl [1007–1026], aryl [1027], acyl [1028–1030], vinyl [1024], alkynyl, and cyano groups [1024, 1031, 1032]. Rearrangements of heterocyclic cyclopentadiene-like ring systems are also known [1033, 1034].

At 250 °C, (*S*)-(*Z*,*E*)-5-methyl-2-deuteroocta-2,4-diene (**535**) is rearranged into a 1.5 : 1 mixture of (*R*)-(*E*,*Z*)-3-methyl-7-deuteroocta-3,5-diene (**536**) and (*S*)-(*Z*,*Z*)-3-methyl-7-deuteroocta-3,5-diene (**537**). Both rearrangements are suprafacial (Scheme 5.87). As not a trace of isomeric products could be detected, one estimates the antarafacial mode to have a free energy barrier at least $8\,\text{kcal mol}^{-1}$ higher than that

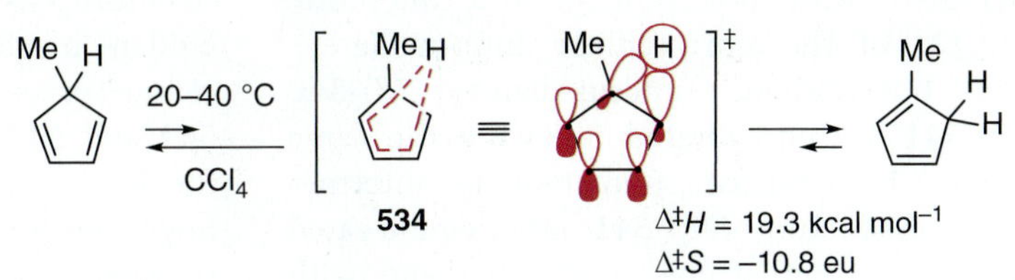

(5.55)

Scheme 5.86 Facile suprafacial (1,5)-migration of hydrogen in cyclopentadienes

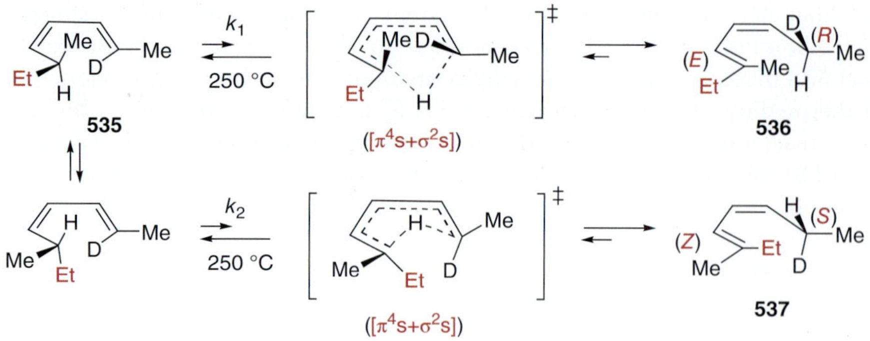

Scheme 5.87 Suprafacial (1,5)-sigmatropic migrations of hydrogen in an acyclic diene.

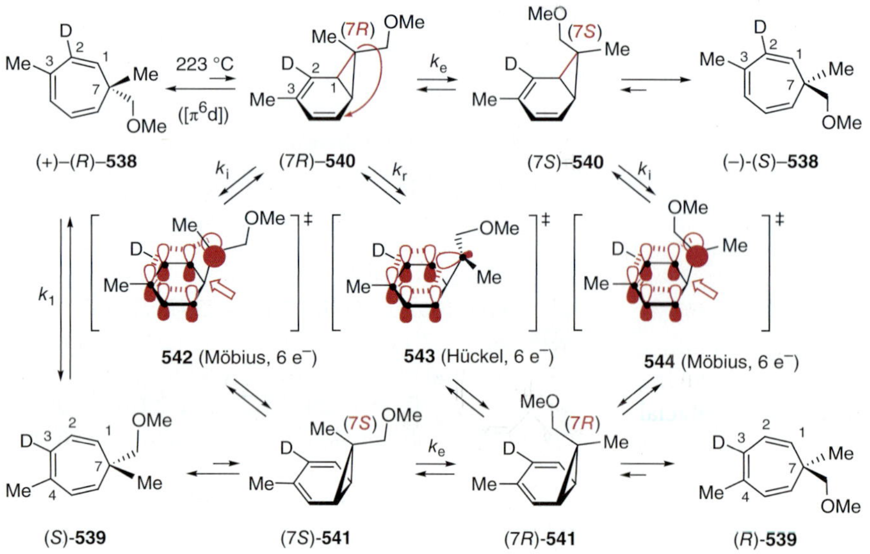

Scheme 5.88 Thermal suprafacial (1,5)-sigmatropic rearrangements of norcaradienes occur preferentially with inversion of the migrating center (in contradiction to the Woodward–Hoffmann rules).

of the suprafacial rearrangement (see Section 5.2.8, Scheme 5.16 for other (1,5)-hydrogen shift).

On heating to 223 °C, the 2-deuteriocycloheptatriene derivative (+)-(*R*)-**538** is isomerized into the 3-deuteriocycloheptatriene derivative (*S*)-**539** (Scheme 5.88) with $k_1 = (6.13 \pm 0.2) \times 10^{-6}\ s^{-1}$ (first-order rate law). Competitively with this isomerization, the racemization of (+)-(*R*)-**538** occurs with $k_r = (6.07 \pm 0.3) \times 10^{-6}\ s^{-1}$ (first-order rate law for the isomerization (+)-(*R*)-**538** → (−)-(*S*)-**538**). These isomerizations imply the disrotatory electrocyclic ring closure ([π^6d], Section 5.2.8) of the cycloheptatrienes into their norcaradiene (bicyclo[4.1.0]hepta-2,4-diene) isomers. Norcaradiene intermediates (7*R*)-**540** and (7*S*)-**541** are equilibrated through a reversible (1,5)-sigmatropic rearrangement with inversion of the migrating center (C(7) of the norcaradiene intermediates). Similarly, norcaradiene intermediates (7*S*)-**540** and (7*R*)-**541** are equilibrated following the same mechanism. In contrast, norcaradiene intermediates (7*R*)-**540** and (7*R*)-**541** are equilibrated through a (1,5)-sigmatropic rearrangement with retention of the configuration of the migrating center. The same occurs with (7*S*)-**540** → (7*S*)-**541**. By comparing the rate of appearance of (−)-(*S*)-**528** and of (*S*)-**539**, Baldwin and Broline measured the rate constant $k_i = (2.96 \pm 0.09) \times 10^{-6}\ s^{-1}$ for the norcaradiene isomerization with inversion of configuration of the migrating center, the rate constant $k_r = (0.08 \pm 0.17) \times 10^{-6}\ s^{-1}$ for the migration with retention, and the rate constant $k_e = (0.10 \pm 0.13) \times 10^{-6}\ s^{-1}$ for the rearrangement of norcaradiene intermediates following other mechanisms (e.g. (1,3)-sigmatropic rearrangement of their cycloheptatriene isomers). Thus, the mechanism favored by the Woddward–Hoffmann rules, the (1,5)-sigmatropic rearrangement with retention of the migrating center ([π^4s+ω^2s] = [π^5s+ω^1s], transition structure **543**) is at least 30 times slower than the "forbidden" mechanism with inversion of the migrating center ([π^4s+ω^2a] = [π^5s+ω^1a]: transition structures **542** and **544**). The latter mechanism is favored by the least motion principle (avoids rotation of the migration center). Quantum mechanical calculations suggested that both modes of (1,5)-sigmatropic rearrangements of norcaradienes have transition states that are diradical in character with similar relative stabilities [992].

Problem 5.57 Explain (a) why two successive (1,3)-migrations of hydrogen around a cyclopentadiene system are much slower than a one-step, concerted (1,5)-sigmatropic rearrangement and (b) demonstrate that the suprafacial (1,5)-sigmatropic shift of hydrogen in conjugated dienes is faster than the antarafacial mode.

Problem 5.58 Cyclopentadiene has a pK_a = 15.0 in water (Table 1.24) and 18.0 in DMSO (Table 1.25). Is it possible that the isomerization of 5-alkylcyclopentadiene into 1- and 2-alkylcyclopentadiene can occur via deprotonation and reprotonation in CCl_4?

5.5.7 (1,7)-Sigmatropic rearrangements

The active form of vitamin D_3 is 1α,25-dihydroxyvitamin D_3 (**549**) [1035, 1036]. It is the result of the photometabolism of 7,8-dehydrocholesterol (**545**). UV-B light (skin, or food) induces the conrotatory ring opening of the cyclohexadiene moiety giving triene **546** (Scheme 5.89). Thermal antarafacial (1,7)-migration of an hydrogen atom converts **546** into triene **548** [15, 88, 1037, 1038]. The latter is then oxidized at C(25) in the liver first and then at C(1) in the kidney to give **549**. In the case of isomerization **546** → **548**, the Woodward–Hoffmann rules ([π^6a+ω^2s]) are obeyed: the transition state **547** realizes a Möbius system with eight electrons that is electronically stabilized (Section 4.9).

Several examples of antarafacial (1,7)-migrations of hydrogen have been reported (see also Section 5.2.8, Scheme 5.16) [1039, 1040]. In the simpler case of equilibrium (5.57) that uses triene **550**, activation parameters and a primary deuterium kinetic isotopic effect k_H/k_D(60 °C) = 7 are consistent with a concerted one-step mechanism [1041].

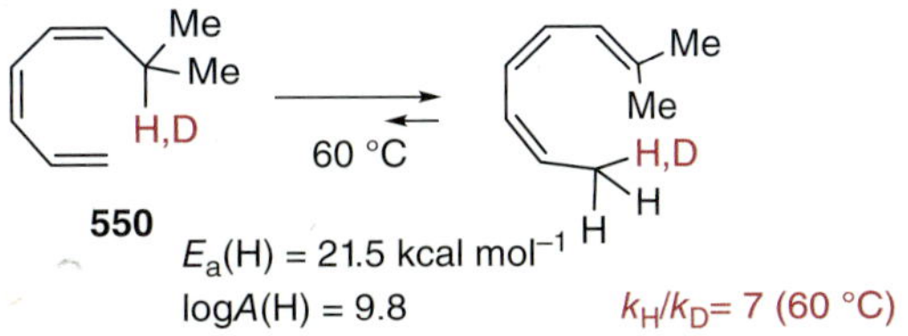

(5.57)

Bicyclo[6.1.0]nona-2,4,6-trienes undergo thermal (1,7)-alkyl shifts in so-called walk rearrangements. The suprafacial reactions proceed generally with inversion of the migrating group as shown with isomerization **551** → **552** [1042].

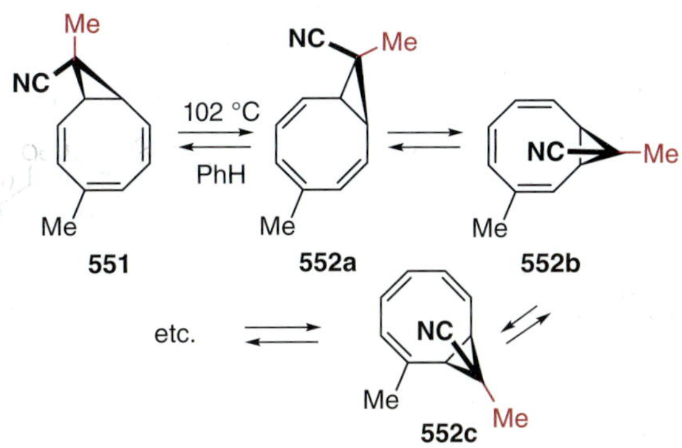

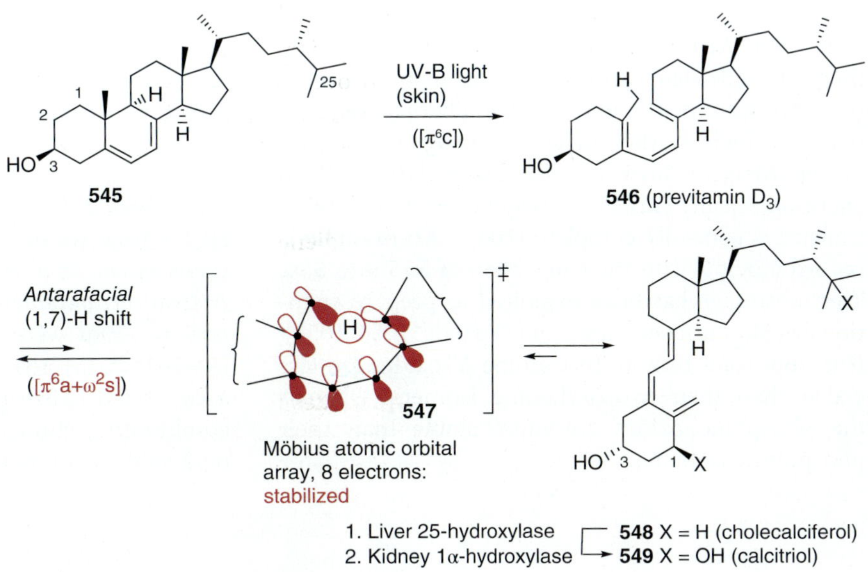

Scheme 5.89 Conversion of 7,8-dehydrocholesterol into the active form of vitamin D_3.

Problem 5.59 Give a mechanism for the isomerization **A** ⇄ **B** [1043].

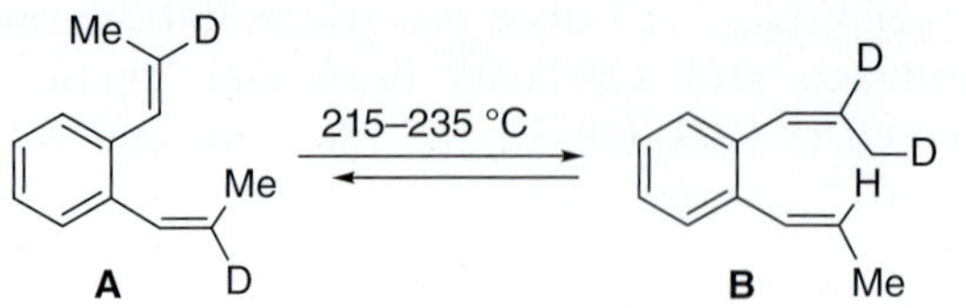

5.5.8 (2,3)-Sigmatropic rearrangements

In 1919, Meisenheimer et al. reported an early example of (2,3)-sigmatropic rearrangement that isomerizes allylic tertiary amines-*N*-oxides **553** into *O*-allyl hydroxylamines **554**. The reaction with crotylmethylaniline-*N*-oxide occurs with inversion of the crotyl group, in agreement with a thermal intramolecular (2,3)-rearrangement [1044–1046]. Reductive cleavage of the N—O bonds of hydroxylamines with $Mo(CO)_5$, e.g., delivers the corresponding allylic alcohols and secondary amines [1047]. Using enantiomerically pure amines, the *O*-allyl hydroxylamines are obtained with modest stereoselectivity because of the low stereoselectivity of the oxidation (with *m*CPBA) of the tertiary amines into their *N*-oxides and because of incomplete stereoselectivity in the subsequent (2,3)-sigmatropic rearrangement [1048–1051].

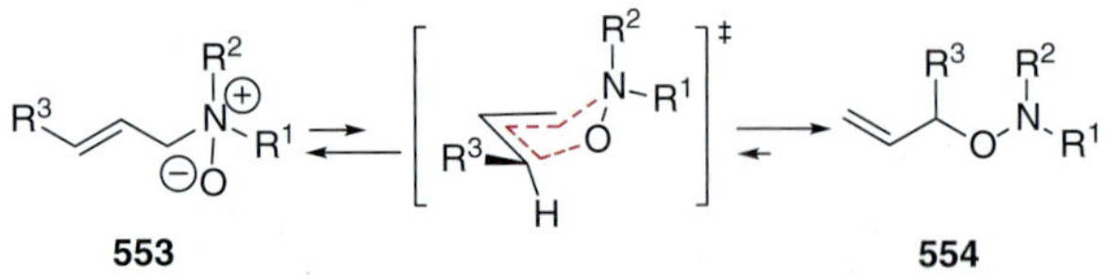

In 1948, Arbuzov and Nikoronov discovered the thermal (2,3)-sigmatropic rearrangement of allyldiarylphosphinites into allyldiphenylphosphine oxides, which is the phospha analogous reaction of the Meisenheimer rearrangement [1052]. With enantiomerically enriched phosphinites, the chirality transfer is generally complete [1053]. An example is reported below with the conversion of **555** into **556**. This advantage has been exploited to prepare enantiomerically enriched diphospine ligands [1054, 1055]. It is important to note that amine *N*-oxides are less stable than their hydroxylamine isomers, whereas the phosphine oxides are more stable than their phosphinite isomers.

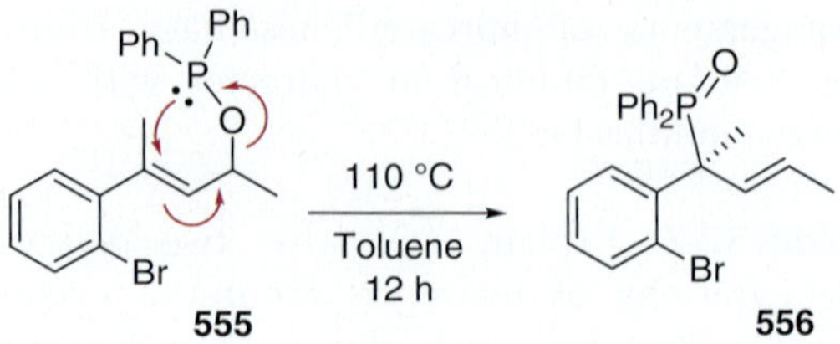

In 1950, Cope et al. found that allyl arenesulfinates [1056] are isomerized below 100 °C into the corresponding sulfones, whereas alkyl arenesulfinates are stable at this temperature [1057]. Nevertheless, the latter are also rearranged at higher temperature into sulfones usually via ion pair formation, as reported first in 1917 by Hinsberg [1058]. An application of the (2,3)-sigmatropic rearrangement of propargyl sulfinates is given below [1059]. In this case, the sulfones are more stable than their sulfinate isomers (see Figure 5.50 for other sulfur and selenium derivatives).

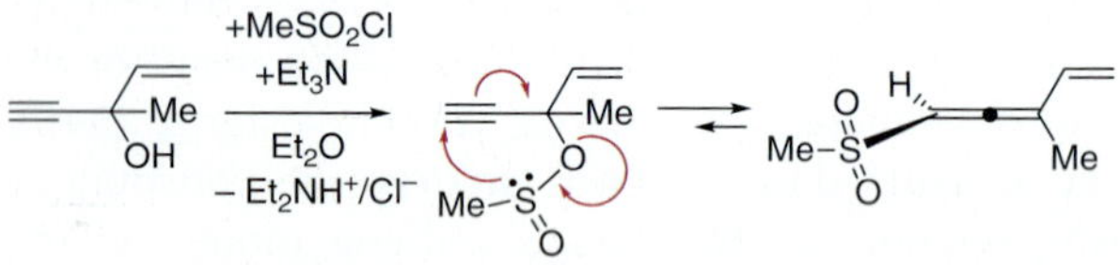

In 1951, Challenger and coworkers reacted diallyl sulfide with chloramine T and obtained sulfilimine **557**, which isomerized readily into *N*-allylsulfenamide **558** [1060]. They also found that sulfilimine **559** was isomerized into **560**, in agreement with a (2,3)-sigmatropic rearrangement (involves an allylic transposition) [1061].

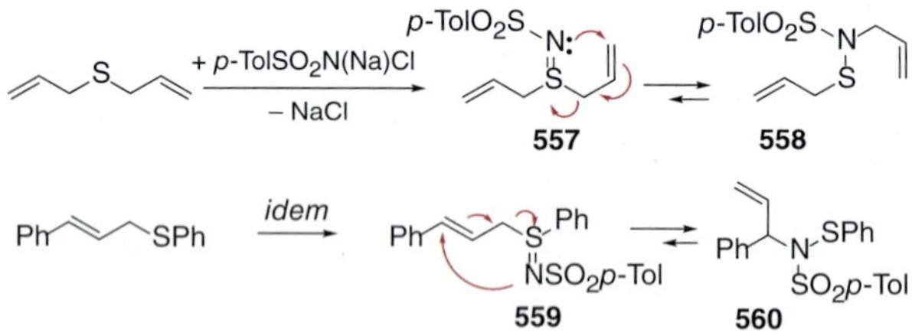

In 1966, Mislow and coworkers found that the allyl *p*-tolyl sulfoxide (*R*)-**561** is racemized 560 000 times as fast as phenyl tolyl sulfoxide (Scheme 5.90). Activation parameters $\Delta^{\ddagger}H = 22\,\text{kcal mol}^{-1}$ and $\Delta^{\ddagger}S = -9$ eu were measured for the racemization (*R*)-**561** ⇄ (*S*)-**561**. These data are consistent with a concerted (2,3)-sigmatropic rearrangement [1062] equilibrating chiral allyl sulfoxides (pyramidal trigonal sulfur center) with achiral allyl *p*-tolylsulfenate

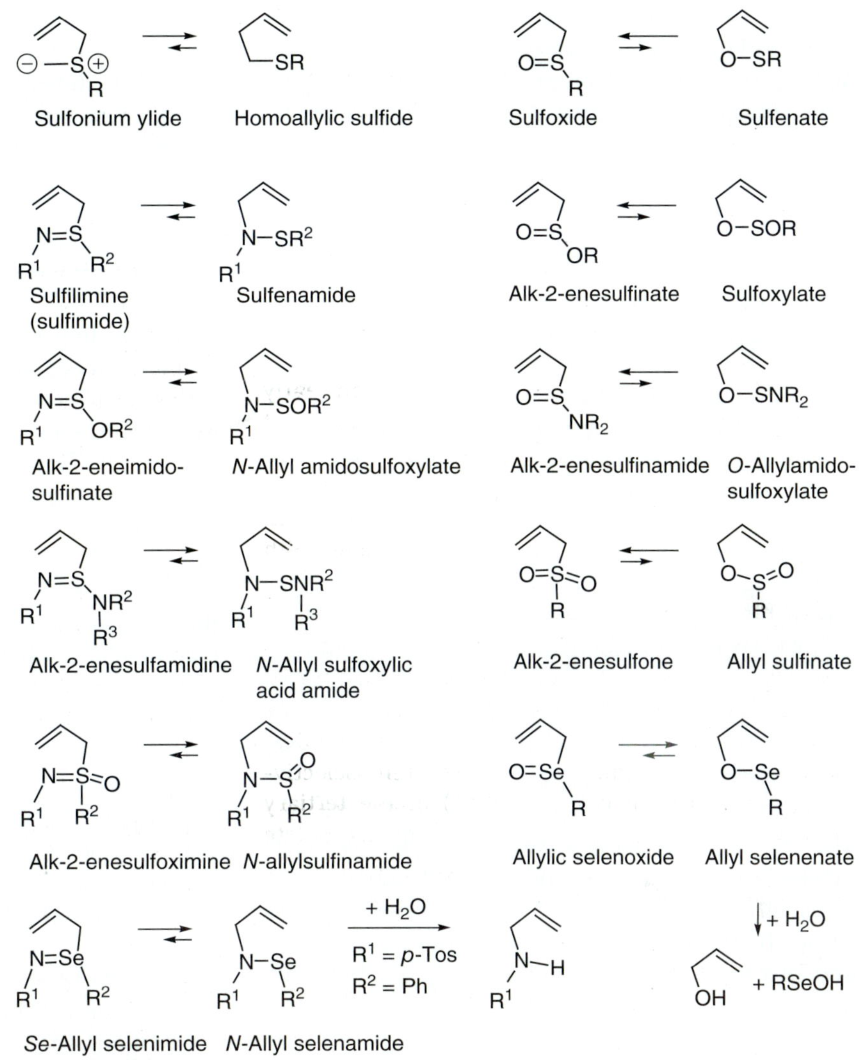

Figure 5.50 (2,3)-Sigmatropic rearrangements of allylic sulfur and selenium compounds and the usual equilibrium positions.

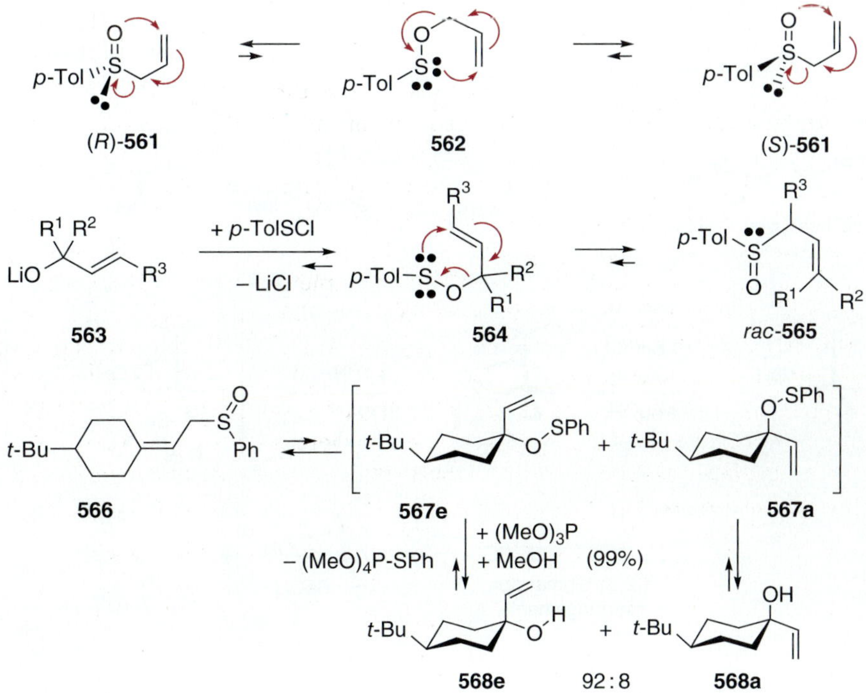

Scheme 5.90 Mislow–Braverman–Evans sulfoxide/sulfinate rearrangements.

(digonal planar sulfur center) **562**, an unstable intermediate of the racemization process. Allylic alkoxides **563** react with *p*-toluenesulfinyl chloride to give the corresponding sulfinates **564** that are immediately isomerized into the more stable allylsulfoxides (±)-**565** [1063]. In 1967, Braverman and Stabinsky showed that cinnamyl trichloromethanesulfenate, that is prepared at −70 °C, isomerizes at low temperature into cynnamyl trichloromethyl sulfoxide [1064]. By quenching the sulfinates with thiophilic agents, Evans and coworkers developed a useful method that converts allyl sulfoxides into allylic alcohols (Scheme 5.90) [1065, 1066]. The method has been applied to the enantioselective and stereoselective synthesis of steroids [1067, 1068], 1,3-dienes [1069], prostaglandins [1070], *cis*-retinoids [1071], gabosines [1072], and other products of biological interest [1073–1075]. Propargylic sulfinates are rearranged into allenic sulfoxides, useful dienophiles [1059].

Other allylic sulfur compounds such as sulfilimines and sulfoxides can undergo (2,3)-sigmatropic rearrangements. Examples are listed in Figure 5.50 [1076] together with the rearrangements of selenium analogs [1073, 1077, 1078].

The (3,2)-sigmatropic rearrangements of selenoxides into allylic selenates are exergonic and are much quicker than the corresponding rearrangements of allylsulfoxides into sulfinates that are usually endergonic [1079]. Allylic selenates are hydrolyzed readily and provide the corresponding allylic alcohols. An example of enantioselective conversion of achiral allyl selenides into enantiomerically enriched allylic alcohols is shown in Scheme 5.91 [1080]. The method relies upon the enantioselective oxidation of the selenide precursor with the Katzuki–Sharpless catalyst [1081–1084].

Treatment of allylic selenides with *t*-BuOCl gives the corresponding allylic chloroselenuranes as only products. Their subsequent nucleophilic displacement with lithium *N*-protected amides generates the corresponding allylic selenimides that are rearranged quickly into *N*-allyl selenamides. Hydrolysis of the latter produces corresponding allyl amines. Starting with enantiomerically pure 2-*exo*-hydroxy-10-bornyl derivatives, enantiomerically enriched allyl amines have been obtained (Scheme 5.92) [1085].

Allylic oxidation of alkenes with selenium dioxide produces (*E*)-allylic alcohols predominantly [1086, 1087]. As proposed by Sharpless and Lauer [1088], the process involves first an oxaseleno ene-reaction (Section 5.7.3) that give a β,γ-unsaturated seleninic acid intermediate which, on its turn, undergoes a quick (2,3)-sigmatropic rearrangement forming an allyl selenoxylic acid [1089, 1090]. The latter can be hydrolyzed into the corresponding allylic alcohol and $Se(OH)_2$ or oxidized by H_2O_2 (or by t-BuO_2H) as illustrated in Scheme 5.93. The ene-reaction of SeO_2 is sensitive to the bulk of the alkene, which explains the chemoselectivity observed [1091].

The (2,3)-Wittig rearrangement of metallated allylic ethers was discovered in 1949 [955, 1092].

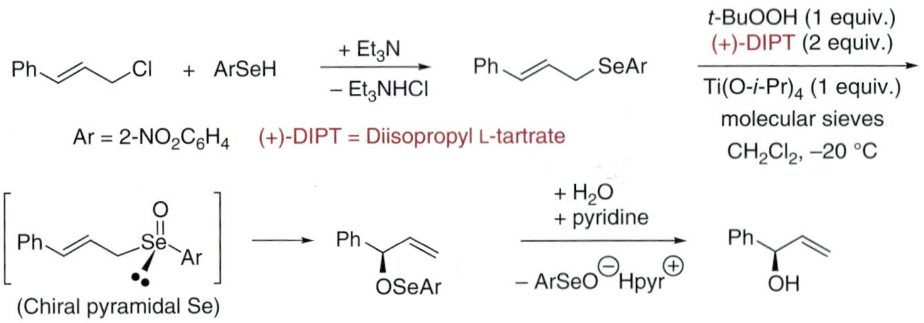

Scheme 5.91 Enantioselective synthesis of an allylic alcohol through enantioselective oxidation of an allylic selenide and subsequent fast (2,3)-sigmatropic rearrangement selenoxide → selenate.

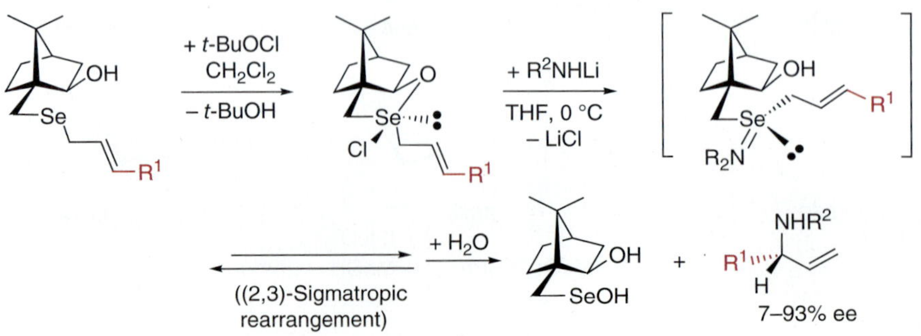

Scheme 5.92 Asymmetric synthesis of allyl amines via the (2,3)-sigmatropic rearrangement Se-allyl selenimide → *N*-allyl selenamide.

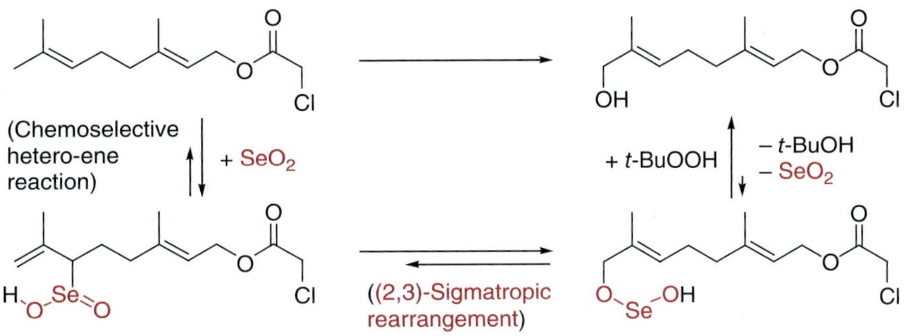

Scheme 5.93 Chemoselective allylic oxidation catalyzed by selenium dioxide.

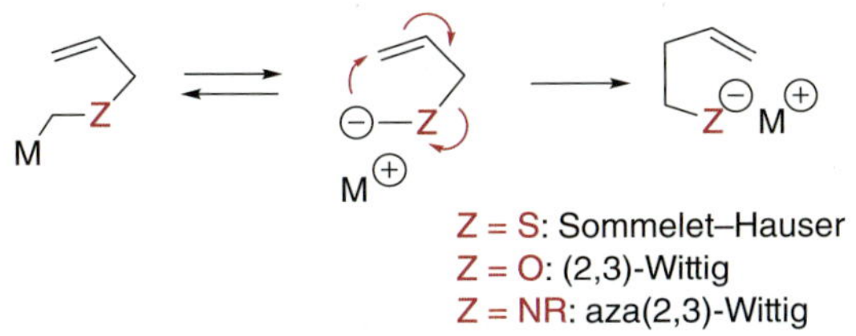

It can be stereoselective [1093] as shown in Scheme 5.94 for (*Z*)-allyl ethers [1094]. It is usually less stereoselective for (*E*)-allyl ethers. The diastereoselectivity of the reactions can be explained by considering five-membered transition structures in which gauche interactions are minimized [1095–1097]. The (2,3)-Wittig rearrangement has now become a powerful tool in synthetic chemistry [1098–1107].

There are several options for the asymmetric (2,3)-Wittig rearrangement [1108]. If the starting allylic ether is enantiomerically enriched, chirality transfer (diastereoselective reaction) leads to enantiomerically enriched products. Alternatively, α-metallation of an achiral ether can use an enantiomerically pure chiral base that leads to enantioselectivity as illustrated (Scheme 5.95) with the total asymmetric synthesis of (+)-eupomatilone 2, a natural lignan isolated from the Australian shrub *Eupomatia bennettii* [1108b]. In this case, the favored transition state involves coordination of the organolithium intermediate with an enantiomerically pure chiral bis(oxazoline) ligand.

Aza (2,3)-Wittig rearrangements of allyl alkyl amines are more difficult than for their ether analogs. This is due to the lower exothermicity of the isomerization aminoalkylmetal → metal amide than of the corresponding oxyalkylmetal → metal alcoholate isomerization. Nevertheless, substituted vinyl aziridines are excellent substrates for the aza-(2,3)-Wittig rearrangement [1109] as illustrated with the synthesis of indolizidine 209D (Scheme 5.96), a noncompetitive blocker of neuromuscular transmission isolated in minute amount from dendrobatid frogs. The synthesis starts from the enantiomerically enriched allylaziridine **569**, which is metallated with LDA at low temperature and is immediately isomerized into lithium amide **570**. After aqueous work-up, the alkene is hydrogenated and the carboxylic ester reduced with $LiAlH_4$ into the corresponding primary alcohol. Swern oxidation of the latter ($DMSO + (ClCO)_2 \rightarrow [Me_2S{-}OCOCOCl]^+ + Cl^- \rightarrow [Me_2S{-}Cl]^+Cl^- + CO + CO_2$; $R^1R^2CHOH + [Me_2S{-}Cl]^+Cl^- + 2Et_3N \rightarrow [Me_2S{-}O{-}C(H)R^1R^2]^+ + 2Cl^- + [Et_3NH]^+ + Et_3N \rightarrow Me_2S + 2[Et_3NH]^+ + 2Cl^- + R^1COR^2$) gives the corresponding aldehyde that is olefinated (Wittig olefination with a stabilized Wittig reagent, Section 5.3.7) into acrylate derivative

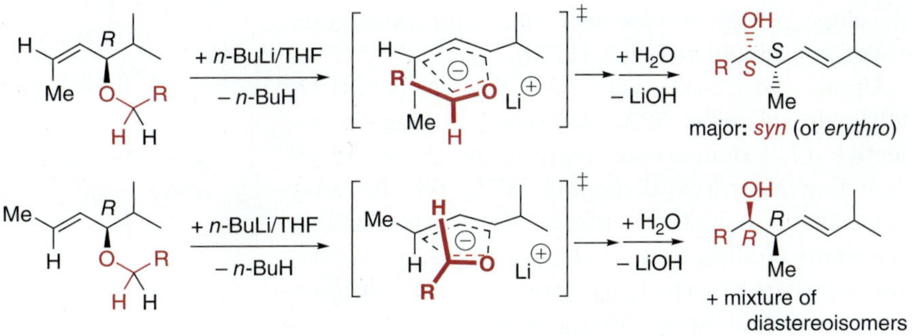

Scheme 5.94 Chirality transfer and diastereoselectivity in the (2,3)-Wittig rearrangement.

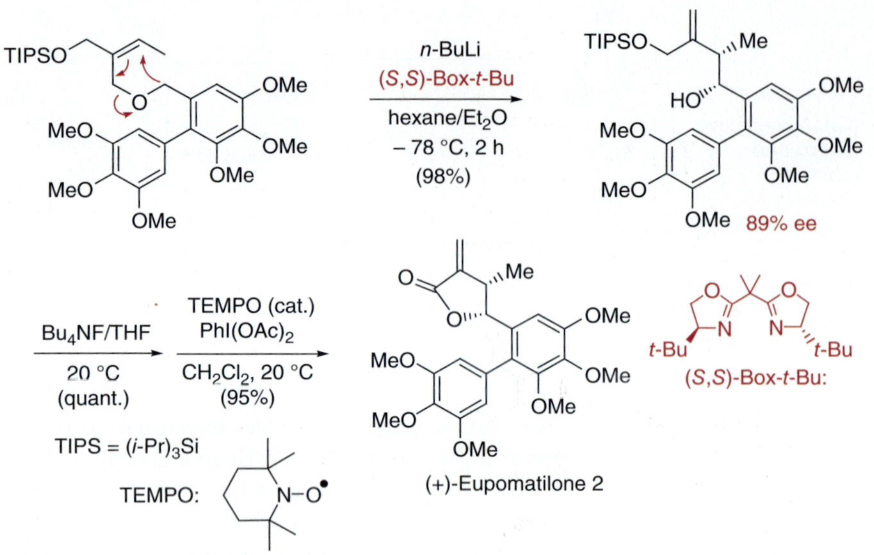

Scheme 5.95 Asymmetric synthesis of natural (+)-eupomatilone 2 applying an enantioselective metallation of a benzyl allyl ether and subsequent stereoselective (2,3)-Wittig rearrangement.

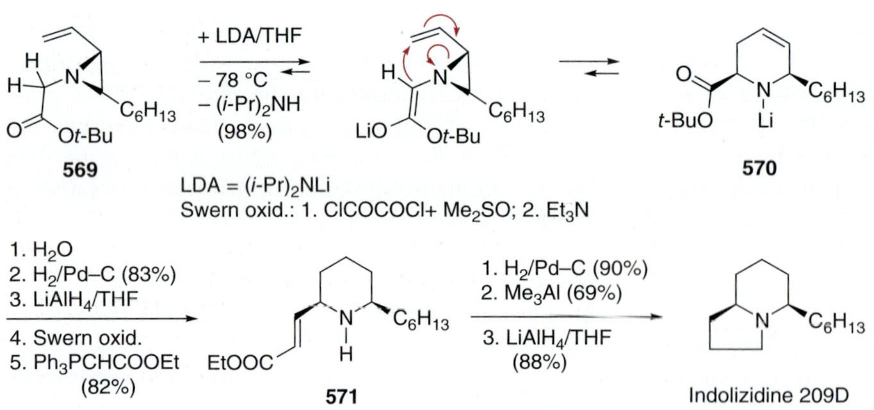

Scheme 5.96 Asymmetric synthesis of indolizidine 209D applying an aza-(2,3)-Wittig rearrangement.

571. Palladium-catalyzed alkene hydrogenation (Section 7.8.1) gives an saturated ester that undergoes intramolecular amidification promoted by Me_3Al (Section 7.2.2). Reduction of the resulting lactam with $LiAlH_4$ provides indolizidine 209D [1110]. Examples of diastereoselective acyclic aza-(2,3)-Wittig rearrangements have been reported [1111, 1112].

BF_3- and BBr_3-mediated (2,3)-sigmatropic rearrangements of allylic α-amino amides have been developed. They provide secondary amines in good yields as exemplified below [1113].

Upon treatment with *n*-BuLi, *N*-benzyl-*O*-allylhydroxylamine **572** undergoes a diastereoselective (2,3)-sigmatropic rearrangement to afford *N*-benzyl-*N*-hydroxyallylamine **573**, which, after subsequent reduction, affords the corresponding *N*-benzyl-*N*-allylamine [1114, 1115]. In this case, the *syn*-diastereoselectivity arises from the lithium coordination to the methoxy group.

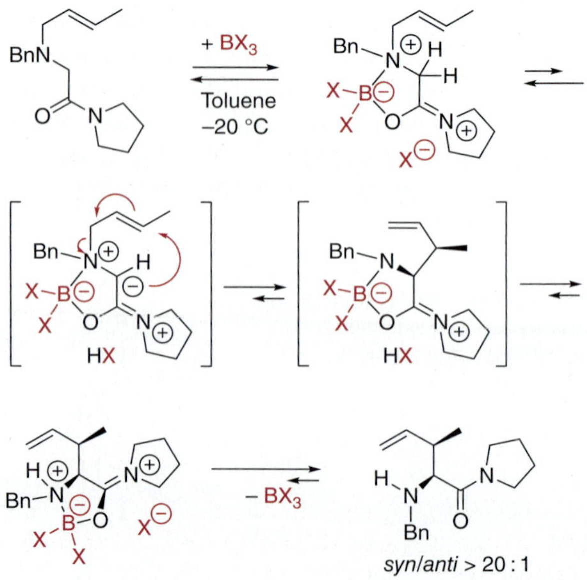

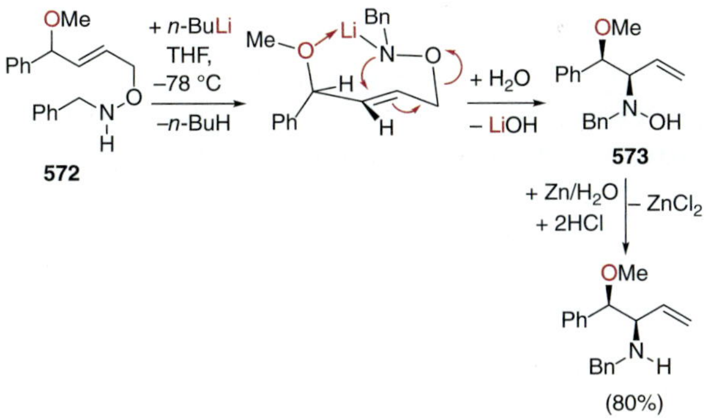

[Allylic ylides](#) of type **575** (obtained by deprotonation of the corresponding onium salts **574**) undergo (2,3)-sigmatropic rearrangements into corresponding homoallyl derivatives **576** [1116]. Ylidic precursors such as amines [1117–1122], ethers [1122–1126], sulfides [1127–1137], and halogens have been investigated. Most of these reactions can be catalyzed and made enantioselective [1138–1140].

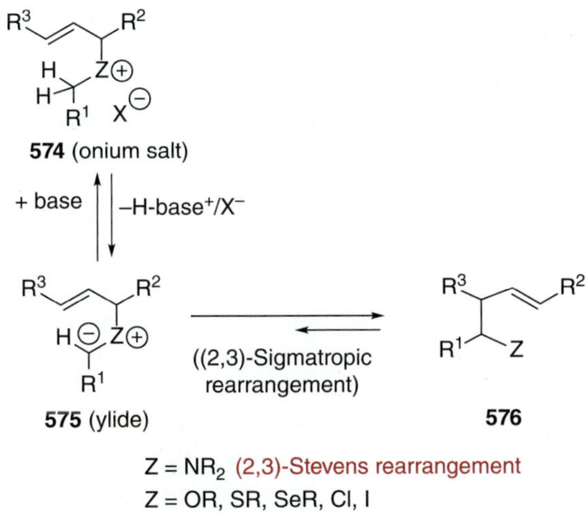

An example of stereoselective (2,3)-sigmatropic rearrangement of an unstabilized nitrogen ylide is given with reaction **577** → **578** [1141].

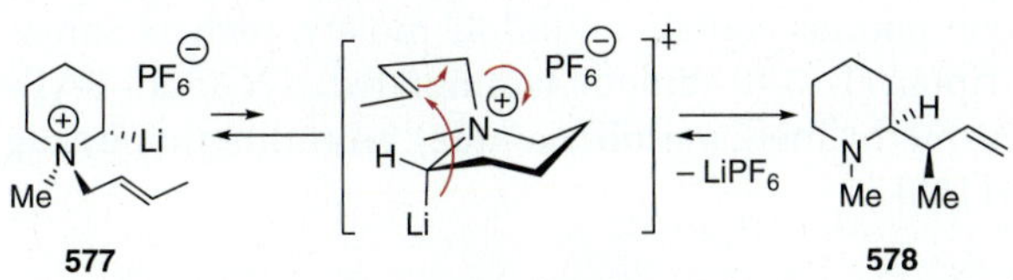

Carbenemetal species (Sections 7.7.2, 7.7.11, and 8.4.2) cyclopropanate alkenes and react with ethers to generate the corresponding oxonium ylides. In the presence of a copper salt α-diazoacetate, **579** engenders a carbenoid intermediate that reacts in an intramolecular manner both with the methoxy group and the (*Z*)-alkene moiety, the formation of ylide **580** being preferred by a factor of 8 : 1 over the cyclopropanation. The ylide is rearranged into the macrolactone **581**. In the presence of an enantiomerically pure bis(oxazoline ligand), the process becomes enantioselective [1142].

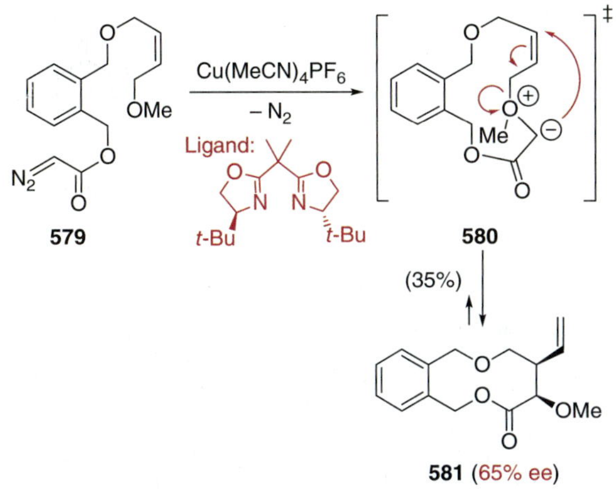

The sulfur ylide mediated ring expansion involving the intramolecular (2,3)-Wittig rearrangement has been developed by Vedejs and coworkers as a useful method for medium ring synthesis [1143, 1144]. Sulfide **582** adds to 2-oxobut-1-yl trifluoromethanesulfonate giving a sulfonium salt that is deprotonated with K_2CO_3 producing the stabilized sulfur ylide **583**. The latter undergoes a (2,3)-sigmatropic rearrangement into **584** [1145].

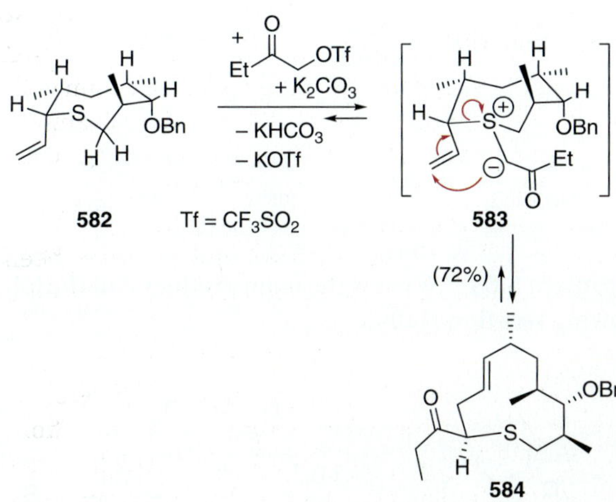

Problem 5.60 Complete the following scheme of the synthesis of Yomogi alcohol.

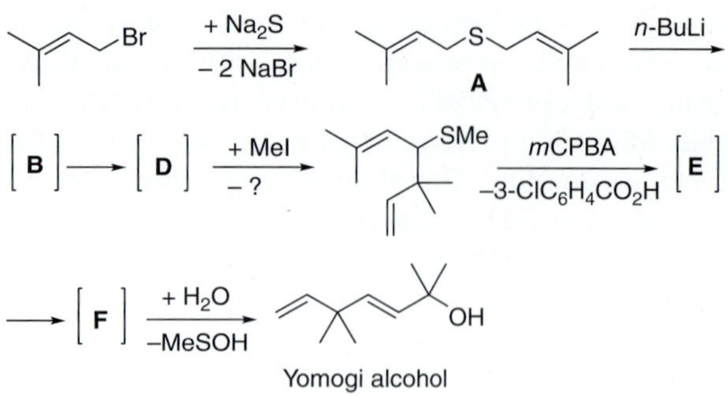

Problem 5.61 Explain the following transformations.

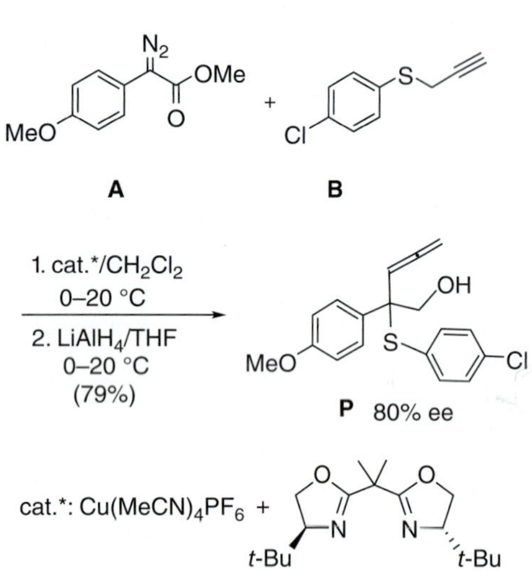

Problem 5.62 Complete the following scheme of the synthesis of **P**.

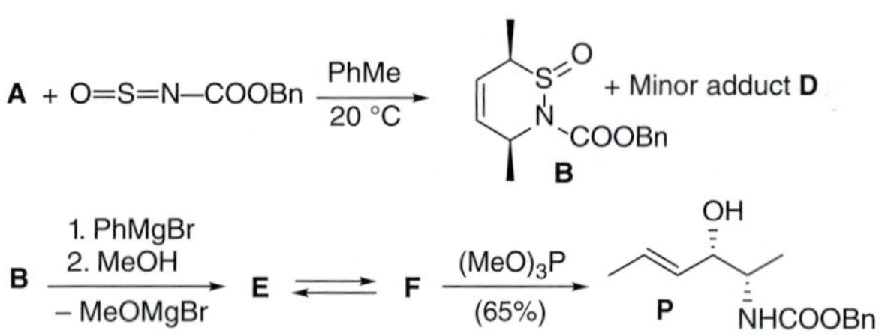

Problem 5.63 What is the main product **P** of the following reaction [1106].

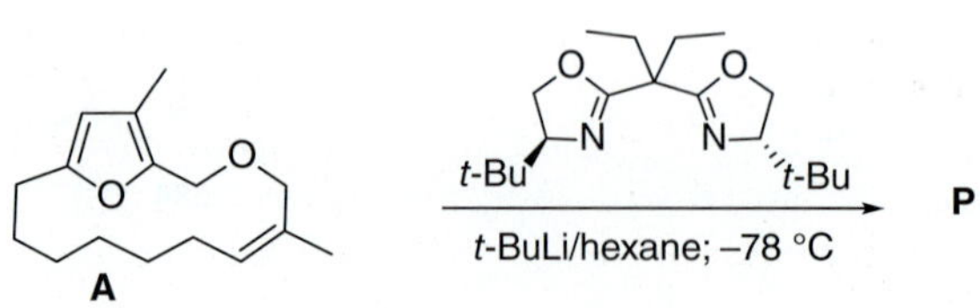

5.5.9 (3,3)-Sigmatropic rearrangements

(3,3)-Sigmatropic rearrangements are defined as processes to migrate a σ bond of two connected allyl systems from the C(1)-position to the C(3)-position. Described the first time in 1940, the Cope rearrangement (5.58) is the prototype of such process as only C—C bonds are reorganized during the course of the reaction (see Section 3.4.3) [1146]. Other synthetically useful (3,3)-sigmatropic rearrangements are the Claisen (5.59), aza-Claisen (5.60), and thia-Claisen rearrangements (5.61). They represent examples of hetero-Cope rearrangements [1147–1150].

(5.58) $Z = CH_2$: Cope rearrangement
(5.59) Z = O: Claisen rearrangement
(5.60) Z = RN: aza-Claisen rearrangement
(5.61) Z = S: thia-Claisen rearrangement

5.5.9.1 Fischer indole synthesis (3,4-diaza-Cope rearrangement)

The first example of (3,3)-sigmatropic rearrangement came with the Fischer indole synthesis discovered by Fischer and Jourdan [1151, 1152]. It can be seen as a 3,4-diaza-Cope rearrangement (Scheme 5.97). Under acidic conditions (e.g. HCl, H_2SO_4, polyphosphoric acid, *p*-toluenesulfonic acid, $ZnCl_2$, $FeCl_3$, and $AlCl_3$), an arylhydrazine reacts with an aldehyde or a ketone to generate the corresponding hydrazone, which isomerizes into the respective enamine (or ene-hydrazine). After protonation, a cyclic (3,3)-sigmatropic rearrangement occurs producing an imine that equilibrates with an aminoacetal that, on its turn, loses an equivalent of NH_3 producing the corresponding indole [1153, 1154]. The intermediate hydrazones can be obtained by palladium-catalyzed nucleophilic displacement of aryl bromides (Section 7.12.2) with hydrazones (Buchwald modification of the Fischer indole synthesis) [1155–1157]. Several important bioactive compounds contain an indole moiety, such as Sumatriptan (1-[3-(2-dimethylaminoethyl)-1*H*-indol-5-yl]-*N*-methylmethanesulfonamide), an antimigraine drug [1158].

5.5.9.2 Claisen rearrangement and its variants (3-oxa-Cope rearrangements)

Discovered in 1912, the *ortho*-aryl-Claisen rearrangement is a (3,3)-sigmatropic rearrangement that

Scheme 5.97 Fischer indole synthesis

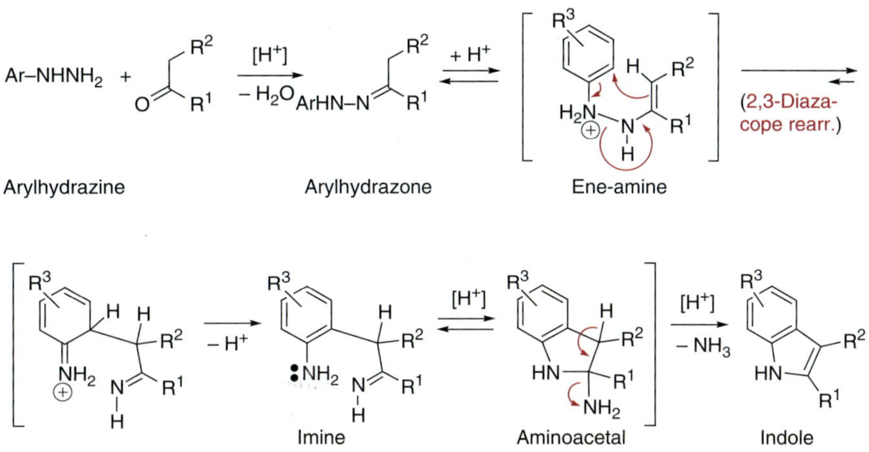

Figure 5.51 Examples of (3,3)-sigmatropic rearrangements: (a) ortho-Claisen, (b) para-Claisen, and (c) aliphatic Claisen rearrangement.

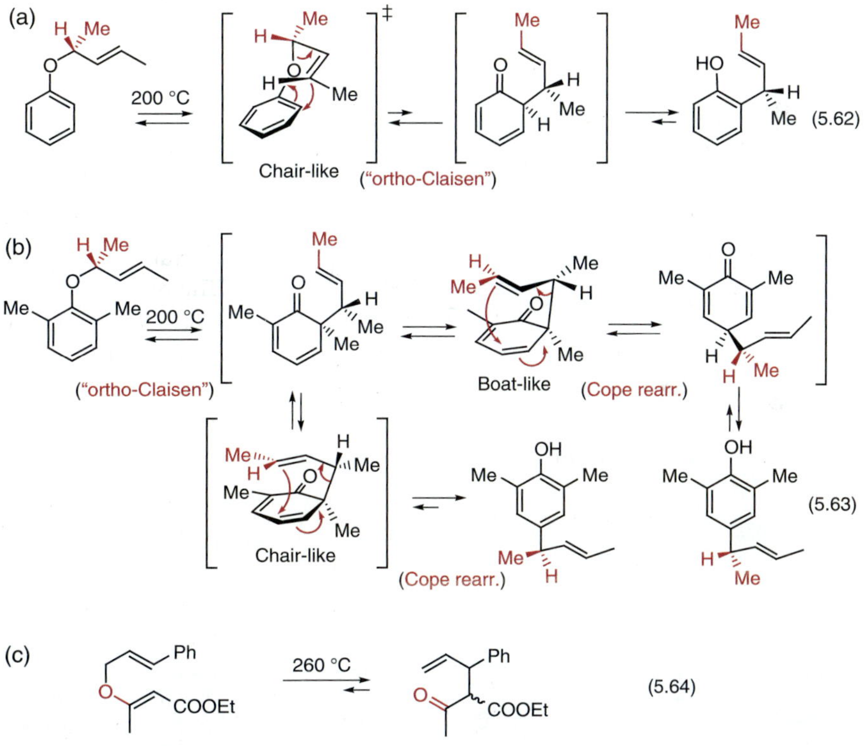

585 **586**

converts allyl aryl ethers into *ortho*-allylphenols on heating (e.g. reaction (5.62), Figure 5.51a) [1159, 1160]. When both *ortho* positions of the aryl group are substituted, *para*-aryl-Claisen rearrangements are observed [1161]. The latter reactions imply two successive (3,3)-sigmatropic rearrangements (e.g. reaction (5.63), Figure 5.51b). Both the *ortho*-aryl- and *para*-aryl-Claisen rearrangements are monomolecular processes that involve six-membered cyclic transition states, as confirmed by the observation that enantiomerically enriched (−)-α,γ-dimethylallyl phenyl ether is isomerized into optically active (+)-2-(α,γ-dimethylallyl)-phenol, and (−)-α,γ-dimethylallyl-2,6-xylyl ether is isomerized into optically active (−)-4-(α,γ-dimethylallyl)-2,6-xylol [1162]. These results are consistent with chair or boat-shaped six-membered transition structures. Quantum mechanical calculations find that the chair-like transition structures are lower in energy than their respective boat-like counterparts for the *ortho*-aryl Claisen rearrangements of 3-phenoxy-(*E*)- and -(*Z*)-but-2-ene [1163].

Bergmann and Corte [1164] and Lauer and Kilburn [1165] reported the isomerization (5.64) of ethyl β-cinnamyloxycrotonate (**585**) into ethyl 2-oxo-4-phenylhex-5-ene-3-carboxylate (**586**) (Figure 5.51c), the first example of an aliphatic-Claisen rearrangement.

In 1938, Hurd and Pollack reported the archetypical rearrangement (5.59) of allyl vinyl ether into pent-4-enal that requires heating to 200 °C [1166]. In general, the thermal aliphatic Claisen rearrangements are one-step, monomolecular processes involving cyclic transition states [1167] as confirmed by their negative entropies of activation (for the gas phase isomerization (5.59): $\Delta^{\ddagger}H = 25.7 \pm 0.3$ kcal mol^{-1}; $\Delta^{\ddagger}S = -14.1 \pm 0.7$ eu) [1168, 1169]. For the rearrangement of allyl vinyl ether into pent-4-enal (reaction (5.59)), transition state spectroscopy (femtochemistry) and quantum mechanical calculations suggest that the allylic C—O bond breaking and new C—C bond forming processes are not synchronous [1170], in agreement with the diradicaloid model for the transition structure of this reaction (Figure 5.52, Scheme 5.111).

Nature utilizes the aliphatic Claisen rearrangement in the biosynthesis of aromatic α-amino acids (the shikimate pathway). Chorismate **587**

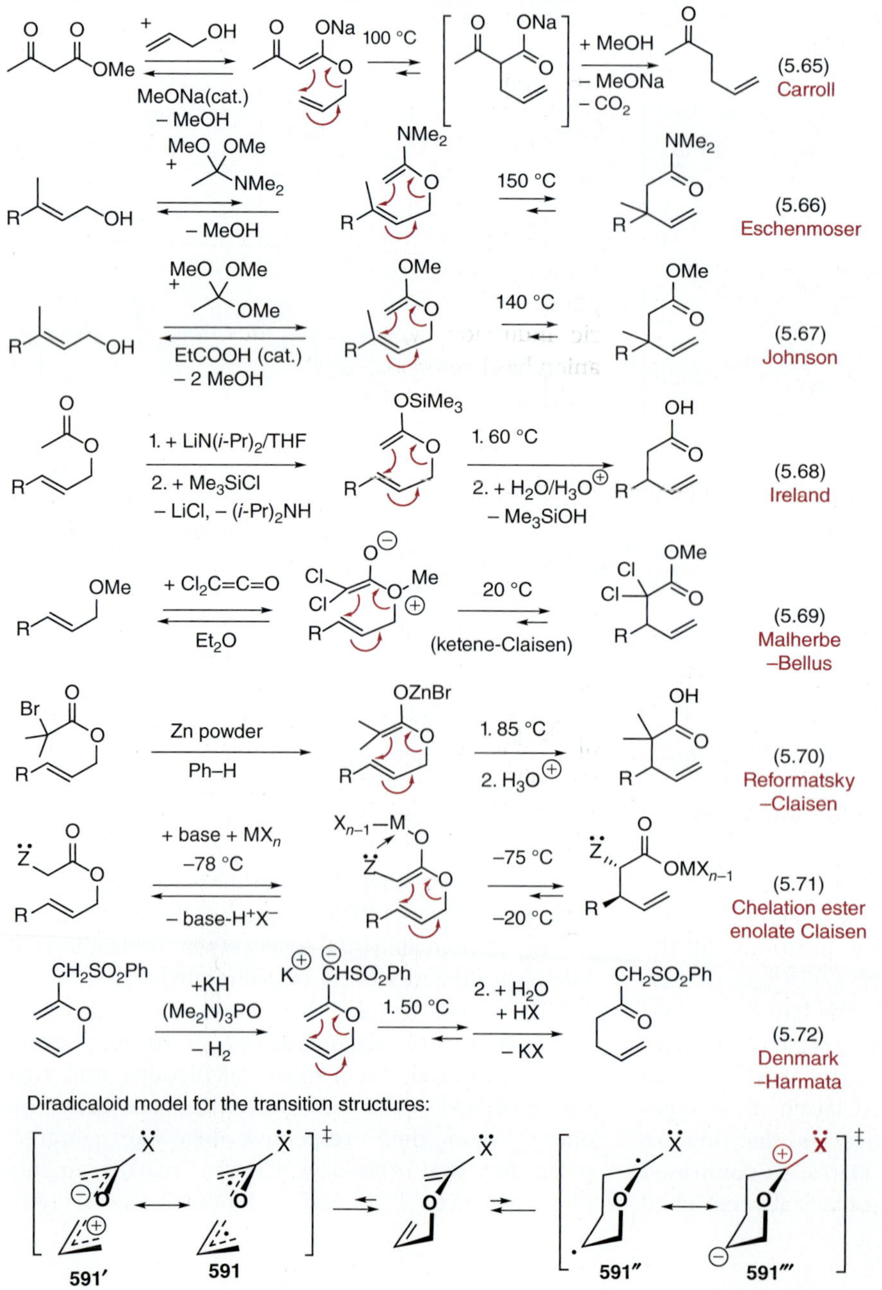

Figure 5.52 Variants (5.65)–(5.72) of the aliphatic Claisen rearrangement.

is isomerized stereospecifically into prephenate **589** via the chair-shaped transition structure **588** [1171–1173]. In the absence of the enzyme (chorismate mutase), reaction **587** → **589** also occurs with the same high stereoselectivity, but 10^6 times more slowly [1174]. This (3,3)-sigmatropic rearrangement has been catalyzed by monoclonal antibodies raising against haptene **590**, which mimics transition structure **588** [1175]. The Claisen rearrangement and its variants (Figure 5.52, Schemes 5.98–5.102) are a very important reaction for the stereoselective construction of C—C bonds [1176–1178].

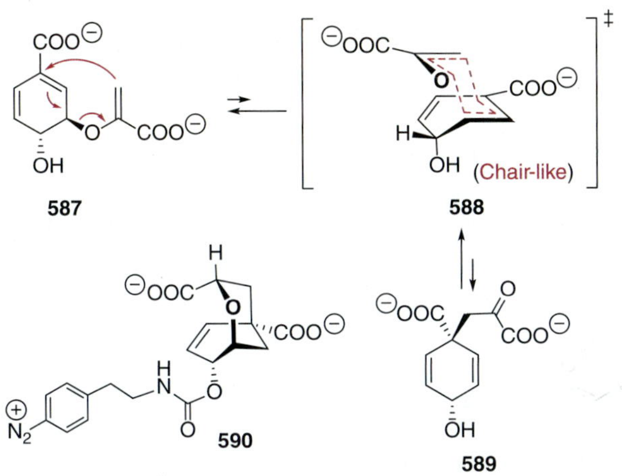

The rate of the Claisen rearrangement depends on the nature of the substituents. In general, radical stabilizing groups accelerate the reaction [1170], in agreement with the diradicaloid model **591** ↔ **591′** ↔ **591″** ↔ **591‴** (Figure 5.52). Formally, there are two limiting mechanisms for the (3,3)-sigmatropic rearrangements: (i) a dissociative mechanism that generates an allyl/oxaallyl radical pair **591**, which maintains a strong interaction between the two radicals through electron exchange and for which the charge transfer configuration **591′** can be written, and (ii) an associative mechanism that involves the formation of the 2-oxacyclohexa-1,4-diyl diradical **591″**, which can be stabilized by electron exchange and for which the zwitterionic limiting structure **591‴** can be written. This model is supported by kinetic deuterium isotopic effects (Figure 5.54) [1179].

It explains the large acceleration effect observed for the (3,3)-sigmatropic rearrangements of allyl 1-oxyalkenyl and allyl/1-dialkylaminoalkenyl systems (Figure 5.52) compared with the thermal Claisen rearrangement of allyl vinyl ether and of its alkyl substituted derivatives [1180, 1181]. The oxy and dialkylamino substituents stabilize both the diradical form **591″** and zwitterionic form **591‴** in the Claisen variants reported by Carroll (reaction (5.65)) [1182], Eschenmoser (reaction (5.66)) [1183, 1184], Johnson (reaction (5.67)) [1185], Ireland (reaction (5.68)) [1186–1188], Malherbe and Bellus (reaction (5.69)) [1189, 1190], and Baldwin (the Reformatsky–Claisen rearrangement (6.70)) [1191, 1192]. The enolates and their Lewis acid complexes of alk-2-enyl glycinates undergo fast ester-enolate-Claisen rearrangements (reaction (5.71)) that permit the enantioselective synthesis of substituted γ,δ-unsaturated α-amino acids when the Lewis acid (MX_n) is enantiomerically enriched [1193–1196]. A significant acceleration is also observed with the Denmark–Harmata variant (5.72) that involves a sulfonyl-stabilized allyl anions [1197], phosphine oxide, phosphate, or phosphamide-stabilized anions [1198]. Asymmetric induction with chiral phosphorous stabilized anion has been realized [1199].

Claisen rearrangements are accelerated by 1-substitution of the alkenyl moiety of the allyl alkenyl ether by electron-releasing groups as predicted by the diradicaloid model **591″** ↔ **591‴**. Model **591** ↔ **591′** predicts rate enhancement of the Claisen rearrangement for allyl alkenyl ethers bearing an electron-releasing group at C(2) of their alkenyl unit (Figure 5.52). This was indeed observed by Koreeda and Luengo with the Claisen variant called anionic oxy-Claisen rearrangement shown in Scheme 5.98 [1200]. Depending on solvent and on the counterion of the enolate, tight ion pair (**592**), solvent-separated ion pair (**592′**), or free ion pair forms. The larger the cation M^+ and the more polar the solvent, the more M^+ dissociates from the enolate anion, the more the negative charge becomes available to the transition state of the reaction. This explains why the anionic oxy-Claisen rearrangement (Scheme 5.98) is faster in THF (polar, solvates M^+) than in toluene (nonpolar) and for large counterion K^+ (less associated with

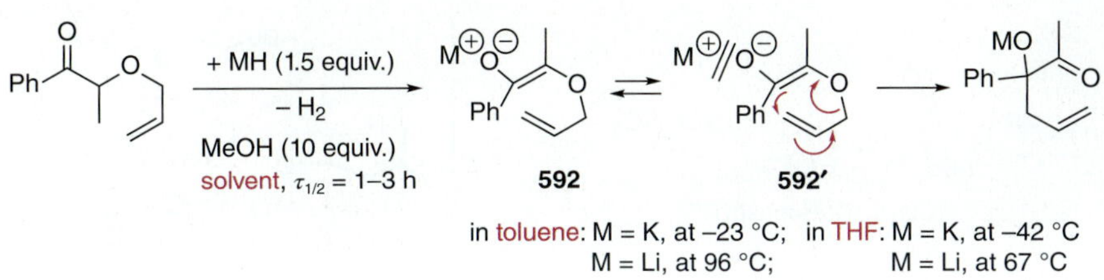

Scheme 5.98 Example of an anionic oxy-Claisen rearrangement. It is faster when the counterion dissociates from the enolate anion.

the enolate) than for the small Li^+ cation (strongly associated with the oxygen atom of the enolate) [1201]. The large rate enhancements observed for oxy-Claisen rearrangements compared with the rearrangements of alkyl substituted allyl vinyl ethers are similar to the accelerations observed for the oxy-Cope rearrangements (Figure 5.52, Section 5.5.9.7).

Rychnovsky and Lee reported that geranyl 1-methylvinyl ether is rearranged already at 0 °C in the presence of triisobutylaluminum [1202]. The latter forms an oxonium Lewis complex **593** that undergoes a Claisen rearrangement into **594**. Intramolecular hydride transfer reduces the ketone so-obtained (Scheme 5.99). The high-rate enhancement of the Claisen rearrangement upon formation of the oxonium species is an example of charge-accelerated Claisen rearrangement. The rate enhancement is explained by the diradicaloid model for the transition state of this reaction that admits considerable contribution of the allyl/oxoalkyl radical pair and ion pair **595** ↔ **595′**. The Lewis acid coordination to the oxygen atom stabilizes the enolate-limiting structure **595′** significantly. Charge-accelerated sulfonium (3,3)-sigmatropic rearrangement has also been reported [1203].

Using an enantiomerically enriched chiral Lewis acid, enantioselective Claisen rearrangements [1203–1209] can be realized as illustrated in Scheme 5.100 [1210, 1211].

Hydrogen bond donors such as protic solvents [1212, 1213], ureas, and thioureas [1214, 1215] accelerate the Claisen rearrangements. Water itself accelerates the Claisen rearrangements (on-water acceleration) [691]. Enantiomerically enriched chiral guanidinium salts promote the Claisen rearrangement of *O*-allyl α-ketoesters and induce enantioselectivity (Scheme 5.101) [1216]. This is ascribed to the stabilization of the developing negative charge on the oxyallyl fragment (see the diradicaloid model **591** ↔ **591′**) in the transition state **596** of the pericyclic reaction by N—H hydrogen bridging. Secondary interactions are also involved between the 2-phenylpyrrol ring of the catalyst and the substrate [1217].

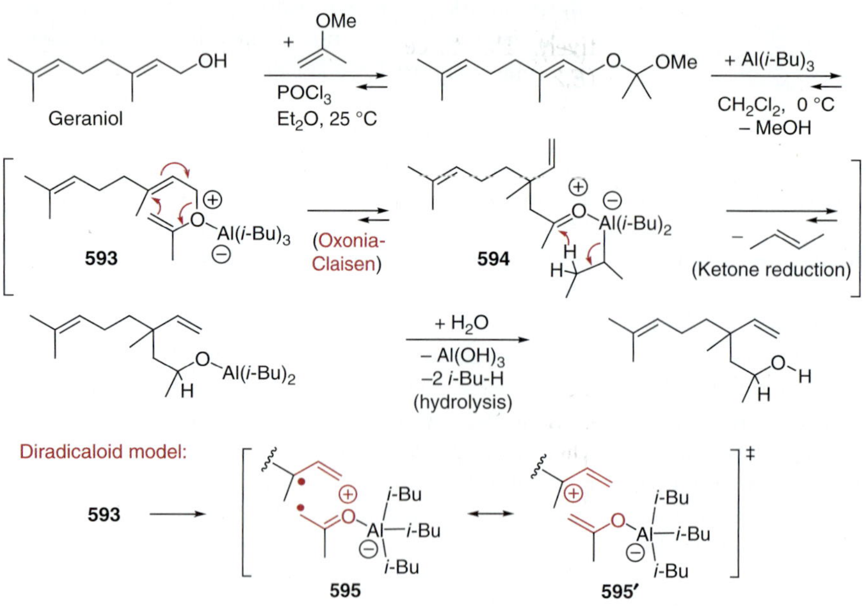

Scheme 5.99 Claisen rearrangement is strongly accelerated by coordination of the ethereal moiety to a Lewis acid.

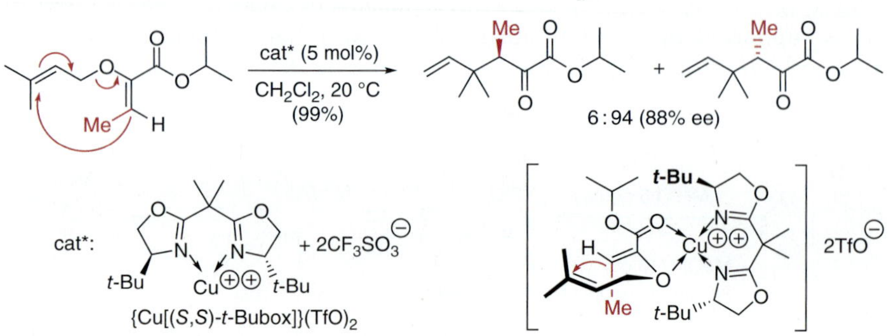

Scheme 5.100 Example of a catalytic enantioselective Claisen rearrangement.

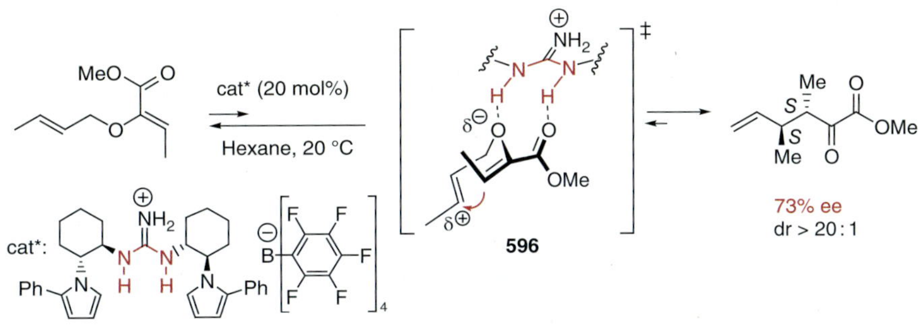

Scheme 5.101 Example of a guanidinium-catalyzed enantioselective Claisen rearrangement.

The Coates–Claisen rearrangement (5.73) of enols and acetals of unsaturated aldehydes produces 5-alkoxy γ,δ-unsaturated carbonyl compounds [1218, 1219].

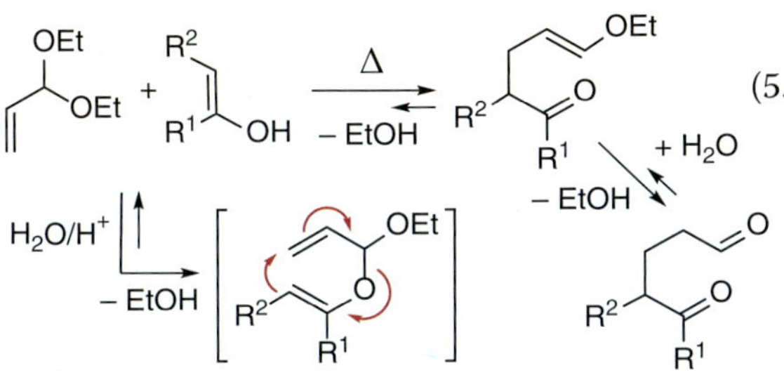

(5.73)

The Saucy–Marbet–Claisen rearrangement (5.74) converts propargyl ethers into β-substituted allenyl carbonyls [1220]. An example of asymmetric synthesis of allenyl oxindoles using an enantiomerically enriched Pd(II) catalyst is shown in Scheme 5.102 [1221]. As for the enantioselective aliphatic Claisen rearrangements of Schemes 5.100 and 5.103, the chiral cationic palladium complex coordinates both the carbonyl group of the ester substituent and the ethereal group of the propargyl vinyl ether moiety. This activates the (3,3)-sigmatropic rearrangement and induces the enantioselectivity. The oxoindoles are central building blocks for the construction of indole alkaloids and are attractive templates for pharmaceutical agents [1222–1228].

The aliphatic Claisen rearrangement is a suprafacial, concerted, asynchronous (Scheme 5.103) pericyclic process that may be seen as an intramolecular S_N2' alkylation. In the case of (*E*,*E*)-but-2-enyl propenyl ether (*E*,*E*)-**597**, a 95.9 : 4.1 mixture of racemic *syn*- and *anti*-2,3-dimethylpent-4-enal is obtained at 142 °C (Scheme 5.103). Two enantiomeric chair-like transition states **Tc** and *ent*-**Tc** leading to (*S*,*R*)-**598** and (*R*,*S*)-**598**, respectively, are preferred over the two enantiomeric boat-like transition states **Tb** and *ent*-**Tb** that lead to (*S*,*S*)-**598** and (*R*,*R*)-**598**, respectively. The three other isomeric ethers (*Z*,*Z*)-**597**, (*E*,*Z*)-**597**, and (*Z*,*E*)-**597** are also isomerized into the mixture of racemic *syn*- and *anti*-**298** showing a c. 95 : 5 preference for chair-like transition states [1181, 1229, 1230]. If gauche interactions between the substituents of the alkenyl allyl ether destabilize the chair-like more than the boat-like transition structures, boat-like transition states might be preferred as illustrated with rearrangement **597′** → **599** [1231, 1232].

5.5.9.3 Aza-Claisen rearrangements (3-aza-Cope rearrangements)

The thermal aromatic aza-Claisen rearrangements are difficult reactions as they require heating to

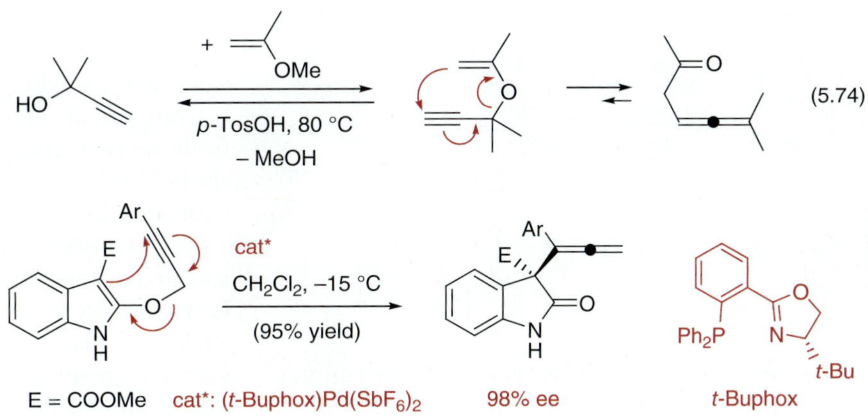

(5.74)

Scheme 5.102 Catalytic enantioselective Saucy–Marbet–Claisen rearrangement.

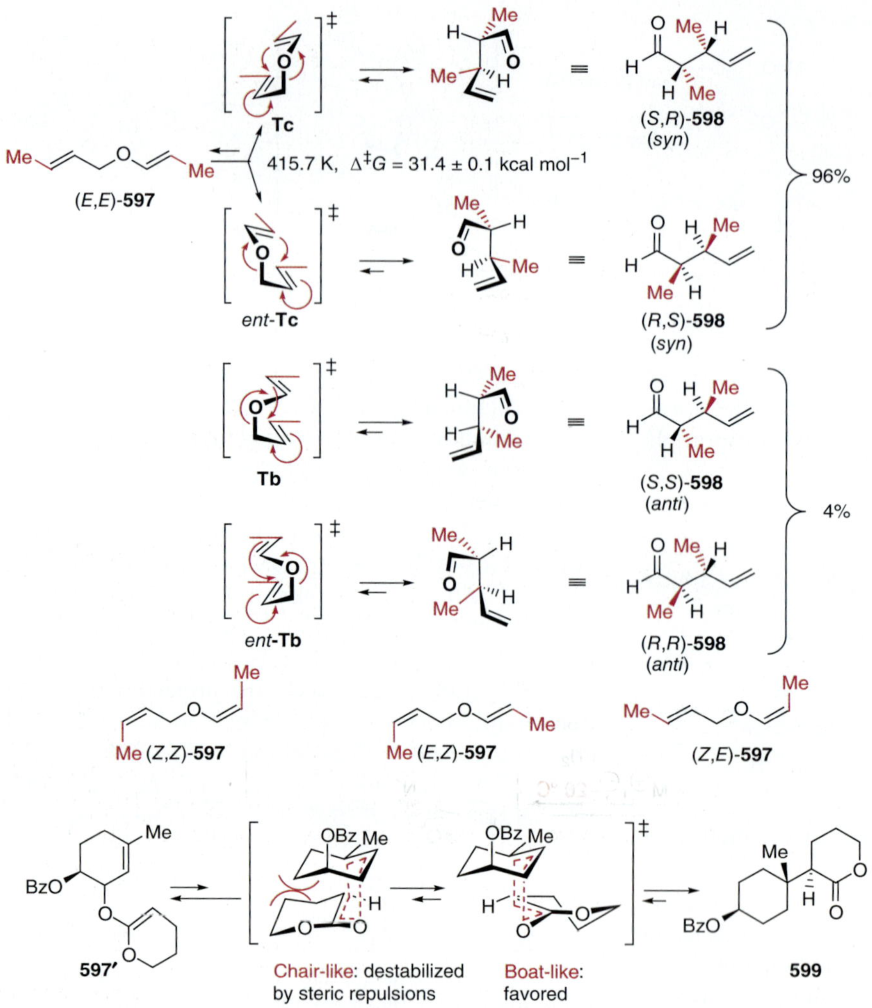

Scheme 5.103 Chair-like transition states are usually preferred in acyclic Claisen rearrangement unless the chair structures are destabilized more than the boat transition structures by gauche interactions between the substituents.

200–350 °C [1233]. Nevertheless, they are catalyzed by CF_3CO_2H [1234, 1235]. In the case of reaction (5.75), the reaction occurs at 80 °C in water [1236]. Formation of ammonium ions by protonation transforms the aza-Claisen rearrangements into charge-accelerated aza-Claisen rearrangements [1237].

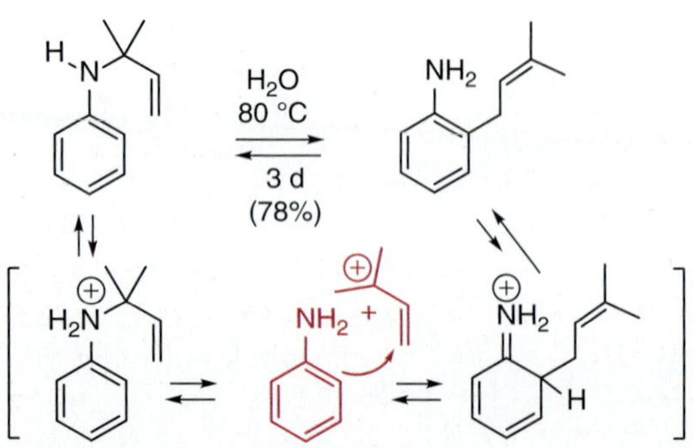

(5.75)

The rearrangements of *N*-alkenyl *N*-allylamines (e.g. aliphatic 3-aza-Claisen rearrangement (5.76), Scheme 5.104) are very difficult thermally. When an electron-releasing oxy [1238] or amino group [1239] substitutes the N(3) position, the thermal reaction is somewhat easier. In the presence of a Lewis acid such as $TiCl_4$, $BF_3{\bullet}Et_2O$, $AlCl_3$, Me_2AlCl, $MeAlCl_2$, Me_3Al, etc., the rearrangements are strongly accelerated as illustrated in Scheme 5.104 with reaction (5.77) [1150, 1240]. The aza-Claisen (or 3-aza-Cope) rearrangements are also catalyzed by Pd(0) complexes [1241]. Aza-Claisen rearrangements of tertiary ammonium derivatives are facile thermal reactions and have been used extensively [1242] in the alkaloid synthesis as illustrated with reaction (5.78) [1243–1246], an example of alkyne-carbonester-Claisen rearrangement [1247]. A similar rearrangement occurs at room temperature for the zwitterions resulting from the addition tertiary allyl amines and cyclic α-vinylamines with acetylenic sulfones (reaction (5.79)) [1248]. Water elimination from *N*-allyl amides using $Ph_3P/CCl_4/Et_3N$ or

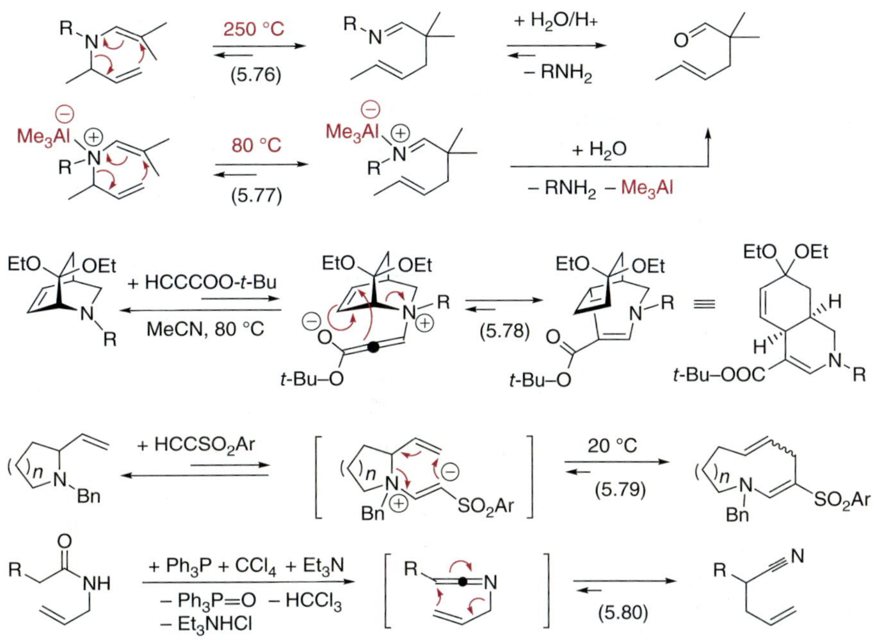

Scheme 5.104 Example of aza-Claisen and charge-accelerated aza-Claisen rearrangements.

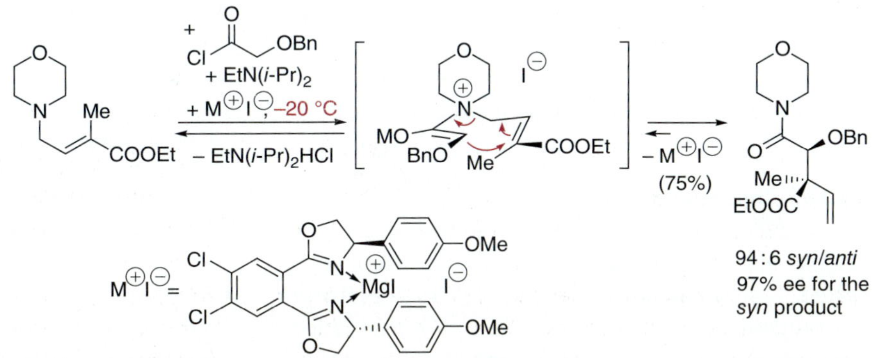

Scheme 5.105 Example of an asymmetric acyl-aza-Claisen rearrangement.

$I_2/P(OEt)_3/Et_3N$ generates 3-azahexa-1,2,5-triene that are rearranged at 20 °C into the corresponding γ,δ-unsaturated carbonitriles (reaction (5.80)) [1249].

The reaction of a tertiary amine with an acyl chloride generates a first intermediate acyl ammonium chloride that eliminates an equivalent of HCl in the presence of a base giving a zwitterionic species. The latter can be coordinated to a Lewis acid that promotes the subsequent (3,3)-sigmatropic rearrangement into the corresponding γ,δ-unsaturated amide. An asymmetric version of this rearrangement called acyl-aza-Claisen rearrangement is presented in Scheme 5.105 [1207].

5.5.9.4 Overman rearrangement (1-oxa-3-aza-Cope rearrangement)

In 1974, Overman reported that trichloroacetamidates derived from allylic alcohols undergo (3,3)-sigmatropic rearrangements (for instance, reaction (5.81)) at 120 °C [1250] (Scheme 5.106a). The rearrangements are strongly accelerated (by a factor up to 10^{12}) by $Hg(O_2CCF_3)_2$. In the latter case, the reaction follows a two-step mechanism with the formation of an intermediate of type **600** resulting from an intramolecular mercurioimination of the allyl ether moiety (Scheme 5.106b). Subsequent elimination generates the final trichloroacetamides that are more stable than the initial trichloroacetamidate tautomers. After hydrolysis of the latter, allyl amines are obtained that are important synthetic intermediates [1251, 1252]. The Overman rearrangement is also catalyzed by Pd(II) complexes [1253]. In 1997, Overman and coworkers found the first enantioselective catalyst able to induce enantioselective Overman rearrangements [1254]. Since then, a large variety of chiral catalysts have been developed that permit the obtainment of allyl amines with high ee's [1255–1260].

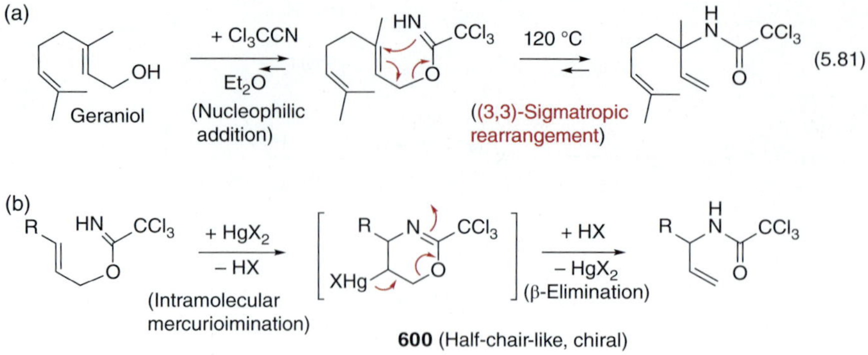

Scheme 5.106 Overman aza-Claisen rearrangements: (a) uncatalyzed, thermal reaction and (b) HgX_2-catalyzed. When HgX_2 is replaced by an enantiomerically pure Pd(II) complex, enantioselective rearrangements are possible.

5.5.9.5 Thia-Claisen rearrangement (3-thia-Cope rearrangement)

Hruby and coworkers have developed the thia-Claisen rearrangement for the synthesis of *anti*-β-functionalized γ,δ-unsaturated α-amino acids (Scheme 5.107) [1261]. As for the Claisen and aza-Claisen rearrangements that are strongly accelerated by formation of oxonium and ammonium ions, respectively (see above), the formation of sulfonium species also leads to charge-accelerated thia-Claisen rearrangements [1262]. This can be realized using a thiophilic Lewis acid catalyst such as $FeBr_3$.

5.5.9.6 Cope rearrangements

The Cope rearrangements of simple hexa-1,5-dienes [1146] are considered to follow one-step, concerted suprafacial/suprafacial (3,3)-sigmatropic rearrangements (noted $[\pi^2s+\sigma^2s+\pi^2s]$) that adopt six-membered ring chair-like transition structures (Scheme 5.108) [338, 1023, 1263, 1264]. Labeled hexa-1,5-diene **601** equilibrates with **602** exclusively. No isomeric product **602′** + **602″** resulting from formal (1,3)-sigmatropic rearrangements or of recombination of two allyl radical is observed below 300 °C.

As shown in Section 3.4.3, the Cope rearrangement (5.58) of unsubstituted hexa-1,5-diene has activation parameters $\Delta^{\ddagger}H = 33.5 \pm 0.5\,\text{kcal mol}^{-1}$ and $\Delta^{\ddagger}S = -13.8 \pm 1.0$ eu (210–258 °C) [1265]. An alternative two-step mechanism involving cyclohexa-1,4-diyl diradical intermediate such as **604** (Scheme 5.108, Figure 5.53) would require $\Delta^{\ddagger}H > 36.6\,\text{kcal mol}^{-1}$ [1266]. Pyrolysis of hexa-1,5-diene into two allyl radicals gives $DH°(\text{allyl}^\bullet/\text{allyl}^\bullet) = 56.1 \pm 1.6\,\text{kcal mol}^{-1}$ (625–900 K) [1267]. Thus, one can estimate a degree of concert for the Cope rearrangement (5.58) of $(56.1 \pm 1.6) - (33.5 \pm 0.5) = 22.6 \pm 2.1\,\text{kcal mol}^{-1}$. As shown in Figure 5.53, the FMO theory (see also Figure 4.29, Section 4.7.5) predicts a strongly stabilizing interaction between two allyl radicals that are maintained in six-membered ring structure (like

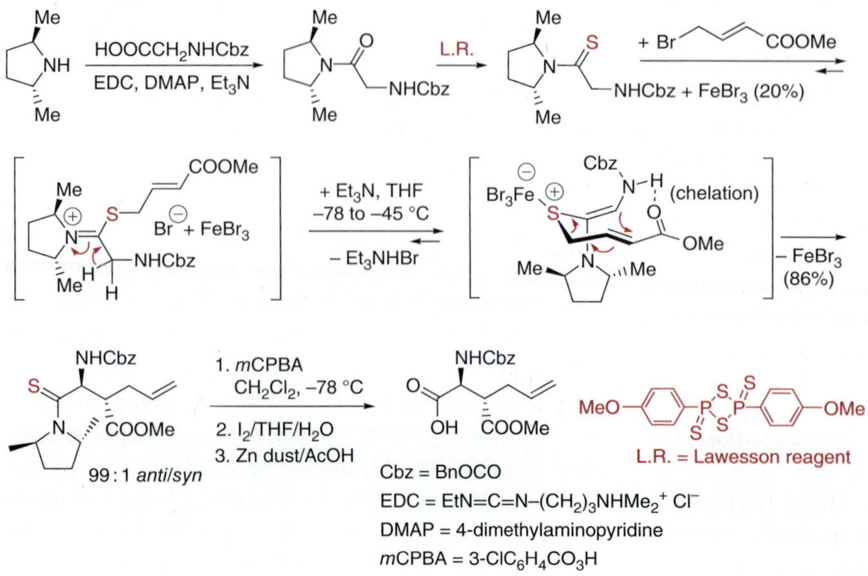

Scheme 5.107 Example of a charge-accelerated thia-Claisen rearrangement applied to the asymmetric synthesis of *anti*-β-functionalized, γ,δ-unsaturated α-amino acids.

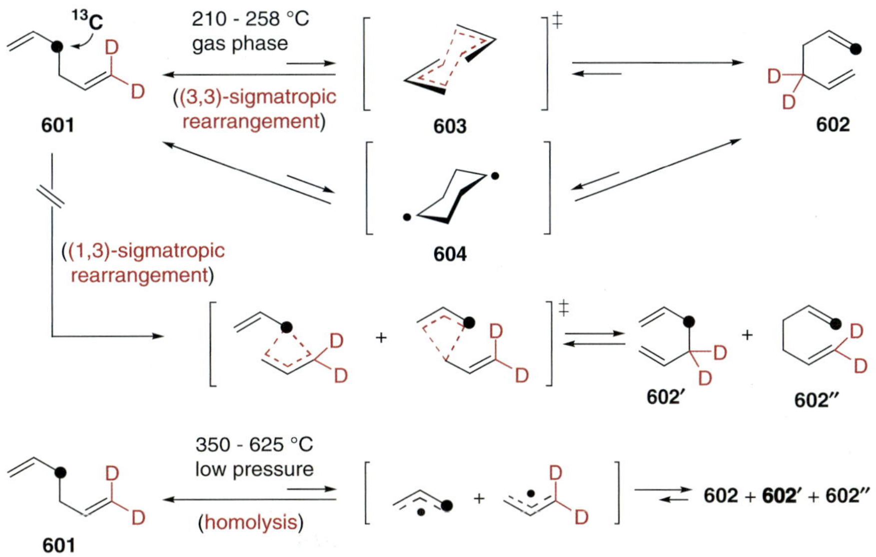

Scheme 5.108 The Cope rearrangement (reaction below 300 °C) follows an associative mechanism with the formation of a six-membered transition state **603** or cyclohexa-1,4-diyl diradical intermediate **604**. Thermochemical calculations, kinetic deuterium isotopic effects, as well as quantum mechanical calculations support the one-step concerted mechanism via chair-like transition structure **603**.

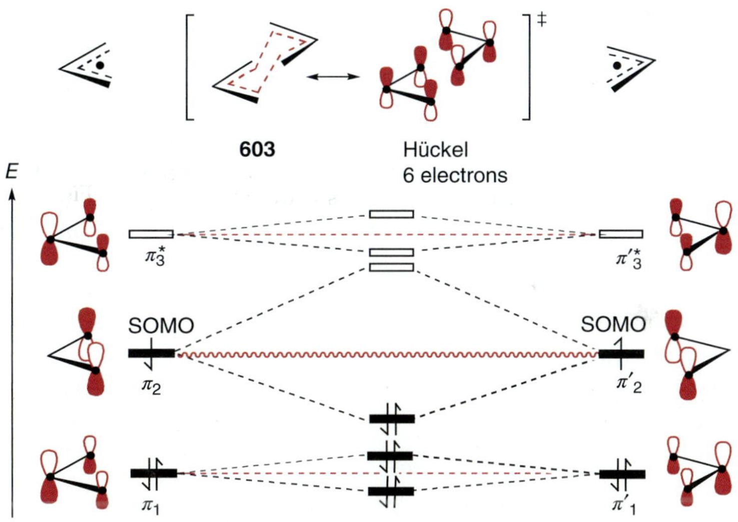

Figure 5.53 FMO diagram showing the origin of the energy of concert in the Cope rearrangement of simple hexa-1,5-dienes. As for cyclohexane, the chair-like conformation of the transition state is more stable than the boat-like conformation.

a suprafacial/suprafacial [π^3s+π^3s]-cycloaddition). Most of the stabilization arises from the overlap between the two allyl radical SOMOs.

On heating, enantiomerically enriched (*R*,*E*)-3-methyl-3-phenylhepta-1,5-diene is isomerized exclusively into (*S*,*E*)- and (*R*,*Z*)-3-methyl-6-phenylhepta-1,5-diene. The nonformation of (*S*,*Z*) and (*R*,*E*)-3-methyl-6-phenylhepta-1,5-diene confirms a chair-like transition state. If 2,5-dimethyl-2-phenylcyclohexa-1,4-diyl radicals in their chair conformation should have been formed as intermediates, the results (nearly complete transfer of chirality) demonstrate that these hypothetical intermediates do not have the time to undergo chair/chair interconversion [1268]. In contrast, Wessel and Berson have found that the gas phase pyrolysis of (*R*,*E*)-5-methylocta-1,2,6-triene (an allene–alkene system) gives (*R*,*E*)-4-methyl-3-methylidenehepta-1,5-diene with 68% retention of configuration, which implies the formation of 16% of the enantiomer (*S*,*E*)-4-methyl-3-metylidenehepta-1,5-diene [1269]. This is interpreted in terms of the formation a *cis*-2,6-dimethyl-3-methylidenecyclohexa-1,4-diyl diradical intermediate that has the time to undergo chair/chair interconversion and to generate two possible enantiomeric products. On their side, Roth and coworkers have found that heating hepta-1,2,6-triene (**605**) generates 3-methylidenehexa-1,5-diene (**607**). The same product is obtained by pyrolysis of **610**, **611**, **612**, or **613** (Scheme 5.109). The results are consistent with the formation of 2-methylidenecyclohexa-1,4-diyl diradical (**606**), which lives long enough (>10^{-10} s) for the bimolecular reaction with SO_2, giving **608**, and with O_2, giving hydroperoxide **609**, to be observed

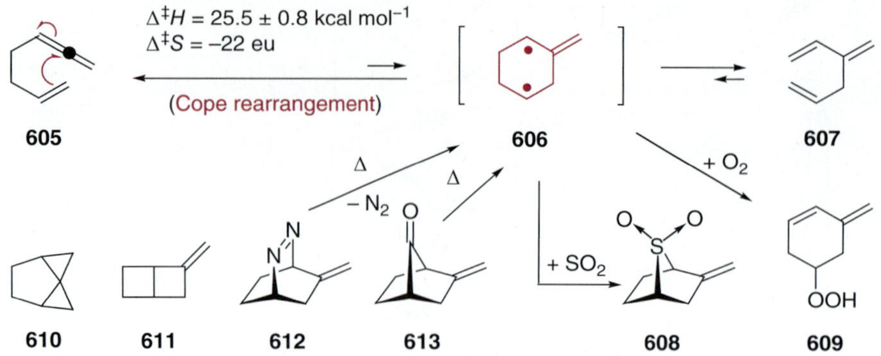

Scheme 5.109 Bimolecular quenching of 2-methylidenecyclohexa-1,4-diyl diradical in the Cope rearrangement of hepta-1,2,6-triene.

[1270, 1271]. Compared with cyclohexa-1,4-diyl diradical (**604**), diradical **606** benefits allylic stabilization and from the lower endothermicity arising from the fact that an allene instead of an alkene moiety has added to an alkene unit when going from **605** to **606**. This makes **606** only 16.6 kcal mol^{-1} above **605**, which is significantly less than the enthalpy of activation of 25.5 ± 0.8 kcal mol^{-1} measured for isomerization **605** ⇄ **607**.

Dewar and Wade proposed that the Cope rearrangement (5.82) of 2,5-diphenylhexa-1,5-diene involves the intermediacy of diradical **614**, which is stabilized by conjugative stabilization by the two phenyl groups [1272, 1273]. The experimental $\Delta^{\ddagger}H$ = 21.3 ± 0.3 kcal mol^{-1} measured [1274, 1275] for this degenerate rearrangement is compatible with the latter hypothesis as one estimates $\Delta_r H°$(2,5-diphenylhexa-1,5-diene ⇄ diradical **614**) = 20.8 ± 4.2 kcal mol^{-1}.

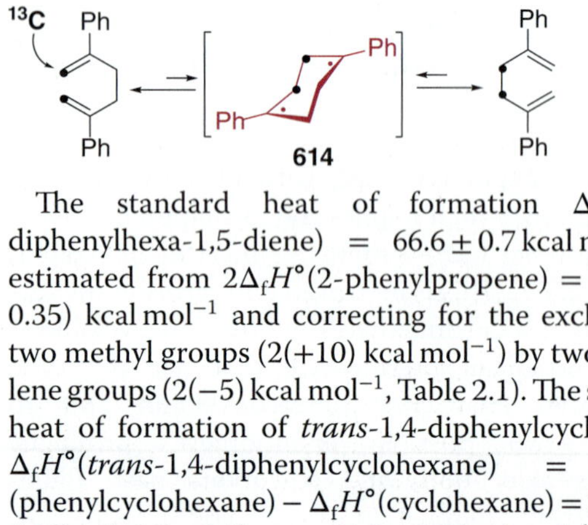

(5.82)

The standard heat of formation $\Delta_f H°$(2,5-diphenylhexa-1,5-diene) = 66.6 ± 0.7 kcal mol^{-1} is estimated from 2$\Delta_f H°$(2-phenylpropene) = 2(28.3 ± 0.35) kcal mol^{-1} and correcting for the exchange of two methyl groups (2(+10) kcal mol^{-1}) by two methylene groups (2(−5) kcal mol^{-1}, Table 2.1). The standard heat of formation of *trans*-1,4-diphenylcyclohexane $\Delta_f H°$(*trans*-1,4-diphenylcyclohexane) = 2$\Delta_f H°$ (phenylcyclohexane) − $\Delta_f H°$(cyclohexane) = 2(−4.0 ± 0.35) − (29.8 ± 0.2) = 21.8 ± 0.9 kcal mol^{-1}. Thus, $\Delta_f H°$(*trans*-1,4-diphenylcyclohexa-1,4-diyl diradical: **614**) = $\Delta_f H°$(*trans*-1,4-diphenylcyclohexane) + 2*DH*° (PhCMe$_2$•/H•) − *DH*°(H•/H•) = (21.8 ± 0.7) + 2(84.9 ± 1.3) − (104.2 ± 0.2) = 87.4 ± 3.5 kcal mol^{-1}. This gives $\Delta_r H°$(2,5-diphenylhexa-1,5-diene ⇄ diradical **614**) = (87.4 ± 3.5)−(66.6±0.7) = 20.8 ± 4.2 kcal mol^{-1} (with *DH*°(PhCMe$_2$•/H•) = 84.9 ± 1.3 kcal mol^{-1} taken from Ref. [1276]). Quantum mechanical calculations agree with a two-step mechanism involving diradical intermediate **614** [1277].

Tetraphenyl substituted hexa-1,5-diene *meso*-**615** is equilibrated at 77–115 °C with the racemic *threo*-**615** (1 : 1 mixture of (*S*,*S*)-**615** and (*R*,*R*)-**615**). The phenyl substituent effect on the relative stability of benzyl radical compared with methyl radical amounts to *DH*°(CH$_3$•/H•)–*DH*°(Ph–CH$_2$•/H•) = (104.7 ± 0.2) − (89.7 ± 1.3) = 15.0 ± 1.5 kcal mol^{-1}. The phenyl substituent effect on allyl radical must not be larger than 7.5 kcal mol^{-1} as the spin is delocalized equally onto two carbon centers. Thus, the four phenyl groups of diene **615** are expected to make *DH*°(**616•**/**616•**) = *DH*°(allyl•/allyl•) = 56.2 − 4(7.5) + correction for the stabilization of **615** by π-conjugation = 26.2 + 4 kcal mol^{-1} (Table 2.6) ≈ 30.2 kcal mol^{-1}, a value similar to the enthalpy of activation = 30.7 ± 0.2 kcal mol^{-1} measured for the thermal isomerization *meso*-**615** ⇄ *threo*-**615**. Thus, this reaction might generate two (*E*,*E*)-1,3-diphenylallyl radicals **616•** in its rate-determining step. This hypothesis that explains the non-Cope rearrangement is supported by the observation (ESR) of free radical **616•** on heating *meso*-**615** in diphenyl ether to 251 °C. Enantiomerically enriched (+)-*threo*-**615** is racemized rapidly at 40–65 °C (Scheme 5.110) with activation parameters consistent with a concerted one-step mechanism implying the C_s-symmetrical transition, chair-like structure **617** [1278].

In the More O'Ferrall–Jencks diagram of Figure 3.17 (Section 3.4.4), the trajectory of the Cope rearrangement (5.58) of unsubstituted hexa-1,5-diene from reactant to product (see also Figure 5.54) tells us that the transition structure **603** is in between two interacting allyl radicals and the cyclohexa-1,4-diyl diradical. It does not tells us whether the degree of C(3)—C(4) bond breaking is the same as the degree of bond forming between C(1) and C(6). The transition structure could have a C_{2h} chair structure (synchronous mechanism) or be of lower C_1-symmetry (asynchronous mechanism) in which the extent on σ(C—C) bond breaking and σ(C—C) bond making are not the same. Quantum mechanical calculations

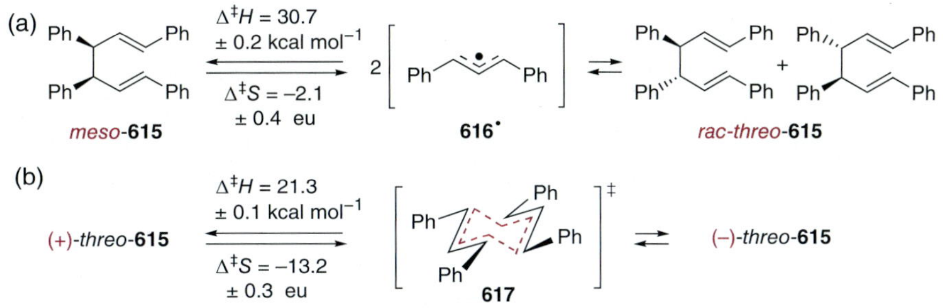

Scheme 5.110 (a) Non-Cope and (b) Cope rearrangement of (*E*,*E*)-1,3,4,6-tetraphenylhexa-1,5-dienes.

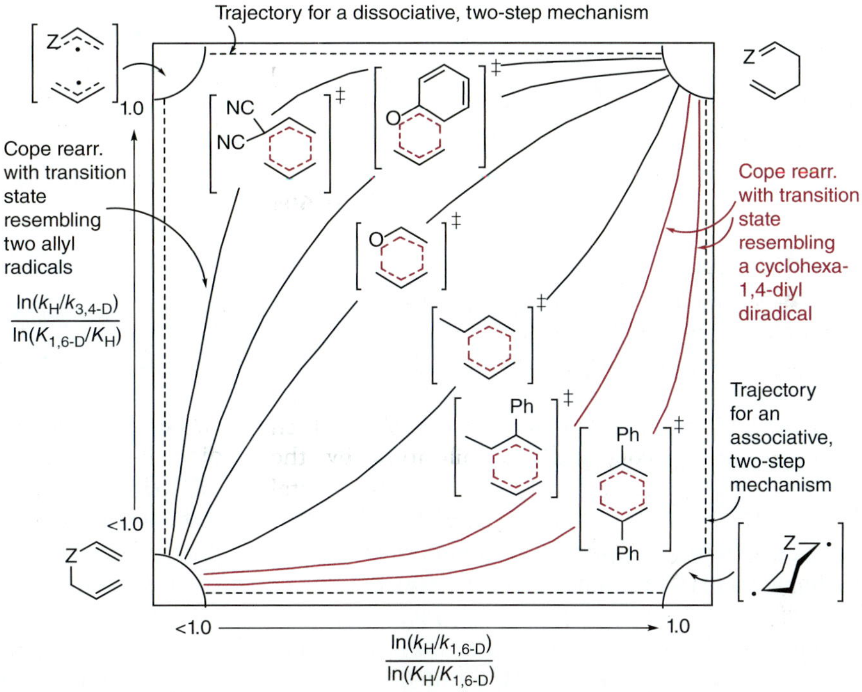

Figure 5.54 More O'Ferrall–Jencks type of diagram for Cope and Claisen rearrangements. The horizontal coordinate represents the shortening of the C(1)—C(6) distance, the vertical coordinate represents the lengthening of the σ(C(3)—C(4)) bond, the enthalpy is perpendicular to the plane made by the two former coordinates. The position of the transition states are deduced from the comparison of kinetic and equilibrium deuterium isotopic effects.

predict for **603** a C_{2h} chair structure with 1.965 Å for the bond length of σ(C—C)bond that forms and the bond that is broken, and 1.407 Å (C—C bond length in allyl radical: 1.386 Å [1279]) for the four other C—C bonds of the two interacting allyl moieties [1280, 1281]. This is not the same for other thermal (3,3)-sigmatropic rearrangements such as the Cope rearrangement of substituted hexa-1,5-dienes and of hetero-Cope rearrangements. Their transition states will resemble either the cyclohexa-1,4-diyl diradical or two allyl radicals with lower symmetry. A transition structure resembling two interacting allyl radicals will be favored by substituents at C(1), C(3), C(4), and C(6), whereas a transition structure that resembles cyclohexa-1,4-diyl diradical is favored by substituents at C(2) and C(5). Doering has suggested the term "chameleonic" to describe such variable transition structures [1282]. On comparing secondary kinetic deuterium isotopic effects with the secondary equilibrium isotopic effects (Section 3.9), one can decide whether a transition state is early (resembles the reactant) or late (resembles the product) in a concerted, one-step reaction (Bell–Evans–Polanyi theory). For instance, if the rate constant ratio $k_H/k_{1,6\text{-D}}$ of the Cope rearrangement of 3-methylhexa-1,5-diene (reaction (5.83)) and of 1,1,6,6-tetradeuterio-3-methylhexa-1,5-diene (reaction (5.84)) should be similar to the equilibrium constant ratio $K_H/K_{1,6\text{-D}}$ for equilibrium (5.83) and (5.84), the transition structure would resemble the cyclohexa-1,4-diyl diradical (the C—H/C—D bond strength are changed in the same way in the transition structure and in the product as one goes from sp^2- to sp^3-hybridized carbon centers). For the reverse reactions, one expects the secondary deuterium isotopic effect $k'_H/k'_{3,4-D}$ to be close to unity, which is different from the equilibrium isotopic effect $1/(K_H/K_{1,6\text{-D}})$. If the transition state should resemble two allyl radicals, the forward reaction would show a kinetic isotopic effect $k_H/k_{1,6\text{-D}}$ close to unity and different from the equilibrium isotopic effect $K_H/K_{1,6\text{-D}}$ (C—H/C—D bonds do not change their strength significantly as sp^2-hybridized carbon atoms are involved in the reactant and the transition state). On the contrary, the kinetic deuterium isotopic effect for the reverse reaction, $k'_H/k'_{3,4-D}$, will be different from unity and similar to $1/(K_{\mathrm{H}}/K_{1,6\text{-D}})$. Thus, if one compares $k_{\mathrm{H}}/k_{1,6\text{-D}}$ with $K_{\mathrm{H}}/K_{1,6\text{-D}}$ and $k'_H/k'_{3,4-D}$ with $1/(K_{\mathrm{H}}/K_{1,6\text{-D}})$, one has the possibility to define the location of the transition state in the More O'Ferrall–Jencks diagram. This has been done by Gajewski and Conrad (Figure 5.54) for substituted hexa-1,5-dienes (Cope rearrangements) and allyl phenyl and allyl vinyl ether (Claisen rearrangements) [1283–1285].

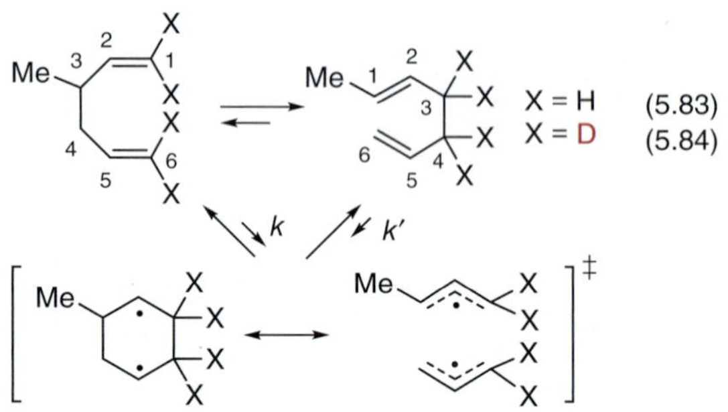

$k_{\mathrm{H}}/k_{1,6\text{-D}} = k(5.83)/k(5.84)$
$K_{\mathrm{H}}/K_{1,6\text{-D}} = K(5.83)/K(5.84)$
$k'_{\mathrm{H}}/k'_{3,4\text{-D}} = k'(\text{reverse of } (5.83))/k'(\text{reverse of } (5.84))$
$K_{\mathrm{H}}/K_{3,4\text{-D}} = 1/(K_{\mathrm{H}}/K_{1,6\text{-D}}) = K_{1,6\text{-D}}/K_{\mathrm{H}}$

These studies support the diradicaloid model (Scheme 5.111) for the "chameleonic" transition structures of the (3,3)-sigmatropic rearrangements and which has already been applied to explain substituent effects (Figure 5.52: **591** ↔ **591′** ↔ **591″** ↔ **591‴**) and catalytic effects (Scheme 5.99: **595** ↔ **595′**) on the rates of Claisen rearrangements.

The Cope rearrangement of *cis*-1,2-alkenylcycloalkanes represents a very efficient synthesis of (*Z*,*Z*)-cycloalkadienes [1286]. In 1960, Vogel first reported the rearrangement (5.85) of 1,2-*cis*-divinylcyclopropane into cyclohepta-1,4-diene [977]. Because of the higher ring strain of cyclopropane compared with that of cyclohepta-1,4-diene (Table 2.3), the reaction is exothermic and occurs much faster (already at room temperature) than the parent rearrangement (5.58) that requires heating above 200 °C (Scheme 5.108) [1287]. The Cope rearrangement (5.86) of *cis*-1,2-divinyl cyclobutane into (*E*,*E*)-cycloocta-1,5-diene occurs at 120 °C [1288]. These rearrangements proceed in a concerted manner via boat-like transition structures (suprafacial/suprafacial (3,3)-sigmatropic rearrangements, also noted $[\pi^2\mathrm{s}+\sigma^2\mathrm{s}+\pi^2\mathrm{s}]$) [1023, 1263, 1289]. This is also the case for rearrangements (5.87) [1290] and (5.88) [1291, 1292]. Semibullvalene undergoes a very fast Cope rearrangement of its *cis*-1,2-divinylcyclopropane unit. Even though the rearrangement of semibullvalene is degenerate ($\Delta_r H = 0$), it is extremely fast ($\Delta^{\ddagger}H = 6\,\mathrm{kcal\,mol^{-1}}$) because of the proximity of the two alkenyl moieties that are forced to approach to each other in the ground state of the reactant, i.e. toward the geometry of the boat-like transition structure (see Problem 3.17, Section 3.6.5).

Z = CH_2 $\Delta^{\ddagger}H$ = 19–20 kcal mol^{-1} (20°C) (5.85)

Z = CH_2CH_2 $\Delta^{\ddagger}H$ = 23–24 kcal mol^{-1} (100°C) (5.86)

Z = NMe_2 $\Delta^{\ddagger}H$ = 18–19 kcal mol^{-1} (20°C) (5.87)

Z = O $\Delta^{\ddagger}H$ = 22–24 kcal mol^{-1} (80°C) (5.88)

Z = H,H $\Delta^{\ddagger}H$ = 33–34 kcal mol^{-1} (210°C) (5.58)

The diradicaloid model (Scheme 5.111) permits to evaluate the substituent effects on the rates of (3,3)-sigmatropic rearrangements. As an illustration, we present (Scheme 5.112) the synthesis of β-sinensal reported by Thomas [1293]. The key synthetic intermediate is **618**, a compound containing an allyl alkenyl ether that can undergo a Claisen rearrangement and a hexa-1,5-diene unit that can undergo a Cope rearrangement. Aliphatic Claisen rearrangements are usually more exothermic than Cope rearrangements. Because of the Dimroth principle or Bell–Evans–Polanyi theory ($\Delta^{\ddagger}H = \alpha\Delta_r H + \beta$), the former rearrangements are expected to be faster than the latter for systems having the same type of substitution. Moreover, because of the favorable charge transfer in the limiting structure allyl radical + oxoethyl radical ↔ allyl cation + enolate anion of the transition state, Claisen rearrangements (Figure 5.54) are also expected to be faster than analogous Cope rearrangements. In **618**, two terminal alkyl groups substitute the allyl vinyl ether moiety. More important is the terminal vinyl substitution that is expected to stabilize significantly the allyl + oxoethyl diradical ↔ ion pair limiting structure **619** ↔ **619′**. For the hexa-1,5-dienyl unit, there is also a vinyl substitution but not at a terminal carbon center. The latter is expected to stabilize the cyclohexa-1,4-diyl diradical limiting structure **620** of the Cope transition state. In this structure, the two alkyl substituents do not contribute to its stability. In agreement with

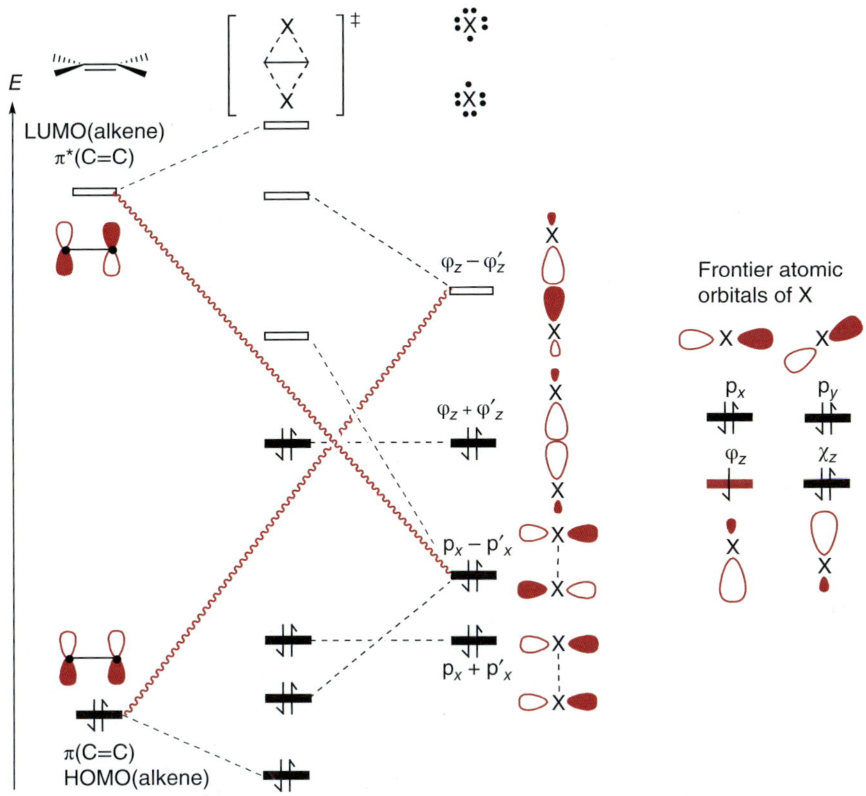

Figure 5.55 FMO diagram representing the transition structure of a concerted 1,2-dyotropic rearrangement in a 1,2-dihalogenoalkane (vicinal dihalide).

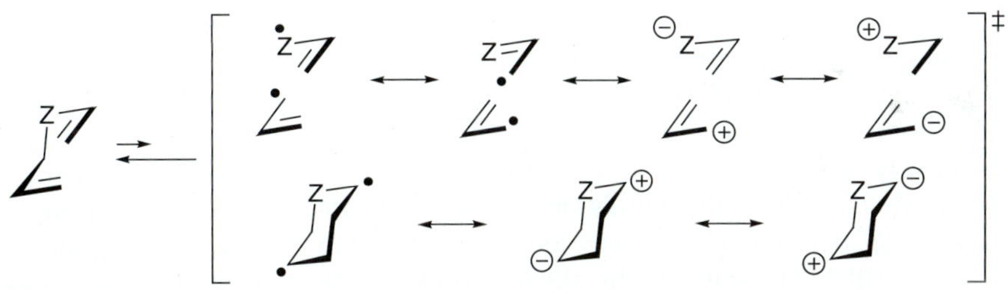

Scheme 5.111 Diradicaloid model for the variable transition structures of the (3,3)-sigmatropic rearrangements.

this prediction, heating of **618** to 100 °C generates intermediate **621** via a Claisen rearrangement which, on its turn, undergoes a Cope rearrangement at 100 °C producing β-sinensal. The latter rearrangement is relatively fast because it is activated by the 3-carbaldehyde and 4-alkyl groups as suggested by the limiting structures **622** ↔ **622′** of the transition state of this rearrangement.

Beginning with Overman's report of a $PdCl_2(NCPh)_2$-catalyzed Cope rearrangement of 1,5-dienes [1294], a number of studies showed that electrophilic metals can mediate this reaction through 4-metallacyclohexyl cation intermediates that are fragmented to the corresponding Cope products under milder conditions than the thermal rearrangements [1295–1297]. When substituents are chosen to render the Cope rearrangement irreversible, enantioselective reaction induced by enantiomerically enriched gold catalysts is possible [1298].

5.5.9.7 Facile anionic oxy-Cope rearrangements

The gas phase Cope rearrangement of dienol **623** produces enone **624** as major product [1299, 1300]. This reaction, called oxy-Cope rearrangement, requires heating to 320 °C ($E_a = 41.8 \pm 0.4$ kcal mol^{-1}, $\log A = 12.5 \pm 0.2$) [1301]. Quantum mechanical calculations predicted a slightly asynchronous but nevertheless concerted transition state for the Cope rearrangement of 3-hydroxyhexa-1,5-diene [1302]. The methoxy-substituted derivative **625** is more reactive as its rearrangement into **626** occurs already at 175 °C ($E_a = 35.9 \pm 1.8$ kcal mol^{-1}, $\log A = 12.6 \pm 0.6$) [1303]. This is expected from the diradicaloid model of Scheme 5.111 that predicts stabilization of the allyl limiting structures (radical and ion pair) by 3-methoxy substitution (Scheme 5.113). In agreement with the diradicaloid model, Evans and Golob found that the potassium alcoholate of **625** undergoes its Cope

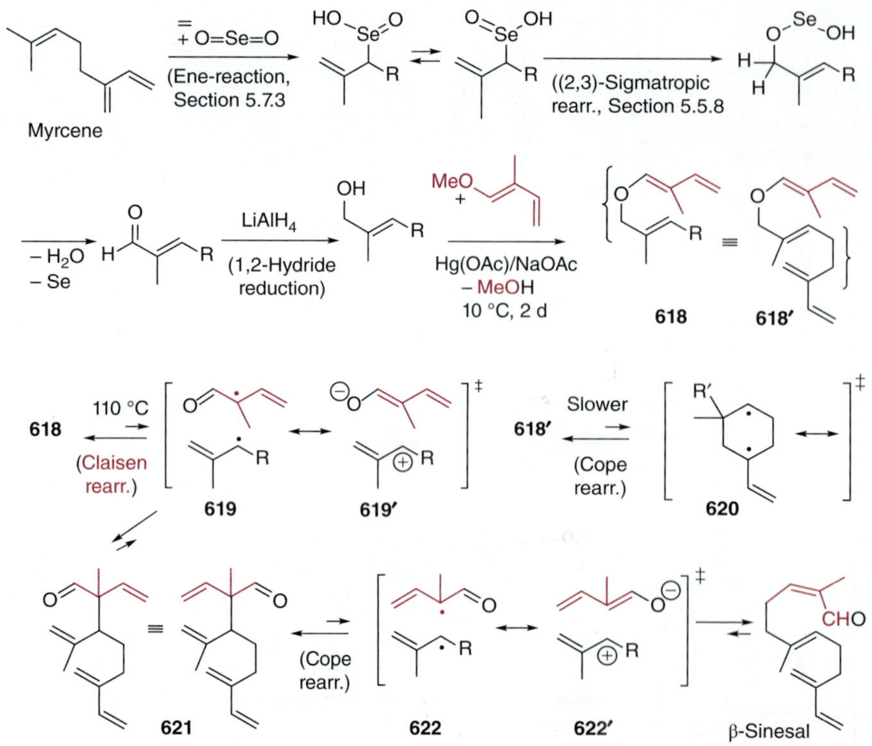

Scheme 5.112 Thomas synthesis of β-sinensal, the fragrance of the Chinese orange.

rearrangement already at room temperature (acceleration factor of c. 10^{12}). Further acceleration (10^{17} overall) is observed when the potassium counterion is encapsulated with 18-crown-6 ether (encapsulation of K^+ increases significantly the distance between the negatively charged centers and the metallic cation). Thus, when the alkoxide is fully dissociated from its counterion ("naked" alcoholate), its negative charge is fully available to stabilize the allylic cation moiety of the limiting structure **627′** of the transition state. Quantum mechanical calculations on the anionic oxy-Cope rearrangement supports a concerted, one-step, asynchronous mechanism [1304].

Because of their ease, anionic oxy-Cope rearrangements have found a large number of applications in the synthesis of natural products and analogs of biological interest [1286, 1305–1309]. An interesting example is the thermal reaction of cyclonona-2,4,7-trien-1-ol that undergoes a (1,5)-hydrogen shift to generate cyclonona-3,7-dienone and its anionic oxy-Cope rearrangement at 25 °C, which produces aldehyde **628** [1310], a synthetic precursor of prostaglandins [1311], and of the algal sperm attractant multifidene [1312].

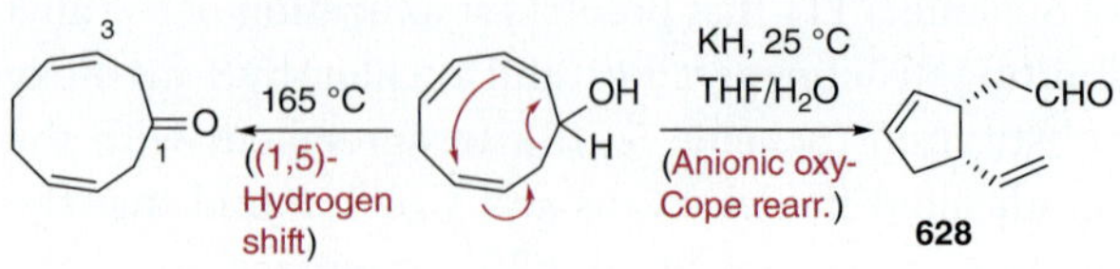

Additions of vinyllithium to 1,2-diketones give the corresponding dialkoxides that can undergo dianionic oxy-Cope rearrangements at room temperature or below [1313]. Evans et al. observed that anionic oxy-Cope rearrangements of 1,5-diene-3-alkoxides with 4-thiophenoxy substituents are subject to a large rate acceleration, but fragmentation can also occur [1314]. A similar acceleration effect was found by Paquette for 6-thiophenoxy substitution [1315]. Macdonald and coworkers have reported anionic 3-amino-Cope rearrangements and suggested that they occur by a allyl/allyl cleavage and recombination mechanism [1316]. Quantum mechanical calculations confirmed that strongly electron-releasing substituents at C(4) and/or C(6) of the hexa-1,5-diene-3-alkoxide moiety may render the two-step mechanism preferred to the concerted, one-step rearrangement [1317], as predicted by the diradicaloid model. Indeed, when the allyl cation/allyl anion–ion pair limiting structure is highly stabilized, it is not just a limiting structure of the concerted reaction, but it relaxes into an intermediate of the pericyclic process. In agreement with the diradicaloid model, electron-withdrawing substituents can also accelerate the Cope rearrangement. For instance, the rearrangement of hexa-1,5-diene-2-carboxaldehydes is accelerated in the presence of acylhydrazides and CF_3SO_3H because of the formation of iminium salt intermediates of type

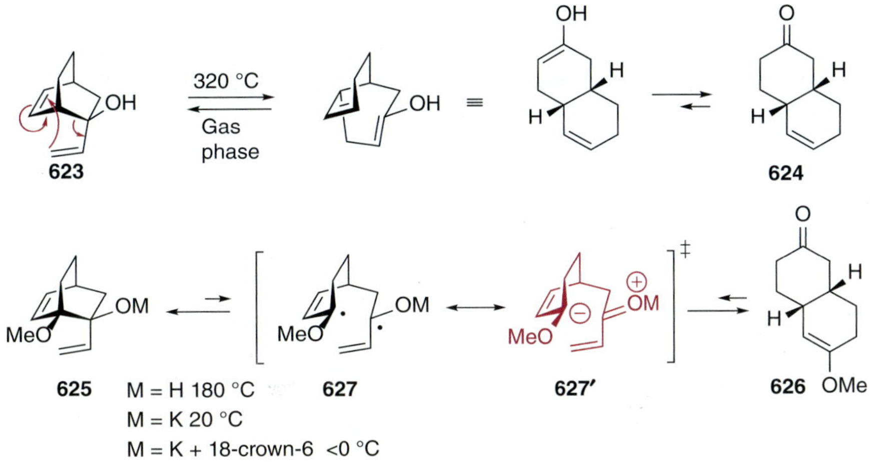

Scheme 5.113 The oxy-Cope rearrangement is strongly accelerated on converting the alcohol into its alcoholate. The more the counterion is dissociated from the alcoholate anion, the faster is the rearrangement.

R′CH=CHCH$_2$CH$_2$C(=CH$_2$)—CH=N$^{(+)}$R$_2$/TfO$^-$ [1318].

In 1994, Schultz and coworkers reported antibody catalysts that accelerate the oxy-Cope rearrangement of allylic alcohol **629** to enol **630** by factors of 10^3–10^5 (k_{cat}/k_{uncat}) [1319]. The catalytic antibodies were raised to hapten **631**, which mimics the chair-like transition structure expected for this rearrangement. Kinetic studies showed that antibody AZ28 lowers $\Delta^{\ddagger}H$ from 27.4 kcal mol^{-1} for the uncatalyzed reaction to 15.4 ± 2.4 kcal mol^{-1} for the catalyzed process. This favorable change in enthalpy offsets a very unfavorable change in $\Delta^{\ddagger}S$, which is −3 eu for the uncatalyzed reaction and −23 ± 8 eu for the antibody-catalyzed process [1320]. Quantum mechanics and flexible docking modeling suggested a stepwise mechanism for the latter reaction involving a 2-hydroxycyclohexa-1,4-diyl diradical intermediate that is more strongly coordinated by the antibody than the starting diene. The binding implies, among several factors, hydrogen bonding of the hydroxy group of **629** with the carboxylate anion of an aspartate unit of the antibody [1321].

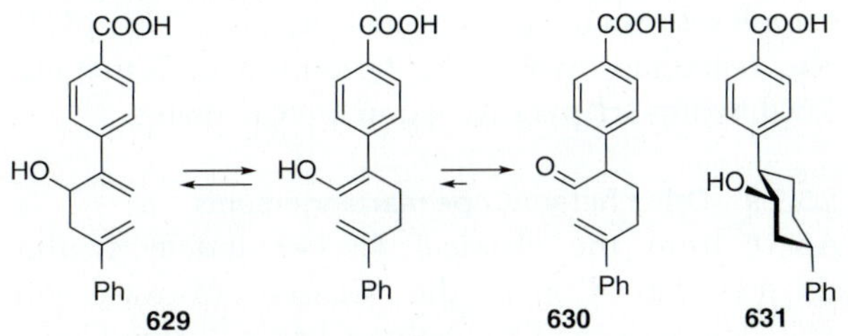

5.5.9.8 Acetylenic Cope rearrangements

The first examples of acetylenic Cope rearrangements were reported by Black and Landor (reaction (5.89)) [1322] and by Hunstman et al. (reactions (5.90) and (5.91)) during the mid-1960s [1323]. Above 200 °C, hexa-1-en-5-yne is equilibrated with hexa-1,2,5-triene with $\Delta^{\ddagger}H(5.91) \approx 32$ kcal mol^{-1}, which is lower by a 3–5 kcal mol^{-1} than the estimated enthalpy difference $\Delta_f H°$(cyclohex-1-ene-1,4-diyl diradical: **632**) − $\Delta_f H°$(hexa-1-en-5-yne). This suggests a concerted mechanism for the acetylene-Cope rearrangement (5.91).

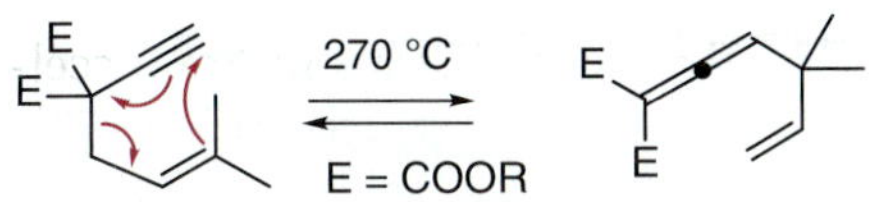

(5.89)

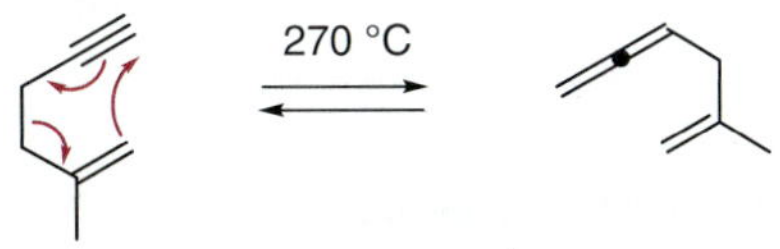

(5.90)

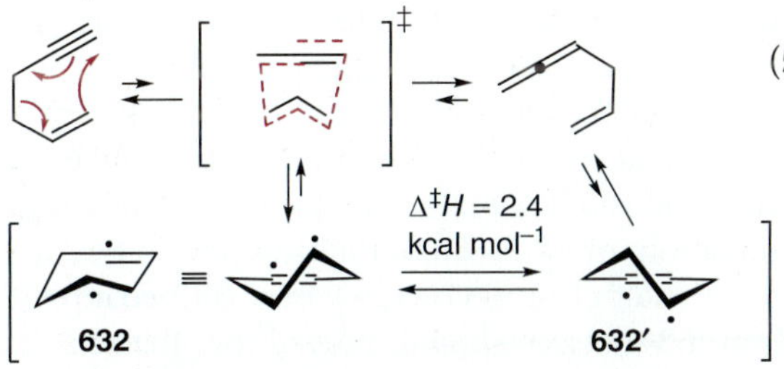

(5.91)

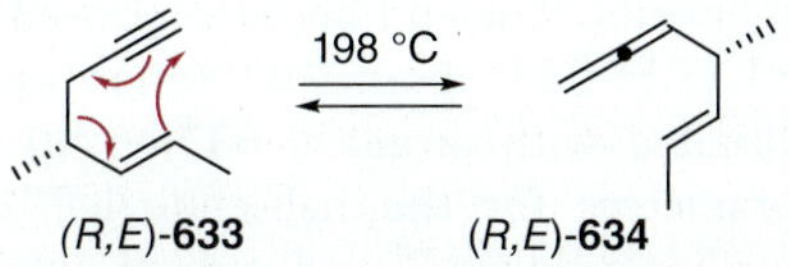

An experimental test for the preferred concerted mechanism (Scheme 5.114) was realized by Owens and Berson who found that optically active acetylene (*R*,*E*)-**633** is equilibrated (after several half-lives) in the gas phase at 198 °C with allene (*R*,*E*)-**634** with essentially complete preservation of enantiomeric purity [1324]. Racemization of (*R*,*E*)-**633** should imply the formation of 3,5-dimethylcyclohex-1-ene-1,4-diyl diradicals capable to undergo half-chair/half-chair

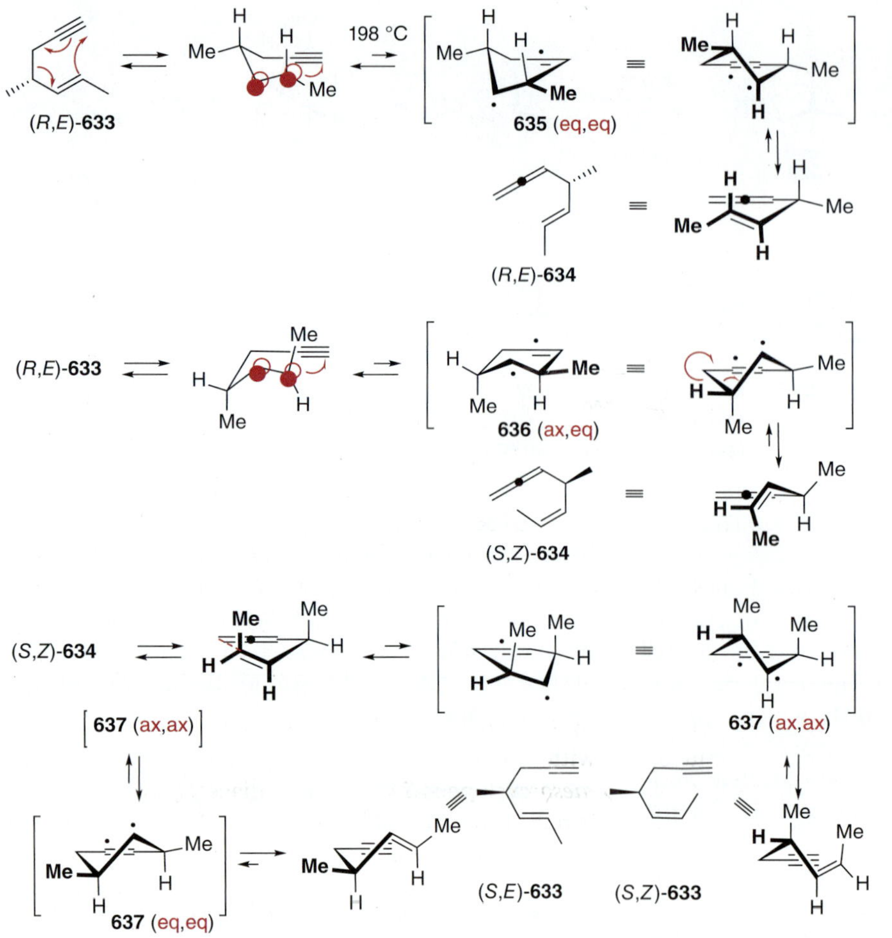

Scheme 5.114 Acetylenic Cope rearrangement racemization of (*R*,*E*)-**633** that would require the intermediacy of conformationally flexible cyclohex-1-ene1,4-diyl diradicals does not occur.

interconversion. Formally, acetylene (*R*,*E*)-**633** can equilibrate with diradicals **635** and **636** that are isomerized exclusively into allenes (*R*,*E*)-**634** and (*S*,*Z*)-**634**, respectively. The latter allenes can be equilibrated with acetylene (*S*,*Z*)-**633** (not into (*S*,*E*)-**633**) via diradical **637**(ax,ax). Formation of (*S*,*E*)-**633** (racemization of (*R*,*E*)-**633**) requires the interconversion of half-chair **637**(ax,ax) into **637**(eq,eq). If diradical **637**(ax,ax) should indeed be formed in these rearrangements, it would be decomposed into (*S*,*Z*)-**634** and (*S*,*E*)-**633** about 100 times as fast as it is equilibrated with diradical **637**(eq,eq). The activation parameters for the half-chair/half-chair interconversion in cyclohex-3-enyl radical amounts to $E_a = 2.4\,\text{kcal mol}^{-1}$ and $\log A = 12.6$ [1325] ($E_a = 1.3\,\text{kcal mol}^{-1}$ for cyclohex-3-enone). Assuming the same activation parameters for the interconversion **637**(ax,ax) ⇄ **637**(eq,eq), one calculates a rate constant of $3 \times 10^{11}\,\text{s}^{-1}$ for the process at 198 °C, the temperature of the isomerization (*R*,*E*)-**633** ⇄ (*R*,*E*)-**634**. This would mean that the isomerization of diradical **637**(ax,ax) into these compounds should have a rate constant of about $(1\text{–}4) \times 10^{13}\,\text{s}^{-1}$. This rate exceeds the upper limit imposed by transition state theory (Section 3.3) for the decomposition of an intermediate, which is the rate of the transition state itself, about $10^{13}\,\text{s}^{-1}$ at 198 °C. Thus, diradicals of type **635**, **636**, **637**(ax,ax), and **637**(eq,eq) do not occur significantly in this rearrangement. A concerted mechanism best accounts for the experimental results [344].

Acetylene oxy-Cope rearrangements have found applications in organic synthesis [1326]. The reaction can be catalyzed by silver salts [1327] and/or by gold(I) complexes such as [(*t*-Bu)$_3$P]AuCl [1328–1331]. The reactions involve the formation of 5-metalla-3,6-dihydro-2*H*-pyranyl cation intermediates.

5.5.9.9 Other hetero-Cope rearrangements

Apart from the classical Fischer indole synthesis (3,4-diaza-Cope), the Claisen (3-oxa-Cope), aza-Claisen (3-aza-Cope), thia-Claisen (3-thia-Cope), and Overman (1-aza,3-oxa-Cope) rearrangements, several other hetero-Cope rearrangements have been described. Among them, a very useful reaction is the 2-azonia-Cope rearrangement (5.92) [1332]. As for the Cope rearrangements, the 2-azonia-Cope rearrangements of simple 2-azonia-1,5-dienes are reversible. Several approaches have been applied to drive irreversible rearrangements: (i) aryl conjugation

Scheme 5.115 Facile 2-azonia-Cope rearrangement of 4-hydroxy-2-azonia-1,5-dienes applied to the enantioselective synthesis of (−)-strychnine.

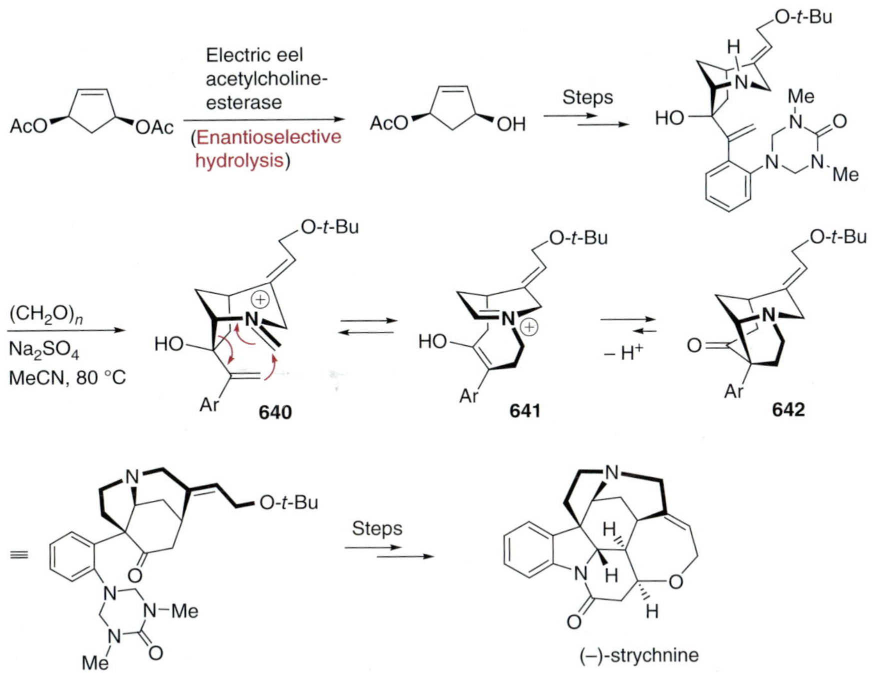

of the iminium ion [1333, 1334], (ii) trapping of the iminium ion in a subsequent nucleophile-induced ene-iminium cyclization [1335], (iii) selective cleavage of the iminium ion of one of the equilibrating 2-azonia-1,5-diene isomers [1336–1339], and (iv) a tandem Mannich cyclization process as established by Overman and Kakimoto [1340] by using 4-hydroxy-2-azonia-1,5-dienes **638** derived from the acid-catalyzed condensation of 1-aminomethylallyic alcohols with aldehydes [1341, 1342]. The rearrangements of 4-hydroxy-2-azonia-1,5-dienes are facile, as their transition state **639** ↔ **639′** is stabilized by a facile electron exchange that does not generate further charges.

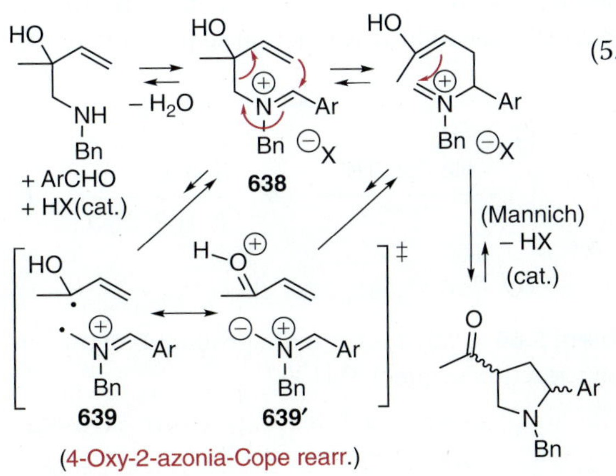

(5.92)

The reaction has been applied as a key step in a number of alkaloid total synthesis [1343–1347] as exemplified with the first asymmetric synthesis of (−)-strychnine (Scheme 5.115) presented by Overman and coworkers [1343, 1347]. It starts with the enzymatic enantioselective hydrolysis of *meso*-cyclopent-2-en-1,4-diyl diacetate. The key step is rearrangement **640** ⇄ **641**. On its turn, enol **641** is isomerized into the more stable ketone **642**. Strychnine has played an important role in the development of natural product chemistry [1348, 1349]. It was first isolated in 1818 from *Strychnos ignatii* by Pelletier and Caventou. It is the first plant alkaloid obtained in pure form. Its structure was established in 1946 [1350] and a first total synthesis of it was realized by Woodward et al. [1351].

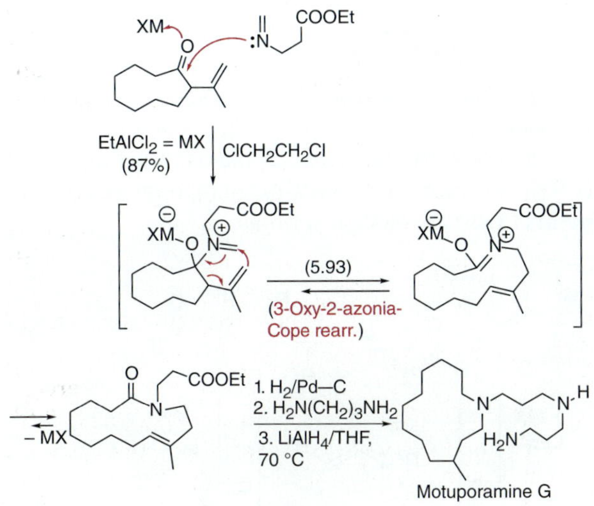

In the presence of a Lewis acid, β,γ-unsaturated ketones and oximes or imines undergo nucleophilic addition to produce zwitterion intermediates, and subsequent exergonic 3-oxy-2-azonia-Cope rearrangements (5.93) give homoallylic amides. An example of application of this reaction is given below with a total synthesis of Motuporamine G [1346c].

In the presence of a protic or Lewis acid promoter, homoallylic alcohols add to aldehydes forming hemiacetals that can be heterolyzed into the corresponding oxycarbenium ions that realize 2-oxania-1,5-dienes. The latter intermediates undergo facile 2-oxania-Cope rearrangements [1352]. An example (5.94) is shown with oxaniadiene **643** for which tethering an oxygen nucleophile drives the oxania-Cope to product [1353].

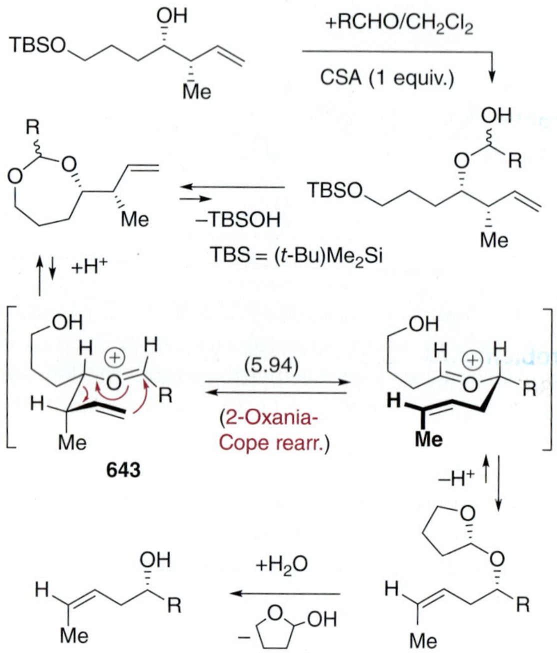

CSA:

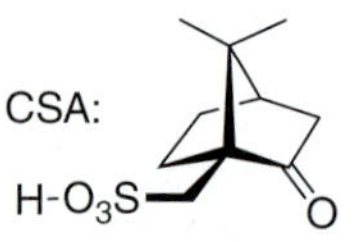

Camphorsulfonic acid

Analogous to the 3-oxy-2-azonia rearrangement (5.93), exergonic 3-oxy-2-oxonia-Cope rearrangements (5.95) are facile reactions.

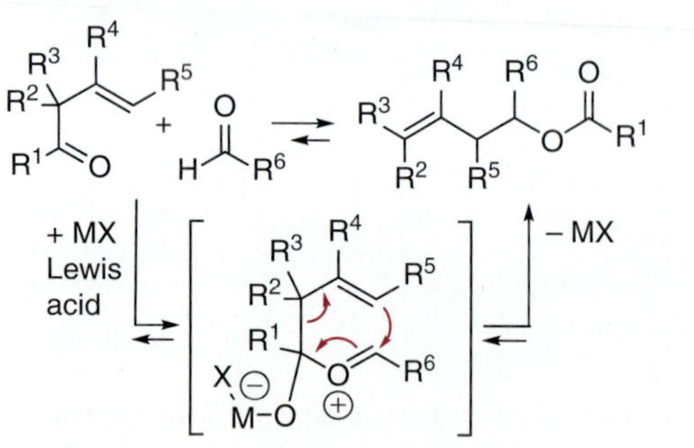

(5.95)

Enantiomerically enriched (99% ee) β,γ-unsaturated ketone (*R*)-**644** reacts with benzaldehyde in the presence of 10% $BF_3{\cdot}OEt_2$ (=MX) to give a 4 : 1 mixture of lactones (*S,E*)-**646** and (*R,Z*)-**646** with complete transfer of chirality (Scheme 5.116). The results can be interpreted in terms of chair-like transition structure **645**, that is inscribed in a trans-decalin-like species, for the favored reaction (*R*)-**644** → (*S,E*)-**646**, and of chair-like transition structure **645″**, that is inscribed in a *cis*-decalin-like structure, for the minor reaction (*R*)-**644** → (*R,Z*)-**646**. Alternative transition structures **645′** and **645‴** with a pseudo-axial phenyl substituent can be ruled out [1354].

An example of thermal 1-oxa-2-aza-Cope rearrangement is given with reaction (5.96) reported by Zakarian and Lu [1355]. Quantum mechanical calculations support a concerted (3,3)-sigmatropic rearrangement involving a boat-like transition structure. Lewis acids such as SbF_5 and $TiCl_4$ preferentially coordinate to nitrogen of the nitroso group (and the carbonyl of the ester moiety) and substantially decrease the activation barrier of the reaction [1356].

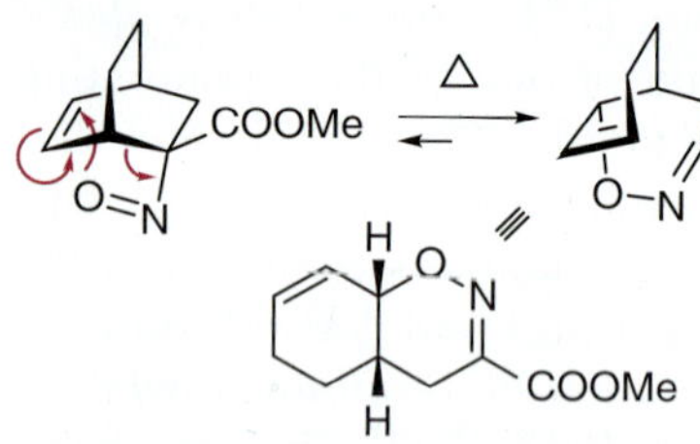

(5.96)

Problem 5.64 What is the heat of reaction allyl vinyl ether ⇄ pent-4-enal?

Problem 5.65 What is the intermediate **B** and the product **P** of the following reaction sequence? [1200]

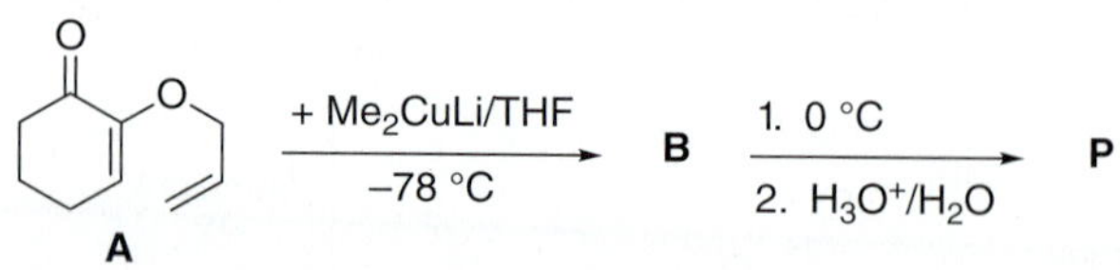

Problem 5.66 What is the main product **P** of the following reaction sequence? [1357]

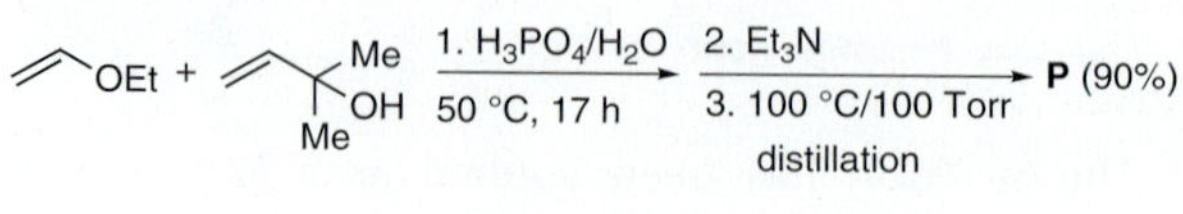

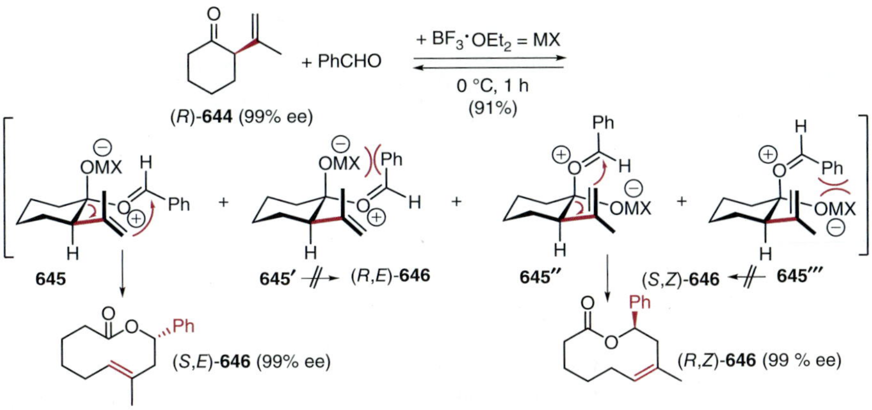

Scheme 5.116 Chirality transfer in the 3-oxa-2-oxania-Cope rearrangement. Application to the synthesis of 10-membered-ring lactones.

Problem 5.67 What is the main product **P** of the following reaction? [1358]

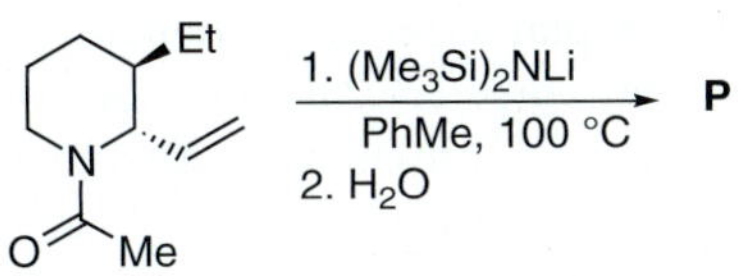

Problem 5.68 What is intermediate **B** and the major product **P** of the following reaction? [1359]

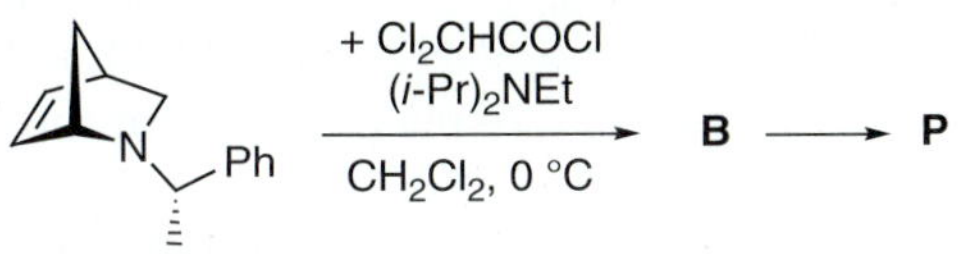

Problem 5.69 What is the major product **P** of the following reactions? [1360]

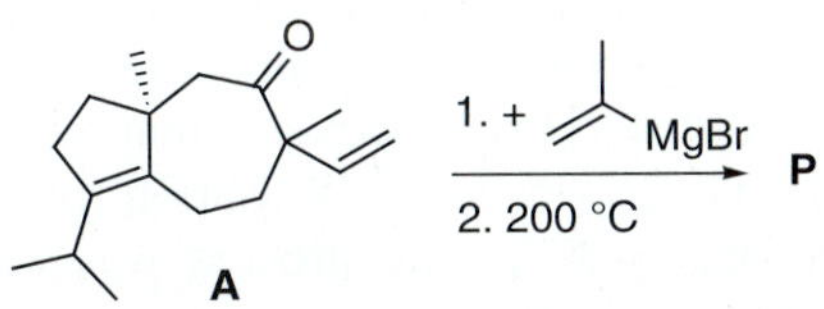

Problem 5.70 What is the product **P** of the following reactions? [1361]

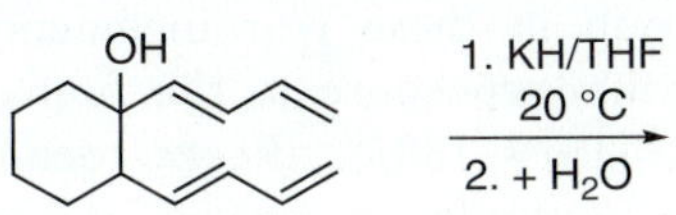

Problem 5.71 What are major products **P** and **Q** of the following reactions? [1313]

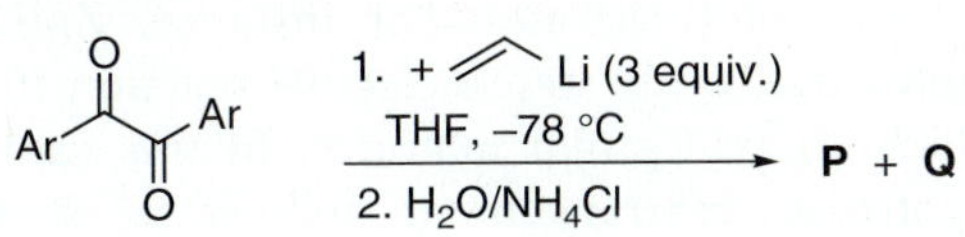

Problem 5.72 Explain the following reaction [1346c].

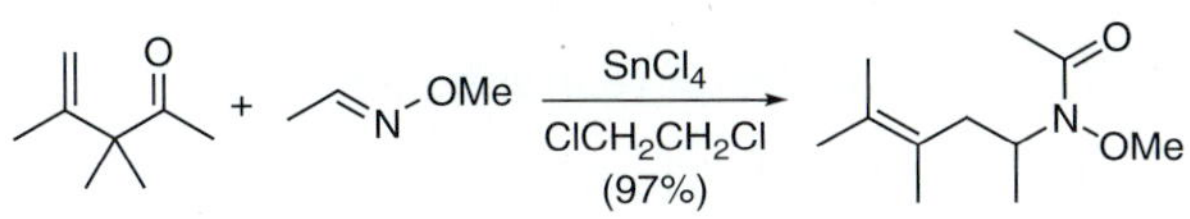

Problem 5.73 Explain the following reactions [1362].

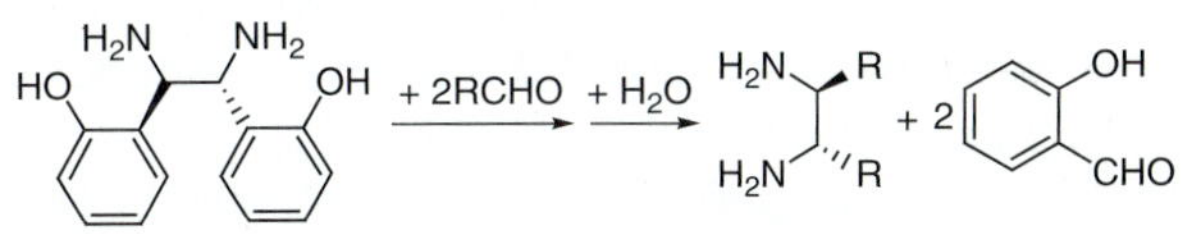

5.6 Dyotropic rearrangements and transfers

In 1972, Reetz [1363, 1364] defined dyotropic (from Greek *dyo*, meaning two) rearrangements as a new class of pericyclic valence isomerizations in which two σ-bonds migrate intramolecularly (two bond shifts) [1365, 1366]. Reactions in which the two migrating groups interchange their relative positions were designated as type I of dyotropic rearrangements (Scheme 5.117a), whereas those of type II involve migration to new bonding sites without position interchange (Scheme 5.117b). We can add to the type II of dyotropic rearrangements the intermolecular dyotropic transfer of dihydrogen in the alkene and alkyne reductions with diimide and related (rare) intermolecular dihydrogen (H—H) transfer from two vicinal conjugated allylic C—H moieties to alkenes (Scheme 5.117c). Mechanical quantum calculations suggested that type II transfer of H_2 transfer cannot be excluded for hydrogenation of carbonyl

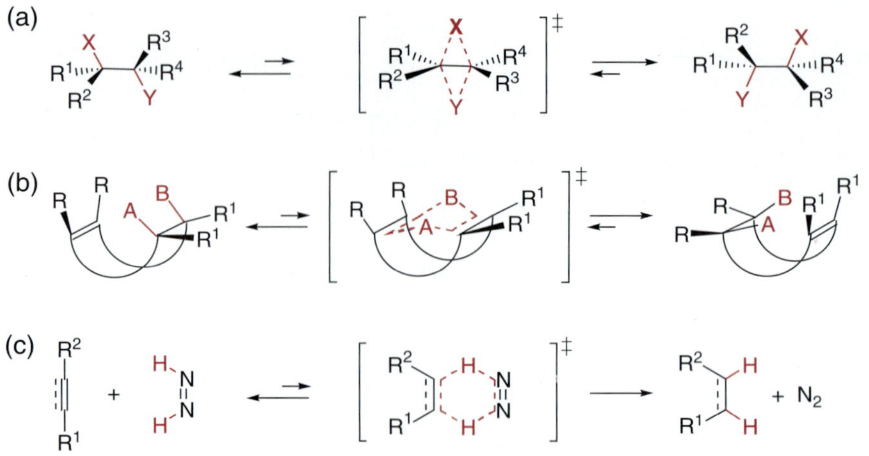

Scheme 5.117 (a) Type I, (b) intramolecular type II, and (c) intermolecular type II dyotropic rearrangements.

compounds by transition metal complexes of type H–[Ru]–NH_2–R [1367].

5.6.1 Type I dyotropic rearrangements

In 1952, Grob and Winstein [1368] elucidated the mechanism of the mutarotation of 5α,6β-dibromocholestane (the product of bromination of cholest-5-ene) into 5β,6α-dibromocholestane and proposed that the two bromides migrate simultaneously and intramolecularly [1369]. Similar rearrangements have been reported for 1,2-dibromocyclohexanes and other alkyl dibromides [1370, 1371]. Computational studies support the generality of such vicinal dihalide reactions [1372]. Heating either (*R*,*R*)-**647** or (*S*,*S*)-**647** in benzene to 80 °C led to their racemization (reaction (5.97)) with first-order kinetics ($k = 0.0015\,\text{s}^{-1}$, $\Delta^{\ddagger}G = 25.3\,\text{kcal mol}^{-1}$) without any formation of *meso*-dibromide (*R*,*S*)-**647** after prolonged heating, thus demonstrating the stereospecificity of the 1,2-dyotropic rearrangement. The results rule out a dissociation/recombination mechanism. Quantum mechanical calculations support a one-step, concerted 1,2-dyotropic rearrangement with a transition structure **648**, which has a strong diatropic ring current (what is not the case for B_2H_6 that lacks two electrons compared with **648**) [1373].

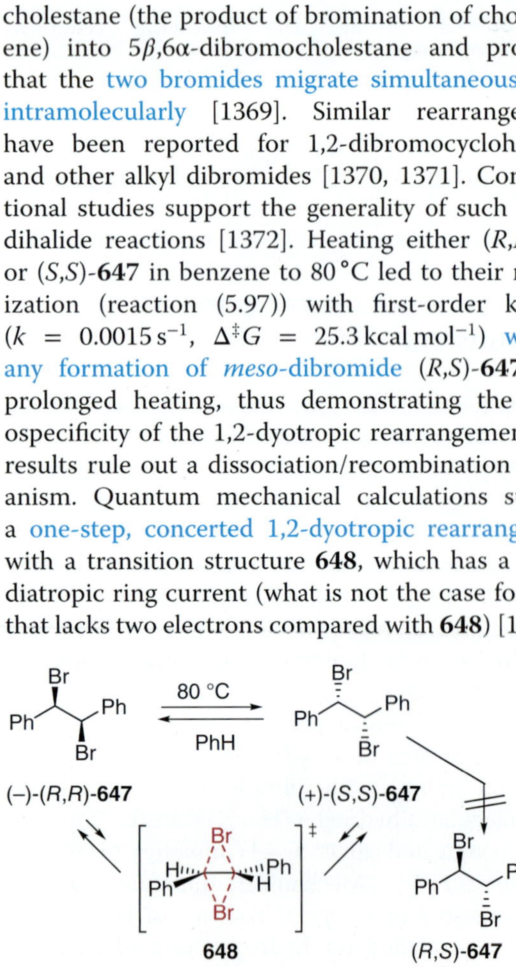

(5.97)

The transition structure **648** can be represented as the result of the interaction of an alkene with two halide radicals residing above, for one of them, and below its π-plane, for the other one. The alkene contributes with one HOMO (π) and one LUMO (π^*); each radical $X^{\bullet}$ contributes with one φ_z (SOMO) and one p_x (HOMO) type of FMOs, the two other nonbonding valence electron pairs are in π_y and χ_z orbitals that cannot interact with the alkene π MOs for reasons of symmetry and geometry, respectively. Stabilization arises from the overlaps HOMO(alkene: π)/LUMO(diradical: $\varphi_z - \phi'_z$) and LUMO(alkene: π^*)/HOMO(diradical: $p_x - p'_x$) as shown in the FMO diagram of Figure 5.55.

The isomerization of four-substituted β-lactones into butyrolactones is one of the most used 1,2-dyotropic rearrangements in organic synthesis [1374, 1375]. In 1979, Mulzer and Brüntrup reported a first example (5.98) that is promoted by $MgBr_2$ [1376]. This procedure permits the preparation of 3-substituted butyrolactones that cannot be prepared via the standard alkylation of the lactones enolates [1377]. The process was named as Wagner–Meerwein-type dyotropic rearrangement by analogy with the carbenium ion rearrangements (Section 5.5.1). Black and Fields found with reaction (5.99) that hydrogen migrates faster than alkyl group in these rearrangements [1378]. They applied this procedure to the preparation of spirolactones [1379–1381]. Reetz demonstrated the migratory aptitude of acyclic amino groups in β-lactone dyotropic rearrangements to generate 3-amino-γ-lactones [1382]. Other Lewis acids than $MgBr_2$ can be used to induce the rearrangements of β-lactones. For instance, $Zn(OTf)_2$ promotes the highly stereoselective reaction (5.100) in which an acyl group migrates. In this case, the 1,2-dyotropic rearrangement follows a stepwise mechanism: the Lewis acid promotes the heterolysis of the lactone giving a carbenium ion intermediate that undergoes a subsequent Wagner–Meerwein rearrangement [1383]. In the presence of montmorillonite clay, tetraazafulvalenes are isomerized into 1,4,5,8-tetraazanaphthalenes through 1,2-dyotropic rearrangements (5.101) [1384, 1385]. 1,2-Dyotropic rearrangements are also catalyzed in biological systems by a family of isomerases that use adenosyl cobalamin (coenzyme B_{12}) as cofactor [1386]. Examples are glutamate mutase [1387], 2-methyl glutarate mutase [1388], methyl malonyl-coenzyme A mutase [1389], isobutyryl-coenzyme A mutase [1390], diol dehydratase [1391], glycerol dehydratase [1392], ethanolamine-ammonia lyase [1393], L-β-lysine-5,6-aminomutase [1394], L-β-lysine-2,3-aminomutase [1395], and D-ornithine 4,5-aminomutase [1396]. These enzyme-catalyzed rearrangements often imply radical intermediates [1397]. Terpenes are natural products that can have thousands of distinct structurally and stereochemically complex molecular architectures. The core hydrocarbon frameworks of these molecules are constructed via carbocation rearrangements promoted by terpene synthase (cyclase) enzymes. When the transformation requires formally two successive Wagner–Meerwein rearrangements, the latter might occur in two distinct steps or in a concerted and stereoselective way that corresponds to a type I dyotropic transfer [1398]. The migrating group can be hydrogen, alkyl groups, and/or anionic leaving groups (counteranion) [1399].

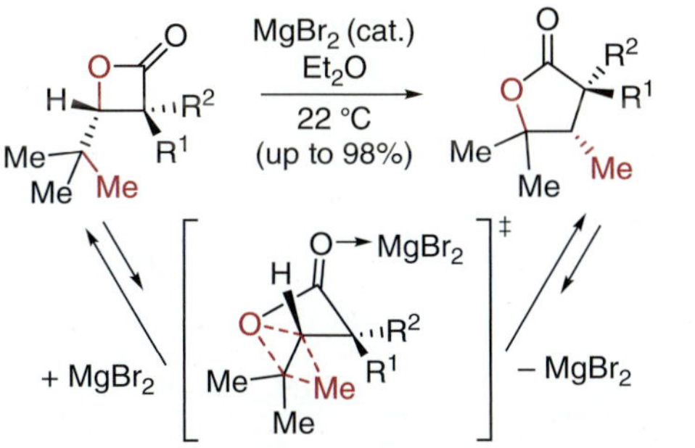
(5.98)

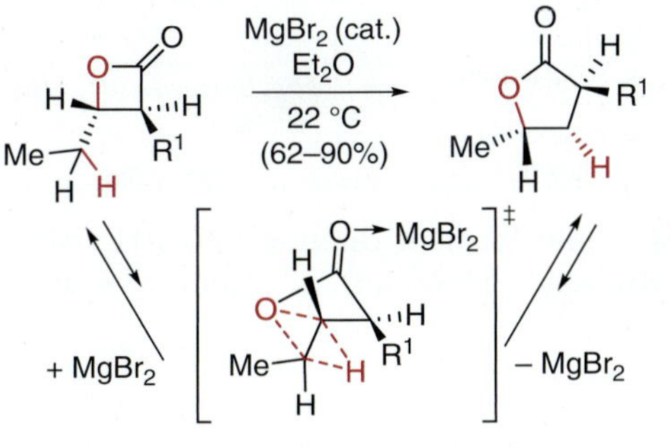
(5.99)

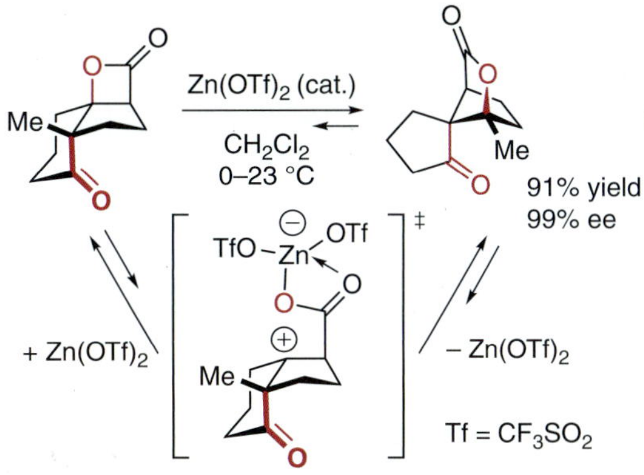
(5.100)

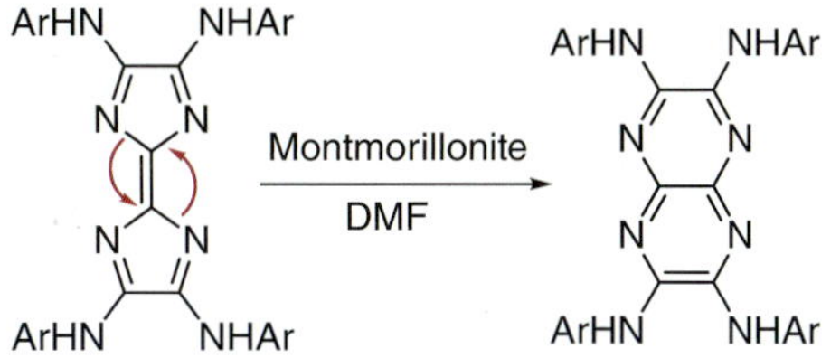
(5.101)

In 1973, Frainnet and coworkers described an exchange of trialkylsilyl groups in bis(organosilyl) hydroxylamines [1400]. Later, West and coworkers found a reversible intramolecular rearrangement that they suggested proceeds via a concerted 1,2-dyotropic rearrangement with simultaneous migration of two trialkylsilyl groups [1401–1403]. If a dialkylfluorosilyl group is incorporated in the hydroxylamine, the rearrangement is irreversible (Scheme 5.118) due to more favorable enthalpic anomeric effects (Section 2.7.9) for the F–Si–N + O–Si–C moieties than for the F–Si–O + N–Si–C moieties [1404].

Mintz and coworkers reported the 1,2-dyotropic rearrangements (5.103) of α-zirconocenyl thioethers. The reactions follow first-order kinetics, and labeling experiments showed that the rearrangements are intramolecular as confirmed by their negative entropy of activation ($\Delta^{\ddagger}S \approx -19$ eu) [1405]. Erker and Petrenz have described a similar rearrangement (5.104) [1406].

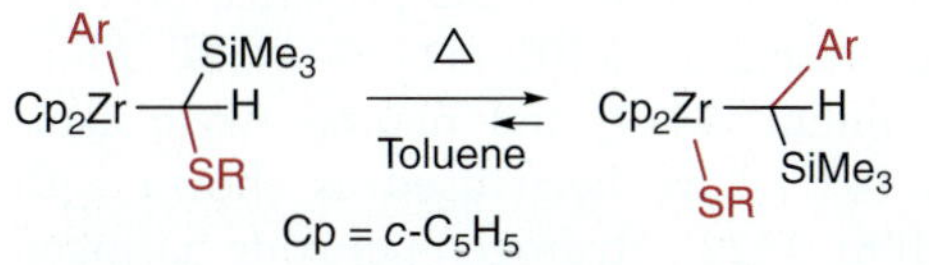
(5.103)

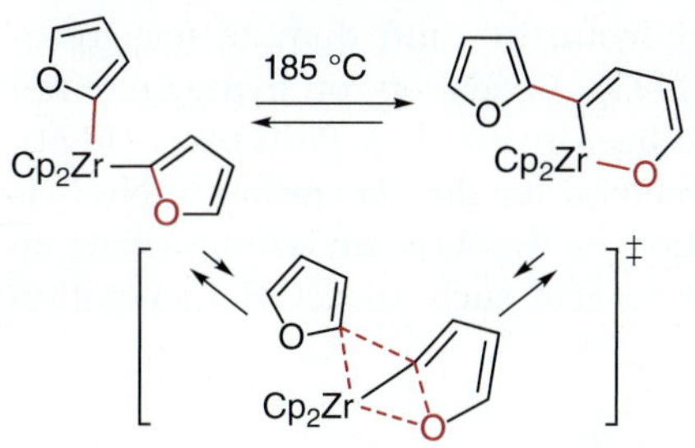
(5.104)

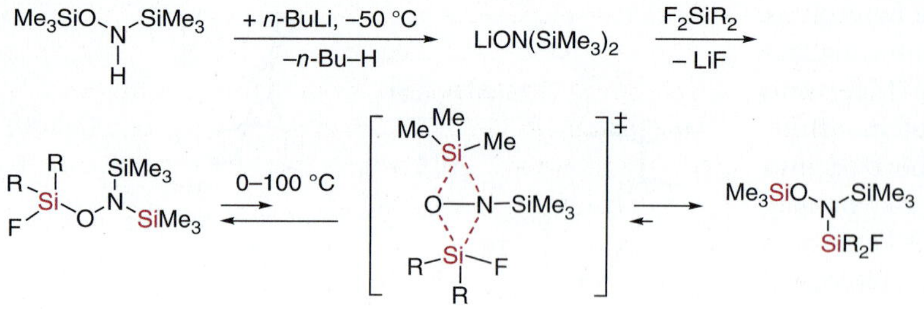

(5.102)

Scheme 5.118 Dyotropic rearrangement involving the 1,2-shifts of silyl groups.

Problem 5.74 What is product **P** in the following reaction? [1407]

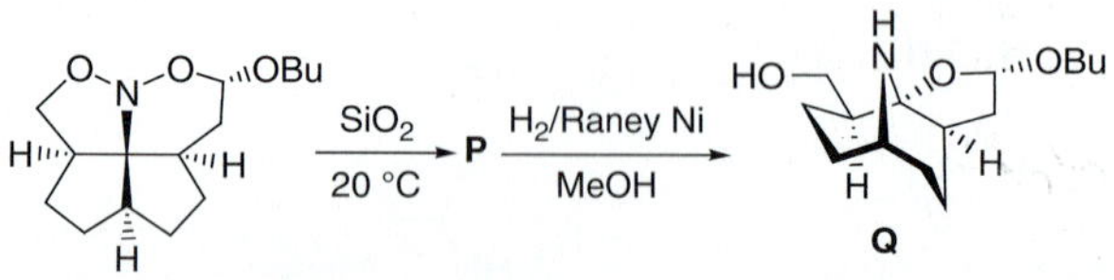

Problem 5.75 What is the isomer of diphenyl (trimethylsilyl)methyl acetate obtained on heating to 190 °C? [1408]

5.6.2 Alkene and alkyne reductions with diimide

Diimide, or diazene (HN=NH), is an unstable compound that undergoes disproportion into hydrazine ($H_2N—NH_2$) and N_2 [1409] and decomposes into H_2 and N_2 [1410]. Nevertheless, it reduces alkenes, alkynes, and azo compounds efficiently [1411, 1412], and not so much polarized double bonds such as carbonyl [1413], nitro, sulfoxyl, and cyano groups. Compared with transition-metal-catalyzed hydrogenation (Section 7.8), diimide does not hydrogenolyze benzyl ethers, benzyl amines, N—O bond, and O—O bond [1414, 1415]. One of the simplest method to generate diimide is the air or O_2 oxidation of hydrazine in the presence of a Cu catalyst (e.g. reaction (5.105) [1416–1418]. More recently, flavinium [1419] and neutral flavin catalysts [1420, 1421] have been used as shown with reaction (5.106) [1422]. Hydrogen peroxide has been used to convert hydrazine into diimide (e.g. reaction (5.107)) [1423]. Oxidation of hydrazine into diimide has also been realized by $PhI(OAc)_2$ [1424]. An alternative method for the generation of N_2H_2 is the decomposition of dipotassium azodiformate in the presence of an acid such as AcOH (generation of HOOC—N=N—COOH that decomposes at 0 °C into H—N=N—H + $2CO_2$) [1425–1427]. An example (reaction (5.108)) is given with the *Z*-stereoselective monoreduction of a iodoacetylene [1428]. Diimide is also formed by β-elimination of arenesulfinic acid from the corresponding sulfonylhydrazide ($ArSO_2NHNH_2 \rightarrow ArSO_2H + N_2H_2$) [1429, 1430] as exemplified with reaction (5.109) [1431]. On heating, 1-amino-2,2,-diphenylaziridine is fragmented into 1,1-diphenylethylene and diimide [1432].

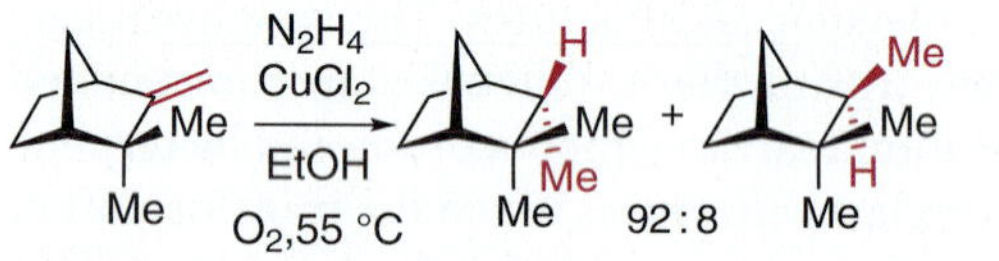

(5.105)

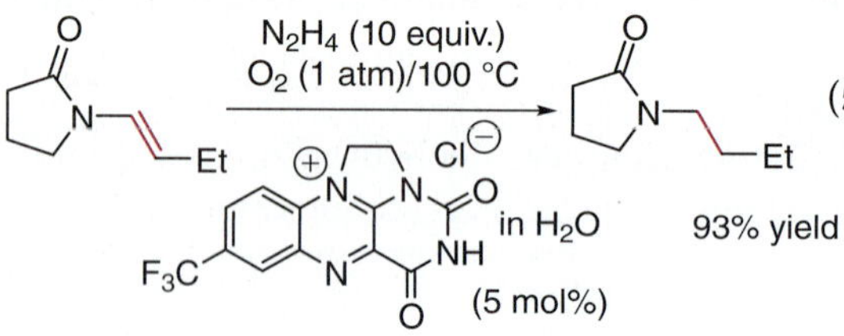

(5.106)

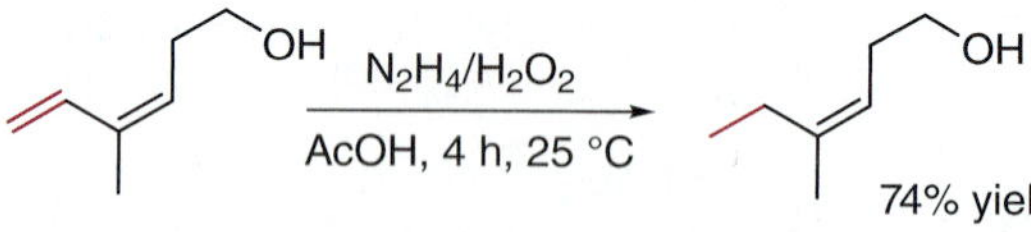

(5.107)

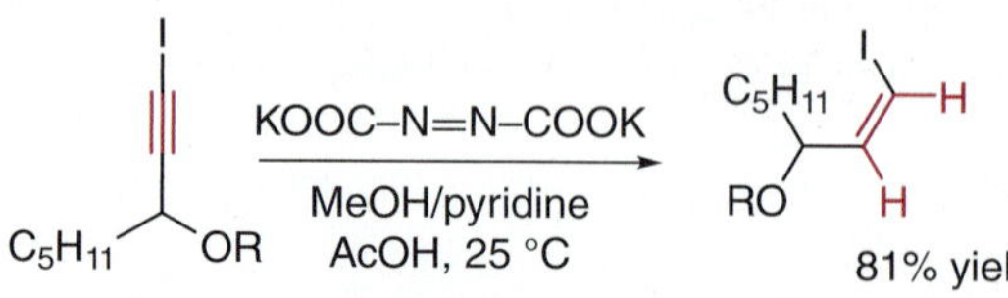

(5.108)

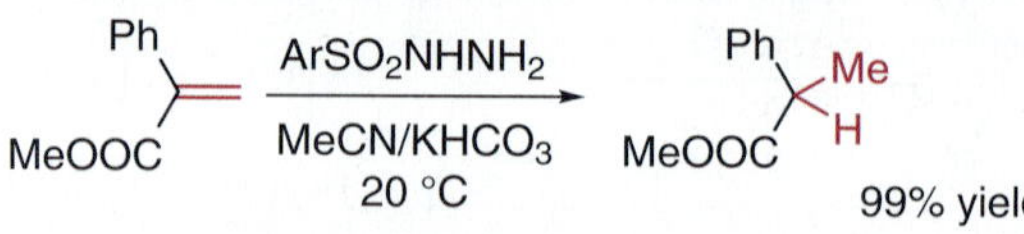

(5.109)

The most stable form of diimide is (*E*)-HN=NH (UV spectroscopy). However, it equilibrates with

(*Z*)-HN=NH in a fast process that is solvent catalyzed [1433]. The rate-limiting step of the alkene or alkyne reduction is the hydrogen transfer from (*Z*)-HN=NH to the unsaturated compound (Scheme 5.119). Quantum mechanical calculations [1434] support a one-step, concerted mechanism that can be noted [π^2s+σ^2s+σ^2s], which is thermally allowed according to the Woodward–Hoffmann rules [1, 2] (analogy with the Diels–Alder reaction). The reduction of ethylene by diimide is highly exothermic (c. −83 kcal mol^{-1}). Therefore, it is not a surprise that the barrier for the transfer is largely entropic. For the reduction of fumaric acid into succinic acid, a rate constant $k = (1.32 \pm 0.07) \times 10^2\ \mathrm{M^{-1}\,s^{-1}}$ has been measured at 25 °C [1435].

The diimide reductions are the fastest for the least substituted alkenes (steric effect) and for the most strained alkenes (increase of the exothermicity) [1436]. Even though the barrier is mostly entropic, monosubstituted ethylenes with the lowest ionization energies (EI) react the fastest [1437], suggesting a non-negligible enthalpic contribution ($\Delta^{\ddagger}H = \alpha\Delta_r H + \beta$) to the free enthalpy of activation ($\Delta^{\ddagger}G = \Delta^{\ddagger}H - T\Delta^{\ddagger}S$).

Problem 5.76 What is the product of the following reaction?

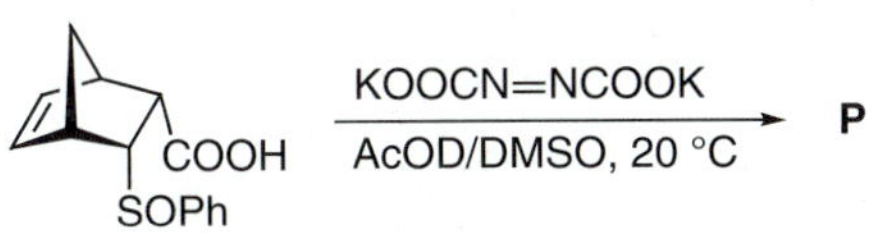

5.6.3 Type II dyotropic rearrangements

In 1967, Doering and Rosenthal reported the intermolecular suprafacial/suprafacial dihydrogen transfer (5.110) that reduces 1,2-dimethylcyclohex-1-ene into *cis*-1,2-dimethylcyclohexane with *cis*-9,10-dihydronaphthalene. The reaction (noted [π^2s+σ^2s+σ^2s]) is highly stereoselective as no trace of *trans*-1,2-dimethylcyclohexane is formed [1438]. An earlier example of intramolecular dihydrogen transfer (5.111) was presented by McKenzie [1439]. This initial finding was further extended to similar frameworks containing a cyclohexa-1,3-diene or a 2-pyrazoline moiety, which ensure a relatively high exothermicity to these type II dyotropic rearrangements due to aromatization [1440–1443].

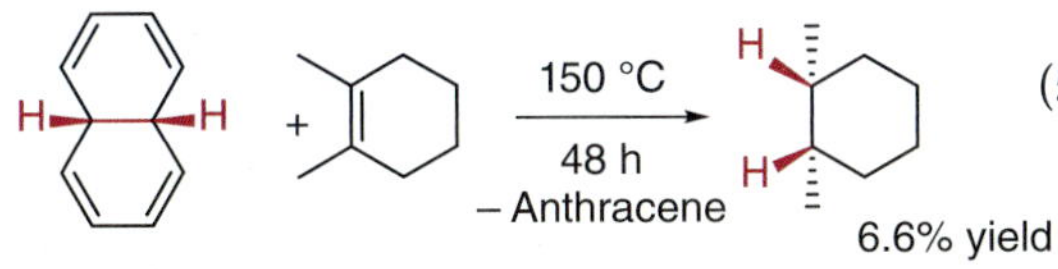

(5.110)

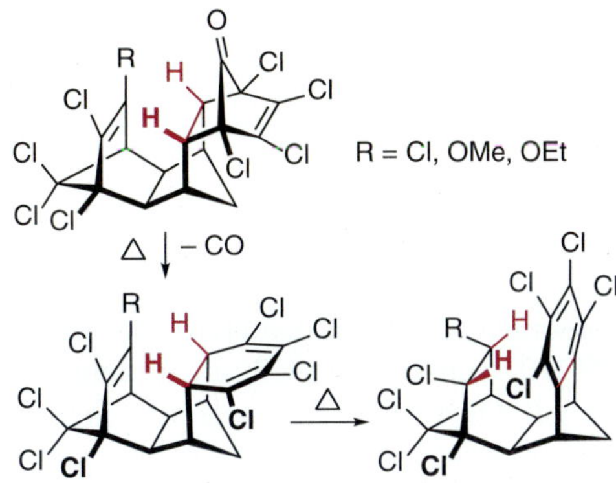

(5.111)

The addition of dimethyl acetylenedicarboxylate to 2-norborneno[*c*]furan (**255**) is highly *endo* face selective (Section 5.3.14) and give the *syn*-sesquinorbornene derivative **649** that is hydrogenated chemo- and stereoselectively into **650** (Scheme 5.120). When heated in C_6D_6 to 130–160 °C, **650** equilibrates with **651** ($k = (2.0 \pm 0.1) \times 10^{-6}\ \mathrm{s^{-1}}$ at 130 °C, K = [**651**]/[**650**] = 1.1–1.2) through a dyotropic transfer of H—H (5.112) that is approximatey 30 times faster than cycloreversion of **650** into norbornene and dimethyl furan-3,4-dicarboxylate. Cycloreversion of **651** into cyclopentadiene and dimethyl 7-oxanorborna-2,5-diene-2,3-dicarboxylate is not observed. Equilibrium **650** ⇄ **651** represents the first thermoneutral dyotropic [π^2s+σ^2s+σ^2s] rearrangement ever reported. Its rate is not affected by concentration, solvent, SiO_2, and water. Exclusive migration

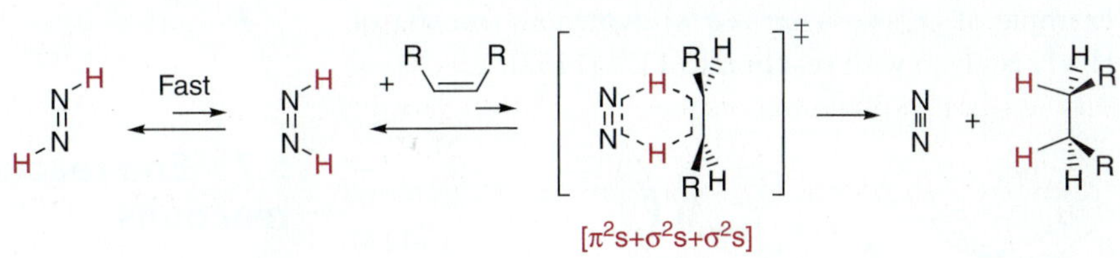

Scheme 5.119 Mechanism of the reduction of alkene by diimide.

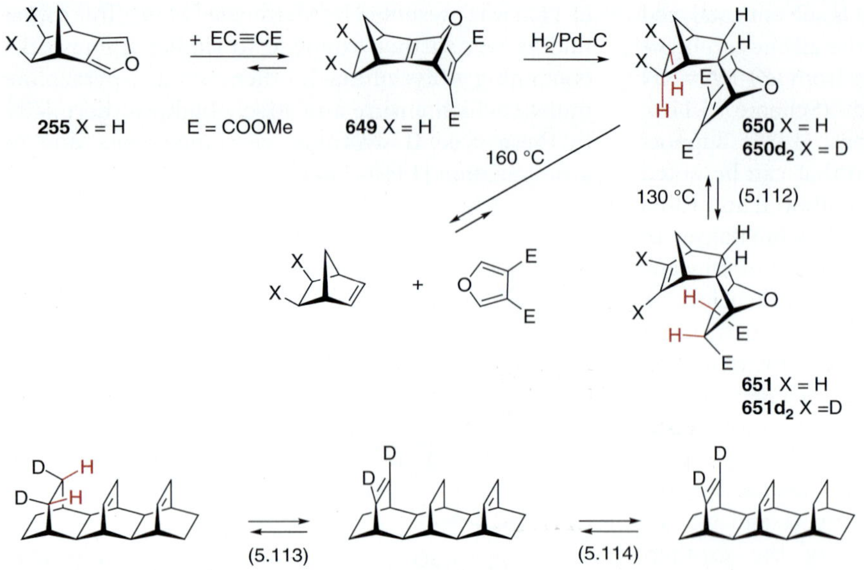

Scheme 5.120 Examples of thermoneutral type II dyotropic rearrangements.

of H—H is observed with the dideuterated derivative **650-d$_2$** ⇄ **651-d$_2$** consistently with an intramolecular transfer of dihydrogen. The reaction follows a concerted pericyclic mechanism. An alternative two-step process involving a single hydrogen transfer that generates a diradical intermediate would require $\Delta^{\ddagger}H \geq 52\,\text{kcal mol}^{-1}$, which is higher than the measured activation enthalpy $\Delta^{\ddagger}H = 35\text{–}39\,\text{kcal mol}^{-1}$. The slightly positive activation entropy measured ($\Delta^{\ddagger}S = 2\text{–}11$ eu) is also consistent with an intramolecular concerted mechanism [1444]. Paquette et al. [1445–1447] reported related type II dyotropic hydrogen transfer in *syn*-sesquinorbornenedisulfones and, in agreement with findings by Mackenzie, found that the proximity of the ethano and ethylene moieties undergoing the dyotropic transfer is a prerequisite for a smooth reaction (see Section 3.6.4) [1448]. Other examples of thermoneutral dyotropic transfers are equilibria (5.113) and (5.114) described by Grimme et al. (Scheme 5.120) [1449].

Further examples of exothermic [π^2s+σ^2s+σ^2s] dyotropic transfers of H—H are presented with reactions (5.115) [1450] and (5.116) [1451]. A rare example of [π^4s+σ^2s+π^2s+σ^2s] dyotropic transfer of H—H is given with reaction (5.117) [1452].

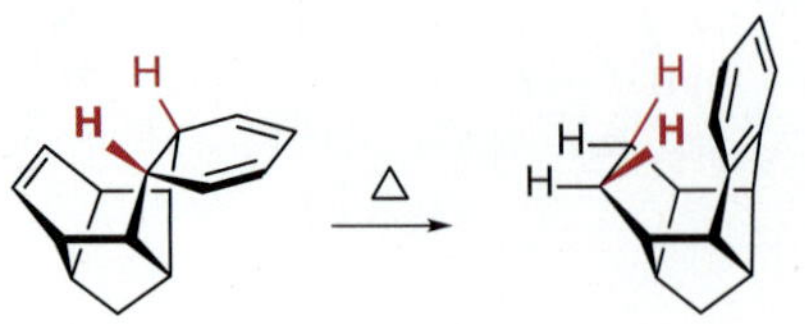

(5.115)

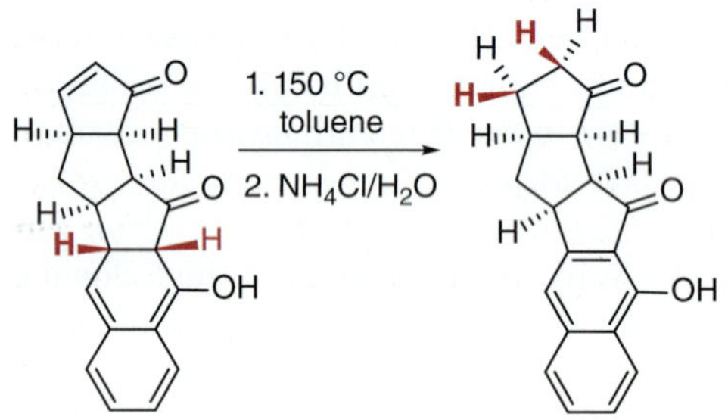

(5.116)

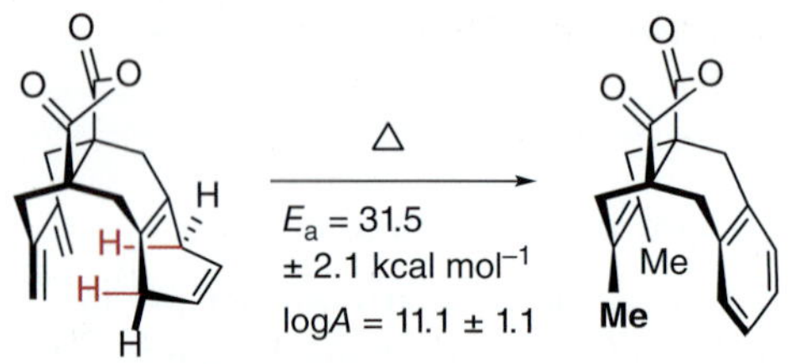

(5.117)

Problem 5.77 What are the intermediate compounds **B**, **D** and the final product **P** of the following reaction sequence?

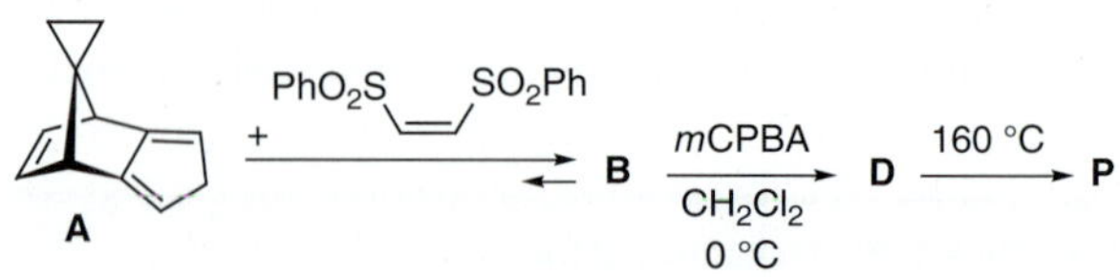

5.7 Ene-reactions and related reactions

The ene-reactions combine allyl systems (A=B—U—G) or progargyl systems (A≡B—U—G) (enes) with an unsaturated compound (X=Y, W≡Z: enophiles) forming adducts resulting from the formation of a new single bond and a transfer of the allylic or propargylic group G. The reactions construct C—C or C-heteroatom bonds in an atom economical way as they do not produce any coproduct [630]. When the enophile is a Lewis base (e.g. RCH=CH—COX, RCH=CHCN, RC≡C—COX, RC≡C—CN, RCH=O; RCH=NR′, R—N=NCOX, R—N=O, SO_2, etc.), Brønsted or Lewis acids can be used as catalysts (Sections 7.6.1–7.6.5). When the reactants are hydrocarbons, transition metal complexes can catalyze their ene-reactions (Section 8.3.2). The reactions can be seen as additions of two unsaturated compounds followed by a group transfer (Scheme 5.121). The reverse reactions are called retro-ene reactions: they have been introduced in Section 3.4.4. The reactions can follow stepwise mechanisms such as the formation of a 1,4-diradical or 1,4-zwitterion intermediates, followed by transfer of the allylic group G (as a radical, cation, or anion), or follow a concerted, one-step mechanism in which the transfer of the G group is assisted by the formation of the new single bond. The G group can be H, R_3Si, any metallic species, any electrofugal group. The overall processes are pericyclic reactions that can be noted $[\pi^2s+\sigma^2s+\pi^2s]$ for those showing "thermally allowed" stereoselectivity [1453–1455]. All carbon enes (A=B—U—G; e.g. CH_2=CH—C(R^1)(R^2)—H, G=H) and enophiles (X=Y; e.g. maleic anhydride, acrylic derivatives) undergo Alder ene-reactions that have been described first in 1943 [1453].

5.7.1 Thermal Alder ene-reactions

Pyrolysis of ethene + propene at 410–481 °C under 33–300 Torr produces pent-1-ene (reaction (5.118), $E_a = 37.0 \pm 1.0\,\text{kcal mol}^{-1}$, $\log A = 7.45 \pm 0.3$), hex-1-ene (reaction (5.119), $E_a = 37.0 \pm 1.0\,\text{kcal mol}^{-1}$, $\log A = 6.55 \pm 0.3$), and 4-methylpent-1-ene (reaction (5.120), $E_a = 37.0 \pm 1.0\,\text{kcal mol}^{-1}$, $\log A = 6.55 \pm 0.3$). Methane and ethane are also formed as the result of radical reactions of ethene and propene themselves. The rate of formation of pent-1-ene is first order in both the concentrations of ethene and propene, and the rate of formation of hex-1-ene and 4-methylpent-1-ene are second order in the concentration of propene. Formation of methane and ethane is accelerated on adding O_2, but not the ene-reactions (5.118)–(5.120), in agreement with molecular concerted pericyclic processes [1456]. At a somewhat lower temperature, the reaction of isobutene with acetylene in a flow reactor (350 °C, 170 atm) produces 2-methylpenta-1,4-diene (reaction (5.121)) [1457]. Quantum mechanical calculations predicted that the barrier of the ene reaction of propene decreases along the enophiles in the order $H_2C=CH_2 > HC\equiv CH$, $H_2C=NH > H_2C=CH(COOMe) > H_2C=O > H_2=CPH > H_2C=S$ [1458].

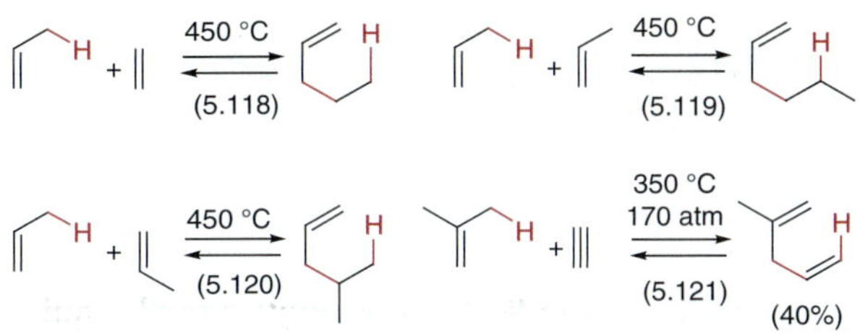

As seen for cycloadditions (Section 5.3), the condensation of two unsaturated compounds may generate 1,4-diradical or 1,4-zwitterion intermediates that evolve to the cycloadducts [135]. Alternatively, the latter species are limiting structures of the transition states of cycloadditions following concerted mechanisms, as admitted by the diradicaloid model. Similarly, the diradicaloid model **652** ↔ **652′** ↔ **652″** can be used to represent the transition state of concerted Alder ene-reaction. Thus, any substituent that stabilizes the diradical **652** and/or the zwitterion **652′** will accelerate the reaction. This is indeed the case when the alkene or alkyne (enophile) is conjugated with a π-electron-withdrawing group such as

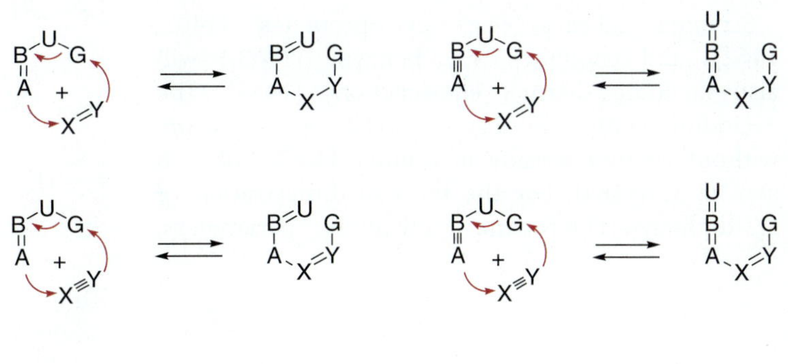

Scheme 5.121 Ene-reactions and retro-ene reactions.

COCl, COOH, COOR, COR, CHO, CN, SO_2R, etc. (activated enophile), and the allylic system (ene) is 2-substituted by electron-releasing groups such an alkyl group. For instance, naturally occurring β-pinene adds to acryloyl chloride (CH_2=CHCOCl) already at 70 °C (reaction (5.122)). As predicted by the diradicaloid model, the ene-reaction of α,β-unsaturated esters, ketones, and aldehydes can be accelerated by coordination of their carbonyl groups to Lewis acids (Sections 7.6.2–7.6.5), as for the Diels–Alder reactions (Sections 7.6.6 and 7.6.7) and other cycloadditions (Sections 7.6.8–7.6.10) [1459]. An interesting example is the chemo- and stereoselective reaction (5.123) reported by Roche chemists and which controls the stereoselectivity of the side chain in a steroid [1460].

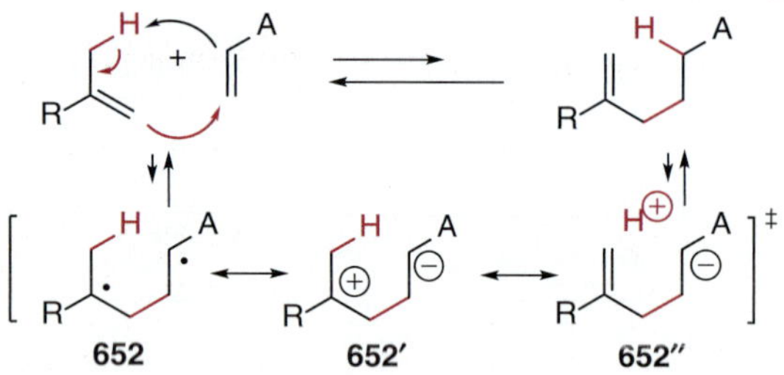

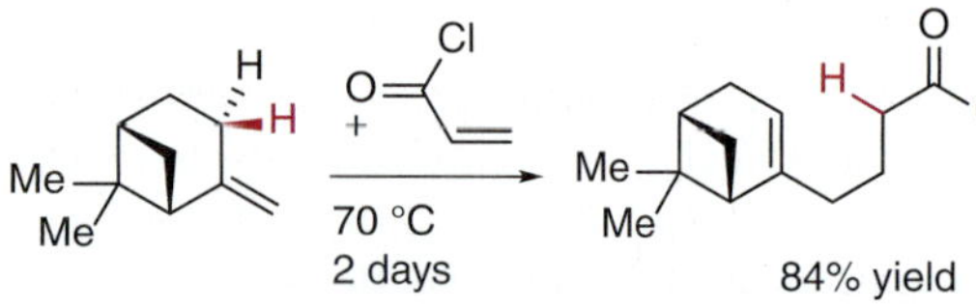

(5.122)

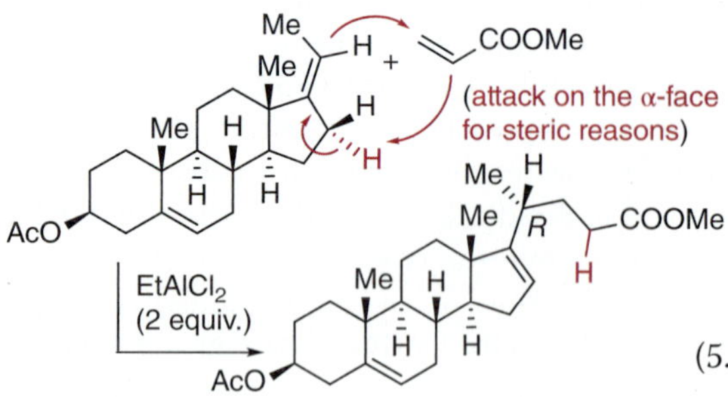

(5.123)

Strained alkenes (e.g. cyclopropenes [1461, 1462]) and enophiles (e.g. benzyne [1463]) will also facilitate thermal ene-reactions [1455]. The cyclodimerization (5.124) of cyclopropene occurs without catalyst already in dilute CH_2Cl_2 solution at −25 °C [1464]. For the thermal dimerization of 1,3-diphenylcyclopropene activation parameters, $\Delta^{\ddagger}H$ = 9.7 kcal mol^{-1} and $\Delta^{\ddagger}S$ = −32.8 eu have been measured. The primary deuterium kinetic isotopic effect for the dimerization of 3-deuterio-1,3-diphenylcyclopropene is k_H/k_D = 3.1 [1465]. This is similar to k_H/k_D for the transfer of hydrogen in other ene-reactions [1455]. A rare case of CN-Alder ene-reaction has been reported by Shamov and coworkers [1466]. On heating 3-methylcyclopropene-3-carbonitrile, a first cyclodimerization (5.125) occurs, which implies the migration of a cyano group. Then, the 1-methylcyclopropene moiety of the dimer so-obtained acts as ene and undergoes two successive H-Alder ene-reactions (5.126) with two molecules of 3-methylcyclopropene-3-carbonitrile (enophile) giving a tetrameric compound (reaction (5.126)).

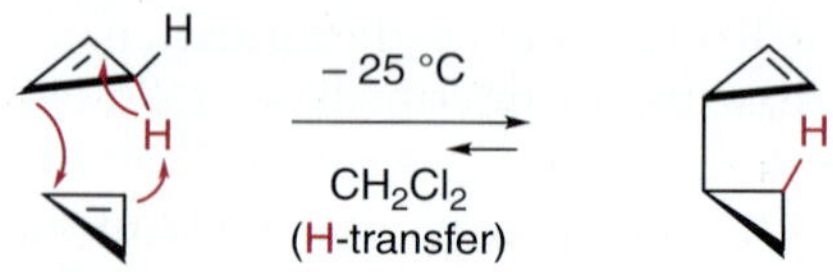

(5.124)

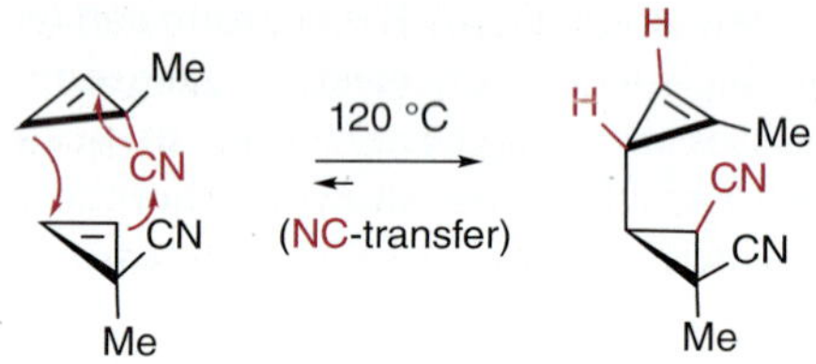

(5.125)

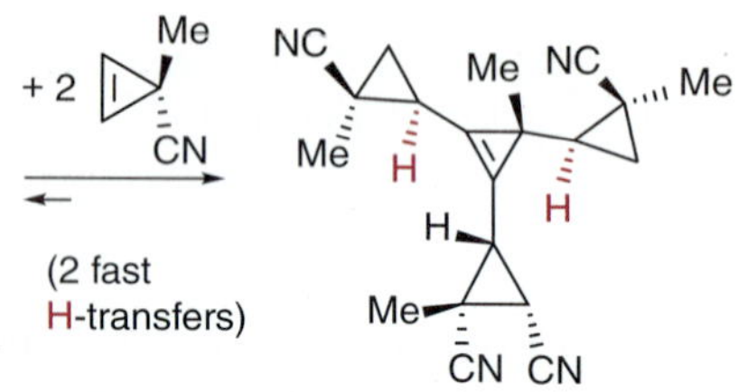

(5.126)

Thermal intramolecular ene-reactions might be easier than corresponding intermolecular processes and can be highly stereoselective [1467]. An example is given with reactions (5.127) that engage an allene moiety as enophile. In this case, the thermal ene-reactions compete with the corresponding intramolecular $[\pi^2+\pi^2]$-cycloadditions [1468].

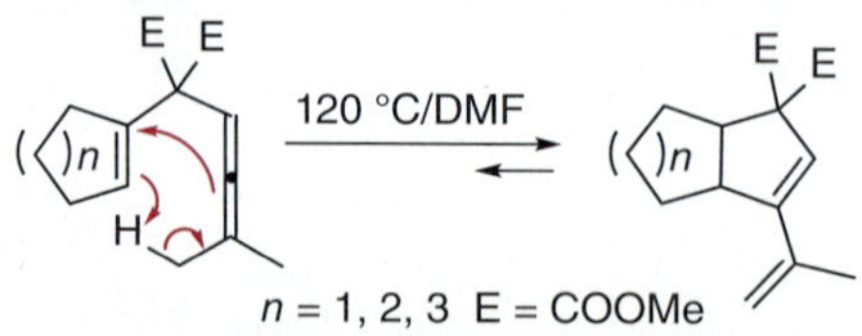

(5.127)

As an illustration of the utility of an intramolecular acetylene ene-reaction, the synthesis of enantio-enriched acetic acid (*R*)-CHTDCOOH developed by Arigoni and coworkers is out-lined in Scheme 5.122 [1454]. The synthesis features a first ene-reaction of the doubly labeled enyne **653** into **654**. Then, a retro-carbonyl-ene reaction (see Sections 3.4.4 and 5.7.2) converts **654** into **655**, the face selectivity for the deuterium transfer being determined by the configuration of the allylic moiety. A final Kuhn–Roth oxidation of alkene **655** with $KMnO_4$ provides (*R*)-CHTDCOOH.

Intramolecular ene-reactions (ene cyclizations) are very useful in organic synthesis [1469]. When the enophiles (e.g. α,β-unsaturated esters, amides, carbamates, ketones, aldehydes, etc.) can be coordinated to an enantiomerically enriched Lewis acid, enantioselective reactions can be realized [295, 1470]. Conia and coworkers have cyclized a great variety of ethylenic ketones to the corresponding cycloalkyl ketones [1471–1474]. The thermal Conia reactions (5.128) require elevated temperatures; they involve pre-equilibrium with enol intermediates that undergo oxa-ene-reaction with an intramolecular hydrogen atom transfer from the enol (ene) to the ethylenic or acetylenic moiety (enophile). The Conia reactions are accelerated by traces of water. They can be catalyzed by bases or acids. Alkene and alkyne allylations can be catalyzed by complexes of Pd(0), Ni(0), Rh(I), and Zn(0). When enantiomerically enriched phosphine ligands are employed, enantioselective reactions can be realized (Section 8.3.2) [1475].

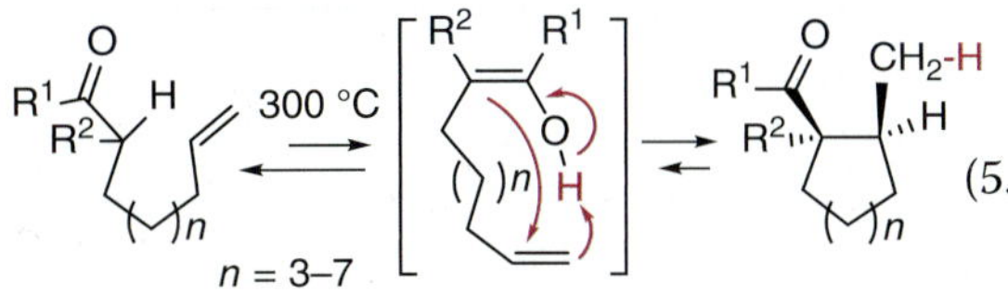

(5.128)

Problem 5.78 On heating methylidenecyclobutane with an excess of maleic anhydride to 195 °C for 30 hours, a 1:2 adduct is obtained. Explain [1476].

Problem 5.79 Explain the following reaction [673].

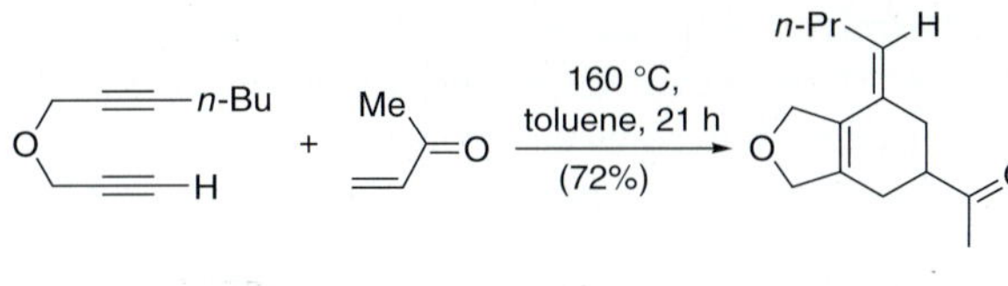

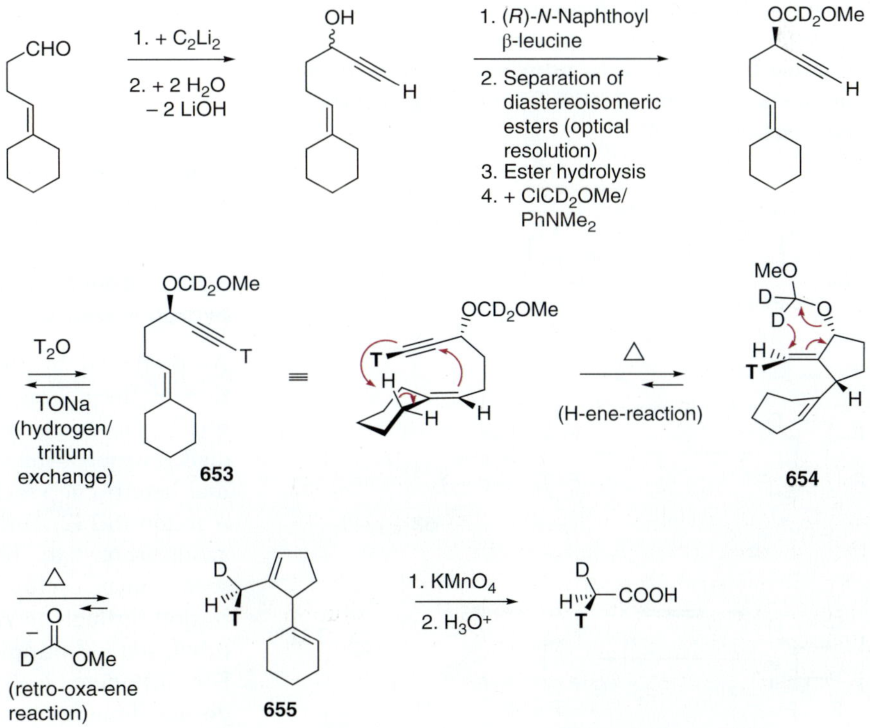

Scheme 5.122 Arigoni's enantioselective synthesis of chiral acetic acid.

5.7.2 Carbonyl ene-reactions

From a synthetic point of view, the carbonyl ene-reaction [294, 295, 1455, 1477] constitutes a more efficient alternative (more atom economical [630]) than the common additions of allylmetals to carbonyl compounds [1478, 1479]. The thermal intermolecular formaldehyde-ene reaction, the Prins reaction, has been very much studied [1480]. It requires very reactive 1,1-di- and trisubstituted alkenes (enes) and high temperatures (180–220 °C) [1481, 1482]. An example is given with reaction (5.129) [1481] that converts β-pinene into nopol, a valuable compound in the perfumery [1483]. Blomquist found the reaction to work more smoothly in Ac_2O/AcOH mixtures [1484, 1485]. Chloral (Cl_3CCHO) [1486, 1487] and glyoxalates (ROOC—CHO) are more reactive enophiles than fomaldehyde (destabilization of the carbonyl moiety by the adjacent electron-withdrawing groups: increase of the electrophilicity of the carbonyl group and of the exothermicity of the carbonyl ene-reaction) and react with the same activated alkenes already at 90–130 °C. Ketones destabilized by electron-withdrawing groups such as oxomalonic esters (ROOC—CO—COOR), carbonyl cyanide (N≡C—CO—C≡N), pyruvic esters (ROOC—CO—Me), pyruvonitrile (N≡C—CO—Me), hexafluoroacetone (F_3C—CO—CF_3), and 1,1,1-trifluoromethyl ketones (F_3C—CO—R) can be engaged in thermal, noncatalyzed carbonyl ene-reactions. Hexafluorocyclobutanone is an excellent enophile that reacts with propyne (reaction (5.130)) and allene (reaction (5.131)) in uncatalyzed carbonyl ene-reactions [1488].

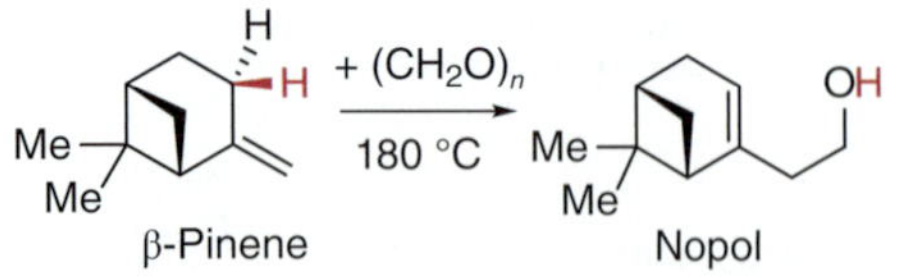

(5.129)

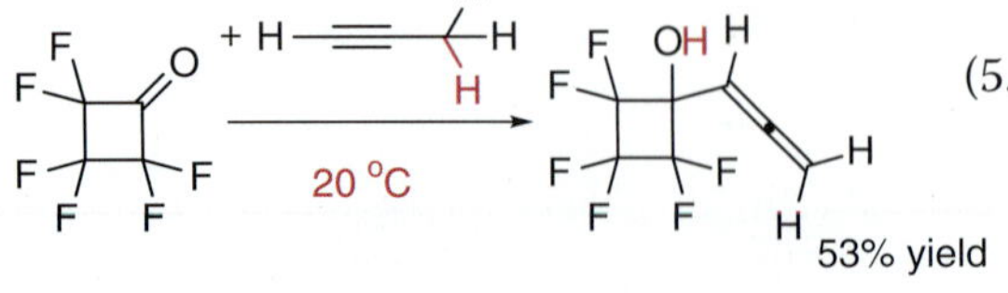

(5.130)

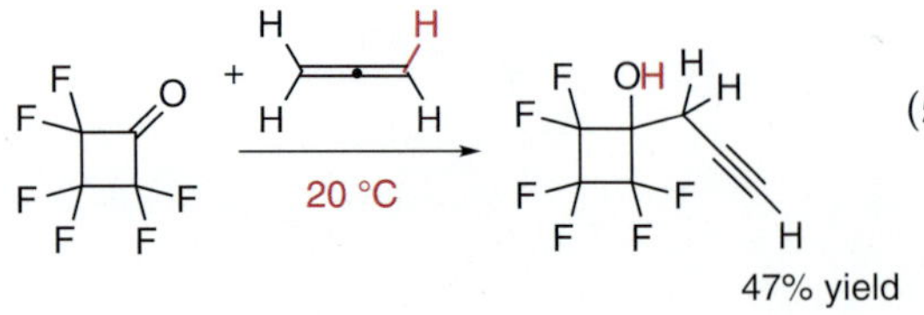

(5.131)

Thermal intramolecular carbonyl ene-reactions of unsaturated aldehydes and ketones have been very much explored. On heating (+)-β-citronellal to 180 °C, Ohloff obtained (−)-isopulegol (60%) as major product together with the three other possible diastereoisomers [1489, 1490]. With α-ketoester **656**, a 63 : 14 mixture of *trans*-**657** and *cis*-**657** is obtained (Scheme 5.123). The major product arises from transition structure **656′**, which minimizes $A^{1,3}$ allylic strain (Section 2.6.10) [1491]. This relatively high stereoselectivity is attributed to a concerted mechanism for thermal reaction **656** → **667**. This might not be the case with the Prins reaction (5.129), which is catalyzed by traces of acids and the reactor walls (formation of carbenium ion intermediates). Carbonyl ene-reactions are catalyzed by Lewis or protic acids. The latter reactions usually follow multistep mechanisms [1492]. When the carbonyl compounds is coordinated to an enantiomerically enriched Lewis or protic acid catalyst, very useful enantioselective carbonyl ene-reactions have been realized (Mikami ene-reaction: Section 7.6.3) [1493–1497]. For thermal retro-carbonyl ene-reactions, see Section 3.4.4.

Problem 5.80 Explain the following reaction [1498].

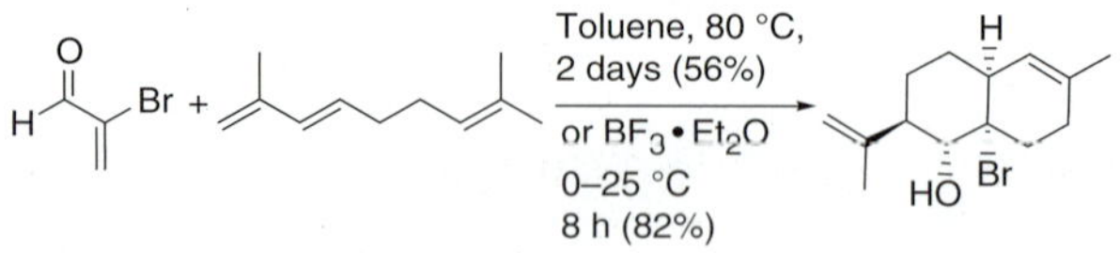

Problem 5.81 Can one use the carbonyl ene-reaction to capture CO_2 with an alkene?

5.7.3 Other hetero-ene reactions involving hydrogen transfers

As predicted by the diradicaloid model (stabilization of the charge transfer configuration **652′** and **652″**, the negative charge resides on NCOOR), dialkyl azodicarboxylate are good enophiles in thermal hetero-ene reactions. An example is given with reaction (5.132) [1455]. The ene-reactions of dialkyl azodicarboxylate are accelerated in water [1499]. As seen above, cyclopropene undergoes quick dimerization through an Alder ene-reaction. When substituted, steric hindrance slows down the dimerization. For instance, 1-alkyl-2-trimethylsilylcyclopropenes do not dimerize at 25 °C and can react with dialkyl azodicarboxylates in hetero-ene reactions at this

Scheme 5.123 Examples of thermal, intramolecular carbonyl ene-reactions.

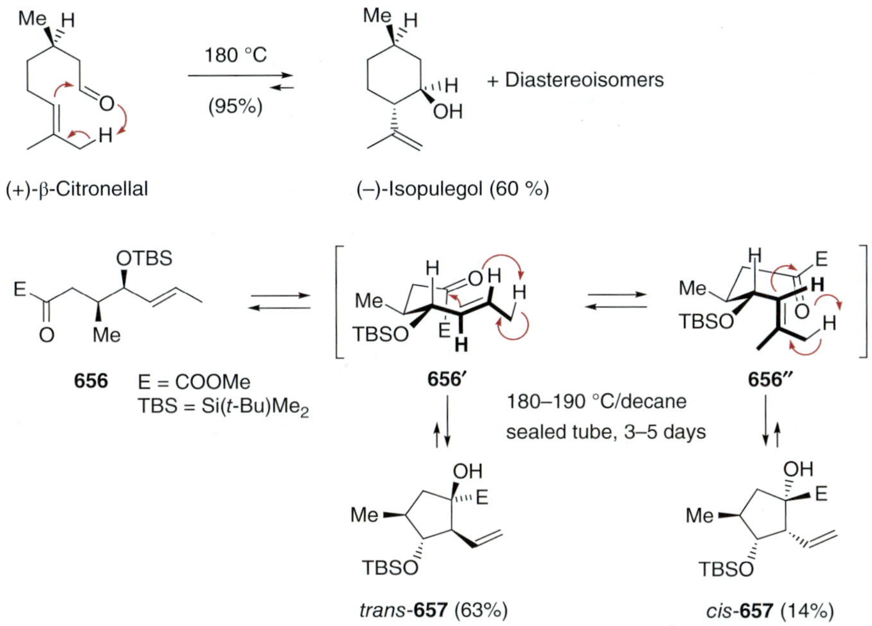

temperature [1500, 1501]. The dye-sensitized photo-oxidation (Section 6.9) of alkenes that have an allylic C—H bond generated the corresponding allylic hydroperoxides that can be reduced into the corresponding alcohols, for instance with $LiAlH_4$ or Na_2SO_3. An example is given with the photooxidation (5.133) of limonene [1502]. Singlet dioxygen ($^1O{=}O$) reacts with the alkene in a dioxa ene-reaction as shown by Foote [1503, 1504]. The mechanism of the reaction will be discussed in Section 6.9.4. It does not involve the formation of allylic radical **658** as the products obtained are not racemized. We have seen in Section 5.5.8 (Scheme 5.93) that SeO_2 undergoes quick, concerted oxaseleno ene-reaction (5.134) with alkenes having one allylic C—H bond [1089, 1505]. A subsequent (3,2)-sigmatropic rearrangement and hydrolysis generate the corresponding allylic alcohol. The analogous oxathia ene-reaction with sulfur dioxide does not occur with simple alkenes as the reaction is endergonic at room temperature, unless one stabilizes the forming β,γ-unsaturated sulfinic acid as a polymeric complex with BCl_3 (reaction (5.135) [1506]. With alkylallenes, the ene-reaction with SO_2 is possible as illustrated with reaction (5.136) [1507]. In this case, the isomerization of an allene derivative into a butadiene system brings an extra exothermicity of c. $-12\,\text{kcal mol}^{-1}$ ($\Delta_f H°(CH_2{=}C{=}CH{-}CH_3) = 38.8\,\text{kcal mol}^{-1}$, $\Delta_f H°(CH_2{=}CH{-}CH{=}CH_2) = 26.0\,\text{kcal mol}^{-1}$, Table 1.A.2), which is not available for the ene-reaction of SO_2 with a simple alkene. No ene-reaction has been reported yet for sulfur monoxide (S=O), although it undergoes (2+1)-cheletropic additions with alkenes and alkynes, (4+1)-cheletropic additions with 1,3-dienes, and (6+1)-cheletropic additions with conjugated trienes [1508–1510]. Sulfur trioxide (SO_3) usually reacts with alkenes at low

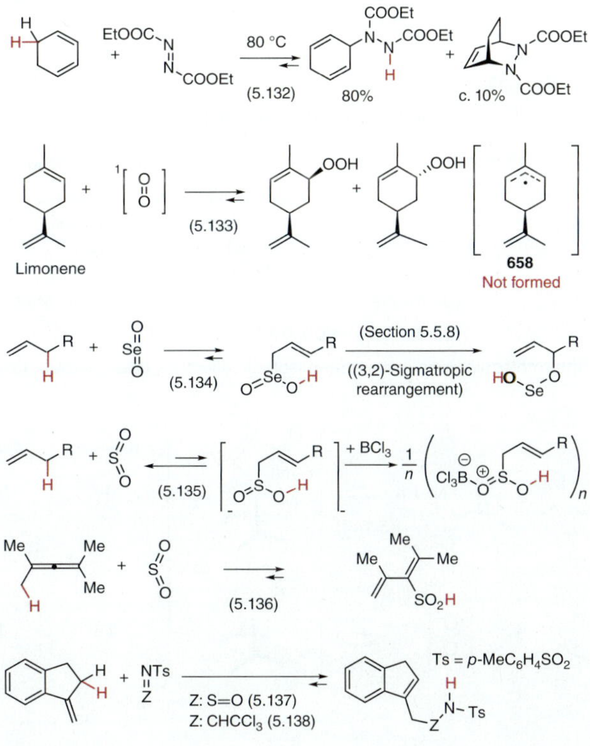

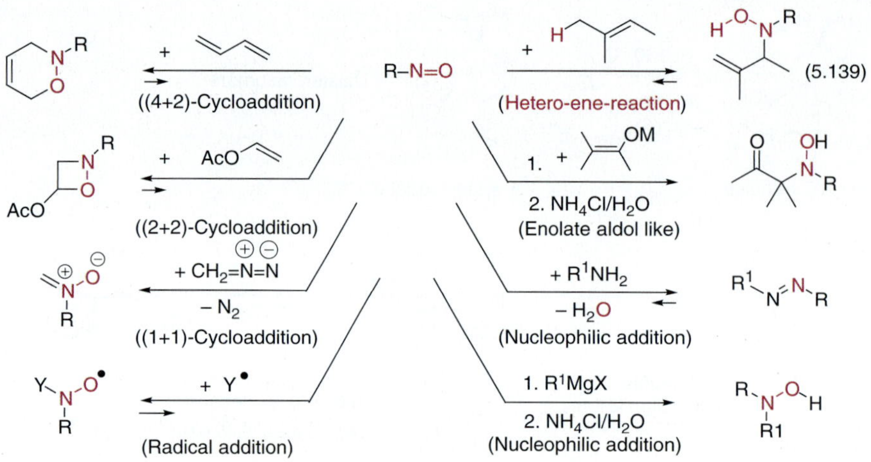

Scheme 5.124 The most common reactions of nitroso compounds.

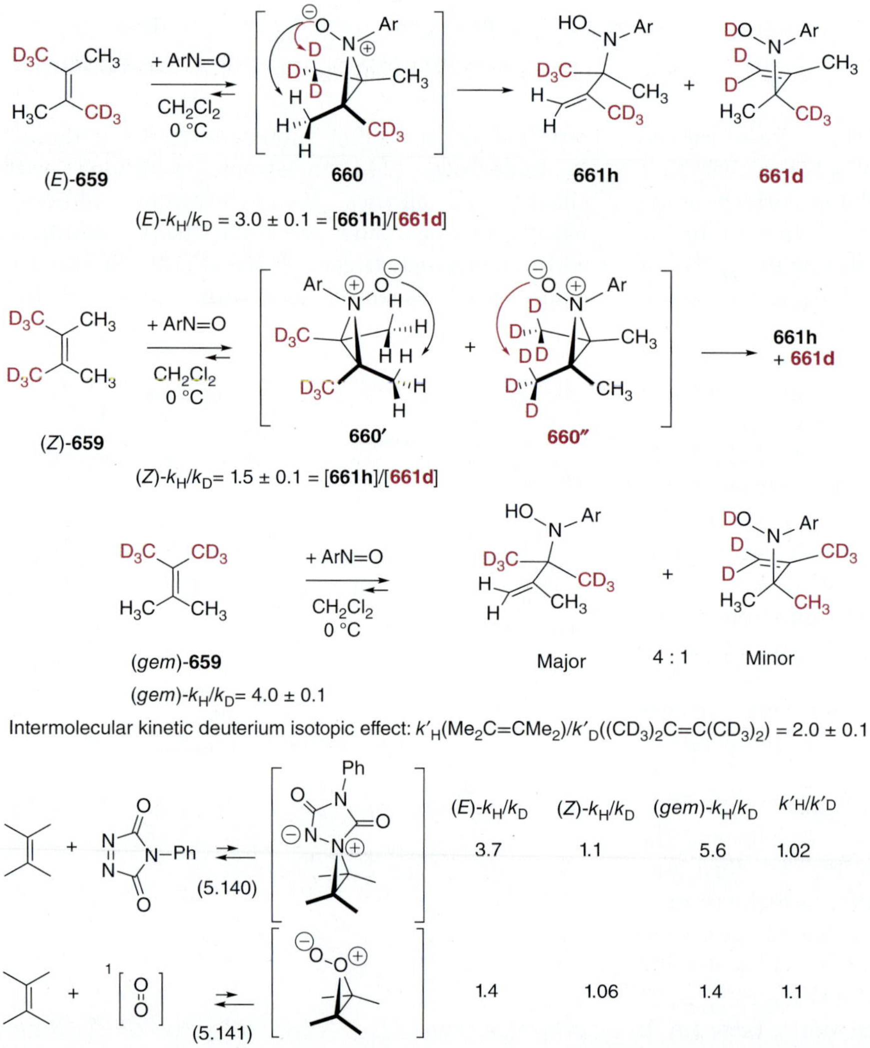

Scheme 5.125 Intramolecular (k_H/k_D) and intermolecular (k'_H/k'_D) kinetic deuterium isotope effects in the nitroso ene-reaction of 4-O_2N-C_6H_4NO with 2,3-dimethylbut-2-ene. Comparison with the diaza ene-reaction of *N*-phenyltriazolinedione and the dioxa ene-reaction of $^1O_2$.

temperature in (2+2)-cycloadditions. The β-sultones so-obtained may rearrange into the corresponding β,γ-unsaturated sulfonic acids [1511, 1512]. Hetero ene-reactions (5.137) and (5.138) have been reported for N-sulfinyl-p-toluenesulfonamide [1513–1515] and p-toluenesulfonylimines such as N-tosylchloral imine [1516, 1517]. Lewis acid induced hetero-ene-reactions of $(MeO)_2S{=}O/BF_3{\cdot}Et_2O$, $SOCl_2/Et_2AlCl$, [1518, 1519] and $ArS(NH_2){=}O/Yb(OTf)_3/Me_3SiCl$ have been reported [1520].

The nitroso compounds undergo a variety of reactions, which include nucleophilic additions, radical additions, reductions, isomerization, hetero-Diels–Alder reactions, $[\pi^2+\pi^2]$-cycloadditions, and the formation of nitrogen–carbon ylides with carbenes (Scheme 5.124) [1521]. The nitroso ene-reaction (5.139) was discovered in 1965 [1522]. It constitutes a mild methodology for the direct regioselective and stereoselective allylic nitrogen functionalization of alkenes. Acyl-nitroso compounds are the most reactive nitroso enophiles [311].

Applying the Stephenson's isotope test (Section 3.9.5), [1523] Greene and coworkers found that the reaction of C_6F_5NO with (Z)- and (E)- 2,3-(CD_3)-but-2-ene ((E)-**659** and (Z)-**659**) is a two-step process, the first one, the rate-determining-step, is the irreversible formation of an aziridine N-oxide intermediate of type **660** that undergoes a subsequent fast hydrogen transfer (Scheme 5.125) [1524]. With the less reactive $4\text{-}O_2N{-}C_6H_4{-}N{=}O$ (=ArNO), the same test suggested that the formation of the aziridine N-oxide **660**, or similar polarized diradical intermediate, is reversible [1525–1527]. For reaction (E)-**659** + ArNO, which generates intermediate **660**, the ratio of products **661h** and **661d** depends on the relative rates of hydrogen vs. deuterium migration from the allylic position to the oxygen center of the aziridine N-oxide. This gives an intramolecular kinetic isotope effect (E)-k_H/k_D = 3.0 ± 0.1 that is much larger than unity, as expected for a primary kinetic deuterium isotope effect (Section 3.9.1). The same hypothesis explains the intramolecular kinetic deuterium isotope effect (gem)-k_H/k_D = 4.0 ± 0.1 measured for the ene-reactions of ArNO with $(CD_3)_2C{=}CMe_2$ and with $Me_2C{=}CMe_2$. For reaction (Z)-**659** + ArNO → **661h** + **661d**, two aziridine N-oxide intermediates **660′** and **660″** form. In **660′**, only hydrogen can migrate, whereas in **660″**, only deuterium can migrate. Thus, the proportion of products formed [**661h**]/[**661d**] gives the proportion of intermediates **660′** vs. **660″**. This proportion is expected to be close to unity if the reaction is irreversible, and larger than unity if intermediates **660′** and **660″** should equilibrate with (Z)-**659** (**660′** undergoes hydrogen transfer faster than **660″** undergoes deuterium transfer). The later hypothesis can be retained as the intermolecular kinetic deuterium isotope effect k'_H/k'_D = 2.0 ± 0.1 given by the rate ratio of the reactions of ArNO with the nonlabeled 2,3-dimethylbut-2-ene and the perdeuterated 2,3-dimethylbut-2-ene is also significantly larger than unity. Related kinetic deuterium isotope effects are reported in Scheme 5.125 for the diaza ene-reaction (5.140) of N-phenyltriazolinedione and the dioxa ene-reaction (5.141) of singlet dioxygen ($^1O_2$) with 2,3-dimethylbut-2-ene. In these two latter reactions, the formation of the three-membered zwitterionic intermediate is less reversible.

Hetero-enes may react with all carbon enophiles. An example is given with reaction (5.142) that combines aldehyde-derived hydrazones with methyl acrylate or acrylonitrile. The diazenes so-obtained

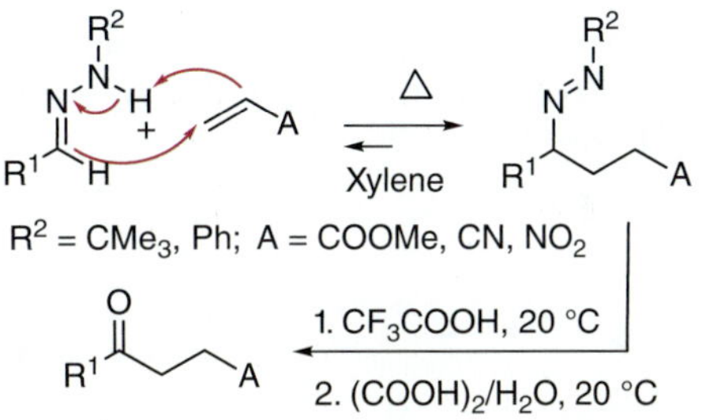

(5.142)

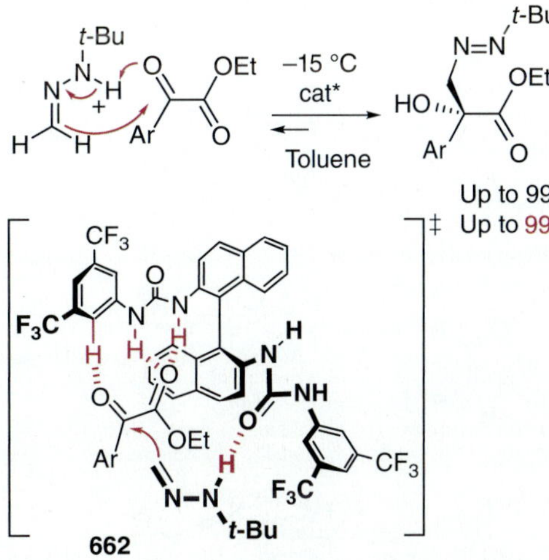

(5.143)

Up to 99% yield
Up to 99% ee

662

can be converted into ketones after acid-promoted isomerization into the corresponding hydrazones and acid-promoted hydrolysis of the latter [1528]. Similar diaza ene-reactions reacting formaldehyde *N-tert*-butyl hydrazone with nitroalkenes have been catalyzed by a chiral bis-thiourea to afford the corresponding diazenes with moderate ee's [1529]. High enantioselectivity has been reported for the carbonyl diaza ene-reaction (5.143) in the presence of an enantiomerically pure axially chiral bis-urea catalyst that permits dual coordination of both the ene and enophile through hydrogen bridging as outlined with the hypothetical transition structure **662** [1530].

The aminoborane $(F_3C)_2B{=}NMe_2$ matches the electron-donating and electron-withdrawing characteristics of the peripheral groups and makes this aminoborane olefin-like in terms of reactivity toward unsaturated systems. It undergoes cycloaddition reactions as well as ene-reactions with alkenes either in the mode (5.144) or (5.145). Similar reactions are also observed with alkynes. With carbonyl compounds having an α-hydrogen atom, ene-reactions of type (5.146) are observed. Similarly, reactions of type (5.147) occur with carbonitriles. Carbonyl compounds and nitriles devoid of α-hydrogen atom react with $(F_3C)_2B{=}NMe_2$ in ene-reaction of type (5.145) with hydrogen transfer from the dimethylamino group [1531, 1532].

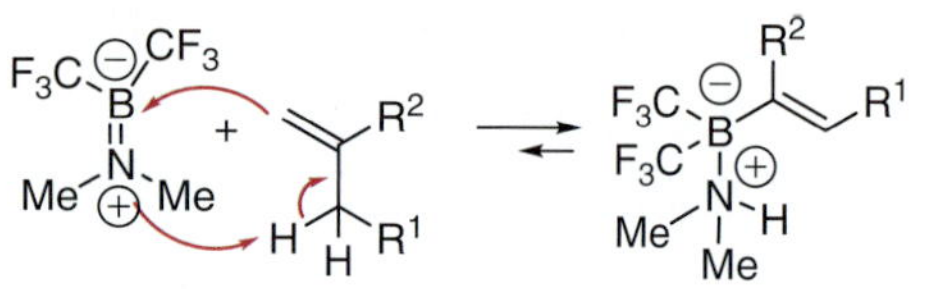

(5.144)

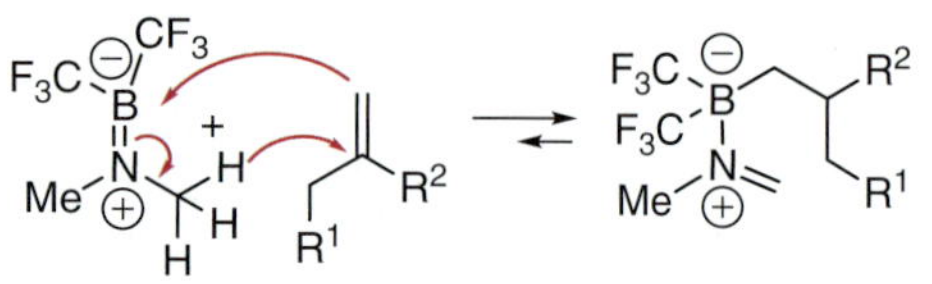

(5.145)

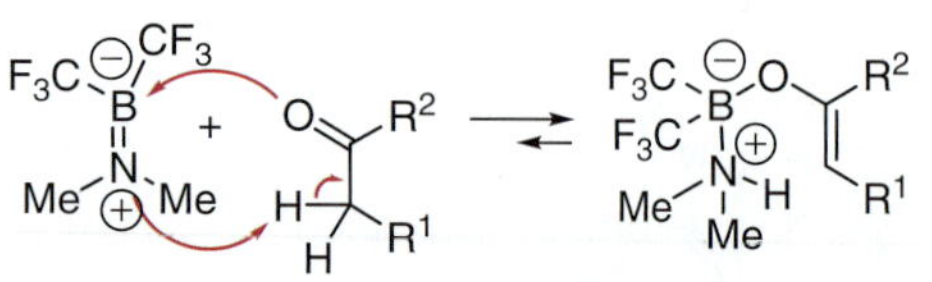

(5.146)

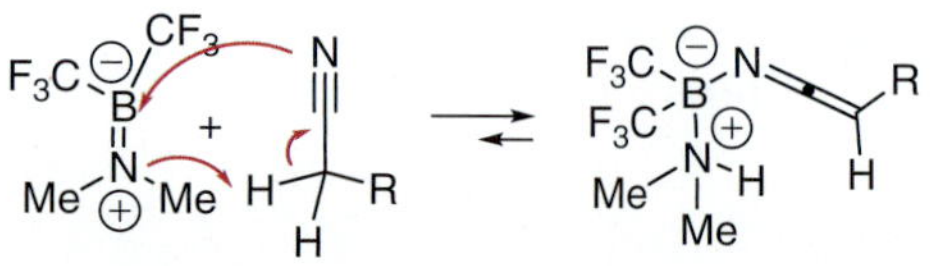

(5.147)

Problem 5.82 Explain the following transformations [1533, 1534].

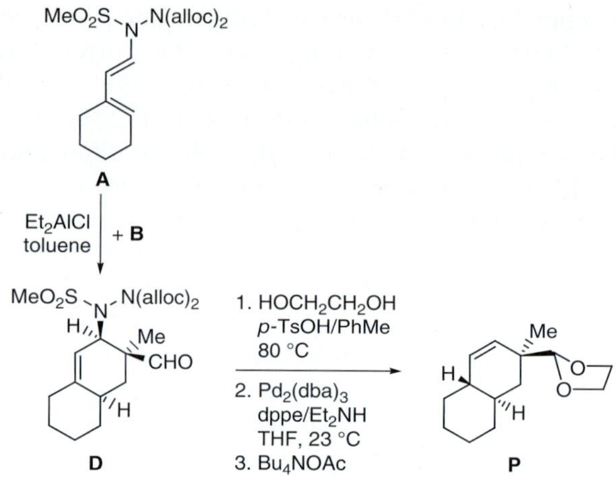

alloc, $CH_2{=}CHCH_2OCO$; dba, (*E*,*E*)-PhCH=CHCOCH=CHPh dppe, 1,2-bis(diphenylphosphino)ethane, *p*-TsOH, *p*-toluenesulfonic acid

5.7.4 Metallo-ene-reactions

In all the ene-reactions presented above, an atom of hydrogen (G = H) is transferred. Allylic metallic reagents of type $RCH{=}CH{-}CHRML_n$ (metallo-enes) are prepared readily and are very common. They react with aldehydes, ketones, esters, and imines (enophiles) in facile, noncatalyzed metalla ene-reactions with transfer of the metallic entity ML_n from the carbon center of the allylmetal reagent to the heteroatom (oxygen and nitrogen) of the enophile. They may also react with unpolar unsaturated systems such as alkenes and alkynes as shown first by Lehmkuhl and Reinehr in 1970 for allyl Grignard and allylaluminum compounds [1535, 1536]. Evidence for a concerted suprafacial mechanism (noted $[\pi^2s{+}\sigma^2s{+}\pi^2s]$) was provided by the addition (5.148), which combines allylmagnesium reagents with 3,3-dimethylcyclopropene at room temperature producing exclusively cis-products [1537]. Negative entropies of activation ($\Delta^{\ddagger}S$ = −18 to −24 eu) have been measured for these metallo-ene-reactions [1538]. Quantum mechanical calculations for the reactions of allyl-MgH and allyl-MgCl with ethylene also support concerted mechanisms (pericyclic reactions) [1539]. Since then, the metallo-ene-reaction of allyllithium [1540], allylboron [1541–1544], allylzinc [1545–1549], and more examples of allylmagnesium [1550, 1551] and allylaluminum [1552] to alkenes, alkynes, allenes, and enol ethers have been reported. In 1971, Gaudemar reported the condensation (5.149) of allylzinc with alkenylmagnesium reagents in apparent metallo-ene-reactions that generate the corresponding alk-4-ene-1,1-dimetal species [1553]. The latter can be reacted with two different electrophiles (E^1X and E^2X) in two successive reactions, which make the process (now called the Gaudemar–Normant reaction) quite useful [1554–1557]. Quantum mechanical calculations have suggested that the mechanism of the Gaudemar–Normant reaction might not be a one-step metalla ene-reaction but be a two-step process forming first allyl(alkenyl)zinc intermediates that undergo metalla-Claisen rearrangements [1558]. The addition of allylzinc reagents to alkenyboronates generates *gem*-zincio(boryl) compounds [1559].

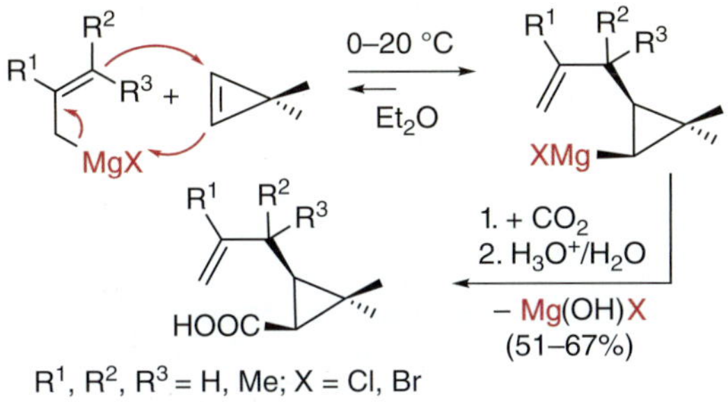

(5.148)

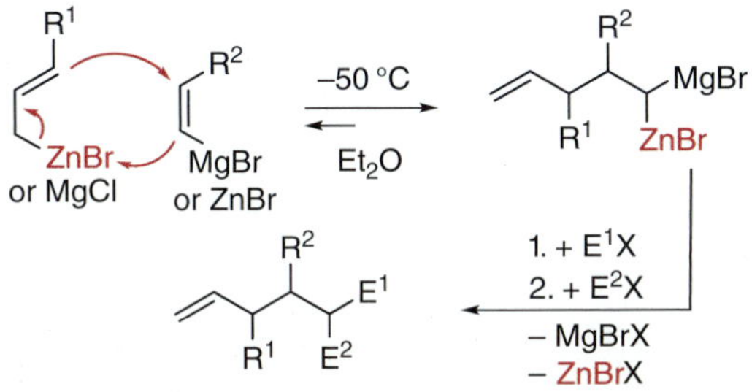

(5.149)

Metallo-ene-reactions with transfer of palladium and nickel are known since the 1960s and have been evidenced by studying the mechanism of the polymerization of butadiene induced by these metals [1560–1563]. A synthetic application of the pallado-ene-reaction is given in Scheme 5.126 with the synthesis of a prostaglandin endoperoxide analog [1564]. A large number of intramolecular catalytic in Ni, Pd, and Pt metallo-ene-reactions have been applied in organic synthesis [1470, 1565, 1566].

5.7.5 Carbonyl allylation with allylmetals: carbonyl metallo-ene-reactions

Like aldol and aldol-like reactions (Section 5.7.6), the allylation of aldehydes and ketones is a powerful tool of organic synthesis as it permits the stereocontrolled formation of carbon–carbon bonds in the homoallylic alcohols so-obtained (reaction (5.150)) [1478, 1479, 1567–1570]. Allylboranes and alk-2-ene-1-boronates are the most popular allylation agents because they are readily available, nontoxic, and lead to the best stereoselectivities, and the best enantioselectivities using enantiomerically enriched allylboron reagents. All other possible allylmetals and allylmetalloids have been used, for instance, M = Mg, Li, K, Zn, B, Si (Hosomi–Sakurai allylation), Sn, Al, In, Cr, Ti, and Zr. The metallo-ene reagents can also be propargyl and allenylmetal species that permit the preparation of allenyl alcohols (reactions (5.151)) and propargyl alcohols (reactions (5.152)), respectively [1571–1574].

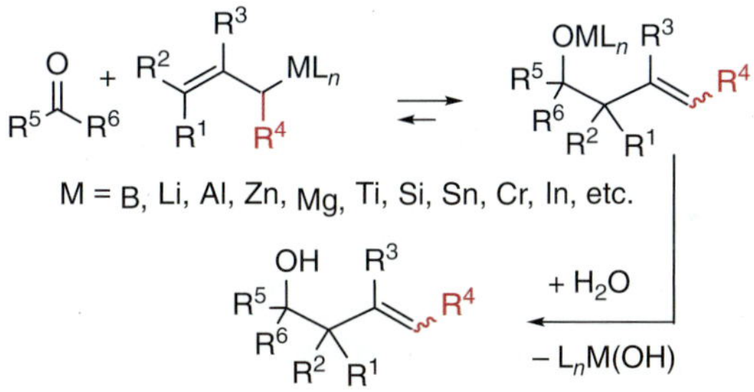

(5.150)

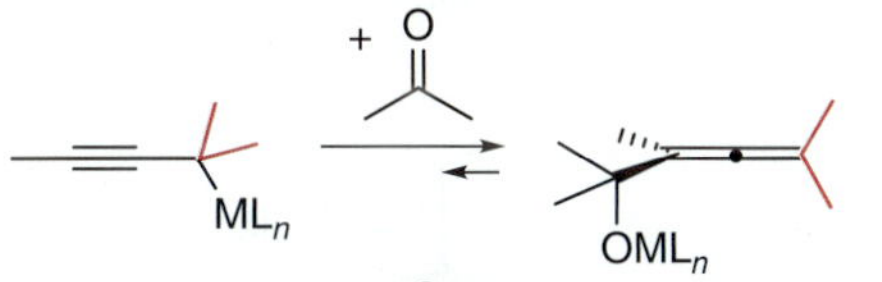

(5.151)

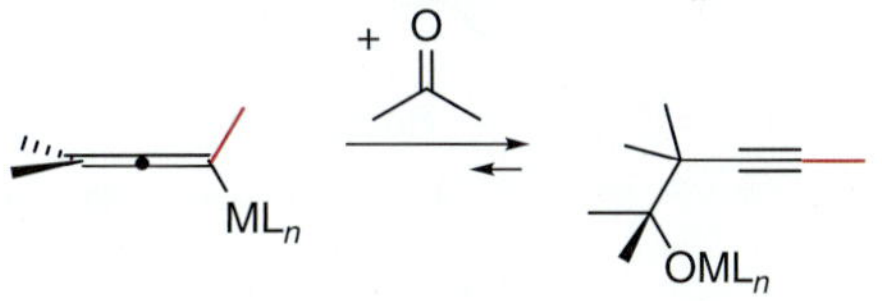

(5.152)

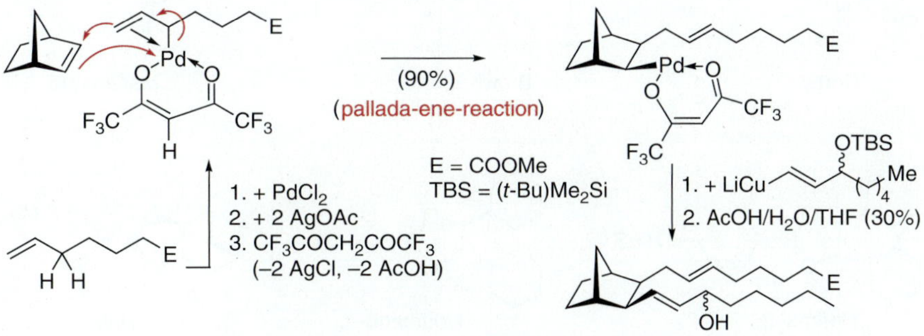

Scheme 5.126 Example of a pallado-ene-reaction

Denmark and Weber have proposed a classification system for carbonyl allylations based on the nature of their mechanisms of reaction [1575]. Allylmetals of type I have a strong Lewis acid character and are not sterically hindered about their metallic centers such as M = B, Al, $SiCl_3$, Ti, Zr, and Li. They form Lewis acid/base complexes with the carbonyl compounds that evolve via six-membered cyclic (closed) transition structures (Scheme 5.127a) as expected for one-step pericyclic processes. Allylmetals of type II such as allyltrialkylsilanes and allyltrialkylstannanes (M = Si, Sn) have a weak Lewis acid character and do not interact strongly with the carbonyl compounds. In a first step, they add to aldehydes and ketones in noncyclic (extended, open) transition states generating zwitterionic intermediates **663** that evolve to the final alcoholates in a second step (Scheme 5.127b). Usually, these nucleophilic additions require the presence of an activator such as a Lewis acid catalyst that increases the electrophilicity of the carbonyl compound by coordination to it and stabilizes the alkoxy moiety of the zwitterionic intermediate **663**. Stereoselectivity is generally better for type I allylmetals. The more rigid, the chair-like six-membered cyclic transition structure, the better the stereoselectivity and this is realized the best for bora-ene reagents as they make the shortest possible M—C and M—O bonds [1571, 1573, 1574, 1576–1587]. The diastereoselectivity is a consequence of the balance between gauche effects and other steric interactions between the substituents of the cyclic transition structures.

The addition of crotylboronic esters to aldehydes proceeds with high diastereoselectivities: (*Z*)-crotylboronates give *syn* adducts whereas (*E*)-crotylboronates give *anti* adducts [1588]. The transition states adopt six-membered cyclic chair-type conformations [1589]. When the ligand is chiral, the transition structures are diastereomeric and the products form in unequal amounts under conditions of kinetic control [1590, 1591]. Figure 5.56 gives several

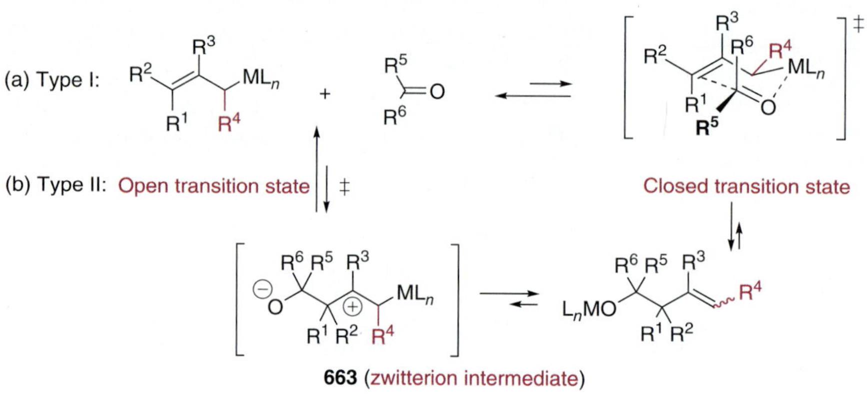

Scheme 5.127 Mechanistic classification for allylic metal reagents of (a) type I, and (b) type II.

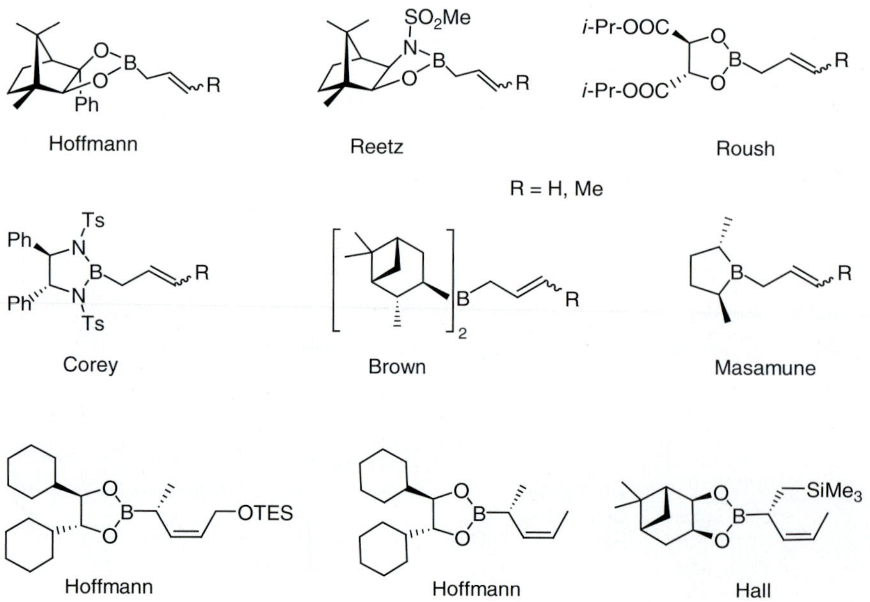

Figure 5.56 Commonly used enantiomerically enriched chiral allyl- and crotylboron reagents for the asymmetric allylation and crotylation of aldehydes and ketones.

chiral allyl/crotylboron reagents proposed by Hoffmann and coworkers [1592–1594], Reetz et al. [1595], Roush and coworkers [1585, 1586, 1597], Corey et al. [1582], Brown and coworkers, [1598, 1599] and Masamune [1580, 1581] and Hall [1570, 1587].

Enantiomerically pure tartaric ester derived allylboronate **664**, introduced first by Roush and coworkers [1583, 1600], allows one to prepare homoallylic alcohols with high ee's. The diastereoselectivity observed (relative configuration between the (*S*,*S*)-tartarate or (*R*,*R*)-tartarate and the newly created chiral center of the homoallyl alcohol) seems to be controlled by an attractive electrostatic interaction between the carbonyl group of the aldehyde (or ketone) and that of the carboxylate moiety of the tartaric system that can be the best approach than the former, as depicted in Scheme 5.128 with the diastereomeric transition structures **665a** and **665b** [1601, 1602].

Addition of (*E*)- and (*Z*)-crotylboranes and analogs (X = alkyl, RO, RS) to aldehydes proceeds with high simple diastereoselection (Scheme 5.129). These results have been rationalized in terms of a chair-like cyclic six-membered transition states. However, the energy difference is not so large compared with boat-like transition states, and both possibilities must be considered in a discussion of the diastereoselection. Nevertheless, in general, (*E*)-allylboranes give rise mainly to the *anti* diastereoisomers, whereas (*Z*)-allylboranes lead to the corresponding *syn* diastereoisomers [1567, 1570, 1603].

As other nucleophilic reagents, allylmetals add diastereoselectively to aldehydes and ketones with α-chiral centers. As summarized in Scheme 5.130, the additions follow the Cram, Felkin–Anh model when the carbonyl compound does not have a polar or a chelating *α*-substituent (in this model, the large substituent is also the group that can hyperconjugate the best, i.e. exchange σ-electrons with the antiperiplanar C—C bond that forms), the polar-Felkin–Anh model when a polar substituent that is not a good Lewis base, and by chelation-controlled additions if an *α*-substituent is a Lewis base that can cocoordinate the Lewis acid (the metal atom of the metalla-ene reagent or an added Lewis acid) that activates the carbonyl group. In the three cases, the C—C bond that forms in the ene-reaction follows the path that is the least sterically hindered [1604, 1605].

When the ligands about the boron center and the aldehyde are chiral, the inherent stereoselectivities of each partner may be either matched or mismatched. In many cases, the chiral allylboron reagents can be used to control the relative and absolute configuration of the newly formed stereogenic centers of the resulting homoallylic alcohols independently of the configuration of the aldehyde [1606]. This is

Scheme 5.128 Diastereoselective allylboration of acetaldehyde: enantioselective synthesis of pent-5-en-2-ol.

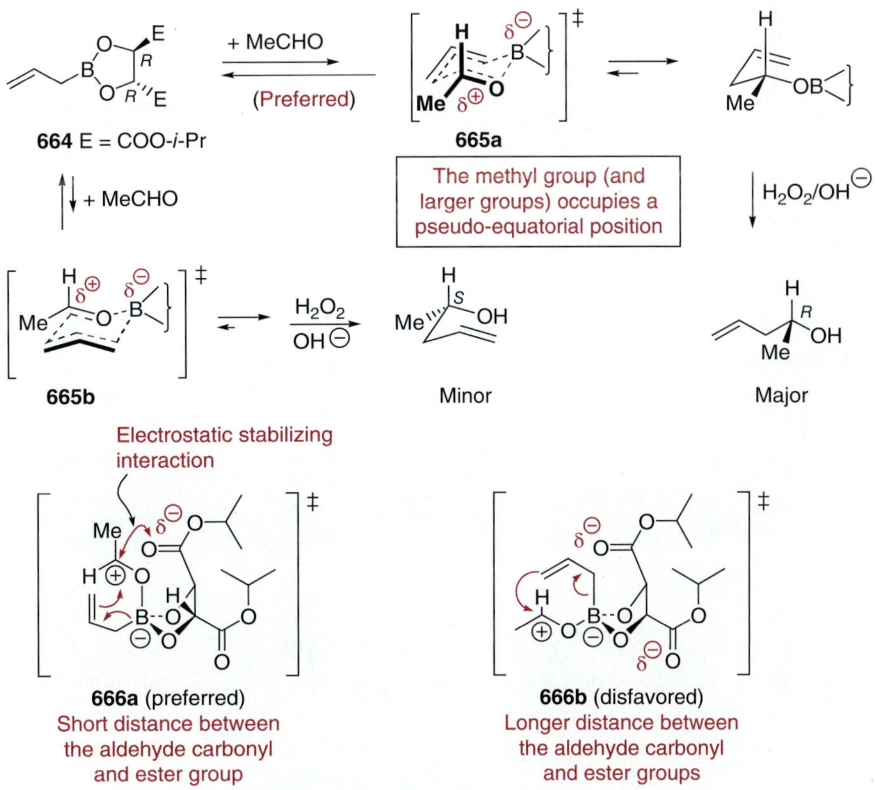

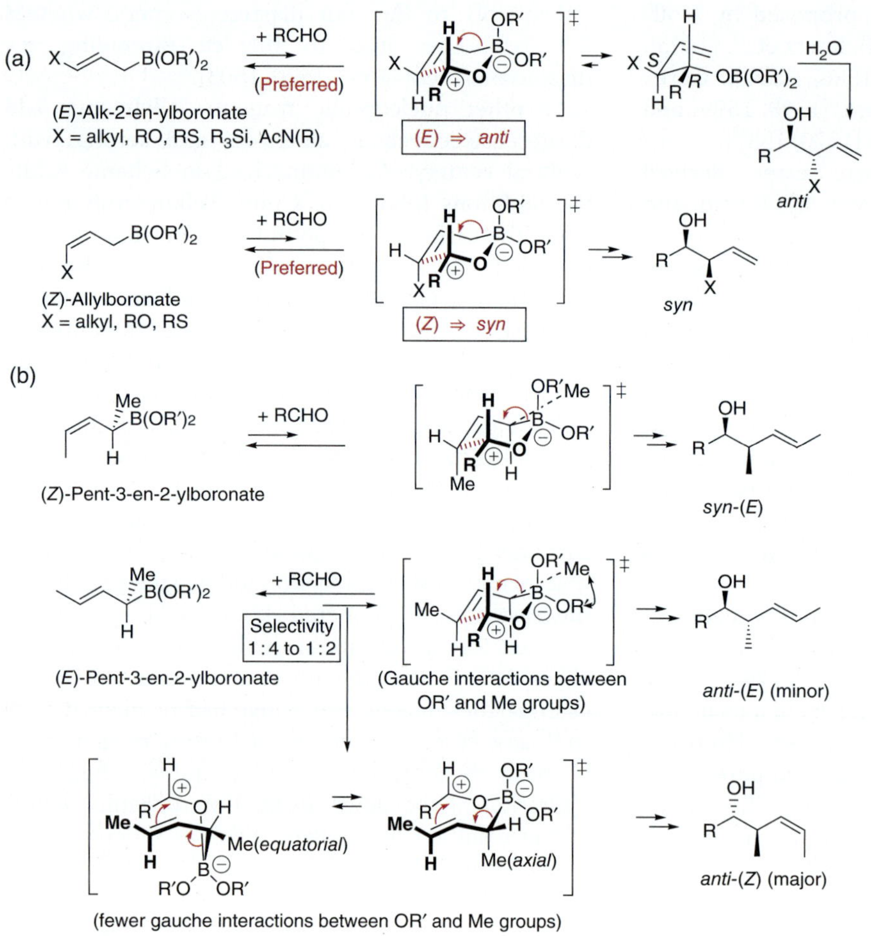

Scheme 5.129 Diastereoselectivity of the (a) allylation of aldehydes with 3-substituted alk-2-en-1-ylboronates and (b) with 1,3-dimethylalk-2-en-1-ylboronates. When the boronic ester derives from an enantiomerically pure diol, enantiomerically enriched homoallylic alcohols are obtained. The transition structures of these carbonyl boro-ene-reactions are represented as zwitterion resulting from the formation of boron ← aldehyde Lewis acid–base complexes, which interact intramolecularly with the allylic C=C double bond.

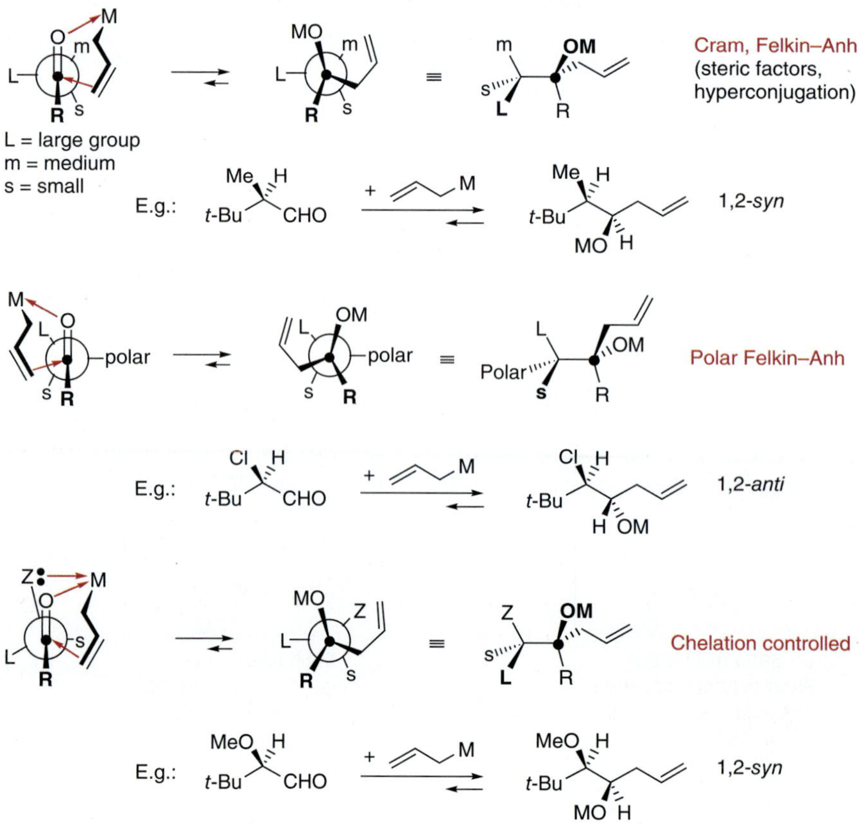

Scheme 5.130 Preferred facial selectivities for the allylations of aldehydes and ketones with an α-chiral center.

illustrated with the two crotylborations (5.153) and (5.154) reported by Roush and coworkers for their synthesis of (−)-Bafilomycin A_1 [1607, 1608]. Chelation control cannot be invoked in these reactions as the alk-2-en-1-ylboronates cannot coordinate the aldehyde **667** and its β-etheral group simultaneously. Thus, the facial selectivity of the addition to the aldehyde is given by the Cram, Felkin–Anh rule (steric factor) and must lead to preferred 1,2-*syn* product (Scheme 5.130). This forces the homoallylic alcohol that forms to have the (*S*) configuration. This is also expected (Scheme 5.128) when aldehyde **667** reacts with *ent*-**664** (derived from (*S*,*S*)-tartaric acid). Thus, stereochemical matching is realized in reaction (5.153). In the case of reaction (5.154), aldehyde **668** has the same absolute configuration for its α-center than aldehyde **667**. Because of the Cram, Felkin–Anh rule, the allylboration should give a 1,2-*syn* product with (*S*) configuration for the homoallylic alcohol center. This is opposite to what is expected for the allylation with **664** that prefers the formation of (*R*)-alcohol. Thus, reaction (5.154) is mismatched in terms of the stereochemistry and should be less stereoselective than reaction (5.153), as reported.

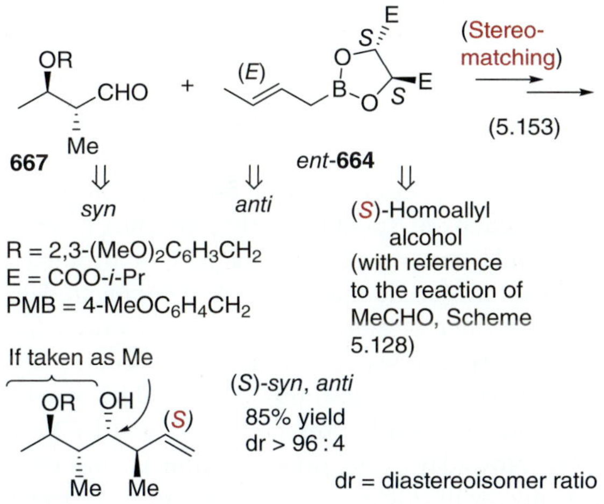

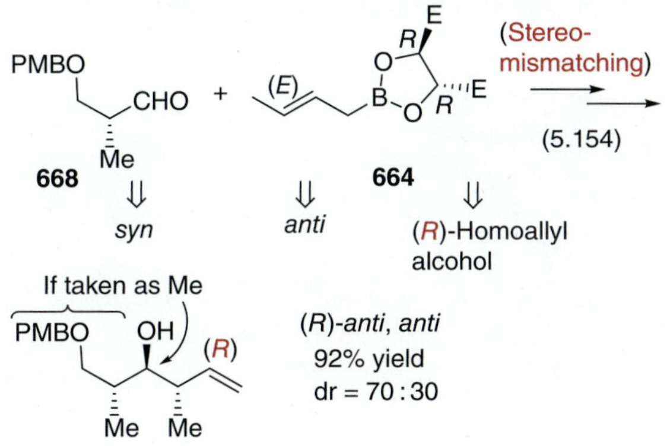

Allyl and crotyl complexes of titanium have also found useful applications in the stereoselective allylation and crotylation of aldehydes and ketones [1609–1611]. In the case of the reaction of *meso*-dialdehyde **669**, the absolute configuration of the titanium reagent **670** controls the stereoselectivity (Zimmerman–Traxler, cyclic transition states) and gives the *anti*-Cram product **671**. The reaction *ent*-**670** + **669** produces *ent*-**671** [1612].

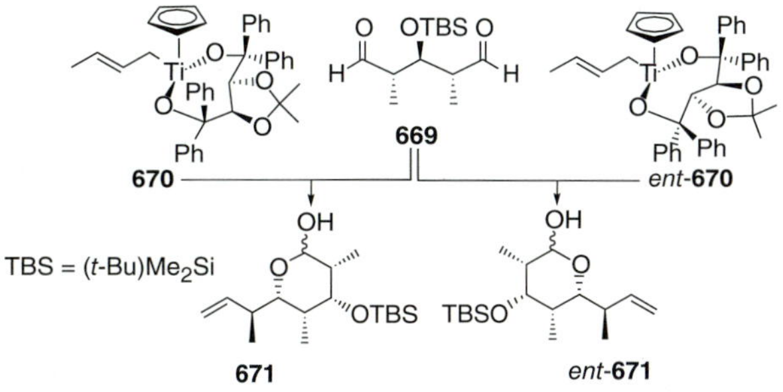

In contrast, reaction of aldehyde (*S*)-**672** (derived from the Roche ester) with the allenylstannane **673** promoted with $BF_3{\cdot}Et_2O$ gives the expected Cram, Felkin–Anh product **674** (2,3-*syn*) with high diastereoselectivity. In this case, the reaction adopts an open transition state without chelation. When allenyl stannane **675** and $InBr_3$ are reacted with aldehyde (*S*)-**672**, Sn/In exchange occurs first. The allenylindium intermediate so-obtained reacts with (*S*)-**672** adopting a cyclic transition state that follows the Cram, Felkin–Anh rule and leads selectively to **676**, the 4-epimer of **674** [1613–1616].

Allylboration of aldehydes with alk-2-en-1-ylboronates can be accelerated by Lewis acids such as $Sc(OTf)_2$, $Cu(OTf)_2$ [1617], BF_3, $TiCl_4$, and $AlCl_3$ [1618] and by protic acids. An enantioselective allylation using Yamamoto's Lewis-acid-assisted activation [1619] has been developed by Hall and Rauniyar [1620–1623]. In the presence of a catalytical amount of a BINOL-derived phosphoric acid, the reaction of $CH_2{=}C{=}CHCH_2B(pin) + RCHO$ provides the corresponding buta-1,3-diene-2-yl carbinols with high yields and ee values [1624]. Activation of the ene-reaction might be realized through hydrogen bridging of the acid catalyst to aldehyde or to one or two pinacol oxygen centers of the alk-2-en-1-ylboronate. This permits the asymmetric allylboration of ketones [1625] and acyl imines [1626] catalyzed by enantiomerically enriched diols. A boron-based catalyst, generated in situ from a readily accessible valine-derived aminophenol and a *Z*- or an *E*-γ-substituted boronic acid pinacol ester, permits the enantioselective allylation of ketones [1627].

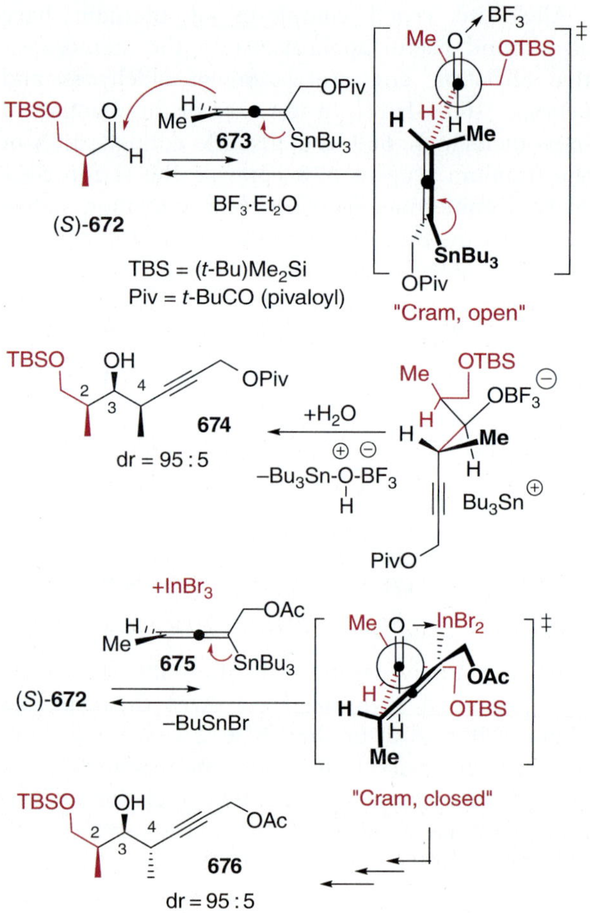

Asymmetric allyl- and allenylboration of aldehydes with allyl- and propargylboronic acid pinacol ester have been catalyzed by a chiral hydroxy–carboxylic acid [1628]. Efficient catalytic allylation of ketones, imines, and hydrazones with alk-2-en-1-ylboronates using a catalytic amount of zinc amide such as $[(Me_3Si)_2N]_2Zn$ has been reported. In this process, the boron/zinc exchange is the rate-determining step that leads to the corresponding allylzinc amides [1629].

Problem 5.83 What are the products of reaction of allylsilanes with sulfur dioxide? [1630]

Problem 5.84 What is the major product **P** of the following reaction? [1631]

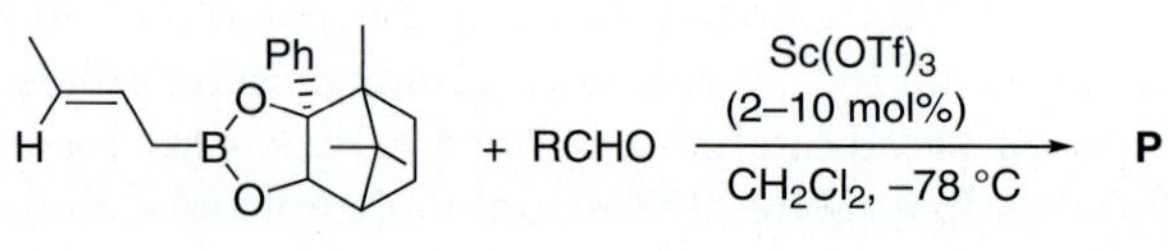

Problem 5.85 Explain the following reactions. Are the stereoselectivities expected? [1594]

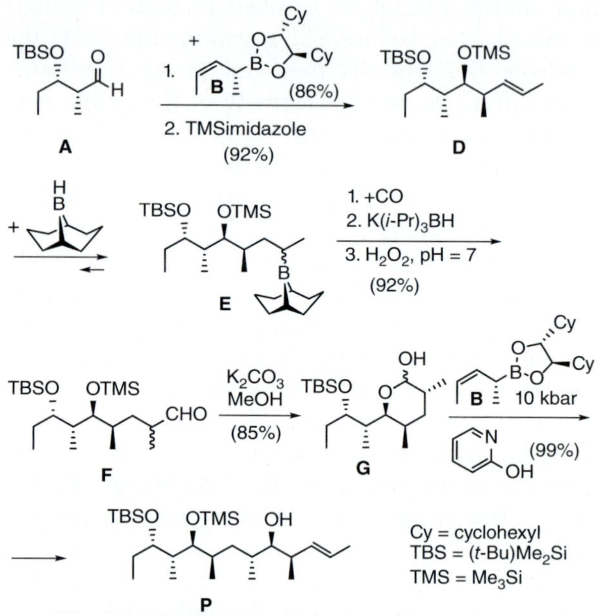

5.7.6 Aldol reaction

In 1858, Wurtz found that glycol is dehydrated with $ZnCl_2$ to acetaldehyde, which reacts further to an undefined compound of the same composition. In 1864, Borodin (Russian music composer, surgeon, and chemist) [1632] reported the action of sodium metal on valeraldehyde (pentanal) that produces several compounds, among them aldol **677**, a product of aldol reaction (or aldol addition) and conjugated enal **678**, a product of aldol condensation (results from the elimination of water from **677**, see Section 2.9.1) [1633]. Further contributions on the aldol reaction were reported then by Kekulé [1634], and by Wurtz [1635]. Nowadays, the aldol reaction is one of the most powerful means of forming carbon–carbon bonds in organic chemistry [1636–1638].

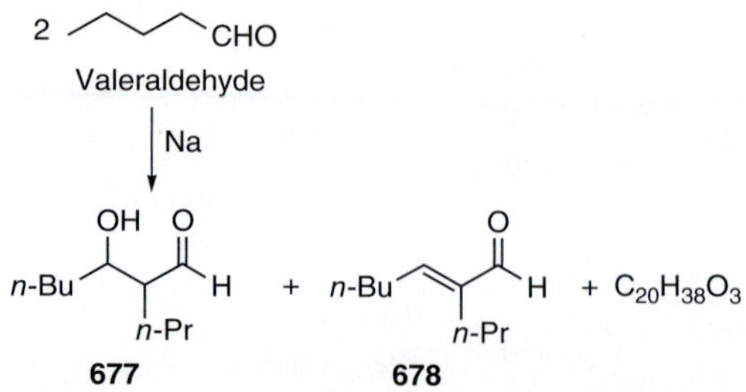

Aldol reactions (e.g. $RCH_2COR^1 + R^2CHO \rightleftarrows R^2CH(OH)—CH(R)—COR^1$) may proceed via two different mechanisms, both can be considered as carbonyl oxa ene-reactions. Like the Alder ene-reactions, they are atom economical (no coproduct), aldol reactions constitute hydrocarbations of aldehydes, and ketones (addition of a C—H moiety across a C=O double bond). In the first mechanism, aldehydes and ketones with at least one α-hydrogen atom are equilibrated with their enols or converted into enol ethers. The latter are nucleophiles that add onto the carbonyl group of an aldehyde, for example, usually activated by protonation or by coordination with a Lewis acid. This is the "enol mechanism" that generates the aldol in two steps, usually through an open transition state (1,2-addition of the nucleophile to the carbonyl compound) giving an hydroxycarbenium or alkoxycarbenium ion intermediate, which is then deprotonated (Scheme 5.131a). In theory, reaction enol + carbonyl compound ⇄ aldol could follow a concerted, one-step process (with a closed transition structure: Scheme 5.131b). In this mechanism, the C—C bond formation (nucleophilic addition) is assisted by the proton transfer from the enol to the carbonyl compound. In the second route, enolizable carbonyl compounds can be converted into their enolates with base B: giving the corresponding ion pairs enolate$^-$/BH^+, or by hydrogen/metal exchange with R–M giving enolate$^-$/M^+ + RH. This is the "enolate mechanism" in which the enolates add then to the carbonyl group of aldehydes or ketones. Here also, the aldol formation may go through an open (Scheme 5.131c) or a closed transition state (Scheme 5.131d). Aldol reactions (have negative entropies of reaction) are reversible (Table 2.7). On heating, aldols, and aldolates, undergo retro-aldol reactions and/or may lose 1 equiv. of water (β-elimination) to give the corresponding product of aldol condensation (crotonalization, with positive entropy of reaction: Section 2.9.1).

Aldolases are enzymes that catalyze the aldol reaction in Nature, a very important biochemical reaction that is highly enantioselective. Some aldolases can accept a variety of substrates (little substrate specific). They are two main types of aldolases. Class I aldolases catalyze aldol condensations converting first one of the two carbonyl compounds into an enamine. The latter is more nucleophilic than the corresponding enol. The enzyme then forces the enamine to add to the second carbonyl compound, the electrophile (Scheme 5.132a). Class II aldolases catalyze aldol reactions by converting first one of the carbonyl compound into a metallic enolate, usually zinc, and then force it to add to the second carbonyl compound that is activated by the Lewis acid Zn^{2+} (Scheme 5.132b) [1639, 1640]. As many "chemical" aldol reactions, the aldolase-catalyzed aldol reactions are formal carbonyl oxa ene-reactions that follow multistep mechanisms (are not concerted pericyclic reactions).

Simple amino acids such as L-proline are class I aldolase mimics as reported first in the 1970s by Wiechert and coworkers [1641] and Hajos and Parrish [1642] and coworkers for intramolecular aldol reactions. In the 2000s, several enantioselective intermolecular aldol reactions catalyzed by L-proline or related cyclic amino acids have been developed principally by List et al. [1643], Barbas and coworkers, [1644] and MacMillan and coworker [1645]. They are exemplified in Scheme 5.133a for the aldol reaction of propanal that gives 3-hydroxy-2-methylpentanal with 4 : 1 *anti* vs. *syn* diastereoselectivity and 99% ee

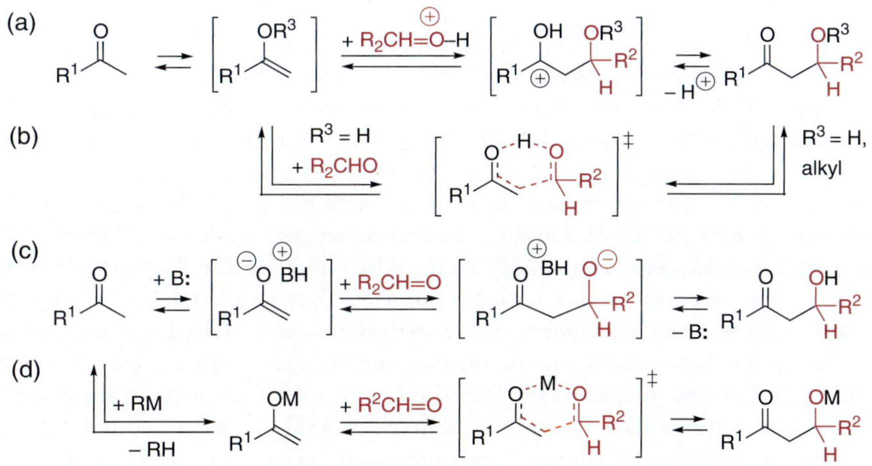

Scheme 5.131 Possible mechanisms for aldol reactions. Mechanisms (a) and (b) are two possible routes for the "enol mechanism" and (c) and (d) are two possible routes for the "enolate mechanism."

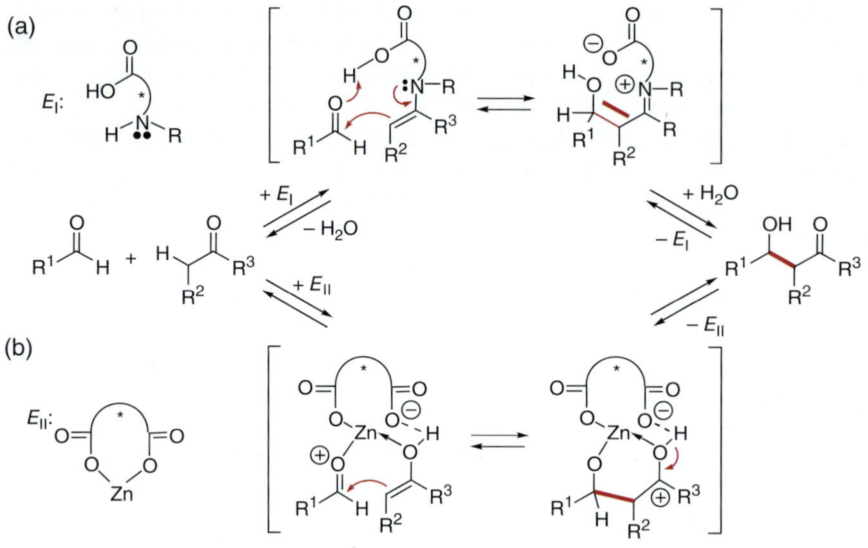

Scheme 5.132 General mechanisms of the aldol reactions catalyzed by class I (a) and class II (b) aldolase. Note: the reactions are reversible; two new stereogenic centers are formed in the aldol, thus four possible products can form. Under conditions of kinetic control (relatively low temperature), the enzyme-catalyzed aldol reactions can be highly diastereoselective and enantioselective.

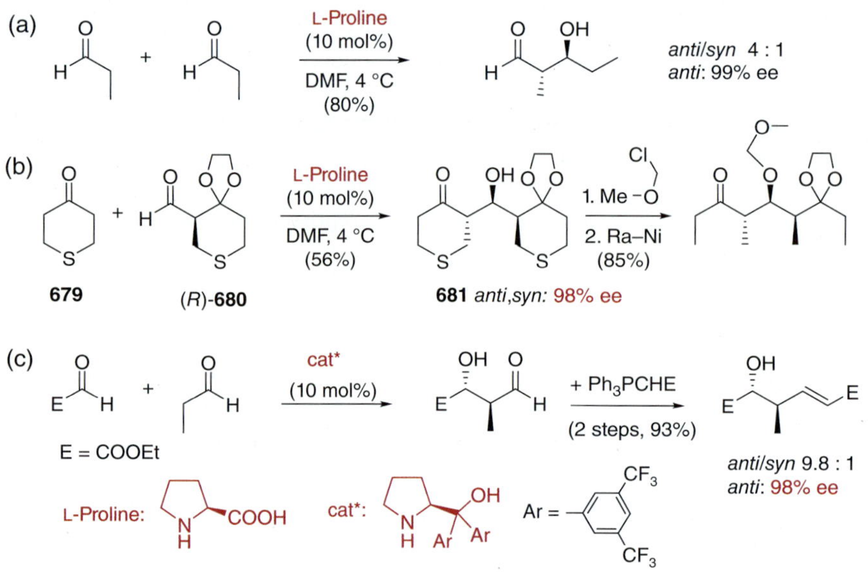

Scheme 5.133 Examples of enantioselective aldol reactions catalyzed by L-proline (a, b) or by a L-prolinol derivative (c) that mimics class I aldolase-catalyzed reactions.

for the major *anti* aldol. With the example shown in Scheme 5.133b, the enamine derived from the achiral ketone **679** adds to the chiral aldehyde (*R*)-**680** following the Cram, Felkin–Anh rule as the *α*,β-*anti*,*β*,*γ*-*syn* product is the major product isolated in 56% yield with 98% ee [1646]. The same result is obtained with the reaction of **679** and *rac*-(*R*,S)-**680**. L-Proline catalyzes racemization (*S*)-**680** ⇄ (*R*)-**680** much faster than it catalyzes its condensation with ketone **679**. Thus, the L-proline-catalyzed aldol reaction permits us to realize a dynamic kinetic resolution (Section 3.7.3): aldehyde (*R*)-**680** reacts much faster than (*S*)-**680** with the intermediate enamine (Scheme 5.133). L-Prolinol derivatives have also been found to induce asymmetry in the aldol reaction of activated aldehydes (Scheme 5.133c) [1647].

The admitted mechanism of the L-proline-catalyzed aldol reaction is given in Scheme 5.134. It is based on quantum mechanical calculations [1648] and experiments [1649–1652] such as the crystal structures of isolated proline-derived enamines [1653]. At 25 °C and in water $\log K(\mathrm{MeCHO} + \mathrm{Me_2CO} \rightleftarrows \mathrm{MeCH(OH)CH_2COMe}) = 1.59$ (Table 2.4, Section 2.9.1). This corresponds to $K = 39 = k_1/k_{-1}$ the rate constant ratio for the forward and reverse reaction (aldol and retro-aldol reaction). The reaction is exergonic, but reversible. Thus, good enantioselectivity can be realized only when the reaction is stopped before completion (one does not let time enough for the retro-aldol reaction to intervene) or by lowering the temperature of reaction (4 °C in DMF, instead of 25 °C). Being a condensation of two reactants

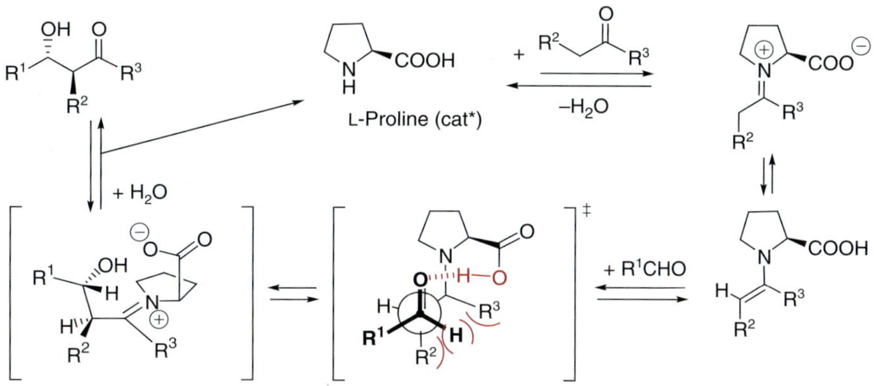

Scheme 5.134 Proposed mechanism for the enantioselective aldol reaction catalyzed by L-proline.

into one product, its entropy variation is negative (Section 2.9). The latter becomes more negative, the larger the aldehyde and ketone undergoing the intermolecular aldol reaction. As a consequence, good kinetic enantioselectivity can be reached only with small aldehydes and ketones, unless the exothermicity of the reaction is increased by using an activated aldehyde (e.g. α-substituted by an electron-withdrawing group) or by making the entropy of reaction less negative (e.g. on realizing intramolecular aldol reactions) [1654]. Proline-catalyzed cross-aldol reactions are accelerated by bifunctional urea $PhNHCONHCH_2CH_2NMe_2$ [1655].

Class II aldolase mimics have been developed for enantioselective aldol reactions. In 1997, Shibazaki and coworkers proposed the α,α'-binaphthol-derived tetrametallic complex **682** [1656]. In 2000, Trost and coworkers [1657] developed the dizinc complex **683**. On his side, Mahrwald proposed the amphoteric titanium salts resulting from the combination of *rac*-**684** and enantiomerically pure (*R*)-mandelic acid [1658]. In the latter case, the aldol reaction of pentan-3-one with propanal gives the *syn*-aldol as major product with 72% ee. Shair and coworkers have developed a biomimetic Cu(II)-catalyzed decarboxylative thioester aldol reaction using catalyst **685** [1659, 1660].

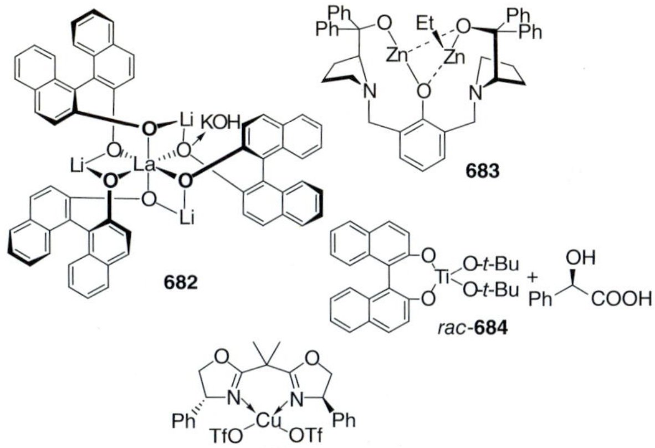

682 **683** *rac*-**684** **685**

Shibazaki and coworkers proposed the four-component catalyst (cat*) shown in Scheme 5.135 for the direct catalytic enantio- and diastereoselective aldol reaction of thionamides [1661]. This is an example of catalyzed carbonyl thia-ene reaction in which a thioenolate is the nucleophile, and the aldehyde the electrophile. Because of the high affinity of copper for sulfur, the Cu(II) thioenolate formation is preferred to the formation of the Cu(II) enolate from the aldehyde. Many more asymmetric direct aldol reactions catalyzed by enantiomerically enriched Lewis acids have been presented [1662].

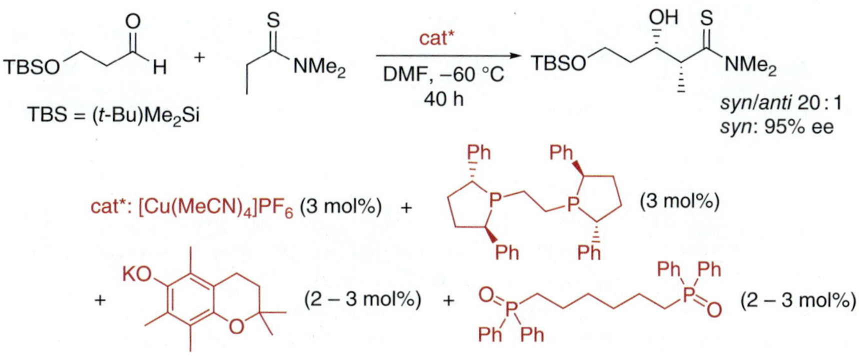

Scheme 5.135 A four-component catalyst for the enantioselective aldol reaction of thioamides and enolizable aldehydes.

5.7.7 Reactions of metal enolates with carbonyl compounds

In 1957, Zimmermann and Traxler proposed that some aldol reactions such as the Ivanov reaction undergo through six-membered transition states having a chair conformation [1663]. (*E*)-Enolates give rise to *anti* products, whereas (*Z*)-enolates generate *syn* products (Scheme 5.136; see also Schemes 5.54, 5.127–5.129) [1664]. Steric factors make the preferred transition structures to place their substituents equatorially and avoiding *syn*-pentane interactions [1665]. In reality, only a few metallic enolates such as lithium and boron enolates follow the Zimmerman–Traxler model (metals that make the shortest C—O bonds). Thus, in some cases, the stereoselectivity of the aldol reactions of metallic enolates may be unpredictable as boat-like transition states may compete. Furthermore, alkali enolates, like other polar organometallic compounds, are strongly solvated and have the tendency to aggregate [1666, 1667], which complicates the study of their reactivity. Nevertheless, lithium enolates have been very much applied in aldol reactions [1668].

Methods have been developed to generate either (*Z*)- or (*E*)-enolates with good (*E*) vs. (*Z*) stereoselectivity as illustrated in Scheme 5.137 [1665] and in Table 5.6.

The mechanism of formation of the lithium (*E*)-enolate **687** resulting from the reaction of ethyl isopropyl ketone (**686**) with LiHMDS in toluene at −78 °C in the presence of Et_3N is outlined in Scheme 5.138. At low concentration of Et_3N complex, unsolvated complex **688** is the major form and enolization follows a rate law with first-order $[Et_3N]$ dependence. At high concentration of Et_3N, fully solvated complex **689** becomes the dominant form, and a zeroth-order $[Et_3N]$ dependence is observed for the formation (k_{obs}) of enolate (*E*)-**687**. A plot of k_{obs} vs. [LiHMDS] at elevated concentration of Et_3N reveals a zeroth-order dependence. A kinetic deuterium isotopic effect $k_H/k_D > 10$ is evaluated for this reaction, confirming a proton transfer in the rate-determining step with transition state **690**. In agreement with the Zimmerman–Traxler model (Scheme 5.136), the lithium (*E*)-enolate (*E*)-**687** reacts with isobutyraldehyde giving the *anti*-α-methylaldol as major product in 72% yield. Reaction **686** + LiHMDS in pure THF at −78 °C generates the isomeric (*Z*)-enolate (*Z*)-**687** that reacts with isobutyraldehyde giving the *syn*-α-methylaldol as major product [1675].

In general, additions of lithium enolates to aldehydes with chiral α-center follow the Cram, Felkin–Anh rule (Section 5.7.5, Scheme 5.130). Several efficient protocols are available for the stereoselective and enantioselective aldol reactions of metal enolates. Among them, the boron-mediated aldol reaction has demonstrated its power in the field of total synthesis of complex natural products, especially those containing 1,3-dioxygen moieties as in polyketides and polypropionates, two important classes of products of the metabolism in plants, bacteria, insects, fungi, and marine organisms [1680–1682]. These compounds can inhibit the growth of bacteria, viruses, fungi, parasites, or human tumor cells with biological and pharmaceutical significance [1638, 1683]. The characteristic features of the boron-mediated aldol reactions compared to those of lithium-mediated ones are (i) the reacting boron enolate species in solution appears to be homogeneous and uncomplicated in terms of aggregation while lithium enolates exist as aggregates; (ii) the B—O bond and B—C bond are shorter than the Li—O and Li—C bonds, which make the transition structures (Zimmerman–Traxler model) more compact; (iii) the nucleophilicity and basicity of the boron enolate are much less pronounced than for lithium enolates, which reduces the possible side reactions such as proton transfer, crotonalization, and β-elimination of β-alkoxy groups, etc.

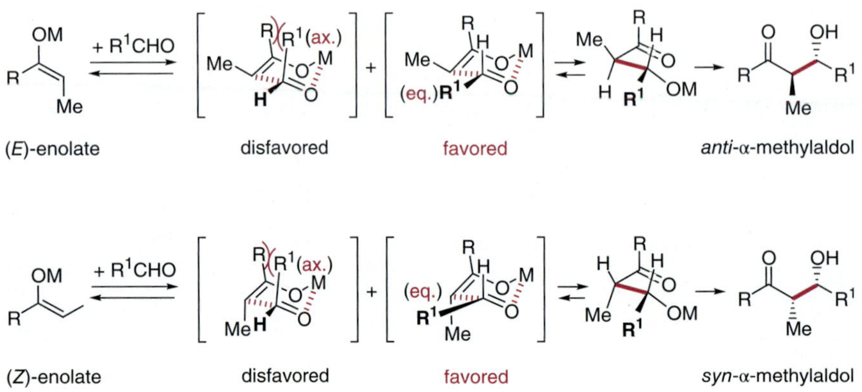

Scheme 5.136 Zimmermann–Traxler model for the transitions states of aldol reactions of metal enolates (carbonyl metalla-oxy-ene reactions).

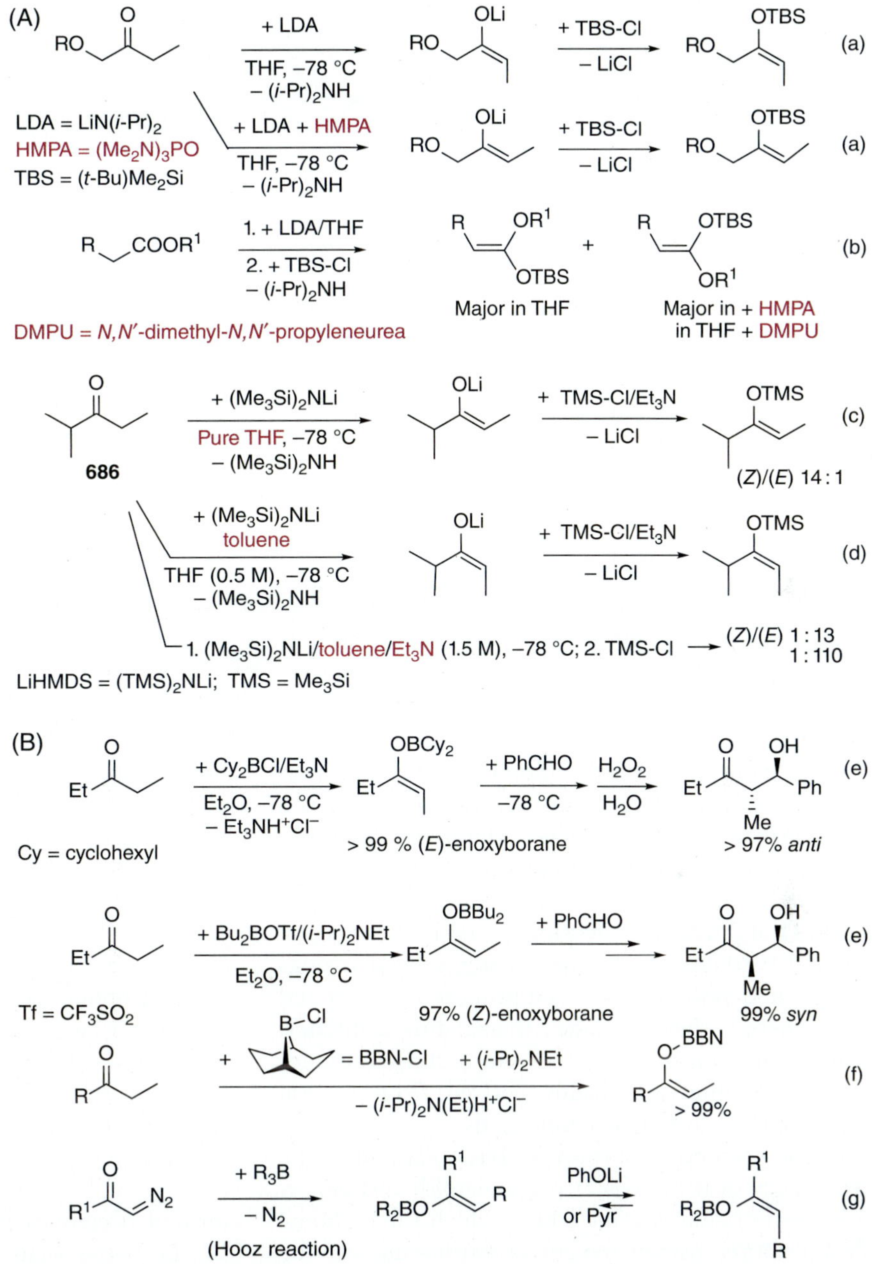

Scheme 5.137 Selected methods for the stereoselective preparation of achiral (A) lithium and silyl (*E*)- or (*Z*)-enolates and (B) of (*E*)- and (*Z*)-enoxyboranes. (a) [1669–1672], (b) [1673, 1674], (c) [1675], (d) [1675], (e) [1676], (f) [1677, 1678], and (g) [1679].

Table 5.6 Solvent-dependent (*E*)/(*Z*) selectivities for the formation of lithium enolates from carbonyl compounds (RCOEt, 0.05 M) in toluene/Et_3N (1.5 M) and neat THF at −78 °C using LiHMDS ($(Me_3Si)_2NLi$, 0.15 M), as given by the (*E*)/(*Z*) ratio of the enoxysilanes (RC(OTMS)=CHMe) (OTMS, Me_3SiO) resulting from their reaction with Me_3SiCl/Et_3N [1675].

	R = Et	*i*-Pr	*c*-C_6H_{11}	Ph	MeO	*syn*-$Me_2CCH(OTBDMS)CH(Me)$
THF	1 : 4	1 : 14	1 : 30	1 :>100	1 : 12	1 : 8
Et_3N/PhMe	140 : 1	100 : 1	80 : 1	3.5:1	22 : 1	50 : 1

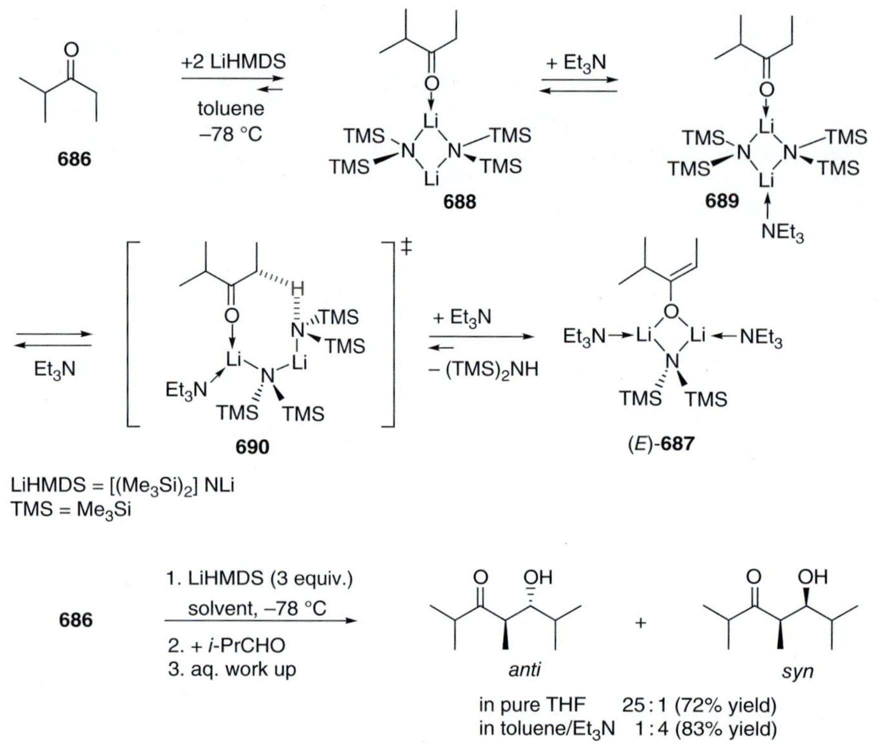

Scheme 5.138 Proposed mechanism for the enolization with LiHMDS in toluene/Et_3N and examples of stereoselective aldol reactions.

One of the most used method for the asymmetric synthesis of aldols is the reaction of aldehydes with enantiomerically enriched chiral enoxyboranes (boron enolates) derived from ester surrogates that are amides of *N*-acyloxazolidin-2-ones. Compounds like **691** and **692** were developed first by Evans et al. around 1980 [202, 1684–1686]. Later, several analogs and related systems have been proposed by him and by others. They include thiazolidine-2-thiones, 2-imidazilidinones, and oxazolidine-2-thiones (Table 5.7) that are readily derived from natural enantiomerically pure α-amino acids.

In the presence of 1 equiv. of Bu_2BOTf and of the base (Et_3N or $(i\text{-}Pr)_2NMe$), propionamide **691** generates the (*Z*)-enoxyborane **694** in which the carbonyl group of its carbamate moiety is coordinating with the boron Lewis acid (Scheme 5.139). Upon addition of aldehyde R^1CHO, the latter Lewis base competes with the former and generates several fast equilibrating Lewis acid–base complexes including **695**, **695′**, and **695″**. For steric reasons, **695′** is less stable than **695**. Because of dipole/dipole interactions between the C=O and enolate C—O bonds in **695**, facile rotation about the acyclic N—C bond isomerizes **695** into **695″**, which undergoes the carbonyl bora-oxa-ene reaction (cyclic transition state, product-determining step) giving the major boron aldolate **696**. After aqueous work-up, the "Evans *syn*" product **697** is obtained with *syn*/*anti* selectivity that can be better than 98 : 2. This amide reacts with MeON(Me)H·HCl and $AlCl_3$ and forms the Weinreb amide **698**, with recovery of the chiral auxiliary **699**, the acylation of which with propionyl chloride gives **691**. After conversion of alcohol **698** into a stable silyl ether, reduction of the Weinreb amide gives aldehyde **700**, which can then be engaged in another cross-aldol reaction as electrophilic partner. Alternatively, the Weinreb amide can be reacted with a Grignard reagent that gives the corresponding enantiomerically enriched *syn*-α-methyl-β-silyloxyketone that can be used in subsequent aldol reactions as enolate, for instance. Interestingly, under Heathcock's conditions that use 2 equiv. of Bu_2BOTf with **691**, the *anti*-isomer is formed preferentially with *anti*/*syn* selectivity that can be better than 98 : 2 [1702]. This is explained by invoking an open transition state **701**, the aldehyde is coordinated to the second equivalent of Bu_2BOTf and ignores the boron atom of the enoxyborane formed with the first equivalent of Bu_2BOTf. Aromatic, heteroaromatic aldehydes and acrolein derivatives react with the benzyl analog of **691** in the presence of Et_3N (2 equiv.), Me_3SiCl (1.5 equiv.), and $MgCl_2·Et_2O$ in EtOAc at 20 °C giving preferentially the corresponding *anti*-products [1703].

The thiazolidine-2-thiones **693** (Table 5.7) introduced first by Fujita in 1985 have been developed later by the groups of Crimmins [1689, 1704, 1705] and Evans et al. [1706]. Depending on reaction

Table 5.7 Commonly used enantiomerically enriched precursors of metal enolates in diastereoselective aldol reactions.

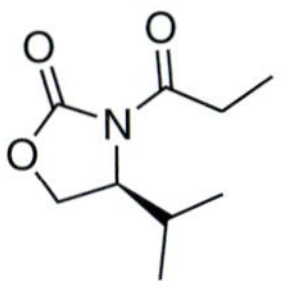 **691** (Evans 1980) [1687]	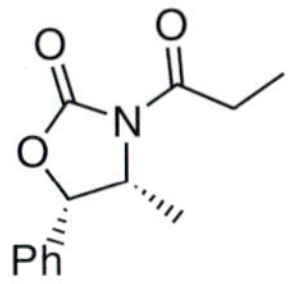 **692** (Evans 1981) [1687]	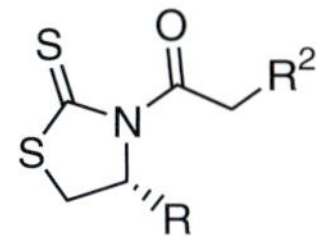 **693** (Fujita/Nagao 1985) [1688] (Crimmins 1997) [1689] (Sammakia 2005) [1690]	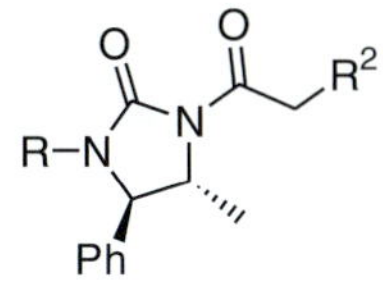 (Cardillo 1988) [1691] (Roos 1991) [1692]
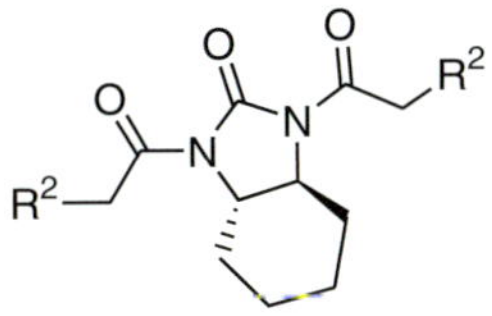 (Davies 1991) [1693]	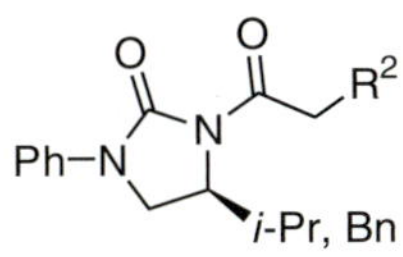 (Kim 2000) [1694]	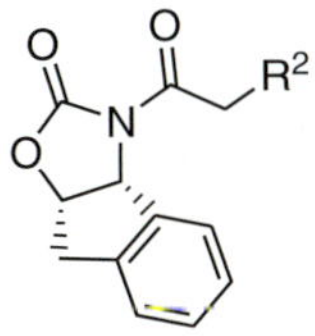 (Ghosh 1992) [1695]	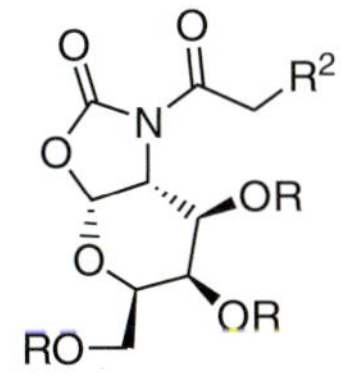 (Kunz 1992) [1696]
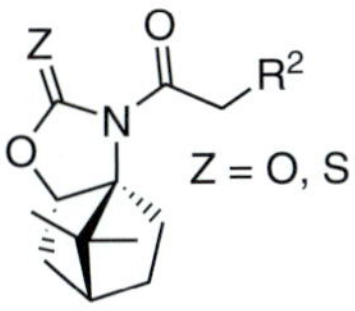 (Yan 1993) [1697]	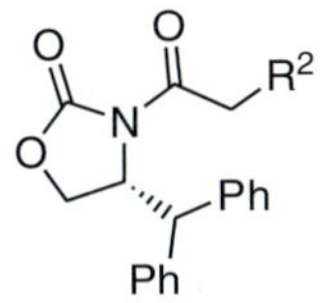 (Sibi 1995) [1698]	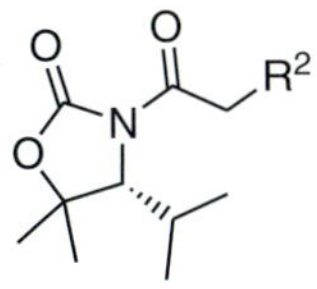 (Seebach 1998) [1699] (Davies 1999) [1700]	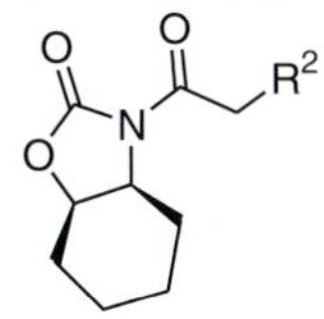 (Parrodi 2001) [1701]

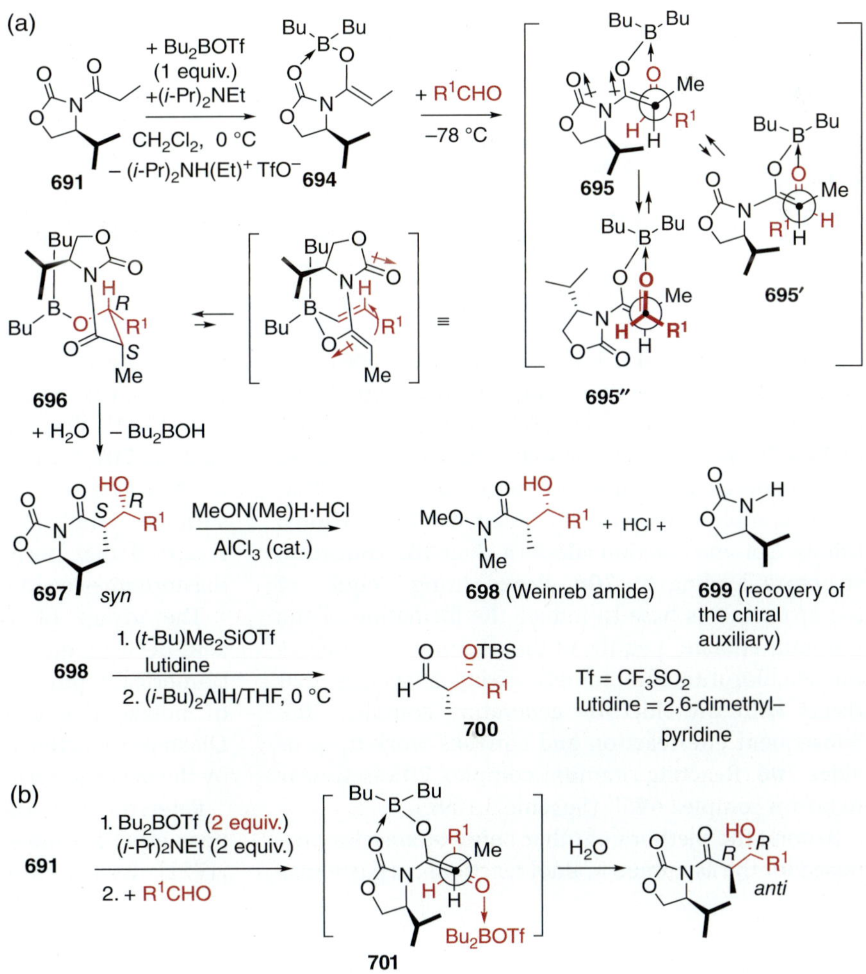

Scheme 5.139 (a) Evans boron-induced aldol reaction uses 1 equiv. of dibutylboronyl triflate giving selectively the *syn*-product, whereas (b) under Heathcock conditions, 2 equiv. of dibutylboronyl triflate are used, which leads selectively to the corresponding *anti*-product.

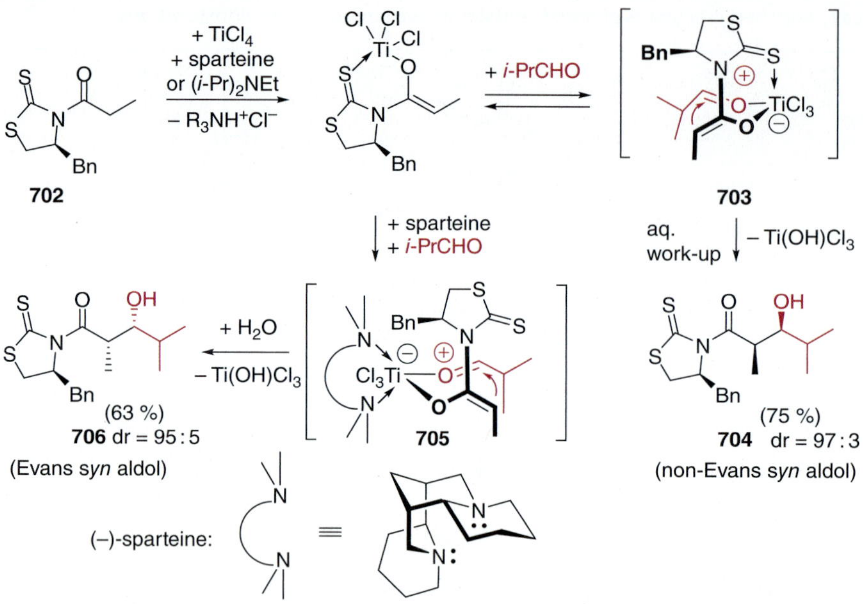

Scheme 5.140 Depending on the conditions, Evans-*syn* and non-Evans-*syn* aldols can be obtained.

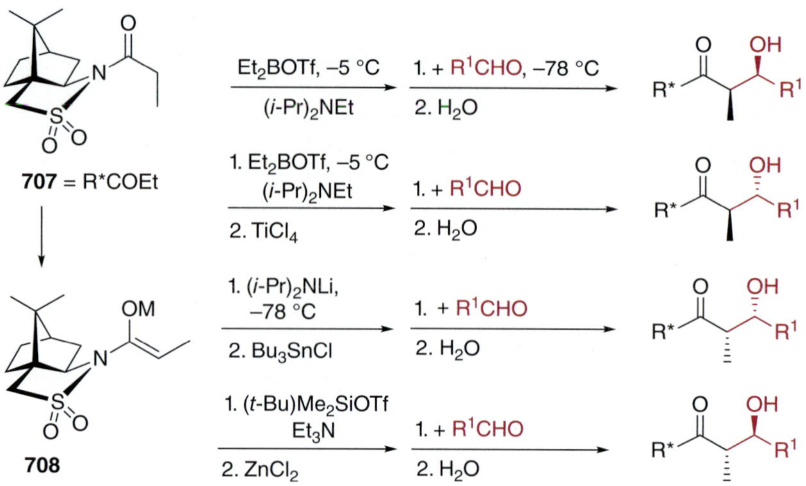

Scheme 5.141 The use of Oppolzer's sultams in asymmetric aldol reactions.

conditions, with **702**, both "*Evans syn*" (**706**) and "*non-Evans syn*" diastereoisomeric products (**704**) can be obtained selectively (Scheme 5.140). The use of 1 equiv. of $(i\text{-Pr})_2NEt$ or sparteine with 1 equiv. of $TiCl_4$ generates a trichlorotitanium (*Z*)-enolate that forms complex **705** with the aldehyde. The carbonyl titania-oxa-ene reaction adopts a chair-like transition structure leading to **706**. When using 2 equiv. of (−)-sparteine as base to induce the formation of the titanium enolate, 1 equiv. of the diamine coordinates the trichlorotitanium moiety giving a species that reacts with the aldehyde generating complex **705**. Subsequent ene-reaction and aqueous work-up provides **706**. Reacting titanium complex **705** is similar to boron complex **695″** (Scheme 5.139).

Among the plethora of other chiral auxiliaries proposed for the asymmetric aldol reactions, in particular by Heathcock et al. [1707, 1708], Masamune et al. [1709–1712], Myers et al. [1713], and Ghosh and Liu [1714], Oppolzer's *N*-acylbornane-10,2-sultams such as **707** [1715, 1716] have become very popular as they give solid aldols that are purified readily by simple crystallization [1717]. Conditions have been reported that permit us to obtain all four possible diastereomeric aldols (Scheme 5.141).

The group of Paterson has used the D-lactic acid-derived ethyl ketones **709** and **710** as enantiomerically pure templates [1718] in their synthesis of polyketides and polypropionates [1719, 1720]. Diastereoselectivity of the aldol reactions is controlled by the α-oxy substituent as shown in Scheme 5.142.

Paterson et al. also developed the enantiomerically pure ethyl ketone **711** (and its enantiomer *ent*-**711**) [1721]. Its (*E*)-enoxydicyclohexylborane reacts with

Scheme 5.142 Diastereoselectivity of boron aldol reactions of Paterson's lactate-derived ethyl ketones.

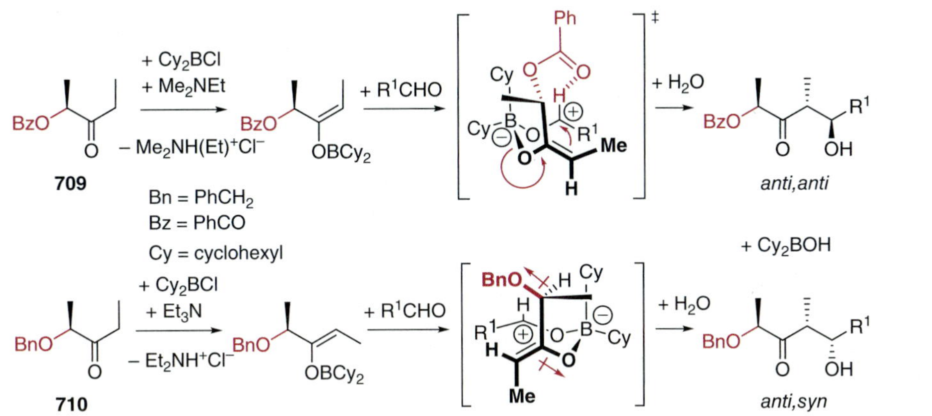

Scheme 5.143 Diastereoselective boron aldol reactions of 1-benzyloxy-2-methylpentan-3-one.

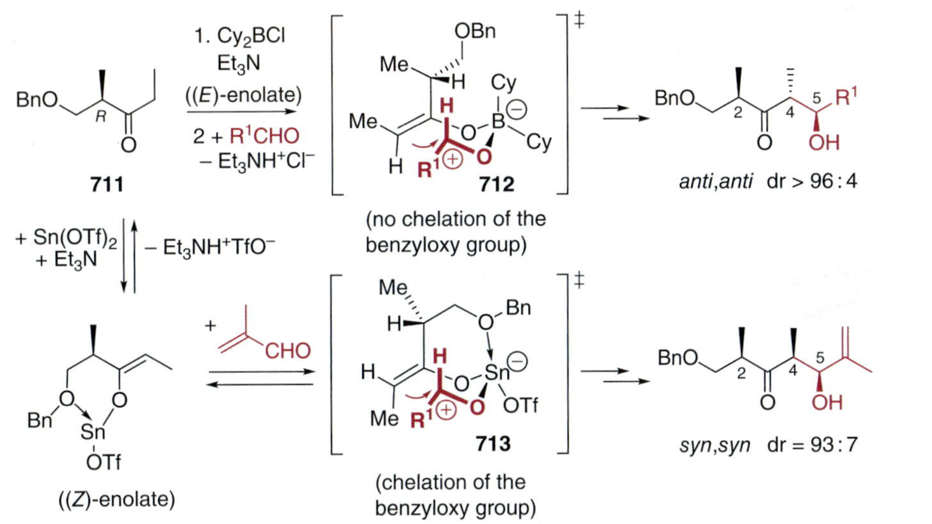

aldehydes through transition states of type **712** giving the corresponding 2,4-*anti*,4,5-*anti*-aldols. These transition states minimize $A^{1,3}$-allylic strain and electrostatic repulsions between the lone pairs of the benzyloxy and enolate oxygen atoms. The β-oxy substituent does not interact with the boron Lewis acid. When using tin triflate and Et_3N for the (*Z*)-enolization of **711**, transition state **713** forms and leads to the 2,4-*syn*,4,5-*syn*-product. In this case, the β-benzyloxy group cocoordinates with the Sn(II) Lewis acid and controls the diastereoselectivity (Scheme 5.143). When the β-benzyloxy (or another oxy substituent) is on a chiral carbon center, one speaks of 1,5-stereoinduction [1722]. Early examples have been reported by Masamune and coworkers in 1989 for enoxyboranes [1723]. Important contributions involving β-alkoxyketones have been made by Paterson and coworkers [1724, 1725], Evans et al. [1726–1728], Dias et al. [1722, 1729, 1730], and others [1731–1735], and for the addition of enoxysilanes (Mukaiyama aldol reaction) by Evans et al. [1736] and by Denmark and coworkers [1737–1739].

The readily available β-ketoimide **714** has been converted by Evans et al. into three of the four possible diastereomeric aldols with good diastereoselectivities (Scheme 5.144) [1740–1742]. The cross-aldol via the (*Z*)-enoxytitanium trichloride probably involves the transition state **715** that leads to the 2,4-*syn*-4,5-*syn*-product. When using $Sn(OTf)_2$ and Et_3N, the (*Z*)-enoxystannyl triflate forms and reacts according to transition state **716** that leads to the 2,4-*anti*-4,5-*syn* product preferentially. This difference in diastereoselectivity arises from the fact that the Ti(IV) Lewis acid is capable of being hexacoordinated, whereas the Sn(II) Lewis acid prefers to be tetracoordinated. Instead of the cyclic transition structure **716**, the tin (*E*)-enolate could be formed and the aldehyde would then react with it in an open transition structure. The (*E*)-enoxyborane derived from $(c\text{-Hex})_2BCl$ and **714** produces the 2,4-*anti*-4,5-*anti* product selectively via transition state **717**. The aldols can be reduced selectively either into *syn*- or *anti*-1,3-diols [1743–1750].

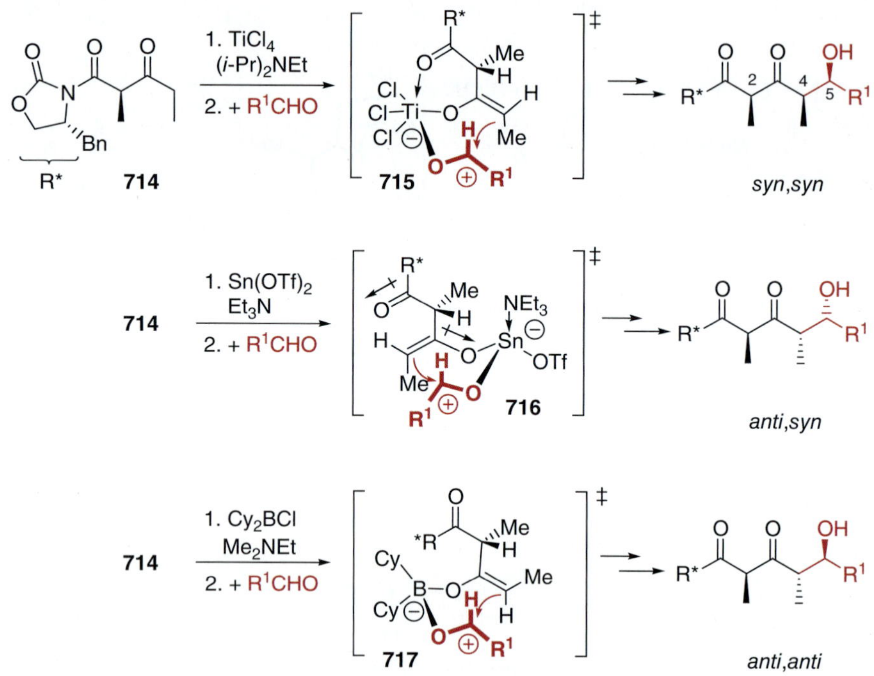

Scheme 5.144 Depending on the reaction conditions, three different diastereomeric aldols can be obtained selectively from the same β-ketoimide.

Carboxylic esters are less readily enolized than ketones, but with special reagents and under adequate reaction conditions, their enoxyboranes can be obtained and have been used in cross-aldol reactions by the groups of Corey and Kim [1751], Brown and coworker [1752], Masamune and coworkers [1753], and Abiko [1754]. When employing propionic esters derived from enantiomerically enriched chiral alcohols, asymmetric ester aldol reactions have been realized. Interestingly, the boron enolate generated by reaction of propionate **718** with (c-Hex)$_2$BOTf (2 equiv.) and Et$_3$N (2.4 equiv.) reacts at −78 °C with all kinds of aldehydes giving the corresponding *anti*-aldols **719** with high diastereoselectivity (95 : 5 to >99 : 1) [1711]. In contrast, under the same conditions, propionate **718′** leads to the corresponding *syn*-aldol **720** as major products (diastereoselectivity 95 : 5 to >97 : 3) [1755].

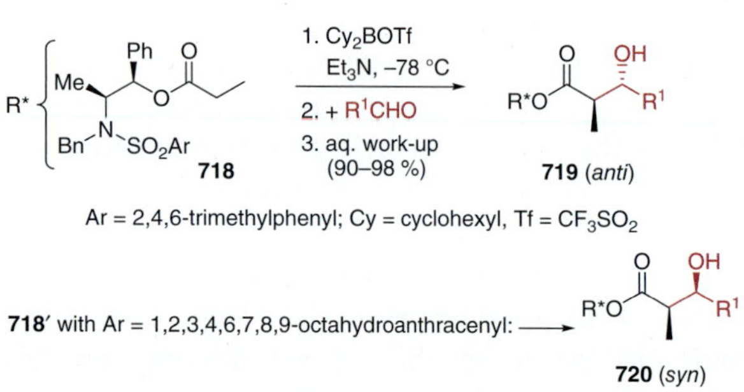

The preceding asymmetric aldol reactions employ enantiomerically enriched chiral aldehydes and/or enolates. In the next example, Paterson et al. have used achiral aldehyde and enolizable ketone and introduced the chirality with the boron reagent [1756]. Thus, the (*Z*)-enoxyborane **721** formed by reaction of pentan-3-one with bis((−)-isopinocampheyl)boryl triflate and Hünig base ((i-Pr)$_2$NEt) reacts with ethanal through transition state **722** giving the *syn*-aldol **723** as major product and with 82% ee.

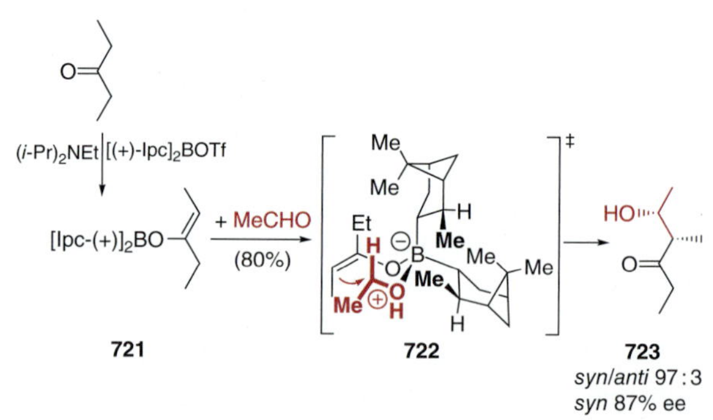

The reaction of enoxysilanes with aldehydes (Mukaiyama aldol reaction) [1757] has been catalyzed originally by stoichiometric amounts of strong Lewis acid such as TiCl$_4$, SnCl$_4$, AlCl$_{13}$, BF$_3$·OEt$_2$, and ZnCl$_2$ (Section 7.6.1). Later on, catalytical versions were developed using Lewis acids like

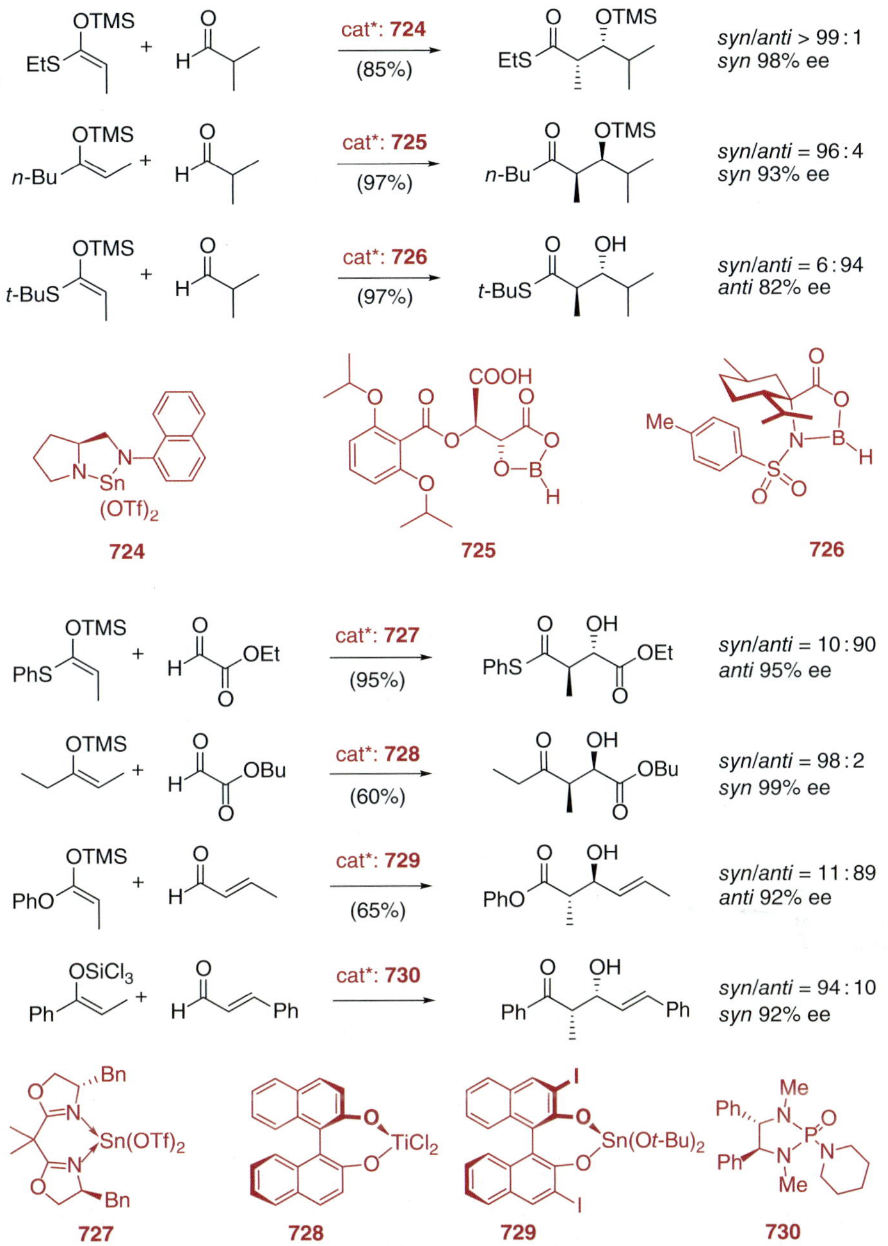

Scheme 5.145 Examples of asymmetric catalyzed Mukaiyama aldol reactions.

Sn(IV), Sn(II), Mg(II), Zn(II), Li(I), Bi(III), Ln(II), Pd(II), Ti(IV), Zr(IV), Ru(II), Rh(I), Fe(II), Au(III), Cu(II), Au(I), R_3SiX, and Ar_3C^+/X^- as catalysts [1758]. Methods using F^- sources and Lewis bases were also developed [1759]. The mechanism of the Mukaiyama aldol reactions depends on the nature of the substrates, the promoter, and the reaction conditions. Quite often, they adopt open transition structures [1595, 1760, 1761]. In 1986, Reetz et al. showed that asymmetric Mukayiama aldol reaction can be induced by substoichiometric amount of an enantiomerically pure chiral Lewis acid complex [1762]. There is now a plethora of chiral catalysts that have been proposed. Examples are given in Scheme 5.145 with Lewis acids **724** [1763, 1764], **725** [1765], **726** [1766], **727** [1767], **728** [1768], and **729** [1769] for the asymmetric reactions of trimethylsilyl enol ethers, and with a chiral base **730** (analogous of HMPA) for the aldol reaction of a trichlorosilyl enolate [1770, 1771]. Further discussion on acid-catalyzed Mukaiyama aldol reaction will be given in Section 7.6.1 [1638c].

Problem 5.86 Using propionamide *ent*-**696**, prepare the polypropionate fragment **P** [1772].

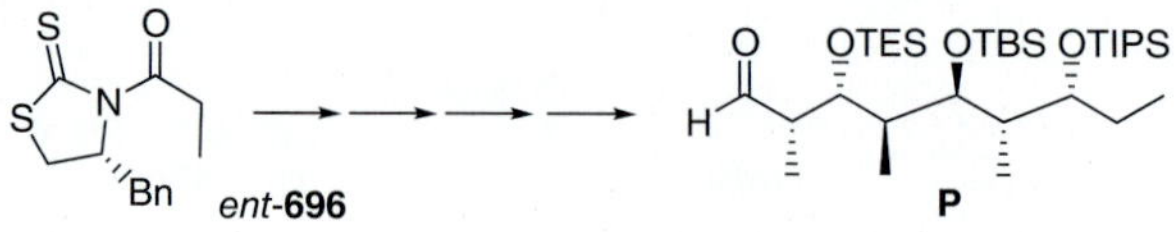

TES = Et_3Si
TBS = $(t\text{-Bu})Me_2Si$
TIPS = $(i\text{-Pr})_3Si$

Problem 5.87 What is the major product **P** of the following reaction? [1685]

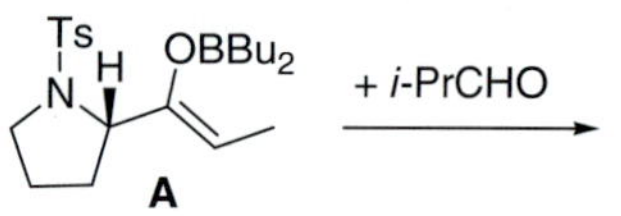

Problem 5.88 What are the major products of the following cross-aldol reactions all starting with **A** and finishing with a work-up using $H_2O/H_2O_2/KOH$? [1773]

A: 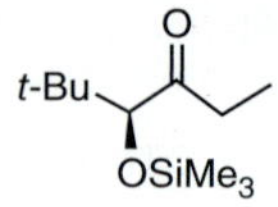

a) 1. $(i\text{-Pr})_2NLi/THF$, –78 °C; 2. TMEDA, then + PhCHO.
b) $Bu_2BOTf/(i\text{-Pr})_2NEt$, then PhCHO.
c) MTMP/THF, –5 °C, then PhCHO.
d) MTMP/THF/HMPA/dioxane, then $(i\text{-PrO})_3TiCl$ and PhCHO.

TMEDA = $Me_2NCH_2CH_2NMe_2$;
HMPA = $(Me_2N)_3P{=}O$; Tf = CF_3SO_2; MTMP = 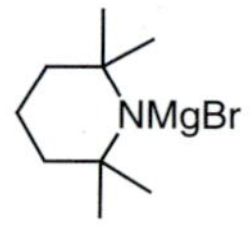

References

1 Woodward, R.B. and Hoffmann, R. (1969). Conservation of orbital symmetry. *Angew. Chem. Int. Ed. Engl.* 8 (11): 781–853.

2 Woodward, R.B. and Hoffmann, R. (1970). *The Conservation of Orbital Symmetry*, 1–184. New York: Academic Press; 3rd printing, New York: Wiley-VCH, Weinheim, and Academic Press, 2014.

3 (a) Woodward, R.B. and Hoffmann, R. (1965). Stereochemistry of electrocyclic reactions. *J. Am. Chem. Soc.* 87 (2): 395–397. (b) Breulet, J. and Schaeffer, H.F. III (1984). Conrotatory and disrotatory stationary points for the electrocyclic isomerization of cyclobutene to *cis*-butadiene. *J. Am. Chem. Soc.* 106: 1221–1226. (c) Oliva, J.M., Gerratt, J., and Lardakov, P.B. (1997). Conrotatory and disrotatory pathways in the electrocyclic isomerization of cyclobutene to *cis*-butadiene: the spin-coupled viewpoint. *J. Chem. Phys.* 107: 8917–8926. (d) Misale, A., Niyomchon, S., and Maulike, N. (2016). Cyclobutenes: at a crossroad between diastereoselective syntheses of dienes and unique Pd-Catalyzed asymmetric allylic substitutions. *Acc. Chem. Res.* 49 (11): 2444–2458.

4 Bauld, N.L. (1992). Hole and electron transfer catalyzed pericyclic reactions. *Adv. Electron Transfer Chem.* 2: 1–66.

5 Bauld, N.L., Cessac, J., Chang, C.S. et al. (1976). Cyclobutene-butadiene anion-radical electrocyclic reaction. *J. Am. Chem. Soc.* 98 (15): 4561–4567.

6 Willstäter, R. and von Schmaedel, W. (1905). Über einige Derivate des Cyclobutanes. *Liebigs Ann. Chem.* 38: 1992–1999.

7 Coke, J.L. (1972). Stereochemistry of Hofmann eliminations. *Sel. Org. Transform.* 2: 269–307.

8 Vogel, E. (1954). Über die Stabilität des Ungesattigten Kohlenstoff-Vierringes. *Angew. Chem.* 66 (20): 640–641.

9 Criegee, R. and Noll, K. (1959). Umsetzungen in der Reihe des 1,2,3,4-Tetramethyl-Cyclobutans. *Justus Liebigs Ann. Chem.* 627 (1–3): 1–14.

10 Branton, G.R., Frey, H.M., and Skinner, R.F. (1966). Thermal isomerization of cyclobutenes. 8. *Cis*- and *trans*-1,2,3,4-tetramethylcyclobutene and bicyclo[4.2.0]oct-7-ene. *Trans. Faraday Soc.* 62 (522P): 1546–1552.

11 Brauman, J.I. and Archie, W.C. (1972). Energies of alternate electrocyclic pathways - pyrolysis of *cis*-3,4-dimethylcyclobutene. *J. Am. Chem. Soc.* 94 (12): 4262–4265.

12 Srinivasan, R. (1969). Thermal and photochemical isomerization of *cis*-3,4-dimethycyclobutene. *J. Am. Chem. Soc.* 91 (27): 7557–7561.

13 Stephenson, L.M., Brauman, J.I., and Gemmer, R.V. (1972). Thermal *cis-trans* isomerization of butadiene. *J. Am. Chem. Soc.* 94 (24): 8620–8622.

14 Douglas, J.E., Rabinovitch, B.S., and Looney, F.S. (1955). Kinetics of the thermal *cis-trans* isomerization of dideuteroethylene. *J. Chem. Phys.* 23 (2): 315–323.

15 Havinga, E. and Schlatmann, J.L.M.A. (1961). Remarks on specificities of photochemical and

thermal transformations in Vitamin D field. *Tetrahedron* 16 (1–4): 146–152.
16 Longuet-Higgins, H.C. and Abrahamson, E.W. (1965). The electronic mechanism of electrocyclic reactions. *J. Am. Chem. Soc.* 87: 2045–2046.
17 Vollmer, J.J. and Servis, K.L. (1968). Woodward–Hoffmann rules - electrocyclic reactions. *J. Chem. Educ.* 45 (4): 214–220.
18 (a) Epiotis, N.D. (1973). Configuration interaction and organic reactivity. 2. Electrocyclic reactions. *J. Am. Chem. Soc.* 95 (4): 1200–1206. (b) Shi, H., Roettger, D.C., and East, A.L.L. (2008). Photochemistry studied with ab initio orbital-correlation and state-correlation plots: classic cyclobutene ring opening, and the reaction of N_2 with photoexcited O_2. *J. Comput. Chem.* 29 (6): 883–891.
19 Cooper, W. and Walters, W.D. (1958). The thermal isomerization of cyclobutene. *J. Am. Chem. Soc.* 80 (16): 4220–4224.
20 Hoffmann, R. and Woodward, R.B. (1968). Conservation of orbital symmetry. *Acc. Chem. Res.* 1 (1): 17–22.
21 Dewar, M.J.S. (1966). A molecular orbital theory of organic chemistry. 8. Aromaticity and electrocyclic reactions. *Tetrahedron* 22: 75–92.
22 Dewar, M.J.S. (1971). Aromaticity and pericyclic reactions. *Angew. Chem. Int. Ed. Engl.* 10 (11): 761–776.
23 Zimmerman, H.E. (1966). Molecular orbital correlation diagrams Möbius systems and factors controlling ground- and excited-state reactions. 2. *J. Am. Chem. Soc.* 88 (7): 1566–1567.
24 Zimmerman, H.E. (1971). Möbius-Hückel concept in organic chemistry - application to organic molecules and reactions. *Acc. Chem. Res.* 4 (8): 272–280.
25 Dewar, M.J.S. and Dougherty, R.G. (1976). *The PMO Theory of Organic Chemistry*, 1–576. New York: Plenum.
26 (a) Day, A.C. (1975). Aromaticity and generalized Woodward–Hoffmann rules for pericyclic reactions. *J. Am. Chem. Soc.* 97 (9): 2431–2438. (b) Geerlings, P., Ayers, P.W., Toro-Labbé, A. et al. (2012). The Woodward–Hoffmann rules reinterpreted by conceptual density functional theory. *Acc. Chem. Res.* 45 (5): 683–695. (c) Schleyer, P.v.R., Wu, J.I., Cossio, F.P., and Fernández, I. (2014). Aromaticity in transition structures. *Chem. Soc. Rev.* 43 (14): 4909–4921.
27 Branton, G.R., Frey, H.M., Montague, D.C., and Stevens, I.D.R. (1966). Thermal unimolecular isomerization of cyclobutenes. 7. Bicyclo[3.2.0]hept-6-ene. *Trans. Faraday Soc.* 62 (519P): 659–663.
28 (a) Radlick, P. and Fenical, W. (1967). Thermal rearrangements of *cis*-bicyclo[6.2.0]deca-9-enes and a new Cope rearrangement. *Tetrahedron Lett.* (48): 4901–4904. (b) Ralph, M.J., Harrowen, D.C., Gaulier, S. et al. (2015). The profound effect of the ring size in the electrocyclic opening of cyclobutene-fused bicyclic systems. *Angew. Chem. Int. Ed.* 54 (5): 1527–1531.
29 Bajorek, T. and Werstiuk, N.H. (2002). Pyrolysis and UV photoelectron spectroscopy of bicyclo[3.2.0]hept-6-en-2-one; preparation and detection of cyclohepta-2(*Z*),4(*E*)-dien-1-one. *J. Chem. Soc., Chem. Commun.* (6): 648–649.
30 Steiner, R.P. and Michl, J. (1978). π,π-Biradicaloid hydrocarbons - pleiadene family. 3. Facile symmetry-forbidden thermal conversion of a polycyclic butadiene moiety to a cyclobutene. *J. Am. Chem. Soc.* 100 (20): 6413–6415.
31 Frey, H.M. and Solly, R.K. (1969). Thermal isomerization of cyclobutenes. 12. 3,3-Diethylcyclobutene and 3-ethyl-3-methylcyclobutene. *Trans. Faraday Soc.* 65 (554P): 448–452.
32 Curry, M.J. and Stevens, I.D.R. (1980). The thermal ring-opening of 3,3-disubstituted cyclobutenes. *J. Chem. Soc., Perkin Trans.* 2 (10): 1391–1398.
33 Dolbier, W.R., Koroniak, H., Houk, K.N., and Sheu, C.M. (1996). Electronic control of stereoselectivities of electrocyclic reactions of cyclobutenes: a triumph of theory in the prediction of organic reactions. *Acc. Chem. Res.* 29 (10): 471–477.
34 (a) Kirmse, W., Rondan, N.G., and Houk, K.N. (1984). Stereoselective substituent effects on conrotatory electrocyclic reactions of cyclobutenes. *J. Am. Chem. Soc.* 106 (25): 7989–7991. (b) Saal, A. and Merzoud, L. (2016). NBO analysis of the torquoselectivity of the electrocyclic reactions of mono-substituted cyclobutenes. *J. Comput. Methods Sci. Eng.* 16 (2): 437–447. (c) Rudolf, K., Spellmeyer, D.C., and Houk, K.N. (1987). Prediction and Experimental Verification of the Stereoselective Electrocyclization of 3-Formylcyclobutene. *J. Org. Chem.* 52 (16): 3708–3710. (d) Houk, K.N., Spellmeyer, D.C., Jefford, C.W., Rimbault, C.G., Wang, Y. and Miller, R.D. (1988). Electronic Control of the Stereoselectivities of Electrocyclic Reactions of Cyclobutenes Against

Incredible Steric Odds. *J. Org. Chem.* 53 (9): 2125–2127.

35 Murakami, M., Miyamoto, Y., and Ito, Y. (2001). A silyl substituent can dictate a concerted electrocyclic pathway: inward torquoselectivity in the ring opening of 3-silyl-1-cyclobutene. *Angew. Chem. Int. Ed.* 40 (1): 189–190.

36 Lee, P.S., Zhang, X.Y., and Houk, K.N. (2003). Origins of inward torquoselectivity by silyl groups and other sigma-acceptors in electrocyclic reactions of cyclobutenes. *J. Am. Chem. Soc.* 125 (17): 5072–5079.

37 Niwayama, S., Kallel, E.A., Spellmeyer, D.C. et al. (1996). Substituent effects on rates and stereoselectivities of conrotatory electrocyclic reactions of cyclobutenes. A theoretical study. *J. Org. Chem.* 61 (8): 2813–2825.

38 (a) Charlton, J.L. and Alauddin, M.M. (1987). Orthoquinodimethanes. *Tetrahedron* 43 (13): 2873–2889. (b) Chapman, O.L., Tsou, U.P.E., and Johnson, J.W. (1987). Thermal isomerization of benzocyclobutene. *J. Am. Chem. Soc.* 109 (2): 553–559.

39 Kametani, T., Nemoto, H., Ishikawa, H. et al. (1976). Formal regioselective and stereoselective total synthesis of Estrone - convenient synthesis of D-homoestrone. *J. Am. Chem. Soc.* 98 (11): 3378–3379.

40 Kametani, T., Nemoto, H., Ishikawa, H. et al. (1977). A stereoselective total synthesis of Estrone by an intramolecular cycloaddition reaction of olefinic *o*-quinodimethane. *J. Am. Chem. Soc.* 99 (10): 3461–3466.

41 Petrow, V. (1970). Contraceptive progestagens. *Chem. Rev.* 70 (6): 713–726.

42 Kametani, T., Matsumoto, H., Nemoto, H., and Fukumoto, K. (1978). Asymmetric total synthesis of Estradiol by an intramolecular cycloaddition of benzocyclobutene derivative. *J. Am. Chem. Soc.* 100 (19): 6218–6220.

43 Jefford, C.W., Bernardinelli, G., Wang, Y. et al. (1992). Torquoselectivity in the electrocyclic conversion of benzocyclobutenes to *ortho*-xylylenes. *J. Am. Chem. Soc.* 114 (4): 1157–1165.

44 Murakami, M., Usui, I., Hasegawa, M., and Matsuda, T. (2005). Contrasteric stereochemical dictation of the cyclobutene ring-opening reaction by a vacant boron p orbital. *J. Am. Chem. Soc.* 127 (5): 1366–1367.

45 (a) Frontier, A.J. and Collison, C. (2005). The Nazarov cyclization in organic synthesis. Recent advances. *Tetrahedron* 61 (32): 7577–7606. (b) Simeonov, S.P., Nunes, J.P.M., Guerra, K. et al. (2016). Synthesis of chiral cyclopentenones. *Chem. Rev.* 116 (10): 5744–5893. (c) Vinogradov, M.G., Turova, O.V., and Zlotin, S.G. (2017). Nazarov reaction: current trends and recent advances in the synthesis of natural compounds and their analogs. *Org. Biomol. Chem.* 15: 8245–8269.

46 (a) Pellissier, H. (2005). Recent developments in the Nazarov process. *Tetrahedron* 61 (27): 6479–6517. (b) Hoffmann, M., Weibel, J.-M., de Frémont, P. et al. (2014). Gold(I)/(III)-catalyzed rearrangement of divinyl ketones and acyloxyalkynyloxiranes into cyclopentenones. *Org. Lett.* 16: 908–911. (c) Ondet, P., Lemière, G., and Duñach, E. (2017). Cyclisations catalysed by bismuth(III) triflate. *Eur. J. Org. Chem.* (4): 761–780.

47 Tius, M.A. (2005). Some new Nazarov chemistry. *Eur. J. Org. Chem.* (11): 2193–2206.

48 Shoppee, C.W. and Lack, R.E. (1969). Intramolecular electrocyclic reactions. I. Structure of bromohydroxyphorone - 3-bromo-5-hydroxy-4,4,5,5-tetramethylcyclopent-2-enone. *J. Chem. Soc. C* (10): 1346–1349.

49 Shoppee, C.W. and Cooke, B.J.A. (1972). Intramolecular electrocyclic reactions. 2. Reactions of 1,5-diphenylpenta-1,4-dien-3-one. *J. Chem. Soc., Perkin Trans. 1* (18): 2271–2276.

50 (a) Vaidya, T., Eisenberg, R., and Frontier, A.J. (2011). Catalytic Nazarov cyclization: the state of the art. *ChemCatChem* 3 (10): 1531–1548. (b) Jolit, A., Dickinson, C.F., Kitamura, K. et al. (2017). Catalytic enantioselective Nazarov cyclization. *Eur. J. Org. Chem.* (40): 6067–6076.

51 Williams, D.R., Robinson, L.A., Nevill, C., and Reddy, J.P. (2007). Strategies for the synthesis of fusicoccanes by Nazarov reactions of dolabelladienones: total synthesis of (+)-Fusicoauritone. *Angew. Chem. Int. Ed.* 46 (6): 915–918.

52 Grant, T.N., Rieder, C.J., and West, F.G. (2009). Interrupting the Nazarov reaction: domino and cascade processes utilizing cyclopentenyl cations. *J. Chem. Soc., Chem. Commun.* (38): 5676–5688.

53 Nakanishi, W. and West, F.G. (2009). Advances in the Nazarov cyclization. *Curr. Opin. Drug Discovery Dev.* 12 (6): 732–751.

54 Aggarwal, V.K. and Belfield, A.J. (2003). Catalytic asymmetric Nazarov reactions promoted by chiral Lewis acid complexes. *Org. Lett.* 5 (26): 5075–5078.

55 Liao, S., Sun, X.L., and Tang, Y. (2014). Side arm strategy for catalyst design: modifying

bisoxazolines for remote control of enantioselection and related. *Acc. Chem. Res.* 47 (8): 2260–2272.

56 Liang, G.X. and Trauner, D. (2004). Enantioselective Nazarov reactions through catalytic asymmetric proton transfer. *J. Am. Chem. Soc.* 126 (31): 9544–9545.

57 Shimada, N., Stewart, C., and Tius, M.A. (2011). Asymmetric Nazarov cyclizations. *Tetrahedron* 67 (33): 5851–5870.

58 Denmark, S.E. and Jones, T.K. (1982). Silicon-directed Nazarov cyclization. *J. Am. Chem. Soc.* 104 (9): 2642–2645.

59 (a) Denmark, S.E. and Klix, R.C. (1988). Silicon-directed Nazarov cyclizations. 7. Linearly-fused tricyclics. *Tetrahedron* 44 (13): 4043–4060. (b) Denmark, S.E., Wallace, M.A., and Walker, C.B. Jr. (1990). Silicon-directed Nazarov cyclizations. 8. Stereoelectronic control of torquoselectivity. *J. Org. Chem.* 55 (21): 5543–5545.

60 Smith, D.A. and Ulmer, C.W. (1993). Theoretical studies of the Nazarov cyclization. 3. Torquoselectivity and hyperconjugation in the Nazarov cyclization - the effects of inner versus outer β-methyl and β-silyl groups. *J. Org. Chem.* 58 (15): 4118–4121.

61 Williams, C.W., Shenje, R., and France, S. (2016). Catalytic, interrupted formal homo-Nazarov cyclization with (hetero)arenes: access to α-(hetero)aryl cyclohexanones. *J. Org. Chem.* 81 (18): 8253–8267.

62 He, W., Sun, X.F., and Frontier, A.J. (2003). Polarizing the Nazarov cyclization: efficient catalysis under mild conditions. *J. Am. Chem. Soc.* 125 (47): 14278–14279.

63 He, W., Herrick, I.R., Atesin, T.A. et al. (2008). Polarizing the Nazarov cyclization: the impact of dienone substitution pattern on reactivity and selectivity. *J. Am. Chem. Soc.* 130 (3): 1003–1011.

64 Leboeuf, D., Huang, J., Gandon, V., and Frontier, A.J. (2011). Using Nazarov electrocyclization to stage chemoselective (1,2)-migrations: stereoselective synthesis of functionalized cyclopentenones. *Angew. Chem. Int. Ed.* 50 (46): 10981–10985.

65 (a) Grant, T.N. and West, F.G. (2006). A new approach to the Nazarov reaction via sequential electrocyclic ring opening and ring closure. *J. Am. Chem. Soc.* 128 (29): 9348–9349. (b) Grant, T.N. and West, F.G. (2007). Interrupted Nazarov reactions using dichlorocyclopropanes: a novel mode of arene trapping. *Org. Lett.* 9 (19): 3789–3792.

66 Wu, Y.K. and West, F. (2010). Brønsted acid-mediated Nazarov cyclization of vinylallenes. *J. Org. Chem.* 75 (15): 5410–5413.

67 Rueping, M., Leawsuwan, W., Antonchick, A.P., and Nachtsheim, B.J. (2007). Chiral Brønsted acids in the catalytic asymmetric Nazarov cyclization - the first enantioselective organocatalytic electrocyclic reaction. *Angew. Chem. Int. Ed.* 46 (12): 2097–2100.

68 Basak, A.K., Shimada, N., Bow, W.F. et al. (2010). An organocatalytic asymmetric Nazarov cyclization. *J. Am. Chem. Soc.* 132 (24): 8266–8267.

69 Cao, P., Deng, C., Zhou, Y.Y. et al. (2010). Asymmetric Nazarov reaction catalyzed by chiral tris(oxazoline)/Cu(II). *Angew. Chem. Int. Ed.* 49 (26): 4463–4466.

70 Hastings, C.J., Bergman, R.G., and Raymond, K.N. (2014). Origins of large rate enhancements in the Nazarov cyclization catalyzed by supramolecular encapsulation. *Chem. Eur. J.* 20 (14): 3966–3973.

71 Saito, A., Umakoshi, M., Yagyu, N., and Hanzawa, Y. (2008). Novel one-pot approach to synthesis of indanones through Sb(V)-catalyzed reaction of phenylalkynes with aldehydes. *Org. Lett.* 10 (9): 1783–1785.

72 Giese, S. and West, F.G. (2000). Ionic hydrogenation of oxyallyl intermediates: the reductive Nazarov cyclization. *Tetrahedron* 56 (52): 10221–10228.

73 Boche, G., Buckl, K., Martens, D. et al. (1979). Ringöffnung von 2,3-Diphenylcyclopropyl- zu 1,3-Diphenylallyllithium-Verbindungen. *Chem. Ber.* 112: 2961–2996.

74 Liu, K., Zhao, H.M., Ma, S.Y., and Li, Z.H. (2004). Computational study of the thermal ring opening of cyclopropyl radical, cation and anion. *J. Mol. Struct. THEOCHEM* 672 (1–3): 209–213.

75 Nieto-Faza, O., Silva, L.C., González-Pérez, A. et al. (2009). Torquoselectivity in the electrocyclic ring-opening of cyclopropyl anions. *J. Phys. Org. Chem.* 22: 378–385.

76 (a) Kauffmann, T., Koppelma, E., and Habersaa, K. (1972). Ring-opening of *N*-Lithio-2, 3(*cis*)-diphenylaziridine to 1,3(*cis,trans*)-diphenyl-2-azaallyllithium and its rearrangement to *trans,trans* form. *Angew. Chem.*

Int. Ed. Engl. 11 (4): 291–292. (b) Pearson, W.H., Walters, M.A., and Oswell, K.D. (1986). Intramolecular 2-azaallyl anion cycloadditions. Application to the synthesis of fused bicyclic pyrrolidines. *J. Am. Chem. Soc.* 108: 2769–2771. (c) Pearson, W.H., Walters, M.A., Rosen, M.K., and Harter, W.G. (2002). Generation and cycloaddition of semi-stabilized 2-azaallyllithiums via electrocyclic ring-opening or cycloreversion reactions. *ARKIVOC* (Part 8): 91–111.

77 Fowler, J.E., Alberts, I.L., and Schaefer, H.F. (1991). Mechanistic study of the electrocyclic ring-opening reaction of thiirane. *J. Am. Chem.Soc.* 113 (13): 4768–4776.

78 Mloston, G., Huisgen, R., and Polborn, K. (1999). Cycloadditions of adamantanethione S-methylide to heteromultiple bonds. *Tetrahedron* 55 (38): 11475–11494.

79 (a) Huisgen, R., Scheer, W., and Huber, H. (1967). Stereospecific conversion of *cis-trans* isomeric aziridines to open-chain azomethine ylides. *J. Am. Chem. Soc.* 89 (7): 1753–1755. (b) Coldham, I. and Hufton, R. (2005). Intramolecular dipolar cycloaddition reactions of azomethine ylides. *Chem. Rev.* 105: 2765–2810. (c) Pankova, A.S., Sorokina, M.V., and Kuznetsov, M.A. (2015). Thermal rearrangement of 2,3-diaryl-1-phthalimidoaziridines. *Tetrahedron Lett.* 56 (40): 5381–5385.

80 Crawford, R.J., Vukov, V., and Tokunaga, H. (1973). Kinetics of gas phase reactions of (+)-*trans*-2,3-divinyloxirane. *Can. J. Chem.* 51 (22): 3718–3725.

81 Stogryn, E.L., Passannante, A.J., and Gianni, M.H. (1964). Valence tautomerism of vinyl-substituted three-membered heterocycles. 1. Conversion of 1,2-divinylethylene oxide to 4,5-dihydrooxepine. *J. Org. Chem.* 29 (5): 1275–1276.

82 (a) Vogel, E. and Günther, H. (1967). Benzene oxide-oxepin valence tautomerism. *Angew. Chem. Int. Ed. Engl.* 6 (5): 385–401. (b) Günther, H. and Jikeli, G. (1973). Application of ^{13}C resonance spectroscopy. 4. Spectra of oxepines and benzene oxides and influence of oxygen on chemical shift of ^{13}C. *Chem. Ber.* 106 (6): 1863–1880. (c) Kassaee, M.Z., Arshadi, S., and Ahmadi-Taheri, N. (2005). Substituent effects on tautomerization of oxepine to benzene oxide: a Hammett study via ab initio. *J. Mol. Struct. THEOCHEM* 715: 107–115.

83 Huisgen, R. (1977). Electrocyclic ring-opening reactions of ethylene oxides. *Angew. Chem. Int. Ed. Engl.* 16 (9): 572–585.

84 (a) Jean, Y., Volatron, F., and Anh, N.T. (1983). Theoretical study of electrocyclic ring-opening of oxirane. *J. Mol. Struct. THEOCHEM* 10: 167–173. (b) Joshi, A., You, X.Q., Barckholtz, T.A., and Wang, H. (2005). Thermal decomposition of ethylene oxide: potential energy surface, master equation analysis, and detailed kinetic modeling. *J. Phys. Chem. A* 103 (35): 8016–8027.

85 Olivella, S., Sole, A., and Bofill, J.M. (1990). A theoretical investigation of the thermal ring-opening of cyclopropyl radical into allyl radical - evidence for a highly nonsymmetric transition state. *J. Am. Chem. Soc.* 112 (6): 2160–2167.

86 Arnold, P.A. and Carpenter, B.K. (2000). Computational studies on the ring openings of cyclopropyl radical and cyclopropyl cation. *Chem. Phys. Lett.* 328 (1–2): 90–96.

87 Mann, D.J. and Hase, W.L. (2002). Ab initio direct dynamics study of cyclopropyl radical ring-opening. *J. Am. Chem. Soc.* 124 (13): 3208–3209.

88 Verloop, A., Koevoet, A.L., and Havinga, E. (1957). Studies on Vitamin-D and related compounds. 7. Some remarks on the stereochemistry of the trienoic systems of Vitamin-D and related compounds. *Recl. Trav. Chim. Pays-Bas* 76 (8): 689–702.

89 Marvell, E.N., Caple, G., Schatz, B., and Pippin, W. (1973). Electrocyclic reactions - stereochemistry of triene electrocyclization. *Tetrahedron* 29 (23): 3781–3789.

90 Marvell, E.N., Caple, G., Delphey, C. et al. (1973). Triene electrocyclizations - some structure reactivity relations. *Tetrahedron* 29 (23): 3797–3806.

91 (a) Sünnemann, H.W., Banwell, M.G., and de Meijere, A. (2007). Synthesis and use of new substituted 1,3,5-hexatrienes in studying thermally induced 6 π-electrocyclizations. *Eur. J. Org. Chem.* (23): 3879–3893. (b) Detert, H., Lenoir, D., and Zipse, H. (2009). Thermal rearrangements of perchlorohexatrienes: structures and experimental and theoretical evaluation of pathways to isomerization and cyclization. *J. Org. Chem.* (8): 1181–1190.

92 Reisman, S.E., Nani, R.R., and Levin, S. (2011). Büchner and beyond: arene cyclopropanation as applied to natural product total synthesis. *Synlett* (17): 2437–2442.

93 Anciaux, A.J., Demonceau, A., Noels, A.F. et al. (1981). Transition-metal-catalyzed reactions of diazo-compounds. 2. Addition to aromatic-molecules - catalysis of Büchner synthesis of cycloheptatrienes. *J. Org. Chem.* 46 (5): 873–876.

94 (a) McNamara, O.A. and Maguire, A.R. (2011). The norcaradiene-cycloheptatriene equilibrium. *Tetrahedron* 67 (1): 9–40. (b) Bateman, L.M., McNamara, O.A., Buckley, N.R. et al. (2015). A study of the norcaradiene-cycloheptatriene equilibrium in a series of azulenones by NMR spectroscopy; the impact of substitution on the position of equilibrium. *Org. Biomol. Chem.* 13: 11026–11038.

95 (a) Paquette, L.A. (1971). Valence isomerism in selected heterocyclic systems. *Angew. Chem. Int. Ed. Engl.* 10: 11–20. (b) Kassaee, M.Z., Arshadi, S., Haerizade, B.N., and Vessally, E. (2005). Electronic effects on 1*H*-azepines valence tautomerization: an ab initio comparative study. *J. Mol. Struct. THEOCHEM* 731 (1–3): 29–37.

96 Tang, T.H., Lew, C.S.Q., Cui, Y.P. et al. (1994). A theoretical study of substituted norcaradiene with some strong electron-withdrawing groups at position-7. *J. Mol. Struct. THEOCHEM* 111: 149–164.

97 Rudi, A. and Kashman, Y. (1993). Aaptosine - a new cytotoxic 5,8-diazabenz[cd]azulene alkaloid from the Red Sea sponge *Aaptos aaptos*. *Tetrahedron Lett.* 34 (29): 4683–4684.

98 Prinzbach, H. and Vogel, P. (1969). Photochemical conversions. 27. A stable benzene oxide. *Helv. Chim. Acta* 52 (2): 396–401.

99 (a) Suzuki, T., Li, Q., Khemani, K.C. et al. (1991). Systematic inflation of Buckminsterfullerene C_{60} - synthesis of diphenyl fulleroids C_{61} to C_{66}. *Science* 254 (5035): 1186–1188. (b) Bingel, C. and Schiffer, H. (1995). Biscyclopropanation of C_{70}. *Liebigs Ann. Chem.* (8): 1551–1553. (c) Jin, B., Shen, J., Peng, R. et al. (2015). DMSO: An efficient catalyst for the cyclopropanation of C_{60}, C_{70}, SWNTs, and graphene through the Bingel reaction. *Ind. Eng. Chem. Res.* 54 (11): 2879–2885. (d) Nikolaev, D.N., Davidovich, P.B., and Piotrovskii, I.B. (2016). Synthesis of fullerene C_{60} monoadducts. Cyclopropanation of C_{60} with sulfonium ylides. *Russ. J. Org. Chem.* 52 (7): 1050–1053. (e) Wang, J.-J., Lin, H.-S., Niu, C., and Wang, G.-W. (2017). The cyclopropanation of [60]fullerobenzofurans via electrosynthesis. *Org. Biomol. Chem.* 15 (15): 3248–3254.

100 Warner, P.M. (1994). Fullerenes vs fulleroids - understanding their relative energies. *J. Am. Chem. Soc.* 116 (24): 11059–11066.

101 de Lera, A.R., Alvarez, R., Lecea, B. et al. (2001). On the aromatic character of electrocyclic and pseudopericyclic reactions: thermal cyclization of (2*Z*)-hexa-2,4-5-trienals and their Schiff bases. *Angew. Chem. Int. Ed.* 40 (3): 557–561.

102 Fadnavis, N.W., Sharfuddin, M., and Vadivel, S.K. (1999). Resolution of racemic 2-amino-1-butanol with immobilised penicillin G acylase. *Tetrahedron: Asymmetry* 10 (23): 4495–4500.

103 Ross, J.A., Seiders, R.P., and Lemal, D.M. (1976). Extraordinarily facile sulfoxide automerization. *J. Am. Chem. Soc.* 98 (14): 4325–4327.

104 Birney, D.M. (2010). Theory, experiment and unusual features of potential energy surfaces of pericyclic and pseudopericyclic reactions with sequential transition structures. *Curr. Org. Chem.* 14 (15): 1658–1668.

105 Birney, D.M. and Wagenseller, P.E. (1994). An ab-initio study of the reactivity of formylketene - pseudopericyclic reactions revisited. *J. Am. Chem. Soc.* 116 (14): 6262–6270.

106 Birney, D.M., Ham, S., and Unruh, G.R. (1997). Pericyclic and pseudopericyclic thermal cheletropic decarbonylations: when can a pericyclic reaction have a planar, pseudopericyclic transition state? *J. Am. Chem. Soc.* 119 (19): 4509–4517.

107 Birney, D.M., Xu, X.L., and Ham, S. (1999). (1,3), (3,3), and (3,5)-sigmatropic rearrangements of esters are pseudopericyclic. *Angew. Chem. Int. Ed.* 38 (1–2): 189–193.

108 Fabian, W.M.F., Kappe, C.O., and Bakulev, V.A. (2000). Ab initio and density functional calculations on the pericyclic vs pseudopericyclic mode of conjugated nitrile ylide 1,5-electrocyclizations. *J. Org. Chem.* 65 (1): 47–53.

109 Chamorro, E., Notario, R., Santos, J., and Perez, P. (2007). A theoretical scale for pericyclic and pseudopericyclic reactions. *Chem. Phys. Lett.* 443 (1–3): 136–140.

110 Bates, R.B. and Mccombs, D.A. (1969). Pentadienyl-cyclopentenyl anion rearrangement. *Tetrahedron Lett.* (12): 977–978.

111 Bates, R.B., Deines, W.H., Mccombs, D.A., and Potter, D.E. (1969). Preparation of heptatrienyl anions - heptatrienyl-cycloheptadienyl anion rearrangement. *J. Am. Chem. Soc.* 91 (16): 4608–4608.

112 Huisgen, R. (1980). 1,5,-Electrocyclizations - an important principle of heterocyclic chemistry. *Angew. Chem. Int. Ed. Engl.* 19 (12): 947–973.

113 Auwers, K. and Heimke, P. (1927). Pyrazoline. *Liebigs Ann. Chem.* 458: 186–220.

114 Pommelet, J.C. and Chuche, J. (1976). Comparative thermal rearrangement of *cis* and *trans* vinylaziridines. *Can. J. Chem.* 54 (10): 1571–1581.

115 Mloston, G. and Heimgartner, H. (2011). Synthesis of five-membered sulfur-heterocycles via 1,5-dipolar electrocyclization of thiocarbonyl ylides and related processes. *Curr. Org. Chem.* 15 (5): 675–693.

116 Huisgen, R., Dahmen, A., and Huber, H. (1967). Stereospecific conrotatory valence isomerization of octatetraenes to cycloocta-1,3,5-trienes. *J. Am. Chem. Soc.* 89 (26): 7130–7131.

117 Thomas, B.E., Evanseck, J.D., and Houk, K.N. (1993). Electrocyclic reactions of 1-substituted 1,3,5,7-octatetraenes - an ab initio molecular orbital study of torquoselectivity in 8-electron electrocyclizations. *J. Am. Chem. Soc.* 115 (10): 4165–4169.

118 Thomas, B.E., Evanseck, J.D., and Houk, K.N. (1993). Transition structures of the electrocyclic reactions of *cis,cis,cis*-1,3,5-cyclooctatriene. *Isr. J. Chem.* 33 (3): 287–293.

119 Lecea, B., Arrieta, A., and Cossio, F.P. (2005). Substituent effects in eight-electron electrocyclic reactions. *J. Org. Chem.* 70 (3): 1035–1041.

120 Hulot, C., Amiri, S., Blond, G. et al. (2009). Understanding the torquoselectivity in 8 π-electrocyclic cascade reactions: synthesis of fenestradienes versus cyclooctatrienes. *J. Am. Chem. Soc.* 131 (37): 13387–13398.

121 Mauksch, M. and Tsogoeva, S.B. (2009). A preferred disrotatory 4n electron Möbius aromatic transition state for a thermal electrocyclic reaction. *Angew. Chem. Int. Ed.* 48 (16): 2959–2963.

122 Nicolaou, K.C., Petasis, N.A., and Zipkin, R.E. (1982). The Endiandric acid cascade - electrocyclizations in organic synthesis. 4. Biomimetic approach to Endiandric acids-A-acid-G - total synthesis and thermal studies. *J. Am. Chem. Soc.* 104 (20): 5560–5562.

123 Beaudry, C.M., Malerich, J.P., and Trauner, D. (2005). Biosynthetic and biomimetic electrocyclizations. *Chem. Rev.* 105 (12): 4757–4778.

124 Charpenay, M., Boudhat, A., Hulot, C. et al. (2013). Expeditious cascade reactions: controlled syntheses of fenestradienes and cyclooctatrienes under Pd catalysis. *Tetrahedron* 69 (36): 7568–7591.

125 Skropeta, D. and Rickards, R.W. (2007). Domino pericyclic reactions of acyclic conjugated (*E,Z,E,E*)-tetraenes. *Tetrahedron Lett.* 48 (18): 3281–3284.

126 (a) Lawrence, A.L., Wegner, H.A., Jacobsen, M.F. et al. (2006). Investigations towards the synthesis of (−)-coprinolone, via a thermal 8 π-6-π electrocyclization cascade of 1,5,7-trien-4-ones. *Tetrahedron Lett.* 47 (49): 8717–8720. (b) Staley, S.W. and Heyn, A.S. (1975). Electrocyclic reaction of cyclodecatrienyl anion. *J. Am. Chem. Soc.* 97 (13): 3852–3854.

127 Huisgen, R. (1968). Cycloadditions - definition classification and characterization. *Angew. Chem. Int. Ed. Engl.* 7: 321–328.

128 Epiotis, N.D. (1974). Theory of pericyclic reactions. *Angew. Chem. Int. Ed. Engl.* 13 (12): 751–780.

129 Diels, O. and Alder, K. (1928). Synthesen in der hydroaromatischen Reihe. I Mitteilung: Anlagerung von "Di-en" kohlenwasserstoffen. *Liebigs Ann. Chem.* 460: 98–122.

130 Evans, M.G. (1939). The activation energies of reactions involving conjugated systems. *Trans. Faraday Soc.* 35 (2): 0824–0834.

131 Zhao, H.M. and Wu, D.X. (1986). Mo approach to the aromaticity of transition states of the generalized pericyclic reactions. *Tetrahedron* 42 (2): 515–522.

132 Bernardi, F., Celani, P., Olivucci, M. et al. (1995). Theoretical study of the aromatic character of the transition states of allowed and forbidden cycloadditions. *J. Am. Chem. Soc.* 117 (42): 10531–10536.

133 Ponec, R. and Strnad, M. (1991). Topological aspects of chemical reactivity – Evans–Dewar principle in terms of molecular similarity approach. *J. Phys. Org. Chem.* 4 (11): 701–705.

134 Zimmerman, H.E. (1966). On molecular orbital correlation diagrams occurrence of Möbius systems in cyclization reactions and factors controlling ground- and excited-state reactions. I. *J. Am. Chem. Soc.* 88 (7): 1564–1565.

135 Bartlett, P.D. (1968). 1,2- and 1,4-cycloaddition to conjugated dienes. *Science* 159 (3817): 833–838.

136 Reed, H.W.B. (1951). The formation of *trans*-1-2-divinylcyclobutane from butadiene. *J. Chem. Soc.* 685–687.

137 Lebedev, S.V. and Skavronskii, N.A. (1913). Polymerization of some hydrocarbons of

the divinyl series. *J. Soc. Chem. Ind. London* 33: 1224–1224.

138 Ziegler, K. and Wilms, H. (1947). Ein neuer Weg in die Cyclooctanreihe. *Angew. Chem.* 59 (5–6): 177–177.

139 Vogel, E. (1956). Über den Mechanismus der Butadien-Dimerisation Zum 8-Ring. *Angew. Chem.* 68 (12): 413–414.

140 Shuraeva, V.N. and Kastsobashvili, V.Y. (1977). Kinetics of cyclodimerization of acrylonitrile and styrene. *React. Kinet. Catal. Lett.* 7 (3): 267–272.

141 Doering, W.v.E. and Guyton, C.A. (1978). Stereochemistry of acrylonitrile dimerization. *J. Am. Chem. Soc.* 100 (10): 3229–3230.

142 Wiberg, K.B., Matturro, M.G., Okarma, P.J. et al. (1986). Bicyclo[2.2.0]hex-1(4)-ene. *Tetrahedron* 42 (6): 1895–1902.

143 Borden, W.T. (1989). Pyramidalized alkenes. *Chem. Rev.* 89 (5): 1095–1109.

144 Shelton, G., Hrovat, D.A., Wei, H., and Borden, W.T. (2006). Why does perfluorination render bicyclo[2.2.0]hex-1(4)-ene stable toward dimerization? Calculations provide the answers. *J. Am. Chem. Soc.* 128 (36): 12020–12027.

145 Shtarov, A.B., Krusic, P.J., Smart, B.E., and Dolbier, W.R. (2001). Bimolecular kinetic studies with high-temperature gas-phase ^{19}F NMR: cycloaddition reactions of fluoroolefins. *J. Am. Chem. Soc.* 123 (41): 9956–9962.

146 Huisgen, R. and Plumet-Ortega, J. (1978). Evidence for a zwitterionic intermediate in (2+2)-cycloaddition of tetracyanoethylene to a *trans*-fixed 1,3-diene. *Tetrahedron Lett.* 19 (41): 3975–3978.

147 Huisgen, R. (1980). Cycloaddition mechanism and the solvent dependence of rate. *Pure Appl. Chem.* 52 (10): 2283–2302.

148 Hosoya, T., Hasegawa, T., Kuriyama, Y., and Suzuki, K. (1995). Stereospecificity in (2+2)-cycloaddition of benzyne and ketene silyl acetal. *Tetrahedron Lett.* 36 (19): 3377–3380.

149 Bruce, M.I., Rogers, J.R., Snow, M.R., and Swincer, A.G. (1981). (Cyclopentadienyl)Ru and -Os chemistry. Cleavage of tetracyanoethylene under mild conditions. *J. Chem. Soc., Chem. Commun.* 271–272.

150 Kivala, M., Boudon, C., Gisselbrecht, J.P. et al. (2009). Organic super-acceptors with efficient intramolecular charge-transfer interactions by (2+2)-cycloadditions of TCNE, TCNQ, and F_4-TCNQ to donor-substituted cyanoalkynes. *Chem. Eur. J.* 15 (16): 4111–4123.

151 Onuma, K., Kai, Y., Yasuoka, N., and Kasai, N. (1975). Crystal and molecular structure of [*trans*-bis(trimethylphosphine)propynyl-1-(4′-dicyanomethylenecyclohexa-2′,5′-dien-1-yliden)-3,3-dicyano-2-methylprop-2-en-1-yl]Pt, a reaction product of *trans*-bis (trimethylphosphine)bis-(propynyl)Pt and 7,7,8,8-tetracyanoquinodimethane. *Bull. Chem. Soc. Jpn.* 48 (6): 1696–1700.

152 Michinobu, T., Boudon, C., Gisselbrecht, J.P. et al. (2006). Donor-substituted 1,1,4,4-tetracyanobutadienes (TCBDs): new chromophores with efficient intramolecular charge-transfer interactions by atom-economic synthesis. *Chem. Eur. J.* 12 (7): 1889–1905.

153 Jarowski, P.D., Wu, Y.L., Boudon, C. et al. (2009). New donor-acceptor chromophores by formal (2+2)-cycloaddition of donor-substituted alkynes to dicyanovinyl derivatives. *Org. Biomol. Chem.* 7 (7): 1312–1322.

154 Hayashi, Y., Niihata, S., and Narasaka, K. (1990). (2+2)-Cycloaddition reaction between allenyl sulfides and electron deficient olefins promoted by Lewis acids. *Chem. Lett.* 19 (11): 2091–2094.

155 Narasaka, K., Hayashi, Y., Shimadzu, H., and Niihata, S. (1992). Asymmetric (2+2)-cycloaddition reaction catalyzed by a chiral titanium reagent. *J. Am. Chem. Soc.* 114 (23): 8869–8885.

156 Engler, T.A., Letavic, M.A., Iyengar, R. et al. (1999). Asymmetric reactions of 2-methoxy-1,4-benzoquinones with styrenyl systems: enantioselective syntheses of 8-aryl-3-methoxybicyclo[4.2.0]oct-3-en-2,5-diones, 7-aryl-3-hydroxybicyclo[3.2.1]oct-3-en-2,8-diones, 2-aryl-6-methoxy-2,3-dihydrobenzofuran-5-ols, and pterocarpans. *J. Org. Chem.* 64 (7): 2391–2405.

157 Canales, E. and Corey, E.J. (2007). Highly enantioselective (2+2)-cycloaddition reactions catalyzed by a chiral aluminum bromide complex. *J. Am. Chem. Soc.* 129 (42): 12686–12687.

158 Alcaide, B., Almendros, P., and Aragoncillo, C. (2010). Exploiting (2+2)-cycloaddition chemistry: achievements with allenes. *Chem. Soc. Rev.* 39 (2): 783–816.

159 Applequist, D.E. and Roberts, J.D. (1956). Small-ring compounds. 15. Methylenecyclobutene and related substances. *J. Am. Chem. Soc.* 78 (16): 4012–4022.

160 Pasto, D.J., Sugi, K.D., and Alonso, D.E. (1992). Stereochemical features of the (2+2)-cycloaddition reactions of chiral allenes. 4. Cycloaddition of enantioenriched

1,3-dimethylallene with methyl propiolate. *J. Org. Chem.* 57: 1146–1150.

161 Pasto, D.J. and Sugi, K.D. (1992). Stereochemical features of the (2+2)-cycloaddition reactions of chiral allenes. 3. The cycloaddition of enantioenriched 1,3-dimethylallene with the symmetrically 1,1-disubstituted alkenes 1,1-dichloro-2,2-difluoroethene and 1,1-diphenylethene. *J. Org. Chem.* 57 (1): 12–17.

162 Brummond, K.M. and Chen, D.T. (2005). Microwave-assisted intramolecular (2+2)-allenic-cycloaddition reaction for the rapid assembly of bicyclo[4.2.0]octa-1,6-dienes and bicyclo[5.2.0]nona-1,7-dienes. *Org. Lett.* 7 (16): 3473–3475.

163 Siebert, M.R., Osbourn, J.M., Brummond, K.M., and Tantillo, D.J. (2010). Differentiating mechanistic possibilities for the thermal, intramolecular (2+2)-cycloaddition of allene-ynes. *J. Am. Chem. Soc.* 132 (34): 11952–11966.

164 Staudinger, H. (1905). Ketene, eine neue Körperklasse. *Ber. Dtsch. Chem. Ges.* 38: 1735–1739.

165 Mitzel, T.M. (2001). Ketene. In: *e-EROS Encyclopedia of Reagents for Organic Synthesis.* Wiley.

166 Hyatt, J.A. and Raynolds, P.W. (1994). Ketene cycloadditions. In: *Organic Reactions*, vol. 45, 159–646. Hoboken, NJ, USA: Wiley.

167 Tidwell, T.T. (2006). Ketene chemistry after 100 years: ready for a new century. *Eur. J. Org. Chem.* (3): 563–576.

168 Meier, H. and Zeller, K.P. (1975). Wolff rearrangement of alpha-diazo carbonyl-compounds. *Angew. Chem. Int. Ed. Engl.* 14 (1): 32–43.

169 Kirmse, W. (2002). 100 years of the Wolff rearrangement. *Eur. J. Org. Chem.* (14): 2193–2256.

170 Candeias, N.R., Trindade, A.F., Gois, P.M.P., and Afonso, C.A.M. (2014). The Wolff rearrangement. In: *Comprehensive Organic Synthesis*, 2e, vol. 3 (ed. P. Knochel and G.A. Molander), 944–991. Amsterdam, The Netherlands: Elsevier B.V.

171 Ye, T. and McKervey, M.A. (1994). Organic synthesis with α-diazocarbonyl compounds. *Chem. Rev.* 94 (4): 1091–1160.

172 Schmidlin, J. and Bergman, M. (1910). Darstellung des Ketens aus Aceton. *Ber. Dtsch. Chem. Ges.* 43: 2821–2823.

173 Zhang, Z., Zhang, Y., and Wang, J. (2011). Carbonylation of metal carbene with carbon monoxide: generation of ketene. *ACS Catal.* 1 (11): 1621–1630.

174 Staudinger, H. (1907). Zur Kenntniss der Ketene. Diphenylketene. *Liebigs Ann. Chem.* 356 (1/3): 51–123.

175 Yamabe, S., Dai, T.S., Minato, T. et al. (1996). Ketene is a dienophile for (4+2) (Diels–Alder) reactions across its C=O bond. *J. Am. Chem. Soc.* 118 (27): 6518–6519.

176 Machiguchi, T., Hasegawa, T., Ishiwata, A. et al. (1999). Ketene recognizes 1,3-dienes in their *s-cis* forms through $[\pi^4+\pi^2]$ (Diels–Alder) and $[\pi^2+\pi^2]$ (Staudinger) reactions. An innovation of ketene chemistry. *J. Am. Chem. Soc.* 121 (20): 4771–4786.

177 Tidwell, T.T. (1995). *Ketenes.* New York: Wiley.

178 Machiguchi, T., Okamoto, J., Takachi, J. et al. (2003). Exclusive formation of α-methyleneoxetanes in ketene-alkene cycloadditions. Evidence for intervention of both an α-methyleneoxetane and the subsequent 1,4-zwitterion. *J. Am. Chem. Soc.* 125 (47): 14446–14448.

179 Wang, X.B. and Houk, K.N. (1990). Carbenoid character in transition structures for reactions of ketenes with alkenes. *J. Am. Chem. Soc.* 112 (5): 1754–1756.

180 Ma, N.L. and Wong, M.W. (2000). A theoretical study of the properties and reactivities of ketene, thioketene, and selenoketene. *Eur. J. Org. Chem.* (8): 1411–1421.

181 Yamabe, S., Minato, T., and Osamura, Y. (1993). Dual one-center frontier orbital interactions in (2+2)-cycloadditions of ketenes. *J. Chem. Soc., Chem. Commun.* (5): 450–452.

182 Rice, F.O. and Greenberg, J. (1934). Ketene. II. Rate of polymerization. *J. Am. Chem. Soc.* 56 (7): 2132–2134.

183 Rice, F.O. and Roberts, R. (1943). The structure of diketene. *J. Am. Chem. Soc.* 65: 1677–1681.

184 Clemens, R.J. (1986). Diketene. *Chem. Rev.* 86 (2): 241–318.

185 Huisgen, R. and Otto, P. (1968). Mechanism of dimerization of dimethylketene. *J. Am. Chem. Soc.* 90 (19): 5342–5343.

186 Seidl, E.T. and Schaefer, H.F. (1991). Theoretical investigation of the dimerization of ketene - does the $[\pi 2s+\pi 2a]$-cycloaddition reaction exist? *J. Am. Chem. Soc.* 113 (14): 5195–5200.

187 Pascal, R.A., Baum, M.W., Wagner, C.K. et al. (1986). Measurement of deuterium kinetic isotope effects in organic and biochemical reactions by natural abundance deuterium NMR-spectroscopy. *J. Am. Chem. Soc.* 108 (21): 6477–6482.

188 Schaad, L.J., Gutman, I., Hess, B.A., and Hu, J. (1991). Ab initio calculation of secondary deuterium-isotope ratios in the formation of diketene. *J. Am. Chem. Soc.* 113 (14): 5200–5203.

189 Ogasawara, M., Okada, A., Nakajima, K., and Takahashi, T. (2009). Palladium-catalyzed synthesis of endocyclic allenes and their application in stereoselective (2+2)-cycloaddition with ketenes. *Org. Lett.* 11 (1): 177–180.

190 Greene, A.E., Charbonnier, F., Luche, M.J., and Moyano, A. (1987). Enantioselective ring construction through asymmetric olefin-ketene cycloaddition - a highly enantio-controlled approach to (−)-α-cuparenone and (+)-β-Cuparenone. *J. Am. Chem. Soc.* 109 (15): 4752–4753.

191 Tidwell, T.T. (2008). Hugo (ugo) Schiff, Schiff bases, and a century of β-lactam synthesis. *Angew. Chem. Int. Ed.* 47 (6): 1016–1020.

192 Fu, N. and Tidwell, T.T. (2008). Preparation of β-lactams by (2+2)-cycloaddition of ketenes and imines. *Tetrahedron* 64 (46): 10465–10496.

193 Staudinger, H. (1907). Über Ketene. 4. Mitteilung. Reaktionen des Diphenylketens. *Ber. Dtsch. Chem. Ges.* 40: 1145–1148.

194 Staudinger, H. (1907). Über Ketene. 5. Mitteilung. Reaktionen des Dimethylketens. *Ber. Dtsch. Chem. Ges.* 40: 1149–1153.

195 Miller, M.J. (2000). Preface - recent aspects of the chemistry of β-lactams. II. *Tetrahedron* 56 (31): IX–IX.

196 Singh, G.S. (2003). Recent progress in the synthesis and chemistry of azetidinones. *Tetrahedron* 59 (39): 7631–7649.

197 Palomo, C., Aizpurua, J.M., Ganboa, I., and Oiarbide, M. (1999). Asymmetric synthesis of β-lactams by Staudinger ketene-imine cycloaddition reaction. *Eur. J. Org. Chem.* (12): 3223–3235.

198 (a) Jiao, L., Liang, Y., and Xu, J. (2006). Origin of the relative stereoselectivity of the β-lactam formation in the Staudinger reaction. *J. Am. Chem. Soc.* 128 (18): 6060–6069. (b) Domingo, L.R., Rios-Gutiérrez, M., and Sáez, J.A. (2015). Unravelling the mechanism of the ketene-imine Staudinger reaction. An ELF quantum topological analysis. *RSC Adv.* 5 (47): 37119–37129.

199 Moore, H.W. and Gheorghiu, M.D. (1981). Cyanoketenes - synthesis and cycloadditions. *Chem. Soc. Rev.* 10 (3): 289–328.

200 Sheehan, J.C. and Ryan, J.J. (1951). The synthesis of substituted penicillins and simpler structural analogs. 1. α-Amino monocyclic β-lactams. *J. Am. Chem. Soc.* 73 (3): 1204–1206.

201 Evans, D.A. and Sjogren, E.B. (1985). The asymmetric synthesis of β-lactam antibiotics. 1. Application of chiral oxazolidones in the Staudinger reaction. *Tetrahedron Lett.* 26 (32): 3783–3786.

202 Evans, D.A. and Sjogren, E.B. (1985). The asymmetric synthesis of β-lactam antibiotics .2. The 1st enantioselective synthesis of the carbacephalosporin nucleus. *Tetrahedron Lett.* 26 (32): 3787–3790.

203 Ojima, I. and Delaloge, F. (1997). Asymmetric synthesis of building-blocks for peptides and peptidomimetics by means of the β-lactam synthon method. *Chem. Soc. Rev.* 26 (5): 377–386.

204 Ruhland, B., Bhandari, A., Gordon, E.M., and Gallop, M.A. (1996). Solid-supported combinatorial synthesis of structurally diverse β-lactams. *J. Am. Chem. Soc.* 118 (1): 253–254.

205 France, S., Weatherwax, A., Taggi, A.E., and Lectka, T. (2004). Advances in the catalytic, asymmetric synthesis of β-lactams. *Acc. Chem. Res.* 37 (8): 592–600.

206 Zhang, Y.R., He, L., Wu, X. et al. (2008). Chiral N-heterocyclic carbene catalyzed Staudinger reaction of ketenes with imines: highly enantioselective synthesis of *N*-boc β-lactams. *Org. Lett.* 10 (2): 277–280.

207 Marchand, J. and Ghosez, L. (1972). Cycloadditions of keteneimmonium cations to olefins and dienes - new synthesis of 4-membered rings. *J. Am. Chem. Soc.* 94 (8): 2870–2872.

208 Ghosez, L., Bogdan, S., Ceresiat, M. et al. (1987). Syntheses and uses of azetidiniminium salts. *Pure Appl. Chem.* 59 (3): 393–398.

209 O'Brien, J.M. and Kingsbury, J.S. (2011). A Practical synthesis of 3-acyl cyclobutanones by (2+2) annulation. Mechanism and utility of the Zn(II)-catalyzed condensation of α-chloroenamines with electron-deficient alkenes. *J. Org. Chem.* 76 (6): 1662–1672.

210 (a) Houge, C., Frisquehesbain, A.M., Mockel, A. et al. (1982). Models for asymmetric (2+2)-cycloadditions. *J. Am. Chem. Soc.* 104 (10): 2920–2921. (b) Lachia, M., Jung, P.M.J., and De Mesmaeker, A. (2012). A novel approach toward the synthesis of strigolactones through intramolecular (2+2)-cycloaddition of ketenes and ketene-iminiums to olefins. Application to the asymmetric synthesis of GR-24. *Tetrahedron Lett.* 53 (34): 4514–4517.

211 (a) Madelaine, C., Valerio, V., and Maulide, N. (2011). Revisiting keteniminium

salts: more than the nitrogen analogs of ketenes. *Chem. Asian J.* 6 (9): 2224–2239. (b) Kolleth, A., Lumbroso, A., Tanriver, G. et al. (2016). Synthesis of aminocyclobutanes via (2+2)-cycloadditions involving keteniminium intermediates. *Tetrahedron Lett.* 57 (25): 2697–2702. (c) Kolleth, A., Lumbroso, A., Tanriver, G. et al. (2017). Synthesis of 4-membered ring alkaloid analogues via intramolecular (2+2)-cycloaddition involving keteniminium salt intermediates. *Tetrahedron Lett.* 58 (30): 2904–2909.

212 Barbaro, G., Battaglia, A., Bruno, C. et al. (1996). Synthesis of 2-azetidiniminium salts. 1. Diastereoselectivity in keteniminium triflate/imine cycloadditions. *J. Org. Chem.* 61 (24): 8480–8488.

213 Arrieta, A., Cossio, F.P., and Lecea, B. (1999). Competitive mechanisms and origins of stereocontrol in the (2+2) thermal cycloaddition between imines and keteniminium cations. A complementary entry to 2-azetidinones (β-lactams) and related compounds. *J. Org. Chem.* 64 (6): 1831–1842.

214 Alajarin, M., Vidal, A., Tovar, F. et al. (1999). Surpassing torquoelectronic effects in conrotatory ring closures: origins of stereocontrol in intramolecular ketenimine-imine(2+2)-cycloadditions. *Chem. Eur. J.* 5 (3): 1106–1117.

215 Alhusaini, A.H., Muqtar, M., and Ali, S.A. (1991). Study of ketene-alkene cycloadditions and cycloreversions of cyclobutanones. *Tetrahedron* 47 (36): 7719–7726.

216 Wittig, G. and Schöllkopf, U. (1954). Über Triphenyl-Phosphinmethylene als Olefinbildende Reagenzien. 1. *Chem. Ber.* 87 (9): 1318–1330.

217 Hoffmann, R.W. (2001). Georg Wittig and his accomplishments: still relevant beyond his 100th birthday. *Angew. Chem. Int. Ed.* 40 (8): 1411–1416.

218 (a) Nicolaou, K.C., Harter, M.W., Gunzner, J.L., and Nadin, A. (1997). The Wittig and related reactions in natural product synthesis. *Liebigs Ann. Chem.* 7: 1283–1301. (b) Parkavatkar, P.T., Torray, P.S., and Tiwe, S.G. (2013). Recent developments of Wittig reaction in organic synthesis through tandem or sequential processes. *Curr. Org. Chem.* 10 (2): 288–317. (c) Lao, Z. and Toy, P.H. (2016). Catalytic Wittig and aza-Wittig reactions. Beilstein. *J. Org. Chem.* 12: 2577–2587.

219 Wittig, G., Eggers, H., and Duffner, P. (1958). Benzocycloheptatrien, Benzocyclooctatrien und Benzocyclooctatetraen. 6. Mitteilung über neuartige Synthesen von Cyclopolyenen. *Liebigs Ann. Chem.* 619 (1–3): 10–27.

220 Walker, B. (1996). Ylides and related compounds. *Organophosphorus Chem.* 27: 264–307.

221 Pommer, H. and Nürrenbach, A. (1975). Industrial synthesis of terpene compounds. *Pure Appl. Chem.* 43 (3–4): 527–551.

222 O'Brien, C.J., Tellez, J.L., Nixon, Z.S. et al. (2009). Recycling the waste: the development of a catalytic Wittig reaction. *Angew. Chem. Int. Ed.* 48 (37): 6836–6839.

223 Murphy, J.A., Commeureuc, A.G.J., Snaddon, T.N. et al. (2005). Direct conversion of *N*-methoxy-*N*-methylamides (Weinreb amides) to ketones via a nonclassical Wittig reaction. *Org. Lett.* 7 (7): 1427–1429.

224 Aggarwal, V.K., Fulton, J.R., Sheldon, C.G., and de Vicente, J. (2003). Generation of phosphoranes derived from phosphites. A new class of phosphorus ylides leading to high *E* selectivity with semi-stabilizing groups in Wittig olefinations. *J. Am. Chem. Soc.* 125 (20): 6034–6035.

225 López-Ortíz, F., García-López, J., Álvarez-Manzaneda, R., and Pérez-Álvarez, I.J. (2004). Isolable 1,2-oxaphosphetanes: from curiosities to starting materials for the synthesis of olefins. *Mini-Rev. Org. Chem.* 1: 65–76.

226 Vedejs, E. and Marth, C.F. (1990). Mechanism of Wittig reaction - evidence against betaine intermediates. *J. Am. Chem. Soc.* 112 (10): 3905–3909.

227 Vedejs, E. and Peterson, M.J. (1994). *Stereochemistry and Mechanism of the Wittig Reaction*, Topics in Stereochemistry, vol. 21 (ed. E.L. Eliel and S.H. Wilen), 1–157. Hoboken, NJ: Wiley.

228 Alagona, G. and Ghio, C. (2010). Dependence of the Wittig reaction mechanism on the environment and on the substituents at the aldehyde group and at the phosphonium ylide. *Int. J. Quantum Chem.* 110 (3): 765–776.

229 Robiette, R., Richardson, J., Aggarwal, V.K., and Harvey, J.N. (2006). Reactivity and selectivity in the Wittig reaction: a computational study. *J. Am. Chem. Soc.* 128 (7): 2394–2409.

230 Schlosser, M. and Christmann, K.F. (1967). Olefinations with phosphorus-ylides. I. Mechanism and stereochemistry of Wittig reaction. *Liebigs Ann. Chem.* 708: 1–35.

231 Bergelson, L.D. and Shemyakin, M.M. (1964). New methods in preparative organic chemistry. 4. Synthesis of naturally occurring unsaturated fatty acids by sterically controlled carbonyl olefination. *Angew. Chem. Int. Ed. Engl.* 3 (4): 250–260.

232 (a) Schlosser, M. (1970). The stereochemistry of the Wittig reaction. In: *Topics in Stereochemistry*, vol. 5 (ed. E.L. Eliel and N.L. Allinger), 1–30. New York: Wiley. (b) Vedejs, E. and Peterson, M.J. (1996). The Wittig reaction: stereoselectivity and a history of mechanistic ideas (1953–1995). In: *Advances in Carbanion Chemistry*, vol. 2, 1–85. Greenwich, CT, USA: JAI Press. (c) Pashley, A. (2013). Making alkenes: the Wittig reaction. *Chemistry Review* 23 (2): 30–33.

233 Giese, B., Schoch, J., and Rüchardt, C. (1978). Reactivity-selectivity principle. 3. Application of reactivity-selectivity principle to Wittig reactions. *Chem. Ber.* 111 (4): 1395–1403.

234 Rüchardt, C., Panse, P., and Eichler, S. (1967). Zum Mechanismus der Wittig-Reaktion mit Triarylalkoxycarbonylmethylen-Phosphoranen. *Chem. Ber.* 100 (4): 1144–1164.

235 El-Batta, A., Jiang, C., Zhao, W. et al. (2007). Wittig reactions in water media employing stabilized ylides with aldehydes. Synthesis of α,β-unsaturated esters from mixing aldehydes, α-bromoesters, and Ph_3P in aqueous $NaHCO_3$. *J. Org. Chem.* 72 (14): 5244–5259.

236 Orsini, F., Sello, G., and Fumagalli, T. (2006). One-pot Wittig reactions in water and in the presence of a surfactant. *Synlett* (11): 1717–1718.

237 Reitz, A.B., Mutter, M.S., and Maryanoff, B.E. (1984). Observation of *cis* and *trans* oxaphosphetanes in the Wittig reaction by high-field P-31 NMR-spectroscopy. *J. Am. Chem. Soc.* 106 (6): 1873–1875.

238 Kawashima, T., Kato, K., and Okazaki, R. (1993). Synthesis, structure, and thermolysis of a 3-methoxycarbonyl-1,2-λ(5)-oxaphosphetane. *Angew. Chem. Int. Ed. Engl.* 32 (6): 869–870.

239 Kawashima, T., Kato, K., and Okazaki, R. (1992). Novel synthetic route to isolable pentacoordinate 1,2-oxaphosphetanes and mechanism of their thermolysis, the 2nd step of the Wittig reaction. *J. Am. Chem. Soc.* 114 (10): 4008–4010.

240 Kojima, S., Sugino, M., Matsukawa, S. et al. (2002). First isolation and characterization of an anti-apicophilic spirophosphorane bearing an oxaphosphetane ring: a model for the possible reactive intermediate in the Wittig reaction. *J. Am. Chem. Soc.* 124 (26): 7674–7675.

241 Teo, B.K. and Calabrese, J.C. (1976). Stereochemical systematics of metal clusters - structural characterization of tetrameric triphenylphosphine silver chloride - analysis of bonded vs nonbonded interactions in cubane-like (R3Y)4M4X4 species. *Inorg. Chem.* 15 (10): 2467–2474.

242 Castaneda, F., Silva, P., Garland, M. et al. (2011). Molecular structures and preferred conformations of stabilized keto diester phosphonium ylides. *J. Mol. Struct.* 1004 (1–3): 284–291.

243 Castaneda, F., Silva, P., Bunton, C.A. et al. (2011). Conformations of diester triphenylphosphonium ylides with an ylidic ester or keto and ester ylidic groups. *Acta Crystallogr., Sect. C: Cryst. Struct. Commun.* 67: O319–O323.

244 Schlosser, M. and Schaub, B. (1982). *cis*-selectivity of salt-free Wittig reactions - a Leeward approach of the aldehyde at the origin. *J. Am. Chem. Soc.* 104 (21): 5821–5823.

245 (a) Maryanoff, B.E. and Reitz, A.B. (1989). The Wittig olefination reaction and modifications involving phosphoryl-stabilized carbanions – stereochemistry, mechanism, and selected synthetic aspects. *Chem. Rev.* 89 (4): 863–927. (b) McNulty, J., McLeod, D., Das, P., and Zepeda-Velazquez, C. (2015). Wittig reactions of trialkylphosphine-derived ylides: new directions and applications in organic synthesis. *Phosphorus Sulfur Silicon Relat. Elem.* (Taylor & Francis Ltd.) 190 (5–6): 619–632.

246 Schlosser, M. and Christmann, K.F. (1966). *trans*-selective olefin syntheses. *Angew. Chem. Int. Ed. Engl.* 5 (1): 126–126.

247 Corey, E.J. and Yamamoto, H. (1970). Modification of Wittig reaction to permit stereospecific synthesis of certain trisubstituted olefins. Stereospecific synthesis of α-Santalol. *J. Am. Chem. Soc.* 92 (1): 226–228.

248 Wilker, S., Laurent, C., Sarter, C. et al. (1995). Detection of a thiaphosphetane during the reaction of the ylide $Ph_3P{=}CH_2$ with thiobenzophenone. *J. Am. Chem. Soc.* 117 (27): 7293–7294.

249 Ernst, H. (2002). Recent advances in industrial carotenoid synthesis. *Pure Appl. Chem.* 74 (11): 2213–2226.

250 Ramirez, F., Mckelvie, N., and Desai, N.B. (1962). New synthesis of 1,1-dibromoolefins via phosphine-dibromomethylenes - reaction of triphenylphosphine with carbon tetrabromide. *J. Am. Chem. Soc.* 84 (9): 1745–1747.

251 Corey, E.J. and Fuchs, P.L. (1972). Synthetic method for formyl→ethynyl conversion (RCHO→RC≡CH or RC≡CR′). *Tetrahedron Lett.* 13 (36): 3769–3772.

252 Chalifoux, W.A. and Tykinski, R.R. (2006). Polyyne synthesis using carbene/carbenoid rearrangements. *Chem. Rec.* 6 (4): 169–182.

253 Horner, L., Hoffmann, H., and Wippel, H.G. (1958). Phosphororganische Verbindungen. 12. Phosphinoxyde als Olefinierungsreagenzien. *Chem. Ber.* 91 (1): 61–63.

254 Horner, L., Hoffmann, H., Wippel, H.G., and Klahre, G. (1959). Phosphororganische Verbindungen. 20. Phosphinoxyde als Olefinierungsreagenzien. *Chem. Ber.* 92 (10): 2499–2505.

255 (a) Wadsworth, W. and Emmons, W.D. (1961). Utility of phosphonate carbanions in olefin synthesis. *J. Am. Chem. Soc.* 83 (7): 1733–1738. (b) Gu, Y. and Tian, S.-K. (2012). Olefination reactions of phosphorus-stabilized carbon nucleophiles. In: *Stereoselective Alkene Synthesis*, Topics in Current Chemistry, vol. 327 (ed. J. Wang), 197–238. Berlin, Heidelberg: Springer-Verlag. (c) Bisceglia, J.A. and Orelli, L.R. (2015). Recent progress in the Horner–Wadsworth–Emmons reaction. *Curr. Org. Chem.* 19 (9): 744–775.

256 Blanchette, M.A., Choy, W., Davis, J.T. et al. (1984). Horner–Wadsworth–Emmons reaction - use of lithium-chloride and an amine for base sensitive compounds. *Tetrahedron Lett.* 25 (21): 2183–2186.

257 Rathke, M.W. and Nowak, M. (1985). The Horner–Wadsworth–Emmons modification of the Wittig reaction using triethylamine and lithium or magnesium salts. *J. Org. Chem.* 50 (15): 2624–2626.

258 Larsen, R.O. and Aksnes, G. (1983). Kinetic study of the Horner-Reaction. 1. *Phosphorus, Sulfur Silicon Relat. Elem.* 15 (2): 219–228.

259 Larsen, R.O. and Aksnes, G. (1983). Kinetic study of the Horner-Reaction. 2. *Phosphorus, Sulfur Silicon Relat. Elem.* 15 (2): 229–237.

260 Still, W.C. and Gennari, C. (1983). Direct synthesis of *Z*-unsaturated esters - a useful modification of the Horner-Emmons olefination. *Tetrahedron Lett.* 24 (41): 4405–4408.

261 Patois, C., Savignac, P., About-Jaudet, E., and Collignon, N. (1996). Bis(trifluoroethyl) (carboethoxymethyl)phosphonate - (acetic acid, [bis(2,2,2-trifluoroethoxy)phosphinyl]-, ethyl ester). *Org. Synth.* 73: 152–158.

262 Peterson, D.J.A. (1968). Carbonyl olefination reaction using silyl-substituted organometallic compounds. *J. Org. Chem.* 33 (2): 780–784.

263 Ager, D.J. (1984). The Peterson reaction. *Synthesis* (5): 384–398.

264 (a) van Staden, L.F., Gravestock, D., and Ager, D.J. (2002). New developments in the Peterson olefination reaction. *Chem. Soc. Rev.* 31 (3): 195–200. (b) Romanenko, V.D., Shevchuk, M.V., and Kukhar, V.P. (2011). Application of silicon-based methodologies for the synthesis of functionalized mono- and bisphosphonic acids. *Curr. Org. Chem.* 15 (16): 2774–2801.

265 Beadham, I. and Micklefield, J. (2005). Reagents for carbonyl methylenation in organic synthesis. *Curr. Org. Synth. (Bentham)* 2 (2): 231–259.

266 Julia, M. and Paris, J.M. (1973). Syntheses using sulfones. 5. Method for general synthesis of double bonds. *Tetrahedron Lett.* (49): 4833–4836.

267 Lythgoe, B. and Waterhouse, I. (1977). New 1,2-elimination reaction with a radical mechanism. *Tetrahedron Lett.* 18 (48): 4223–4226.

268 Kende, A.S. and Mendoza, J.S. (1990). An improved variant of the Julia olefin synthesis - reductive elimination of β-hydroxy-imidazolyl sulfones by SmI_2. *Tetrahedron Lett.* 31 (49): 7105–7108.

269 Baudin, J.B., Hareau, G., Julia, S.A., and Ruel, O. (1991). A direct synthesis of olefins by reaction of carbonyl compounds with lithio derivatives of 2-[alkyl-sulfonyl or (2′-alkenyl)-sulfonyl or benzyl-sulfonyl]-benzothiazoles. *Tetrahedron Lett.* 32 (9): 1175–1178.

270 Warren, L.A. and Smiles, S. (1930). Dehydro-2-naphtholsulphone. *J. Chem. Soc.* 1327–1331.

271 Baudin, J.B., Hareau, G., Julia, S.A., and Ruel, O. (1993). Stereochemistry of the olefin formation from *anti* and *syn* heterocyclic β-hydroxysulfones. *Bull. Soc. Chim. Fr.* 130 (3): 336–357.

272 Baudin, J.B., Hareau, G., Julia, S.A. et al. (1993). Stereochemistry of direct olefin formation from carbonyl compounds and lithiated heterocyclic sulfones. *Bull. Soc. Chim. Fr.* 130 (6): 856–878.

273 Blakemore, P.R., Cole, W.J., Kocienski, P.J., and Morley, A. (1998). A stereoselective synthesis of *trans*-1,2-disubstituted alkenes based on the condensation of aldehydes with metallated 1-phenyl-1*H*-tetrazol-5-yl sulfones. *Synlett* (1): 26–28.

274 Kocienski, P.J., Bell, A., and Blakemore, P.R. (2000). 1-*tert*-Butyl-1*H*-tetrazol-5-yl sulfones in the modified Julia olefination. *Synlett* (3): 365–366.

275 Blakemore, P.R. (2002). The modified Julia olefination: alkene synthesis via the condensation of metallated heteroarylalkylsulfones with carbonyl compounds. *J. Chem. Soc., Perkin Trans. 1* (23): 2563–2585.

276 Boggian, D.B. and Mata, E.G. (2010). Efficient alkene synthesis on solid support using the Julia–Kocienski coupling. *Mol. Diversity* 14 (4): 847–853.

277 Aissa, C. (2009). Mechanistic manifold and new developments of the Julia–Kocienski reaction. *Eur. J. Org. Chem.* (12): 1831–1844.

278 Claridge, T.D., Davies, S.G., Lee, J.A. et al. (2008). Highly (*E*)-selective Wadsworth–Emmons reactions promoted by methylmagnesium bromide. *Org. Lett.* 10 (23): 5437–5440.

279 Liu, Y.K., Ma, C., Jians, K. et al. (2009). Asymmetric tandem Michael addition-Wittig reaction to cyclohexenone annulation. *Org. Lett.* 11: 2848–2851.

280 Carruthers, W. (1990). *Cycloaddition Reactions in Organic Synthesis*, Tetrahedon Organic Chemistry Series, 1e, vol. 8, 1–373. Oxford: Pergamon Press; 2nd impression, Amsterdam, Netherlands: Elsevier, 2013, 1–382.

281 Kobayashi, S. and Jørgensen, K.A. (eds.) (2001). *Cycloaddition Reactions in Organic Synthesis*, 1–332. Weinheim: Wiley-VCH.

282 Nishiwaki, N. (ed.) (2014). *Methods and Applications of Cycloaddition Reactions in Organic Synthesis*, 1–659. Hoboken, NJ: Wiley.

283 Takao, K.-I., Munakata, R., and Tadano, K.-I. (2005). Recent advances in natural product synthesis by using intramolecular Diels–Alder reactions. *Chem. Rev.* 105: 4779–4807.

284 (a) Juhl, M. and Tanner, D. (2009). Recent applications of intramolecular Diels–Alder reactions to natural product synthesis. *Chem. Soc. Rev.* 38 (11): 2983–2992. (b) Nawrat, C.C. and Moody, C.J. (2014). Quinones as dienophiles in the Diels–Alder reaction: history and applications in total synthesis. *Angew. Chem. Int. Ed.* 53: 2056–2077. (c) Klier, L., Turr, F., Poulsen, P.H., and Jørgensen, K.A. (2017). Asymmetric cycloaddition reactions catalysed by diarylprolinol silyl ethers. *Chem. Soc. Rev.* 46 (4): 1080–1102. (d) Cao, M.-H., Green, N.J., and Xu, S.-Z. (2017). Application of the aza-Diels–Alder reaction in the synthesis of natural products. *Org. Biomol. Chem.* 15: 3105–3129.

285 Woodward, R.B., Sondheimer, F., and Taub, D. (1951). The total synthesis of Cholesterol. *J. Am. Chem. Soc.* 73 (7): 3548–3548.

286 Woodward, R.B., Sondheimer, F., and Taub, D. (1951). The total synthesis of Cortisone. *J. Am. Chem. Soc.* 73 (8): 4057–4057.

287 Sarett, L.H., Arth, G.E., Lukes, R.M. et al. (1952). Stereospecific total synthesis of Cortisone. *J. Am. Chem. Soc.* 74 (19): 4974–4976.

288 Stork, G., van Tamelen, E.E., Friedman, L.J., and Burgstahler, A.W. (1951). Cantharidin - a stereospecific total synthesis. *J. Am. Chem. Soc.* 73 (9): 4501–4501.

289 Nicolaou, K.C., Snyder, S.A., Montagnon, T., and Vassilikogiannakis, G. (2002). The Diels–Alder reaction in total synthesis. *Angew. Chem. Int. Ed.* 41 (10): 1668–1698.

290 Kagan, H.B. and Riant, O. (1992). Catalytic asymmetric Diels–Alder reactions. *Chem. Rev.* 92 (5): 1007–1019.

291 Ahrendt, K.A., Borths, C.J., and MacMillan, D.W.C. (2000). New strategies for organic catalysis: the first highly enantioselective organocatalytic Diels–Alder reaction. *J. Am. Chem. Soc.* 122 (17): 4243–4244.

292 Pellissier, H. (2009). Asymmetric hetero-Diels–Alder reactions of carbonyl compounds. *Tetrahedron* 65 (15): 2839–2877.

293 Pellissier, H. (2012). Asymmetric organocatalytic cycloadditions. *Tetrahedron* 68 (10): 2197–2232.

294 Mikami, K. and Shimizu, M. (1992). Asymmetric ene reactions in organic synthesis. *Chem. Rev.* 92 (5): 1021–1050.

295 Mikami, K. and Aikawa, K. (2010). Asymmetric ene reactions and cycloadditions. In: *Catalytic Asymmetric Synthesis*, 3e (ed. I. Ojima), 683–737. New York: Wiley.

296 Ishihara, K. and Sakakura, A. (2011). $[\pi^4+\pi^2]$-Cycloaddition reactions. In: *Science of Synthesis, Stereoselective Synthesis*, vol. 3, 67–123. Stuttgart: Thieme Verlag.

297 Merino, P., Marqués-López, E., Tejero, T., and Herrera, R.P. (2010). Enantioselective organocatalytic Diels–Alder reactions. *Synthesis* (1): 1–26.

298 Xu, Z. and Wang, H. (2011). Asymmetric inverse-electron-demand hetero-Diels–Alder reaction via enamine-metal Lewis acid bifunctional catalysis. *Synlett* (20): 2907–2912.

299 Bernardi, L., Fochi, M., Comes Franchini, M., and Ricci, A. (2012). Bioinspired organocatalytic asymmetric reactions. *Org. Biomol. Chem.* 10 (15): 2911–2922.

300 Needleman, S.B. and Changkuo, M.C. (1962). Diels–Alder syntheses with heteroatomic compounds. *Chem. Rev.* 62 (5): 405–431. An experimental and theoretical study of

the hetero-Diels–Alder reactions between (*E*)-2-aryl-1-cyano-1-nitroethenes and ethyl vinyl ether: one.

301 (a) Tietze, L.F. and Kettschau, G. (1997). Hetero Diels–Alder reactions in organic chemistry. In: *Stereoselective Heterocyclic Synthesis I*, Topics in Current Chem, vol. 189 (ed. P. Metz), 1–120. Heidelberg: Springer-Verlag. (b) Jasiniski, R., Kubik, M., Lapczuk-Krygier, A. et al. (2014). An experimental and theoretical study of the hetero-Diels–Alder reactions between (*E*)-2-aryl-1-cyano-1-nitroethenes and ethyl vinyl ether: one-step or zwitterionic, two-step mechanism? *React. Kinet., Mech. Catal.* 113 (2): 333–345.

302 Weinreb, S.M. and Staib, R.R. (1982). Synthetic aspects of Diels–Alder cycloadditions with heterodienophiles. *Tetrahedron* 38 (21): 3087–3128.

303 Danishefsky, S., Kato, N., Askin, D., and Kerwin, J.F. (1982). Stereochemical consequences of the Lewis acid-catalyzed cyclocondensation of oxygenated dienes with aldehydes - a rapid and stereoselective entry to various natural products derived from propionate. *J. Am. Chem. Soc.* 104 (1): 360–362.

304 Danishefsky, S., Kerwin, J.F., and Kobayashi, S. (1982). Lewis acid-catalyzed cyclocondensations of functionalized dienes with aldehydes. *J. Am. Chem. Soc.* 104 (1): 358–360.

305 Barone, V., Arnaud, R., Chavant, P.Y., and Vallee, Y. (1996). Substituent effects in the hetero-Diels–Alder reaction of thiocarbonyl compounds with butadiene. *J. Org. Chem.* 61 (15): 5121–5129.

306 Rohr, U., Schatz, J., and Sauer, J. (1998). Thio- and selenocarbonyl compounds as "Superdienophiles" in [π4+π2]-cycloadditions. *Eur. J. Org. Chem.* (12): 2875–2883.

307 Girling, P., Kiyoi, T., and Whiting, A. (2011). Mannich–Michael versus formal aza-Diels–Alder approaches to piperidine derivatives. *Org. Biomol. Chem.* 9 (9): 3105–3121.

308 Yao, S.L., Saaby, S., Hazell, R.G., and Jørgensen, K.A. (2000). Catalytic enantioselective aza-Diels–Alder reactions of imines - an approach to optically active nonproteinogenic alpha-amino acids. *Chem. Eur. J.* 6 (13): 2435–2448.

309 Jørgensen, K.A. (2000). Catalytic asymmetric hetero-Diels–Alder reactions of carbonyl compounds and imines. *Angew. Chem. Int. Ed.* 39 (20): 3558–3588.

310 (a) Bodnar, B.S. and Miller, M.J. (2011). The nitrosocarbonyl hetero-Diels–Alder reaction as a useful tool for organic syntheses. *Angew. Chem. Int. Ed.* 50 (25): 5629–5646. (b) Tripoteau, F., Eberlin, L., Fox, M.A. et al. (2013). A novel, efficient synthesis of *N*-aryl pyrroles *via* reaction of 1-boronodienes with arylnitroso compounds. *J. Chem. Soc., Chem. Commun.* 49: 5414–5416. (c) de los Santos, J.M., Ignacio, R., Es Sbai, Z. et al. (2014). Hetero-Diels–Alder reaction of phosphorylated nitroso alkenes with enol ethers on water: a clean approach toward 1,2-oxazine derivatives. *J. Org. Chem.* 79: 7607–7615. (d) Sandoval, D., Samoshin, A.V., and Read de Alaniz, J. (2015). Asymmetric electrophilic α-amination of silyl enol ether derivatives via the nitrosocarbonyl hetero-ene reaction. *Org. Lett.* 17 (18): 4514–4517. (e) Huang, J., Chen, Z., Yuan, J., and Peng, Y. (2016). Recent advances in highly selective applications of nitroso compounds. *Asian J. Org. Chem.* 5 (8): 951–960. (f) Lewis, R.B. and Read de Alaniz, J. (2017). Nitrosocarbonyl hetero-Diels–Alder cycloaddition with 2-substituted 1,3-butadienes. *Tetrahedron* 73 (29): 4045–4051.

311 (a) Johannsen, M. and Jørgensen, K.A. (1998). Allylic amination. *Chem. Rev.* 98 (4): 1689–1708. (b) Quadrelli, P., Romana, S., Piccanello, A., and Caramella, P. (2009). The remarkable *cis* effect in the ene reactions of nitrosocarbonyl intermediates. *J. Org. Chem.* 74 (6): 2301–2310. (c) Huang, S.-H., Hong, H.-X., Li, W.-H., and Hong, R. (2012). Research progress on nitroso-ene reaction. *Chin. J. Org. Chem.* 32 (10): 1776–1791. (d) Memeo, M.G. and Quadrelli, P. (2017). Generation and trapping of nitrosocarbonyl intermediates. *Chem. Rev.* 117 (3): 2108–2200. (e) Ouyang, J., Mi, X.-W., Wang, Y., and Hong, R. (2017). Crossing the ring: a journey of the nitroso-ene reaction. *Synlett* 28 (7): 762–772. (f) Carosso, S., Memeo, M.G., Bovio, B. et al. (2017). *N,O-n*ucleosides from ene reaction of (nitrosocarbonyl)mesitylene with crotyl alcohol: selectivity, scope, and limitations. *Synthesis* 49 (9): 1972–1982. (g) Zhai, L., Tian, X.-C., Wang, C. et al. (2017). Construction of morphan derivatives by nitroso-ene cyclization: mechanistic insight and total synthesis of (+/-)-Kopsone. *Angew. Chem. Int. Ed.* 56 (38): 11599–11603.

312 Bansal, R.K. and Kumawat, S.K. (2008). Diels–Alder reactions involving C=P- functionality. *Tetrahedron* 64 (49): 10945–10976.

313 Kresze, G. and Wucherpfennig, W. (1967). Organic synthesis with imides of sulfur dioxide. *Angew. Chem. Int. Ed. Engl.* 6 (2): 149–167.

314 El-Sayed, I. (2005). Carbon vs sulfur addition of nucleophiles to sulfines. Part 1. The case of sulfinates. *Monatsh. Chem.* 136 (4): 543–551.

315 Deguin, B. and Vogel, P. (1992). Hetero-Diels–Alder addition of sulfur dioxide to 1,3-dienes - suprafaciality, regioselectivity, and stereoselectivity. *J. Am. Chem. Soc.* 114 (23): 9210–9211.

316 Vogel, P. and Sordo, J.A. (2006). Hetero-Diels–Alder and cheletropic additions of sulfur dioxide to conjugated dienes. Experimental facts and theoretical analysis. *Curr. Org. Chem.* 10 (16): 2007–2036.

317 Vogel, P., Turks, M., Bouchez, L. et al. (2007). New organic chemistry of sulfur dioxide. *Acc. Chem. Res.* 40 (10): 931–942.

318 Tietze, L.F., Kettschau, G., Gewert, J.A., and Schuffenhauer, A. (1998). Hetero–Diels–Alder reactions of 1-oxa-1,3-butadienes. *Curr. Org. Chem.* 2 (1): 19–62.

319 Boger, D.L. (1986). Diels–Alder reactions of heterocyclic azadienes - scope and applications. *Chem. Rev.* 86 (5): 781–793.

320 Boger, D.L. (1996). Azadiene Diels–Alder reactions: scope and applications. Total synthesis of natural and *ent*-Fredericamycin A. *J. Heterocycl. Chem.* 33 (5): 1519–1531.

321 Boger, D.L. (1998). Heterocyclic and acyclic azadiene Diels–Alder reactions: total synthesis of Nothapodytine B. *J. Heterocycl. Chem.* 35 (5): 1003–1011.

322 Capozzi, G., Giacomelli, S., Menichetti, S., and Nativi, C. (2001). Generation and trapping of α,β-unsaturated thioketones. Synthesis of 5,6-dihydrothiopyranes. *Synthesis* (3): 409–412.

323 Weinreb, S.M. and Scola, P.M. (1989). *N*-acyl imines and related hetero dienes in (4+2)-cycloaddition reactions. *Chem. Rev.* 89 (7): 1525–1534.

324 (a) Thalhammer, F., Wallfahrer, U., and Sauer, J. (1990). Reactivity of simple open-chain and cyclic dienophiles in inverse-type Diels–Alder reactions. *Tetrahedron Lett.* 31 (47): 6851–6854. (b) Liu, F., Liang, Y., and Houk, K.N. (2014). Theoretical elucidation of the origins of substituent and strain effects on the rates of Diels–Alder reactions of 1,2,4,5-tetrazines. *J. Am. Chem. Soc.* 136 (32): 11483–11493.

325 Sharma, P., Kumar, A., Sahu, V., and Singh, J. (2008). Frontier orbital interactions in the hetero Diels–Alder cycloaddition of diazadienes. *Can. J. Chem.* 86 (5): 384–394.

326 Schoch, J., Wiessler, M., and Jaeschke, A. (2010). Post-synthetic modification of DNA by inverse-electron-demand Diels–Alder reaction. *J. Am. Chem. Soc.* 132 (26): 8846–8847.

327 Boccardo, G., Capozzi, G., Giuntini, M. et al. (1997). α,α'-dioxothiones. 2. Asymmetric Diels–Alder reactions of chiral non-racemic α,α'-dioxothiones. *Tetrahedron* 53 (51): 17383–17394.

328 Berson, J.A. (1992). Discoveries missed, discoveries made - creativity, influence, and fame in chemistry. *Tetrahedron* 48 (1): 3–17.

329 Stobbe, H. and Reuss, F. (1912). Polymerisation of the cyclopentadienes. *Liebigs Ann. Chem.* 391 (1/3): 151–168.

330 Krämer, G. and Spilker, A. (1896). Über das Cyclopentadien im Steinhohlentheer, Das Inden der Fettreihe. *Ber. Dtsch. Chem. Ges.* 29: 552–561.

331 Kistiakowsky, G.B. and Mears, W.H. (1936). The kinetics of cyclopentadiene. *J. Am. Chem. Soc.* 58: 1060–1060.

332 Harkness, J.B., Kistiakowsky, G.B., and Mears, W.H. (1937). Studies in gaseous polymerizations. *J. Chem. Phys.* 5 (9): 682–694.

333 Bergel, F. and Widmann, E. (1928). Zur Konstitution des Dicyclopentadienes. *Liebigs Ann. Chem.* 467: 76–91.

334 Alder, K. and Stein, G. (1933). The polymerisation of cyclical hydrocarbons. VI. The stereochemistry of cyclopentadiene polymerisation. *Liebigs Ann. Chem.* 504: 216–257.

335 von Euler, H. and Josephson, K.O. (1920). Condensation of double formations. I.: The condensation from isoprene with benzochinon. *Ber. Dtsch. Chem. Ges.* 53: 822–826.

336 Diels, O., Blom, J.H., and Koll, W. (1925). Endo-methylenepiperidazine resulting from cyclopentadiene and azo ester and its transformation into 1,3-diaminocyclopentane. *Liebigs Ann. Chem.* 443: 242–262.

337 Littmann, E.R. (1936). The mechanism of the diene synthesis. *J. Am. Chem. Soc.* 58: 1316–1317.

338 Houk, K.N., González, J., and Li, Y. (1995). Pericyclic reaction transition states - passions and punctilios, 1935–1995. *Acc. Chem. Res.* 28 (2): 81–90.

339 Woodward, R.B. and Katz, T.J. (1959). The mechanism of the Diels–Alder reaction. *Tetrahedron* 5 (1): 70–89.
340 Dewar, M.J.S. (1959). Mechanism of the Diels–Alder reaction. *Tetrahedron Lett.* 4: 16–18.
341 Dewar, M.J.S. and Pyron, R.S. (1970). Nature of transition state in some Diels–Alder reactions. *J. Am. Chem. Soc.* 92 (10): 3098–3103.
342 Horn, B.A., Herek, J.L., and Zewail, A.H. (1996). Retro–Diels–Alder femtosecond reaction dynamics. *J. Am. Chem. Soc.* 118 (36): 8755–8756.
343 Bartlett, P.D. and Schueller, K.E. (1968). Cycloaddition. 8. Ethylene as a dienophile. A minute amount of 1,2-cycloaddition of ethylene to butadiene. *J. Am. Chem. Soc.* 90 (22): 6071–6077.
344 Houk, K.N., Li, Y., and Evanseck, J.D. (1992). Transition structures of hydrocarbon pericyclic reactions. *Angew. Chem. Int. Ed. Engl.* 31 (6): 682–708.
345 Coffman, D.D., Barrick, P.L., Cramer, R.D., and Raasch, M.S. (1949). Synthesis of tetrafluorocyclobutanes by cycloalkylation. *J. Am. Chem. Soc.* 71 (2): 490–496.
346 Getty, S.J. and Borden, W.T. (1991). Why does tetrafluoroethylene not undergo Diels–Alder reaction with 1,3-butadiene - an ab initio investigation. *J. Am. Chem. Soc.* 113 (11): 4334–4335.
347 Northrop, B.H. and Houk, K.N. (2006). Vinylcyclobutane-cyclohexene rearrangement: theoretical exploration of mechanism and relationship to the Diels–Alder potential surface. *J. Org. Chem.* 71 (1): 3–13.
348 Li, Y. and Houk, K.N. (1993). Diels–Alder dimerization of 1,3-butadiene - an ab initio CASSCF study of the concerted and stepwise mechanisms and butadiene ethylene revisited. *J. Am. Chem. Soc.* 115 (16): 7478–7485.
349 Goldstein, E., Beno, B., and Houk, K.N. (1996). Density functional theory prediction of the relative energies and isotope effects for the concerted and stepwise mechanisms of the Diels–Alder reaction of butadiene and ethylene. *J. Am. Chem. Soc.* 118 (25): 6036–6043.
350 Alder, K. and Stein, G. (1937). Untersuchungen über den Verlauf der Diensynthese. *Angew. Chem.* 50: 0510–0519.
351 Houk, K.N., Lin, Y.T., and Brown, F.K. (1986). Evidence for the concerted mechanism of the Diels–Alder reaction of butadiene with ethylene. *J. Am. Chem. Soc.* 108 (3): 554–556.
352 Klärner, F.G., Krawczyk, B., Ruster, V., and Deiters, U.K. (1994). Evidence for pericyclic and stepwise processes in the cyclodimerization of chloroprene and 1,3-butadiene from pressure-dependence and stereochemistry - experimental and theoretical volumes of activation and reaction. *J. Am. Chem. Soc.* 116 (17): 7646–7657.
353 Stewart, C.A. (1972). Competing diradical and electrocyclic reactions - difference in activation volumes. *J. Am. Chem. Soc.* 94 (2): 635–637.
354 Goldstein, M.J. and Thayer, G.L. (1965). Mechanism of a Diels–Alder reaction. 2. Structure of transition state. *J. Am. Chem. Soc.* 87 (9): 1933–1941.
355 Taagepera, M. and Thornton, E.R. (1972). Secondary deuterium-isotope effects - transition state in reverse Diels–Alder reaction of 9,10-dihydro-9,10-ethanoanthracene - potentially general method for experimentally determining transition state symmetry and distinguishing concerted from stepwise mechanisms. *J. Am. Chem. Soc.* 94 (4): 1168–1177.
356 Gajewski, J.J., Peterson, K.B., Kagel, J.R., and Huang, Y.C.J. (1989). Transition state structure variation in the Diels–Alder reaction from secondary deuterium kinetic isotope effects - the reaction of nearly symmetrical dienes and dienophiles is nearly synchronous. *J. Am. Chem. Soc.* 111 (25): 9078–9081.
357 Storer, J.W., Raimondi, L., and Houk, K.N. (1994). Theoretical secondary kinetic isotope effects and the interpretation of transition state geometries. 2. The Diels–Alder reaction transition state geometry. *J. Am. Chem. Soc.* 116 (21): 9675–9683.
358 (a) Houk, K.N., Li, Y., Storer, J. et al. (1994). Concerted and stepwise mechanisms in cycloaddition reactions - potential surfaces and isotope effects. *J. Chem. Soc., Faraday Trans.* 90 (12): 1599–1604. (b) Pham, H.V. and Houk, K.N. (2014). Diels–Alder reactions of allene with benzene and butadiene: concerted, stepwise and ambimodal transition states. *J. Org. Chem.* 79 (19): 8968–8976.
359 (a) Beno, B.R., Houk, K.N., and Singleton, D.A. (1996). Synchronous or asynchronous? An "experimental" transition state from a direct comparison of experimental and theoretical kinetic isotope effects for a Diels–Alder reaction. *J. Am. Chem. Soc.* 118 (41): 9984–9985. (b) Wang, T., Niu, D.W., and Hoye, T.R. (2016). The hexadehydro-Diels–Alder cycloisomerization reaction proceeds by a stepwise mechanism. *J. Am. Chem. Soc.* 138 (25): 7832–7835. (c) Mloston, G., Grzelak, P., Linden, A., and

Heimgartner, H. (2017). Thia-Diels–Alder reactions of hetaryl thioketones with nonactivated 1,3-dienes leading to 3,6-dihydro-2*H*-pyrans: evidence for a diradical mechanism. *Chem. Heterocycl. Compd.* 53 (5): 518–525. (d) Rana, A., Paul, I., and Schmittel, M. (2017). Dynamic effects in the didehydro-Diels–Alder (DDDA) reaction of enyne-ketoenes: 50% stepwise bond formation in spite of concerted transition state. *J. Phys.Org. Chem.* 30 (9): Article No. e3732 (pp. 1–7).

360 Singleton, D.A., Schulmeister, B.E., Hang, C. et al. (2001). Isotope effects and the distinction between synchronous, asynchronous, and stepwise Diels–Alder reactions. *Tetrahedron* 57: 5149–5160.

361 Kwiatkowska, M., Jasinski, R., Mikulska, M., and Baranski, A. (2010). Secondary α-deuterium kinetic isotope effects in (2+4)-cycloaddition of (*E*)-2-phenylnitroethene to cyclopentadiene. *Monatsh. Chem.* 141 (5): 545–548.

362 Pross, A. and Shaik, S.S. (1983). A qualitative valence-bond approach to organic reactivity. *Acc. Chem. Res.* 16 (10): 363–370.

363 Bachmann, W.E. and Deno, N.C. (1949). The Diels–Alder reaction of 1-vinylnaphthalene with α,β-unsaturated and α,β,γ,δ-unsaturated acids and derivatives. *J. Am. Chem. Soc.* 71 (9): 3062–3072.

364 Koopmans, T. (1934). The classification of wave functions and eigen-values to the single electrons of an atom. *Physica* 1: 104–113.

365 Eibler, E., Hocht, P., Prantl, B. et al. (1997). Transitions of electron demand in pericyclic reactions: normal, neutral, and inverse Diels–Alder reactions of polyhalogenated cyclopentadienes. *Liebigs Ann. Chem.* 1997 (12): 2471–2484.

366 Ess, D.H., Jones, G.O., and Houk, K.N. (2006). Conceptual, qualitative, and quantitative theories of 1,3-dipolar and Diels–Alder cycloadditions used in synthesis. *Adv. Synth. Catal.* 348 (16–17): 2337–2361.

367 Dewar, M.J.S. and Pierini, A.B. (1984). Mechanism of the Diels–Alder reaction - studies of the addition of maleic anhydride to furan and methylfurans. *J. Am. Chem. Soc.* 106 (1): 203–208.

368 Dewar, M.J.S., Olivella, S., and Stewart, J.J.P. (1986). Mechanism of the Diels–Alder reaction - reactions of butadiene with ethylene and cyanoethylenes. *J. Am. Chem. Soc.* 108 (19): 5771–5779.

369 Dewar, M.J.S., Olivella, S., and Rzepa, H.S. (1978). Ground states of molecules. 49. MINDO-3 study of retro-Diels–Alder reaction of cyclohexene. *J. Am. Chem. Soc.* 100 (18): 5650–5659.

370 (a) Bernard-Henriet, C. and Chanon, M. (1996). The Diels–Alder additions of phenylcyclopentadienes: tests for the PMO theory and the diradicaloid model. *Tetrahedron Lett.* 37 (14): 2417–2420. (b) Monnat, F., Vogel, P., Meana, R., and Sordo, J.A. (2003). Equilibrium and kinetic deuterium isotope effects on the hetero-Diels–Alder addition of sulfur dioxide. *Angew. Chem. Int. Ed.* 42 (33): 3924–3927. (c) Silva López, C., Nieto Faza, O., and de Lera, A.R. (2008). Characterization of the switch in the mechanism of an intramolecular Diels–Alder reaction. *J. Org. Chem.* 73 (2): 467–473. (d) Ess, D.H., Hayden, A.E., Klärner, F.-G., and Houk, K.N. (2008). Transition states for the dimerization of 1,3-cyclohexadiene: A DFT, CASPT2, and CBS-QB3 quantum mechanical investigation. *J. Org. Chem.* 73 (19): 7586–7592. (e) Deslongchamps, G. and Deslongchamps, P. (2013). Bent bonds and the antiperiplanar hypothesis as a simple model to predict Diels–Alder reactivity: retrospective or perspective? *Tetrahedron* 69 (30): 6022–6033.

371 Fringuelli, F. and Taticchi, A. (1990). *Dienes in the Diels–Alder reaction*, 1–368. New York: Wiley.

372 Kataoka, F., Shimizu, N., and Nishida, S. (1980). Cycloaddition reaction of some representative 1-cyclopropyl-1,3-butadienes with tetracyanoethylene and reaction of the resultant vinylcyclobutanes - easy vinylcyclobutane cyclohexene rearrangement. *J. Am. Chem. Soc.* 102 (2): 711–716.

373 Bargagna, A., Schenone, P., Bondavalli, F., and Longobardi, M. (1980). Reaction of sulfene and dichloroketene with open-chain *N,N*-disubstituted α-aminomethyleneketones - synthesis of *N,N*-disubstituted 4-amino-3,4-dihydro-(5-methyl-6-phenyl)(5,6-diphenyl)-1,2-oxathiin 2,2-dioxides and of 4-amino-3-chloro-5-methyl-6-phenyl-2*H*-pyran-2-ones. *J. Heterocycl. Chem.* 17 (6): 1201–1206.

374 Clever, H.A., Wang, G.Y., Mollberg, W.C. et al. (1992). Spontaneous homopolymerization competes with Diels–Alder cycloaddition of 1-aryl-1,3-butadienes to dienophiles containing a leaving group. *J. Org. Chem.* 57 (25): 6837–6842.

375 Lakhdar, S., Terrier, F., Vichard, D. et al. (2010). The Diels–Alder reaction of 4,6-dinitrobenzofuroxan with 1-trimethylsilyloxybuta-1,3-diene: a case example of a stepwise cycloaddition. *Chem. Eur. J.* 16 (19): 5681–5690.

376 Wijnen, J.W., Zavarise, S., Engberts, J.B.F.N., and Charton, M. (1996). Substituent effects on an inverse electron demand hetero Diels–Alder reaction in aqueous solution and organic solvents: cycloaddition of substituted styrenes to di(2-pyridyl)-1,2,4,5-tetrazine. *J. Org. Chem.* 61 (6): 2001–2005.

377 Boger, D.L. and Weinreb, S.N. (1987). *Hetero-Diels–Alder Methodology in Organic Synthesis*, 1–366. New York: Academic Press.

378 Rücker, C., Lang, D., Sauer, J. et al. (1980). Reactivity of substituted 1,3-butadienes in Diels–Alder reactions. *Chem. Ber.* 113 (5): 1663–1690.

379 Bischof, P. and Heilbronner, E. (1970). Photoelectron spectra of cycloalkenes and cycloalkadienes. *Helv. Chim. Acta* 53 (7): 1677–1682.

380 Sustmann, R., Böhm, M., and Sauer, J. (1979). Significance of diene-1,4-distance on the reactivity in Diels–Alder reactions. *Chem. Ber.* 112 (3): 883–889.

381 Asmus, P. and Klessing, M. (1974). Photoelectron spectra of organic compounds. 6. Exocyclic methylene compounds. *Tetrahedron* 30 (15): 2477–2483.

382 Degroot, A., Evenhuis, B., and Wynberg, H. (1968). Syntheses and properties of sterically hindered butadienes. A modification of Chugaev reaction. *J. Org. Chem.* 33 (6): 2214–2218.

383 Paton, R.S., Kim, S., Ross, A.G. et al. (2011). Experimental Diels–Alder reactivities of cycloalkenones and cyclic dienes explained through transition state distortion energies. *Angew. Chem. Int. Ed.* 50 (44): 10366–10368.

384 Hayden, A.E. and Houk, K.N. (2009). Transition state distortion energies correlate with activation energies of 1,4-dihydrogenations and Diels–Alder cycloadditions of aromatic molecules. *J. Am. Chem. Soc.* 131 (11): 4084–4089.

385 Osuna, S. and Houk, K.N. (2009). Cycloaddition reactions of butadiene and 1,3-dipoles to curved arenes, fullerenes, and nanotubes: theoretical evaluation of the role of distortion energies on activation barriers. *Chem. Eur. J.* 15 (47): 13219–13231.

386 Pilet, O., Birbaum, J.L., and Vogel, P. (1983). Interaction between non-conjugated chromophores. 18. Preparation and Diels–Alder reactivity of 2,3,5,6,7,8-hexamethylidenebicyclo[2.2.2]octane ([2.2.2]hericene) - force-field calculations of exocyclic dienes as a moiety of bicyclic skeletons. *Helv. Chim. Acta* 66 (1): 19–34.

387 Mohraz, M., Wang, J.Q., Heilbronner, E. et al. (1980). Radical-cation states of 2,3,5,6,7,8-hexamethylidenebicyclo[2.2.2]octane. *Helv. Chim. Acta* 63 (3): 568–570.

388 Mohraz, M., Batich, C., Heilbronner, E. et al. (1979). Radical-cation states of 2,3,5,6-tetramethylenenorbornane, 2,3,5,6-tetramethylenebicyclo[2.2.2]octane and of related compounds. *Recl. Trav. Chim. Pays-Bas* 98 (5): 361–367.

389 Pfeffer, H.U., Klessinger, M., Erker, G., and Roth, W.R. (1975). Photoelectron spectra of organic compounds. 8. 7,8-dimethylenebicyclo[2.2.2]octa-2,5-diene and related compounds. *Chem. Ber.* 108 (9): 2923–2933.

390 Birbaum, J.L. and Vogel, P. (1988). Tandem cheletropic additions of sulfur dioxide to [2.2.2]hericene - the barrelene effect. *Helv. Chim. Acta* 71 (6): 1474–1482.

391 Trabert, L. and Hopf, H. (1980). Alicyclic compounds. 3. New method of preparation and reactions of [4]radialene. *Liebigs Ann. Chem.* (11): 1786–1800.

392 Vogel, P. (2008). Combinatorial synthesis of linearly condensed polycyclic compounds, including anthracyclinones, through tandem Diels–Alder additions. In: *Anthracycline Chemistry and Biology*, Topics in Current Chemistry, vol. 282 (ed. H. Krohn), 187–214. Heidelberg: Springer-Verlag.

393 Chung, Y.S., Dürr, B.F., Nanjappan, P., and Czarnik, A.W. (1988). Diene substituent effects on the rate of the retro-Diels–Alder reaction - cycloreversion reactivity varying over a range of 5 powers of 10. *J. Org. Chem.* 53 (6): 1334–1338.

394 Boger, D.L., Corbett, W.L., Curran, T.T., and Kasper, A.M. (1991). Inverse electron demand Diels–Alder reactions of *N*-sulfonyl α,β-unsaturated imines - a general approach to implementation of the 4-π participation of 1-aza-1,3-butadienes in Diels–Alder reactions. *J. Am. Chem. Soc.* 113 (5): 1713–1729.

395 Bunnage, M.E. and Nicolaou, K.C. (1997). The oxide anion accelerated retro-Diels–Alder reaction. *Chem. Eur. J.* 3 (2): 187–192.

396 Fukui, K. (1966). A MO-theoretical illumination for principle of stereoselection. *Bull. Chem. Soc. Jpn.* 39 (3): 498–503.

397 Fukui, K. (1971). Recognition of stereochemical paths by orbital interaction. *Acc. Chem. Res.* 4 (2): 57–64.

398 Fukui, K. (1982). The role of Frontier Orbitals in chemical reactions (Nobel lecture). *Angew. Chem. Inte. Ed. Engl.* 21 (11): 801–809.

399 Fukui, K. (1982). Role of Frontier Orbitals in chemical reactions. *Science* 218 (4574): 747–754.

400 Inagaki, S., Fujimoto, H., and Fukui, K. (1976). Orbital mixing rule. *J. Am. Chem. Soc.* 98 (14): 4054–4061.

401 Eisenstein, O., Lefour, J.M., Anh, N.T., and Hudson, R.F. (1977). Simple prediction of cycloaddition orientation. 1. Diels–Alder reactions. *Tetrahedron* 33 (5): 523–531.

402 Alston, P.V., Ottenbrite, R.M., and Shillady, D.D. (1973). Secondary orbital interactions determining regioselectivity in Diels–Alder reaction. *J. Org. Chem.* 38 (23): 4075–4077.

403 Alston, P.V., Ottenbrite, R.M., and Cohen, T. (1978). Secondary orbital interactions determining regioselectivity in Diels–Alder reaction. 3. Disubstituted dienes. *J. Org. Chem.* 43 (10): 1864–1867.

404 Cohen, T., Ruffner, R.J., Shull, D.W. et al. (1978). Secondary orbital interactions determining regioselectivity in Diels–Alder reaction. 4. Experimental and theoretical examination of reaction of acrylonitrile with 1-(phenylthio)-2-methoxy-1,3-butadiene - determination of conformations of 4 cyclohexene adducts by H-1 NMR-spectroscopy. *J. Org. Chem.* 43 (21): 4052–4057.

405 Kahn, S.D., Pau, C.F., Overman, L.E., and Hehre, W.J. (1986). Modeling chemical reactivity. 1. Regioselectivity of Diels–Alder cycloadditions of electron-rich dienes with electron-deficient dienophiles. *J. Am. Chem. Soc.* 108 (23): 7381–7396.

406 Ussing, B.R., Hang, C., and Singleton, D.A. (2006). Dynamic effects on the periselectivity, rate, isotope effects, and mechanism of cycloaddions of ketenes with cyclopentadiene. *J. Am. Chem. Soc.* 128 (23): 7594–7607.

407 Hammond, G.S. (1955). A correlation of reaction rates. *J. Am. Chem. Soc.* 77 (2): 334–338.

408 Yarnell, A. (2003). Hammond Postulate: 1955 paper used transition-state theory to explain structure-reactivity relationships. *Chem. Eng. News* 81 (20): 42.

409 Evans, D.A., Sims, C.L., and Bryan, C.A. (1972). Complementarity of (4+2)-cycloaddition reactions and (2,3)-sigmatropic rearrangements in synthesis - new synthesis of functionalized Hasubanan derivatives. *J. Am. Chem. Soc.* 94 (8): 2891–2892.

410 Sauer, J. and Sustmann, R. (1980). Mechanistic aspects of Diels–Alder reactions - a critical survey. *Angew. Chem. Int. Ed. Engl.* 19 (10): 779–807.

411 Harada, M., Asaba, K.N., Iwai, M. et al. (2012). Asymmetric total synthesis of an Iboga-type indole alkaloid, Voacangalactone, newly isolated from *Voacanga africana*. *Org. Lett.* 14 (22): 5800–5803.

412 Lutz, E.F. and Bailey, G.M. (1964). Regulation of structural isomerism in simple Diels–Alder adducts. *J. Am. Chem. Soc.* 86 (18): 3899–3901.

413 Singleton, D.A., Merrigan, S.R., Beno, B.R., and Houk, K.N. (1999). Isotope effects for Lewis acid catalyzed Diels–Alder reactions. The experimental transition state. *Tetrahedron Lett.* 40 (32): 5817–5821.

414 Fringuelli, F., Pizzo, F., Taticchi, A., and Wenkert, E. (1983). Diels–Alder reactions of cycloalkenones. 3. Effects of specific reaction parameters. *J. Org. Chem.* 48 (17): 2802–2808.

415 Fringuelli, F., Taticchi, A., and Wenkert, E. (1990). Diels–Alder reactions of cycloalkenones in organic synthesis - a review. *Org. Prep. Proced. Int.* 22 (2): 131–165.

416 Jung, M.E. and Gaede, B. (1979). Synthesis and Diels–Alder reactions of *E*-1-trimethylsilylbuta-1,3-diene. *Tetrahedron* 35 (5): 621–625.

417 Carter, M.J., Fleming, I., and Percival, A. (1981). The Diels–Alder route to allylsilanes from 1-trimethylsilylbutadienes. *J. Chem. Soc., Perkin Trans. 1* (9): 2415–2434.

418 Fu, Y., Liu, L., and Guo, Q.X. (2005). Extraordinary β-silyl effects on C—H, N—H, and O—H bond dissociation energies. *J. Mol. Struct. THEOCHEM* 725 (1–3): 1–4.

419 Zutterman, F. and Krief, A. (1983). Synthesis and Diels–Alder reactions of allylidenecyclopropane. *J. Org. Chem.* 48 (7): 1135–1137.

420 Koreeda, M. and Ciufolini, M.A. (1982). Natural product synthesis via allylsilanes. 1. Synthesis and reactions of (1*E*,3*E*)-4-acetoxy-1-(trimethylsilyl)-1,3-butadiene and

its use in the total synthesis of (+/-)-shikimic acid. *J. Am. Chem. Soc.* 104 (8): 2308–2310.

421 Laugraud, S., Guingant, A., Chassagnard, C., and D'Angelo, J. (1988). Regioselective synthesis of 2-(phenylthio)juglone and 3-(phenylthio)juglone derivatives. *J. Org. Chem.* 53 (7): 1557–1560.

422 Trost, B.M., Vladuchick, W.C., and Bridges, A.J. (1980). Sulfur as a regiochemical control element - cycloadditions of 2-alkoxy(acyloxy)-3-alkyl(aryl)thiobuta-1,3-dienes. *J. Am. Chem. Soc.* 102 (10): 3554–3572.

423 Trost, B.M., Vladuchick, W.C., and Bridges, A.J. (1980). Sulfur as a regiochemical control element - synthesis of 2-alkoxy(acyloxy)-3-alkyl(aryl)thiobuta-1,3-dienes. *J. Am. Chem. Soc.* 102: 3548–3554.

424 Jefferson, A., Nicovich, J.M., and Wine, P.H. (1994). Temperature-dependent kinetics studies of the reactions Br(P-2(3/2)) + CH_3SCH_3=CH_3SCH_2 + HBr - heat of formation of the CH_3SCH_2 radical. *J. Phys. Chem.* 98 (29): 7128–7135.

425 Bordwell, F.G. and Liu, W.Z. (1998). Effects of sulfenyl, sulfinyl and sulfonyl groups on acidities and homolytic bond dissociation energies of adjacent C—H and N—H bonds. *J. Phys. Org. Chem.* 11 (6): 397–406.

426 Birch, A.J. and Powell, V.H. (1970). Synthesis of some polycyclic quinones through 1-methoxycyclohexa-1,3-dienes. *Tetrahedron Lett.* (40): 3467–3470.

427 Kelly, T.R., Gillard, J.W., Goerner, R.N., and Lyding, J.M. (1977). Regiochemical control in Diels–Alder reactions of substituted naphthoquinones - model studies on a regiospecific approach to adriamycinone. *J. Am. Chem. Soc.* 99 (16): 5513–5514.

428 Muxfeldt, H. (1962). Synthese eines Terramycin-Bausteins. *Angew. Chem.* 74 (21): 825–828.

429 Kelly, T.R., Parekh, N.D., and Trachtenberg, E.N. (1982). Regiochemical control in the Diels–Alder reaction of substituted naphthoquinones - the directing effects of C-6 oxygen substituents. *J. Org. Chem.* 47 (25): 5009–5013.

430 Tamariz, J. and Vogel, P. (1984). A doubly-convergent and regioselective synthesis of (+/-)-daunomycinone. *Tetrahedron* 40 (22): 4549–4560.

431 Avenati, M., Carrupt, P.A., Quarroz, D., and Vogel, P. (1982). Interaction between non-conjugated chromophores. 14. Regioselectivity of the Diels–Alder additions of 2-substituted 5,6-dimethylidenenorbornanes and dimethylidenebicyclo[2.2.2]octanes. *Helv. Chim. Acta* 65 (1): 188–203.

432 Kraus, G.A. and Taschner, M.J. (1980). Timed Diels–Alder reactions. *J. Am. Chem. Soc.* 102 (6): 1974–1977.

433 Anderson, E.D. and Boger, D.L. (2011). Inverse electron demand Diels–Alder reactions of 1,2,3-triazines: pronounced substituent effects on reactivity and cycloaddition scope. *J. Am. Chem. Soc.* 133 (31): 12285–12292.

434 Hayashi, Y., Samanta, S., Gotoh, H., and Ishikawa, H. (2008). Asymmetric Diels–Alder reactions of α,β-unsaturated aldehydes catalyzed by a diarylprolinol silyl ether salt in the presence of water. *Angew. Chem. Int. Ed.* 47 (35): 6634–6637.

435 Alder, K. and Stein, G. (1934). Over the steric course of addition reactions and substitution reactions. I. Regarding the stereochemistry of the serving synthesis. *Liebigs Ann. Chem.* 514: 1–33.

436 Alder, K. and Stein, G. (1937). Examinations into the process of diene synthesis. *Angew. Chem.* 50: 0510–0519.

437 Alder, K. and Mölls, H.H. (1956). Über den Sterischen Verlauf der Dien-Synthese von Cycloheptadien-(1.3) mit Maleinsaure Anhydrid - Ein Beitrag zur Kenntnis Bicyclischer Systeme mit Brücken. *Chem. Ber.* 89 (8): 1960–1971.

438 Stephenson, L.M., Smith, D.E., and Current, S.P. (1982). *Endo* preference in the Diels–Alder cycloaddition of butadiene and maleic anhydride. *J. Org. Chem.* 47 (21): 4170–4171.

439 Breslow, R. and Guo, T. (1988). Diels–Alder reactions in nonaqueous polar solvents - kinetic effects of chaotropic and antichaotropic agents and of β-cyclodextrin. *J. Am. Chem. Soc.* 110 (17): 5613–5617.

440 (a) García, J.I., Mayoral, J.A., and Salvatella, L. (2004). The source of the *endo* rule in the Diels–Alder reaction: are secondary orbital interactions really necessary? *Eur.J. Org. Chem.* (1): 85–90. (b) Levandowski, B.J. and Houk, K.N. (2016). Hyperconjugative, secondary orbital, electrostatic, and steric effects on the reactivities and *endo* and *exo* stereoselectivities of cyclopropene Diels–Alder reactions. *J. Am. Chem. Soc.* 138 (51): 16731–16736.

441 Wassermann, A. (1965). *Diels–Alder Reactions. Organic Background and Physico-Chemical Aspects*, 1e, 1–109. Amsterdam: Elsevier Publ. Comp.

442 Woodward, R.B. and Baer, H. (1944). Studies on diene-addition reactions. II. The reaction of 6,6-pentamethylenefulvene with maleic anhydride. *J. Am. Chem. Soc.* 66: 645–649.

443 Inukai, T. and Kojima, T. (1966). Aluminum chloride catalyzed diene condensation. 2. Stronger adherence to Alder *endo* rule. *J. Org. Chem.* 31 (6): 2032–2033.

444 Sauer, J. and Kredel, J. (1966). Einfluss von Lewis–Säuren auf das *endo/exo*-Verhaltnis bei Diels–Alder additionen des cyclopentadiens. *Tetrahedron Lett.* 7 (7): 731–736.

445 Corey, E.J. and Ensley, H.E. (1975). Preparation of an optically active prostaglandin intermediate via asymmetric induction. *J. Am. Chem. Soc.* 97 (23): 6908–6909.

446 Corey, E.J., Albonico, S.M., Koellike, U. et al. (1971). Reagents for stereoselective carbonyl reduction - improved synthetic route to primary prostaglandins. *J. Am. Chem. Soc.* 93 (6): 1491–1492.

447 Corey, E.J., Weinshenker, N.M., Schaaf, T.K., and Huber, W. (1969). Stereo-controlled synthesis of prostaglandins $F_{2\alpha}$ and E_2. *J. Am. Chem. Soc.* 91 (20): 5675–5677.

448 Corey, E.J., Schaaf, T.K., Huber, W. et al. (1970). Total synthesis of prostaglandins $F_{2\alpha}$ and E_2 as naturally occurring forms. *J. Am. Chem. Soc.* 92 (2): 397–398.

449 Corey, E.J., Imwinkelried, R., Pikul, S., and Xiang, Y.B. (1989). Practical enantioselective Diels–Alder and aldol reactions using a new chiral controller system. *J. Am. Chem. Soc.* 111 (14): 5493–5495.

450 Corey, E.J. (2002). Catalytic enantioselective Diels–Alder reactions: methods, mechanistic fundamentals, pathways, and applications. *Angew. Chem. Int. Ed.* 41 (10): 1650–1667.

451 Hashimoto, S., Komeshima, N., and Koga, K. (1979). Asymmetric Diels–Alder reaction catalyzed by chiral alkoxyaluminium dichloride. *J. Chem. Soc., Chem. Commun.* (10): 437–438.

452 Takemura, H., Komeshima, N., Takahashi, I. et al. (1987). Stereochemical aspects of asymmetric Diels–Alder reaction catalyzed by chiral alkoxyaluminum dichlorides. *Tetrahedron Lett.* 28 (46): 5687–5690.

453 Bao, J.M., Wulff, W.D., and Rheingold, A.L. (1993). Vaulted biaryls as chiral ligands for asymmetric catalytic Diels–Alder reactions. *J. Am. Chem. Soc.* 115 (9): 3814–3815.

454 Kobayashi, S., Murakami, M., Harada, T., and Mukaiyama, T. (1991). The asymmetric Diels–Alder reaction of α,β-unsaturated aldehydes with dienes using a chiral boron reagent as a catalyst. *Chem. Lett.* (8): 1341–1344.

455 Evans, D.A., Miller, S.J., Lectka, T., and von Matt, P. (1999). Chiral bis(oxazoline)Cu(II) complexes as Lewis acid catalysts for the enantioselective Diels–Alder reaction. *J. Am. Chem. Soc.* 121 (33): 7559–7573.

456 Corey, E.J., Sarshar, S., and Lee, D.H. (1994). First example of a highly enantioselective catalytic Diels–Alder reaction of an achiral C_{2v}-symmetrical dienophile and an achiral diene. *J. Am. Chem. Soc.* 116 (26): 12089–12090.

457 Singh, R.P., Bartelson, K., Wang, Y. et al. (2008). Enantioselective Diels–Alder reaction of simple α,β-unsaturated ketones with a cinchona alkaloid catalyst. *J. Am. Chem. Soc.* 130 (8): 2422–2423.

458 Shen, J. and Tan, C.H. (2008). Brønsted-acid and Brønsted-base catalyzed Diels–Aider reactions. *Org. Biomol. Chem.* 6 (18): 3229–3236.

459 Payette, J.N. and Yamamoto, H. (2007). Regioselective and asymmetric Diels–Alder reaction of 1-and 2-substituted cyclopentadienes catalyzed by a Brønsted acid activated chiral oxazaborolidine. *J. Am. Chem. Soc.* 129 (31): 9536–9537.

460 Winstein, S., Shatavsky, M., Norton, C., and Woodward, R.B. (1955). 7-Norbornenyl and 7-norbornyl cations. *J. Am. Chem. Soc.* 77 (15): 4183–4184.

461 Ishida, M. and Inagaki, S. (2009). *Orbitals in Chemistry*, vol. 289 (ed. S. Inagaki), 183–218. Berlin, Heidelberg: Springer-Verlag.

462 Macaulay, J.B. and Fallis, A.G. (1990). Heteroatom-directed π-facial diastereoselection in Diels–Alder cycloadditions of plane-nonsymmetric cyclopentadienes. *J. Am. Chem. Soc.* 112 (3): 1136–1144.

463 Datta, S.C., Franck, R.W., Tripathy, R. et al. (1990). Effect of allylic substituents on the face selectivity of Diels–Alder reactions of semicyclic dienes. *J. Am. Chem. Soc.* 112 (23): 8472–8478.

464 (a) Cieplak, A.S. (1999). Inductive and resonance effects of substituents on π-face selection. *Chem. Rev.* 99 (5): 1265–1336.
(b) Levandowski, B.J., Zou, L.F., and Houk, K.N. (2016). Schleyer hyperconjugative aromaticity and Diels–Alder reactivity of 5-substituted cyclopentadienes. *J. Comput. Chem.* 37 (1): 117–123.

465 Magnus, P., Cairns, P.M., and Moursounidis, J. (1987). Trimethylsilyl accelerated retro-Diels–Alder reaction - a quantitative measure of the β-effect. *J. Am. Chem. Soc.* 109 (8): 2469–2471.

466 Watson, W.H., Galloy, J., Bartlett, P.D., and Roof, A.A.M. (1981). Double-bond deformation in two crystalline derivatives of *syn*-sesquinorbornene (Δ-4A,8A-octahydro-1,4,5,8-dimethanonaphthalene). *J. Am. Chem. Soc.* 103 (8): 2022–2031.

467 Sugimoto, T., Kobuke, Y., and Furukawa, J. (1976). *syn*-Selectivity in Diels–Alder reactions of isodicyclopentadiene. *J. Org. Chem.* 41 (8): 1457–1459.

468 Gleiter, R. and Paquette, L.A. (1983). σ,π-Interaction as a controlling factor in the stereoselectivity of addition reactions. *Acc. Chem. Res.* 16 (9): 328–334.

469 Brown, F.K. and Houk, K.N. (1985). Torsional and steric control of stereoselectivity in isodicyclopentadiene cycloadditions. *J. Am. Chem. Soc.* 107 (7): 1971–1978.

470 Hagenbuch, J.P., Vogel, P., Pinkerton, A.A., and Schwarzenbach, D. (1981). The π-anisotropy of the double-bond of a *syn*-11-oxasesquinorbornene derivative. (1) Stereoselectivity of the Diels–Alder additions of (2-norborneno)furan. (2) Crystal-structure of 11-oxa-*endo*-tetracyclo[6.2.1.^{13,6}.0^{2,7}]dodec-2(7)-ene-9,10-*exo*-dicarboxylic anhydride. *Helv. Chim. Acta* 64 (6): 1818–1832.

471 Carrupt, P.A., Berchier, F., and Vogel, P. (1985). Face selectivity in the Diels–Alder additions and chelotropic addition of sulfur dioxide to 2-(D)methylidene-3-methylidenebicyclo[2.2.1] heptane. *Helv. Chim. Acta* 68 (6): 1716–1729.

472 Inagaki, S. and Fukui, K. (1974). Origin and direction of nonequivalent orbital extension and stereochemical behaviors of plane-asymmetric olefins - *exo*-electrophilic addition to norbornene. *Chem. Lett.* (5): 509–514.

473 Bickelhaupt, F.M. (1999). Understanding reactivity with Kohn–Sham molecular orbital theory: E2–S_N2 mechanistic spectrum and other concepts. *J. Comput. Chem.* 20: 114–128.

474 (a) Ohwada, T. (2009). Orbital phase environments and stereoselectivities. In: *Orbitals in Chemistry*, vol. 289 (ed. S. Inagaki), 129–181. Heildelberg, Berlin: Springer-Verlag. (b) Liu, P.Y., Wu, Y.J., Pye, C.C., and Thornton, P.D. (2012). Facial selectivity in the Diels–Alder reactions of 2,2-disubstituted cyclopent-4-ene-1,3-dione derivatives and a computational examination of the facial selectivity of the Diels–Alder reactions of structurally related dienes and dienophiles. *Eur. J. Org. Chem.* (6): 1186–1194.

475 Kahn, S.D. and Hehre, W.J. (1987). Modeling chemical reactivity. 5. Facial selectivity in Diels–Alder cycloadditions. *J. Am. Chem. Soc.* 109 (3): 663–666.

476 Naperstkow, A.M., Macaulay, J.B., Newlands, M.J., and Fallis, A.G. (1989). π-Facial diastereoselectivity in Diels–Alder reactions of 2,5-dimethylthiophene oxide. *Tetrahedron Lett.* 30 (38): 5077–5080.

477 Treiber, A., Dansette, P.M., El Amri, H. et al. (1997). Chemical and biological oxidation of thiophene: preparation and complete characterization of thiophene *S*-oxide dimers and evidence for thiophene S-oxide as an intermediate in thiophene metabolism in vivo and in vitro. *J. Am. Chem. Soc.* 119 (7): 1565–1571.

478 Li, Y.Q., Thiemann, T., Sawada, T. et al. (1997). Lewis acid catalysis in the oxidative cycloaddition of thiophenes. *J. Org. Chem.* 62 (23): 7926–7936.

479 Furukawa, N., Zhang, S.Z., Horn, E. et al. (1998). X-ray crystallographic analysis of 2,5-bis(diphenylmethylsilyl)thiophene monooxide and the Diels–Alder reaction of thiophene monooxide with dienophiles. *Heterocycles* 47 (2): 793–809.

480 Takayama, J., Sugihara, Y., Takayanagi, T., and Nakayama, J. (2005). *syn*-π-Face- and *endo*-selective, inverse electron-demand Diels–Alder reactions of 3,4-di-*tert*-butylthiophene 1-oxide with electron-rich dienophiles. *Tetrahedron Lett.* 46 (24): 4165–4169.

481 Danishefsky, S. and Kitahara, T. (1974). Useful diene for Diels–Alder reaction. *J. Am. Chem. Soc.* 96 (25): 7807–7808.

482 Laclef, S., Exner, C.J., Turks, M. et al. (2009). Synthesis of (*E,Z*)-1-alkoxy-3-acyloxy-2-methylpenta-1,3-dienes via Danishefsky-type dienes or *O*-acylation of enones. *J. Org. Chem.* 74 (22): 8882–8885.

483 Savard, J. and Brassard, P. (1979). Regiospecific syntheses of quinones using vinylketene acetals derived from unsaturated esters. *Tetrahedron Lett.* (51): 4911–4914.

484 Huang, Y. and Rawal, V.H. (2000). Hetero-Diels–Alder reactions of 1-amino-3-siloxy-1,3-butadienes under strictly thermal conditions. *Org. Lett.* 2 (21): 3321–3323.

485 Dane, E. and Schmitt, J. (1938). Syntheses in the hydromatic range. IV. (1) The condensation of 1-methylcyclopentenedione-2,3 with 1-vinyl-6-methoxy-3,4-dihydro-naphthalene. (2) Presentation and diene syntheses of 3-oxy-2,6-dimethyl-benzoquinone-(1,4). *Liebigs Ann. Chem.* 536: 196–203.

486 Kumar, A. (2001). Salt effects on Diels–Alder reaction kinetics. *Chem. Rev.* 101 (1): 1–19.

487 Tiwari, S. and Kumar, A. (2006). Diels–Alder reactions are faster in water than in ionic liquids at room temperature. *Angew. Chem. Int. Ed.* 45 (29): 4824–4825.

488 Mccarrick, M.A., Wu, Y.D., and Houk, K.N. (1992). *exo*-Lone-pair effect on hetero-Diels–Alder cycloaddition stereochemistry. *J. Am. Chem. Soc.* 114 (4): 1499–1500.

489 Jørgensen, K.A. (2004). Hetero-Diels–Alder reactions of ketones - a challenge for chemists. *Eur. J. Org. Chem.* (10): 2093–2102.

490 Danishefsky, S.J., Larson, E., Askin, D., and Kato, N. (1985). On the scope, mechanism, and stereochemistry of the Lewis acid-catalyzed cyclocondensation of activated dienes with aldehydes - an application to the erythronolide problem. *J. Am. Chem. Soc.* 107 (5): 1246–1255.

491 Maruoka, K., Itoh, T., Shirasaka, T., and Yamamoto, H. (1988). Asymmetric hetero-Diels–Alder reaction catalyzed by chiral organo-aluminum reagent. *J. Am. Chem. Soc.* 110 (1): 310–312.

492 Thompson, C.F., Jamison, T.F., and Jacobsen, E.N. (2001). FR901464: total synthesis, proof of structure, and evaluation of synthetic analogues. *J. Am. Chem. Soc.* 123 (41): 9974–9983.

493 Du, H.F., Zhang, X., Wang, Z., and Ding, K.L. (2005). One catalyst for two distinct reactions: sequential asymmetric hetero Diels–Alder reaction and diethylzinc addition. *Tetrahedron* 61 (40): 9465–9477.

494 Berkessel, A., Erturk, E., and Laporte, C. (2006). Chiral Cr(III) porphyrins as highly enantioselective catalysts for hetero-Diels–Alder reactions between aldehydes and dienes. *Adv. Synth. Catal.* 348 (1–2): 223–228.

495 Watanabe, Y., Washio, T., Shimada, N. et al. (2009). Highly enantioselective hetero-Diels–Alder reactions between Rawal's diene and aldehydes catalyzed by chiral dirhodium(II) carboxamidates. *J. Chem. Soc., Chem. Commun.* (47): 7294–7296.

496 Du, H., Zhang, X., Wang, Z. et al. (2008). BINOLate - magnesium catalysts for enantioselective hetero-Diels–Alder reaction of Danishefsky's diene with aldehydes. *Eur. J. Org. Chem.* (13): 2248–2254.

497 Miesowicz, S., Chaladaj, W., and Jurczak, J. (2010). Oxo-Diels–Alder reaction of Danishefsky's diene with aldehydes, catalyzed by chiral tridentate Cr(III)-Schiff base complexes. *Synlett* (9): 1421–1425.

498 Lin, L., Kuang, Y., Liu, X., and Feng, X. (2011). Indium(III)-catalyzed asymmetric hetero-Diels–Alder reaction of Brassard-type diene with aliphatic aldehydes. *Org. Lett.* 13 (15): 3868–3871.

499 Giguere, D., Martel, J., Shiao, T.C., and Roy, R. (2011). Enantioselective de novo synthesis of 4-deoxy-D-hexopyranoses via hetero-Diels–Alder cycloadditions: total synthesis of ezoaminuroic acid-l and neosidomycin. *J. Org. Chem.* 76 (23): 9687–9698.

500 Kozmin, S.A., Janey, J.M., and Rawal, V.H. (1999). 1-Amino-3-siloxy-1,3-butadienes: highly reactive dienes for the Diels–Alder reaction. *J. Org. Chem.* 64 (9): 3039–3052.

501 Kozmin, S.A. and Rawal, V.H. (1999). Chiral amino siloxy dienes in the Diels–Alder reaction: applications to the asymmetric synthesis of 4-substituted and 4,5-disubstituted cyclohexenones and the total synthesis of (−)-α-elemene. *J. Am. Chem. Soc.* 121 (41): 9562–9573.

502 Huang, Y. and Rawal, V.H. (2002). Hydrogen-bond-promoted hetero-Diels–Alder reactions of unactivated ketones. *J. Am. Chem. Soc.* 124 (33): 9662–9663.

503 Gordillo, R., Dudding, T., Anderson, C.D., and Houk, K.N. (2007). Hydrogen bonding catalysis operates by charge stabilization in highly polar Diels–Alder reactions. *Org. Lett.* 9 (3): 501–503.

504 Huang, Y., Unni, A.K., Thadani, A.N., and Rawal, V.H. (2003). Hydrogen bonding: single enantiomers from a chiral-alcohol catalyst. *Nature* 424 (6945): 146–146.

505 Unni, A.K., Takenaka, N., Yamamoto, H., and Rawal, V.H. (2005). Axially chiral biaryl diols catalyze highly enantioselective hetero-Diels–Alder reactions through hydrogen bonding. *J. Am. Chem. Soc.* 127 (5): 1336–1337.

506 Rajaram, S. and Sigman, M.S. (2005). Design of hydrogen bond catalysts based on a modular oxazoline template: application to an enantioselective hetero Diels–Alder reaction. *Org. Lett.* 7 (24): 5473–5475.

507 Orlova, G. and Goddard, J.D. (2001). Competition between diradical stepwise and concerted

mechanisms in chalcogeno-Diels–Alder reactions: a density functional study. *J. Org. Chem.* 66 (11): 4026–4035.
508 Kerwin, J.F. and Danishefsky, S. (1982). On the Lewis acid-catalyzed cyclocondensation of imines with a siloxydiene. *Tetrahedron Lett.* 23 (37): 3739–3742.
509 Danishefsky, S. and Kerwin, J.F. (1982). On the Lewis acid-catalyzed cyclocondensation of silyloxy dienes with α,β-unsaturated aldehydes. *J. Org. Chem.* 47 (16): 3183–3184.
510 Danishefsky, S., Langer, M.E., and Vogel, C. (1985). On the use of the imine-diene cyclocondensation reaction in the synthesis of Yohimbine congeners. *Tetrahedron Lett.* 26 (48): 5983–5986.
511 Sunden, H., Ibrahem, I., Eriksson, L., and Cordova, A. (2005). Direct catalytic enantioselective aza-Diels–Alder reactions. *Angew. Chem. Int. Ed.* 44 (31): 4877–4880.
512 Kobayashi, S., Komiyama, S., and Ishitani, H. (1998). The first enantioselective aza-Diels–Alder reactions of imino dienophiles on use of a chiral zirconium catalyst. *Angew. Chem. Int. Ed. Engl.* 37 (7): 979–981.
513 Itoh, J., Fuchibe, K., and Akiyama, T. (2006). Chiral Brønsted acid catalyzed enantioselective aza-Diels–Alder reaction of Brassard's diene with imines. *Angew. Chem. Int. Ed.* 45 (29): 4796–4798.
514 Shang, D., Xin, J., Liu, Y. et al. (2008). Enantioselective aza-Diels–Alder reaction of aldimines with "Danishefsky-type diene" catalyzed by chiral Sc(III)-*N,N'*-dioxide complexes. *J. Org. Chem.* 73 (2): 630–637.
515 Jurcik, V., Arai, K., Salter, M. et al. (2008). Niobium-catalyzed highly enantioselective aza-Diels–Alder reactions. *Adv. Synth. Catal.* 350 (5): 647–651.
516 Weinreb, S.M. (1988). Synthetic methodology based upon *N*-sulfinyl dienophile (4+2)-cycloaddition reactions. *Acc. Chem. Res.* 21 (8): 313–318.
517 Bayer, A., Endeshaw, M.M., and Gautun, O.R. (2004). Asymmetric hetero-Diels–Alder reactions of *N*-sulfinyl dienophiles using chiral bis(oxazoline)Cu(II) and -Zn(II) triflates. *J. Org. Chem.* 69 (21): 7198–7205.
518 Gademann, K., Chavez, D.E., and Jacobsen, E.N. (2002). Highly enantioselective inverse-electron-demand hetero-Diels–Alder reactions of α,β-unsaturated aldehydes. *Angew. Chem. Int. Ed.* 41 (16): 3059–3061.
519 He, M., Uc, G.J., and Bode, J.W. (2006). Chiral N-heterocyclic carbene catalyzed, enantioselective oxodiene Diels–Alder reactions with low catalyst loadings. *J. Am. Chem. Soc.* 128 (47): 15088–15089.
520 Evans, D.A., Olhava, E.J., Johnson, J.S., and Janey, J.M. (1998). Chiral C_2-symmetric Cu(II) complexes as catalysts for enantioselective hetero-Diels–Alder reactions. *Angew. Chem. Int. Ed. Engl.* 37 (24): 3372–3375.
521 Evans, D.A., Johnson, J.S., and Olhava, E.J. (2000). Enantioselective synthesis of dihydropyrans. Catalysis of hetero-Diels–Alder reactions by bis(oxazoline)Cu(II) complexes. *J. Am. Chem. Soc.* 122 (8): 1635–1649.
522 Fang, X., Chen, X., and Chi, Y.R. (2011). Enantioselective Diels–Alder reactions of enals and alkylidene diketones catalyzed by N-heterocyclic carbenes. *Org. Lett.* 13 (17): 4708–4711.
523 Buonora, P., Olsen, J.C., and Oh, T. (2001). Recent developments in imino Diels–Alder reactions. *Tetrahedron* 57 (29): 6099–6138.
524 Esquivias, J., Gómez Arrayás, R., and Carretero, J.C. (2007). Catalytic asymmetric inverse-electron-demand Diels–Alder reaction of *N*-sulfonyl-1-aza-1,3-dienes. *J. Am. Chem. Soc.* 129 (6): 1480–1481.
525 Esquivias, J., Alonso, I., Gómez Arrayás, R., and Carretero, J.C. (2009). Inverse-electron-demand Diels–Alder reactions of *N*-(heteroarylsulfonyl)-1-aza-1,3-dienes catalyzed by chiral Lewis acids. *Synthesis* (1): 113–126.
526 He, M., Struble, J.R., and Bode, J.W. (2006). Highly enantioselective azadiene Diels–Alder reactions catalyzed by chiral N-heterocyclic carbenes. *J. Am. Chem. Soc.* 128 (26): 8418–8420.
527 Glushkov, V. and Tolstikov, A. (2008). Synthesis of substituted 1,2,3,4-tetrahydroquinolines by the Povarov reaction. New potential of a classical reaction. *Usp. Khim.* 77 (2): 138–160.
528 Smith, C.D., Gavrilyuk, J.I., Lough, A.J., and Batey, R.A. (2010). Lewis acid catalyzed three-component hetero-Diels–Alder (Povarov) reaction of *N*-arylimines with strained norbornene-derived dienophiles. *J. Org. Chem.* 75 (3): 702–715.
529 Wichterle, O. (1947). Addition of nitroso compound to a conjugated system. *Collect. Czech. Chem. Commun.* 12: 292–304.
530 Kresze, G. and Firl, J. (1965). Orientierung und Mechanismus bei der Diensynthese von Nitrosoverbindungen (Additionsreaktionen

der Nitrosogruppe. 8). *Tetrahedron Lett.* (17): 1163–1170.

531 Leach, A.G. and Houk, K.N. (2001). Transition states and mechanisms of the hetero-Diels–Alder reactions of hyponitrous acid, nitrosoalkanes, nitrosoarenes, and nitrosocarbonyl compounds. *J. Org. Chem.* 66 (15): 5192–5200.

532 Leach, A.G. and Houk, K.N. (2002). Diels–Alder and ene reactions of singlet oxygen, nitroso compounds and triazolinediones: transition states and mechanisms from contemporary theory. *J. Chem. Soc., Chem. Commun.* (12): 1243–1255.

533 Kresze, G. and Firl, J. (1968). Addition reactions of nitroso group. 9. Diels–Alder reaction of 1-mono- or of 1,4-disubstituted dienes with nitroso compounds. *Tetrahedron* 24 (3): 1043–1050.

534 Vogt, P.F. and Miller, M.J. (1998). Development and applications of amino acid derived chiral acylnitroso hetero Diels–Alder reactions. *Tetrahedron* 54 (8): 1317–1348.

535 Felber, H., Kresze, G., Prewo, R., and Vasella, A. (1986). Diastereoselectivity and reactivity in the Diels–Alder reactions of α-chloronitroso ethers. *Helv. Chim. Acta* 69 (5): 1137–1146.

536 (a) Yamamoto, Y. and Yamamoto, H. (2004). Catalytic, highly enantio, and diastereoselective nitroso Diels–Alder reaction. *J. Am. Chem. Soc.* 126 (13): 4128–4129. (b) Pous, J., Courant, T., Bernadat, G. et al. (2015). Regio-, diastereo-, and enantioselective nitroso-Diels–Alder reaction of 1,3-diene-1-carbamates catalyzed by chiral phosphoric acids. *J. Am. Chem. Soc.* 137 (37): 11950–11953.

537 Wichterle, O. and Rocek, J. (1953). Adice Dienu Na Thionylaminy Priprava Heterocyklickych Sloucenin 1,2-Thiazinove Rady. *Chem. Listy* 47 (12): 1768–1780.

538 Macaluso, A. and Hamer, J. (1966). 1,4-Cycloaddition reaction between *N*-sulfinylbenzenesulfonamide and 1,3-cyclopentadiene. An equilibrium study. *J. Org. Chem.* 31 (9): 3049–3050.

539 de Bruin, G. (1914). A crystallized compound of isoprene with sulphur dioxide. In: *Proceedings of the Koninklijke Akademie Van Wetenschappen Te Amsterdam*, vol. 17, 585–586.

540 Heldeweg, R.F. and Hogeveen, H. (1976). (2+4)($\pi+\pi,\pi$) and (2+4)(n+π-π) Modes of addition in reaction between SO_2 and a diene - kinetic vs thermodynamic control. *J. Am. Chem. Soc.* 98 (8): 2341–2342.

541 Durst, T. and Tetreault-Ryan, L. (1978). Reaction of *ortho*-quinodimethane with sulfur dioxide competition between $\pi + \pi$-π and n + π-π cycloadditions. *Tetrahedron Lett.* (26): 2353–2354.

542 Markovic, D., Roversi, E., Scoppelliti, R. et al. (2003). The hetero-Diels–Alder addition of sulfur dioxide: the pseudo-chair conformation of a 4,5-dialkylsultine. *Chem. Eur. J.* 9 (20): 4911–4915.

543 Fernández, T., Sordo, J.A., Monnat, F. et al. (1998). Sulfur dioxide promotes its hetero-Diels–Alder and cheletropic additions to 1,2-dimethylidenecyclohexane. *J. Am. Chem. Soc.* 120 (50): 13276–13277.

544 Exner, C.J., Turks, M., Fonquerne, F., and Vogel, P. (2011). Concise synthesis of complicated polypropionates through one-pot disymmetrical two-directional chain elongation. *Chem. Eur. J.* 17: 4246–4253.

545 Turks, M., Exner, C.J., Hamel, C., and Vogel, P. (2009). Umpolung with sulfur dioxide: carbon-carbon cross-coupling of electron-rich 1,3-dienes and alkenes; application to the enantioselective synthesis of long-chain polyketide fragments. *Synthesis* (7): 1065–1074.

546 Laclef, S., Turks, M., and Vogel, P. (2010). Total synthesis and determination of the absolute configuration of (−)-dolabriferol. *Angew. Chem. Int. Ed.* 49 (45): 8525–8527.

547 Turks, M., Murcía, M.C., Scopelliti, R., and Vogel, P. (2004). First asymmetric synthesis of the cyclohexanone subunit of baconipyrones A and B. Revision of its structure. *Org. Lett.* 6 (18): 3031–3034.

548 Turks, M., Huang, X.G., and Vogel, P. (2005). Expeditious asymmetric synthesis of a stereoheptad corresponding to the C(19)-C(27)-ansa chain of Rifamycins: formal total synthesis of Rifamycin S. *Chem. Eur. J.* 11 (2): 465–476.

549 Craita, C., Didier, C., and Vogel, P. (2007). Short synthesis of the C_{16}–C_{28} polyketide fragment of apoptolidin A aglycone. *J. Chem. Soc., Chem. Commun.* (23): 2411–2413.

550 Turks, M., Fairweather, K.A., Scopelliti, R., and Vogel, P. (2011). Efficient asymmetric synthesis of long-chain polyketides containing up to ten contiguous stereogenic centres by double-chain elongation. *Eur. J. Org. Chem.* (18): 3317–3328.

551 Geisler, A. and Würthwein, E.U. (1994). Diels–Alder reaction and ene-reaction of 2-azaallenium salts. *Tetrahedron Lett.* 35 (1): 77–80.

552 Dubey, S.K. and Knaus, E.E. (1985). Some regiochemical and stereochemical aspects of the Diels–Alder reaction of nitrosocarbonyl compounds with *N*-substituted 1,2-dihydropyridines. *J. Org. Chem.* 50 (12): 2080–2086.

553 Padwa, A. and Pearson, W.H. (eds.) (2002). Synthetic Applications of 1,3-Dipolar Cycloaddition Chemistry Toward Heterocycles and Natural Products., 1–952. Chichester: Wiley.

554 (a) Nájera, C. and Sansano, J.M. (2003). Azomethine ylides in organic synthesis. *Curr. Org. Chem.* 7 (11): 1105–1150. (b) Adrio, J. and Carretero, J.C. (2014). Recent advances in the catalytic asymmetric 1,3-dipolar cycloaddition of azomethine ylides. *J. Chem. Soc., Chem. Commun.* 50 (83): 12434–12446. (c) Feng, J.-J. and Zhang, J. (2016). Synthesis of unsaturated N-heterocycles by cycloadditions of aziridines and alkynes. *ACS Catal.* 6 (10): 6651–6661. (d) Zhang, X., Yuan, C., Zhang, C. et al. (2016). Thermal (3+2)-cycloaddition of *N*-iminoquiazoline ylides with allenoates for synthesis of tetrahydropyrazoloquinazoline derivatives. *Tetrahedron* 72 (50): 82745–88281. (e) Döndas, H.A., de Gracia Retamosa, M., and Sansano, J.M. (2017). Current trends towards the synthesis of bioactive heterocycles and natural products using 1,3-dipolar cycloadditions (1,3-DC) with azomethine ylides. *Synthesis* 49 (13): 2819–2851. (f) Otero-Fraga, J., Suárez-Pantiga, S., Montesinos-Magraner, M., and Mendoza, A. (2017). Direct and stereospecific (3+2) synthesis of pyrrolidines from simple unactivated alkenes. *Angew. Chem. Int. Ed.* 56 (42): 12962–12966.

555 Büchner, E. (1888). Einwirkung von Diazoessigäther au die Aether ungesättigter Säuren. *Ber. Dtsch. Chem. Ges.* 21: 2637–2647.

556 Michael, A. (1893). Über die Einwirkung von Diazobenzolimid auf Acetylenedicarbonsäuremethylester. *J. Prakt. Chem.* (48): 2, 94–95.

557 Pechmann, H.v. (1898). Pyrazol aus Acetylen und Diazomethan. *Ber. Dtsch. Chem. Ges.* 31 (3): 2950–2951.

558 Huisgen, R. (1963). 1,3-Dipolar Cycloadditions. Past and Future. *Angew. Chem. Int. Ed. Engl.* 2 (10): 565–598.

559 (a) Rüchhardt, C., Sauer, J., and Sustmann, R. (2005). "Rolf Huisgen": some highlights of his contributions to organic chemistry. *Helv. Chim. Acta* 88 (8): 1154–1184. (b) Houk, K.N. (2010). Rolf Huisgen's profound adventures in chemistry. *Helv. Chim. Acta* 93 (7): 1241–1260.

560 Wentrup, C. and Kambouris, P. (1991). N-sulfides - dinitrogen sulfide, thiofulminic acid, and nitrile sulfides. *Chem. Rev.* 91 (3): 363–373.

561 Bender, H., Carnovale, F., Peel, J.B., and Wentrup, C. (1988). Dinitrogen sulfide, N_2S, revealed by photoelectron-spectroscopy. *J. Am. Chem. Soc.* 110 (11): 3458–3461.

562 Asinger, F. and Gentz, F. (1965). Zur Kenntnis des 2-Methyl-2,4-Dineopentyl-Imidazolin-3-thions-(5) aus Methyl-Neopentylketon (Über die Gemeinsame Einwirkung von Elementarem Schwefel und Gasformigem Ammoniak auf Ketone. 51). *Monatsh. Chem. Verw. Teile Anderer Wiss.* 96 (5): 1474–1484.

563 Huisgen, R., Fulka, C., Kalwinsch, I. et al. (1984). Recent developments of the chemistry of thiocarbonyl ylides. *Bull. Soc. Chim. Belg.* 93 (7): 511–532.

564 Mloston, G. and Heimgartner, H. (2000). Generation and typical reactions of thiocarbonyl ylides. *Pol. J. Chem.* 74 (11): 1503–1532.

565 Mloston, G. and Heimgartner, H. (2002). Thiocarbonyl ylides. In: *Synthetic Applications of 1,3-Dipolar Cycloaddition Chemistry Toward Heterocycles and Natural Products*, vol. 59 (ed. A. Padwa and W.H. Pearson), 315–360. New York: Wiley.

566 Zwanenburg, B. (1982). The chemistry of sulfines. *Recl. Trav. Chim. Pays-Bas* 101 (1): 1–27.

567 Block, E. (1992). The organosulfur chemistry of the genus *Allium* - implications for the organic chemistry of sulfur. *Angew. Chem. Int. Ed. Engl.* 31 (9): 1135–1178.

568 Huisgen, R., Mloston, G., and Polborn, K. (1996). 1,3-Dipolar activity in cycloadditions of an aliphatic sulfine. *J. Org. Chem.* 61 (19): 6570–6574.

569 Huisgen, R., Mloston, G., Polborn, K. et al. (1997). Cycloadditions of alicyclic thione *S*-oxides with alicyclic thiones. *Liebigs Ann. Chem.* (1): 179–185.

570 Huisgen, R. and Rapp, J. (1987). Thiocarbonyl *S*-sulfides, A new class of 1,3-dipoles. *J. Am. Chem. Soc.* 109 (3): 902–903.

571 Okuma, K., Shigetomi, T., Shibata, S., and Shioji, K. (2004). Synthesis of 1,2,4-trithiolanes from thione *S*-oxides and Lawesson reagent at room temperature. *Bull. Chem. Soc. Jpn.* 77 (1): 187–188.

572 Lan, Y., Zou, L., Cao, Y., and Houk, K. (2011). Computational methods to calculate accurate activation and reaction energies of 1,3-dipolar cycloadditions of 24 1,3-dipoles. *J. Phys. Chem. A* 115 (47): 13906–13920.

573 Senning, A. (1991). *N,N′*-Di-*tert*-butylsulfur diimides. *Sulfur Lett.* 13 (97): 100.

574 Huisgen, R. and Peng, X. (1986). 1,3-Cycloadditions of a thionitroso *S*-sulfide. *Tetrahedron Lett.* 27 (50): 6063–6066.

575 Lan, Y., Wheeler, S.E., and Houk, K. (2011). Extraordinary difference in reactivity of ozone (OOO) and sulfur dioxide (OSO): a theoretical study. *J. Chem. Theory Comput.* 7 (7): 2104–2111.

576 Dodson, R.M., Srinivas, V., Sauers, R.F., and Sharma, K.S. (1972). Disulfur monoxide - reaction with dienes. *J. Org. Chem.* 37 (15): 2367–2372.

577 Fulcher, B.C., Hunter, M.L., and Welker, M.E. (1993). A simple synthesis of the S_2O source, 4,5-diphenyl-1,2-dithiin-1-oxide. *Synth. Commun.* 23 (2): 217–222.

578 (a) Schenk, P.W. and Steudel, R. (1964). Disulfurmonoxide. *Angew. Chem.* 76 (2): 97–97. (b) Ishii, A., Kawai, T., Tekura, K. et al. (2001). A convenient method for the generation of a disulfur monoxide equivalent and its reaction with diazoalkanes to yield dithiirane 1-oxides. *Angew. Chem. Int. Ed.* 40 (10): 1924–1926.

579 Nakayama, J., Aoki, S., Takayama, J. et al. (2004). Reversible disulfur monoxide (S_2O)-forming retro-Diels–Alder reaction. Disproportionation of S_2O to trithio-ozone (S_3) and sulfur dioxide (SO_2) and reactivities of S_2O and S_3. *J. Am. Chem. Soc.* 126 (29): 9085–9093.

580 Padwa, A., Burgess, E.M., Gingrich, H.L., and Roush, D.M. (1982). On the problem of regioselectivity in the 1,3-dipolar cycloaddition reaction of munchnones and sydnones with acetylenic dipolarophiles. *J. Org. Chem.* 47 (5): 786–791.

581 Padwa, A., Sheehan, S.M., and Straub, C.S. (1999). An isomunchnone-based method for the synthesis of highly substituted 2(1*H*)-pyridones. *J. Org. Chem.* 64 (23): 8648–8659.

582 Dennis, N., Katritzky, A.R., and Takeuchi, Y. (1976). 1,3-Dipolar character of 6-membered aromatic rings. 17. Synthetic applications of heteroaromatic betaines with 6-membered rings. *Angew. Chem. Int. Ed. Engl.* 15 (1): 1–9.

583 Katritzky, A.R. and Dennis, N. (1989). Cycloaddition reactions of heteroaromatic 6-membered rings. *Chem. Rev.* 89 (4): 827–861.

584 Sammes, P.G. (1986). Recent studies on 3-oxidopyrylium and its derivatives. *Gazz. Chim. Ital.* 116 (3): 109–114.

585 Radhakrishnan, K.V. (2008). Heterocycles via pyrylium and pyridinium betaines. *Top. Heterocycl. Chem.* 13: 71–98.

586 Singh, V., Krishna, U.M., Vikrant, and Trivedi, G.K. (2008). Cycloaddition of oxidopyrylium species in organic synthesis. *Tetrahedron* 64 (16): 3405–3428.

587 Huisgen, R. (1963). 1.3-Dipolare Cycloadditionen - Ruckschau und Ausblick. *Angew. Chem.* 75 (13): 604–637.

588 Huisgen, R. (1968). On mechanism of 1,3-dipolar cycloadditions. A reply. *J. Org. Chem.* 33 (6): 2291–2297.

589 Mcdouall, J.J.W., Robb, M.A., Niazi, U. et al. (1987). An MO-SCF study of the mechanisms for 1,3-dipolar cycloadditions. *J. Am. Chem. Soc.* 109 (15): 4642–4648.

590 Bihlmaier, W., Geittner, J., Huisgen, R., and Reissig, H.U. (1978). Stereospecificity of diazomethane cycloadditions. *Heterocycles* 10: 147–152.

591 Firestone, R.A. (1968). On mechanism of 1,3-dipolar cycloadditions. *J. Org. Chem.* 33 (6): 2285–2290.

592 Huisgen, R., Mloston, G., and Langhals, E. (1986). The 1st two-step 1,3-dipolar cycloadditions - interception of an intermediate. *J. Org. Chem.* 51 (21): 4085–4087.

593 Huisgen, R., Mloston, G., and Langhals, E. (1986). 1st two-step 1,3-dipolar cycloadditions - nonstereospecificity. *J. Am. Chem. Soc.* 108 (20): 6401–6402.

594 Huisgen, R., Giera, H., and Polborn, K. (2005). 2,2,6,6-Tetramethylcyclohexanethione *S*-methylide, a highly hindered thiocarbonyl ylide: two-step cycloadditions. *Tetrahedron* 61 (25): 6143–6153.

595 Karadakov, P.B., Cooper, D.L., and Gerratt, J. (1998). Modern valence-bond description of chemical reaction mechanisms: the 1,3-dipolar addition of fulminic acid to ethyne. *Theor. Chem. Acc.* 100 (1–4): 222–229.

596 Nguyen, M.T., Chandra, A.K., Sakai, S., and Morokuma, K. (1999). Another look at the mechanism of the concerted 1,3-dipolar cycloaddition of fulminic acid to acetylene. *J. Org. Chem.* 64 (1): 65–69.

597 Xu, L., Doubleday, C.E., and Houk, K. (2009). Dynamics of 1,3-dipolar cycloaddition reactions of diazonium betaines to acetylene and

ethylene: bending vibrations facilitate reaction. *Angew. Chem. Int. Ed.* 48 (15): 2746–2748.
598 Xu, L., Doubleday, C.E., and Houk, K. (2010). Dynamics of 1,3-dipolar cycloadditions: energy partitioning of reactants and quantitation of synchronicity. *J. Am. Chem. Soc.* 132 (9): 3029–3037.
599 (a) Kavitha, K. and Venuvanalingam, P. (2002). 1,3-Dipolar additions involving allenes: a density functional study of concerted and stepwise mechanisms. *J. Chem. Soc., Perkin Trans.* 2 (12): 2130–2139. (b) Lan, Y. and Houk, K. (2010). Mechanism and stereoselectivity of the stepwise 1,3-dipolar cycloadditions between a thiocarbonyl ylide and electron-deficient dipolarophiles: a computational investigation. *J. Am. Chem. Soc.* 132 (50): 17921–17927. (c) Siadati, S.A. (2016). Beyond the alternatives that switch the mechanism of the 1,3-dipolar cycloadditions from concerted to stepwise or vice versa: a literature review. *Prog. React. Kinet. Mech.* 41 (4): 331–344.
600 Houk, K.N., Firestone, R.A., Munchausen, L.L. et al. (1985). Stereospecificity of 1,3-dipolar cycloadditions of *para*-nitrobenzonitrile oxide to *cis*-dideuterioethylene and *trans*-dideuterioethylene. *J. Am. Chem. Soc.* 107 (24): 7227–7228.
601 Houk, K.N. (1975). The Frontier molecular orbital theory of cycloaddition reactions. *Acc. Chem. Res.* 11: 361–369.
602 Fleming, I. (1976). *Frontier Orbitals and Organic Chemical Reactions*, 24–109. London: Wiley.
603 Sustmann, R. and Trill, H. (1972). Substituent effects in 1,3-dipolar cycloadditions of phenyl azide. *Angew. Chem. Int. Ed. Engl.* 11 (9): 838–840.
604 Sustmann, R., Wenning, E., and Huisgen, R. (1977). Perturbational analysis of 1,3-dipolar cycloaddition reactions of diazomethane. *Tetrahedron Lett.* 18 (10): 877–880.
605 Geittner, J., Huisgen, R., and Sustmann, R. (1977). Kinetics of 1,3-dipolar cycloaddition reactions of diazomethane - correlation with HOMO-LUMO energies. *Tetrahedron Lett.* 18 (10): 881–884.
606 Jikyo, T., Schatz, J., and Maas, G. (2003). Reactivity of heterophospholes toward 1,3-dipolar cycloaddition of diazo compounds - an FMO analysis. *J. Phys. Org. Chem.* 16 (8): 504–512.
607 Benchouk, W. and Mekelleche, S. (2008). Theoretical study of the mechanism and regioselectivity of the 1,3-dipolar cycloaddition of diazomethane with methyl acrylate using theoretical approaches. *J. Mol. Struct. THEOCHEM* 862 (1–3): 1–6.
608 Ess, D.H. and Houk, K. (2008). Theory of 1,3-dipolar cycloadditions: distortion/interaction and frontier molecular orbital models. *J. Am. Chem. Soc.* 130 (31): 10187–10198.
609 Harcourt, R.D. (1972). Valence formulas and 1,3-dipolar cycloaddition reactions. *J. Mol. Struct.* 12 (3): 351–366.
610 Harcourt, R.D. (1978). Re-statement of concerted-diradical mechanism for 1,3-dipolar cycloaddition reactions. *Tetrahedron* 34 (20): 3125–3127.
611 Harcourt, R.D. (2001). Reply to comment "Electronic reorganization in 1,3-dipolar cycloaddition of fulminic acid to acetylene". *J. Phys. Chem. A* 105 (48): 10947–10948.
612 Braida, B., Walter, C., Engels, B., and Hiberty, P.C. (2010). A clear correlation between the diradical character of 1,3-dipoles and their reactivity toward ethylene or acetylene. *J. Am. Chem. Soc.* 132 (22): 7631–7637.
613 Simonsen, K.B., Bayon, P., Hazell, R.G. et al. (1999). Catalytic enantioselective inverse-electron demand 1,3-dipolar cycloaddition reactions of nitrones with alkenes. *J. Am. Chem. Soc.* 121 (16): 3845–3853.
614 Karlsson, S. and Hogberg, H.E. (2001). Asymmetric 1,3-dipolar cycloadditions for the construction of enantiomerically pure heterocycles. A review. *Org. Prep. Proced. Int.* 33 (2–3): 105–172.
615 Huang, Z.Z., Kang, Y.B., Zhou, J. et al. (2004). Diastereoselectivity-switchable and highly enantioselective 1,3-dipolar cycloaddition of nitrones to alkylidene malonates. *Org. Lett.* 6 (10): 1677–1679.
616 Phomkeona, K., Takemoto, T., Ishima, Y. et al. (2008). Chiral bis(2-oxazolinyl)xanthene (xabox)/transition-metal complexes catalyzed 1,3-dipolar cycloaddition reactions and Diels–Alder reactions. *Tetrahedron* 64 (8): 1813–1822.
617 Carmona, D. and Oro, L.A. (2011). Iridium-catalyzed 1,3-dipolar cycloadditions. In: *Iridium Catalysis*, Topics in Organometallic Chemistry, vol. 34 (ed. P.G. Andersson), 209–229. Berlin, Heidelberg: Springer-Verlag.
618 Pellissier, H. (2007). Asymmetric 1,3-dipolar cycloadditions. *Tetrahedron* 63 (16): 3235–3285.
619 Stanley, L.M. and Sibi, M.P. (2008). Enantioselective Cu-catalyzed 1,3-dipolar cycloadditions. *Chem. Rev.* 108 (8): 2887–2902.

620 de Cozar, A. and Cossio, F.P. (2011). Stereocontrolled (3+2)-cycloadditions between azomethine ylides and dipolarophiles: a fruitful interplay between theory and experiment. *Phys. Chem. Chem. Phys.* 13 (23): 10858–10868.

621 Nájera, C., de Gracia Retamosa, M., Martín-Rodríguez, M. et al. (2009). Synthesis of prolines by enantioselective 1,3-dipolar cycloaddition of azomethine ylides and alkenes catalyzed by chiral phosphoramidite-Ag(I) complexes. *Eur. J. Org. Chem.* (32): 5622–5634.

622 Danielsson, J., Toom, L., and Somfai, P. (2011). 1,3-Dipolar cycloaddition of azomethine ylides to aldehydes: synthesis of *anti* α-amino-β-hydroxy esters. *Eur. J. Org. Chem.* (3): 607–613.

623 Grubert, L., Jugelt, W., Bress, H.J. et al. (1992). Reactions of 1,3-dipoles with heterocycles. 8. Synthesis of 1,8a-dihydro[1,2,4] triazolo[4,3-a]pyridines and benzologs. *Liebigs Ann. Chem.* (9): 885–894.

624 Grubert, L., Patzel, M., Jugelt, W. et al. (1994). Reactions of 1,3-dipoles with heterocycles. 9. Reactions of nitrilimines with pyrazines, pyrimidines and 1*H*-pyrimidinethiones. *Liebigs Ann. Chem.* (10): 1005–1011.

625 Huisgen, R., Fisera, L., Giera, H., and Sustmann, R. (1995). Thiones as superdipolarophiles - rates and equilibria of nitrone cycloadditions to thioketones. *J. Am. Chem. Soc.* 117 (38): 9671–9678.

626 Huisgen, R., Mloston, G., Polborn, K., and Sustmann, R. (2003). 1,3-Dithiolanes from cycloadditions of alicyclic and aliphatic thiocarbonyl ylides with thiones: regioselectivity. *Chem. Eur. J.* 9 (10): 2256–2263.

627 Sustmann, R., Sicking, W., and Huisgen, R. (2003). Thioformaldehyde *S*-methylide and thioacetone *S*-methylide: an ab initio MO study of structure and cycloaddition reactivity. *Chem. Eur. J.* 9 (10): 2245–2255.

628 Vedejs, E. and Wilde, R.G. (1986). (2+3)-Dipolar cycloadditions of a monomeric thioaldehyde. *J. Org. Chem.* 51 (1): 117–119.

629 Kolb, H.C., Finn, M.G., and Sharpless, K.B. (2001). Click chemistry: diverse chemical function from a few good reactions. *Angew. Chem. Int. Ed.* 40 (11): 2004–2021.

630 Trost, B.M. (1995). Atom economy - a challenge for organic synthesis - homogeneous catalysis leads the way. *Angew. Chem. Int. Ed. Engl.* 34 (3): 259–281.

631 Ostrovskis, P., Volla, C.M., Turks, M., and Markovic, D. (2013). Application of metal free click chemistry in biological studies. *Curr. Org. Chem.* 17 (6): 610–640.

632 Rostovtsev, V.V., Green, L.G., Fokin, V.V., and Sharpless, K.B. (2002). A stepwise Huisgen cycloaddition process: copper(I)-catalyzed regioselective "ligation" of azides and terminal alkynes. *Angew. Chem. Int. Ed.* 41 (14): 2596–2599.

633 Tornoe, C.W., Christensen, C., and Meldal, M. (2002). Peptidotriazoles on solid phase: [1,2,3]-triazoles by regiospecific Cu(I)-catalyzed 1,3-dipolar cycloadditions of terminal alkynes to azides. *J. Org. Chem.* 67 (9): 3057–3064.

634 (a) Spiteri, C. and Moses, J.E. (2010). Copper-catalyzed azide-alkyne cycloaddition: regioselective synthesis of 1,4,5-trisubstituted 1,2,3-triazoles. *Angew. Chem. Int. Ed.* 49 (1): 31–33. (b) Haldon, E., Nicasio, M.C., and Pérez, P.J. (2015). Copper-catalysed azide-alkyne cycloadditions (CuAAC): an up-date. *Org. Biomol, Chem.* 13 (37): 9528–9550. (c) Brittain, W.D.G., Buckley, B.R., and Fossey, J.S. (2016). Asymmetric Cu-catalyzed azide–alkyne cycloadditions. *ACS Catal.* 6 (6): 3629–3636. (d) Zhao, Z., Yao, Z.H., and Xu, X.W. (2017). Recent advances in Cu-catalyzed click reaction and its applications. *Curr. Org. Chem.* 21 (22): 2240–2248.

635 Liang, L. and Astruc, D. (2011). The Cu(I)-catalyzed alkyne-azide cycloaddition (CuAAC) "click" reaction and its applications. An overview. *Coord. Chem. Rev.* 255 (23–24): 2933–2945.

636 (a) Fatiadi, A.J. (1987). The classical permanganate ion - still a novel oxidant in organic-chemistry. *Synthesis* (2): 85–127. (b) Bataille, C.J.R. and Donohoe, T.J. (2011). Osmium-free direct *syn*-dihydroxylation of alkenes. *Chem. Soc. Rev.* 40 (1): 114–128.

637 Schröder, M. (1980). Osmium tetraoxide *cis*-hydroxylation of unsaturated substrates. *Chem. Rev.* 80 (2): 187–213.

638 Criegee, R. (1936). Osmium acid ester as an intermediate product in oxidation. *Liebigs Ann. Chem.* 522: 75–96.

639 Hofmann, K.A. (1912). Oxygen transmission by mean of osmiumtetroxide and activation of chlorate solutions. *Ber. Dtsch. Chem. Ges.* 45: 3329–3336.

640 Milas, N.A. and Sussman, S. (1936). The hydroxylation of the double bond. *J. Am. Chem. Soc.* 58: 1302–1304.

641 Milas, N.A., Trepagnier, J.H., Nolan, J.T., and Iliopulos, M.I. (1959). A study of the hydroxylation of olefins and the reaction of osmium tetroxide with 1,2-glycols. *J. Am. Chem. Soc.* 81 (17): 4730–4733.

642 Schneider, W.P. and McIntosch, A.V.J. (1956). Hydroxylation of D17(20)-pregnenes. US Patent 2769824; 19561106.

643 (a) Sundermeier, U., Dobler, C., and Beller, M. (2004). Recent developments in the Os-catalyzed dihydroxylation of olefins. In: *Modern Oxidation Methods* (ed. J.E. Bäckvall), 1–20. Weinheim: Wiley-VCH. (b) Zaitsev, A. and Adolfsson, H. (2006). Recent development in asymmetric dihydroxylation. *Synthesis* (11): 1725–1756. (c) Neumeyer, M. and Brückner, R. (2016). Nonracemic γ-lactones from the Sharpless asymmetric dihydroxylation of *β*,*γ*-unsaturated carboxylic esters. *Eur. J. Org. Chem.* (30): 5060–5087.

644 Sharpless, K.B. and Akashi, K. (1976). Osmium catalyzed vicinal hydroxylation of olefins by *tert*-butyl hydroperoxide under alkaline conditions. *J. Am. Chem. Soc.* 98 (7): 1986–1987.

645 Akashi, K., Palermo, R.E., and Sharpless, K.B. (1978). Major improvement in Os-catalyzed vicinal hydroxylation of olefins by *tert*-butyl hydroperoxide. *J. Org. Chem.* 43 (10): 2063–2066.

646 Minato, M., Yamamoto, K., and Tsuji, J. (1990). Osmium tetraoxide catalyzed vicinal hydroxylation of higher olefins by using hexacyanoferrate(III) ion as a co-oxidant. *J. Org. Chem.* 55 (2): 766–768.

647 Imada, Y., Saito, T., Kawakami, T., and Murahashi, S.I. (1992). Osmium-catalyzed asymmetric dihydroxylation of olefins using chiral isoxazolidine ligands. *Tetrahedron Lett.* 33 (35): 5081–5084.

648 Torii, S., Liu, P., and Tanaka, H. (1995). Electrochemical Os-catalyzed asymmetric dihydroxylation of olefins with Sharpless ligand. *Chem. Lett.* (4): 319–320.

649 Hentges, S.G. and Sharpless, K.B. (1980). Asymmetric induction in the reaction of osmium-tetroxide with olefins. *J. Am. Chem. Soc.* 102 (12): 4263–4265.

650 Kolb, H.C., Van Nieuwenhze, M.S., and Sharpless, K.B. (1994). Catalytic asymmetric dihydroxylation. *Chem. Rev.* 94: 2483–2547.

651 Becker, H. and Sharpless, K.B. (1996). A new ligand class for the asymmetric dihydroxylation of olefins. *Angew. Chem. Int. Ed. Engl.* 35 (4): 448–451.

652 DelMonte, A.J., Haller, J., Houk, K.N. et al. (1997). Experimental and theoretical kinetic isotope effects for asymmetric dihydroxylation. Evidence supporting a rate-limiting "(3+2)" cycloaddition. *J. Am. Chem. Soc.* 119 (41): 9907–9908.

653 Gotthard, H. and Huisgen, R. (1968). 1,3-Dipolar cycloadditions. 35. Δ2-Pyrazolines from sydnones and alkenes. *Chem. Ber.* 101 (2): 552–563.

654 Brunn, E., Funke, E., Gotthard, H., and Huisgen, R. (1971). 1.3-Dipolar cycloadditions. 61. Cycloadditions of *N*-substituted oxazolium-5-olates to nitriles, nitro, nitroso and azo compounds. *Chem. Ber.* 104 (5): 1562–1572.

655 Nguyen, T.B., Martel, A., Gaulon-Nourry, C. et al. (2012). 1,3-Dipolar cycloadditions of nitrones to hetero-substituted alkenes. Part 2: Sila-, thia-, phospha- and halo-substituted alkenes. *Org. Prep. Proced. Int.* 44 (1): 1–81.

656 Schore, N.E. (1988). Transition-metal-mediated cycloaddition reactions of alkynes in organic synthesis. *Chem. Rev.* 88 (7): 1081–1119.

657 (a) Saito, S. and Yamamoto, Y. (2000). Recent advances in the transition-metal-catalyzed regioselective approaches to polysubstituted benzene derivatives. *Chem. Rev.* 100 (8): 2901–2915. (b) Shaaban, M.R., El-Sayed, R., and Elwahy, A.H.M. (2011). Construction of fused heterocycles by metal-mediated (2+2+2)-cyclotrimerization of alkynes and/or nitriles. *Tetrahedron* 67 (34): 6095–6130. (c) Broere, D.L.J. and Ruijter, E. (2012). Recent advances in transition metal catalyzed (2+2+2)-cyclo(co)trimerization reactions. *Synthesis* 44 (17): 2639–2672.

658 Galan, B.R. and Rovis, T. (2009). Beyond Reppe: building substituted arenes by (2+2+2)-cycloadditions of alkynes. *Angew. Chem. Int. Ed.* 48 (16): 2830–2834.

659 Maitlis, P.M. (1976). Palladium(II)-induced oligomerization of acetylenes - organometallic detective story. *Acc. Chem. Res.* 9 (3): 93–99.

660 Jankova, S., Dracinsky, M., Cisarova, I., and Kotora, M. (2008). Synthesis and rearrangement of Dewar benzenes into biaryls: experimental evidence for conrotatory ring opening. *Eur. J. Org. Chem.* (1): 47–51.

661 Balasubramanian, K.K., Selvaraj, S., and Venkataramani, P.S. (1980). Trimerization of aryl ethynyl ketones. *Synthesis* (1): 29–30.

662 Matsuda, K., Nakamura, N., and Iwamura, H. (1994). Exploratory and mechanistic studies of the trimerization reaction of benzoylacetylenes

in the presence of a secondary amine. *Chem. Lett.* (9): 1765–1768.

663 Yang, J.C. and Verkade, J.G. (1998). Disilane-catalyzed cyclotrimerization of acetylenes. *J. Am. Chem. Soc.* 120 (27): 6834–6835.

664 Yang, J.C. and Verkade, J.G. (2000). Disilane-catalyzed and thermally induced oligomerizations of alkynes: a comparison. *Organometallics* 19 (5): 893–900.

665 Berthelot, M. (1866). Action de la chaleur sur quelques carbures d'hydrogène. *C.R. Hebd. Seances Acad. Sci.* 62: 905–909.

666 Wagenseller, P.E., Birney, D.M., and Roy, D. (1995). On the development of aromaticity in cycloadditions - ab-initio transition structures for the trimerization of acetylene and for the addition of ethylene and acetylene to formylketene. *J. Org. Chem.* 60 (9): 2853–2859.

667 Jiao, H.J. and Schleyer, P.v.R. (1998). Aromaticity of pericyclic reaction transition structures: magnetic evidence. *J. Phys. Org. Chem.* 11 (8–9): 655–662.

668 Viehe, H.G., Merényi, R., Oth, J.F.M. et al. (1964). Valence-bond isomers of (substituted) benzene - simultaneous formation of substituted tetracyclohexane tricyclohexene + bicyclohexadiene by spontaneous trimerization of *t*-butylfluoroacetylene. *Angew. Chem. Int. Ed. Engl.* 3 (11): 755–756.

669 Viehe, H.G., Merényi, R., Oth, J.F.M., and Valange, P. (1964). Formation of 1,2,3-tri-t-butyltrifluorobenzene by spontaneous trimerization of t-butyfluoroacetylene (1). *Angew. Chem. Int. Ed. Engl.* 3 (11): 746.

670 Ballester, M., Castaner, J., Riera, J., and Tabernero, I. (1986). Synthesis and chemical behavior of perchlorophenylacetylene. *J. Org. Chem.* 51 (9): 1413–1419.

671 Hopf, H. and Witulski, B. (1993). Cyanoalkynes - magic wands for the preparation of novel aromatic compounds. *Pure Appl. Chem.* 65 (1): 47–56.

672 Yao, Z.K. and Yu, Z.X. (2011). Mechanisms of the thermal cyclotrimerizations of fluoro- and chloroacetylenes: density functional theory investigation and intermediate trapping experiments. *J. Am. Chem. Soc.* 133 (28): 10864–10877.

673 Robinson, J.M., Sakai, T., Okano, K. et al. (2010). Formal (2+2+2)-cycloaddition strategy based on an intramolecular propargylic ene reaction/Diels–Alder cycloaddition cascade. *J. Am. Chem. Soc.* 132 (32): 11039–11041.

674 Sakai, T. and Danheiser, R.L. (2010). Cyano Diels–Alder and cyano ene reactions. Applications in a formal (2+2+2)-cycloaddition strategy for the synthesis of pyridines. *J. Am. Chem. Soc.* 132 (38): 13203–13205.

675 Lan, Y., Danheiser, R.L., and Houk, K.N. (2012). Why Nature eschews the concerted (2+2+2)-cycloaddition of a nonconjugated cyanodiyne. Computational study of a pyridine synthesis involving an ene-Diels–Alder-bimolecular hydrogen-transfer mechanism. *J. Org. Chem.* 77 (3): 1533–1538.

676 (a) Ullman, E.F. (1958). Diels–Alder reaction of an unconjugated diene. *Chem. Ind. (London)* (36): 1173–1174. (b) Vasil'ev, N.V., Truskanova, T.D., Buzaev, A.V. et al. (2005). Cycloaddition of norbornadiene to fluorine-containing heteroaddends. *Russ. Chem. Bull.* 54 (4): 1038–1040.

677 Blomquist, A.T. and Meinwald, Y.C. (1959). Synthesis of some conjugated cyclobutane polyolefins and their 1,2-cycloaddition to tetracyanoethylene. *J. Am. Chem. Soc.* 81 (3): 667–672.

678 Sargeant, P.B. (1969). Fluorocyclopropanes. 3. 1,4-Cycloaddition reactions of perfluorocyclopropene and 1,2-bis(trifluoromethyl)-3,3-difluorocyclopropene. *J. Am. Chem. Soc.* 91 (11): 3061–3068.

679 Matsumoto, K., Taketsuna, H., Ikemi, Y. et al. (1998). Cycloaddition reactions of oxaquadricyclane, oxanorbornadiene, and related compounds with cyclooctyne. *Heterocycles* 49: 79–83.

680 Rieber, N., Alberts, J., Lipsky, J.A., and Lemal, D.M. (1969). Δ-1,2-Diazetines. *J. Am. Chem. Soc.* 91 (20): 5668–5669.

681 (a) Smith, C.D. (1966). Cycloaddition reactions of quadricyclanes. *J. Am. Chem. Soc.* 88 (18): 4273–4274. (b) Petrov, V.A., Davidson, F., and Smart, B.E. (2004). Quadricyclane-thermal cycloaddition to polyfluorinated carbonyl compounds - a simple synthesis of polyfluorinated 3-oxatricyclo[4.2.1.0(2,5)]non-7-enes. *J. Fluorine Chem.* 125 (10): 1543–1552.

682 Tabushi, I., Yamamura, K., and Yoshida, Z. (1972). Regiospecific and stereospecific $[2^2+\sigma^2+\sigma^2]$-cycloaddition reaction of quadricyclane. *J. Am. Chem. Soc.* 94 (3): 787–792.

683 (a) Papadopoulos, M., Jost, R., and Jenner, G. (1983). Synthesis of annelated oxetans via cycloaddition reactions of quadricyclane under thermal and high pressure conditions. *J. Chem. Soc., Chem. Commun.* (5): 221–222. (b) Hatsui, T. and Takeshita, H.

(1993). Rose-Bengal-sensitized photooxidation of quadricyclane. A [σ^2+σ^2+π^2]-cycloaddition of singlet oxygen. *Chem. Lett.* (1): 129–132. (c) Barlow, M.G., Suliman, N.N.E., and Tipping, A.E. (1995). Unsaturated nitrogen compounds containing fluorine. Part 18. Novel cycloaddition products from the reaction of quadricyclane with fluorinated azines and the azomethine imide 2-hexafluoroisopropylidene-3-methyl-3-phenyl-5,5-bis(trifluoromethyl) tetrahydropyrazol-2-ium-1-ide. *J. Fluorine Chem.* 72 (1): 147–156.

684 Prinzbach, H. and Rivier, J. (1967). Cycloadditions to derivatives of methylenequadricyclane and of quadricyclanone (bishomocyclobutadiene). *Angew. Chem. Int. Ed. Engl.* 6 (12): 1069–1070.

685 Delucchi, O. and Lucchini, V. (1982). Reactions of quadricyclane with sulfur dioxide - formation of a stable B-sultine and catalyzed isomerization to norbornadiene. *J. Chem. Soc., Chem. Commun.* (8): 464–465.

686 Prato, M., Maggini, M., Scorrano, G., and Lucchini, V. (1993). Addition of quadricyclane to C_{60} - easy access to fullerene derivatives bearing a reactive double-bond in the side-chain. *J. Org. Chem.* 58 (14): 3613–3615.

687 Landis, M.E. and Mitchell, J.C. (1979). Cycloaditions with quadricyclane - synthesis of fused-ring 1,2-diazetidines. *J. Org. Chem.* 44 (13): 2288–2290.

688 Papadopoulos, M. and Jenner, G. (1983). Pressure effect and estimation of concertedness in some [π^2+σ^2+σ^2]-cycloaddition reactions of quadricyclane. *New J. Chem.* 7 (8–9): 463–464.

689 Paquette, L.A., Kesselmayer, M.A., and Künzer, H. (1988). Comparative analysis via secondary deuterium isotope effects of transition state structural features associated with cycloadditions to norbornadiene [π^2+π^2+π^2] and quadricyclane [π^2+σ^2+σ^2]. *J. Org. Chem.* 53 (21): 5183–5185.

690 Jones, G.A., Shephard, M.J., Paddon-Row, M.N. et al. (1999). Unexpected complexity in the thermal [π^2+σ^2+σ^2]- cycloaddition reactions of quadricyclane: theory and isotope effects. *J. Am. Chem. Soc.* 121 (18): 4334–4339.

691 (a) Narayan, S., Muldoon, J., Finn, M.G. et al. (2005). "On water": Unique reactivity of organic compounds in aqueous suspension. *Angew. Chem. Int. Ed.* 44 (21): 3275–3279. (b) Bain, R.N., Sathyamoorthi, S., and Zare, R.N. (2017). "On-Droplet" chemistry: the cycloaddition of diethyl azodicarboxylate and quadricyclane. *Angew. Chem. Int. Ed.* 56 (47): 15083–15087.

692 Jung, Y.S. and Marcus, R. (2007). On the theory of organic catalysis on water. *J. Am. Chem. Soc.* 129 (17): 5492–5502.

693 (a) Domingo, L.R., Saez, J.A., Zaragoza, R.J., and Arno, M. (2008). Understanding the participation of quadricyclane as nucleophile in polar [π^2+σ^2+σ^2]-cycloadditions toward electrophilic π molecules. *J. Org. Chem.* 73 (22): 8791–8799. (b) Kiselev, V.D., Kornilov, D.A., Anikin, O.V. et al. (2018). Kinetics and thermochemistry of the unusual [π^2+σ^2+σ^2]-cycloaddition of quadricyclane with some dienophiles. *J. Phys. Org. Chem.* 31 (1): e3737.

694 Prinzbach, H., Argüelles, M., Vogel, P., and Eberbach, W. (1967). Cycloadditions to derivatives of oxaquadricyclanes (bishomofuran). *Angew. Chem. Int. Ed. Engl.* 6 (12): 1070–1071.

695 Prinzbach, H., Bingmann, H., Markert, J. et al. (1986). Photochemical transformations. 64. The [π^3+σ^3] route to oxepines-benzene oxides. *Chem. Ber.* 119 (2): 589–615.

696 Cramer, C.J., Harmata, M., and Rashatasakhon, P. (2001). Intramolecular (4+3)-cycloadditions. Theoretical and experimental evaluation of *endo/exo* preferences of a cyclopentenyl cation. *J. Org. Chem.* 66 (16): 5641–5644.

697 Harmata, M. (1997). Intramolecular (4+3)-cycloaddition reactions. *Adv. Cycloaddit.* 4: 41–86.

698 (a) Harmata, M. and Schreiner, P.R. (2001). Intramolecular (4+3)-cycloadditions. A theoretical analysis of simple diastereoselectivity in reactions of alkoxyallylic cations and furans. *Org. Lett.* 3 (23): 3663–3665. (b) Hassan, A.H.E., Lee, J.K., Pae, A.N. et al. (2015). Synthesis of the tricyclic ring structure of Daphnanes via intramolecular (4+3)-cycloaddition/SmI2-pinacol coupling. *Org. Lett.* 17 (11): 2672–2675.

699 (a) Harmata, M. (2001). Exploration of fundamental and synthetic aspects of the intramolecular (4+3)-cycloaddition reaction. *Acc. Chem. Res.* 34 (7): 595–605. (b) Ling, J., Lam, S., Low, K.-H., and Chiu, P. (2017). Dearomative intramolecular (4+3)-cycloadditions of arenes with epoxy and aziridinyl enolsilanes. *Angew. Chem. Int. Ed.* 56 (30): 8879–8882.

700 Harmata, M. (2006). Asymmetric catalytic (4+3)-cycloaddition reactions. *Adv. Synth. Catal.* 348 (16–17): 2297–2306.

701 Hartung, I.V. and Hoffmann, H.M.R. (2004). 8-Oxabicyclo[3.2.1]oct-6-en-3-ones: application

to the asymmetric synthesis of polyoxygenated building blocks. *Angew. Chem. Int. Ed.* 43 (15): 1934–1949.

702 (a) Harmata, M. (2010). The (4+3)-cycloaddition reaction: simple allylic cations as dienophiles. *J. Chem. Soc., Chem. Commun.* 46 (47): 8886–8903. (b) Lohse, A.G. and Hsung, R.P. (2011). (4+3)-Cycloaddition reactions of nitrogen-stabilized oxyallyl cations. *Chem. Eur. J.* 17 (14): 3812–3822. (c) Alcázar, P., Cruz, I., González-Romero, C. et al. (2015). Synthesis and structural studies of novel fused seven-membered carbocycles derived from exo-2-oxazolidinone dienes through (4+3)-cycloadditions. *Tetrahedron* 71 (10): 1604–1615.

703 (a) Zhi, Y., Zhao, K., Shu, T., and Enders, D. (2016). Synthesis of benzotriazepine derivatives via (4+3)-cycloaddition of aza-*o*-quinone methide intermediates and azomethine imines. *Synthesis* 48 (2): 238–244. (b) Zhao, H.-W., Pang, H.-L., Tian, T. et al. (2016). Construction of 2,3,4,7-tetrahydro-1,2,4,5-oxatriazepines via (4+3)-cycloadditions of -halogeno hydrazones with nitrones. *Adv. Synth. Catal.* 358 (11): 1826–1832. (c) Wang, W., Yuan, C., Liu, Y. et al. (2016). (4+3)-Cycloaddition of phthalazinium dicyanomethanides with azoalkenes formed in situ: Synthesis of triazepine derivatives. *J. Org. Chem.* 81 (17): 7597–7603. (d) Mei, G.-J., Zhu, Z.-Q., Zhao, J.-J. et al. (2017). Brønsted acid-catalyzed stereoselective (4+3)-cycloadditions of *ortho*-hydroxybenzyl alcohols with *N,N′*-cyclic azomethine imines. *J. Chem. Soc., Chem. Commun.* 53 (18): 2768–2771.

704 (a) Fort, A.W. (1962). Evidence for a delocalized intermediate in Favorskii rearrangement - 2,6-Lutidine-promoted methanolysis of A-chlorodibenzyl ketone. *J. Am. Chem. Soc.* 84 (13): 2620–2625. (b) Hamblin, G.D., Jimenez, R.P., and Sorensen, T.S. (2007). A theoretical study of Favorskii reaction stereochemistry. Lessons in torquoselectivity. *J. Org. Chem.* 72 (21): 8033–8045. (c) Tsuchida, N., Yamazaki, S., and Yamabe, S. (2008). A new mechanism for the Favorskii rearrangement. *Org. Biomol. Chem.* 6 (17): 3109–3117.

705 Föhlisch, B., Gottstein, W., Kaiser, R., and Wanner, I. (1980). Preparation and cycloaddition of allylium intermediates from α-haloketones under solvolysis conditions. *Tetrahedron Lett.* 21 (31): 3005–3008.

706 Föhlisch, B., Franz, T., and Kreiselmeier, G. (2005). Based-induced trifluoroethanolysis of acyclic di- and tri-halogeno ketones; Favorskii rearrangement and (4+3)-cycloaddition. *Eur. J. Org. Chem.* 4687–4698.

707 Föhlisch, B. and Reiner, S. (2004). Hexachloroacetone as a precursor for a tetrachloro-substituted oxyallyl intermediate: (4+3)-Cycloaddition to 1,3-dienes. *Molecules* 9: 1–10.

708 Perkow, W. (1954). Umsetzungen mit Alkylphosphiten. 1. Umlagerungen bei der Reaktion mit Chloral und Bromal. *Chem. Ber.* 87 (5): 755–758.

709 Harmata, M. and Rashatasakhon, P. (2000). Intramolecular (4+3)-cycloadditions. Aspects of stereocontrol in the synthesis of cyclooctanoids. A synthesis of (+)-Dactylol. *Org. Lett.* 2 (18): 2913–2915.

710 Lautens, M., Aspiotis, R., and Colucci, J. (1996). Diastereoselective intermolecular (4+3)-cycloadditions via an extended transition state: a route to enantiomerically enriched cycloadducts. *J. Am. Chem. Soc.* 118 (44): 10930–10931.

711 Murray, D.H. and Albizati, K.F. (1990). Ambiphilic reactivity of 1,1-dimethoxyacetone. *Tetrahedron Lett.* 31 (29): 4109–4112.

712 Stark, C.B.W., Pierau, S., Wartchow, R., and Hoffmann, H.M.R. (2000). Chiral allyl cations are captured by furan with 100 % stereoselectivity: synthesis of enantiopure 2-alkoxy-8-oxabicyclo[3.2.1]oct-6-en-3-ones by low-temperature (4+3)-cycloaddition. *Chem. Eur. J.* 6 (4): 684–691.

713 Krenske, E.H., Houk, K., and Harmata, M. (2010). Origin of stereoselectivity in the (4+3)-cycloadditions of chiral alkoxy siloxyallyl cations with furan. *Org. Lett.* 12 (3): 444–447.

714 Antoline, J.E., Krenske, E.H., Lohse, A.G. et al. (2011). Stereoselectivities and regioselectivities of (4+3)-cycloadditions between allenamide-derived chiral oxazolidinone-stabilized oxyallyls and furans: experiment and theory. *J. Am. Chem. Soc.* 133 (36): 14443–14451.

715 Huang, J. and Hsung, R.P. (2005). Chiral Lewis acid-catalyzed highly enantioselective (4+3)-cycloaddition reactions of nitrogen-stabilized oxyallyl cations derived from allenamides. *J. Am. Chem. Soc.* 127 (1): 50–51.

716 Hendrickson, J.B. and Farina, J.S. (1980). A new 7-ring cycloaddition reaction. *J. Org. Chem.* 45 (16): 3359–3361.

717 Hendrickson, J.B. and Farina, J.S. (1980). A simple synthesis of 8-membered and 10-membered carbocyclic rings. *J. Org. Chem.* 45 (16): 3361–3363.

718 Magnus, P., Waring, M.J., Ollivier, C., and Lynch, V. (2001). Stereoselective synthesis of the bicyclo[5.3.0]decane portion of the diterpene antibiotic Guanacastepene using a pyrylium-ylide (5+2)-cycloaddition reaction. *Tetrahedron Lett.* 42 (30): 4947–4950.

719 (a) Pellissier, H. (2011). Recent developments in the (5+2)-cycloaddition. *Adv. Synth. Catal.* 353 (2–3): 189–218. (b) Mei, G.J., Yuan, H., Gu, Y. et al. (2014). Dearomative indole (5+2)-cycloaddition reactions: stereoselective synthesis of highly functionalized cyclohepta[*b*]indoles. *Angew. Chem. Int. Ed.* 53 (41): 11051–11055. (c) Mei, G., Liu, X., Qiao, C. et al. (2015). Type II Intramolecular (5+2)-cycloaddition: facile synthesis of highly functionalized bridged ring systems. *Angew. Chem. Int. Ed.* 54: 1754–1758. (d) Krüger, S. and Gaich, T. (2016). Catalytic enantioselective intermolecular (5+2)-dipolar cycloadditions of a 3-hydroxy-4-pyrone-derived oxidopyrylium ylide. *Eur. J. Org. Chem.* 4893–4899. (e) Fuhr, K.N., Hisch, D.R., Murelli, R.P., and Brenner-Moyer, S.E. (2017). Catalytic enantioselective intermolecular (5+2)-dipolar cycloadditions of a 3-hydroxy-4-pyrone-derived oxidopyrylium ylide. *Org. Lett.* 19 (23): 6356–6359.

720 Hassner, A., D'Costa, R., McPhail, A.T., and Butler, W. (1981). Cycloaddition of vinyl aziridines with unsaturated substrates - a novel rearrangement of an unsaturated nitro compound. *Tetrahedron Lett.* 22 (38): 3691–3694.

721 Kanno, E., Yamanoi, K., Koya, S. et al. (2012). (5+2)-cycloaddition reaction of 2-vinylaziridines and sulfonyl isocyanates. Synthesis of seven-membered cyclic ureas. *J. Org. Chem.* 77 (5): 2142–2148.

722 Koya, S., Yamanoi, K., Yamasaki, R. et al. (2009). Selective synthesis of eight-membered cyclic ureas by the (6+2)-cycloaddition reaction of 2-vinylazetidines and electron-deficient isocyanates. *Org. Lett.* 11 (23): 5438–5441.

723 Harmata, M., Jones, D.E., Kahraman, M. et al. (1999). Chiral cations in (4+3)-cycloadditions. *Tetrahedron Lett.* 40 (10): 1831–1834.

724 Kende, A.S. and Huang, H. (1997). Asymmetric (4+3)-cycloadditions from chiral α-chloro imines. *Tetrahedron Lett.* 38 (19): 3353–3356.

725 (a) Sieburth, S.M. and Cunard, N.T. (1996). The (4+4)-cycloaddition and its strategic application in natural product synthesis. *Tetrahedron* 52 (18): 6251–6282. (b) Yu, Z.-X., Wang, Y., and Wang, Y.-Y. (2010). Transition-metal-catalyzed cycloadditions for the synthesis of eight-membered carbocycles. *Chem. Asian J.* 5 (5): 1072–1088. (c) Xie, M.-S., Liu, X.-H., Wu, X.-X. et al. (2013). Catalytic Asymmetric (8+2)-cycloaddition: Synthesis of cycloheptatriene-fused pyrrole derivatives. *Angew. Chem. Int. Ed.* 52 (21): 5604–5607. (d) Xie, M.-S., Wu, X.-X., Wang, G. et al. (2014). Catalytic asymmetric (8+2)-cycloaddition for the construction of cycloheptatriene-fused pyrrolidin-3,3′-oxindoles. *Acta Chim. Sinica* 72 (7): 856–8621.

726 Mose, R., Preegel, G., Larsen, J. et al. (2017). Organocatalytic stereoselective (8+2) and (6+4)-cycloadditions. *Nat. Chem.* 9 (5): 487–492.

727 (a) Houk, K.N. and Woodward, R.B. (1970). Cycloaddition reactions of cycloheptatriene and 2,5-dimethyl-3,4-diphenylcyclopentadienone. *J. Am. Chem. Soc.* 92 (13): 4143–4145. (b) Houk, K.N. and Woodward, R.B. (1970). Cycloaddition reactions of tropone and 2,5-dimethyl-3,4-diphenylcyclopentadienone. *J. Am. Chem. Soc.* 92 (13): 4145–4147.

728 Mukherjee, D., Watts, C.R., and Houk, K.N. (1978). Periselectivity in (4+2) and (6+4)-cycloadditions of diphenylnitrilimine to tropone. *J. Org. Chem.* 43 (5): 817–821.

729 Houk, K.N., Sims, J., Watts, C.R., and Luskus, L.J. (1973). Origin of reactivity, regioselectivity, and periselectivity in 1,3-dipolar cycloadditions. *J. Am. Chem. Soc.* 95 (22): 7301–7315.

730 Leach, A.G., Goldstein, E., and Houk, K.N. (2003). A cornucopia of cycloadducts: theoretical predictions of the mechanisms and products of the reactions of cyclopentadiene with cycloheptatriene. *J. Am. Chem. Soc.* 125 (27): 8330–8339.

731 Cope, A.C. and Schmitz, W.R. (1950). Cyclic polyolefins. 7. Structure of the 8-membered cyclic dimer of chloroprene. *J. Am. Chem. Soc.* 72 (7): 3056–3062.

732 Trahanovsky, W.S., Cassady, T.J., and Woods, T.L. (1981). Preparation of 2,3-dimethylene-2,3-dihydrofurans by the Flash Vacuum Pyrolysis of substituted furylmethyl esters. *J. Am. Chem. Soc.* 103 (22): 6691–6695.

733 Leung, M.K. and Trahanovsky, W.S. (1995). The effects of A-methyl group substitution

on the dimerization products of furan-based *o*-quinodimethanes. *J. Am. Chem. Soc.* 117 (3): 841–851.

734 Krishna, U.M., Deodhar, K.D., Trivedi, G.K., and Mobin, S.M. (2004). Novel route to functionalized cyclooctanoids via (5+3)-cycloaddition. *J. Org. Chem.* 69 (3): 967–969.

735 Hutton, J. and Waters, W.A. (1966). A 1,6-addition of nitrosobenzene to cycloheptatriene. *J. Chem. Soc., Chem. Commun.* (18): 634–635.

736 Murphy, W.S. and McCarthy, J.P. (1968). 1,6-Cycloaddition to *N*-ethoxycarbonylazepine. *J. Chem. Soc., Chem. Commun.* (19): 1155–1156.

737 Askani, R., Chesick, J.P., and Chesick, J.P. (1973). (6+2)-Cycloadditions with homofulvenes. *Chem. Ber.* 106 (1): 8–19.

738 Imafuku, K. and Arai, K. (1989). A simple and convenient synthesis of 5-alkyl-substituted 3-isopropenyl-acetyltropolones and 3-acetyltropolones. *Synthesis* (7): 501–505.

739 Harre, M., Raddatz, P., Walenta, R., and Winterfeldt, E. (1982). Cyclopentenone derivatives. 6. 4-Oxo-2-cyclopentenyl acetate - a synthetic intermediate. *Angew. Chem. Int. Ed. Engl.* 21 (7): 480–492.

740 Himeda, Y., Yamataka, H., Ueda, I., and Hatanaka, M. (1997). (3+2)-Annulation of allylidenetriphenylphosphorane with 1,2-diacylethylenes and 1,2-diacylacetylenes: a one-step synthesis of tri- and tetrasubstituted cyclopentadienes and fulvenes. *J. Org. Chem.* 62 (19): 6529–6538.

741 Nair, V., Anilkumar, G., Radhakrishnan, K.V. et al. (1997). Diels–Alder reactions of a 6-arenyl fulvene with dienes and dienophiles and related chemistry. *Tetrahedron* 53 (51): 17361–17372.

742 Rawson, D.I., Carpenter, B.K., and Hoffmann, H.M.R. (1979). Allyl cation route to 7-membered rings. 19. Loss of allyl cation configuration in cycloadditions to electron-rich conjugated dienes. *J. Am. Chem. Soc.* 101 (7): 1786–1793.

743 Hong, B.C., Shr, Y.J., Wu, J.L. et al. (2002). Novel (6+2)-cycloaddition of fulvenes with alkenes: a facile synthesis of the anislactone and hirsutane framework. *Org. Lett.* 4 (13): 2249–2252.

744 Krishnan, K., Sajisha, V., Anas, S. et al. (2006). (6+3)-Cycloaddition of pentafulvenes with 3-oxidopyrylium betaine: a novel methodology toward the synthesis of 5-8 fused oxabridged cyclooctanoids. *Tetrahedron* 62 (25): 5952–5961.

745 Gupta, Y.N., Doa, M.J., and Houk, K.N. (1982). Intramolecular (6+4)-cycloadditions - intramolecular control of periselectivity. *J. Am. Chem. Soc.* 104 (25): 7336–7338.

746 Yoshida, Z., Shibata, M., Ogino, E., and Sugimoto, T. (1985). A novel transformation of a calicene to an azulenofulvene. *Angew. Chem. Int. Ed. Engl.* 24 (1): 60–61.

747 Wu, T.C. and Houk, K.N. (1985). Construction of linearly fused tricyclopentanoids by intramolecular (6+2)-cycloadditions of fulvenes with enamines. *J. Am. Chem. Soc.* 107 (18): 5308–5309.

748 Coskun, N., Ma, J., Azimi, S. et al. (2011). 1,2-Dihydropentalenes from fulvenes by(6+2)-cycloadditions with 1-isopropenylpyrrolidine. *Org. Lett.* 13 (22): 5952–5955.

749 Sung, M.J., Lee, H.I., Chong, Y., and Cha, J.K. (1999). Synthetic studies on (4+3)-cycloadditions of oxyallyls. Part 11. Facile synthesis of the tricyclic core of Sarain A. 3-Oxidopyridinium betaine cycloaddition approach. *Org. Lett.* 1 (12): 2017–2019.

750 Trost, B.M. and Seoane, P.R. (1987). (6+3)-Cycloaddition to 9-membered ring carbocycles. *J. Am. Chem. Soc.* 109 (2): 615–617.

751 Hong, B.C., Gupta, A.K., Wu, M.F. et al. (2003). Hetero (6+3)-cycloaddition of fulvenes with *N*-alkylidene glycine esters: a facile synthesis of the delavayine and incarvillateine framework. *Org. Lett.* 5 (10): 1689–1692.

752 Doering, W.v.E. and Wiley, D.W. (1960). Heptafulvene (methylenecycloheptatriene). *Tetrahedron* 11 (3): 183–198.

753 Nair, V. and Anilkumar, G. (1998). Cycloaddition reactions of heptafulvenes: an overview. *Synlett* (9): 950–960.

754 Nair, V. and Abhilash, K. (2008). (8+2)-Cycloaddition reactions in organic synthesis. *Synlett* (3): 301–312.

755 Nair, V., Abhilash, K., Biju, A.T., and Suresh, E. (2007). Efficient higher-order cycloaddition reactions of 8,8-dicyanoheptafulvene with dienes. *Synthesis* (12): 1833–1836.

756 Nair, V., Abhilash, K., and Suresh, E. (2006). A novel protocol for the generation of tropothione and its trapping with electron deficient dienophiles. *Tetrahedron Lett.* 47 (52): 9329–9331.

757 Rigby, J.H., Ateeq, H.S., Charles, N.R. et al. (1993). Metal-promoted higher-order cycloaddition reactions - stereochemical, regiochemical,

and mechanistic aspects of the [π6+π4] reaction. *J. Am. Chem. Soc.* 115 (4): 1382–1396.

758 Rigby, J.H. (1993). Transition-metal promoted higher-order cycloaddition reactions in organic synthesis. *Acc. Chem. Res.* 26 (11): 579–585.

759 Funk, R.L. and Bolton, G.L. (1986). Intramolecular (6+4) and (4+2) tropone olefin cycloaddition reactions - facile synthesis of the ingenane ring-system. *J. Am. Chem. Soc.* 108 (15): 4655–4657.

760 Rigby, J.H. and Fleming, M. (2002). Construction of the Ingenane core using an Fe(III) or Ti(IV) Lewis acid-catalyzed intramolecular (6+4)-cycloaddition. *Tetrahedron Lett.* 43 (48): 8643–8646.

761 Simig, G. and Schlosser, M. (1994). The spontaneous cyclodimerization of 2,3-dihydroisoquinolines after base-promoted elimination of methanol from 4-methoxy-1,2,3,4-tetrahydroisoquinolines. *Tetrahedron Lett.* 35 (19): 3081–3082.

762 (a) Stark, B.P. and Duke, A.J. (1967). *Extrusion Reactions*, 1–190. Oxford: Pergamon Press. (b) Rzepa, H.S. (1988). Cheletropic elimination of CO and N_2 - a comparison of the MNDO, AM1, and ab initio SCF-MO methods. *J. Chem. Res. S* (7): 224–225. (c) Zahedi, E., Mozaffari, M., Shahsavar, F. et al. (2017). Understanding the kinetics and mechanism of thermal cheletropic elimination of N_2 from (2,5-dihydro-1H-pyrrol-1-ium-1-ylidene) amide using RRKM and ELF theories. *Res. Chem. Intermed.* 43 (3): 1575–1590. (d) Zahedi, E., Shaabani, S., and Shiroudi, A. (2017). Following the molecular mechanism of decarbonylation of unsaturated cyclic ketones using bonding evolution theory coupled with NCI analysis. *J. Phys. Chem. A* 121 (44): 8504–8517.

763 Grigg, R., Hayes, R., and Jackson, J.L. (1969). Thermal elimination reactions of nitrogen and sulphur heterocycles. *J. Chem. Soc., Chem. Commun.* (20): 1167–1168.

764 Korth, H.G., Sustmann, R., Lommes, P. et al. (1994). Nitric-oxide cheletropic Traps (Nocts) with improved thermal stability and water solubility. *J. Am. Chem. Soc.* 116 (7): 2767–2777.

765 Lauricella, R., Triquigneaux, M., André-Barrès, C. et al. (2010). 5-Hydroxy-2,2,6,6-tetramethyl-4-(2-methylprop-1-en-yl)cyclohex-4-ene-1,3-dione, A novel cheletropic trap for nitric oxide EPR detection. *J. Chem. Soc., Chem. Commun.* 46 (21): 3675–3677.

766 Roberts, V.A. and Garst, M.E. (1985). Synthesis of 4-substituted cycloheptatrienones by oxidative cheletropic elimination of nitrosobenzene from 6-substituted 8-phenyl-8-azabicyclo[3.2.1]octa-3,6-dien-1-ones. *J. Org. Chem.* 50 (6): 893–894.

767 Shustov, G.V. and Rauk, A. (2000). Cheletropic decomposition of cyclic nitrosoamines revisited: the nature of the transition states and a critical role of the ring strain. *J. Org. Chem.* 65 (12): 3612–3619.

768 Bumgardner, C., Freeman, J.P., and McCallum, K.S. (1961). Small ring heterocyclic nitrosoamines. *J. Am. Chem. Soc.* 83 (21): 4417–4419.

769 Martinu, T. and Dailey, W.P. (2000). Facile one-pot preparation of 3-chloro-2-(chloromethyl)propene and an ab initio study of the deamination reaction of nitrosoaziridine. *J. Org. Chem.* 65 (20): 6784–6786.

770 Weber, L. (2007). Phospha- and arsaalkenes $RE{=}C(NMe_2)_2$ (E=P, As) as novel phosphinidene- and arsinidene-transfer reagents. *Eur.J. Inorg. Chem.* (26): 4095–4117.

771 Roversi, E., Monnat, F., Schenk, K. et al. (2000). O-17 NMR spectroscopy of sulfolenes (2,5-dihydrothiophene-1,1-dioxides) and sultines (3,6-dihydro-1,2-oxathiin-2-oxides)-experiment and quantum calculations: synthesis of 4,9-dioxo-1,2-oxathiacyclodecane-2-oxide, a new heterocycle. *Chem. Eur. J.* 6 (10): 1858–1864.

772 Asay, M., Jones, C., and Driess, M. (2011). N-Heterocyclic carbene analogues with low-valent Group 13 and Group 14 elements: syntheses, structures, and reactivities of a new generation of multitalented ligands. *Chem. Rev.* 111 (2): 354–396.

773 Tokitoh, N. and Okazaki, R. (2000). Recent topics in the chemistry of heavier congeners of carbenes. *Coord. Chem. Rev.* 210: 251–277.

774 Becerra, R. and Walsh, R. (2007). What have we learnt about heavy carbenes through laser flash photolysis studies? *Phys. Chem. Chem. Phys.* 9 (22): 2817–2835.

775 (a) Salaün, J. (2000). Cyclopropane derivatives and their diverse biological activities. In: *Small Ring Compounds in Organic Synthesis VI*, Topics in Current Chemistry, vol. 207 (ed. A. de Meijere), 1–67. Heidelberg, Berlin: Springer-Verlag. (b) Chen, D.Y.-K., Pouwer, R.H., and Richard, J.-A. (2012). Recent advances in the total synthesis of cyclopropane-containing natural products. *Chem. Soc. Rev.* 41 (13): 4631–4642. (c) Keglevich, P., Keglevich, A., Hazai, L.

et al. (2014). Natural compounds containing a condensed cyclopropane ring. Natural and synthetic aspects. *Curr. Org. Chem.* 18 (15): 2037–2042. (d) Taylor, R.D., MacCross, M., and Lawson, A.D.G. (2014). Rings in drugs. *J. Med. Chem.* 57: 5845–5859. (e) Tatale, T.T. (2016). The "cyclopropyl fragment" is a versatile player that frequently appears in preclinical/clinical drug molecules. *J. Med. Chem.* 59: 8712–8756.

776 Donaldson, W.A. (2001). Synthesis of cyclopropane containing natural products. *Tetrahedron* 57 (41): 8589–8627.

777 Faust, R. (2001). Fascinating natural and artificial cyclopropane architectures. *Angew. Chem. Int. Ed.* 40 (12): 2251–2253.

778 (a) Reissig, H.U. and Zimmer, R. (2003). Donor-acceptor-substituted cyclopropane derivatives and their application in organic synthesis. *Chem. Rev.* 103 (4): 1151–1196. (b) O'Connor, N.R., Wood, J.L., and Stoltz, B.M. (2016). Synthetic applications and methodological developments of donor-acceptor cyclopropanes and related compounds. *Isr. J. Chem.* (6–7): 437–444. (c) Wang, L.-J. and Tang, Y. (2016). Asymmetric ring-opening reactions of donor-acceptor cyclopropanes and cyclobutanes. *Isr. J. Chem.* 56 (6–7): 463–475. (d) Rassadin, V.A. and Six, Y. (2016). Ring-opening, cycloaddition and rearrangement reactions of nitrogen-substituted cyclopropane derivatives. *Tetrahedron* 72 (32): 4701–4757. (e) Budynina, E.M., Ivanov, K.L., Sorokin, I.D., and Melkinov, M.Y. (2017). Ring opening of donor-acceptor cyclopropanes with N-nucleophiles. *Synthesis* 49 (14): 3035–3068.

779 (a) Kulinkovich, O.G. (2003). The chemistry of cyclopropanols. *Chem. Rev.* 103 (7): 2597–2632. (b) Nikolaev, A. and Orellana, A. (2016). Transition-metal-catalyzed C—C and C—X bond forming reactions using cyclopropanols. *Synthesis* 48 (12): 1741–1768.

780 Hudlicky, T. and Reed, J.W. (2010). From discovery to application: 50 Years of the vinylcyclopropane-cyclopentene rearrangement and its impact on the synthesis of natural products. *Angew. Chem. Int. Ed.* 49 (29): 4864–4876.

781 Pellissier, H. (2010). Recent developments in the reactivity of methylene- and alkylidenecyclopropane derivatives. *Tetrahedron* 66 (43): 8341–8375.

782 (a) Pellissier, H. (2008). Recent developments in asymmetric cyclopropanation. *Tetrahedron* 64 (30–31): 7041–7095. (b) Bartoli, G., Bencivenni, G., and Dalpozzo, R. (2014). Asymmetric cyclopropanation reactions. *Synthesis* 46 (8): 979–1029.

783 Kim, H.Y., Salvi, L., Carroll, P.J., and Walsh, P.J. (2010). One-pot catalytic enantio- and diastereoselective syntheses of *anti*-, *syn-cis*-disubstituted, and *syn*-vinyl cyclopropyl alcohols. *J. Am. Chem. Soc.* 132 (1): 402–412.

784 Nozaki, H., Moriuti, S., Takaya, H., and Noyori, R. (1966). Asymmetric induction in carbenoid reaction by means of a dissymmetric Cu chelate. *Tetrahedron Lett.* (43): 5239–5244.

785 Doyle, M.P. and Forbes, D.C. (1998). Recent advances in asymmetric catalytic metal carbene transformations. *Chem. Rev.* 98 (2): 911–935.

786 (a) Lebel, H., Marcoux, J.F., Molinaro, C., and Charette, A.B. (2003). Stereoselective cyclopropanation reactions. *Chem. Rev.* 103 (4): 977–1050. (b) Ebner, C. and Carreira, E.M. (2017). Cyclopropanation strategies in recent total synthesis. *Chem. Rev.* 117 (18): 11651–11679.

787 Bellemin-Laponnaz, S. and Gade, L.H. (2002). A modular approach to C_1 and C_3 chiral N-tripodal ligands for asymmetric catalysis. *Angew. Chem. Int. Ed.* 41 (18): 3473–3475.

788 Li, J., Liao, S.-H., Xiong, H. et al. (2012). Highly diastereo- and enantioselective cyclopropanation of 1,2-disubstituted alkenes. *Angew. Chem. Int. Ed.* 51 (35): 8838–8841.

789 Deng, C., Wang, L.-J., Zhu, J., and Tang, Y. (2012). A chiral cagelike Cu(I) catalyst for the highly enantioselective synthesis of 1,1-cyclopropane diesters. *Angew. Chem. Int. Ed.* 51 (46): 11620–11623.

790 Skell, P.S. and Garner, A.Y. (1956). The stereochemistry of carbene-olefin reactions - reactions of dibromocarbene with the *cis*-2-butenes and *trans*-2-butenes. *J. Am. Chem. Soc.* 78 (14): 3409–3411.

791 Moore, W.R., Moser, W.R., and Laprade, J.E. (1963). Reactions of dibromocarbene adducts of bicyclo[2.2.1]heptene and bicyclo[2.2.1]heptadiene. *J. Org. Chem.* 28 (9): 2200–2205.

792 Hoffmann, R. (1968). Trimethylene and addition of methylene to ethylene. *J. Am. Chem. Soc.* 90 (6): 1475–1485.

793 Bodor, N., Wasson, J.S., and Dewar, M.J.S. (1972). Ground-states of molecules. 19. Carbene and its reactions. *J. Am. Chem. Soc.* 94 (26): 9095–9102.

794 Nagase, S. and Fueno, T. (1976). Localized molecular-orbital studies of chemical reaction

.2. Abstraction and addition reactions of triplet methylene. *Bull. Chem. Soc. Jpn.* 49 (11): 2920, 2926.

795 Zurawski, B. and Kutzelnigg, W. (1978). Ab initio calculation of reaction path for addition of singlet (1A1) methylene to ethylene. *J. Am. Chem. Soc.* 100 (9): 2654–2659.

796 Houk, K.N., Rondan, N.G., and Mareda, J. (1985). Theoretical studies of halocarbene cycloaddition selectivities - a new interpretation of negative activation energies and entropy control of selectivity. *Tetrahedron* 41 (8): 1555–1563.

797 Bernardi, F., Bottoni, A., Canepa, C. et al. (1997). MCSCF/MP_2 study of the cheletropic addition of singlet and triplet CF_2 and $C(OH)_2$ to the ethene double bond. *J. Org. Chem.* 62 (7): 2018–2025.

798 Blavins, J.J., Cooper, D.L., and Karadakov, P.B. (2004). Spin-coupled study of addition reactions of singlet dihalocarbenes with ethene. *Int. J. Quantum Chem.* 98 (5): 465–472.

799 Keating, A.E., Merrigan, S.R., Singleton, D.A., and Houk, K.N. (1999). Experimental proof of the non-least-motion cycloadditions of dichlorocarbene to alkenes: kinetic isotope effects and quantum mechanical transition states. *J. Am. Chem. Soc.* 121 (16): 3933–3938.

800 Xu, L., Doubleday, C.E., and Houk, K.N. (2011). Dynamics of carbene cycloadditions. *J. Am. Chem. Soc.* 133 (44): 17848–17854.

801 Baceiredo, A., Bertrand, G., and Sicard, G. (1985). Synthesis of the 1st A-diazo phosphines - phosphorus carbon multiple-bond character of phosphinocarbenes. *J. Am. Chem. Soc.* 107 (16): 4781–4783.

802 Igau, A., Baceiredo, A., Trinquier, G., and Bertrand, G. (1989). [Bis(diisopropylamino) phosphino]trimethylsilylcarbene - a stable nucleophilic carbene. *Angew. Chem. Int. Ed. Engl.* 28 (5): 621–622.

803 Lavallo, V., Mafhouz, J., Canac, Y. et al. (2004). Synthesis, reactivity, and ligand properties of a stable alkyl carbene. *J. Am. Chem. Soc.* 126 (28): 8670–8671.

804 Khramov, D.M., Rosen, E.L., Lynch, V.M., and Bielawski, C.W. (2008). Diaminocarbene[3]ferrocenophanes and their transition-metal complexes. *Angew. Chem. Int. Ed.* 47 (12): 2267–2270.

805 Moerdyk, J.P. and Bielawski, C.W. (2012). Diamidocarbenes as versatile and reversible (2+1)-cycloaddition reagents. *Nat. Chem.* 4 (4): 275–280.

806 Goumri-Magnet, S., Kato, T., Gornitzka, H. et al. (2000). Stereoselectivity and stereospecificity of cyclopropanation reactions with stable (phosphanyl)(silyl)carbenes. *J. Am. Chem. Soc.* 122 (18): 4464–4470.

807 Lecea, B., Ayerbe, M., Arrieta, A. et al. (2007). Theoretical study on the mechanism of the (2+1) thermal cycloaddition between alkenes and stable singlet (phosphino)(silyl)carbenes. *J. Org. Chem.* 72 (2): 357–366.

808 Kraakman, P.A., Dewolf, W.H., and Bickelhaupt, F. (1989). The stereochemistry of the 1,4-addition of dibromocarbene to 1,2-dimethylenecycloheptane. *J. Am. Chem. Soc.* 111 (22): 8534–8535.

809 Sweeney, J.B. (2002). Aziridines: Epoxides' ugly cousins? *Chem. Soc. Rev.* 31 (5): 247–258.

810 Yudin, A.K. (ed.) (2006). *Aziridines an Epoxides in Organic Synthesis*, 1–492. Weinheim: Wiley-VCH.

811 Takeuchi, H., Igura, T., Mitani, M. et al. (1978). Stabilization of singlet ethoxycarbonylnitrene by 1,4-dioxan. 3. Stereospecificity for addition to 4-methylpent-2-ene of nitrene generated from ethyl azidoformate. *J. Chem. Soc., Perkin Trans. 2* (8): 783–786.

812 Mansuy, D., Mahy, J.P., Dureault, A. et al. (1984). Iron-porphyrin and Mn-porphyrin catalyzed aziridination of alkenes by tosyl-iminoiodobenzene and acyl-iminoiodobenzene. *J. Chem. Soc., Chem. Commun.* (17): 1161–1163.

813 Han, H., Park, S.B., Kim, S.K., and Chang, S. (2008). Copper-nitrenoid formation and transfer in catalytic olefin aziridination utilizing chelating 2-pyridylsulfonyl moieties. *J. Org. Chem.* 73 (7): 2862–2870.

814 Yamada, Y., Yamamoto, T., and Okawara, M. (1975). Synthesis and reaction of new type I-*N* ylide, *N*-Tosyliminoiodinane. *Chem. Lett.* (4): 361–362.

815 Fantauzzi, S., Caselli, A., and Gallo, E. (2009). Nitrene transfer reactions mediated by metallo-porphyrin complexes. *J. Chem. Soc., Dalton Trans.* (28): 5434–5443.

816 Evans, D.A., Faul, M.M., Bilodeau, M.T. et al. (1993). Bis(oxazoline) Cu-complexes as chiral catalysts for the enantioselective aziridination of olefins. *J. Am. Chem. Soc.* 115 (12): 5328–5329.

817 Li, Z., Conser, K.R., and Jacobsen, E.N. (1993). Asymmetric alkene aziridination with readily available chiral diimine-based catalysts. *J. Am. Chem. Soc.* 115 (12): 5326–5327.

818 Müller, P. and Fruit, C. (2003). Enantioselective catalytic aziridinations and asymmetric nitrene insertions into CH bonds. *Chem. Rev.* 103 (8): 2905–2919.

819 Pellissier, H. (2010). Recent developments in asymmetric aziridination. *Tetrahedron* 66 (8): 1509–1555.

820 (a) Jenkins, D.M. (2012). Atom-economical C_2+N_1 aziridination: progress towards catalytic intermolecular reactions using alkenes and aryl azides. *Synlett* (9): 1267–1270. (b) Chandrachud, P.P., Bass, H.M., and Jenkins, D.M. (2016). Synthesis of fully aliphatic aziridines with a macrocyclic tetracarbene Fe catalyst. *Organometallics* 35 (11): 1652–1657. (c) Damiano, C., Intrieri, D.M., and Gallo, E. (2018). Aziridination of alkenes promoted by Fe or Ru complexes. *Inorg. Chim. Acta* 470: 51–67.

821 Moncada, S., Palmer, R.M.J., and Higgs, E.A. (1991). Nitric-oxide - physiology, pathophysiology, and pharmacology. *Pharmacol. Rev.* 43 (2): 109–142.

822 Palmer, R.M.J., Ashton, D.S., and Moncada, S. (1988). Vascular endothelial-cells synthesize nitric oxide from L-arginine. *Nature* 333 (6174): 664–666.

823 Marletta, M.A., Yoon, P.S., Iyengar, R. et al. (1988). Macrophage oxidation of L-arginine to nitrite and nitrate - nitric oxide is an intermediate. *Biochemistry* 27 (24): 8706–8711.

824 Hibbs, J.B., Taintor, R.R., Vavrin, Z., and Rachlin, E.M. (1988). Nitric oxide - a cytotoxic activated macrophage effector molecule. *Biochem. Biophys. Res. Commun.* 157 (1): 87–94.

825 Snyder, S.H. (1992). Nitric oxide - 1st in a new class of neurotransmitters. *Science* 257 (5069): 494–496.

826 Culotta, E. and Koshland, D.E. (1992). NO news is good news. *Science* 258 (5090): 1862–1865.

827 Schuman, E.M. and Madison, D.V. (1991). A requirement for the intercellular messenger nitric oxide in long term potentiation. *Science* 254 (5037): 1503–1506.

828 Adam, W., Bottle, S.E., Grice, I.D. et al. (1996). The cheletropic trapping of nitric oxide by the bis-ketene 1,2-dicarbonylcyclohexa-3,5-diene and the diene 3,4-diphenyl-2,5-dimethyl-2,4-hexadiene. *Tetrahedron Lett.* 37 (12): 2113–2116.

829 Turro, N.J. and Hammond, W.B. (1966). Cyclopropanone. *J. Am. Chem. Soc.* 88 (15): 3672–3972.

830 Hammond, W.B. and Turro, N.J. (1966). Cyclopropanones. 3. 2,2-Dimethylcyclopropanone. *J. Am. Chem. Soc.* 88 (12): 2880–2881.

831 Sclove, D.B., Pazos, J.F., Camp, R.L., and Greene, F.D. (1970). Thermal reactions of a cyclopropanone - racemization and decarbonylation of *trans*-2,3-di-*tert*-butylcyclopropanone. *J. Am. Chem. Soc.* 92 (25): 7488–7488.

832 Pazos, J.F., Pacifici, J.G., Pierson, G. et al. (1974). *Trans*-di-*tert*-butylcyclopropanone - preparation, properties, resolution, and reaction with nucleophiles. *J. Org. Chem.* 39 (14): 1990–1995.

833 Chamorro, E. (2003). The nature of bonding in pericyclic and pseudopericyclic transition states: thermal chelotropic decarbonylations. *J. Chem. Phys.* 118 (19): 8687–8698.

834 Dolbier, W.R. and Frey, H.M. (1974). Thermal decomposition of cyclopent-3-enone. *J. Chem. Soc., Perkin Trans. 2* (14): 1674–1676.

835 Buxton, J.P. and Simpson, C.J.S.M. (1986). Thermal decarbonylations of unsaturated cyclic ketones - kinetics and dynamics. *Chem. Phys.* 105 (3): 307–316.

836 Birney, D.M. and Berson, J.A. (1985). Norborna-2,5-dien-7-one - a covalent benzene carbon monoxide adduct - a new point on the cycloreversion structure reactivity correlation curve. *J. Am. Chem. Soc.* 107 (15): 4553–4554.

837 Rubio, M., Hernández, A.G., Daudey, J.P. et al. (1980). Thermal decarbonylation of some strained ketones. *J. Org. Chem.* 45 (1): 150–154.

838 Zhu, Q., Liu, M.-F., Wang, B., and Cheng, Y. (2007). An unprecedented tandem 1,3-dipolar cycloaddition-cheletropic elimination: a facial approach to novel push-pull olefins. *Org. Biomol. Chem.* 5 (8): 1282–1286.

839 (a) Backer, H.J. and Strating, J. (1934). Cyclical sulphones, derivatives and butadienes. *Recl. Trav. Chim. Pays-Bas* 53: 525–543. (b) Deeming, A.S., Emmett, E.J., Richards-Taylor, C.S., and Willis, M.C. (2014). Rediscovering the chemistry of sulfur dioxide: New developments in synthesis and catalysis. *Synthesis* 46 (20): 2701–2710. (c) Skillinghaug, B., Rydfjord, J., and Odell, L.R. (2016). Synthesis of sodium aryl sulfinates from aryl bromides employing 1,4-diazabicyclo[2.2.2]octane bis(sulfur dioxide) adduct (DABSO) as a bench-stable, gas-free alternative to SO_2. *Tetrahedron Lett.* 57 (5): 533–536. (d) Van Mileghem, S. and De Borggraeve, W.M. (2017). A convenient multigram synthesis of DABSO using sodium sulfite as SO_2 source. *Org. Process Res. Dev.* 21 (5): 785–787.

840 Isaacs, N.S. and Laila, A.A.R. (1976). Reaction of 1,3-dienes with sulfur dioxide. 1. Thermal decomposition of 2,5-dihydrothiophen 1,1-dioxides. *J. Chem. Soc., Perkin Trans. 2* (13): 1470–1475.

841 Monnat, F., Vogel, P., and Sordo, J.A. (2002). Hetero-Diels–Alder and cheletropic additions of sulfur dioxide to 1,2-dimethylidenecycloalkanes. Determination of thermochemical and kinetics parameters for reactions in solution and comparison with estimates from quantum calculations. *Helv. Chim. Acta* 85 (3): 712–732.

842 Monnat, F., Vogel, P., Rayon, V.M., and Sordo, J.A. (2002). Ab initio and experimental studies on the hetero-Diels–Alder and cheletropic additions of sulfur dioxide to (*E*)-1-methoxybutadiene: a mechanism involving three molecules of SO_2. *J. Org. Chem.* 67 (6): 1882–1889.

843 Nicolaou, K.C., Barnette, W.E., and Ma, P. (1980). A remarkably simple, highly efficient, and stereoselective synthesis of steroids and other polycyclic systems - total synthesis of estra-1,3,5(10)-trien-17-one via intramolecular capture of *ortho*-quinodimethanes generated by cheletropic elimination of SO_2. *J. Org. Chem.* 45 (8): 1463–1470.

844 Wojciechowski, K. (2001). Aza-*ortho*-xylylenes in organic synthesis. *Eur. J. Org. Chem.* (19): 3587–3605.

845 De Lucchi, O. and Lucchini, V. (1982). Homocheletropic addition of sulfur dioxide to norbornadiene. *J. Chem. Soc., Chem. Commun.* (19): 1105–1106.

846 Roulet, J.M., Deguin, B., and Vogel, P. (1994). Homocheletropic addition of sulfur dioxide to 1,4-dienes - competition between cheletropic and homocheletropic additions of polyenes. *J. Am. Chem. Soc.* 116 (8): 3639–3640.

847 Roulet, J.M., Vogel, P., Wiesemann, F., and Pinkerton, A.A. (1995). Competition between cheletropic and homocheletropic additions of sulfur dioxide to 2,3,5,6-tetrakis(methylene)bicyclo[2.2.n]alkanes - crystal and molecular structures of two 3-thiabicyclo[3.1.0]hexane 3,3-dioxide derivatives. *Tetrahedron* 51 (6): 1685–1696.

848 Staudinger, H. and Pfenninger, F. (1916). The effect of sulphur dioxide on diphenyl-diazomethane. *Ber. Dtsch. Chem. Ges.* 49: 1941–1951.

849 Hesse, G., Reichold, E., and Majmudar, S. (1957). Über die Einwirkung von Schwefeldioxyd auf Diazoverbindungen. 2. Äthylensulfon. *Chem. Ber.* 90 (10): 2106–2113.

850 Neureiter, N.P. and Bordwell, F.G. (1963). Mechanism for formation of *cis*-alkenes from A-chloro sulfones. *J. Am. Chem. Soc.* 85 (8): 1209–1210.

851 Tokura, N., Nagai, T., and Matsumur, S. (1966). Stereospecific formation and decomposition of a *cis*-episulfone from phenyldiazomethane and sulfur dioxide. *J. Org. Chem.* 31 (1): 349–350.

852 Opitz, G. and Fischer, K. (1965). Synthesis of olefins from sulfenes and diazoalkanes via episulfones. *Angew. Chem. Int. Ed. Engl.* 4 (1): 70–70.

853 Markovic, D. and Vogel, P. (2004). Polysulfones: Catalysts for alkene isomerization. *Angew. Chem. Int. Ed.* 43 (22): 2928–2930.

854 Markovic, D., Steunenberg, P., Ekstrand, M., and Vogel, P. (2004). Polysulfones: solid organic catalysts for the chemoselective cleavage of methyl-substituted allyl ethers under neutral conditions. *Chem. Commun.* (21): 2444–2445.

855 Bordwell, F.G. and Williams, J.M. (1968). Ramberg–Bäcklund reaction of benzyl A-halobenzyl and halomethyl sulfones. *J. Am. Chem. Soc.* 90 (2): 435–439.

856 Meyers, C.Y., Malte, A.M., and Matthews, W.S. (1969). Ionic reactions of carbon tetrachloride. Survey of reactions with ketones, alcohols, and sulfones. *J. Am. Chem. Soc.* 91 (26): 7510–7512.

857 (a) Paterson, D.E., Griffin, F.K., Alcaraz, M.L., and Taylor, R.J.K. (2002). A Ramberg–Bäcklund approach to the synthesis of C-glycosides, C-linked disaccharides, and C-glycosyl amino acids. *Eur. J. Org. Chem.* (7): 1323–1336. (b) Taylor, R.J.K., McAllister, G.D., and Franck, R.W. (2006). The Ramberg–Bäcklund reaction for the synthesis of C-glycosides, C-linked-disaccharides and related compounds. *Carbohydr. Res.* 341 (10): 1298–1311.

858 (a) Chan, T.L., Fong, S., Li, Y. et al. (1994). A new one-flask Ramberg–Bäcklund reaction. *J. Chem. Soc., Chem. Commun.* (15): 1771–1772. (b) Lou, X. (2015). The recent advance of Ramberg–Bäcklund reaction. *Mini-Rev. Org. Chem.* 12 (5): 449–454.

859 Souto, J.A., López, C.S., Faza, O.N. et al. (2005). 2-Alkylidenesulfol-3-enes by (regio- and) stereoselective cheletropic addition of SO_2 to (di)vinylallenes. *Org. Lett.* 7 (8): 1565–1568.

860 Wojciechowski, K. (1993). Reactions of aza-*ortho*-xylylenes generated from

2,1-benzisothiazoline 2,2-dioxides. *Tetrahedron* 49 (33): 7277–7286.

861 Bouchez, L.C. and Vogel, P. (2005). Synthesis of the C(1)-C(11) polyene fragment of Apoptolidin with a new sulfur dioxide-based organic chemistry. *Chem. Eur. J.* 11 (16): 4609–4620.

862 (a) Hill, N.J. and West, R. (2004). Recent development in the chemistry of stable silylenes. *J. Organomet. Chem.* 689 (24): 4165–4183. (b) Lee, V.Y. and Sekiguchi, A. (eds.) (2010). *Organometallic Compounds of Low-Coordinate Si, Ge, Sn, and Pb. From Phantom Species to Stable Compounds*, 1–448. Hoboken, NJ: Wiley. (c) Raoufrnoghaddam, S., Zhou, Y.P., Wang, Y.W., and Driess, M. (2017). N-heterocyclic silylenes as powerful steering ligands in catalysis. *J. Organomet. Chem.* 829: 2–10.

863 Mizuhata, Y., Sasamori, T., and Tokitoh, N. (2009). Stable heavier carbene analogues. *Chem. Rev.* 109 (8): 3479–3511.

864 Weidenbruch, M. (1995). Cyclotrisilanes. *Chem. Rev.* 95 (5): 1479–1493.

865 Gilman, H., Cottis, S.G., and Atwell, W.H. (1964). Preparation + reactions of some 7-silanorbornadienes. I. Approach to dimethylsilylene. *J. Am. Chem. Soc.* 86 (8): 1596–1599.

866 Atwell, W.H. (2009). The reaction of silylenes with acetylenes: in search of the reaction pathways. *Organometallics* 28 (13): 3573–3586.

867 (a) Seyferth, D. and Annarelli, D.C. (1975). Generation of dimethylsilylene under mild conditions by thermolysis of hexamethylsilirane. *J. Am. Chem. Soc.* 97 (24): 7162–7163. (b) Becerra, R. and Walsh, R. (2010). Kinetic studies of reactions of organosilylenes: what have they taught us? *J. Chem. Soc., Dalton Trans.* 39 (39): 9217–9228.

868 Seyferth, D., Annarelli, D.C., and Vick, S.C. (1976). 1,1-Dimethyl-2,3-bis(trimethylsilyl)-1-silirene, A stable silacyclopropene. *J. Am. Chem. Soc.* 98 (20): 6382–6384.

869 Conlin, R.T. and Gaspar, P.P. (1976). Tetramethylsilacyclopropene. *J. Am. Chem. Soc.* 98 (12): 3715–3716.

870 Himmel, H.J. and Schnockel, H. (2003). An alternative way of characterising the bonding in compounds featuring main-group elements and with the potential for multiple bonding: on the dissociation of binary main-group hydrides. *Chem. Eur. J.* 9 (3): 748–755.

871 Suzuki, H., Tokitoh, N., and Okazaki, R. (1995). Thermal dissociation of disilenes into silylenes. *Bull. Chem. Soc. Jpn.* 68 (9): 2471–2481.

872 Hermanns, J. and Schmidt, B. (1998). Five- and six-membered silicon-carbon heterocycles. Part 1. Synthetic methods for the construction of silacycles. *J. Chem. Soc., Perkin Trans. 1* (14): 2209–2230.

873 Jutzi, P., Kanne, D., and Krüger, C. (1986). Decamethylsilicocene - synthesis and structure. *Angew. Chem. Int. Ed. Engl.* 25 (2): 164–164.

874 Denk, M., Lennon, R., Hayashi, R. et al. (1994). Synthesis and structure of a stable silylene. *J. Am. Chem. Soc.* 116 (6): 2691–2692.

875 Kira, N., Ishida, S., Iwamoto, T., and Kabuto, C. (1999). The first isolable dialkylsilylene. *J. Am. Chem. Soc.* 121 (41): 9722–9723.

876 Kira, M., Iwamoto, T., and Ishida, S. (2007). A helmeted dialkylsilylene. *Bull. Chem. Soc. Jpn.* 80 (2): 258–275.

877 Rodriguez, R., Gau, D., Contie, Y. et al. (2011). Synthesis of a phosphine-stabilized silicon(II) hydride and its addition to olefins: a catalyst-free hydrosilylation reaction. *Angew. Chem. Int. Ed.* 50 (48): 11492–11495.

878 Rodriguez, R., Troadec, T., Kato, T. et al. (2012). Synthesis and characterization of an isolable base-stabilized silacycloprop-1-ylidene. *Angew. Chem. Int. Ed.* 51 (29): 7158–7161.

879 Rekken, B.D., Brown, T.M., Fettinger, J.C. et al. (2012). Isolation of a stable, acyclic, two-coordinate silylene. *J. Am. Chem. Soc.* 134 (15): 6504–6507.

880 Protchenko, A.V., Birjkumar, K.H., Dange, D. et al. (2012). A stable two-coordinate acyclic silylene. *J. Am. Chem. Soc.* 134 (15): 6500–6503.

881 Sun, L., Jin, G., Feng, W. et al. (2011). Cycloaddition reactions of N-heterocyclic stable silylenes with ethylene and formaldehyde. *J. Organomet. Chem.* 696 (4): 841–845.

882 Nag, M. and Gaspar, P.P.A. (2009). Computational study of the mechanism of addition of singlet carbene analogues to 1,3-butadiene to form 1,1-dimethylmetallacyclopent-3-enes [$MMe_2C_4H_6$, M=Si, Ge, Sn] and their reverse retro-addition reactions. *Organometallics* 28 (19): 5612–5622.

883 Schriewer, M. and Neumann, W.P. (1983). Chemistry of heavy carbene analogs: R_2Si,R_2Ge,R_2Sn. 8. Germylenes - singlets or triplets – (1+4)-cheletropic cycloadditions of Me_2Ge and GeI_2 to conjugated dienes. *J. Am. Chem. Soc.* 105 (4): 897–901.

884 Marx, R., Neumann, W.P., and Hillner, K. (1984). (1+4)-cheletropic cycloadditions of stannylenes (alkyl)2Sn, (amino)2Sn, and (halogen)2Sn. *Tetrahedron Lett.* 25 (6): 625–628.

885 Chrostowska, A., Lemierre, V., Dargelos, A. et al. (2009). Gas phase synthesis and characterization of heteroleptic divalent Ge compounds by FVT/UV-PES. *J. Organomet. Chem.* 694 (1): 43–51.

886 Harris, D.H. and Lappert, M.F. (1974). Monomeric, volatile bivalent amides of Group-4B elements, $M(NR^1R^2)_2$ and $M(NR^1R^2)_2$ (M=Ge, Sn, or M=PbR^1, M=$SiMe_3$, M=CMe_3). *J. Chem. Soc., Chem. Commun.* (21): 895–896.

887 Davidson, P.J., Harris, D.H., and Lappert, M.F. (1976). Sub-valent Group 4B metal alkyls and amides. 1. Synthesis and physical properties of kinetically stable bis[bis(trimethysilyl)methyl] Ge(II), bis[bis(trimethysilyl)methyl]Sn(II), and bis[bis(trimethsilyl)methyl]Pb(II). *J. Chem. Soc., Dalton Trans.* (21): 2268–2274.

888 Bucher, G., Borst, M.L.G., Ehlers, A.W. et al. (2005). Infrared, UV/Vis, and W-band EPR spectroscopic characterization and photochemistry of triplet mesitylphosphinidene. *Angew. Chem. Int. Ed.* 44 (21): 3289–3293.

889 Harrison, J.J. and Williamson, B.E. (2005). Magnetic circular dichroism and absorption spectra of phosphinidene in noble-gas matrices. *J. Phys. Chem. A* 109 (7): 1343–1347.

890 Cowley, A.H., Gabbai, F., Schluter, R., and Atwood, D. (1992). New approaches to the generation of phosphinidenes. *J. Am. Chem. Soc.* 114 (8): 3142–3144.

891 Mathey, F. (2007). Developing the chemistry of monovalent phosphorus. *Dalton Trans.* (19): 1861–1868.

892 Aktas, H., Slootweg, J.C., and Lammertsma, K. (2010). Nucleophilic phosphinidene complexes: access and applicability. *Angew. Chem. Int. Ed.* 49 (12): 2102–2113.

893 Marinetti, A., Mathey, F., Fischer, J., and Mitschler, A. (1982). Generation and trapping of terminal phosphinidene complexes - synthesis and X-ray crystal structure of stable phosphirene complexes. *J. Am. Chem. Soc.* 104 (16): 4484–4485.

894 Marinetti, A., Mathey, F., Fischer, J., and Mitschler, A. (1982). Stabilization of 7-phosphanorbornadienes by complexation - X-ray crystal structure of 2,3-bis(methoxycarbonyl)-5,6-dimethyl-7-phenyl-7-phosphanorbornadiene$(CO)_5Cr$. *J. Chem. Soc., Chem. Commun.* (12): 667–668.

895 Hitchcock, P.B., Lappert, M.F., and Leung, W.P. (1987). The 1st stable transition-metal (Mo or W) complexes having a metal phosphorus(III) double-bond - the phosphorus analogs of metal aryl-imides and alkyl-imides - X-ray structure of $[Mo(E^5\text{-}C_5H_5)_2]=P[2,4,6\text{-tris}(\textit{tert}\text{-Bu})C_6H_2]$. *J. Chem. Soc., Chem. Commun.* (17): 1282–1283.

896 Bohra, R., Hitchcock, P.B., Lappert, M.F., and Leung, W.P. (1989). Synthetic and structural studies on some bis(cyclopentadienyl)Mo and W complexes containing doubly bonded Sn or P. *Polyhedron* 8 (13–14): 1884–1884.

897 Borst, M.L.G., Bulo, R.E., Winkel, C.W. et al. (2005). Phosphepines: convenient access to phosphinidene complexes. *J. Am. Chem. Soc.* 127 (16): 5800–5801.

898 Hillner, K. and Neumann, W.P. (1986). Additions of a stable stannylene R_2Sn to vinyl carbonyl compounds and 1,2-diketones. *Tetrahedron Lett.* 27 (44): 5347–5350.

899 Bulo, R.E., Trion, L., Ehlers, A.W. et al. (2004). C=C Bond insertion of a complexed phosphinidene into 1,6-methano[10]annulene. *Chem. Eur. J.* 10 (21): 5332–5337.

900 Fittig, R. (1860). Über einige Derivate des Acetons. *Liebigs Ann. Chem.* 114: 54–63.

901 Butlerow, A. (1873). Über trimethylessigsaüre. *Liebigs Ann. Chem.* 170: 151–162.

902 Bunton, C.A., Hadwick, T., Llewellyn, D.R., and Pocker, Y. (1956). The mechanism of the pinacol-pinacolone rearrangement. *Chem. Ind. (London)* (24): 547–547.

903 (a) Itoh, S. and Yamataka, H. (2010). Reaction pathway of aliphatic pinacol-type rearrangement re-examined. *J. Phys. Org. Chem.* 23 (8): 789–795. (b) Mora, J.R., Márquez, E., Lezama, J. et al. (2011). The reaction mechanism of the gas phase thermal decomposition kinetics of neopentyl halides: A DFT study. *Int. J. Quantum Chem.* 111 (15): 4011–4019. (c) Drosos, N., Cheng, G.J., Ozkai, E. et al. (2017). Catalytic reductive pinacol-type rearrangement of unactivated 1,2-diols through a concerted, stereoinvertive mechanism. *Angew. Chem. Int. Ed.* 56 (43): 13377–13381.

904 Whitmore, F.C. (1932). Common basis of intramolecular rearrangements. *J. Am. Chem. Soc.* 54: 3274–3283.

905 Beggs, J.J. and Meyers, M.B. (1970). Stereochemical fate of an asymmetric migrating group in pinacol rearrangement. *J. Chem. Soc. B* (5): 930–934.

906 Saunders, M. and Kates, M.R. (1978). Rates of degenerate 1,2-hydride and 1,2-methide shifts from ^{13}C nuclear magnetic resonance spectra of tertiary alkyl cations. *J. Am. Chem. Soc.* 100 (22): 7082–7083.

907 (a) Saunders, M. and Vogel, P. (1971). Equilibrium deuterium isotope effects in systems undergoing rapid rearrangements - dimethyl-*tert*-butylcarbonium ion and cyclopentyl cation. *J. Am. Chem. Soc.* 93 (10): 2559–2561. (b) Sandbeck, D.J.S., Markewich, D.J., and East, A.L.L. (2016). The carbocation rearrangement mechanism, clarified. *J. Org. Chem.* 81 (4): 1410–1415.

908 Meot-Ner, M., Solomon, J.J., and Field, F.H. (1976). Stability of some C7 tertiary alkyl carbonium ions. *J. Am. Chem. Soc.* 98 (4): 1025–1026.

909 Dewar, M.J.S. (1951). A review of the Π-complex theory. *Bull. Soc. Chim. Fr.* 18 (3–4): C71–C79.

910 Floss, H.G. and Lee, S.S. (1993). Chiral methyl groups - small is beautiful. *Acc. Chem. Res.* 26 (3): 116–122.

911 Cannizzaro, S. (1853). Über den der Benzolsäure entsprechenden Alkohol. *Liebigs Ann. Chem.* 88: 129–130.

912 Geissman, T.A. (1944). The Cannizzaro reaction. *Org. React.* 2: 94–113.

913 Abaee, M.S., Sharifi, R., and Mojtahedi, M.M. (2005). Room-temperature Cannizzaro reaction under mild conditions facilitated by magnesium bromide ethyl etherate and triethylamine. *Org. Lett.* 7 (26): 5893–5895.

914 Basavaiah, D., Sharada, D.S., and Veerendhar, A. (2006). Organo-base mediated Cannizzaro reaction. *Tetrahedron Lett.* 47 (32): 5771–5774.

915 Schmitt, E., Schiffers, I., and Bolm, C. (2009). Organocatalytic conversion of arylglyoxals into optically active mandelic acid derivatives. *Tetrahedron Lett.* 50 (26): 3185–3188.

916 Franzen, V. (1957). Asymmetrische Katalyse der Intramolekularen Cannizzaro-Reaktion. *Chem. Ber.* 90 (9): 2036–2039.

917 Maruyama, K., Murakami, Y., Yoda, K. et al. (1992). Catalysis by Co Schiffs base complexes in highly selective conversion of arylglyoxals to A-aryl-A-hydroxyacetic esters. *J. Chem. Soc., Chem. Commun.* (21): 1617–1618.

918 Russell, A.E., Miller, S.P., and Morken, J.P. (2000). Efficient Lewis acid catalyzed intramolecular Cannizzaro reaction. *J. Org. Chem.* 65 (24): 8381–8383.

919 Wang, P., Tao, W.J., Sun, X.-L. et al. (2013). A highly efficient and enantioselective intramolecular Cannizzaro reaction under TOX/Cu(II) catalysis. *J. Am. Chem. Soc.* 135 (45): 16849–16852.

920 Wieland, H. (1911). Triphenylmethyl-peroxide. An article on the chemistry of the free radicals. *Ber. Dtsch. Chem. Ges.* 44: 2550–2556.

921 Berson, J.A., Walia, J.S., and Olsen, C.J. (1962). Reactions of 2-bornyl radical. 2. Free radical Wagner–Meerwein rearrangement. *J. Am. Chem. Soc.* 84 (17): 3337–3348.

922 Davis, A.C. and Francisco, J.S. (2011). Ab Initio study of hydrogen migration across *n*-alkyl radicals. *J. Phys. Chem. A* 115 (14): 2966–2977.

923 Fossey, J. and Nedelec, J.Y. (1981). Theoretical study of free radical migrations. *Tetrahedron* 37 (17): 2967–2976.

924 Mariano, D., Vera, A., and Pierini, A.B. (2004). Species with negative electron affinity and standard DFT methods. *Phys. Chem. Chem. Phys.* 6 (11): 2899–2903.

925 Pauling, L. (1960). *The Nature of the Chemical Bond and the Structure of Molecules and Crystals*, 3e, 1–664. Ithaca, NY: Cornwell University Press.

926 Bartlett, P.D. and Cotman, J.D. (1950). Migration aptitude as a criterion of ionic mechanism in the rearrangement of mono-*para*-nitrotriphenylmethyl hydroperoxide. *J. Am. Chem. Soc.* 72 (7): 3095–3095.

927 Asensio, A. and Dannenberg, J.J. (2001). Phenyl-bridging in the 2-phenylethyl radical. A molecular orbital study. *J. Org. Chem.* 66 (18): 5996–5999.

928 Maillard, B. and Ingold, K.U. (1976). Kinetic applications of electron paramagnetic resonance spectroscopy. 24. Neophyl rearrangements. *J. Am. Chem. Soc.* 98 (5): 1224–1226.

929 Franz, J.A., Barrows, R.D., and Camaioni, D.M. (1984). Arrhenius parameters for rearrangements of the neophyl, 1-indanylmethyl, 2-allylbenzyl, and 2-(2-vinylphenyl)ethyl radicals relative to hydrogen abstraction from tributylstannane. *J. Am. Chem. Soc.* 106 (14): 3964–3967.

930 Surzur, J.M. and Teissier, P. (1967). Addition radicalaire d' esters sur les alcools éthyléniques. *C.R. Seances Acad. Sci., Ser. C* 264 (24): 1981–1984.

931 Tanner, D.D. and Law, F.C.P. (1969). Free radical acetoxy group migration. *J. Am. Chem. Soc.* 91 (26): 7535–7537.

932 (a) Beckwith, A.L. and Thomas, C.B. (1973). Mechanism of rearrangement of B-acyloxyalkyl radicals. *J. Chem. Soc., Perkin Trans.* 2 (6): 861–872. (b) Lacote, E. and Renaud, P. (1998). Rate enhancement of the radical 1,2-acyloxy shift (Surzur–Tanner rearrangement) by complexation with Lewis acids. *Angew. Chem. Int. Ed.* 37 (16): 2259–2262.

933 (a) Beckwith, A.L.J. and Duggan, P.J. (1996). The mechanism of the β-(acyloxy)alkyl radical rearrangement: substituent and solvent effects. *J. Am. Chem. Soc.* 118 (50): 12838–12839. (b) Wiile, U. (2006). Computational study on the 1,2-rearrangement in beta-(nitroxy)vinyl and β-(acetoxy) vinyl radicals. *J. Org. Chem.* 71 (11): 4040–4046.

934 Korth, H.G., Sustmann, R., Gröninger, K.S. et al. (1988). Electron-spin resonance spectroscopic investigation of carbohydrate radicals. 4. 1,2-Acyloxyl migration in pyranosyl radicals. *J. Org. Chem.* 53 (18): 4364–4369.

935 (a) Crich, D., Huang, X.H., and Beckwith, A.L.J. (1999). Free-radical ring contraction of six-, seven-, and eight-membered lactones by a 1,2-shift mechanism. A kinetic and ^{17}O NMR spectroscopic study. *J. Org. Chem.* 64 (5): 1762–1764. (b) Choi, S.Y., Crich, D., Horner, J.H. et al. (1999). β-(Phosphatoxy)alkyl and β-(acyloxy)alkyl radical migrations studied by laser flash photolysis. *Tetrahedron* 55 (11): 3317–3326.

936 Grovenstein, E.J. and Wentwort, G. (1967). Carbanions. 9. Evidence for a duality of mechanism in rearrangement of carbanions. *J. Am. Chem. Soc.* 89 (8): 1852–1862.

937 Grovenstein, E., Singh, J., Patil, B.P., and Vanderveer, D. (1994). Carbanions-27 - rearrangements of (9-alkyl-9-fluorenyl) methyllithium (or cesium) and 2,2-diphenyl-3, 3-dimethylbutyllithium. *Tetrahedron* 50 (20): 5971–5998.

938 Zimmerman, H. and Zweig, A. (1961). Carbanion rearrangements. 2. *J. Am. Chem. Soc.* 83 (5): 1196–1213.

939 Grovenstein, E. (1978). Skeletal rearrangements of organo-alkali metal compounds. *Angew. Chem. Int. Ed. Engl.* 17 (5): 313–332.

940 Grovenstein, E., Black, K.W., Goel, S.C. et al. (1989). Carbanions. 24. Rearrangements of (*E*)- and (*Z*)-2,2-diphenyl-3-pentenyl alkali metal compounds. *J. Org. Chem.* 54 (7): 1671–1679.

941 Grovenstein, E., Chiu, K.W., and Patil, B.B. (1980). Carbanions. 20. Rearrangements in reactions of 5-chloro-4,4-diphenyl-2-pentyne, 1-chloro-5,5-dimethyl-2,2-diphenyl-3-hexyne, and 5,5-dimethyl-2,2-diphenyl-3-hexyne with alkali metals - cation effects on migratory aptitudes of groups. *J. Am. Chem. Soc.* 102 (18): 5848–5859.

942 Grovenstein, E. and Cottingham, A.B. (1977). Carbanions. 17. Rearrangements of 2,2-diphenyl-4-pentenyl alkali metal compounds. *J. Am. Chem. Soc.* 99 (6): 1881–1889.

943 Baldwin, J.E. and Urban, F.J. (1970). Anion rearrangements . A case of a non-degenerate all-carbon (3,2)-sigmatropic process. *J. Chem. Soc., Chem. Commun.* (3): 165–166.

944 Stevens, T.S., Creighton, E.M., Gordon, A.B., and MacNicol, M. (1928). Degradation of quaternary ammonium salts Part I. *J. Chem. Soc.* 3193–3197.

945 (a) Thomson, T. and Stevens, T.S. (1932). Degradation of quaternary ammonium salts. Part V. Molecular rearrangement in related sulphur compounds. *J. Chem. Soc.* 69–73. (b) Vanecko, J.A. and West, F.G. (2002). A novel, stereoselective silyl-directed Stevens (1,2)-shift of ammonium ylides. *Org. Lett.* 4 (17): 2813–2816. (c) Bott, T.M., Vanecko, J.A., and West, F.G. (2009). One-carbon ring expansion of azetidines via ammonium ylide (1,2)-shifts: a simple route to substituted pyrrolidines. *J. Org. Chem.* 74 (7): 2832–2836. (d) Lahm, G., Pacheco, J.C.O., and Opatz, T. (2014). Rearrangements of nitrile-stabilized ammonium ylides. *Synthesis* (18): 2413–1421.

946 (a) Okazaki, Y., Asai, T., Ando, F., and Koketsu, J. (2006). The Stevens rearrangement of sulfur ylide generated by electrochemical reduction of sulfonium salt. *Chem. Lett.* 35 (1): 98–99. (b) Murphy, G.K., Marmsaeter, F.P., and West, F.G. (2006). Polycyclic oxonium ylides - use of cyclic acetals as convenient scaffolds in the construction of fused bicyclic compounds containing a medium ring. *Can. J. Chem.* 84 (10): 1470–1486. (c) Hosseini, S.N., Johnston, R.R., and West, F.G. (2017). Evidence for heterolytic cleavage of a cyclic oxonium ylide: implications for the mechanism of the Stevens (1,2)-shift. *J. Chem. Soc., Chem. Commun.* 53 (94): 12654–12656.

947 (a) Sommelet, M. (1937). A particular method of intramolecular rearrangement. *C.R. Hebd. Seances Acad. Sci.* 205: 56–58. (b) Sweeney, J.B. (2009). Sigmatropic rearrangements of "onium" ylids. *Chem. Soc. Rev.* 38 (4): 1027–1038. (c) Ghigo, G., Cagnina, S., Maranzana, A., and Tonachini, G. (2010). The mechanism of the Stevens and Sommelet–Hauser rearrangements. A theoretical study. *J. Org. Chem.* 75 (11): 3608–3617. (d) Tayama, E., Watanabe, K., and Matano, Y. (2016). Ring-strain effects in base-induced Sommelet–Hauser rearrangement:

application to successive stereocontrolled transformations. *Eur. J. Org. Chem.* (21): 3631–3641.

948 (a) Kantor, S.W. and Hauser, C.R. (1951). Rearrangements of benzyltrimethylammonium ion and related quaternary ammonium ions by sodium amide involving migration into the ring. *J. Am. Chem. Soc.* 73 (9): 4122–4131. (b) Tayama, E., Watanabe, K., and Sotome, S. (2017). Structural and mechanistic studies of the base-induced Sommelet–Hauser rearrangement of *N*-α-branched benzylic azetidine-2-carboxylic acid-derived ammonium salts. *Org. Biomol. Chem.* 15 (31): 6668–6678.

949 Biellman, J.F. and Schmitt, J.L. (1973). Solvation and rearrangement of a carbanion. *Tetrahedron Lett.* 14 (46): 4615–4618.

950 Schlenk, W. and Bergmann, E. (1928). Forschungen auf dem Gebiete der alkaliorganischen Verbindungen. VI. Über Alkali-Additionsprodukte aus Diarylketonen. *Liebigs Ann. Chem.* 464: 22–34.

951 Wittig, G. and Lohmann, L. (1942). About the cation isomerisation of certain benzyl ethers through the effect of phenyl lithium. *Liebigs Ann. Chem.* 550: 260–268.

952 Wittig, G. (1954). Fortschritte auf dem Gebiet der Organischen Aniono-Chemie. *Angew. Chem.* 66 (1): 10–17.

953 (a) Baldwin, J.E., DeBernardis, J., and Patrick, J.E. (1970). Anion rearrangements. Duality of mechanism in decomposition of allylic ether anions and synthetic applications. *Tetrahedron Lett.* 11 (5): 353–356. (b) Strunk, S. and Schlosser, M. (2006). Wittig rearrangement of lithiated allyl aryl ethers: A mechanistic study. *Eur. J. Org. Chem.* (19): 4393–4397.

954 Hauser, C.R. and Kantor, S.W. (1951). Rearrangement of benzyl ethers to carbinols by potassium amide - mechanism of isomerization of carbanions involving 1,2-shifts. *J. Am. Chem. Soc.* 73 (4): 1437–1441.

955 (a) Cast, J., Stevens, T.S., and Holmes, J. (1960). Molecular rearrangement and fission of ethers by alkaline reagents. *J. Chem. Soc.* 3521–3527. (b) Isobe, M., Chang, W.C., Tsou, P.K. et al. (2015). Stereochemical course of Wittig rearrangements of dihydropyran allyl propargyl ethers. *J. Org. Chem.* 80 (12): 6222–6237. (c) Velasco, R., López, C.S., Faza, O.N., and Sanz, R. (2016). Exploring the reactivity of α-lithiated aryl benzyl ethers: inhibition of the (1,2)-Wittig rearrangement and the mechanistic proposal revisited. *Chem. Eur. J.* 22 (42): 15958–15068.

956 Lansbury, P.T., Pattison, V.A., Sidler, J.D., and Bieber, J.B. (1966). Mechanistic aspects of rearrangement and elimination reactions of A-metalated benzyl alkyl ethers. *J. Am. Chem. Soc.* 88 (1): 78–84.

957 Schäfer, H., Schöllkopf, U., and Walter, D. (1968). Migration tendencies of alkyl groups in Wittig rearrangement. *Tetrahedron Lett.* 9 (24): 2809–2814.

958 Weiske, T. and Schwarz, H. (1986). Intrinsic alkyl radical properties inferred from the study of unimolecular dissociations of gaseous carboxylic acid cation-radicals. *Tetrahedron* 42 (22): 6245–6251.

959 Schöllkopf, U. (1970). Recent results in carbanion chemistry. *Angew. Chem. Int. Ed. Engl.* 9 (10): 763–773.

960 Garst, J.F. and Smith, C.D. (1976). Wittig rearrangements of aralkyl alkyl ethers. *J. Am. Chem. Soc.* 98 (6): 1526–1537.

961 Eichinger, P.C.H., Bowie, J.H., and Blumenthal, T. (1986). Gas phase Wittig rearrangement of carbanions derived from benzyl ethers. *J. Org. Chem.* 51 (26): 5078–5082.

962 Sheldon, J.C., Taylor, M.S., Bowie, J.H. et al. (1999). The gas phase 1,2-Wittig rearrangement is an anion reaction. A joint experimental and theoretical study. *J. Chem. Soc., Perkin Trans. 2* (2): 333–340.

963 Fritsch, P. (1894). Über die Darstellung von Diphenylacetaldehyd und eine neue Synthese von Tolanderivaten. *Liebigs Ann. Chem.* 279: 319–323.

964 Buttenberg, W.P. (1894). Condensation des dichloracetals mit phenol und toluol. *Liebigs Ann. Chem.* 279 (3): 324–337.

965 Wiechell, H. (1894). Condensation des dichloracetals mit anisol und phenetol. *Liebigs Ann. Chem.* 279 (3): 337–344.

966 Du, Z.M. and Erickson, K.L. (2010). Carbanionic rearrangements of halomethylenecyclobutanes. Stereochemistry of the migrating group. *J. Org. Chem.* 75 (21): 7129–7140.

967 Brook, A.G. (1958). Isomerism of some A-hydroxysilanes to silyl ethers. *J. Am. Chem. Soc.* 80 (8): 1886–1889.

968 (a) Brook, A.G. (1974). Some molecular rearrangements of organosilicon compounds. *Acc. Chem. Res.* 7 (3): 77–84. (b) Kirschning, A., Luiken, S., Migliorini, A., and Loreto, M.A. (2009). Studies on the 1,2-Brook rearrangement of bissilyl ketones. *Synlett* (3): 439–432. (c) Matsuya, Y., Wada, K., Minato, D., and Sugimoto, K. (2016). Highly efficient access to both geometric isomers of silyl enol ethers: sequential (1,2)-Brook/Wittig reactions. *Angew. Chem. Int. Ed.* 55 (34): 10079–10082. (d) Han, M.-Y., Yang, F.-Y., Zhou, D., and Xu, Z.-G. (2017). Organobase-catalyzed (1,2)-Brook rearrangement of silyl glyoxylates. *Org. Biomol. Chem.* 15 (6): 1418–1425.

969 Speier, J.L. (1952). The preparation and properties of (hydroxyorgano)-silanes and related compounds. *J. Am. Chem. Soc.* 74 (4): 1003–1010.

970 Jiang, X.L. and Bailey, W.F. (1995). Facile retro-(1,4)-Brook rearrangement of a [(2-siloxycyclopentyl)methyl]lithium species. *Organometallics* 14 (12): 5704–5707.

971 Kawashima, T., Naganuma, K., and Okazaki, R. (1998). Generation and decomposition of a pentacoordinate spirobis[1,2-oxasiletanide]. *Organometallics* 17 (3): 367–372.

972 Jankowski, P., Raubo, P., and Wicha, J. (1994). Tandem transformations initiated by the migration of a silyl group - some new synthetic applications of silyloxiranes. *Synlett* (12): 985–992.

973 Lautens, M., Delanghe, P.H.M., Goh, J.B., and Zhang, C.H. (1995). Studies in the transmetalation of cyclopropyl, vinyl, and epoxy stannanes. *J. Org. Chem.* 60 (13): 4213–4227.

974 Moser, W.H. (2001). The Brook rearrangement in tandem bond formation strategies. *Tetrahedron* 57 (11): 2065–2084.

975 Brook, A.G., Warner, C.M., and Limburg, W.W. (1967). Stereochemistry of silylcarbinol to silyl ether rearrangement. *Can. J. Chem.* 45 (11): 1231–1246.

976 Pasto, D.J. and Brophy, J.E. (1991). (1,3)-Hydrogen sigmatropic rearrangements in alkyl-substituted allenes. *J. Org. Chem.* 56 (14): 4554–4556.

977 Vogel, E. (1960). Kleine Kohlenstoff-Ringe. *Angew. Chem.* 72 (1): 4–26.

978 Overberger, C.G. and Borchert, A.E. (1960). Novel thermal rearrangements accompanying acetate pyrolysis in small ring systems. *J. Am. Chem. Soc.* 82 (4): 1007–1008.

979 Flowers, M.C. and Frey, H.M. (1961). Thermal unimolecular isomerisation of vinylcyclopropane to cyclopentene. *J. Chem. Soc.* 3547–3548.

980 Wellington, C.A. (1962). Thermal isomerization of vinylcyclopropane. *J. Phys. Chem.* 66 (9): 1671–1674.

981 (a) Baldwin, J.E. (2003). Thermal rearrangements of vinylcyclopropanes to cyclopentenes. *Chem. Rev.* 103 (4): 1197–1212. (b) von E. Doering, W. and Barsa, E.A. (2004). Fate of diradicals in the caldera: stereochemistry of thermal stereomutation and ring enlargement in *cis*- and *trans*-1-cyano-2(*E*)-propenylcyclopropanes. *J. Am. Chem. Soc.* 126: 12353–12362.

982 Ellis, R.J. and Frey, H.M. (1963). Thermal unimolecular decomposition of isopropenylcyclobutane. *Trans. Faraday Soc.* 59 (489): 2076–2079.

983 (a) Leber, P.A. and Baldwin, J.E. (2002). Thermal (1,3) carbon sigmatropic rearrangements of vinylcyclobutanes. *Acc. Chem. Res.* 35 (5): 279–287. (b) Baldwin, J.E. and Fedé, J.M. (2006). Thermal isomerizations of 2-d-1-(*E*)-propenylcyclobutanes to 4-d-3-methylcyclohexenes. *J. Am. Chem. Soc.* 128 (17): 5608–5609. (c) Leber, P.A., Bogdan, A.R., Powers, D.C., and Baldwin, J.E. (2007). Thermal isomerizations of *cis,anti,cis*-tricyclo[6.4.0.0^{2,7}]dodec-3-ene to *trans*- and *cis,endo*-tricyclo[6.2.2.0^{2,7}]dodec-9-ene: diradical conformations and stereochemical outcomes in (1,3) carbon shifts. *Tetrahedron* 63 (27): 6331–7338. (d) Leber, P.A., Mann, G.R. III, Hancock-Cerutti, W. et al. (2012). Thermal isomerizations of *cis,anti,cis*-tricyclo[7.4.0.0^{2,8}]tridec-10-ene. *J. Org. Chem.* 77 (7): 3468–3474.

984 Baldwin, J.E., Kiemle, D.J., and Kostikov, A.P. (2009). Quantitative analyses of stereoisomeric 3,4-d_2-cyclohexenes in the presence of 3,6-d_2-cyclohexenes. *J. Org. Chem.* 74 (10): 3866–3874.

985 Baldwin, J.E. and Kostikov, A.P. (2010). On the stereochemical characteristic of the thermal reactions of vinylcyclobutane. *J. Org. Chem.* 75 (9): 2767–2775.

986 Baldwin, J.E. and Keliher, E.J. (2002). Activation parameters for three reactions interconverting isomeric 4-and 6-deuteriobicyclo [3.1.0]hex-2-enes. *J. Am. Chem. Soc.* 124 (3): 380–381.

987 Suhrada, C.P. and Houk, K.N. (2002). Potential surface for the quadruply degenerate rearrangement of bicyclo[3.1.0]hex-2-ene. *J. Am. Chem. Soc.* 124 (30): 8796–8797.

988 (a) Doubleday, C., Suhrada, C.P., and Houk, K.N. (2006). Dynamics of the degenerate rearrangement of bicyclo[3.1.0]hex-2-ene. *J. Am. Chem. Soc.* 128 (1): 90–94. (b) Tao, H.-R. and Fang, C.-C. (2008). Thermal (1,3)-sigmatropic rearrangements of bicyclic and tricyclic vinylcyclobutanes: a gray zone between the concerted and stepwise extremes. *Theor. Chem. Acc.* 121 (1–2): 91–101.

989 Deslongchamps, G. and Deslongchamps, P. (2016). Bent bonds and the antiperiplanar hypothesis - a simple model to rationalize (1,3)-sigmatropic alkyl shifts. *Org. Biomol. Chem.* 14 (32): 7754–7767.

990 Berson, J.A. and Nelson, G.L. (1967). Inversion of configuration in migrating group of a thermal (1,3)-sigmatropic rearrangement. *J. Am. Chem. Soc.* 89 (21): 5503–5504.

991 Beno, B.R., Wilsey, S., and Houk, K.N. (1999). The C_7H_{10} potential energy landscape: concerted transition states and diradical intermediates for the retro-Diels–Alder reaction and (1,3)-sigmatropic shifts of norbornene. *J. Am. Chem. Soc.* 121 (20): 4816–4826.

992 Kless, A., Nendel, M., Wilsey, S., and Houk, K.N. (1999). Origin of the preference for the orbital symmetry forbidden stereochemistry of the (1,5)-sigmatropic shift of substituted norcaradienes. *J. Am. Chem. Soc.* 121 (18): 4524–4525.

993 Roth, W.R. and Friedrich, A. (1969). Stereochemistry of (1,3)-sigmatropic rearrangement of bicyclo[2.1.1]hex-2-ene. *Tetrahedron Lett.* 10 (31): 2607–2910.

994 Carpenter, B.K. (1992). Anatomy of a bond cleavage - the dynamics of rearrangement of a strained ring hydrocarbon. *J. Org. Chem.* 57 (17): 4645–4648.

995 Carpenter, B.K. (1992). Intramolecular dynamics for the organic chemist. *Acc. Chem. Res.* 25 (11): 520–528.

996 Pikulin, S. and Berson, J.A. (1988). Static structure of a regular intermediate controls the course of the thermal (1,3)-sigmatropic rearrangement of 6-methylenebicyclo[3.1.0] hex-2-enyl derivatives. *J. Am. Chem. Soc.* 110 (25): 8500–8512.

997 Ellis, R.J. and Frey, H.M. (1964). Thermal unimolecular isomerizations of *cis*- and *trans*-1-methyl-2-vinylcyclopropanes. *J. Chem. Soc.* 5578–5583.

998 Roth, W.R. and König, J. (1965). Wasserstoff-Verschiebungen. 3. Thermische Isomerisierung von *cis*-Hexadien-(1,4). *Liebigs Ann. Chem.* 688: 28–39.

999 Simpson, J.M. and Richey, H.G. (1973). Effects of methoxyl and phenyl substituents on thermal rearrangements of vinylcyclopropane. *Tetrahedron Lett.* 14 (27): 2545–2548.

1000 McGaffin, G., Grimm, B., Heinecke, U. et al. (2001). On the substituent effects of the thermal ethenyleyclopropane-to-cyclopentene rearrangement: gas-phase kinetics of ethoxy-, methylthio- and trimethylsilyl-substituted ethenylcyclopropanes. *Eur. J. Org. Chem.* (18): 3559–3573.

1001 Childs, R.F. and Winstein, S. (1968). Ring opening and fivefold degenerate scrambling in hexa-and heptamethylbicyclo[3.1.0]hexenyl cations. *J. Am. Chem. Soc.* 90 (25): 7146–7147.

1002 Genaev, A., Sal'nikov, G., and Shubin, V. (2007). Energy barriers to carousel rearrangements of carbocations: quantum-chemical calculations vs. experiment. *Russ. J. Org. Chem.* 43 (8): 1134–1138.

1003 Vogel, P., Saunders, M., Hasty, N.M. Jr., and Berson, J.A. (1971). Bicyclo[3.1.0]hex-3-hn-2-yl cation. *J. Am. Chem. Soc.* 93 (6): 1551–1552.

1004 Sustmann, R. and Lübbe, F. (1979). Rearrangements of free-radicals. 7. Electrocyclic and sigmatropic rearrangements of bicyclo[3.1.0]hexenyl radical. *Chem. Ber.* 112 (1): 42–56.

1005 Ingold, K.U. and Walton, J.C. (1986). Small, strained bicycloalkyl radicals and some homolytic reactions involving their parent bicycloalkanes. *Acc. Chem. Res.* 19 (3): 72–77.

1006 McLean, S. and Haynes, P. (1965). Hydrogen migration in cyclopentadienes. *Tetrahedron* 21 (9): 2329–2342.

1007 Spangler, C.W. (1976). Thermal (1,j)-sigmatropic rearrangements. *Chem. Rev.* 76 (2): 187–217.

1008 Davison, A. and Rakita, P.E. (1970). Fluxional behavior of cyclopentadienyl, methylcyclopentadienyl, and pentamethylcyclopentadienyl compounds of Si, Ge, and Sn. *Inorg. Chem.* 9 (2): 289–294.

1009 Johnson, H.D., Hartford, T.W., and Spangler, C.W. (1978). Fluxional cyclopentadienylboranes. *J. Chem. Soc., Chem. Commun.* (6): 242–242.

1010 Schöller, W.W. (1984). The 1,5-sigmatropic shift of a BR_2 unit (R=H, F, OH, NH_2, Cl, or SH) over a cyclopentadiene system. *J. Chem. Soc., Dalton Trans.* (10): 2233–2236.

1011 Skattebøl, L. (1969). Thermal rearrangement of 5-methyl-1,2,4-hexatriene - A facile 1,5-hydrogen migration. *Tetrahedron* 25 (20): 4933–4938.

1012 Ma, S.M. and Gu, Z.H. (2006). Sequential rearrangement of 1,2,4*Z*,7-tetraenes involving (1,5)-hydrogen migration and electrocyclization: an efficient synthesis of eight-membered cyclic compounds. *J. Am. Chem. Soc.* 128 (15): 4942–4943.

1013 Wolinsky, J., Baird, M.D., and Chollar, B. (1962). Thermal rearrangement of 1,3-dienes. *J. Am. Chem. Soc.* 84 (14): 2775–2779.

1014 Roth, W.R. and Konig, J. (1966). Wasserstoffverschiebungen. 4. Kinetischer Isotopeneffekt der 1,5-Wasserstoffverschiebung im *cis*-Pentadien-(1,3). *Liebigs Ann. Chem.* 699: 24–32.

1015 Liu, Y.P., Lynch, G.C., Truong, T.N. et al. (1993). Molecular modeling of the kinetic isotope effect for the (1,5)-sigmatropic rearrangement of *cis*-1,3-pentadiene. *J. Am. Chem. Soc.* 115 (6): 2408–2415.

1016 Hess, B.A. and Schaad, L.J. (1983). Ab initio study of the transition structure of the (1,5)-sigmatropic hydrogen transfer in *cis*-1,3-pentadiene. *J. Am. Chem. Soc.* 105 (24): 7185–7186.

1017 Dondoni, A., Rossini, G., Mossa, G., and Caglioti, L. (1968). Azo-hydrazo conversion. Kinetics of isomerization of 1-tosylazocyclohexene to cyclohex-2-enone tosylhydrazone. *J. Chem. Soc. B* (12): 1404–1407.

1018 Saettel, N.J. and Wiest, O. (2000). Ab initio studies of (1,5)-H shifts: pentadiene and beyond. *J. Org. Chem.* 65 (8): 2331–2336.

1019 Xing, C.G. and Skibo, E.B. (2000). Sigmatropic reactions of the aziridinyl semiquinone species. Why aziridinyl benzoquinones are metabolically more stable than aziridinyl indoloquinones. *Biochemistry* 39 (35): 10770–10780.

1020 Alajarin, M., Sánchez-Andrada, P., López-Leonardo, C., and Álvarez, A. (2005). On the mechanism of phthalimidine formation via *o*-phthalaldehyde monoimines. New (1,5)-H sigmatropic rearrangements in molecules with the 5-aza-2,4-pentadienal. *J. Org. Chem.* 70 (19): 7617–7623.

1021 Gao, X., Gaddam, V., Altenhofer, E. et al. (2012). C—H activation in *S*-alkenyl sulfoximines: an *endo* 1,5-hydrogen migration. *Angew. Chem. Int. Ed.* 51 (28): 7016–7019.

1022 Brasil, E.M., Borges, R.S., Romero, O.A. et al. (2012). Azo-hydrazo conversion via (1,5)-hydrogen shifts. A combined experimental and theoretical study. *Tetrahedron* 68 (34): 6902–6907.

1023 Gajewski, J.J. (2003). *Hydrocarbon Thermal Isomerizations*, 2e, 1–440. Amsterdam: Elsevier Science.

1024 Shen, G.Y., Tapia, R., and Okamura, W.H. (1987). Substituent effect studies on the thermal (1,5)-sigmatropic hydrogen shifts of vinylallenes. *J. Am. Chem. Soc.* 109 (24): 7499–7506.

1025 Baldwin, J.E. and Broline, B.M. (1982). Stereochemistry of thermal (1,5)-carbon shifts in norcaradienes - complete kinetic analysis of the degenerate thermal isomerization of (+)-2-deuterio-3,7-dimethyl-7-methoxymethylcycloheptatriene. *J. Am. Chem. Soc.* 104 (10): 2857–2865.

1026 Borden, W.T., Lee, J.G., and Young, S.D. (1980). Stereochemistry of the thermal and photochemical rearrangement of 7,7-dimethylbicyclo[4.1.1]octa-2,4-diene to 8,8-dimethylbicyclo[5.1.0]octa-2,4-diene. *J. Am. Chem. Soc.* 102 (14): 4840–4843.

1027 Youssef, A.K. and Ogliarus, M.A. (1972). Correction of literature concerning reactions of polyarylated carbinols - novel suprafacial (1,5)-sigmatropic rearrangement. *J. Org. Chem.* 37 (16): 2601–2604.

1028 Nelsen, S.F., Gillespie, J.P., Hintz, P.J., and Seppanen, E.D. (1973). Formation and decomposition reactions of 5-(3′,3′-dimethylallenyl)-1,5-bis(carbomethoxy)cyclopentadiene. *J. Am. Chem. Soc.* 95 (25): 8380–8384.

1029 Martin, M. and Regitz, M. (1974). Investigations on diazo-compounds and azides. 21. Sigmatropic (1,5)-acyl shifts in cycloaddition of 2-diazocycloalkanones to activated acetylenes. *Liebigs Ann. Chem.* 1974 (10): 1702–1708.

1030 Jones, D.W. and Kneen, G. (1975). *ortho*-Quinonoid compounds. 8. Photodecarbonylation of 1,3-diphenylinden-2-one adducts - (1,5)-sigmatropic shifts in sterically stabilized *ortho*-quinodimethanes. *J. Chem. Soc., Perkin Trans. 1* (2): 171–175.

1031 Collett, M.J., Jones, D.W., and Renyard, S.J. (1986). *ortho*-Quinonoid compounds. 19. Relative migratory aptitudes for doubly and triply bonded groups in 1,3-dimethyldibenz[e,g]indenes. *J. Chem. Soc., Perkin Trans. 1* (8): 1471–1477.

1032 Battye, P.J. and Jones, D.W. (1986). Relative migratory aptitudes of doubly and triply bonded groups in a cycloheptatriene system. *J. Chem. Soc., Perkin Trans. 1* (8): 1479–1489.

1033 Leigh, W.J. and Arnold, D.R. (1979). Photochemical and thermal rearrangements of some pyrazoles-*H*-3. *Can. J. Chem.* 57 (10): 1186–1200.

1034 Majchrzak, M.W., Jefferson, E., and Warkentin, J. (1990). Stepwise mechanism of formal (1,5)-sigmatropic rearrangement of dimethyl 3,3-dialkyl-3*H*-pyrazole-4,5-dicarboxylates. *J. Am. Chem. Soc.* 112 (6): 2449–2451.

1035 Miller, W.L. and Portale, A.A. (2003). Vitamin D biosynthesis and vitamin D 1 α-hydroxylase deficiency. *Endocr. Dev.* 6: 156–174.

1036 Zhu, J. and DeLuca, H.F. (2012). Vitamin D 25-hydroxylase - four decades of searching, are we there yet? *Arch. Biochem. Biophys.* 523 (1): 30–36.

1037 Velluz, L., Amiard, G., and Petit, A. (1949). Le preécalciférol - Ses relations d'équilibre avec le calciférol. *Bull. Soc. Chim. Fr.* 16 (5–6): 501–507.

1038 (a) Schlatmann, J.L.M., Pot, J., and Havinga, E. (1964). Studies on Vitamin D + related compounds. 18. Investigation into interconversion of precalciferol + calciferol + of analogous compounds. *Recl. Trav. Chim. Pays-Bas* 83 (9–1): 1173–1184. (b) Kulesza, U., Sigüeiro, R., Mouriño, A., and Sicinski, R.R. (2013). Synthesis of 9-alkylated calcitriol and two 1α,25-dihydroxy-9-methylene-10,19-dihydrovitamin D_3 Analogues with a non-natural triene system by thermal sigmatropic rearrangements. *J. Org. Chem.* 78: 1444–1450.

1039 (a) Okamura, W.H., Hoeger, C.A., Miller, K.J., and Reischl, W. (1988). Studies of Vitamin-D(calciferol) and its analogs. 33. A study of the (1,7)-sigmatropic shift of a 1-hydroxylated 3-desoxy previtamin-D to Vitamin-D - observation of a modest primary deuterium kinetic isotope effect. *J. Am. Chem. Soc.* 110 (3): 973–974. (b) Okamura, W.H., Elnagar, H.Y., Ruther, M., and Dobreff, S. (1993). Thermal (1,7)-sigmatropic shift of previtamin-D_3 to Vitamin-D_3 – synthesis and study of pentadeurerio derivatives. *J. Org. Chem.* 58 (3): 600–610. (c) Meana-Pañeda, R. and Fernández-Ramos, A. (2012). Tunneling and conformational flexibility play critical roles in the isomerization mechanism of Vitamin D. *J. Am. Chem. Soc.* 134 (1): 346–354.

1040 Dmitrenko, O., Bach, R.D., Sicinski, R.R., and Reischl, W. (2003). Computational insight into the effect of C(19) substituents on (1,7)-hydrogen shift in previtamin D. *Theor. Chem. Acc.* 109 (4): 170–175.

1041 (a) Baldwin, J.E. and Reddy, V.P. (1988). Primary deuterium kinetic isotope effects for the thermal (1,7)-sigmatropic rearrangement of 7-methylocta-1,3(*Z*),5(*Z*)-triene. *J. Am. Chem. Soc.* 110 (24): 8223–8228. (b) Gurskii, M.E., Gridnev, I.D., Il'ichev, Y.V. et al. (1992). Investigation of the (1,7)-sigmatropic hydrogen shift in a parent compound (*Z,Z*)-1,3,5-heptatriene and its monodeuterium analog. *Angew. Chem. Int. Ed.* 31 (6): 781–783. (c) Gridnev, I.D., Gurskii, M.E., Ignatenko, A.V., and Bubnov, Y.N. (1993). A series of sigmatropic rearrangements in 2,4,6-heptatrienylborane – kinetic study of a (1,7)-hydrogen-shift facilitated by a (1,3)-boron-shift. *Organometallics* 12 (7): 2487–2495. (d) Jiao, H. and von Ragué Schleyer, P. (1993). A detailed theoretical analysis of the (1,7)-sigmatropic hydrogen shift: the Möbius character of the eight-electron transition structure. *Angew. Chem. Int. Ed.* 32 (12): 1763–1765. (e) Mousavipour, S.H., Fernández-Ramos, A., Meana-Pañeda, R. et al. (2007). Direct-dynamics VTST study of the (1,7)-hydrogen-shift in 7-methylocta-1,3(*Z*),5(*Z*)-triene. A model system for the hydrogen transfer reaction in previtamin D_3. *J. Phys. Chem. A.* 111 (4): 719–725. (f) Kryvohuz, M. (2014). Calculation of kinetic isotope effects for intramolecular hydrogen shift reactions using semiclassical instant on approach. *J. Phys. Chem. A* 118 (3): 535–544.

1042 Klärner, F.G. (1972). Degenerate rearrangement of *cis*-bicyclo[6.1.0]nona-2,4,6-triene. *Angew. Chem. Int. Ed. Engl.* 11 (9): 832–833.

1043 Heimgartner, H., Schmid, H., and Hansen, H.J. (1972). Thermal behavior of 1,2-dipropenylbenzenes. *Helv. Chim. Acta* 55 (5): 1385–1403.

1044 Meisenheimer, J., Greeske, H., and Willmersdorf, A. (1922). The behaviour of allyl- and benzyl- against caustic soda. *Ber. Dtsch. Chem. Ges.* 55: 513–522.

1045 Kleinschmidt, R.F. and Cope, A.C. (1944). Rearrangement of allyl groups in dyad systems - amine oxides. *J. Am. Chem. Soc.* 66: 1929–1933.

1046 Cope, A.C. and Towle, P.H. (1949). Rearrangement of allyldialkylamine oxides and benzyldimethylamine oxide. *J. Am. Chem. Soc.* 71 (10): 3423–3428.

1047 Guarna, A., Occhiato, E.G., Pizzetti, M. et al. (2000). Stereoselective Meisenheimer rearrangement using BTAa's as chiral auxiliaries. *Tetrahedron: Asymmetry* 11 (20): 4227–4238.

1048 Yamamoto, Y., Oda, J., and Inouye, Y. (1976). Self-immolative asymmetric synthesis. 2. Transfer of chirality from tetrahedral carbon to trigonal carbon in chiral amine oxide rearrangement. *J. Org. Chem.* 41 (2): 303–306.

1049 Enders, D. and Kempen, H. (1994). Enantioselective synthesis of allylic alcohols via asymmetric (2,3)-sigmatropic Meisenheimer rearrangement. *Synlett* (11): 969–971.

1050 Buston, J.E.H., Coldham, I., and Mulholland, K.R. (1998). Studies into the asymmetric Meisenheimer rearrangement. *Tetrahedron: Asymmetry* 9 (11): 1995–2009.

1051 (a) Davies, S.G. and Smyth, G.D. (1996). Asymmetric synthesis of (*R*)-hexane-1,5-diol and (*R*)-hex-3-ene-1,5-diol via a tandem asymmetric conjugate addition stereospecific Meisenheimer rearrangement protocol. *Tetrahedron: Asymmetry* 7 (4): 1001–1004. (b) Bao, H.-L., Qi, X.-B., and Tambar, U.K. (2011). Stereoselective (2,3)-rearrangements of amine *N*-oxides. *Synlett* (13): 1789–1792.

1052 Arbuzov, A.E. and Nikonorov, K.V. (1948). O Poluchenii Nekotorykh Efirov Difenilfosfinistoi Kisloty. *Zh. Obshch. Khim.* 18 (11): 2008–2015.

1053 Demay, S., Harms, K., and Knochel, P. (1999). Stereoselective preparation of phosphine oxides via a (2,3)-sigmatropic shift of allylic diphenylphosphinites. *Tetrahedron Lett.* 40 (27): 4981–4984.

1054 Gavryushin, A., Polborn, K., and Knochel, P. (2004). Novel chiral diphosphine ligands with a pinene core obtained via an allylphosphinite-allylphosphine oxide rearrangement. *Tetrahedron: Asymmetry* 15 (14): 2279–2288.

1055 Busacca, C.A., Qu, B., Farber, E. et al. (2013). (2,3)-sigmatropic rearrangements of 2-phosphineborane 2-propen-1-ols: rapid access to enantioenriched diphosphine monoxide derivatives. *Org. Lett.* 15 (5): 1136–1139.

1056 Cope, A.C., Morrison, D.E., and Field, L. (1950). Thermal rearrangement of allyl-type sulfoxides, sulfones and sulfinates. *J. Am. Chem. Soc.* 72 (1): 59–67.

1057 (a) Braverman, S. (1990). Rearrangements. In: *The Chemistry of Sulphinic Acids, Esters and their Derivatives*, Patai's Chemistry of Functional Groups, Chapter 11 (ed. S. Patai), 297–349. Chichester: Wiley. (b) Maeno, M., Shibata, N., and Cahard, D. (2015). Trifluoromethyl sulfoxides from allylic alcohols and electrophilic SCF_3 donor by (2,3)-sigmatropic rearrangement. *Org. Lett.* 17 (8): 1990–1993.

1058 Hinsberg, O. (1917). The reaction between aromatic sulphinc acid and di- and triphenyl-carbinol. *Ber. Dtsch. Chem. Ges.* 50: 468–473.

1059 (a) Braverman, S. and Cherkinsky, M. (2007). (2,3)-Sigmatropic rearrangements of propargylic and allenic systems. In: *Sulfur-Mediated Rearrangement II*, Topics in Current Chemistry, vol. 275 (ed. E. Schaumann), 67–101. Heidelberg, Berlin: Springer-Verlag. (b) Hampton, C.S. and Harmata, M. (2015). (2,3)-Sigmatropic rearrangement of propargylic sulfinates by a pale Ag-1-USY zeolite and silver salts. *Adv. Synth. Catal.* 357 (2–3): 549–552. (c) Tata, R.R., Hampton, C.S., and Harmata, M. (2017). Preparation of propargylic sulfinates and their (2,3)-sigmatropic rearrangement to allenic sulfones. *Adv. Synth. Catal.* 359 (7): 1232–1241.

1060 Ash, A.S.F., Challenger, F., and Greenwood, D. (1951). The isomerisation of diallyl sulphilimines. *J. Chem. Soc.* 1877–1882.

1061 Briscoe, P.A., Challenger, F., and Duckworth, P.S. (1956). The isomerisation of sulphidimines. 3. The action of chloramine-T on cinnamyl phenyl sulphide, and the hydrolysis and oxidation of *N*-substituted toluene-*para*-sulphonamides. *J. Chem. Soc.* 1755–1768.

1062 Freeman, F., Bathala, R.M., Cavillo, J.E. et al. (2006). (2,3)-Sigmatropic rearrangements of hydrogen and alkyl 3-propenyl sulfoxides: A computational study. *Int. J. Quantum Chem.* 106 (11): 2390–2397.

1063 Miller, E.G., Rayner, D.R., and Mislow, K. (1966). Thermal rearrangement of sulfenates to sulfoxides. *J. Am. Chem. Soc.* 88 (13): 3139–3140.

1064 Braverman, S. and Stabinsky, Y. (1967). Rearrangement of allylic trichloromethanesulphenates. *J. Chem. Soc., Chem. Commun.* (6): 270–271.

1065 Evans, D.A., Andrews, G.C., and Sims, C.L. (1971). Reversible (1,3)-transposition of sulfoxide and alcohol functions - potential synthetic utility. *J. Am. Chem. Soc.* 93 (19): 4956–4957.

1066 Evans, D.A. and Andrews, G.C. (1974). Allylic sulfoxides - useful intermediates in organic synthesis. *Acc. Chem. Res.* 7 (5): 147–155.

1067 Morera, E. and Ortar, G. (1983). Preparative and stereochemical features of the sulfoxide-sulfenate (2,3)-sigmatropic rearrangement in 17-vinyl-17-hydroxy steroids. *J. Org. Chem.* 48 (1): 119–121.

1068 Sato, T., Otera, J., and Nozaki, H. (1989). Efficient stereocontrol in the (2,3)-sigmatropic rearrangement of allylic sulfoxides. *J. Org. Chem.* 54 (12): 2779–2780.

1069 Reich, H.J. and Wollowitz, S. (1982). Conversion of allyl alcohols to 1,3-dienes by sequential sulfenate sulfoxide (2,3)-sigmatropic rearrangement and *syn*-elimination. *J. Am. Chem. Soc.* 104 (25): 7051–7059.

1070 Miller, J.G., Kurz, W., Untch, K.G., and Stork, G. (1974). Highly stereoselective total syntheses of prostaglandins via stereospecific sulfenate-sulfoxide transformations - 13-*cis*-15-B-Prostaglandins-E1 to prostaglandins-E1. *J. Am. Chem. Soc.* 96 (21): 6774–6775.

1071 Iglesias, B., Torrado, A., de Lera, A.R., and López, S. (2000). A pericyclic cascade to the stereocontrolled synthesis of 9-*cis*-retinoids. *J. Org. Chem.* 65 (9): 2696–2705.

1072 Shinada, T., Fuji, T., Ohtani, Y. et al. (2002). Syntheses of Gabosine A, B, D, and E from allyl sulfide derived from (−)-quinic acid. *Synlett* (8): 1341–1343.

1073 (a) Koprowski, M., Krawczyk, E., Skowronska, A. et al. (2001). Stereospecific (2,3)-sigmatropic rearrangement of allylic sulfoxides and selenoxides. Synthesis of novel polycyclic allylic alcohols and α-hydroxy ketones. *Tetrahedron* 57 (6): 1105–1118. (b) Colomer, I., Velado, M., Fernández de la Pradilla, R., and Viso, A. (2017). From allylic sulfoxides to allylic sulfenates: fifty years of a never-ending (2,3)-sigmatropic rearrangement. *Chem. Rev.* 117 (24): 14291–14243.

1074 Brebion, F., Nájera, F., Delovrie, B. et al. (2007). Addition of bis-sulfinyl anions to ketones: stereoselective synthesis of allylic alcohols through Evans–Mislow rearrangement based domino reactions. *Synthesis* (15): 2273–2278.

1075 de la Pradilla, R.F., Lwoff, N., del Águila, M.A. et al. (2008). (2,3)-Sigmatropic rearrangements of 3-sulfinyl dihydropyrans: application to the syntheses of the cores of *ent*-dysiherbaine and deoxymalayamicin A. *J. Org. Chem.* 73 (22): 8929–8941.

1076 Reggelin, M. (2007). (2,3)-Sigmatropic rearrangements of allylic sulfur compounds. In: *Sulfur-Mediated Rearrangement II*, Topics in Current Chemistry, vol. 275 (ed. E. Schaumann), 1–65. Heidelberg, Berlin: Springer-Verlag.

1077 Nishibayashi, Y. and Uemura, S. (2000). Selenoxide elimination and (2,3)-sigmatropic rearrangement. In: *Organoselenium Chemistry: Modern Developments in Organic Synthesis*, Topics in Current Chemistry, vol. 208 (ed. T. Wirth), 201–233.

1078 Bayse, C.A. and Antony, S. (2009). Theoretical studies of (2,3)-sigmatropic rearrangements of allylic selenoxides and selenimides. *Molecules* 14 (9): 3229–3236.

1079 Reich, H.J., Yelm, K.E., and Wollowitz, S. (1983). Kinetics, thermodynamics, and stereochemistry of the allyl sulfoxide sulfenate and selenoxide selenenate (2,3)-sigmatropic rearrangements. *J. Am. Chem. Soc.* 105 (8): 2503–2504.

1080 Komatsu, N., Nishibayashi, Y., and Uemura, S. (1993). Asymmetric (2,3)-sigmatropic rearrangement via chiral selenoxide with Sharpless oxidants. *Tetrahedron Lett.* 34 (14): 2339–2342.

1081 Katsuki, T. and Sharpless, K.B. (1980). The 1st practical method for asymmetric epoxidation. *J. Am. Chem. Soc.* 102 (18): 5974–5976.

1082 Katsuki, T., Lee, A.W.M., Ma, P. et al. (1982). Synthesis of saccharides and related polyhydroxylated natural products. 1. Simple alditols. *J. Org. Chem.* 47 (7): 1373–1378.

1083 Sharpless, K.B., Behrens, C.H., Katsuki, T. et al. (1983). Stereo and regioselective openings of chiral 2,3-epoxy alcohols - versatile routes to optically pure natural-products and drugs - unusual kinetic resolutions. *Pure Appl. Chem.* 55 (4): 589–604.

1084 Sharpless, K.B., Woodard, S.S., and Finn, M.G. (1983). On the mechanism of titanium-tartrate catalyzed asymmetric epoxidation. *Pure Appl. Chem.* 55 (11): 1823–1836.

1085 (a) Kurose, N., Takahashi, T., and Koizumi, T. (1996). Asymmetric (2,3)-sigmatropic rearrangement of chiral allylic selenimides. *J. Org. Chem.* 61 (9): 2932–2933. (b) Armstrong, A. and Emmerson, D.P.G. (2011). Enantioselective synthesis of α-alkyl, α-vinyl amino acids via (2,3)-sigmatropic rearrangement of selenimides. *Org. Lett.* 13 (5): 1040–1043. (c) Armstrong, A., Emmerson, D.P.G., Milner, H.J., and Sheppard, R.J. (2014). (2,3)-Sigmatropic rearrangement of allylic selenimides: strategy for the synthesis of peptides, peptidomimetics, and *N*-aryl vinyl glycines. *J. Org. Chem.* 79 (9): 3895–3907.

1086 Paulmier, C. (1986). *Selenium Reagents and Intermediates in Organic Synthesis*, 1–463. Oxford: Pergamon Press.

1087 Bulman Page, P.C. and McCarthy, T.J. (1991). Oxidation adjacent to C=C bonds. In: *Comprehensive Organic Synthesis*, vol. 7 (Oxidation) (ed. B.M. Trost and I. Fleming), 83–117. Oxford: Pergamon Press.

1088 Sharpless, K.B. and Lauer, R.F. (1972). Selenium dioxide oxidation of olefins - evidence for intermediacy of allylseleninic acids. *J. Am. Chem. Soc.* 94 (20): 7154–7155.

1089 Ra, C.S. and Park, G. (2003). Ab initio studies of the allylic hydroxylation: DFT calculation on the reaction of 2-methyl-2-butene with selenium dioxide. *Tetrahedron Lett.* 44 (5): 1099–1102.

1090 Ra, C.S., Park, E.M., and Park, G. (2008). DFT calculation on the stereochemistry of the allylic oxidation. Selenium dioxide-mediated oxidation of an exocyclic olefinic hydrindane compound. *Bull. Korean Chem. Soc.* 29 (12): 2513–2516.

1091 Fairlamb, I.J.S., Dickinson, J.M., and Pegg, M. (2001). Selenium dioxide E-methyl oxidation of suitably protected geranyl derivatives - synthesis of farnesyl mimics. *Tetrahedron Lett.* 42 (11): 2205–2208.

1092 Wittig, G., Doser, H., and Lorenz, I. (1949). Über die Isomerisierbarkeit Metallierter Fluorenylather. *Liebigs Ann. Chem.* 562 (3): 192–205.

1093 Still, W.C. and Mitra, A. (1978). Highly stereoselective synthesis of *Z*-trisubstituted olefins via (2,3)-sigmatropic rearrangement - preference for a pseudoaxially substituted transition state. *J. Am. Chem. Soc.* 100 (6): 1927–1928.

1094 Tsai, D.J.S. and Midland, M.M. (1984). Acyclic stereocontrol through diastereoselective and enantioselective (2,3)-sigmatropic Wittig rearrangements. *J. Org. Chem.* 49 (10): 1842–1843.

1095 Mikami, K. and Nakai, T. (1991). Acyclic stereocontrol via (2,3)-Wittig sigmatropic rearrangement. *Synthesis* (8): 594–604.

1096 Nakai, K. and Mikami, K. (1994). The (2,3)-Wittig rearrangement. *Org. React.* 46: 105–209.

1097 Nakai, T. and Mikami, K. (1986). (2,3)-Wittig sigmatropic rearrangements in organic synthesis. *Chem. Rev.* 86 (5): 885–902.

1098 Marshall, J.A. and Jenson, T.M. (1984). Chirality transfer in the (2,3)-Wittig rearrangement. *J. Org. Chem.* 49 (10): 1707–1712.

1099 Brückner, R. (1989). Asymmetric induction and simple diastereoselectivity in the (2,3)-Wittig rearrangement of ester enolates. *Chem. Ber.* 122 (4): 703–710.

1100 Brückner, R. (1989). Asymmetric induction in the (2,3)-Wittig rearrangement - the stereoselective synthesis of unsaturated alcohols with three contiguous stereogenic centers. *Chem. Ber.* 122 (1): 193–198.

1101 Marshall, J.A. and Wang, X.J. (1990). (2,3)-Wittig rearrangement of nonracemic propargyloxyacetic acids and esters - synthesis of optically active 2,5-dihydrofurans. *J. Org. Chem.* 55 (10): 2995–2996.

1102 Marshall, J.A. and Wang, X.J. (1991). Synthesis of enantioenriched α-hydroxy-α-allenylacetic acids by (2,3)-Wittig rearrangement of α-(propargyloxy)acetates. *J. Org. Chem.* 56 (16): 4913–4918.

1103 Crawley, J.E., Kaye, A.D., Pattenden, G., and Roberts, S.M. (1993). Stereoselective alkene synthesis via silicon-directed (2,3)-sigmatropic rearrangements - an approach to leukotrienes. *J. Chem. Soc., Perkin Trans. 1* (17): 2001–2007.

1104 (a) Denmark, S.E. and Miller, P.C. (1995). Asymmetric (2,3)-Wittig rearrangements with chiral, phosphorus anion-stabilizing groups. *Tetrahedron Lett.* 36 (37): 6631–6634. (b) Denmark, S.E. and Cullen, L.R. (2015). Development of a phase-transfer-catalyzed, (2,3)-Wittig rearrangement. *J. Org. Chem.* 80 (23): 11818–11848.

1105 Lay, L.G., Meldal, M., Nicotra, F. et al. (1997). Stereoselective synthesis of the C-analogue of β-D-glucopyranosyl serine. *J. Chem. Soc., Chem. Commun.* (15): 1469–1470.

1106 Tsubuki, M., Takahashi, K., and Honda, T. (2003). Stereoselective construction of β-isopropenyl alcohol moiety at the C(2) and (3) of Kallolide A and Pinnatin A using a (2,3)-Wittig rearrangement of cyclic furfuryl ethers. *J. Org. Chem.* 68 (26): 10183–10186.

1107 (a) Sasaki, M., Ikemoto, H., Kawahata, M. et al. (2009). (2,3)-Wittig rearrangement of enantiomerically enriched 3-substituted 1-propenyloxy-1-phenyl-2-propen-1-yl carbanions: effect of heteroatoms and conjugating groups on planarization of an α-oxy-benzylcarbanion through a double bond. *Chem. Eur. J.* 15 (18): 4663–4666. (b) Rycek, L. and Hudlicky, T. (2017). Applications of the Wittig–Still rearrangement in organic synthesis. *Angew. Chem. Int. Ed.* 56 (22): 6022–6066.

1108 (a) Nakai, T. and Tomooka, K. (1997). Asymmetric (2,3)-Wittig rearrangement as a general tool for asymmetric synthesis. *Pure Appl. Chem.* 69 (3): 595–600. (b) Hirokawa, Y., Kitamura, M., Mizubayashi, M. et al. (2013). Total synthesis of eupomatilones 1, 2, and 5 by enantioselective (2,3)-Wittig rearrangement. *Eur. J. Org. Chem.* (4): 721–727. (c) Kennedy, C.R., Guidera, J.A., and Jacobsen, E.N. (2016). Synergistic ion-binding catalysis demonstrated via an enantioselective, catalytic (2,3)-Wittig rearrangement. *ACS Cent. Sci.* 2 (6): 416–423.

1109 (a) Ahman, J. and Somfai, P. (1994). Aza-(2,3)-Wittig rearrangements of vinylaziridines. *J. Am. Chem. Soc.* 116 (21): 9781–9782. (b) Somfai, P. and Panknin, O. (2007). Investigations of the (2,3)-sigmatropic rearrangements of vinylaziridines and allylic amines. *Synlett* (8): 1190–1202. (c) Blid, J., Panknin, O., Tuzina, P., and Somfai, P.

(2007). Lewis acid mediated asymmetric (2,3)-sigmatropic rearrangement of allylic amines. Scope and mechanistic investigation. *J. Org. Chem.* 71 (4): 1294–1300.

1110 Ahman, J. and Somfai, P. (1995). Enantioselective total synthesis of (−)-indolizidines 209B and 209D via a highly efficient aza-(2,3)-Wittig rearrangement of vinylaziridines. *Tetrahedron* 51 (35): 9747–9756.

1111 Anderson, J.C., Smith, S.C., and Swarbrick, M.E. (1997). Diastereoselective acyclic aza-(2,3)-Wittig sigmatropic rearrangements. *J. Chem. Soc., Perkin Trans. 1* (10): 1517–1521.

1112 (a) Anderson, J.C. and Whiting, M. (2003). Total synthesis of (+/-)-kainic acid with an aza-(2,3)-Wittig sigmatropic rearrangement as the key stereochemical determining. *J. Org. Chem.* 68 (16): 6160–6163. (b) Drouillat, B., Wright, K., Quinodoz, P. et al. (2015). Aza-(2,3)-Wittig sigmatropic rearrangement of allylic tertiary amines: a successful example with high chirality transfer. *J. Org. Chem.* 80 (13): 6936–6940. (c) Everett, R.X. and Wolfe, J.P. (2015). Aza-Wittig rearrangements of *N*-benzyl and *N*-allyl glycine methyl esters. Discovery of a surprising cascade aza-Wittig rearrangement/hydroboration reaction. *J. Org. Chem.* 80 (18): 9041–9056.

1113 Blid, J., Brandt, P., and Somfai, P. (2004). Lewis acid mediated (2,3)-sigmatropic rearrangement of allylic α-amino amides. *J. Org. Chem.* 69 (9): 3043–3049.

1114 Davies, S.G., Jones, S., Sanz, M.A. et al. (1998). A novel (2,3) intramolecular rearrangement of *N*-benzyl-*O*-allylhydroxylamines. *J. Chem. Soc., Chem. Commun.* (20): 2235–2236.

1115 Bull, S.D., Davies, S.G., Domingez, S.H. et al. (2002). Diastereoselective (2,3)-sigmatropic rearrangements of lithium *N*-benzyl-*O*-allylhydroxylamides bearing a stereogenic centre adjacent to the migration terminus. *J. Chem. Soc., Perkin Trans. 1* (19): 2141–2150.

1116 (a) Markó, I.E. (1991). The Stevens and related rearrangements. In: *Comprehensive Organic Synthesis*, vol. 3 (ed. B.M. Trost and I. Flemming), 913–974. Oxford: Pergamon Press. (b) Hock, K. and Koenigs, R.M. (2017). Enantioselective (2,3)-sigmatropic rearrangements: metal-bound or free ylides as reaction intermediates? *Angew. Chem. Int. Ed.* 56 (44): 13566–13568.

1117 (a) Glaeske, K.W. and West, F.G. (1999). Chirality transfer from carbon to nitrogen to carbon via cyclic ammonium ylides. *Org. Lett.* 1 (1): 31–33. (b) Biswas, B., Collins, S.C., and Singleton, D.A. (2014). Dynamics and a unified understanding of competitive (2,3)- and (1,2)-sigmatropic rearrangements based on a study of ammonium ylides. *J. Am. Chem. Soc.* 136 (10): 3740–3743. (c) Zhang, J., Chen, Z.-X., Du, T. et al. (2016). Aryne-mediated (2,3)-sigmatropic rearrangement of tertiary allylic amines. *Org. Lett.* 18 (19): 4872–4875. (d) Kasten, K., Slawin, A.M.Z., and Smith, A.D. (2017). Enantioselective synthesis of β-fluoro-β-aryl-α-aminopentenamides by organocatalytic (2,3)-sigmatropic rearrangement. *Org. Lett.* 19 (19): 5182–5185.

1118 Arbore, A.P.A., Cane-Honeysett, D.J., Coldham, I., and Middleton, M.L. (2000). A rapid approach to amino-acid derivatives by (2,3)-Stevens rearrangement. *Synlett* (2): 236–238.

1119 Smith, S.C. and Bentley, P.D. (2002). Tandem cyclisation and (2,3)-Stevens rearrangement to 2-substituted pyrrolidines. *Tetrahedron Lett.* 43 (5): 899–902.

1120 Sweeney, J.B., Tavassoli, A., Carter, N.B., and Hayes, J.F. (2002). (2,3)-Sigmatropic rearrangements of didehydropiperidinium ylids. *Tetrahedron* 58 (51): 10113–10126.

1121 Sweeney, J., Tavassoli, A., and Workman, J.A. (2006). Asymmetric ammonium ylide rearrangements: the effect of nitrogen asymmetry. *Tetrahedron* 62 (49): 11506–11512.

1122 Workman, J.A., Garrido, N.P., Sancon, J. et al. (2005). Asymmetric (2,3)-rearrangement of glycine-derived allyl ammonium ylids. *J. Am. Chem. Soc.* 127 (4): 1066–1067.

1123 Doyle, M.P., Griffin, J.H., Chinn, M.S., and Vanleusen, D. (1984). Rearrangements of ylides generated from reactions of diazo-compounds with allyl acetals and thioketals by catalytic methods - heteroatom acceleration of the (2,3)-sigmatropic rearrangement. *J. Org. Chem.* 49 (11): 1917–1925.

1124 Pirrung, M.C. and Werner, J.A. (1986). Intramolecular generation and (2,3)-sigmatropic rearrangement of oxonium ylides. *J. Am. Chem. Soc.* 108 (19): 6060–6062.

1125 (a) Roskamp, E.J. and Johnson, C.R. (1986). Generation and rearrangements of oxonium ylides. *J. Am. Chem. Soc.* 108 (19): 6062–6063. (b) Li, Z.-J., Parr, B.T., and Davies, H.M.L. (2012). Highly stereoselective C—C bond formation by Rh-catalyzed tandem ylide formation/(2,3)-sigmatropic rearrangement between donor/acceptor carbenoids and chiral allylic alcohols. *J. Am. Chem. Soc.* 134 (26): 10942–10946.

1126 (a) Hodgson, D.M., Pierard, F.Y.T.M., and Stupple, P.A. (2001). Catalytic enantioselective rearrangements and cycloadditions involving ylides from diazo compounds. *Chem. Soc. Rev.* 30 (1): 50–61. (b) Hodgson, D.M., Man, S., Powell, K.J. et al. (2014). Intramolecular oxonium ylide formation (2,3)-sigmatropic rearrangement of diazocarbonyl-substituted cyclic unsaturated acetals: a formal synthesis of hyperolactone C. *J. Org. Chem.* 79 (20): 9728–9734. (c) Rao, S. and Prabhu, K.R. (2017). Gold-catalyzed (2,3)-sigmatropic rearrangement: reaction of aryl allyl alcohols with diazo compounds. *Org. Lett.* 19 (4): 846–849.

1127 (a) Blackburn, G.M., Ollis, W.D., Plackett, J.D. et al. (1968). Sigmatropic reactions of allylic sulphur ylids. *J. Chem. Soc., Chem. Commun.* (4): 186–188. (b) Li, Y.-Y., Huang, Z.-X., Wu, X.-H. et al. (2012). Rh(II)-catalyzed (2,3)-sigmatropic rearrangement of sulfur ylides derived from *N*-tosylhydrazones and sulfides. *Tetrahedron* 68 (26): 5234–5240. (c) Zhang, H., Wang, B., Yi, H. et al. (2015). Rh(II)-catalyzed (2,3)-sigmatropic rearrangement of sulfur ylides derived from cyclopropenes and sulfides. *Org. Lett.* 17 (13): 3322–3325. (d) Zhang, Z., Sheng, Z., Yu, W. et al. (2017). Catalytic asymmetric trifluoromethylthiolation via enantioselective (2,3)-sigmatropic rearrangement of sulfonium ylides. *Nat. Chem.* 9 (10): 970–976.

1128 Baldwin, J.E., Hackler, R.E., and Kelly, D.P. (1968). Rearrangement of vinylsulphonium ylids. An alternative mechanism for squalene synthetase. *J. Chem. Soc., Chem. Commun.* (10): 537–538.

1129 Gulea, M., Marchand, P., Saquet, M. et al. (1999). Generation and reactivity of α-phosphorylated sulfur stabilised carbanions and ylides; selective syntheses of α-thiosubstituted phosphonates. *Phosphorus, Sulfur Silicon Relat. Elem.* 153: 327–328.

1130 Aggarwal, V.K., Ferrara, M., Hainz, R., and Spey, S.E. (1999). (2,3)-Sigmatropic rearrangement of allylic sulfur ylides derived from trimethylsilyldiazomethane (TMSD). *Tetrahedron Lett.* 40 (50): 8923–8927.

1131 Hiroi, K. and Sato, S. (1985). Asymmetric induction reactions. 2. Stereochemical studies on asymmetric (2,3)-sigmatropic rearrangements using chiral ketenimines. *Chem. Pharm. Bull.* 33 (11): 4691–4700.

1132 Grieco, P.A., Meyers, M., and Finkelho, R.S. (1974). (2,3)-Sigmatropic rearrangements of acetylenic and allenic sulfonium ylides - synthesis of allenes and conjugated dienes. *J. Org. Chem.* 39 (1): 119–120.

1133 (a) Cagle, P.C., Meyer, O., Weickhardt, K. et al. (1995). Enantioselective syntheses of organosulfur compounds via (2,3)-rearrangements of ylides derived from di(allyl) and di(propargyl) sulfide complexes - control of carbon configuration by an easily resolved and recycled chiral transition-metal auxiliary. *J. Am. Chem. Soc.* 117 (47): 11730–11744. (b) Zhang, Y. and Wang, J.B. (2010). Catalytic (2,3)-sigmatropic rearrangement of sulfur ylide derived from metal carbene. *Coord. Chem. Rev.* 254 (9–10): 941–953.

1134 Braverman, S., Zafrani, Y., and Rahimipour, S. (2001). Base catalyzed (2,3)-sigmatropic rearrangements of propargylic sulfonium and selenonium salts. *Tetrahedron Lett.* 42 (15): 2911–2914.

1135 Gassman, P.G., Miura, T., and Mossman, A. (1982). (2,3)-Sigmatropic rearrangement of ylides derived from benzylic selenonium salts. *J. Org. Chem.* 47 (6): 954–959.

1136 Nishibayashi, Y., Ohe, K., and Uemura, S. (1995). The first example of enantioselective carbenoid addition to organochalcogen atoms - application to (2,3)-sigmatropic rearrangement of allylic chalcogen ylides. *J. Chem. Soc., Chem. Commun.* (12): 1245–1246.

1137 Kurose, N., Takahashi, T., and Koizumi, T. (1997). Asymmetric (2,3)-sigmatropic rearrangement of chiral allylic selenonium ylides. *J. Org. Chem.* 62 (14): 4562–4563.

1138 (a) Doyle, M.P., Tamblyn, W.H., and Bagheri, V. (1981). Highly effective catalytic methods for ylide generation from diazo-compounds - mechanism of the Rh-catalyzed and Cu-catalyzed reactions with allylic compounds. *J. Org. Chem.* 46 (25): 5094–5102. (b) Doyle, M.P., Forbes, D.C., Vasbinder, M.M., and Peterson, C.S. (1998). Enantiocontrol in the generation and diastereoselective reactions of catalytically generated oxonium and iodonium ylides. Metal-stabilized ylides as reaction intermediates. *J. Am. Chem. Soc.* 120 (30): 7653–7654. (c) Xu, B. and Tambar, U.K. (2017). Copper-catalyzed enantio-, diastereo-, and regioselective (2,3)-rearrangements of iodonium ylides. *Angew. Chem. Int. Ed.* 56 (33): 9868–9871.

1139 Simonneaux, G., Galardon, E., Paul-Roth, C. et al. (2001). Ru-porphyrin-catalyzed carbenoid addition to allylic compounds: application to (2,3)-sigmatropic rearrangements of ylides. *J. Organomet. Chem.* 617 (1): 360–363.

1140 (a) West, T.M., Spoehrle, S.S.M., Kasten, K. et al. (2015). Catalytic stereoselective (2,3)-rearrangement reactions. *ACS Catal.* 5 (12): 7446–7479. (b) Sheng, Z., Zhang, Z., Chu, C. et al. (2017). Transition metal-catalyzed (2,3)-sigmatropic rearrangements of ylides: an update of the most recent advances. *Tetrahedron* 73 (29): 4011–4022. (c) West, T.H., Spoehrle, S.S.M., and Smith, A.D. (2017). Isothiourea-catalysed chemo- and enantioselective (2,3)-sigmatropic rearrangements of *N,N*-diallyl allylic ammonium ylides. *Tetrahedron* 73 (29): 4138–4149.

1141 Gawley, R.E. and Moon, K. (2007). Stereoselective (2,3)-sigmatropic rearrangements of unstabilized nitrogen ylides. *Org. Lett.* 9 (16): 3093–3096.

1142 Doyle, M.P. and Peterson, C.S. (1997). Macrocyclic oxonium ylide formation and internal (2,3)-sigmatropic rearrangement. Catalyst influence on selectivity. *Tetrahedron Lett.* 38 (30): 5265–5268.

1143 Vedejs, E. and Hagen, J.P. (1975). Macrocycle synthesis by repeatable (2,3)-sigmatropic shifts - ring-growing reactions. *J. Am. Chem. Soc.* 97 (23): 6878–6880.

1144 Vedejs, E. (1984). Sulfur-mediated ring expansions in total synthesis. *Acc. Chem. Res.* 17 (10): 358–364.

1145 Vedejs, E., Buchanan, R.A., Conrad, P.C. et al. (1989). Total synthesis of D,L-methynolide - medium-ring sulfides by ylide ring expansion. *J. Am. Chem. Soc.* 111 (22): 8421–8430.

1146 (a) Cope, A.C. and Hardy, E.M. (1940). The introduction of substituted vinyl groups. V. A rearrangement involving the migration of an allyl group in a three-carbon system. *J. Am. Chem. Soc.* 62: 441–444. (b) Graulich, N. (2011). The Cope rearrangement – the first born of a great family. *WIREs Comput. Mol. Sci.* 1: 172–190.

1147 (a) Blechert, S. (1989). The hetero-Cope rearrangement in organic synthesis. *Synthesis* (2): 71–82. (b) Jung, J.-W., Kim, S.H., and Suh, Y.-G. (2017). Advances in aza-Claisen rearrangement induced ring-expansion strategies. *Asian J. Org. Chem.* 6 (9): 1117–1129.

1148 (a) Allin, S.M. and Baird, R.D. (2001). Development and synthetic applications of asymmetric (3,3)-sigmatropic rearrangements. *Curr. Org. Chem.* 5 (4): 395–415. (b) Touchet, S., Macé, A., Roisnel, T. et al. (2013). (3,3)-Sigmatropic rearrangement of boronated allylcyanates: a new route to α-aminoboronate derivatives and trisubstituted tetrahydrofurans. *Org. Lett.* 15 (11): 2512–2715. (c) Henrion, S., Carboni, B., Cossio, F.P. et al. (2016). Stereospecific synthesis of α-amino allylsilane derivatives through a (3,3)-allyl cyanate rearrangement. Mild formation of functionalized disiloxanes. *J. Org. Chem.* 81 (11): 4633–4644. (d) Ilardi, E., Stivala, C.E., and Zakarian, A. (2009). (3,3)-Sigmatropic rearrangements: recent applications in the total synthesis of natural products. *Chem. Soc. Rev.* 38 (11): 3133–3148. (e) Kaldre, D., Maryasin, B., Kaiser, D. et al. (2017). An asymmetric redox arylation: chirality from sulfur to carbon through a sulfonium (3,3)-sigmatropic rearrangement. *Angew. Chem. Int. Ed.* 56 (8): 2212–2215.

1149 Palko, R., Egyed, O., Bombicz, P. et al. (2008). Formation of new pyridyl substituted enamines. Observation of a diaza-Cope rearrangement. *Tetrahedron* 64 (45): 10375–10380.

1150 Maity, P., Pemberton, R.P., Tantillo, D.J., and Tambar, U.K. (2013). Brønsted acid catalyzed enantioselective indole aza-Claisen rearrangement mediated by an arene CH—O interaction. *J. Am. Chem. Soc.* 135 (44): 16380–16383.

1151 Fischer, E. and Jourdan, F. (1883). Über die Hydrazine der Brenztraubensäure. *Ber. Dtsch. Chem. Ges.* 16 (2): 2241–2245.

1152 Fischer, E. and Hess, O. (1884). Synthese von Indolderivaten. *Ber. Dtsch. Chem. Ges.* 17: 559–568.

1153 (a) Robinson, B. (1963). Fischer indole synthesis. *Chem. Rev.* 63 (4): 373–401. (b) Robinson, B. (1969). Recent studies on Fischer indole synthesis. *Chem. Rev.* 69 (2): 227–250.

1154 (a) Zhang, J.-Z., Yin, Z.-W., Leonard, P. et al. (2013). A variation of the Fischer indolization involving condensation of quinone monoketals and aliphatic hydrazines. *Angew. Chem. Int. Ed.* 52 (6): 1753–1757. (b) Satoh, H., Ueda, H., and Tokuyama, H. (2013). Divergent total synthesis of (−)-aspidophytine and its congeners via Fischer indole synthesis. *Tetrahedron* 69 (1): 89–95. (c) Martínez, A., Webber, M.J., Müller, S., and List, B. (2013). Versatile access to chiral indolines by catalytic asymmetric Fischer indolization. *Angew. Chem. Int. Ed.* 52 (36): 9486–9490. (d) Lim, B.-Y., Jung, B.-E., and Cho, C.-G. (2014). Ene-hydrazide from enol triflate for the regioselective Fischer indole synthesis. *Org. Lett.* 16 (17): 4492–4495. (e) Matcha, K. and Antochick, A.P. (2014). Cascade multicomponent synthesis of indoles, pyrazoles, and pyridazinones by functionalization of alkenes. *Angew. Chem. Int. Ed.* 53 (44): 11960–11964. (f) Xu, K., Gilles, T., and Breit, B. (2015). Asymmetric synthesis of *N*-allylic indoles via regio- and enantioselective allylation of aryl hydrazines. *Nat. Commun.* 6: 7616. (g) Smith, J.M., Moreno, J., Boal, B.W., and Garg, N.K. (2015). Fischer indolizations as a strategic platform for the total synthesis of Picrinine. *J. Org. Chem.* 80 (18): 8954–8967. (h) Shmatova, O.I., Shevchenko, N.E., and Nenajdenko, V.G. (2015). Fischer reaction with 2-perfluoroalkylated cyclic imines ? An efficient route to 2-perfluoroalkyl-substituted tryptamines and their derivatives and homologues. *Eur. J. Org. Chem.* (29): 6479–6488. (i) Park, J., Kim, D.-H., Das, T., and Cho, C.-G. (2016). Intramolecular Fischer indole synthesis for the direct synthesis of 3,4-fused tricyclic indole and application to the total synthesis of (−)-aurantioclavine. *Org. Lett.* 18 (19): 5098–5101. (j) Susik, R.B., Morrill, L.A., PIcazo, E., and Garg, N.K. (2017). Pardon the interruption: a modification of Fischer's venerable reaction for the synthesis of heterocycles and natural products. *Synlett* 28 (1): 1–11. (k) Ghumro, S.A., Saleem, S., al-Rashida, M. et al. (2017). *N,N*-Dimethylpyridin-4-amine (DMAP) based ionic liquids: evaluation of physical properties via molecular dynamics simulations and application as a catalyst for Fisher indole and 1*H*-tetrazole synthesis. *RSC Adv.* 7 (54): 34197–34207. (l) Heravi, M.M., Rohani, S., Zadsirjan, V., and Zahedi, N. (2017). Fischer indole synthesis applied to the total synthesis of natural products. *RSC Adv.* 7 (83): 52852–52887.

1155 Wagaw, S., Yang, B.H., and Buchwald, S.L.A. (1998). Pd-catalyzed strategy for the preparation of indoles: a novel entry into the Fischer indole synthesis. *J. Am. Chem. Soc.* 120 (26): 6621–6622.

1156 Chae, J.Y. and Buchwald, S.L. (2004). Palladium-catalyzed regioselective hydrodebromination of dibromoindoles: application to the enantioselective synthesis of indolodioxane U86192A. *J. Org. Chem.* 69 (10): 3336–3339.

1157 (a) Wagaw, S., Yang, B.H., and Buchwald, S.L. (1999). A palladium-catalyzed method for the preparation of indoles via the Fischer indole synthesis. *J. Am. Chem. Soc.* 121 (44): 10251–10263. (b) DeAngelis, A., Wang, D.H., and Buchwald, S.L. (2013). Mild and rapid Pd-catalyzed cross-coupling with hydrazine in continuous flow: application to the synthesis of functionalized heterocycles. *Angew. Chem. Int. Ed.* 52 (12): 3434–3437.

1158 (a) Kaushik, N.K., Kaushik, N., Attri, P. et al. (2013). Biomedical importance of indoles. *Molecules* 18 (6): 6620–6662. (b) Inman, M. and Moody, C.J. (2013). Indole synthesis - something old, something new. *Chem. Sci.* 4 (1): 29–41. (c) Imran, S., Taha, M., and Ismail, N.H. (2015). A review of bisindolylmethane as an important scaffold for drug discovery. *Curr. Med. Chem.* 22 (38): 4412–4433. (d) Zhang, M.-Z., Chen, Q., and Yang, G.-F. (2015). A review on recent developments of indole-containing antiviral agents. *Eur. J. Med. Chem.* 89: 421–441. (e) Chadha, N. and Silakari, O. (2017). Indoles as therapeutics of interest in medicinal chemistry: bird's eye view. *Eur. J. Med. Chem. (Elsevier)* 134: 159–184.

1159 (a) Claisen, L. (1912). The rearrangement of phenol-allyl-ather in C-allyl-phenole. *Ber. Dtsch. Chem. Ges.* 45: 3157–3166. (b) Claisen, L. and Tietze, E. (1926). On the mechanism of the rearrangement of phenol allyl ethers (2 announcement). *Ber. Dtsch. Chem. Ges.* 59: 2344–2351.

1160 (a) Nubbemeyer, U. (2003). *Synthesis* (7): 961–1008. (b) Hiratani, K. and Albrecht, M. (2008). The tandem Claisen rearrangement in the construction of building blocks for supramolecular chemistry. *Chem. Soc. Rev.* 37 (11): 2413–2421.

1161 Rhoads, S.J. and Raulins, N.R. (1975). The Claisen and Cope rearrangements. *Org. React.* 22: 1–252.

1162 Alexander, E.R. and Kluiber, R.W. (1951). Studies on the mechanism of the Claisen rearrangement - the *ortho* rearrangement of (−)-α,γ-dimethylallyl phenyl ether and the *para* rearrangement of (−)-α,γ-dimethylallyl 2,6-xylyl ether. *J. Am. Chem. Soc.* 73 (9): 4304–4306.

1163 Ramadhar, T.R. and Batey, R.A. (2011). Resolving the mechanistic origins of *E/Z*-selectivity differences for the *ortho*-aryl-Claisen (3,3)-sigmatropic rearrangement through DFT calculations. *Comput. Theor. Chem.* 974 (1–3): 76–78.

1164 Bergmann, E. and Corte, H. (1935). Migration of allyl groups in the ethyl acetoacetate series. *J. Chem. Soc.* 1363–1365.

1165 Lauer, W.M. and Kilburn, E.I. (1937). The rearrangement of some β-allyloxycrotonic esters. *J. Am. Chem. Soc.* 59: 2586–2588.

1166 Hurd, C.D. and Pollack, M.A. (1938). The rearrangement of vinyl allyl ethers. *J. Am. Chem. Soc.* 60: 1905–1911.

1167 (a) Ralls, J.W., Lundin, R.E., and Bailey, G.F. (1963). Preparation and thermal rearrangement of alkenyl 3-alkenyloxy-2-butenoates - catalysis of aliphatic Claisen rearrangement by ammonium chloride. *J. Org. Chem.* 28 (12): 3521–3526. (b) Hill, R.K. and Edwards, A.G. (1964). Stereochemistry of no-mechanism reactions. 2. Asymmetric induction in the Claisen rearrangement. *Tetrahedron Lett.* 5 (44): 3239–3243.

1168 Schuler, F.W. and Murphy, G.W. (1950). The kinetics of the rearrangement of vinyl allyl ether. *J. Am. Chem. Soc.* 72 (7): 3155–3159.

1169 Gajewski, J.J., Gee, K.R., and Jurayj, J. (1990). Energetic and rate effects of the trifluoromethyl group at C-2 and C-4 on the aliphatic Claisen rearrangement. *J. Org. Chem.* 55 (6): 1813–1822.

1170 (a) Iwakura, I., Kaneko, Y., Hayashi, S. et al. (2013). The reaction mechanism of Claisen rearrangement obtained by transition state spectroscopy and single direct-dynamics trajectory. *Molecules* 18 (2): 1995–2004. (b) González-Navarrete, P., Andrés, J., and Safont, V.S. (2018). A bonding evolution analysis for the thermal Claisen rearrangement: an experimental and theoretical exercise for testing the electron density flow. *Phys. Chem. Chem. Phys.* 20 (1): 535–541.

1171 Gorisch, H. (1978). Mechanism of Chorismate mutase reaction. *Biochemistry* 17 (18): 3700–3705.

1172 Kast, P., Tewari, Y.B., Wiest, O. et al. (1997). Thermodynamics of the conversion of chorismate to prephenate: experimental results and theoretical predictions. *J. Phys. Chem. B* 101 (50): 10976–10982.

1173 Marti, S., Andres, J., Moliner, V. et al. (2004). A comparative study of Claisen and Cope rearrangements catalyzed by chorismate mutase. An insight into enzymatic efficiency: transition state stabilization or substrate preorganization? *J. Am. Chem. Soc.* 126 (1): 311–319.

1174 Copley, S.D. and Knowles, J.R. (1985). The uncatalyzed Claisen rearrangement of chorismate to prephenate prefers a transition state of chairlike geometry. *J. Am. Chem. Soc.* 107 (18): 5306–5308.

1175 Jackson, D.Y., Jacobs, J.W., Sugasawara, R. et al. (1988). An antibody-catalyzed Claisen rearrangement. *J. Am. Chem. Soc.* 110 (14): 4841–4842.

1176 Ziegler, F.E. (1977). Stereochemistry and regiochemistry of Claisen rearrangement - applications to natural products synthesis. *Acc. Chem. Res.* 10 (6): 227–232.

1177 Ziegler, F.E. (1988). The thermal, aliphatic Claisen rearrangement. *Chem. Rev.* 88 (8): 1423–1452.

1178 Castro, A.M.M. (2004). Claisen rearrangement over the past nine decades. *Chem. Rev.* 104 (6): 2939–3002.

1179 McMichael, K.D. and Korver, G.L. (1979). Secondary deuterium-isotope effects and transition state structure in the aromatic Claisen rearrangement. *J. Am. Chem. Soc.* 101 (10): 2746–2747.

1180 Frey, H.M. and Montague, D.C. (1968). Thermal unimolecular isomerization of 1-methylallyl vinyl ether in gas phase and in solution. *Trans. Faraday Soc.* 64 (549P): 2369–2374.

1181 Ramadhar, T.R. and Batey, R.A. (2011). Accurate prediction of experimental free energy of activation barriers for the aliphatic-Claisen rearrangement through DFT calculations. *Comput. Theor. Chem.* 976 (1–3): 167–182.

1182 (a) Carroll, M.F. (1940). Addition of α,β-unsaturated alcohols to the active methylene group. Part I. The action of ethyl acetoacetate on linalool and geraniol. *J. Chem. Soc.* 704–706. (b) Defossey, M., Blanchard, N., Meyer, C., and Cossy, J. (2004). Total synthesis of Zincophorin and its methyl ester. *J. Org. Chem.* 69 (14): 4626–4647. (c) Skytte, D.M., Rasmussen, H.B., and Christensen, S.B. (2009). Carroll rearrangement to construct the norneolignan skeleton. *Nat. Prod. Res.* 22 (2): 116–123.

1183 (a) Felix, D., Gschwend, K., Wick, A.E., and Eschenmoser, A. (1969). Claisen rearrangement of allyl and benzyl alcohols with 1-dimethylamino-1-methoxyethene. *Helv. Chim. Acta* 52 (4): 1030–1042. (b) Walkowiak, J., Tomas-Szwaczyk, M., Haufe, G., and Koroniak, H. (2012). Synthesis of α-trifluoromethylated amides by Eschenmoser–Claisen-type rearrangement of allylic alcohols. *J. Fluorine Chem.* 143: 189–197. (c) Yoshida, M., Kasai, T., Mizuguchi, T., and Namba, K. (2014). Total synthesis of (−)-HM-3 and (−)-HM-4 utilizing a Pd-catalyzed addition of an arylboronic acid to an allenic alcohol followed by

Eschenmoser-Claisen rearrangement. *Synlett* 25 (8): 1160–1162.

1184 (a) Wick, A.E., Steen, K., Felix, D., and Eschenmoser, A. (1964). Claisensche Umlagerungen bei Allyl- + Benzylalkoholen mit Hilfe von Acetalen des *N,N*-Dimethylacetamids. *Helv. Chim. Acta* 47 (8): 2425–2429. (b) Olson, J.A. and Shea, K.M. (2011). Critical perspective: named reactions discovered and developed by women. *Acc. Chem. Res.* 44 (5): 311–321.

1185 (a) Johnson, W.S., Werthemann, L., Bartlett, W.R. et al. (1970). Simple stereoselective version of Claisen rearrangement leading to *trans*-trisubstituted olefinic bonds. Synthesis of squalene. *J. Am. Chem. Soc.* 92 (3): 741–743. (b) Meza-Aviña, M.-E., Ordoñez, M., Fernández-Zertuche, M. et al. (2005). Martínez de los Ríos-Corsino, A. A. Synthesis of some monocyclic analogues of mycophenolic acid via the Johnson ortho ester Claisen rearrangement. *Bioorg. Med. Chem.* 13 (23): 6521–6528. (c) Ermolovich, Y.V., Zhabinskii, V.N., and Khripach, V.A. (2013). 1,4-Chirality transfer via the ester enolate Claisen rearrangements in the preparation of (25R)- and (25S)-cholestenoic acids. *Steroids* 78 (7): 683–692.

1186 Ireland, R.E. and Mueller, R.H. (1972). Claisen rearrangement of allyl esters. *J. Am. Chem. Soc.* 94 (16): 5897–5898.

1187 (a) Ireland, R.E., Mueller, R.H., and Willard, A.K. (1976). Ester enolate Claisen rearrangement - construction of prostanoid skeleton. *J. Org. Chem.* 41 (6): 986–996. (b) Ireland, R.E., Mueller, R.H., and Willard, A.K. (1976). Ester enolate Claisen rearrangement - stereochemical control through stereoselective enolate formation. *J. Am. Chem. Soc.* 98 (10): 2868–2877.

1188 (a) Bunte, J.O., Cuzzupe, A.N., Daly, A.N., and Rizzacasa, M.A. (2006). Formal total synthesis of (+)-Zaragozic acid C through an Ireland-Claisen rearrangement. *Angew. Chem. Int. Ed.* 45 (38): 6376–6380. (b) Stivala, C.E. and Zakarian, A. (2008). Total synthesis of (+)-pinnatoxin A. *J. Am. Chem. Soc.* 130 (12): 3774–3776. (c) Araoz, R., Servent, D., Molgó, J. et al. (2011). Total synthesis of pinnatoxins A and G and revision of the mode of action of pinnatoxin A. *J. Am. Chem. Soc.* 133 (27): 10499–10511. (d) Ernouf, G., Brayer, J.L., Folleas, B. et al. (2017). Synthesis of alkylidene(*gem*-difluorocyclopropanes) from propargyl glycolates by a one-pot difluorocyclopropenation/Ireland–Claisen rearrangement sequence. *J. Org. Chem.* 82 (7): 3965–3975. (e) Kuilya, T.K. and Goswami, R.K. (2017). Stereoselective total synthesis of carolacton. *Org. Lett.* 19 (9): 2366–2369. (f) Seizert, C.A. and Ferreira, E.M. (2017). A boron-based Ireland–Claisen approach to the synthesis of pordamacrine A. *Tetrahedron* 73 (29): 4186–4194. (g) Murakata, M. and Ikeda, T. (2017). Stereoselective synthesis of the viridiofungin analogue NA808 from a chiral tetrahydrofuran-carboxylic acid. *Org. Biomol. Chem.* 15 (31): 6632–6639. (h) Xu, B., Xun, W., Wang, T.-Z., and Qiu, F.G. (2017). Total synthesis of (+)-aplykurodinone-1. *Org. Lett.* 19 (18): 4861–4863.

1189 (a) Malherbe, R. and Bellus, D. (1978). New type of Claisen rearrangement involving 1,3-dipolar intermediates - preliminary communication. *Helv. Chim. Acta* 61 (8): 3096–3099. (b) Malherbe, R., Rist, G., and Bellus, D. (1983). Reactions of haloketenes with allyl ethers and thioethers - a new type of Claisen rearrangement. *J. Org. Chem.* 48 (6): 860–869.

1190 (a) Kling, M.R., McNaughton, G.A., and Taylor, J.K. (1993). The stereoselective synthesis of unsaturated nine-membered lactones using the Malherbe–Bellus variant of the Claisen rearrangement. *J. Chem. Soc., Chem. Commun.* (21): 1593–1595. (b) Mahmood, A. (2017). The steering pathway: Ketene–Claisen rearrangement (KCR)-1978–2016. *Tetrahedron* 73 (16): 2173–2190.

1191 Baldwin, J.E. and Walker, J.A. (1973). Reformatsky–Claisen reaction - new synthetically useful sigmatropic process. *J. Chem. Soc., Chem. Commun.* (4): 117–118.

1192 (a) Greuter, H., Lang, R.W., and Romann, A.J. (1988). Fluorine-containing organozinc reagents. 4. The Reformatskii–Claisen reaction of chlorodifluoroacetic acid-derivatives. *Tetrahedron Lett.* 29 (27): 3291–3294. (b) Brandange, S., Holmgren, E., Leijonmarck, H., and Rodríguez, B. (1995). On the preparation of *N*-acylpyrroles and their use in the synthesis of ketones. *Acta Chem. Scand.* 49 (12): 922–928. (c) Ishihara, J., Koyama, N., Nishino, Y. et al. (2009). A new variant of Reformatsky–Claisen rearrangement mediated by indium chloride. *Synlett* (14): 2351–2355. (d) Ishihara, J., Watanabe, Y., Koyama, N. et al. (2011). Indium-mediated Reformatsky–Claisen rearrangement. *Tetrahedron* 67 (20): 3659–3667. (e) Ishihara, J. and Hatakeyama, S. (2012). Recent developments in the Reformatsky–Claisen rearrangement. *Molecules* 17 (12): 14249–14259.

1193 Kazmaier, U. (1994). Synthesis of unsaturated amino-acids by (3,3)-sigmatropic rearrangement of chelate-bridged glycine ester enolates. *Angew. Chem. Int. Ed. Engl.* 33 (9): 998–999.

1194 Kazmaier, U., Mues, H., and Krebs, A. (2002). Asymmetric chelated Claisen rearrangements in the presence of chiral ligands - scope and limitations. *Chem. Eur. J.* 8 (8): 1850–1855.

1195 Sakaguchi, K., Suzuki, H., and Ohfune, Y. (2001). Chirality transferring (3,3)-sigmatropic rearrangement of (1-acyloxy-2-alkenyl)trialkylsilane: synthesis of optically active vinylsilane-containing α-amino acid. *Chirality* 13 (7): 357–365.

1196 Sakaguchi, K., Yamamoto, M., Kawamoto, T. et al. (2004). Synthesis of optically active β-alkyl aspartate via (3,3)-sigmatropic rearrangement of α-acyloxytrialkylsilane. *Tetrahedron Lett.* 45 (30): 5869–5872.

1197 (a) Denmark, S.E. and Harmata, M.A. (1982). Carbanion-accelerated Claisen rearrangements. *J. Am. Chem. Soc.* 104 (18): 4972–4974. (b) Denmark, S.E. and Harmata, M.A. (1983). Carbanion-accelerated Claisen rearrangements. 2. Studies on internal asymmetric induction. *J. Org. Chem.* 48 (19): 3369–3370. (c) Denmark, S.E. and Harmata, M.A. (1984). Carbanion-accelerated Claisen rearrangements. 3. Vicinal quaternary centers. *Tetrahedron Lett.* 25 (15): 1543–1546.

1198 (a) Denmark, S.E. and Marlin, J.E. (1991). Carbanion-accelerated Claisen rearrangements. 7. Phosphine oxide and phosphonate anion stabilizing groups. *J. Org. Chem.* 56 (3): 1003–1013. (b) Denmark, S.E., Stadler, H., Dorow, R.L., and Kim, J.H. (1991). Carbanion-accelerated Claisen rearrangements. 8. Phosphoramide anion-stabilizing groups. *J. Org. Chem.* 56 (17): 5063–5079.

1199 Denmark, S.E., Marlin, J.E., and Rajendra, G. (2013). Carbanion-accelerated Claisen rearrangements: asymmetric induction with chiral phosphorus-stabilized anions. *J. Org. Chem.* 78 (1): 66–82.

1200 (a) Koreeda, M. and Luengo, J.I. (1985). Anionic oxy-Claisen rearrangement of enolates of α-allyloxy ketones - a remarkable rate-accelerating effect exhibited by the nature of the counterion. *J. Am. Chem. Soc.* 107 (19): 5572–5573. (b) see also: Kirchner, J.J., Pratt, D.V., and Hopkins, P.B. (1988). Anionic oxy-Claisen rearrangement of a tricyclic α-allyloxy ketone. *Tetrahedron Lett.* 29 (34): 4229–4232.

1201 Jansma, M.J. and Hoye, T.R. (2012). Oxyanionic sigmatropic rearrangements relevant to cyclooctadienone formation in penostatins I and F. *Org. Lett.* 14 (18): 4738–4741.

1202 Rychnovsky, S.D. and Lee, J.L. (1995). A mild, regioselective ketal Claisen rearrangement promoted by triisobutylaluminum. *J. Org. Chem.* 60 (14): 4318–4319.

1203 Huang, X.L., Klimczyk, S., and Maulide, N. (2012). Charge-accelerated sulfonium (3,3)-sigmatropic rearrangements. *Synthesis* 44 (2): 175–183.

1204 (a) Maruoka, K., Banno, H., and Yamamoto, H. (1990). Asymmetric Claisen rearrangement catalyzed by chiral organoaluminum reagent. *J. Am. Chem. Soc.* 112 (21): 7791–7793. (b) Maruoka, K., Banno, H., and Yamamoto, H. (1991). Enantioselective activation of ethers by chiral organoaluminum reagents - application to asymmetric Claisen rearrangement. *Tetrahedron: Asymmetry* 2 (7): 647–662.

1205 Enders, D., Knopp, M., and Schiffers, R. (1996). Asymmetric (3,3)-sigmatropic rearrangements in organic synthesis. *Tetrahedron: Asymmetry* 7 (7): 1847–1882.

1206 (a) Ito, H. and Taguchi, T. (1999). Asymmetric Claisen rearrangement. *Chem. Soc. Rev.* 28 (1): 43–50. (b) Rodrigues, A.F., Costa, T., Wender-Alves, S. et al. (2015). Recent advances in the asymmetric Claisen rearrangement promoted by chiral organometallic Lewis acids or organic Brønsted–Lowry acids. *Curr. Org. Synth. (Bentham)* 12 (6): 795–805.

1207 Yoon, T.P. and MacMillan, D.W.C. (2001). Enantioselective Claisen rearrangements: development of a first generation asymmetric acyl-Claisen reaction. *J. Am. Chem. Soc.* 123 (12): 2911–2912.

1208 Tan, J., Cheon, C.H., and Yamamoto, H. (2012). Catalytic asymmetric Claisen rearrangement of enolphosphonates: construction of vicinal tertiary and all-carbon quaternary centers. *Angew. Chem. Int. Ed.* 51 (33): 8264–8267.

1209 Trung, C., Linton, E.C., Deitch, J. et al. (2012). Copper(II)- and Pd(II)-catalyzed enantioselective Claisen rearrangement of allyloxy- and propargyloxy-indoles to quaternary oxindoles and spirocyclic lactones. *J. Org. Chem.* 77 (24): 11034–11055.

1210 Abraham, L., Czerwonka, R., and Hiersemann, M. (2001). The catalytic enantioselective Claisen rearrangement of an allyl vinyl ether. *Angew. Chem. Int. Ed.* 40 (24): 4700–4703.

1211 Abraham, L., Körner, M., Schwab, P., and Hiersemann, M. (2004). The catalytic diastereo- and enantioselective Claisen rearrangement of 2-alkoxycarbonyl-substituted allyl vinyl ether. *Adv. Synth. Catal.* 346 (11): 1281–1294.

1212 White, W.N. and Wolfarth, E.F. (1970). *Ortho* Claisen rearrangement. 9. Effect of solvent on substituent effect. *J. Org. Chem.* 35 (10): 3585–3585.

1213 White, W.N. and Wolfarth, E.F. (1970). *Ortho* Claisen rearrangement. 8. Solvent effects. *J. Org. Chem.* 35 (7): 2196–2199.

1214 Curran, D.P. and Kuo, L.H. (1995). Acceleration of a dipolar Claisen rearrangement by hydrogen-bonding to a soluble diaryl urea. *Tetrahedron Lett.* 36 (37): 6647–6650.

1215 Kirsten, M., Rehbein, J., Hiersemann, M., and Strassner, T. (2007). Organocatalytic Claisen rearrangement: theory and experiment. *J. Org. Chem.* 72 (11): 4001–4011.

1216 Uyeda, C. and Jacobsen, E.N. (2008). Enantioselective Claisen rearrangements with a hydrogen-bond donor catalyst. *J. Am. Chem. Soc.* 130 (29): 9228–9229.

1217 Uyeda, C. and Jacobsen, E.N. (2011). Transition state charge stabilization through multiple non-covalent interactions in the guanidinium-catalyzed enantioselective Claisen rearrangement. *J. Am. Chem. Soc.* 133 (13): 5062–5075.

1218 (a) Coates, R.M., Shah, S.K., and Mason, R.W. (1982). Stereoselective total synthesis of (+/-)-gymnomitrol via reduction-alkylation of α-cyano ketones. *J. Am. Chem. Soc.* 104 (8): 2198–2208. (b) Coates, R.M., Rogers, B.D., Hobbs, S.J. et al. (1987). Synthesis and Claisen rearrangement of alkoxyallyl enol ethers - evidence for a dipolar transition state. *J. Am. Chem. Soc.* 109 (4): 1160–1170.

1219 (a) Mahatthananchai, J., Kaeobamrung, J., and Bode, J.W. (2012). Chiral N-heterocyclic carbene-catalyzed annulations of enals and ynals with stable enols: a highly enantioselective Coates–Claisen rearrangement. *ACS Catal.* 2 (4): 494–503. (b) Candish, L. and Lupton, D.W. (2013). N-heterocyclic carbene-catalyzed Ireland–Coates Claisen rearrangement: synthesis of functionalized β-lactones. *J. Am. Chem. Soc.* 135 (1): 58–61.

1220 (a) Saucy, G. and Marbet, R. (1967). Über eine neuartige Synthese von β-Ketoallenen durch Reaktion von tertiaren Acetylencarbinolen mit Vinyläthern eine ergiebige Methode zur Darstellung des Pseudojonons und Verwandter Verbindungen. *Helv. Chim. Acta* 50 (4): 1158–1167. (b) Tejedor, D., Méndez-Abt, G., Cotos, L., and Carcía-Telado, F. (2013). Propargyl-Claisen rearrangement: allene synthesis and beyond. *Chem. Soc. Rev.* 42 (2): 458–471.

1221 Cao, T., Deitch, J., Linton, E.C., and Kozlowski, M.C. (2012). Asymmetric synthesis of allenyl oxindoles and spirooxindoles by a catalytic enantioselective Saucy–Marbet Claisen rearrangement. *Angew. Chem. Int. Ed.* 51 (10): 2448–2451.

1222 Wang, H.J., Gloer, J.B., Wicklow, D.T., and Dowd, P.F. (1998). Mollenines A and B: new dioxomorpholines from the ascostromata of *Eupenicillium molle. J. Nat. Prod.* 61 (6): 804–807.

1223 Kato, H., Yoshida, T., Tokue, T. et al. (2007). Notoamides A–D: prenylated indole alkaloids isolated from a marine-derived fungus, *Aspergillus sp. Angew. Chem. Int. Ed.* 46 (13): 2254–2256.

1224 Gupta, L., Talwar, A., and Chauhan, P.M. (2007). Bis and tris indole alkaloids from marine organisms: new leads for drug discovery. *Curr. Med. Chem.* 14 (16): 1789–1803.

1225 Miller, K.A. and Williams, R.M. (2009). Synthetic approaches to the bicyclo[2.2.2]diazaoctane ring system common to the paraherquamides, stephacidins and related prenylated indole alkaloids. *Chem. Soc. Rev.* 38 (11): 3160–3174.

1226 Kochanowska-Karamyan, A.J. and Hamann, M.T. (2010). Marine indole alkaloids: potential new drug leads for the control of depression and anxiety. *Chem. Rev.* 110 (8): 4489–4497.

1227 Shiri, M. (2012). Indoles in multicomponent processes (MCPs). *Chem. Rev.* 112 (6): 3508–3549.

1228 Lindel, T., Marsch, N., and Adla, S.K. (2012). Indole prenylation in alkaloid synthesis. In: *Alkaloid Synthesis*, Topics in Current Chemistry, vol. 309 (ed. J.-J. Knölker), 67–129. Heidelberg, Berlin: Springer-Verlag.

1229 Vittorelli, P., Winkler, T., Hansen, H.J., and Schmid, H. (1968). Stereochemistry of transition states in aliphatic Claisen rearrangements. *Helv. Chim. Acta* 51 (6): 1457–1461.

1230 Hansen, H.J. and Schmid, H. (1974). Stereochemistry of sigmatropic (3,3)-rearrangement and sigmatropic (5,5)-rearrangement. *Tetrahedron* 30 (13): 1959–1969.

1231 Chapleo, C.B., Hallett, P., Lythgoe, B. et al. (1977). Calciferol and its relatives. 19. Synthetic applications of cyclic orthoesters - stereospecific synthesis of a bicyclic alcohol related to Vitamins D^2. *J. Chem. Soc., Perkin Trans. 1* (10): 1211–1218.

1232 Cave, R.J., Lythgoe, B., Metcalfe, D.A., and Waterhouse, I. (1977). Stereochemical aspects of some Claisen rearrangements with cyclic orthoesters. *J. Chem. Soc., Perkin Trans. 1* (10): 1218–1228.

1233 Mumm, O. and Moller, F. (1937). Tests on the theory of allyl redistribution. *Ber. Dtsch. Chem. Ges.* 70: 2214–2227.

1234 Khusnitdinov, R.N., Fakhretdinov, R.N., Abdrakhmanov, I.B. et al. (1983). Synthesis of *ortho*-substituted unsaturated anilines. *Zh. Org. Khim.* 19 (5): 1034–1039.

1235 Roe, J.M., Webster, R.A.B., and Ganesan, A. (2003). Total synthesis of (+)-okaramine J featuring an exceptionally facile N-reverse-prenyl to C-prenyl aza-Claisen rearrangement. *Org. Lett.* 5 (16): 2825–2827.

1236 Beare, K.D. and McErlean, C.S. (2013). Revitalizing the aromatic aza-Claisen rearrangement: implications for the mechanism of 'on-water' catalysis. *Org. Biomol. Chem.* 11 (15): 2452–2459.

1237 Nubbemeyer, U. (2005). Recent advances in charge-accelerated aza-Claisen rearrangements. In: *Natural Product Synthesis II*, Topics in Current Chemistry, vol. 244 (ed. J. Mulzer), 149–213. Heidelberg, Berlin: Springer-Verlag.

1238 Gomes, M.J.S., Sharma, L., Prabhakar, S. et al. (2002). Studies in 3-oxy-assisted 3-aza Cope rearrangements. *J. Chem. Soc., Chem. Commun.* (7): 746–747.

1239 Glória, P.M., Prabhakar, S., and Lobo, A.M. (2008). Studies in *N*-amino-3-aza Cope rearrangements. *Tetrahedron Lett.* 49 (52): 7355–7357.

1240 (a) Barta, N.S., Cook, G.R., Landis, M.S., and Stille, J.R. (1992). Studies of the regiospecific 3-aza-Cope rearrangement promoted by electrophilic reagents. *J. Org. Chem.* 57 (26): 7188–7194. (b) Cook, G.R., Barta, N.S., and Stille, J.R. (1992). Lewis acid-promoted 3-aza-Cope rearrangement of *N*-alkyl-*N*-allylenamines. *J. Org. Chem.* 57 (2): 461–467.

1241 Murahashi, S., Makabe, Y., and Kunita, K. (1988). Palladium(0)-catalyzed rearrangement of *N*-allylenamines - synthesis of δ,ε-unsaturated imines and γ,δ-unsaturated carbonyl compounds. *J. Org. Chem.* 53 (19): 4489–4495.

1242 Baxter, E.W., Labaree, D., Chao, S., and Mariano, P.S. (1989). Model studies probing the amino-Claisen rearrangement approach to hydroisoquinoline synthesis - development of methods for stereocontrolled introduction of Reserpine-e ring-type functionality. *J. Org. Chem.* 54 (12): 2893–2904.

1243 Mariano, P.S., Dunawaymariano, D., Huesmann, P.L., and Beamer, R.L. (1977). Novel approaches to synthesis of hydroisoquinolines using aza-Claisen rearrangements of *N*-vinylisoquinuclidenes. *Tetrahedron Lett.* (49): 4299–4302.

1244 Mariano, P.S., Dunawaymariano, D., and Huesmann, P.L. (1978). Amino-Claisen rearrangements of *N*-vinylisoquinuclidenes in novel approaches to the synthesis of hydroisoquinolines and hydrophenanthridines. *J. Org. Chem.* 44 (1): 124–133.

1245 Kunng, F.A., Gu, J.M., Chao, S.C. et al. (1983). A novel synthetic approach to reserpine based upon amino-Claisen rearrangements of zwitterionic *N*-vinylisoquinuclidenes. *J. Org. Chem.* 48 (23): 4262–4266.

1246 Chao, S., Kunng, F.A., Gu, J.M. et al. (1984). Rauwolfia alkaloid synthesis approach employing the zwitterionic amino-Claisen rearrangement - improvements in the efficiency for yohimbane ring construction and unambiguous de-ring-fusion stereochemical assignments. *J. Org. Chem.* 49 (15): 2708–2711.

1247 Baxter, E.W., Labaree, D., Ammon, H.L., and Mariano, P.S. (1990). Formal total synthesis of deserpidine demonstrating a versatile amino-Claisen rearrangement: Wenkert cyclization strategy for the preparation of functionalized yohimbane ring-systems. *J. Am. Chem. Soc.* 112 (21): 7682–7692.

1248 Weston, M.H., Nakajima, K., and Back, T.G. (2008). Tandem conjugate additions and 3-aza-Cope rearrangements of tertiary allyl amines and cyclic α-vinylamines with acetylenic sulfones. Applications to simple and iterative ring expansions leading to medium and large-ring nitrogen heterocycles. *J. Org. Chem.* 73 (12): 4630–4637.

1249 Walters, M.A., Hoem, A.B., Arcand, H.R. et al. (1993). An extremely mild 3-aza-Claisen reaction. 2. New conditions and the rearrangement of α-heteroatom substituted amides. *Tetrahedron Lett.* 34 (9): 1453–1456.

1250 Overman, L.E. (1974). Thermal and mercuric ion catalyzed (3,3)-sigmatropic rearrangement of allylic trichloroacetimidates - (1,3)-transposition of alcohol and amine functions. *J. Am. Chem. Soc.* 96 (2): 597–599.

1251 (a) Overman, L.E. (1976). General method for synthesis of amines by rearrangement of allylic trichloroacetimidates - (1,3)-transposition of alcohol and amine functions. *J. Am. Chem. Soc.* 98 (10): 2901–2910. (b) Overman, L.E. (1984). Mercury(II)-catalyzed and Pd(II)-catalyzed (3,3)-sigmatropic rearrangements. *Angew. Chem. Int. Ed. Engl.* 23 (8): 579–586.

1252 (a) Tsuzaki, S., Usui, S., Oishi, H. et al. (2015). Total synthesis of sphingofungin F by orthoamide-type Overman rearrangement of an unsaturated ester. *Org. Lett.* 17 (7): 1704–1707. (b) Fernandes, R.A., Kattanguru, P., Gholap, S.P., and Chaudhari, D.A. (2017). Recent advances in the Overman rearrangement: synthesis of natural products and valuable compounds. *Org. Biomol. Chem.* 15 (13): 2672–2710. (c) Martínez-Alsina, L.A., Murray, J.C., Buzon, L.M. et al. (2017). Spiropiperidine sultam and lactam templates: diastereoselective Overman rearrangement and metathesis followed by NH arylation. *J. Org. Chem.* 82 (23): 12246–12256.

1253 Overman, L.E. and Carpenter, N.E. (2005). The allylic trihaloacetimidate rearrangement. *Org. React.* 66: 1–107.

1254 Calter, M., Hollis, T.K., Overman, L.E. et al. (1997). First enantioselective catalyst for the rearrangement of allylic imidates to allylic amides. *J. Org. Chem.* 62 (5): 1449–1456.

1255 Uozumi, Y., Kato, K., and Hayashi, T. (1998). Asymmetric aza-Claisen rearrangement of allyl imidates catalyzed by homochiral cationic Pd(II) complexes. *Tetrahedron: Asymmetry* 9 (6): 1065–1072.

1256 Anderson, C.E. and Overman, L.E. (2003). Catalytic asymmetric rearrangement of allylic trichloroacetimidates. A practical method for preparing allylic amines and congeners of high enantiomeric purity. *J. Am. Chem. Soc.* 125 (41): 12412–12413.

1257 Overman, L.E., Owen, C.E., Pavan, M.M., and Richards, C.J. (2003). Catalytic asymmetric rearrangement of allylic *N*-aryl trifluoroacetimidates. A useful method for transforming prochiral allylic alcohols to chiral allylic amines. *Org. Lett.* 5 (11): 1809–1812.

1258 Prasad, R.S., Anderson, C.E., Richards, C.J., and Overman, L.E. (2005). Synthesis of *tert*-leucine-derived Co oxazoline palladacycles. Reversal of palladation diastereoselectivity and application to the asymmetric rearrangement of *N*-aryl trifluoroacetimidates. *Organometallics* 24 (1): 77–81.

1259 Jautze, S., Seiler, P., and Peters, R. (2007). Macrocyclic ferrocenyl-bisimidazoline palladacycle dimers as highly active and enantioselective catalysts for the aza-Claisen rearrangement of *Z*-configured *N*-*para*-methoxyphenyl trifluoroacetimidates. *Angew. Chem. Int. Ed.* 46 (8): 1260–1264.

1260 (a) Jautze, S., Seiler, P., and Peters, R. (2008). Synthesis of nearly enantiopure allylic amines by aza-Claisen rearrangement of *Z*-configured allylic trifluoroacetimidates catalyzed by highly active ferrocenylbispalladacycles. *Chem. Eur. J.* 14 (5): 1430–1444. (b) Fischer, D.F., Barakat, A., Xin, Z.q. et al. (2009). The asymmetric aza-Claisen rearrangement: development of widely applicable pentaphenylferrocenyl palladacycle catalysts. *Chem. Eur. J.* 15 (35): 8722–8741. (c) Weiss, M., Frey, W., and Peters, R. (2012). Asymmetric synthesis of heterobimetallic planar chiral ferrocene pallada-/platinacycles and their application to enantioselective aza-Claisen rearrangements. *Organometallics* 31 (17): 6365–6372.

1261 Liu, Z., Qu, H., Gu, X. et al. (2008). Enantioselective synthesis of *anti*-β-substituted γ,δ-unsaturated amino acids: a highly selective asymmetric thio-Claisen rearrangement. *Org. Lett.* 10 (18): 4105–4108.

1262 (a) Ellwood, A.R., Mortimer, A.J.P., Tocher, D.A., and Porter, M.J. (2008). Diastereoselective thia-Claisen rearrangement of pyrrolidinone-derived ketene N,S-acetals. *Synlett* (14): 2199–2203. (b) Ellwood, A.R., Mortimer, A.J.P., Goodman, J.M., and Porter, M.J. (2013). Reversal of facial selectivity in a thia-Claisen rearrangement by incorporation of a vinylic bromine substituent. *Org. Biomol. Chem.* 11 (43): 7530–7539.

1263 (a) Gajewski, J.J. and Conrad, N.D. (1978). Mechanism of Cope rearrangement. *J. Am. Chem. Soc.* 100 (19): 6268–6269. (b) Mackey, J.L., Yang, Z.-Y., and Houk, K.N. (2017). Dynamically concerted and stepwise trajectories of the Cope rearrangement of 1,5-hexadiene. *Chem. Phys. Lett.* 683: 253–257.

1264 Wiest, O., Black, K.A., and Houk, K.N. (1994). Density-functional theory isotope effects and activation energies for the Cope and Claisen rearrangements. *J. Am. Chem. Soc.* 116 (22): 10336–10337.

1265 Doering, W.v.E., Toscano, V.G., and Beasley, G.H. (1971). Kinetics of Cope rearrangement of 1,1-dideuteriohexa-1,5-diene. *Tetrahedron* 27 (22): 5299–5306.

1266 McGuire, M.J. and Piecuch, P. (2005). Balancing dynamic and nondynamic correlation for diradical and aromatic transition states: a renormalized coupled-cluster study of the Cope rearrangement of 1,5-hexadiene. *J. Am. Chem. Soc.* 127 (8): 2608–2614.

1267 Rossi, M., King, K.D., and Golden, D.M. (1979). Equilibrium constant and rate constant for allyl radical recombination in the gas phase. *J. Am. Chem. Soc.* 101 (5): 1223–1230.

1268 Hill, R.K. and Gilman, N.W. (1967). Asymmetric induction in Cope rearrangement. *J. Chem. Soc., Chem. Commun.* (13): 619–620.

1269 Wessel, T.E. and Berson, J.A. (1994). Stabilization of a putative cyclohexane-1,4-diyl intermediate elicits an antarafacial Cope rearrangement via a stepwise mechanism - pyrolysis of (*R*,*E*)-5-methyl-1,2,6-octatriene to 4-methyl-3-methylene-1,5-heptadiene. *J. Am. Chem. Soc.* 116 (2): 495–505.

1270 Roth, W.R., Wollweber, D., Offerhaus, R. et al. (1993). The energy-well of diradicals. 4. 2-Methylene-1,4-cyclohexadiyl. *Chem. Ber.* 126 (12): 2701–2715.

1271 Debbert, S.L., Carpenter, B.K., Hrovat, D.A., and Borden, W.T. (2002). The iconoclastic dynamics of the 1,2,6-heptatriene rearrangement. *J. Am. Chem. Soc.* 124 (27): 7896–7897.

1272 Dewar, M.J.S. and Wade, L.E. (1977). Study of mechanism of Cope rearrangement. *J. Am. Chem. Soc.* 99 (13): 4417–4424.

1273 Dewar, M.J.S. and Jie, C.X. (1987). Mechanism of the Cope rearrangement. *J. Am. Chem. Soc.* 109 (20): 5893–5899.

1274 Roth, W.R., Lennartz, H.W., Doering, W.V. et al. (1990). A frustrated Cope rearrangement - thermal interconversion of 2,6-diphenylhepta-1,6-diene and 1,5-diphenylbicyclo[3.2.0]heptane. *J. Am. Chem. Soc.* 112 (5): 1722–1732.

1275 Roth, W.R., Gleiter, R., Paschmann, V. et al. (1998). Non-concerted Cope rearrangement. *Eur. J. Org. Chem.* (6): 961–967.

1276 Muralha, V.S.F., dos Santos, R.M.B., and Simões, J.A.M. (2004). Energetics of alkylbenzyl radicals: a time-resolved photoacoustic calorimetry study. *J. Phys. Chem. A* 108 (6): 936–942.

1277 Hrovat, D.A., Chen, J.G., Houk, K.N., and Borden, W.T. (2000). Cooperative and competitive substituent effects on the Cope rearrangements of phenyl substituted 1,5-hexadienes elucidated by Becke3LYP/6-31G* calculations. *J. Am. Chem. Soc.* 122 (31): 7456–7460.

1278 von Doering, W., Birladeanu, L., Sarma, K. et al. (2000). A non-Cope among the Cope rearrangements of 1,3,4,6-tetraphenylhexa-1,5-dienes. *J. Am. Chem. Soc.* 122 (2): 193–203.

1279 Hirota, E., Yamada, C., and Okunishi, M. (1992). Infrared diode-laser spectroscopy of the allyl radical - the ν-11 band. *J. Chem. Phys.* 97 (5): 2963–2970.

1280 Hrovat, D.A., Beno, B.R., Lange, H. et al. (1999). A Becke3LYP/6-31G* study of the Cope rearrangements of substituted 1,5-hexadienes provides computational evidence for a chameleonic transition state. *J. Am. Chem. Soc.* 121 (45): 10529–10537.

1281 Blavins, J.J., Cooper, D.L., and Karadakov, P.B. (2004). Aromatic vs diradical character in the transition states of the cope rearrangements of 1,5-hexadiene and its cyano derivatives. *J. Phys. Chem. A* 108 (1): 194–202.

1282 Doering, W.V. and Wang, Y.H. (1999). Perturbation of Cope's rearrangement: 1,3,5-Triphenylhexa-1,5-diene. Chameleonic or centauric transition region? *J. Am. Chem. Soc.* 121 (43): 10112–10118.

1283 Gajewski, J.J. and Conrad, N.D. (1978). Variable transition state structure in Cope rearrangement as deduced from secondary deuterium kinetic isotope effects. *J. Am. Chem. Soc.* 100 (19): 6269–6270.

1284 Gajewski, J.J. and Conrad, N.D. (1979). Variable transition state structure in (3,3)-sigmatropic shifts from α-secondary deuterium isotope effects. *J. Am. Chem. Soc.* 101 (22): 6693–6704.

1285 Gajewski, J.J. and Conrad, N.D. (1979). Aliphatic Claisen rearrangement transition state structure from secondary α-deuterium isotope effects. *J. Am. Chem. Soc.* 101 (10): 2747–2748.

1286 Hill, R.K. (1991). Cope, oxy-Cope and anionic oxy-Cope rearrangements. In: *Comprehensive Organic Synthesis*, vol. 5 (Combining C—C π bonds) (ed. B.M. Trost and I. Fleming), 785–826. Oxford: Pergamon Press.

1287 Brown, J.M., Golding, B.T., and Stofko, J.J. (1973). Isolation and characterization of *cis*-divinylcyclopropane. *J. Chem. Soc., Chem. Commun.* (9): 319–320.

1288 Berson, J.A. and Dervan, P.B. (1972). Incremental substituent effects leading to steric blockade of boat-like six-center Cope rearrangement of *cis*-1,2-dialkenylcyclobutanes. *J. Am. Chem. Soc.* 94 (21): 7597–7598.

1289 Ozkan, I. and Zora, M. (2003). Transition structures, energetics, and secondary kinetic isotope effects for Cope rearrangements of *cis*-1,2-divinylcyclobutane and *cis*-1,2-divinylcyclopropane: a DFT study. *J. Org. Chem.* 68 (25): 9635–9642.

1290 Pommelet, J.C. and Chuche, J. (1974). Synthesis and thermal isomerization of *cis* *N*-methyl-2,3-divinyl aziridine. *Tetrahedron Lett.* 15 (44): 3897–3898.

1291 Braun, R.A. (1963). Preparation of 4,5-dihydrooxepine and 1,2-divinylethylene oxide. *J. Org. Chem.* 28: 1383–1384.

1292 Zora, M. (2005). A comparison of the Cope rearrangements of *cis*-1,2-divinyleyelopropane, *cis*-2,3-divinylaziridine, *cis*-2,3-divinyloxirane, *cis*-2,3-divinylphosphirane, and *cis*-2,3-divinylthiirane: a DFT study. *J. Org. Chem.* 70 (15): 6018–6026.

1293 Thomas, A.F. (1969). Addition of a functionalized isoprene unit to an allyl alcohol. I. Synthesis of β-Sinensal and related topics. *J. Am. Chem. Soc.* 91 (12): 3281–3289.

1294 Overman, L.E. and Renaldo, A.F. (1990). Catalyzed sigmatropic rearrangements. 10. Mechanism of the Pd dichloride catalyzed Cope rearrangement of acyclic dienes - a substituent effect study. *J. Am. Chem. Soc.* 112 (10): 3945–3949.

1295 Nakamura, H., Iwama, H., Ito, M., and Yamamoto, Y. (1999). Palladium(0)-catalyzed Cope rearrangement of acyclic 1,5-dienes. Bis(π-allyl)Pd(II) intermediate. *J. Am. Chem. Soc.* 121 (46): 10850–10851.

1296 Fanning, K.N., Jamieson, A.G., and Sutherland, A. (2006). Palladium(II)-catalysed rearrangement reactions. *Curr. Org. Chem.* 10 (9): 1007–1020.

1297 Siebert, M.R. and Tantillo, D.J. (2007). Transition state complexation in Pd-promoted (3,3)-sigmatropic shifts. *J. Am. Chem. Soc.* 129 (28): 8686–8687.

1298 Felix, R.J., Weber, D., Gutierrez, O. et al. (2012). A gold-catalysed enantioselective Cope rearrangement of achiral 1,5-dienes. *Nat. Chem.* 4 (5): 405–409.

1299 Berson, J.A. and Jones, M. (1964). Stepwise mechanisms in oxy-Cope rearrangement. *J. Am. Chem. Soc.* 86 (22): 5017–5019.

1300 Berson, J.A. and Jones, M. (1964). Synthesis of ketones by thermal isomerization of 3-hydroxy-1,5-hexadienes. Oxy-Cope rearrangement. *J. Am. Chem. Soc.* 86 (22): 5019–5020.

1301 Berson, J.A. and Walsh, E.J. (1968). Allylic and α-oxygen resonance energy effects on strengths of carbon-carbon bonds. Activation parameters in stepwise biallyl rearrangements. *J. Am. Chem. Soc.* 90 (17): 4730–4732.

1302 Baumann, H. and Chen, P. (2001). Density functional study of the oxy-Cope rearrangement. *Helv. Chim. Acta* 84 (1): 124–140.

1303 Evans, D.A. and Golob, A.M. (1975). (3,3)-Sigmatropic rearrangements of 1,5-diene alkoxides - powerful accelerating effects of alkoxide substituent. *J. Am. Chem. Soc.* 97 (16): 4765–4766.

1304 Yoo, H.Y., Houk, K.N., Lee, J.K. et al. (1998). New paradigm for anionic heteroatom Cope rearrangements. *J. Am. Chem. Soc.* 120 (1): 205–206.

1305 Evans, D.A. and Nelson, J.V. (1980). Stereochemical study of the (3,3)-sigmatropic rearrangement of 1,5-diene-3-alkoxides - application to the stereoselective synthesis of (+/-)-Juvabione. *J. Am. Chem. Soc.* 102 (2): 774–782.

1306 Paquette, L.A. (1990). Stereocontrolled construction of complex cyclic ketones via oxy-Cope rearrangement. *Angew. Chem. Int. Ed. Engl.* 29 (6): 609–626.

1307 Paquette, L.A. (1995). Centenary lecture - bridgehead unsaturation in compounds of nature - a proper forum for unleashing the potential of organic synthesis. *Chem. Soc. Rev.* 24 (1): 9–17.

1308 Paquette, L.A. (1997). Recent applications of anionic oxy-Cope rearrangements. *Tetrahedron* 53 (41): 13971–14020.

1309 (a) Gentric, L., Hanna, I., and Ricard, L. (2003). Synthesis of the complete carbocyclic skeleton of vinigrol. *Org. Lett.* 5 (7): 1139–1142. (b) Gauvreau, D. and Barriault, L. (2005). Conservation of the planar chiral information in the tandem oxy-Cope/ene reaction. *J. Org. Chem.* 70 (4): 1382–1388. (c) Chiang, P.C., Kaeobamrung, J., and Bode, J.W. (2007). Enantioselective, cyclopentene-forming annulations via NHC-catalyzed benzoin-oxy-Cope reactions. *J. Am. Chem. Soc.* 129 (12): 3520–3521. (d) Smith, A.B. III, Bosanac, T., and Basu, K. (2009). Evolution of the total synthesis of (−)-Okilactomycin exploiting a tandem oxy-Cope rearrangement/oxidation, a Petasis–Ferrier union/rearrangement, and ring-closing metathesis. *J. Am. Chem. Soc.*

131 (6): 2348–2358. (e) Li, F., Tartakoff, S.S., and Castle, S.F. (2009). Total synthesis of (−)-acutumine. *J. Am. Chem. Soc.* 131 (19): 6674–6675. (f) Roosen, P.C. and Vanderwal, C.D. (2014). Investigations into an anionic oxy-Cope/transannular conjugate addition approach to 7,20-diisocyanoadociane. *Org. Lett.* 16 (17): 4368–4371. (g) Fujimoto, Y., Yanai, H., and Matsumoto, T. (2016). Concise total synthesis of elliptoxanthone A by utilizing aromatic oxy-Cope rearrangement for efficient C-isoprenylation of xanthone skeleton. *Synlett* 27 (15): 2229–2232.

1310 Paquette, L.A., Crouse, G.D., and Sharma, A.K. (1982). Relationship of the anionic behavior of unsaturated medium-ring alcohols to structure - generation and antarafacial cyclization of coiled 8-π-electron carbanions. *J. Am. Chem. Soc.* 104 (16): 4411–4423.

1311 Paquette, L.A. and Crouse, G.D. (1981). Stereocontrolled preparation of precursors to all primary prostaglandins from butadiene. *Tetrahedron* 37: 281–287.

1312 Paquette, L.A., Coghlan, M.J., and Hayes, P.C. (1984). Absolute configuration of multifidene as deduced by total synthesis of the unnatural levorotatory enantiomer. *J. Org. Chem.* 49 (23): 4516–4518.

1313 Clausen, C., Wartchow, R., and Butenschon, H. (2001). Reactions of 1,2-diketones with vinyllithium: addition reactions and dianionic oxy-Cope rearrangements of cyclic and acyclic substrates. *Eur. J. Org. Chem.* (1): 93–113.

1314 Evans, D.A., Baillargeon, D.J., and Nelson, J.V. (1978). General approach to synthesis of 1,6-dicarbonyl substrates - new applications of base-accelerated oxy-Cope rearrangements. *J. Am. Chem. Soc.* 100 (7): 2242–2244.

1315 Paquette, L.A., Wang, H.L., Zeng, Q.B., and Shih, T.L. (1998). Heteroatomic modulation of oxyanionic cope rearrangement rates. Consequences on competing nucleophilic cleavage of an oxetane ring in precursors to paclitaxel. *J. Org. Chem.* 63 (19): 6432–6433.

1316 Sprules, T.J., Galpin, J.D., and Macdonald, D. (1993). Charge-accelerated Cope rearrangements of 3-amino-1,5-dienes. *Tetrahedron Lett.* 34 (2): 247–250.

1317 Haeffner, F., Houk, K.N., Reddy, Y.R., and Paquette, L.A. (1999). Mechanistic variations and rate effects of alkoxy and thioalkoxy substituents on anionic oxy-Cope rearrangements. *J. Am. Chem. Soc.* 121 (50): 11880–11884.

1318 Kaldre, D. and Gleason, J.L. (2016). An organocatalytic Cope rearrangement. *Angew. Chem. Int. Ed.* 55 (38): 11557–11561.

1319 (a) Braisted, A.C. and Schultz, P.G. (1994). An antibody-catalyzed oxy-Cope rearrangement. *J. Am. Chem. Soc.* 116 (5): 2211–2212. (b) Ulrich, H.D., Mundroff, E., Santarsiero, B.D. et al. (1997). The interplay between binding energy and catalysis in the evolution of a catalytic antibody. *Nature* 389 (6648): 271–275.

1320 Driggers, E.M., Cho, H.S., Liu, C.W. et al. (1998). Mechanistic studies of an antibody-catalyzed pericyclic rearrangement. *J. Am. Chem. Soc.* 120 (9): 1945–1958.

1321 Black, K.A., Leach, A.G., Kalani, M.Y.S., and Houk, K.N. (2004). Antibody-catalyzed oxy-Cope rearrangement: mechanism and origins of catalysis and stereoselectivity from DFT quantum mechanics and flexible docking. *J. Am. Chem. Soc.* 126 (31): 9695–9708.

1322 Black, D.K. and Landor, S.R. (1965). Allenes. X. Claisen-Cope rearrangement of propargyl vinyl systems. *J. Chem. Soc.* 6784–6788.

1323 Huntsman, W.D., Deboer, J.A., and Woosley, M.H. (1966). Thermal rearrangement of 1-alken-5-ynes and 1,2,5-alkatrienes. *J. Am. Chem. Soc.* 88 (24): 5846–5350.

1324 Owens, K.A. and Berson, J.A. (1990). Stereochemistry of the thermal acetylenic Cope rearrangement - experimental test for a 1,4-cyclohexenediyl as a mechanistic intermediate. *J. Am. Chem. Soc.* 112 (16): 5973–5985.

1325 Berson, J.A., Griller, D., Owens, K., and Wayner, D.D.M. (1987). Ring-inversion barriers for the 3-cyclohexenyl and 4-cyclohexenyl radicals in solution. *J. Org. Chem.* 52 (15): 3316–3319.

1326 (a) Jacobi, P.A., Armacost, L.M., Brielmann, H.L. et al. (1994). Enynones in organic synthesis. 6. Synthesis of spirocyclic methylenecyclopentenones and analogs of the methylenomycin class of antibiotics - Mechanism of phenol catalysis. *J. Org. Chem.* 59 (18): 5292–5304. (b) Gupta, A.K., Samal, S.K., Ila, H., and Junjappa, H. (1996). Trimethyl borate induced thermal cycloaromatization of 1-aryl-1-(prop-2-ynyl)-3,3-bis(alkylthio)-2-propen-1-ols through acetylenic oxy-Cope rearrangement. *Tetrahedron Lett.* 37 (16): 2817–2820.

1327 Bluthe, N., Goré, J., and Malacria, M. (1986). Silver mediated acetylenic oxy-Cope rearrangement. *Tetrahedron* 42 (5): 1333–1344.

1328 Sherry, B.D. and Toste, F.D. (2004). Gold(I)-catalyzed propargyl Claisen rearrangement. *J. Am. Chem. Soc.* 126 (49): 15978–15979.

1329 Suhre, M.H., Reif, M., and Kirsch, S.F. (2005). Gold(I)-catalyzed synthesis of highly substituted furans. *Org. Lett.* 7 (18): 3925–3927.

1330 Binder, J.T. and Kirsch, S.F. (2006). Synthesis of highly substituted pyrroles via a multimetal-catalyzed rearrangement-condensation-cyclization domino approach. *Org. Lett.* 8 (10): 2151–2153.

1331 Sherry, B.D., Maus, L., Laforteza, B.N., and Toste, F.D. (2006). Gold(I)-catalyzed synthesis of dihydropyrans. *J. Am. Chem. Soc.* 128 (25): 8132–8133.

1332 (a) Horowitz, R.M. and Geissman, T.A. (1950). A cleavage reaction of α-allylbenzylamines. *J. Am. Chem. Soc.* 72 (4): 1518–1522. (b) Ent, H., De Koning, H., and Speckamp, W.N. (1986). 2-Azonia-Cope rearrangement in *N*-acyliminium cyclization. *J. Org. Chem.* 51 (10): 1687–1691.

1333 Geisel, M., Grob, C.A., and Wohl, R.A. (1969). Structure and mechanism in fragmentation reactions. 1. Stereoisomeric 10-chlorodecahydroisoquinolines. *Helv. Chim. Acta* 52 (8): 2206–2215.

1334 Sugiura, M., Mori, C., and Kobayashi, S. (2006). Enantioselective transfer aminoallylation: synthesis of optically active homoallylic primary amines. *J. Am. Chem. Soc.* 128 (34): 11038–11039.

1335 Gelas-Mialh, Y., Gramain, J.-C., Louvet, A., and Remuson, R. (1992). Diastereoselective cyclization of allysilanes on α-acyliminium ions - a new approach to (+/-)-myrtine and (+/-)-epimyrtine. *Tetrahedron Lett.* 33 (1): 73–76.

1336 Agami, C., Couty, F., and Poursoulis, M. (1992). Aza-Cope rearrangement - asymmetric syntheses of α-amino-acids - (*R*)-Homoserine lactone and (*R,R*)-4-(1-hydroxy-1-methylethyl)proline. *Synlett* (10): 847–848.

1337 Bennett, D.J. and Hamilton, N.M. (2000). A facile synthesis of *N*-benzylallylglycine. *Tetrahedron Lett.* 41 (41): 7961–7964.

1338 Aron, Z.D. and Overman, L.E. (2005). Stereocontrolled synthesis of angularly substituted 1-azabicyclic rings by cationic 2-aza-Cope rearrangements. *Org. Lett.* 7 (5): 913–916.

1339 Ito, T., Overman, L.E., and Wang, J. (2010). Enantioselective synthesis of angularly substituted 1-azabicyclic ring systems: dynamic kinetic resolution using aza-Cope rearrangements. *J. Am. Chem. Soc.* 132 (10): 3272–3273.

1340 Overman, L.E. and Kakimoto, M. (1979). Carbon–carbon bond formation via directed 2-azonia-(3,3)-sigmatropic rearrangements - new pyrrolidine synthesis. *J. Am. Chem. Soc.* 101 (5): 1310–1312.

1341 Overman, L.E. and Mendelson, L.T. (1981). Synthesis applications of aza-Cope rearrangements. 5. A stereoselective synthesis of *cis*-3a-aryloctahydroindoles and a new short entry to amaryllidaceae alkaloids. *J. Am. Chem. Soc.* 103 (18): 5579–5581.

1342 Overman, L.E. (1992). Charge as a key component in reaction design - the invention of cationic cyclization reactions of importance in synthesis. *Acc. Chem. Res.* 25 (8): 352–359.

1343 Knight, S.D., Overman, L.E., and Pairaudeau, G. (1993). Enantioselective total synthesis of (−)-strychnine. *J. Am. Chem. Soc.* 115 (20): 9293–9294.

1344 Overman, L.E. and Shim, J. (1993). Synthesis applications of cationic aza-Cope rearrangements. 25. Total synthesis of amaryllidaceae alkaloids of the 5,11-methanomorphanthridine type - efficient total syntheses of (−)-pancracine and (+/-)-pancracine. *J. Org. Chem.* 58 (17): 4662–4672.

1345 Brummond, K.M. and Lu, J.L. (2001). Tandem cationic aza-Cope rearrangement-Mannich cyclization approach to the core structure of FR901483 via a bridgehead iminium ion. *Org. Lett.* 3 (9): 1347–1349.

1346 (a) Hanessian, S., Tremblay, M., and Petersen, J.F.W. (2004). The *N*-acyloxyiminium ion aza-Prins route to octahydroindoles: total synthesis and structural confirmation of the antithrombotic marine natural product oscillarin. *J. Am. Chem. Soc.* 126 (19): 6064–6071. (b) Fürstner, A. and Korte, A. (2008). Total synthesis of epohelmin B and its analogues. *Chem. Asian J.* 3 (2): 310–318. (c) Zhou, L.-J., Li, Z.-M., Zou, Y. et al. (2012). Tandem nucleophilic addition/oxy-2-azonia-Cope rearrangement for the formation of homoallylic amides and lactams: total synthesis and structural verification of motuporamine G. *J. Am. Chem. Soc.* 134 (49): 20009–20012. (d) Mu, W.-B., Zhou, L.-J., Zou, Y. et al. (2014). Irreversible oxy-2-azonia-Cope rearrangements for the synthesis of functionalized allyl α-amino acid derivatives. *Eur. J. Org. Chem.* (11): 2379–2385. (e) Mu, W.-B., Zhou, L.-J., Zou, Y. et al. (2015). Removable silyl group as a

"Masked proton" in oxy-2-oxonia(azonia)-Cope rearrangement: applications in stereoselective total synthesis of natural macrolides. *Eur. J. Org. Chem.* (22): 4982–4959.

1347 Carballo, R.M., Purino, M., Ramírez, M.A. et al. (2010). Iron(III)-catalyzed consecutive aza-Cope-Mannich cyclization: synthesis of *trans*-3,5-dialkyl pyrrolidines and 3,5-dialkyl-2,5-dihydro-1*H*-pyrroles. *Org. Lett.* 12 (22): 5334–5337.

1348 Mori, M. (2010). Total synthesis of strychnine. *Heterocycles* 81 (2): 259–292.

1349 Cannon, J.S. and Overman, L.E. (2012). Is there no end to the total syntheses of Strychnine? Lessons learned in strategy and tactics in total synthesis. *Angew. Chem. Int. Ed.* 51 (18): 4288–4311.

1350 Briggs, L.H., Openshaw, H.T., and Robinson, R. (1946). Strychnine and Brucine. 42. Constitution of the neo-series of bases and their oxidation products. *J. Chem. Soc.* 903–908.

1351 Woodward, R.B., Cava, M.P., Ollis, W.D. et al. (1954). The total synthesis of strychnine. *J. Am. Chem. Soc.* 76 (18): 4749–4751.

1352 (a) Nokami, J., Anthony, L., and Sumida, S. (2000). New route to α-adducts of homoallylic alcohols by an acid-catalyzed stereospecific allyl-transfer reaction from γ-adducts. *Chem. Eur. J.* 6 (16): 2909–2913. (b) Yan, L., He, G.-L., Yin, R.-F. et al. (2014). Synthesis of polyketide stereoarrays enabled by a traceless oxonia-Cope rearrangement. *Angew. Chem. Int. Ed.* 53 (43): 11600–11604.

1353 McDonald, F.E., Pereira, C.L., and Chen, Y.H. (2011). Fumonisin: a template for methodology development and drug discovery. *Pure Appl. Chem.* 83 (3): 445–459.

1354 Zou, Y., Ding, C., Zhou, L. et al. (2012). Tandem cross-dimerisation/oxonia-Cope reaction of carbonyl compounds to homoallylic esters and lactones. *Angew. Chem. Int. Ed.* 51 (23): 5647–5651.

1355 Zakarian, A. and Lu, C.D. (2006). Development of the 1,2-oxaza-Cope rearrangement. *J. Am. Chem. Soc.* 128 (16): 5356–5357.

1356 Jabbari, A. and Houk, K.N. (2006). Hetero-Cope rearrangements of nitrosobutenes. DFT studies of thermal and acid-catalyzed reactions. *Org. Lett.* 8 (26): 5975–5978.

1357 Marbet, R. and Saucy, G. (1967). Über die Reaktion von tertiaren Vinylcarbinolen mit Vinyläthern - Eine neue Methode zur Herstellung von γ,δ-ungesattigten Aldehyden. *Helv. Chim. Acta* 50 (7): 2095–2100.

1358 Suh, Y.G., Kim, S.A., Jung, J.K. et al. (1999). Asymmetric total synthesis of fluvirucinine A_1. *Angew. Chem. Int. Ed.* 38 (23): 3545–3547.

1359 Cid, M.M. and Pombovillar, E. (1993). Enantioselective synthesis of 3-azabicyclo[4.3.0]nonane alkaloids. *Helv. Chim. Acta* 76 (4): 1591–1607.

1360 Mehta, G., Karra, S.R., and Krishnamurthy, N. (1994). An oxy-Cope rearrangement route for the enantioselective construction of 5,11-fused framework of dolabellane diterpenes. *Tetrahedron Lett.* 35 (17): 2761–2762.

1361 Wender, P.A., Sieburth, S.M., Petraitis, J.J., and Singh, S.K. (1981). Macroexpansion methodology - medium ring synthesis based on an eight unit ring expansion process. *Tetrahedron* 37 (23): 3967–3975.

1362 So, S.M., Mui, L., Kim, H., and Chin, J. (2012). Understanding the interplay of weak forces in (3,3)-sigmatropic rearrangement for stereospecific synthesis of diamines. *Acc. Chem. Res.* 45 (8): 1345–1355.

1363 Reetz, M.T. (1972). Dyotropic rearrangements, a new class of orbital-symmetry controlled reactions - Type-I. *Angew. Chem. Int. Ed. Engl.* 11 (2): 129–130.

1364 Reetz, M.T. (1972). Dyotropic rearrangements - new class of orbital-symmetry controlled reactions - Type-II. *Angew. Chem. Int. Ed.* 11 (2): 130–131.

1365 Reetz, M.T. (1973). Primary and secondary orbital effects in dyotropic rearrangements. *Tetrahedron* 29 (14): 2189–2194.

1366 (a) Fernández, I., Cossio, F.P., and Sierra, M.A. (2009). Dyotropic reactions: mechanisms and synthetic applications. *Chem. Rev.* 109 (12): 6687–6711. (b) Davis, R.L. and Tantillo, D.J. (2010). Dissecting a dyotropic rearrangement. *J. Org. Chem.* 75 (5): 1693–1700.

1367 Faza, O.N., López, C.S., and Fernández, I. (2013). Noyori hydrogenation: aromaticity, synchronicity, and activation strain analysis. *J. Org, Chem.* 78 (11): 5669–5676.

1368 Grob, C.A. and Winstein, S. (1952). Mechanismus der Mutarotation von 5,6-Dibromcholestan. *Helv. Chim. Acta* 35 (3): 780–802.

1369 Barton, D.H.R. and Head, A.J. (1956). Long-range effects in alicyclic Systems. 1. The rates of rearrangement of some steroidal dibromides. *J. Chem. Soc.* 932–937.

1370 Barili, P.L., Bellucci, G., Berti, G. et al. (1970). Substituent effects on formation and equilibration of *trans*-1,2-dibromocyclohexanes. *J. Chem. Soc., Chem. Commun.* (21): 1437–1438.

1371 Bellucci, G., Marsili, A., Mastrori, E. et al. (1974). Kinetics of thermal racemization of some 1,2-dihalides. *J. Chem. Soc., Perkin Trans. 2* (2): 201–204.

1372 Fernández, I., Sierra, M.A., and Cossio, F.P. (2006). Stereoelectronic effects on type I 1,2-dyotropic rearrangements in vicinal dibromides. *Chem. Eur. J.* 12 (24): 6323–6330.

1373 Braddock, D., Roy, D., Lenoir, D. et al. (2012). Verification of stereospecific dyotropic racemisation of enantiopure D and L-1,2-dibromo-1,2-diphenylethane in non-polar media. *J. Chem. Soc., Chem. Commun.* 48 (71): 8943–8945.

1374 Leverett, C.A., Purohit, V.C., Johnson, A.G. et al. (2012). Dyotropic rearrangements of fused tricyclic β-lactones: application to the synthesis of (-)-curcumanolide A and (-)-curcumalactone. *J. Am. Chem. Soc.* 134 (32): 13348–13356.

1375 Ren, W., Bian, Y., Zhang, Z. et al. (2012). Enantioselective and collective syntheses of xanthanolides involving a controllable dyotropic rearrangement of *cis*-β-lactones. *Angew. Chem. Int. Ed.* 51 (28): 6984–6988.

1376 Mulzer, J. and Brüntrup, G. (1979). Stereospecific synthesis of γ-lactones by dyotropic Wagner–Meerwein rearrangement. *Angew. Chem. Int. Ed. Engl.* 18 (10): 793–794.

1377 Zenk, P.C. and Wiley, R.A. (1984). A general synthesis for 4-alkyl-2,3-dihydrofurans. *Synthesis* (8): 695–697.

1378 Black, T.H. and Fields, J.D. (1988). A versatile synthesis of 3-substituted 5-alkyl butyrolactones via dyotropic rearrangement. *Synth. Commun.* 18 (2): 125–130.

1379 Black, T.H., Dubay, W.J., and Tully, P.S. (1988). Dyotropic rearrangement of cycloalkyl β-lactones - formation of spiro vs fused butyrolactones as a function of ring size. *J. Org. Chem.* 53 (25): 5922–5927.

1380 Black, T.H. and Dubay, W.J. (1988). Lactone synthesis via dyotropic rearrangement - stereospecific construction of fused butyrolactones with three contiguous asymmetric centers. *Tetrahedron Lett.* 29 (15): 1747–1750.

1381 Black, T.H. and Dubay, W.J. (1987). A new synthesis of substituted spiro butyrolactones via dyotropic rearrangement. *Tetrahedron Lett.* 28 (41): 4787–4788.

1382 Reetz, M.T., Schmitz, A., and Holdgrün, X. (1989). Tandem aldolization lactonization dyotropic rearrangement of α-amino-aldehydes. *Tetrahedron Lett.* 30 (40): 5421–5424.

1383 Davis, R.L., Leverett, C.A., Romo, D., and Tantillo, D.J. (2011). Switching between concerted and stepwise mechanisms for dyotropic rearrangements of β-lactones leading to spirocyclic, bridged γ-butyrolactones. *J. Org. Chem.* 76 (17): 7167–7174.

1384 Stöckner, F., Beckert, R., Gleich, D. et al. (2007). Polyazaacenes - on the way to stable, fluorescent and redox-active derivatives. *Eur. J. Org. Chem.* (8): 1237–1243.

1385 Hill, J.H.M. (1963). Oxidative dimerization of benzimidazole. *J. Org. Chem.* 28 (7): 1932–1933.

1386 Banerjee, R. (2003). Radical carbon skeleton rearrangements: catalysis by coenzyme B-12-dependent mutases. *Chem. Rev.* 103 (6): 2083–2094.

1387 Buckel, W. and Golding, B.T. (1996). Glutamate and 2-methyleneglutarate mutase: from microbial curiosities to paradigms for coenzyme B-12-dependent enzymes. *Chem. Soc. Rev.* 25 (5): 329–337.

1388 Reitzer, R., Gruber, K., Jogl, G. et al. (1999). Glutamate mutase from *Clostridium cochlearium*: the structure of a coenzyme B_{12}-dependent enzyme provides new mechanistic insights. *Structure* 7 (8): 891–902.

1389 Mancia, F. and Evans, P.R. (1998). Conformational changes on substrate binding to methylmalonyl CoA mutase and new insights into the free radical mechanism. *Structure* 6 (6): 711–720.

1390 Zerbe-Burkhardt, K., Ratnatilleke, A., Philippon, N. et al. (1998). Cloning, sequencing, expression, and insertional inactivation of the gene for the large subunit of the coenzyme B_{12}-dependent isobutyryl-CoA mutase from *Streptomyces cinnamonensis*. *J. Biol. Chem.* 273 (11): 6508–6517.

1391 Abeles, R.H. and Dolphin, D. (1976). Vitamin-B_{12} coenzyme. *Acc. Chem. Res.* 9 (3): 114–120.

1392 Toraya, T. and Fukui, S. (1977). Coenzyme-B_{12}-dependent diol dehydrase. 9. Immunochemical evidence for difference between coenzyme-B_{12}-dependent diol dehydratase and glycerol dehydratase. *Eur. J. Biochemistry* 76 (1): 285–289.

1393 Babior, B.M. (1970). Mechanism of action of ethanolamine ammonia-lyase, A B_{12}-dependent enzyme. 7. Mechanism of hydrogen transfer. *J. Biol. Chem.* 245 (22): 6125–6133.

1394 Stadtman, T.C. (1973). Lysine metabolism by *Clostridia*. *Adv. Enzymol. Relat. Areas Mol. Biol.* 38: 413–448.

1395 (a) Kunz, F., Retey, J., Arigoni, D. et al. (1978). Absolute-configuration of 3,5-diaminohexanoic acid produced in β-lysine mutase reaction. *Helv. Chim. Acta* 61 (3): 1139–1145. (b) Retey, J., Arigoni, D., Stadtman, T.C., and Kunz, F. (1978). Investigations on β-lysine mutase reaction - mechanism and steric course. *Helv. Chim. Acta* 61 (8): 2989–2998.

1396 Chen, H.P., Wu, S.H., Lin, Y.L. et al. (2001). Cloning, sequencing, heterologous expression, purification, and characterization of adenosylcobalamin-dependent D-ornithine aminomutase from *Clostridium sticklandii*. *J. Biol. Chem.* 276 (48): 44744–44750.

1397 (a) Smith, D.M., Golding, B.T., and Radom, L. (1999). Understanding the mechanism of B_{12}-dependent methylmalonyl-CoA mutase: partial proton transfer in action. *J. Am. Chem. Soc.* 121 (40): 9388–9399. (b) Kozlowski, P.M., Kamachi, T., Toraya, T., and Yoshizawa, K. (2007). Does Cob(II)alamin act as a conductor in coenzyme B_{12} dependent mutases? *Angew. Chem. Int. Ed.* 46 (6): 980–983.

1398 Hong, Y.J. and Tantillo, D.J. (2009). A potential energy surface bifurcation in terpene biosynthesis. *Nat. Chem.* 1 (5): 384–389.

1399 Santalla, H., Faza, O.N., Gómez, G. et al. (2017). From hydrindane to decalin: a mild transformation through a dyotropic ring expansion. *Org. Lett.* 19 (13): 3648–3651.

1400 Duboudin, F., Frainnet, E., Vincon, G., and Dabescat, F. (1973). Synthesis and new type of rearrangement of disilicon hydroxylamines. *C. R. Seances Acad. Sci., Ser. C* 276 (18): 1469–1472.

1401 Nowakowski, P. and West, R. (1976). Rearrangement of organosilyl groups between oxygen and nitrogen in tris(organosilyl)hydroxylamines - possible dyotropic rearrangement. *J. Am. Chem. Soc.* 98 (18): 5616–5620.

1402 West, R., Nowakowski, P., and Boudjouk, P. (1976). Irreversible thermal rearrangement of tris(organosilyl)hydroxylamines. *J. Am. Chem. Soc.* 98 (18): 5620–5627.

1403 Reetz, M.T. (1977). Dyotropic rearrangements. 11. Mechanism of thermal rearrangement of allyl (silylmethyl) ethers. *Chem. Ber.* 110 (3): 965–978.

1404 Wolfgramm, R., Müller, T., and Klingebiel, U. (1998). Irreversible rearrangement of silyl groups between oxygen and nitrogen in tris(silyl)hydroxylamines. *Organometallics* 17 (15): 3222–3226.

1405 Ward, A.S., Mintz, E.A., and Kramer, M.P. (1988). Thermally induced rearrangement of α-zirconocenyl thioethers – carbon–carbon bond formation via a transition-metal hydroxymethyl or metalloxirane analog. *Organometallics* 7 (1): 8–12.

1406 Erker, G. and Petrenz, R. (1989). Formation of a conjugated oxametallacyclohexadiene by dyotropic rearrangement of bis(2-furyl)zirconocene. *J. Chem. Soc., Chem. Commun.* (6): 345–346.

1407 Denmark, S.E. and Montgomery, J.I. (2005). Synthesis of *cis,cis,cis,cis*-[5.5.5.4]-1-azafenestrane with discovery of an unexpected dyotropic rearrangement. *Angew. Chem. Int. Ed.* 44 (24): 3732–3736.

1408 Reetz, M.T. and Greif, N. (1977). Dyotropic rearrangements. 14. Thermal rearrangement of silymethyl acetates. *Angew. Chem. Int. Ed. Engl.* 16 (10): 712–713.

1409 Tang, H.R. and Stanbury, D.M. (1994). Direct-detection of aqueous diazene - its UV spectrum and concerted dismutation. *Inorg. Chem.* 33 (7): 1388–1391.

1410 Foner, S.N. and Hudson, R.L. (1958). Diimide - identification and study by mass spectrometry. *J. Chem. Phys.* 28 (4): 719–720.

1411 Hünig, S., Müller, H.R., and Thier, W. (1965). Chemistry of diimine. *Angew. Chem. Int. Ed. Engl.* 4 (4): 271–280.

1412 Pasto, D.J. and Taylor, R.T. (1991). Reduction with diimide. *Org. React.* 40: 91–155.

1413 Van Tamelen, E.E., Davis, M., and Deem, M.F. (1965). Selectivity in diimide reductions of aldehydes, ketones and hetero-substituted olefins. *J. Chem. Soc., Chem. Commun.* (4): 71–72.

1414 Adam, W. and Eggelte, H.J. (1977). Prostanoid endoperoxide model compounds - 2,3-dioxabicyclo[2.2.1]heptane via selective diimide reduction. *J. Org. Chem.* 42 (24): 3987–3988.

1415 Adam, W. and Eggelte, H.J. (1977). Cyclic peroxides. 58. 2,3-Dioxabicyclo[2.2.2]octane by selective reduction of double-bonds with azodicarboxylate. *Angew. Chem. Int. Ed. Engl.* 16 (10): 713–713.

1416 Corey, E.J., Mock, W.L., and Pasto, D.J. (1961). Chemistry of diimide - some new systems for the hydrogenation of multiple bonds. *Tetrahedron Lett.* (11): 347–352.

1417 Corey, E.J., Pasto, D.J., and Mock, W.L. (1961). Chemistry of diimide. 2. Stereochemistry of hydrogen transfer to carbon-carbon multiple bonds. *J. Am. Chem. Soc.* 83 (13): 2957–2958.

1418 Hünig, S., Müller, H.R., and Thier, W. (1961). Reduktionen mit Diimid. *Tetrahedron Lett.* 2 (11): 353–357.

1419 Imada, Y., Iida, H., and Naota, T. (2005). Flavin-catalyzed generation of diimide: an environmentally friendly method for the aerobic hydrogenation of olefins. *J. Am. Chem. Soc.* 127 (42): 14544–14545.

1420 Smit, C., Fraaije, M.W., and Minnaard, A.J. (2008). Reduction of carbon–carbon double bonds using organocatalytically generated diimide. *J. Org. Chem.* 73 (23): 9482–9485.

1421 Imada, Y., Kitagawa, T., Ohno, T. et al. (2010). Neutral flavins: green and robust organocatalysts for aerobic hydrogenation of olefins. *Org. Lett.* 12 (1): 32–35.

1422 Marsh, B.J., Heath, E.L., and Carbery, D.R. (2011). Organocatalytic diimide reduction of enamides in water. *J. Chem. Soc., Chem. Commun.* 47 (1): 280–282.

1423 Nagendrappa, G. and Devaprabhakara, D. (1970). Diimide reduction of allenes. *Tetrahedron Lett.* 11 (49): 4243–4244.

1424 Moriarty, R.M., Vaid, R.K., and Duncan, M.P. (1987). Hypervalent iodine oxidation of hydrazine hydrate in dichloromethane - a new method for diimide generation. *Synth. Commun.* 17 (6): 703–708.

1425 King, C.V. (1940). The catalytic decomposition of azodicarbonate ion. *J. Am. Chem. Soc.* 62: 379–385.

1426 Van Tamelen, E., Dewey, R.S., and Timmons, R.J. (1961). Reduction of olefins by means of azodicarboxylic acid in situ. *J. Am. Chem. Soc.* 83 (17): 3725–3726.

1427 Hamersma, J.W. and Snyder, E.I. (1965). Diimide reductions using potassium azodicarboxylate. *J. Org. Chem.* 30 (11): 3985–3988.

1428 Lüthy, C., Konstantin, P., and Untch, K.G. (1978). Total synthesis of dl-19-hydroxyprostaglandin-E1 and dl-13-*cis*-15-*epi*-19-hydroxyprostaglandin-E1. *J. Am. Chem. Soc.* 100 (19): 6211–6217.

1429 Mori, K., Sato, A., Ohki, M., and Matsui, M. (1972). Synthesis of compounds with Juvenile–Hormone activity. 11. New routes for stereocontrolled construction of trisubstituted *cis* double-bond portion of *Cecropia* Juvenile Hormones. *Tetrahedron* 28 (14): 3739–3745.

1430 Buszek, K.R. and Brown, N. (2007). Improved method for the diimide reduction of multiple bonds on solid-supported substrates. *J. Org. Chem.* 72 (8): 3125–3128.

1431 Marsh, B.J. and Carbery, D.R. (2009). One-pot *o*-nitrobenzenesulfonylhydrazide (NBSH) formation - diimide alkene reduction protocol. *J. Org. Chem.* 74 (8): 3186–3188.

1432 Annunziata, R., Fornasier, R., and Montanari, F. (1974). Evidence for formation of diimide in thermal fragmentation of 1-amino-2,2-diphenylaziridine. *J. Org. Chem.* 39 (22): 3195–3197.

1433 McKee, M.L. (1993). Catalyzed *cis-trans* isomerization of diazene - a computational study in the gas and aqueous phases. *J. Phys. Chem.* 97 (51): 13608–13614.

1434 McKee, M.L., Squillacote, M.E., and Stanbury, D.M. (1992). Ab initio investigation of dihydrogen transfer from *cis*-1,2-diazene. *J. Phys. Chem.* 96 (8): 3266–3272.

1435 Tang, H.R., McKee, M.L., and Stanbury, D.M. (1995). Absolute rate constants in the concerted reduction of olefins by diazene. *J. Am. Chem. Soc.* 117 (35): 8967–8973.

1436 Garbisch, E.W. Jr., Schildcrout, S.M., Patterson, D.B., and Sprecher, C.M. (1965). Strain effects. 2. Diimide reductions of olefins. *J. Am. Chem. Soc.* 87 (13): 2932–2944.

1437 Nelson, D.J., Henley, R.L., Yao, Z.Y., and Smith, T.D. (1993). Diimide reduction of representative alkenes and correlation of their relative reaction rates with corresponding ionization potentials. *Tetrahedron Lett.* 34 (37): 5835–5838.

1438 Doering, W.v.E. and Rosenthal, J.W. (1967). Hydrogenation by *cis*-9,10-dihydronaphthalene. *J. Am. Chem. Soc.* 89 (17): 4534–4535.

1439 Mackenzie, K. (1965). Isodrin in Diels–Alder reaction. *J. Chem. Soc.* 4646–4653.

1440 Mackenzie, K., Proctor, G., and Woodnutt, D.J. (1987). Intramolecular 2H group-transfer (dyotropic rearrangements) in alicyclic and heterocyclic bridged-ring systems. *Tetrahedron* 43 (24): 5981–5993.

1441 Mackenzie, K., Howard, J.A.K., Mason, S. et al. (1993). On intramolecular dyotropy - structural effects on reaction rates, crystal structure molecular mechanics correlations and primary deuterium kinetic isotope effects - (parameters for intramolecular recognition). *J. Chem. Soc., Perkin Trans. 2* (6): 1211–1228.

1442 Mackenzie, K., Astin, K.B., Gravett, E.C. et al. (1998). Intramolecular thermal ($\sigma4+\pi2$) dyotropy: primary H_2 kinetic isotope effects. Experimental limiting barrier parameters for quantum tunnelling in 2H transfer processes and mechanistically significant substituent effects. *J. Phys. Org. Chem.* 11 (12): 879–886.

1443 Mackenzie, K. (2008). Intramolecular ($\sigma 4+\pi 2$) thermal dyotropy of pre-stressed molecules. The correlation of reactivity with experimental crystallographic and calculated structural features, substituent and kinetic isotope effects. *Recent Res. Dev. Org. Chem.* 4: 295–326.

1444 Hagenbuch, J.P., Stampfli, B., and Vogel, P. (1981). Thermoneutral dyotropic transfer of hydrogen in a hydrocarbon. *J. Am. Chem. Soc.* 103 (13): 3934–3935.

1445 Paquette, L.A., Kesselmayer, M.A., and Rogers, R.D. (1990). Quantitation of proximity effects on rate - a case-study involving dyotropic hydrogen migration within *syn*-sesquinorbornene disulfones carrying central substituents having different spatial demands. *J. Am. Chem. Soc.* 112 (1): 284–291.

1446 Paquette, L.A., O'Doherty, G.A., and Rogers, R.D. (1991). Intramolecular reaction rate is not determined exclusively by the distance separating reaction centers - the kinetic consequences of modulated ground-state strain on dyotropic hydrogen migration in systems of very similar geometric disposition. *J. Am. Chem. Soc.* 113 (20): 7761–7762.

1447 O'Doherty, G.A., Rogers, R.D., and Paquette, L.A. (1994). Consequences of modulated precompression along reaction coordinates - synthesis, crystallographic structural studies, and rate of intramolecular dyotropy in an extended series of *syn*-sesquinorbornene disulfones. *J. Am. Chem. Soc.* 116 (24): 10883–10894.

1448 Houk, K.N., Li, Y., McAllister, M.A. et al. (1994). Mechanistic analysis of double hydrogen dyotropy in *syn*-sesquinorbornene disulfones - a combined kinetic and theoretical evaluation of primary deuterium isotope effects. *J. Am. Chem. Soc.* 116 (24): 10895–10913.

1449 Grimme, W., Pohl, K., Wortmann, J., and Frowein, D. (1996). Dyotropic hydrogen transfer in dihydrosesqui- and in dihydro-*syn*-sesterbicyclo[2.2.2]octenes. *Liebigs Ann. Chem.* 1996 (11): 1905–1916.

1450 Chow, T.J. and Ding, M.F. (1986). Dyotropic hydrogen migration in pentacyclo [7.6.0.02,13.03,8.010,14]pentadeca-4,6,11-triene. *Angew. Chem. Int. Ed. Engl.* 25 (12): 1121–1122.

1451 Marchand, A.P., Annapurna, P., Watson, W.H., and Nagl, A. (1989). Cycloaddition of the anion derived from homophthalic anhydride to *cis,cisoid,cis*-tricyclo[6.3.0.02,6]undeca-4,9-diene-3,11-dione - aromatization as a driving force for intramolecular sigmatropic hydrogen transfer. *J. Chem. Soc., Chem. Commun.* (5): 281–282.

1452 Geich, H., Grimme, W., and Proske, K. (1992). Dyotropic (6+4)-hydrogen migration in a 2,3-bis(methylene)decahydroanthracene. *J. Am. Chem. Soc.* 114 (4): 1492–1493.

1453 Alder, K., Pascher, F., and Schmitz, A. (1943). Absorption of maleic acid-anhydride and azodicarbon acid-ester in simple unsaturated carbohydrogens. Information on the substitution processes in the allyl position. *Chem. Ber. Dtsch. Chem. Ges.* 76: 27–53.

1454 Townsend, C.A., Scholl, T., and Arigoni, D. (1975). New synthesis of chiral acetic acid. *J. Chem. Soc., Chem. Commun.* (22): 921–922.

1455 (a) Hoffmann, H.M.R. (1969). The ene reaction. *Angew. Chem. Int. Ed. Engl* 8: 556–557. (b) Domingo, L.R., Aurell, M.J., and Pérez, P. (2014). Understanding the polar mechanism of the ene reaction. A DFT study. *Org. Biomol. Chem.* 12 (38): 7581–7590.

1456 Richard, C. and Back, M.H. (1978). Ene reactions of olefins. 2. Addition of ethylene to propylene and to isobutene and addition of propylene to propylene. *Int. J. Chem. Kinet.* 10 (4): 389–405.

1457 Cywinski, N.F. (1965). Synthesis of 1,4-diolefins. *J. Org. Chem.* 30 (2): 361–364.

1458 Fernández, I. and Bickelhaupt, F.M. (2012). Alder-ene reaction: aromaticity and activation-strain analysis. *J. Comput. Chem.* 33 (5): 509–516.

1459 Snider, B.B. (1980). Lewis acid-catalyzed ene reactions. *Acc. Chem. Res.* 13 (11): 426–432.

1460 Batcho, A.D., Berger, D.E., Davoust, S.G. et al. (1981). Stereoselective introduction of steroid side-chains at C(17) and C(20). *Helv. Chim. Acta* 64 (5): 1682–1687.

1461 (a) Breslow, R. and Dowd, P. (1963). Dimerization of triphenylcyclopropene. *J. Am. Chem. Soc.* 85 (18): 2729–2735. (b) Sakai, S. (2006). Theoretical studies on the ene reaction mechanisms of propene and cyclopropene with ethylene and cyclopropene: concerted or stepwise. *J. Phys. Chem. A* 110 (47): 12891–12899.

1462 (a) Baird, M.S. (1988). Functionalized cyclopropenes as synthetic intermediates. In: *Small Ring Compounds in Organic Synthesis III*, Topics in Current Chemistry, vol. 144 (ed. A. de Meijere), 137–209. Heildelberg, Berlin, New York: Springer-Verlag. (b) Deng, Q.L., Thomas, B.E., Houk, K.N., and Dowd, P. (1997). Transition structures of the ene reactions of cyclopropene. *J. Am. Chem. Soc.* 119 (29): 6902–6908.

1463 (a) Candito, D.A., Paneleev, J., and Lautens, M. (2011). Intramolecular aryne-ene reaction: Synthetic and mechanistic studies. *J. Am. Chem. Soc.* 133 (36): 14200–14203. (b) Candito, D.A., Dobrovolsky, D., and Lautens, M. (2012). Development of an intramolecular aryne ene reaction and application to the formal synthesis of (+/-)-Crinine. *J. Am. Chem. Soc.* 134 (37): 15577–15580. (c) Karmakar, R., Mamidipalli, P., Yun, S.Y., and Lee, D. (2013). Alder-ene reactions of arynes. *Org. Lett.* 15 (8): 1938–1941. (d) Trinchera, P., Sun, W., Smith, J.E., and Palomas, D. (2017). Intermolecular aryne ene reaction of Hantzsch esters: stable covalent ene adducts from a 1,4-dihydropyridine reaction. *Org. Lett.* 19 (17): 4644–4647.

1464 Dowd, P. and Gold, A. (1969). Thermal dimerization of cyclopropene. *Tetrahedron Lett.* 10 (2): 85–86.

1465 Komatsu, K., Niwa, T., Akari, H., and Okamoto, K. (1985). 1,3-Diarylcyclopropenes - syntheses and a facile ene dimerization. *J. Chem. Res., Synop.* (8): 252–253.

1466 Ashirov, R.V., Shamov, G.A., Lodochnikova, O.A. et al. (2008). Tetramerization of 3-methyl-cyclopropene-3-carbonitrile: a novel CN-Alder-ene reaction. *J. Org. Chem.* 73 (15): 5985–5988.

1467 (a) Resek, J.E. (2008). Intramolecular ene reaction of a chiral bicyclic lactam. *J. Org. Chem.* 73 (24): 9792–9794. (b) Buisone, O., Gandon, V., Fensterbank, L. et al. (2008). Thermal intramolecular Alder-ene cycloisomerization of 1,6-allenynes. *Synlett* (5): 751–754.

1468 Närhi, K., Franzen, J., and Bäckvall, J.E. (2006). An unexpectedly mild thermal Alder-ene-type cyclization of enallenes. *J. Org. Chem.* 71 (7): 2914–2917.

1469 Oppolzer, W. and Snieckus, V. (1978). Intramolecular ene reactions in organic synthesis. *Angew. Chem. Int. Ed. Engl.* 17 (7): 476–486.

1470 Oppolzer, W. and Stammen, B. (1997). On the faciality of intramolecular Pd(0)-catalysed "metallo-ene-type" cyclisations. *Tetrahedron* 53 (10): 3577–3586.

1471 Rouessac, F., Leperchec, P., and Conia, J.M. (1967). Thermolysis and photolysis of un-saturated ketones. V. Thermal cyclisation of aliphatic ε,ζ-ethylenic ketones. *Bull. Soc. Chim. Fr.* (3): 818–821.

1472 Leperchec, P., Conia, J.M., and Rouessac, F. (1967). Thermolysis and photolysis of unsaturated ketones. 6. Thermal cyclisation of ε,ζ-ethylenic potential ketones - β-ketoesters enol acetates ketals enol ethers. *Bull. Soc. Chim. Fr.* (3): 822–825.

1473 Leperchec, P., Rouessac, F., and Conia, J.M. (1967). Thermolysis and photolysis of unsaturated ketones. 7. Stereochemistry and mechanism of ε,ζ-unsaturated ketone thermal cyclisation. *Bull. Soc. Chim. Fr.* (3): 826–829.

1474 Bloch, R. and Conia, J.M. (1967). La cyclisation thermique des aldéhydes ε,ζ-éthyléniques. *Tetrahedron Lett.* (35): 3409–3412.

1475 Mikami, K. and Terada, M. (1999). Ene-type reactions. In: *Comprehensive Asymmetric Catalysis III* (ed. E.N. Jacobsen, A. Pfaltz and H. Yamamoto), 1143–1174. Heidelberg, Berlin: Springer-Verlag.

1476 Alder, K. and Dortmann, H.A. (1952). Über Substituierende Addition und Dien-Synthese beim Methylen-Cyclobutan. 9. Über Substituierende Additionen. *Chem. Ber.* 85 (6): 556–565.

1477 Clarke, M.L. and France, M.B. (2008). The carbonyl ene reaction. *Tetrahedron* 64 (38): 9003–9031.

1478 Denmark, S.E. and Almstead, N.G. (2000). Allylation of carbonyls: methodology and stereochemistry. In: *Modern Carbonyl Chemistry* (ed. J. Otera), 299–402. Weinheim: Wiley-VCH.

1479 Denmark, S.E. and Fu, J.P. (2003). Catalytic enantioselective addition of allylic organometallic reagents to aldehydes and ketones. *Chem. Rev.* 103 (8): 2763–2793.

1480 Adams, D.R. and Bhatnagar, S.P. (1977). Prins reaction. *Synthesis* (10): 661–672.

1481 Bain, J.P. (1946). Nopol. 1. The reaction of β-pinene with formaldehyde. *J. Am. Chem. Soc.* 68 (4): 638–641.

1482 Arnold, R.T. and Dowdall, J.F. (1948). Preparation and reactions of methylenecyclohexane. *J. Am. Chem. Soc.* 70 (7): 2590–2591.

1483 Opanasenko, M., Dhakshinamoorthy, A., Hwang, Y.K. et al. (2013). Superior performance of metalorganic frameworks over zeolites as solid acid catalysts in the Prins reaction: green synthesis of Nopol. *ChemSusChem* 6 (5): 865–871.

1484 Blomquist, A.T. and Verdol, J.A. (1955). The thermal isobutylene-formaldehyde condensation. *J. Am. Chem. Soc.* 77 (1): 78–80.

1485 Blomquist, A.T., Verdol, J., Adami, C.L. et al. (1957). Thermal condensation of cyclic olefins with formaldehyde. *J. Am. Chem. Soc.* 79 (18): 4976–4980.

1486 Colonge, J. and Perrot, A. (1957). Etude sur les trichlorométhylcarbinols secondaires éthyléniques et saturés. 1. Preparation. *Bull. Soc. Chim. Fr.* 1957 (1): 204–208.

1487 Colonge, J. and Perrot, A. (1957). Etude sur les trichlorométhylcarbinols secondaires saturés et éthyléniques. 2. Action des réactifs alcalins. *Bull. Soc. Chim. Fr.* 1957 (4): 658–662.

1488 England, D.C. (1961). Perfluorocyclobutanones. *J. Am. Chem. Soc.* 83 (9): 2205–2206.

1489 Ohloff, G. (1960). Zur Thermischen Isomerisation von Citral. *Tetrahedron Lett.* 1 (32): 10–14.

1490 Schulte-Elte, K.H. and Ohloff, G. (1967). Über eine aussergewöhnliche Stereospezifität bei der Hydroborierung der Diastereomeren (1*R*)-Isopulegole mit Diboran. *Helv. Chim. Acta* 50 (1): 153–165.

1491 Helmboldt, H. and Hiersemann, M. (2009). Synthetic studies toward Jatrophane diterpenes from *Euphorbia characias*. Enantioselective synthesis of (-)-15-*O*-acetyl-3-*O*-propionyl-17-norcharaciol. *J. Org. Chem.* 74 (4): 1698–1708.

1492 Achmatowicz, O. and Bialecka-Florjanczyk, E. (1996). Mechanism of the carbonyl-ene reaction. *Tetrahedron* 52 (26): 8827–8834.

1493 Mikami, K., Terada, M., and Nakai, T. (1990). Catalytic asymmetric glyoxylate ene reaction - a practical access to α-hydroxy esters in high enantiomeric purities. *J. Am. Chem. Soc* 112 (10): 3949–3954.

1494 Corey, E.J., Barnes-Seeman, D., Lee, T.W., and Goodman, S.N. (1997). A transition-state model for the Mikami enantioselective ene reaction. *Tetrahedron Lett.* 38 (37): 6513–6516.

1495 Johnson, J.S. and Evans, D.A. (2000). Chiral bis(oxazoline)Cu(II) complexes: versatile catalysts for enantioselective cycloaddition, aldol, Michael, and carbonyl ene reactions. *Acc. Chem. Res.* 33 (6): 325–335.

1496 Evans, D.A., Kvaerno, L., Dunn, T.B. et al. (2008). Total synthesis of (+)-azaspiracid-1. An exhibition of the intricacies of complex molecule synthesis. *J. Am. Chem. Soc.* 130 (48): 16295–16309.

1497 Evans, D.A. and Wu, J. (2005). Enantioselective *syn*-selective Sc-catalyzed ene reactions. *J. Am. Chem. Soc.* 127 (22): 8006–8007.

1498 Kraus, G.A. and Kim, J. (2004). Tandem Diels–Alder/ene reactions. *Org. Lett.* 6 (18): 3115–3117.

1499 Pirrung, M.C., Sarma, K.D., and Wang, J. (2008). Hydrophobicity and mixing effects on select heterogeneous, water-accelerated synthetic reactions. *J. Org. Chem.* 73 (22): 8723–8730.

1500 Sun, C., Li, J., Lee, D. et al. (2012). Formal C-H amination of cyclopropenes. *J. Chem. Soc., Chem. Commun.* 48 (89): 10990–10992.

1501 Huang, G., Xia, Y., Sun, C. et al. (2013). Theoretical studies on the mechanism of the C—H amination of silyl cyclopropenes by azodicarboxylates. *J. Org. Chem.* 78 (3): 988–995.

1502 Schenck, G.O., Gollnick, K., Schröter, S. et al. (1964). Zur Chemischen + Sterischen Selektivität der Photosensibilisierten O_2-Übertragung auf (+)-Limonen =(+)-Carvomenthen. *Liebigs Ann. Chem.* 674: 93–117.

1503 Foote, C.S. (1968). Photosensitized oxygenations and role of singlet oxygen. *Acc. Chem. Res.* 1 (4): 104–110.

1504 Alberti, M.N. and Orfanopoulos, M. (2010). Unraveling the mechanism of the singlet oxygen ene reaction: recent computational and experimental approaches. *Chem. Eur. J.* 16 (31): 9414–9421.

1505 Singleton, D.A. and Hang, C. (2000). Isotope effects and the mechanism of allylic hydroxylation of alkenes with selenium dioxide. *J. Org. Chem.* 65 (22): 7554–7560.

1506 Markovic, D., Volla, C.M., Vogel, P. et al. (2010). BCl_3-mediated ene reaction of sulfur dioxide and unfunctionalized alkenes. *Chem. Eur. J.* 16 (20): 5969–5975.

1507 Capozzi, G., Lucchini, V., Marcuzzi, F., and Melloni, G. (1980). Stable sulfinic acids from the ene reaction of methyl-substituted allenes with sulfur dioxide. *Tetrahedron Lett.* 21 (34): 3289–3292.

1508 Dodson, R.M. and Sauers, R.F. (1967). Sulphur monoxide - reaction with dienes. *J. Chem. Soc., Chem. Commun.* (22): 1189–1190.

1509 Nakayama, J., Tajima, Y., Xue-Hua, P., and Sugihara, Y. (2007). (1+2)-Cycloadditions of sulfur monoxide (SO) to alkenes and alkynes and (1+4)-cycloadditions to dienes (polyenes). Generation and reactions of singlet SO? *J. Am. Chem. Soc.* 129 (23): 7250–7251.

1510 Dodson, R.M. and Nelson, J.P. (1969). Sulphur monoxide - reaction with trienes. *J. Chem. Soc., Chem. Commun.* (19): 1159–1160.

1511 Nagayama, M., Okumura, O., Noda, S., and Mori, A. (1973). Stereospecific reaction of olefins with sulfur trioxide-dioxan. *J. Chem. Soc., Chem. Commun.* (21): 841–841.

1512 Cerfontain, H., Kramer, J.B., Schonk, R.M., and Bakker, B.H. (1995). Reactivity and regioselectivity in the reaction of allylsilanes and some analogous all-carbon alkenes with sulfur trioxide. *Recl. Trav. Chim. Pays-Bas* 114 (9–10): 410–420.

1513 Deleris, G., Kowalski, J., Dunogues, J., and Calas, R. (1977). *N*-sulfinyl benzene sulfonamide - new enophile offering many possibilities for synthesis. *Tetrahedron Lett.* 18 (48): 4211–4214.

1514 Münsterer, H., Kresze, G., Brechbiel, M., and Kwart, H. (1982). Transition state geometry in superene reactions. *J. Org. Chem.* 47 (13): 2677–2679.

1515 Deleris, G., Gadras, A., and Dunogues, J. (1990). Selective alkylation on least nucleophilic nitrogen atom of *para*-aminoarylsufonamides. *Tetrahedron* 46 (17): 5999–6010.

1516 Braxmeier, H. and Kresze, G. (1985). The use of *N*-nonafluorobutanesulfonyl imines as enophiles - a new synthesis of allylglycine derivatives. *Synthesis* (6–7): 683–684.

1517 Starflinger, W., Kresze, G., and Huss, K. (1986). Ene reaction mechanisms. 3. Intermolecular and intramolecular kinetic isotope effects (KIE) for some ene reactions of hetero enophiles. *J. Org. Chem.* 51 (1): 37–40.

1518 Dragan, V.A. and Moiseenkov, A.M. (1993). Ene-type reaction of trisubstituted monoterpene alkenes with methyl sulfite and thionyl chloride. *Russ. Chem. Bull.* 42 (1): 95–97.

1519 Dragan, V.A. and Moiseenkov, A.M. (1992). An ene-type reaction of thionyl chloride with linear isoprenoids. *Mendeleev Commun.* 2 (4): 150–151.

1520 Alajarin, M., Pastor, A., and Cabrera, J. (2004). Arenesulfinamides as new reagents for the synthesis of allylic sulfoxides. *Synlett* (6): 995–998.

1521 Adam, W. and Krebs, O. (2003). The nitroso ene reaction: a regioselective and stereoselective allylic nitrogen functionalization of mechanistic delight and synthetic potential. *Chem. Rev.* 103 (10): 4131–4146.

1522 Banks, R.E., Barlow, M.G., and Haszeldine, R.N. (1965). Perfluoroalkyl derivatives of nitrogen. 16. Reaction of trifluoronitrosomethane with butadiene and with isobutene. *J. Chem. Soc.* 4714–4718.

1523 Grdina, S.B., Orfanopoulos, M., and Stephenson, L.M. (1979). Stereochemical dependence of isotope effects in the singlet oxygen-olefin reaction. *J. Am. Chem. Soc.* 101 (11): 3111–3112.

1524 Seymour, C.A. and Greene, F.D. (1982). On the reaction of the nitroso group with olefins - mechanisms of ene reactions. *J. Org. Chem.* 47 (26): 5226–5227.

1525 Adam, W., Griesbeck, A., and Staab, E. (1986). A simple access to 2-epoxy alcohols: Ti(IV)-catalyzed oxygen transfer from allylic hydroperoxides. *Angew. Chem. Int. Ed. Engl.* 98: 269–270.

1526 Adam, W. and Nestler, B. (1993). Direct epoxy hydroxylation of hydroperoxy homoallylic alcohols: multidentate oxygen donor and oxygen acceptor substrates in Ti(IV)-catalyzed epoxidations. *J. Am. Chem. Soc.* 115: 7226–7231.

1527 Adam, W., Krebs, O., Orfanopoulos, M. et al. (2003). Intramolecular and intermolecular kinetic isotope effects (KIE) in the nitrosoarene ene reaction: experimental evidence for reversible intermediate formation. *J. Org. Chem.* 68 (6): 2420–2425.

1528 Baldwin, J.E., Adlington, R.M., Jain, A.U. et al. (1986). Applications of the thermal ene reaction of aldehyde *tert*-butyl-hydrazones and phenylhydrazones. *Tetrahedron* 42 (15): 4247–4252.

1529 Monge, D., Daza, S., Bernal, P. et al. (2013). Synthesis of enantioenriched azo compounds: organocatalytic Michael addition of formaldehyde *N-tert*-butyl hydrazone to nitroalkenes. *Org. Biomol. Chem.* 11 (2): 326–335.

1530 Crespo-Pelna, A., Monge, D., Martín-Zamora, E. et al. (2012). Asymmetric formal carbonyl-ene reactions of formaldehyde *tert*-butyl hydrazone with α-keto esters: dual activation by bis-urea catalysts. *J. Am. Chem. Soc.* 134 (31): 12912–12915.

1531 Pawelke, G. and Bürger, H. (1996). Trifluoromethyl-substituted aminoboranes and amine boranes revealing alkene and alkane chemistry. *Appl. Organomet. Chem.* 10 (3–4): 147–174.

1532 Dutmer, B.C. and Gilbert, M. (2011). Computational studies of pericyclic reactions of aminoborane $(F_3C)_2B{=}N(CH_3)_2$: ene reactions vs hydrogen transfers and regiochemical and conformational preferences. *Organometallics* 30 (4): 778–791.

1533 Sammis, G.M., Flamme, E.M., Xie, H. et al. (2005). Design, synthesis, and reactivity of 1-hydrazinodienes for use in organic synthesis. *J. Am. Chem. Soc.* 127 (24): 8612–8613.

1534 Jabbari, A., Sorensen, E.J., and Houk, K.N. (2006). Transition states of the retro-ene reactions of allylic diazenes. *Org. Lett.* 8 (14): 3105–3107.

1535 Lehmkuhl, H. and Reinehr, D. (1970). Addition of allylmagnesium halides on 1-alkenes. *J. Organomet. Chem.* 25 (2): C47–C50.

1536 Lehmkuhl, H. and Reinehr, D. (1970). Allyl aluminum compounds. *J. Organomet. Chem.* 23 (2): C25–C29.

1537 Lehmkuhl, H. and Mehler, K. (1978). Addition of organomagnesium halides to C=C bonds. 12. Addition of alkylmagnesium halides to 3,3-dimethylcyclopropene - new route to *cis*-chrysanthemic or *trans*-chrysanthemic acid. *Liebigs Ann. Chem.* 1978 (11): 1841–1853.

1538 Lehmkuhl, H. and Janssen, E. (1978). Addition of organomagnesium halides to C=C bonds. 13. Kinetic investigations of addition of 2-alkenylmagnesium compounds to olefins. *Liebigs Ann. Chem.* 1978 (11): 1854–1875.

1539 Yamada, T., Udagawa, T., and Sakai, S. (2010). Driving force of metallo (Mg–H and Mg–Cl)-ene reaction mechanisms. *Phys. Chem. Chem. Phys.* 12 (15): 3799–3805.

1540 (a) Cheng, D., Zhu, S.R., Liu, X.J. et al. (1999). The lithium-ene cyclization and the purported carbanion-induced ene cyclization. *J. Am. Chem. Soc.* 121 (43): 10241–10242. (b) Sakai, S. and Hikita, T. (2008). Theoretical studies of metallo (Li and Na)-ene reaction mechanisms. *J. Phys. Chem. A* 112 (43): 10985–10992.

1541 Bubnov, Y.N., Nesmeyanova, O.A., Rudashevskaya, T.Y. et al. (1971). Reaction of allylboranes with methylopropene. *Tetrahedron Lett.* 12 (24): 2153–2160.

1542 Mikhailov, B.M. and Bubnov, Y.N. (1971). Method for preparation of 1,4-dienes via allylboration of vinyl ethers. *Tetrahedron Lett.* 12 (24): 2127–2130.

1543 Mikhailov, B.M. (1972). Allylboron compounds. *Organomet. Chem. Rev. Sect. A: Subj. Rev.* 8 (1): 1–65.

1544 Frangin, Y. and Gaudemar, M. (1976). New method for synthesis of 1,4- and 1,6-dienes. *Bull. Soc. Chim. France; Partie II-Chim. Mol. Org. Biol.* (7–8): 1173–1177.

1545 Lehmkuhl, H. and Nehl, H. (1973). Additions of 2-alkenylzinc to unsaturated-hydrocarbons. *J. Organomet. Chem.* 60 (1): 1–10.

1546 Frangin, Y., Favre, E., and Gaudemar, M. (1976). New property of propargylic organozincs - addition to acetylenic compounds and their organomagnesiums. *C.R. Seances Acad. Sci., Ser. C* 282 (5): 277–280.

1547 (a) Bernadou, F. and Miginiac, L. (1976). New example of reversible reaction in synthesis of organometallic compounds - addition of α-ethylenic organozinc compounds to true α-acetylenic compounds. *Tetrahedron Lett.* 17 (35): 3083–3086. (b) Negishi, E. and Miller, J.A. (1983). Selective carbon-carbon bond formation via transition-metal catalysis. 37. Controlled carbometalation. 16. Novel syntheses of α,β-unsaturated cyclopentenones via allylzincation of alkynes. *J. Am. Chem. Soc.* 105 (22): 6761–6763.

1548 Molander, G.A. (1983). Allylmetalations - the allylzincation of 1-(trimethylsilyl)-1-alkynes. *J. Org. Chem.* 48 (26): 5409–5411.

1549 Lichtenberg, C., Spaniol, T.P., and Okuda, J. (2012). Dimerization of the allylzinc cation: selective coupling of allyl anions in a metallo-ene reaction. *Angew. Chem. Int. Ed.* 51 (32): 8101–8105.

1550 Nakamura, E. and Kubota, K. (1997). Carbometalation of cyclopropene. Diastereoselective *cis*-addition of zincated amides, esters, and hydrazones. *J. Org. Chem.* 62 (4): 792–793.

1551 Lorthiois, E., Marek, I., and Normant, J.F. (1998). Diastereoselective and enantioselective intramolecular amino-zinc-enolate carbometalation reactions. A new polysubstituted pyrrolidines synthesis. *J. Org. Chem.* 63 (8): 2442–2450.

1552 Stefani, A. (1973). Alkylaluminum promoted stereospecific addition to vinyl olefins. *Helv. Chim. Acta* 56 (4): 1192–1198.

1553 Gaudemar, M. (1971). New property of allyl organozinc compounds - addition to acetylenic zinc and organomagnesium compounds and to vinyl organomagnesium compounds. *C. R. Seances Acad. Sci., Ser. C* 273 (23): 1669–1672.

1554 Knochel, P. and Normant, J.F. (1986). Synthesis and reactivity of *gem*-dimetallic organic compounds. 1. *Tetrahedron Lett.* 27 (9): 1039–1042.

1555 Knochel, P. and Normant, J.F. (1986). Mixed *gem*-dimetallic organic compounds - a new class of multi-coupling reagents. 2. *Tetrahedron Lett.* 27 (9): 1043–1046.

1556 Normant, J.F., Marek, I., and Lefrancois, J.M. (1992). Organobismetallic zinc reagents - their preparation and use in diastereoselective reactions. *Pure Appl. Chem.* 64 (12): 1857–1864.

1557 Marek, I., Lefrançois, J.M., and Normant, J.F. (1994). Diastereoselective carbometalation of vinylmetals. *J. Org. Chem.* 59 (15): 4154–4161.

1558 Hirai, A., Nakamura, M., and Nakamura, E. (2000). Mechanism of addition of allylmetal to vinylmetal. Dichotomy between metallo-ene reaction and metalla-Claisen rearrangement. *J. Am. Chem. Soc.* 122 (48): 11791–11798.

1559 Nakamura, M., Hara, K., Hatakeyama, T., and Nakamura, E. (2001). Regioselective allylzincation of alkenylboronate. *Org. Lett.* 3 (20): 3137–3140.

1560 Hughes, R.P. and Powell, J. (1972). Transition-metal promoted reactions of unsaturated hydrocarbons. 1. Mechanism of 1,3-diene insertion into (allyl)Pd bonds. *J. Am. Chem. Soc.* 94 (22): 7723–7732.

1561 Wilke, G., Schneider, J., Müller, H. et al. (1963). Cyclooligomerisation von Butadien und Übergangsmetall-π-Komplexe. *Angew. Chem.* 75 (1): 10–19.

1562 (a) Wilke, G. (1975). Karl Ziegler, In Memoriam. In: *Coordination Polymerization* (ed. J.C.W. Chien), 1–14. New York: Academic Press. (b) Wilke, G. (2003). Fifty years of Ziegler catalysts: consequences and development of an invention. *Angew. Chem. Int. Ed.* 42 (41): 5000–5008.

1563 Wilke, G. (1988). Contributions to organo-nickel chemistry. *Angew. Chem. Int. Ed. Engl.* 27 (1): 185–206.

1564 Larock, R.C., Burkhart, J.P., and Oertle, K. (1982). Organopalladium approaches to prostaglandins. 2. Synthesis of prostaglandin endoperoxide analogs via π-(allyl)Pd additions to bicyclic olefins. *Tetrahedron Lett.* 23 (10): 1071–1074.

1565 (a) Oppolzer, W. (1989). Intramolecular, stoichiometric (Li, Mg, Zn) and catalytic (Ni, Pd, Pt) metallo-ene reactions in organic synthesis. *Angew. Chem. Int. Ed. Engl.* 28 (1): 38–52. (b) Oppolzer, W., Pimm, A., Stammen, B., and Hume, W.E. (1997). Palladium-catalysed intramolecular cyclisations of olefinic propargylic carbonates and application to the diastereoselective synthesis of enantiomerically pure (-)-α-thujone. *Helv. Chim. Acta* 80 (3): 623–639. (c) Cardenas, D.J., Alcami, M., Cossio, F. et al. (2003). The final steps of the Oppolzer cyclization: mechanism of the insertion of alkenes into (allyl)Pd(II) complexes. *Chem. Eur. J.* 9 (1): 96–105.

1566 (a) Shvo, Y. and Arisha, A. (2001). Intermolecular metallo-ene reaction catalyzed by Pd. *J. Org. Chem.* 66 (14): 4921–4922. (b) Zhu, G.-G., Tong, X.-F., Cheng, J. et al. (2005). Palladium-catalyzed tandem cyclization/Suzuki coupling of 1,6-enynes: reaction scope and mechanism. *J. Org. Chem.* 70 (5): 1712–1717. (c) Liang, H.-B., Yan, F.-C., Dong, X. et al. (2017). Palladium-catalyzed cascade metallo-ene/Suzuki coupling reaction of allenamides. *J. Chem. Soc., Chem. Commun.* 53 (21): 3138–3141.

1567 Yamamoto, Y. and Asao, N. (1993). Selective reactions using allylic metals. *Chem. Rev.* 93 (6): 2207–2293.

1568 Chemler, S.R. and Roush, W.R. (2000). Recent applications of allylation reaction to the synthesis of natural products. In: *Modern Carbonyl Chemistry* (ed. J. Otera), 403–490. Weinheim: Wiley-VCH.

1569 Kennedy, J.W.J. and Hall, D.G. (2003). Recent advances in the activation of boron and silicon reagents for stereocontrolled allylation reactions. *Angew. Chem. Int. Ed.* 42 (39): 4732–4739.

1570 Hall, D.G. (2007). Lewis and Brønsted acid catalyzed allylboration of carbonyl compounds: from discovery to mechanism and applications. *Synlett* (11): 1644–1655.

1571 Corey, E.J., Yu, C.M., and Lee, D.H. (1990). A Practical and general enantioselective synthesis of chiral propa-1,2-dienyl and propargyl carbinols. *J. Am. Chem. Soc.* 112 (2): 878–879.

1572 Hernández, E. and Soderquist, J.A. (2005). Nonracemic α-allenyl carbinols from asymmetric propargylation with the 10-trimethylsilyl-9-borabicyclo[3.3.2]decanes. *Org. Lett.* 7 (24): 5397–5400.

1573 Burgos, C.H., Canales, E., Matos, K., and Soderquist, J.A. (2005). Asymmetric allyl- and crotylboration with the robust, versatile, and recyclable 10-TMS-9-borabicyclo[3.3.2]decanes. *J. Am. Chem. Soc.* 127 (22): 8044–8049.

1574 Canales, E., Prasad, K.G., and Soderquist, J.A. (2005). B-allyl-10-Ph-9-borabicyclo[3.3.2]decanes: strategically designed for the asymmetric allylboration of ketones. *J. Am. Chem. Soc.* 127 (33): 11572–11573.

1575 Denmark, S.E. and Weber, E.J. (1983). On the stereochemistry of allylmetal-aldehyde condensations. *Helv. Chim. Acta* 66 (6): 1655–1660.

1576 Hoffmann, R.W. and Zeiss, H.J. (1981). Stereoselective synthesis of alcohols. 8. Diastereoselective synthesis of β-methylhomoallyl alcohols via crotylboronates. *J. Org. Chem.* 46 (7): 1309–1314.

1577 Hoffmann, R.W. and Zeiss, H.J. (1979). Diastereoselective synthesis of β-methyl homoallyl alcohols. *Angew. Chem. Int. Ed. Engl.* 18 (4): 306–307.

1578 Brown, H.C. and Jadhav, P.K. (1983). Asymmetric carbon–carbon bond formation via β-allyldiisopinocampheylborane - simple synthesis of secondary homoallylic alcohols with excellent enantiomeric purities. *J. Am. Chem. Soc.* 105 (7): 2092–2093.

1579 Jadhav, P.K., Bhat, K.S., Perumal, P.T., and Brown, H.C. (1986). Chiral synthesis via organoboranes. 5. Asymmetric allylboration via chiral allyldialkylboranes - synthesis of homoallylic alcohols with exceptionally high enantiomeric excess. *J. Org. Chem.* 51 (4): 432–439.

1580 García, J., Kim, B.M., and Masamune, S. (1987). Asymmetric addition of (*E*)-crotyl-*trans*-2,5-dimethylborolanes and (*Z*)-crotyl-*trans*-2,5-dimethylborolanes to aldehydes. *J. Org. Chem.* 52 (21): 4831–4832.

1581 Short, R.P. and Masamune, S. (1989). Asymmetric allylboration with *B*-allyl-2-(trimethylsilyl)borolane. *J. Am. Chem. Soc.* 111 (5): 1892–1894.

1582 Corey, E.J., Yu, C.M., and Kim, S.S. (1989). A practical and efficient method for enantioselective allylation of aldehydes. *J. Am. Chem. Soc.* 111 (14): 5495–5496.

1583 Roush, W.R., Walts, A.E., and Hoong, L.K. (1985). Diastereoselective and enantioselective aldehyde addition reactions of 2-allyl-1,3,2-dioxaborolane-4,5-dicarboxylic esters, a useful class of tartrate ester modified allylboronates. *J. Am. Chem. Soc.* 107 (26): 8186–8190.

1584 Roush, W.R., Hoong, L.K., Palmer, M.A.J. et al. (1990). Asymmetric synthesis using tartrate ester modified allylboronates. 2. Single and double asymmetric reactions with alkoxy substituted aldehydes. *J. Org. Chem.* 55 (13): 4117–4126.

1585 Roush, W.R., Ando, K., Powers, D.B. et al. (1990). Asymmetric synthesis using diisopropyl tartrate modified (*E*)-crotylboronates and (*Z*)-crotylboronates - preparation of the chiral crotylboronates and reactions with achiral aldehydes. *J. Am. Chem. Soc.* 112 (17): 6339–6348.

1586 Roush, W.R., Palkowitz, A.D., and Ando, K. (1990). Acyclic diastereoselective synthesis using tartrate ester modified crotylboronates - double asymmetric reactions with α-methyl chiral aldehydes and synthesis of the C(19)–C(29) segment of rifamycin-S. *J. Am. Chem. Soc.* 112 (17): 6348–6359.

1587 Hall, D.G. (2008). New preparative methods for allylic boronates and their application in stereoselective catalytic allylborations. *Pure Appl. Chem.* 80 (5): 913–927.

1588 (a) Poh, J.-S., Lau, S.-H., Dykes, I.G. et al. (2016). A multicomponent approach for the preparation of homoallylic alcohols. *Chem. Sci.* 7 (11): 6803–6807. (b) Diner, C. and Czabo, K.J. (2017). Recent advances in the preparation and application of allylboron species in organic synthesis. *J. Am. Chem. Soc.* 139 (1): 2–14.

1589 Hoffmann, R.W. (1988). α-Chiral allylboronates - reagents for asymmetric synthesis. *Pure Appl. Chem.* 60 (1): 123–130.

1590 (a) Matteson, D.S. (1988). Asymmetric synthesis with boronic esters. *Acc. Chem. Res.* 21 (8): 294–300. (b) Wohlfahrt, M., Harms, K., and Koert, U. (2011). Asymmetric allylboration of *vic*-tricarbonyl compounds: total synthesis of (+)-awajanomycin. *Angew. Chem. Int. Ed.* 50 (36): 8404–8406.

1591 Chen, M. and Roush, W.R. (2013). Crotylboron-based synthesis of the polypropionate units of chaxamycins A/D, Salinisporamycin, and Rifamycin S. *J. Org. Chem.* 78 (1): 3–8.

1592 Hoffmann, R.W. and Zeiss, H.J. (1980). Enhancement of 1,2-asymmetric induction on C—C linkage by use of chiral 2-butenylboronic esters. *Angew. Chem. Int. Ed. Engl.* 19 (3): 218–219.

1593 Ditrich, K., Bube, T., Stürmer, R., and Hoffmann, R.W. (1986). Total synthesis of mycinolide-V, the aglycone of a macrolide antibiotic of the mycinamycin series. *Angew. Chem. Int. Ed. Engl.* 25 (11): 1028–1030.

1594 Hoffmann, R.W. and Rolle, U. (1994). Synthesis of a C-15-C-27 aegment of Venturicidine. *Tetrahedron Lett.* 35 (27): 4751–4754.

1595 Reetz, M.T., Kesseler, K., and Jung, A. (1984). Aldol additions to α-alkoxy and β-alkoxy aldehydes - the effect of chelation on simple diastereoselectivity. *Tetrahedron* 40 (21): 4327–4336.

1596 (a) Chen, M., Handa, M., and Roush, W.R. (2009). Enantioselective synthesis of 2-methyl-1,2-*syn*- and 2-methyl-1,2-*anti*-3-butenediols via allene hydroboration-aldehyde allylboration reaction sequences. *J. Am. Chem. Soc.* 131: 14602–14603. (b) Kister, J., DeBaillie, A.C., Lira, R., and Roush, W.R. (2009). Stereoselective synthesis of γ-substituted (*Z*)-allylic boranes via kinetically controlled hydroboration of allenes with

10-TMS-9-borabicyclo[3.3.2]decane. *J. Am. Chem. Soc.* 131: 14174–14175.

1597 Wang, Z. (2010). Roush crotylation. In: *Comprehensive Organic Name Reactions and Reagents*, vol. 549, 2435–2438. Hoboken, NJ: Wiley.

1598 Brown, H.C. and Bhat, K.S. (1986). Chiral synthesis via organoboranes. 7. Diastereoselective and enantioselective synthesis of *erythro*-β-methylomoallyl and *threo*-β-methylhomoallyl alchols via enantiomeric (*Z*)-crotylborane and (*E*)-crotylborane. *J. Am. Chem. Soc.* 108 (19): 5919–5923.

1599 Racherla, U.S. and Brown, H.C. (1991). Chiral synthesis via organoboranes. 27. Remarkably rapid and exceptionally enantioselective (approaching 100 % ee) allylboration of representative aldehydes at 100 °C under new salt-free conditions. *J. Org. Chem.* 56 (1): 401–404.

1600 Roush, W.R. and Halterman, R.L. (1986). Diisopropyl tartrate modified (*E*)-crotylboronates - highly enantioselective propionate (*E*)-enolate equivalents. *J. Am. Chem. Soc.* 108 (2): 294–296.

1601 Gung, B.W., Xue, X.W., and Roush, W.R. (2002). The origin of diastereofacial control in allylboration reactions using tartrate ester derived allylboronates: attractive interactions between the Lewis acid coordinated aldehyde carbonyl group and an ester carbonyl oxygen. *J. Am. Chem. Soc.* 124 (36): 10692–10697.

1602 Gung, B.W. and Xue, X.W. (2001). Asymmetric transition states of allylation reaction: an ab initio molecular orbital study. *Tetrahedron: Asymmetry* 12 (21): 2955–2959.

1603 (a) Kennedy, J.W.J. and Hall, D.G. (2005). Recent advances in the preparation of allylboronates and their use in tandem reactions with carbonyl compounds. In: *Boronic Acids* (ed. D.G. Hall), 241–274. Weinheim: Wiley-VCH. (b) Lachance, H. and Hall, D.G. (2008). Allylboration of carbonyl compounds. In: *Organic Reactions*, vol. 73 (ed. S.E. Denmark), 1–573. Hoboken, NJ: Wiley.

1604 (a) Bartlett, P.A. (1980). Stereocontrol in the synthesis of acyclic systems - applications to natural product synthesis. *Tetrahedron* 36 (1): 3–72. (b) Christoffers, J., Scharl, H., Frey, W., and Baro, A. (2004). Transformation of an optically active decahydro-6-isoquinolone scaffold: perfect Felkin–Anh diastereoselectivity. *Org. Lett.* 6 (7): 1171–1173. (c) Díaz-Oltra, S., Carda, M., Murga, J. et al. (2008). Aldol reactions between L-erythrulose derivatives and chiral α-amino and α-fluoro aldehydes: competition between Felkin–Anh and Cornforth transition states. *Chem. Eur. J.* 14 (30): 9240–9254. (d) Stanton, G.R., Johnson, C.N., and Walsh, P.J. (2010). Overriding Felkin control: a general method for highly diastereoselective chelation-controlled additions to α-silyloxy aldehydes. *J. Am. Chem. Soc.* 132 (12): 4399–4408. (e) Raffier, L., Stanton, G.R., and Walsh, P.J. (2013). Chelation-controlled additions to α-silyloxy aldehydes: an autocatalytic approach. *Org. Lett.* 15 (24): 6174–6177. (f) Rosenberg, R.E. and Kelly, W.J. (2015). Felkin–Anh is not enough. *J. Phys. Org. Chem.* 28 (1): 47–56.

1605 Paddon-Row, M.N., Rondan, N.G., and Houk, K.N. (1982). Staggered models for asymmetric induction - attack trajectories and conformations of allylic bonds from ab initio transition structures of addition reactions. *J. Am. Chem. Soc.* 104 (25): 7162–7166.

1606 Kolodiazhnyi, O.I. (2003). Multiple stereoselectivity and its application in organic synthesis. *Tetrahedron* 59 (32): 5953–6018.

1607 Roush, W.R. and Bannister, T.D. (1992). Stereoselective synthesis of the C(13)-C(25) segment of bafilomycin-A_1. *Tetrahedron Lett.* 33 (25): 3587–3590.

1608 Scheidt, K.A., Tasaka, A., Bannister, T.D. et al. (1999). Total synthesis of (-)-Bafilomycin A_1: application of diastereoselective crotylboration and methyl ketone aldol reactions. *Angew. Chem. Int. Ed.* 38 (11): 1652–1655.

1609 Seebach, D., Beck, A.K., Schiess, M. et al. (1983). Some recent advances in the use of Ti reagents for organic synthesis. *Pure Appl. Chem.* 55 (11): 1807–1822.

1610 Cossy, J., Willis, C., Bellosta, V., and BouzBouz, S. (2002). Enantioselective allyltitanations and metathesis reactions. Application to the synthesis of piperidine alkaloids (+)-sedamine and (-)-prosophylline. *J. Org. Chem.* 67 (7): 1982–1992.

1611 Duthaler, R.O. and Hafner, A. (1992). Chiral Ti complexes for enantioselective addition of nucleophiles to carbonyl groups. *Chem. Rev.* 92 (5): 807–832.

1612 Cossy, J., BouzBouz, S., Pradaux, F. et al. (2002). Chiral Ti complexes. Synthesis of optically active unsaturated alcohols, diols, polypropionates and their use in the synthesis of biologically active compounds. *Synlett* (10): 1595–1606.

1613 Marshall, J.A. (1996). Chiral allylic and allenic stannanes as reagents for asymmetric synthesis. *Chem. Rev.* 96: 31–47.

1614 Marshall, J.A., Yu, R.H., and Perkins, J.F. (1995). Diastereoselective and enantioselective synthesis of allenylcarbinols through S_E2' addition of transient nonracemic propargylic stannanes to aldehydes. *J. Org. Chem.* 60 (17): 5550–5555.

1615 Marshall, J.A., Perkins, J.F., and Wolf, M.A. (1995). Synthesis of *syn,syn,anti,syn syn,anti* and *anti,anti* stereotriads from a single pair of enantiomeric reagents. *J. Org. Chem.* 60 (17): 5556–5559.

1616 Marshall, J.A., Lu, Z.H., and Johns, B.A. (1998). Synthesis of discodermolide subunits by S_E2' addition of nonracemic allenylstannanes to aldehydes. *J. Org. Chem.* 63 (3): 817–823.

1617 Kennedy, J.W.J. and Hall, D.G. (2002). Dramatic rate enhancement with preservation of stereospecificity in the first metal-catalyzed additions of allylboronates. *J. Am. Chem. Soc.* 124 (39): 11586–11587.

1618 Ishiyama, T., Ahiko, T., and Miyaura, N. (2002). Acceleration effect of Lewis acid in allylboration of aldehydes: catalytic, regiospecific, diastereospecific, and enantioselective synthesis of homoallyl alcohols. *J. Am. Chem. Soc.* 124 (42): 12414–12415.

1619 Yamamoto, H. and Futatsugi, K. (2005). "Designer acids": combined acid catalysis for asymmetric synthesis. *Angew. Chem. Int. Ed.* 44 (13): 1924–1942.

1620 Penner, M., Rauniyar, V., Kaspar, L.T., and Hall, D.G. (2009). Catalytic asymmetric synthesis of palmerolide A via organoboron methodology. *J. Am. Chem. Soc.* 131 (40): 14216.

1621 Rauniyar, V. and Hall, D.G. (2004). Lewis acids catalyze the addition of allylboronates to aldehydes by electrophilic activation of the dioxaborolane in a closed transition structure. *J. Am. Chem. Soc.* 126 (14): 4518–4519.

1622 Rauniyar, V. and Hall, D.G. (2006). Catalytic enantioselective and catalyst-controlled diastereofacial-selective additions of allyl- and crotylboronates to aldehydes using chiral bronsted acids. *Angew. Chem. Int. Ed.* 45 (15): 2426–2428.

1623 (a) Rauniyar, V. and Hall, D.G. (2007). Preparation of a C_2-symmetric binol-derived diol and its application in the catalytic enantioselective and catalyst-controlled diastereoselective allylboration of aldehydes. *Synthesis* (21): 3421–3426. (b) Rauniyar, V. and Hall, D.G. (2009). Rationally improved chiral Brønsted acid for catalytic enantioselective allylboration of aldehydes with an expanded reagent scope. *J. Org. Chem.* 74 (11): 4236–4241.

1624 (a) Huang, Y.-Y., Yang, X., Lv, Z.-C. et al. (2015). Asymmetric synthesis of 1,3-butadienyl-2-carbinols by the homoallenylboration of aldehydes with a chiral phosphoric acid catalyst. *Angew. Chem. Int. Ed.* 54 (25): 7299–7302. (b) Wang, H., Jain, P., Antilla, J.C., and Houk, K.N. (2013). Origins of stereoselectivities in chiral phosphoric acid catalyzed allylborations and propargylations of aldehydes. *J. Org. Chem.* 78 (3): 1208–1215. (c) Jain, P. and Antilla, J.C. (2010). Chiral Brønsted acid-catalyzed allylboration of aldehydes. *J. Am. Chem. Soc.* 132 (34): 11884–11886. (d) Chen, M. and Roush, W.R. (2012). Enantioselective synthesis of *anti*- and *syn*-homopropargyl alcohols via chiral Brønsted acid catalyzed asymmetric allenylboration reactions. *J. Am. Chem. Soc.* 134 (6): 10947–10952.

1625 (a) Barnett, D.S., Moquist, P.N., and Schaus, S.E. (2009). The mechanism and an improved asymmetric allylboration of ketones catalyzed by chiral biphenols. *Angew. Chem. Int. Ed.* 48 (46): 8679–8682. (b) Grayson, M.N., Pellegrinet, S.C., and Goodman, J.M. (2012). Mechanistic insights into the BINOL-derived phosphoric acid-catalyzed asymmetric allylboration of aldehydes. *J. Am. Chem. Soc.* 134 (5): 2716–2722. (c) Xing, C.H., Liao, Y.X., Zhang, Y. et al. (2012). Asymmetric allylboration of aldehydes with pinacol allylboronates catalyzed by 1,1′-spirobiindane-7,7′-diol (SPINOL) based phosphoric acids. *Eur. J. Org. Chem.* (6): 1115–1118.

1626 (a) Lou, S., Moquist, P.N., and Schaus, S.E. (2007). Asymmetric allylboration of acyl imines catalyzed by chiral diols. *J. Am. Chem. Soc.* 129 (49): 15398–15404. (b) Chen, J.L.-Y. and Aggarwal, V.K. (2014). Highly diastereoselective and enantiospecific allylation of ketones and imines using borinic esters: contiguous quaternary stereogenic centers. *Angew. Chem. Int. Ed.* 53 (41): 10992–10996. (c) Tan, Q.-Y., Wang, X.-Q., Xiong, Y. et al. (2017). Chiral amino alcohol accelerated and stereocontrolled allylboration of iminoisatins: highly efficient construction of adjacent quaternary stereogenic centers. *Angew. Chem. Int. Ed.* 56 (17): 4829–4833.

1627 (a) Robbins, D.W., Lee, K.A., Silverio, D.L. et al. (2016). Practical and broadly applicable catalytic enantioselective additions of allyl-B(pin) compounds to ketones and α-ketoesters. *Angew. Chem. Int. Ed.* 55 (33): 9610–9614. (b) van der Mei, F.W., Qin, G.-M., Morrison, R.J., and Hoveyda, A.H. (2017). Practical, broadly applicable, α-selective, *Z*-selective, diastereoselective, and enantioselective addition of allylboron compounds to mono-, di-, tri-, and polyfluoroalkyl ketones. *J. Am. Chem. Soc.* 139 (26): 9053–9065.

1628 Ota, Y., Kawato, Y., Egami, H., and Hamashima, Y. (2017). Enantioselective allyl-, and allenylboration of aldehydes catalyzed by chiral hydroxyl carboxylic acid. *Synlett* 28 (8): 976–980.

1629 Cui, Y., Li, W., Sato, T. et al. (2013). Catalytic use of zinc amide for transmetalation with allylboronates: general and efficient catalytic allylation of carbonyl compounds, imines, and hydrazones. *Adv. Synth. Catal.* 355 (6): 1193–1205.

1630 Bouchez, L. and Vogel, P. (2002). One-pot, three-component synthesis of open-chain, polyfunctional sulfones. *Synthesis* (2): 225–231.

1631 Lachance, H., Lu, X.S., Gravel, M., and Hall, D.G. (2003). Scandium-catalyzed allylboration of aldehydes as a practical method for highly diastereo- and enantioselective construction of homoallylic alcohols. *J. Am. Chem. Soc.* 125 (34): 10160–10161.

1632 Podlech, J. (2010). "Try and fall sick ..."-the composer, chemist, and surgeon Aleksandr Borodin. *Angew. Chem. Int. Ed.* 49 (37): 6490–6495.

1633 Borodin, A. (1864). Über die Einwirkung des Natriums auf Valeraldehyd. *J. Prakt. Chem.* 93: 413–425.

1634 Kekulé, A. (1869). Condensationsprodukte des Aldehyds: Crotonaldehyd. *Ber. Dtsch. Chem. Ges.* 2: 365–368.

1635 Wurtz, C.A. (1872). Über einen Aldehyd-Alkohol. *J. Prakt. Chem.* 5: 457–464.

1636 (a) Smith, M.B. (2013). *March's Advanced Organic Chemistry: Reactions, Mechanisms, and Structure*, 7e, Chapter 16, 1067–1252. New York: Wiley. (b) Mandal, S., Mandal, S., Ghosh, S.K. et al. (2016). Review of the aldol reaction. *Synth. Commun.* 46 (16): 1327–1342.

1637 (a) Palomo, C., Oiarbide, M., and García, J.M. (2004). Current progress in the asymmetric aldol addition reaction. *Chem. Soc. Rev.* 33 (2): 65–75. (b) Braun, M. (2004). Fundamentals and transition-state models. Aldol additions of group 1 and 2 enolates. In: *Modern Aldol Reactions*, vol. 1 (ed. R. Mahrwald), 1–62. Weinheim: Wiley-VCH. (c) Mlynarski, J. and Bas, S. (2014). Catalytic asymmetric aldol reactions in aqueous media - a five year update. *Chem. Soc. Rev.* 43 (2): 577–587. (d) Mokgweetsi, P., Botlhole, L., and Singh, G.S. (2016). Recent progress in asymmetric aldol reactions under solvent-free conditions or in aqueous medium. *Curr. Green Chem.* 3 (4): 318–327. (e) Heravi, M.M., Zadsirjan, V., Dehghani, M., and Hosseintash, N. (2017). Current applications of organocatalysts in asymmetric aldol reactions: an update. *Tetrahedron: Asymmetry* 28 (5): 587–707.

1638 (a) Schetter, B. and Mahrwald, R. (2006). Modern aldol methods for the total synthesis of polyketides. *Angew. Chem. Int. Ed.* 45 (45): 7506–7525. (b) Li, J. and Menche, D. (2009). Direct methods for stereoselective polypropionate synthesis. A survey. *Synthesis* (14): 2293–2315. (c) Gafi, W. and Yamamoto, H. (2016). Second generation of aldol reaction. *Acc. Chem. Res.* 49 (9): 1757–1768.

1639 (a) Greenberg, W.A., Varvak, A., Hanson, S.R. et al. (2004). Development of an efficient, scalable, aldolase-catalyzed process for enantioselective synthesis of statin intermediates. *Proc. Natl. Acad. Sci. U.S.A.* 101 (16): 5788–5793. (b) Samland, A.:.K. and Sprenger, G.A. (2006). Microbial aldolases as C—C bonding enzymes - unknown treasures and new developments. *Appl. Microbiol. Biotechnol.* 71 (3): 253–264. (c) Dean, S.M., Greenberg, W.A., and Wong, C.H. (2007). Recent advances in aldolase-catalyzed asymmetric synthesis. *Adv. Synth. Catal.* 349 (8–9): 1308–1320.

1640 (a) Gutierrez, M.L., Garrabou, X., Agosta, E. et al. (2008). Serine hydroxymethyl transferase from Streptococcus thermophilus and L-threonine aldolase from *Escherichia coli* as stereocomplementary biocatalysts for the synthesis of β-hydroxy-α,ω-diamino acid derivatives. *Chem. Eur. J.* 14 (15): 4647–4656. (b) Li, Y.-H., Yu, H., Cao, H.-Z. et al. (2008). Pasteurella multocida sialic acid aldolase: a promising biocatalyst. *Appl. Microbiol. Biotechnol.* 79 (6): 963–970. (c) Aragon, C.C., Palomo, J.M., Filice, M., and Mateo, C. (2016). Immobilization of aldolase for C—C bond formation. *Curr. Org. Chem.* 20 (11): 1243–1251. (d) Bisterfeld, C.M., Classen, T., Kuberl, I. et al. (2016). Redesigning aldolase stereoselectivity by homologous grafting. *PLoS One* 11 (6): e0156525. (e) Rahmi, M., Geertsema, E.M., Miao, Y.-F. et al. (2017). Inter-and intramolecular aldol reactions promiscuously catalyzed by a proline-based tautomerase. *Org. Biomol. Chem.*

15 (13): 2809–2816. (f) Chen, Q.-J., Chen, X., Cui, Y.-F. et al. (2017). A new D-threonine aldolase as a promising biocatalyst for highly stereoselective preparation of chiral aromatic β-hydroxy-α-amino acids. *Catal. Sci. Technol.* 7 (4): 5964–5973.

1641 Eder, U., Sauer, G., and Weichert, R. (1971). Total synthesis of optically active steroids. 6. New type of asymmetric cyclization to optically active steroid CD partial structures. *Angew. Chem. Inter. Ed. Engl.* 10 (7): 496–497.

1642 Hajos, Z.G. and Parrish, D.R. (1974). Asymmetric synthesis of bicyclic intermediates of natural product chemistry. *J. Org. Chem.* 39 (12): 1615–1621.

1643 List, B., Lerner, R.A., and Barbas, C.F.I. (2000). Proline-catalyzed direct asymmetric aldol reactions. *J. Am. Chem. Soc.* 122 (10): 2395–2396.

1644 Sakthivel, K., Notz, W., Bui, T., and Barbas, C.F.I. (2001). Amino acid catalyzed direct asymmetric aldol reactions: a bioorganic approach to catalytic asymmetric carbon-carbon bond-forming reactions. *J. Am. Chem. Soc.* 123 (22): 5260–5267.

1645 Northrup, A.B. and MacMillan, D.W.C. (2002). The first direct and enantioselective cross-aldol reaction of aldehydes. *J. Am. Chem. Soc.* 124 (24): 6798–6799.

1646 Ward, D.E., Jheengut, V., and Akinnusi, O.T. (2005). Enantioselective direct intermolecular aldol reactions with enantiotopic group selectivity and dynamic kinetic resolution. *Org. Lett.* 7 (6): 1181–1184.

1647 Urushima, T., Yasui, Y., Ishikawa, H., and Hayashi, Y. (2010). Polymeric ethyl glyoxylate in an asymmetric aldol reaction catalyzed by diarylprolinol. *Org. Lett.* 12 (13): 2966–2969.

1648 (a) Bahmanyar, S., Houk, K.N., Martin, H.J., and List, B. (2003). Quantum mechanical predictions of the stereoselectivities of proline-catalyzed asymmetric intermolecular aldol reactions. *J. Am. Chem. Soc.* 125 (9): 2475–2479. (b) Allemann, C., Gordillo, R., Clemente, F.R. et al. (2004). Theory of asymmetric organocatalysis of aldol and related reactions: rationalizations and predictions. *Acc. Chem. Res.* 37 (8): 558–569. (c) Renzi, P., Hioe, J., and Gschwind, R.M. (2017). Enamine/dienamine and Brønsted acid catalysis: Elusive intermediates, reaction mechanisms, and stereoinduction modes based on in situ NMR spectroscopy and computational studies. *Acc. Chem. Res.* 50 (12): 2936–2948.

1649 Notz, W., Tanaka, F., and Barbas, C.F. (2004). Enamine-based organocatalysis with proline and diamines: the development of direct catalytic asymmetric aldol, Mannich, Michael, and Diels–Alder reactions. *Acc. Chem. Res.* 37 (8): 580–591.

1650 Mukherjee, S., Yang, J.W., Hoffmann, S., and List, B. (2007). Asymmetric enamine catalysis. *Chem. Rev.* 107 (12): 5471–5569.

1651 Schmid, M.B., Zeitler, K., and Gschwind, R.M. (2010). The elusive enamine intermediate in proline-catalyzed aldol reactions: NMR detection, formation pathway, and stabilization trends. *Angew. Chem. Int. Ed.* 49 (29): 4997–5003.

1652 Sharma, A.K. and Sunoj, R.B. (2010). Enamine versus oxazolidinone: what controls stereoselectivity in proline-catalyzed asymmetric aldol reactions? *Angew. Chem. Int. Ed.* 49 (36): 6373–6377.

1653 Bock, D.A., Lehmann, C.W., and List, B. (2010). Crystal structures of proline-derived enamines. *Proc. Natl. Acad. Sci. U.S.A.* 107 (48): 20636–20641.

1654 (a) Enders, D., Fronert, J., Bisschops, T., and Boeck, F. (2012). Asymmetric total synthesis of smyrindiol employing an organocatalytic aldol key step. *Beilstein J. Org. Chem.* 8: 1112–1117. (b) Fanton, J., Camps, F., Castillo, J.A. et al. (2012). Enzymatic and organocatalyzed asymmetric aldolization reactions for the synthesis of thiosugar scaffolds. *Eur. J. Org. Chem.* (1): 203–210. (c) Palathi, S., Das, J., Samanta, M., and Dash, J. (2014). Asymmetric aldol reaction of thiazole-carbaldehydes: regio- and stereoselective synthesis of tubuvalin analogues. *Adv. Synth. Catal.* 356 (16): 3370–3376. (d) Dibello, E., Gamenara, D., and Seoane, G.A. (2017). Efficient synthesis of orthogonally protected rare L-hexoses and derivatives. *Synthesis* 49 (5): 1087–1092.

1655 (a) Ishikawa, R., Yoshida, N., Akao, Y. et al. (2014). Total syntheses of (+)-sesamin and (+)-sesaminol. *Chem. Lett.* 43 (10): 1572–1574. (b) Inai, M., Ishikawa, R., Yoshida, N. et al. (2015). Stereocontrolled total syntheses of optically active furofuran lignans. *Synthesis* 47 (22): 3513–3521.

1656 Yamada, Y.M.A., Yoshikawa, N., Sasai, H., and Shibasaki, M. (1997). Direct catalytic asymmetric aldol reactions of aldehydes with unmodified ketones. *Angew. Chem. Int. Ed. Engl.* 36 (17): 1871–1873.

1657 (a) Trost, B.M. and Ito, H. (2000). A direct catalytic enantioselective aldol reaction via a novel catalyst design. *J. Am. Chem. Soc.* 122 (48): 12003–12004. (b) Trost, B.M., Shin, S.H., and Sclafani, J.A. (2005). Direct asymmetric Zn-aldol reaction of methyl vinyl ketone and its synthetic applications. *J. Am. Chem. Soc.* 127 (24): 8602–8603.

1658 Mahrwald, R. (2000). Titanium(IV) alkoxide ligand exchange with α-hydroxy acids: the enantioselective aldol addition. *Org. Lett.* 2 (25): 4011–4012.

1659 Magdziak, D., Lalic, G., Lee, H.M. et al. (2005). Catalytic enantioselective thioester aldol reactions that are compatible with protic functional groups. *J. Am. Chem. Soc.* 127 (20): 7284–7285.

1660 Fortner, K.C. and Shair, M.D. (2007). Stereoelectronic effects dictate mechanistic dichotomy between Cu(II)-catalyzed and enzyme-catalyzed reactions of malonic acid half thioesters. *J. Am. Chem. Soc.* 129 (5): 1032–1033.

1661 (a) Iwata, M., Yazaki, R., Chen, I.H. et al. (2011). Direct catalytic enantio- and diastereoselective aldol reaction of thioamides. *J. Am. Chem. Soc.* 133 (14): 5554–5560. (b) Kawato, Y., Iwata, M., Yazaki, R. et al. (2011). A simplified catalytic system for direct catalytic asymmetric aldol reaction of thioamides; application to an enantioselective synthesis of atorvastatin. *Tetrahedron* 67 (35): 6539–6546.

1662 (a) Paradowska, J., Stodulski, M., and Mlynarski, J. (2007). Direct catalytic asymmetric aldol reactions assisted by zinc complex in the presence of water. *Adv. Synth. Catal.* 348 (7): 1041–1046. (b) Li, H., Da, C.-S., Xiao, Y.-H. et al. (2008). Direct asymmetric aldol reaction of aryl ketones with aryl aldehydes catalyzed by chiral BINOL-derived zincate catalyst. *J. Org. Chem.* 73 (18): 7398–7401. (c) Chen, X.-H., Zhu, Y., Qiao, Z. et al. (2010). Efficient synthesis of β-hydroxy-α-amino acid derivatives via direct catalytic asymmetric aldol reaction of α-isothiocyanato imide with aldehydes. *Chem. Eur. J.* 16 (33): 10124–10129. (d) Lu, Z.-J., Mei, H.-B., Zhang, G.-Q. et al. (2010). Zinc-prolinamide complex catalyzed direct asymmetric aldol reactions in the presence of water. *Sci. China: Chem.* 53 (11): 2291–2296. (e) Kotani, S., Aoki, S., Sugiura, M., and Nakajima, M. (2011). Trichlorosilyl triflate for enantioselective direct-type aldol reaction with chiral phosphine oxide. *Tetrahedron Lett.* 52 (22): 2834–2836. (f) Zhang, Q.-Y., Cui, X.-L., Zhang, L. et al. (2015). Redox tuning of a direct asymmetric aldol reaction. *Angew. Chem. Int. Ed.* 54 (17): 5210–5213. (g) Wiedenhoeft, D., Benoit, A.:.R., Wu, Y.-B. et al. (2016). Multifunctional heterocyclic scaffolds for hybrid Lewis acid/Lewis base catalysis of carbon–carbon bond formation. *Tetrahedron* 72: 2905–3916. (h) Liu, Z., Takeuchi, T., Pluta, R. et al. (2017). Direct catalytic asymmetric aldol reaction of α-alkylamides. *Org. Lett.* 19 (3): 710–713. (i) Noda, N., Amemiya, F., Weidner, K. et al. (2017). Catalytic asymmetric synthesis of CF3-substituted tertiary propargylic alcohols via direct aldol reaction of α-N-3 amide. *Chem. Sci.* 8 (4): 3260–3269. (j) Matsuzawa, A., Noda, H., Kumagai, N., and Shibazaki, M. (2017). Direct catalytic asymmetric aldol addition of an α-CF3 amide to arylglyoxal hydrates. *J. Org. Chem.* 82 (15): 8304–8308.

1663 Zimmerman, H.E. and Traxler, M.D. (1957). The stereochemistry of the Ivanov and Reformatsky reactions. 1. *J. Am. Chem. Soc.* 79 (8): 1920–1923.

1664 (a) Ivanoff, D. and Spassoff, A. (1931). A method for the preparation of phenyl-, *o*-chlorophenyl-, and *p*-chlorophenylmalonic acid. *Bull. Soc. Chim. Fr.* 49: 19–23. (b) Ivanov, D., Vassilev, G., and Panayotov, I. (1975). Syntheses and reactions of organolithium reagents derived from weakly acidic C—H compounds. *Synthesis* (2): 83–98.

1665 (a) Heathcock, C.H., Buse, C.T., Kleschick, W.A. et al. (1980). Acyclic stereoselection. 7. Stereoselective synthesis of 2-alkyl-3-hydroxy carbonyl compounds by aldol condensation. *J. Org. Chem.* 45 (6): 1066–1081. (b) Heathcock, C.H. (1992). Modern enolate chemistry: regio- and stereoselective formation of enolate configuration on subsequent reactions. In: *Modern Synthetic Methods* (ed. R. Scheffold), 1–102. Basel, Switzerland, Weinheim: Helvetica Chimica Acta-VCH. (c) Liu, C.M., Smith, W.J., Gustin, D.J., and Roush, W.R. (2005). Experimental evidence for chair-like transition states in aldol reactions of methyl ketone lithium enolates: stereoselective synthesis and utilization of a deuterium-labeled enolate as a probe of reaction stereochemistry. *J. Am. Chem. Soc.* 127 (16): 5770–5771.

1666 (a) Streitwieser, A. (2006). Ion pair aggregates and reactions; experiment and theory. *J. Mol. Model.* 12 (5): 673–680. (b) Gruver, J.M., Liou, L.R., McNeil, A.J. et al. (2008). Solution structures of lithium enolates, phenolates, carboxylates, and alkoxides in the presence of $N,N,N'N'$-tetramethylethylenediamine: a prevalence of cyclic dimers. *J. Org. Chem.* 73 (19): 7743–7747. (c) Kolonko, K.J., Wheritt, D.J., and Reich, H.J. (2011). Mechanistic studies of the lithium enolate of 4-fluoroacetophenone: rapid-injection NMR study of enolate formation, dynamics, and aldol reactivity. *J. Am. Chem. Soc.* 133 (42): 16774–16777. (d) Larrañaga, O., de Lózar, A., Bickelhaupt, F.M. et al. (2013). Aggregation and cooperative effects in the aldol reactions of lithium enolates. *Chem. Eur. J.* 19 (41): 13761–13773. (e) Tomasevich, L.L. and Collum, D.B. (2014). Method of continuous variation: characterization of alkali metal enolates using ^{1}H and ^{19}F NMR spectroscopies. *J. Am. Chem. Soc.* 136 (27): 9710–9718. (f) Tallmadge, E.H. and Collum, D.B. (2015). Evans enolates: solution structures of lithiated oxazolidinone-derived enolates. *J. Am. Chem. Soc.* 137 (40): 13087–13095. (g) Tallmadge, E.H., Jermaks, J., and Collum, D.B. (2016). Structure-reactivity relationships in lithiated Evans enolates: influence of aggregation and solvation on the stereochemistry and mechanism of aldol additions. *J. Am. Chem. Soc.* 138 (1): 345–355.

1667 Braun, M. (2015). Lithium enolates: 'Capricious' structures - reliable reagents for synthesis. *Helv. Chim. Acta* 98 (1): 1–31.

1668 (a) Ager, D.J. and East, M.B. (1996). Enolate reactions of carbonyl compounds. In: *Asymmetric Synthetic Methodology*, 97–147. Boca Raton, FL, New York, London, Tokyo: CRC Press, Inc. (b) Corral-Bautista, F. and Mayr, H. (2015). Quantification of the nucleophilic reactivities of cyclic β-keto ester anions. *Eur. J. Org. Chem.* (34): 7594–7601.

1669 Ireland, R.E. and Willard, A.K. (1975). Stereoselective generation of ester enolates. *Tetrahedron Lett.* (46): 3975–3978.

1670 Ireland, R.E., Wipf, P., and Xiang, J.N. (1991). Stereochemical control in the ester enolate Claisen rearrangement. 2. Chairlike vs boatlike transition state selection. *J. Org. Chem.* 56 (11): 3572–3582.

1671 Ireland, R.E., Wipf, P., and Armstrong, J.D. (1991). Stereochemical control in the ester enolate Claisen rearrangement. 1. Stereoselectivity in silyl ketene acetal formation. *J. Org. Chem.* 56 (2): 650–657.

1672 Xie, L.F., Isenberger, K.M., Held, G., and Dahl, L.M. (1997). Highly stereoselective kinetic enolate formation: steric vs electronic effects. *J. Org. Chem.* 62 (21): 7516–7519.

1673 Seebach, D., Amstutz, R., Laube, T. et al. (1985). Structures of three lithium ester enolates by X-ray-diffraction - derivation of reaction-path for cleavage into ketene and alcoholate. *J. Am. Chem. Soc.* 107 (19): 5403–5409.

1674 Mukhopadhyay, T. and Seebach, D. (1982). Substitution of HMPT by the cyclic urea DMPU as a cosolvent for highly reactive nucleophiles and bases. *Helv. Chim. Acta* 65 (1): 385–391.

1675 Godenschwager, P.F. and Collum, D.B. (2008). Lithium hexamethyldisilazide-mediated enolizations: influence of triethylamine on *E/Z* selectivities and enolate reactivities. *J. Am. Chem. Soc.* 130 (27): 8726–8732.

1676 Brown, H.C., Dhar, R.K., Bakshi, R.K. et al. (1989). Major effect of the leaving group in dialkylboron chlorides and triflates in controlling the stereospecific conversion of ketones into either (*E*)-enol or (*Z*)-enol borinates. *J. Am. Chem. Soc.* 111 (9): 3441–3442.

1677 Van Horn, D.E. and Masamune, S. (1979). Stereoselective synthesis of β-hydroxy-α-methylketones via *Z*-vinyloxyboranes and *E*-vinyloxyboranes generated directly from cyclohexyl ethyl ketone. *Tetrahedron Lett.* 20 (24): 2229–2232.

1678 Brown, H.C., Ganesan, K., and Dhar, R.K. (1993). Enolboration. 4. An examination of the effect of the leaving group-(X) on the stereoselective enolboration of ketones with various R2BX/Et3N - new reagents for the selective generation of either *Z*-enol or *E*-enol borinates from representative ketones. *J. Org. Chem.* 58 (1): 147–153.

1679 Masamune, S., Mori, S., Van Horn, D., and Brooks, D.W. (1979). *E*-vinyloxyboranes and *Z*-vinyloxyboranes (alkenyl borinates) - stereoselective formation and aldol condensation. *Tetrahedron Lett.* 20 (19): 1665–1668.

1680 Davies-Coleman, M.T. and Garson, M.J. (1998). Marine polypropionates. *Nat. Prod. Rep.* 15 (5): 477–493.

1681 O'Hagan, D. (1993). *The Polyketide Metabolites*, Ellis Harwood Series in Organic Chemistry, 1–250. Harlow: Ellis Horwood Ltd.

1682 O'Hagan, D. (1995). Biosynthesis of fatty-acid and polyketide metabolites. *Nat. Prod. Rep.* 12 (1): 1–32.

1683 Mahrwald, R. (2009). *Aldol Reactions*, 1–214. Heildelberg, Berlin: Springer-Verlag.

1684 Evans, D.A., Vogel, E., and Nelson, J.V. (1979). Stereoselective aldol condensations via boron enolates. *J. Am. Chem. Soc.* 101 (20): 6120–6123.

1685 Evans, D.A. and Taber, T.R. (1980). Enantioselective aldol condensations via boron enolates - a steric model for asymmetric induction. *Tetrahedron Lett.* 21 (49): 4675–4678.

1686 Evans, D.A., Bartroli, J., and Shih, T.L. (1981). Enantioselective aldol condensations. 2. *erythro*-Selective chiral aldol condensations via boron enolates. *J. Am. Chem. Soc.* 103 (8): 2127–2129.

1687 Evans, D.A. (1982). Studies in asymmetric synthesis. The development of practical chiral enolate synthons. *Aldrichim. Acta* 15: 23–32.

1688 Nagao, Y., Inoue, T., Hashimoto, K. et al. (1985). A facile chiral synthesis of (+)-Prelog–Djerassi lactonic acid methyl ester using 5-membered heterocyclic chiral reagents. *J. Chem. Soc., Chem. Commun.* (20): 1419–1420.

1689 Crimmins, M.T., King, B.W., and Tabet, E.A. (1997). Asymmetric aldol additions with Ti enolates of acyloxazolidinethiones: dependence of selectivity on amine base and Lewis acid stoichiometry. *J. Am. Chem. Soc.* 119 (33): 7883–7884.

1690 Zhang, Y.C., Phillips, A.J., and Sammakia, T. (2004). Highly selective asymmetric acetate aldol reactions of an *N*-acetyl thiazolidinethione reagent. *Org. Lett.* 6 (1): 23–25.

1691 Cardillo, G., Damico, A., Orena, M., and Sandri, S. (1988). Diastereoselective alkylation of 3-acylimidazolidin-2-ones - synthesis of (*R*)-lavandulol and (*S*)-lavandulol. *J. Org. Chem.* 53 (10): 2354–2356.

1692 Drewes, S.E., Malissar, D.G.S., and Roos, G.H.P. (1991). Crystalline, enantiomerically pure aldols from a (-)-ephedrine-derived *N*-acylimidazolidin-2-one. *Chem. Ber.* 124 (12): 2913–2914.

1693 Davies, S.G. and Mortlock, A.A. (1991). Bifunctional chiral auxiliaries. 1. The aldol reaction between dialkylboron enolates of 1,3-dipropionyl-*trans*-4,5-tetramethyleneimidazolidin-2-one and aldehydes. *Tetrahedron Lett.* 32 (36): 4787–4790.

1694 Kim, T.H. and Lee, G.J. (2000). L-valinol and L-phenylalaninol-derived 2-imidazolidinones as chiral auxiliaries in asymmetric aldol reactions. *Tetrahedron Lett.* 41 (10): 1505–1508.

1695 Ghosh, A.K., Duong, T.T., and Mckee, S.P. (1992). Highly enantioselective aldol reaction - development of a new chiral auxiliary from *cis*-1-amino-2-hydroxyindan. *J. Chem. Soc., Chem. Commun.* (22): 1673–1674.

1696 Rück, K. and Kunz, H. (1992). Bicyclic carbohydrate oxazolidinone template for stereoselective 1,4-additions of organoaluminum chlorides to unsaturated carbocylic acid derivatives. *Synlett* 343–344.

1697 Yan, T.H., Tan, C.W., Lee, H.C. et al. (1993). Asymmetric aldol reactions - a novel model for switching between chelation-controlled and non-chelation-controlled aldol reactions. *J. Am. Chem. Soc.* 115 (7): 2613–2621.

1698 Sibi, M.P., Deshpande, P.K., and Ji, J.G. (1995). A new chiral oxazolidinone derived from diphenylalaninol - aldol, alkylation, and Diels–Alder reactions. *Tetrahedron Lett.* 36 (49): 8965–8968.

1699 Hintermann, T. and Seebach, D. (1998). A useful modification of the Evans auxiliary: 4-isopropyl-5,5-diphenyloxazolidin-2-one. *Helv. Chim. Acta* 81 (11): 2093–2126.

1700 Bach, J., Bull, S.D., Davies, S.G. et al. (1999). *N*-acyl-5,5-dimethyl-oxazolidin-2-ones as latent aldehyde equivalents. *Tetrahedron Lett.* 40 (36): 6677–6680.

1701 de Parrodi, C.A., Clara-Sosa, A., Pérez, L. et al. (2001). Highly diastereoselective alkylation, acylation and aldol condensation of *cis*- and *trans*-(*N*-acyloyl)hexahydrobenzoxazolidin-2-ones. *Tetrahedron: Asymmetry* 12 (1): 69–79.

1702 Danda, H., Hansen, M.M., and Heathcock, C.H. (1990). Acyclic stereoselection. 48. Reversal of stereochemistry in the aldol reactions of a chiral boron enolate. *J. Org. Chem.* 55 (1): 173–181.

1703 Evans, D.A., Tedrow, J.S., Shaw, J.T., and Downey, C.W. (2002). Diastereoselective magnesium halide-catalyzed *anti*-aldol reactions of chiral *N*-acyloxazolidinones. *J. Am. Chem. Soc.* 124 (3): 392–393.

1704 Crimmins, M.T., King, B.W., Tabet, E.A., and Chaudhary, K. (2001). Asymmetric aldol additions: use of $TiCl_4$ and (-)-sparteine for the soft enolization of *N*-acyl oxazolidinones, oxazolidinethiones, and thiazolidinethiones. *J. Org. Chem.* 66 (3): 894–902.

1705 Crimmins, M.T. and Dechert, A.M. (2009). Enantioselective total synthesis of (-)-pironetin: iterative aldol reactions of thiazolidinethiones. *Org. Lett.* 11 (7): 1635–1638.

1706 Evans, D.A., Downey, C.W., Shaw, J.T., and Tedrow, J.S. (2002). Magnesium halide-catalyzed *anti*-aldol reactions of chiral *N*-acylthiazolidinethiones. *Org. Lett.* 4 (7): 1127–1130.

1707 Heathcock, C.H., Pirrung, M.C., Buse, C.T. et al. (1979). Acyclic stereoselection. 6. Reagent for achieving high 1,2 diastereoselection in the aldol conversion of chiral aldehydes into 3-hydroxy-2-methylcarboxylic acids. *J. Am. Chem. Soc.* 101 (23): 7077–7079.

1708 Vandraanen, N.A., Arseniyadis, S., Crimmins, M.T., and Heathcock, C.H. (1991). Acyclic stereoselection. 53. Protocols for the preparation of each of the four possible stereoisomeric α-alkyl-β-hydroxy carboxylic acids from a single chiral aldol reagent. *J. Org. Chem.* 56 (7): 2499–2506.

1709 Masamune, S., Choy, W., Kerdesky, F.A.J., and Imperiali, B. (1981). Stereoselective aldol condensation - use of chiral boron enolates. *J. Am. Chem. Soc.* 103 (6): 1566–1568.

1710 Masamune, S., Hirama, M., Mori, S. et al. (1981). Total synthesis of 6-deoxyerythronolide-B. *J. Am. Chem. Soc.* 103 (6): 1568–1571.

1711 Abiko, A., Liu, J.F., and Masamune, S. (1997). The *anti*-selective boron-mediated asymmetric aldol reaction of carboxylic esters. *J. Am. Chem. Soc.* 119 (10): 2586–2587.

1712 Abiko, A., Liu, J.F., Wang, G.Q., and Masamune, S. (1997). New isoxazolidine-based chiral auxiliaries for asymmetric syntheses. *Tetrahedron Lett.* 38 (18): 3261–3264.

1713 Myers, A.G., Widdowson, K.L., and Kukkola, P.J. (1992). Silicon-directed aldol condensation - evidence for a pseudorotational mechanism. *J. Am. Chem. Soc.* 114 (7): 2765–2767.

1714 Ghosh, A.K. and Liu, C.F. (2003). Enantioselective total synthesis of (+)-amphidinolide T1. *J. Am. Chem. Soc.* 125 (9): 2374–2375.

1715 Oppolzer, W., Blagg, J., Rodriguez, I., and Walther, E. (1990). Bornanesultam-directed asymmetric synthesis of crystalline, enantiomerically pure *syn*-aldols. *J. Am. Chem. Soc.* 112 (7): 2767–2772.

1716 Oppolzer, W. (1990). Camphor as a natural source of chirality in asymmetric synthesis. *Pure Appl. Chem.* 62 (7): 1241–1250.

1717 Storer, R.I., Takemoto, T., Jackson, P.S. et al. (2004). Multi-step application of immobilized reagents and scavengers: a total synthesis of Epothilone C. *Chem. Eur. J.* 10 (10): 2529–2547.

1718 Paterson, I., Wallace, D.J., and Velazquez, S.M. (1994). Studies in polypropionate synthesis - high π-face selectivity in *syn*- and *anti*-aldol reactions of chiral boron enolates of lactate-derived ketones. *Tetrahedron Lett.* 35 (48): 9083–9086.

1719 Paterson, I. and Wallace, D.J. (1994). *anti*-Aldol reactions of lactate-derived ketones - application to the total synthesis of (-)-ACRL Toxin-IIIB. *Tetrahedron Lett.* 35 (50): 9477–9480.

1720 Paterson, I., Wallace, D.J., and Cowden, C.J. (1998). Polyketide synthesis using the boron-mediated, *anti*-aldol reactions of lactate-derived ketones: total synthesis of (-)-ACRL, toxin IIIB. *Synthesis* 639–652.

1721 Paterson, I., Goodman, J.M., and Isaka, M. (1989). Aldol reactions in polypropionate synthesis - high π-face selectivity of enol borinates from α-chiral methyl and ethyl ketones under substrate control. *Tetrahedron Lett.* 30 (50): 7121–7124.

1722 Dias, L.C., Polo, E.C., Ferreira, M.A., and Tormena, C.F. (2012). 1,5-Stereoinduction in boron-mediated aldol reactions of *β*,δ-bisalkoxy methylketones containing cyclic protecting groups. *J. Org. Chem.* 77 (8): 3766–3792.

1723 Blanchette, M.A., Malamas, M.S., Nantz, M.H. et al. (1989). Synthesis of Bryostatins. 1. Construction of the C(1)-C(16) fragment. *J. Org. Chem.* 54 (12): 2817–2825.

1724 Paterson, I., Gibson, K.R., and Oballa, R.M. (1996). Remote 1,5-*anti* stereoinduction in the boron-mediated aldol reactions of β-oxygenated methyl ketones. *Tetrahedron Lett.* 37 (47): 8585–8588.

1725 Paterson, I. and Tudge, M. (2003). A fully stereocontrolled total synthesis of (+)-leucascandrolide A. *Tetrahedron* 59 (35): 6833–6849.

1726 Evans, D.A., Coleman, P.J., and Cote, B. (1997). 1,5-Asymmetric induction in methyl ketone aldol addition reactions. *J. Org. Chem.* 62 (4): 788–789.

1727 Evans, D.A., Cote, B., Coleman, P.J., and Connell, B.T. (2003). 1,5-Asymmetric induction in boron-mediated β-alkoxy methyl ketone aldol addition reactions. *J. Am. Chem. Soc.* 125 (36): 10893–10898.

1728 Evans, D.A., Welch, D.S., Speed, A.W. et al. (2009). An aldol-based synthesis of (+)-peloruside A, a potent microtubule stabilizing agent. *J. Am. Chem. Soc.* 131 (11): 3840–3841.

1729 Dias, L.C., Bau, R.Z., de Sousa, M.A., and Zukerman-Schpector, J. (2002). High 1,5-*anti* stereoinduction in boron-mediated aldol reactions of methyl ketones. *Org. Lett.* 4 (24): 4325–4327.

1730 Dias, L.C., Pinheiro, S.M., de Oliveira, V.M. et al. (2009). Addition of kinetic boron enolates generated from β-alkoxy methyl ketones to aldehydes. Density functional theory calculations on the transition structures. *Tetrahedron* 65 (42): 8714–8721.

1731 Arefolov, A. and Panek, J.S. (2002). Studies directed toward the total synthesis of discodermolide: asymmetric synthesis of the C1—C14 fragment. *Org. Lett.* 4 (14): 2397–2400.

1732 Park, P.K., O'Malley, S.J., Schmidt, D.R., and Leighton, J.L. (2006). Total synthesis of dolabelide D. *J. Am. Chem. Soc.* 128 (9): 2796–2797.

1733 Li, P., Li, J., Arikan, F. et al. (2009). Total synthesis of etnangien. *J. Am. Chem. Soc.* 131 (33): 11678–11679.

1734 Li, P., Li, J., Arikan, F. et al. (2010). Stereoselective total synthesis of etnangien and etnangien methyl ester. *J. Org. Chem.* 75 (8): 2429–2444.

1735 Zhang, Y., Arpin, C.C., Cullen, A.J. et al. (2011). Total synthesis of dermostatin A. *J. Org. Chem.* 76 (19): 7641–7653.

1736 Evans, D.A., Dart, M.J., Duffy, J.L., and Yang, M.G. (1996). A stereochemical model for merged 1,2- and 1,3-asymmetric induction in diastereoselective Mukaiyama aldol addition reactions and related processes. *J. Am. Chem. Soc.* 118 (18): 4322–4343.

1737 Denmark, S.E. and Fujimori, S. (2001). Diastereoselective aldol addition reactions of a chiral methyl ketone trichlorosilyl enolate under Lewis base catalysis. *Synlett* 1024–1029.

1738 Denmark, S.E., Fujimori, S., and Pham, S.M. (2005). Lewis base catalyzed aldol additions of chiral trichlorosilyl enolates and silyl enol ethers. *J. Org. Chem.* 70 (26): 10823–10840.

1739 Denmark, S.E. and Fujimori, S. (2005). Total synthesis of RK-397. *J. Am. Chem. Soc.* 127 (25): 8971–8973.

1740 Evans, D.A., Ennis, M.D., Le, T. et al. (1984). Asymmetric acylation reactions of chiral imide enolates - the 1st direct approach to the construction of chiral β-dicarbonyl synthons. *J. Am. Chem. Soc.* 106 (4): 1154–1156.

1741 Evans, D.A., Ng, H.P., Clark, J.S., and Rieger, D.L. (1992). Diastereoselective *anti*-aldol reactions of chiral ethyl ketones - enantioselective processes for the synthesis of polypropionate natural products. *Tetrahedron* 48 (11): 2127–2142.

1742 Evans, D.A., Clark, J.S., Metternich, R. et al. (1990). Diastereoselective aldol reactions using β-keto imide derived enolates - a versatile approach to the assemblage of polypropionate systems. *J. Am. Chem. Soc.* 112 (2): 866–868.

1743 Narasaka, K. and Pai, F.C. (1984). Stereoselective reduction of β-hydroxyketones to 1,3-diols – highly selective 1,3-asymmetric induction via boron chelates. *Tetrahedron* 40 (12): 2233–2238.

1744 Chen, K.M., Hardtmann, G.E., Prasad, K. et al. (1987). 1,3-*syn* diastereoselective reduction of β-hydroxyketones utilizing alkoxydialkylboranes. *Tetrahedron Lett.* 28 (2): 155–158.

1745 Keck, G.E., Wager, C.A., Sell, T., and Wager, T.T. (1999). An especially convenient stereoselective reduction of β-hydroxy ketones to *anti*-1,3-diols using samarium diiodide. *J. Org. Chem.* 64 (7): 2172–2173.

1746 Evans, D.A. and Hoveyda, A.H. (1990). Reduction of β-hydroxy ketones with catecholborane - a stereoselective approach to the synthesis of *syn*-1,3-diols. *J. Org. Chem.* 55 (18): 5190–5192.

1747 Evans, D.A., Chapman, K.T., and Carreira, E.M. (1988). Directed reduction of β-hydroxy ketones employing tetramethylammonium triacetoxyborohydride. *J. Am. Chem. Soc.* 110 (11): 3560–3578.

1748 Paterson, I. and Temal-Laib, T. (2002). Toward the combinatorial synthesis of polyketide libraries: asymmetric aldol reactions with α-chiral aldehydes on solid support. *Org. Lett.* 4 (15): 2473–2476.

1749 Bode, S.E., Wolberg, M., and Müller, M. (2006). Stereoselective synthesis of 1,3-diols. *Synthesis* (4): 557–588.

1750 Dieckmann, M. and Menche, D. (2013). Stereoselective synthesis of 1,3-*anti*-diols by an IPC-mediated domino aldol-coupling/reduction sequence. *Org. Lett.* 15 (1): 228–231.

1751 Corey, E.J. and Kim, S.S. (1990). Versatile chiral reagent for the highly enantioselective synthesis of either *anti* or *syn* ester aldols. *J. Am. Chem. Soc.* 112 (12): 4976–4977.

1752 Ganesan, K. and Brown, H.C. (1994). Enolboration. 6. Dicyclohexyliodoborane, a versatile reagent for the stereoselective synthesis of either *Z*-enolates or *E*-enolates from representative esters. *J. Org. Chem.* 59 (9): 2336–2340.

1753 Abiko, A., Liu, J.F., and Masamune, S. (1996). Concerning the boron-mediated aldol reaction of carboxylic esters. *J. Org. Chem.* 61 (8): 2590–2591.

1754 Abiko, A. (2004). Boron-mediated aldol reaction of carboxylic esters. *Acc. Chem. Res.* 37 (6): 387–395.

1755 Inoue, T., Liu, J.F., Buske, D.C., and Abiko, A. (2002). Boron-mediated aldol reaction of carboxylic esters: complementary *anti-* and *syn-*selective asymmetric aldol reactions. *J. Org. Chem.* 67 (15): 5250–5256.

1756 Paterson, I., Lister, M.A., and McClure, C.K. (1986). Enantioselective aldol condensations - the use of ketone boron enolates with chiral ligands attached to boron. *Tetrahedron Lett.* 27 (39): 4787–4790.

1757 (a) Mukaiyama, T., Banno, K., and Narasaka, K. (1974). New cross-aldol reactions - reactions of silyl enol ethers with carbonyl compounds activated by $TiCl_4$. *J. Am. Chem. Soc.* 96 (24): 7503–7509. (b) Mohamed, S.H., Trabelsi, M.M., and Champagne, B. (2016). Unraveling the concerted reaction mechanism of the non-catalyzed Mukaiyama reaction between *C,O,O*-tris(trimethylsilyl)ketene acetal and aldehydes using density functional theory. *J. Phys. Chem. A* 120 (28): 5649–5657.

1758 Mahrwald, R. (1998). Lewis acid catalysts in enantioselective aldol addition. *Recent Res. Dev. Synth. Org. Chem.* 1: 123–150.

1759 Denmark, S.E. and Stavenger, R.A. (2000). Asymmetric catalysis of aldol reactions with chiral Lewis bases. *Acc. Chem. Res.* 33 (6): 432–440.

1760 Reetz, M.T. (1985). Selective reactions of organotitanium reagents. *Pure Appl. Chem.* 57 (12): 1781–1788.

1761 Reetz, M.T. (1993). Structural, mechanistic, and theoretical aspects of chelation-controlled carbonyl addition reactions. *Acc. Chem. Res.* 26 (9): 462–468.

1762 Reetz, M.T., Kyung, S.H., Bolm, C., and Zierke, T. (1986). Enantioselective C—C bond formation with chiral Lewis acids. *Chem. Ind. (London)* 23: 824–824.

1763 Kobayashi, S. and Mukaiyama, T. (1989). Asymmetric aldol reaction of silyl enol ethers with aldehydes promoted by the combined use of chiral diamine coordinated Sn(II) triflate and tributyltin fluoride. *Chem. Lett.* (2): 297–300.

1764 Kobayashi, S., Uchiro, H., Fujishita, Y. et al. (1991). Asymmetric aldol reaction between achiral silyl enol ethers and achiral aldehydes by use of a chiral promoter system. *J. Am. Chem. Soc.* 113 (11): 4247–4252.

1765 Furuta, K., Maruyama, T., and Yamamoto, H. (1991). Catalytic asymmetric aldol reactions - use of a chiral (acyloxy)borane complex as a versatile Lewis acid catalyst. *J. Am. Chem. Soc.* 113 (3): 1041–1042.

1766 Parmee, E.R., Tempkin, O., Masamune, S., and Abiko, A. (1991). New catalysts for the asymmetric aldol reaction - chiral boranes prepared from α,α-disubstituted glycine arenesulfonamides. *J. Am. Chem. Soc.* 113 (24): 9365–9366.

1767 Evans, D.A., MacMillan, D.W.C., and Campos, K.R. (1997). C_2-symmetric Sn(II) complexes as chiral Lewis acids. Catalytic enantioselective *anti*-aldol additions of enolsilanes to glyoxylate and pyruvate esters. *J. Am. Chem. Soc.* 119 (44): 10859–10860.

1768 Mikami, K. and Matsukawa, S. (1993). Enantioselective and diastereoselective catalysis of the Mukaiyama aldol reaction - ene mechanism in Ti-catalyzed aldol reactions of silyl enol ethers. *J. Am. Chem. Soc.* 115 (15): 7039–7040.

1769 Yamashita, Y., Ishitani, H., Shimizu, H., and Kobayashi, S. (2002). Highly *anti*-selective asymmetric aldol reactions using chiral Zr catalysts. Improvement of activities, structure of the novel Zr complexes, and effect of a small amount of water for the preparation of the catalysts. *J. Am. Chem. Soc.* 124 (13): 3292–3302.

1770 Denmark, S.E., Stavenger, R.A., Wong, K.T., and Su, X.P. (1999). Chiral phosphoramide-catalyzed aldol additions of ketone enolates. Preparative aspects. *J. Am. Chem. Soc.* 121 (21): 4982–4991.

1771 (a) Denmark, S.E., Wynn, T., and Beutner, G.L. (2002). Lewis base activation of Lewis acids. Addition of silyl ketene acetals to aldehydes. *J. Am. Chem. Soc.* 124 (45): 13405–13407. (b) Denmark, S.E., Beutner, G.L., Wynn, T., and Eastgate, M.D. (2005). Lewis base activation of Lewis acids: catalytic, enantioselective addition of silyl ketene acetals to aldehydes. *J. Am. Chem. Soc.* 127 (11): 3774–3789.

1772 Crimmins, M.T. and Slade, D.J. (2006). Formal synthesis of 6-deoxyerythronolide B. *Org. Lett.* 8 (10): 2191–2194.

1773 Heathcock, C.H. (1990). Understanding and controlling diastereofacial selectivity in carbon-carbon forming reactions. *Aldrichim. Acta* 23: 99–111.

6

Organic photochemistry

6.1 Introduction

Organic photochemistry plays a very important role in preparative chemistry, biology, medicine, material sciences, and chemical mechanistic studies [1, 2]. Photochemical reactions are often applied as key steps in the synthesis of complicated organic compounds [3–9]. Special flow and microflow reactors have been developed that have the advantage over conventional batch reactors to permit more efficient light penetration, better temperature control, and easy removal of photoproducts from the irradiated area where photodecomposition might otherwise occur [10–13].

Absorption of light produces an electronically excited state with a greater energy content corresponding to the quantum of energy in the absorbed photon. The electronic energy can be restored to the medium in different forms without net structural change of the compound; these are called photophysical processes. Alternatively, structural changes may occur, and these are called photochemical processes [14–17]. Photochemical activation often occurs without additional reagents, which reduces formation of byproducts, an advantage in the context of green chemistry [18–20]. More and more reactions can be run with sunlight as a renewable energy source [21, 22]. The latter are particularly interesting when they are photocatalytical, i.e. require less than one photon per molecule to be transformed (quantum yield > 100%, e.g. visible light redox photocatalysis) [23–27]. Over 100 years ago, in 1912, Ciamician suggested that one day chemical factories made of glass would be installed in the desert, without smokestacks and without smoke [28, 29].

In 1825, Faraday found that sunlight promoted chemical reactions of chlorine (Cl_2), including the addition to ethylene to form 1,2-dichloroethane and the addition to benzene to give 1,2,3,4,5,6-hexachlorocyclohexane [30]. Later, Draper and coworkers showed that light initiates a violent reaction between H_2 ad Cl_2 producing HCl. Together with Grotthuss, he established the Grotthuss–Draper law which states that light must be absorbed by a substance in order for a photochemical reaction to take place [31]. In 1877, Downes and Blunt showed that bacteria could be killed by sunlight and that the effects varied when different filters were used [32]. They also found that oxalic acid was destroyed by sunlight and that O_2 plays an active role in the photodecomposition process. They demonstrated the wavelength dependence of organic photochemistry [33]. In 1879, they reported the sunlight decomposition of H_2O_2 and interpreted the results by invoking the formation of hydroxyl radical intermediate, $HO^•$ [33], as was demonstrated to be the case in 1952 [34]. The first organic photochemical reaction was reported by Trommsdorff in 1834 when he described how crystals of α-santonin (**1**) turn yellow and "burst" when exposed to sunlight [35, 36]. After absorption of a photon, a formal type I dyotropic rearrangement **1** → **4** occurs (Section 5.6.1) that implies the 1,2-shifts of an alkyl group and of a hydrogen atom (Scheme 6.1). This rearrangement can be visualized as a stepwise process **1** → **2** → **3** → **4** → **5**. In the crystal, the yellow, antiaromatic cyclopentadienone **5** waits for another molecule **5** to be formed and brought nearby. These two cyclopentadienones **5** undergo a quick Diels–Alder reaction producing **6**. Absorption of another photon induces a $[\pi^2s+\pi^2s]$-cycloaddition giving **7** that distorts the crystals and leads to their fracture, often quite dramatically [37].

6.2 Photophysical processes of organic compounds

The color of a compound in solution results from the absorption of part of the visible light (wavelength $\lambda = 400–800$ nm, wavenumber $1/\lambda = 25\,000$ to $12\,500\ cm^{-1}$, $\Delta E = 71.4$ to $35.7\ kcal\ mol^{-1}$) spectrum. Absorption of violet (λ = 400 nm), blue (450 nm),

Organic Chemistry: Theory, Reactivity and Mechanisms in Modern Synthesis, First Edition. Pierre Vogel and Kendall N. Houk.

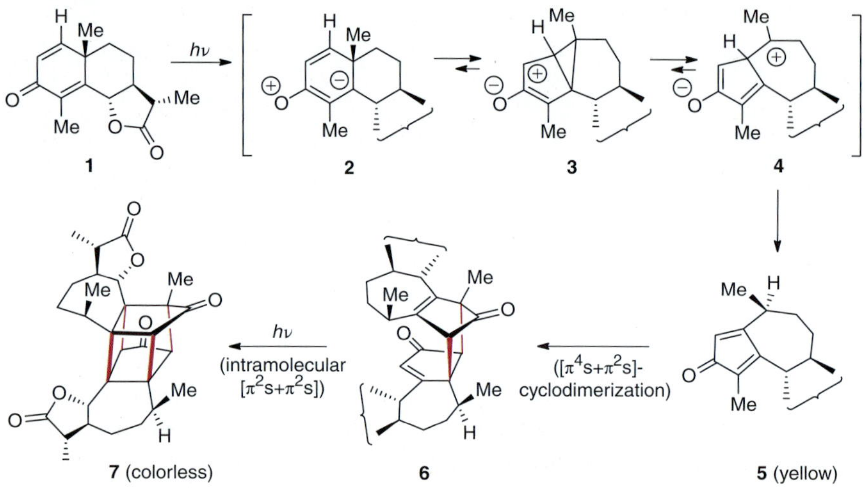

Scheme 6.1 The sunlight-induced photoreaction of crystalline α-santonin.

Table 6.1 Radiation spectrum (1 kcal mol^{-1} = 4.18 kJ mol^{-1} ≅ 0.0434 eV/molecule = 350 cm^{-1} = 1.05 × 10^{13} Hz).

γ-Rays	X-rays	Vacuum UV	Near UV	Visible	Near infrared	Infrared	Microwaves	Radio
λ (nm)	0.01	100	200	400	800	5000	10^7	10^9
ν (Hz)	3×10^{19}	3×10^{15}	1.5×10^{15}	7.5×10^{14}	3.75×10^{14}	6×10^{13}	3×10^{10}	3×10^{8}
$1/\lambda$ (cm^{-1})		10^5	5×10^4	2.5×10^4	12.5×10^4	2000	1	0.01
ΔE (kcal mol^{-1})		286	143	71.4	35.7	5.7	2.9×10^{-3}	2.9×10^{-5}

blue-green (500 nm), yellow (550 nm), orange (600 nm), or red (700 nm) light renders a substance yellow, orange, red, violet, blue-green, or green, respectively. The visible spectrum constitutes only a small part of the total radiation spectrum (Table 6.1). A compound absorbing UV light ($\lambda < 400$ nm) is colorless. Organic photochemistry uses UV light ($\lambda > 180$ nm: quartz vessel; $\lambda > 210$ nm: Vycor vessel; $\lambda > 280$ nm: Pyrex (borosilicate) vessel) and visible light. Alkanes as well as simple fluoro- and chloroalkanes absorb with $\lambda < 180$ nm. They can be used as solvents of photochemical processes using monochromatic (e.g. low-pressure quartz Hg arc lamps and He lamps) or polychromatic UV light (e.g. high-pressure Hg burners) and quartz vessels [38].

6.2.1 UV–visible spectroscopy: electronic transitions

The absorbance of sample solution at a particular wavelength λ is $A(\lambda) = \log I_0 - \log I$, where I_0 is the intensity of the reference beam (entering into the sample solution) and I the intensity of the beam that comes out of the sample solution. This is measured by UV–visible spectrometers. For diluted solutions, $A(\lambda)$ is proportional to the concentration [M] of solute **M** and length ℓ of the solution crossed by the beam (Beer–Lambert law). One defines the molar absorptivity or molar extinction coefficient as $\varepsilon = A/[\mathrm{M}]\ell$, where [M] is in mol l^{-1} (M^{-1}) and ℓ in centimeter (cm). In methanol solution, 2,5-dimethylhexa-2,4-diene has a maximum of absorbance at $\lambda_{max} = 242.5$ nm and molar extinction coefficient $\varepsilon = 13\,000$. The λ_{max} and ε values define the electronic transition and characterize the chromophore [39, 40]. They are associated with the electronic excitation of the compound from its ground state ψ_0 to an electronically excited state ψ^*. Although the energy absorption by a molecule is quantized and corresponds to $\Delta E = h\nu$ (one photon) with h the Planck constant and ν the light frequency, a UV or visible spectrum does not consist of sharp lines but consists of broad bands (with a Franck–Condon contour) over a wide range of wavelengths (Figure 6.1). At room temperature, most molecules are in their lowest vibrational level v_0; very few are distributed (Boltzmann relationship (1.18), Section 1.4) among excited vibrational states. As the time required for the electronic transition is c. 10^{-16} seconds, much shorter than the time required for a geometrical change of the molecules ($>10^{-13}$ seconds), the electronic excitation converts the molecules in their ground state ψ_0 into molecules that reside in several vibrational levels,

v'_0, v'_1, v'_2, v'_3, v'_4, ... (and rotational levels), of the electronically excited state ψ^*. All these excited molecules have the same geometries as those of the molecules in their initial ground state ψ_0, just before the photon absorption. This is the Franck–Condon principle (Section 1.8) [41–43]. The electronic transition of minimal energy corresponds to the excitation of molecules from the v_0 level of ψ_0 to the v'_0 level of ψ^*. This transition, called the adiabatic transition, is not always available as the geometries of these two states may not be the same (do not coincide: Figure 6.1b). The molar extinction coefficient ε is proportional to the probability of the transition that is given by the product $\psi_0 \cdot \psi^*$. The latter can be essentially zero if these two functions are orthogonal for reasons of symmetry and/or geometry. One speaks of "allowed" transition when it is not the case and of "forbidden transition" when it is the case (see below). The wavy lines drawn with the vibrational levels (Figure 6.1) give the probability to find a molecule with the geometries available for this level.

The vacuum UV spectra of alkanes such as ethane and cyclopropane start at $\lambda < 152$ nm and $\lambda < 172$ nm, respectively. The electronic transitions are associated with valence $\sigma \to \sigma^*$ and Rydberg $\sigma \to 3s$(carbon) transitions [44, 45]. Fluoro- and chloroalkanes absorb below 200 nm and their electronic transitions are assigned to $n \to \sigma^*$, $n \to 3s$(F) or $n \to 4s$(Cl), $n \to 4p,3d$, and $\sigma \to \sigma^*$ transitions [46, 47]. The $n \to \sigma^*$ transition is of lower energy for iodoalkanes than for bromoalkanes, chloroalkanes, and fluoroalkanes [48, 49]. This is explained by considering the C—X bond strength that decreases from C—F to C—I, and thus, the LUMO(C—X) decreases from C—F to C—I. The lowest energy electronic transition of alcohols, ethers, and amines (Figure 6.2) also correspond to $n \to \sigma^*$ transitions. The vacuum UV absorption spectrum of ethylene starts at $\lambda = 180$ nm with $\lambda_{max} = 165$ nm ($\Delta E = 173.3$ kcal mol^{-1}) [50]. Calculations show that this spectrum is not due uniquely to the $\pi \to \pi^*$ excitation (valence bands) but is overridden by Rydberg bands that arise from the $\pi \to 3s$(carbon) excitation [50, 51].

As seen for the ground state and electronically excited states of cyclobutene (Figure 5.2) and of *s-cis*-butadiene (Figure 5.3, Section 5.2.2), these functions can be approximated by combinations (variation theorem, Section 4.3, Eq. (4.5)) of electronic configurations [52]. The function that describes the ground state of ethylene, $\psi_0(CH_2{=}CH_2)$, can be approximated by the ground-state configuration:

$$\phi_0 = [\sigma_1(C-H)]^2[\sigma_2(C-H)]^2[\sigma_3(C-H)]^2 [\sigma_4(C-H)]^2[\sigma(C-C)]^2[\pi(C=C)]^2$$

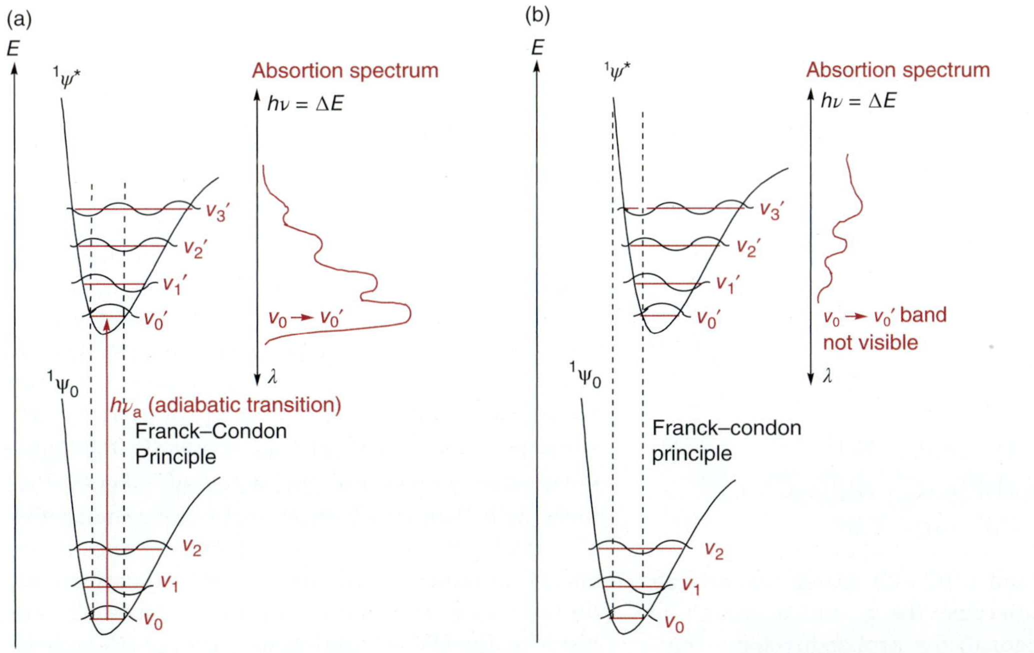

Figure 6.1 The Franck–Condon principle explained for a diatomic molecule: (a) absorption spectrum for a molecule for which the geometries of its ground state and its electronically excited states coincide: λ_{max} (with the largest ε value) corresponds to the adiabatic transition $\psi_0(v_0) \to \psi^*(v'_0)$ and (b) absorption spectrum of a molecule for which the geometries of ψ_0 and ψ^* are different: the band of maximal λ (the lowest energy band) is not the most intense and corresponds to a vertical transition $\psi_0(v_0) \to \psi^*(v'_{i>0})$.

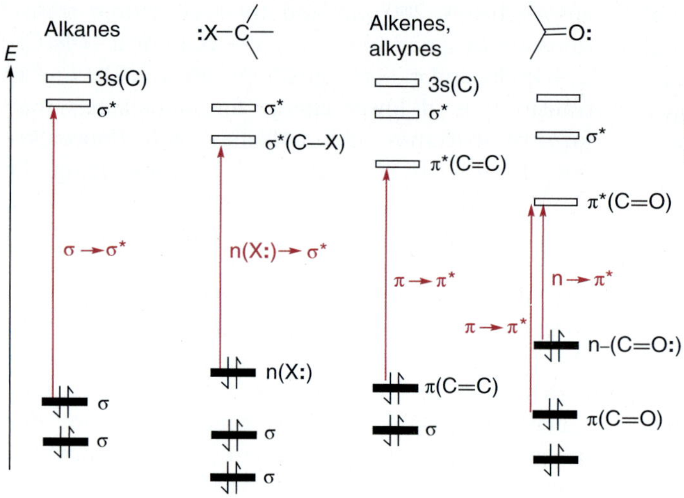

Examples of chromophores	λ_{max} (nm)	ε ($M^{-1}\,cm^{-1}$)
CH_3—CH_3	133	10 000
CH_2=CH_2	165	1 600
CH_3Cl	173	200
CH_3OH	177	200
n-HexCH=CH_2	177	12 600
n-HexC≡CH	185	2 000
Et_3N:	199	3 950
Et—Br	208	300
Me—I	259	400
Me_2C=O	280 187	12

Figure 6.2 FMO model for the representation of the valence electronic transitions of alkanes, haloalkanes, alcohols, ethers, amines, alkenes, alkynes, and ketones.

where σ_1(C—H), σ_2(C—H), σ_3(C—H), and σ_4(C—H) are C—H bond localized orbitals distributing eight electrons of the four σ(C—H) bonds, σ(C—C) is the C—C bond localized orbital distributing two electrons (mixes with the σ(C—H) localized orbitals), and π(C=C) is the highest occupied molecular orbital (HOMO) of ethylene (does not mix with the σ-orbitals as they are orthogonal). When a photon is absorbed by ethylene, one of its electrons is promoted into an empty orbital. The lowest energy configuration of the first excited states $\psi^*_1(CH_2{=}CH_2)$ is

$$\phi_1 = [\sigma_1(C-H)]^2[\sigma_2(C-H)]^2[\sigma_3(C-H)]^2 [\sigma_4(C-H)]^2[\sigma(C-C)]^2[\pi(C=C)]^1 [\pi^*(C=C)]^1$$

where π^*(C=C) is the lowest unoccupied molecular orbital (LUMO) of ethylene. Other excited configurations have similar energy, e.g.

$$\phi_2 = [\sigma_1(C-H)]^2[\sigma_2(C-H)]^2[\sigma_3(C-H)]^2 [\sigma_4(C-H)]^2[\sigma(C-C)]^2[\pi(C=C)]^1 [\sigma^*_5(C-H)]^1 \text{and}$$

$$\phi_3 = [\sigma_1(C-H)]^2[\sigma_2(C-H)]^2 [\sigma_3(C-H)]^2[\sigma_4(C-H)]^2[\sigma(C-C)]^2 [\pi(C=C)]^1[\sigma^*(C-C)]^1$$

where σ^*_5(C—H) and σ^*(C—C) are empty orbitals of the σ-skeleton. Because the π- and σ-orbitals of ethylene are orthogonal, the probability for a transition doing the promotion of an electron from the HOMO(ethylene) into a σ^* empty orbital (transition noted $\pi \rightarrow \sigma^*$) is much weaker than for the transition π(C=C) → π^*(C=C) (noted $\pi \rightarrow \pi^*$). Thus, configuration ϕ_1 is a good approximation of its first electronically excited state $\psi^*_1(CH_2{=}CH_2)$. As a consequence, the energy difference between π(C=C) and π^*(C=C) ($E_{LUMO} - E_{HOMO}$) is correlated with energy $\Delta E = h\nu$ of the electronic transition $\psi_0(\nu_0) \rightarrow \psi^*(\nu'_0)$ in ethylene. Excited configurations involving 3s(C) orbitals have to be added to the list of excited configurations constructed above from the bond orbitals for a more accurate description of the electronic transitions of ethylene.

On substituting ethylene by a methyl group, as in propene (λ_{max} = 170 nm, ε = 20 000, ΔE = 168.2 kcal mol^{-1}), and by two methyl groups, as in (*E*)-but-2-ene (λ_{max} = 174 nm, ε = 24 000, ΔE = 164.4 kcal mol^{-1}), the energy of the electronic transition decreases by 4–5 kcal/mol per methyl group (Table 6.2) [53]. This effect is called a "bathochromic shift." The opposite effect is called a "hypsochromic shift." Hyperconjugation π(C=C)/σ(CH_3) rises the HOMO(alkene) energy with respect to HOMO(CH_2=CH_2), and the interaction π^*(C=C)/σ^*(CH_3) lowers the energy of LUMO (alkene) with respect to that of LUMO(CH_2=CH_2). With this simple model, one explains the highly polarizable singlet excited states of alkenes [54] (Scheme 6.2) and the bathochromic shifts induced by alkyl and other substituents (stabilizing substituent effects on radicals, carbenium ions, and carbanions) on the $\pi \rightarrow \pi^*$ transition of ethylene (Table 6.2). For instance, the UV absorption spectrum of (*E*)-stilbene shows λ_{max} = 295 nm (ΔE = 97 kcal mol^{-1}). The UV absorption spectrum of (*Z*)-stilbene shows λ_{max} = 280 nm (ΔE = 102 kcal mol^{-1}). This is due to the fact that conjugation between the phenyl

Table 6.2 UV–visible absorption spectra of selected unsaturated compounds.

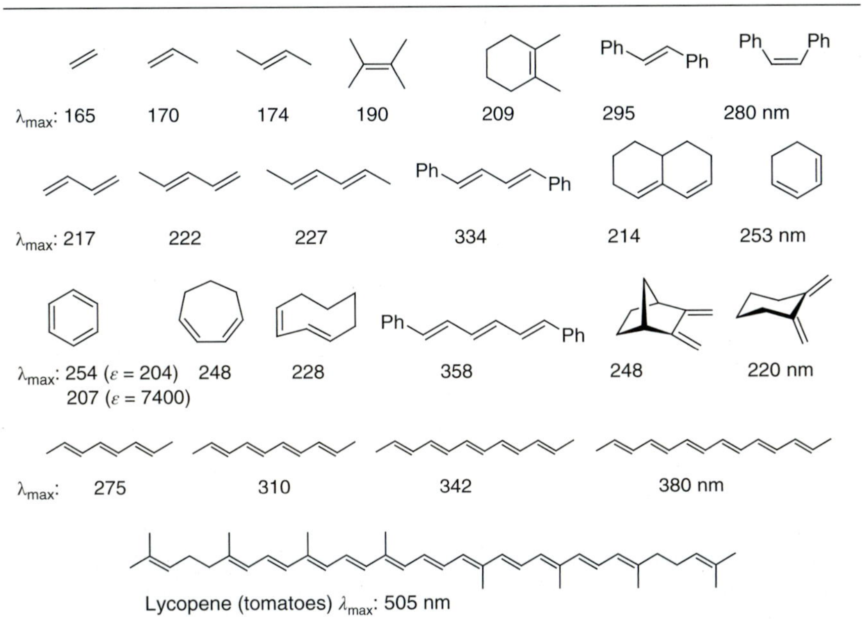

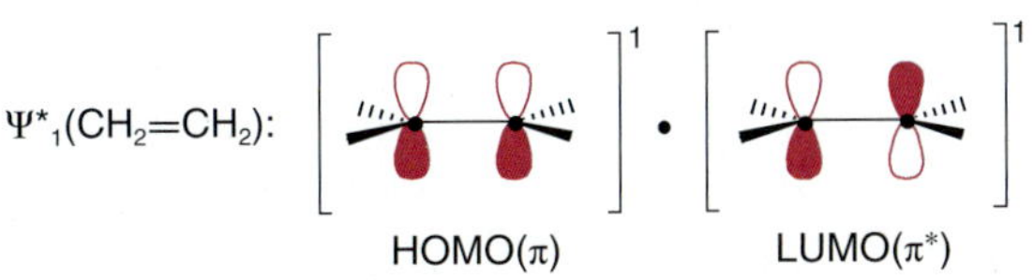

Scheme 6.2 Simple model for the valence electronically excited state of alkenes.

substituents and the ethylene moiety is reduced in the partially twisted (*Z*)-stilbene (gauche interactions between the Ph rings) compared with planar (*E*)-stilbene. This simple model can be applied to all other chromophores as shown below with the Woodward–Fieser rules (Table 6.4).

Simple aliphatic aldehydes and ketones display two absorptions [55]. The lowest energy one in the near-UV region corresponds to a n(CO) → π*(CO) transition ($\lambda_{max} \approx 290$ nm in RCHO, R = Me, Et, *i*-Pr, *n*-Bu; 278 nm in MeCOMe; 285 nm in EtCOEt) and the second one of type π(CO) → π*(CO) is at $\lambda_{max} < 200$ nm (187 nm for MeCOMe). The extinction coefficient ε is usually small for the n(CO) → π*(CO) transitions ("forbidden" transitions) as they imply orthogonal frontier molecular orbitals (FMOs), which is not the case with the higher energy π(CO) → π*(CO) transitions (Figure 6.2), which are "allowed" transitions. A similar situation exists for haloalkanes, alcohols, ethers, and amines for which the low-energy n(X:) → σ* transitions are associated with small ε values, whereas the high energy $\sigma \rightarrow \sigma^*$ transitions have much larger ε values.

The UV absorption spectra of acetaldehyde, acetone, acetyl chloride, acetic anhydride, acetamide, methyl acetate, and acetic acid show increasing energies (lower λ_{max} values) for their n(CO) → π*(CO) transitions (Figure 6.3). Hyperconjugation Me/CO and n(CO)/Me can explain the increase of the LUMO/HOMO energy gap going from acetaldehyde to acetone. The same interpretation can be retained for the other carbonyl compounds.

The UV absorption spectrum of butadiene displays a $\lambda_{max} = 217$ nm ($\varepsilon = 21\,000$ M^{-1} cm^{-1}), which corresponds to a stabilization of the excited state of ethylene by c. 40 kcal mol^{-1} for its vinyl substitution. This result is explained by considering the lower energy gap between LUMO(butadiene)/HOMO(butadiene) compared with LUMO(ethylene)/HOMO(ethylene) (Sections 4.5.2 and 4.5.5). As expected, substitution of butadiene as in (*E*)- and (*Z*)-penta-1,3-diene induces a bathochromic shift of c. 5 nm (Table 6.2). (*E,E*)-Hexa-2,4-diene shows λ_{max} = 227 nm (ε = 24 000). The $\pi \rightarrow \pi^*$ valence excitation bands of conjugated *s-cis*-dienes usually requires less energy than for similarly substituted *s-trans*-dienes (compare the spectra of (*E,E*)-hexa-2,4-diene (λ_{max} = 227 nm) and (*E,Z*)-cycloocta-1,3-diene (λ_{max} = 228 nm)) with those of cyclohexa-1,3-diene (λ_{max} = 253 nm) and cyclohepta-1,3-diene (λ_{max} = 248 nm). If the *s-cis*-butadiene moiety deviates from planarity, this leads to a reduced conjugation between two ethylene units, and thus to a higher LUMO/HOMO energy gap, and thus to a higher $\pi \rightarrow \pi^*$ valence

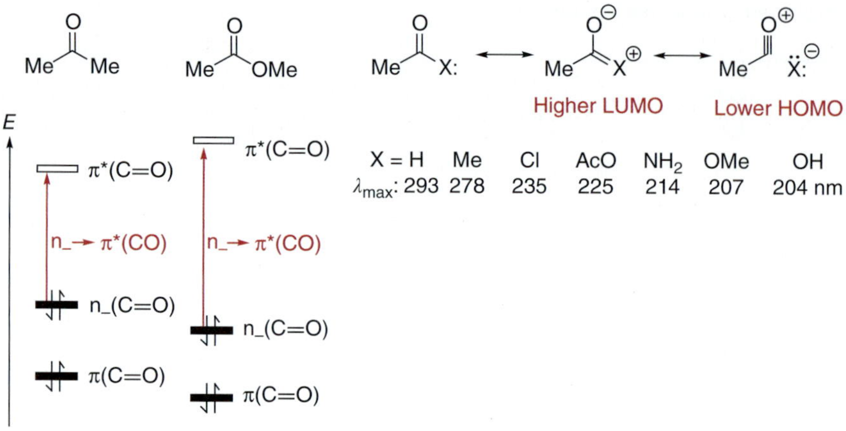

Figure 6.3 UV absorption spectra of various carbonyl compounds.

excitation energy (smaller λ_{max} values). For instance, 1,2-dimethylidenecyclohexane that adopts a chair conformation for its six-membered ring has λ_{max} = 220 nm, whereas 2,3-dimethylidenebicyclo[2.2.1]heptane that maintains its butadiene moiety planar shows λ_{max} = 248 nm. The more olefinic units are conjugated, the higher the λ_{max} value in their UV–visible absorption spectra. For (*E,E,E*)-octa-2,4,6-triene, λ_{max} = 275 nm (ε = 50 000). Lycopene, the compound responsible for the red color of tomatoes, and which contains 11 conjugated ethylenic moieties, has λ_{max} = 505 nm [56]. Methyl vinyl ketone (Table 6.3), a conjugated enone, shows two maxima in its UV absorption spectrum associated with the "forbidden" n(CO) → π*(CO) transition (λ_{max} = 320 nm (ε = 24)) and the "allowed" π(CO) → π*(CO) (λ_{max} = 219 nm (ε = 3600)). Because the corresponding n(CO) → π*(CO) transition in (*E*)-but-2-enal has a very small ε value, its UV absorption spectrum displays only one maximum at λ_{max} = 217 nm (compare with the spectrum of butadiene and (*E*)-penta-1,3-diene; Table 6.2). The UV absorption spectra of (*E*)-but-2-enal, *all-trans*-hexa-2,4-dienal, octa-2,4,6-trienal, deca-2,4,6,8-tetraenal, and dodeca-2,4,6,8,10-pentaenal (Table 6.3) see their λ_{max} values increase with the number of conjugated ethylene units, in analogy with $\pi \rightarrow \pi$* valence excitation bands of conjugated polyenes (Table 6.2).

Table 6.3 UV absorption spectra of conjugated enones and enals.

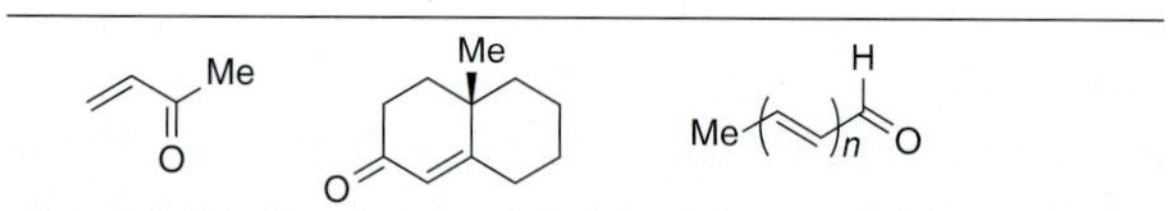

	Methyl vinyl ketone	Bicyclic enone	n = 1	2	3	4	5
λ_{max}:	320 (ε = 24)	310 (ε = 100)					
	219 (3600)	242 (18000)	217	270	312	343	370 nm

Benzene and other aromatic compounds exhibit more complex UV absorption spectra that are not explained by simple $\pi \rightarrow \pi$* transitions. The complexity arises from the existence of several low-lying excited states. For benzene, one moderate intensity band occurs at λ_{max} = 204 nm (ε = 7400) and a less intense band appears at λ_{max} = 254 nm (ε = 204) (Table 6.4). Annulation of benzene with one benzene ring generates naphthalene (λ_{max} = 280 nm) and then anthracene (λ_{max} = 375 nm) or phenanthrene (λ_{max} = 350 nm). Linear annulation of anthracene with benzene generates naphthacene (λ_{max} = 375 nm), which is yellow, and then pentacene (λ_{max} = 575 nm), which is blue. Triple annulation of benzene with benzo groups generates coronene (λ_{max} = 400 nm), which is yellow.

Woodward and several authors have found that the substituent effects on the λ_{max} of a given chromophore are additive. This has led to the Woodward–Fieser rules for calculating the λ_{max} values as summarized in Table 6.4 [57–60].

6.2.2 Fluorescence and phosphorescence: singlet and triplet excited states

After UV–visible light absorption, the compound in its electronically and vibrationally excited state $\psi^*(v'_1, v'_2, v'_3, v'_4, \ldots)$ will relax into $\psi^*(v'_0)$ in a few picoseconds (10^{-12} seconds). From here, the compound can give off its excess of energy as heat to the surroundings, (internal conversion) with typical rates of 10^{11}–10^{12} s^{-1}, or emit light. There are two types of emissions: fluorescence (photons $h\nu_f$) with rate constants k_f in the range 10^{11} to 10^6 s^{-1} and phosphorescence (photons $h\nu_p < h\nu_f$) with rate constant k_p in the range 10^3 to 10^{-1} s^{-1} (measured by time-resolved emission spectroscopy: this technique permits one to measure processes that occur on time scales as short as 10^{-15} s) [61, 62].

Table 6.4 Woodward–Fieser rules for the calculation of the λ_{max} of conjugated π-systems: λ_{max} = base value + sum of substituent increments.

Chromophore	Base value	Substituent increments	
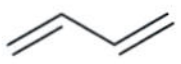	217 nm	extended C=C:	+ 30
		alkyl	5
		Cl, Br	5
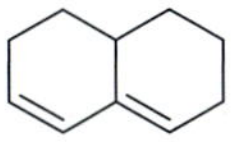	214	OR (ether)	6
		OCOR (ester)	0
		SR (thioether)	30
	253	NR_2 (dialkylamino)	60

Chromophore	X	Base value	position:	α	β	γ	δ
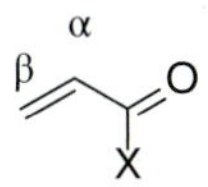	X = H	210 b)	alkyl	+ 10	12	18	18
	X = alkyl	215	Cl	15	12	12	12
	X = OR	195	Br	25	30	30	30
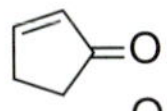		202	OH	35	30	50	
			OR	35	30	30	
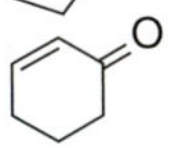		215	OCOR	6	6	6	
			SR		85		
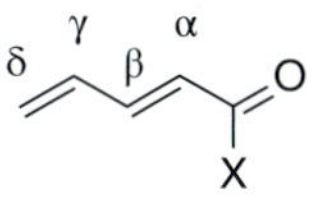	X = H	240	NR_2		95		
	X = alkyl	245	vinyl				30
	X = OR	225	phenyl				60

Chromophore	Y = H	Me	Cl	Br	OH	OMe	NH_2	CN	CO_2H	CHO	NO_2
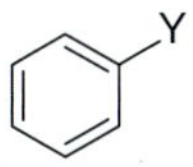	204	207	210	210	211	217	230	224	230	250	269
	254	261	264	261	270	269	280	271	273		

Chromophore	Y = H		position: Y = alkyl,	OH,	OR,	Cl,	Br,	NH_2,	NMe_2
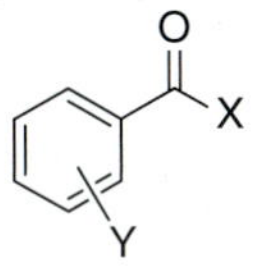	X = H	250	*ortho* + 3	7	0	2	13	20	20
	X = alkyl	246	*meta* 3	7	0	2	13	20	20
	X = OH,OR	230	*para* 10	25	10	15	58	85	45

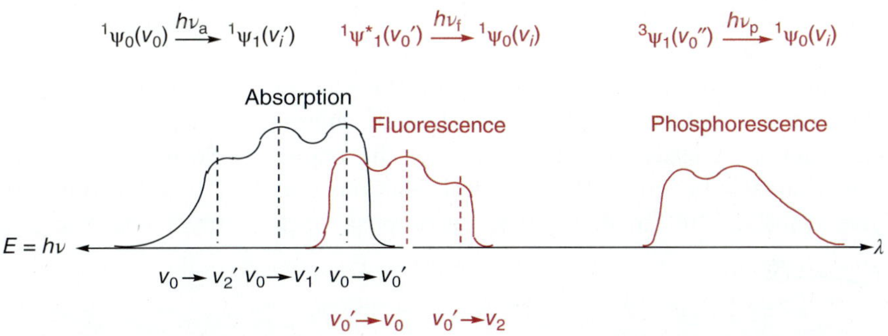

Figure 6.4 UV–visible light absorption, fluorescence, and phosphorescence.

Fluorescence is a spin "allowed" electronic transition, therefore quick, between an excited and a ground state having the same spin state (singlet or triplet). Phosphorescence is a spin "forbidden" electronic transition, therefore slow, between an excited and a ground state with different spin states. Typically, for an organic compound, absorption of a photon ($h\nu_a$) converts singlet $^1\psi_0(\nu_0)$ into $^1\psi^*_1(\nu'_0, \nu'_1, \nu'_2, \nu'_3, \nu'_4, \ldots)$ (superscript 1 for singlet spin state, all electron pairs have opposite or antiparallel spins). The latter relaxes into $^1\psi^*_1(\nu'_0)$ and then it decays into $^1\psi_0$ producing heat, or by fluorescence emitting a photon $h\nu_f$, usually of lower energy than photon $h\nu_a$. The lifetime of the first singlet excited state depends on the rate (k_{ISC}) of its intersystem crossing (ISC) $^1\psi^*_1(\nu'_0) \rightarrow {}^3\psi^*_1(\nu''_i)$ (see below), on the rate of fluorescence (k_f) and the rate of nonradiative decay (internal conversion) into the ground state of the molecule (k_{IC}). The latter can be affected by the medium and by compounds that can dissipate its energy through different mechanisms such as electron transfer (ET: Section 6.10) and/or formation of transient intermediates that decompose back into the starting compounds (mono- or bimolecular processes: nonproductive photochemistry) or into products of photoisomerization, photocondensation, or photofragmentation (Sections 6.3–6.9). The absorption and fluorescence spectra might have one band in common if the lowest energy transition ${}^1\psi_0(\nu_0) \rightarrow {}^1\psi^*{}_1(\nu'_0)$ is available (Franck–Condon principle); thus, transition ${}^1\psi^*{}_1(\nu'_0) \rightarrow {}^1\psi_0(\nu_0)$ can occur (Figure 6.4). If not, all bands $h\nu_f < h\nu_a$. In electronically excited states $\psi^*{}_I$, two electrons occupy two different MOs. For the lowest energy transition ${}^1\psi_0(\nu'_0) \rightarrow {}^1\psi^*{}_1(\nu'_0, \nu'_1, \nu'_2, \nu'_3, \nu'_4, \ldots)$, these are the HOMO and LUMO (Figures 6.2 and 6.3). Thus, according to the Pauli exclusion principle [63], the electronically excited states can also exist as triplet spin state (the two unshared electrons have their spin parallel). The lowest energy triplet excited state ${}^3\psi^*{}_1(\nu''_0)$ (superscript ^{3} for triplet spin state) is always lower in energy than the corresponding singlet excited state ${}^1\psi^*{}_1(\nu'_0)$ as a consequence of the Pauli exclusion principle that keeps the electrons of ${}^3\psi^*{}_1(\nu''_0)$ apart and leads to the Fermi–Coulomb hole [40]. Thus, the phosphorescence photons have energies $h\nu_p < h\nu_f$ and the rate of their emission is low because the electronic transition ${}^3\psi^*{}_1(\nu''_0) \rightarrow {}^1\psi_0(\text{v}_i)$ is spin "forbidden." The lifetime of an excited triplet state depends on the structure of the compound and on the medium that affect the rate (k_P) of phosphorescence, the rate (k_{NR}) of nonradiative decay into the ground state of the molecule, and the rate (k_c) of chemical reaction it leads to. Deactivation of the triplet states can also be induced by chemical reactions that give transient intermediates that decompose back into the starting compounds (nonproductive photochemistry) or a photochemical reaction [64, 65]. The fluorescence and phosphorescence quantum yields are defined as Φ_f = number of $h\nu_f$ photons emitted divided by the number of molecules electronically excited, and Φ_p = number of $h\nu_p$ photons emitted divided by the number of molecules electronically excited. For many organic compounds, spin flipping ${}^1\psi^*{}_1(\nu'_0) \rightarrow {}^3\psi^*{}_1(\nu''_i)$ is a slow process, as ISC is quantum mechanically forbidden to first order because singlet and triplet wave functions are orthogonal. Converting ${}^1\psi^*{}_1(\nu'_0)$ into ${}^3\psi^*{}_1(\nu''_i)$ involves a change in spin angular momentum. This is made easier if the spin momentum can be coupled with the orbital angular momentum. Spin–orbit coupling (soc) [66] is possible when heavy atoms are present in the molecules (e.g. rate constant k_{ISC} = c. 10^6 s^{-1} for naphthalene, 10^9 s^{-1} for 1-bromonaphthalene) [67]. If not, like in compounds comprising only C, H, O, and N atoms, ISC is usually difficult. According to the El-Sayed's rules, ISC is slow ("forbidden") when the states being interconverted are both of the π,π^* type (e.g. alkenes, polyenes, and arenes; Figure 6.2) or when both are of the n,π^* type (e.g. ketones, imines, and O, N, S-heterocyclic compounds, Figure 6.2) [68–71]. However, when a π,π^* state interconverts into a n,π^* state, or *vice versa*, ISC is more favorable because the orbital momentum has changed. In the case of a carbonyl chromophore (Figure 6.5), the first singlet excited state ${}^1\psi^*{}_1$ is a n,π^* state with one electron in the n(CO) HOMO and one in the π^*(CO) LUMO. For alkyl ketones, the lowest energy singlet (${}^1\psi^*{}_1$) and triplet state (${}^3\psi^*{}_1$) are both n,π^* states (noted ${}^1(\text{n},\pi^*)_1$ and ${}^3(\text{n},\pi^*)_1$); thus, ISC is not favorable. In contrast, for diaryl ketones such as benzophenone, the second excited triplet state ${}^3\psi^*{}_2$ is a π,π^* state (noted: ${}^3(\pi,\pi^*)_2$), which is close in energy to the first singlet excited state ${}^1\psi^*{}_1$ (${}^1(\text{n},\pi^*)_1$). Thus, ISC ${}^1(\text{n},\pi^*)_1 \rightarrow {}^3(\pi,\pi^*)_2$ becomes now much easier (e.g. k_{ISC} c. 10^8 s^{-1} for acetone, 10^{11} s^{-1}

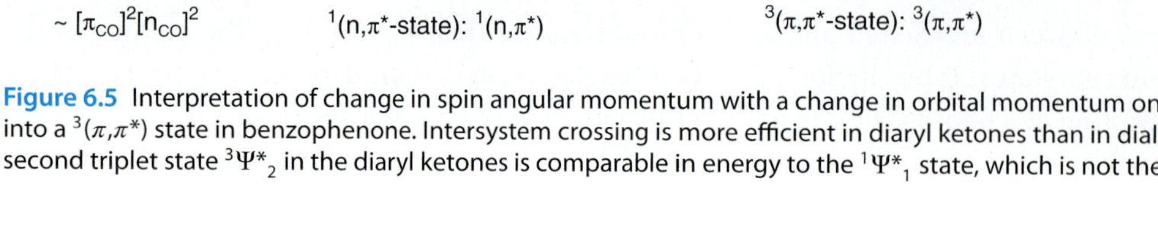

Figure 6.5 Interpretation of change in spin angular momentum with a change in orbital momentum on converting a ${}^1(\text{n},\pi^*)$ state into a ${}^3(\pi,\pi^*)$ state in benzophenone. Intersystem crossing is more efficient in diaryl ketones than in dialkyl ketones because the second triplet state ${}^3\Psi^*{}_2$ in the diaryl ketones is comparable in energy to the ${}^1\Psi^*{}_1$ state, which is not the case for dialkyl ketones.

for benzophenone, and 10^{12} s^{-1} for benzaldehyde). Spin flipping in these two orbitals occurs simultaneously by moving one electron in a 2p orbital to the n(C=O) orbital. This changes the orbital angular momentum and compensates to some extent for the change in spin angular momentum, making the process more favorable. When singlet state $^1\psi^*_1$ (^{1}(n,π^*)$_1$) is reached, it quickly relaxes into $^3\psi^*_1$ (^{3}(π,π^*)$_1$) [72]. Because of the very efficient ISC in benzophenone, this compound plays a very important role as triplet photosensitizer (or sensitizer) in organic photochemistry (Sections 6.3–6.8). The energy difference between ^{1}(π,π^*)$_1$ and ^{3}(π,π^*)$_1$ states ($S_1 - T_1$ splitting) is large for π-systems like ethylene (70 kcal mol^{-1}), benzene (40 kcal mol^{-1}), naphthalene (35 kcal mol^{-1}), anthracene (30 kcal mol^{-1}), (*E*)-stilbene (40 kcal mol^{-1}), (*Z*)-stilbene (30 kcal mol^{-1}), and much less for the energy difference between ^{1}(n,π^*)$_1$ and ^{3}(n,π^*)$_1$ states of carbonyl compounds like formaldehyde (10 kcal mol^{-1}), acetone (8 kcal mol^{-1}), and benzophenone (7 kcal mol^{-1}). This is due to the fact that the Fermi hole that separates electrons in the triplet state has a larger stabilizing effect in ^{3}(π,π^*)$_1$ than in ^{3}(n,π^*)$_1$ states. In ^{3}(π,π^*)$_1$ states, the electron circulating in the π(HOMO) orbital occupies the same space as the electron circulating in the π^*(LUMO) orbital. In contrast, in ^{3}(n,π^*)$_1$ states, the electron circulating in the n_(HOMO) orbital occupies another space than the electron circulating in the π^*(LUMO) orbital. In the ^{1}(n,π^*)$_1$ state, the repulsion between the two unpaired electrons is nearly the same as the ^{3}(n,π^*)$_1$ state [73]. All the photophysical processes described above are summarized in the Jablonski diagram (Figure 6.6).

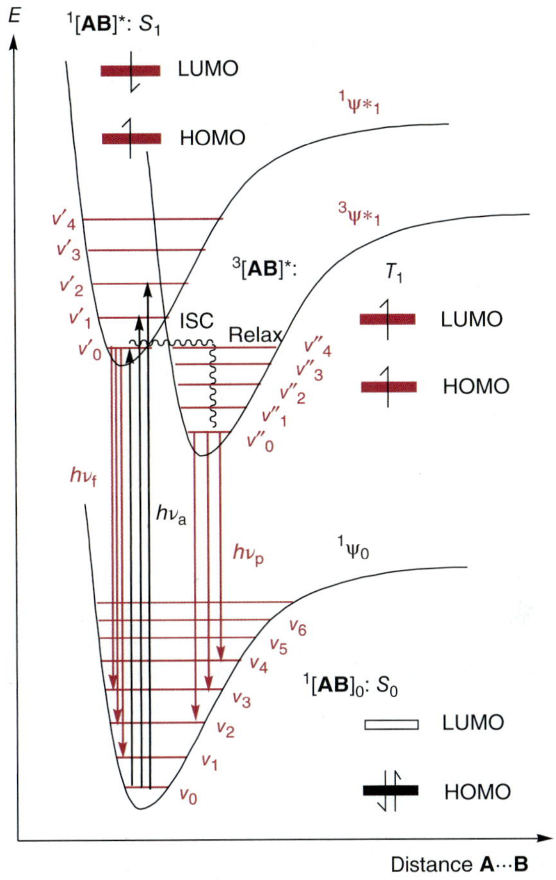

Figure 6.6 The Jablonski diagram for a diatomic molecule AB.

6.2.3 Bimolecular photophysical processes

In solution, an excited molecule ^{1}**A*** or ^{3}**A*** resulting from the excitation $\mathbf{A} + h\nu_a \rightarrow {}^1\mathbf{A}^* \rightarrow {}^3\mathbf{A}^*$ is constantly colliding with solvent molecules **M** with bimolecular diffusion rate constant, k_{diff}, c. 10^{10} M^{-1} s^{-1}. This facilitates the several radiationless relaxation processes of the electronically excited molecules. The collisions compete with fluorescence, which is a unimolecular process with rate constant, k_f, of the order 10^8 to 10^5 s^{-1}. Both temperature and viscosity of the medium contributes to quantum yield of fluorescence. For instance, the fluorescence efficiency of (*E*)-stilbene increases by a factor of 3 in going from ethanol ($\phi_f \approx 0.05$) to viscous glycerol ($\phi_f \approx 0.15$). Most substituted ethylenes and polyenes do not display fluorescence or phosphorescence even at 77 K. Compounds **Q** that have a chromophore (not like solvents used in photochemistry) or can interact specifically with the excited molecule ^{1}**A*** or ^{3}**A*** can induce radiationless relaxation more efficiently than solvent molecules. Such a compound **Q** is called a quencher and leads to collision-induced relaxation $\mathbf{A}^* + \mathbf{Q} \rightarrow \mathbf{A} + \mathbf{Q}^*$ (quenching). This process is an energy transfer. In some cases, it does not require a collision: it occurs when the excited molecule **A*** emits a photon that is absorbed by **Q** (see below). In the collision-induced process, **Q*** can be in its electronic ground state but vibrationally excited or be an electronically excited state. Compounds that are efficient quenchers of singlet excited ^{1}**A*** are called singlet quenchers, and compounds that are efficient quenchers of triplet excited ^{3}**A*** are called triplet quenchers. Kinetics of the fluorescence or phosphorescence quenching generally follow the Stern–Volmer analysis (Scheme 6.3). The maximum value of the quenching rate constant, k_q, is the bimolecular diffusion rate, k_{diff}. In general, quenching requires an intimate contact between **A*** and **Q**. Many olefins and simple dienes are efficient quenchers for which a relationship between the ionization energies

$$\mathbf{A} \xrightarrow{h\nu_a} \mathbf{A}^* \xrightarrow{k_1} \mathbf{A} \quad \text{Lifetime of } \mathbf{A}^* \text{ without quencher } \mathbf{Q}\text{: } \tau_1 = 1/k_1$$

$$\mathbf{Q} + \mathbf{A}^* \xrightarrow{k_q} \mathbf{A} + \mathbf{Q}^* \quad \text{Lifetime of } \mathbf{A}^* \text{ with quencher } \mathbf{Q}\text{: } \tau_q$$

$$\mathbf{Q}^* \longrightarrow \mathbf{Q} + \text{heat;} \quad \text{or } \mathbf{Q}^* \longrightarrow \mathbf{Q} + h\nu_q$$

$$1/\tau_q = k_1 + k_q[\mathbf{Q}] = 1/\tau_1 + k_q[\mathbf{Q}]$$

Scheme 6.3 Stern–Volmer quenching kinetics for compound **A**.

(Section 1.8, Table 1.A.21) of the quencher (IE(**Q**)) and k_d has been observed. Both electron-rich (low IE(**Q**) value, low −EA(**Q**) value) and electron-poor quenchers (high IE(**Q**) value) are effective, whereas alkyl-substituted alkenes and dienes are less effective. The mechanism of quenching can be viewed as a photochemical reaction that does not produce any new product, but return to starting materials.

Quenching can be a radiationless ($\mathbf{Q}^* \rightarrow \mathbf{Q} + \text{heat}$) or a radiative process ($\mathbf{Q}^* \rightarrow \mathbf{Q} + h\nu$). There are also radiative processes involving two molecules. Two molecules can equilibrate with a noncovalent complex (e.g. $\mathbf{A} + \mathbf{A} \rightarrow (\mathbf{A})_2$ or $\mathbf{A} + \mathbf{B} \rightarrow (\mathbf{A}{\cdot}\mathbf{B})$), and this complex can absorb light. This is referred to an absorption complex. If two molecules **A** and **B** act together to emit a photon, an exciplex $[(\mathbf{A}{\cdot}\mathbf{B})]^*$ is involved. If the two molecules are the same, the exciplex is called an excimer (for excited dimer). Usually, absorption complexes (**A**·**B**) absorb light of lower energy than the isolated monomers **A** and **B**. This is responsible for the changes observed in the UV–visible absorption spectra as a function of concentration (deviation from the Beer–Lambert law, appearance of lower energy bands, and broadening of the spectra compared with the spectra measured in the gas phase). The most common absorption complexes involve polyaromatic compounds (e.g. anthracene, naphthacene, pyrene and their substituted derivatives) or charge transfer complexes that are associated with charge transfer absorption, also known as donor–acceptor absorption. The donor **D** is a molecule with low IE(**D**) value (Section 1.8, Table 1.A.21), whereas the acceptor **A** has a relatively high IE(**A**) value and a high electron affinity value, −EA(**A**) (Section 1.8, Table 1.A.22). Light absorption of complex (**D**·**A**) ($(\mathbf{D}{\cdot}\mathbf{A}) + h\nu_a \rightarrow [(\mathbf{D}{\cdot}\mathbf{A})]^*$) is usually accompanied by an electron transfer leading to an electronically excited state ψ^*_1, (noted: $^1[(\mathbf{D}{\cdot}\mathbf{A})]^*$) for which the charge-transferred configuration ϕ^*_{CT} (noted: $\mathbf{D}^{+\bullet}\,\mathbf{A}^{-\bullet}$) contributes significantly. The charge transfer transitions are usually of lower energies than the absorption transitions of isolated **D** and **A**. Because of their highly polar character, charge transfer absorptions are extremely sensitive to solvent polarity, moving to longer wavelength as the solvent polarity increases. The electronically excited complexes $[(\mathbf{D}{\cdot}\mathbf{A})]^*$ commonly undergo relaxation by return-electron transfer (BET). The emission of an exciplex $[(\mathbf{A}{\cdot}\mathbf{B})]^*$ resulting from the collision of an excited molecule $\mathbf{A}^*$ and molecule **B** is usually of lower energy than emission from isolated $\mathbf{A}^*$ and $\mathbf{B}^*$. Exciplex $[(\mathbf{A}{\cdot}\mathbf{B})]^*$ is expected to be a highly polarizable species as an electron from a bonding MO has been put in one antibonding MO. This enhanced polarizability promotes the formation of weak complex via dipole-induced–dipole interactions (London dispersion forces). Commonly, exciplexes remain in their singlet state before relaxation, but sometimes, they undergo ISC, and then the first excited triplet state $^3\mathbf{A}^*$ or $^3\mathbf{B}^*$ is created. Thus, exciplexes and excimers can emit from both their singlet (fluorescence) and triplet state (phosphorescence). Several photochemical processes (Section 6.10) involve a photoinduced electron transfer (PET). The two limiting cases of such transfers are summarized in Figure 6.7.

If the energy transfer $\mathbf{A}^* + \mathbf{Q} \rightarrow \mathbf{A} + \mathbf{Q}^*$ produces an electronically excited state of **Q**, one speaks of photosensitization of **Q** by $\mathbf{A}^*$ [74]. This process of photosensitization, or sensitization, obeys the Wigner's spin conservation rule that states that total spin must be conserved during the energy transfer. Thus, singlet/singlet $^1\mathbf{A}^* + {}^1\mathbf{Q}_0 \rightarrow {}^1\mathbf{A}_0 + {}^1\mathbf{Q}^*$ and triplet/triplet $^3\mathbf{A}^* + {}^1\mathbf{Q}_0 \rightarrow {}^1\mathbf{A}_0 + {}^3\mathbf{Q}^*$ sensitizations are both allowed. Also known are energy transfers of type $^3\mathbf{A}^* + {}^3\mathbf{Q}^* \rightarrow {}^1\mathbf{A}_0 + {}^1\mathbf{Q}^*$. Two major mechanisms are recognized for these energy transfers. In the Förster mechanism [75, 76] associated with nonradiative singlet/singlet energy transfers (Figure 6.8a), there is a coupling of the two electronic transitions $^1\mathbf{A}^* \rightarrow {}^1\mathbf{A}_0$ and $^1\mathbf{Q}_0 \rightarrow {}^1\mathbf{Q}^*$ [77]. The probability for this transfer is proportional to the transition dipoles D_A and D_Q. The transition dipole arises from the electron displacement between the ground and electronically excited states, and the change in distance between positive and negative regions of the molecule upon excitation. In this mechanism, a direct overlap of the wave functions of $^1\mathbf{A}^*$ and $^1\mathbf{Q}_0$ is not required. It is simply a coupling of transition dipoles D_A and D_Q. The probability of the transfer depends on the distance r_{AQ} separating **A** from **Q**. The rate constant of the energy transfer is given by $k_{dc} \approx (D_A)^2{\cdot}(D_Q)^2/(r_{AQ})^6$. The requirement for the energy transfer is an energy matching between the initial and final states. **Q** is excited by the energy equal to the energy lost from $\mathbf{A}^*$. Thus, for this mechanism to operate, part of the emission spectra of **A** must overlap the absorption spectrum of **Q**. If **A** and **Q** are attached to a large molecule such as a biomolecule,

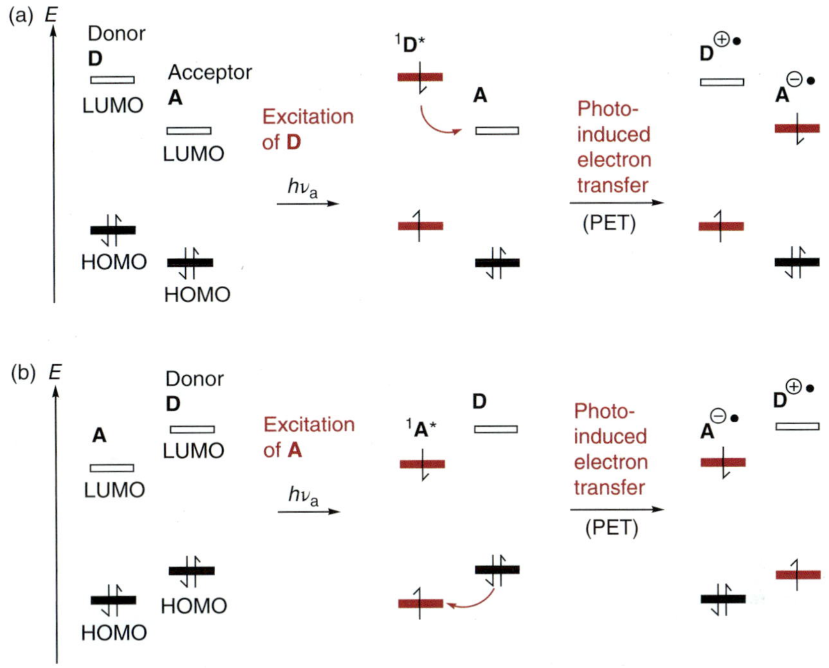

Figure 6.7 Photoinduced electron transfer: (a) the electron flows from HOMO(donor **D**) to LUMO(**D**) and then to LUMO(acceptor **A**) and (b) the electron flows from HOMO(**A**) to LUMO(**A**) and then an electron from HOMO(**D**) fills HOMO(**A**).

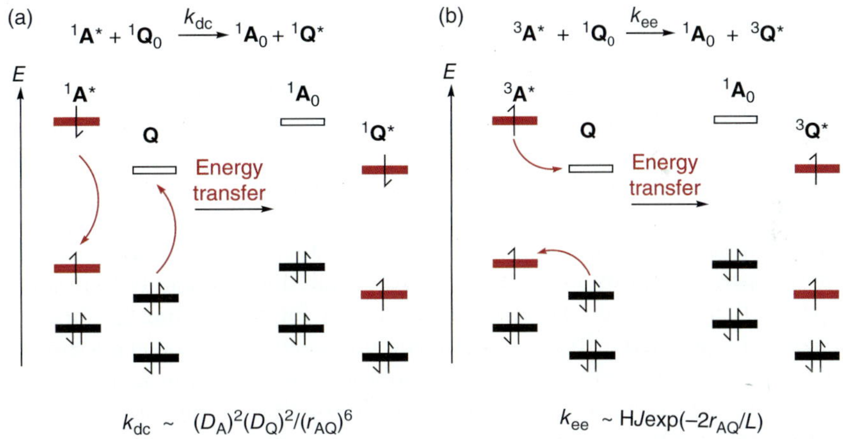

Figure 6.8 Simple models for energy transfers (sensitization); (a) the Förster mechanism (electronic transition dipole coupling); (b) the Dexter mechanism (electron exchange).

the quantum yield of emission from **Q*** can be used to measure the distance separating **A** from **Q** (up to 80 Å = 8 nm) in the biomolecule. This is the basis of FRET (Förster resonance energy transfer or fluorescence resonance energy transfer), an extremely useful research tool in chemistry, material sciences [78–80], and biology [81–86].

In the Dexter (or electron exchange) mechanism (Figure 6.8b) [87] occurring for collisional energy transfer in triplet/triplet sensitization, there is a direct interaction of the wave functions of $^3\mathbf{A}^*$ and $^1\mathbf{Q}_0$ that exchange one electron. The rate constant for the electron exchange is given by $k_{ee} \approx HJ \exp(-2r_{AQ}/L)$, where H is related to the magnitude of the specific orbital interaction that promotes the electron exchange, J is related to the spectral overlap between the absorption spectrum of **A** and the absorption spectrum of **Q**, and L is the sum of the van der Waals radii of **A** and **Q**. As k_{ee} decreases exponentially with the distance r_{AQ} between **A** and **Q**, the efficiency of the energy transfer falls off deeply as the separation between **A** and **Q** increases (up to 10 Å). Such triplet/triplet sensitization plays an important role in organic photochemistry.

Many photochemical reactions (reactant **A**) rely upon photosensitization (or sensitization) in which a sensitizer **S** absorbs light of lower energy than the energy necessary to excite reactant **A** directly. Because triplet excited state $^3\mathbf{A}^*$ (arising from singlet ground-state molecule $^1\mathbf{A}_0$) has a much longer lifetime than singlet excited state $^1\mathbf{A}^*$, the sensitization usually involves the energy transfer $^3S^* + {}^1A_0 \rightarrow {}^3A^* + {}^1S_0$. Then, $^3A^*$ evolves into product(s) **P**. As seen above, efficiency of the energy transfer depends on the lifetime of $^3S^*$, on the concentration of **A**, on the viscosity of the medium, and on the intervention of competitive processes capable of wasting the excitation energy of $^3S^*$ and/or of $^3A^*$. One of them is electron transfer (ET) in which the triplet excited state of the reactant gives an electron to the sensitizer ($^3A^* + {}^1S_0 \rightarrow A^{\bullet+}$ (cation radical) + $S^{\bullet-}$ (anion radical)) or takes an electron from the sensitizer ($^3A^* + {}^1S_0 \rightarrow A^{\bullet-}$ (anion radical) + $S^{\bullet+}$ (cation-radical)). The charge transfer complexes that result relax into the starting $^1A_0 + {}^1S_0$ through return electron transfer (RET), or undergo chemical reactions [88].

6.3 Unimolecular photochemical reactions of unsaturated hydrocarbons

In general, a photochemical reaction pathway involves two hypersurfaces: one is the excited state and the other one the ground state. These two hypersurfaces might be connected at a point (or an ensemble of points: intersection space, joined hypersurfaces: seam) where their potential energy is the same and, if the states have the same spin multiplicity, this region is called a conical intersection (CI), or "conical intersection seam" or simply a "seam" [89–95]. After photon absorption ($^1A_0 + h\nu \rightarrow {}^1A^*_i$), the excited state of reactant **A** relaxes into $^1A^*_1(\nu'_0)$ (state function $^1\psi^*_1(\nu'_0)$). If the latter survives long enough for a change in geometry (>10^{-13} seconds), one recognizes at least four types of possible photochemical reactions converting **A** into product(s) **P**. In the first case (Figure 6.9a), $^1A^*_1(\nu'_0)$ evolves to an energy minimum, which is close in energy and geometry (transient species **F**) to the ground-state hypersurface of the thermal reaction **A** ⇄ **P**. Species **F** (funnel that can be a Franck–Condon minimum, an avoided crossing, or a conical intersection: a seam) is an electronically excited conformer of $^1A^*_1$, which can drop quickly onto the ground-state hypersurface $^1\psi_0(A \rightleftarrows P)$ and then evolves competitively into **A** and **P**. Note that the Born–Oppenheimer approximation can break down during this process as nuclei can move rapidly when $^1\psi^*_1(A \rightleftarrows P)$ approaches the funnel that connects it with $^1\psi_0(A \rightleftarrows P)$. This may lead to special dynamic effects that affect the proportion of **A** and **P** produced (we do not treat these effects in this book). In a second case (Figure 6.9b), $^1A^*_1(\nu'_0)$ evolves to the ground state $^1I_0(\nu'''_i)$ of intermediate **I** of the thermal reaction $^1\psi_0(A \rightleftarrows I \rightleftarrows P)$ and then produces **A** and **P** competitively. In a third case (Figure 6.9c), the electronically excited state $^1I^*_1(\nu^{iv}_0)$ of intermediate **I** is reached. This species decays into $^1I_0(\nu'''_i)$, which evolves competitively into **A** and **P**. In the fourth case (Figure 6.9d), $^1A^*_1(\nu'_0)$ evolves to the electronically excited state $^1P^*_1(\nu^v_0)$ of product **P**, which then decays into $^1P_0$. Alternatively, $^1P^*_1(\nu^v_0)$ can emit light (fluorescence) or undergo ISC into $^3P^*_1$ that decays into $^1P_0$ + heat, or emit light (phosphorescence). The quantum photochemical yield is defined as $\phi_{(A,P)}$ = number of molecules of **P** produced/number of photons absorbed by the system. Because of competition between the photochemical processes seen above, many photochemical reactions have quantum yields lower than 100%. It can be larger than 100% in the case of photocatalysis such as radical chain reactions (see below) [96].

6.3.1 Photoinduced (*E*)/(*Z*)-isomerization of alkenes

The lowest excited states of substituted alkenes are mostly of $^1(\pi,\pi^*)$ and $^3(\pi,\pi^*)$ types, for which the π-bond order is significantly reduced, thus rendering facile the rotation about the π(C=C) double bond. This process dominates the photochemistry of alkenes for both singlet and triplet excited states [97]. The vertical and twisted singlet excited states of alkenes possess a zwitterionic character, whereas the corresponding triplet states are diradical in character [98]. The triplet states of alkenes are obtained most readily through sensitized photoreactions. As any other diradical, the triplet intermediates can undergo typical reactions of radicals such as hydrogen abstraction from the medium or from alkyl groups, and/or alkene polymerization. The direct photoexcitation of alkyl-substituted alkenes ((*E*)-**A**, (*Z*)-**A**) uses light with wavelengths ~200 nm (ΔE = 143 kcal mol^{-1}), which requires a quartz vessel and a medium- or a high-pressure Hg lamp. Aryl-substituted alkenes require less drastic conditions and their photochemistry can be carried out in Pyrex vessels. After direct excitation, the $^1(\pi,\pi^*)_1$ state is obtained, which does not undergo ISC readily but starts to rotate about the C=C double bond until a nearly perpendicular excited 1,2-diradical is formed **I**(D)* in an electronically excited state that funnels onto the thermal isomerization hypersurface $^1\psi_0((E)\text{-}A \rightleftarrows (Z)\text{-}A)$ (see e.g. Figure 6.10). Both the vertical (Franck–Condon principle) and twisted singlet excited state possess a zwitterionic character. In the case of the thermal (*Z*) ⇄ (*E*) isomerization of but-2-ene, the transition state of the reaction is c. 60 kcal mol^{-1} above the alkene [99, 100]. This leads to partial isomerization

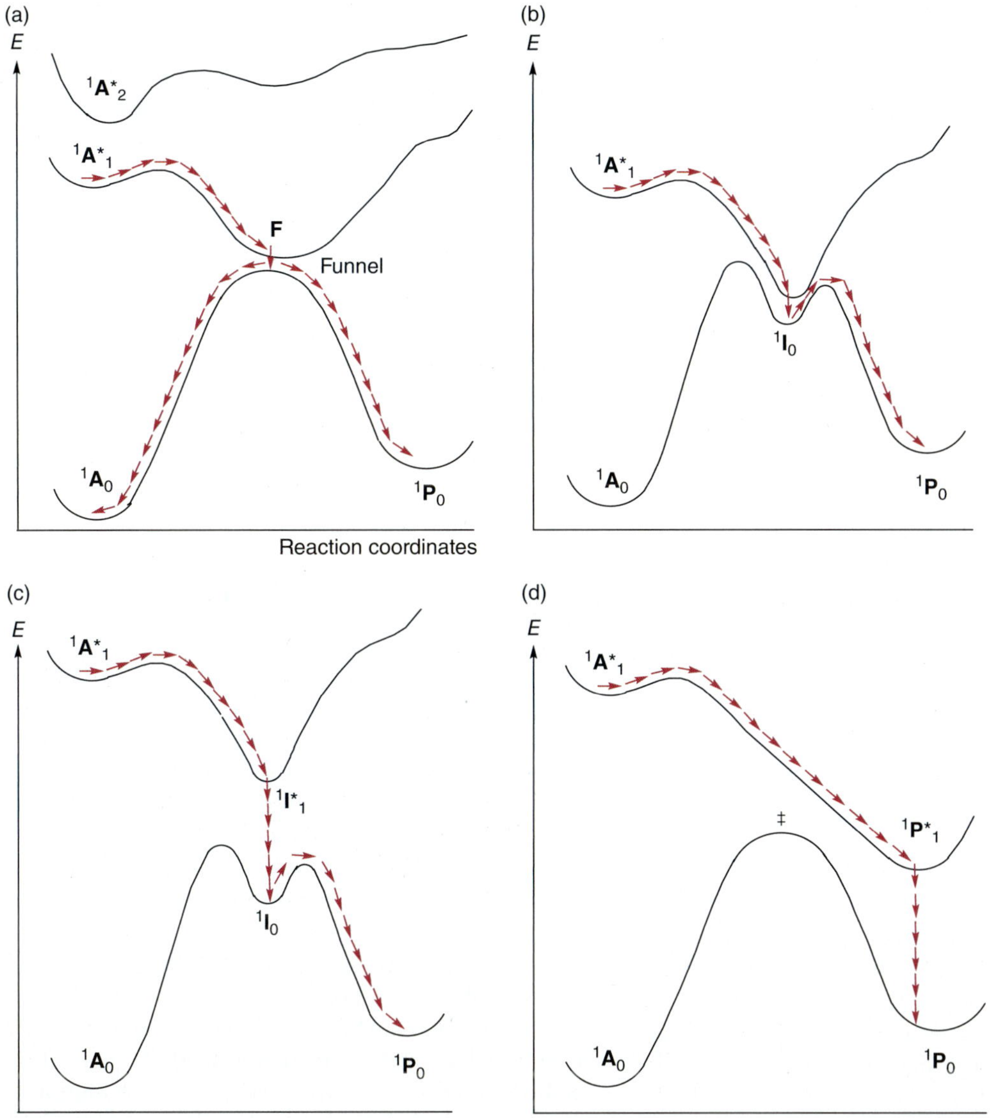

Figure 6.9 Examples of potential energy surfaces for photochemical reactions. (a) A diabatic photoreaction with the formation of an intermediate **F** (singlet or triplet state) that "funnels" onto the ground-state energy hypersurface; (b) the reactant in its excited state (singlet or triplet) evolves into an intermediate of the ground-state hypersurface ($\mathbf{I}_0$); (c) the reactant in its excited state evolves into an excited state of the intermediate (**I***); the latter then decays to its ground state, (d) an adiabatic photoreaction in which the excited state of reactant evolves to the excited state of the product (singlet or triplet state).

of (*E*)-**A** into (*Z*)-**A** and of (*Z*)-**A** into (*E*)-**A**. The proportion of (*E*)- and (*Z*)-alkene depends on the photolysis conditions, but not on the thermodynamic equilibrium constant $K((E)\text{-}\mathbf{A} \rightleftarrows (Z)\text{-}\mathbf{A})$. At a given irradiation wavelength, λ_{irr}, isomeric (*E*)- and (*Z*)-alkenes have different efficiencies of light absorption ε_E and ε_Z, and different quantum yields for their interconversion $\phi_E(E \rightarrow Z)$ and $\phi_Z(Z \rightarrow E)$, respectively. The photostationary state is given by $[(E)\text{-}\mathbf{A}]/[(Z)\text{-}\mathbf{A}] = \varepsilon_Z\, \phi_Z/\varepsilon_E\phi_E$.

At λ_{irr} = 313 nm, the direct photoisomerization of (*E*)-stilbene into (*Z*)-stilbene, and *vice versa*, has quantum yield $\phi_E(E \rightarrow Z) \approx 0.50$ and $\phi_Z(Z \rightarrow E) \approx 0.35$ (Scheme 6.4) [101, 102]. Direct photoexcitation of (*Z*)-stilbene results in the emission from (*E*)-stilbene. This is consistent with an adiabatic process that converts a fraction of $^1(\pi,\pi^*)_1$ state of (*Z*)-stilbene directly into $^1(\pi,\pi^*)_1$ state of (*E*)-stilbene (see Figure 6.9d) [103]. Using picosecond laser spectroscopy, Hochstrasser and coworkers demonstrated that irradiation of (*E*)-stilbene with a pulse at 265 nm produces an intermediate absorbing visible light at λ_{max} = 584 nm and which has a lifetime of 68 ps. This intermediate is the twisted

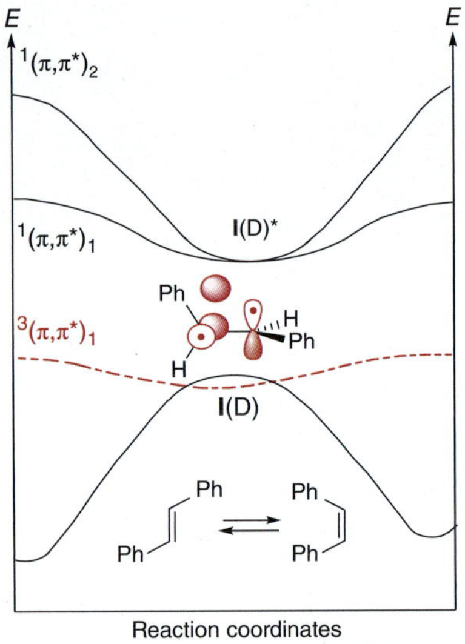

Figure 6.10 Prototype hypersurface diagram for the thermal and photochemical (*E*) ⇄ (*Z*)-isomerization of alkenes.

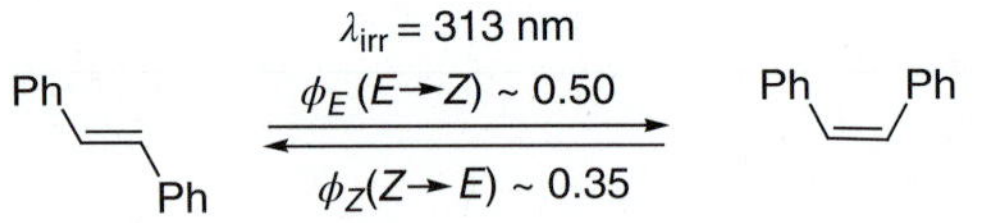

Scheme 6.4 Photoinduced (*E*) ⇄ (*Z*)-isomerization of stilbene (1,2-diphenylethylene).

1,2-diphenylethane-1,2-diyl diradical residing in its $^1(\pi,\pi^*)_1$ state [104]. In contrast, and using femtosecond laser spectroscopy, they found that irradiation of (*Z*)-stilbene produces an intermediate that is not intercepted on the 150 fs time scale [105, 106]. Apparently, there is an energy barrier to rotation on the excited state for the (*E*)-isomer, with a much lower barrier, or no barrier for the (*Z*)-isomer.

The triplet excited state of alkenes is much lower in energy than the corresponding excited singlet state. The triplet has a diradical character. It can be reached through sensitization upon irradiating mixtures containing triplet sensitizers such as benzene ($E(^3(\pi,\pi^*)_1) - E(^1\psi_0) = E_T \sim 80$ kcal mol^{-1}) or acetone ($E_T \sim 78$–80 kcal mol^{-1}). In the case of triplet sensitization of (*E*)- and (*Z*)-stilbene hypersurface, $^3(\pi,\pi^*)_1$ is reached. There is a small energy barrier separating $^3(\pi,\pi^*)_1$ state of (*E*)-stilbene and the twisted 1,2-diradical intermediate. The energies of the $^3(\pi,\pi^*)_1$ states of (*E*)-stilbene and (*Z*)-stilbene are 49 and 57 kcal mol^{-1}, respectively. Therefore, with a proper choice of sensitizer, one can control the photostationary state composition.

The small-ring (*Z*)-cycloalkenes (cyclopropene, cyclobutene, or cyclopentene) are not photoisomerized into their much less stable (*E*)-isomers. Laser flash irradiation of 1-phenylcyclohexene ((*E*)-**8**, (*E*) because of the Ph substituent, Cahn–Ingold–Prelog priority rules [107, 108]) produces (*Z*)-**8** (with λ_{max} = 380 nm and a lifetime of 9 μs) that can be observed by fast absorption spectroscopy (scheme 6.5). At −75 °C, this intermediate quickly reacts in a Diels–Alder reaction with a nearby molecule of (*E*)-**8** to provide a dimeric product **9**. A subsequent (1,3)-hydrogen shift gives the final product **10** [109]. Using photoacoustic calorimetry, Peters and coworkers measured a strain energy of 48.6 kcal mol^{-1} for (*Z*)-1-phenylcyclohexene ((*Z*)-**8**) [110].

There are direct spectroscopic and chemical trapping experiments that confirm the formation of strained (*E*)-cyclohexene, (*E*)-cycloheptene, and (*E*)-cyclooctene in the triplet sensitization of the corresponding (*Z*)-cycloalkenes [111, 112]. The (*E*)-cycloalkenes are highly polarizable and are readily protonated by weak acids such as alcohols, generating cationic intermediates that form ethers or eliminate a proton to produce isomeric alkenes [113, 114]. Flash photolysis of (*Z*)-cyclohept-2-enone in alkane solvent generates the transient (*E*)-cyclohept-2-enone [115, 116].

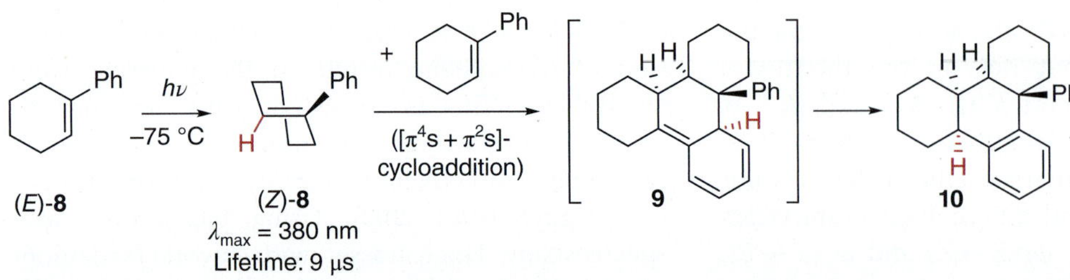

Scheme 6.5 Formation and chemical trapping of (*Z*)-1-phenylcyclohexene.

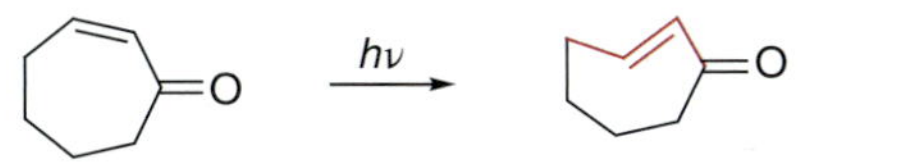

(E)-Cycloalkenes have C_2-symmetry, and they are, therefore, chiral. Chirality transfer from an enantiomerically pure sensitizer to a (E)-cycloalkene generated by photolysis of the corresponding (Z)-cycloalkene constitutes an elegant approach to asymmetric synthesis. The method is called chiral photochemistry or photochirogenesis [117–121]. Enantiodifferentiation in the photoisomerization of cyclooctene has been realized by photosensitization with enantiomerically pure arenecarboxylates [122, 123], aroyl-β-cyclodextrins [124], and planar chiral paracyclophanes [125]. An example is shown (Scheme 6.6) for the enantioselective synthesis of (R)-(−)-(E;Z)-cycloocta-1,5-diene ((R)-(−)-**11**). The (E,Z)-cycloocta-1,5-diene is extracted from the reaction mixture by formation of a Ag^+ complex with aqueous $AgNO_3$ solution (forms a more stable π-complex with Ag^+ than (Z,Z)-cycloocta-1,5-diene). Treatment of this solution with aqueous NH_3 liberates (R)-(−)-**11**, which is extracted at 0 °C with pentane.

The photochemical (Z) → (E) isomerization of an alkene group in rhodopsin is the primary event in vision. Rhodopsin is a membrane-bound protein that combines with (11Z)-retinal by making an imine with one lysine side chain. This imine is the chromophore of vision with a relatively broad absorption spectrum and with λ_{max} = 500 nm (ε = 40 000 M^{-1} cm^{-1}). Absorption of a photon induces the isomerization to the *all*-(E)-retinal, a process that occurs on the picosecond time scale (Scheme 6.7). Within the cavity of the enzyme, this structure change induces a signaling cascade in the cell that ultimately reaches the visual cortex of the brain.

Similar to stilbene, azobenzene (Ph—N=N—Ph) undergoes geometrical photoisomerization. (Z)-Azobenzene absorbs visible light of λ = 450 nm and is isomerized into (E)-azobenzene quantitatively as the latter does not absorb visible light but absorbs UV light. When (E)-azobenzene is irradiated at λ = 350 nm, the (Z)-isomer forms. This reaction is not accompanied by secondary reactions, as those observed for stilbene (photoisomerization into phenanthrene, (2+2)-cycloaddition, see below) [126]. Azobenzene and its derivatives have been used as photoswitches for photoelectronics and biomaterials [127–129], as well as for in vivo red-light-induced devices [130].

Scheme 6.6 Enantiodifferentiating photoisomerization of cycloocta-1,5-diene.

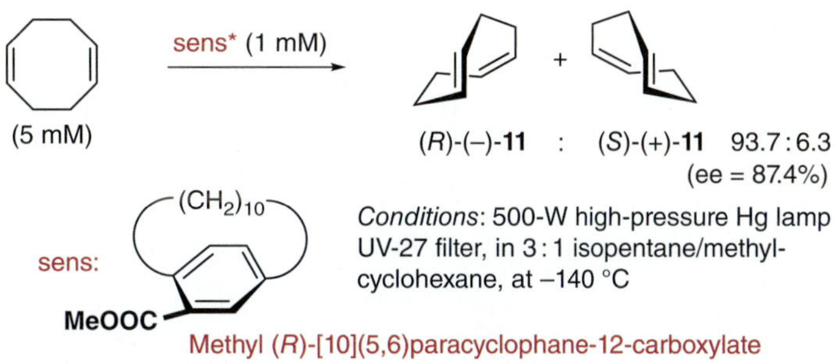

Scheme 6.7 (Z) → (E)-Photoisomerization of rhodopsin-retinyl imine in the vision process.

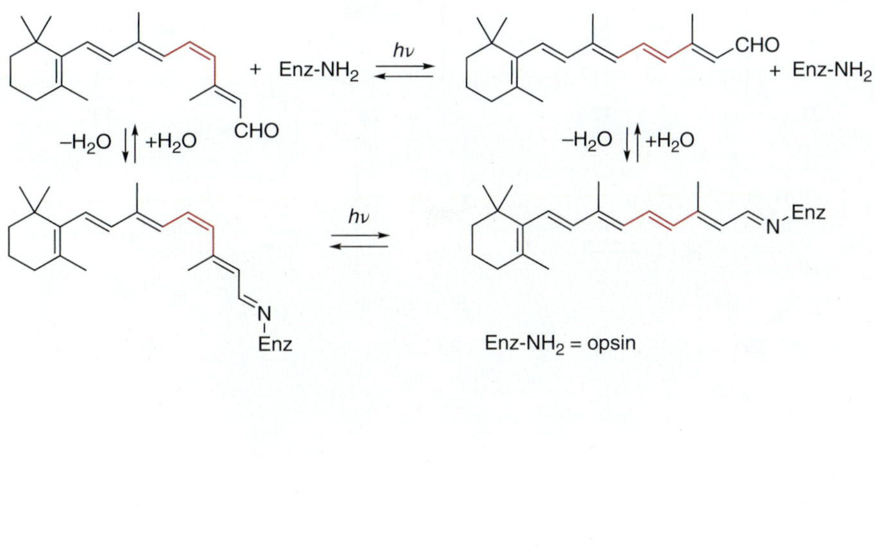

Problem 6.1 Direct photoexcitation of (*E*,*E*)-1-deuteropenta-1,3-diene results in the selective isomerization of the CH=C(D)H double bond giving (*Z*,*E*)-1-deuteropenta-1,3-diene. Is this observation compatible with the formation of diradical intermediates or not? [16]

Problem 6.2 What are the products formed upon irradiation of 1-methylcyclohexene in toluene containing methanol? [114]

Problem 6.3 On direct photoirradiation of cyclooct-2-enone (**A**) that shows a λ_{max} at 321 nm, one obtains an isomer **B** with a λ_{max} = 283 nm. What is isomer **B**? [131]

6.3.2 Photochemistry of cyclopropenes, allenes, and alkynes

Photolysis of cyclopropene in an Ar matrix at 8 K (reaction (6.1)) yields allene and propyne [132]. Thermolysis of cyclopropene at 190–240 °C also produces allene (2%) and propyne (98%) [133]. Photolysis of spiro[2.4]hept-1-ene (**12**) in pentane solution (reaction (6.2)) with unfiltered low-pressure Hg lamp results in rearrangement to the isomeric allene **13**, cyclopentylethyne (**14**), and 1-vinylcyclopentene (**15**) [134]. Photolysis of allene **13** in pentane produces **12** + **14** + **15** and 3-methylidenecyclohexene (**16**) [135, 136].

(6.1)

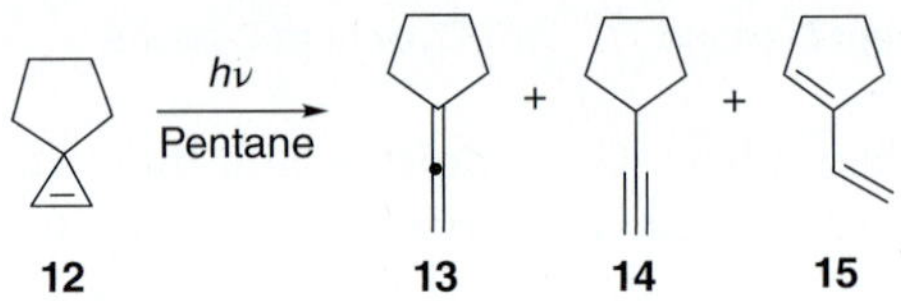

(6.2)

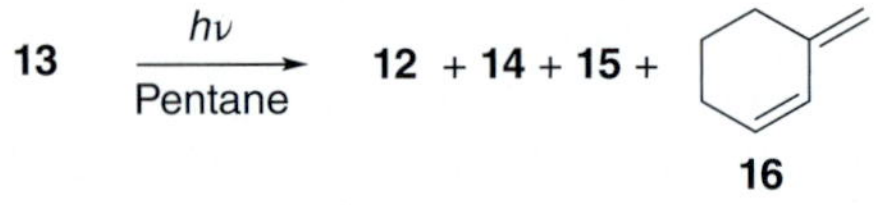

(6.3)

Experimental and theoretical studies on the photoisomerization of cyclopropenes **17**, allenes **18**, and alkynes **19** (Scheme 6.8) support the intermediacy of 1,3-diradicals of type **20** ⇄ **20′** ⇄ **20″** and of alkylidene carbenes of type **21**. In the case of the photoexcitation of 1-alkylcyclopropenes, conjugated dienes are also formed. The process involves (1,2)-hydrogen shift of type **22** → **23** → **24**.

In general, alkyl-substituted alkynes do not rearrange into the corresponding allenes or cyclopropenes upon direct irradiation in solution. They are usually reduced into the corresponding alkene by hydrogen atom abstraction from the medium or from the alkyl substituents [137]. Photolysis of cyclononyne

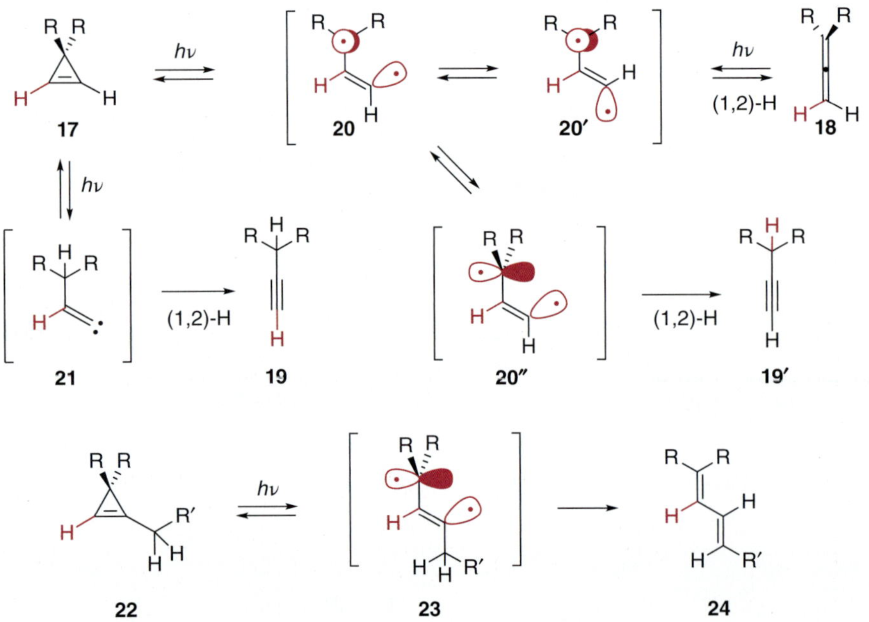

Scheme 6.8 The photoinduced interconversion of cyclopropenes, allenes, and alkynes involves carbene and diradical intermediates.

(**25**) in pentane solution gives bicyclo[4.3.0]non-1-ene (**27**) and (*Z*)-cyclononene (**29**). They result from the competing intramolecular hydrogen atom transfer (HAT) **25** → **26** → **27** and the intermolecular reduction of **25** → **28** → **29**.

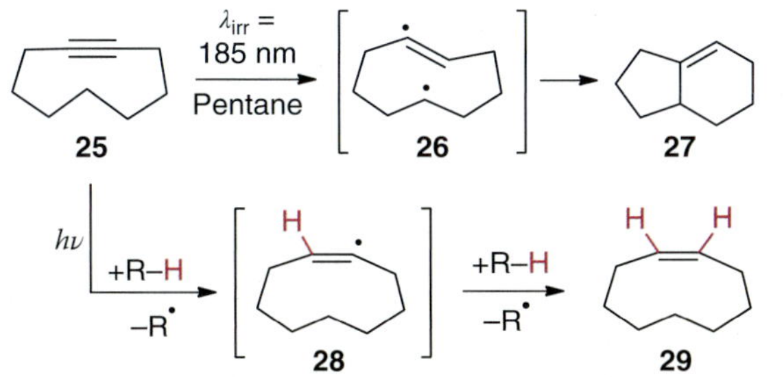

Upon sensitized photoactivation, alkyl-substituted cyclopropenes that contain γ or δ-C—H bonds on their side chains can undergo HATs giving diradical intermediates that can disproportionate or cyclize as illustrated in Scheme 6.9 [138]. In the case of the sensitized photoreaction (sens = thioxanthone = 9-oxothiazanthene, in PhH, Pyrex, Hg-burner) of cyclopropene **30**, a mixture of cyclopropane derivatives (*Z*)- and (*E*)-**32** is formed that involves the formation of diradical intermediate **31**. A quantum yield of 25% is measured for this photoreaction. With the deuterated compound **30**-d, products (*Z*) and (*E*)-**32**-d are formed with a quantum yield of 8.5%. This corresponds to a deuterium isotopic effect of $k_H/k_D \approx 3$, which is typical for a primary kinetic deuterium isotopic effect (Section 3.9) with an early transition state. The rate-determining step is the hydrogen (deuterium) transfer in the triplet excited state of cyclopropenes **30** and **30-d**, leading to intermediate **31**.

6.3.3 Electrocyclic ring closures of conjugated dienes and ring opening of cyclobutenes

Apart from their (*Z*) ⇄ (*E*) isomerization under direct or sensitized photoirradiation, conjugated dienes can undergo photoinduced isomerizations into the corresponding cyclobutenes and/or bicyclo[1.1.0]butanes. On comparing the first excited singlet electronic configuration of *s-cis*-butadiene with the first excited singlet electronic configuration of cyclobutene, Longuet-Higgins (Section 5.2.2) found that these two electronic configurations share the same symmetry only if they evolve on a hypersurface maintaining the C_s-symmetry (a mirror plane of symmetry) for a concerted reaction that would interconvert them (isomerization (6.4)). This implies a favored disrotatory mode of what Woodward and Hoffmann named a photochemical electrocyclization and noted [π^4d]. The same analysis concludes that the conrotatory mode (noted [π^4c]) is favored for the concerted thermal ring opening of cyclobutenes into 1,3-dienes:

Scheme 6.9 Examples of photoinduced hydrogen transfer reactions in alkyl-substituted cyclopropenes.

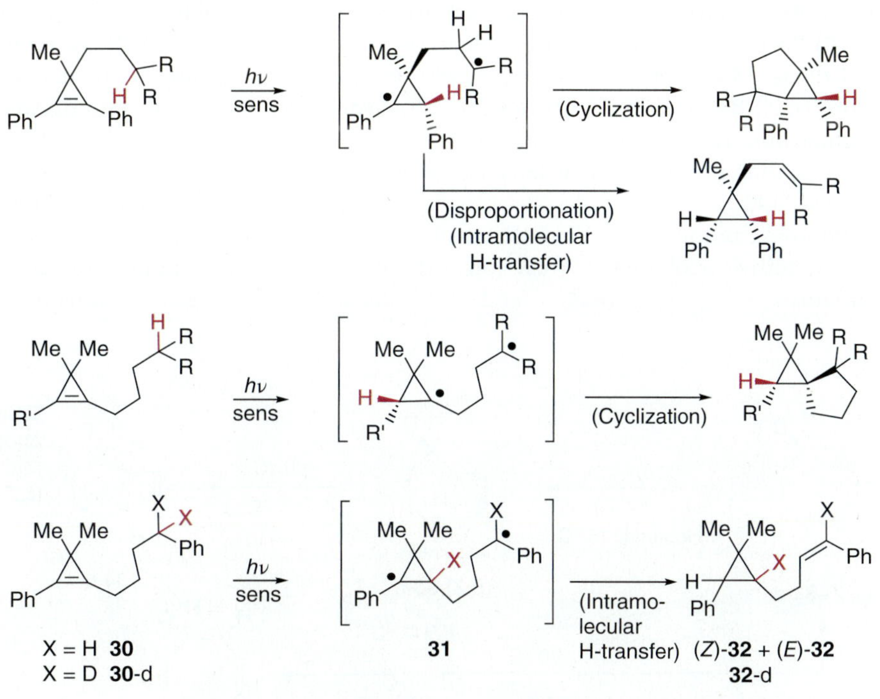

see also the discussion of the Woodward–Hoffmann rules (Section 5.2.3). This is verified with the photochemical ring closure (6.4) of (*E*,*E*)-hexa-2,4-diene into *cis*-3,4-dimethylcyclobutene at low conversion (<10%), when only small amounts of (*E*,*Z*)- and (*Z*,*Z*)-stereoisomers have been formed. Direct irradiation (quartz vessel, Et_2O solvent) of (*Z*,*Z*)-cycloocta-1,3-diene ((*Z*,*Z*)-**33**) gives bicyclo[4.2.0]oct-7-ene (**34**) (reaction (6.5)) [139, 140] in apparent agreement with the Longuet-Higgins analysis and the Woodward–Hoffmann rules for concerted reactions in the first singlet excited state.

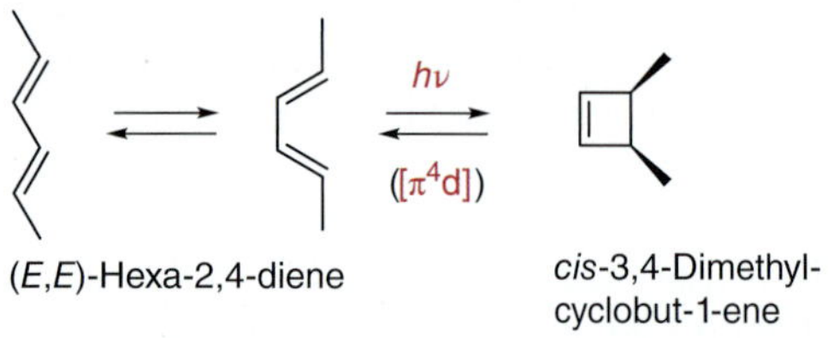

(6.4)

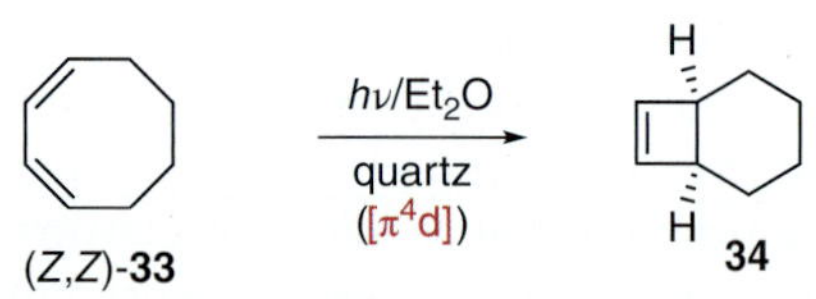

(6.5)

At 80 °C, the photosensitized isomerization of the deuterated derivative (*Z*,*Z*)-**33**-d produces **34**-d (Scheme 6.10). At lower temperature (15 °C), the sensitized photoisomerization gives first (*E*,*Z*)-**33**-d, which, above 50 °C, is isomerized into (*Z*,*Z*)-**33**-d and **34**-d (76%) in an "allowed" thermal [π^4c] electrocyclization, competitively with its isomerization into (*Z*,*Z*)-**33**-d (24%) [141]. As no 6-deuterocycloocta-1,3-diene (**35**) is formed, the thermal isomerization (*E*,*Z*)-**33** → (*Z*,*Z*)-**33** is not a (1,5)-hydrogen sigmatropic rearrangement, but a (*E*) → (*Z*) isomerization that becomes facile because of ring strain release [142].

Photoexcitation of *cis*- (reaction (6.6)) and *trans*-3,4-dimethylcyclobutene (reaction (6.7)) gives mixtures of (*E*,*E*)-, (*E*,*Z*)-, and (*Z*,*Z*)-hexa-2,4-diene (products of ring opening) and products of [2+2]-cycloreversion [38]. The photoinduced ring openings are not stereoselective, contrary to the corresponding thermal isomerization (Sections 5.2.1–5.2.4) [143]. Similarly, the direct irradiation of bicyclo[4.2.0]oct-7-ene and of its 7-methyl and 7-trifluoromethyl substituted derivatives are not stereoselective, as mixtures of (*Z*,*Z*)- and (*E*,*Z*)-cycloocta-1,3-dienes are formed ([π^4d] and [π^4c] electrocyclic ring openings) together with the formation of acetylenes and cyclohexene ((2+2)-cycloreversion) [144–146].

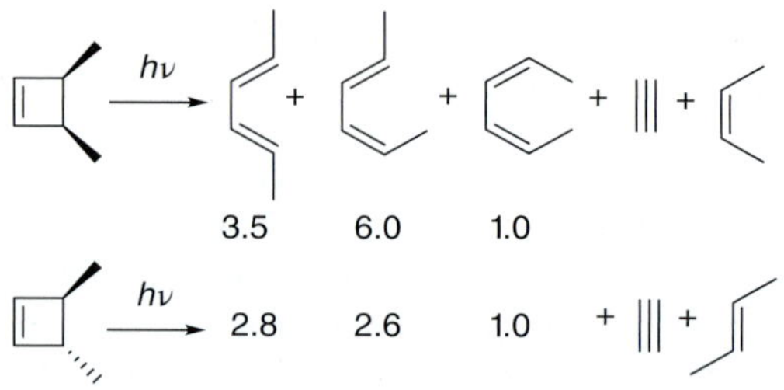

(6.6)

(6.7)

Apart from (*E*) ⇄ (*Z*) isomerizations, direct irradiation of 1,3-dienes that cannot adopt the *s-cis* conformation are isomerized into their corresponding bicyclo[1.1.0]butanes in their singlet excited states as illustrated with photoreactions (6.8) and (6.9) [38, 147]. Direct irradiation of butadiene yields mostly cyclobutene and a small amount of bicyclo[1.1.0]butane [148]. In contrast, direct irradiation of 2,3-bis(*t*-butyl)butadiene yields 1,3-bis(*t*-butyl)bicyclo[1.1.0]butane (reaction (6.10)) as only photoproduct [149]. Although mono and dialkyl substituted cyclobutadienes are not photoisomerized into their tetrahedrane (tricyclo[1.1.0.0^{2,4}]butane) derivatives, tris-and tetrakis(*t*-butyl)cyclobutadiene generate the corresponding tetrahedranes (reaction (6.11)) under direct irradiation [150]. Woodward and Hoffmann consider the disrotatory photoisomerization butadiene → cyclobutene to be a concerted

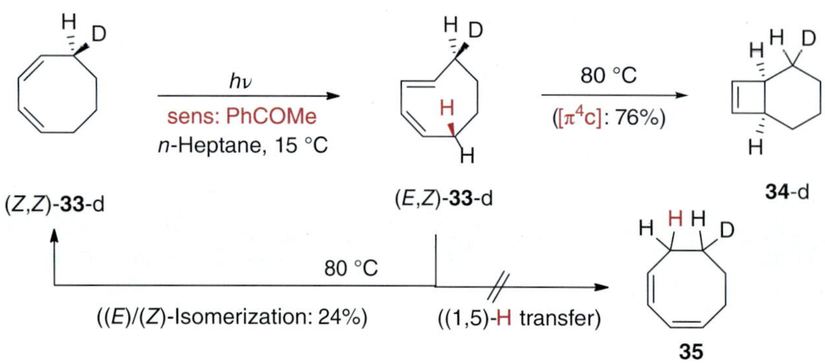

Scheme 6.10 The sensitized photoisomerization of (*Z*,*Z*)-cycloocta-1,3-diene generates (*E*,*Z*)-cycloocta-1,3-diene.

and allowed process. They see the photoinduced isomerization butadiene → bicyclo[1.1.0]butane as an "allowed" process if one considers it to be an intramolecular antarafacial/antarafacial $[\pi^2a+\pi^2a]$-cycloaddition. Quantum mechanical computations support a concerted disrotatory ring closure mechanism for the singlet excited state of butadiene into cyclobutene. Similarly, a concerted and synchronous path characterized by a conrotatory rotation of the two terminal methylene groups is predicted for the direct photoisomerization of butadienes into bicyclo[1.1.0]butanes [151]. The thermal isomerization of bicyclo[1.1.0]butanes into butadienes follows the same mechanism [152–154]. It is supported by quantum mechanical calculations [155].

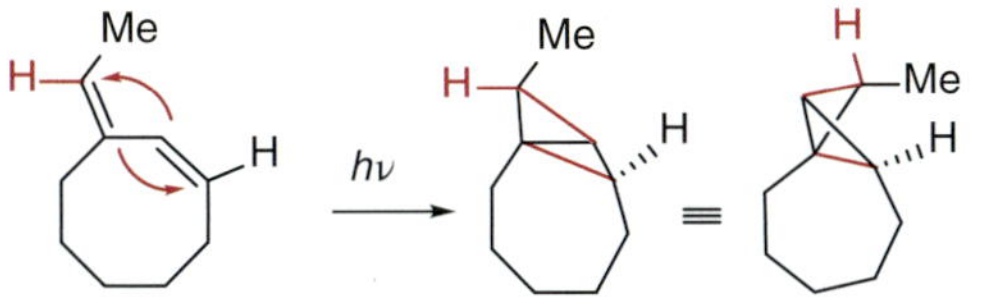

(6.8)

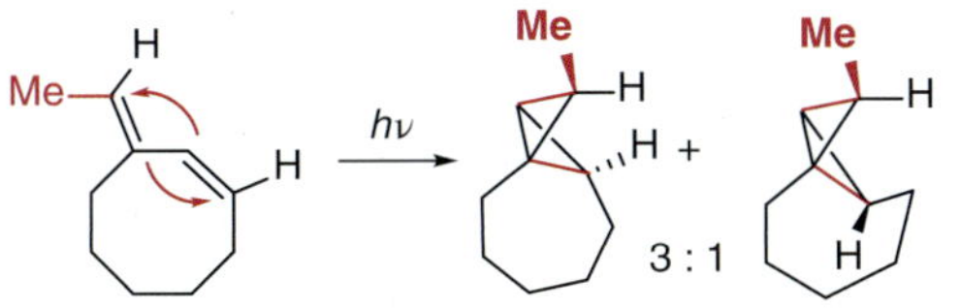

(6.9)

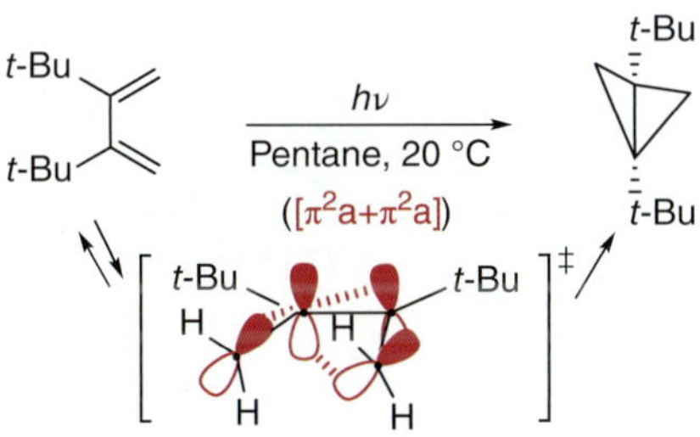

(6.10)

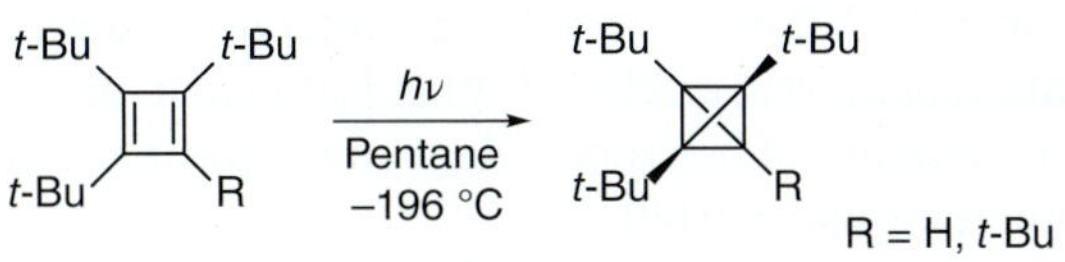

(6.11)

Problem 6.4 Does the transition state of the thermal isomerization of bicyclo[1.1.0]butane benefit from some aromatic stabilization?

6.3.4 The di-π-methane (Zimmerman) rearrangement of 1,4-dienes

One of the most important photochemical reaction of 1,4-dienes is their rearrangement into vinylcyclopropanes, the di-π-methane (DPM) rearrangement (6.12). The reaction discovered by Gary Griffin in 1965 (with the photorearrangement of 1,3-diphenylpropene into *cis*- and *trans*-1,2-diphenylcyclopropane) [156] has been extensively investigated by Zimmerman and coworkers [157–159]. Depending on the structure of the 1,4-diene and the reaction conditions, the intramolecular (2+2)-cycloaddition (6.13) can compete (Section 6.8.1). The second example of a DPM rearrangement reported in 1966 was Zimmerman's report on the photolysis of barrelene (1–2% pentane solution containing 3–8% acetone as sensitizer) that affords 25–40% semibullvalene (Scheme 6.11a) and 1–2% cyclooctatetraene [160, 161].

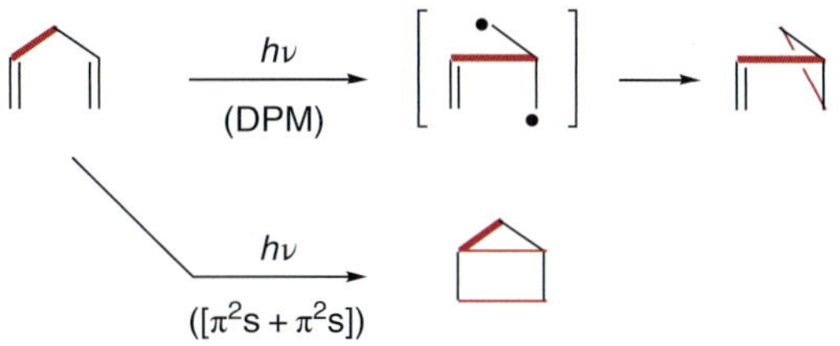

(6.12)

(6.13)

The sensitized photoisomerization of barrelene into semibullvalene suggests that there are two diradical intermediates **36** and **37** on the triplet excited hypersurface (Scheme 6.11a) [162]. The DPM rearrangement of acyclic diene **38** proceeds in the absence of triplet sensitizer. The reaction occurs via diradical intermediates of type **39** and **40** that leads to vinylcyclopropane **41** (Scheme 6.11b). Similarly, with alkene **42** in which a phenyl ring is one of the two vinyl components of the DPM rearrangement, the reaction involves intermediates **43** and **44** (Scheme 6.11c). As an example taken from Nature, one can cite the conversion of erythrolide B into erythrolide A by sunlight in a glass vessel [163].

The direct irradiation of barrelene mostly generates cyclooctatetraene in a three step process (reaction (6.14)). First, a photoinduced $[\pi^2s+\pi^2s]$-cycloaddition gives a tetracyclic compound that undergoes a thermal $[\pi^4s+\pi^2s]$-cycloreversion into bicyclo[4.2.0]octa-2,4,7-triene. Finally, a disrotatory ring opening $([\pi^4d])$ of the cyclobutene moiety of the latter that occurs simultaneously with the $[\pi^6d]$-ring opening of its cyclohexadiene moiety provides cyclooctatetraene [164].

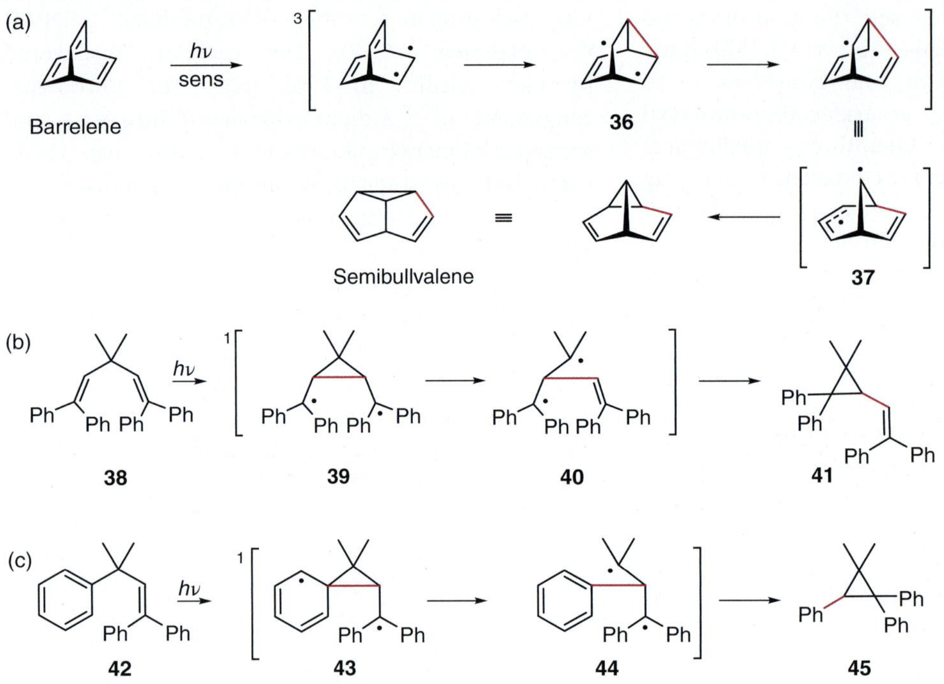

Scheme 6.11 Examples of DPM rearrangements of (a) a bicyclic homoconjugated diene, (b) an acyclic homoconjugated diene, and (c) an allylbenzene system in which the benzene ring is one of the π-partners of the photoreaction.

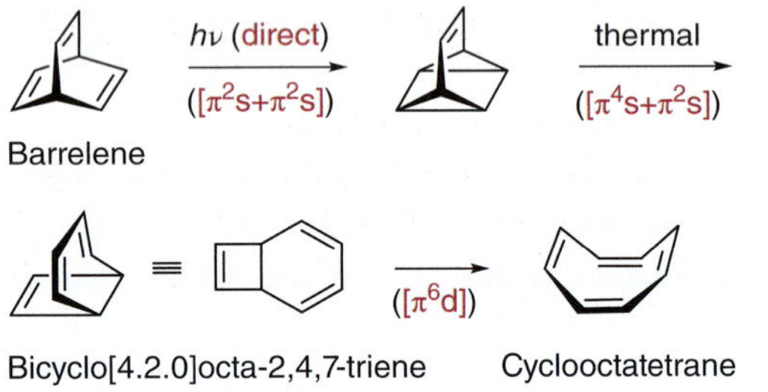

(6.14)

Like racemic sodium ammonium tartrate that crystallizes below 28 °C into mixtures of enantiomeric crystals that can be differentiated by their morphology and separated by hand, as done by Pasteur (Section 1.14.3), dibenzobarrelene derivative **46** (obtained by Diels–Alder reaction of anthracene with diisopropyl acetylenedicarboxylate) that is achiral in solution crystallizes from ethanol into two enantiomeric crystals of the space group $P2_12_12_1$. Irradiation of one of the two forms of these crystals give dibenzosemibullvalene derivative **47**, the product of DPM rearrangement (6.15), with 100% ee [165, 166]. This realizes an absolute asymmetric synthesis from achiral molecules in the chiral crystalline environment [167]. The dynamics of the triplet surface reaction have been explored for dibenzobarrelene and the selectivities of reactions related to the conversion of **46** to **47** have been studied [165, 166].

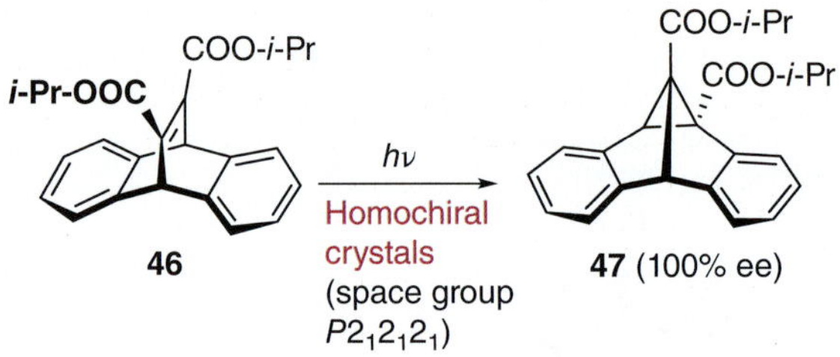

(6.15)

Problem 6.5 What is the major product of the direct irradiation of benzobarrelene? What is the main product of irradiation of benzobarrelene in the presence of acetophenone? [168, 169]

Problem 6.6 What is the major product of the sensitized irradiation of benzobarrelene-5-carbonitrile? [170]

Problem 6.7 What is the major product of the photoexcitation of 1,4-methano-1,4-dihydronaphthalene-5-carbonitrile? [171–173]

Problem 6.8 Propose a mechanism for the photoisomerization of 1,3-diphenylpropene into *cis*- and *trans*-1,2-diphenylcyclopropane and 1-phenylindane [156].

6.3.5 Electrocyclic interconversions of cyclohexa-1,3-dienes and hexa-1,3,5-trienes

Vitamin D plays a universal function as a signaling molecule in several biological processes [174, 175]. Besides controlling calcium uptake, there is evidence that it can prevent cancer [176] and autoimmune diseases [177]. The major source of vitamin D is exposure to sunlight, which induces the conrotatory ring opening ([π^6c]) of ergosterol (provitamin D = Pro) into previtamin D (Pre) in the skin (reaction (6.16), Scheme 6.12) [178, 179]. The latter equilibrates with vitamin D (Vita) through a thermal (1,7)-hydrogen shift (Section 5.5.7) [180, 181]. In solution, the photochemistry of previtamin D is complicated by a number of side reactions such as the (Z) ⇄ (E) isomerization of previtamin D into tachysterol (Tachy) [182], the ring closure into lumisterol (Lumi) [183, 184], and the formation of toxic compounds (Toxi) such as **48** resulting from a disrotatory [π^4d] photoelectrocyclization (Scheme 6.12) [185, 186]. Photoinduced conrotatory electrocyclic [π^6c] ring openings have been applied in the total synthesis of natural products [187].

Photoinduced electrocyclic ring opening of cyclohexa-1,3-diene into hexa-1,3,5-triene has been investigated extensively by time-resolved spectroscopy [188, 189] and by quantum mechanical calculations [190, 191], including real-time simulations [192]. Absorption of UV light promotes cyclohexa-1,3-diene into its first excited singlet state $^1(\pi,\pi^*)_1$. This species undergoes twisting about its C_2-axis and stretching of the σ(C(5)—C(6)) bond. This renders possible the overlap between the π-system at C(1),C(2),C(3),C(4), and the σ(C(5)—C(6)) bond. Thus, excited cyclohexa-1,3-diene evolves on a hypersurface (called 1B) that maintains an axis of symmetry C_2. This corresponds to a [π^6c] conrotatory ring opening in agreement with the Woodward–Hoffmann rules (reaction (6.17), Figure 6.11). During this rearrangement, the character of the excited state changes adiabatically from 1B to 2A. There is an avoided crossing between the $^1(\pi,\pi^*)_1$ hypersurface and the ground state. Symmetry breaking is thought to be necessary to reach the conical intersection. The entire excited state process terminates within 200 ps [193]. In dilute solution, the direct irradiation of cyclohexa-1,3-diene produces hexa-1,3,5-triene and benzene by elimination of H_2. Sensitized photoexcitation leads to cyclodimerization ((2+2) and (4+2)-cycloadditions, Section 6.8) [194].

The direct irradiation of hexa-1,3,5-triene and its substituted derivatives produces mixtures of products: (Z) ⇄ (E) isomerization, [π^4d]-electrocyclization, and intramolecular (2+2)-cycloadditions

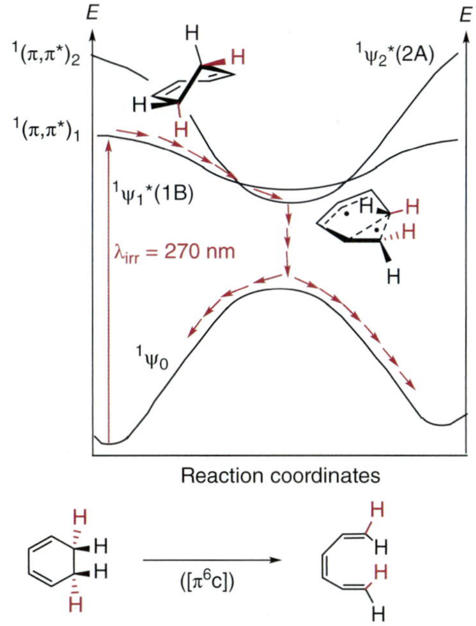

(6.17)

Figure 6.11 Conrotatory photoinduced electrocyclic ring opening of cyclohexa-1,3-diene into *s-cis*-hexa-1,3,5-triene.

compete with the electrocyclic [π^6c] ring closure [195, 196]. Sensitized triplet photochemistry of hexa-1,3,5-trienes mostly leads to their (E) ⇄ (Z) isomerization [197, 198]. Upon irradiation of (E)-stilbene in cyclohexane, its (Z)-isomer forms (Section 6.3.1) [199], which undergoes a photoinduced [π^6c]-ring closure into an unstable dihydophenanthrene *trans*-**49**, which is then converted thermally or photochemically back to (Z)-stilbene [200]. In the presence of an oxidizing agent, such as O_2 or I_2, *trans*-**49** is converted irreversibly into phenanthrene [201–203]. In the vapor phase, *cis*-**49** is formed instead of *trans*-**49** (Scheme 6.13), and *cis*-**49** then undergoes the thermal elimination of H_2 and formation of phenanthrene. Intermediate *cis*-**49** does not react with O_2 [204]. In the gas phase, *trans*-**49** also forms, but it cannot eliminate H_2. At 170 °C, it is immediately converted back into (Z)-stilbene and its concentration remains too low to react with O_2. This hypothesis is supported by the fact that the quantum yield (ϕ) for the formation of phenanthrene is c. 10 times higher in solution than in the vapor phase.

Irradiation of (Z)-1,2-diarylethylenes **50** and **51** in the presence of I_2 (the Mallory reaction [205]) produces [5]-helicene and [6]-helicene, respectively. Because of gauche interactions between the terminal benzene rings, these helicenes are not planar and adopt C_2-symmetrical structures. They are thus chiral [206]. In 1874, Le Bel pointed out that circularly polarized light must, in principle, induce "absolute asymmetric synthesis" [207]. In 1971, Kagan and coworkers showed that the irradiation of **51** with circularly polarized light generates [6]-helicene with not more than 0.20% ee [208]. Not much better optical yields were obtained for the irradiation with circularly polarized light of other (*Z*)-1,2-diarylethylenes generating [6]-, [7]-, [8]-, and [9]-helicenes [209, 210]. Irradiation of alkene **52** in a nematic mesophase (1 : 1 mixture of *p*-cyanophenyl *p*-butylbenzoate and *p*-cyanophenyl *p*-heptylbenzoate) rendered chiral by mechanical twisting gave [8]-helicene with an optical yield of c. 0.21% [211]. Higher optical purity has been realized for the synthesis of [5]-helicene derivatives by oxidative irradiation (I_2 + propylene oxide for the quenching of HI) of the macrocyclic diether made of (*S*,*S*)-pentane-2,4-diol and (*Z*,*Z*)-1,4-bis[2-(4-hydroxyphenyl)ethenyl]benzene [212a]. There are related reactions of many *cis*-1,2-diarylethylenes and numerous applications. For example, Irie et al. have shown the great utility of *cis*-1,2-dithienylethylenes, which undergo wavelength-dependent electrocyclic ring closure and opening [212b].

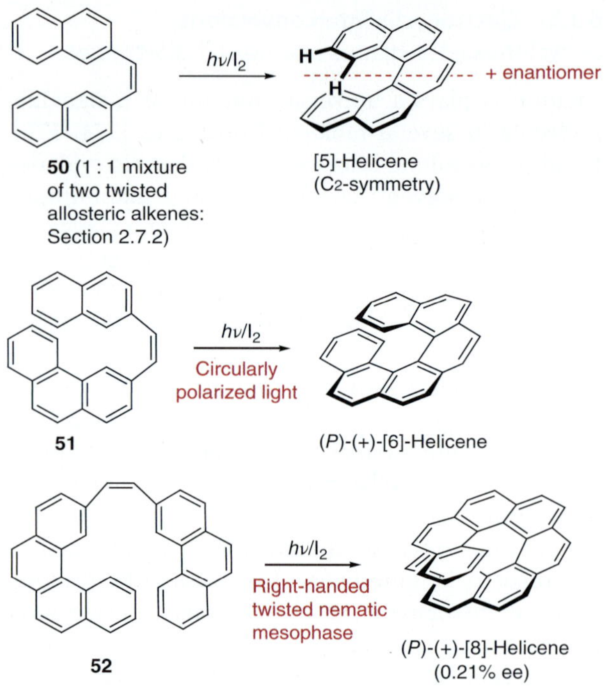

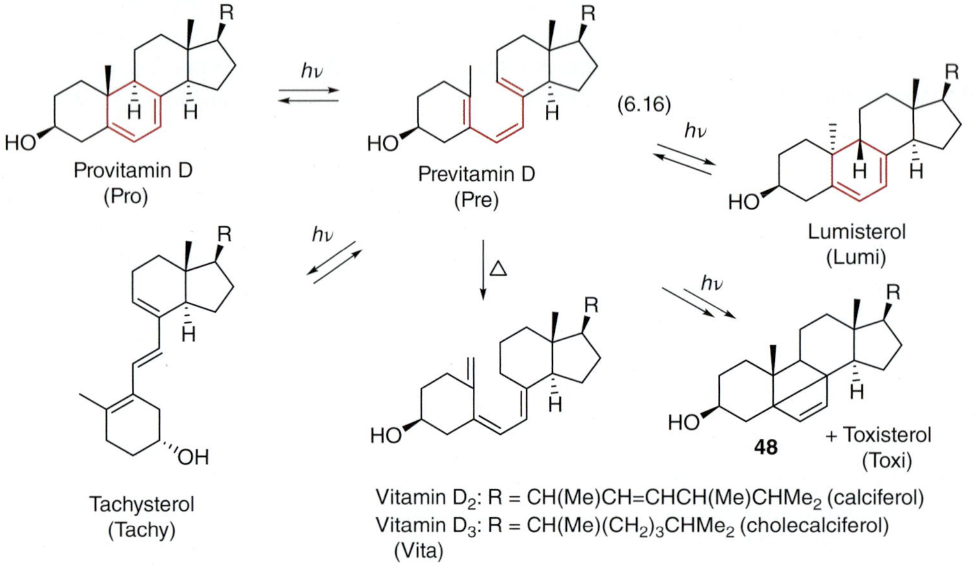

Scheme 6.12 Photochemical and thermal reactions occurring in the in vitro synthesis of vitamin D.

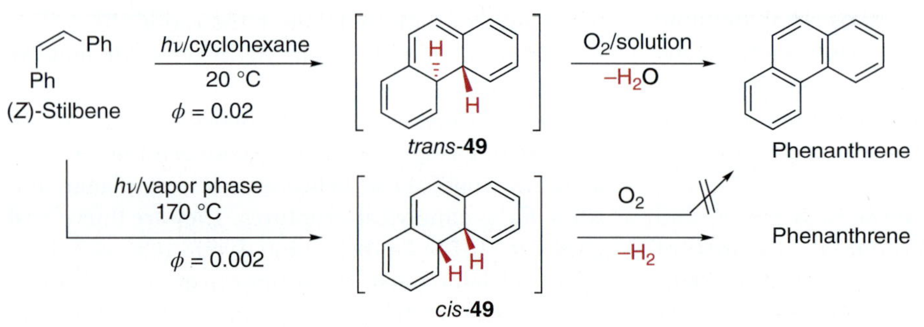

Scheme 6.13 Electrocyclic ring closures of (*Z*)-stilbene.

Problem 6.9 What are products **B** and **P** of the following photoreaction? [213]

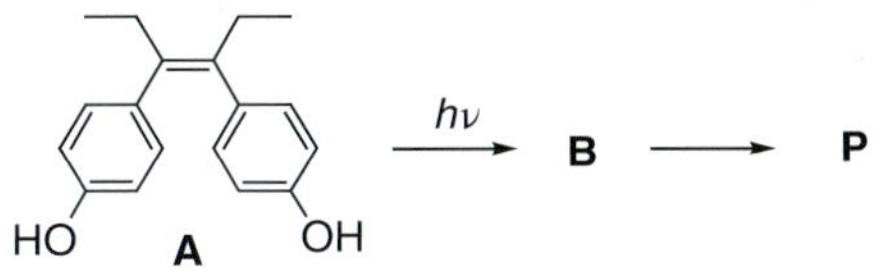

Problem 6.10 What is the product formed upon irradiation of diphenyl ether in the presence of I_2? [214]

6.4 Unimolecular photochemical reactions of carbonyl compounds

The photochemistry of aldehydes and ketones has been studied and used more thoroughly than the photochemistry of carboxylic acids, esters, and amides. Aldehydes and ketones (λ_{max} > 260 nm) require light of longer wavelength and lesser energy than carboxylic derivatives (λ_{max} < 225 nm, Figure 6.3) for their direct electronic excitation (n(CO) → π*(CO) transition) and thus lead to more chemoselective processes.

Upon excitation, the singlet $^1(n,\pi^*)_1$ state gives the corresponding triplet $^3(n,\pi^*)_1$ state through ISC. Both excited states can be represented as 1,2-diradicals. Compared with the ground state of R(X)C=O for which the oxygen center is electron rich and the carbon atom of the carbonyl is electron poor (Section 4.5.15), the excited $^1(n,\pi^*)_1$ and $^3(n,\pi^*)_1$ states have an electrophilic oxygen center and a nucleophilic carbon center (umpolung). There is one electron less in the π-plane of the carbonyl group (one electron less in the n_(CO) orbital) and one electron in excess perpendicular to it (in the $2p_C$ and $2p_O$ orbitals). This weakens the $\sigma(C_\alpha$—CO) bonds (like hyperconjugation in a carbenium ion (Sections 4.8.6 and 5.5.1)) and makes the oxygen center electrophilic through its half-occupied orbital. It may abstract a radical such as H• (intra- or intermolecular hydrogen abstraction) or accept an electron from an electron-rich moiety (intra- or intermolecular photoreduction). Apart from photocycloadditions (Section 6.8.5), the most important photochemical reactions of carbonyl compounds are summarized in Scheme 6.14 [215, 216].

HAT to the triplet excited state of carbonyl compounds also occurs in intermolecular processes. The most studied case is the photoreduction of ketones and aldehydes by secondary alcohols according to $R^1(R^2)C{=}O + R^3(R^4)CHOH + h\nu \rightarrow R^1(R^2)CHOH + R^3(R^4)C{=}O$ (+ pinacols) [217]. In a similar manner, triplet excited state of imines are reduced into amines according to $R^1(R^2)C{=}NR^3 + Me_2CHOH + h\nu \rightarrow R^1(R^2)CH{-}NH(R^3) + Me_2CO$ + (vicinal diamines) [218].

6.4.1 Norrish type I reaction (α-cleavage)

In 1907, Ciamician and Silber [219] reported the photoinduced α-cleavage of carbonyl compounds, a reaction called the Norrish type I reaction [220, 221]. Gas phase irradiation of acetone (Scheme 6.15) at λ_{irr} > 266 nm induces the C_α—CO cleavage (6.18) giving acetyl and methyl radical [222]. Under these conditions, the fragmentation of acetyl radical into CO and methyl radical does not occur as the acetyl radical formed does not have enough energy to surmount the barrier for further dissociation. However, at high energies (λ_{irr} < 193 nm), where Rydberg states are excited, both C_α—CO bonds of acetone are cleaved (reaction (6.19)), giving two methyl radicals and CO in a nonconcerted manner [223]. Femtosecond elementary dynamics of the dissociation give a time scale of 50 fs for the primary dissociation $(Me_2CO)^* \rightarrow MeCO^\bullet + Me^\bullet$, and 500 ns for the secondary breakage $MeCO^\bullet \rightarrow Me^\bullet + CO$. [224, 225] A similar conclusion has been reached for the gas phase photochemistry of butanone, pentan-3-one [226] cyclopentanone, and other dialkyl ketones [227]. In these cases, no intermediates such as $MeCOCH_2^\bullet$ or $EtCOCH_2^\bullet$ due to C_β—Me cleavage have been found to compete with the C_α—CO bond cleavage. The acetyl and alkyl radicals formed in this way can recombine or undergo disproportionation through intra- or intermolecular HAT as illustrated below. Ultrafast electron diffraction studies have permitted Zewail and coworkers to establish the structures and chemistries of excited aromatic carbonyl compounds. Following femtosecond excitation, the first excited singlet states are obtained that evolve to molecular dissociation products. For benzaldehyde (PhCHO), benzene and CO are formed, and for acetophenone (PhCOMe), benzoyl (PhCO•) and methyl (Me•) radicals are generated, along with ISC into the first excited triplet states. For the latter, the excitation is of $^3(\pi,\pi^*)$ type and is localized in the phenyl ring [228].

In the case of enantiomerically enriched ketone (*S*)-**53**, irradiation leads to racemization (6.20) concurrently with the formation of benzaldehyde and styrene resulting from the disproportionation of benzoyl radical and 1-phenylethyl radical. Racemization is slow compared with the fragmentation process. It can be stopped by adding radical scavengers such as dodecanethiol or a stable nitroxide radical such as TEMPO ((2,2,6,6-tetramethylpiperidin-1-yl)oxyl). This indicates that recombination with racemization

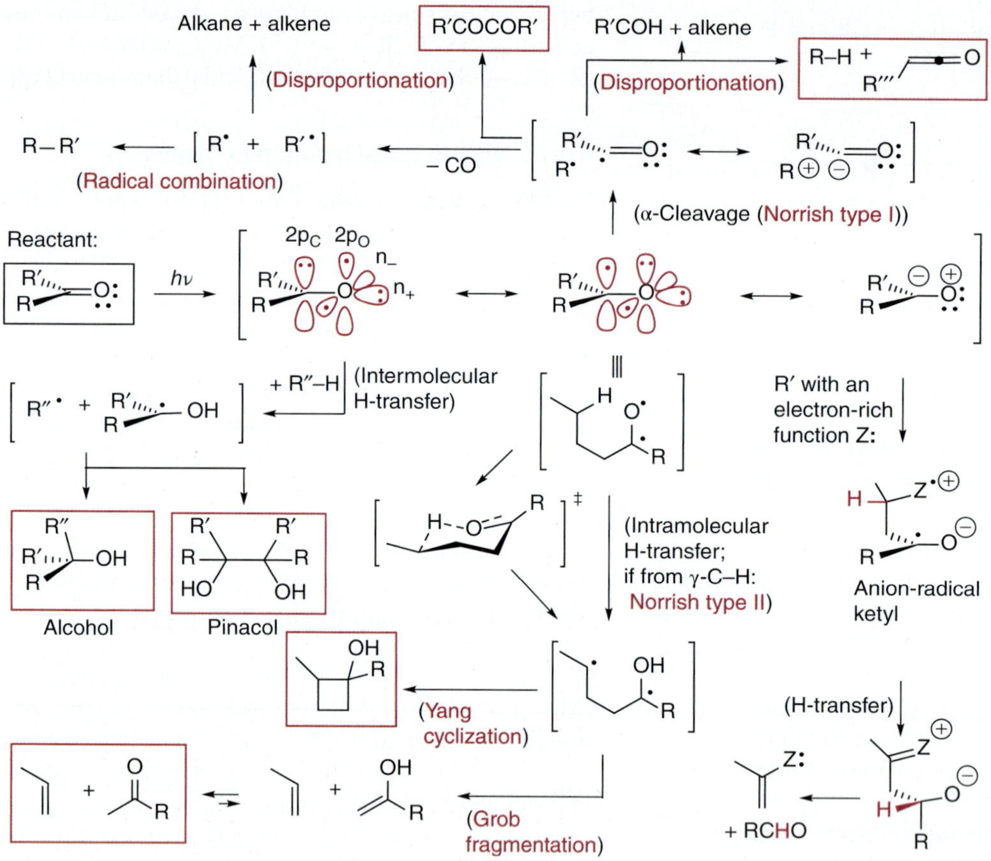

Scheme 6.14 Reactivity of the excited states of carbonyl compounds, excluding intermolecular reactions such as intermolecular hydrogen transfers and cycloadditions (see Section 6.8.5) and photoinduced electron transfers (see Section 6.10). The possible product of photolysis is shown in red boxes.

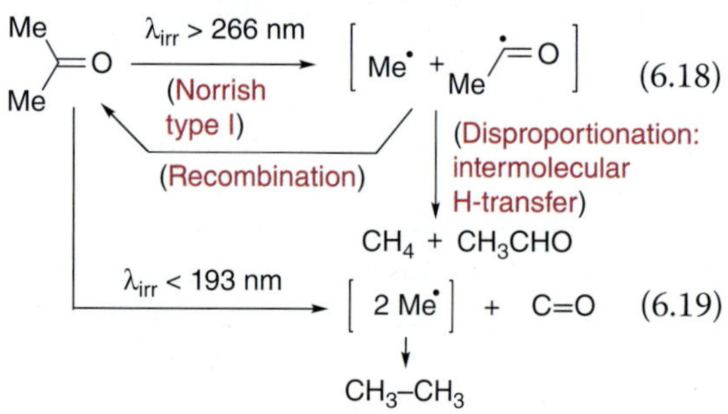

Scheme 6.15 Gas phase photochemistry of acetone.

involves free radicals that have escaped from the solvent cage, not between the initial radical pair. Such recombination of free radicals is consistent with the expectation that ISC of the triplet state radical pair to the singlet state requires much more time ($>10^{-8}$ seconds) than the lifetime of a solvent cage ($<10^{-10}$ seconds) in solution [229, 230].

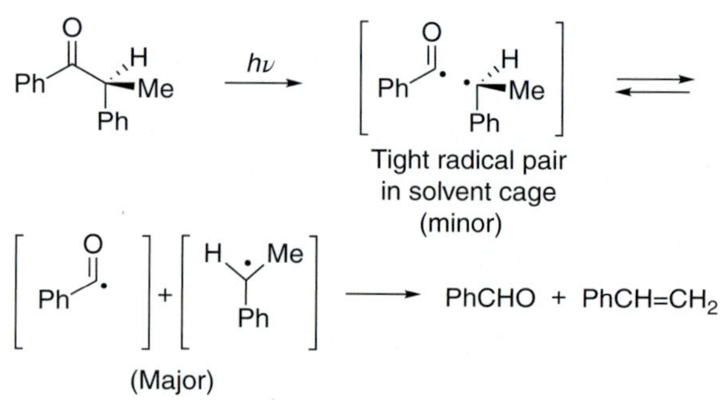

(6.20)

Similarly, irradiation of cyclopentanone **54** leads to its α-epimerization (6.21) [231, 232]. When relatively stable alkyl radicals can form, decarbonylation becomes facile as shown with photoreaction (6.22) that generates two benzyl radicals that cannot undergo disproportionation reactions and thus are forced to combine into 1,2-diphenylethane [233].

Photoreaction (6.23) of cyclohexanone produces a mixture of hex-5-enal and *n*-butylketene arising from intramolecular (1,5)-hydrogen shifts [234–236]. In this case, the decarbonylation is a slow process because of the intrinsic instability of the primary alkyl radical that must be formed in such a fragmentation. When HATs are difficult or impossible for structural reasons, the photodecarbonylation becomes the favored process [237] as illustrated with reactions (6.24) [238] and (6.25) [239] that generate highly strained cycloalkynes.

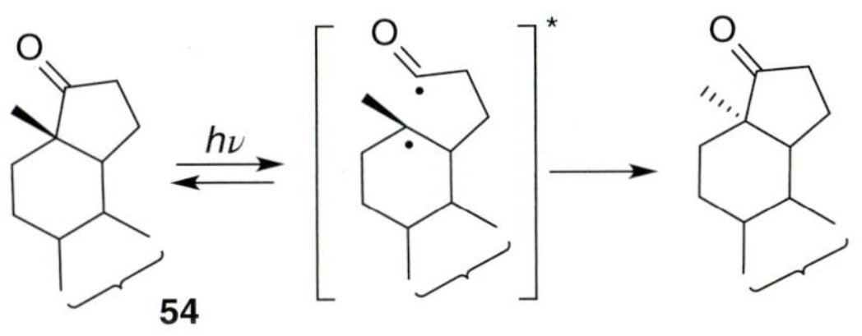

(6.21)

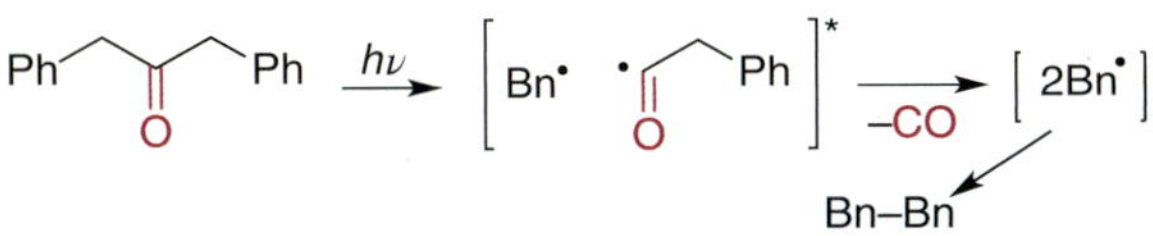

(6.22)

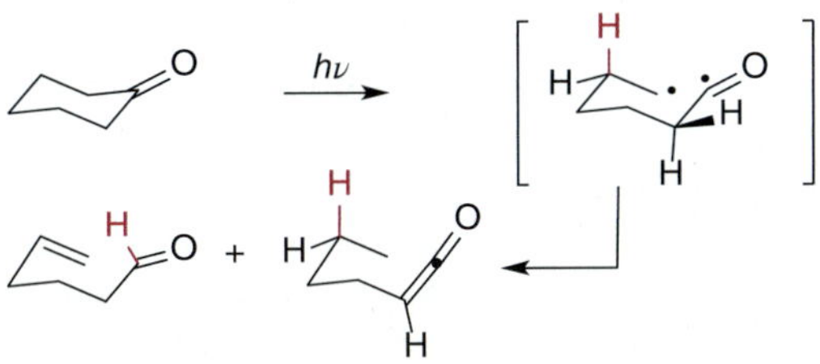

(6.23)

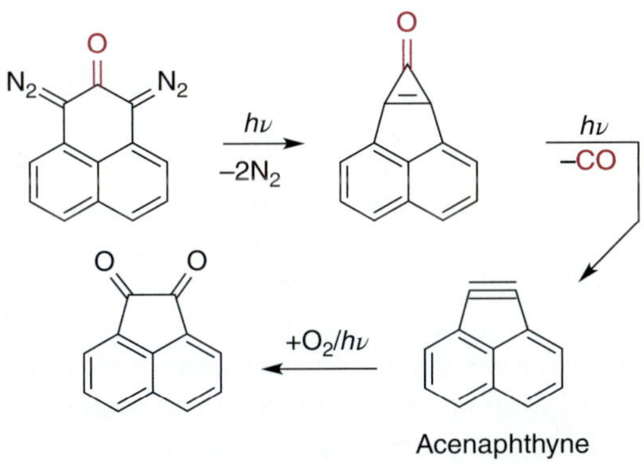

(6.24)

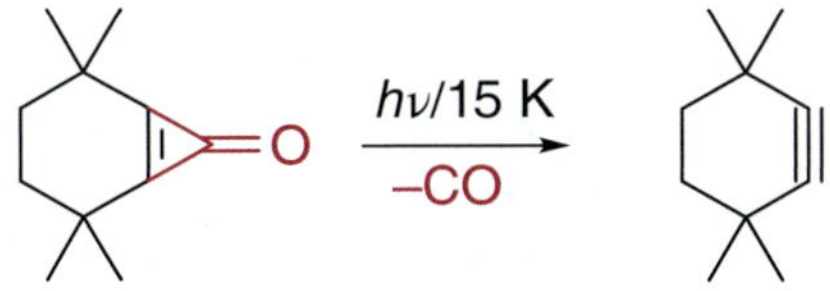

(6.25)

Norrish type I photodecarbonylation (6.26) is the key step in Maier's synthesis of tetrakis(*t*-butyl)tetrahedrane (Scheme 6.16) [240].

The photochemistry of cyclobutanone and its derivatives is completely dominated by the primary photochemical process of α-cleavage [241, 242]. The subsequent chemistry is special because it occurs via the singlet $^1(n,\pi^*)$ excited state, rather than the triplet $^3(n,\pi^*)$ state. Apart from its decarbonylation into cyclopropane and fragmentation into ketene and ethylene, the initial 1-oxobuta-1,4-diyl diradical undergoes rearrangement into tetrahydrofur-2-ylidene, an oxycarbene that is evidenced by its quenching with nucleophiles such as methanol [243] and by low-temperature irradiation in a matrix [244]. An oxycarbene is a relatively stable carbene due to $n(O:) \rightarrow 2p_C$ donation. This is exemplified with cyclobutanone **55** (Scheme 6.17). Oxycarbene intermediates have been identified in the photoinduced rearrangements of other cyclic ketones [245–247].

Photochemical double decarbonylation of α-diketones of the type bis(areno)bicyclo[2.2.2]octa-5,7-diene-2,3-dione, known as the Strating–Zwanenburg reaction [248], has been applied to the synthesis of oligoacenes (n linearly condensed benzene rings) such as pentacene, hexacene, and heptacene [249, 250]. Photolysis of racemic camphor with circularly polarized light decomposes camphor in an enantioselective manner. After 99% of degradation of the ketone, the recovered camphor has an ee of c. 20% [251, 252].

6.4.2 Norrish type II reaction and other intramolecular hydrogen transfers

Irradiation of optically active ketone (*R*)-**56** leads to its racemization, to fragmentation into acetophenone plus 2-methylbut-1-ene resulting from a Norrish type

Scheme 6.16 Maier's synthesis of tetrakis(*t*-butyl)tetrahedrane.

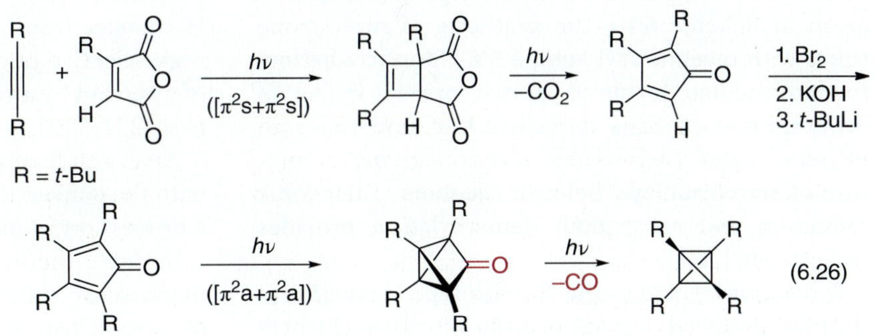

(6.26)

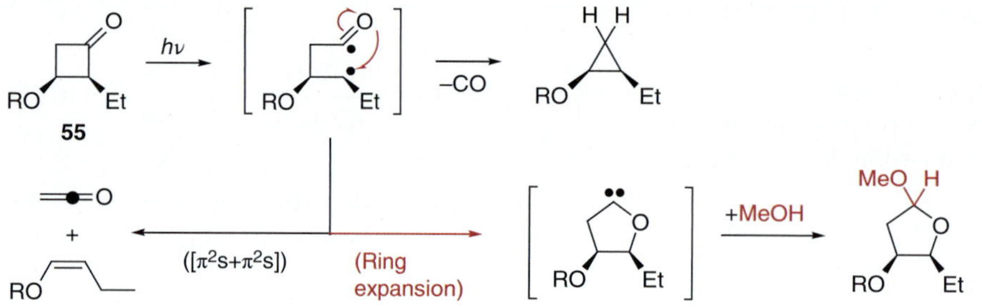

Scheme 6.17 Stereospecific photodecarbonylation of a cyclobutanone derivative, its cycloreversion into ketene and an alkene, and its rearrangement into an oxycarbene.

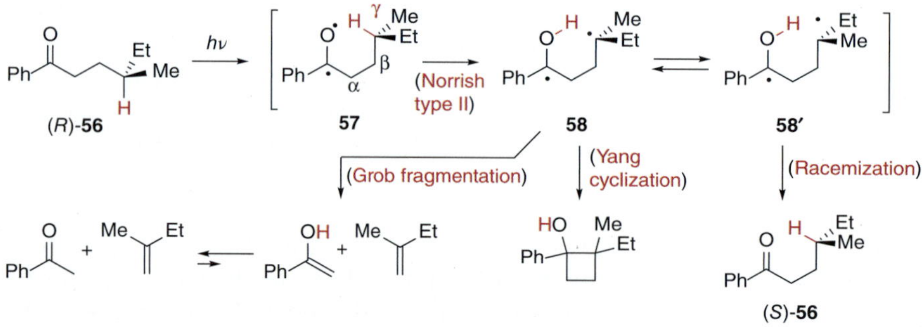

Scheme 6.18 Example of a Norrish type II reaction ((1,5)-hydrogen atom transfer).

II reaction, and to rearrangement into 2-ethyl-2-methyl-1-phenylcyclobutanol (Scheme 6.18) [253]. These observations are interpreted in terms of the formation of 1,4-diradical **58** resulting from an intramolecular (1,5)-hydrogen shift (transfer of a hydrogen atom from the γ-C—H center) in the initially formed $n\pi^*$ excited state **57**. The γ-H transfer proceeds most likely through a six-membered chair-like transition state [254, 255]. As these reactions occur on a triplet hypersurface (ISC requires > 10^{-9} seconds), many rotations about the $\sigma(C_\beta—C\gamma)$ bond have the time to occur (one rotation requires about 10^{-13} seconds). Direct spectroscopic detection of 1,4-diradical of type **58** was possible, and lifetimes in the range of 100–1000 ns have been measured for these species in solution at 20 °C [256–258]. Triplet diradicals **58** have several reaction channels [258]. The Grob fragmentation gives an alkene and an enol that equilibrates with the corresponding ketone [259]. A cyclization (Yang cyclization [260]) produces cyclobutanol derivatives [261–264].

Another example of a Norrish type II reaction is given in Scheme 6.19; the synthesis of *rac*-estrone starts with racemic aryl ketone **59**. Upon irradiation, the intermediate 1,4-diradical that forms is in fact an *ortho*-quinodimethane derivative that undergoes an intramolecular Diels–Alder reaction giving a mixture of stereoisomeric benzylic alcohols. Their ionic reduction and subsequent demethylation provides *rac*-estrone [265].

A reaction related to the Norrish type II reaction is the photoinduced cleavage of *ortho*-nitrobenzyl ethers into the corresponding *ortho*-nitrosobenzaldehyde and alcohols (Scheme 6.20). The *ortho*-nitrobenzyl group is a very useful protective group of alcohols and other moieties as it can be removed under neutral conditions by simple irradiation in the visible spectrum as the nitrophenyl chromophore absorbs in the 300–380 nm region. Sometimes, biochemists refer to this as "decaging" when it is used to release a biochemically active compound in a living system [266–269]. The acinitro intermediate has been detected by flash photolysis studies, with an absorption maximum of 408 nm. Its lifetime depends on pH (milliseconds at pH 5–6, seconds at pH > 8).

The Norrish type II reaction involves an intramolecular hydrogen transfer from a γ-C—H center, but abstraction of a hydrogen atom from another position is possible if the migration from the γ-C—H position is prevented for structural or geometrical reasons. Weigel and Wagner have shown that migration from a β-C—H position occurs in the photolysis of β-arylketones that leads to the formation of cyclopropanols [270]. An example of intermolecular H-transfer from a remote position is given with the regioselective photoinduced oxidation (Scheme 6.21) of a steroid by attachment of a benzophenone moiety to it [271, 272].

An example of a (1,6)-HAT is shown (Scheme 6.22) with the conversion of ketone **60** into dodecahedrane, a process developed by Paquette [273a].

In fact, there is some evidence from quantum mechanical calculations that a seven-membered hydrogen abstraction transition state is preferred over

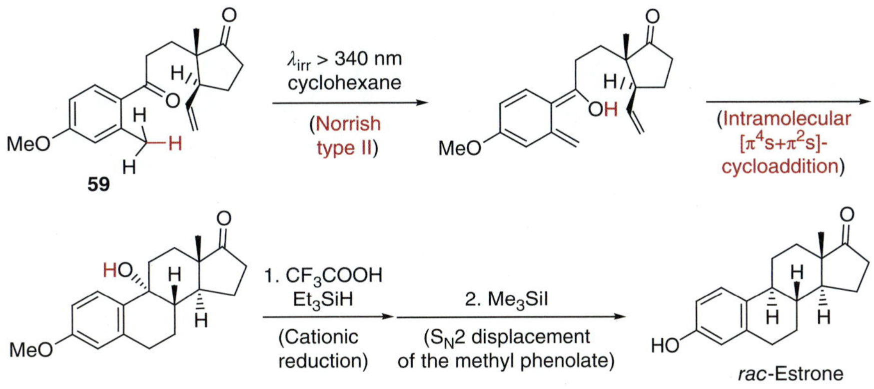

Scheme 6.19 Synthesis of (±)-estrone applying a Norrish type II reaction.

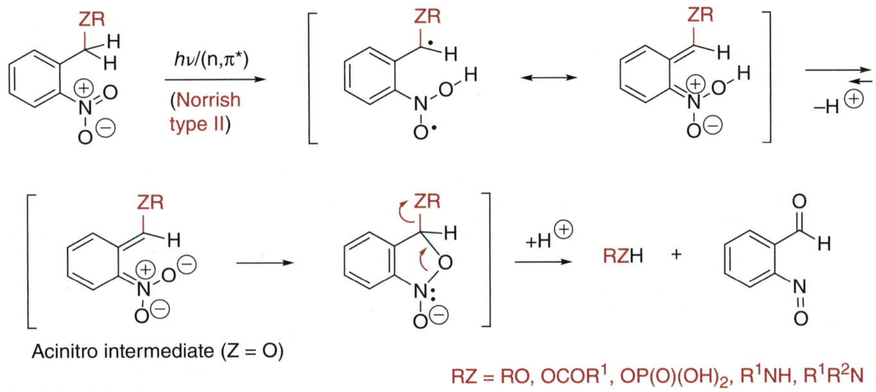

Scheme 6.20 The photocleavage of *ortho*-nitrobenzyl ethers, carboxylates, phosphates, and amines.

six because the C···H—C or O···H—C are rather short and the heavy atoms can form a nearly ideal staggered cyclohexane chair-like structure [273b].

Intermolecular HATs are common for electronically excited ketones [16]. For instance, irradiation of a mixture of acetone and isopropanol generates two 2-hydroxyprop-2-yl radicals that dimerize into pinacol or undergo disproportionation back into acetone and isopropanol (Scheme 6.14). Rate constants for the intermolecular H-atom abstractions ^{3}[PhCOMe]* + $Me_2C(OH)$-H, $PhCH_2$-H, $PhCD_2$-D, 4-$MeC_6H_4CH_2$-H → PhC$^\bullet$(OH)Me + $Me_2C^\bullet$OH, $PhCH_2^\bullet$, $PhCD_2^\bullet$, and 4-$MeC_6H_4CH_2^\bullet$ are $\approx 1\times10^6, 1\times10^5$, 0.2×10^5, and 7×10^5 M^{-1} s^{-1}, respectively, at 25 °C [274].

Problem 6.11 What are the major products of the irradiation (λ_{irr} > 300 nm) of 2,2,4,4-tetramethylcyclobuta-1,3-dione in (i) degassed benzene and (ii) in MeOH at 20 °C? [275]

Problem 6.12 Photolysis of *trans*-1-benzoyl-1-methyl-4-*t*-butylcyclohexane (**A**) gives benzaldehyde and **B**, the α-epimer of **A**. Photolysis of **B** gives an isomeric cyclobutanol **P**. What is **P**? Explain these reactions [276].

Problem 6.13 What products are formed upon irradiation (λ_{irr} = 275–380 nm) of gaseous heptanal (mixed with air, 1 atm, 23 °C)? [277]

Problem 6.14 Irradiation (in MeCN, Pyrex vessel, medium pressure Hg lamp, λ_{irr} > 290 nm) of 2,5-dimethylbenzoyloxirane results in 2-hydroxymethyl-6-methylindan-1-one. Give a mechanism for this photochemical isomerization [278].

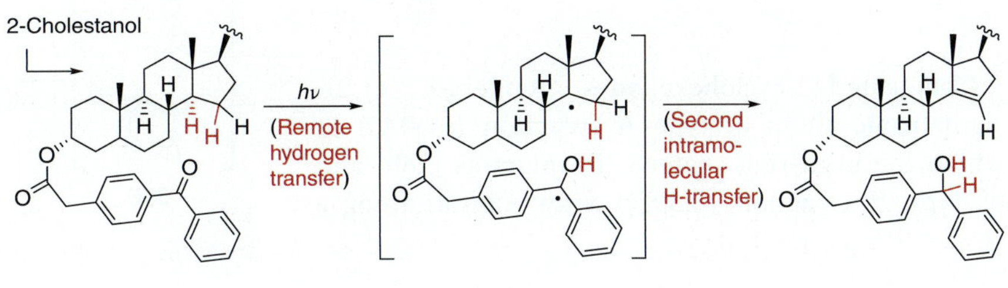

Scheme 6.21 Breslow's remote and regioselective oxidation of a steroid by intermolecular photoinduced hydrogen transfer.

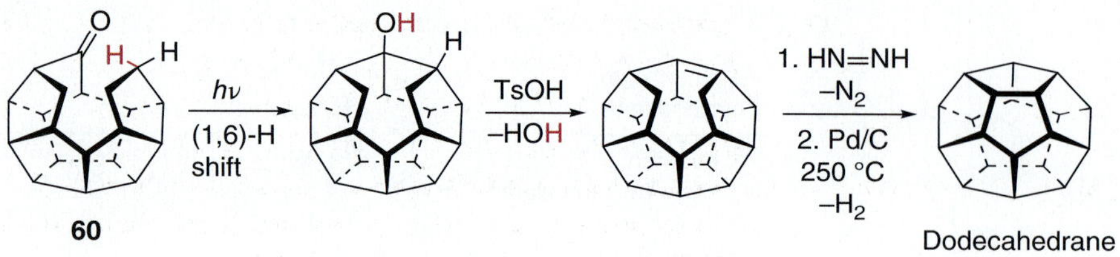

Scheme 6.22 Last steps of the Paquette's synthesis of dodecahedrane.

6.4.3 Unimolecular photochemistry of enones and dienones

Because they absorb UV light filtered by Pyrex vessels, the photochemistry of enones has been very much studied [279]. Just as with other alkenes, acyclic α,β- and β,γ-enones undergo $(E) \rightleftarrows (Z)$ isomerization upon photoexcitation. Cyclic enones can also isomerize irradiation of (*Z*)-cyclohept-2-enone leads to photoisomerization into the unstable (*E*)-cyclohept-2-enone that can be trapped in a Diels–Alder reaction producing the *trans*-cycloadduct **61** (reaction (6.27)) [280, 281]. The photochemical decarbonylation of cyclopropenones into the corresponding alkynes (reaction (6.28)) [233, 282, 283] follows a two-step mechanism [284–286]. Photolysis of simple cyclopent-2-enones generates cyclopropylketenes that are recognized by their products of addition of alcohols (e.g. reaction (6.29)) [287].

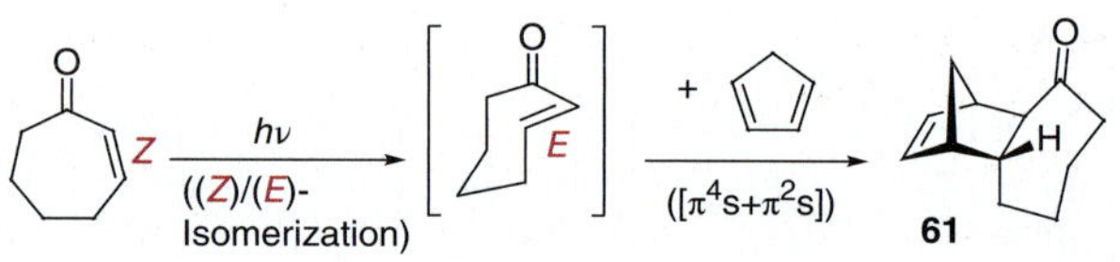

(6.27)

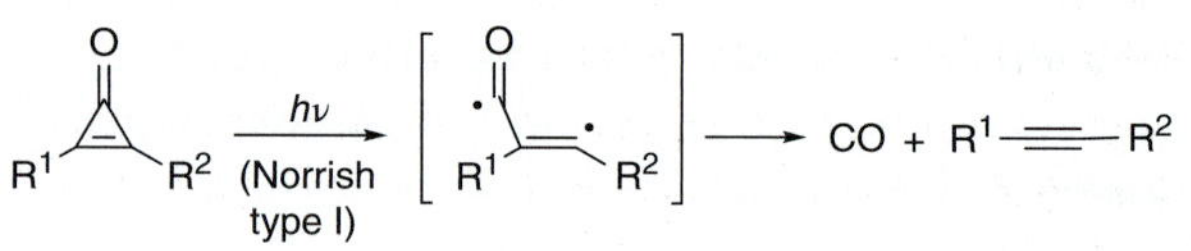

(6.28)

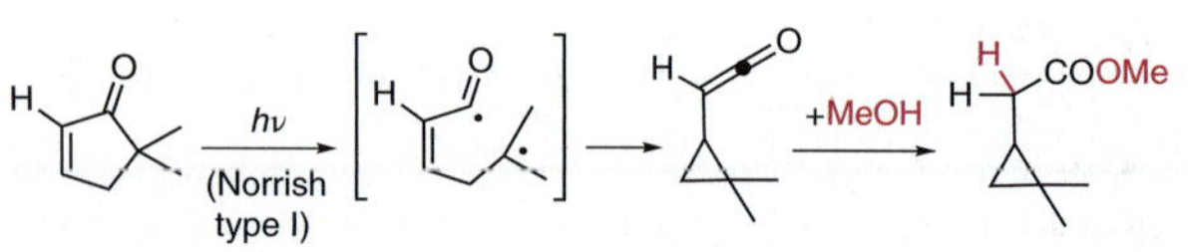

(6.29)

Conjugated cyclohexenones undergo (1,2)-sigmatropic shifts of type A (reaction (6.30A)) in which the intra-ring center C(5) migrates [288] and of type B (reaction (6.30B)) with migration of a 4-substituent [289]. Both rearrangements originate from the triplet $^3(n,\pi^*)$ state [290, 291] and generate bicyclo[3.1.0]hexan-2-one derivatives [292, 293]. They can be explained in terms of diradical rearrangements **62** → **63A** and **62** → **63B**, respectively. Photorearrangement (6.31) [294] and several similar photoreactions of enantiomerically enriched cyclohexenones [295, 296] are found to proceed with >95% retention of optical activity. Thus, the stereoselectivities of (1,2)-shifts of type A imply slow σ(C—C) bond rotation in diradical intermediates of type **63A** compared with their ring closure. Alternatively, the rearrangements might be concerted processes that can be seen as intramolecular $[\sigma^2a+\pi^2a]$-cycloadditions that are "allowed" photochemically by the Woodward–Hoffmann rules. They involve conrotatory breaking of the (1,2)-migrating σ-bond and antarafacial combination of the alkene moiety as shown with transition state **64**. Another explanation invokes the initial photoisomerization of the (*Z*)-cyclohex-2-enones into their (*E*)-stereoisomers that precedes the (1,2)-alkyl shift. Smaller (*Z*)-cycloenones do not undergo similar rearrangements as they cannot be photo-isomerized into (*E*)-cycloenones. This brings support to the latter hypothesis [297].

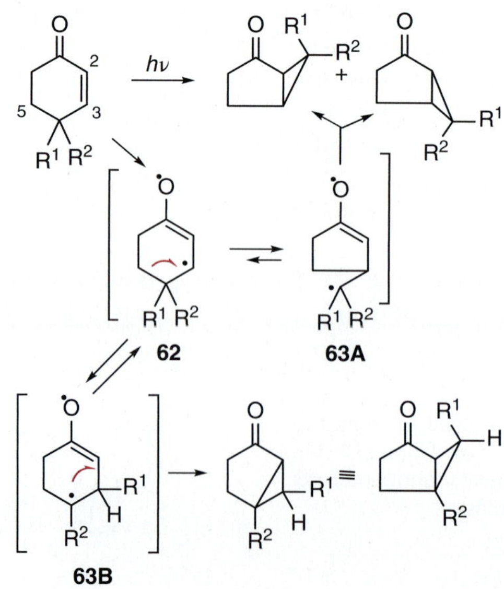

(6.30A)

(6.30B)

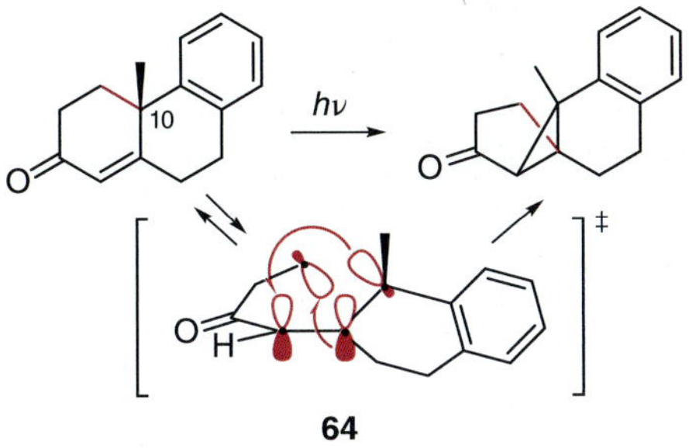

(6.31)

Direct irradiation of *β,γ*-unsaturated ketones generally results in (1,3)-sigmatropic rearrangements with (1,3)-acyl shifts [298–302], whereas sensitized photoexcitation induces oxa-di-π-methane rearrangements, also considered as (1,2)-acyl shifts [303]. The examples shown in Scheme 6.23 illustrate the usefulness of spin selectivity in organic photochemistry [304–306].

Cyclohexa-2,4-dienones undergo two typical photorearrangements (Scheme 6.24) that are the pseudo 6-electron electrocyclic ring opening to form linear ketenes when the $^1(n,\pi^*)$ excited state is involved [307–309] and a (1,2)-sigmatropic rearrangement to form bicyclo[3.1.0]hexenones when the $^1(\pi,\pi^*)$ excited state is involved [310–312].

The photochemistry of cross-conjugated dienones such as cyclohexa-2,5-dienones began in 1830 with the observation of the light-induced isomerization of α-santonin in solution into lumisantonin (see also the solid state photoreaction of α-santonin, Scheme 6.1). The latter is photoisomerized into mazdasantonin, a cyclohexa-2,4-dienone that undergoes a photoinduced electrocyclic ring opening into a ketene that reacts with water to give photosantonic acid [313, 314].

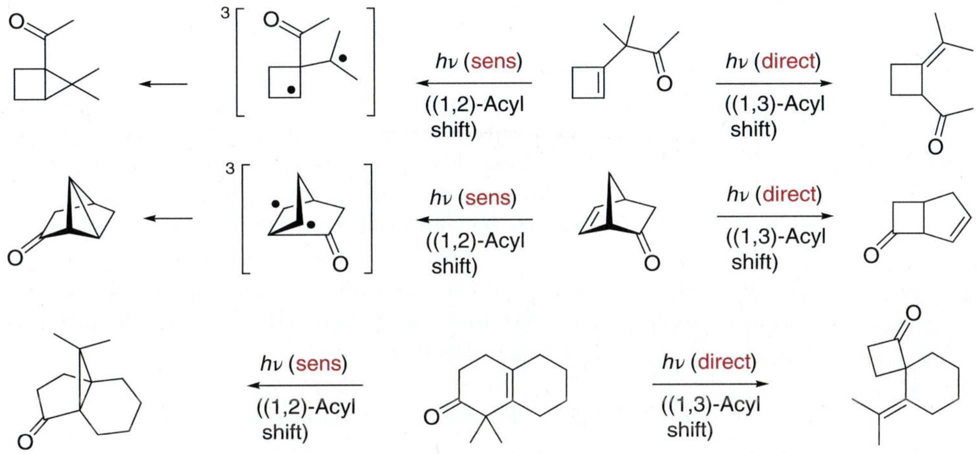

Scheme 6.23 Homoconjugated enones undergo (1,3)-acyl shift in their first singlet excited state, and (1,2)-acyl shifts (oxa-di-π-methane rearrangements) in their first triplet excited state.

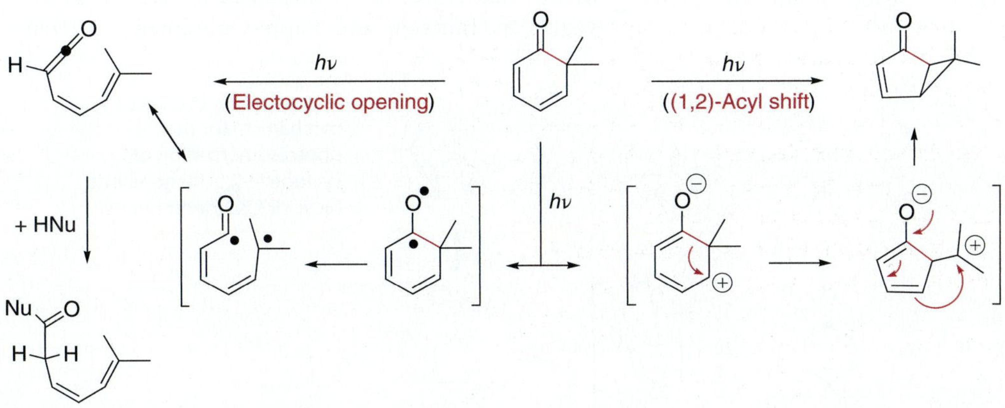

Scheme 6.24 Cyclohexa-2,4-dienones undergo photochemical electrocyclic openings and (1,2)-sigmatropic rearrangements.

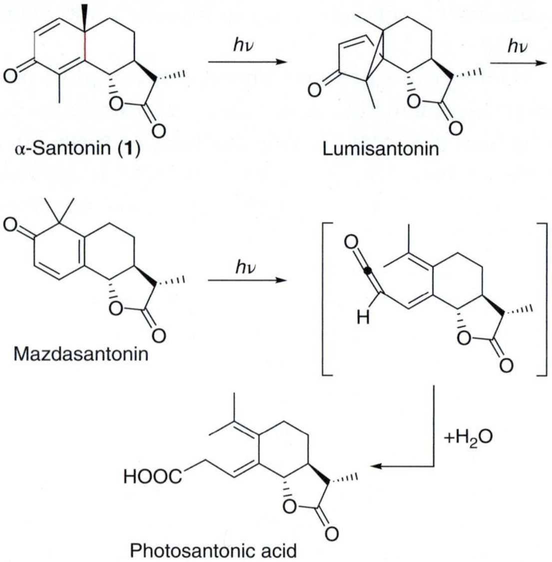

The mechanism proposed for the photoisomerization of cyclohexa-2,5-dienones into bicyclo[3.1.0]hex-3-en-2-ones (Scheme 6.25) involves a triplet $^3(n,\pi^*)$ excited state [315]. It is reminiscent of the DPM rearrangement (Section 6.3.4) and corresponds to a (1,2)-sigmatropic rearrangement [316, 317]. After light absorption and ISC, 1,4-diradical **65** forms that equilibrates with the 1,3-diradical **65′**. Because of the presence of the cross-conjugated carbonyl group, the limiting zwitterionic structure **65″** is relatively important. Like cyclopropyl alkyl cations that undergo facile cyclopropylalkyl ⇄ cyclopropylalkyl rearrangements, **65″** is isomerized into **66**, which is a limiting structure of the final bicyclo[3.1.0]hexenone.

Problem 6.15 Photolysis of 4-methyl-4-(3-methylbutyl)cyclopent-2-en-1-one leads to three products involving a diradical intermediate resulting from an intramolecular hydrogen transfer. What are these three products? [318]

Problem 6.16 Explain the following photoisomerization [296].

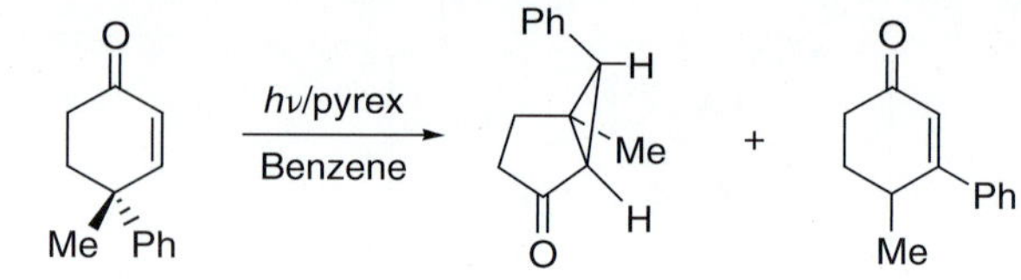

(*R*)-(+)-**A** (−)-**B**

Problem 6.17 What are the products of direct and sensitized photolysis of bicyclo[2.2.2]oct-5-en-2-one? [319]

Problem 6.18 What is the main product of photolysis of 5,6-dimethylidenebicyclo[2.2.1]heptan-2-one? [320]

6.5 Unimolecular photoreactions of benzene and heteroaromatic analogs

Thermal pericyclic reactions of benzene and aromatic analogs are difficult reactions because they lead to the destruction of aromatic electronic stabilization. In contrast, benzene and aromatic analogs are quite reactive in their electronically excited states and undergo facile $[\pi^4d]$-electrocyclizations into Dewar benzenes, and intramolecular "crossed" (2+2)-cycloadditions into benzvalenes.

6.5.1 Photoisomerization of benzene

Originally mentioned by Sir James Dewar in 1867 as one of several alternatives to the Kekulé formulation for benzene, Dewar benzene (bicyclo[2.2.0]hexa-2,5-diene) had never been observed before 1962. That year, van Tamelen and Pappas obtained 1,2,5-tri(*t*-

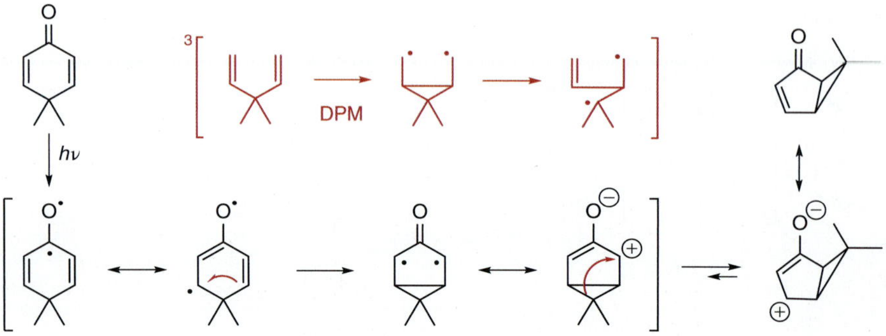

65 65′ 65″ 66

Scheme 6.25 Proposed mechanism for the photoisomerization of cyclohexa-2,5-dienone into bicyclo[3.2.0]heptenones.

butyl)bicyclo[2.2.0]hexa-2,5-diene by direct irradiation (Vycor filter, Hg burner) of 1,2,5-tri(*t*-butyl) benzene in ether solution (reaction (6.32)) [321]. In 1964, Viehe and coworkers isolated the corresponding hexasubstituted Dewar benzene, benzvalene, and prismane resulting from the spontaneous cyclotrimerization of *t*-butylfluoroacetylene (Section 5.3.19, Scheme 5.62) [322]. In 1965, Kaplan and coworkers identified tri(*t*-butyl)benzvalene and tri(*t*-butyl)prismane during the photoinduced rearrangement of tri-(*t*-butyl)benzene [323]. In 1967, they reported that unsubstituted benzene is photoisomerized into benzvalene by irradiation at λ_{irr} = 254 nm (reaction (6.33)) [324] and into fulvene upon irradiation in the gas phase at λ_{irr} = 185 nm [325]. It was then found that (*Z*)-hexa-1,3-diene-5-yne also forms under these conditions [326]. Photolysis (λ_{irr} = 254 nm) of benzene in argon matrix gives a mixture of Dewar benzene (reaction (6.35)), benzvalene (reaction (6.33)), and fulvene (reaction (6.34)) [327]. Prismane derivatives are formed during the photochemical interconversion 1,2,4,5- into 1,2,3,5-tetra(t-butyl)benzene [328] and of perfluoro-1,3,5- and 1,2,4-trimethylbenzene [329]. Direct irradiation of Dewar benzene gives benzene plus prismane (reaction (6.36)) [330], a compound made first by Katz and Acton [331]. Direct irradiation of polysubstituted Dewar benzenes has provided stable prismane derivatives (intramolecular [π^2s+π^2s]-cycloadditions) [332–334].

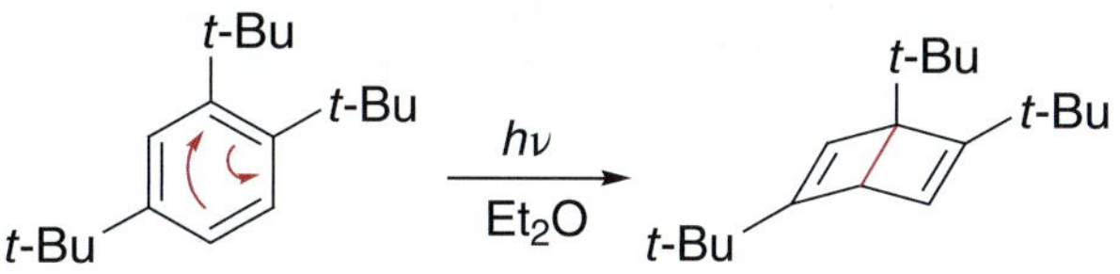

(6.32)

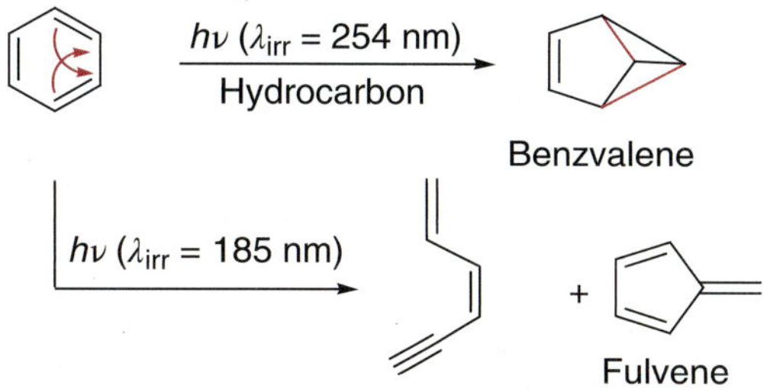

(6.33)

(6.34)

Quantum mechanical calculations support the mechanism presented in Scheme 6.26 for the photoisomerization of benzene. In its first excited $^1(\pi,\pi^*)$ state (λ_{irr} = 254 nm), benzene evolves to prefulvene diradicaloid **67s** ⇄ **67a** [335] and to cyclopentadienylcarbene **68**. Almost without energy barrier, prefulvene diradicaloids **67a** ⇄ **67s** form benzvalene and fulvene [336]. In its second singlet excited state, benzene finds a path for its electrocyclization into

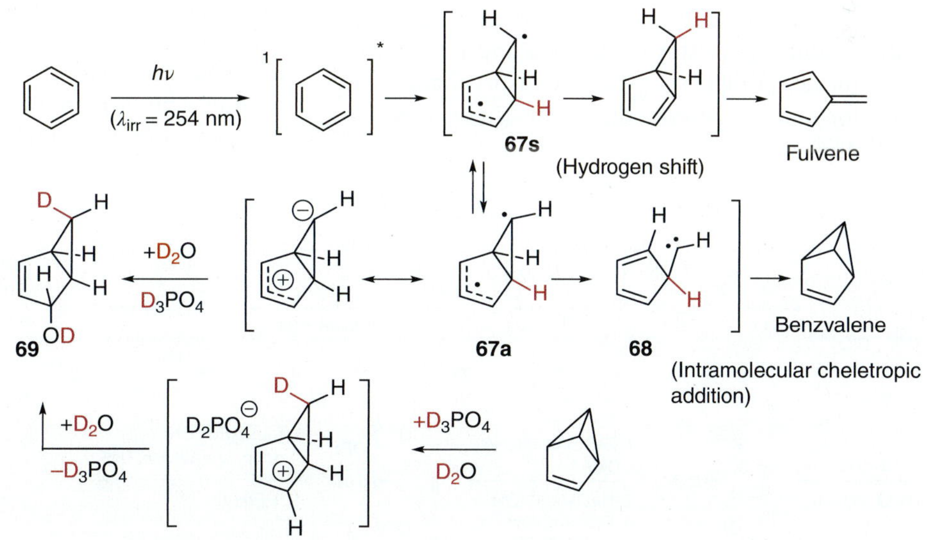

Scheme 6.26 The photoisomerizations of benzene.

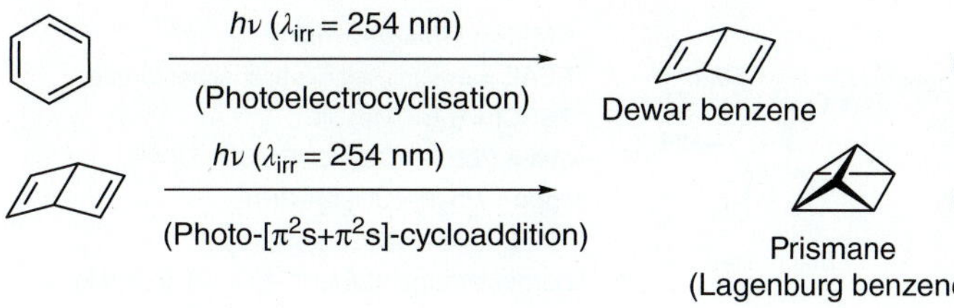

(6.35)

(6.36)

Dewar benzene (reaction (6.35)) [337]. Irradiation of benzene in CF_3CH_2OH gives a 2 : 1 mixture of 2-(2,2,2-trifluoroethyloxy)bicyclo[3.1.0]hex-3-ene and 6-(2,2,2-trifluoroethyloxy)bicyclo[3.1.0]hex-2-ene [338, 339]. This can be taken as a confirmation of the formation of prefulvene diradicaloids **67a** ⇄ **67s**. Irradiation of benzene in D_3PO_4/D_2O gives **69** with high stereoselectivity. This is consistent with the acid-catalyzed hydrolysis of benzvalene that is protonated into bicyclo[3.1.0]hex-3-en-2-yl cation intermediate [340, 341].

Problem 6.19 Give a possible mechanism for the photoisomerization of 1,3,4-trideuteriobenzene into its isomeric trideuteriobenzenes [342].

6.5.2 Photoisomerizations of pyridines, pyridinium salts, and diazines

In 1970, Wilzbach and Rausch reported the photoisomerization (6.37) of pyridine into 2-azabicyclo[2.2.0] hexa-2,5-diene [343]. At 20 °C, the latter compound is isomerized in 15 minutes into pyridine. In the presence of $NaBH_4$ in water, it is reduced into 2-azabicyclo[2.2.0]hex-5-ene. In pure water, its hydration provides 5-aminopenta-2,4-dienal. Chapman et al. found that direct irradiation of pyridine at 8 K in an Ar matrix produces only cyclobutadiene and HCN as observable products [344]. The fragmentation involves the formation of 2-azabicyclo[2.2.0]hexa-2,5-diene [327]. Irradiation of perfluorotetra- and -trialkylpyridines generates stable 2-aza and 1-azabicyclo [2.2.0]hexa-2,5-dienes [345]. Aza-prismanes are also formed as intermediates in these reactions [346, 347].

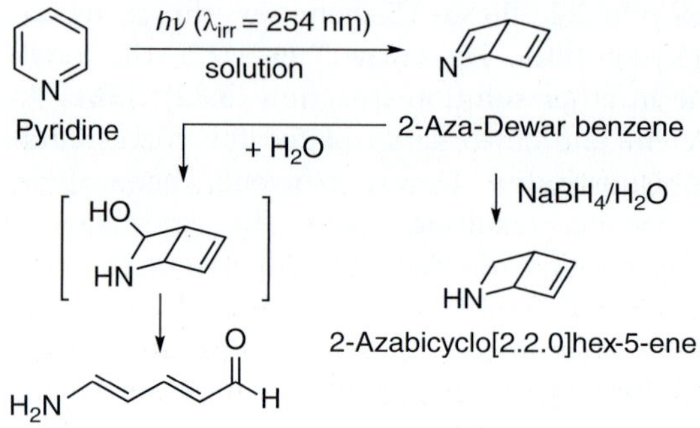

(6.37)

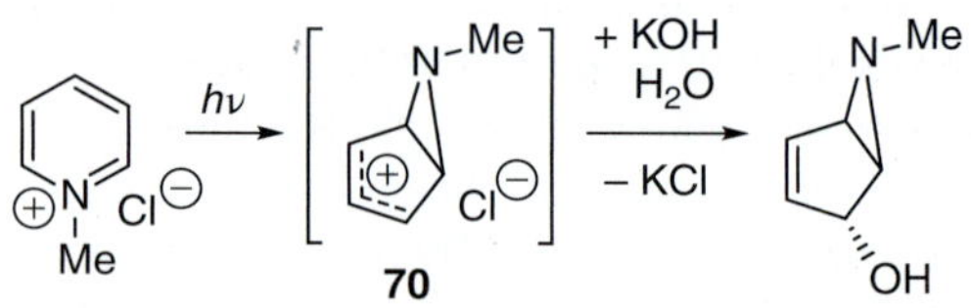

(6.38)

In 1972, Kaplan et al. reported the photosolvolysis of pyridinium salts such as *N*-methylpyridinium chloride under basic aqueous conditions that produce 6-methyl-6-azabicyclo[3.1.0]hex-3-en-2-*exo*-ol (reaction (6.38)) [348]. Irradiation of substituted *N*-alkylpyridinium salts gives mixtures of products resulting from the (1,4)-sigmatropic rearrangement (walk of the aziridine moiety) of the 6-azabicyclo[3.1.0]hexenyl cation intermediate **70** (like bicyclo[3.1.0]hex-3-en-2-yl cation, Section 5.5.5) [349]. The products of photohydration of pyridinium salts are very useful synthetic intermediates [350] as illustrated with the synthesis of mannostatin A (a powerful α-mannosidase inhibitor) reported by Mariano and Ling (Scheme 6.27) [351]. Photolysis

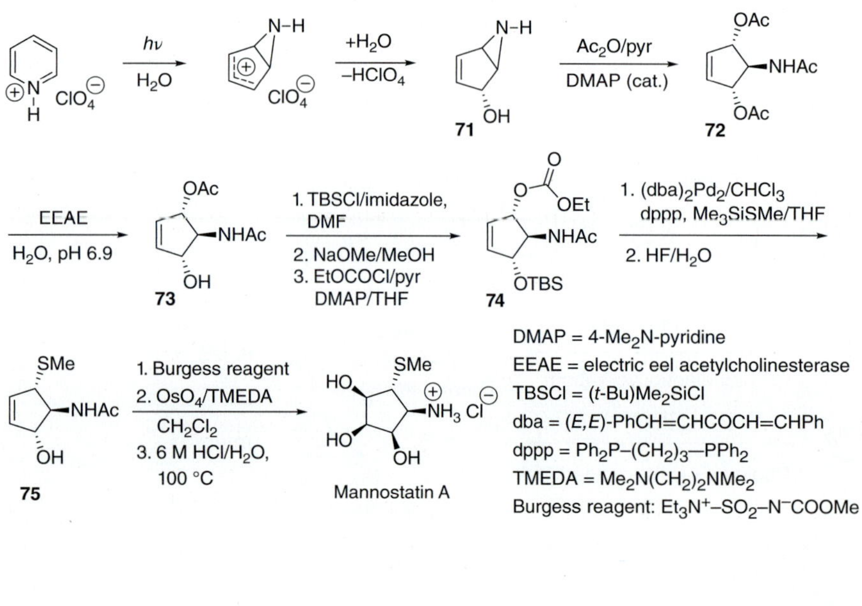

Scheme 6.27 Total asymmetric synthesis of mannostatin A according to Ling and Mariano.

of pyridinium perchlorate under acidic conditions provides 6-azabicyclo[3.1.0]hex-3-en-2-*exo*-ol (**71**). This compound is hydrolyzed into a diol, which is polyacetylated into *meso*-diacetate **72**. Enzymatic desymmetrization (enantioselective hydrolysis of the diacetate) of **72** using electric eel acetylcholinesterase as catalyst affords the key intermediate **73** with 80% ee. After silylation of the alcohol and alkaline methanolysis of the acetate, the corresponding ethyl carbonate **74** is obtained. Palladium-catalyzed substitution of carbonate **74** leads to the methyl thio-ether **75** after desilylation of the silyl ether in aqueous HF. Inversion of the configuration of alcohol **75** follows the Wipf's procedure [352] using the Burgess reagent. Alkene dihydroxylation is *syn*-selective because of the lateral control by the alcoholic moiety. After boiling in 6 M aqueous HCl, the hydrochloride salt of mannostatin A is obtained in 11% overall yield.

Pyrazine (1,4-diazine) photoequilibrates with pyrimidine (1,3-diazine) as a result of the interchange of adjacent ring atoms (photoreaction (6.39)) [353, 354]. In the gas phase, irradiation of 2-methylpyrazine leads to a mixture of 2-, 4-, and 5-methylpyrimidines. A similar photochemical valence isomerization is observed with dimethylpyrazine [355].

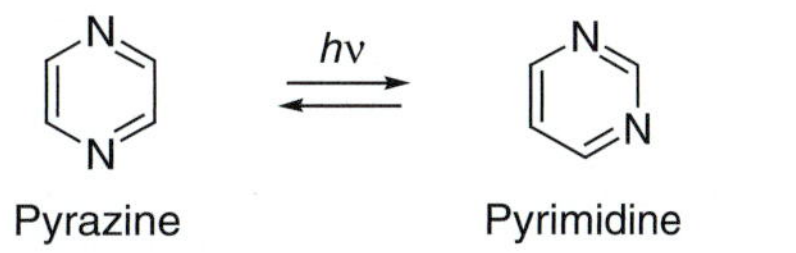

(6.39)

Problem 6.20 Photohydration of 1,4-dimethylpyrimidium chloride in KOH/H_2O gives three products. What are they? [348]

Problem 6.21 What products are formed in the alkaline photohydration of 3-alkoxypyridinium salts in water? [356]

Problem 6.22 What is the product of photoisomerization of 3-oxido-1-phenylpyridinium betaine? [357]

6.5.3 Photolysis of five-membered ring heteroaromatic compounds

Low-pressure gas phase photolysis of furan (254 nm) gives CO, cyclopropene, propyne, and allene (reaction (6.40)) [358]. At higher pressure, products of Diels–Alder addition of furan to cyclopropene and to cyclopropene-3-carbaldehyde have been observed (reaction (6.41)) [359]. Photolysis of furan at 10 K in a Ar matrix gives Dewar furan (5-oxabicyclo[2.1.0]

Scheme 6.28 Examples of photoreactions of furan and derivatives.

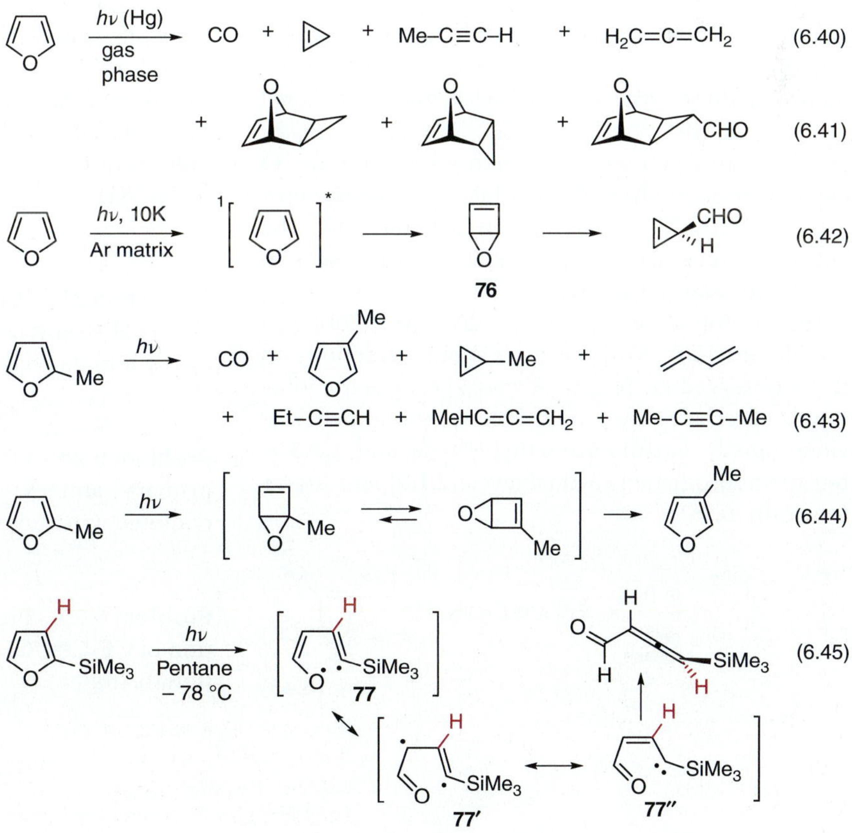

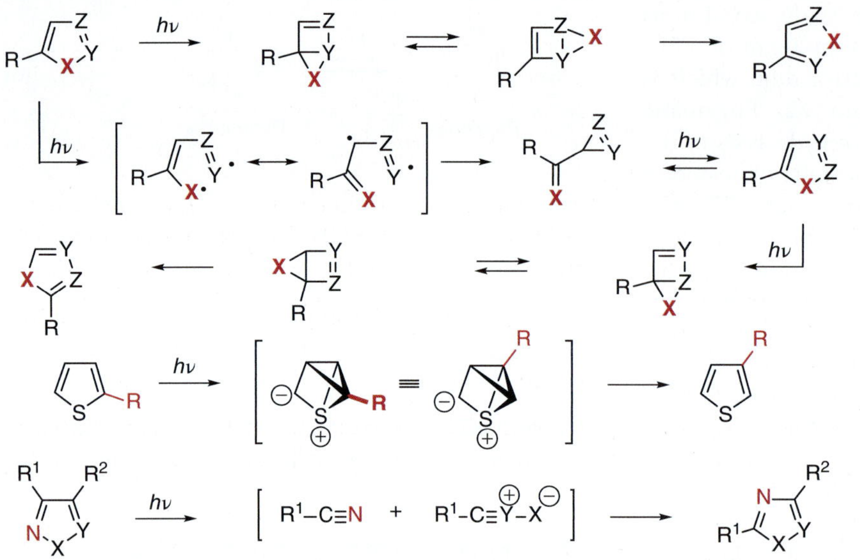

Scheme 6.29 Main photochemical reactions of five-atom heteroaromatic compounds.

pent-2-ene: **76**) [360]. Its rearrangement generates cyclopropene-3-carbaldehyde (reaction (6.42)). Photolysis of 2-methylfuran gives several products including 3-methylfuran [361, 362]. The latter is explained by the formation of 1-methyl-5-oxabicyclo[2.1.0]pent-2-ene that undergoes (1,3)-shift of the oxa bridge into 2-methyl-5-oxabicyclo[2.1.0]pent-2-ene. Disrotatory ring opening of the latter produces 3-methylfuran (reaction (6.44)). Irradiation of 2-trimethylsilylfuran gives 4-trimethylsilylbuta-2,3-dienal in 68% yield (reaction (6.45)) [363]. A possible mechanism for this photoisomerization implies the C—O cleavage into diradical **77** for which the carbene limiting structure **77″** can be written (Scheme 6.28). The latter undergoes facile intramolecular C—H insertion ((1,2)-hydrogen shift) [364].

The photolysis of pyrroles (azoles) [365, 366] and thiophenes [367–370] leads to similar reactions as those observed for furans. A special case is the photolysis of tetra(trifluoromethyl)thiophene, which provides mostly di(trifluoromethyl)ethyne and 1,2,3,4-tetra(trifluoromethyl)-5-thiabicyclo[2.1.0]pent-2-ene (reaction (6.46)) [371].

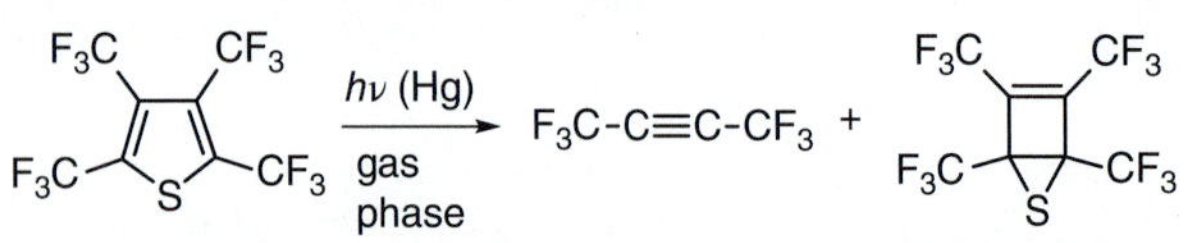

(6.46)

The various routes for the photochemical reactions of five-center heteroaromatic compounds are summarized in Scheme 6.29 [372]. The quantum yields of these processes are usually low as the heterocyclic compounds possess efficient mechanisms for the relaxation of their excited states. For instance, imidazole in histidine and DNA adenine base relax from their electronically excited states on fs time scales, providing an inherent self-protection mechanism against potentially harmful radiation [373–376]. UV excitation of pyrrole, imidazole, pyrazole, and triazole induces fast N—H cleavage as the main photoreaction [377–381].

Problem 6.23 Propose a mechanism for the photochemical isomerization of 5-methylpyrrole-2-carbonitrile into its regioisomers [382].

Problem 6.24 On irradiating thiophenes with primary amines, the corresponding pyrroles are obtained. Give an explanation [368].

Problem 6.25 Propose a mechanism for the formation of $CF_3C{\equiv}CCF_3$ in the photochemical fragmentation (6.46).

6.6 Photocleavage of carbon–heteroatom bonds

6.6.1 Photo-Fries, photo-Claisen, and related rearrangements

UV light irradiation of aryl esters generates mixtures of 2-hydroxy-, 4-hydroxyphenones and the corresponding phenols (e.g. reaction (6.47)) [383]. The first example was observed by Anderson and Reese in 1960 for the rearrangement of catechol monoacetate [384]. Irradiation of *N*-arylamides generates mixtures of 2-amino-, 4-aminophenones and the corresponding anilines (e.g. reaction (6.48)) [385, 386]. The reactions are called photo-Fries rearrangements and are useful synthetic tools [387, 388]. They play an important role in the photodegradation of polycarbonates, polyesters, and polyamides [389–393]. They involve radical pair intermediates [394]. Photo-Fries rearrangements of *N*-arylsulfonamides to (aminoaryl)sulfone derivatives are also useful [395].

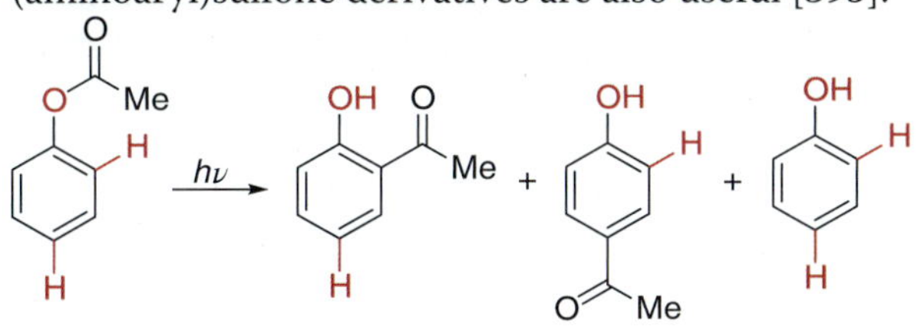

(6.47)

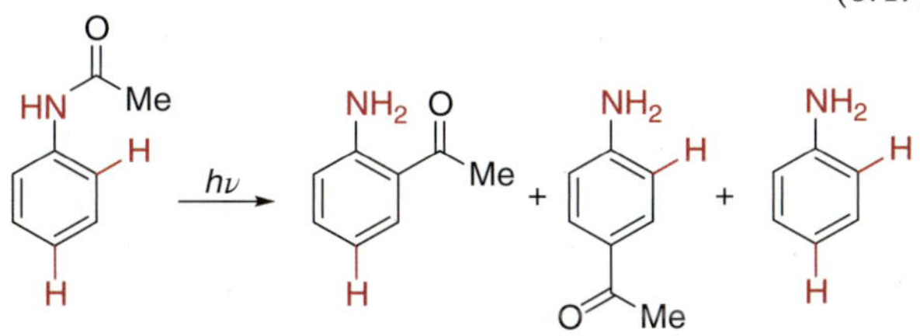

(6.48)

Lochbrunner et al. have investigated the photo-Fries rearrangement of 4-(*tert*-butyl)phenyl acetate (**78**) by two-color femtosecond pump probe spectroscopy. The spectral transmission changes in the visible and ultraviolet spectral region; this change makes it possible to establish that the photoinduced homolytic cleavage (Scheme 6.30) occurs within 2 ps, which indicates a very small energy barrier for the dissociation of the first excited singlet state of **78** into the singlet radical pair **79**. The subsequent geminate recombination of **79** into intermediate cyclohexadienone **80** takes place in c. 13 ps [396]. This is much shorter than the time necessary for spin conversion (1–10 ns). The results support the hypothesis that the initial reaction proceeds from an aromatic ring $^1(\pi,\pi^*)_1$ excited state, not from a $^1(n,\pi^*)$ ester excited state, and evolves to a dissociative $^1(\pi,\sigma^*)$ state. Finally, **80** equilibrates relatively slowly with 2-hydroxy-5-(*t*-butyl)phenone (**81**). The latter reaction is catalyzed by protic acids. The concurrent formation of phenol in reaction (6.47) and of aniline in reaction (6.48) results from the diffusion of the radicals out of the initial solvent cage [397].

Aryl carbonates, [398] aryl carbamates (e.g.: reaction (6.49)) and arenesulfonanilides (e.g.: reaction (6.52)) [399] also undergo photo-Fries rearrangements. Photo-rearrangements of alkyl aryloxyacetate (e.g.: reaction (6.50)), [400] of alkyl arylaminoacetate, of benzylamines (e.g.: reaction (6.51)), [401] of benzyl enol ethers, [402] of benzyl alkyl ketones, [403] of aryloxyacetones [404] and a-benzyloxystyrenes [405] all involve homolytic cleavage and the formation singlet radical pair intermediates that recombine into transposed products analogously to photo-Fries rearrangements.

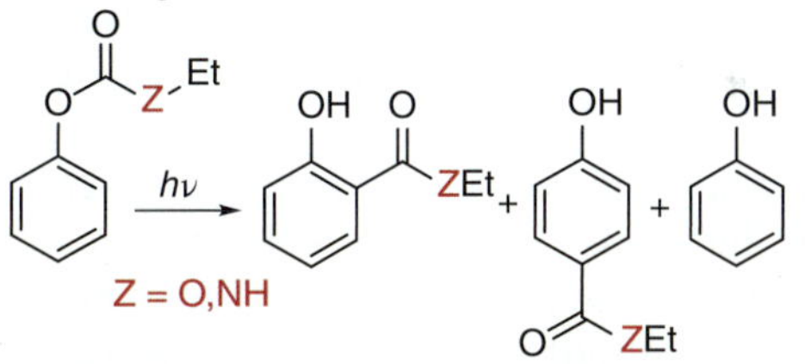

(6.49)

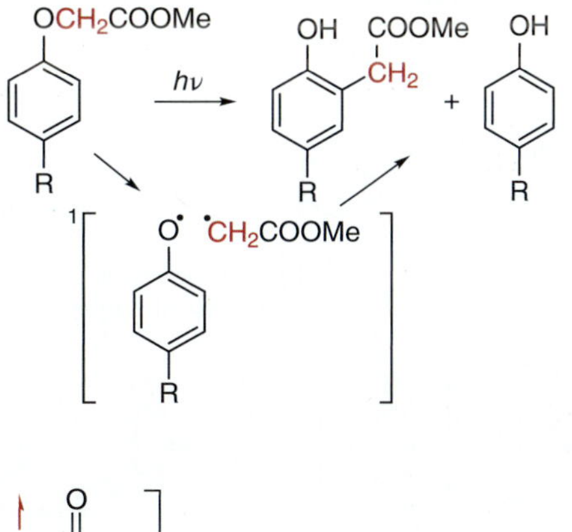

(6.50)

Scheme 6.30 The mechanism of the photo-Fries rearrangement involves the formation of singlet radical pair intermediates.

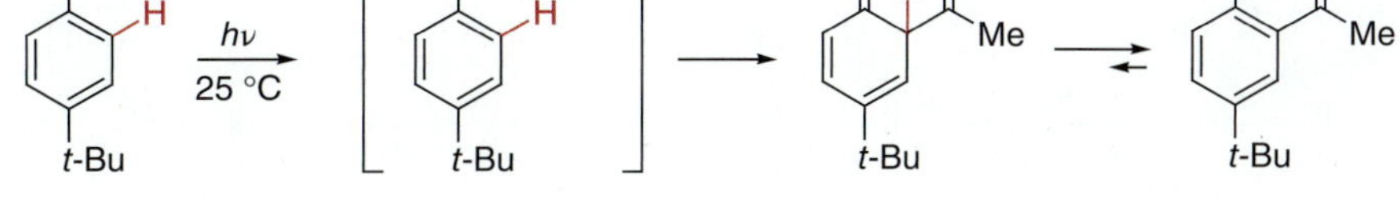

78 **79** ($\tau_{1/2}$ = 13 ps) **80** **81**

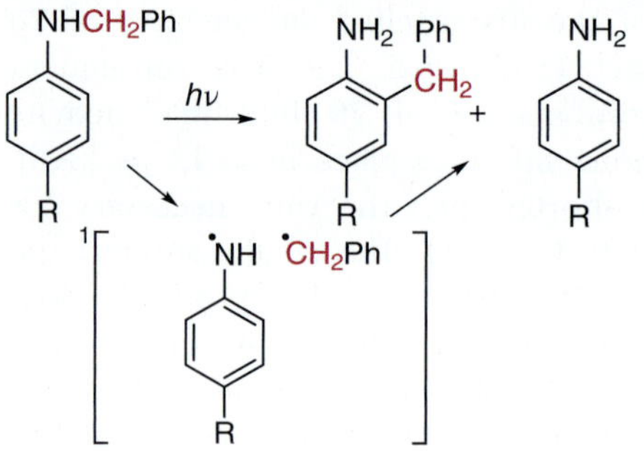

(6.51)

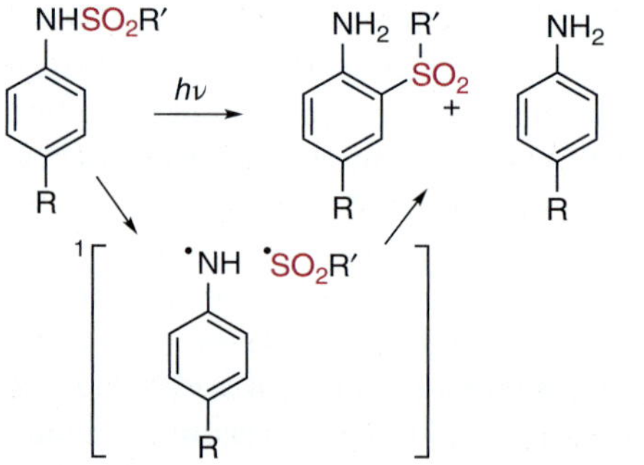

(6.52)

The thermal reaction of aryl allyl ethers to afford *ortho* rearranged products, the classical Claisen rearrangement (section 5.5.9.2) has its photochemical equivalent, the photo-Claisen rearrangement [406] that was described first by Kharasch et al. for allyl phenyl ether (Scheme 6.31) and benzyl phenyl ether [407]. Early evidence for the formation of radical pairs and their recombination was found with the rearrangement of 3-methyl-3-phenoxybut-1-ene into a mixture of *ortho*- and *para*-(3-methylbut-2-en-1-yl) and (2-methylbut-3-en-2-yl)phenol [408]. The quantum yield of the photorearrangement of allyl phenyl ether (benzene solution) is not affected by a triplet quencher such as penta-1,3-diene [409]. In contrast, triplet sensitization with acetophenone in benzene causes a slight increase of the phenol yield and decrease in the quantum yield of the rearranged product 2-allyl- and 4-allylphenol, confirming that the homolysis of the allyl ether bond occurs in the first singlet excited state [410]. Transient cyclohexa-2,4- and 2,5-dienones are necessary intermediates in the photo-Claisen rearrangement (Scheme 6.31). In the case of the photolysis of benzyl phenyl ether (in benzene, 20 °C) (Scheme 6.31b), cyclohexadienone **82** has a lifetime of c. five seconds [411]. It cannot be seen when the solvent is acetic acid, as acids catalyze the exothermic cyclohexadienone → phenol rearrangement. Cyclohexadienone **82** was found to give photoinduced ring opening into ketene **83** upon applying two-laser-two-color flash photolysis [412]. The concurrent formation of phenol arises from the competitive diffusion of the radicals out of the initial solvent cage. When irradiating allyl phenyl ether or benzyl phenyl ether as guests in β-cyclodextrin (cycloheptaamylose), the formation of phenol can be stopped [413, 414].

Problem 6.26 Explain the following one-pot, two-step preparation of quinolines **Q** [415].

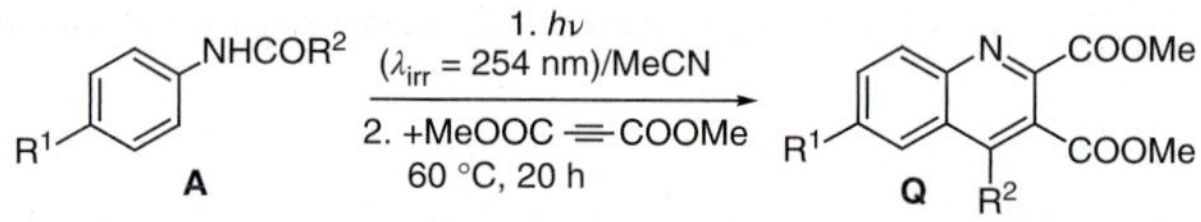

Problem 6.27 What is the main product of the photolysis of benzyl 2,4,6-trimethylphenyl ether? [411]

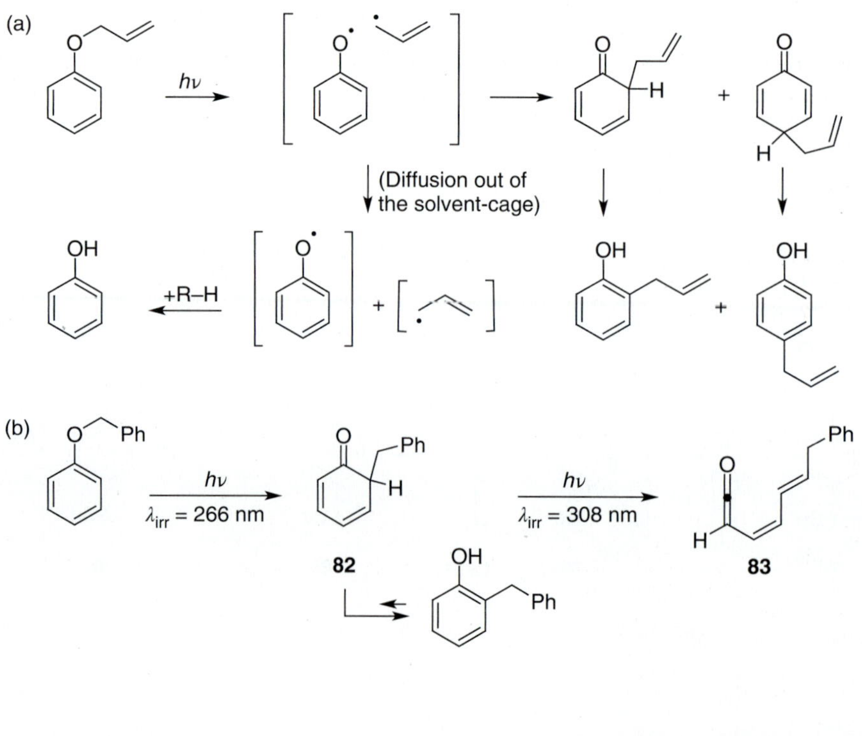

Scheme 6.31 Photo-Claisen rearrangements involve singlet radical pair intermediates that recombine into transient cyclohexadienones; (a) migration of allyl radical, (b) migration of benzyl radical.

6.6.2 Photolysis of 1,2-diazenes, 3*H*-diazirines, and diazo compounds

Thermolysis and photolysis of acylic tertiary azoalkanes (1,2-diazenes) are very useful methods to generate radical pairs and radicals. Photolysis of cyclic azoalkanes generates diradical intermediates. Photolysis of diazirines, diazoalkanes, and, in some cases, *gem*-diazides (see however Section 6.7.3) generates carbene intermediates. The photolysis of monoazides generate the highly reactive nitrene intermediates (Section 6.7.1). As seen above (Section 6.3.1), azobenzene and derivatives (1,2-diraryl-1,2-diazenes) undergo quantitative (*E*) ⇄ (*Z*) photoisomerization. (*E*)-Diazobenzene ((*E*)-Ph—N=N—Ph) does not expel N_2 readily (forming the relatively unstable phenyl radical, see Table 1.A.7) and is stable up to 600 °C. In contrast, azotriphenylmethane (1,2-trityl-1,2-diazene) is a transient species even at −40 °C as it loses N_2 readily to generate the highly stable trityl ($Ph_3C^{\bullet}$) radical (Section 1.9) [416]. The activation energy for the thermal loss of N_2 depends on the relative stability of the radicals formed in agreement with the Dimroth principle and the Bell–Evans–Polanyi theory ($\Delta^{\ddagger}H = \alpha\Delta_r H + \beta$) [417].

Azoalkanes have been known since 1909 [418]. Irradiation of (*E*)-azomethane ((*E*)-Me—N=N—Me) at 25 °C in benzene produces (*Z*)-azomethane with a quantum yield of 0.42. The reverse reaction proceeds with a quantum yield of 0.45 [419, 420]. Femtosecond laser excitation of (*E*)-MeN=NMe induces partial rotation about the N=N double bond leading to a twisted species that fragments within a few femtoseconds into the methyldiazenyl (Me—N=N$^{\bullet}$) and methyl radicals. The second C—N bond cleavage occurs consecutively with a small energy barrier within 100 fs [420]. With primary and secondary azoalkanes, the (*Z*)-isomer can be isolated. This is not the case with tertiary azoalkanes for which the (*Z*)-isomers decompose rapidly into N_2 and alkyl radicals that combine into alkanes or disproportionate into alkanes + alkenes. For instance (Scheme 6.32a), irradiation of (*E*)-**84** at −126 °C produces (*Z*)-**84**, which decomposes at −36 °C into N_2 [421]. The gas phase photodissociation of **85** was studied with time-resolved spectroscopy (Scheme 6.32b). The results indicate a two-step mechanism [422] for the dediazoniation in which the formation of 1,1-dimethylallyl radical + methyldiazenyl radical is formed within 2 ns after the excitation, whereas methyl radical + N_2 are formed within c. 12 ns [423].

Photolysis of AIBN (azoisobutyronitrile: $NC(Me)_2C$—N=N—$C(Me)_2CN$) leads to only 5% of radical–radical reaction of 2-cyanoisopropyl radical, but to 95% in the crystal [424]. Persistent triplet radical pairs have been seen by low-temperature photolysis of the corresponding azoalkane in a matrix [425]. Photolysis of 5-*exo*,6-*exo*-dideutero-2,3-diazanorborn-2-ene (**86**) gives a 4 : 1 mixture of housanes *exo*-**88** and *endo*-**88** [426, 427]. The major product *exo*-**88** is formed in a two-step mechanism involving diradical intermediate **87**, which undergoes a S_H2 process (second-order homolytic substitution). If the C_2-symmetrical diradical intermediate **89** would form as a unique intermediate, it would lead to a 1 : 1 mixture of *exo*-**88** and *endo*-**88** (Scheme 6.33) [428]. The two-step mechanism is confirmed by measuring the quantum yield of the N_2 formation as a function of viscosity. The more viscous the medium, the lower the quantum yield. In a more viscous medium, the dediazoniation of diradical **87** is retarded [429]. Photolysis of 2,3-diazanorbornene in an argon matrix at 5 K allowed the observation of triplet cyclopenta-1,3-diyl diradical **89** [430].

Thermolysis of azo compound **90** gives barrelene in 100% yield. Direct irradiation of **90** gives a 24 : 73

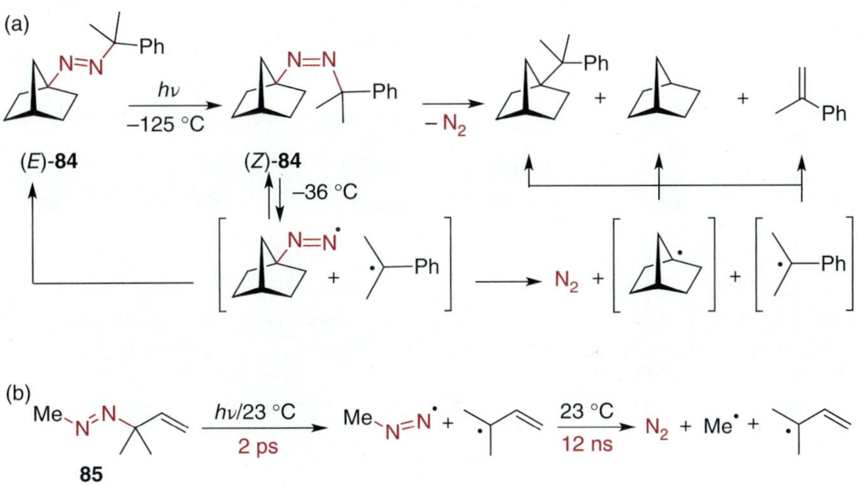

Scheme 6.32 (a) (*E*)/(*Z*)-Photoisomerization of an azoalkane and the two-step thermal decomposition of a its (*Z*)-isomer into N_2 and alkyl radicals. (b) Stepwise mechanism for the photodediazoniation of an acyclic azoalkane.

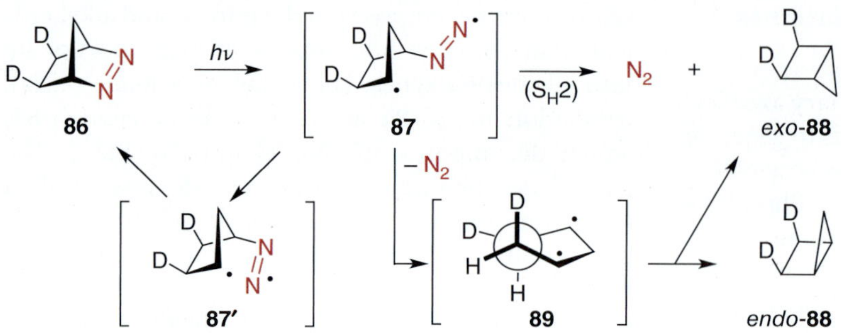

Scheme 6.33 The two-step mechanism for the photodenitrogenation of 2,3-diazanorbornene.

mixture of barrelene and semibullvalene. Sensitized photolysis of **90** produces semibullvalene only, in agreement with the theory presented above for the DPM rearrangement (Section 6.3.4) that occurs exclusively on the triplet energy hypersurface. The singlet excited state is responsible for the formation of barrelene [164].

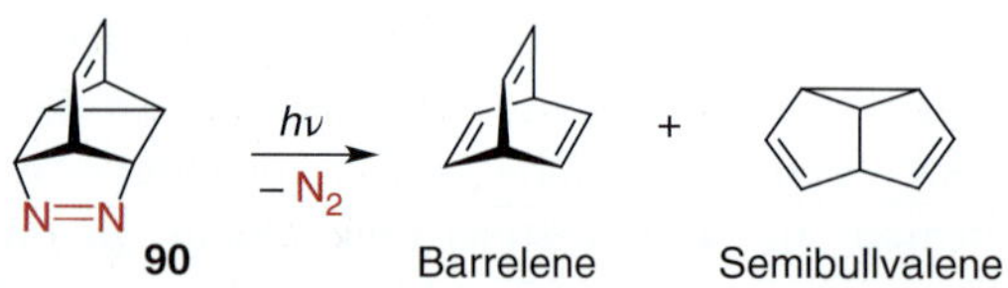

The first examples of 3*H*-diazirines were reported by Paulsen [431] and by Schmitz and Ohme [432]. Their thermolysis or photolysis realizes a powerful method for the generation of carbene intermediates [433, 434]. 3*H*-Diazirines are widely used as photoaffinity reagents for labeling biological receptor sites [435–437]. The initial product of gas phase irradiation (λ_{irr} = 320 nm) of unsubstituted 3*H*-diazirine is diazomethane ($CH_2{=}N_2$) with a quantum yield of 0.2 [438]. The nearest homologs, 3-methyl-, 3-ethyl-, and 3,3-dimethyl-3*H*-diazirine do not form isomeric diazoalkanes when irradiated in the gas phase, but lose N_2 instead giving alkenes resulting formally from (1,2)-H shifts in the corresponding carbene intermediates. Irradiation of the same 3*H*-diazirines in solution results in diazoalkanes. Photolysis of 3-methyl-3-phenyl-3*H*-diazirine (**91**) in ether in the presence of *N*-methylmaleimide (**92**) produces a mixture of products of cyclopropanation *trans*-**93** (23%) and *cis*-**93** (20%) and small amounts of products **94** (8%, resulting from the reaction of diazoalkane **96** + carbene **97**) and **95** (reaction of two carbene intermediates **97**) [438, 439]. When the photolysis is carried out with the gum obtained by evaporation of the solvent (ether), two further products *trans*-**98** and *cis*-**98** are formed that result from the quenching of intermediate diazoalkane **96** by *N*-methylmaleimide (**92**) in dipolar cycloadditions (Scheme 6.34). This suggests that irradiation of **91** generates diazoalkane **96** in parallel with carbene **97** [440].

Laser flash photolysis of 3-chloro-3-isopropyldiazirine, 3-chloro-3-propyldiazirine, and 3,3-(adamant-2-yl)-3*H*-diazirine (25 °C in isooctane, λ_{irr} = 355 nm, 200 ps pulse) generates transient species appearing 1 ms after irradiation and detected by their UV absorption spectra ($\lambda_{max} \approx 240$ nm). The latter are diazoalkanes that decompose into N_2 and the corresponding carbenes. These carbenes rearrange into alkenes through 1,2-hydrogen migrations [441]. Applying femtosecond time-resolved UV–visible spectroscopy to the photolysis (λ_{irr} = 350 nm, in

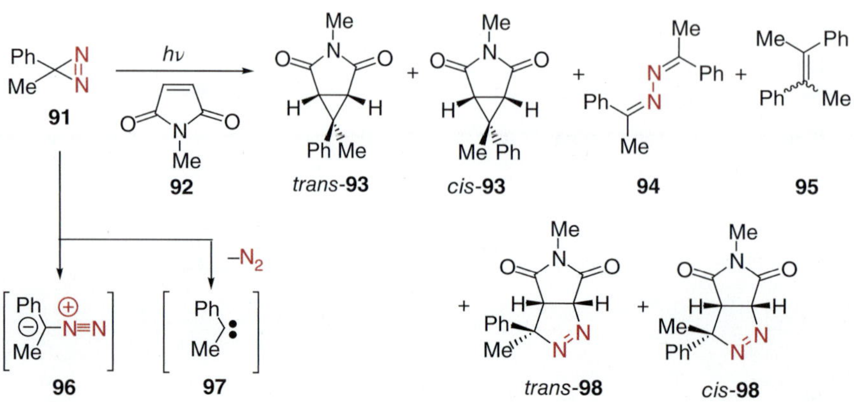

Scheme 6.34 Photolysis of diazirines generates isomeric diazoalkanes competitively with the formation of carbenes by denitrogenation.

cyclohexane) of 3-alkyl-3-phenyl-3*H*-diazirines **99**, Platz and coworkers demonstrated that diazirines **99** are converted into the corresponding alkenes **100** via (1,2)-H shifts that occur in concert with N_2 extrusion. The reactions evolve in the first singlet excited states [442].

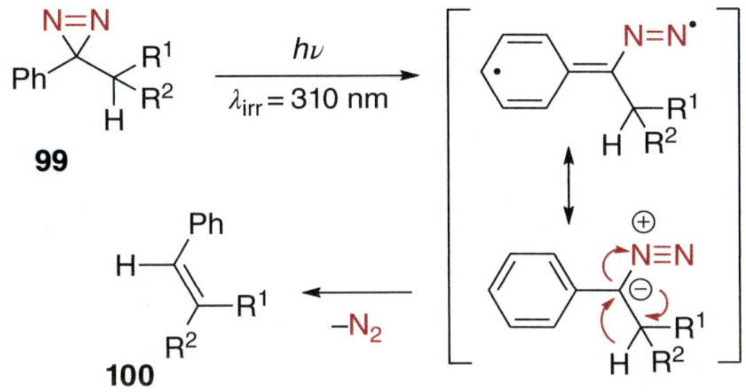

Photoisomerizations of diazomethane and its alkyl derivatives into 3*H*-diazirines have not been observed yet. Only fluorinated alkyl or silyl derivatives form the isomeric 3*H*-diazirines upon long-wavelength irradiation in low yield. Prolonged irradiation or short-wavelength excitation leads to loss of N_2 with formation of products resulting formally from the corresponding carbene intermediates as illustrated with the photolyses of **101** and **105** in Ar matrices [443, 444]. Photolysis of (trimethylsilyl)diazomethane (**101**) at $\lambda_{irr} > 360$ nm leads to a photostationary mixture of **101** and **102**. Prolonged irradiation induces dediazoniation and formation of triplet carbene **103** that has been characterized by its electron spin resonance (ESR) spectrum. Above 45 K, carbene **103** undergoes a (1,2)-methyl shift with formation of the silene **104** [445]. Irradiation of bis(diazomethyl)silane **105** with blue light gives bis(diazirine) **106** and a small amount of silirene **107**. Shorter wavelength irradiation (λ_{irr} = 305 nm) produces **107** as the only product [446, 447].

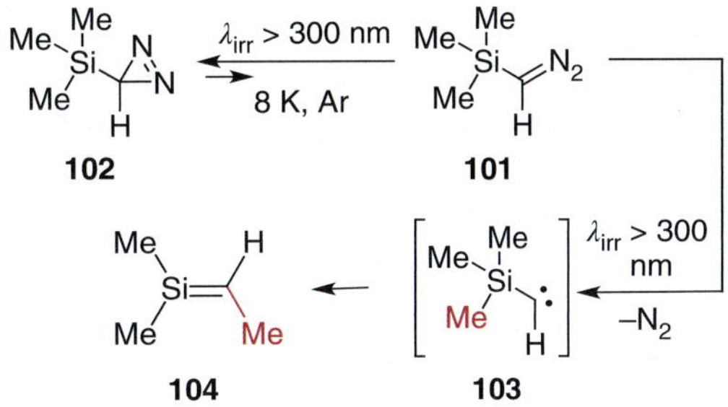

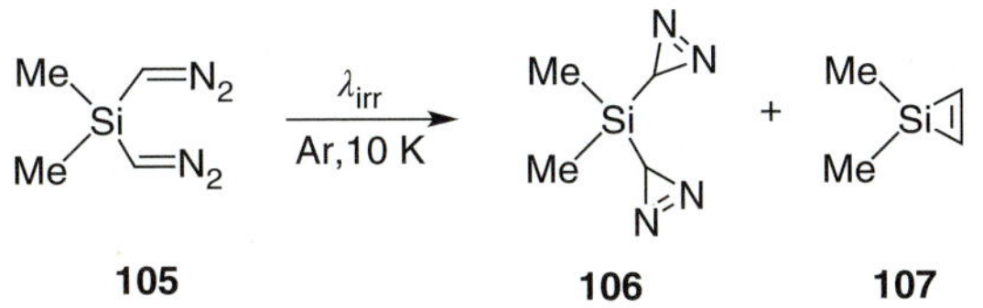

	106	107
$\lambda_{irr} > 405$ nm	Major	Minor
$\lambda_{irr} = 305$ nm	0%	100%

α-Diazocarboxamides are isomerized into their 3*H*-diazirine isomers by visible light [448]. Argon matrix photolysis of *N*,*N*-dialkyldiazoacetamides at 7–10 K results in either the formation of C—H insertion products or Wolff rearrangement to ketenes [449]. Depending on the structure and the light wavelength, photolysis of cyclic α-diazoketones and diazodicarboxylic esters generates the corresponding isomeric 3*H*-diazirines concurrently with products resulting from the reactions of the carbene intermediate (rearrangement and/or reaction with the medium) [439]. α-Diazocarbonyl compounds **108** undergo heat or light-induced dediazoniation with the formation of a variety of reactive intermediates. When ketenes are formed, the reaction is the Wolff rearrangement discovered in 1902 [450, 451]. Two mechanisms have been proposed for the photoinduced Wolff rearrangement (Scheme 6.35). In 1966,

Scheme 6.35 The photoinduced Wolff rearrangement follows a one-step, concerted process for *syn*-α-diazoketones, and a two-step mechanism for *anti*-α-diazoketones, with the formation of a carbene intermediate.

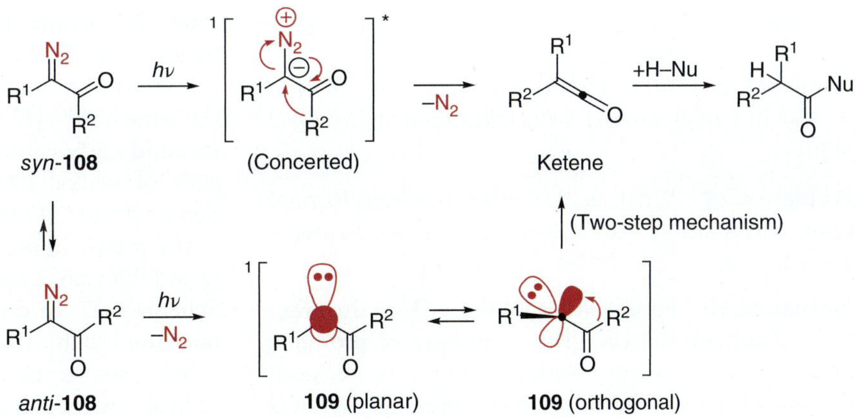

Kaplan and Meloy proposed that one-step formation of ketene is favored upon decomposition of the first excited state of α-diazocarbonyl compounds in their *syn* conformation, whereas a two-step mechanism via the formation of a carbene intermediate of type **109** prevails when the photoexcited α-diazocarbonyl compound adopts the *anti*-conformation [452]. The initial carbene is expected to be planar. This is the case for aryl derivatives for which the excited states decompose within 300 ps [453]. The carbene then rotates into an orthogonal conformer that permits overlap between the empty 2p orbital of the carbene with the σ(C—R^2) bond of the group that migrates in the last step. Diphenylketene from the photoinduced (λ_{irr} = 270 nm, MeCN) Wolff rearrangement of azibenzyl (PhCOC(=N_2)Ph) is formed by two parallel pathways: a stepwise mechanism with the formation of benzoylphenyl carbene intermediate with a slow rise time constant of c. 650 ps and directly in the azibenzyl excited state leading to a vibrationally hot ketene. Photolysis (λ_{irr} = 270 nm, $CHCl_3$) of diazoacetone (MeCOC(=N_2)H) that exists mostly in the *syn* conformation leads to methylketene through a concerted process [454, 455].

The photoinduced Wolff rearrangements of α-diazoketones can be accompanied by the formation of isomeric 3*H*-diazirines as exemplified with the irradiation of diazocamphor (**110**) [456] and hexafluoro-3-diazobutanone (**111**) [457, 458].

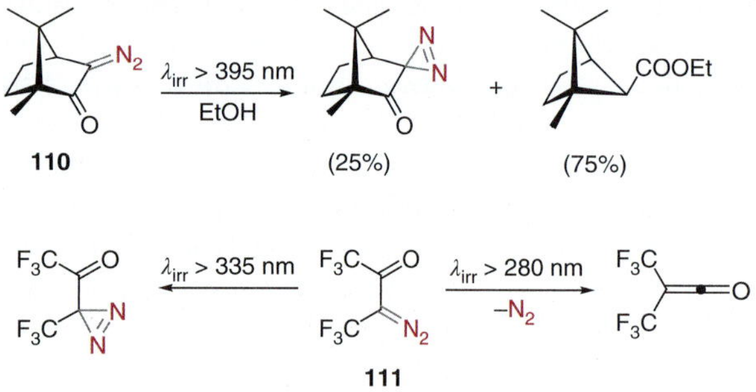

Problem 6.28 What are the products of irradiation at 350 nm of 9-diazo-1,8-diazafluorene in MeOH? [459]

Problem 6.29 What is the main product formed upon irradiation of 2,3-diazabicyclo[2.2.1]hept-2-ene?

Problem 6.30 Photolysis of **A** [460, 461] in the presence of fumarodinitrile gives a mixture of products including a 2*H*-azirine resulting from a rare case of carbene cheletropic addition to a carbonitrile. What is this product and the other products of reaction? [462]

A: 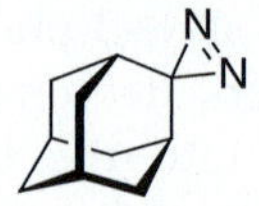

6.6.3 Photolysis of alkyl halides

Early studies on the photolysis of alkyl halides were conducted mainly in the gas phase in which UV light induces homolysis R—X → R$^\bullet$ + X$^\bullet$. Alkyl bromides and iodides absorb UV light of lower energy (transitions n(X:) → σ^*(C—X)), whereas simple R—F and R—Cl absorb below 200 nm in the vacuum UV (Figure 6.2). Alkyl bromides and iodides are thus more prone to selective photochemical reactions than alkyl chlorides and fluorides, except for benzylic fluorides and chlorides that undergo C—F and C—Cl bond cleavage readily upon irradiation at 254 nm. Gas phase irradiation of iodoalkanes leads to their C—I bond homolysis and, for secondary and tertiary derivatives, to competitive β-elimination of HI. With 193 nm excitation, C—C and C—H cleavage compete with C—I homolysis [463]. Alkenyl and aryl halides have a richer photochemistry as many more excited states than (n,σ^*) states are available, including dissociative (π,σ^*) states and bound (π,π^*) states [464–468].

Irradiation of alkyl bromides and iodides in solution leads to competitive radical and ionic reactions. The initially formed radical pair can escape from the solvent cage to participate in radical processes or undergo electron transfer to generate first a tight ion pair and then a solvent-separated ion pair and, finally, a pair of free carbocation and halide anion intermediates. The type of products obtained depends strongly on the solvent polarity and on the nature of the leaving group (nucleofugal group). Alkyl iodides are more prone than their corresponding bromides to ionic photoreactions (Scheme 6.36a) [469, 470]. Irradiation of primary and secondary alkyl halides also leads to competitive α-eliminations of HX with the formation of carbene intermediates (Scheme 6.36b) [471]. Simultaneous formation of radical and carbocation intermediates from the excited state of 1-alkyl-1,1-diarylmethyl chlorides has been demonstrated [472].

The photochemical behavior of 1-iodonorbornane (1-iodobicyclo[2.2.1]heptane) is summarized in Scheme 6.37. In diethyl ether, the formation of norbornane results from the intermolecular transfer of a hydrogen atom from the solvent to 1-norbornyl radical intermediate; 1-ethoxynorbornane results from the quenching of the 1-norbornyl cation (an

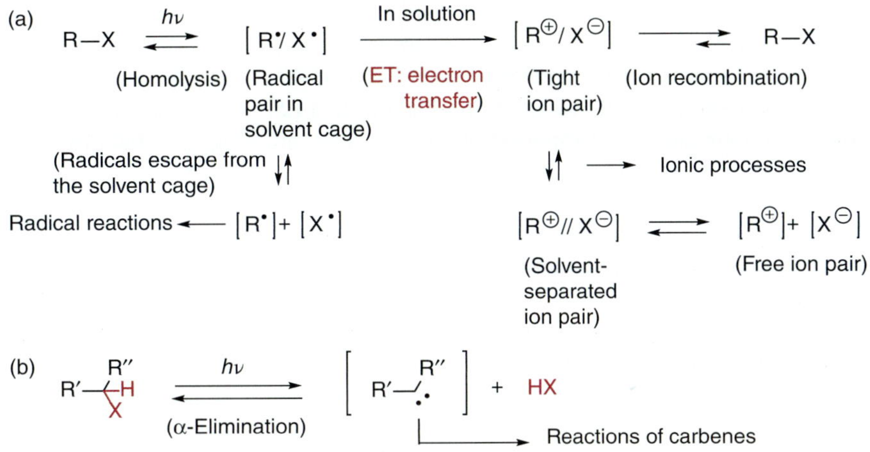

Scheme 6.36 In solution, the photochemistry of alkyl bromides and iodides involves (a) radical and ionic intermediates and (b) carbene intermediates for systems able to undergo α-elimination of HX.

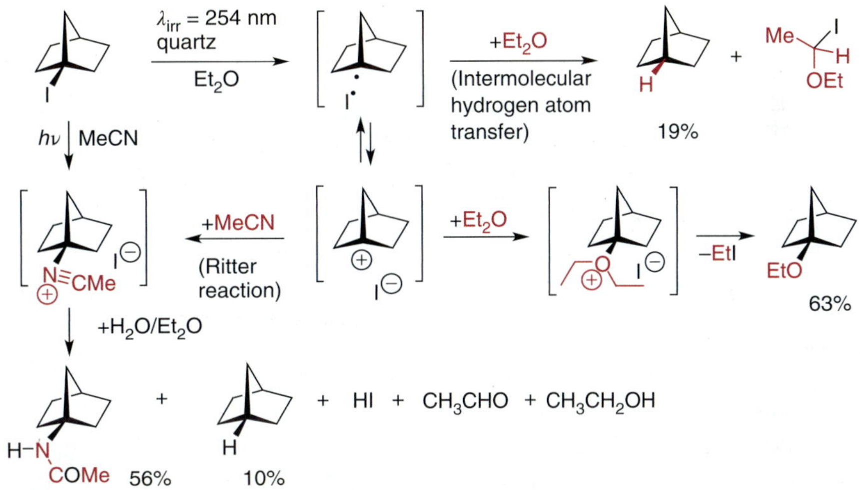

Scheme 6.37 Photoreactions of 1-iodonorbornane in ether and in acetonitrile.

unstable tertiary carbenium ion that cannot adopt a planar geometry) by Et_2O that gives a relatively stable oxonium ion, which, on its turn, reacts with iodide anion. In acetonitrile, the Ritter reaction of 1-norbornyl cation is observed [473, 474].

Photolysis of dihalomethanes (X—CH_2—Y) in an Ar matrix at 12 K leads to the photostationary state (6.53) [475]. In the case of CH_2I_2, Br—CH_2—I, and Cl—CH_2—I, the colored ion pairs $CH_2{=}I^+I^-$ (λ_{max} = 370, 545 nm), $CH_2{=}Br^+I^-$ (λ_{max} = 403, 660 nm), and $CH_2{=}Cl^+I^-$ (λ_{max} = 428, 745 nm), respectively, are observed. F—CH_2—I does not afford the corresponding ion pair. Upon warming the matrix to 26–30 K, the ion pairs recombine into the corresponding dihalomethanes [476]. Upon excitation at 268 nm in acetonitrile or hexane, $ICH_2^•$ radical is formed within <200 fs. This radical combines with $I^•$ in c. 500 fs to form CH_2I_2. Then, on the picosecond time scale, the molecule/radical $CH_2I_2/I^•$ forms. The latter then undergoes charge transfer into $CH_2I_2{}^+I^-$. On the ns time scale, I_3^- is then generated after the decay of ion pair $CH_2I_2{}^+I^-$ [477].

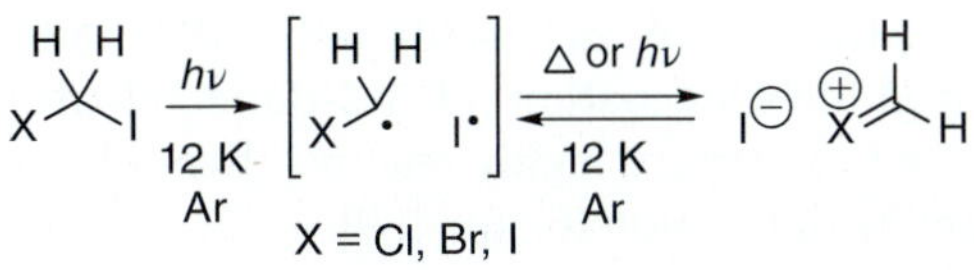

(6.53)

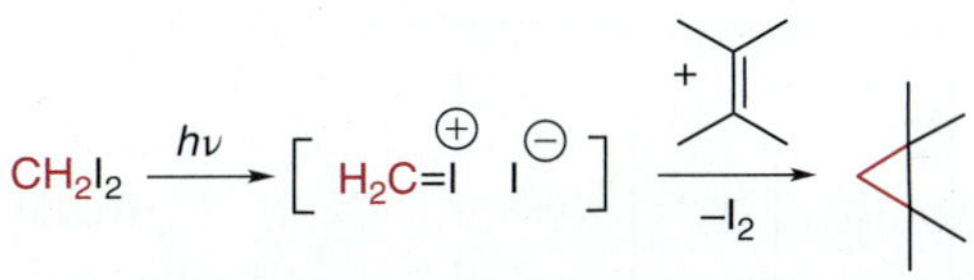

(6.54)

Irradiation of CH_2I_2 in the presence of alkenes generates cyclopropanes according to reaction (6.54). CH_2Br_2 can also be used for the photochemical cyclopropanation of alkenes. The electrophilic nature of the reactive intermediate $CH_2{=}I^+I^-$ is confirmed by the observation that trapping in the presence of LiBr affords Br—CH_2—I and that the reaction rates increase with the alkene nucleophilicity. Cyclopropanation (6.54) is less sensitive to steric factors than the Simmons–Smith method ($CH_2I_2 + Zn/Cu \rightarrow IZnCH_2—I$), therefore being particularly useful for the cyclopropanation of congested alkenes [478, 479]. Irradiation of diiodide **112** affords the product of monohydrogenation **113** in cyclohexane, dimethyl acetal **114** in methanol, and 2-methylbut-2-ene in 1,2-dichloroethane [480, 481]. Irradiation of CH_2I_2 in the gas phase generates first the radical pair $ICH_2^•/I^•$, which then absorbs a photon to dissociate into methylene (CH_2:) and I_2 [482, 483]. Irradiation of 1,1-dichloro-1,1-diphenylmethane in solution produces diphenylcarbene (Ph_2C:), which has been detected by ESR spectroscopy [484].

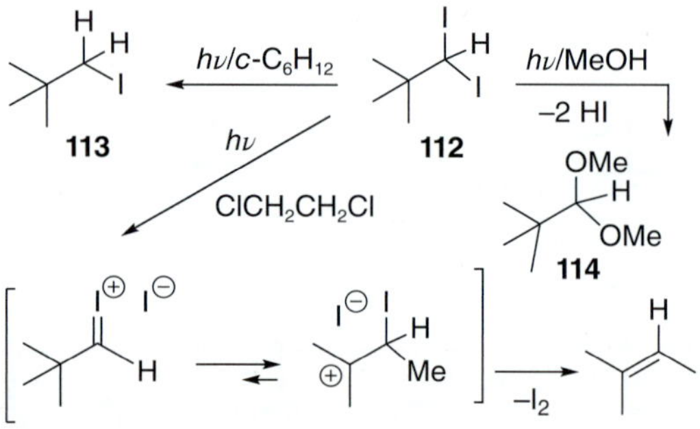

Photolysis of 1,2-dihaloalkanes gives rise to β-haloalkyl radicals **115** that exist as equilibria with their bridged structures **116** (reaction (6.55)). Bridged structures are preferred for $X—CH_2—CH_2^•$ with X = I and Br [485]. In the case of X = Cl, both the open and bridged structures exist, whereas for X = F, only the open structure $F—CH_2—CH_2^•$ is present [486]. In benzene, or acetonitrile, β-bromoalkyl radicals lose $Br^•$ thermally forming the corresponding alkenes + Br_2. In the presence of a H-atom donor, hydrogen abstraction by $Br^•$ yields HBr [487, 488]. Photoinduced dehalogenation of 1,2-diiodoalkanes has been used to prepare highly reactive dienophiles as illustrated with reaction (6.56) [489].

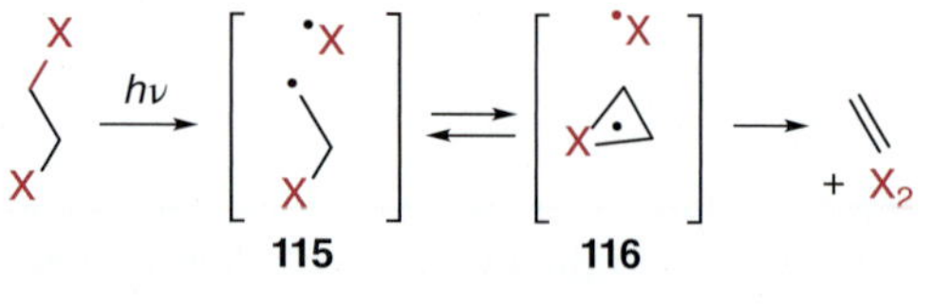

(6.55)

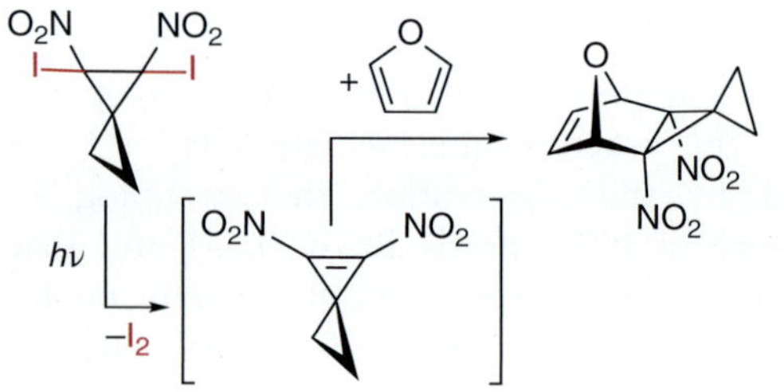

(6.56)

Laser flash photolysis (λ_{irr} = 266 nm, cyclohexane) of 1,2-bis(chloromethyl)benzene generates first 2-(chloromethyl)benzyl radical (λ_{max} = 330 nm) and then *ortho*-xylylene (*ortho*-quinodimethane; λ_{max} = 360 nm) (reaction (6.57)) [490]. Similarly, laser flash photolysis of 1,4-bis(chloromethyl)benzene generates first 4-(chloromethyl)benzyl radical (λ_{max} = 320 nm) and then *para*-xylylene or *para*-quinodimethane (λ_{max} = 290 nm) (reaction (6.58)) [491]. The non-Kekulé hydrocarbon *meta*-xylylene diradical is obtained by photolysis (λ_{irr} = 254 nm) of 1,3-bis(chloromethyl)benzene in glassy ethanol (reaction (6.59)) [492]. Naphthoquinodimethane diradicals have been detected in the photolysis (λ_{irr} = 254 nm) of 1,3-, 1,6-, and 2,7-bis(chloromethyl)naphthalenes in low-temperature matrix media, or by laser flash irradiation at room temperature [493, 494].

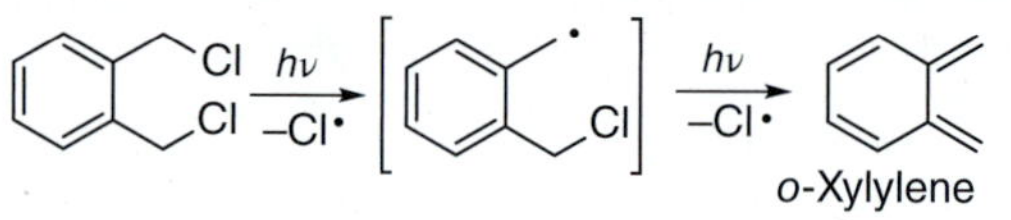

(6.57)

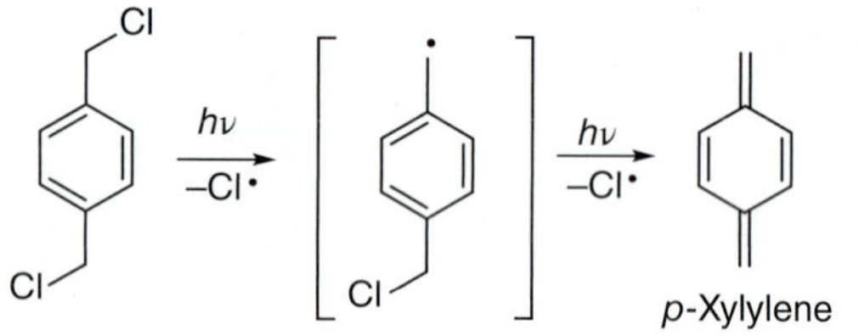

(6.58)

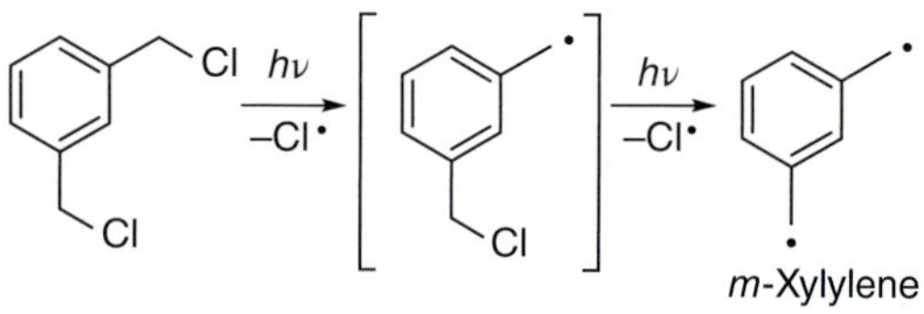

(6.59)

Photodissociations of the C—Br and C—Cl bonds in *p*-bromo- and *p*-chloromethyl-benzophenone in solution have been investigated by time-resolved ESR spectroscopy and laser flash photolysis techniques at 295 K. Both reactions generate *p*-benzoylbenzyl radical ($ArCH_2^•$). The C—Cl bond photodissociation occurs only in the triplet state $^3[ArCH_2—Cl]^*$, whereas the C—Br photodissociation occurs only in the singlet excited state $^1[ArCH_2—Br]^*$ upon direct irradiation. The C—Br bond also dissociates in the triplet excited state $^3[ArCH_2—Br]^*$ of the benzyl bromide upon sensitized (acetone) photolysis. The C—Br dissociation is faster than $^1[ArCH_2—Br]^* \rightarrow {}^3[ArCH_2—Br]^*$ ISC. This is not the case for the corresponding benzyl chloride [495].

In analogy with alkyl bromides and iodides, the direct irradiation (λ_{irr} = 254 nm, quartz vessel) of phenylthio and phenylseleno ethers in solution generates radical and ionic intermediates. A possible mechanism involves initial homolytic cleavage of the

C—S or C—Se bond followed by electron transfer within the resulting radical pair cage [496].

Problem 6.31 What are the products formed upon photolysis of 1,3-dichloro-1,3-diphenylpropane in cyclohexane? [497]

6.6.4 Solution photochemistry of aryl and alkenyl halides

In a non-nucleophilic solvent, photolysis (λ_{irr} = 254 nm) of aryl halides (Ar—X) [498] leads to their reduction and formation of corresponding arenes, Ar—H. The solvent transfers a hydrogen atom to the aryl radical intermediate (photohomolysis, Scheme 6.38a) [499, 500]. In nucleophilic solvents, Nu—H products of substitution resulting from quenching of the ion pair intermediates Ar^+/X^- (photoheterolysis, Scheme 6.38b) are obtained. The latter mechanism is favored for aryl halides substituted with electron-donating groups. In the triplet states (^{3}[Ar—X]*), aryl halides are cleaved in polar solvents into the corresponding triplet aryl cation intermediates ($^3[Ar^+]$) [501–503]. Formation of 2-pyridyl cation intermediates have been observed during the photolysis of 2-halopyridines [504]. With PhCl, photohomolysis is the primary process under most reaction conditions, except in CF_3CH_2OH, where photoheterolysis competes [505]. A third reaction path is a PET from the medium (M) generating the corresponding radical cation $PhX^{•+}$ and radical anion $M^{•-}$. The radical cation reacts with the nucleophile

Scheme 6.38 The main photoreactions of aryl halides: (a) photoinduced homolysis, (b) photoinduced heterolysis, (c) intermolecular photoinduced electron transfer to a catalyst M, (d) photoinduced electron transfer to a radical, (e) photoinduced electron transfer from the neutral nucleophile, initiation of a chain process ($S_{RN}1$ mechanism), (f) photoinduced electron transfer from the anionic nucleophile to the aryl halide.

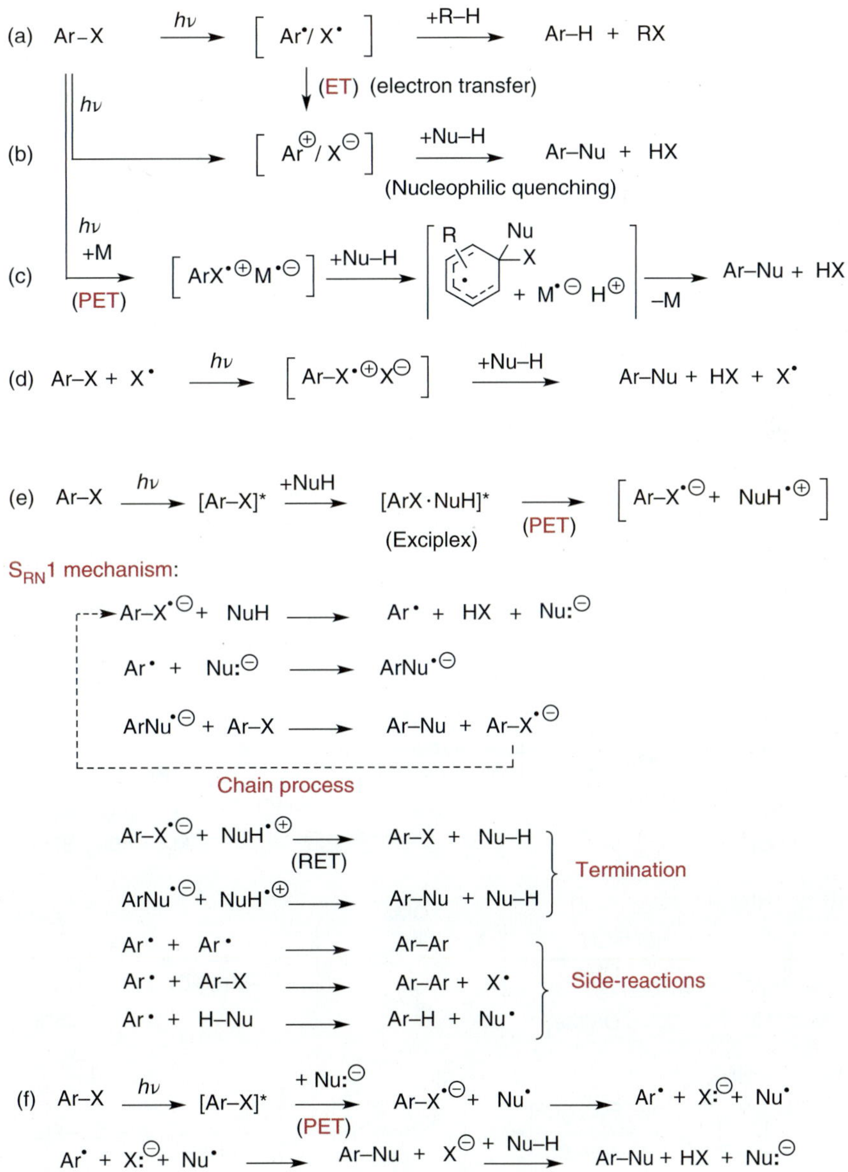

Nu—H to form the products of nucleophilic substitution Ph—Nu + HX (Scheme 6.38c). Under certain conditions, the initially formed X$^{\bullet}$ radical, which has a relatively high electronegativity, may intervene in the PET (Scheme 6.38d). Light can also promote the electron transfer from the nucleophile Nu—H to the aryl halides (Scheme 6.38e). This is the $S_{RN}1$ mechanism. Alternatively, the conjugate base Nu:$^{-}$ undergoes PET (Scheme 6.38f). As indicated in Scheme 6.38e, several side reactions may compete with the displacement reaction Ar—X + Nu—H → Ar—Nu + HX. Most of them are the result of radical reactions, such as radical combination (for relatively long-lived radicals), radical addition to unsaturated compounds, and H or X atom transfer.

Irradiation at 254 nm of a 1 : 1 mixture of chlorobenzene and cyclopentene at 20 °C gives a mixture containing *cis-* and *trans-*2-phenylcyclopentyl chloride (60%; products of C—Cl bond insertion) and small amounts of 1-phenylcyclopentene (Scheme 6.39a) and chlorobiphenyls (Scheme 6.39b). Similar products are observed in the photolysis of bromobenzene and iodobenzene in cyclopentene. In contrast, fluorobenzene gives only a small amount of products of C—F insertion and undergoes mostly photoinduced [π^4s+π^2s]-cycloadditions with the alkene (Section 6.8.4) [506]. Gas phase laser flash photolysis (λ_{irr} = 193 nm) of fluorobenzene produces HF and products of decomposition of C_6H_4 ($C_2H_2 + C_4H_2$) formed via secondary photolysis [507]. Photolysis of electron-rich aryl chlorides (and mesylates, triflates, and phosphates) [508] leads to aryl cation intermediates that can add to alkenes giving carbenium ion intermediates that lose a proton (Scheme 6.39c: kind of nonmetal Mizoroki–Heck coupling (see Section 8.8.7), Scheme 6.39d) or another electrofugal group such as R_3Si or R_3Sn (Scheme 6.39e,f). They also add to phenylacetylene giving 2-aryl-1-phenylvinyl cation

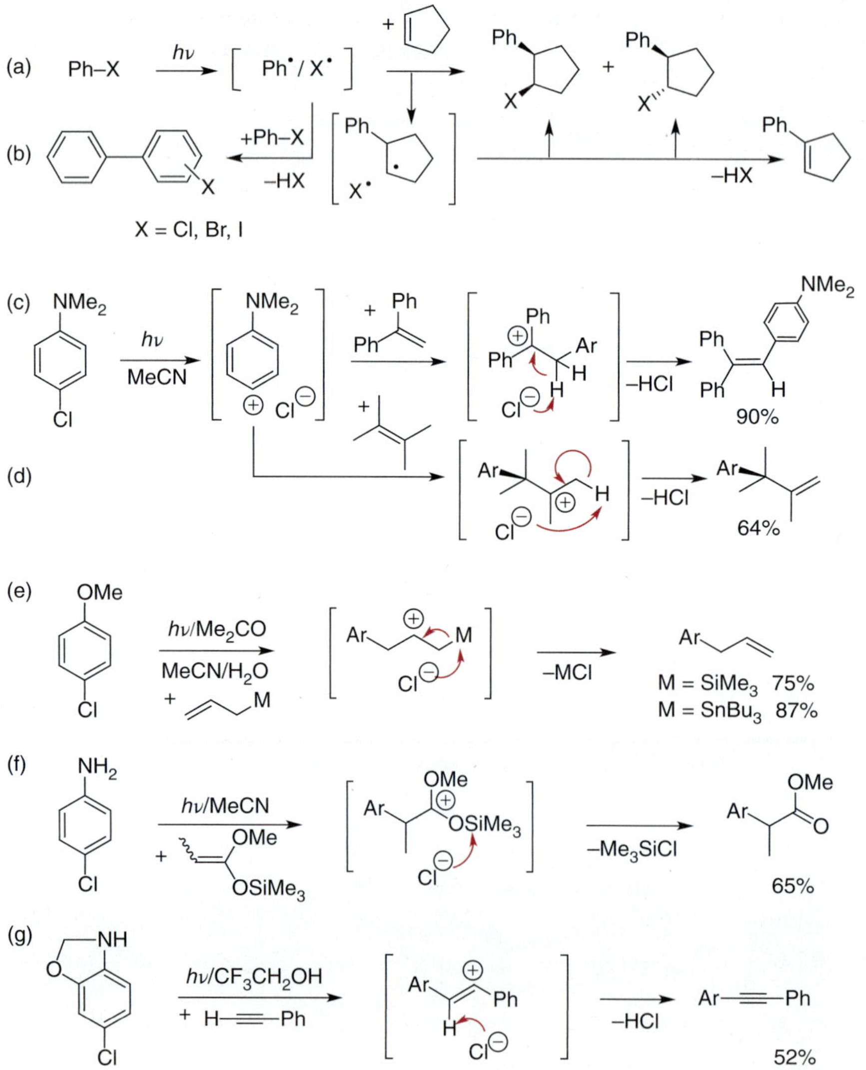

Scheme 6.39 Photoinduced arylations of alkenes and alkynes: (a) photo-induced homolysis, addition of the radicals, (b) radical arylation of alkenes, (c) and (d) photo-induced heterolysis and electrophilic substitution of the alkene (addition/elimination mechanism), (e) the same using nucleophilic allylmetals (nucleophilic allylation of the aryl halide), (f) electrophilic arylation of ketene acetals, formal α-arylation of esters (g) electrophilic arylation of terminal alkynes (addition/elimination mechanism).

intermediates that eliminate a proton (Scheme 6.39g) [509, 510].

Irradiation of iodoaryl azides [511] and of iodopyridyl azides generates first the corresponding iodoarylnitrenes + N_2 (Section 6.7.1). Arylnitrenyl and pyridylnitrenyl radicals are then formed that have been characterized by IR and ESR spectroscopy [512].

UV photolysis of (*E*)-bromostyrene ((*E*)-PhCH=CHBr) gives products of substitution and elimination resulting from vinyl radical and vinyl cation intermediates. Photolysis of deuterated (*E*)-PhCD=CHBr and (*E*)-PhCH=CDBr results in the scrambling of the deuterium label in the elimination products, which demonstrates that alkenyl cation intermediates are formed; deuterium scrambling occurs via vinylene benzenium ion (6,6-(ethene-1,2-diyl)cyclohexadienyl cation) [513].

Problem 6.32 The photolysis (quartz, high-pressure Hg burner) of iodobenzene in anisole gives a 71 : 15 : 14 mixture of *ortho-*, *meta-*, and *para*-phenylanisole. Similarly, photolysis of 2-iodothiazole in anisole gives a 63 : 14 : 23 mixture of *ortho-*, *meta-*, and *para*-(2-thiazolyl)anisole. Similar product distributions are obtained in the photolysis of iodobenzene and 1-iodothiazole in benzonitrile. Propose a mechanism for these photoreactions [514].

Problem 6.33 What are the major products formed upon irradiation of 1,2-diiodobenzene in cyclohexane, benzene, and benzene in the presence of tetraphenylcyclopentadienone (tetracyclone)? [515]

6.6.5 Photolysis of phenyliodonium salts: formation of aryl and alkenyl cation intermediates

Diphenyliodonium salts exist as tight ion pairs in acetonitrile. Their direct irradiation (λ_{irr} = 254 nm, quartz) produces mostly PhI by a photohomolytic cleavage reaction from a charge transfer excited state, whereas in aqueous acetonitrile, solvent-separated ion pairs form and give substantial amounts of 2-, 3-, and 4-iodobiphenyls, acetanilide, benzene, and HX by an initial photoheterolytic cleavage pathway (Scheme 6.40) [516, 517]. In the presence of colored transition metal complexes (sensitizer), radical arylations of arenes and aromatic heterocyclic compounds can be induced by visible light [518].

The nonphotochemical thermal solvolysis (solvent SOH) of (*E*)- and (*Z*)-2-phenyl-1-propenyl(phenyl)

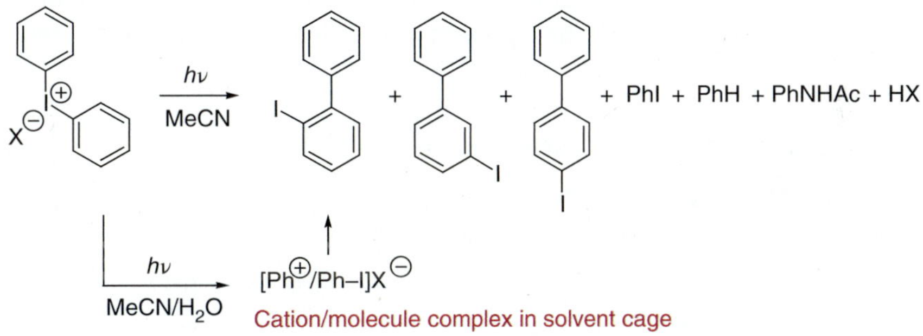

Scheme 6.40 Photolysis of diphenyliodonium salts induces homolytic cleavage in pure MeCN and heterolytic cleavage in aqueous MeCN.

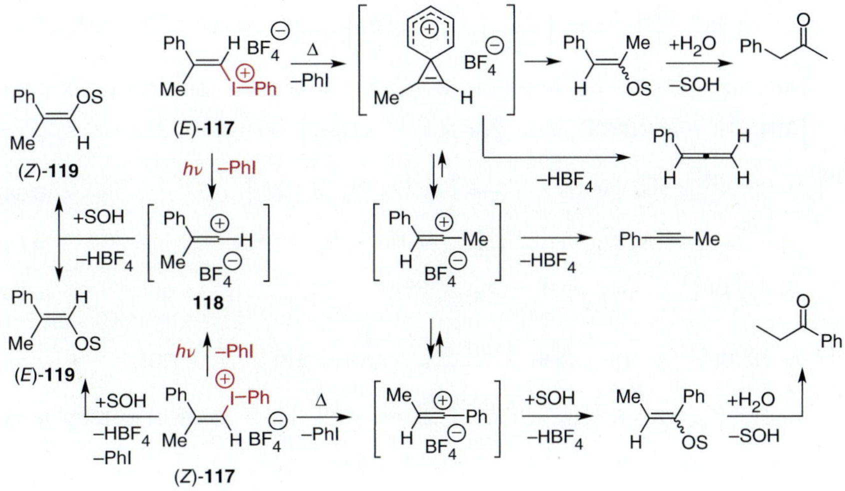

Scheme 6.41 Although thermal solvolysis favors the formation of rearranged products, photoinduced solvolysis of 2-phenyl-1-propenyl (phenyl)iodonium tetrafluoroborate gives un-rearranged products.

iodonium tetrafluoroborate ((*E*)-**117**, (*Z*)-**117**) gives mostly products of substitution and elimination involving migration of the phenyl group, whereas photolysis (λ_{irr} = 248 nm, quartz) of (*E*)- and (*Z*)-**117** generates the primary vinylic cation intermediate **118** without migration of the phenyl group (Scheme 6.41). The latter reacts with the solvent giving a mixture of un-rearranged ethers (*E*)- and (*Z*)-**119** [519, 520]. Strained cyclic vinyl cation intermediates such as cyclopent-1-en-1-yl and cyclohex-1-en-1-yl cation are formed upon UV irradiation of the corresponding phenyliodonium tetrafluoroborates [521].

6.6.6 Photolytic decomposition of arenediazonium salts in solution

Arene- and heteroaromatic diazonium salts (Ar—$N_2^+X^-$, readily prepared by diazotization of the corresponding primary amines Ar—NH_2) are extremely useful synthetic intermediates [522, 523]. The classical and less classical (photochemical, transition metal) catalyzed reactions of Ar—$N_2^+X^-$ are summarized in Scheme 6.42. On heating, they generally decompose into the corresponding Ar—X compounds or products of solvolysis (solvent SOH): Ar—OS + HX + N_2 (reaction (6.60)). When X^- = BF_4^-, PF_6^-, SbF_6^-, or AsF_6^-, the fluoroarenes ArF are formed (Balz–Schiemann reaction (6.61)) [524]. The arenediazonium salts represent an excellent source of aryl radicals (Ar•) because of their relatively high reduction potentials [525–527]. Dediazoniation can be induced thermally, for instance, for X^- = I^- (reaction (6.62)) [528], or through single-electron transfer from a copper(I) salt (CuCl and CuBr) as in the Sandmeyer reaction (6.63) [529] or with Cu + HCl or HBr as in the Gatterman reaction [530]. The aryl radicals can be quenched with alkenes (Meerwein arylation of olefins (6.64)) [531, 532] or by benzene and derivatives in intramolecular arene arylations as in the Pschorr reaction [533–536], or intermolecular arene arylations (Gomberg–Bachmann reaction (6.65)) [537–540]. The reactions are accompanied by reduction into arenes Ar—H in medium that can provide a hydrogen atom like alcohols (e.g. reactions (6.66) and (6.67)) [541, 542].

Aryldiazenyl radical (Ar—N=N•) can be formed upon direct or sensitized irradiation of the arenediazonium salts [543]. Usually, they decompose quickly (reaction (6.68)) into N_2 and the corresponding aryl radical (Ar•), which can be used as above in C—C bond forming reactions. Alternative photoinduced dediazoniation can generate aryl cation Ar^+, which can be reacted with all kinds of nucleophiles (reaction (6.69)) [544–548] including carbon nucleophiles such as benzene (electrophilic arylation of benzene) and allyltrimethylsilane (electrophilic allylation) [549]. In 1971, Cohen and coworker reported the photochemical Balz–Schiemann fluorination (6.70) [550] and have applied it to the synthesis of important bioactive

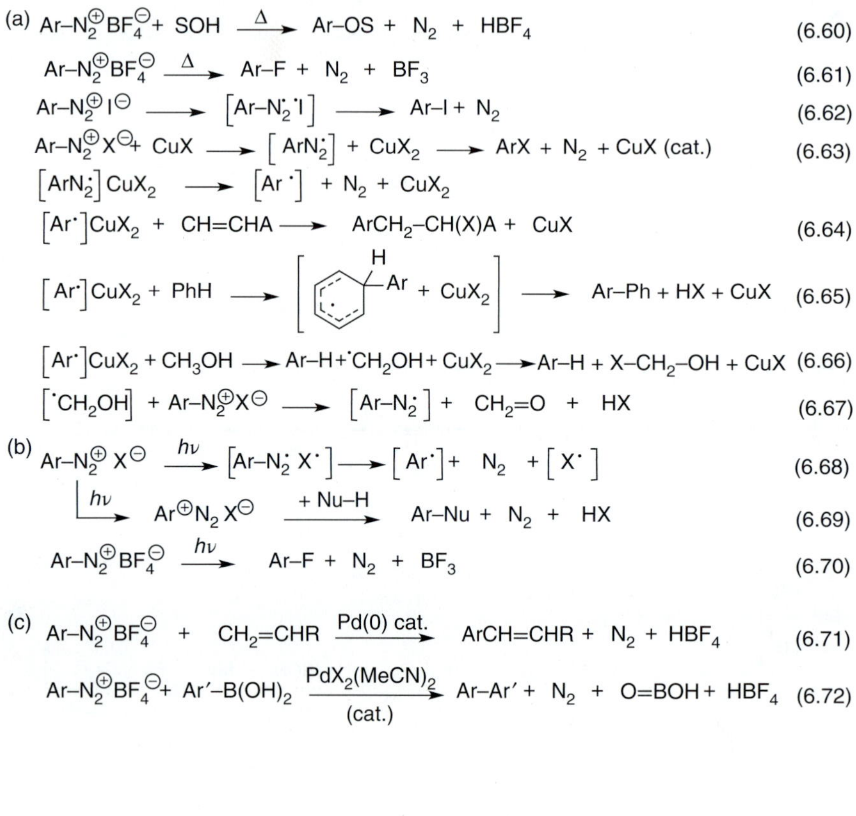

Scheme 6.42 (a) Classical reactions of arenediazonium salts, (b) photochemical formation of aryl radicals and cations from arenediazonium salts, and (c) palladium-catalyzed C-X cross-coupling reactions with alkenes (Mizoroki–Heck reaction, see Section 8.8.7) and with areneboronic acids (Suzuki–Miyaura reaction, see Section 8.8.4).

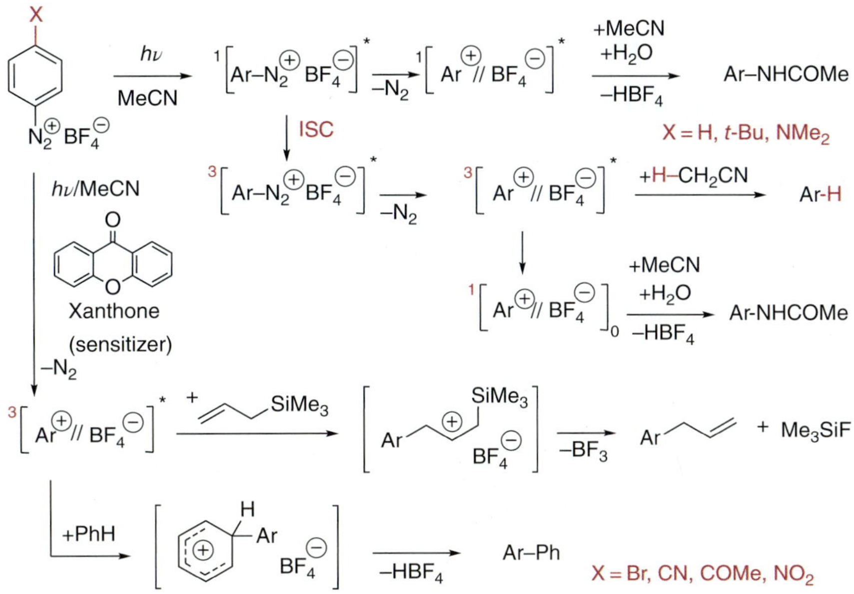

Scheme 6.43 Sensitized photolysis of diazonium salts generates triplet aryl cation intermediates that can be trapped with allylsilanes (C-allylation) and benzene (benzene arylation).

fluoroheteroaromatic compounds [551–558]. When run in HF/pyridine, excellent yields of the fluoro compounds are obtained [559]. The direct irradiation (λ_{irr} = 310 nm, 20 °C) of 4-substituted benzenediazonium tetrafluoroborates 4-X—$C_6H_4N_2^+BF_4^-$ with X = H, *t*-Bu, and Me_2N leads to singlet aryl cations that react with MeCN in Ritter reactions (Scheme 6.43). With other substituents such as X = Br, CN, COMe, and NO_2, ISC competes and triplet aryl cations form competitively from the excited arenediazonium salts. Triplet aryl cations abstract an H-atom from the solvent (MeCN) competitively with their relaxation into the corresponding singlet aryl cations that react as above in Ritter reactions with MeCN. In the presence of π-nucleophiles such as allyltrimethylsilane and benzene, the triplet aryl cations give allylarene and biphenyl derivatives, respectively. Sensitized irradiation of the arenediazonium salts also produces the triplet aryl cations, thus realizing an efficient alternative for these arylations (Scheme 6.43) [549].

Visible light along with 1% eosin Y induces the direct C—H bond arylation of heteroarenes with arenediazonium tetrafluoroborates (reaction (6.73)) [560]. Arenediazonium anthracenesulfonates photodecompose under visible light as the anion absorbs it [561].

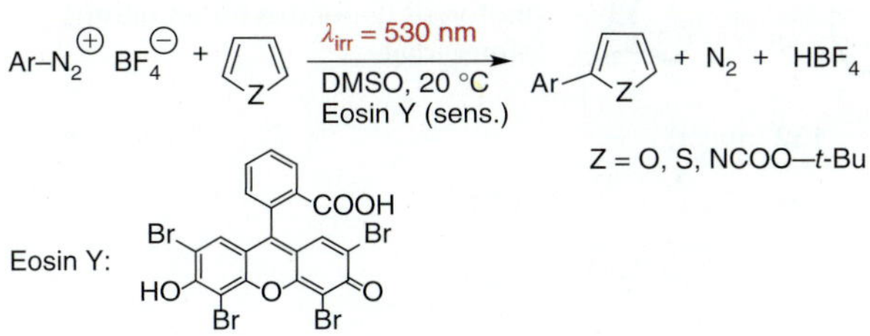

(6.73)

Unstable *O*-(trifluoromethyl)dibenzofuranium salts of type **121** have been prepared by low-temperature photolysis of arenediazonium salts **120**. Salts **121** are efficient reagents for the *O*- and *N*-trifluoromethylation of alcohols, phenols, amines, anilines, and pyridines under very mild conditions [562].

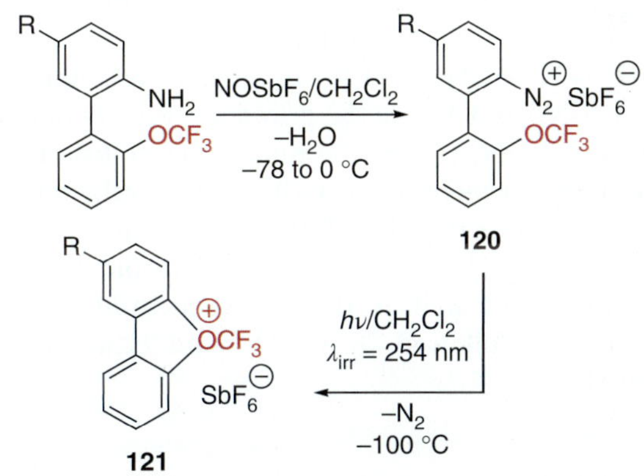

6.7 Photocleavage of nitrogen—nitrogen bonds

Compounds with a N=N double bond like the azoalkanes (1,2-diazenes), 3*H*-diazirines, azo compounds (Section 6.6.2), and the arenediazonium salts (Section 6.6.6) can lose a very stable N_2 molecule either thermally or under photolysis. The process involves cleavage of C—N bonds. We present here photochemical reactions in which N—N bonds are cleaved and form N_2. The substrates are azides, diazides, 1,2,3-triazoles, and tetrazoles.

6.7.1 Photolysis of azides

Azides represent an important class of organic compounds widely used as masked amines, as dipolar reagents in thermal and catalyzed 1,3-dipolar cycloadditions (Section 5.3.16), as starting materials of iminophosphoranes in aza-Wittig reactions to prepare various heterocyclic compounds, and as a source of nitrene intermediates [563–569]. On heating, alkyl azides **122** lose N_2 and generate imines through a one-step process that involves hydrogen or alkyl group migrations from the α-carbon center to the nearby nitrogen center (reaction (6.74)) [570]. Direct irradiation of alkyl azides also generates imines without the formation of nitrene intermediates [571]. For instance, direct photolysis of 1-azidoadamantane (**123**) produces imine **124**, a relatively strained compound [572] that dimerizes into the (2+2)-cycloadduct **125**, or reacts with methanol to give **126** [573]. Similarly, direct irradiation of α-methylbenzyl azide (**127**) produces imines **128–130** that result from the competitive migration of a hydrogen, a phenyl, and a methyl group, respectively [571]. Studies on the direct photolysis (λ_{irr} = 313 nm) of methyl azide (^{1}[Me—N_3]) suggested a concerted hydrogen atom migration and N_2 elimination to form methanimine (CH_2=NH) [574], a mechanism analogous to that reported for the thermal reaction $MeN_3 \rightarrow N_2 + CH_2{=}NH$ [575]. Singlet methylnitrene ($^1CH_3$—N:) and N_2 are the primary products of photolysis of MeN_3 (λ_{irr} = 254 nm) under collision-free conditions (synchrotron radiation). Methylnitrene then isomerizes into "hot" CH_2=NH that decomposes into H_2 + H—N=C: through a cheletropic elimination of H_2, as well as through sequential H-atom loss (N—H followed by C—H bond cleavage), to form HNC.

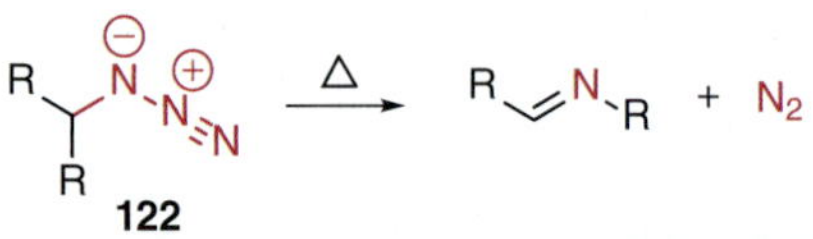

(6.74)

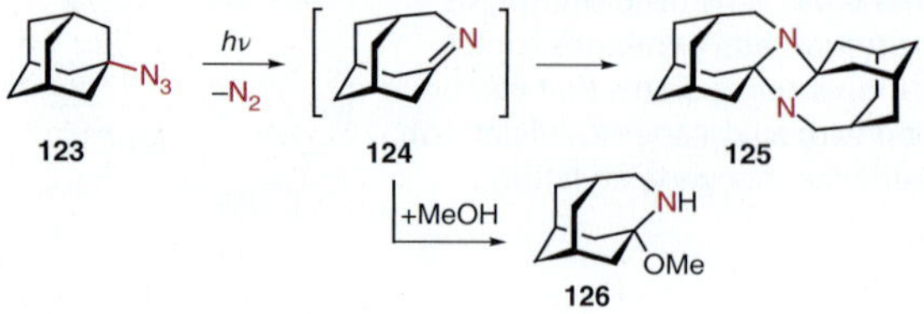

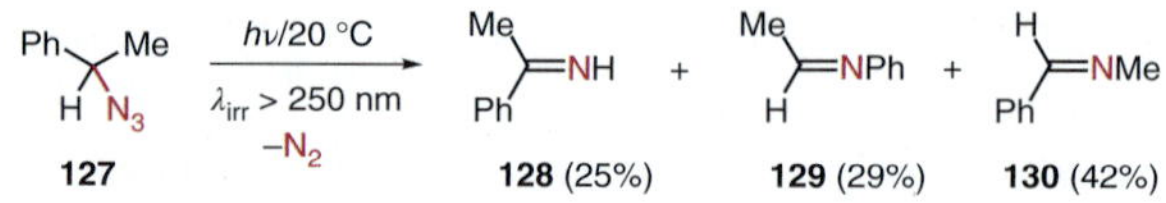

Triplet-sensitized photolysis of 1-azidoadamantane (**123**) with acetone, acetophenone, or benzophenone (Scheme 6.44) leads to di(adamant-1-yl)-1,2-diazene (**132**) via dimerization of (adamant-1-yl)nitrene (**131**). Triplet nitrene **131** also abstracts hydrogen atoms from toluene (solvent) to form 1-aminoadamantane, 1-(benzylamino)adamantane, and the imine resulting from 1-aminoadamantane and benzaldehyde [576].

The photochemistry of aromatic azides has attracted wide interest for practical applications such as photoresists, surface modification of nanomaterials, and affinity labeling of biomolecules [577–582]. Ultrafast time-resolved studies of the photochemistry of aryl azides have permitted the establishment of detailed mechanisms of the photochemical processes observed [583]. UV irradiation of phenyl azide (PhN_3) yields singlet phenylnitrene (^{1}[PhN:]) [584], which, above 165 K, rapidly ring expands to 1-azacyclohepta-1,2,4,6-tetraene (**134**) [585] via benzazirine **133**. The latter intermediate has been intercepted with ethanethiol (to generate 2-ethylthioaniline) before it opens to **134** [586]. Ketenimine **134** is trapped by diethylamine producing azepine **135**, or by water giving lactam **136** (Scheme 6.45a). In the absence of nucleophile, 1-azacyclohepta-1,2,4,6-tetraene reverts to singlet phenylnitrene under conditions of high dilution; otherwise, it reacts with PhN_3 to form a polymeric tar [587]. Under photochemical conditions, phenylnitrene interconverts into isomeric pyridyl-

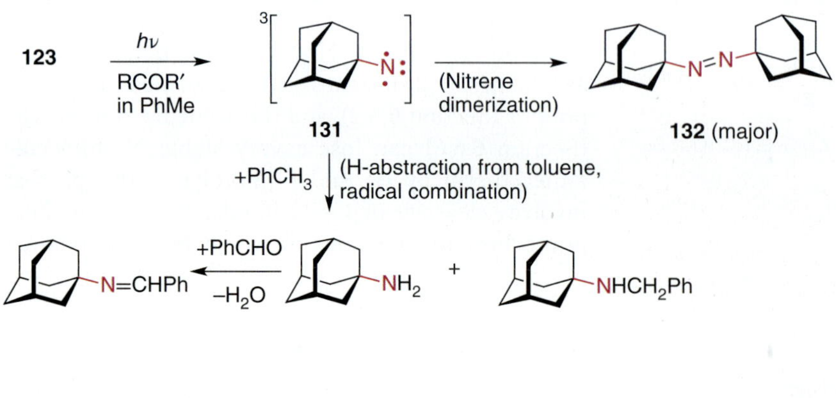

Scheme 6.44 Triplet-sensitized photolysis generates triplet nitrene intermediates.

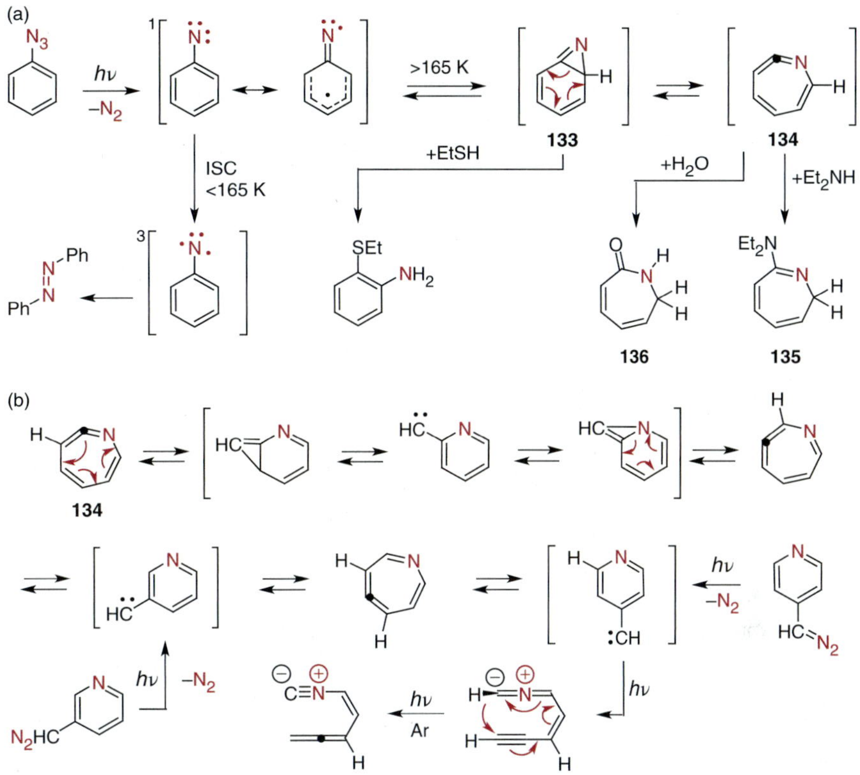

Scheme 6.45 (a) Photolysis of phenyl azide and (b) nitrene/carbene rearrangement of phenylnitrene.

methylenes via 1-azacyclohepta-1,2,4,6-tetraene (Scheme 6.45b) [588], a reaction that also occurs under thermal conditions as described by Wentrup and Crow [589–591]. Contrary to the behavior of phenylcarbene, singlet phenylnitrene does not react readily with alkenes ((2+1)-cheletropic reactions forming aziridines) or alkanes (C—H insertion producing *N*-alkylanilines) [592]. Laser flash (λ_{irr} = 266 nm, 150 ps) photolysis of PhN_3 in pentane at 20 °C generates a transient species absorbing at 350 nm, which decays into **134** with activation parameters E_a = 5.3 ± 0.3 kcal mol^{-1} and log A = 13.6 ± 0.4. This transient is assigned to singlet phenylnitrene [593, 594]. Similar experiments with substituted phenyl azides lead to similar energy barriers for the conversion of the corresponding singlet nitrenes into azacycloheptatetraenes [595], except for *ortho*-fluorosubstituted derivatives that are less sensitive toward rearrangements [511, 596]. By means of laser flash photolysis at room temperature [597], or by photolysis in Ar matrices, substituted benzazirines have been characterized by UV and IR spectroscopy [598–600]. Below 165 K, ISC to triplet phenylnitrene (^{3}[PhN:]) dominates and it eventually dimerizes into azobenzene [601]. It is not very prone to abstracting a hydrogen atom from the medium [602, 603]. UV light irradiation at 80 K of single crystals of aryl azides generates the corresponding triplet arylnitrenes that have been characterized by X-ray diffraction studies [604]. Ultrafast photolysis of PhN_3 (λ_{irr} = 308 nm) in 100% formic acid (HCOOH) produces singlet excited state of PhN_3 that decomposes within 300 fs into ^{1}[PhN:]. Within 10–20 ps, phenylnitrene is protonated into phenylnitrenium ion ($PhNH^+$) (λ_{max} = 500 nm) that survives c. 100–200 ps in HCOOH giving PhNHOOCH [605].

Photolysis of 2-methoxy-6-methoxycarbonylphenyl azide (**137**) produces the corresponding singlet arylnitrene **138** that has a lifetime of 800 ps in MeCN at 20 °C. It reacts with water at 20 °C with a rate constant of 1.9×10^8 M^{-1} s^{-1} giving hydroxylamine **139** [606]. In the absence of water, it forms ketenimine **140** and isoxazole **141** [607].

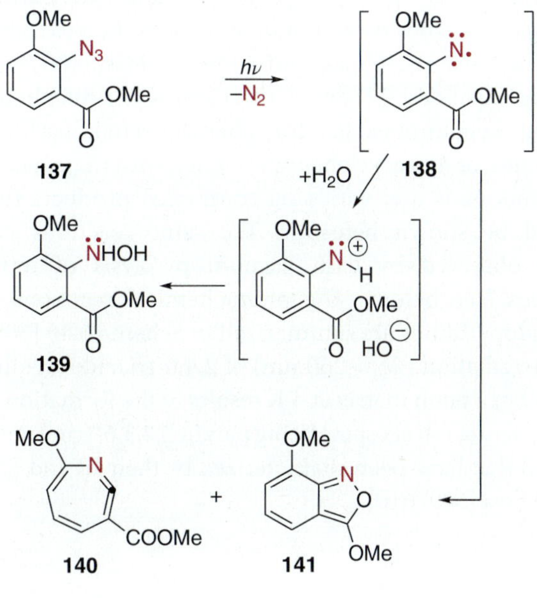

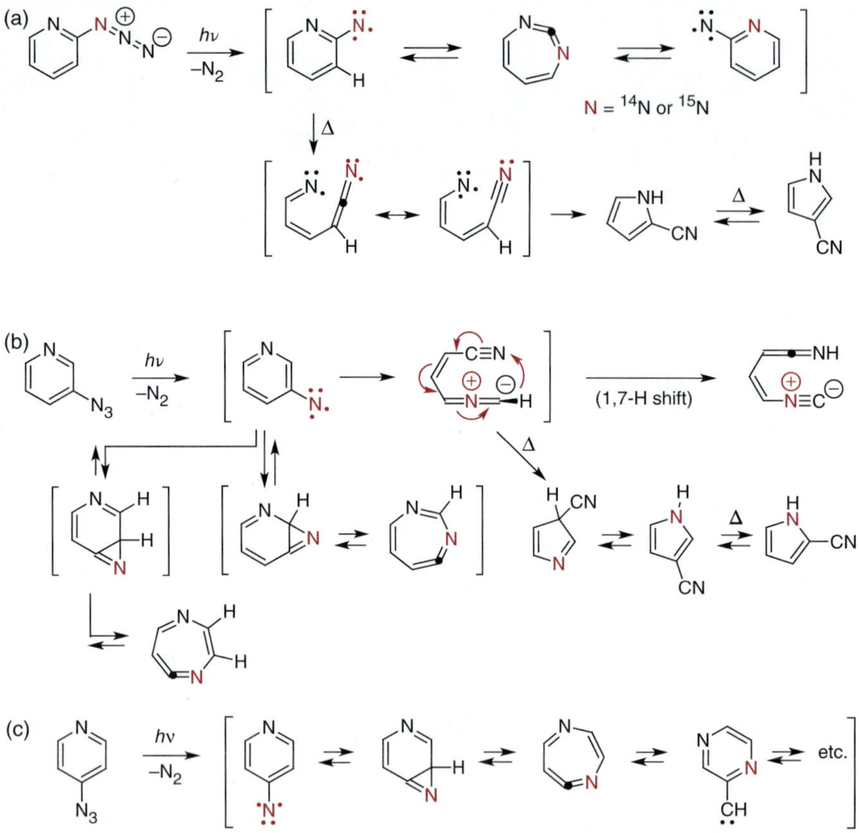

Scheme 6.46 The gas phase photolysis of (a) 2-azidopyridine, (b) 3-azidopyridine, and (c) 4-azidopyridine generates the corresponding singlet nitrenes that undergo reversible isomerization into diazacycloheptatetraene intermediates that can give ring-opened and/or ring-contracted products.

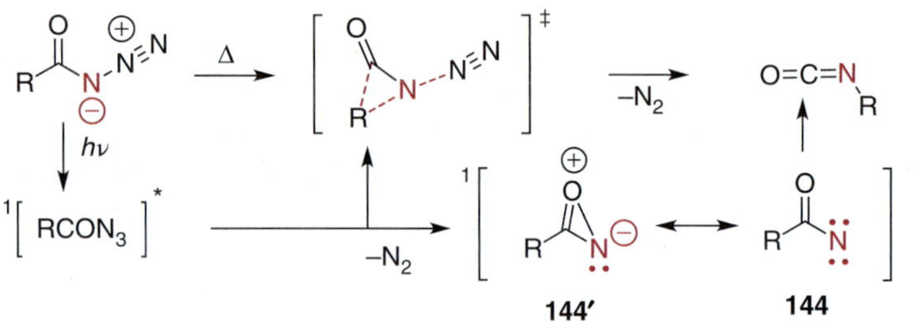

(6.75)

Scheme 6.47 The photo-Curtius reactions of acyl azides derive mostly from the singlet excited state of the acyl azides and from the corresponding singlet acylnitrene intermediates.

Photolysis of azidopyridines and derivatives generates the corresponding singlet heteroarylnitrenes that equilibrate with heteroarylcarbenes, with ring-expanded products that are seven-membered ring ketenimines (as for phenylnitrene), carbodiimides, or diazacycloheptatetraenes, with ring-opened compounds and with ring-contracted products that will be shown hereafter. The same reactions can be observed by flash vacuum pyrolysis of tetrazoles (see below). The photochemical reactions of azidopyridines are summarized in Scheme 6.46 [608].

Irradiation ($\lambda_{irr} > 260$ nm) of 2,4,6-triazidopyridine **142** in a neon matrix at 4 K results in the formation of the septet (all six spins being parallel) 2,4,6-trinitrenes **143** that have been characterized by their IR and ESR spectra [609, 610].

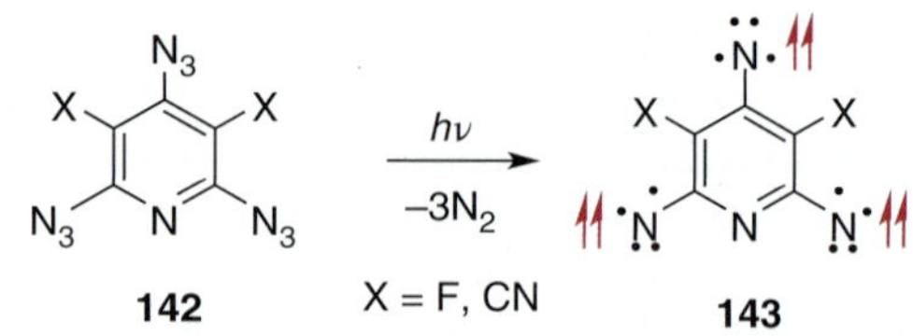

Problem 6.34 Propose a sequence of reactions that convert 1,3-diaminobenzene into (*E*,*E*)-3-azahexa-3,5-dien-1-yne-6-carbonitrile (NC—CH=CH—CH=N—C≡CH) [611].

6.7.2 Photo-Curtius rearrangement

The Curtius rearrangement (reaction (6.75)) [612] of acyl azides to isocyanates + N_2 can take place either thermally or photochemically [613]. Thermal Curtius

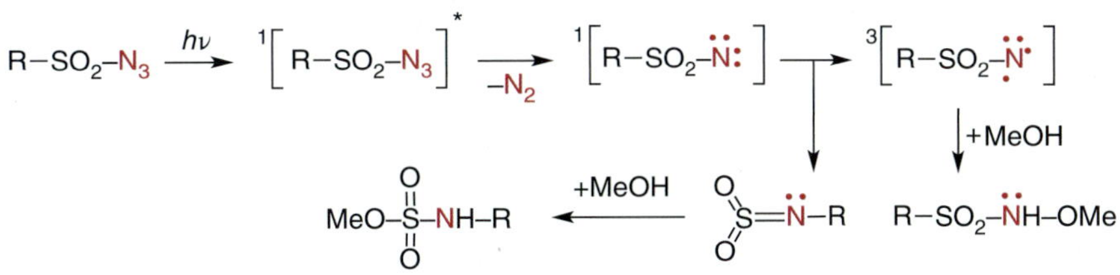

Scheme 6.48 The photochemical reactions of sulfonyl azides involve the relatively long-lived triplet sulfonylnitrene intermediates.

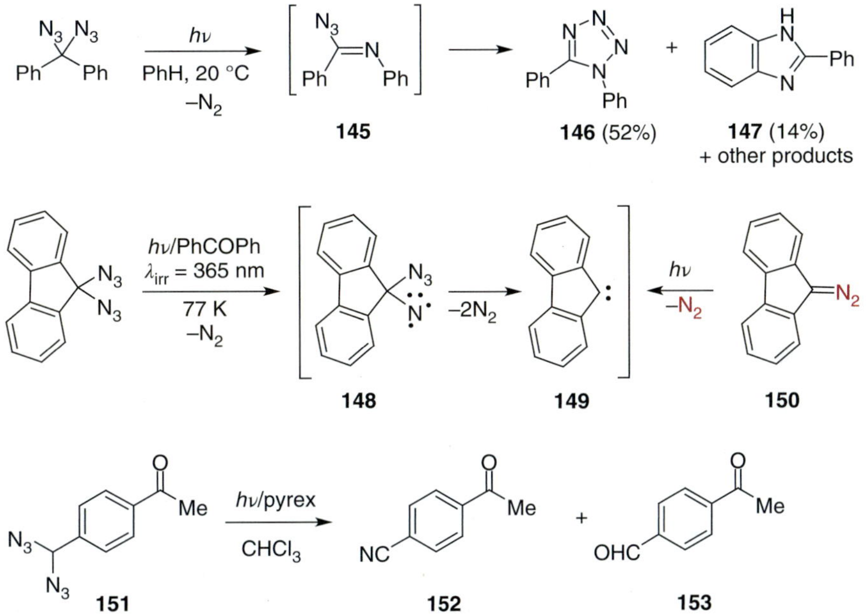

Scheme 6.49 Examples of photochemical reactions of *gem*-diazides.

rearrangements are one-step, concerted processes (Scheme 6.47). The photo-Curtius rearrangement can follow a mechanism in which the acyl azide in its singlet excited state is fragmentized into N_2 + R—N=C=O in a concerted manner, like the thermal reaction, or in a two-step mechanism involving the formation of singlet acylnitrene intermediate (^{1}[R—CO—N:]) that can be reacted in cheletropic reactions with electron-rich alkenes and fullerene C_{60} [614]. In the case of photolysis (Ar matrix, 12 K, λ_{irr} = 308 nm) of benzoyl azide ($PhCON_3$), a small amount of phenyl cyanate (PhOCN) forms next to phenyl isocyanate (PhNCO) [615]. In agreement with Hund's rule, alkyl and arylnitrenes have triplet ground states, but acylnitrenes do not. This is because of a bonding interaction between the oxygen atom of the carbonyl group and the electrophilic nitrogen atom of the nitrene, and this interaction (**144** ↔ **144′**) selectively stabilizes the singlet state over the triplet state as a closed-shell singlet state is realized [616–618].

Schuster and coworkers have found that the triplet states of acyl azides do not produce isocyanates [619, 620]. Ultrafast time-resolved infrared and UV–visible spectroscopy as well as computational studies by Platz and coworkers confirm that singlet acyl azide excited state must be the precursor to the corresponding isocyanate [621–623]. In contrast to acyl azides, IR and ESR spectroscopy demonstrate that the triplet states of sulfonylnitrenes are the ground states responsible of the products observed for the photolysis of sulfonyl azides [624]. In the case of naphthalene-2-sulfonyl azide, the singlet sulfonylnitrene is a short-lived species (c. 70 ps in CCl_4) that decays into the longer lived triplet sulfonylnitrene [625]. Benzenesulfonylnitrene adds methanol to give the corresponding *N*-methoxysulfonamide, concurrently with its pseudo-Curtius rearrangement giving *N*-sulfonylphenylamine, which adds methanol producing methyl *N*-phenylsulfamate (Scheme 6.48).

6.7.3 Photolysis of geminal diazides

In 1967, Moriarty and Kliegman reported that the direct irradiation of diazidodiphenylmethane results in the formation of 1,5-diphenyltetrazole (**146**) and a small amount of 2-phenylbenzimidazole (**147**) [626].

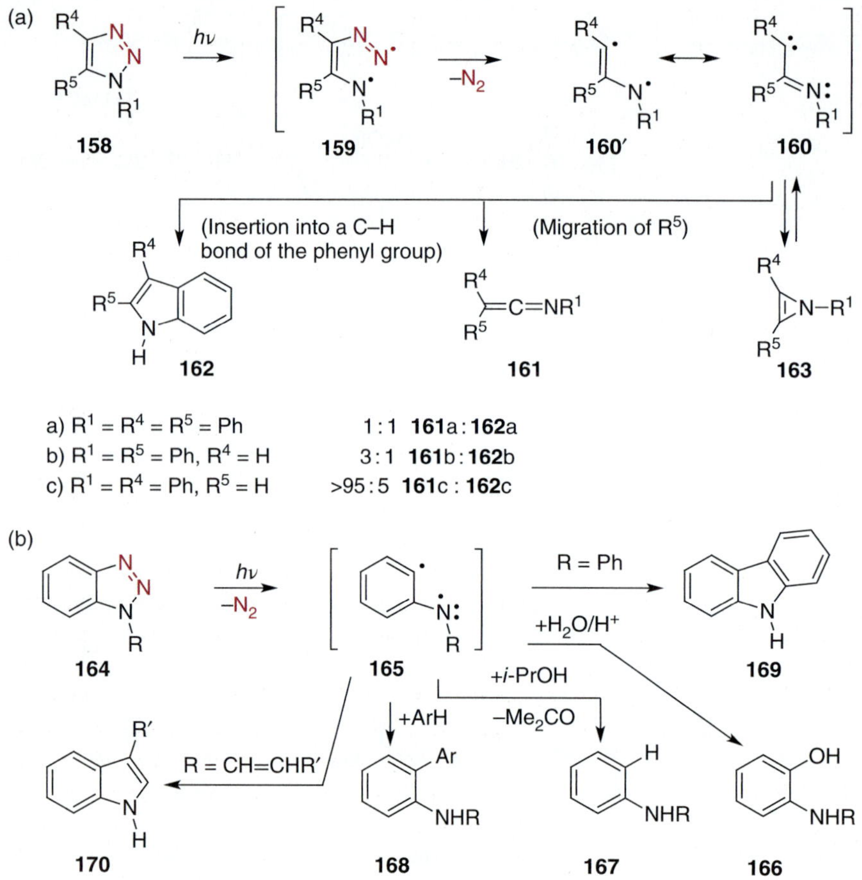

Scheme 6.50 Photolysis of (a) 1*H*-1,2,3-triazole and (b) benzotriazole derivatives generate diradical intermediates by dediazoniation.

Dediazonation generates the azidoimine intermediate **145**, which undergoes an electrocyclic cyclization into **146** and a concurrent loss of N_2 producing **147**. The same year, Barash et al. found that the sensitized photolysis of 9,9-diazido-9*H*-fluorene in a glass leads to its dediazoniation forming first nitrene **148**. Upon prolonged irradiation, **148** generates triplet carbene **149** (ESR spectroscopy) [627]. The same carbene is obtained upon irradiation of diazo compound **150** (Scheme 6.49). Photolysis of diazides **151** in $CHCl_3$ gives benzonitrile **152** (major) and benzaldehyde **153** (minor) [628].

Laser flash photolysis of **151** at 20 °C showed that the first excited triplet state of the ketone is formed initially and leads to C—N homolysis and formation of radical **154** (λ_{max} = 380 nm, lifetime of c. 2 μs). Radical **154** loses 1 equiv. of N_2 yielding imine radical **155** (λ_{max} = 300 nm). Disproportionation of **155** produces carbonitrile**152** and an imine that adds H_2O (present in traces) to give aldehyde **153**. Photolysis of diazide **151** in Ar matrix at 14 K results in triplet nitrene **156**. Irradiation of **156** yields triplet imine-nitrene **157**.

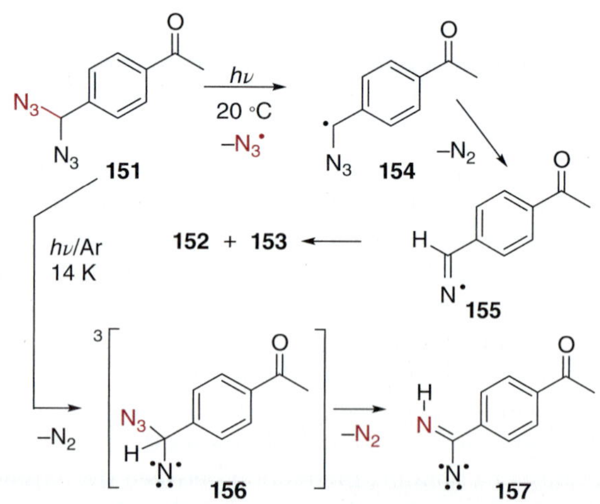

6.7.4 Photolysis of 1,2,3-triazoles and of tetrazoles

Photolysis (high-pressure Hg burner, quartz) of 1-phenyl-1*H*-1,2,3-triazoles **158** generates mixtures of iminoketenes **161** and indoles **162**. A possible mechanism (Scheme 6.50a) implies C—N cleavage into diradical **159** that loses N_2 to form iminocarbenes **160** [629]. The latter intermediates can equilibrate with the corresponding 1*H*-azirine **163** [630]. Irradiation (λ_{irr} = 254 nm, quartz) of 4-phenyl-1,2,3-triazole in methanol produces phenylacetonitrile in 35% yield [631, 632]. UV irradiation of *N*-substituted benzotriazoles **164** gives rise to diradical intermediates **165** that show seven-line ESR spectra, typical of triplet species, when the irradiation is carried out in a matrix at 77 K. This triplet undergoes hydrolysis into aminophenols **166** in aqueous H_2SO_4, reduction into anilines **167** in isopropanol, and to products of arylation **168** in aromatic solvents. Photolysis of 1-phenyl-1*H*-benzotriazole produces carbazole **169** as the main product (Scheme 6.50b) [633]. With 1-alkenyl-1*H*-benzotriazole, the corresponding indoles **170** are obtained [634, 635]. In a similar way, photolysis of 1-alkenyl-1*H*-1,2,3-triazoles furnish the corresponding pyrrole derivatives [636].

Photolysis of 5-phenyl-2*H*-tetrazole (**171a**) in Ar matrix (λ_{irr} = 254 nm, 12 K) causes elimination of N_2 and formation of benzonitrile imine (**172**). Prolonged irradiation converts **172** into phenylcarbodimide (PhN=C=NH) probably via 1*H*-diazirine **173**. A second route to PhC=N=NH is via the 5-phenyl-1*H*-tetrazole (**171b**, tautomer of **171a**) and imidoylnitrene intermediate **174** [637].

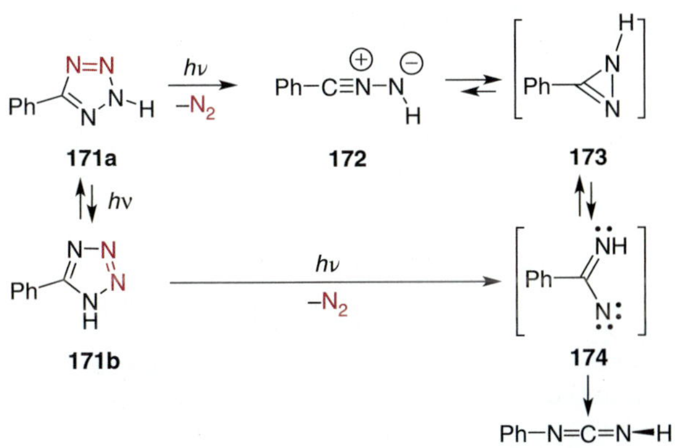

Problem 6.35 What products result from the irradiation (Hg burner, Pyrex vessel) of 1,1-diazido-2,3,4,5,6-penta-*O*-benzyl-D-glucose in CH_2Cl_2 [638] and of 2,3,4,6-tetra-*O*-acetyl-D-glucopyranosylidene diazide in the presence of acrylonitrile in quartz tubes? [639]

Problem 6.36 Propose a mechanism for the photolysis of 1-methyl-4-phenyl-1*H*-tetrazole-5(4*H*)-thione that produces methylphenylcarbodiimide [640].

Problem 6.37 What are the products formed upon irradiation (254 nm) of 2,4-diphenyl-2*H*-tetrazole? [637]

6.8 Photochemical cycloadditions of unsaturated compounds

Direct light absorption by an alkene (RCH=CHR′) in the gas phase or in dilute solution produces the first excited state, $^1(\pi,\pi^*)$ state, for which the π-bond order is significantly reduced, thus favoring rotation about the π(C=C) double bond (Section 6.3.1). This process dominates the uni- and bimolecular photochemistry of acyclic alkenes and medium-size ring cycloalkenes for both their singlet and triplet excited states [97]. The vertical and twisted singlet excited states of alkenes possess zwitterionic character, whereas the corresponding triplet states $^3(\pi,\pi^*)$ are diradical in character [98]. The latter has a much longer lifetime than its singlet excited state because decay from $^3(\pi,\pi^*)$ to the ground state is spin forbidden (Section 6.2.2). Because bimolecular reactions (photochemical condensations) involve the encounter of two molecules, these processes will be more frequent with triplet than with singlet excited states, at least for dilute solutions. Obviously, for compounds with two π-functions in the same molecule, intramolecular photochemical reactions involving these two functions have a great chance to occur in their singlet excited state. The triplet state of an alkene is obtained advantageously through sensitized photoreaction (requires UV light of lower energy than direct irradiation). As diradicals, the triplet intermediates can undergo typical reactions of radicals such as hydrogen abstraction from the medium or from alkyl side chains or can react with another alkene molecule in its ground state leading to dimeric products. These latter reactions usually occur through a nonstereoselective (2+2)-cycloaddition that involves diradical intermediates (Scheme 6.51a). Another possible reaction is the polymerization of the alkene. Depending on the substituents, the electronically excited alkene may react with another compound present in solution rather than with itself and undergo a photoinduced condensation (Scheme 6.51b). Alternatively, the other reacting partner (e.g. an electron-poor alkene, Scheme 6.51c, or an electron-rich compound such a tertiary amine, Scheme 6.51d) may involve electron transfer with an electron with the electronically excited alkene initiating various possible photochemical processes. For concentrated solutions, excimers (absorption complex of two identical alkene molecules) or exciplexes (absorption complex of two molecules of different alkenes) can form upon light absorption (Section 6.2.3). These excited complexes might have their own photochemistry; for instance, they may

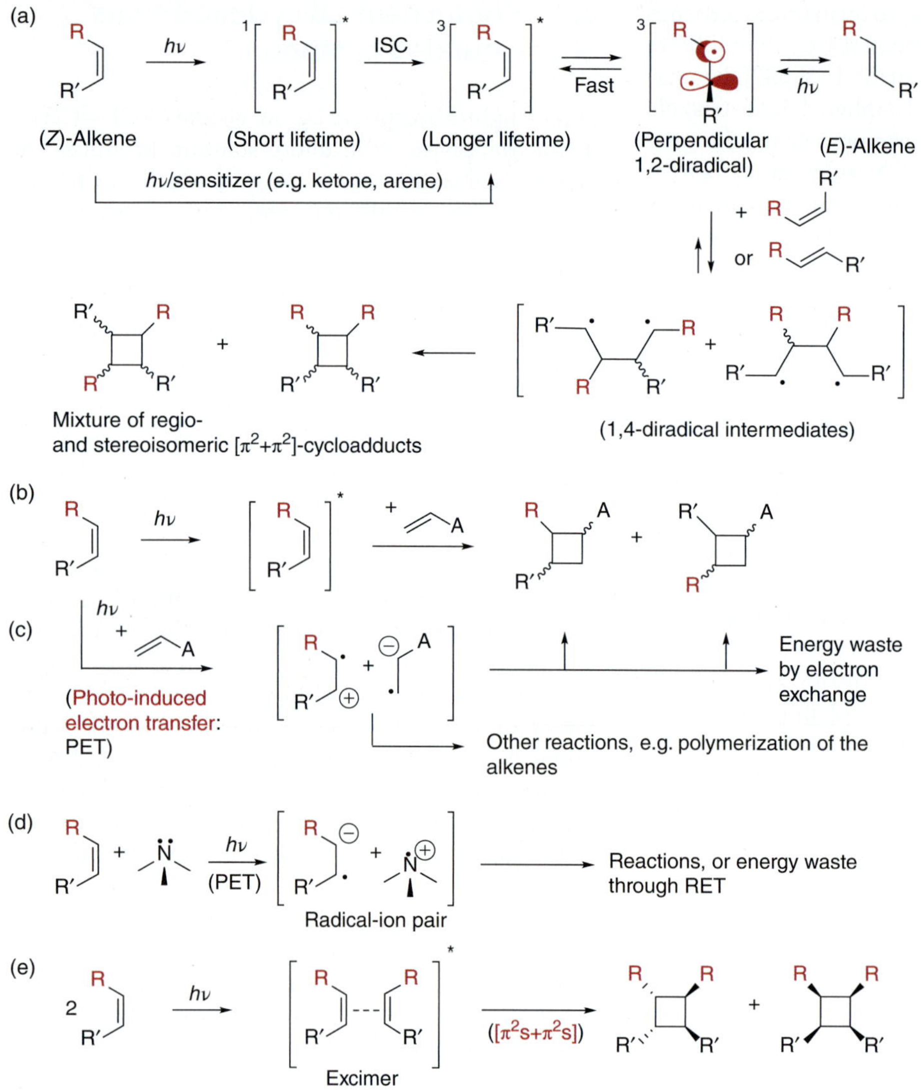

Scheme 6.51 Bimolecular photochemical reactions of alkenes: (a) non-stereoselective (2+2)-cycloaddition *via* 1,4-diradical intermediates, (b) and (c) non-stereoselective (2+2)-cycloaddition implying photo-induced electron transfer between the cycloaddends and formation of radical-cation/radical-anion pair intermediates, (d) photo-induced electron transfer from an electron rich compound to the alkene, (e) stereoselective $[\pi^2s+\pi^2s]$-cycloaddition.

undergo stereoselective $[\pi^2s+\pi^2s]$-cycloadditions in their first singlet excited states (Scheme 6.51e) as predicted by the Longuet-Higgins model (Section 5.3.1, Figure 5.18) and the Woodward–Hoffmann rules for concerted reactions [641]. As we shall see below, the photochemical (2+2)-cycloaddition is a powerful method for the construction of cyclobutane derivatives, especially when combining readily available cyclic α,β-unsaturated ketones or esters with alkenes, alkynes, or allenes [642–648].

Nonconjugated π-functions will undergo intramolecular cycloadditions. These intramolecular photochemical $[\pi^2+\pi^2]$-cycloadditions are presented in this section. The tethering of two π-functions drastically reduces the number of possible orientations for cycloaddition and renders these photocycloadditions much more regio- and stereoselective than intermolecular cycloadditions. These intramolecular photocycloadditions have found wide applications in the synthesis of complicated natural and unnatural organic compounds [348, 643, 649].

6.8.1 Photochemical intramolecular (2+2)-cycloadditions of alkenes

In 1908, Ciamician and Silber reported that sunlight affects a reaction of carvone that leads to the formation of carvone camphor, a substance with a smell similar to that of camphor [650]. This is an early example of photochemical intramolecular $[\pi^2+\pi^2]$-cycloaddition (reaction (6.76)) of two alkene moieties [651].

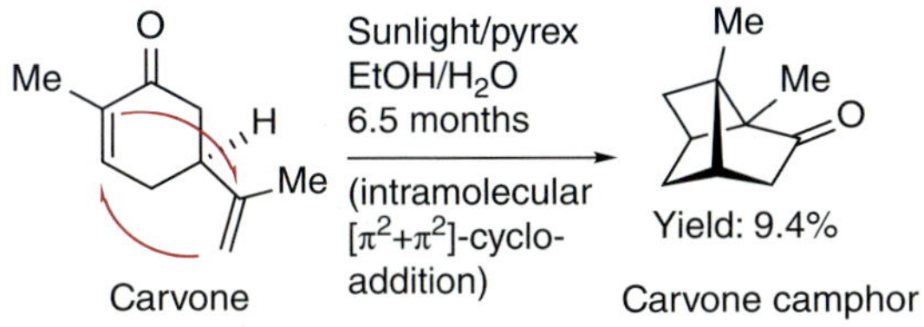

(6.76)

In the gas phase, the mercury-sensitized excitation of hexa-1,5-diene gives bicyclo[2.1.1]hexane that results from a "crossed" $[\pi^2+\pi^2]$-cycloaddition (6.77) together with several other products resulting from radical formation, including allylcyclopropane. Bicyclo[2.2.0]hexane that would have resulted from an intramolecular "straight" $[\pi^2+\pi^2]$-cycloaddition is not observed [652].

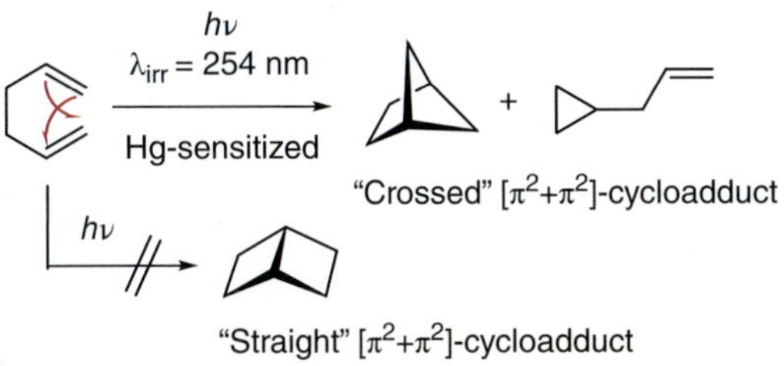

(6.77)

Direct irradiation of butadiene gives a mixture of cyclobutene arising from an electrocyclic ring closing reaction, and bicyclo[1.1.0]butane resulting from the "crossed" $[\pi^2+\pi^2]$-cycloaddition of its *s-trans*-conformer (Section 6.3.3) [148]. Mercury-sensitized excitation of hepta-1,6-diene gives both possible "straight" (bicyclo[3.2.0]heptane) and "crossed" (bicyclo[3.1.1]heptane) (2+2)-cycloadducts together with several products resulting from hydrogen atom migration and free radical processes [653]. Direct irradiation of myrcene ($\lambda_{irr} > 220$ nm) produces cyclobutene **175** (electrocyclic ring closing of the butadiene moiety) as the major product, together with small amounts of β-pinene and bicyclo[2.1.1]hexane derivative **176** (intramolecular "crossed" $[\pi^2+\pi^2]$-cycloadditions) and several products derived from radical intermediates [654]. Photosensitized (ketone) excitation of myrcene gives **176** as the only product [655].

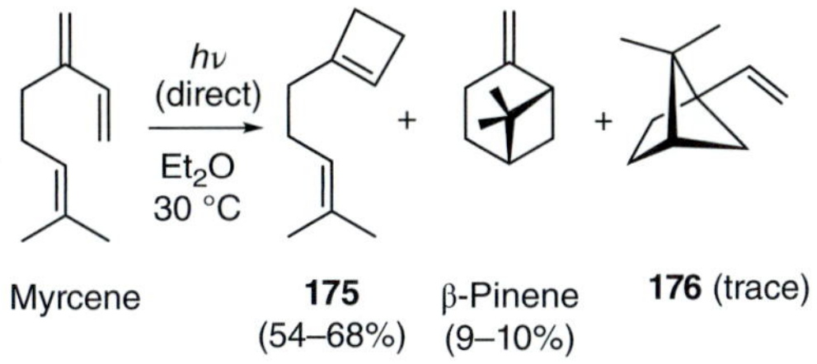

(6.78)

The mercury-sensitized excitation of cycloocta-1,5-diene in the gas phase gives mostly a polymeric material and small amounts of tricyclo[3.3.0.0^{2,6}]octane resulting from a intramolecular "crossed" $[\pi^2+\pi^2]$-cycloaddition. None of the isomeric tricyclo[4.2.0.0^{2,5}]octane resulting from a "straight" cycloaddition is detected [656–658]. The ratio between the products of "straight" (or parallel) and "crossed" cycloaddition is rationalized by "the rule of five" proposed in 1967 by Srinivasan and Carlough [659] for the intramolecular alkene + alkene cycloadditions and by Hammond and coworkers for the intramolecular alkene + diene cycloadditions [660]. Accordingly, the preferred photocycloadditions imply the formation of diradical intermediates that contain a five-membered ring, even if other diradicals that do not contain a five-membered ring should be more stable (Scheme 6.52) [661]. The formation of diradicals has been confirmed by quenching them with H_2Se [662].

Depending on substitution, the intramolecular $[\pi^2+\pi^2]$-cycloadditions follow (e.g. photoreactions (6.79) and (6.82)) or do not follow the "rule of five" (e.g. photoreactions (6.80) and (6.81)) [663, 664]. The chemoselectivity depends on factors such as reversibility or irreversibility of the diradical formation and on their conformational properties [665]. However, this complexity has not impeded organic chemists from the application of the photochemical intramolecular (2+2)-cycloadditions to the construction of complex natural products and analogs of biological interest [3, 642, 666–669].

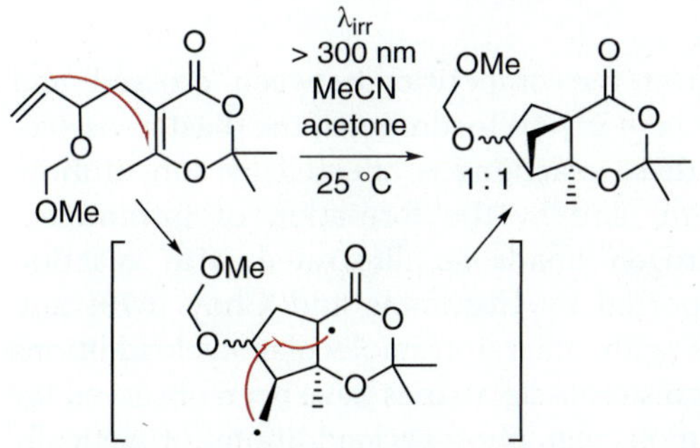

(6.79)

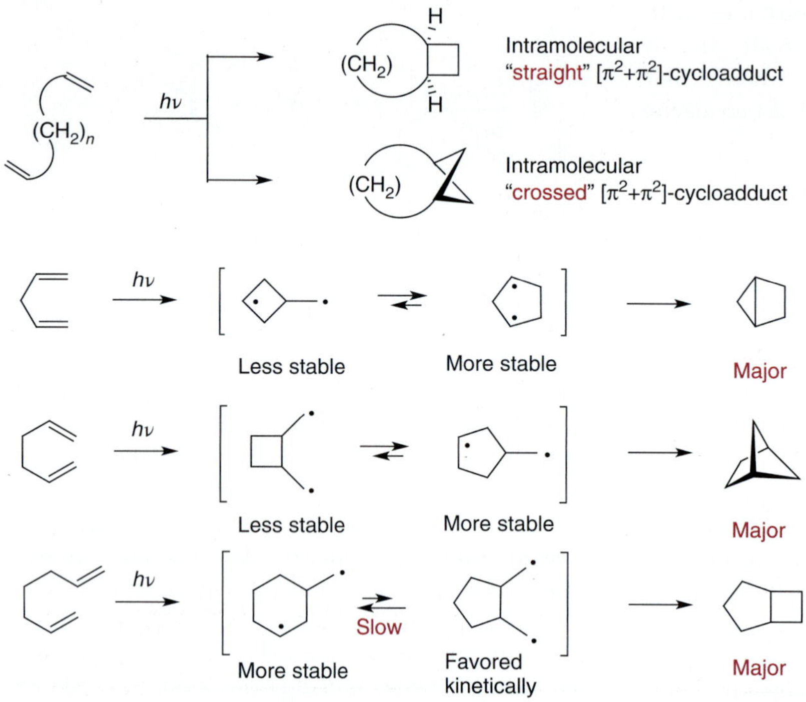

Scheme 6.52 The "rule of five" for the intramolecular photoinduced $[\pi^2+\pi^2]$-cycloadditions.

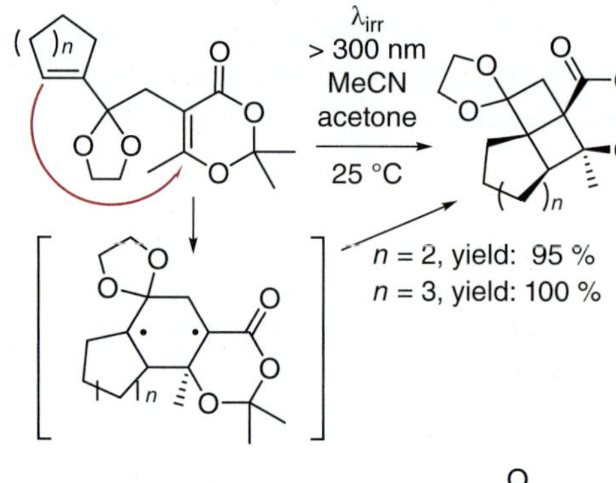

(6.80)
(6.81)

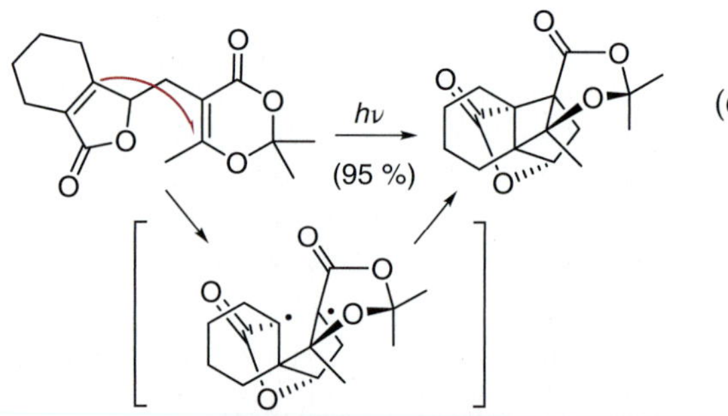

(6.82)

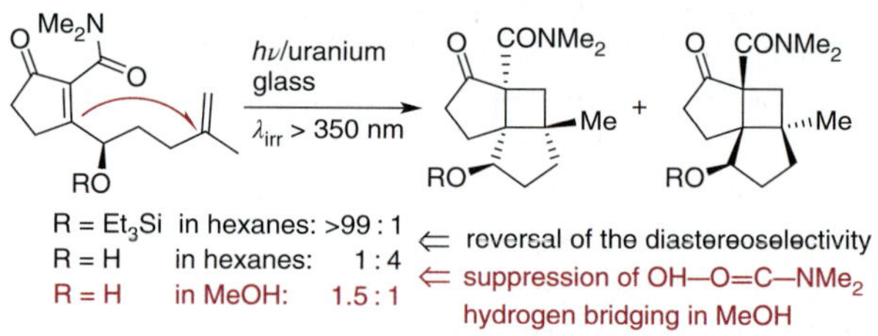

(6.83)

Apart from the competition between "crossed" and "straight" $[\pi^2+\pi^2]$-cycloadditions, the diastereoselectivity of these reactions is affected by substitution, the solvent, and by the formation of intramolecular hydrogen bonds as illustrated with reaction (6.83) reported by Crimmins and Choy [670] and by others with other intramolecular cycloadditions [671]. High stereoselectivities have been observed for the intramolecular photocycloadditions of optically active substrates [672–674].

Following the "rule of five," irradiation of vinylogous allenamides **177** can lead to either "straight" (**179**) or "crossed" cycloadducts **181** (Scheme 6.53). The latter are rearranged into pyrrole derivatives **180** and bicyclic systems **182**, respectively. The chemoselectivity depends on substitution. For amines **177** with R = H, Me pyrroles are obtained as major products, whereas with amides **177** containing R = MeCO, **182** is the major product of photoreaction [675].

Direct irradiation of norbornadiene (**N**) (Hg burner, quartz vessel, and ether) gives quadricyclane (**Q**) in 67% yield (reaction (6.84)) [676]. The direct irradiation of barrelene, or of its benzo-annulated derivatives (Section 6.3.4) and Dewar benzene derivatives (Section 6.5.1), reaction (6.84), leads to a "straight" intramolecular $[\pi^2 s+\pi^2 s]$-cycloaddition. For geometrical reasons, the "crossed" $[\pi^2+\pi^2]$-cycloaddition cannot be realized. Other products are cyclopentadiene + acetylene (retro-Diels–Alder reaction) and toluene. Norbornadiene absorbs UV light up to λ = 226 nm (final absorption), at somewhat lower

Scheme 6.53 The photocyclization of vinylogous allenamides.

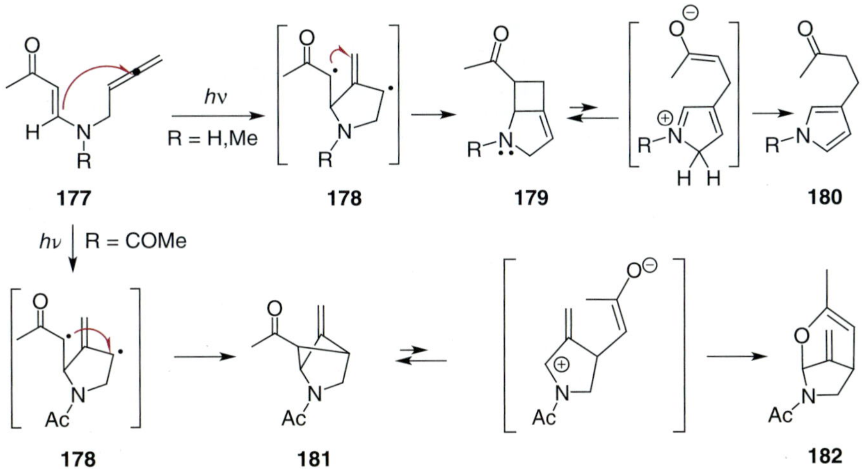

energy than monocyclic alkenes such as norbornene (final absorption at $\lambda = 210$ nm), indicating through space and through bond interactions between the two alkene units of norbornadiene [677]. Quadricyclane is isolated in up to 95% yield upon sensitized irradiation (acetone, acetophenone or benzophenone as sensitizer) of norbornadiene, with quantum yields approaching 100%. A photostationary equilibrium is reached, and the relative amount of norbornadiene (**N**) and quadricyclane (**Q**) depends on the triplet energies of the sensitizer **S** [678, 679]. The mechanism of photosensitization by organic carbonyl compounds such as acetophenone occurs through the lowest triplet states: (i) $^1\mathbf{S}_0 + h\nu \rightarrow {}^3\mathbf{S}^*$; (ii) $^1\mathbf{N}_0 + {}^3\mathbf{S}^* \rightarrow {}^3\mathbf{N}^* + {}^1\mathbf{S}_0$; (iii) $^3\mathbf{N}^* \rightarrow {}^3\mathbf{Q}^*$; and (iv) $^3\mathbf{Q}^* \rightarrow {}^1\mathbf{Q}_0$ [680, 681]. Contrary to the behavior of other bicyclic 1,4-dienes, the triplet sensitized photoreaction of norbornadiene does not give the corresponding products of DPM rearrangement (Section 6.3.4). Using chloranil, benzoquinone, or 2,5-dichlorobenzoquinone as triplet sensitizers, the triplet state of norbornadiene ($^3\mathbf{N}^*$) and quadricyclane ($^3\mathbf{Q}^*$) cannot be attained because of efficient return electron transfer (Section 6.2.3); consequently, the photoisomerization $\mathbf{N} \rightarrow \mathbf{Q}$ does not occur. Interestingly, with 3,3′,4,4′-benzophenonetetracarboxylic dianhydride as triplet sensitizer (**S**), the triplet state of norbornadiene is not attained, but the triplet state of ion pair $^3(\mathbf{N}^{+\bullet}/\mathbf{S}^{-\bullet})$ is realized. Through bond-coupled electron transfer (BCET), the latter permits the attainment of $^3\mathbf{Q}^*$, which produces quadricyclane [682]. (Arylphosphine)copper(I) halides such as $(MePh_2P)_3CuCl$ that absorb visible light are efficient sensitizers (quantum yield ≅ 100% for ISC) for the photoisomerization of norbornadiene into quadricyclane. Its triplet state has a lifetime of 3.3 μs in benzene at 25 °C and it transfers its triplet energy to norbornadiene with a rate constant $k = 8 \times 10^7$ M^{-1} s^{-1} [683]. Thus, the visible light photoisomerization $\mathbf{N} \rightarrow \mathbf{Q}$ has been considered as a solar energy storing device ($\Delta_r H°(\mathbf{Q} \rightarrow \mathbf{N}) \cong -23$ kcal mol^{-1} or -250 kcal kg^{-1}; $\Delta_f H°(\mathbf{N}) = 57 \pm 6$ kcal mol^{-1}, $\Delta_f H°(\mathbf{Q}) = 80.4$ kcal mol^{-1}) (Table 1.A.3) compared with the combustion of gaseous ethanol ($\Delta_r H°$ ($EtOH + 3O_2 \rightarrow 2CO_2 + 3H_2O$) $= -305.44$ kcal mol^{-1} or -6630 kcal kg^{-1}) [684–687].

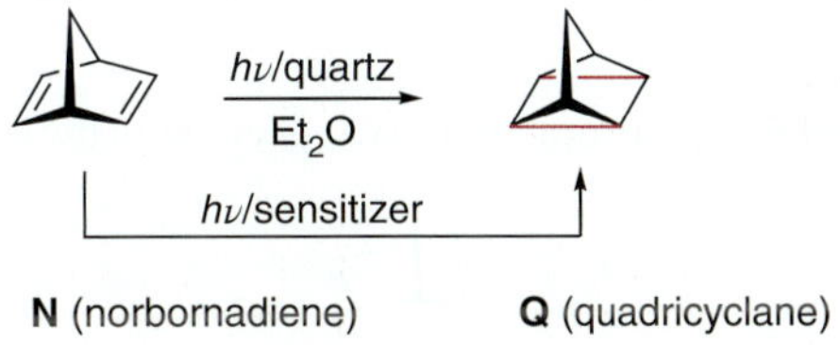

(6.84)

A landmark application of intramolecular [π^2s+π^2s]-cycloadditions was Eaton's [688, 689] synthesis of cubane (Scheme 6.54) [690]. Cubane has a rich chemistry and several derivatives of it have been prepared, including octanitrocubane, a powerful explosive and propellant [691]. Cubane can be rearranged into valence isomers such as *syn*-tricyclo[4.2.0.0^{2,5}]octa-3,7-diene [692], cuneane, and semibullvalene in the presence of transition metal catalysts [693].

Enantioselective intramolecular (2+2)-photocycloadditions have been observed in the presence of complexing agents that force the initial substrates to adopt a preferred chiral conformation. An example is given with the photochemical transformation **183** to **185** in the presence of 2.5 equiv. of the U-shaped compound **184** derived from Kemp's triacid (*cis,cis*-1,3,5-trimethylcyclohexane-1,3,5-tricarboxylic acid) [694].

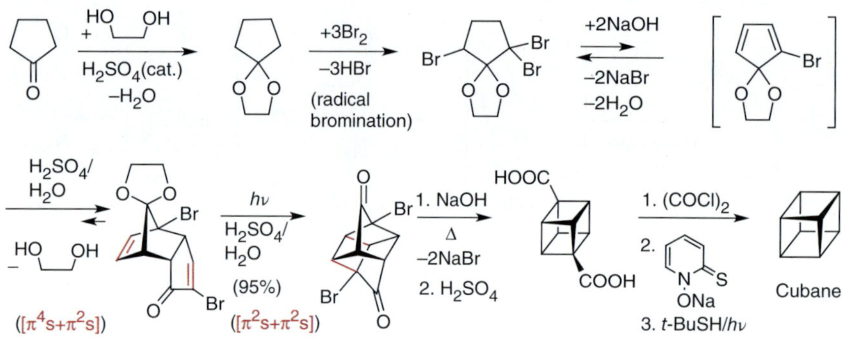

Scheme 6.54 The Eaton synthesis of cubane.

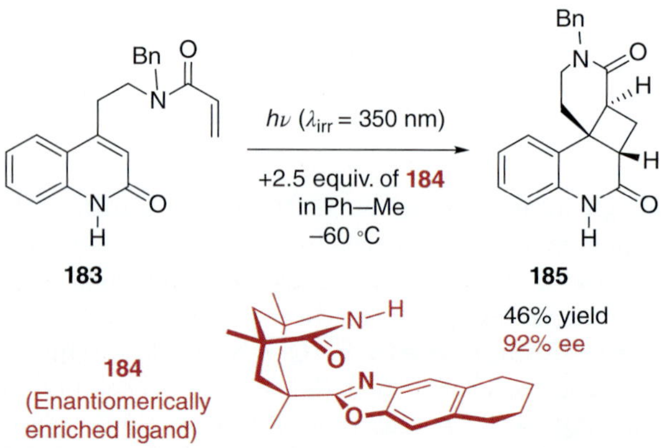

183 185

46% yield 92% ee

184 (Enantiomerically enriched ligand)

Problem 6.38 What are the major products of photochemical intramolecular cycloadditions of cycloocta-1,4-diene [658] and of cyclo-deca-2,6-dienone? [695]

Problem 6.39 Propose a two-step synthesis of **P** [696].

P: 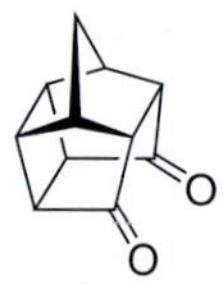

Problem 6.40 What is the product of UV irradiation of 4-(*t*-butyl)-2-(5-methylhex-4-en-1-yl)cyclohex-2-en-1-one? [697]

Problem 6.41 What is the major product **P** of irradiation of **A**? [674]

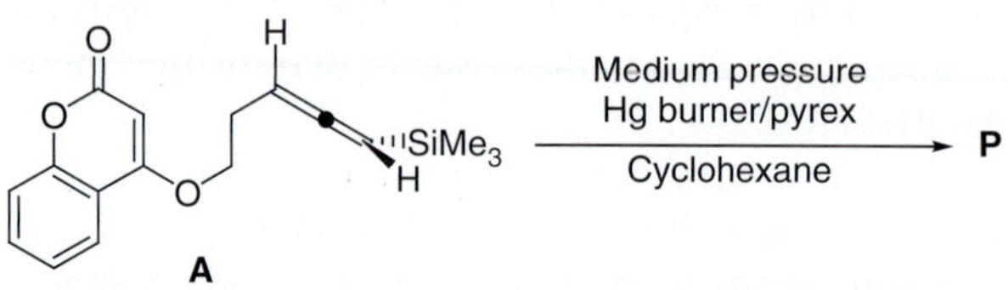

Problem 6.42 What are the intermediate products formed in the sequence of reactions that converts **A** into **P**? [698]

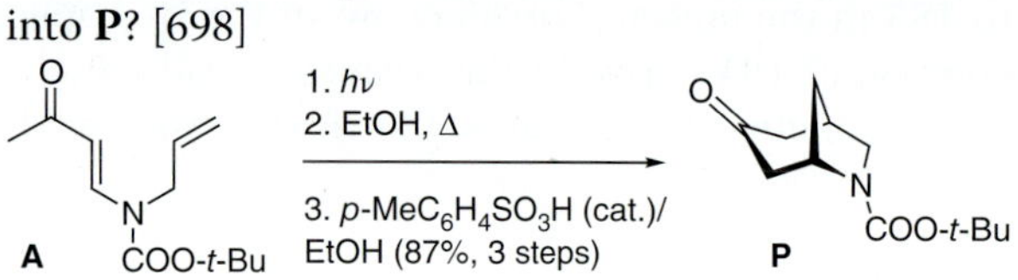

Problem 6.43 What are the main products of triplet-sensitized irradiation of (i) dimethyl octa-2,6-diene-1,8-dioate and (ii) dimethyl nona-2,7-diene-1,9-dioate? [651]

6.8.2 Photochemical intermolecular (2+2)-cycloadditions of alkenes

In 1969, Yamazaki and Cvetanovic found that direct irradiation (λ_{irr} = 214 nm) of (*Z*)-but-2-ene mostly leads to its isomerization into (*E*)-but-2-ene and but-1-ene. At low conversion (c. 0.1%), some cycloadducts **186a** and **186b** resulting from [π^2s+π^2s]-cycloadditions are formed with quantum yields representing only 3–4% of the quantum yield of the formation of (*E*)-but-2-ene. At higher conversion, cycloadducts **186c** and **186d** also form. Cycloadduct **186c** results from the [π^2s+π^2s]-cycloaddition of (*E*)-but-2-ene onto itself, and stereoisomer **186d** results from the [π^2s+π^2s]-cycloaddition of (*Z*)- and (*E*)-but-2-ene (Scheme 6.55). When diluting (*Z*)-but-2-ene with neopentane, the ratio cycloadducts/(*E*)-but-2-ene decreases sharply. Close proximity of the but-2-ene molecules in the liquid phase is essential for their photochemical cyclodimerization [699]. In the gas phase, the relative amount of cyclodimers is much lower than in solution. No cyclodimerization is observed in the gas phase mercury-photosensitized reaction [700].

Small-size cycloalkenes that cannot be photoisomerized into their (*E*)-isomers are more prone than acyclic alkenes to undergo intermolecular photochemical [π^2+π^2]-cycloadditions. These reactions involve triplet excited states. An early example is the photodimerization of norbornene in benzene (λ_{irr} = 254 nm) [701]. Benzene triplet photosensitizes the cyclodimerization of norbornene into [π^2s+π^2s]-cycloadducts **187** and **188**. In its triplet state, benzene can add concurrently to norbornene giving 1,3-cycloadducts **189** (see Section 6.8.4) [702]. The sensitized photocycloaddition of norbornene to maleic anhydride gives the two [π^2s+π^2s]-cycloadducts **190** and **191** [703].

Scheme 6.55 Direct irradiation of (*Z*)-but-2-ene: the $[\pi^2+\pi^2]$-cyclodimers are minor compounds in the pure liquid. The major photoreaction is the ($Z \rightleftarrows E$) isomerization.

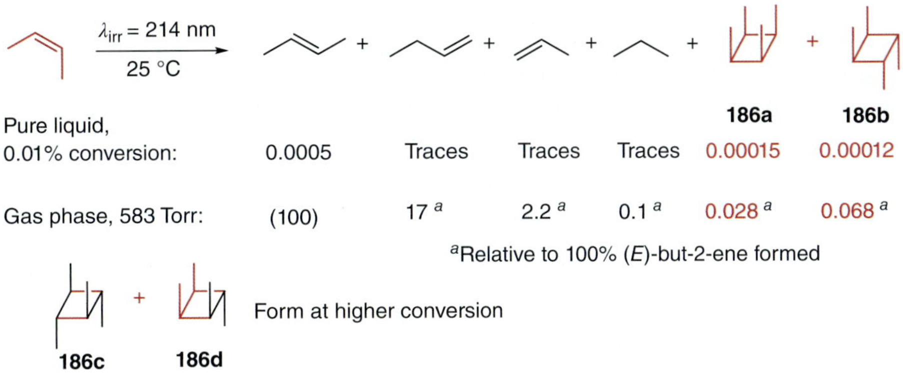

Scheme 6.56 Alkene triplets as 1,2-diradicals.

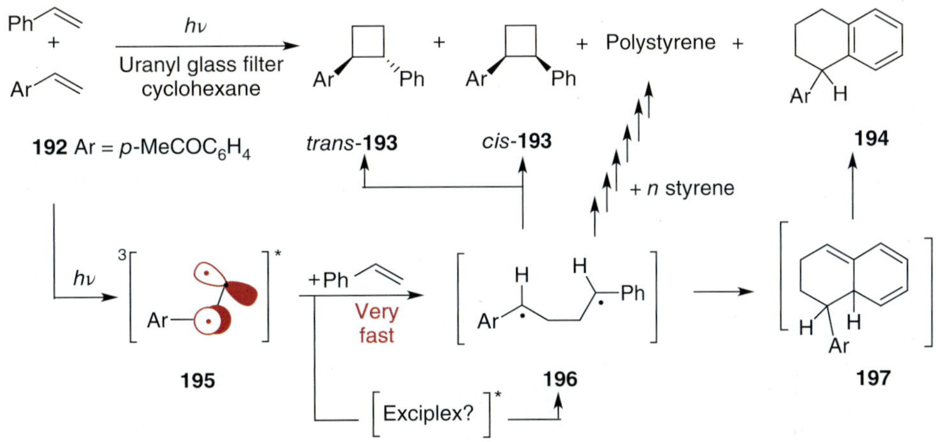

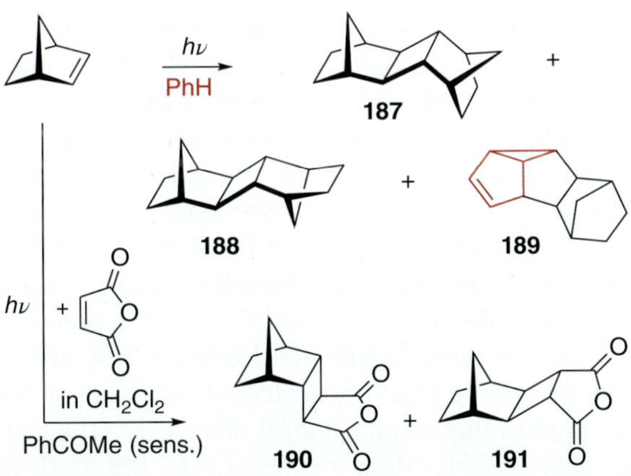

Irradiation of *p*-acetylstyrene (**192**) in the presence of styrene gives a 20 : 6 : 1 mixture of *trans*-**193**, *cis*-**193**, and **194** together with 7–11% of polystyrene. After light absorption, the triplet state of **192** undergoes rapid twisting of its double bond forming 1,2-diradical **195**, which adds to styrene generating 1,4-diradical intermediate **196**. The latter cyclizes into cyclobutanes *trans*- and *cis*-**193** or starts the polymerization of styrene. It also equilibrates with **197** that isomerizes into **194** (Scheme 6.56). The rate constant for the disappearance of ^{3}[**192**]* has been determined by laser flash photolysis to 32 M^{-1} s^{-1}. The rate of its reaction with styrene is one order of magnitude higher that the rate estimated for a model reaction that adds a primary alkyl radical to styrene. This suggests that the 1,2-diradical **195** is more reactive than a simple primary alkyl radical. It may form first an exciplex with styrene responsible for the fast cycloaddition [98].

The De Mayo reaction (6.85) is a photochemical reaction in which the enol of a 1,3-diketone reacts with an alkene (or another species with a C=C bond), and the resulting 2-acylbutanol ring undergoes a spontaneous retro-aldol reaction to yield 1,5-diketone [704].

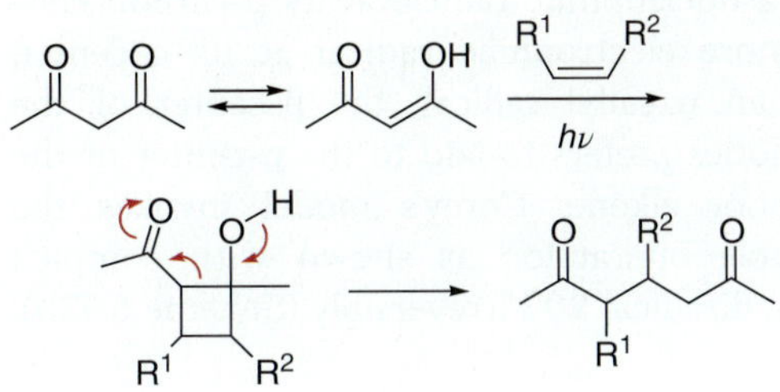

(6.85)

An example of application of the De Mayo reaction is given below with the synthesis of (±)-hirsutene starting with dimedone (5,5-dimethylcyclohexa-1,3-dione) and 2-methylcyclopent-2-enol (**198**). The photochemical (2+2)-cycloaddition is regioselective giving the head-to-tail (the carbonyl and methyl groups point toward opposite directions) cycloadduct **199** as the major product (see below). Retro-aldol of **199** gives **200** that is silylated. McMurry olefination (K, $TiCl_3$) provides alkene **201**. After desilylation, Jones oxidation of the resulting alcohol mixture into the corresponding ketone, face-selective catalytic hydrogenation of the cycloalkene moiety, and Wittig methylidenation (Section 5.3.6) (±)-hirsutene is obtained [705].

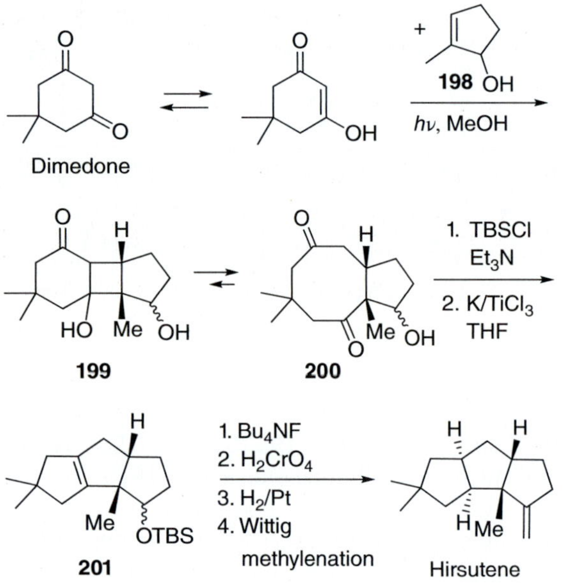

Photocycloadditions of cyclohex-2-enones to electron-deficient alkenes generally occur with head-to-head regioselectivity [279, 642, 645, 706]. This can be rationalized assuming that the most stable 1,4-diradical **202** that can form involves bond formation between the β-carbon centers of both reactants (Scheme 6.57a). Alternatively, the triplet state of the enone is predicted by quantum calculations [707–709], (postulated by Corey et al. in 1964), [710] to be $^3(\pi\pi^*)$ in nature, nonplanar, and with a nucleophilic radical at its β-carbon center and more electrophilic radical at its α-center, which is an oxyallyl radical; the β-center of the excited enones prefers to add to the β-center of the electron-poor alkene. Corey's model invokes the dipole/dipole interaction as shown with complex **203** giving diradical **202** irreversibly (Scheme 6.57b), but the selectivity is better explained by the intrinsic nature of the α and β radicals of the relaxed excited states. Head-to-head diradicals **204** might also form with electron-rich alkenes. However, the less stable regioisomeric diradicals **205** are also formed. They cyclize onto the head-to-tail cycloadducts more rapidly than their cleavage into the starting reactants. In contrast, diradicals **204** fragmentize into the cycloaddends more rapidly than they cyclize into head-to-head cycloadducts (Scheme 6.57c). The LUMO of the triplet excited state of cyclohex-2-enone (see Figure 4.25, Section 4.5.15) is populated by one electron, which is responsible of the polarization shown and that leads to the formation of complexes **203** (Scheme 6.57b) and **206** (Scheme 6.57d) with Michael acceptors and electron-rich alkenes, respectively. Relatively low quantum yields are indicative of reversible diradical formation. The regioselectivity of the photocycloaddition of cyclohex-2-enone with 1-cycloalkenecarboxylates changes from head-to-head to head-to-tail depending on the ring size of the acrylic ester. This change of regioselectivity has been explained invoking competition with back-reactions [711].

The photochemical cycloaddition of allene to cyclopent-2-enone yields methylidenecyclobutane with preferential head-to-head regiochemistry (ratio of cycloadducts **207**/**208**: 88 : 12) in contrast to the preferential head-to-tail orientation of the photocycloadditions of cyclic enones with 1-alkylalkenes [712]. In the presence of high concentration of H_2Se, the four possible 1,4-diradical intermediates **209′**–**212′** are reduced into the corresponding products **209**–**212** [713]. The results demonstrate that the most stable diradical **212′** (stabilized by allyl and carbonyl conjugation) is not the major intermediate formed. The dominant initial diradical is **209′** and results from the bonding between C(2) of the enone and C(1) of allene. This diradical is not the precursor of the major head-to-head cycloadduct **207**, but the precursor of the minor head-to-tail cycloadduct **208**. This implies that diradical **209′** reverts into the starting material faster than it cyclizes into **208**. Both the less dominant diradicals **210′** and **211′** cyclize into the major cycloadduct **207** (Scheme 6.58). The proportion of photocycloadducts **207** and **208** depends on the rate by which the diradical intermediates **209′**–**211′** are formed, upon their conformation, spin/orbit coupling, and singlet–triplet energy spitting that influence the competition between their reversion to starting materials and their cyclization [714].

Cyclopent-2-enone is not photoisomerized into its (*E*)-stereoisomer and gives only *cis*-$[\pi^2+\pi^2]$-cyclo-

Scheme 6.57 Regioselectivity of the photochemical $[\pi^2+\pi^2]$-cycloadditions of cyclohex-2-enone: (a) and (b) with electron-poor alkenes (Michael acceptors) head-to-head cycloadducts are generally preferred; (c) and (d) with electron-rich alkenes, head-to-tail cycloadducts are formed.

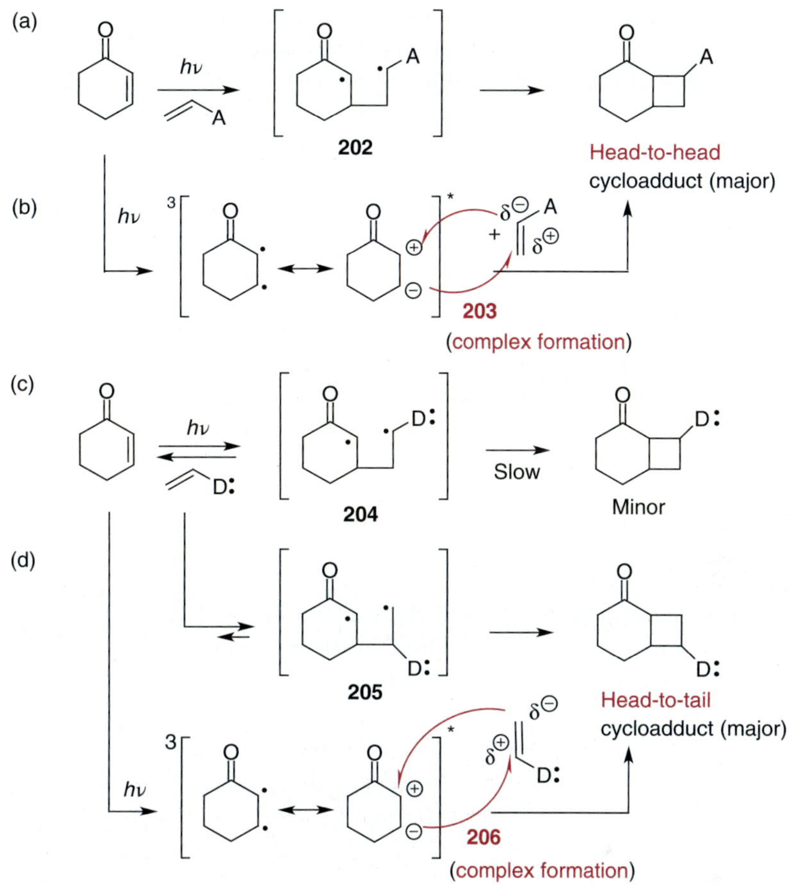

Scheme 6.58 Trapping 1,4-diradical intermediates with hydrogen selenide. The most stable diradical **212′** resulting from the reaction of cyclopent-2-enone with allene does not lead to the major head-to-head photocycloadduct.

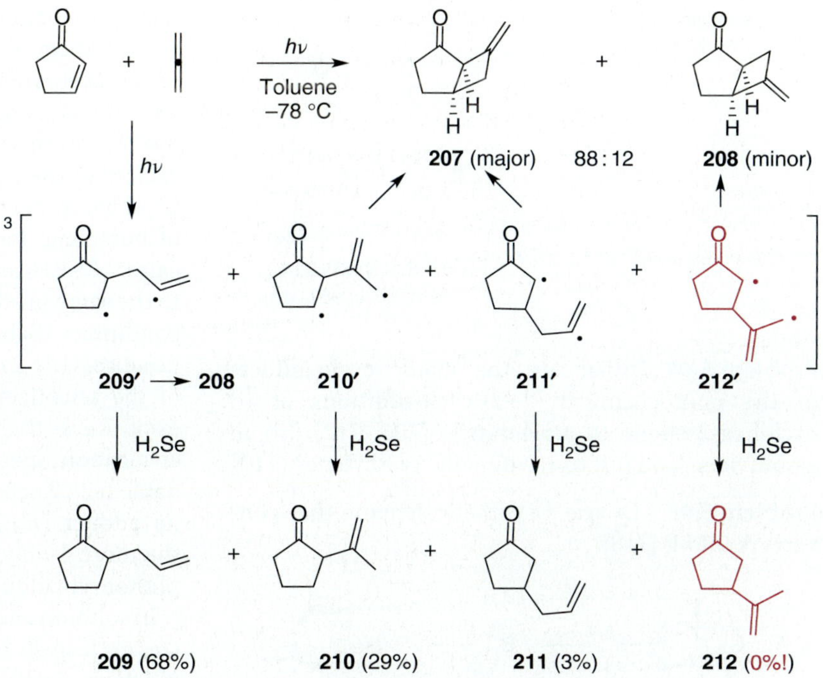

adducts. In contrast, cyclohex-2-enone is photoisomerized into (*E*)-cyclohex-2-enone. Thus, the photochemical cycloadditions of cyclohex-2-enone to alkenes give mixtures of *cis*- and *trans*-(2+2)-cycloadduct as illustrated with reaction (6.86) [710].

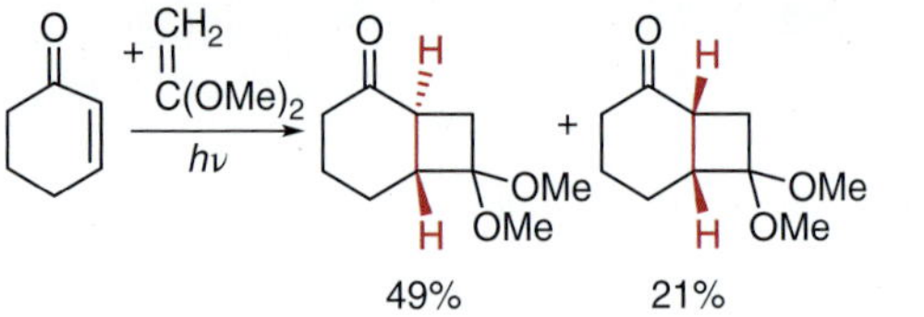

(6.86)

High enantiomeric excesses in the intermolecular photochemical $[\pi^2+\pi^2]$-cyclodimerization of coumarin derivatives have been observed when the reaction is carried out in homochiral crystals [715]. In cyclohexane solution and in the presence of 1 equiv. of host **213**, irradiation of 6-methylcoumarin gives a 2 : 1 inclusion complex $[\mathbf{213}]_2 \subset \mathbf{214}$ from which (*S*,*S*,*S*,*S*)-(+)-**214** is isolated in 60% yield and with 95% ee [716].

Photochemical $[\pi^2+\pi^2]$ cycloaddition has become the most frequently applied method for the construction of cyclobutane derivatives [717].

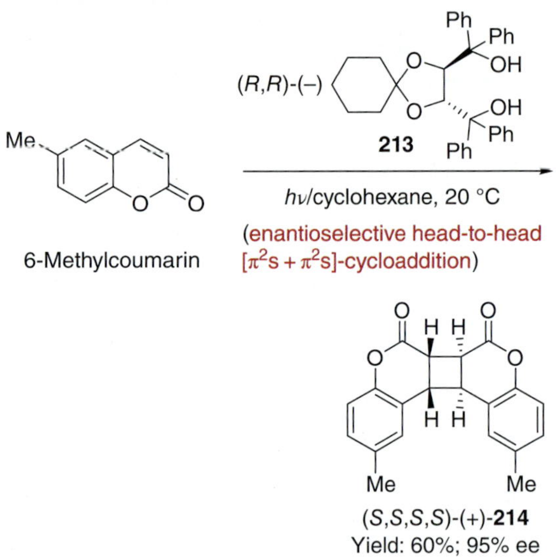

Problem 6.44 What are the major cycloadducts of the photochemical (2+2)-cycloadditions of (i) cyclohex-2-enone to acrylonitrile [706, 712], (ii) to isobutylene, and (iii) to 1,1-dimethoxyethylene? [710]

Problem 6.45 Devise a synthetic scheme that converts **A** into **P** [718].

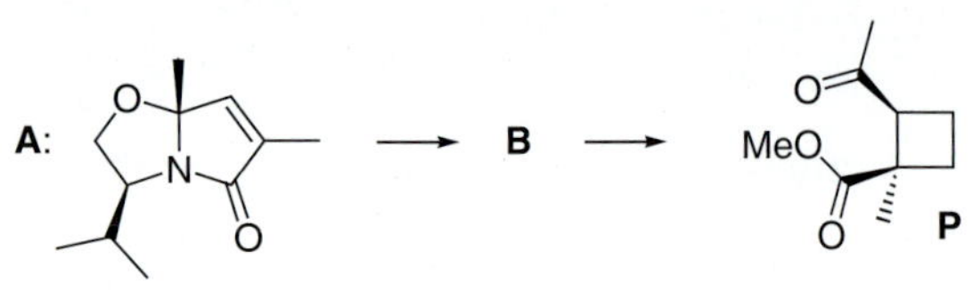

Problem 6.46 What is the major product **P** of UV irradiation of **A** + **B**? [719]

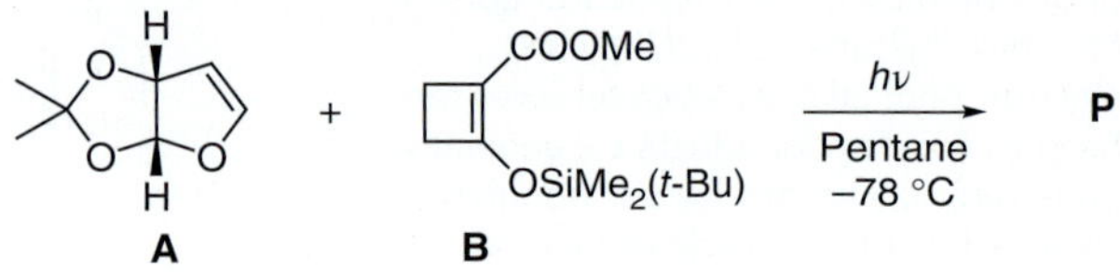

Problem 6.47 What is the major product **P** of the photolysis of **A** with acetylene? [720]

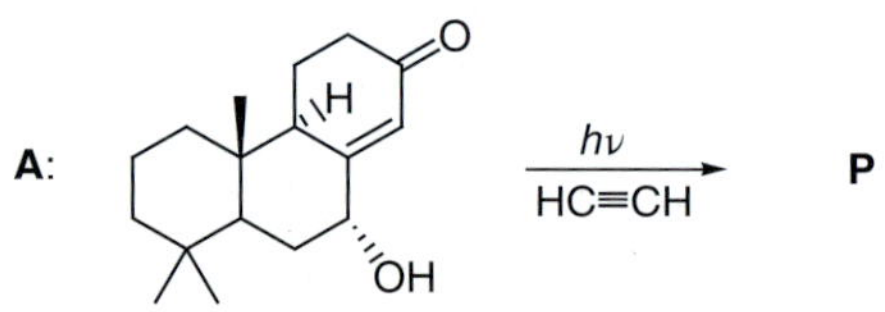

6.8.3 Photochemical intermolecular (4+2)-cycloadditions of dienes and alkenes

On heating (autoclave, 150 °C), butadiene forms mostly the Diels–Alder cycloadduct 4-vinylcyclohexene (**217**), c. 5% of *trans*-1,2-divinylcyclobutane (*trans*-**218**) and a small amount of cycloocta-1,5-diene that arises from the Cope rearrangement of *cis*-1,2-divinylcyclobutane (*cis*-**218**) (Section 5.3.9). The two latter compounds result from the reaction of two molecules of *s-trans*-butadiene into (*E*,*E*)-octa-1,7-diene-3,6-diyl diradical ((*E*,*E*)-**216**). The triplet sensitized photoreaction of butadiene at −10 °C gives mixtures of (*E*,*E*)-**216**, *cis*-**218**, and *trans*-**218** and their proportion varies with the triplet energy (E_T) of the sensitizer [678]. For instance, the ratio is 3 : 19 : 78 using acetophenone (E_T = 73.6 kcal mol^{-1}) and 45 : 10 : 44 using benzil (PhCOCOPh, E_T = 53.7 kcal mol^{-1}). Both the *s-trans* and *s-cis*-butadiene are excited in their first triplet excited states (*E*)-**215** and (*Z*)-**215**, respectively. These add to another molecule of butadiene forming octa-1,7-diene-3,6-diyl diradicals (*E*,*E*)-**216** and (*E*,*Z*)-**216**, respectively, that evolve to the same products as those obtained under thermal conditions (Scheme 6.59) [721]. The proportion of diradicals (*E*)-**215** and (*Z*)-**215** depends on the nature of the sensitizer because less energy is required to excite *s-cis*- than *s-trans*-butadiene (compare the UV absorption spectra of *s-cis*- and *s-trans*-1,3-dienes, Table 6.2). Accordingly, the less diradical (*E*)-**215** and its adduct (*E*,*E*)-**216** are formed as intermediates, the more 4-vinylcyclohexene (**217**) is obtained as the photocycloadduct.

In solution and at 0 °C, cyclopentadiene cyclodimerizes exclusively into the *endo* Diels–Alder cycloadduct (Section 5.3.8). By contrast, the photosensitized cyclodimerization gives a c. 1 : 1 : 1 mixture of *endo*-,

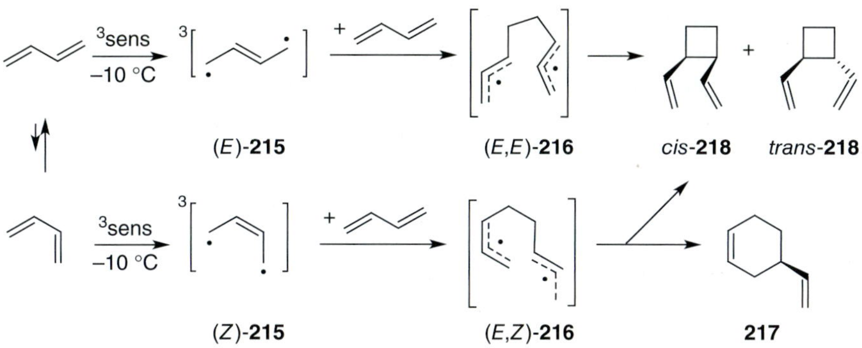

Scheme 6.59 The triplet-sensitized photolysis at −10 °C of butadiene gives the same (2+2)- and (4+2)-cycloadducts as those obtained under heating at 150 °C.

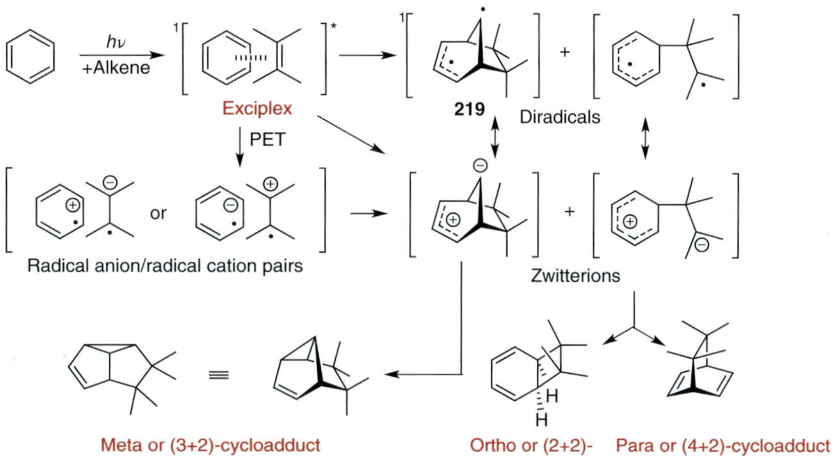

Scheme 6.60 Three possible modes of photochemical cycloadditions of benzene with alkenes and hypothetical stepwise mechanisms.

exo-dicyclopentadiene ((4+2)-photocycloadducts) and *trans*-tricyclo[5.3.0.0^{2,6}]deca-3,9-diene ((2+2)-photocycloadduct) [722]. Thermal dimerization (200 °C) of cyclohexa-1,3-diene gives a 4 : 1 mixture of *endo*- and *exo*-dicyclohexadiene (Diels–Alder cycloadducts). The photosensitized dimerization of cyclohexa-1,3-diene gives a 1 : 3 : 1 mixture of *exo*-dicyclohexadiene ((4+2)-photocycloadduct), *trans, cis–trans-*, and *cis,cis,cis*-tricyclo[6.4.0.0^{2,7}] dodeca-3,11-diene ((2+2)-photocycloadducts) [194].

6.8.4 Photochemical cycloadditions of benzene and derivatives to alkenes

Three types of photochemical cycloadditions are often observed with arenes and alkenes: the ortho, or (2+2)-cycloaddition, the meta, or (3+2)-cycloaddition, and the para, or (4+2)-cycloaddition (Scheme 6.60) [723–727]. The meta photocycloadditions are the most studied and permit the quick construction of complicated structures that have found wide applications in organic synthesis [5, 668, 726–729]. They imply the formation of diradical intermediates of type **219**.

The first examples of ortho photochemical cycloadditions were reported in 1957 by Büchi and coworkers [730] and, in 1959, by Angus and Bryce-Smith with the photocycloaddition of maleic anhydride to benzene. The photochemical (2+2)-cycloaddition is the preferred mode of cycloadditions of arenes with alkynes [731, 732]. This reaction is also observed when there is a relatively large difference between the electron-donor and electron-acceptor properties of the arene and alkene [733–738]. In contrast, the photochemical (3+2)-cycloaddition is the preferred mode of reaction when this difference is small [737]. It was discovered in 1966 by Kaplan and coworkers [739] and by Bryce-Smith et al. [740]. Photochemical (4+2)-cycloadditions take place in a few cases where the steric factors are important or when the alkene is an allene [741–743] or a diene [744, 745]. This mode of photocycloaddition is more frequent with anthracene than with naphthalene and benzene derivatives [746, 747]. In the case of benzene + ethylene, the photoinduced cycloaddition involves the first singlet excited state of benzene that reacts with ethylene giving a 1 : 1 mixture of ortho and meta photocycloadducts [748, 749].

As a rule, photochemical (2+2)-cycloadditions of arenes + alkenes are favored when electron transfer is possible between the first singlet excited state of the arene and the other cycloaddend. For instance, in the case of benzene reacting with the electron-poor acrylonitrile (reaction (6.87)), the charge transfer complex **220** can form. When benzene reacts with electron-rich ethyl vinyl ether, a mixture of ortho and meta cycloadducts is obtained (reaction (6.88)). The charge transfer complex **221** can be invoked as intermediate [750–752]. Photochemical (2+2)-cycloadditions of 1-cyanonaphthalene to substituted pyridines have been reported [753].

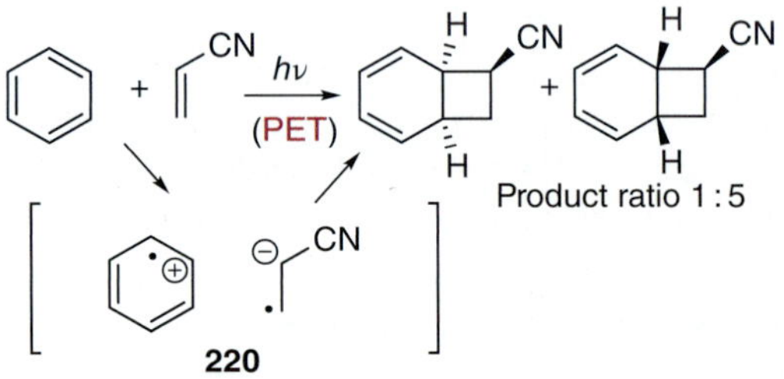

(6.87)

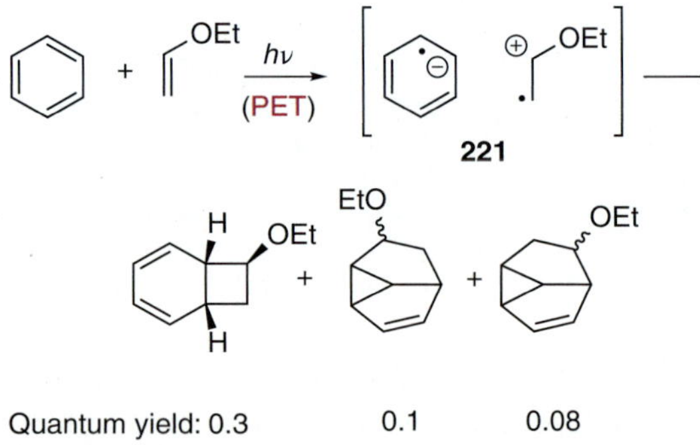

(6.88)

Irradiation of benzene in the presence of unactivated acyclic alkenes or cycloalkenes gives mostly meta cycloadducts. The reaction involves the first singlet excited state $^1(\pi,\pi^*)$ of benzene and an exciplex (Scheme 6.60). The occurrence of such an exciplex has been detected by emission spectroscopy [754–756]. In the presence of Xe, singlet–triplet ISC is accelerated (heavy atom effect). This leads to a decrease of the quantum yield for the formation of (3+2)-cycloadduct, thus supporting the hypothesis that the photochemical process occurs on the singlet hypersurface [757]. The exciplex evolves to a singlet diradical of type **219** (Scheme 6.60), a short-lived intermediate that has never been trapped chemically. Being a singlet diradical, it undergoes quick cyclization into the cycloadducts as the process is spin allowed. The intermediacy of diradicals of type **219** is supported by the regioselectivity observed for the photochemical (3+2)-cycloaddition of 1,3-dimethylbenzene with cyclopentene (Scheme 6.61), which is the same as that observed for the photochemical fragmentations of azo compounds **223** and **224** that generate diradical intermediate **222** [758, 759].

Monosubstituted benzene derivatives undergo regioselective meta photocycloadditions. The regioselectivity can be explained by considering the zwitterionic limiting structures of the initial diradicals of type **219** and **222**, as shown with reactions (6.89) and (6.90).

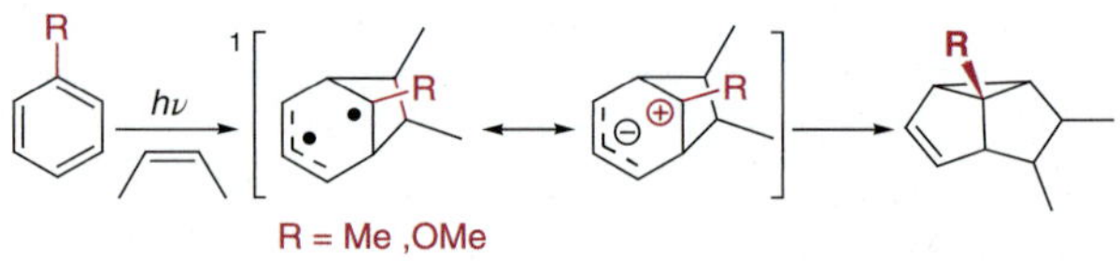

(6.89)

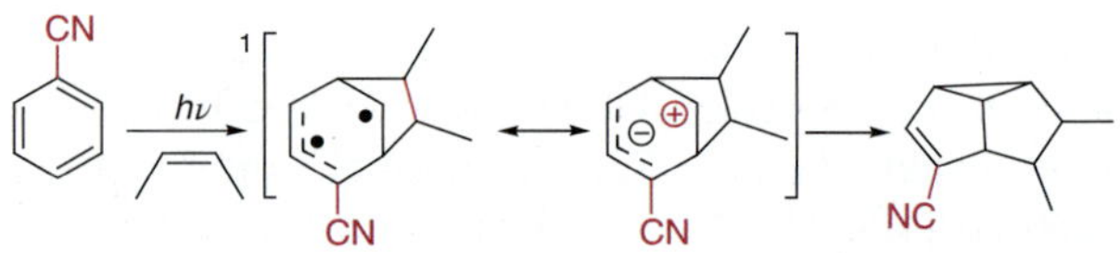

(6.90)

With (*Z*)-1,2-disubstituted alkenes, *endo*- and *exo*-meta-cycloadducts can form. The photocycloadditions are *endo*-stereoselective with cyclopentene (e.g. reaction (6.91) [760], see also Scheme 6.61) and *exo*-stereoselective with 1,3-dioxole (e.g. reaction (6.92)). Competing attractive secondary frontier orbital interactions and repulsive gauche and electrostatic interactions have been invoked to explain the

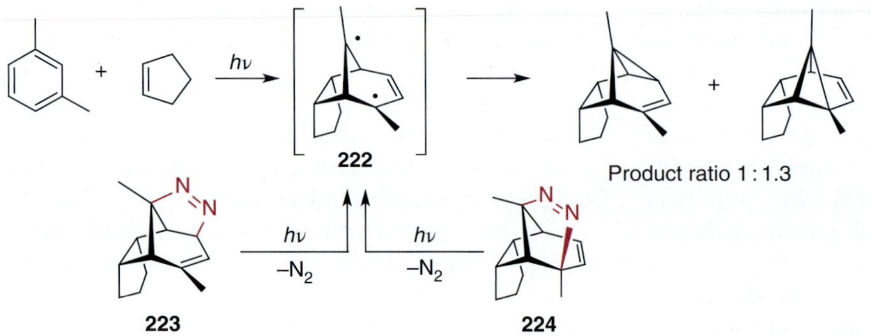

Scheme 6.61 Mechanism of the photochemical (3+2)-cycloaddition of 1,3-dimethyl benzene with cyclopentene involves a diradical intermediate.

results [668, 728].

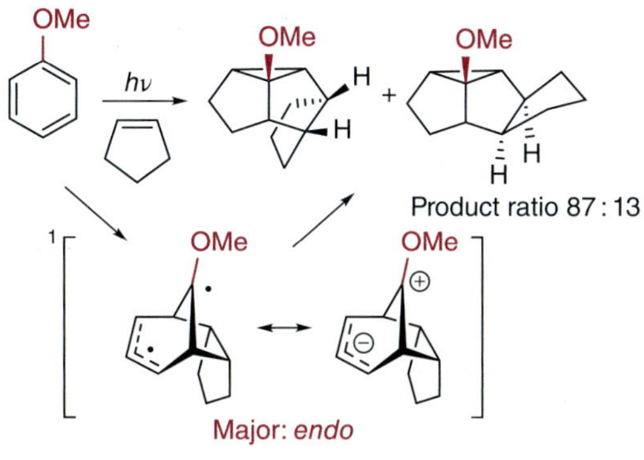

(6.91)

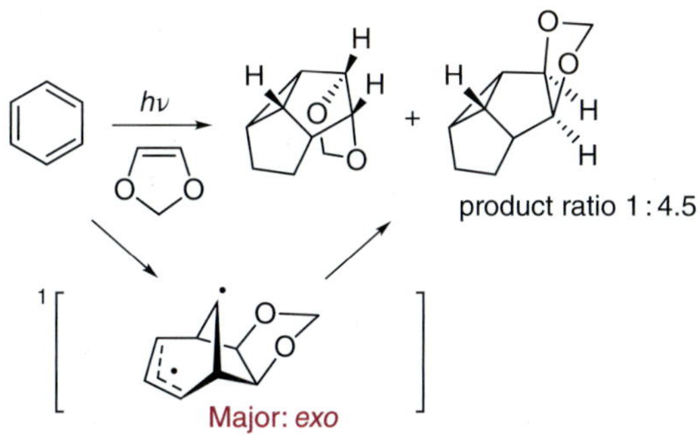

(6.92)

Irradiation of 1,4-dicyanobenzene with isobutylene in acetonitrile gives product **225** resulting from a (2+2+2)-cycloaddition of 1,4-dicyanobenzene + isobutylene + MeCN and a subsequent elimination of HCN (Scheme 6.62). After photon absorption, exciplex **226** is formed that has undergone electron transfer from the alkene to the electron-poor benzene derivative. The isobutylene radical cation reacts with a molecule of acetonitrile (solvent) on its least sterically hindered center to give a radical anion/radical cation pair **227**. The latter collapses into zwitterion **228** that undergoes cyclization into **229**. Elimination of HCN restores aromaticity and produces the final product **225** [761].

Upon irradiation of 1,4-dicyanonaphthalene with isobutylene, Kubo et al. obtained a product of 1,8-cycloaddition, a formal photochemical (3+2)-cycloaddition across carbon centers C(1) and C(8), followed by a HCN elimination (reaction (6.93)) [762, 763]. With 1,2-dicyanonaphthalene, the photochemical reaction of isobutylene in acetonitrile gives a mixture of three products, the first one, as in reaction (6.93), results from a 1,8-cycloaddition followed by HCN elimination, the second one is a (2+2)-cycloadduct across C(1a)—C(8) of 1,2-dicyanonaphthalene, and the third one results from a (2+2+2)-cycloaddition arene + alkene + MeCN, analogous to the process that generates **225** (Scheme 6.62) [764, 765].

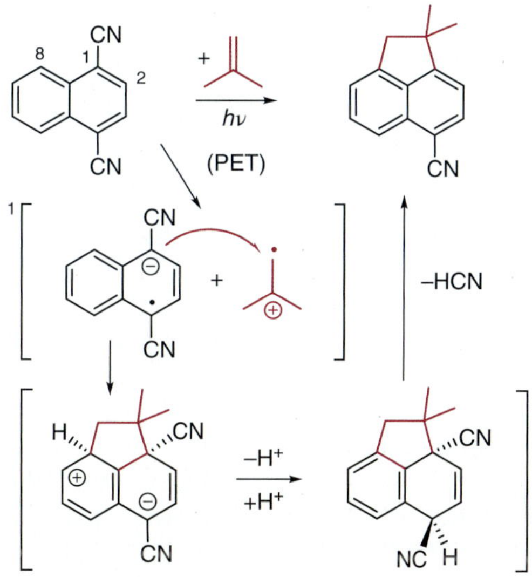

(6.93)

Irradiation of naphthalene with 2,5-dimethylhexa-2,4-diene at −5 °C gives a (4+4)-cycloadduct (reaction (6.94)) that rearranged quickly at 35 °C into **230** via a Cope rearrangement (Scheme 6.63) [766]. Photochemical (4+4)-cycloaddition (6.95)

Scheme 6.62 Example of a photochemical (2+2+2)-cycloaddition of an arene, alkene, and acetonitrile.

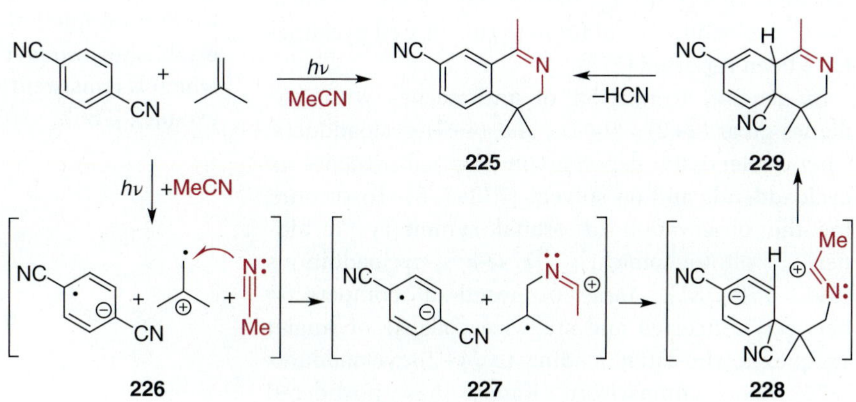

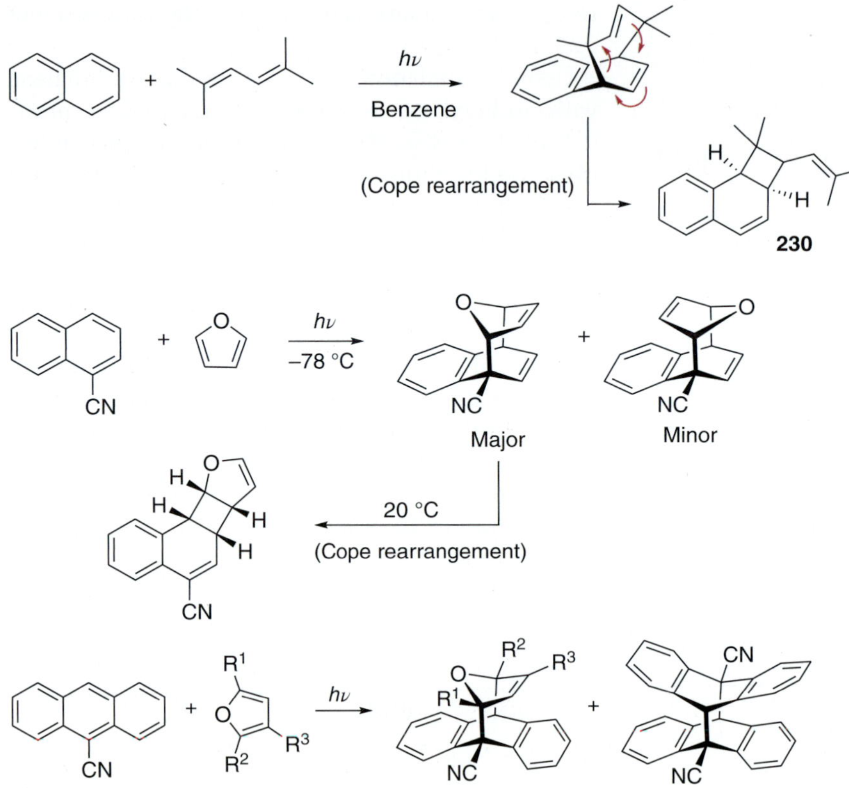

(6.94)

(6.95)

(6.96)

Scheme 6.63 Examples of photochemical $[\pi^4s+\pi^4s]$-cycloadditions of arenes.

produces two stereoisomeric cycloadducts, and the major one also undergoes a facile Cope rearrangement (Scheme 6.63) [767, 768]. Irradiation of 9-cyanoanthracene with furans gives mixtures of the cross-(4+4)-cycloadducts and the (4+4)-cyclodimer of 9-cyanoanthracene (reaction (6.96), Scheme 6.63) [769]. Intramolecular photochemical (4+4)-cycloaddition of 1,2-di(9-anthryl) ethanol in benzene occurs in the singlet excited state with a quantum yield of 0.34 [770]. The photochemical (4+4)-cyclodimerization of anthracene, which involves C(9) and C(10) centers [771], has been reported already in 1909 and proposed to be a potential solar energy storage device [684]. Photochemical (4+4)-cycloadditions of furan to substituted pyridines have been reported [772].

In general, irradiation of anthracenes with 1,3-dienes gives (2+2)-, (4+2)-, and (4+4)-cycloadducts. Chemoselectivity depends on the substituents of cycloaddends and on solvent [773–778]. To account for the observation of orbital symmetry "forbidden" photochemical $[\pi^4s+\pi^2s]$-cycloadditions [641, 779–781], Yang proposed a competition between concerted and stepwise collapse of singlet exciplexes, the latter leading to (4+2)-cycloadducts [775]. For comparison, Kaupp has postulated direct competing stepwise diradical pathways that lead to "allowed" photochemical $[\pi^2s+\pi^2s]$- and $[\pi^4s+{}^4\pi s]$-cycloadditions and "forbidden" photochemical $[\pi^4s+\pi^2s]$-cycloadditions [782–785].

Irradiation of 9,10-dichloroanthracene in the presence of 2,5-dimethylhexa-2,4-diene in benzene at 25 °C gives the single (4+2)-cycloadduct **231** (reaction (6.97)). The quantum yields of the fluorescence of the first excited state of 9,10-dichloroanthracene and of the exciplex it forms with 2,5-dimethylhexa-2,4-diene have been measured as a function of MeI concentration (singlet quencher: heavy atom effect). The quantum yield of **231** depends on the concentration of 9,10-dichloroanthracene and it is proportional to the exciplex fluorescence quantum yield. This establishes a singlet mechanism for adduct formation, which is consistent with a stepwise collapse of a polar exciplex [786].

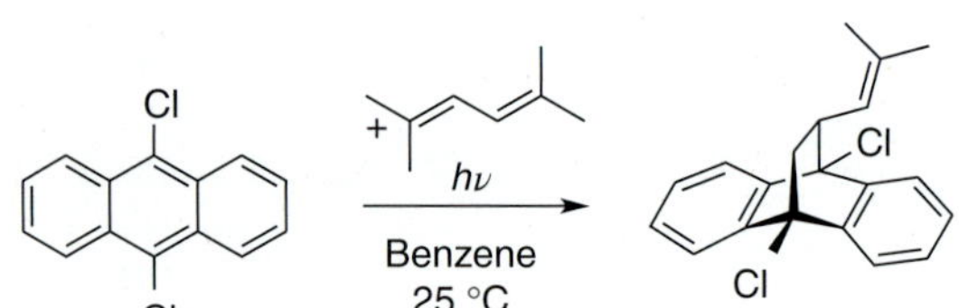

231

(6.97)

Problem 6.48 What are the major photocycloadducts of the irradiation of cyclopentene with the methyl carbamate of isoindoline? [787]

Problem 6.49 What is the major photocycloadduct of acrylonitrile with 2,6-dimethoxybenzonitrile? [788]

Problem 6.50 What products are formed upon irradiation of benz[a]anthracene in the presence of cyclohexa-1,3-diene? [789–791]

6.8.5 Photochemical cycloadditions of carbonyl compounds

The Paternó–Büchi reaction is one of the first organic photochemical reactions (Scheme 6.64). In 1909, Paternó and Chieffi described the formation of oxetanes by light-induced (2+2)-cycloadditions of alkenes to aldehydes and ketones. Under sunlight irradiation, benzaldehyde adds to 2-methylbut-2-ene giving a major isomeric oxetane **232** (reaction (6.98)) [792]. In 1954, Büchi et al. confirmed the structures of the products obtained by Paternó and pointed out the synthetic power of this photochemical reaction, [793] which is more advantageous in several aspects than conventional thermal reactions [794–798]. The regioselectivity of the photochemical (2+2)-cycloadditions of carbonyl compounds to electron-rich alkenes is generally explained as follows: (i) The excited ketone undergoes ISC within picoseconds ($^1(n,\pi^*) \rightarrow {}^3(n,\pi^*)$) [799]. (ii) The triplet adds to the alkene forming 1,4-diradical intermediates; usually, the most stable one is formed preferentially

Scheme 6.64 Paternó–Büchi reactions to electron-rich alkenes involve the $^3(n,\pi^*)$ triplet excited state of the carbonyl compounds and the formation of 1,4-diradical intermediates that cyclize into oxetanes, or undergo other reactions typical of diradicals.

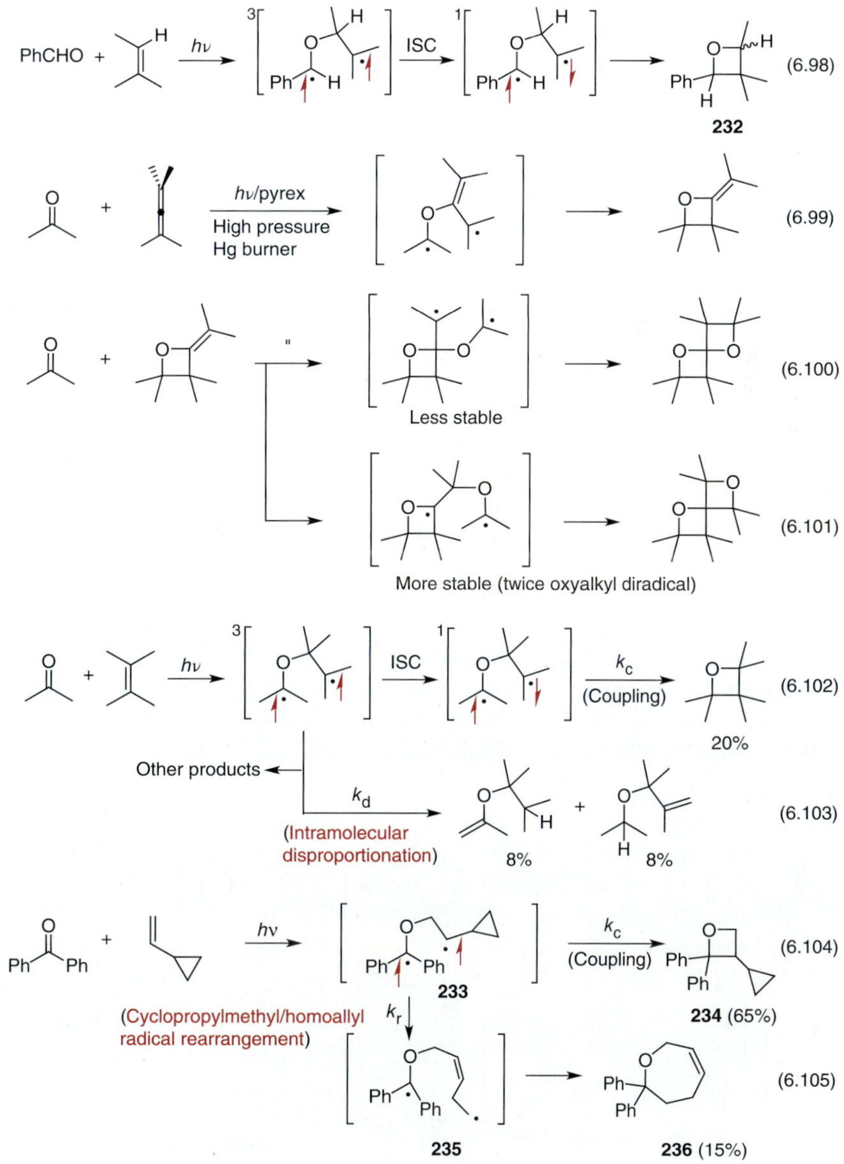

and evolves to products after spin inversion [800, 801] as illustrated with the two successive cycloadditions (6.99) and (6.101) of acetone to tetramethylallene (Scheme 6.64) [802], or it can be cleaved into the starting cycloaddends. As we shall see below, singlet first excited states of the carbonyl compounds can also lead to the oxetanes [803, 804]. The formation of 1,4-diradical intermediates was suggested in early studies by the observation that the photochemical (2+2)-cycloadditions of benzophenone to (*E*)- and (*Z*)-but-2-ene give the same 1 : 9 mixture of the *cis*- and *trans*-2,2-diphenyl-3,4-dimethyldioxetane regardless of the configuration of the starting alkene. Evidence for the formation of 1,4-diradicals is given by the observation of side products resulting from other reactions than the cyclization of the 1,4-diradicals into the corresponding oxetanes. This is illustrated with the photochemical reaction (6.102) of acetone with tetramethylethylene that gives the expected oxetane concurrently with products of intramolecular disproportionation (reaction (6.103)) and other products [805, 806] and with the Paternó–Büchi reaction (6.104) of benzophenone with vinylcyclopropane, which gives oxetane **234** next to the rearranged product **236** resulting from the cyclopropylmethyl/homoallyl radical rearrangement **233** ⇄ **235** (reaction (6.105)) [807, 808].

Picosecond dynamics confirm that the triplet state of benzophenone adds directly to the electron-rich 1,4-dioxene to form the triplet 1,4-diradical intermediate without the intermediacy of exciplex or charge transfer species (Scheme 6.65) [801]. In the absence of quenchers, the triplet state of benzophenone ($^3[Ph_2CO]^*$) in MeCN has a half-life greater than 1 μs, appearing within 25 ps following irradiation at 355 nm [809]. In the presence of one molar dioxene, $^3[Ph_2CO]^*$ has a half-life of c. 175 ps only and generates diradical ^{3}[**237**] with an absorption maximum λ_{max} = 535 nm. The same observations are made in DMF (*N*,*N*-dimethylformamide), DME ($MeOCH_2CH_2OMe$), cyclohexane, EtOH, and dimethyl sulfoxide (DMSO) (Me_2SO). In MeCN, the triplet diradical ^{3}[**237**] decays with a rate constant of c. 6×10^8 s^{-1}. It undergoes ring closing into oxetane **238** and β-scission into the starting cycloaddends. Because the 1,4-diradical is formed in its triplet state, ISC into ^{1}[**237**] must take place before any bond-forming reaction can occur. As the 1,4-diradical decays, a new species appears, absorbing at λ_{max} = 690 nm in MeCN and at λ_{max} = 726 nm in DMSO, concurrently with the formation of oxetane **238**. This species is identified as the radical anion of benzophenone. Its formation is independent of the initial benzophenone concentration, which eliminates the possibility of a reaction between Ph_2CO and the 1,4-diradical, but confirms that it arises from the heterolysis of the 1,4-diradical intermediate.

Generally, the photochemical cycloadditions of carbonyl compounds to electron-poor alkenes follow different mechanisms than those operating in the (2+2)-cycloadditions to electron-rich alkenes [810–815]. Ultraviolet irradiation of solutions of α,β-unsaturated nitriles (acrylonitrile and derivatives) with dialkyl ketones give mixtures of oxetanes that result from the reactions of the $^1(n,\pi^*)$ singlet states of the ketones with the alkenes [816] and of alkene cyclodimers resulting from their quenching of the $^3(n,\pi^*)$ triplet states of the ketones. The regioselectivity of the formation of oxetane is opposite to that expected for the formation of the most stable 1,4-diradical intermediates. Thus, as for the photochemical (2+2)-cycloadditions of electron-rich and electron-poor alkenes to conjugated enones (Section 6.8.2, Scheme 6.57), one needs to consider the polarization of the ketone in its first excited state that renders its oxygen center electrophilic (Umpolung). Accordingly, the excited state of the ketone looks for the most nucleophilic center of the acrylonitrile system, which is the α-carbon center as shown with reaction (6.106). Because ISC is rapid in excited ketones, it competes with the (2+2)-cycloaddition

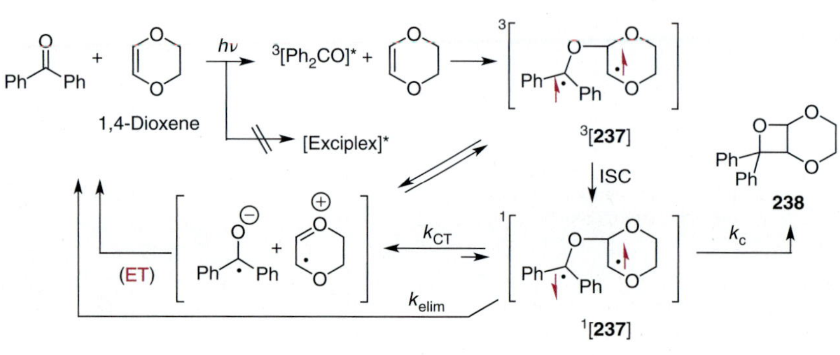

Scheme 6.65 The Paternó–Büchi reaction of benzophenone and 1,4-dioxene engenders a triplet 1,4-diradical and a benzophenone ketyl radical anion detected by their visible absorption spectra.

of its singlet state. Electron-poor alkenes are efficient quenchers of the triplet state of ketones. This promotes the cyclodimerization of the alkenes as exemplified with reaction (6.107).

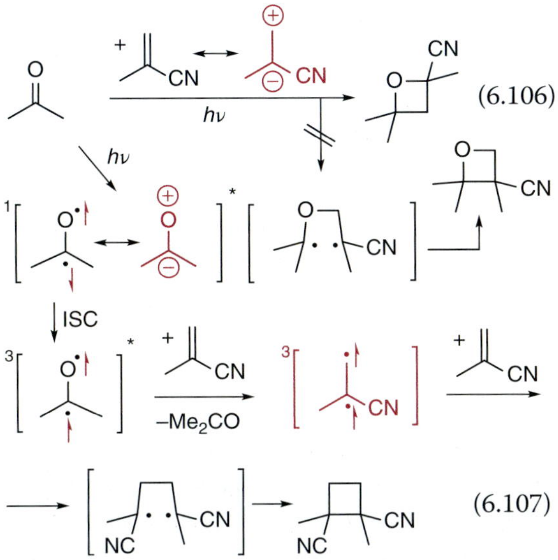

(6.106)

(6.107)

The Paternó–Büchi reaction of benzaldehyde with 2,3-dihydrofuran (Scheme 6.66) in *n*-hexane gives a 88 : 12 mixture of the *endo*- and *exo*-oxetane **240**. The high regioselectivity of this photochemical (2+2)-cycloaddition is explained invoking the favored formation of 1,4-diradical **239**, the most stable diradical out of the four possible isomeric diradicals that can be formed in this condensation. As for Diels–Alder reactions, the *endo* stereoisomer is favored, probably because of stabilizing secondary interactions between the two cycloaddends (contrasteric attractive interactions). On varying the concentration of the 1 : 1 mixture of benzaldehyde + 2,3-dihydrofuran from 0.001 to 2 M, the *endo*/*exo* product ratio does not vary. At low concentration (<0.01 M), the Paternó–Büchi reaction of propionaldehyde with 2,3-dihydrofuran gives the corresponding oxetane with a 85 : 15 *endo*/*exo* stereoselectivity, similarly to the reaction of benzaldehyde. However, on increasing the concentration of the 1 : 1 mixture of cycloaddends, the *endo*/*exo* product ratio diminishes to 48 : 52 when the concentration reaches 0.5 M. A similar concentration effect on the *endo*/*exo* stereoselectivity is observed for the photochemical reaction of acetaldehyde with 2,3-dihydrofuran (Scheme 6.66) [803] and for the Paternó–Büchi reactions of propionaldehyde and acetaldehyde with 2,3-dihydropyran [817]. Because excited benzaldehyde undergoes very fast ISC ($k_{ISC} = 10^{12}\ s^{-1}$, Section 6.2.2), it has no chance to react in a bimolecular process in its singlet excited state; only its triplet state encounters the alkenes. In contrast, aliphatic aldehydes have longer singlet excited states, as their ISC is slower. Thus, their singlet excited states encounter the alkenes at high concentrations; at low concentrations, the bimolecular processes (second-order rate law) are slower than ISC (first-order rate law) and, as a consequence, triplet states of aldehydes react with alkenes. The change in product stereoselectivity arises from the fact that the reaction of the triplet states favors *endo*-cycloadducts, whereas singlet excited states of aldehydes form *exo*-cycloadducts preferentially.

Photocycloaddition (6.108) of enantiomerically enriched glyoxaxate **241** to 2,2-dimethyl-1,3-dioxolene is highly diastereoselective as one of the two faces of the ketone moiety is impeded to react with the alkene [818]. Stereoselective Paternó–Büchi reactions of silyl enol ethers with aromatic aldehydes have been reported [819, 820]. As for enol ethers, photochemical cycloadditions of *N*-acyl-enamines to aldehydes are regio- and stereoselective [821]. The 3-*N*-acetylaminooxetane so obtained can be converted into diastereomerically pure 1,2-aminoalcohols [822].

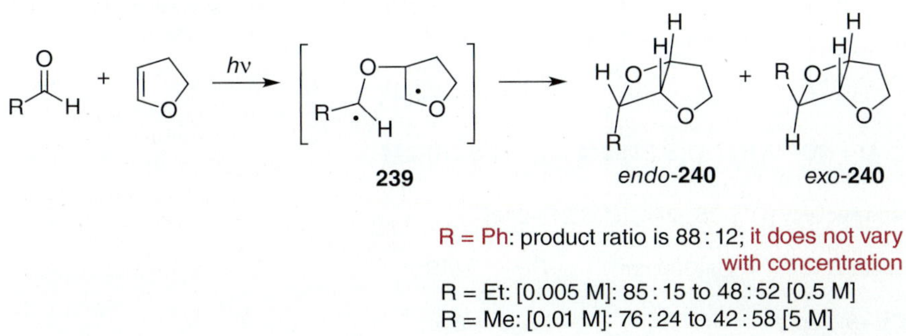

Scheme 6.66 Spin-directed stereoselectivity of the Paternó–Büchi reaction of aldehydes with 2,3-dihydrofuran. PhCHO with large k_{isc} reacts in its excited triplet state exclusively, whereas propionaldehyde and acetaldehyde with smaller k_{ISC} react through their singlet excited states at high concentrations and through their triplet excited states at low concentrations.

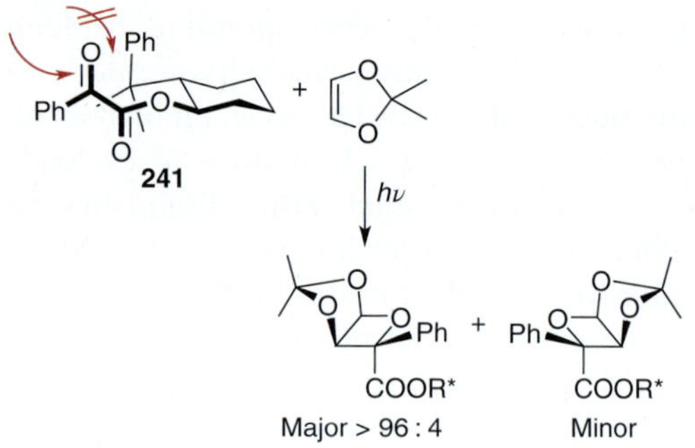

(6.108)

Enantiomerically pure chiral cyanobenzoate (*S*)-**242** (electron-poor) undergoes photochemical (2+2)-cycloaddition with 1,1-diphenylethylene (**243**: electron-rich) giving mixtures of oxetanes (1*S*′,2*S*)-**244** and (1*S*′,2*R*)-**244** in proportion (diastereoselectivity) that varies with temperature, concentration, solvent, and wavelength of irradiation (λ_{irr}). Yield is not affected by air (O_2), thus confirming that the product-forming process does not involve triplet states. The diastereomeric ratio dr = [(1*S*′,2*S*)-**244**]/[(1*S*′,2*R*)-**244**] = 82 : 18 (diastereomeric excess, de = 64%) upon irradiation (in methylcyclohexane, 25 °C, 0.2 M in (*S*)-**242**, 1 M in **243**) at 254 and 290 nm, but gradually decreases to 72 : 28 (de = 44%) at λ_{irr} = 310 nm and reaches 52 : 48 (de = 4%) at λ_{irr} = 330 nm. Short-wavelength irradiation ($\lambda_{irr} \leq 290$ nm) excites ester (*S*)-**242** to its first excited state ^{1}[(*S*)-**242**]*, which generates an exciplex with alkene **243** and eventually produces oxetanes **244**. Long-wavelength irradiation (λ_{irr} = 330 nm) excites the charge transfer complex (**242·243**) into its singlet state ^{1}[(**242·243**)]* that evolves to oxetanes **244** with lower, or opposite diastereoselectivity (see below). The charge transfer complex (**242·243**) has a binding constant K_{CT} = 0.04 M^{-1} in methylcyclohexane, K_{CT} = 0.8 in toluene, K_{CT} = 0.2 in MeCN at 25 °C. In apolar solvents, these binding constants increase at lower temperatures and decrease at higher temperature (e.g. K_{CT} = 0.03 in methylcyclohexane at 50 °C). Direct irradiation in methylcyclohexane (λ_{irr} = 290 nm) at −50 °C leads to dr = 67 : 33 (de = 34%), whereas irradiation of the charge transfer complex (λ_{irr} = 330 nm) leads to dr = 63 : 37 (de = 26%) in the same solvent at −50 °C. Thus, depending on the solvent, concentration of cycloaddends, and temperature, the proportion of charge transfer complex varies, and as a consequence, the diastereoselectivity of the photochemical cycloaddition varies with the wavelength of irradiation. As we shall see, the diastereoselectivity for direct irradiation (λ_{irr} = 290 nm) and for irradiation of the charge complex (λ_{irr} = 330 nm) depends on solvent and temperature. The diastereomer ratio (dr) depends on the difference of enthalpies of activation of the two parallel reactions: $\ln(\text{dr}) = \Delta\Delta^{\ddagger}G/RT$ (dr = (rate of formation of (1*S*′,2*S*)-**244**)/(rate of formation of (1*S*′,2*R*)-**244**) taken as rate constant ratio; one uses $\ln k = -\Delta^{\ddagger}G/RT + \ln T + 23.76$, Section 3.3). As $\Delta^{\ddagger}G = \Delta^{\ddagger}H - T\Delta^{\ddagger}S$, $\Delta\Delta^{\ddagger}G = \Delta\Delta^{\ddagger}H - T\Delta\Delta^{\ddagger}S$; by measuring dr as a function of temperature and solvent, the activation parameter differences $\Delta\Delta^{\ddagger}H$ and $\Delta\Delta^{\ddagger}S$ are obtained (Eyring analysis) for the diastereoselectivity of the photocycloaddition (*S*)-**242** + **243** (Scheme 6.67). Except for the values

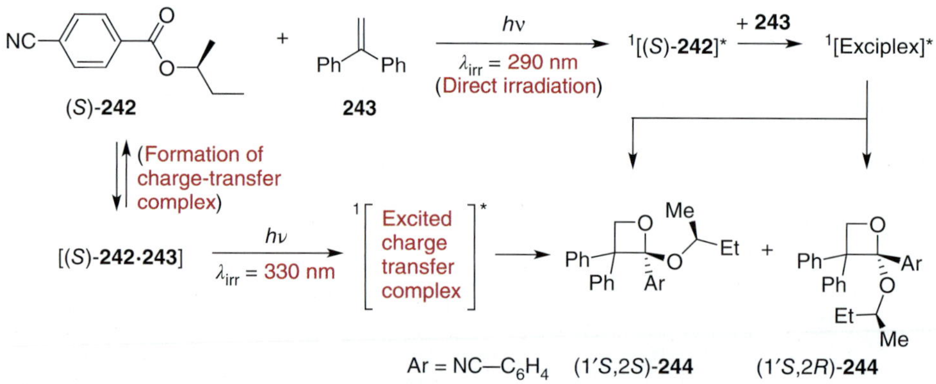

Scheme 6.67 Solvent and temperature effects on the diastereoselectivity (dr) of the Paternó–Büchi reaction of a chiral alkyl 4-cyanobenzoate with 1,1-diphenylethylene upon direct or charge-transfer complex irradiation (1 cal = 4.184 J).

Activation parameters for the diastereoselectivity [(1′*S*,2*S*)-**244**]/[(1′*S*,2*R*)-**244**]:

λ_{irr} = 290 nm	$\Delta\Delta^{\ddagger}H$	$\Delta\Delta^{\ddagger}S$	λ_{irr} = 330 nm	$\Delta\Delta^{\ddagger}H$	$\Delta\Delta^{\ddagger}S$
In methylcyclohexane:	+1.5 kcal mol^{-1},	+9.3.eu		−0.8	−2.5
In toluene:	+1.5	+8.3		−0.9	−3.1
In MeCN:	−0.05	−0.6		+0.23	+0.6

Scheme 6.68 Photochemical reactions of 1,2-diketones with silyl ketene acetals.

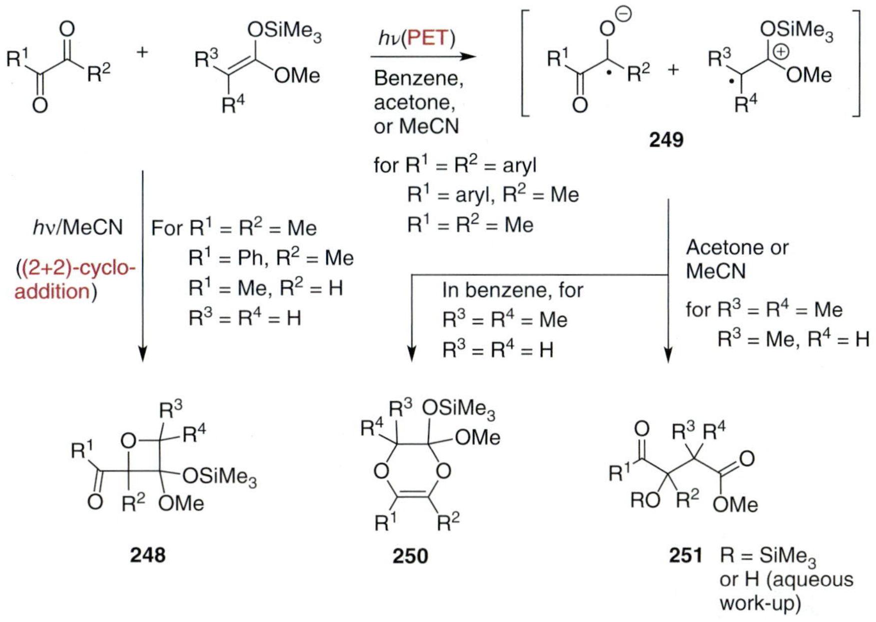

measured in MeCN, a polar solvent, those measured in nonpolar solvents such as methylcyclohexane and toluene are larger and of different signs for the reactions involving the classical exciplex intermediate than the values collected for the irradiation of the charge transfer complex. The direct irradiation (λ_{irr} = 290 nm) involves a $^1(\pi,\pi^*)$ singlet excited state of the cyanobenzoate (*S*)-**242**, which borrows some intensity from the $^1(n,\pi^*)$ state of its carbonyl group. This renders the carbonyl group electrophilic on its oxygen center and thus able to interact strongly with the electron-rich alkene **243**. For that reason, the chiral center C(2′) can more critically control the product diastereoselectivity upon direct irradiation, rather than irradiation of the charge transfer complex for which the singlet excited state maintains a looser interaction between the two cycloaddends, as indicated by the smaller $\Delta\Delta^{\ddagger}H$ and $\Delta\Delta^{\ddagger}S$ values measured for the irradiation at λ_{irr} = 330 nm [823, 824]. In the previous case of bimolecular (2+2)-cycloaddition, the diastereoselectivity, or face selectivity of the carbonyl cycloaddend, arises from steric factors (methyl vs. ethyl group). Face selectivity and regioselectivity of the bimolecular Paternó–Büchi reaction can be controlled by an allylic hydroxy group of the alkene cycloaddend [825, 826].

For obvious reasons, regio- and stereoselectivity can be very high for intramolecular Paternó–Büchi reactions [827–837]. An example is given with photoreaction (6.109) [838].

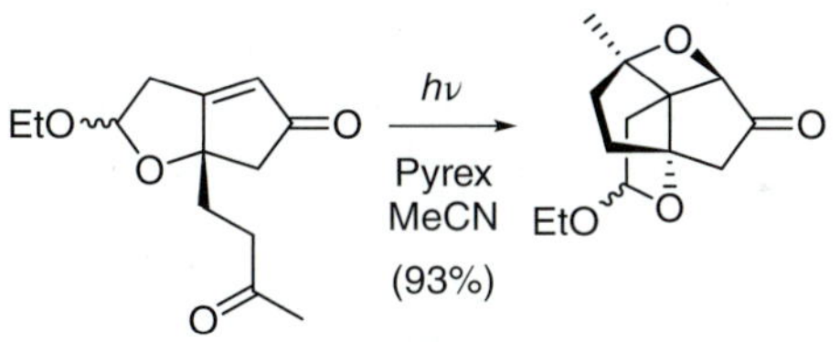

(6.109)

The far-UV component (280–320 nm) of solar light has mutagenic, carcinogenic, and lethal effects on living organisms. DNA is the most significant cellular target for far-UV light. The two major DNA lesions are the cyclobutane pyrimidine dimers **245** (*cis-syn* form for double-stranded DNA, and, to a smaller extent, *trans-syn* forms for single-stranded DNA) and pyrimidine (6-4) pyrimidone photoproducts **247** that arise from initial Paternó–Büchi reactions forming oxetanes **246**. The two types of photoproducts **245** and **247** are responsible for harmful effects of UV irradiation on organisms, such as growth delay, mutagenesis, and death [839–843]. The two modes of photochemical (2+2)-cycloadditions can be observed in solution with pyrimidines and ketones. UV irradiation of pyrimidines in acetone produces mostly cyclobutane products [844], whereas, in the presence of benzophenone, the corresponding oxetanes are the main photoadducts [845–848].

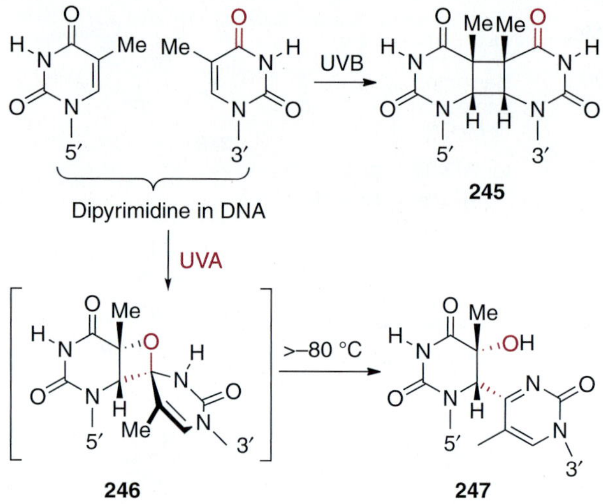

245 246 247

Irradiation of 1,2-diketones and silyl ketene acetals solutions leads to three types of reactions depending on substituents and solvent (Scheme 6.68). With trimethylsilyl ketene acetals derived from methyl acetate and methyl propionate, oxetanes **248** form in MeCN through Paternó–Büchi reactions. With the more electron-rich silyl ketene acetals, PET generates the corresponding radical anion/radical cation pairs **249** that evolve either to 1,4-dioxenes **250** (formal (4+2)-cycloadditions) or to aldol-type products **251** [849]. Photochemical (4+2)-cycloadditions of benzo[b]thiophene-2,3-dione with alkenes have been reported [850].

Problem 6.51 What are the products of Paternó–Büchi reaction of but-2-yne with benzaldehyde at −45 °C? [851]

Problem 6.52 What is the major product of photochemical reaction of furan with benzaldehyde? [852–855]

Problem 6.53 What is the major product of irradiation (λ_{irr} = 300 nm) of 3,4,5-trimethylisoxazole with benzaldehyde in MeCN at −10 °C? [856]

6.8.6 Photochemical cycloadditions of imines and related C=N double-bonded compounds

The UV absorption spectra of nonconjugated imines ($R^1(R^2)C{=}NR^3$) and *O*-alkyl oximes ($R^1(R^2)C{=}NOR^3$) generally show weak bands ($\varepsilon \sim 100$) at λ_{max} = 230–235 nm (compare with acetaldehyde: λ_{max} = 293 nm, Figure 6.3, and acetone: λ_{max} = 280 nm, Figure 6.2) that are considered to be due to $n \rightarrow \pi^*$ transitions [857, 858]. Direct irradiation requires UV light or higher energy than the direct excitation of aliphatic aldehydes and ketones (quartz vessel and high-pressure Hg burner), whereas imines and oximes conjugated with arene or alkene systems can be excited by longer wavelength irradiation. Acyclic imines and *O*-alkyl oximes have low photochemical reactivity because of their fast (*E*) ⇄ (*Z*) photoisomerization that wastes their excited energy efficiently [859, 860]. For instance, the excited state of Ph(Me)C=NOMe exhibits neither bond fission nor emission [861, 862]. In contrast, 3-phenyl-2-oxazoline and its 4- and 5-disubstituted derivatives are photolyzed into various products through reactions starting from N—O bond fission [863–865]. The 3-aryl-2-oxazolines **252** undergo photochemical (2+2)-cycloaddition (6.110) with benzene and (6.111) with indene. Photochemical (2+2)-cycloadditions of **252** to furan and thiophene have also been described [866]. UV-light-induced $[\pi^2+\pi^2]$-cycloaddition of *N*-arylsulfonylimines to styrenes or benzofurans have been reported [867]. Cyclic *N*-acylimines (aza-analogs of α,β-unsaturated ketones) undergo photochemical $[\pi^2+\pi^2]$-cycloadditions as exemplified with reaction (6.112) [868, 869]. The acetone-sensitized photoreaction of 1,3-dimethyl-6-azauracil (**253**) in the presence of alkenes gives the corresponding (2+2)-cycloadducts. With vinyl acetate (reaction (6.113)), the regioselectivity is complete [870].

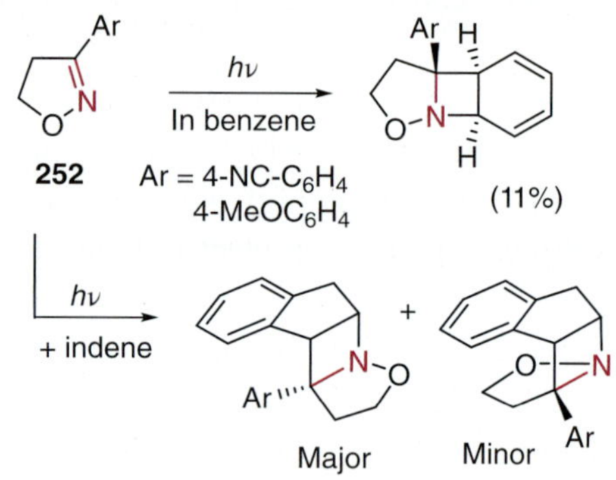

(6.110)

(6.111)

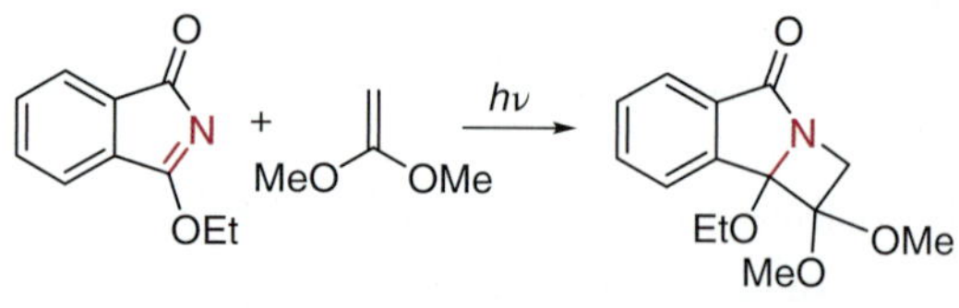

(6.112)

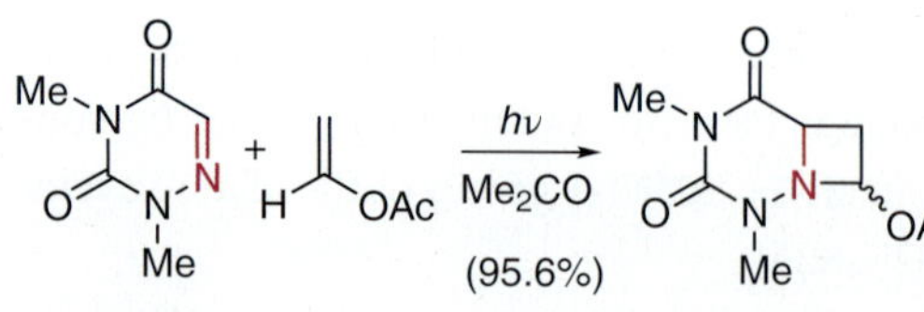

(6.113)

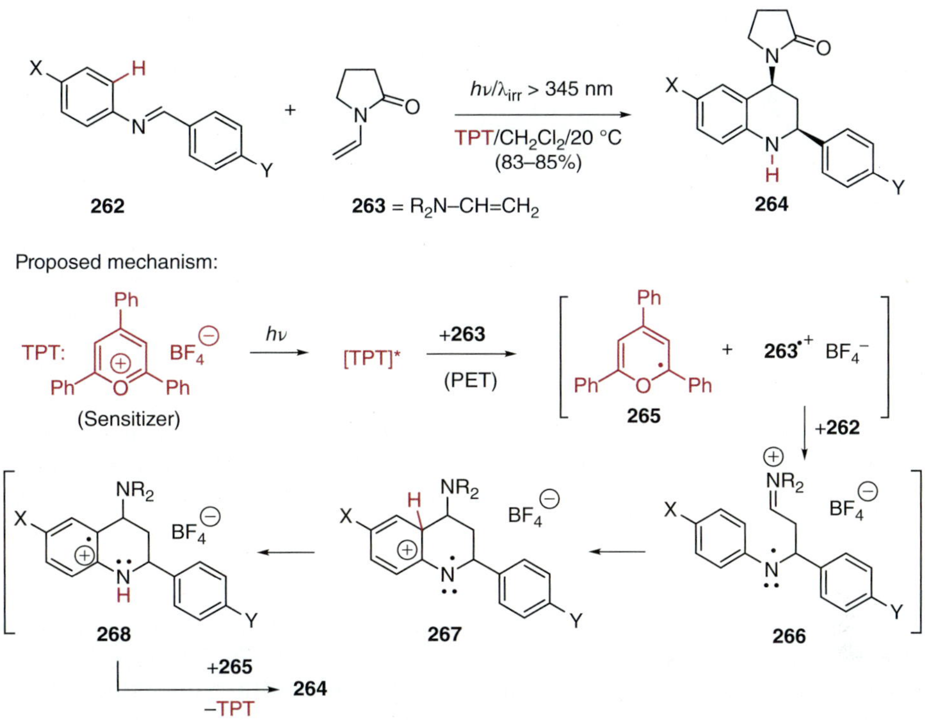

Scheme 6.69 Photosensitized Diels–Alder reaction of arylimines with *N*-vinylpyrrolidinone involving electron transfer to the excited sensitizer.

The photochemical (2+2)-cycloadditions of aromatic thiones **254** with ketenimines **255** have been reported to yield the corresponding cycloadducts **257** resulting from the preferential reaction of the C=C double bond of the ketenimines [871]. This can be interpreted in terms of the formation and fast cyclization of the most stable 1,4-diradical intermediates **256** that can be generated in this case. Photochemical (2+2)-cycloadditions of aromatic aldehydes, ketones, [872] and quinones to ketenimines give the corresponding 2-imino- and 3-iminooxetanes, also with preferential cycloaddition across the C=C bond of the ketenimines. In some cases, the 2-iminooxetanes are rearranged photochemically into their β-lactam isomers [873].

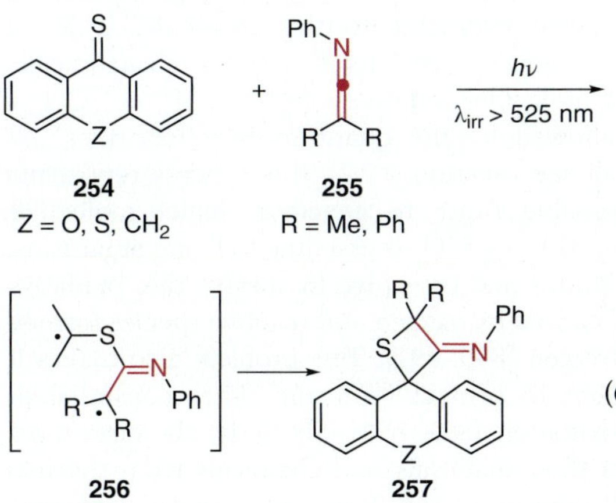

(6.114)

Direct irradiation (Pyrex) in *t*-BuOH or PhCOMe-sensitized excitation (in benzene) of imine **258** produces imine **260** through an "aza-di-π-methane" rearrangement. The isomeric aziridine **259** and the corresponding product **261** of intramolecular (2+2)-cycloaddition are not observed [874].

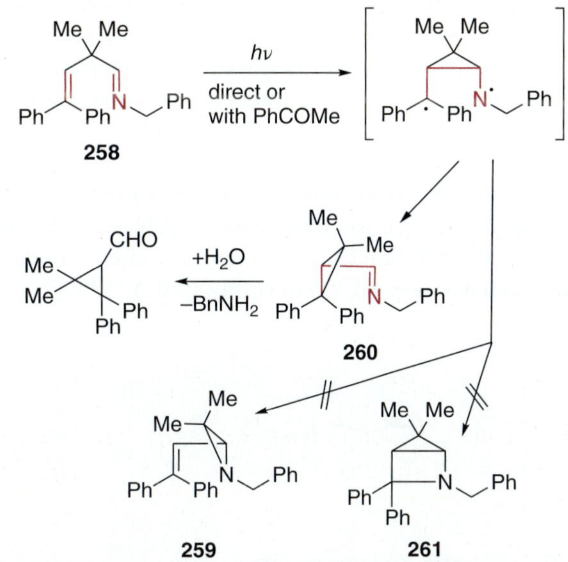

The "aza-di-π-methane rearrangement" (6.115) of a β,γ-unsaturated iminium triflate has been reported [875].

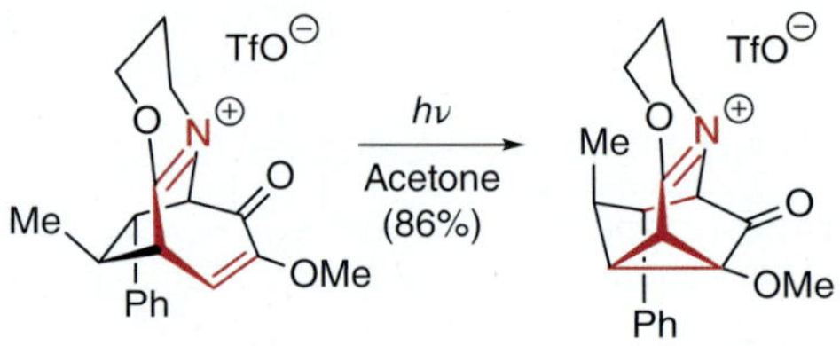

(6.115)

Formal (4+2)-cycloadditions of *N*-arylimines **262** with *N*-vinylpyrrolidinone (**263**) have been realized by using 2,4,6-triphenylpyrylium tetrafluoroborate (**TPT**) as an electron transfer photosensitizer [876] to produce the corresponding 2-oxopyrrolidin-1-yltetrahydroquinolines *cis*-**264** in good yield and with high stereoselectivity (Scheme 6.69) [877]. This is an example of photochemical cycloaddition that involves a PET (Section 6.10) from one of the two reactants to a sensitizer. Irradiation ($\lambda_{irr} \geq 345$ nm) of the colored salt **TPT** leads to its excited state [**TPT**]* that abstracts an electron from the electron-rich alkene **263** producing radical cation $\mathbf{263}^{\bullet+}/BF_4^-$ and radical **265** [878]. Radical cation $\mathbf{263}^{\bullet+}/BF_4^-$ adds to the C=N double bond of **262** to give iminium-aminyl radical cation **266**, which undergoes cyclization forming the aminyl-cyclohexadienyl radical cation **267**. A formal hydrogen atom migration gives the cyclohexadienyl radical cation **268** that takes an electron from radical **265** giving the final product **264** and the sensitizer **TPT** in its ground state. If the radical cation intermediate **268** was able to oxidize the starting alkene **263** into radical cation $\mathbf{263}^{\bullet+}$, the process would be a photocatalyzed reaction. Similar photochemical Diels–Alder reactions have been reported for *N*-vinylcarbazole + imines **262**, and for vinylogues of **262** [879].

Problem 6.54 What is the product of direct irradiation (λ_{irr} = 254 nm, quartz vessel, MeCN, −20 °C) or sensitized excitation (high-pressure Hg lamp, Solidex filter, and acetone, 60 °C) of compound **A**? [880]

A: 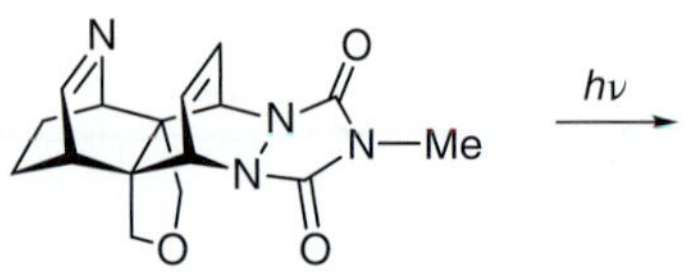

6.9 Photo-oxygenation

In the presence of a dye sensitizer and visible light, molecular oxygen (which has a triplet ground state noted: $^3O_2(^3\Sigma^-{}_g)$) is excited into singlet oxygen, $^1O_2(^1\Delta_g)$ or $^1O_2(^1\Sigma^+{}_g)$ (photosensitization) [74]. The lowest energy $^1O_2(^1\Delta_g)$ (22.4 kcal mol^{-1} above $^3O_2$) is relatively long-lived and has become a powerful reagent in organic synthesis [881]. Singlet oxygen undergoes Diels–Alder reactions [882] with naphthalenes, anthracenes, naphthacenes, and heteroaromatic compounds and with aliphatic conjugated dienes giving endoperoxides (Section 6.9.3). Alkenes possessing an allylic C—H bond react with $^1O_2(^1\Delta_g)$ in dioxa-ene reactions giving the corresponding allylic hydroperoxides [883]. With electron-rich enoxysilanes, these ene reactions may complete with (2+2)-cycloadditions. With allyltin compounds, stanna-dioxa-ene reactions are observed (Section 6.9.4). Endoperoxides and allylic hydroperoxides are important synthetic intermediates [884–888]. Singlet oxygen undergoes (2+2)-cycloadditions with electron-rich alkenes or alkenes that do not have allylic C—H bonds forming 1,2-dioxetanes. Sometimes, epoxides are formed concurrently (Section 6.9.5). 1,3-Dipolar cycloadditions have been observed with azomethine ylides (that equilibrate with aziridine) and with diazoalkanes (Section 6.9.6). Singlet oxygen also reacts with electron-rich aromatic compounds, with amines and with thioethers giving all sorts of products depending on the structure of the substrates (Section 6.9.7). The reactions may involve radical, diradical and zwitterion intermediates and/or electron transfer with the formation of superoxide radical anion ($O_2^{\bullet-}$) [889]. Singlet oxygen $^1O_2(^1\Delta_g)$ is highly electrophilic: it is sometimes represented (see Figure 6.12) as a 1,2-zwitterion $^1[O{=}O]^* \leftrightarrow {}^-O{-}O^+$, in contrast with triplet oxygen that is a 1,2-diradical and can be represented as $^3O_2 \leftrightarrow {}^\bullet O{-}O^\bullet$.

6.9.1 Reactions of ground-state molecular oxygen with hydrocarbons

The open-shell electronic structure of molecular oxygen in its ground state ($^3O_2(^3\Sigma^-{}_g)$) confers a 1,2-diradical character to this molecule. It is an efficient quencher of triplet states of electronically excited molecules, a process ($^3O_2 + {}^3[M]^* \rightarrow {}^1[O_2]^* + {}^1M_0$) allowed by the quantum selection rules. As we shall see (Section 6.9.2), this process represents one possible route to generate singlet molecular oxygen. Although $^3O_2$ is essential to living organisms, living things had to evolve to survive the “oxidative stress” caused by oxygen and reactive species formed from oxygen [890, 891]. This problem also makes it important to protect food and other consumables from oxidation [892, 893]. Similarly, chemists must protect their materials and chemicals from thermal and photoinduced reactions with molecular oxygen.

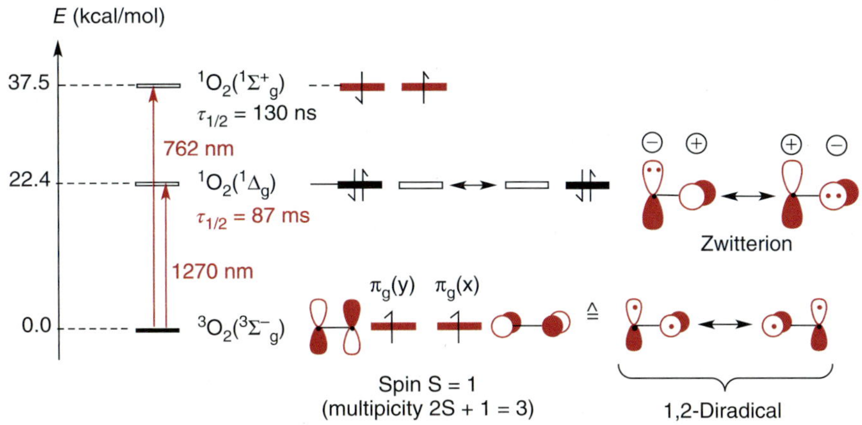

Figure 6.12 Electronic states of molecular oxygen and their representations. Lifetime of singlet excited states ($\tau_{1/2}$) depends very much on solvent: they are given here for CCl_4.

When an organic compound absorbs light, its excited states may react with $^3O_2$ and generate reactive oxygen species such as singlet oxygen ($^1O_2$), hydroxyl radical ($HO^\bullet$), hydroperoxide radical ($HOO^\bullet$), superoxide anion ($O_2^{\bullet-}$), and hydrogen peroxide (H_2O_2). These species are responsible for the degradation of living and nonliving materials ("oxidative stress"). Antioxidants are thus added to almost all commercial products; they are radical scavenging agents, an important business for the chemical industry. Nature also protects living species with radical scavenging agents.

Oxidation of *t*-Bu—H with O_2 in the gas phase into *t*-BuOOH is exothermic by −24.1 kcal mol^{-1}. Direct, uncatalyzed oxidation of *t*-Bu—H by $^3O_2$ does not occur at room temperature because the formation of *t*-butyl radical (*t*-Bu$^\bullet$) + hydroperoxide radical ($HOO^\bullet$) is endergonic, with $\Delta_rG°(t\text{-Bu—H} + {}^3O_2 \rightleftarrows t\text{-Bu}^\bullet + HOO^\bullet) \cong 43.5$ kcal mol^{-1} (endothermicity $\Delta_rH° = 43.5$ kcal mol^{-1}, entropy of reaction $\Delta_rS° \cong 0$ eu, the same number of molecules for the reactants and products, Section 2.9). This reaction involves a transition state combining two molecules, so that the activation entropy can be estimated: $\Delta^\ddagger S \cong -33$ eu and $-T\Delta^\ddagger S = 0$ eu at 25 °C. This gives $\Delta^\ddagger G(t\text{-Bu-H} + {}^3O_2 \rightarrow t\text{-Bu}^\bullet + HOO^\bullet) \geq 53.5$ kcal mol^{-1}. Thus, this reaction does not occur at room temperature (otherwise self-serving of gasoline at a gas station would be a very dangerous activity). A catalyst and/or a hot spot is required to ignite a gaseous mixture of hydrocarbon and $^3O_2$ such as in spark-ignition (e.g. Otto) or compression-ignition (Diesel) engines. Radicals *t*-Bu$^\bullet$ and $HOO^\bullet$ combine into *t*-BuOOH in a highly exothermic reaction with $\Delta_rH°(t\text{-Bu}^\bullet\text{+}HOO^\bullet \rightleftarrows t\text{-BuOOH}) \cong -67.6$ kcal mol^{-1}. Hydroperoxide radical can also abstract a H-atom from isobutane and generate *t*-butyl radical and H_2O_2, an endothermic and endergonic process of only 10 kcal mol^{-1} ($\Delta_rG°(t\text{-Bu—H} + HOO^\bullet \rightleftarrows t\text{-Bu}^\bullet + H_2O_2) = \Delta_rH° - T\Delta_rS° \cong 10$ kcal mol^{-1}). Tert-butyl radical can react with $^3O_2$ giving *t*-BuOO$^\bullet$, an exothermic condensation ($\Delta_rH° = -34.1$ kcal mol^{-1}) and exergonic reaction ($\Delta_rS° \cong -33$ eu, $\Delta_rG°(t\text{-Bu}^\bullet + {}^3O_2 \rightleftarrows t\text{-BuOO}^\bullet) \cong \Delta_rH° - T\Delta_rS° \cong -24$ kcal mol^{-1}). Then, $HOO^\bullet$ and *t*-BuOO$^\bullet$ can abstract a hydrogen atom from isobutane to give *t*-BuOOH and *t*-Bu$^\bullet$ in endothermic and endergonic processes with $\Delta_rG° \cong \Delta_rH° = 10$ kcal mol^{-1}. Tert-butyl peroxide equilibrates with 2 *t*-BuO$^\bullet$, with $\Delta H°(t\text{-BuO}^\bullet/t\text{-BuO}^\bullet) = 38.2$ kcal mol^{-1}, a fragmentation that is favored by the entropy ($\Delta_rS° \geq 35$ eu) and for which $\Delta^\ddagger G(t\text{-Bu—OO—}t\text{-Bu} \rightleftarrows 2\,t\text{-BuO}^\bullet) \cong 38.2 - T\Delta^\ddagger S \geq 28$ kcal mol^{-1}. Tert-butyl hydroperoxide equilibrates with *t*-BuO$^\bullet$ and $HO^\bullet$ radicals with $\Delta_rH°(t\text{-BuOOH} \rightleftarrows t\text{-BuO}^\bullet + HO^\bullet) \cong 42.3$ kcal mol^{-1}, a process also favored by the entropy. Isobutane reacts with *t*-BuO$^\bullet$ and generates *t*-Bu$^\bullet$ + *t*-BuOH and with $HO^\bullet$ to generate *t*-Bu$^\bullet$ + H_2O, in two exothermic and exergonic reactions (Scheme 6.70a). The slowest reaction is the direct reaction of isobutane with $^3O_2$ to form *t*-Bu$^\bullet$ + $HOO^\bullet$. However, once $HOO^\bullet$ is formed, all subsequent reactions are faster and generate more reactive oxygen species ($HO^\bullet$, *t*-BuOO$^\bullet$, and *t*-BuO$^\bullet$). Uncontrolled by radical scavengers, the oxidation reaction is explosive.

There are many other species that can abstract one H-atom from an alkane. In trace amounts, radicals (e.g. X$^\bullet$ = $HOO^\bullet$, $HO^\bullet$, $Cl^\bullet$, $Br^\bullet$, etc., radicals formed in thermal or photochemical processes) can catalyze the oxidation of alkanes and other hydrocarbons (Scheme 6.70b). The chemical industry has developed a large number of catalytical systems for the controlled aerobic oxidation of hydrocarbons into useful chemicals such as hydroperoxides, alcohols, alkenes, ketones, and carboxylic acids (Section

(a) $t\text{-Bu-H} + {}^3O_2 \rightleftarrows t\text{-BuOOH}$ $\Delta_r H° = -24.1$ kcal mol^{-1}

$\Delta_f H°$: −32.0 0.0 −56.1 kcal mol^{-1}

$t\text{-Bu-H} + {}^3O_2 \rightleftarrows t\text{-Bu}^\bullet + HOO^\bullet$ $\Delta_r H° = 43.5$ kcal mol^{-1}, $\Delta^{\ddagger}G° > 53.5$ kcal mol^{-1}

$\Delta_f H°$: −32.0 0.0 11 0.5 kcal mol^{-1}

$t\text{-Bu}^\bullet + HOO^\bullet \rightleftarrows t\text{-BuOOH}$ $\Delta_r H° = -67.6$ kcal mol^{-1}

$t\text{-Bu-H} + HOO^\bullet \rightleftarrows t\text{-Bu}^\bullet + H_2O_2$ $\Delta_r H° \approx 10$ kcal/mol

$\Delta_f H°$: −32.0 0.5 11 −32.5 kcal mol^{-1}

$t\text{-Bu-H} + t\text{-BuOO}^\bullet \rightleftarrows t\text{-Bu}^\bullet + t\text{-BuOOH}$ $\Delta_r H° \approx 10$ kcal mol^{-1}

$t\text{-Bu}^\bullet + {}^3O_2 \rightleftarrows t\text{-BuOO}^\bullet$ $\Delta_r H° = -34.1$ kcal mol^{-1}

$\Delta_f H°$: 11 0.0 −23.1 (est.)

$t\text{-Bu}^\bullet + t\text{-BuOO}^\bullet \rightleftarrows t\text{-Bu-OO-}t\text{-Bu}$ $\Delta_r H° = -69.4$ kcal mol^{-1}

$\Delta_f H°$: 11 −23.1 (est.) −81.5 kcal mol^{-1}

$t\text{-BuOOH} \rightleftarrows t\text{-BuO}^\bullet + HO^\bullet$ $\Delta_r H° = 43.8$ kcal mol^{-1}

$\Delta_f H°$: −56.1 −21.6 (est.) 9.3 kcal mol^{-1}

$t\text{-Bu-H} + t\text{-BuO}^\bullet \rightleftarrows t\text{-Bu}^\bullet + t\text{-BuOH}$ $\Delta_r H° = -10.1$ kcal mol^{-1}

$\Delta_f H°$: −32.0 −21.6 11 −74.7 kcal mol

$t\text{-Bu-H} + HO^\bullet \rightleftarrows t\text{-Bu}^\bullet + H_2O$ $\Delta_r H° = -24.7$ kcal mol^{-1}

$\Delta_f H°$: −32.0 9.3 11 −57.8 kcal mol^{-1}

(b) $t\text{-Bu-H} + X^\bullet\text{ (cat.)} \rightleftarrows t\text{-Bu}^\bullet + HX$ (initiation)

$t\text{-Bu}^\bullet + {}^3O_2 \rightleftarrows t\text{-BuOO}^\bullet$ (oxigenation)

$t\text{-Bu-H} + t\text{-BuOO}^\bullet \rightleftarrows t\text{-Bu}^\bullet + t\text{-BuOOH}$ (propagation: chain process)

$t\text{-BuOO}^\bullet + HX \rightleftarrows t\text{-BuOOH} + X^\bullet\text{ (cat.)}$ (catalyst regeneration)

$t\text{-Bu}^\bullet + X^\bullet \rightleftarrows t\text{-Bu-X}$ (termination)

Scheme 6.70 (a) Thermochemical data for uncatalyzed gas phase reactions of isobutane with ${}^3O_2$ that produces *t*-butyl hydroperoxide and *t*-butanol + water. $\Delta_f H°(t\text{-BuOO}^\bullet) \cong -23.1$ kcal mol^{-1} is estimated from $DH°(t\text{-BuOO}^\bullet/H^\bullet) \cong DH°(HOO^\bullet/H^\bullet) = 85.1$ kcal mol^{-1} (calculated from $\Delta_f H°(H^\bullet) = 52.1$, $\Delta_f H°(HOO^\bullet) = 0.5$, $\Delta_f H°(HOOH) = -32.5$ kcal mol^{-1}; $\Delta_f H°(t\text{-BuO}^\bullet) \cong -21.6$ kcal mol^{-1} is estimated from $DH°(t\text{-BuO}^\bullet/t\text{-BuO}^\bullet) \cong DH°(MeO^\bullet/MeO^\bullet) = 2\,\Delta_f H°(MeO^\bullet) - \Delta_f H°(MeOOMe) = 2(4.1) + 30.0 = 38.2$ kcal mol^{-1} and from $\Delta_f H°(t\text{-BuOO-}t\text{-Bu}) = -81.5$ kcal mol^{-1} [NIST WebBook of Chemistry]) and (b) possible mechanism for the catalyzed oxidation of isobutylene into *t*-butyl hydroperoxide.

7.11). Examples of catalytic systems include transition metal salts, clusters and complexes [894–898], metal surfaces [899, 900], activated carbons, [901] and *N*-hydroxyphtalimide (generates $HO^\bullet$) [902–905].

In the gas phase, the activation barrier of the direct, uncatalyzed allylic oxidation of alkenes with ${}^3O_2$ is lowered by about 10 kcal mol^{-1} compared with ${}^3O_2$ oxidation of alkanes (stabilization by π-conjugation of the allyl radicals). In the case of reaction of isobutylene to give 2-methylallyl (methallyl) hydroperoxide, the activation free energy is estimated to be $\Delta^{\ddagger}G(Me_2C{=}CH_2 + {}^3O_2 \rightarrow H_2C{=}C(Me)CH_2^\bullet + HOO^\bullet) \geq 43$ kcal mol^{-1} (Scheme 6.71a). Another route to methallyl hydroperoxide is the dioxa-ene reaction (Scheme 6.71b). In principle, the condensation of isobutylene with ${}^3O_2$ generates a 1,4-diradical intermediate. In this case, the triplet diradical ^{3}[**269**] forms with an endothermicity

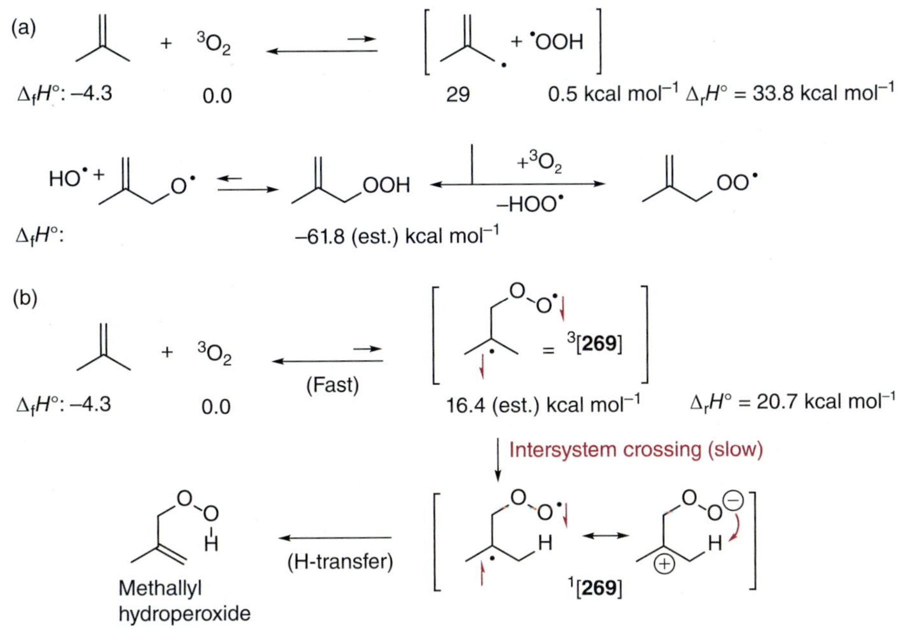

Scheme 6.71 In the absence of light, $^3O_2$ oxidizes alkenes relatively quickly through the formation of 1,4-diradical intermediates. Estimates of the thermochemical data for the direct oxidation of isobutylene by molecular oxygen into methallyl hydroperoxide (a) through direct allylic oxidation and (b) through a two-step dioxa-ene reaction. Estimates (see Tables 1.A.4 and 1.A.7): $\Delta_f H°(i\text{-PrCH}_2\text{OOH}) = \Delta_f H°(i\text{-PrCH}_2\text{OH}) + \Delta_f H°(\text{EtOOH}) - \Delta_f H°(\text{EtOH}) = -67.8 - 50 + 56 = -61.8$ kcal mol^{-1}; $\Delta_f H°(\mathbf{269}) = \Delta_f H°(i\text{-PrCH}_2\text{OOH}) + DH°(t\text{-Bu}^\bullet/\text{H}^\bullet) + DH°(\text{HOO}^\bullet/\text{H}^\bullet) - DH°(\text{H}^\bullet/\text{H}^\bullet) = -61.8 + 95.2 + 87.2 - 104.2 = 16.4$ kcal mol^{-1}.

$\Delta_r H°(\text{Me}_2\text{C=CH}_2 + {}^3\text{O}_2 \rightleftarrows {}^3[\mathbf{269}]) \cong 20.7$ kcal mol^{-1}. Considering an entropy of condensation of c. −33 eu, an entropy cost $-T\Delta_r S° \cong 10$ kcal mol^{-1} is implied in this oxidation at 25 °C. Thus, its activation free energy is estimated to be $\Delta^{\ddagger}G(\text{Me}_2\text{C=CH}_2 + {}^3\text{O}_2 \rightarrow {}^3[\mathbf{269}]) \geq 20.7$ kcal mol^{-1} at 25 °C. This gives a lower limit for the rate constant for the formation of diradical **269**: $k(\text{Me}_2\text{C=CH}_2 + {}^3\text{O}_2 \rightarrow {}^3[\mathbf{269}])] \leq 8 \times 10^{-9}$ M^{-1} s^{-1} at 25 °C under one atmosphere. This corresponds to a half-life $\tau_{1/2} > 145$ days. This is less than the lower limit for the rate constant $k'(\text{Me}_2\text{C=CH}_2 + {}^3\text{O}_2 \rightarrow \text{H}_2\text{C=C(Me)CH}_2\text{OOH})$ of the oxidation of isobutylene into methallyl hydroperoxide, as the triplet intermediate $^3[\mathbf{269}]$ must undergo ISC into $^1[\mathbf{269}]$ before it can undergo a hydrogen transfer generating methallyl hydroperoxide, a process that competes with the fragmentation of diradical $^3[\mathbf{269}]$ into the starting reactants. However, the thermochemical data demonstrate that alkenes have a relatively easy route available for oxidation in the absence of light. It is clear that alkenes (and all conjugated π-systems) must be stored in the absence of air. Traces of air may lead to their polymerization, as diradical intermediates such as **269** can initiate such a reaction concurrently with the formation of the corresponding hydroperoxides. The polymerization can be retarded by adding radical scavenging agents.

6.9.2 Singlet molecular oxygen

In 1928, Mulliken [906, 907] predicted that molecular oxygen exists in two singlet electronically excited states noted $^1\text{O}_2(^1\Delta_g)$ and $^1\text{O}_2(^1\Sigma^+{}_g)$ that are much lower in energy than the first triplet excited state $^3\text{O}_2(^3\Sigma^-{}_g)$ evidenced by the intense Schumann UV absorption band at $\lambda \cong 180$ nm. The relative energy of $^1\text{O}_2(^1\Sigma^+{}_g)$ (with respect to its ground state $^3\text{O}_2(^3\Sigma^-{}_g)$) has been evaluated first by Childs and Mecke who measured the very weak absorption band at $\lambda = 762$ nm for gaseous O_2 (the spin forbidden "atmospheric oxygen band") [908]. Ellis and Kneser discovered another transition of very low intensity at $\lambda = 1261$ nm with liquid O_2 [909], thus identifying the lower lying $^1\text{O}_2(^1\Delta_g)$ state (Figure 6.12) [910]. Direct irradiation of $^3\text{O}_2$ dissolved in Freon-113 ($\text{Cl}_2\text{CF—CF}_2\text{Cl}$) by a He and Ne laser ($\lambda_{irr} = 632.8$ nm) promotes the transition $2\ \text{O}_2(^3\Sigma^-{}_g) \rightarrow 2\ ^1\text{O}_2(^1\Delta_g)$ and direct irradiation with neodyme laser (YAG, $\lambda_{irr} = 1064$ nm) promotes the transition $^3\text{O}_2(^1\Sigma^-{}_g) \rightarrow {}^1\text{O}_2(^1\Delta_g)$ [911]. The lifetime of $^1\text{O}_2(^1\Delta_g)$ in solution has been determined by its emission spectrum (phosphorescence: radiative transition $^1\text{O}_2(^1\Delta_g) \rightarrow {}^3\text{O}_2(^3\Sigma^-{}_g)$) through time-resolved laser spectroscopy [912]. Quantum yield for the phosphorescence ($\lambda \cong 1275$ nm) of $^1\text{O}_2(^1\Delta_g)$ in benzene generated by photosensitization with anthracene-9,10-dicarbonitrile amounts to c. 2.5×10^{-4} [913].

$$\text{H}_2\text{O}_2 + \text{OCl}^- \rightarrow {}^1\text{O}_2 + \text{H}_2\text{O} + \text{Cl}^- \qquad (6.32)$$

In 1927, Mallet reported a red luminescence accompanying reaction (6.32) of hydrogen peroxide with sodium hypochlorite in water that generates molecular oxygen [914]. The same observation was made later for the reaction of Br_2 in alkaline H_2O_2. In 1942, Groh and Kirrmann measured the prominent visible bands associated with this chemiluminescence as occurring at $\lambda_{max} = 632$ nm and $\lambda_{max} = 578$ nm corresponding to ${}^1O_2({}^1\Sigma^+{}_g) \rightarrow {}^3O_2({}^3\Sigma^-{}_g)$ transitions [915]. Later studies by Khan and Kasha identified emission bands arising not only from singlet ${}^1O_2({}^1\Sigma^+{}_g)$ molecules at $\lambda_{max} = 762$ nm and from singlet ${}^1O_2({}^1\Delta_g)$ molecules at $\lambda_{max} = 1270$ nm [916] but also from dimeric complexes of singlet oxygen. The red pair at $\lambda_{max} = 633.4$ nm and 703.2 nm corresponds to the ${}^1O_2({}^1\Delta_g)\cdot{}^1O_2({}^1\Delta_g) \rightarrow {}^3O_2({}^3\Sigma^-{}_g)\cdot{}^3O_2({}^3\Sigma^-{}_g)$ transition, and a green band at $\lambda_{max} = 478$ nm corresponds to the ${}^1O_2({}^1\Delta_g)\cdot{}^1O_2({}^1\Sigma^+{}_g) \rightarrow {}^3O_2({}^3\Sigma^-{}_g)\cdot{}^3O_2({}^3\Sigma^-{}_g)$ transition. These bands correlate with the high-pressure (150 atm) gaseous absorption bands. No emission corresponding to ${}^1O_2({}^1\Sigma^+{}_g)\cdot{}^1O_2({}^1\Sigma^+{}_g) \rightarrow {}^3O_2({}^3\Sigma^-{}_g)\cdot{}^3O_2({}^3\Sigma^-{}_g)$ transition is observed because the lifetime of ${}^1O_2({}^1\Sigma^+{}_g)$ is much shorter than that of ${}^1O_2({}^1\Delta_g)$, which leaves little chance for two molecules of ${}^1O_2({}^1\Sigma^+{}_g)$ to encounter [917, 918]. Decomposition of H_2O_2 with hypervalent iodine compound $PhI(OCOCF_3)_2$ also produces ${}^1O_2({}^1\Delta_g)$ [919]. Aubry and Cazin have shown that singlet oxygen forms during the decomposition of H_2O_2 catalyzed by sodium molybdate. It also forms during the decomposition of $MoO_6{}^{2-}$ and $MoO_8{}^{2-}$ generated by reaction of $MoO_4{}^{2-}$ with H_2O_2 ($MoO_4{}^{2-} + 2\,H_2O_2 \rightarrow MoO_6{}^{2-} + 2\,H_2O$; $MoO_6{}^{2-} \rightarrow MoO_4{}^{2-} + {}^1O_2$, and $MoO_6{}^{2-} + 2\,H_2O_2 \rightarrow MoO_8{}^{2-} + 2\,H_2O$; $MoO_8{}^{2-} \rightarrow MoO_6{}^{2-} + {}^1O_2$) [920, 921]. The reduction of Cu^{2+} by H_2O_2 in alkaline solution produces $O_2{}^{\bullet-}$, which reduces Cu^{2+} into Cu^+ and ${}^1O_2$ ($Cu^{2+} + H_2O_2 \rightarrow Cu^+ + O_2{}^{\bullet-} + H^+$; $Cu^{2+} + O_2{}^{\bullet-} \rightarrow Cu^+ + {}^1O_2$). $La(NO_3)_3$ in alkaline solution decomposes H_2O_2 into ${}^1O_2$ [922]. $CaO_2\cdot(H_2O_2)_2$ decomposes into ${}^1O_2$ upon heating [923]. The reaction of H_2O_2 + MeCN in alkaline solution is another source of ${}^1O_2$ (MeCN + $HOO^- \rightarrow$ MeC(=NH)OO^- + $H_2O \rightleftarrows$ MeC(N=H)OOH + HO^-; MeC(N=H)OOH + MeC(=NH)$OO^- \rightarrow$ $MeCONH_2$ + MeC(=NH)O^- + ${}^1O_2$) [924, 925]. Hydrogen peroxide can be decomposed into other species than ${}^1O_2$ for instance through Fenton's chemistry [926] that produces $HO^\bullet$ together with $HOO^\bullet \rightleftarrows H^+ + O_2{}^{\bullet-}$, a process that involves reactions $Fe^{2+} + H_2O_2 + H^+ \rightarrow Fe^{3+} + HO^\bullet + H_2O$; $Fe^{3+} + H_2O_2 \rightarrow Fe^{2+} + HOO^\bullet + H^+$ as well as the Haber–Weiss reaction $O_2{}^{\bullet-} + H_2O_2 \rightarrow O_2 + HO^- + HO^\bullet$ [927–930]. Singlet oxygen ${}^1O_2({}^1\Delta_g)$ has been observed directly in the gas phase by spontaneous Raman scattering spectroscopy [931]. In 1964, Foote and Wexler [932, 933] have shown that oxygenation products obtained with reaction (6.32) are the same as those arising from photosensitized photooxygenation [934, 935].

There are more methods to generate singlet molecular oxygen (see below). The most frequently used procedure in organic synthesis is bubbling O_2 or air in an illuminated solution of the compound to oxidize in the presence of a dye, the sensitizer (e.g. Figure 6.13). After having washed their bed linens and white shirts with soap and water, our ancestors were drying them on grass in the sunshine. This renders the cotton whiter. This is still practiced in many places in the world. Chlorophyll of the grass (Figure 6.13) is the dye that absorbs visible light and sensitizes the formation of singlet oxygen, the oxidant that cleans (bleaches) the cotton by "cold burning" stains and dirt. Self-cleaning cotton fabrics impregnated with sensitized (anthraquinone-2-carboxylic acid = sensitizer) titanium oxide have been developed [936]. In the early 1900s, the toxicity of O_2 in the presence of a dye and light was recognized: it is called the "photodynamic effect" [937, 938]. In some cases, photodynamic effects have therapeutic applications. Annals over 3000 years old by Egyptians and Indians report the use of topically applied vegetable and plant substances to induce photoreactions to treat skin diseases. Today, psoralens are the dyes (Figure 6.13) used in the photodynamic therapy of neurodermatitis, eczema, cutaneous T-cell, and *lymphomalichen ruber planus* [939, 940]. Porphyrinoid compounds, including chlorins, bacteriochlorins, and phthalocyanines [941], are mostly used in the photodynamic therapy of cancers [942–945]. Examples are porfimer sodium (Photofrin®), a mixture of monomeric, dimeric and oligomeric esters derived from hematoporphyrin, and aluminum phthalocyanine tetrasulfonate (Photosens®) (Figure 6.13). In the chemical laboratory, soluble sensitizers commonly used are rose bengal, eosin yellow, erythrosin B, methylene blue, and zinc tetraphenylphorphyrin. Colorless compounds absorbing UV light such as aromatic compounds (e.g. anthracene, benzophenone, and 9,10-anthraquinone) have been used as sensitizers for the generation of ${}^1O_2({}^1\Delta_g)$. Dyes immobilized on solid supports have proven to be especially convenient [946–949], and polymer-bound dyes such as Sensitox® (rose bengal on Merrifield resin) are commercially available [950, 951]. Fullerene C_{60} has also been used as a sensitizer [952]. UV irradiation of aerated aqueous solution of C_{60} generates ${}^1O_2$ and/or $O_2{}^{\bullet-}$depending on the aggregation state of C_{60} [953]. Since the physical decay ${}^1O_2({}^1\Sigma^+{}_g) \rightarrow {}^1O_2({}^1\Delta_g)$ + heat is very fast, only the lowest energy singlet molecular

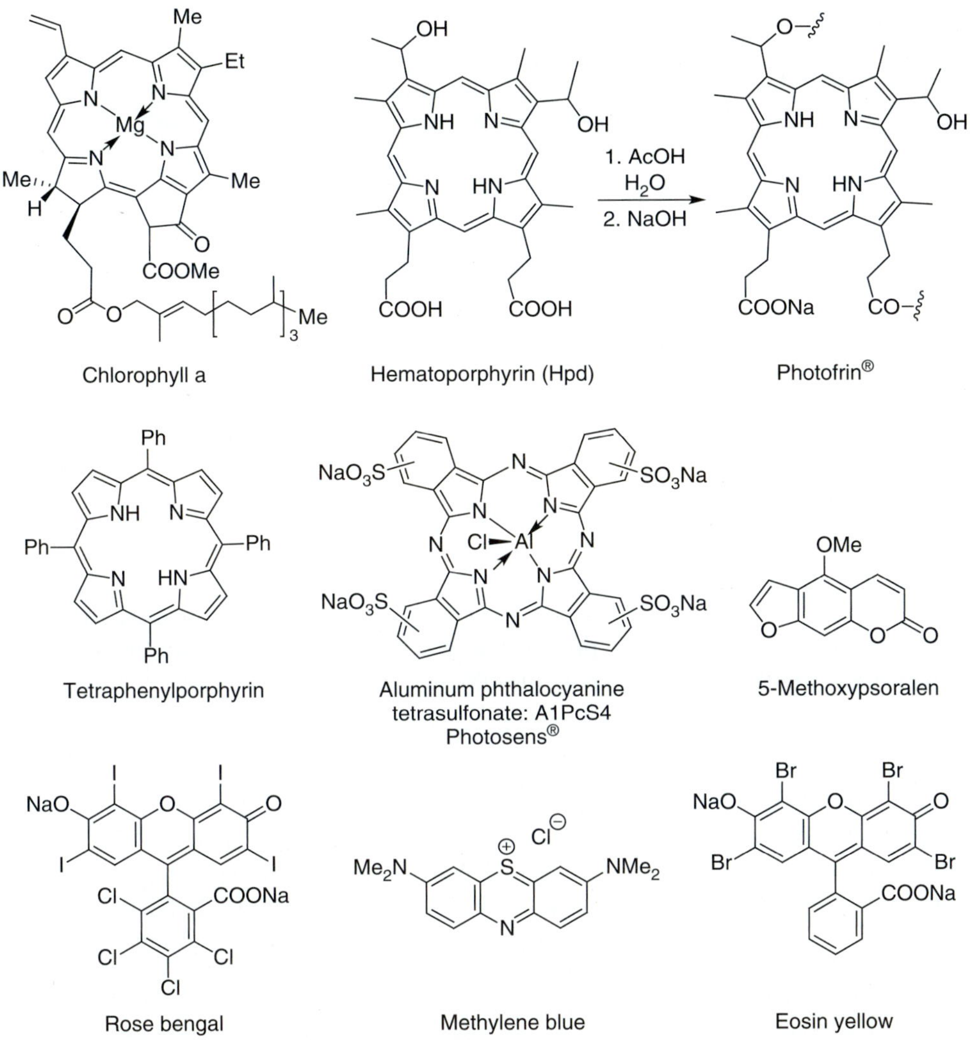

Figure 6.13 Examples of sensitizers used to generate singlet molecular oxygen.

oxygen undergoes reactions with organic compounds [954, 955].

In 1931, Kautsky and Hirsch first proposed a mechanism (Scheme 6.72a) for the photosensitized formation of $^1O_2$ in which the triplet excited sensitizer (^{3}[sens(T_1)]*) transfers its energy directly to $^3O_2$ producing $^1O_2$ and the sensitizer (^{1}[sens(S_0)]) in its ground state (reaction (6.117)) [956–958]. This is a spin allowed energy transfer that is the most common mechanism for the photosensitized formation of $^1O_2(^1\Delta_g)$. When the sensitizer is anthracene in benzene, the fraction of triplet state ^{3}[sens(T_1)]* that results in the formation of $^1O_2(^1\Delta_g)$ is about 0.8, and the rate constant of quenching $k_q(^3[\text{sens}(T_1)]^* + {}^3O_2 \rightarrow [^1\text{sens}(S_0)] + {}^1O_2) \cong 4 \times 10^9\ M^{-1}\ s^{-1}$ [959]. $^1O_2$ subsequently reacts with acceptor (or substrate) **A** to form the products of photo-oxygenation (Section 6.9.3). In 1935, Schönberg proposed an alternative mechanism (Scheme 6.72b) in which the triplet excited sensitizer forms a diradical with $^3O_2$ (reaction (6.118)), and the diradical reacts with **A** giving the products of photo-oxygenation and the sensitizer in its ground state [960]. Ground-state $^3O_2$ interacts with singlet excited molecules ^{1}[**M**]* in two possible ways (Scheme 6.72c). In the first way, spin allowed energy transfer produces $^1O_2(^1\Delta_g)$ and triplet excited ^{3}[**M**]* (reaction (6.117′)). This implies that the singlet (S_1)/triplet(T_1) energy gap in **M** is greater than 22.4 kcal mol^{-1} (Figure 6.12). This mechanism is observed for **M** = linearly fused benzene rings and is nearly diffusion controlled (rate constant of $^1O_2(^1\Delta_g)$ formation from singlet excited anthracene in MeCN: $k_q = 30 \times 10^9\ M^{-1}\ s^{-1}$) [961, 962]. For a compound **M** capable of fluorescence and phosphorescence, a determination of whether the singlet or triplet excited

(a) $^{1}\text{sens}(S_o) + hn \rightarrow {}^{1}[\text{sens}(S_1)]^* \rightarrow {}^{3}[\text{sens}(T_1)]^*$

$$^{3}[\text{sens}(T_1)]^* + {}^{3}O_2 \rightarrow {}^{1}\text{sens}(S_o) + {}^{1}O_2(^{1}D_g) \qquad (6.117)$$

$$^{1}O_2(^{1}D_g) + \mathbf{A} \rightarrow \mathbf{A}\text{–}O_2 \text{ (products of photo-oxidation)}$$

(b) $$^{3}[\text{sens}(T_1)]^* + {}^{3}O_2 \rightarrow {}^{\bullet}\text{sens-}O_2^{\bullet} \qquad (6.118)$$

$$^{\bullet}\text{sens-}O_2^{\bullet} + \mathbf{A} \rightarrow \mathbf{A}\text{–}O_2 + {}^{1}\text{sens}(S_o)$$

(c) $$^{1}\mathbf{M}_o + hn \rightarrow {}^{1}[\mathbf{M}]^* + {}^{3}O_2 \rightarrow {}^{3}[\mathbf{M}]^* + {}^{1}O_2(^{1}D_g) \qquad (6.117')$$

(d) $$^{1}[\mathbf{M}]^* + {}^{3}O_2 \rightarrow \mathbf{M}^{\bullet+} + O_2^{\bullet-} \qquad (6.119)$$

(e) $$^{1}O_2(^{1}D_g) + \mathbf{M} \rightarrow O_2^{\bullet-} + \mathbf{M}^{\bullet+} \qquad (6.120)$$

Scheme 6.72 Photosensitized photo-oxygenation: (a) via singlet molecular oxygen $^{1}O_2(^{1}\Delta_g)$ generated by triplet sensitization (Kautsky's mechanism); (b) by formation of diradical (Schönberg's mechanism); (c) via $^{1}O_2(^{1}\Delta_g)$ generated by singlet sensitization; (d) via photo-induced electron-transfer (PET, Section 6.10) to triplet oxygen that generates the radical-cation of the substrate (or reactant) and superoxide radical-anion; and (e) via singlet oxygen that abstracts an electron (PET; Section 6.10) from the substrate (reactant).

state reacts can be determined by measuring the quantum yield of fluorescence and phosphorescence as a function of the concentration of $^{3}O_2$. The second way of quenching singlet excited state $^{1}[\mathbf{M}]^*$ by $^{3}O_2$ is an electron transfer (6.119) producing superoxide anion radical ($O_2^{\bullet-}$) that reacts with radical cation $\mathbf{M}^{\bullet+}$ or with other electrophilic compounds present in the medium (Scheme 6.72d). Electron transfer might also be involved in the reaction of $^{1}O_2(^{1}\Delta_g)$ with reactant (or substrate) **M** (Scheme 6.72e; reaction (6.120); see Sections 6.9.2, 6.9.5, and 6.9.7, for examples). The lifetime of $^{1}O_2(^{1}\Delta_g)$ depends very much on solvent that promotes its physical (radiationless) deactivation [963]. Singlet molecular oxygen is quenched strongly by interaction with C—H and O—H bonds ($\tau_{1/2} \cong 5$ μs in H_2O, 10 μs in MeOH, 30 μs in hexane and PhH, and 75 μs in MeCN). The longest lifetimes are in CCl_4 (87 ms) and Freon-113 ($Cl_2CF\text{-}CF_2Cl$; c.100 ms) [964]. Tertiary aliphatic amines deactivate $^{1}O_2(^{1}\Delta_g)$ efficiently (singlet oxygen quenchers) [910, 959, 965, 966]. They form exciplexes with $^{1}O_2(^{1}\Delta_g)$ with substantial charge transfer; the exciplexes then proceed to the starting amine + $^{3}O_2$ (physical quenching) [888]. Alternatively, an electron transfer $R_3N{:} + {}^{1}O_2 \rightarrow R_3N^{\bullet+} + O_2^{\bullet-} \rightarrow R_3N{:} + {}^{3}O_2 + \text{heat}$ can intervene (chemical quenching) [967–971] as with DABCO (1,4-diazabicyclo[2.2.2]octane) [967] and substituted *N*,*N*-dimethylanilines [889] for instance. Sodium azide is a very efficient quencher of $^{1}O_2(^{1}\Delta_g)$ [966], with a rate constant $k_q \cong 2.2 \times 10^{8}$ M^{-1} s^{-1} in MeOH [972]. It has generally been assumed that natural antioxidants such as carotenoids are efficient quenchers of singlet oxygen [973]. Studies in mammalian cells applying microscope-based time-resolved spectroscopy have shown this not to be the case for β-carotene [974]. Deactivation of $^{1}O_2(^{1}\Delta_g)$ by 2,2,6,6-tetramethylpiperidine leads to the formation of TEMPO, a persistent nitroxide radical that can be analyzed by ESR [975–977]. This reaction is inhibited by proline [978].

Under UV irradiation, ozone in the gas phase produces singlet oxygen. Electrical discharge also produces singlet oxygen as shown with the following reactions [979].

$$O_3 \rightarrow {}^{1}O_2(^{1}\Delta_g) + O(^{1}D)$$

$$O(^{1}D) + {}^{3}O_2(^{3}\Sigma^{-}{}_g) \rightarrow {}^{1}O_2(^{1}\Sigma^{+}{}_g) + O(^{3}P)$$

Ozone reacts at low temperature (−70 °C) with trialkyl phosphites giving the corresponding phosphitozonides. The latter decompose at −35 °C into the corresponding phosphate esters and singlet oxygen [980–986].

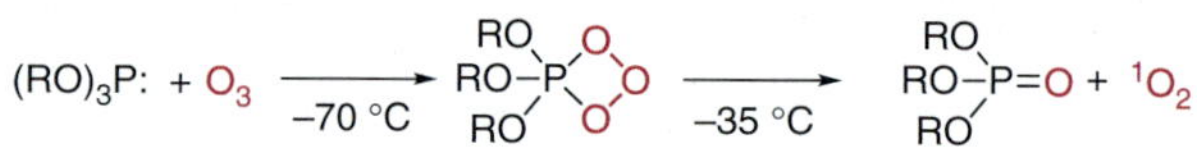

((2+2)-Cycloreversion)

Peroxynitrite ((dioxido)oxidonitrate anion: O=NOO⁻) is a strong oxidant produced by the reaction of superoxide radical anion and nitrogen monoxide ($NO^{\bullet} + O_2^{\bullet-} \rightarrow O{=}NOO^{-}$). At neutral to alkaline pH, O=NOOH decomposes into $^{1}O_2(^{1}\Delta_g)$ with a 2–10% yield [987]. Organic oxides obtained by ozonization of secondary alcohols are decomposed producing singlet oxygen [988, 989]. Fragmentation of 1,1-dihydroperoxide monoperacetates is a source of $^{1}O_2$ ($R_2C(OOH)OOAc \rightarrow R_2C{=}O + AcOH + {}^{1}O_2$) [990]. Sonication (ultrasound irradiation) of water at 1.56 MHz produces hydroxyl and hydrogen radical. In the presence of $^{3}O_2$, hydroxyl radical generates $HOOO^{\bullet}$ that decomposes immediately into $HO^{\bullet}$ and singlet oxygen $^{1}O_2$ [991].

In 1867, Fritsche reported that, in the presence of light, naphthacene reacts with molecular oxygen giving a material that regenerates oxygen on heating [992]. In the meantime, a large number of endoperoxides have been characterized [993]. In 1926, Dufraisse and coworkers discovered that red solutions of rubrene become colorless upon exposure to air and light. The endoperoxide **270** that forms gives up to an 80% yield of molecular oxygen at 140–150 °C [994, 995].

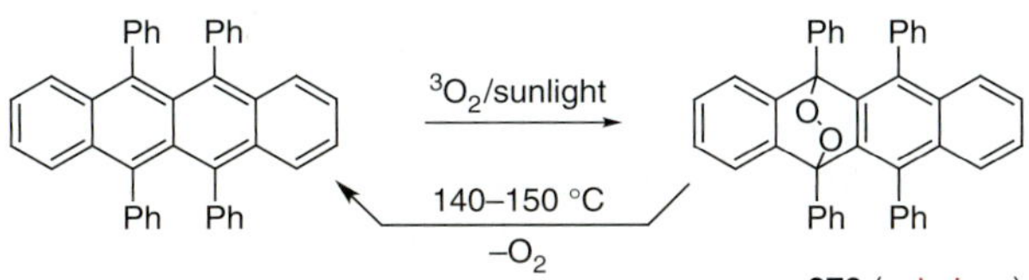

Rubrene (red) **270** (colorless)

The 9,10-endoperoxide of 9,10-diphenylanthracene (DPA) dissociates at 80 °C with a half-life of 16 hours (reaction (6.121)) [996, 997]. Singlet oxygen $^1O_2(^1\Delta_g)$ is formed (spin allowed process), which leads to the same reaction products as photosensitized oxygenation [998, 999]. Heat and activation parameters of retro-Diels–Alder reactions (6.121) and (6.122) have been reported. By measuring the chemiluminescence (light emitted by singlet oxygen), the yield in the formation of $^1O_2(^1\Delta_g)$ approaches 95% for the thermolysis (6.122), an endothermic reaction with no activation entropy. Thus, the bulk of the energy for the formation of $^1O_2(^1\Delta_g)$ comes from activation energy and not from reaction exothermicity. In the case of the thermolysis (6.121), the yield in $^1O_2(^1\Delta_g)$ is only 35% and a positive activation entropy is measured. A late transition state is proposed in the case of the thermolysis (6.121) (no spin/orbit coupling, the system cannot leave the singlet energy hypersurface) and of an early transition state in the case of the thermolysis (6.122) (spin/orbit coupling is efficient because of the diradical character reached in the transition state, see Section 6.9.3 for the mechanism of the $[\pi^4+\pi^2]$-cycloaddition of $^1O_2(^1\Delta_g)$) [1000, 1001]. *N*-Substituted α-pyridones undergo reversible $[\pi^4+\pi^2]$-cycloadditions (6.123) with $^1O_2(^1\Delta_g)$ [1002].

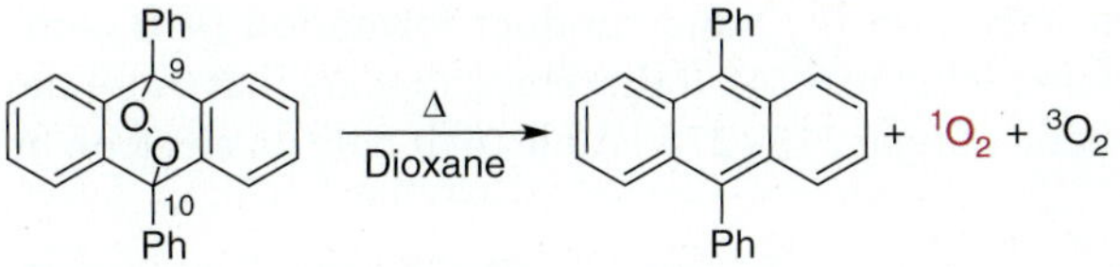

Δ_rH^o = 5 kcal mol^{-1} 35% 65%
$\Delta^{\ddagger}H$ = 32.5 ± 0.2 kcal mol^{-1}; $\Delta^{\ddagger}S$ = 9.5 ± 0.5 eu

(6.121)

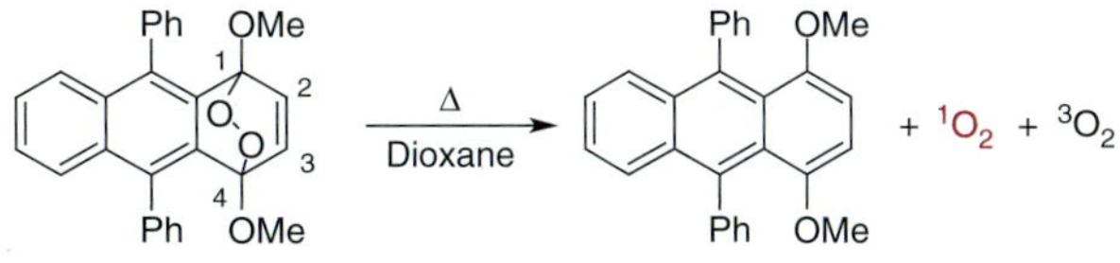

Δ_rH^o = 1 kcal mol^{-1} 95% 5%
$\Delta^{\ddagger}H$ = 24.2 ± 0.2 kcal mol^{-1}; $\Delta^{\ddagger}S$= −0.3 ± 0.7 eu

(6.122)

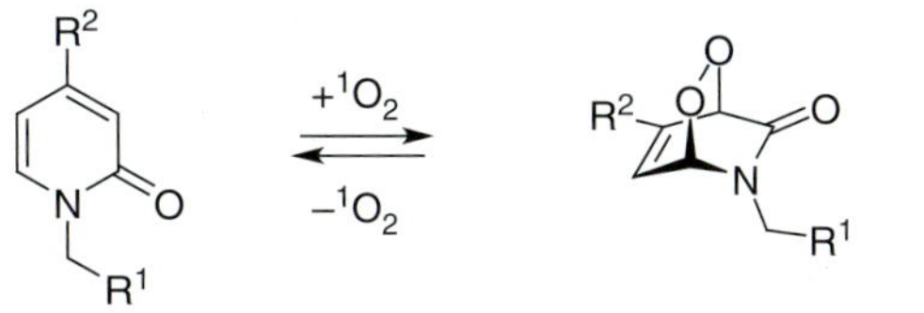

(6.123)

Problem 6.55 Why is the lifetime of $^1O_2(^1\Sigma^+{}_g)$ much shorter than that of $^1O_2(^1\Delta_g)$?

6.9.3 Diels–Alder reactions of singlet oxygen

In 1928, Windaus and Brunken reported on the photochemical oxidation of Ergosterine (or Ergosterol), a natural product. Its cyclohexa-1,3-diene moiety undergoes a $[\pi^4+\pi^2]$-cycloaddition of dioxygen [1003]. In 1951, Schenck and Ziegler reported on the photosensitized oxygenation of cyclohexadiene derivatives that generate the corresponding endoperoxides through $[\pi^4+\pi^2]$-cycloadditions [1004]. Photo-oxygenation of α-terpinene opened a short synthesis of (±)-ascaridole (Scheme 6.73) [1005–1007], an anthelmintic (toxic) drug that expels

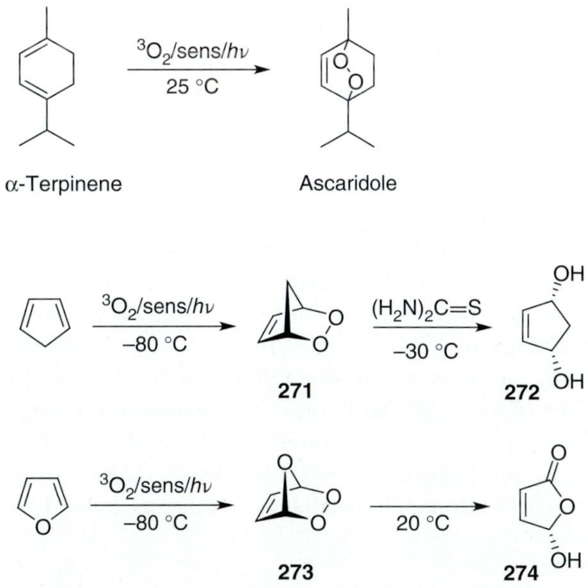

Scheme 6.73 Early examples by Schenck of Diels–Alder reactions of singlet oxygen with cyclic aliphatic 1,3-dienes.

Table 6.5 Relative rate constants for the formation of endoperoxides of dienes **275–278** in CH_2Cl_2, −20 °C, sensitizer: tetraphenylporphyrin (Figure 6.13); rate constants of their Diels–Alder reactions with TCNE in toluene at 25 °C, and their gas phase ionization energies (IE).

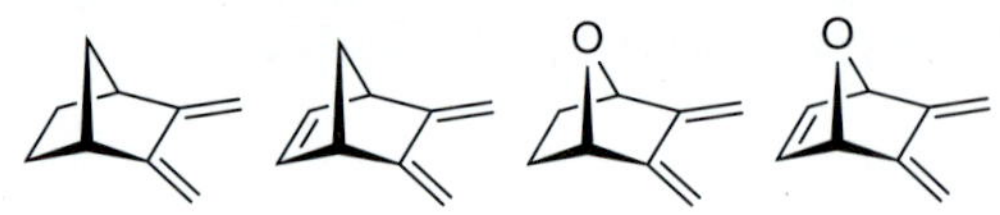

	275	**276**	**277**	**278**
k_{rel} ($^1O_2$)	(1.0)	0.12	0.07	<0.001
k^{II} (TCNE)	690	53	1.5	0.004×10^{-4} $M^{-1}\,s^{-1}$
IE (gas phase)	8.41	8.48	8.60	8.87 eV

parasitic worms from plants, animals, and the human body. At low temperatures, cyclopentadiene and furan undergo Diels–Alder reactions during low temperature-photosensitized (rose bengal) oxygenation, giving the unstable endoperoxides **271** and **273**, respectively. Reduction of **271** with thiourea at −30 °C gives *cis*-cyclopent-4-ene-1,3-diol (**272**). Unstable endoperoxide **273** is rapidly isomerized into **274** at room temperature (Scheme 6.73). It can be reduced with SO_2 into succindialdehyde [1008].

The mechanisms (one-step concerted pathways vs. multistep pathways, which intermediates?) of photo-oxygenation of unsaturated compounds is still a matter of debate. Thermochemical data (Scheme 6.71) gives an estimate of 20.7 kcal mol^{-1} for the endothermicity of the formation of the alkyl/peroxyl 1,4-diradical **269**, the adduct of $^3O_2$ to isobutylene. As $^1O_2(^1\Delta_g)$ is 22.4 kcal mol^{-1} higher in energy than ground-state $^3O_2$ (Figure 6.12), the formation of **269** by addition of $^1O_2(^1\Delta_g)$ to isobutylene (Scheme 6.71) must be exothermic. This must also be the case when a conjugated diene adds to $^1O_2(^1\Delta_g)$ to give the corresponding allyl/peroxyl diradical (see Scheme 6.75: **281** is relatively more stable than **269** because of allylic π-conjugation). Thus, it is not a surprise that the gas phase kinetics for the Diels–Alder reactions of $^1O_2(^1\Delta_g)$ (generated by discharge flow using a microwave generator) to cyclopentadiene (E_a = 3.9 kcal mol^{-1}), cyclohexa-1,3-diene (E_a = 5.5 kcal mol^{-1}), furan (E_a = 5.3 kcal mol^{-1}), and 2-methylfuran (E_a = 3.9 kcal mol^{-1}) have Arrhenius activation energies that are similar and very low. The Arrhenius *A* factor of these four reactions are comprised between 1.5×10^8 and 2×10^8 M^{-1} s^{-1}, typical for bimolecular processes [1009]. For other kinetic studies in solution, it is found that the reactions of $^1O_2(^1\Delta_g)$ with several conjugated dienes (e.g. $k = 7 \times 10^6$ $M^{-1}\,s^{-1}$ for cyclohexa-1,3-diene in MeOH) have rate constants that do not vary significantly with temperature and do not depend on the ionization potential of the dienes. This suggests that the reactivity of singlet oxygen is entropy controlled in these cases. This is consistent either with a one-step concerted mechanism (does not have to be synchronous) or with a two-step mechanism with the rate-determining step leading to an intermediate such as a vinyl-1,4-diradical (**281**) or a vinylperepoxide (**282**, Scheme 6.75) [1010], which would rearrange rapidly into products [1011–1015]. With exocyclic dienes **275–278** (Table 6.5) that are very similar in terms of geometry, flexibility, and bulk, the relative rates of formation of the corresponding endoperoxides vary significantly and correlate with the second-order rate constants measured for the Diels–Alder reactions of ethylenetetracarbonitrile (tetracyanoethylene: TCNE) with these dienes. The relative rate constants correlate with the IEs of the dienes, i.e. the lower the IE, the faster the cycloadditions of both TCNE and singlet oxygen, which is expected for cycloadditions (Section 5.3.10) with "normal electron-demand" ($^1O_2(^1\Delta_g)$ is highly electrophilic, like TCNE). It was verified that none of the dienes **275–278** inhibits the photo-oxidation of the others. Because the *s-cis*-butadiene moiety in these dienes are all planar and have the same distance separating their terminal CH_2 groups, one cannot invoke steric and/or geometric factors that would affect their reactivity. If the photo-oxygenations should follow one-step concerted mechanisms, one would expect an energy barrier $\Delta^{\ddagger}H > 0$ for these reactions [1016]. Alternatively, the encounter complex or exciplex diene + $^1O_2(^1\Delta_g)$ undergoes competing physical deactivation into $^3O_2$ + diene (rate constant k_{q1}) and electron transfer (k_{ET}) into radical cation/radical anion intermediate of type diene$^{•+}$/$O_2^{•-}$. The latter undergoes competitive chemical deactivation into diene + $^3O_2$ (k_{q2}) and leads to the formation of products of photo-oxygenation directly (k_{r1}), or via an intermediate (k_i, e.g. a diradical) for which chemical deactivation (k_{q3}) and product formation (k_{r2}) compete (Scheme 6.74). It therefore appears that more the diene of type **275–278** is able to donate an electron or

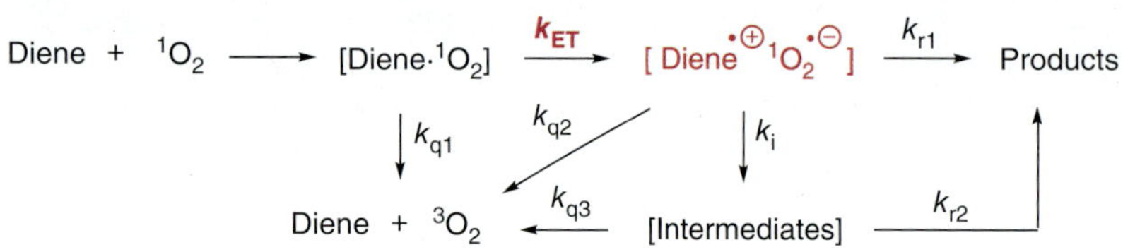

Scheme 6.74 Possible general mechanism for the photo-oxygenation of a 1,3-diene (see below for the nature of the possible intermediates).

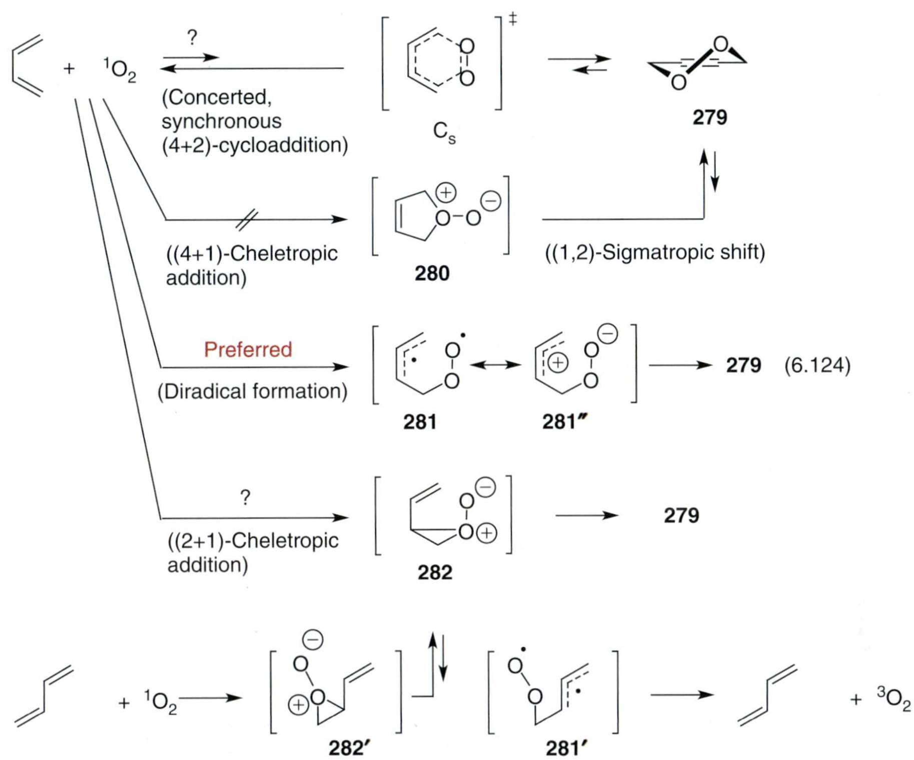

Scheme 6.75 Possible mechanisms for the (4+2)-cycloaddition of singlet oxygen to butadiene in the gas phase.

stabilize the diene$^{\bullet+}$/$O_2^{\bullet-}$ intermediate, the larger is k_{ET} and the more endoperoxide formation competes ($k_{r1}+k_{r2}$) with physical and chemical deactivation ($k_{q1}+k_{q2}+k_{q3}$).

Acyclic conjugated dienes that cannot exist in their *s-cis*-conformation react with $^1O_2(^1\Delta_g)$ in $[\pi^2+\pi^2]$-cycloadditions (Section 6.9.5) giving 1,2-dioxetanes and in dioxa-ene reactions (Section 6.9.4) producing hydroperoxides.

Quantum mechanical calculations by Liwo and coworkers on the reaction of butadiene + $^1O_2(^1\Delta_g) \rightarrow$ 3,6-dihydro-1,2-dioxin (**279**) in the gas phase have suggested that the concerted, synchronous $[\pi^4+\pi^2]$-cycloaddition and a two-step mechanism involving the (4+1)-cheletropic addition forming zwitterionic 2,5-dihydrofuran 1-oxide (**280**) are unlikely (Scheme 6.75). The preferred mechanism (6.124) appears to be the formation of diradical **281** from *s-cis*-butadiene in the rate-determining step. The calculated transition state for this process is only 1 kcal mol^{-1} lower in energy than the calculated C_s-symmetrical transition state for a concerted, synchronous Diels–Alder reaction [1017]. Earlier calculations by Houk and coworkers predicted the latter mechanism to be possible [1018, 1019], and others by Dewar and Thiel predicted a preferred mechanism involving vinylperepoxide intermediate **282** arising from the (2+1)-cheletropic addition of singlet oxygen to *s-cis*-butadiene, and **282′** from *s-trans*-butadiene. The latter intermediates can equilibrate, which is not as easy for corresponding diradicals **281** and **281′**. Thus, the perepoxide mechanism could explain the formation of endoperoxides from *s-trans*-1,3-dienes that cannot equilibrate readily with their *s-cis* conformers [1020]. The diradical mechanism (6.124) is supported by the observation of nonstereospecific reactions of singlet oxygen with acyclic 1,3-dienes (see Scheme 6.76) [1021, 1022]. The formation of diradical intermediate **281′** that cannot be cyclized readily into endoperoxide **279** is one possible mechanism for the chemical deactivation of $^1O_2(^1\Delta_g)$ (Scheme 6.75). Quantum calculations on the gas phase reaction $^1O_2(^1\Delta_g)$ + cyclohexa-1,3-diene → 2,3-dioxabicyclo[2.2.2]oct-5-ene also predicted a two-step mechanism with the formation of an allyl/peroxyl diradical intermediate similar to **281**. One calculates an activation enthalpy of 6.5 kcal mol^{-1} for this reaction, which is nearly the same (E_a = 5.5 kcal mol^{-1}) as the experimental value (see above) [1023].

The photo-oxygenations of (*E,E*)-, (*E,Z*)-, and (*Z,Z*)-2,4-dimethylhexa-2,4-diene (**283**) in CD_2Cl_2 give mixtures of endoperoxides *cis*-**284** and *trans*-**284** (nonstereoselective (4+2)-cycloaddition of $^1O_2(^1\Delta_g)$), allylic hydroperoxides **286** (products of dioxa-ene reactions of $^1O_2(^1\Delta_g)$), and acetaldehyde + (*E*)- and (*Z*)-crotonaldehyde resulting from concurrent (2+2)-cycloadditions of $^1O_2(^1\Delta_g)$ forming the unstable dioxetanes **285** that undergo (2+2)-cyloreversions

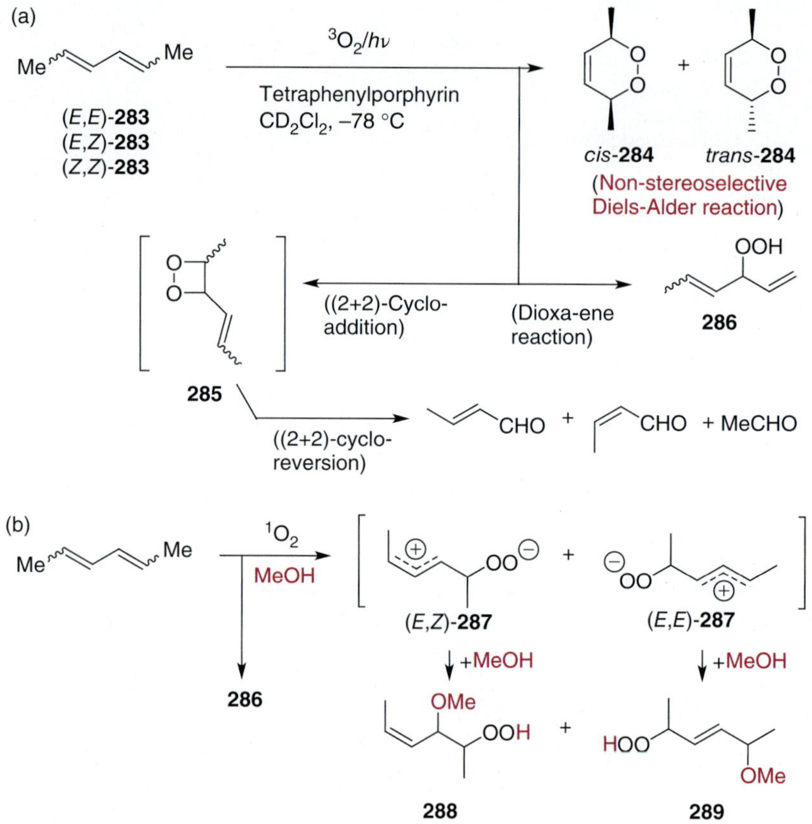

Scheme 6.76 Photo-oxygenations of acyclic 1,3-dienes are not stereoselective and give several products resulting from (a) concurrent (4+2)-, (2+2)-cycloadditions and dioxa-ene reactions; (b) in polar solvent, allyl/peroxyl diradical intermediates equilibrate with their zwitterions.

(Scheme 6.76a). When run in MeOH, the photo-oxygenations of dienes **283** give further products **288** and **289** of 1,2- and 1,4-addition of MeOH to the dienes resulting from the nucleophilic quenching of zwitterionic intermediates **287** (Scheme 6.76b) [1021, 1024–1028]. In the case of the photo-oxygenation of 2,4-dimethylpenta-1,3-diene, the ratio of hydroperoxide (resulting from a dioxa-ene reaction) and of endoperoxide (resulting from a Diels–Alder reaction of singlet oxygen) increases with the solvent polarity [1029].

The importance of charge transfer during photo-oxygenation is shown in the (4+2)-cycloaddition (6.125) of 1,4-dimethylnaphthalene (**290**) to $^1O_2(^1\Delta_g)$ that gives endoperoxide **291** (with rate constant k_r) concurrently with singlet oxygen deactivation (rate constant k_q). In THF (nonpolar solvent), $k_r = 0.3 \times 10^4\ M^{-1}\ s^{-1}$ and $k_r+k_q = 2.4 \times 10^4\ M^{-1}\ s^{-1}$. In EtOH, CF_3CH_2OH, and formamide (more polar solvents), k_r = 1.5, 24, $124 \times 10^4\ M^{-1}\ s^{-1}$, and $k_r + k_q$ = 6.6, 90, $324 \times 10^4\ M^{-1}\ s^{-1}$, respectively, are measured. This is consistent with a two-stage mechanism implying the formation of solvated exciplex [**290**·$^1O_2$] with charge transfer character [1030].

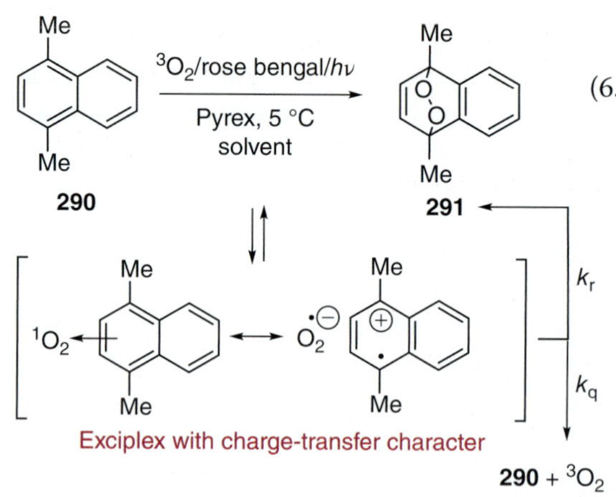

(6.125)

The latter hypothesis is corroborated by the stereoselectivity observed for the photo-oxygenations (6.126) in which hydroxy and hydroxymethyl groups lead to contrasteric lateral control of the (4+2)-cycloadditions. These groups stabilize the *syn* approach of $^1O_2(^1\Delta_g)$ by hydrogen bridging with the negatively charged oxygen centers of the exciplex and/or charge transfer complex, whereas other groups impede the *syn* approach of $^1O_2(^1\Delta_g)$ for steric reasons [1031].

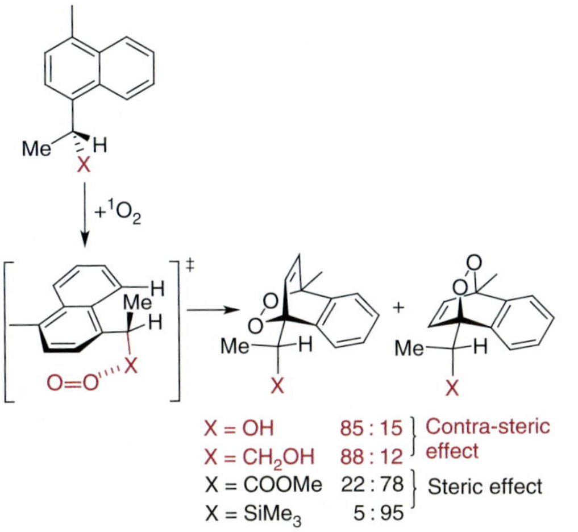

(6.126)

An example of synthetic application of endoperoxides is given with the synthesis of (±)-crotepoxide, a natural compound with anticancer properties. The key reaction (6.127) is the thermal isomerization of the endoperoxide into a *cis*-diepoxide [1032].

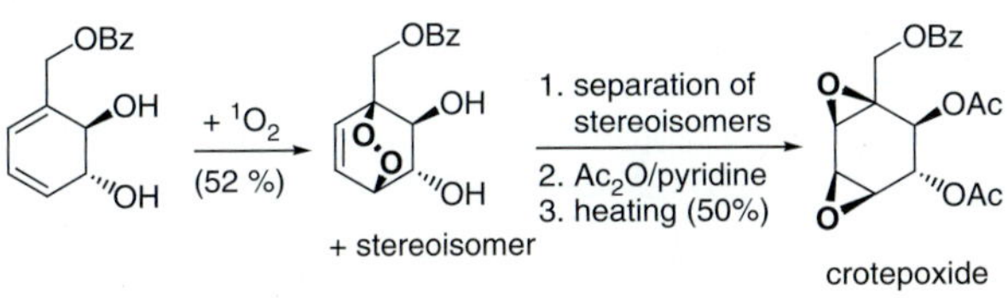

(6.127)

As seen above, photo-oxygenation of furan (and derivatives) at low temperature provides the corresponding endoperoxides. At room temperature and in nonprotic apolar solvents, the latter undergoes isomerization into the corresponding *cis*-diepoxides and epoxylactones as shown with **273** → **292** + **293** (Scheme 6.77). In alcohols and at room temperature, the endoperoxides give lactone acetals as exemplified with **273** + MeOH → **294** [1033]. Triphenylphosphine reduces **273** into malealdehyde ((*Z*)-but-2-enedial). In the case of monocylic endoperoxide **295** resulting from the photo-oxygenation of diene **275** [1034], the thermal isomerization is slower than isomerization **273** → **292** and produces *trans*-diepoxide **297**. No *cis*-diepoxide is formed, probably because of electrostatic repulsion between the two oxyradical moieties in intermediate **296**. In the presence of tertiary amines, endoperoxides undergo Kornblum–DeLaMare rearrangements giving corresponding γ-hydroxyenones [1035, 1036]. The same isomerizations are catalyzed by $[Rh(CO)_2Cl]_2$, as shown with the conversion of **295** into **298**. Treatment of **298** with acetic acid induces elimination of water and formation of furan **299** (Scheme 6.77).

Using an enantiomerically pure chiral base, Kornblum–DeLaMare rearrangements can be enantioselective. The first examples were reported in 1980 by Hagenbuch and Vogel with the rearrangements of **295** and **300**, compounds in which the stereogenic centers of the *meso*-endoperoxides (bridgehead center C(1)) are three bonds away from the enantiotopic protons the chiral base catalyst has

Scheme 6.77 Examples of reactions of endoperoxides.

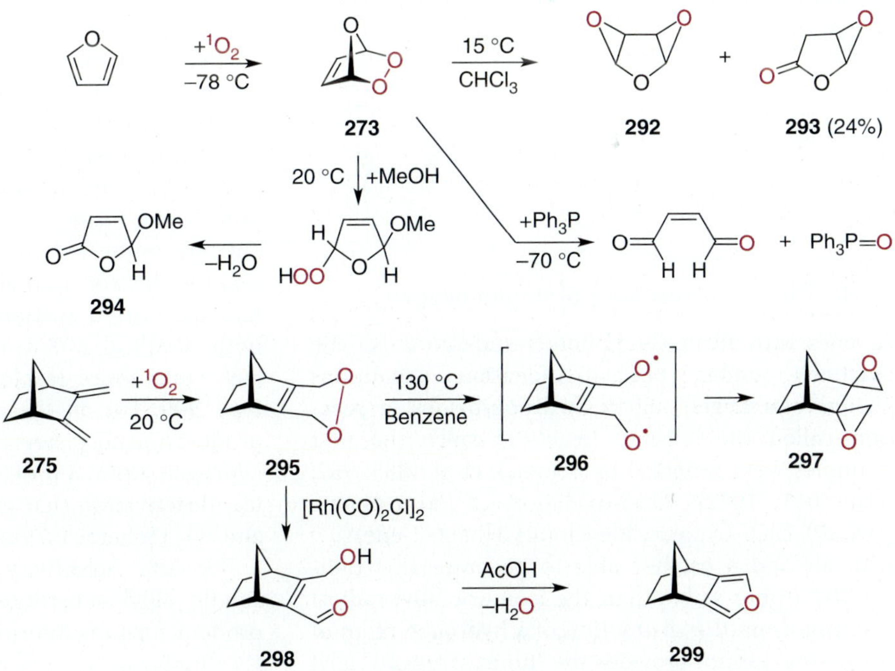

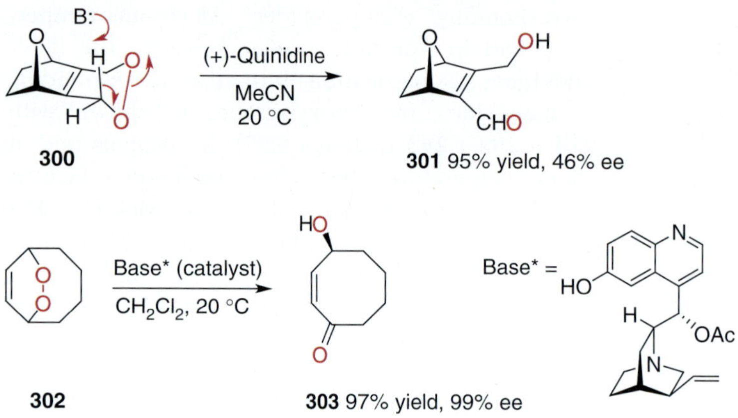

Scheme 6.78 Examples of enantioselective Kornblum–DeLaMare rearrangements.

to recognize [1037]. In 2006, Toste and coworkers reported excellent yields and enantioselectivities for the Kornblum–DeLaMare rearrangements of bicyclic endoperoxides as illustrated with **302** → **303** (Scheme 6.78) [1038].

Problem 6.56 What is the product of photo-oxygenation of 2,3,5,6-tetramethylidene-7-oxabicyclo [2.2.1]heptane? [1016]

Problem 6.57 What is the major product of the following reaction sequence? [1039]

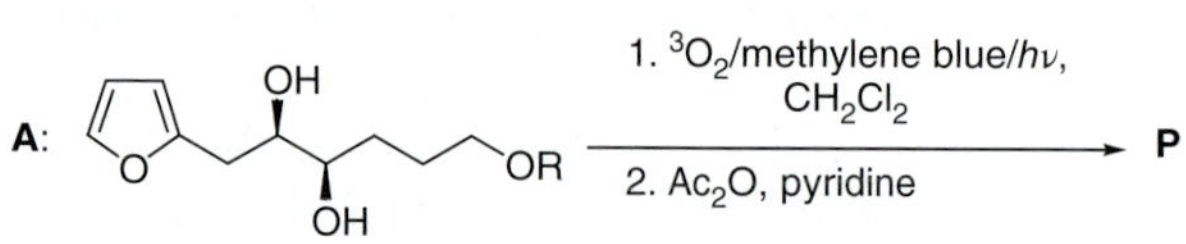

Problem 6.58 Propose a synthetic sequence to convert **A** into **P** [1040].

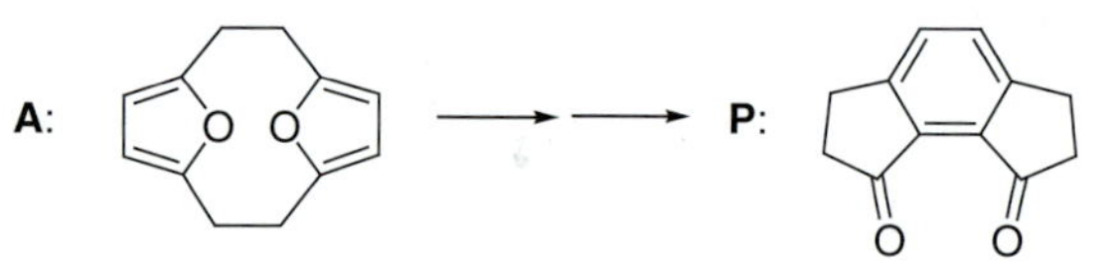

6.9.4 Dioxa-ene reactions of singlet oxygen

Alkenes with allylic C—H bonds undergo dioxa-ene reactions under photo-oxygenation conditions giving "rearranged" allylic hydroperoxides, a reaction called the Schenck reaction, since the first examples were reported by Schenck et al. [883, 932, 933, 1041, 1042]. Auto-oxidation of (+)-α-pinene ((1*R*,5*R*)-2,6,6-trimethylbicyclo[3.1.1]hept-2-ene) with air and a catalyst able to generate allyl radical **304** (more stable than the isomeric allyl radical resulting from the abstraction of a hydrogen atom of the methyl group) provides the "nonrearranged" allyl hydroperoxide **305**. In contrast, under photochemical sensitization conditions, (+)-α-pinene is oxidized into the "rearranged" allyl hydroperoxide **306** [1008].

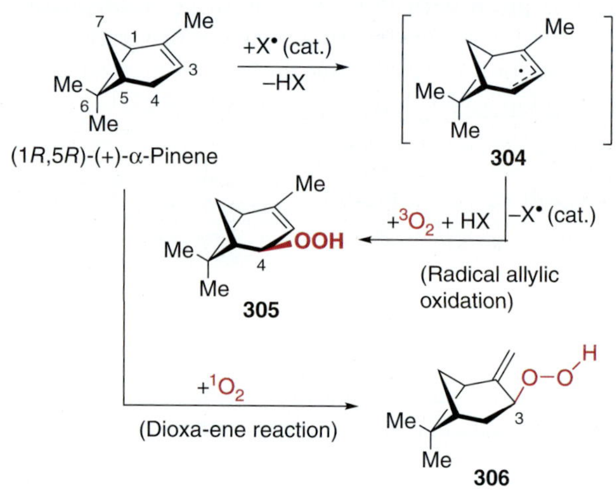

The initially proposed concerted mechanism (transition state of type **307**) [1043] has been challenged by mechanisms involving intermediates on the route to the hydroperoxides. As suggested by thermochemical estimations, the 1,4-alkyl/peroxyl diradicals of type **308** might form in an exothermic process that is entropy controlled [1044, 1045]. In a polar medium and for electron-rich alkenes, this intermediate may take the form a zwitterion **308′** [1046]. At the limit, both diradical **308** and zwitterion **308′** could collapse into more stable perepoxide intermediate of type **309**. For all these possible intermediates, the product-forming step is the migration of an allylic hydrogen atom, a process that competes with chemical deactivation that gives back the starting alkene and $^3O_2$ (Scheme 6.79).

The site selectivity of photo-oxygenations of simple alkyl substituted alkenes suggest that the product-forming intermediates are not freely rotating 1,4-diradicals or zwitterionic species for which

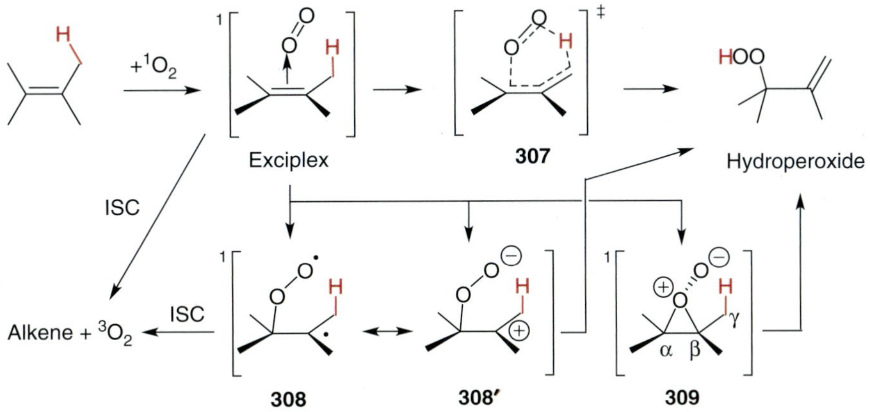

Scheme 6.79 Possible mechanisms for the dioxa-ene reaction of alkenes.

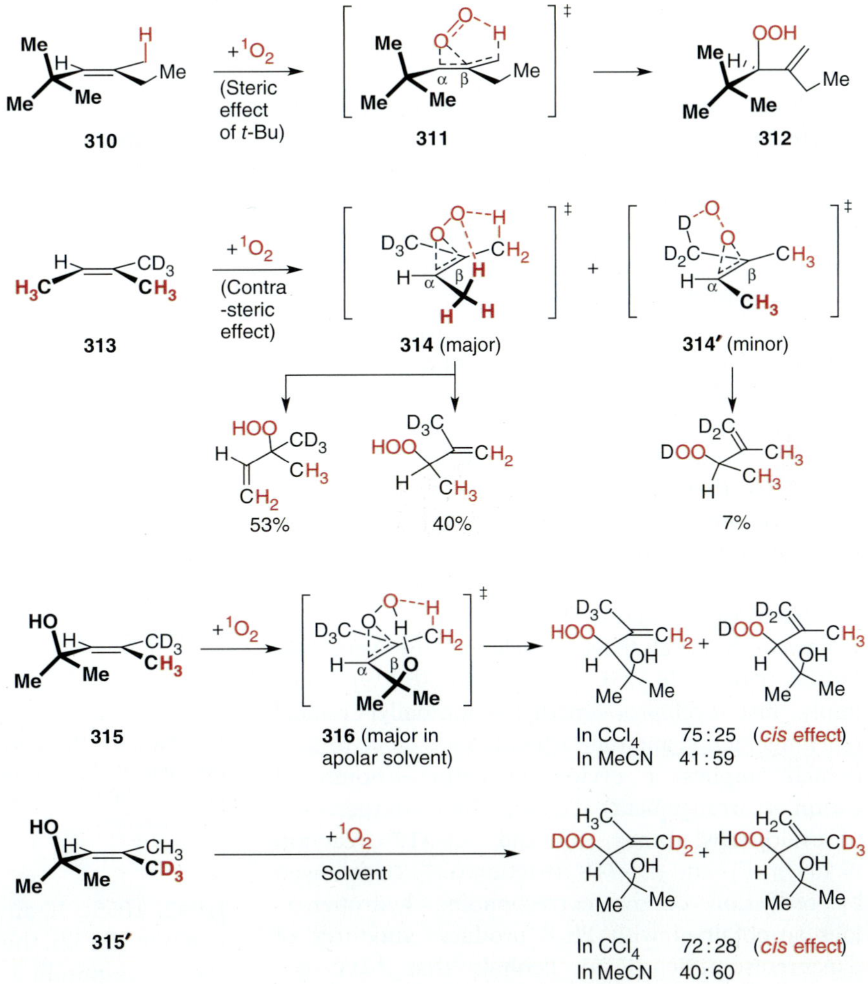

Scheme 6.80 The "*cis* effect" confirms that intermediates or transition states preceding formation of the hydroperoxides are bridged species that can be represented as "symmetrical" or distorted perepoxides that possess an important partial negative charge at the terminal oxygen center, rendering possible contrasteric hydrogen-bridging interactions.

the most stable tertiary alkyl radical or carbenium ion moiety would be formed. For instance, the ene-reaction **310** + $^1O_2 \rightarrow$ **312** implies exclusively migration of a H-atom from the methyl group and not from the ethyl group. The bulk of the *t*-butyl group favors the product-determining intermediate or transition state **311**. In contrast, reaction **313** + $^1O_2$ involves an intermediate or transition state **314** that is favored over **314′**. The polar species **314** that possesses a relatively important negative charge at its terminal oxygen center is stabilized by two O···H—C bridging interactions, whereas for **314′**, only one such hydrogen bridging is possible. The contrasteric effect that stabilizes **314** is called the "*cis*-effect." Species **311, 314**, and **314′** can be either "symmetrical" perepoxides with two C—O bonds having nearly the same bond length, or not. At the limit, the perepoxides can be distorted becoming partially bridged species resembling diradical ↔ zwitterion intermediates or transition states of type **308** ↔ **308′** in which the terminal oxygen center is not in the C(β)—C(α)—O plane. The partial bridging between the internal oxygen center and the β-carbon center avoids the formation of less stable secondary carbon-centered radical or cation and permits the two O···H—C stabilizing interactions. As found for the Diels–Alder reactions (6.126) of singlet oxygen with naphthalene derivatives, the dioxa-ene reactions of allylic alcohols **315** and **315′** are *syn*-selective with respect to the alcoholic moiety as they favor the "*cis* effect" by hydrogen bridging of the intermediate or transition state **316** that produces the allylic hydroperoxides (Scheme 6.80) [1047–1049]. Aryl groups and heteroatom-based moieties (including those involving S, Si, Sn, and Ge) can influence the regioselectivity of the dioxa-ene reaction [1049–1052].

Newcomb et al. have developed the very fast radical clocks (6.128) [1053] and (6.129) (they imply fast cyclopropylmethyl → homoallyl radical rearrangements) and the carbenium ion clock (6.130) (which implies a cyclopropylmethyl → homoallyl cation rearrangement) [1054]. Photo-oxygenation of cyclopropylalkenes (*E*)- and (*Z*)-**317** (reaction (6.131)), (*E*)- and (*Z*)-**318** (reaction (6.132)) followed by reduction of the corresponding hydroperoxides so-obtained with Ph_3P produces mixtures of diastereoisomeric allylic alcohols that have not undergone any cyclopropyl/homoallyl radical or cation rearrangement (Scheme 6.81). These results are best explained by invoking symmetrical or distorted perepoxides as transition states, rather than intermediates, in the product-determining steps

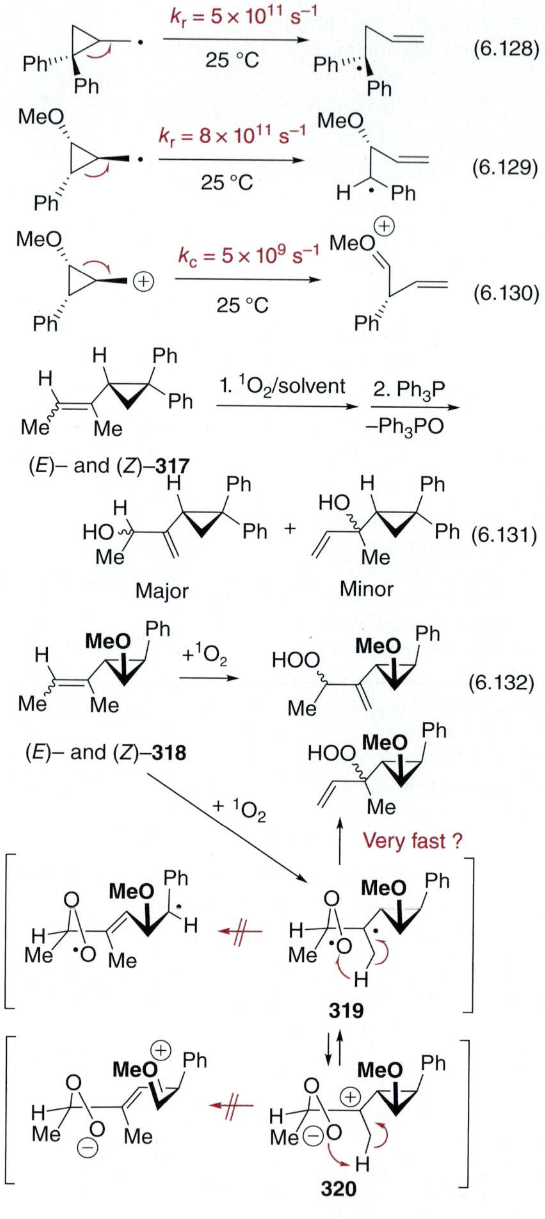

Scheme 6.81 The dioxa-ene reaction of singlet oxygen does not generate freely rotating diradical intermediates but it involves either "symmetrical" or "distorted" perepoxide intermediates.

[1042, 1055]. If diradical **319** should be formed in reaction (6.132), the H-atom migration that produces the corresponding hydroperoxide must be about 100 times faster (limit of detection of rearranged product) than the cyclopropyl → homoallyl radical rearrangement and would have a rate constant reaching 8×10^{13} s^{-1}, which is almost as fast as a C—H bond vibration (infrared spectroscopy: $1/\lambda$ = 3000 cm^{-1},

$\nu = 9 \times 10^{13}\ s^{-1}$). If zwitterion **320** were to be formed in reaction (6.132), migration of the proton to its terminal oxygen center must reach a rate constant of $5 \times 10^{11}\ s^{-1}$, which is possible. Thus, intermediate **320** can be seen as a distorted perepoxide that undergoes fast intramolecular proton migration.

Quantum mechanical calculations (gas phase) and kinetic deuterium isotopic effects on the dioxa-ene reaction of tetramethylethylene support a concerted "two-step no-intermediate" mechanism [1056]. In this mechanism, there is a single transition state that resembles the perepoxide geometrically, but the forming CO bonds are much longer than expected in an intermediate. The product ratios and isotope effects are determined after the transition state and can be understood by molecular dynamics simulations. Such "ambimodal" transition states have been observed for a variety of reactions [1056].

Other calculations for the same reaction in water, DMSO, and cyclohexane, using a similar method, but including molecular mechanics (MM) and molecular dynamics (MD) averaging for solvation energies, support the nature of the rate-determining transition state but predict a two-step mechanism with the formation of a perepoxide intermediate that gives rapid hydrogen abstraction to form products [1057]. In the case of the photo-oxygenation of (*E*)-cyclooctene, quantum mechanical calculations also find a perepoxide intermediate **322**. The (*E*)-cyclooctene imposes a substantial strain in the transition state for allylic hydrogen migration, so that the perepoxide is now a minimum. Intermediate **322** is formed through a polarized diradical intermediate **321** ↔ **321′** that leads to the alkene isomerization (Scheme 6.82) [1058]. The photo-oxidation (Na lamp, tetraphenylporphyrin (TPP) as sensitizer, in $CHCl_3$, in the presence of 2,6-di(*t*-butyl)phenol + pyridine as radical scavenger) of (*E*)-cyclooctene produces a mixture composed of the expected hydroperoxide **324**, (*Z*)-cyclooctene, *cis*- and *trans*-9-oxabicyclo[6.1.0]nonane (*cis*-**323**, *trans*-**323**), and cyclooct-2-enol (**325**). Fragmentation of intermediate **321** ↔ **321′** produces (*Z*)-cyclooctene and $^3O_2$. Epoxides *cis*- and *trans*-**323** result probably from oxygen transfer from **321** to (*Z*)- and (*E*)-cyclooctene, respectively, with liberation of allylic alcohol **325**. In the presence of $P(OPh)_3$, perepoxide **322** is reduced into *trans*-**323**, and hydroperoxide **324** is reduced into alcohol **325**. A small amount of *n*-octadial also forms that results from the competitive (2+2)-cycloreversion of dioxetane **326** arising from the concurrent (2+2)-cycloaddition of $^1O_2$ to (*E*)-cyclooctene [1059, 1060].

Dragoco in Germany and Firmenich in Switzerland produce Rose Oxide, an important fragrance, by photooxygenation of (−)-β-citronellol (Scheme 6.83) [1061]. The reaction can be done in the dark by catalytic (Na_2MoO_4) decomposition of H_2O_2 [1062].

Allylic hydroperoxides can be isomerized into their epoxy alcohols under Lewis acid catalysis. An example is given with the enantioselective photo-oxygenation (TPP as sensitizer, CH_2Cl_2, 0 °C) of achiral alkene **327** into epoxyalcohol (*S*)-**328** [1063]. On mixing (+)-diethyl tartrate ((+)-DET) with Ti(O-*t*-Bu)$_4$, a homochiral Lewis acid catalyst forms that induces enantioselective epoxidation of the alkene moiety (Katsuki–Sharpless asymmetric epoxidation of allylic alcohols [1064, 1065] described in Section 3.7.7).

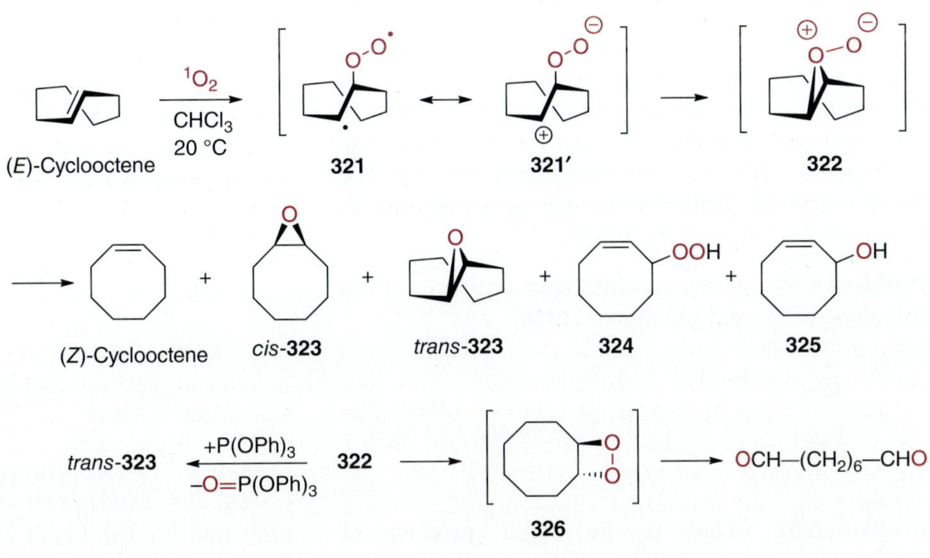

Scheme 6.82 The photo-oxygenation of (*E*)-cyclooctene involves a polarized 1,4-diradical of the type alkyl/peroxyl and a perepoxide as intermediates.

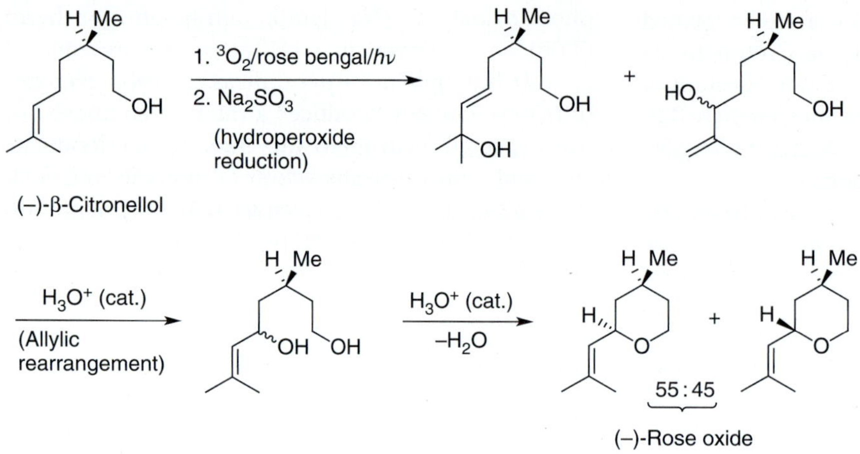

Scheme 6.83 Industrial synthesis of rose oxide.

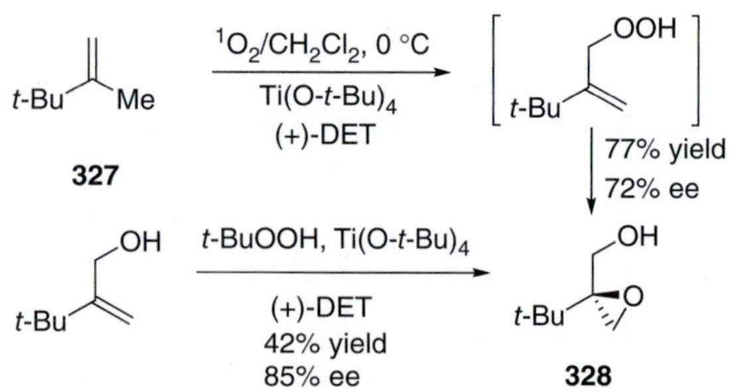

Problem 6.59 What are the products of photo-oxygenation of adamantylideneadamantane (**A**) in the presence of PhSOMe? [1066]

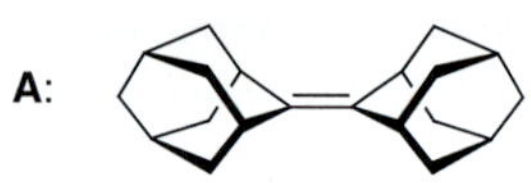

Problem 6.60 Photo-oxygenation of 1,2,3,4,5,6,7,8-octahydronaphthalene (**A**) gives a product **B** that is esterified with PhCOCl/pyridine into **C**. Product **C** undergoes a Criegee's rearrangement into **Q**. The latter reacts in alkaline water giving a diketone **P**. What are **B**, **C**, **Q**, and **P**? [1067]

Problem 6.61 What are the major products of the following photo-oxygenations? [1048]

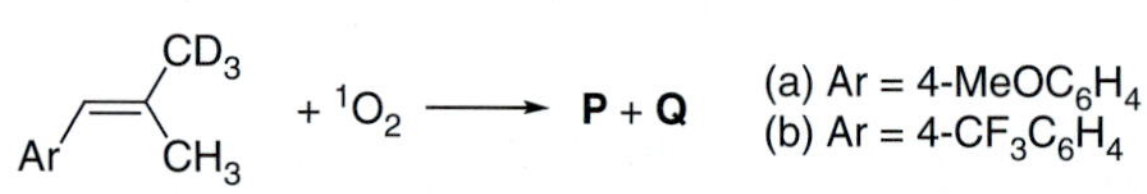

Problem 6.62 What is the major product of photo-oxygenation of the enol trimethylsilyl ether derived from methyl isobutyrate? [1068]

Problem 6.63 Explain the following reaction sequence [1069].

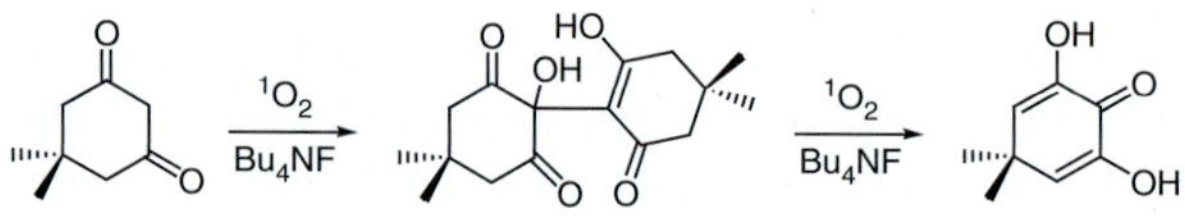

6.9.5 (2+2)-Cycloadditions of singlet oxygen

In 1969, Kopecky and Mumford reported the first example of 1,2-dioxetane that was obtained by base-induced cyclization of a β-bromohydroperoxide. They obtained 3,3,4-trimethyl-1,2-dioxetane, a compound that decomposes thermally with light emission (chemiluminescence, Section 6.11.3) [1070]. In 1970, Bartlett and Schaap (reaction (6.133)) [1071] and Mazur and Foote (reaction (6.134)) [1072] independently reported the isolation of 1,2-dioxetanes from the (2+2)-cycloadditions of $^1O_2$ to alkenes. As seen above (Scheme 6.76 and 6.82), and depending on the nature of the unsaturated organic compound, the (2+2)-cycloaddition may compete with other reactions of singlet oxygen. It is especially favored with electron-rich alkenes [1073–1079] and dienes [1028, 1080], or with alkenes that cannot undergo ene-reactions (no allylic C—H or C—metal) [1081].

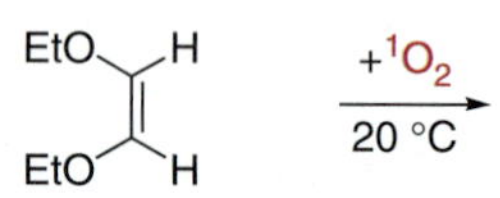

(6.133)

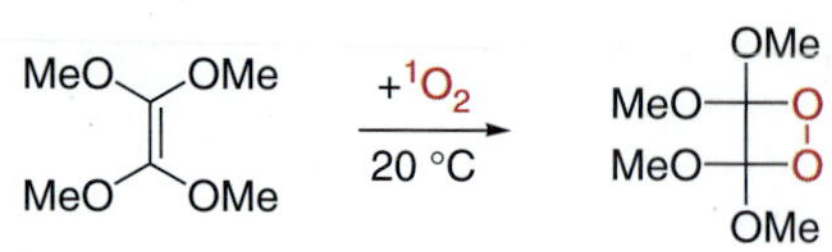

(6.134)

As with cycloadditions and ene-reactions, concerted and nonconcerted mechanisms have been proposed for the (2+2)-cycloadditions of $^1O_2(^1\Delta_g)$. For ethylene and vinyl ether, quantum mechanical calculations favor the formation of 1,4-diradical intermediates of type **308** (see Scheme 6.79) in the rate-determining step that involves ring closures into the corresponding 1,2-dioxetanes (reaction (6.135)) faster than formation of the corresponding perepoxides. If formed, the latter equilibrate back into the initial diradical intermediates, as they are separated by a large energy barrier to their isomerization into the 1,2-dioxetanes [1055, 1082]. In the case of butadiene, the same quantum mechanical calculations also predict the formation of an allyl/peroxyl diradical (**281**, see Scheme 6.75) that undergoes fast ring closure into 3-vinyl-1,2-dioxetane (**329**): the 2-vinylperepoxide (**282**) [966] is not an intermediate [1083]. Kinetic deuterium isotopic effects of the competitive Diels–Alder reaction, ene-reaction, and (2+2)-cycloaddition of singlet oxygen to 2,5-dimethylhexa-2,4-diene have been interpreted in terms of the formation of a common perepoxide intermediate [1084].

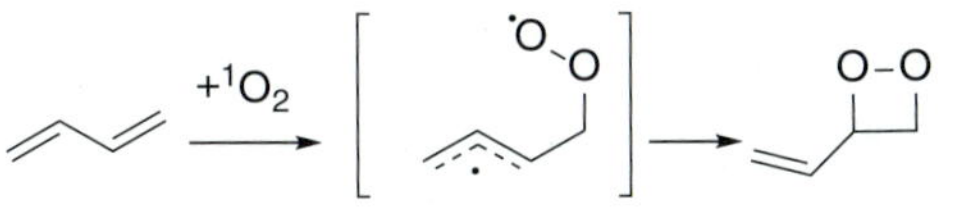

281 **329** (6.135)

A mechanism (6.136) involving electron transfer from the alkene to $^1O_2(^1\Delta_g)$ with the formation of a radical cation and $O_2^{\bullet-}$ has been proposed [1072] and subsequently refuted for the photo-oxidation of 2,3-diaryl-1,4-diox-2-enes [1076]. With electron-rich alkenes such as enamines [1085] and indoles [1086], it may occur and involve the formation of zwitterionic intermediates [1055, 1087] that can be viewed as charge-transferred limiting structures (e.g. **308′**, see Scheme 6.79) of corresponding alkyl/peroxyl 1,4-diradicals. For the photo-oxygenation of electron-rich substrates (M = alkenes, 1,3-dienes, sulfides) using 5,10-dicyanoanthracene as a sensitizer, a nonsinglet oxygen mechanism (6.137) has been proposed [1088]. Formation of radical anion $DCA^{\bullet-}$ has been proven by ESR spectroscopy [1089].

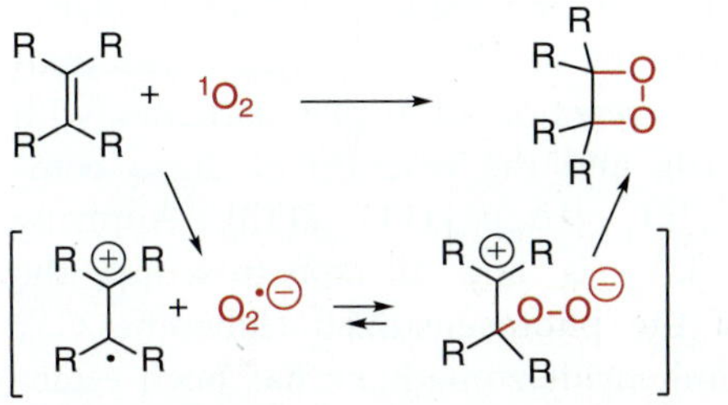
(6.136)

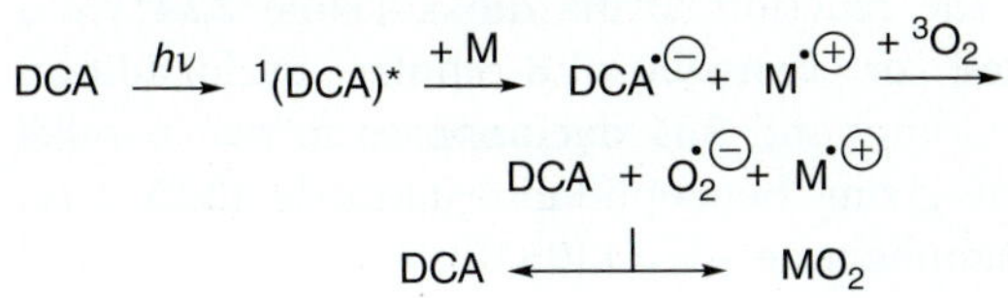

(6.137)

Pyrroles and indoles react with singlet oxygen in competitive (2+2)- and (4+2)-cycloadditions giving adducts that are converted into all kinds of synthetic intermediates [1090]. Photo-oxygenation of ene-carbamate (*Z*)-**330** gives a 80 : 20 mixture of 1,2-dioxetane (1′*S*,2′*S*,4*R*)- and (1′*R*,2′*R*,4*R*)-**331** resulting from a (2+2)-cycloaddition as the major product. A concurrent ene-reaction produces a hydroperoxide as the minor product [1091, 1092].

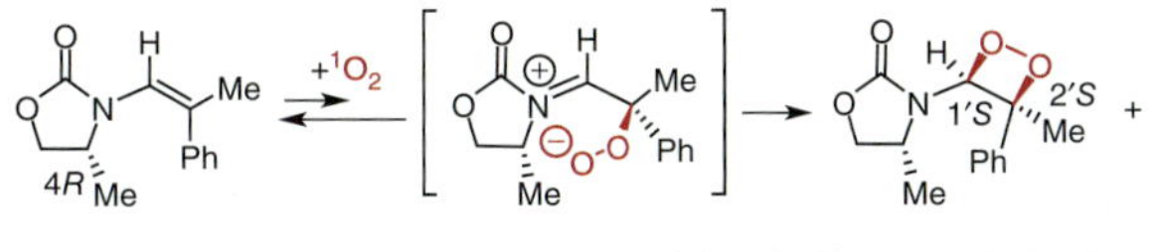
(*Z*)-**330** (1′*S*,2′*S*,4*R*)-**331** + (1′*R*,2′*R*,4*R*)-**331** 80 : 20

Problem 6.64 What are the main products of photo-oxygenation of **A**? [1093]

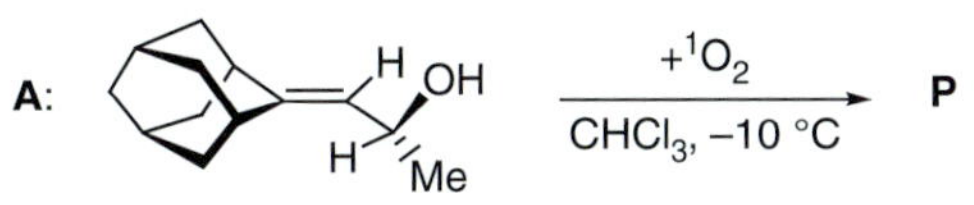

Problem 6.65 *Trans*-resveratrol (**A**) is an antioxidant agent found in several plants and wine, with several beneficial health effects such as cardiovascular modulating effect and antiaging and antitumor effects. The photo-oxygenation of **A** in CD_3CN using methylene blue as a sensitizer gives two aldehydes (**P1** and **P2**) and moracin M after heating the reaction mixture to 80 °C. Explain these transformations [1094].

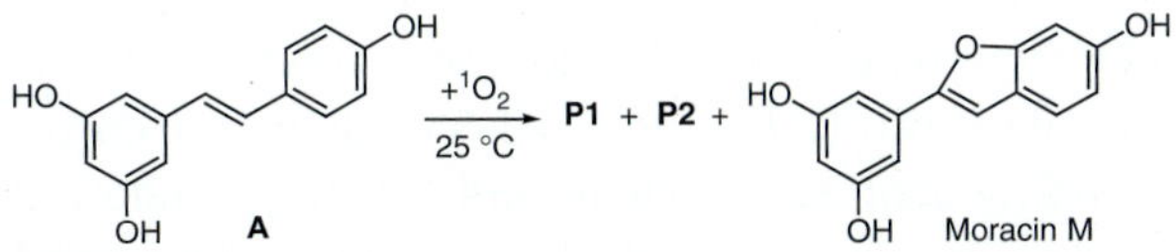

6.9.6 1,3-Dipolar cycloadditions of singlet oxygen

Nitrones [$R^1R^2C^{(-)}$—$N^{(+)}(R^3)$=O] react with $^1O_2(^1\Delta_g)$ to give addition products [1095]. Aziridines are photoisomerized into azomethine ylides (Section 5.3.16, Table 5.5) and react with $^1O_2(^1\Delta_g)$ (rose bengal as the sensitizer) forming 1,2,4-dioxazolidines (e.g. reactions (6.138) and (6.139)) [1090]. The latter are decomposed into a variety of products as illustrated in Scheme 6.84 [1096–1098].

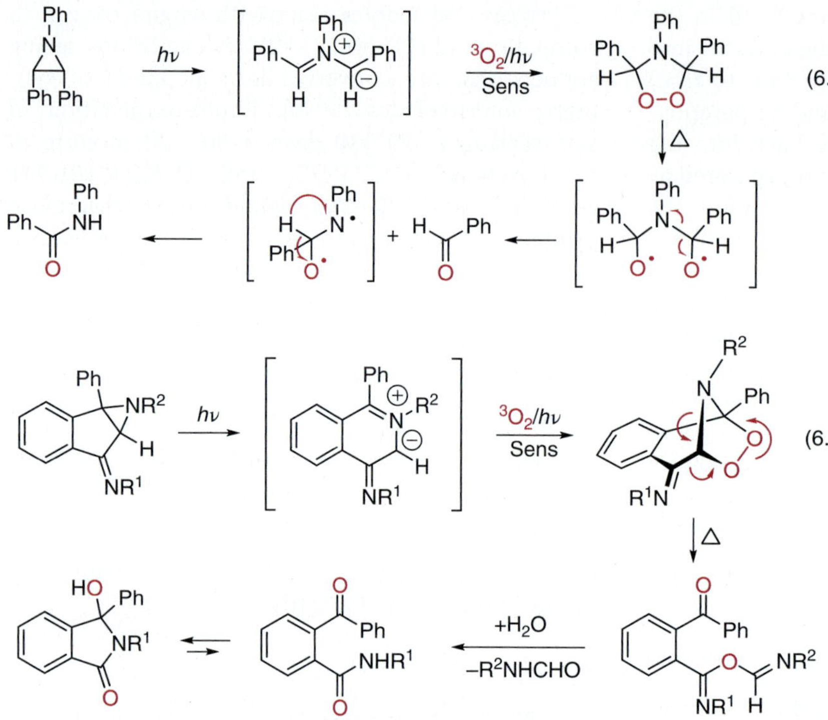

(6.138)

(6.139)

Scheme 6.84 Example of photo-oxygenation of aziridines involving singlet oxygen dipolar addition to azomethine ylides.

DCA (5,10-dicyanoanthracene)-sensitized photo-oxygenation of *cis*- and *trans*-2,3-diphenylaziridine in MeCN gives exclusively *cis*-3,5-diphenyl-1,2,4-dioxazolidine. Photo-oxidation of *N*-alkyl-substituted 2,3-diphenylaziridines provides both isomeric peroxides *cis*- and *trans*-**333**. The mechanism proposed for these reactions (Scheme 6.85) implies electron transfer (6.140) from the aziridines to the excited sensitizer with the formation of radical cations (*E*,*E*)- and (*E*,*Z*)-**332** (electrocyclic ring openings are easier for radical cations than for the neutral aziridines) and radical anion $DCA^{\bullet -}$. The latter reduces $^3O_2$ into superoxide radical anion that reacts with the radical cations (*E*,*E*)- and (*E*,*Z*)-**332** giving the corresponding azomethine ylides and $^1O_2(^1\Delta_g)$. Followed by 1,3-dipolar cycloadditions producing *cis*- and *trans*-**333** [1099, 1100].

With the exception of 2,3-diphenylindene oxide (reaction (6.141)) that is photo-oxidized into the corresponding ozonide in the presence of DCA as unique sensitizer, 2,3-diphenyloxiranes are photo-oxidized into the corresponding ozonides only when another sensitizer like biphenyl is present together with DCA (cosensitized electron transfer photo-oxygenation, e.g. reaction (6.142)) [1101, 1102]. Both *cis*- and *trans*-1,2-diaryloxiranes give the corresponding *cis*-ozonides [1103]. Under similar conditions, photo-oxygenation of 1,2-diphenylcyclopropanes gives isomeric 1,2-dioxolanes [1104].

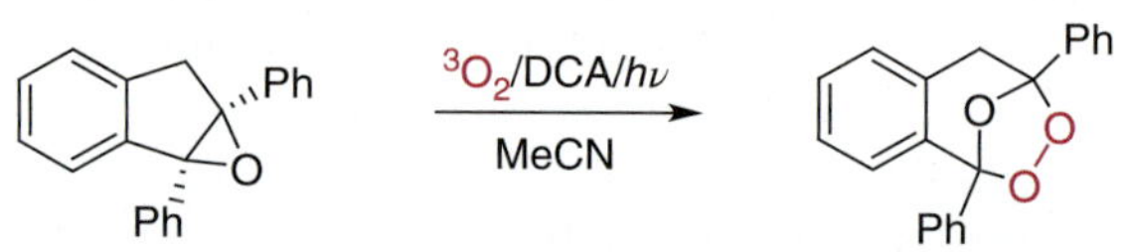

(6.141)

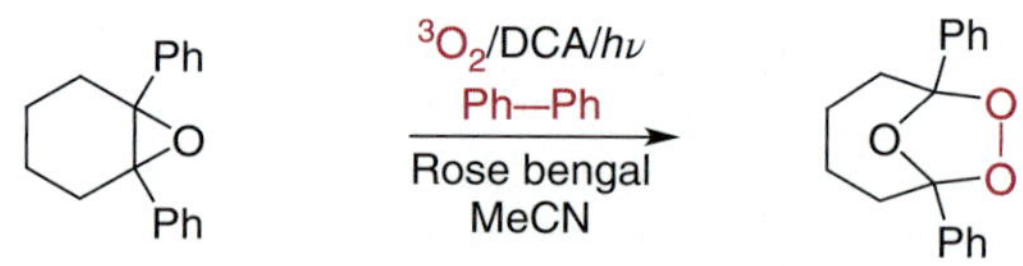

(6.142)

There are several routes for the synthesis of carbonyl oxides, including the conversion of 1,2,3-trioxolanes during the ozonolyis of alkenes (Criegee reaction) [1105–1107], the reaction of triplet carbenes with $^3O_2$ [1108–1110], and the reaction of diazo compounds with $^1O_2$ ($^1\Delta_g$) [1111–1113]. Applying time-resolved UV–vis and IR spectroscopy, the mechanism of the photosensitized (fullerene C_{60}) oxidation of diphenyldiazomethane has been established. The reaction forms dioxadiazole **334** via a concerted or two-step 1,3-dipolar cycloaddition [1114]. Compound **334** decomposes in two parallel reactions giving benzophenone *O*-oxide (**335**) + N_2 and benzophenone + N_2O [952].

Scheme 6.85 Photo-oxygenation of aziridines involving photoelectron transfer (DCA = anthracene-9,10-dicarbonitrile).

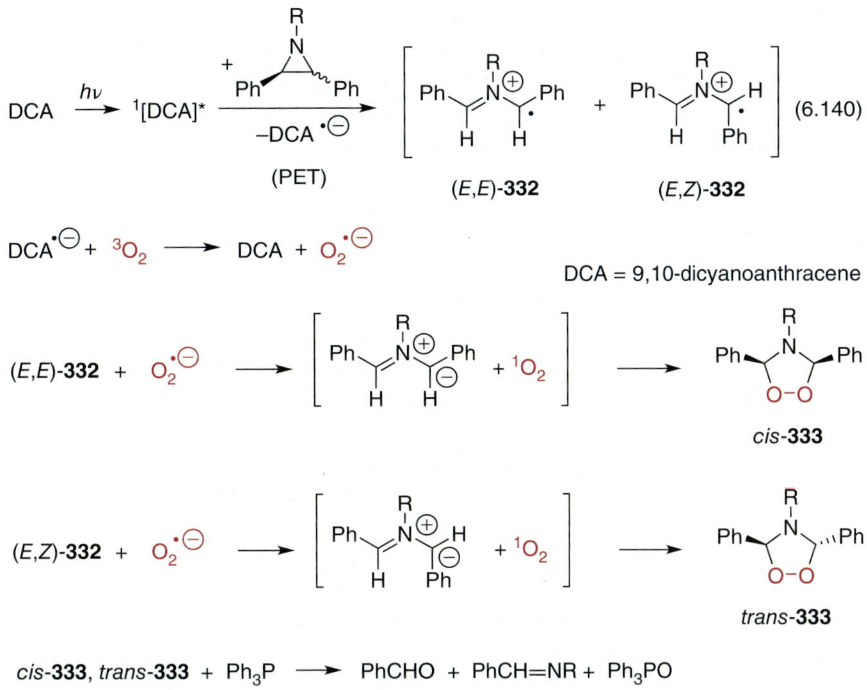

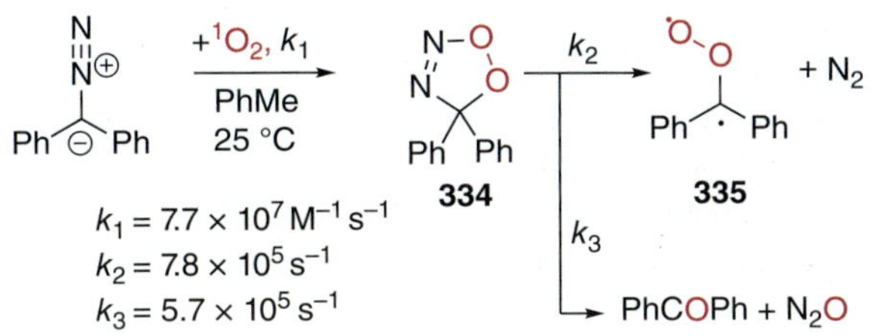

6.9.7 Nonpericyclic reactions of singlet oxygen

The photo-oxidation of phenols and naphthols is an important mode of degradations of these pollutants (e.g. phenol is a general disinfectant, a general reagent for the chemical and pharmaceutical industry; 2,4-dichlorophenoxyacetic acid is a widely applied herbicide; bisphenol A (2,2-di(4-hydroxyphenyl)propane) is a common plasticizer) in water [1115–1118]. They are converted into quinones that are readily degraded into less harmful products. Several 1,4-naphthoquinones are important synthetic intermediates that can be derived from 1-naphthols by oxidation applying all kinds of dangerous oxidants. A greener approach is to use air and solar light [1119]. Singlet oxygen (obtained by visible light and a dye sensitizer) reacts with electron-rich phenols ArOH through electron transfer giving radical cation $ArOH^{\bullet+}$ and superoxide radical anion ($O_2^{\bullet-}$). They react to form *ortho-* and *para-*hydroperoxides **336** and **337** that eliminate 1 equiv. of water providing the corresponding ortho- and parabenzoquinones, respectively (Scheme 6.86a) [1118, 1120]. Quantum mechanical calculations do not support the formation of endoperoxide through Diels–Alder reaction of $^1O_2$ as an alternative mechanism [1121]. In the absence of a dye sensitizer, compounds that absorb UV light such as aryl ketones are sensitizers for the photo-oxidation of phenols into their quinones (Scheme 6.86b) [1122–1124]. In this case, the ketone in its triplet state abstracts a H-atom from the phenol giving the corresponding aryloxyl and ketyl radical (reaction (6.143)). The ketyl radical reacts with $^3O_2$ through hydrogen transfer that liberates the starting ketone (sensitizer) and produces the hydroperoxyl radical ($HOO^\bullet$), the conjugate acid of superoxide radical anion. The latter reacts with the aryloxyl radical generating hydroperoxides **336** and **337** that eliminate 1 equiv. of water producing ortho- and parabenzoquinone, respectively. Alternatively (Scheme 6.86b′), $^3O_2$ reacts with the aryloxyl radical leading finally to the same products of photo-oxidation. In the absence of any sensitizer and in deoxygenated solution, UV irradiation of phenols generates aryloxyl radicals and solvated electrons. The latter process is easier in alkaline media in which the phenols are phenolate salts. In the case of the UV irradiation of phenol between pH 11 and 12, only *p*-benzosemiquinone is observed by ESR spectroscopy (Scheme 6.86c). Phenoxyl radical forms according to reaction (6.144). It is an electrophilic species that adds hydroxide anion quickly giving adduct **338**. A second equivalent

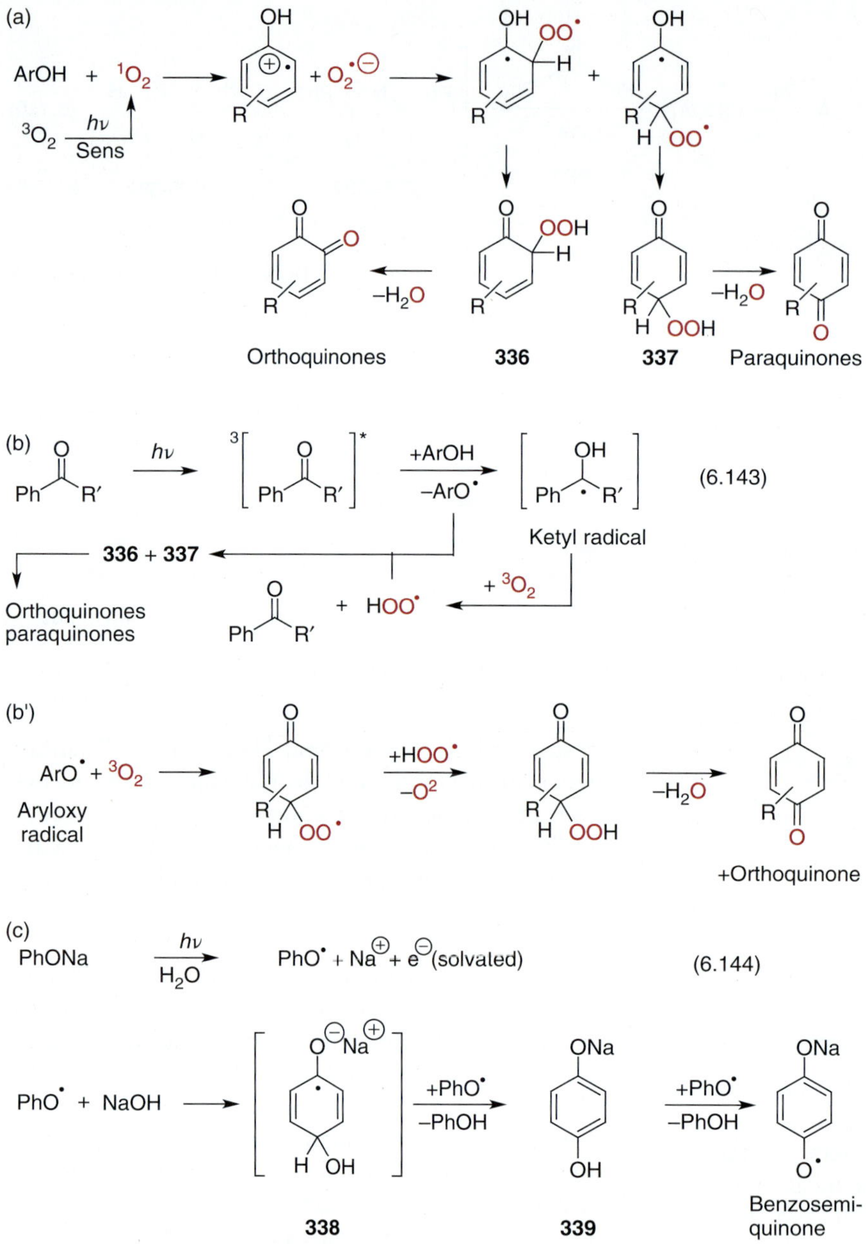

Scheme 6.86 Possible mechanisms of the photo-oxygenation of phenols: (a) under visible-light and with a dye-sensitizer; (b) under UV light in the presence a triplet sensitizer; (c) formation of solvated electrons by UV irradiation of phenolates.

of phenoxyl radical abstracts a hydrogen atom from **338** giving phenol + hydroquinolate **339**. A third equivalent of phenoxyl radical finally oxidizes **339** into benzosemiquinone [1125, 1126]. Photo-oxidation of 1-naphthol in aerated aqueous solution generates 1,4-napthoquinone, 2-hydroxy-1,4-naphthoquinone, and 6-hydroxy-1,4-naphthoquinone as main products. Traces of 1,2-naphthoquinones and 5-hydroxyl-1,4-naphthoquinone have also been detected. Solvated electrons have been detected by laser flash photolysis [1127]. The solar photocatalytic decomposition of phenol in aerated aqueous suspension of reduced anatase (e.g. Cu deposited TiO_2 powder) gives first hydroquinone, pyrocatechol, 1,2,4-benzenetriol, pyrogallol, and 2-hydroxy-1,4-benzoquinone at pH 3.5. These compounds are then decomposed photocatalytically into CO_2 and H_2O [1128, 1129]. Photo-oxidation of bisphenol A under UV irradiation and in aqueous TiO_2 dispersion starts with the methyl group oxidation by $HO^•$ and $HOO^•$ radicals [1130]. Gallic acid in water is degraded by visible light at pH 7 where $^1O_2(^1\Delta_g)$, $O_2^{•-}$, and H_2O_2 are formed from $^3O_2$ and triplet excited riboflavin (vitamin B_2) [1131].

The industry produces phenol in large amount in three steps from benzene (Ph—H + CH_2=CH—Me → Ph-*i*-Pr (cumene), cumene + air → Ph—CMe_2—OOH → PhOH + Me_2CO) [1132]. A one-step process has

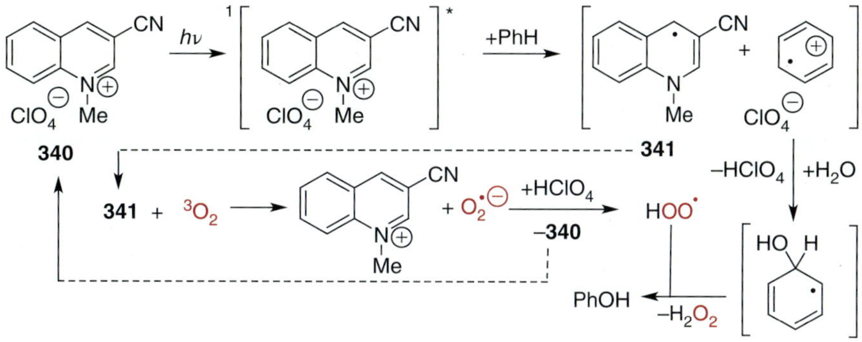

Scheme 6.87 Direct photo-oxidation of benzene into phenol.

been reported, which is the direct photo-oxidation of benzene ($Ph—H + H_2O + O_2 \rightarrow PhOH + H_2O_2$) using 3-cyano-1-methylquinolium perchlorate (**340**) as the photocatalyst (Scheme 6.87). The proposed reaction mechanism implies the formation of radical **341** + benzene radical cation. The radical cation reacts with H_2O giving 6-hydroxycyclohexadienyl radical. The latter is then oxidized through H-atom transfer to $HOO^•$ that results from the reaction of $^3O_2$ with radical **341** and protonation. This produces phenol and hydrogen peroxide [1133].

Dialkyl and alkyl aryl sulfides (but not diaryl sulfides) are photo-oxidized into sulfoxides according to mechanism (6.145), which implies persulfoxide intermediates [1134–1136]. In aprotic solvents, the photo-oxygenations are less efficient than in methanol. The photo-oxygenation of ethyl benzyl sulfide produces (Scheme 6.88) mostly benzaldehyde in benzene and MeCN. In methanol, only 25% of benzaldehyde forms in addition to the corresponding sulfoxide and sulfone (reaction (6.146)) [1137–1140].

Scheme 6.88 Sensitized photo-oxygenation of sulfides.

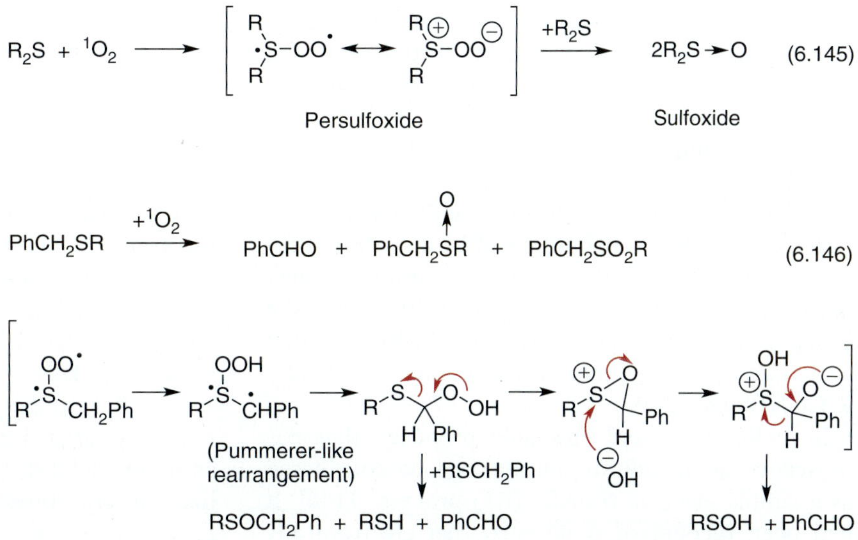

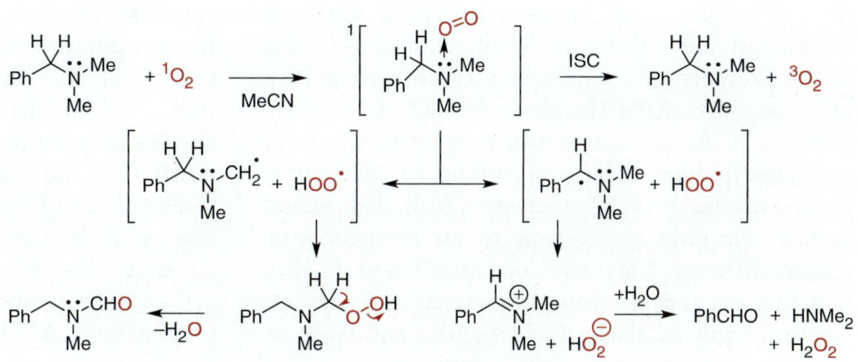

Scheme 6.89 Possible mechanism for the reaction of $^1O_2(^1\Delta_g)$ with *N*,*N*-dimethylbenzyl amine.

Singlet oxygen is quenched efficiently by tertiary amines. The process (physical quenching) involves the formation of an exciplex (or encounter complex: $^1O_2$·amine) with partial charge transfer from amine to $^1O_2$ (Section 6.9.2). In some cases, chemical quenching accompanies physical quenching and leads to products of photo-oxidation as illustrated in Scheme 6.89 for the reaction of *N*,*N*-dimethylbenzylamine that gives *N*-benzyl-*N*-methylformamide and benzaldehyde [1141, 1142].

6.10 Photoinduced electron transfers

PETs (see Figure 6.7, Section 6.2.3) may compete with other processes in bimolecular reactions of electronically excited molecules [1143]. The phenomenon permits the generation of highly reactive species such as radical cations, radical anions, and radicals that are exploited in photocatalysis (Sections 6.10.2–6.10.6), in dye-sensitized solar cells (DSSCs) (Section 6.10.7), and in light-controlled polymerization processes [1144]. A recent example of a reaction exploiting a PET is the visible-light-assisted dealkylative amide formation from tertiary amines and carboxylic acids. Irradiation of a tertiary amine of type $R^1CH_2N(R^2)R^3$ + in CCl_4 leads to the formation of radical ion pair $[R^1CH_2N(R^2)R^3]^{\bullet+}/CCl_4^{\bullet-}$ that decomposes into iminium salt $R^1CH{=}N^{(+)}(R^2)R^3/Cl^-$ and $CHCl_3$. In the presence of a carboxylic acid R^4COOH, the iminium salt forms adduct $R^3(R^2)N$—$CH(R^1)$—$OCOR^4 + HCl$, followed by an intramolecular nucleophilic addition of the amine moiety of $R^3(R^2)N$—$CH(R^1)$—$OCOR^4$ onto its carboxylic group giving a four-membered ring zwitterion that undergoes $[\sigma^2+\sigma^2]$-cycloreversion into $R^1CH{=}O + R^4CON(R^2)R^3$ [1145].

In 1936, Weiss and Fischgold proposed that fluorescence quenching in solution can be considered as a simple electron transfer (ET) process [1146]. It had been recognized quite early that electronically excited molecules have an enhanced tendency to take or to give away an electron [1147]. We have mentioned above that the photolysis of aryl halides in the presence of a nucleophile can involve a PET of type $[Ar—X]^* + H—Nu\colon \rightarrow Ar—X^{\bullet-} + H—Nu^{\bullet+}$ (Scheme 6.38e, Section 6.6.4), that tertiary amines can photoreduce alkenes according to $(alkene)^* + R_3N\colon \rightarrow alkene^{\bullet-} + R_3N^{\bullet+}$ (Scheme 6.51d), that excited arenes can give an electron to an electron-poor alkene (Scheme 6.60, reactions (6.87) and (6.93)), and that excited electron-poor arenes can take an electron from an alkene (Scheme 6.62 and 6.69) or from an aziridine (Scheme 6.85). Benzene is oxidized into $PhH^{\bullet+}$ by irradiation in the presence of *N*-methylquinolium salt as a sensitizer (Scheme 6.87). Singlet oxygen is deactivated by trialkylamines through a mechanism that generally involves electron transfer: $R_3N\colon + {}^1O_2 \rightarrow R_3N^{\bullet+} + O_2^{\bullet-}$ (Section 6.9.2). The dye-sensitized photo-oxidation of phenols implies ET: $ArOH + {}^1O_2 \rightarrow ArOH^{\bullet+} + O_2^{\bullet-}$(Section 6.9.7). In general, the most important interaction in an exciplex (encounter complex of an excited molecule $^1\mathbf{A}^*$ or $^3\mathbf{A}^*$ that lives longer than 10^{-9} seconds, the sensitizer, and another molecule **B**, the quencher) is the energy transfer (6.147) (see also Figure 6.8, Section 6.2.2). One recognizes the reductive electron transfer (6.148) in which **A*** is the acceptor and **B** the donor, and the oxidative electron transfer (6.149) in which **A*** is the donor and **B** the acceptor. The transferred electron, in principle, can go back to the donor as in reaction (6.150), a process called back electron transfer or return electron transfer (RET), which contributes to the deactivation (quenching) of **A*** (energy wasting) [1148].

$$\mathbf{A}^* + \mathbf{B} \rightarrow \mathbf{A} + \mathbf{B}^* \quad (6.147)$$

$$\mathbf{A}^* + \mathbf{B} \rightarrow \mathbf{A}^{\bullet-} + \mathbf{B}^{\bullet+} \quad (6.148)$$

$$\mathbf{A}^* + \mathbf{B} \rightarrow \mathbf{A}^{\bullet+} + \mathbf{B}^{\bullet-} \quad (6.149)$$

$$\mathbf{A}^{\bullet-} + \mathbf{B}^{\bullet+} \rightarrow \mathbf{A} + \mathbf{B} \quad (6.150)$$

In condensed phases, vibrational relaxation is fast (10^{-12} to 10^{-13} seconds) compared with exciplex formation. As a consequence, the molecules involved in bimolecular processes are thermally equilibrated species. In the gas phase, **A** and **B** are separated by a relatively large distance. The charge transfer state $\mathbf{A}^{\bullet-} + \mathbf{B}^{\bullet+}$ is higher in energy by $\Delta E = \text{IE}(\mathbf{B}) + (-\text{EA}(\mathbf{A}))$ compared with molecules **A** and **B** in their ground state and separated by a relatively large distance. When they get in contact as in an exciplex, the system gains coulombic stabilization. The coulombic potential energy amounts to $E_c \approx -Q_A \cdot Q_B / 4\pi\varepsilon_0 r$, where Q_A is the negative charge on radical anion $\mathbf{A}^{\bullet-}$,Q_B the positive charge on radical cation $\mathbf{B}^{\bullet+}$, ε_0 is the electric constant (permittivity) of the gas phase $= 8.854 \times 10^{-12}$ F m^{-1} (farads/meter), and r is the distance separating $\mathbf{A}^{\bullet-}$and $\mathbf{B}^{\bullet+}$, thus $\Delta E \approx \text{IE}(\mathbf{B}) + (-\text{EA}(\mathbf{A})) - Q_A \cdot Q_B / 4\pi\varepsilon_0 r$. For species in solution, IE becomes oxidation potential $\Delta E(\mathbf{B}^{\bullet+}/\mathbf{B})$ and electron affinity becomes reduction potential $\Delta E(\mathbf{A}^{\bullet-}/\mathbf{A})$. Thus, in solution, the free energy of the thermal electron transfer is given by $\Delta_{ET}G(\mathbf{A} + \mathbf{B} \rightarrow \mathbf{A}^{\bullet-} + \mathbf{B}^{\bullet+}) \approx \Delta E(\mathbf{B}^{\bullet+}/\mathbf{B}) - \Delta E(\mathbf{A}^{\bullet-}/\mathbf{A}) - Q_A \cdot Q_B / 4\pi\varepsilon r$, with ε = permittivity of the medium. One defines $\Delta E(\mathbf{A} \rightarrow \mathbf{A}^*)_{0,0}$ as the energy for the

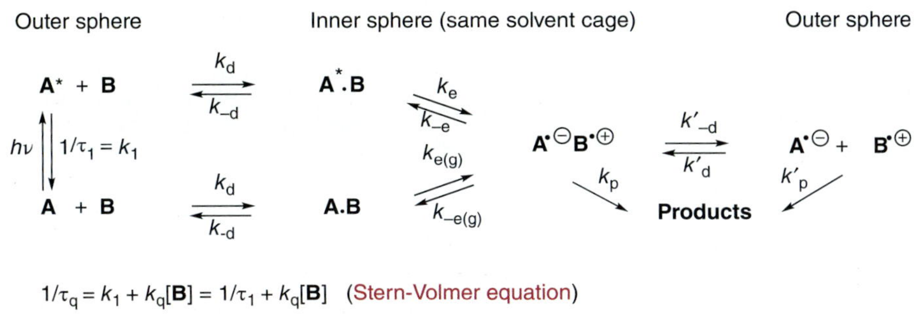

Figure 6.14 Kinetic mechanism for the photoinduced electron transfer reactions. In the absence of **B**, the deactivation of **A*** has the rate constant $k_1 = 1/\tau_1$; k_d, k_{-d}, k'_d, and k'_{-d} are diffusional rate constants for molecules and ions evolving between outer (two different solvent cages for **A** and **B** and their charge-transferred species) and inner sphere (**A** and **B** and their charge-transferred species are in the same solvent cage).

adiabatic transition (difference in energy between the vibrational level ν'_0 of **A*** and ν_0 of **A**, see Section 6.2.1). The free energy of the PET is given by the Rehm–Weller effective free energy (6.151) [1149–1153].

$$\Delta_{PET}G(\mathbf{A}^* + \mathbf{B} \rightarrow \mathbf{A}^{\bullet-} + \mathbf{B}^{\bullet+}) \approx \Delta E(\mathbf{B}^{\bullet+}/\mathbf{B}) - \Delta E(\mathbf{A}^{\bullet-}/\mathbf{A}) - Q_A \cdot Q_B/4\pi\varepsilon r - \Delta E(\mathbf{A}/\mathbf{A}^*)_{0,0} \quad (6.151)$$

The plot of the logarithm of the quenching rate constant k_q (see Scheme 6.3, Section 6.2.3 for a definition of k_q) vs. $\Delta_{PET}G$ leads to the Rehm–Weller empirical equation (6.152), where k_d and k_{-d} are the rate constants for diffusional encounter (see Figure 6.14 for a general kinetic mechanism for PET) and separation, k_0 is an Arrhenius frequency factor for the electron transfer, $\Delta^{\ddagger}G$ represents the free energy of activation for the electron transfer, and $\Delta^{\ddagger}G_0$ is the intrinsic value for $\Delta_{PET}G = 0$ (corresponds to the β term in the Bell–Evans–Polanyi equation: $\Delta^{\ddagger}H = \alpha\Delta_rH + \beta$). Experimental deviations to Eq. (6.152) have been reported [1154].

$$k_q = k_d/\{1 + (k_{-d}/k_0)(e^{\Delta\ddagger G/RT} + e^{\Delta_{PET}G/RT})\} \quad (6.152)$$

with $\Delta^{\ddagger}G = \{(\Delta_{PET}G/2)^2 + (\Delta^{\ddagger}G_0)^2\}^{1/2} + \Delta_{PET}G/2$

6.10.1 Marcus model

The Marcus model (6.153) applies successfully to the rate constant $k_{e(g)}$ of electron transfer (ET) involving ground-state molecules, as well as to the rate constant k_e of electron transfer involving electronically excited molecules (Figure 6.14):

$$k_e = \nu_N \kappa_e e^{(\Delta^{\ddagger}G/RT)} \quad (6.153)$$

with $\Delta^{\ddagger}G = (\Delta_{PET}G + \lambda)^2/4\lambda$ and where ν_N is the average nuclear frequency factor, κ_e is the electronic transmission coefficient, and λ is the nuclear reorganizational energy [1155–1159]. From what precedes the quantum yield of photoreactions involving an electron transfer will be the highest for reactions with $\Delta_{PET}G = -\lambda$ and for those where RET is retarded, i.e. with $k_{-e(g)}$ small or nil. In order to retard RET, **A** and **B** must be large molecules (e.g. electron-accepting sensitizers **A** such as polycyanoaromatic compounds, anthraquinones, benzophenones, organometallic complexes, etc., or electron-donating sensitizers **B** such as tertiary amines, phenanthrene, 1,5-dimethoxynaphthalene, 1,6-bis(dimethylamino)pyrene, and organometallic complexes) for which the charges in $\mathbf{A}^{\bullet-}{\cdot}\mathbf{B}^{\bullet+}$ are delocalized, or when either **A** or **B** is a cation or an anion; e.g. cation $\mathbf{A}^+$ is excited into $[\mathbf{A}^+]^*$ that gives exciplex $[\mathbf{A}^+]^*{\cdot}\mathbf{B}$ undergoing ET into $\mathbf{A}{\cdot}\mathbf{B}^{\bullet+}$, a process that does not involve charge separation and recombination. One approaches this ideal situation when $\mathbf{A}^+$ belongs to a solvent-separated ion pair $\mathbf{A}^+//\mathbf{X}^-$ or to a free ion pair. Examples (TPT, *N*-methylquinolium hexafluorophosphate, dimethyl (MV^{2+}) and dialkyl (RV^{2+}) viologen (*N*,*N*′-dialkyl-4,4′-bipyridinium) salts, 9-alkylacridinium salts, and organometallic cationic complexes) will be given in the next chapters. Reactions involving PET have found a large number of applications in organic synthesis [24, 25, 1160–1164].

6.10.2 Catalysis through photoreduction

Light accelerates radical anion substitution processes. An early example is the reaction of *p*-nitrobenzyl chloride with the potassium salt of 2-nitropropane, which proved to follow a chain mechanism (Scheme 6.90a) [1165]. Other examples are reactions (6.154) and (6.155) for which quantum yields of 570 and 6000 have been measured [1166]. Photolysis of DDT (1,1,1-trichloro-2,2-bis(*p*-chlorophenyl)ethane, a pesticide that is degraded very slowly in nature) and $PhNEt_2$ at λ_{irr} = 310 nm yields 1,1-dichloro-2,2-bis

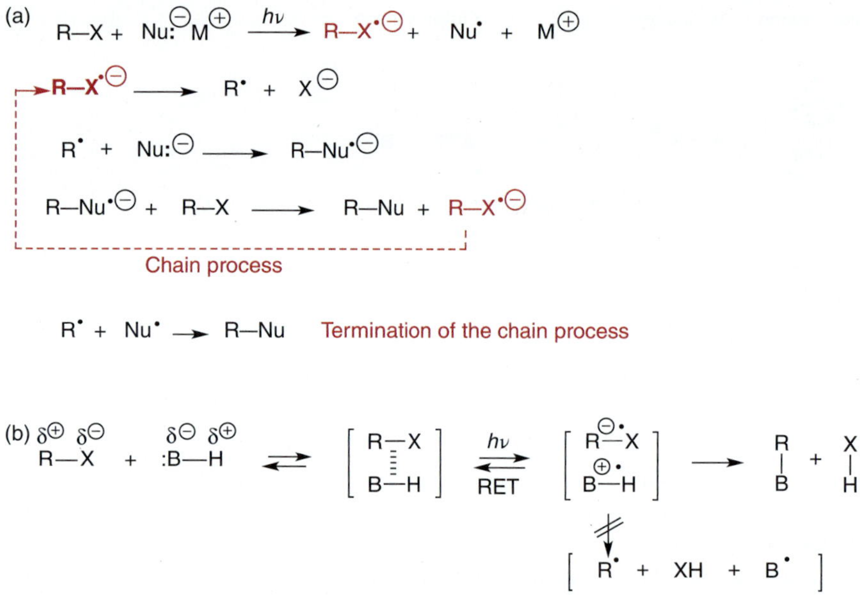

Scheme 6.90 Photoinduced nucleophilic substitutions: (a) with anionic nucleophiles Nu:⁻ that imply PET and (b) with Lewis bases :B-H that imply charge transfer complexes, the irradiation of which gives radical cation/radical anion pair intermediates, not free radicals.

(p-chlorophenyl)ethylene (DDE), 1,1-dichloro-2,2-bis(p-chlorophenyl)ethane (DDD), p,p'-dichlorobenzophenone, and HCl [1167]. The photoinduced substitutions of halogenated alkanes that cannot undergo S_N2 (e.g. 1-haloadamantane) or S_N1 (e.g. 1-halonorbornane) reactions for geometrical reasons, or when using nonpolar solvents (e.g. pentane), might occur through a charge transfer complex, the irradiation of which generates a polarized species (radical cation/radical anion pair intermediate) that does not permit free radicals to diffuse out of the solvent cage, as they are not quenched by $^3O_2$ and other radical scavenging agents such as TEMPO and benzoquinone (Scheme 6.90b). An example is given with the photoinduced methanolysis of (1R,2S,5R)-menthyl chloride into the methyl ether of (1R,2S,5R)-menthol (reaction (6.156)) in pentane/MeOH, which occurs with complete retention of configuration [1168].

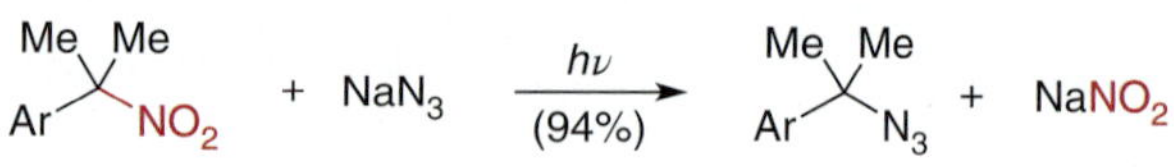

(6.154)

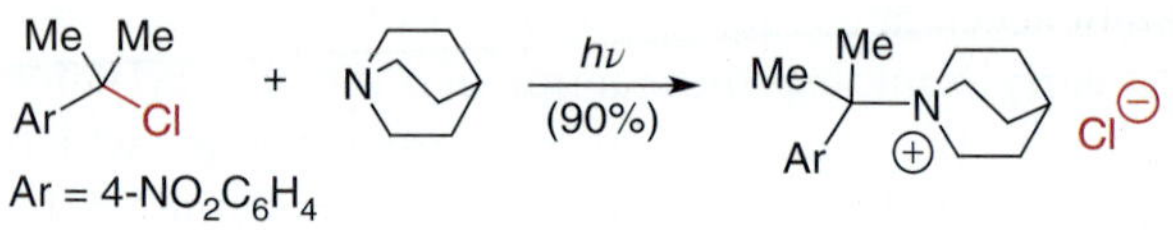

(6.155)

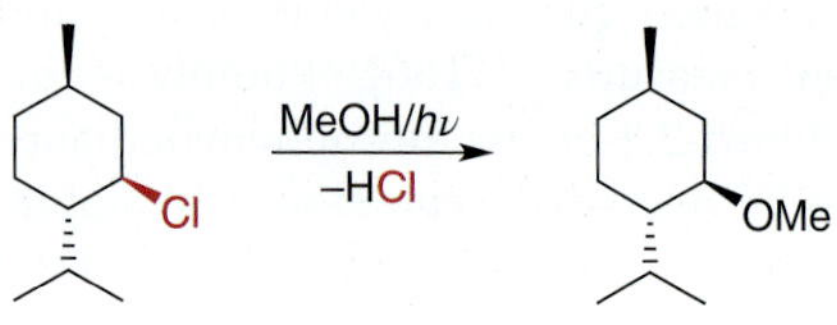

(6.156)

Visible light photoredox catalysis with transition metal complexes is finding many applications in organic synthesis [24, 1169–1173]. Yoon and coworkers reported that the intramolecular (2+2)-cycloaddition of bisenone **342** into cycloadduct **343** can be photocatalyzed with visible light and tris(2,2′-bipyridine)ruthenium dichloride ($Ru(bpy)_3Cl_2$) as a sensitizer (photoredox catalyst) in the presence of (i-Pr)$_2$NEt (Hünig base) and $LiBF_4$ [1174]. The proposed mechanism (Scheme 6.91) involves radical anion intermediate **344** resulting from the reduction of **342** by the photoreduced form of the sensitizer ($Ru(bpy)_2Cl$). Intermediate **344** undergoes quick cyclization into cis-**345** and then into the ketyl radical cis-**346**. The latter is then oxidized into **343** + $LiBF_4$ by the amine salt (i-Pr)$_2$NEt$^{\bullet+}BF_4^-$ resulting from the reduction of the sensitizer in its excited state ([$Ru(bpy)_3Cl_2$]*) by (i-Pr)$_2$NEt [1175]. This mechanism is corroborated by earlier reports of Krische and coworkers who found that the same reaction **342** → **343** can be promoted by single-electron transfer from a cobalt catalyst [1176] or by cathode reduction [1177]. Examples of intramolecular photocatalyzed hetero-Diels–Alder reactions of bisenones have also been reported by Yoon and coworkers using similar photocatalyzed conditions [1178]. In the presence of formic acid (reductant), (i-Pr)$_2$NEt and $Ru(bpy)_3Cl_2$ visible light induces the reductive cyclization of **342** into **347** [1179]. In this case, $trans$-**345** forms, which is reduced by the amine salt (i-Pr)$_2$NEt$^{\bullet+}/BF_4^-$ giving the iminium salt (i-Pr)$_2$N$^{(+)}$=CHMe/BF_4^- as the coproduct. The latter then reacts with HCOOH to give (i-Pr)$_2$N$^{(+)}$H(Et)/BF_4^- + CO_2.

Scheme 6.91 Example of a visible-light-photocatalyzed intramolecular (2+2)-cycloaddition.

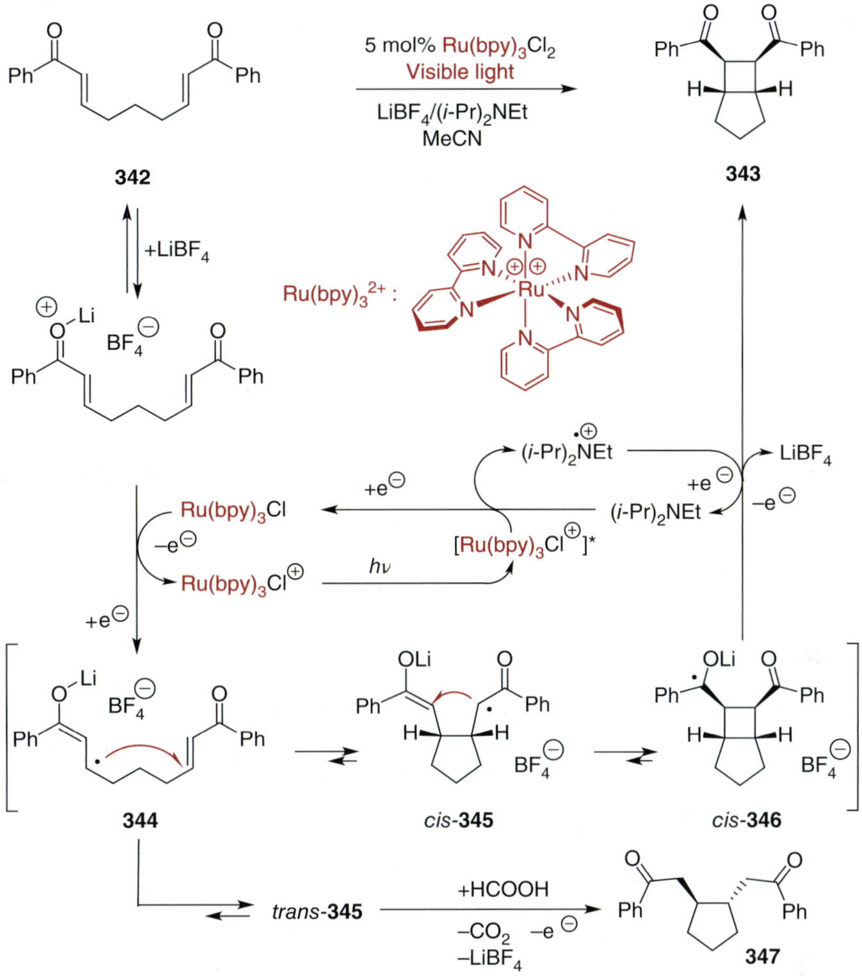

The N-alkoxyphthalimide group has been used as a photoredox-labile protective group for aldehydes as shown in Scheme 6.92 [1180]. This reaction can be seen as a photocatalyzed β-elimination of type $PhCH_2—O—N(COR)_2 \rightarrow PhCHO + HN(COR)_2$.

Under similar conditions ($Ru(bpy)_3Cl_2$ as a photoredox catalyst, secondary amine $HN(R^3)R^4$ as a cocatalyst, or a organocatalyst), MacMillan and coworkers realized a formal S_N2 displacement reaction $R^1CH(H)CHO + BrCH_2COOR^2 \rightarrow R^1CH(CH_2COOR^2)CHO + HBr$ (aldehyde α-alkylation). The reaction involves the formation of an enamine $R^1CH{=}CH—N(R^3)R^4$ that adds the α-stabilized alkyl radical $^{\bullet}CH_2COOR^2$ (combination of organocatalysis with photoredox catalysis). When the amine is enantiomerically pure, the addition of the latter radical onto the enamine intermediate becomes face selective (diastereoselective), which permits the obtainment of enantiomerically enriched products of aldehyde α-alkylation [1181a]. An application of this concept is presented for the photoredox catalytic enantioselective α-trifluoromethylation of aldehydes, which uses an iridium(III) complex as the sensitizer (Scheme 6.93a). In this reaction, $CF_3—I$ is a formal source of CF_3^+, which is converted into a trifluoromethyl radical in the photoredox catalytical process [1181c]. Other sources of electrophilic trifluoromethyl are available, such as CF_3SO_2Cl and the Umemoto's and Togni's reagents (Scheme 6.93b). Under visible light irradiation and in the presence of 1 mol% of $Ru(bpy)_3Cl_2$ and 20 mol% CuOAc, the trifluoromethylation of arylboronic acids ($ArB(OH)_2$) with CF_3I to give the corresponding $Ar—CF_3$ can be realized [1181d]. MacMillan and coworkers showed that the radical responsible of the α-alkylation of the aldehyde can be generated in situ by HAT from thiophenol to an alkene under visible-light-induced photoreduction of an Ir(III) complex (combining three catalytic processes: photoredox, enamine, and HAT catalysis) [1182]. The oxytrifluoromethylation of alkenes $CF_3—X + ROH + R^1(R^2)C{=}C(R^3)H \rightarrow R^1(R^2)(OR)—CH(R^3)CF_3 + HX$ can also be photocatalyzed in

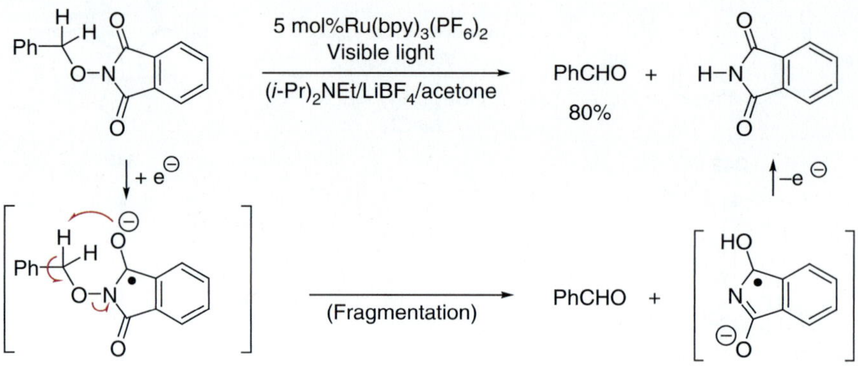

Scheme 6.92 *N*-Alkoxyphthalimides as masking groups of aldehydes.

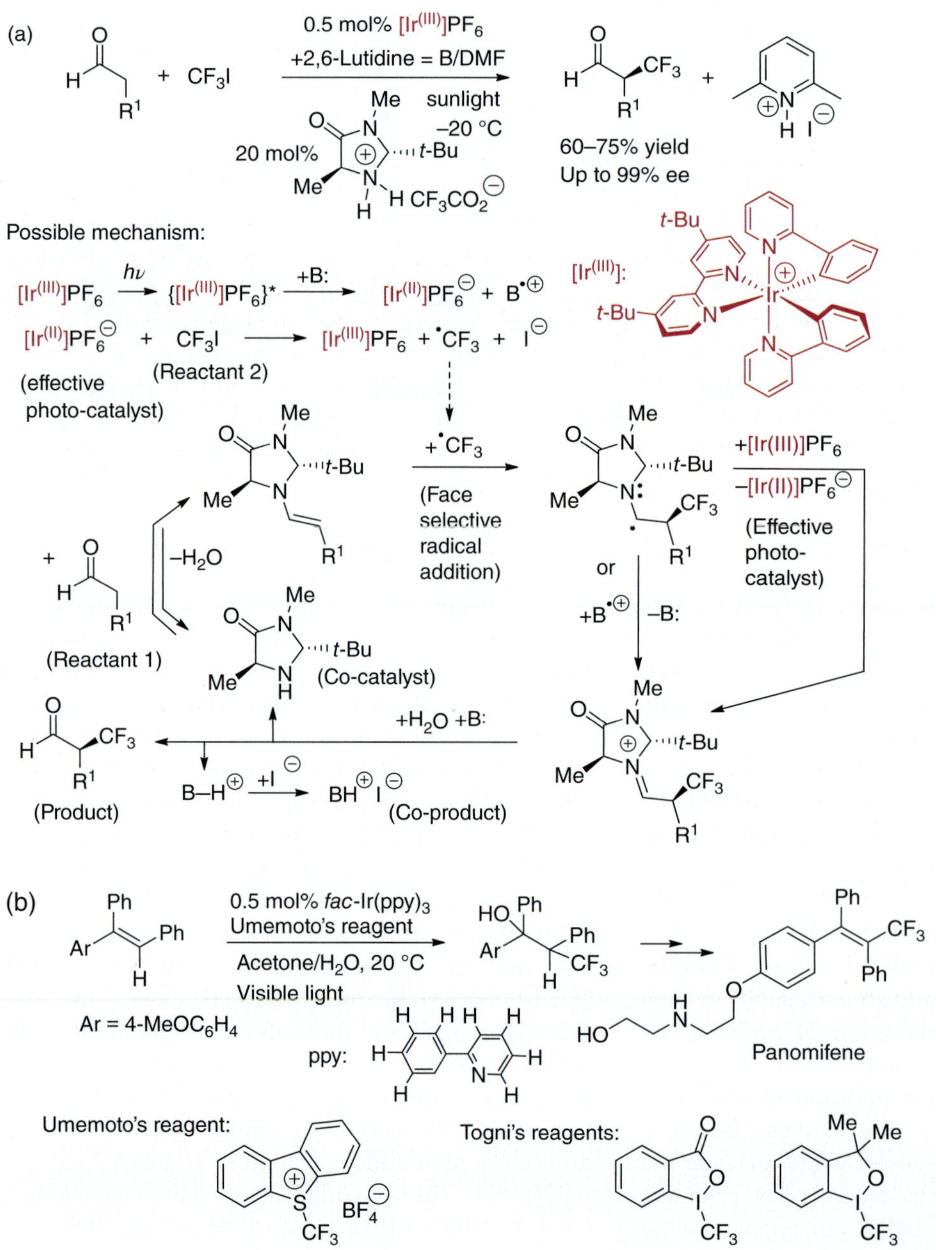

Scheme 6.93 (a) Enantioselective α-trifluoromethylation of aldehydes via photoredox organocatalysis and possible mechanisms and (b) example of alkene oxytrifluoromethylation.

the presence of iridium(III) complexes as sensitizers (Scheme 6.93b) [1183]. PET is a very useful means to generate intrinsically unstable free radical intermediates [1184] (Section 6.10.6) as exemplified above for the trifluoromethyl radical ($DH°(CF_3^•/H^•) = 108$ kcal mol^{-1} to be compared with $DH°(CH_3^•/H^•) = 104.7$ kcal mol^{-1}). Other unstable alkyl ($DH°(Et^•/H^•) = 100.7$ kcal mol^{-1}) as well as alkenyl ($DH°(CH_2{=}CH^•/H^•) = 110.6$ kcal mol^{-1}) and aryl ($DH°(Ph^•/H^•) = 113.2$ kcal mol^{-1}, Table 1.A.7) free radicals can be generated by visible light photolysis (Sections 6.10.3 and 6.10.6, reaction (6.154)). Formal aromatic electrophilic substitutions of indole, pyrroles, and furans with diethyl bromomalonate (Ar—H + $(EtOOC)_2CHBr$ → Ar—CH$(COOEt)_2$ + HBr) have been mediated by visible light PET using $Ru(bpy)_3Cl_2$ as the photocatalyst and Et_3N or 4-$MeOC_6H_4NPh_2$ as the base. Again, the reactions are photocatalyzed and generate $(EtOOC)_2CH^•$ radical intermediate [1184a].

Direct aldehyde C—H arylation (RCH=O + Ar—Br + K_2CO_3 → R(Ar)C=O + $KHCO_3$ + KBr) and alkylation (RCHO + R′—Br + K_2CO_3 → R(R′)C=O + $KHCO_3$ + KBr) have been realized by one-pot triple catalysis. Visible-light-excited [Ir(III)]* complex (redox catalyst) abstracts an electron from quinuclidine (organic catalyst of type $(R'')_3N$:) giving a radical cation of type $(R'')_3N^{•+}$, which abstracts a H-atom from RCO producing radical RCO$^•$ + quinuclidinium ion ($(R'')_3NH^+$), and a reduced [Ir(II)] complex. The latter reduces a nickel catalyst of type $L_2Ni(II)Br_2$ into $L_nNi(I)Br + Br^-$ first and then into $L_nNi(0)$ that undergoes oxidative addition (Section 7.7.8) onto Ar—Br and R′—Br giving species of type $L_nNi(II)(Ar,R')Br$ that react with radicals RCO$^•$ generating corresponding adducts $L_nNi(III)(RCO)(Ar,R')Br$. The latter are cleaved (reductive elimination, Section 7.7.8) into the final product (RCO)(Ar,R′) and $L_nNi(I)Br$, which return into the catalytic cycle oxidizing complex [Ir(II)] into complex [Ir(III)] and regenerating $L_nNi(0)$. Meanwhile, bromide anion became the counterion of the quinuclidinium ion. The latter salt is neutralized by K_2CO_3. This liberates quinuclidine (catalyst), which returns to the catalytic system by reducing the photoexcited [Ir(III)]* complex [1185].

Problem 6.66 Propose a mechanism for the visible-light-photocatalyzed conversion of **A** into **P**, a formal intramolecular (3+2)-cycloaddition [1186].

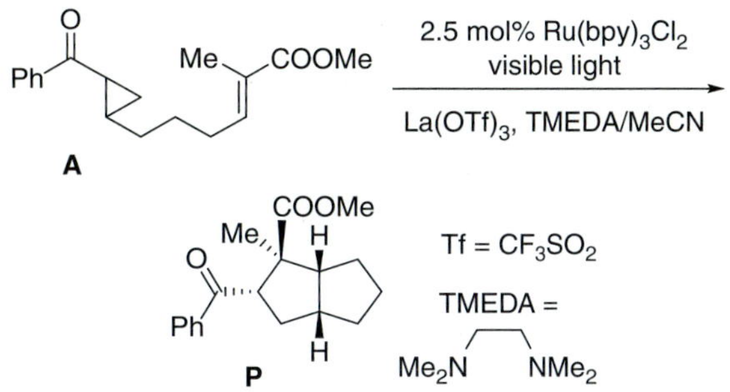

6.10.3 Photoinduced net reductions

The reduction of phenacyl bromide (**348**) into acetophenone (Scheme 6.94) uses visible light (λ_{irr} = 452 nm), tris(2,2′-bipyridine)ruthenium dichloride ($Ru(bpy)_3Cl_2$) as the photocatalyst, and 9,10-dihydro-10-methylacridine (**349**) as the stoichiometric reductant. The latter is oxidized into *N*-methylacridinium bromide (**350**). The reaction is initiated by a PET from **349** to $[Ru(bpy)_3Cl_2]^*$ (lifetime of c. 1 ms) to give the radical cation **351** and the reduced species $Ru(bpy)_3Cl$. The latter then reduces **348** into radical anion **348**$^{•-}$ returning catalyst $Ru(bpy)_3Cl_2$. Radical anion **348**$^{•-}$ undergoes mesolysis, the fragmentation into bromide anion (Br^-) and phenacyl radical ($PhCOCH_2^•$). The latter radical then abstracts an electron and a proton (formal HAT) from radical cation **351** to furnish acetophenone and **350**. The mechanism proposed is supported by the observation that light emission of excited $Ru(bpy)_3Cl_2$ is quenched by **349** and not by **348** [1187].

Using Hünig's base as a primary reductant ($EtN(i\text{-}Pr)_2 \rightarrow EtN^{(+)}(i\text{-}Pr){=}CMe_2 + 2e^- + H^+$) and formic acid (HCOOH), Hantzsch ester [1188] (diethyl 1,4-dihydro-2,6-dimethylpyridine-3,5-dicarboxylate), or isopropanol as a secondary stoichiometric reductant ($EtN^{(+)}(i\text{-}Pr){=}CMe_2 + HCOOH \rightarrow EtN(i\text{-}Pr)_2 + CO_2 + H^+$), a general method has been developed for the photocatalytic reductive dehalogenation [1189]. The photocatalyzed (sensitizer: $Ru(bpy)_3Cl_2$) reduction of phenacetyl sulfonium salts by 1,4-dihydropyridines occurs under visible light [1190]. Under similar conditions, β-ketosulfones are reduced into the corresponding ketones [1191]. Using iridium complex, *fac*-$Ir(ppy)_3$(ppy=2-phenylpyridine) as a photocatalyst, tributylamine as a primary reductant, and either HCOOH or Hantzsch ester as a secondary reductant, the Stephenson group has reduced alkyl as well as alkenyl and aryl iodides under

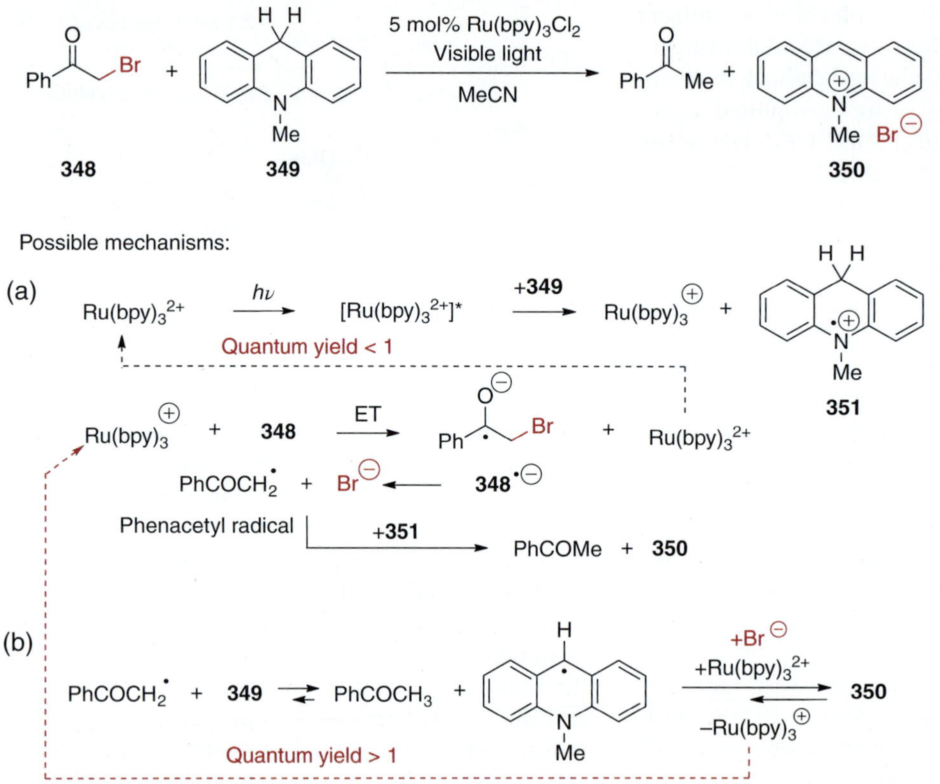

Scheme 6.94 Reductive dehalogenation of phenacyl bromide using visible light and a ruthenium(II) complex as a photocatalyst.

visible light. Interestingly, these conditions allow for the selective dehalogenation of aryl iodides in the presence of bromides and chlorides as illustrated with reaction (6.157) [1192]. Remarkably, this process has been applied to the one-pot deoxygenation of alcohols [1193].

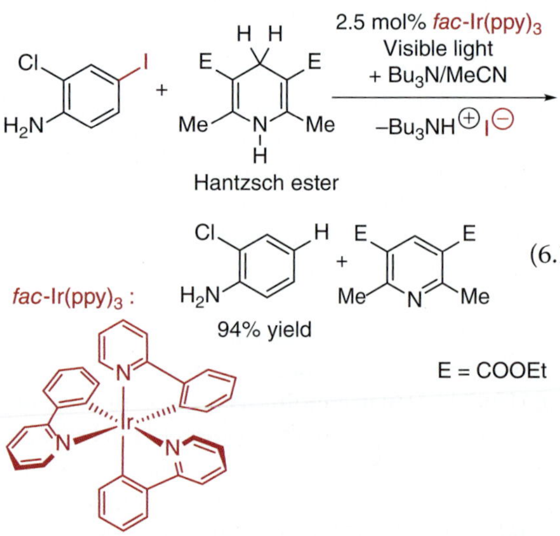

(6.157)

Photoreductions (cat.: $Ru(bpy)_3Cl_2$, visible light) of α-ketoepoxides and α-ketoaziridines with Hantzsch ester into the corresponding β-hydroxy and β-aminoketones, respectively, have been realized [1194]. The reactions imply the formation of ketyl radical anion intermediates that undergo ring openings into α-ketoalkyl radicals: the latter are reduced by 1,4-dihydropyridine or can be allylated with ethyl 3-tosylprop-1-ene-2-carboxylate, as illustrated with reaction (6.158) [1195].

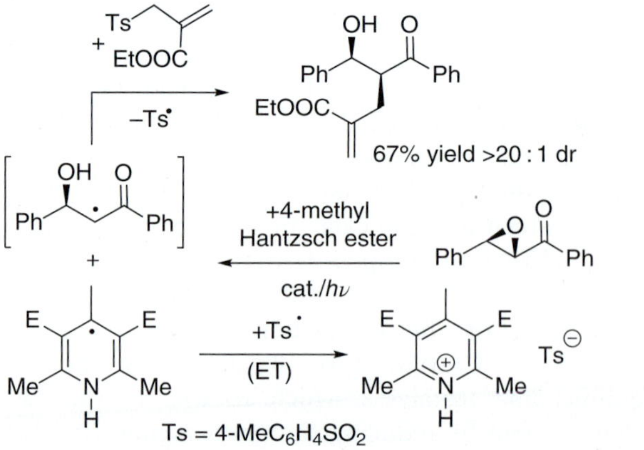

(6.158)

The reductive photocyclization of mandelic acid derivative **352** into 1,4-dioxane **353** uses anthracene-9,10-dicarbonitrile (DCA) as a sensitizer (0.3 equiv.) and 1,4-dimethoxynaphthalene (DMN) as a cosensitizer (0.15 equiv.) and ascorbic acid (H_2A) as a reductant (2.6 equiv.) (Scheme 6.95). With λ_{irr} = 405 nm, only DCA is excited and thus

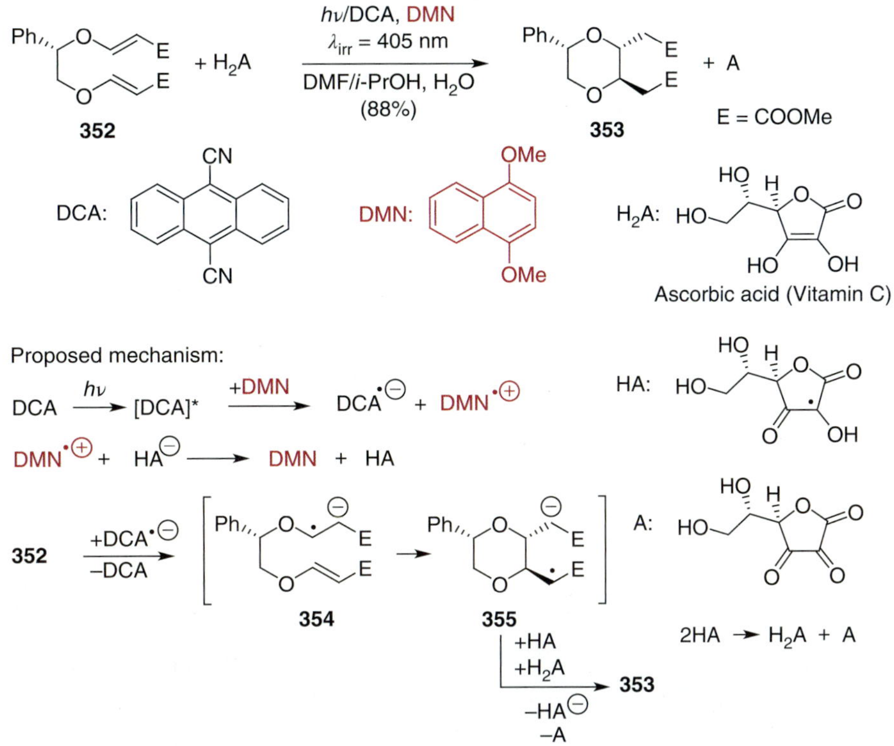

Scheme 6.95 Example of photoreductive stereoselective cyclization using a sensitizer and a cosensitizer, and ascorbic acid as a reductant.

DMN cannot act as a primary photoelectron donor. Thus, a PET generates radical anion $DCA^{\bullet-}$and radical cation $DMN^{\bullet+}$. The latter abstracts an electron from the conjugate base of ascorbic acid (HA^-) and DMN is liberated as a cosensitizer. Meanwhile, radical anion $DCA^{\bullet-}$ reduces the acrylic derivative **352** giving a radical anion **354** that undergoes cyclization into radical anion **355**. Protonation by ascorbic acid (H_2A) and hydrogen transfer from semidehydroascorbic acid (HA) produce product **353** [1196].

Problem 6.67 What are the possible products of irradiation (λ_{irr} = 254 nm) of 7-oxabicyclo[2.2.1]heptan-2-one in MeCN containing 5 equiv. of Et_3N? [1197]

Problem 6.68 Explain the following transformations [1198]:

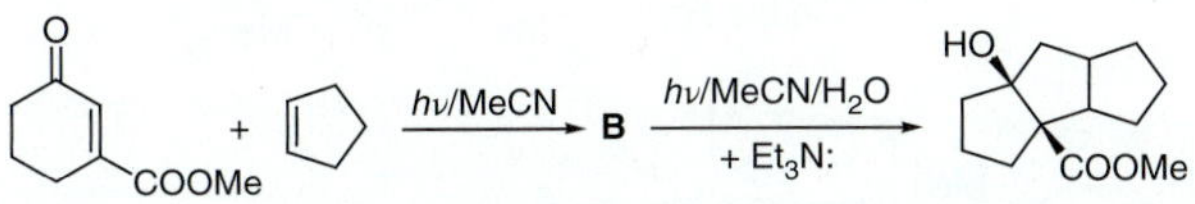

Problem 6.69 What are the main products **P** (94%) + **Q** of the following photoreaction (Ts = tosyl = *p*-toluenesulfonyl = 4-$MeC_6H_4SO_2$)? [1194]

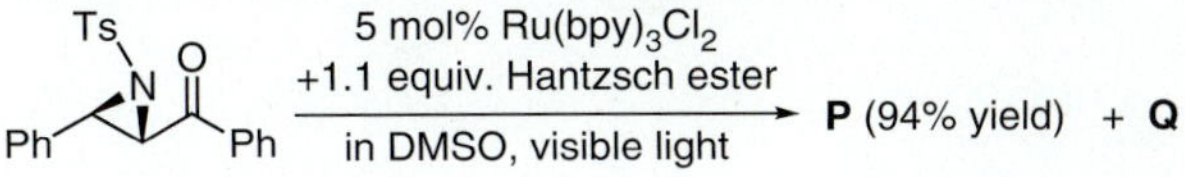

Problem 6.70 What is the major product **P** (69%) of the following photoreaction (E = COOMe)? [1199]

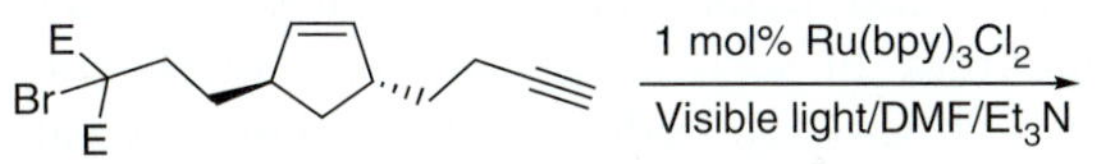

Problem 6.71 Propose a mechanism for the photocatalyzed reductive cleavage of carbamate **B** (protected form of amines) using ascorbic acid as a stoichiometric reductant [1200].

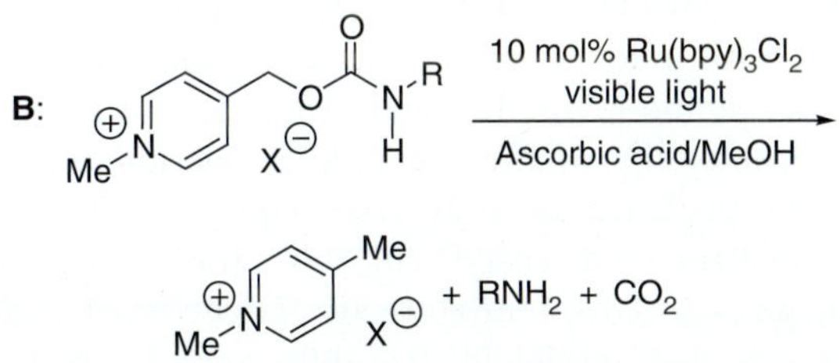

6.10.4 Catalysis through photo-oxidation

Irradiation of methyl *p*-cyanobenzoate (**356**) and 1,1-diphenylethylene in a nonpolar solvent such as benzene mostly gives oxetane **357** (Paternó–Büchi reaction, Section 6.8.5). In a more polar solvent such as MeCN, another product **358** forms in 70% yield, and, in methanol, the product of addition **359** is the main product of photoreaction. An exciplex forms that generates the radical cation $Ph_2C^{(+)}$-$CH_2^{\bullet}$ and radical anion $\mathbf{356}^{\bullet-}$. In the nonpolar solvent, they

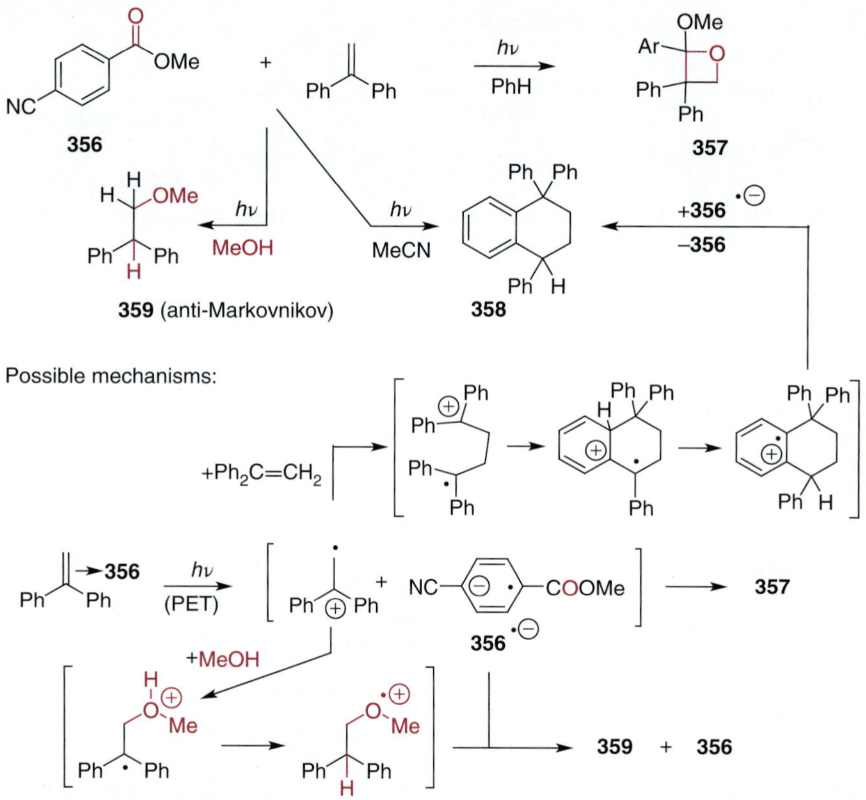

Scheme 6.96 Depending on solvent polarity, the radical anion/radical cation pair form initially in a PET can give different products.

combined into oxetane **357**. In a polar solvent, they escape from the solvent cage and the radical cation adds to $Ph_2C{=}CH_2$, leading finally to **358** after capture of an electron from **356**$^{\bullet-}$. In the nucleophilic solvent MeOH, initial radical cation $Ph_2C^{(+)}{-}CH_2^{\bullet}$ adds 1 equiv. of MeOH giving another radical cation that evolves into **359** after capture of an electron from radical anion **356**$^{\bullet-}$(Scheme 6.96) [1201].

The thermal Diels–Alder reaction of cyclohexa-1,3-diene with indole is very difficult especially because of the geometry of the diene (Section 5.3.12) and the aromaticity of the dienophile (Section 5.3.11). In the presence of only 0.05 equiv. of TPT as a sensitizer and acetyl chloride, the photoreaction gives a 3.3 : 1 mixture of *N*-acetylated *endo*- and *exo*-(4+2)-cycloadducts **360e** and **360x**. PET generates the radical cation of indole, which is an electron-poor dienophile that has lost most of the aromatic stabilization of its azole moiety and, thus, can add quickly to cyclohexa-1,3-diene giving a radical cation that abstracts an electron from an indole molecule, starting a chain process (Scheme 6.97) [1202].

Intramolecular and intermolecular Diels–Alder reactions of electron-rich alkenes and dienes can be accelerated by single-electron transfer from either the diene or dienophiles. The process involves the formation of a radical cation intermediate able to induce a chain reaction analogous to that presented in Scheme 6.97. The formation of the initial radical cation can be induced photochemically, for instance, with visible light when a ruthenium(II) complex is used as a photocatalyst. An illustration is given below with reaction (6.159) where the sensitizer and photocatalyst is the Ru(II)(2,2′-bipyrazine)$_3$ salt **361** [1203]. By varying the electron demand of the ligands of the transition metal complex, one can modify its redox properties until the most efficient photocatalyst is obtained [25].

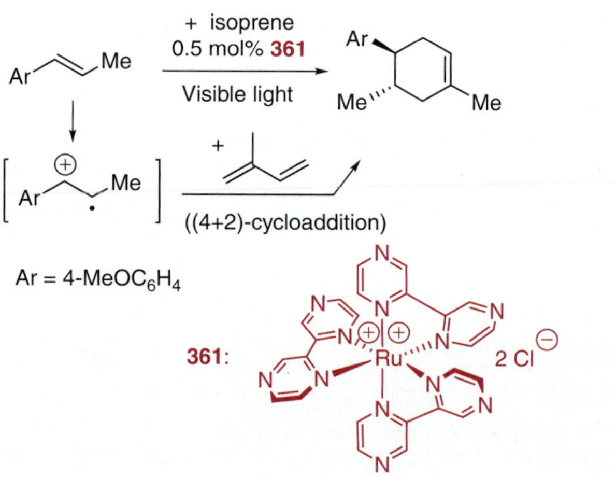

(6.159)

Scheme 6.97 Difficult Diels–Alder reactions can be photocatalyzed by PET: the electron-rich dienophile is made electron poor by photo-oxidation into its radical cation (Umpolung by PET).

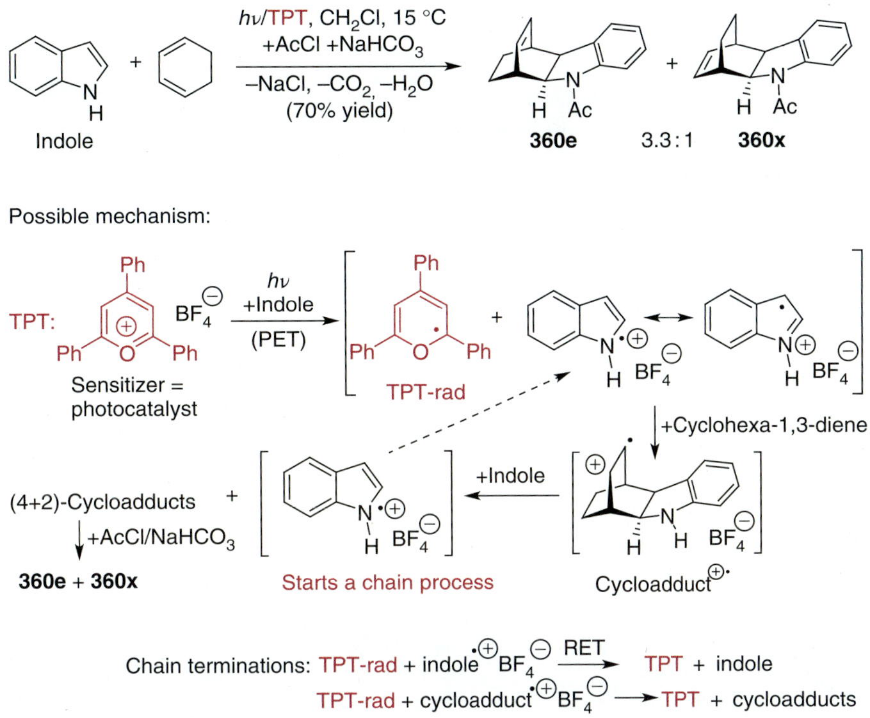

The intramolecular (2+2)-cycloaddition (6.160) is photocatalyzed by visible light in the presence of $Ru(bpy)_3(BF_4)_2$ as a sensitizer and dimethyl viologen bis(tetrafluoroborate) $[MV(BF_4)_2]$ as an oxidant (Scheme 6.98) [1204]. The latter oxidizes (PET) the Ru(II) complex in its excited state into a Ru(III) species that, on its turn, abstracts an electron from the starting diene **362**. This converts it into radical cation **362**$^{\bullet+}$, which undergoes fast cyclization into radical cation **363**$^{\bullet+}$. The latter takes an electron from another molecule of diene **362**, producing cycloadduct **363** and radical cation **362**$^{\bullet+}$, which starts a chain process.

An example of photocatalytic activation of amines is given with reaction (6.161) that can be seen as a formal nucleophilic substitution in which the nucleophile is an amino-carbanion displacing arenecarbonitrile, the cyanide anion being the nucleofugal group. The sensitizer and photocatalyst *fac*-$Ir(ppy)_3$ in its excited state abstracts an electron from the tertiary amine giving a radical cation that loses a proton with the generation of aminoalkyl radical **365** (Scheme 6.99). The Ir(II) complex reduces the arenecarbonitrile into its radical anion **364**. The aminoalkyl radical **365** adds to the radical anion **364** giving an intermediate **366** that loses the cyanide anion (forms HCN with the proton provided by the oxidation of the initial amine). An alternative mechanism has been proposed in which the arenecarbonitrile is reduced into radical anion **364** by the excited sensitizer. The Ir(IV) complex obtained in this PET oxidizes the amine into its radical cation, which then loses a proton generating the aminoalkyl radical **365** [1205].

Problem 6.72 Explain the following reaction (TPT = 2,4,6-triphenylpyrylium tetrafluoroborate) [879].

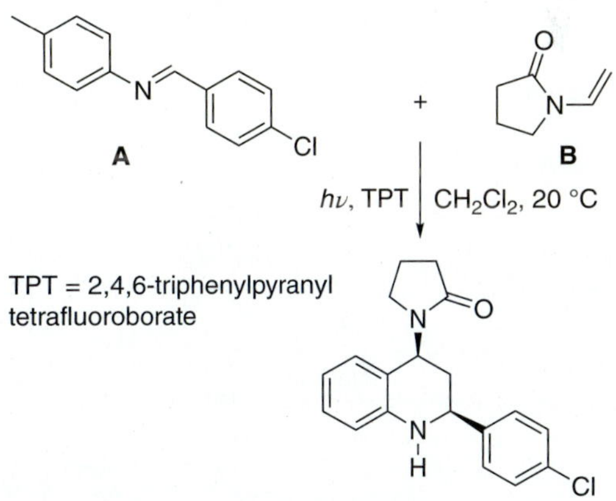

Problem 6.73 Propose a mechanism (5 mol% $Ru(bpy)_3Cl_2$ as a sensitizer, visible light) for the following reaction [1171, 1206, 1207].

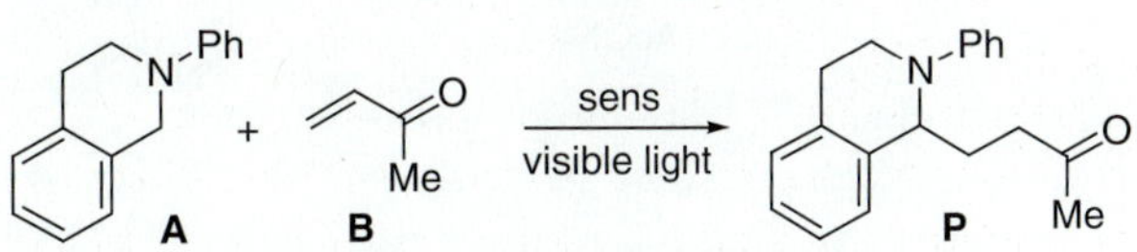

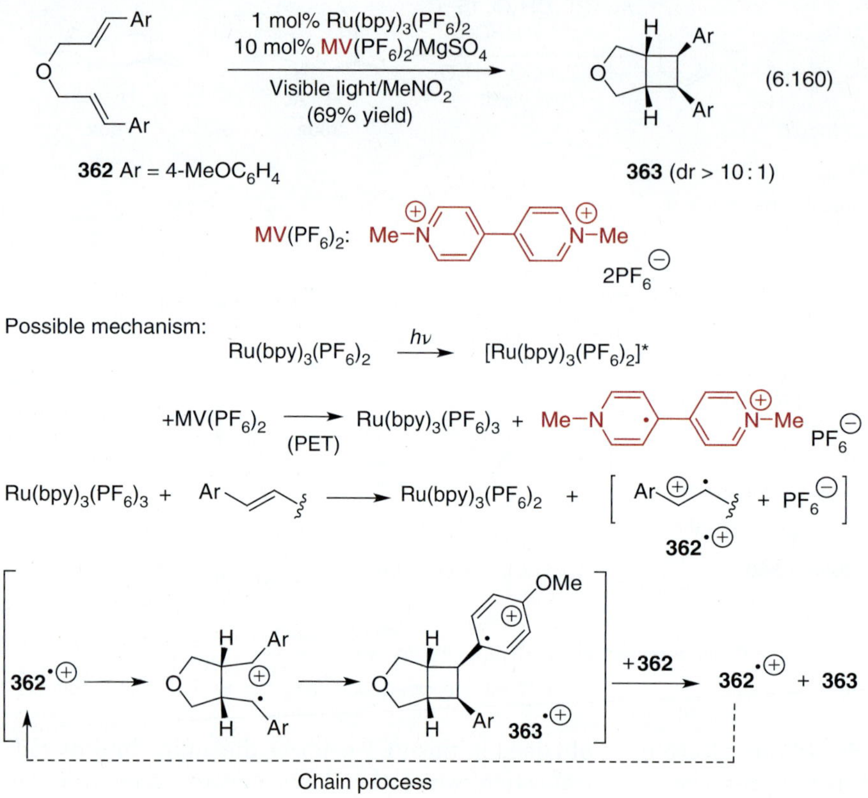

Scheme 6.98 Example of a (2+2)-cycloaddition by oxidative visible light photocatalysis.

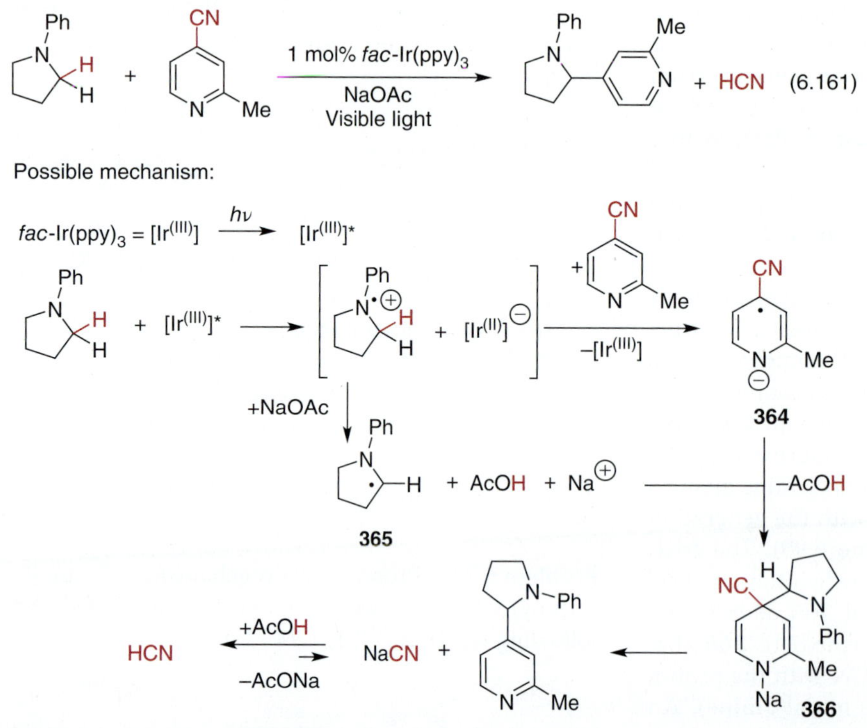

Scheme 6.99 Example of visible light photocatalytic activation of α-C—H bonds of tertiary amines.

Problem 6.74 What is product **P** of the visible-light-induced cyclization of **A** (sensitizer: 10-mesityl-9-methylacridinium perchlorate)? [1208]

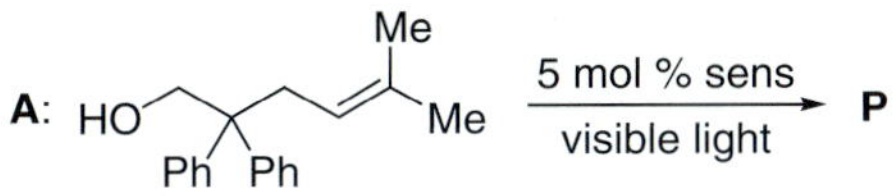

6.10.5 Photoinduced net oxidations

Acetal radical cations $R_2C(OR')_2^{\bullet+}$ that can be obtained by PET have the tendency to fragment into alkyl radical $R^\bullet$ and the relatively stable dialkoxycarbenium ions $RC^{(+)}(OR')_2$, which can be quenched by nucleophiles. If the nucleophile is an alcohol, orthoesters are obtained as illustrated with the conversion of 1,2:5,6-di-*O*-isopropylidene-D-mannitol (**367**) into orthoester **368** and bisorthoester **369** upon irradiation in the presence of benzene-1,2,4,5-tetracarbonitrile (TCB) as a sensitizer [1209].

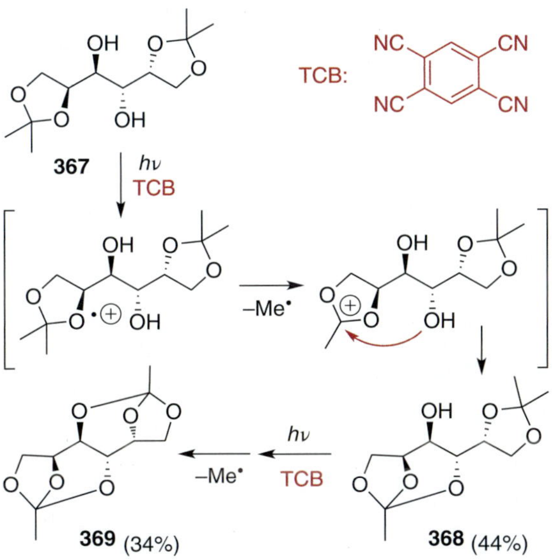

Irradiation of pyridinium salts in MeOH generates the corresponding radicals **370** (detected by ESR) and $[MeOH]^{\bullet+}$ radical cation (reaction (6.162)) through an intermolecular PET [1210]. The Kosower's salt (**371**) and related pyridinium salts undergo intramolecular PET (reaction (6.163)) with the formation of diradical **372** [1211, 1212]. Similarly, iminium salts can undergo PET generating aminoalkyl radicals that are quenched by the radical cations formed during the PET, as illustrated with reaction (6.164) [1213].

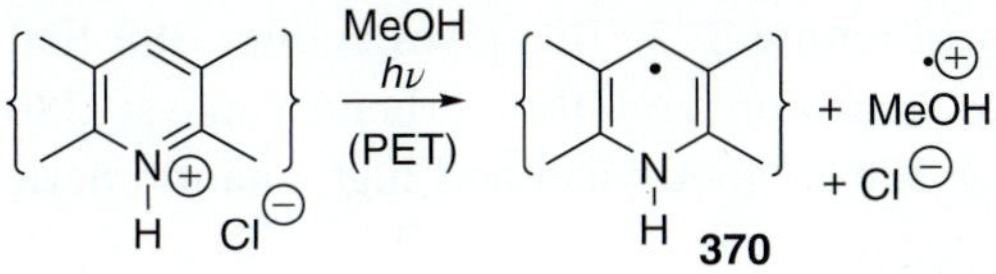

(6.162)

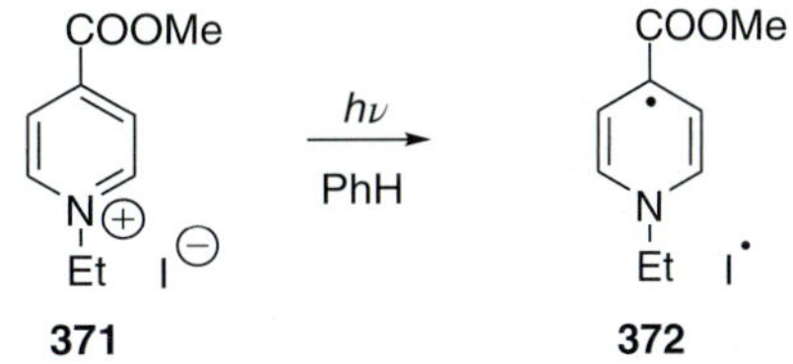

(6.163)

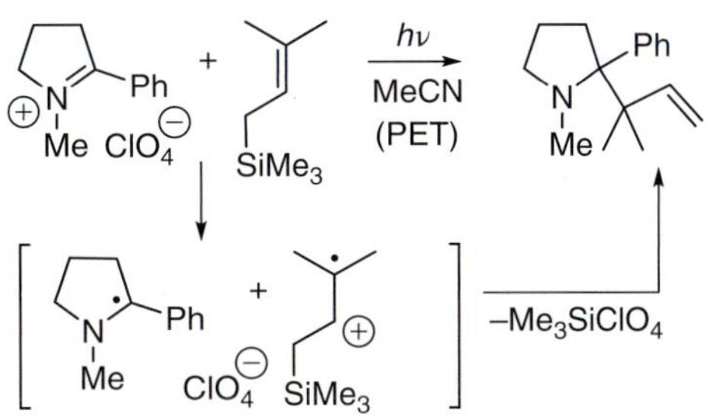

(6.164)

In 1984, Cano-Yelo and Deronzier reported that visible light irradiation of the ruthenium(II) complex of type RuL_3Cl_2 induces the oxidation of benzylic alcohols into the corresponding aldehydes using aryldiazonium salt **373** as an oxidant (Scheme 6.100) [1214]. In its photoexcited state, the Ru(II) complex reduces the aryldiazonium salt **373** into the corresponding radical + X^- that immediately decomposes into aryl radical **374** + N_2 + X^-. The Ru(III) complex that is formed in this photoredox process then oxidizes the benzylic alcohol into the corresponding radical cation **375**, with the recovery of the initial Ru(II) complex, the photocatalyst. HAT from radical cation **375** to radical **374** provides **376**, the conjugate acid of the final aldehyde. Benzophenone + HX (neutralized by collidine) + N_2 are the coproducts. Yield is quantitative for the oxidation of *p*-$MeOC_6H_4CH_2OH$, but only 24% for the less electron-rich $PhCH_2OH$.

The oxidative hydroxylation of arylboronic acids ($ArB(OH)_2$) into the corresponding phenols (ArOH) can be carried out with $^3O_2$ as a stoichiometric oxidant, with $Ru(bpy)_3Cl_2$ as a sensitizer and a photocatalyst, and with Hünig's base as a sacrificial reductant under visible light irradiation (Scheme 6.101). The key reactive species in the reaction is superoxide ($O_2^{\bullet-}$), which is generated by reduction of $^3O_2$ by the Ru(I) species generated by PET from the tertiary amine [1215].

Photoredox catalysis for the formation of C—C bonds between tertiary *N*-arylamines and nitroalkanes via an oxidative aza-Henry reaction has been reported first by Stephenson and coworkers [1216]. In the presence of a catalytic amount of $Ir(ppy)_2(dtbbpy)PF_6$ (**378**), visible light irradiation induced the

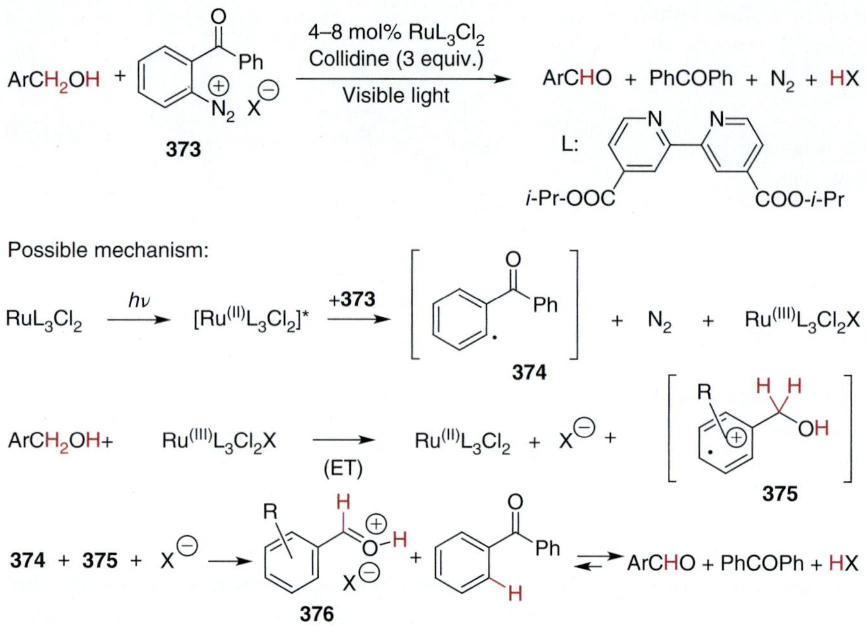

Scheme 6.100 Visible-light-photocatalyzed oxidation of benzylic alcohols into aldehydes.

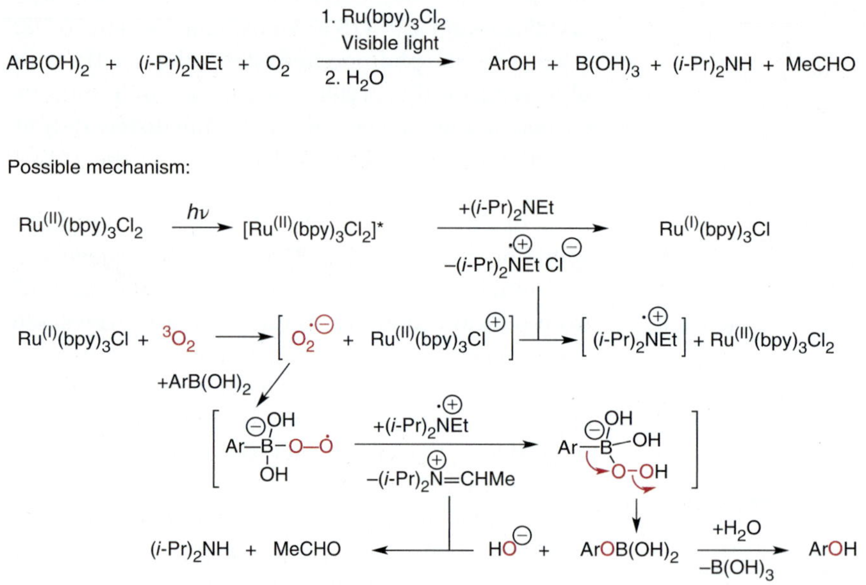

Scheme 6.101 Photoinduced aerobic oxidative hydroxylation of arylboronic acids.

coupling of nitroalkanes with tertiary amine **377** via the in situ-generated iminium ion (Scheme 6.102). The reductive quenching of the Ir(III) excited state by **377** leads to the ammonium radical cation, with subsequent catalyst turnover (Ir(II) → Ir(III)) that is affected by atmospheric oxygen and/or $MeNO_2$. Other nucleophiles (e.g. KCN, acetone, silyl enol ethers, allylsilanes, indole, phenylacetylene, and Me_3Si-CF_3 (Ruppert–Prakash reagent)) than $MeNO_2$ have been C—C coupled oxidatively with **377** and other *N*-arylated tetrahydroisoquinolines [24].

DiRocco and Rovis have combined photoredox catalysis with *N*-heterocyclic carbene (NHC) catalysis (Sections 7.2.5, 7.3.3, and 7.5.1–7.5.3) to realize an enantioselective oxidative α-acylation of amines with aldehydes (Scheme 6.103) [1217]. The stoichiometric oxidant is 1,3-dinitrobenzene (*m*-DNB). The enantiomerically pure carbene **380** adds to butanal giving electron-rich enol **381** (Breslow intermediate, Section 7.5.1, Scheme 7.15). The latter then reacts with the iminium ion **382**, resulting from the photoinduced oxidation of amine **377**. This gives adduct **383** that eliminates a proton and the carbene catalyst **380** producing **379** in good yield and high enantiomeric excess.

Scheme 6.102 Example of a photoinduced oxidative aza-Henry reaction combining a tertiary amine with nitromethane.

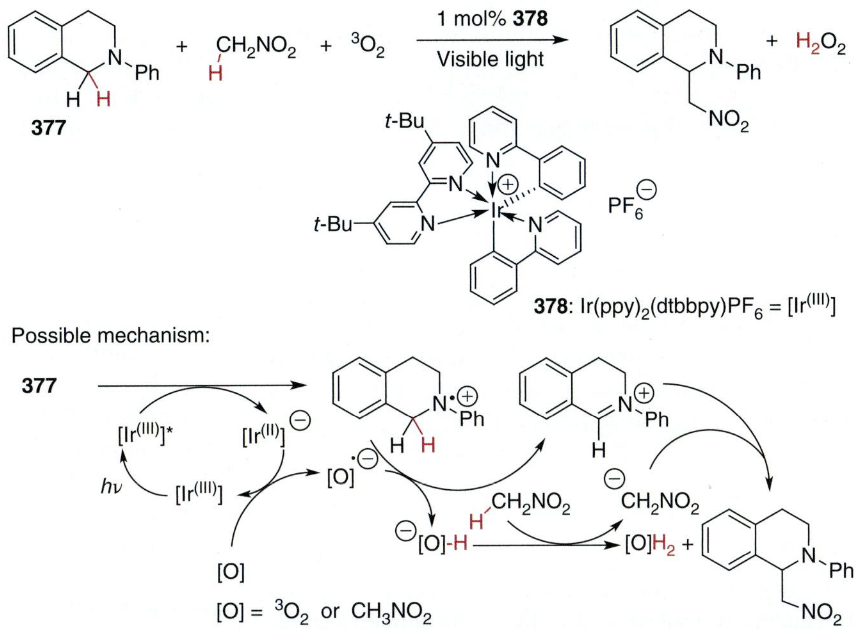

Scheme 6.103 Example of a photocatalyzed enantioselective oxidative α-acylation of an amine with an aldehyde.

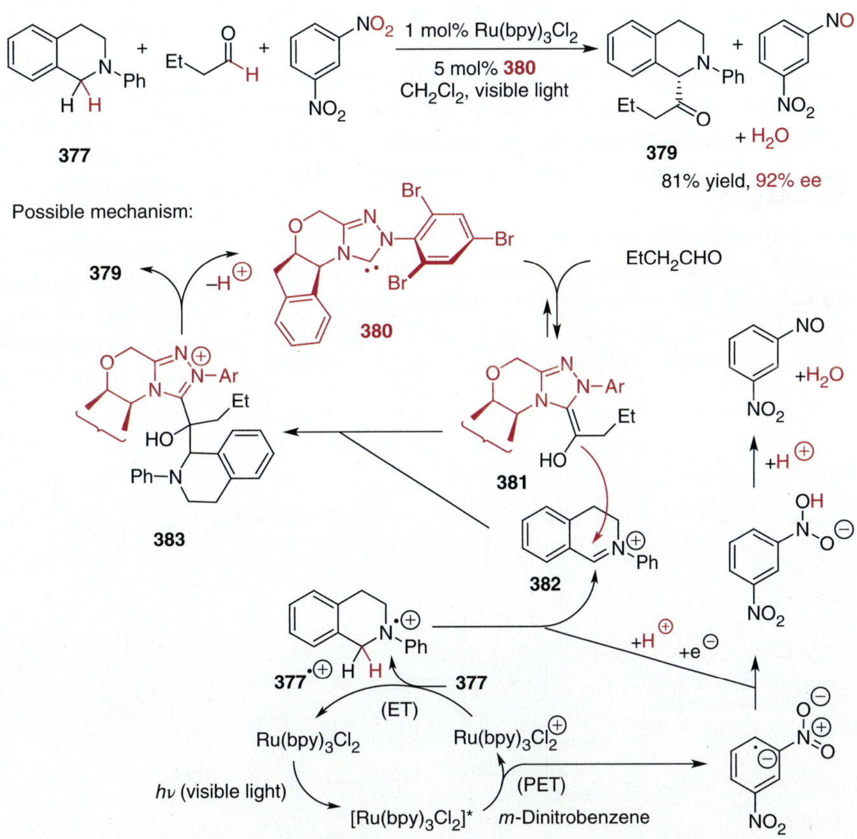

Problem 6.75 Propose a mechanism for the oxidative conversion of thiobenzamide **A** into benzothiazole **P** [1218].

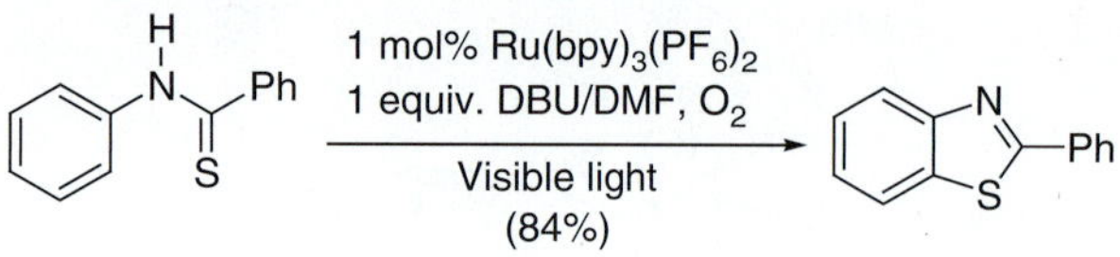

Problem 6.76 Oxidative 1,1′-biaryl coupling of β-naphthols can be achieved under visible light irradiation in the presence of a Ru(II) complex as a sensitizer and $Co(acac)_3$ as an oxidant. Propose a mechanism for this reaction [1219, 1220].

6.10.6 Generation of radical intermediates by PET

A large variety of radical intermediates can be formed by thermal homolysis and/or by photoinduced homolysis (Section 6.6.3 and 6.6.6). In principle compounds of type R—X can be reduced by PET into the corresponding radical anions $RX^{\bullet-}$ that may fragment into $R^\bullet + X^-$ or into $R^- + X^\bullet$ [1221]. These intermediates can enter in a large variety of monomolecular or bimolecular reactions. Aromatic compounds with electron-withdrawing substituents (Ar—Y), typically the arenepolynitriles, are oxidants in their electronically excited states and can interact with electron-rich compounds R—H giving $ArY^{\bullet-} + RH^{\bullet+}$. Radical cations are highly acidic and are deprotonated with the solvent giving neutral radical $R^\bullet$, which adds to the radical anion $ArY^{\bullet-}$ giving a carbanion intermediate of type cyclohexadienyl. The latter can either eliminate the nucleofugal group Y^- or be protonated to give 1,2- and 1,4-dihydroarenes **384** (Scheme 6.104a). In this case, the electrofugal group is H^+ that originates from R—H. Reactant R—H can be alkylbenzenes, alkylalkenes, amines, ethers, alkylsilanes, and allystannanes that generate α-stabilized carbon-centered radicals. With a strong oxidant such as TCB, unstabilized alkyl radicals can be formed from reactants R—E with E = electrofugal that can be Me_3Si, Bu_3Sn, or COOH (Scheme 6.104b) [1209, 1222].

The addition of tertiary amines across alkenes can be realized photochemically using various sensitizers [1207]. An example is given with the photoinduced addition of *N*-methylpyrrolidine to the chiral menthyloxyfuranone **385** (Scheme 6.105a) [1223]. The triplet excited state of the ketone sensitizer (^{3}[ArCO]*) abstracts an electron from *N*-methylpyrrolidine giving the corresponding ketyl radical anion ($Ar_2C^\bullet—O^-$), which then abstracts an α-hydrogen atom from the amine radical cation giving hydroxyalkyl radical **387** and the electron-rich aminoalkyl radical **388**. The latter then adds to the electron-poor alkene **385** giving radical **389**. Electron-rich radicals add the fastest to electron-poor alkenes; electron-poor radicals add the fastest to electron-rich alkenes (*β* term (intrinsic barrier depends on steric factors and the ease by which the reacting partners can exchange an electron) in the Bell–Evans–Polanyi relationship: $\Delta^{\ddagger}H = \alpha\Delta_r H + \beta$). Hydrogen transfer from *N*-methylpyrrolidine to **389** provides the final adducts **386a** + **386b** and generates radical **388**, which can start a chain process. The efficiency of the process depends on the occurrence of side reactions that are the combination of two radicals

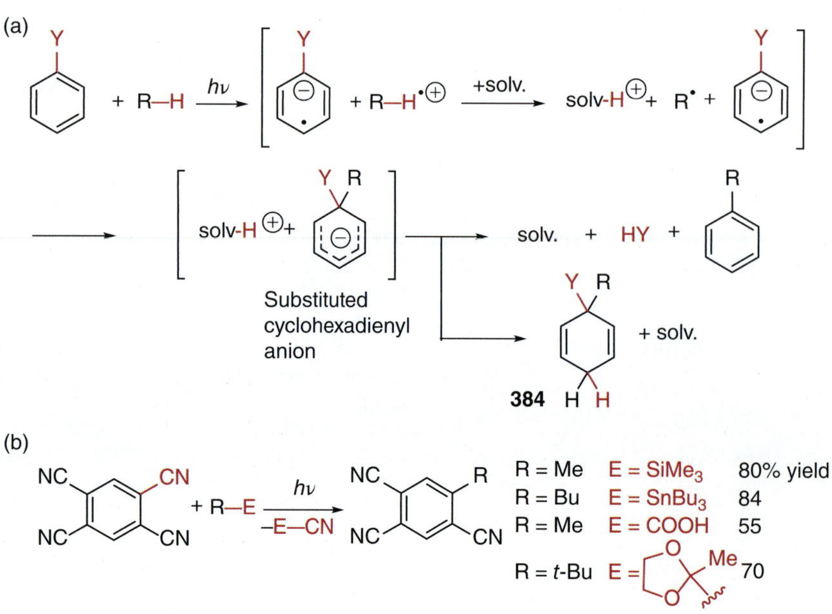

Scheme 6.104 PET oxidative formation of neutral radicals and their reactions with aromatic radical anions: (a) formal electrophilic arylation of alkanes, (b) metathetical reaction exchanging a cyanide nucleofugal group of a cyanoarene with an alkyl nucleophile.

Scheme 6.105 (a) Example of a photoinduced addition of an amine to an alkene and (b) possible mechanism for the recovery of the ketone sensitizer.

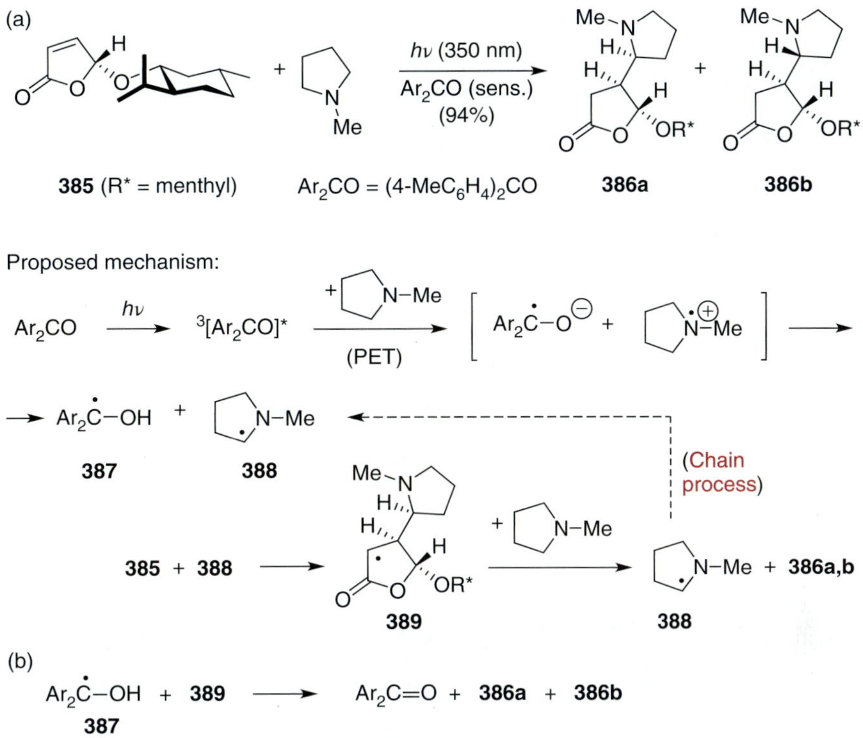

387 giving the corresponding pinacol derivative, the disproportionation of **387** into $Ar_2CO + Ar_2CHOH$, the combination of **387** + **388**, and the disproportionation of radical **388**, to mention the most obvious reactions. With 4,4′-dimethoxybenzophenone, these reactions are slower than those involving other triplet sensitizers such as benzophenone (Ph_2CO) or acetophenone (PhCOMe), probably because of the enhanced stability of radical **387** compared with that of the hydroxyalkyl radicals derived from the latter ketones. Furthermore, **387** interacts efficiently with radical **389** through an electron transfer generating the stable hydroxy-carbenium ion $Ar_2C{=}O^{(+)}{-}H$ (Scheme 6.105b), which is relatively more stable than $Ph_2C{=}O^{(+)}{-}H$ and $Ph(Me)C{=}O^{(+)}{-}H$ [1162].

When the enantiomerically pure diaryl ketone **391** is used as a sensitizer, an enantioselective intramolecular addition of the tertiary amine and alkene moieties of **390** giving **392** has been realized [1224].

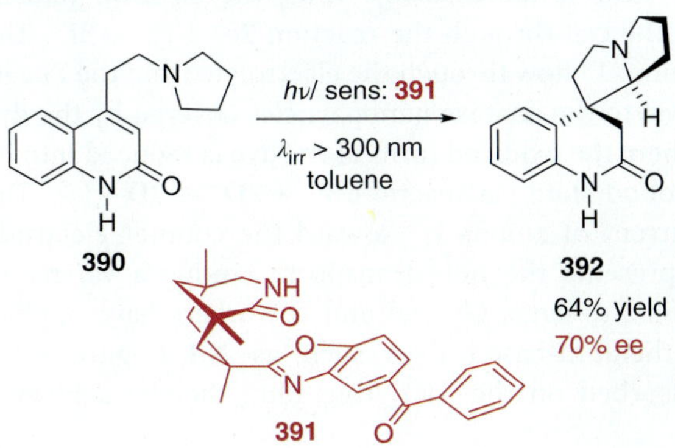

Neutral alkyl radicals R• can be generated under visible light irradiation of *N*-(acyloxy)phthalimides **393** in the presence of $Ru(bpy)_3Cl_2$ as a photocatalyst and 1-benzyl-1,4-dihydronicotinamide (**394**) as a reductant. This produces coproducts phthalimide and CO_2 next to radical **397**. A possible mechanism for the formation of R• is shown in Scheme 6.106 [1225–1227]. PET oxidizes amine **394** into radical cation **395** and the Ru(I) complex so-obtained reduces **393** into radical anion **393**•−. Protonation of the latter gives radical **396** that is rapidly fragmented into CO_2, phthalimide, and radical R•.

Problem 6.77 UV irradiation of **A** in methanol gives two products **P** and **Q**. What are they? [1164]

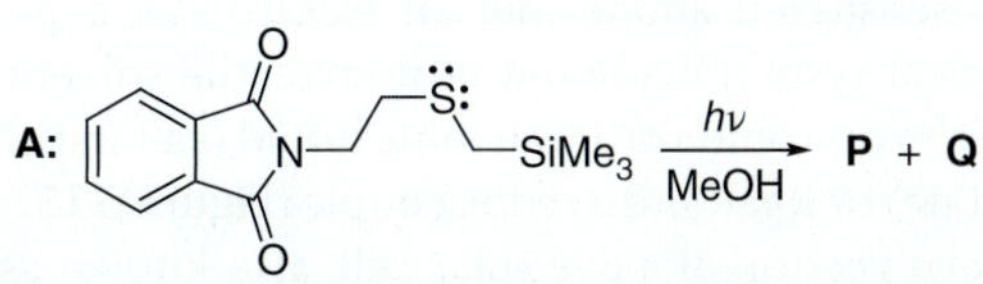

Problem 6.78 What are the main products of the following photoreactions that use visible light, 2.5 mol% *fac*-$Ir(ppy)_3$ as a sensitizer, 10 equiv. of Bu_3N, and 10 equiv. of formic acid? [1192]

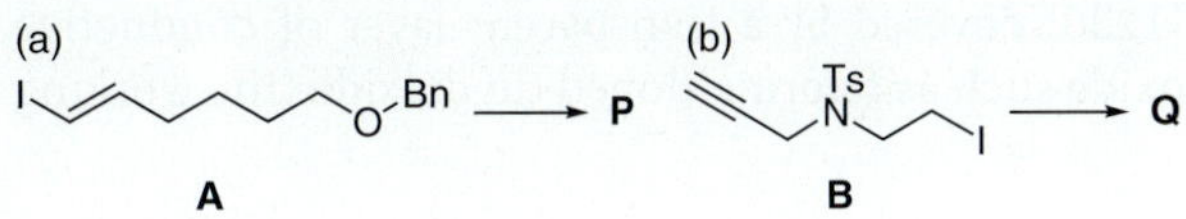

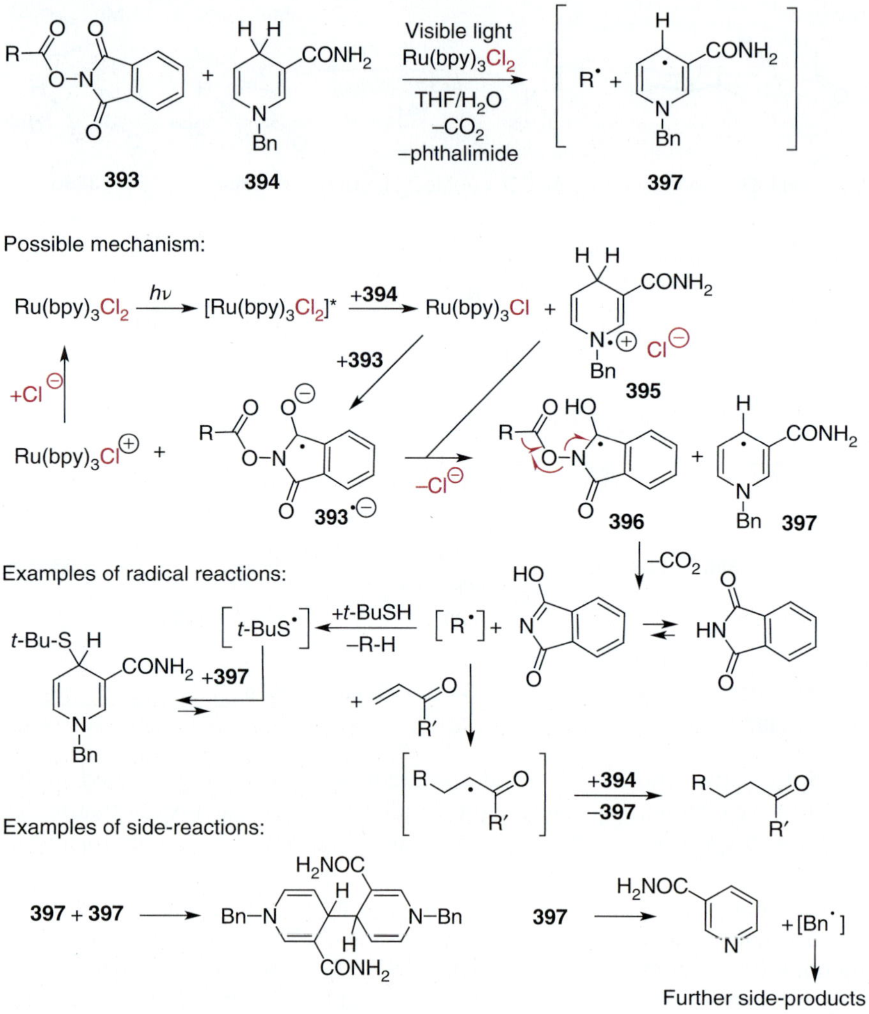

Scheme 6.106 *N*-(Acyloxy) phthalimides as masking groups of alkyl radicals.

6.10.7 Dye-sensitized solar cells

DSSCs are now used to convert sunlight into electricity. Because of their lower cost of production, they are expected to compete favorably with the silicon-based photovoltaic cells (PVs) [1228]. They are based on a semiconductor formed between a photosensitized anode and an electrolyte, a photochemical system made out of nonvolatile solvent like 3-methoxypropionitrile, an ionic liquid (salt in the liquid state) or a gel, and a redox couple (Figure 6.15). The modern version of a dye solar cell, also known as the Grätzel cell, was originally coinvented by O'Reagan and Grätzel [1229]. Dyes (sensitizer = **D**) blended with nanoparticles of anatase (**A**), a crystalline form of TiO_2, harvest the light. A PET from the excited dye (**D***) to the mesoporous oxide generates $\mathbf{A}^+$ and an electron that reaches the collecting electrode (anode) made out of glass (or another transparent material) [1230] covered by a transparent layer of conductive oxide such as fluorine-doped tin dioxide (the working electrode: WE). This electron is conducted through an outer circuit to reach the counter electrode (CE), performing electrical work on the way. The counter electrode can be made out of glass or another material covered by a layer of conductive oxide, itself coated by a layer of Pt nanoparticles, the catalyst onto which the electron reduces the oxidized form of the redox couple. One redox couple that has been used a lot is I_3^- (the oxidized form: I_2 + LiI, or KI, or Bu_4NI) and $3I^-$ (the reduced form). Thus, the electron returns to the cell through the reaction $2e^- + I_3^- \rightarrow 3I^-$. The anions I^- flow through the electrolyte from the Pt catalyst to the anatase nanoparticles covered by the dye where the oxidized form of the dye is reduced into its ground state via reaction $3I^- + 2\mathbf{D}^+ \rightarrow 2\mathbf{D} + I_3^-$. The current of anions I_3^- toward the counter electrode represents the hole transport. Among a variety of DSSC designs, Grätzel and coworkers have applied ruthenium-based dyes such as **N3** (Figure 6.15) adsorbed on the TiO_2 thin film, thereby achieving

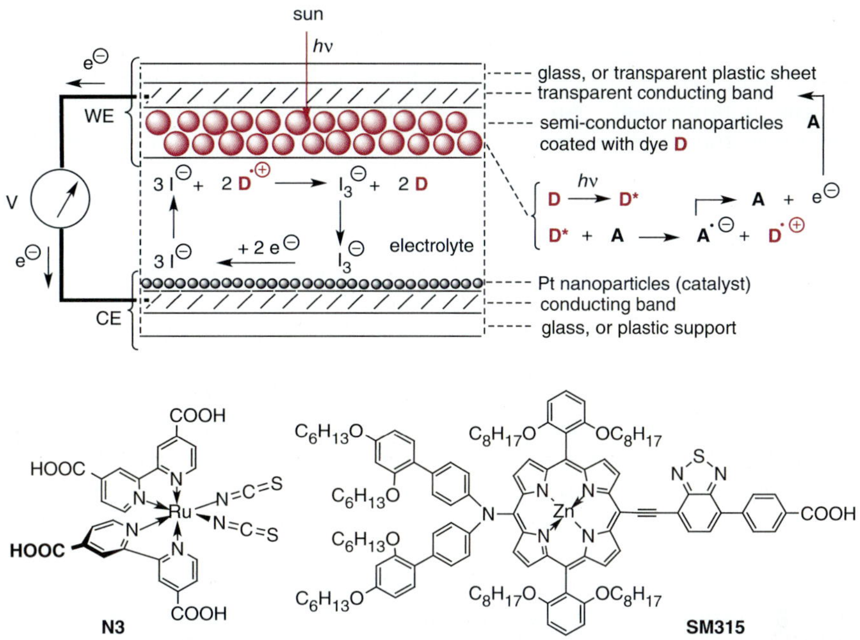

Figure 6.15 Schematic representation of a dye-sensitized solar cell (WE = anode or working electrode, CE = counter electrode) and two examples of dye-sensitizers.

power conversion efficiency >10% [1231, 1232]. For coumarin or polyene-type organic dyes, Ushida and coworkers reported up to 9.2% solar efficiency [1233]. Chlorophyll (Section 6.9.2, Figure 6.13) is the pigment responsible for light absorption in photosynthesis. It consists of a magnesium–porphyrin ring structure linked to a hydrocarbon tail [1234]. With the zinc–porphyrin-containing dye **SM315** (Figure 6.15), Grätzel and coworkers obtained 13% solar power conversion efficiency [1235, 1236].

Among light absorbers other than organometallic dyes, the perovskites of type $MeNH_3{\cdot}PbX_3$ (X = Cl, Br, I) have been very much studied in the recent years. They have shown solar energy conversion efficiency up to 22.1% [1237]. Unfortunately, they suffer from instability and represent a problem to the environment (Pb). However, in 2017, the Grätzel group presented a device based on $(HOOC(CH_2)_4NH_3)_2{\cdot}PbI_4/MeNH_3{\cdot}PbI_3$ that maintains nearly the same power conversion efficiency for one year [1238].

6.11 Chemiluminescence and bioluminescence

Chemiluminescence (CL) [1239–1243] and bioluminescence (BL) [1244–1246] are electromagnetic radiations (UV, visible, or IR light emission) produced when a chemical or biochemical reaction yields electronically excited intermediates or products, which either emit light (direct CL) or donate their energy to other molecules responsible for the emission (indirect or sensitized CL). Chemiluminescence at room temperature and in the absence of a flame has attracted man since he discovered fireflies and other luminescent (bioluminescent) organisms including bacteria, marine organisms, and insects. Several man-invented reactions engaging organic and/or inorganic compounds are chemiluminescent, and some of them with quantum yields of luminescence greater than those of bioluminescent living species [1247]. Firefly emission has a quantum yield of 0.41 [1248]. In principle, one molecule of reactant can yield one quantum $h\nu$ of light. In a 100% efficient chemiluminescent reaction, 1 mol of reactant produces 1 einstein (1 mol of quanta) of light. This corresponds to 67 000 lm·hour of light, the output of a 40-W incandescent bulb operating for six days [1249]. Modern instruments (photomultiplier associated with electronics that count photons) can detect very low intensity light emissions; thus, chemiluminescence has become a very powerful tool of analytical chemistry (e.g. trace analysis [1250]) and biology (e.g. protein detection) [1251–1258]. One of the oldest known chemiluminescent reactions is that of elemental phosphorous burning in moist air that produces a green glow above the solid. Vapor of phosphorous reacts with $^3O_2$ producing $(PO)_2$ and HPO in their electronically excited states [1249, 1259, 1260]. Hydrogen peroxide reacts with hypohalides and with carbonitriles forming singlet oxygen that emits light (Section 6.9.2) [1261]. Acidified $KMnO_4$ oxidizes organic and inorganic compounds often accompanied with chemiluminescence [1262], reactions useful in analytical chemistry [1263–1265]. Oxygenation of Grignard reagents can be luminescent [1266]. Auto-oxidation of hydrocarbons can also be chemiluminescent. For instance, ethylbenzene generates the triplet excited state of acetophenone (lifetime 10^{-6} to 10^{-7} seconds) detected by its phosphorescence [1242]. A blue luminescence is observed by the reaction of oxalyl chloride and hydrogen peroxide that generates monoperoxyoxalic chloride ($ClCOCOCl + H_2O_2 \rightarrow ClCOCO_3H$) as an intermediate that decomposes according to $ClCOCO_3H \rightarrow HCl + 2CO_2 + O_2 + h\nu$ (Section 6.11.4) [1267]. Instruments testing air quality detect the light emitted by the reaction of ozone with nitric oxide: $O_3 + {}^{\bullet}N{=}O \rightarrow O_2 + NO_2{}^{\bullet} + h\nu$ [1268, 1269]. Nitric oxide plays an important role as a signaling molecule in smooth muscle tonus regulation, neurotransmission, and the immune response. It can also be analyzed in biological samples through its chemiluminescent reaction with ozone [1259, 1270]. The reactions of ozone with organic compounds can also be chemiluminescent [1271]. In the case of furan + O_3, fluorescence from the first singlet states of *cis*- and *trans*-glyoxal is observed [1272].

6.11.1 Thermal isomerization of Dewar benzene into benzene

When a reaction becomes exothermic enough, its energy hypersurface may intersect or get close in energy to that of the electronically excited state of the product, or of a reaction intermediate, and the latter may emit light from its singlet (fluorescence) or triplet excited state (phosphorescence). Quite often, the quantum yield for emission (number of photons emitted/number of molecules of reactant converted) is low and no luminescence can be detected. However, one can add a suitable quencher (or activator (ACT), see below) to the reaction mixture that quenches the excitation energy of the excited product or reaction intermediate. Then, the singlet or triplet excited state of the quencher can emit light. A suitable quencher is a compound efficient in the excitation energy transfer (reaction (6.147)) and that has a good quantum yield for light emission (see the Jablonski diagram, Figure 6.6, Section 6.2.2). This is illustrated (Figure 6.16) with the thermal isomerization of Dewar benzene into benzene [1273], a reaction with an exothermicity of −60 kcal mol^{-1} [1274, 1275] and that requires an activation enthalpy of 23 kcal mol^{-1} [1276, 1277]. In its transition state, the energy reaches c. 83 kcal mol^{-1} above the ground state of benzene, $[PhH]_0$, which is enough (at 25 °C) to populate its triplet excited state, $^3[PhH]^*$, but not its singlet excited state, $^1[PhH]^*$. Quantum yield for the phosphorescence of benzene ($^3[PhH]^* \rightarrow [PhH]_0 + h\nu$) is extremely low (c. 10^{-9}) [1278]; thus, in the absence of any quencher, the thermal isomerization of Dewar benzene into benzene does not emit visible light. In the presence of DPA or 9,10-diphenylethynylanthracene (DPEA) that are singlet–singlet quenchers of donors with singlet excited state higher than 74 kcal mol^{-1}, a very weak emission is detected, which corresponds to the fluorescence of these two quenchers. Thus, $^1[PhH]^*$ is little populated during the thermal isomerization of Dewar benzene, consistent with its high energy (115 kcal mol^{-1} above the ground state of benzene).

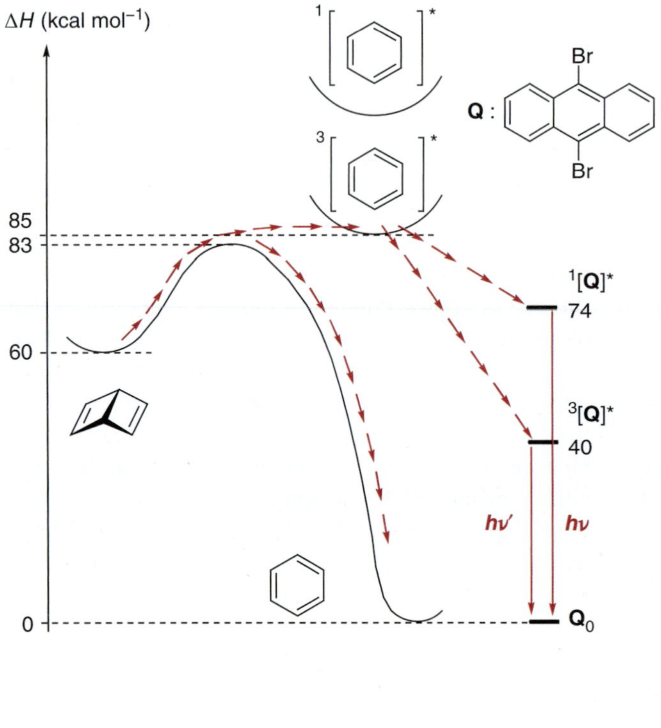

Figure 6.16 Thermal isomerization of Dewar benzene into benzene generates a small amount of benzene in its triplet state that is quenched by 9,10-dibromoanthracene, the light emitter.

However, in the presence of 9,10-dibromoanthracene (DBA), which is able to quench excited molecules in their singlet and triplet states as well [1279], an intense luminescence accompanies the thermal isomerization of Dewar benzene into benzene. It is estimated that only one molecule in 10^3–10^4 of Dewar benzene produces a triplet benzene [1273].

Under comparable conditions, benzvalene and prismane are also isomerized thermally into benzene, but they do not emit light in the presence of DBA. Interestingly, the thermal isomerization at 63 °C of 3,3′-Di(3-methylcyclopropenyl) into 1,2-, 1,3-, and 1,4-dimethylbenzene ($E_a = 37 \pm 2$ kcal mol^{-1}) is luminescent in the presence of DBA. At 63 °C, the luminescence is constant for hours. When the sample is heated quickly to 200 °C for 50 seconds and then quickly cooled back to 63 °C (temperature jump), the increased initial light emission is no longer constant, but it decays with a first-order rate constant of c. 2×10^{-4} s^{-1}. This observation is consistent with the formation of one or several intermediates that are responsible for the chemiluminescence. They are proposed to be dimethyl Dewar benzenes that rearrange into xylenes [1280]. The latter experiment shows that chemiluminescence can be used to detect unstable intermediates or products of reaction formed at a very low concentration.

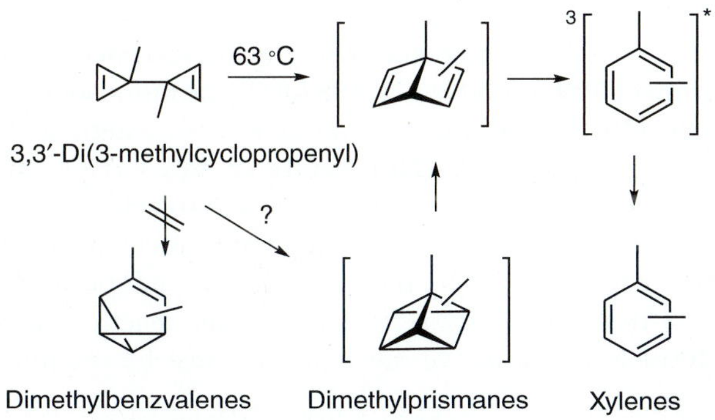

6.11.2 Oxygenation of electron-rich organic compounds

The first discovered chemiluminescent organic compound is probably lophine (**398**, Ar = Ph) [1281]. It reacts at 25 °C with $^3O_2$ giving mostly the corresponding phenylamidine salt **401** with emission of a blue–green light (Scheme 6.107) [1282]. The reactions of 2-aryl-4,5-diphenylimidazoles **398** with $^3O_2$ involve the formation of acyclic peroxides **399** that equilibrate with the corresponding 1,2-dioxetane intermediates **400**. The latter undergo formal [σ^2+σ^2]-cycloreversion producing salts **401** and light [1283, 1284]. With 2-phenyl-4-(4-fluorophenyl)-5-(4-trifluorophenyl) imidazole, it was shown that chemiluminescence may also arise from acyclic peroxides of type **402** [1285]. A concurrent reaction for acyclic peroxides **399** is the cyclization into the endoperoxides **402** that decompose into salts **403** and singlet oxygen, as evidenced by its reaction with 2,5-diphenylbenzo[c]furan [1286]. The dioxetane path, which is chemiluminescent, is favored for electron-donating 2-aryl groups such as 4-(Me_2N)C_6H_4, whereas with electron-withdrawing 2-aryl groups such as 4-(NO_2)C_6H_4, the nonchemiluminescent path forming $^1O_2$ is favored [1287]. Auto-oxidation of lophine and derivatives represent a useful tool in analytical chemistry [1288].

Auto-oxidation of tetrakis(dimethylamino)ethylene (**404**) produces mostly tetramethylurea, tetramethyloxamide (**405**), and tetramethylhydrazine accompanied with green light emission [1289], which corresponds to the fluorescence spectrum of **404** [1290]. Alkene **404** has become an oxygen probe in analytical chemistry [1291]. Product analysis and kinetics studies in solution containing an alcohol ROH [1292, 1293], and in the gas phase [1294], led to the mechanism outlined in Scheme 6.108. The electron-rich alkene **404** reacts with $^3O_2$ through an ET giving the pair $404^{\bullet+} + O_2^{\bullet-}$ that undergoes cyclization into dioxetane **406** concurrently with

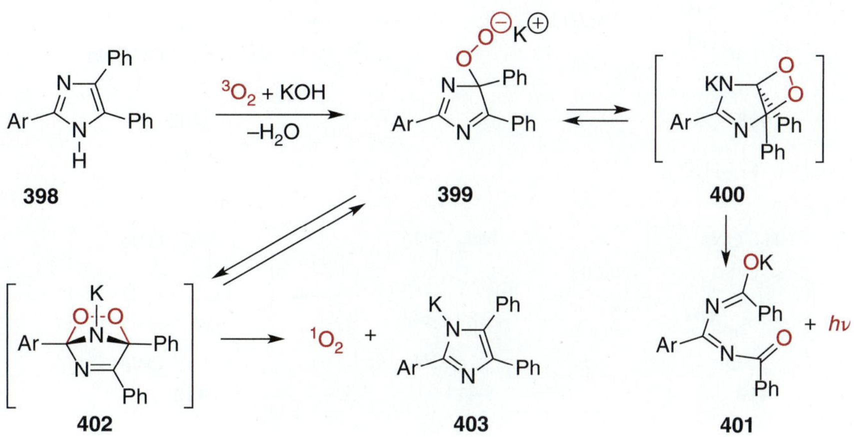

Scheme 6.107 Auto-oxidation of lophine and derivatives.

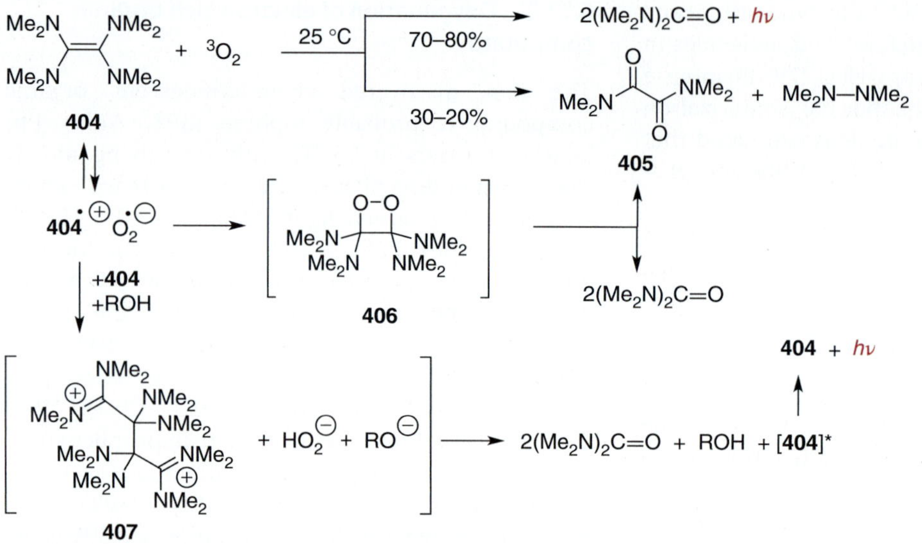

Scheme 6.108 Auto-oxidation of tetrakis(dimethylamino) ethylene.

its reaction with the alcohol ROH producing **407** + HO_2^- + RO^-. Radical cation **404**$^{\bullet+}$ dimerizes into dication **407**, which then reacts with HO_2^- giving 2 $(Me_2N)_2C{=}O$, ROH and the electronically excited ethylene [**404**]*. The latter emits light (fluorescence). In parallel, dioxetane **406** fragments into two different ways: one of them ((2+2)-cycloreversion) produces tetramethylurea and the other one involves the elimination of tetramethylhydrazine giving a 1,2-dioxetene that is immediately isomerized (electrocyclic ring opening) into oxamide **405**. In the gas phase, the initial product of reaction of $^3O_2$ with alkene **404** is not a zwitterion. It is a 1,4-diradical [$^\bullet$O—O—$(Me_2N)_2$C—C$^\bullet$$(NMe)_2$]. Compared with the diradical $^\bullet$O—O—CH_2—C$^\bullet$$Me_2$ arising from the reaction of isobutylene + $^3O_2$, which is 18.6 kcal mol^{-1} higher in enthalpy than the reactants, this diradical enjoys stabilization by two dimethylamino substituents (one Me_2N group stabilizes the methyl radical by c. 20 kcal mol^{-1}, Table 1.A.9).

Since the discovery in 1928 by Albrecht [1295] of the chemiluminescent oxidation of luminol (**408**: 5-amino-1,2,3,4-tetrahydrophthalazin-1,4-dione) with H_2O_2 in the presence of metallic ions (catalyze the decomposition of H_2O_2 into $O_2 + H_2O$), several hydrazides have been found to undergo chemiluminescent oxidations. White et al. established that the auto-oxidation of **408** in DMSO or DMF is almost quantitative in the formation of the 3-aminophthalate salt **409** + N_2 and requires 2 equiv. of NaOH and 1 equiv. of $^3O_2$. The spectrum of the emitted light (λ_{max} = 424 nm) coincides with the fluorescence spectrum of **409**. With NaOH, **408** forms the monosodium salt **410**. Kinetic measurements are consistent with pre-equilibrium of **410** + NaOH ⇄ **411** + H_2O and give a rate of light emission = k[**410**][HO^-][$^3O_2$] (Scheme 6.109) [1296, 1297]. A possible mechanism of the reaction could be the formation of endoperoxide **412** that undergoes retro-Diels–Alder reaction into **409** + N_2. Alternative mechanisms must be retained

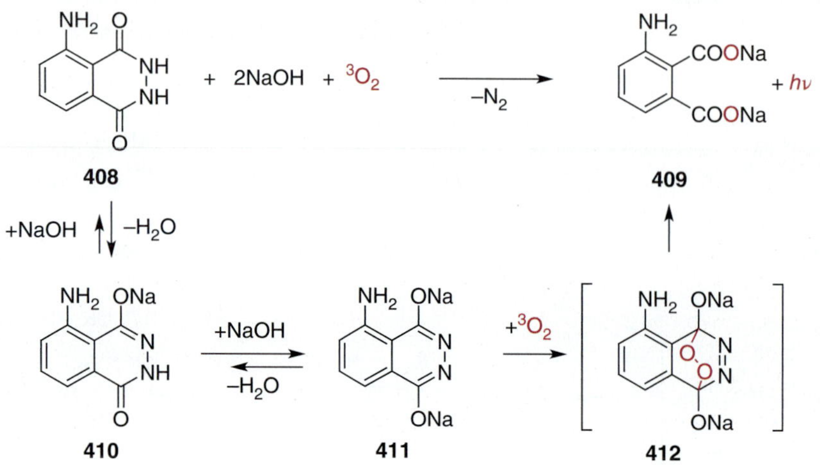

Scheme 6.109 Auto-oxidation of luminol in alkaline medium.

Scheme 6.110 Reactions of cycloalkynes with $^3O_2$ are chemiluminescent. Heats of formation are from the NIST Chemistry WebBook and Table 1.A.2. The heat of didehydrogenation (c. 31 kcal mol^{-1}) of 1,2-dioxetane into dioxetene is taken as $\Delta_f H°$(cyclobutene) $-\Delta_f H°$(cyclobutane) = 37.5 (±0.4) − 6.6 kcal mol^{-1}.

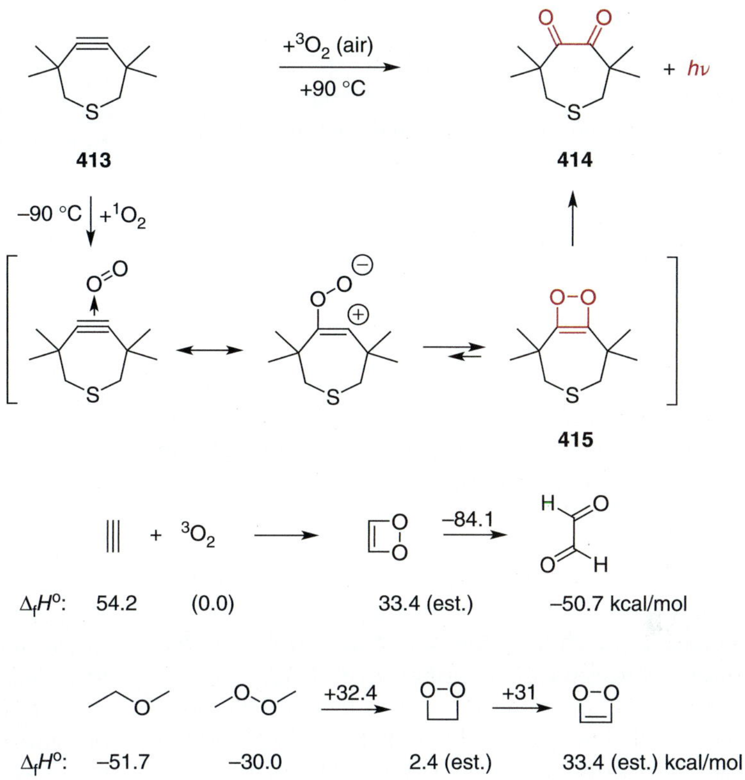

as acyclic hydrazides also undergo chemiluminescent auto-oxidations [1298–1300].

Subnanomolar detection for H_2O_2 using luminol chemiluminescence and cobalt(II) salts as a catalyst has been established [1301]. Luminol is used to detect blood traces at crime scenes, as iron found in hemoglobin catalyzes its oxidation by H_2O_2. Biologists use chemiluminescence of luminol in cellular assays for the detection of copper, iron, and cyanides, as well as of specific proteins by western blot [1302]. Oxidation of hydrazine and its derivatives can also be chemiluminescent [1303, 1304]. Chemiluminescence of luminol can be enhanced by additives such as phenylhydrazine [1305] and hydrazine [1306].

The strained cycloalkyne **413** reacts with $^3O_2$ at 90 °C giving 1,2-diketone **414** with light emission that corresponds to the fluorescence of the latter, showing that the singlet excited state of dione **414** forms in this reaction. The same chemiluminescent reaction is observed at −90 °C when **413** reacts with $^1O_2$. The reaction generates dioxetene **415** as an intermediate that undergoes quick electrocyclic ring opening into ^{1}[**414**]* [1307]. Thermochemical data give an exothermicity of c. −22 kcal mol^{-1} for the (2+2)-cycloaddition of $^3O_2$ to acetylene and an exothermicity of −84.1 kcal mol^{-1} for the isomerization of dioxetene into glyoxal (Scheme 6.110). If one considers the extra ring strain imposed by the annulation of the dioxetene unit of **415** with the thiacycloheptene moiety, the exothermicity of reaction **415** → **414** is surely high enough (>85 kcal mol^{-1}) to populate ^{1}[**414**]*. For *s-cis*-glyoxal, the absorption spectrum gives λ = 487.5 nm for the ground state (ν_0) → singlet excited state (ν'_0) transition [1308], which corresponds to ΔE = 58.6 kcal mol^{-1} (Figure 6.1). The same electronic transition in dione **414** is not expected to be higher in energy than 85 kcal mol^{-1} even if the two carbonyls should deviate somewhat from planarity.

Problem 6.79 Lucigenin (dimethyl diacridylium salt **A**) and its derivatives react with H_2O_2 in alkaline solution with emission of green to blue light [1309, 1310]. Propose a mechanism for this reaction.

A:

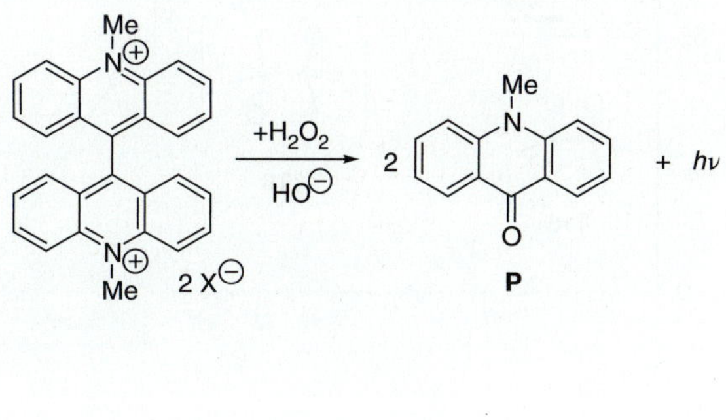

6.11.3 Thermal fragmentation of 1,2-dioxetanes

Before 1,2-dioxetanes had been prepared, they had been proposed to be intermediates in the reactions responsible for the bioluminescence of fireflies [1311, 1312]. In 1969, Kopecky and Mumford prepared 3,3,4-trimethyl-1,2-dioxetane and showed that it fragments into acetaldehyde + acetone on heating to 60–70 °C in benzene solution [1070]. The reaction is chemiluminescent with λ_{max} of emission between 430 and 440 nm depending on the initial concentration (emission from monomeric carbonyl compounds and their excimers). Similarly, heating tetramethyl-1,2-dioxetane (**416**) fragmentizes into acetone (reaction (6.162)) and is chemiluminescent (Figure 6.17) [1313–1315]. With an exothermicity of −61 kcal mol^{-1} and an enthalpy of activation of 25 kcal mol^{-1}, reaction (6.165) provides a total of 86 kcal mol^{-1} of energy in its transition state that is enough to reach both the singlet and triplet excited state of the product (2 acetone) [1316, 1317].

Quantum mechanical calculations [1318–1320] suggest that **416** undergoes first breakage of the O—O bond into diradical **418** via twisted transition state **417**. The process goes from the lowest singlet potential hypersurface (S_0) of the 1,2-dioxetane to the lowest triplet hypersurface (T_1) of products; the T_1 hypersurface and the S_0 hypersurface cross each other twice in the region of the diradical **418**. Spin–orbit interaction should be effective at the first crossing point because of the orthogonality between the π- and σ-localized orbitals at the oxygen centers. Thus, the initially formed diradical transfers from the S_0 to the T_1 hypersurface and then undergoes C—C bond cleavage to give a triplet excited acetone + a ground-state acetone molecule. There is a small energy barrier that separates diradical **418** from the triplet excited products (Figure 6.17). With alkyl-substituted 1,2-dioxetane, the quantum yield of chemiluminescence is generally low (<0.01%) [1321] as they are preferentially deactivated by nonradiative pathways or are quenched by traces of dissolved $^3O_2$. However, when dioxetane is substituted by an aromatic or heteroaromatic ring with low oxidation potential, the decomposition rate, the emission intensity, and the chemiluminescence quantum yield are significantly enhanced [1322, 1323]. In 1982, Schaap and Gagnon reported that the thermal decomposition of dioxetane **420** (obtained by (2+2)-cycloaddition of $^1O_2$ to alkene **419**) is strongly accelerated upon deprotonation of its phenolic moiety (Scheme 6.111) [1324]. The same group reported quantum yield of chemiluminescence (fluorescence, emission from singlet excited states of products) reaching 50% (0.50 einstein mol^{-1}). Heating dioxetane **422** in toluene to 80 °C ($\Delta^{\ddagger}H \approx 25$ kcal mol^{-1}, $\Delta^{\ddagger}S \approx 0$ eu, rate constant c. 0.003 s^{-1}) produces aldehyde **423** and butanone with a weak chemiluminescence. Using DPA as a singlet quencher and DBA as a triplet quencher, chemiexcitation quantum yields of emission $\varphi^{T} \approx 0.07$ (from triplet exited state of **423**) and $\varphi^{S} \approx 0.004$ (from the singlet excited state of **423**) have been determined. Desilylation of **422** in THF by Bu_4NF leads to a visible yellow flash with λ_{max} = 550 nm corresponding to the emission from the singlet excited sate of the phenolate salt of aldehyde **423**. The phenolate salt of dioxetane **422** is decomposed ($\Delta^{\ddagger}H \approx 20.5$ kcal mol^{-1},

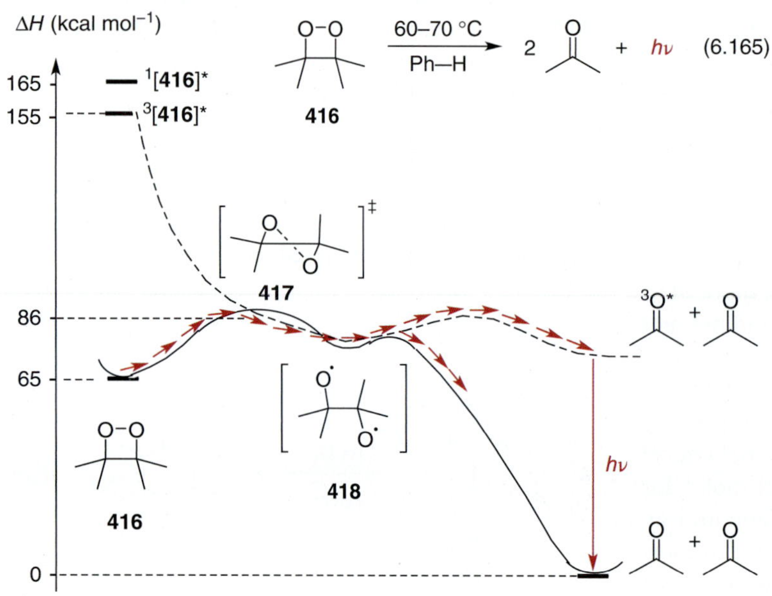

Figure 6.17 Representation of the chemiluminescent thermal fragmentation of tetramethyl-1,2-dioxetane into acetone.

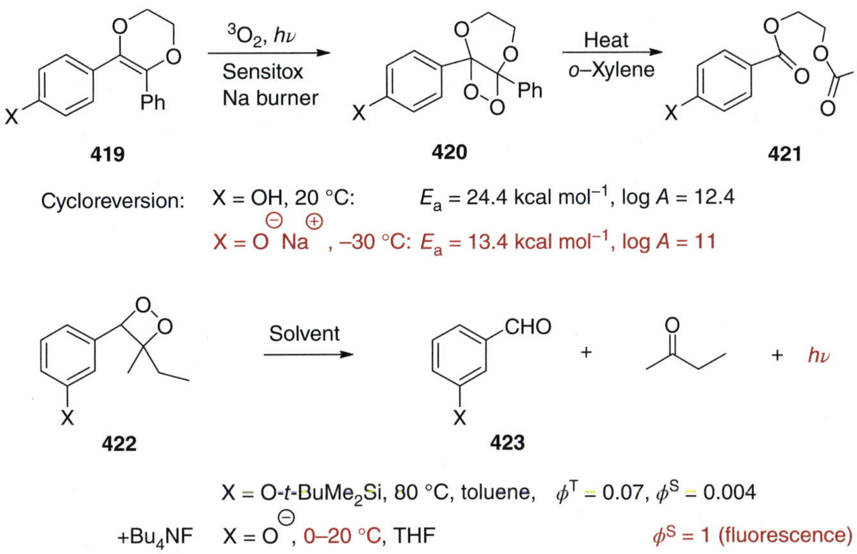

Scheme 6.111 An electron-rich aryl substituent accelerates the thermal fragmentation of 1,2-dioxetanes.

$\Delta^{\ddagger}S \approx 5$ eu) c. 10^4 times as quick as the silyl ether **422**, with a quantum yield of fluorescence reaching 100% [1325]. Chemiluminescence efficiency depends on the conformation of the oxidoaryl group in aryl substituted 1,2-dioxetanes [1326].

In the presence of electron-rich aromatic compounds such as *N*,*N*-dimethyldihydrophenazine, rubrene, Ph_3N, pyrene, or anthracene (intermolecular activator: ACT), the decomposition of dimethylperoxylactone (**424**) is strongly accelerated. Schuster and coworkers formulated a chemiexcitation mechanism involving a rate-determining electron transfer from the ACT to the peroxide, called the CIEEL (chemically initiated electron exchange luminescence) mechanism [1327–1332]. The intermolecular CIEEL process (Scheme 6.112) is initiated by the formation of a charge transfer complex **425**. In this complex, the O—O bond is weakened by electron transfer from the electron-rich ACT to the σ*(O—O) LUMO. This facilitates the O—O bond breakage and generates radical anion/radical cation pair **426**. The fragmentation of **426** generates either CO_2 + acetone + ACT without light production, or $CO_2 + ACT^{\bullet+}$ + ketyl radical anion $Me_2CO^{\bullet-}$. Return electron transfer produces the singlet excited ACT which fluoresces. Similar mechanisms (intramolecular CIEEL) apply to the thermal decomposition of aryl-substituted dioxetanes like **420**.

The thermal unimolecular decomposition of 1,2-dioxetanes substituted with an acridinium moiety in a spiro fashion like in **427** have a high quantum yield of chemiluminescence (up to 15%: 0.15 einstein mol^{-1}). They generate ketone **428** and aldehydes **429** [1333]. The decomposition rate constants of **427** show linear free-energy correlation for electron-withdrawing substituents X, with Hammett reaction constant of $\rho = 1.3 \pm 0.1$. This is consistent with a mechanism involving an endothermic intramolecular ET from the nitrogen lone pair of the acridinium moiety to the 1,2-dioxetane system with simultaneous O—O bond cleavage in the rate-determining step as implied for the intramolecular CIEEL mechanism [1334]. Methyl substitution of the acridine moiety of acridine-containing

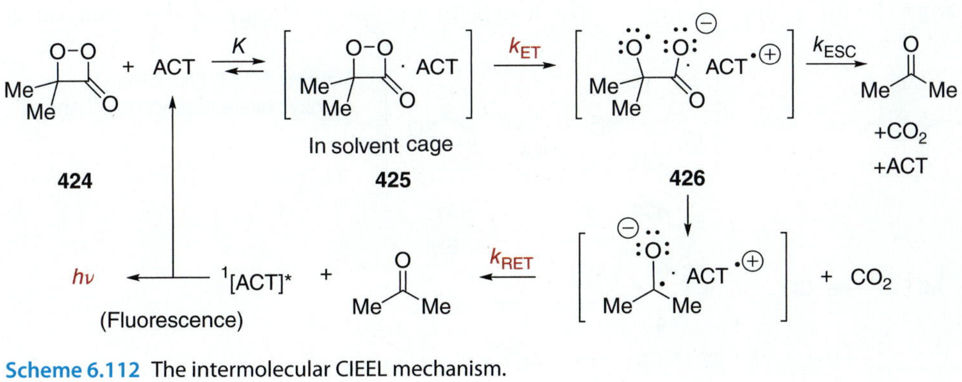

Scheme 6.112 The intermolecular CIEEL mechanism.

1,2-dioxetanes accelerates their chemiluminescent fragmentations. The more electron-rich the acridine ring, the faster the intramolecular electron transfer [1323].

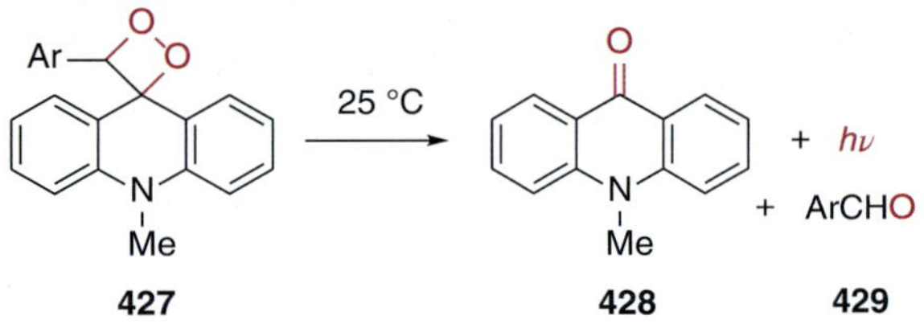

6.11.4 Peroxylate chemiluminescence

In 1963, Chandross discovered the blue chemiluminescent reaction of oxalyl chloride with hydrogen peroxide in the presence of DPA [1267]. Aromatic oxalic esters, like bis(2,4,6-trichlorophenyl) oxalate, react under basic conditions with hydrogen peroxide, producing light with high quantum yield if an appropriate ACT is present, for instance rubrene ($\varphi^S \approx 0.68$) [1322, 1323]. It has been proposed that high-energy intermediates (HEIs) such as **430** and **431** are formed during the elementary peroxyoxalate reaction (Scheme 6.113).

The reaction of $ClCOCOCl + H_2O_2$ in the presence of DPA (=ACT) at 20 °C results in low-intensity chemiluminescence emission that corresponds to the fluorescence of the ACT. However, delayed addition of ACT to the reaction mixture gives rise to an intense flash of light, which decays within a few seconds. The observed rate constant k_{obs} obtained upon delayed ACT addition depends on the final concentration of the latter, consistently with a bimolecular interaction between HEI and ACT and the occurrence of a CIEEL mechanism in the chemiexcitation step. This excludes an alternative mechanism in which the ACT would be excited by energy transfer (reaction (6.147)). The HEI decomposition rate constant in the absence of ACT (k_D) and the catalytic rate constant (k_{CAT}) can be calculated from the linear dependence of the observed k_{obs} on the concentration ACT added in delay ($k_D \approx 0.096\ s^{-1}$, $k_{CAT} \approx 3740\ M^{-1}\,s^{-1}$, 20 °C). Activation parameters for the unimolecular decomposition of HEI ($\Delta^{\ddagger}H \approx 11.2$ kcal mol^{-1}, $\Delta^{\ddagger}S \approx -23$ eu) and for the bimolecular reaction with the ACT ($\Delta^{\ddagger}H \approx 4.2$ kcal mol^{-1}, $\Delta^{\ddagger}S \approx -27$ eu) show that HEI is much less stable than 1,2-dioxetanes and 1,2-dioxetanones and demonstrate the highly favored interaction with the ACT. A linear free-energy correlation between k_{CAT} and the oxidation potentials of several ACTs has been observed, which confirms the intervention of a CIEEL mechanism [1335].

6.11.5 Firefly bioluminescence

Bioluminescence is a natural phenomenon that emits visible light produced by a chemical reaction within a living organism. The North American firefly (*Photinus pyralis*) is one of the most luminous and the most studied species [1336]. The enzyme, luciferase, catalyzes a two-step reaction, by which electronically excited oxyluciferin (OLH) is formed (Scheme 6.114). ^{1}[OLH]* and its conjugate base ^{1}[OL]* are the light emitters. For monomeric OLH in a water solution, an absorption band at 380 nm is observed under acidic/neutral pH and at 420 nm under basic pH. OLH can exist in six different species that arise from the keto/enol equilibrium, and from reversible deprotonation of the phenol or enol moieties. As OLH is unstable in a basic medium, one does not know much about the structure of its deprotonated forms. Nevertheless, it is believed that under bioluminescence conditions, OLH exists in its anionic keto from, whereas in water solution, it exists as an enol species [1337]. Bioluminescence ranges from λ_{max} = 530–640 nm depending on the pH, polarity of the medium [1338], microenvironment of the enzyme [1339], and several other factors [1340, 1341]. OLH forms π/π stacking complexes both in the ground state and excited states. Depending on pH, these complexes may adopt different conformations, which explain the lower energy of the emission at

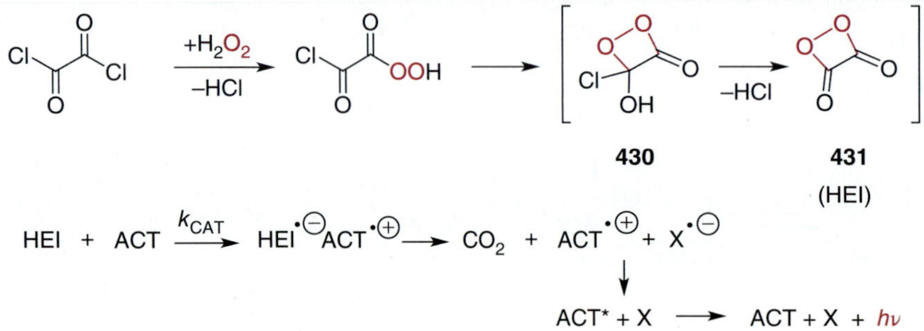

Scheme 6.113 Elementary peroxyoxalate reaction mechanism.

Scheme 6.114 Mechanism of the firefly bioluminescence.

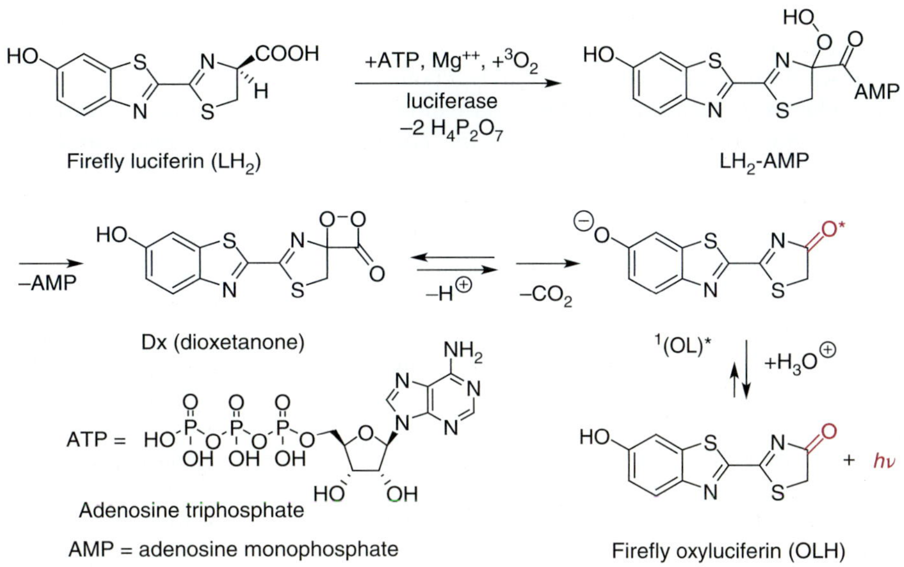

acidic/neutral pH, in comparison with the emission at basic pH [1342]. The process requires 1 equiv. of molecular oxygen, ATP (adenosine triphosphate), and Mg^{++} as cofactors (cocatalysts). The first step corresponds to the trapping by $^3O_2$ of the carbanion α to the carboxylic group giving LH_2-AMP. Then, an intramolecular substitution generates a dioxetanone (Dx) + AMP. Deprotonation of the phenol group of the dioxetanone triggers the chemiluminescent reaction by forming the dioxetanone anion that fragments into CO_2 and the singlet excited state of OLH. Several mechanisms have been proposed to explain the chemiluminescence of dioxetanone Dx. As the latter compound is analogous to dioxetanes substituted by electron-rich aromatic groups (Scheme 6.114), an intermolecular CIEEL mechanism is quite plausible, but other mechanisms cannot be rejected [1343–1347].

Bioluminescent imaging (BLI) has been applied in the imaging of all kinds of processes in the life sciences such as monitoring enzyme activity, detecting bioactive small molecules, evaluating the properties of molecular transporters, etc. [1348] For most BLI techniques, the firefly luciferin/luciferase system is used in vivo and in vitro. Simple synthetic firefly luciferin analogs emitting blue, green, and near-infrared light have been developed and are expected to expand the field of BLI applications [1349, 1350].

References

1 Horspool, W.M. and Lenci, F. (eds.) (2003). *CRC Handbook of Organic Photochemistry and Photobiology*, 2e, vol. 1 and 2, 1–2904. Boca Raton, FL: CRC Press.

2 Griesbeck, A., Oelgemüller, M., and Ghetti, F. (eds.) (2012). *CRC Handbook of Organic Photochemistry and Photobiology*, 3e, vol. 1 and 2, 1–1694. Boca Raton, FL: CRC Press.

3 (a) Hoffmann, N. (2008). Photochemical reactions as key steps in organic synthesis. *Chem. Rev.* 108 (3): 1052–1103. (b) Karkas, M.D., Porco, J.A., and Stephenson, C.R.J. (2016). Photochemical approaches to complex chemotypes: applications in natural product synthesis. *Chem. Rev.* 116 (117): 9683–9747.

4 Ralph, M.J. (2011). Organic photochemistry in the construction of heterocyclic compounds. *Curr. Org. Chem.* 15 (15): 2658–2672.

5 Hoffmann, N. (2012). Photochemical reactions of aromatic compounds and the concept of the photon as a traceless reagent. *Photochem. Photobiol. Sci.* 11 (11): 1613–1641.

6 Nicolaou, K., Becker, J., Lim, Y.H. et al. (2009). Total synthesis and biological evaluation of (+)- and (−)-bisanthraquinone antibiotic BE-43472B and related compounds. *J. Am. Chem. Soc.* 131 (41): 14812–14826.

7 Hu, W., Qin, H., Cui, Y., and Jia, Y. (2013). Total synthesis of (+)- and (−)-decursivine and (+/-)-serotobenine through a cascade Witkop photocyclization/elimination/addition sequence:

scope and mechanistic insights. *Chem. Eur. J.* 19 (9): 3139–3147.

8 (a) Bach, T. and Hehn, J.-P. (2011). Photochemical reactions as key steps in natural product synthesis. *Angew. Chem. Int. Ed.* 50 (5): 1000–1045. (b) Liu, W. and Li, C.-J. (2017). Recent synthetic applications of catalyst-free photochemistry. *Synlett* 28 (20): 2714–2754.

9 Hoffmann, N. (2012). Photochemical key steps in organic synthesis. In: *CRC Handbook of Organic Chemistry and Photobiology*, 3e (ed. A. Griesbeck, M. Oelgemöller and F. Ghetti), 95–124. Boca Raton, FL: CRC Press.

10 Wegner, J., Ceylan, S., and Kirschning, A. (2012). Flow chemistry – a key enabling technology for (multistep) organic synthesis. *Adv. Synth. Catal.* 354 (1): 17–57.

11 Oelgemöller, M. and Shvydkiv, O. (2011). Recent advances in microflow photochemistry. *Molecules* 16 (9): 7522–7550.

12 (a) Oelgemöller, M. (2012). Highlights of photochemical reactions in microflow reactors. *Chem. Eng. Technol.* 35 (7): 1144–1152. (b) Politano, F. and Oksdath-Mansilla, G. (2018). Light on the horizon: current research and future perspectives in flow photochemistry. *Org. Proc. Res. Dev.* 22 (9): 1045–1062.

13 Conradi, M. and Junkers, T. (2013). Efficient $[\pi^2+\pi^2]$-photocycloadditions under equimolar conditions by employing a continuous UV-flow reactor. *J. Photochem. Photobiol., A* 259: 41–46.

14 Braslavsky, S.E. and Houk, K.N. (1988). Glossary of terms used in photochemistry. *Pure Appl. Chem.* 60 (7): 1055–1106.

15 Anslyn, E.V. and Dougherty, D.A. (2006). Photochemistry. In: *Modern Physical Organic Chemistry*, 935–1000. Sausalito, CA: University Science Books.

16 Turro, N.J., Ramamurphy, V., and Scaiono, J.C. (2010). *Modern Molecular Photochemistry of Organic Molecules*, 1–1110. Sausalito, CA: University Science Books.

17 Turro, N.J. (1978). *Modern Molecular Photochemistry*, 1–628. Menlo Park, CA: The Benjamin/Cummings Publishing Company.

18 Poliakoff, M. and Licence, P. (2007). Sustainable technology: green chemistry. *Nature* 450: 810–812.

19 Sheldon, R.A. (2007). The E factor: fifteen years on. *Green Chem.* 9: 1273–1283.

20 Dichiarante, V. and Protti, S. (2012). Photochemistry in ecosustainable syntheses. In: *CRC Handbook of Organic Photochemistry and Photobiology*, 3e (ed. A. Griesbeck, M. Oelgemöller and F. Lenci), 213–236. Boca Raton, FL: CRC Press.

21 (a) Protti, S. and Fagnoni, M. (2009). The sunny side of chemistry: green synthesis by solar light. *Photochem. Photobiol. Sci.* 8 (11): 1499–1516. (b) Chen, Y., Lu, L.-Q. and Yu, D.-G. et al. (2019). Visible light-driven organic photochemical synthesis in China. *Sci. China Chem.* 62 (1): 24–57.

22 Coyle, E.E. (2012). Solar photochemistry from the beginnings of organic photochemistry to the solar production of chemicals. In: *CRC Handbook of Organic Photochemistry and Photobiology*, 3e (ed. A. Griesbeck, M. Oelgemöller and F. Lenci), 237–248. Boca Raton, FL: CRC Press.

23 Yoon, T.P., Ischay, M.A., and Du, J. (2010). Visible light photocatalysis as a greener approach to photochemical synthesis. *Nat. Chem.* 2 (7): 527–532.

24 (a) Prier, C.K., Rankic, D.A., and MacMillan, D.W. (2013). Visible light photoredox catalysis with transition metal complexes: applications in organic synthesis. *Chem. Rev.* 113 (7): 5322–5363. (b) Koenig, B. (2017). Photocatalysis in organic synthesis - past, present, and future. *Eur. J. Org. Chem.* 2017 (15): 1979–1981.

25 Schultz, D.M. and Yoon, T.P. (2014). Solar synthesis: prospects in visible light photocatalysis. *Science* 343 (6174): 985–993.

26 Koike, T. and Akita, M. (2013). Visible light-induced photoredox catalysis: an easy access to green radical chemistry. *Synlett* 24 (19): 2492–2505.

27 Narayanam, J.M. and Stephenson, C.R.J. (2011). Visible light photoredox catalysis: applications in organic synthesis. *Chem. Soc. Rev.* 40 (1): 102–113.

28 Ciamician, G. (1912). The photochemistry of the future. *Science* 36: 385–394.

29 Albini, A. and Fagnoni, M. (2004). Green chemistry and photochemistry were born at the same time. *Green Chem.* 6: 1–6.

30 Faraday, M. (1821). On two new compounds of chlorine and carbon, and on new compound of iodine and hydrogen. *Philos. Trans. R. Soc. London* 111: 47–74.

31 (a) Draper, J.W. (1861). Chemical action of light. In: *Text-book on Chemistry*, 99–105. New York, NY, USA, Chap. 23–24: Harper & Brothers, Publ. (b) Tidwell, T. (2013). Sunlight and free radicals. *Nature Chemistry* 5: 637–639.

32 Downes, A. and Blunt, T.P. (1877). Researches on the effect of light upon bacteria and

other organisms. *Proc. R. Soc. London* 26: 488–500.

33 Downes, A. and Blunt, T.P. (1879). Behavior of oxalic acid and the oxalates of the alkalies, and of potassium iodide, in sunlight. *Proc. R. Soc. London* 29: 219–221.

34 Hunt, J.P. and Taube, H. (1952). The photochemical decomposition of hydrogen peroxide. Quantum yields, tracer and fractionation effects. *J. Am. Chem. Soc.* 74 (23): 5999–6002.

35 Trommsdorff, H. (1834). Über Santonin. *Ann. Pharm. (Leipzig)* 11: 190–207.

36 Roth, H.D. (1989). The beginning of organic photochemistry. *Angew. Chem. Int. Ed. Engl.* 28 (9): 1193–1207.

37 Natarajan, A., Tsai, C.K., Khan, S.I. et al. (2007). The photorearrangement of α-santonin is a single-crystal-to-single crystal reaction: a long kept secret in solid state organic chemistry revealed. *J. Am. Chem. Soc* 129: 9846–9847.

38 Leigh, W.J. (1993). Techniques and applications of far-UV photochemistry in solution – the photochemistry of the C_3H_4 and C_4H_6 hydrocarbons. *Chem. Rev.* 93 (1): 487–505.

39 (a) Bernath, P.F. (2005). *Spectra of Atoms and Molecules*, Topics in Physical Chemistry, 2e, 1–454. Oxford: Oxford University Press. (b) Bernath, P.F. (2015). *Spectra of Atoms and Molecules*, Topics in Physical Chemistry, 3e, 1–504. Oxford: Oxford University Press.

40 Atkins, P.W. and Friedman, R.S. (2010). *Molecular Quantum Mechanics*, 5e, 1–537. Oxford: Oxford University Press.

41 Franck, J. (1926). Elementary processes of photochemical reactions. *Trans. Faraday Soc.* 21: 536–542.

42 Condon, E. (1926). A theory of intensity distribution in band systems. *Phys. Rev.* 28: 1182–1201.

43 Condon, E. (1928). Nuclear motions associated with electron transitions in diatomic molecules. *Phys. Rev.* 32: 858–872.

44 Raymonda, J.W. and Simpson, W.T. (1967). Experimental and theoretical study of σ-bond electronic transitions in alkanes. *J. Chem. Phys.* 47 (2): 430–447.

45 Sandorfy, C. (1981). Rydberg states in chemistry. *Int. J. Quantum Chem.* 19 (6): 1147–1156.

46 Russell, B.R., Edwards, L.O., and Raymonda, J.W. (1973). Vacuum ultraviolet absorption spectra of the chloromethanes. *J. Am. Chem. Soc.* 95 (7): 2129–2133.

47 Sandorfy, C. (1976). UV absorption of fluorocarbons. *Atmos. Environ.* 10 (5): 343–351.

48 Doucet, J., Sauvageau, P., and Sandorfy, C. (1975). Photoelectron and far ultraviolet absorption spectra of chlorofluoro derivatives of ethane. *J. Chem. Phys.* 62 (2): 355–359.

49 Doucet, J., Gilbert, R., Sauvageau, P., and Sandorfy, C. (1975). Photoelectron and far ultraviolet spectra of CF_3Br, CF_2BrCl, and CF_2Br_2. *J. Chem. Phys.* 62 (2): 366–369.

50 Robin, M.B., Basch, H., Kuebler, N.A. et al. (1968). Assignments in the ultraviolet spectra of olefins. *J. Chem. Phys.* 48 (11): 5037–5047.

51 Hazra, A., Chang, H.H., and Nooijen, M. (2004). First principles simulation of the UV absorption spectrum of ethylene using the vertical Franck–Condon approach. *J. Chem. Phys.* 121 (5): 2125–2136.

52 (a) Robin, M.B. (1974). *Higher Excited States of Polyatomic Molecules*, vol. 1, 1–390. New York: Academic Press. (b) Robin, M.B. (1985). *Higher Excited States of Polyatomic Molecules*, vol. 3, 1–465. Amsterdam: Elsevier Inc.

53 Merer, A.J. and Mulliken, R.S. (1969). Ultraviolet spectra and excited states of ethylene and its alkyl derivatives. *Chem. Rev.* 69: 639–656.

54 Wulfman, C.E. and Kumei, S. (1971). Highly polarizable singlet excited states of alkenes. *Science* 172: 1061–1061.

55 Martinez, R.B., Buitrago, A.A., Howell, N.W. et al. (1992). The near UV absorption spectra of several aliphatic aldehydes and ketones at 300 K. *Atmos. Environ.* 26A: 785–792.

56 Suhai, S. (1986). On the excitonic nature of the first UV absorption peak in polyenes. *Int. J. Quantum Chem.* 29: 469–476.

57 Woodward, R.B. (1942). Structure and absorption spectra III. Normal conjugated dienes. *J. Am. Chem. Soc.* 64: 72–75.

58 Woodward, R.B. (1942). Structure and absorption spectra, IV. Further observations on α,β-unsaturated ketones. *J. Am. Chem. Soc.* 64 (1): 76–77.

59 Fieser, L.F., Fieser, M., and Rajagopalan, S. (1948). Absorption spectroscopy and the structure of diosterols. *J. Org. Chem.* 13: 800–806.

60 Kalsi, P.S. (2007). *Spectroscopy of Organic Compounds*, 6e, 1–652. New Delhi: New Age International Publishers.

61 Zewail, A.H. (2000). Femtochemistry. Past, present, and future. *Pure Appl. Chem.* 72 (12): 2219–2231.

62 Zewail, A.H. (2000). Femtochemistry: atomic-scale dynamics of the chemical bond using ultrafast lasers – (Nobel lecture). *Angew. Chem. Int. Ed.* 39 (15): 2586–2631.

63 Pauli, W. (1925). Über den Zusammenhang des Abschusses des Elektronengruppen im Atom mit der Komplexstructur der Spektren. *Z. Angew. Phys.* 31: 765–783.
64 Rice, S.A. (1975). Dynamics of primary photochemical processes. In: *Excited States*, vol. 2 (ed. E.C. Lim), 111–320. Amsterdam, The Netherlands: Elsevier B.V.
65 Zalesskaya, G.A., Yakovlev, D.L., Sambor, E.G., and Bely, N.N. (2002). The triplet lifetime dependence on vibrational energy for gas-phase carbonyl aromatic compounds: benzophenone and fluorenone. *Phys. Chem. Chem. Phys.* 4 (22): 5634–5639.
66 Marian, C.M. (2012). Spin-orbit coupling and intersystem crossing in molecules. *Wiley Interdiscip. Rev.: Comput. Mol. Sci.* 2 (2): 187–203.
67 El-Sayed, M.A. and Pavlopoulos, T. (1963). Intramolecular heavy-atom effect on polarization of naphthalene phosphorescence. *J. Chem. Phys.* 39 (7): 1899–1900.
68 El-Sayed, M.A. (1963). Spin-orbit coupling and radiationless processes in nitrogen heterocyclics. *J. Chem. Phys.* 38 (12): 2834–2838.
69 El-Sayed, M.A. (1968). Triplet state – its radiative and nonradiative properties. *Acc. Chem. Res.* 1 (1): 8–16.
70 Shimakura, N., Fujimura, Y., and Nakajima, T. (1977). Theory of intersystem crossing in aromatic compounds – extension of El-Sayed rule. *Chem. Phys.* 19 (2): 155–163.
71 Fujimura, Y., Shimakura, N., and Nakajima, T. (1977). Mechanisms of intersystem crossing in aromatic hydrocarbons. *J. Chem. Phys.* 66 (8): 3530–3537.
72 Baba, M. (2011). Intersystem crossing in the $^1n\pi^*$ and $^1\pi\pi^*$ states. *J. Phys. Chem. A* 115 (34): 9514–9519.
73 Cundall, R.B. and Davies, A.S. (1967). Primary processes in the gas phase photochemistry of carbonyl compounds. *Prog. React. Kinet.* 4: 149–213.
74 Jori, G. (2012). Photosensitization. In: *CRC Handbook of Organic Photochemistry and Photobiology*, 3e (ed. A. Griesbeck, M. Oelgemöller and F. Lenci), 1105–1116. Boca Raton, FL: CRC Press.
75 Förster, T. (1948). Zwischenmolekulare Energiewanderung und Fluoreszenz. *Ann. Phys.* 2 (1–2): 55–75.
76 Förster, T. (1951). *Fluorenzenz Organische Verbindungen*, 1–312. Göttingen: Vandenhoech und Ruprecht.
77 Valeur, B. and Berberan-Santos, M.N. (2012). Excitation, energy transfer. In: *Molecular Fluorescence: Principles and Applications*, 2e, 213–261. Weinheim: Wiley-VCH.
78 Alex, S. and Tiwari, A. (2015). Functionalized gold nanoparticles: synthesis, properties and applications – a review. *J. Nanosci. Nanotechnol.* 15 (3): 1869–1894.
79 Rehor, I. and Cigler, P. (2014). Precise estimation of HPHT nanodiamond size distribution based on transmission electron microscopy image analysis. *Diamond Relat. Mater.* 46: 21–24.
80 Li, Z., He, M., Xu, D., and Liu, Z. (2014). Graphene materials-based energy acceptor systems and sensors. *J. Photochem. Photobiol., C* 18: 1–17.
81 Truong, K. and Ikura, M. (2001). The use of FRET imaging microscopy to detect protein-protein interactions and protein conformational changes in vivo. *Curr. Opin. Struct. Biol.* 11 (5): 573–578.
82 Sekar, R.B. and Periasamy, A. (2003). Fluorescence resonance energy transfer (FRET) microscopy imaging of live cell protein localizations. *J. Cell Biol.* 160 (5): 629–633.
83 Silvius, J.R. and Nabi, I.R. (2006). Fluorescence-quenching and resonance energy transfer studies of lipid microdomains in model and biological membranes (review). *Mol. Membr. Biol.* 23 (1): 5–16.
84 Ni, Q. and Zhang, J. (2010). Dynamic visualization of cellular signaling. *Adv. Biochem. Eng./Biotechnol.* 119: 79–97.
85 Yuan, L., Lin, W., Zheng, K., and Zhu, S. (2013). FRET-based small-molecule fluorescent probes: rational design and bioimaging applications. *Acc. Chem. Res.* 46 (7): 1462–1473.
86 Jones, A.M., Grossmann, G., Danielson, J.A. et al. (2013). In vivo biochemistry: applications for small molecule biosensors in plant biology. *Curr. Opin. Plant Biol.* 16 (3): 389–395.
87 Dexter, D.L. (1953). A theory of sensitized luminescence in solids. *J. Chem. Phys.* 21 (5): 836–850.
88 Gould, I.R. and Farid, S. (1996). Dynamics of bimolecular photoinduced electron-transfer reactions. *Acc. Chem. Res.* 29 (11): 522–528.
89 Bernardi, F., Olivucci, M., and Robb, M.A. (1995). Following reaction paths in organic photochemistry – the special role of surface crossings. *Pure Appl. Chem.* 67 (1): 17–24.
90 Robb, M.A., Bernardi, F., and Olivucci, M. (1995). Conical intersections as a mechanistic

feature of organic photochemistry. *Pure Appl. Chem.* 67 (5): 783–789.

91 Bernardi, F., Olivucci, M., and Robb, M.A. (1996). Potential energy surface crossings in organic photochemistry. *Chem. Soc. Rev.* 25 (5): 321–328.

92 Klessinger, M. (1995). Conical intersections and the mechanism of singlet photoreactions. *Angew. Chem. Int. Ed. Engl.* 34 (5): 549–551.

93 Martinez, T.J. (2010). Physical chemistry seaming is believing. *Nature* 467 (7314): 412–413.

94 Sicilia, F., Bearpark, M., Blancafort, L., and Robb, M.A. (2007). An analytical second-order description of the S_0/S_1 intersection seam: fulvene revisited. *Theor. Chem. Acc.* 118 (1): 241–251.

95 Sicilia, F., Blancafort, L., Bearpark, M.J., and Robb, M.A. (2007). Quadratic description of conical intersections: characterization of critical points on the extended seam. *J. Phys. Chem. A* 111 (11): 2182–2192.

96 Turro, N.J., Ramamurphy, V., and Scaino, J.C. (2009). *Principles of Molecular Photochemistry: An Introduction*, 1–495. Sausalito, CA: University Science Books.

97 Kropp, P.J. (2003). Photorearrangement and fragmentation of alkenes. In: *CRC Handbook of Organic Photochemistry and Photobiology*, Chapter 13, 2e (ed. F. Lenci and W. Horspool). Boca Raton, FL: CRC Press.

98 Caldwell, R.A., Diaz, J.F., Hrncir, D.C., and Unett, D.J. (1994). Alkene triplets as 1,2-biradicals – the photoaddition of p-acetylstyrene to styrene. *J. Am. Chem. Soc.* 116 (18): 8138–8145.

99 (a) Gajewski, J.J. (1981). *Hydrocarbon Thermal Isomerizations*. New York: Academic Press. (b) Gajewski, J.J. (2004). *Hydrocarbon Thermal Isomerizations*, 2e, 1–442. Amsterdam: Elsevier Inc.

100 Ni, T., Caldwell, R.A., and Melton, L.A. (1989). The relaxed and spectroscopic energies of olefin triplets. *J. Am. Chem. Soc.* 111 (2): 457–464.

101 Hammond, G.S., Saltiel, J., Bradshaw, J.S. et al. (1964). Mechanisms of photochemical reactions in solution. 22. Photochemical *cis*-trans isomerization. *J. Am. Chem. Soc.* 86 (16): 3197–3217.

102 Waldeck, D.H. (1991). Photoisomerization dynamics of stilbenes. *Chem. Rev.* 91 (3): 415–436.

103 Saltiel, J., Waller, A.S., and Sears, D.F. (1992). Dynamics of *cis*-stilbene photoisomerization – the adiabatic pathway to excited trans-stilbene. *J. Photochem. Photobiol., A* 65 (1–2): 29–40.

104 Doany, F.E., Heilweil, E.J., Moore, R., and Hochstrasser, R.M. (1984). Picosecond study of an intermediate in the trans to *cis* isomerization pathway of stiff stilbene. *J. Chem. Phys.* 80 (1): 201–206.

105 Abrash, S., Repinec, S., and Hochstrasser, R.M. (1990). The viscosity dependence and reaction coordinate for isomerization of *cis*-stilbene. *J. Chem. Phys.* 93 (2): 1041–1053.

106 Sension, R.J., Repinec, S.T., Szarka, A.Z., and Hochstrasser, R.M. (1993). Femtosecond laser studies of the *cis*-stilbene photoisomerization reactions. *J. Chem. Phys.* 98 (8): 6291–6315.

107 Cahn, R.S., Ingold, C., and Prelog, V. (1966). Specification of molecular chirality. *Angew. Chem. Int. Ed. Engl.* 5 (4): 385–415.

108 Prelog, V. and Helmchen, G. (1982). Basic principles of the CIP-system and proposals for a revision. *Angew. Chem. Int. Ed. Engl.* 21 (8): 567–583.

109 Dauben, W.G., Vanriel, H.C.H.A., Hauw, C. et al. (1979). Photochemical formation of trans-1-phenylcyclohexene – chemical proof of structure. *J. Am. Chem. Soc.* 101 (7): 1901–1903.

110 Goodman, J.L., Peters, K.S., Misawa, H., and Caldwell, R.A. (1986). Use of pulsed time-resolved photoacoustic calorimetry to determine the strain energy of trans-1-phenylcyclohexene and the energy of the relaxed 1-phenylcyclohexene triplet. *J. Am. Chem. Soc.* 108 (21): 6803–6805.

111 Dunkin, I.R. (2003). Matrix photochemistry of small ring compounds. In: *CRC Handbook of Organic Photochemistry and Photobiology*, Chapter 15, 2e (ed. W.M. Horspool and F. Lenci). Boca Raton, FL: CRC Press.

112 (a) Inoue, Y., Ueoka, T., Kuroda, T., and Hakushi, T. (1983). Singlet photosensitization of simple alkenes. Part 4. *cis*-trans-Photoisomerization of cycloheptene sensitized by aromatic esters. Some aspects of the chemistry of trans-cycloheptene. *J. Chem. Soc., Perkin Trans. 2* 983–988. (b) Mori, T. and Inoue, Y. (2003). Photochemical isomerization of cycloalkenes. In: *CRC Handbook of Organic Photochemistry and Photobiology*, Chapter 16, 2e (ed. W.M. Horspool and F. Lenci). Boca Raton, FL: CRC Press.

113 Marshall, J.A. (1970). Photochemically induced ionic reactions of cycloalkenes. *Science* 170 (3954): 137–141.

114 Kropp, P.J. (2003). Photoreactions of alkenes in protic media. In: *CRC Handbook of Organic Photochemistry and Photobiology*, Chapter 9, 2e (ed. W.M. Horspool and F. Lenci). Boca Raton, FL: CRC Press.

115 Bonneau, R., Fornier de Violet, P., and Joussot-Dubien, J. (1977). trans-Cycloheptenone – flash photolytic study of reactive intermediate in photochemistry of 2-cycloheptenone. *New J. Chem.* 1 (1): 31–34.

116 Goldfarb, T.D. (1978). Kinetic studies of transient photochemical isomers of 2-cycloheptenone, 1-acetylcyclohexene and 2-cyclohexenone. *J. Photochem.* 8 (1): 29–38.

117 Inoue, Y. (1992). Asymmetric photochemical reactions in solution. *Chem. Rev.* 92 (5): 741–770.

118 Griesbeck, A.G. and Meierhenrich, U.L. (2002). Asymmetric photochemistry and photochirogenesis. *Angew. Chem. Int. Ed.* 41 (17): 3147–3154.

119 Inoue, Y. and Ramamurphy, V. (eds.) (2004). Chiral photochemistry. In: *Molecular and Supramolecular Photochemistry*, vol. 11, 1–686. New York: Marcel Dekker.

120 Yang, C. and Inoue, Y. (2012). Photochirogenesis. In: *CRC Handbook of Organic Photochemistry and Photobiology*, 3e (ed. A. Griesbeck, M. Oelgemöller and F. Ghetti), 125–176. Boca Raton, FL: CRC Press.

121 Inoue, Y., Wada, T., Asaoka, S. et al. (2000). Photochirogenesis: multidimensional control of asymmetric photochemistry. *J. Chem. Soc., Chem. Commun.* (4): 251–259.

122 Inoue, Y., Yokoyama, T., Yamasaki, N., and Tai, A. (1989). Temperature switching of product chirality upon photosensitized enantiodifferentiating *cis*-trans isomerization of cyclooctene. *J. Am. Chem. Soc.* 111 (16): 6480–6482.

123 Inoue, Y., Yamasaki, N., Yokoyama, T., and Tai, A. (1993). Highly enantiodifferentiating photoisomerization of cyclooctene by congested and or triplex-forming chiral sensitizers. *J. Org. Chem.* 58 (5): 1011–1018.

124 Lu, R., Yang, C., Cao, Y. et al. (2008). Enantiodifferentiating photoisomerization of cyclooctene included and sensitized by aroyl-β-cyclodextrins: a critical enantioselectivity control by substituents. *J. Org. Chem.* 73 (19): 7695–7701.

125 Maeda, R., Wada, T., Mori, T. et al. (2011). Planar-to-planar chirality transfer in the excited state. Enantiodifferentiating photoisomerization of cyclooctenes sensitized by planar-chiral paracyclophane. *J. Am. Chem. Soc.* 133 (27): 10379–10381.

126 Gilbert, A. (2003). Cyclization of stilbene and its derivatives. In: *CRC Handbook of Organic Photochemistry and Photobiology*, Chapter 33, 2e (ed. W.M. Horspool and F. Lenci). Boca Raton, FL: CRC Press.

127 Willner, I. (1997). Photoswitchable biomaterials: en route to optobioelectronic systems. *Acc. Chem. Res.* 30 (9): 347–356.

128 Willner, I. and Rubin, S. (1996). Control of the structure and functions of biomaterials by light. *Angew. Chem. Int. Ed. Eng.* 35 (4): 367–385.

129 Hong, F.T. (2003). Molecular electronic switches in photobiology. In: *CRC Handbook of Organic Photochemistry and Photobiology*, Chapter 134, 2e (ed. W.M. Horspool and F. Lenci). Boca Raton, FL: CRC Press.

130 Samanta, S., Beharry, A.A., Sadovski, O. et al. (2013). Photoswitching azo compounds in vivo with red light. *J. Am. Chem. Soc.* 135 (26): 9777–9784.

131 Beccalli, E.M. and Marchesini, A. (1984). Synthesis of strained cycloalkene derivatives. *Gazz. Chim. Ital.* 114 (5–6): 287–288.

132 Chapman, O.L. (1974). Photochemical studies. 56. Low-temperature photochemical studies. *Pure Appl. Chem.* 40 (4): 511–523.

133 Bailey, I.M. and Walsh, R. (1978). Gas phase pyrolysis of cyclopropene. 1. Kinetics and mechanism. *J. Chem. Soc., Faraday Trans. 1* 74: 1146–1158.

134 Steinmetz, M.G., Yen, Y.P., and Poch, G.K. (1983). Comparison between ground and excited state reactivity of cyclopropenes – thermolysis and far-UV (185 nm) photolysis of spiro[2.4]hept-1-ene. *J. Chem. Soc., Chem. Commun.* (24): 1504–1505.

135 Steinmetz, M.G., Stark, E.J., Yen, Y.P. et al. (1983). Far-UV (185-nm) photochemistry of allenes – photo-isomerization and cycloreversion reactions of vinylidenecyclobutane. *J. Am. Chem. Soc.* 105 (24): 7209–7210.

136 Shimizu, T. (2003). Photochemistry of allenes. In: *CRC Handbook of Organic Photochemistry and Photobiology*, Chapter 24, 2e (ed. W.M. Horspool and F. Lenci). Boca Raton, FL: CRC Press.

137 (a) Barltrop, J.A. and Giles, D. (1969). Studies in organic photochemistry. X. Photochemical addition of cyclohexene-1,4-diones to olefins and acetylenes. *J. Chem. Soc. C* 105–114.

(b) Guillemein, J.C., Breneman, C.M., Joseph, J.C., and Ferris, J.P. (1998). Regioselectivity of the photochemical addition of ammonia, phosphine, and silane to olefinic and acetylenic nitriles. *Chem. Eur. J.* 4 (6): 1074–1082. (c) Zeidan, T.A., Kovalenko, S.V., Manoharan, M. et al. (2005). Triplet acetylenes as synthetic equivalents of 1,2-bicarbenes: phantom n,π^* state controls reactivity in triplet photocycloaddition. *J. Am. Chem. Soc.* 127: 4270–4285.

138 Padwa, A., Chou, C.S., Rosenthal, R.J., and Terry, L.W. (1988). Photochemistry of cyclopropene derivatives – intramolecular hydrogen transfer reaction of some 1-(alkyl-substituted)cyclopropenes. *J. Org. Chem.* 53 (18): 4193–4201.

139 Srinivasan, R. (1962). Kinetics of photochemical dimerization of olefins to cyclobutane derivatives. 1. Intramolecular addition. *J. Am. Chem. Soc.* 84 (21): 4141–4145.

140 Dauben, W.G. and Cargill, R.L. (1962). Photochemical transformations. 11. Isomerization of 1,3-cyclooctadiene. *J. Org. Chem.* 27 (5): 1910.

141 Liu, R.S.H. (1967). Photosensitized isomerization of 1,3-cyclooctadienes and conversion to bicyclo[4.2.0]oct-7-ene. *J. Am. Chem. Soc.* 89 (1): 112–114.

142 Bramham, J. and Samuel, C.J. (1989). Thermal *E*-isomerization to *Z*-isomerization of (*E*,*Z*)-cycloocta-1,3-diene. *J. Chem. Soc., Chem. Commun.* (1): 29–30.

143 Lawless, M.K., Wickham, S.D., and Mathies, R.A. (1995). Resonance Raman view of pericyclic photochemical ring-opening reactions: beyond the Woodward–Hoffmann rules. *Acc. Chem. Res.* 28 (12): 493–502.

144 Leigh, W.J., Zheng, K., Nguyen, N. et al. (1991). Cyclobutene photochemistry. 5. Identification of the excited states responsible for the ring opening and cycloreversion reactions of alkylcyclobutenes. *J. Am. Chem. Soc.* 113 (13): 4993–4999.

145 Leigh, W.J. and Zheng, K.C. (1991). Cyclobutene photochemistry. 5. Reinvestigation of the photochemistry of *cis*-tricyclo[6.4.0.0(2,7)]dodec-1-ene and trans-tricyclo[6.4.0.0(2,7)]dodec-1-ene. *J. Am. Chem. Soc.* 113 (6): 2163–2166.

146 Leigh, W.J., Zheng, K.C., and Clark, K.B. (1991). Cyclobutene photochemistry. 4. The photochemistry of *cis*-bicyclo[5.2.0]non-8-ene and trans-bicyclo[5.2.0]non-8-ene. *J. Org. Chem.* 56 (4): 1574–1580.

147 Dauben, W.G. and Ritscher, J.S. (1970). Photochemistry of ethylidenecyclooctense – mechanism of bicyclobutane formation. *J. Am. Chem. Soc.* 92 (9): 2925–2926.

148 Srinivasan, R. (1963). A simple synthesis of bicyclo[1.1.0]butane and its relation to internal conversion of electronic energy in 1,3-butadiene. *J. Am. Chem. Soc.* 85 (24): 4045–4046.

149 Hopf, H., Lipka, H., and Traetteberg, M. (1994). Sterically hindered double-bond system. 3. Photoisomerization of highly alkylated butadienes. *Angew. Chem. Int. Ed. Engl.* 33 (2): 204–205.

150 Maier, G. (1988). Tetrahedrane and cyclobutadiene. *Angew. Chem. Int. Ed. Engl.* 27 (3): 309–333.

151 Garavelli, M., Frabboni, B., Fato, M. et al. (1999). Photochemistry of highly alkylated dienes: computational evidence for a concerted formation of bicyclobutane. *J. Am. Chem. Soc.* 121 (7): 1537–1545.

152 Wiberg, K.B. and Lavanish, J.M. (1966). Formation and thermal decomposition of bicyclo[1.1.0]butane-2-exo-D1. *J. Am. Chem. Soc.* 88 (22): 5272–5275.

153 Blanchard, E.P. and Cairncross, A. (1966). Bicyclo[1.1.0]butane chemistry. I. Synthesis and reactions of 3-methylbicyclo[1.1.0]butanecarbonitriles. *J. Am. Chem. Soc.* 88 (3): 487–495.

154 Closs, G.L. and Pfeffer, P.E. (1968). Steric course of thermal rearrangements of methylbicyclobutanes. *J. Am. Chem. Soc.* 90 (9): 2452–2453.

155 Nguyen, K.A. and Gordon, M.S. (1995). Isomerization of bicyclo[1.1.0]butane to butadiene. *J. Am. Chem. Soc.* 117 (13): 3835–3847.

156 Griffin, G.W., Covell, J., Petterson, R.C. et al. (1965). Photoisomerization of cyclopropane derivatives. Photointerconversion of propenes and cyclopropanes. *J. Am. Chem. Soc.* 87 (6): 1410–1411.

157 Zimmerman, H.E. (2012). Di-π-methane rearrangement. In: *CRC Handbook of Organic Photochemistry and Photobiology*, 3e (ed. A. Griesbeck, M. Oelgemüller and F. Ghetti), 511–526. Boca Taton, FL: CRC Press.

158 Zimmerman, H.E. and Armesto, D. (1996). Synthetic aspects of the di-π-methane rearrangement. *Chem. Rev.* 96 (8): 3065–3112.

159 Hixson, S.S., Mariano, P.S., and Zimmerman, H.E. (1973). Di-π-methane and oxa-di-π-methane rearrangements. *Chem. Rev.* 73 (5): 531–551.

160 Zimmerman, H.E. and Grunewald, G.L. (1966). Chemistry of barrelene. 3. A unique photoisomerization to semibullvalene. *J. Am. Chem. Soc.* 88 (1): 183–184.

161 Zimmerman, H.E., Binkley, R.W., Givens, R.S., and Sherwin, M.A. (1967). Mechanistic organic photochemistry. 24. Mechanism of conversion of barrelene to semibullvalene. A general photochemical process. *J. Am. Chem. Soc.* 89 (15): 3932–3933.

162 Zimmerman, H.E. and Kutateladze, A.G. (1996). New approaches to singlet-triplet spin-orbit coupling in photochemistry and diradical chemistry. Mechanistic and exploratory organic photochemistry. *J. Am. Chem. Soc.* 118 (1): 249–250.

163 Look, S.A., Fenical, W., Vanengen, D., and Clardy, J. (1984). Erythrolides – unique marine diterpenoids interrelated by a naturally occurring di-π-methane rearrangement. *J. Am. Chem. Soc.* 106 (17): 5026–5027.

164 Zimmerman, H.E., Boettcher, R.J., Buehler, N.E. et al. (1976). Independent generation of cyclopropyldicarbinyl diradical species of di-π-methane rearrangement – excited singlet, triplet, and ground-state hypersurfaces of barrelene photochemistry – mechanistic and exploratory organic photochemistry. *J. Am. Chem. Soc.* 98 (24): 7680–7689.

165 Evans, S.V., García-Garibay, M., Omkaram, N. et al. (1986). Use of chiral single crystals to convert achiral reactants to chiral products in high optical yield – application to the di-π-methane and Norrish type-II photorearrangements. *J. Am. Chem. Soc.* 108 (18): 5648–5650.

166 Fu, T.Y., Liu, Z.Q., Scheffer, J.R., and Trotter, J. (1993). Supramolecular photochemistry of crystalline host-guest assemblies – absolute asymmetric photorearrangement of the host component. *J. Am. Chem. Soc.* 115 (25): 12202–12203.

167 Sakamoto, M. (1997). Absolute asymmetric synthesis from achiral molecules in the chiral crystalline environment. *Chem. Eur. J.* 3 (5): 684–689.

168 Zimmerman, H.E., Givens, R.S., and Pagni, R.M. (1968). Control of photochemical reaction pathways by excited state multiplicity. Mechanistic and exploratory organic photochemistry. 34. *J. Am. Chem. Soc.* 90 (15): 4191–4193.

169 Zimmerman, H.E., Givens, R.S., and Pagni, R.M. (1968). Photochemistry of benzobarrelene. Mechanistic and exploratory photochemistry. 35. *J. Am. Chem. Soc.* 90 (22): 6096–6108.

170 Bender, C.O., Brooks, D.W., Cheng, W. et al. (1978). Polar substituents in barrelene photochemistry – mechanistic aspects of photochemistry of 2-cyanobenzobarrelene. *Can. J. Chem.* 56 (23): 3027–3037.

171 Paquette, L.A., Cottrell, D.M., Snow, R.A. et al. (1975). Substituent control of regiospecific pathways in di-π-methane photorearrangements which utilize benzo-vinyl bonding schemes. *J. Am. Chem. Soc.* 97 (11): 3275–3276.

172 Paquette, L.A., Cottrell, D.M., and Snow, R.A. (1977). Control by meta substituents of benzo-vinyl bonding options during triplet sensitized photorearrangement of benzonorbornadienes and anti-7,8-benzotricyclo[4.2.2.$0^{2,5}$]deca-3,7,9-trienes. *J. Am. Chem. Soc.* 99 (11): 3723–3733.

173 Santiago, C., Houk, K.N., Snow, R.A., and Paquette, L.A. (1976). Bridging regioselectivity in triplet-sensitized di-π-methane photorearrangements of ortho-substituted benzonorbornadienes – case for importance of benzene HOMO and LUMO polarization. *J. Am. Chem. Soc.* 98 (23): 7443–7445.

174 Beissert, S. and Granstein, R.D. (1996). UV-induced cutaneous photobiology. *Crit. Rev. Biochem. Mol. Biol.* 31 (5–6): 381–404.

175 Holick, M.F. (2003). Vitamin D deficiency: what a pain it is. *Mayo Clin. Proc.* 78 (12): 1457–1459.

176 Garland, C.F., Gorham, E.D., Mohr, S.B., and Garland, F.C. (2009). Vitamin D for cancer prevention: global perspective. *Ann. Epidemiol.* 19 (7): 468–483.

177 Wang, S. (2009). Epidemiology of vitamin D in health and disease. *Nutr. Res. Rev.* 22 (2): 188–203.

178 Rappoldt, M.P. and Havinga, E. (1960). Studies on vitamin-D and related compounds. 11. Investigations on sterols. 13. The photoisomerisation of ergosterol. *Recl. Trav. Chim. Pays-Bas* 79 (4): 369–381.

179 Tapavicza, E., Meyer, A.M., and Furche, F. (2011). Unravelling the details of vitamin D photosynthesis by non-adiabatic molecular dynamics simulations. *Phys. Chem. Chem. Phys.* 13 (47): 20986–20998.

180 Tian, X.Q., Chen, T.C., Matsuoka, L.Y. et al. (1993). Kinetic and thermodynamic studies of the conversion of previtamin-D_3 to vitamin-D_3 in human skin. *J. Biol. Chem.* 268 (20): 14888–14892.

181 Tian, X.Q. and Holick, M.F. (1999). A liposomal model that mimics the cutaneous production of

vitamin D_3 – studies of the mechanism of the membrane-enhanced thermal isomerization of previtamin D_3 to vitamin D_3. *J. Biol. Chem.* 274 (7): 4174–4179.

182 Redwood, C., Bayda, M., and Saltiel, J. (2013). Photoisomerization of pre- and provitamin D_3 in EPA at 77 K: one-bond-twist, not hula-twist. *J. Phys. Chem. Lett.* 4 (5): 716–721.

183 Windaus, A., Dimroth, K., and Breywisch, W. (1940). The photochemical process in the formation of photopyrocalciferol. *Liebigs Ann. Chem.* 543: 240–247.

184 Havinga, E., Dekock, R.J., and Rappoldt, M.P. (1960). The photochemical interconversions of provitamin-D, lumisterol, previtamin-D and tachysterol. *Tetrahedron* 11 (4): 276–284.

185 Boomsma, F., Jacobs, H.J.C., Havinga, E., and Vandergen, A. (1977). Studies on vitamin-D and related compounds. 26. Over-irradiation products of previtamin-D and tachysterol – toxisterols. *Recl. Trav. Chim. Pays-Bas* 96 (4): 104–112.

186 Barrett, A.G.M., Barton, D.H.R., Russell, R.A., and Widdowson, D.A. (1977). Photochemical-transformations. 34. Structures of toxisterols-2. *J. Chem. Soc., Perkin Trans. 1* (6): 631–643.

187 Corey, E.J. and Hortmann, A.G. (1963). Total synthesis of dihydrocostunolide. *J. Am. Chem. Soc.* 85 (24): 4033–4034.

188 Reid, P.J., Doig, S.J., Wickham, S.D., and Mathies, R.A. (1993). Photochemical ring-opening reactions are complete in picoseconds – a time-resolved UV resonance Raman-study of 1,3-cyclohexadiene. *J. Am. Chem. Soc.* 115 (11): 4754–4763.

189 Kosma, K., Trushin, S., Fuss, W., and Schmid, W. (2009). Cyclohexadiene ring opening observed with 13 fs resolution: coherent oscillations confirm the reaction path. *Phys. Chem. Chem. Phys.* 11 (1): 172–181.

190 Celani, P., Bernardi, F., Robb, M.A., and Olivucci, M. (1996). Do photochemical ring-openings occur in the spectroscopic state? B-1(2) pathways for the cyclohexadiene/hexatriene photochemical interconversion. *J. Phys. Chem.* 100 (50): 19364–19366.

191 Nenov, A., Koelle, P., Robb, M.A., and de Vivie-Riedle, R. (2010). Beyond the van der Lugt/Oosterhoff model: when the conical intersection seam and the S_1 minimum energy path do not cross. *J. Org. Chem.* 75 (1): 123–129.

192 Li, A., Yuan, S., Dou, Y. et al. (2009). Semi-classical dynamic simulation of photon-induced ring-opening of cyclohexadiene to hexatriene. *Chem. Phys. Lett.* 478 (1–3): 28–32.

193 Garavelli, M., Page, C.S., Celani, P. et al. (2001). Reaction path of a sub-200 fs photochemical electrocyclic reaction. *J. Phys. Chem. A* 105 (18): 4458–4469.

194 Valentine, D., Hammond, G.S., and Turro, N.J. (1964). Thermal + photosensitized dimerizations of cyclohexadiene. *J. Am. Chem. Soc.* 86 (23): 5202–5208.

195 Vroegop, P.J., Lugtenburg, J., and Havinga, E. (1973). Conformational equilibrium and photochemistry of hexa-1,3,5-trienes. *Tetrahedron* 29 (10): 1393–1398.

196 Whitesell, J.K. and Minton, M.A. (1987). A novel synthesis of Ikarugamycin – the carbocyclic portion. *J. Am. Chem. Soc.* 109 (21): 6403–6408.

197 Minnaard, N.G. and Havinga, E. (1973). Some aspects of solution photochemistry of 1,3-cyclohexadiene, (*Z*)-1,3,5-hexatriene and (*E*)-1,3,5-hexatriene. *Recl. Trav. Chim. Pays-Bas* 92 (12): 1315–1320.

198 Møller, S., Langkilde, F.W., and Wilbrandt, R. (1991). Sensitized triplet photochemistry of (*E*)-1,3,5-hexatriene and (*Z*)-1,3,5-hexatriene. *J. Photochem. Photobiol., A* 62 (1): 93–106.

199 Smakula, A. (1934). The photochemical transformation of trans-stilbene. *Z. Phys. Chem.* B25: 90–98.

200 Buckles, R.E. (1955). Illumination of *cis*-stilbenes and trans-stilbenes in dilute solutions. *J. Am. Chem. Soc.* 77 (4): 1040–1041.

201 Mallory, F.B., Gordon, J.T., Wood, C.S. et al. (1962). Photochemistry of stilbenes. 1. *J. Am. Chem. Soc.* 84 (22): 4361–4362.

202 Wood, C.S. and Mallory, F.B. (1964). Photochemistry of stilbenes. 4. Preparation of substituted phenanthrenes. *J. Org. Chem.* 29 (11): 3373–3377.

203 Mallory, F.B., Wood, C.S., and Gordon, J.T. (1964). Photochemistry of stilbenes. 3. Some aspects of mechanism of photocyclization to phenanthrenes. *J. Am. Chem. Soc.* 86 (15): 3094–3102.

204 Srinivasan, R. and Powers, J.C. (1963). Mechanism of photochemical formation of phenanthrene from *cis*-stilbene in vapor phase. *J. Am. Chem. Soc.* 85 (9): 1355–1356.

205 Jørgensen, K.B. (2010). Photochemical oxidative cyclisation of stilbenes and stilbenoids-the Mallory-reaction. *Molecules* 15 (6): 4334–4358.

206 Martin, R.H. (1974). Helicenes. *Angew. Chem. Int. Ed. Engl.* 13 (10): 649–659.

207 Le Bel, J.A. (1874). Sur les relations qui existent entre les formules atomiques des corps organiques et le pouvoir rotatoire de leur dissolution. *Bull. Soc. Chim. Fr.* 22 (337): 347.

208 Moradpour, A., Nicoud, J.F., Balavoine, G. et al. (1971). Photochemistry with circularly polarized light – synthesis of optically active hexahelicene. *J. Am. Chem. Soc.* 93 (9): 2353–2354.

209 Bernstein, W.J., Calvin, M., and Buchardt, O. (1972). Absolute asymmetric synthesis. 1. On mechanism of photochemical synthesis of non-racemic helicenes with circularly polarized-light – wavelength dependence of optical yield of octahelicene. *J. Am. Chem. Soc.* 94 (2): 494–498.

210 Bernstein, W.J., Calvin, M., and Buchardt, O. (1973). Absolute asymmetric synthesis. 3. Hindered rotation about aryl-ethylene bonds in excited-states of diaryl ethylenes – structural effects on asymmetric synthesis of 2-substituted and 4-substituted hexahelicenes. *J. Am. Chem. Soc.* 95 (2): 527–532.

211 Nakazaki, M., Yamamoto, K., Fujiwara, K., and Maeda, M. (1979). Mechanically directed absolute asymmetric syntheses of helicenes in a twisted nematic mesophase. *J. Chem. Soc., Chem. Commun.* (23): 1086–1087.

212 (a) El Abed, R., Ben Hassine, B., Genêt, J.P. et al. (2004). Synthesis of a configurationally locked [5]helicene derivative. *Synthesis* (15): 2513–2516. (b) Irie, M. (2000). Diarylethenes for memories and switches. *Chem. Rev.* 100: 1685–1716.

213 Doyle, T.D., Filipescu, N., Benson, W.R., and Banes, D. (1970). Photocyclization of α,α-diethyl-4,4′-stilbenediol – isolation of a stable tautomer of elusive dihydrophenanthrenes. *J. Am. Chem. Soc.* 92 (21): 6371–6372.

214 Pollard, R. and Wan, P. (1993). Synthetic applications of diaryl ether photochemistry – a review. *Org. Prep. Proced. Int.* 25 (1): 1–13.

215 Moore, W.M., Hammond, G.S., and Foss, R.P. (1961). Mechanisms of photoreactions in solutions. 1. Reduction of benzophenone by benzhydrol. *J. Am. Chem. Soc.* 83 (13): 2789–2794.

216 Turro, N.J., Schore, N., Hautala, R. et al. (1972). Molecular photochemistry of alkanones in solution – α-cleavage, hydrogen abstraction, cycloaddition, and sensitization reactions. *Acc. Chem. Res.* 5 (3): 92–101.

217 Schuster, D.I. and Karp, P.B. (1980). Photochemistry of hetones in solution. 58. Mechanism of photo-reduction of benzophenone by benzhydrol. *J. Photochem.* 12 (4): 333–344.

218 Ortega, M., Rodríguez, M.A., and Campos, P.J. (2005). Photoreduction of imines. An environmentally friendly approach to obtain amines. *Tetrahedron* 61 (49): 11686–11691.

219 Ciamician, G. and Silber, P. (1907). Chemische Lichtwirkungen. *Ber. Dtsch. Chem. Ges.* 40: 2415–2424.

220 Laue, T. and Plagens, A. (eds.) (2005). Norrish type I reaction. In: *Named Organic Reactions*, 2e, 320–324. Chichester: Wiley.

221 Weiss, D.S. (1981). The Norrish type I reaction in cycloalkanone photochemistry. *Org. Photochem.* 5: 347–420.

222 Cundall, R.B. and Davies, A.S. (1966). Mechanism of gas phase photolysis of acetone. *Proc. R. Soc. London, Ser. A* 290 (1423): 563–582.

223 Lightfoot, P.D., Kirwan, S.P., and Pilling, M.J. (1988). Photolysis of acetone at 193.3 nm. *J. Phys. Chem.* 92 (17): 4938–4946.

224 Kim, S.K., Pedersen, S., and Zewail, A.H. (1995). Direct femtosecond observation of the transient intermediate in the α-cleavage reaction of $(CH_3)_2CO$ to $2CH_3 + CO$ – resolving the issue of concertedness. *J. Chem. Phys.* 103 (1): 477–480.

225 Diau, E.W.G., Kotting, C., and Zewail, A.H. (2001). Femtochemistry of Norrish type-1 reactions. I. Experimental and theoretical studies of acetone and related ketones on the S-1 surface. *ChemPhysChem* 2 (5): 273–293.

226 Kim, S.K. and Zewail, A.H. (1996). Femtosecond elementary dynamics of transition states and asymmetric alpha-cleavage in Norrish reactions. *Chem. Phys. Lett.* 250 (3–4): 279–286.

227 Cundall, R.B. and Davies, A.S. (1966). Role of triplet state in gas phase photolysis of aliphatic ketones and aldehydes. *Trans. Faraday Soc.* 62 (525P): 2444–2451.

228 Park, S.T., Feenstra, J.S., and Zewail, A.H. (2006). Ultrafast electron diffraction: excited state structures and chemistries of aromatic carbonyls. *J. Chem. Phys.* 124 (17): 174707-1–174707-23.

229 Lewis, F.D. and Magyar, J.G. (1973). Cage effects in photochemistry of (S)-(+)-2-phenylpropiophenone. *J. Am. Chem. Soc.* 95 (18): 5973–5976.

230 Step, E.N., Buchachenko, A.L., and Turro, N.J. (1992). The cage effect in the photolysis of (S)-(+)-α-methyldeoxybenzoin – can triplet radical pairs undergo germinate recombination in non-viscous homogeneous solution. *J. Org. Chem.* 57 (26): 7018–7024.

231 Wehrli, H. and Schaffner, K. (1962). Photochemische Reaktionen. 12. Zur UV-Bestrahlung von 17-Oxo-Steroiden. *Helv. Chim. Acta* 45 (1): 385–389.

232 Rickborn, B., Alumbaugh, R.L., and Pritchard, G.O. (1964). Photolysis of *cis* and trans-2,6-dimethylcyclohexanone. *Chem. Ind. (London)* (47): 1951–1953.

233 Quinkert, G., Opitz, K., Wiersdorff, W.W., and Weinlich, J. (1963). Lichtinduzierte Decarbonylierung geloster Ketone bei Raumtemperatur. *Tetrahedron Lett.* 4 (27): 1863–1868.

234 Srinivasan, R. (1959). Photoisomerization processes in cyclic ketones. 2. Cyclohexanone and 2-methylcyclohexanone. *J. Am. Chem. Soc.* 81 (11): 2601–2604.

235 Dalton, J.C., Dawes, K., Turro, N.J. et al. (1971). Molecular chemistry. 48. Type-I and type-II photochemical reactions of some 5-membered and 6-membered cycloalkanones. *J. Am. Chem. Soc.* 93 (26): 7213–7221.

236 Coyle, J.D. (1971). Product ratios in ring-opening photoreactions of cycloalkanones. *J. Chem. Soc. B* (9): 1736–1740.

237 Starr, J.E. and Eastman, R.H. (1966). Structural features facilitating photodecarbonylation of cyclic ketones. *J. Org. Chem.* 31 (5): 1393–1402.

238 Krebs, A., Cholcha, W., Müller, M. et al. (1984). Generation of 7-membered and 6-membered alkynes through protolysis and thermolysis of cyclopropenones. *Tetrahedron Lett.* 25 (44): 5027–5030.

239 Chapman, O.L., Gano, J., West, P.R. et al. (1981). Acenaphthyne. *J. Am. Chem. Soc.* 103 (23): 7033–7036.

240 Maier, G., Pfriem, S., Schafer, U., and Matusch, R. (1978). Small rings. 25. Tetra-tert-butyltetrahedrane. *Angew. Chem. Int. Ed. Engl.* 17 (7): 520–521.

241 Benson, S.W. and Kistiakowsky, G.B. (1942). The photochemical decomposition of cyclic ketones. *J. Am. Chem. Soc.* 64: 80–86.

242 Blacet, F.E. and Miller, A. (1957). The photochemical decomposition of cyclohexanone, cyclopentanone and cyclobutanone. *J. Am. Chem. Soc.* 79 (16): 4327–4329.

243 Turro, N.J. and Southam, R.M. (1967). Molecular photochemistry. 4. Solution photochemistry of cyclobutanone and some derivatives. *Tetrahedron Lett.* 8 (6): 545–551.

244 Quinkert, G., Kaiser, K.H., and Stohrer, W.D. (1974). Low-temperature electronic spectroscopic observations on light-induced cyclobutanone-tetrahydrofurylidene rearrangement. *Angew. Chem. Int. Ed. Engl.* 13 (3): 198–199.

245 Agosta, W.C. and Herron, D.K. (1968). Solution photolysis of camphor. *J. Am. Chem. Soc.* 90 (25): 7025–7030.

246 Herz, W. and Iyer, V.S. (1977). Influence of substitution on photochemistry of rigid cyclopentenones. *J. Org. Chem.* 42 (9): 1573–1578.

247 Roberts, S.M. (2003). Carbene formation in the photochemistry of cyclic ketones. In: *CRC Handbook of Organic Photochemistry & Photobiology*, Chapter 49, 2e (ed. W.M. Horspool and F. Lenci). Boca Raton, FL: CRC Press.

248 Strating, J., Zwanenburg, B., Wagenaar, A., and Udding, A.C. (1969). Evidence for expulsion of *bis*-CO from bridged α-diketones. *Tetrahedron Lett.* (3): 125–128.

249 Mondal, R., Adhikari, R.M., Shah, B.K., and Neckers, D.C. (2007). Revisiting the stability of hexacenes. *Org. Lett.* 9 (13): 2505–2508.

250 Mondal, R., Toenshoff, C., Khon, D. et al. (2009). Synthesis, stability, and photochemistry of pentacene, hexacene, and heptacene: a matrix isolation study. *J. Am. Chem. Soc.* 131 (40): 14281–14289.

251 Kagan, H.B., Balavoine, G., and Moradpour, A. (1974). Can circularly polarized-light be used to obtain chiral compounds of high optical purity. *J. Mol. Evol.* 4 (1): 41–48.

252 Balavoine, G., Moradpour, A., and Kagan, H.B. (1974). Preparation of chiral compounds with high optical purity by irradiation with circularly polarized-light, a model reaction for prebiotic generation of optical-activity. *J. Am. Chem. Soc.* 96 (16): 5152–5158.

253 Wagner, P.J., Zepp, R.G., and Kelso, P.A. (1972). Type-II photoprocesses of phenyl ketones – evidence for a biradical intermediate. *J. Am. Chem. Soc.* 94 (21): 7480–7488.

254 Dorigo, A.E., Mccarrick, M.A., Loncharich, R.J., and Houk, K.N. (1990). Transition structures for hydrogen-atom transfers to oxygen – comparisons of intermolecular and intramolecular processes and open-shell and closed-shell systems. *J. Am. Chem. Soc.* 112 (21): 7508–7514.

255 Sauers, R.R. and Edberg, L.A. (1994). Modeling of Norrish type-II reactions by semiempirical and ab initio methodology. *J. Org. Chem.* 59 (23): 7061–7066.

256 Scaiano, J.C. (1982). Laser flash-photolysis studies of the reactions of some 1,4-biradicals. *Acc. Chem. Res.* 15 (8): 252–258.

257 Johnston, L.J. and Scaiano, J.C. (1989). Time-resolved studies of biradical reactions in solution. *Chem. Rev.* 89 (3): 521–547.

258 Moorthy, J.N., Koner, A.L., Samanta, S. et al. (2006). Diastereomeric discrimination in the lifetimes of Norrish type II triplet 1,4-biradicals and stereocontrolled partitioning of their reactivity (Yang cyclization versus type II fragmentation). *Chem. Eur. J.* 12 (34): 8744–8749.

259 Hasegawa, T. (2003). Norrish type II processes of ketones: influence of environment. In: *CRC Handbook of Organic Photochemistry and Photobiology*, Chapter 55, 2e (ed. W.M. Horspool and F. Lenci). Boca Raton, FL: CRC Press.

260 Yang, N.C. and Yang, D.D.H. (1958). Photochemical reactions of ketones in solution. *J. Am. Chem. Soc.* 80 (11): 2913–2914.

261 Ignatenko, V.A. and Tochtrop, G.P. (2013). Approach for expanding triterpenoid complexity via divergent Norrish–Yang photocyclization. *J. Org. Chem.* 78 (8): 3821–3831.

262 Wagner, P.J. (2003). Yang photocyclization: coupling of biradicals formed by intramolecular hydrogen abstraction of ketones. In: *CRC Handbook of Organic Photochemistry and Photobiology*, Chapter 58, 2e (ed. W.M. Horspool and F. Lenci). Boca Raton, FL: CRC Press.

263 Griesbeck, A.G., Heckroth, H., and Lex, J. (1999). Stereoselective Yang cyclizations of α-amido ketones. *J. Chem. Soc., Chem. Commun.* (12): 1109–1110.

264 Wu, J., Zhang, W., and Wang, C. (2009). Synthesis of benzopyrrolizidinones by Norrish–Yang cyclization of 3-(acylmethyl)-2-alkylisoindol-1-ones. *Synthesis* (11): 1821–1828.

265 Quinkert, G. and Stark, H. (1983). Stereoselective synthesis of enantiomerically pure natural-products – estrone as example. *Angew. Chem. Int. Ed. Engl.* 22 (9): 637–655.

266 Bochet, C.G. and Blanc, A. (2012). Photolabile protecting groups in organic synthesis. In: *CRC Handbook of Organic Photochemistry and Photobiology*, 3e (ed. A. Griesbeck, M. Oelgemöller and F. Ghetti), 73–94. Boca Raton, FL: CRC Press.

267 Greenberg, M.M. and Gilmore, J.L. (1994). Cleavage of oligonucleotides from solid-phase supports using o-nitrobenzyl photochemistry. *J. Org. Chem.* 59 (4): 746–753.

268 Holmes, C.P. (1997). Model studies for new o-nitrobenzyl photolabile linkers: substituent effects on the rates of photochemical cleavage. *J. Org. Chem.* 62 (8): 2370–2380.

269 Aujard, I., Benbrahim, C., Gouget, M. et al. (2006). o-Nitrobenzyl photolabile protecting groups with red-shifted absorption: syntheses and uncaging cross-sections for one- and two-photon excitation. *Chem. Eur. J.* 12 (26): 6865–6879.

270 Weigel, W. and Wagner, P.J. (1996). Photocyclization to cyclopropanols and exciplex emission of β-arylpropiophenones. *J. Am. Chem. Soc.* 118 (50): 12858–12859.

271 Breslow, R., Baldwin, S., Flechtne, T. et al. (1973). Remote oxidation of steroids by photolysis of attached benzophenone groups. *J. Am. Chem. Soc.* 95 (10): 3251–3262.

272 Breslow, R. (1972). Centenary lecture – biomimetic chemistry. *Chem. Soc. Rev.* 1 (4): 553–580.

273 (a) Paquette, L.A. (1982). Dodecahedrane – the chemical transliteration of Platos Universe (a review). Proceedings of the National Academy of Sciences of the United States of America-Physical. *Sciences* 79 (14): 4495–4500. (b) Dorigo, A.E. and Houk, K.N. (1987). Transition structures for intramolecular hydrogen atom transfers: the energetic advantage of seven-membered over six-membered transition structures. *J. Am. Chem. Soc.* 109: 2195–2197.

274 Wagner, P.J., Truman, R.J., Puchalski, A.E., and Wake, R. (1986). Extent of charge-transfer in the photoreduction of phenyl ketones by alkylbenzenes. *J. Am. Chem. Soc.* 108 (24): 7727–7738.

275 Turro, N.J., Leermake, P.A., Wilson, H.R. et al. (1965). Photochemistry of 1,3-cyclobutanediones. Decomposition modes and chemical intermediates. *J. Am. Chem. Soc.* 87 (12): 2613–2615.

276 Lewis, F.D., Johnson, R.W., and Johnson, D.E. (1974). Conformational control of photochemical behavior – competitive α-cleavage and γ-hydrogen abstraction of alkyl phenyl ketones. *J. Am. Chem. Soc.* 96 (19): 6090–6099.

277 Paulson, S.E., Liu, D.L., Orzechowska, G.E. et al. (2006). Photolysis of heptanal. *J. Org. Chem.* 71 (17): 6403–6408.

278 Solomek, T., Stacko, P., Veetil, A.T. et al. (2010). Photoenolization-induced oxirane ring opening in 2,5-dimethylbenzoyl oxiranes to form pharmaceutically promising indanone derivatives. *J. Org. Chem.* 75 (21): 7300–7309.

279 Schuster, D.I. (1989). The photochemistry of enones. In: *Chemistry of Enones* (ed. S. Patai and Z. Rappoport), 623–756. Chichester: Wiley.

280 Corey, E.J., Tada, M., Lamahieu, R., and Libit, L. (1965). Trans-2-cycloheptenone. *J. Am. Chem. Soc.* 87 (9): 2051–2052.

281 Eaton, P.E. and Lin, K. (1965). Trans-2-cycloheptenone. *J. Am. Chem. Soc.* 87 (9): 2052–2054.

282 Dehmlow, E.V., Neuhaus, R., and Schell, H.G. (1988). Cyclopropenone chemistry. 10. 2-Alkoxy-3-alkylcyclopropenones. *Chem. Ber.* 121 (3): 569–571.

283 Vladimir, A.P. and Popik, V.V. (2003). Highly efficient photochemical generation of a triple bond: synthesis, properties, and photodecarbonylation of cyclopropenones. *J. Org. Chem.* 68 (20): 7833–7840.

284 Nguyen, L.T., De Proft, F., Nguyen, M.T., and Geerlings, P. (2001). Theoretical study of (2+1)-cycloaddition of CO and CS to acetylenes forming cyclopropenones and cyclopropenethiones. *J. Org. Chem.* 66 (12): 4316–4326.

285 Nguyen, L.T., De Proft, F., Nguyen, M.T., and Geerlings, P. (2001). Theoretical study of cyclopropenones and cyclopropenethiones: decomposition via intermediates. *J. Chem. Soc., Perkin Trans. 2* (6): 898–905.

286 Poloukhtine, A. and Popik, V.V. (2006). Mechanism of the cyclopropenone decarbonylation reaction. A density functional theory and transient spectroscopy study. *J. Phys. Chem. A* 110 (5): 1749–1757.

287 Agosta, W.C. and Smith, A.B. (1971). Photochemical reactions of simple cyclopentenones. *J. Am. Chem. Soc.* 93 (21): 5513–5520.

288 Kwie, W.W., Shoulders, B.A., and Gardner, P.D. (1962). Photoisomerization of Δ-Cholesten-3-one. *J. Am. Chem. Soc.* 84 (11): 2268–2269.

289 Zimmerman, H.E. and Wilson, J.W. (1964). Mechanistic & exploratory organic photochemistry. 9. Phenyl migration in irradiation of 4,4-diphenylcyclohexenone. *J. Am. Chem. Soc.* 86 (19): 4036–4037.

290 Schuster, D.I., Bonneau, R., Dunn, D.A. et al. (1984). Photochemistry of ketones in solution. 72. Characterization of transient intermediates in the photochemistry of cyclohexenones. *J. Am. Chem. Soc.* 106 (9): 2706–2707.

291 Schuster, D.I., Dunn, D.A., Heibel, G.E. et al. (1991). Enone photochemistry – dynamic properties of triplet excited-states of cyclic conjugated enones as revealed by transient absorption-spectroscopy. *J. Am. Chem. Soc.* 113 (16): 6245–6255.

292 Schuster, D.I. (1980). Photochemical rearrangements of enones. In: *Rearrangements in Ground and Excited States*, vol. 3 (ed. P. De Mayo), 167–279. New York: Academic Press.

293 Margaretha, P. (2003). Photorearrangement reactions of cyclohex-2-enones. In: *CRC Handbook of Organic Photochemistry and Photobiology*, Chapter 76, 2e (ed. W.M. Horspool and F. Lenci). Boca Raton, FL: CRC Press.

294 Chapman, O.L., Sieja, J.B., and Welstead, W.J. (1966). Stereospecific rearrangement of an excited triplet ketone. *J. Am. Chem. Soc.* 88 (1): 161–162.

295 Bellus, D., Kearns, D.R., and Schaffner, K. (1969). Photochemical reactions. 52. Photochemistry of α,β-unsaturated cyclic ketones. Specific reactions of n,π^* and π,π^*-triplet states of O-acetyltestosterone and 10-methyl-D-1,9-octalone-2. *Helv. Chim. Acta* 52 (4): 971–1009.

296 Schuster, D.I., Brown, R.H., and Resnick, B.M. (1978). Photochemistry of ketones in solution. 53. Stereospecific triplet-state photorearrangements of chiral 2-cyclohexenones – type-A lumiketone rearrangement and phenyl migrations. *J. Am. Chem. Soc.* 100 (14): 4504–4512.

297 Schuster, D.I. and Hussain, S. (1980). Photochemistry of ketones in solution. 57. Synthesis and photochemistry of a constrained 2-cyclohexenone. *J. Am. Chem. Soc.* 102 (1): 409–410.

298 Singh, W. (2003). 1,3-Acyl migrations in β,γ-unsaturated ketones. In: *CRC Handbook of Organic Photochemistry and Photobiology*, Chapter 79, 2e (ed. W.M. Horspool and F. Lenci). Boca Raton, FL: CRC Press.

299 Engel, P.S. and Schexnayder, M.A. (1975). Systematic structural modifications in photochemistry of β,γ-unsaturated ketones. 1. Cyclic olefins. *J. Am. Chem. Soc.* 97 (1): 145–153.

300 Schexnayder, M.A. and Engel, P.S. (1975). Systematic structural modifications in photochemistry of β,γ-unsaturated ketones. 2. Acyclic olefins and acetylenes. *J. Am. Chem. Soc.* 97 (17): 4825–4836.

301 Armesto, D., Ortíz, M.J., and Agarrabeita, A.R. (2003). Recent advances in Di-π-methane processes. Novel reactions of 1,4-unsaturated compounds promoted by triplet sensitization and photoelectron transfer. *J. Photosci.* 10 (1): 9–20.

302 Houk, K.N. (1976). Photochemistry and spectroscopy of β,γ-unsaturated carbonyl-compounds. *Chem. Rev.* 76 (1): 1–74.

303 Rao, V.J. and Sriniivas, K. (2012). Oxa-di-π-methane rearrangement of β,γ-unsaturated ketones. In: *CRC Handbook of Organic Photochemistry and Photobiology*, 3e (ed. A. Griesbeck, M. Oelgemöller and F. Ghetti), 527–548. Boca Raton, FL: CRC Press.

304 Griesbeck, A.G. (2003). Spin-selectivity in photochemistry: a tool for organic synthesis. *Synlett* (4): 451–472.

305 Uyehara, T., Kabasawa, Y., and Kato, T. (1986). Rearrangement approaches to cyclic skeletons. 4. The total synthesis of (+/-)-pinguisone and (+/-)-deoxopinguisone based on photochemical (1,3)-acyl migration of a bicyclo[3.2.2]non-6-en-2-one. *Bull. Chem. Soc. Jpn.* 59 (8): 2521–2528.

306 Shin, J.H. and Fenical, W. (1991). New diterpenoids from the Caribbean Gorgonian Eunicea calyculata – photochemical Interconversion of the cembrene and cubitene skeletons. *J. Org. Chem.* 56 (3): 1227–1233.

307 Barton, D.H.R. and Quinkert, G. (1960). Photochemical transformations. 6. Photochemical cleavage of cyclohexadienones. *J. Chem. Soc.* 1–9.

308 Chapman, O.L. and Lassila, J.D. (1968). Direct observation of ketene intermediates in photochemical reactions. *J. Am. Chem. Soc.* 90 (9): 2449–2450.

309 Quinkert, G. (1965). Light-induced formation of acids from cyclic ketones. *Angew. Chem. Int. Ed. Engl.* 4 (3): 211–222.

310 Barton, D.H.R. (1959). Some photochemical rearrangements. *Helv. Chim. Acta* 42 (7): 2604–2616.

311 Quinkert, G., Bronster, B., and Schmieder, K.R. (1972). Light-induced reactions. 14. Three isomerization routes originating from different electronic states of a linearly-conjugated cyclohexadienone. *Angew. Chem. Int. Ed. Engl.* 11 (7): 637–640.

312 Quinkert, G., Scherer, S., Reichert, D. et al. (1997). Stereoselective ring opening of electronically excited cyclohexa-2,4-dienones: cause and effect. *Helv. Chim. Acta* 80 (6): 1683–1772.

313 Barton, D.H.R., DeMayo, P., and Shafiq, M. (1957). Photochemical transformations. 1. Some preliminary investigations. *J. Chem. Soc.* 929–935.

314 Birladeanu, L. (2003). The stories of Santonin and Santonic acid. *Angew. Chem. Int. Ed.* 42 (11): 1202–1208.

315 Zimmerman, H.E. and Swenton, J.S. (1967). Mechanistic organic photochemistry. 21. Electronic details of 2,5-cyclohexadienone rearrangement. *J. Am. Chem. Soc.* 89 (4): 906–912.

316 Zimmerman, H.E. and Schuster, D.I. (1961). Photochemical rearrangement of 4,4-diphenylcyclohexadienone – paper I on a general theory of photochemical reactions. *J. Am. Chem. Soc.* 83 (21): 4486–4488.

317 Zimmerman, H.E. and Schuster, D.I. (1962). A new approach to mechanistic organic photochemistry. 4. Photochemical rearrangements of 4,4-diphenylcyclohexadienone. *J. Am. Chem. Soc.* 84 (23): 4527–4540.

318 Wolff, S., Smith, A.B., Schreiber, W.L., and Agosta, W.C. (1972). Photochemistry of cyclopentenones – hydrogen abstraction by β-Carbon atom. *J. Am. Chem. Soc.* 94 (22): 7797–7806.

319 Givens, R.S., Oettle, W.F., Coffin, R.L., and Carlson, R.G. (1971). Mechanistic studies on organic photochemistry. 3. Photochemistry of bicyclo[2.2.2]octenone and benzobicyclo[2.2.2]octadienone. *J. Am. Chem. Soc.* 93 (16): 3957–3962.

320 Schwager, L. and Vogel, P. (1980). Interactions between non-conjugated chromophores. 12. The photochemistry of 5,6-dimethylidene-2-morbornanone – synthesis and Diels–Alder reactivity of 2,3-dimethylidenebicyclo[2.1.1]hexane. *Helv. Chim. Acta* 63 (5): 1176–1180.

321 Van Tamelen, E.E. and Pappas, S.P. (1962). Chemistry of Dewar-benzene – 1,2,5-Tri-*t*-butylbicyclo[2.2.0]hexa-2,5-diene. *J. Am. Chem. Soc.* 84 (19): 3789–3791.

322 Viehe, H.G., Merényi, R., Oth, J.F.M. et al. (1964). Valence-bond isomers of (substituted)benzene. *Angew. Chem. Int. Ed.* 3: 755–756.

323 Wilzbach, K.E. and Kaplan, L. (1965). Photoisomerization of tri-*t*-butylbenzenes. Prismane and benzvalene isomers. *J. Am. Chem. Soc.* 87 (17): 4004–4006.

324 (a) Wilzbach, K.E., Ritscher, J.S., and Kaplan, L. (1967). Benzvalene tricyclic valence isomer of benzene. *J. Am. Chem. Soc.* 89 (4): 1031–1032. (b) Kaplan, L. and Wilzbach, K.E. (1968). Photolysis of benzene vapor. Benzvalene formation at wavelengths 2537–2370 A. *J. Am. Chem. Soc.* 90 (12): 3291–3292.

325 Kaplan, L. and Wilzbach, K.E. (1967). Photolysis of benzene vapor at 1849 A. *J. Am. Chem. Soc.* 89 (4): 1030–1031.

326 Kaplan, L., Walch, S.P., and Wilzbach, K.E. (1968). Photolysis of benzene vapor at 1849 A. Formation of *cis*-1,3-hexadien-5-yne. *J. Am. Chem. Soc.* 90 (20): 5646–5647.

327 Johnstone, D.E. and Sodeau, J.R. (1991). Matrix-controlled photochemistry of benzene and pyridine. *J. Phys. Chem.* 95 (1): 165–169.

328 Arnett, E.M. and Bollinger, J.M. (1964). Evidence for a Ladenburg benzene intermediate in the photoisomerization of 1,2,4,5-tetra-t-butylbenzene to 1,2,3,5-tetra-t-butylbenzene. *Tetrahedron Lett.* (50): 3803–3808.

329 Barlow, M.G., Haszeldi, R.N., and Kershaw, M.J. (1974). Valence-bond isomer chemistry. 6. Photochemical interconversion of perfluoro-1,3,5-1,2,4-trimethylbenzenes and perfluoro-1,3,5-1,2,4-trimethylbenzenes via para-bonded benzene and prismane isomers. *J. Chem. Soc., Perkin Trans. 1* (14): 1736–1740.

330 Turro, N.J., Ramamurthy, V., and Katz, T.J. (1977). Energy-storage and release – direct and sensitized photoreactions of Dewar-benzene and prismane. *New J. Chem.* 1 (5): 363–365.

331 Katz, T.J. and Acton, N. (1973). Synthesis of prismane. *J. Am. Chem. Soc.* 95 (8): 2738–2739.

332 Wingert, H. and Regitz, M. (1986). Syntheses with cyclobutadienes. 14. Tetra-tert-butylprismanes by Dewar-benzene isomerization. *Z. Naturforsch., B: Chem. Sci.* 41 (10). 1306–1310.

333 Wingert, H., Maas, G., and Regitz, M. (1986). Syntheses with cyclobutadienes. 15. Prismane/Dewar-benzene isomerization X-ray crystal-structure of tert-butyl 3,4,5-tri-tert-butyl-2,6-bis(trifluoromethyl)prismane-1-carboxylate. *Tetrahedron* 42 (19): 5341–5353.

334 Maier, G., Bauer, I., Huberpatz, U. et al. (1986). Small Rings. 59. – unusual preparation as well as crystal and molecular-structure of trimethyl 3,4,5-tri-tert-butylprismane-1,2,6-tricarboxylate. *Chem. Ber.* 119 (4): 1111–1116.

335 Palmer, I.J., Ragazos, I.N., Bernardi, F. et al. (1993). An MO-SCF study of the S_1 and S_2 photochemical reactions of benzene. *J. Am. Chem. Soc.* 115 (2): 673–682.

336 Dreyer, J. and Klessinger, M. (1996). The photochemical formation of fulvene from benzene via prefulvene – a theoretical study. *Chem. Eur. J.* 2 (3): 335–341.

337 Bryce-Smith, D., Gilbert, A., and Robinson, D.A. (1971). Direct transformation of second excited singlet state of benzene into Dewar-benzene. *Angew. Chem. Int. Ed. Engl.* 10 (10): 745–746.

338 Kaplan, L., Ritscher, J.S., and Wilzbach, K.E. (1966). Photochemical 1,3 addition of alcohols to benzenes. *J. Am. Chem. Soc.* 88 (12): 2881–2882.

339 Farenhorst, E. and Bickel, A.F. (1966). Photochemical reactions of benzene with acids. *Tetrahedron Lett.* (47): 5911–5913.

340 Berson, J.A. and Hasty, N.M. Jr., (1971). Solvolytic behavior of bicyclo[3.1.0]hex-3-en-2-yl derivatives – some observations on mechanism of photolysis of benzene in hydroxylic media. *J. Am. Chem. Soc.* 93 (6): 1549–1551.

341 Kaplan, L., Wilzbach, K.E., and Rausch, D.J. (1972). Photosolvation of benzene – mechanism of formation of bicyclo[3.1.0]hex-3-en-2-yl and of bicyclo[3.1.0]hex-2-en-6-yl derivatives. *J. Am. Chem. Soc.* 94 (24): 8638–8640.

342 Wilzbach, K.E., Harkness, A.L., and Kaplan, L. (1968). Photochemical rearrangement of benzene-1,3,5-D3. *J. Am. Chem. Soc.* 90 (5): 1116–1118.

343 Wilzbach, K.E. and Rausch, D.J. (1970). Photochemistry of nitrogen heterocycles. Dewar pyridine and its intermediacy in photoreduction and photohydration of pyridine. *J. Am. Chem. Soc* 92: 2178–2179.

344 Chapman, O.L., Mcintosh, C.L., and Pacansky, J. (1973). Photochemical transformations. 48. Cyclobutadiene. *J. Am. Chem. Soc.* 95 (2): 614–617.

345 Chambers, R.D. and Middleton, R. (1977). Stable 2-azabicyclo[2.2.0]hexa-2,5-diene derivatives. *J. Chem. Soc., Chem. Commun.* (5): 154–155.

346 Chambers, R.D. and Middleton, R. (1977). Photochemistry of halogenocarbon compounds. 3. Rearrangements involving azaprismanes. *J. Chem. Soc., Perkin Trans. 1* (12): 1500–1504.

347 Kobayashi, Y. and Kumadaki, I. (1982). Dewar heterocycles and related compounds. *Adv. Heterocycl. Chem.* 31: 169–206.

348 Kaplan, L., Wilzbach, K.E., and Pavlik, J.W. (1972). Photohydration of pyridinium ions. *J. Am. Chem. Soc.* 94 (9): 3283–3283.

349 King, R.A., Lüthi, H.P., Schaefer, H.F. et al. (2001). The photohydration of N-alkylypyridinium salts: theory and experiment. *Chem. Eur. J.* 7 (8): 1734–1742.

350 Damiano, T., Morton, D., and Nelson, A. (2007). Photochemical transformations of

pyridinium salts: mechanistic studies and applications in synthesis. *Org. Biomol. Chem.* 5 (17): 2735–2752.

351 Ling, R. and Mariano, P.S. (1998). A demonstration of the synthetic potential of pyridinium salt photochemistry by its application to a stereocontrolled synthesis of (+)-Mannostatin A. *J. Org. Chem.* 63 (17): 6072–6076.

352 Wipf, P. and Miller, C.P. (1993). Stereospecific synthesis of peptide analogs with Allo-Threonine and D-Allo-threonine residues. *J. Org. Chem.* 58 (6): 1575–1578.

353 Lahmani, F. and Ivanoff, N. (1967). Photoisomerization of pyrazine and of its methyl derivatives. *Tetrahedron Lett.* (40): 3913–3916.

354 Lahmani, F. and Ivanoff, N. (1972). Mercury sensitization of isomerization of diazines. *J. Phys. Chem.* 76 (16): 2245–2248.

355 Su, M.D. (2006). CASCSF study on the photochemical transposition reactions of pyrazines. *J. Phys. Chem. A* 110 (30): 9420–9428.

356 Penkett, C.S. and Simpson, I.D. (1999). Photosolvolysis reactions of 3-alkoxypyridinium tetrafluoroborate salts. *Tetrahedron* 55 (19): 6183–6204.

357 Dennis, N., Katritzky, A.R., and Wilde, H. (1976). 1,3-Dipolar character of 6-membered aromatic rings. 27. Photochemically induced valence bond tautomerism and dimerization of 3-oxido-1-phenylpyridinium. *J. Chem. Soc., Perkin Trans. 1* (21): 2338–2343.

358 Srinivasan, R. (1967). Photosensitized decarbonylation of furans. *J. Am. Chem. Soc.* 89 (7): 1758–1758.

359 Srinivasan, R. (1967). Structures of excited states of furan produced on Hg($^3P_1$) photosensitization. *J. Am. Chem. Soc.* 89 (18): 4812–4813.

360 Rendall, W.A., Clement, A., Torres, M., and Strausz, O.P. (1986). Dewar-furan and Dewar-thiophene – low-temperature matrix photolysis of furan and thiophene. *J. Am. Chem. Soc.* 108 (7): 1691–1692.

361 Hiraoka, H. and Srinivasan, R. (1968). Photoisomerization of 2-alkylfurans to 3-alkylfurans. *J. Am. Chem. Soc.* 90 (10): 2720–2721.

362 Hiraoka, H. (1970). Photochemistry of methylfurans. Selectivities of ring contraction reactions. *J. Phys. Chem.* 74 (3): 574–581.

363 Barton, T.J. and Hussmann, G.P. (1983). Extraordinary photo-rearrangement of silylfurans and subsequent thermal rearrangements. *J. Am. Chem. Soc.* 105 (20): 6316–6318.

364 D'Auria, M. (2000). Ab initio study on the photochemical isomerization of furan derivatives. *J. Org. Chem.* 65 (8): 2494–2498.

365 Hiraoka, H. (1970). Photoisomerization of 2-cyanopyrrole. *J. Chem. Soc., Chem. Commun.* (20): 1306–1306.

366 Barltrop, J., Day, A.C., Moxon, P.D., and Ward, R.R. (1975). Permutation patterns and phototranspositions of 2-cyanopyrroles. *J. Chem. Soc., Chem. Commun.* (19): 786–787.

367 Kellogg, R.M., Dik, J.K., Vandriel, H., and Wynberg, H. (1970). Photochemistry of thiophenes. 9. Rearrangements of alkylthiophenes and dithienyls. *J. Org. Chem.* 35 (8): 2737–2741.

368 Couture, A. and Lablache, A. (1971). Photochemical study. 3. Photochemistry of thiophene. *Tetrahedron* 27 (5): 1059–1068.

369 D'Auria, M. (2011). Photochemical and photophysical behavior of thiophene. In: *Advances in Heterocyclic Chemistry*, vol. 104 (ed. A.R. Katritzky), 127–390. Elsevier B.V.

370 D'Auria, M. and Racioppi, R. (2009). On the photoisomerization of thiophene and thiazole derivatives. *Heterocycles* 97 (3): 737–748.

371 Wiebe, H.A., Heicklen, J., and Braslavsky, S. (1972). Photolysis of tetra(trifluoromethyl)thiophene vapor. *Can. J. Chem.* 50 (16): 2721–2724.

372 D'Auria, M. (2001). Photochemical isomerization of pentaatomic heterocycles. In: *Advances in Heterocyclic Chemistry*, vol. 79 (ed. A.R. Katritzky), 41–88. Elsevier B.V.

373 Crespo-Hernández, C.E., Cohen, B., Hare, P.M., and Kohler, B. (2004). Ultrafast excited-state dynamics in nucleic acids. *Chem. Rev.* 104 (4): 1977–2019.

374 Yu, H., Evans, N.L., Stavros, V.G., and Ullrich, S. (2012). Investigation of multiple electronic excited state relaxation pathways following 200 nm photolysis of gas-phase imidazole. *Phys. Chem. Chem. Phys.* 14 (18): 6266–6272.

375 Montero, R., Peralta Conde, A., Ovejas, V. et al. (2012). Ultrafast evolution of imidazole after electronic excitation. *J. Phys. Chem. A* 116 (44): 10752–10758.

376 Slavicek, P. and Farnik, M. (2011). Photochemistry of hydrogen bonded heterocycles probed by photodissociation experiments and ab initio methods. *Phys. Chem. Chem. Phys.* 13 (26): 12123–12137.

377 Ashfold, M.N.R., Cronin, B., Devine, A.L. et al. (2006). The role of π,σ^* excited states in the photodissociation of heteroaromatic molecules. *Science* 312 (5780): 1637–1640.

378 Devine, A.L., Cronin, B., Nix, M.G., and Ashfold, M.N. (2006). High resolution photofragment translational spectroscopy studies of the near ultraviolet photolysis of imidazole. *J. Chem. Phys.* 125 (18).

379 Cronin, B., Devine, A.L., Nix, M.G., and Ashfold, M.N. (2006). Near ultraviolet photolysis of deuterated pyrrole. *Phys. Chem. Chem. Phys.* 8 (29): 3440–3445.

380 Williams, C.A., Roberts, G.M., Yu, H. et al. (2012). Exploring ultrafast H-atom elimination versus photofragmentation pathways in pyrazole following 200 nm excitation. *J. Phys. Chem. A* 116 (11): 2600–2609.

381 King, G.A., Oliver, T.A., Nix, M.G., and Ashfold, M.N. (2010). Exploring the mechanisms of H atom loss in simple azoles: ultraviolet photolysis of pyrazole and triazole. *J. Chem. Phys.* 132 (6).

382 Su, M.-D. (2006). Photochemical isomerization reactions of cyanopyrroles: a theoretical study. *J. Phys. Chem. A* 110 (46): 12653–12661.

383 Kobsa, H. (1962). Rearrangement of aromatic esters by ultraviolet radiation. *J. Org. Chem.* 27 (7): 2293–2298.

384 Anderson, J.C. and Reese, C.B. (1960). Photo-induced Fries rearrangements. *Proc. Chem. Soc., London* (6): 217–217.

385 Shizuka, H. and Tanaka, I. (1969). Photochemistry of acetanilide. 4. Photochemical reactions in vapor phase and rigid matrixes. *Bull. Chem. Soc. Jpn.* 42 (4): 909–913.

386 Bellus, D. (1971). Photo-Fries rearrangement and related photochemical (1,j)-shifts of carbonyl and sulfonyl groups. *Adv. Photochem.* 8: 109–159.

387 Magnus, P. and Lescop, C. (2001). Photo-Fries rearrangement for the synthesis of the diazonamide macrocycle. *Tetrahedron Lett.* 42 (41): 7193–7196.

388 Magauer, T., Martin, H.J., and Mulzer, J. (2009). Total synthesis of the antibiotic Kendomycin by macrocyclization using photo-Fries rearrangement and ring-closing metathesis. *Angew. Chem. Int. Ed.* 48 (33): 6032–6036.

389 Mehr, M., van Driel, W., Jansen, K. et al. (2013). Photodegradation of bisphenol A polycarbonate under blue light radiation and its effect on optical properties. *Opt. Mater.* 35 (3): 504–508.

390 Arrieta, C., David, E., Dolez, P., and Toan, V. (2011). Hydrolytic and photochemical aging studies of a Kevlar (R)-PBI blend. *Polym. Degrad. Stab.* 96 (8): 1411–1419.

391 Diepens, M. and Gijsman, P. (2011). Outdoor and accelerated weathering studies of bisphenol A polycarbonate. *Polym. Degrad. Stab.* 96 (4): 649–652.

392 mine-Khodja, A., Boulkamh, A., and Boule, P. (2004). Photochemical behaviour of phenylurea herbicides. *Photochem. Photobiol. Sci.* 3 (2): 145–156.

393 Pankasem, S., Kuczynski, J., and Thomas, J.K. (1994). Photochemistry and photodegradation of polycarbonate. *Macromolecules* 27 (14): 3773–3781.

394 Bellus, D. and Hrdlovic, P. (1967). Photochemical rearrangement of aryl vinyl and substituted vinyl esters and amides of carboxylic acids. *Chem. Rev.* 67 (6): 599–609.

395 Park, K.K., Lee, J.J., and Ryu, J. (2003). Photo-Fries rearrangement of N-arylsulfonamides to aminoaryl sulfone derivatives. *Tetrahedron* 59 (39): 7651–7659.

396 Lochbrunner, S., Zissler, M., Piel, J. et al. (2004). Real time observation of the photo-Fries rearrangement. *J. Chem. Phys.* 120 (24): 11634–11639.

397 Gohdo, M., Takamasu, T., and Wakasa, M. (2011). Photochemical primary process of photo-Fries rearrangement reaction of 1-naphthyl acetate as studied by MFE probe. *Phys. Chem. Chem. Phys.* 13 (2): 755–761.

398 Pac, C. and Tsutsumi, S. (1964). The photochemical rearrangement of ethyl phenyl carbonate. *Bull. Chem. Soc. Jpn.* 37 (9): 1392–1392.

399 Nozaki, H., Okada, T., Noyori, R., and Kawanisi, M. (1966). Photochemical rearrangement of arenesulphonanilides to P-aminodiarylsulphones. *Tetrahedron* 22 (7): 2177–2180.

400 Kelly, D.P. and Pinhey, J.T. (1964). The photochemical rearrangement of phenoxyacetic acids. *Tetrahedron Lett.* (46): 3427–3429.

401 Fischer, M. (1968). Photochemical reactions. 2. Photolytic fragmentation of benzhydryl and benzyl amines. *Chem. Ber.* 101 (2): 731–737.

402 Pinhey, J.T. and Schaffner, K. (1968). Photochemical rearrangement of benzylenol ethers. *Aust. J. Chem.* 21 (9): 2265–2276.

403 Ogata, Y., Takagi, K., and Izawa, Y. (1968). Photoinduced reactions of benzyl methyl ketone. *Tetrahedron* 24 (4): 1617–1621.

404 Dirania, M.K.M. and Hill, J. (1968). Photolysis of aryloxyacetones. *J. Chem. Soc. C* (11): 1311–1316.

405 Izawa, Y. and Ogata, Y. (1970). Photochemical rearrangements of α-benzyloxystyrenes to β-phenylpropiophenones. *J. Org. Chem.* 35 (9): 3192–3195.

406 Galindo, F. (2005). The photochemical rearrangement of aromatic ethers – a review of the photo-Claisen reaction. *J. Photochem. Photobiol., C* 6 (2–3): 123–138.
407 Kharasch, M.S., Stampa, G., and Nudenberg, W. (1952). Photochemical para-rearrangement of phenyl ethers. *Science* 116 (3012): 309–309.
408 Carroll, F.A. and Hammond, G.S. (1972). Mechanisms of photochemical reactions in solution. 72. Electronic pathways in photodissociation of 3-methyl-1-phenoxybut-2-ene. *J. Am. Chem. Soc.* 94 (20): 7151–7152.
409 Shimamura, N. and Sugimori, A. (1971). Direct and sensitized photoreactions of allyl phenyl ether. *Bull. Chem. Soc. Jpn.* 44 (1): 281–283.
410 Grimme, S. (1992). MO-Theoretical investigation on the photodissociation of C—O bonds in aromatic compounds. *Chem. Phys.* 163 (3): 313–330.
411 Benn, R., Dreeskamp, H., Schuchmann, H.P., and Vonsonntag, C. (1979). Photolysis of aromatic ethers. *Z. Naturforsch., B: Chem. Sci.* 34B (7): 1002–1009.
412 Jiménez, M.C., Miranda, M.A., Scaiano, J.C., and Tormos, R. (1997). Two-photon processes in the photo-Claisen and photo-Fries rearrangements. Direct observation of dienic ketenes generated by photolysis of transient cyclohexa-2,4-dienones. *J. Chem. Soc., Chem. Commun.* (16): 1487–1488.
413 Syamala, M.S. and Ramamurthy, V. (1988). Modification of photochemical reactivity by cyclodextrin complexation – selectivity in photo-Claisen rearrangement. *Tetrahedron* 44 (23): 7223–7233.
414 Pitchumani, K., Devanathan, S., and Ramamurthy, V. (1992). Modification of photochemical reactivity on formation of inclusion complexes – photorearrangement of benzyl phenyl ethers and methyl phenoxyacetates. *J. Photochem. Photobiol., A* 69 (2): 201–208.
415 Guerrini, G., Taddei, M., and Ponticelli, F. (2011). Synthesis of functionalized quinolines and benzo[c][2,7]naphthyridines based on a photo-Fries rearrangement. *J. Org. Chem.* 76 (18): 7597–7601.
416 Engel, P.S. (1980). Mechanism of the thermal and photochemical decomposition of azoalkanes. *Chem. Rev.* 80 (2): 99–150.
417 Al-Sader, B.H. and Crawford, R.J. (1970). Allyl radicals from 3,3′-azo-1-propene. *Can. J. Chem.* 48 (17): 2745–2754.
418 Thiele, J. (1909). On hydrazo and azomethane. *Ber. Dtsch. Chem. Ges.* 42: 2575–2580.
419 Thompson, A.M., Goswami, P.C., and Zimmerman, G.L. (1979). Kinetic analysis of the photochemistry of alkyldiazenes in hydrocarbon solution – quasi-steady state. *J. Phys. Chem.* 83 (3): 314–320.
420 Diau, E.W.G. and Zewail, A.H. (2003). Femtochemistry of trans-azomethane: a combined experimental and theoretical study. *ChemPhysChem* 4 (5): 445–456.
421 Green, J.G., Dubay, G.R., and Porter, N.A. (1977). N-15 CIDNP Investigation of dialkyl azo-compounds – evidence for tert-alkyl diazenyl radicals. *J. Am. Chem. Soc.* 99 (4): 1264–1265.
422 Porter, N.A. and Marnett, L.J. (1973). Photolysis of unsymmetric azo-compounds – *cis*-Azo compound intermediates. *J. Am. Chem. Soc.* 95 (13): 4361–4367.
423 Adams, J.S., Burton, K.A., Andrews, B.K. et al. (1986). Time-resolved observation of sequential bond-cleavage in a gas phase azoalkane. *J. Am. Chem. Soc.* 108 (25): 7935–7938.
424 McBride, J.M. (1971). Rotational diffusion control of radical disproportionation in solid-state photolysis of azobisisobutyronitrile. *J. Am. Chem. Soc.* 93 (23): 6302–6303.
425 Bartlett, P.D. and McBride, J.M. (1967). Configuration, conformation and spin in radical pairs. *Pure App. Chem.* 15 (1): 89–108.
426 Roth, W.R. and Martin, M. (1967). Zur Stereochemie des thermischen und photochemischen Zerfalls von 2,3-Diazabicyclo[2.2.1]hept-2-en. *Liebigs Ann. Chem.* 702 (1): –5.
427 Adam, W., Oppenlander, T., and Zang, G. (1985). Photochemistry of the azoalkanes 2,3-diazabicyclo[2.2.1]hept-2-ene and spiro[cyclopropane-1,7′-[2,3]diazabicyclo[2.2.1]hept-2-ene] – on the questions of one-bond vs 2-bond cleavage during the denitrogenation, cyclization vs rearrangement of the 1,3-diradicals, and double inversion. *J. Org. Chem.* 50 (18): 3303–3312.
428 Adam, W., García, H., Diedering, M. et al. (2002). Stereochemical memory in the temperature-dependent photodenitrogenation of bridgehead-substituted DBH-type azoalkanes: inhibition of inverted-housane formation in the diazenyl diradical through the mass effect (inertia) and steric hindrance. *J. Am. Chem. Soc.* 124 (41): 12192–12199.
429 Adam, W. and Trofimov, A.V. (2003). Viscosity as an experimental tool in studying the mechanism of stereoselective and product-selective

chemical transformations through frictional impositions. *Acc. Chem. Res.* 36 (8): 571–579.

430 Buchwalter, S.L. and Closs, G.L. (1975). Electron-spin resonance study of matrix-isolated 1,3-cyclopentadiyl, a localized 1,3-carbon biradical. *J. Am. Chem. Soc.* 97 (13): 3857–3858.

431 Paulsen, S.R. (1960). 3,3-Dialkyl-diazacycloprop-1-en. *Angew. Chem.* 72 (21): 781–782.

432 Schmitz, E. and Ohme, R. (1961). Neue Diaziridin Synthese. *Angew. Chem.* 73 (6): 220–220.

433 Liu, M.T.H. (1982). The thermolysis and photolysis of diazirines. *Chem. Soc. Rev.* 11 (2): 127–140.

434 (a) Kaczor, A., Gómez-Zavaglia, A., Cardoso, A.L. et al. (2006). Methyl 3-methyl-2H-azirine-2-carboxylate photochemistry studied by matrix isolation. FTIR and DFT calculations. *J. Chem. Phys. A.* 110: 10742–10749. (b) Inui, H. and Murata, S. (2005). Photochemistry of 2-(1-naphthyl)-2H-azirines in matrixes and in solutions: wavelength-dependent C—C and C—N bond cleavage of the azirine ring. *J. Am. Chem. Soc.* 127 (8): 2628–2636.

435 Brunner, J., Senn, H., and Richards, F.M. (1980). 3-Trifluoromethyl-3-phenyldiazirine – a new carbene generating group for photolabeling reagents. *J. Biol. Chem.* 255 (8): 3313–3318.

436 Winnacker, M., Breeger, S., Strasser, R., and Carell, T. (2009). Novel diazirine-containing DNA photoaffinity probes for the investigation of DNA-protein-interactions. *ChemBioChem* 10 (1): 109–118.

437 Das, J. (2011). Aliphatic diazirines as photoaffinity probes for proteins: recent developments. *Chem. Rev.* 111 (8): 4405–4417.

438 Amrich, M.J. and Bell, J.A. (1964). Photoisomerization of diazirine. *J. Am. Chem. Soc.* 86 (2): 292–293.

439 Korneev, S.M. (2011). Valence isomerization between diazo compounds and diazirines. *Eur. J. Org. Chem.* (31): 6153–6175.

440 Bennett, P., Harwood, L.M., Macro, J., and Mcgregor, A. (1994). Medium-dependent trapping of diazoalkane intermediates during photolysis of diazirines. *J. Chem. Soc., Chem. Commun.* (8): 961–962.

441 Bonneau, R. and Liu, M.T.H. (1996). Quantum yield of formation of diazo compounds from the photolysis of diazirines. *J. Am. Chem. Soc.* 118 (30): 7229–7230.

442 Zhang, Y., Kubicki, J., and Platz, M.S. (2010). Evidence of hydrogen migration in an alkylphenyldiazirine excited state. *Org. Lett.* 12 (14): 3182–3184.

443 Chapman, O.L., Chang, C.C., Kolc, J. et al. (1976). 1,1,2-Trimethylsilaethylene. *J. Am. Chem. Soc.* 98 (24): 7844–7846.

444 Erni, B. and Khorana, H.G. (1980). Fatty-acids containing photoactivable carbene precursors – synthesis and photochemical properties of 3,3-bis(1,1-difluorohexyl)diazirine and 3-(1,1-difluorooctyl)-3H-diazirine. *J. Am. Chem. Soc.* 102, 11: 3888–3896.

445 Chedekel, M.R., Skoglund, M., Kreeger, R.L., and Shechter, H. (1976). Solid-state chemistry – discrete trimethylsilylmethylene. *J. Am. Chem. Soc.* 98 (24): 7846–7848.

446 Trommer, M., Sander, W., and Marquard, C. (1994). 1,1-Dimethyl-1H-silirene. *Angew. Chem. Int. Ed. Engl.* 33 (7): 766–769.

447 Zielinski, M., Trommer, M., and Sander, W. (1999). Photochemistry of bis(diazomethyl)silanes: a novel approach to 1H-silirenes. *Organometallics* 18 (15): 2791–2800.

448 Lowe, G. and Parker, J. (1971). Photochemical isomerisation of some diazo-compounds into diazirines. *J. Chem. Soc., Chem. Commun.* (18): 1135–1136.

449 Wentrup, C., Bibas, H., Kuhn, A. et al. (2013). Matrix-IR spectroscopic investigations of the thermolysis and photolysis of diazoamides. *J. Org. Chem.* 78 (21): 10705–10717.

450 Wolff, L. (1902). Announcement from the chemistry laboratory of the University of Jena diazo anhydrides. *Liebigs Ann. Chem.* 325: 129–133.

451 Kirmse, W. (2002). 100 years of the Wolff rearrangement. *Eur. J. Org. Chem.* (14): 2193–2256.

452 Kaplan, F. and Meloy, G.K. (1966). Structure of diazoketones. A study of hindered internal rotation. *J. Am. Chem. Soc.* 88 (5): 950–956.

453 Burdzinski, G. and Platz, M.S. (2010). Ultrafast time-resolved studies of the photochemistry of diazo carbonyl compounds. *J. Phys. Org. Chem.* 23 (4): 308–314.

454 Burdzinski, G., Zhang, Y., Wang, J., and Platz, M.S. (2010). Concerted Wolff rearrangement in two simple acyclic diazocarbonyl compounds. *J. Phys. Chem. A* 114 (50): 13065–13068.

455 Burdzinski, G., Zhang, Y., Selvaraj, P. et al. (2010). Direct observation of 1,2-hydrogen migration in the excited states of alkyl diazo esters by ultrafast time resolved IR spectroscopy. *J. Am. Chem. Soc.* 132 (7): 2126–2130.

456 Rau, H. and Bokel, M. (1990). Photochemistry of bichromophoric molecules with Camphor structure. 2. The photochemistry of 3-diazocamphor. *J. Photochem. Photobiol., A* 53 (3): 311–322.

457 Laganis, E.D., Janik, D.S., Curphey, T.J., and Lemal, D.M. (1983). Photochemistry of perfluoro-3-diazo-2-butanone. *J. Am. Chem. Soc.* 105 (25): 7457–7459.

458 Torres, M., Raghunathan, P., Bourdelande, J.L. et al. (1986). Low-temperature photolysis of perfluorodiazopentanones – the formation of perfluoropropionylmethylmethylene, perfluoroacetylethylmethyle, perfluoroacetylethyl and perfluoropropionylmethyl diazirenes. *Chem. Phys. Lett.* 127 (3): 205–209.

459 Li, Y.Z. and Schuster, G.B. (1987). Photochemistry of 9-diazo-3,6-diazafluorene – through-space or through-bond transmission of electronic effects. *J. Org. Chem.* 52 (18): 3975–3979.

460 Bally, T., Matzinger, S., Truttmann, L. et al. (1994). Matrix spectroscopy of 2-adamantylidene, A dialkylcarbene with singlet ground-state. *Angew. Chem. Int. Ed. Engl.* 33 (19): 1964–1966.

461 Bonneau, R., Hellrung, B., Liu, M.T.H., and Wirz, J. (1998). Adamantylidene revisited: flash photolysis of adamantanediazirine. *J. Photochem. Photobiol., A* 116 (1): 9–19.

462 Knoll, W., Mieusset, J.-L., Arion, V.B. et al. (2010). 2*H*-Azirines from a concerted addition of alkylcarbenes to nitrile groups. *Org. Lett.* 12 (10): 2366–2369.

463 Ross, P.L. and Johnston, M.V. (1995). Excited-state photochemistry of iodoalkanes. *J. Phys. Chem.* 99 (12): 4078–4085.

464 Sammes, J.P. (1973). Photochemistry of the C—X group. In: *Chemistry of the Carbon–Halogen Bond* (ed. S. Patai), 747–794. New York: Wiley.

465 Saplay, K.M. and Damodaran, N.P. (1983). Photochemistry of organic halogen compounds. *J. Sci. Ind. Res.* 42 (11): 602–614.

466 Lodder, G. (1983). Recent advances in the photochemistry of the carbo-halogen bond. In: *Chemistry of Halides, Pseudo-Halides Azides*, vol. 2 (ed. S. Patai and Z. Rappoport), 1605–1683. Chichester: Wiley.

467 Sage, A.G., Oliver, T.A.A., Murdock, D. et al. (2011). n/σ^* and π/σ^* excited states in aryl halide photochemistry: a comprehensive study of the UV photodissociation dynamics of iodobenzene. *Phys. Chem. Chem. Phys.* 13 (18): 8075–8093.

468 Kadi, M., Davidsson, J., Tarnovsky, A.N. et al. (2001). Photodissociation of aryl halides in the gas phase studied with femtosecond pump-probe spectroscopy. *Chem. Phys. Lett.* 350 (1–2): 93–98.

469 Kropp, P.J. (1984). Photobehavior of alkyl-halides in solution – radical, carbocation, and carbene intermediates. *Acc. Chem. Res.* 17 (4): 131–137.

470 Kosmrlj, B. and Sket, B. (2000). Photolysis of bromo- and chloro-substituted benzyl derivatives. Competition between ionic and radical pathways. *J. Org. Chem.* 65 (21): 6890–6896.

471 Kropp, P.J., Sawyer, J.A., and Snyder, J.J. (1984). Photochemistry of alkyl-halides. 11. Competing reaction via carbene and ccarbocationic Intermediates. *J. Org. Chem.* 49 (9): 1583–1589.

472 Lee, S.G. and Park, Y.T. (2003). Selective photosubstitution reactions of tertiary 1-alkyl-1,1-diarylmethyl chloride: simultaneous formation of radicals and carbocations from the excited state of 1-alkyl-1,1-diarylmethyl chloride. *Bull. Korean Chem. Soc.* 24 (9): 1241–1242.

473 Kropp, P.J., Poindexter, G.S., Pienta, N.J., and Hamilton, D.C. (1976). Photochemistry of alkyl halides. 4. 1-Norbornyl, 1-norbornylmethyl, 1-adamantyl and 2-adamantyl, and 1-octyl bromides and iodides. *J. Am. Chem. Soc.* 98 (25): 8135–8144.

474 Kropp, P.J. and Adkins, R.L. (1991). Photochemistry of alkyl halides. 12. Bromides vs. iodides. *J. Am. Chem. Soc.* 113 (7): 2709–2717.

475 Miranda, M.A., Pérez-Prieto, J., Font-Sanchis, E., and Scaiano, J.C. (2001). One- vs two-photon processes in the photochemistry of 1,n-dihaloalkanes. *Acc. Chem. Res.* 34 (9): 717–726.

476 Maier, G., Reisenauer, H.P., Hu, J. et al. (1990). Photochemical isomerization of dihalomethanes in argon matrices. *J. Am. Chem. Soc.* 112 (13): 5117–5122.

477 Saitow, K.-i., Naitoh, Y., Tominaga, K., and Yoshihara, K. (2008). Photodissociation of CH_2I_2 and subsequent electron transfer in solution. *Chem. Asian J.* 3 (4): 696–709.

478 Kropp, P.J., Pienta, N.J., Sawyer, J.A., and Polniaszek, R.P. (1981). Photochemistry of alkyl halides. 7. Cyclopropanation of alkenes. *Tetrahedron* 37 (19): 3229–3236.

479 Kropp, P.J. and Pienta, N.J. (1983). Photochemistry of alkyl halides. 9. Geminal dihalides. *J. Org. Chem.* 48 (12): 2084–2090.

480 Neuman, R.C. and Wolcott, R.G. (1966). Photochemistry of 1,1-diiodoalkanes. *Tetrahedron Lett.* (50): 6267–6272.

481 Moret, E., Jones, C.R., and Grant, B. (1983). Photochemistry of organic geminal diiodides. *J. Org. Chem.* 48 (12): 2090–2092.

482 Kroger, P.M., Demou, P.C., and Riley, S.J. (1976). Polyhalide photofragment spectra. 1. 2-photon 2-step photodissociation of methylene iodide. *J. Chem. Phys.* 65 (5): 1823–1834.

483 Schmitt, G. and Comes, F.J. (1980). Spectroscopic investigations of the photolysis of diiodomethane. *J. Mol. Struct.* 61: 51–54.

484 Haider, K., Platz, M.S., Després, A. et al. (1988). An exceptionally simple method of preparing matrix-isolated biradicals, biradicaloids, and carbenes. *J. Am. Chem. Soc.* 110 (7): 2318–2320.

485 Skell, P.S. and Traynham, J.G. (1984). Radical brominations of alkyl bromides and the nature of β-bromoalkyl radicals. *Acc. Chem. Res.* 17 (5): 160–166.

486 Ihee, H., Zewail, A.H., and Goddard, W.A. (1999). Conformations and barriers of haloethyl radicals (CH_2XCH_2, X=F, Cl, Br, I): ab initio studies. *J. Phys. Chem. A* 103 (33): 6638–6649.

487 Scaiano, J.C., Barra, M., Calabrese, G., and Sinta, R. (1992). Photochemistry of 1,2-dibromoethane in solution – a model for the generation of hydrogen bromide. *J. Chem. Soc., Chem. Commun.* (19): 1418–1419.

488 Scaiano, J.C., Barra, M., Krzywinski, M. et al. (1993). Laser flash-photolysis determination of absolute rate constants for reactions of bromine atoms in solution. *J. Am. Chem. Soc.* 115 (18): 8340–8344.

489 Wade, P.A. and Kondracki, P.A. (1994). Trapping evidence for 1,2-dinitrospiropentene. *J. Chem. Soc., Chem. Commun.* (10): 1263–1264.

490 Fujiwara, M., Mishima, K., Tamai, K. et al. (1997). Spectroscopic studies on photochemical formation of o-xylylene in solution. *J. Phys. Chem. A* 101 (27): 4912–4915.

491 Miranda, M.A., Pérez-Prieto, J., Font-Sanchis, E., and Scaiano, J.C. (1998). Mechanistic studies on the photogeneration of o- and p-xylylenes from α,α'-dichloroxylenes. *J. Chem. Soc., Chem. Commun.* (15): 1541–1542.

492 Haider, K.W., Migirdicyan, E., Platz, M.S. et al. (1990). The mechanism of formation of meta-xylylene type biradicals produced by photolysis of polymethyl benzenes or dihalomethyl benzenes. *J. Am. Chem. Soc.* 112 (2): 733–738.

493 Biewer, M.C., Biehn, C.R., Platz, M.S. et al. (1991). An exceptionally simple method of preparation of biradicals. 2. Low-temperature fluorescence spectra and ambient-temperature laser-induced fluorescence spectra of 1,3-naphthoquinodimethane, 1,6-naphthoquinodimethane, 2,6-naphthoquinodimethane, and 2,7-naphthoquinodimethane. *J. Am. Chem. Soc.* 113 (2): 616–620.

494 Biewer, M.C., Platz, M.S., Roth, M., and Wirz, J. (1991). A carbene to biradical rearrangement – reaction paths from (8-methyl-1-naphthyl)carbene to acenaphthene. *J. Am. Chem. Soc.* 113 (21): 8069–8073.

495 Yamaji, M., Suzuki, A., Ito, F. et al. (2005). A new-type photoreaction of a carbonyl compound – part 2. Photoinduced ω-bond dissociation in p-halomethylbenzophenone studied by time-resolved EPR technique and laser flash photolysis. *J. Photochem. Photobiol., A* 170 (3): 253–259.

496 Kropp, P.J., Fryxell, G.E., Tubergen, M.W. et al. (1991). Photochemistry of phenyl thioethers and phenyl selenoethers – radical vs. ionic behavior. *J. Am. Chem. Soc.* 113 (19): 7300–7310.

497 Miranda, M.A., Pérez-Prieto, J., Font-Sanchis, E. et al. (1997). Laser flash, laser-drop, and lamp photolysis of 1,3-dichloro-1,3-diphenylpropane. One- versus two-photon reaction pathways. *J. Org. Chem.* 62 (17): 5713–5719.

498 Han, K.L. and He, G.Z. (2007). Photochemistry of aryl halides: photodissociation dynamics. *J. Photochem. Photobiol., C* 8 (2): 55–66.

499 Bunce, N.J., Bergsma, J.P., Bergsma, M.D. et al. (1980). Structure and mechanism in the photo-reduction of aryl chlorides in alkane solvents. *J. Org. Chem.* 45 (18): 3708–3713.

500 Da Silva, J.P., Jockusch, S., and Turro, N.J. (2009). Probing the photoreactivity of aryl chlorides with oxygen. *Photochem. Photobiol. Sci.* 8 (2): 210–216.

501 (a) Miller, G.C., Miille, M.J., Crosby, D.G. et al. (1979). Photosolvolysis of 3,4-dichloroaniline in water – evidence for an aryl cation intermediate. *Tetrahedron* 35 (15): 1797–1800. (b) Majek, M. and Jacobi von Wangelin, A. (2016). Mechanistic perspectives on organic photoredox catalysis for aromatic substitutions. *Acc. Chem. Res.* 49 (10): 2316–2327.

502 Protti, S., Fagnoni, M., Mella, M., and Albini, A. (2004). Aryl cations from aromatic halides. Photogeneration and reactivity of 4-hydroxy(methoxy)phenyl cation. *J. Org. Chem.* 69 (10): 3465–3473.

503 Manet, I., Monti, S., Grabner, G. et al. (2008). Revealing phenylium, phenonium,

vinylenephenonium, and benzenium ions in solution. *Chem. Eur. J.* 14 (3): 1029–1039.

504 Ohkura, K., Seki, K., Terashima, M., and Kanaoka, Y. (1991). Direct photolysis of halopyridines in solutions – generation of the 2-pyridyl cation. *Chem. Pharm. Bull.* 39 (12): 3168–3169.

505 Lazzaroni, S., Protti, S., Fagnoni, M., and Albini, A. (2010). Participation of a heterolytic path in the photochemistry of chlorobenzene. *J. Photochem. Photobiol., A* 210 (2–3): 140–144.

506 Bryce-Smith, D., Dadson, W.M., and Gilbert, A. (1980). Photoreactions of halogenobenzenes with cyclopentene – insertion into carbon-halogen bonds. *J. Chem. Soc., Chem. Commun.* (3): 112–113.

507 Wu, C.Y., Wu, Y.J., and Lee, Y.P. (2004). Molecular elimination in photolysis of fluorobenzene at 193 nm: internal energy of HF determined with time-resolved Fourier-transform spectroscopy. *J. Chem. Phys.* 121 (18): 8792–8799.

508 De Carolis, M., Protti, S., Fagnoni, M., and Albini, A. (2005). Metal-free cross-coupling reactions of aryl sulfonates and phosphates through photoheterolysis of aryl-oxygen bonds. *Angew. Chem. Int. Ed.* 44 (8): 1232–1236.

509 Fagnoni, M. and Albini, A. (2005). Arylation reactions: the photo-S_N1 path via phenyl cation as an alternative to metal catalysis. *Acc. Chem. Res.* 38 (9): 713–721.

510 Protti, S., Dichiarante, V., Dondi, D. et al. (2012). Singlet/triplet phenyl cations and benzyne from the photodehalogenation of some silylated and stannylated phenyl halides. *Chem. Sci.* 3 (4): 1330–1337.

511 Grote, D. and Sander, W. (2009). Photochemistry of fluorinated 4-iodophenylnitrenes: matrix isolation and spectroscopic characterization of phenylnitrene-4-yls. *J. Org. Chem.* 74 (19): 7370–7382.

512 Grote, D., Finke, C., Neuhaus, P., and Sander, W. (2012). Matrix isolation and spectroscopic characterization of 2,5,6-trifluoropyridylnitren-3-yl. *Eur. J. Org. Chem.* (17): 3229–3236.

513 Gronheid, R., Zuilhof, H., Hellings, M.G. et al. (2003). Photochemical generation of a primary vinyl cation from (*E*)-bromostyrene: mechanisms of formation and reaction. *J. Org. Chem.* 68 (8): 3205–3215.

514 Vernin, G., Jauffred, R., Ricard, C. et al. (1972). Homolytic aromatic substitution by heterocyclic free-radicals. 2. Reaction of thiazol-2-yl and benzothiazol-2-yl radicals with aromatic compounds. *J. Chem. Soc., Perkin Trans. 2* (9): 1145–1150.

515 Kampmeier, J.A. and Hoffmeister, E. (1962). Photolysis of 1,2-diiodobenzene – a photochemical source of benzyne. *J. Am. Chem. Soc.* 84 (19): 3787–3788.

516 Dektar, J.L. and Hacker, N.P. (1990). Photochemistry of diaryliodonium salts. *J. Org. Chem.* 55 (2): 639–647.

517 Hacker, N.P., Leff, D.V., and Dektar, J.L. (1991). The photochemistry of diphenyliodonium halides – evidence for reactions from solvent-separated and tight ion-pairs. *J. Org. Chem.* 56 (7): 2280–2282.

518 Tobisu, M., Furukawa, T., and Chatani, N. (2013). Visible light-mediated direct arylation of arenes and heteroarenes using diaryliodonium salts in the presence and absence of a photocatalyst. *Chem. Lett.* 42 (10): 1203–1205.

519 Gronheid, R., Lodder, G., Ochiai, M. et al. (2001). Thermal and photochemical solvolysis of (*E*)- and (*Z*)-2-phenyl-1-propenyl(phenyl)iodonium tetrafluoroborate: benzenium and primary vinylic cation intermediates. *J. Am. Chem. Soc.* 123 (36): 8760–8765.

520 Gronheid, R., Lodder, G., and Okuyama, T. (2002). Photosolvolysis of (*E*)-styryl(phenyl)iodonium tetrafluoroborate. Generation and reactivity of a primary vinyl cation. *J. Org. Chem.* 67 (3): 693–702.

521 Slegt, M., Gronheid, R., van der Vlugt, D. et al. (2006). Photochemical generation of six- and five-membered cyclic vinyl cations. *J. Org. Chem.* 71 (6): 2227–2235.

522 Zollinger, H. (1994). *Diazo Chemistry I. Aromatic and Heteroaromatic Compounds*, 1–453. Weinheim: Wiley-VCH.

523 Zollinger, H. (1995). *Diazo Chemistry II. Aliphatic, Inorganic and Organometallic Compounds*, vol. 2, 1–522. Weinheim: Wiley-VCH.

524 Sellers, C. and Suschitzky, H. (1968). Use of arenediazonium hexafluoro-antimonates and -arsenates in preparation of aryl fluorides. *J. Chem. Soc. C* (18): 2317–2319.

525 Zollinger, H. (1973). Reactivity and stability of arenediazonium ions. *Acc. Chem. Res.* 6 (10): 335–341.

526 Galli, C. (1988). Radical reactions of arenediazonium ions – an easy entry into the chemistry of the aryl radical. *Chem. Rev.* 88 (5): 765–792.

527 Allongue, P., Delamar, M., Desbat, B. et al. (1997). Covalent modification of carbon surfaces by aryl radicals generated from the electrochemical reduction of diazonium salts. *J. Am. Chem. Soc.* 119 (1): 201–207.

528 Carey, J.G. and Millar, I.T. (1960). Formation and stability of aryldiazonium iodides and tri-iodides. *Chem. Ind. (London)* (4): 97–98.

529 Hanson, P., Hammond, R.C., Goodacre, P.R. et al. (1994). Sandmeyer reactions. 2. Estimation of absolute rate constants for some hydrogen-transfer reactions and for the transfer of water ligands on Cu(II) to aryl radicals by use of a Pschorr radical clock. *J. Chem. Soc., Perkin Trans. 2* (4): 691–696.

530 Beletskaya, I.P., Sigeev, A.S., Peregudov, A.S., and Petrovskii, P.V. (2007). Catalytic Sandmeyer bromination. *Synthesis* (16): 2534–2538.

531 Meerwein, H., Büchner, E., and van Emster, K. (1939). Memorandum from the Chemistry Institute, University of Marburg. The action of aromatic diazo compound on α,β-unsaturated carbonyl bonds. *J. Prakt. Chem.* 152 (7/10): 237–266.

532 Heinrich, M.R. (2008). Intermolecular olefin functionalisation involving aryl radicals generated from arenediazonium salts. *Chem. Eur. J.* 15: 820–833.

533 Pschorr, P. (1896). Neue Synthese des Phenanthrenes und seiner Derivate. *Ber. Dtsch. Chem. Ges.* 29: 496–501.

534 Leake, P.H. (1956). The Pschorr synthesis. *Chem. Rev.* 56 (1): 27–48.

535 Hanson, P., Lovenich, P.W., Rowell, S.C. et al. (1999). Sandmeyer reactions. Part 4. An investigation into the cyclisation modes of Pschorr phenanthrene synthesis. *J. Chem. Soc., Perkin Trans. 2* (1): 49–63.

536 Laali, K.K. and Shokouhimehr, M. (2009). The Pschorr reaction, a fresh look at a classical transformation. *Curr. Org. Synth.* 6 (2): 193–202.

537 Hassan, J., Sévignon, M., Gozzi, C. et al. (2002). Aryl-aryl bond formation one century after the discovery of the Ullmann reaction. *Chem. Rev.* 102 (5): 1359–1469.

538 Gomberg, M. and Bachmann, W.E. (1924). The synthesis of biaryl compounds by means of the diazo reaction. *J. Am. Chem. Soc.* 46: 2339–2343.

539 Bolton, R. and Williams, G.H. (1986). Homolytic arylation of aromatic and polyfluoroaromatic compounds. *Chem. Soc. Rev.* 15 (2): 261–289.

540 Pratsch, G., Wallaschkowski, T., and Heinrich, M.R. (2012). The Gomberg–Bachmann reaction for the arylation of anilines with aryl diazotates. *Chem. Eur. J.* 18 (37): 11555–11559.

541 Zollinger, H. (1978). Nitrogen as leaving group – dediazoniations of aromatic diazonium ions. *Angew. Chem. Int. Ed. Engl.* 17 (3): 141–150.

542 Scaiano, J.C. and Kimthuan, N. (1983). Diazonium salts in photochemistry. 2. Scavenging of Norrish type-II biradicals. *J. Photochem.* 21 (2): 167–170.

543 Canning, P.S.J., Maskill, H., McCrudden, K., and Sexton, B. (2002). A product analytical study of the thermal and photolytic decomposition of some arenediazonium salts in solution. *Bull. Chem. Soc. Jpn.* 75 (4): 789–800.

544 Detar, D.F. and Turetzky, M.N. (1955). The mechanisms of diazonium salt reactions. 1. The products of the reactions of benzenediazonium salts with methanol. *J. Am. Chem. Soc.* 77 (7): 1745–1750.

545 Swain, C.G., Sheats, J.E., and Harbison, K.G. (1975). Evidence for phenyl cation as an intermediate in reactions of benzenediazonium salts in solution. *J. Am. Chem. Soc.* 97 (4): 783–790.

546 Detar, D.F. and Turetzky, M.N. (1956). The mechanisms of diazonium salt reactions. 4. A kinetic study of the reactions of benzenediazonium fluoroborate with methanol in the absence of oxygen. *J. Am. Chem. Soc.* 78 (16): 3925–3928.

547 Pazo-Llorente, R., Bravo-Díaz, C., and González-Romero, E. (2003). Solvolysis of some arenediazonium salts in binary $EtOH/H_2O$ mixtures under acidic conditions. *Eur. J. Org. Chem.* (17): 3421–3428.

548 Canning, P.S.J., McCrudden, K., Maskill, H., and Sexton, B. (1998). Dediazoniation reactions of arenediazonium ions under solvolytic conditions: fluoride anion abstraction from trifluoroethanol and α-hydrogen atom abstraction from ethanol. *J. Chem. Soc., Chem. Commun.* (18): 1971–1972.

549 Milanesi, S., Fagnoni, M., and Albini, A. (2005). (Sensitized) photolysis of diazonium salts as a mild general method for the generation of aryl cations. Chemoselectivity of the singlet and triplet 4-substituted phenyl cations. *J. Org. Chem.* 70 (2): 603–610.

550 Kirk, K.L. and Cohen, L.A. (1971). Photochemical decomposition of diazonium fluoroborates – application to synthesis of ring-fluorinated imidazoles. *J. Am. Chem. Soc.* 93 (12): 3060–3061.

551 Nagai, W., Kirk, K.L., and Cohen, L.A. (1973). Synthesis of 2-amino-L-histidine and

2-aminohistamine. *J. Org. Chem.* 38 (11): 1971–1974.
552 Kirk, K.L. and Cohen, L.A. (1973). Photochemistry of diazonium salts. 3. New and facile synthesis of 4-fluoroimidazoles. *J. Org. Chem.* 38 (20): 3647–3648.
553 Kirk, K.L. and Cohen, L.A. (1973). Photochemistry of diazonium salts. 1. Synthesis of 4-fluoroimidazoles, 4-fluorohistamine, and 4-fluorohistidine. *J. Am. Chem. Soc.* 95 (14): 4619–4624.
554 Kirk, K.L., Nagai, W., and Cohen, L.A. (1973). Photochemistry of diazonium salts. 2. Synthesis of 2-fluoro-L-histidine and 2-fluorohistamine, and halogen lability of 2-fluoroimidazoles. *J. Am. Chem. Soc.* 95 (25): 8389–8392.
555 Declercq, E., Luczak, M., Reepmeyer, J.C. et al. (1975). Fluoro-imidazoles as antiviral agents and inhibitors of polynucleotide biosynthesis. *Life Sci.* 17 (2): 187–194.
556 Klee, C.B., Kirk, K.L., Cohen, L.A., and Mcphie, P. (1975). Histidine ammonia-lyase – use of 4-fluorohistidine in identification of rate-determining step. *J. Biol. Chem.* 250 (13): 5033–5040.
557 Klee, C.B., Lajohn, L.E., Kirk, K.L., and Cohen, L.A. (1977). 2-Fluorourocanic acid, a Potent reversible inhibitor of urocanase. *Biochem. Biophys. Res. Commun.* 75 (3): 674–681.
558 Takahashi, K., Kirk, K.L., and Cohen, L.A. (1984). Photochemistry of diazonium salts. 5. Syntheses of 2,4-difluoroimidazole-5-carboxylic acid and related-compounds. *J. Org. Chem.* 49 (11): 1951–1954.
559 Sawaguchi, M., Fukuhara, T., and Yoneda, N. (1999). Preparation of aromatic fluorides: facile photo-induced fluorinative decomposition of arenediazonium salts and their related compounds using pyridine-nHF. *J. Fluorine Chem.* 97 (1–2): 127–133.
560 Hari, D.P., Schroll, P., and Koenig, B. (2012). Metal-free, visible-light-mediated direct C—H arylation of heteroarenes with aryl diazonium salts. *J. Am. Chem. Soc.* 134 (6): 2958–2961.
561 Tamaoki, N., Takahashi, Y., and Yamaoka, T. (1994). Photochemistry of benzenediazonium anthracenesulfonates – photolysis of benzenediazonium salts by excitation of the anion. *J. Chem. Soc., Chem. Commun.* (15): 1749–1750.
562 Umemoto, T., Adachi, K., and Ishihara, S. (2007). CF_3 oxonium salts, O-(trifluoromethyl)dibenzofuranium salts: in situ synthesis, properties, and application as a real CF_3^+ species reagent. *J. Org. Chem.* 72 (18): 6905–6917.
563 L'Abbé, G. (1969). Decomposition and addition reactions of organic azides. *Chem. Rev.* 69 (3): 345–363.
564 L'Abbé, G. (1975). Reactions of vinyl azides. *Angew. Chem. Int. Ed. Engl.* 14 (12): 775–782.
565 Scriven, E.F.V. and Turnbull, K. (1988). Azides – their preparation and synthetic uses. *Chem. Rev.* 88 (2): 297–368.
566 Bräse, S., Gil, C., Knepper, K., and Zimmermann, V. (2005). Organic azides: an exploding diversity of a unique class of compounds. *Angew. Chem. Int. Ed.* 44 (33): 5188–5240.
567 Patonay, T., Konya, K., and Juhasz-Toth, E. (2011). Syntheses and transformations of α-azido ketones and related derivatives. *Chem. Soc. Rev.* 40 (5): 2797–2847.
568 Bräse, S. and Banert, K. (eds.) (2010). *Organic Azides-Synthesis and Applications*, 1–536. Chichester: Wiley.
569 Falvey, D.E. and Gudmundsdottir, A. (eds.) (2013). *Nitrenes and Nitrenium ions*, 1–606. Hoboken, NJ: Wiley.
570 Wentrup, C. (2013). Chemical activation in azide and nitrene chemistry: methyl azide, phenyl azide, naphthyl azides, pyridyl azides, benzotriazoles, and triazolopyridines. *Aust. J. Chem.* 66 (8): 852–863.
571 Abramovitch, R.A. and Kyba, E.P. (1971). Photodecomposition of alkyl azides – absence of freedom of choice and non-nitrene mechanism. *J. Am. Chem. Soc.* 93 (6): 1537–1538.
572 Wayne, G.S., Snyder, G.J., and Rogers, D.W. (1993). Measurement of the strain energy of a transient bridgehead imine, 4-azahomoadamant-3-ene, by photoacoustic calorimetry. *J. Am. Chem. Soc.* 115 (21): 9860–9861.
573 Radziszewski, J.G., Downing, J.W., Jawdosiuk, M. et al. (1985). 4-Azahomoadamant-3-ene – spectroscopic characterization and photoresolution of a highly reactive strained bridgehead imine. *J. Am. Chem. Soc.* 107 (3): 594–603.
574 Shang, H.R., Yu, C., Ying, L.M., and Zhao, X.S. (1995). Dissociation dynamics of methyl azide on the first excited singlet surface. *Chem. Phys. Lett.* 236 (3): 318–323.
575 Bock, H. and Dammel, R. (1987). Molecular properties and models. 15. The pyrolysis of azides in the gas phase. *Angew. Chem. Int. Ed. Engl.* 26 (6): 504–526.

576 Klima, R.F. and Gudmundsdottir, A.D. (2004). Intermolecular triplet-sensitized photolysis of alkyl azides. Trapping of triplet alkyl nitrenes. *J. Photochem. Photobiol., A* 162 (2–3): 239–247.

577 Platz, M.S. (1997). Photoaffinity labeling. *Photochem. Photobiol.* 65 (2): 193–194.

578 Breslow, D.S. (1984). Industrial applications. In: *Azides and Nitrenes* (ed. E.F.V. Scriven), 491–521. Orlando, FL: Academic Press.

579 Moghaddam, M.J., Yang, W., Bojarski, B. et al. (2012). Azide photochemistry for facile modification of graphitic surfaces: preparation of DNA-coated carbon nanotubes for biosensing. *Nanotechnology* 23 (42): 425503.

580 Buchmüller, K.L., Hill, B.T., Platz, M.S., and Weeks, K.M. (2003). RNA-tethered phenyl azide photocrosslinking via a short-lived indiscriminant electrophile. *J. Am. Chem. Soc.* 125 (36): 10850–10861.

581 Mecomber, J.S., Murthy, R.S., Rajam, S. et al. (2008). Photochemical functionalization of polymer surfaces for microfabricated devices. *Langmuir* 24 (7): 3645–3653.

582 Morris, J.L., Reddington, S.C., Murphy, D.M. et al. (2013). Aryl azide photochemistry in defined protein environments. *Org. Lett.* 15 (4): 728–731.

583 Wang, J., Burdzinski, G., and Platz, M.S. (2013). Ultrafast time-resolved studies of the photochemistry of aryl azides. In: *Nitrenes and Nitrenium Ions* (ed. D.E. Falvey and A.D. Gudmunsdottir), 1–32. Hoboken, NJ: Wiley.

584 Schrock, A.K. and Schuster, G.B. (1984). Photochemistry of phenyl azide – chemical-properties of the transient intermediates. *J. Am. Chem. Soc.* 106 (18): 5228–5234.

585 Chapman, O.L. and Leroux, J.P. (1978). 1-Aza-1,2,4,6-ccloheptatetraene. *J. Am. Chem. Soc.* 100 (1): 282–285.

586 Carroll, S.E., Nay, B., Scriven, E.F.V. et al. (1977). Decomposition of aromatic azides in ethanethiol. *Tetrahedron Lett.* (36): 3175–3178.

587 Meijer, E.W., Nijhuis, S., and Vanvroonhoven, F.C.B.M. (1988). Poly-1,2-azepines by the photopolymerization of phenyl azides – precursors for conducting polymer-films. *J. Am. Chem. Soc.* 110 (21): 7209–7210.

588 Chapman, O.L., Sheridan, R.S., and Leroux, J.P. (1978). Photochemical interconversion of phenylnitrene and isomeric pyridylmethylenes. *J. Am. Chem. Soc.* 100 (19): 6245–6247.

589 Wentrup, C. (1969). Aromatic nitrene-carbene interconversion. *J. Chem. Soc., Chem. Commun.* (23): 1386–1387.

590 Crow, W.D. and Wentrup, C. (1968). Reactions of excited molecules. 7. Thermal interconversion of 2-pyridylcarbene and phenyl nitrene. *Tetrahedron Lett.* (59): 6149–6152.

591 Wentrup, C. (2011). Nitrenes, carbenes, diradicals, and ylides. Interconversions of reactive intermediates. *Acc. Chem. Res.* 44 (6): 393–404.

592 Platz, M.S. (1995). Comparison of phenylcarbene and phenylnitrene. *Acc. Chem. Res.* 28 (12): 487–492.

593 Leyva, E., Platz, M.S., Persy, G., and Wirz, J. (1986). Photochemistry of phenyl azide – the role of singlet and triplet phenylnitrene as transient intermediates. *J. Am. Chem. Soc.* 108 (13): 3783–3790.

594 Gritsan, N.P., Yuzawa, T., and Platz, M.S. (1997). Direct observation of singlet phenylnitrene and measurement of its rate of rearrangement. *J. Am. Chem. Soc.* 119 (21): 5059–5060.

595 Gritsan, N.P., Gudmundsdottir, A.D., Tigelaar, D. et al. (2001). A laser flash photolysis and quantum chemical study of the fluorinated derivatives of singlet phenylnitrene. *J. Am. Chem. Soc.* 123 (9): 1951–1962.

596 Gritsan, N. and Platz, M. (2006). Kinetics, spectroscopy, and computational chemistry of arylnitrenes. *Chem. Rev.* 106 (9): 3844–3867.

597 Tsao, M.L. and Platz, M.S. (2003). Photochemistry of ortho,ortho′-dialkyl phenyl azides. *J. Am. Chem. Soc.* 125 (39): 12014–12025.

598 Morawietz, J. and Sander, W. (1996). Photochemistry of fluorinated phenyl nitrenes: matrix isolation of fluorinated azirines. *J. Org. Chem.* 61 (13): 4351–4354.

599 Carra, C., Nussbaum, R., and Bally, T. (2006). Experimental and theoretical study of 2,6-difluorophenylnitrene, its radical cation, and their rearrangement products in argon matrices. *ChemPhysChem* 7 (6): 1268–1275.

600 Inui, H., Sawada, K., Oishi, S. et al. (2013). Aryl nitrene rearrangements: spectroscopic observation of a benzazirine and its ring expansion to a ketenimine by heavy-atom tunneling. *J. Am. Chem. Soc.* 135 (28): 10246–10249.

601 Gritsan, N.P., Zhu, Z.D., Hadad, C.M., and Platz, M.S. (1999). Laser flash photolysis and computational study of singlet phenylnitrene. *J. Am. Chem. Soc.* 121 (6): 1202–1207.

602 Sankaranarayanan, J., Rajam, S., Hadad, C.M., and Gudmundsdottir, A.D. (2010). The ability of triplet nitrenes to abstract hydrogen atoms. *J. Phys. Org. Chem.* 23 (4): 370–375.

603 Singh, P.N., Mandel, S.M., Sankaranarayanan, J. et al. (2007). Selective formation of triplet alkyl nitrenes from photolysis of β-azido-propiophenone and, their reactivity. *J. Am. Chem. Soc.* 129 (51): 16263–16272.

604 Takayama, T., Mitsumori, T., Kawano, M. et al. (2010). Direct observation of arylnitrene formation in the photoreaction of arylazide crystals. *Acta Crystallogr., Sect. B: Struct. Sci.* 66: 639–646.

605 Wang, J., Kubicki, J., and Platz, M.S. (2007). An ultrafast study of phenyl azide: the direct observation of phenylnitrenium ion. *Org. Lett.* 9 (20): 3973–3976.

606 Novak, M. and Wang, Y.-T. (2013). Aryl nitrenium ions in aqueous solution. In: *Nitrenes and Nitrenium Ions* (ed. D.E. Falvey and A.D. Gudmunsdottir), 167–190. Hoboken, NJ: Wiley.

607 Xue, J., Luk, H.L., Eswaran, S.V. et al. (2012). Ultrafast infrared and UV–Vis studies of the photochemistry of methoxycarbonylphenyl azides in solution. *J. Phys. Chem. A* 116 (22): 5325–5336.

608 Wentrup, C., Nguyen, M.L., Lukosch, A. et al. (2013). 3-Pyridylnitrene, 2-and 4-pyrimidinylcarbenes, 3-quinolylnitrenes, and 4-quinazolinylcarbenes. Interconversion, ring expansion to diazacycloheptatetraenes, ring opening to nitrile ylides, and ring contraction to cyanopyrroles and cyanoindoles. *Beilstein J. Org. Chem.* 9: 743–753.

609 Finke, C., Grote, D., Seidel, R.W. et al. (2012). Matrix isolation and IR spectroscopic characterization of 3,5-difluoropyridyl-2,4,6-trinitrene. *J. Phys. Org. Chem.* 25 (6): 486–492.

610 Chapyshev, S.V., Neuhaus, P., Grote, D., and Sander, W. (2010). Matrix isolation and magnetic parameters of septet 3,5-dicyanopyridyl-2,4,6-trinitrene. *J. Phys. Org. Chem.* 23 (4): 340–346.

611 Chapyshev, S.V. and Tomioka, H. (2003). Photochemical transformations of quintet m-phenylenedinitrenes. *Bull. Chem. Soc. Jpn.* 76 (11): 2075–2089.

612 Curtius, T. (1894). Hydrazide und Azide organischer Säure I. Abhandlung. *J. Prakt. Chem.* 50: 275–294.

613 Platz, M.S. (2004). Nitrenes. In: *Reactive Intermediate Chemistry*, Chapter 11 (ed. R.A. Moss, M.S. Platz and M. Jones Jr.,). Hoboken, NJ: Wiley.

614 Clauss, K.U., Buck, K., and Abraham, W. (1995). The photochemistry of acyl azides. 9. Direct and sensitized photolytic generation of acylnitrenes for cycloaddition reactions. *Tetrahedron* 51 (26): 7181–7192.

615 Wentrup, C. and Bornemann, H. (2005). The Curtius rearrangement of acyl azides revisited – formation of cyanate (R-O-CN). *Eur. J. Org. Chem.* (21): 4521–4524.

616 Gritsan, N.P. and Pritchina, E.A. (2001). Are aroylnitrenes species with a singlet ground state? *Mendeleev Commun.* 11 (3): 94–95.

617 Pritchina, E.A., Gritsan, N.P., Maltsev, A. et al. (2003). Matrix isolation, time-resolved IR, and computational study of the photochemistry of benzoyl azide. *Phys. Chem. Chem. Phys.* 5 (6): 1010–1018.

618 Pritchina, E.A., Gritsan, N.P., and Bally, T. (2005). Ground-state multiplicity of acylnitrenes: computational and experimental studies. *Russ. Chem. Bull.* 54 (3): 525–532.

619 Sigman, M.E., Autrey, T., and Schuster, G.B. (1988). Aroylnitrenes with singlet ground-states – photochemistry of acetyl-substituted aroyl and aryloxycarbonyl azides. *J. Am. Chem. Soc.* 110 (13): 4297–4305.

620 Autrey, T. and Schuster, G.B. (1987). Are aroylnitrenes ground-state singlets – photochemistry of β-naphthoyl azide. *J. Am. Chem. Soc.* 109 (19): 5814–5820.

621 Kubicki, J., Zhang, Y., Wang, J. et al. (2009). Direct observation of acyl azide excited states and their decay processes by ultrafast time resolved infrared spectroscopy. *J. Am. Chem. Soc.* 131 (12): 4212–4213.

622 Vyas, S., Kubicki, J., Luk, H.L. et al. (2012). An ultrafast time-resolved infrared and UV–Vis spectroscopic and computational study of the photochemistry of acyl azides. *J. Phys. Org. Chem.* 25 (8): 693–703.

623 Kubicki, J., Zhang, Y., Vyasi, S. et al. (2011). Photochemistry of 2-naphthoyl azide. An ultrafast time-resolved UV–Vis and IR spectroscopic and computational study. *J. Am. Chem. Soc.* 133 (25): 9751–9761.

624 Smolinsky, G., Wasserman, E., and Yager, W.A. (1962). EPR of ground state triplet nitrenes. *J. Am. Chem. Soc.* 84 (16): 3220–3221.

625 Kubicki, J., Luk, H.L., Zhang, Y. et al. (2012). Direct observation of a sulfonyl azide excited state and its decay processes by ultrafast time-resolved IR spectroscopy. *J. Am. Chem. Soc.* 134 (16): 7036–7044.

626 Moriarty, R.M. and Kliegman, J.M. (1967). Photochemical decomposition of geminal diazides. 2. Solution photolysis of benzophenone diazide. *J. Am. Chem. Soc.* 89 (23): 5959–5960.

627 Barash, L., Wasserman, E., and Yager, W.A. (1967). Generation of methylenes from geminal diazides via excited nitrenes. *J. Am. Chem. Soc.* 89 (15): 3931–3932.

628 Ranaweera, R.A., Sankaranarayanan, J., Casey, L. et al. (2011). Triplet-sensitized photoreactivity of a geminal diazidoalkane. *J. Org. Chem.* 76 (20): 8177–8188.

629 Burgess, E.M., Carithers, B.R., and McCullagh, L. (1968). Photochemical decomposition of 1H-1,2,3-triazole derivatives. *J. Am. Chem. Soc.* 90 (7): 1923–1924.

630 Mitchell, G. and Rees, C.W. (1987). Photolysis of 1-aryl-1,2,3-triazoles – rearrangement via 1H-azirines. *J. Chem. Soc., Perkin Trans. 1* (2): 413–422.

631 Boyer, J.H. and Selvaraj, R. (1969). Photolysis of vic-triazoles. *Tetrahedron Lett.* (1): 47–50.

632 Selvarajan, R. and Boyer, J.H. (1972). Photo and thermal elimination of nitrogen from 4-phenyl-1,2,3-triazole and 4,5-diphenyl-1,2,3-triazole. *J. Heterocycl. Chem.* 9 (1): 87–90.

633 Maerky, M., Schmid, H., and Hansen, H.J. (1979). Photoreactions of 1-alkylbenztriazoles. *Helv. Chim. Acta* 62 (7): 2129–2153.

634 Wender, P.A. and Cooper, C.B. (1986). The photochemistry of 1-alkenylbenzotriazoles – methodology for the synthesis of indoles. *Tetrahedron* 42 (11): 2985–2991.

635 Wender, P.A. and Cooper, C.B. (1987). Indole synthesis based on triazole photochemistry – total synthesis of 7-methoxymitosene. *Tetrahedron Lett.* 28 (49): 6125–6128.

636 Al-Jalal, N.A., Al-Awadi, N.A., brahim, M.R., and lnagdi, M.H. (2013). The photochemistry of 1-alkenyl-substituted-1,2,3-triazoles leading to formation of pyrrole derivatives. *Int. J. Chem.* 5 (4): 80–89.

637 Bégué, D., Qiao, G.G.-H., and Wentrup, C. (2012). Nitrile imines: matrix isolation, IR spectra, structures, and rearrangement to carbodiimides. *J. Am. Chem. Soc.* 134 (11): 5339–5350.

638 Yokoyama, M., Hirano, S., Hachiya, T., and Togo, H. (1994). Thermolysis and photolysis of sugar diazides. *Chem. Lett.* (6): 1107–1110.

639 Praly, J.P., Distefano, C., Descotes, G., and Faure, R. (1994). Photolysis of sugar anomeric diazides – sugar-derived tetrazoles as evidences for a major nitrene decomposition pathway. *Tetrahedron Lett.* 35 (1): 89–92.

640 Alawode, O.E., Robinson, C., and Rayat, S. (2011). Clean photodecomposition of 1-methyl-4-phenyl-1H-tetrazole-5(4H)-thiones to carbodiimides proceeds via a biradical. *J. Org. Chem.* 76 (1): 216–222.

641 Woodward, R.B. and Hoffmann, R. (1971). *The Conservation of Orbital Symmetry*, 1–184. New York: Academic Press.

642 Schuster, D.I., Lem, G., and Kaprinidis, N.A. (1993). New insights into an old mechanism – $[\pi^2+\pi^2]$-photocycloaddition of enones to alkenes. *Chem. Rev.* 93 (1): 3–22.

643 Winkler, J.D., Bowen, C.M., and Liotta, F. (1995). $[\pi^2+\pi^2]$-Photocycloaddition/fragmentation strategies for the synthesis of natural and unnatural products. *Chem. Rev.* 95 (6): 2003–2020.

644 Namyslo, J.C. and Kaufmann, D.E. (2003). The application of cyclobutane derivatives in organic synthesis. *Chem. Rev.* 103 (4): 1485–1537.

645 Meyer, L., Alouane, N., Schmidt, K., and Margaretha, P. (2003). Photocycloaddition of cyclohex-2-enones to acrylonitrile. *Can. J. Chem.* 81 (6): 417–422.

646 Lohmeyer, B. and Margaretha, P. (2005). Photocycloaddition of cyclohex-2-enones to alkylidenemalononitriles (1,1-dicyanoalkenes). *Photochem. Photobiol. Sci.* 4 (8): 637–640.

647 Margaretha, P. (2005). Photocycloaddition of cycloalk-2-enones to alkenes. In: *Synthetic Organic Photochemistry*, vol. 12 (ed. A.G. Griesbeck and J. Mattay), 211–237.

648 Lee-Ruff, E. and Mladenova, G. (2003). Enantiomerically pure cyclobutane derivatives and their use in organic synthesis. *Chem. Rev.* 103 (4): 1449–1483.

649 Alcaide, B., Almendros, P., and Aragoncillo, C. (2010). Exploiting $[\pi^2+\pi^2]$-cycloaddition chemistry: achievements with allenes. *Chem. Soc. Rev.* 39 (2): 783–816.

650 Ciamician, G. and Silber, P. (1908). Chemische Lichtwirkungen. XIII. *Ber. Dtsch. Chem. Ges.* 41: 1928–1935.

651 Dilling, W.L. (1966). Intramolecular photochemical cycloaddition reactions of nonconjugated olefins. *Chem. Rev.* 66 (4): 373–393.

652 Srinivasan, R. (1963). Mercury photosensitized isomerization of 1,5-hexadiene. *J. Phys. Chem.* 67 (6): 1367–1368.

653 Srinivasan, R. and Hill, K.A. (1965). Photoisomerization of 1,6-heptadiene sensitized by $Hg_6(^3P_1)$ atoms. *J. Am. Chem. Soc.* 87 (22): 4988–4992.

654 Crowley, K.J. (1965). Photoisomerizations. 6. Cyclobutene formation and diene migration in simple 1,3-dienes. *Tetrahedron* 21 (5): 1001–1014.

655 Liu, R.S.H. and Hammond, G.S. (1964). Photosensitized cyclization of myrcene – problem of addition of dienes to alkenes. *J. Am. Chem. Soc.* 86 (9): 1892–1893.

656 Srinivasan, R. (1963). Mercury photosensitized isomerization of 1,5-cyclooctadiene to tricyclo[3.3.0.0(2,6)]octane. *J. Am. Chem. Soc.* 85 (6): 819–820.

657 Srinivasan, R. (1964). Photochemical transformations of 1,5-cyclooctadiene. *J. Am. Chem. Soc.* 86 (16): 3318–3321.

658 Moon, S. and Ganz, C.R. (1968). Photochemistry of homoconjugated cyclic dienes. I. photoisomerization of 1,4-cyclooctadiene. *Tetrahedron Lett.* (60): 6275–6277.

659 Srinivasan, R. and Carlough, K.H. (1967). Mercury (3P1) photosensitized internal cycloaddition reactions in 1,4-, 1,5-, and 1,6-dienes. *J. Am. Chem. Soc.* 89 (19): 4932–4936.

660 Liu, R.S.H. and Hammond, G.S. (1967). Photosensitized internal addition of dienes to olefins. *J. Am. Chem. Soc.* 89 (19): 4936–4944.

661 Gleiter, R. and Sander, W. (1985). Light-induced [π^2+π^2]-biycloaddition reactions of nonconjugated dienes – the effect of through-bond interaction. *Angew. Chem. Int. Ed. Engl.* 24 (7): 566–568.

662 Maradyn, D.J. and Weedon, A.C. (1995). Trapping of triplet 1,4-biradicals with hydrogen selenide in the intramolecular photochemical cycloaddition reaction of 3-(4′-pentenyl)cycloalk-2-enones – verification of the rule-of-5. *J. Am. Chem. Soc.* 117 (19): 5359–5360.

663 Becker, D., Nagler, M., Sahali, Y., and Haddad, N. (1991). Regiochemistry and stereochemistry of intramolecular [π^2+π^2]-photocycloaddition of C=C double-bonds to cyclohexenones. *J. Org. Chem.* 56 (14): 4537–4543.

664 Blaauw, R.H., Briere, J.F., de Jong, R. et al. (2001). Intramolecular photochemical dioxenone-alkene [π^2+π^2]-cycloadditions as an approach to the bicyclo[2.1.1]hexane moiety of Solanoeclepin A. *J. Org. Chem.* 66 (1): 233–242.

665 Bradley, S.A., Bresnan, B.J., Draper, S.M. et al. (2011). Photochemical [π^2+π^2]-cycloaddition reactions of 6-alkenyl-3-phenylcyclohex-2-en-1-ones: using biradical conformation control to account for exceptions to the "rule-of-five". *Org. Biomol. Chem.* 9 (8): 2959–2968.

666 Schreiber, S.L. (1985). [π^2+π^2]-Photocycloadditions in the synthesis of chiral molecules. *Science* 227 (4689): 857–863.

667 Crimmins, M.T. (1988). Synthetic applications of intramolecular enone + olefin photocycloadditions. *Chem. Rev.* 88 (8): 1453–1473.

668 Dekeukeleire, D. and He, S.L. (1993). Photochemical strategies for the construction of polycyclic molecules. *Chem. Rev.* 93 (1): 359–380.

669 Hue, B.T.B., Dijkink, J., Kuiper, S. et al. (2006). Synthesis of the tricyclic core of solanoeclepin A through intramolecular [π^2+π^2]-photocycloaddition of an allene butenolide. *Eur. J. Org. Chem.* (1): 127–137.

670 Crimmins, M.T. and Choy, A.L. (1997). Solvent effects on diastereoselective intramolecular [π^2+π^2]-photocycloadditions: reversal of selectivity through intramolecular hydrogen bonding. *J. Am. Chem. Soc.* 119 (42): 10237–10238.

671 Ng, S.M., Bader, S.J., and Snapper, M.L. (2006). Solvent-controlled intramolecular [π^2+π^2]-photocycloadditions of α-substituted enones. *J. Am. Chem. Soc.* 128 (22): 7315–7319.

672 Carreira, E.M., Hastings, C.A., Shepard, M.S. et al. (1994). Asymmetric induction in intramolecular [π^2+π^2]-photocycloadditions of 1,3-disubstituted allenes with enones and enoates. *J. Am. Chem. Soc.* 116 (15): 6622–6630.

673 Shepard, M.S. and Carreira, E.M. (1997). Asymmetric photocycloadditions with an optically active allenylsilane: trimethylsilyl as a removable stereocontrolling group for the enantioselective synthesis of exo-methylenecyclobutanes. *J. Am. Chem. Soc.* 119 (11): 2597–2605.

674 Shepard, M.S. and Carreira, E.M. (1997). Enantioselective allene/enone photocycloadditions: the use of an inexpensive optically active 1,3-disubstituted allene. *Tetrahedron* 53 (48): 16253–16276.

675 Winkler, J.D. and Ragains, J.R. (2006). Intramolecular photoaddition of vinylogous amides with allenes: a novel approach to the synthesis of pyrroles. *Org. Lett.* 8 (18): 4031–4033.

676 Dauben, W.G. and Cargill, R.L. (1961). Photochemical transformations. 8. Isomerization of bicyclo[2.2.1]hepta-2,5-diene to tetracyclo[2.2.1.0(2,6).0(3,5)]heptane (quadricyclane). *Tetrahedron* 15 (1–4): 197–201.

677 Wilcox, C.F., Winstein, S., and Mcmillan, W.G. (1960). Neighboring carbon and hydrogen. 34. Interaction of non-conjugated chromophores. *J. Am. Chem. Soc.* 82 (20): 5450–5454.

678 Hammond, G.S., Turro, N.J., and Fischer, A. (1961). Photosensitized cycloaddition reactions. *J. Am. Chem. Soc.* 83 (22): 4674–4675.

679 Hammond, G.S., Turro, N.J., Wyatt, P., and Deboer, C.D. (1964). Photosensitized isomerization involving saturated centers. *J. Am. Chem. Soc.* 86 (12): 2532–2533.

680 Turro, N.J., Cherry, W.R., Mirbach, M.F., and Mirbach, M.J. (1977). Energy acquisition, storage, and release – photochemistry of cyclic azoalkanes as alternate entries to energy surfaces interconnecting norbornadiene and quadricyclane. *J. Am. Chem. Soc.* 99 (22): 7388–7390.

681 Barwise, A.J.G., Gorman, A.A., Leyland, R.L. et al. (1978). Pulse-radiolysis study of quenching of aromatic carbonyl triplets by norbornadienes and quadricyclanes – mechanism of interconversion. *J. Am. Chem. Soc.* 100 (6): 1814–1820.

682 Cuppoletti, A., Dinnocenzo, J.P., Goodman, J.L., and Gould, I.R. (1999). Bond-coupled electron transfer reactions: photoisomerization of norbornadiene to quadricyclane. *J. Phys. Chem. A* 103 (51): 11253–11256.

683 Fife, D.J., Moore, W.M., and Morse, K.W. (1985). Photosensitized isomerization of norbornadiene to quadricyclane with (arylphosphine)Cu(I) halides. *J. Am. Chem. Soc.* 107 (24): 7077–7083.

684 Jones, G., Chiang, S.H., and Xuan, P.T. (1979). Energy storage in organic photoisomers. *J. Photochem.* 10 (1): 1–18.

685 Hautala, R.R., Little, J., and Sweet, E. (1977). Use of functionalized polymers as photosensitizers in an energy-storage reaction. *Sol. Energy* 19 (5): 503–508.

686 Bren, V.A., Dubonosov, A.D., Minkin, V.I., and Chernoivanov, V.A. (1991). Norbornadiene-quadricyclane as efficient molecular system of accumulation of solar-energy. *Usp. Khim.* 60 (5): 913–948.

687 Kucharski, T.J., Tian, Y., Akbulatov, S., and Boulatov, R. (2011). Chemical solutions for the closed-cycle storage of solar energy. *Energy & Energy Environ. Sci.* 4 (11): 4449–4472.

688 Eaton, P.E. and Cole, T.W. (1964). Cubane system. *J. Am. Chem. Soc.* 86 (5): 962–964.

689 Eaton, P.E. and Cole, T.W. (1964). Cubane. *J. Am. Chem. Soc.* 86 (15): 3157–3158.

690 Eaton, P.E., Nordari, N., Tsanaktsidis, J., and Upadhyaya, S.P. (1995). Barton decarboxylation of cubane-1,4-dicarboxylic acid – optimized procedures for cubanecarboxylic acid and cubane. *Synthesis* (5): 501–502.

691 Eaton, P.E., Zhang, M.X., Gilardi, R. et al. (2002). octanitrocubane: a new nitrocarbon. *Propellants, Explos., Pyrotech.* 27 (1): 1–6.

692 Cassar, L., Eaton, P.E., and Halpern, J. (1970). Catalysis of symmetry-restricted reactions by transition metal compounds – valence isomerization of cubane. *J. Am. Chem. Soc.* 92 (11): 3515–3518.

693 Cassar, L., Eaton, P.E., and Halpern, J. (1970). Silver(I)-catalyzed and Pd(II)-catalyzed isomerizations of cubane – synthesis and characterization of cuneane. *J. Am. Chem. Soc.* 92 (21): 6366–6368.

694 Selig, P. and Bach, T. (2006). Photochemistry of 4-(2′-aminoethyl)quinolones: enantioselective synthesis of tetracyclic tetrahydro-1aH-pyrido[4′,3′,2,3]-cyclobuta[1,2-c] quinoline-2,11(3H,8H)-diones by intra- and intermolecular [π^2+π^2]-photocycloaddition reactions in solution. *J. Org. Chem.* 71 (15): 5662–5673.

695 Gleiter, R., Sander, W., Irngartinger, H., and Lenz, A. (1982). A simple path to tricyclo[5.3.0.0^{2,8}]decane and its derivatives. *Tetrahedron Lett.* 23 (26): 2647–2650.

696 Cookson, R.C., Crundwell, E., Hill, R.R., and Hudec, J. (1964). Photochemical cyclisation of Diels–Alder adducts. *J. Chem. Soc.* 3062–3075.

697 Becker, D. and Haddad, N. (1986). About the stereochemistry of intramolecular [π^2+π^2]-photocycloadditions. *Tetrahedron Lett.* 27 (52): 6393–6396.

698 Kwak, Y.S. and Winkler, J.D. (2001). Synthesis of 6-azabicyclo[3.2.1]octan-3-ones via vinylogous imide photochemistry: an approach to the synthesis of the Hetisine alkaloids. *J. Am. Chem. Soc.* 123 (30): 7429–7430.

699 Yamazaki, H. and Cvetanovic, R.J. (1969). Stereospecific photochemical cyclodimerization of 2-butene in liquid phase. *J. Am. Chem. Soc.* 91 (2): 521–522.

700 Yamazaki, H., Cvetanovic, R.J., and Irwin, R.S. (1976). Kinetics of stereospecific photochemical cyclodimerization of 2-butene in liquid phase. *J. Am. Chem. Soc.* 98 (8): 2198–2205.

701 Arnold, D.R., Trecker, D.J., and Whipple, E.B. (1965). Stereochemistry of pentacyclo[8.2.1.14,7.02,9.03,8]tetradecanes and -dienes. Norbornene and norbornadiene dimers. *J. Am. Chem. Soc.* 87 (12): 2596–2602.

702 Srinivasan, R. (1972). Photochemical addition of benzene to bicyclo[2.2.1]hept-2-ene (norbornene). *J. Phys. Chem.* 76 (1): 15–19.

703 Cargill, R.L. and Willcott, M.R. (1966). Photochemical cycloaddition reactions with norbornene. *J. Org. Chem.* 31 (12): 3938–3940.

704 De Mayo, P. and Takeshita, H. (1963). Photochemical syntheses. 6. Formation of heptandiones from acetylacetone and alkenes. *Can. J. Chem.* 41 (2): 440–449.

705 Disanayaka, B.W. and Weedon, A.C. (1987). Application of the De Mayo reaction to the preparation of tricyclo[6.3.0.0^{2,6}]undecanes – a photochemical synthesis of (+/-)-Hirsutene. *J. Org. Chem.* 52 (13): 2905–2910.

706 Eaton, P.E. (1968). Photochemical reactions of simple alicyclic enones. *Acc. Chem. Res.* 1 (2): 50–57.

707 Broeker, J.L., Eksterowicz, J.E., Belk, A.J., and Houk, K.N. (1995). On the regioselectivity of photocycloadditions of triplet cyclohexenones to alkenes. *J. Am. Chem. Soc.* 117 (6): 1847–1848.

708 Morell, C., Labet, V., Grand, A. et al. (2009). Characterization of the chemical behavior of the low excited states through a local chemical potential. *J. Chem. Theory Comput.* 5 (9): 2274–2283.

709 D'Auria, M. (2012). Regio- and stereochemistry of the [$\pi^2+\pi^2$]-cycloaddition reaction between enones and alkenes. A DFT study. *Tetrahedron* 68 (42): 8699–8703.

710 Corey, E.J., Bass, J.D., Le Mathieu, R., and Mitra, R.B. (1964). Study of photochemical reactions of 2-cyclohexenones with substituted olefins. *J. Am. Chem. Soc.* 86 (24): 5570–5583.

711 Odo, Y., Shimo, T., Hori, K., and Somekawa, K. (2004). Origin of regioselectivity in photocycloaddition reactions of 2-cyclohexenone with cycloalkenecarboxylates. *Bull. Chem. Soc. Jpn.* 77 (6): 1209–1215.

712 Eaton, P.E. (1964). Photocondensation reactions of unsaturated ketones. *Tetrahedron Lett.* (48): 3695–3698.

713 Maradyn, D.J., Sydnes, L.K., and Weedon, A.C. (1993). Origin of the regiochemistry in the photochemical cycloaddition reaction of 2-cyclopentenone with allene – trapping of triplet 1,4-biradical intermediates with hydrogen selenide. *Tetrahedron Lett.* 34 (15): 2413–2416.

714 Froese, R.D.J., Lange, G.L., and Goddard, J.D. (1996). Theoretical studies of regioselectivity in the photochemical cycloaddition of allene to cyclopentenone. *J. Org. Chem.* 61 (3): 952–961.

715 (a) Tanaka, K., Mochizuki, E., Yasui, N. et al. (2000). Single-crystal-to-single-crystal enantioselective [$\pi^2+\pi^2$]-photodimerization of coumarin, thiocoumarin and cyclohex-2-enone in the inclusion complexes with chiral host compounds. *Tetrahedron* 56 (36): 6853–6865. (b) Tanaka, K. and Toda, F. (2000). Solvent-free organic synthesis. *Chem. Rev.* 100 (3): 1025–1074.

716 Tanaka, K. and Fujiwara, T. (2005). Enantioselective [$\pi^2+\pi^2$]-photodimerization reactions of coumarins in solution. *Org. Lett.* 7 (8): 1501–1503.

717 Poplata, S., Troster, A., Zou, Q.-Q., and Bach, T. (2016). Recent advances in the synthesis of cyclobutanes by olefin [$\pi2+\pi2$]-photocycloaddition reactions. *Chem. Rev.* 116 (17): 9745–9815.

718 Meyers, A.I. and Fleming, S.A. (1986). Efficient asymmetric (2+2)-photocycloaddition leading to chiral cyclobutanes – application to the total synthesis of (−)-grandisol. *J. Am. Chem. Soc.* 108 (2): 306–307.

719 Sulikowski, M.M., Davies, G.E.R.E., and Smith, A.B. (1992). Avermectin Milbemycin synthetic studies. 7. An approach to the southern-hemisphere of Milbemycin α-1. *J. Chem. Soc., Perkin Trans. 1* (8): 979–989.

720 Abad, A., Arno, M., Marin, M.L., and Zaragoza, R.J. (1993). Spongian pentacyclic diterpenes – stereoselective synthesis of aplyroseol-1, aplyroseol-2 and deacetylaplyroseol-2. *J. Chem. Soc., Perkin Trans. 1* (16): 1861–1867.

721 Liu, R.S.H., Turro, N.J., and Hammond, G.S. (1965). Mechanisms of photochemical reactions in solution. 31. Activation and deactivation of conjugated dienes by energy transfer. *J. Am. Chem. Soc.* 87 (15): 3406–3412.

722 Turro, N.J. and Hammond, G.S. (1962). Photosensitited dimerization of cyclopentadiene. *J. Am. Chem. Soc.* 84 (14): 2841–2842.

723 Mccullough, J.J. (1987). Photoadditions of aromatic compounds. *Chem. Rev.* 87 (4): 811–860.

724 Wender, P.A. and Dore, T.M. (1995). Intra- and intermolecular cycloadditions of benzene derivatives. In: *CRC Handbook of Organic Photochemistry and Photobiology* (ed. W.M. Horspool and P.-S. Song), 280–290. Boca Raton, FL: CRC Press.

725 Gilbert, A. (2003). Intra- and intermolecular cycloadditions of benzene derivatives. In: *CRC Handbook of Organic Photochemistry and Photobiology*, Chapter 41, 2e, vol. 1 and 2 (ed. W.

Horspool and F. Lenci). Boca Raton, FL: CRC Press.

726 Mattay, J. (2007). Photochemistry of arenes-reloaded. *Angew. Chem. Int. Ed.* 46 (5): 663–665.

727 Streit, U. and Bochet, C.G. (2011). The arene-alkene photocycloaddition. *Beilstein J. Org. Chem.* 7: 525–542.

728 Cornelisse, J. (1993). The meta photocycloaddition of arenes to alkenes. *Chem. Rev.* 93 (2): 615–669.

729 Hoffmann, N. (2004). Photochemical cycloaddition between benzene derivatives and alkenes. *Synthesis* (4): 481–495.

730 Atkinson, J.G., Ayer, D.E., Büchi, G., and Robb, E.W. (1963). Photochemical reactions. XII. Addition reactions of olefins and acetylenes with benzonitrile. *J. Am. Chem. Soc.* 85: 2257–2263.

731 Hanzawa, Y. and Paquette, L. (1982). Photochemical preparation of 5-tert-butyl-4-methoxycarbonyltetracyclo[3.3.0.0(2,4).0(3,6)]oct-7-ene. *Synthesis* 8: 661–662.

732 Pirrung, M.C. (1987). Intramolecular arene alkyne photocycloadditions. *J. Org. Chem.* 52 (8): 1635–1637.

733 Bowman, R.M., Calvo, C., Mccullou, J.J., and Miller, R.C. (1973). Singh, I. Photoaddition of naphthalene and acrylonitrile – X-ray and Nuclear Magnetic-Resonance study of cycloadducts. *Can. J. Chem.* 51 (7): 1060–1067.

734 Bowman, R.M., Chamberl, T.R., Huang, C.W., and Mccullough, J.J. (1974). Medium effects and quantum yields in photoaddition of naphthalene and acrylonitrile – chemical evidence on an exciplex structure. *J. Am. Chem. Soc.* 96 (3): 692–700.

735 Mccullough, J.J., Miller, R.C., Fung, D., and Wu, W.S. (1975). Exciplex and electron-transfer chemistry – reactions of naphthonitrile S_1 states with tetramethylethylene. *J. Am. Chem. Soc.* 97 (20): 5942–5943.

736 Akhtar, I.A. and Mccullough, J.J. (1981). Photoadditions of 1-naphthols and 2-naphthols and derivatives to acrylonitrile – conversion of the cyclobutane adducts to cyanoethylnaphthols. *J. Org. Chem.* 46 (7): 1447–1450.

737 Mattay, J. (1987). Selectivities in photocycloaditions of arenes to olefins. *J. Photochem.* 37 (1): 167–183.

738 Mccullough, J.J., Macinnis, W.K., Lock, C.J.L., and Faggiani, R. (1982). Naphthonitrile alkene exciplexes – comparison of bimolecular and bichromophoric cases – effects of linking on fluorescence and photochemistry. *J. Am. Chem. Soc.* 104 (17): 4644–4658.

739 Wilzbach, K.E. and Kaplan, L. (1966). A photochemical 1,3-cycloaddition of olefins to benzene. *J. Am. Chem. Soc.* 88 (9): 2066–2967.

740 Bryce-Smith, D., Gilbert, A., and Orger, B.H. (1966). Photochemical 1,3-cycloaddition of olefins to aromatic compounds. *J. Chem. Soc., Chem. Commun.* (15): 512–514.

741 Bryce-Smith, D., Gilbert, A., and Foulger, B.E. (1972). Photoaddition of allenes to benzene. *J. Chem. Soc., Chem. Commun.* (11): 664–665.

742 Haddaway, K., Somekawa, K., Fleming, P. et al. (1987). chemistry of allene cation-radicals probed by the use of theoretical and electron-transfer photochemical methods. *J. Org. Chem.* 52 (19): 4239–4253.

743 Birbaum, F., Neels, A., and Bochet, C.G. (2008). Photochemistry of allenyl salicylaldehydes. *Org. Lett.* 10 (15): 3175–3178.

744 Berridge, J.C., Forrester, J., Foulger, B.E., and Gilbert, A. (1980). The photochemical-reactions of benzene with 1,2-dienes, 1,3-dienes, and 1,4-dienes. *J. Chem. Soc., Perkin Trans. 1* (11): 2425–2434.

745 Berridge, J.C., Gilbert, A., and Taylor, G.N. (1980). The photochemical cycloaddition of furan and benzene. *J. Chem. Soc., Perkin Trans. 1* (10): 2174–2178.

746 Scharf, H.D., Leismann, H., Erb, W. et al. (1975). Photocycloaddition of dichlorovinylene carbonate to benzene and naphthalene. *Pure Appl. Chem.* 41 (4): 581–600.

747 Dopp, D., Krüger, C., Memarian, H.R., and Tsay, Y.H. (1985). 1,4-Photocycloaddition of α-morpholinoacrylonitrile to 1-acylnaphthalenes. *Angew. Chem. Int. Ed. Engl.* 24 (12): 1048–1049.

748 Clifford, S., Bearpark, M.J., Bernardi, F. et al. (1996). Conical intersection pathways in the photocycloaddition of ethene and benzene: a CASSCF study with MMVB dynamics. *J. Am. Chem. Soc.* 118 (31): 7353–7360.

749 Serrano-Pérez, J.J., de Vleeschouwer, F., De Proft, F. et al. (2013). How the conical intersection seam controls chemical selectivity in the photocycloaddition of ethylene and benzene. *J. Org. Chem.* 78 (5): 1874–1886.

750 Bryce-Smith, D., Gilbert, A., Orger, B., and Tyrrell, H. (1974). Polar and stereochemical aspects of 1,2-photoaddition of ethylenes to benzene. *J. Chem. Soc., Chem. Commun.* (9): 334–336.

751 Gilbert, A., Taylor, G.N., and Binsamsudin, M.W. (1980). The photocycloaddition of enol ethers to benzene. *J. Chem. Soc., Perkin Trans. 1* (4): 869–876.

752 Gilbert, A., Taylor, G.N., and Collins, A. (1980). Photochemical reactions of benzene derivatives with electron-donor ethylenes. *J. Chem. Soc., Perkin Trans. 1* (6): 1218–1224.

753 Sakamoto, M., Yagi, T., Mino, T. et al. (2000). A novel photochemical cycloaddition of 1-cyanonaphthalene to substituted pyridines. *J. Am. Chem. Soc.* 122 (34): 8141–8144.

754 Leismann, H. and Mattay, J. (1978). Formation of benzene exciplex in presence of enediol ethers and triethylamine. *Tetrahedron Lett.* (44): 4265–4268.

755 Leismann, H., Mattay, J., and Scharf, H.D. (1978). Intermolecular exciplex formation between benzene and some olefinic compounds. *J. Photochem.* 9 (4): 338–339.

756 Leismann, H., Mattay, J., and Scharf, H.D. (1984). Photochemical cycloaddition of olefins to aromatic compounds. 5. Formation and deactivation of exciplexes from singlet benzene or toluene and 1,3-dioxoles. *J. Am. Chem. Soc.* 106 (14): 3985–3991.

757 Mattay, J., Leismann, H., and Scharf, H.D. (1978). Photochemical cycloaddition of benzene to 1,3-dioxoles and 1,4-dioxene. 2. Mechanistic results. *Mol. Photochem.* 9 (2): 119–156.

758 Reedich, D.E. and Sheridan, R.S. (1985). Independent generation of arene meta photoaddition biradicals. *J. Am. Chem. Soc.* 107 (11): 3360–3361.

759 Reedich, D.E. and Sheridan, R.S. (1988). Direct evidence for both thermal and photochemical stepwise cleavage in a pair of isomeric azo-compounds. *J. Am. Chem. Soc.* 110 (11): 3697–3698.

760 De Vaal, P., Osselton, E.M., Krijnen, E.S. et al. (1988). The meta photocycloaddition of alkyl-substituted and alkoxy-substituted arenes to cyclopentene. *Recl. Trav. Chim. Pays-Bas* 107 (6): 407–411.

761 de Lijser, H.J.P. and Arnold, D.R. (1997). Radical-ions in photochemistry. 44. The photo-NOCAS reaction with acetonitrile as the nucleophile. *J. Org. Chem.* 62 (24): 8432–8438.

762 Kubo, Y., Noguchi, T., and Inoue, T. (1992). A novel 1,8-photoaddition of methyl 4-cyanonaphthalenecarboxylate and 1,4-naphthalenedicarbonitrile to alkenes. *Chem. Lett.* (10): 2027–2030.

763 Kubo, Y., Inoue, T., and Sakai, H. (1992). A novel 1,8-photoaddition of dimethyl 1,4-naphthalenedicarboxylate to alkenes. *J. Am. Chem. Soc.* 114 (20): 7660–7663.

764 Kubo, Y., Kiuchi, K., and Inamura, I. (1999). Photochemical reactions of 1,2-dicyanonaphthalene with alkenes and dienes. *Bull. Chem. Soc. Jpn.* 72 (5): 1101–1108.

765 Kubo, Y., Yoshioka, M., Kiuchi, K. et al. (1999). (3+2)-Photocycloadditions of dicyanonaphthalenes to alkenes. *Tetrahedron Lett.* 40 (3): 527–530.

766 Yang, N.C., Libman, J., and Savitzky, M.F. (1972). Photochemical additions of acyclic 1,3-dienes to naphthalene. *J. Am. Chem. Soc.* 94 (26): 9226–9227.

767 Pac, C., Sakurai, H., and Sugioka, T. (1972). Photochemical (4+4)-cycloaddition of furan to α-naphthonitrile. *Chem. Lett.* (1): 39–42.

768 Noh, T. and Kim, D. (1996). Reinvestigation in the photoreaction of 1-naphthalenecarbonitrile and furan. *Tetrahedron Lett.* 37 (52): 9329–9332.

769 Mizuno, K., Pac, C., and Sakurai, H. (1974). Photochemical reactions of aromatic compounds. 20. Photocycloaddition of 9-cyanoanthracene to furan and mono-furans and dimethylfurans. *J. Chem. Soc., Perkin Trans. 1* (20): 2360–2364.

770 Becker, H.D., Hansen, L., and Andersson, K. (1986). Synthesis and photochemical isomerization of 1,2-di-9-anthrylethanol and 1,2-di-9-anthrylethanone. *J. Org. Chem.* 51 (15): 2956–2961.

771 Greene, F.D., Misrock, S.L., and Wolfe, J.R. (1955). The structure of anthracene photodimers. *J. Am. Chem. Soc.* 77 (14): 3852–3855.

772 Sakamoto, M., Yagi, T., Kobaru, S. et al. (2004). A novel photochemical cycloaddition of 2-alkoxy-3-cyanopyridines to 2-cyanofuran. *Tetrahedron Lett.* 45 (23): 4437–4440.

773 Yang, N.C. and Libman, J. (1972). Photochemical additions of conjugated dienes to anthracene. *J. Am. Chem. Soc.* 94 (4): 1405–1406.

774 Yang, N.C., Libman, J., Loeschen, R.L. et al. (1972). Photochemical addition of acyclic 1,3-dienes to 9-cyanoanthracene and 9-anthraldehyde, a $[\pi^4+\pi^2]$ stereospecific photochemical cycloaddition. *J. Am. Chem. Soc.* 94 (4): 1406–1407.

775 Yang, N.C., Srinivasachar, K., Kim, B., and Libman, J. (1975). Chemistry of exciplexes. 4. Orientation of photocycloaddition of 1,3-dienes to anthracenes. *J. Am. Chem. Soc.* 97 (17): 5006–5008.

776 Yang, N.C., Yates, R.L., Masnovi, J. et al. (1979). Chemistry of exciplexes – photocycloadditions of anthracenes to conjugated polyenes. *Pure Appl. Chem.* 51 (2): 173–180.

777 Yang, N.C., Shou, H., Wang, T., and Masnovi, J. (1980). Chemistry of exciplexes. 10. Role of excimers in photocycloadditions of arenes to 1,3-dienes. *J. Am. Chem. Soc.* 102 (21): 6652–6654.

778 Caldwell, R.A. (1980). A predictor of reactivity in allowed photodimerizations and photocycloadditions. *J. Am. Chem. Soc.* 102 (12): 4004–4007.

779 Hoffmann, R. and Woodward, R.B. (1968). Conservation of orbital symmetry. *Acc. Chem. Res.* 1 (1): 17–22.

780 Woodward, R.B. and Hoffmann, R. (1969). Conservation of orbital symmetry. *Angew. Chem. Int. Ed. Engl.* 8 (11): 781–853.

781 Zimmerman, H.E. (1966). On molecular orbital correlation diagrams occurrence of Möbius systems in cyclization reactions and factors controlling ground- and excited-state reactions. I. *J. Am. Chem. Soc.* 88 (7): 1564–1565.

782 Kaupp, G. (1972). [π^2+π^4]-Photoaddition and [π^4+π^4]-photoaddition of cyclopentadiene to anthracene. *Angew. Chem. Int. Ed. Engl.* 11 (8): 718–719.

783 Kaupp, G., Dyllickbrenzinger, R., and Zimmermann, I. (1975). Photo-Diels–Alder reactions of anthracene. *Angew. Chem. Int. Ed. Engl.* 14 (7): 491–492.

784 Kaupp, G. and Grüter, H.W. (1979). Quantitative field desorption mass-spectrometry. 18. Polar [π^4+π^2]-photocycloadditions and [π^4+π^4]-photocycloadditions – effects of hybridization in diradicals. *Angew. Chem. Int. Ed. Engl.* 18 (11): 881–882.

785 Kaupp, G., Grüter, H.W., and Teufel, E. (1983). Photoadditions of anthracenes to 2,4-hexadiene and 1,3,5-cycloheptatriene. *Chem. Ber.* 116 (2): 630–644.

786 Saltiel, J., Dabestani, R., Sears, D.F. et al. (1995). Photocycloaddition of 9,10-dichloroanthracene to 2,5-dimethyl-2,4-hexadiene – the singlet pathway for [π^4+π^2]-adduct formation in benzene. *J. Am. Chem. Soc.* 117 (36): 9129–9138.

787 Piet, D., de Bruijn, S., Sum, J., and Lodder, G. (2005). New molecular entities via intermolecular meta photocycloaddition. *J. Heterocycl. Chem.* 42 (2): 227–231.

788 Al-Jalal, N. and Gilbert, A. (1990). Substituent effects on the photocycloaddition reactions of anisoles to acrylonitriles. *Recl. Trav. Chim. Pays-Bas* 109 (1): 21–25.

789 Yang, N.C., Masnovi, J., and Chiang, W. (1979). Chemistry of exciplexes. 8. Role of conservation of orbital symmetry in cycloadditions of olefins to aromatic hydrocarbons. *J. Am. Chem. Soc.* 101 (21): 6465–6466.

790 Yang, N.C., Masnovi, J., Chiang, W. et al. (1981). Chemistry of exciplexes. 11. Photocycloaddition of 1,3-cyclohexadiene to polynuclear aromatic hydrocarbons. *Tetrahedron* 37 (19): 3285–3300.

791 Yang, N.C.C., Gan, H., Kim, S.S. et al. (1990). Stereoselectivity in π^4/σ+π^4/σ photocycloaddition of 1,3-cyclohexadiene to arenes. *Tetrahedron Lett.* 31 (27): 3825–3828.

792 Paternó, E. and Chieffi, G. (1909). Sintesi in chimica organica per mezzo della luce. Nota II. Composti degli idrocarboni non saturi con aldeidi e chetoni. *Gazz. Chim. Ital.* 39: 341–361.

793 Büchi, G., Inman, C.G., and Lipinsky, E.S. (1954). Light-catalyzed organic reactions. 1. The reaction of carbonyl compounds with 2-methyl-2-butene in the presence of ultraviolet light. *J. Am. Chem. Soc.* 76 (17): 4327–4331.

794 Bach, T. (1998). Stereoselective intermolecular (2+2)-photocycloaddition reactions and their application in synthesis. *Synthesis* (5): 683–703.

795 Griesbeck, A.G. and Bondock, S.B. (2004). Oxetane formation: stereocontrol. In: *CRC Handbook of Organic Photochemistry and Photobiology*, 2e (ed. W. Horspool and F. Lenci), 1–19. Boca Raton, FL: CRC Press.

796 Griesbeck, A.G. and Bondock, S.B. (2004). Oxetane formation: intermolecular additions. In: *CRS Handbookof Organic Photochemistry and Photobiology*, 2e (ed. W. Horspool and F. Lenci), 1–21. Boca Raton, FL: CRC Press.

797 Abe, M. (2010). Formation of four-membered ring. oxetanes. In: *Handbook Synthetic Photochemistry* (ed. A. Albini and M. Fagnoni), 217–239. Weinheim: Wiley-VCH.

798 D'Auria, M. (2012). Paternó–Büchi Reaction. In: *CRC Handbook of Organic Photochemistry and Photobiology*, 3e (ed. A. Griesbeck, M. Oelgemöller and F. Ghetti), 653–681. Boca Raton, FL: CRC Press.

799 Brogaard, R.Y., Schalk, O., Boguslavskiy, A.E. et al. (2012). The Paternó–Büchi reaction: importance of triplet states in the excited-state reaction pathway. *Phys. Chem. Chem. Phys.* 14 (24): 8572–8580.

800 Freilich, S.C. and Peters, K.S. (1981). Observation of the 1,4-biradical in the Paternó–Büchi reaction. *J. Am. Chem. Soc.* 103 (20): 6255–6257.

801 Freilich, S.C. and Peters, K.S. (1985). Picosecond dynamics of the Paternó–Büchi reaction. *J. Am. Chem. Soc.* 107 (13): 3819–3822.

802 Arnold, D.R. and Glick, A.H. (1966). Photocycloaddition of carbonyl compounds to allenes. *J. Chem. Soc., Chem. Commun.* (22): 813–814.

803 Griesbeck, A.G., Fiege, M., Bondock, S., and Gudipati, M.S. (2000). Spin-directed stereoselectivity of carbonyl-alkene photocycloadditions. *Org. Lett.* 2 (23): 3623–3625.

804 Griesbeck, A.G., Bondock, S., and Cygon, P. (2003). Substantial ^{2}H-magnetic isotope effects on the diastereoselectivity of triplet photocycloaddition reactions. *J. Am. Chem. Soc.* 125 (30): 9016–9017.

805 Japar, S.M., Abrahams, E.W., and Davidson, J.A. (1972). Path of triplet excitation energy in simple carbonyl monoolefins system in liquid phase. *J. Phys. Chem.* 76 (4): 478–487.

806 Carless, H.A.J. (1974). Photochemical reactions of acetone with 2,3-dimethylbut-2-ene. *J. Chem. Soc., Perkin Trans. 2* (7): 834–842.

807 Shimizu, N., Ishikawa, M., Ishikura, K., and Nishida, S. (1974). New Substrate to investigate radical cycloadditions. 3. Photocycloaddition of aromatic carbonyl-compounds to vinylcyclopropane and Its derivatives. *J. Am. Chem. Soc.* 96 (20): 6456–6462.

808 Hu, S.K. and Neckers, D.C. (1997). Lifetime of the 1,4-biradical derived from alkyl phenylglyoxylate triplets: an estimation using the cyclopropylmethyl radical clock. *J. Org. Chem.* 62 (3): 755–757.

809 Porter, G. and Wilkinson, F. (1961). Primary photochemical processes in aromatic molecules. 5. Flash photolysis of benzophenone in solution. *Trans. Faraday Soc.* 57 (10): 1686–1691.

810 Turro, N.J., Wriede, P., Dalton, J.C. et al. (1967). Molecular photochemistry. V. Photocycloaddition of alkyl ketones to electron-deficient double bonds. *J. Am. Chem. Soc.* 89 (15): 3950–3951.

811 Dalton, J.C. and Turro, N.J. (1970). Photoreactivity of n/π* excited states of alkyl ketones. *Annu. Rev. Phys. Chem.* 21: 499–560.

812 Coyle, J.D. and Carless, H.A.J. (1972). Selected aspects of photochemistry. 1. Photochemistry of carbonyl compounds. *Chem. Soc. Rev.* 1 (4): 465–480.

813 Barltrop, J.A. and Carless, H.A.J. (1972). Organic photochemistry. Photocycloaddition of aliphatic ketones to α,β-unsaturated nitriles. *J. Am. Chem. Soc.* 94 (6): 1951–1959.

814 Turro, N.J. and Farrington, G.L. (1980). Quenching of the fluorescence of 2-norbornanone and derivatives by electron-rich and electron-poor ethylenes. *J. Am. Chem. Soc.* 102 (19): 6051–6055.

815 Turro, N.J. and Farrington, G.L. (1980). Photoinduced oxetane formation between 2-norbornanone and derivatives with electron-poor ethylenes. *J. Am. Chem. Soc.* 102 (19): 6056–6063.

816 Turro, N.J., Wriede, P.A., and Dalton, J.C. (1968). Molecular photochemistry. 8. Evidence for a singlet-state complex in photocycloaddition of acetone to trans-1,2-dicyanoethylene. *J. Am. Chem. Soc.* 90 (12): 3274–3275.

817 Griesbeck, A.G., Abe, M., and Bondock, S. (2004). Selectivity control in electron spin inversion processes: regio- and stereochemistry of Paternó–Büchi photocycloadditions as a powerful tool for mapping intersystem crossing processes. *Acc. Chem. Res.* 37 (12): 919–928.

818 Nehrings, A., Scharf, H.D., and Runsink, J. (1985). Chiral induction in photochemical-reactions. 3. Photochemical synthesis of an L-erythrose building block and its use in the preparation of methyl 2,3-O-isopropylidene-β-L-apio-L-furanoside. *Angew. Chem. Int. Ed. Engl.* 24 (10): 877–878.

819 Bach, T. (1995). Silyl enol ethers in Paternó–Büchi reactions – Functional group tolerance and the effect of β-alkyl substitution. *Liebigs Ann. Chem.* (5): 855–865.

820 D'Auria, M. and Racioppi, R. (2010). A DFT study of 1,4-biradical intermediates involved in stereoselective Paternó–Büchi reactions. *Eur. J. Org. Chem.* (20): 3831–3836.

821 Vogt, F., Joedicke, K., Schröder, J., and Bach, T. (2009). Paternó–Büchi reactions of silyl enol ethers and enamides. *Synthesis* (24): 4268–4273.

822 Bach, T. and Schröder, J. (1999). Photocycloaddition of N-acyl enamines to aldehydes and its application to the synthesis of diastereomerically pure 1,2-amino alcohols. *J. Org. Chem.* 64 (4): 1265–1273.

823 Matsumura, K., Mori, T., and Inoue, Y. (2009). Wavelength control of diastereodifferentiating Paternó–Büchi reaction of chiral cyanobenzoates with diphenylethene through direct versus charge-transfer excitation. *J. Am. Chem. Soc.* 131 (47): 17076–17077.

824 Matsumura, K., Mori, T., and Inoue, Y. (2010). Solvent and temperature effects on diastereodifferentiating Paternó–Büchi reaction of chiral alkyl cyanobenzoates with diphenylethene upon direct versus charge-transfer excitation. *J. Org. Chem.* 75 (16): 5461–5469.

825 D'Auria, M., Emanuele, L., Racioppi, R., and Valente, A. (2008). Paternó–Büchi reaction between aromatic carbonyl compounds and

1-(3-furyl) alkanols. *Photochem. Photobiol. Sci.* 7 (1): 98–103.

826 Yabuno, Y., Hiraga, Y., Takagi, R., and Abe, M. (2011). Concentration and temperature dependency of regio- and stereoselectivity in a photochemical $[\pi^2+\pi^2]$-cycloaddition reaction (the Paternó–Büchi reaction): origin of the hydroxy-group directivity. *J. Am. Chem. Soc.* 133 (8): 2592–2604.

827 Meinwald, J. and Hamner, A.T. (1969). 7-Oxatricyclo[4,3,0,0(5,8)]nonane and exo-bicyclo[3.2.1]octan-8-ol via an intramolecular Paternó–Büchi reaction. *J. Chem. Soc., Chem. Commun.* (22): 1302–1302.

828 Guenard, D. and Beugelmans, R. (1975). Studies on photochemical reactions. 27. Competition between intramolecular Paternó–Büchi reaction and Norrish types I and II reactions – formation of a di-norsteroids. *Bull. Soc. Chim. Fr.* (9–10): 2202–2206.

829 Sasaki, T., Eguchi, S., and Suzuki, T. (1980). Synthesis of adamantane derivatives. 50. Facile synthesis of 2,4-oxa-bridged protoadamantanes and their conversions to 2-substituted and 2,4-disubstituted protoadamantanes and a 2,4-disubstituted adamantane. *J. Org. Chem.* 45 (19): 3824–3827.

830 Gleiter, R. and Sigwart, C. (1994). Synthesis and properties of 2,4,6-trimethylenetricyclo[3.3.0.0^{3,7}]octane (stellatriene) and its alkyl derivatives – candidates for a 2-step Cope rearrangement. *J. Org. Chem.* 59 (5): 1027–1038.

831 Meier, R.M. and Ganem, B. (1994). An approach to 6-oxygenated Shikimic acid-derivatives by photolysis of the pyruvate ester of dimethyl epoxychorismate. *J. Org. Chem.* 59 (18): 5436–5438.

832 Heckendorn, R., Fuhrer, H., Kalvoda, J. et al. (1995). Synthesis, structure, and reactivity of secosteroids containing a medium-sized ring. 36. Transannular photocyclization of some unsaturated 5,10-secosteroidal ketones – a reevaluation. *Helv. Chim. Acta* 78 (5): 1291–1297.

833 Hu, S.K. and Neckers, D.C. (1997). Rapid regio- and diastereoselective Paternó–Büchi reaction of alkyl phenylglyoxylates. *J. Org. Chem.* 62 (3): 564–567.

834 Hammaecher, C. and Portella, C. (2008). New 6-oxa-2-silabicyclo[2.2.0]hexanes by photochemical conversion of acyl(allyl)(dimethyl) silanes. *J. Chem. Soc., Chem. Commun.* (44): 5833–5835.

835 Valiulin, R.A., Dressen, D.G., Riggs, J.R. et al. (2009). Effect of intramolecular Paternó–Büchi reaction on the thermodynamics and kinetics of nearly degenerate (3,3)-sigmatropic shift in fluxional polycycles. *J. Org. Chem.* 74 (9): 3484–3490.

836 Pepper, H.P., Kuan, K.K., and George, J.H. (2012). Synthesis of a Liphagal–Frondosin C hybrid and speculation on the biosynthesis of the frondosins. *Org. Lett.* 14 (6): 1524–1527.

837 Raghunathan, R., Kumarasamy, E., Iyer, A. et al. (2013). Intramolecular Paternó–Büchi reaction of atropisomeric α-oxoamides in solution and in the solid-state. *J. Chem. Soc., Chem. Commun.* 49 (77): 8713–8715.

838 Iriondo-Alberdi, J., Perea-Buceta, J.E., and Greaney, M.F. (2005). A Paternó–Büchi approach to the synthesis of merrilactone A. *Org. Lett.* 7 (18): 3969–3971.

839 Cadet, J., Voituriez, L., Grand, A. et al. (1985). Recent aspects of the photochemistry of nucleic-acids and related model compounds. *Biochimie* 67 (3–4): 277–292.

840 Cadet, J., Berger, M., Decarroz, C. et al. (1986). Photosensitized reactions of nucleic-acids. *Biochimie* 68 (6): 813–834.

841 Taylor, J.S. (1995). Dna, sunlight and skin-cancer. *Pure Appl. Chem.* 67 (1): 183–190.

842 Cadet, J., Berger, M., Douki, T. et al. (1997). Effects of UV and visible radiation on DNA – final base damage. *Biol. Chem.* 378 (11): 1275–1286.

843 Middleton, C.T., de La Harpe, K., Su, C. et al. (2009). DNA excited-state dynamics: from single bases to the double helix. *Ann. Rev. Phys. Chem.* 60: 217–239.

844 Varghese, A.J. (1975). Photocycloaddition of acetone to uracil and cytosine. *Photochem. Photobiol.* 21 (3): 147–151.

845 Hei, X.M., Song, Q.H., Li, X.B. et al. (2005). Origin of a large temperature dependence of regioselectivity observed for $[\pi^2+\pi^2]$-photocycloaddition (Paternó–Büchi reaction) of 1,3-dimethylthymine with benzophenone and its derivatives: conformational property of the intermediary triplet 1,4-diradicals. *J. Org. Chem.* 70 (7): 2522–2527.

846 Song, Q.H., Wang, H.B., Li, X.B. et al. (2006). Notable substituent and temperature effects on the regioselectivity and the efficiency in Paternó–Büchi reaction of 4,4′-disubstituted benzophenones with 1,3-dimethyluracil and 1,3-dimethylthymine. *J. Photochem. Photobiol., A* 183 (1–2): 198–204.

847 Kong, F.F., Wang, J.B., and Song, Q.H. (2011). Heavy atom effects in the Paternó–Büchi reaction of pyrimidine derivatives with 4,4′-disubstituted benzophenones. *Beilstein J. Org. Chem.* 7: 113–118.

848 Liu, X.L., Wang, J.B., Tong, Y., and Song, Q.H. (2013). Regioselectivity and competition of the Paternó–Büchi reaction and triplet-triplet energy transfer between triplet benzophenones and pyrimidines: control by triplet energy levels. *Chem. Eur. J.* 19 (39): 13216–13223.

849 Cho, D.W., Lee, H.Y., Oh, S.W. et al. (2008). Photoaddition reactions of 1,2-diketones with silyl ketene acetals. Formation of β-hydroxy-γ-ketoesters. *J. Org. Chem.* 73 (12): 4539–4547.

850 Verma, S., Srivastava, B., Sharma, K. et al. (2012). Photoinduced $[\pi^4+\pi^2]$-cycloaddition reactions of benzo[b]thiophene-2,3-dione with alkenes. *Res. Chem. Intermed.* 38 (3–5): 723–731.

851 Friedrich, L.E. and Bower, J.D. (1973). Detection of an oxetene intermediate in photoreaction of benzaldehyde with 2-butyne. *J. Am. Chem. Soc.* 95 (20): 6869–6870.

852 Toki, S., Shima, K., and Sakurai, H. (1965). Organic photochemical reactions. I. Synthesis of substituted oxetanes by photoaddition of aldehydes to furans. *Bull. Chem. Soc. Jpn.* 38 (5): 760–762.

853 Shima, K. and Sakurai, H. (1966). Organic photochemical reactions. 4. Photoaddition reactions of various carbonyl compounds to furan. *Bull. Chem. Soc. Jpn.* 39 (8): 1806–1808.

854 D'Auria, M., Emanuele, L., and Racioppi, R. (2004). Regio- and stereoselectivity in the Paternó–Büchi reaction on furan derivatives. *J. Photochem. Photobiol., A* 163 (1–2): 103–111.

855 D'Auria, M., Racioppi, R., and Viggiani, L. (2010). Paternó–Büchi reaction between furan and heterocyclic aldehydes: oxetane formation vs. metathesis. *Photochem. Photobiol. Sci.* 9 (8): 1134–1138.

856 Griesbeck, A.G., Franke, M., Neudoerfl, J., and Kotaka, H. (2011). Photocycloaddition of aromatic and aliphatic aldehydes to isoxazoles: cycloaddition reactivity and stability studies. *Beilstein J. Org. Chem.* 7: 127–134.

857 Dargelos, A. and Sandorfy, C. (1977). Photoelectron and far ultraviolet absorption-spectra of simple oximes. *J. Chem. Phys.* 67 (7): 3011–3013.

858 Sandorfy, C. (1981). Spectroscopy of non-aromatic Schiff-bases. *J. Photochem.* 17 (3–4): 297–302.

859 Padwa, A. (1977). Photochemistry of carbon–nitrogen double-bond. *Chem. Rev.* 77 (1): 37–68.

860 Pratt, A.C. (1977). Photochemistry of imines. *Chem. Soc. Rev.* 6 (1): 63–81.

861 Padwa, A. and Albrecht, F. (1974). Concentration effects in photochemical syn-anti isomerization of an oxime ether. *J. Org. Chem.* 39 (16): 2361–2366.

862 Padwa, A. and Albrecht, F. (1974). Photochemical syn-anti isomerization about carbon-nitrogen double-bond. *J. Am. Chem. Soc.* 96 (15): 4849–4857.

863 Mukai, T. and Sukawa, H. (1973). Organic photochemistry. 28. Synthesis of 2-phenyl-1,3-oxazepine by irradiation of 4-phenyl-2, 3-oxazabicyclo[3.2.0]hepta-3,6-diene. *Tetrahedron Lett.* (21): 1835–1838.

864 Giezendanner, H., Rosenkranz, H.J., Hansen, H.J., and Schmid, H. (1973). Photoreactions. 30. Photochemistry of 3,5-diaryl-2-isoxazolines. *Helv. Chim. Acta* 56 (7): 2588–2611.

865 Ito, Y. and Matsuura, T. (1975). Photoinduced reactions. 80. Photochemistry of 2-isoxazolines. *Tetrahedron* 31 (11–12): 1373–1380.

866 (a) Kumagai, T., Shimizu, K., Kawamura, Y., and Mukai, T. (1981). Organic photochemistry. 50. Photochemistry of 3-aryl-2-isoxazoline. *Tetrahedron* 37 (19): 3365–3376.
(b) Sampedro, D., Soldevilla, A., Campos, P.J. et al. (2008). Regio- and stereochemistry of $[\pi^2+\pi^2]$-photocycloadditions of imines to alkenes: a computational and experimental study. *J. Org. Chem.* 73 (21): 8331–8336.

867 Sakamoto, R., Inada, T., Sakurai, S., and Maruoa, K. (2016). $[\pi^2+\pi^2]$-Photocycloadditions between the carbon–nitrogen double bonds of imines and carbon–carbon double bonds. *Org. Lett.* 18 (24): 6252–6255.

868 Howard, K.A. and Koch, T.H. (1975). Photochemical reactivity of keto imino ethers. 5. (2+2)-Photocycloaddition to carbon–nitrogen double-bond of 3-ethoxyisoindolone. *J. Am. Chem. Soc.* 97 (25): 7288–7298.

869 Rodehorst, R.M. and Koch, T.H. (1975). Photochemical reactivity of keto imino ethers. 6. Type-1 rearrangement and (2+2)-photocycloaddition to carbon-nitrogen double-bond of 2-oxazolin-4-ones. *J. Am. Chem. Soc.* 97 (25): 7298–7304.

870 Swenton, J.S. and Hyatt, J.A. (1974). Photosensitized cycloadditions to 1,3-dimethyl-6-azauracil and 1,3-dimethyl-6-azathymine – imine linkage unusually reactive toward photocycloaddition. *J. Am. Chem. Soc.* 96 (15): 4879–4885.

871 Visser, R.G., Baaij, J.P.B., Brouwer, A.C., and Bos, H.J.T. (1977). Photochemical

(2+2)-cycloaddition of aromatic thiones to ketenimines – acid-catalyzed rearrangement of 2-iminothietanes. *Tetrahedron Lett.* (49): 4343–4344.

872 Singer, L.A. and Bartlett, P.D. (1964). The photocycloaddition of aromatic aldehydes and ketones to dimethyl-N(2-cyano-2-propyl)ketenimine. *Tetrahedron Lett.* 5 (28): 1887–1892.

873 Ogino, K., Yamashina, S., Matsumoto, T., and Kozuka, S. (1979). Photocycloadditions of para-quinones to ketenimines. *J. Chem. Soc., Perkin Trans. 1* (6): 1552–1559.

874 Armesto, D., Martin, J.A.F., Perezossorio, R., and Horspool, W.M. (1982). A Novel aza-di-π-methane rearrangement the photoreaction of 4,4-dimethyl-1,6,6-triphenyl-2-aza-hexa-2,5-diene. *Tetrahedron Lett.* 23 (20): 2149–2152.

875 Goodell, J.R., Poole, J.L., Beeler, A.B. et al. (2011). Synthesis and reactivity of bicyclo[3.2.1]octanoid-derived cyclopropanes. *J. Org. Chem.* 76 (23): 9792–9800.

876 Miranda, M.A. and García, H. (1994). 2,4,6-Triphenylpyrylium tetrafluoroborate as an electron-transfer photosensitizer. *Chem. Rev.* 94 (4): 1063–1089.

877 Jia, X., Han, B., Zhang, W. et al. (2006). Cation-radical aza-Diels–Alder reaction between N-arylimines and N-vinyllactams: a facile synthesis of 4-lactam-N-yl tetrahydroquinolines. *Synthesis* (17): 2831–2836.

878 Pérez-Ruiz, R., Domingo, L.R., Jiménez, M.C., and Miranda, M.A. (2011). Experimental and theoretical studies on the radical-cation-mediated imino-Diels–Alder reaction. *Org. Lett.* 13 (19): 5116–5119.

879 Zhang, W., Guo, Y.P., Liu, Z.G. et al. (2005). Photochemically catalyzed Diels–Alder reaction of arylimines with N-vinylpyrrolidinone and N-vinylcarbazole by 2,4,6-triphenylpyrylium salt: synthesis of 4-heterocycle-substituted tetrahydroquinoline derivatives. *Tetrahedron* 61 (5): 1325–1333.

880 Fischer, G., Fritz, H., and Prinzbach, H. (1986). Photochemical transformations. 66. An intramolecular imine ene photo-[π^2+π^2]-cycloaddition reaction. *Tetrahedron Lett.* 27 (11): 1269–1272.

881 Zamadar, M. and Greer, A. (2010). Singlet oxygen as a reagent in organic synthesis. In: *Handbook of Synthetic Photochemistry* (ed. A. Albini and F. Fagnoni), 353–386. Weinheim: Wiley-VCH.

882 Iesce, M.R. and Cermola, F. (2012). Photooxygenation, [π^2+π^2] and [π^4+π^2]. In: *CRC Handbook of Organic Photochemistry and Photobiology*, 3e (ed. A. Griesbeck, M. Oelgemöller and F. Ghetti), 727–764. Boca Raton, FL: CRC Press.

883 Alberti, M.N. and Orfanopoulos, M. (2012). Singlet oxygen-mediated allylic oxidation. In: *CRC Handbook of Organic Photochemistry and Photobiology*, 3e (ed. A. Griesbeck, M. Oegemöller and F. Ghetti), 765–787. Boca Raton, FL: CRC Press.

884 Ohloff, G. (1975). Singlet oxygen – reagent in organic synthesis. *Pure Appl. Chem.* 43 (3–4): 481–502.

885 Frimer, A.A. (1979). Reaction of singlet oxygen with olefins – question of mechanism. *Chem. Rev.* 79 (5): 359–387.

886 Nonell, S. and Flors, C. (eds.) (2016). *Singlet Oxygen: "Applications in Biosciences and Nanosciences"*, vol. 1, 1–472. Royal Society of Chemistry.

887 Clennan, E.L. (2000). New mechanistic and synthetic aspects of singlet oxygen chemistry. *Tetrahedron* 56 (47): 9151–9179.

888 Clennan, E.L. and Pace, A. (2005). Advances in singlet oxygen chemistry. *Tetrahedron* 61 (28): 6665–6691.

889 Saito, I., Matsuura, T., and Inoue, K. (1983). Formation of superoxide ion via one-electron transfer from electron-donors to singlet oxygen. *J. Am. Chem. Soc.* 105 (10): 3200–3206.

890 Cadet, J., Douki, T., and Ravanat, J.-L. (2008). Oxidatively generated damage to the guanine moiety of DNA: mechanistic aspects and formation in cells. *Acc. Chem. Res.* 41 (8): 1075–1083.

891 Cadet, J., Douki, T., and Ravanat, J.-L. (2010). Oxidatively generated base damage to cellular DNA. *Free Radical Biol. Med.* 49 (1): 9–21.

892 Choe, E. and Min, D.B. (2006). Mechanisms and factors for edible oil oxidation. *Compr. Rev. Food Sci. Food Saf.* 5 (4): 169–186.

893 Choe, E. and Min, D.B. (2006). Chemistry and reactions of reactive oxygen species in foods. *Crit. Rev. Food Sci. Nutr.* 46 (1): 1–22.

894 Fan, L., Nakayama, Y., and Fujimoto, K. (1997). Air oxidation of supercritical phase isobutane to tert-butyl alcohol. *J. Chem. Soc., Chem. Commun.* (13): 1179–1180.

895 Das, B.K. and Chakrabarty, R. (2011). Cobalt(III)-oxo cubane clusters as catalysts for oxidation of organic substrates. *J. Chem. Sci.* 123 (2): 163–173.

896 Boisvert, L. and Goldberg, K.I. (2012). Reactions of late transition metal complexes with molecular oxygen. *Acc. Chem. Res.* 45 (6): 899–910.

897 Sakamoto, T., Takagaki, T., Sakakura, A. et al. (2008). Hydroxylation of benzene to phenol under air and carbon monoxide catalyzed by molybdovanadophosphates. *J. Mol. Catal. A: Chem.* 288 (1–2): 19–22.

898 Tani, M., Sakamoto, T., Mita, S. et al. (2005). Hydroxylation of benzene to phenol under air and carbon monoxide catalyzed by molybdovanadophosphoric acid. *Angew. Chem. Int. Ed.* 44 (17): 2586–2588.

899 Veser, G. and Schmidt, L.D. (1996). Ignition and extinction in the catalytic oxidation of hydrocarbons over platinum. *AIChE J.* 42 (4): 1077–1087.

900 Davis, M.B., Pawson, M.D., Veser, G. et al. (2000). Methane oxidation over noble metal gauzes: an LIF study. *Combust. Flame* 123 (1–2): 159–174.

901 Gniot, I., Kirszensztejn, P., and Kozlowski, M. (2009). Oxidative dehydrogenation of isobutane using modified activated carbons as catalysts. *Appl. Catal., A* 362 (1–2): 67–74.

902 Ishii, Y. and Sakaguchi, S. (1999). A new strategy for alkane oxidation with O_2 using N-hydroxyphthalimide (NHPI) as a radical catalyst. *Catal. Surv. Jpn.* 3 (1): 27–35.

903 Ishii, Y., Sakaguchi, S., and Iwahama, T. (2001). Innovation of hydrocarbon oxidation with molecular oxygen and related reactions. *Adv. Synth. Catal.* 343 (5): 393–427.

904 Sawatari, N., Yokota, T., Sakaguchi, S., and Ishii, Y. (2001). Alkane oxidation with air catalyzed by lipophilic N-hydroxyphthalimides without any solvent. *J. Org. Chem.* 66 (23): 7889–7891.

905 Ishii, Y. and Sakaguchi, S. (2006). Recent progress in aerobic oxidation of hydrocarbons by N-hydroxyimides. *Catal. Today* 117 (1–3): 105–113.

906 Mulliken, R.S. (1928). The assignment of quantum numbers for electrons in molecules I. *Phys. Rev.* 32 (2): 0186–0222.

907 Mulliken, R.S. (1928). Interpretation of the atmospheric oxygen bands, electronic levels of the oxygen molecule. *Nature* 122: 505–505.

908 Childs, W. and Mecke, R. (1931). Intensity measurements in the atmospheric oxygen band at 7600 A. *Z. Angew. Phys.* 68 (5–6): 344–361.

909 Ellis, J. and Kneser, H. (1933). Combination relationship in the absorption spectrum of the liquid oxygen. *Z. Angew. Phys.* 86 (9–10): 583–591.

910 Schweitzer, C. and Schmidt, R. (2003). Physical mechanisms of generation and deactivation of singlet oxygen. *Chem. Rev.* 103 (5): 1685–1757.

911 Evans, D.F. and Tucker, J.N. (1976). Reactivity of $(^1\Delta_g)_2$ and $(^1\Delta_g)$ states of oxygen produced by direct laser excitation. *J. Chem. Soc., Faraday Trans. 2* 72: 1661–1666.

912 Hurst, J.R., Mcdonald, J.D., and Schuster, G.B. (1982). Lifetime of singlet oxygen in solution directly determined by Laser spectroscopy. *J. Am. Chem. Soc.* 104 (7): 2065–2067.

913 Schmidt, R., Brauer, H.D., and Drexhage, K.H. (1987). Determination of the phosphorescence quantum yield of singlet molecular-oxygen $(^1\Delta_g)$ in benzene. *Chem. Phys. Lett.* 138 (1): 18–22.

914 Mallet, L. (1927). Phénomènes de luminescence au cours de réactions oxydantes en solutions aqueuses. *C.R. Hebd. Séances Acad. Sci., Paris* 185: 352–354.

915 Groh, P. and Kirrmann, A. (1942). Spectral study of a luminescent reaction. *C.R. Hebd. Séances Acad. Sci.* 215: 275–276.

916 Khan, A.U. and Kasha, M. (1966). Physical theory of chemiluminescence in systems evolving molecular oxygen. *J. Am. Chem. Soc.* 88 (7): 1574–1576.

917 Khan, A.U. and Kasha, M. (1970). Chemiluminescence arising from simultaneous transitions in pairs of singlet oxygen molecules. *J. Am. Chem. Soc.* 92 (11): 3293–3300.

918 Shakhashiri, B.Z. and Williams, L.G. (1976). Singlet oxygen in aqueous-solution – lecture demonstration. *J. Chem. Educ.* 53 (6): 358–361.

919 Catir, M., Kilic, H., Nardello-Rataj, V. et al. (2009). Singlet oxygen generation from [bis(trifluoroacetoxy)iodo]benzene and hydrogen peroxide. *J. Org. Chem.* 74 (12): 4560–4564.

920 Aubry, J.M. (1985). Search for singlet oxygen in the decomposition of hydrogen peroxide by mineral compounds in aqueous solutions. *J. Am. Chem. Soc.* 107 (21): 5844–5849.

921 Aubry, J.M. and Cazin, B. (1988). Chemical sources of singlet oxygen. 2. Quantitative generation of singlet oxygen from hydrogen peroxide disproportionation catalyzed by molybdate ions. *Inorg. Chem.* 27 (12): 2013–2014.

922 Hanaoka, S., Lin, J.M., and Yamada, M. (2000). Chemiluminescence behavior of the decomposition of hydrogen peroxide catalyzed by Cu(II)-amino acid complexes and its application to the determination of tryptophan and phenylalanine. *Anal. Chim. Acta* 409 (1–2): 65–73.

923 Pierlot, C., Nardello, V., Schrive, J. et al. (2002). Calcium peroxide diperoxohydrate as a storable chemical generator of singlet oxygen for organic synthesis. *J. Org. Chem.* 67 (8): 2418–2423.

924 Brauer, H.D., Eilers, B., and Lange, A. (2002). Formation of singlet molecular oxygen by the Radziszewski reaction between acetonitrile and hydrogen peroxide in the absence and presence of ketones. *J. Chem. Soc., Perkin Trans. 2* (7): 1288–1295.

925 Almeida, E.A., Miyamoto, S., Martinez, G.R. et al. (2003). Direct evidence of singlet molecular oxygen [O_2 ($^1\Delta\gamma$)] production in the reaction of acetonitrile with hydrogen peroxide in alkaline solutions. *Anal. Chim. Acta* 482 (1): 99–104.

926 Fenton, H.J.H. (1894). Oxidation of tartaric acid in presence of iron. *J. Chem. Soc. Trans.* 65: 899–911.

927 Haber, F. and Weiss, J. (1932). Über die Katalyse des Hydroperoxydes. *Naturwissenschaften* 20: 948–950.

928 Koppenol, W.H. (2001). The Haber–Weiss cycle – 70 years later. *Redox Rep.* 6 (4): 229–234.

929 Sun, H.H., Sun, S.P., Fan, M.H. et al. (2007). A kinetic study on the degradation of p-nitroaniline by Fenton oxidation process. *J. Hazard. Mater.* 148 (1–2): 172–177.

930 Kaczmarek, M. and Lis, S. (2012). Chemiluminescence determination of ibuprofen and ketoprofen using the Fenton system in the presence of Eu(III) ions. *Anal. Methods* 4 (7): 1964–1967.

931 Tian, W., Shi, W., Yang, H. et al. (2012). Production of singlet oxygen by the reaction of non-basic hydrogen peroxide with chlorine gas. *Phys. Chem. Chem. Phys.* 14 (38): 13344–13349.

932 Foote, C.S. and Wexler, S. (1964). Olefin oxidations with excited singlet molecular oxygen. *J. Am. Chem. Soc.* 86 (18): 3879–3880.

933 Foote, C.S. and Wexler, S. (1964). Singlet oxygen. Probable intermediate in photosensitized autoxidations. *J. Am. Chem. Soc.* 86 (18): 3880–3881.

934 Foote, C.S., Wexler, S., Ando, W., and Higgins, R. (1968). Chemistry of singlet oxygen. 4. Oxygenations with hypochlorite-hydrogen peroxide. *J. Am. Chem. Soc.* 90 (4): 975–981.

935 Foote, C.S. (1968). Photosensitized oxygenations and role of singlet oxygen. *Acc. Chem. Res.* 1 (4): 104–110.

936 Rahal, R., Pigot, T., Foix, D., and Lacombe, S. (2011). Photocatalytic efficiency and self-cleaning properties under visible light of cotton fabrics coated with sensitized TiO_2. *Appl. Catal., B* 104 (3–4): 361–372.

937 Raab, O. (1904). Über die Wirkung Fluorescierenden Stoffe auf Infusionen. *Z. Biol.* 39: 524–546.

938 Hausmann, W. and Kolmer, W. (1909). On the sensitive effect of plant and animal pigments on paramecia. *Biochem. Z.* 15: 12–18.

939 Bethea, D., Fullmer, B., Syed, S. et al. (1999). Psoralen photobiology and photochemotherapy: 50 years of science and medicine. *J. Dermatol. Sci.* 19 (2): 78–88.

940 Tamesis, M.E. and Morelli, J.G. (2010). Vitiligo treatment in childhood: a state of the art review. *Pediatr. Dermatol.* 27 (5): 437–445.

941 Ormond, A.B. and Freeman, H.S. (2013). Dye sensitizers for photodynamic therapy. *Materials* 6 (3): 817–840.

942 (a) Dougherty, T.J., Kaufman, J.E., Goldfarb, A. et al. (1978). Photoradiation therapy for treatment of malignant tumors. *Cancer Res.* 38 (8): 2628–2635. (b) Chilakamarthi, U. and Giribabu, L. (2017). Photodynamic therapy: past, present and future. *Chem. Rec.* 17 (8): 775–802.

943 Moan, J. and Peng, Q. (2003). An outline of the hundred year history of PDT. *Anticancer Res.* 23 (5A): 3591–3600.

944 Taquet, J.-P., Frochot, C., Manneville, V., and Barberi-Heyob, M. (2007). Phthalocyanines covalently bound to biomolecules for a targeted photodynamic therapy. *Curr. Med. Chem.* 14 (15): 1673–1687.

945 (a) Yuan, A., Wu, J., Tang, X. et al. (2013). Application of near-infrared dyes for tumor imaging, photothermal, and photodynamic therapies. *J. Pharm. Sci.* 102 (1): 6–28. (b) Deng, K.-R., Li, C.-X., Huang, S.-S. et al. (2017). Recent progress in near infrared light triggered photodynamic therapy. *Small* 13 (4): 1702299.

946 Williams, J.R., Orton, G., and Unger, L.R. (1973). Preparation of singlet oxygen by heterogeneous photosensitization. *Tetrahedron Lett.* (46): 4603–4606.

947 Nilsson, R. and Kearns, D.R. (1974). Some useful heterogeneous systems for photosensitized generation of singlet oxygen. *Photochem. Photobiol.* 19 (2): 181–184.

948 Lacombe, S. and Pigot, T. (2011). New materials for sensitized photo-oxygenation. *Photochemistry* 38: 307–329.

949 Ronzani, F., Costarramone, N., Blanc, S. et al. (2013). Visible-light photosensitized oxidation of α-terpinene using novel silica-supported sensitizers: photooxygenation vs. photodehydrogenation. *J. Catal.* 303: 164–174.
950 Blossey, E.C. and Neckers, D.C. (1973). Polymer-based sensitizers for photooxidations. *J. Am. Chem. Soc.* 95 (17): 5820–5822.
951 Paczkowski, J. and Neckers, D.C. (1985). Photochemical properties of Rose-Bengal. 11. Fundamental studies in heterogeneous energy transfer. *Macromolecules* 18 (12): 2412–2418.
952 Torres-Alacan, J. and Sander, W. (2008). Reaction of diphenyldiazomethane with singlet oxygen studied by time-resolved IR spectroscopy. *J. Org. Chem.* 73 (18): 7118–7123.
953 Lee, J., Fortner, J.D., Hughes, J.B., and Kim, J.H. (2007). Photochemical production of reactive oxygen species by C_{60} in the aqueous phase during UV irradiation. *Environ. Sci. Technol.* 41 (7): 2529–2535.
954 (a) Scurlock, R.D., Wang, B.J., and Ogilby, P.R. (1996). Chemical reactivity of singlet sigma oxygen ($^1\Sigma^+{}_g$) in solution. *J. Am. Chem. Soc.* 118 (2): 388–392. (b) Sivéry, A., Barras, A., Boukherroub, R. et al. (2014). Production rate and reactivity of singlet oxygen $^1O_2(^1\Sigma^+{}_g)$ directly photoactivated at 1270 nm in lipid nanocapsules dispersed in water. *J. Phys. Chem. C* 118 (5): 2885–2893.
955 Ogilby, P.R. (2010). Singlet oxygen: there is indeed something new under the sun. *Chem. Soc. Rev.* 39 (8): 3181–3209.
956 Kautsky, H. and Hirsch, A. (1931). Energy transformations in boundary layers, IV. Announcement: interaction between excited pigment molecules and oxygen. *Ber. Dtsch. Chem. Ges.* 64: 2677–2683.
957 Kautsky, H. (1937). The interaction between sensitisers and oxygen in light. *Biochem. Z.* 291 (4/6): 271–284.
958 Kautsky, H. (1939). Quenching of luminescence by oxygen. *Trans. Faraday Soc.* 35 (1): 0216–0218.
959 Wilkinson, F., Helman, W.P., and Ross, A.B. (1995). Rate constants for the decay and reactions of the lowest electronically excited singlet state of molecular oxygen in solution – an expanded and revised compilation. *J. Phys. Chem. Ref. Data* 24 (2): 663–1021.
960 Schönberg, A. (1935). Notes of the photochemical formation of biradicals. *Justus Liebigs Ann. Chem.* 518: 299–302.
961 Wilkinson, F., McGarvey, D.J., and Olea, A.F. (1993). Factors governing the efficiency of singlet oxygen production during oxygen quenching of singlet and triplet states of anthracene derivatives in cyclohexane solution. *J. Am. Chem. Soc.* 115 (25): 12144–12151.
962 Abdel-Shafi, A.A. and Wilkinson, F. (2000). Charge transfer effects on the efficiency of singlet oxygen production following oxygen quenching of excited singlet and triplet states of aromatic hydrocarbons in acetonitrile. *J. Phys. Chem. A* 104 (24): 5747–5757.
963 Schmidt, R. and Brauer, H.D. (1987). Radiationless deactivation of singlet oxygen ($^1\Delta_g$) by solvent molecules. *J. Am. Chem. Soc.* 109 (23): 6976–6981.
964 Schmidt, R. and Bodesheim, M. (1995). Collision-induced radiative transitions $b^1\Sigma_g{}^+ \rightarrow a^1\Delta_g$, $b^1\Sigma_g{}^+ \rightarrow X^3\Sigma_g{}^-$, and $a^1\Delta_g \rightarrow X^3\Sigma_g{}^-$ of O_2. *J. Phys. Chem.* 99 (43): 15919–15924.
965 Ogryzlo, E.A. and Tang, C.W. (1970). Quenching of oxygen ($^1\Delta_g$) by amines. *J. Am. Chem. Soc.* 92 (17): 5034–5036.
966 Kearns, D.R. (1971). Physical and chemical properties of singlet molecular oxygen. *Chem. Rev.* 71 (4): 395–427.
967 (a) Smith, W.F. (1972). Kinetic evidence for both quenching and reaction of singlet oxygen with triethylamine in pyridine solution. *J. Am. Chem. Soc.* 94 (1): 186–190. (b) Seikel, K. and Brauer, H.D. (1991). Pressure effect on the radiationless deactivation of singlet oxygen by 1,4-diazobicyclo[2.2.2]octane in six different solvents. *Ber. Bunsen Ges. Phys. Chem.* 95 (8): 900–904.
968 Tsubomur, H., Yagishit, T., and Toi, H. (1973). Intermediates and mechanism of photooxygenation reaction of triethylamine. *Bull. Chem. Soc. Jpn.* 46 (10): 3051–3055.
969 Bernstein, R. and Foote, C.S. (1999). Singlet oxygen involvement in the photochemical reaction of C_{60} and amines. Synthesis of an alkyne-containing fullerene. *J. Phys. Chem. A* 103 (36): 7244–7247.
970 Cocquet, G., Rool, P., and Ferroud, C. (2000). Photosensitized oxidation, by single-electron transfer, of catharanthine and vindoline: a highly regio- and diastereoselective photocyanation reaction. *J. Chem. Soc., Perkin Trans. 1* (14): 2277–2281.
971 (a) Liu, Q., Li, Y.-N., Zhang, H.-H. et al. (2012). Reactivity and mechanistic insight into visible-light-induced aerobic

cross-dehydrogenative coupling reaction by organophotocatalysts. *Chem. Eur. J.* 18 (2): 620–627. (b) Ghosh, T., Das, A., and König, B. (2017). Photocatalytic N-formylation of amines via a reductive quenching cycle in the presence of air. *Org. Biomol. Chem.* 15 (12): 2536–2540.

972 Potter, S.S., Newbold, J.E., Hutchison, C.A., and Edgell, M.H. (1975). Specific cleavage analysis of mammalian mitochondrial-DNA. *Proc. Natl. Acad. Sci. U.S.A.* 72 (11): 4496–4500.

973 (a) Foote, C.S., Chang, Y.C., and Denny, R.W. (1970). Chemistry of singlet oxygen. 10. Carotenoid quenching parallels biological protection. *J. Am. Chem. Soc.* 92 (17): 5216–5218. (b) Cicero, A.F.G. and Colletti, A. (2017). Effects of carotenoids on health: are all the same? Results from clinical trials. *Curr. Pharm. Des.* 23 (17): 2422–2427.

974 Bosio, G.N., Breitenbach, T., Parisi, J. et al. (2013). Antioxidant β-Carotene does not quench singlet oxygen in mammalian cells. *J. Am. Chem. Soc.* 135 (1): 272–279.

975 Ivanov, V.B., Shlyapintokh, V.Y., Shapiro, A.B. et al. (1974). Formation of nitroxyl radicals in reaction of singlet oxygen with amines. *Bull. Acad. Sci. USSR, Div. Chem. Sci.* 23 (8): 1845–1845.

976 Lion, Y., Delmelle, M., and Vandevorst, A. (1976). New method of detecting singlet oxygen production. *Nature* 263 (5576): 442–443.

977 Moan, J. and Wold, E. (1979). Detection of singlet oxygen production by ESR. *Nature* 279 (5712): 450–451.

978 Alia, Mohanty, P., and Matysik, J. (2001). Effect of proline on the production of singlet oxygen. *Amino Acids* 21 (2): 195–200.

979 Dumas, J.L. and Garnier, B. (1975). Singlet oxygen production by an HF discharge in gaseous phase and determination of this excited species concentration and half-life time. *J. Chim. Phys. Phys.- Chim. Biol.* 72 (9): 1045–1051.

980 Thompson, Q.E. (1961). Ozone oxidation of nucleophilic substances. 1. Tertiary phosphite esters. *J. Am. Chem. Soc.* 83 (4): 845–851.

981 Murray, R.W. and Kaplan, M.L. (1968). Singlet oxygen sources in ozone chemistry. *J. Am. Chem. Soc.* 90 (2): 537–538.

982 Wasserman, E., Murray, R.W., Kaplan, M.L., and Yager, W.A. (1968). Electron paramagnetic resonance of $^1\Delta$ oxygen from a phosphite-ozone complex. *J. Am. Chem. Soc.* 90 (15): 4160–4161.

983 Murray, R.W. and Kaplan, M.L. (1968). Gas-phase reactions of singlet oxygen from a chemical source. *J. Am. Chem. Soc.* 90 (15): 4161–4162.

984 Schaap, A.P., Kees, K., and Thayer, A.L. (1975). 1-Phospha-2,8,9-trioxaadamantane ozonide – convenient source of singlet molecular oxygen. *J. Org. Chem.* 40 (8): 1185–1186.

985 Bartlett, P.D. and Chu, H.K. (1980). Mechanism of the direct reaction of phosphite ozonides with olefins. *J. Org. Chem.* 45 (15): 3000–3004.

986 Caminade, A.M., Elkhatib, F., Koenig, M., and Aubry, J.M. (1985). Phosphite ozonides with an oxygen singlet source – yield, mechanism. *Can. J. Chem.* 63 (11): 3203–3209.

987 Miyamoto, S., Ronsein, G.E., Correa, T.C. et al. (2009). Direct evidence of singlet molecular oxygen generation from peroxynitrate, a decomposition product of peroxynitrite. *Dalton Trans.* (29): 5720–5729.

988 Deslongchamps, P., Atlani, P., Frehel, D. et al. (1974). Oxidation of acetals by ozone. *Can. J. Chem.* 52 (21): 3651–3664.

989 Abdrakhmanova, A., Khalitova, L., Spirikhin, L.V. et al. (2007). Synthesis and thermal decomposition of hydrotrioxide obtained by ozonization of exo-bicyclo[2.2.1]heptan-2-ol. *Russ. Chem. Bull.* 56 (2): 271–275.

990 Hang, J., Ghorai, P., Finkenstaedt-Quinn, S.A. et al. (2012). Generation of singlet oxygen from fragmentation of monoactivated 1,1-dihydroperoxides. *J. Org. Chem.* 77 (3): 1233–1243.

991 Matsumura, Y., Iwasawa, A., Kobayashi, T. et al. (2013). Detection of high-frequency ultrasound-induced singlet oxygen by the ESR spin-trapping method. *Chem. Lett.* 42 (10): 1291–1293.

992 Fritzsche, M. (1867). Note sur les carbures d'hydrogène solides, tirés du goudron de houille. *C.R. Hebd. Séances Acad. Sci., Paris* 64: 1035–1037.

993 Fudickar, W. and Linker, T. (2012). Why triple bonds protect acenes from oxidation and decomposition. *J. Am. Chem. Soc.* 134 (36): 15071–15082.

994 Moureu, C., Dufraisse, C., and Dean, P.M. (1926). Un peroxide organique dissociable: le peroxide de rubrène. *C.R. Hebd. Séances Acad. Sci., Paris* 182: 1584–1587.

995 Bergmann, W. and Mclean, M.J. (1941). Transannular peroxides. *Chem. Rev.* 28 (2): 367–394.

996 Dufraisse, C. and Enderlin, L. (1930). Research on the susceptible structure of presenting reversable oxidability. *C.R. Hebd. Séances Acad. Sci., Paris* 190: 1229–1230.

997 Dufraisse, C. and Houpiliart, J. (1937). Research on the dissociable organic oxides – hydrogenation of photooxides. *C.R. Hebd. Séances Acad. Sci., Paris* 205: 740–743.

998 Wasserman, H.H. and Scheffer, J.R. (1967). Singlet oxygen reactions from photoperoxides. *J. Am. Chem. Soc.* 89 (12): 3073–3075.

999 Wasserman, H.H., Scheffer, J.R., and Cooper, J.L. (1972). Singlet oxygen reactions with 9,10-diphenylanthracene peroxide. *J. Am. Chem. Soc.* 94 (14): 4991–4996.

1000 Turro, N.J., Chow, M.F., and Rigaudy, J. (1979). Thermolysis of anthracene endoperoxides – concerted vs diradical mechanisms – microscopic reversibility in endothermic chemiluminescent reactions. *J. Am. Chem. Soc.* 101 (5): 1300–1302.

1001 Aubry, J.M., Pierlot, C., Rigaudy, J., and Schmidt, R. (2003). Reversible binding of oxygen to aromatic compounds. *Acc. Chem. Res.* 36 (9): 668–675.

1002 Matsumoto, M., Yamada, M., and Watanabe, N. (2005). Reversible 1,4-cycloaddition of singlet oxygen to N-substituted 2-pyridones: 1,4-Endoperoxide as a versatile chemical source of singlet oxygen. *J. Chem. Soc., Chem. Commun.* (4): 483–485.

1003 Windaus, A. and Brunken, J. (1928). About the photochemical oxidation of ergosterine. *Liebigs Ann. Chem.* 460: 225–235.

1004 Schenck, G.O. and Ziegler, H. (1951). Endoperoxyde der Norcaradienreihe. *Naturwissenschaften* 38 (15): 356–356.

1005 Schenck, G.O. and Ziegler, K. (1944). Die Synthese des Ascaridols. *Naturwissenschaften* 32: 157–157.

1006 Schenck, G.O. (1948). Über Askaridol und seine Synthese. Süddeutsche Apotheker-Zeitung. *Zeitschrift für Apotheker und Vertreter verwandter Berufszweige* 88: 3–6.

1007 Schenck, G.O. (1954). Zur Reaktion Photochemisch Angeregter Molekeln mit O_2. *Naturwissenschaften* 41 (19): 452–453.

1008 Schenck, G.O. (1957). Aufgaben und Möglichkeiten der Preparativen Strahlenchemie. *Angew. Chem.* 69 (1–2): 63–63.

1009 Ashford, R.D. and Ogryzlo, E.A. (1974). Arrhenius parameter for some gas phase cycloaddition reactions of singlet molecular oxygen. *Can. J. Chem.* 52 (20): 3544–3548.

1010 Hasty, N.M. and Kearns, D.R. (1973). Mechanisms of singlet oxygen reactions – intermediates in reaction of singlet oxygen with dienes. *J. Am. Chem. Soc.* 95 (10): 3380–3381.

1011 Gorman, A.A., Lovering, G., and Rodgers, M.A.J. (1979). Entropy-controlled reactivity of singlet oxygen ($^1\Delta_g$) toward furans and indoles in toluene – variable temperature study by pulse radiolysis. *J. Am. Chem. Soc.* 101 (11): 3050–3055.

1012 Monroe, B.M. (1981). Rate constants for the reaction of singlet oxygen with conjugated dienes. *J. Am. Chem. Soc.* 103 (24): 7253–7256.

1013 (a) Clennan, E.L. and Mehrsheikhmohammadi, M.E. (1984). Mechanism of endoperoxide formation. 2. Possibility of exciplexes on the reaction coordinates. *J. Org. Chem.* 49 (7): 1321–1322. (b) Clennan, E.L. and Mehrsheikhmohammadi, M.E. (1984). Mechanism of endoperoxide formation. 3. Utilization of the Young and Carlsson kinetic techniques. *J. Am. Chem. Soc.* 106 (23): 7112–7118.

1014 (a) Fudickar, W. and Linker, T. (2014). Intermediates in the formation and thermolysis of peroxides from oxidations with singlet oxygen. *Aust. J. Chem.* 67 (3): 520–327. (b) Tzirakis, M.D. and Alberti, N.M. (2015). On the mechanism of the (4+2)-cycloaddition of $^1O_2$ with 1,3-dienes: identifying the putative biradical/dipolar intermediate by employing the gem-diphenylcyclopropylcarbinyl radical clock as a mechanistic probe. *ARKIVOC* 83–100.

1015 (a) Adam, W., Bosio, S., Bartoschek, A., and Griesbeck, A.G. (2004). Photooxygenation of 1,3-dienes. In: *CRC Handbook of Organic Photochemistry and Photobiology*, 2e (ed. W. Horspool and F. Lenci), 25–125/19. Boca Raton, FL: CRC Press. (b) Clennan, E.L. (2016). Overview of the chemical reactions of singlet oxygen. In: *Singlet Oxygen, Volume 1*, Comprehensive Series in Photochemical & Photobiological Sciences, vol. 13 (ed. E. Sage, L. Rhodes and M. Trotta), 353–367. Royal Society of Chemistry.

1016 Hagenbuch, J.P., Birbaum, J.L., Métral, J.L., and Vogel, P. (1982). Interaction between non-conjugated chromophores. 17. Effects of remote substitution on the photooxidation of exocyclic *s-cis*-butadienes grafted onto bicyclo[2.2.1]heptanes – thermal and Rh-catalyzed rearrangements of 3,6-dihydro-1,2-dioxines. *Helv. Chim. Acta* 65 (3): 887–895.

1017 Bobrowski, M., Liwo, A., Oldziej, S. et al. (2000). CAS MCSCF/CAS MCQDPT2 study of the mechanism of singlet oxygen addition to 1,3-butadiene and benzene. *J. Am. Chem. Soc.* 122 (34): 8112–8119.

1018 McCarrick, M.A., Wu, Y.D., and Houk, K.N. (1993). Hetero-Diels–Alder reaction transition structures – reactivity, stereoselectivity, catalysis, solvent effects, and the exo-lone-pair effect. *J. Org. Chem.* 58 (12): 3330–3343.

1019 Li, Y. and Houk, K.N. (1993). Diels–Alder dimerization of 1,3-butadiene – an ab-initio CASSCF study of the concerted and stepwise mechanisms and butadiene ethylene revisited. *J. Am. Chem. Soc.* 115 (16): 7478–7485.

1020 Dewar, M.J.S. and Thiel, W. (1975). Ground states of molecules. 30. MINDO-3 study of reactions of singlet ($^1\Delta_g$) oxygen with carbon-carbon double bonds. *J. Am. Chem. Soc.* 97 (14): 3978–3986.

1021 Clennan, E.L. (1991). Synthetic and mechanistic aspects of 1,3-diene photooxidation. *Tetrahedron* 47 (8): 1343–1382.

1022 Lissi, E.A., Encinas, M.V., Lemp, E., and Rubio, M.A. (1993). Singlet oxygen $O_2(^1\Delta_g)$ bimolecular processes – solvent and compartmentalization effects. *Chem. Rev.* 93 (2): 699–723.

1023 Sevin, F. and McKee, M.L. (2001). Reactions of 1,3-cyclohexadiene with singlet oxygen. A theoretical study. *J. Am. Chem. Soc.* 123 (19): 4591–4600.

1024 Manring, L.E., Kanner, R.C., and Foote, C.S. (1983). Chemistry of singlet oxygen. 43. Quenching by conjugated olefins. *J. Am. Chem. Soc.* 105 (14): 4707–4710.

1025 Manring, L.E. and Foote, C.S. (1983). Chemistry of singlet oxygen. 44. Mechanism of photooxidation of 2,5-dimethylhexa-2,4-diene and 2-methyl-2-pentene. *J. Am. Chem. Soc.* 105 (14): 4710–4717.

1026 Gollnick, K. and Griesbeck, A. (1984). Interactions of singlet oxygen with 2,5-dimethyl-2,4-hexadiene in polar and non-polar solvents evidence for a vinylog ene-reaction. *Tetrahedron* 40 (17): 3235–3250.

1027 Gollnick, K. and Griesbeck, A. (1984). Solvent dependence of singlet oxygen substrate interactions in ene-reactions, (4+2)-cycloaddition and (2+2)-cycloaddition reactions. *Tetrahedron Lett.* 25 (7): 725–728.

1028 O'Shea, K.E. and Foote, C.S. (1988). Chemistry of singlet oxygen. 51. Zwitterionic Intermediates from 2,4-Hexadienes. *J. Am. Chem. Soc.* 110 (21): 7167–7170.

1029 Griesbeck, A.G., Fiege, M., Gudipati, M.S., and Wagner, R. (1998). Photooxygenation of 2,4-dimethyl-1,3-pentadiene: solvent dependence of the chemical (ene-reaction and (4+2)-cycloaddition) and physical quenching of singlet oxygen. *Eur. J. Org. Chem.* (12): 2833–2838.

1030 Aubry, J.M., Mandardcazin, B., Rougee, M., and Bensasson, R.V. (1995). Kinetic studies of singlet oxygen $[\pi^4+\pi^2]$-cycloadditions with cyclic 1,3-dienes in 28 solvents. *J. Am. Chem. Soc.* 117 (36): 9159–9164.

1031 Adam, W. and Prein, M. (1996). π-Facial diastereoselectivity in the (4+2)-cycloaddition of singlet oxygen as a mechanistic probe. *Acc. Chem. Res.* 29 (6): 275–283.

1032 Demuth, M.R., Garrett, P.E., and White, J.D. (1976). Synthesis of (+/-)-crotepoxide, (+/-)-epicrotepoxide, and (+/-)-isocrotepoxide. *J. Am. Chem. Soc.* 98 (2): 634–635.

1033 (a) Gollnick, K. and Griesbeck, A. (1985). Singlet oxygen photooxygenation of furans – isolation and reactions of (4+2)-cycloaddition products (unsaturated sec-ozonides). *Tetrahedron* 41 (11): 2057–2068. (b) Kalaitzakis, D., Montagnon, T., Antonatou, E. et al. (2013). From simple furans to complex nitrogen-bearing aromatic polycycles by means of a flexible and general reaction sequence initiated by singlet oxygen. *Chem. Eur. J.* 19 (31): 10119–10123. (c) Kalaitzakis, D., Montagnon, T., Ionnou, G.I. et al. (2015). *ARKIVOC* (3): 154–166.

1034 Hagenbuch, J.P. and Vogel, P. (1979). Conformational analysis and the rearrangements of (norborn-2-eno)[d](3,6-dihydro-1,2-iioxine). *Tetrahedron Lett.* (6): 561–564.

1035 Kornblum, N. and Delamare, H.E. (1951). The base catalyzed decomposition of a dialkyl peroxide. *J. Am. Chem. Soc.* 73 (2): 880–881.

1036 (a) Kelly, D.R., Bansal, H., and Morgan, J.J.G. (2002). The mechanism of the tertiary amine catalysed isomerisation of endoperoxides to hydroxyketones: synthesis and chemistry of the intermediate postulated in the peroxide attack mechanism. *Tetrahedron Lett.* 43 (51): 9331–9333. (b) Palframan, M.J., Kociok-Kohn, G., and Lewis, S.E. (2012). Photooxygenation of a microbial arene oxidation product and regioselective Kornblum–DeLaMare rearrangement: total synthesis of zeylenols and zeylenones. *Chem. Eur. J.* 18 (15): 4766–4774. (c) Zheng, X.-J., Lu, S.-L., and Li, Z.-P. (2013). The rearrangement of tert-butylperoxides for the construction of polysubstituted furans. *Org. Lett.*

15 (21): 5432–5435. (d) Yaremenko, I.A., Vil, V.A., Demchuk, D.V., and Terent'ev, A.D. (2016). Rearrangements of organic peroxides and related processes. *Beilstein J. Org. Chem.* 12: 1647–1748.

1037 Hagenbuch, J.P. and Vogel, P. (1980). Asymmetric induction in the rearrangement of monocyclic endoperoxides into γ-hydroxy-α,β-unsaturated aldehydes. *J. Chem. Soc., Chem. Commun.* (22): 1062–1063.

1038 Staben, S.T., Linghu, X., and Toste, F.D. (1996). Enantioselective synthesis of γ-hydroxyenones by chiral based-catalyzed Kornblum DeLaMare rearrangement. *J. Am. Chem. Soc.* 128: 12658–12659.

1039 Montagnon, T., Tofi, M., and Vassilikogiannakis, G. (2008). Using singlet oxygen to synthesize polyoxygenated natural products from furans. *Acc. Chem. Res.* 41 (8): 1001–1011.

1040 Wasserman, H.H. and Doumaux, A.R. (1962). An unusual intramolecular Diels–Alder reaction in photooxidation of a (furano) heterocyclophane. *J. Am. Chem. Soc.* 84 (23): 4611–4612.

1041 Schenck, G.O., Eggert, H., and Denk, W. (1953). Photochemische reaktionen. 3. Über die Bildung von Hydroperoxyden bei Photosensibilisierten Reaktionen von O_2 mit geeigneten Akzeptoren, insbesondere mit α-pinen und β-pinen. *Liebigs Ann. Chem.* 584 (2–3): 177–198.

1042 (a) Alberti, M.N. and Orfanopoulos, M. (2010). Recent mechanistic iinsights in the singlet oxygen ene-reaction. *Synlett* (7): 999–1026. (b) Alberti, M.N. and Orfanopoulos, M. (2010). Unraveling the mechanism of the singlet oxygen ene reaction: recent computational and experimental approaches. *Chem. Eur. J.* 16 (31): 9414–9421. (c) Griesbeck, A.G., Goldfuss, B., Jäger, C. et al. (2017). Strong asymmetry in the perepoxide bifurcation mechanism: the large-group effect in the singlet oxygen ene-reaction with allylic alcohols. *ChemPhotoChem* 1 (5): 213–221.

1043 Gorman, A.A. and Rodgers, M.A.J. (1981). Singlet molecular oxygen. *Chem. Soc. Rev.* 10 (2): 205–231.

1044 Harding, L.B. and Goddard, W.A. (1980). Mechanism of the ene reaction of singlet oxygen with olefins. *J. Am. Chem. Soc.* 102 (2): 439–449.

1045 Hurst, J.R. and Schuster, G.B. (1982). Ene reaction of singlet oxygen – an entropy-controlled process determines the reaction rate. *J. Am. Chem. Soc.* 104 (24): 6854–6856.

1046 Jefford, C.W. (1979). Role of zwitterionic peroxides in controlling hydroperoxidation. *Tetrahedron Lett.* (11): 985–988.

1047 Stratakis, M. and Orfanopoulos, M. (2000). Regioselectivity in the ene-reaction of singlet oxygen with alkenes. *Tetrahedron* 56 (12): 1595–1615.

1048 Alberti, M.N. and Orfanopoulos, M. (2006). Stereoelectronic and solvent effects on the allylic oxyfunctionalization of alkenes with singlet oxygen. *Tetrahedron* 62 (46): 10660–10675.

1049 Yardimci, S.D., Kaya, N., and Balci, M. (2006). Regioselectivity in the ene-reaction of singlet oxygen with cyclic alkenes: photooxygenation of methyl-substituted 1,4-cyclohexadiene derivatives. *Tetrahedron* 62 (46): 10633–10638.

1050 Dang, H.S. and Davies, A.G. (1992). Ene-reactions of allylically stannylated cholestenes – singlet oxygenation of 7-α-triphenylstannylcholest-5-en-3-β-ol, and of 7-α-triphenylstannyl-cholest-5-ene-3-one and 7-α-tributylstannyl-cholest-5-ene-3-one, and the rearrangement of 5-α-tributylstannylperoxy-3-β-benzoyloxycholest-6-ene and of 7-α-tributylstannylperoxy-3-β-benzoyloxycholest-5-ene. *J. Chem. Soc., Perkin Trans. 2* (7): 1095–1101.

1051 Dang, H.S. and Davies, A.G. (1992). The effect of ligands, solvent and temperature on the reactions of allyltin(IV) compounds with singlet oxygen. *J. Organomet. Chem.* 430 (3): 287–298.

1052 Adam, W. and Klug, P. (1993). Photooxygenation of vinylstannanes – tin-substituted allylic hydroperoxides through the regioselective and diastereoselective ene-reaction with singlet oxygen. *J. Org. Chem.* 58 (12): 3416–3420.

1053 Newcomb, M., Johnson, C.C., Manek, M.B., and Varick, T.R. (1992). Picosecond radical kinetics – ring openings of phenyl substituted cyclopropylcarbinyl radicals. *J. Am. Chem. Soc.* 114 (27): 10915–10921.

1054 Newcomb, M. and Chestney, D.L. (1994). A hypersensitive mechanistic probe for distinguishing between radical and carbocation intermediates. *J. Am. Chem. Soc.* 116 (21): 9753–9754.

1055 Yoshioka, Y., Yamada, S., Kawakami, T. et al. (1996). Ab initio molecular orbital studies of singlet oxygen reactions of olefins, enol ethers, and enamines. *Bull. Chem. Soc. Jpn.* 69 (10): 2683–2699.

1056 Singleton, D.A., Hang, C., Szymanski, M.J. et al. (2003). Mechanism of ene-reactions of singlet oxygen. A two-step no-intermediate mechanism. *J. Am. Chem. Soc.* 125 (5): 1319–1328.

1057 Sheppard, A.N. and Acevedo, O. (2009). Multidimensional exploration of valley-ridge inflection points on potential-energy surfaces. *J. Am. Chem. Soc.* 131 (7): 2530–2540.

1058 Leach, A.G., Houk, K.N., and Foote, C.S. (2008). Theoretical prediction of a perepoxide intermediate for the reaction of singlet oxygen with trans-cyclooctene contrasts with the two-step no-intermediate ene-reaction for acyclic alkenes. *J. Org. Chem.* 73 (21): 8511–8519.

1059 Inoue, Y. and Turro, N.J. (1980). Reaction of singlet oxygen with trans-cyclooctene – evidence for a 1,2-dioxetane intermediate. *Tetrahedron Lett.* 21 (45): 4327–4330.

1060 Poon, T.H.W., Pringle, K., and Foote, C.S. (1995). Reaction of cyclooctenes with singlet oxygen – trapping of a perepoxide intermediate. *J. Am. Chem. Soc.* 117 (29): 7611–7618.

1061 Ohloff, G., Klein, E., and Schenck, G.O. (1961). Darstellung von Rosenoxyden und anderen Hydropyran-Derivaten über Photohydroperoxyde. *Angew. Chem.* 73 (16): 578–578.

1062 Alsters, P.L., Jary, W., Nardello-Rataj, V., and Aubry, J.-M. (2010). "Dark" singlet oxygenation of β-citronellol: a key step in the manufacture of rose oxide. *Org. Process Res. Dev.* 14 (1): 259–262.

1063 Adam, W., Griesbeck, A., and Staab, E. (1986). A convenient "one-pot" synthesis of epoxy alcohols via photooxygenation of olefins in the presence of Ti(IV) catalyst. *Tetrahedron Lett.* 27: 2839–2842.

1064 (a) Katsuki, T. and Sharpless, K.B. (1980). The 1st practical method for asymmetric epoxidation. *J. Am. Chem. Soc.* 102 (18): 5974–5976. (b) Katsuki, T. (1999). Epoxidation of allylic alcohols. In: *Comprehensive asymmetric catalysis* (ed. E.N. Jackobsen, A. Pfaltz and H. Yamamoto), 621–648. Heidelberg, Berlin: Springer-Verlag.

1065 (a) Huber, S., Cokoja, M., and Kuhn, F.E. (2014). Historical landmarks of the application of molecular transition metal catalysts for olefin epoxidation. *J. Organomet. Chem.* 751: 25–32. (b) Noji, M., Kobayashi, T., Uechi, Y. et al. (2015). Asymmetric epoxidation of allylic alcohols catalyzed by vanadium-binaphthylbishydroxamic acid complex. *J. Org. Chem.* 80 (6): 3203–3210. (c) Heravi, M.-M., Lashaki, T.B., and Poorahmad, N. (2015). Applications of sharpless asymmetric epoxidation in total synthesis. *Tetrahedron: Asymmetry* 26 (8–9): 405–495. (d) Chen, C., Yuan, H.-Y., Wang, H.-F. et al. (2016). Highly efficient epoxidation of allylic alcohols with hydrogen peroxide catalyzed by peroxoniobate-based ionic liquids. *ACS Catal.* 6 (5): 3354–3364. (e) Botulol-Ares, J.M., Hanson, J.R., Hernández-Galán, R., and Collado, I.G. (2017). Mild epoxidation of allylic alcohols catalyzed by Ti(III) complexes: selectivity and mechanism. *ACS Omega* 2 (7): 2083–3090.

1066 Schaap, A.P., Recher, S.G., Faler, G.R., and Villasenor, S.R. (1983). Evidence for a perepoxide intermediate in the 1,2-cycloaddition of singlet oxygen to adamantylideneadamantane – nucleophilic oxygen atom transfer to sulfoxides. *J. Am. Chem. Soc.* 105 (6): 1691–1693.

1067 Schenck, G.O. and Schulte-Elte, K.H. (1958). Photosensibilisierte synthese tertiarer Allylhydroperoxyde aus symmetrisch Tetrasubstituierten Äthylenen. *Liebigs Ann. Chem.* 618 (1–3): 185–193.

1068 Adam, W. and Delfierro, J. (1978). Cyclic peroxide series. 61. Synthesis of α-hydroperoxy esters by singlet oxygenation of ketene acetals. *J. Org. Chem.* 43 (6): 1159–1161.

1069 Wasserman, H.H. and Pickett, J.E. (1985). The fluoride ion effect in the reactions of singlet oxygen with enols. *Tetrahedron* 41 (11): 2155–2162.

1070 Kopecky, K.R. and Mumford, C. (1969). Luminescence in thermal decomposition of 3,3,4-trimethyl-1,2-dioxetane. *Can. J. Chem.* 47 (4): 709–711.

1071 Bartlett, P.D. and Schaap, A.P. (1970). Stereospecific formation of 1,2-dioxetanes from *cis*-diethoxyethylenes and trans-diethoxyethylenes by singlet oxygen. *J. Am. Chem. Soc.* 92 (10): 3223–3225.

1072 Mazur, S. and Foote, C.S. (1970). Chemistry of singlet oxygen. 9. A stable dioxetane from photooxygenation of tetramethoxyethylene. *J. Am. Chem. Soc.* 92 (10): 3225–3226.

1073 Huber, J.E. (1968). Photooxygenation of enamines – a partial synthesis of progesterone. *Tetrahedron Lett.* (29): 3271–3272.

1074 Adam, W., Fierro, J.D., Quiroz, F., and Yany, F. (1980). Singlet oxygenation of ketene acetals – formation of 1,2-dioxetanes and their thermal rearrangement to α-peroxy esters. *J. Am. Chem. Soc.* 102 (6): 2127–2128.

1075 Wasserman, H.H., Lipshutz, B.H., Tremper, A.W., and Wu, J.S. (1981). Preparation of β-lactams from azetidine-2-carboxylic acids and esters. *J. Org. Chem.* 46 (15): 2991–2999.

1076 Zaklika, K.A., Kaskar, B., and Schaap, A.P. (1980). Mechanisms of photooxygenation. 1.

Substituent effects on the (2+2)-cycloaddition of singlet oxygen to vinyl ethers. *J. Am. Chem. Soc.* 102 (1): 386–389.

1077 Gollnick, K. and Knutzenmies, K. (1991). Dye-sensitized photooxygenation of 2,3-dihydrofurans – competing (2+2)-cycloadditions and ene-reactions of singlet oxygen with a rigid cyclic enol ether system. *J. Org. Chem.* 56 (12): 4017–4027.

1078 Gollnick, K. and Knutzenmies, K. (1991). Dye-sensitized photooxygenation of 2,3-dihydrothiophenes – formation of stable 1,2-dioxetanes from 4,5-dialkyl-substituted derivatives. *J. Org. Chem.* 56 (12): 4027–4031.

1079 Adam, W., Kumar, A.S., and Sahamoller, C.R. (1995). Photooxygenation of vinyl sulfides – substituent effects on the (2+2)-cycloaddition versus Schenck ene-reaction modes. *Tetrahedron Lett.* 36 (43): 7853–7854.

1080 Clennan, E.L. and Nagraba, K. (1988). Additions of singlet oxygen to alkoxy-substituted dienes – the mechanism of the singlet oxygen 1,2-cycloaddition reaction. *J. Am. Chem. Soc.* 110 (13): 4312–4318.

1081 Cermola, F. and Iesce, M.R. (2002). Substituent and solvent effects on the photosensitized oxygenation of 5,6-dihydro-1,4-oxathiins. Intramolecular oxygen transfer vs normal cleavage of the dioxetane intermediates. *J. Org. Chem.* 67 (14): 4937–4944.

1082 Harding, L.B. and Goddard, W.A. (1977). Intermediates in chemiluminescent reaction of singlet oxygen with ethylene – Ab initio studies. *J. Am. Chem. Soc.* 99 (13): 4520–4523.

1083 Maranzana, A., Ghigo, G., and Tonachini, G. (2000). Diradical and peroxirane pathways in the [π^2+p^2]-cycloaddition reactions of $^1\Delta_g$ dioxygen with ethene, methyl vinyl ether, and butadiene: a density functional and multireference perturbation theory study. *J. Am. Chem. Soc.* 122 (7): 1414–1423.

1084 Vassilikogiannakis, G., Stratakis, M., and Orfanopoulos, M. (1998). Primary and secondary isotope effects in the photooxidation of 2,5-dimethyl-2,4-hexadiene. Elucidation of the reaction energy profile. *J. Org. Chem.* 63 (18): 6390–6393.

1085 Foote, C.S., Dzakpasu, A.A., and Lin, J.W.P. (1975). Chemistry of singlet oxygen. 20. Mechanism of sensitized photooxidation of enamines. *Tetrahedron Lett.* (14): 1247–1250.

1086 Saito, I., Matsugo, S., and Matsuura, T. (1979). 1,2-Dioxetane formation in an indole system. *J. Am. Chem. Soc.* 101 (16): 4757–4759.

1087 Saito, I., Matsugo, S., and Matsuura, T. (1979). Photoinduced reactions. 110. Mechanism of indole-singlet oxygen reactions – interception of zwitterionic intermediates and ene-reaction. *J. Am. Chem. Soc.* 101 (24): 7332–7338.

1088 Eriksen, J., Foote, C.S., and Parker, T.L. (1977). Photosensitized oxygenation of alkenes and sulfides via a non-singlet-oxygen mechanism. *J. Am. Chem Soc.* 99 (19): 6455–6456.

1089 Schaap, A.P., Zaklika, K.A., Kaskar, B., and Fung, L.W.M. (1980). Mechanisms of photooxygenation. 2. Formation of 1,2-dioxetanes via 9,10-dicyanoanthracene-sensitized electron-transfer processes. *J. Am. Chem. Soc.* 102 (1): 389–391.

1090 George, M.V. and Bhat, V. (1979). Photooxygenations of nitrogen-heterocycles. *Chem. Rev.* 79 (5): 447–478.

1091 Sivaguru, J., Solomon, M.R., Poon, T. et al. (2008). The reaction of singlet oxygen with enecarbamates: a mechanistic playground for investigating chemoselectivity, stereoselectivity, and vibratioselectivity of photooxidations. *Acc. Chem. Res.* 41 (3): 387–400.

1092 Rajeev, R. and Sunoj, R.B. (2012). On the origin of regio- and stereoselectivity in singlet oxygen addition to enecarbamates. *J. Org. Chem.* 77 (5): 2474–2485.

1093 Adam, W., Saha-Möller, C.R., and Schambony, S.B. (1999). A highly diastereoselective dioxetane formation by the hydroxy-directed (2+2)-cycloaddition of singlet oxygen to a chiral allylic alcohol. *J. Am. Chem. Soc.* 121 (9): 1834–1838.

1094 Celaje, J.A., Zhang, D., Guerrero, A.M., and Selke, M. (2011). Chemistry of trans-Resveratrol with singlet oxygen: (2+2)-addition, (4+2)-addition, and formation of the phytoalexin moracin M. *Org. Lett.* 13 (18): 4846–4849.

1095 Ching, T.Y. and Foote, C.S. (1975). Chemistry of singlet oxygen. 22. Photooxidation of nitrones. *Tetrahedron Lett.* (44): 3771–3774.

1096 Bhat, V. and George, M.V. (1977). Dye-sensitized photooxygenation of aziridines. *Tetrahedron Lett.* (47): 4133–4136.

1097 Bhat, V. and George, M.V. (1979). Photooxygenation of aziridines and some potential azomethine ylides. *J. Org. Chem.* 44 (19): 3288–3292.

1098 Bhat, V., Dixit, V.M., Ugarker, B.G. et al. (1979). Photo-oxygenations of sydnones and azomethine imines. *J. Org. Chem.* 44 (17): 2957–2961.

1099 Schaap, A.P., Prasad, G., and Gagnon, S.D. (1983). Formation of a 1,2,4-dioxazolidine by electron-transfer photooxygenation of 1-butyl-2,3-diphenylaziridine. *Tetrahedron Lett.* 24 (30): 3047–3050.

1100 Schaap, A.P., Prasad, G., and Siddiqui, S. (1984). Formation of 1,2,4-dioxazolidines by electron-transfer photooxygenation of aziridines. *Tetrahedron Lett.* 25 (29): 3035–3038.

1101 Schaap, A.P., López, L., and Gagnon, S.D. (1983). Formation of an ozonide by electron-transfer photooxygenation of tetraphenyloxirane – cosensitization by 9,10-dicyanoanthracene and biphenyl. *J. Am. Chem. Soc.* 105 (3): 663–664.

1102 Schaap, A.P., Siddiqui, S., Gagnon, S.D., and López, L. (1983). Stereoselective formation of *cis*-stilbene ozonide from the cosensitized electron-transfer photooxygenation of *cis*-2,3-diphenyloxirane and trans-2,3-diphenyloxirane. *J. Am. Chem. Soc.* 105 (15): 5149–5150.

1103 Schaap, A.P., Siddiqui, S., Prasad, G. et al. (1984). Stereoselective formation of *cis* ozonides by electron-transfer photooxygenation of naphthyl-substituted epoxides – stereochemical assignments of ozonides by X-ray crystallography and chromatographic resolution. *J. Am. Chem. Soc.* 106 (20): 6087–6088.

1104 Schaap, A.P., Siddiqui, S., Prasad, G. et al. (1984). Cosensitized electron-transfer photooxygenation – the photochemical preparation of 1,2,4-trioxolanes, 1,2-dioxolanes and 1,2,4-dioxazolidines. *J. Photochem.* 25 (2–4): 167–181.

1105 Criegee, R. and Wenner, G. (1949). Die Ozonisierung des 9,10-Oktalins. *Liebigs Ann. Chem.* 564 (1): 9–15.

1106 Story, P.R. and Burgess, J.R. (1967). Ozonolysis. Evidence for carbonyl oxide tautomerization and for 1,3-dipolar addition to olefins. *J. Am. Chem. Soc.* 89 (22): 5726–5727.

1107 Criegee, R. (1975). Mechanism of ozonolysis. *Angew. Chem. Int. Ed. Engl.* 14 (11): 745–752.

1108 Bartlett, P.D. and Traylor, R.G. (1962). Reaction of diphenyldiazomethane with oxygen – Criegee carbonyl oxide. *J. Am. Chem. Soc.* 84 (17): 3408–3809.

1109 Casal, H.L., Tanner, M., Werstiuk, N.H., and Scaiano, J.C. (1985). Fluorenone oxide – transient spectroscopy and kinetics of its formation and reactions. *J. Am. Chem. Soc.* 107 (16): 4616–4620.

1110 Sander, W., Marquardt, R., Bucher, G., and Wandel, H. (1996). Chemistry and spectroscopy of aromatic diradicals in cryogenic matrices. *Pure Appl. Chem.* 68 (2): 353–356.

1111 Casal, H.L., Sugamori, S.E., and Scaiano, J.C. (1984). Study of carbonyl oxide formation in the reaction of singlet oxygen with diphenyldiazomethane. *J. Am. Chem. Soc.* 106 (24): 7623–7624.

1112 Scaiano, J.C., McGimpsey, W.G., and Casal, H.L. (1989). Generation and transient spectroscopy of substituted diaryl carbonyl oxides. *J. Org. Chem.* 54 (7): 1612–1616.

1113 Nojima, T., Ishiguro, K., and Sawaki, Y. (1995). Formation and cycloreversion of dioxadiazole intermediates in the reaction of diazo-compounds with singlet oxygen. *Chem. Lett.* (7): 545–546.

1114 Bethell, D. and Mckeivor, R. (1977). Intermediates in decomposition of aliphatic diazo-compounds. 13. Mechanistic studies on reaction of diaryldiazomethanes with singlet molecular-oxygen. *J. Chem. Soc., Perkin Trans. 2* (3): 327–333.

1115 Juretic, D., Kusic, H., Dionysiou, D.D., and Bozic, A.L. (2013). Environmental aspects of photooxidative treatment of phenolic compounds. *J. Hazard. Mater.* 262: 377–386.

1116 Montano, M., Gutleb, A.C., and Murk, A.J. (2013). Persistent toxic burdens of halogenated phenolic compounds in humans and wildlife. *Environ. Sci. Technol.* 47 (12): 6071–6081.

1117 Barbieri, Y., Massad, W.A., Díaz, D.J. et al. (2008). Photodegradation of bisphenol A and related compounds under natural-like conditions in the presence of riboflavin: kinetics, mechanism and photoproducts. *Chemosphere* 73 (4): 564–571.

1118 Martire, D.O. and Gónzalez, M.C. (2000). Electron transfer reactions of singlet molecular oxygen with phenols. *J. Phys. Org. Chem.* 13 (4): 208–212.

1119 Suchard, O., Kane, R., Roe, B.J. et al. (2006). Photooxygenations of 1-naphthols: an environmentally friendly access to 1,4-naphthoquinones. *Tetrahedron* 62 (7): 1467–1473.

1120 Thomas, M.J. and Foote, C.S. (1978). Chemistry of singlet oxygen. 26. Photooxygenation of phenols. *Photochem. Photobiol.* 27 (6): 683–693.

1121 Bobrowski, M., Liwo, A., Oldziej, S. et al. (2002). Ab initio study of the mechanism of singlet-dioxygen addition to hydroxyaromatic compounds: negative evidence for the involvement of peroxa and endoperoxide intermediates. *J. Comput. Chem.* 23 (11): 1076–1089.

1122 Canonica, S., Jans, U., Stemmler, K., and Hoigne, J. (1995). Transformation kinetics of phenols in water – photosensitization by dissolved natural organic material and aromatic ketones. *Environ. Sci. Technol.* 29 (7): 1822–1831.

1123 Canonica, S., Hellrung, B., and Wirz, J. (2000). Oxidation of phenols by triplet aromatic ketones in aqueous solution. *J. Phys. Chem. A* 104 (6): 1226–1232.

1124 Huvaere, K., Olsen, K., and Skibsted, L.H. (2009). Quenching of triplet-excited flavins by flavonoids. Structural assessment of antioxidative activity. *J. Org. Chem.* 74 (19): 7283–7293.

1125 Tomkiewicz, M., Groen, A., and Cocivera, M. (1971). Electron paramagnetic resonance spectra of semiquinone intermediates observed during photooxidation of phenol in water. *J. Am. Chem. Soc.* 93 (25): 7102–7103.

1126 Cocivera, M., Groen, A., and Tomkiewi, M. (1972). Electron-paramagnetic resonance and nuclear spin polarization study of photooxidation of phenol in water. *J. Am. Chem. Soc.* 94 (19): 6598–6604.

1127 Brahmia, O. and Richard, C. (2005). Photochemical transformation of 1-naphthol in aerated aqueous solution. *Photochem. Photobiol. Sci.* 4 (6): 454–458.

1128 Okamoto, K., Yamamoto, Y., Tanaka, H. et al. (1985). Heterogeneous photocatalytic decomposition of phenol over TiO_2 powder. *Bull. Chem. Soc. Jpn.* 58 (7): 2015–2022.

1129 Devi, L. and Kavitha, R. (2014). Enhanced photocatalytic activity of sulfur doped TiO_2 for the decomposition of phenol: a new insight into the bulk and surface modification. *Mater. Chem. Phys.* 143 (3): 1300–1308.

1130 Watanabe, N., Horikoshi, S., Kawabe, H. et al. (2003). Photodegradation mechanism for bisphenol A at the TiO_2/H_2O interfaces. *Chemosphere* 52 (5): 851–859.

1131 Pajares, A., Bregliani, M., Montaña, M. et al. (2010). Visible-light promoted photoprocesses on aqueous gallic acid in the presence of riboflavin. Kinetics and mechanism. *J. Photochem. Photobiol., A* 209 (2–3): 89–94.

1132 Schmidt, R.J. (2005). Industrial catalytic processes – phenol production. *Appl. Catal., A* 280 (1): 89–103.

1133 Ohkubo, K., Kobayashi, T., and Fukuzumi, S. (2011). Direct oxygenation of benzene to phenol using quinolinium ions as homogeneous photocatalysts. *Angew. Chem. Int. Ed.* 50 (37): 8652–8655.

1134 Foote, C.S., Denny, R.W., Weaver, L. et al. (1970). Quenching of singlet oxygen. *Ann. N.Y. Acad. Sci.* 171 (1): 139–148.

1135 Foote, C.S. and Peters, J.W. (1971). Chemistry of singlet oxygen. 14. Reactive intermediate in sulfide photooxidation. *J. Am. Chem. Soc.* 93 (15): 3795–3796.

1136 Waluk, J. (2012). Photooxygenations of sulfur compounds. In: *CRC Handbook of Organic Photochemistry and Photobiology*, 3e (ed. A. Griesbeck, M. Oelgemöller and F. Ghetti), 789–808. Boca Raton, FL: CRC Press.

1137 Corey, E.J. and Ouannes, C. (1976). Intermediary in oxidation of sulfureted benzyl alkyl-compounds by singlet oxygen. *Tetrahedron Lett.* (47): 4263–4266.

1138 Bonesi, S.M., Mella, M., d'Alessandro, N. et al. (1998). Photosensitized oxygenation of benzyl ethyl sulfide. *J. Org. Chem.* 63 (26): 9946–9955.

1139 Toutchkine, A. and Clennan, E.L. (2000). A new mechanism for oxidative C—S bond cleavage during reactions of singlet oxygen with organic sulfides: electronically dictated reaction selectivity in the persulfoxide intermediate. *J. Am. Chem. Soc.* 122 (8): 1834–1835.

1140 Baciocchi, E., Del Giacco, T., Lanzalunga, O., and Lapi, A. (2007). Singlet oxygen promoted carbon–heteroatom bond cleavage in dibenzyl sulfides and tertiary dibenzylamines. Structural effects and the role of exciplexes. *J. Org. Chem.* 72 (25): 9582–9589.

1141 Baciocchi, E., Del Giacco, T., and Lapi, A. (2004). Oxygenation of benzyldimethylamine by singlet oxygen. Products and mechanism. *Org. Lett.* 6 (25): 4791–4794.

1142 Baciocchi, E., Del Giacco, T., and Lapi, A. (2006). Quenching of singlet oxygen by tertiary aliphatic amines. Structural effects on rates and products. *Helv. Chim. Acta* 89 (10): 2273–2280.

1143 (a) Weller, A. (1982). Exciplex and radical pairs in photochemical electron-transfer. *Pure Appl. Chem.* 54 (10): 1885–1888. (b) Julliard, M. and Chanon, M. (1983). Photoelectron-transfer catalysis – its connections with thermal and electrochemical analogs. *Chem. Rev.* 83 (4): 425–506. (c) Kavarnos, G.J. (1993). *Fundamentals of Photoinduced Electron Transfer*, 1–359. Weinheim: Wiley-VCH. (d) Festov, S.V., Mikhailova, V.A., and Ivanov, A.I. (2016). Non-equilibrium effects in ultrafast photoinduced charge transfer kinetics. *J. Photochem. Photobiol., C* 29: 48–72. (e) Dereka, B., Koch, M., and Vauthey, E. (2017). Looking at photoinduced charge transfer processes in the IR:

answers to several long-standing questions. *Acc. Chem. Res.* 50 (2): 426–434.

1144 (a) Chen, M., Zhong, M.J., and Johnson, J.A. (2016). Light-controlled radical polymerization: mechanisms, methods, and applications. *Chem. Rev.* 116 (17): 10167–10211. (b) Zhang, X.-P., Xi, W.-X., Huang, S.-J. et al. (2017). Wavelength-selective sequential polymer network formation controlled with a two-color responsive initiation system. *Macromolecules* 50 (15): 5652–5660.

1145 Cohen, I., Mishra, A.K., Parvari, G. et al. (2017). Sunlight assisted direct amide fromation via a charge-transfer complex. *J. Chem. Soc., Chem. Commun.* 53: 10128–10131.

1146 Weiss, J. and Fischgold, H. (1936). The mechanisms of the fluorescence obliteration in solutions. *Z. Phys. Chem. Abt. B:* 32 (2): 135–138.

1147 (a) Bryant, D.A. and Frigaard, N.U. (2006). Prokaryotic photosynthesis and phototrophy illuminated. *Trends Microbiol.* 14 (11): 488–496. (b) Johnson, M.P. (2016). Photosynthesis. *Essays Biochem.* 60: 255–273.

1148 (a) Benniston, A.C. and Harriman, A. (1994). Dynamics of charge-transfer and recombination in a covalently-linked, face-to-face electron donor-acceptor complex. *J. Am. Chem. Soc.* 116 (23): 11531–11537. (b) Roth, H.D. (2008). Biradicals by triplet recombination of radical ion pairs. *Photochem. Photobiol. Sci.* 7 (5): 540–546. (c) Sasaki, S., Drummen, G.P.C., and Konishi, G. (2016). Recent advances in twisted intramolecular charge transfer (TICT) fluorescence and related phenomena in materials chemistry. *J. Mater. Chem. C* 4 (14): 2731–2743.

1149 Rehm, D. and Weller, A. (1969). Kinetics and mechanics of electron transfer during fluorescence quenching in acetonitrile. *Ber. Bunsen Ges. Phys. Chem.* 73 (8–9): 834–839.

1150 Rehm, D. and Weller, A. (1970). Kinetics of fluorescence quenching by electron and H-atom transfer. *Isr. J. Chem.* 8 (2): 259–271.

1151 Balzani, V., Bolletta, F., and Scandola, F. (1980). Vertical and nonvertical energy-transfer processes – a general classical treatment. *J. Am. Chem. Soc.* 102 (7): 2152–2163.

1152 Weller, A. (1982). Photoinduced electron-transfer in solution – exciplex and radical-ion-pair formation free enthalpies and their solvent dependence. *Z. Phys. Chem.* 133 (1): 93–98.

1153 Weller, A. (1982). Mechanism and spin-dynamics of photoinduced electron-transfer reactions. *Z. Phys. Chem.* 130 (2): 129–138.

1154 (a) Pan, Y., Fu, Y., Liu, S.-X. et al. (2006). Studies on photoinduced H-atom and electron transfer reactions of o-naphthoquinones by laser flash photolysis. *J. Phys. Chem. A* 110 (23): 7316–7322. (b) Farid, S., Dinnocenzo, J.P., Merkel, P.B. et al. (2011). Reexamination of the Rehm–Weller data set reveals electron transfer quenching that follows a Sandros–Boltzmann dependence on free energy. *J. Am. Chem. Soc.* 133 (30): 11580–11587.

1155 Marcus, R.A. (1956). Electrostatic free energy and other properties of states having nonequilibrium polarization. 1. *J. Chem. Phys.* 24 (5): 979–989.

1156 Marcus, R.A. (1956). On the theory of oxidation-reduction reactions involving electron transfer. 1. *J. Chem. Phys.* 24 (5): 966–978.

1157 Marcus, R.A. (1964). Chemical & electrochemical electron-transfer theory. *Annu. Rev. Phys. Chem.* 15: 155–196.

1158 Marcus, R.A. (1993). Electron-transfer reactions in chemistry – theory and experiment (Nobel lecture). *Angew. Chem. Int. Ed. Engl.* 32 (8): 1111–1121.

1159 Marcus, R.A. and Sutin, N. (1985). Electron transfers in chemistry and biology. *Biochim. Biophys. Acta* 811 (3): 265–322.

1160 Mattay, J. (1989). Photoinduced electron transfer in organic synthesis. *Synthesis* (4): 233–252.

1161 Pandey, G. (1993). Photoinduced electron-transfer (PET) in organic synthesis. In: *Photoinduced Electron Transfer V*, Topics in Current Chemistry, vol. 168 (ed. J. Mattay), 175–221. Berlin, Heidelberg, Germany: Springer Verlag.

1162 Hoffmann, N. (2008). Efficient photochemical electron transfer sensitization of homogeneous organic reactions. *J. Photochem. Photobiol., C* 9 (2): 43–60.

1163 Griesbeck, A.G., Hoffmann, N., and Warzecha, K.D. (2007). Photoinduced-electron-transfer chemistry: from studies on PET processes to applications in natural product synthesis. *Acc. Chem. Res.* 40 (2): 128–140.

1164 Cho, D.W., Yoon, U.C., and Mariano, P.S. (2011). Studies leading to the development of a single-electron transfer (SET) photochemical strategy for syntheses of macrocyclic polyethers, polythioethers, and polyamides. *Acc. Chem. Res.* 44 (3): 204–215.

1165 Russell, G.A. and Danen, W.C. (1966). Coupling reactions of 2-nitro-2-propyl anion. *J. Am. Chem. Soc.* 88 (23): 5663–5663.

1166 Kornblum, N. (1975). Substitution reactions which proceed via radical-anion intermediates. *Angew. Chem. Int. Ed. Engl.* 14 (11): 734–745.

1167 Miller, L.L. and Narang, R.S. (1970). Induced photolysis of DDT. *Science* 169 (3943): 368–370.

1168 (a) Miller, J.B. and Salvador, J.R. (2002). Photoinduced electron-transfer substitution reactions via unusual charge-transfer intermediates. *J. Org. Chem.* 67 (2): 435–442. (b) Romero, N.A. and Nicewicz, D.A. (2016). Organic photoredox catalysis. *Chem. Rev.* 116 (17): 10075–10166.

1169 (a) Cismesia, M.A., Ischay, M.A., and Yoon, T.P. (2013). Reductive cyclizations of nitroarenes to hydroxamic acids by visible light photoredox catalysis. *Synthesis* 45 (19): 2699–2705. (b) Yoon, T.P. (2013). Visible light photocatalysis: the development of photocatalytic radical-ion cycloadditions. *ACS Catal.* 3 (5): 895–902. (c) Espelt, L.R., Wiensch, E.M., and Yoon, T.P. (2013). Brønsted acid co-catalysts in photocatalytic radical addition of α-amino C—H bonds across Michael acceptors. *J. Org. Chem.* 78 (8): 4107–4114. (d) Tyson, E.L., Niemeyer, Z.L., and Yoon, T.P. (2014). Redox mediators in visible light photocatalysis: photocatalytic radical thiol-ene additions. *J. Org. Chem.* 79 (3): 1427–1436. (e) Farney, E.P. and Yoon, T.P. (2014). Visible-light sensitization of vinyl azides by transition-metal photocatalysis. *Angew. Chem. Int. Ed.* 53 (3): 793–797. (f) Skubi, K.L., Blum, T.R., and Yoon, T.P. (2016). Dual catalysis strategies in photochemical synthesis. *Chem. Rev.* 116 (17): 10035–10074. (g) Savateev, A., Ghosh, I., Koenig, B., and Antonietti, M. (2018). Photoredox catalytic organic transformations using heterogeneous carbon nitrides. *Angew. Chem. Int. Ed.* 57 (49): 15936–15947.

1170 (a) Twilton, J., Le, C., Zhang, P. et al. (2017). The merger of transition metal and photocatalysis. *Nat. Rev. Chem.* 1 (7): UNSP 0052. (b) Xie, J., Jin, H.-M., and Hasmi, A.S.K. (2017). The recent achievements of redox-neutral radical C—C cross-coupling enabled by visible-light. *Chem. Soc. Rev.* 46 (17): 5193–5203. (c) Parasram, M. and Gevorgyan, V. (2017). Visible light-induced transition metal-catalyzed transformations: beyond conventional photosensitizers. *Chem. Soc. Rev.* 46 (20): 6227–6240.

1171 (a) Wu, J., LI, J.-W., LI, H., and Zhu, C.-Y. (2017). Visible light photoredox catalysis mediated elementary steps in organometallic reaction. *Chin. J. Org. Chem.* 37 (9): 2203–2210. (b) Ruan, L.-H., Dong, Z.-C., Chen, C.-X. et al. (2017). Recent progress on the nickel/photoredox dual catalysis. *Chin. J. Org. Chem.* 37 (10): 2544–2554.

1172 Lee, K.N. and Ngai, M.-Y. (2017). Recent developments in transition-metal photoredox-catalysed reactions of carbonyl derivatives. *J. Chem. Soc., Chem. Commun.* (98): 53, 13093–13112.

1173 Eivgi, O. and Lemcoff, N.G. (2018). Turning the light on: recent developments in photoinduced olefin metathesis. *Synthesis* 50 (1): 49–63.

1174 Ischay, M.A., Anzovino, M.E., Du, J., and Yoon, T.P. (2008). Efficient visible light photocatalysis of (2+2) enone cycloadditions. *J. Am. Chem. Soc.* 130 (39): 12886–12887.

1175 Du, J. and Yoon, T.P. (2009). Crossed intermolecular (2+2)-cycloadditions of acyclic enones via visible light photocatalysis. *J. Am. Chem. Soc.* 131 (41): 14604–14605.

1176 Roh, Y., Jang, H.Y., Lynch, V. et al. (2002). Anion-radical chain cycloaddition of tethered enones: intramolecular cyclobutanation and Diels–Alder cycloaddition. *Org. Lett.* 4 (4): 611–613.

1177 Yang, J.K., Felton, G.A.N., Bauld, N.L., and Krische, M.J. (2004). Chemically induced anion-radical cycloadditions: intramolecular cyclobutanation of bis(enones) via homogeneous electron transfer. *J. Am. Chem. Soc.* 126 (6): 1634–1635.

1178 Hurtley, A.E., Cismesia, M.A., Ischay, M.A., and Yoon, T.R. (2011). Visible light photocatalysis of radical anion hetero-Diels–Alder cycloadditions. *Tetrahedron* 67 (24): 4442–4448.

1179 Du, J., Espelt, L.R., Guzei, I.A., and Yoon, T.P. (2011). Photocatalytic reductive cyclizations of enones: divergent reactivity of photogenerated radical and radical-anion intermediates. *Chem. Sci.* 2 (11): 2115–2119.

1180 Zlotorzynska, M. and Sammis, G.M. (2011). Photoinduced electron-transfer promoted redox fragmentation of N-alkoxyphthalimides. *Org. Lett.* 13 (23): 6264–6267.

1181 (a) Nicewicz, D.A. and MacMillan, D.W. (2008). Merging photoredox catalysis with organocatalysis: the direct asymmetric alkylation of aldehydes. *Science* 322 (5898): 77–80. (b) Nagib, D.A., Scott, M.E., and MacMillan, D.W. (2009). Enantioselective α-trifluoromethylation of aldehydes via photoredox organocatalysis. *J. Am. Chem. Soc.* 131 (31): 10875–10877. (c) Welin, E.R., Warkentin, A.A., Conrad, J.C., and MacMillan, D.W.C. (2015). Enantioselective α-alkylation of aldehydes by photoredox organocatalysis: rapid access to pharmacophore fragments from β-cyanoaldehydes. *Angew. Chem. Int. Ed.* 54 (33): 9668–9672. (d) Ye, Y. and Sanford, M.S. (2012). Merging visible-light photocatalysis and transition-metal catalysis in the Cu-catalyzed trifluoromethylation of boronic acids with CF_3I. *J. Am. Chem. Soc.* 134 (22): 9034–9037. (e) see also: Gualandi, A., Marchini, M., Mengozzi, L. et al. (2015). Organocatalytic enantioselective alkylation of aldehydes with $[Fe(bpy)_3]Br_2$ catalyst and visible light. *ACS Catal.* 5 (10): 5927–5931. (f) Yerien, D.E., Cooke, M.V., Garcia Vior, M.C. et al. (2019). Radical fluoroalkylation reactions of (hetero)arenes and sulfides under red light photocatalysis. *Org. Biomol. Chem.* 17 (15): 3741–3746.

1182 Capacci, A.-G., Malinowski, J.T., McAlpine, N.J. et al. (2017). Direct, enantioselective α-alkylation of aldehydes using simple olefins. *Nat. Chem.* 9: 1073–1077.

1183 Yasu, Y., Koike, T., and Akita, M. (2012). Three-component oxytrifluoromethylation of alkenes: highly efficient and regioselective difunctionalization of C=C bonds mediated by photoredox catalysts. *Angew. Chem. Int. Ed.* 51 (38): 9567–9571.

1184 (a) Furst, L., Matsuura, B.S., Narayanam, J.M. et al. (2010). Visible light-mediated intermolecular C—H functionalization of electron-rich heterocycles with malonates. *Org. Lett.* 12 (13): 3104–3107. (b) Huo, H.-H., Shen, X.-D., Wang, C.-Y. et al. (2014). Asymmetric photoredox transition-metal catalysis activated by visible light. *Nature* 515 (7525): 100–103. (c) Telitel, S., Dumur, F., Telitel, S. et al. (2015). Photoredox catalysis using a new iridium complex as an efficient toolbox for radical, cationic and controlled polymerizations under soft blue to green lights. *Polym. Chem.* 6 (4): 613–624. (d) Wang, C.-Y., Qin, J., Shen, X.-D. et al. (2016). Asymmetric radical-radical cross-coupling through visible-light-activated Ir catalysis. *Angew. Chem. Int. Ed.* 55 (2): 685–688. (e) Huo, H.-H., Huang, X.-Q., Shen, X.-D. et al. (2016). Visible-light-activated enantioselective perfluoroalkylation with a chiral iridium photoredox catalyst. *Synlett* 27 (5): 749–753. (f) Bahamonde, A. and Melchiorre, P. (2016). Mechanism of the stereoselective α-alkylation of aldehydes driven by the photochemical activity of enamines. *J. Am. Chem. Soc.* 138 (25): 8019–8030. (g) Franchino, A., Rinaldi, A., and Dixon, D.J. (2017). α-Alkylation of ketimines using visible light photoredox catalysis. *RSC Adv.* 7 (69): 43655–43659. (h) Zhang, H.-H. and Yu, S.-Y. (2017). Radical alkylation of imines with 4-alkyl-1,4-dihydropyridines enabled by photoredox/Brønsted acid cocatalysis. *J. Org. Chem.* 82 (19): 9995–10006. (i) Wang, D.-H., Zhang, L., and Luo, S.-Z. (2017). Photo-induced catalytic asymmetric free radical reactions. *Acta Chim. Sinica* 75 (1): 22–33.

1185 Zhang, X.-H. and MacMillan, D.W.C. (2017). Direct aldehyde C—H arylation and alkylation via the combination of nickel, hydrogen atom transfer, and photoredox catalysis. *J. Am. Chem. Soc.* 139 (33): 11353–11356.

1186 Lu, Z., Shen, M., and Yoon, T.P. (2011). (3+2)-Cycloadditions of aryl cyclopropyl ketones by visible light photocatalysis. *J. Am. Chem. Soc.* 133 (5): 1162–1164.

1187 Fukuzumi, S., Mochizuki, S., and Tanaka, T. (1990). Photocatalytic reduction of phenacyl halides by 9,10-dihydro-10-methylacridine – control between the reductive and oxidative quenching pathways of tris(bipyridine)Ru complex utilizing an acid catalysis. *J. Phys. Chem.* 94 (2): 722–726.

1188 Zheng, C. and You, S.L. (2012). Transfer hydrogenation with Hantzsch esters and related organic hydride donors. *Chem. Soc. Rev.* 41 (6): 2498–2518.

1189 (a) Narayanam, J.M.R., Tucker, J.W., and Stephenson, C.R.J. (2009). Electron-transfer photoredox catalysis: development of a tin-free reductive dehalogenation reaction. *J. Am. Chem. Soc.* 131 (25): 8756–8757. (b) McTieman, C.D., Pitre, S.P., Ismaili, H., and Scaiano, J.C. (2014). Heterogeneous light-mediated reductive dehalogenations and cyclizations utilizing Pt nanoparticles on titania ($PtNP@TiO_2$). *Adv. Synth. Catal.* 356 (13): 2819–2824. (c) McTieman, C.D., Pitre, S.P., and Scaiano, J.C. (2014). Photocatalytic dehalogenation of vicinal dibromo compounds utilizing sexithiophene and visible-light irradiation. *ACS Catal.* 4 (11): 4034–4039. (d) Martinez-Haya, R., Miranda, M.A., and Marin, M.L. (2017). Metal-free photocatalytic reductive dehalogenation using visible-light: a time-resolved mechanistic study. *Eur. J. Org. Chem.* (15): 2164–2169. (e) Marin, M., Miranda, M.A., and Marin, M.L. (2017). A comprehensive mechanistic study on the visible-light photocatalytic reductive dehalogenation of haloaromatics mediated by $Ru(bpy)_3Cl_2$. *Catal. Sci. Technol.* 7: 4852–4858.

1190 van Bergen, T.J., Hedstrand, D.M., Kruizinga, W.H., and Kellogg, R.M. (1979). Chemistry of dihydropyridines. 9. Hydride transfer from 1,4-dihydropyridines to sp^3-hybridized carbon in sulfonium salts and activated halides – studies with NAD(P)H models. *J. Org. Chem.* 44 (26): 4953–4962.

1191 Nakamura, K., Fujii, M., Mekata, H. et al. (1986). NAD(P)+-NAD(P)H model. 58. Desulfonylation of β-keto sulfones with the Hantzsch ester, an NAD(P)H model. *Chem. Lett.* (1): 87–88.

1192 (a) Nguyen, J.D., D'Amato, E.M., Narayanam, J.M., and Stephenson, C.R. (2012). Engaging unactivated alkyl, alkenyl and aryl iodides in visible-light-mediated free radical reactions. *Nat. Chem.* 4 (10): 854–859. (b) Staveness, D.M., Bosque, I., and Stephenson, C.R.J. (2016). Free radical chemistry enabled by visible light-induced electron transfer. *Acc. Chem. Res.* 49 (10): 2295–2306.

1193 Nguyen, J.D., Reiss, B., Dai, C., and Stephenson, C.R.J. (2013). Batch to flow deoxygenation using visible light photoredox catalysis. *J. Chem. Soc., Chem. Commun.* 49 (39): 4352–4354.

1194 Larraufie, M.H., Pellet, R., Fensterbank, L. et al. (2011). Visible-light-induced photoreductive generation of radicals from epoxides and aziridines. *Angew. Chem. Int. Ed.* 50 (19): 4463–4466.

1195 Guindon, Y., Jung, G., Guerin, B., and Ogilvie, W.W. (1998). Hydrogen and allylation transfer reactions in acyclic free radicals. *Synlett* (3): 213–220.

1196 Pandey, G., Gaikwad, A.L., and Gadre, S.R. (2006). Chiral 6-phenyl-2,3-bismethylenemethoxycarbonyl-[1,4]-dioxane as a designer synthon for an efficient and short synthesis of optically pure 2,6-dioxabicyclo[3.3.0] octane-3,7-dione. *Tetrahedron Lett.* 47 (5): 701–703.

1197 Cossy, J., Ranaivosata, J.L., Bellosta, V. et al. (1995). Reductive oxa ring opening of 7-oxabicyclo[2.2.1]heptan-2-ones. Synthesis of C-α-galactosides of carbapentopyranoses. *J. Org. Chem.* 60 (26): 8351–8359.

1198 Lange, G.L., Decicco, C.P., Willson, J., and Strickland, L.A. (1989). Ring expansions of (2+2)-photoadducts – potential applications in the synthesis of triquinane and taxane skeletons. *J. Org. Chem.* 54 (8): 1805–1810.

1199 Tucker, J.W., Nguyen, J.D., Narayanam, J.M. et al. (2010). Tin-free radical cyclization reactions initiated by visible light photoredox catalysis. *J. Chem. Soc., Chem. Commun.* 46 (27): 4985–4987.

1200 Edson, J.B., Spencer, L.P., and Boncella, J.M. (2011). Photorelease of primary aliphatic and aromatic amines by visible-light-induced electron transfer. *Org. Lett.* 13 (23): 6156–6159.

1201 Neunteuf, R.A. and Arnold, D.R. (1973). Radical-ions in photochemistry. 1. 1,1-Diphenylethylene cation-radical. *J. Am. Chem. Soc.* 95 (12): 4080–4081.

1202 (a) Gieseler, A., Steckhan, E., Wiest, O., and Knoch, F. (1991). Photochemically induced radical-cation Diels–Alder reaction of indole and electron-rich dienes. *J. Org. Chem.* 56 (4): 1405–1411. (b) Wiest, O., Steckhan, E., and Grein, F. (1992). Selectivity in radical-cation Diels–Alder reactions of índole and electron-rich dienes – a semiempirical approach. *J. Org. Chem.* 57 (14): 4034–4037.

1203 (a) Lin, S., Ischay, M.A., Fry, C.G., and Yoon, T.P. (2011). Radical-cation Diels–Alder cycloadditions by visible light photocatalysis. *J. Am. Chem. Soc.* 133 (48): 19350–19353. (b) Pitre, S.P., Scaiano, J.C., and Yoon, T.P. (2017). Photocatalytic indole Diels–Alder cycloadditions mediated by heterogeneous Pt-modified titanium dioxide. *ACS Catal.* 7 (10): 6440–6444.

1204 Ischay, M.A., Lu, Z., and Yoon, T.P. (2010). $[\pi^2+\pi^2]$-Cycloadditions by oxidative visible light photocatalysis. *J. Am. Chem. Soc.* 132 (25): 8572–8574.

1205 McNally, A., Prier, C.K., and MacMillan, D.W.C. (2011). Discovery of an α-amino C—H arylation reaction using the strategy of accelerated serendipity. *Science* 334 (6059): 1114–1117.

1206 Kohls, P., Jadhav, D., Pandey, G., and Reiser, O. (2012). Visible light photoredox catalysis: generation and addition of N-aryltetrahydroisoquinoline-derived α-amino radicals to Michael acceptors. *Org. Lett.* 14 (3): 672–675.

1207 (a) Marinkovic, S. and Hoffmann, N. (2001). Efficient radical addition of tertiary amines to electron-deficient alkenes using semiconductors as photochemical sensitisers. *J. Chem. Soc., Chem. Commun.* (17): 1576–1577. (b) Hoffmann, N., Bertrand, S., Marinkovic, S., and Pesch, J. (2006). Efficient radical addition of tertiary amines to alkenes using photochemical electron transfer. *Pure Appl. Chem.* 78 (12): 2227–2246. (c) Miyake, Y., Nakajima, K., and Nishibayashi, Y. (2012). Visible-light-mediated utilization of α-aminoalkyl radicals: addition to electron-deficient alkenes using photoredox catalysts. *J. Am. Chem. Soc.* 134 (7): 3338–3341. (d) Nakajima, K., Miyake, Y., and Nishibayashi, Y. (2016). Synthetic utilization of α-aminoalkyl radicals and related species in visible light photoredox patalysis. *Acc. Chem. Res.* 49 (9): 1946–1956.

1208 Hamilton, D.S. and Nicewicz, D.A. (2012). Direct catalytic anti-Markovnikov hydroetherification of alkenols. *J. Am. Chem. Soc.* 134 (45): 18577–18580.

1209 Mella, M., Fagnoni, M., Freccero, M. et al. (1998). New synthetic methods via radical-cation fragmentation. *Chem. Soc. Rev.* 27 (1): 81–89.

1210 Castellano, A., Catteau, J.P., and Lablache-Combier, A. (1975). Mechanism of photochemical substitution of six-membered ring monoazaaromatic compounds by methanol in neutral and HCl acidified medium. *Tetrahedron* 31 (18): 2255–2261.

1211 Kosower, E.M. (1965). Flash photolysis of a pyridinium iodine through charge-transfer band. *Tetrahedron Lett.* (50): 4481–4485.

1212 Cozzens, R.F. and Gover, T.A. (1970). Flash photolysis of charge-transfer band of 1-methylpyridinium, 1-methylcollodinium and 1-methylquinolinium iodides. *J. Phys. Chem.* 74 (15): 3003–3006.

1213 Mariano, P.S. (1983). Electron-transfer mechanisms in photochemical transformations of iminium salts. *Acc. Chem. Res.* 16 (4): 130–137.

1214 Cano-Yelo, H. and Deronzier, A. (1984). Photo-oxidation of some carbinols by the Ru(II) polypyridyl complex aryl diazonium salt system. *Tetrahedron Lett.* 25 (48): 5517–5520.

1215 Zou, Y.-Q., Chen, J.-R., Liu, X.-P. et al. (2012). Highly efficient aerobic oxidative hydroxylation of arylboronic acids: photoredox catalysis using visible light. *Angew. Chem. Int. Ed.* 51 (3): 784–788.

1216 (a) Condie, A.G., González-Gómez, J.C., and Stephenson, C.R.J. (2010). Visible-light photoredox catalysis: Aza-Henry reactions via C—H functionalization. *J. Am. Chem. Soc.* 132 (5): 1464–1465. (b) see also: Bartling, H., Eisenhofer, A., König, B., and Gschwind, R.M. (2016). The photocatalyzed Aza-Henry reaction of N-aryltetrahydroisoquinolines: comprehensive mechanism, $H^{\bullet}$ versus H^{+}-abstraction, and background reactions. *J. Am. Chem. Soc.* 138 (36): 11860–11871.

1217 DiRocco, D.A. and Rovis, T. (2012). Catalytic asymmetric α-acylation of tertiary amines mediated by a dual catalysis mode: *N*-heterocyclic carbene and photoredox catalysis. *J. Am. Chem. Soc.* 134 (19): 8094–8097.

1218 Cheng, Y., Yang, J., Qu, Y., and Li, P. (2012). Aerobic visible-light photoredox radical C—H functionalization: catalytic synthesis of 2-substituted benzothiazoles. *Org. Lett.* 14 (1): 98–101.

1219 Hamada, T., Ishida, H., Usui, S. et al. (1993). A novel photocatalytic asymmetric synthesis of (R)-(+)-1,1′-bi-2-naphthol derivatives by oxidative coupling of 3-substituted-2-naphthol with Δ-$[Ru(Menbpy)_3]_2^{+}$ [Menbpy = 4,4′-Di(1R,2S,5R)-(−)-menthoxycarbonyl-2,2′-bipyridine], which possesses molecular helicity. *J. Chem. Soc., Chem. Commun.* (11): 909–911.

1220 Hamada, T., Ishida, H., Usui, S. et al. (1994). Enantioselective and photocatalytic oxidation of 1,1′-bi-2-naphthol with a chiral Ru complex which includes molecular helicity. *J. Mol. Catal.* 88 (1): L1–L5.

1221 (a) Eaton, D.F. (1984). Electron-transfer induced photofragmentation as a route to free-radicals. *Pure Appl. Chem.* 56 (9): 1191–1202. (b) Staveness, D., Bosque, I., and Stephenson, C.R.J. (2016). Free radical chemistry enabled by visible light-induced electron transfer. *Acc. Chem. Res.* 49 (10): 2295–2306. (c) Camussi, I., Mannucci, B., Speltini, A. et al. (2019). g-C_3N_4 – Singlet oxygen made easy for organic synthesis: scope and limitations. *ACS Sust. Chem. Eng.* 7 (9): 8176–8182.

1222 Albini, A., Mella, M., and Freccero, M. (1994). A New method in radical chemistry – generation of radicals by photoinduced electron-transfer and fragmentation of the radical-cation. *Tetrahedron* 50 (3): 575–607.

1223 Bertrand, S., Hoffmann, N., and Pete, J.P. (2000). Highly efficient and stereoselective radical addition of tertiary amines to electron-deficient alkenes – application to the enantioselective synthesis of necine bases. *Eur. J. Org. Chem.* (12): 2227–2238.

1224 Bauer, A., Westkamper, F., Grimme, S., and Bach, T. (2005). Catalytic enantioselective reactions driven by photoinduced electron transfer. *Nature* 436 (7054): 1139–1140.

1225 (a) Okada, K., Okamoto, K., Morita, N. et al. (1991). Photosensitized decarboxylative Michael addition through *N*-(acyloxy)phthalimides via an electron-transfer mechanism. *J. Am. Chem. Soc.* 113 (24): 9401–9402. (b) Okada, K., Okubo, K., Morita, N., and Oda, M. (1992). Reductive decarboxylation of *N*-(acyloxy)phthalimides via redox-initiated radical chain mechanism. *Tetrahedron Lett.* 33 (48): 7377–7380. (c) Okada, K., Okubo, K., Morita, N., and Oda, M. (1993). Redox-mediated decarboxylative photo-phenylselenenylation of N-acyloxyphthalimides. *Chem. Lett.* (12): 2021–2024.

1226 (a) Tang, Q., Liu, X.-B., Liu, S. et al. (2015). *N*-(Acyloxy)phthalimides as tertiary alkyl radical precursors in the visible light photocatalyzed tandem radical cyclization of N-arylacrylamides to 3,3-dialkyl substituted oxindoles. *RSC Adv.* 5 (108): 89009–89014. (b) Pratsch, G., Lackner, G.L., and Overman, L.E. (2015). Constructing quaternary carbons from *N*-(acyloxy)phthalimide precursors of tertiary radicals using visible-light photocatalysis. *J. Org. Chem.* 80 (12): 6025–6036. (c) Chen, W.-M., Shang, R., Fu, M.-C., and Fu, Y. (2017). Photoredox-catalysed decarboxylative alkylation of *N*-heteroarenes with *N*-(acyloxy)phthalimides. *Chem. Eur. J.* 23 (11): 2537–2541.

1227 Schwarz, J. and König, B. (2017). Decarboxylative alkynylation of biomass-derived compounds by metal-free visible light photocatalysis. *ChemPhotoChem* 1 (5): 237–242.

1228 (a) Goncalves, L.M., Bermudez, V.d.Z., Ribeiro, H.A., and Mendes, A.M. (2008). Dye-sensitized solar cells: a safe bet for the future. *Energy Environ. Sci.* 1 (6): 655–667. (b) Sugathan, V., John, E., and Sudhakar, K. (2015). Recent improvements in dye sensitized solar cells: a review. *Renewable Sustainable Energy Rev.* 52: 54–64. (c) Wang, J.-Y., Liu, K., Ma, L.-C., and Zhan, X.-W. (2016). Triarylamine: versatile platform for organic, dye-sensitized, and perovskite solar cells. *Chem. Rev.* 116 (23): 14675–14725. (d) Gong, J.-W., Sumathy, K., Qiao, Q.-Q., and Zhou, Z.-P. (2017). Review on dye-sensitized solar cells (DSSCs): advanced techniques and research trends. *Renewable Sustainable Energy Rev.* 68 (Part 1): 234–246. (e) Kumara, N.T.R.N., Lim, A., Lim, C.M. et al. (2017). Recent progress and utilization of natural pigments in dye sensitized solar cells: a review. *Renewable Sustainable Energy Rev.* 78: 301–317. (f) Wu, J.-H., Lan, Z., Lin, J.-M. et al. (2017). Counter electrodes in dye-sensitized solar cells. *Chem. Soc. Rev.* 46 (19): 5975–6023. (g) Yeoh, M.E. and Chan, K.Y. (2017). Recent advances in photo-anode for dye-sensitized solar cells: a review. *Int. J. Energy Res.* 41 (15): 2446–2467.

1229 O'Regan, B. and Grätzel, M. (1991). A low-cost, high-efficiency solar-cell based on dye-sensitized colloidal TiO_2 films. *Nature* 353 (6346): 737–740.

1230 (a) Peter, L.M. (2011). The Grätzel cell: where next? *J. Phys. Chem. Lett.* 2 (15): 1861–1867. (b) Weerasinghe, H.C., Huang, F., and Cheng, Y.-B. (2013). Fabrication of flexible dye sensitized solar cells on plastic substrates. *Nano Energy* 2 (2): 174–189. (c) Parisi, M.L., Maranghi, S., and Basosi, R. (2014). The evolution of the dye sensitized solar cells from Grätzel prototype to up-scaled solar applications: a life cycle assessment approach. *Renewable Sustainable Energy Rev.* 39: 124–138. (d) Higashino, T. and Imahori, H. (2015). Porphyrins as excellent dyes for dye-sensitized solar cells: recent developments and insights. *J. Chem. Soc., Dalton Trans.* 44 (2): 448–463. (e) Galliano, S., Bella, F., Gerbaldi, C. et al. (2017). Photoanode/electrolyte interface stability in aqueous dye-sensitized solar cells. *Energy Technol.* 5 (2): 300–311.

1231 Nazeeruddin, M.K., Kay, A., Rodicio, I. et al. (1993). Conversion of light to electricity by *cis*-X_2bis(2,2′-bipyridyl-4,4′-dicarboxylate)Ru(II) charge-transfer sensitizers (X=Cl-, Br-, I-, CN-, and SCN-) on nanocrystalline TiO_2 electrodes. *J. Am. Chem. Soc.* 115 (14): 6382–6390.

1232 Nazeeruddin, M.K., Péchy, P., and Grätzel, M. (1997). Efficient panchromatic sensitization of nanocrystalline TiO_2 films by a black dye based on a trithiocyanato-Ru complex. *J. Chem. Soc., Chem. Commun.* (18): 1705–1706.

1233 (a) Horiuchi, T., Miura, H., Sumioka, K., and Uchida, S. (2004). High efficiency of dye-sensitized solar cells based on metal-free indoline dyes. *J. Am. Chem. Soc.* 126 (39): 12218–12219. (b) Kim, B.-G., Chung, K., and Kim, J. (2013). Molecular design principle of all-organic dyes for dye-sensitized solar cells. *Chem. Eur. J.* 19 (17): 5220–5230. (c) Wu, Y. and Zhu, W. (2013). Organic sensitizers from D-π-A to D-A-π-A: effect of the internal electron-withdrawing units on molecular absorption, energy levels and photovoltaic performances. *Chem. Soc. Rev.* 42 (5): 2039–2058.

1234 Fleming, I. (1967). Absolute configuration and structure of chlorophyll. *Nature* 216 (5111): 151–152.

1235 Mathew, S., Yella, A., Gao, P. et al. (2014). Dye-sensitized solar cells with 13% efficiency achieved through the molecular engineering of porphyrin sensitizers. *Nat. Chem.* 6 (3): 242–247.

1236 Li, L.-L. and Diau, E.W.-G. (2013). Porphyrin-sensitized solar cells. *Chem. Soc. Rev.* 42 (1): 291–304.

1237 (a) Correa-Baena, J.P., Salita, M., Buonassisi, T. et al. (2017). Promises and challenges of perovskite solar cells. *Science* 358 (6364): 739–744. (b) Assadi, M.K., Bakhoda, S., Saldur, R., and Hanael, H. (2018). Recent progress in perovskite solar cells. *Renewable Sustainable Energy Rev.* 81 (Part 2): 2812–2822. (c) Mesquita, I., Andrade, L., and Mendes, A. (2018). Perovskite solar cells: materials, configurations and stability. *Renewable Sustainable Energy Rev.* 82 (Part 3): 2471–2487. (d) Wang, K., Zheng, L., and Zhu, T. et al. (2019). Efficient perovskite solar cells by hybrid perovskites incorporated with heterovalent neodymium cations. *Nano Energy* 61: 352–360.

1238 Grancini, G., Roldán-Carmona, C., Zimmermann, I. et al. (2017). One-year stable perovskite solar cells by 2D/3D interface engineering. *Nat. Commun.* 8: 15684.

1239 Trautz, M. (1905). Studies on chemiluminescence. *Z. Phys. Chem. Stoechiom. Verwandtschafts.* 53 (1): 1–111.

1240 Wegscheider, R. (1923). Chemiluminescence. *Recl. Trav. Chim. Pays-Bas* 42: 585–587.

1241 Rideal, E.K. (1929). Chemiluminescence. *Nature* 123: 417–419.

1242 Gundermann, K.-D. (1965). Chemilumineszenz bei organischen Verbindungen. *Angew. Chem. Int. Ed. Engl.* 4 (7): 566–580.

1243 (a) McCapra, F. (1966). Chemiluminescence of organic compounds. *Q. Rev. Chem. Soc.* 20 (4): 485–510. (b) Welsch, E. (2009). Chemiluminescence. *Chem. Rev. (Deddington, U.K.)* 19 (1): 3–7.

1244 Hastings, J.W. (1983). Biological diversity, chemical mechanisms, and the evolutionary origins of bioluminescent systems. *J. Mol. Evol.* 19 (5): 309–321.

1245 Haddock, S.H., Moline, M.A., and Case, J.F. (2010). Bioluminescence in the sea. *Ann. Rev. Mar. Sci.* 2: 443–493.

1246 Vrsansky, P., Chorvat, D., Fritzsche, I. et al. (2012). Light-mimicking cockroaches indicate tertiary origin of recent terrestrial luminescence. *Naturwissenschaften* 99 (9): 739–749.

1247 Matsumoto, M. (2004). Advanced chemistry of dioxetane-based chemiluminescent substrates originating from bioluminescence. *J. Photochem. Photobiol., C* 5 (1): 27–53.

1248 Ando, Y., Niwa, K., Yamada, N. et al. (2008). Firefly bioluminescence quantum yield and colour change by pH-sensitive green emission. *Nat. Photonics* 2 (1): 44–47.

1249 Rauhut, M.M. (1969). Chemiluminescence from concerted peroxide decomposition reactions. *Acc. Chem. Res.* 2 (3): 80–87.

1250 (a) Ocaña-González, J.A., Ramos-Payán, M., Fernández-Torres, R. et al. (2014). Application of chemiluminescence in the analysis of wastewaters – a review. *Talanta* 122: 214–222. (b) Seidel, M. and Niessner, R. (2014). Chemiluminescence microarrays in analytical chemistry: a critical review. *Anal. Bioanal. Chem.* 406 (23): 5589–5612. (c) Arakama, H. (2017). Development of highly sensitive analytical methods for biologically relevant materials and their pharmaceutical applications. *Chem. Pharm. Bull.* 65 (12): 1099–1112. (d) Tiwari, A. and Dhoble, S.J. (2018). Recent advances and developments on integrating nanotechnology with chemiluminescence assays. *Talanta* 180: 1–11.

1251 Baeyens, W.R.G., Schulman, S.G., Calokerinos, A.C. et al. (1998). Chemiluminescence-based detection: principles and analytical applications in flowing streams and in immunoassays. *J. Pharm. Biomed. Anal.* 17 (6–7): 941–953.

1252 (a) Kricka, L.J., Joos, T., and Fortina, P. (2003). Protein microarrays: a literature survey. *Clin. Chem.* 49 (12): 2109–2109. (b) Kricka, L.J. (2003). Clinical applications of chemiluminescence. *Anal. Chim. Acta* 500 (1–2): 279–286.

1253 Kricka, L.J. and Wilding, P. (2003). Microchip PCR. *Anal. Bioanal. Chem.* 377 (5): 820–825.

1254 Roda, A., Guardigli, M., Pasini, P., and Mirasoli, M. (2003). Bioluminescence and chemiluminescence in drug screening. *Anal. Bioanal. Chem.* 377 (5): 826–833.

1255 Roda, A., Guardigli, M., Michelini, E. et al. (2003). Analytical bioluminescence and chemiluminescence. *Anal. Chem.* 75 (21): 462A–470A.

1256 Gámiz-Gracía, L., García-Campaña, A.M., Soto-Chinchilla, J.J. et al. (2005). Analysis of pesticides by chemiluminescence detection in the liquid phase. *TrAC, Trends Anal. Chem.* 24 (11): 927–942.

1257 García-Campaña, A.M. and Lara, F.J. (2007). Trends in the analytical applications of chemiluminescence in the liquid phase. *Anal. Bioanal. Chem.* 387 (1): 165–169.

1258 Kricka, L.J., Voyta, J.C., and Bronstein, I. (2000). Chemiluminescent methods for detecting and quantitating enzyme activity. In: *Bioluminescence and Chemiluminescence, Part C*, vol. 305 (Ser. ed. M. Ziegler, T. Baldwin, Chief-ed. J. Abelson, and M. Simon), 370–390. New York: Academic Press.

1259 Woldman, Y.Y., Sun, J., Zweier, J.L., and Khramtsov, V.V. (2009). Direct chemiluminescence detection of nitric oxide in aqueous solutions using the natural nitric oxide target soluble guanylyl cyclase. *Free Radical Biol. Med.* 47 (10): 1339–1345.

1260 Bowen, E.J. and Pells, E.G. (1927). A study of the chemiluminescence of phosphorus vapour. *J. Chem. Soc.* 1096–1099.

1261 Mckeown, E. and Waters, W.A. (1964). Chemiluminescence as diagnostic feature of heterolytic reactions which produce oxygen. *Nature* 203 (494): 1063–1063.

1262 Kautsky, H. and Neitzke, O. (1925). Emission matter spectra by excitation through light and through chemical reactions. *Z. Angew. Phys.* 31: 60–71.

1263 Kricka, L.J. (2000). Application of bioluminescence and chemiluminescence in biomedical sciences. In: *Bioluminescence & Chemiluminescence, Part C*, vol. 305, 333–345. Elsevier.

1264 Hindson, B.J. and Barnett, N.W. (2001). Analytical applications of acidic potassium permanganate as a chemiluminescence reagent. *Anal. Chim. Acta* 445 (1): 1–19.

1265 Kang, J., Zhang, Y., Huang, Z. et al. (2011). Directly oxidized chemiluminescence of 2-substituted-4,5-di(2-furyl)-1H-imidazole by acidic potassium permanganate and its analytical application for determination of albumin. *J. Fluoresc.* 21 (4): 1607–1615.

1266 Thomas, C.D. and Dufford, R.T. (1933). Efficiency of the chemiluminescence accompanying oxidation of Grignard compounds. *J. Opt. Soc. Am.* 23 (7): 253–257.

1267 (a) Chandross, E.A. (1963). A new chemiluminescent system. *Tetrahedron Lett.* (12): 761–765. (b) Bardsley, R.L. and Hercules, D.M. (1968). Chemiluminescence of Grignard reagents. Some spectroscopic and mechanistic studies. *J. Am. Chem. Soc.* 90 (17): 4545–4549. (c) Bolton, P.H. and Kearns, D.R. (1974). Molecular oxygen induced chemiluminescence of aryl Grignard reagents. *J. Am. Chem. Soc.* 96 (14): 4651–4654. (d) Bolton, P.H. and Kearns, D.R. (1974). Anaerobic chemiluminescent reaction of aryl Grignard reagents with aryl peroxides. *J. Phys. Chem.* 78 (19): 1896–1899.

1268 Buzanovskii, V.A. and Bulaev, A.A. (2008). Chemiluminescent gas analyzers. *Chem. Pet. Eng.* 44: 514–518.

1269 Buzanovskii, V.A. and Bulaev, A.A. (2008). Gas analyzers for monitoring air of workplace. *Med. Tr. Prom. Ekol.* 2008 (10): 37–45.

1270 Pinder, A.G., Rogers, S.C., Khalatbari, A. et al. (2008). The measurment of nitric oxide and its metabolites in biological samples by ozone-based chemilumiscence. *Methods Mol. Biol.* 476: 11–25.

1271 Kummer, W.A., Pitts, J.N., and Steer, R.P. (1971). Chemiluminescent reactions of ozone with olefins and sulfides. *Environ. Sci. Technol.* 5 (10): 1045–1047.

1272 Arora, P.K. (1984). Gas phase chemiluminescent reaction of ozone with furan. *J. Photochem.* 27 (1): 1–8.

1273 Lechtken, P., Breslow, R., Schmidt, A.H., and Turro, N.J. (1973). Thermal rearrangement of Dewar-benzenes to benzene triplet states – examples of spin forbidden nonadiabatic pericyclic reactions. *J. Am. Chem. Soc.* 95 (9): 3025–3027.

1274 Oth, J.F.M. (1968). Kinetics and thermochemistry of thermal rearrangement of hexamethyl(Dewar-benzene) and of hexamethylprismane. *Angew. Chem. Int. Ed. Engl.* 7 (8): 646–647.

1275 Li, Z.R., Rogers, D.W., McLafferty, F.J. et al. (1999). Ab initio calculations of enthalpies of hydrogenation, isomerization, and formation of cyclic C-6 hydrocarbons. Benzene isomers. *J. Phys. Chem. A* 103 (3): 426–430.

1276 Breslow, R., Napierski, J., and Schmidt, A.H. (1972). Unsymmetrical substituent effects on antiaromatic transition state for rearrangement of bicyclo[2.2.0]hexadiene to benzene. *J. Am. Chem. Soc.* 94 (16): 5906–5907.

1277 Ferrar, L., Mis, M., and Robello, D.R. (2008). Thermal isomerization of Dewar-benzene derivatives. *Tetrahedron Lett.* 49 (26): 4130–4133.

1278 Cundall, R.B., Evans, G.B., Griffith, P.A., and Keene, J.P. (1968). Pulse radiolysis of benzene-biacetyl solutions. *J. Phys. Chem.* 72 (11): 3871–3877.

1279 Belyakov, V.A. and Vassilev, R.F. (1967). On some problems concerning investigation of liquid-phase chemiluminescence. *Photochem. Photobiol.* 6 (1): 35–40.

1280 Turro, N.J., Schuster, G.B., Bergman, R.G. et al. (1975). Mechanisms of energy-storage and release – chemiluminescent valence isomerizations – evidence for a Dewar-benzene intermediate in a biscyclopropenyl to benzene rearrangement. *J. Am. Chem. Soc.* 97 (16): 4758–4760.

1281 Radziszewski, B. (1877). Untersuchungen über Hydrobenzamid. Amarin in Lophin. *Ber. Dtsch. Chem. Ges.* 10: 70–75.

1282 Philbrook, G.E. and Maxwell, M.A. (1964). Chemiluminescent intensities from substituted lophines. *Tetrahedron Lett.* 5 (19): 1111–1116.

1283 Kimura, M., Nishikawa, H., Kura, H. et al. (1993). Maximization of the chemiluminescence efficiency of 1,4,5-triarylhydroperoxy-4H-isoimidazoles. *Chem. Lett.* (3): 505–508.

1284 Kimura, M., Lu, G., Iga, H. et al. (2007). The stereoselective thermal rearrangement of chiral lophine peroxides. *Tetrahedron Lett.* 48 (17): 3109–3113.

1285 Lu, G., Wada, J., Kimoto, T. et al. (2014). The chemiexcitation of the chemiluminescence of lophine peroxide anions via a partially cyclic transition state. *Eur. J. Org. Chem.* (6): 1212–1219.

1286 Frandsen, P.L., Hakansson, K., Holm, A., and Harrit, N. (1991). Peroxide chemistry of triaryl-substituted imidazoles – fenflumizole, a nonsteroidal, antiinflammatory agent. *Acta Chem. Scand.* 45 (6): 627–631.

1287 Kimura, M., Lu, G., Nishigawa, H. et al. (2007). Singlet oxygen generation from lophine hydroperoxides. *Luminescence* 22 (2): 72–76.

1288 Nakashima, K. (2003). Lophine derivatives as versatile analytical tools. *Biomed. Chromatogr.* 17 (2–3): 83–95.

1289 Pruett, R.L., Barr, J.T., Rapp, K.E. et al. (1950). Reactions of polyfluoro olefins. 2. Reactions with primary and secondary amines. *J. Am. Chem. Soc.* 72 (8): 3646–3650.

1290 Winberg, H.E., Downing, J.R., and Coffman, D.D. (1965). Chemiluminescence of tetrakis(dimethylamino)ethylene. *J. Am. Chem. Soc.* 87 (9): 2054–2055.

1291 Freeman, T.M. and Seitz, W.R. (1981). Oxygen probe based on tetrakis(alkylamino)ethylene chemiluminescence. *Anal. Chem.* 53 (1): 98–102.

1292 Heller, C.A. and Fletcher, A.N. (1965). Oxidation and chemiluminescence of tetrakis(dimethylamino)ethylene. I. Reversible reactions of oxygen with tetrakis(dimethylamino)ethylene and n-decane. *J. Phys. Chem.* 69 (10): 3313–3317.

1293 Fletcher, A.N. and Heller, C.A. (1967). Oxidation and chemiluminescence of tetrakis(dimethylamino)ethylene. 3. Kinetics quantum yield and mechanism of luminescence. *J. Phys. Chem.* 71 (5): 1507–1518.

1294 Toby, S., Astheimer, P.A., and Toby, F.S. (1992). Chemiluminescence in the gas phase reaction between retrakis(dimethylamino)ethylene and oxygen. *J. Photochem. Photobiol., A* 67 (1): 1–12.

1295 Albrecht, H.O. (1928). Über die Chemiluminescenz des Aminophthalsaürehydrazyds. *Z. Phys. Chem.* 136: 321–330.

1296 White, E.H. and Bursey, M.M. (1964). Chemiluminescence of luminol & related hydrazides – light emission step. *J. Am. Chem. Soc.* 86 (5): 941–942.

1297 White, E.H., Kagi, H.H., Hill, J.H.M., and Zafiriou, O. (1964). Chemiluminescence of luminol – chemical reaction. *J. Am. Chem. Soc.* 86 (5): 940–941.

1298 Kautsky, H. and Kaiser, K.H. (1950). Lumineszenzanalyse Chemilumineszierender Losungen. 1. *Z. Naturforsch., B: Chem. Sci.* 5 (7): 353–361.

1299 Ojima, H. (1961). Chemiluminescence of non-cyclic hydrazides (relationships between structure of hydrazides and chemiluminescent power). *Naturwissenschaften* 48 (18): 600–603.

1300 White, E.H., Bursey, M.M., Roswell, D.F., and Hill, J.H.M. (1967). Chemiluminescence of some monoacylhydrazides. *J. Org. Chem.* 32 (4): 1198–1202.

1301 Yuan, J.C. and Shiller, A.M. (1999). Determination of subnanomolar levels of hydrogen peroxide in seawater by reagent-injection chemiluminescence detection. *Anal. Chem.* 71 (10): 1975–1980.

1302 Nieman, T. (1989). Detection based on solution-phase chemiluminescence systems. In: *Chemiluminescence & Photochemical Reaction Detection in Chromatography* (ed. J.W. Birks), 99–123. Weinheim and New York: Wiley-VCH.

1303 Townshend, A. and Wheatley, R.A. (1998). Oxidative chemiluminescence of some nitrogen nucleophiles in the presence of formic acid as an ancillary reductant. *Analyst* 123 (2): 267–272.

1304 Townshend, A. and Wheatley, R.A. (1998). Oxidative chemiluminescence assay of 2,4-dinitrophenylhydrazine. *Analyst* 123 (5): 1041–1046.

1305 Chandel, A., Khan, S., Kher, R., and Tiwari, A. (2012). Investigation on the chemiluminescence reaction of the phenylhydrazine-Luminol-peroxide system. *Luminescence* 27 (6): 455–458.

1306 Shukla, M., Tiwari, A., Brahme, N. et al. (2013). Enhancing effect of hydrazine on chemiluminescence of luminol-H_2O_2 system. *J. Appl. Spectrosc.* 80 (2): 305–307.

1307 Turro, N.J., Ramamurthy, V., Liu, K.C. et al. (1976). Reaction of strained acetylenes with molecular oxygen – novel chemiluminescent reaction, evidence for a dioxetene, and a mechanism for thermal generation of singlet oxygen. *J. Am. Chem. Soc.* 98 (21): 6758–6761.

1308 Beyer, R.A. and Lineberg, W.C. (1973). Time and energy resolved fluorescence of *cis*-glyoxal. *Chem. Phys. Lett.* 20 (6): 600–604.

1309 Gleu, K. and Petsch, W. (1935). Chemoluminescence of dimethyl-diacridylium salts. *Angew. Chem.* 48: 0057–0059.

1310 McCapra, F. and Richardson, D.G. (1964). The mechanism of chemiluminescence – a new chemiluminescent reaction. *Tetrahedron Lett.* 5 (43): 3167–3172.

1311 McCapra, F. (1968). An application of theory of electrocyclic reactions to bioluminescence. *Chem. Commun.* (3): 155–156.

1312 McCapra, F., Chang, Y.C., and François, V.P. (1968). Chemiluminescence of a firefly Luciferin analogue. *J. Chem. Soc., Chem. Commun.* (1): 22–23.

1313 Turro, N.J., Lechtken, P., Schore, N.E. et al. (1974). Tetramethyl-1,2-dioxetane – experiments in chemiexcitation, chemiluminescence, photochemistry, chemical dynamics, and spectroscopy. *Acc. Chem. Res.* 7 (4): 97–105.

1314 Adam, W. (1980). Thermal generation of electronic excitation with hyperenergetic molecules. *Pure Appl. Chem.* 52 (12): 2591–2608.

1315 Schuster, G.B. and Schmidt, S.P. (1982). Chemiluminescence of organic compounds. *Adv. Phys. Org. Chem.* 18: 187–238.

1316 (a) Adam, W. (1983). Four-membered ring peroxides: 1,2-Dioxetanes and α-peroxylactones. In: *The Chemistry of Peroxides* (ed. S. Patai), 829–920. New York: Wiley. (b) Matsumoto, M. and Watanabe, N. (2005). Structural aspects of 1,2-dioxetanes active toward intramolecular charge-transfer-induced chemiluminescent decomposition. *Bull. Chem. Soc. Jpn.* 78 (11): 1899–1920.

1317 (a) Andronico, L.A., Quintavalla, A., Lombardo, M. et al. (2016). Synthesis of 1,2-dioxetanes as thermochemiluminescent labels for ultrasensitive bioassays: rational prediction of olefin photooxygenation outcome by using a chemometric approach. *Chem. Eur. J.* 22 (50): 18156–18168. (b) Umehara, Y., Son, A., Kondo, T., and Tanabe, K. (2017). Dioxetane formation and chemiluminescent emission upon the combination of a vinylphenol derivative with naphthalene endoperoxide. *RSC Adv.* 7 (16): 9472–9475.

1318 Takano, Y., Tsunesada, T., Isobe, H. et al. (1999). Theoretical studies of decomposition reactions of dioxetane, dioxetanone, and related species. CT induced luminescence mechanism revisited. *Bull. Chem. Soc. Jpn.* 72 (2): 213–225.

1319 Wilsey, S., Bernardi, F., Olivucci, M. et al. (1999). The thermal decomposition of 1,2-dioxetane revisited. *J. Phys. Chem. A* 103 (11): 1669–1677.

1320 Tanaka, C. and Tanaka, J. (2000). Ab initio molecular orbital studies on the chemiluminescence of 1,2-dioxetanes. *J. Phys. Chem. A* 104 (10): 2078–2090.

1321 Adam, W. and Baader, W.J. (1985). Effects of methylation on the thermal stability and chemiluminescence properties of 1,2-dioxetanes. *J. Am. Chem. Soc.* 107 (2): 410–416.

1322 Augusto, F.A., de Souza, G.A., de Souza, S.P. Jr., et al. (2013). Efficiency of electron-transfer initiated chemiluminescence. *Photochem. Photobiol.* 89 (6): 1299–1317.

1323 (a) Di Fusco, M., Quintavalla, A., Trombini, C. et al. (2013). Preparation and characterization of thermochemiluminescent acridine-containing 1,2-dioxetanes as promising ultrasensitive labels in bioanalysis. *J. Org. Chem.* 78 (22): 11238–11246. (b) Augusto, F.A., Francés-Monerris, A., Fernández-Galván, I. et al. (2017). Mechanism of activated chemiluminescence of cyclic peroxides: 1,2-dioxetanes and 1,2-dioxetanones. *Phys. Chem. Chem. Phys.* 19 (5): 3955–3962.

1324 Schaap, A.P. and Gagnon, S.D. (1982). Chemiluminescence from a phenoxide substituted 1,2-dioxetane – a model for firefly bioluminescence. *J. Am. Chem. Soc.* 104 (12): 3504–3506.

1325 Nery, A.L.P., Weiss, D., Catalani, L.H., and Baader, W.J. (2000). Studies on the intramolecular electron transfer catalyzed thermolysis of 1,2-dioxetanes. *Tetrahedron* 56 (30): 5317–5327.

1326 Matsumoto, M., Suzuki, H., Watanabe, N. et al. (2011). Crucial dependence of chemiluminescence efficiency on the syn/anti conformation for intramolecular charge-transfer-induced decomposition of bicyclic dioxetanes bearing an oxidoaryl group. *J. Org. Chem.* 76 (12): 5006–5017.

1327 Schmidt, S.P. and Schuster, G.B. (1978). Dioxethanone chemiluminescence by chemically-initiated electron exchange

pathway – efficient generation of excited singlet states. *J. Am. Chem. Soc.* 100 (6): 1966–1968.
1328 Schmidt, S.P. and Schuster, G.B. (1980). Chemiluminescence of dimethyldioxetanone – unimolecular generation on exchange luminescence, the primary light generating reaction. *J. Am. Chem. Soc.* 102 (1): 306–314.
1329 Adam, W. and Cueto, O. (1979). Cyclic peroxide. 81. Fluorescer-enhanced chemiluminescence of α-peroxylactones via electron exchange. *J. Am. Chem. Soc.* 101 (22): 6511–6515.
1330 Turro, N.J. and Chow, M.F. (1980). Chemiluminescent thermolysis of α-peroxylactones. *J. Am. Chem. Soc.* 102 (15): 5058–5064.
1331 Schuster, G.B. (1979). Chemiluminescence of organic peroxides – conversion of ground-state reactants to excited-state products by the chemically-initiated electron-exchange luminescence mechanism. *Acc. Chem. Res.* 12 (10): 366–373.
1332 Wilson, T. (1995). Comments on the mechanisms of chemiluminescence and bioluminescence. *Photochem. Photobiol.* 62 (4): 601–606.
1333 Lee, C. and Singer, L.A. (1980). Structural effects on the intramolecular electron-transfer induced decomposition of a series of 1,2-dioxetanes derived from 9-alkylidene-10-methylacridans. *J. Am. Chem. Soc.* 102 (11): 3823–3829.
1334 Ciscato, L.F.M.L., Bartoloni, F.H., Weiss, D. et al. (2010). Experimental evidence of the occurrence of intramolecular electron-transfer in catalyzed 1,2-dioxetane decomposition. *J. Org. Chem.* 75 (19): 6574–6580.
1335 Ciscato, L.F.M.L., Bartoloni, F.H., Bastos, E.L., and Baader, W.J. (2009). Direct kinetic observation of the chemiexcitation step in peroxyoxalate chemiluminescence. *J. Org. Chem.* 74 (23): 8974–8979.
1336 (a) Inouye, S. (2010). Firefly luciferase: an adenylate-forming enzyme for multicatalytic functions. *Cell. Mol. Life Sci.* 67 (3): 387–404. (b) Pinto da Silva, L. and Esteves da Silva, J.C.G. (2011). Computational investigation of the effect of pH on the color of firefly bioluminescence by DFT. *ChemPhysChem* 12 (5): 951–960. (c) Niwa, K. (2016). Quantum yield determination based on photon number measurement, protocols for firefly bioluminescence reactions. *Methods Mol. Biol.* 1461: 55–61.
1337 (a) Solntev, K.M., Laptenok, S.P., and Naumov, P. (2012). Photoinduced dynamics of oxyluciferin analogues: unusual enol "super" photoacidity and evidence for keto-enol isomerization. *J. Am. Chem. Soc.* 134 (40): 16452–16455. (b) Maltsev, O.V., Nath, N.K., Naumov, P., and Hintermann, L. (2014). Why is firefly oxyluciferin a notoriously labile substance? *Angew. Chem. Int. Ed.* 53 (3): 847–850.
1338 Naumov, P., Ozawa, Y., Ohkubo, K., and Fukuzumi, S. (2009). Structure and spectroscopy of oxyluciferin, the light emitter of the firefly bioluminescence. *J. Am. Chem. Soc.* 131 (32): 11590–11605.
1339 Nakatsu, T., Ichiyama, S., Hiratake, J. et al. (2006). Structural basis for the spectral difference in luciferase bioluminescence. *Nature* 440 (7082): 372–376.
1340 Seliger, H.H. and McElroy, W.D. (1964). Colors of firefly bioluminescence – enzyme configuration & species specificity. *Proc. Natl. Acad. Sci. U.S.A.* 52 (1): 75–81.
1341 (a) Hosseinkhani, S. (2011). Molecular enigma of multicolor bioluminescence of firefly luciferase. *Cell. Mol. Life Sci.* 68 (7): 1167–1182. (b) Cheng, Y.-Y. and Li, Y.-J. (2015). What exactly is the light emitter of a firefly? *J. Chem. Theory Comput.* 11 (11): 5360–5370. (c) Hirano, T. (2016). Molecular origin of color variation in firefly (beetle) bioluminescence: a chemical basis for biological imaging. *Curr. Top. Med. Chem.* 16 (24): 2638–2647.
1342 Pinto da Silva, L., Simkovitch, R., Huppert, D., and Esteves da Silva, J.C.G. (2013). Theoretical photodynamic study of the photoprotolytic cycle of firefly oxyluciferin. *ChemPhysChem* 14 (12): 2711–2716.
1343 Koo, J.Y., Schmidt, S.P., and Schuster, G.B. (1978). Bioluminescence of firefly – key steps in formation of electronically excited-state for model systems. *Proc. Natl. Acad. Sci. U.S.A.* 75 (1): 30–33.
1344 Isobe, H., Takano, Y., Okumura, M. et al. (2005). Mechanistic insights in charge-transfer-induced luminescence of 1,2-dioxetanones with a substituent of low oxidation potential. *J. Am. Chem. Soc.* 127 (24): 8667–8679.
1345 Yue, L., Liu, Y.J., and Fang, W.H. (2012). Mechanistic Insight into the chemiluminescent decomposition of firefly dioxetanone. *J. Am. Chem. Soc.* 134 (28): 11632–11639.
1346 Pinto da Silva, L. and Esteves da Silva, J.C.G. (2013). Interstate crossing-induced chemiexcitation as the reason for the chemiluminescence of dioxetanones. *ChemPhysChem* 14 (5): 1071–1079.
1347 da Pinto, Silva, L., Santos, A.J., and da Esteves, Silva, J.C.G. (2013). Efficient firefly

chemi/bioluminescence: evidence for chemiexcitation resulting from the decomposition of a neutral firefly dioxetanone molecule. *J. Phys. Chem. A* 117 (1): 94–100.

1348 Li, J., Chen, L., Du, L., and Li, M. (2013). Cage the firefly Luciferin! – a strategy for developing bioluminescent probes. *Chem. Soc. Rev.* 42 (2): 662–676.

1349 Miura, C., Kiyama, M., Iwano, S. et al. (2013). Synthesis and luminescence properties of biphenyl-type firefly luciferin analogs with a new, near-infrared light-emitting bioluminophore. *Tetrahedron* 69 (46): 9726–9734.

1350 Iwano, S., Obata, R., Miura, C. et al. (2013). Development of simple firefly luciferin analogs emitting blue, green, red, and near-infrared biological window light. *Tetrahedron* 69 (19): 3847–3856.

7

Catalytic reactions

7.1 Introduction

In 1746, Roebuck developed the lead chamber for sulfuric acid production in which nitrogen oxides catalyze the oxidation of SO_2 into SO_3. Since then, catalysts have been in use in a large portion of the chemical industry. Catalysts can be inorganic, organic, organometallic compounds, enzymes, or microorganisms [1–3]. Organocatalysis refers to a form of catalysis, where the rate of a chemical reaction is increased by an organic compound referred to as an "organocatalyst" consisting of carbon, hydrogen, sulfur, and other nonmetal elements. Most common are acids and bases. In the past two decades, organocatalysis has become a pillar of asymmetric organic synthesis [4–10]. The term catalyst was coined by Berzelius in 1835 [11], and the term "organocatalyst" was first used by MacMillan in 2000 [6, 12]. A catalyst is a compound, generally used in substoichiometric amount, that takes part in a reaction, resulting in an increased rate, but that is not consumed. It does not affect the equilibrium constant of the reaction. A catalytic (or catalytical) reaction is a reaction that is catalyzed. Some reactions are promoted by an additive. In these cases, the additive speeds up the reactions and is converted to another species during the reaction. Most reactions can be catalyzed, usually by changing their mechanism, which is in fact always the case if the catalyst interacts with one or more species of the reaction (reactants, intermediates, transition states, and products). Catalysts are classified as heterogeneous (most frequent in the industry, bulk chemicals, reactions at surfaces, Section 3.2.10) or homogeneous (Sections 3.2.8 and 3.2.9). A homogeneous catalyst dissolves in the reaction mixture (completely or partially), whereas a heterogeneous catalyst does not dissolve and acts with the reaction partners through physisorption (-2 to $-10\,kcal\,mol^{-1}$: binding energies, attraction between permanent or induced dipoles or quadrupoles, attraction due to van der Waals-London forces) or through chemisorption (covalent/electrostatic forces such as hydrogen bridging, electron transfer). Selected examples of homogeneous catalytical reactions have been presented in the preceding chapters. A yeast enzyme called invertase catalyzes the hydrolysis of sucrose into D-fructose and D-glucose, whose kinetics follow the Michaelis–Menten mechanism (Section 3.2.8). A phosphatase-catalyzed hydrolysis is given in Scheme 3.5 (Section 3.5.1). Kinetic optical resolutions have been discussed in Section 3.7.1. They are catalyzed by enzymes (e.g. hydrolases, esterases, proteases, and nitralases) or man-made enantiomerically pure catalysts (see the Noyori's (*S*)-(BINAP)Ru$(OAc)_2$-catalyzed hydrogenation of enones [BINAP: 2,2′-bis(diphenylphosphino)-1,1′-binaphthyl]; the enantioselective hydrolysis of epoxides catalyzed by Jacobsen's (salen)Co-OAc catalysts; the enantioselective esterification of alcohols catalyzed by the Fu's catalyst, a chiral 4-dimethylaminopyridine (DMAP) analog; the enantioselective esterification of α-arylalkanoic acids using a benzotetramisole type of catalyst (Scheme 3.13); the enantioselective dehydrogenation of alcohols catalyzed by ruthenium catalysts; the enantioselective ring-closing alkene metathesis by the Hoveyda–Schrock catalysts; the silver-catalyzed isomerization of α-allenic alcohols into 2,4-dihydrofurans catalyzed by an enantiomerically pure silver phosphate; the asymmetric epoxidation by the Katsuki–Sharpless catalysts (see also Section 3.7.6), by the Jacobsen's catalysts, or tetraphenylphosphonium monoperoxysulfate in the presence of nonracemic chiral iminium salts; and the enantioselective α-allylation of ketones catalyzed by palladium and (*S*)-P-PHOS). Dynamic kinetic resolution (kinetic deracemization, Section 3.7.3) based on hydantoinase-catalyzed hydrolytic ring opening of hydantoins to form the corresponding *N*-carbamoyl α-amino acids has been described in Scheme 3.15. Racemic secondary alcohols have been converted into enantiomerically pure acetates by Bäckvall utilizing a lipase for enantioselective esterification and

Organic Chemistry: Theory, Reactivity and Mechanisms in Modern Synthesis, First Edition. Pierre Vogel and Kendall N. Houk.

a ruthenium catalyst for alcohol enantiomerization. Noyori's enantioselective hydrogenation of ketones has been presented in Scheme 3.16. Deracemization of 2-aryl-3*H*-indolines has been realized by Toste utilizing achiral phosphoric acid (Section 3.7.3). Desymmetrization of achiral compounds has been presented in Section 3.7.6 and illustrated with several examples based on enantioselective catalysis utilizing bacteria (e.g. *Bacillus proteus*-catalyzed addition of HCN to PhCHO), lipase-catalyzed acetylation, or baker's yeast-catalyzed enantioselective reduction of ketones into secondary alcohols. The industrial synthesis of (−)-ephedrine based on microbial-mediated acyloin condensation is shown in Scheme 3.22. Aldol reactions can be catalyzed by rabbit muscle aldolase (Scheme 3.23) or by L-proline and derivatives. Enantioselective oxidation of sulfides into sulfoxides has been catalyzed by modified Sharpless titanium catalysts or the Juliá–Colonna and other enantioselective alkene epoxidations (Scheme 3.24). An example of asymmetric Baeyer–Villiger oxidation has been described in Section 3.7.6. Nonlinear effects in enantioselective catalyzed reactions have been presented in Section 3.7.7. An example of enantioselective S_N2' displacement reaction catalyzed by a chiral Pd complex is given in Scheme 5.25 (Section 5.3.5). Nazarov cyclizations can be catalyzed by various Brønsted (Scheme 5.7) or Lewis acids (Schemes 5.9 and 5.10). An example of enantioselective Nazarov cyclization catalyzed by a chiral N—H acid is given in Scheme 5.12. Cycloadditions can be catalyzed by protic or Lewis acids. Examples are given for $[\pi^2+\pi^2]$-cycloadditions in Section 5.3.3. Scheme 5.28 presents an example of enantioselective β-lactam synthesis catalyzed by a chiral diaminocarbene. Diels–Alder reactions of acrylic derivatives are catalyzed by protic (Brønsted) and Lewis acids (see Scheme 5.45 for a catalytical enantioselective Diels–Alder reaction utilized by Corey in its total synthesis of prostaglandins). Hetero-Diels–Alder reaction can also be catalyzed by acids, and several examples of catalytic enantioselective reactions have been presented (Section 5.3.16). Examples of regio- and enantioselective 1,3-dipolar cycloadditions catalyzed by chiral Lewis acids have been illustrated for the Sharpless asymmetric dihydroxylation of alkenes (Section 5.3.17). More examples of catalyzed cycloadditions will be given in this chapter. Claisen rearrangement is accelerated by coordination of their ethereal moiety to a Lewis acid (Scheme 5.99). Examples of catalytic enantioselective Claisen rearrangements utilizing a chiral Cu^{++} complex (Scheme 5.100), a chiral guanidinium salt (Scheme 5.101), or a chiral Pd^{++} complex (Scheme 5.102) have been given. Aza-Claisen rearrangements are accelerated by protic and Lewis acids. A catalytic enantioselective acyl-aza-Claisen rearrangement is given in Scheme 5.105. Anionic oxy-Cope rearrangements (Section 5.5.9.7) are catalyzed by crown ethers (Scheme 5.113) of antibody catalysts. Alder ene reactions of acrylic and propiolic acid derivatives are catalyzed, as Diels–Alder reactions, by Brønsted and Lewis acids (Section 5.7.1). Aldol reactions are catalyzed by acids and bases. Simple amino acids such as L-proline (typical organocatalysts) are class I aldolase mimics that permit catalytic enantioselective aldol reactions (see e.g. Scheme 5.133). Class II aldolase mimics that are metallic complexes have also been developed (Section 5.7.6). Light may catalyze reactions that involve radical intermediates and follow a radical chain mechanism. These reactions are not treated here. However, UV–visible irradiation of aryl and alkenyl halides (Section 6.6.4) may induce $S_{RN}1$ reactions. For instance, a photoinduced electron transfer (PET) generates the radical initiator $Ar^\bullet$ according to $Ar{—}X + H{—}Nu + h\nu \rightarrow Ar{—}X^{\bullet-} + H{—}Nu^{\bullet+} \rightarrow Ar^\bullet + HX + Nu^-$; then $Ar^\bullet + Nu^- \rightarrow Ar{—}Nu^{\bullet-}$ and $Ar{—}Nu^{\bullet-} + Ar{—}X \rightarrow Ar{—}Nu + Ar{—}X^{\bullet-}$. The same reaction can be catalyzed by an electron-rich organic, organometallic, or metallic compound able to transfer an electron (ET) to Ar—X according to $Ar{—}X + M \rightarrow Ar{—}X^{\bullet-} + M^{\bullet+}$ (Scheme 6.38). There are many more examples of catalytical processes initiated by PET or ET. Some of them have been described in Section 6.10.2 such as intramolecular (2+2)-cycloadditions (Scheme 6.91), β-elimination of *N*-alkoxyphtalimides (Scheme 6.92), enantioselective α-trifluoromethylation of aldehydes (Scheme 6.93a), and alkene oxytrifluoromethylation (Scheme 6.93b) photocatalyzed by visible light in the presence of a Ru(II) or Ir(III) complex. The latter reactions are examples of catalysis through photoreduction. Catalysis through photo-oxidation has also been described (Section 6.10.4). The catalysts, or sensitizers, can be organic compounds such as terephthalic derivatives, pyrylium salts, or transition metal complexes.

A catalyst of a given reaction is characterized by its turnover number (TON), which is the maximum number of moles of reactants that 1 mol of catalyst is able to convert into products before it loses its activity. One defines the lifetime of the catalyst for given reaction conditions and the turnover frequency (TOF [time^{-1}]), which quantifies the specific activity of the catalyst under the conditions used. For most relevant industrial applications, TOF is in the range between 10^{-2} and $10^2\ s^{-1}$ and for enzymes between 10^3 and

$10^7\ s^{-1}$. The turnover rate, often used, is synonymous for TOF.

In general, any elementary process will be either accelerated (decrease of activation free energy $\Delta^{\ddagger}G$) or retarded (increase of $\Delta^{\ddagger}G$) if the reactants and/or the transition structures (activated complex) are associated (make complexes or adducts) with one or more than one (cluster) spectator molecule (solvent, additive, catalyst, inhibitor, and a combination of them). This is represented in Figure 7.1 for a bimolecular reaction involving reactants **A** and **B** occurring (a) in the gas phase, without any other molecule, (b) in the presence of an additive **C** that stabilizes the transition structure better than the reactants ($\Delta^{\ddagger}G(\text{B}) < \Delta^{\ddagger}G(\text{A})$), and (c) in the presence of an additive **D** that stabilizes the reactants better than the transition structure ($\Delta^{\ddagger}G(\text{C}) > \Delta^{\ddagger}G(\text{A})$). This representation does not involve dynamic effects, i.e. the possibility to have a collection of possible transition structures in the transition state. A catalytic effect results from the combination of an enthalpic effect and an entropy effect.

Water introduces many unusual features into the energetics of chemical and biological reactions [13, 14]. Beyond the properties of bulk water, a single molecule of H_2O or an aggregate may influence the rate of a reaction [15, 16]. For instance, water forms complexes with radicals and polar molecules in thermal reactions [17, 18], and such complexation alters the reaction activation-free energies. Kinetic studies as well as quantum chemical mechanical calculations show that one molecule of water accelerates the hydrogen abstraction from acetaldehyde by hydroxyl radical, to produce acetyl radical and a water molecule, as represented in Figure 7.2 [19].

A homogeneous catalyst is a molecule that adds intermediates and transition structures on the pathway from reactants to products (Figure 7.3). A catalyst makes a first complex with one reactant and then labilizes the bonds that must be broken. If the process is bimolecular, the first complex must be able to bind with little activation-free energy to the second reactant (template effect) and bring both reactants into a stabilized activated complex (transition state). A good catalyst is a compound that does not bind too

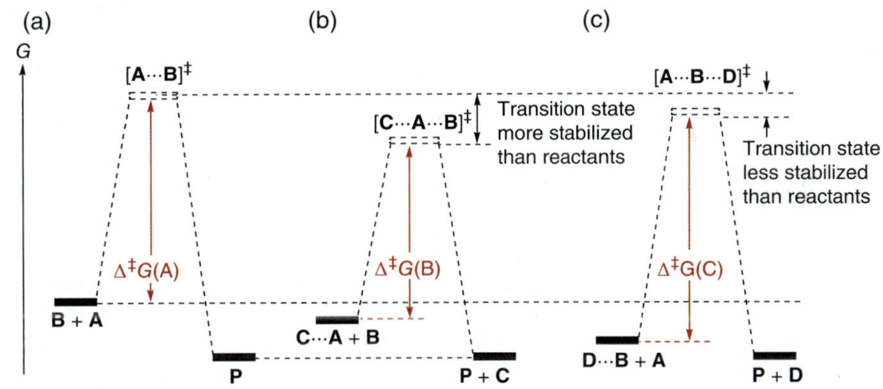

Figure 7.1 (a) Free energy diagram for an uncatalyzed gas-phase bimolecular reaction converting **A** + **B** → **P**; (b) free energy diagram for a catalyzed (catalyst **C**) bimolecular reaction; (c) free energy diagram for a bimolecular reaction inhibited by additive **D**. If **D** is in substoichiometric concentration, the uncatalyzed reaction (a) will compete with (c). The retardation effect of **D** increases with its concentration.

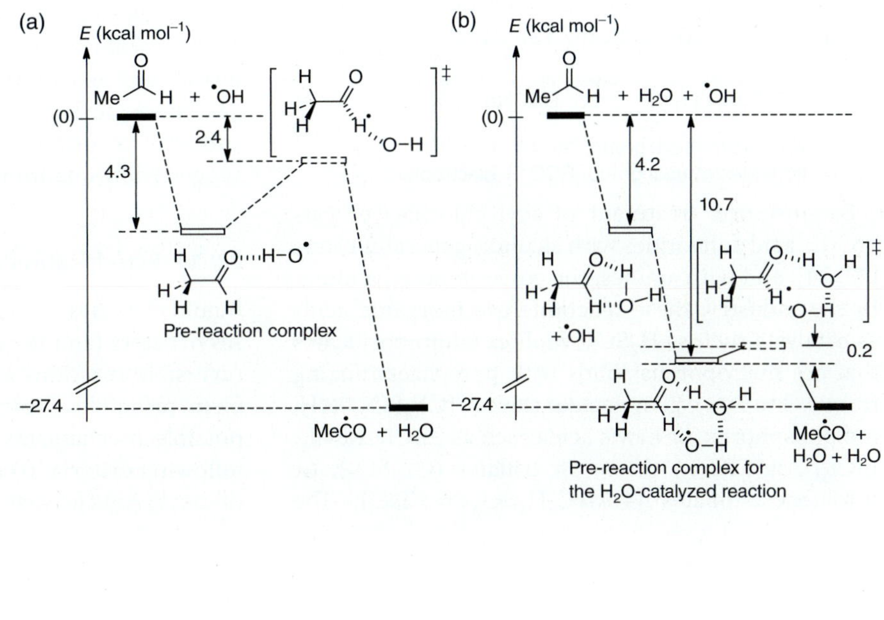

Figure 7.2 Calculated relative energies ($\Delta E \cong \Delta H$) for the reaction $MeCHO + {}^{\bullet}OH \rightarrow MeC^{\bullet}{=}O + H_2O$, (a) in the absence of H_2O and (b) in the presence of H_2O.

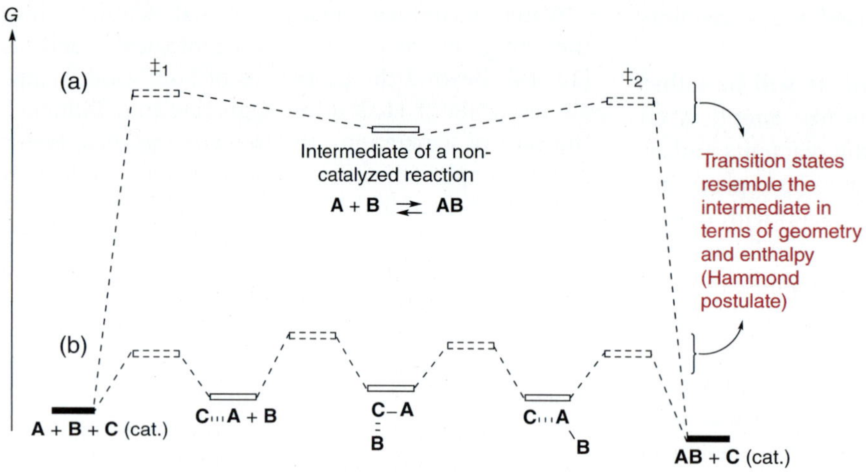

Figure 7.3 (a) Free energy diagram for an uncatalyzed reaction converting reactants **A** + **B** into product **AB** and (b) for a reaction catalyzed by catalyst **C**.

strongly to the reactants, does not form intermediates that are more stable than the products, and that lowers the activation-free energies of all the elementary processes. It combines with the reactants to equilibrate with the initial complex first, then converts this complex into the next intermediate, and so on. A good catalyst should not bind to the product. Otherwise, the product becomes an inhibitor of the catalyzed reaction.

7.2 Acyl group transfers

Acylation (reaction (7.1)) is one of the most important reactions of preparative chemistry and in biochemistry. It exchanges a proton (electrofugal group) from H—Nu for an acyl cation (electrophile). It can also be seen as a nucleophilic displacement: exchange of X^- in RCOX for Nu^- [20]. The most applied reactions nowadays in pharmaceutical chemistry are (i) amide bond formation, (ii) Suzuki–Miyaura coupling (Section 8.8.4), and (iii) nucleophilic aromatic substitution (S_NAr, Section 7.12) [21, 22].

$$\text{RCOX} + \text{H–Nu} \xrightleftharpoons{\text{cat.}} \text{RCONu} + \text{HX} \tag{7.1}$$

X^- = nucleofugal group; Nu^- = nucleophile
H^+ = electrofugal group; RCO^+ = electrophile

For instance, treatment of acyl chlorides or carboxylic acid anhydrides with alcohols generates esters [23, 24], and with amines, the corresponding amides. In the industry, these reactions use inorganic acids as catalysts such as H_2SO_4, zeolites (aluminosilicates that are microporous solids with pore sizes ranging from c. 3 to 7 Å; e.g. molecular sieves), $(CF_3SO_2)_2NH$, red phosphorus, or Lewis acids such as $ZnCl_2$, $RuCl_3$, $InCl_3$, $Mg(ClO_4)_2$, Cu or Bi triflates (CF_3SO_3), or *p*-toluenesulfonates ($4\text{-MeC}_6H_4SO_3$ = TsO). The same reactions are also base-catalyzed (Section 7.2.1, Table 7.1). Transesterification (7.2), a reaction producing esters by catalytical alcoholysis of other esters, also plays an important role in the industry, for instance, for the manufacture of glycerol esters in fat chemistry (e.g. biodiesel), the transesterification of dimethyl carbonate to diethyl carbonate, or the production of polyethylene terephthalate from dimethyl terephthalate.

$$\text{R}^1\text{COOR}^2 + \text{R}^3\text{OH} \xrightleftharpoons{\text{cat.}} \text{R}^1\text{COOR}^3 + \text{R}^2\text{OH} \tag{7.2}$$

For these reactions, usual catalysts are Brønsted mineral acids (H_3PO_4, H_2SO_4, HCl) [25–27], and organic acids such as $MeSO_3H$ and *p*-toluenesulfonic acid (TsOH) [28]; alkoxides such as NaOR, KOR, and ROMgBr [29–33]; Lewis bases such as 4-DMAP [34, 35]; 1,8-diazabicyclo[5.4.0]undec-7-ene (DBU) [36]; Lewis acids such as BX_3 [37] or $AlCl_3$ [38]; and amphoteric compounds (compounds able to react either as acids or as bases) such as Bu_3SnOR [39–42], perfluorotin oxides [43], $Al(OR)_3$ [44, 45], titanium oxides chlorides [46–49], or palladium oxides [50, 51]. More recently, diaminocarbenes [52] have been introduced as catalysts [53–56]. Mechanisms of the gas-phase acyl group transfers involve tetrahedral adducts as intermediates or transition states, as suggested by quantum mechanical calculations [57].

7.2.1 Esterification and ester hydrolysis

Ingold classifies the acid- and base-catalyzed hydrolyses of esters (and the formation of esters, as these are reversible reactions and thus have the same mechanism: principle of microscopic reversibility) into eight possible mechanisms (Table 7.1), depending on the following criteria: (i) acid (A)- or base (B)-catalyzed, (ii) acyl (Ac) cleavage or alkyl (AL) cleavage, and (iii)

Table 7.1 Ingold's classification of mechanisms for ester hydrolysis and formation.

1) Acid-catalyzed acyl cleavage $A_{AC}1$ (IUPAC: $A_N + D_N + A_N + D_N$) (S_N1 like at the acyl group)
$RCOOR' + H_3O^+ + H_2O \rightleftarrows [RCO(O^+H)R'(H_2O)_2 \rightleftarrows RCO^+ + R'OH + 2H_2O]$ (*the slowest step*) $\rightleftarrows [RCO(O^+H_2) + R'OH + H_2O \rightleftarrows RC^+(OH)_2 + R'OH + H_2O] \rightleftarrows RCOOH + R'OH + H_3O^+$
2) Acid-catalyzed acyl cleavage $A_{AC}2$ ($A_h + A_N + A_hD_h + D_h$) (add./elim.: adduct intermediate)
$RCOOR' + H_3O^+ + H_2O \rightleftarrows [RC^+(OR')OH(OH_2) + H_2O \rightleftarrows RC(OR')(OH)(O^+H_2...OH_2)]$ (*slow*) $\rightleftarrows [RC(OH)_2(O^+HR')...OH_2 \rightleftarrows RC^+(OH)_2...OH_2 + R'OH]$ (*slow*) $\rightleftarrows RCOOH + R'OH + H_3O^+$
3) Acid-catalyzed alkyl cleavage $A_{AL}1$ ($A_h + A_N + A_N + D_h$) (S_N1 like at the alkyl group)
$RCOOR' + H_3O^+ + H_2O \rightleftarrows [RC^+(OH)(OR')(H_2O)_2 \rightleftarrows RCOOH + [R'^+] + 2H_2O]$ (*the slowest step*) $\rightleftarrows [RCOOH + R'O^+H_2(OH_2)] \rightleftarrows RCOOH + R'OH + H_3O^+$
4) Acid-catalyzed alkyl cleavage $A_{AL}2$ ($A_h + A_ND_N + D_N$) (S_N2 like at the alkyl group)
$RCOOR' + H_3O^+ + H_2O \rightleftarrows [RCO(O^+HR')(OH_2)_2] \rightleftarrows [RCOOH + R'O^+H_2(OH_2)]$ (slow) $\rightleftarrows RCOOH + R'OH + H_3O^+$
5) Base-catalyzed acyl cleavage $B_{AC}1$ ($D_N + A_N + A_{xh}D_h$) (S_N1 like at the acyl group)
$RCOOR' + HO^- \rightleftarrows [RCO^+ + R'O^- + HO^-]$ (*slow*) $\rightleftarrows RCOOH + R'O^- \rightleftarrows RCOO^- + R'OH$
6) Base-catalyzed acyl cleavage $B_{AC}2$ ($A_N + D_N + A_{xh}D_h$) (adduct intermediate)
$RCOOR' + HO^- \rightleftarrows [R(OH)(OR')(O^-)$ (*slow*) $\rightleftarrows RCOOH + R'O^-] \rightleftarrows RCOO^- + R'OH$
7) Base-catalyzed alkyl cleavage $B_{AL}1$ ($D_N + A_N + A_{xh}D_h$) (S_N1 like at the alkyl group)
$RCOOR' + HO^- \rightleftarrows [RCOO^- + R'^+ + HO^-]$ (*slow*) $\rightleftarrows RCOO^- + R'OH$
8) Base-catalyzed alkyl cleavage $B_{AL}2$ (A_ND_N) (S_N2 like at the alkyl group)
$RCOOR' + HO^- \rightleftarrows RCOO^- + R'OH$
9) Ketene formation ($D_h + D_N + A_N + A_h$) (elim./add.)
$RCH_2COOR' + HO^- \rightleftarrows [RCH{=}C{=}O + R'O^- + H_2O]$ (*slow*) $\rightleftarrows [RCH{=}C{=}O + R'OH + HO^-] \rightleftarrows [RCH{=}C(OH)O^- + R'OH] \rightleftarrows RCH_2COO^- + R'OH$

unimolecular (1) or bimolecular (2) [58]. All eight of these are S_N1, S_N2, or tetrahedral (add./elim.) mechanisms. A ninth mechanism implies the base-catalyzed elimination of alcohol and the formation of a ketene intermediate that adds HO^- to generate an enolate that is finally protonated into the final carboxylic acid (E_{1cb} pathway) [59]. The most common mechanisms are $B_{AC}2$ for base-catalyzed acyl transfers and $A_{AC}2$ for acid-catalyzed acyl transfers. Both involve acyl-oxygen cleavage as evidenced by hydrolysis in $H_2{}^{18}O$ that results in ^{18}O appearing in the acid and not in the alcohol, and for esters with chiral R′ groups, alcohols with retention of configuration are formed. Usually, two molecules of H_2O are required in the transition state of the $A_{AC}2$ mechanism.

7.2.2 Acid or base-catalyzed acyl transfers

In the absence of any catalyst and in a solvent that cannot play neither the role of an acid nor that of a base (e.g. PhMe), reaction (7.1) can follow three limiting mechanisms represented in the More O'Ferrall–Jencks type of diagram [60, 61] given in Figure 7.4 [62, 63].

With this hypothetical reaction, one sees that the two-step nucleophilic addition route implies the

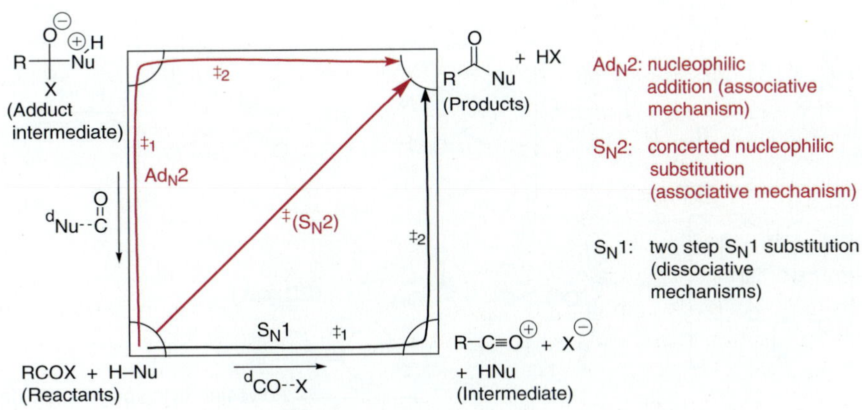

Figure 7.4 Limiting mechanisms for an acyl transfer reaction neither catalyzed nor assisted by the solvent (*x*-axis describes distance CO· · ·X, *y*-axis distance Nu· · ·C=O, and the free energy is perpendicular to the plane).

formation of a zwitterion from uncharged reactants. It is obvious that such a process will be facilitated by polar solvent and by any additive that can take a proton from it (base-catalyzed addition: $B_{AC}2$) or protonate the alkoxide moiety (acid-catalyzed addition: $A_{AC}2$). Any compound able to give a proton and take a proton from the zwitterionic, tetrahedral intermediate will catalyze the addition reaction. This is the case with carboxylic acids such as benzoic acid (Scheme 7.1), which catalyzes the formation of poly(ethylene)terephthalate by transesterification of methyl terephthalate with ethylene glycol [64].

In an analogous way to benzoic acid that is a bifunctional catalyst (OH: acid; C=O: base), 1,5,7-triazabicyclo[4.4.0]dec-5-ene (TBD, $pK_a = 26$ in MeCN; $pK_a = 21$ in tetrahydrofuran [THF]) is an efficient catalyst of the ring-opening polymerization of cyclic esters (lactones), the key step being the formation of *N*-acyl intermediates that undergo facile intermolecular reactions with the hydroxyl moiety (Scheme 7.2) [65].

Stabilization of the alkoxide moiety of the tetrahedral intermediate resulting from the addition of HNu + RCOX (Figure 7.4) can be realized by oxyphilic Lewis acids (see Table 7.3) [67]. Thus, acetylation of alcohols with acetic acid can be carried out in the presence of triflates (trifluoromethanesulfonates) such as $Yb(OTf)_3$, $La(OTf)_3$, $Ce(OTf)_3$, $Cu(OTf)_2$, $Sc(OTf)_3$, $Er(OTf)_3$, $In(OTf)_3$, $Bi(OTf)_3$, or ytterbium *p*-toluenesulfonate ($Yb(OTs)_3$) [68]. These Lewis acids are stable in carboxylic acids, alcohols, and water. Direct condensation of carboxylic acids with alcohols has been catalyzed by hafnium(IV) salts such as $HfCl_4(THF)_2$ and $Hf(O\text{-}t\text{-}Bu)_4$, by titanium(IV) salts such as $Ti(O\text{-}t\text{-}Bu)_4$ [48], and by $ZrOCl_2(H_2O)_8$ or $HfOCl_2(H_2O)_8$ that are water-tolerant and reusable homogenous catalysts [69]. Synergism in catalytic activity with the combined use of $Hf(O\text{-}i\text{-}Pr)_4$ or $Zr(O\text{-}i\text{-}Pr)_4$ and $Fe(O\text{-}i\text{-}Pr)_3$ has been demonstrated. These catalysts are environment benign and can be extracted with ionic liquids for their recovery and reuse [70]. Stabilization of the alkoxide moiety of the tetrahedral intermediate resulting from the addition of H—Nu + RCO—X can be realized by a neighboring hydroxy group of the leaving group X. For instance, in aminolyses of adenosine 3′-ester, derivatives are faster than aminolyses of corresponding 3′-esters of 2′-deoxyadenosine [71]. This property is exploited by Nature in the ribosome (peptidyl *t*-RNA). Esterification and transesterification [72] have been catalyzed by aluminum dodecatungstophosphate ($AlPW_{12}O_{40}$) [73] and $K_5CoW_{12}O_{14}(H_2O)_3$ [74] that are water-tolerant and recyclable, nontoxic Lewis acids. Heterogeneous catalysts are preferred in the industry because their recovery requires simple filtration. A large number of these catalysts are available and contain supported Lewis acid or amphoteric compounds (see below, Section 7.2.3). Montmorillonites that are constituted of two-dimensional silicate sheets, with alternating layers that are negatively charged and

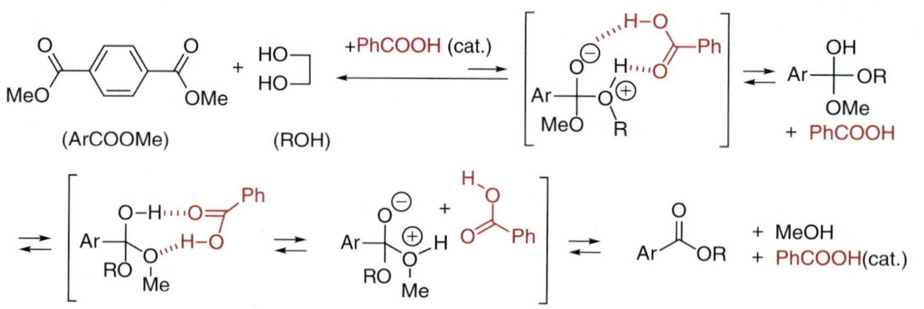

Scheme 7.1 Benzoic acid is a bifunctional (acid/base) catalyst for transesterification. It can give and take a proton, both processes can be concerted.

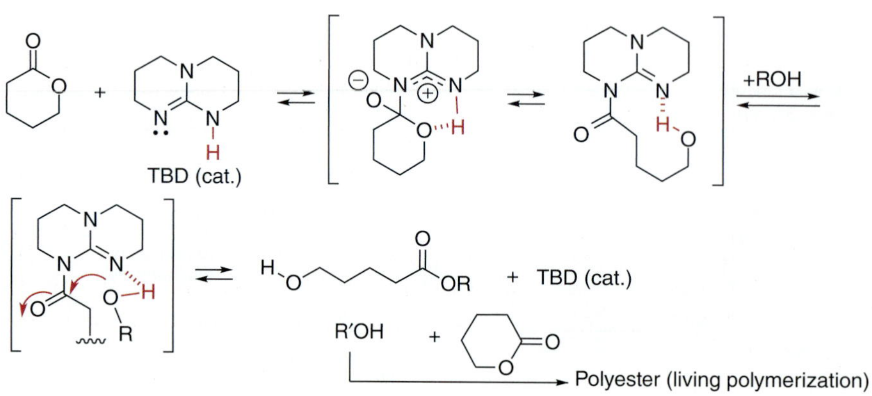

Scheme 7.2 TBD-catalyzed alcoholysis of δ-lactone (polyester formation).

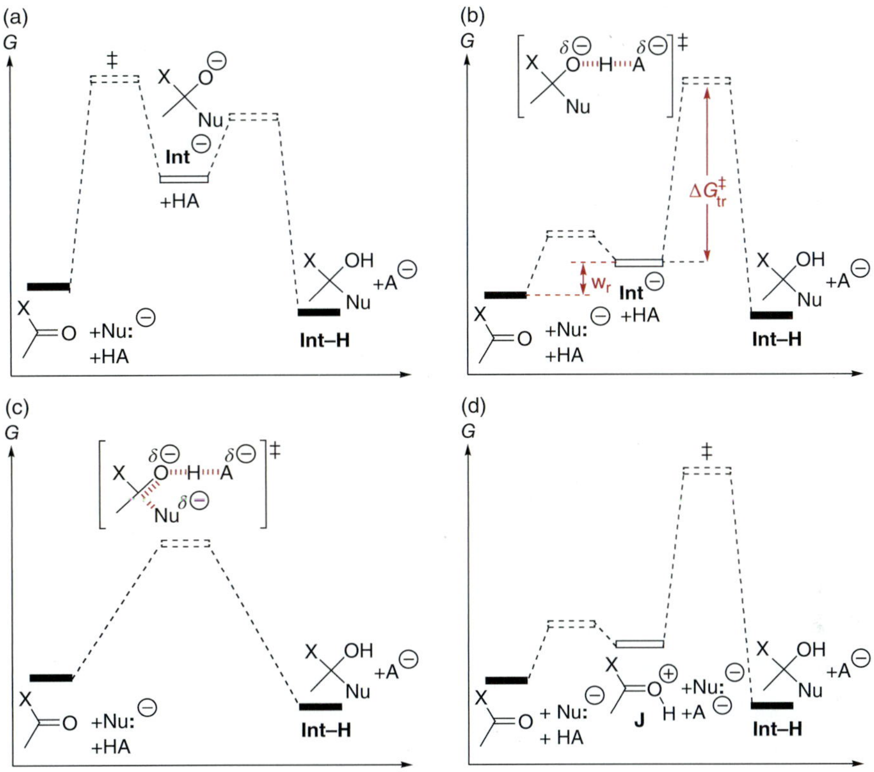

Figure 7.5 (a) The nucleophilic addition is the rate-determining step; (b) the proton transfer from the acid catalyst is the rate-determining step; (c) the nucleophile addition and proton transfer are concerted; and (d) the nucleophile addition onto the conjugate acid of the carbonyl compound is the rate-determining step.

positively charged, are suitable for impregnation of all kinds of catalysts, for instance $TiCl_4 + H_2O$, and represent efficient catalysts for esterification of carboxylic acids with alcohols [75]. Other heterogeneous catalysts can be silica-supported sulfonic acids [76] or graphite bisulfate [77]. Figure 7.5 represents the free energy diagrams of four possible mechanisms of the acid (HA)-catalyzed nucleophilic (Nu:$^-$) addition to a carbonyl compound (ester, carboxylic acid, acyl halide, carboxamide, aldehyde, and ketone). In the case of Figure 7.5a, the rate-determining step is the one-step addition of Nu:$^-$ generating the tetrahedral alcoholate intermediate **Int**$^-$, which is then protonated in a second step. In this case, the reaction is not acid-catalyzed. It could be base-catalyzed if the conjugate acid Nu—H was not completely dissociated (weak acid). In the case of Figure 7.5b, a fast pre-equilibrium exists between reactants and intermediate **Int**$^-$; the rate-determining step is the proton transfer from HA to **Int**$^-$ generating **Int–H** and A$^-$. The energy barrier of this process is discussed hereafter. In Figure 7.5c, a one-step process converts reactants and catalyst HA into a neutral adduct **Int–H** and the conjugate base A$^-$ of the catalyst **HA**. The addition of anionic nucleophile Nu:$^-$ is assisted by a proton transfer to the forming alkoxide moiety. In a fourth mechanism (Figure 7.5d), the catalyst protonates the carbonyl compound, but not the anionic nucleophile, in a fast pre-equilibrium that generates a hydroxycarbenium ion intermediate **J**. The rate-determining step is the nucleophilic addition Nu:$^-$ + **J** → **Int–H**. This mechanism is followed for reactant H—Nu with pK_a lower than pK_a of the conjugate acid of the carbonyl compound: R(X)C=O$^{(+)}$H. The energy barrier of such a reaction arises mostly from solvent reorganization (desolvation of Nu:$^-$, solvation of A$^-$, and desolvation of cationic intermediate **J**).

We now examine mechanism (b) (Figure 7.5b). The energy barrier is the sum of the free energy difference W_r between intermediate **Int**$^-$ and reactants and the activation free enthalpy $\Delta^{\ddagger}G_{tr}$ for the one-step proton transfer **Int**$^-$ + HA → [**Int**$^-$···H··A]‡ → **Int–H** + A$^-$. In this reaction, bond formation **Int**···**H** assists bond breaking A···H. In the 1920s, Brønsted observed that strongly exothermic acid- and base-catalyzed reactions have a low activation energy [78]. In 1933, Dimroth showed for a series of analogous reactions that a relationship exists between the rates and the heats of these reactions [79–83]. For comparable reactions, the more exothermic the reaction, the faster it occurs. For instance, in the case of reactions $Na^{\bullet} + R{-}X \rightleftarrows NaX + R^{\bullet}$ in which sodium atoms react with alkyl halides in the gas phase to generate the corresponding alkyl radicals and sodium halides, Evans and Polanyi [84, 85] found the relationship $\Delta^{\ddagger}H = \alpha\Delta_r H^{\circ} + \beta$ between the activation enthalpies and the heats of the reactions, where α and β are constants typical of the type of reactions under consideration [86]. Bell found the same relationship for proton transfers [87, 88]. For the exchange of proton in solution (acid/base equilibrium), the coefficient α varies between 0 and 1, and this is generally known as the Brønsted relationship [89]. Developed first for electron transfer reactions, the Marcus theory proposes relationship (7.3) for concerted proton transfer reactions [80, 90, 91].

$$\Delta^{\ddagger}G_{\text{tr}} = \left(1 + \frac{(\Delta_{\text{r}}G)^2}{\lambda}\right)\left(\frac{\lambda}{4}\right) \tag{7.3}$$

where $\Delta_r G$ is the free energy difference between $\mathbf{Int}^- + \text{AH}$ and $\mathbf{Int{-}H} + \text{A}^-$ $(-1.36\ [\text{p}K_a(\mathbf{Int{-}H}) - \text{p}K_a(\text{HA})]$ at 25 °C) and λ is the solvent reorganization term related to solvation changes between $\mathbf{Int}^- + \text{HA}$ and $\mathbf{Int{-}H} + \text{A}^-$.

7.2.3 Amphoteric compounds are good catalysts for acyl transfers

The tetranuclear zinc cluster $Zn_4(OCOCF_3)_6O$ catalyzes the direct conversion of esters, lactones, and carboxylic acids to oxazolines with remarkable chemoselectivity. The possible mechanism for these reactions is outlined in Scheme 7.3 for methyl esters **1** reacting with amino alcohols **2** giving oxazolines **3**.

The zinc cluster $Zn_4(OCOCF_3)O$ is an amphoteric compound (the metallic centers are Lewis acids and the oxygen centers are Lewis bases). This is also the case with distannoxane catalysts of type **4** [39–42]. Other oxometallic species catalyze the nucleophilic acyl substitution reactions with protic nucleophiles. Examples are TiO(acetylacetonate)$_2$, vanadyl dichloride $VOCl_2(THF)_n$ [49], and $Y_5(O\text{-}i\text{-}Pr)_{13}O$ [92], all being water-tolerant catalysts.

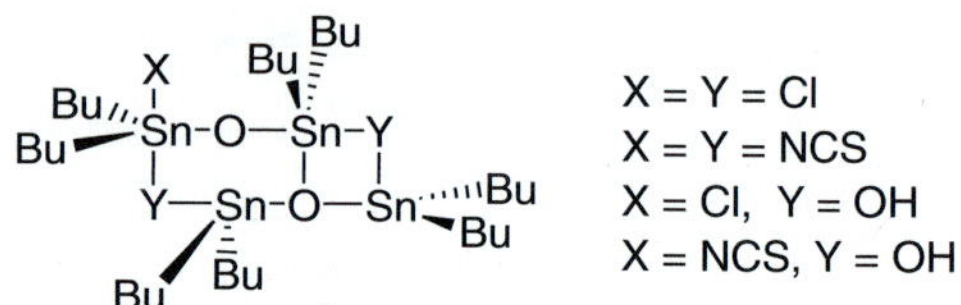

4

7.2.4 Catalysis by nucleofugal group substitution

In 1967, Litvinenko and Kirichenko established that 4-DMAP provides a 10^4-fold rate enhancement vs. pyridine in the benzoylation of 3-chloroaniline [93, 94]. In 1969, Steglich and Höfle [95] found that the difficult acetylation of tertiary alcohols [96] with acetic anhydride does not occur in pure pyridine, but only in the presence of DMAP as a catalyst [97]. Hassner and Alexanian proposed a room temperature direct esterification of carboxylic acids with alcohols catalyzed by DMAP or better, by 4-pyrrolidinopyridine (PYP) in the presence of N,N'-dicyclohexylcarbodiimide (DCC: drying agent) (reaction (7.4)) [98].

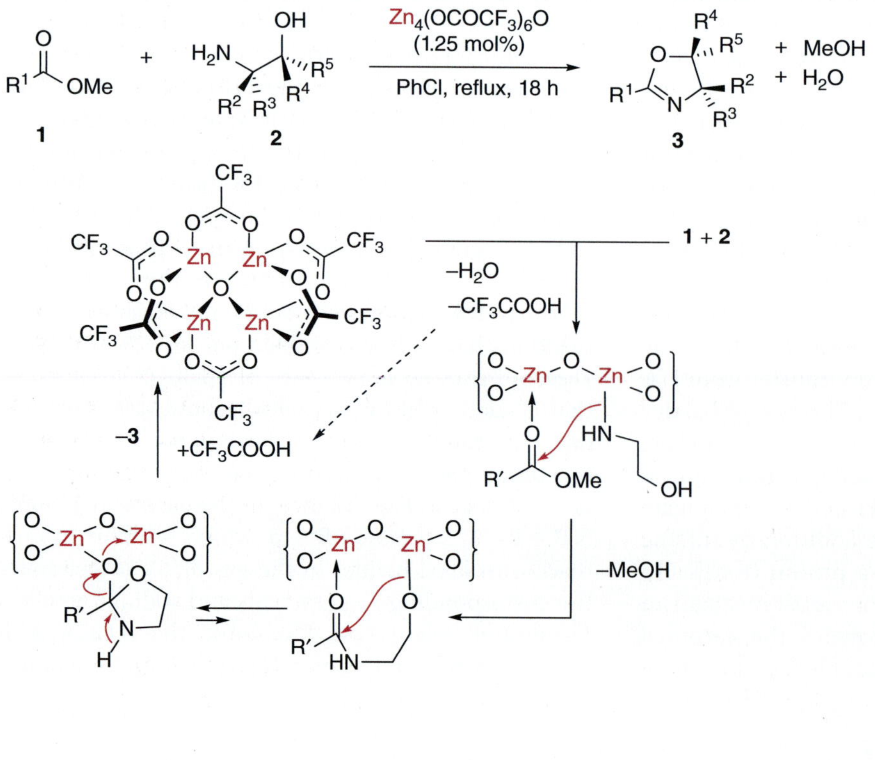

Scheme 7.3 Postulated catalytic cycle of zinc cluster-catalyzed transamination of esters by 2-aminoalcohols yielding oxazolines.

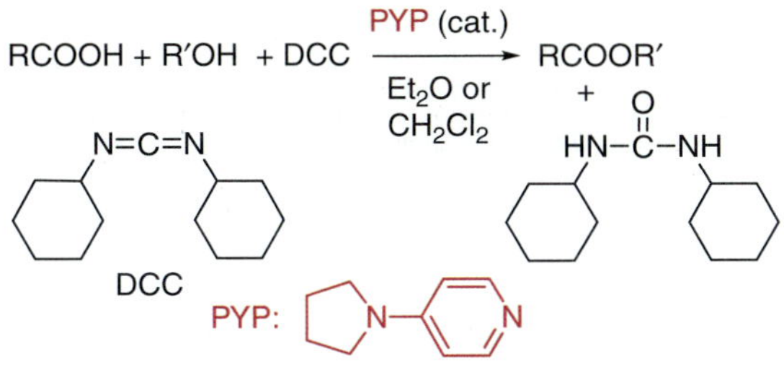

(7.4)

DMAP and other 4-aminopyridines have been used to catalyze acyl or sulfonyl transfer reactions, including the reactions of $(t\text{-BuO}_2\text{CO})_2\text{O}$ (Boc_2O) with amines, alcohols, diols, aminoalcohols, and aminothiols [99] and the chemoselective trifluoroacetylation of anilines using ethyl trifluoroacetate [100]. The catalytical effect of DMAP and analogous 4-aminopyridines results from their high nucleophilicity, favoring a fast addition/elimination process that generates *N*-acylpyridinium intermediates **5**. The latter react rapidly with the protic nucleophiles H—Nu: and generates the corresponding products of acyl transfer RCONu (reaction (7.5)).

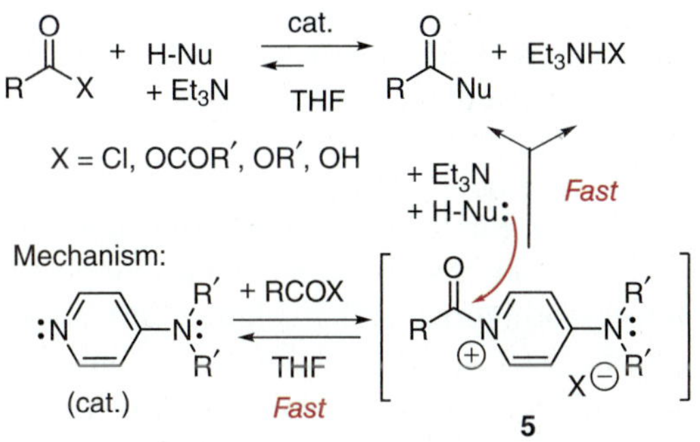

(7.5)

Electron-rich pyridines listed here below are proposed to be better than DMAP as nucleophilic catalysts [101–103]. In 1996, Vedejs and Chen introduced **6** as the first member of the "chiral DMAP" family that was used in stoichiometric amount with 2 equiv. of a Lewis acid to carry out the resolution of 1-arylethanols [104]. Enantiomerically pure DMAP derivatives such as PPY* and DMAPC$_5$Ph$_5$ (Fu's catalysts also introduced in 1996) [105, 106] are commercially available and can be used as chiral catalysts in kinetic resolution of alcohols, amines, and to induce asymmetric cycloadditions, protonations, C-acylation, halogenations, and Michael additions [107]. A large number of other enantiomerically pure *p*-dialkylaminopyridines have been proposed for enantioselective acyl transfers [108].

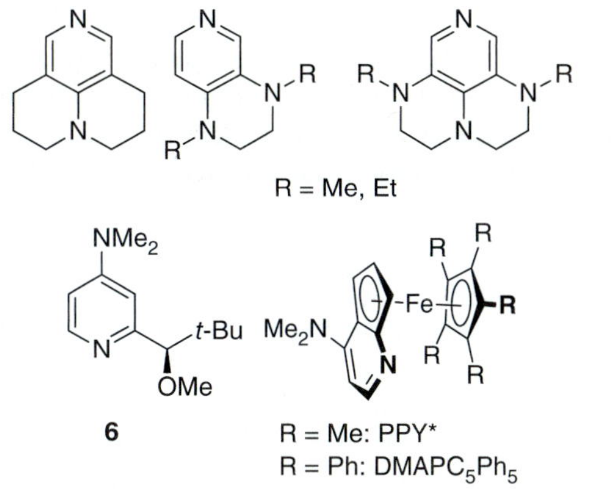

When the nucleophilic catalyst is part of a compound able to recognize a polyfunctional substrate, regioselective acyl transfer can be realized. The classical conversion of *n*-octyl β-D-glucopyranoside (**7**) into its 4-*O*-isobutyryl derivative **8** (46% overall yield) requires five synthetic steps and a final separation of regioisomeric esters. With 10 mol% catalyst **9**, this conversion is realized in isobutyryl anhydride at

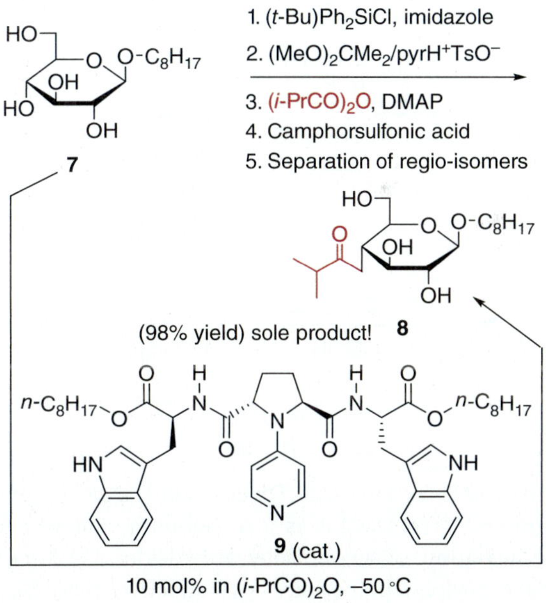

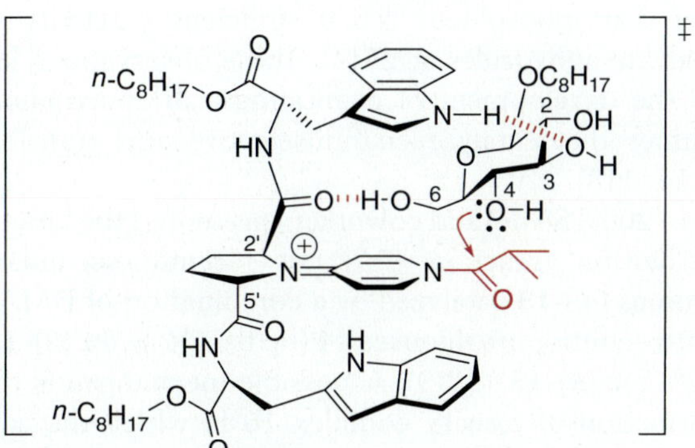

10

−50 °C giving a sole product in 98% yield. Transition structure **10** is proposed for this highly regioselective monoacylation [109]. Compound **9** is polyfunctional like amphoteric salts (Section 7.2.3). It is a nucleophile (pyridine) that activates the acyl transfer reagent, equipped with arms that are hydrogen bond donors (N—H of amides and indole moieties) to realize the required template effect with the alcoholic substrate. Many catalysts for asymmetric synthesis are of this type.

Transesterification of phosphonoacetates catalyzed with DMPA exchanges only the carboxylic esters. Similarly, reactions of phosphonoacetates with primary and secondary amines in the presence of DMAP produce the corresponding phosphonoacetamides in good yields [110]. Instead of DMAP, the cyanide anion can catalyze acyl transfer reactions. In a study comparing the catalytical activities of DMAP, 2-hydroxypyridine, imidazole, and NaCN in the difficult aminolyses of ethyl (*S*)-1-ethyl-2-pyrrolidinecarboxylate (**11**) in MeOH (reaction (7.6)), NaCN was a superior catalyst that does not induce racemization of products **12** [111].

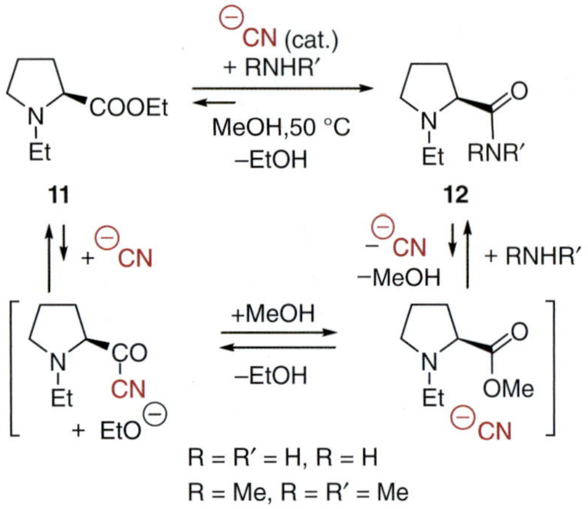

(7.6)

In 1993, Vedejs and Diver found that Bu_3P is comparable to DMAP as a nucleophilic catalyst for the acylation of alcohols by anhydrides [112, 113]. The *P*-arylphosphines are less effective than Bu_3P, but diarylphosphines retain sufficient reactivity to activate anhydrides at 25 °C. These observations led to the development of enantiomerically pure phosphines that catalyze enantioselective acyl transfers [114–118].

In 2009, Seidel and coworkers presented the kinetic resolution (reaction (7.7)) of racemic secondary amines (±)-**13** catalyzed by a combination of DMAP with enantiopure thiourea **14** (up to 77% ee for (*S*)-**13**, 82% for (*R*)-**15**)) [119]. A possible mechanism is the formation of ternary complex **16** in which the acyl cation counterion (benzoate anion) binds the chiral thiourea and controls the face selectivity of the amine addition to the acyl cation [120–122].

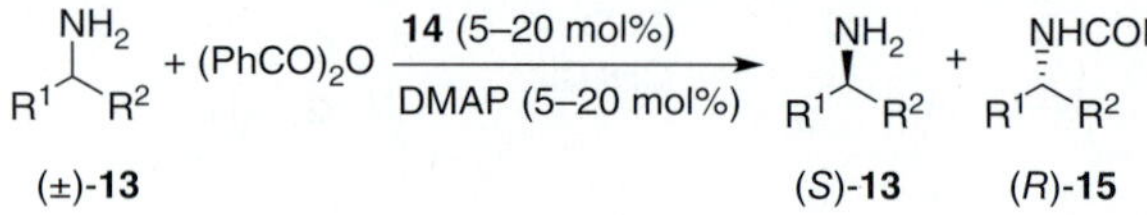

(7.7)

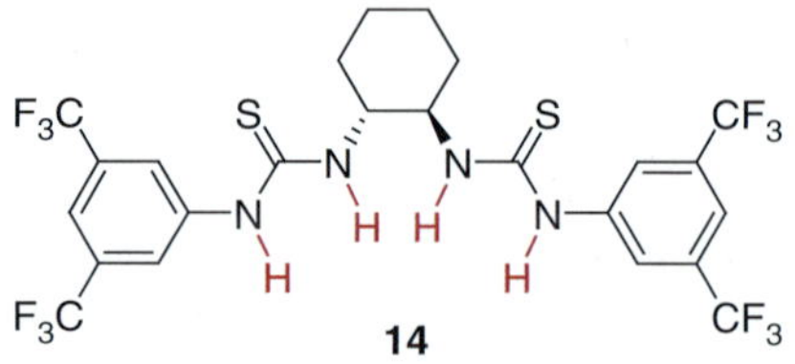

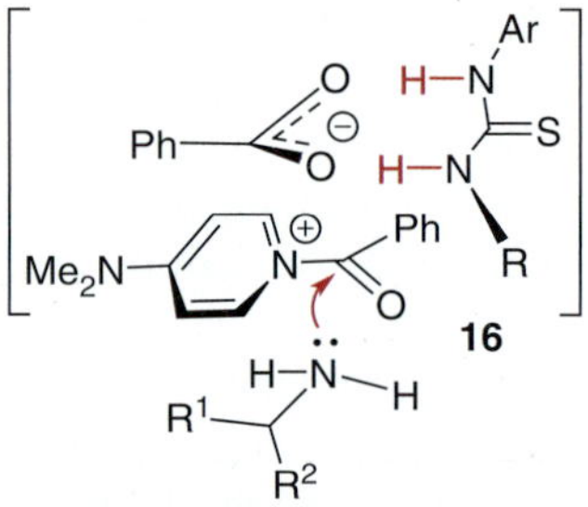

Amidines (e.g. DBU, 1,5-diazabicyclo[4.3.0]non-5-ene, DBN) [123, 124], isothiourea (e.g. 2,3,6,7-tetrahydro-5H-thiazolo[3,2-a]pyrimidine: THTP, 3,4-dihydro-2*H*-pyrimido[2,1-*b*]benzothiazole, DHPH), and guadinines (e.g. TBD) can also form acylium salts with carboxylic anhydrides that make them catalyze acyl transfers through nucleophilic substitution in analogy with electron-rich pyridines (nucleophilic catalysts), not just as bases. Several enantiomerically pure amidine, isothiourea, and guadinine derivatives have been prepared and tested for their ability to catalyze enantioselective acyl transfers as shown first by Birman et al. [125]. Examples of such enantiopure catalysts are given below [126, 127].

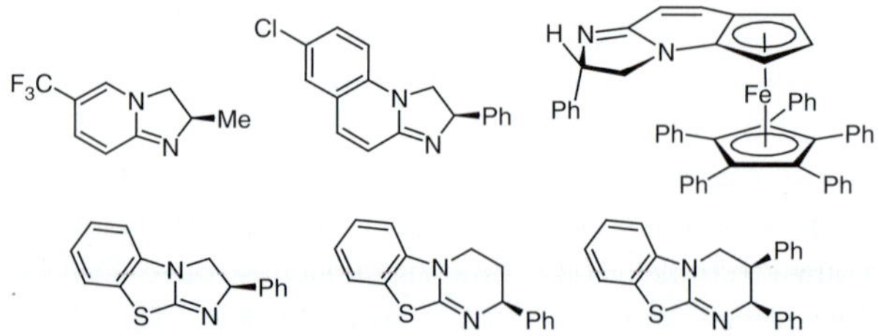

7.2.5 N-heterocyclic carbene-catalyzed transesterifications

In 1988, Bertrand and coworkers reported the first example of long-lived carbene: $Me_3SiC(:)P[N(i\text{-}Pr)_2]_2$ [128]. The first example of stable, crystalline N-heterocyclic carbene (*N*,*N*′-di(1-adamantyl) imidazol-2-ylidene) was described by Arduengo et al. [129]. In 1994, Bakhtiar and Smith found that acyl transfers can be promoted by stoichiometric amounts of carbenes [130]. In 2002, Nolan's [53] and Hedrick's [54] research groups reported that N-heterocyclic carbenes (NHC) such as **17** and **18** (imidazol-2-ylidenes) are efficient catalysts in the transesterification of esters and alcohols [55, 131], as exemplified with reaction (7.8).

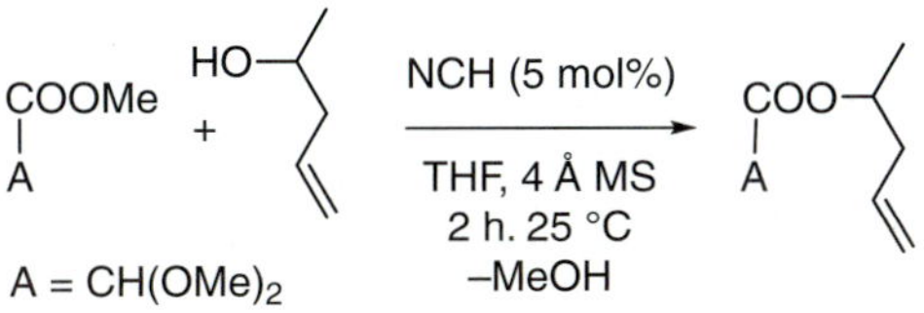

(7.8)

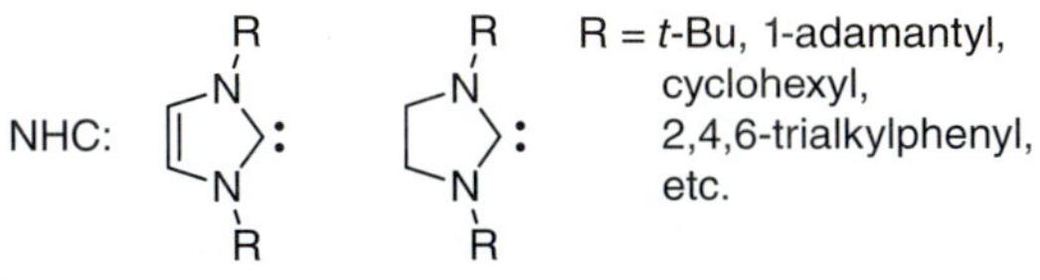

17 **18**

N-heterocyclic carbenes act as catalysts for the living ring-opening polymerization of lactones that produces polyesters (see e.g. Scheme 7.2) [132, 133]. The precatalyst can be an alcohol adduct of the NHC, e.g. **19**, which, upon heating (α-elimination), generates 1,3,4-triphenyl-1,4-dihydro-1*H*-1,2,4-triazol-5-ylidene (**20**). Today NHCs are used as organocatalysts for a large variety of reactions [134, 135].

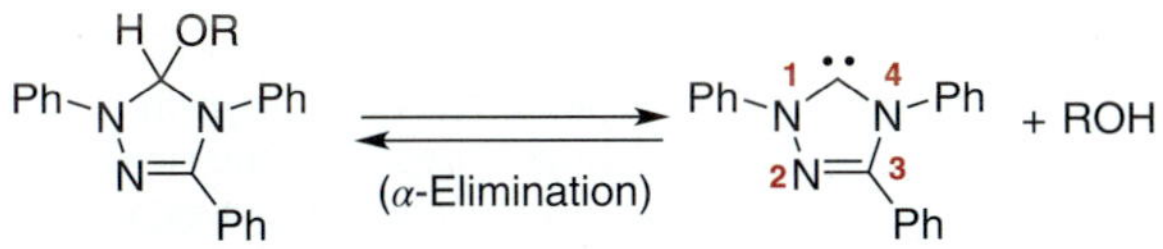

19 (dormant catalyst: pre-catalyst) **20** (active catalyst)

Quantum mechanical calculations led Hu and coworkers to propose the mechanism presented in Scheme 7.4 for the NHC-catalyzed transesterification [136]. The role of NHC is to assist proton transfer from alcohol to the carbonyl group of the ester, forming the tetrahedral intermediate **21**, which then decomposes into the acylated product.

Singh and Nolan also found that the nucleophilic NHC can catalyze the transesterification of phosphorous esters under mild conditions utilizing imidazolium salts as precatalysts [137]. NHCs catalyze the CO_2 fixation as a vinyl carbonate (reaction (7.9)). For instance, the carboxylative cyclization of 2-methylprop-3-yn-2-ol with 5 mol% 1,3-diisopropyl-4,5-dimethylimidazol-2-ylidene (100 °C, 10 MPa, 15 hours) gives 5-methylidene-4,4-dimethyl-1,3-dioxolan-2-one in 80% yield [138].

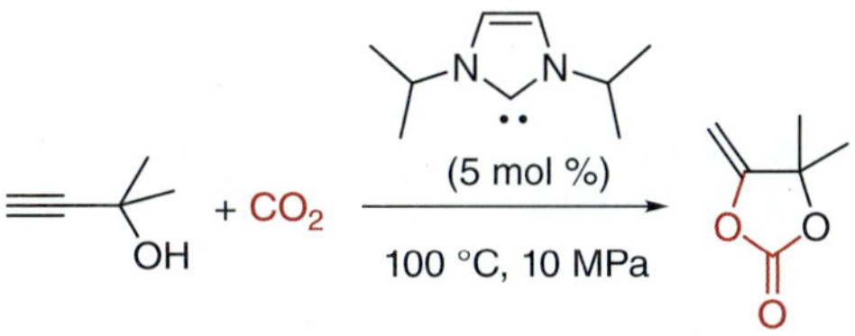

(7.9)

In 2004, Suzuki et al. reported the first examples of NHC of type **23** (obtained by in situ α-elimination from salts **22**) as catalysts for enantioselective acyl transfers [139]. They applied them to the kinetic resolution of racemic secondary aryl alkyl alcohols [140]. Further examples were reported then by the Maruoka's group [141]. In this case, one postulated the formation of 2-acylimidazolium ion intermediates of type **24**, the NHC acting as a nuleophilic catalyst. The formation of 2-acylimidazolium ion intermediates has also been postulated in the NHC-catalyzed (4+2)-cycloaddition/decarboxylation of silyl dienol ethers with *α*,β-unsaturated acyl fluorides [142, 143] and for several other NHC-catalyzed reactions involving carboxylic moieties [144–146].

Scheme 7.4 One possible mechanism (quantum mechanical calculations) of the NHC-catalyzed transesterification.

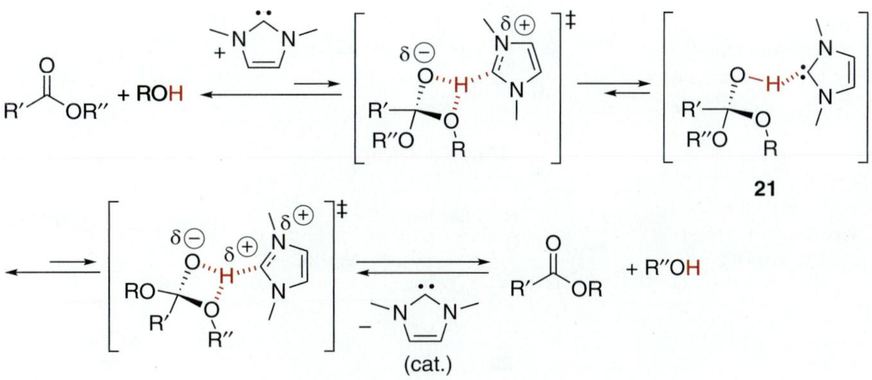

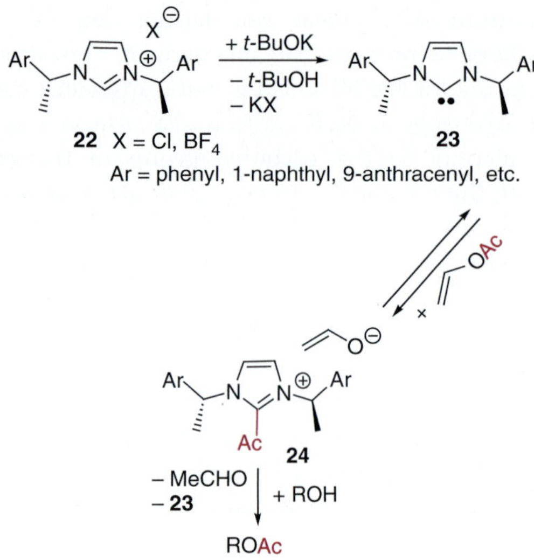

7.2.6 Enzyme-catalyzed acyl transfers

One important biological acyl transfer is the hydrolysis of the amide linkage in a protein (e.g. **25**). Chymotrypsin is one of the most well-studied peptidases (enzymes that catalyze the cleavage of peptide bonds) [147–149]. Instead of using water as a nucleophile, the alcohol side chain of serine-195 is used (Scheme 7.5). The addition of this alcoholic moiety is base-catalyzed by the side chain of histidine-57, an imidazole (step 1). Ion pairing of the resulting imidazolium ion with asparatate-102 enhances the basicity of histidine-57. The combination of serine, histidine, and aspartate constitutes the catalytic triad. This generates a tetrahedral intermediate **26** (as for mechanism $B_{AC}2$, Table 7.1) that is stabilized by hydrogen bonding in a pocket called the oxyanion hole. The amide NH bonds from the enyzme backbone contribute to the hydrogen bonding donors. After reorientation of the tetrahedral intermediate **26**, the leaving group departure is acid-catalyzed by the imidazolium ion created in step 1. This forms an acyl-enzyme intermediate **27** (end of step 2) and one of the products, the amine $R'NH_2$. Then, water enters the active site and attacks (step 3) the carbonyl group of **28**, a base-catalyzed addition promoted, as for the serine-OH addition in step 1, by the imidazole unit of histidine-57. This produces tetrahedral intermediate **29**, which then liberates serine-198 and the carboxylic acids RCOOH (**30**), the second product.

Another amide bond cleavage mechanism is followed by the carboxypeptidase A catalyzed hydrolysis [150, 151]. In this case, activation of the nucleophilic addition of water implies co-coordination of water (the nucleophile) and the carbonyl group (the electrophile) of the amide to a zinc cation (Lewis acid, template effect: transition state **31**). Furthermore, the nucleophilicity of water is enhanced by hydrogen bridging with glutamate-270 and by coordination to Zn(II). Its pK_a is expected to be decreased significantly because it coordinates Zn(II), as shown for complex **33** [152, 153]. Further electrophilic activation of the carboxamide moiety is realized by hydrogen bridging of its amino group with the phenolic moiety of tyrosine-248 and of its carbonyl group with arginium cation Arg-127 (step 1, Scheme 7.6). This leads to the formation of the tetrahedral intermediate **32**. In a second step, tyrosine- and glutamic acid catalyzed elimination of the amine $R'NH_2$ occurs, followed by the liberation of the carboxylic acid, RCOOH.

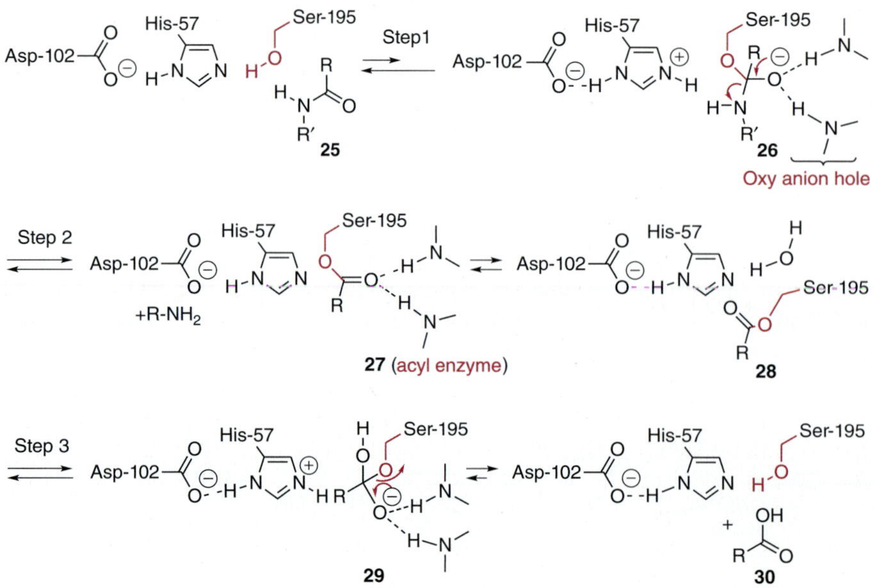

Scheme 7.5 Chymotrypsin-catalyzed peptide bond cleavage: an example of covalent catalysis by amide alcoholysis through serine (serine protease).

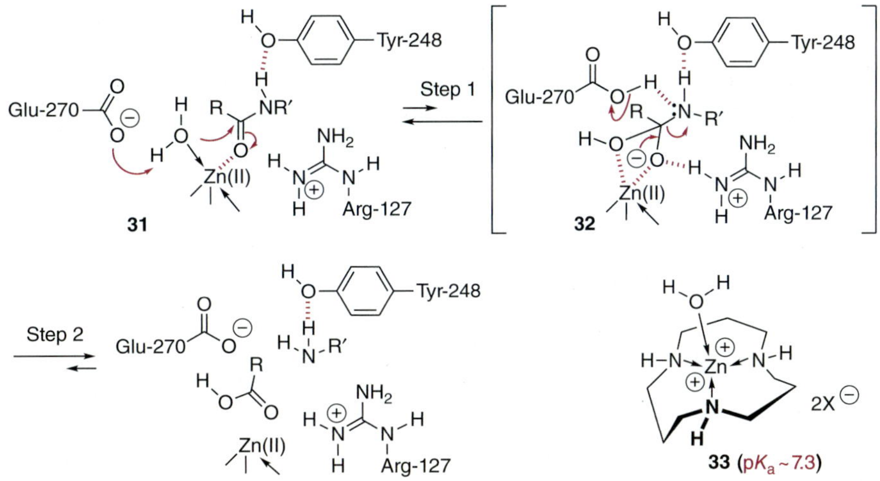

Scheme 7.6 Carboxypeptidase A (prototypical metalloprotease) catalyzed peptide bond cleavage. Example of zinc(II) electrophilic activation.

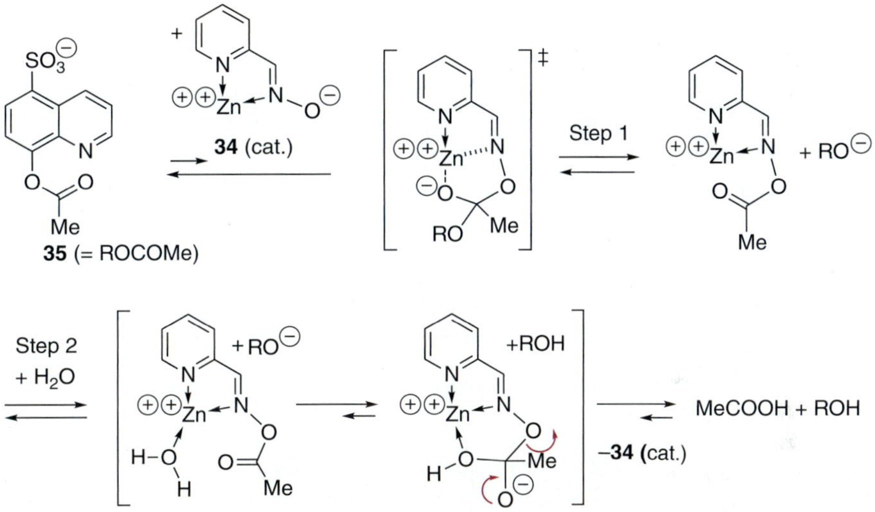

Scheme 7.7 Ester hydrolysis catalyzed by a zinc(II) aldoxime complex.

7.2.7 Mimics of carboxypeptidase A

In 1965, Breslow and Chipman [154] reported that zinc pyridinecarboxaldoxime anion complex **34** is an effective catalyst in the hydrolysis of 8-acetoxyquinoline-5-sulfonate (**35**), a weakly complexing substrate. The reactions occur in two steps involving acyl transfer within a catalyst–substrate complex (Scheme 7.7).

When **34** is coordinated with β-cyclodextrin derivative **36** containing a linking bipyridyl group, complex **37** so-obtained catalyzes the hydrolysis of ester **38** with a rate enhancement of 1 700 000-fold over the background reaction (pH = 7 at 37 °C) [155, 156]. The two β-cyclodextrin units offer a cavity to the substrate like an enzyme [157]. The adamantyl moiety of **38** enters the lyophilic cavity of one cyclodextrin unit and the leaving group (*p*-nitrophenolate) enters the other cyclodextrin unit, repelling water molecules [158]. The substrate is thus forced to approach the zinc pyridinecarboxaldoxime anion that adds (Scheme 7.7) onto its carboxylic moiety, generating an acyl-catalyst intermediate that is then hydrolyzed into *p*-nitrophenol and the carboxylic product **39** (Scheme 7.8). Several applications of cyclodextrins as catalysts have been reported [159].

7.2.8 Direct amide bond formation from amines and carboxylic acids

Amides represent one of the most important chemical building blocks found in Nature, in pharmaceuticals [160–162], in agrochemicals, and in materials [163, 164]. Using gas-phase standard heats of formation [165], one calculates standard heats of reaction $\Delta_r H°(\text{AcOH} + \text{Me}_2\text{NH} \rightleftarrows \text{AcONMe}_2 + \text{H}_2\text{O}) = -4.4\ \text{kcal mol}^{-1}$ and $\Delta_r H°(\text{AcOH} + \text{PhNH}_2 \rightleftarrows \text{AcONHPh} + \text{H}_2\text{O}) = -8.0\ \text{kcal mol}^{-1}$. In water and at 25 °C, the equilibrium constant $K°(\text{RCOOR} + \text{R}'\text{NH}_2 \rightleftarrows \text{RCONHR}' + \text{H}_2\text{O}) = 10^3\text{–}10^{4.4}$ for the formation

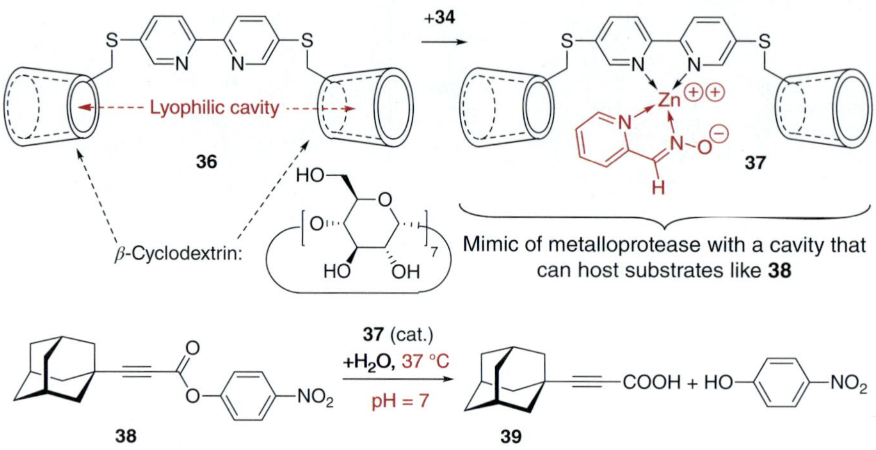

Scheme 7.8 The interior of cyclodextrins imitates the hydrophobic part of the cavity of enzymes.

Table 7.2 Examples of coupling reagents for the one-pot amide (peptide) bond formation from carboxylic acids and amines.

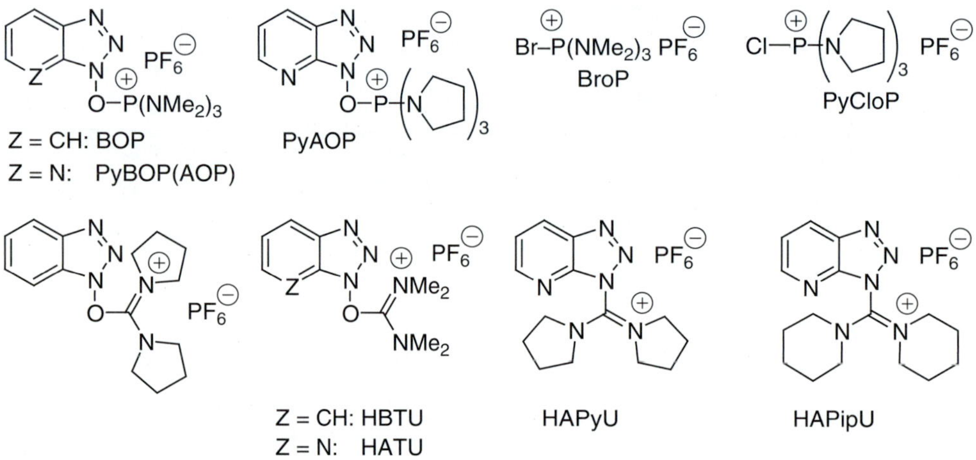

of protected and unprotected di- and tripeptides [166]. Penicillin acylase-catalyzed amidifications of phenyl acetic acid with primary amines are exergonic in diluted water for not too basic amines ($pK_a(RNH_3^+) < 8.5$) [167]. Although the direct formation of amides from amines and carboxylic acids (reaction (7.10)) have been known for 150 years, this process has little synthetic utility as it requires too harsh conditions (heating that leads to secondary products, epimerization, and polymers) [168]. Amide bonds are generally formed from carboxylic acid derivatives such as acyl halides (mostly chlorides and fluorides; the Schotten–Baumann reaction) or anhydrides, including mixed anhydrides, or active esters as those obtained by reaction of carboxylic acids with coupling reagents such as Castro's reagent

BOP (benzotriazol-1-yl-oxy-tris(dimethylamino) phosphonium hexafluorophosphate) and analogous phosphonium-based coupling reagents ByBOP, PyAOP, BroP, PyCloP, or the uronium/guanidinum-based coupling reagents BCC, HBTU (*O*-(1*H*-benzotriazol-1-yl)-*N,N,N′,N′*-tetramethyluronium hexafluorophosphate), HATU, HAPyCl, and TAPipU (Table 7.2) [169]. The latter reagents are suitable for peptide synthesis requiring smooth reaction conditions to reduce concurrent epimerization of the α-amino acid moieties. For instance, HATU is added to the carboxylic acids at room temperature to generate adducts **40** that evolve to the corresponding intermediates **43** and the coproduct **44**. Intermediates **43** are activated carboxylic derivatives that add primary amines readily producing the final carboxamides and a second coproduct **41**. Unfortunately, the one-pot method produces at least a further coproduct because the amine may also react with HATU and lead to **41** + **42** (Scheme 7.9). In terms of atomic

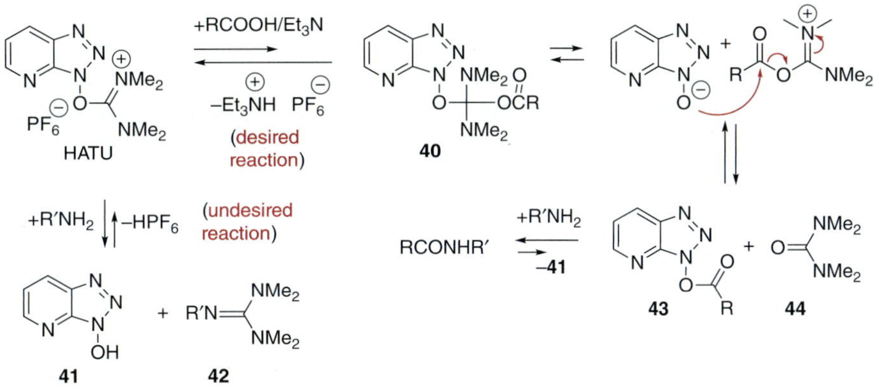

Scheme 7.9 HATU-promoted one-pot amide bond formation produces several coproducts.

economy [170], these reactions are poorly behaved and thus efficient catalysts permitting the direct amide formation (reaction (7.10)) under smooth reaction conditions are most welcome [171–173]. Nature makes amides using enzymes as catalysts. In the laboratory, mostly lipases (e.g. peptide amidase extracted from orange peel, *Candida antarctica* lipase B, *Candida cylindracea* lipase, Novozym, and *porcine pancreas* lipase) and microorganisms (e.g. *Bacillus cereus*, *Streptomyces halstedii*, and *Bacillus subtilis*) have been used as biocatalysts for direct amidification of carboxylic acids [174]. As the biocatalysts cannot accept any amines and carboxylic acids as reagents, chemical catalysts are highly desired.

$$\mathrm{RCOO^{(-)}H_3N^{(+)}R' \rightleftarrows RCOOR + R'NH_2 \rightleftarrows RCONHR' + H_2O} \qquad (7.10)$$

Since 1965, stoichiometric amounts of boron-based compounds have been used to promote the direct amidification of carboxylic acids with amines [175]. For instance, in the presence of 2 equiv. of $B(OCH_2CF_3)_3$, carboxylic acids and amines are condensed into the corresponding amides upon heating for several hours at 80–100 °C [176, 177]. The boron reagents assist amide formation directly by formation of acyloxy boron intermediates = mixed anhydrides [178–183]. In 1996, Yamamoto and coworkers reported that 3,4,5-trifluorobenzeneboronic acid catalyzes (Soxhlet refluxing toluene, xylene or mesitylene, and molecular sieve for the drying) the direct amide condensation reactions [180]. They later reported that *N*-alkyl-4-boronopyridinium halides (**45**, **46**) are more effective than boric acid and areneboronic acids [184]. The proposed mechanism of these catalytical amidifications is shown with reaction (7.11) [185]. Activated alumina balls have been found to promote the direct amide formation from carboxylic acids and amines upon heating to 140 °C for three hours [186].

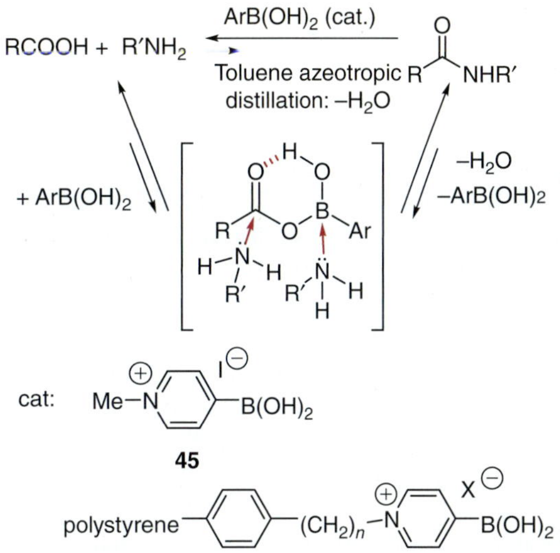

(7.11)

46 n = 1, 2, 4; X = Cl, Br, NTf_2 (Tf = CF_3SO_2)

In 2012, Hall and coworkers found that 5-methoxy-2-iodobenzeneborinic acid (MIBA) catalyst leads to higher yields for a wider range of aliphatic and heteroaromatic carboxylic acids and aliphatic amines [187]. The methods can be applied to the formation of α-peptides (4 Å molecular sieves, boiling CH_2Cl_2, 48 hours) and β-peptides as illustrated with reaction (7.12) [188].

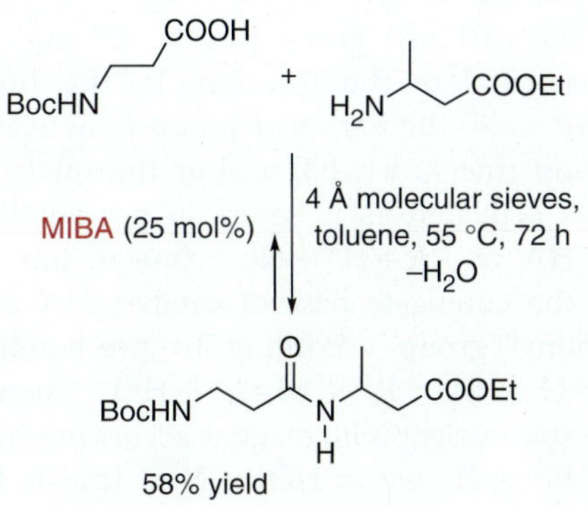

(7.12)

A large number of metallic compounds have been proposed as catalysts for the direct amidification of carboxylic acids with amines. Most of them require heating to above 100 °C such as (cyclopentadienyl)$_2$ZrCl$_2$, ZrCl$_4$ [189], or Nb$_2$O$_5$ [190]. The combination of Ph$_3$P + CCl$_4$ has been found to catalyze direct amidifications [191]. Without solvent, direct amide formation $R^1COOH + R^2(R^3)NH_2Cl + 1,1'$-carbonyldiimidazole→$R^1CON(R^3)R^2 + HCl + CO_2 + 2$ imidazole can be achieved at room temperature in a ball-mill [192].

7.3 Catalysis of nucleophilic additions

Additions of carbon nucleophiles and heteronucleophiles (E—Nu ↔ E$^+$/Nu:$^-$ ⇄ E$^+$ + Nu:$^-$) to polar unsaturated compounds such as aldehydes (RCH=O), ketones R(R′)C=O, imines (R(R′)C=NR″), and carbonitriles (R—C≡N) (1,2-additions: E—Nu + R$_2$C=Z ⇄ Nu—C(R$_2$)—Z—E), to their α,β-unsaturated derivatives (competition between 1,2- and 1,4-additions: E—Nu + R$_2$C=CR′—CR″=Z ⇄ Nu—C(R$_2$)—CR′=CR″—Z—E), and to alkenes and alkynes, bearing electron-withdrawing substituents (E—Nu + R$_2$C=C(R′)—A ⇄ Nu—CR$_2$—C(R′)(E)—A, e.g. A = SOR, SO$_2$R; CF$_3$), represent the most important reactions of synthetic interest. All these reactions can be catalyzed as illustrated in the next chapters by selected examples.

7.3.1 Catalysis of nucleophilic additions to aldehydes, ketones and imines

As for the acyl transfer reactions, nucleophilic additions to aldehydes and ketones of reagent E—Nu (Scheme 7.10) can be catalyzed by a protic (Brønsted) acid (HA), a Lewis base (B:), or a Lewis acid (MX ↔ M$^+$/X$^-$). In the absence of a catalyst, the electrofugal moiety of the reagent E—Nu can activate the carbonyl group (mech. a2) and, eventually, autocatalyze the reaction. In the presence of a protic acid, the carbonyl group is activated by protonation (mech. b1, b3) and/or the nucleophilic reagent is activated as it enters in pre-equilibrium E—Nu + HA ⇄ EA + H$^+$ + Nu:$^-$ (mech. b2). Alternatively, the conjugate base of catalyst HA adds to the carbonyl group according to pre-equilibrium R^1(R^2)C=O + HA ⇄ R^1(R^2)C=A$^+$ + HO$^-$. Then, HO$^-$ activates the nucleophilic reagent according to equilibrium HO$^-$ + E—Nu ⇄ HOE + Nu:$^-$ (mech. b3). In the presence of a Lewis base B:, the latter activates E—Nu by deprotonation when E = H, or by entering into pre-equilibrium B: + E—Nu ⇄ E—B$^+$ + Nu:$^-$ when E ≠ H (mech. c1). Alternatively, the base acts as a nucleophilic catalyst by adding first to the carbonyl moiety and equilibrating with an oxycarbenium ion intermediate of type R^1(R^2)C=O$^{(+)}$—E, which then adds the negatively charged nucleophile Nu:$^-$ rapidly (mech. c2). In the presence of a Lewis acid MX, the carbonyl group is activated by forming an ion pair of type R^1(R^2)C=O$^{(+)}$—M/X$^-$ (mech. d1). Alternatively, E—Nu can be activated by entering into pre-equilibrium E—Nu + MX ⇄ EX + M$^+$/Nu:$^-$ (mech. d2). In these mechanistic limits, the activation of the carbonyl group implies the formation of ion pairs of type R^1(R^2)C=O$^{(+)}$—E/Nu:$^-$ (mech. a2, b2, R^1(R^2)C=O$^{(+)}$—H/A:$^-$ (mech. b1, b3) or R^1(R^2)C=O$^{(+)}$—M/X:$^-$ (mech. d1), and the activation of the nucleophilic reagent E—Nu implies the formation of ion pairs of type H$^+$/Nu:$^-$ (mech. b2), E—B$^+$/Nu:$^-$ (mech. c1) or M$^+$/Nu:$^-$ (mech. d2). As we shall see, in many cases, these species do not have to be formed for activation; instead, complexes of type R^1(R^2)C=O → E—Nu and R^1(R^2)C=O…H—A (hydrogen-bonded carbonyl) or Lewis acid–base complexes of type R^1(R^2)C=O → MX react with the nucleophile (E—Nu, Nu:$^-$, or Nu:$^-$···H—A). The same mechanistic limits can be retained for the nucleophilic additions of imines (Scheme 7.10) [193].

From the above discussion, one predicts that the best catalysts of the nucleophilic additions (and of the reverse elimination reaction) possess on the same template functional groups capable of electrophilic activation of the carbonyl or imine moieties and nucleophilic assistance to the dissociation of reagent E—Nu as shown below with reaction (7.13) and as illustrated already in Schemes 7.1 and 7.2 (Section 7.2.2) where benzoic acid and TBD act as bifunctional catalysts (i.e. as proton donor and proton acceptor, both proton transfers being concerted or not). Further illustration is with the asymmetric Strecker reaction (7.14) catalyzed by the chiral, enantiomerically pure bicyclic guanidine **47** reported by Corey and Grogan [194], and, in the Section 7.3.2, with the mutarotation of 2,3,4,6-tetra-*O*-methyl-α-D-glucopyranose catalyzed by 2-pyridinone.

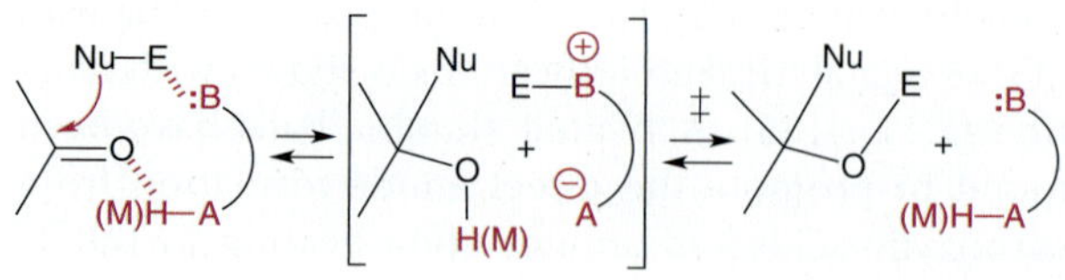

(7.13)

Scheme 7.10 Possible mechanism limits for the nucleophilic additions of carbonyl compounds: (a) uncatalyzed reaction, (b) protic acid catalyzed, (c) Lewis base catalyzed, and (d) Lewis acid catalyzed.

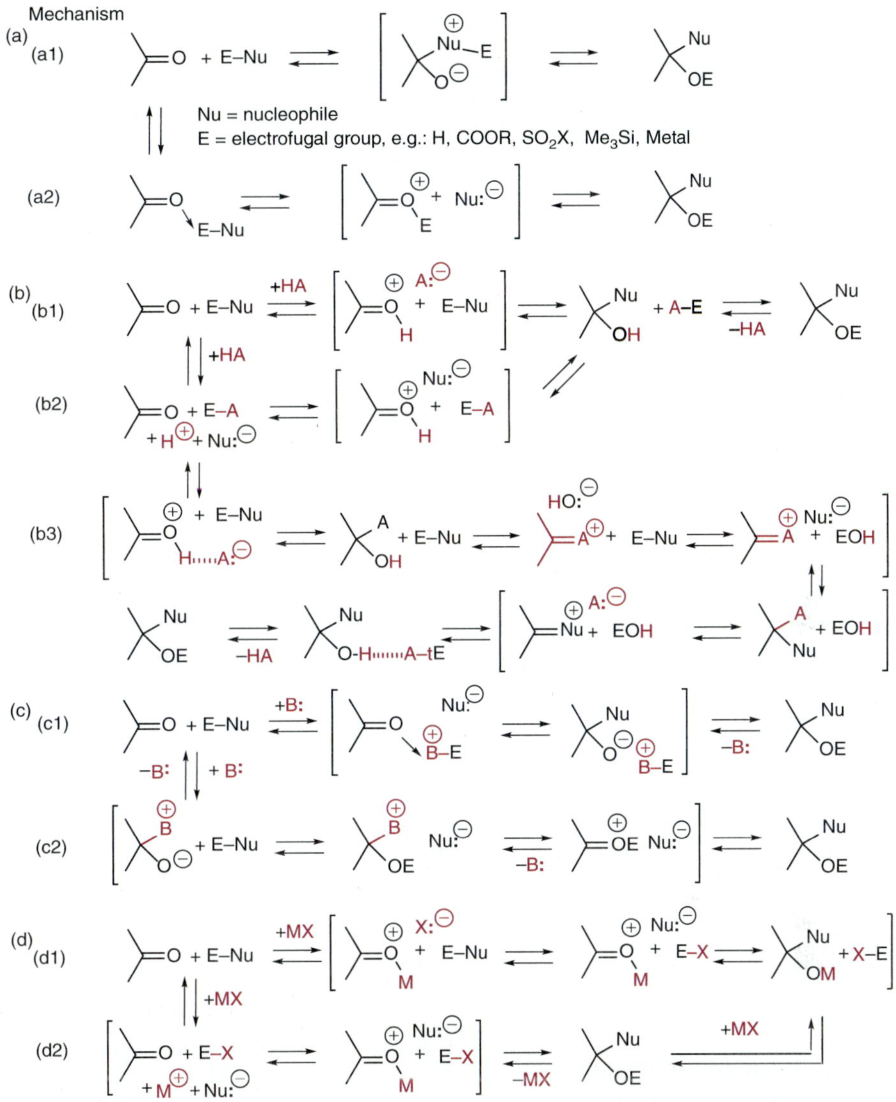

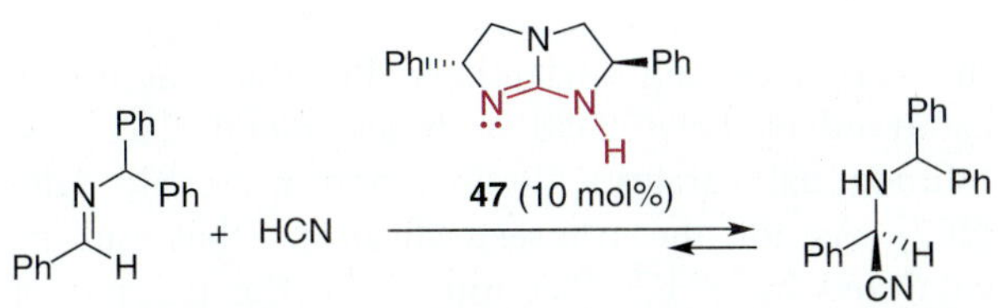

(7.14)

7.3.2 Bifunctional catalysts for nucleophilic addition/elimination

In 1952, Swain and Brown [195, 196] reported that neutral 2-pyridinone (**48**) is a better catalyst of the mutarotation (epimerization) of 2,3,4,6-tetra-*O*-methyl-α-D-glucopyranose (α-**49**) into its β-anomer (β-**49**) in benzene than in pyridine and phenol alone, and than in a 1 : 1 mixture of pyridine and phenol. This was taken as a proof of a rate-controlling step involving two-proton transfers (formation of complex **51**) coupled with the C(1)…O bond cleavage of the glucose moiety as shown with transition state **52** (Scheme 7.11).

In the gas phase and in dilute solution of benzene, α-pyridone (**48**) and its tautomer 2-hydroxypyridine (**48′**) have similar stability [197], each of them may catalyze the mutarotation [198]. The mechanism proposed in Scheme 7.11 implies that α-pyridone (equilibrates with dimers) forms a precomplex **51** with α-**49**. The two proton transfers occur in concert with the breaking of the C(1)—O bond, leading to the hydroxyaldehyde **50** in one step. Fast rotation about the C(1)—C(2) bond engenders rotameric hydroxyaldehyde **50′** that undergoes intramolecular alcohol addition furnishing

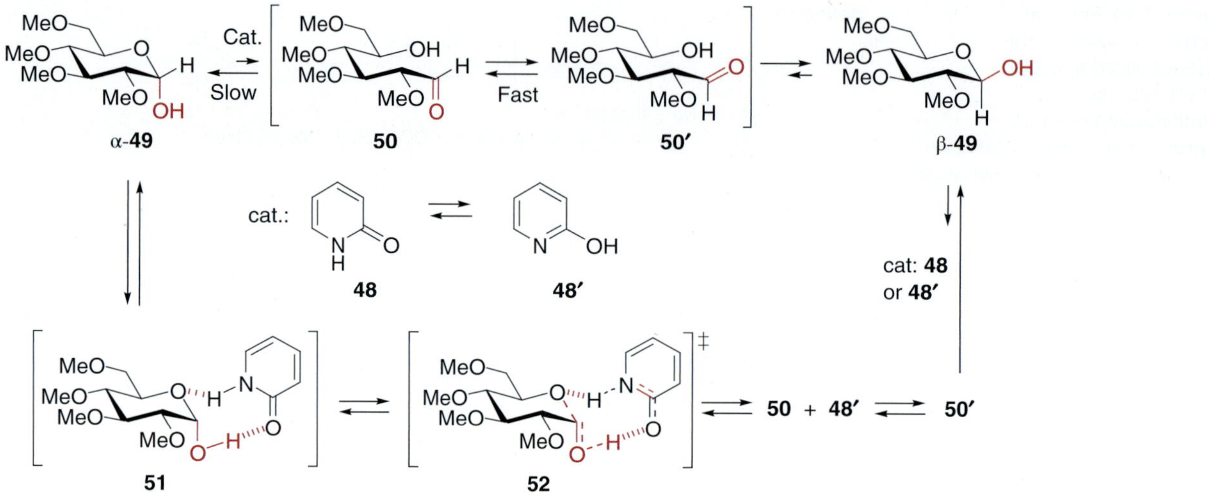

Scheme 7.11 Mutarotation of tetramethylglucose catalyzed by the difunctional catalyst α-pyridone ⇄ 2-hydroxypyridine in benzene (apolar solvent).

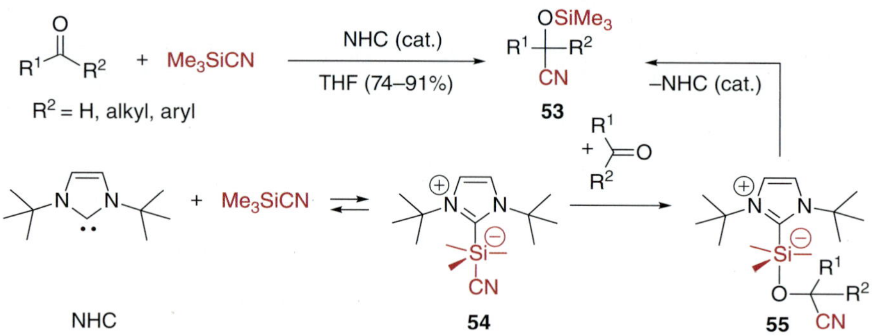

Scheme 7.12 NHC-catalyzed cyanosilylation of aldehydes and ketones.

hemiacetal β-**49**. This addition is also catalyzed by **48** or **48′** [199, 200]. Kinetic deuterium isotope effects determined with (*N*-D)-2-pyridinone and (1-*O*-D)-2,3,4,6-tetra-*O*-methyl-α-D-glucopyranose agree with the above hypothesis, although they could not tell whether the two proton transfers are synchronous or not [201]. Analogs of α-pyridinone such as benzoic acid, formamide/formamidic [202] acid tautomeric couple, or 2-hydroxy-4-methylquinoline; 2-aminopyridine; imidazole [203]; and 2′,3′,5′-tri-*O*-protected RNA nucleosides [204] can also act as bifunctional catalysts. Sodium 2-pyridinolate has been found to catalyze aminolysis of esters [205].

7.3.3 σ- and π-Nucleophiles as catalysts for nucleophilic additions to aldehydes and ketones

Enantio-enriched cyanohydrins are important building blocks for the preparation of α-hydroxy acids, α-hydroxy aldehydes, 1,2-diols, and α-amino alcohols [206]. Cyanohydrins can be obtained by cyanosilylation of aldehydes and ketones using Me_3SiCN (=E—Nu in Scheme 7.10) that produces the corresponding silyl ethers **53**. The reaction is catalyzed by Lewis acid or bases, metal alkoxides, bifunctional catalysts, iodine, and inorganic salts [207]. As for the transesterifications that can be catalyzed by NHCs (Section 7.2.5), the transfer of cyano group from Me_3SiCN to a carbonyl compound has been catalyzed by diaminocarbenes [208]. It is proposed (Scheme 7.12) that the NHC-catalyzed cyanosilylation follows a nucleophilic catalysis pathway (mech. c1, Scheme 7.10) implying intermediate **54** with a pentacoordinated silicon center [209–211]. The latter transfers the cyano group to the aldehyde or ketone generating adduct **55**, which then dissociates into product **53** and the catalyst (NHC).

NHCs are neutral σ-nucleophilic catalysts. According to Wang and Tian [212], electron-rich π-systems should also be catalysts for the cyanosilylation of carbonyl compounds. Indeed, they found that the reaction of PhCHO + Me_3SiCN in the presence of $Me_2C{=}C(OSiMe_3)OMe$ (10%) is at least 100-fold faster than without catalyst at 25 °C.

Aniline acts as a nucleophilic catalyst (mech. b3, Scheme 7.10) in oxime ligation in aqueous solution through formation of intermediate iminium salt, followed by transimination by the oxime (reaction (7.15)) [213–215].

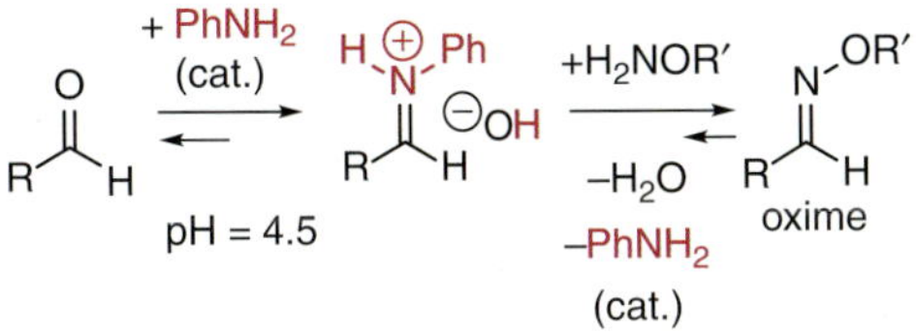
(7.15)

Since the pioneering work of Oguni and coworkers in 1993 that used the enantiopure chiral Schiff base **56** and $Ti(O\text{-}i\text{-}Pr)_4$ as catalyst (e.g. reaction (7.16)) [216], a large number of enantioselective syntheses of cyanohydrins have been developed [217–220]. The transition structure **57** has been proposed to interpret the results (mech. d1, Scheme 7.10).

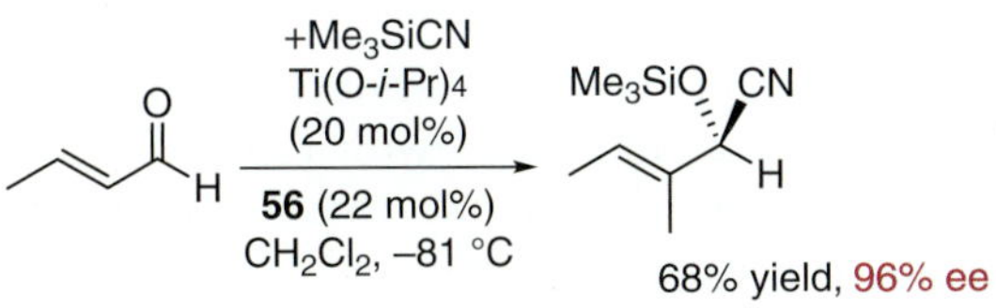
(7.16)

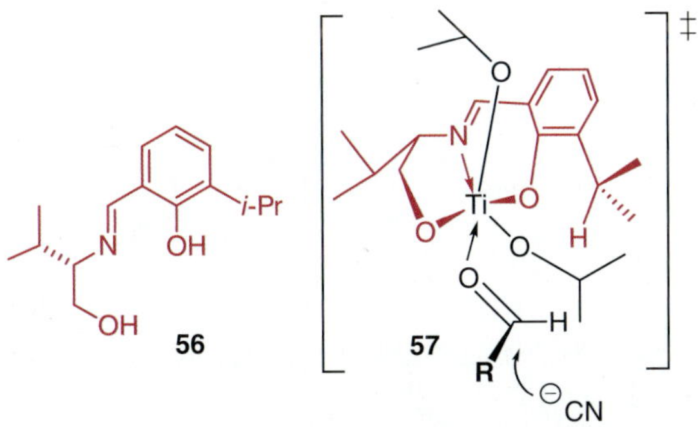
56 **57**

An example of enantioselective cyanation of an aldimine [221] with the quinine-derived thiourea catalyst **58** is given with reaction (7.17) for which transition structure **59** is proposed to explain the high yield and enantioselectivity (mech. b1, Scheme 7.10).

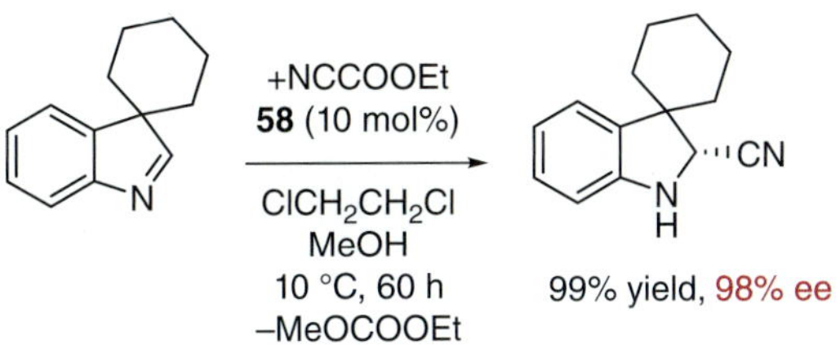
(7.17)

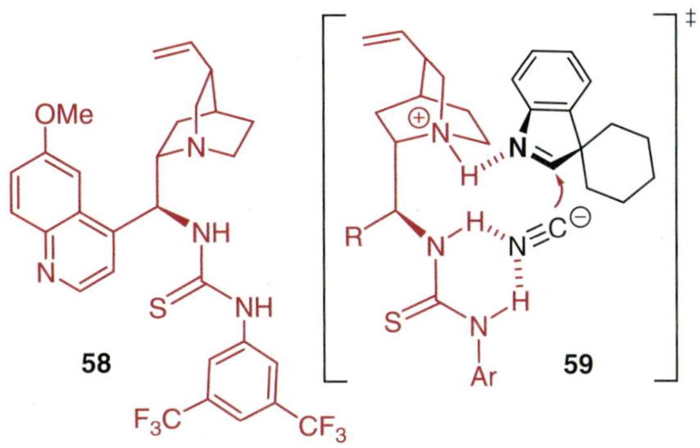
58 **59**

7.3.4 Catalysis by self-assembled encapsulation

Self-assembly in water of 6 equiv. of ligand **60** (L) with 4 equiv. of $Gd(acac)_3$ generates the negatively charged supramolecular assembly $Gd_4L_6^{12-}$ that can host all kinds of small monocationic guests in its cavity. The naphthalene walls render the hydrophobic interior and thus able to promote acid-catalyzed reaction on encapsulation. This was suggested by the observation that *N,N,N′,N′*-tetramethyl-1,4-diaminobutane or *N,N,N′,N′*-tetramethyl-1,2-diaminoethane are doubly protonated when encapsulated with $Gd_4L_6^{12-}$. Orthoesters require normally acidic conditions to be hydrolyzed (acid-catalyzed elimination, reverse of nucleophilic addition to a carbonyl compound for one step of the overall process). Under neutral or alkaline conditions, their hydrolysis is very slow. However, when orthoesters $HC(OR)_3$ (R = Me, Et, Pr, *i*-Pr, Bu, *i*-Bu) are mixed in D_2O (22 °C, pH = 11, K_2CO_3) with small amounts of $Gd_4L_6^{12-}$, hydrolysis proceeds. The phenomenon is not observed with larger orthoesters (e.g. with R = pentyl). Thus, upon encapsulation, the orthoesters are protonated into **61** and then dissociated into two alcohol moieties and the conjugate acids **62** of the corresponding formates. The latter diffuse out of the host and are saponified by the alkaline medium (Figure 7.6). Importantly, the catalysis by the host of type $Gd_4L_6^{12-}$ obeys Michaelis–Menten kinetics (Section 3.2.8) and exhibits competitive inhibition. When compared to the background hydrolysis reactions under the same conditions, rate accelerations (k_{cat}/k_{uncat}) for triethyl orthoformate and triisopropyl formate are 560 and 890, respectively (pH = 11, 22 °C, D_2O) [222, 223].

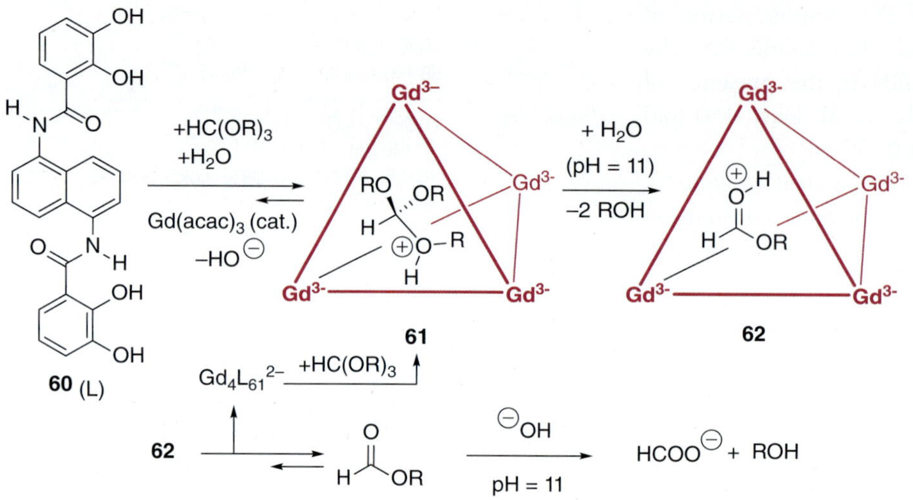

Figure 7.6 Catalysis by supramolecular assembly. Hydrolysis of orthoesters under alkaline conditions.

7.3.5 Catalysis of 1,4-additions (conjugate additions)

Conjugate nucleophilic additions to electron-deficient alkenes [224] and dienes [225] are one of the most useful tools in organic synthesis. In 1977, Wynberg and coworkers [226] reported that cinchona alkaloids catalyze the asymmetric conjugate addition of aromatic thiols to cycloalkenones (e.g. reaction (7.18)) [227]. Mechanical quantum calculations proposed transition state of type **63** for these reactions [228].

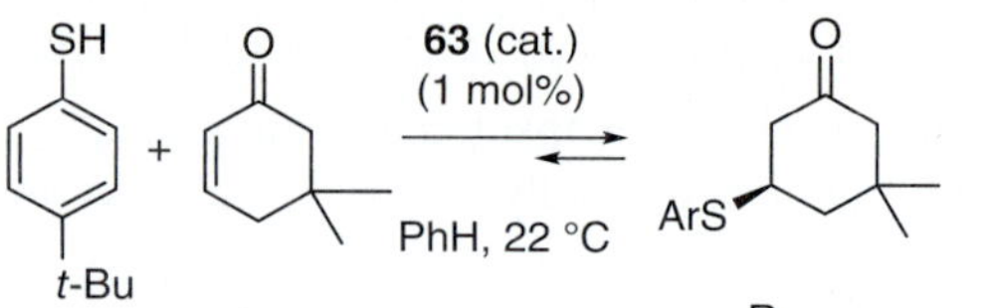

(7.18)

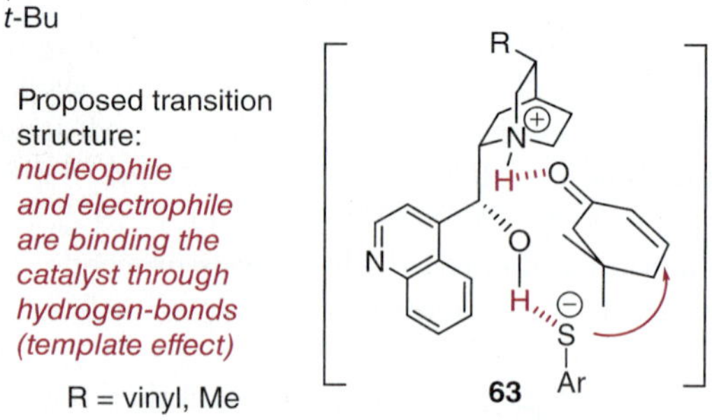

This has been the starting point of several enantioselective conjugate additions using chiral catalysts [229–233]. For instance, the enantioselective Michael addition (7.19) of dimethyl malonate to cyclohex-2-enone is carried out on the kilogram scale. It uses the bifunctional Lewis acid/Brønsted base catalyst **64** developed by Shibasaki and coworkers [207, 234–236]. Adduct **65** is the starting material in the synthesis of (−)-strychnine.

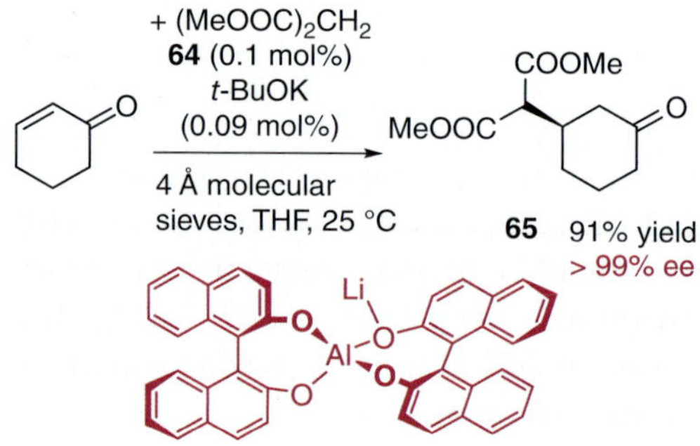

(7.19)

The bifunctional thiourea catalyst **66** developed by Takemoto and coworkers in 2003 [237, 238] activates both the nucleophile and the carbonyl group of the Michael acceptor as illustrated with reaction (7.20) [239]. In a similar manner, the chiral squaramide catalyst **67** reported by Rawal and coworkers in 2008 catalyzes the enantioselective conjugate additions (8.21) of 1,3-diketones and 3-ketoesters to (*E*)-2-nitrostyrene [240].

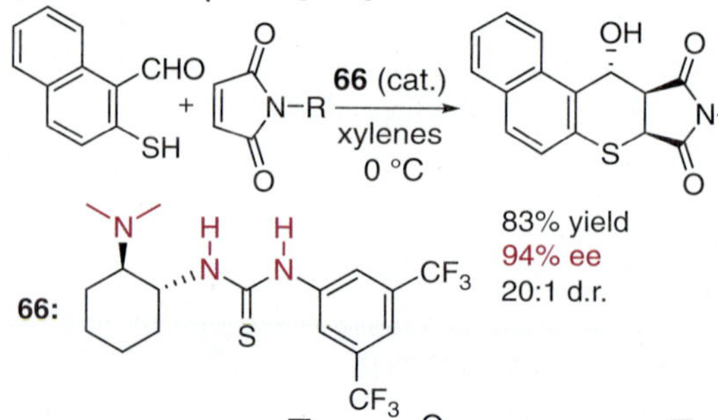

(7.20)

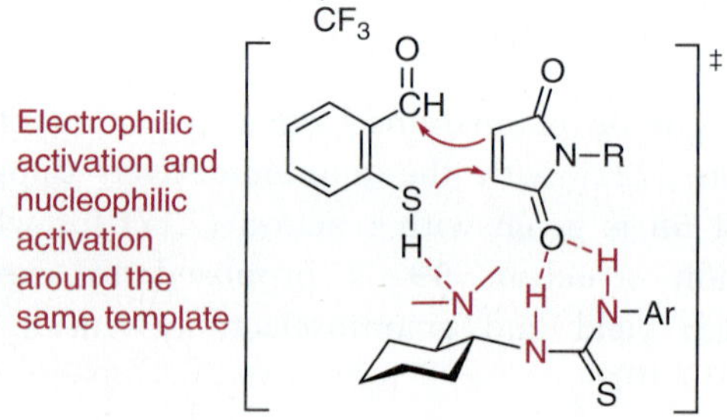

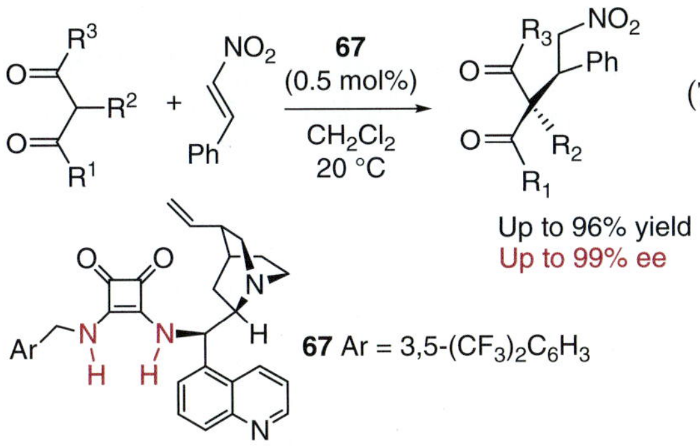

(7.21)

7.4 Anionic nucleophilic displacement reactions

Nucleophilic substitution is one of the most important reactions to construct C—C and C–heteroatom bonds. There are four classes of nucleophilic displacement reactions: reaction (7.22) involves neutral electrophile R—X and nucleophile Nu: and generates a salt (e.g. Menschutkin reaction: quaternization of amines and phosphines, formation of oxonium and sulfonium ions); reaction (7.23) is like the reverse reaction of equilibrium (7.22) in which a positively charged nucleofugal group is substituted by a negatively charged nucleophile Y:⁻; reaction (7.24) engages a neutral electrophile R—X and an anionic nucleophile Y:⁻ with a proton or cation as a counterion; and finally, in reaction (7.25), a neutral nucleophile attacks a cationic electrophile.

$$\mathrm{Nu:} + \mathrm{R-X} \rightleftarrows \mathrm{R-Nu^+} + \mathrm{X:^-} \tag{7.22}$$

$$\mathrm{Y:^-} + \mathrm{R-Z^+} \rightleftarrows \mathrm{R-Y} + \mathrm{Z:} \tag{7.23}$$

$$\mathrm{M^+/Y:^-} + \mathrm{R-X} \rightleftarrows \mathrm{R-Y} + \mathrm{M^+/X:^-} \tag{7.24}$$

$$\mathrm{Nu:} + \mathrm{R-Z^+} \rightleftarrows \mathrm{R-Nu^+} + \mathrm{Z:} \tag{7.25}$$

The S_N2 reaction (7.24) is very important for organic synthesis and we discuss now how it can be accelerated and/or catalyzed. In the gas phase, a naked anion Y:⁻ and a neutral electrophile R—X equilibrate without activation barrier with complex [Y:• •R—X]⁻ [108, 241–245]. The substrate R—X solvates the anionic nucleophile. A subsequent bond reorganization gives another complex [Y—R• •X:]⁻ with an energy barrier that depends on the exothermicity of the reaction and the polarizability of the reactants (Bell–Evans–Polanyi theory: $\Delta^{\ddagger}H = \alpha\Delta_r H + \beta$). Polarizability depends above all on the ionization energy (IE) of Y:⁻ (the easier it can lose an electron, the more nucleophilic it is) and on the electron affinity (−EA) of R—X (the easier R—X can accept an electron, the more stabilized is the corresponding radical anion [R—X]$^{\bullet-}$; the more electrophilic is R—X, the faster is the nucleophilic substitution). The Marcus equation $\Delta^{\ddagger}G = (\Delta_r G°)^2/(16\Delta^{\ddagger}G_0) + \Delta^{\ddagger}G_0 + 0.5\Delta_r G°$, originally developed to model electron transfer reactions [80], has been found to apply to methyl transfer (S_N2) reactions in the gas phase [15, 246]. In this theory, $\Delta_r G°$, $\Delta^{\ddagger}G$, and $\Delta^{\ddagger}G_0$ are the free energy of reaction, activation free energy, and "intrinsic" free energy of activation (corresponds to β in the Bell–Evans–Polanyi theory: $\Delta^{\ddagger}H = \alpha\Delta_r H + \beta$), respectively. Shaik and Pross, and others, have described the S_N2 reaction in terms of valence bond theory [247–255]. The transition state of the gas-phase S_N2 reaction can be represented by the limiting structures: $[\mathrm{Y:^-/R^\bullet/X^\bullet} \leftrightarrow \mathrm{Y:^-/R^+/X^-} \leftrightarrow \mathrm{Y^\bullet/R^\bullet/X:^-} \leftrightarrow \mathrm{Y^\bullet/R:^-/X^\bullet}]^{\ddagger}$. This representation allows one to predict that the weaker the R—X bond ($DH°(\mathrm{R^\bullet/X^\bullet})$), the faster the reaction. Furthermore, the higher is the stability of cation R⁺ and of nucleofugal group X:⁻, and/or the higher is the stability of the carbanion R:⁻, the faster is the reaction.

7.4.1 Pulling on the leaving group

The electron affinity of substrates R—X can be enhanced by coordination to an electrophilic catalyst, such as a metallic cation M′⁺. For it to act as a catalyst of the S_N2 reaction, the binding energy in M—Y must be lower than that in M′—X [256]. This is the case with soluble Ti⁺ and Ag⁺ salts (e.g. $AgNO_3$, $AgClO_4$, $AgOSO_2Cl$, and AgOTf) in MeCN and alkyl halides (R—Hal) as reactants (reaction (7.26)) [257, 258].

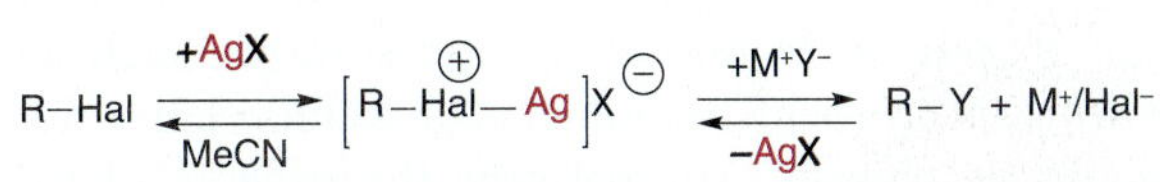

(7.26)

In the case of S_N2 reactions of sulfonic esters, the counter cation M⁺ of M⁺/Y:⁻ can assist the departure of the nucleofugal group (sulfonate anion) by coordinating the S=O groups [259, 260]. Alternatively, the displacement of a sulfonic ester can be activated through hydrogen donors making strong hydrogen bond with the S=O moieties. Similarly, a sulfone is an electrophilic reagent when the departure of the nucleofugal group (sulfinate anion) is activated by hydrogen bridging of the SO_2 moiety. Applying this feature, an asymmetric Mannich-type reaction (7.27) of aromatic α-amido sulfone with malonate using guanidine–thiourea bifunctional organocatalyst **68** has been proposed [261].

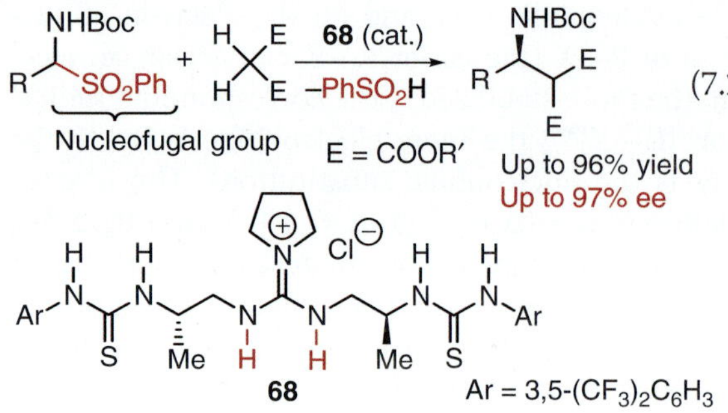

(7.27)

7.4.2 Phase transfer catalysis

In solution, a large extent of the energy barrier of a S_N2 reaction of type (7.24) implies dissociation of $M^+/Y:^-$ and desolvation of the anionic nucleophile $Y:^-$. Thus, such a reaction will be faster in an aprotic, polar solvent, such as MeCN, DMSO, DMF, or *N*-methylpyrrolidone or sulfolane, than in a protic polar solvent such as H_2O, alcohols, AcOH, or CF_3COOH. The latter solvents stabilize the nucleophile by specific solvation, hydrogen bridging, thus making it less nucleophilic. A "naked" anionic nucleophile $Y:^-$ will be much more reactive than a nucleophile dressed up by the solvent. The nature of the counterion M^+ is very important for the rate of the substitution reaction. The larger it is, the better it dissociates from $Y:^-$ and allows for equilibria with free ion pairs to form. When $Y:^-$ is strongly associated with M^+ (formation of tight ion pair), it is less reactive as nucleophile than as free ion, completely dissociated from M^+. Any additive to the reaction mixture that takes away M^+ and leaves $Y:^-$ "naked" will catalyze S_N2 displacement reactions of type (7.24). In some solvents, $Y:^-$ is more "naked" (less solvated) than in others. Nonpolar, aprotic solvents such as hydrocarbons (pentane, toluene, and chlorobenzene) should be chosen. Alternatively, the organic phase can be the reactant RX itself. However, in most cases, salts $M^+/Y:^-$ are not soluble in such a medium. The salts dissolve better in water. Thus, under such conditions, we have a two phase system (R—X in the organic phase, or realizing the organic phase, and MY a solid phase, or as an aqueous, nonmiscible solution). In order to bring the anionic nucleophile $Y:^-$ in contact with the electrophilic reagent R—X, it must be transported into the organic phase. As pioneered by Makosza [262–265], this can be done using a phase transfer catalyst (PTC) [266–269] that is a large ammonium or phosphonium ion (with large alkyl groups making them soluble in the organic phase) [270–277] or by encapsulation (formation of stable complexes or crytaplexes) of M^+ with a suitable crown ether [278, 279] or cryptand [280]. For example, the displacement reaction (7.28) of 1-chlorooctane with aqueous NaCN is accelerated > 1000-folds by the addition of hexadecyltributylphosphonium bromide (**68**) as a PTC (Scheme 7.13). The high rate of displacement is mainly due to the high lipophilicity of the phosphonium salt Q^+/CN^- and the large ionic radius of Q^+, which reduces the electrostatic interaction between Q^+ and CN^- and thus leaves the anionic nucleophile dissociated. The use of enantiomerically pure ammonium or phosphonium salts as PTCs allows enantioselective C—C bond formation [274, 276, 277, 281, 282], in particular for the industrial synthesis of amino acids [283].

Nicolas and Lacour have prepared the triazatriangulenium cations **70** from trityl salt **69** (Scheme 7.14) and shown that they can be used as PTCs for several organic reactions, thanks to their intrinsic stability in strongly basic and nucleophilic conditions ($pK_R+ > 20$) [284].

Crown ethers and cryptands make stable complexes with metal ions M^+ and thus increase dissociation of MY into ions pairs and provide highly reactive, unsolvated ("naked") anions $Y:^-$ [285–287]. The nature of crown ether or cryptand can be chosen

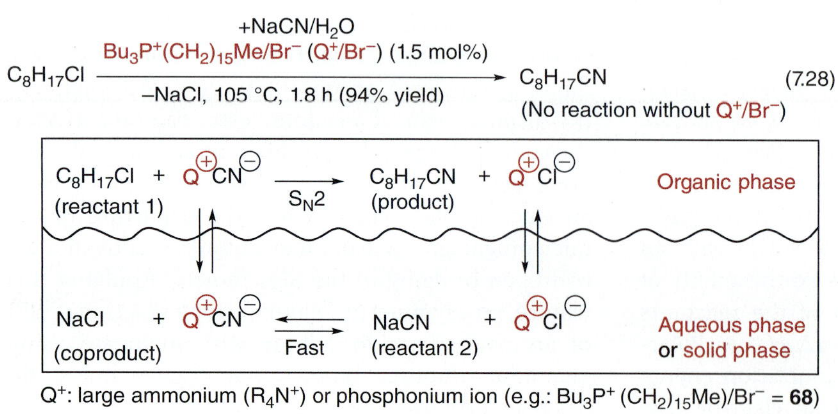

Scheme 7.13 Example of phase transfer catalysis (PTC).

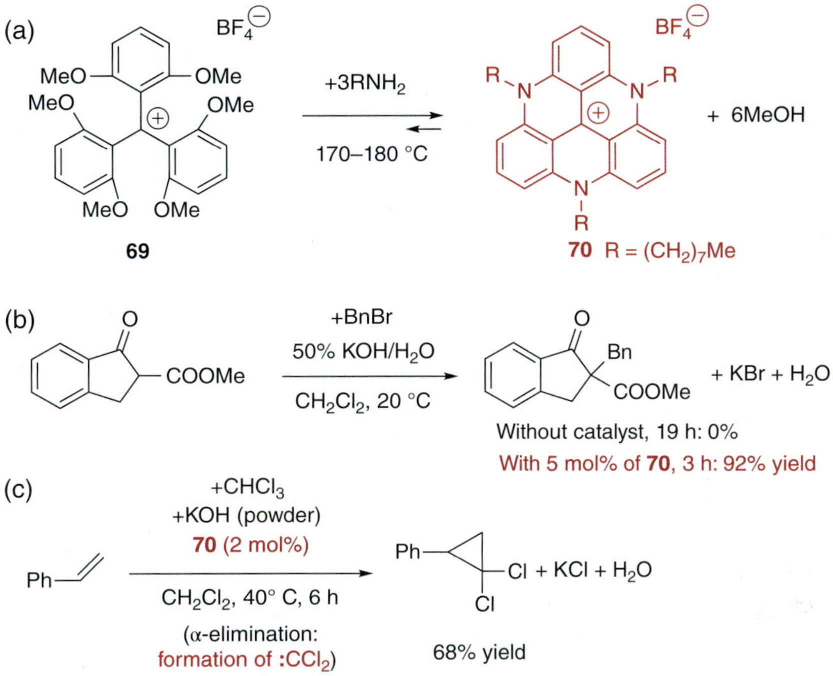

Scheme 7.14 Lacour's transfer catalyst (trityl cation): (a) preparation; (b) examples of nucleophilic displacement; and (c) example of 1,1-elimination and cyclopropanation.

to fit M^+ the best and for an optimal catalytical effect. For instance, 12-crown-4 binds Li^+ but not K^+, whereas 18-crown-6 binds K^+ not Li^+ [288]. Crown ethers having aliphatic chains, e.g. dicyclohexyl-18-crown-6 (**71**), can be used as PTCs in anion-promoted two-phase reactions [289, 290]. An example of such processes is shown with reaction (7.29). *n*-Octyl methanesulfonate is converted into n-C_8H_{17}-I with an aqueous solution of NaI containing 5 mol% of **71**. Crown ether **71** also catalyzes the oxidation of oct-1-ene in benzene (insoluble in H_2O) with an aqueous solution of $KMnO_4$ into *n*-heptanoic acid. Similarly, alkyl-substituted aza-macrobicyclic polyethers such as **72** are highly efficient catalysts of S_N2 displacement reactions of type (7.24) forming C—C or C–heteroatom bonds [291]. Crown ethers derived from monosaccharides allow asymmetric phase transfer reactions [292–295].

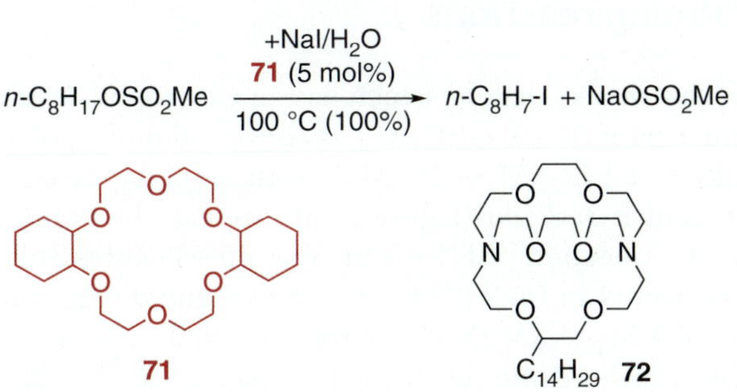

(7.29)

Organofluorine compounds are very important in material sciences, in crop protection, and in medicine [296–299]. One of the most economically viable methods for their preparation is nucleophilic fluorination of alkyl halides or sulfonate esters using alkaline fluorides. Because the fluoride anion makes strong hydrogen bonds, it is a strong base in the gas phase: $\Delta_r H°(X{:}^- + MeOH \rightleftarrows [X\cdots H\cdots OMe]^-) = -29.6$, -14.2, $-11.0\,kcal\,mol^{-1}$ for X = F, Cl, I, respectively [300, 301]; $\Delta_r H°(R_4NX(solid) + HX(gas) \rightleftarrows R_4N^+[X\cdots H\cdots X]^-) = -37, -12, -9.4, -7.3\,kcal\,mol^{-1}$ for X = F, Cl, Br, I, respectively [302–305]. It promotes eliminations through E_2H mechanism that compete with S_N2 displacements for reactions in polar aprotic solvent or in ionic liquids [306–308]. Liotta and Harris have reported the use of 18-crown-6 ether for solubilizing KF in acetonitrile and benzene solutions [309]. For benzene solution at 90 °C, 2-bromooctane reacts to give 32% of 2-fluorooctane and 68% of octenes. With 1-bromooctane, the proportion of products of elimination is reduced to 8%. Cox et al. found that Bu_4NF containing 0.1–0.3 equiv. of H_2O reacts at 25 °C with 2-bromooctane without solvent giving a 1 : 9 mixture of 2-fluorooctane and alkenes [310]. For the reaction of 1-bromooctane under the same conditions, the product of S_N2 substitution, 1-fluorooctane, is formed in 48% yield, the other products being oct-1-ene and octan-1-ol (see also Ref. [311]). Competition between the S_N2 and E_2H processes can be affected by temperature, solvent, and the amount of water present in the reaction mixture. Although water makes stable clusters $F^-(H_2O)_n$ with much less reactivity than naked F^- [312], the proportion of S_N2 vs. E_2H product might increase [313, 314]. By using insoluble PTCs that are fixed to either an insoluble resin [315] or an inorganic support, the problem of waste can be resolved as the catalyst can be recovered after the reaction by simple filtration. Unfortunately, most of these insoluble catalysts are less efficient than the conventional soluble PTCs. Fluorous tetraalkylphosphonium iodides catalyze the synthesis of propylene carbonate in supercritical CO_2 [316] and recyclable fluorous chiral ammonium salts have been used to promote the asymmetric alkylation of a protected glycine derivative [317]. The N,N'-dialkyl-4,13-diaza-18-crown-6 lariat ethers **73** and **74** with polyfluorinated side arms are efficient PTCs under solid–liquid conditions. In the case of reaction (7.30) and using **73**, the catalyst has been recycled several times by fluorous solid-phase extraction without loss in activity [318].

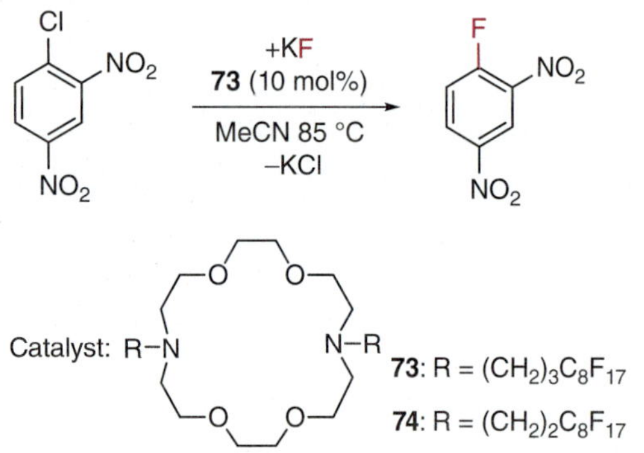

(7.30)

When chiral ammonium, phosphonium, or crown ether are used for enantioselective reactions, the nucleophilic reagents are associated with chiral, enantiopure cationic counterions (asymmetric ion pairing catalysis) [319]. Asymmetric electrophilic reactions have been realized in which the electrophile is associated with a chiral, enantiopure anion responsible for the phase-transfer-catalyzed process (chiral anion phase transfer [CAPT] catalysis) [320–324]. An example is given with the enantioselective electrophilic fluorination (7.31) of alkenes developed by Toste and coworkers [325]. In a similar way, electrophilic enantioselective fluorolactonization has been realized by Hamashima and coworkers [326]. Chiral anion transfer of arenediazonium cations has permitted the enantioselective synthesis of C(3)-diazenated pyrroloindolines in a similar way [327].

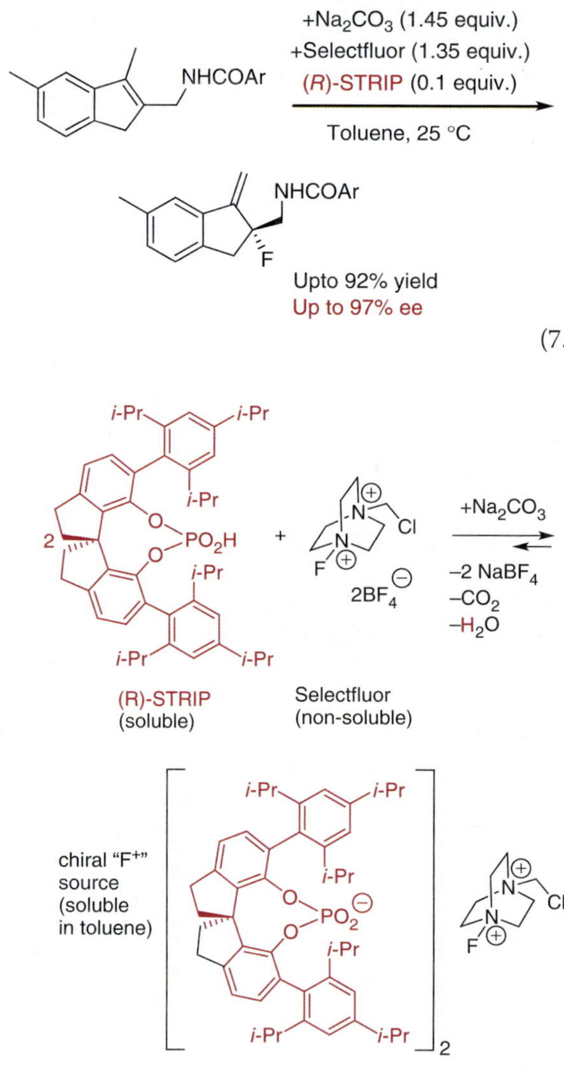

(7.31)

7.5 Catalytical Umpolung C—C bond forming reactions

When an alkyl halide, which has an electrophilic carbon center (R—X ↔ R^+/X^-), is converted into a polar alkylmetal (R—M ↔ R^-/M^+), a nucleophilic carbon reagent is realized (Figure 7.7a) (see e.g. the Frankland zincation $3EtI + 3Zn \rightarrow Et_2Zn + EtZnI + ZnI_2$ discovered in 1849 [328–330], the Grignard reagents R—X + Mg → RMgX discovered in 1900 [331, 332], the Ziegler–Colonius direct lithiation R—X + 2Li → R—Li + LiX [333, 334], and the Wittig–Gilman

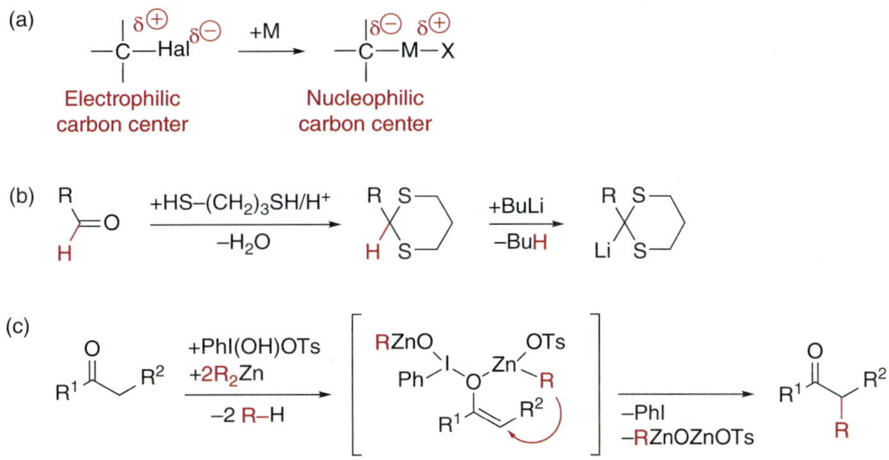

Figure 7.7 Examples of stoichiometric Umpolung: (a) halogen/metal exchange; (b) metallation of dithianes derivatives of aldehydes; and (c) oxidative Umpolung α-alkylation of ketones.

halogen/metal exchange reaction MeI + BuLi → MeLi + BuI) [335–343]. One speaks of Umpolung (inversion of polarity) of the carbon center. Unsaturated compounds such as aldehydes, alkenes, and alkynes conjugated with aldehyde, ketone, ester, carbonitrile, carboxamide, sulfonyl, and phosphonyl moieties are electrophiles undergoing 1,2 and/or 1,4-additions with nucleophiles. They can be converted into nucleophilic carbon reagents upon suitable modifications. A well-known case is the conversion of aldehydes into dithianes that can be metalated by hydrogen/metal exchange to generate nucleophilic carbon reagents that are acyl anion equivalents (Figure 7.7b) [344]. Alternatively, an electron-poor (electrophilic) π-system can be converted into a nucleophile by addition of a nucleophilic species or by capture of an electron (single-electron transfer, SET). Similarly, an electron-rich (nucleophilic) π-system can be converted into an electrophilic reagent by protonation, by addition of an electrophile [345, 346] (see e.g. direct α-hydroxylation of ketones with Ph—I=O [347], Smoliakova's C-glycosidation using glycals [348, 349], and oxidative α-alkylation of ketones (Figure 7.7c)), or by oxidation (SET) into a radical cation (Section 7.5.8) [350]. In this chapter, we shall be concerned by C—C bond forming cross-coupling reactions involving catalytical Umpolung.

7.5.1 Benzoin condensation: Umpolung of aldehydes

In 1832, Wöhler and Liebig discovered the so-called benzoin condensation catalyzed by cyanide anion (reaction (7.32)) [351]. In this reaction, one aldehyde molecule does the hydrocarbation of another aldehyde molecule. In 1903, Lapworth proposed the mechanism given below for this reaction [352] (see however Ref. [353]).

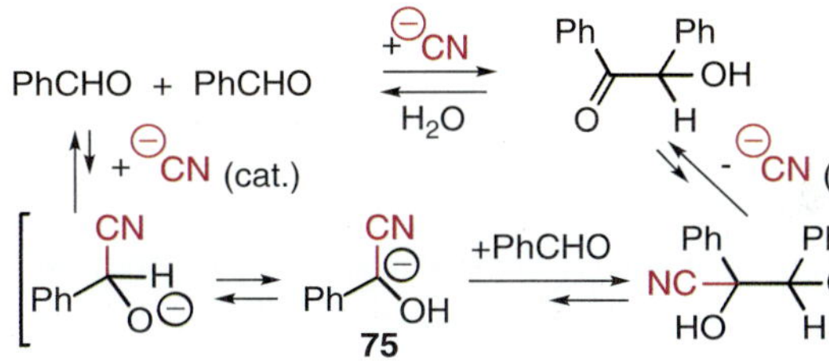

(7.32)

An intermediate carbanion **75** represents an active aldehyde with inverted reactivity of the carbonyl carbon atom (i.e. nucleophilic) [354]. In 1943, Ukai et al. found that thiazolium salts can be used as catalysts for the benzoin condensation as well as cyanide ion [355]. A few years later, Mizuhara and Handler described the catalytical activity of natural thiamine (**76**) which, in the form of his pyrophosphate (**77**), is the coenzyme for a number of important biochemical reactions, including decarboxylation of pyruvic acid into acetaldehyde, the transketolase-catalyzed reaction, and the oxidative decarboxylation of pyruvic acid into "active acetate" [356]. In 1958, Breslow presented the mechanism shown in Scheme 7.15 for the thiazolium ion-catalyzed benzoin condensation. Deuterium exchange studies demonstrated the easy deprotonation of thiazolium salts **78** into corresponding zwitterions **79′** (α-elimination of HX). They can be seen as thiazol-2-ylidenes **79** that are relatively stable carbenes with a strong σ-nucleophilic character. Intermediate carbenes **79** are the catalytically active species [357, 358].

They add to the carbonyl group of benzaldehyde giving adducts **80** that equilibrate (probably by protonation/deprotonation) with the electron-rich alkenes **81**. The latter add onto benzaldehyde giving zwitterions **82** that equilibrate with **83**. Finally, β-elimination furnishes benzoin and the catalyst **79**. In the case of benzoin condensation catalyzed by 3-benzyl-5-(2-hydroxyethyl)-4-methylthiazolium bromide in MeOH buffered with $Et_3N/Et_3NH^+Cl^-$,

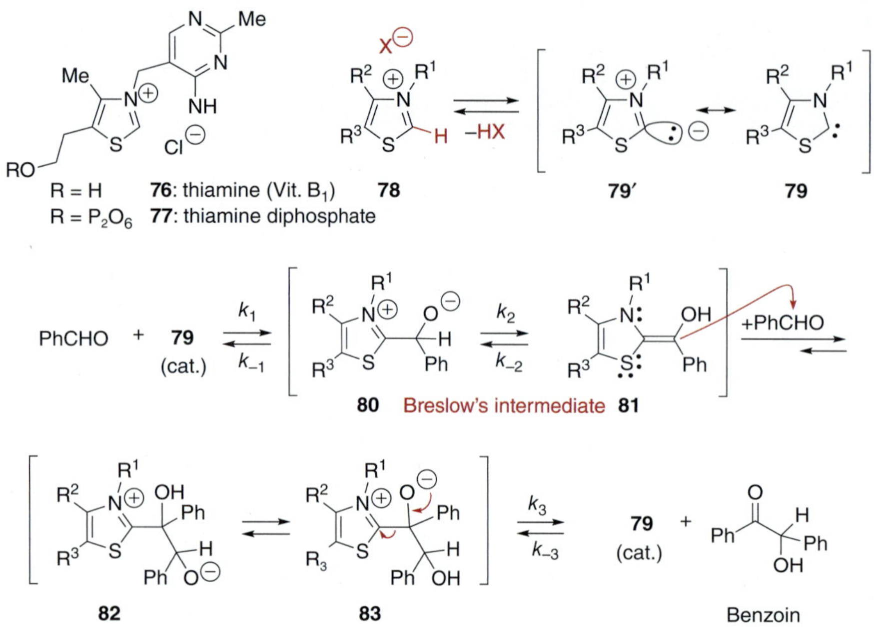

Scheme 7.15 The thiazolium ion-catalyzed benzoin condensation: Breslow's mechanism involving thiazol-2-ylidene intermediates **79**.

White and Leeper found that all the three steps k_1, k_2, and k_3 are partially rate determining. A normal kinetic deuterium isotope effect for the overall reaction $k_H/k_D \cong 3.4$ is observed using PhCDO, and a large inverse solvent isotope effect $k_D/k_H \cong 5.9$ is observed using CD_3OD consistently with the mechanism shown in Scheme 7.15 [358–360]. As thiazol-2-ylidenes **79**, 1,3,5-triphenyl-1,3,4-triazol-2-ylidene (**20**, Section 7.2.5) also catalyzes the formoin condensation (7.33) converting formaldehyde into glycoaldehyde (**84**) [361, 362]. This reaction is the first step of the "formose reaction" discovered in 1861 by Butlerow and that generates mixtures of racemic aldoses and ketoses (monosaccharides) by oligomerization of formaldehyde in the presence of $Ca(OH)_2$ [363, 364]. When carried out in 8 : 1 dimethylformamide (DMF)/water with Et_3N and thiamine hydrochloride (catalyst) at 75 °C for one hour, DL-2-*C*-hydroxymethyl-3-pentulose (**85**) is obtained in 28% yield [365, 366].

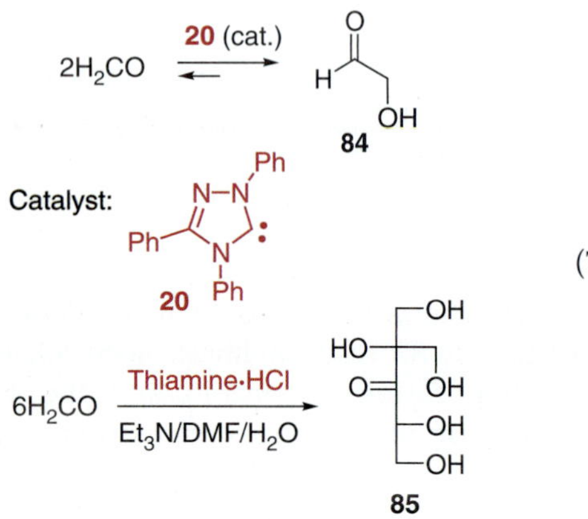

(7.33)

In the presence of secondary amines such as piperidine and morpholine, which combine with benzaldehyde to give iminium ion intermediates, modest yields of α-aminoketones ($PhCOCH(NR_2)Ph$) have been obtained in boiling EtOH in the presence of 3-methylthiazolium tetrafluoroborate and Et_3N [367]. However, aromatic aldehydes or benzoins react with unactivated imines in the presence of NHC affording α-aminoketones in good yields [368]. Benzoin condensations have been catalyzed by diaminocarbenes resulting from the Wanzlick equilibrium [369] and by NHC obtained by α-elimination of benzimidazonium salts [370]. NaCN promotes the intramolecular cyclization of dialdimines in DMF, MeOH, or CH_2Cl_2/H_2O (phase transfer conditions) yielding various heterocyclic systems [371]. On attaching thiazolium salts to carbon center C(6) of a D-glucose unit of γ-cyclodextrin, as in **86**, Breslow and Kool realized "artificial enzymes" that catalyze the benzoin condensation of benzaldehyde more efficiently than the corresponding "free" thiazolium ions. This is explained by invoking that two benzaldehyde molecules can reside simultaneously in the hydrophobic interior of γ-cyclodextrin (cyclomaltooctaose). Analogs to **86** made of β-cyclodextrin (cyclomaltoheptaose) do not catalyze the benzoin condensation as well [372, 373].

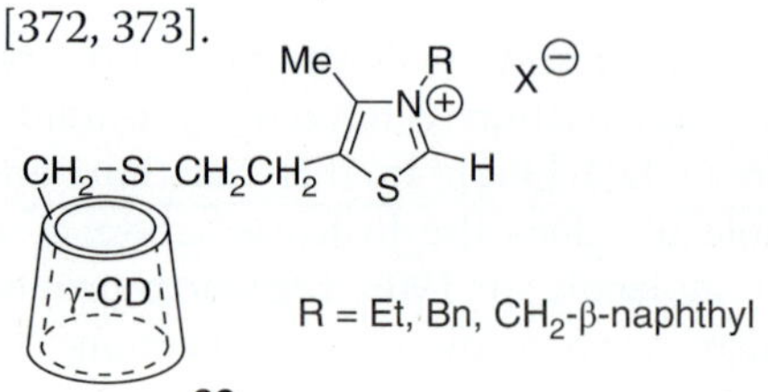

86

The α-hydroxycarbonyl compounds are valuable synthetic intermediates. Cross-condensation of two different aldehydes through the benzoin reaction represents the most direct route to these systems. As all the steps in the benzoin condensation are reversible, selectivity for the product of cross-coupling depends on their relative stability with respect to the products of homocoupling. In most cases, it remains small; thus, selectivity in the product of cross-coupling can be obtained only by using an excess of one of the two aldehydes. Using cyanide-catalyzed benzoin reaction involving mixtures of acylsilanes **87** and aldehydes **90**, Johnson and coworkers found conditions under which high yields of products of cross-coupling **94** are obtained (an example of carbasilylation of a C=O bond). Using $La(CN)_3$ as a catalyst, the scope of the cross-silyl benzoin addition can be extended to aryl, heteroaryl, alkyl, and α,β-unsaturated aldehydes [374]. The proposed mechanism (Scheme 7.16) implies a cyanide-promoted (1,2)-Brook rearrangement **88** → **89** [375, 376]. Acyl phosphonates also undergo cyanide-catalyzed additions to aldehydes [377]. A strategy utilizing NHCs derived from thiazolium salts has been developed for the generation of equivalents of carbonyl anions from acylsilanes [378].

Enantioselective benzoin syntheses have been realized using enantiopure, chiral NHCs as catalysts [134, 135]. An example is given in Scheme 7.17a. Enantioselective cross-benzoin reactions of aliphatic aldehydes and α-ketoesters (Scheme 7.17b) and aza-benzoin reactions of enals with activated ketimines (Scheme 7.17c) have also been realized. In these reactions, α-eliminations of HBF_4 from the precatalysts **95**, **96**, and **97** generate the corresponding NHCs that are chiral nucleophiles. They add to the carbonyl groups as cyanide anion or thiazol-2-ilydenes (carbenes of Breslow), leading to the Umpolung of aldehydes. Asymmetric catalyzed Umpolung of imines have also been described [379].

7.5.2 Stetter reaction: Umpolung of aldehydes

Since its discovery in 1973 [380], the Stetter reaction (7.34) has become one of the most important Umpolung procedures for the synthesis of 1,4-dicarbonyl

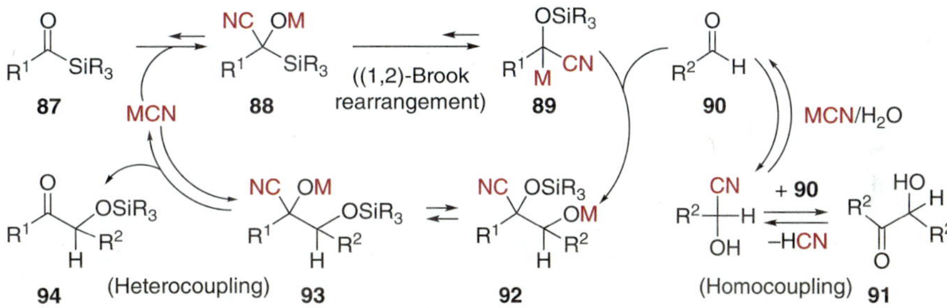

Scheme 7.16 Proposed mechanism for the cyanide-catalyzed cross-silyl benzoin addition. In this case (MCN = $La(CN)_3$), homocoupling **90** → **91** is slower than cross-coupling **87** + **90** → **94**.

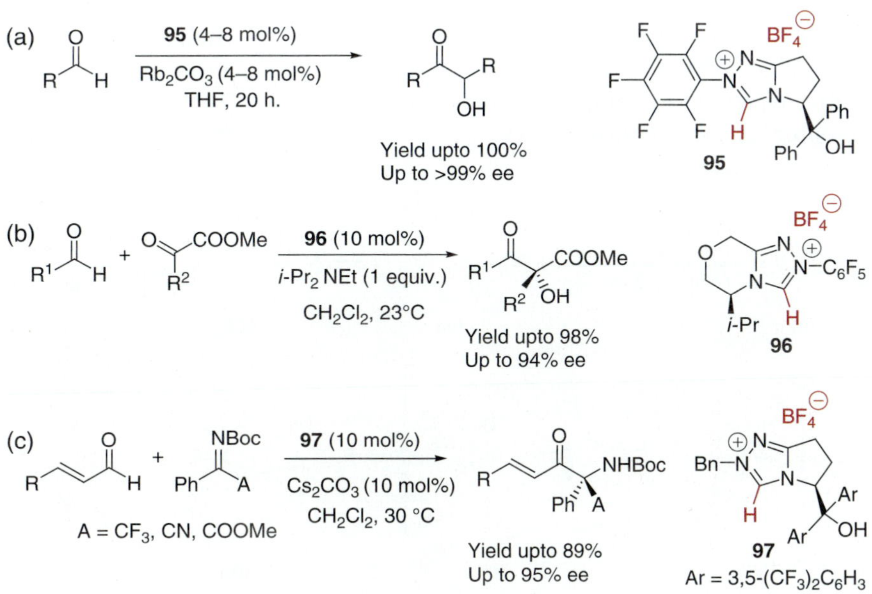

Scheme 7.17 Examples of enantioselective, intermolecular benzoin condensations. (a) Homobenzoin reaction, (b) cross-benzoin reaction, and (c) aza-benzoin reaction.

compounds [381, 382]. In the presence of Et_3N, thiazolium salts of type **78** generate the corresponding electron-rich carbenes **79** that react with aldehydes to generate enol intermediates (see **81** in Scheme 7.15) that, on their turn, add to α,β-unsaturated carbonyl compounds **98** in a 1,4-fashion producing the final 1,4-dicarbonyl compounds **99**.

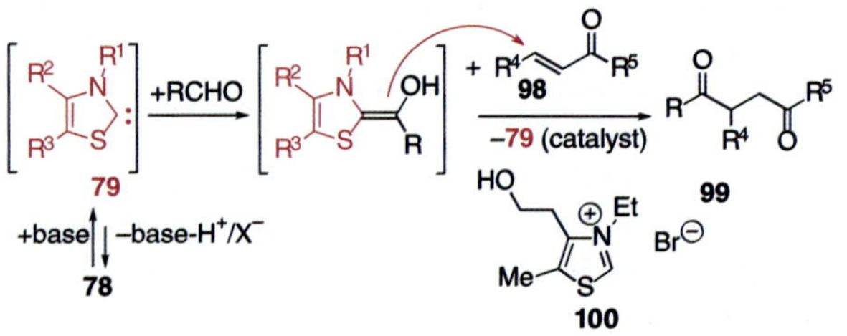

(7.34)

Because aldehydes RCHO can also react with Breslow's intermediates **81** and give products of benzoin homocoupling, the conjugated enals or enones **98** must react faster with **81** than with RCHO for successful intermolecular Stetter reactions. Thus, intramolecular Stetter reactions have found a faster development than intermolecular Stetter reactions [383–385], including for asymmetric versions [386, 387]. The Stetter reaction is a two-step process in which the rate-determining step C—C coupling of the Breslow's intermediate and the Michael acceptor (1,4-nucleophilic addition) precedes final proton transfer [388]. Using acylsilanes **87** (Scheme 7.16) and enones of type **98** in the presence of DBU (base), THF (solvent), and thiazolium salt **100** (precatalyst), good yields are obtained for the intermolecular Stetter reaction [389]. An example of catalytic enantioselective reaction is given with reaction (7.35) [390]. A NHC-catalyzed intramolecular Stetter reaction involving an aldimine and a Michael acceptor has been reported [391].

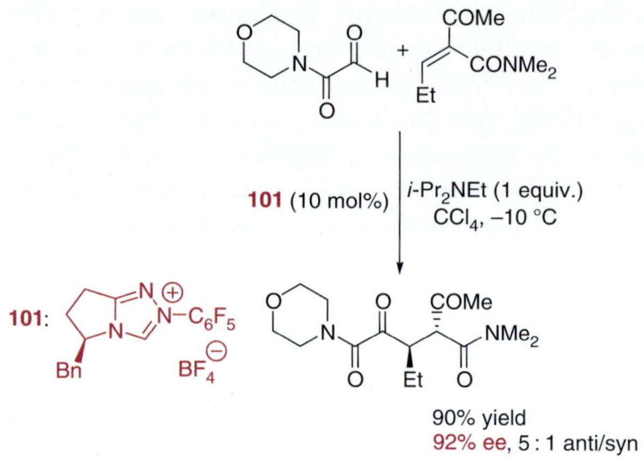

(7.35)

7.5.3 Umpolung of enals

In 2004, the groups of Glorius [392] and Bode [393] reported the first examples of NHC-catalyzed annulations of enals and aldehydes into lactones. On combining α,β-unsaturated aldehydes **102** with 2,2,2-trifluoro-1-phenylethanone (**103**) in the presence of 5 mol% of (IMes·HCl) and 10 mol% *t*-BuOK in THF, a mixture of *cis*/*trans* γ-butyrolactones **107** was obtained (Scheme 7.18). In this case, the aldehyde reacts with diaminocarbene (IMes: Herrmann's diaminocarbene [394]) generated in situ giving the nucleophilic dienol **104**, which can be seen as a homoenolate. It adds to ketone **103** producing intermediates **105** ⇄ **106** that generate product **107** and liberate the catalyst (IMes). Enantioselective lactonization of cinnamaldehyde with $PhCOCF_3$ has been realized applying a chiral diaminocarbene catalyst [395].

Nair et al. uncovered a IMes-catalyzed cross-condensation of enals and enones producing 1,3,4-trisubstituted cyclopentenes (Scheme 7.19a) [396]. The diaminocarbene induces the Umpolung of the enal through the corresponding homoenol,

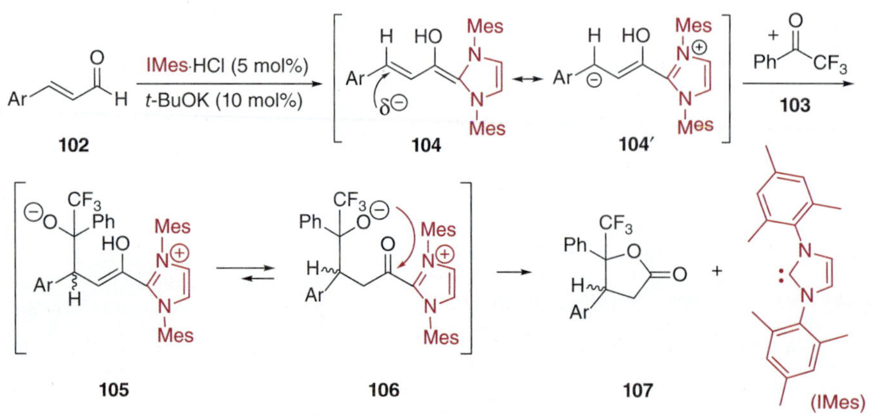

Scheme 7.18 Synthesis of γ-butyrolactones by Umpolung of an enal with a diaminocarbene catalyst (IMes).

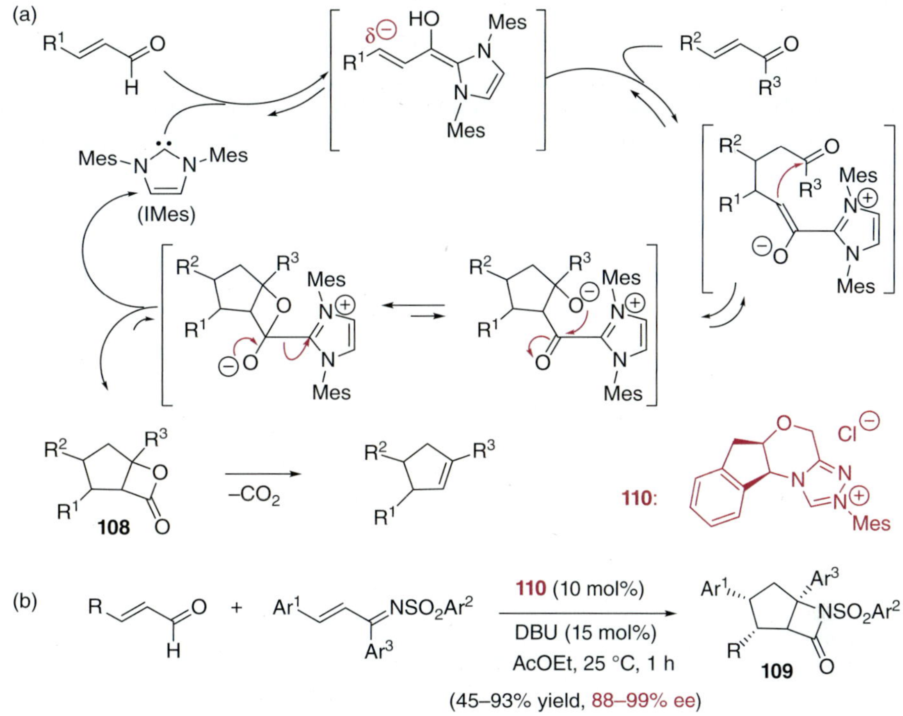

Scheme 7.19 (a) Nair's synthesis of 1,3,4-trisubstituted cyclopentenes; (b) Bode's asymmetric synthesis of β-lactams.

which undergoes conjugate addition to the enone. The intermediate so obtained is cyclized into the corresponding β-lactones **108** that undergo (2+2)-cycloreversion with liberation of CO_2. When the enones are replaced by α,β-unsaturated imines, the corresponding lactams **109** form, as shown first by He and Bode. Using an enantiopure diaminocarbene precursor such as **110**, catalytic enantioselective lactamizations have been realized (Scheme 7.19b) [397]. The group of Scheidt has found that enantioselective β-protonation of homoenolates of type **104** ↔ **104′** can be realized in alcohols using an enantiopure NHC precursor as a catalyst. Thus, enals of type $R^1(R^2)C{=}CHCHO$ react with alcohols R^3OH producing enantiomerically enriched esters of type $R^1(R^2)CH{-}CH_2{-}COOR^3$ [398].

Problem 7.1 Give a possible mechanism for the thiazolium-catalyzed acylsilane additions to α,β-unsaturated ketones that produce 1,4-diketones [389].

Problem 7.2 What are the main products **P** of the following reactions? [399].

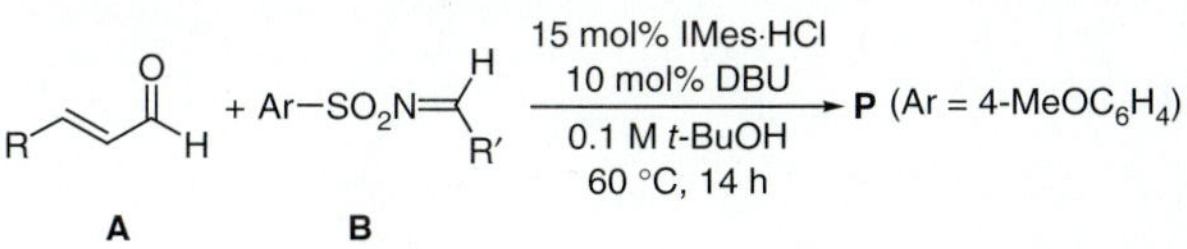

Problem 7.3 Explain the following reaction catalyzed by IMes [400].

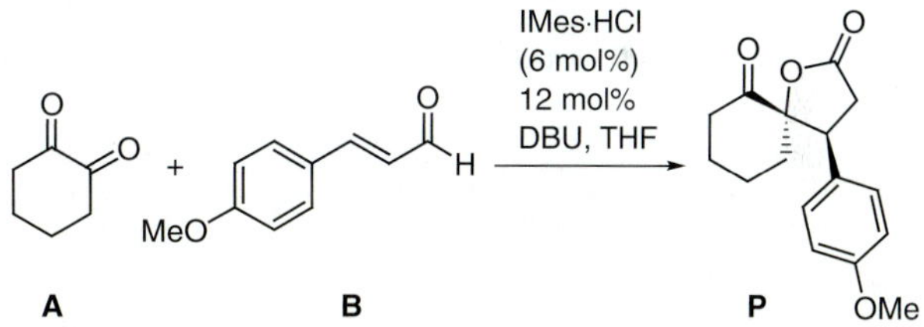

7.5.4 Umpolung of Michael acceptors

Huisgen et al. showed that dimethyl azodicarboxylate adds to triphenylphosphine forming zwitterion **111** (E = COOMe). The latter can be trapped with 1,2-diketones as illustrated with reaction (7.36) [401].

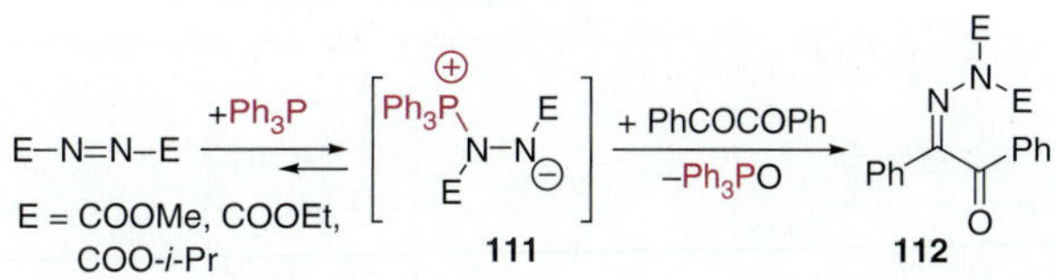

(7.36)

In this process, the electrophilic nitrogen center of the azo compound becomes nucleophilic after addition of the phosphine catalyst; zwitterion **111** is formed that adds to electrophilic carbonyl moieties. In 1973, White and Baizer at Monsanto reported a phosphine-catalyzed Michael addition of 2-nitropropane to ethyl acrylate, acrylonitrile, and methyl vinyl ketone ($CH_2{=}CHCOOMe + Me_2CHNO_2 \rightarrow Me_2C(NO_2)CH_2CH_2COOMe$). The weak basicity of the catalysts, Bu_3P, Ph_3P, $PhMe_2P$, and Ph_2MeP, suggested a mechanism in which the phosphines behave as nucleophiles rather than as bases. In this hypothesis, zwitterionic intermediates of type $R_3P^{(+)}CH_2CH^{(-)}COOMe_3$ are formed that deprotonate the carbon acid (Me_2CHNO_2) generating the corresponding conjugate base $Me_2C^{(-)}NO_2$. The latter adds to the activated alkenes [402]. The reaction has been extended to other carbon acids than 2-nitropropane [403]. In 1992, Trost and Kazmaier reported that Ph_3P catalyzes the isomerization of conjugated ynones, ynoates, and ynecarboxamides (**113**) into conjugated dienes **121** (Scheme 7.20) [404].

The proposed mechanism implies the formation of carbanionic intermediates (e.g. **114** ⇄ **115**; **117** ⇄ **118** ↔ **119** ⇄ **120**) resulting from protonation and deprotonation of zwitterionic phosphonium intermediates [116]. Methyl but-2-ynoate (**122**) adds to Ph_3P generating zwitterion **123**. In the presence of carbon acids such as malonic esters and analogs with $pK_a < 16$, proton transfer occurs giving ion pair **124** (AcOH/AcONa). Collapse of ion pair **124**

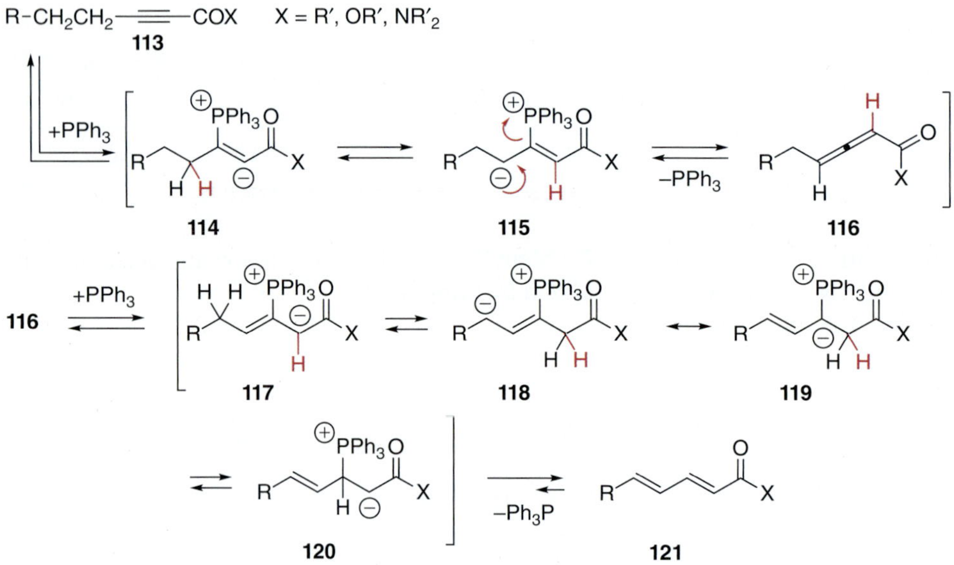

Scheme 7.20 Isomerization of conjugated alkynes into conjugated dienes.

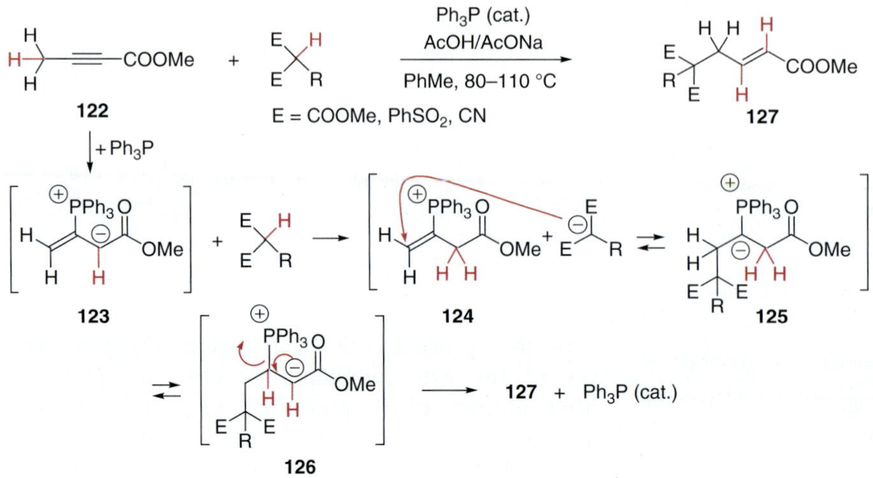

Scheme 7.21 Trost's Umpolung γ-alkylation of an ynoate by triphenylphosphine.

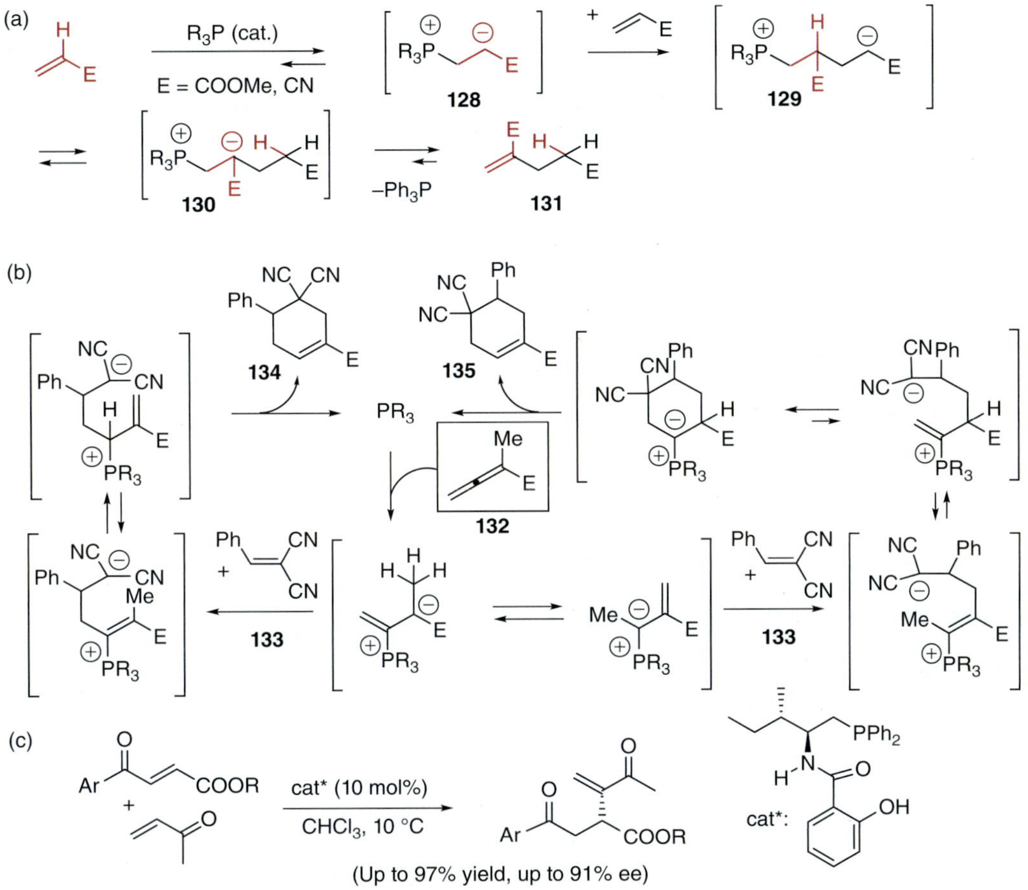

Scheme 7.22 The addition of phosphine to Michael acceptors engenders nucleophilic intermediates. (a) The Rauhut–Currier dimerization of acrylic derivatives; (b) proposed mechanisms for the phosphine-catalyzed (4+2)-annulations of alleneoate and electron-poor alkenes; (c) example of catalytic enantioselective cross-Rauhut–Currier reaction.

produces intermediate **125** that isomerizes into its tautomer **126**. β-Elimination of Ph_3P furnishes product of cross-coupling **127**. This can be viewed as Umpolung of the γ-position of activated alkyne **122** from nucleophilic to electrophilic because of its acidity, as demonstrated by Trost and Li in 1994 (Scheme 7.21) [405].

Oxygen nucleophiles (alcohols and carboxylates) [406] and acidic nitrogen nucleophiles such as *p*-toluenesulfonamides have also been utilized as Umpolung catalysts in the nucleophilic γ-additions of conjugated yne-ketones and yne-esters [407]. Fu and coworkers [408] have optimized the NHC-catalyzed mode [409] of β-alkylation of enolates, as shown with reaction (7.37). Ph_3P has been used as nucleophilic catalyst for the Umpolung addition of azoles to electron-deficient allenes [410].

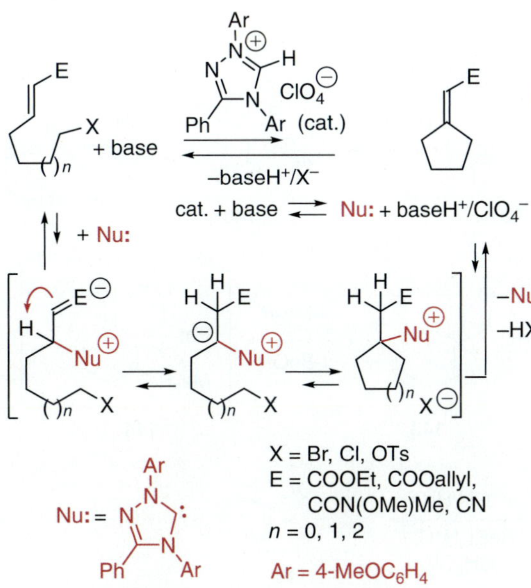

(7.37)

7.5.5 Rauhut–Currier reaction

In 1963, Rauhut and Currier reported phosphine-catalyzed dimerizations of acrylonitrile and acrylic esters (Scheme 7.22a) [411]. These transformations were then investigated by McClure [412] and Baizer and Anderson [413]. They are believed to involve reversible phosphine conjugate additions to electron-poor alkenes giving zwitterions **128**, followed by Michael additions of the intermediate enolates to a second alkene moiety producing **129**. A prototropic shift gives zwitterion **130**. Finally, elimination forms the dimeric compound **131**. The Rauhut–Currier reaction is a hydrocarbation of an activated alkene with the α-C—H moiety of an activated alkene. Small trialkylphosphines are superior to amine nucleophiles such as DABCO, DBU, Et_2NH, and DMAP (4-Me_2N-pyridine) [414, 415]. The Rauhut–Currier applied to allene esters and 2-phenylethylene-1,1-dicarbonitrile generates cyclohexene derivatives. For instance, the reaction of alleneoate **132** and electron-poor alkene **133** in the presence of $P(NMe_2)_3$ catalyst produces 98% yield of **134**, whereas the same reaction in the presence of (4-$ClC_6H_4)_3P$ as catalyst leads to 92% yield of isomeric cyclohexene **135** (Scheme 7.22b) [416]. Related to the Rauhut–Currier reaction is the formal (4+1)-cycloaddition of sulfur (S_8) with 2 equiv. of acetylenedicarboxylic esters producing tetraalkyl thiophene-2,3,4,5-tetracarboxylate, a reaction catalyzed by nucleophilic bases such as triethyl phosphite, isoquinoline, pyridine, and *N*-methylimidazole [417]. An example of catalytic enantioselective intermolecular cross-Rauhut–Currier reaction is given in Scheme 7.22c [418]. Many more enantioselective catalyzed condensations of two Michael acceptors in the Rauhut–Currier mode have been reported [419].

Problem 7.4 What is the product **P** when **A** is treated with Me_3P and *t*-amyl-OH? [420].

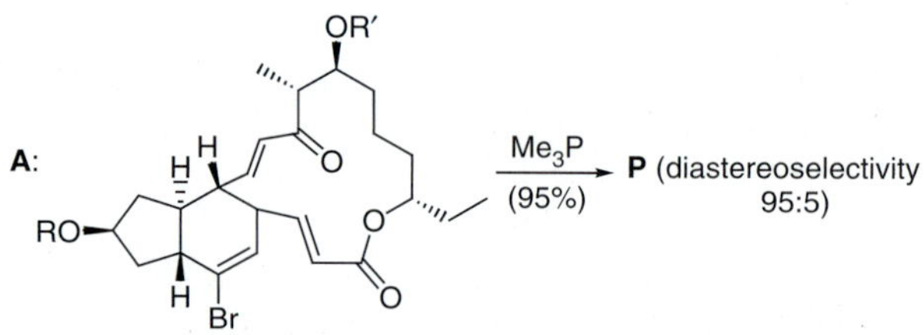

7.5.6 Morita–Baylis–Hillman reaction

The Morita–Baylis–Hillman (MBH) reaction is an atom economical coupling reaction of activated alkenes **136** and aldehydes **137** in the presence of a nucleophilic catalyst (Scheme 7.23a). It corresponds to the hydrocarbation of the aldehyde with the α-C—H bond of the activated alkene. The reaction involves formally a sequence of Michael addition, aldol reaction, and β-elimination. The reaction was first reported by Morita et al. in 1968 [421] and in 1972

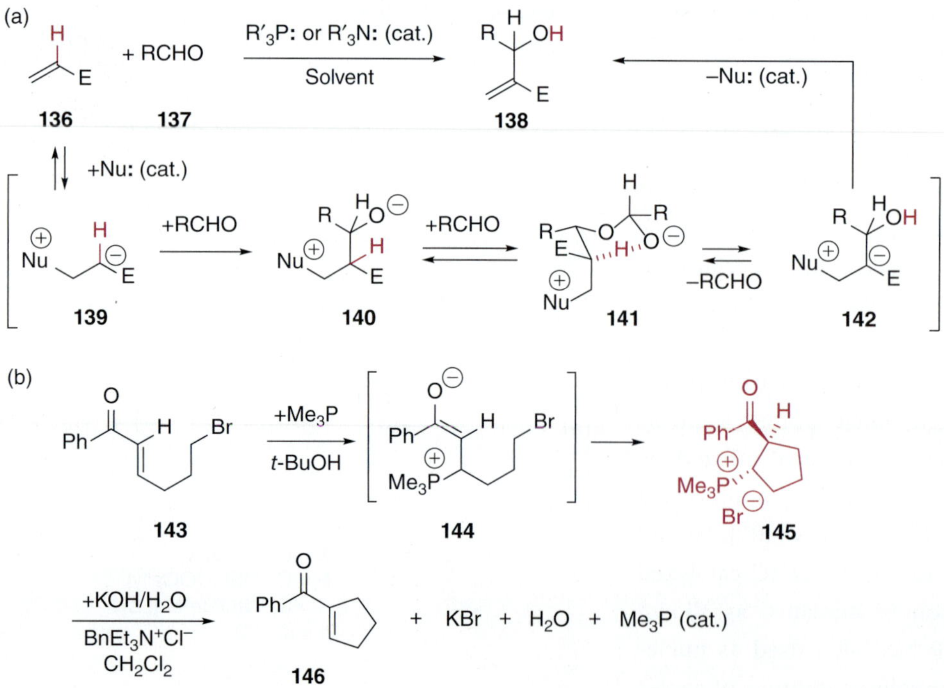

Scheme 7.23 The Morita–Baylis–Hillman (MBH) reaction: (a) an intermolecular version, (b) an intramolecular version.

by Baylis and Hillman [422, 423]. The nucleophilic catalyst can be a phosphine or a tertiary amine such as DABCO [424, 425]. Activated alkenes include acrylic esters, acrylonitriles, vinyl ketones, phenyl vinyl sulfones, phenyl vinyl sulfonates, vinyl phosphates, and acrolein. β-Substituted alkenes require more forcing conditions because of the slow addition of the nucleophilic catalyst. A variety of electrophiles can be used such as aldehydes, α-ketoesters, 1,2-diketones, aldimines, methyl α-bromoesters [426, 427], and arenes [428].

The mechanism of the MBH reaction is proposed to involve three steps: (i) nucleophilic additions to the activated alkene **136** generating a zwitterionic **139** that adds to aldehyde **137** (or another electrophiles) to give zwitterion **140**, (ii) follows a protonation and deprotonation step **140** → **141** → **142**, and (iii) β-elimination with recovery of the nucleophilic catalyst (Scheme 7.23a). Quantum mechanical calculations for a reaction in aprotic solvent suggested that deprotonation of the α-position **140** → **142** is the rate-determining step and that it proceeds though a cyclic transition state **141**, with proton transfer to a hemiacetal alkoxide formed by addition of a second equivalent of aldehyde to intermediate **140**. This mechanism is supported by the observation of second-order kinetics with respect to RCHO concentration in the absence of protic solvent. In contrast, in MeOH, the alcohol serves as a shuttle to transfer the proton from carbon to oxygen (**140** → **142**) [429]. In the case of the intramolecular MBH alkylation of **143** → **146**, intermediate **145** has been isolated and characterized (Scheme 7.23b) [430]. Reaction **143** → **146** corresponds to a δ-elimination of HBr involving the α-C—H moiety of the conjugated enone **143**, HBr being neutralized by KOH/H_2O buffered by $BnEt_3N^+Cl^-$. Treatment of acetates **148** (resulting from the MBH reactions **147** + Ph_3P) with α-silyloxyfuran leads to an efficient and stereoselective synthesis of γ-butenolides **150** (reaction (7.38)) [431].

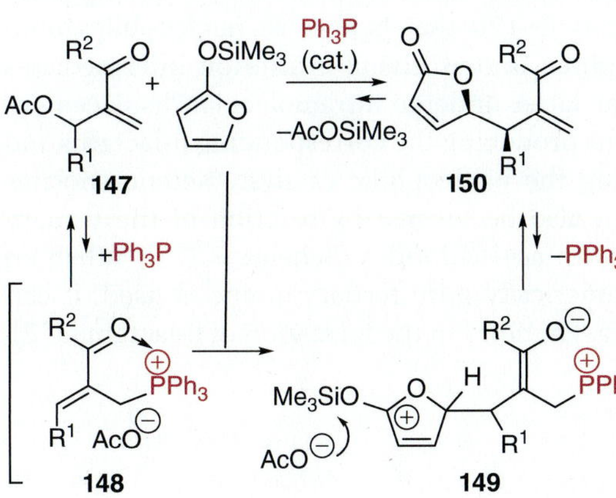

(7.38)

Much effort has been devoted to the development of catalytic enantioselective MBH reactions [432–435]. The catalyst can be a chiral Lewis base (e.g. **151** [436] and **152** [437]), a chiral tertiary phosphine (e.g. **153** [438] and **154** [439]), a chiral Lewis acid (e.g. **155** [234, 440]), or a chiral Brønsted acid (e.g. **156**) [441]. An example of aza-Morita–Baylis–Hillman (aza-MBH) reaction is given with reaction (7.39), which applies the *P*-chirogenic organocatalyst **157** [442]. Many more examples of catalytic enantioselective aza-MBH reactions have been reported [443].

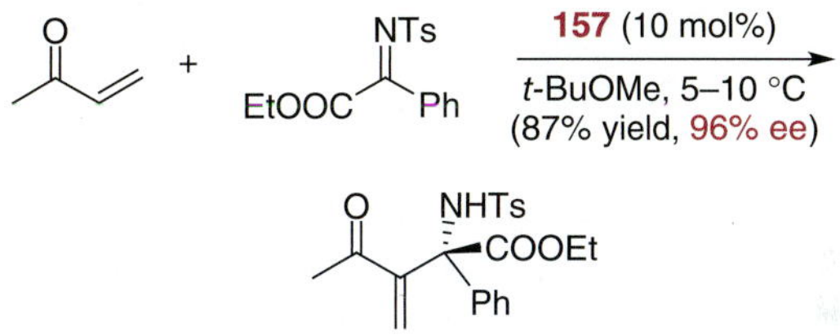

(7.39)

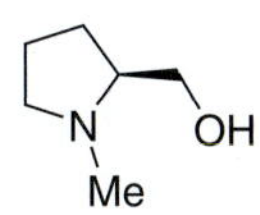

151

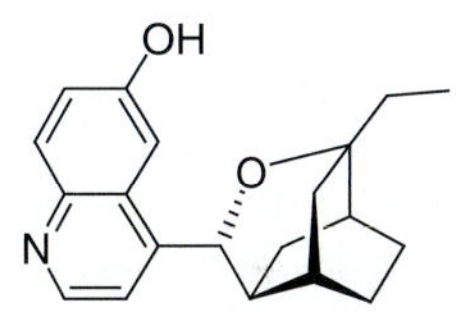

152: β-Isocupreidine

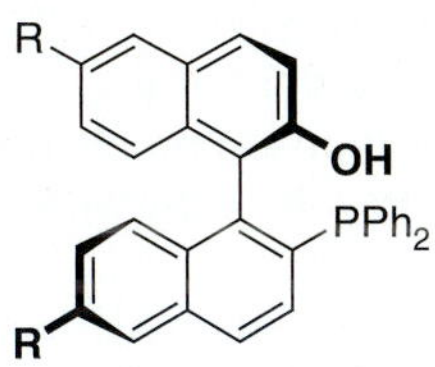

153 R = $(CH_2)_2C_6F_{13}$

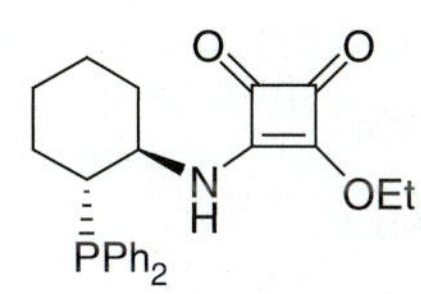

154

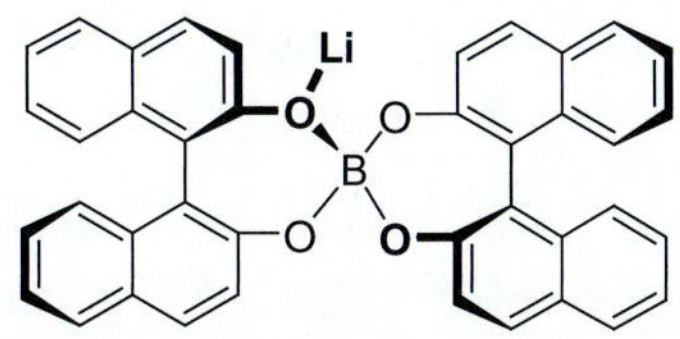

155

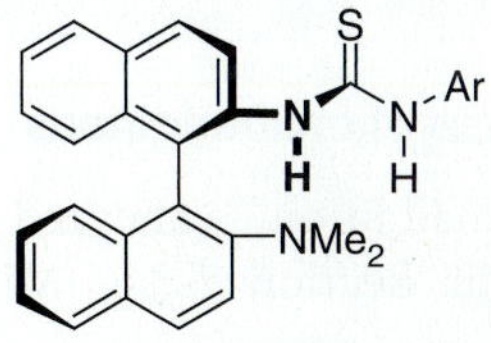

156 Ar = 3,5-$(CF_3)_2C_6H_3$

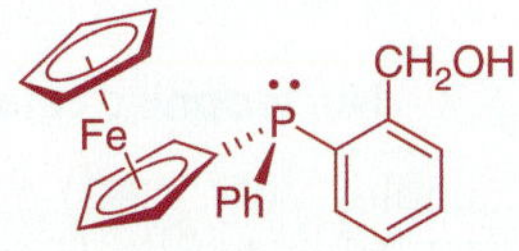

157

Problem 7.5 Propose a mechanism for the following condensations catalyzed by pyridine [444].

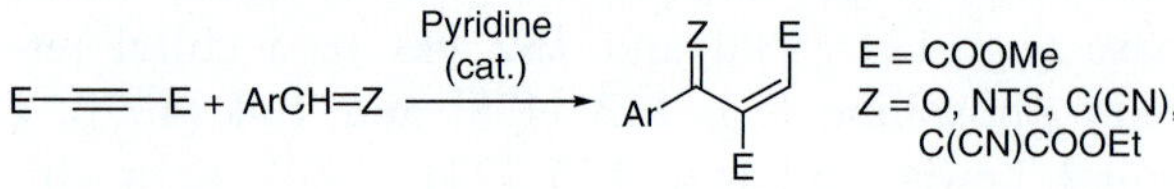

Problem 7.6 Propose a mechanism for the DABCO (1,4-diazabicyclo[2.2.2]octane)-catalyzed chromene synthesis **A** + **B** → **C** [445].

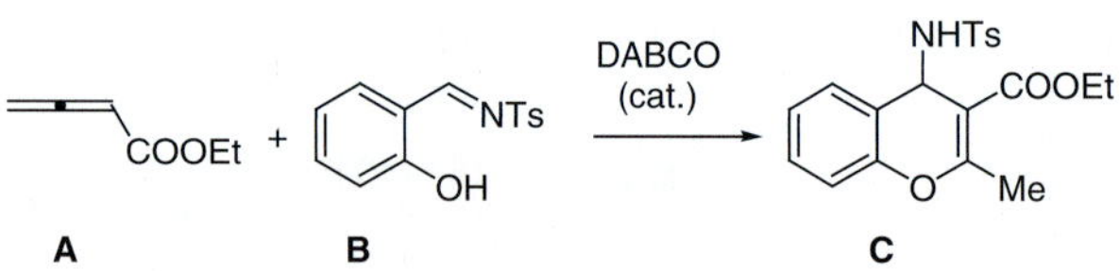

Problem 7.7 Propose a mechanism for the Ph_3P-catalyzed double Michael addition **A** + **B** → **C** [446, 447].

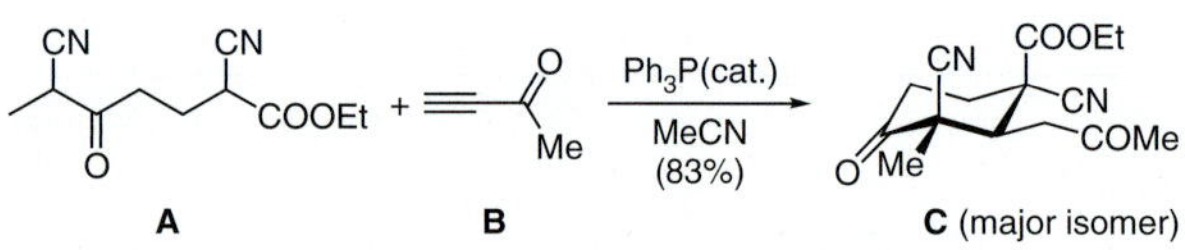

Problem 7.8 Lu and coworkers [448, 449] have pioneered the catalytical (3+2)-cycloaddition of allenoate esters and conjugated enones to give cyclopentane adducts in the presence of Ph_3P. Cowen and Miller [450] have developed an enantioselective version of the reaction using phosphine-containing α-amino acid derivatives as catalysts. Propose a mechanism for the (3+2)-cycloaddition **A** + **B** → **P** in the presence of catalyst **C**.

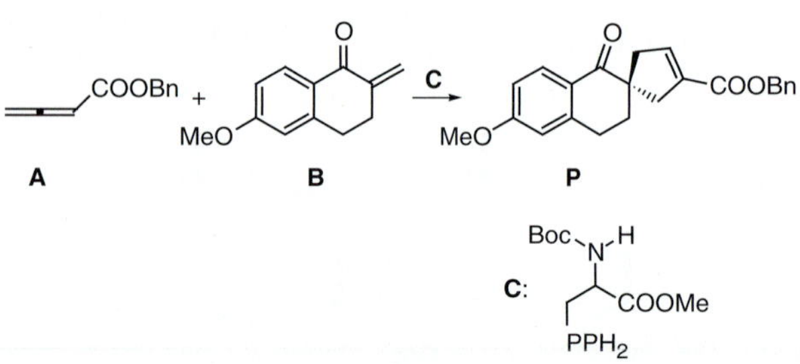

7.5.7 Nucleophilic catalysis of cycloadditions

Examples of (2+2)-cycloaddition catalyzed by NHC have been given in Section 5.3.6. Mixing 2-alkyl-2,3-butadienoates **158** with *N*-tosylimines **159** in the presence of 20 mol% Bu_3P results in the diastereoselective formation of tetrahydropyridines **160**. The butadienoates **158** act as 1,4-dipole synthons giving products of (4+2)-cycloaddition (Scheme 7.24). A γ-addition to imine **159** by the zwitterionic phosphonium adduct **161** gives **162**. The successive proton transfer step giving **163** sets up for an *N*-tosyl Michael addition to close the six-membered ring [451]. Several related examples of phosphine-catalyzed (4+2)-cycloadditions have been reported [417, 452–455], including enantioselective reactions utilizing enantiopure phosphines as catalysts [456–458]. Enantioselective (4+2)-cycloadditions of α,β-unsaturated imines catalyzed by enantiopure phosphines have also been reported [459–462].

L-Proline catalyzes the enantioselective aza-Diels–Alder reaction of arylimines with conjugated enones as illustrated in Scheme 7.25. L-Proline and the enone equilibrate with the corresponding 2-aminodiene **164** that adds to the imine equilibrating with the arylamine and formaldehyde. The reaction is assumed to involve the formation of the zwitterionic intermediate **165** that cyclizes quickly and irreversibly into the corresponding Diels–Alder cycloadduct [463]. For other examples of catalyzed enantioselective aza-Diels–Alder reactions, see Refs. [464–468].

In 2006, Bode and coworkers reported a stereoselective hetero-Diels–Alder reaction proceeding via a NHC-bound azolium enolate intermediate of type **168** derived from ethyl 4-oxobut-2-enoate and enantiomerically pure diaminocarbene **166** (Scheme 7.26) [469]. Enolate **168** is the dienophile that adds to the azadiene giving zwitterion **169**. The latter undergoes facile β-elimination providing cycloadduct **167** and the starting catalyst (**166**). Yields up to 90% with 97–99% ee were obtained. In the meantime, more examples of nucleophilic catalyzed (4+2)-cycloadditions have been reported [470, 471].

Ketenes react with tertiary amines giving enolates of type **170** that behave as nucleophiles onto electrophilic imines giving zwitterion intermediates **171**. The latter undergo intramolecular displacement reactions producing the corresponding β-lactams and liberating the nucleophilic catalyst. Ketene-enolates **170** can also be formed by reaction of the tertiary amine with acyl chlorides (Scheme 7.27a). When an enantiomerically pure tertiary amine is used, it can induce asymmetry in the formation of β-lactam [472].

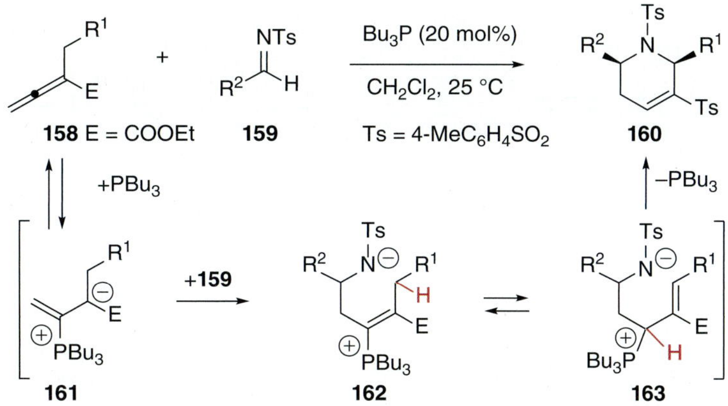

Scheme 7.24 A phosphine-catalyzed (4+2)-annulation.

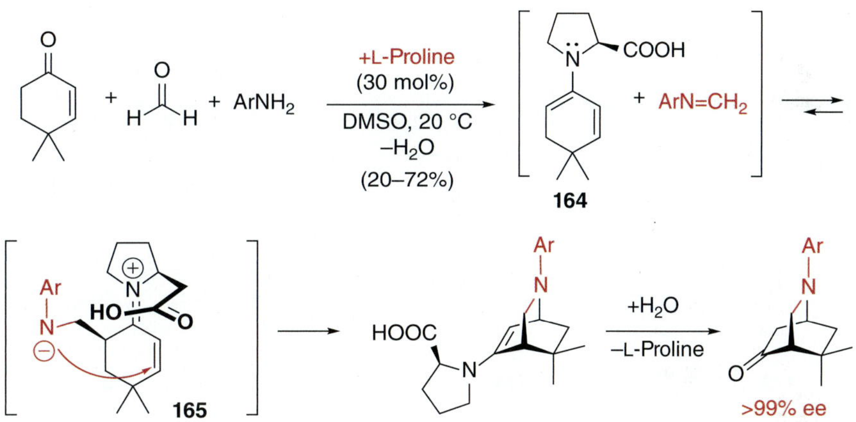

Scheme 7.25 Enantioselective aza-Diels–Alder reaction catalyzed by L-proline.

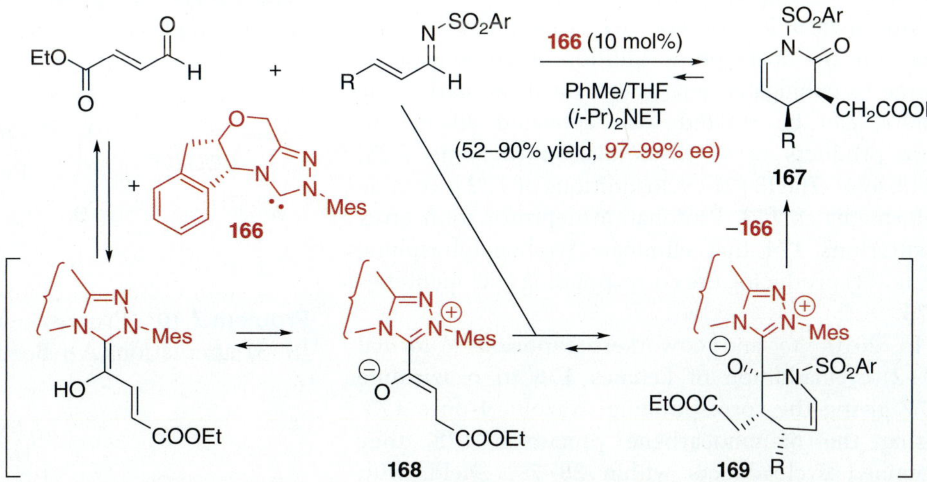

Scheme 7.26 Bode's enantioselective hetero-Diels–Alder reaction.

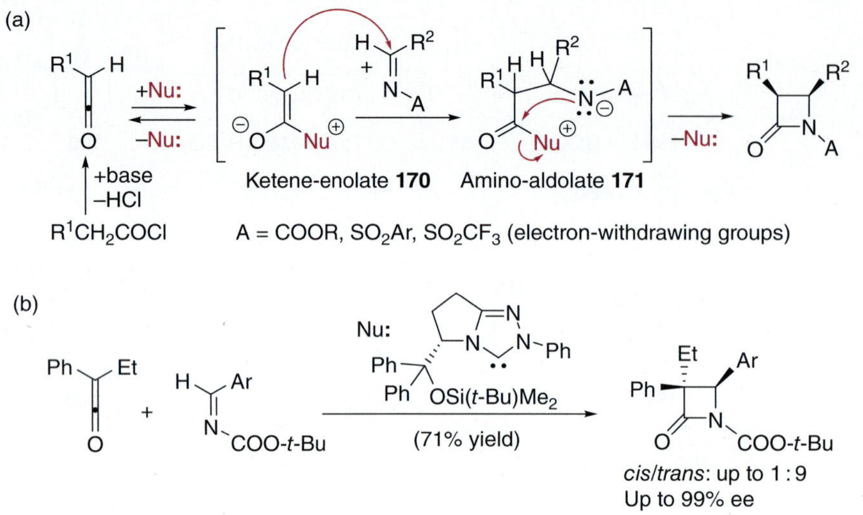

Scheme 7.27 (a) Nucleophile-catalyzed (2+2)-cycloaddition. (b) Asymmetric synthesis of β-lactams via enantiomerically pure ketene-enolate intermediates (NHC-catalyzed Staudinger reaction).

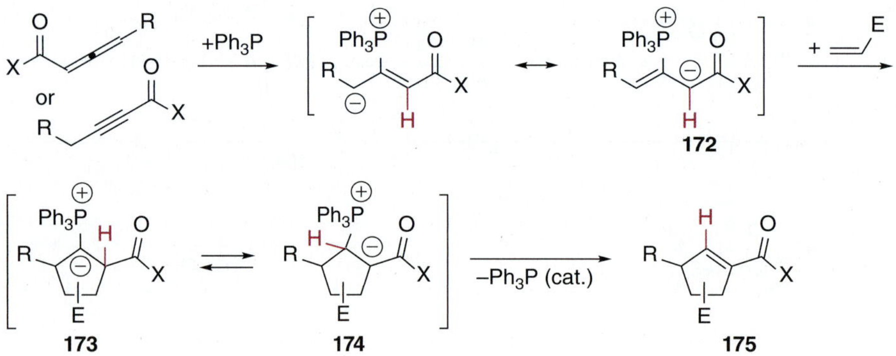

Scheme 7.28 Phosphine-catalyzed (3+2)-annulation.

Instead of a tertiary amine, one can employ an enantiomerically pure chiral NHC as a nucleophilic catalyst (Scheme 7.27b) [473].

The allylic anions of type **161** (Scheme 7.24) resulting from the nucleophilic addition of triphenylphosphine to conjugated yne-esters, or allene carboxylic esters, can be reacted with activated alkenes to give products of (3+2)-annulation (Scheme 7.28) [448, 474]. The (3+2)-cycloadditions of **172** give ylide intermediates **173**. Protonation/deprotonation gives zwitterions **174** that eliminate triphenylphosphine (catalyst) producing the corresponding cyclopentenes **175**.

In 2010, Ye and coworkers reported a formal (3+2)-cycloaddition of ketenes **176** to oxaziridine **177** giving the corresponding oxazolin-4-ones **179**. Using the diaminocarbene precursor **178**, they obtained cycloadducts within 38–78% yield, with high diastereoselectivity (dr up to 15 : 1) and enantioselectivity (up to 95% ee). The proposed reaction mechanism is shown in Scheme 7.29 [475].

In the presence of a chiral phoshine as a catalyst, malonic esters or equivalent carbon acids react with forming enantio-enriched cyclopentene derivatives resulting from (4+1)-annulations [476].

Problem 7.9 Propose a mechanism for reaction **A** + **B** → **C** [477].

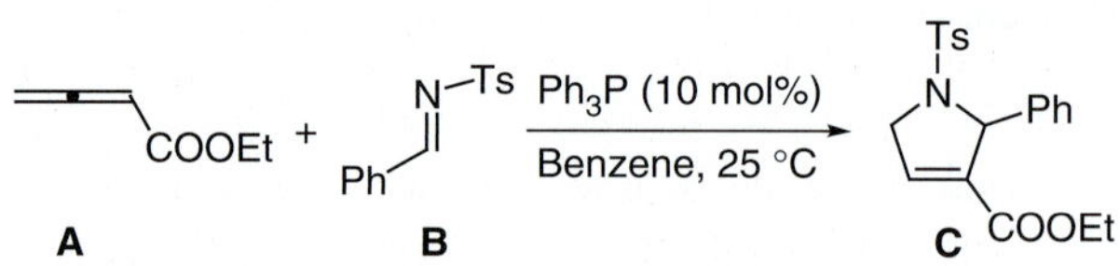

Problem 7.10 Propose a mechanism for the (8+3)-annulations **A** + **B** → **C** [478].

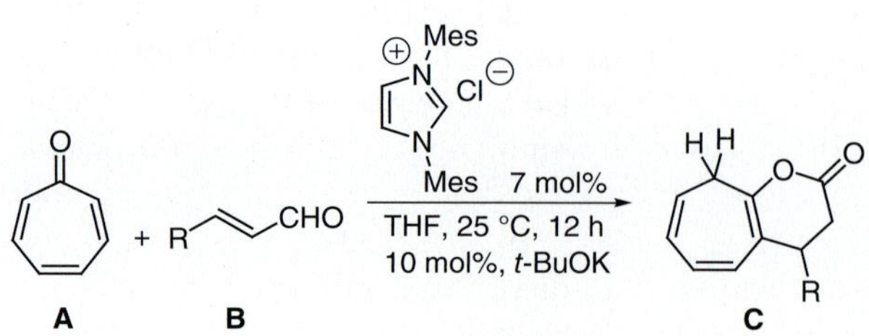

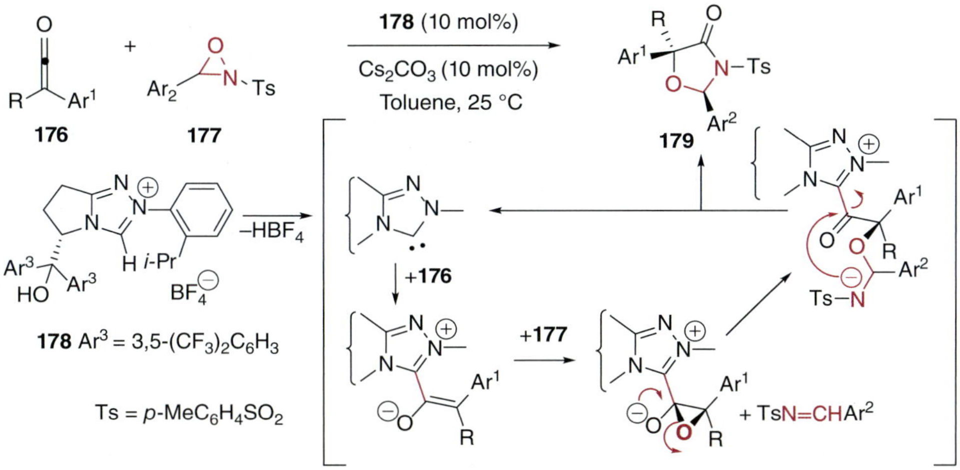

Scheme 7.29 NHC-catalyzed (3+2)-cycloaddition of ketenes and oxaziridines.

7.5.8 Catalysis through electron-transfer: hole-catalyzed reactions

In 1963, Scott et al. proposed that the polymerization of *N*-vinylcarbazole (NVC) in MeCN induced by chloranil (CA), ethylenetetracarbonitrile (TCNE), or Wurster's salts (e.g. $Me_2NC_6H_4NMe_2^{•+}/Br_3^-$) is initiated by the radical cation NVC$^{•+}$ [479]. In 1964, Ellinger found that in MeOH and in the presence of daylight, the polymerization of NVC is diverted into a (2+2)-cyclodimerization [480]. This is an early example of a PET process that catalyzes a pericyclic reaction. In 1965, Ledwith and coworkers found that NVC can be cyclodimerized in MeOH in the presence of Fe(III) or Ce(IV) salts [481]. They also reported that other enamines can undergo (2+2)-cyclodimerization under similar conditions [482] or in the presence of tris(*p*-bromophenyl)ammonium perchlorate [483]. The cation radical chain nature of the reaction (Scheme 7.30a) was established by this group in 1970 [484]. In this mechanism, the radical cation of the enamine adds to another molecule of enamine giving the ammonio-alkyl radical **181** that is cyclized into amine radical cation **182**. The latter is reduced into the cycloadduct **180** by the starting enamine (the chain process starts) or by the reduced form of the oxidant (catalyst), which terminates the chain process. Ledwith and coworkers also found that one electron oxidation (SET) of the cycloadduct **180** in MeOH gives the open-chain methanolysis product **183**. The latter is also formed by oxidative coupling of NVC in MeOH, demonstrating the interconvertability of cation radicals **181** ⇄ **182** bearing carbazolyl substituents [485]. In 1964, the ^{60}Co γ-radiolysis of cyclohexa-1,3-diene was reported to produce primarily Diels–Alder cyclodimerization [486], a reaction very difficult on heating (Section 5.3.12). In 1969, Schutte and Freeman established the radical cation chain mechanism of this reaction [487]. UV irradiation of 1,1-diphenylethylene in the presence of methyl 4-cyanobenzoate induces its formal (4+2)-cyclodimerization (Scheme 7.30b) [488].

Unsymmetrically substituted cyclobutanes have been prepared via crossed intermolecular [π^2+π^2]-cycloadditions of styrenes induced by visible light photocatalysis (Scheme 7.31a). The photocatalyst must be chosen for its photoredox potential that makes its excited state capable to oxide selectively the electron richest of the two styrenes [489], **184** in our case (Scheme 7.31a). Radical cation **184**$^{•+}$ so obtained reacts with the other styrene **185** faster than with **184** for steric reasons. This generates radical cation cycloadduct **186**$^{•+}$, which is reduced by the electron-richest styrene **184** into product **187** and radical cation **184**$^{•+}$, the latter starting the chain process. Regeneration of the photoredox catalyst is insured by a SET from O_2. If radical cation **186**$^{•+}$ oxidizes the reduced form of the photocatalyst, it terminates the chain process. When Ar^1 = *p*-methoxyphenyl and Ar^2 = *p*-toluyl, cross-cyclodimerization (**184** + **185**) gives the corresponding (2+2)-cycloadduct **187** in 86% yield and with less than 8% of the product of cyclodimerization of **184** [490]. An example of visible light photocatalytic Diels–Alder addition between an electron-rich diene and an electron-rich dienophile is given in Scheme 7.31b [491]. The [π^2+π^2]-cyclodimerization of aryl vinyl ethers (**188**) are catalyzed by PET. On irradiating the sensitizer (e.g. sens = 1,4-dicyanobenzene, DCB), its excited triplet state abstracts an electron from one vinyl ether unit generating the radical cation intermediate **188**$^{•+}$

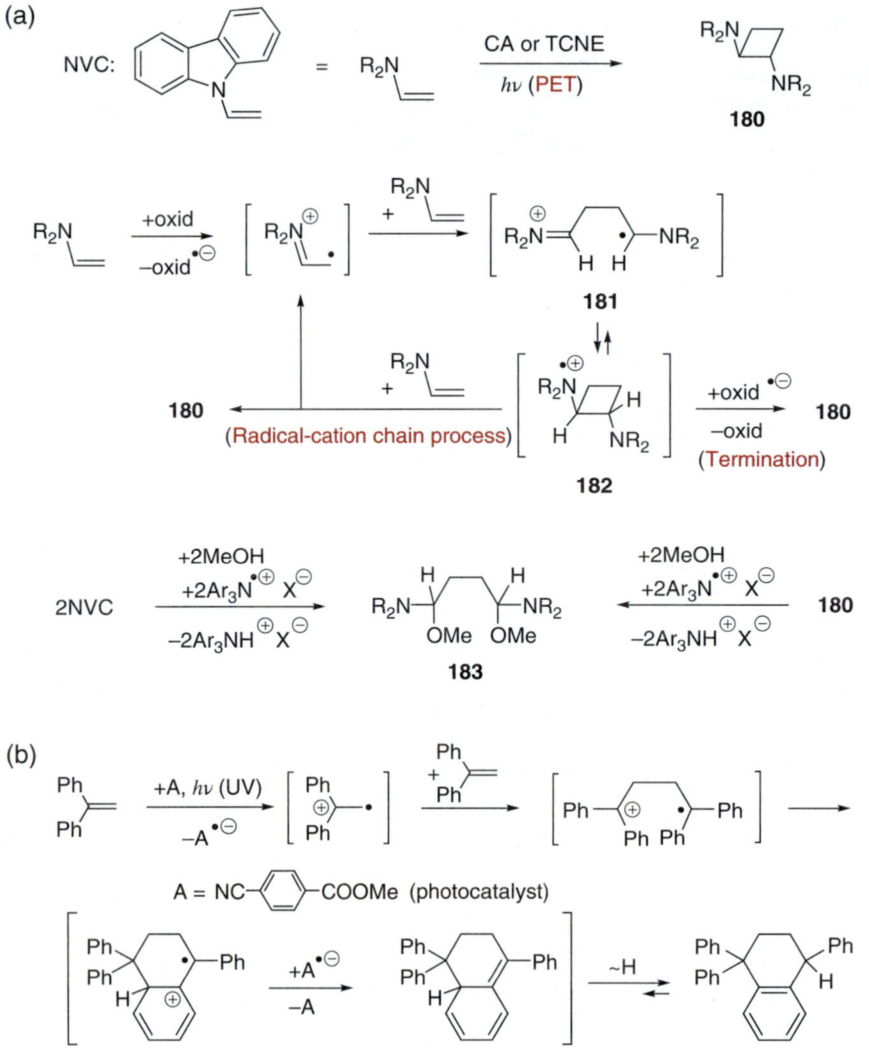

Scheme 7.30 Early examples of cycloadditions involving radical cation intermediates. (a) Ledwith's mechanism for the hole-catalyzed $[\pi^2+\pi^2]$-cyclodimerization of enamines induced by SET or PET from the electron-rich alkene to the oxidant; (b) PET-induced cyclodimerization of 1,1-diphenylethylene.

(+ sens$^{\bullet-}$). The latter is electron poor and thus adds rapidly to an electron-rich vinyl ether with formation of 1,4-radical cation **189**. Cyclization gives radical cation **190**, which abstracts an electron from vinyl ether **188** to form the product of (2+2)-cycloaddition and radical cation **188**$^{\bullet+}$, which starts another cycle of reaction (Scheme 7.31c) [492]. The regioselectivity of the cycloaddition is controlled by the relative stability of isomeric 1,4-radical cation intermediates that can be formed. In this case, **189** is the most stable radical cation that can arise from the addition **188** + **188**$^{\bullet+}$.

Bauld and coworkers have reported the cyclodimerization of cyclohexa-1,3-diene catalyzed by a triarylaminium salt (Scheme 7.32a) [493–496]. The latter abstracts an electron from the diene generating a radical cation intermediate that becomes an electron-poor dienophile able to add quickly to another molecule of cyclohexa-1,3-diene furnishing a (4+2)-adduct that is a radical cation. The latter takes an electron from a diene with the delivery of the Diels–Alder adduct and regeneration of the diene-radical cation that starts another cycloaddition cycle (Scheme 7.32c). The process can be applied to catalyze the cycloaddition of cyclohexa-1,3-diene to (*E*,*E*)-hexa-2,4-diene (Scheme 7.32a). In 2002, Singleton and coworkers demonstrated with ^{13}C kinetic isotope effect measurements and quantum mechanical calculations that the $(p\text{-}BrC_6H_4)N^{\bullet+}/SbCl_6^-$-catalyzed Diels–Alder reaction of cyclohexadiene and indole follows a stepwise mechanism initiated by attack of the diene radical cation into the C(3)-position of indole, which is the rate-determining step of the reaction (Scheme 7.32b) [497].

Under thermal conditions and applying catalyst of type $Ar_3N^{\bullet+}/SbCl_6^-$, $[\pi^4+\pi^2]$-, $[\pi^2+\pi^2]$-cycloadditions, vinylcyclobutane rearrangement and alkene polymerization compete [498]. $SnCl_2$ and $CuSO_4$ catalyze the $[\pi^2s+\pi^2s]$-cycloreversion of

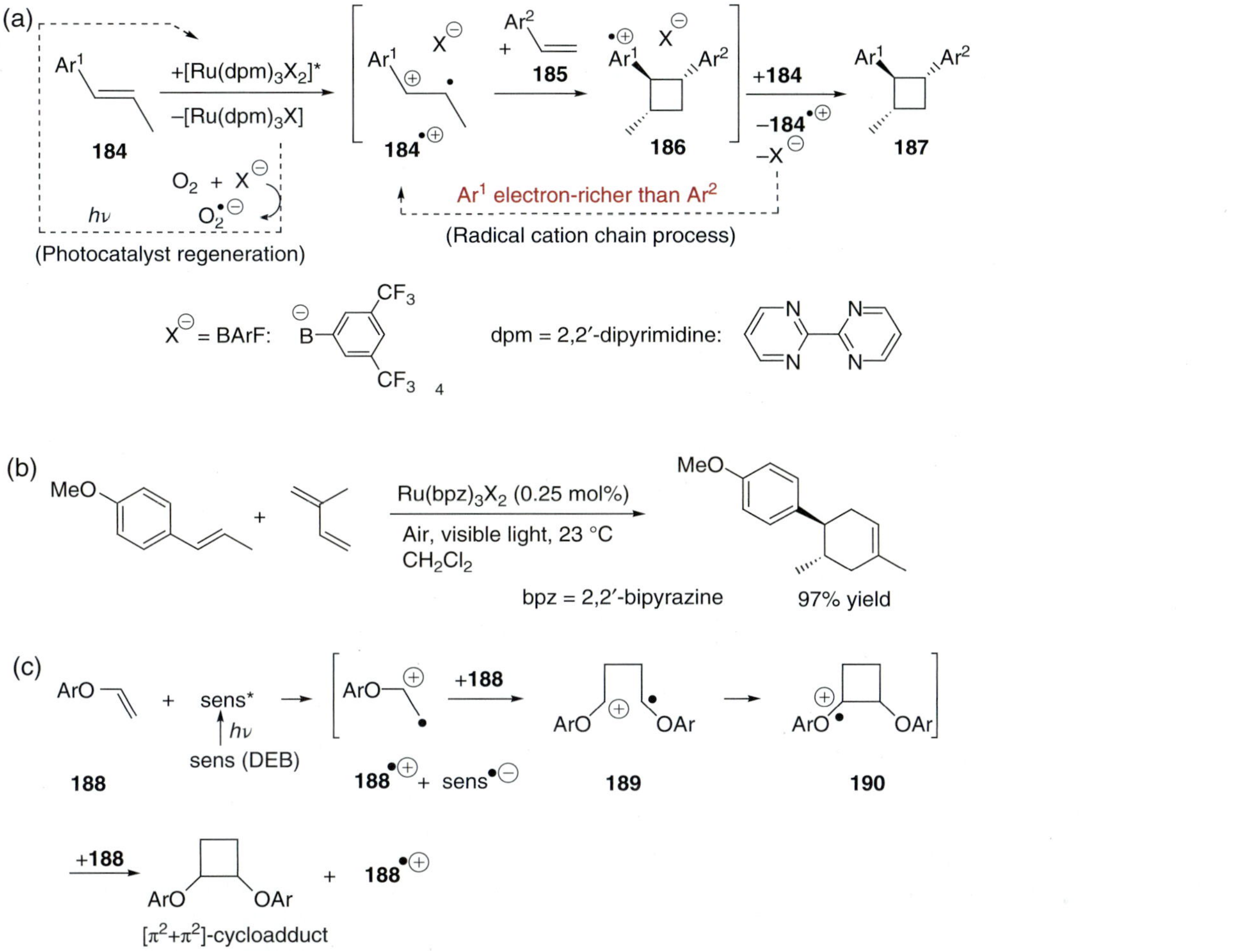

Scheme 7.31 Examples of photocatalytic radical cation intermolecular cross-cycloadditions. (a) Formal [π^2+π^2]-cycloaddition of two different styrenes; (b) formal [π^4+π^2]-cycloaddition of isoprene with a styrene; (c) [π^2+π^2]-cyclodimerization of aryl enol ethers photocatalyzed by benzene-1,4-dicarbonitrile (1,4-dicyanobenzene, DCB).

quadricyclane into norbornadiene, via a hole-catalyzed process [499]. An example of SET-catalyzed [π^2a+π^2a]-cycloaddition has been given with the isomerization of tetrakis(trimethylsilyl)cyclobutadiene into tetrakis(trimethylsilyl)tetrahedrane (Section 2.5.2) [500]. Cyclopropanation of alkenes with N_2CH_2COOEt [501], as well as epoxidation with SeO_2, has been catalyzed by $(p\text{-}BrC_6H_4)_3N^{\bullet+}/SbCl_6^-$ in CH_2Cl_2 at 0 °C [502]. An example of photoassisted Cope rearrangement on zeolite has been reported [503]. [π^2+π^2]-Cycloadditions of aryl alkenyl ethers to alkenes giving cyclobutanes have been realized by anodic oxidation in $LiClO_4$/nitromethane solution [504]. Hole-catalyzed (3+2)-cycloadditions of cyclopropanes **191** to vinyl ethers **192** via PET (reaction (7.40)) produce the corresponding cyclopentanes **193** [505].

(7.40)

DCB = p-NC-C_6H_4-CN

Problem 7.11 $Fe(ClO_4)_3$ catalyzes the (2+2)-cyclodimerization of *trans*-anethole and gives a *trans, anti, trans*-cyclobutane derivative stereospecifically in good yield. Alumina-supported $Fe(ClO_4)_3$ is an especially efficient catalyst. Propose a mechanism for this reaction [506].

Problem 7.12 What is the major product **P** of the following visible-light-induced reaction (bpz = 2,2-bipyrazine)? [491].

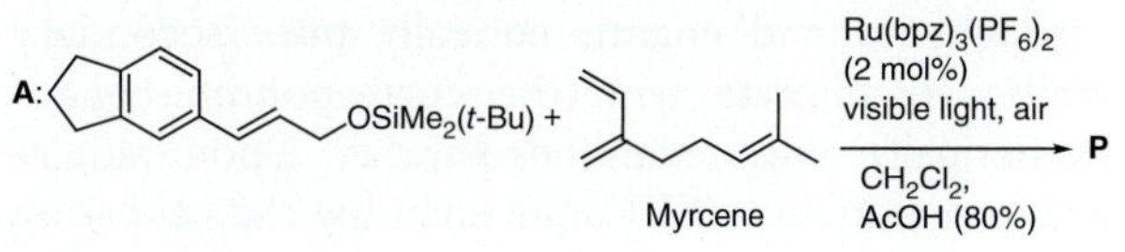

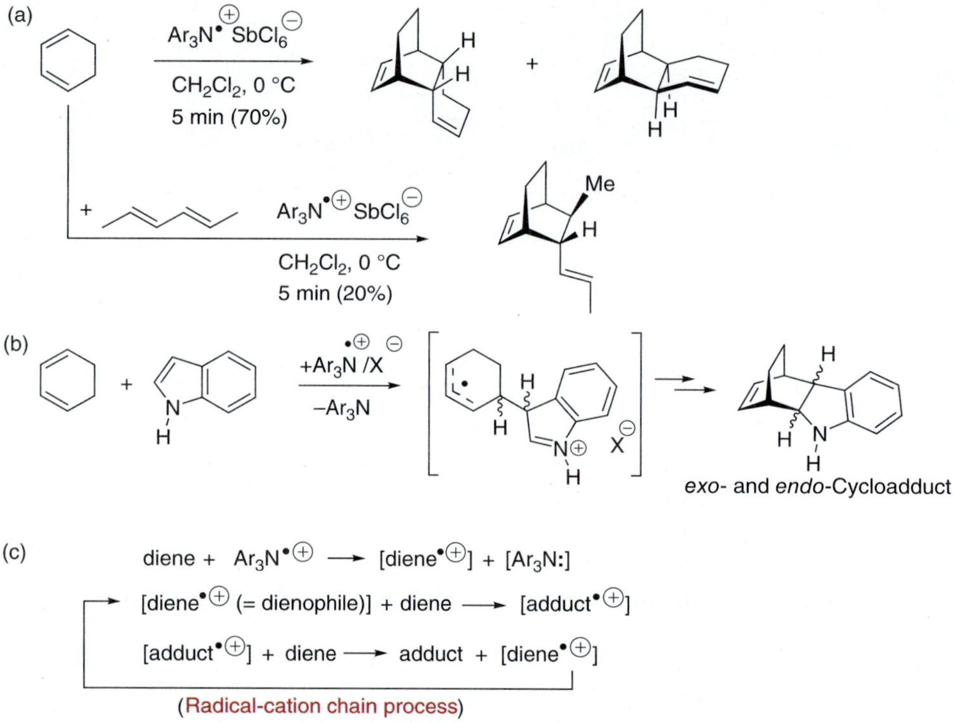

Scheme 7.32 (a,b) Difficult thermal Diels–Alder reactions of cyclohexadiene are quick at 0 °C upon SET oxidation; (c) proposed mechanism for the hole-catalyzed $[\pi^4+\pi^2]$-cycloadditions.

Problem 7.13 Propose a mechanism for the $(p\text{-BrC}_6H_4)_3N^{\bullet+}/SbCl_6^-$-catalyzed epoxidation of 1-vinylcyclohexene with benzeneselenic anhydride. What is the main product or reaction? [502].

7.5.9 Umpolung of enamines

Enamines are electron-rich alkenes that react as nucleophiles [507–510]. They generate relatively stable carbocation intermediates (α-aminocarbenium ions ↔ iminium ions) when they react with electrophilic reagents. They are also radicophilic as they add radicals forming amino-substituted alkyl radicals that are highly stabilized by the amino substituent, especially by *N*,*N*-dialkylamino groups (Tables 1.A.7 and 1.A.9). Enamines derived from enantiomerically pure secondary amines and aldehydes play a crucial role in asymmetric organocatalysis [511–513]. As for any electron-rich alkenes (enolates, enols ethers, enol thioethers, arylalkenes, etc.), they are readily oxidized into their radical cations that are electrophilic. Thus, the single-electron oxidation of enamines realizes their Umpolung.

Aldehydes and enantiomerically pure secondary amines equilibrate with the corresponding enantiomerically pure enamines + water. Upon single-electron oxidation (SET), the enamine radical cation is formed: it can react with all kinds of carbon nucleophiles inter- and intramolecularly, generating alkyl radicals that, on their turn, can be oxidized into the corresponding carbocation intermediates that are quenched by the medium, or a neighboring nucleophilic moiety and then the medium, liberating the secondary amine and 1 equiv. of water. Overall, the process realizes an oxidative C—C-coupling in which the secondary amine can be used in a catalytic amount. With aldehydes of type RCH_2CHO, in the presence 20 mol% of enantiomerically pure amine **194** (Scheme 7.33), using allylsilane as nucleophile and 2 equiv. of CAN as oxidant, MacMillan and coworkers have realized the enantioselective oxidative α-allylation of aldehydes ($RCH_2CHO + 2Ce^{(IV)}(NH_4)_2(NO_3)_6 + base + CH_2{=}CH{-}CH_2SiMe_3 \rightarrow RCH(allyl)CHO + 2Ce^{(III)}(NH_4)_2(NO_3)_5 + baseH^+/NO_3^- + Me_3SiONO_2$) in excellent yields and ee's [514]. Using enoxysilanes as nucleophiles under similar conditions, enantioselective α-enolation of aldehydes has been carried out (e.g. **195** + $CH_2{=}C(Ph)OSiMe_3$ → **197**, Scheme 7.33a). In the case of aldehyde **195**, and in the absence of enoxysilane, oxidative cyclization **195** → **198** occurs (Scheme 7.33b) [515]. Similar reaction with 4-(aryl or heteroaryl)-substituted aldehydes leads to enantioselective intramolecular oxidative α-arylation

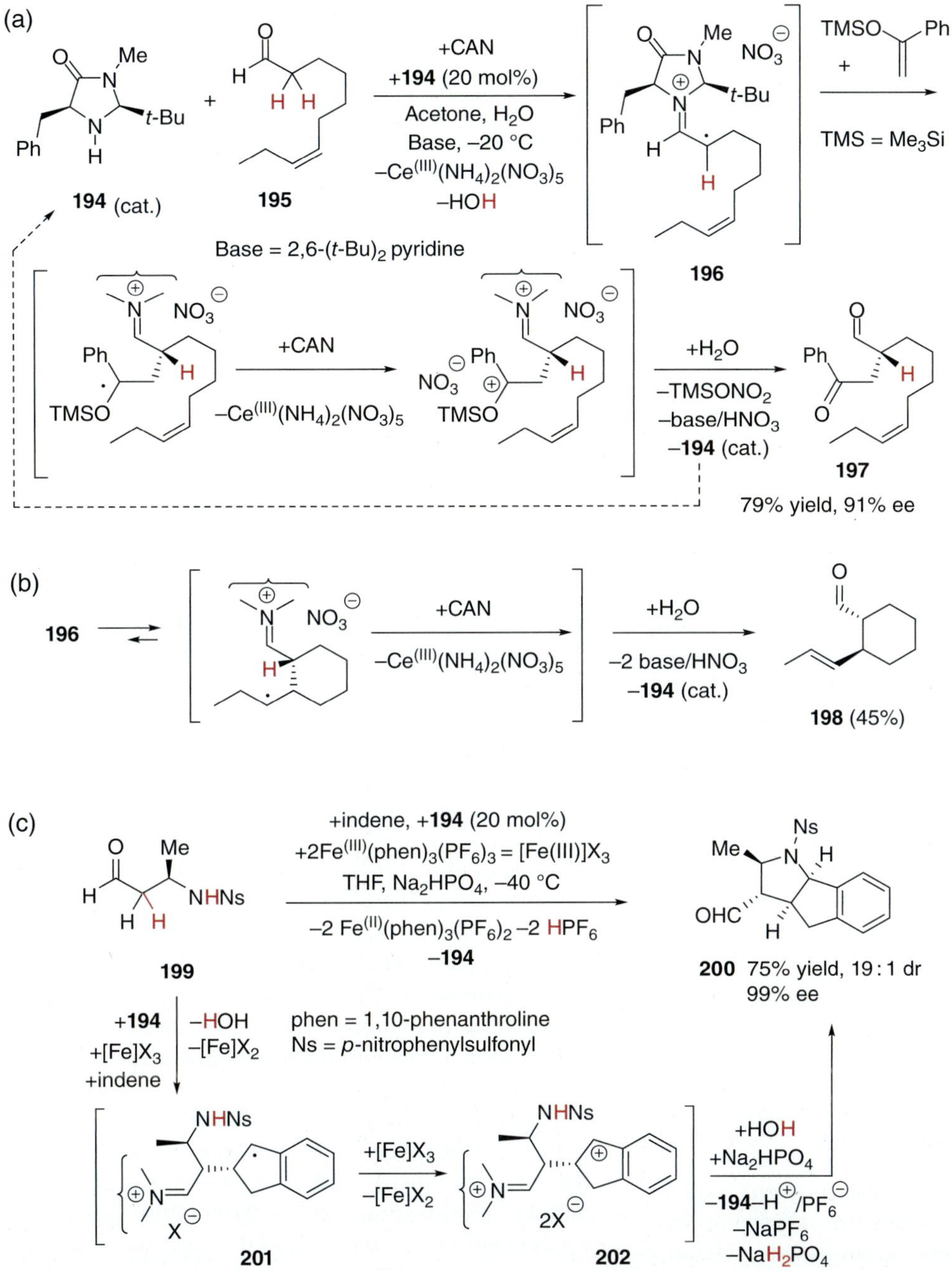

Scheme 7.33 Examples of enantioselective organo-SOMO catalyzed reactions generating α-substituted aldehydes via the Umpolung of enantiomerically pure enamine intermediates. (a) Oxidative α-enolation; (b) oxidative cyclizations; and (c) oxidative cycloaddition.

of the aldehydes [516]. Using alkenylpotassium trifluoroborate salts as nucleophiles, enantioselective α-alkenylation of aldehydes has been realized ($RCH_2CHO + Ar(R)C{=}CHBF_3K + 2CAN + NaHCO_3 \rightarrow RCH[CH{=}C(R)Ar]\text{-}CHO + 2Ce^{(III)}(NH_4)_2(NO_3)_5 + NaNO_3 + H_2O + CO_2 + BF_3 + KNO_3$) [517]. With silyl nitronate as nucleophiles ($R''CH{=}N^{(+)}(—O^{(-)})—OSiR'_3$), enantioselective oxidative aldehyde α-nitroalkylation has been developed [518]. An enantioselective synthesis of pyrrolidines has been proposed based on enamine Umpolung. For instance, in the presence of a catalytic amount of **194**, sulfonamide **199** and indene undergo oxidative cycloaddition into the tricyclic pyrrolidine derivative **200** in the presence 2 equiv. of iron(III)(phenanthroline)$_3$PF$_6$, the oxidant. The

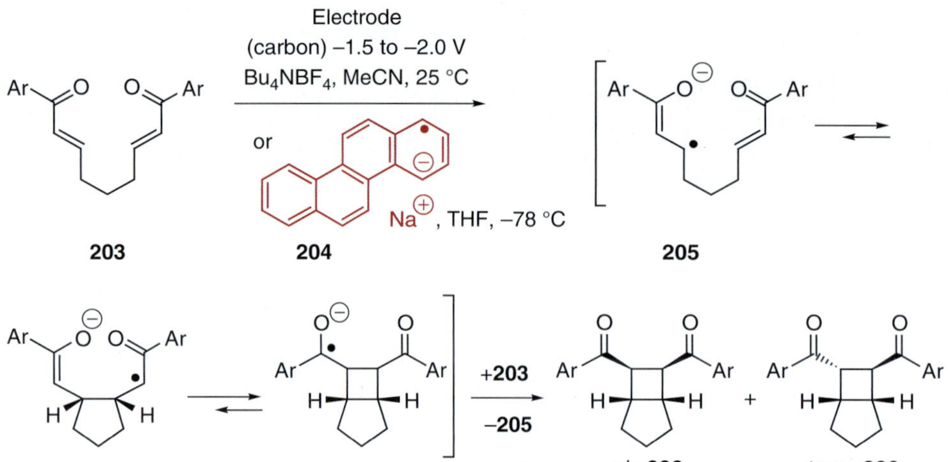

Scheme 7.34 Electrochemically or chemically induced radical anion $[\pi^2+\pi^2]$-cycloaddition.

intermediate enamine that equilibrates with aldehyde **199** and the chiral catalyst **194** is oxidized into its radical cation that adds to the alkene (indene) generating radical intermediate **201**. The latter is then oxidized into benzylic cation intermediate **202** that is quenched intramolecularly by the sulfonamide moiety, giving, after hydrolysis, product **200**, and recovery of the chiral catalyst (Scheme 7.33c). The reaction is highly diastereo- and enantioselective [519]. If the sulfonylamido group is exchanged for an arene or heteroarene group as in 2-aryl(heteroaryl)aldehydes, their reactions with styrenes realize enantioselective oxidative cycloadditions [520]. MacMillan calls the oxidative processes based on the Umpolung of enantiomerically pure enamine intermediates as enantioselective organo-SOMO catalyzed reactions (SOMO catalysis) [514].

7.5.10 Catalysis through electron-transfer: Umpolung through electron capture

Electrochemical reactions have proven valuable for the Umpolung of alkenes by triggering radical anion-based or radical cation-based cyclization reactions [521–524].

Cathodic reduction of 2- and 4-vinylpyridine and of 2-vinylquinoline gives *trans*-1,2-di(heteroaryl) cyclobutanes as major products. They arise via radical anion + substrate cycloaddition realizing probably a chain process [525]. Similar cyclobutanations induced electrochemically have been reported [526, 527]. In the case of bis(enone) **203**, the intramolecular $[\pi^2+\pi^2]$-cycloaddition is induced electrochemically or by a substoichiometric amount of sodium chrysene radical anion (**204**). The major product is *trans*-**206** under electrochemical conditions (MeCN, 25 °C) and *cis*-**206** (61.7%) under radical anion-induced conditions (THF, −78 °C) [528]. The reaction proceeds via radical anion **205** as shown in Scheme 7.34.

Exposure of aromatic bis(enones) **207** to the methyl Gilman reagent $Me_2CuLi{\cdot}LiI$ at 0 °C in tetrahydrofuran solvent provides the products of tandem conjugate addition/Michael cyclization **209**, along with the products of $[\pi^2+\pi^2]$-cycloaddition **212** (Scheme 7.35). Products **209** are favored by rapid addition of the Gilman reagent in excess at higher concentrations, whereas **212** are favored by slow addition of $Me_2CuLi{\cdot}LiI$ at lower concentrations and loadings. Notably, $[\pi^2+\pi^2]$-cycloaddition to form **212** is catalytic in Gilman reagent. Kinetic data revealed that the ratio [**209**]/[**212**] changes such that the cycloaddition pathway becomes dominant upon increased consumption of Gilman reagent. These data are consistent with a concentration-dependent speciation of the Gilman reagent and differential reactivity of the aggregates present at higher and lower concentrations [529]. The dependence of the competition between the SET chemistry (**207** → **210** → **211** → **212**) and Gilman methylation (**207** → **210** → **209**) upon the concentration of $Me_2CuLi{\cdot}LiI$ suggests that the species responsible for methylation is probably a dimeric species, whereas the species responsible for SET is probably the monomeric Gilman reagent, which is present in tetrahydrofuran solutions as solvent-separated ion pair [530].

7.6 Brønsted and Lewis acids as catalysts in C—C bond forming reactions

Two acid–base definitions are used in organic chemistry: the Brønsted–Lowry definition [531] and

Scheme 7.35 Competitive Michael addition and [$\pi^2+\pi^2$]-cycloaddition during the reaction of the methyl Gilman reagent with a bis(enone).

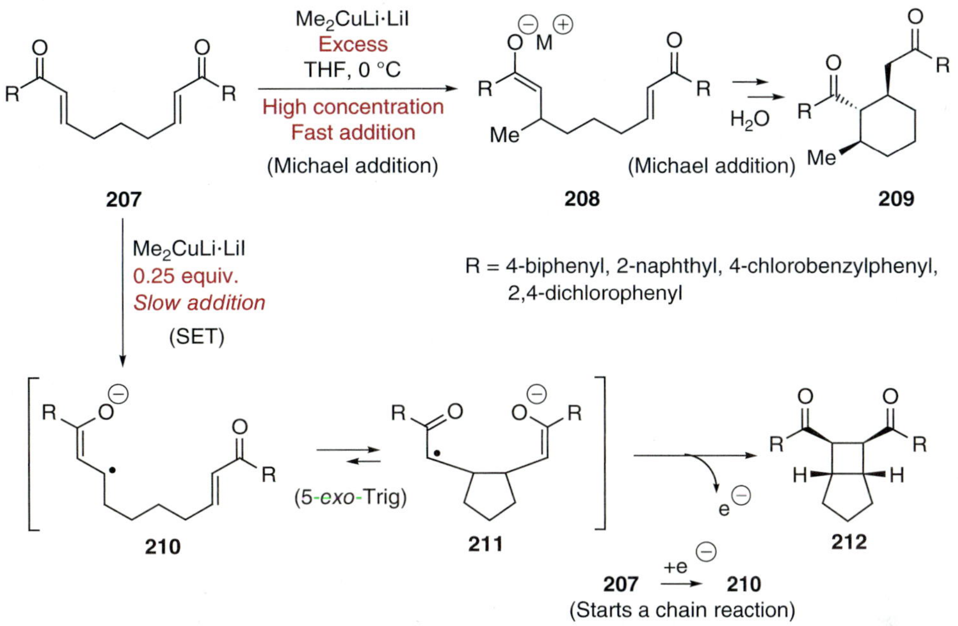

the Lewis definition [532]. A Brønsted acid HA is defined as a proton donor ($HA \rightleftarrows H^+ + A^-$) and a base B: is a proton acceptor ($B: + H^+ \rightleftarrows BH^+$). Carbonyl and imine compounds equilibrate with protic (Brønsted) acids HA giving salts **213**, singly bonded complex **214** or doubly bonded complexes **215** [533]. Brønsted acids of types RCOOH, $(RO)_2P(=O)OH$, RSO_3H, $(RSO_2)_2NH$, $(RO)_2P(=O)NHSO_2CF_3$, and $(RSO_2)_3CH$ are becoming efficient catalysts for a range of C—C bond forming reactions [534]. A large collection of enantiopure Brønsted and Lewis acids is now available and can be used to catalyze enantioselective nucleophilic additions (e.g. asymmetric cyanation of carbonyl compounds, imines, α,β-unsaturated carbonyl compounds with HCN, Me_3SiCN, EtOCOCN, MeCOCN, acetone cyanohydrin [221, 535], asymmetric addition of terminal acetylenes to aldehydes [536], and asymmetric addition of alkylboron reagents to aldehyde catalyzed by enantiopure Ti complexes [537]).

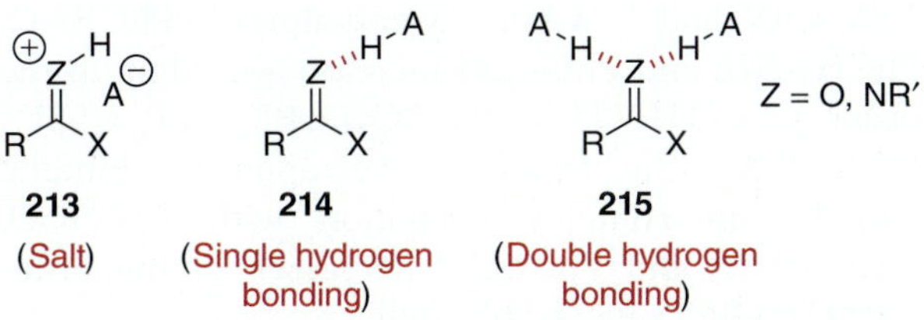

Lewis acid–base theory [538] is more comprehensive than the Brønsted–Lowry theory. Lewis acids are neutral or charged electron-deficient species with a vacant orbital responsible of the electrophilic character of the acid. Lewis acids include Brønsted acids, metal ions, metal atoms, neutral complexes (e.g. BF_3, $AlCl_3$, and methylaluminoxane (MAO) = product of the partial hydrolysis of Me_3Al: $—[Al(Me)—O—]_n$), various ionic liquids (salts in the liquid state) [539–542], and superacids (nonaqueous acids more acidic than H_3O^+) [543]. Pearson distinguishes hard and soft acids, and hard and soft bases (Section 1.13) [544, 545]. A hard acid is positively charged and concentrates the charge on one or few atoms, a soft acid is neutral or negatively charged with a low-lying empty orbital (high electron affinity). The classical boron Lewis acids BX_3, RBX_2, and R_2BX (X = F, Cl, Br, I, OSO_2CF_3 = OTf) act as Lewis acids because of their empty 2p(B) orbital on boron [546–548]. Enthalpies for their combination with pyridine as the reference base show the following Lewis acidity order for group IIIB halides: $AlX_3 > BX_3 > GaX_3$. The Lewis acidity of BX_3 toward Me_3N and Et_2O increases in the order F < Cl < Br < I, i.e. the reverse of the order expected on the basis of relative σ-donor strength of the halogen atom [549, 550]. Quantum mechanical calculations on the relative acid strengths of BF_3 and BCl_3 toward Lewis bases such as NH_3, Me_3N, H_2O, and Me_2O show that the BX bond length increases during complex $X_3B \leftarrow$ base ($\leftrightarrow X_3B^{(-)}$–base$^{(+)}$) formation. It takes more energy to lengthen the short B—F bonds than the longer and weaker B—Cl bonds. It is for that reason that BCl_3 is a stronger Lewis acid than BF_3 toward the above strong bases. An energy partitioning analysis of $Cl_3B \leftarrow :NH_3$ and $F_3B \leftarrow :NH_3$ shows that the stronger bond in the former complex comes from enhanced covalent interactions between the Lewis acid and the Lewis base, which can be related to the energetically lower lying lowest unoccupied molecular orbital (LUMO) of BCl_3 [551]. In contrast, in the formation of a complex with a weak base such as CO, HCN, MeCN, and CH_3F, BX_3 is barely distorted from planarity and so the acidity of BF_3 is greater than that of BCl_3 because the charge on boron atom is greater in BF_3 than in BCl_3 [552]. The BF_3 and BCl_3 complexes of Et_2O are less stable than those of Me_2O, and the same trend is observed for complexes of Et_2S and Me_2S. This is a manifestation of F-strain (Section 2.5.1), which destabilizes $Et_2Z{:}\rightarrow BX_3$ complexes more than $Me_2Z{:}\rightarrow BX_3$ complexes. This phenomenon (F-strain) diminishes from boron to aluminum and from oxygen to sulfur. According to Corey and Lee, hydrogen bridging of the C—H bond of the CHO group of aldehydes with the σ-ligand X in BX_3 and related Lewis acids can be significant and play an important role in the addition stereoselectivity of aldehydes [553].

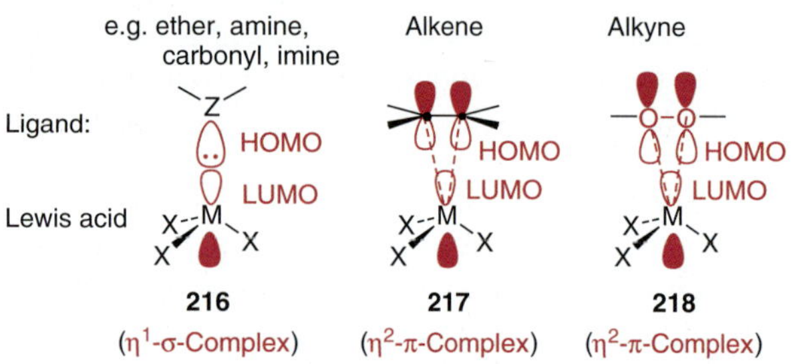

Lewis acids MX_n make η^1-σ-complexes **216** (dative complexes) with bases having a heteroelement bearing a nonbonded electron pair. With π-systems such as alkenes and alkynes, they generate η^2-π-complexes **217** and **218**, respectively. For light metallic atoms M, bonding in these complexes results mostly from the HOMO(ligand) → LUMO(MX_n) interaction (donation). Back-donation is weak because MX_n does not have a high-lying highest occupied molecular orbital (HOMO) to retrocede electrons into the LUMO of the ligand. With heavier metallic atoms M, especially with transition metals, high-lying HOMOs are available and they overlap with the ligand LUMO. Computed enthalpies of binding of BCl_3, $MgCl_2$, $AlCl_3$, CuCl, $CuCl_2$, AgCl, AuCl, $AuCl_3$, and $PtCl_2$ for benzaldehyde, benzylimine, and phenylacetylene are given in Table 7.3. The data show that BCl_3 makes complexes with aldehydes and imines that are much more stable than complexes with acetylenes. This is also the case with $MgCl_2$ and $AlCl_3$ but to a much smaller extent. In fact, one can state that BCl_3 does not form η^2-π-complexes, whereas $AlCl_3$ does form π-complexes with alkenes and alkynes. The $AlCl_3$-catalyzed cyclotrimerization (7.41) of but-2-yne into hexamethylbicydo[2.2.0]hexa-2,5-diene illustrates this point [554, 555].

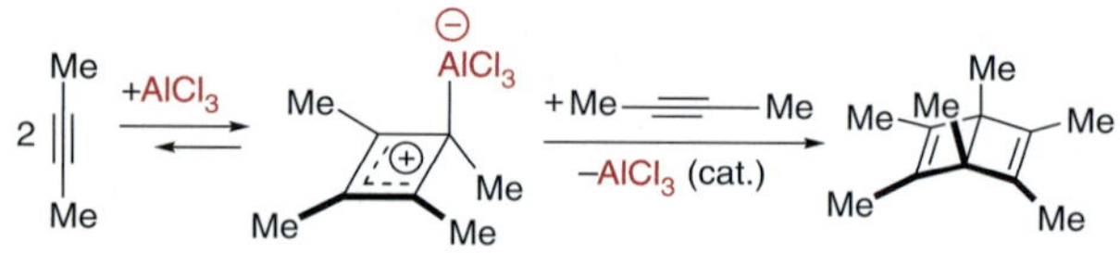

(7.41)

Quantum mechanical calculations give for the binding energies in complexes $H_2C{=}O{:}\rightarrow AlCl_3$, $H(Cl)C{=}O{:}\rightarrow AlCl_3$, $H(CO){-}Cl\rightarrow AlCl_3$, π-($C_2H_4$)$AlCl_3$, $\pi(CH{\equiv}CH)AlCl_3$, and $H_3CCl\rightarrow AlCl_3$: 26, 21, 9, 15, 14, and 14 kcal mol^{-1}, respectively [556]. These data explain why $AlCl_3$ and other trivalent aluminum compounds play important roles as catalysts in several reactions such as the Friedel–Crafts alkylation and acylation of benzene derivatives [557], cycloadditions [558–562], Alder-ene reactions (Section 5.7.1), hetero-ene reactions (Sections 5.7.2 and 5.7.3), aminoalkylations with imines [563, 564], and Ziegler–Natta polymerization reactions. The transition metal chlorides make much more stable π-complexes than $MgCl_2$ and $AlCl_3$ in agreement with their ability to back-donate electrons from their high-lying HOMOs into the LUMOs of alkenes and alkynes (Section 7.7). It is not surprising that the Zeise's salt $K(C_2H_4PtCl_3]\cdot H_2O$ has been the first stable π-complex of an alkene and a transition metal to have been crystallized, already in 1827 [565, 566]. One also notices that CuCl is not much worse than AuCl in its ability to interact with alkynes and alkenes and that transition metal chlorides CuCl, AgCl, AuCl, and $PtCl_2$ do not discriminate between aldehydes, imines, alkenes, and alkynes in their complexation as do $AlCl_3$, $MgCl_2$, and especially BCl_3. Halides of Ag, Cu, Au, and Pt follow the same trend. The heats of complex formation increase from X = Br to X = F for binding to PhCH=O, PhCH=NH, and PhC≡CH. For AgX, CuX, and AuX binding enthalpies to PhCH=O, PhCH=NH, and PhC≡CH increase according to the order Cl < OTf (Tf = CF_3SO_2) < BF_4 < PF_6 ≈ SbF_6. With PtX_2, the trend is the opposite, binding enthalpy in complex formation with PhCH=O, PhCH=NH, and PhC≡CH decreases in the order F ≈ OTf > BF_4 > PF_6 ≈ SbF_6 [66].

As MX_n Lewis acids often form aggregates or polymers, or decompose (e.g. AuF), they can be ligated for instance to Ph_3P. This increases their stability but reduces their acidity. Thus, the binding

Table 7.3 DFT-computed binding energies in kcal mol^{-1} of complexes of Lewis acids with PhCH=O, PhCH=NH forming η^1-σ-complexes, and with PhCH=CH$_2$ and PhC≡CH forming η^2-π-complexes.[a)]

	BCl_3	$MgCl_2$	$AlCl_3$	CuCl	$CuCl_2$	AgCl	AuCl	$AuCl_3$	$PtCl_2$
PhCH=O	18.9	34.5	40.7	37.4	26.4	26.4	33.1	35.9	46.9
PhCH=NH	42.1	44.2	55.1	51.8	41.2	39.6	53.6	60.3	71.5
PhC≡CH	0.9	15.2	19.1	33.1	14.3	22.6	34.7	32.5	49.4
PhCH=CH_2	0.4	15.7	19.2	33.6	18.1	24.4	37.5	36.8	53.9

	AgBr	AgF	AgOTf [b)]	$AgBF_4$	$AgPF_6$	$AgSbF_6$
PhCH=O	25	28	26	28	29	28
PhCH=NH	39	43	40	41	43	42
PhC≡CH	21	26	25	26	27	26

	CuBr	CuF	CuOTf	$CuBF_4$	$CuPF_6$	$CuSbF_6$
PhCH=O	36	41	40	43	42	42.5
PhCH=NH	50	55	56	60	68	69
PhC≡CH	31	37	39	42	41	42

	AuBr	AuF	AuOTf	$AuBF_4$	$AuPF_6$	$AuSbF_6$
PhCH=O	31	35	41	43	42	42.5
PhCH=NH	50	60	62	67	66	66
PhC≡CH	32	42	44	45	47	48

	$PtBr_2$	PtF_2	$Pt(OTf)_2$	$Pt(BF_4)_2$	$Pt(PF_6)_2$	$Pt(SbF_6)_2$
PhCH=O	43	49	48	42	38	40
PhCH=NH	64	80	84	72	67	66
PhC≡CH	46	58	54	45	28	30

1 cal = 4.184 J.
a) Taken from [66]; binding energies. $\Delta_r H°(A + MX_n \rightleftarrows A{\cdot}MX_n)$.
b) Tf = CF_3SO_2.

of $(Ph_3P)AuCl$ with PhC≡CH or styrene becomes endothermic (4.9 and 5.5 kcal mol^{-1}, respectively). In the case of $(Ph_3P)AuOTf$, slightly exothermic binding is found with the PhC≡CH and PhCH=CH_2 (c. −0.5 and −3.3 kcal mol^{-1}, respectively). $(Ph_3P)AuOTf$ and PhCH=NH make a stable complex. Gold Lewis acids such as $AuCl_3$, Au_2Cl_6, $(Ph_3P)AuX$ (X = Cl, SbF_6), BF_4, and OTf are excellent homogeneous catalysts for Michael additions and several other C—C bond forming reactions (see e.g. Scheme 7.45) [567–575]. The free energy changes of ligand association have been measured in the gas phase for Cu^+ [576] and Ag^+ [577]. Selected data are given in Table 7.4. On the basis of the absolute hardness, η, scale developed by Pearson [578] (Section 1.13), one finds only a limited agreement of the hard and soft acid–base (HSAB) concept for these systems. The data show that Cu^+ is a better σ-catalyst than Ag^+.

7.6.1 Mukaiyama aldol reactions

At room temperature, aldol reactions of type 2 $RCH_2CHO \rightleftarrows RCH_2CH(OH)CH(R)$ generally proceeds under thermodynamic control (Section 2.9.1): they are reversible reactions. They are catalyzed either by acids or bases (Section 5.7.6) [579–581]. As a consequence of this reversibility, intermolecular direct aldol reactions with two different carbonyl compounds suffer from lack of stereoselectivity, except when using promoters such as L- and D-proline [582–585] or other α-amino acids [586, 587] and amines [588] in DMSO or DMF below room temperature. Thus, several alternative methods have been developed that utilize preformed magnesium or lithium enolates (Hauser, 1949) [589–595], boron enolates (Mukaiyama, 1971) [596], enoxysilanes + Lewis acids (Mukaiyama aldol reaction, 1973) [597–603], tin

Table 7.4 Free energies ($\Delta_r G°$) in kcal mol^{-1} at 298 K for $CuL_2^+ \rightleftarrows Cu^+ + 2L$[a] and $AgL_2^+ \rightleftarrows Ag^+ + 2L$,[b] in kcal mol^{-1}, for L = σ-donors (1 cal = 4.184 J).

L	CuL_2^+	AgL_2^+	$\eta(L)$[c]	L	CuL_2^+	AgL_2^+	$\eta(L)$[c]
1-Methylimidazole	116.4			Et_2O	77.5	61.8	
Bu_3N:	101.8			AcOMe	77.7	59.2	
$Me_2(O{=})S$:	95.5	c. 75		*i*-PrBr	66.7	56.4	
Me_2S:	85.5	72.4	~6	EtBr	62.3	52.3	
$CH_3C{\equiv}N$:	90.9	70.8	7.5	EtOH	70.3	55.1	
:NH_3	88.2	70.4	8.0	MeOH	66.6	50.0	~7.5
$Me_2C{=}O$:	79.3	60.1	5.6	H_2O	59.3	43.9	9

a) Taken from [576].
b) Taken from [577].
c) Absolute hardness $\eta = [IE(L) + EA(L))]/2$ (Section 1.13).

enolates (Mukaiyama, 1982) [604, 605], and titanium enolates (De Kimpe, 1980) (Section 5.7.7) [606, 607]. Originally, the Mukaiyama aldol reaction (7.42) used $TiCl_4$ in stoichiometric amount for its promotion in nonpolar solvent.

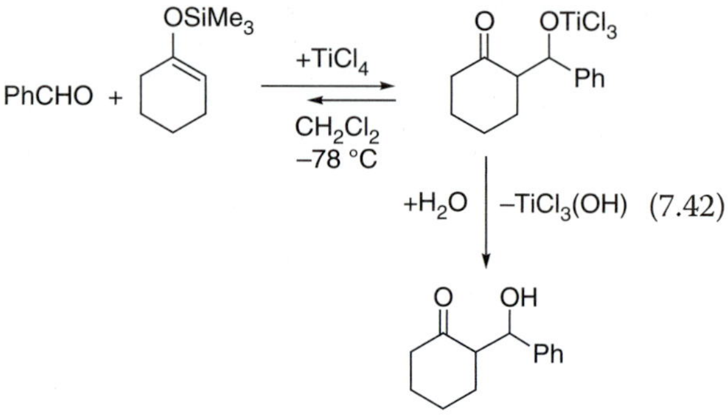

(7.42)

Thus, $TiCl_4$ is a promoter rather than a catalyst in this reaction. For the reaction of $Me_2C{=}C(OMe)OSiMe_3$ and PhCOCOPh, the corresponding titanium aldolate complex $[(PhCO)PC(O)C(Me_2)C(OMe)OTi(Cl)_2]_2(\mu Cl)_2$ has been crystallized and characterized by X-ray diffraction studies [608]. Myers et al. [609, 610], Denmark et al. [611, 612], and Miura et al. [613] have shown that the Mukaiyama aldol reaction can proceed without catalyst by increasing the Lewis acidity of silicon of the silyl enol ether (e.g. by using enoxysilacyclobutanes) and by increasing the acidity of enoxysilanes by substitution (e.g. enoxysilanes derived from *N,N*-dialkylamides: *O*-silyl ketene *N,O*-acetals, or from esters, *O*-silyl *O,O*-ketene acetals). For instance, reactions of enantiomerically pure dimethylsiloxane **219** with aldehydes in CH_2Cl_2 at 23 °C are highly *anti*-selective giving products **221** (reaction (7.43)), in agreement with one-step concerted mechanisms implying boat six-membered transition structures of type **220** in which pentavalent silicon is trigonal bipyramidal, the aldehyde is equatorially oriented, coordinated through the lone pair *trans* to the R group, and the silicon atom is bound to the enol through an oxygen 2p-type orbital. The boat-like transition structure ensures a better overlap between the bonding carbons than in a chair-like transition structure, thus explaining the *anti*-stereoselectivity. This mechanism is supported by quantum mechanical calculations that favor twist-boat six-membered ring transition state over stepwise pathways involving four-membered ring oxetane intermediate [614]. It is analogous to a concerted sila carbonyl ene-reaction (Section 5.7.7).

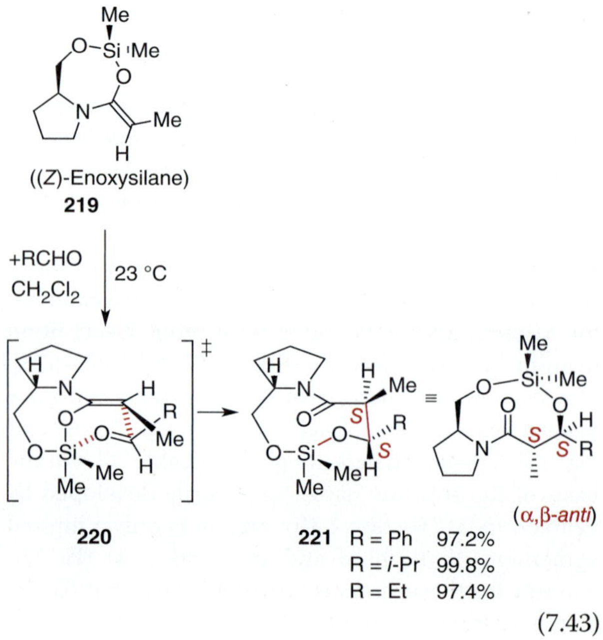

(7.43)

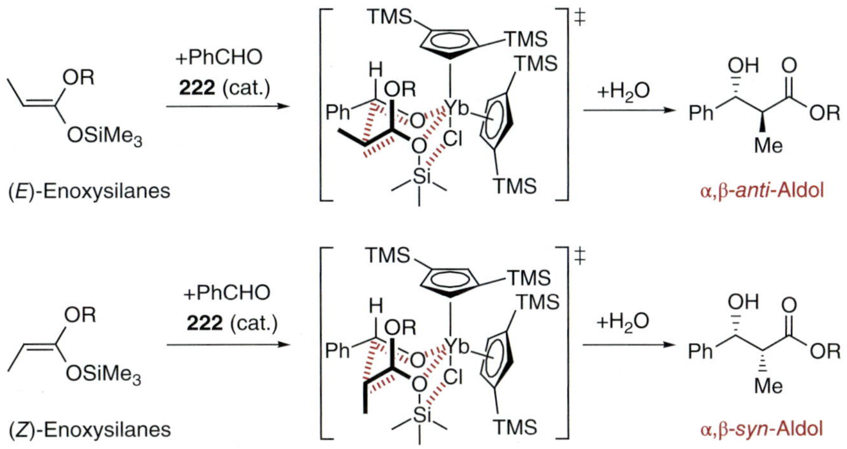

Scheme 7.36 Streitwieser's ytterbium-catalyzed Mukaiyama reactions (OR substituent prefers pseudo-axial positions in the six-membered chair-like transition states because of the conformational anomeric effect, Section 2.7.10).

The *O*-(silacyclobutyl)ketene acetals derived from esters, thiol esters, and amides with (*E*)-configuration undergo facile uncatalyzed aldol reactions with aliphatic and conjugated aldehydes. Using *O*,*O*-ketene acetals of (*E*)-configuration, the *α*,β-*syn* aldols are obtained with excellent diastereoselectivity [611].

The Mukaiyama aldol reaction can be catalyzed by Bu_4NF [615], or by trityl salts, including enantiopure derivatives [616]. Bis[bis(1,3-trimethylsilyl)cyclopentadienyl]ytterbium(III) chloride (**222**) is an efficient catalyst in the presence of additional Me_3SiCl for the reaction of silyl ester enolates with aliphatic and aromatic aldehydes, to yield β-silyloxy esters (Scheme 7.36). Interestingly, although (*E*)-enoxysilanes produce *anti*-aldols as major products, (*Z*)-enoxysilanes generate *syn* stereoisomers preferentially [617].

Other catalysts such as [(cyclopentadienyl)$_2$Zr(O-*t*-Bu)THF][BPh_4] [618], 1-phenyl-2:3,6:7-dibenzosuberyl salts [619], (cyclopentadienyl)$_2$Zr(OTf)$_2$·THF or (cyclopentadienyl)$_2$Zr(OTf)$_2$ [620], $AgPF_6$ coordinated to diphosphines [621], $BiCl_3$ [622], and Bi(OTf)$_3$ [623], have been applied as catalysts, but they did not lead to better diastereoselectivities. When BF_3·OEt_2, $LiClO_4$, Yb(OTf)$_3$, Sn(OTf)$_2$, Zn(OTf)$_2$ [624], Ph_3COTf, Ph_3CClO_4, and (cyclopentadienyl)$_2$Ti(OTf)$_2$ [625] are used as catalysts, trace amounts of water hydrolyze them with the generation of protic acids such as $HClO_4$ and TfOH. The latter react with the enoxysilanes, generating Me_3SiClO_4 and Me_3SiOTf, respectively, both of which are the actual catalysts of the Mukaiyama aldol reactions. The following sequence of increasing acidic activity has been reported: $Me_3SiCl < Me_3SiOSO_2Me < Me_3SiOTs < Me_3SiBr \ll Me_3SiOTf \approx Me_3SiOSO_2F < Me_3SiI < Me_3SiClO_4 < Me_3SiN(SO_2F)_2 < Me_3SiNTf_2$ [626]. R_3SiX mixed with other Lewis acids generate complexes that can be good catalysts in Mukaiyama aldol reactions (e.g. $Me_3SiCl + SnCl_2$ [627], $Me_3SiCl + InCl_3$ [628] or $Me_3SiOTf + B(OTf)_3$ [629]). Quantum mechanical calculations on $TiCl_4$-, BCl_3-, $AlCl_3$-, and $GaCl_3$-catalyzed reaction of formaldehyde and vinyloxysilane (CH_2=$CHOSiH_3$) suggested multistep mechanisms in which H_2C=O is strongly activated by coordination to the Lewis acid, making complexes H_2C=O:→MCl_n that form stable intermolecular π/π complexes with the enoxysilane. The rate-determining step corresponds to the C—C bond formation and a chlorine atom shift as shown in transition state **223** of reaction (7.44) [630].

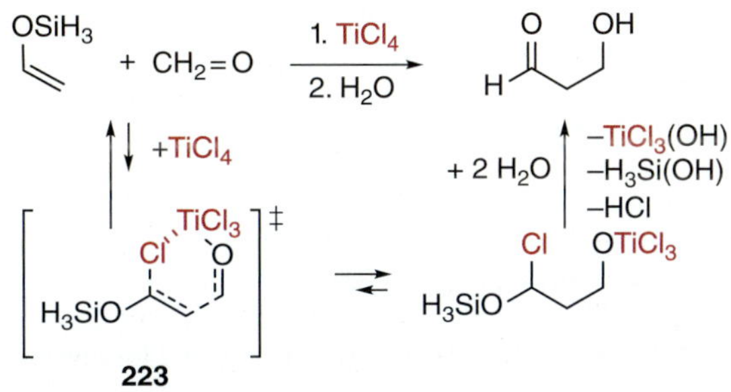

(7.44)

Ghosez and Mathieu, on one hand, and Mikami et al., on the other hand [632], have introduced Me_3SiNTf_2 (prepared from $Me_3SiCl + AgNTf_2$) as a much more powerful carbonyl-activating reagent than Me_3SiOTf (Tf = CF_3SO_2). Yamamoto and coworkers have demonstrated the high activity of Me_3SiNTf_2 and of Me_3SiCTf_3 as catalysts for the Mukaiyama aldol reactions [633]. They also showed that Me_3SiX-catalyzed reactions proceed through each catalytic cycle under the influence of X:$^-$. The silyl group of Me_3SiNTf_2 and Me_3SiCTf_3 is not released; the enoxysilane transfers intermolecularly to the product. In the case of Me_3SiOTf, the silyl group remains in the product and that of the enoxysilane becomes the catalyst of the next catalytic cycle (Scheme 7.37) [634, 635]. In the case of reaction of benzaldehyde with enoxysilane **224** in the presence of Me_3SiNTf_3 and Me_3SiCTf_3, intermediates **227** resulting from the C–C coupling do not transfer $Tf_2N{:}^-$ and $Tf_3C{:}^-$ anions to the neighboring silyl group, but react with PhCHO to give ion pair **228/229**. Cation **228** is the activated electrophile that adds to enoxysilane **224** giving ion pair **230/229**. This ion pair is the catalyst for the next catalytical cycles, generating product **231** on reacting with PhCHO and producing ion pair **228/229**. A possible termination reaction is the relatively slow transfer of a silyl group from cation **230** to its counterion **229** with liberation of Me_3SiX ($X = Tf_2N$, Tf_3C).

Using tris(trimethylsilyl)silyl (TTMSS) enol ethers and 0.05 mol% $HNTf_2$ or $TTMSSNTf_2$ as catalysts, Mukaiyama aldol reactions with aldehydes generate silyl aldols (e.g. **232**) in high yield (Scheme 7.38). These aldols can react in situ with a second equivalent of TTMSS enol ether giving aldol (e.g. **233**) with high yield and *syn*-diastereoselectivity [636]. The TTMSS group substitutes the TBS and Me_3Si (TMS) groups in Scheme 7.37. It is a much stronger Lewis acid than TBS and TMS. It confers higher reactivity to the enoxysilane and large steric bulk for superior *syn*-diastereoselectivity, allowing one-pot reaction cascades that can create important synthetic intermediates with high diastereoselectivity.

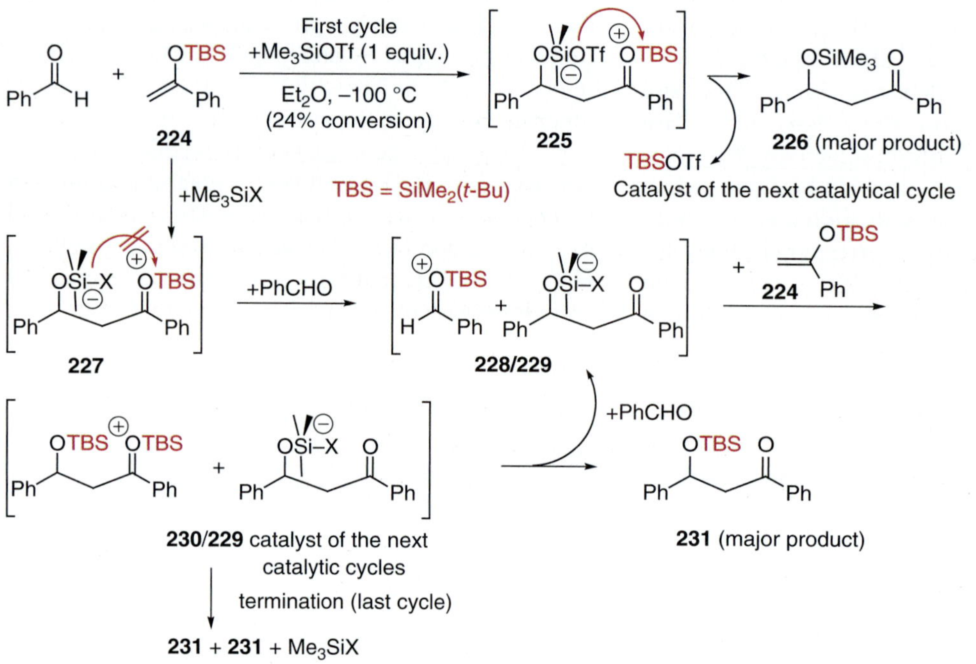

Scheme 7.37 Crucial role of the conjugate base X:⁻ for silyl Lewis acid induced Mukaiyama aldol reactions.

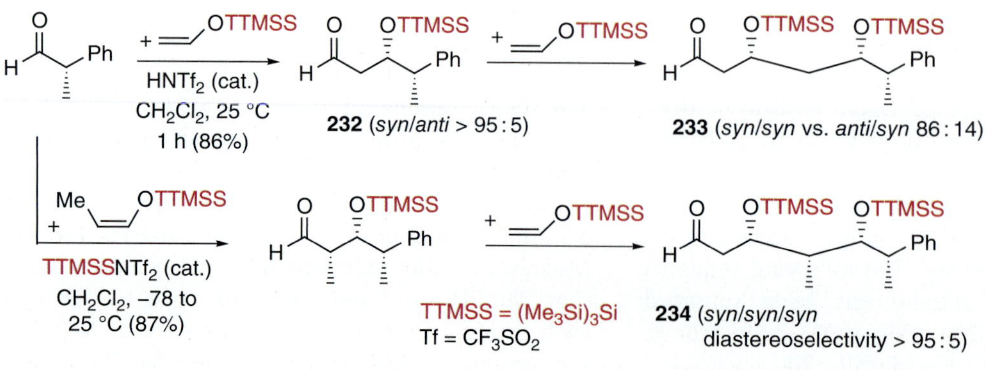

Scheme 7.38 Yamamoto's tris(trimethylsilyl)silyl-governed cross-aldol reaction.

Acid $HCTf_3$ is one of the strongest neutral Brønsted acids in the gas phase ($\Delta_r G$(X—CH_2—H ⇄X—CH_2^- + H^+) = 408.5 kcal mol^{-1} for X = H, 364 (for X = CN), 362 (X = COMe), 350 (X = NO_2), 340 (X = SO_2CF_3), and 289 for X = $CF_3SO_2)_3$C—H)) [637, 638]. Yamamoto and coworkers have developed strong carbon acids such as pentafluorophenyl-bis(triflyl)methane and polystyrene bond tetrafluorophenyl-bis(triflyl)methane [639–641]. The latter is a highly acidic heterogeneous Brønsted acid catalyst that is effectively swollen by nonpolar organic solvents, and its catalytical activities are superior to those of Nafion® SAC-13 [642]. Examples of enantioselective Mukaiyama aldol reactions catalyzed by enantiomerically pure Lewis acids have been presented in Scheme 5.145. An example of asymmetric Mukaiyama aldol reaction catalyzed by chiral disulfonimide **235** (N—H Brønsted acid) has been developed by List and coworkers (Scheme 7.39a) [643]. Enantioselective vinylogous Mukaiyama–Mannich reaction (using an imine as the electrophilic partner) catalyzed by **235** (Scheme 7.39b) has been reported by the same group [644].

Problem 7.14 Explain the high *syn*-diastereoselectivity of reaction **232** + CH_2=CHOSi($SiMe_3)_3$ → **233**.

7.6.2 Metallo-carbonyl-ene reactions

The Mukaiyama aldol reaction (e.g. reaction (7.43)) can be seen as a metallo-oxa-carbonyl-ene reaction (Section 5.7.7). In the absence of Lewis or Brønsted acid, the transition state of thermal reaction (7.45) that would follow a concerted mechanism can be represented by a diradicaloid (1,4-diradical ↔ 1,4-zwitterion) of type **236** ↔ **236′**. The role of the acid catalyst A is to stabilize the zwitterionic limiting structure **236′** by equilibrating with zwitterion **237**. As a consequence, the reaction may follow a multistep mechanism in which **237** is one reaction intermediate that can adopt quasi-cyclic or acyclic conformations. In 1976, Hosomi and Sakurai [645] reported that $TiCl_4$ promotes allylation of aldehydes and ketones with allylsilanes (reaction (7.46)). The reaction generates homoallylic alcohols after aqueous work-up [646, 647]. This is an example of metalla-carbonyl-ene reaction (Section 5.7.5) promoted by a Lewis acid. Very electrophilic ketones such as hexafluoroacetone react with allylsilanes without catalyst [648]. The Hosomi–Sakurai reaction is not only regiospecific but also stereospecifically *anti* with respect to the silyl group, which leads to the formation of a (*E*)-homoallylic alcohol (reaction (7.47)) [649]. It can also use acetals as electrophilic partner [650]. An important synthetic application of allylsilanes

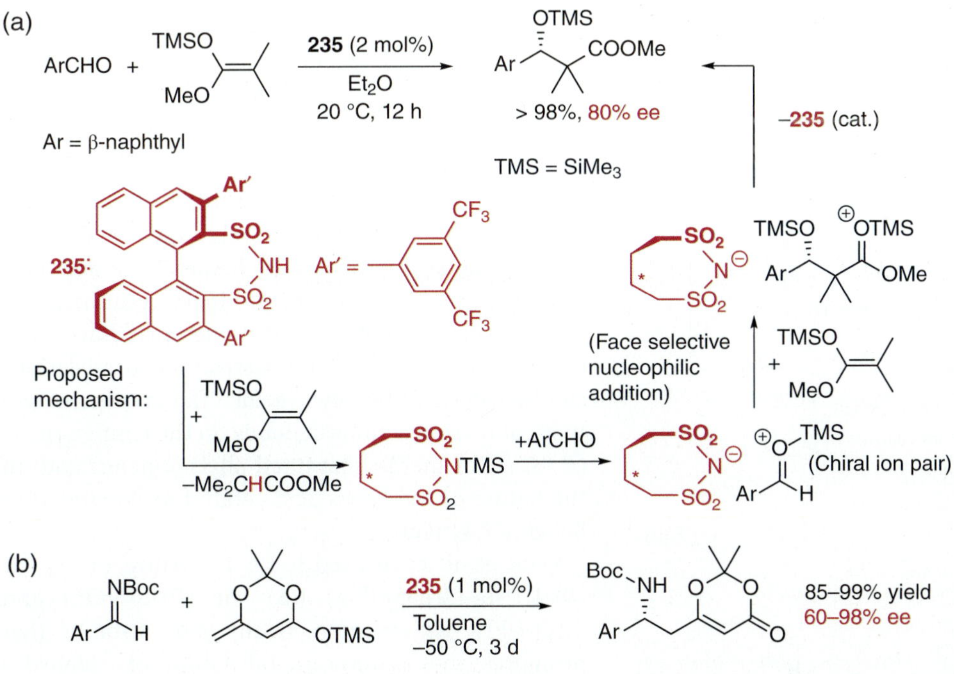

Scheme 7.39 (a) Catalytic enantioselective Mukaiyama aldol reaction catalyzed by a disulfonimide; (b) enantioselective vinylogous Mukaiyama–Mannich reactions.

involves the reactions of (*E*)-crotytrimethylsilane and (*E*)-cinnamyltrimethylsilane with aldehydes in the presence of $TiCl_4$ to give homoallyl alcohols with over 93% *syn-selectivity* (reaction (7.48)). Lower *syn*-selectivity is observed for the reactions of (*Z*)-crotylsilanes [651, 652]. These results can be compared with those of allylic organometals containing lithium, boron, aluminum, titanium, zirconium, or chromium that add to aldehydes and ketones for which the (*Z*)-allylic stereoisomers lead to *syn* and (*E*)-isomers to *anti*-homoallyl alcohols after aqueous work-up (see Scheme 5.127). In these cases, the stereoselectivity is generally accepted to be determined by chair-like six-membered ("closed") transition states [653–658].

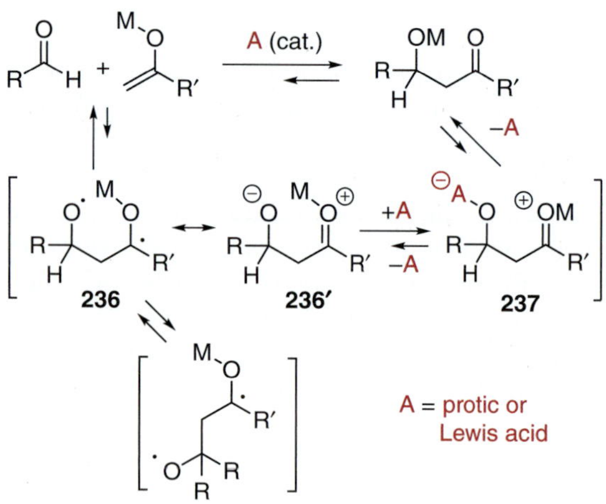

(7.45)

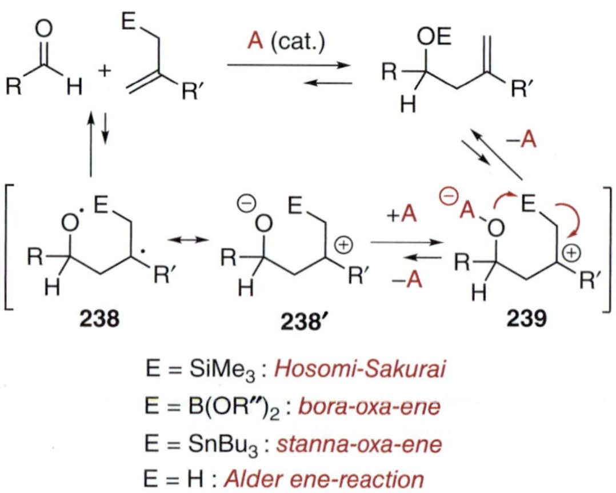

(7.46)

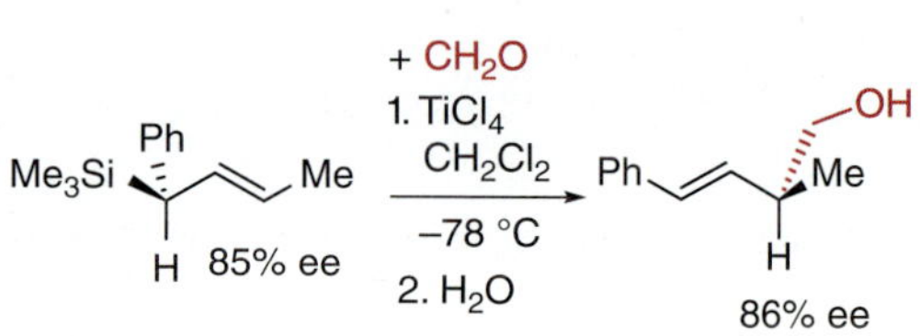

(7.47)

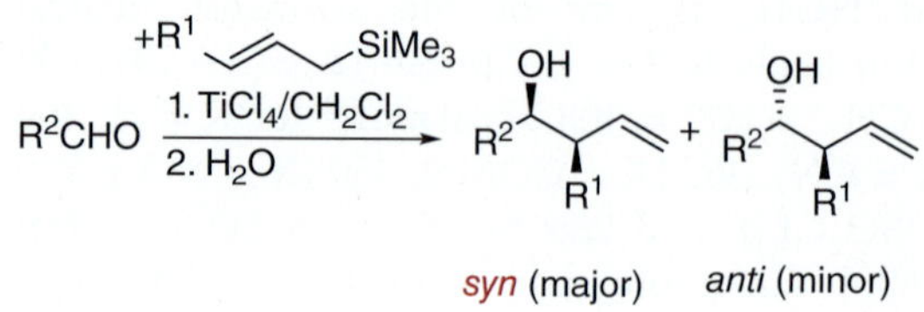

(7.48)

Crotylsilanes (Scheme 7.40) and crotyl(trialkyl)tins with aldehydes in the presence of $BF_3{\cdot}Et_2O$ react via acyclic ("open") transition states to give *syn*-products regardless of the configuration ((*E*) or (*Z*)) of the crotyl double bond [659–663].

In a similar way as the Mukaiyama aldol reaction (Scheme 7.39), the catalytic asymmetric Hosomi–Sakurai reaction is possible when catalyzed by N—H acids analogous to **235** [664]. An example applying chiral *N*-(perfluorooctanesulfonyl)thiophosphoramide **240** as a catalyst is shown with reaction (7.49). Highly diastereo- and enantioselective Mukaiyama aldol reactions have been observed with the same catalyst [665].

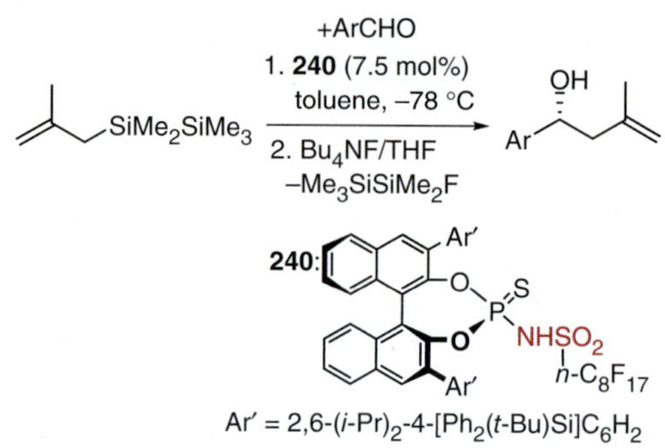

(7.49)

Crotylstannanes and 3-substituted allylstannanes react with aldehydes in a similar way as the corresponding allylboronates (Scheme 7.41) and give mixtures of *syn*- and *anti*-products resulting, formally, from metalla-carbonyl-ene reactions with cyclic ("closed") transition states. These allylations can be catalyzed by Lewis acids. The diastereoselectivity of these reactions depends on the configuration (*E* vs. *Z*) of the 3-substituted allylstannanes and on the nature of the Lewis acid catalyst as illustrated in Scheme 7.41 [666].

Keck et al. have used a 1 : 1 mixture of (*S*)- or (*R*)-1,1′-bi-(2-naphthol) ((*S*)- or (*R*)-BINOL) and $Ti(O\text{-}i\text{-}Pr)_4$ as enantiomerically pure catalyst that promotes the asymmetric allylation of aldehydes with achiral allyltin compounds (reaction (7.50)) [667–672].

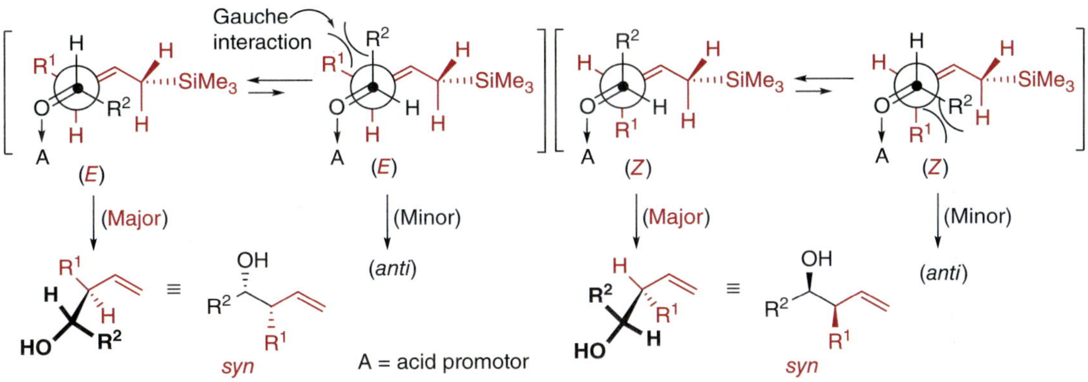

Scheme 7.40 Open transition states for the Hosomi–Sakurai reaction explain the *syn* selectivity regardless of the (*E*) or (*Z*)-configuration of the crotylsilane.

Scheme 7.41 The diastereoselectivity of the allylations of alkoxy-substituted aldehydes depends on the nature of the Lewis promoter.

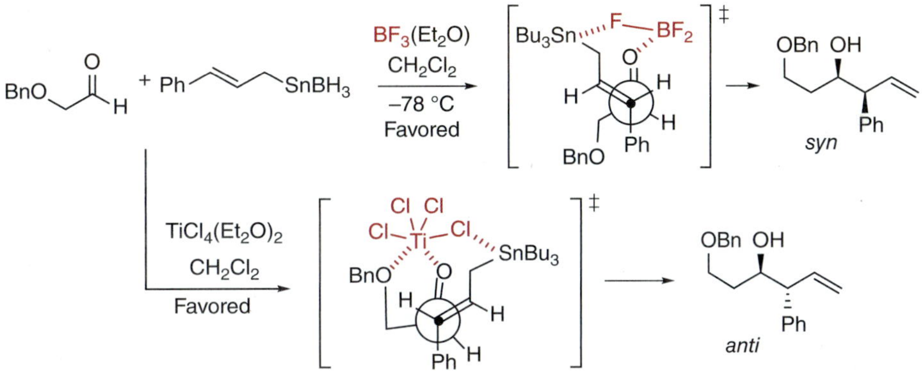

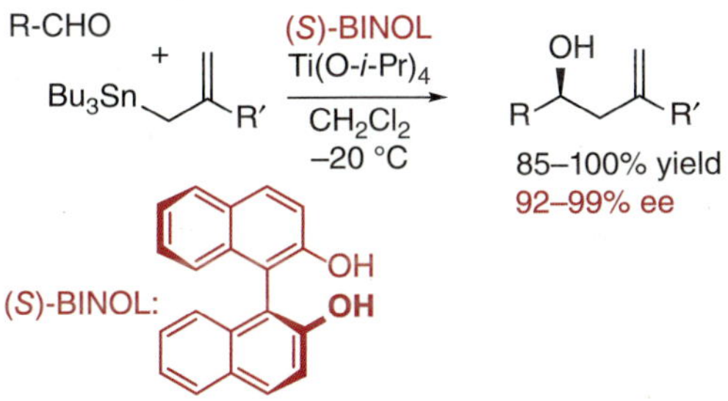

(7.50)

Allylic boron compounds are used for the allylation of aldehydes and ketones (bora-carbonyl-ene reactions, Section 5.7.5) [673]. These reactions can be run in ionic liquids [674]. A characteristic feature of these boron derivatives is their ability to undergo metallotropic rearrangement ((1,3)-sigmatropic rearrangements, or allylic rearrangements) at a rate depending on substituents and temperature. For example, attempts to prepare the (*E*)- or (*Z*)-isomers of tricrotylborane give the equilibrium mixture rather than a pure product. However, dialkylallylboranes can be used in diastereoselective reactions at low temperature. By contrast, allylboronic esters are relatively stable toward isomerization. The reactions of alkenyldialkoxy boranes with aldehydes have been reported first by Mikhailov in 1972 [675]. With two amino substituents, allylboranes are particularly resistant to boratropy [676, 677]. Because of the Lewis acid character of the boron center in allylboranes and allylboronates, strong Lewis acid/base complexes form between both the reactants and then six-membered chair-like ("closed") transition states are realized. This hypothesis allows one to explain the stereoselectivity of these allylations [678]. The concerted character of the reaction is confirmed by kinetic secondary deuterium isotope effects ($k_H/k_D = 0.85$, toluene, 25 °C) [679] measured for the reaction of benzaldehyde (PhCHO, PhCDO) with allylboronate derived from di(isopropyl) (*R*,*R*)-tartrate (the Roush allylboronation [680]).

In 1981, Hoffmann and coworkers have developed camphor-based allylboronates that were the first chiral allylboron reagents ever reported [681]. The chiral reagent **241** reacts with benzaldehyde to give the corresponding allylic alcohol **242** with only 11% ee after work-up (Scheme 7.42a). Hall and coworkers have found that Lewis acids catalyze the allylboration of aldehydes, and, in the case of reaction **241** + PhCHO → **242**, 90% yield and 92% ee can be obtained (CH_2Cl_2, −78 °C) in the presence of

10 mol% of $Sc(OTf)_3$ [682–686]. The discovery that Lewis acids can catalyze the allylboration of aldehydes opened the door to the use of achiral allylboranes and allylboronates for the enantioselective synthesis of homoallyl alcohols, where the asymmetry is induced by enantiomerically pure Lewis acid catalysts [687] as shown with reactions reported by Shibasaki and coworkers (Scheme 7.42b), on the one hand [688], and Hall and coworker (Scheme 7.42c), on the other hand [689]. On their side, Hoveyda and coworkers have used a mixture of $Zn(OMe)_2$ and a chiral 2-aminomethyl phenol to catalyze the diastereoselective and enantioselective allylation of flurorinated ketones [690].

7.6.3 Carbonyl-ene reactions

Carbonyl compounds can be used as enophiles in thermal ene-reactions (Section 5.7.3) [691] or in carbonyl ene-reactions promoted with Lewis acid [692–696] or Ni(cyclooctadiene)$_2$ + $EtOPPh_2$ [697]. For the $SnCl_4$-catalyzed ene-reaction of diethyl oxomalonate with 1-arylcyclopentenes (reaction (7.51)), Salomon and coworkers measured $\rho = -3.9$ (positive electron-demand on the aryl group) indicative of a positive charge build up in the central carbon atom of the ene moiety in the transition state of the reaction [698] (see the diradicaloid model presented in Section 5.7.1).

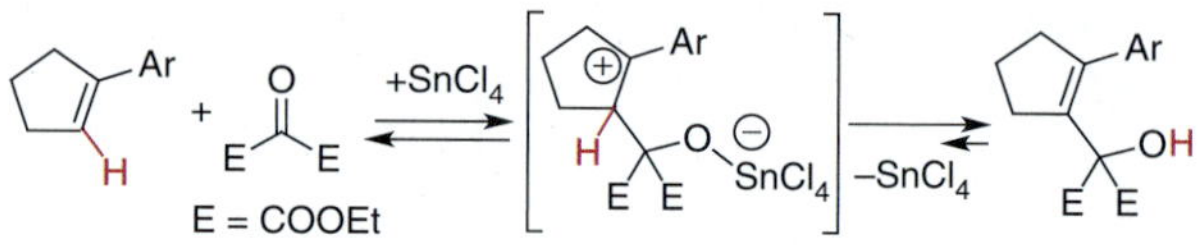

(7.51)

On the other hand, kinetic isotope effect for the $SnCl_4$-catalyzed reaction of diethyl oxomalonate with Ph—CH$_2$(D$_2$)—C)Me)=CH$_2$ is relatively small ($k_H/k_D < 1.2$) [699]. In the case of R_2AlCl-catalyzed reactions of formaldehyde with tetramethylethylene,

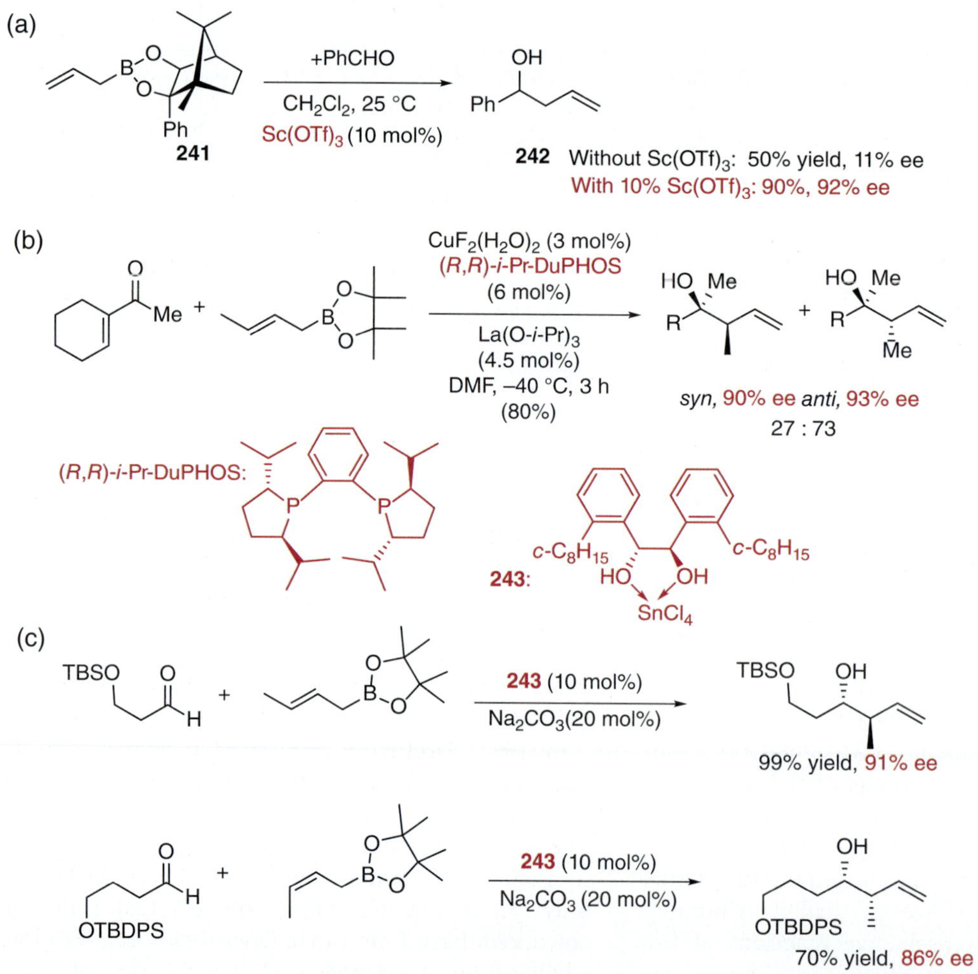

Scheme 7.42 Lewis acid catalyzed allylborations: (a) with a chiral allylboronate and scandium triflate; (b) Shibasaki's enantioselective crotonylboration of a ketone; and (c) Hall's enantioselective crotonylboration of aldehydes.

the deuterium kinetic isotope effects are indicative of a multistep mechanism with a product-determining step that follows the rate-determining step. For the Et_2AlCl-catalyzed carbonyl ene-reaction of formaldehyde with 2-methylbut-2-ene (reaction (7.52)), Singleton and Hang measured the kinetic isotope effects $k_{^{12}C}/k_{^{13}C}$ and k_H/k_D shown below [700].

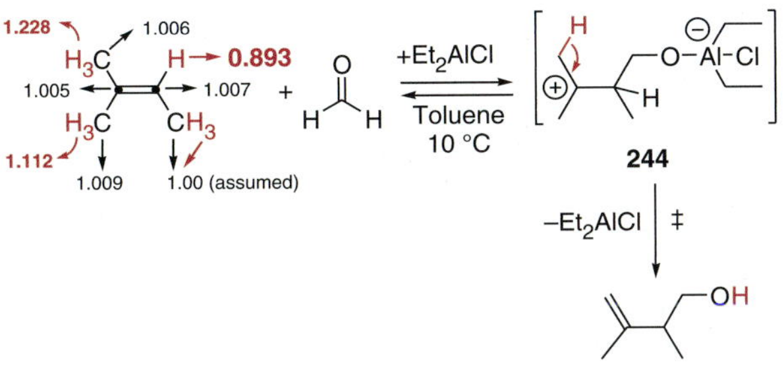

(7.52)

The reactive methyl group exhibits a $k_{^{12}C}/k_{^{13}C}$ = 1.006 and a k_H/k_D = 1.22–1.23. The latter represents a combination of primary and secondary isotope effects and is consistent with a significant primary deuterium isotope effect, in agreement with a mechanism involving the reversible formation of a zwitterion of type **244**, followed by rate-limiting proton transfer. Enantioselective carbonyl-ene reaction (Mikami ene-reaction [701, 702]) has been demonstrated several times [691, 703–710]. An example is given with reaction (7.53) that applies a strongly acidic *N*-trifluoromethanesulfonylphosphoramide catalyst (cat) [711]. Because homoallylic alcohols are very important synthetic intermediates, a large number of asymmetric carbonyl-ene reactions catalyzed by homochiral Lewis of Brønsted acids have been reported [712–720].

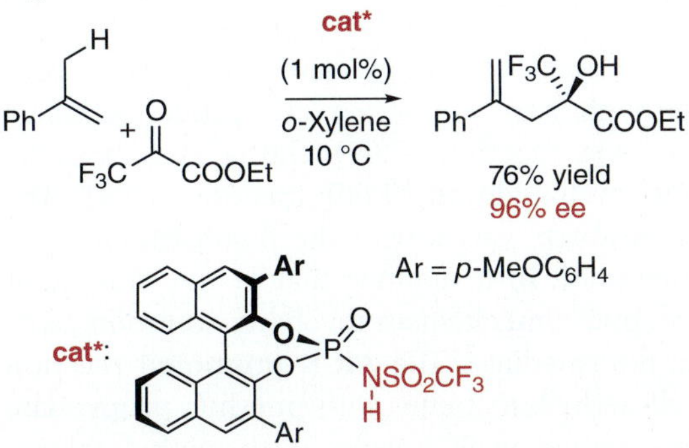

(7.53)

Problem 7.15 What is the major product of reaction of dimethyl oxomalonate with 2-methylhepta-2,6-diene in the presence of 0.2 equiv. of $SnCl_4$?

Problem 7.16 Ene-reaction of formaldehyde with methyl hept-6-ynoate using $Me_2AlCl + MeAlCl_2$ as catalyst produces an antibiotic isolated from Sapium japonicum. What is this product? [721].

Problem 7.17 What is the major product of reaction of isoprene with 3-methylbutanal in the presence of Me_2AlCl? [722].

7.6.4 Imine-ene reactions

The imine ene-reaction (7.54) is an atom economical way to generate homoallyl amines. Under thermal conditions, the reaction is limited to intramolecular reactions, or to reactive imines such as glyoxalate imines [723–726]. Quantum mechanical calculations find a concerted mechanism for the uncatalyzed reaction of CH_3—CH=CH_2 + CH_2=NH in which the H-atom is transferred in concert with the C—C bond formation [727].

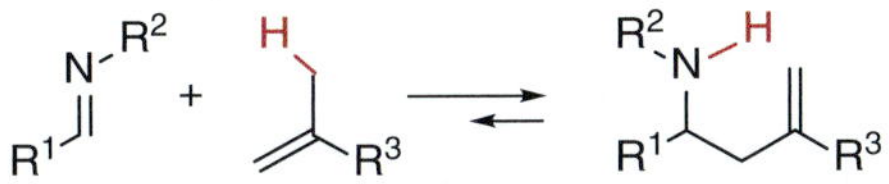

(7.54)

Lanthanide Lewis acids have been shown to effectively activate imino groups [728–730]. Nakagawa and coworkers reported that $Yb(OTf)_3$ with Me_3SiCl or Me_3SiOTf catalyzes imine-ene reactions as shown with the three-component reaction of aldehydes, tosylamide, and α-methylstyrene (reaction (7.55)) [731]. Brønsted acids can also promote imino-ene reactions [732].

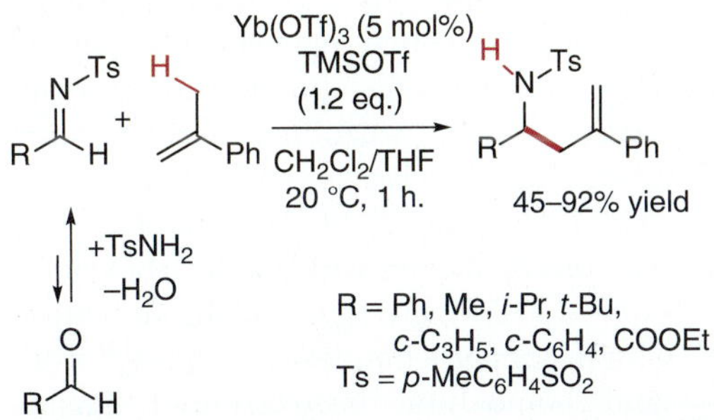

(7.55)

When using deuterated alkene **245** and tosylimine **246** derived from benzaldehyde, the reaction gives **247** with nonspecific deuterium labeling. A control experiment in which **245** is treated with the above catalyst alone leads to a 12 : 88 mixture of (*Z*)- and (*E*)-2-phenylbut-2-ene with the 1-methyl and 3-CH_2 moieties partially deuterated (Scheme 7.43). This demonstrates that $Yb(OTf)_3/Me_3SiOTf$ catalyzes the isomerization of the double C=C bond (via **248**) more rapidly than the imine-ene reaction [733].

Using enantiomerically pure chiral Lewis acid catalyst, enantioselective imine ene-reactions have

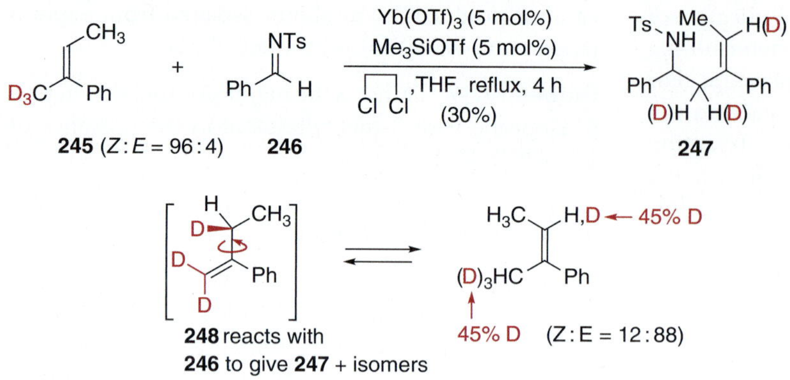

Scheme 7.43 Alkene isomerization competes with imine-ene-reaction.

been realized [734–738]. For instance, in the presence of (*R*)-Tol-BINAP-Cu-bis(acetonitrile) perchlorate (**249** + $CuClO_4(MeCN)_2$) as the catalyst, enantioselective imino-ene reaction (7.56) of alkenes with α-imino ester EtOCOCH=NTs has been reported [739].

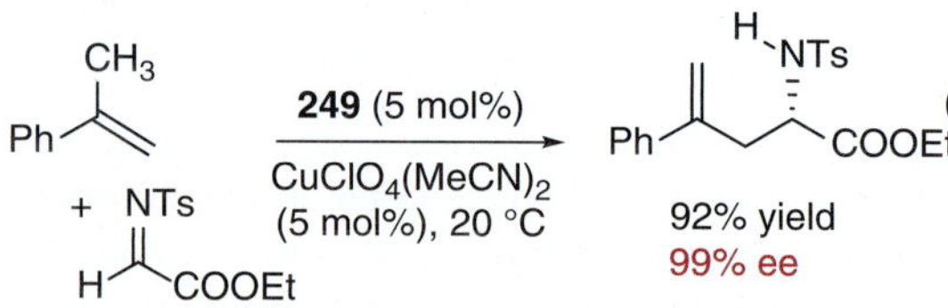

(7.56)

249: 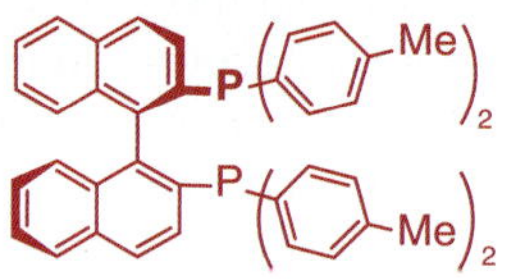

Ts = p-$MeC_6H_4SO_2$

A simple pathway for the synthesis of α-hydroxy and α-amino carbonyl compounds with high enantioselectivity is provided by the heterogeneously catalyzed enantioselective carbonyl-ene and imino-ene reactions. The catalyst is a bis(oxazoline)-modified CuH Zeolite Y system [735]. Lewis acids such as $Cu(OTf)_2$, $Zn(OTf)_2$, $Yb(OTf)_3$, and $Nd(OTf)_3$ catalyze the diaza-ene reactions of alkenes with azodicarboxylates (ROOC—N=N—COOR), giving the allylic amination products. Chiral copper complexes prepared from $Cu(OTf)_2$ and homochiral bis(oxazoline) ligands catalyze the enantioselective diaza-ene reactions of alkenes with azodicarboxylates [740].

7.6.5 Alder-ene reaction

The reaction of an alkene having an allylic hydrogen (an all carbon ene reaction) with an alkene or an alkyne (enophile) to form a new C—C bond with migration of the ene bond and an 1,5-hydrogen shift is referred to as the Alder-ene reaction (reaction (7.57), Section 5.7.1). It shows analogy with the Diels-Alder cycloaddition, originally known to Diels and Alder as the 'diene addition'. It provides an atom economical process for C—C bond formation (no side product formed [171]). Without an efficient catalyst, the intermolecular ene-reactions require extreme conditions and exhibit lack of selectivity [741, 742]. In 1974, Snider found that Alder-ene reactions of acrylate and propynoate esters can be catalyzed by $AlCl_3$ as illustrated with reaction (7.58) [743]. Later, it was found that alkylaluminum halides, which are proton scavengers, lead to better yields and extend the scope of the Alder-ene reaction [722, 744–746], including Lewis acid-catalyzed asymmetric version of the Alder ene-reaction [747].

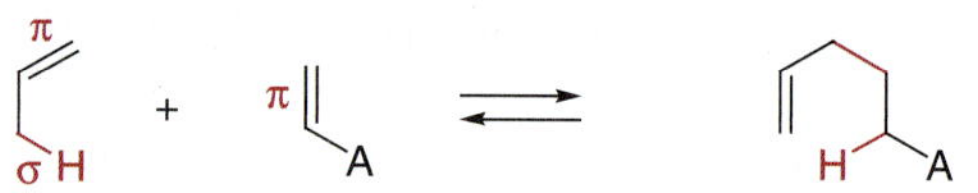

(7.57)

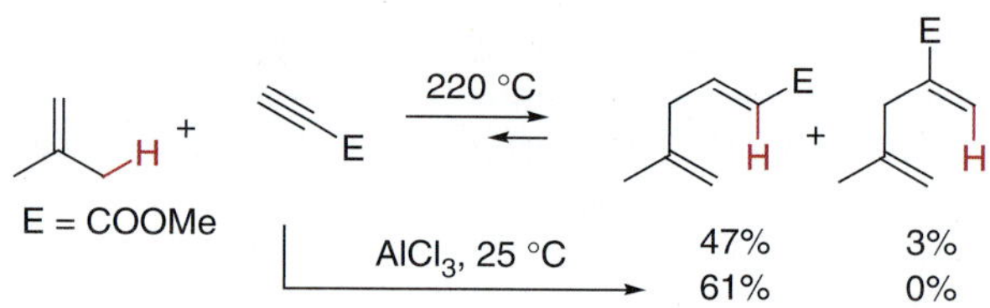

(7.58)

The reactions of propynoate derivatives proceed through two competing concerted reactions: an Alder-ene reaction (7.59) and a stereospecific [$\pi^2s+\pi^2s$]-cycloaddition (7.60) (Section 5.3.4). The ratio of products varies with the β-substituent X of the propynoate in a manner that is not consistent with a stepwise mechanism involving common zwitterionic intermediate. The $AlCl_3$-promoted reaction of 2,3-dimethylbut-2-ene with methyl propynoate gives ene adduct in 90% yield. With methyl chloropropynoate, it gives a cyclobutene adduct in 76% yield. The [$\pi^2s+\pi^2s$]-cycloaddition of propynoate esters has a less polar transition state than the Alder-ene reaction. This is indicated by the reactions of 1-alkenes (IUPAC: alk-1-enes), which give almost 1 : 1 mixtures of regioisomeric cyclobutenes. Except where steric effects interfere, ene reactions predominate with alkenes containing disubstituted vinylic carbon center, which can lead to a transition state with partial tertiary carbenium ion character. With monosubstituted ethylenes, ene reactions are much slower, allowing cycloadditions to compete, whereas with 1,2-disubstituted ethylenes, steric hindrance to ene reactions makes cycloadditions exclusive. The kinetic isotope effect $k_H/k_D = 2.64$ obtained for the $AlCl_3$-promoted Alder-ene reaction of methyl propynoate with $(CD_3)_2C{=}C(Me)_2$ suggests a concerted mechanism involving hydrogen transfer in the rate-determining step. In a stepwise mechanism, the isotope effect would be smaller [741]. The $AlCl_3$-catalyzed reaction of methyl propynoate to cyclohexene gives a mixture of $[\pi^2s+\pi^2s]$-cycloadduct and product of Alder-ene reaction, both reactions having the same activation volume $\Delta^{\ddagger}V = -20\,cm^2\ mol^{-1}$ (the proportion of the two products does not change between 1 and 2420 bars). This suggests dipolar acyclic transition states with common intermediates for the catalyzed Alder-ene reactions and $[\pi^2s+\pi^2s]$-cycloadditions [748, 749]. Alder-ene reactions of unactivated alkenes and alkynes can be catalyzed by transition metal complexes (Section 8.3.2).

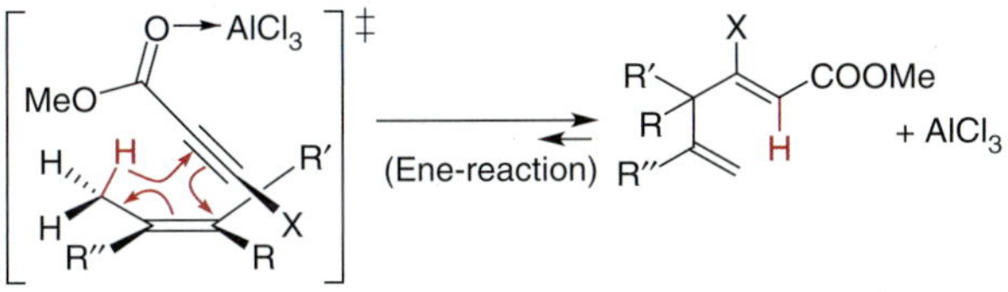

(7.59)

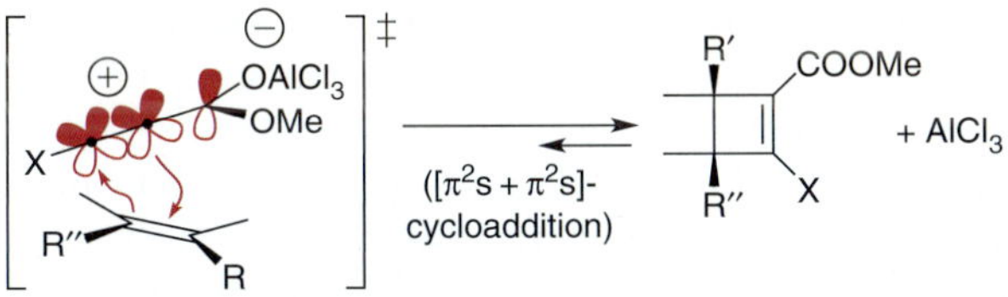

(7.60)

7.6.6 Diels–Alder reaction

Examples of acid-catalyzed cycloadditions have been given for the Diels–Alder (Sections 5.3.13 and 5.3.14) and hetero-Diels–Alder reaction (Section 5.3.16), including for asymmetric versions [750, 751] and for 1,3-dipolar cycloadditions (Section 5.3.7) [752]. The diradicaloid model developed for these reactions (Section 5.3.4 for $[\pi^2+\pi^2]$-cycloadditions; Sections 5.3.11, 5.3.13, 5.3.14, and 5.3.16 for $[\pi^4+\pi^2]$-cycloadditions; Section 5.3.17 for 1,3-dipolar cycloadditions) predicts that two unsaturated compounds may combine into a diradical ↔ zwitterion species (equilibria (7.61)–(7.63)) that is associated with the transition state of a concerted, one-step cycloaddition, or an intermediate of a two-step or multistep reaction [753]. If B: (e.g. COR′, COOR′, and CN) and Z: (e.g. O:, NR′) are Lewis bases, they can be protonated or form stable Lewis acid/base complexes of types **250–252**. The result is to stabilize the zwitterionic form of these transition structures, or intermediates, and, as a consequence, to accelerate the cycloadditions. Quite often, in parallel with their rate increase, the reactions have increased stereo- and regioselectivities. When the catalyst is chiral and enantio-enriched, enantioselective cycloadditions are possible.

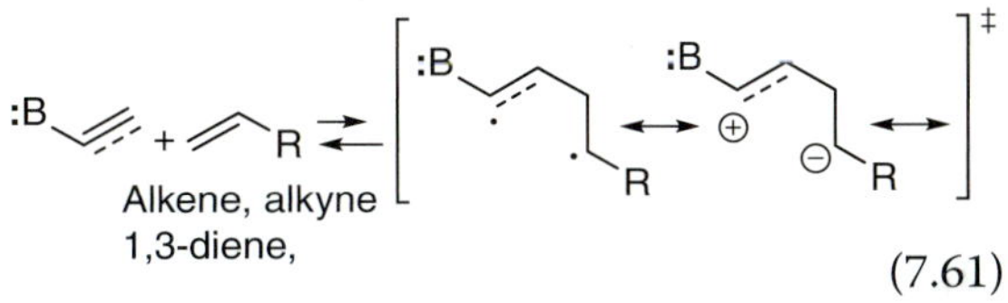

(7.61)

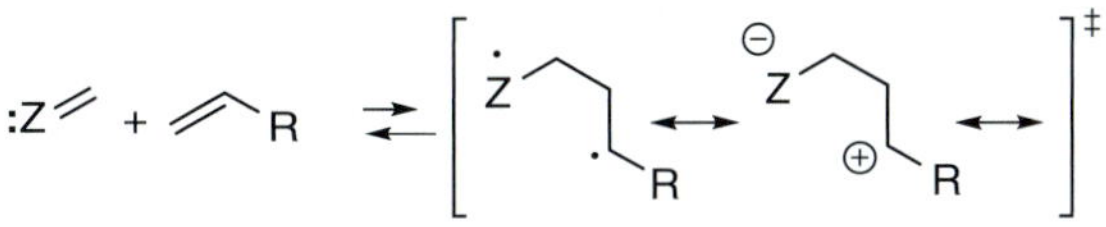

(7.62)

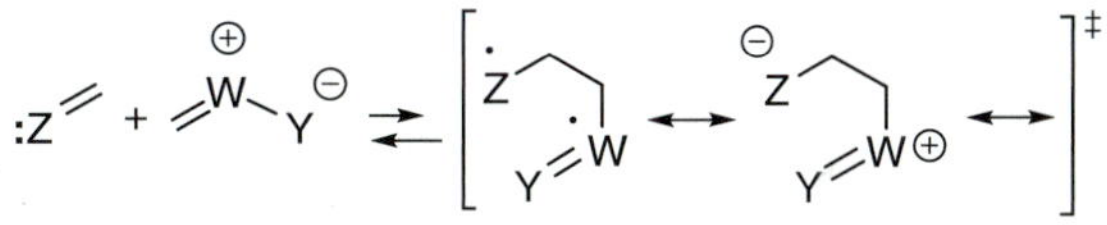

(7.63)

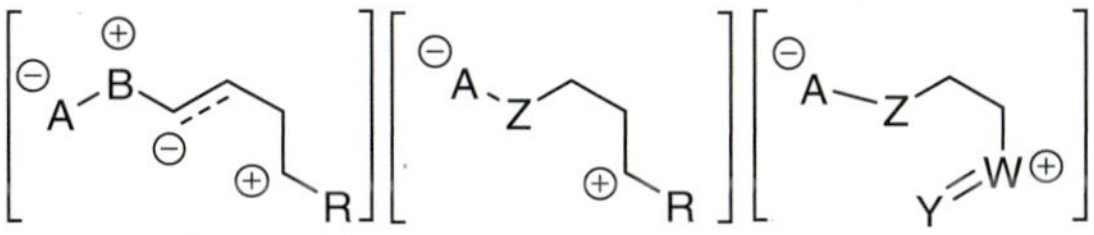

250 **251** **252**

In 1942, Wassermann observed that Diels–Alder additions of 1,3-dienes to α,β-unsaturated carbonyl compounds are accelerated in the presence of protic acids [754, 755]. In 1960, Yates and Eaton reported that the Diels–Alder additions of maleic anhydride, *p*-benzoquinone, and dimethyl fumarate are highly accelerated in the presence of $AlCl_3$ [559]. This discovery made possible the $[\pi^4s+\pi^2s]$-cycloadditions of little reactive α,β-unsaturated aldehydes and ketones. For instance, whereas the cycloaddition of butadiene to cyclohex-2-enone requires heating to 110 °C for 228 hours to furnish the corresponding Diels–Alder adduct in only 29% yield, in the presence of $AlCl_3$, the reaction is complete (95% yield) at 70 °C after five minutes only [756, 757]. Polar dienophiles such as conjugated enals, enones, ene-esters, ynals, ynones, yne-esters, α,β-unsaturated nitroalkanes, sulfones, phosphonic esters, etc., are Lewis bases forming complexes with Brønsted and Lewis acids in which the electrophilicity of their alkene or alkyne units is enhanced. According to the MO theory (Section 5.3.3), the LUMO of these complexed dienophiles is lowered in energy and thus permits a faster concerted cycloaddition. Organic solvent soluble Lewis acids such as $EtAlCl_2$ [758], $MeAlCl_2$ [759], Et_2AlCl, $ZnCl_2$, $ZnBr_2$, ZnI_2, $AlBr_3$, $B(OAc)_3$, $BF_3{\cdot}Et_2O$, $SnCl_4$, $TiCl_4$, $Yb(OTf)_3$, and Me_3SiOTf; cationic complexes of Co, Fe, Mo, Ni, Cu, Ru, V, W, Yb, Ti, Pd, and Pt [760, 761], and $MeReO_3$; [762] or solids such as silica gel [763, 764], magnesium silicate, alumina [765–767], zeolites [768, 769], clays [770–773], silica, and alumina modified by Lewis acids (Et_2AlCl, $TiCl_4$, Me_3Al, $TiCl_2(O\text{-}i\text{-}Pr)_2$, $Ti(O\text{-}i\text{-}Pr)_4$ [774], $Mg(ClO_4)_2$, $Ba(ClO_4)_2$, and $LiClO_4$) [775] have been used as catalysts [776]. Using zeolite catalysts, the Diels–Alder addition (7.64) of myrcene and methacrolein gives Precyclemone®, a fragrance used in perfumery, with high yield and regioselectivity [777]. The reaction gives a mixture of two possible regioisomers under uncatalyzed conditions (160 °C, six hours) [778]. Acyclic (*Z*)-1,3-dienes are poor substrates for Diels–Alder additions. However, reaction (7.65) occurs already at −78 °C when promoted with $SnCl_4$ and gives a 96 : 4 *endo* + *exo*-product mixture in 90% yield [759].

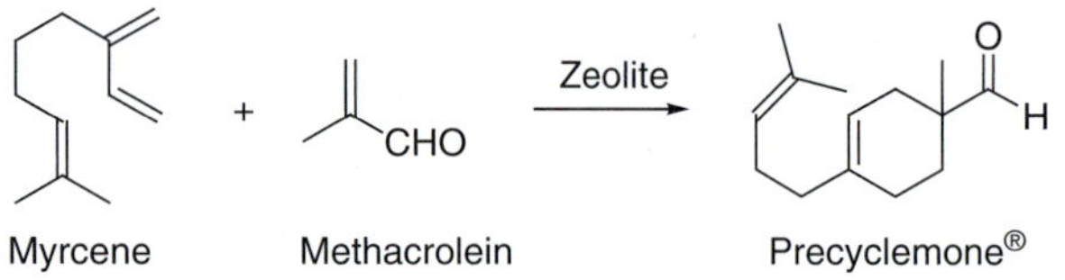

(7.64)

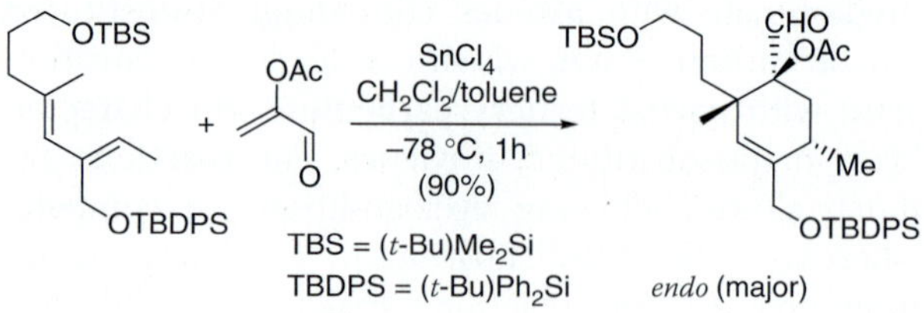

(7.65)

α,β-Unsaturated lactones such 5*H*-furan-2-one and their alkyl-substituted derivatives are poor dienophiles, even in the presence of common Lewis acid promoters. However, the doubly aluminated triflic amides such as $TfN[Al(Me)Cl]_2$ and $TfN[Al(i\text{-}Bu)_2]_2$ that are derived from $TfNH_2 + 2Me_2AlCl$ and $TfNH_2 + 2(i\text{-}Bu)_3Al$, respectively, are excellent promoters of Diels–Alder reactions as illustrated with the highly regioselective reaction (7.66) [779].

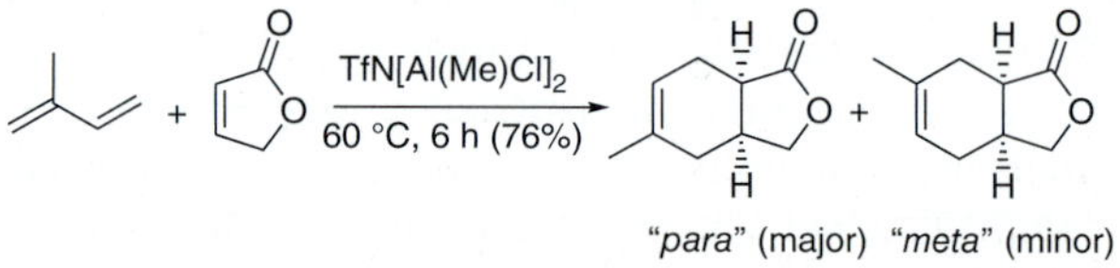

(7.66)

Cycloaddends (alkenes and alkynes) π-conjugated with functions such as CHO, COR, C≡N, or C=NR are activated by hydrogen bond donors [780, 781]. A large number of asymmetric cycloadditions, including Diels–Alder and hetero-Diels–Alder reactions; formal (3+2)-, (5+1)-, and (5+2)-cycloadditions; and 1,3-dipolar cycloadditions have been catalyzed by enantiopure chiral thiourea or squaramides (Figure 7.8; see also Takemoto's thiourea **66** and Rawal's squaramide **67**, Section 7.3.5) [787, 788]. Enantiopure chiral trityl salts prepared in situ by mixing trityl chloride derivatives ($Ar^1Ar^2Ar^3C$—Cl) with chiral phosphonate, phosphoramide, and

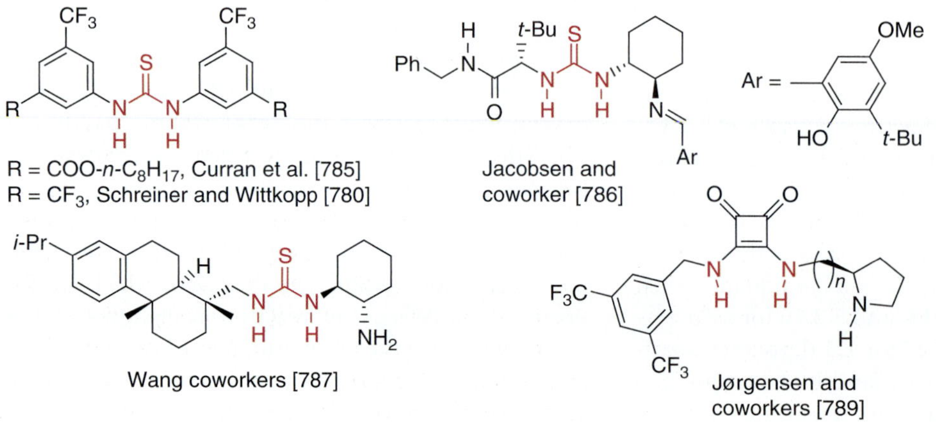

Figure 7.8 Examples of hydrogen bond donor catalysts that activate cycloaddends such as aldehydes, ketones, imines, and alkenes (and alkynes) conjugated with these functions [780, 782–786].

bis(sulfuryl)amide silver or sodium salts catalyze enantioselective Diels–Alder reactions [789]. Transition metal-catalyzed (4+2)-cycloadditions of nonactivated cycloaddends will be presented in Section 8.4.9.

7.6.7 Brønsted and Lewis acid-catalyzed hetero-Diels-Alder reactions

The carbonyl group of aldehydes and of activated ketones acts as dienophiles in $[\pi^4+\pi^2]$-cycloadditions leading to dihydropyrans [790–794]. These hetero-Diels–Alder reactions have been catalyzed by protic acids [795]. An efficient asymmetric version using only 0.2 mol% of a chiral binaphthol-derived imidodiphosphorimidate (CF_3SO_2N—H acid) and aldehyde dienophiles has been reported by List and coworkers [796]. The oxa-Diels–Alder reactions have also been catalyzed by Lewis acids such as cationic Pd(II) complexes [797], cationic Fe(III) complexes [798, 799], $Sc(n\text{-}C_8F_{17}SO_3)_3$ [800], $TiCl_4$ [801], or $AgBF_4$ [802]. In the case of cycloaddition (7.67), the diastereoselectivity *syn* vs. *anti* depends on the nature of the Lewis acid catalyst. With $BF_3{\cdot}Et_2O$, the *anti*-cycloadduct is the major product isolated [803], whereas, using a Yamamoto–Saito catalyst ($Al(OAr)_2Me$) [804], the *syn*-cycloadduct is the major product. Enantioselective hetero-Diels–Alder reaction of Brassard-type diene ((*E,E*)-1,3-dimethoxy-1-trimethylsilyloxypenta-1,3-diene) to aliphatic aldehydes and isatins has been catalyzed by enantiopure chiral In(III) [805] and Mg(II) complexes of *N,N′*-dioxides, respectively [806].

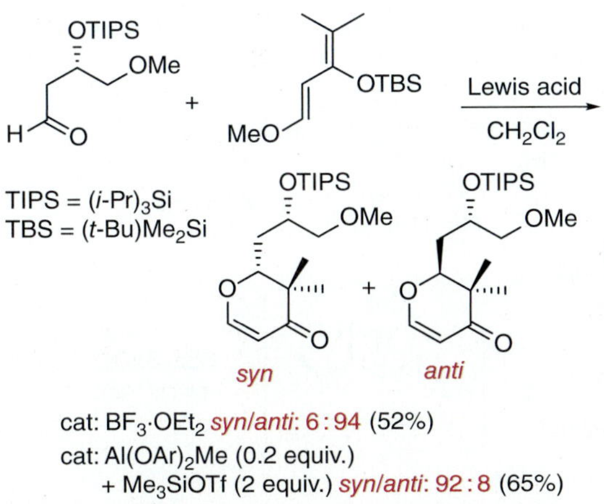

(7.67)

Imino-Diels–Alder reactions (7.68) use imines as dienophiles [794]. They are catalyzed by metal triflates such as $Yb(OTf)_3$, $Sm(OTf)_3$, or $In(OTf)_3$ [807]. These catalysts can tolerate the presence of small amounts of water [808–811]. The choice of Lewis acid can have a strong influence on the *endo* vs. *exo* diastereoselectivity as illustrated with reaction (7.69) [812].

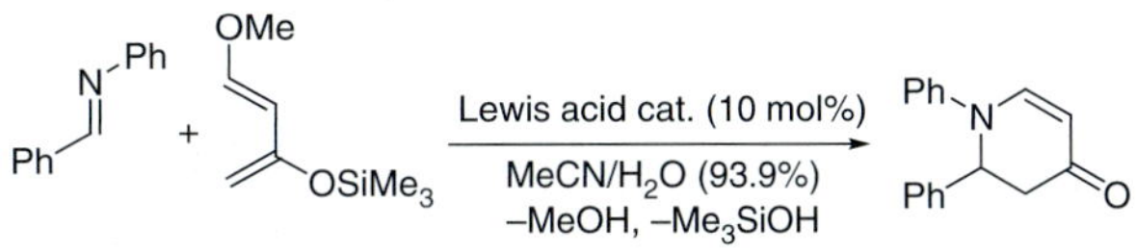

(7.68)

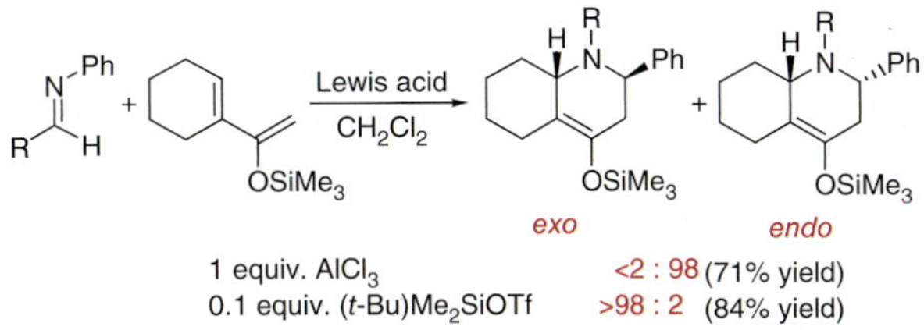

(7.69)

The complex $[(PPh_3)Ag(CB_{11}H_6Br_6)]$ (**253**) is an effective, selective (0.1 mol% loading), and recyclable catalyst for Diels–Alder reactions [813]. A rather "naked," thus activated Ag^+ species makes **253** a strongly nitrophilic Lewis acid. In comparison, $(Ph_3P)AgOTf$ is a weaker catalyst for this reaction. Enantioselective Diels–Alder reactions of imines catalyzed by a chiral phosphoric acid have been reported [813]. The zinc salt resulting from the reaction of Et_2Zn with (*R*)-3,3′-di(9-anthracenyl)-1,1′-bi-2-napththol catalyzes enantioselective inverse electron demand imino-Diels–Alder reactions [814].

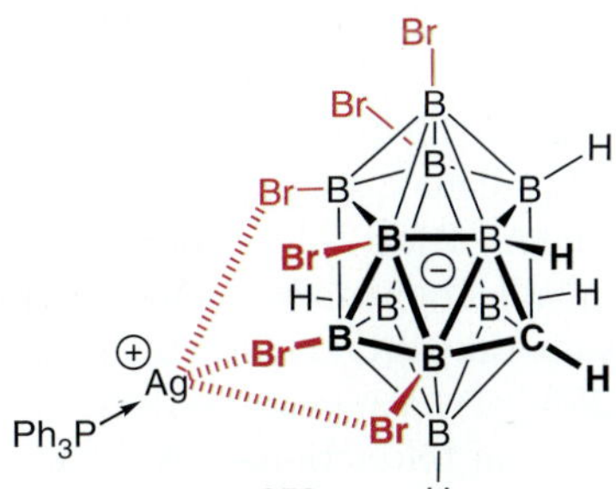

253

The $[\pi^4+\pi^2]$-cycloadditions of α,β-unsaturated aldehydes and ketones (1-oxa-1,3-dienes) to enol ethers (Diels–Alder additions with inverse electron demand, Section 5.3.11) is an attractive route to 3,4-dihydro-2*H*-pyranes and hexose derivatives. An illustration is given with the Tietze's syntheses of L- and D-mannopyranosides (Scheme 7.44a). Interestingly, depending on the nature of the Lewis acid promoter, different diastereomeric adducts

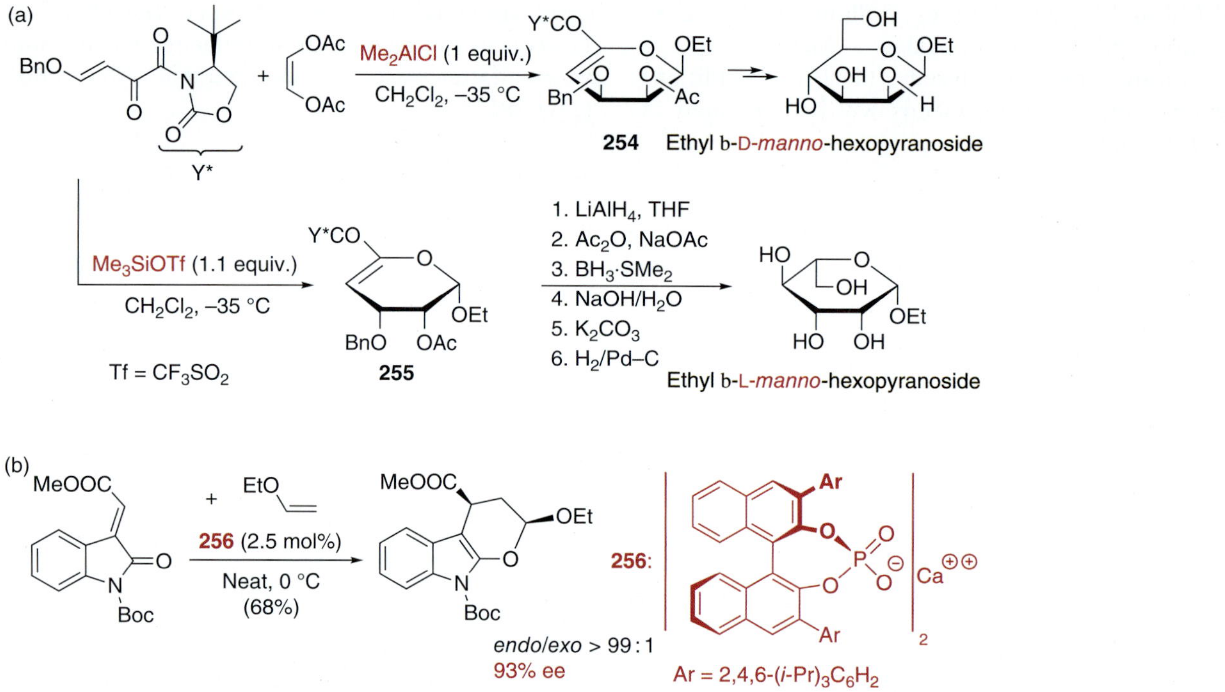

Scheme 7.44 Examples of hetero-Diels–Alder reaction of 1-oxa-1,3-dienes promoted by Lewis acids. (a) Tietze's syntheses of ethyl D- and L-*manno*-hexopyranosides; (b) Cheng's enantioselective synthesis of dihydropyran-fused indoles.

254 and **255** are obtained [815]. An enantioselective oxa-hetero-Diels–Alder reaction catalyzed by the chiral phosphoric acid derivative **256** has been presented by the Cheng's group (Scheme 7.44b) [816]. Enantioselective catalyzed hetero-Diels–Alder reactions of *o*-quinone methides and azlactones to generate dihydrocoumarins have been catalyzed by a chiral *N,N′*-dioxide–Sc(III) complex [817]. Unactivated alkenes add enantioselectively to *o*-quinone methides activated by enantiopure chiral BINOL-based *N*-triflylphosphoramide [818].

As 1-oxa-1,3-dienes, simple 1-aza-1,3-dienes tend to undergo hetero-Diels–Alder additions with inverse electron demand. They can be catalyzed by nitrophilic Lewis acids such as $Cu(OTf)_2$ and $BiCl_3$ as shown with reaction (7.70) [819]. An enantioselective version catalyzed by a chiral bifunctional protic acid has been reported [819c].

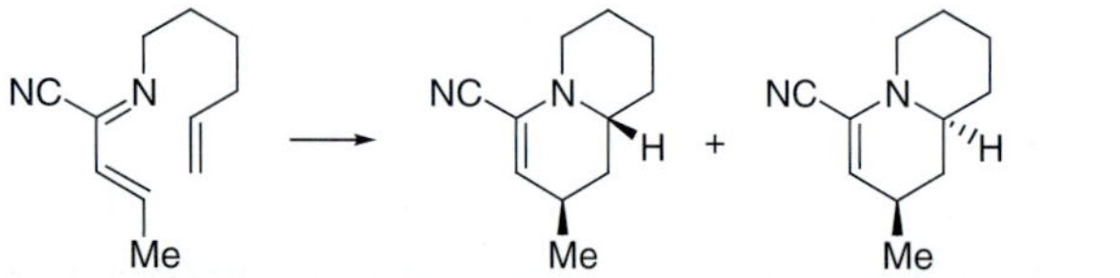

Conditions: 110 °C, 24 h	1	:	4	
$Cu(OTf)_2$, –20 °C, 38 h	1	:	3.7	(60%)
$BiCl_3$, 0 °C, 72 h	1	:	4.8	(69%)

(7.70)

Like 1-aza-1,3-dienes, 2-aza-1,3-dienes have proven to be useful reagents for the Diels–Alder-based synthesis of pyridones, isoquinolones, and pyrimidones. They can be catalyzed by Lewis acids such as $BF_3{\cdot}Et_2O$, $TiCl_4$, $SnCl_4$, and $InCl_3$. In reaction (7.51), $Dy(OTf_3)$ catalyzes in situ generation of *N*-arylimines and their subsequent $[\pi^4+\pi^2]$-cycloadditions to the *N*-protected enamine **257** [820]. $KHSO_4$ has been found to catalyze similar one-pot three component coupling of aldehydes, anilines, and dienophiles such as dihydropyrane, cyclopentadiene, and ethyl vinyl ether [821].

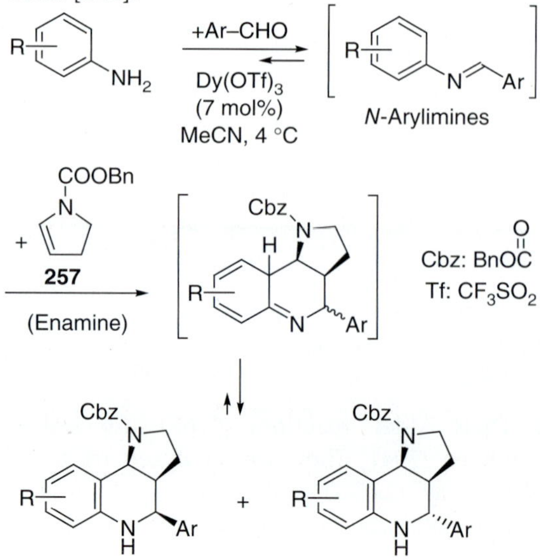

(7.71)

2-Aza-1,3-dienes substituted at C(1) and C(3) with electron-releasing groups such as Me_3SiO, MeS, and Me_2N undergo $[\pi^4+\pi^2]$-cycloadditions with α,β-conjugated enones and ene-esters (Diels–Alder reactions with normal electron demand). They can be catalyzed by oxyphilic Lewis acids such as R_3SiOTf, $Eu(fod)_3$, Et_2AlCl, or $TiCl_4$ [807]. In the presence of $AlCl_3$ or $EtAlCl_2$ (1–3 equiv.), conjugated carbodiimides react as 2-aza-dienes with various dienophiles such as acetylenic compounds, enamines, vinyl ethers, and aldehyde to give nitrogen heterocycles as cycloadducts [822]. Examples are given with reactions (7.72) and (7.73).

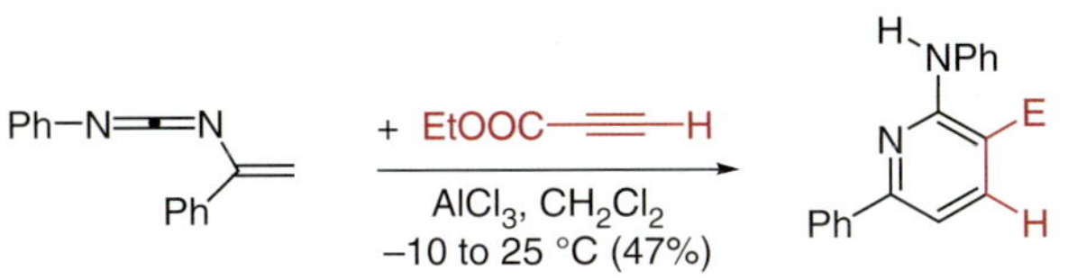

(7.72)

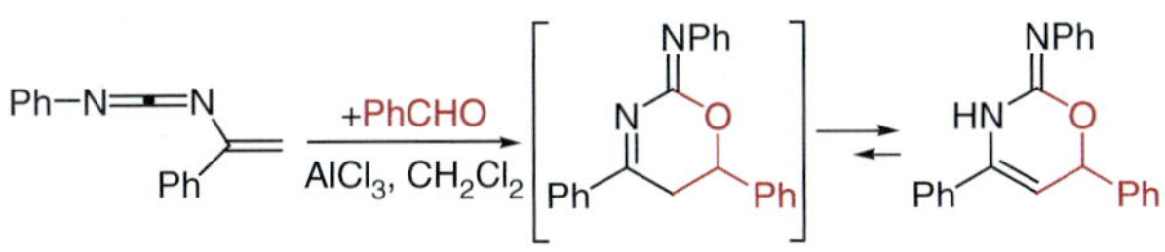

(7.73)

Enantioselective hetero-Diels–Alder reactions of 1,3-dienes with 1,2-diazene of type Boc—N=N—C(=O)—NAr have been catalyzed by Au(I) complexes of chiral phosphoramidites [823]. Enantioselective hetero-Diels–Alder reactions of electron-rich alkenes with diazenes of type PhCO—N=N—R have been catalyzed by Cu(I) complexes of (*S,S*)-bisoxazolines [824]. Complexes made of $Cu(MeCN)_2BF_4$ and chiral diphosphine ligands or other ligands catalyze the enantioselective hetero-Diels–Alder reactions of 1,3-dienes with nitroso compounds [825].

7.6.8 Acid-catalyzed (2+2)-cycloadditions

Cyclobutanes can be obtained by photochemical $[\pi^2s+\pi^2s]$-cycloaddition of alkenes (Sections 6.8.1 and 6.8.2) [826–829]. Unfortunately, this method generally fails for acyclic enones, except for 1,3-dicarbonyl compounds (de Mayo reaction (6.85), Section 6.8.2) [830]. Thermal $[\pi^2+\pi^2]$-cycloadditions of alkenes are possible with fluorinated alkenes, highly strained alkenes, and alkynes and for electron-rich alkenes reacting with electron-poor alkenes (Section 5.3.4) [831–839]. For example, cyclobutanes result from the reaction of Michael acceptors and alkenes substituted by strongly electron-releasing group such as NR_2, OR, $OSiR_3$, and SR [840–843]. Lewis acid promoted $[\pi^2+\pi^2]$-cycloaddition of alkenes with alkynes or allenyl esters and ketones has been used widely [741, 844–846]. Examples are given with reactions (7.74) [847] and (7.75) [848].

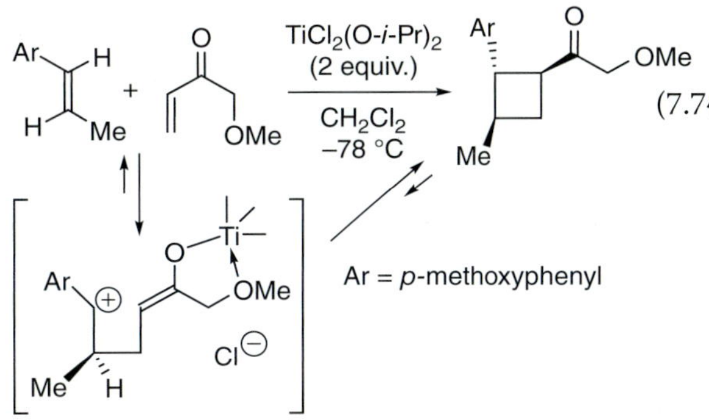

(7.74)

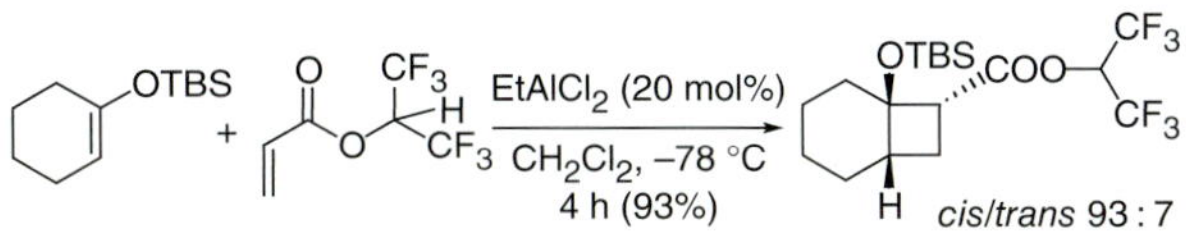

(7.75)

$AgNTf_2$ catalyzes $[\pi^2+\pi^2]$-cycloadditions of alkynyl silyl ethers and of silyl enol ethers to α,β-unsaturated carbonyl compounds [849]. Tf_2NH serves as an efficient catalyst for $[\pi^2+\pi^2]$-cycloadditions of silyl enol ethers to methyl acrylate [850] and allylsilanes [848–851]. An example of asymmetric Lewis acid catalyzed $[\pi^2+\pi^2]$-cycloaddition of a conjugated enone and an enol ether presented by Canales and Corey has been given in Section 5.3.4 [852]. The oxyphilic catalyst is an aluminum bromide that activates the carbonyl group of the enone (Table 7.3). An other example is shown in Scheme 7.45a, which uses oxaborolidine **258** as the enantiopure chiral catalyst, which also an oxophilic Lewis acid [853]. A second example applies a zinc complex resulting from the combination of $Zn(NTf_2)_2$ with enantiopure chiral *N,N*′-dioxide **259**, another oxyphilic Lewis acid (Scheme 7.45b) [854]. Salts of transition metals are less oxyphilic and can be highly carbophilic (Table 7.3). They interact with the cycloaddends by reacting with their C=C of C≡C bonds rather than with their polar substituents (Section 8.4). This point is illustrated in Scheme 7.45c with a formal $[\pi^2+\pi^2]$-cycloaddition of a 2,3-dimethylindole derivative with an allenamide catalyzed by a gold(I) complex resulting from the combination of $AuCl(Me_2S)$ and chiral diphosphine **260** [855]. The first example of enantioselective (phosphine)gold(I)-catalyzed $[\pi^2+\pi^2]$-cycloaddition was reported by Toste and coworkers in 2007 [856]. Since then, many more examples of enantioselective gold-catalyzed (2+2)-cycloaddition have been reported [857–862]. With a mixture of $Cu(ClO_4)_2{\cdot}6H_2O$ and a chiral bis-oxazolidine ligand as catalysts, enantioselective (2+2)-cycloadditions of dimethyl methylidenemalonate to arylsubstituted alkenes have been realized (THF, −70 °C) [863]. For more transition metal catalyzed $[\pi^2+\pi^2]$-cycloadditions, see Section 8.4.4.

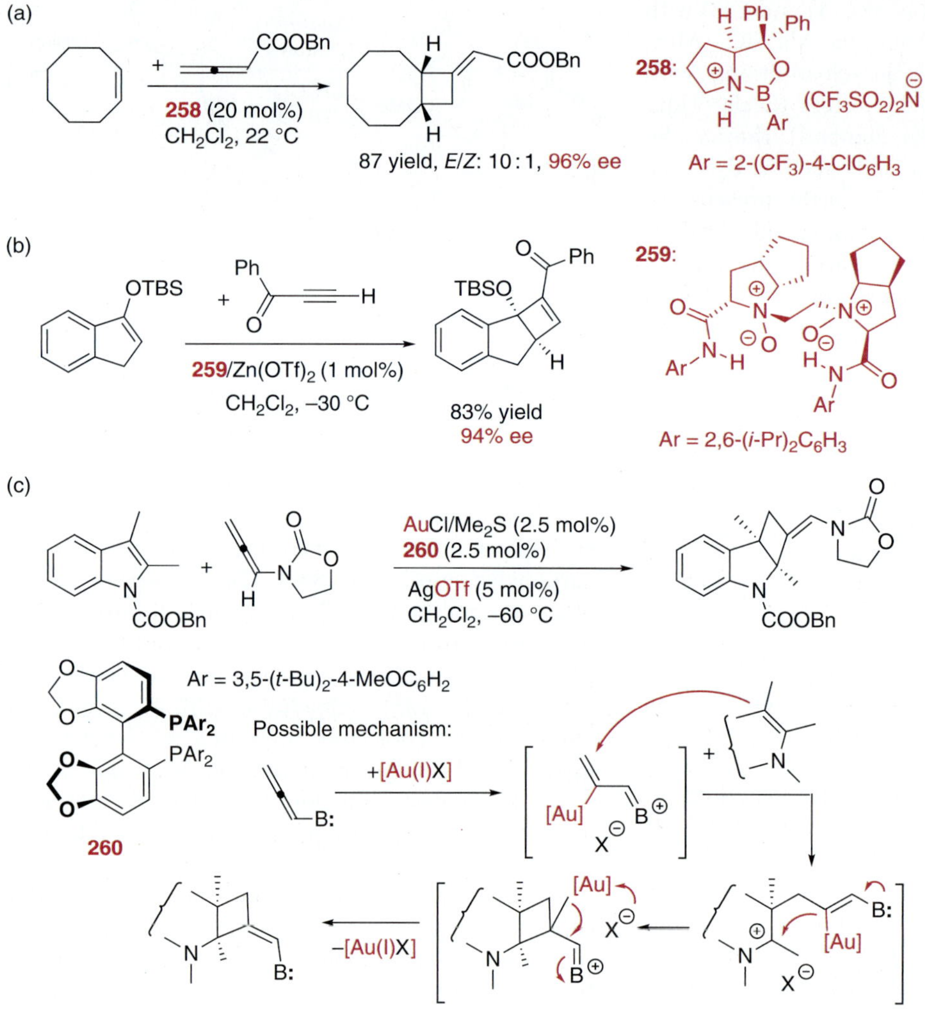

Scheme 7.45 Examples of catalytic enantioselective $[\pi^2+\pi^2]$-cycloadditions: (a) applying an oxyphilic Lewis acid as the catalyst; (b) cycloaddition of an alkynone with an enoxysilane catalyzed by a *N,N′*-dioxide-zinc complex; and (c) cycloaddition of an indole derivative with an allenamide catalyzed by a (diphosphine) gold(I) complex: a carbophilic Lewis acid.

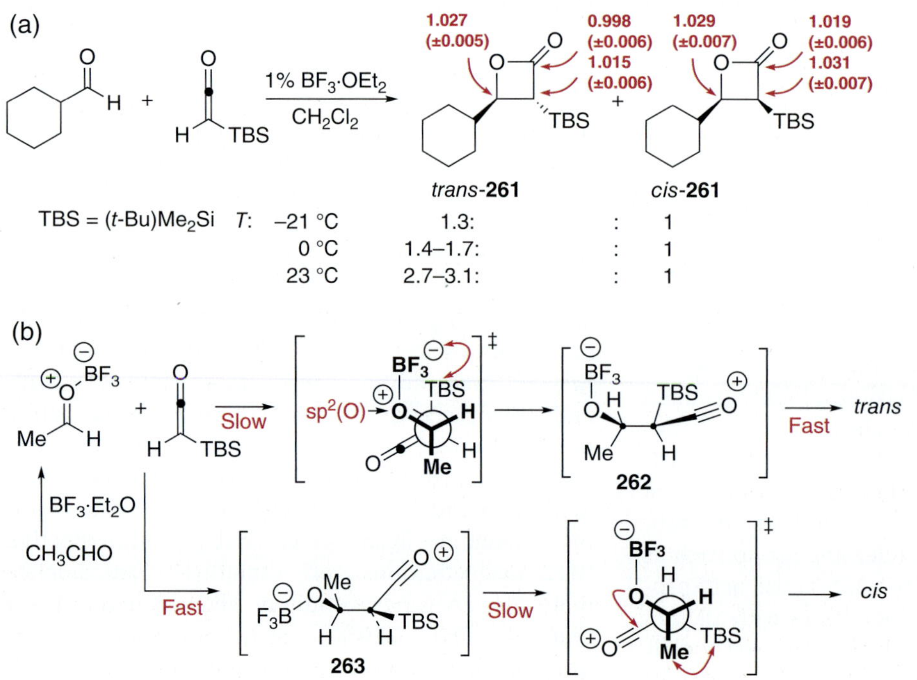

Scheme 7.46 (a) ^{13}C-kinetic isotope effects for a BF_3/Et_2O-catalyzed $[\pi^2+\pi^2]$-cycloaddition of an aldehyde with a ketene and (b) proposed mechanisms.

β-Lactones are important synthetic intermediates [864–869] that can be obtained from direct [$\pi^2+\pi^2$]-cycloaddition of ketenes with carbonyl compounds [870–875]. For the uncatalyzed process, quantum mechanical calculations favor a concerted suprafacial–suprafacial cycloadditions [876–878]. For the reaction of $CH_3CHO{\cdot}BH_3 + Cl_2C{=}C{=}O$, a concerted, asynchronous mechanism is predicted [877]. The $BF_3{\cdot}Et_2O$-catalyzed cycloaddition of cyclohexanecarboxaldehyde with *tert*-butyldimethylsilylketene gives a mixture of adducts *trans*-**261** and *cis*-**261** (are not isomerized under the conditions of their formation) in high yield and with proportion that varies with temperature (reaction (7.76)). The ^{13}C-kinetic isotope effects ($k_{^{13}C}/k_{^{12}C}$, standard deviations given in parenthesis) for this reaction at 0 °C were determined at natural abundance by NMR spectroscopy (Section 3.9.5). They differ significantly for the formation of diastereomeric trans- and cis-cycloadduct. This is indicative of differing mechanisms or rate-limiting steps in the formation of these products. The increased preference for trans-cycloadduct at higher temperatures indicates that it is "entropically favored." High-level quantum mechanical calculations employing an Onsager-type continuum solvent model predicted stepwise cycloaddition mechanisms (acetaldehyde as model aldehyde) [879]. In the case of the formation of the trans-cycloadduct, the first step generating zwitterionic intermediate **262** is rate-limiting, and, in the case of the formation of the cis-cycloadduct, the second step, the ring closure of zwitterionic intermediate **263**, is the rate-determining step. A pathway involving *syn*-acetaldehyde/BF_3 complex is calculated to have an energy barrier higher by c. 5.3 kcal mol^{-1} (Scheme 7.46).

Several methods of catalytic asymmetric ketene [$\pi^2+\pi^2$]-cycloadditions are available; most of them apply nucleophilic catalysis [880, 881]. In 2014, the Feng's group has reported that metallic complexes of chiral *N,N′*-dioxide analogous to **259** (Scheme 7.45) catalyze [$\pi^2+\pi^2$]-cycloadditions of disubstituted ketenes for the asymmetric synthesis of β-lactones [882].

7.6.9 Lewis acid catalyzed (3+2)- and (3+3)-cycloadditions

In the presence of a Lewis acid, cyclopropanes with electron-releasing and electron-withdrawing substituents on two vicinal carbon centers can equilibrate with zwitterions as shown with equilibrium **264** + $TiCl_4$ ⇄ **265**. The zwitterions can be quenched with unsaturated compounds and generate five-membered ring systems. In the presence of stoichiometric amount of Me_2AlCl, methyl 2-methylthiocyclopropyl ketone and silyl enol ethers of acetophenone afford methyl cyclopentyl ketone derivatives [883]. Similarly, $TiCl_4$ promotes the (3+2)-cycloaddition of substituted methylidenecyclopropyl ketones to allyltrimethylsilane [884], as illustrated with reaction (7.77) catalyzed by $TiCl_4$ [885].

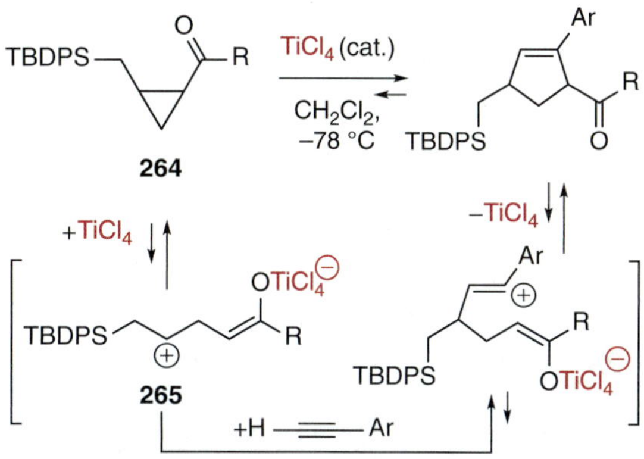

265 (cation stabilized by β-silicon effect)
TBDPS = (*t*-Bu)Ph_2Si; R = Ph, Bu, *t*-Bu
Ar = Ph, 4-ClC_6H_4, 4-MeOC_6H_4

(7.77)

2-Phenylcyclopropane-1,1-dicarboxylic esters (**266**, R′ = Ph) react with imines in the presence of 5 mol% $Sc(OTf)_3$ to generate the corresponding pyrrolidines [886]. Similarly, aryl isocyanide undergo formal (3+2)-cycloaddition catalyzed by $Pr(OTf)_3$ generating 3-arylamino-2(arylimino)xyclopent-3-ene-1,1-dicarboxylates [887]. $GaCl_3$ catalyzes the insertion of 2-arylcyclopropane-1,1-carboxylic diesters that produces pyrazolidines [888]. One-step diastereoselective synthesis of *cis*-2,5-disubstituted tetrahydrofurans **268** via Lewis acid catalyzed (3+2)-cycloadditions of donor–acceptor cyclopropanes **266** to aldehydes has been developed by Johnson and coworkers (reaction (7.78)) [889]. Chiral complexes of $Cu(OTf)_2$ catalyze the enantioselective (3+2)-cycloadditions of donor–acceptor cyclopropanes and enamines producing amino-substituted cyclopentanes [890]. The reactions involve the formation of zwitterion intermediates **267** and proceeds with inversion at the cyclopropane donor site (substituted by R′). They allow chirality transfer to the product with high stereoselectivity. Transition metal complexes catalyze (3+2)- (Section 8.4.6) [891], (3+2+1)- (Section 8.4.12), and (3+4)-cycloadditions (Section 8.4.1) of cyclopropanes and three-membered heterocycles.

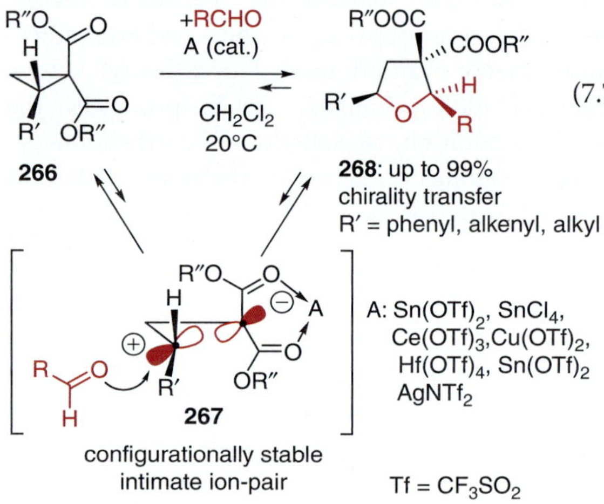

(7.78)

In the presence of various Lewis acids such as $BF_3{\cdot}Et_2O$, Me_3SiOTf, $TiCl_4$, or $Zn(OTf)_2$, $CpW(CO)_3$-propargyl compounds bearing a tethered epoxides undergo formal (3+3)-cycloadditions (Scheme 7.47a) [892].

Lewis acids such as $BF_3{\cdot}Et_2O$, $TiCl_4$, $In(OTf)_3$, $Zn(OTf)_2$, $SnCl_4$, $AlCl_3$, $FeCl_3$, $Mg(OTf)_2$, and CuCl catalyze the formal (3+3)-cycloadditions of α,β-unsaturated aldehydes with 4-hydroxy-2-pyrone, diketones, and vinylogous esters (Scheme 7.47b) [893, 894]. Highly site-, regio-, and stereoselective synthesis of arylidene oxazinc diesters has been described that apply the Lewis acid promoted (3+3)-cycloaddition of methylidenecyclopropane 1,1-dicarboxylic esters with *C*,*N*-diarylnitrones [895]. Enantioselective formal (3+3)-cycloaddition of nitrones to cyclopropane-1,1-dicarboxylic esters catalyzed by chiral Lewis acids (e.g. **269**) have been reported (Scheme 7.47c) [896, 897]. In the presence of a Lewis acid cyclopropylamines add to tropones (Scheme 7.47d) [898, 899]. For transition metal catalyzed (3+3)-cycloadditions, see Section 8.4.11.

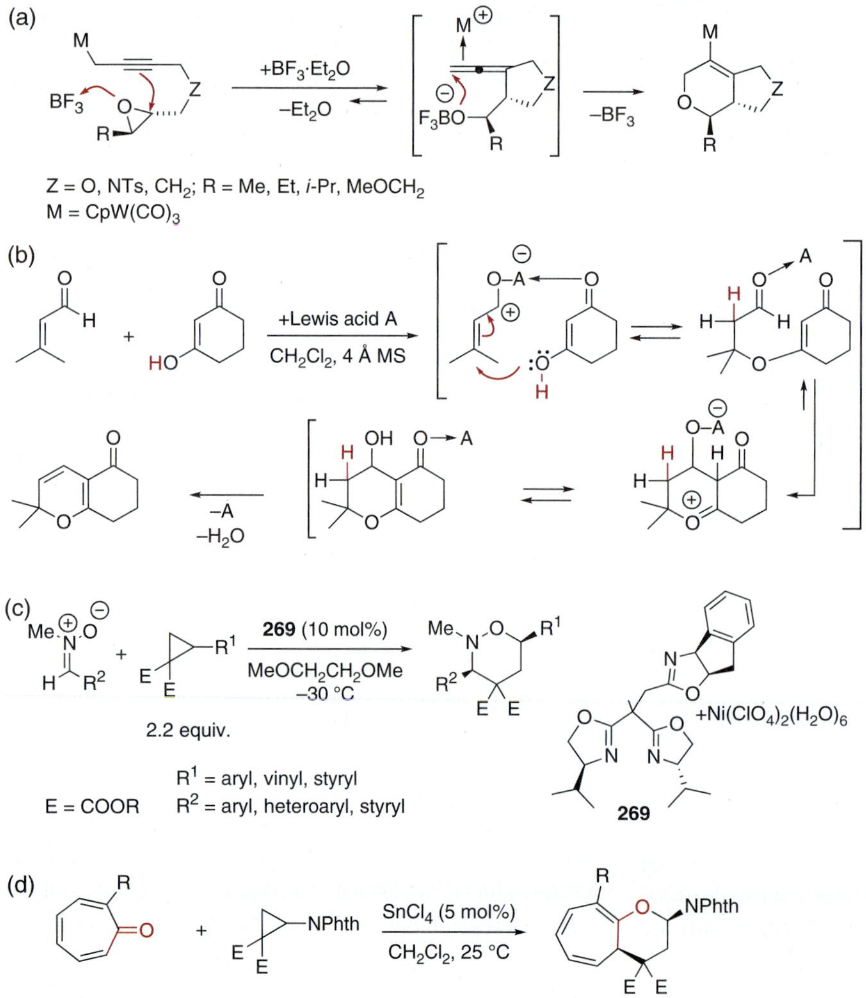

Scheme 7.47 Example of Lewis acid promoted (3+3)-cycloadditions: (a) oxirane + allene; (b) enol + enone; (c) enantioselective addition of nitrone to polar cyclopropanes that realized a kinetic optical resolution of chiral cyclopropanes; and (d) polar cyclopropanes + tropone can be seen as (8+3)-cycloadditions.

7.6.10 Lewis acid promoted (5+2)-cycloadditions

Pipitzol has been proposed to result from the thermal decomposition (reaction (7.79)) of Perezone and coworker [900–902], a reaction proposed to involve an intramolecular (5+2)-cycloaddition of the conjugate acid of the benzoquinone moiety (a 3-hydroxypentadienyl cation) to the alkene unit [903].

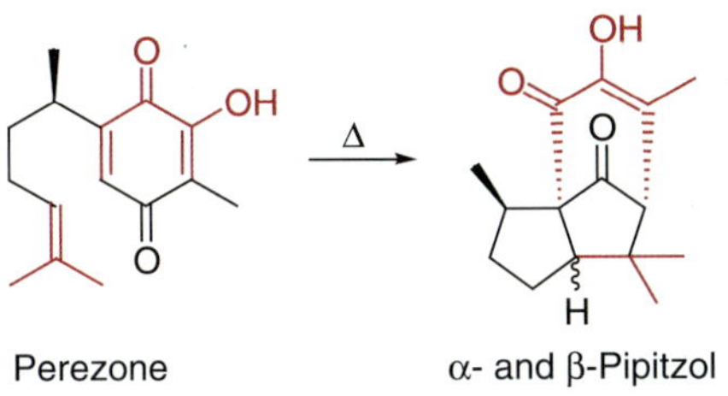

(7.79)

This reaction has been applied by Büchi and Mak in their synthesis of neolignans (guianin and burchellin [904]) and the sesquiterpene gymnomitrol [905]. Grieco and coworkers found that trimethylsilyl trifluoromethanesulfonate (TMSOTf) is an effective promoter of intramolecular and intermolecular cationic (5+2)-cycloadditions in 3.0 M $LiClO_4$(EtOAc) at −23 °C, as illustrated with reaction (7.80) [906, 907].

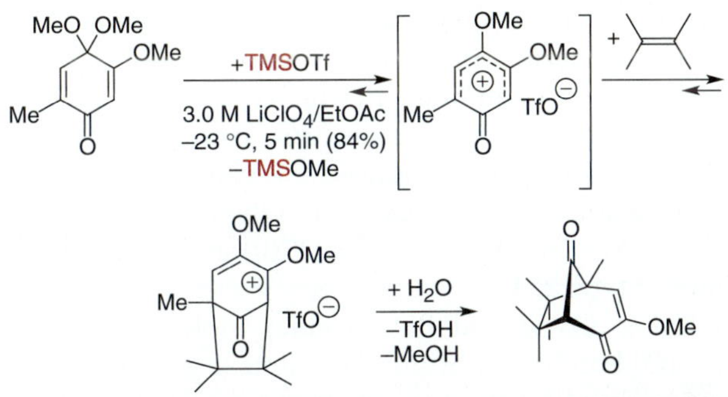

(7.80)

Under the influence of $EtAlCl_2$, [5-benzoyloxy-2-(triisopropylsiloxy)pent-1-en-3-yne)](hexacarbonyl) dicobalt is ionized into a relatively stable (hexacarbonyl)(propargyl)dicobalt cation intermediate that reacts with silyl enol ethers to yield the corresponding seven-membered acetylenedicobalt complexes as illustrated with reaction (7.81) [908].

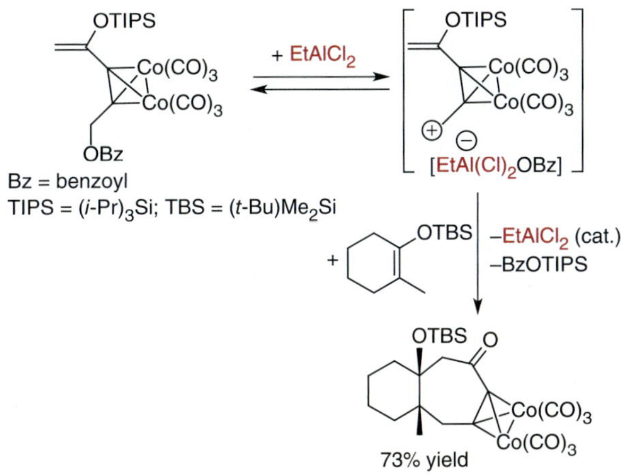

(7.81)

Reaction of penta-1,4-dien-3-ol with Cp*Co(ethylene)$_2$/HBF_4·OEt_2 at −78 °C, followed by addition of KPF_6/H_2O generates the corresponding stable Cp*Co(pentadienyl) tetrafluoroborate that undergoes intermolecular (5+2)-cycloaddition with alkynes to generate the corresponding Cp*Co(cycloheptadienyl) salts [909]. Six-substituted TpMo(CO)$_2$ (η-2,3,4-pyranyl)- and TpMo(CO)$_2$(η-2,3,4-pyridinyl) complexes (Tp = hydridotrispyrazolylborate) add in a (5+2)-manner under the influence of $EtAlCl_2$ to Michael acceptors giving the corresponding 2-substituted 8-oxa- and 8-azabicyclo[3.2.1]octenes complexes as illustrated with reaction (7.82) [910].

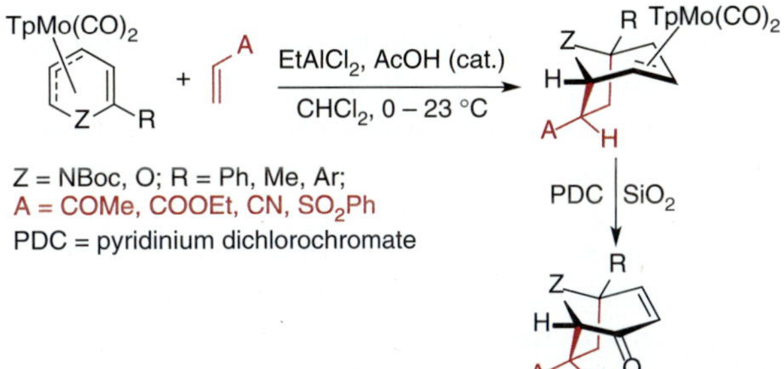

(7.82)

Using their bifunctional primary amine/thiourea catalysts (Figure 7.8, Section 7.6.6), Jacobsen and coworkers realized highly enantioselective (5+2)-cycloadditions of oxidopyrylium ylide dipoles to enol ethers [911, 912]. Vicario and coworkers [913] reported enantioselective (5+2)-cycloadditions of oxidopyrylium ylides with enals catalyzed by a chiral bifunctional secondary amine-squaramide, catalyst developed originally by Jørgensen and coworkers (Figure 7.8) for asymmetric $[\pi^2+\pi^2]$-cycloadditions [914]. High yields and enantioselectivities result from the cooperation between nucleophilic and

Brønsted acid catalysis made possible by the bifunctional catalysts. For transition metal catalyzed (5+2)-cycloadditions, see Section 8.4.14.

Problem 7.18 Explain the regioselectivity of Lewis acid catalyzed cycloadditions (7.69) and (7.75).

Problem 7.19 What is the main product **P** of reaction **A** + **B** [915].

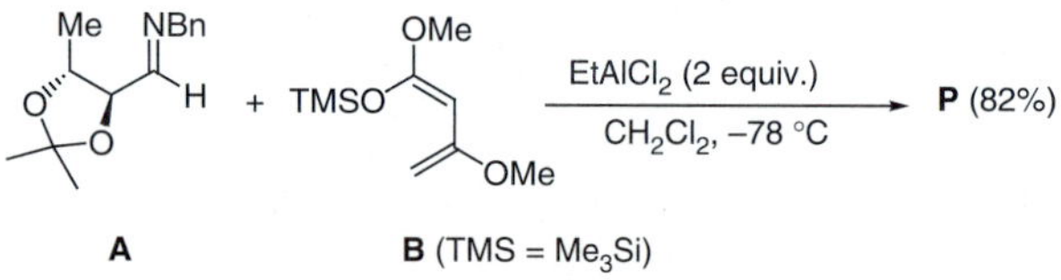

Problem 7.20 Give a possible mechanism for reaction (7.72).

Problem 7.21 What are the successive products **B** and **C** produced in the following reaction? [916, 917].

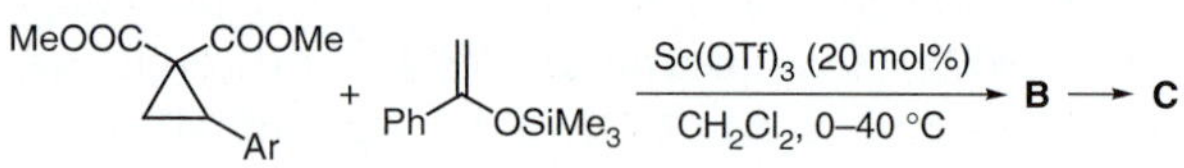

7.7 Bonding in transition metal complexes and their reactions

In the next chapters, we shall describe the most useful organic reactions catalyzed by transition metals and their complexes, either as heterogeneous or homogeneous catalysts. Some of these catalysts are expected to contribute to a more sustainable industrial development through "greener" chemistry [918]. We present here simple models of the bonding and reactivity of transition organometallic complexes. Excellent introductions to this topic and to transition organometallic chemistry can be found in the textbook of Hartwig [919]. The majority of simple transition metal–ligand bonding can be accommodated by a molecular orbital or "ligand field" treatment [920, 921]. As the difference in electronegativity between transition metals and carbon is not large, covalency is important in explaining the metal–ligand bond. Ionic character increases for metal–ligand bonds involving more electronegative atoms than carbon. Even for nonpolar carbon–metal (M—C) bonds, their reactivity is high because of their relatively weak bond energies and the high polarizability of the metals (the electrons in the M—C bond can be polarized easily). The number of valence electrons in transition metal complexes is crucial to understand the structures and the reactivity of these species. This is described by the 18-electron rule [922–924], which is by no means absolute but provides a useful framework for interpretation [925–930]. The octet rule for second-row elements arises from the use of four valence orbitals (one 2s and three 2p), therefore requiring eight valence electrons to attain the atomic number of the next heaviest noble gas (Ne) [931–933]. With the transition elements, nine orbitals are available (one s, three p's, and five d's) and full use of them requires 18 valence electrons. As for carbon species such as stable diaminocarbenes (Section 7.2.5) that have a carbon center with formally only a sextet of electrons, there are several stable transition metal complexes with only 16 valence electrons. As for the stable carbenes, steric factors make the attachment of an ultimate ligand or dimerization thermodynamically unfavored.

7.7.1 The π-complex theory

In 1945, Dewar proposed that alkenes and any unsaturated organic compounds with π-electrons can form π-complexes with electron-deficient partners in the same way as water and ammonia that possess nonbonding electron pairs (Lewis bases) [934–936]. Thus, the bromonium ion intermediates formed during the bromine additions to alkenes are π-complexes combining the alkenes (Lewis bases) and Br^+ (Lewis acid). Bonding arises from electron transfer (donation) from the HOMO of the alkene (π_{CC} orbital) to the LUMO of Br^+, an unsaturated species as it lacks two electrons to reach the electronic configuration of Kr. Alkenes are thus able to coordinate unsaturated systems as the cyanide anion can coordinate Fe^{++} to form the hexacyanoferrate(II) anion, $[Fe(CN)_6]^{4-}$, a coordination compound, in which the Lewis acid Fe^{++} forms bonds by sharing electron pairs provided by the NC^- ions. Dewar noticed the similarity between coordination compounds such as the Lewis acid/base complexes (e.g. $H_3N{:} \rightarrow BF_3$ (**270**)) and bromonium ions of type **271**.

Using alkene and Br^+, frontier molecular orbital (FMO) bonding in bromonium ions (e.g. **271**) can be represented by the perturbational molecular orbital (PMO) diagram (Section 4.4.2) of Figure 7.9. Most of the bonding in this bridged cation, or π-complex, arises from the electron transfer from the HOMO of the alkene to the LUMO of Br^+ (donation). A closer analysis reveals that Br^+ has a HOMO that shares the same symmetry as the LUMO of the alkene. Thus, a HOMO (Br^+) → LUMO (alkene) interaction is not excluded (back-donation) and it contributes somewhat to the stability of **271** as proposed by Dewar in 1951 [923, 937–939].

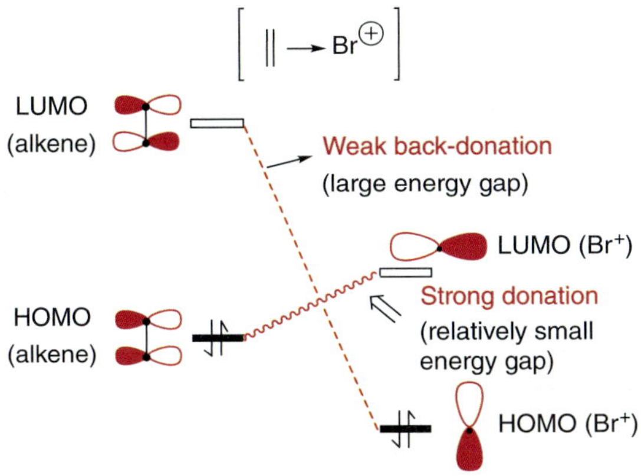

Figure 7.9 MO diagram representing bonding in bromonium ions (alkene + Br^+).

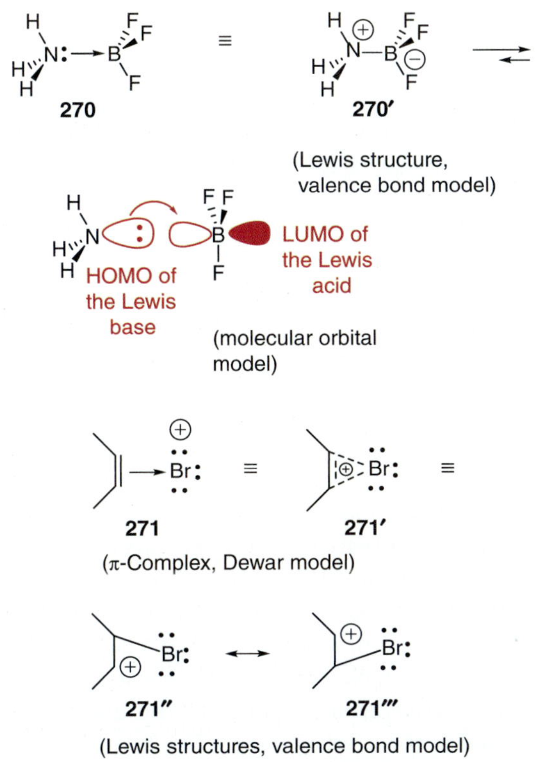

Dewar also recognized that corner-protonated cyclopropanes, intermediates in Wagner–Meerwein rearrangements of carbenium ions, can be represented as π-complexes of alkenes and carbenium ions (Figure 7.10a) [940, 941]. He also noticed that cyclopropane itself can be represented as the complex of ethylene and methylene (carbene). Difference in bonding between cyclopropane and corner-protonated cyclopropane resides in the relative importance of HOMO (alkene)/LUMO (unsaturated species) interaction (donation) and HOMO (unsaturated species)/LUMO (alkene) interaction (back-donation).

In the case of corner-protonated cyclopropane, donation is much more important than back-donation for the stability of this cation, whereas for cyclopropane, donation and back-donation both contribute significantly to the stability of cyclopropane (Figure 7.10b). An excited state of $:CH_2$ is used to represent bonding in cyclopropane in which the two electrons do not reside in the lowest energy sp^2-hybridized orbital but in the 2p(C) atomic orbital. In fact, this is not necessary to represent bonding in cyclopropane. When π^*(ethylene) overlaps with the sp^2 orbital of $:CH_2$, two molecular orbitals of cyclopropane are generated, one bonding MO and one antibonding MO. Similarly, when π(ethylene) overlaps with the 2p(C) orbital of $:CH_2$, two MOs of cyclopropane are generated, one bonding and one antibonding MO. One thus generates a total of two bonding MOs and two nonbonding MOs. The four electrons brought by the two fragments that combine are stabilized as they occupy two bonding MOs.

In 1947, Walsh recognize that Ag^+ complexes of alkenes can also be represented as π-complexes (Figure 7.11a), as bromonium ions (**271**, Figure 7.9) and corner-protonated cyclopropanes (Figure 7.10a) [942]. In 1953, Chatt and Duncanson adapted the Dewar π-complex theory to alkene complexes of platinum [943] and especially to the Zeise's salt $K[(CH_2{=}CH_2)PtCl_3]\cdot H_2O$ [944]. In the latter complex, the ethylene ligand is perpendicular to the plane containing Pt and the three Cl atoms [945, 946]. There is a donation HOMO(ethylene) → LUMO(Pt(II)). The LUMO(Pt(II)) is an orbital hybrid $5d6s6p^2$. Back-donation involves a nonbonding HOMO of the metal and the LUMO of ethylene. The HOMO(Pt(II)) is an orbital hybrid 5d6p (Figure 7.11b). The relative importance of donation and back-donation as defined above depends on the metal electropositivity and on the other ligands it bears. In the case of a metal/alkene complex, back-donation implies electron transfer from the metal to the empty antibonding orbital π^*. As a consequence, the C—C bond is elongated. The less electropositive is the metallic moiety, the longer is the C—C bond of the π-complex (see e.g. [945, 946]).

The π-complex theory of Dewar establishes a link between organometallic complexes and organic compounds with three-membered ring structures. These systems are seen to arise from the combination of an alkene with a monoatomic (e.g. Br^+, Ag^+, and $\ddot{\underset{..}{O}}$:) or polyatomic (e.g. Me^+, $:CH_2$, $Fe(CO)_4$, $Ni(PPh_3)_3$, $Pt(PPh_3)_3$, $Mn(CO)_5{}^+$, RhCl, and $PtCl_3{}^-$) species generating three-membered ring structures. Bonding arises mostly from the overlap of two pairs of

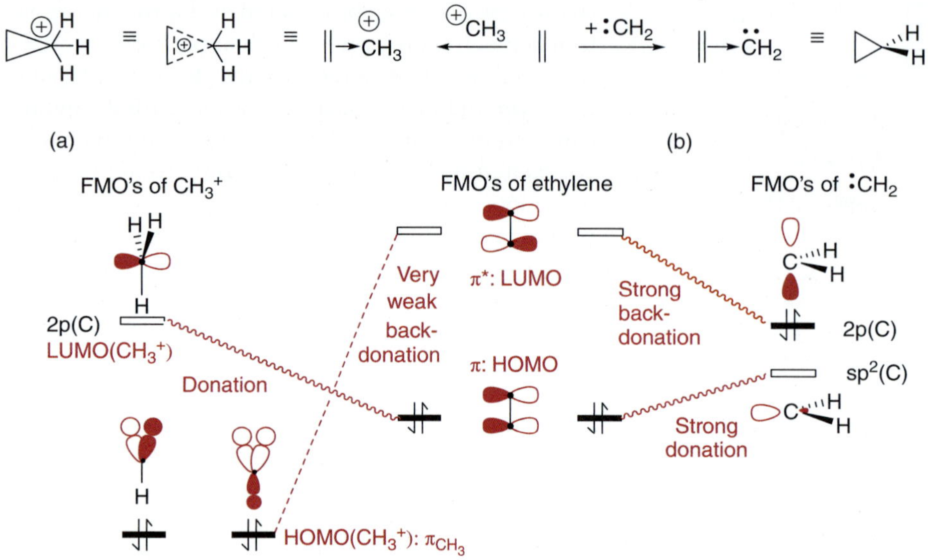

Figure 7.10 MO diagrams representing bonding (a) in c-$C_3H_7^+$ and (b) in cyclopropane.

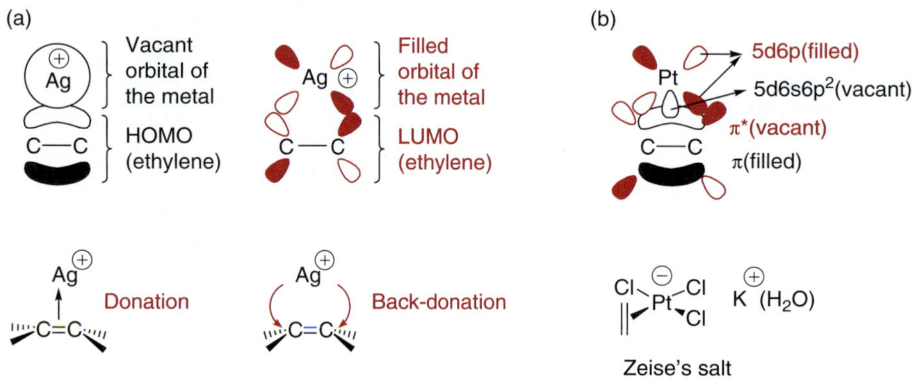

Figure 7.11 (a) Dewar's representation of the complex of Ag^+ with ethylene and (b) representation of Zeise's salt according to Chatt and Duncanson.

MOs sharing a total of four electrons. Each fragment provides two FMOs [938].

7.7.2 The isolobal formalism

Hoffmann has extended the Dewar theory on π-complexes and related structures to all organic, organometallic, and inorganic compounds [947–949]. In this formalism, any compound can be viewed as arising from the combination of two or several fragments that share electrons as the result of the overlap of suitable FMOs of these fragments. For instance, H_2 arises from the combination of two $H^•$ radicals (Figure 4.3, Section 4.4.1). In analogy, methyl and hydrogen radicals form methane (Me—H) as the result of the overlap of the FMOs $sp^3(Me^•)$ and $1s(H^•)$. As radicals $Me^•$ and $H^•$ have each an FMO that can overlap, these two fragments are said isolobal (noted: ⟵o⟶). The energy gained on combining two radicals depends on the square of the overlap between their SOMOs and the energy difference between these two SOMOs (PMO theory, Section 4.4.2). A similar treatment considering methane arising from the combination of Me^+ and H^- shows that these species are also isolobal, but their FMOs are not populated in the same way. Two methyl cations are isolobal, but their head/head combination (forming $(H_3C—CH_3)^{++}$ dication) does not generate a stable compound because they do not provide any electron to populate the bonding MO generated by the overlap of their FMOs. Furthermore, the dication is destabilized by electrostatic repulsion between the two positive charges. Note that BH_3 which is neutral, isoelectronic, and isolobal with methyl cation combines into a more stable dimeric structure B_2H_6 (diborane) that is a double σ-complex (coordination of B of one BH_3 with a B—H bond of the other BH_3; Section 7.7.4). Two carbanions (e.g. Me^-) do not give a stable compound because four electrons will have to be distributed in the $\sigma_{(C—C)}$ and $\sigma^*_{(C—C)}$ MOs and this leads to a net destabilization (Figure 4.4b, Section 4.3.2, equations (4.27) & (4.28)) that adds to the electrostatic repulsion between the two negative charges of the dianion. All alkyl, alkenyl, aryl, hetaryl, alkynyl, and halide radicals are isolobal. They all can be combined with radicals to form stable compounds. Depending on the electronegativity difference between two combining fragments, the σ-bond is more or less polar. Lewis bases, such as amines, and Lewis acids, such as BF_3, are also isolobal and are combined into stable Lewis complexes (e.g. **270**) as they have only two electrons to share. The FMOs of these species are a HOMO for the base and a LUMO for the Lewis acid.

Octacarbonyldicobalt, $Co_2(CO)_8$, first prepared in 1910 by Mond et al. by reaction of cobalt metal with CO (at 150 °C, under 30–40 atm) is a stable complex [950]. The metal used in this preparation was obtained by reduction of cobalt oxalate with H_2 (300 °C, 5 atm). A convenient laboratory synthesis of $Co_2(CO)_8$ that does not require high pressure starts with the preparation of $K[Co(CO)_4]$. The latter compound results from the passing of CO through an alkaline (KOH) suspension of $Co(CN)_2$. On acidifying, $H[Co(CO)_4]$ is obtained, which is decomposed on heating, producing $Co_2(CO)_8$ and H_2 [951, 952]. Mononuclear $Co(CO)_4$ is an unstable complex as it has 17 valence electrons, making this species analogous to hydrogen (with one electron) or alkyl radicals (with seven valence electrons at one carbon center). Thus, $Co(CO)_4$ is isolobal with these radicals and will combine with them and with itself to give stable compounds. Alkyl(tetracarbonyl)cobalt are obtained by reaction of $K[Co(CO)_4]$ or $Na[Co(CO)_4]$ with the corresponding alkyl iodides (e.g. $K[Co(CO)_4] + R{-}I \rightarrow R{-}Co(CO)_4 + KI$). Acyl(tetracarbonyl)cobalt can be obtained by a similar route (e.g. $K[Co(CO)_4] + R{-}COX \rightarrow RCO{-}Co(CO)_4 + KX$). The crystal structure of $Co_2(CO)_8$ shows two bridging carbonyl groups [953]. The IR spectra of isolated $Co_2(CO)_8$ molecules condensed from the gas phase into an argon matrix show the presence of the three forms **272–274**. The order of their relative stabilities below 77 K is **272** > **273** > **274**, with an estimated $\Delta_r G^{77\,K}$(**272** ⇄ **273**) = 6.3 kcal mol^{-1} [954].

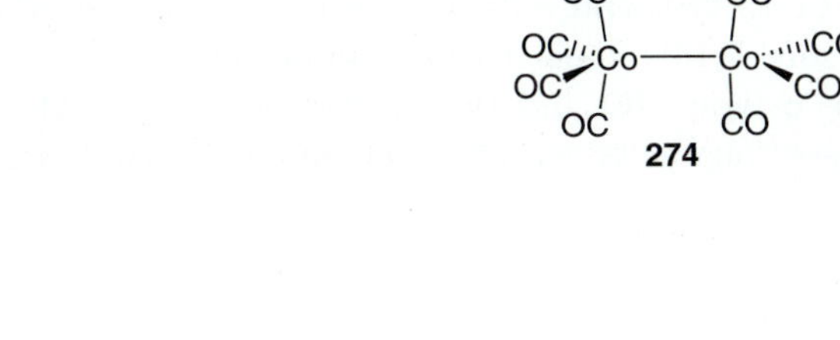

272 **273** **274**

Decacarbonyldimanganese, $Mn_2(CO)_{10}$, is the first organometallic compound of manganese to have been isolated and characterized [955, 956]. It was obtained as a volatile product from the reaction of $MnI_2 + RMgX$ under a CO atmosphere. Mononuclear complex $Mn(CO)_5$ is a 17 valence electron metallic species; thus, it is unstable and prone to combine with molecular fragments isolobal with it, i.e. with itself and with radicals. Like other radicals, it is paramagnetic. It can be generated upon UV irradiation of $Mn_2(CO)_{10}$ and be combined with H_2 ($Mn_2(CO)_{10} + H_2 \rightarrow 2HMn(CO)_5$) or with a halogen ($Mn_2(CO)_{10} + X_2 \rightarrow 2[MnCO)_5]X$). In 1927, Job and Cassel obtained hexacarbonylchromium ($Cr(CO)_6$) by reaction of $CrCl_3$ with PhMgBr under CO atmosphere. In contrast with $Co(CO)_4$ and $Mn(CO)_5$ that react quickly with air, water, and protic solvents, $Cr(CO)_6$ is diamagnetic, an octahedric compound, colorless and stable in the air. This arises from the fact that its metal atom is "saturated" with 18 valence electrons.

As we have seen (Section 7.7.1), molecular fragments such as carbenes (alkylidenes), nitrene (:NR), and molecules such as alkenes have the possibility to offer more than one FMO that can overlap with the FMOs of other species. Thus, alkylidenes have a cylindro-symmetrical sp^2 and a π-symmetrical 2p(C) FMO. Any molecular fragment or molecule that can offer FMOs capable to overlap with the FMOs of alkylidenes is said isolobal with alkylidenes (Figure 7.12). Thus, two carbenes combine into stable alkenes. Carbenes combine with carbon monoxide to form the corresponding ketenes ($R_2C{:} + {:}C{=}O \rightarrow R_2C{=}C{=}O$) and carbenes combine with alkenes to form cyclopropanes (Section 5.4.1).

The difference between carbon monoxide (:C=O) and alkylidene ligands (:CR_2) is that the former has two degenerate LUMOs that are π*(C=O), whereas the latter have one 2p(C) LUMO and a much higher lying supra-LUMO that is the antisymmetrical combination of two antibonding σ*(C—R) bond orbitals. Thus, back-donation from a carbene ligand will involve mostly the 2p-orbital which is a low-lying LUMO, lower lying than π*(C=O) of CO, and thus quite efficient for its interaction with a π-type HOMO of a partner. Any Lewis bases such as carbanions (R^-, RCO^-), anions RZ^- (Z = O, S, NR′, PR′), ethers, alcohols, water (R(R′)O:), thiols, thioethers (R(R′)S:), amines (R(R′)(R″)N:), phosphines (R_3P:), arsines (R_3As:), phosphites ($(RO)_2(O{=})P$:), carbonyl compounds (RCOR′), and carbonitriles (RC≡N:) are isoelectronic and isolobal. They combine with unsaturated transition metal species of type ML_n

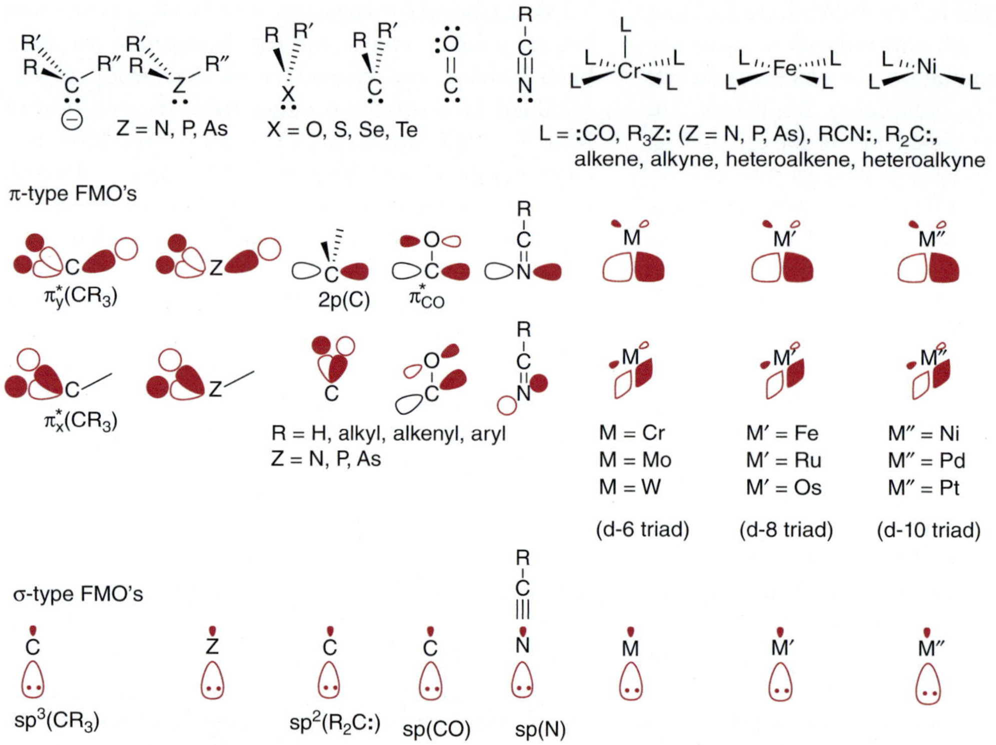

Figure 7.12 Isolobal fragments with one σ-type (cylindro-symmetrical) and one π-type (planar symmetrical) FMO.

(with 16, 14, or less valence electrons) to form complexes of type R—ML_n^-, R—CO—ML_n^-, RZ—ML_n^-, R(R′)O: → ML_n, R(R′)S: → ML_n, R(R′)(R″)Z: → ML_n ↔ R(R′)(R″)$Z^{(+)}$—$ML_n^{(-)}$ (Z = N, P, As), $(RO)_2$(O=)P: → ML_n, R(R′)C=O: → ML_n, and RCN: → ML_n, respectively. All these ligands have a σ-type HOMO and a π-type LUMO. In the case of R(R′)X: and R(R′)(R″)Z:, their LUMOs are combinations of type σ^* and are relatively high in energy. Thus, back-donation in complexes bearing these ligands will not be as important as in complexes containing alkylidene or carbonyl ligands. Carbonitriles (e.g. MeC≡N:) are analogous to :C=O as ligands. In the former, the nitrogen atom bears the σ-type HOMO, whereas in the latter, the highest lying HOMO resides mostly on the carbon center. This can be understood by considering the limiting structures :C=O ↔ $^-$C≡O$^+$. Thus, carbonitriles bind to transition metals through their nitrogen atom, whereas carbon monoxide prefers to bind the metal atom through its carbon center. Carbonyl compounds (RCHO, RCOR′, RCOOR′, RCONHR′, RCOCl, etc.,) and imines (R(R′)C=N—R″) can bind to transition metals. They are Lewis bases like amines and phosphines. They can also form π-complexes. Species such as R_3C^-, R_3N:, R_3P:, and R_3As: will not combine with alkenes or alkynes to form stable compounds as the HOMO of these partners overlap (four electron interaction leads to destabilization). There are neither LUMO(ligand)/HOMO(alkene) nor LUMO(alkene)/HOMO(ligand) stabilizing interaction that can compensate for the repulsive HOMO(ligand)/HOMO(alkene) interaction with metallic species such as ML_5, $M'L_4$, and $M''L_3$ (Figure 7.12). The σ-type orbitals of the metal are the LUMOs that interact with the HOMOs of these ligands. When alkenes and alkynes are coordinating these metallic species, donation implies HOMO(alkene, alkyne, π_{CC}) → LUMO(metal) interactions, and back-donation HOMO(metal) → LUMO (alkene, alkyne) as discussed above (Dewar–Chatt–Duncanson model, Figure 7.11).

When $Co(CO)_4$ and $Mn(CO)_5$ lose a carbonyl ligand, the unsaturated complexes $Co(CO)_3$ and $Mn(CO)_4$ are obtained, respectively. These have 15 valence electrons and can be seen as 1,1,1-triradicals like the methyne fragment that results theoretically from the homolytical cleavage of the triple C≡C bond of acetylene. Fragments $Co(CO)_3$, $Mn(CO)_4$ (with 15 VE), and :C$^•$—H (with 5 VE) are isolobal (Figure 7.13). Their FMOs can be represented by one σ-type MO and two orthogonal π-type MOs. These fragments can all be combined pairwise into

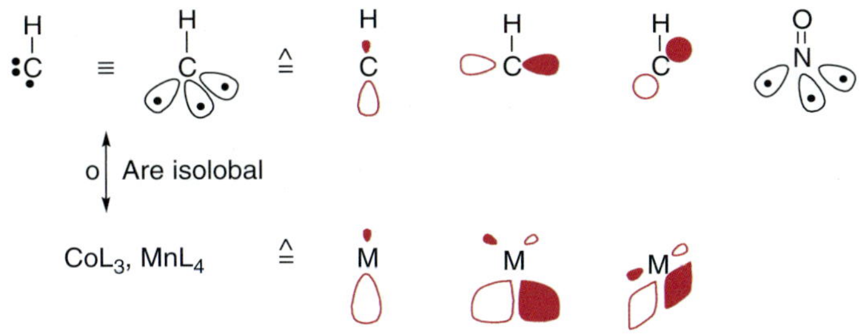

Figure 7.13 FMOs of isolobal (noted ←o→) carbyne, CoL_3, and MnL_4 fragments (L = CO, R_3N:, R_3P:, :R_3As:, RC≡N:, X_2C:, etc.).

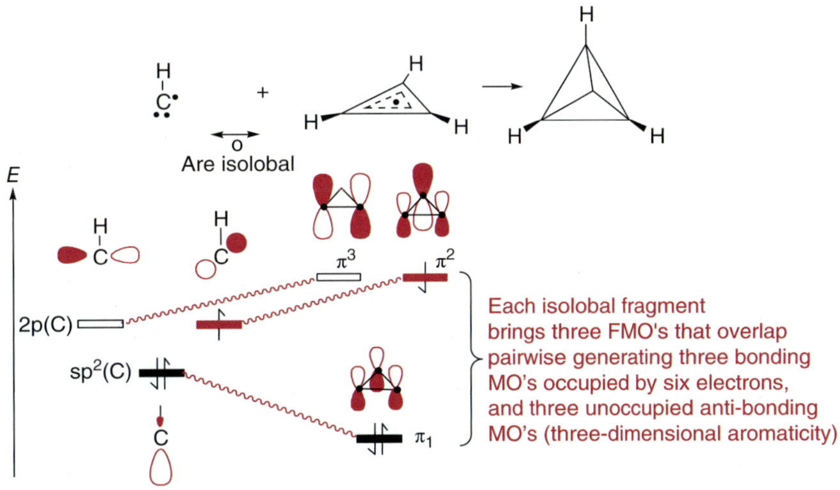

Figure 7.14 PMO diagram showing the favorable combination of a carbyne fragment with cyclopropenyl radical to give tetrahedrane.

triply bonded species, and with cyclopropenyl radical (which is isolobal with :C•—H, Figure 7.14) to generate tetrahedrical structures. Tetrahedrane is an unstable hydrocarbon that isomerizes into cyclobutadiene unless it is tetrasubstituted by large alkyl or trialkylsilyl groups (Sections 2.4.2 and 6.2.3). The same type of combinations involving $Co_2(CO)_6$ and alkynes will generate stable tetrahedrical complexes instead of 1,2-dicobaltocyclobutadiene compounds. Nitric oxide (or nitrogen oxide, or nitrogen monoxide: :N•=O) is a free radical with three unshared electron at nitrogen. It is isolobal with carbyne and thus forms stable nitrosyl complexes with metallic complexes having only 15 VE, such as $Co(CO)_3$ (e.g. $Co_2(CO)_8$ + 2NO → $2Co(CO)_3NO$ + 2CO; $Na[Mn(CO)_5]$ + NOCl → $Mn(CO)_4NO$ + NaCl) [957, 958].

Not considering ring strain in tetrahedrane, a stable species is engendered as the six electrons shared between the two molecular fragments occupy the three bonding orbitals of the tetrahedrical product. Bond angle deformations responsible of the strain in hydrocarbons almost vanish in transition metal complexes, a property arising from the more dispersed metal orbitals (s, p, d) and their larger number (9 instead of 4) compared with carbon [959–963].

7.7.3 σ-Complexes of dihydrogen

Dihydrogen, H_2, can be considered to be isolobal with ethylene or a carbene. Its FMOs σ_{H-H} and σ^*_{H-H} overlap with the $\sigma(sp^2)$ and 2p(C) FMOs of methylene, respectively, giving two binding MOs and two antibonding MOs. The four electrons available (two from H_2, two from :CH_2) occupy the two bonding MOs, thus realizing a stable compound, i.e. methane (Figure 7.15a). Similarly, H_2 forms stable complexes with 16-VE transition metal fragments generating either σ-complex noted η^2-H_2 or dihydrides (oxidative addition of the metal, Figure 7.15b) [964–968]. If donation HOMO(H_2) → LUMO(metal) dominates in the binding, the complex will be a σ-complex in which the two H atoms maintain some binding (relatively short H···H distance). Alternatively, if back-donation HOMO(metal) → LUMO(H_2) is important, this will separate both H atoms as the antibonding σ^*_{H-H} orbital gets filled by electrons, and the complex will be a dihydride as illustrated with the Vaska's dihydride resulting from the reaction of H_2 with $Ir(CO)(Ph_3P)_2Cl$ (Figure 7.15) [969–973].

In 1984, however, Kubas et al. obtained the first example of a dihydrogen σ-complex $W(CO)_3(Cy_3P)_2(\eta^2$-$H_2)$ (Kubas' complex, Figure 7.17a) from the reversible reaction of H_2 with $W(CO)_3(Cy_3P)_2$

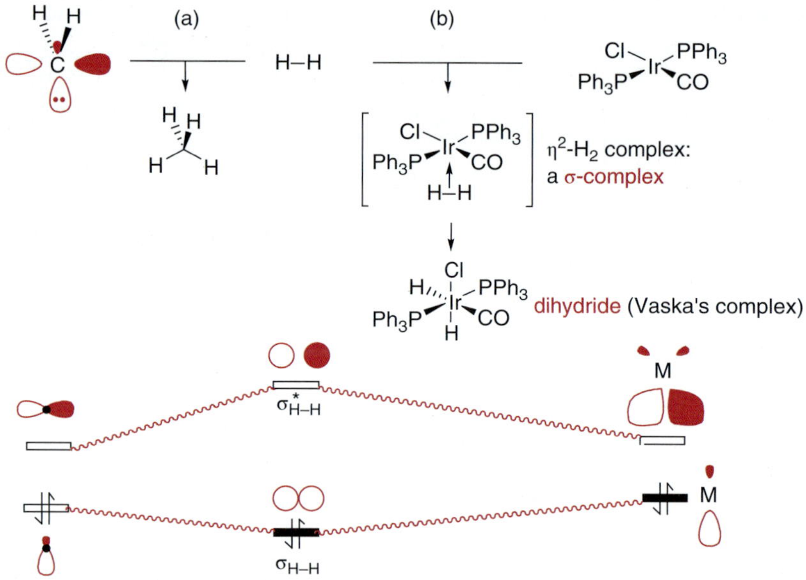

Figure 7.15 (a) H_2 is isolobal with methylene (:CH_2) and combines with it to form methane; (b) H_2 reacts with Ir(CO)(PPh_3)$_2$Cl (with 16 VE) generating the Vaska's complex probably via an intermediate σ-complex of H_2.

(a 16 VE complex, Cy = cyclohexyl). The complex retains a bond between the two hydrogen atoms (r_{H-H} = 0.84 Å [0.74 Å in H_2]; r_{W-H} = 1.75 Å) [974]. In this case, the oxidative addition that generates the corresponding dihydride W(H)$_2$ is arrested! Since this discovery, a large number of dihydrogen (η^2-H$_2$) σ-complexes have been described [975–978]. In 1985, Crabtree and Lavin obtained complex **276** by displacement of the labile aqua ligand of **275** with H_2 [979, 980]. Complex **276** contains a (η^2-H$_2$) σ-complex as well as a terminal hydride. At 300 K, the ^{1}H NMR spectrum of **276** shows a broad resonance for the two types of protons (hydride and η^2-H$_2$). At 240 K, a sharp resonance for Ir(hydride) and a broad one for Ir(η^2-H$_2$) are seen. The H–D analog of **276** showed ^1J(H,D) = 29.5 Hz (43 Hz for uncoordinated H–D). In a dihydride complex of type L$_n$M(H)(D), $^2J_{H,D} \approx 1$ Hz. Complex **276** can be deprotonated by tertiary amines or by BuLi.

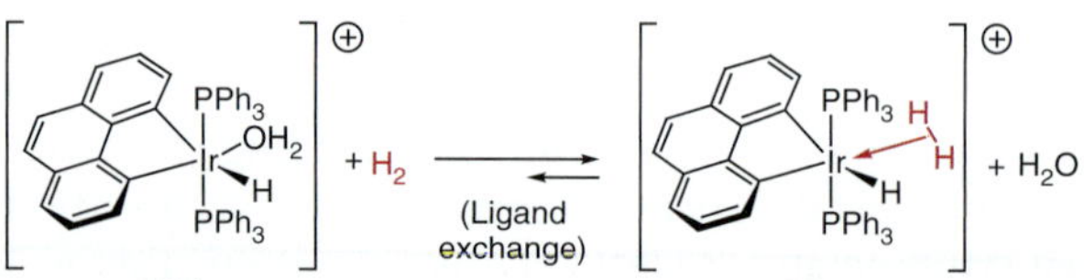

275 **276**

Distinction between σ-complexes (η^2-H$_2$) and classical dihydrides can be done by measuring T_1 relaxation times of the protons in the ^{1}H NMR spectra for complexes in solution [979, 980] or by single-crystal neutron diffraction (e.g. r_{H-H} in *trans*-Cr(CO)$_4$(Me$_3$P)(η^2-H$_2$): 0.88 Å; in Os(NH$_2$)$_4$(AcO)(η^2-H$_2$): 1.31 Å [981]. Photolysis of solutions of M(CO)$_6$ (M = Cr, W) at 195 K in the presence of H_2 gas affords the corresponding σ-complexes M(CO)$_5$(η_2-H$_2$) [979, 980]. Typical σ-complexes of H_2 have H···H distance in the range 0.8–1.0 Å, whereas dihydrides and polyhydrides have H···H distances larger than 1.5 Å. A new class of dihydrogen complexes has been recognized where the H···H distance falls between these limits. Such complexes are termed "stretched" or "elongated" dihydrogen complexes [982]. Examples are ReH$_7$[P(*p*-tolyl)$_3$]$_2$ [983] and Os(NH$_2$)$_4$(AcO)(η^2-H$_2$) [984, 985]. The study of various dihydrogen complexes and their isomerization into dihydrides is of prime importance to understand catalytic hydrogenations (Section 7.8).

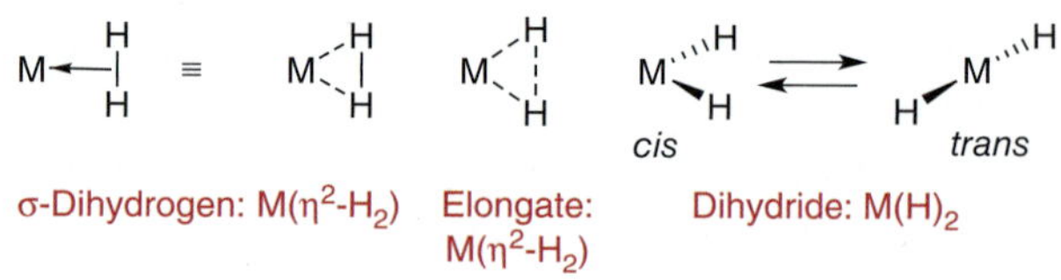

Dihydrogen in σ-complexes can be displaced by all kinds of ligands (e.g. R$_3$P:, R$_2$O:, RCN:, alkene, alkyne) including N_2. At 24 °C, the ^{1}H NMR spectrum of the Kubas' complex **277** shows a broad singlet at δ_H = −4.35 ppm for H_2, meaning that the coupling $^3J_{H,P}$ is small. A triplet at δ_H = −3.55 ppm is also seen, which witnesses of a larger $^3J_{H,P}$ coupling between this proton and the two phosphorous-31 centers of the two Cy$_3$P ligands. This signal was assigned to the dihydride **278** resulting from the oxidative addition of η^2-H$_2$ of **277**. The reaction is reversible. We are in the presence of tautomerism of the type

$M(\eta^2\text{-}H_2) \rightleftarrows M(H)_2$ [974]. Following the Tolman conventions, [922] σ-complex **277** is a saturated complex with 18 VE and with an oxidation number (or degree, sometimes called state) of 0 as it does not bear any σ-bonded ligands such as halide, alkoxide, hydride, or alkyl. In contrast, dihydride **278** bears two σ-bonded ligands. By convention, each hydrogen atom in the dihydride takes formally an electron from the metallic atom, thus conferring to it the oxidation number of +2. For that reason, reaction **277** → **278** is called an oxidative addition, even though hydrogenation is a reduction in classical chemistry. The reverse reaction **278** → **277** is called a reductive elimination (Figure 7.16a). By convention, the π-Lewis bases and ligands (e.g. alkenes and alkynes) that make π-complexes do not change the oxidation number of the metal. Similarly, η^2-coordination of σ-bonds such H—H, R—H, X—H, C—C, C—X, C—M, etc. (Figure 7.16a) do not change the oxidation number of the metal. Ligands such as acyl, alkyl, allyl, cyclopentadienyl, heptatrienyl, halide, alkoxy, and triaalkylsilyl take one electron from the metal and increase its oxidation number by +1. Ligands such as butadiene, benzene, and cyclooctatetraene are considered to be two, three, and four alkene moieties, respectively, and do not change the oxidation number of the metal. An organic model reaction for the reversible hydrogenation $W(CO)_3(Cy_3P)_2 + H_2 \rightleftarrows$ **277** $\rightleftarrows$ **278** is the reaction of methyl cation with D_2, which exchanges hydrogen and deuterium (Figure 7.16b). In the gas phase, methonium ion (CH_5^+, saturated, with an octet of VE) [986] has been represented by Olah as a σ-complex of H_2 (a species possessing formally a three-atom two-electron bond and three "classical" two-atom two-electron C—H bonds) with methyl cation adopting C_s symmetry [987, 988]. However, structures with other symmetries have similar stabilities [989] and permit the H/D exchange with D_2 [990–993]. In the gas phase, methonium ion (carbonium ion, carbon with an octet) is much more stable than methyl cation (carbenium ion, unsaturated carbon center, with a sextet of valence electrons): $\Delta_f H°(CH_5^+) = 216.3\,\text{kcal mol}^{-1}$; $\Delta_f H°(CH_3^+) = 261.3\,\text{kcal mol}^{-1}$ (Table 1.A.14),

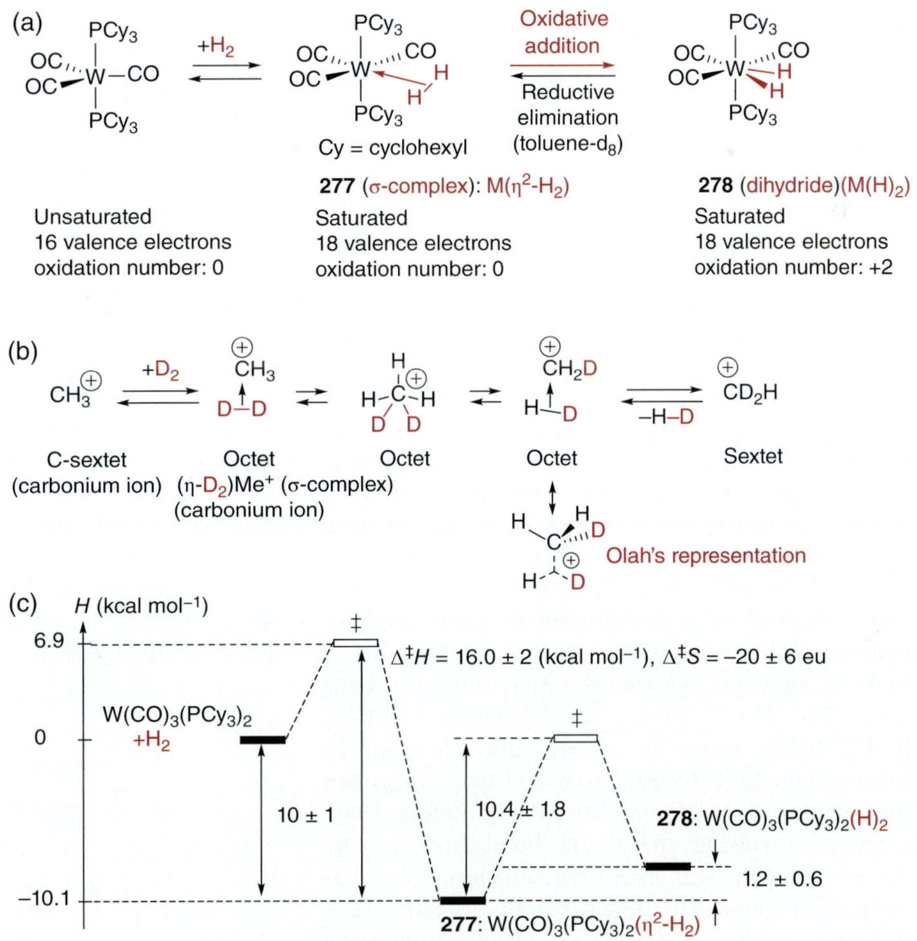

Figure 7.16 (a) Example of a reversible addition of H_2 to an unsaturated transition metal; (b) the reversible exchange of $D_2 + Me^+$ (C-sextet) $\rightleftarrows$ $CH_2D^+ + H—D$ in the gas phase implies the formation of methonium ion $CH_3D_2^+$ (C-octet); it models the hydrogenation of transition metal complexes. (c) Enthalpy diagram for the hydrogenation of $W(CO)_3(PCy_3)_2$.

$\Delta_r H°(CH_3^+ + H_2 \rightleftarrows CH_5^+) = -45$ kcal mol^{-1}; and $PA(CH_4) = 131.6$ kcal mol^{-1} (Table 1.A.15) [994, 995]. The enthalpy diagram of Figure 7.16c has been evaluated for the equilibria $W(CO)_3(Cy_3P)_2 + H_2 \rightleftarrows$ **277** $\rightleftarrows$ **278**. Complexes in which both η^2-H_2 and dihydrides reside simultaneously on the same metallic center have been reported for compounds such as $(Ph_2PEt)_3Fe(\eta^2_2\text{-}H_2)(H)_2$, and $(Ph_3P)_3Ru(\eta^2\text{-}H_2)(H)_2$ [996–998].

7.7.4 σ-Complexes of C—H bonds and agostic bonding

Hydrides such as BH_3 [999, 1000], AlH_3 [1001–1003], GaH_3 [1003], and InH_3 [1004] are Lewis acids that exist in equilibria with dimers and oligomers in which the M—H bond makes σ-donor complexes with the metallic center of another hydride [1005, 1006]. For instance:

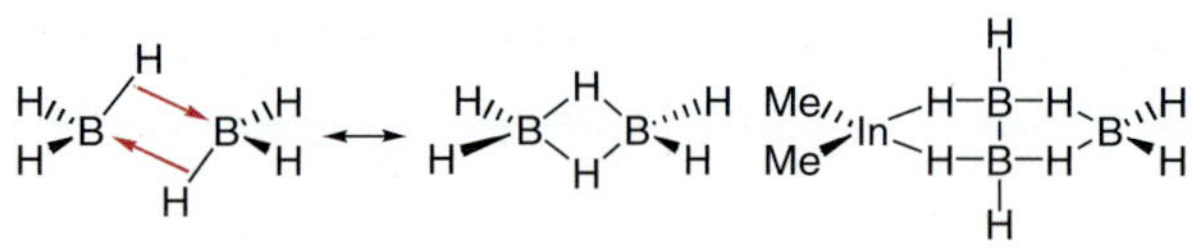

The B—H bond in borane does not differ significantly from the C—H bond; thus, C—H bonds are expected to form σ-complexes with coordinately unsaturated transition metal complexes. This make possible the functionalization and/or substitution of unactivated C—H bonds of alkanes [1007–1013]. Alkanes are called "paraffins," meaning not enough affinity. These compounds are relatively inert because their C—C and C—H bonds are strong and highly localized; they have no empty orbitals of low energy, or filled orbitals of high energy that could readily participate in a chemical reaction, as for alkenes and alkynes. In classical chemistry, the breaking of a C—H bond requires the attack by a reactive species such as radicals, carbenes, superacids, or carbenium ions. Activated C—H bonds [1014, 1015] can undergo hydrogen/metal exchange with Grignard or organolithium reagents (Gilman hydrogen/lithium exchange reactions) [340, 1016]. In 1898, Dimroth reported the mercuration of benzene ($Ph\text{—}H + Hg(OAc)_2 \rightleftarrows PhHgOAc + AcOH$) [1017, 1018], which is an aromatic electrophilic substitution. In 1932, Kharasch and Isbell extended the reaction to gold reagents [1019]. Before 1980, reactions involving the C—H bond breakage by transition metal complexes were known only for aromatic C—H bonds [1020] (e.g. reaction (7.83) [1021]; (Cp = η^5-cyclopentadienyl) or for intramolecular reactions where the group containing the reacting C—H bond is attached to the metal (e.g. reaction (7.84) [1022] and (7.85) [1023]). In 1969, Shilov and coworkers showed that H/D exchange is possible with alkanes in D_2O/AcOD solution, including CH_4, in the presence of K_2PtCl_4 catalyst [1024, 1025].

$$Cp_2Ti(H)_2 + PhH \xrightarrow{h\nu} Cp_2Ti(Ph)(H) + H_2 \tag{7.83}$$

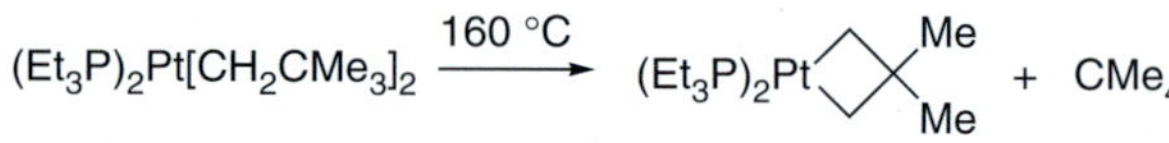

(7.84)

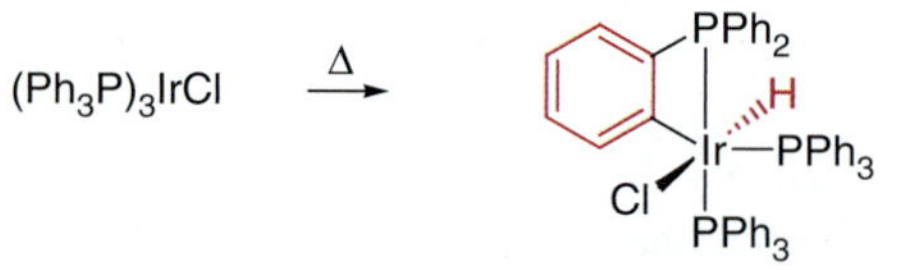

(7.85)

As for the oxidative addition of H_2 with a coordinately unsaturated transition metal complex that involves a pre-equilibrium forming a σ-complex, the oxidative addition of a C—H bond might also imply a more or less stable σ-complex with the metal. The first direct observation of σ-complexes of alkanes was made by Turner and coworkers with photochemically generated $M(CO)_5$ (M = Cr, W, Mo) fragments in alkanes and alkane/rare gas matrices [1026–1031]. Any C—H bond is in fact isolobal and isoelectronic with the H—H bond and thus must generate σ-complexes. The HOMO(C—H) → LUMO(metal) donation and HOMO(metal) → LUMO(C—H) back-donation are not highly stabilizing because of the relatively large energy gap between these FMOs. Nevertheless, these complexes should be observable at low temperature. One of the earliest examples of such σ-complex is **279** in which the β-C—H bond from a coordinated ethyl group interacts with the titanium center [1032]. For this type of intramolecular σ-complex, one speaks of "agostic" bonding [1033–1037] (for α-C—H agostic interactions, see [1038]). This suggested that intermolecular C—H activation of alkanes might be possible [1039, 1040].

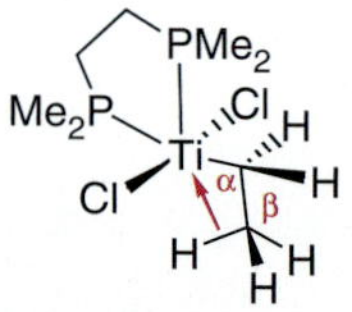

279: agostic bonding; i.e.: intramolecular σ-complex with a β C–H bond

In 1982, Janowicz and Bergman reported a first case with reaction (7.86). Irradiation of dihydride **280** generated a 16 VE iridium complex intermediate [1041] that is able to undergo intermolecular oxidative addition with one C—H bond of cyclohexane producing the corresponding cyclohexyl(hydride) complex **281** [1042].

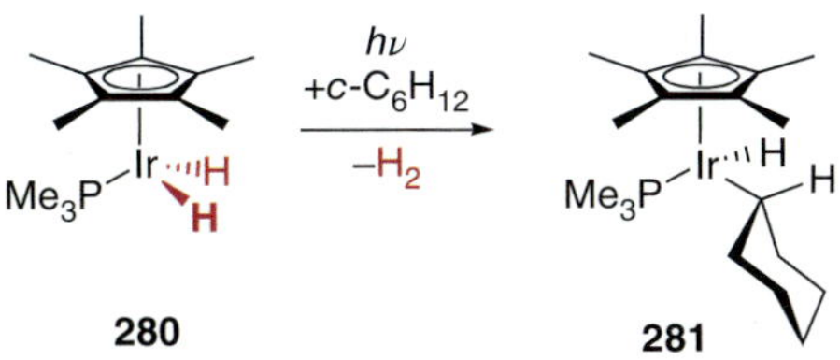

(7.86)

The same year, Hoyano and Graham reported the photoreaction $Cp^*Ir(CO)_2(H)_2$ + neopentane → $Cp^*Ir(CO)(H)CH_2CMe_3 + H_2$ [1043]. This was followed by many more similar discoveries of intermolecular oxidative additions of transition metal complexes onto C—H bonds of alkanes [1044–1047]. For instance, irradiation of $Cp^*Ir(CO)_2$ (Cp^* = pentamethylcyclopentadienyl) in liquid krypton at 193 K in the presence of neopentane generates first $Cp^*Ir(CO)(Kr)$ characterized by its IR spectrum [1048]. The latter complex reacts with neopentane forming the σ-(C—H)-complex $Cp^*Ir(CO)(\eta^2\text{-}Me_4C)$. This complex is slowly transformed into (alkyl)hydride isomer $Cp^*Ir(CO)(H)CH_2CMe_3$ [1049]. The same sequence of reactions has been observed with propane and cyclohexane. In this latter case, the energy barrier for the oxidative addition converting the σ_{C-H} complex $Cp^*Rh(CO)(\eta^2\text{-}c\text{-}C_6H_{12})$ into the corresponding (cyclohexyl)hydride isomer amounts to 4.6 kcal mol^{-1} only. Contrary to hydrogen abstractions by radicals ($R—H + X^\bullet \rightleftarrows R^\bullet + HX$) for which tertiary alkane C—H is more reactive than primary alkane C—H and sp^2(C)—H bonds, the oxidative additions follow the reactivity sequence Ar—H > H—CH_2R > H–cycloalkyl [1050, 1051]. However, in the case of photolysis of [(tris-3,5-dimethylpyrazolyl)borate] $Rh(CNCH_2CMe_3)(PhN{=}C{=}NCH_2CMe_3)$ in alkane mixtures, it was found that additions are about 1.5 times faster for secondary than for primary C—H bonds [1052]. Isotope exchange provides evidence for the role of the metastable C—H σ-complexes as shown with equilibria (7.87) [1053].

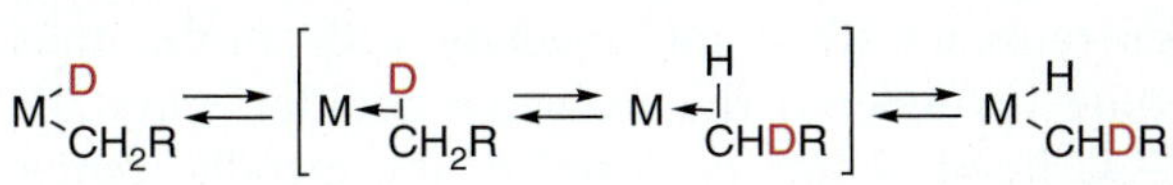

(7.87)

Because it has been shown that complete loss of alkane from the metallic center and subsequent readdition does not take place, exchange must proceed through σ-complexes [1026, 1027]. Bond strength in C—H σ-complexes varies between 5 and 11 kcal mol^{-1}. It depends on the ionization energy of the alkane. For the irradiation of $Cp^*Rh(CO)_2$ in neopentane/Kr mixtures, Bergman and coworkers established that $\Delta_r G^{165\,K}(Cp^*Rh(CO)(Kr) + Me_4C \rightleftarrows Cp^*Rh(CO)(\eta^2\text{-}Me_4C) + Ar) = -2.5$ kcal mol^{-1} and $\Delta^\ddagger G^{165K}(Cp^*Rh(CO)(\eta^2\text{-}Me_4C)$ $Cp^*Ir(CO)(H)(CH_2CMe_3)) = 5.9$ kcal mol^{-1} [1049, 1054]. George and coworkers have studied the σ-complex $CpRe(CO)_2$(*n*-heptane) by time-resolved infrared spectroscopy at 25 °C and have found this complex to have the longest lifetime measured for any known σ-alkane complex at room temperature [1055]. The photolysis of $CpRe(CO)_3$ in neat cyclopentane at −80 °C resulted in the formation of $CpRe(CO)_2(c\text{-}C_5H_{10})$ [1056]. Apart from the "side-on" C—H σ-complexes considered above that are called η^2-C,H-complexes, other types of σ-complexes can be formed between an alkane and a metallic species. A metal can coordinate two C—H bonds, giving η^2-H,H σ-complexes, or three C—H bonds, giving η^3-H,H,H σ-complexes. Alternatively, if the alkane binds in an end-on manner with only one hydrogen atom, η^1-H σ-complexes are formed. The study of C—H complexes of transition metal is essential to understand the mechanisms of reactions that involve C—H activation by transition metal catalysts (Section 8.3). Silane Si—H bonds can also form σ-complexes with transition metallic species [1057].

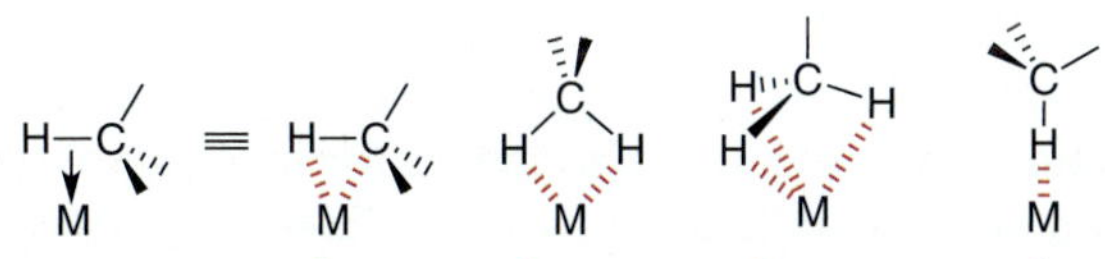

7.7.5 σ-Complexes of C—C bonds and C—C bond activation

The σ(CC) bond in alkanes is isoelectronic and isolobal to the σ(H—H) and σ(C—H) bonds; thus, unsaturated transition metallic species are expected to form σ-complexes with single C—C bonds. These σ-complexes may then undergo oxidative addition to generate the corresponding dialkyl complexes, a possible route for the C—C bond activation in alkanes by transition metals (reaction (7.88)) [1058–1065].

Table 7.5 Homolytic bond dissociation enthalpies of selected transition metal complexes $DH°(M^•/X^•) = \Delta_r H°(MX \rightleftarrows M^• + X^•)$, in kcal mol^{-1} (1 cal = 4.184 J).[a)]

M	X	$DH°(M^•/X^•)$	M	X	$DH°(M^•/X^•)$
$Ti(CH_2CMe_3)_3$	CH_2CMe_3	47.3	$Mo(Cp)_2Me$	Me	39.7 ± 2
$Ti(CH_2SiMe_3)_3$	CH_2SiMe_3	60.5	$Mo(Cp)_2Bu$	Bu	36.8 ± 3
$Ti(Bn)_3$	Bn	48.0	$Mo(Cp)(CO)_3$	H	67.4 ± 1
$Ti(Cp)_2Me$	Me	71.2	$Mo(Cp)(CO)_3$	Et	44.2 ± 3
$Ti(Cp)_2Ph$	Ph	79.3	$W(Cp)_2H$	H	74.3 ± 1
$Ti(Cp^*)_2Me$	Me	67.2 ± 2	$W(Cp)_2Me$	Me	52.8 ± 1
$Ti(Cp^*)_2Ph$	Ph	66.9 ± 5	$Re(CO)_5$	Me	52.6
$Zr(CH_2CMe_3)_3$	CH_2CMe_3	59.5	$Mn(CO)_5$	H	58.6 ± 2
$Hf(CH_2CMe_3)_3$	CH_2CMe_3	63.6	$Mn(CO)_5$	Me	44.7 ± 1
$Zr(Cp^*)_2H$	H	81.0	$Mn(CO)_5$	Ph	49.5 ± 3
$Zr(Cp^*)_2Me$	Me	67.9	$Fe(CO)_4H$	H	65.0
$Zr(Cp^*)_2Ph$	Ph	74.6	$Fe(CO)_4$	C_2H_4	36.3 ± 4
$Hf(Cp^*)_2H$	H	82.7	$Co(CO)_4$	H	54.3
$Hf(Cp^*)_2Me$	Me	73.1	$CpCr(CO)_3$	H	61.7
$Hf(Cp^*)_2Bu$	Bu	65.5	$CpMo(CO)_3$	H	65.2
$Mo(Cp)_2H$	H	61.4 ± 2	$CpW(CO)_3$	H	81.0

Cp = η^5-*c*-C_5H_5, Cp* = η^5-*c*-Me_5C_5.
a) Taken from Ref. [1066–1068].

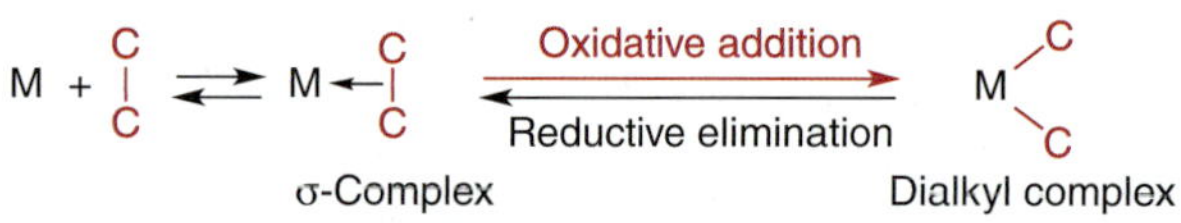

(7.88)

Differences in bond dissociation energies for C—C and C—H bonds in alkanes favor the initial cleavage of C—C bonds under thermal conditions (e.g. $DH°(Me^•/Me^•) \cong 89.6$ kcal mol^{-1}, $DH°(Et^•/H^•) \cong 100.7$ kcal mol^{-1}, Table 1.A.7). Similarly, bond dissociation energies are much smaller for M–alkyl bonds than for metal–hydride bonds (Table 7.5) [1066–1068]. The directional nature of the sp^3-hybridized orbitals of the SOMO of the alkyl radicals prevents optimal overlap with the FMOs of the transition metal, whereas the 1s SOMO ($H^•$) ensures a much more isotropic electron distribution and thus a more favorable SOMO ($H^•$)/FMO(metal) overlap [1069–1071].

Kinetics favor C—H vs. C—C activation with transition metals [1072]. The product distribution of the dissociative chemisorption of propane with close-packed Ir(111) surface at 600 K is 18.5% activation via C—C bond cleavage to yield chemisorbed methyl and ethyl groups, 34.8% activation via secondary C—H bond cleavage to form isopropyl intermediates, and 46.7% activation via primary C—H bond cleavage to form *n*-propyl-Ir products [1073]. The unfavorable thermodynamics of the C—C bond oxidative additions can be overcome by choosing strained cycloalkanes such as cyclopropane and cyclobutane [1065]. For instance, UV irradiation at −60 °C of dihydride $Cp^*Rh(H_2)PMe_3$ in cyclopropane gives first Cp*Rh(H)(cyclopropyl)PMe_3, which is isomerized into rhodiacyclobutane $Cp^*Rh(CH_2—CH_2—CH_2)PMe_3$ at −10 °C [1074]. In this case, the direct coordination of a σ(C—C) bond by the unsaturated rhodium species does not occur and the C—H oxidative addition of cyclopropane is favored kinetically (and thermodynamically). Direct C—C bond attack by a rhodium complex is the initial step of the Rh(I)-catalyzed rearrangement of (cyclopropylmethoxy)trimethylsilane into (2-methylallyloxy)trimethlysilane [1075]. As we shall see, this type of C—C activation is applied in transition metal catalyzed reaction (Section 8.4.1). Milstein and coworkers developed pincer-type chelating ligands to cleave strong aryl C—C bonds under mild conditions (e.g. reaction (7.89)) [1076–1081]. Ligand exchange between the bidentate ligand substituting two ethylene ligands in $[Rh(CH_2{=}CH_2)_2Cl]_2$ generates an Rh(I) intermediate with 14 VE, thus prone to undergo the oxidative addition onto the C—C bond, which is closed to the metallic center [1080, 1082].

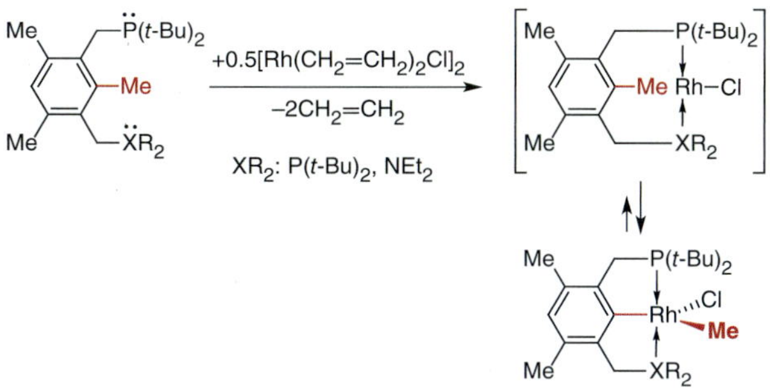

(7.89)

The C—C bond activation involves probably pre-equilibria with the formation of σ-complexes. Such complexes have been identified. They imply C—C agostic interactions [1064] as illustrated with **282** [1083] and **283** [1084]. In the latter complex, rhodium makes an agostic interaction with one C—C bond of a cyclopropane moiety, and an agostic interaction with a γ-C—H bond of one of the isopropyl groups. Upon coordination, the C—C bond is elongated.

282: 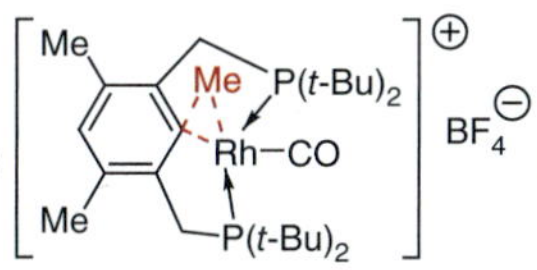

283: 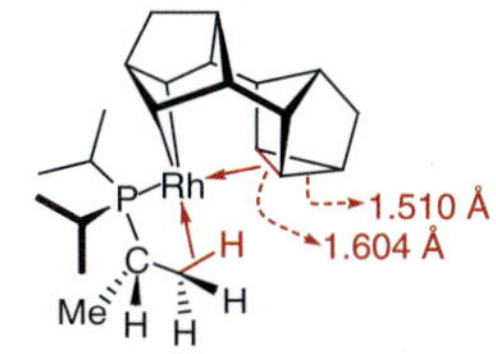

Suresh and Koga suggested that two C—C agostic bonds and a carbene-like C(3) carbon to be present in the ruthanacyclobutane intermediate of the migratory insertion step in the dissociative olefin metathesis reaction catalyzed by Grubb's catalysts (Section 8.5) [1085]. This hypothesis is supported by quantum mechanical calculations [1086] and NMR studies on titanium complexes [1087]. This is reminiscent to the structure of cyclobutyl cation that must be seen as bicyclobutonium ion equilibrating with cyclopropylmethyl cation [1088–1093]. The unsaturated carbenium center of the cyclobutyl cation coordinates the C(3) carbon just like the unsaturated ruthenium metal in rutheniacyclobutane interacts with the two σ(C—C) bonds joining the C(3) center of the ruthenacyclobutane. A LUMO(Ru)/HOMO[σ(C(2)—C(3)) − σ(C(3)—C(4))] interaction is responsible of this C,C,C-agostic bonding.

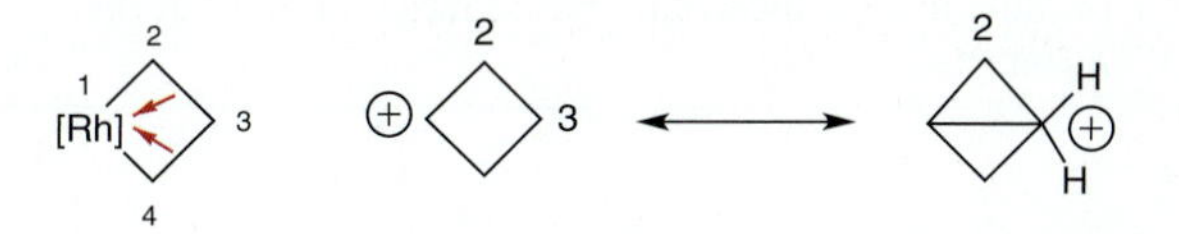

Rhodiacyclobutane Cyclobutyl cation ⟷ 1H-bicyclobutonium ion

7.7.6 Reactions of transition metal complexes are modeled by reactions of organic chemistry

In Sections 7.7.7–7.7.14, we shall present the fundamental reactions of transition metal complexes and show how they are modeled by reactions in organic chemistry [941]. A summary is collected in Table 7.6. These reactions will be illustrated with important chemical processes using transition metal complexes as catalysts. Chapter 8 is devoted to transition metal catalyzed C—C bond forming reactions in bulk and fine chemistry. Knowledge of the fundamental reactions described below permits rationalization of these catalyzed reactions at the molecular level.

7.7.7 Ligand exchange reactions

Usually, ligand exchange is the first step in the synthesis of a transition metal complex. It is also the beginning and the end of a process catalyzed by a transition metal complex. Ligand exchange (or ligand substitution) compares with substitutions at the carbon atom in organic chemistry. They can be catalyzed by Lewis acids [1094]. There are two limiting mechanisms: the dissociative (D) mechanism that is analogous to the S_N1 heterolysis and the associative (A) mechanism that resembles the S_N2 nucleophilic substitution. As for the nucleophilic displacements in organic chemistry, further mechanisms involving thermal SET or photochemical electron transfer (PET) must also be considered. The SET mechanism can be catalyzed by additives ready to take or give an electron to the reacting complex (e.g. Al_2Br_6 is an electron acceptor; $Me_3N \rightarrow O = Me_3N^{(+)}—O^{(-)}$, DMSO = $Me_2S{=}O$ = $Me_2S^{(+)}—O^{(-)}$ are electron donors). Redox effects at electrodes must also be considered [1095]. Complexes of Pt(II), Pd(II), Ni(II), Au(III), Rh(I), and Ir(I) have been studied abundantly for their ligand exchange reactions. In general, those with 16 valence electrons follow the associative mechanism [1096, 1097]. In some cases, saturated transition metal complexes (18 valence electrons) might also be substituted through an associative mechanism [1098, 1099]. For instance, complexes containing ligands that can change their hapticity by ligand slippage undergo ligand substitution following associative mechanisms [1100–1102]. Three typical examples are given with reactions (7.90–7.92) [1103, 1104].

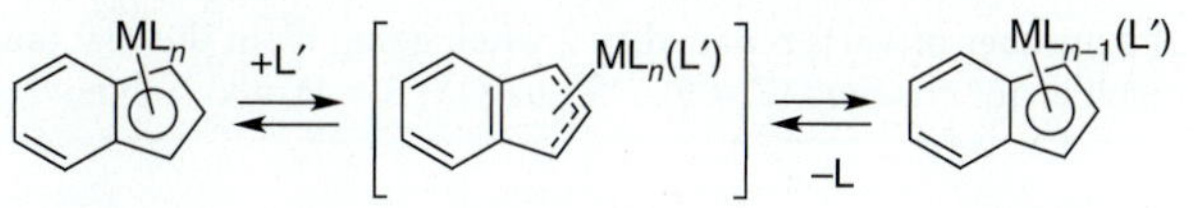

η5-Indenyl η3-Indenyl η5-Indenyl

(7.90)

Table 7.6 Fundamental reactions of transition metal complexes and comparison with organic reactions. Tolman's rules are applied to define (conventionally) the oxidation number *x* of the metal, M (VE: number of valence electrons).

1. Ligand exchange (Section 7.7.7)

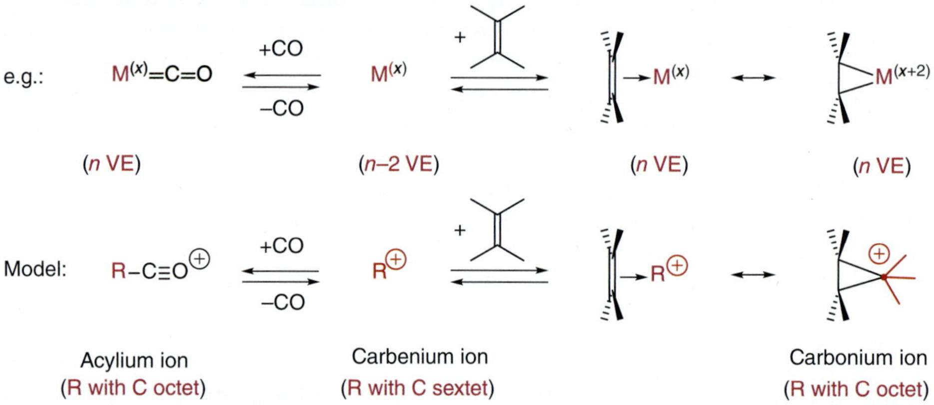

No change of valence electron count and oxidation number between the metal center of reactant [M]=CO and product [M](π-alkene) by convention

2. Oxidative addition/reductive elimination (Section 7.7.8)

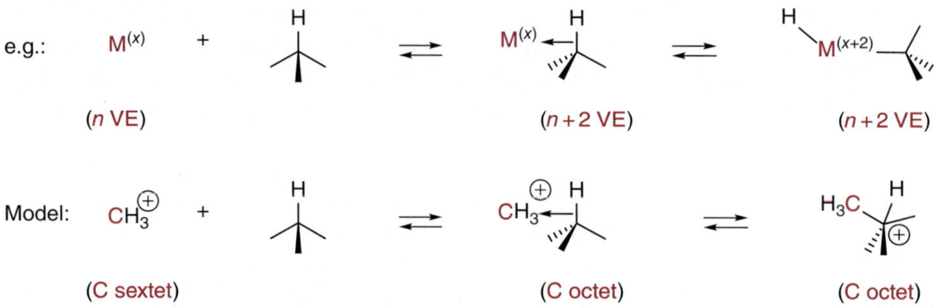

Oxidative addition requires an unsaturated metallic species (with 16, 14 VE). The metal oxidation number is increased by +2 and the metal of adduct (double σ complex) has two more VE than the metal in the starting complex. Note that unsaturated [M] can add to any kinds of X—Y bonds reversibly. Addition of carbenes R_2C: (C with a sextet) to X—H bonds generates stable compounds $R_2C(X)$—H with an octet carbon center. The reverse reaction is an α-elimination

3. α-Insertion/α-elimination (Section 7.7.9)

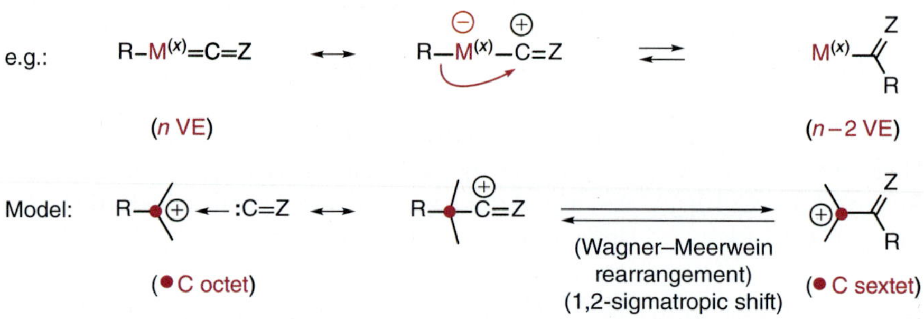

The number of VE is reduced by 2 when going from the reactant to the product in an α-insertion; the oxidation number of the metal is not changed (Z = O, NR′, C(X)Y; R = H, alkyl, alkenyl, aryl, R′O, Hal, etc.)

Table 7.6 (Continued)

4. β-Insertion/β-elimination (Section 7.7.10)

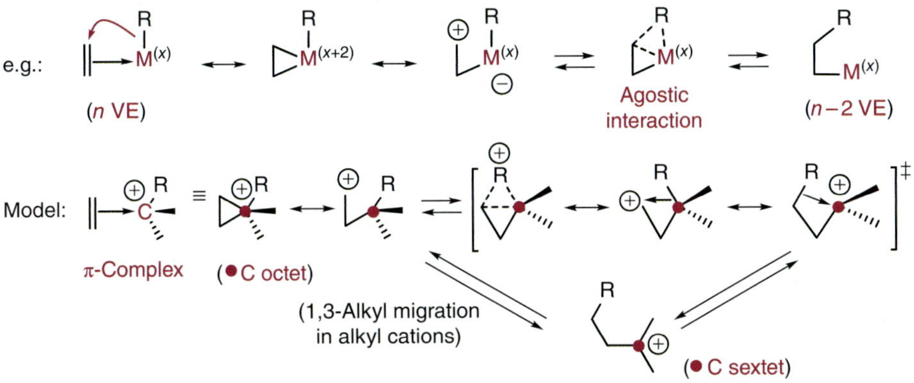

The number of VE is reduced by 2 when going from reactant to product in the β-insertion; the oxidation number of the metal is not changed (R = H, alkyl, alkenyl, aryl, R′O, Hal, etc.)

5. α-cycloinsertion/α-cycloelimination (Section 7.7.11)

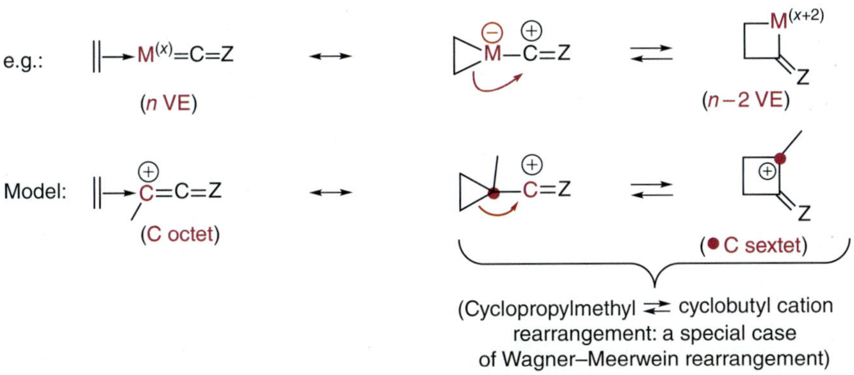

The number of VE is reduced by 2 when going from reactant to product in the α-cycloinsertion; the oxidation number of the metal is increased by +2

6. Oxidative cyclization/reductive fragmentation (β-cycloinsertion/β-cycloelimination: Section 7.7.14)

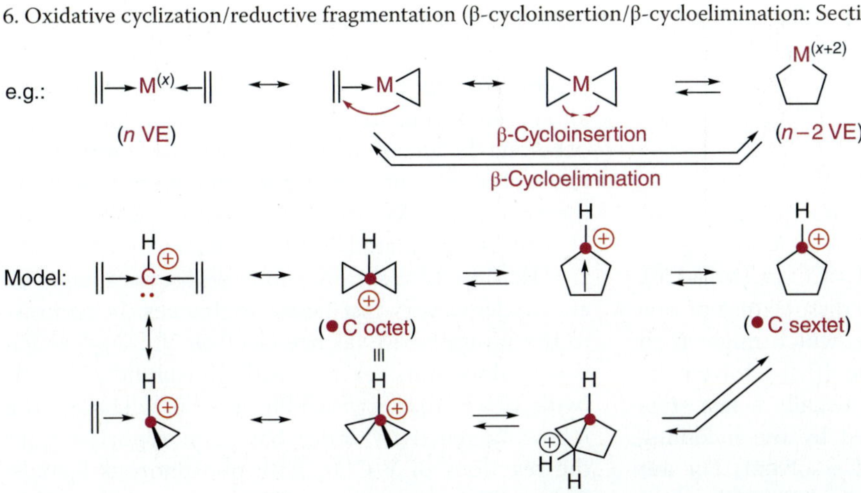

The number of VE is reduced by 2 when going from reactant to product in the oxidative cyclization and the oxidation number of the metal increases by +2. The metal-catalyzed cycloinsertions can combine other unsaturated reactants than alkenes

(continued)

Table 7.6 (Continued)

7. Reactions at the coordinated ligands

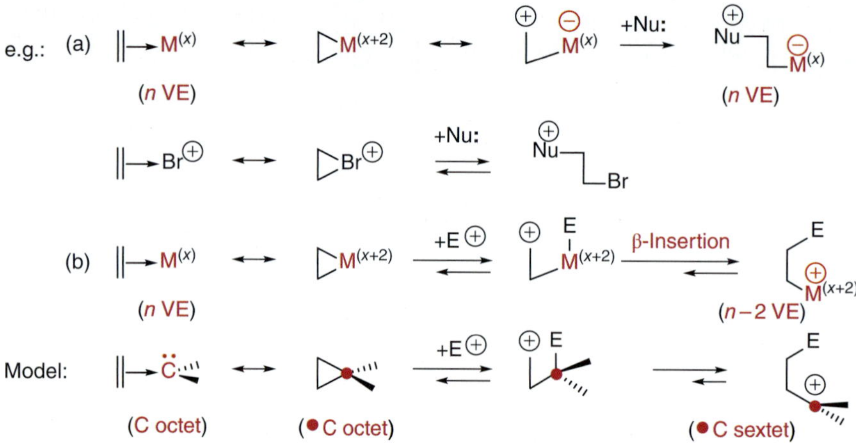

$[M] = M^{(x)}$ for a metal bearing eventually other ligands, including solvent molecules.

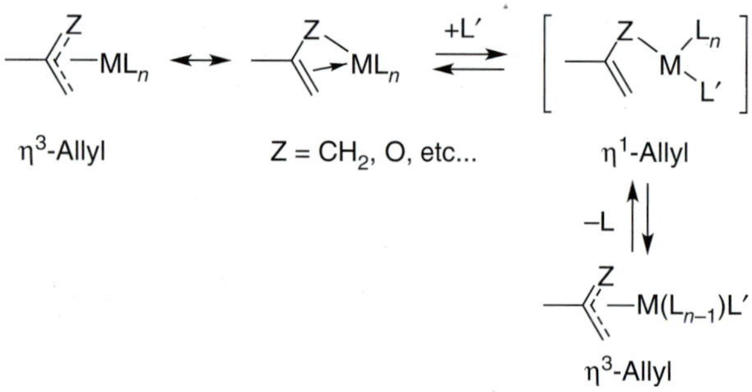

(7.91)

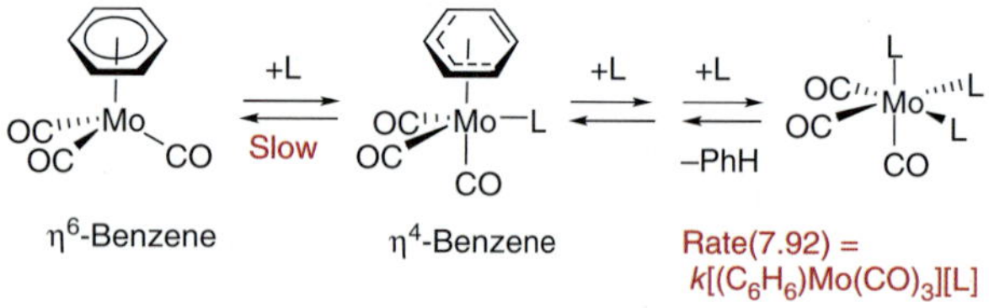

(7.92)

Saturated complexes such as $Ni(CO)_4$ (more general: [M]=CO) undergo ligand exchange following dissociative mechanisms of type $[M]{=}CO \rightleftarrows [M] + CO$. The rate-determining step is the dissociation of one CO giving $Ni(CO)_3$ intermediate, which might then coordinate one solvent molecule ($[M] + \text{solvent} \rightleftarrows [M](\text{solvent})$). In a second step, usually a fast step, the solvent molecule is displaced by the incoming ligand L ($[M](\text{solvent}) + L \rightleftarrows [M]L + \text{solvent}$). For stable complexes such as $Mo(CO)_6$ and $W(CO)_6$, their dissociation activation enthalpies ($\Delta^{\ddagger}H$) are 31.7 and 39.9 kcal mol^{-1}, respectively, to be compared to the heats of dissociation $\Delta_r H°(Mo(CO)_6 \rightleftarrows Mo(CO)_5 + CO) = 41 \pm 3$ kcal mol^{-1} and $\Delta_r H°(W(CO)_6 \rightleftarrows W(CO)_5 + CO) = 46 \pm 3$ kcal mol^{-1}. The rate law of these ligand exchanges is given by $-d[M(CO)_6]/dt = (k_1 + k_2[L])[M(CO)_6]$. It implies that the incoming ligand L activates the predissociation of the saturated $M(CO)_6$ complexes, either through an associative mechanism or through an SET mechanism. Rates of substitution of CO in $M(CO)_6$ depend considerably on the metal. Reaction $Cr(CO)_6 + PPh_3 \rightarrow Cr(CO)_5(PPh_3) + CO$ requires heating to 130 °C. The 18 VE anion $V(CO)_6^-$ resists to substitution even in molten PPh_3 [1105]. For $M(CO)_6$ where M = Cr, Mo, and W, the observed order of reactivity does not parallel the order of M—C bond energies, which is W > Mo > Cr. Ligand exchange can be induced photochemically [1106, 1107]. In the case of $M(CO)_n$, the photodissociation is analogous to the Norrish type I photoreaction of ketone that generates CO and two alkyl radicals and to the photolysis of ketene that generates H_2C: and CO (Section 6.4.1). As already seen with reaction (7.86), H_2 can be eliminated from dihydrides by photolysis [1041]. Most 17-electron complexes are paramagnetic and undergo very fast ligand exchanges. In contrast to the saturated 18 valence electron $V(CO)_6^-$ which is very slow in its reaction with phosphines, $V(CO)_6$ (with 17 VE) reacts with PPh_3 at −70 °C [1108]. The following reactivity order has been reported from the reactions of $V(CO)_6$ with phosphorous ligands: $PMe_3 > PBu_3 > P(OMe)_3 > PPh_3$. As negative activation entropy (c. −28 eu) is found for ligand exchange reaction $V(CO)_6 + Ph_3P \rightarrow V(CO)_5(Ph_3P) + CO$, an associative mechanism must be followed [1109–

Scheme 7.48 Ligand exchange reaction catalyzed by a radical ($X^\bullet$).

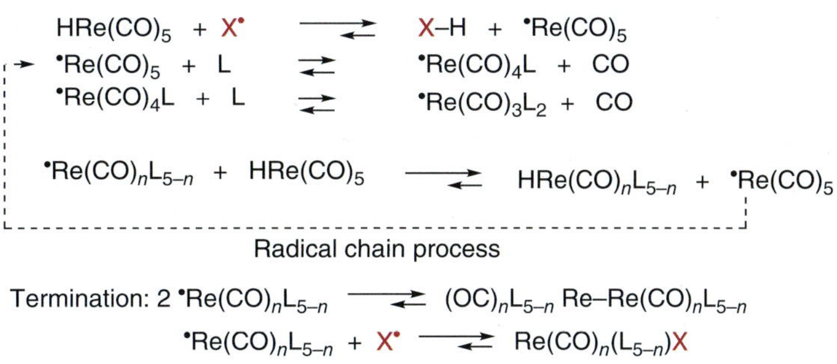

1111]. This reaction is about 10^{10} times as fast as substitution $Cr(CO)_6 + Ph_3P \rightarrow Cr(CO)_5(PPh_3) + CO$ in which $Cr(CO)_6$ is a 18 VE complex. Ligand exchanges can be strongly accelerated by SET. The substitution of MeCN ligand for PPh_3 is catalyzed at the platinum electrode [1112]. Some ligand exchange reactions involve radical intermediates (chain process). An example is given in Scheme 7.48 for substitution of carbonyl group in $HRe(CO)_5$ by ligands L.

7.7.8 Oxidative additions and reductive eliminations

As seen for the reactions of H_2 and alkanes, an unsaturated (14, 16 VE) transition metal complex ML_n (= [M]) can add across single H—H (Section 7.7.3), C—H (Section 7.7.4), and C—C bonds (Section 7.7.5). This can be generalized to any reagent A–B that reacts with unsaturated complex [M] to form a σ-complex first and then an adduct [M](A)(B) (reaction (7.93)). According to Tolman's rules, [922] M—A and M—B bonds take each one electron from the metal, thus increasing its oxidation number by +2. As each σ-ligand A and B in [M](A)(B) can be considered as an hydride, alkyl anion, or halide anion, each contribute by two VE in the final complex [M](A)(B). In the latter reactions, associative mechanisms are followed. They imply retention of configuration at the centers of A and B attached to the metal [1113, 1114].

$$\underset{(14,\,16\ \text{VE})}{[\text{M}] + \text{A–B}} \underset{\text{Reductive elimination}}{\overset{\text{Oxidative addition}}{\rightleftharpoons}} \underset{(16,\,18\ \text{VE})}{[\text{M}]\langle^{\text{A}}_{\text{B}}} \qquad (7.93)$$

Kinetic studies on the formation of Vaska's complex $(Ir(CO)(Ph_3P)Cl + H_2\ Ir(CO)(Ph_3P)(H)_2Cl)$ (Figure 7.15b, Section 7.7.3) in several aromatic solvents gave second-order rate laws: rate = $k_{obs}[Ir(CO)(PPh_3)_2(Cl)][H_2]$ with $\Delta^{\ddagger}H$ = 10–12 kcal mol^{-1} and $\Delta^{\ddagger}S$ = −20 to −30 eu in agreement with early transition states ($\Delta_r H° \approx -16$ kcal mol^{-1}) [1115].

Stoutland and Bergman have shown that heating of $Cp^*Ir(PMe_3)$(cyclohexyl)(H) to 130 °C generates a 16 VE iridium species that reacts with ethylene. Reaction (7.94) involves first an oxidative addition of one of the C—H bonds of ethylene giving a metal(vinyl)hydride adduct, which then undergoes a thermal rearrangement into the π-complex [1116]. This observation contrasts with the report of Jones and Feher who found that the reaction of benzene with intermediate complex $Cp^*Ir(Me_3P)$ generates first a η^2-complex with benzene. In a second step, the latter is rearranged into the product of C—H oxidative addition (reaction (7.95)) [1117].

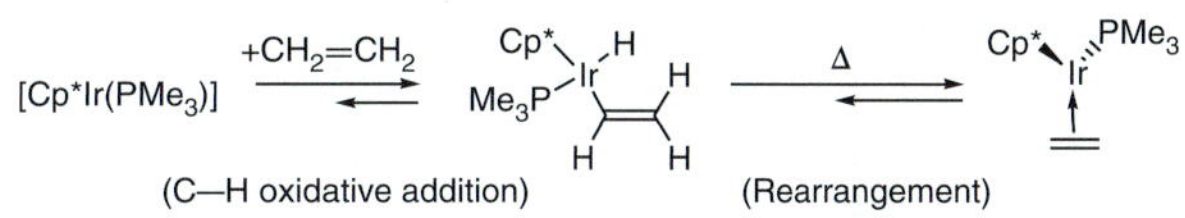

(7.94)

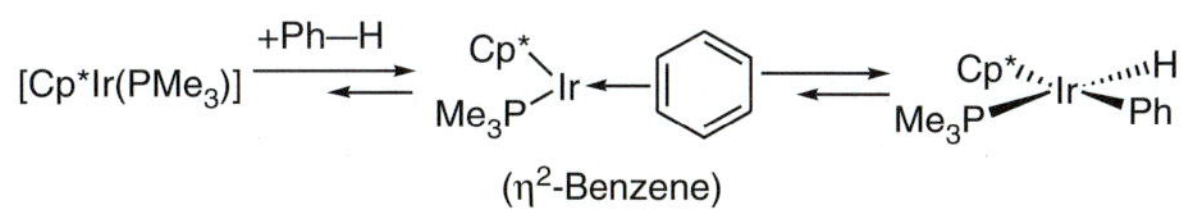

(7.95)

The catalytic addition of NH_3 to alkenes and the coupling of NH_3 with arenes are considered to be two challenges for catalytic chemistry [1118]. To accomplish the oxidative addition of the N—H bond of NH_3, one must overcome the usually more favored reaction that coordinates the transition metal to NH_3 by its nonbonding electron pair (classical Werner's coordination chemistry) [1119–1125]. Terminal amido complexes of late transition metals have been prepared by other routes than direct N–H oxidative additions [1126–1128]. Hartwig and coworkers have found that the electron-rich pincer iridium complex **284** reacts with an excess of propene giving the π-complex **285** and 2 equiv. of propane [1129]. Treatment of **285** with NH_3 at 25 °C in cyclohexane generates the monomeric amido(hydride) complex **286** and liberates propene.

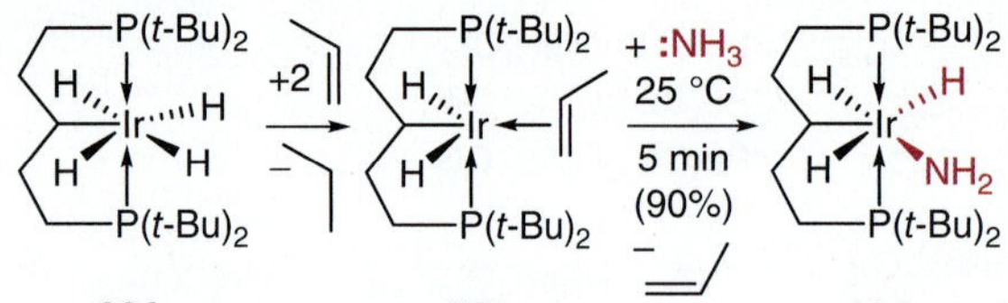

284 **285** **286**

Alkyl, allyl, acyl, and benzyl halides undergo oxidative additions with electron-rich transition metal complexes following S_N2 mechanisms. These reactions (7.96) are characterized by negative activation entropies ($\Delta^{\ddagger}S = -40$ to -50 eu) and are accelerated by polar solvents, as S_N2 Mentschukin amine quaternization reactions. The more nucleophilic is the reacting saturated complex L[M], the faster is the oxidative addition [1095, 1130].

$$\mathrm{L[M]} + \mathrm{RX} \rightleftharpoons \{\mathrm{L}\overset{\oplus}{\mathrm{[M]}}\mathrm{R} + \mathrm{X}^{\ominus}\} \underset{-\mathrm{L}}{\rightleftharpoons} \mathrm{[M](R)(X)} \qquad (7.96)$$

Changing Ph_3P ligand for Et_3P or Me_3P increases the rate of the reaction. The following reactivity order is observed for the reactions of halides RX with various nickel complexes: $Ni(PAr_3)_4 > Ni(PR_3)_3(alkene) > Ni(PAr_3)_3(alkene) > Ni(cycloocta\text{-}1,5\text{-}diene)_2$.

The reaction rate is strongly affected by steric factors. For the oxidative additions of alkyl iodides, the following reactivity order can be retained: MeI > EtI > *i*-PrI. Furthermore, the better is the nucleofugal group, the faster is the reaction: ROTs > RI > RBr > RCl [1131]. In certain cases, ionic intermediates such as **287** have been isolated (reaction analogous to the amine quaternization) [1132, 1133].

287

The first demonstration of stereochemical inversion at carbon in the oxidative addition of benzyl halides to palladium was provided by two sequential reactions, in which the stereochemistry of only the oxidative addition reaction was unknown (Scheme 7.49). Optically active deuterated benzyl chloride (*S*)-**288** was reacted with $Pd(Ph_3P)_4$ giving the stable complex **289**. In the presence of CO, **289** undergoes ligand exchange and α-insertion (Section 7.7.8) giving **290**. Methanolysis of **290** provides the methyl ester (*R*)-**291**. Complete inversion of configuration occurs at the benzylic center in this sequence of reactions for which it was established that the carbonyl α-insertion proceeds with complete retention of configuration (Section 7.7.8). The same product (*R*)-**291** is obtained by reaction of (*S*)-**288** with $Pd(PPh_3)_3(CO)$ in MeOH. Similarly, reaction of (*S*)-**288** with $(Et_3P)_4Pd$ first and then with CO and subsequent methanolysis also furnishes (*R*)-**291**, but with only 72% of inversion [1134].

Palladium complexes are efficient catalysts in nucleophilic substitution of allylic acetates or derivatives (Tsuji–Trost reaction, Section 8.8.6) [1135–1138]. In DMF (*N*,*N*-dimethylformamide), the oxidative addition of Pd(0) complexes generated from $Pd(dba)_2$ (dba = *trans*,*trans*-dibenzylideneacetone) and 2 equiv. of PPh_3 or from $Pd(dba)_2$, bidentate ligand dppb (1,4-bis(diphenylphosphanyl)butane), or dppf (1,1′-bis(diphenylphosphanyl)ferrocene) gives cationic (η^3-allyl)palladium(II) complexes **292** with AcO^- as counterion, as shown by Amatore et al. [1139]. The reactions are reversible and proceed through two successive equilibria (Scheme 7.50) [1140, 1141]. The overall equilibrium constants depend on the type of phosphine ligands. The stability order for **292** is dppb > dppf > PPh_3.

The reactivity of allylpalladium cations **292** toward nucleophiles is affected by the presence of other counterions than acetate. Chloride anion can modify not only reactivity but also regioselectivity and enantioselectivity [1142]. On the basis of ^{31}P NMR spectroscopy and voltammetric studies, Amatore et al. established that $Pd(dba)(PPh_3)_2$ and PPh_3 equilibrate with $Pd(PPh_3)_3$ and dba (reaction (7.97)) with an

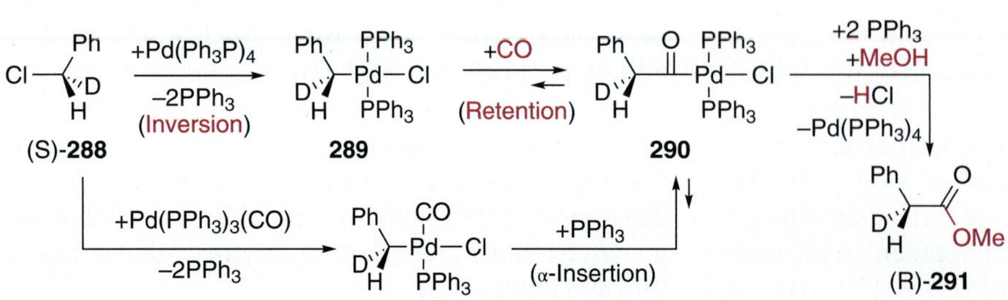

Scheme 7.49 Stille's demonstration of S_N2-type displacement in the oxidative addition of benzyl chloride to palladium complexes.

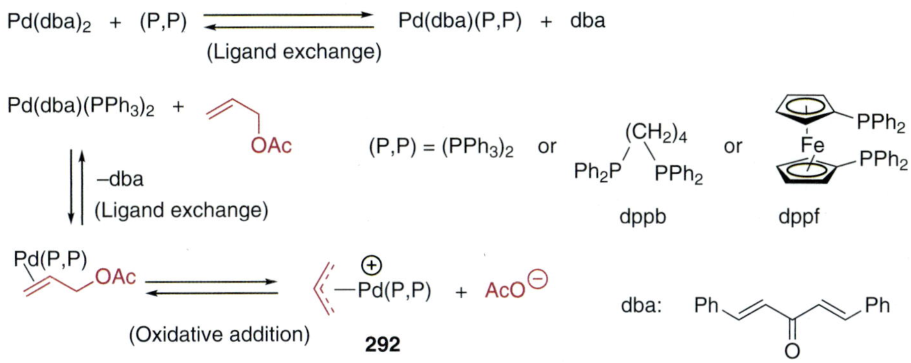

Scheme 7.50 Mechanism of the oxidative addition of Pd(0) complexes to allyl acetate.

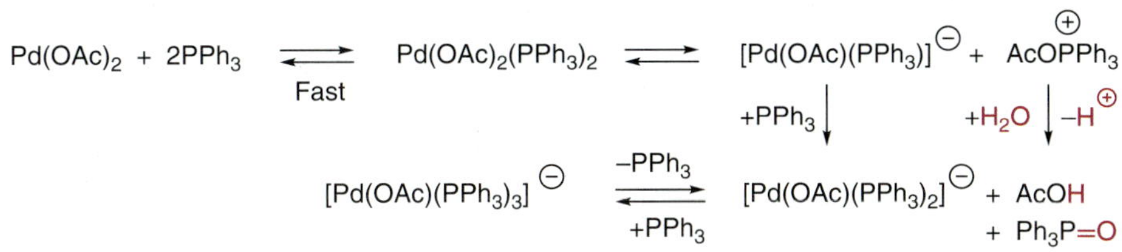

Scheme 7.51 Formation of Pd(0) species in the mixture $Pd(OAc)_2/PPh_3$.

equilibrium constant $K(7.97) = 0.14$ in THF at 25 °C [1143–1145].

$$[Pd(dba)(PPh_3)_2] + PPh_3 \underset{-dba}{\overset{K}{\rightleftharpoons}} [Pd(PPh_3)_3] \quad (7.97)$$

Thus, dba is a better ligand than PPh_3. In monomeric $[Pd(dba)_2L_2]$ complexes, dba coordinates to the Pd center as a monodentate ligand. Similar equilibria were found for tris(2-furyl)phosphine [1146] and $AsPh_3$ [1147]. $[Pd(dba)(AsPh_3)_2]$ was found to be more stable than $[Pd(dba)(PPh_3)_2]$. The oxidative addition of PhI to a 1 : 2 mixture of $[Pd_2(dba)_2]/PPh_3$ is 10 times slower than the addition of PhI to $Pd(PPh_3)_4$. Using bidentate diphosphines instead of PPh_3, the oxidative addition of PhI is the slowest for the diphosphine with the smallest bite angle [1143, 1148]. Coordinately saturated $[Pd(P,P)_2]$ complexes were found unreactive toward PhI. Amatore and Jutand established the important role played by anionic ligands in oxidative additions [1149]. Tricoordinated anionic complexes $[Pd(OAc)L_2]^-$ or $[PdClL_2]^-$ are proposed to be the actual species that undergo oxidative addition, rather than the neutral $[PdL_2]$ complexes. These anions are present in reaction mixtures when $Pd(OAc)_2$ or $[PdCl_2L_2]$ (L = CH_3CN, PhCN, PPh_3, etc.) are used. When mixed together, $Pd(OAc)_2$ and PPh_3 generate $[Pd(OAc)_2(PPh_3)_2]$ quickly [1150]. The latter complex undergoes slow intramolecular reduction to form a Pd(0) complex (Scheme 7.51). This reduction is promoted by electron-withdrawing substituents at the aryl group of the phosphine ligand [1151].

In THF, $[PdCl_2(PPh_3)_2]$ equilibrates with the three anionic complexes **293–295** [1152]. The oxidative addition **294** + PhI affords **296**, a 18 VE complex [1153].

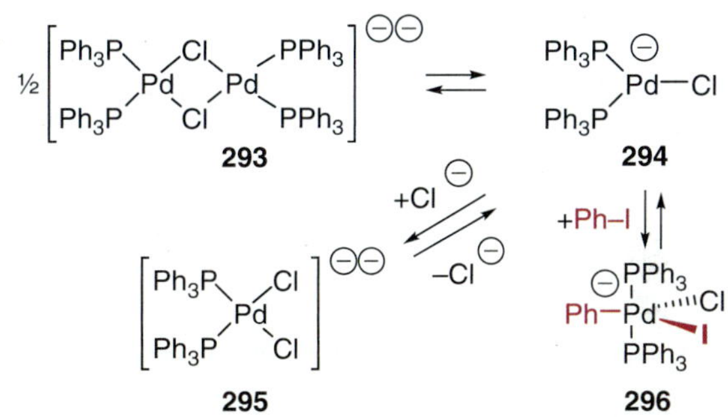

Similarly, $[Pd(L)_2OAc]^-$ reacts with aryl iodides giving five-coordinate Pd complexes of type $[Pd(L)_2(Ar)(OAc)I]^-$. This hypothesis has been challenged by Thiel and coworkers [1154] who notice that five-coordinate Pd species are unstable [1082] unless they bear polydentate ligands [1155]. Based on quantum mechanical calculations, they propose the mechanism shown in Scheme 7.52. It implies the formation of the relatively stable intermediate **297** in which the aryl halide (e.g. PhI) coordinates to the Pd metal via the halide atom (η^1-complex) [1156]. The calculations suggested that **297** rearranges to the η^2-arene complexes **298** and **299**, with a rate-determining step corresponding to the displacement of halide anion by the metallic center that leads first to *cis*-**300**. Then, the latter is isomerized slowly into *trans*-**300** [1157].

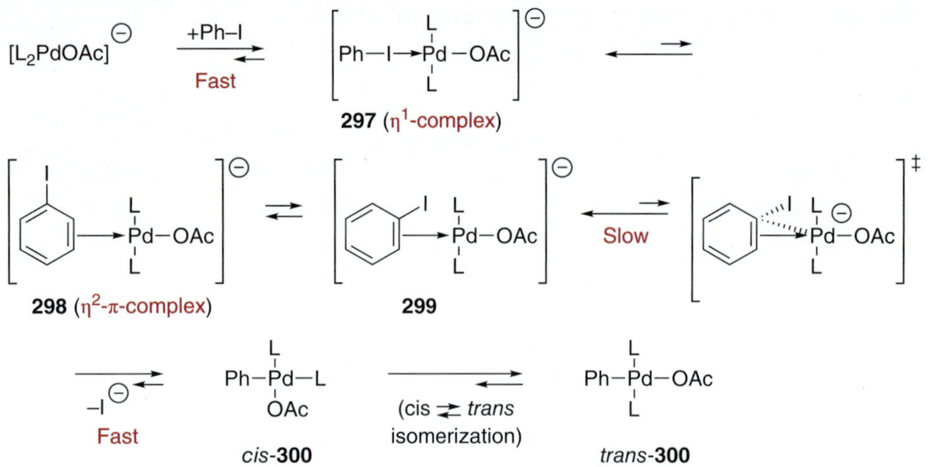

Scheme 7.52 Thiel's mechanism for the oxidative addition of PhI to $Pd(L)_2OAc^-$ (L = phosphine ligand).

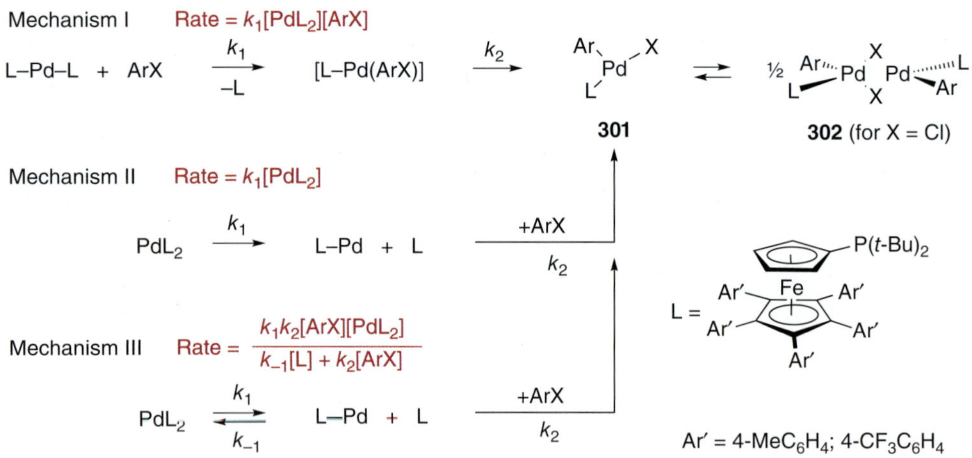

Scheme 7.53 Hartwig's mechanisms of the oxidative additions of ArX to Pd(0) complexes bearing bulky ligands.

Barrios-Landeros and Hartwig have studied the oxidative additions of PhI, PhBr, and PhCl to (bisphosphine)Pd(0) complex PdL_2 with hindered ligands that give adducts **301** [1158–1160]. They propose that different mechanisms (Scheme 7.53) are followed depending on the nature of the aryl halide. With PhI, mechanism I (S_N2 like; similar to that of Scheme 7.52) is preferred. The rate for this mechanism depends only on the concentration of PhI. With PhBr, mechanism II is preferred. It involves irreversible, rate-determining dissociation of one of the phosphine ligands L from **301** (S_N1 like, without return), followed by oxidative addition to form **301**, which is stabilized by solvent coordination. With PhCl, a mechanism III similar to mechanism II is followed. In the latter, the phosphine dissociation is reversible (S_N1 like with return). The rates of reactions following mechanism II are independent of concentrations [ArX] and [L], whereas reactions following mechanism III have rates depending on [ArX] and [L].

The oxidative addition of C—F bonds of aromatic and heteroaromatic fluorocompounds requires electron-rich complexes of Pd, Pt, and Ni [1161]. The oxidative additions of C_6F_5OTf (Tf = CF_3SO_2) to PdL_4 complexes proceed smoothly for L = PPh_3 but require the addition of LiCl for L = $AsPh_3$ [1162]. Similarly, the oxidative additions of C_6F_5OTf to [Pd(dppe)(alkene)] complexes are very slow but can be accelerated by addition of LiCl. This promotes the formation of $[Pd(C_6F_5Cl)(dppe)]$ (dppe = 1,2-bis(diphenylphosphanyl)ethane) [1163]. The oxidative addition of alkenyl triflates to $Pd(PPh_3)_4$ in DMF gives $[Pd(\eta^1\text{-alkenyl})(PPh_3)_2(DMF)](TfO)$ [1163]. The following reactivity order was established:

alkenyl-OTf ≫ vinyl-Br > Ph-Br and vinyl-OTf > PhOTf. Roy and Hartwig have studied the oxidative additions of aryl tosylates (ArOTs) to Pd(0) complexes **303** in benzene at 25 °C [1164, 1165]. PhOTs gives first **304** that then reacts with $(n\text{-oct})_4$NBr to give **305**. The oxidative addition of PhOTs is faster if $(n\text{-oct})_4$NBr is present from the beginning of the reaction. In the presence of Bu_4NPF_6 or in polar solvents, the reaction PhOTs + **303** → **305** is even faster. Thus, the addition of $(n\text{-oct})_4$NBr to the benzene solution increases its polarity, which accelerates the oxidative addition.

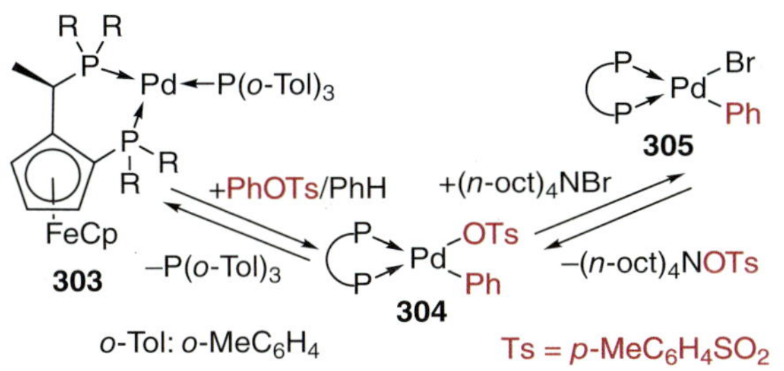

The oxidative additions of halogens X_2 (X = Cl, Br, I) to electron-rich transition metal complexes such as **306** imply the capture of X^+ in a first step giving **307**, followed by loss of a neutral ligand, and finally collapse of the outer sphere halide, X^-, to give the corresponding adducts **308** [1166].

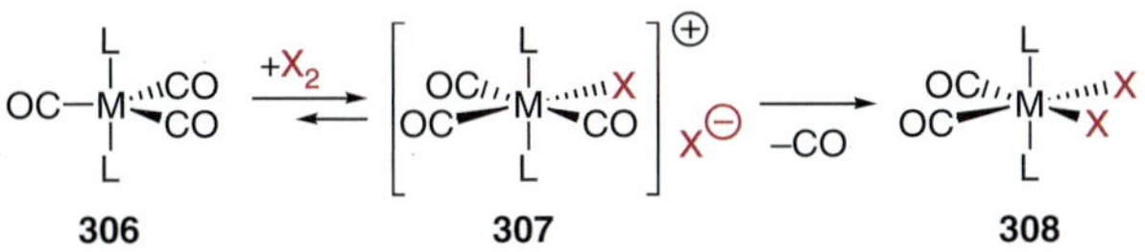

The oxidative additions of EtBr or BnBr on analogs of Vaska's complex follow radical chain mechanisms (Scheme 7.54). They imply initiation by a paramagnetic impurity, In•, or triplet $^3O_2$ (air). Kinetics of these reactions shows induction periods [1167, 1168]. The mechanism is supported by the observation of product of radical intermediate rearrangements; for instance, for R = cyclopropylmethyl, products of type $CH_2{=}CH{-}CH_2CH_2ML_{11}X$ are formed concurrently. This would not be observed in the case of net S_N2 process.

The oxidative addition of a strong protic acid can follow mechanism (7.98) and gives hydrides (reactions (7.99)–(7.101)) [1169, 1170].

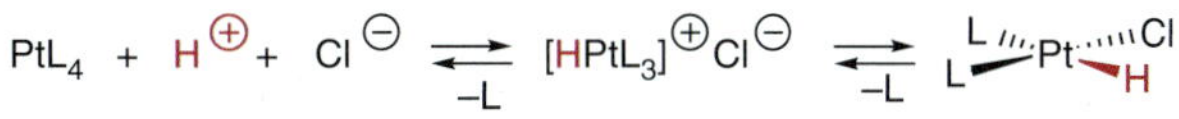

(7.98)

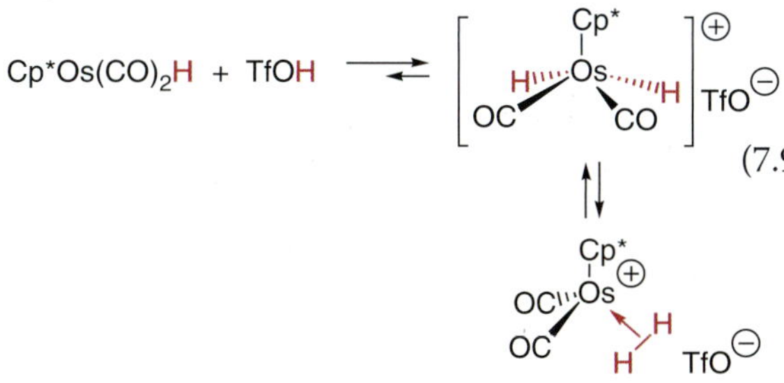

(7.99)

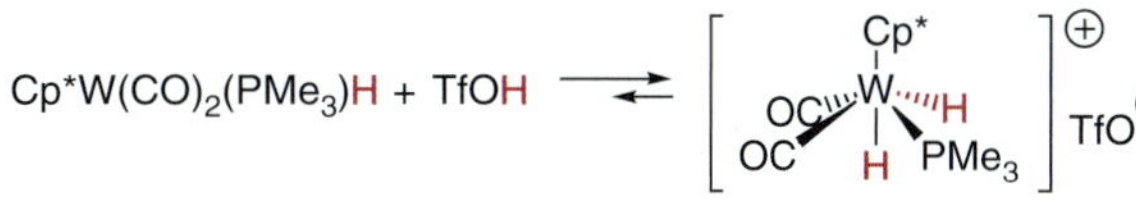

(7.100)

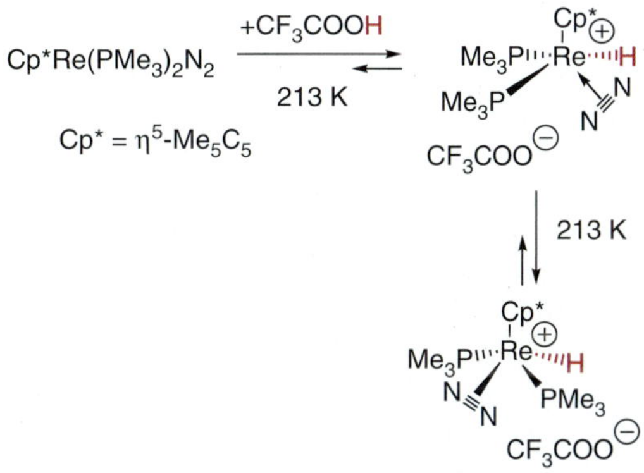

(7.101)

Reductive elimination is the reverse of oxidative addition whereby oxidation state, coordination number, and VE count all decrease by

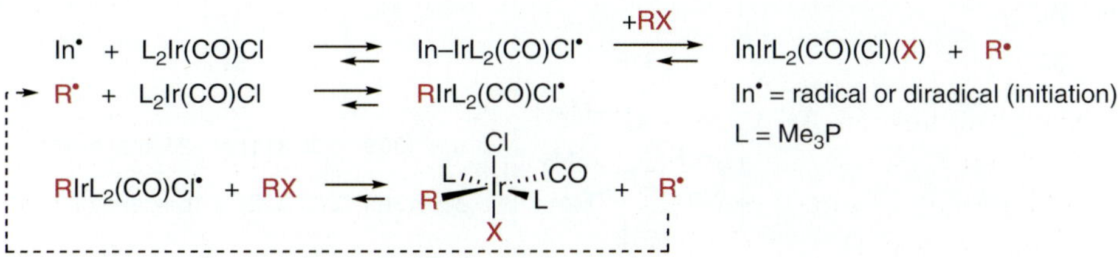

Scheme 7.54 Radical mechanism for the oxidative additions of Vaska's iridium complexes to alkyl halides.

two units. A large number of catalytical processes (Chapter 8) imply reductive eliminations in the product-forming step [1171, 1172]. Mechanisms of reductive eliminations are the same as for the reverse oxidative addition (principle of microscopic reversibility) as already seen with reactions $(R_3P)_2W(CO)_3(H)_2 \rightleftarrows H_2 + (R_3P)_2W(CO)_3$ (Figure 7.16a, Section 7.7.3). Reaction (7.102) generates substituted alkenes in which the methyl group is attached to the sp^2 center of the η^1-alkenyl ligand (retention of configuration).

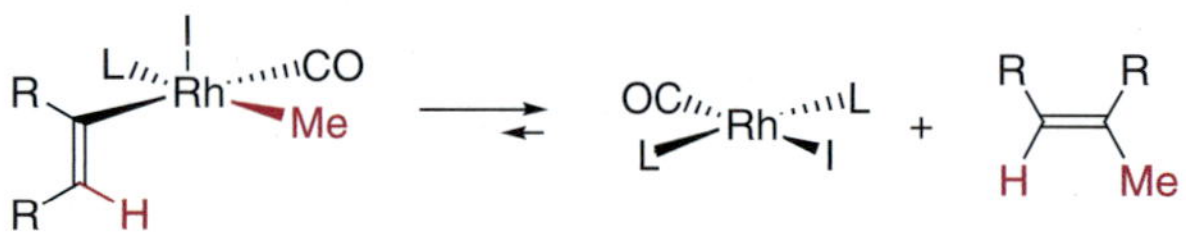

(7.102)

Reductive eliminations are favored in coordinately saturated complexes with bulky ligands (relief of B-strain), with complexes of metal with high oxidation state, and the presence of groups (e.g. CO, $P(OR)_3$) that stabilize the reduced metal fragment upon loss of ligands. The reactions are rapid for complexes of the late transition metals in groups 8–10, particularly d^8-nickel(II) and palladium(II), as well as for d^6-platinum(IV), palladium(IV), rhodium(III), and iridium(III). Copper triad (Co, Ag, and Au) complexes also have a tendency to undergo reductive eliminations [1173–1176]. Most reductive eliminations forming H—H, C—H, or C—C bonds occur by a concerted three-centered process that imply that the two leaving groups X and Y (reaction (7.103)) are cis to each other [1177, 1178]. Crossover experiments (reaction (7.104)) showed that equimolar mixture of nondeuterated and hexadeuterated dimethyl complexes produced exclusively CH_3—CH_3 and CD_3—CD_3. The absence of CD_3—CH_3 product strongly supports the monomolecular mechanism retained for reaction (7.103)) [1179].

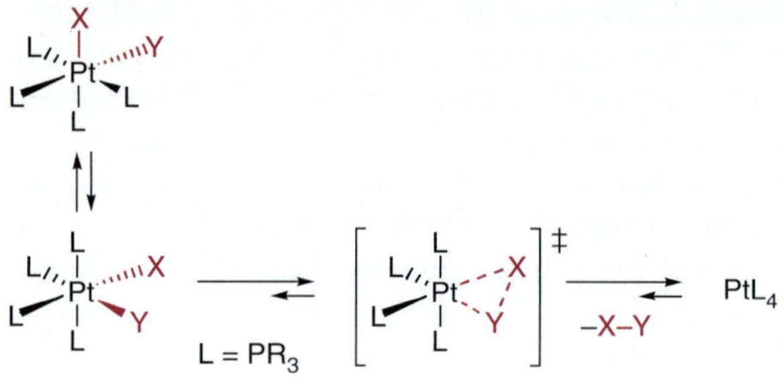

(7.103)

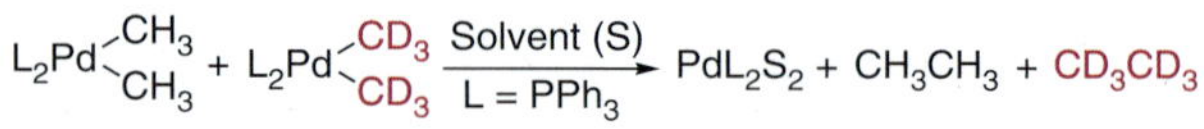

(7.104)

As demonstrated by Stille and coworkers for the oxidative addition of Pd(0) complexes, the reductive elimination from Pd(II) complexes occur with retention of configuration of the leaving groups (Scheme 7.49). Thermolysis of platinum complex **309** gives two competitive reductive eliminations (Scheme 7.55). The first reaction ($\Delta_r H$(**309** $\rightleftarrows$ **311** + Me—Me) $\simeq -25$ kcal mol^{-1}) produces ethane and the second is endothermic ($\Delta_r H$(**309** $\rightleftarrows$ **312** + MeI) $\simeq 16$ kcal mol^{-1}) and furnishes methyl iodide [1180]. On adding NaI in acetone, only complex **312** and MeI are formed. In the presence of $AgBF_4$ (quenches I^-), only reaction **309** → **311** + CH_3CH_3 is observed. In agreement with kinetic data, these two reductive eliminations involve formation of cationic intermediate **310** (S_N1 like process), the expected adduct for the oxidative addition MeI + **312**. In cationic complex **310**, a relatively low-lying LUMO confers a higher polarizability to this complex compared with neutral **309**. This explains the lower activation barriers for the former than for the latter reductive eliminations.

Quantum mechanical calculations on the C—C coupling reactions $M(PH_3)_2(R)_2 \rightarrow M(PH_3)_2 + R$—$R$ for M = Pd and Pt complexes suggested that the energy barrier of the reductive elimination decreases in the

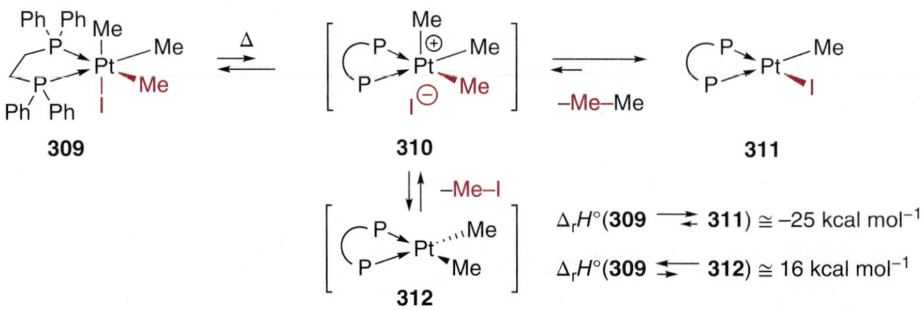

Scheme 7.55 Example of reductive eliminations involving a coordinately unsaturated complex intermediate.

order R = Me > CH≡C > Ph > CH_2=CH. Me—Me coupling is the most energy demanding, whereas the vinyl–vinyl coupling is the least demanding. All reductive eliminations forming C—C bonds are exothermic, and their exothermicity changes in the following order: CH_2=CH—CH=CH_2 ≈ Ph—Ph ≈ Me—Me > CH≡C—C≡CH. For reactions with similar exothermicity, the rate constants of the C—C bond formation is determined by the orientation of the two M—C bonds in $M(PH_3)_2(R)_2$ and by steric interactions between R and the coligands. For reactions with high exothermicity, the activation barrier is strongly influenced by the thermodynamics, in agreement with the Dimroth principle or Bell–Evans–Polanyi theory for concerted processes ($\Delta^{\ddagger}H = \alpha\Delta_r H + \beta$). The combination of these factors explains the overall trend in the ease of C—C bond formation in reactions $M(PH_3)_2(R)_2 \rightarrow M(PH_3)_2$ + R—R: vinyl–vinyl > Ph—Ph > CH≡C—C≡CH > Me—Me. For reactions $M(PH_3)_2(R)(R') \rightarrow M(PH_3)_2$ + R—R′ (R and R′ are different), the energy barriers and exothermicities are very close to the averages of the corresponding values of the symmetrical R—R and R′—R′ coupling reactions. The reductive eliminations are easier for Pd than for Pt complexes because the Pd—R(R′) bond energies are smaller than the corresponding Pt—R(R′) bond energies [1181]. The reductive eliminations of complexes **313** bearing bidentate diphosphine ligands (n = 1–4) showed that only the complex with $Cy_2PCH_2PCy_2$, the ligand with the smallest bite angle, leads to smooth elimination of CH_3—CH_3. The reaction is probably preceded by the dissociation of one of the phosphine moieties of the bidentate ligand [1182]. The resulting Pd(0) complex dimerizes into **314** (n = 1). Complexes **313** (n = 2, 3, 4) do not eliminate ethane under mild conditions.

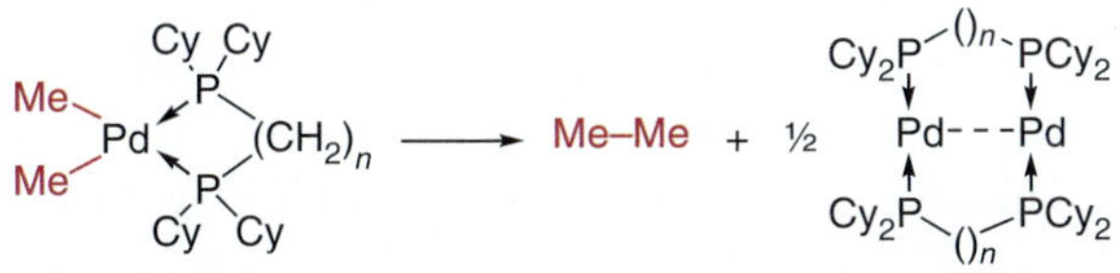

313 n = 1 Cy = cyclohexyl **314**

On the other hand, the fastest reductive elimination for a series of complexes Pd(L—L)Me_2 with L—L = dppp (1,3-(diphenylphosphanyl)propane), dppf(1,1′-bis(diphenylphosphanyl)ferrocene), and dppr (1,1′-bis(diphenylphosphanyl)ruthenocene) was observed with the ligand with the largest bite angle [1183–1191]. Schwartz and coworker have shown that C—C coupling of allylic halides with allylic organometallic reagents does not proceed unless an electron-deficient alkene such as maleic anhydride is added [1192–1194]. Bis(η^3-allyl)Pd complexes do not eliminate hexa-1,5-diene readily [1195, 1196]. In the presence of phosphine ligands, (η^1-allyl)(η^3-allyl)Pd complexes are formed [1197, 1198], which can decompose to form dinuclear Pd derivatives [1199]. If diphosphine ligands are added, bis(η^1-allyl)Pd species are formed that undergo fast reductive eliminations [1200]. Quantum mechanical calculations suggested that complexes *syn*-**315** and *anti*-**315** undergo reductive elimination into **316** with E_a = 11.1 and 8.5 kcal mol^{-1}, respectively. Both reactions form a C—C bond between C(3)—C(3′) of the two allyl moieties. The calculations also suggested much higher energy barriers for the reductive eliminations combining C(1)—C(1′) (E_a = 20.9 kcal mol^{-1}) or C(1)—C(3′) (E_a = 22.8 kcal mol^{-1}) [1201].

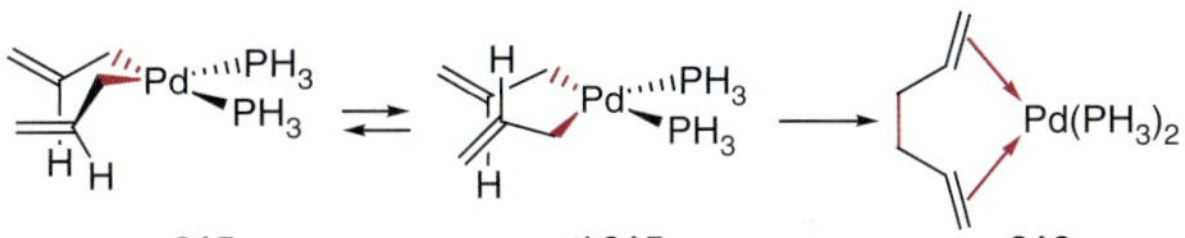

syn-**315** *anti*-**315** **316**

Norton has shown that reductive elimination of H_2 from $Os(CO)_4(H)_2$ ($k = 6 \times 10^{-5}$ s^{-1} at 126 °C) implies a first dissociative step (rate-determining step) liberating CO and $Os(CO)_3(H)_2$ (Scheme 7.56) [1202]. The latter complex coordinates a molecule of $Os(CO)_4(H)_2$ and then furnishes H_2 and the dinuclear complex $Os_2(CO)_7(H)_2$. The latter combines with CO and produces $Os_2(CO)_8(H)_2$.

Because of the high exothermicity of the reductive eliminations [M](alkyl)(hydride) → [M] + alkyl—H,

Scheme 7.56 Reductive elimination of H_2 from $Os(CO)_4(H)_2$.

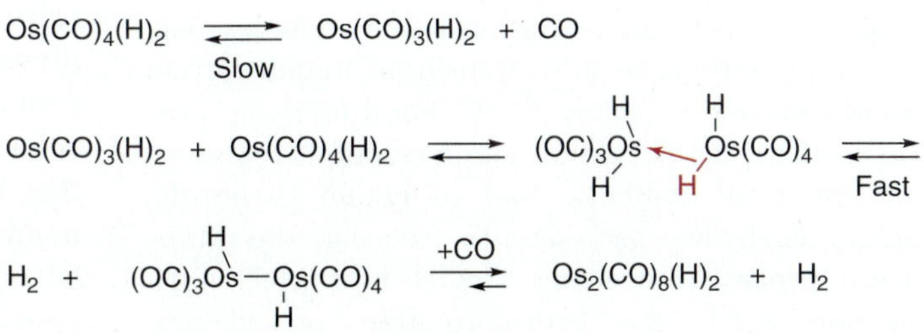

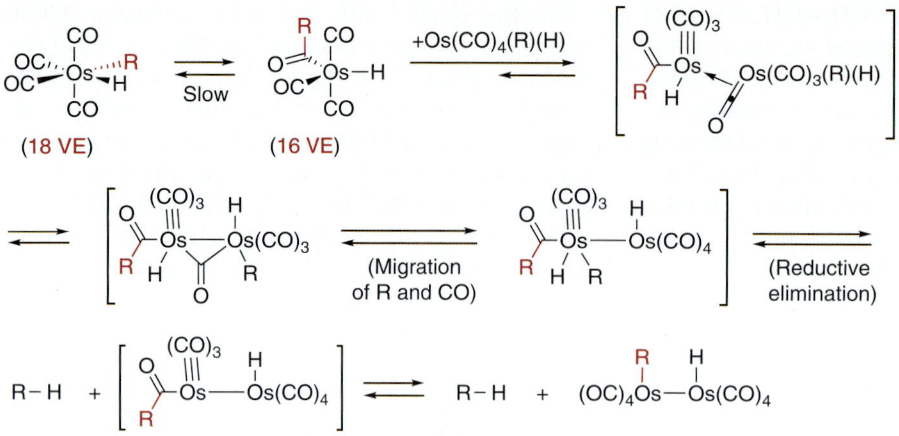

Scheme 7.57 Norton's mechanism for the reductive elimination of RH from $Os(CO)_4(R)(H)$.

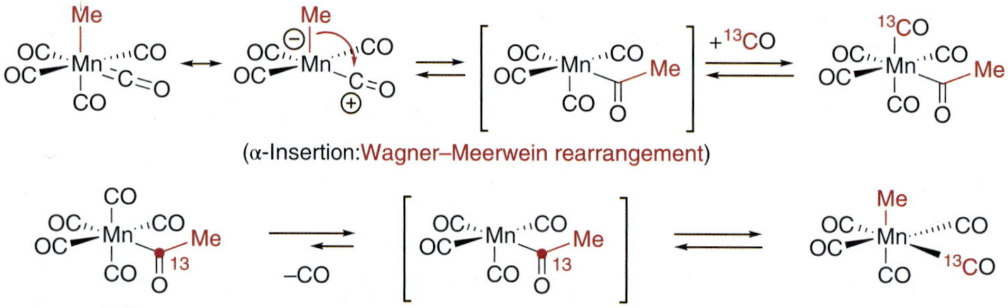

317 318

Scheme 7.58 Mechanism of the α-insertion $MeMn(CO)_5 \rightleftarrows (MeCO)Mn(CO)_4$. The CO ligand exchange is slower than α-insertion of the methyl group into a carbonyl ligand.

these reactions are generally fast. In the case of *cis*-$Os(CO)_4(R)(H)$, the mechanism shown in Scheme 7.57 has been proposed by Norton. The rate-determining step of the process is the α-insertion of the alkyl group into a carbonyl ligand giving the 16 VE complex $(RCO)Os(CO)_3(H)$. This coordinately unsaturated species combines with $Os(CO)_4(R)(H)$, an hypothesis confirmed by the observation of the formation of CDH_3 and CHD_3 when mixtures of $Os(CO)_4(Me)(H)$ and $Os(CO)_4(CD_3)(D)$ are allowed to react. A dinuclear species undergoes the reductive elimination giving product R—H.

7.7.9 α-Insertions/α-eliminations

Migratory insertion of carbon-based moieties (e.g. CO and carbenes) into transition metal–carbon bonds occurs in many C—C bond-forming processes (Chapter 8). Carbonylation reactions represent an important synthetic tool to obtain carboxylic acids, aldehydes, and ketones (see e.g. the Monsanto process converting MeOH into MeCOOH, Section 8.2.2; the hydroformation of alkenes, Section 8.2.3) [1203–1205]. Mechanistic studies indicate as the essential step the insertion of a carbon monoxide molecule into a metal–carbon bond, a process in which the alkyl group actually migrates from the metal to the CO ligand (O=C=M—R ⇄ M—C(=O)—R) [1206]. The same is recognized for alkyl insertion into (carbene) metal moieties (X(R′)C=M—R ⇄ M—C(X)(R′)—R) [1207]. Under CO atmosphere, $Me—Mn(CO)_5$ equilibrates with acyl complex $(MeCO)Mn(CO)_5$ (Scheme 7.58). The first step of the reaction is the migration of the methyl group to one of the carbonyl groups. This generates a 16 valence electron complex $(MeCO)Mn(CO)_4$ that evolves to product $(MeCO)Mn(CO)_5$ under CO pressure [1208]. In the presence of ^{13}CO, only one of the five carbonyl groups contains the ^{13}C-label; no ^{13}C-label is found in the acetyl group of $(MeCO)Mn(CO)_5$. This demonstrates that CO (and ^{13}CO) does not attack the Me—Mn bond and proves the migration mechanism. The latter process is analogous to a Wagner–Meerwein rearrangement (Table 7.6).

On heating complex **317** with the ^{13}C-label at the carbonyl moiety of the acetyl group, complex **318** is obtained. Ligand exchange in these complexes is slower than methyl group migration. As for suprafacial (1,2)-sigmatropic alkyl shifts (Wagner–Meerwein rearrangements, Section 5.5.1), retention of configuration is observed at the migrating carbon center. This is illustrated with reaction (7.105), which inserts a η^1-alkenyl group into a carbonyl ligand, and with α-elimination (7.106) catalyzed by Wilkinson's complex ($RhCl(Ph_3P)_3$) that converts the cyclopropanecarbaldehyde **319** into CO and cyclopropane **320** stereospecifically [1209].

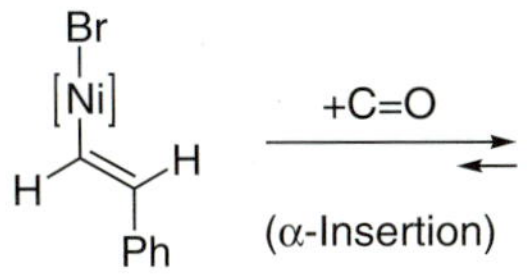

(7.105)

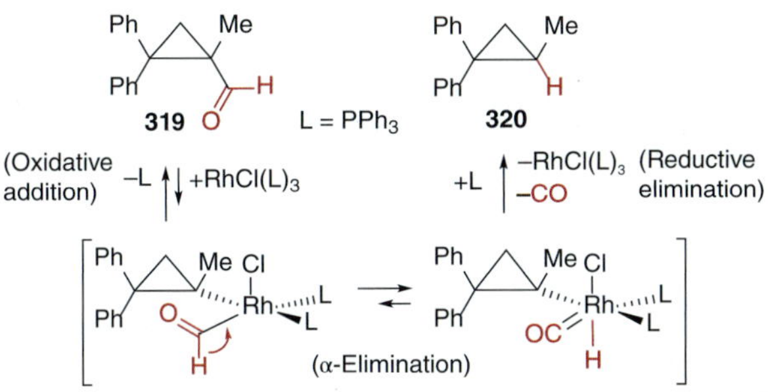

(7.106)

A further insight into the α-insertion process has been obtained with the stereoselective CO insertion. It implies ethyl group migration into the *cis*-carbonyl group. If migration of the carbonyl group into the Fe—C bond of Fe—Et would have occurred, *ent*-**322** would have been formed instead of **322**.

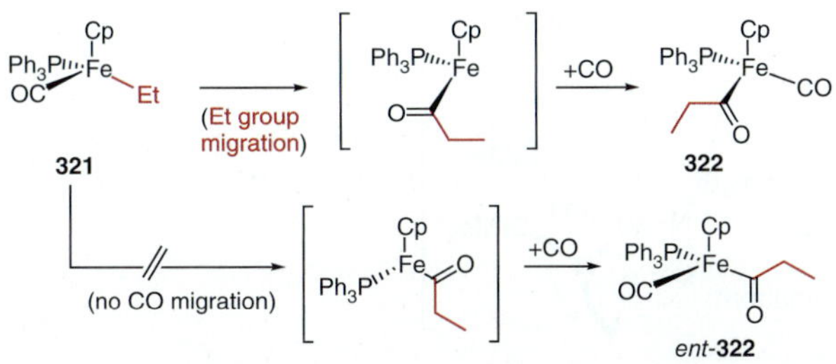

Reactions of α-insertions in alkyl(polycarbonyl) metal complexes can be accelerated by Brønsted or Lewis acid catalysts (e.g. reaction (7.107)) [1094]. These reactions can be catalyzed by SET, for instance, using $(NH_4)_2Ce(NO_3)_6$ or by trityl salts (reaction (7.108)) as electron acceptors [1210]. Anodic oxidation is an alternative way to catalyze α-insertion of carbon monoxide.

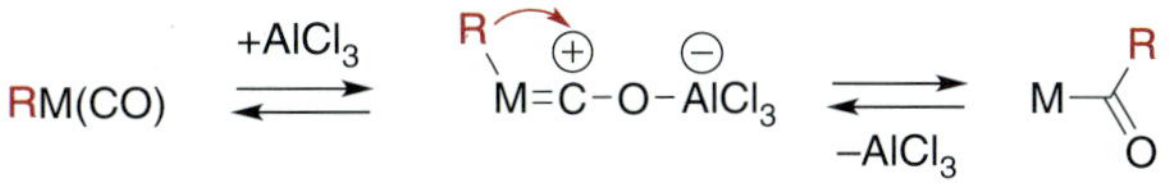

(7.107)

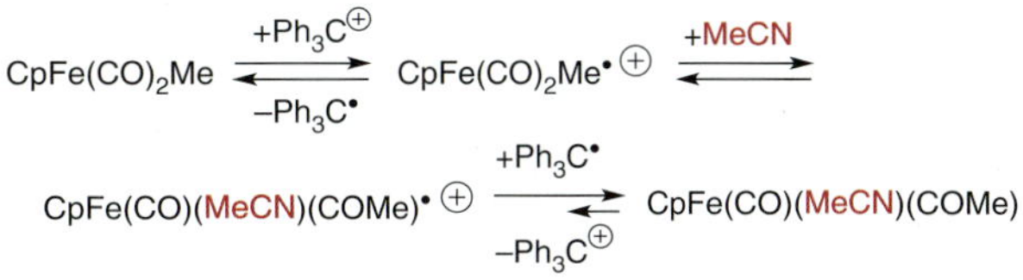

(7.108)

Electron-withdrawing (alkylidene)metal complexes also undergo α-insertions: an example is given with reaction (7.109). $Cp_2W(Me)I$ reacts with PhMgBr to give $Cp_2W(Me)(Ph)$. SET induced by ferrocenium hexafluorophosphate in CH_2Cl_2 generates ferrocene and radical cation **323**. The latter transfers one hydrogen atom to the trityl radical furnishing the cationic (methylene)metal complex **324** and Ph_3CH. The unstable complex **324** undergoes fast α-insertion of the phenyl group (Wagner–Meerwein rearrangement) providing **325** in the presence of acetonitrile. The same complex **325** is obtained on heating $[Cp_2W(CH_2{=}CH_2)CH_2Ph]^+/PF_6^-$ in CH_3CN [1211]. Another example reported by Werner and coworkers is shown with reaction (7.110) [1212]. When **326** is treated with a trityl salt, a hydride transfer generates methyleneruthenium complex **327**. The latter undergoes a Wagner–Meerwein methyl group migration (α-insertion) giving coordinately unsaturated complex **328** that undergoes a β-elimination (Section 7.7.7) with the formation of the stable (ethylene)-(hydrido)ruthenium complex **329**.

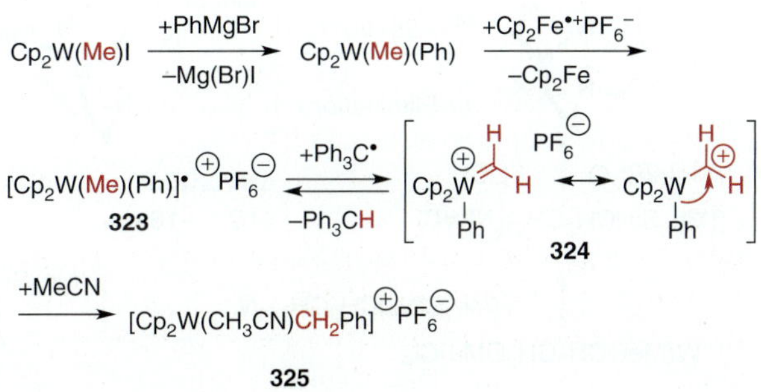

(7.109)

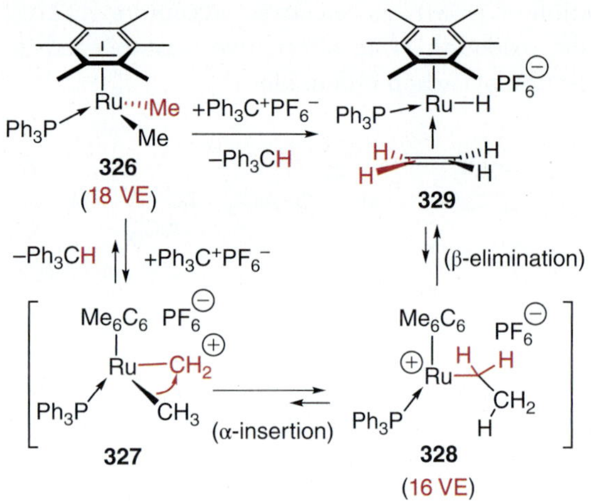

(7.110)

Schrock and coworkers have shown that rapid equilibria can exist between complexes resulting from α-insertion or β-insertion of hydride [1213]. Treatment of $Ta(Np)_2Cl_3$ with Et_2Zn generates $Ta(Np)_2(Et)Cl_2$, (Ta^{+5}, with 10 VE). This unsaturated complex coordinates a trimethylphosphine and gives an intermediate **330** resulting from a fast α-elimination of a hydride from one of its neopentyl groups. Complex **330** then undergoes a fast reductive elimination of neopentane after coordinating a second molecule of trimethylphosphine providing **331**. This ethylmetal complex undergoes a facile β-elimination giving intermediate **332**, a metalhydride that equilibrates with **333**, via an α-insertion of hydrogen into the alkylidene moiety (Scheme 7.59a). In the case of alkylmolybdenum and alkyltungsten complexes, α-elimination can be faster than β-elimination [1214]. An example of double α-elimination of hydrogen is shown in Scheme 7.59b [1215].

The ^{13}C NMR spectrum of complex **333** shows a resonance at $\delta_c = 139$ ppm for the methylene carbon atom (CH_2 of t-Bu-CH_2) attached to the metallic center and $^1J(C,H) = 98$ Hz, which is a low value for a C—H bond. This suggests that the σ-(C—H) bond is weakened by α-agostic interaction with the metal atom represented by limiting structures **333** ↔ **333′** ↔ **333″**.

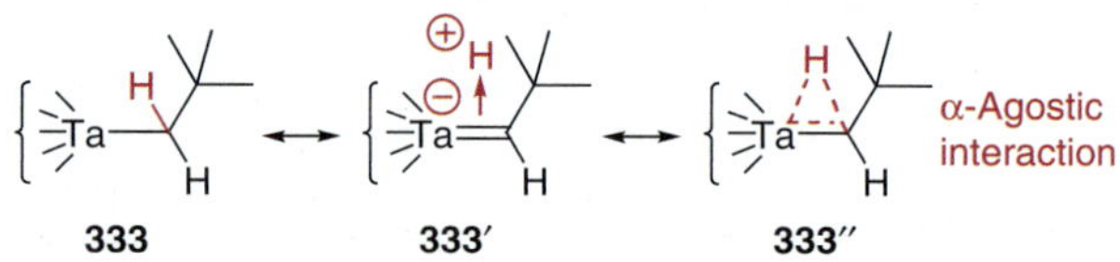

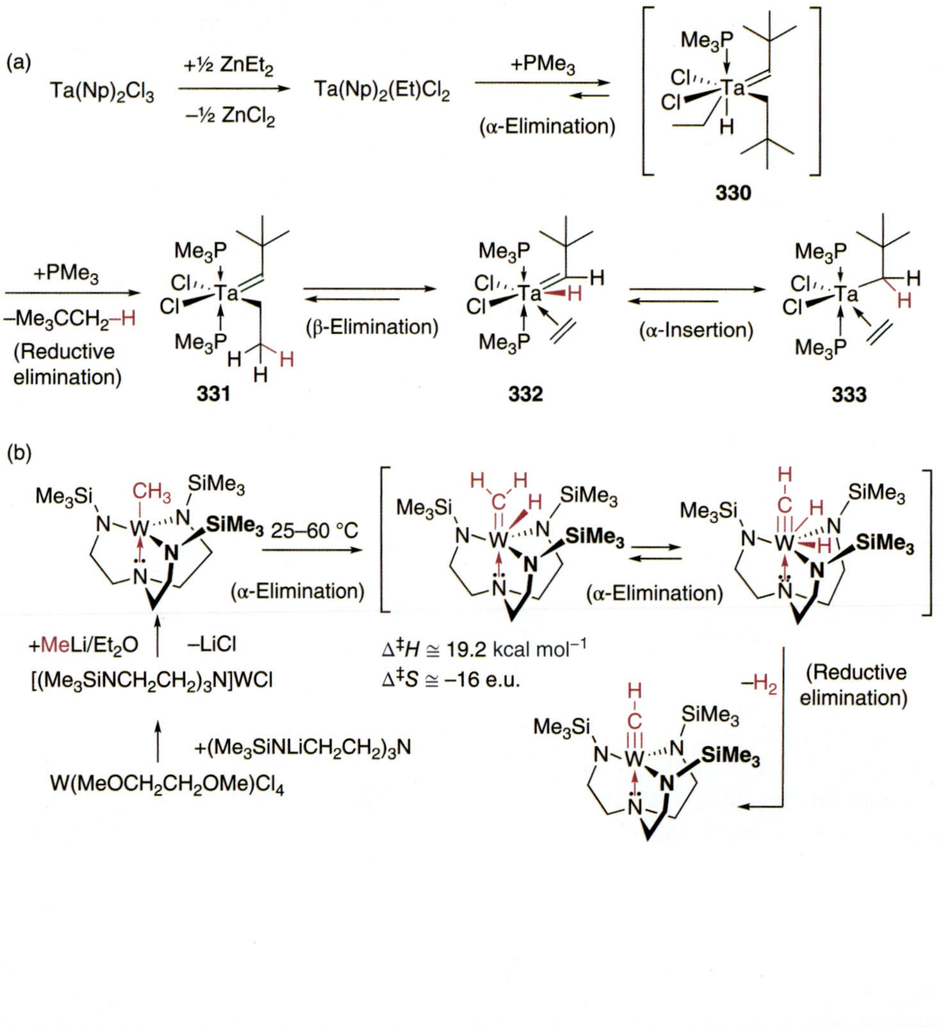

Scheme 7.59 (a) Equilibria between complexes resulting from α-insertion and β-insertion of hydride. (b) Example of double α-elimination of hydrogen and formation of an alkylidynemetal complex (HC≡M).

Problem 7.22 Explain the following reaction [1216].

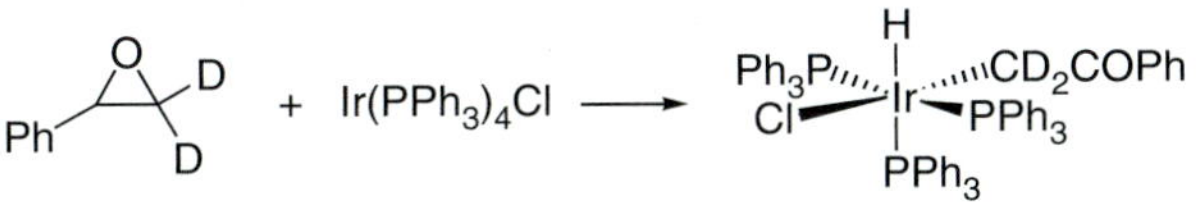

Problem 7.23 What is the structure of product **P** in the following reaction?

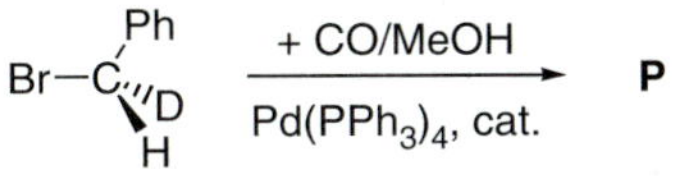

Problem 7.24 Explain the following transformation: $Cp_2Mo(PR_3)Me^+/PF_6^- \rightarrow Cp_2Mo(H)CH_2P^+R_3/PF_6^-$ [1217].

Problem 7.25 $Ir(PPh_3)_2(Cl)_2(COCH_2Ph)$ ($\nu_{C=O}$ = 1670 cm^{-1}) rearranges into **P** ($\nu_{C=O}$ = 2040 cm^{-1}) at 30 °C in benzene. What is the structure of **P**?

7.7.10 β-Insertions/β-eliminations

Hydroborations of alkenes and alkynes are typical 1,2-insertions, or β-insertions, into metal–H bonds. Similarly, alkenes and alkynes can be β-inserted into transition metalhydride bonds, or more generally into M—R bonds (R-alkyl, aryl, halides, alkoxides, amides, etc.). The reverse of the β-insertions are the β-eliminations (e.g. equilibria (7.111) and (7.112)).

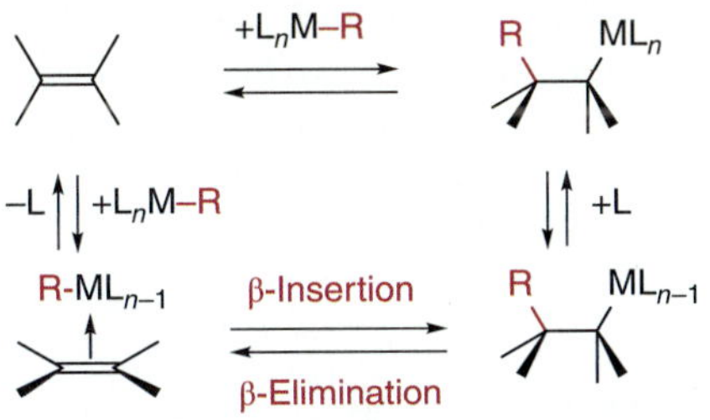

(7.111)

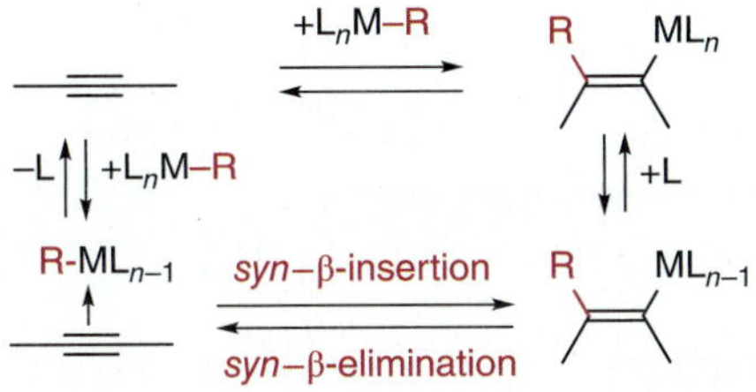

(7.112)

β-Insertions operate in many reactions catalyzed by transition metal complexes such as alkene and alkyne hydrogenation, hydroformylation of alkenes, alkene oligomerization, and polymerization [1218, 1219]. The β-elimination often terminates a catalytical process that produces alkenes or alkynes. In the Ziegler–Natta polymerization of alkenes catalyzed by $TiCl_3/Et_3Al$ in bulk $MgCl_2$, L_nTi—Et intermediates are formed that coordinate a propene unit forming a $L_{n-1}Ti$(propene)(Et) complex **334**. β-Insertion generates the coordinately unsaturated intermediate **336**, which, on its turn, coordinates a second unit of propene, starting the polymerization process. The chain growth is interrupted when the last $L_{n-1}Ti$(long chain alkyl) complex **338** undergoes a β-elimination of a hydride, generating a π-complex of the long-chain alkene, and eventually liberates the monoalkene **339**. This is the Cossee's mechanism (Scheme 7.60) [1179, 1180, 1220–1223].

Chain transfer occurs between intermediates **338** and alkenes **339**. This implies that the chain growth from **338** does not involve coordination of the monomer, but an oligomer of type **339**. The β-insertion of alkyl group into an alkene unit, as illustrated with **334** → **336**, corresponds to corner-to-corner alkyl migrations in corner-protonated or corner-alkylated cyclopropanes (Table 7.6). The π-complex **334** can be seen as a metallacyclopropane **334′** in which the metal has a high valence electron count. The transition state **335** of β-insertion can be seen as an intramolecular σ-complex **335′**, i.e. a species involving agostic interaction with a β-(C—C) bond. β-Insertion of hydride is generally more facile than β-insertion of an alkyl group in late transition state complexes, as illustrated (Scheme 7.61) for reactions **340** → **341** and **342** → **343** [1224, 1225].

Eisch et al. showed for the first time with reaction (7.113) that an alkyne can also undergo β-insertion of an alkyl group in a titanium complex [1226]. The reaction is *syn* selective. Another example is giving with the hydroplatination (7.114) of hexafluorobut-2-yne [1227].

(7.113)

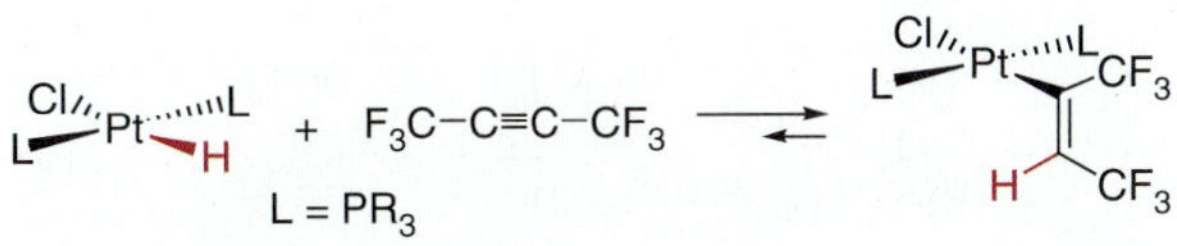

(7.114)

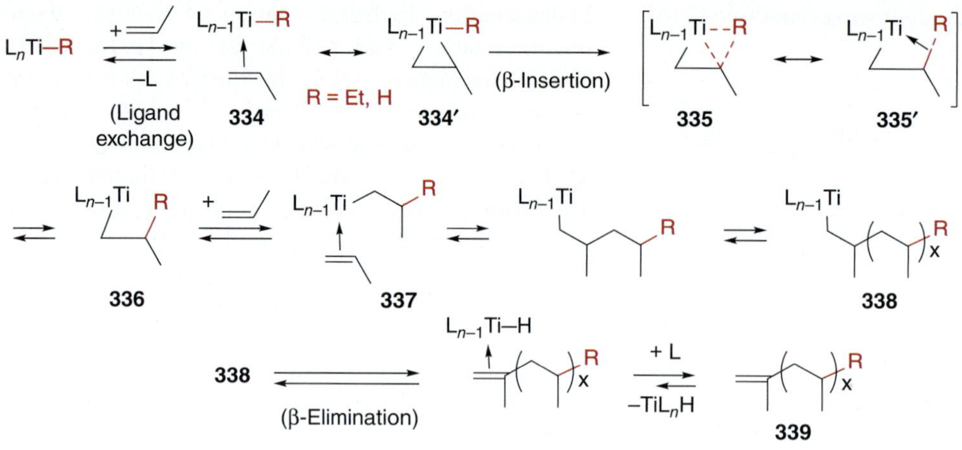

Scheme 7.60 Cossee's mechanism of the Ziegler–Natta polymerization of propylene.

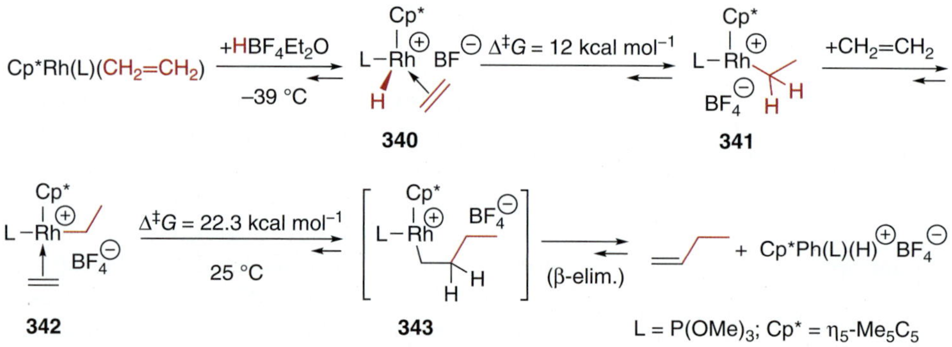

Scheme 7.61 β-Insertion of hydride is faster than β-insertion of ethyl group.

The *syn* selectivity of the β-insertion/β-elimination arises from β-agostic interactions. UV irradiation of $CpM(CO)_3Et$ (M = Mo, W) at 75 K in alkane matrices generates **344** that shows β-H agostic interactions. At 195 K, complex **344** is isomerized ($\Delta^{\ddagger}G \cong 10\,kcal\,mol^{-1}$) into hydride **345**, which is not a transition structure but a relatively stable intermediate (corresponds to edge-protonated cyclopropane). The 18 VE (ethylene)metal hydride complex **345** undergoes β-insertion of hydrogen on heating in the presence of PPh_3, giving ethylmetal complex **346**.

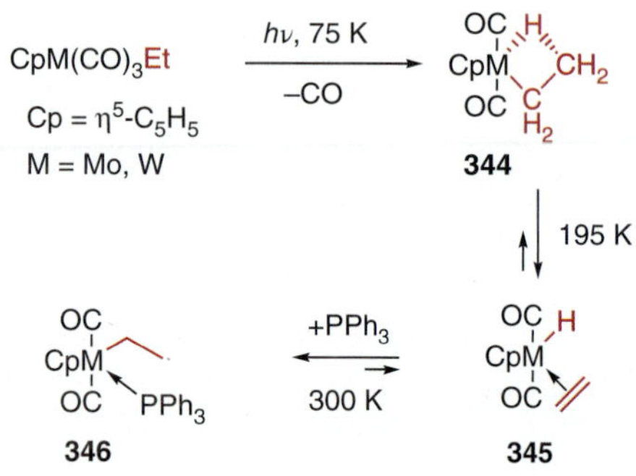

A model for the transition state of β-insertion/β-elimination reaction is given with salt **347** obtained on treatment of $Cp^*Co(ethylene)PR_3$ with HBF_4. The ^1H NMR spectrum of **347** shows resonances at $\delta_H = -12.6, -1.78, -0.44, 1.655$, and 2.86 ppm. This is an example of stable edge-protonated metallacyclopropane [1228]. Another example is cation **349** resulting from the treatment of diferracyclopropane **348** with HBF_4 or CF_3SO_3H [1229].

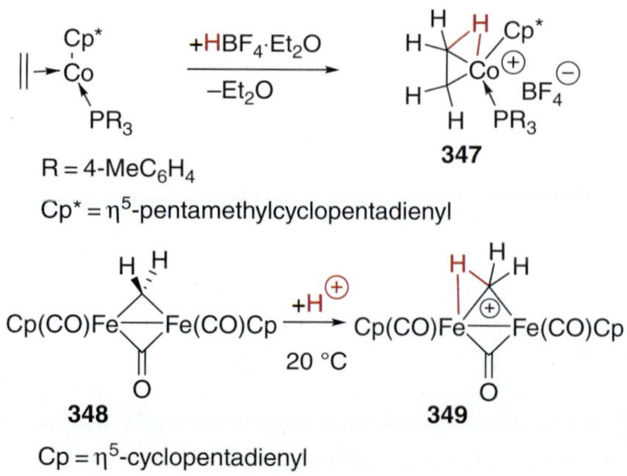

Alkene and alkyne hydrozirconations with the Schwartz's reagent (reaction (7.115)) are typical β-insertions [1230]. Utility of Schwartz's reagent is demonstrated with the conversion of alkynes into corresponding (*E*)-1-iodoalk-1-enes with high stereoselectivity (reaction (7.116)) [1231].

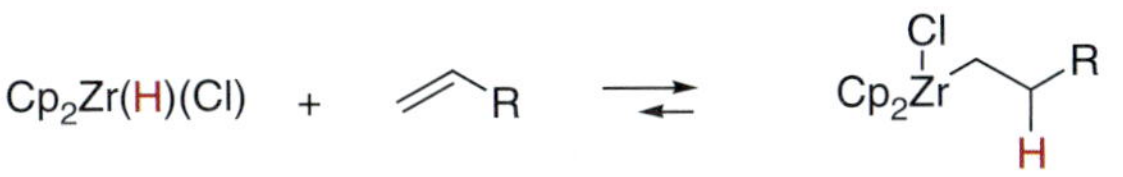

(7.115)

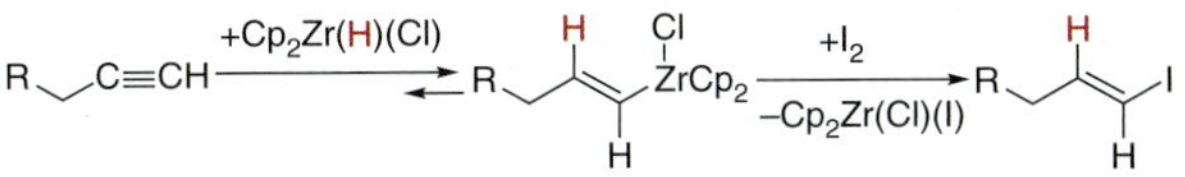

(7.116)

Comparison of the reactivities of neutral and cationic (alkyl)Pd and (alkyl)Pt complexes revealed that the creation of a vacant site adjacent to the alkyl ligand attached to the metallic center induces marked rate enhancement of the β-hydrogen elimination. This is illustrated with ethylpalladium complexes **350** and **352**. Decomposition of **350** (X = Cl, AcO) occurs at 40 °C in CH_2Cl_2 and follows first-order rate laws with respect to the concentration of **350**. The reactions are retarded on adding acetate anion to the mixtures, which confirms the mechanism involving the formation of the ion pair intermediate **351** in the rate-determining step (S_N1 with return). When the salt **352** is used as the starting material, the β-elimination occurs already at −30 °C [1232].

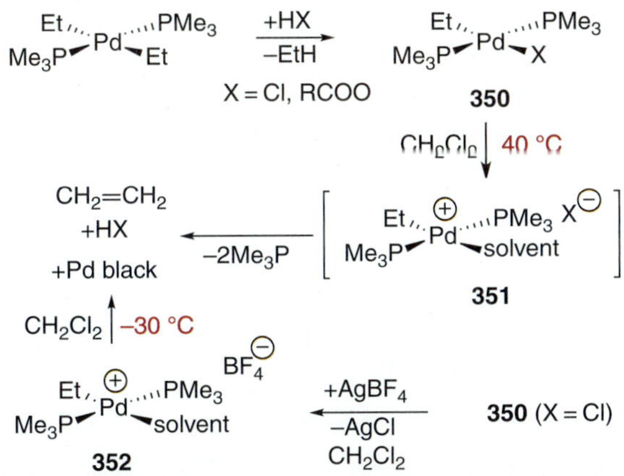

Iridium(III) complexes such as $Ir(H)_2(COMe)(PPh_3)^+$ [1233, 1234], $Cp^*Ir(PMe_3)(Me)^+$, [1235–1237] or **353** [1238] activate alkanes and generate π-complexes of the corresponding alkenes (e.g. alkane dehydrogenation (7.117)). Stable complexes of benzyne (1,2-didehydrobenzene) and derivatives have been prepared through β-elimination from arylmetal complexes as illustrated by reactions (7.118)–(7.120) [1239]. Steric factors strongly affects the thermodynamics (equilibria between β-insertion and β-elimination) and the kinetics of β-insertions. For instance, for β-hydrogen insertions (7.121), the relative rates are pent-1-ene > styrene ≫ (*Z*)-but-2-ene > cyclopentene > (*E*)-but-2-ene > isobutylene.

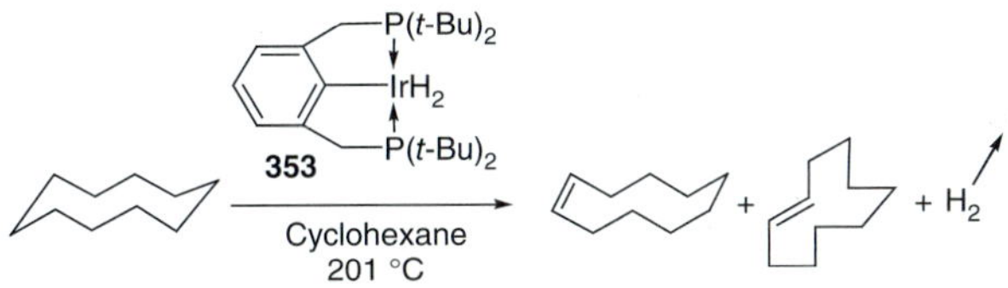

(7.117)

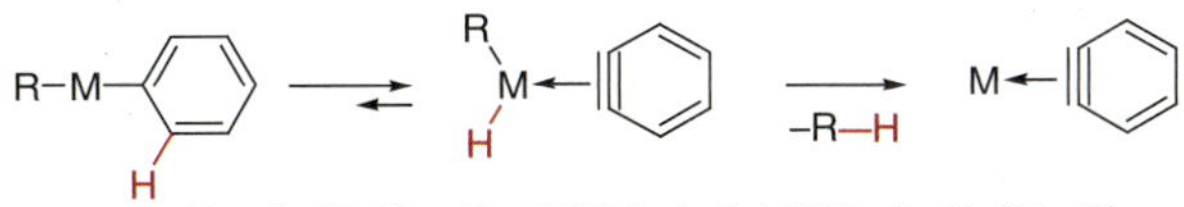

M = Cp*TaMe$_2$, Cp$_2$Zr(PMe$_3$), CpV(PMe$_3$)$_2$, Ru(Me$_3$P)$_4$

(7.118)

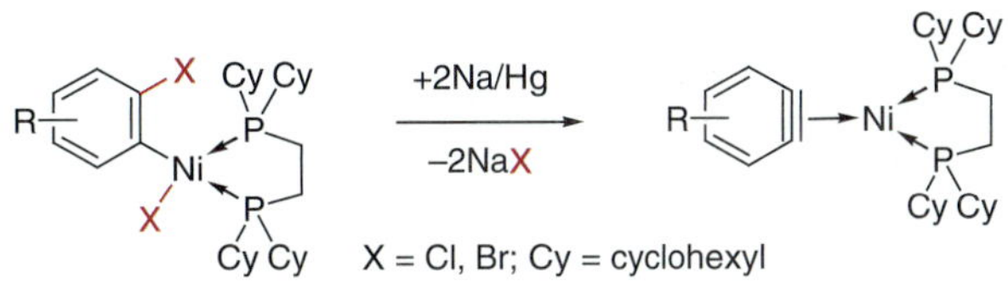

(7.119)

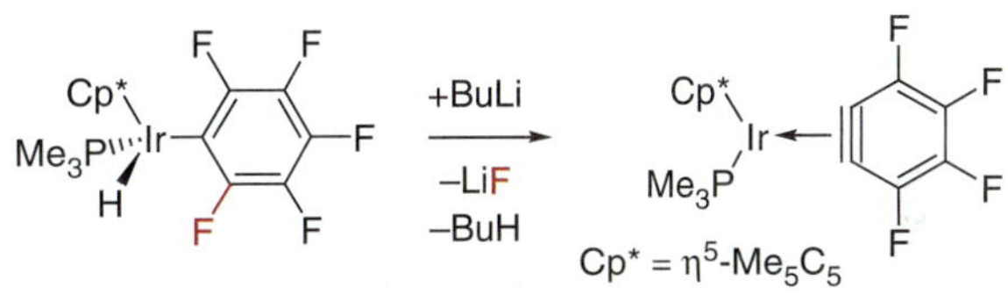

(7.120)

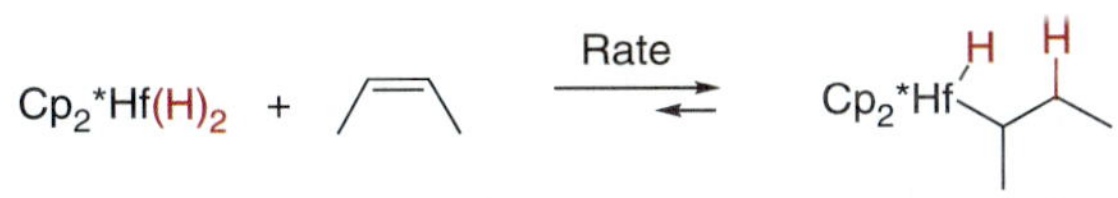

(7.121)

The rate of isobutylene insertion into $Cp^*(\eta^5\text{-}Me_4C_5H)Zr(H)_2$ is 3.8×10^3 times greater than that for $(Cp^*)_2Ir(H)_2$ at −63 °C, demonstrating the striking steric effect for isobutylene insertion imposed by a tenth methyl substituent on the two pentamethylcyclopentadienyl ligands (B-strain build up in the complex formation and subsequent hydrogen β-insertion). A primary kinetic deuterium isotopic effect k_H/k_D = 2.4 at 23 °C is found for the insertions of isobutylene into $Cp_2^*Zr(H)_2$ and $Cp_2^*Zr(D)_2$(Cp^* = pentamethylcyclopentadienyl). Thus, the rate-determining step of this reaction implies the four-center transition state **354**.

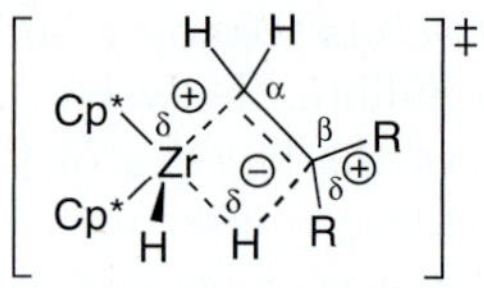

354

Rates of β-insertion of $Cp^*Zr(H)_2$ into para-substituted α-methylstyrenes afforded a free linear energy correlation of the observed rate constants with the Hammett constants σ giving $\rho = -0.46$. This relative small ρ value indicates cyclic transition states **354** with little charge separation, with only a slight positive charge development at the β-carbon center of the coordinated alkene. Also consistent with transition states **354** are the highly negative activation entropies (−30 to −48 eu) [1209, 1240, 1241].

Problem 7.26 What is product **P** of reaction C_4H_9—C≡C—D + $Cp_2Zr(H)(Cl)$ → **P**?

Problem 7.27 Internal alkenes and alkynes isomerize to 1-alkyl and alk-1-en-1-yl complexes, respectively, when reacted with the Schwartz's reagent. Explain.

Problem 7.28 Explain the stereoselectivity of the following reactions [1242].

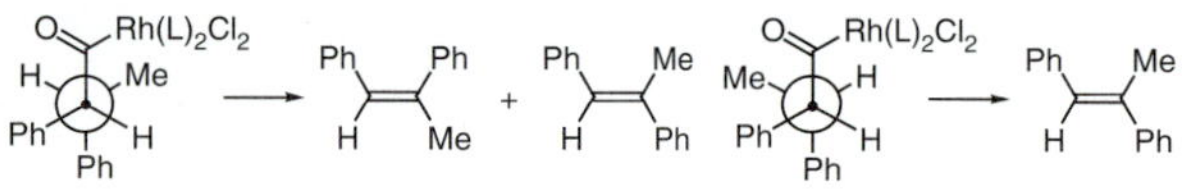

7.7.11 α-Cycloinsertions/α-cycloeliminations: metallacyclobutanes, metallacyclobutenes

In 1966, Jolly and Pettit prepared the (carbene)metal complex $CpFe(CO)_2{=}CH_2^+/BF_4^-$ and showed that it reacts with cyclohexene giving norcarane [1243]. In 1971, Green et al. prepared the first (carbene)ruthenium complex CpRu[=C(Me)(OMe)] $(CO)(PCy_3)^+/PF_6^-$ also capable of alkene cyclopropanation [1244]. These reactions were considered to proceed via coordination of the alkene cis to the carbene ligand, using complexes of type **355**, followed by α-cycloinsertion. The overall process can be viewed as (2+2)-cycloaddition of the C=Z and M=C double bond. In fact, the α-cycloinsertion **355** → **356** corresponds to a cyclopropylmethyl → cyclobutyl cation rearrangement (Table 7.6). The resulting metallocyclobutanes **356** undergo reductive elimination producing three-membered ring compounds (cyclopropane derivatives with Z = CR_2, C=O, C=NR, C=CR_2; aziridine derivatives with Z = NR; and oxirane derivatives with Z = O) (Scheme 7.62).

Depending on the nature of the metallic complex, the metallacyclobutanes can undergo concurrently other reactions such as alkene metathesis (Section 8.5.2) and β-hydrogen eliminations (Scheme 7.62c). The latter reaction generates η^3-allylmetal complexes of type **362** that equilibrate through β-insertions of hydrogen with alkene complexes of type **363**. In 1978, Green and coworkers [1217, 1245] proposed an alternative mechanism to the Cossee's mechanism (Scheme 7.60) for the Ziegler–Natta polymerization of alkenes. Instead of relying on β-insertions and β-eliminations, they suggested that the initial alkylmetal catalyst undergoes an α-elimination with the formation of an (alkylidene)metal species. The latter coordinates an alkene unit and, after α-cycloinsertion, gives a metallacyclobutane of type **356** (Scheme 7.62d). On its turn, **356** undergoes α-elimination of a hydrogen giving an intermediate of type **364**, which produces another (alkylidene)metal species of type **365** through a reductive elimination. When another alkene coordinates **365**, the polymerization process is initiated [1246]. There are (alkylidene)metal complexes that catalyze the oligomerization of olefins [1247–1249]. Others catalyze the polymerization of cyclic olefins [1250] and diolefins [1251]. Using alkyltungsten complexes that cannot undergo β-elimination, Schrock et al. have prepared (alkylidene)tungsten complex **366** and shown that it reacts with 2.4 equiv. of ethylene giving *t*-butylethylene and tungstanacyclobutane **367** (Scheme 7.63) [1251].

A large number of metallacyclobutanes have been characterized. The first example (**368**) was described by Tipper [1252]. He found that cyclopropane, like ethylene, can form complexes with platinum. Complex **368** was obtained on treating cyclopropane with H_2PtCl_4/AcOH [1253].

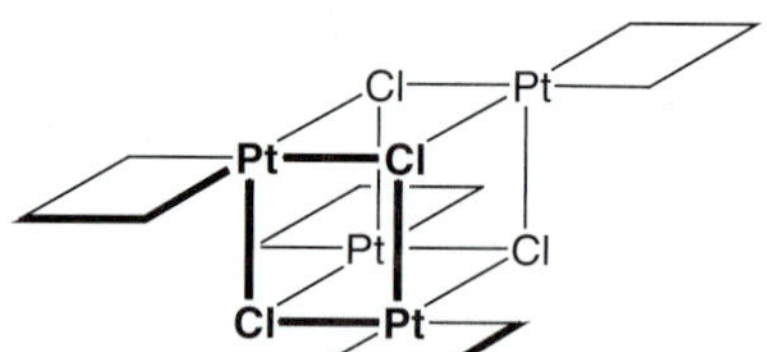

368

Problem 7.29 Snyder and Hours reported the $PtCl_2$-catalyzed rearrangement **A** → **P** [1254]. With the deuterated derivatives **B**, **C**, and **D**, products **Q**, **T**, and **W**, respectively, were obtained with the following kinetic isotopic effects: $k(\mathbf{A})/k(\mathbf{B}) = 1.54$, $k(\mathbf{A})/k(\mathbf{C}) = 5.44$, and $k(\mathbf{A})/k(\mathbf{D}) = 11.3$ at 140 °C [1255]. Propose a mechanism for these catalyzed isomerizations.

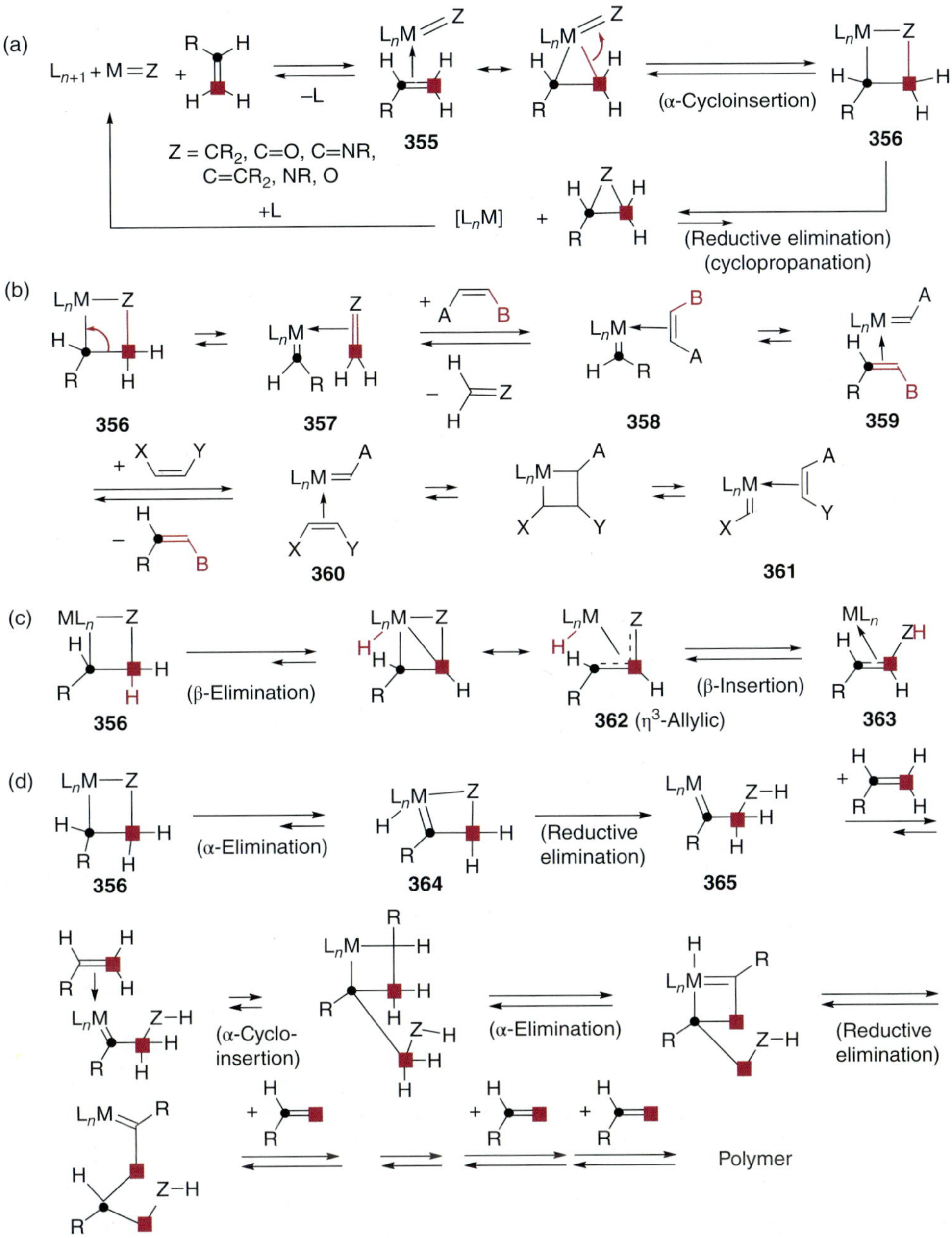

Scheme 7.62 Metallocyclobutane arising from the α-cycloinsertion of (alkene)(alkylidene)metal complexes intermediates in (a) cyclopropanation reaction of alkenes by carbene–metal complexes, (b) in Chauvin's mechanism of alkene metathesis, (c) in the formation of η^3-allyl complexes, and (d) in the formation of η^1-alkylidene complexes involved in the polymerization of olefins following the "carbene-to-metallocycle" mechanism of Green.

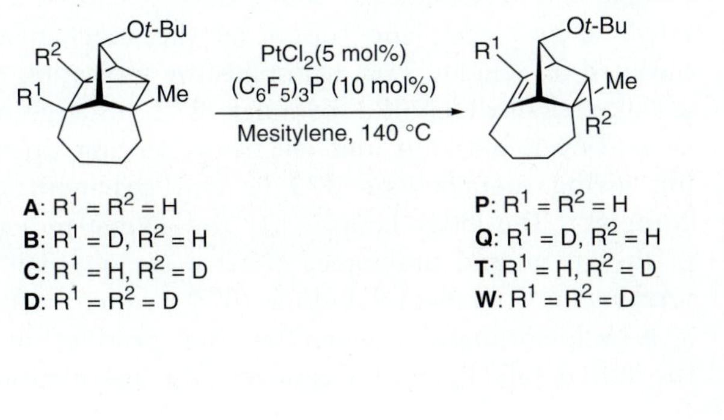

7.7.12 Metallacyclobutenes: alkyne polymerization, enyne metathesis, cyclopentadiene synthesis

The platinacyclobutenone **369** was obtained by reaction of diphenylcyclopropenone with $Pt(PPh_3)_4$ at 20 °C (reaction (7.122)). In this case, the oxidative addition of the metal onto the σ(C—C) bond giving

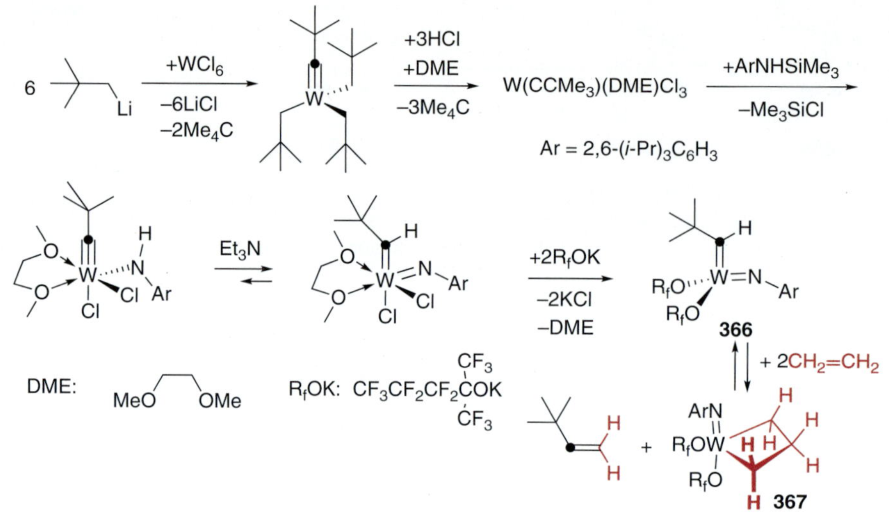

Scheme 7.63 Schrock's demonstration of the formation of a tungstanacyclobutane in the reaction of (alkylidene)W complex with ethylene.

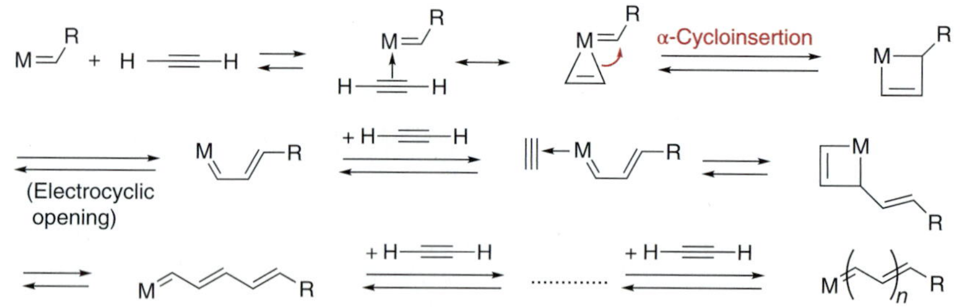

Scheme 7.64 Schrock's polymerization of acetylene catalyzed by an alkenylidenemetal complex $W(CHCMe_3)(NAr)(O\text{-}t\text{-}Bu)_2$ (analog of complex **366**). It implies repetitive α-cycloinsertion/metallacyclobutene electrocyclic opening.

369 (relief of cyclic strain) is more favorable than formation of a π-complex with the alkene unit [1256]. Tebbe and Harlow have shown that titanacyclobutene **370** can exchange its bis(trimethylsilyl)ethyne moiety for diphenylacetylene giving complex **371** (reaction (7.123)) [1257].

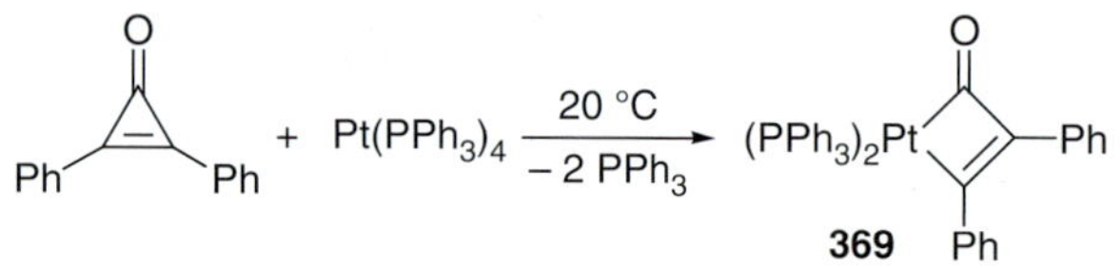

(7.122)

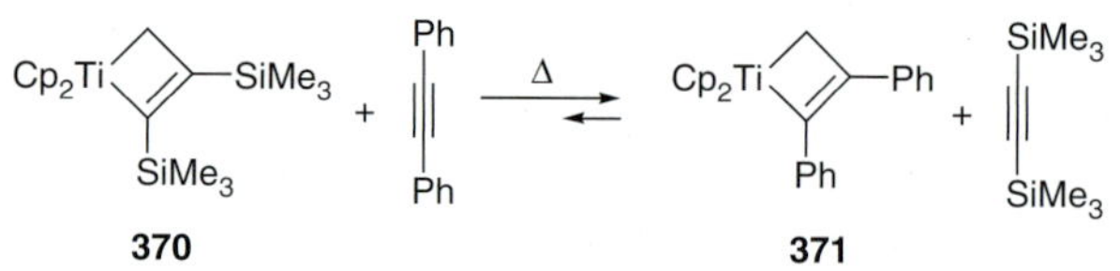

(7.123)

Schrock and coworkers have shown that metallacyclobutene complexes are catalysts of the polymerization of alkynes into polyacetylenes (Scheme 7.64) [1258, 1259], important material used to fabricate organic conductors [1260]. For example, they found that [*o*-(trimethylsilyl)phenyl]acetylene undergoes living polymerization with $W(NR^1)(CHR^2)(OR^3)_2$ catalysts giving soluble and highly conjugated polyenes [1261, 1262].

Enyne metathesis (or enyne bond reorganization) combines an alkene and an alkyne to produce a 1,3-diene (e.g. reaction (7.124)). Formally, the two methylidene fragments of the alkene add across the alkyne triple bond, converting it into a sp^2–sp^2 single bond of the conjugated diene product. Common catalysts for the enyne metathesis are $PdCl_2$, Fischer's carbene complexes, or Grubbs ruthenium alkylidene catalysts (Section 8.6.2) [1263–1265].

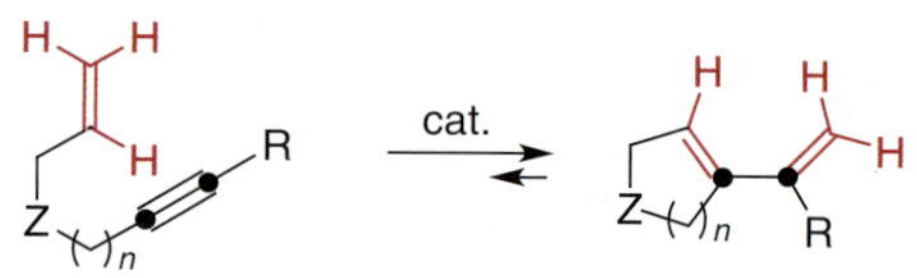

(7.124)

For the (alkylidene)ruthenium catalyst, quantum mechanical calculations for $(Me_3P)_2Cl_2Ru{=}CH_2$, ethylene and acetylene supported the mechanism outlined in Scheme 7.65. It implies the formation of a (alkyne)(alkylidene)Ru complex **372** undergoing slow α-cycloinsertion and electrocyclic ring opening giving metalladiene **373** (a vinylcarbenemetal complex). The latter coordinates the alkene moiety of the enyne and undergoes a fast α-cycloinsertion forming a metallacyclobutane **374**, which, after a α-cycloelimination, generates the product and the initial (alkylidene)Ru catalyst. The calculations

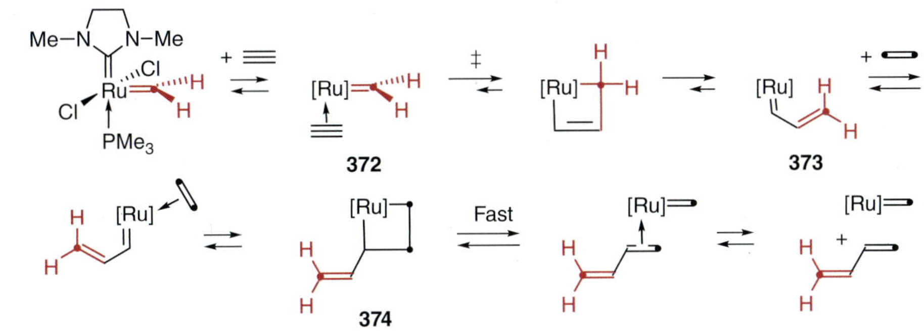

Scheme 7.65 Schematized mechanism for the enyne metathesis as calculated for a simpler analog of the Grubbs second-generation catalyst.

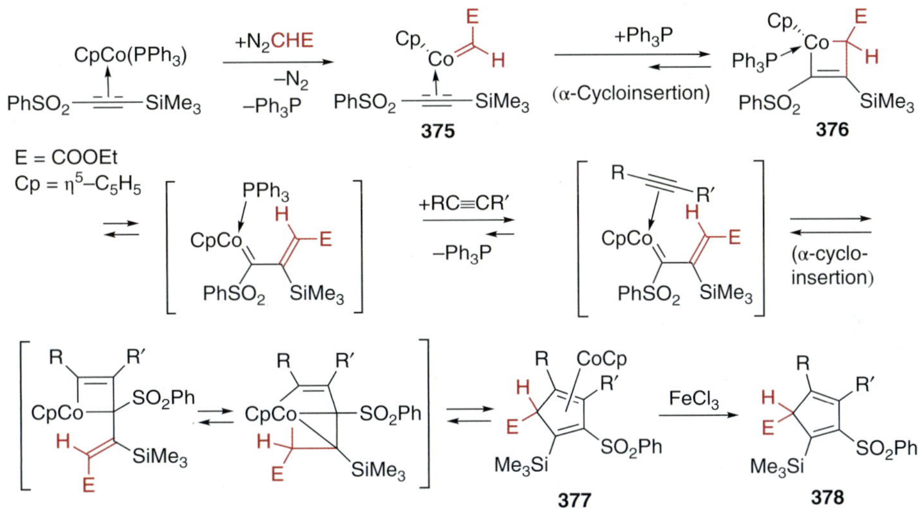

Scheme 7.66 O'Connor's synthesis of cyclopentadienes implying the cocondensation of two alkynes and a diazoacetate.

suggested that the ruthenacyclobutene is not an intermediate between **372** and **373** in the catalytic cycle. Instead, **373** is formed directly from **372** [1266].

O'Connor et al. [1267] have observed that cobaltacyclobutene **376** resulting from the (alkylidene)(alkyne)cobalt complex **375** reacts with an alkyne R—C≡C—R′ giving cyclopentadiene **378** after oxidation of complex **377** (Scheme 7.66).

Problem 7.30 Propose a mechanism for the reaction of **A** with the (alkylidene)chromium complex **B** that gives **P**. How can one prepare **B**? [1268].

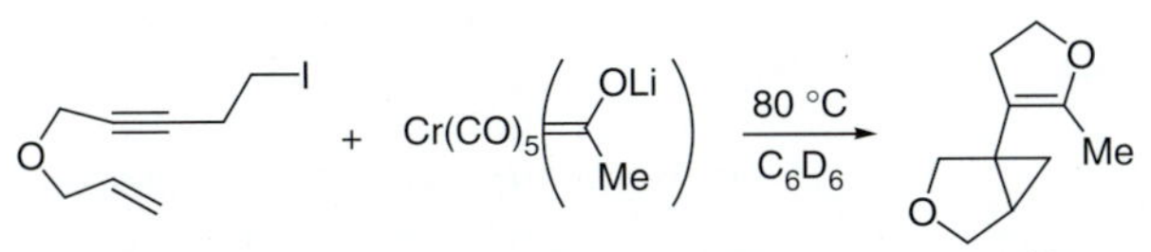

A **B** **P**

Problem 7.31 Propose mechanisms for all the synthetic steps given in Scheme 7.63.

7.7.13 Metallacyclobutadiene: alkyne metathesis

In 1968, Pennella et al. [1269] reported the first catalyst consisting of a heterogeneous mixture of tungsten oxide and silica that promotes alkyne metathesis at 200–450 °C (reaction (7.125)) [1270]. The second active catalytic system was introduced by Mortreux and Blanchard in 1974 [1271]. It consisted of $Mo(CO)_6$ and phenolic additives (e.g. 4-chlorophenol and resorcinol) and promotes the alkyne metathesis already at 60–130 °C [1272]. A major breakthrough in rational catalyst design for alkyne metathesis came with the (alkylidyne)tungsten complexes $(RO)_3W{\equiv}CR$ developed by Schrock's group [1251, 1273]. On his side, Fürstner et al. introduced a highly active monochloromolybdenum complex obtained by activation of Mo[N(*t*-Bu)(3,5-diMeC_6H_3)] with CH_2Cl_2 as chlorine source [1274, 1275]. Cummins and coworkers [1276, 1277] and Moore and coworkers have used Mo(≡CR)$(OAr)_3$ species as efficient catalysts [1278–1280]. The complex W(≡C-*t*-Bu)$(O\text{-}t\text{-}Bu)_3$ is active for the metathesis of internal alkynes, whereas the Mo analog and W(≡C-*t*-Bu)$(NR_2)_3$ are almost inactive [1281, 1282]. The (alkylidyne)molybdenum and tungsten complexes of type **379** containing fluoroalkoxide or phenoxide ligands react with internal alkynes to give isolable metallacyclobutadiene complexes **380** and are good catalysts for alkyne metathesis (Section 8.6.4) [1282].

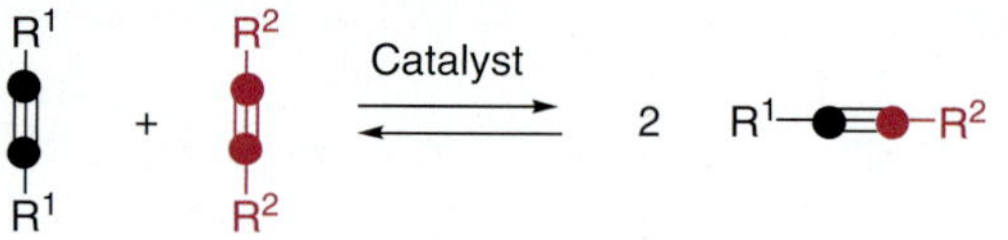

(7.125)

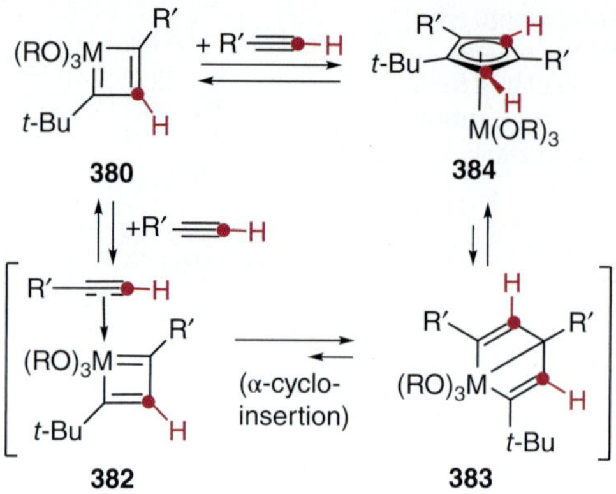

(7.126)

Molybdenum complexes in general are less reactive than their tungsten analogs. Molybdenacyclobutadienes of type **380** are much more prone to loose alkyne (⇒alkyne metathesis) than tungstenacyclobutadiene analogs. Molybdenum complexes of type **379** are more likely to catalyze the polymerization of alkynes, a reaction that competes with metathesis [1283]. The formation of **380** can be viewed as a two-step process in which **379** coordinates the alkynes to make a (alkylidyne)(alkyne)metal complex that undergoes a subsequent α-insertion.

Quantum mechanical calculations on the reactions of MeC≡CMe with W(≡CMe)$(OMe)_3$ suggested a one-step, direct (2+2)-cycloaddition [1284]. One interesting property of metallacyclobutadienes of type **380** is that they undergo facile base-induced β-elimination of 1 equiv. of alcohol ROH giving the "deprotiometallacyclobutadiene" complexes **381** (Scheme 7.67). Apart from (alkylidyne)Mo and W complexes of type **379**, a rhenium complex Re(≡C-*t*-Bu)(NAr)$(OCMe(CF_3)_2)$ was found to be an active catalyst for the alkyne metathesis of internal alkynes. The metallacyclobutadienes **380** are coordinately unsaturated and thus can coordinate with another unit of alkyne giving **382**. An α-cycloinsertion equilibrates **380** with metalla-Dewar-benzene **383**. This process is impeded using large alkoxy ligands. Further insertion of alkynes into **383** may lead to alkyne polymerization [1285]. Alternatively, **383** can isomerize into cyclopentadienyl complexes of type **384** (reaction (7.126)).

Problem 7.32 Propose a mechanism for the oligomerization and polymerization of terminal alkynes induced by an (alkylidyne)metal catalyst.

7.7.14 Metallacyclopentanes, metallacyclopentenes, metallacyclopentadienes: oxidative cyclizations (β-cycloinsertions) and reductive fragmentations (β-cycloeliminations)

Irradiation of $Fe(CO)_5$ at −30 °C in the presence of methyl acrylate generates the (dialkene)(tricarbonyl) iron complex **385** [1286]. At 20 °C and in the presence of CO, complex **385** undergoes an isomerization (β-cycloinsertion) into the ferracyclopentane **386**. According to Tolman's convention, the oxidation number of the iron atom is increased by two units going from **385** to **386**; therefore, one speaks of an oxidative cyclization (reaction (7.127)). If one considers the limiting structure shown below for **385**, isomerization **385** → **386** corresponds to a β-insertion of an alkyl moiety of the ferracyclopropane part into the alkene ligand. The reverse process **386** → **385** is obviously a reductive fragmentation of a metallacyclopentane into a dialkene complex, or a β-cycloelimination.

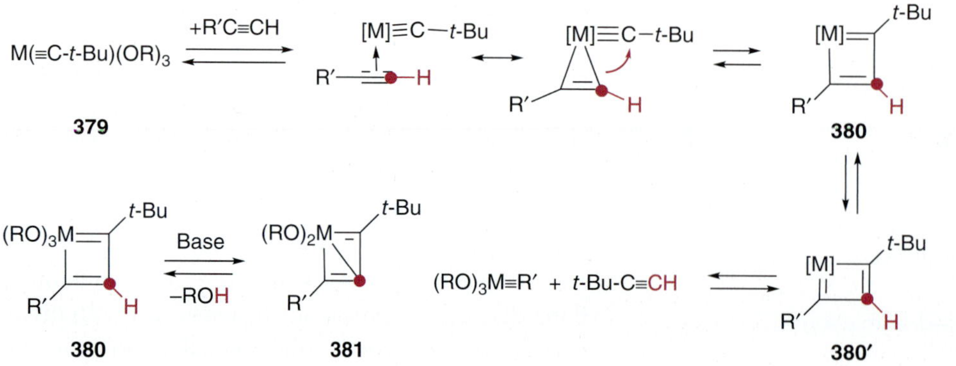

Scheme 7.67 Possible mechanism of alkyne metathesis involving a formal (2+2)-cycloaddition of an alkyne and alkylidynemetal complex and formation of a metallacyclobutadiene intermediate that undergoes valence bond isomerization.

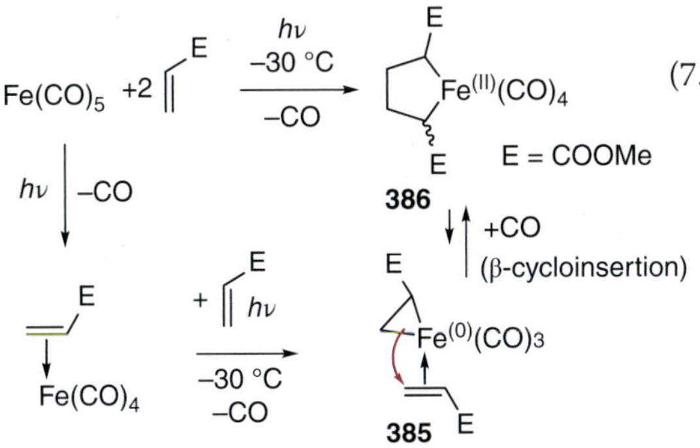

(7.127)

Isomerization **385** → **386** is modeled by the isomerization of protonated spirocyclopentane into cyclopentyl cation (Table 7.6) and the reverse reaction **386** → **385** is modeled by the gas-phase fragmentation of cyclopentyl cation into ethylene and allyl cation [1287–1290]. Metallacyclopentanes are intermediates (e.g. **387** and **388**) in processes that isomerize strained compounds possessing four-membered ring cycles such as cubane (reaction (7.128)) [1291, 1292] and 1,3-bishomocubane (reaction (7.129)) [1293].

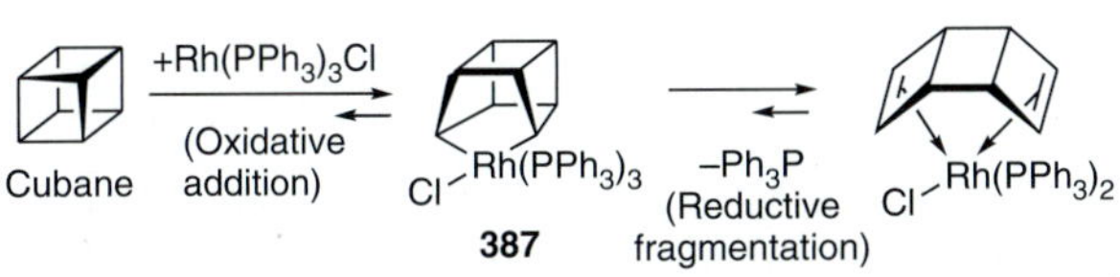

(7.128)

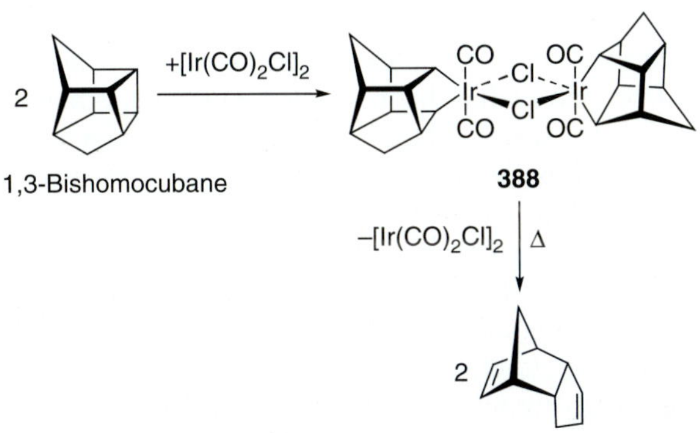

(7.129)

Depending on the nature of the metal, the colligands, and the valence electron count about the metal, metallacyclopentanes can undergo various reactions such as the formation of cyclobutanes via a reductive elimination, fragmentation into two alkenes via β-cycloelimination (reductive fragmentation), or β-elimination liberating an alkene [1294]. Metallacyclopentadienes are obtained by treatment of alkynes with coordinately unsaturated transition metal complexes. For instance, reduction of $FeCl_2(PMe_3)_3$ in the presence of but-2-yne produces a ferra(2,3,4,5-tetramethylcyclopentadiene) complex [1295]. They can react with a third molecule of alkyne and generate benzene derivatives (see Berthelot's cyclotrimerization of acetylene into benzene, Section 8.4.10). Metallacyclopentadienes can be intermediates in the formation of (cyclobutadiene)metal complexes starting with alkynes, in the tetracyclomerization of alkynes into cyclooctatetraenes (Section 8.4.18), and in alkyne metathesis (Section 8.6.4) [1296].

Problem 7.33 What are the products of reaction of Cp_2Zr(propene) with (a) deca-2,8-diyne; (b) with $Me_3SiCH{=}CH_2 + PhC{\equiv}CPh$? [1297].

7.8 Catalytic hydrogenation

One of the first references to the use of catalytic hydrogenation appears in 1863 when Debus produced methylamine by passing the vapor of HCN mixed with H_2 over Pt black. von Wilde in 1874 reported the reduction of acetylene and of ethylene to ethane in the presence or Pt black [1298]. However, the widespread use of catalytic hydrogenation did not start until 1897 when Sabatier and Senerens in Toulouse found that ethylene and hydrogen when passed over cobalt, iron, copper, or platinum at 300 °C gave ethane and that benzene and H_2 over nickel gave cyclohexane [1299]. They also discovered that the fluid consistency of vegetable oils is due to their lower hydrogenation content than solid fats such as butter or lard. In 1903, they reported that aldehydes and ketones can be reduced into alcohols with H_2 in the presence of nickel catalyst [1300, 1301] (Sabatier won the Nobel Prize together with Grignard in 1912). The first high-pressure catalytic hydrogenations were carried out over iron and nickel oxide by Ipatieff [1302, 1303]. Nowadays, hydrogenation of unsaturated compounds including alkenes, alkynes, arenes, heteroarenes, imines, aldehydes, and ketones is one of the most important processes utilized by the industry for the production of petrochemicals, fine chemicals, pharmaceuticals, materials, and food. The reactions are carried out on heterogeneous catalysts or with homogeneous catalysts that are generally based on expensive noble metals such as Pd, Pt, Rh, and Ir or toxic metals such as Ni and Co. When combined with enantio-enriched chiral ligands or used in chiral environment, they permit enantioselective catalytic hydrogenations. In 2001, for their work on asymmetric catalytic hydrogenation, Knowles [1304–1306] and Noyori [1307, 1308] received the Nobel Prize that they shared with Sharpless for his work on chirally catalyzed oxidation reactions [1309]. Kagan and coworkers also pioneered asymmetric catalytic hydrogenation and contributed significantly to asymmetric catalysis with his readily available tartaric acid (from renewable source) derived diphosphine ligands (e.g. (−)-2,3-*O*-isopropylidene-2,3-dihydroxy-1,4-bis (diphenylphosphino)butane, DIOP) [1310, 1311]. Efficient hydrogenation catalysts (working at room temperature under 1 atm of H_2, including asymmetric versions) based on inexpensive and nontoxic iron complexes [1312–1318] as well as nontransition metal based systems such as the frustrated Lewis pairs [1319–1322] are emerging and render hydrogenation reactions less costly and more environment friendly. In 2006, Stephan and coworkers discovered that *p*-$(Ar_2PH)^+$-C_6F_4-$BH(C_6F_5)_2$-H^- (Ar = 2,4,6-$Me_3C_6H_2$) liberates H_2 to yield *p*-$(Mes)_2P$-C_6F_4-$B(C_6F_5)_2$ reversibly [1323]. In 2007, they found that a bulky Lewis base such as *t*-Bu_3P and a bulky Lewis acid such as $B(C_6F_5)_3$ (a frustrated Lewis pair) is capable to activate H_2 through heterolytic cleavage [1324] and permits hydrogenation of imine, enamines, enol ethers, alkenes, alkynes, and anthracene [1325–1327]. In 1996, Parks and Piers discovered that $B(C_6F_5)_3$ catalyzes carbonyl hydrosilylation [1328, 1329]. Nowadays, hydrosilylation is a promising alternative to the catalytic hydrogenation of alkenes, alkynes, carbonyl compounds, carboxylic esters, carboxamides, and CO_2, especially when inexpensive PMBS (polymethylhydrosiloxane: $[-(Me(H)SiO)_n-]$ can be used [1330, 1331]. These reactions can be catalyzed by iron complexes [1332]. Enantioselective iron-catalyzed hydrosilylations of ketones and imines are realized in the presence of enantio-enriched chiral ligands [1333–1338].

7.8.1 Heterogeneous catalysts for alkene, alkyne, and arene hydrogenation

The now “classical” hydrogenation catalysts are supported noble metals, Raney nickel, and supported Ni and Cu, all of which are able to activate H_2 under mild conditions (1 atm, 25 °C) [1339–1341]. Because only surface atoms are active, the metal is used as very small particles in order to give the highest possible specific surface area. Quite often, the metallic particles are deposited on other materials or dispersed in a polymer. The trend today is to use metallic nanoparticles [1342–1348] and “green solvent” such as water [1345, 1349, 1350]. Nanoparticles have the tendency to aggregate into less active microparticles. One thus searches to stabilize them by additives such as organic surfactants, heteroatom ligands, ionic liquids, and polymers [1349, 1351]. Although a large number of studies have been carried out [1343, 1344, 1352, 1353], one does not have a precise molecular picture of the heterogeneous catalytic hydrogenation mechanism. Catalysts are therefore still chosen on an empirical basis by trial and error (combinatorial assaying) [1354–1356]. The addition of a second metal often improves the properties of a metallic catalyst. Classical examples are the Lindlar catalyst (Pd–Pb/$CaCO_3$) where the addition of Pb allows the selective monohydrogenation of acetylenes [1357], or the Nishimura catalyst (Rh–Pt oxide) for the mild hydrogenation of aromatic rings [1340]. For instance, in the presence of the latter catalyst, hydrogenation of chiral α-amino and α-hydroxyacids to the corresponding aminoalcohols and diols, respectively, occurs with high yields at 25 °C and with no racemization [1358]. Another example of bimetallic catalyst is Pt–Co/Al_2O_3, which promotes the hydrogenation of hex-2-enal to the corresponding allylic alcohol with 92% selectivity at 90% conversion [1359, 1360]. In 1934, Horiuti and Polanyi proposed a mechanism that consists of three steps: (i) alkene adsorption on the metallic surface of the hydrogenated metal catalyst, (ii) hydrogen migration to the β-carbon of the alkene with formation of a σ-bonded alkyl group forming an alkylmetal species, and (iii) reductive elimination of the free alkane [1361, 1362]. The second step is reversible and permits deuterium/hydrogen exchange when D_2 is used. Today, this mechanism is still debated. Most of the more recent work on alkene hydrogenation has been carried out with ethylene on Pt(111) surfaces [1363–1365] and has been aided by the preparation and characterization of related hydrocarbon intermediates [1366, 1367]. Several conclusions have been reached from those studies. First, thermal activation of C_2H_4 chemisorbed on transition metal surfaces leads to a number of competing reactions, including molecular desorption, dehydrogenation to other strongly bonded surface species, including ethylidyne (Pt≡C—CH_3; Scheme 7.68), hydrogenation to ethane, and H–D exchange. Second, the formation of Pt≡C—CH_3 involves more than one step, and most likely goes through the formation of ethylidene (Pt=CH—CH_3) intermediate. Third, ethylene hydrogenation under vacuum is stepwise via the rate-determining formation of an ethyl (Pt—CH_2CH_3) intermediate. Fourth, the catalytic hydrogenation of CH_2=CH_2 under atmospheric pressure takes place in the presence of strongly bonded carbonaceous deposits, a layer of

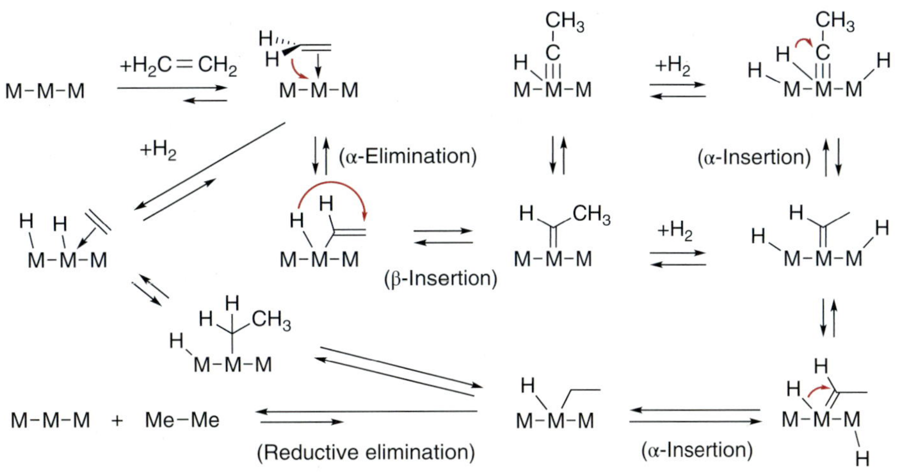

Scheme 7.68 Possible mechanism of the Pt(111) surface catalyzed hydrogenation of ethylene.

ethylidyne (M≡C—CH_3) in the case of Pt, Pd, and Rh surfaces; the role of those absorbed species in the mechanism of hydrogenation is still unclear. Finally, H–D exchange also takes place via ethyl–metal intermediates but occurs at temperatures much lower than those needed for any other reactions. The thermal chemistry (280 K) of propene on Pt(111) single-crystal surfaces produces H_2 from propene dehydrogenation (probably via Pt≡C—CH_2CH_3) and propane from self-hydrogenation of the olefin. Hydrogen coadsorption weakens the adsorption of propene and enhances the production of propane. With D_2, multiple H–D exchanges are observed, including formation of $CD_3CD{=}CH_2$ and $CD_3CD_2CD_3$. This result rules out any mechanism relying solely on the intermediacy of allyl–metal intermediates [1368, 1369].

Essentially monodisperse, small ruthenium nanoparticles (1–2 nm) are obtained by hydrogenation of [Ru(cod)(cot)] in THF at 25 °C (cod = cycloocta-1,5-diene, cot = cycloocta-1,3,5-triene). The nanoparticles can be stabilized by diphosphine ligands such as dppb (1,4-bis(diphenylphosphino)butane) and dppd (1,10-bis(diphenylphosphino)decane) or amines such as hexadecylamine. These nanoparticles are very active catalysts for hydrogenation. They contain hydrides on their surface as shown by NMR [1370] and titration by reaction with alkenes (1–1.3 hydrides *per* surface Ru atom). When ethylene is reacted at 25 °C with those nanoparticles in the absence of H_2, NMR demonstrates the presence of methyl-Ru moieties at their surface. They probably form according to the mechanism shown in Scheme 7.69 [1371].

Most metallic catalysts lead to *syn*-addition of H_2 across alkyne and alkene moieties. However, hydrogenation of pent-1-ene over Raney nickel has been found to be accompanied by some isomerization to both (*E*)- and (*Z*)-pent-2-ene [1372, 1373]. Exchange of hydrogen atoms between reactants and absorbed H_2 can be detected by isotopic exchange. Allylic positions undergo such exchange readily [1374]. Polar groups that can interact with the surface of the metallic catalyst might induce face-selective alkene *cis*-hydrogenations as illustrated with the Raney nickel catalyzed hydrogenation (7.130) [1375].

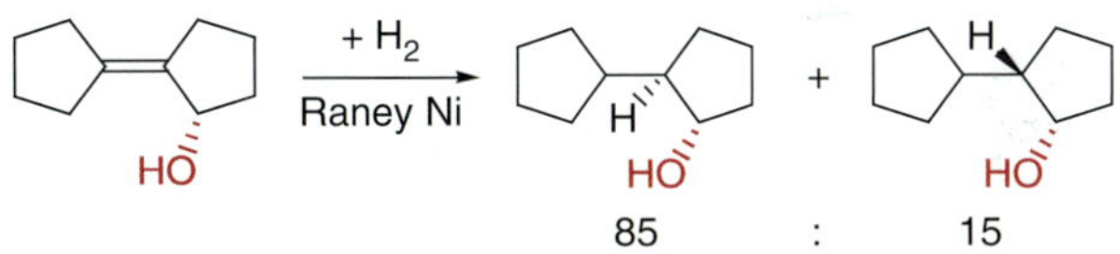

(7.130)

Preferred catalysts for the hydrogenation of carbocyclic and heterocyclic aromatic compounds are Rh > Ru > Pt > Ni > Pd, but this order of activity can change depending on the substitution of the substrates. Most functional groups, except carboxyl functions are usually hydrogenated before the aromatic rings. Sometimes, bimetallic catalysts (e.g. 4.5% Pd + 0.5% Rh/C) are more active than monometallic catalysts (e.g. 5% Rh/C) [1376]. Complexes of Nb and Ta alkoxide are also able to catalyze the hydrogenation of nonfunctional aromatic rings [1377]. Wilkinson and coworkers also found that $RuH(PPh_3)Cl$ (prepared form $RuCl_2(PPh_3)_3$ by base-promoted hydrogenolysis at 25 °C, 1 atm) is a more active hydrogenation catalyst for terminal alkene than $RhCl(PPh_3)_3$ in benzene [1378].

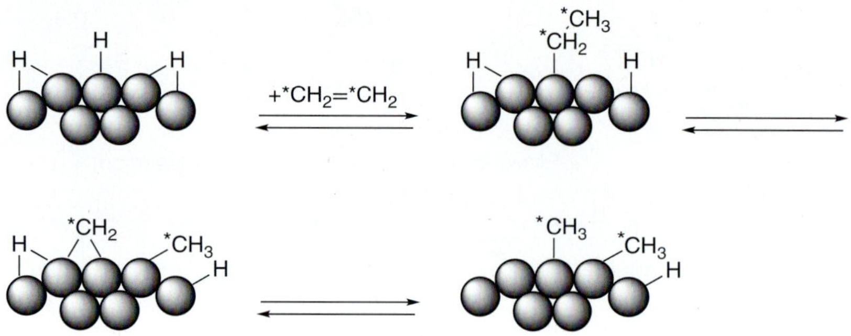

Scheme 7.69 Reaction of ethylene with hydrides at the surface of Ru nanoparticles.

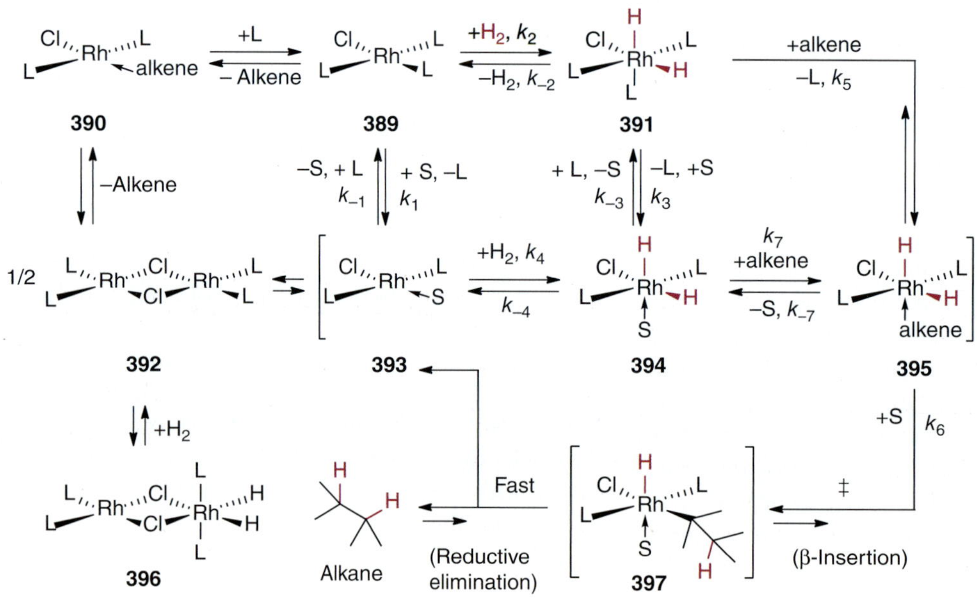

Scheme 7.70 Halpern's ("hydride route") mechanism for the catalytic hydrogenation of alkenes catalyzed by Wilkinson's catalyst **389** (S, solvent; alkene, cyclohexene; alkane, cyclohexane; L, PPh_3; k_6, rate-determining step).

7.8.2 Homogeneous catalysts for alkene and alkyne hydrogenation

In 1938, Calvin reported that CuOAc and cuprous salicylaldehyde in quinoline solution are homogeneous catalysts for the hydrogenation of cupric compounds and quinone [1379]. In 1939, Iguchi reported that $[Rh(NH_4)_5(H_2O)]Cl_3$ catalyzes the hydrogenation of benzoquinone into hydroquinone [1380]. He also found that $RhCl_3{\cdot}4H_2O$ catalyzes the hydrogenation of alkene moieties of carotene, fumaric acid, maleic acid, and the N=N double bond of azobenzene. In 1961, Halpern et al. reported the homogeneous hydrogenation of maleic acid to succinic acid in the presence of a Ru(II) species obtained by reaction of $TiCl_3$ on $(NH_4)_2RuCl_6$ [1381]. In 1966, the discovery of $RhCl(PPh_3)_3$ (**389**), commonly referred to as "Wilkinson's catalyst," marked the entry of modern homogeneous catalysis [1382]. It catalyzes the homogeneous hydrogenation of alkenes and alkynes to alkanes and of unsaturated aldehydes to saturated aldehydes [1383]. In 1971, Schrock and Osborn reported the cationic complexes of type $[Rh(diene)L_2]^+$ (diene = cycloocta-1,5-diene, norbornadiene; L = PPh_3, $AsPh_3$) that react with H_2 to generate cationic complexes of type $[Rh(PPh_3)_2(H_2)S_2]^+$ (S = solvent) with the elimination of the corresponding cycloalkane [1384, 1385]. The latter complexes react with Ph_3P to give cations of type $[RhL_4H_2]^+$, or alternatively with CO yielding complexes such as $[Rh(PPh_3)_3(CO)_3]^+$. This sets the basis for the subsequent development in asymmetric hydrogenation. Combining kinetic studies, equilibrium parameters of individual reactions, and spectroscopic measurements (NMR), Halpern et al. proposed the mechanism of Scheme 7.70 for the hydrogenation of alkenes catalyzed by $RhCl(PPh_3)_3$. The rate law derived from this mechanism ("hydride route") is given by Eq. (7.131) where alkene = cyclohexene and L = PPh_3 [1386, 1387]. Species **389–392** and **396** have been

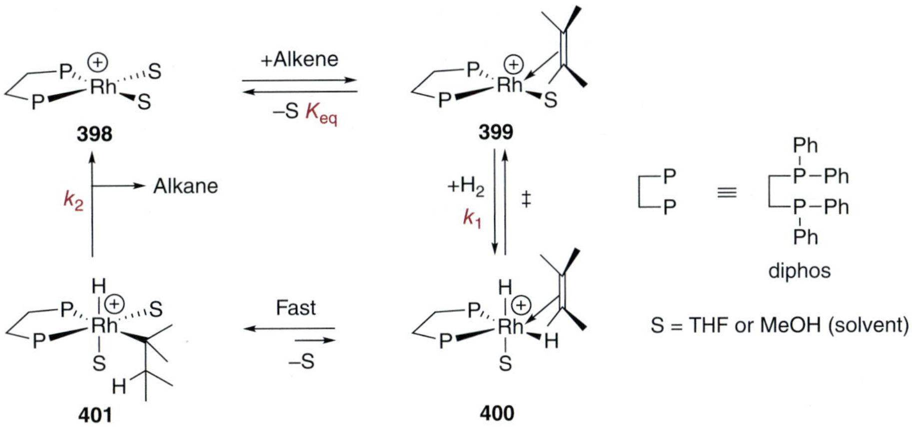

Scheme 7.71 Halpern's "olefin path" mechanism for the hydrogenation of alkenes catalyzed by $[Rh(diphos)(solvent)_2]^+$.

characterized in the solid state and/or in solution.

$$[k_{cat}]^{-1} = \frac{[Rh]}{rate} = \frac{1}{k_6} + \frac{[L]}{K_5k_6[alkene]} + \frac{[L]}{K_4k_4[H_2]} \tag{7.131}$$

Oxidative addition (Sections 7.7.6 and 7.7.8) of H_2 onto Wilkinson's catalyst (**389**) gives dihydride **391**, which can exchange one PPh_3 ligand by a molecule of solvent, giving **394**, or by one alkene unit, equilibrating (equilibrium constant K_5) with **395** [1388]. A parallel pathway implies first exchange of a PPh_3 ligand by a solvent molecule giving intermediate **393**, the latter adds H_2 furnishing dihydride **394**, which then exchanges its solvent ligand for an alkene molecule. In **395**, the template effect of the metallic complex plays its role in bringing together both reactants (alkene + H_2) about the same metallic center. The rate-determining step (k_6) is the β-insertion (Sections 7.7.6 and 7.7.10) **395** → **397**. A reductive elimination generates the product (alkane) and the unsaturated Rh(I) species **393** that starts another catalytic cycle. When using a more electron-rich alkene such as styrene, a slightly different mechanism in which a second, parallel path involves the coordination of two styrene molecules [1389]. In this case, olefin binding to the rhodium precedes H_2 addition ("olefin path"). Competition between the "hydride route" and "olefin path" depends not only on the nature of the alkene but also on that of the phosphine ligands.

With cationic rhodium catalysts of type $[RhL_2(S)_2]^+$ (e.g. **398**) (S = THF, MeOH; L_2 = two tertiary phosphine chelating ligands such as diphos = $Ph_2PCH_2CH_2PPh_2$), the mechanism shown in Scheme 7.71 has been proposed [1390, 1391], with rate equation (7.132). It involves an olefin rather than a hydride path. The kinetically important sequences is (i) alkene coordination **398** + alkene → **399**, (ii) H_2 oxidative addition **399** → **400**, (iii) hydrogen β-insertion **400** → **401**, and (iv) reductive elimination **401** → **398** + alkene. Complexes **398**, **399**, and **401** have been detected and characterized by spectroscopy. Equilibrium constant K_{eq}(**398** + alkene ⇄ **399**) depends strongly on the structure of the alkene. Even aromatic compounds such as benzene coordinate to **398**, but are not reduced. They are competitive inhibitors of the alkene hydrogenation.

$$\text{rate} = k_{cat}[Rh][alkene][H_2] \quad k_{cat} = \frac{k_1K_{eq}}{1 + K_{eq}[alkene]} \tag{7.132}$$

A mechanism similar to that of Scheme 7.71 has been proposed for the $RuH(PPh_3)_3Cl$ [1392] and $[RuH(CO)(MeCN)_2(PPh_3)_2]BF_4$-catalyzed hydrogenation of alkenes [1393]. Marks and coworkers have provided organolanthanide complexes of type $(Cp^*_2MH)_2$ (Cp^* = η^5-Me_5C_5, M = La, Nd, Sm, Lu; $(Cp^*)_2$ = $Me_2Si(C_5Me_4)_2$, M = Nd, Sm, Lu) that are extremely active as alkene hydrogenation catalysts (25 °C, 1 atm H_2) [1394]. Kinetic analysis, together with substrate selectivities, labeling experiments, and stoichiometric model reactions supported the mechanism shown in Scheme 7.72, which involves

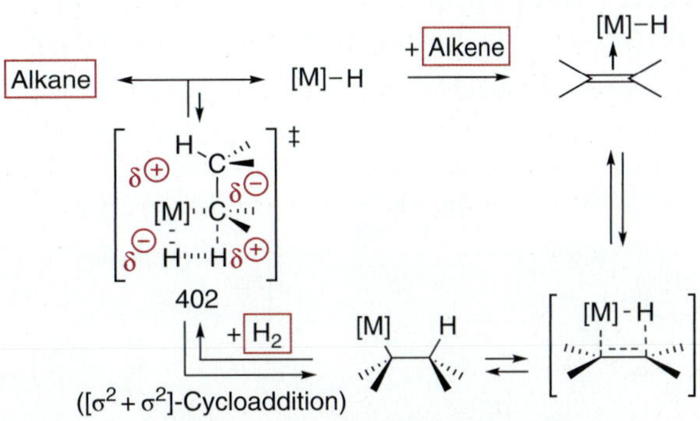

Scheme 7.72 The mechanism for the hydrogenation of alkenes catalyzed by organolanthanides.

olefin/hydride insertion followed by a four-center hydrogenolysis (transition structure **402**) rather than via classical oxidative addition/reductive elimination sequence.

Complexes of palladium have been used to catalyze alkene and alkyne hydrogenation. For instance, $Pd(OAc)_2 + P(t\text{-}Bu)_3$ is an efficient catalyst for selective reduction of alkenes under hydrogen transfer conditions [1395]. The source of H_2 is HCOOH, thus avoiding the hazards associated with H_2 gas (reaction (7.133)) [1396].

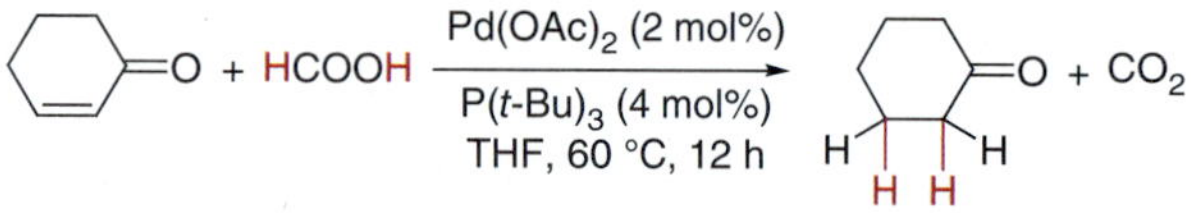

(7.133)

The active palladium species is probably $Pd[P(t\text{-}Bu)_3]_2$ that forms upon reduction of $Pd(OAc)_2$ by the phosphine ($Pd(OAc)_2 + 3R_3P + H_2O \rightarrow Pd(R_3P)_2 + R_3P{=}O + 2AcOH$) [1143, 1150, 1397, 1398]. This species reacts with HCOOH forming a dihydride and CO_2. Olefin coordination may precede or succeed this step as shown in Scheme 7.73.

Hydrogen transfer from $Me_2NH{\cdot}BH_3$ to alkenes is catalyzed by $Re^{(I)}$ complexes of type $(R_3P)_2(L)Re(NO)Br_2$ (R = *i*-Pr, *c*-C_6H_{11}; L = H_2, CH_3CN, $CH_2{=}CH_2$). The corresponding alkanes are formed in high yield (dioxane, 85 °C) together with $(H_2B{-}NH_2)_2$ [1399]. The mechanism of catalytic hydrogenation with homogeneous Rh and Pd catalysts is strongly influenced by substrate polarity. It is generally assumed that the rate-determining step (e.g. k_6: β-insertion **395** → **397**, Scheme 7.70) implies a nucleophilic transfer of hydrogen from the Rh–H to the coordinated alkene. It appears that Rh–H is polarized into Rh^+/H^- rather than into Rh^-/H^+. In the case of hydrogenation catalyzed by $Pd(diphos)_2$ (bis[1,2-bis(diphenylphosphino)ethane]palladium(0)), both modes of polarization Pd–H ↔ Pd^+/H^- and Pd–H ↔ Pd^-/H^+ intervene. This viewpoint [1400] is based on the observation that hydrogen atoms of alkenes can be exchanged by deuterium with homogeneous catalysts [1401–1403] and that deuteration of (*Z*)-crotonic acid is accompanied by the formation of deuterated (*E*)-crotonic acid. In the case of reaction (7.134) catalyzed by $Pd(diphos)_2$, the recovered (*E*)-cinnamic acid contains 37% D at the β-carbon center and 6% at the α-carbon center. When catalyzed by Wilkinson's catalyst ($RhCl(PPh_3)_3$), 22% of D resides on the β-carbon center, and none is found on the α-carbon center (reaction (7.135)). In contrast, although the $Pd(diphos)_2$-catalyzed deuteration (7.136) of (*E*)-β-methoxystyrene (**403**) leads to H/D exchange at both C(1) and C(2) positions, the $RhCl(PPh_3)$-catalyzed deuteration generates only the alkene deuterated at the C(2) position (reaction (7.137)). In support to the above hypothesis, it is observed that homogeneous hydrogenations catalyzed by the Wilkinson's catalyst are faster for electron-poor alkenes than for electron-rich alkenes [1404].

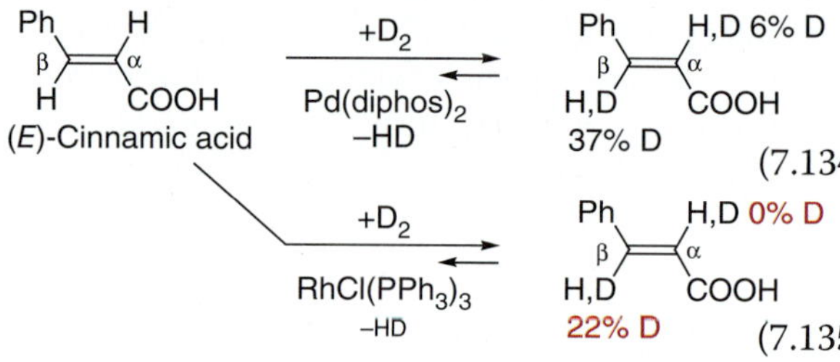

(7.134)

(7.135)

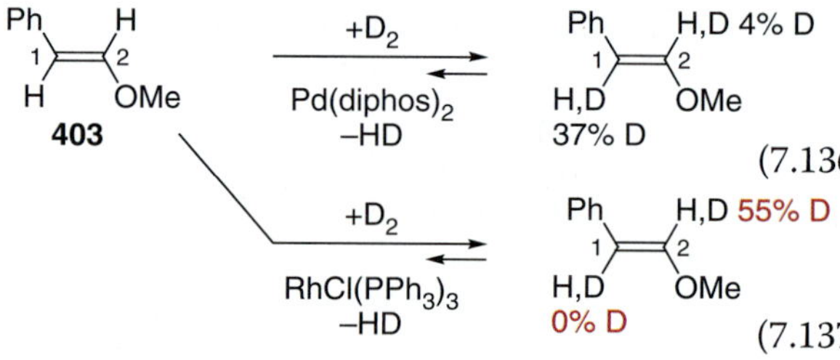

(7.136)

(7.137)

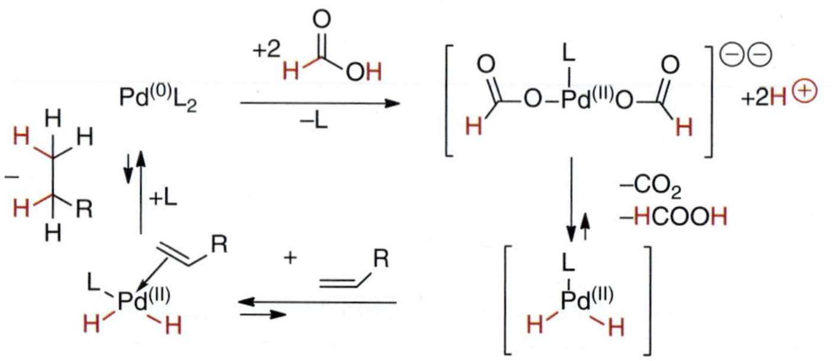

Scheme 7.73 Possible mechanism for the Pd(0)-catalyzed hydrogenation of alkenes using HCOOH as the hydrogen source (hydrogen transfer reaction).

Hydridoruthenium complex $RuH(CO)(P\text{-}Cy_3)_2Cl$ and $RuH(CO)(NHC)(PPh_3)Cl$ (NHC: N-heterocyclic carbene) are good catalysts for hydrogenation of alkenes [1405, 1406]. Substitution of Ph_3P in Wilkinson's catalysts ($Rh(Ph_3P)_3X$) by NHCs and phosphites (e.g. $P(OPh)_3$) gives a good homogeneous catalyst for alkene hydrogenation [1407, 1408].

7.8.3 Dehydrogenation of alkanes

Oxidative dehydrogenation of alkanes to give alkenes and H_2O is catalyzed by finely dispersed platinum metal [1409, 1410], vanadium–magnesium oxides [1411], or vanadium supported on mesoporous silica [1412]. Neutral atoms of Rh, Pd, Ir, Pt, lanthanides, and thorium react in the gas phase with CH_4. Most transition metal cations react with CH_4 and larger alkanes in the gas phase [1413]. Cationic platinum clusters, Pt_n^+ ($1 \leq n \leq 21$), also react with alkanes in the gas phase with reaction rate trend: $k(CH_4) < k(C_2H_6) < k(C_3H_8)$. Methane dehydrogenation occurs according to a 1,1-elimination (or α-elimination) mechanism. Ethane dehydrogenation takes place according to both a 1,1- and a 1,2-mechanism, both leading to $[Pt_nC_2H_4]^+ + H_2$. In the presence of D_2, all hydrogen atoms in $[Pt_nC_2H_4]^+$ are exchanged with deuterium atoms [1414]. Stoichiometric dehydrogenation of alkanes was reported by Crabtree et al. in 1979 [1233]. For instance, cyclopentane reacts with $[IrH_2(acetone)(PPh_3)_2]^+/BF_4^-$ to give $[(cyclopentadienyl)IrH(PPh_3)_2]^+/BF_4^-$ with t-BuCH=CH$_2$ as "hydrogen acceptor." Oxidative addition of an alkane C—H bond is the initial step of the pathway as observed by Janowicz and Bergman in 1982, via photogeneration of $Cp^*Ir(PMe_3)$ from $Cp^*Ir(H)_2(PMe_3)$ (Cp = cyclopentadienyl, Cp* = pentamethylcyclopentadienyl) and reaction from the alkane solvent, RH, to give corresponding $Cp^*Ir(R)(H)(PMe_3)$ products [1042, 1415]. Felkin and coworkers found selective conversion of cycloalkanes into cycloalkenes using $ReH_7(PR_3)_2$ catalyst and t-Bu—CH=CH$_2$ as the hydrogen acceptor [1416]. TON up to 1000 were observed for reaction (7.138) catalyzed by the pincer complexes **404** [1417] or **405** [1418].

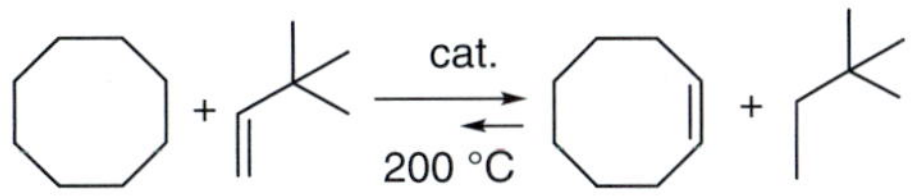

(7.138)

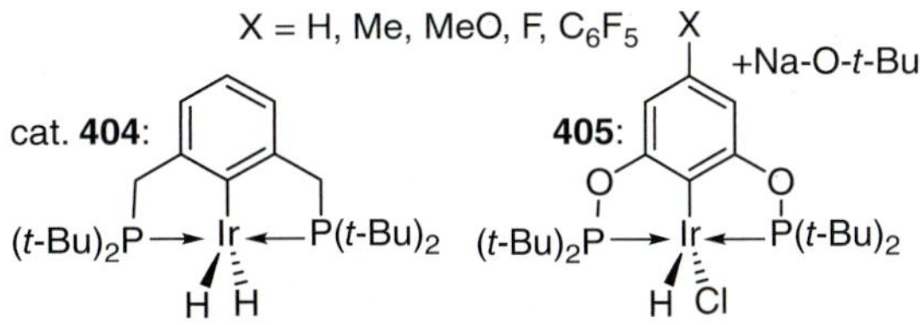

Methane steam reforming (MSR) is an established process for large-scale production of H_2 in industry ($CH_4 + H_2O \rightleftarrows CO + 3H_2$; $CO + H_2O \rightleftarrows CO_2 + H_2$) [1419–1421]. At 800 °C and in the presence of supported Ni-catalyst incorporating CeO_2, high level of H_2 production with no significant carbon deposition has been demonstrated [1422].

7.8.4 Hydrogenation of alkynes into alkenes

A number of complexes are known to reduce alkynes selectively into alkenes [1423, 1424]. For instance, $OsH(CO)(PR_3)Cl$ (PR_3 = $PMe(t\text{-}Bu)_2$ or $P(i\text{-}Pr)_3$: the "Zaragoza–Würzburg catalysts" [1425]) catalyze the sequential hydrogenation of phenylacetylene in 2-propanol solution at 60 °C. Selectivities close to 100% were obtained for the hydrogenation of alkyne to alkene. The kinetics give a rate = k_{cat} [Os][PhC≡CH][H_2]. The slow step is the reaction of vinylosmium intermediates $Os(CH{=}CHPh)Cl(CO)(PR_3)$ with H_2 to yield styrene (Scheme 7.74). In the absence of PhC≡CH, the

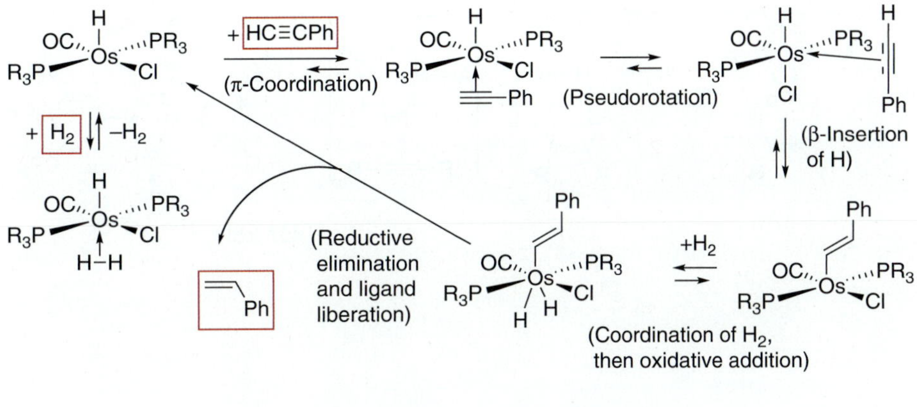

Scheme 7.74 Possible mechanism for the hydrogenation of phenylacetylene into styrene catalyzed by $OsH(CO)(PR_3)Cl$ complexes.

same osmium complex catalyzes the hydrogenation of styrene to ethylbenzene at rates about three times faster than those of the alkyne hydrogenation, although the reaction is zero order with respect to $[H_2]$ (rate = k_{cat}[Os][alkyne]) and so, in this case, H_2 addition is not the rate-determining step. Hydrogenation of styrene is the kinetically favored reaction, and thus, the high selectivity for reducing the alkyne to the alkene must have a thermodynamic origin, which is the formation of the stable vinylosmium intermediates. Thus, the alkyne is an efficient inhibitor of the hydrogenation of the alkene [1426].

Cationic complex [Ir(cycloocta-1,5-diene)(η^2-$Pr_2PCH_2CH_2OMe$)]BF_4 [1427] and *cis*-dihydride {[$P(CH_2CH_2PPh_2)_3$]Fe(H)(H_2)}BPh_4 [1428] have been shown to be good catalysts for the hydrogenation of alkynes into alkenes [1408, 1429]. With catalyst **406**, a diimine–palladium complex, the yield of hydrogenation (7.139) of internal alkynes into corresponding (*Z*)-alkenes is very high (>99%) and occurs under mild conditions (25 °C, 1 bar H_2, TOF = 100–200 h^{-1}). The chemoselectivity is also remarkable as other reducible functions such as alkenes in conjugated enynes, carboxylic and nitro groups are not affected [1430, 1431]. Palladium NHC complex also catalyzes the semihydrogenation of alkynes [1432]. Milstein and coworkers have reported on the (*E*)-selective semihydrogenation of alkynes catalyzed by an iron pincer complex derived from 1,8-bis(diphenylphosphino)acridine [1433].

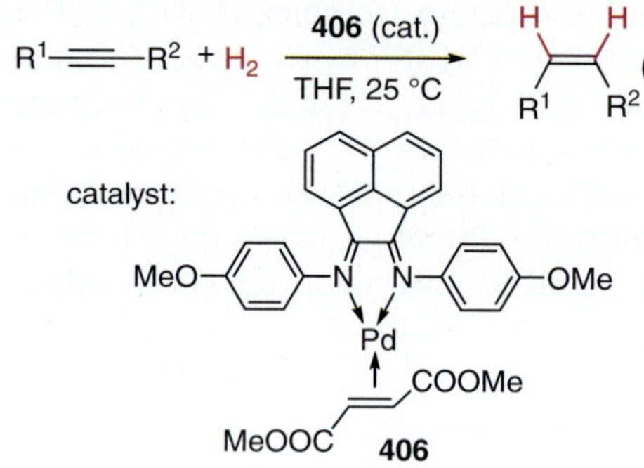

(7.139)

Kinetics of the stereoselective semihydrogenation of oct-4-yne by catalyst **407** are consistent with the mechanism shown in Scheme 7.75 with rate law (7.140). The first step of the process is the alkyne coordination and then heterolytic H_2 activation (hydrogenolysis of one Pd—N bond). Then the alkyne is hydropalladed, followed by addition of N–H to Pd and reductive coupling of the vinyl and hydride moieties. This generates an alkene complex that is finally displaced by the alkyne liberating the product (alkene). Under hydrogen-limiting conditions, side reactions occur such as formation of catalytically inactive palladacycle **408** by oxidative alkyne coupling. It is also found that (*Z*)-oct-4-ene is the primary product of reaction, from which the minor product (*E*)-oct-4-ene is formed by an H_2-assisted, palladium-catalyzed isomerization reaction [1408, 1434].

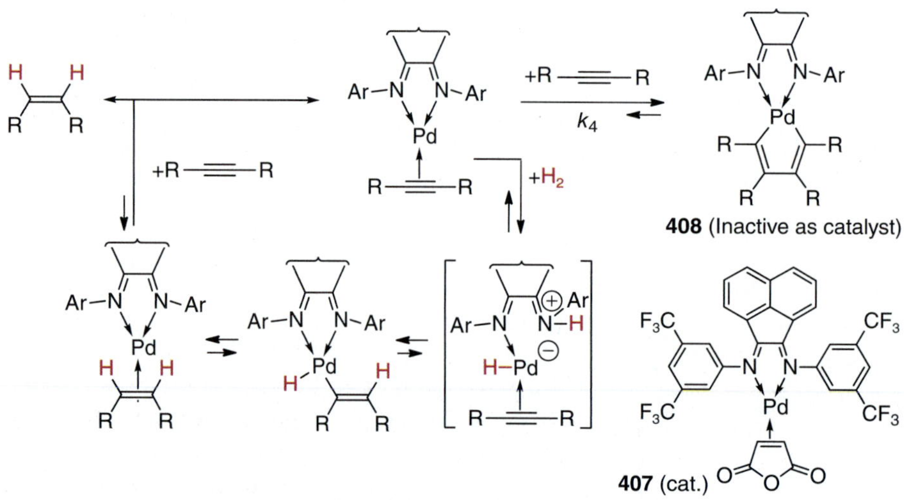

Scheme 7.75 Catalytic cycle of the (*Z*)-selective semihydrogenation of oct-4-yne. In the absence of H_2 pallacycle, **408** forms.

$$-\mathrm{d}[\mathrm{oct}-4-\mathrm{yne}]/\mathrm{d}t = \mathrm{k_{cat}} \times [\mathrm{oct}-4-\mathrm{yne}]^{0.65}[\mathbf{407}][\mathrm{H_2}] \quad (7.140)$$

7.8.5 Catalytic hydrogenation of arenes and heteroarenes

Catalytic hydrogenation of arenes and heteroarenes uses catalysts such as dispersed Rh, Pt, Pd, Ni, Ru, or Os metal (or mixtures of the latter) on charcoal or alumina under pressurized H_2, at room temperature or above [1435–1438]. For instance, 1-and 2-substituted indanes [1439, 1440] and 1,2,3,4-tetrahydronaphthalenes [1436, 1441] are hydrogenated into product mixtures in which the corresponding substituted *cis,cis*-bicyclo[4.3.0]nonanes and bicyclo[4.4.0]decanes, respectively, are the major products [1442]. Catalytic hydrogenation of Equilin in acidic media using PtO_2 (Adams catalyst) as a catalyst led to complete reduction of its aromatic rings with concomitant C—O bond hydrogenolysis and reduction of the carbonyl group. However, when using Ni and Ru catalysts, chemoselective reductions (7.141) and (7.142) are possible. (−)-1-(2-Pyridyl)-2-propanol is hydrogenated over PtO_2 in AcOH [1443] or EtOH [1444] giving (−)-sedridine in high yield and high diastereoselectivity (reaction (7.143)).

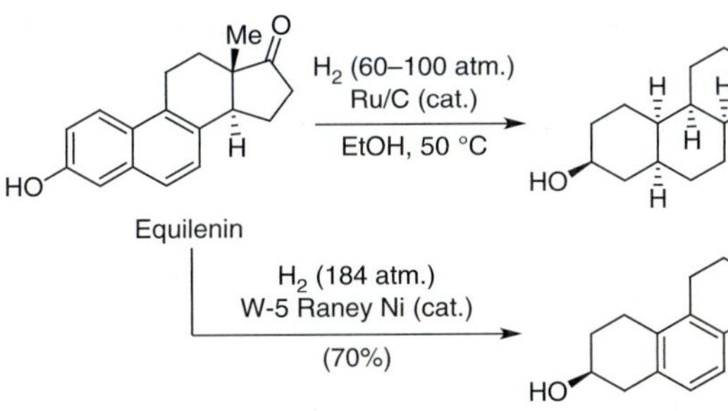

(7.141)

(7.142)

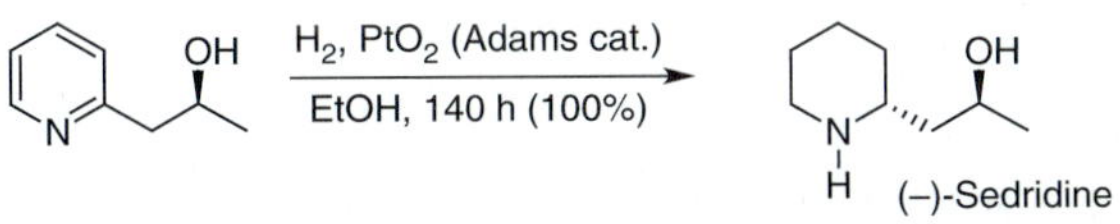

(7.143)

In 2000, Kuwano et al. reported a highly enantioselective hydrogenation of indoles by using the enantiomerically pure rhodium complex made of $[\mathrm{Rh(norbornadiene)_2}]\mathrm{SbF_6}$ and diphosphine ligand (*S,S*)-(*R,R*)-PhTRAP ((R,R)-2,2″-bis[(*S*)-1-(diphenylphosphino)ethyl]-1,1″-biferrocene; reaction (7.144)) [1445]. Since then, a large number of enantioselective catalytic hydrogenations of arenes and heteroarenes have been described [1446, 1447]. Reactions (7.145) [1447] and (7.146) [1448] are illustrations. Silyl enol ethers can be hydrogenated enantioselectively by using a "frustrated Lewis pair" as the catalyst that is a combination of $(t\text{-Bu})_3P$ and of a chiral borane generated in situ by the hydroboration of an enantio-enriched chiral diene with $HB(C_6F_5)_2$ [1321].

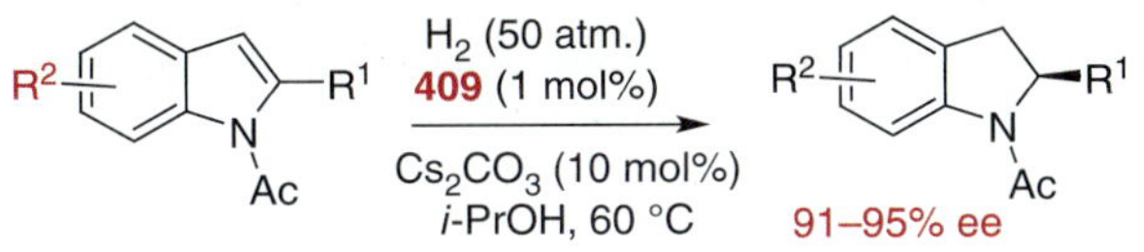

(7.144)

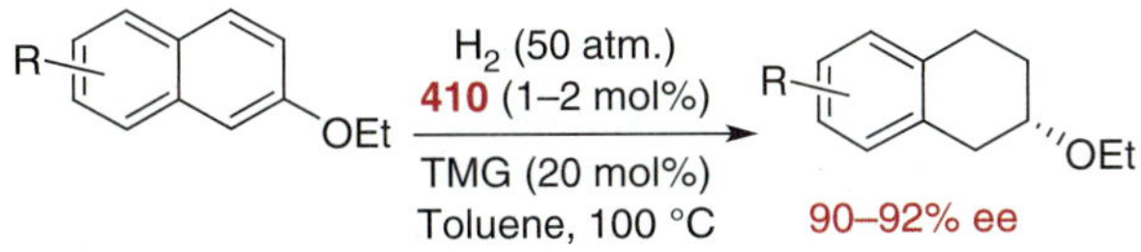

TMG = tetramethylguadinine = $(Me_2N)_2C{=}NH$

(7.145)

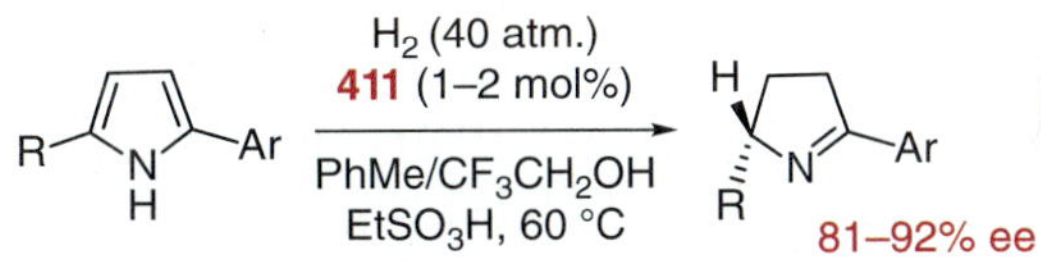

(7.146)

409: $[\mathrm{Rd(norbornadiene)_2}]\mathrm{PF_6}$ + (*S,S*)-(*R,R*)-PhTRAP

410: {Ru(*p*-cymene)[(*S,S*)-(*R,R*)-PhTRAP]}Cl

411: $\mathrm{Pd(OCOCF_3)_2}$ + C4-TunePhos

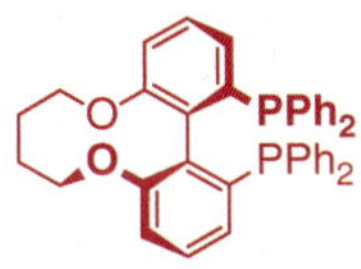

C4-TunePhos

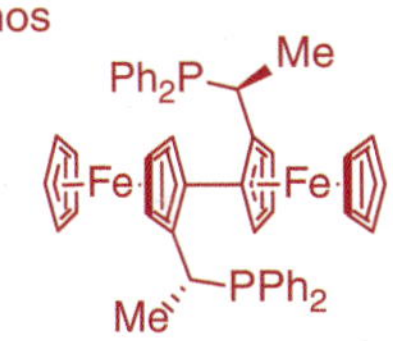

(*S,S*)-(*R,R*)-PhTRAP

7.8.6 Catalytic hydrogenation of ketones and aldehydes

For the heterogeneous hydrogenation of carbonyl compounds into alcohols and imines into amines, the preferred catalysts are highly dispersed (e.g. nanoparticles supported or not) of Pd, Pt, Ni, Ru, and Au. Chemo-, regio-, and stereoselectivity can be controlled by catalyst, solvent, pH, modifiers, and reaction conditions [1449]. Homogeneous Rh, Ru, Ir, and Fe complexes are used for hydrogenation with H_2 or transfer hydrogenation (the hydrogen sources being usually isopropanol, HCOOH, $NH_4^+/HCOO^-$, or $Et_3NH^+/HCOO^-$) [1395, 1450–1452]. Transfer hydrogenation of ketones and imines is well known and gives corresponding alcohols and amines [1453–1457]. One of the first examples of catalytic hydrogenation of carbonyl compounds was reported by Coffey in 1967 [1458]. Butanal was reduced into

1-butanol at 50 °C and under 1 atm of H_2 in the presence of $[IrH_3(PPh_3)_3]$ in acetic acid. The catalytic cycles can be divided into two parts: (i) the reaction of the ruthenium hydride with the unsaturated compound and (ii) the regeneration of the hydride from H_2 or from the hydrogen donor in transfer hydrogenations. Ruthenium homogeneous hydrogenation catalysts have been the most studied and often surpass rhodium and iridium complexes [1459, 1460]. The original Meerwein–Ponndorf–Verley mechanism for transfer hydrogenation (7.147) catalyzed by aluminum alkoxides might have a parallel in Ru-catalyzed reactions [1461–1465]. In this case, the substrate and a hydrogen donor coordinate to the metal and the hydride transfers directly from the donor to the carbonyl compound without formation of a ruthenium hydride. The Meerwein–Ponndorf–Verley reaction has been carried out traditionally using homogeneous catalysts such as Al alkoxides [1466], and more recently with SmI_2 [1467], Ti, Zr alkoxides [1468], or Li isoproxide [1469]. Recoverable heterogeneous catalysts such as activated Mg–Al hydrotalcite [1470] and hydrous zirconia supported by silica gel have also been applied [1471, 1472].

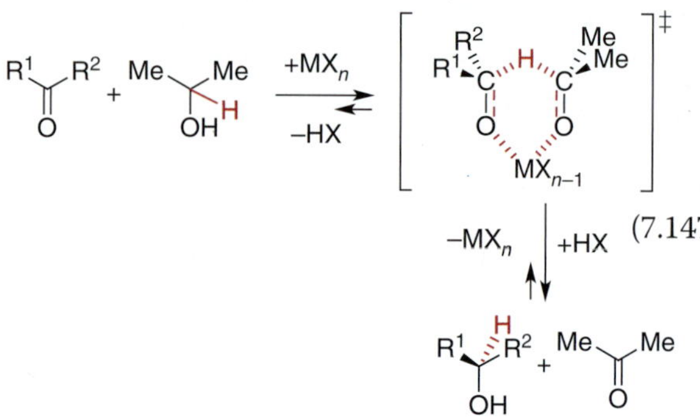

(7.147)

In the case of Ru-catalyzed hydrogenations, H_2 coordinates first to the metal as a η^2-H_2 ligand. A heterolytic splitting occurs and gives the ruthenium hydride and a protonated base (Scheme 7.76, **412** → **413**). In certain cases, the intramolecular heterolytic splitting can occur to give a hydride and a protonated ancillary ligand (**412** → **414**). The oxidative addition **412** + H_2 → **415** is a third option.

One possible mechanism for the homogeneous H_2-hydrogenation of carbonyl compounds is the inner sphere hydride transfer (Scheme 7.77a) and that is probably followed for the $Ru(H)(H_2)(PPh_3)_3$-catalyzed hydrogenation of cyclohexanone into cyclohexanol in toluene and by several other related reductions [1473]. One possible mechanism of $RuCl_2(PPh_3)_3$-catalyzed reduction of ketones by isopropanol is shown in Scheme 7.77b [1474, 1475].

The selective hydrogenation of α,β-unsaturated aldehydes into allylic alcohols can be carried out with H_2 in the presence of platinum group metals combined with a second more electropositive metal, a metal oxide, and/or an organic modifier [1476]. The use of supercritical CO_2 increases the chemoselectivity of the hydrogenation of cinnamaldehyde [1477]. Hex-2-enal is hydrogenated with $Pt–Co/Al_2O_3$ catalyst into hex-2-enol with 92% selectivity [1359]. A wide range of α,β-unsaturated ketones have been hydrogenated into the corresponding allylic alcohols using the Noyori catalysts Ru(binap)(diamine) (e.g. **416**) with high chemoselectivities and high enantiomeric excesses [1478]. An example is given with reduction (7.148). A very efficient chiral catalyst for the enantioselective hydrogenation of acetophenones derivatives (e.g. reaction (7.149)) is the iridium complex Ir-(*S*)-SpiroPAP-3-Me, which has shown TON up to 4 550 000 [1479–1481].

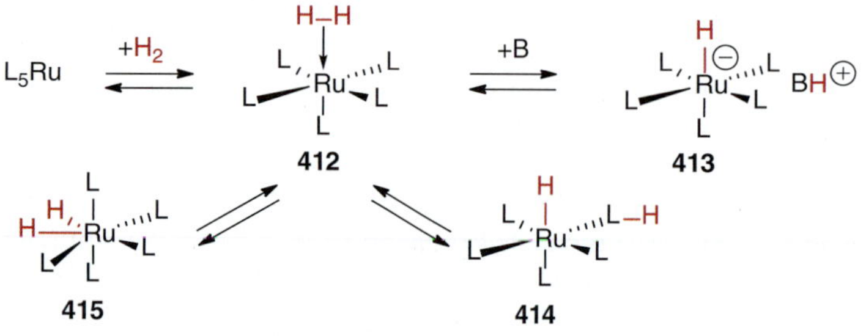

Scheme 7.76 Formation of ruthenium hydrides by reaction with H_2.

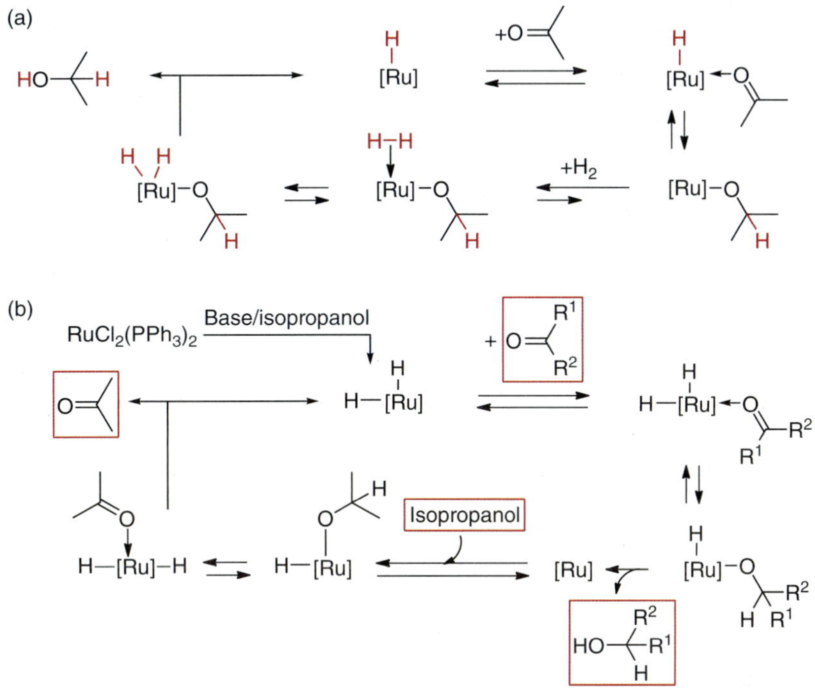

Scheme 7.77 (a) Catalytic cycle for the inner sphere H_2-hydrogenation of carbonyl compounds; (b) hydrogenation of ketones by transfer from isopropanol catalyzed by $RuCl_2(PPh_3)_3$/base ([Ru] = $Ru(PPh_3)_3$).

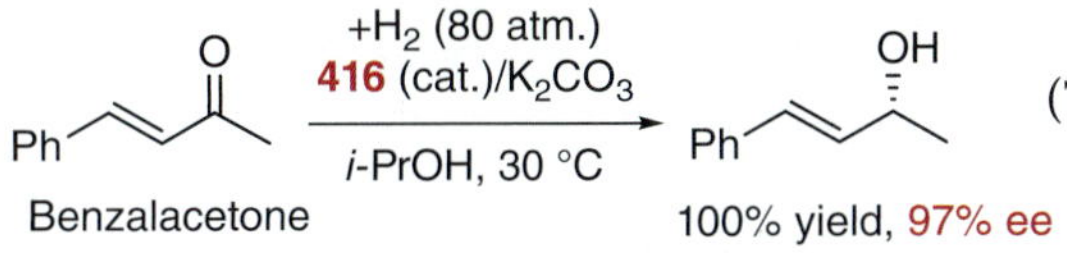

(7.148)

416:

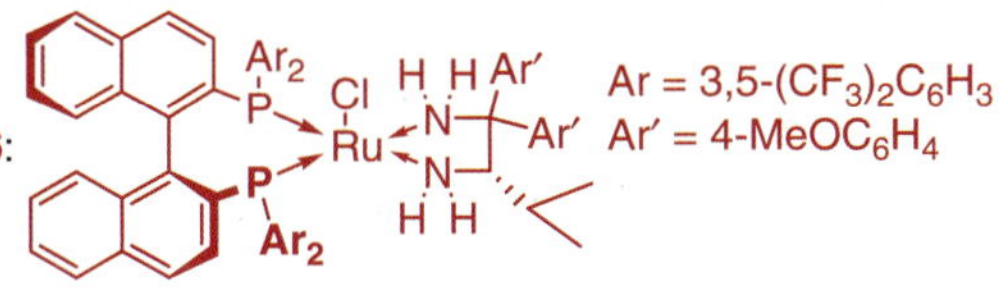

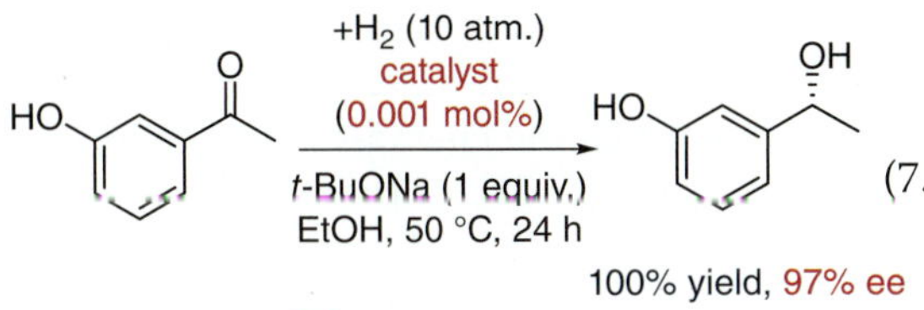

(7.149)

catalyst:

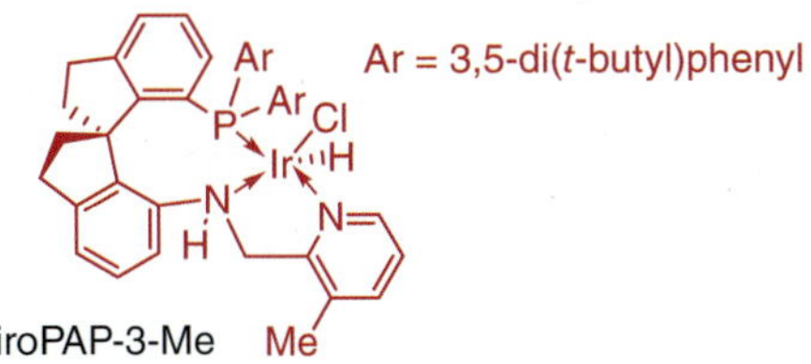

Ir-(S)-SpiroPAP-3-Me

Gold nanocolloids as recyclable quasi-homogeneous catalysts have been used in the chemoselective hydrogenation of α,β-unsaturated aldehydes and ketones into allylic alcohols [1482]. Casey and Guan have presented iron catalyst **417** for the hydrogenation of ketones and chemoselective hydrogenation of unsaturated ketones into unsaturated alcohols (reaction (7.150)) [1483]. Similar 1,2-selectivity has been found for other Fe-catalyzed hydrogenations [1315, 1484].

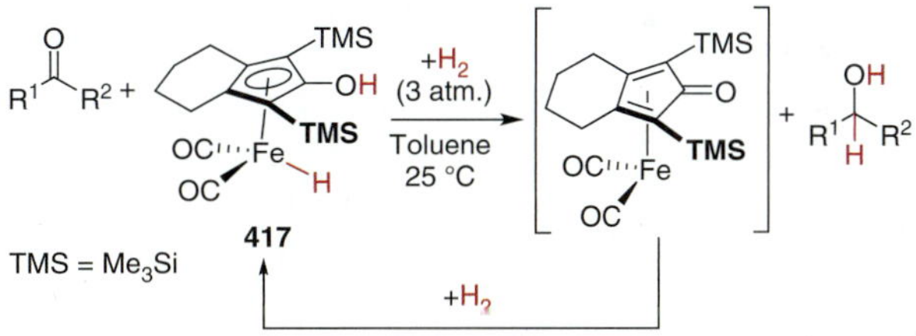

(7.150)

Hydrogenation of nonenolizable ketones without transition metal catalyst is possible under harsh conditions (200 °C, > 100 bar H_2, t-BuOK as a base) [1485–1487]. Alcohols can be temporarily converted into carbonyl compounds by metal-catalyzed (e.g. $RuCl_2(PPh_3)_3$, $CpRuCl(PPh_3)_2$, $[Cp^*IrCl_2]_2$, and $[Ir(cod)Cl]_2$) removal of H_2 (Cp = η^5-C_5H_5, Cp* = η^5-Me_5C_5, and cod = cycloocta-1,5-diene). Carbonyl compounds are reacted in situ with amines to give corresponding imines, or with enolates or ylides to give corresponding alkenes that are hydrogenated by the metal dihydride resulting from H_2 removal from the alcohol (borrowing H_2 in the activation of alcohols: see Section 8.3.10, direct hydrocarbation via hydrogen transfer) [1455]. An example is given with reaction (7.151).

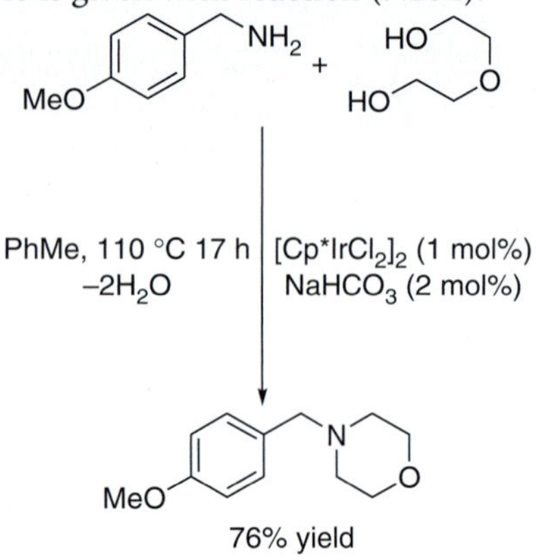

(7.151)

Problem 7.34 Give a possible mechanism for the chemoselective hydrogenation of cinnamaldehyde into (*E*)-3-phenylprop-2-en-1-ol catalyzed by the cationic iron complex $[P(o\text{-}Ph_2PC_6H_4)_3FeF]^+/BF_4^-$ in *i*-PrOH (CF_3COOH at 140 °C and under H_2 [20 bar]) [1315].

7.8.7 Catalytic hydrogenation of carboxylic acids, their esters and amides

The hydrogenation of carboxylic acids ($RCOOH + 2H_2 \rightarrow RCH_2OH + H_2O$) ($\Delta_rH°(AcOH + 2H_2 \rightleftarrows EtOH + H_2O) = -10.3 \pm 1.1$ kcal mol^{-1}) and their esters ($RCOOR' + 2H_2 \rightarrow RCH_2OH + R'OH$) ($\Delta_rH°(AcOEt + 2H_2 \rightleftarrows 2EtOH) = -5.5 \pm 1.2$ kcal mol^{-1}) is an important method of upgrading of bio-based starting materials such as seed and vegetable oils. The process generates fatty alcohols and other bulk chemicals [1488–1490]. Important materials such as polyesters and polyurethanes can be prepared using diols obtained via the hydrogenation of dicarboxylic acids and their esters. Compared with stoichiometric methods (Bouveault–Blanc reduction with Na/EtOH [1491, 1492], reduction with $LiAlH_4$ or $NaBH_4$) [1493], the catalytic hydrogenation is cheaper and produces much less waste. The catalytic hydrogenation of carboxamides into alcohols and amines ($RCONR'R'' + 2H_2 \rightarrow RCH_2OH + R'R''NH$) ($\Delta_rH°(n\text{-}C_3H_7CONH_2 + 2H_2 \rightleftarrows n\text{-}BuOH + NH_3) = -10.3 \pm 1.3$ kcal mol^{-1}) are more difficult but can be carried out [1494, 1495]. Acid anhydrides are more reactive toward reduction, followed by imides, esters, lactones, and carboxamides. Early work by Adkins et al. focused on Cu/Cr/Ba oxide-containing catalysts [1496]. In 1934, they reported that butyl oleate is hydrogenated over Cu/Zn oxide at 300 °C to give the corresponding unsaturated primary alcohol in 60% yield [1494]. Conventional catalysts are Cu + Zn chromite [1497, 1498], and Raney Ni [1499, 1500]. Metal oxides such as ZrO_2, CeO_2, ZnO, and combination of metal oxides such as SnO_2/Al_2O_3, TiO_2/Cr_2O_3, rare earth oxides, and CuO/Cr_2O_3 catalyze the selective hydrogenation of carboxylic acids into aldehydes ($RCOOH + H_2 \rightarrow RCHO + H_2O$). The process usually requires high temperature (300–500 °C) [1501]. Heterogeneous hydrogenation of some fatty esters is done in the industry at relatively high temperature (200–300 °C) and high H_2 pressures (200–300 atm) [1502]. An efficient homogeneous catalyst **418** for the catalytic hydrogenation of esters into alcohols at 115 °C in dioxane (H_2, 5.3 atm) has been developed by Milstein and coworkers [1503]. Catalyst **419** + NaOMe permits the chemoselective reduction of unsaturated esters [1504]. Milstein and coworkers

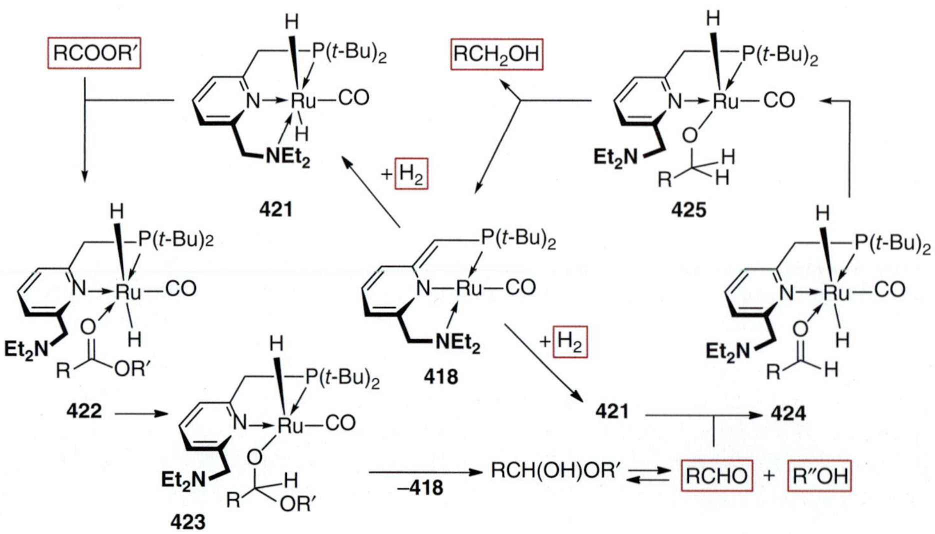

Scheme 7.78 Homogeneous catalytic hydrogenation of esters into alcohols.

also demonstrated that dearomatized bipyridyl-based PNN–Ru(II) (PNN: phosphine-diamine ligand) pincer complex **420** (Scheme 7.78) can serve as a homogeneous catalyst for the hydrogenations $ArCH_2NH–COR + 2H_2 \rightarrow ArCH_2NH_2 + RCH_2OH$ [1505, 1506]. The reverse reactions $RCH_2OH + RCH_2OH \rightarrow RCOOCH_2R + 2H_2$ [1507], $RCH_2OH + R'CH_2NH_2 \rightarrow RCONHCH_2R' + 2H_2$ [1508], and $RCOOH + HO^- \rightarrow RCOO^- + 2H_2$ are also catalyzed by the same pincer-type catalysts [1509].

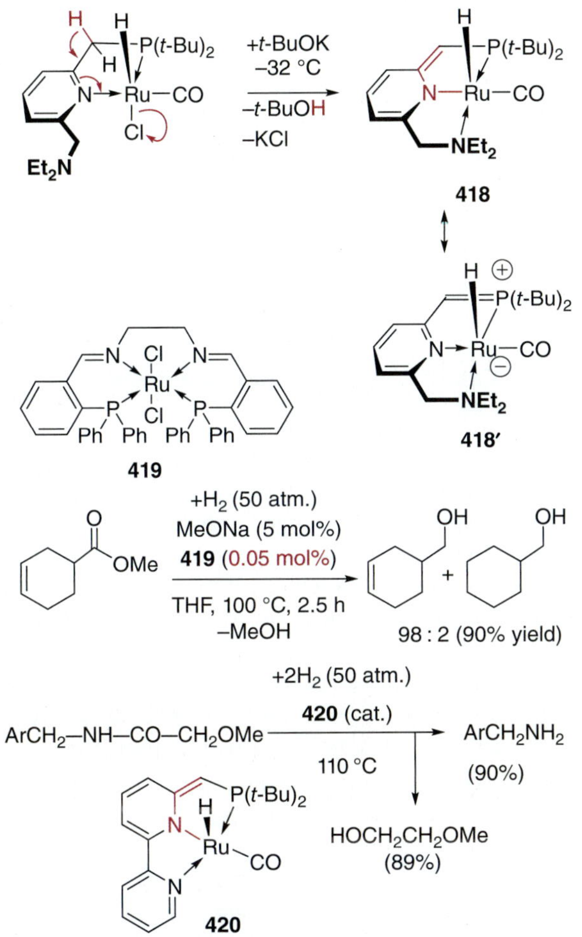

A possible cycle for the catalysis of ester hydrogenation by **418** is shown in Scheme 7.78. Initially, H_2 adds to **418** restituting aromaticity and forming a coordinately saturated ruthenium *trans*-dihydride complex **421**, as observed. Dissociation of the amine "arm" provides a site for ester coordination, giving intermediate **422**. Hydride insertion from the metal to the carbonyl moiety is a β-insertion furnishing a ruthenium alkoxide **423**. It dissociates with the formation of aldehyde RCHO and alcohol R′OH. Aldehyde RCHO then coordinates **418** giving **424**. Hydride insertion (β-insertion of hydrogen atom) gives alkoxide **425**. Reductive elimination provides alcohol RCH_2OH and the catalyst **418**. The success of complex **418** is probably due to the pincer ligand hemilability [1510–1512].

7.8.8 Hydrogenation of carbon dioxide

Atmospheric CO_2 is a major contributor to the greenhouse effect, and various strategies are being developed for lowering its concentration in the atmosphere [1513, 1514]. CO_2 is an interesting chemical feedstock of low cost and is not toxic. A variety of useful products can be prepared from it [1515–1518], and among them, formic acid and derivatives are prepared through catalyzed hydrogenation, a slightly exothermic ($\Delta_r H°(CO_2 + H_2 \rightleftarrows HCOOH) = -3.5$ kcal mol^{-1}) (Table 1.A.1) but endergonic ($\Delta_r S°(CO_2 + H_2 \rightleftarrows HCOOH) = -22.9$ eu) [165] (http://webbook.nist.gov) reaction under standard conditions. Thus, high pressure and neutralization of formic acid into formate salts are required to achieve the hydrogenation of CO_2 [1519–1523]. Obviously, plants are the best in converting CO_2 into a large variety of useful products through photosynthesis [1524], and one day, an artificial leaf will be created by chemists [1525]. Research on CO_2 capture also intends to use solar energy in several ways [1526]. Artificial photosynthesis uses reduction with photoinduced electrons, bulk electrolysis of CO_2-saturated solution with electricity produced by photovoltaics, CO_2 hydrogenation using solar produced H_2, and the thermochemical reaction of CO_2 catalyzed by metal oxides at extremely high temperature reached by solar collectors [1527]. Hydrogenation of CO_2 using Ni/SiO_2 heterogeneous catalyst (1.4 bar, 500–600 K) produces mainly CO and H_2O (the reverse water shift reaction) [1528, 1529]. High selectivity for CO formation was also observed with calcium-, magnesium-, and Pt-based catalysts dispersed on charcoal [1530]. Hydrogenation of CO_2 over Cu/ZrO_2 catalysts generates MeOH, an exothermic reaction under standard conditions ($\Delta_r H°(CO_2 + 3H_2 \rightleftarrows MeOH + H_2O) = -12.8$ kcal mol^{-1} but with a devaforable reaction entropy ($\Delta_r S°(CO_2 + 3H_2 \rightleftarrows MeOH + H_2O) = -42.3$ kcal mol^{-1}, Tables 1.1 and 1.4), which imposes to work under pressure, or at atmospheric pressure not above room temperature [1531–1533]. Low-pressure CO_2 hydrogenation into MeOH has been catalyzed by gold nanoparticles activated by CeO_x/TiO_2 [1534]. Methane is the main product formed by hydrogenation of CO_2 (a highly exothermic reaction that can pay for the entropy cost of condensation; $\Delta_r H°(CO_2 + 4H_2 \rightleftarrows CH_4 + 2H_2O) = -39.4$ kcal mol^{-1}, $\Delta_r S°(CO_2 + 4H_2 \rightleftarrows CH_4 + 2H_2O) = -40.7$ eu), Tables 1.A.1 and 1.A.4) on Rh-zeolithe catalysts, whereas CO is formed on Li-doped Rh-zeolithe catalysts [1535]. In imidazolium-based ionic liquid using Ru nanoparticles as catalysts, the CO_2 reduction into CH_4 has reached 69% yield using 0.24% Ru catalyst at 40 bar of H_2 plus 40 bar of CO_2 and at 150 °C [1536]. Synthesis of formic acid and derivative utilizing CO_2 and H_2 with mainly homogeneous transition metal catalysts have been reported since the 1970s [1537]. A breakthrough was made in 1994 by Noyori and coworkers who obtained high turnover frequencies using supercritical CO_2 (*sc*CO_2) as both reactant and solvent [1538]. Above 31.1 °C and 73.8 bar, CO_2 reaches a supercritical state, a fluid capable to dissolve large amounts of other gases such as H_2, CO, $CH_2{=}CH_2$, and CH_4, light alkanes, and transition metal catalysts [1539]. Dissolving H_2, a tertiary amine, a catalyst precursor such as $RuH_2(PMe_3)_4$ or $RuCl_2(PMe_3)_4$, and a promoting additive such as H_2O, MeOH, or DMSO in *sc*CO_2 at 50 °C leads to the generation of HCOOH with TOF > 4000 h^{-1}. The high rate of reaction is attributed to rapid diffusion, weak catalyst solvation, and the high miscibility of H_2 in *sc*CO_2. The HCOOH synthesis can be coupled with subsequent reactions of HCOOH, for example, with alcohols or amines to give highly efficient routes to formate esters or formamides. The possible mechanism of CO_2 hydrogenation is shown in Scheme 7.79 [1540].

In a study using $Ru(dmpe)_2H_2$ (dmpe = $Me_2PCH_2CH_2PMe_2$) as a catalyst, it was observed that both CO_2 and HCOOH are inserted rapidly to form ruthenium formates *cis*-[$Ru(dmpe)_2OCHO)_2$] and *trans*-[$Ru(dmpe)_2H(OCHO)$] at low CO_2 pressure (<5 bar). Quantum mechanical calculations suggested that H_2 insertion into intermediate of type **426** via Ru(η^2-H_2)-like complexes is the rate-determining step [1537]. Nozaki and coworkers reported that the pincer iridium complex **427** catalyzes the hydrogenation of CO_2 into HCOOK in aqueous KOH/THF with TOF and TON reaching 150 000 h^{-1} and 3 500 000, respectively, at 120 °C and under 60 bars [1541–1543]. Milstein and coworkers found that the iron complex **428** catalyzes reaction $NaHCO_3 + H_2 \rightarrow HCOONa + H_2O$ in H_2O/THF at 80 °C and under 1 atm, with TOF and TON reaching 150 h^{-1} and 800, respectively [1544]. Complex **428** catalyzes the hydrogenation $CF_3COOR + 2H_2 \rightarrow CF_3CH_2OH + ROH$ (25 bar H_2, 1,4-dioxane, MeONa, 40 °C) [1545]. In the presence of a reducing agent (Et_3Al furnishing the equivalent of H_2), allenes react with CO_2 to give β,γ-unsaturated carboxylic acids. The reactions are catalyzed by a silyl pincer-type Pd complex [1546].

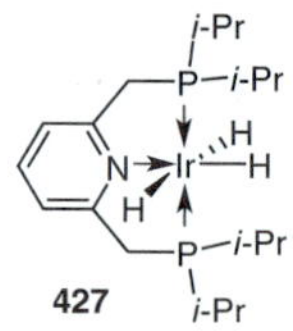

427

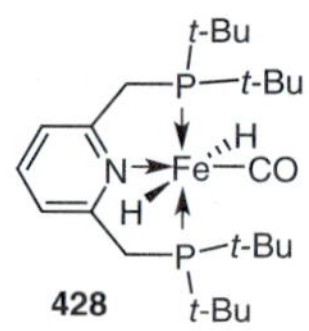

428

$$NaHCO_3 \xrightarrow[-H_2O]{+H_2\ \mathbf{428}\ (0.1\ mol\%)} HCOONa$$

Problem 7.35 Give a plausible mechanism for the hydrogenation of CO_2 into HCOOH catalyzed by the iron complex **428** [1544].

7.8.9 Catalytic hydrogenation of nitriles and imines

Hydrogenation of nitriles is one of the most important methods to obtain primary amines, and especially diamines that are of high industrial importance [1547, 1548]. Heterogeneous catalysts commonly used are Raney Ni, Raney Co, Pd/C, Pt/C, Ru/C, and Rh/C. The same metals can also be supported on SiO_2, Al_2O_3, or carbon nanoparticles [1549]. A large variety of homogeneous transition metal complexes have been proposed to catalyze nitrile hydrogenation [1550–1554]. The simple cobalt complex **429** is an example [1555].

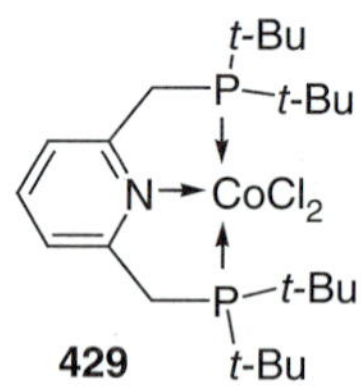

429

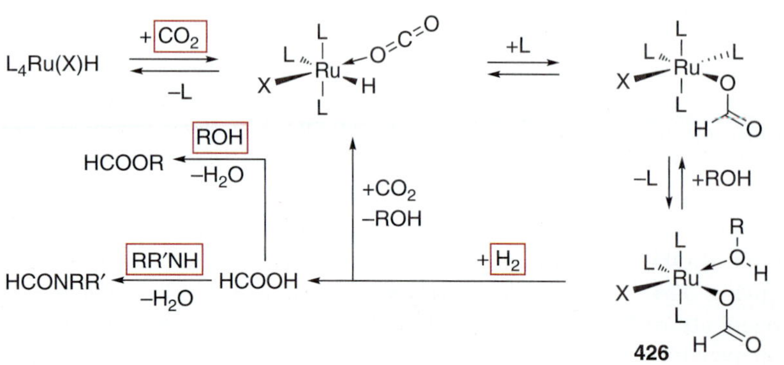

Scheme 7.79 Possible mechanism for the ruthenium-catalyzed hydrogenation of carbon dioxide (X = H, Cl, R = H, Me, L = PMe_3).

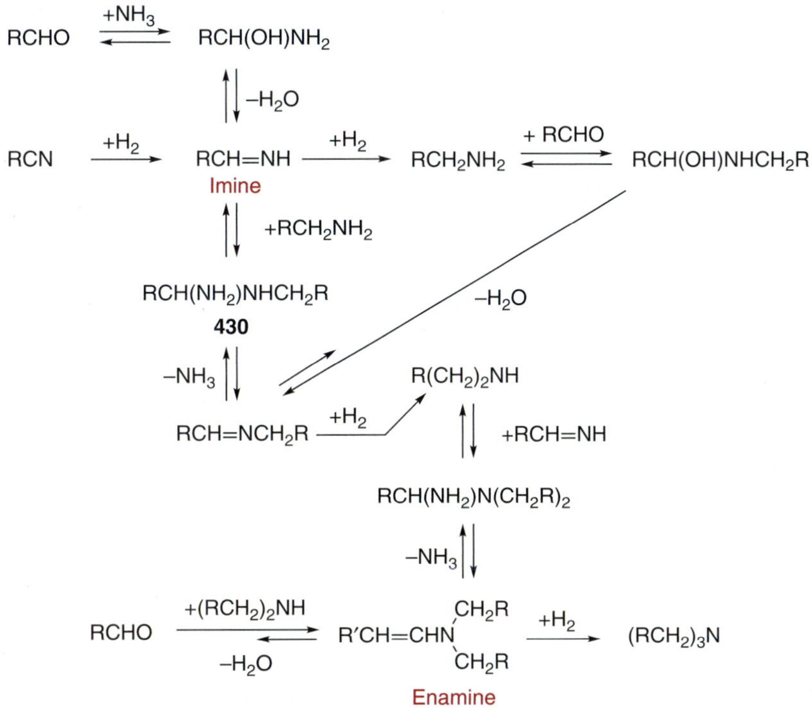

Scheme 7.80 Catalytic hydrogenation of nitriles and the reductive amination of aldehydes with NH_3.

Nitriles are hydrogenated first to the imine and then to the primary amine. As explained first by von Braun et al. in 1923 (Scheme 7.80) [1556], the primary amine can react with the imine and form secondary amine via a *gem*-diamine of type **430**. The latter undergoes either direct hydrogenolysis or elimination of NH_3, followed by hydrogenation. Tertiary amines can also be formed by addition of secondary amines to the imine intermediates via a similar route [1557]. Formation of secondary and tertiary amines can be avoided by addition of NH_3 to the reaction mixture [1558, 1559]. Alternatively, the addition of NaOH [1560] and LiOH is also effective to improve the selectivity for primary amines [1561]. The HO^- ions prevent heterogeneous catalyst deactivation by avoiding polyamine formation. Pretreatment of Ni and Co catalyst with CO, CO_2, aldehydes, or ketones also gives significantly less secondary amines.

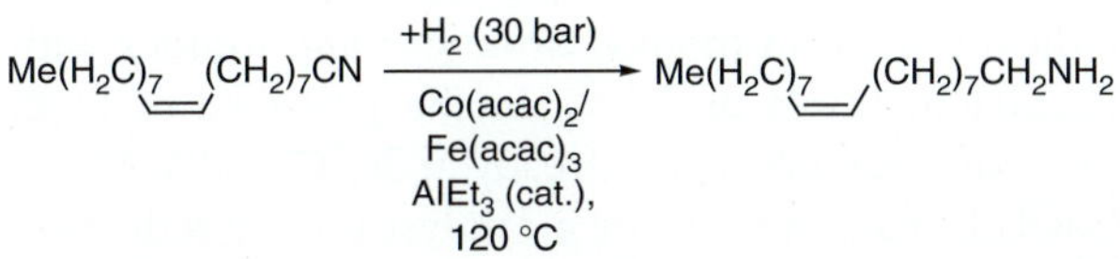

(7.152)

Chemoselective hydrogenation of unsaturated nitriles into unsaturated primary amines is possible with high chemoselectivity applying a Ziegler-type Co–Fe catalyst even in the absence of NH_3 (reaction (7.152)) [1562]. In 2006, Stephan and coworkers reported the first nonmetal compound to reversibly activate and liberate H_2 [1323]. The phosphonium-borate **432** is formed by reaction of phosphine-boran species *p*-$(2,4,6\text{-}Me_3C_6H_2)_2P\text{-}C_6F_4\text{-}B(C_6F_5)_2$ (**431**) with H_2, while heating zwitterion **432** above 100 °C liberates H_2 and regenerates **431** (equilibrium (7.153)). The latter compound, and related systems (R = *t*-Bu), are able to catalyze the hydrogenation of nitriles, imines (e.g. reaction (7.154)), and aziridines to produce primary and secondary amines in high yields under relatively mild reaction conditions (120 °C, 5 atm H_2, toluene) [1326, 1563]. Compound **431** is a "frustrated Lewis pair" in which the base and acid moieties cannot form a Lewis acid/base pairs for steric reasons. The combination of $(t\text{-}Bu)_3P$ and $(C_6F_5)_3B$ also forms a "frustrated Lewis pair" able to activate H_2 at 25 °C [1319–1322, 1564].

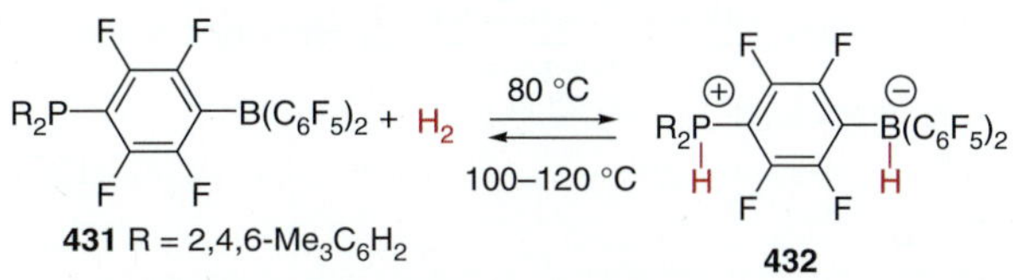

(7.153)

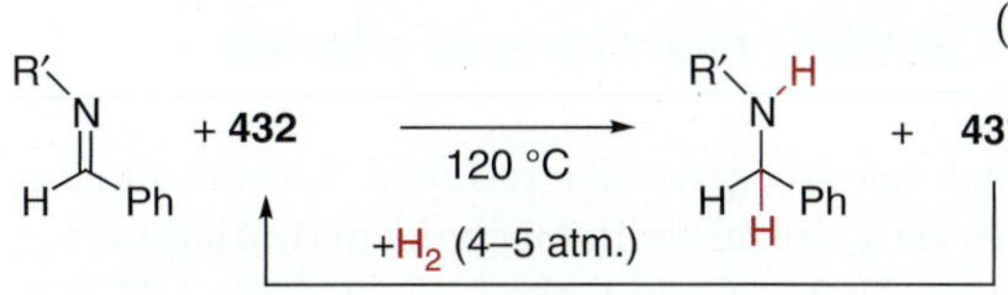

(7.154)

7.8.10 Catalytic hydrogenolysis of C–halogen and C–chalcogen bonds

There are five common general methods for the direct reduction of alkyl halides to the corresponding alkanes (R–Hal + H_2 → R–H + H–Hal): (i) radical reduction employing tin or silicon hydride reagent, (ii) low-valent metal reduction (e.g. Zn/AcOH, Al–Hg/H_2O/MeCN, SmI_2/THF, Li/NH_3/THF, Na/*t*-BuOH, Ni/EtOH, and $NiCl_2$/Zn/H_2O), (iii) metal hydride reduction, usually by an ionic mechanism (e.g. $LiAlH_4$/THF, $LiEt_3BH$/THF, $NaBH_4$/THF, $NaBH_3CN/ZnI_2$, Et_3SiH/CF_3COOH, $Et_3SiH/(C_6F_5)_3B$, $(Me_3Si)_3SiH/Et_3B$/toluene, Bu_3SnH/MgBr/Et_3B/CH_2Cl_2, and $NaBH_4/InCl_3/H_2O$), (iv) catalytic hydrogenolysis with transition metal catalysis (e.g. H_2/Pd–C, H_2/Cp_2ZrH_2, $NH_4^+HCOO^-$/Pd – C, H_2/Ni/Et_3N/MeOH, and Ni/THF), and (v) electrochemical reduction [1565]. An alternative, indirect method is the base-induced elimination to generate an alkene that is subsequently hydrogenated. Hydrogenation over Pd/C, typically in the presence of a base, remains a useful method for the reduction of chloroalkane, bromoalkanes, and iodoalkanes into the corresponding alkanes. Catalytic dehalogenation of aryl chlorides under H_2 atmosphere has been catalyzed by Ru/phosphine complexes [1566], by Rh/Pt/C-based catalysts [1567], by silica gel-supported Ni catalysts [1568], and by molybdenum sulfide clusters [1569]. Raney nickel or combination of zinc and $NiCl_2$ are useful catalysts for the hydrogenation of alkyl bromides and iodides. Hydrogenation over Pd/C remains the method of choice for the reduction of benzylic ethers. Hydrogenation over Pd/C is a good method for the reduction $ArCHO \rightarrow ArCH_3$. Sulfur- and selenium-containing compounds are almost universally reduced by Raney nickel/H_2/EtOH. Wilkinson's catalyst ($RhCl(PPh_3)_3$) catalyzes the reductive coupling of disulfides and diselenides with alkyl halides, using H_2 as a reducing agent (reaction (7.155)) [1570].

$$(R^1Y)_2 + 2R^2X \xrightarrow[\text{THF/Et}_3\text{N, 50 °C, 24 h}]{+\text{H}_2\text{ (1 atm.), RhCl(PPh}_3)_3\text{ (0.03 equiv.)}} 2R^1\text{-}Y\text{-}R^2 + 2HX/Et_3N \tag{7.155}$$

Y = Se, S; X = Br, Cl

7.9 Catalytic reactions of silanes

Alkyl halides are generally reduced with trialkyltin hydrides via a radical chain mechanism (initiation; e.g. $NCCMe_2$—N=N—CMe_2CN → N_2 + $2Me_2(NC)C^\bullet$, $Me_2(NC)C^\bullet$ + Bu_3SnH → Me_2(NCCH + $Bu_3Sn^\bullet$, radical chain propagation: $Bu_3Sn^\bullet$ + R—X→Bu_3SnX + $R^\bullet$, $R^\bullet$ + Bu_3SnH → R—H + $Bu_3Sn^\bullet$, and termination; e.g. $2R^\bullet$ → R—R) [1571–1574]. Alternative reduction procedures are desired owing to the toxicity of the tin reagents and problems separating tin byproducts from reaction products. Silanes are thermodynamically capable of such reductions, but the Si—H bond is too strong to support a radical chain reaction [1575, 1576]. Use of $(Me_3Si)_3SiH$ possessing a weak Si—H bond circumvents this problem but is not economically attractive [1575, 1576].

7.9.1 Reduction of alkyl halides

High loadings of Pd(II) salts have been shown to induce reduction of certain classes of alkyl halides by Et_3SiH [1577]. Strong Lewis acids (e.g. $AlCl_3$) have been used in combination with Et_3SiH; skeletal rearrangements accompanying reduction support a mechanism involving carbocation-like intermediates [1578]. The combination of $Et_3SiH + Ph_3C^+/BAr_4^-$ (Ar = C_6F_5) can reduce alkyl fluorides via a mechanism involving generation of carbocations through fluoride anion abstraction by Et_3Si^+ (formed from Ph_3C^+ and Et_3SiH) followed by hydride transfer to the carbocation by Et_3SiH [1579]. Similar chemistry has been described using a bissilylated onium salts [1580]. Yang and Brookhart have found that the cationic Ir(III) hydride complex **434** is a versatile and highly active catalyst (used in 0.5 mol%) for the reduction of a large variety of alkyl halides by Et_3SiH (Scheme 7.81) [1581].

The catalytic cycle appears to operate by a unique process in which an electrophilic iridium silane complex **435** acts as a silylating reagent to produce a silyl-substituted halonium ion **436**, which is then readily reduced by the nucleophilic dihydride of type $[Ir]H_2$ formed following silyl transfer. This mechanism parallels that proposed by Piers for hydrosilylation of ketones using $(C_6F_5)_3B/Et_3SiH$ where the silane is activated by $(C_6H_5)_3B$ and transfers Et_3Si^+ to the ketone [1329, 1582–1584]. $In(OAc)_3$ catalyzes the radical reduction of alkyl bromides and iodides with Ph_3SiH in EtOH at 70 °C. In the presence of Et_3B, the reduction proceeds already at 30 °C [1585, 1586]. $GaCl_3$ is also found to be a catalyst for the reduction of haloalkanes with poly(methylhydrosilane) (PMHS). A possible mechanism for these reductions is given below with reactions (7.156)–(7.158) [1587].

$$PhSiH_3 + MX_3 \rightleftharpoons PhSiH_2X + H\text{—}MX_2 \tag{7.156}$$

Scheme 7.81 Synthesis of catalyst **434** and possible mechanism for the catalyzed reduction R—X + Et_3SiH → R—H + Et_3SiX (X = F, Cl, Br, I).

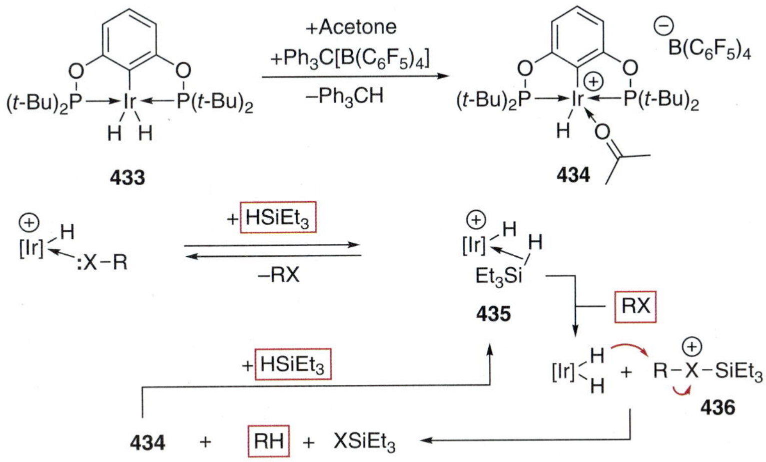

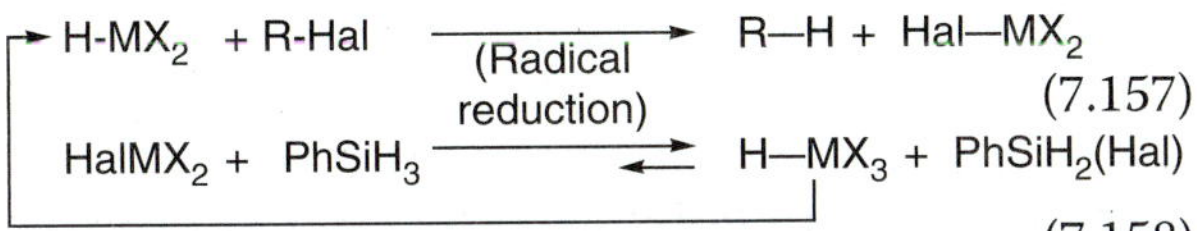

(7.157)

(7.158)

7.9.2 Reduction of carbonyl compounds

In 1972, Ojima et al. reported that hydrosilylations of carbonyl compounds can be catalyzed by Wilkinson's catalyst ($Rh(Ph_3P)_3Cl$) [1588, 1589]. The less expensive copper hydride $[(Ph_3P)CuH]_6$ reported first by Osborn and coworkers in 1971 [1590] was shown by Stryker and coworkers in 1988 to catalyze the conjugated reductions of α,β-unsaturated carbonyl derivatives [1591]. Using hydrosilanes, Lipshutz et al. found that $[(Ph_3P)CuH]_6$ catalyzed the conjugate hydrosilylation under mild reaction conditions [1592, 1593]. In the presence of enantio-enriched chiral ligands (e.g. diphosphines) and Cu salts, asymmetric catalytic conjugate hydrosilylations (1,4-additions) of enones and hydrosilylations of ketones (1,2-additions) have been realized [1594, 1595]. Hydrosilylation of aldehydes and ketones can be catalyzed by iron [1596–1602], nickel [1603–1605], cobalt, or manganese salts and complexes [1606]. In 2007, Nishiyama and Furuta first reported on iron-catalyzed enantioselective hydrosilylation of ketones [1607]. They were followed by reports by several other groups [1333, 1608–1611]. An example is given with hydrosilylations (7.159) catalyzed by the iminopyridine-oxazoline iron complex **437** [1610]. Cobalt-catalyzed enantioselective hydrosilylation of ketones have also been reported [1612]. Hydrosilylation of imines in a protic medium produces the corresponding amines, a reaction also catalyzed by iron complexes [1613–1615]. An example of asymmetric imine hydrosilylation is given with the zinc-catalyzed reaction (7.160) [1616].

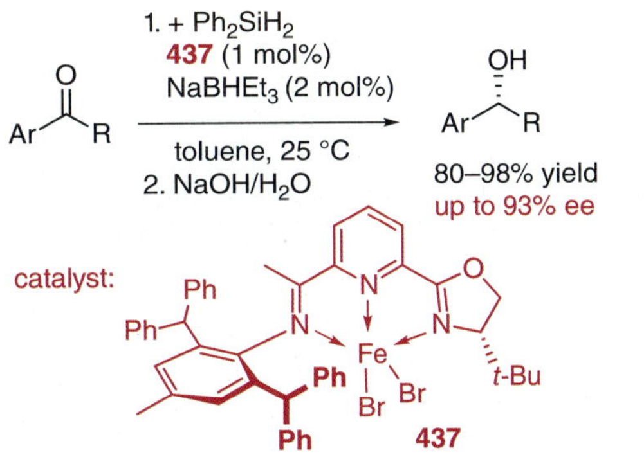

(7.159)

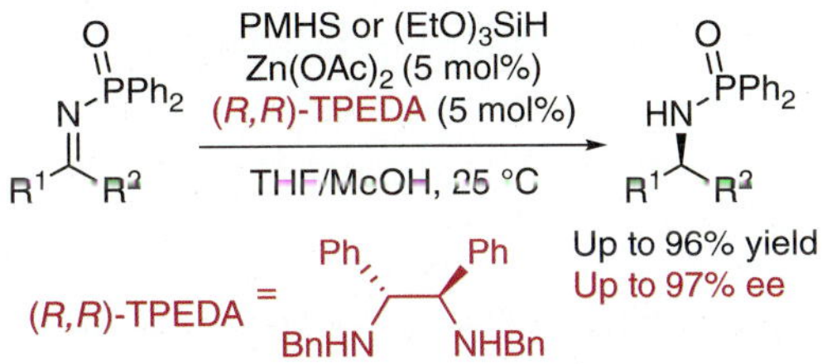

(7.160)

Hydrosilylation of carbonyl compounds and imines has been catalyzed by cationic aluminum alkyl complexes such as **438** [1617], by electrophilic phosphonium salt **439** [1618], and by neutral silicon compound **440** [1619]. Metal-free enantioselective hydrosilylation has been catalyzed by "frustrated Lewis pairs" made from the combination of tricyclohexylphosphine and an enantio-enriched chiral alkenylborane derived in situ from a diyne (e.g. reaction (7.161)) [1620].

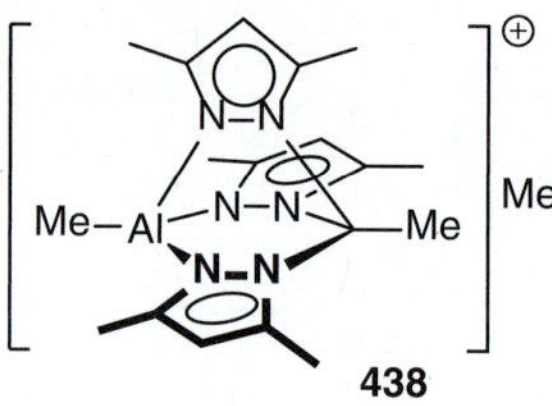

438

$[(C_6F_5)_3PF]^{\oplus}/[B(C_6F_5)_4]^{\ominus}$

439

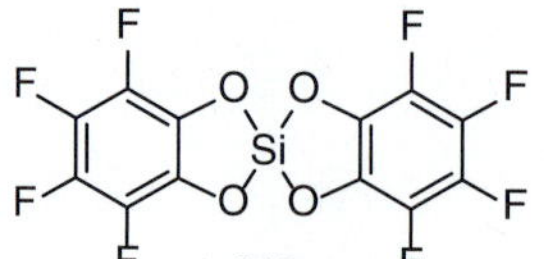

440

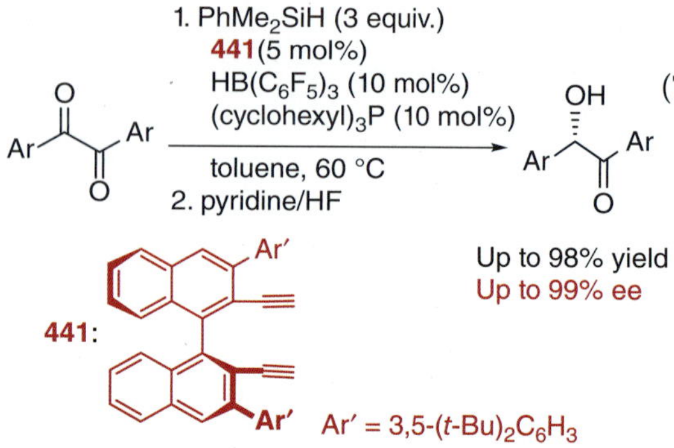

(7.161)

Carboxylic esters can be reduced into primary alcohols via hydrosilylation catalyzed by Rh [1621, 1622] Ir [1623], Ru [1624, 1625], or Fe complexes [1314, 1626, 1627], by MoO_2Cl_2 [1628], V(O)(O-*i*-Pr)$_3$ [1629], or by diethylzinc [1630]. Similarly, secondary and tertiary amines can be obtained via catalyzed hydrosilylations of carboxamides. The reaction have been catalyzed by Rh [1622, 1631], Pt [1632, 1633], Ir [1634], Mo [1635–1637], Ru [1624, 1638], Fe [1639], Zn [1640, 1641], and Co complexes [1642], and by MnO_2Cl_2 [1643], InI_3 [1644], Et_2Zn [1645], and metal-free catalysts such as $B(C_6F_5)_3$ [1646–1648]. The following standard homolytic bond dissociation enthalpies $DH°(H_3Si^•/H^•) = 92.2$, $DH°(H^•/H^•) = 104.2$, $DH°(HO^•/H^•) = 119.2$ (Table 1.A.13) and $DH°(H_3Si^•/HO^•) = 124.1\ \text{kcal mol}^{-1}$ [1649] suggest that the hydrosilylation of CO_2 into a silyl formate must be more exothermic than CO_2 hydrogenation into HCOOH. The reacting Si—H bond is weaker than the H—H bond, and the forming Si—O bond is stronger than the corresponding H—O bond. Transition metal complexes containing Ru, Co, Rh, Ir, Nb, Cu, and main group Zn catalyze the hydrosilylation of CO_2 producing a variety of products [1650]. In 2010, Piers and coworkers reported that the "frustrated Lewis pair" consisting of 2 equiv. of 2,2,6,6-tetramethylpiperidine (TMP) and tris(pentafluorophenyl)borane activates CO_2 to form a boratocarbamate/ammonium pair **442** that reacts with Et_3SiH giving a silyl carbamate **443** and ion pair $TMPH^+/HB(C_6F_5)_3^-$. The latter reacts with CO_2 via hydride transfer to form the formatoborate salt **444**. In the presence of an excess of $B(C_6F_5)_3$ and Et_3SiH, **444** is hydrosilylated to produce formatosilane **445** and to regenerate catalyst **442**. The formatosilane **445** is hydrosilylated by the $Et_3Si/B(C_6F_5)_3$ system into methane with formation of 2 equiv. of $(Et_3Si)_2O$ as a coproduct (Scheme 7.82) [1651]. Selective reduction of CO_2 to CH_4 by tandem hydrosilylation with Et_3SiH has been catalyzed with mixed Lewis acids $Al(C_6F_5)_3 + B(C_6F_5)_3$. In this case, the highly oxyphilic aluminum compound catalyzes the first reduction $CO_2 + Et_3SiH \rightarrow HCOOSiEt_3$, whereas the subsequent reductions are catalyzed by $B(C_6F_5)_3$. With 1% of $Al(C_6F_5)_3$ and 5% of $B(C_6F_5)_3$, CH_4 has been produced in 94% yield [1652].

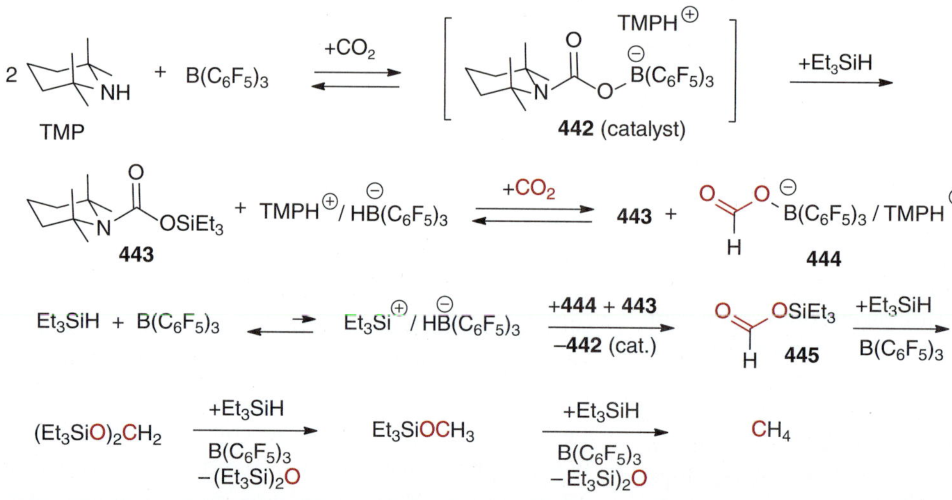

Scheme 7.82 Deoxygenative hydrosilylation of carbon dioxide catalyzed by a "frustrated Lewis pair."

7.9.3 Alkene hydrosilylation

Hydrosilylation of alkenes produces corresponding alkylsilanes that are important raw materials in the manufacture of silicon rubbers, molding implants, releasing coatings, and adhesives [1653]. In 1947, Sommer et al. reported the first hydrosilylation reaction of oct-1-ene with trichlorosilane ($HSiCl_3$) in the presence of diacetyl peroxide. The reaction selectivity was low due to the radical mechanism [1654]. In the meantime several transition metal complexes have been found to catalyze alkene and alkyne hydrosilylations [1655–1660]. In 1962, Nesmeyanov et al. found that $Fe(CO)_5$ catalyzed alkene hydrosilylations between 100 and 150 °C [1661]. More efficient catalysts such as the [bis(imino)pyridine]iron complex **447** were reported by Chirik's group in 2004 [1662] and by Nakazawa's group more recently [1663]. The iron complex exhibits TOF of up to 100 000 h^{-1} [1664, 1665], but are not compatible with carbonyl groups unless the ligands are chosen to render them electron richer and less oxyphilic [1666]. With the Ni pincer complex **448** as the catalyst, Hu and coworkers have realized highly chemoselective anti-Markovnikov hydrosilylations as illustrated with reaction (7.162) [1667].

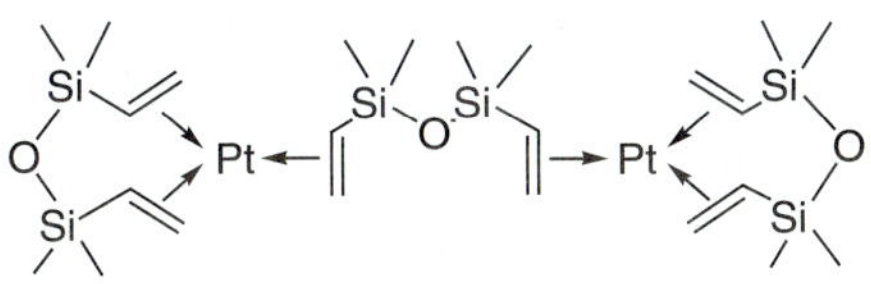

446

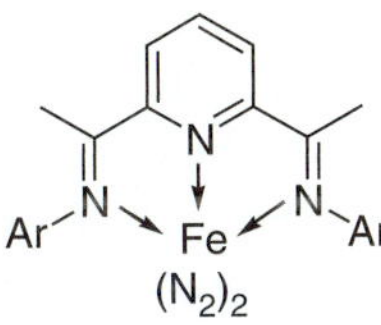

447

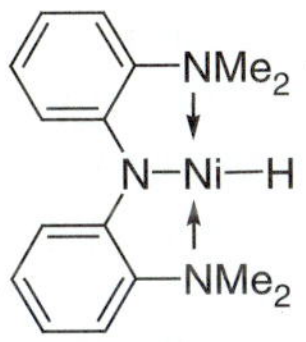

448

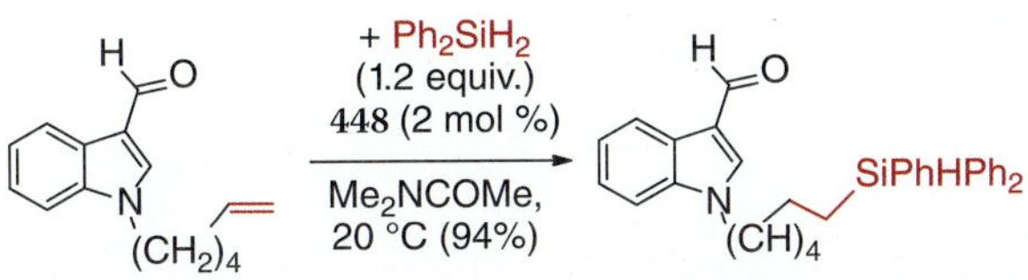

(7.162)

In 2016, Huang and coworkers reported that the regioselectivity of terminal alkene hydrosilylation depends on the metal of phosphine–iminopyridine complexes. With the iron complex **449**, they find highly regioselective anti-Markovnikov additions, whereas with the cobalt catalyst **450**, Markovnikov addition are highly favored [1668]. In 1976, Kumada and coworkers reported a first example of enantioselective anti-Markovnikov hydrosilylation of 1,1-disubstituted alkenes utilizing Pt or Ni complexes bearing a chiral phosphine or diphosphine. Yields and ee's were modest. Much better results (reaction (7.163)) have been reported recently by Lu and coworkers applying the iron catalyst **451** [1669]. Asymmetric Markonikov hydrosilylations have been realized utilizing Rh- or Pd-based enantio-enriched catalysts [1670, 1671].

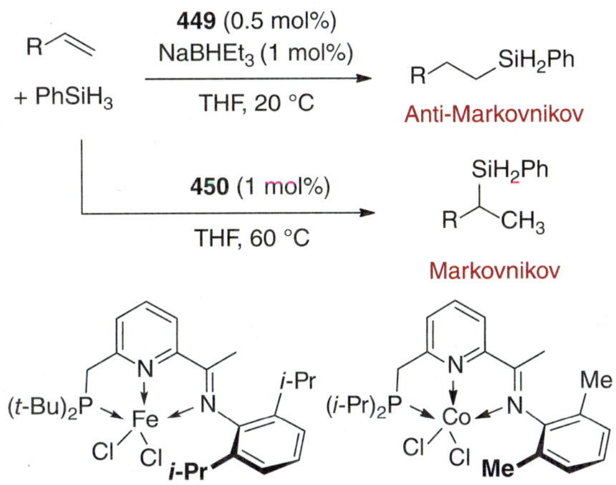

449 450

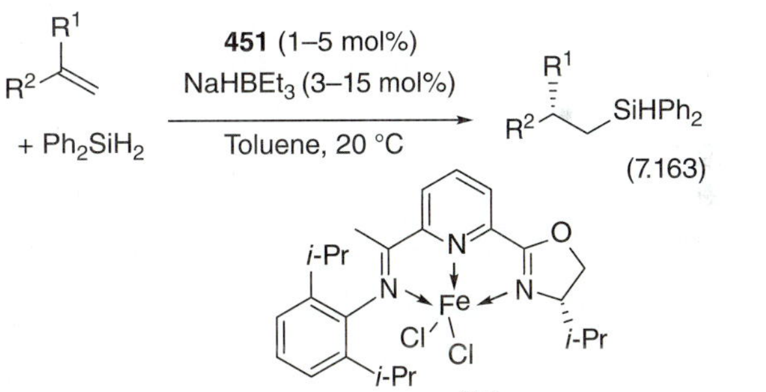

451

(7.163)

Problem 7.36 Propose a mechanism for the hydrosilylation of ketones with $(EtO)_2MeSiH$ in toluene at 65 °C catalyzed by iron(II) complex of type **452** [1602, 1672].

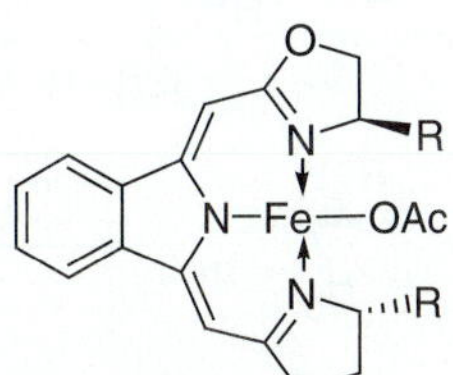

452

Problem 7.37 Propose a mechanism for the bis(tributylphosphine)dichloroplatinum(II)-catalyzed alkene hydrosilylation [1673].

Problem 7.38 Propose a mechanism for the hydrosilylation of alkenes with $PhSiH_3$ catalyzed by the cationic ruthenium complex $[Cp^*(i\text{-}Pr_3P)(H)_2Ru{=}Si(H)Ph{\cdot}Et_2O]^+/[B(C_6F_5)_4]^-$ [1674].

7.10 Hydrogenolysis of C—C single bonds

Hydrogenolysis of alkanes occurs on heating them with H_2 in the presence of Pt black [1675, 1676], of Pt dispersed on Al_2O_2 [1677] or on SiO_2 [1678, 1679], of Ti/Mo [1680], or of Ir, Rh, and Ru blacks [1681, 1682]. In 1992, Basset and coworkers obtained silica-supported zirconium monohydride $(\equiv SiO)_3ZrH$ by reaction of $ZrNp_4$ (Np = neopentyl: CH_2CMe_3) with silica dehydroxylated at 500 °C under vacuum ($\equiv$SiOH). This gives first $\equiv SiOZrNp_3 + NpH$, then hydrogenation of $\equiv SiOZrNp_3$ generates $\equiv SiOZrH_3 +$ 3NpH [1683]. The reaction of $\equiv SiOZrH_3$ with 2 $\equiv$Si—O—Si$\equiv$ gives $(\equiv SiO)_3ZrH + 2\ \equiv SiH$ [1684]. These silica-supported zirconium hydrides catalyze olefin polymerization [1685, 1686], C—C bond hydrogenolysis in alkanes, [1687] and H/D isotope exchange in alkanes [1688, 1689]. They catalyze the hydrogenolysis of propane, butanes, and pentanes, but not of ethane, under mild conditions (25–150 °C) [1690]. To render the supported zirconium hydride even more electrophilic than on silica, $ZrNp_4$ was reacted with partially dehydroxylated silica–alumina (500 °C, vacuum). After hydrogenation at 150 °C, silica–alumina-supported zirconium hydride is obtained with concomitant formation of silica hydride and aluminum hydride (**453**), which is capable to catalyze hydrogenolysis of polyethylene and polypropylene to Diesels and lower alkanes at low H_2 pressure (1 atm) and at 150 °C. A possible mechanism for this process is shown in Scheme 7.83. The first step is an unselective C—H activation of the paraffinic chain, which occurs by σ-bond metathesis with the liberation of H_2. Then follows a β-alkyl transfer giving an (alkene)(alkyl)zirconium complex. Hydrogenation of the latter produces alkane R′–H and another (allyl)zirconium species that, on its turn, is also hydrogenated into alkane R–H with liberation of the initial catalyst **453** [1691].

In the absence of H_2, **453** catalyzes the polymerization of ethylene and of propylene (200 atm, 25 °C). Yermakov considered grafted zirconium dihydrides $(\equiv Si{—}O)_2ZrH_2$ as the catalytic sites for the olefin polymerization and H/D exchange reactions [1685, 1686]. Quantum mechanical calculations suggested that dihydrides can to catalyze alkane hydrogenolysis [1692, 1693]. C—C bond activation of strained hydrocarbons is always easier than for acyclic alkanes or unstrained cycloalkanes. Hydrogenation of cyclopropane over heterogeneous Ni, Pd, and Pt catalysts differ fundamentally from alkene hydrogenation in being first order with respect to hydrocarbon and of zero order with respect to H_2. The surface is covered with chemisorbed H and the rate-determining step of the reaction involves physically absorbed cyclopropane [1694, 1695]. With Ni and Co dispersed on SiO_2, methane and ethane form in addition to propane [1696, 1697]. Kinetics and H/D exchange studies suggested that the supported Ni-catalyzed hydrogenation and hydrocracking of cyclopropane involve the rate-determining oxidative of a Ni metal into a C—C bond with the formation of nickelacyclobutane intermediate [1698]. Catalytic hydrogenation of cyclopropane on the Pt(111) surface at 350 K gives propane exclusively. No methane or ethane products are observed. Soft X-ray methods allowed in situ characterization of the C—H bonds. At

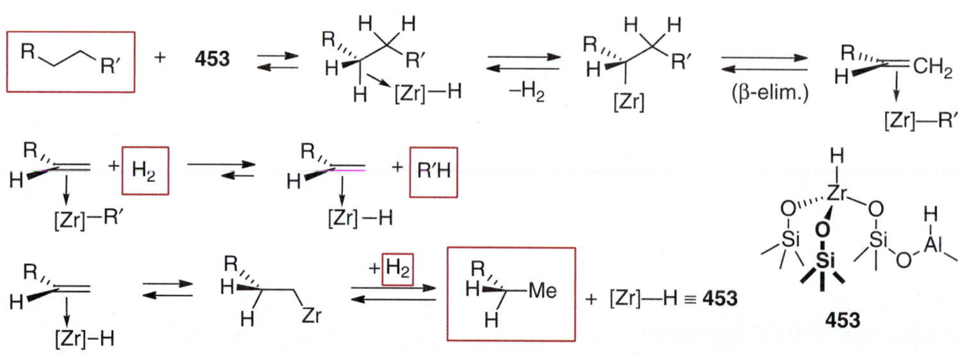

Scheme 7.83 Proposed mechanism for the hydrogenolytic degradation of polyethylene.

Scheme 7.84 Selective catalytic isomerization of a cyclopropane derivatives and its subsequent hydrogenation.

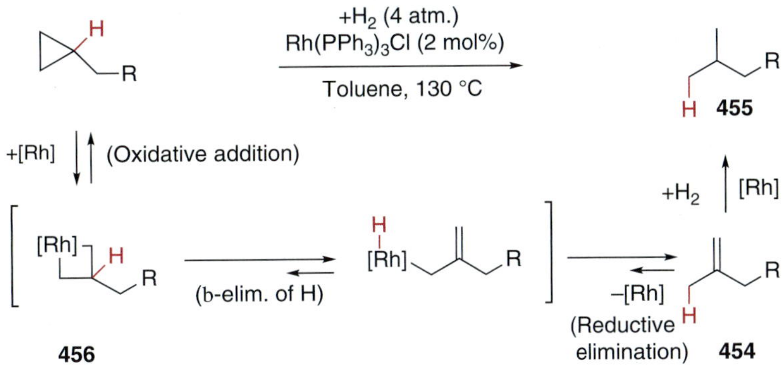

320 K, platinacyclobutane intermediate absorbed in a upright configuration relative to the surface has been characterized. Increasing temperature to 350 K under the reaction conditions increases the average C—H stoichiometry to C_3H_7. Thus, the platinacyclobutane undergoes hydrogen atom insertion with the formation of a (propyl)Pt intermediate that inserts itself another hydrogen atom to produce propane [1699]. Hydrogenation of 1,1-disubstituted cyclopropane derivatives using Pd/C in MeOH or PtO_2 in AcOH gives preferentially Me_2CR_2 if R = alkyl, CH_2OH, alkoxy (electron-releasing groups), and $EtCHR_2$ if R = carboxy, acyl, CN (electron-withdrawing groups) [1700, 1701]. Under similar conditions, ethyl bicyclo[1.1.0]butane-2-carboxylate gives 95% EtCH-MeCOOEt and 5% $BuCO_2Et$, whereas ethyl bicyclo [1.1.0]butane-1-carboxylate gives 99% EtCHMe-COOEt and <1% BuCOOEt [1702]. Pd/C-catalyzed hydrogenation of dimethyl bicyclo[1.1.0]butane-2,2-dicarboxylate gives 94% $EtMeC(COO_2Me)_2$, 5% PrCH $(CO_2Me)_2$, 1% cyclopropane, and no cyclobutane-1,1-dicarboxylic ester [1703]. Using Wilkinson's catalyst, Bart and Chirik found a selective C—C bond activation of monosubstituted cyclopropane (Scheme 7.84) [1075]. The reaction proceeds without H_2 to produce alkene **454**. In the presence of H_2, the latter is hydrogenated into alkane **455**. The process probably involves oxidative addition of the Rh(I) complex to generate the rhodiacyclobutane **456**. Then follows a β-elimination of hydrogen followed by a reductive elimination liberating alkene **454**.

Catalytic hydrogenation of cyclobutane derivatives over platinized carbon catalyst occurs above 150 °C [1704, 1705]. Cyclopentane is hydrogenated into *n*-pentane under similar conditions [1706]. Murakami et al. reported a highly regio- and stereoselective hydrogenation of cyclobutanone **457** into alcohol **458** [1707]. In the presence of $[Rh(cod)_2Cl]_2$ as the catalyst, the pincer diphosphine ligand **459** is hydrogenolyzed into **460** and methane [1708].

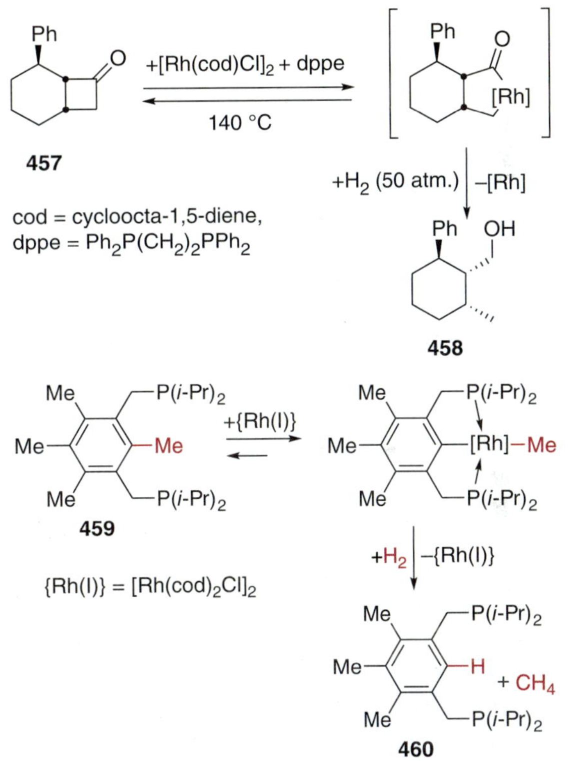

7.11 Catalytic oxidations with molecular oxygen

Oxygenation with O_2 is an essential transformation in chemical and biological processes, most of which are associated with transition metal catalysts. Nature uses metalloproteins such as cytochrome P450 (Section 7.11.1). Air or pure oxygen is the ideal oxidant for "green chemistry," as it is inexpensive, abundant, noncorrosive, and environmentally benign. There are several industrial processes using catalytic oxidation of hydrocarbons such as the production of maleic anhydride by the classical V_2O_5-catalyzed oxidation of benzene [1709, 1710], or the P_2O_5/V_2O_5-catalyzed oxidation of *n*-butane (Monsanto process) [1711], the production of phthalic anhydride by V_2O_5-catalyzed oxidation of naphthalene, *o*-xylene and tar containing them [1712, 1713], the oxidation of cyclohexane into cyclohexanol (intermediates in the production of polyamide fibers and plastics such as nylon-6 and nylon-6,6 [1714]), and cyclohexanone catalyzed by cobalt [1715] or gold catalysts [1716, 1717]. Selective aerobic oxidation of propylene to propylene oxide has been catalyzed by silver-supported tungsten oxide nanostructures [1718]. Acidic dehydration of hexoses in biomass furnishes 5-hydroxymethyfurfural (HMF) [1719], an inexpensive starting material for "green" chemistry [1720–1725]. Applying vanadium-based catalyst, the aerobic oxidation of HMF generates maleic anyhydride and acid in 64% yield [1726]. Benzene can be oxidized into phenol under similar conditions [1727]. With nonprecious Fe–Co-based catalysts, aerobic oxidation of HMF provides furan-2,5-dicarbaldehyde [1728]. With Pd/FeO_3 nanoparticles, furan-2,5-dicarboxylic acid is obtained [1729, 1730]. As Haruta's discovery of the exceptional reactivity of gold nanoclusters [1731], there has been an exponential growth in the use of metal-oxide-supported gold particles as catalysts [1732], including for low-temperature CO oxidation [1733–1736], direct propylene epoxidation [1737–1739], and the water–gas shift reaction [1740, 1741]. Gold monolayers and bilayers on TiO_x are found to be electron rich and to have significantly different electronic properties compared with bulk Au [1734]. Water activates the oxidation of CO by O_2 into CO_2 catalyzed by Au(111) surfaces. This is interpreted in terms of CO reacting directly with activated H_2O or OH groups formed from H_2O interacting with atomic oxygen on Au(111) [1733]. Hutchings and coworkers found that gold supported on graphite catalyzes the selective aerobic oxidation of cyclohexane to cyclohexanol and cyclohexanone under mild conditions (70 °C, 3 atm O_2) [1742]. Shul'pin et al. reported the oxidation of cyclooctane using $NaAuCl_4$ as a catalyst and H_2O_2 as an oxidant [1743]. Shi and coworkers found a selective oxidation of primary alcohols to aldehydes using gold catalyst with O_2 as an oxidant in toluene at 90 °C [1744]. Heterogeneous gold catalyst has been used to catalyze the aerobic oxidation of alcohols [1745–1764]. Ionic liquids are used more and more frequently as a solvent or a cocatalyst in selective aerobic oxidations [1765].

7.11.1 Heme-dependent monooxygenase oxidations

The cytochrome P450 enzymes (e.g. **461**) constitute a large family of cysteinato-heme proteins that catalyze a large diversity of biotransformations, the most common of which are hydroxylation ($C{-}H + ½\ O_2 \rightarrow C{-}OH$), epoxidation, and heteroatom oxidation. They are common in all forms of life and play a key role in the oxidative transformation of endogeneous and exogeneous compounds (xenobiotics) [1766, 1767]. Terpenoids form the largest group of plant-specialized metabolites and exhibit essential functions in plant metabolism, propagation, and defense. The general mechanisms of terpene biosynthesis involve enzymes catalyzing the formation of the isoprenoid basic carbon skeletons. In the subsequent pathway steps, it is mainly cytochrome P450 that catalyzes the decoration of these basic skeletons through oxidation [1768, 1769]. Cytochrome P450 catalyzes the oxidation of food, drugs, and other chemicals we ingest. It is the P450 enzymes that protect us from the toxins in the air we breathe [1770–1774].

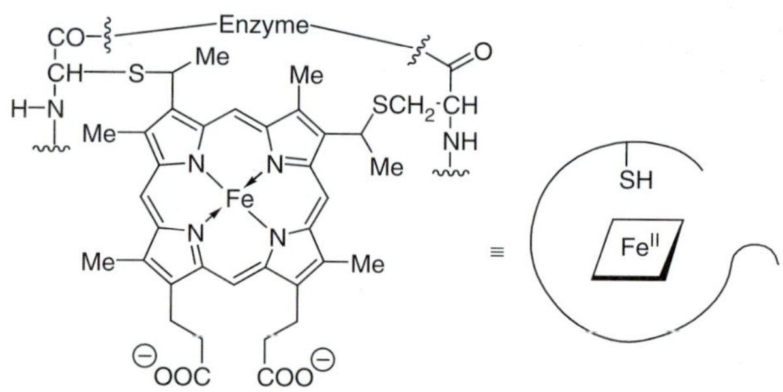

461: Heme *c* (cytochromes *c*, c_1)

The function of heme is to activate molecular oxygen to a species that can insert into various molecules. The proposed mechanism for cytochrome P450-catalyzed oxidation is shown in Scheme 7.85 [1766, 1775, 1776]. Early calculations by Löew et al. [1777] supported the oxygen-centered radical intermediate **468** as the active oxygen species. Intermediates **463–466** have been isolated in a stable form and are well characterized [1778, 1779].

The electrophilic nature of the activated heme species **468** ↔ **468′** ↔ **468″** is evident when substrates are utilized in which carbon atoms have equal access to the active-site heme. In these cases, reaction occurs preferentially at the more substituted (tertiary > secondary > primary) carbon center [1780]. For the benzylic hydroxylation, a large kinetic deuterium isotopic effect ($k_H/k_D > 11$) has been found, suggesting a hydrogen atom abstraction in the rate-determining

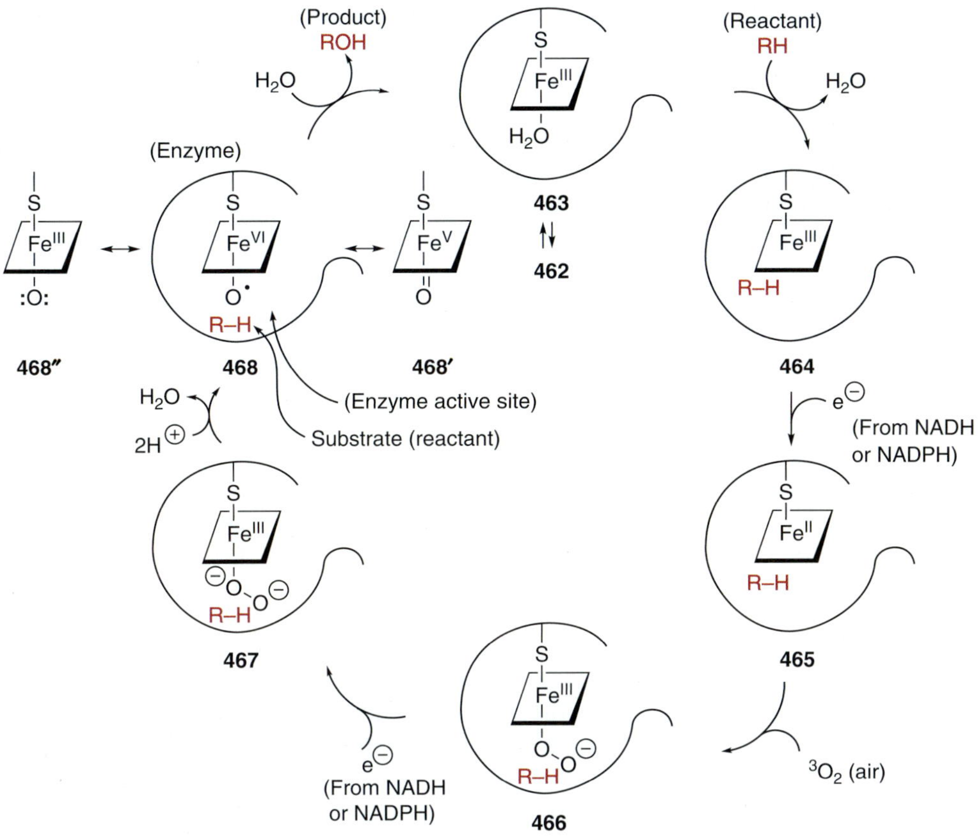

Scheme 7.85 Molecular oxygen activation by heme-dependent monoxygenases.

Scheme 7.86 Cytochrome P450-catalyzed oxidation of *trans*-1-methyl-2-phenylcyclopropane.

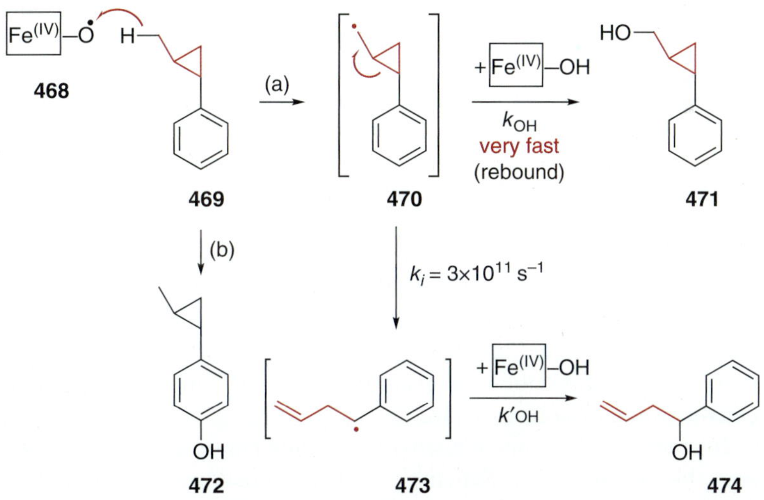

step [1781, 1782]. A two-step mechanism is evidenced by several experiments [1776, 1783, 1784]. In one of them, support for a radical intermediate was obtained with the ultrafast radical clock probe, *trans*-1-methyl-2-phenylcyclopropane (**469**, Scheme 7.86) [1785]. The cytochrome P450-catalyzed oxidation of **469** generates cyclopropylmethanol **471** and homoallyl alcohol **474** (pathway (a)). The cyclopropylmethyl radical intermediate **470**, arising from the H-atom abstraction from **469** by **468**, rearranges at the remarkable rate of $k_i = 10^{11}\ s^{-1}$ into the homoallylic radical intermediate **473**. Both radical intermediates are quenched in the catalytic site of the enzyme, a process called "oxygen rebound," and give **471** and **474**, respectively. Formation of rearranged alcohol **474** is strong evidence for radical intermediates. Pathway (b) accounts for the formation of the third product **472** resulting from the concurrent aromatic hydroxylation of **469**. Further mechanistic details are abundant in the literature [1785–1799]. If steric factors should retard the oxygen rebound, outer-shell electron transfer might oxidize the radical intermediate into a carbenium ion that undergoes quick Wagner–Meerwein rearrangement or other isomerization and subsequent deprotonation or nucleophilic quenching [1800]. If the carbenium ion intermediate is a primary alkyl cation, the Wagner–Meerwein rearrangement is expected to be concerted with electron transfer.

7.11.2 Chemical aerobic C—H oxidations

Synthetic metalloporphyrins have been searched to mimic cytochrome P450 as a catalyst of aerobic oxidations of organic compounds in water [1801–1803]. In 1979, Groves et al. developed a first (porphyrin)iron complex that catalyzes hydroxylations and epoxidations via the oxygen transfer from PhI=O (iodosylbenzene) [1804]. Since then, systems have been found the use of O_2 as the oxidant instead of PhI=O, $PhI(OAc)_2$, H_2O_2, HOCl, or peroxides. For instance, aerobic oxidation of cyclohexane into a mixture of cyclohexanol and cyclohexanone has been catalyzed by $[(5,10,15,20\text{-tetraphenylporphyrin})Fe(II)]_2O$ with a TON = 90 000 [1805]. Ton-scale air oxidation of cyclohexane (20 °C) into cyclohexanol (87% yield) is catalyzed by (5,10,15,20-tetraphenylporphyrin)Co(II) [1806]. $Co(OAc)_3$ has long been recognized for its high oxidizing power [1807]. An electron transfer mechanism (Scheme 7.87) has been proposed for the oxidation of cyclohexane involving initial interaction with a C—H bond with Co(III) ions, producing cyclohexyl acetate, 2-acetoxycyclohexanone, and cyclohexylidene diacetate. In the presence of O_2, mostly adipic acid is formed [1808–1810]. Catalytic oxidation of cyclohexane with O_2 in the presence of Co_3O_4 nanocrystals (30–50 nm) leads to 89% selectivity to cyclohexanol and cyclohexanone at 7.6% conversion of cyclohexane [1811].

The radical cation mechanism (the rate-determining step being the SET; Scheme 7.87) is consistent

$$\text{R–H} + Co(OAc)_3 \rightleftharpoons RH^{\bullet+}Co(OAc) + AcO^-$$
$$\text{R–H}^{\bullet+} \rightleftharpoons R^{\bullet} + H^+$$
$$R^{\bullet} + Co(OAc)_3 \longrightarrow ROAc + Co(OAc)_2$$
$$R^{\bullet} + O_2 \longrightarrow ROO^{\bullet} \longrightarrow \text{Products}$$

Scheme 7.87 Initial steps in the $Co(OAc)_3$-catalyzed O_2 oxidation of hydrocarbons.

with the nonobservation of primary k_H/k_D kinetic isotopic effect. In the case of the oxidation of methyl-substituted benzene derivatives, a relatively high negative ρ value of −2.4 (correlated with σ^+ at 65 °C) was observed. Radical cation intermediates were detected by ESR flow techniques. The addition of LiCl leads to an increased activity of $Co(OAc)_3$ and to ρ value of −1.35 (correlated with σ^+ at 24 °C) suggesting that $Co(OAc)_2Cl$ has a higher oxidation potential and lower selectivity than $Co(OAc)_3$ [1812]. Once the hydrocarbon radical cation $RH^{\bullet+}$ is formed, it undergoes deprotonation forming the corresponding alkyl radical $R^{\bullet}$. In the case of toluene oxidation into benzoic acid [1813], the Morimoto's mechanism shown in Scheme 7.88 explains the oxidation of benzyl radical [1814].

Aerobic oxidation of cyclohexanone (0.5 M) in 9 : 1 $AcOH/H_2O$ catalyzed by $Co(OAc)_2{\cdot}4H_2O$ (0.002 M) and NaBr (0.01 M) gives a number of products of α-oxidation and then adipic acid. The probable mechanism of these oxidations is shown in Scheme 7.89 [1815, 1816].

The kinetics (redox potential: $\Delta E = 0.074\log([Co^{3+}]/[Co^{2+}])$, concentration of dissolved O_2, and concentration of Br^- show an oscillatory behavior [1817–1819]. This is due to a relatively slow autooxidation of cyclohexanone (RH) that leads to gradual increase of the concentration of hydroperoxide (ROOH) and α-hydroxyketone ROH. When ROH builds to moderate concentrations, a second stage begins, and a facile oxidation of ROH by Co^{3+} occurs to give the open acyl radical $A\text{–}CO^{\bullet}$. The subsequent steps lead to rapid generation of peracid $A\text{-}CO_3H$, which rapidly converts Co^{2+} to Co^{3+}. Co^{3+}, in turn, then oxidizes ROH even more rapidly into radical $A\text{–}CO^{\bullet}$. Another oscillating observable is chemiluminescence due to light emission from the electronically excited state of cyclohexa-1,2-dione ($\mathbf{D}^* \rightarrow \mathbf{D} + h\nu$) generated by dimerization of α-ketocyclohexylperoxy radicals ($ROO^{\bullet}$). Oscillatory kinetic behaviors are known for enzyme-catalyzed reactions [1820, 1821] and metal-catalyzed alkane oxidations [1822]. In the mechanisms shown in Schemes 7.88 and 7.89, the hydroperoxides are formed via multistep processes implying the reactions of intermediate alkyl radicals ($R^{\bullet}$) with triplet oxygen ($^3O_2$). Direct electrophilic attack of the alkane C—H bond by $Co(OAc)_3$ is also a possibility by analogy with the $PtCl_2$-catalyzed H/D-exchange of alkane in CH_3COOD/D_2O [1024]. The $Pt(II)Cl_4{}^{2-}$-catalyzed H/D exchange of alkanes in $AcOD/D_2O$ follows Shilov's mechanism: $PtCl_4{}^{2-} + R\text{—}H \rightleftharpoons Cl_3Pt(R\text{—}H)^- + Cl^- \rightleftharpoons Cl_3Pt(H)(R)^- + Cl^-$; $Cl_3Pt(H)(R)^- + Cl^- + AcOD \rightleftharpoons Cl_3Pt(D)(R)^- + Cl^- + AcOH \rightleftharpoons Cl_3Pt(R\text{—}D)^- + Cl^- + AcOH \rightleftharpoons$

Scheme 7.88 Morimoto's mechanism for the Co(III)-catalyzed autooxidation of toluene into benzaldehyde and benzoic acid.

$$PhCH_2^{\bullet} + O_2 \longrightarrow PhCH_2O_2^{\bullet}$$
$$PhCH_2O_2^{\bullet} + Co^{2+} \longrightarrow PhCHO + Co^{3+} + HO^-$$
$$PhCHO + Co^{3+} \longrightarrow PhC^{\bullet}{=}O + Co^{2+} + H^+$$
$$(HO^- + H^+ \longrightarrow H_2O)$$
$$PhC^{\bullet}{=}O + O_2 \longrightarrow PhC({=}O){-}O{-}O^{\bullet}$$
$$PhCOOO^{\bullet} + PhCH_3 \longrightarrow PhCH_2^{\bullet} + PhCO_3H$$
$$PhCO_3^{\bullet} + PhCHO \longrightarrow PhC^{\bullet}{=}O + PhCO_3H$$
$$PhCO_3H + Co^{2+} \longrightarrow PhCO_2^{\bullet} + HO^- + Co^{3+}$$
$$PhCO_2^{\bullet} + PhCH_3 \longrightarrow PhCH_2^{\bullet} + PhCO_2H$$
$$PhCO_2^{\bullet} + PhCHO \longrightarrow PhC^{\bullet}{=}O + PhCO_2H$$
Competing reaction: $PhCO_3^{\bullet} + PhCHO \longrightarrow PhCO_2H + PhO^{\bullet} + CO$
$$PhCO^{\bullet} \longrightarrow Ph^{\bullet} + CO$$
$$Ph^{\bullet} + PhCO_3^{\bullet} \longrightarrow PhO^{\bullet} + PhCOO^{\bullet}$$
$$PhCOO^{\bullet} + PhCHO \longrightarrow PhCOOH + PhC^{\bullet}O$$
$$PhO^{\bullet} + PhCH_3 \longrightarrow PhOH + PhCH_2^{\bullet}$$

$R{-}D + PtCl_4^{2-} + AcOH$. In Shilov's oxidation of alkanes, $Pt(IV)Cl_6^{2-}$ is the stoichiometric oxidant: $Cl_3Pt(H)(R)^- + Cl^- \rightleftarrows Cl_3Pt(R)^{2-} + HCl$, $Cl_3Pt(R)^{2-} + Pt(IV)Cl_6^{2-} \rightleftarrows Cl_5Pt(R)^{2-} + Pt(II)Cl_4^{2-}$, and $Cl_5Pt(R)^{2-} + H{-}Nu\ (H_2O, HCl) \rightarrow Cl_4Pt^{2-} + R{-}Nu + HCl$ [1823, 1824]. This suggests that a suitable transition metal complex [M] that undergoes oxidative addition of a C—H bond should generate the corresponding adduct, e.g. $R_3C{-}[M]{-}H$. Then, O_2 should react with it to produce the corresponding alcohol and release the initial catalyst [M] according to $R_3C{-}[M]{-}H + ½\ O_2 \rightarrow R^3COH + [M]$. Several efforts are done to find such conditions. In 1996, Yoneyama and Crabtree found that $Pd(OAc)_2$ catalyzed the arene oxidation using $PhI(OAc)_2$ as the stoichiometric oxidant [1825]. ($Ar{-}H + Pd(OAc)_2 \rightarrow Ar{-}Pd(OAc) + AcOH$; $Ar{-}Pd(OAc) + PhI(OAc)_2 \rightarrow ArOAc + Pd(OAc)_2 + PhI$). Several strategies have been explored by Sanford and coworkers to render these arene acetylations regioselective. Using *N*-halogenosuccimides as oxidants, the corresponding chlorides, bromides, and iodides are obtained. Interestingly, with *N*-fluoropyridinium oxidants direct C—H fluorination is possible. Furthermore, with $[Ph_2I]BF_4$, $[Mes{-}I{-}Ar]BF_4$, and $[Ar_2I]BF_4$, arene phenylation and arylations have been realized [1826–1828].

Salicylaldehyde (2-hydroxybenzaldehyde) has been prepared by O_2 oxidation of *ortho*-cresol (2-methylphenol) catalyzed by an heterogeneous Cu/Co/C catalyst in the presence of pyridine and methanol [1829]. Sterically hindered complex *cis*-$[Ru(dmp)_2(S)_2](PF_6)_2$ ($S = H_2O$, MeCN, dmp = 2,9-dimethyl-1,10-phenanthroline) catalyzes the epoxidation of alkenes with O_2 and of alkanes into alcohols with H_2O_2 [1830]. Aerobic oxidation of methane in the presence of CF_3COOH gives methyl trifluoroacetate at 80 °C. The reaction involves an electron transfer chain, which is a combination of the three redox couples: $Pd^{2+}/Pd°$, quinone/hydroquinone, and NO_2/NO (Scheme 7.90) [1831, 1832]. The direct selective oxidation of methane into methanol with wet O_2 over Cu-exchanged zeolites at 110–125 °C has been realized [1833].

Labile Fe(II) and Fe(III) complexes $[Fe(bpy)_2^{2+}]$ and $[Fe(bpy)_2^{3+}]$ in MeCN (bpy = 2,2′-bipyridine) activate O_2 for the oxidation of limonene to produce mainly carvone, carveol, limonene oxide, and perillaldehyde [1834].

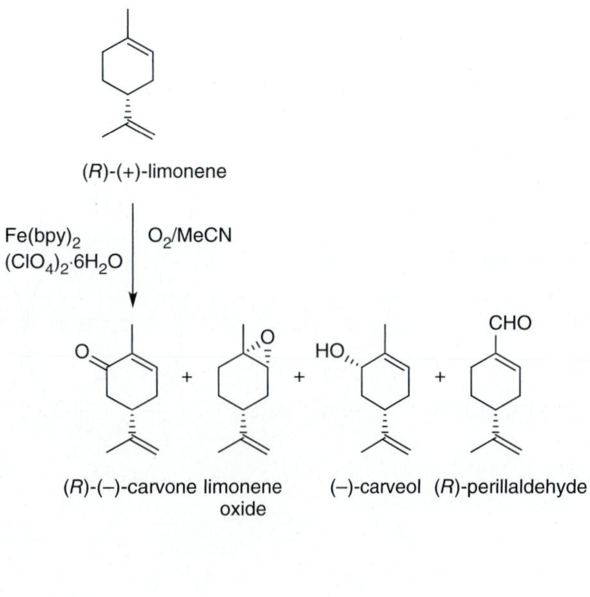

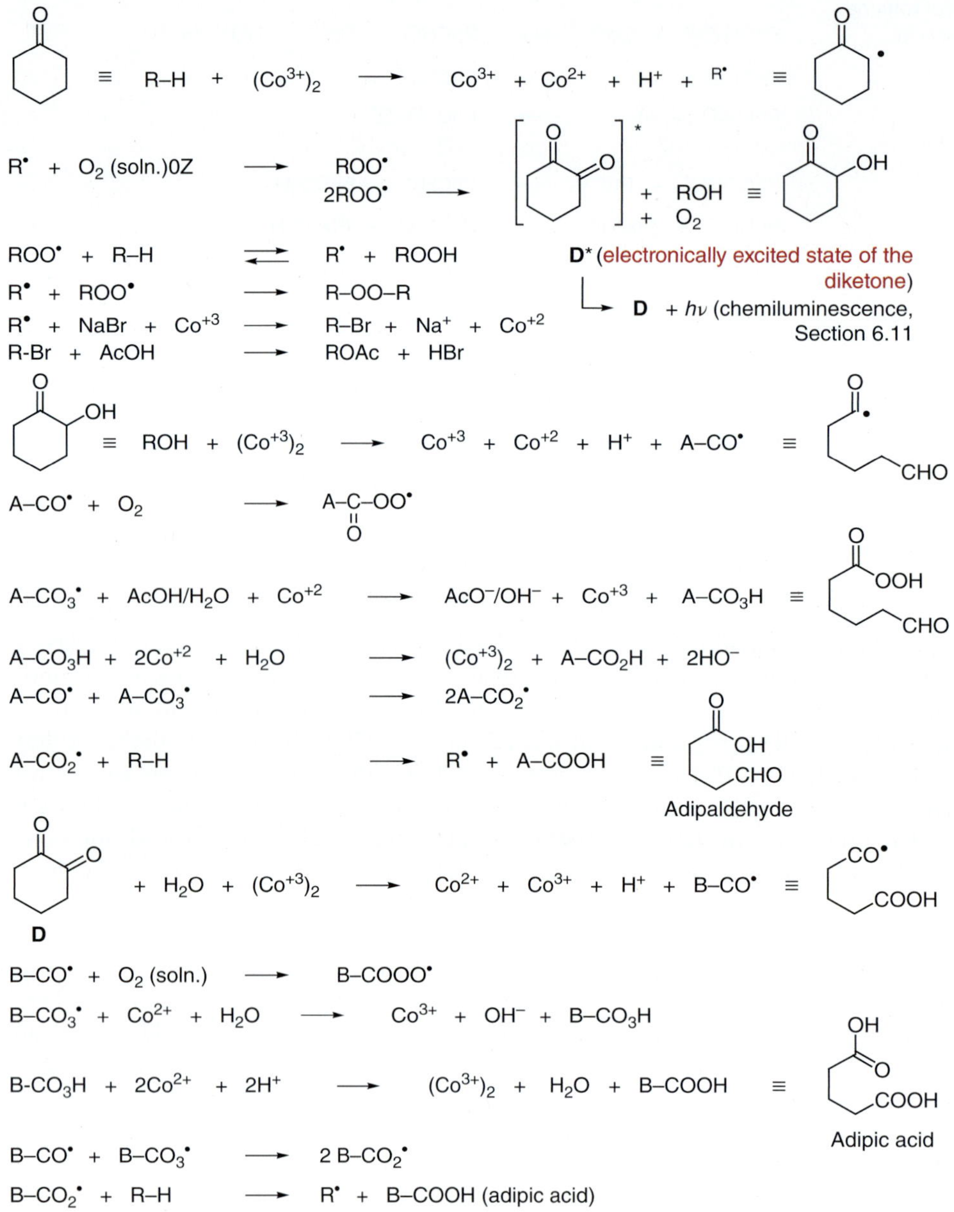

Scheme 7.89 Wasserman's mechanism for the $Co(OAc)_3$-catalyzed O_2 oxidation of cyclohexanone (O_2(soln) = oxygen in solution).

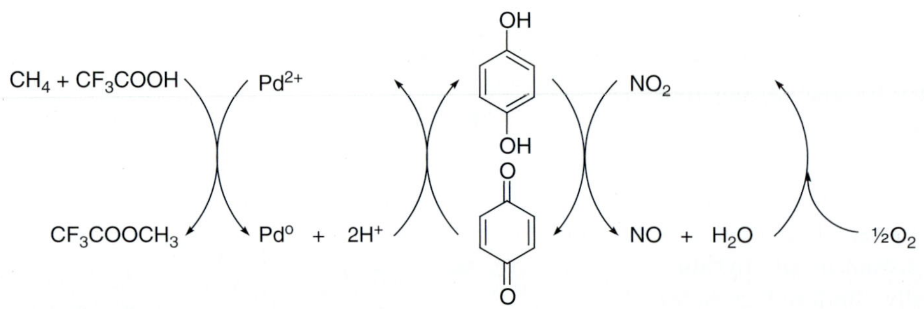

Scheme 7.90 Catalytic O_2 oxidation of methane in CF_3COOH through an electron transfer chain.

The polyoxometalates such as phosphovanado-molybdates of the Keggin structure, $[PV_xMo_{12-x}O_{40}]^{((3+x)-)}$, catalyze dehydrogenations through electron transfer oxidations by the polyoxometalate that is then reoxidized by O_2. These aerobic oxidations are useful for the aromatization of cyclohexadienes, oxidative phenol homocoupling, and alcohol oxidation [1835]. Hydrogen peroxide (H_2O_2) is a convenient oxidant that can be used to convert methane and ethane into methanol and ethanol, respectively, in the presence of a suitable catalyst. Among transition metal chlorides, $OsCl_3$ exhibits the best TOF for reaction occurring at 90 °C [1836].

7.11.3 Reductive activation of molecular oxygen

In general, the term "reductive oxidation" stands for the process: $O_2 + 2H^+ + 2e^- \rightarrow [O] + H_2O$. One speaks of reductive activation of molecular oxygen [1837] when the metallic catalyst activates O_2 by an electron transfer. The enzymes called "monooxygenases" employs this process with the coordination sphere of an active metal site for the generation of highly reactive species (see e.g. Scheme 7.85, Section 7.11.1). Barton and coworkers established an aerobic oxidation using Zn powder as the reducing agent under acidic conditions [1838–1840]. The reaction system composed of an iron complex, pyridine, AcOH, and Zn (powder) is called "Gif system." It oxidizes saturated hydrocarbons to alcohols or ketones by aerobic oxidation at room temperature. Moro-Oka showed that catalyst **475** that mimics the active site of methanemonooxygenase can oxidize alkanes, alkenes, and benzene derivatives into the corresponding alcohols, epoxides, and phenols at 25 °C (reactions (7.164)–(7.166)) [1841].

$$\text{R–H} \xrightarrow[\text{CF}_3\text{COOH/CH}_2\text{Cl}_2]{\text{O}_2\text{/Zn, } \mathbf{475} \text{ (cat.)}} \text{ROH} \qquad (7.164)$$

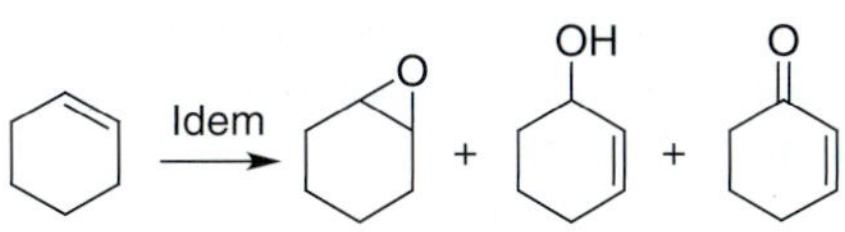

(7.165)

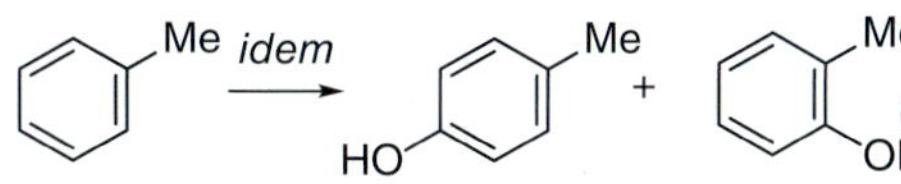

(7.166)

catalyst:

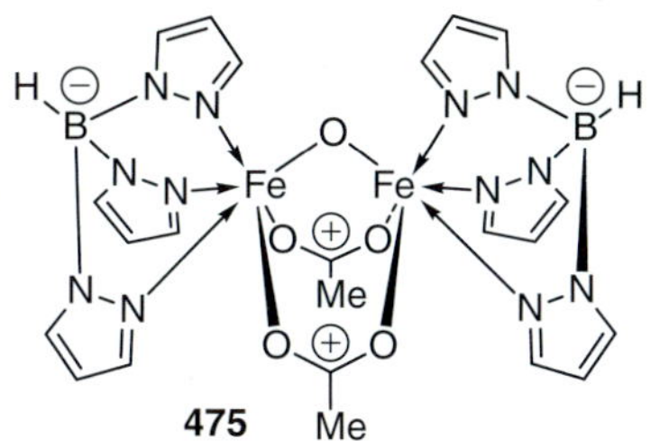

475

Cheaper reducing agents than zinc are H_2, CO, and aldehydes. Aerobic epoxidation of alkenes with H_2 gas was first demonstrated by Tabushi and coworkers with the epoxidation (7.167) of cyclohexene with O_2, H_2, colloidal platinum, and **476** as cocatalysts [1842, 1843]. Colloidal Pt catalyzes electron transfer from H_2 to **476** to form a Mn(II) species that quenches O_2 to generate peroxide TPP—Mn(II)—O_2 (TPP = 5,10,15,20-tetraphenylporphyrinate). The latter corresponds closely to the native NADH–hemoprotein–O_2 system.

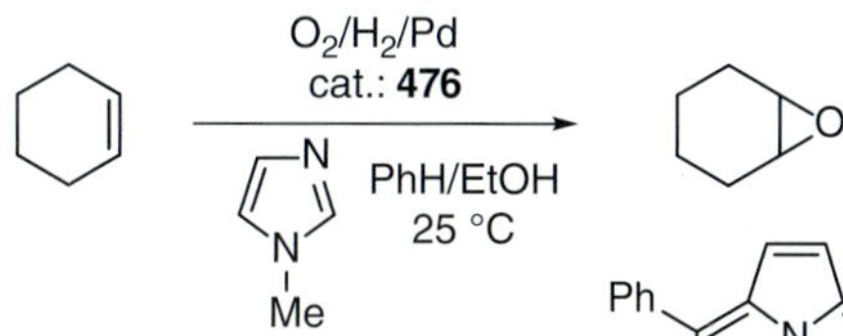

(7.167)

catalyst: **476** (TPPMnCl):

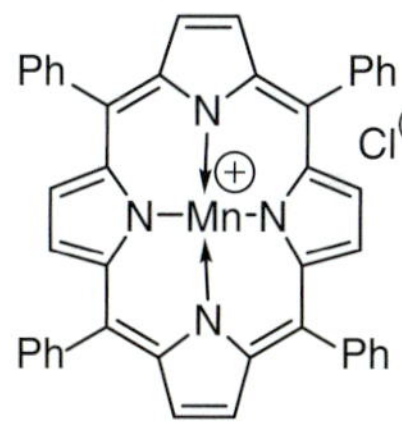

A bicomponent catalyst containing Pt/Mo/P permitted the oxidation with O_2/H_2 of cyclohexane into cyclohexanol and cyclohexanone (in MeCN, 35 °C) [1844]. A possible mechanism is given in Scheme 7.91. Reduction of absorbed O_2 on Pt by H_2 results in the formation of peroxide species. Pt—O_2 interacts with Mo^{5+} ions included in the polyatomic catalyst to

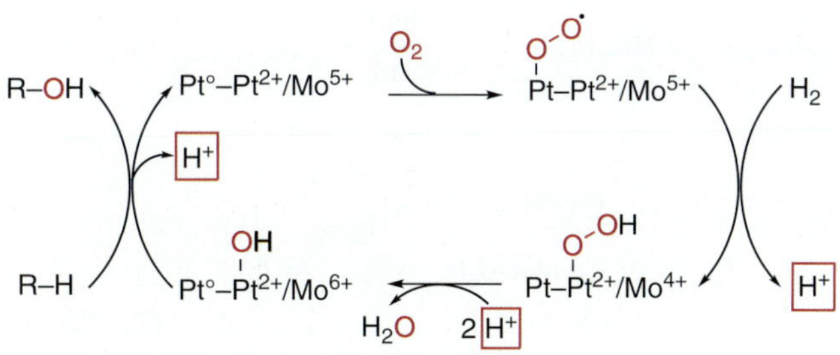

Scheme 7.91 Possible mechanism for the reductive activation of O_2 in catalytic systems including Pt and heteropolymetallic compounds.

generate hydroxyl radicals, which oxidize cyclohexane into cyclohexanol, and then into cyclohexanone. Metal complexes and *p*-benzoquinones are commonly used as catalysts in aerobic oxidation reactions that operate via an "oxidase" type mechanism involving two independent half reactions (see e.g. Scheme 7.90): (i) oxidation of the organic substrate mediated by the oxidized catalyst and (ii) oxidation of the reduced catalysts by O_2 [1845–1849].

Direct hydroxylation of benzene derivatives to phenols with O_2 occurs in $Cu(I)/O_2/H_2SO_4$ system. Cu(I) is consumed stoichiometrically. On adding Pd as a catalyst, and using H_2 as the reductant, Cu(II) is reduced back to Cu(I), and thus copper is used as the cocatalyst [1850, 1851]. The aerobic oxidation of benzene to phenol is possible at 180 °C using CO as the reductant and $Pd(OAc)_2$ as the catalyst (reaction (7.168)) [1852–1854]. The most effective hydroxylation of aromatic compounds with O_2 has been presented by Miyake et al. employing H_2 as the reductant. Reaction (7.169) proceeds at 20–60 °C in AcOH [1855]. Sodium formate has been used as reductant in the reductive activation of O_2 catalyzed by a two-phase system TPP—Mn(III) + [Rh(III) (η^5-Me_5C_5)(bipyridine)Cl_2]. The system permits epoxidation of alkenes [1856]. Mukaiyama and Yamada have found that aerobic epoxidation of alkenes can be catalyzed by Ni(II), Fe(III), V(IV), and Mn(II) complexes in the presence of MeCHO as the reductant [1857]. An example is given with reaction (7.170).

$$\text{Ph–H} + O_2 + CO \xrightarrow[180\,^\circ\text{C, 1 h}]{Pd(OAc)_2\text{(cat.)}} \text{PhOH} + O{=}C{=}O \tag{7.168}$$

$$\text{Ph–H} + O_2 + H_2 \xrightarrow[\text{Cu(I), AcOH}]{\text{Pt/Pd(cat.)}} \text{PhOH} + H_2O \tag{7.169}$$

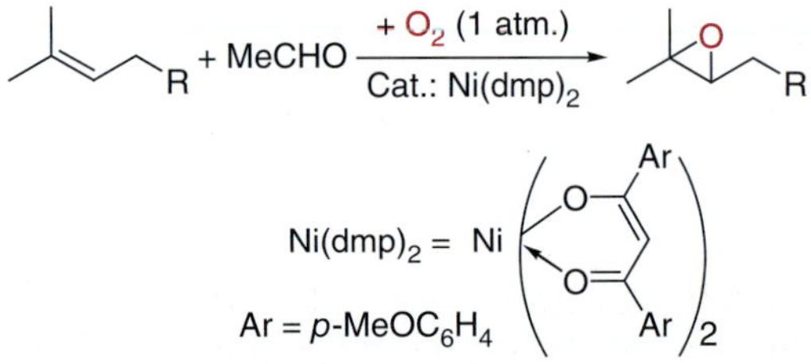

(7.170)

7.11.4 Oxidation of alcohols with molecular oxygen

The oxidation of primary alcohols into the corresponding aldehydes or carboxylic acids and of secondary alcohols into the corresponding ketones is an important organic synthetic method using, for instance, Jone's reagent (CrO_3/H_2SO_4/acetone/H_2O) [1858], Collins' reagent (CrO_3/pyridine) [1859], MnO_2 [1860], Fétizon's Ag_2CO_3/Celite [1861], Corey's PCC (pyridinium chlorochromate) [1862], Corey's PDC (pyridinium dichromate) [1863, 1864], Corey's chromate catalyst (2,4-dimethylpenta-2,4-diyl chromate), and acetic peracid [1865] or Moffatt's oxidation (DMSO (dimethylsulfoxide) + DCC [1866]). DCC can be replaced by Ac_2O [1867], P_4O_{10} [1868], SO_3 [1869], or $(CF_3CO)_2O$ [1870] or better by ClCOCOCl (Swern oxidation) [1871, 1872]. Dess–Martin periodinane is a very mild oxidizing agent that converts primary alcohols into aldehydes and secondary alcohols into ketones [1873, 1874]. Sodium hypochlorite can serve as oxidant in the presence of TEMPO ((2,2,6,6-tetramethylpiperidin-1-yl)oxyl) or its derivatives as catalysts [1875]. TEMPO is a stable radical discovered in 1960 by Lebedev and Kazarnowskii [1876]. The process based on the Anelli–Montanari protocol [1877] is applied in the industry but present limitations due to the concurrent formation chlorides [1878]. Less expensive methods producing less waste and secondary products uses air, O_2, or H_2O_2 as oxidants in the presence of a reusable catalyst [1879]. Oxidation of secondary alcohols into ketone by O_2 has been catalyzed by Pt [1880–1882], Rh, [1883–1885] Ru compounds [1886–1893], and Pd complexes with various ligands [1894, 1895] and by monodispersed Au nanoparticles embedded in a polyometalate self-assembly matrix [1896]. Yamada and Mukaiyama found that $Co[(CF_3CO)_2CH]_2$ catalyzes the O_2 oxidation of secondary alcohols into ketones (reaction (7.171)) [1897]. A "green" route to cyclohexanone from cyclohexanol uses H_2O_2 as oxidant and nonprecious h-WO_3 nanorods as catalyst (80 °C, 1 atm) [1898].

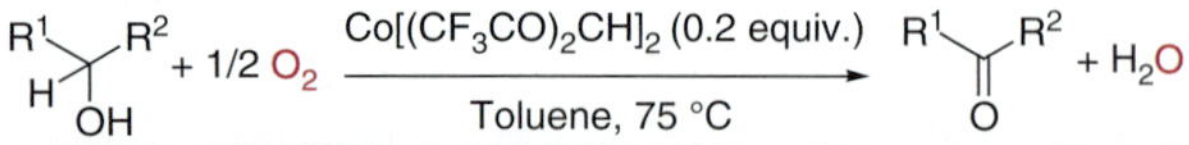

(7.171)

Markó et al. have developed a copper-based catalyst that catalyzes the aerobic oxidation of primary alcohol into aldehydes and secondary alcohols into corresponding ketones (Scheme 7.92) [1899, 1900]. In the presence of H_2O, the phenanthroline-CuCl and bis(*t*-butyl) azodicarboxylate (DBAD) complex

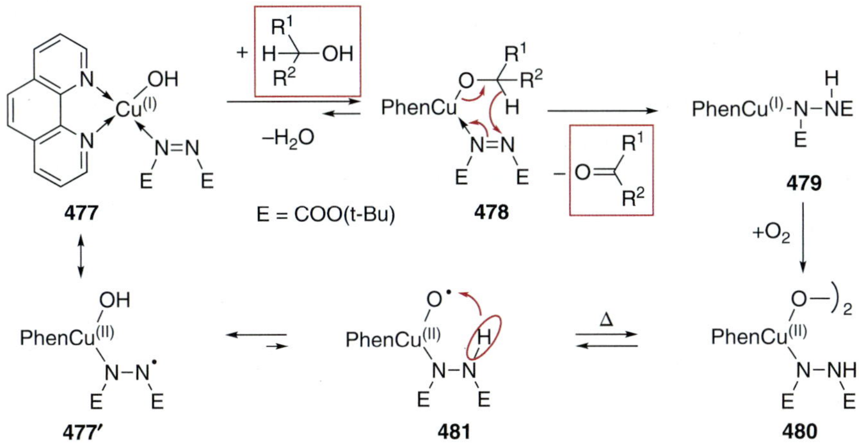

Scheme 7.92 Markó's copper-catalyzed aerobic oxidation of alcohols into aldehydes and ketones (Phen ≡ 1.10-phenanthroline; solvent: toluene or fluorobenzene, K_2CO_3 (2 equiv.), 80 °C; initial catalyst: phenCuCl (5 mol%) + DBAD (5 mol%)).

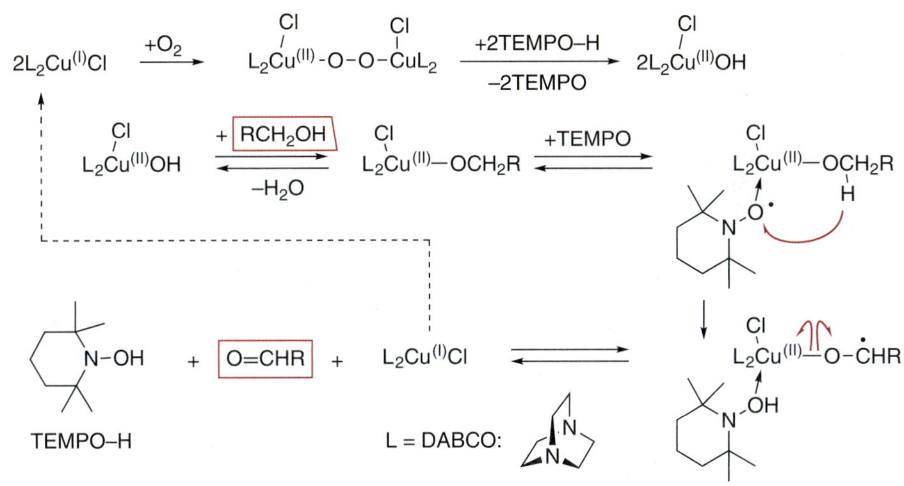

Scheme 7.93 Possible mechanism for the catalytic oxidation of alcohols by O_2 and TEMPO/Cu(II) (diamine) catalyst.

intermediate **477** forms. In the presence of the alcohol, **478** is obtained. A kind of retro-ene reaction with transfer of the hydrogen atom to the nitrogen center of **478** leads to product (aldehyde and ketone) and generates intermediate **479**. The latter is also formed by the reaction of ENH–NHE, K_2CO_3, and initial 1,10-phenanthroline-CuCl complex. The Cu(I) complex **479** reacts with Cu(II) oxide **481**, which undergoes fast γ-hydrogen atom transfer forming the more stable nitrogen-centered radical **477′**, which is the limiting structure of complex **477**.

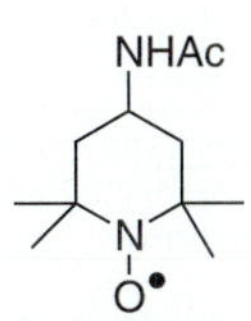

482

Jiang and Raganskas have proposed a four-component system consisting of 4-acetamido-TEMPO (**482**) (4 mol%)/$Cu(ClO_4)_2 \cdot 6H_2O$ (4 mol%) and 4-DMAP for room-temperature aerobic alcohol oxidation in the ionic liquid [BMPy]PF_6 (BMPy = *N*-*n*-butyl-*N*-methylpyrrolidinium) [1901] or in DMSO (dimethylsulfoxide) [1902]. Repo and coworkers have proposed $CuSO_4$/phenanthroline and TEMPO as catalytic system for the O_2 oxidation of alcohols at 80–100 °C and 10 bar O_2 [1903]. With CuCl/DABCO (5 mol%), the O_2 oxidation occurs already at 25 °C (DABCO = 1,4-diazabicyclo[2.2.2]octane) [1904]. A possible mechanism is shown in Scheme 7.93 [1905–1907].

Selective aerobic oxidation of primary alcohols into carboxylic acids has been found for the aerobic oxidation catalyzed by $VO(acac)_2$ (2 mol%) (acac = acetylacetonate), Cu(II) 2-ethylhexanoate (2 mol%), and DABCO (6 mol%) in ionic liquid [HMIN]OTf ([HMIN]OTf = 1-hexyl-3-methylimidazolium trifluoromethanesulfonate) at 95 °C, whereas the corresponding aldehydes are obtained when the Cu(II) salt is omitted [1908]. Palladium nanoparticles in ionic liquids are reusable catalysts for aerobic oxidation of alcohols [1909]. The oxidation of allylic and benzylic alcohols to the corresponding aldehydes can be carried out at 25 °C with O_2 in the presence of n-$Pr_4N^+/RuO_4^- + CuCl$ [1910], by supported Ru/Al_2O_3 catalyst [1911], by trimetallic $Ru/CeO_2/CoO(OH)$, [1912] and $VOSO_4$ + pyridine ligands in water [1913]. Carbon nanotubes are metal-free catalysts for the selective aerobic oxidation of benzyl alcohol to benzaldehyde [1914]. More metal-free catalysts for aerobic oxidation of alcohols are being developed [1915].

7.11.5 Wacker process

The Wacker olefin oxidation via $PdCl_2$ catalyst and air-recyclable $CuCl_2$ co-oxidant was discovered in 1959 by Smidt et al. at the Wacker Chemie company in Germany (also called the Hoechst-Wacker process) [1916]. It was based on the observation reported by Phillips in 1894 that ethylene and stoichiometric amount of $PdCl_2$ in water solution produces ethanal and palladium metal (reaction (7.172)) [1917]. In the presence of $CuCl_2$, Pd(0) is oxidized into Pd(II) (reaction (7.173)), thus enabling the use of Pd(II) salt in catalytic amounts. CuCl obtained can be oxidized by O_2 (reaction (7.174)).The oxygen atom incorporated in the product does not come from O_2 but from water. One speaks here of Pd oxidase catalysis [1918]. Overall, the Wacker reaction is $CH_2{=}CH_2 + \frac{1}{2}\ O_2 \rightarrow CH_3CHO$ (with $\Delta_r H° = -53.3 \pm 0.45\ \mathrm{kcal\ mol^{-1}}$, Tables 1.A.2 and 1.A.4). Its mechanism is still a matter of debate [1919–1921]. It is one of the first industrial processes using homogeneous palladium catalysis. Inexpensive and environmentally friendly $Fe(SO_4)_3$ has been proposed as terminal oxidant for the Wacker process [1922]. Using only O_2 as oxidant (1 atm) and CF_3COOH (1 equiv.) and $Pd(AcO)_2$ as catalysts (10 mol%) in 10 : 1 DMSO/H_2O, terminal olefins have been oxidized into corresponding methyl ketones, and 4-arylbut-1-enes into corresponding α,β-unsaturated methyl ketones [1923].

$$CH_2 = CH_2 + Pd(II)Cl_2 + H_2O \rightarrow CH_3CHO + Pd(0) + 2HCl \quad (7.172)$$

$$Pd(0) + 2\,CuCl_2 \rightarrow PdCl_2 + 2\,CuCl \quad (7.173)$$

$$2\,CuCl + \tfrac{1}{2}\,O_2 + 2\,HCl \rightarrow 2\,CuCl_2 + H_2O \quad (7.174)$$

Under industrial conditions (LL: low [Cl⁻] (<1 M) and low [$CuCl_2$] (<1 M)), ethylene is oxidized into acetaldehyde via internal *syn* addition of water. Henry suggested an inner-sphere process involving two fast ligand exchange reactions $PdCl_4^{2-} + CH_2{=}CH_2 \rightleftarrows$ **484** + Cl^- and **484** + $H_2O \rightleftarrows$ **485** + Cl^- (Scheme 7.94), followed by pre-equilibrium deprotonation **485** ⇄ **486**, and then rate-determining *syn* hydroxypalladation **486** → **487** to yield the rate law (7.175) [1924].

$$\mathrm{rate}_{LL} = -\frac{d[C_2H_4]}{dt} = \frac{k[PdCl_4^{2-}][C_2H_4]}{[H^+][Cl^-]^2} \quad (7.175)$$

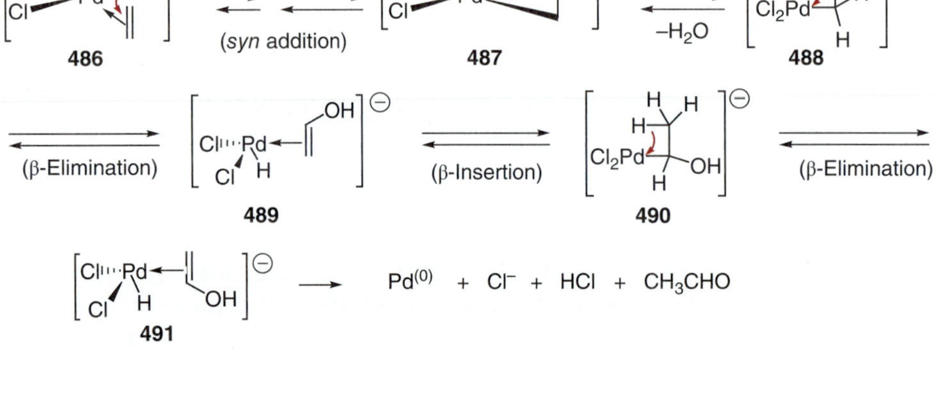

Scheme 7.94 Henry's mechanism of the Wacker process under LL conditions (low concentrations of Cl⁻ and $CuCl_2$): the rate-determining step is the *syn*-hydroxypalladation of the alkene.

Experiments using $CD_2{=}CD_2$ in H_2O generate CD_3CHO whereas oxidation of $CH_2{=}CH_2$ in D_2O generates CH_3CHO. Thus, the keto/enol tautomerization is not a mechanistic step. For the reaction of $CH_2{=}CH_2/CD_2{=}CD_2$, a relatively small kinetic deuterium isotope effect is observed ($k_H/k_D = 1.07$). This excludes hydride transfer in the rate-determining step. However, $k_H/k_D = 1.9$ is observed for the oxidation of $CD_2{=}CH_2$, suggesting that the rate-determining step precedes the formation of oxidized products. Applying quantum mechanical calculations, Goddard and coworkers found that the rate-determining step in the mechanism shown in Scheme 7.94 is not hydroxypalladation as proposed by Henry, but isomerization **492** ⇄ **488** (reaction (7.176)), a step necessary for subsequent hydride transfer (β-elimination of H) [1925].

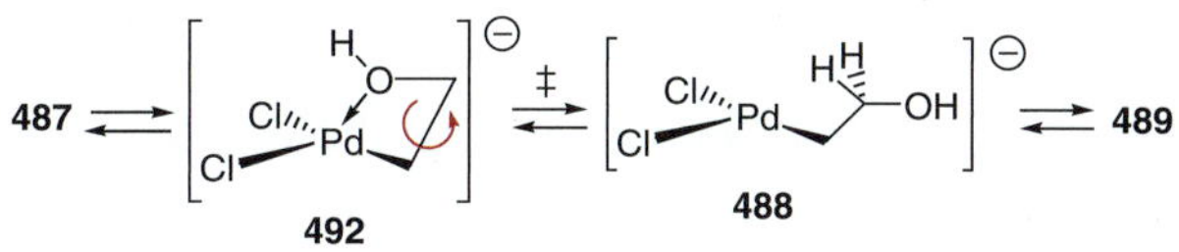

(7.176)

At high concentrations of Cl^- and $CuCl_2$ (> 2.5 M) (HH conditions), both acetaldehyde and chlorohydrin are obtained. The ratio of chlorohydrin to aldehyde increases with increasing [$CuCl_2$] [1926], implying a second (or higher)-order dependence of chlorohydrin formation on $CuCl_2$, thus leading to rate equation (7.177),

$$\text{rate}_{HH} = -\frac{d[C_2H_4]}{dt} = \frac{k_1[PdCl_4^-][C_2H_4][CuCl_2]}{[Cl^-]} + \frac{k_2[PdCl_4^-][C_2H_4][CuCl_2]^2}{[Cl^-]} \quad (7.177)$$

where k_1 and k_2 are the rate constants associated with the formation of CH_3CHO and chlorohydrin products, respectively. Mechanism (7.178) has been proposed for the formation of chlorohydrin. It implies outer-sphere *anti*-addition of H_2O to the (alkene)Pd(II) complex **484** with the formation of intermediate **493** and subsequent reductive elimination [1927]. The rate-determining step is the nucleophilic attack by H_2O, which is of low energy in the presence of $CuCl_2$.

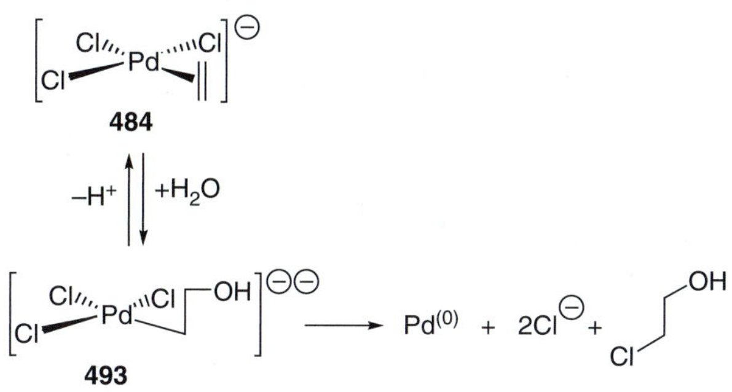

(7.178)

The main mechanistic feature in Wacker process is that $PtCl_2$ (electrophilic activator) promotes the nucleophilic addition of water to the alkene, followed by an hydride shift or β-elimination (nucleofugal group) that reduces Pd(II) into Pd(0). When the reaction is carried out in pure AcOH, vinyl acetate forms. These considerations led Tsuji et al. in the earlier 1960s [1928–1933] and then others (see e.g. the Moritani–Fujiwara reaction) to invent new C—C bond forming reactions in which soft carbon nucleophiles such as CO or malonates are used (Section 8.5.2) [1934]. Under standard conditions (O_2, $PdCl_2$, $CuCl_2$, DMF/H_2O, 60 °C), terminal alkenes are oxidized into methyl ketones (Markovnikov products), cycloalkenes into cycloalkanones; this is the Tsuji–Wacker reaction [1935, 1936]. Allyl alcohol is oxidized into acrolein [1937, 1938]. Internal olefins give low yield of oxidation products unless lateral control is possible by the presence of a nitrogen nucleophile [1939, 1940], The reaction requires high pressure of O_2 and is limited to alkyl- and arylalkenes [1941]. In 2013, Grubbs and coworkers found new conditions of solvent (dimethylacetamide/MeCN/H_2O) and cocatalysts that permit the conversion of internal olefins into ketones at room temperature and under 1 atm of O_2. The reaction can be applied to alkenes containing alcohols, phenols, carboxylic acids, esters, and amide moieties [1942]. They also reported conditions that convert terminal olefins into aldehydes (anti-Markovnikov products) as shown with reaction (7.179). Silver nitrite favors probably the addition of oxygen-centered radical of type [M]–$NO_2^\bullet$ to the alkene (selective formation of secondary instead of primary alkyl radical intermediates) [1943–1945].

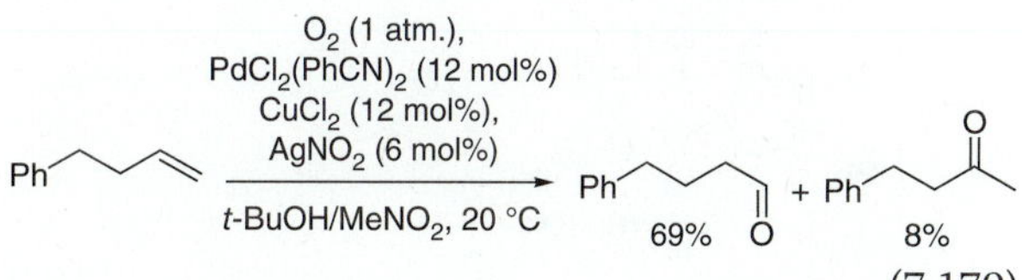

(7.179)

Pd-catalyzed aerobic oxidation of alkenes can provide other compounds than aldehydes and ketones [1845–1947], as illustrated with the formation of acetals (reaction (7.180)) and of vicinal diacetates (reaction (7.181)).

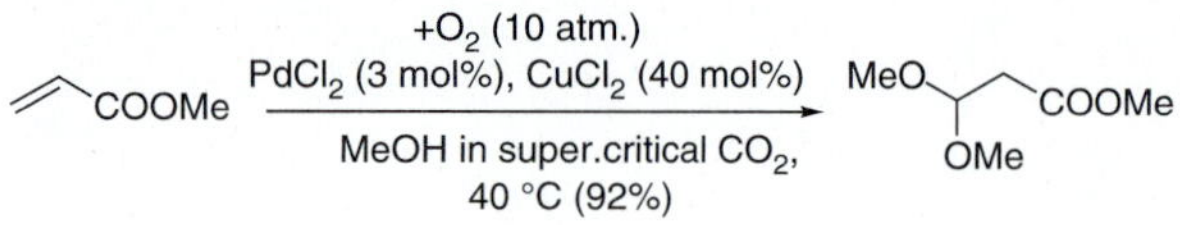

(7.180)

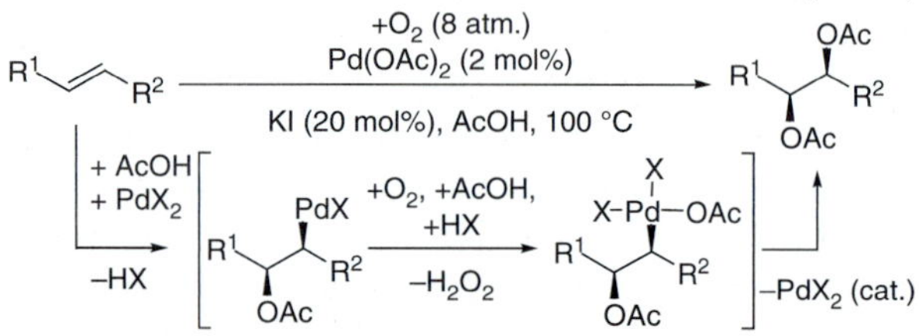

(7.181)

Alkynes can be cleaved into their carboxylic esters (e.g. Ar—C≡C—Ar + 2MeOH + O_2 → 2ArCOOMe + H_2O_2) or produce 2,3,4,5-tetrakis(aryl)furans via the formation of intermediate 1,2,3,4-tetrakis(aryl) butane-1,4-diones. Other C—C bond forming reactions have been reported for the Pd-catalyzed aerobic oxidation of unsaturated compounds such as the formation of 1,3,5-trisubstiuted benzene by oxidative trimerization of terminal alkenes and other benzene derivatives by the oxidative cocondensation of alkenes and alkynes [1947]. In the absence of Pd reoxidant, oxygen atom in the products comes from O_2. One speaks here of oxygenase catalysis [1948]. An example is given with the oxime-assisted intramolecular deoxygenation of alkenes (e.g. reaction (7.182)) with air (1 atm) as the sole oxidant [1949]. A Pd(II) dinuclear complex incorporating four bridging hexahydro-2*H*-pyrimido[1,2-*a*]pyrimidine ligands catalyzes the α-hydroxylation of ketones ($R^1COCH(R^2)R^3 + O_2 \rightarrow R^1COC(OH)(R^2)R^3$) by air (1 atm, 0–6 °C), a reaction that also incorporates oxygen from O_2 into the products [1950].

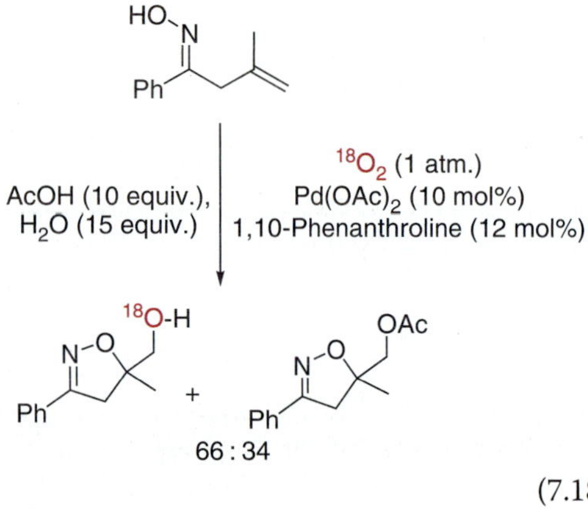

(7.182)

Two pathways have been proposed for the air oxidation of Pd(0) into Pd(II) species. The Pd(II)hydride pathway implies first oxidative addition of HX + $L_nPd(0) \rightarrow L_nPd(H)(X)$ and subsequent coordination of O_2 and β-insertion of hydride to generate hydroperoxide $L_nPd(X)(OOH)$. The latter reacts with HX liberating $L_nPd(II)X_2$ and H_2O_2. The second pathway starts with the formation of a peroxo complex $L_nPd(\eta^2\text{-O=O})$. The latter reacts then with HX giving hydroperoxide $L_nPd(X)(OOH)$ [1918]. Several peroxo complexes of Pd and Pt have been characterized by radiocrystallography [1948]. Pd(1,3-bis(2,6-di(isopropyl)phenyl)imidazole-2-ylidene)$_2$ (=L_2Pd) reacts with O_2 forming the crystalline bis-superoxo complex $L_2Pd(—O—O^\bullet)_2$ [1951].

Problem 7.39 Propose a mechanism for the formation of methyl 3,3-dimethoxypropanoate by oxidation of methyl acrylate by O_2 in MeOH-catalyzed $PdCl_2/CuCl_2$ [1947].

Problem 7.40 Propose a mechanism for the following three-component synthesis of 1,4-dihydopyridines.

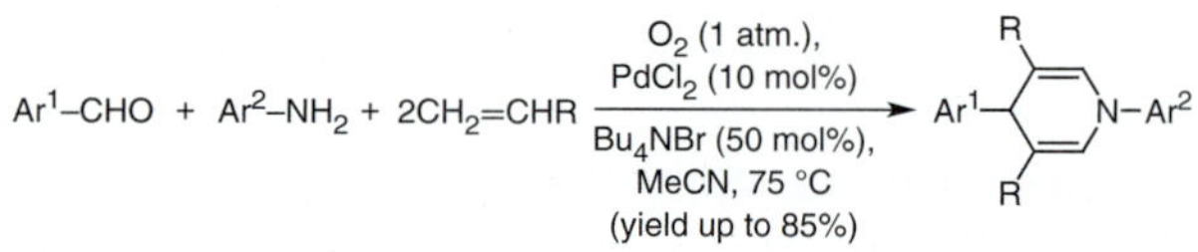

7.12 Catalyzed nucleophilic aromatic substitutions

Uncatalyzed, direct aromatic and heteroaromatic nucleophilic substitutions Ar–X + H(M)–Nu ⇄ Ar–Nu + H(M)X (S_NAr) may follow different mechanisms depending on the nature of the aryl (heteroaryl) group Ar, the nucleofugal group X, and the nucleophile H(M)–Nu. Two-step dissociative S_N1 mechanism is possible only with extremely good leaving groups such as X = N_2^+/X^-. It involves the formation of tight $(Ar^+/N_2/X^-)$ ⇄ solvent-separated $(Ar^+//N_2/X^-)$ molecule–ion pairs as intermediates (Zollinger's mechanism) [1952–1956]. Arylphenyliodonium salts $[Ar\text{-}I\text{-}Ph]^+X^-$ may also undergo S_N1 displacement reactions [1957]. Because the S_N2 mechanism (associative, one-step) requires inversion of configuration, it is never followed by S_NAr. With anionic nucleophiles (M^+/Nu^-), associative addition/elimination mechanism is possible that involves the intermediacy of Meisenheimer adduct (cyclohexadienyl anion) [1958, 1959]. With electron-poor aryl halides or pseudo-halides, neutral nucleophiles H—Nu or Z:

add to them readily and lead to the products of substitution (e.g. 2,4-$(NO_2)_2C_6H_3$–F (Sanger's reagent) + H_2N–peptide→2,4-$(NO_2)_2C_6H_3$–NH–peptide + HF) [1960–1962]. When the addition/elimination mechanism implies the generation of nonstabilized cyclohexadienyl anion, an alternative elimination/addition mechanism can be followed, which implies the generation of benzyne (aryne) intermediates (mechanism of Roberts and Wittig [1963, 1964]; e.g. Ph—Cl + $NaNH_2$ ⇄ [benzyne = 1,2-didehydrobenzene] + NaCl + NH_3 → Ph—NH_2 + NaCl; phenol synthesis (Dow process, 350 °C, 310 atm): PhCl + 2NaOH → PhONa + NaCl + H_2O) and the formation of products of *ipso* and *ortho*-substitution [1965, 1966]. As for any bimolecular process (addition, cycloaddition, and substitution), SET between H(M)–Nu and Ar–X can occur as shown by Kim and Bunnett [1967, 1968]. Radical anions of type $[Ar{-}X]^{\bullet-}$ are generated in the first step in which the C(sp^2)—X bonds are weakened with respect to Ar—X. This permits their dissociation into stable X^- anion and unstable aryl (heteroaryl) radicals $Ar^\bullet$ [1969]. The latter react either with $[H(M){-}Nu]^{\bullet+}$ or with H(M)–Nu to produce the corresponding products of *ipso*-substitution (radical chain mechanism, e.g. $Ar^\bullet + Nu^-$ ⇄ $Ar{-}Nu^{\bullet-}$; $Ar{-}Nu^{\bullet-}$ + Ar–X ⇄ Ar–Nu + $[Ar{-}X]^{\bullet-}$). This is the $S_{RN}1$ mechanism of Kornblum et al. [1970–1972] and Russell et al. [1973], which is favored for Ar–X with high electron affinities (–EA) and nucleophiles with low ionization energies (IE) [1974–1978]. Reactions following the $S_{RN}1$ mechanism can be catalyzed by alkali metals, other metal complexes, and organic compounds with low IEs [1967, 1968, 1979], light [1980–1983], microwave heating [1984–1990], mechanical grinding (electrostriction), and electrochemically [1991–1994]. Complexes of Cu, Pd, or Ni are commonly used to catalyze S_NAr. As we shall see, several mechanisms have been proposed for these catalyzed displacement reactions (Scheme 7.95) [1995].

7.12.1 Ullmann–Goldberg reactions

In 1901, Ullmann and Bielecki reported the formation of diphenyl derivatives from the reductive homocoupling of aryl halides induced by metallic copper (Ullmann coupling: 2Ar–X + 2Cu → Ar–Ar + 2CuX) [1996]. In 1903, Ullmann reported the Cu-mediated *N*-arylation of amines [1997] and, in 1905, the Cu-mediated *O*-arylation of phenols [1998]. These reactions required large amounts of Cu, a base, and elevated temperatures (>200 °C). In 1906, Goldberg reported the first Cu-catalyzed *N*-arylation of amines and carboxamides (Cu in catalytic amount, NaOAc, in $PhNO_2$, 210 °C) [1999]. Although these reactions require harsh conditions, they have found numerous industrial applications such as the synthesis of intermediates in pharmaceutical, agrochemical, fine, and polymer chemistry [2000, 2001]. Without ligand, Cu(I) halides catalyze the *O*-arylation of phenols and the *N*-arylation of benzyl amine and imidazole in *N*-methylpyrrolidone at 160 °C [2002]. Arylation of amines is possible under much milder conditions applying Pd- (Buchwald–Hartwig reaction, Section 7.12.2), Cu-, [2001, 2003] or Ni-based catalysts [2004–2008]. In 1998, Ma et al. devised the first protocol for a mild Cu-catalyzed coupling of aryl bromides with chelating α-amino acids (e.g. reaction (7.183)) [2009]. It was then extended to β-amino acids [2010]. The amino acid acts as the ligand (N,O-chelator) of the Cu precatalyst, thus increasing its solubility and, as a consequence, permitting to run the reaction at lower temperature (80–120 °C). Almost any kind of Cu source can be used, even metallic Cu. In the latter case, Cu_2O that covers the metallic surface is taken up by the ligand into solution [1995]. In 1999, Goodbrand and Hu reported the use of CuCl + 1,10-phenanthroline (N,N-chelator) in toluene and KOH at 125 °C in their syntheses of triarylamines ($Ar^1(Ar^2)NH + Ar^3{-}I + KOH \rightarrow Ar^1(Ar^2)NAr^3 + KI + H_2O$; and $Ar^1{-}NH_2 + 2Ar^3{-}I + 2KOH \rightarrow Ar^1N(Ar^3)_2 + 2KI + 2H_2O$) [2011]. The combination of CuI + 1,10-phenanthroline catalyzes amination of bromoarylguanidines to give 2-aminobenzimidazoles [2012]. In 2001, Lang et al. found that ethylene glycol can be used as a solvent and O,O-chelator [2013, 2014]. (e.g. reaction (7.184) reported by Buchwald and coworkers) [2014]. The same year, Buchwald and coworkers reported the Cu-catalyzed amidation of aryl and heteroaryl halides using CuI and 1,2-diamines such as *trans*-1,2-di(methylamino)cyclohexane in combination with stoichiometric amounts of K_3PO_4, K_2CO_3, or Cs_2CO_3 (e.g. reaction (7.185)) [2015–2017]. The latter method has been applied to *N*-arylation of indoles [2018], pyrroles, pyrazoles, imidazoles, indazoles, and triazoles [2019]. Kinetic studies suggested that the diamine serves to prevent multiple ligation of the amide to Cu(I) and to activate the σ(C—I) bond in Ar—I [2020, 2021]. *N*-Arylations of amines by aryl iodides and bromides have been catalyzed by CuI with diethylsalicyclamide as ligand (N,O-chelator) [2022]. In the presence of 2-isobutyrylcyclohexanone (O,O-chelator), these Cu-catalyzed reactions occur at room temperature [2023]. Cu-catalyzed *N*-arylation of imidazoles and benzimidazoles are high yielded using 4,7-dimethoxy-1,10-phenanthroline as ligand

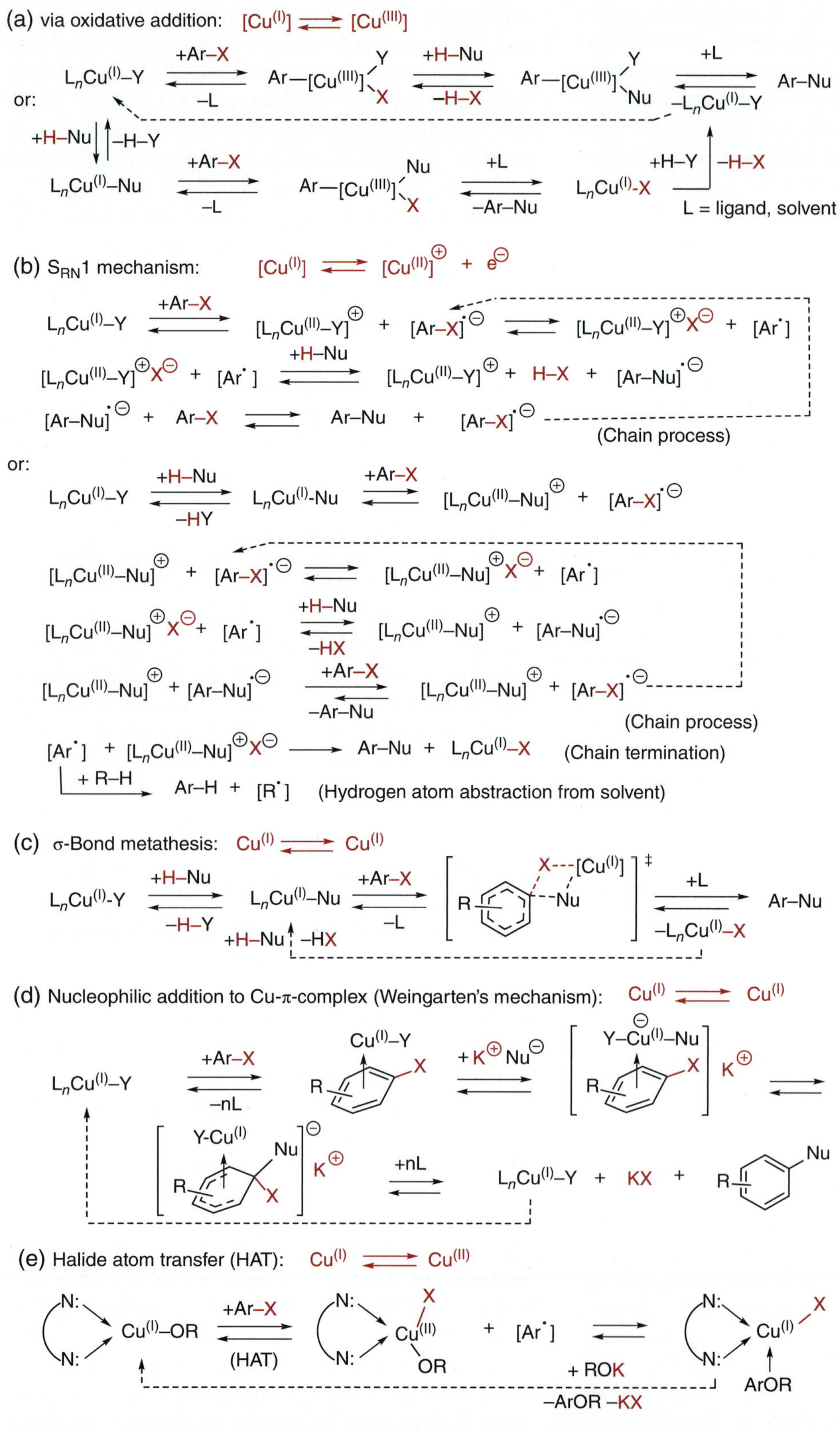

Scheme 7.95 Possible mechanisms for the Cu-catalyzed Ullmann–Goldberg reactions.

[2024, 2025]. By using a combination of CuI and proline, primary and secondary amines are N-arylated by aryl iodides at room temperature [2026]. O-Arylations of phenols by aryl iodides and bromides have been catalyzed by CuI + 2-picolinic acid (reaction (7.186)). Aminoarenethiolate Cu(I) complexes catalyze the N-arylation of benzylamine and imidazole [2027], and the O-arylation of phenols with PhBr [2028, 2029]. C-Arylation of malonic esters by aryl iodides has been catalyzed by CuI + 2-phenylphenol (THF, 70 °C) [2030] and with Cu(I) salts in the presence of 2-picolinic acid [2031] or L-proline (examples of the Ullmann–Hurtley condensation, Section 8.8.10) [2032]. In 2009, Buchwald and coworker demonstrated that unprotected aminophenols can be O-arylated selectively in the presence of a Cu-catalyst and a suitable ligand, whereas N-arylation is favored in the presence of a Pd-catalyst [2033]. This exemplifies the complementarities of Cu- and Pd-catalyzed Ullmann–Goldberg reactions [2034].

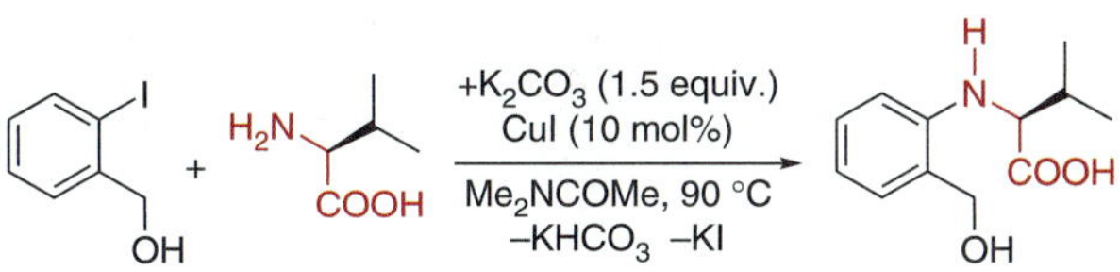

(7.183)

$$\text{Ar–I} + \text{HN(R}^1\text{)R}^2 \xrightarrow[\substack{\text{HOCH}_2\text{CH}_2\text{OH (2 equiv.),}\\ i\text{-PrOH, 80 °C}\\ -\text{K}_2\text{HPO}_4\ -\text{KI}}]{+\text{K}_3\text{PO}_4\text{ (2 equiv.), CuI (5 mol\%)}} \text{Ar–N(R}^1\text{)R}^2 \quad (7.184)$$

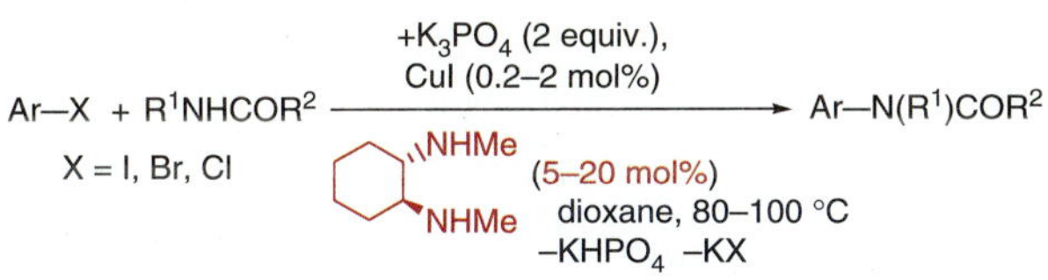

(7.185)

$$\underset{X = \text{I, Br}}{\text{Ar}^1\text{–X}} + \text{Ar}^2\text{–OH} \xrightarrow[\substack{\text{2-Picolinic acid (10 mol\%),}\\ \text{DMSO, 90 °C}\\ -\text{K}_2\text{HPO}_4\ -\text{KX}}]{\substack{+\text{K}_3\text{PO}_4\text{ (2 equiv.),}\\ \text{CuI (5 mol\%)}}} \text{Ar}^1\text{–O–Ar}^2 \quad (7.186)$$

Five types of possible mechanisms for the Cu(I)-catalyzed Ullmann–Goldberg reactions have been proposed (Scheme 7.95):

A) Oxidative addition of L_nCu(I)–Y to Ar–X generates the corresponding adduct Ar–[Cu(III)](Y)–X that is then displaced by H–Nu giving Ar–[Cu](Y)–Nu + HX (or Ar–[Cu](X)–Nu + HY). A final reductive elimination produces Ar–Nu and [Cu(I)]–Y (or [Cu(I)]–X that equilibrates with HY giving the initial catalyst [Cu(I)]–Y + HX (coproduct)). Alternatively, the first step is the exchange L_nCu(I)–Y + H–Nu ⇄ L_nCu(I)–Nu + HY. It is followed by the oxidative addition to Ar–X that generates intermediate Ar–[Cu(III)](X)–Nu. A final reductive elimination produces Ar–Nu (product) and [Cu(I)]–X (Scheme 7.95a). This mechanism was first proposed by Cohen et al. [2035] and has been retained in several instances [2028, 2036]. The groups of Buchwald and coworkers [2020, 2021], Hartwig and coworkers, [2037, 2038] and Zhang et al. have characterized LCu(I)(amidate) intermediates (L = ancillary ligand) and shown that they activate the C—X bonds of substrates Ar—X [2039–2041]. Similarly, Taillefer and coworkers suggested that LCu(I)(pyrazolate) complexes are the active Cu species in the Cu(I)-catalyzed coupling of aryl halides with pyrazoles [2042]. In 2010, Ribas and coworkers reported direct evidence (in situ ^{1}H NMR, UV–visible spectroscopic studies) for the Cu(I)/Cu(III) redox step relevant to the Ullmann–Goldberg reaction [2043] and for the formation of aryl-Cu(III) species [2044].
B) Either L_nCu(I)–Y or L_nCu(I)–Nu complex transfers an electron (SET) to Ar–X and initiates a S_{RN}Ar process (Scheme 7.95b). This mechanism was proposed first by Hey and Waters [2045]. In 1978, Arai et al. found that the CuBr-catalyzed reaction of 1-bromoanthraquinone (Ar–Br) with 2-aminoethanol (H–Nu) giving 1-(2-hydroxyethylamino)anthraquinone (Ar–Nu) involves the formation of 1-bromoanthraquinone radical anion ($[\text{Ar–Br}]^{\bullet-}$) and radical cation $\text{Cu(II)Br}^{\bullet+}$ resulting from the reduction by the Cu(I) catalyst. As anthraquinone (Ar–H) is also formed as the coproduct, it indicates the formation of anthraquinone-1-yl radical intermediate ($\text{Ar}^{\bullet}$) resulting from the fragmentation of 1-bromoanthraquinone radical anion [2046–2048]. Other studies support S_{RN}1 mechanisms [2049, 2050].
C) Complex L_n[Cu(I)]–Nu and Ar–X undergo σ-bond metathesis (four-center intermediates: Scheme 7.95c) [2051–2053]. This formal $[\sigma^2 s+\sigma^2 s]$-cycloaddition may involve SET with the formation of diradical-zwitterion intermediates [1995, 2050, 2054].

D) The catalyst forms a π-complex with the aryl halide and follows a nucleophilic addition of H–Nu generating a Cu-stabilized cyclohexadienyl anion intermediate that expels the nucleofugal group X and produces Ar–Nu (Weingarten's mechanism) [2055].

E) Quantum mechanical calculations by Houk and coworkers on the *N*-arylation of $MeNH_2$ and *O*-arylation of MeOH by Ph-I catalyzed by Cu(I) complexes containing either a β-diketone or a 1,10-phenanthroline complex suggested that the $C(sp^2)$—I bond activation is neither an oxidative addition nor a σ-metathesis. In agreement with experimental results, *N*-arylation is calculated to be faster than *O*-arylation if a β-diketone is used as the ligand and the reaction follows a $S_{RN}1$ mechanism. In contrast, *O*-arylation is calculated to be faster than *N*-arylation when the ligand is 1,10-phenanthroline and the reaction mechanism implies halide atom transfer with the formation of phenyl radical (Scheme 7.95e) [2056]. Calculations by Zhang et al. on the *O*-arylation of phenol with aryl halides using CuBr + 1,10-phenanthroline as catalyst also support the halide atom transfer (HAT) mechanism [2057].

Problem 7.41 Propose a Cu-catalyzed synthesis of aryl thioethers [2058, 2059].

7.12.2 Buchwald–Hartwig reactions

Pd-catalyzed Ullmann–Goldberg reactions are the Buchwald–Hartwig cross-coupling reactions or Buchwald–Hartwig reactions [2060–2065]. The Pd-catalyzed displacement reactions have been developed for *N*-arylations (e.g. reactions (7.187) and (7.188) [2066]), *N*-alkenylations [2067–2069] (e.g. reaction (7.189) [2070]), *O*-arylations [2071–2076] (e.g. reaction (7.190) [2077, 2078]), *S*-arylations [2065, 2079–2083] (e.g. reaction (7.191) [2084]), and for α-*C*-arylations of carbonyl compounds, malonic diesters, and carbonitriles (see Section 8.8.10) [2085]. Electrophilic partners can be aryl or heteroaryl halides (Ar–X, X = Cl, Br, I), triflates [2086], nonaflates ($Ar–OSO_2C_4F_9$) [2087], mesylates (ArOMs) [2088], tosylates (Ar–OTs) [2089, 2090], imidazolsulfonates, or sulfamates ($ArOSO_2NMe_2$). The reactions require 1 equiv. of a base to neutralize the coproduct HX. A large number of Pd ligands have been developed. Electron-rich ones permit the activation (oxidative addition of Pd) of the strongest σ(C—X) bonds such C—Cl bonds [2088, 2091–2094]. Ligands BrettPhos and RuPhos have found the widest number of applications [2095]. NHC ligands have also contributed to the development of efficient catalysts [2096–2098], including recyclable catalysts [2099]. Catalysts for reactions in aqueous media [2100, 2101] or solvent-free conditions have been presented [2102, 2103].

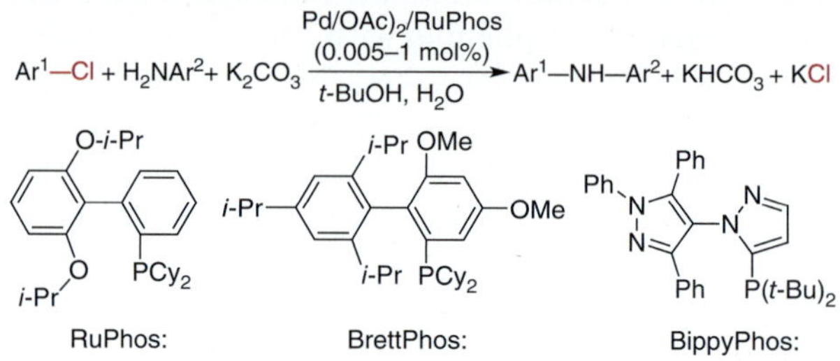

(7.187)

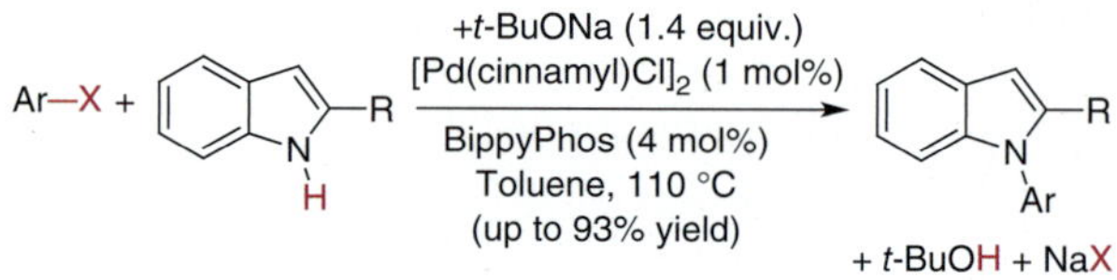

(7.188)

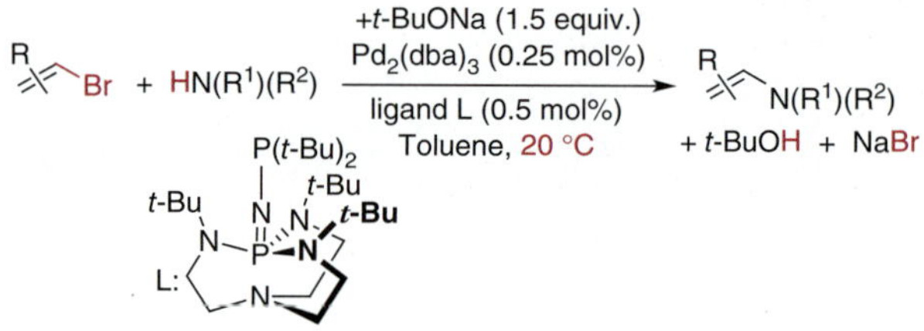

(7.189)

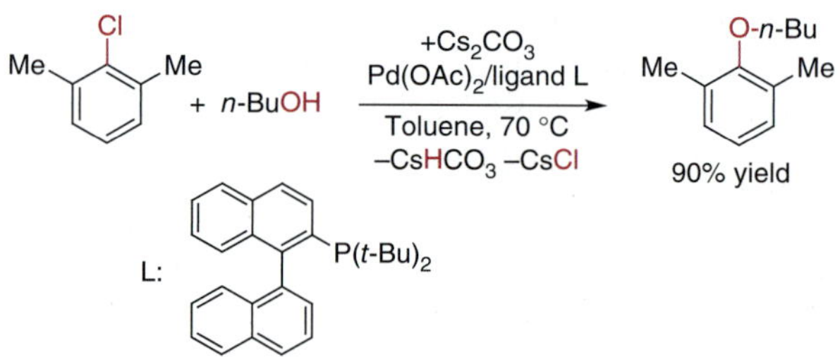

(7.190)

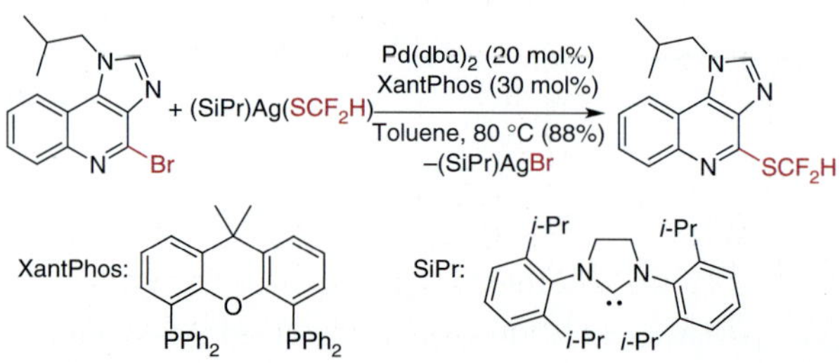

(7.191)

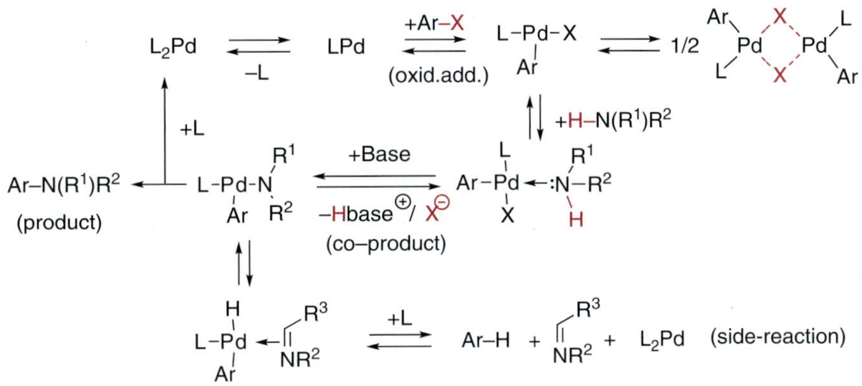

Scheme 7.96 Mechanism of the Pd-catalyzed Ullmann–Goldberg reaction: Buchwald–Hartwig amination.

The first example of Pd-catalyzed *N*-arylation was reported by Migita and coworkers in 1983 (Ar–Br + $Bu_3Sn–NEt_2 \rightarrow Ar–NEt_2 + Bu_3SnBr$, toluene, 100 °C) [2104]. In 1984, Boger and Panek found a first example of intramolecular *N*-arylation catalyzed by $Pd(PPh_3)_4$ (THF, 80 °C) [2105]. In 1994, Hartwig and coworkers reported a systematic study of the Pd complexes involved in the original Migita report and concluded that $Pd[P(o\text{-tolyl})_3]_2$ is the active catalyst that adds to Ar–Br equilibrating with dimeric complex $\{Ar\text{-}Pd[P(o\text{-tolyl})_3]\text{-}Br\}_2$ (Scheme 7.96) [2106]. The same year, Buchwald and coworker reported on the transamination of $Bu_3SnEt_2 + HNR_2 \rightleftarrows HNEt_2$ (volatile) $+ Bu_3SnNR_2$. The latter was used in Pd-catalyzed *N*-arylations [2107]. In 1995, the groups of Buchwald (using NaO-*t*-Bu as the base) [2108] and Hartwig (using $(Me_3Si)_2NLi$ (LiHMDS) as the base) [2109] reported that $PdCl_2[P(o\text{-toluyl})_3]_2$ or $Pd(dba)_2/P(o\text{-toluyl})_3$ catalyzed *N*-arylations of amine (organotin-free coupling) by aryl bromides at 55–100 °C. The accepted mechanism of the Buchwald–Hartwig amination is shown in Scheme 7.96. As for all other Pd-catalyzed arylations, the first step is the oxidative addition of an unsaturated Pd(0) complex to the aryl halide or pseudo-halide (Ar–X); then, the amine ($R^1(R^2)NH$) coordinates the adduct. The base promotes the irreversible elimination of HBr generating $LPd(II)(Ar)NR^1R^2$. A final reductive elimination liberates the product of *N*-arylation and the catalytically active Pd(0) complex. An unproductive side reaction is the β-hydride elimination from the metallic amide forming Ar–H and an imine [2110–2113].

Problem 7.42 What is the product **P** of the following reaction? [2114].

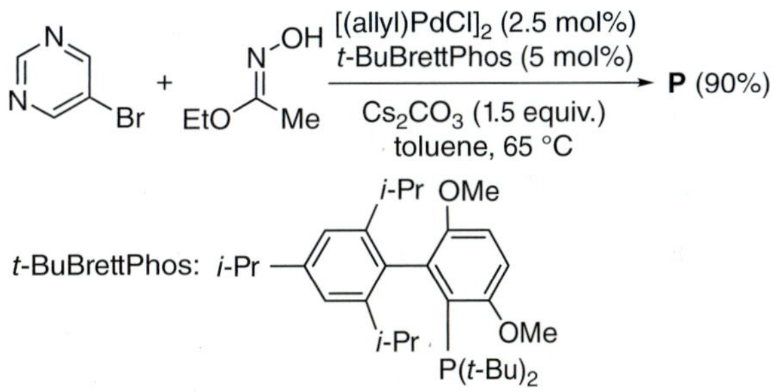

References

1 (a) Haber, J. (1994). Catalysis – where science and industry meet. *Pure Appl. Chem.* 66: 1597–1620. (b) Blaser, H.-U. (2015). Looking back on 35 years of industrial catalysis. *Chimia* 69 (7–8): 393–406.

2 (a) Hagen, J. (2006). *Industrial Catalysis: A Practical Approach*, 2e, 1–507. Weinheim: Wiley-VCH. (b) Gröger, H. (2016). Shibasaki catalysts and their use for asymmetric synthetic applications by the chemical industry. *Eur. J. Org. Chem.* 24: 4116–4123. (c) Marion, P., Bernela, B., Piccirilli, A. et al. (2017). Sustainable chemistry: how to produce better and more from less? *Green Chem.* 19 (21): 4973–4989.

3 (a) Lloyd, L. (2011). Fundamental and applied catalysis. In: *Handbook of Industrial Catalysts*, 1–490. New York: Springer. (b) Pugin, B. and Blaser, H.-U. (2010). Immobilized complexes for enantioselective catalysis: when will they be used in industry? *Top. Catal.* 53 (143–14): 953–962. (c) Angelici, C., Weckhuysen, B.M.,

and Bruijnincx, P.C.A. (2013). Chemocatalytic conversion of ethanol into butadiene and other bulk chemicals. *ChemSusChem* 6 (9): 1595–1614. (d) Hübner, S., de Vries, J.G., and Farina, V. (2016). Why does industry not use immobilized transition metal complexes as catalysts? *Adv. Synth. Catal.* 358 (1): 3–25. (e) Rinaldi, R., Jastrzebski, R., Clough, M.T. et al. (2016). Paving the way for lignin valorisation: recent advances in bioengineering, biorefining and catalysis. *Angew. Chem. Int. Ed.* 55 (29): 8164–8215.

4 (a) Dalko, P.I. and Moisan, L. (2004). In the golden age of organocatalysis. *Angew. Chem. Int. Ed.* 43: 5138–5175. (b) Dalko, P.I. (2007). Do we need asymmetric organocatalysis? *Chimia* 61 (5): 213–218. (c) Lofti, T.-T., Hamid, D., and Dalko, P.I. (2011). Asymmetric α-alkylation of aldehydes: efficiency with elegance. *Angew. Chem. Int. Ed.* 50 (51): 12146–12147. (d) Yan, Q., Lihui, Z., and Sanzhong, L. (2017). Organocatalysis in inert C—H bond functionalization. *Chem. Rev.* 117 (13): 9433–9520.

5 List, B. (2007). Introduction: organocatalysis. *Chem. Rev.* 107: 5413–5415.

6 MacMillan, D.W.C. (2008). The advent and development of organocatalysis. *Nature* 455: 304–308.

7 Bertelsen, S. and Jørgensen, K.A. (2009). Organocatalysis-after the gold rush. *Chem. Soc. Rev.* 38: 2178–2189.

8 List, B. (2010). Emil Knoevenagel and the roots of aminocatalysis. *Angew. Chem. Int. Ed.* 49: 1730–1734.

9 Albrecht, L., Jiang, H., and Jørgensen, K.A. (2011). A simple recipe for sophisticated cocktails: organocatalytic one-pot reactions-concept, nomenclature, and future perspectives. *Angew. Chem. Int. Ed.* 50: 8492–8509.

10 (a) Sun, B.F. (2015). Total synthesis of natural and pharmaceutical products powered by organocatalytic reactions. *Tetrahedron Lett.* 56: 2133–2140. (b) Porta, R., Benaglia, M., and Puglisi, A. (2016). Flow chemistry: recent developments in the synthesis of pharmaceutical products. *Org. Process Res. Dev.* 20 (1): 2–25.

11 (a) Berzelius, J.J. (1835). *Årsberättelsen om framsteg i fysik och kemi.* Stockholm: Royal Swedish Academy of Sciences. (b) Berzelius, J.J. (1835). *Jahres-Bericht über Fortschritte der Physischen Wissenschaften*, vol. 15. Tübingen: Laupp Verlag. (c) Berzelius, J.J. (1836). Quelques idées sur une nouvelle force agissant dans les combinaisons des corps organiques. *Ann. Chim. Phys.* (journal founded in 1789 by Lavoisier) 61: 146–151.

12 Ahrendt, K.A., Borths, C.J., and MacMillan, D.W.C. (2000). New strategies for organic catalysis: the first highly enantioselective organocatalytic Diels–Alder reaction. *J. Am. Chem. Soc.* 122 (17): 4243–4244.

13 Garrett, B.C. (2004). Ions at the air/water interface. *Science* 303 (5661): 1146–1147.

14 Tait, M.J. and Franks, F. (1971). Water in biological systems. *Nature* 230 (5289): 91–94.

15 Chabinyc, M.L., Craig, S.L., Regan, C.K., and Brauman, J.I. (1998). Gas phase ionic reactions: dynamics and mechanism of nucleophilic displacements. *Science* 279 (5358): 1882–1886.

16 Aloisio, S. and Francisco, J.S. (2000). Radical-water complexes in Earth's atmosphere. *Acc. Chem. Res.* 33 (12): 825–830.

17 Smith, I.W.M. and Ravishankara, A.R. (2002). Role of hydrogen bonded intermediates in the bimolecular reactions of the hydroxyl radical. *J. Phys. Chem. A* 106 (19): 4798–4807.

18 Hansen, J.C. and Francisco, J.S. (2002). Radical-molecule complexes: changing our perspective on the molecular mechanisms of radical-molecule reactions and their impact on atmospheric chemistry. *ChemPhysChem* 3 (10): 833–840.

19 Vöhringer-Martinez, E., Hansmann, B., Hernández-Soto, H. et al. (2007). Water catalysis of a radical-molecule gas-phase reaction. *Science* 315 (5811): 497–501.

20 Vogel, P., Lam, Y.-H., Simon, A., and Houk, K.N. (2016). Organocatalysis: fundamentals and comparisons to metal and enzyme catalysis. *Catalysts* 6 (9): 128.

21 Brown, D.G.B.J. (2016). Analysis of past and present synthetic methodologies on medicinal chemistry: where have all the new reactions gone? *J. Med. Chem.* 59: 4443–4458.

22 Boström, J. and Brown, D.G. (2016). Stuck in a rut with old chemistry. *Drug Discov. Today* 21: 701–703.

23 Otera, J. (1993). Transesterification. *Chem. Rev.* 93 (4): 1449–1470.

24 (a) Otera, J. (2001). In search of practical esterification. *Angew. Chem. Int. Ed.* 40 (11): 2044–2045. (b) Otera, J. (2003). *Esterification: Methods, Reactions and Applications*, 1–303. Weinheim: Wiley-VCH. (c) Otera, J. and Nishikido, J. (2009). *Esterification: Methods, Reactions and Applications*, 2e, 1–303. Weinheim: Wiley-VCH.

25 Rehberg, C.E. and Fisher, C.H. (1944). Preparation and properties of the *n*-alkyl acrylates. *J. Am. Chem. Soc.* 66: 1203–1207.

26 Rehberg, C.E. (1955). *n*-Butyl acrylate. *Org. Synth.* 3: 146–147.

27 Dewolfe, R.H. (1974). Synthesis of carboxylic and carbonic ortho esters. *Synthesis* 3: 153–172.

28 Rothman, E.S., Silbert, L.S., Pfeffer, P.E., and Hecht, S.S. (1972). Enol esters. 15. Synthesis of highly hindered esters via isopropenyl ester intermediates. *J. Org. Chem.* 37 (22): 3551–3552.

29 Taft, R.W. Jr.,, Newman, M.S., and Verhoek, F.H. (1950). The kinetics of the base-catalyzed methanolysis of *ortho*-substituted, *meta*-substituted and *para*-substituted L-menthyl benzoates. *J. Am. Chem. Soc.* 72 (10): 4511–4519.

30 Billman, J.H., Smith, W.T., and Rendall, J.L. (1947). Anticonvulsants. 5. Esters of γ-diethylamino-α-phenylbutyric acid. *J. Am. Chem. Soc.* 69 (8): 2058–2059.

31 Reimer, M. and Downes, H.R. (1921). Preparation of esters by direct replacement of alkoxyl groups. *J. Am. Chem. Soc.* 43: 945–951.

32 Rossi, R.A. and De Rossi, R.H. (1974). Preparation of benzoate esters of tertiary alcohols by transesterification. *J. Org. Chem.* 39 (6): 855–856.

33 Frank, R.L., Davis, H.R., Drake, S.S., and McPherson, J.B. (1944). Ester interchange by means of the Grignard complex. *J. Am. Chem. Soc.* 66: 1509–1510.

34 Taber, D.F., Amedio, J.C., and Patel, Y.K. (1985). Preparation of β-keto-esters by 4-DMAP-catalyzed ester exchange. *J. Org. Chem.* 50 (19): 3618–3619.

35 Mottet, C., Hamelin, O., Garavel, G. et al. (1999). A simple and efficient preparation of propargylic β-keto esters through transesterification. *J. Org. Chem.* 64 (4): 1380–1382.

36 Seebach, D., Thaler, A., Blaser, D., and Ko, S.Y. (1991). Transesterifications with 1,8-diazabicyclo[5.4.0]undec-7-ene/lithium bromide (DBU/LiBr) – also applicable to cleavage of peptides from resins in Merrifield syntheses. *Helv. Chim. Acta* 74 (5): 1102–1118.

37 Yazawa, H., Tanaka, K., and Kariyone, K. (1974). Reaction of carboxylic esters with boron tribromide – convenient method for synthesis of amides and transesterification. *Tetrahedron Lett.* 46: 3995–3996.

38 Blossey, E.C., Turner, L.M., and Neckers, D.C. (1973). Polymer protected reagents. 2. Esterifications with P-$AlCl_3$. *Tetrahedron Lett.* (21): 1823–1826.

39 Otera, J., Yano, T., Kawabata, A., and Nozaki, H. (1986). Novel distannoxane-catalyzed transesterification and a new entry to α,β-unsaturated carboxylic acids. *Tetrahedron Lett.* 27 (21): 2383–2386.

40 Otera, J., Dan-oh, N., and Nozaki, H. (1993). Unique template effects of distannoxane catalysts in transesterification of diol esters. *Tetrahedron* 49 (15): 3065–3074.

41 Pereyre, M., Colin, G., and Delvigne, J.P. (1969). Alkoxytrialkyltin compounds as transesterification reaction catalysts. *Bull. Soc. Chim. Fr.* 1: 262–263.

42 Otera, J., Dan-oh, N., and Nozaki, H. (1991). Novel template effects of distannoxane catalysts in highly efficient transesterification and esterification. *J. Org. Chem.* 56 (18): 5307–5311.

43 Xiang, J.N., Toyoshima, S., Orita, A., and Otera, J. (2001). A practical and green chemical process: fluoroalkyldistannoxane-catalyzed biphasic transesterification. *Angew. Chem. Int. Ed.* 40 (19): 3670–3672.

44 Kunz, H. and Waldmann, H. (1983). 1,3-Dithian-2-yl methyl ester as 2-step protecting group for the carboxy function in peptide synthesis. *Angew. Chem. Int. Ed. Engl.* 22 (1): 62–62.

45 Waldmann, H. and Kunz, H. (1988). 1,3-Dithian-2-ylmethyl esters as 2-step carboxy-protecting groups in the synthesis of *N*-glycopeptides. *J. Org. Chem.* 53 (18): 4172–4175.

46 Seebach, D., Hungerbühler, E., Näf, R. et al. (1982). Titanate-mediated transesterifications with functionalized substrates. *Synthesis* 2: 138–141.

47 Imwinkelried, R., Schiess, M., and Seebach, D. (1987). Diisopropyl (2*S*,3*S*)-2,3-*O*-isopropylidenetartrate [1,3-dioxolane-4,5-dicarboxylic acid, 2,2-dimethyl-, bis(1-methylethyl)ester, (4*R*-*trans*)-]. *Org. Synth.* 65: 230–235.

48 Ishihara, K., Ohara, S., and Yamamoto, H. (2000). Direct condensation of carboxylic acids with alcohols catalyzed by hafnium(IV) salts. *Science* 290 (5494): 1140–1142.

49 Chen, C.T., Kuo, J.H., Ku, C.H. et al. (2005). Nucleophilic acyl substitutions of esters with protic nucleophiles mediated by amphoteric, oxotitanium, and vanadyl species. *J. Org. Chem.* 70 (4): 1328–1339.

50 Kubota, M., Yamamoto, T., and Yamamoto, A. (1979). Trans-esterification reaction catalyzed by alkoxo(triphenylphosphine)Cu(I) complexes. *Bull. Chem. Soc. Jpn.* 52 (1): 146–150.

51 Kim, Y.J., Osakada, K., Takenaka, A., and Yamamoto, A. (1990). Alkylnickel and alkylpalladium alkoxides associated with alcohols through hydrogen bonding. *J. Am. Chem. Soc.* 112 (3): 1096–1104.

52 Hopkinson, M.N., Richter, C., Schedler, M., and Glorius, F. (2014). An overview of N-heterocyclic carbenes. *Nature* 510 (7506): 485–496.

53 Grasa, G.A., Kissling, R.M., and Nolan, S.P. (2002). N-heterocyclic carbenes as versatile nucleophilic catalysts for transesterification/acylation reactions. *Org. Lett.* 4 (21): 3583–3586.

54 Nyce, G.W., Lamboy, J.A., Connor, E.F. et al. (2002). Expanding the catalytic activity of nucleophilic N-heterocyclic carbenes for transesterification reactions. *Org. Lett.* 4 (21): 3587–3590.

55 (a) Grasa, G.A., Gueveli, T., Singh, R., and Nolan, S.P. (2003). Efficient transesterification/acylation reactions mediated by N-heterocyclic carbene catalysts. *J. Org. Chem.* 68 (7): 2812–2819. (b) Singh, R., Kissling, R.M., Letellier, M.A., and Nolan, S.P. (2004). Transesterification/acylation of secondary alcohols mediated by N-heterocyclic carbene catalysts. *J. Org. Chem.* 69 (1): 209–212. (c) Zeng, T.-G., Song, G.-G., and Li, C.-J. (2009). Separation, recovery and reuse of N-heterocyclic carbene catalysts in transesterification reactions. *J. Chem. Soc., Chem. Commun.* (41): 6249–6251.

56 (a) Samanta, R.C., De Sarkar, S., Frölich, R. et al. (2013). N-Heterocyclic carbene (NHC) catalyzed chemoselective acylation of alcohols in the presence of amines with various acylating reagents. *Chem. Sci.* 4 (5): 2177–2184. (b) He, L., Guo, H., Ma, X.-W. et al. (2014). N-Heterocyclic carbene-catalyzed synthesis of acetyltributylcitrate via a transesterification reaction. *Chin. Chem. Lett.* 25 (2): 215–217. (c) Asadi, M., Hooper, J.F., and Lupton, D.W. (2016). Biodiesel synthesis using integrated acid and base catalysis in continuous flow. *Tetrahedron* 72 (26): 3729–3733.

57 Lee, I., Kim, C.K., Li, H.G. et al. (2000). Acyl-transfer mechanisms involving various acyl functional groups: >X—Y with X=C, S, P and Y=O, S. *J. Am. Chem. Soc.* 122 (45): 11162–11172.

58 (a) Day, J.W.C. and Ingold, C.K. (1941). Mechanism and kinetics of carboxylic ester hydrolysis and carboxyl esterification. *Trans. Faraday Soc.* 37: 686–705. (b) Ingold, C.K. (1969). *Structure and Mechanism in Organic Chemistry*, 2e, 1–1266. Ithaca, NY: Cornell University Press. (c) Hori, K., Ikenaga, Y., Arata, K. et al. (2007). Theoretical study on the reaction mechanism for the hydrolysis of esters and amides under acidic conditions. *Tetrahedron* 63 (5): 1264–1269.

59 Rao, G.V., Balakrishnan, M., and Venkatasubramanian, N. (1978). E1cB pathway of acyl transfer. *J. Sci. Ind. Res.* 37 (10): 547–557.

60 (a) More O'Ferrall, R.A. (1970). Relationships between E2 and E1cB mechanism of β-elimination. *J. Chem. Soc. B* (2): 274–277. (b) see also: Scudder, P.H. (1990). Use of reaction cubes for generation and display of multiple mechanistic pathways. *J. Org. Chem.* 55 (14): 4238–4240.

61 Jencks, W.P. (1972). General acid-base catalysis of complex reactions in water. *Chem. Rev.* 72: 705–718.

62 Williams, A.F. (1989). Concerted mechanisms of acyl group transfer reactions in solution. *Acc. Chem. Res.* 22: 387–392.

63 Anslyn, E.V. and Dougherty, D.A. (2006). Considering multiple reaction coordinates. In: *Modern Physical Organic Chemistry* (ed. J. Murzek), 407–413. Sausalito, CA: University Science Books.

64 Otton, J. and Ratton, S. (1988). Investigation of the formation of poly(ethylene-terephthalate) with model molecules – kinetics and mechanism of the catalytic esterification and alcoholysis reactions. 1. Carboxylic acid catalysis (monofunctional reactants). *J. Polym. Sci., Part A: Polym. Chem.* 26 (8): 2183–2197.

65 (a) Pratt, R.C., Lohmeijer, B.G.G., Long, D.A. et al. (2006). Triazabicyclodecene: a simple bifunctional organocatalyst for acyl transfer and ring-opening polymerization of cyclic esters. *J. Am. Chem. Soc.* 128 (14): 4556–4557. (b) Simon, L. and Goodman, J.M. (2007). The mechanism of TBD-catalyzed ring-opening polymerization of cyclic esters. *J. Org. Chem.* 72 (25): 9656–9662. (c) Nifant'ev, I., Shlyakhtin, A., Bagrov, V. et al. (2016). Theoretical and experimental studies of 1,5,7-triazabicyclo[4.4.0]dec-5-ene-catalyzed ring opening/ring closure reaction mechanism for 5-, 6- and 7-membered cyclic esters and carbonates. *React. Kinet. Mech. Catal.* 117 (2): 447–476.

66 Yamamoto, Y. (2007). From σ- to π-electrophilic Lewis acids. Application to selective organic transformations. *J. Org. Chem.* 72 (21): 7817–7831.

67 Yatsimirsky, A.K. (2005). Metal ion catalysis in acyl and phosphoryl transfer: transition states as ligands. *Coord. Chem. Rev.* 249 (17–18): 1997–2011.

68 Parac-Vogt, T.N., Deleersnyder, K., and Binnemans, K. (2005). Lanthanide(III) tosylates as new acylation catalysts. *Eur. J. Org. Chem.* 9: 1810–1815.

69 Nakayama, M., Sato, A., Ishihara, K., and Yamamoto, H. (2004). Water-tolerant and reusable catalysts for direct ester condensation between equimolar amounts of carboxylic acids and alcohols. *Adv. Synth. Catal.* 346 (11): 1275–1279.

70 Sato, A., Nakamura, Y., Maki, T. et al. (2005). Zr(IV)-Fe(III), -Ga(III), and -Sn(IV) binary metal complexes as synergistic and reusable esterification catalysts. *Adv. Synth. Catal.* 347 (10): 1337–1340.

71 Bayryamov, S.G., Rangelov, M.A., Mladjova, A.P. et al. (2007). Unambiguous evidence for efficient chemical catalysis of adenosine ester aminolysis by its 2′/3′-OH. *J. Am. Chem. Soc.* 129 (18): 5790–5791.

72 Hoydonckx, H.E., De Vos, D.E., Chavan, S.A., and Jacobs, P.A. (2004). Esterification and transesterification of renewable chemicals. *Top. Catal.* 27 (1–4): 83–96.

73 Firouzabadi, H., Iranpoo, N., and Jafari, A.A. (2006). Facile and selective preparation of esters from carboxylic acids catalyzed by aluminum dodecatungstophosphate (AlPW12O40) as a versatile, recyclable and a highly water tolerant green Lewis acid catalyst. *Lett. Org. Chem.* 3 (1): 25–28.

74 Bose, D.S., Satyender, A., Das, A.P.R., and Mereyala, H.B. (2006). A facile, catalytic and environmentally benign method for esterification of carboxylic acids and transesterification of carboxylic esters with nearly equimolar amounts of alcohols. *Synthesis* 14: 2392–2396.

75 Kawabata, T., Mizugaki, T., Ebitani, K., and Kaneda, K. (2003). Highly efficient esterification of carboxylic acids with alcohols by montmorillonite-enwrapped titanium as a heterogeneous acid catalyst. *Tetrahedron Lett.* 44 (51): 9205–9208.

76 Bossaert, W.D., De Vos, D.E., Van Rhijn, W.M. et al. (1999). Mesoporous sulfonic acids as selective heterogeneous catalysts for the synthesis of monoglycerides. *J. Catal.* 182 (1): 156–164.

77 Bertin, J., Kagan, H.B., Luche, J.L., and Setton, R. (1974). Graphite electrolytic lamellar reagents in organic chemistry – esterifications in presence of graphite bisulfate. *J. Am. Chem. Soc.* 96 (26): 8113–8115.

78 Brønsted, J.N. (1928). Acid and basic catalysis. *Chem. Rev.* 5 (3): 231–338.

79 Dimroth, O. (1933). Relationships between affinity and reactions speed. *Angew. Chem.* 46: 571–576.

80 Marcus, R.A. (1964). Chemical + electrochemical electron-transfer theory. *Annu. Rev. Phys. Chem.* 15: 155–196.

81 Cohen, A.O. and Marcus, R.A. (1968). On slope of free energy plots in chemical kinetics. *J. Phys. Chem.* 72 (12): 4249–4256.

82 Marcus, R.A. (1969). Unusual slopes of free energy plots in kinetics. *J. Am. Chem. Soc.* 91 (26): 7224–7225.

83 Guthrie, J.P. (1996). Multidimensional Marcus theory: an analysis of concerted reactions. *J. Am. Chem. Soc.* 118 (51): 12878–12885.

84 Evans, M.G. and Polanyi, M. (1936). Further considerations on the thermodynamics of chemical equilibria and reaction rates. *Trans. Faraday Soc.* 32 (2): 1333–1359.

85 Evans, M.G. and Polanyi, M. (1938). Inertia and driving force of chemical reactions. *Trans. Faraday Soc.* 34 (1): 11–24.

86 Dill, K.A. and Bromberg, S. (2010). *Molecular Driving Forces: Statistical Thermodynamics in Biology, Chemistry, Physics, and Nanoscience*, 2e, 1–784. New York: Garland Science.

87 Bell, R.P. (1936). The theory of reactions involving proton transfers. *Proc. R. Soc. London, Ser. A* 154 (882): 414–429.

88 Bell, R.P. (1938). The kinetics of proton transfer reactions. *Trans. Faraday Soc.* 34 (1): 0229–0236.

89 Brønsted, J.N. and Pedersen, K. (1924). The catalytic disintegration of nitramide and its physical-chemical relevance. *Z. Phys. Chem.* 108 (3/4): 185–235.

90 Marcus, R.A. (1993). Electron-transfer reactions in chemistry – theory and experiment (Nobel lecture). *Angew. Chem. Int. Ed. Engl.* 32 (8): 1111–1121.

91 Albery, W.J. (1980). The application of the Marcus relation to reactions in solution. *Annu. Rev. Phys. Chem.* 31: 227–263.

92 Lin, M.H. and Rajanbabu, T.V. (2000). Metal-catalyzed acyl transfer reactions of enol esters: role of $Y_5((OPr)\text{-}Pr\text{-}i)_{13}O$ and $(thd)_2Y((OPr)\text{-}Pr\text{-}i)$ as transesterification catalysts. *Org. Lett.* 2 (7): 997–1000.

93 Litvinenko, L.M. and Kirichenko, A.I. (1967). Basicity and stereospecificity in nucleophile catalysis by tertiary amines. *Dokl. Akad. Nauk* 176 (1): 97–100.

94 Litvinenko, L.M. and Kirichenko, A.I. (1968). Basicity and stereospecificity in nucleophile catalysis by tertiary amines. *Chem. Abstr.* 68: 68325u.

95 Steglich, W. and Höfle, G. (1969). *N,N*-dimethyl-4-pyridinamine, a very effective acylation catalyst. *Angew. Chem. Int. Ed. Engl.* 8 (12): 981–981.

96 Haslam, E. (1980). Recent developments in methods for the esterification and protection of the carboxyl group. *Tetrahedron* 36 (17): 2409–2433.

97 Höfle, G., Steglich, W., and Vorbrüggen, H. (1978). 4-dialkylaminopyridines as acylation catalysts. 4. 4-dialkylaminopyridines as highly active acylation catalysts. *Angew. Chem. Int. Ed. Engl.* 17 (8): 569–583.

98 Hassner, A. and Alexanian, V. (1978). Synthetic methods. 12. Direct room-temperature esterification of carboxylic acids. *Tetrahedron Lett.* 46: 4475–4478.

99 Basel, Y. and Hassner, A. (2000). Di-*tert*-butyl dicarbonate and 4-(dimethylamino)pyridine revisited. Their reactions with amines and alcohols. *J. Org. Chem.* 65 (20): 6368–6380.

100 Prashad, M., Hu, B., Har, D. et al. (2000). A new, convenient and selective 4-dimethylaminopyridine-catalyzed trifluoroacetylation of anilines with ethyl trifluoroacetate. *Tetrahedron Lett.* 41 (51): 9957–9961.

101 Singh, S., Das, G., Singh, O.V., and Han, H. (2007). Development of more potent 4-dimethylaminopyridine analogues. *Org. Lett.* 9 (3): 401–404.

102 Held, I., Xu, S.J., and Zipse, H. (2007). Modular design of pyridine-based acyl-transfer catalysts. *Synthesis* 8: 1185–1196.

103 (a) Baidya, M., Kobayashi, S., Brotzel, F. et al. (2007). DABCO and DMAP – why are they different in organocatalysis? *Angew. Chem. Int. Ed.* 46 (32): 6176–6179. (b) Candish, L., Nakano, Y., and Lupton, D.W. (2014). Lewis base catalysis of three n-□* mediated reactions with N-heterocyclic carbenes (NHCs), isothioureas, bicyclic tertiary amines, and electron-rich pyridyls. *Synthesis* 46 (14): 1823–1835.

104 Vedejs, E. and Chen, X.H. (1996). Kinetic resolution of secondary alcohols. Enantioselective acylation mediated by a chiral (dimethylamino)pyridine derivative. *J. Am. Chem. Soc.* (118): 1809–1810.

105 Ruble, J.C. and Fu, G.C. (1996). Chiral π-complexes of heterocycles with transition metals: a versatile new family of nucleophilic catalysts. *J. Org. Chem.* 61 (21): 7230–7231.

106 Wurz, R.P., Lee, E.C., Ruble, J.C., and Fu, G.C. (2007). Synthesis and resolution of planar-chiral derivatives of 4-(dimethylamino)pyridine. *Adv. Synth. Catal.* 349 (14–15): 2345–2352.

107 Wurz, R.P. (2007). Chiral dialkylaminopyridine catalysts in asymmetric synthesis. *Chem. Rev.* 107 (12): 5570–5595.

108 (a) Müller, C.E. and Schreiner, P.R. (2011). Organocatalytic enantioselective acyl transfer onto racemic as well as *meso* alcohols, amines, and thiols. *Angew. Chem. Int. Ed.* 50: 6012–6042. (b) Larionov, E., Mahesh, M., Spivey, A.C. et al. (2012). Theoretical prediction of selectivity in kinetic resolution of secondary alcohols catalyzed by chiral DMAP derivatives. *J. Am. Chem. Soc.* 134 (22): 9390–9399. (c) Mandai, H., Fujii, K., and Suga, S. (2018). Recent topics in enantioselective acyl transfer reactions with dialkylaminopyridine-based nucleophilic catalysts. *Tetrahedron Lett.* 59 (19): 1787–1803.

109 Kawabata, T., Muramatsu, W., Nishio, T. et al. (2007). A catalytic one-step process for the chemo- and regioselective acylation of monosaccharides. *J. Am. Chem. Soc.* 129 (42): 12890–12895.

110 Hatakeyama, S., Satoh, K., Sakurai, K., and Takano, S. (1987). Efficient ester exchange-reaction of phosphonoacetates. *Tetrahedron Lett.* 28 (24): 2713–2716.

111 Högberg, T., Strom, P., Ebner, M., and Raemsby, S. (1987). Cyanide as an efficient and mild catalyst in the aminolysis of esters. *J. Org. Chem.* 52 (10): 2033–2036.

112 Vedejs, E. and Diver, S.T. (1993). Tributylphosphine – a remarkable acylation catalyst. *J. Am. Chem. Soc.* 115 (8): 3358–3359.

113 Vedejs, E., Bennett, N.S., Conn, L.M. et al. (1993). Tributylphosphine-catalyzed acylations of alcohols – scope and related reactions. *J. Org. Chem.* 58 (25): 7286–7288.

114 (a) Vedejs, E., Daugulis, O., Mackay, J.A., and Rozners, E. (2001). Enantioselective acyl transfer using chiral phosphine catalysts. *Synlett*

10: 1499–1505. (b) Mackay, J.A. and Vedejs, E. (2006). Synthesis and reactivity of new chiral bicyclic phospholanes as acyl-transfer catalysts. *J. Org. Chem.* 71 (2): 498–503.

115 Methot, J.L. and Roush, W.R. (2004). Nucleophilic phosphine organocatalysis. *Adv. Synth. Catal.* 346 (9–10): 1035–1050.

116 Holz, J.G., Gensow, M.N., Zayas, O., and Boerner, A. (2007). Synthesis of chiral heterocyclic phosphines for application in asymmetric catalysis. *Curr. Org. Chem.* 11: 61–106.

117 Spivey, A.C. and Aseniyadis, S. (2010). Amine, alcohol and phosphine catalysts for acyl transfer reactions. In: *Asymmetric Organocatalysis*, Topics in Current Chemistry, vol. 291 (ed. B. List), 233–280.

118 (a) Wei, Y. and Shi, M. (2014). Applications of chiral phosphine-based organocatalysts in catalytic asymmetric reactions. *Chem. Asian J.* 10: 2720–2734. (b) Xiao, Y.-M., Sun, Z.-H., Guo, H.-C., and Kwon, O. (2014). Chiral phosphines in nucleophilic organocatalysis. *Beilstein J. Org. Chem.* 10: 2089–2121.

119 Kanta De, C., Klauber, E.G., and Seidel, D. (2009). Merging nucleophilic and hydrogen bonding catalysis: an anion binding approach to the kinetic resolution of amines. *J. Am. Chem. Soc.* 131: 17060–17061.

120 (a) Klauber, E.G., Kanta De, C., Shah, T.K., and Seidel, D. (2010). Merging nucleophilic and hydrogen bonding catalysis: an anion binding approach to the kinetic resolution of propargylic amines. *J. Am. Chem. Soc.* 132: 13624–13626. (b) Klauber, E.G., Mittal, N., Shah, T.K., and Seidel, D.A. (2011). Dual-catalysis/anion-binding approach to the kinetic resolution of allylic amines. *Org. Lett.* 13: 2464–2467. (c) Mittal, N., Lippert, K.M., Kanta De, C. et al. (2015). A dual-catalysis anion-binding approach to the kinetic resolution of amines: insights into the mechanism via a combined experimental and computational study. *J. Am. Chem. Soc.* 137: 5748–5758.

121 Takemoto, Y. (2010). Development of chiral thiourea catalysts and its application to asymmetric catalytic reactions. *Chem. Pharm. Bull.* 58 (5): 593–601.

122 Sun, Y.-L., Wei, Y., and Shi, M. (2017). Applications of chiral thiourea-amine/phosphine organocatalysts in catalytic asymmetric reactions. *ChemCatChem* 9 (5): 718–727.

123 Reed, R., Reau, R., Dahan, F., and Bertrand, G. (1993). DBU and DBN are strong nucleophiles – X-ray crystal structures of onio-substituted and dionio-substituted phosphanes. *Angew. Chem. Int. Ed. Engl.* 32: 399–401.

124 (a) Birman, V.B., Li, X., and Han, Z. (2007). Nonaromatic amidine derivatives as acylation catalysts. *Org. Lett.* 9: 37–40. (b) Ahlemeyer, N.A., Streff, E.V., and Muthupandi, P. (2017). Dramatic acceleration of an acyl transfer-initiated cascade by using electron-rich amidine-based catalysts. *Org. Lett.* 19 (24): 6486–6489.

125 Birman, V.B., Uffman, E.W., Hui, J. et al. (2004). 2,3-Dihydroimidazo[1,2-a]pyridines: a new class of enantioselective acyl transfer catalysts and their use in kinetic resolution of alcohols. *J. Am. Chem. Soc.* 126: 12226–12227.

126 Yang, X., Bumbu, V.D., Liu, P. et al. (2012). Catalytic, enantioselective *N*-acylation of lactams and thiolactams using amidine-based catalysts. *J. Am. Chem. Soc.* 134: 17605–17612.

127 (a) Taylor, J.E., Bull, S.D., and Williams, J.M.J. (2012). Amidines, isothioureas, and guanidines as nucleophilic catalysts. *Chem. Soc. Rev.* 41: 2109–2121. (b) Merad, J., Borkar, P., Bouyon Yenda, T. et al. (2015). Highly enantioselective acylation of acyclic *meso*-1,3-diols through synergistic isothiourea-catalyzed desymmetrization/chiroablative kinetic resolution. *Org. Lett.* 17 (9): 2118–2121.

128 Igau, A., Grützmacher, H., Baceiredo, A., and Bertrand, G. (1988). Analogous α,α'-bis-carbenoid triply bonded species – Synthesis of a stable λ-3-phosphinocarbene λ-5-phosphaacetylene. *J. Am. Chem. Soc.* 110 (19): 6463–6466.

129 Arduengo, A.J.I., Harlow, R.L., and Kline, M. (1991). A stable crystalline carbene. *J. Am. Chem. Soc.* 113 (1): 361–363.

130 Bakhtiar, C. and Smith, E.H. (1994). Transfer of alkoxycarbonyl from alkyl imidazolium-2-carboxylates to benzyl alcohol, a cyclohexanone enamine and diethylamine. *J. Chem. Soc., Perkin Trans. 1* 239–243.

131 Grasa, G.A., Singh, R., and Nolan, S.P. (2004). Transesterification/acylation reactions catalyzed by molecular catalysts. *Synthesis* 7: 971–985.

132 Connor, E.F., Nyce, G.W., Myers, M. et al. (2002). First example of N-heterocyclic carbenes as catalysts for living polymerization: organocatalytic ring-opening polymerization of cyclic esters. *J. Am. Chem. Soc.* 124 (6): 914–915.

133 Coulembier, O., Lohmeijer, B.G.G., Dove, A.P. et al. (2006). Alcohol adducts of N-heterocyclic carbenes: latent catalysts for the thermally-controlled living polymerization of cyclic esters. *Macromolecules* 39 (17): 5617–5628.

134 Enders, D., Niemeier, O., and Henseler, A. (2007). Organocatalysis by N-heterocyclic, carbenes. *Chem. Rev.* 107 (12): 5606–5655.

135 (a) Flanigan, D., Michailidis, M.F.R., White, N.A., and Rovis, T. (2015). Organocatalytic reactions enabled by N-heterocyclic carbenes. *Chem. Rev.* 115: 9307–9387. (b) Xia, F. and Ye, S. (2016). *N-Heterocyclic Carbenes as Asymmetric Organocatalysts*, RSC Green Chemistry Series, Part 2, vol. 41, 256–316. Royal Society of Chemistry. (c) Levens, A., An, F., Breugst, M. et al. (2016). Influence of the N-substituents on the nucleophilicity and Lewis basicity of N-heterocyclic carbenes. *Org. Lett.* 18 (15): 5366–3569. (d) Zhao, M., Zhang, Y.-T., Chen, J., and Zhou, L. (2018). Enantioselective reactions catalyzed by N-heterocyclic carbenes. *Asian J. Org. Chem.* 7 (1): 54–69. (e) Chen, X.-Y., Li, S., Vetica, F. et al. (2018). *N*-Heterocyclic-carbene-catalyzed domino reactions via two or more activation modes. *iScience* 2: 1–26. (f) Wang, W., Cui, L., Sun, P. et al. (2018). Reusable *N*-heterocyclic carbene complex catalysts and beyond: a perspective on recycling strategies. *Chem. Rev.* 118 (19): 9843–9929.

136 Lai, C.L., Lee, H.M., and Hu, C.H. (2005). Theoretical study on the mechanism of N-heterocyclic carbene catalyzed transesterification reactions. *Tetrahedron Lett.* 46 (37): 6265–6270.

137 Singh, R. and Nolan, S.P. (2005). Synthesis of phosphorus esters by transesterification mediated by N-heterocyclic carbenes (NHCs). *J. Chem. Soc., Chem. Commun.* (43): 5456–5458.

138 Kayaki, Y., Yamamoto, M., and Ikariya, T. (2009). N-Heterocyclic carbenes as efficient organocatalysts for CO_2 fixation reactions. *Angew. Chem. Int. Ed.* 48 (23): 4194–4197.

139 Suzuki, Y., Yamauchi, K., Muramatsu, K., and Sato, M. (2004). First example of chiral N-heterocyclic carbenes as catalysts for kinetic resolution. *J. Chem. Soc., Chem. Commun.* 23: 2770–2771.

140 Suzuki, Y., Muramatsu, K., Yamauchi, K. et al. (2006). Chiral N-heterocyclic carbenes as asymmetric acylation catalysts. *Tetrahedron* 62: 302–310.

141 Kano, T., Sasaki, K., and Maruoka, K. (2005). Enantioselective acylation of secondary alcohols catalyzed by chiral N-heterocyclic carbenes. *Org. Lett.* 7: 1347–1349.

142 Ryan, S.J., Candish, L., and Lupton, D.W. (2011). N-Heterocyclic carbene-catalyzed (4+2)-cycloaddition/decarboxylation of silyl dienol ethers with α,β-unsaturated acid fluorides. *J. Am. Chem. Soc.* 133: 4694–4697.

143 Ryan, S.J., Stasch, A., Paddon-Row, M.N., and Lupton, D.W. (2012). Synthetic and quantum mechanical studies into the N-heterocyclic carbene catalyzed (4+2)-cycloaddition. *J. Org. Chem.* 77: 1113–1124.

144 Candish, L., Forsyth, C.M., and Lupton, D.W. (2013). *N-tert*-Butyl triazolylidenes: catalysts for the enantioselective (3+2)-annulation of unsaturated acyl azoliums. *Angew. Chem. Int. Ed.* 52: 9149–9152.

145 Candish, L., Levens, A., and Lupton, D.W. (2015). N-Heterocyclic carbene catalysed redox isomerisation of esters to functionalised benzaldehydes. *Chem. Sci.* 6: 2366–2370.

146 Levens, A., Zhang, C., Candish, L. et al. (2015). Enantioselective N-heterocyclic carbene catalyzed diene regenerative (4+2)-annulation. *Org. Lett.* 17: 5332–5335.

147 Sigler, P.B., Blow, D.M., Matthews, B.W., and Henderson, R. (1968). Structure of crystalline α-chymotrypsin. 2. A preliminary report including a hypothesis for activation mechanism. *J. Mol. Biol.* 35 (1): 143–164.

148 Kasserra, H.P. and Laidler, K.J. (1969). Mechanisms of action of trypsin and chymotrypsin. *Can. J. Chem.* 47 (21): 4031–4039.

149 Hedstrom, L. (2002). Serine protease mechanism and specificity. *Chem. Rev.* 102 (12): 4501–4523.

150 Quiocho, F.A. and Lipsomb, W.N. (1971). Carboxypeptidase A: a protein and an enzyme. *Adv. Protein Chem.* 25: 1–78.

151 Christianson, D.W. and Lipscomb, W.N. (1989). Carboxypeptidase-A. *Acc. Chem. Res.* 22 (2): 62–69.

152 Satoh, T., Imai, T., Kitajyo, Y., and Kakuchi, T. (2005). Novel synthetic method for preparing artificial carbohydrate polymers. In: *Curr. Top. Polym. Res.* (ed. R.K. Bregg), 195–231. Hauppauge: Nova Science Publishers, Inc.

153 Kimura, E., Shiota, T., Koike, T. et al. (1990). Zn(II) complex of 1,5,9-triazacyclododecane ([12]anen-3) as a model for carbonic anhydrase. *J. Am. Chem. Soc.* 112 (15): 5805–5811.

154 Breslow, R. and Chipman, D. (1965). Mixed metal complexes as enzyme models. I. Intracomplex nucleophilic catalysis by an oxime anion. *J. Am. Chem. Soc.* 87 (18): 4195–4196.

155 Zhang, B.L. and Breslow, R. (1997). Ester hydrolysis by a catalytic cyclodextrin dimer enzyme mimic with a metallobipyridyl linking group. *J. Am. Chem. Soc.* 119 (7): 1676–1681.

156 Yan, J.M. and Breslow, R. (2000). An enzyme mimic that hydrolyzes an unactivated ester with catalytic turnover. *Tetrahedron Lett.* 41 (13): 2059–2062.

157 Luzhkov, V. and Aqvist, J. (1999). Free-energy perturbation calculations of binding and transition-state energies: hydrolysis of phenyl esters by β-cyclodextrin. *Chem. Phys. Lett.* 302 (3–4): 267–272.

158 Breslow, R. (2006). The hydrophobic effect in reaction mechanism studies and in catalysis by artificial enzymes. *J. Phys. Org. Chem.* 19 (12): 813–822.

159 (a) Bhosale, S.V. and Bhosale, S.V. (2007). β-Cyclodextrin as a catalyst in organic synthesis. *Mini Rev. Org. Chem.* 4 (3): 231–242. (b) Bai, C.C., Tian, B.R., Zhao, T. et al. (2017). Cyclodextrin-catalyzed organic synthesis: reactions, mechanisms, and applications. *Molecules* 22 (9): 1475/1–1475/17. (c) Macaev, F. and Boldescu, V. (2015). Cyclodextrins in asymmetric and stereospecific synthesis. *Symmetry* 7 (4): 1699–1720. (d) Mishra, N.P., Mohapatra, S., Panda, P., and Nayak, S. (2018). One-pot multicomponent approach towards the synthesis of heterocyclic compounds by using β-cyclodextrin as a catalyst. *Curr. Org. Chem.* 22 (20): 1959–1985.

160 Ghose, A.K., Viswanadhan, V.N., and Wendoloski, J.J. (1999). A knowledge-based approach in designing combinatorial or medicinal chemistry libraries for drug discovery. 1. A qualitative and quantitative characterization of known drug databases. *J. Comb. Chem.* 1: 55–68.

161 Carey, J.S., Laffan, D., Thomson, C., and Williams, M.T. (2006). Analysis of the reactions used for the preparation of drug candidate molecules. *Org. Biomol. Chem.* 4 (12): 2337–2347.

162 (a) Roughley, S.D. and Jordan, A.M. (2011). The medicinal chemist's toolbox: an analysis of reactions used in the pursuit of drug candidates. *J. Med. Chem.* 54: 3451–3479. (b) Narendar, R.T., Beatriz, A., Jayathirtha, R.V., and Pires de Lima, D. (2019). Carbonyl compounds' journey to amide bond formation. *Chem. Asian J.* 14 (3): 344–388.

163 Deming, T.J. (2007). Synthetic polypeptides for biomedical applications. *Prog. Polym. Sci.* 32: 858–875.

164 Holowka, E.P., Sun, V.Z., Kamei, D.T., and Deming, T.J. (2007). Polyarginine segments in block copolypeptides drive both vesicular assembly and intracellular delivery. *Nat. Mater.* 6: 52–57.

165 Linstrom, P.J. and Mallard, W.G. (eds.) (2009). *NIST Chemistry WebBook, NIST Standard Reference Database*. Gaithersburg, MD: National Institute of Standards and Technology.

166 Ulijn, R.V., Moore, B.D., Janssen, A.E.M., and Halling, P.J. (2002). A single aqueous reference equilibrium constant for amide synthesis-hydrolysis. *J. Chem. Soc., Perkin Trans. 2* 1024–1028.

167 Guranda, D.T., Ushakov, G.A., Yolkin, P.G., and Svedas, V.K. (2012). Thermodynamics of phenylacetamides synthesis: linear free energy relationship with the pK of amine. *J. Mol. Catal. B: Enzym.* 74: 48–53.

168 Perreux, L., Loupy, A., and Volatron, F. (2002). Solvent-free preparation of amides from acids and primary amines under microwave irradiation. *Tetrahedron* 58 (11): 2155–2162.

169 (a) Montalbetti, C.A.G.N. and Falque, V. (2005). Amide bond formation and peptide coupling. *Tetrahedron* 61 (46): 10827–10852. (b) Bode, J.W. (2006). Emerging methods in amide- and peptide-bond formation. *Curr. Opin. Drug Discov. Dev.* 9 (6): 765–775. (c) Taylor, J.E. and Bull, S.D. (2014). *N*-Acylation reactions of amines. In: *Comprehensive Organic Synthesis*, 2e, vol. 6 (ed. P. Knochel and G.A. Molander), 427–478. Amsterdam: Elsevier, B. V. (d) Mahe, O., Desroches, J., and Paquin, J.F. (2013). Amide formation using in situ activation of carboxylic acids with $[Et_2NSF_2]BF_4$. *Eur. J. Org. Chem.* (29): 4325–4331. (e) Chen, P.-J., Wang, H.-Y., and Peng, A.-Y. (2015). A mild and efficient amide formation reaction mediated by $P(OEt)_3$ and iodine. *RSC Adv.* 5 (114): 94328–94331. (f) Jaradat, D.M.M. (2018). Thirteen decades of peptide synthesis: key developments in solid phase peptide synthesis and amide bond formation utilized in peptide ligation. *Amino Acids* 50 (1): 39–68.

170 (a) Trost, B.M. (1991). The atom economy – a search for synthetic efficiency. *Science* 254 (5037): 1471–1477. (b) Trost, B.M. (1995). Atom economy – a challenge for organic synthesis – homogeneous catalysis leads the way. *Angew. Chem. Int. Ed. Engl.* 34 (3): 259–281.

171 (a) Pattabiraman, V.R. and Bode, J.W. (2011). Rethinking amide bond synthesis. *Nature* 480:

471–479. (b) Bode, J.W. (2013). Reinventing amide bond formation. In: *Inventing Reactions*, Topics in Organometallic Chemistry, vol. 44 (ed. L.J. Goossen), 13–34. Heidelberg: Springer-Verlag.

172 (a) de Figeiredo, R.M., Suppo, J.-S., and Campagne, J.-M. (2016). Nonclassical routes for amide bond formation. *Chem. Rev.* 116 (19): 12029–12122. (b) Krause, T., Baader, S., Erb, B., and Goossen, L.J. (2016). Atom-economic catalytic amide synthesis from amines and carboxylic acids activated in situ with acetylenes. *Nat. Commun.* 7: 11732.

173 (a) Crochet, P. and Cadierno, V. (2015). Ruthenium-catalyzed amide bond formation. In: *Ruthenium in Catalysis*, Topics in Organometallic Chemistry, vol. 48 (ed. P.H. Dixneuf and C. Bruneau), 81–118. Heidelberg: Springer Verlag. (b) Chen, C., Verpoort, F., and Wu, Q.-G. (2016). Atom-economic dehydrogenative amide synthesis via ruthenium catalysis. *RSC Adv.* 6 (60): 55599–55607. (c) Gusev, D.G. (2017). Rethinking the dehydrogenative amide synthesis. *ACS Catal.* 7 (10): 6656–6662.

174 (a) Lundberg, H., Tinnis, F., Selander, N., and Adolfsson, H. (2014). Catalytic amide formation from non-activated carboxylic acids and amines. *Chem. Soc. Rev.* 43: 2714–2742. (b) Lundberg, H., Tinnis, F., Zhang, J. et al. (2017). Mechanistic elucidation of Zirconium-catalyzed direct amidation. *J. Am. Chem. Soc.* 139 (6): 2286–2295. (c) Hamstra, D.F.J., Lenstra, D.C., Koenders, T.J. et al. (2017). Poly(methylhydrosiloxane) as a green reducing agent in organophosphorus-catalyzed amide bond formation. *Org. Biomol. Chem.* 15 (30): 6426–6432. (d) Wang, Z., Bao, X., Xu, M. et al. (2018). Direct formation of amides from carboxylic acids and amines catalyzed by Niobium(V) oxalate hydrate. *ChemistrySelect* 3 (9): 2599–2603. (e) Rimola, A., Fabbiani, M., Sodupe, M. et al. (2018). How does silica catalyze the amide bond formation under dry conditions? Role of specific surface silanol pairs. *ACS Catal.* 8 (5): 4558–4568. (f) Philpott, H.K., Thomas, P.J., Tew, D. et al. (2018). A versatile biosynthetic approach to amide bond formation. *Green Chem.* 20 (15): 3426–3431. (g) Siddiki, S.M.A.H., Rashed, M.N., Ali, M.A. et al. (2019). Lewis acid catalysis of Nb2O5 for reactions of carboxylic acid derivatives in the presence of basic inhibitors. *ChemCatChem* 11 (1): 383–396.

175 Nelson, P. and Pelter, A. (1965). Trisdialkylaminoboranes – new reagents for synthesis of enamines and amides. *J. Chem. Soc.* 5142–5144.

176 Lanigan, R.M., Sarkov, P., and Sheppard, T.D. (2013). Direct synthesis of amides from carboxylic acids and amines using $B(OCH_2CF_3)_3$. *J. Org. Chem.* 78: 4512–4523.

177 Karaluka, V., Lanigan, R.M., Murray, P.M. et al. (2015). $B(OCH_2CF_3)_3$-mediated direct amidation of pharmaceutically relevant building blocks in cyclopentyl methyl ether. *Org. Biomol. Chem.* 13: 10888–10894.

178 Pelter, A., Levitt, T.E., and Nelson, P. (1970). Some amide forming reactions involving boron reagents. *Tetrahedron* 26 (6): 1539–1544.

179 Latta, R., Springsteen, G., and Wang, B.H. (2001). Development and synthesis of an arylboronic acid-based solid-phase amidation catalyst. *Synthesis* 11: 1611–1613.

180 Ishihara, K., Ohara, S., and Yamamoto, H. (1996). 3,4,5-Trifluorobenzeneboronic acid as an extremely active amidation catalyst. *J. Org. Chem.* 61 (13): 4196–4197.

181 Arnold, K., Davies, B., Giles, R.L. et al. (2006). To catalyze or not to catalyze? Insight into direct amide bond formation from amines and carboxylic acids under thermal and catalyzed conditions. *Adv. Synth. Catal.* 348 (7–8): 813–820.

182 Marcelli, T. (2010). Mechanistic insights into direct amide bond formation catalyzed by boronic acids: halogens as Lewis bases. *Angew. Chem. Int. Ed.* 49: 6840–6843.

183 Wang, C., Yu, H.Z., Fu, Y., and Guo, Q.X. (2013). Mechanism of arylboronic acid-catalyzed amidation reaction between carboxylic acids and amines. *Org. Biomol. Chem.* 11: 2140–2146.

184 (a) Maki, T., Ishihara, K., and Yamamoto, H. (2005). *N*-Alkyl-4-boronopyridinium salts as thermally stable and reusable amide condensation catalysts. *Org. Lett.* 7 (22): 5043–5046. (b) Maki, T., Ishihara, K., and Yamamoto, H. (2005). *N*-Alkyl-4-boronopyridinium halides versus boric acid as catalysts for the esterification of α-hydroxycarboxylic acids. *Org. Lett.* 7 (22): 5047–5050.

185 Maki, T., Ishihara, K., and Yamamoto, H. (2007). New boron(III)-catalyzed amide and ester condensation reactions. *Tetrahedron* 63 (35): 8645–8657.

186 Ghosh, S., Bhaumik, A., Mondal, J. et al. (2012). Direct amide bond formation from carboxylic acids and amines using activated alumina balls as a new, convenient, clean, reusable and low cost heterogeneous catalyst. *Green Chem.* 14 (11): 3220–3229.

187 (a) Gernigon, N., Al-Zoubi, R.M., and Hall, D.G. (2012). Direct amidation of carboxylic acids catalyzed by *o*-iodo-arylboronic acids: catalyst optimization, scope, and preliminary mechanistic study supporting a peculiar halogen acceleration effect. *J. Org. Chem.* 77: 8386–8400. (b) Gernigon, N., Zheng, H., and Hall, D.G. (2013). Solid-supported *o*-iodoarylboronic acid catalyst for direct amidation of carboxylic acids. *Tetrahedron Lett.* 54: 4475–4478. (c) Wang, C., Yu, H.-Z., Fu, Y., and Guo, Q.-X. (2013). Mechanism of arylboronic acid-catalyzed amidation reaction between carboxylic acids and amines. *Org. Biomol. Chem.* 11 (13): 2140–2146.

188 Fatemi, S., Gernigon, N., and Hall, D.G. (2015). A multigram scale lower E-factor procedure for MIBA-catalyzed direct amidation and its application to the coupling of α and β-aminoacids. *Green Chem.* 17: 4016–4028.

189 (a) Lenstra, D.C., Tao Nguyen, D., and Mecinovic, J. (2015). Zirconium-catalyzed direct amide bond formation between carboxylic esters and amines. *Tetrahedron* 71: 5547–5553. (b) Lundberg, H., Tinnis, F., Zhang, J. et al. (2017). Mechanistic elucidation of zirconium-catalyzed direct amidation. *J. Am. Chem. Soc.* 139 (6): 2286–2295.

190 Ali, M.A., Siddiki, S.M.A.H., Onodera, W. et al. (2015). Amidation of carboxylic acids with amines by Nb_2O_5 as a reusable Lewis acid catalyst. *ChemCatChem* 7: 3555–3561.

191 Lenstra, D.C., Rutjes, F.P., and Mecinovic, J. (2014). Triphenylphosphine-catalysed amide bond formation between carboxylic acids and amines. *J. Chem. Soc., Chem. Commun.* 50: 5763–5766.

192 (a) Métro, T.-X., Bonnamour, J., Reidon, T. et al. (2012). Mechanosynthesis of amides in the total absence of organic solvent from reaction to product recovery. *J. Chem. Soc., Chem. Commun.* 48: 11781–11783. (b) see also Duankamol, C., Jaita, S., Wangngae, S. et al. (2015). An efficient mechanochemical synthesis of amides and dipeptides using 2,4,6-trichloro-1,3,5-triazine and PPh3. *RSC Adv.* 5: 52624–52628. (c) Hernandez, J.G., Ardila-Fierro, K.-J., Crawford, D. et al. (2017). Mechanoenzymatic peptide and amide bond formation. *Green Chem.* 19 (11): 2620–2625.

193 Miyabe, H. and Takemoto, Y. (2014). Carbonyl and imine activation. In: *Comprehensive Organic Synthesis*, Chapter 1.19, 2e, vol. 1 (ed. P. Knochel and G.A. Molander), 751–769. Amsterdam: Elsevier Ltd.

194 Corey, E.J. and Grogan, M.J. (1999). Enantioselective synthesis of α-amino nitriles from *N*-benzhydryl imines and HCN with a chiral bicyclic guanidine as catalyst. *Org. Lett.* 1: 157–160.

195 Swain, C.G. and Brown, J.F. (1952). Concerted displacement reactions. 7. The mechanism of acid/base catalysis in non-aqueous solvents. *J. Am. Chem. Soc.* 74 (10): 2534–2537.

196 Swain, C.G. and Brown, J.F. (1952). Concerted displacement reactions. 8. Polyfunctional catalysis. *J. Am. Chem. Soc.* 74 (10): 2538–2543.

197 Beak, P. (1977). Energies and alkylations of tautomeric heterocyclic compounds – old problems, new answers. *Acc. Chem. Res.* 10 (5): 186–192.

198 Morpurgo, S. and Bossa, M. (2003). The epimerisation of 2-tetrahydropyranol catalysed by the tautomeric couples 2-pyridone/2-hydroxypyridine and formamide/formamidic acid as a model for the sugar's mutarotation: a theoretical study. *Phys. Chem. Chem. Phys.* 5 (6): 1181–1189.

199 Rony, P.R. (1968). Polyfunctional catalysis. I. Activation parameters for mutarotation of tetramethyl-D-glucose in benzene. *J. Am. Chem. Soc.* 90 (11): 2824–2831.

200 Arai, K. (1993). Preparation of poly(4-vinyl-2-hydroxypyridine) and its catalytic activity on mutarotation of 2,3,4,6-tetramethyl-α-D-glucose. *J. Polym. Sci., Part A: Polym. Chem.* 31 (1): 193–197.

201 Engdahl, K.A., Bivehed, H., Ahlberg, P., and Saunders, W.H. Jr., (1983). Rate-controlling 2-proton transfer coupled with heavy-atom motion in the 2-pyridinone-catalyzed mutarotation of tetramethylglucose – experimental and calculated deuterium isotope effects. *J. Am. Chem. Soc.* 105 (14): 4767–4774.

202 Kergomard, A. and Renard, M. (1968). Mutarotation of tetramethylglucose catalysed by carboxylic acids in benzene solution. *Tetrahedron* 24 (23): 6643–6651.

203 Breslow, R. (1994). Bifunctional acid-base catalysis by imidazole groups in enzyme mimics. *J. Mol. Catal.* 91 (2): 161–174.

204 Melander, C. and Horne, D.A. (1997). Mutarotation of tetramethylglucose catalyzed by ribonucleosides. *Tetrahedron Lett.* 38 (5): 713–716.

205 Nakamizo, N. (1971). Catalysis in peptide synthesis with active esters. 3. Prominent catalysis of sodium 2-pyridinolate or 4-pyridinolate in aminolysis of esters. *Bull. Chem. Soc. Jpn.* 44 (7): 2006–2006.

206 (a) Khan, N.-u.H., Kureshy, R.I., Abdi, S.H.R. et al. (2008). Metal catalyzed asymmetric cyanation reactions. *Coord. Chem. Rev.* 252: 593–623. (b) Holt, J. and Hanefeld, U. (2009). Enantioselective enzyme-catalysed synthesis of cyanohydrins. *Curr. Org. Synth.* 6 (1): 15–37. (c) Wang, W., Liu, X.H., Lin, L.-L., and Feng, X.-M. (2010). Recent progress in the chemically catalyzed enantioselective synthesis of cyanohydrins. *Eur. J. Org. Chem.* (25): 4751–4769.

207 (a) Chen, F.X. and Feng, X.M. (2005). Synthesis of racemic tertiary cyanohydrins. *Synlett* (6): 892–899. (b) Kanai, M., Kato, N., Ichikawa, E., and Shibasaki, M. (2005). Power of cooperativity: Lewis acid-Lewis base bifunctional asymmetric catalysis. *Synlett* (10): 1491–1508.

208 Song, J.J., Gallou, F., Reeves, J.T. et al. (2006). Activation of TMSCN by N-heterocyclic carbenes for facile cyanosilylation of carbonyl compounds. *J. Org. Chem.* 71 (3): 1273–1276.

209 Brefort, J.L., Corriu, R., Guérin, C., and Henner, B. (1989). Hypervalent silicon hydrides – evidence of their intermediacy in the reaction of optically-active 1-NpPhMeSiH(D) with hydrides. *J. Organomet. Chem.* 370 (1–3): 9–15.

210 Gutsev, G.L. (1992). A theoretical investigation on the structure of the hypervalent carbon and silicon pentahalogenides as well as their singly charged anions. *Chem. Phys.* 166 (1–2): 57–68.

211 Murthy, V.S. and Miller, J.M. (1994). Formation of pentavalent silicon anions, $(CH_3O)_2Si(OEt)_3(-)$ and $CH_3OSiH(OEt)_3(-)$, in the gas phase. *Rapid Commun. Mass Spectrom.* 8 (9): 698–700.

212 Wang, X. and Tian, S.K. (2007). Neutral π-nucleophile-catalyzed cyanation of aldehydes and ketones. *Synlett* (9): 1416–1420.

213 Dirksen, A., Dirksen, S., Hackeng, T.M., and Dawson, P.E. (2006). Nucleophilic catalysis of hydrazone formation and transimination: implications for dynamic covalent chemistry. *J. Am. Chem. Soc.* 128 (49): 15602–15603.

214 Dirksen, A., Hackeng, T.M., and Dawson, P.E. (2006). Nucleophilic catalysis of oxime ligation. *Angew. Chem. Int. Ed.* 45 (45): 7581–7584.

215 Dirksen, A. and Dawson, P.E. (2008). Rapid oxime and hydrazone ligations with aromatic aldehydes for biomolecular labeling. *Bioconjugate Chem.* 19 (12): 2543–2548.

216 Hayashi, M., Miyamoto, Y., Inoue, T., and Oguni, N. (1993). Enantioselective trimethylsilylcyanation of some aldehydes catalyzed by chiral Schiff-base titanium alkoxide complexes. *J. Org. Chem.* 58: 1515–1522.

217 Groger, H. (2003). Catalytic enantioselective Strecker reactions and analogous syntheses. *Chem. Rev.* 103 (8): 2795–2828.

218 North, M., Usanov, D.L., and Young, C. (2008). Lewis acid catalyzed asymmetric cyanohydrin synthesis. *Chem. Rev.* 108: 5146–5226.

219 (a) Gawronski, J., Wascinska, N., and Gajewy, J. (2008). Recent progress in Lewis base activation and control of stereoselectivity in the additions of trimethylsilyl nucleophiles. *Chem. Rev.* 108: 5227–5252. (b) Wang, W., Liu, X., Lin, L., and Feng, X. (2010). Recent progress in the chemically catalyzed enantioselective synthesis of cyanohydrins. *Eur. J. Org. Chem.* 4751–4769.

220 Kurono, N. and Okhuma, T. (2016). Catalytic asymmetric cyanation reactions. *ACS Catal.* 6: 989–1023.

221 (a) Ooi, T., Uematsu, Y., and Maruoka, K. (2006). Asymmetric strecker reaction of aldimines using aqueous potassium cyanide by phase-transfer catalysis of chiral quaternary ammonium salts with a tetranaphthyl backbone. *J. Am. Chem. Soc.* 128 (8): 2548–2549. (b) Rueping, M., Sugiono, E., and Moreth, S.A. (2007). Metal-free, enantioselective Strecker reactions catalyzed by chiral BINOL and TADDOL catalysts. *Adv. Synth. Catal.* 349: 759–764. (c) Negru, M., Schollmeyer, D., and Kunz, H. (2007). Enantioselective Strecker reaction catalyzed by an organocatalyst lacking a hydrogen-bond-donor function. *Angew. Chem. Int. Ed.* 46 (48): 9339–9341. (d) Banphavichit, V., Mansawat, W., Bhantumnavin, W., and Vilaivan, T. (2009). A practical synthesis of optically active arylglycines via catalytic asymmetric Strecker reaction. *Tetrahedron* 65: 5849–5854. (e) Shao, Y.-D. and Tian, S.-K. (2012). A highly enantioselective catalytic Strecker reaction of cyclic (*Z*)-aldimines. *J. Chem. Soc., Chem. Commun.* 48 (40): 4899–4901. (f) Wang, H.-Y., Zheng, C.-W., Chai, Z. et al. (2016). Asymmetric cyanation of imines via dipeptide-derived organophosphine dual-reagent catalysis. *Nat. Commun.* 7: 12720.

222 (a) Pluth, M.D., Bergman, R.G., and Raymond, K.N. (2007). Acid catalysis in basic solution: a supramolecular host promotes orthoformate hydrolysis. *Science* 316 (5821): 85–88. (b) Pluth, M.D., Bergman, R.G., and Raymond, K.N.

(2009). Proton-mediated chemistry and catalysis in a self-assembled supramolecular host. *Acc. Chem. Res.* 42: 1650–1659.

223 Hong, C.M., Kaphan, D.M., Bergman, R.G. et al. (2017). Conformational selection as the mechanism of guest binding in a flexible supramolecular host. *J. Am. Chem. Soc.* 139 (23): 8013–8021.

224 Smith, M.B. (2013). The Michael reaction. In: *March's Advanced Organic Chemistry*, 943–959. Hoboken, NJ: Wiley.

225 Csákÿ, A.G., de la Herrán, G., and Murcia, M.C. (2010). Conjugate addition reactions of carbon nucleophiles to electron-deficient dienes. *Chem. Soc. Rev.* 39: 4080–4102.

226 Helder, R., Arends, R., Bolt, W. et al. (1977). Alkaloid catalyzed asymmetric synthesis. 3. Addition of mercaptans to 2-cyclohexene-1-one – determination of enantiomeric excess using ^{13}C NMR. *Tetrahedron Lett.* 2181–2182.

227 Hiemstra, H. and Wynberg, H. (1981). Addition of aromatic thiols to conjugated cycloalkenones, catalyzed by chiral β-hydroxy amines – a mechanistic study on homogeneous catalytic asymmetric synthesis. *J. Am. Chem. Soc.* 103: 417–430.

228 Grayson, M.N. and Houk, K.N. (2016). Cinchona alkaloid-catalyzed asymmetric conjugate additions: the bifunctional Brønsted acid hydrogen bonding model. *J. Am. Chem. Soc.* 138: 1170–1173.

229 Sibi, M.P. and Manyem, S. (2000). Enantioselective conjugate additions. *Tetrahedron* 56: 8033–8061.

230 (a) Vicario, J.L., Reyes, E., Carrillo, L., and Uria, U. (2014). Organocatalytic asymmetric nucleophilic addition to electron-deficient alkenes. In: *Comprehensive Organic Synthesis*, Chapter 4.03, 2e, vol. 4 (ed. P. Knochel and G.A. Molander), 119–188. Amsterdam: Elsevier. (b) Maudit, M., Baslé, O., Clavier, H. et al. (2014). Metal-catalyzed asymmetric nucleophilic addition to electron-deficient alkenes. In: *Comprehensive Organic Synthesis*, Chapter 4.04, 2e, vol. 4 (ed. P. Knochel and G.A. Molander), 189–341. Amsterdam: Elsevier.

231 Howell, G.P. (2012). Asymmetric and diastereoselective conjugate addition reactions: C—C bond formation at large scale. *Org. Process Res. Dev.* 16: 1258–1272.

232 Melchiorre, P. (2012). Cinchona-based primary amine catalysis in the asymmetric functionalization of carbonyl compounds. *Angew. Chem. Int. Ed.* 51: 9748–9770.

233 (a) Gopula, B., Yang, S.-H., Kuo, T.-S. et al. (2015). Direct synthesis of chiral 3-arylsuccinimides by Rh-catalyzed enantioselective conjugate addition of arylboronic acids to maleimides. *Chem. Eur. J.* 21 (31): 11050–11055. (b) Lu, X., Wen, S., Li, X. et al. (2016). Highly enantioselective conjugate additions of glycine imines to activated alkenes catalyzed by amino acid-derived chiral phosphonium salts. *Eur. J. Org. Chem.* (25): 4298–4301. (c) Drissi-Amraoui, S., Schmid, T.E., Lauberteaux, J. et al. (2016). Copper-catalyzed asymmetric conjugate addition of dimethylzinc to acyl-*N*-methylimidazole Michael acceptors: scope, limitations and iterative reactions. *Adv. Synth. Catal.* 358 (15): 2519–2540. (d) Mahato, C.K., Kundu, M., and Pramanik, A. (2017). Solvent free, fast and asymmetric Michael additions of ketones to nitroolefins using chiral pyrrolidine-pyridone conjugate bases as organocatalysts. *Tetrahedron: Asymmetry* 28 (4): 511–515. (e) Yuan, Y., Yu, B., Bai, X.-F. et al. (2017). Asymmetric synthesis of glutamic acid derivatives by silver-catalyzed conjugate addition-elimination reactions. *Org. Lett.* 19 (18): 4896–4899.

234 Shibasaki, M., Sasai, H., and Arai, T. (1997). Asymmetric catalysis with heterobimetallic compounds. *Angew. Chem. Int. Ed. Engl.* 36: 1237–1256.

235 (a) Shibasaki, M., Kanai, M., Matsunaga, S., and Kumagai, N. (2009). Recent progress in asymmetric bifunctional catalysis using multimetallic systems. *Acc. Chem. Res.* 42: 1117–1127. (b) Yin, L., Takada, H., Lin, S. et al. (2014). Direct catalytic asymmetric vinylogous conjugate addition of unsaturated butyrolactones to α,β-unsaturated thioamides. *Angew. Chem. Int. Ed.* 53: 5327–5331.

236 (a) Li, Y., Deng, P., Zeng, Y.-M. et al. (2016). *anti*-Selective asymmetric Henry reaction catalyzed by a heterobimetallic Cu-Sm-aminophenol sulfonamide complex. *Org. Lett.* 18 (7): 1578–1581. (b) Li, X., Zhang, L., Xiao, Y.-H. et al. (2016). Direct asymmetric aldol reaction of acetophenones with aromatic aldehydes catalyzed by chiral Al/Zn heterobimetallic compounds. *Russ. J. Gen. Chem.* 86 (8): 1922–1930. (c) Karasawa, T., Kumagai, N., and Shibasaki, M. (2018). Heterogeneous heterobimetallic catalysis enabling expeditious access

to CF_3-containing vic-amino alcohols. *Org. Lett.* 20 (1): 308–311.
237 (a) Okino, T., Hoashi, Y., and Takemoto, Y. (2003). Enantioselective Michael reaction of malonates to nitroolefins catalyzed by bifunctional organocatalysts. *J. Am. Chem. Soc.* 125 (42): 12672–12673. (b) Hoashi, Y., Okino, T., and Takemoto, Y. (2005). Enantioselective Michael addition to α,β-unsaturated imides catalyzed by a bifunctional organocatalyst. *Angew. Chem. Int. Ed.* 44: 4032–4035.
238 Zu, L.-S., Xie, H., Li, H. et al. (2010). Chiral amine thiourea-promoted enantioselective domino Michael-aldol reactions between 2-mercaptobenzaldehydes and maleimides. *Adv. Synth. Catal.* 349: 1882–1886.
239 (a) Aleman, J., Milelli, A., Cabrera, S. et al. (2008). Asymmetric 1,4-addition of oxazolones to nitroalkenes by bifunctional cinchona alkaloid thiourea organocatalysts: synthesis of α,α-disubstituted α-amino acids. *Chem. Eur. J.* 14 (35): 10958–10966. (b) Zhang, W., Tan, D., Lee, R. et al. (2012). Highly enantio- and diastereoselective reactions of γ-substituted butenolides through direct vinylogous conjugate additions. *Angew. Chem. Int. Ed.* 51 (40): 10069–10073. (c) Rajkumar, S., Shankland, K., Goodman, J.M., and Cobb, A.J.A. (2013). Organocatalytic domino reaction of cyanosulfones: access to complex cyclohexane systems with quaternary carbon centers. *Org. Lett.* 15 (6): 1386–1389. (d) Anderson, J.C., Barham, J.P., and Rundell, C.D. (2015). Asymmetric intramolecular conjugate addition nitro-Mannich route to *cis*-2-aryl-3-nitrotetrahydroquinolines. *Org. Lett.* 17 (16): 4090–4093. (e) Al-Ani, W., Shankland, K., and Cobb, A.J.A. (2016). Asymmetric organocatalytic synthesis of cyclopentane γ-nitroketones. *Synlett* 27 (1): 17–20. (f) Zhao, H., Xiao, M.-G., Xu, L. et al. (2016). Bifunctional thiourea catalyzed asymmetric Michael addition of anthrone to methyleneindolinones. *RSC Adv.* 6 (45): 38558–38562.
240 Malerich, J.P., Hagihara, K., and Rawal, V.H. (2008). Chiral squaramide derivatives are excellent hydrogen bond donor catalysts. *J. Am. Chem. Soc.* 130: 14416–14417.
241 Bohme, D.K. and Mackay, G.I. (1981). Bridging the gap between the gas phase and solution – transition in the kinetics of nucleophilic displacement reactions. *J. Am. Chem. Soc.* 103 (4): 978–979.
242 Gronert, S. (2001). Mass spectrometric studies of organic ion/molecule reactions. *Chem. Rev.* 101 (2): 329–360.
243 (a) Brauman, J.I., Olmstead, W.N., and Lieder, C.A. (1974). Gas phase nucleophilic displacement reactions. *J. Am. Chem. Soc.* 96 (12): 4030–4031. (b) Wladkowski, B.D., Lim, K.F., Allen, W.D., and Brauman, J.I. (1992). The S_N2 identity exchange reaction $ClCH_2CN + Cl^- \rightleftarrows Cl^- + ClCH_2CN$ – experiment and theory. *J. Am. Chem. Soc.* 114 (23): 9136–9153.
244 Vayner, G., Houk, K.N., Jørgensen, W.L., and Brauman, J.I. (2004). Steric retardation of S_N2 reactions in the gas phase and solution. *J. Am. Chem. Soc.* 126 (29): 9054–9058.
245 (a) Ebrahim, A., Habibi, M., and Amirmijani, A. (2007). The study of counterion effect on the reactivity of nucleophiles in some S_N2 reactions in gas phase and solvent media. *J. Mol. Struct. THEOCHEM* 809: 115–124. (b) Wu, Z., Rong, C.-Y., Lu, T. et al. (2015). Density functional reactivity theory study of S_N2 reactions from the information-theoretic perspective. *Phys. Chem. Chem. Phys.* 17 (40): 27052–27061.
246 Dodd, J.A. and Brauman, J.I. (1984). On the application of the Marcus equation to methyl transfer (S_N2) reactions. *J. Am. Chem. Soc.* 106 (18): 5356–5357.
247 Shaik, S.S. and Pross, A. (1982). S_N2 Reactivity of CH_3X derivatives – a valence bond approach. *J. Am. Chem. Soc.* 104 (10): 2708–2719.
248 Pross, A. (1984). Relationship between rates and equilibria and the mechanistic significance of the Brønsted parameter – a qualitative valence-bond approach. *J. Org. Chem.* 49 (10): 1811–1818.
249 Shaik, S.S., Schlegel, H.B., and Wolfe, S. (1992). *Theoretical Aspects of Physical Organic Chemistry. The S_N2 Mechanism*, 1–285. New York: Wiley.
250 Hoz, S. (1993). Is the transition-state indeed intermediate between reactants and products – the Michael addition reaction as a case study. *Acc. Chem. Res.* 26 (2): 69–74.
251 Shaik, S.S., Ioffe, A., Reddy, A.C., and Pross, A. (1994). Is the avoided crossing state a good approximation for the transition state of a chemical reaction – an analysis of Menschutkin and ionic S_N2 reactions. *J. Am. Chem. Soc.* 116 (1): 262–273.
252 Pross, A. (1995). *Theoretical and Physical Principles of Organic Reactivity*, 1–312. New York: Wiley.

253 Anglada, J.M., Besalú, E., Bofill, J.M., and Crehuet, R. (1999). Prediction of approximate transition states by Bell-Evans-Polanyi principle: I. *J. Comput. Chem.* 20 (11): 1112–1129.

254 Bofill, J.M., Anglada, J.M., Besalú, E., and Crehuet, R. (2001). Quantum chemical reactivity: beyond the study of small molecules. In: *Fundamentals of Molecular Similarity*, 125–141. Springer US.

255 Wester, R., Bragg, A.E., Davis, A.V., and Neumark, D.M. (2003). Time-resolved study of the symmetric S_N2-reaction I-+CH_3I. *J. Chem. Phys.* 119: 10032–10039.

256 (a) Pocker, Y. (1959). Electrophilic catalysis in nucleophilic substitution and elimination. I. Kinetics and mechanism of reaction of *tert*-butyl bromide with alcohols, phenols, and carboxylic acids in nitromethane. *J. Chem. Soc.* 1179–1183. (b) Pocker, Y. (1960). Electrophilic catalysis in nucleophilic substitutions and elimination. II. Kinetics and mechanism of reaction of *tert*-butyl chloride with pyridine, tribenzylamine, radiochloride ions and hydrogen radiochloride in nitromethane. *J. Chem. Soc.* 1972–1977. (c) Lee, S.-S., Kim, H.-S., Hwang, T.-K. et al. (2008). Efficiency of bulky protic solvent for S_N2 reaction. *Org. Lett.* 10 (1): 61–64.

257 Bender, M.L. (1963). Metal ion catalysis of nucleophilic organic reaction in solution. *Adv. Chem. Ser.* 37: 19–36.

258 (a) Pocker, Y. and Kevill, D.N. (1965). Electrophilic catalysis in nucleophilic substitution and elimination. III. Conductances of some silver and tetraethylammonium salts in acetonitrile and kinetics of reaction of 2-octyl bromide with tetraethylammonium and silver nitrates in that solvent. *J. Am. Chem. Soc.* 87 (21): 4760–4770. (b) Pocker, Y. and Kevill, D.N. (1965). Electrophilic catalysis in nucleophilic substitution and elimination. IV. Kinetics of reaction of 2-octyl bromide with tetraethylammonium and silver perchlorates and nitrites in acetonitrile. *J. Am. Chem. Soc.* 87 (21): 4771–4777. (c) Pocker, Y. and Wong, W.H. (1975). Electrophilic catalysis in nucleophilic substitution and elimination. VII. Kinetics and mechanism of reaction of neopentyl iodide with tetraethylammonium and silver nitrates and perchlorates in acetonitrile. *J. Am. Chem. Soc.* 97 (24): 7097–7104.

259 Maia, A., Landini, D., Betti, C. et al. (2005). Catalytic activity and anion activation in S_N2 reactions promoted by complexes of silicon polypodands. Comparison with traditional polyethers. *New J. Chem.* 29 (9): 1195–1198.

260 Buncel, E., Nagelkerke, R., and Thatcher, G.R.J. (2003). Metal ion catalysis and inhibition in nucleophilic displacement reactions at carbon, phosphorous, and sulfur centers part 10 – alkali metal ion catalysis in nucleophilic displacement by ethoxide ion on *p*-nitrophenyl phenylphosphonate: evidence for multiple metal ion catalysis. *Can. J. Chem.* 81 (1): 53–63.

261 Takada, K., Tanaka, S., and Nagasawa, K. (2009). Asymmetric Mannich-type reaction of aromatic α-amido sulfone with malonate using guanidine-thiourea bifunctional organocatalyst. *Synlett* (10): 1643–1646.

262 Makosza, M. and Serafin, B. (1965). Reactions of organic anions. I. Catalytic ethylation of phenylacetonitrile in aqueous medium. *Roczniki Chemii* 39 (9): 1223–1230.

263 Makosza, M. and Jonczyk, A. (1976). Phase-transfer alkylation of nitriles: α-phenylbutyronitrile. *Org. Synth.* 55: 91–95.

264 Makosza, M. and Fedorynski, M. (1996). Thirty years of phase transfer catalysis. *Pol. J. Chem.* 70: 1093–1110.

265 Makosza, M. (2000). Phase-transfer catalysis. A general green methodology in organic synthesis. *Pure Appl. Chem.* 72: 1399–1403.

266 Starks, C.M. (1971). Phase-transfer catalysis. 1. Heterogeneous reactions involving anion transfer by quaternary ammonium and phosphonium salts. *J. Am. Chem. Soc.* 93 (1): 195–199.

267 Starks, C.M. and Liotta, C. (1978). *Phase Transfer Catalysis: Principles and Techniques*, 1–378. New York: Academic Press.

268 Herriott, A.W. and Picker, D. (1975). Phase transfer catalysis – evaluation of catalysts. *J. Am. Chem. Soc.* 97: 2345–2349.

269 (a) Maruoka, K. (2008). Practical aspects of recent asymmetric phase-transfer catalysis. *Org. Process Res. & Dev.* 12 (4): 679–697. (b) Ohmatsu, K., Uraguchi, D., and Ooi, T. (2013). Asymmetric phase-transfer catalysis. In: *Stereoselective Synthesis of Drugs and Natural Products*, vol. 1, (ed. V. Andrushko and N. Andrushko), 119–143. Hoboken: Wiley.

270 Dehmlow, E.V. and Dehmlow, S.S. (1983). *Phase Transfer Catalysis: Principles and Techniques*, 2e, vol. 11, 1–366. Weinheim: Verlag Chemie.

271 Makosza, M. and Fedorynski, M. (1987). Catalysis in 2-phase systems – phase-transfer and related phenomena. *Adv. Catal.* 35: 375–422.

272 Starks, C.M., Liotta, C.l., and Halpern, M. (1994). *Phase-Transfer Catalysis: Fundamentals, Applications, and Industrial Perspectives*, 1–668. New York: Chapman & Hall.

273 Yadav, G.D. (2004). Insight into green phase transfer catalysis. *Top. Catal.* 29 (3–4): 145–161.

274 (a) Shirakawa, S. and Maruoka, K. (2013). Recent developments in asymmetric phase-transfer reactions. *Angew. Chem. Int. Ed.* 52: 4312–4348. (b) Tan, J. and Yasuda, N. (2015). Contemporary asymmetric phase transfer catalysis: large-scale industrial applications. *Org. Process Res. Dev.* 19 (11): 1731–1746. (c) Liu, S., Kumatabara, Y., and Shirakawa, S. (2016). Chiral quaternary phosphonium salts as phase-transfer catalysts for environmentally benign asymmetric transformations. *Green Chem.* 18 (2): 331–341. (d) Kaneko, S., Kumatabara, Y., and Shirakawa, S. (2016). A new generation of chiral phase-transfer catalysts. *Org. Biomol. Chem.* 14 (24): 5367–5376. (e) Ha, M.W. and Park, H.G. (2016). *Cinchona Alkaloid Derivatives as Asymmetric Phase-Transfer Catalysts*, RSC Green Chemistry Series, Part 2, vol. 41, 82–134. Royal Society of Chemistry. (f) Zong, L. and Tan, C.-H. (2017). Phase-transfer and ion-pairing catalysis of pentanidiums and bisguanidiniums. *Acc. Chem. Res.* 50 (4): 842–856.

275 Liu, Y., Shirakawa, S., and Maruoka, K. (2013). Phase transfer catalyzed asymmetric conjugate cyanation of alkylidenemalonates with KCN in the presence of a Brønsted acid additive. *Org. Lett.* 15: 1230–1233.

276 Shirakawa, S., Tokuda, T., and Maruoka, K. (2015). Phase transfer catalyzed asymmetric desymmetrizations of cyclopentanones. *Org. Chem. Front.* 2: 336–339.

277 Shirakawa, S., Koga, K., Tokuda, T. et al. (2014). Catalytic asymmetric synthesis of 3,3′-diaryloxindoles as triarylmethanes with a chiral all-carbon quaternary center: phase transfer catalyzed SNAr reaction. *Angew. Chem. Int. Ed.* 53: 6220–6223.

278 Pedersen, C.J. (1967). Cyclic polyethers and their complexes with metal salts. *J. Am. Chem. Soc.* 89 (26): 7017–7036.

279 Pedersen, C.J. (1967). Cyclic polyethers and their complexes with metal salts. *J. Am. Chem. Soc.* 89 (10): 2495–2496.

280 Newcomb, M., Gokel, G.W., and Cram, D.J. (1974). Pyridyl unit in host compounds. *J. Am. Chem. Soc.* 96 (21): 6810–6811.

281 Ooi, T. and Maruoka, K. (2007). Recent advances in asymmetric phase transfer catalysis. *Angew. Chem. Int. Ed.* 46 (23): 4222–4266.

282 Hashimoto, T. and Maruoka, K. (2007). Recent development and application of chiral phase transfer catalysts. *Chem. Rev.* 107 (12): 5656–5682.

283 Patterson, D.E., Xie, S., Jones, L.A. et al. (2010). Large scale application of asymmetric phase-transfer catalysis for amino acid synthesis. In: *Asymmetric Catalysis on Industrial Scale* (ed. H.-U. Blaser and H.-J. Federsel), 473–484. Weinheim: Wiley-VCH.

284 Nicolas, C. and Lacour, J. (2006). Triazatriangulenium cations: highly stable carbocations for phase-transfer catalysis. *Org. Lett.* 8 (19): 4343–4346.

285 Pedersen, C.J. and Frensdorff, H.K. (1972). Macrocyclic polyethers and their complexes. *Angew. Chem. Int. Ed. Engl.* 11 (1): 16–25.

286 Dibiase, S.A. and Gokel, G.W. (1978). Crown-cation complex effects. 8. Reactions of crown ether activated *tert*-butoxide ion. *J. Org. Chem.* 43 (3): 447–452.

287 Davis, F. and Higson, S. (2001). Crown ethers, cryptands and other compounds. In: *Macrocycles: Construction, Chemistry and Nanotechnology Applications*, Chapter 3, 34–71. Chichester: Wiley.

288 Izatt, R.M., Bradshaw, J.S., Nielsen, S.A. et al. (1985). Thermodynamic and kinetic data for cation macrocycle interaction. *Chem. Rev.* 85 (4): 271–339.

289 Landini, D., Montanari, F., and Pirisi, F.M. (1974). Crown ethers as phase transfer catalysts in two-phase reactions. *J. Chem. Soc., Chem. Commun.* (21): 879–880.

290 Wong, K.H. (1978). Kinetics of esterification of potassium *para*-nitrobenzoate by benzyl bromide using dicyclohexyl-18-crown-6 as phase transfer agent. *J. Chem. Soc., Chem. Commun.* (7): 282–283.

291 Cinquini, M., Montanari, F., and Tundo, P. (1975). Alkyl substituted aza-macrobicyclic polyethers – highly efficient catalysts in 2-phase reactions. *J. Chem. Soc., Chem. Commun.* (10): 393–394.

292 Rapi, Z., Bako, P., Keglevich, G. et al. (2012). Asymmetric phase transfer Darzens reactions catalyzed by D-glucose- and D-mannose-based chiral crown ethers. *Tetrahedron: Asymmetry* 23: 489–496.

293 Bako, P., Keglevich, G., Rapi, Z., and Toke, L. (2012). The enantiomeric differentiation ability of chiral crown ethers based on carbohydrates. *Curr. Org. Chem.* 16: 297–304.

294 Rapi, Z., Demuth, B., Keglevich, G. et al. (2014). Enantioselective Michael addition of malonates to aromatic nitroalkenes catalyzed by

monosaccharide-based chiral crown ethers. *Tetrahedron: Asymmetry* 25: 141–147.

295 Bako, P., Rapi, Z., Gruen, A. et al. (2015). Asymmetric Michael addition of malonates to enones catalyzed by an α-D-glucopyranoside-based crown ether. *Synlett* 26: 1847–1851.

296 Mann, J. (1987). Modern methods for the introduction of fluorine into organic molecules – an approach to compounds with altered chemical and biological activities. *Chem. Soc. Rev.* 16 (4): 381–436.

297 Mikami, K., Itoh, Y., and Yamanaka, M. (2004). Fluorinated carbonyl and olefinic compounds: basic character and asymmetric catalytic reactions. *Chem. Rev.* 104 (1): 1–16.

298 Schlosser, M. (2006). CF_3-bearing aromatic and heterocyclic building blocks. *Angew. Chem. Int. Ed.* 45 (33): 5432–5446.

299 Müller, K., Faeh, C., and Diederich, F. (2007). Fluorine in pharmaceuticals: looking beyond intuition. *Science* 317 (5846): 1881–1886.

300 Caldwell, G. and Kebarle, P. (1984). Binding energies and structural effects in halide anion ROH and RCOOH complexes from gas phase equilibria measurements. *J. Am. Chem. Soc.* 106 (4): 967–969.

301 Weis, P., Kemper, P.R., Bowers, M.T., and Xantheas, S.S. (1999). A new determination of the fluoride ion-water bond energy. *J. Am. Chem. Soc.* 121 (14): 3531–3532.

302 Emsley, J. (1980). Very strong hydrogen bonding. *Chem. Soc. Rev.* 9 (1): 91–124.

303 Gronert, S. and Kass, S.R. (1997). Theoretical studies of eliminations. 6. The regiochemistry and stereochemistry of the gas phase reactions of 3-halocyclohexenes with fluoride. An ab initio study. *J. Org. Chem.* 62 (23): 7991–8000.

304 Pliego, J.R. and Riveros, J.M. (2002). Theoretical study of the gas phase reaction of fluoride and chloride ions with methyl formate. *J. Phys. Chem. A* 106 (2): 371–378.

305 Mugnai, M., Cardini, G., and Schettino, V. (2003). Substitution and elimination reaction of F- with C_2H_5Cl: an ab initio molecular dynamics study. *J. Phys. Chem. A* 107 (14): 2540–2547.

306 Wilkinson, J.A. (1992). Recent advances in the selective formation of the C—F bond. *Chem. Rev.* 92 (4): 505–519.

307 Murray, C.B., Sandford, G., and Korn, S.R. (2003). Ionic liquids as media for nucleophilic fluorination. *J. Fluorine Chem.* 123 (1): 81–84.

308 Kim, D.W., Song, C.E., and Chi, D.Y. (2002). New method of fluorination using potassium fluoride in ionic liquid: significantly enhanced reactivity of fluoride and improved selectivity. *J. Am. Chem. Soc.* 124 (35): 10278–10279.

309 Liotta, C.L. and Harris, H.P. (1974). Chemistry of naked anions. 1. Reactions of 18-crown-6 complex of potassium fluoride with organic substrates in aprotic organic solvents. *J. Am. Chem. Soc.* 96 (7): 2250–2252.

310 Cox, D.P., Terpinski, J., and Lawrynowicz, W. (1984). Anhydrous tetrabutylammonium fluoride – a mild but highly efficient source of nucleophilic fluoride ion. *J. Org. Chem.* 49 (17): 3216–3219.

311 Sun, H.R. and DiMagno, S.G. (2005). Anhydrous tetrabutylammonium fluoride. *J. Am. Chem. Soc.* 127 (7): 2050–2051.

312 Landini, D., Maia, A., and Rampoldi, A. (1989). Dramatic effect of the specific solvation on the reactivity of quaternary ammonium fluorides and poly(hydrogen fluorides), (HF)-normal.F-, in media of low polarity. *J. Org. Chem.* 54 (2): 328–332.

313 Albanese, D., Landini, D., and Penso, M. (1998). Hydrated tetrabutylammonium fluoride as a powerful nucleophilic fluorinating agent. *J. Org. Chem.* 63 (25): 9587–9589.

314 Pliego, J.R. and Pilo-Veloso, D. (2007). Chemoselective nucleophilic fluorination induced by selective solvation of the S_N2 transition state. *J. Phys. Chem. B* 111 (7): 1752–1758.

315 Cinquini, M., Colonna, S., Molinari, H. et al. (1976). Heterogeneous phase transfer catalysts – onium salts, crown ethers, and cryptands immobilized on polymer supports. *J. Chem. Soc., Chem. Commun.* (11): 394–396.

316 He, L.N., Yasuda, H., and Sakakura, T. (2003). New procedure for recycling homogeneous catalyst: propylene carbonate synthesis under supercritical CO_2 conditions. *Green Chem.* 5 (1): 92–94.

317 Shirakawa, S., Tanaka, Y., and Maruoka, K. (2004). Development of a recyclable fluorous chiral phase-transfer catalyst: application to the catalytic asymmetric synthesis of α-amino acids. *Org. Lett.* 6 (9): 1429–1431.

318 Stuart, A.M. and Vidal, J.A. (2007). Perfluoroalkylated 4,13-diaza-18-crown-6 ethers: synthesis, phase transfer catalysis, and recycling studies. *J. Org. Chem.* 72 (10): 3735–3740.

319 Brak, K. and Jacobsen, E.N. (2013). Asymmetric ion-pairing catalysis. *Angew. Chem. Int. Ed.* 52: 534–561.

320 Carter, C., Fletcher, S., and Nelson, A. (2003). Towards phase transfer catalysts with a chiral

anion: inducing asymmetry in the reactions of cations. *Tetrahedron: Asymmetry* 14: 1995–2004.

321 Hamilton, G.L., Kanai, T., and Toste, F.D. (2008). Chiral anion-mediated asymmetric ring opening of *meso*-aziridinium and episulfonium ions. *J. Am. Chem. Soc.* 130: 14984–14986.

322 Phipps, R.J., Hamilton, G.L., and Toste, F.D. (2012). The progression of chiral anions from concepts to applications in asymmetric catalysis. *Nat. Chem.* 4: 603–614.

323 Hennecke, U. (2012). A new approach towards the asymmetric fluorination of alkenes using anionic phase transfer catalysts. *Angew. Chem. Int. Ed.* 51: 4532–4534.

324 Hennecke, U. (2012). New catalytic approaches towards the enantioselective halogenation of alkenes. *Chem. Asian J.* 7: 456–465.

325 (a) Wu, J., Wang, Y.M., Drljevic, A. et al. (2013). A combination of directing groups and chiral anion phase-transfer catalysis for enantioselective fluorination of alkenes. *Proc. Natl. Acad. Sci. U.S.A.* 110: 13729–13733. (b) Yang, X., Wu, T., Phipps, R.J., and Toste, F.D. (2015). Advances in catalytic enantioselective fluorination, mono-, di-, and trifluoromethylation, and trifluoromethylthiolation reactions. *Chem. Rev.* 115: 826–870.

326 Egami, H., Asado, J., Sato, K. et al. (2015). Asymmetric fluorolactonization with a bifunctional hydroxyl carboxylate catalyst. *J. Am. Chem. Soc.* 137 (32): 10132–10135.

327 Nelson, H.M., Reisberg, S.H., Shunatona, H.P. et al. (2014). Chiral anion phase transfer of aryldiazonium cations: an enantioselective synthesis of C3-diazenated pyrroloindolines. *Angew. Chem. Int. Ed.* 53: 5600–5603.

328 Frankland, E. (1849). Isolation of organic radicals. *Liebigs Ann. Chem.* 171–213.

329 Frankland, E. (1853). Über eine neue Reihe organischer Körper, welche Metalle enthalten. *Liebigs Ann. Chem.* 85: 329–373.

330 Seyferth, D. (2001). Zinc alkyls, Edward Frankland, and the beginnings of main-group organometallic chemistry. *Organometallics* 20: 2940–2955.

331 Grignard, V. (1900). Sur quelques nouvelles combinaisons organométalliques du magnésium et leur application à des synthèses d'alcools et d'hydrocarbures. *C.R. Hebd. Seances Acad. Sci.* 130: 1322–1324.

332 Seyferth, D. (2009). The grignard reagents. *Organometallics* 28 (6): 1598–1605.

333 Ziegler, K. and Colonius, H. (1930). Experiments on the alkali-organic compounds. V. A convenient synthesis of a simple lithium alkyle. *Liebigs Ann. Chem.* 479: 135–149.

334 Ziegler, K. (1968). Forty years' stroll through the realms of organometallic chemistry. *Adv. Organomet. Chem.* 6: 1–17.

335 Gilman, H., Langham, W., and Jacoby, A.L. (1939). Metalation as a side reaction in the preparation of organolithium compounds. *J. Am. Chem. Soc.* 61: 106–109.

336 Gilman, H. and Bebb, R.L. (1938). Relative reactivities of organometallic compounds. XX. Metalation. *J. Am. Chem. Soc.* 61: 109–112.

337 Wittig, G. and Pockels, U. (1939). On the exchange of aromatic bonded bromine towards lithium by means of phenyl-lithium (II. Announcement). *Ber. Dtsch. Chem. Ges.* 72: 89–92.

338 Wittig, G. (1940). Syntheses with lithium organic compounds. *Angew. Chem.* 53: 241–247.

339 (a) Frankel, G. (2006). Dynamics of the reorganization behavior of organolithium compounds. In: *The Chemistry of Organolithium Compounds*, Chapter 1, vol. 2 (ed. Z. Rapporport and I. Marek), 1–62. Chichester: Wiley. (b) Vázquez, A.J., de Waisbaum, R.G., and Sbarbati Nudelman, N. (2006). Organolithiums as useful synthetic intermediates for tandem reactions. In: *The Chemistry of Organolithium Compounds*, Chapter 2, vol. 2 (ed. Z. Rappoport and I. Marek), 63–138. Chichester: Wiley. (c) Najera, C. and Yus, M. (2006). The chemistry of acyllithium derivatives. In: *The Chemistry of Organolithium Compounds*, Chapter 3, vol. 2 (ed. Z. Rappoport and I. Marek), 139–294. Chichester: Wiley. (d) Fañanás, F.G. and Sanz, R. (2006). Intramolecular carbolithiation reactions. In: *The Chemistry of Organolithium Compounds*, Chapter 4, vol. 2 (ed. Z. Rappoport and I. Marek), 295–380. Chichester: Wiley.

340 Eisch, J.J. (2002). Henry Gilman: American pioneer in the rise of organometallic chemistry in modern science and technology. *Organometallics* 21 (25): 5439–5463.

341 Wu, G. and Huang, M. (2006). Organolithium reagents in pharmaceutical asymmetric processes. *Chem. Rev.* 106 (7): 2596–2616.

342 Reich, H.J. (2012). What's going on with these lithium reagents? *J. Org. Chem.* 77: 5471–5491.

343 Reich, H.J. (2013). Role of organolithium aggregates and mixed aggregates in organolithium mechanisms. *Chem. Rev.* 113: 7130–7178.

344 Seebach, D. (1979). Methods of reactivity Umpolung. *Angew. Chem. Int. Ed. Engl.* 18 (4): 239–258.

345 Zhdankin, V.V., Mullikin, M., Tykwinski, R. et al. (1989). Carbon–carbon bond formation in reactions of PhIO·HBF_4/silyl enol ether adduct with alkenes or silyl enol ethers. *J. Org. Chem.* 54: 2605–2608.

346 Miyata, O., Miyoshi, T., and Ueda, M. (2013). Umpolung reactions at the α-carbon position of carbonyl compounds. *ARKIVOC* 60–81.

347 Moriarty, R.M., Hu, H., and Gupta, S.C. (1981). Direct α-hydroxylation of ketones using iodosobenzene. *Tetrahedron Lett.* 22: 1283–1286.

348 Smoliakova, I.P., Caple, R., Gregory, D. et al. (1995). Highly selective formation of a β-C-glucosidic bond in the reactions of ArSCl-glucal adducts with silicon-containing nucleophiles. *J. Org. Chem.* 60 (5): 1221–1227.

349 Han, M.M., Smoliakova, I.P., and Koikov, L.N. (2000). Stereoselective synthesis of C-[2-*S*-(*p*-tolyl)-2-thio-β-D-galactopyranosyl] compounds using the reaction of TolSCl adducts of D-galactal with C-nucleophiles. *Carbohydr. Res.* 323 (1–4): 202–207.

350 Shneider, O.S., Pisarevesky, E., Fristrup, P., and Szpilman, A.M. (2015). Oxidative umpolung α-alkylation of ketones. *Org. Lett.* 17: 282–285.

351 Wöhler, F. and Liebig, J. (1832). Untersuchungen über das Radikal der Benzosäure. *Ann. Pharm.* 3: 249–282.

352 Lapworth, A. (1903). Reactions involving the addition of hydrogen cyanide to carbon compounds. *J. Chem. Soc.* 83: 995–1005.

353 Wiberg, K.B. (1954). The deuterium isotope effect of some ionic reactions of benzaldehyde. *J. Am. Chem. Soc.* 76 (21): 5371–5375.

354 Johnson, J.S. (2004). Catalyzed reactions of acyl anion equivalents. *Angew. Chem. Int. Ed.* 43 (11): 1326–1328.

355 Ukai, T., Tanaka, R., and Dokawa, T. (1943). A new catalyst for acyloin condensation. I. *J. Pharm. Soc. Jpn.* 63: 296–300.

356 (a) Mizuhara, S. and Handler, P. (1954). Mechanism of thiamine-catalyzed reactions. *J. Am. Chem. Soc.* 76 (2): 571–573. (b) Kluger, R. (1997). Lessons from thiamin watching. *Pure Appl. Chem.* 69 (9): 1957–1967.

357 (a) Breslow, R. (1958). On the mechanism of thiamine action. 4. Evidence from studies on model systems. *J. Am. Chem. Soc.* 80 (14): 3719–3726. (b) Breslow, R. and Schmuck, C. (1996). The mechanism of thiazolium catalysis. *Tetrahedron Lett.* 37 (45): 8241–8242.

358 (a) Dominguez de Maria, P. and Shamuganathan, S. (2011). Umpolung catalysis in benzoin-type and Stetter-type reactions. From enzymatic performances to bio-mimetic organocatalytic concepts. *Curr. Org. Chem.* 15 (13): 2083–2097. (b) Cardinal-David, B. and Scheidt, K.A. (2013). Carbene catalysis. Beyond the benzoin and Stetter reactions. In: *Inventing Reactions*, Topics in Organometallic Chemistry, vol. 44 (ed. L.J. Goossen), 233–260. Heidelberg: Springer-Verlag.

359 White, M.J. and Leeper, F.J. (2001). Kinetics of the thiazolium ion-catalyzed benzoin condensation. *J. Org. Chem.* 66 (15): 5124–5131.

360 Breslow, R. and Kim, R. (1994). The thiazolium catalyzed benzoin condensation with mild base does not involve a dimer intermediate. *Tetrahedron Lett.* 35 (5): 699–702.

361 Teles, J.H., Melder, J.P., Ebel, K. et al. (1996). The chemistry of stable carbenes. 2. Benzoin-type condensations of formaldehyde catalyzed by stable carbenes. *Helv. Chim. Acta* 79 (1): 61–83.

362 Enders, D. and Balensiefer, T. (2004). Nucleophilic carbenes in asymmetric organocatalysis. *Acc. Chem. Res.* 37 (8): 534–541.

363 Butlerow, M.A. (1861). *Liebigs Ann. Chem.* 120: 295–298.

364 Butlerow, M.A. (1861). Formation synthétique d'une substance sucrée. *C.R. Hebd. Seances Acad. Sci. C. Sci. Chim.* 53: 145–147.

365 Shigemasa, Y., Ueda, T., and Saimoto, H. (1989). Formose reactions. 26. 1st Synthesis of dl-2-C-hydroxymethyl-3-pentulose in the formose reaction. *J. Carbohydr. Res.* 8 (4): 669–673.

366 Tajima, H., Niitsu, T., and Inoue, H. (1999). Polymerization of formaldehyde by an immobilized thiamine catalyst on cation-exchange resin. *J. Chem. Eng. Jpn.* 32: 776–782.

367 López-Calahorra, F., Castro, E., Ochoa, A., and Marti, J. (1996). Further evidences about the role of bis(thiazolin-2-ylidene)s as the actual catalytic species in the generalised benzoin condensation. *Tetrahedron Lett.* 37 (28): 5019–5022.

368 Li, G.Q., Dai, L.X., and You, S.L. (2007). Thiazolium-derived N-heterocyclic carbene-catalyzed cross-coupling of aldehydes with unactivated imines. *J. Chem. Soc., Chem. Commun.* (8): 852–854.

369 Lappert, M.F. and Maskell, R.K. (1982). A new class of benzoin condensation catalyst, the bi-(1,3-dialkylimidazolidin-2-ylidenes). *J. Chem. Soc., Chem. Commun.* (11): 580–581.

370 Iwamoto, K., Hamaya, M., Hashimoto, N. et al. (2006). Benzoin reaction in water as an aqueous medium catalyzed by benzimidazolium salt. *Tetrahedron Lett.* 47 (40): 7175–7177.

371 Reich, B.J.E., Justice, A.K., Beckstead, B.T. et al. (2004). Cyanide-catalyzed cyclizations via aldimine coupling. *J. Org. Chem.* 69 (4): 1357–1359.

372 Breslow, R. and Kool, E. (1988). A γ-cyclodextrin thiazolium salt holoenzyme mimic for the benzoin condensation. *Tetrahedron Lett.* 29 (14): 1635–1638.

373 Ikeda, H., Horimoto, Y., Nakata, M., and Ueno, A. (2000). Artificial holoenzymes for benzoin condensation using thiazolio-appended β-cyclodextrin dimers. *Tetrahedron Lett.* 41 (33): 6483–6487.

374 Xin, L.H., Bausch, C.C., and Johnson, J.S. (2005). Mechanism and scope of the cyanide-catalyzed cross silyl benzoin reaction. *J. Am. Chem. Soc.* 127 (6): 1833–1840.

375 Brook, A.G. (1974). Some molecular rearrangements of organosilicon compounds. *Acc. Chem. Res.* 7 (3): 77–84.

376 Moser, W.H. (2001). The Brook rearrangement in tandem bond formation strategies. *Tetrahedron* 57 (11): 2065–2084.

377 Bausch, C.C. and Johnson, J.S. (2005). Cyanide-catalyzed additions of acyl phosphonates to aldehydes: a new acyl donor for benzoin-type reactions. *Adv. Synth. Catal.* 347 (9): 1207–1211.

378 Mattson, A.E., Bharadwaj, A.R., Zuhl, A.M., and Scheidt, K.A. (2006). Thiazolium-catalyzed additions of acylsilanes: a general strategy for acyl anion addition reactions. *J. Org. Chem.* 71 (15): 5715–5724.

379 (a) Menon, R.S., Biju, A.T., and Nair, V. (2016). Recent advances in N-heterocyclic carbene (NHC)-catalysed benzoin reactions. *Beilstein J. Org. Chem.* 12: 444–461. (b) Wu, Y.-W., Hu, L., Li, Z., and Deng, L. (2015). Catalytic asymmetric umpolung reactions of imines. *Nature* 523 (7561): 445–450. (c) Hu, L., Wu, Y.-W., LI, Z., and Deng, L. (2016). Catalytic asymmetric synthesis of chiral γ-amino ketones via umpolung reactions of omines. *J. Am. Chem. Soc.* 138 (49): 15817–15820. (d) Hu, B. and Deng, L. (2018). Catalytic asymmetric synthesis of trifluoromethylated ©-amino acids through the umpolung addition of trifluoromethyl imines to carboxylic acid derivatives. *Angew. Chem. Int. Ed.* 57 (8): 2233–2237.

380 Stetter, H. and Schreckenberg, M. (1973). Addition of aldehydes to activated double bonds. *Angew. Chem.* 85 (2): 89–89.

381 (a) Stetter, H. (1976). Catalyzed addition of aldehydes to activated double bonds – new synthetic approach. *Angew. Chem. Int. Ed. Engl.* 15 (11): 639–647. (b) Stetter, H. and Kuhlmann, H. (1991). The catalyzed nucleophilic addition of aldehydes to electrophilic double bonds. *Org. React.* 40: 408–496.

382 (a) Izquierdo, J., Hutson, G.E., Cohen, D.T., and Scheidt, K.A. (2012). A continuum of progress: applications of N-heterocyclic carbene catalysis in total synthesis. *Angew. Chem. Int. Ed.* 51 (47): 11686–11698. (b) Yetra, S.R., Patra, A., and Biju, A.T. (2015). Recent advances in the N-heterocyclic carbene (NHC)-organocatalyzed Stetter reaction and related chemistry. *Synthesis* 47 (10): 1357–1378. (c) Haghshenas, P. and Langdon, S.M. (2017). Thieme chemistry journals awardees – where are they now? Carbene organocatalysis for Stetter, benzoin, domino, and ring-expansion reactions. *Synlett* 28 (5): 542–559.

383 Trost, B.M., Shuey, C.D., and Di Ninno, F. (1979). Stereocontrolled total synthesis of (+/-)-hirsutic acid-C. *J. Am. Chem. Soc.* 101 (5): 1284–1285.

384 Ciganek, E. (1995). Esters of 2,3-dihydro-3-oxobenzofuran-2-acetic acid and 3,4-dihydro-4-oxo-2*H*-1-benzopyran-3-acetic acid by intramolecular Stetter reactions. *Synthesis* (10): 1311–1314.

385 Zhou, Z.-Z., Ji, F.-Q., Cao, M., and Yang, G.-F. (2006). An efficient intramolecular stetter reaction in room temperature ionic liquids promoted by microwave irradiation. *Adv. Synth. Catal.* 348: 1826–1830.

386 (a) Enders, D., Breuer, K., Runsink, J., and Teles, J.H. (1996). The first asymmetric intramolecular Stetter reaction. *Helv. Chim. Acta* 79 (7): 1899–1902. (b) Kerr, M.S. and Rovis, T. (2004). Enantioselective synthesis of quaternary stereocenters via a catalytic asymmetric Stetter reaction. *J. Am. Chem. Soc.* 126 (29): 8876–8877. (c) Christmann, M. (2005). New developments in the asymmetric Stetter reaction. *Angew. Chem. Int. Ed.* 44 (18): 2632–2634.

387 (a) Lin, Q.-Q., Li, Y.-Z., Das, D.K. et al. (2016). N-Heterocyclic carbene-catalyzed desymmetrization of functionalized 1,4-dienes via Stetter reaction. *J. Chem. Soc., Chem. Commun.* 52 (38): 6459–6462. (b) Rafinski, Z. (2016).

Novel (-)-β-pinene-derived triazolium salts: synthesis and application in the asymmetric Stetter reaction. *ChemCatChem* 8 (16): 2599–2603. (c) Ema, T., Nanjo, Y., Shiratori, S. et al. (2016). Solvent-free benzoin and Stetter reactions with a small amount of NHC catalyst in the liquid or semisolid state. *Org. Lett.* 18 (21): 5764–5767.

388 Hawkes, K.J. and Yates, B.F. (2008). The mechanism of the Stetter reaction – a DFT study. *Eur. J. Org. Chem.* (33): 5563–5570.

389 Mattson, A.E., Bharadwaj, A.R., and Scheidt, K.A. (2004). The thiazolium-catalyzed sila-Stetter reaction: conjugate addition of acylsilanes to unsaturated esters and ketones. *J. Am. Chem. Soc.* 126 (8): 2314–2315.

390 (a) Liu, Q., Perreault, S., and Rovis, T. (2008). Catalytic asymmetric intermolecular Stetter reaction of glyoxamides with alkylidenemalonates. *J. Am. Chem. Soc.* 130: 14066–14067. (b) DiRocco, D.A., Oberg, K.M., Dalton, D.M., and Rovis, T. (2009). Catalytic asymmetric intermolecular Stetter reaction of heterocyclic aldehydes with nitroalkenes: backbone fluorination improves selectivity. *J. Am. Chem. Soc.* 131: 10872–10874. (c) DiRocco, D.A.; Rovis, T. Carbene-catalyzed Stetter reactions. *Science of Synthesis, Asymmetric Organocatalysis*, List, B.; Maruoka, K., Eds.; Georg Thieme Verlag, Stuttgart, 2012, Vol. 1, pp. 619–637; (d) Wurz, N.E., Daniliuc, C.G., and Glorius, F. (2012). Highly enantioselective intermolecular Stetter reaction of simple acrylates: synthesis of α-chiral γ-ketoesters. *Chem. Eur. J.* 18 (51): 16297–16301.

391 Patra, A., Mukherjee, S., Das, T.K. et al. (2017). N-Heterocyclic-carbene-catalyzed umpolung of imines. *Angew. Chem. Int. Ed.* 56 (10): 2730–2734.

392 Burstein, C. and Glorius, F. (2004). Organocatalyzed conjugate umpolung of α,β-unsaturated aldehydes for the synthesis of γ-butyrolactones. *Angew. Chem. Int. Ed.* 43 (45): 6205–6208.

393 Sohn, S.S., Rosen, E.L., and Bode, J.W. (2004). N-heterocyclic carbene catalyzed generation of homoenolates: γ-butyrolactones by direct annulations of enals and aldehydes. *J. Am. Chem. Soc.* 126 (44): 14370–14371.

394 Herrmann, W.A. (2002). N-Heterocyclic carbenes: a new concept in organometallic catalysis. *Angew. Chem. Int. Ed.* 41 (8): 1290–1309.

395 Burstein, C., Tschan, S., Xie, X., and Glorius, F. (2006). N-heterocyclic carbene catalyzed conjugate umpolung for the synthesis of γ-butyrolactones. *Synthesis* 2418–2439.

396 Nair, V., Vellalath, S., Poonoth, M., and Suresh, E. (2006). N-heterocyclic carbene reaction of chalcones and enals via homoenolate: an efficient synthesis of 1,3,4-trisubstituted cyclopentenes. *J. Am. Chem. Soc.* 128 (27): 8736–8737.

397 He, M. and Bode, J.W. (2008). Enantioselective, NHC-catalyzed bicyclo-β-lactam formation via direct annulations of enals and unsaturated *N*-sulfonyl ketimines. *J. Am. Chem. Soc.* 130: 418–419.

398 Wang, M.H., Barsoum, D., Schwamb, C.B. et al. (2017). Catalytic, enantioselective β-protonation through a cooperative activation strategy. *J. Org. Chem.* 82 (9): 4689–4702.

399 He, M. and Bode, J.W. (2005). Catalytic synthesis of γ-lactams via direct annulations of enals and *N*-sulfonylimines. *Org. Lett.* 7 (14): 3131–3134.

400 Nair, V., Vellalath, S., Poonoth, M. et al. (2006). N-heterocyclic carbene catalyzed reaction of enals and 1,2-dicarbonyl compounds: stereoselective synthesis of spiro γ-butyrolactones. *Org. Lett.* 8 (3): 507–509.

401 Huisgen, R., Blaschke, H., and Brunn, E. (1966). Zur 1.3-dipolaren cycloaddition aromatischer nitriloxide an azodicarbonsaureester. *Tetrahedron Lett.* (4): 405–409.

402 White, D.A. and Baizer, M.M. (1973). Catalysis of Michael reaction by tertiary phosphines. *Tetrahedron Lett.* (37): 3597–3600.

403 (a) Gómez-Bengoa, E., Cuerva, J.M., Mateo, C., and Echavarren, A.M. (1996). Michael reaction of stabilized carbon nucleophiles catalyzed by $[RuH_2(PPh_3)_4]$. *J. Am. Chem. Soc.* 118 (36): 8553–8565. (b) Stewart, I.C., Bergman, R.G., and Toste, F.D. (2003). Phosphine-catalyzed hydration and hydroalkoxylation of activated olefins: use of a strong nucleophile to generate a strong base. *J. Am. Chem. Soc.* 125 (29): 8696–8697.

404 Trost, B.M. and Kazmaier, U. (1992). Internal redox catalyzed by triphenylphosphine. *J. Am. Chem. Soc.* 114 (20): 7933–7935.

405 Trost, B.M. and Li, C.J. (1994). Novel umpolung in C—C bond formation catalyzed by triphenylphosphine. *J. Am. Chem. Soc.* 116 (7): 3167–3168.

406 (a) Trost, B.M. and Li, C.J. (1994). Phosphine-catalyzed isomerization-addition of oxygen nucleophiles to 2-alkynoates. *J. Am. Chem. Soc.* 116 (23): 10819–10820. (b) Alvarez-Ibarra, C., Csákÿ, A.G., de la Oliva, C.G., and Gómez, C. (1999). Carboxylates as pronucleophiles in the phosphine-catalyzed γ-addition reaction. *Tetrahedron Lett.* 40 (48): 8465–8467.

407 Trost, B.M. and Dake, G.R. (1997). Nitrogen pronucleophiles in the phosphine-catalyzed γ-addition reaction. *J. Org. Chem.* 62 (17): 5670–5671.

408 Chen, X.-Y. and Ye, S. (2013). N-heterocyclic carbene-catalyzed reactions of C—C unsaturated bonds. *Org. Biomol. Chem.* 11 (46): 7991–7998.

409 Fischer, C., Smith, S.W., Powell, D.A., and Fu, G.C. (2006). Umpolung of Michael acceptors catalyzed by N-heterocyclic carbenes. *J. Am. Chem. Soc.* 128 (5): 1472–1473.

410 Virieux, D., Guillouzic, A.F., and Cristau, H.J. (2006). Phosphines catalyzed nucleophilic addition of azoles to allenes: synthesis of allylazoles and indolizines. *Tetrahedron* 62 (15): 3710–3720.

411 (a) Rauhut, M.M. and Currier, H. (1963). (American Cyanamid Co.) Dialkyl 2-methylglutarate. US Patent 307,499,91 9,122; *Chem. Abstr*. 1963, 58: 11224a; (b) Aroyan, C.E., Dermenci, A., and Miller, S.J. (2009). The Rauhut-Currier reaction: a history and its synthetic application. *Tetrahedron* 65: 4069–4084.

412 McClure, J.D. (1965). Dimerization of acrylonitrile. US Patent 3,225,083 19651221.

413 Baizer, M.M. and Anderson, J.D. (1965). Electrolytic reductive coupling. 8. Utilization and a new preparation of α-methyleneglutaronitrile. *J. Org. Chem.* 30 (5): 1357–1360.

414 (a) Wang, L.C., Luis, A.L., Agapito, F. et al. (2002). Organocatalytic Michael cycloisomerization of bis(enones): the intramolecular Rauhut-Currier reaction. *J. Am. Chem. Soc.* 124 (11): 2402–2403. (b) Frank, S.A., Mergott, D.J., and Roush, W.R. (2002). The vinylogous intramolecular Morita-Baylis-Hillman reaction: synthesis of functionalized cyclopentenes and cyclohexenes with trialkylphosphines as nucleophilic catalysts. *J. Am. Chem. Soc.* 124 (11): 2404–2405.

415 Krafft, M.E. and Haxell, T.F.N. (2005). Organo-mediated Morita-Baylis-Hillman cyclization reactions. *J. Am. Chem. Soc.* 127 (29): 10168–10169.

416 Tran, Y.S. and Kwon, O. (2007). Phosphine-catalyzed (4+2)-annulation: synthesis of cyclohexenes. *J. Am. Chem. Soc.* 129 (42): 12632–12633.

417 Alizadeh, A. and Hosseinpour, R. (2009). An efficient synthesis of tetraalkyl 2,3,4,5-thiophenetetracarboxylate derivatives. *Synthesis* (12): 1960–1962.

418 (a) Dong, X., Liang, L., Li, E., and Huang, Y. (2015). Highly enantioselective intermolecular cross Rauhut-Currier reaction catalyzed by a multifunctional Lewis base catalyst. *Angew. Chem. Int. Ed.* 54: 1621–1624. (b) see also:Aroyan, C.E. and Miller, S.J. (2007). Enantioselective Rauhut-Currier reactions promoted by protected cysteine. *J. Am. Chem. Soc.* 129: 256–257.

419 (a) Zhang, X.-N. and Shi, M. (2012). A highly nucleophilic multifunctional chiral phosphane-catalyzed asymmetric intramolecular Rauhut-Currier reaction. *Eur. J. Org. Chem.* (31): 6271–6279. (b) Li, S.-H., Liu, Y.-Y., Huang, B. et al. (2017). Phosphine-catalyzed asymmetric intermolecular cross-vinylogous Rauhut-Currier reactions of vinyl ketones with π-quinone methides. *ACS Catal.* 7 (4): 2805–2809. (c) Zhang, X.-Z., Gan, K.-J., Liu, X.-X. et al. (2017). Enantioselective synthesis of functionalized 4-arylhydrocoumarins and 4-arylhydroquinolin-2-ones via intramolecular vinylogous Rauhut-Currier reaction of π-quinone methides. *Org. Lett.* 19 (12): 3207–3210. (d) Tao, M., Zhou, W., and Zhang, J.-L. (2017). Phosphine-catalyzed asymmetric intermolecular cross Rauhut-Currier reaction of β-perfluoroalkyl-substituted enones and vinyl ketones. *Adv. Synth. Catal.* 359 (19): 3347–3353. (e) Wang, H.-G., Wang, K., Man, Y.-Q. et al. (2017). Asymmetric intermolecular Rauhut-Currier reaction for the construction of 3,3-disubstituted oxindoles with quaternary stereogenic centers. *Adv. Synth. Catal.* 359 (22): 3934–3939.

420 Mergott, D.J., Frank, S.A., and Roush, W.R. (2004). Total synthesis of (-)-Spinosyn A. *Proc. Natl. Acad. Sci. U.S.A.* 101 (33): 11955–11959.

421 Morita, K., Suzuki, Z., and Hirose, H. (1968). A tertiary phosphine-catalyzed reaction of acrylic compounds with aldehydes. *Bull. Chem. Soc. Jpn.* 41 (11): 2815–2815.

422 Baylis, A.B. and Hillman, M.E.D. (1972). Acrylic compounds. Ger. Offen 2155113 A 19720510.

423 Baylis, A.B. and Hillman, M.E.D. (1972). Acrylic compounds. *Chem. Abstr.* 77: 34174q.

424 Basavaiah, D., Rao, A.J., and Satyanarayana, T. (2003). Recent advances in the Baylis-Hillman reaction and applications. *Chem. Rev.* 103 (3): 811–891.

425 Langer, P. (2000). New strategies for the development of an asymmetric version of the Baylis-Hillman reaction. *Angew. Chem. Int. Ed.* 39 (17): 3049–3052.

426 Basavaiah, D., Kumaragurubaran, N., and Sharada, D.S. (2001). Baylis-Hillman chemistry: a novel synthesis of functionalized 1,4-pentadienes. *Tetrahedron Lett.* 42 (1): 85–87.

427 Basavaiah, D., Sharada, D.S., Kumaragurubaran, N., and Reddy, R.M. (2002). The Baylis-Hillman reaction: one-pot facile synthesis of 2,4-functionalized 1,4-pentadienes. *J. Org. Chem.* 67 (20): 7135–7137.

428 Köch, P.K. and Krische, M.J. (2004). Phosphine catalyzed α-arylation of enones and enals using hypervalent bismuth reagents: regiospecific enolate arylation via nucleophilic catalysis. *J. Am. Chem. Soc.* 126 (17): 5350–5351.

429 Robiette, R., Aggarwal, V.K., and Harvey, J.N. (2007). Mechanism of the Morita-Baylis-Hillman reaction: a computational investigation. *J. Am. Chem. Soc.* 129 (50): 15513–15525.

430 Krafft, M.E., Haxell, T.F.N., Seibert, K.A., and Abboud, K.A. (2006). Mechanistic implications in the Morita-Baylis-Hillman alkylation: isolation and characterization of an intermediate. *J. Am. Chem. Soc.* 128 (13): 4174–4175.

431 Cho, C.W. and Krische, M.J. (2004). Regio- and stereoselective construction of γ-butenolides through phosphine-catalyzed substitution of Morita-Baylis-Hillman acetates: an organocatalytic allylic alkylation. *Angew. Chem. Int. Ed.* 43 (48): 6689–6691.

432 Masson, G., Housseman, C., and Zhu, J. (2007). The enantioselective Morita-Baylis-Hillman reaction and its aza counterpart. *Angew. Chem. Int. Ed.* 46: 4614–4628.

433 Wei, Y. and Shi, M. (2013). Recent advances in organocatalytic asymmetric Morita-Baylis-Hillman/aza-Morita-Baylis-Hillman reactions. *Chem. Rev.* 113: 6659–6690.

434 (a) Gergelitsova, I., Tauchman, J., Cisarova, I., and Vesely, J. (2015). Bifunctional (thio)urea-phosphine organocatalysts derived from D-glucose and -amino acids and their application to the enantioselective Morita-Baylis-Hillman reaction. *Synlett* 26: 2690–2696. (b) Gao, Y., Xu, Q., and Shi, M. (2015). Enantioselective synthesis of polycyclic indole derivatives based on aza-Morita-Baylis-Hillman reaction. *ACS Catal.* 5: 6608–6614.

435 (a) Isenegger, P.G., Bächle, F., and Pfaltz, A. (2016). Asymmetric Morita-Baylis-Hillman reaction: catalyst development and mechanistic insights based on mass spectrometric back-reaction screening. *Chem. Eur. J.* 22 (49): 17595–17599. (b) Pellissier, H. (2017). Recent developments in the asymmetric organocatalytic Morita-Baylis-Hillman reaction. *Tetrahedron* 73 (20): 2831–2861.

436 Krishna, P.R., Kannan, V., and Reddy, P.V.N. (2004). *N*-methylprolinol catalysed asymmetric Baylis-Hillman reaction. *Adv. Synth. Catal.* 346: 603–606.

437 Liu, Y.-L., Wang, B.-L., Cao, J.-J. et al. (2010). Organocatalytic asymmetric synthesis of substituted 3-hydroxy-2-oxindoles via Morita-Baylis-Hillman reaction. *J. Am. Chem. Soc.* 132: 15176–15178.

438 Shi, M., Chen, L.H., and Teng, W.D. (2005). Asymmetric aza-Morita-Baylis-Hillman reaction of *N*-sulfonated imines with methyl vinyl ketone catalyzed by chiral phosphine Lewis bases bearing perfluoroalkanes as "pony tails". *Adv. Synth. Catal.* 347: 1781–1789.

439 (a) Shi, M., Chen, L.H., and Li, C.Q. (2005). Chiral phosphine lewis bases catalyzed asymmetric aza-Baylis-Hiliman reaction of *N*-sulfonated imines with activated olefins. *J. Am. Chem. Soc.* 127: 3790–3800. (b) Song, H.-L., Yuan, K., and Wu, X.-Y. (2011). Chiral phosphine-squaramides as enantioselective catalysts for the intramolecular Morita-Baylis-Hillman reaction. *J. Chem. Soc., Chem. Commun.* 47: 1012–1014.

440 Matsui, K., Takizawa, S., and Sasai, H. (2005). Bifunctional organocatalysts for enantioselective aza-Morita-Baylis-Hillman reaction. *J. Am. Chem. Soc.* 127: 3680–3681.

441 Wang, J., Li, H., Yu, X.H. et al. (2005). Chiral binaphthyl-derived amine-thiourea organocatalyst-promoted asymmetric Morita-Baylis-Hillman reaction. *Org. Lett.* 7: 4293–4296.

442 Takizawa, S., Rémond, E., Arteaga Arteaga, F. et al. (2013). P-chirogenic organocatalysts: application to the aza-Morita-Baylis-Hillman (aza-MBH) reaction of ketimines. *J. Chem. Soc., Chem. Commun.* 49: 8392–8394.

443 (a) Hu, F.-L. and Shi, M. (2014). The highly enantioselective catalytic aza-Morita-Baylis-Hillman reaction. *Org. Chem. Front.* 1 (5): 587–595. (b) Zhao, X., Li, T.-Z., Qian, J.-Y. et al. (2014). Enantioselective aza-Morita-Baylis-Hillman reaction between acrylates and *N*-Boc isatin ketimines: asymmetric construction of chiral 3-substituted-3-aminooxindoles. *Org. Biomol Chem.* 12 (40): 8072–8078. (c) Gao, Y., Xu, Q., and Shi, M. (2015). Enantioselective synthesis of polycyclic indole derivatives based on aza-Morita-Baylis-Hillman reaction.

ACS Catal. 5 (11): 6608–6614. (d) Takenaga, N., Adachi, S., Furusawa, A. et al. (2016). Planar chiral [2.2]paracyclophane-based phosphine-phenol catalysts: application to the aza-Morita-Baylis-Hillman reaction of *N*-sulfonated imines with various vinyl ketones. *Tetrahedron* 72 (43): 6892–9897.

444 Nair, V., Menon, R.S., Sreekanth, A.R. et al. (2006). Engaging zwitterions in carbon-carbon and carbon-nitrogen bond-forming reactions: a promising synthetic strategy. *Acc. Chem. Res.* 39 (8): 520–530.

445 Shi, Y.L. and Shi, M. (2005). DABCO-catalyzed reaction of allenic esters and ketones with salicyl *N*-tosylimines: synthesis of highly functionalized chromenes. *Org. Lett.* 7 (14): 3057–3060.

446 Grossman, R.B., Pendharkar, D.S., and Patrick, B.O. (1999). (n+1)-Annulation route to highly substituted cyclic ketones with pendant ketone, nitrile, and ester functionality. *J. Org. Chem.* 64 (19): 7178–7183.

447 Grossman, R.B., Comesse, S., Rasne, R.M. et al. (2003). Phosphoramidites are efficient, green organocatalysts for the Michael reaction. Mechanistic insights into the phosphorus-catalyzed Michael reaction of alkynones and implications for asymmetric catalysis. *J. Org. Chem.* 68 (3): 871–874.

448 Zhang, C.M. and Lu, X.Y. (1995). Phosphine-catalyzed cycloaddition of 2,3-butadienoates or 2-butynoates with electron-deficient olefins – a novel (3+2)-annulation approach to cyclopentenes. *J. Org. Chem.* 60 (9): 2906–2908.

449 Du, Y.S., Lu, X.Y., and Yu, Y.H. (2002). Highly regioselective construction of spirocycles via phosphine-catalyzed (3+2)-cycloaddition. *J. Org. Chem.* 67 (25): 8901–8905.

450 Cowen, B.J. and Miller, S.J. (2007). Enantioselective (3+2)-cycloadditions catalyzed by a protected, multifunctional phosphine-containing α-amino acid. *J. Am. Chem. Soc.* 129 (36): 10988–10989.

451 Zhu, X.F., Lan, J., and Kwon, O. (2003). An expedient phosphine-catalyzed (4+2)-annulation: synthesis of highly functionalized tetrahydropyridines. *J. Am. Chem. Soc.* 125 (16): 4716–4717.

452 Tran, Y.S., Martin, T.J., and Kwon, O. (2011). Phosphine-catalyzed (4+2)-annulations of 2-alkylallenoates and olefins: synthesis of multisubstituted cyclohexenes. *Chem. Asian J.* 6: 2100–2105.

453 Dückert, H., Khedkar, V., Waldmann, H., and Kumar, K. (2011). Lewis base catalyzed (4+2)-annulation of electron-deficient chromone-derived heterodienes and acetylenes. *Chem. Eur. J.* 17: 5130–5137.

454 Li, E., Huang, Y., Liang, L., and Xie, P. (2013). Phosphine-catalyzed (4+2)-annulation of γ-substituent allenoates: facile a to functionalized spirocyclic skeletons. *Org. Lett.* 15: 3138–3141.

455 (a) Gomez, C., Gicquel, M., Carry, J.-C. et al. (2013). Phosphine-catalyzed synthesis of 3,3-spirocyclopenteneoxindoles from γ-substituted allenoates: systematic studies and targeted applications. *J. Org. Chem.* 78: 1488–1496. (b) Ni, H., Yu, Z., Yao, W. et al. (2017). Catalyst-controlled regioselectivity in phosphine catalysis: the synthesis of spirocyclic benzofuranones via regiodivergent (3+2)-annulations of aurones and a allenoate. *Chem. Sci.* 8 (8): 5699–5704.

456 Wurz, R.P. and Fu, G.C. (2005). Catalytic asymmetric synthesis of piperidine derivatives through the (4+2)-annulation of imines with allenes. *J. Am. Chem. Soc.* 127: 12234–12235.

457 Xiao, H., Chai, Z., Wang, H.F. et al. (2011). Bifunctional *N*-acyl-aminophosphine-catalyzed asymmetric (4+2)-cycloadditions of allenoates and imines. *Chem. Eur. J.* 17: 10562–10565.

458 (a) Zhong, F., Han, X., Wang, Y., and Lu, Y. (2012). Highly enantioselective (4+2)-annulations catalyzed by amino acid-based phosphines: synthesis of functionalized cyclohexenes and 3-spirocyclohexene-2-oxindoles. *Chem. Sci.* 3: 1231–1234. (b) Xian, Y.M., Gun, H.-C., and Kwon, O. (2016). Nucleophilic chiral phosphines: powerful and versatile catalysts for asymmetric annulations. *Aldrichim. Acta* 49 (1): 1–13.

459 Jin, Z., Yang, R., Du, Y. et al. (2012). Enantioselective intramolecular formal (2+4)-annulation of acrylates and α,β-unsaturated imines catalyzed by amino acid derived phosphines. *Org. Lett.* 14: 3226–3229.

460 Shi, Z., Yu, P., Loh, T.P., and Zhong, G. (2012). Catalytic asymmetric (4+2)-annulation initiated by an aza-Rauhut-Currier reaction: facile entry to highly functionalized tetrahydropyridines. *Angew. Chem. Int. Ed.* 51: 7825–7829.

461 Zhang, X.N., Chen, G.Q., Dong, X. et al. (2013). Phosphine-catalyzed asymmetric (4+2)-annulation of vinyl ketones with oxindole-derived α,β-unsaturated imines: enantioselective syntheses of 2′,3′-Dihydro-1′

H-spiro[indoline-3,4′-pyridin]-2-ones. *Adv. Synth. Catal.* 355: 3351–3357.

462 Wang, G., Rexiti, R., Sha, F., and Wu, X.Y. (2015). Enantioselective (4+2)-cycloaddition reaction of α,β-unsaturated imine and methyl vinyl ketone catalyzed by chiral phosphine. *Tetrahedron* 71: 4255–4262.

463 Sunden, H., Ibrahem, I., Eriksson, L., and Cordova, A. (2005). Direct catalytic enantioselective aza-Diels–Alder reactions. *Angew. Chem. Int. Ed.* 44 (31): 4877–4880.

464 Kobayashi, S., Komiyama, S., and Ishitani, H. (1998). The first enantioselective aza-Diels–Alder reactions of imino dienophiles on use of a chiral zirconium catalyst. *Angew. Chem. Int. Ed.* 37 (7): 979–981.

465 Yao, S.L., Saaby, S., Hazell, R.G., and Jørgensen, K.A. (2000). Catalytic enantioselective aza-Diels–Alder reactions of imines – an approach to optically active nonproteinogenic alpha-amino acids. *Chem. Eur. J.* 6 (13): 2435–2448.

466 Itoh, J., Fuchibe, K., and Akiyama, T. (2006). Chiral Brønsted acid catalyzed enantioselective aza-Diels–Alder reaction of Brassard's diene with imines. *Angew. Chem. Int. Ed.* 45 (29): 4796–4798.

467 Shang, D., Xin, J., Liu, Y. et al. (2008). Enantioselective aza-Diels–Alder reaction of aldimines with "Danishefsky-type diene" catalyzed by chiral scandium(III)-*N*,*N*′-dioxide complexes. *J. Org. Chem.* 73 (2): 630–637.

468 Jurcik, V., Arai, K., Salter, M. et al. (2008). Niobium-catalyzed highly enantioselective aza-Diels–Alder reactions. *Adv. Synth. Catal.* 350 (5): 647–651.

469 He, M., Struble, J.R., and Bode, J.W. (2006). Highly enantioselective azadiene Diels–Alder reactions catalyzed by chiral N-heterocyclic carbenes. *J. Am. Chem. Soc.* 128 (26): 8418–8420.

470 Zhao, X., Ruhl, K.E., and Rovis, T. (2012). N-Heterocyclic carbene catalyzed asymmetric oxidative hetero-Diels–Alder Reactions with simple aliphatic aldehydes. *Angew. Chem. Int. Ed.* 51: 12330–12333.

471 Song, X., Ni, Q., Zhu, C. et al. (2015). Asymmetric N-heterocyclic carbene catalyzed annulation of 2-alkenylbenzothiazoles with α-chloro aldehydes. *Synthesis* 47: 421–428.

472 France, S., Weatherwax, A., Taggi, A.E., and Lectka, T. (2004). Advances in the catalytic, asymmetric synthesis of β-lactams. *Acc. Chem. Res.* 37 (8): 592–600.

473 Zhang, Y.R., He, L., Wu, X. et al. (2008). Chiral N-heterocyclic carbene catalyzed Staudinger reaction of ketenes with imines: highly enantioselective synthesis of *N*-boc β-lactams. *Org. Lett.* 10 (2): 277–280.

474 (a) Pham, T.Q., Pyne, S.G., Skelton, B.W., and Whitet, A.H. (2005). Synthesis of carbocyclic hydantocidins via regioselective and diastereoselective phosphine-catalyzed (3+2)-cycloadditions to 5-methylenehydantoins. *J. Org. Chem.* 70 (16): 6369–6377. (b) Yong, S.R., Williams, M.C., Pyne, S.G. et al. (2005). Synthesis of 2-azaspiro[4.4]nonan-1-ones via phosphine-catalysed (3+2)-cycloadditions. *Tetrahedron* 61 (34): 8120–8129.

475 Shao, P.L., Chen, X.Y., and Ye, S. (2010). Formal (3+2)-cycloaddition of ketenes and oxaziridines catalyzed by chiral Lewis bases: enantioselective synthesis of oxazolin-4-ones. *Angew. Chem. Int. Ed.* 49: 8412–8416.

476 Ziegler, D.T., Riesgo, L., Ikeda, T. et al. (2014). Biphenyl-derived phosphepines as chiral nucleophilic catalysts: enantioselective (4+1) annulations to form functionalized cyclopentenes. *Angew. Chem. Int. Ed.* 53: 13183–13187.

477 Xu, Z.R. and Lu, X.Y. (1998). A novel (3+2)-cycloaddition approach to nitrogen heterocycles via phosphine-catalyzed reactions of 2,3-butadienoates or 2-butynoates and dimethyl acetylenedicarboxylate with imines: a convenient synthesis of pentabromopseudilin. *J. Org. Chem.* 63 (15): 5031–5041.

478 Nair, V., Poonoth, M., Vellalath, S. et al. (2006). An N-heterocyclic carbene-catalyzed (8+3)-annulation of tropone and enals via homoenolate. *J. Org. Chem.* 71 (23): 8964–8965.

479 Scott, H., Miller, G.A., and Labes, M.M. (1963). A radical cation initiated polymerization of *N*-vinylcarbazole. *Tetrahedron Lett.* 1073–1078.

480 Ellinger, L.P. (1964). The polymerization of vinylcarbazole by electron acceptors. 1. *Polymer* 5: 559–578.

481 Bawn, C.E.H., Ledwith, A., and Yang, S.L. (1965). Reaction between *N*-vinyl carbazole and ferric salts. *Chem. Ind.* 769–770.

482 Bell, F.A., Crellin, R.A., Fujii, H., and Ledwith, A. (1969). Cation radicals – metal-catalysed cyclodimerisation of aromatic enamines. *J. Chem. Soc., Chem. Commun.* 251–252.

483 Bell, F.A., Ledwith, A., and Sherring, D.C. (1969). Cation radicals – tris-(*para*-bromophenyl)amminium perchlorate and hexachloroantimonate. *J. Chem. Soc. C* 2719–2720.

484 Crellin, R.A., Lambert, M.C., and Ledwith, A. (1970). Photochemical (2+2)-cycloaddition via a cation radical chain reaction. *J. Chem. Soc., Chem. Commun.* 682–683.

485 Beresford, P., Lambert, M.C., and Ledwith, A. (1970). Cation radicals – ring opening of a cyclobutane by electron transfer. *J. Chem. Soc. C* 2508–2510.

486 Schenck, G.O., Mannsfeld, S.P., Krauch, C.H., and Schomburg, G. (1964). Strahlenchemische Cyclodimerisation von 1.3-Cyclohexadien + Neuartige Bildungsweisen von Diels–Alder–Addukten – Ein Beitrag zum Reaktionsmechanismus der Diels–Alder–Reaktion. *Z. Naturforschung B.* B19: 18–22.

487 Schutte, R. and Freeman, G.R. (1969). Radiation-induced dimerization of 1,3-cyclohexadiene. Solvent effects and formation of Diels–Alder dimers by a cationic chain mechanism. *J. Am. Chem. Soc.* 91: 3715–3720.

488 Neunteuf, R.A. and Arnold, D.R. (1973). Radical ions in photochemistry. 1. 1,1-Diphenylethylene cation radical. *J. Am. Chem. Soc.* 95 (12): 4080–4081.

489 Koike, T.A. (2014). Visible-light radical reaction designed by Ru- and Ir-based photoredox catalysis. *Inorg. Chem. Front.* 1: 562–576.

490 Ischay, M.A., Ament, M.S., and Yoon, T.P. (2012). Crossed intermolecular (2+2)-cycloaddition of styrenes by visible light photocatalysis. *Chem. Sci.* 3: 2807–2811.

491 (a) Yoon, T.P. (2013). Visible light photocatalysis: the development of photocatalytic radical ion cycloadditions. *ACS Catal.* 3 (5): 895–902. (b) Yang, Y., Liu, Q., Zhang, L. et al. (2017). Mechanistic investigation on oxygen-mediated photoredox Diels–Alder reactions with chromium catalysts. *Organometallics* 36: 687–698.

492 Mizuno, K., Ueda, H., and Otsuji, Y. (1981). Photoinduced (2+2)-cycloaddition between aryl vinyl ether and alkyl vinyl ether via electron transfer. *Chem. Lett.* (9): 1237–1240.

493 Bellville, D.J., Wirth, D.D., and Bauld, N.L. (1981). The cation radical catalyzed Diels–Alder reaction. *J. Am. Chem. Soc.* 103 (3): 718–720.

494 Pabon, R.A., Bellville, D.J., and Bauld, N.L. (1983). Cation radical Diels–Alder reactions of electron-rich dienophiles. *J. Am. Chem. Soc.* 105 (15): 5158–5159.

495 Reynolds, D.W. and Bauld, N.L. (1986). The diene component in the cation radical Diels–Alder. *Tetrahedron* 42 (22): 6189–6194.

496 Yueh, W. and Bauld, N.L. (1995). Criteria for the cation radical vs electrophilic mechanistic distinction – the aminium salt-catalyzed Diels–Alder reaction. *J. Chem. Soc., Perkin Trans. 2* (5): 871–873.

497 Saettel, N.J., Wiest, O., Singleton, D.A., and Meyer, M.P. (2002). Isotope effects and the mechanism of an electron-transfer-catalyzed Diels–Alder reaction. *J. Am. Chem. Soc.* 124 (38): 11552–11559.

498 Kim, T., Pye, R.J., and Bauld, N.L. (1990). Competing hole catalyzed Diels–Alder and cyclobutanation vinylcyclobutane rearrangement paths – a mechanistic dichotomy. *J. Am. Chem. Soc.* 112: 6285–6290.

499 Shubina, T.E. and Clark, T. (2010). Catalysis of the quadricyclane to norbornadiene rearrangement by $SnCl_2$ and $CuSO_4$. *Z. Naturforsch. Teil B* 65: 347–356.

500 Nakamoto, M., Inagaki, Y., Ochiai, T. et al. (2011). Cyclobutadiene to tetrahedrane: valence isomerization induced by one-electron oxidation. *Heteroat. Chem.* 22 (3–4): 412–416.

501 Bauld, N.L., Stufflebeme, G.W., and Lorenz, K.T. (1989). Cation radical pericyclic reactions – cyclopropanation. *J. Phys. Org. Chem.* 2 (8): 585–601.

502 Bauld, N.L. and Mirafzal, G.A. (1991). Hole-catalyzed epoxidation. *J. Am. Chem. Soc.* 112: 3613–3614.

503 Lorenz, K.T. and Bauld, N.L. (1985). A photoassisted hole-catalyzed Cope reaction on zeolite. *J. Catal.* 95: 613–616.

504 Arata, M., Miura, T., and Chiba, K. (2007). Electrocatalytic formal (2+2)-cycloaddition reactions between anodically activated enyloxy benzene and alkenes. *Org. Lett.* 9 (21): 4347–4350.

505 Tomioka, H., Kobayashi, D., Hashimoto, A., and Murata, S. (1989). (3+2)-Cycloaddition of cyclopropane with vinyl ether via photoinduced electron transfer. *Tetrahedron Lett.* 30 (35): 4685–4688.

506 Ohara, H., Itoh, T., Nakamura, M., and Nakamura, E. (2001). (2+2)-Cycloaddition reaction of styrene derivatives using an Fe(III) salt catalyst. *Chem. Lett.* (7): 624–625.

507 Stork, G. and Birnbaum, G. (1961). The alkylation of enamines from α,β-unsaturated ketones. *Tetrahedron Lett.* 313–316.

508 Kempf, B., Hampel, N., Ofial, A.R., and Mayr, H. (2003). Structure – nucleophilicity relationships for enamines. *Chem. Eur. J.* 9: 2209–2218.

509 Maji, B., Lakhdar, S., and Mayr, H. (2012). Nucleophilicity parameters of enamides and

their implications for organocatalytic transformations. *Chem. Eur. J.* 18: 5732–5740.

510 Hodgson, D.M. and Charlton, A. (2014). Methods for direct generation of α-alkyl-substituted aldehydes. *Tetrahedron* 70: 2207–2236.

511 Movassaghi, M. and Jacobsen, E.N. (2002). The simplest "enzyme". *Science* 298: 1904–1905.

512 (a) Notz, W., Tanaka, F., and Barbas, C.F. (2004). Enamine-based organocatalysis with proline and diamines: the development of direct catalytic asymmetric Aldol, Mannich, Michael, and Diels–Alder reactions. *Acc. Chem. Res.* 37 (8): 580–591. (b) Kano, T. and Maruoka, K. (2008). Design of chiral bifunctional secondary amine catalysts for asymmetric enamine catalysis. *J. Chem. Soc., Chem. Commun.* (43): 5465–5473. (c) Kano, T. and Maruoka, K. (2013). Unique properties of chiral biaryl-based secondary amine catalysts for asymmetric enamine catalysis. *Chem. Sci.* 4 (3): 907–915. (d) Chaudan, P., Mahajan, S., and Enders, D. (2017). Achieving molecular complexity via stereoselective multiple domino reactions promoted by a secondary amine organocatalyst. *Acc. Chem. Res.* 50 (11): 2809–2821.

513 Mayr, H., Lakhdar, S., Maji, B., and Ofial, A.R. (2012). A quantitative approach to nucleophilic organocatalysis. *Beilstein J. Org. Chem.* 8: 1458–1478.

514 (a) Beeson, T.D., Mastracchio, A., Hong, J.B. et al. (2007). Enantioselective organocatalysis using SOMO activation. *Science* 316 (5824): 582–585. (b) Meciarova, M., Tisovsky, P., and Sebesta, R. (2016). Enantioselective organocatalysis using SOMO activation. *New J. Chem.* 40 (6): 4855–4864.

515 Jang, H.-Y., Hong, J.B., and MacMillan, D.W.C. (2007). Enantioselective organocatalytic singly occupied molecular orbital activation: the enantioselective α-enolation of aldehydes. *J. Am. Chem. Soc.* 129: 7004–7005.

516 Conrad, J.C., Kong, J., Laforteza, B.N., and MacMillan, D.W.C. (2009). Enantioselective α-arylation of aldehydes via organo-SOMO catalysis. An *ortho*-selective arylation reaction based on an open-shell pathway. *J. Am. Chem. Soc.* 131: 11640–11641.

517 Kim, H. and MacMillan, D.W.C. (2008). Enantioselective organo-SOMO catalysis: the α-vinylation of aldehydes. *J. Am. Chem. Soc.* 130: 398–399.

518 Wilson, J.E., Casarez, A.D., and MacMillan, D.W.C. (2009). Enantioselective aldehyde α-nitroalkylation via oxidative organocatalysis. *J. Am. Chem. Soc.* 131: 11332–11334.

519 Jui, N.T., Garber, J.A., Gadini Finelli, F., and MacMillan, D.W.C. (2012). Enantioselective organo-SOMO cycloadditions: a catalytic approach to complex pyrrolidines from olefins and aldehydes. *J. Am. Chem. Soc.* 134: 11400–11403.

520 Jui, N.T., Lee, E.C., and MacMillan, D.W.C. (2010). Enantioselective organo-SOMO cascade cycloadditions: a rapid approach to molecular complexity from simple aldehydes and olefins. *J. Am. Chem. Soc.* 132: 10015–10017.

521 Moeller, K.D. (2000). Synthetic applications of anodic electrochemistry. *Tetrahedron* 56 (49): 9527–9554.

522 (a) Little, D.R. and Schwäbe, M.K. (1997). Electrochemical cyclization at the cathode. In: *Electrochemistry VI: Electroorganic Synthesis: Bond Formation at Anode and Cathode*, Topics in Current Chemistry, vol. 185 (ed. E. Steckhan), 1–48. Berlin, Heidelberg: Springer-Verlag. (b) Utley, J.H.P., Little, R.D., and Nielsen, M.F. (2015). Reductive coupling. In: *Organic Electrochemistry*, Chapter 17, 5 revised and expanded (ed. O. Hammerich and B. Speiser), 621–704. Boca Raton, FL: CRC Press – Taylor & Francis Group.

523 Sperry, J.B., Whitehead, C.R., Ghiviriga, I. et al. (2004). Electrooxidative coupling of furans and silyl enol ethers: application to the synthesis of annulated furans. *J. Org. Chem.* 69 (11): 3726–3734.

524 Mihelcic, J. and Moeller, K.D. (2004). Oxidative cyclizations: the asymmetric synthesis of (-)-allialcol A. *J. Am. Chem. Soc.* 126 (29): 9106–9111.

525 Janssen, R.G., Motevalli, M., and Utley, J.H.P. (1998). Electroreductive coupling of vinylpyridines and vinylquinolines: radical anion-substrate cycloaddition? *J. Chem. Soc., Chem. Commun.* (5): 539–540.

526 Delaunay, J., Mabon, G., Orliac, A., and Simonet, J. (1990). The cyclodimerization of aryl vinyl sulfones – a facile and specific reaction when activated by cathodic electron-transfer. *Tetrahedron Lett.* 31 (5): 667–668.

527 Roh, Y., Jang, H.Y., Lynch, V. et al. (2002). Anion radical chain cycloaddition of tethered enones: intramolecular cyclobutanation and Diels–Alder cycloaddition. *Org. Lett.* 4 (4): 611–613.

528 Yang, J.K., Felton, G.A.N., Bauld, N.L., and Krische, M.J. (2004). Chemically induced anion radical cycloadditions: intramolecular cyclobutanation of bis(enones) via homogeneous electron transfer. *J. Am. Chem. Soc.* 126 (6): 1634–1635.

529 Yang, J.K., Cauble, D.F., Berro, A.J. et al. (2004). Anion radical (2+2)-cycloaddition as a mechanistic probe: stoichiometry- and concentration-dependent partitioning of electron-transfer and alkylation pathways in the reaction of the Gilman reagent $Me_2CuLi{\cdot}LiI$ with bis(enones). *J. Org. Chem.* 69 (23): 7979–7984.

530 John, M., Auel, C., Behrens, C. et al. (2000). The relation between ion pair structures and reactivities of lithium cuprates. *Chem. Eur. J.* 6 (16): 3060–3068.

531 Kauffman, G.B. (1988). The Brønsted–Lowry acid-base concept. *J. Chem. Educ.* 65 (1): 28–31.

532 Santelli, M. and Pons, J.M. (1996). *Lewis Acids and Selectivity in Organic Synthesis*, 1–352. Boca Raton, FL: CRC Press Inc.

533 Finston, H.L. and Rychtman, A.C. (1982). *New View of Current Acid-Base Theories*, 1–216. New York: Wiley-Interscience Publisher.

534 Akiyama, T. and Mori, K. (2015). Stronger Brønsted acids: recent progress. *Chem. Rev.* 115: 9277–9306.

535 Yang, J.Y. and Chen, F.X. (2010). Highly efficient syntheses of β-cyanoketones via conjugate addition of Me_3SiCN to aromatic enones. *Chin. J. Chem.* 28: 981–987.

536 Turlington, M. and Pu, L. (2012). Asymmetric alkyne addition to aldehydes catalyzed by BINOL and its derivatives. *Synlett* 649–684.

537 Kumar, R., Kawasaki, H., and Harada, T. (2013). Enantioselective alkylation of aldehydes using functionalized alkylboron reagents catalyzed by a chiral Ti complex. *Org. Lett.* 15: 4198–4201.

538 Lewis, G.N. (1923). *Valence and the Structure of Atoms and Molecules*, American Chemical Monograph Series, 1–172. New York: The Chemical Catalog Co., Inc.

539 Welton, T. (1999). Room-temperature ionic liquids. Solvents for synthesis and catalysis. *Chem. Rev.* 99: 2071–2083.

540 Jodry, J.J. and Mikami, K. (2005). Ionic liquids. In: *Green Reaction Media in Organic Synthesis* (ed. K.E. Mikami), 9–58. Oxford: Blackwell Publishing.

541 (a) Hallett, J.P. and Welton, T. (2011). Room-temperature ionic liquids: solvents for synthesis and catalysis. 2. *Chem. Rev.* 111: 3508–3576. (b) Ratti, R. (2014). Ionic liquids: synthesis and applications in catalysis. *Adv. Chem.* 729842, pp. 1–16. (c) Claus, J., Sommer, F.O., and Kragl, U. (2018). Ionic liquids in biotechnology and beyond. *Solid State Ionics* 314: 119–128.

542 Kuchenbuch, A. and Giernoth, R. (2015). Ionic liquids beyond simple solvents: glimpses at the state of the art in organic chemistry. *ChemistryOpen* 4: 677–681.

543 (a) Olah, G.A., Surya Prakash, G.K., Sommer, J., and Molnar, A. (2009). *Superacids*, 2e, 1–850. Hoboken, NJ: Wiley. (b) Wiesner, A., Gries, T.W., Steinhauer, S. et al. (2017). Superacids based on pentafluoroorthotellurate derivatives of aluminum. *Angew. Chem. Int. Ed.* 56: 8263–8266.

544 (a) Pearson, R.G. (1963). Hard and soft acids and bases. *J. Am. Chem.Soc.* 85 (22): 3533–3539. (b) Pearson, R.G. (2005). Chemical hardness and density functional theory. *J. Chem. Sci.* 117 (5): 369–377. (c) Pearson, R.G. and Chattaraj, P.K. (2008). The hard-soft acid-base principle. *Chemtracts* 21 (1): 1–7.

545 Reed, C.A. (2013). Myths about the proton. The nature of H+ in condensed media. *Acc. Chem. Res.* 46: 2567–2575.

546 (a) Drago, R.S. (1973). Quantitative evaluation and prediction of donor-acceptor interactions. *Struct. Bond.* 15: 73–139. (b) Drago, R.S. (1980). The coordination model for non-aqueous solvents. *Pure Appl. Chem.* 52 (10): 2261–2274. (c) Drago, R.S. (1980). The interpretation of reactivity in chemical and biological systems with the E-Model and C-Model. *Coord. Chem. Rev.* 33 (3): 251–277.

547 Sivaev, I.B. and Bregadze, V.I. (2014). Lewis acidity of boron compounds. *Coord. Chem. Rev.* 270–271: 75–88.

548 Stephan, D.W. (2017). A tale of two elements: the Lewis acidity/basicity umpolung of boron and phosphorus. *Angew. Chem. Int. Ed.* 56 (22): 5984–5992.

549 (a) Cotton, R.A. and Wilkinson, G. (1980). *Advanced Inorganic Chemistry: A Comprehensive Text*, 4e, 1–1396. New York: Wiley. (b) Cotton, R.A., Wilkinson, G., Murillo, C.A. et al. (1999). *Advanced Inorganic Chemistry*, 6e, 1–1376. New York: Wiley.

550 Jonas, V., Frenking, G., and Reetz, M.T. (1994). Comparative theoretical study of Lewis acid-base complexes of BH_3, BF_3, BCl_3, $AlCl_3$, and SO_2. *J. Am. Chem. Soc.* 116 (19): 8741–8753.

551 (a) Rowsell, B.D., Gillespie, R.J., and Heard, G.L. (1999). Ligand close-packing and the Lewis acidity of BF_3 and BCl_3. *Inorg. Chem.* 38 (21):

4659–4662. (b) Bessac, F. and Frenking, G. (2003). Why is BCl_3 a stronger Lewis acid with respect to strong bases than BF_3? *Inorg. Chem.* 42 (24): 7990–7994.

552 (a) van der Veken, B.J. and Sluyts, E.J. (1997). Reversed Lewis acidity of mixed boron halides: an infrared study of the van der Waals complexes of BF_xCl_y with CH_3F in cryosolution. *J. Am. Chem. Soc.* 119 (47): 11516–11522. (b) Rowsell, B.D., Gillespie, F.J., and Heard, G.L. (1999). Ligand close-packing and the Lewis acidity of BF_3 and BCl_3. *Inorg. Chem.* 38 (21): 4659–4662.

553 Corey, E.J. and Lee, T.W. (2001). The formyl C—H···O hydrogen bond as a critical factor in enantioselective Lewis-acid catalyzed reactions of aldehydes. *J. Chem. Soc., Chem. Commun.* (15): 1321–1329.

554 Schäfer, W. (1966). 1-Stage synthesis of hexamethyl-Dewar-benzene from 2-butyne. *Angew. Chem. Int. Ed. Engl.* 5 (7): 669–669.

555 Schäfer, W. and Hellmann, H. (1967). Hexamethyl(Dewar benzene) (hexamethylbicyclo[2.2.0]hexa-2,5-diene). *Angew. Chem. Int. Ed. Engl.* 6 (6): 518–525.

556 Jasien, P.G. (1992). Lewis acid-base complexes of $AlCl_3$. *J. Phys. Chem.* 96 (23): 9273–9278.

557 (a) Olah, G.A., Krishnamurti, R., and Surya Prakash, G.K. (1991). Friedel-crafts alkylations. In: *Comprehensive Organic Synthesis*, 1e, vol. 3 (ed. B.M. Trost and I. Fleming), 293–334. Oxford: Elsevier Science Ltd. (b) Rueping, M. and Nachtheim, B.J. (2010). A review of new developments in the Friedel–Crafts alkylation – from green chemistry to asymmetric catalysis. *Beilstein J. Org. Chem.* 6 (6): 1–24. https://doi.org/10.3762/bjoc.6.6. (c) El-Hiti, G.A., Smith, K., and Hegazy, A.S. (2015). Catalytic, green and regioselective Friedel–Crafts acylation of simple aromatics and heterocycles over zeolites. *Curr. Org. Chem.* 19 (7): 585–598.

558 Yates, P. and Eaton, P. (1960). Acceleration of the Diels–Alder reaction by aluminum chloride. *J. Am. Chem. Soc.* 82 (16): 4436–4437.

559 (a) Robinson, R. and Fray, G.I. (1962). (Shell Oil Co.) Diels-Alder adducts. US Patents 3,067,244. (b) Inukai, T. and Kasai, M. (1965). Diels–Alder reactions of acrylic acid derivatives catalyzed by aluminum chloride. *J. Org. Chem.* 30 (10): 3567–3569. (c) Inukai, T. and Kojima, T. (1966). Catalytic actions of aluminum chloride on the isoprene-methyl acrylate Diels–Alder reaction. *J. Org. Chem.* 31 (4): 1121–1123. (d) Inukai, T. and Kojima, T. (1966). Aluminum chloride catalyzed diene condensation. II. Stronger adherence to the Alder *endo* rule. *J. Org. Chem.* 31 (6): 2032–2033. (e) McCulloch, A.W. and McInnes, A.G. (1967). Use of Lewis acid accelerated Diels–Alder reactions in phenol synthesis. *Chem. Can.* 19 (9): 51–77.

560 Inukai, T. and Kojima, T. (1971). Aluminum chloride catalyzed diene condensation. 6. Partial rate factors of 2-phenylbutadienes, 2-chlorobutadienes 2-trifluoromethylbutadienes, and 2-cyanobutadienes in reactions with methyl acrylate – differential Hammett correlation. *J. Org. Chem.* 36 (7): 924–928.

561 Chen, Z.Y. and Ortuno, R.M. (1994). Lewis-acid catalyzed Diels–Alder cycloadditions of chiral butenolides to cyclopentadiene – *Endo*-stereoselectivity *exo*-stereoselectivity. *Tetrahedron: Asymmetry* 5 (3): 371–376.

562 Maruoka, K., Akakura, M., Saito, S. et al. (1994). Asymmetric Diels–Alder reaction of unsymmetrical maleates – a chemical access to chiral, unsymmetrical *cis*-cyclohexene-1,2-dicarboxylates. *J. Am. Chem. Soc.* 116 (14): 6153–6158.

563 Corma, A. and García, H. (2003). Lewis acids: from conventional homogeneous to green homogeneous and heterogeneous catalysis. *Chem. Rev.* 103 (11): 4307–4365.

564 Fu, G.C. (2004). From the design of a chiral lewis acid catalyst to metal-catalyzed coupling reactions. *J. Org. Chem.* 69 (10): 3245–3249.

565 Zeise, W.C. (1831). Von der Wirkung zwischen Platinchlorid und Alkohol, und von den dabei enstehenden neuen Substanzen. *Ann. Phys. Chem. (Leipzig).* 21: 497.

566 Seyferth, D. (2001). $[(C_2H_4)PtCl_3](-)$, the anion of Zeise's salt, $K[(C_2H_4)PtCl_3]$center dot H_2O. *Organometallics* 20: 2–6.

567 Hashmi, A.S.K., Salathé, R., Frost, T.M. et al. (2005). Homogeneous catalysis by gold: the current status of C,H activation. *Appl. Catal., A* 291 (1–2): 238–246.

568 LaLonde, R., Wang, Z., Mba, M. et al. (2010). Gold(I)-catalyzed enantioselective synthesis of pyrazolidines, isoxazolidines, and tetrahydrooxazines. *Angew. Chem. Int. Ed.* 49: 598–601.

569 LaLonde, R.L., Brenzovich, W.E., Benitez, D. et al. (2010). Alkylgold complexes by the intramolecular aminoauration of unactivated alkenes. *Chem. Sci.* 1: 226–233.

570 Bandini, M. (2011). Gold-catalyzed decorations of arenes and heteroarenes with C—C multiple bonds. *Chem. Soc. Rev.* 40: 1358–1367.

571 Rudolph, M. and Hashmi, A.S.K. (2012). Gold catalysis in total synthesis – an update. *Chem. Soc. Rev.* 41: 2448–2462.

572 Fürstner, A. (2014). From understanding to prediction: gold- and platinum-based π-acid catalysis for target oriented synthesis. *Acc. Chem. Res.* 47: 925–938.

573 Obradors, C. and Echavarren, A.M. (2014). Gold-catalyzed rearrangements and beyond. *Acc. Chem. Res.* 47: 902–912.

574 Wang, Y.M., Lackner, A.D., and Toste, F.D. (2014). Development of catalysts and ligands for enantioselective gold catalysis. *Acc. Chem. Res.* 47: 889–901.

575 Yang, W. and Hashmi, A.S.K. (2014). Mechanistic insights into the gold chemistry of allenes. *Chem. Soc. Rev.* 43: 2941–2955.

576 Deng, H.T. and Kebarle, P. (1998). Bond energies of copper ion-ligand L complexes CuL_2^+ determined in the gas phase by ion-ligand exchange equilibria measurements. *J. Am. Chem. Soc.* 120 (12): 2925–2931.

577 Deng, H.T. and Kebarle, P. (1998). Binding energies of silver ion ligand, L, complexes AgL_2^+ determined from ligand-exchange equilibria in the gas phase. *J. Phys. Chem. A* 102 (3): 571–579.

578 Pearson, R.G. (1988). Absolute electronegativity and hardness – application to inorganic chemistry. *Inorg. Chem.* 27 (4): 734–740.

579 House, H.O. (1972). *Modern Synthetic Reactions*, 2e, 1–494. Menlo Park, CA: W. A. Benjamin Inc.

580 Wade, L.G. (2005). *Organic Chemistry*, 6e, 1056–1066. Upper Saddle River, NJ: Prentice Hall.

581 Carreira, E.M., Fettes, A., and Marti, C. (2006). Catalytic enantioselective aldol addition reactions. *Org. React.* 67: 1–216.

582 (a) List, B., Lerner, R.A., and Barbas, C.F.I. (2000). Proline-catalyzed direct asymmetric aldol reactions. *J. Am. Chem. Soc.* 122 (10): 2395–2396. (b) List, B. (2001). Asymmetric aminocatalysis. *Synlett* (11): 1675–1686. (c) List, B., Pojarliev, P., and Castello, C. (2001). Proline-catalyzed asymmetric aldol reactions between ketones and α-unsubstituted aldehydes. *Org. Lett.* 3 (4): 573–575.

583 Mokgweetsi, P., Botlhole, L., and Singh, G.S. (2016). Recent progress in asymmetric aldol reactions under solvent-free conditions or in aqueous medium. *Curr. Green Chem.* 3 (4): 318–327.

584 Liu, J. and Wang, L. (2017). Recent advances in asymmetric reactions catalyzed by proline and its derivatives. *Synthesis* (5): 49, 960–972.

585 Robina, I. and Vogel, P. (2007). Synthesis of monosaccharides and analogues – part one: applications of chemo-enzymatic methods and organocatalysis. *Chim. Oggi* 25 (4): 65–67.

586 Hoang, L., Bahmanyar, S., Houk, K.N., and List, B. (2003). Kinetic and stereochemical evidence for the involvement of only one proline molecule in the transition states of proline-catalyzed intra- and intermolecular aldol reactions. *J. Am. Chem. Soc.* 125 (1): 16–17.

587 Allemann, C., Gordillo, R., Clemente, F.R. et al. (2004). Theory of asymmetric organocatalysis of aldol and related reactions: rationalizations and predictions. *Acc. Chem. Res.* 37 (8): 558–569.

588 (a) List, B. (2004). Amine-catalyzed aldol reactions. In: *Modern Aldol Reaction*, Chapter 4, vol. 1 (ed. R. Mahrwald), 161–200. Weinheim: Wiley-VCH. (b) Yiniemela-Sipari, S.M., Pilsola, A., and Pihko, P.M. (2012). Enamine catalysis of intermolecular aldol reactions. In: *Science of Synthesis, Asymmetric Organocatalysis*, vol. 1 (ed. B. List and K. Maruoka), 35–72. Stuttgart: Georg Thieme Verlag. (c) Zhang, L. and Luo, S.-H. (2012). Bio-inspired chiral primary amine catalysis. *Synlett* 23 (11): 1575–1589.

589 Frostick, F.C. and Hauser, C.R. (1949). Condensations of esters by diisopropylaminomagnesium bromide and certain related reagents. *J. Am. Chem. Soc.* 71 (4): 1350–1352.

590 Hamell, M. and Levine, R. (1950). Condensations effected by the alkali amides. IV. The reactions of esters with lithium amide and certain substituted lithium amides. *J. Org. Chem.* 15: 162–168.

591 Rathke, M.W. (1970). Preparation of lithio ethyl acetate – a simple procedure for conversion of aldehydes and ketones to β-hydeoxy esters. *J. Am. Chem. Soc.* 92 (10): 3222–3223.

592 Rathke, M.W. and Lindert, A. (1971). Reaction of lithium *N*-isopropylcyclohexylamide with esters – method for formation and alkylation of ester enolates. *J. Am. Chem. Soc.* 93 (9): 2318–2320.

593 Olofson, R.A. and Doughert, C.M. (1973). Useful synthesis of arylcyclopropanes. *J. Am. Chem. Soc.* 95 (2): 581–582.

594 Olofson, R.A. and Doughert, C.M. (1973). Lithium 2,2,6,6-tetramethylpiperidide and related, strong, proton-specific bases – evaluation in synthesis. *J. Am. Chem. Soc.* 95 (2): 582–584.

595 Braun, M. (2004). Fundamentals and transition state models: aldol addition of group 1 and 2 enolates. In: *Modern Aldol Reactions*, Chapter 1, vol. 1 (ed. R. Mahrwald), 1–62. Weinheim: Wiley-VCH.

596 Mukaiyama, T. and Inomata, K. (1971). Reaction of thioboronite – convenient method for preparation of β-hydroxyalkanethioates. *Bull. Chem. Soc. Jpn.* 44 (11): 3215–3215.

597 Mukaiyama, T., Inomata, K., and Muraki, M. (1973). Vinyloxyboranes as synthetic intermediates. *J. Am. Chem. Soc.* 95 (3): 967–968.

598 Inomata, K., Muraki, M., and Mukaiyama, T. (1973). Reactions of vinyloxyboranes – convenient method for preparation of β-hydroxy thiolesters, esters and ketones. *Bull. Chem. Soc. Jpn.* 46 (6): 1807–1810.

599 Mukaiyama, T., Narasaka, K., and Banno, K. (1973). New aldol type reaction. *Chem. Lett.* (9): 1011–1014.

600 Mukaiyama, T., Banno, K., and Narasaka, K. (1974). New cross-aldol reactions – reactions of silyl enol ethers with carbonyl compounds activated by Ti tetrachloride. *J. Am. Chem. Soc.* 96 (24): 7503–7509.

601 Mukaiyama, T. (2004). Explorations into new reaction chemistry. *Angew. Chem. Int. Ed.* 43 (42): 5590–5614.

602 Mukaiyama, T. and Matsuo, J.-I. (2004). Boron and silicon enolates in crossed aldol reaction. In: *Modern Aldol Reactions*, Chapter 3, vol. 1 (ed. R. Mahrwald), 127–160. Weinheim: Wiley-VCH.

603 Mukaiyama, T. (1982). The directed aldol reaction. *Org. React.* 28: 203–331.

604 Iwasawa, N. and Mukaiyama, T. (1982). Enantioselective cross-aldol reaction via divalent tin enolate. *Chem. Lett.* (9): 1441–1444.

605 Shiima, I. (2004). Tin-promoted aldol reactions and their application to the total syntheses of natural products. In: *Modern Aldol Reactions*, Chapter 4, vol. 2 (ed. R. Mahrwald), 105–166. Weinheim: Wiley-VCH.

606 Verhé, R., De Kimpe, N., De Buyck, L. et al. (1980). Aldol condensation of α-chloroketones in Ti tetrachloride-pyridine. *Bull. Soc. Chim. Belg.* 89 (7): 563–564.

607 Ghosh, A.K. and Shevlin, M. (2004). The development of titanium-based enolate aldol reactions. In: *Modern Aldol Reactions*, Chapter 2, vol. 1 (ed. R. Mahrwald), 63–126. Weinheim: Wiley-VCH.

608 Cozzi, P.G., Floriani, C., Chiesi-Villa, A., and Rizzoli, C. (1994). Mechanism of the Mukaiyama aldol reaction – the 1st solid-state characterization of a trichlorotitanium aldolate. *Organometallics* 13 (5): 2131–2134.

609 Myers, A.G. and Widdowson, K.L. (1990). A Silicon-directed aldol condensation – synthesis of enantiomerically pure *anti*-aldols. *J. Am. Chem. Soc.* 112 (26): 9672–9674.

610 Myers, A.G., Kephart, S.E., and Chen, H. (1992). Silicon-directed aldol reactions – rate acceleration by small rings. *J. Am. Chem. Soc.* 114 (20): 7922–7923.

611 Denmark, S.E., Griedel, B.D., Coe, D.M., and Schnute, M.E. (1994). Chemistry of enoxysilacyclobutanes – highly selective uncatalyzed aldol additions. *J. Am. Chem. Soc.* 116 (16): 7026–7043.

612 Denmark, S.E. and Griedel, B.D. (1994). Chemistry of enoxysilacyclobutanes. 3. Uncatalyzed, *syn*-selective, asymmetric aldol additions. *J. Org. Chem.* 59 (18): 5136–5138.

613 Miura, K., Sato, H., Tamaki, K. et al. (1998). Uncatalyzed aldol reaction using a dimethylsilyl enolate and α-dimethylsilyl ester in *N*,*N*-dimethylformamide. *Tetrahedron Lett.* 39 (17): 2585–2588.

614 (a) Wong, C.T. and Wong, M.W. (2005). Facile uncatalyzed Mukaiyama aldol reactions: an ab initio study of the effects of substituents. *J. Org. Chem.* 70 (1): 124–131. (b) see also Hadj, M.S., Trabelsi, M., and Champagne, B. (2016). Unraveling the concerted reaction mechanism of the non-catalyzed Mukaiyama reaction between C,O,O-tris(trimethylsilyl)ketene acetal and aldehydes using density functional theory. *J. Phys. Chem. A* 120 (28): 5649–5657.

615 (a) Nakamura, E., Shimizu, M., Kuwajima, I. et al. (1983). Fluoride-ion catalyzed aldol reaction between enol silyl ethers and carbonyl compounds. *J. Org. Chem.* 48 (7): 932–945. (b) Maheswaran, H., Joseph, P.J.A., Prasanth, K.L. et al. (2010). Investigations of enantioreversal in both direct and directed enantioselective aldol reactions catalyzed by $CuCl_2$[(-)-sparteine] and $NiCl_2$[(-)-sparteine] complexes. *Tetrahedron: Asymmetry* 21 (17): 2158–2166.

616 (a) Mukaiyama, T., Kobayashi, S., and Murakami, M. (1985). An efficient method for the preparation of *threo* cross-aldols from silyl enol ethers and aldehydes using trityl perchlorate as a catalyst. *Chem. Lett.* 4: 447–450. (b) Chen, C.-T., Chao, S.-D., Yen, K.-C. et al.

(1997). Chiral triarylcarbenium ions in asymmetric Mukaiyama aldol additions. *J. Am. Chem. Soc.* 119 (46): 11341–11342.
617 Gong, L. and Streitwieser, A. Jr., (1990). Organolanthanide catalysis of a Mukaiyama addition-reaction. *J. Org. Chem.* 55 (26): 6235–6236.
618 Hong, Y.P., Norris, D.J., and Collins, S. (1993). Catalysis of the Mukaiyama aldol reaction using cationic zirconocene complexes. *J. Org. Chem.* 58 (13): 3591–3594.
619 Denmark, S.E. and Chen, C.T. (1994). Triarylcarbenium ions as catalysts in the Mukaiyama aldol addition – a mechanistic investigation. *Tetrahedron Lett.* 35 (25): 4327–4330.
620 Lin, S.Q., Bondar, G.V., Levy, C.J., and Collins, S. (1998). Mukaiyama aldol reactions catalyzed by zirconocene bis(triflate) complexes: stereochemistry and mechanisms for C—C bond formation. *J. Org. Chem.* 63 (6): 1885–1892.
621 Ohkouchi, M., Masui, D., Yamaguchi, M., and Yamagishi, T. (2001). Mechanism of Ag(I)-catalyzed Mukaiyama aldol reaction: active species in solution in AgPF6-(*S*)-BINAP versus AgOAc-(*S*)-BINAP systems. *J. Mol. Catal. A: Chem.* 170 (1–2): 1–15.
622 Peidro, L., Le Roux, C., Laporterie, A., and Dubac, J. (1996). $BiCl_3$-catalyzed Mukaiyama-aldol and carbonyl-ene reactions. *J. Organomet. Chem.* 521 (1–2): 397–399.
623 Le Roux, C., Ciliberti, L., Laurent-Robert, H. et al. (1998). Bismuth(III) triflate: a water-stable equivalent of trimethylsilyl triflate for the catalysis of Mukaiyama aldol reactions. *Synlett* 11: 1249–1251.
624 Carreira, E.M. and Singer, R.A. (1994). Metal versus silyl triflate catalysis in the Mukaiyama aldol addition reaction. *Tetrahedron Lett.* 35 (25): 4323–4326.
625 Hollis, T.K. and Bosnich, B. (1995). Homogeneous catalysis – mechanisms of the catalytic Mukaiyama aldol and Sakurai allylation reactions. *J. Am. Chem. Soc.* 117 (16): 4570–4581.
626 Ishihara, K. and Yamamoto, H. (2004). Boron and silicon Lewis acids for Mukiyama aldol reactions. In: *Modern Aldol Reactions*, Chapter 2, vol. 2 (ed. R. Mahrwald), 25–68. Weinheim: Wiley-VCH.
627 Iwasawa, N. and Mukaiyama, T. (1987). An essentially mild and efficient catalyst system for the activation of carbonyl compounds and their derivatives – the combined zse of trimethylsilyl chloride and Sn(II) chloride. *Chem. Lett.* 3: 463–466.
628 Mukaiyama, T., Ohno, T., Han, J.S., and Kobayashi, S. (1991). Reversal phenomena of the preferential activation of aldehydes to the corresponding acetals under non-basic conditions – the chemoselective aldol reaction of aldehydes with *tert*-butyldimethylsilyl enol ethers using TbSCl-$InCl_3$ as a catalyst. *Chem. Lett.* 6: 949–952.
629 Davis, A.P., Muir, J.E., and Plunkett, S.J. (1996). Supersilylating agents from chlorosilanes. *Tetrahedron Lett.* 37 (52): 9401–9402.
630 Wong, C.T. and Wong, M.W. (2007). Mechanism of metal chloride-promoted Mukaiyama aldol reactions. *J. Org. Chem.* 72 (4): 1425–1430.
631 Mathieu, B. and Ghosez, L. (1997). *N*-trimethylsilyl-bis(trifluoromethanesulfonyl)imide: a better carbonyl activator than trimethylsilyl triflate. *Tetrahedron Lett.* 38 (31): 5497–5500.
632 Ishii, A., Kotera, O., Saeki, T., and Mikami, K. (1997). Trimethylsilyl bis(trifluoromethanesulfonyl)amide as an effective catalyst for the Friedel–Crafts alkylation reaction. *Synlett* (10): 1145–1146.
633 Ishihara, K., Hiraiwa, Y., and Yamamoto, H. (2001). A high yield procedure for the Me_3SiNTf_2-induced carbon-carbon bond-forming reactions of silyl nucleophiles with carbonyl compounds: the importance of addition order and solvent effects. *Synlett* (12): 1851–1854.
634 Ishihara, K., Hiraiwa, Y., and Yamamoto, H. (2002). Crucial role of the ligand of silyl Lewis acid in the Mukaiyama aldol reaction. *J. Chem. Soc., Chem. Commun.* 15: 1564–1565.
635 Hiraiwa, Y., Ishihara, K., and Yamamoto, H. (2006). Crucial role of the conjugate base for silyl Lewis acid induced Mukaiyama aldol reactions. *Eur. J. Org. Chem.* (8): 1837–1844.
636 Boxer, M.B. and Yamamoto, H. (2006). Tris(trimethylsilyl)silyl-governed aldehyde cross-aldol cascade reaction. *J. Am. Chem. Soc.* 128 (1): 48–49.
637 Koppel, I.A., Burk, P., Koppel, I., and Leito, I. (2002). Generalized principle of designing neutral superstrong Brønsted acids. *J. Am. Chem. Soc.* 124 (19): 5594–5600.
638 Koppel, I.A., Koppel, J., Pihl, V. et al. (2000). Comparison of Brønsted acidities of neutral CH acids in gas phase and dimethyl sulfoxide. *J. Chem. Soc., Perkin Trans. 2* (6): 1125–1133.
639 Ishihara, K., Hasegawa, A., and Yamamoto, H. (2001). Polystyrene-bound tetrafluorophenylbis(triflyl)methane as an organic-solvent-swellable and strong Brønsted

acid catalyst. *Angew. Chem. Int. Ed.* 40 (21): 4077–4079.

640 Ishihara, K., Hasegawa, A., and Yamamoto, H. (2002). Single-pass reaction column system with super Brønsted acid-loaded resin. *Synlett* (8): 1296–1298.

641 Hasegawa, A., Ishikawa, T., Ishihara, K., and Yamamoto, H. (2005). Facile synthesis of aryl- and alkyl-bis(trifluoromethylsulfonyl)methanes. *Bull. Chem. Soc. Jpn.* 78 (8): 1401–1410.

642 Yamamoto, H. and Boxer, M.B. (2007). Super Brønsted acid catalysis in organic synthesis. *Chimia* 61 (5): 279–281.

643 García-García, P., Lay, F., García-García, P. et al. (2009). A powerful chiral counter-anion motif for asymmetric catalysis. *Angew. Chem. Int. Ed.* 48: 4363–4366.

644 (a) Ratjen, L., García-García, P., Lay, F. et al. (2011). Disulfonimide-catalyzed asymmetric vinylogous and bisvinylogous Mukaiyama aldol reactions. *Angew. Chem. Int. Ed.* 50: 754–758. (b) Wang, Q.-G., van Gemmeren, M., and List, B. (2014). Asymmetric disulfonimide-catalyzed synthesis of δ-amino-β-ketoester derivatives by vinylogous Mukaiyama–Mannich reactions. *Angew. Chem. Int. Ed.* 53 (49): 13592–13595. (c) see also Zhou, F.-T. and Yamamoto, H. (2016). A disulfonimide catalyst for highly enantioselective Mukaiyama–Mannich reaction. *Org. Lett.* 18 (19): 4974–4977. (d) Bai, J.-F., Sasagawa, H., Yurino, T. et al. (2017). In situ generation of *N*-Boc-protected alkenyl imines: controlling the *E/Z* geometry of alkenyl moieties in the Mukaiyama–Mannich reaction. *J. Chem. Soc., Chem. Commun.* 53 (58): 8203–8206. (e) Li, J.-S., Liu, J.-J., Zhang, G.-W., and Ma, J.-A. (2017). Catalytic asymmetric Mukaiyama–Mannich reaction of cyclic C-acylimines with difluoroenoxysilanes: access to difluoroalkylated indolin-3-ones. *Org. Lett.* 19 (23): 6364–6367.

645 Hosomi, A. and Sakurai, H. (1976). Syntheses of γ,δ-unsaturated alcohols from allylsilanes and carbonyl compounds in presence of Ti tetrachloride. *Tetrahedron Lett.* (16): 1295–1298.

646 Hosomi, A. (1988). Characteristics in the reactions of allylsilanes and their applications to versatile synthetic equivalents. *Acc. Chem. Res.* 21: 200–206.

647 Biamonte, M.A. (2009). Sakurai allylation. In: *Name Reactions for Homologation, Part I* (ed. J.J. Li), 539–575. Hoboken, NJ: Wiley.

648 Abel, E.W. and Rowley, R.J. (1975). Reaction of perhalogenoketones with allylic derivatives of Si and Sn. *J. Organomet. Chem.* 84 (2): 199–229.

649 (a) Hayashi, T., Konishi, M., Ito, H., and Kumada, M. (1982). Optically-active allylsilanes. 1. Preparation by Pd-catalyzed asymmetric Grignard cross-coupling and *anti*-stereochemistry in electrophilic substitution reactions. *J. Am. Chem. Soc.* 104 (18): 4962–4963. (b) Chabaud, L., James, P., and Landais, Y. (2004). Allylsilanes in organic synthesis – recent developments. *Eur. J. Org. Chem.* (15): 3173–3199. (c) Ramachandran, P.V., Nicponski, D.R., and Gagare, P.D. (2014). Allylsilanes, allylstannanes, and related compounds. In: *Comprehensive Organic Synthesis*, 2e, vol. 2 (ed. P. Knochel and G.A. Molander), 72–147. Amsterdam: Elsevier B. V.

650 Lade, J.J., Pardeshi, S.D., Vadagaonkar, K.S. et al. (2017). The remarkable journey of catalysts from stoichiometric to catalytic quantity for allyltrimethylsilane inspired allylation of acetals, ketals, aldehydes and ketones. *RSC Adv.* 7 (13): 8011–8033.

651 Hayashi, T., Kabeta, K., Hamachi, I., and Kumada, M. (1983). *Erythro*-selectivity in addition of γ-substituted allylsilanes to aldehydes in the presence of Ti tetrachloride. *Tetrahedron Lett.* 24 (28): 2865–2868.

652 Deleris, G., Dunogues, J., and Calas, R. (1976). Synthesis of γ,δ ethylenic or propargylic alcohols via organosilane compounds. *Tetrahedron Lett.* 17 (28): 2449–2450.

653 Hayashi, T., Fujitaka, N., Oishi, T., and Takeshima, T. (1980). Diastereoselection in the reactions of thioallylic anions with benzaldehyde. *Tetrahedron Lett.* 21 (3): 303–306.

654 Hoffmann, R.W. and Zeiss, H.J. (1981). Stereoselective synthesis of alcohols. 8. Diastereoselective synthesis of β-methylhomoallyl alcohols via crotylboronates. *J. Org. Chem.* 46 (7): 1309–1314.

655 Collum, D.B., McDonald, J.H., and Still, W.C. (1980). Synthesis of the polyether antibiotic Monensin. 2. Preparation of intermediates. *J. Am. Chem.Soc.* 102 (6): 2118–2120.

656 Sato, F., Iida, K., Iijima, S. et al. (1981). *Threo*-selective synthesis of β-methylhomoallyl alcohols via but-2-enyltitanium compounds. *J. Chem. Soc., Chem. Commun.* (21): 1140–1141.

657 Yamamoto, Y. and Maruyama, K. (1981). Crotylzirconium derivatives as a new reagent for the *threo*-selective synthesis of β-methylhomoallyl alcohols. *Tetrahedron Lett.* 22 (30): 2895–2898.

658 Hiyama, T., Kimura, K., and Nozaki, H. (1981). Chromium(II) mediated *threo*-selective synthesis

of homoallyl alcohols. *Tetrahedron Lett.* 22 (11): 1037–1040.

659 Yamamoto, Y., Yatagai, H., Naruta, Y., and Maruyama, K. (1980). *Erythro*-selective addition of crotyltrialkyltins to aldehydes regardless of the geometry of the crotyl unit – stereoselection independent of the stereochemistry of precursors. *J. Am. Chem. Soc.* 102 (23): 7107–7109.

660 Yatagai, H., Yamamoto, Y., and Maruyama, K. (1980). A new procedure for the stereoselective synthesis of (*Z*)-2-alkenylsilanes and alkenyl-tins and their application to *erythro*-selective synthesis of β-alkyl alcohol derivatives. *J. Am. Chem. Soc.* 102 (13): 4548–4550.

661 Koreeda, M. and Tanaka, Y. (1982). Stereoselective acyclic synthesis via allylmetals – an efficient synthesis of (+/-)-4-methylheptan-3-ol, an aggregation pheromone of the smaller european elm bark beetle. *Chem. Lett.* 8: 1297–1298.

662 Koreeda, M. and Tanaka, Y. (1982). Stereoselective acyclic synthesis via allylmetals – structural dependence in a Lewis-acid catalyzed addition of allyltins to aldehydes. *Chem. Lett.* (8): 1299–1302.

663 Pratt, A.J. and Thomas, E.J. (1982). On the use of *E*-1-methoxymethoxybut-2-enyl(tri-*n*-butyl)stannane as a *threo*-selective, homo-enolate equivalent. *J. Chem. Soc., Chem. Commun.* (19): 1115–1117.

664 (a) Mahlau, M., García-García, P., and List, B. (2012). Asymmetric counter-anion-directed catalytic Hosomi–Sakurai reaction. *Chem. Eur. J.* 18: 16283–16287. (b) van Gemmeren, M., Lay, F., and List, B. (2014). Asymmetric catalysis using chiral, enantiopure disulfonimides. *Aldrichimica Acta* 47 (1): 3–13.

665 Sai, M. and Yamamoto, H. (2015). Chiral Brønsted acid as a true catalyst: asymmetric Mukaiyama aldol and Hosomi–Sakurai allylation reactions. *J. Am. Chem. Soc.* 137: 7091–7094.

666 Nishigaichi, Y. and Takuwa, A. (2003). Reversible stereocontrol in the Lewis acid promoted reaction of alkoxyaldehydes toward various allyltins. *Tetrahedron Lett.* 44 (7): 1405–1408.

667 Keck, G.E., Krishnamurthy, D., and Grier, M.C. (1993). Catalytic asymmetric allylation reactions. 3. Extension to methallylstannane, comparison of procedures, and observation of a nonlinear effect. *J. Org. Chem.* 58 (24): 6543–6544.

668 Keck, G.E., Tarbet, K.H., and Geraci, L.S. (1993). Catalytic asymmetric allylation of aldehydes. *J. Am. Chem. Soc.* 115 (18): 8467–8468.

669 Keck, G.E. and Krishnamurthy, D. (1998). Catalytic asymmetric allylation reactions: (*S*)-1-(phenylmethoxy)-4-penten-2-ol. *Org. Synth.* 75: 12–18.

670 Keck, G.E. and Yu, T. (1999). Catalytic asymmetric allylation reactions using BITIP catalysis and 2-substituted allylstannanes as surrogates for β-keto ester dianions. *Org. Lett.* 1 (2): 289–291.

671 Sanchez, C.C. and Keck, G.E. (2005). Total synthesis of (+)-dactylolide. *Org. Lett.* 7 (14): 3053–3056.

672 Keck, G.E. and McLaws, M.D. (2005). Stereoselective synthesis of the C_1—C_{13} segment of dolabelide B. *Tetrahedron Lett.* 46 (29): 4911–4914.

673 Brown, H.C., Racherla, U.S., and Pellechia, P.J. (1990). Organoboranes. 53. A high-field variable-temperature H-1 and B-11 NMR study of the effects of solvent and structure on reactivity in allylboration. *J. Org. Chem.* 55 (6): 1868–1874.

674 Kabalka, G.W., Venkataiah, B., and Das, B.C. (2002). Allylboration of carbonyl compounds in ionic liquids. *Green Chem.* 4 (5): 472–473.

675 Mikhailov, B.M. (1972). Allylboron compounds. *Organomet. Chem. Rev. Sect., A: Subj. Rev.* 8 (1): 1–65.

676 Hofmann, R.W. (1982). Diastereogenic addition of crotylmetal compounds to aldehydes. *Angew. Chem. Int. Ed. Engl.* 21 (8): 555–566.

677 Bubnov, Y.N., Gurskii, M.E., Gridnev, I.D. et al. (1992). Determination of the activation parameters of permanent allylic rearrangement in allylboranes and methallylboranes and in 1,1-bis(dipropylborylmethyl)ethylene by dynamic NMR spectroscopy. *J. Organomet. Chem.* 424 (2): 127–132.

678 (a) Fujita, K. and Schlosser, M. (1982). Selective syntheses with organometallics. 10. A convenient and diastereoselective route to homoallyl alcohols – addition of (*Z*)-alkenyl-dimethoxyborane or (*E*)-alkenyl-dimethoxyborane to aldehydes. *Helv. Chim. Acta* 65 (4): 1258–1263. (b) Diner, C. and Szabo, K.J. (2017). Recent advances in the preparation and application of allylboron species in organic synthesis. *J. Am. Chem. Soc.* 139 (1): 2–14. (c) Li, Y. and Houk, K.N. (1989). Transition structures for the allylboration reactions of formaldehyde by allylborane and allylboronic acid. *J. Am. Chem. Soc.* 111 (4): 1236–1240. (d) Barnette, D.S., Moquist, P.N., and Schaus, S.E. (2009). The mechanism and an improved asymmetric allylboration of ketones catalyzed by chiral biphenols. *Angew. Chem. Int. Ed.* 48 (49): 8679–8682.

679 Gajewski, J.J., Bocian, W., Brichford, N.L., and Henderson, J.L. (2002). Secondary deuterium kinetic isotope effects in irreversible additions of allyl reagents to benzaldehyde. *J. Org. Chem.* 67 (12): 4236–4240.

680 Chemier, S. (2009). Roush allylbornation. In: *Name Reactions for Homologations*, Part 2 (ed. J.J. Li), 613–639. Hoboken, NJ: Wiley.

681 Herold, T., Schrott, U., and Hofmann, R.W. (1981). Stereoselective synthesis of alcohols. 6. Asymmetric synthesis of 4-penten-2-ol via allylboronates of chiral glycols. *Chem. Ber. Recl.* 114 (1): 359–374.

682 Kennedy, J.W.J. and Hall, D.G. (2002). Dramatic rate enhancement with preservation of stereospecificity in the first metal-catalyzed additions of allylboronates. *J. Am. Chem. Soc.* 124 (39): 11586–11587.

683 Lachance, H., Lu, X.S., Gravel, M., and Hall, D.G. (2003). Scandium-catalyzed allylboration of aldehydes as a practical method for highly diastereo- and enantioselective construction of homoallylic alcohols. *J. Am. Chem. Soc.* 125 (34): 10160–10161.

684 Kennedy, J.W.J. and Hall, D.G. (2004). Lewis acid catalyzed allylboration: discovery, optimization, and application to the formation of stereogenic quaternary carbon centers. *J. Org. Chem.* 69 (13): 4412–4428.

685 Rauniyar, V. and Hall, D.G. (2004). Lewis acids catalyze the addition of allylboronates to aldehydes by electrophilic activation of the dioxaborolane in a closed transition structure. *J. Am. Chem. Soc.* 126 (14): 4518–4519.

686 Gravel, M., Lachance, H., Lu, X.S., and Hall, D.G. (2004). Scope and limitations of the scandium-catalyzed enantioselective addition of chiral allylboronates to aldehydes. *Synthesis* (8): 1290–1302.

687 Ishiyama, T., Ahiko, T., and Miyaura, N. (2002). Acceleration effect of Lewis acid in allylboration of aldehydes: catalytic, regiospecific, diastereospecific, and enantioselective synthesis of homoallyl alcohols. *J. Am. Chem. Soc.* 124 (42): 12414–12415.

688 (a) Wada, R., Oisaki, K., Kanai, M., and Shibasaki, M. (2004). Catalytic enantioselective allylboration of ketones. *J. Am. Chem. Soc.* 126 (29): 8910–8911. (b) Oisaki, K., Zhao, D.B., Suto, Y. et al. (2005). New chiral bis(diphenylphospholane) ligands: design, synthesis, and application to catalytic enantioselective aldol reaction to ketones. *Tetrahedron Lett.* 46 (25): 4325–4329.

689 Rauniyar, V. and Hall, D.G. (2009). Rationally improved chiral Brønsted acid for catalytic enantioselective allylboration of aldehydes with an expanded reagent scope. *J. Org. Chem.* 74 (11): 4236–4241.

690 van der Mei, F.W., Qin, C., Morrison, R.J., and Hoveyda, A.H. (2017). Practical, broadly applicable, α-selective, *Z*-selective, diastereoselective, and enantioselective addition of allylboron compounds to mono-, di-, tri-, and polyfluoroalkyl ketones. *J. Am. Chem. Soc.* 139 (26): 9053–9065.

691 Clarke, M.L. and France, M.B. (2008). The carbonyl ene reaction. *Tetrahedron* 64 (38): 9003–9031.

692 Snider, B.B., Rodini, D.J., Kirk, T.C., and Cordova, R. (1982). Dimethylaluminum chloride catalyzed ene reactions of aldehydes. *J. Am. Chem. Soc.* 104 (2): 555–563.

693 Snider, B.B. and Phillips, G.B. (1983). Ethylaluminum dichloride catalyzed ene reactions of aldehydes with non-nucleophilic alkenes. *J. Org. Chem.* 48 (4): 464–469.

694 (a) Aggarwal, V.K. and Vennall, G.P. (1998). $Sc(OTf)_3$, an efficient catalyst for addition of allyltrimethylsilane to aldehydes; chemoselective addition to aldehydes in presence of ketones and in situ acylation (3 component coupling). *Synthesis* (12): 1822–1826. (b) Aggarwal, V.K., Vennall, G.P., Davey, P.N., and Newman, C. (1998). Scandium trifluoromethanesulfonate, an efficient catalyst for the intermolecular carbonyl-ene reaction and the intramolecular cyclisation of citronellal. *Tetrahedron Lett.* 39 (14): 1997–2000.

695 (a) Clarke, M.L. and France, M.B. (2008). The carbonyl ene reaction. *Tetrahedron* 64 (38): 9003–9031. (b) Terada, M. (2011). Ene reactions. In: *Science of Synthesis, Stereoselective Synthesis*, vol. 3 (ed. J.G. de Vries, G.A. Molander and P.A. Evans), 309–346. Stuttgart: Georg Thieme Verlag. (c) Snider, B.B. (2014). *Comprehensive Organic Synthesis*, 2e, vol. 2 (ed. P. Knochel and G.A. Molander), 148–191. Amsterdam: Elsevier B. V.

696 Sultana, S., Bondalapati, S., Indukuri, K. et al. (2013). Scandium(III) triflate catalyzed synthesis of primary homoallylic alcohols via carbonyl-ene reaction. *Tetrahedron Lett.* 54 (12): 1576–1578.

697 Ho, C.Y., Ng, S.S., and Jamison, T.F. (2006). Nickel-catalyzed, carbonyl-ene-type reactions: selective for alpha olefins and more efficient with electron-rich aldehydes. *J. Am. Chem. Soc.* 128 (16): 5362–5363.

698 Salomon, M.F., Pardo, S.N., and Salomon, R.G. (1980). Synthesis of allylcarboxylic acids from olefins with diethyl oxomalonate, an enophilic equivalent of carbon dioxide. *J. Am. Chem. Soc.* 102 (7): 2473–2475.

699 Stephenson, L.M. and Orfanopoulos, M. (1981). Changes in ene reaction mechanisms with Lewis acid catalysis. *J. Org. Chem.* 46 (10): 2200–2201.

700 Singleton, D.A. and Hang, C.C. (2000). ^{13}C and ^{2}H kinetic isotope effects and the mechanism of Lewis acid-catalyzed ene reactions of formaldehyde. *J. Org. Chem.* 65 (3): 895–899.

701 Mikami, K., Terada, M., and Nakai, T. (1990). Catalytic asymmetric glyoxylate ene reaction – a practical access to α-hydroxy esters in high enantiomeric purities. *J. Am. Chem. Soc.* 112 (10): 3949–3954.

702 Corey, E.J., Barnes Seeman, D., Lee, T.W., and Goodman, S.N. (1997). A transition-state model for the Mikami enantioselective ene reaction. *Tetrahedron Lett.* 38 (37): 6513–6516.

703 Mikami, K. and Shimizu, M. (1992). Asymmetric ene reactions in organic synthesis. *Chem. Rev.* 92 (5): 1021–1050.

704 Yuan, Y., Zhang, X., and Ding, K.L. (2003). Quasi solvent-free enantioselective carbonyl-ene reaction with extremely low catalyst loading. *Angew. Chem. Int. Ed.* 42 (44): 5478–5480.

705 Evans, D.A. and Wu, J. (2005). Enantioselective *syn*-selective scandium-catalyzed ene reactions. *J. Am. Chem. Soc.* 127 (22): 8006–8007.

706 Hutson, G.E., Dave, A.H., and Rawal, V.H. (2007). Highly enantioselective carbonyl-ene reactions catalyzed by a hindered silyl-salen-co complex. *Org. Lett.* 9 (20): 3869–3872.

707 Liu, X., Bai, S., Yang, Y. et al. (2012). Enantioselective carbonyl-ene reaction on BINOLate/Ti catalyst encapsulated in magnetic nanoreactors. *J. Chem. Soc.; Chem. Commun.* 48 (26): 3191–3193.

708 Doherty, S., Knight, J.G., and Mehdi-Zodeh, H. (2012). Asymmetric carbonyl-ene and Friedel–Crafts reactions catalysed by Lewis acid Pt group metal complexes of the enantiopure atropisomeric biaryl-like diphosphine (*S*)-Me-2-CATPHOS: a comparison with BINAP. *Tetrahedron: Asymmetry* 23 (3–4): 209–216.

709 Miles, W.H., Fialcowitz, E.J., and Halstead, E.S. (2001). Enantioselective synthesis of (*S*)- and (*R*)-fluoxetine hydrochloride. *Tetrahedron* 57 (50): 9925–9929.

710 Miles, W.H., Connell, K.B., Ulas, G. et al. (2010). Furan approach to vitamin D analogues. Synthesis of the A-ring of calcitriol and 1-α-hydroxy-3-deoxyvitamin D_3. *J. Org. Chem.* 75 (20): 6820–6829.

711 Rueping, M., Theissmann, T., Kuenkel, A., and Koenigs, R.M. (2008). Highly enantioselective organocatalytic carbonyl-ene reaction with strongly acidic, chiral Brønsted acids as efficient catalysts. *Angew. Chem. Int. Ed.* 47 (36): 6798–6801.

712 Zheng, K., Shi, J., Liu, X.H., and Feng, X.M. (2008). Asymmetric carbonyl-ene reaction catalyzed by chiral *N,N′*-dioxide-Ni(II) complex: remarkably broad substrate scope. *J. Am. Chem. Soc.* 130 (47): 15770–15771.

713 Zhao, J.F., Tsui, H.Y., Wu, P.J. et al. (2008). Highly enantioselective carbonyl-ene reactions catalyzed by In(III)-pybox complex. *J. Am. Chem. Soc.* 130 (49): 16492–16493.

714 Zhao, J.F., Tan, B.H., Zhu, M.K. et al. (2010). Enantioselective carbonyl-ene reactions of trifluoropyruvate in ionic liquid via a recyclable In (III)-pybox complex. *Adv. Synth. Catal.* 352 (11–12): 2085–2088.

715 Liu, X.-H., Zheng, K., and Feng, X.-M. (2014). Advancements in catalytic asymmetric intermolecular ene-type reactions. *Synthesis* 46: 2241–2257.

716 Zhang, Y., Zhang, R., Chang, H. et al. (2014). Asymmetric carbonyl-ene reactions promoted by chiral catalysts. *Prog. Chem.* 26: 1492–1505.

717 Xu, X., Wang, X., Liu, Y., and Doyle, M.P. (2014). Enantioselective carbonyl-ene reactions catalyzed by chiral cationic dirhodium(II,III) carboxamidates. *J. Org. Chem.* 79: 12185–12190.

718 Luo, W., Zhao, J., Ji, J. et al. (2015). A catalytic asymmetric carbonyl-ene reaction of *β*,*γ*-unsaturated α-ketoesters with 5-methyleneoxazolines. *J. Chem. Soc., Chem. Commun.* 51: 10042–10045.

719 Zhang, X., Wang, M., Ding, R. et al. (2015). Highly enantioselective and *anti*-diastereoselective catalytic intermolecular glyoxylate-ene reactions: effect of the geometrical isomers of alkenes. *Org. Lett.* 17: 2736–2739.

720 Liu, L., Leutzsch, M., Zheng, Y. et al. (2015). Confined acid-catalyzed asymmetric carbonyl-ene cyclization. *J. Am. Chem. Soc.* 137: 13268–13271.

721 Rodini, D.J. and Snider, B.B. (1980). Dimethylaluminum halide induced reactions of formaldehyde with alkynes – synthesis of α-allenic alcohols and ζ-3-chloroallylic alcohols. *Tetrahedron Lett.* 21 (40): 3857–3860.

722 Snider, B.B. and Rodini, D.J. (1980). Dialkylaluminum chloride catalyzed ene reactions of aldehydes – synthesis of ipsenol. *Tetrahedron Lett.* 21 (19): 1815–1818.

723 Achmatowicz, O. and Pietraszkiewicz, M. (1976). Reaction of active C=N-group with alkenes – synthesis of γ,δ-unsaturated α-amino acids. *J. Chem. Soc., Chem. Commun.* (12): 484–484.

724 Borzilleri, R.M. and Weinreb, S.M. (1995). Imino ene reactions in organic synthesis. *Synthesis* (4): 347–360.

725 Laschat, S. and Grehl, M. (1994). Diastereoselective synthesis of amino-substituted indolizidines and quinolizidines by the intramolecular hetero-ene reaction of prolinal imine and 2-piperidine carbaldimine. *Angew. Chem. Int. Ed. Engl.* 33 (4): 458–461.

726 Mikami, K., Kaneko, M., and Yajima, T. (1993). Asymmetric imine-ene reactions with chiral glyoxylate-derived α-imino esters – an efficient route to asymmetric synthesis of α-amino acids. *Tetrahedron Lett.* 34 (30): 4841–4842.

727 Yang, Q., Liu, Y., and Zhang, W. (2011). A theoretical study of imine-ene reaction influencing factors. *Org. Biomol. Chem.* 9: 6343–6351.

728 Kobayashi, S. and Nagayama, S. (1997). Aldehydes vs aldimines. Unprecedented aldimine-selective nucleophilic additions in the coexistence of aldehydes using a lanthanide salt as a Lewis acid catalyst. *J. Am. Chem. Soc.* 119 (42): 10049–10053.

729 Gröger, H., Saida, Y., Sasai, H. et al. (1998). A new and highly efficient asymmetric route to cyclic α-amino phosphonates: the first catalytic enantioselective hydrophosphonylation of cyclic imines catalyzed by chiral heterobimetallic lanthanoid complexes. *J. Am. Chem. Soc.* 120 (13): 3089–3103.

730 Schlemminger, I., Saida, Y., Gröger, H. et al. (2000). Concept of improved rigidity: how to make enantioselective hydrophosphonylation of cyclic imines catalyzed by chiral heterobimetallic lanthanoid complexes almost perfect. *J. Org. Chem.* 65 (16): 4818–4825.

731 Yamanaka, M., Nishida, A., and Nakagawa, M. (2000). Ytterbium(III) triflate/TMSCI: efficient catalyst for imino ene reaction. *Org. Lett.* 2 (2): 159–161.

732 Oliver, L.H., Puls, L.A., and Tobey, S.L. (2008). Brønsted acid promoted imino-ene reactions. *Tetrahedron Lett.* 49 (31): 4636–4639.

733 Yamanaka, M., Nishida, A., and Nakagawa, M. (2003). Imino ene reaction catalyzed by ytterbium(III) triflate and TMSCl or TMSOTf. *J. Org. Chem.* 68 (8): 3112–3120.

734 Terada, M., Machioka, K., and Sorimachi, K. (2006). High substrate/catalyst organocatalysis by a chiral Brønsted acid for an enantioselective aza-ene-type reaction. *Angew. Chem. Int. Ed.* 45 (14): 2254–2257.

735 Caplan, N.A., Hancock, F.E., Page, P.C.B., and Hutchings, G.J. (2004). Heterogeneous enantioselective catalyzed carbonyl- and imino-ene reactions using Cu bis(oxazoline) zeolite Y. *Angew. Chem. Int. Ed.* 43 (13): 1685–1688.

736 Dias, L.C. (2000). Chiral Lewis acid catalyzed ene-reactions. *Curr. Org. Chem.* 4 (3): 305–342.

737 (a) Yao, S.L., Fang, X.M., and Jørgensen, K.A. (1998). Catalytic enantioselective ene reactions of imines: a simple approach for the formation of optically active α-amino acids. *J. Chem. Soc., Chem. Commun.* 22: 2547–2548. (b) Mikami, K., Yajima, T., and Kaneko, M. (1998). Asymmetric imine-ene reactions: diastereofacial selective reactions with chiral glyoxylate-derived α-imino esters and asymmetric catalysis of enantiofacial selective reactions with prochiral α-imino esters. *Amino Acids* 14 (4): 311–318.

738 Momiyama, N., Okamoto, H., Kikuchi, J. et al. (2016). Perfluorinated aryls in the design of chiral Brønsted acid catalysts: catalysis of enantioselective (4+2)-cycloadditions and ene reactions of imines with alkenes by chiral mono-phosphoric acids with perfluoroaryls. *ACS Catal.* 6 (2): 1198–1204.

739 Drury, W.J. III,, Ferraris, D., Cox, C. et al. (1998). A novel synthesis of α-amino acid derivatives through catalytic, enantioselective ene reactions of α-imino esters. *J. Am. Chem. Soc.* 120 (42): 11006–11007.

740 Aburel, P.S., Zhuang, W., Hazell, R.G., and Jørgensen, K.A. (2005). Catalytic and enantioselective aza-ene and hetero-Diels–Alder reactions of alkenes and dienes with azodicarboxylates. *Org. Biomol. Chem.* 3 (12): 2344–2349.

741 Snider, B.B. (1980). Lewis-acid-catalyzed ene reactions. *Acc. Chem. Res.* 13 (11): 426–432.

742 Trost, B.M., Toste, F.D., and Pinkerton, A.B. (2001). Non-metathesis Ru-catalyzed C—C bond formation. *Chem. Rev.* 101 (7): 2067–2096.

743 Snider, B.B. (1974). Lewis acid catalysis of ene reactions. *J. Org. Chem.* 39 (2): 255–256.

744 Snider, B.B. (1976). Stereospecific aluminum chloride catalyzed (2+2)-cycloaddition of propiolate esters with unactivated alkenes. *J. Org. Chem.* 41 (18): 3061–3062.

745 Snider, B.B., Rodini, D.J., Conn, R.S.E., and Sealfon, S. (1979). Lewis acid-catalyzed reactions of methyl propiolate with unactivated alkenes. *J. Am. Chem. Soc.* 101 (18): 5283–5293.

746 (a) Fienemann, H. and Hoffmann, H.M.R. (1979). Cyclobutenecarboxylic esters via aluminum chloride induced (2+2)-cycloadditions of 2-propynoic esters to cyclic olefins. *J. Org. Chem.* 44 (15): 2802–2804. (b) Snider, B.B. and Duncia, J.V. (1980). Stereoselective and regioselective ene reactions of methyl α-chloroacrylate. *J. Am. Chem. Soc.* 102 (18): 5926–5928.

747 Mikami, K. and Aikama, K. (2010). Asymmetric ene reactions and cycloadditions. In: *Catalytic Asymmetric Synthesis*, 3e (ed. I. Ojima), 683–737. Hoboken, NJ: Wiley.

748 Jenner, G. and Papadopoulos, M. (1996). Pressure study and mechanism of the $AlCl_3$ mediated reactions of cycloalkenes and methyl propynoate. *Tetrahedron Lett.* 37 (9): 1417–1420.

749 Jenner, G. (1997). High-pressure kinetics of Lewis acid-catalyzed cycloadditions and ene reactions. A convincing method for mechanistic delineation. *New J. Chem.* 21 (10): 1085–1090.

750 (a) Miller, J.P. (2013). Recent advances in asymmetric Diels–Alder reactions. *Adv. Chem. Res.* 18: 179–219. (b) Fochi, M. and Caruana, L. (2014). Catalytic asymmetric aza-Diels–Alder reactions: the Povarov cycloaddition reaction. *Synthesis* 46 (2): 135–157. (c) Sarmah, M.M. and Prajapati, D. (2014). Aza-Diels–Alder reaction: an efficient approach for construction of heterocycles. *Curr. Org. Chem.* 18 (12): 1586–1620. (d) Ishihara, K. and Sakakura, A. (2014). Intermolecular Diels–Alder reactions. In: *Comprehensive Organic Synthesis*, 2e, vol. 5 (ed. P. Knochel and G.A. Molander), 351–406. Amsterdam: Elsevier B. V. (e) Ishihara, K. and Sakakura, A. (2014). Hetero-Diels–Alder reactions. In: *Comprehensive Organic Synthesis*, 2e, vol. 5 (ed. P. Knochel and G.A. Molander), 409–465. Amsterdam: Elsevier B. V. (f) Sugani, N. and Bansal, R.K. (2017). Catalytic hetero-Diels–Alder reaction of the carbonyl compounds. *Curr. Catal.* 6 (1): 3–24. (g) Shaykhutdinova, P. and Oestreich, M. (2016). Enantioselective Diels–Alder reactions of cyclohexa-1,3-diene and chalcones catalyzed by intramolecular silicon-sulfur Lewis pairs as chiral Lewis acids. *Organometallics* 35 (16): 2768–2771.

751 Wu, X. and Gong, L.-Z. (2015). Chiral phosphoric acid-catalyzed asymmetric multicomponent reactions. In: *Multicomponent Reactions in Organic Synthesis* (ed. J. Zhi, Q. Wang and M.-X. Wang), 439–470. Weinheim: Wiley-VCH.

752 (a) Kanesawa, S. Cornerstone works for catalytic 1,3-dipolar cycloaddition reactions. (b) Suga, H. and Itoh, K. (2014). Recent advances in catalytic asymmetric 1,3-dipolar cycloadditions of azomethine imines, nitrile oxides, diazoalkanes, and carbonyl ylides. In: *Methods and Applications of Cycloadditions Reactions in Organic Synthesis* (ed. N. Nishiwaki), 175–204. Hoboken, NJ: Wiley. (c) Held, F.E., Grau, D., and Tsogeva, S.B. (2015). Enantioselective cycloaddition reactions catalyzed by BINOL-derived phosphoric acids and *N*-triflyl phosphoramides: recent advances. *Molecules* 20 (9): 16103–16126. (d) Bdiri, B., Zhao, B.-J., and Zhou, Z.-M. (2017). Recent advances in the enantioselective 1,3-dipolar cycloaddition of azomethine ylides and dipolarophiles. *Tetrahedron: Asmmetry* 28 (7): 876–899.

753 Epiotis, N.D. and Shaik, S. (1978). Qualitative potential energy surfaces. 2. Cycloadditions. *J. Am. Chem. Soc.* 100: 9–17.

754 Wassermann, A. (1942). 129. Homogeneous catalysis and solvent effects in a diene synthesis. *J. Chem. Soc.* 623–626.

755 Rubin, W., Steiner, H., and Wassermann, A. (1949). Catalysis of Diels–Alder diene associations. 5. Proton-transfer and electron-transfer processes. *J. Chem. Soc.* 3046–3057.

756 Fringuelli, F., Pizzo, F., Taticchi, A. et al. (1982). Diels–Alder reactions of cycloalkenones. 1. Preparation and structure of the adducts. *J. Org. Chem.* 47 (26): 5056–5065.

757 Angell, E.C., Fringuelli, F., Minuti, L. et al. (1985). Diels–Alder reactions of cycloalkenones. 5. 5-Alkyl-2-cyclohexenones as dienophiles. *J. Org. Chem.* 50 (24): 4686–4690.

758 Haaksma, A.A., Jansen, B.J.M., and De Groot, A. (1992). Lewis acid-catalyzed Diels–Alder reactions of *S*-(+)-carvone with silyloxy dienes – total synthesis of (+)-α-cyperone. *Tetrahedron* 48 (15): 3121–3130.

759 Roush, W.R. and Barda, D.A. (1997). Highly selective Lewis acid-catalyzed Diels–Alder reactions of acyclic (*Z*)-1,3-dienes. *J. Am. Chem. Soc.* 119 (31): 7402–7403.

760 Odenkirk, W., Rheingold, A.L., and Bosnich, B. (1992). Homogeneous catalysis – a Ru-based Lewis acid catalyst for the Diels–Alder reaction. *J. Am. Chem. Soc.* 114 (16): 6392–6398.

761 Pignat, K., Vallotto, J., Pinna, F., and Strukul, G. (2000). Cationic complexes of Pd(II) and Pt(II) as Lewis acid catalysts for the Diels–Alder reaction. *Organometallics* 19 (24): 5160–5167.

762 Zhu, Z.L. and Espenson, J.H. (1997). Aqueous catalysis: methylrhenium trioxide (MTO) as a homogeneous catalyst for the Diels–Alder reaction. *J. Am. Chem. Soc.* 119 (15): 3507–3512.

763 Conrads, M., Mattay, J., and Runsink, J. (1989). Thermal reactions of donor-acceptor systems. 7. Hetero Diels–Alder reactions of allenes on silica-gel surface and under liquid-phase conditions. *Chem. Ber.* 122 (11): 2207–2208.

764 Veselovsky, V.V., Gybin, A.S., Lozanova, A.V. et al. (1988). Dramatic acceleration of the Diels–Alder reaction by adsorption on chromatography adsorbents. *Tetrahedron Lett.* 29 (2): 175–178.

765 Parlar, H. and Baumann, R. (1981). Diels–Alder reaction of cyclopentadiene with acrylic acid derivatives in heterogeneous phases. *Angew. Chem. Int. Ed. Engl.* 20 (12): 1014–1014.

766 Hondrogiannis, G., Pagni, R.M., Kabalka, G.W. et al. (1990). The Diels–Alder reaction of cyclopentadiene and methyl acrylate on γ-alumina. *Tetrahedron Lett.* 31 (38): 5433–5436.

767 Hondrogiannis, G., Pagni, R.M., Kabalka, G.W. et al. (1991). The reaction of optically active menthyl acrylate with cyclopentadiene on γ-alumina. *Tetrahedron Lett.* 32 (21): 2303–2306.

768 Ipaktschi, J. (1986). Diels–Alder reaction in the presence of zeolite. *Z. Naturforsch., B: Chem. Sci.* 41 (4): 496–498.

769 Narayana, M.Y.V.S., Pillai, C.N., and Murthy, Y.V.S. (1991). Diels–Alder reactions catalyzed by zeolites. *Synth. Commun.* 21 (6): 783–791.

770 Laszlo, P. and Lucchetti, J. (1984). Catalysis of the Diels–Alder reaction in the presence of clays. *Tetrahedron Lett.* 25 (15): 1567–1570.

771 Cabral, J. and Laszlo, P. (1989). Product distribution in Diels–Alder addition of *N*-benzylidene aniline and enol ethers. *Tetrahedron Lett.* 30 (51): 7237–7238.

772 Collet, C. and Laszlo, P. (1991). Clay catalysis of the nonaqueous Diels–Alder reaction and the importance of humidity control. *Tetrahedron Lett.* 32 (25): 2905–2908.

773 Cativiela, C., Figueras, F., Fraile, J.M. et al. (1991). Clay-catalyzed asymmetric Diels–Alder reaction of cyclopentadiene with (-)-menthyl acrylate. *Tetrahedron: Asymmetry* 2 (10): 953–956.

774 Cativiela, C., Fraile, J.M., García, J.I. et al. (1993). Silica and alumina modified by Lewis acids as catalysts in Diels–Alder reactions of carbonyl containing dienophiles. *Tetrahedron* 49 (19): 4073–4084.

775 Casaschi, A., Desimoni, G., Faita, G. et al. (1993). Catalysis with inorganic cations. 2. The effect of some perchlorates on Diels–Alder reaction rates. *J. Am. Chem. Soc.* 115 (18): 8002–8007.

776 Cossy, J., Carrupt, P.A., and Vogel, P. (1989). Cycloadditions of enones. In: *The Chemistry of Double-bonded Functional Groups*, Part, 2, Chapter 18, vol. 2 (ed. S. Patai), 1369–1565. Chichester: Wiley.

777 Volcho, K.P., Tatarova, L.E., Korchagina, D.V. et al. (1994). Cycloattachment of carbonyl compounds to olefins on aluminosilicate catalysts. *Zh. Org. Khim.* 30 (5): 641–653.

778 Mousseron-Canet, M. and Mousseron, M. (1956). Sur quelques condensations diéniques du Myrcène. *Bull. Soc. Chim. Fr.* 3: 391–401.

779 (a) Saito, A., Yanai, H., and Taguchi, T. (2004). Bis-aluminated triflic amide promoted Diels–Alder reactions of α,β-unsaturated lactones. *Tetrahedron Lett.* 45 (51): 9439–9442. (b) Yanai, H., Takahashi, A., and Taguchi, T. (2007). Dimethylaluminum methide complex $Tf_2CHAlMe_2$: an effective catalyst for Diels–Alder reaction of α,β-unsaturated lactone derivatives with cyclopentadiene. *Tetrahedron* 63 (49): 12149–12159. (c) Ishihara, J., Nakadachi, S., Watanabe, Y., and Hatakeyama, S. (2015). Lewis acid template-catalyzed asymmetric Diels–Alder reaction. *J. Org. Chem.* 80 (4): 2937–2041.

780 Schreiner, P.R. and Wittkopp, A. (2002). H-bonding additives act like Lewis acid catalysts. *Org. Lett.* 4 (2): 217–220.

781 Wittkopp, A. and Schreiner, P.R. (2003). Metal-free, noncovalent catalysis of Diels–Alder reactions by neutral hydrogen bond donors in organic solvents and in water. *Chem. Eur. J.* 9 (2): 407–414.

782 Curran, D.P. and Kuo, L.H. (1994). Altering the stereochemistry of allylation reactions of cyclic α-sulfinyl radicals with diarylureas. *J. Org. Chem.* 59: 3259–3261.

783 Sigman, M.S. and Jacobsen, E.N. (1998). Schiff base catalysts for the asymmetric Strecker reaction identified and optimized from parallel synthetic libraries. *J. Am. Chem. Soc.* 120: 4901–4902.

784 Jiang, X., Zhang, Y., Liu, X. et al. (2009). Enantio- and diastereoselective asymmetric addition of 1,3-dicarbonyl compounds to

nitroalkenes in a doubly stereocontrolled manner catalyzed by bifunctional rosin-derived amine thiourea catalysts. *J. Org. Chem.* 74: 5562–5567.

785 Jiang, X.X., Zhang, Y.F., Wu, L.P. et al. (2009). Doubly stereocontrolled asymmetric aza-Henry reaction with in situ generation of *N*-Boc-imines catalyzed by novel rosin-derived amine thiourea catalysts. *Adv. Synth. Catal.* 351 (13): 2096–2100.

786 Albrecht, L., Acosta Cruz, F., Fraile, A. et al. (2012). Enantioselective H-bond-directing approach for trienamine-mediated reactions in asymmetric synthesis. *Angew. Chem. Int. Ed.* 51: 9088–9092.

787 (a) Rueping, M., Kuenkel, A., and Atodiresei, I. (2011). Chiral Brønsted acids in enantioselective carbonyl activations – activation modes and applications. *Chem. Soc. Rev.* 40: 4539–4549. (b) Otevrel, J. and Bobal, P. (2017). Biphenyl-based bis(thiourea) organocatalyst for asymmetric and *syn*-selective Henry reaction. *Synthesis* 49 (3): 593–603.

788 Held, F.E. and Tsogoeva, S.B. (2016). Asymmetric cycloaddition reactions catalyzed by bifunctional thiourea and squaramide organocatalysts: recent advances. *Catal. Sci. Technol.* 6: 645–667.

789 (a) Ni, S., Naidu Veluru, R., and Franzen, J. (2016). Chiral anion directed asymmetric carbocation-catalyzed Diels–Alder reaction. *Eur. J. Org. Chem.* 1708–1713. (b) Pommerening, P., Mohr, J., Friebel, J., and Oestreich, M. (2017). Synthesis of a chiral borate counteranion, its trityl salt, and application thereof in Lewis-acid catalysis. *Eur. J. Org. Chem.* (16): 2312–2316.

790 Dale, W.J. and Sisti, A.J. (1954). The reaction of 1,3-dienes with chloral. *J. Am. Chem. Soc.* 76 (1): 81–82.

791 (a) Achmatowicz, O. and Zamojski, A. (1957). The chemistry of carbonyl cyanide. IV. Dienophilic properties of carbonyl cyanide and diethyl mesoxalate. Formation of dihydropyran derivatives. *Bull. Acad. Pol. Sciences*Cl. 3 5 (III): 927–929. (b) Aral, I.H.A. and Ichikizaki, I. (1961). Syntheses of γ-ionone analogs. *Bull. Chem. Soc. Jpn.* 34: 1571–1783.

792 Schönberg, A. and Singer, E. (1971). 1.2.3-Tricarbonyl compounds. 5. Hydrate forming 1,2,3-tricarbonyl compounds as dienophiles in Diels–Alder reactions. *Chem. Ber.* 104 (1): 160–167.

793 Ansell, M.F. and Charalambides, A.A. (1972). Aromatic aldehydes as dienophiles. *J. Chem. Soc., Chem. Commun.* (12): 739–740.

794 Eschenbrenner-Lux, V., Kumar, K., and Waldmann, H. (2014). The asymmetric hetero-Diels–Alder reaction in the syntheses of biologically relevant compounds. *Angew. Chem. Int. Ed.* 53: 11146–11157.

795 (a) Griengl, H. and Geppert, K.P. (1976). Prins reactions of arylaldehydes. 2. Dienes. *Monatsh. Chem.* 107: 675–684. (b) Aggarwal, V.K., Vennall, G.P., Davey, P.N., and Newman, C. (1997). Trifluoromethanesulfonic acid, an efficient catalyst for the hetero Diels–Alder reaction and an improved synthesis of mefrosol. *Tetrahedron Lett.* 38: 2569–2572.

796 Liu, L.-P., Kim, H., Xie, Y.-W. et al. (2017). Catalytic asymmetric (4+2)-cycloaddition of dienes with aldehydes. *J. Am. Chem. Soc.* 139 (39): 13656–13659.

797 Oi, S., Kashiwagi, K., Terada, E. et al. (1996). Cationic Pd(II) complex-catalyzed hetero Diels–Alder reaction of dienes with aldehydes. *Tetrahedron Lett.* 37: 6351–6354.

798 Fujiwara, K., Kurahashi, T., and Matsubara, S. (2012). Cationic Fe(III) porphyrin-catalyzed (4+2)-cycloaddition of unactivated aldehydes with simple dienes. *J. Am. Chem. Soc.* 134: 5512–5515.

799 Kuwano, T., Kurahashi, T., and Matsubara, S. (2013). Iron corrole-catalyzed (4+2)-cycloaddition of dienes and aldehydes. *Chem. Lett.* 42: 1241–1243.

800 Hanamoto, T., Sugimoto, Y., Jin, T.Z., and Inanaga, J. (1997). Scandium(III) perfluorooctanesulfonate [Sc(OPf)$_3$]: a novel catalyst for the hetero Diels–Alder reaction of aldehydes with non-activated dienes. *Bull. Chem. Soc. Jpn.* 70: 1421–1426.

801 Sogani, N., Sinha, P., and Bansal, R.K. (2014). Hetero-Diels–Alder reaction of aromatic aldehydes catalyzed by Ti tetrachloride: computational and experimental results. *Tetrahedron* 70: 735–741.

802 Zou, X., Yang, L., Liu, X. et al. (2015). Silver tetrafluoroborate-catalyzed oxa-Diels–Alder reaction between electrically neutral dienes and aldehydes. *Adv. Synth. Catal.* 357: 3040–3046.

803 Jewett, J.C. and Rawal, V.H. (2007). Total synthesis of pederin. *Angew. Chem. Int. Ed.* 46 (34): 6502–6504.

804 Saito, S. and Yamamoto, H. (1997). Designer Lewis acid catalysts – bulky aluminum reagents for selective organic synthesis. *J. Chem. Soc., Chem. Commun.* (17): 1585–1592.

805 Lin, L., Kuang, Y., Liu, X., and Feng, X. (2011). Indium(III)-catalyzed asymmetric hetero-Diels–Alder reaction of Brassard-type diene with aliphatic aldehydes. *Org. Lett.* 13 (15): 3868–3871.

806 Zheng, J., Lin, L., Kuang, Y. et al. (2014). Magnesium(II)-catalyzed asymmetric hetero-Diels–Alder reaction of Brassard's dienes with isatins. *J. Chem. Soc., Chem. Commun.* 50: 994–996.

807 Buonora, P., Olsen, J.C., and Oh, T. (2001). Recent developments in imino Diels–Alder reactions. *Tetrahedron* 57 (29): 6099–6138.

808 Kobayashi, S., Araki, M., Ishitani, H. et al. (1995). Activation of imines by rare-Earth-metal triflates – $Ln(OTf)_3$- or $Sc(OTf)_3$-catalyzed reactions of imines with silyl enolates and Diels–Alder reactions of imines. *Synlett* (3): 233–234.

809 Ali, T., Chauhan, K.K., and Frost, C.G. (1999). Indium triflate – an efficient catalyst for hetero Diels–Alder reactions. *Tetrahedron Lett.* 40 (30): 5621–5624.

810 (a) Kobayashi, S., Ishitani, H., and Nagayama, S. (1995). $Ln(OTf)_3$-catalyzed or $Sc(OTf)_3$-catalyzed three-components coupling reactions between aldehydes, amines, and dienes or alkenes – efficient syntheses of pyridine and quinoline derivatives. *Chem. Lett.* (6): 423–424. (b) Yu, L.B., Chen, D.P., and Wang, P.G. (1996). Aqueous aza Diels–Alder reactions catalyzed by lanthanide(III) trifluoromethanesulfonates. *Tetrahedron Lett.* 37 (13): 2169–2172.

811 Nogue, D., Paugam, R., and Wartski, L. (1992). High degree of *exo*-selectivity in imino Diels–Alder reactions catalyzed by *tert*-butyldimethylsilyltriflate. *Tetrahedron Lett.* 33 (10): 1265–1268.

812 Hague, C., Patmore, N.J., Frost, C.G. et al. (2001). $[(PPh_3)Ag(CB_{11}H_6Y_6)]$ (Y=H, Br): highly active, selective and recyclable Lewis acids for a hetero-Diels–Alder reaction. *J. Chem. Soc., Chem. Commun.* (21): 2286–2287.

813 (a) Hatanaka, Y., Nantaku, S., Nishimura, Y. et al. (2017). Catalytic enantioselective aza-Diels–Alder reactions of unactivated acyclic 1,3-dienes with aryl-, alkenyl-, and alkyl-substituted imines. *J. Chem. Soc., Chem. Commun.* 53 (64): 8996–8999. (b) Weilbeer, C., Sickert, M., Naumov, S., and Schneider, C. (2017). The Brønsted acid-catalyzed, enantioselective aza-Diels–Alder reaction for the direct synthesis of chiral piperidones. *Chem. Eur. J.* 23 (3): 517–518.

814 Eschenbrenner-Lux, V., Küchler, P., Ziegler, S. et al. (2014). An enantioselective inverse-electron-demand imino Diels–Alder reaction. *Angew. Chem. Int. Ed.* 53: 2134–2137.

815 Tietze, L.F., Schneider, C., and Grote, A. (1996). Stereodivergent hetero-Diels–Alder reactions of chiral 1-oxa-1,3-butadienes through a conformational switch induced by Lewis acids. *Chem. Eur. J.* 2 (2): 139–148.

816 Mao, Z., Li, W., Shi, Y. et al. (2013). Enantioselective construction of dihydropyran-fused indoles through chiral Ca phosphate-catalyzed oxo-hetero-Diels–Alder reactions by using 2-oxoindolin-3-ylidenes as heterodienes. *Chem. Eur. J.* 19: 9754–9759.

817 Hu, H., Liu, Y., Guo, J. et al. (2015). Enantioselective synthesis of dihydrocoumarin derivatives by chiral Sc(III)-complex catalyzed inverse-electron-demand hetero-Diels–Alder reaction. *J. Chem. Soc., Chem. Commun.* 51: 3835–3837.

818 Hsiao, C.C., Raja, S., Liao, H.H. et al. (2015). *Ortho*-Quinone methides as reactive intermediates in asymmetric Brønsted acid-catalyzed cycloadditions with unactivated alkenes by exclusive activation of the electrophile. *Angew. Chem. Int. Ed.* 51: 3835–3837.

819 (a) Motorina, I.A. and Grierson, D.S. (1999). An intramolecular 1-azadiene Diels–Alder approach to the preparation of synthetic equivalents of pyridine. *Tetrahedron Lett.* 40 (40): 7211–7214. (b) Sanap, K.K. and Samant, S.D. (2015). Regiospecific inverse electron demand Diels–Alder reactions of 7-methylcoumarin-4-azadienes. *RSC Adv.* 5 (46): 36696–36706. (c) Laurent, G., Retailleau, P., Folleas, B. et al. (2013). Highly enantioselective aza-Diels–Alder reaction of 1-azadienes with enecarbamates catalyzed by chiral phosphoric acids. *Angew. Cem. Int. Ed.* 52 (42): 11088–11091.

820 Batey, R.A., Simoncic, P.D., Lin, D. et al. (1999). A three-component coupling protocol for the synthesis of substituted hexahydropyrrolo[3,2-c]quinolines. *J. Chem. Soc., Chem. Commun.* (7): 651–652.

821 Kumar, R.S., Nagarajan, R., and Perumal, P.T. (2004). Inverse electron demand Diels–Alder reactions of heterodienes catalyzed by potassium hydrogen sulfate: diastereoselective, one-pot synthesis of pyranobenzopyrans, furanobenzopyrans and tetrahydroquinolines derivatives. *Synthesis* (6): 949–959.

822 Saito, T., Ohkubo, T., Maruyama, K. et al. (1993). Lewis acid-induced hetero-Diels–Alder reaction of conjugated carbodiimides. *Chem. Lett.* (7): 1127–1130.

823 Liu, B., Li, K.N., Luo, S.W. et al. (2013). Chiral Au complex-catalyzed hetero-Diels–Alder reaction of diazenes: highly enantioselective and general for dienes. *J. Am. Chem. Soc.* 135: 3323–3326.

824 Huang, R., Chang, X., Li, J., and Wang, C.J. (2016). Copper(I)-catalyzed asymmetric multicomponent cascade inverse electron-demand aza-Diels Alder/nucleophilic addition/ring-opening reaction involving 2-methoxyfurans as efficient dienophiles. *J. Am. Chem. Soc.* 138: 3998–4001.

825 (a) Maji, B. and Yamamoto, H. (2015). Catalytic enantioselective nitroso Diels–Alder reaction. *J. Am. Chem. Soc.* 137: 15957–15963. (b) Brulikova, L., Harrison, A., Miller, M.J., and Hlavac, J. (2016). Stereo- and regioselectivity of the hetero-Diels–Alder reaction of nitroso derivatives with conjugated dienes. *Beilstein J. Org. Chem.* 12: 1949–1980. (c) Li, J., Tao, H.-Y., and Wang, C.-J. (2017). Copper(I)/TF-BiphamPhos catalyzed asymmetric nitroso Diels–Alder reaction. *J. Chem. Soc., Chem. Commun.* 53 (10): 1657–1659. (d) Lu, Y., Zhou, Y.-H., Zhang, J.-C. et al. (2018). Highly regio- and enantioselective nitroso Diels–Alder reaction of 1, 3-Diene-1-carbamates catalyzed by chiral *N,N′*-dioxide/Copper(II) complex. *Adv. Synth. Catal.* 360 (1): 186–191.

826 Crimmins, M.T. (1988). Synthetic applications of intramolecular enone olefin photocycloadditions. *Chem. Rev.* 88 (8): 1453–1473.

827 Sarkar, N., Nayek, A., and Ghosh, S. (2004). Copper(I)-catalyzed intramolecular asymmetric (2+2)-photocycloaddition. Synthesis of both enantiomers of cyclobutane derivatives. *Org. Lett.* 6 (12): 1903–1905.

828 García-Exposito, E., Bearpark, M.J., Ortuno, R.M. et al. (2002). Theoretical study of the photochemical (2+2)-cycloadditions of cyclic and acyclic α,β-unsaturated carbonyl compounds to ethylene. *J. Org. Chem.* 67 (17): 6070–6077.

829 Lee-Ruff, E. and Mladenova, G. (2003). Enantiomerically pure cyclobutane derivatives and their use in organic synthesis. *Chem. Rev.* 103 (4): 1449–1483.

830 De Mayo, P. (1971). Photochemical syntheses. 37. Enone photoannelation. *Acc. Chem. Res.* 4 (2): 41–47.

831 Gilchrist, T.L. and Storr, R.C. (1979). *Organic Reactions and Orbital Symmetry*, 2e, 1–311. London: Cambridge University Press.

832 Faron, K.L. and Wulff, W.D. (1988). The Cr and W pentacarbonyl groups as reactivity auxiliaries in (2+2)-cycloadditions. *J. Am. Chem. Soc.* 110 (26): 8727–8729.

833 Quendo, A. and Rousseau, G. (1988). Reaction of ketenealkylsilylacetals with ethyl propiolate. *Tetrahedron Lett.* 29 (49): 6443–6446.

834 Fétizon, M., Goulaouic, P., Hanna, I., and Prangé, T. (1988). 1,4-Dioxene in organic synthesis. 6. Substituted 2-vinyl-1,4-dioxenes – useful intermediates for the synthesis of highly functionalized compounds. *J. Org. Chem.* 53 (24): 5672–5679.

835 Nicolaou, K.C., Hwang, C.K., Duggan, M.E., and Reddy, K.B. (1987). Thermal cycloadditions of dimethyl acetylenedicarboxylate with cyclic enol ethers – an entry into medium size oxocyclic systems. *Tetrahedron Lett.* 28 (14): 1501–1502.

836 Clark, R.D. and Untch, K.G. (1979). (2+2)-Cycloaddition of ethyl propiolate and silyl enol ethers. *J. Org. Chem.* 44 (2): 248–253.

837 Semmelhack, M.F., Tomoda, S., Nagaoka, H. et al. (1982). Synthesis of racemic fomannosin and illudol using a biosynthetically patterned common intermediate. *J. Am. Chem. Soc.* 104 (3): 747–759.

838 Majetich, G., Defauw, J., and Ringold, C. (1988). Intramolecular additions of allylsilanes to conjugated dienones – the synthesis of three perforanes. *J. Org. Chem.* 53 (1): 50–68.

839 House, H.O., Gaa, P.C., and Van Derveer, D. (1983). Perhydroazulenes. 3. Conformations of the 4-oxoperhydroazulenes. *J. Org. Chem.* 48 (10): 1661–1670.

840 Belluš, D. and Ernst, B. (1988). Cyclobutanones and cyclobutenones in nature and in synthesis. *Angew. Chem. Int. Ed. Engl.* 27 (6): 797–827.

841 Trost, B.M. (1986). Strain and reactivity – partners for selective synthesis. In: *Small Ring Compounds in Organic Synthesis*, Topics in Current Chemistry, vol. 133 (ed. A. de Meijere), 3–82. Heidelberg: Springer-Verlag.

842 Wong, H.N.C., Lau, K.L., and Tam, K.F. (1986). The application of cyclobutane derivatives in organic synthesis. In: *Small Ring Compounds in Organic Synthesis*, Topics in Current Chemistry, vol. 133 (ed. A. de Meijere), 83–157. Heidelberg: Springer-Verlag.

843 Avenoza, A., Busto, J.H., Canal, N., and Peregrina, J.M. (2003). Reactivity of 2-acylaminoacrylates with ketene diethyl acetal; (2+2)-cycloadditions vs. tandem condensations. *J. Chem. Soc., Chem. Commun.* 12: 1376–1377.

844 Jung, M.E. and Halweg, K.M. (1981). Intramolecular Lewis acid promoted (2+2)-cycloadditions – an efficient total synthesis of (+/-)-Coronafacic acid via an internal Diels–Alder reaction. *Tetrahedron Lett.* 22 (29): 2735–2738.

845 Hoffmann, H.M.R., Ismail, Z.M., and Weber, A. (1981). Synthesis of cyclobutylideneacetic esters via Al chloride promoted (2+2)-cycloadditions of ethyl 2,3-butadienoate to olefins. *Tetrahedron Lett.* 22 (21): 1953–1956.

846 Rosenblum, M. and Scheck, D. (1982). Condensation of propiolic esters with olefins catalyzed by the $C_5H_5Fe(CO)_2$ cation. *Organometallics* 1 (2): 397–399.

847 Engler, T.A., Ali, M.H., and Van der Velde, D. (1989). Non-photochemical (2+2)-cycloadditions of acyclic enones and alkenes. *Tetrahedron Lett.* 30 (14): 1761–1764.

848 Takasu, K., Ueno, M., Inanaga, K., and Ihara, M. (2004). Catalytic (2+2)-cycloaddition reactions of silyl enol ethers. A convenient and stereoselective method for cyclobutane ring formation. *J. Org. Chem.* 69 (2): 517–521.

849 Sweis, R.F., Schramm, M.P., and Kozmin, S.A. (2004). Silver-catalyzed (2+2)-cycloadditions of siloxy alkynes. *J. Am. Chem. Soc.* 126 (24): 7442–7443.

850 Inanaga, K., Takasu, K., and Ihara, M. (2005). A practical catalytic method for preparing highly substituted cyclobutanes and cyclobutenes. *J. Am. Chem. Soc.* 127 (11): 3668–3669.

851 Takasu, K., Hosokawa, N., Inanaga, K., and Ihara, M. (2006). Cyclobutane ring formation by triflic imide catalyzed (2+2)-cycloaddition of allylsilanes. *Tetrahedron Lett.* 47 (34): 6053–6056.

852 Canales, E. and Corey, E.J. (2007). Highly enantioselective (2+2)-cycloaddition reactions catalyzed by a chiral Al bromide complex. *J. Am. Chem. Soc.* 129 (42): 12686–12687.

853 Conner, M.L., Xu, Y., and Brown, M.K. (2015). Catalytic enantioselective allenoate-alkene (2+2)-cycloadditions. *J. Am. Chem. Soc.* 137: 3482–3485.

854 Kang, T., Ge, S., Lin, L. et al. (2016). A Chiral *N,N′*-dioxide-Zn(II) complex catalyzes the enantioselective (2+2)-cycloaddition of alkynones with cyclic enol silyl ethers. *Angew. Chem. Int. Ed.* 55: 5541–5544.

855 Jia, M., Monari, M., Yang, Q.Q., and Bandini, M. (2015). Enantioselective Au-catalyzed dearomative (2+2)-cycloaddition between indoles and allenamides. *J. Chem. Soc., Chem. Commun.* 51: 2320–2323.

856 Luzung, M.R., Mauleón, P., and Toste, F.D. (2007). Gold(I)-catalyzed (2+2)-cycloaddition of allenenes. *J. Am. Chem. Soc.* 129 (41): 12402–12403.

857 Teller, H., Flügge, S., Goddard, R., and Fürstner, A. (2010). Enantioselective gold catalysis: opportunities provided by monodentate phosphoramidite ligands with an acyclic TADDOL backbone. *Angew. Chem. Int. Ed.* 49: 1949–1953.

858 González, A.Z., Benitez, D., Tkatchouk, E. et al. (2011). Phosphoramidite Au(I)-catalyzed diastereo- and enantioselective synthesis of 3,4-substituted pyrrolidines. *J. Am. Chem. Soc.* 133: 5500–5507.

859 Suárez-Pantiga, S., Hernández-Díaz, C., Piedrafita, M. et al. (2012). Phosphite-Au(I)-catalyzed (2+2)-intermolecular cycloaddition of enol ethers with *N*-allenylsulfonamides. *Adv. Synth. Catal.* 354: 1651–1657.

860 Suárez-Pantiga, S., Hernández-Díaz, C., Rubio, E., and González, J.M. (2012). Intermolecular (2+2)-reaction of *N*-allenylsulfonamides with vinylarenes: enantioselective Au(I)-catalyzed synthesis of cyclobutane derivatives. *Angew. Chem. Int. Ed.* 51: 11552–11555.

861 (a) Teller, H., Corbet, M., Mantilli, L. et al. (2012). One-point binding ligands for asymmetric gold catalysis: phosphoramidites with a TADDOL-related but acyclic backbone. *J. Am. Chem. Soc.* 134: 15331–15342. (b) Li, X.X., Zhu, L.L., Zhou, W., and Chen, Z. (2012). Formal intermolecular (2+2)-cycloaddition reaction of alleneamides with alkenes via gold catalysis. *Org. Lett.* 14: 436–439.

862 Ranieri, B., Obradors, C., Mato, M., and Echavarren, A.M. (2016). Synthesis of rumphellaone A and hushinone by a gold-catalyzed (2+2)-cycloaddition. *Org. Lett.* 18: 1614–1617.

863 Hu, J.-L., Feng, L.-W., Wang, L. et al. (2016). Enantioselective construction of cyclobutanes: a new and concise approach to the total synthesis of (+)-Piperarborenine B. *J. Am. Chem. Soc.* 138 (40): 13151–13154.

864 Pommier, A. and Pons, J.M. (1993). Recent advances in β-lactone chemistry. *Synthesis* (5): 441–459.

865 Yang, H.W. and Romo, D. (1999). Methods for the synthesis of optically active β-lactones (2-oxetanones). *Tetrahedron* 55 (21): 6403–6434.

866 Wang, Y.C., Tennyson, R.L., and Romo, D. (2004). β-Lactones: intermediates for natural product total synthesis and new transformations. *Heterocycles* 64: 605–658.

867 Jeong, W., Hedrick, J.L., and Waymouth, R.M. (2007). Organic spirocyclic initiators for the ring-expansion polymerization of β-lactones. *J. Am. Chem. Soc.* 129 (27): 8414–8515.

868 Zhang, W., Matla, A.S., and Romo, D. (2007). Alkyl C—O ring cleavage of bicyclic β-lactones with normant reagents: synthesis of a Merck IND intermediate. *Org. Lett.* 9 (11): 2111–2114.

869 Sawant, K.B. and Jennings, M.P. (2006). Efficient total syntheses and structural verification of both diospongins A and B via a common δ-lactone intermediate. *J. Org. Chem.* 71 (20): 7911–7914.

870 Zaitseva, G.S., Vinokurova, N.G., and Baukov, Y.I. (1975). Reaction of trialkylsilylketenes with aldehydes. *Zh. Obshch. Khim.* 45 (6): 1398–1398.

871 Brady, W.T. and Saidi, K. (1979). Trimethylsilylketene – cycloadditions of ketenes and aldehydes. *J. Org. Chem.* 44 (5): 733–737.

872 Concepcion, A.B., Maruoka, K., and Yamamoto, H. (1995). Organoaluminum-promoted cycloaddition of trialkylsilylketene with aldehydes – a new, stereoselective approach to *cis*-2-oxetanones and 2(*Z*)-alkenoic acids. *Tetrahedron* 51 (14): 4011–4020.

873 Arrastia, I. and Cossio, F.P. (1996). Tandem (2+2)-cycloaddition-cycloreversion reactions in highly polar media: a convergent one-pot entry to substituted alkenes and dienes. *Tetrahedron Lett.* 37 (39): 7143–7146.

874 Hattori, T., Suzuki, Y., Uesugi, O. et al. (2000). Cationic Pd(II) complex-catalyzed (2+2)-cycloaddition and tandem cycloaddition-allylic rearrangement of ketene with aldehydes: an improved synthesis of sorbic acid. *J. Chem. Soc., Chem. Commun.* (1): 73–74.

875 Allen, A.D. and Tidwell, T.T. (2013). Ketenes and other cumulenes as reactive intermediate. *Chem. Rev.* 113: 7287–7342.

876 Lecea, B., Arrieta, A., Roa, G. et al. (1994). Catalytic and solvent effects on the cycloaddition reaction between ketenes and carbonyl compounds to form 2-oxetanones. *J. Am. Chem. Soc.* 116 (21): 9613–9619.

877 Lecea, B., Arrieta, A., López, X. et al. (1995). On the stereochemical outcome of the catalyzed and uncatalyzed cycloaddition reaction between activated ketenes and aldehydes to form *cis*-2-oxetanones and *trans*-2-oxetanones – an ab-initio study. *J. Am. Chem. Soc.* 117 (49): 12314–12321.

878 Lecea, B., Arrieta, A., Arrastia, I., and Cossio, F.P. (1998). Origins of stereocontrol in the (2+2)-cycloaddition between achiral ketenes and chiral α-alkoxy aldehydes. A pericyclic alternative to the aldol reaction. *J. Org. Chem.* 63 (15): 5216–5227.

879 Onsager, L. (1936). Electric moments of molecules in liquids. *J. Am. Chem. Soc.* 58: 1486–1493.

880 Marqués-López, E. and Christmann, M. (2012). β-Lactones through catalytic asymmetric heterodimerization of ketenes. *Angew. Chem. Int. Ed.* 51: 8696–8698.

881 Nelson, S.G., Dura, R.D., and Peelen, T.J. (2013). Catylical asymmetric ketene (2+2) and (4+2)-cycloadditions. In: *Organic Reactions*, vol. 82 (ed. P.A. Evans and S. Weinreb), 471–621. Wiley.

882 Hao, X., Liu, X., Li, W. et al. (2014). Chiral Lewis acid catalyzed asymmetric cycloadditions of disubstituted ketenes for the synthesis of β-lactones and δ-lactones. *Org. Lett.* 16: 134–137.

883 Horiguchi, Y., Suehiro, I., Sasaki, A., and Kuwajima, I. (1993). Diastereoselective (3+2)-cycloaddition of methyl 2-phenlthiocyclopropyl ketone with enol silyl ethers – synthesis of functionalized cyclopentanes. *Tetrahedron Lett.* 34 (38): 6077–6080.

884 Monti, H., Rizzotto, D., Léandri, G., and Monti, J.P. (1994). A novel cyclopentane annulation by (3+2)-cycloaddition of substituted methylenecyclopropyl ketones with allyltrimethylsilane. *Tetrahedron Lett.* 35 (18): 2885–2888.

885 Yadav, V.K. and Sriramurthy, V. (2004). Formal (3+2)-addition of acceptor-substituted cyclopropylmethylsilanes with aryl acetylenes. *Angew. Chem. Int. Ed.* 43 (20): 2669–2671.

886 Kang, Y.B., Tang, Y., and Sun, X.L. (2006). Scandium triflate catalyzed cycloaddition of imines with 1,1-cyclopropanediesters: efficient and diastereoselective synthesis of multisubstituted pyrrolidines. *Org. Biomol. Chem.* 4 (2): 299–301.

887 Korotkov, V.S., Larionov, O.V., and de Meijere, A. (2006). $Ln(OTf)_3$-catalyzed insertion of aryl isocyanides into the cyclopropane ring. *Synthesis* (21): 3542–3546.

888 Korotkov, V.S., Larionov, O.V., Hoftneister, A. et al. (2007). $GaCl_3$-catalyzed insertion of

diazene derivatives into the cyclopropane ring. *J. Org. Chem.* 72 (20): 7504–7510.

889 (a) Pohlhaus, P.D. and Johnson, J.S. (2005). Highly diastereoselective synthesis of tetrahydrofurans via Lewis acid-catalyzed cyclopropane/aldehyde cycloadditions. *J. Org. Chem.* 70 (3): 1057–1059. (b) Pohlhaus, P.D., Sanders, S.D., Parsons, A.T. et al. (2008). Scope and mechanism for Lewis acid-catalyzed cycloadditions of aldehydes and donor-acceptor cyclopropanes: evidence for a stereospecific intimate ion pair pathway. *J. Am. Chem. Soc.* 130 (27): 8642–8650.

890 Verma, K. and Banerjee, P. (2016). Lewis acid-catalyzed (3+2)-cycloaddition of donor-acceptor cyclopropanes and enamines: enantioselective synthesis of nitrogen-functionalized cyclopentane derivatives. *Adv. Synth. Catal.* 358 (13): 2953–2058.

891 Parsons, A.T., Campbell, M.J., and Johnson, J.S. (2008). Diastereoselective synthesis of tetrahydrofurans via Pd(0)-catalyzed (3+2)-cycloaddition of vinylcyclopropaines and aldehydes. *Org. Lett.* 10 (12): 2541–2544.

892 Shen, K.H., Lush, S.F., Chen, T.L., and Liu, R.S. (2001). Synthesis of bicyclic pyrane derivatives via W-mediated (3+3)-cycloaddition of epoxides with tethered alkynes. *J. Org. Chem.* 66 (24): 8106–8111.

893 Kurdyumov, A.V., Lin, N., Hsung, R.P. et al. (2006). A Lewis acid-catalyzed formal (3+3)-cycloaddition of α,β-unsaturated aldehydes with 4-hydroxy-2-pyrone, diketones, and vinylogous esters. *Org. Lett.* 8 (2): 191–193.

894 Hubert, C., Moreau, J., Batany, J. et al. (2008). Brønsted acid-catalyzed synthesis of pyrans via a formal (3+3)-cycloaddition. *Adv. Synth. Catal.* 350 (1): 40–42.

895 Hu, B., Zhu, J.L., Zing, S.Y. et al. (2009). A highly site-, regio-, and stereoselective Lewis acid catalyzed formal (3+3)-cycloaddition of methylenecyclopropane-1,1-diesters with C,N-diarylnitrones. *Chem. Eur. J.* 15 (2): 324–327.

896 Liao, S., Sun, X.L., and Tang, Y. (2014). Side arm strategy for catalyst design: modifying bisoxazolines for remote control of enantioselection and related. *Acc. Chem. Res.* 47 (8): 2260–2272.

897 Kang, Y.-B., Sun, X.-L., and Tang, Y. (2007). Highly enantioselective and diastereoselective cycloaddition of cyclopropanes with nitrones and its application in the kinetic resolution of 2-substituted cyclopropane-1,1-dicarboxylates. *Angew. Chem. Int. Ed.* 46 (21): 3918–3921.

898 Tejero, R., Ponce, A., Adrio, J., and Carretero, J.C. (2013). Ni-catalyzed (8+3)-cycloaddition of tropones with 1,1-cyclopropanediesters. *J. Chem. Soc., Chem. Commun.* 49: 10406–10408.

899 Rivero, A.R., Fernández, I., and Sierra, M.A. (2013). Regio- and diastereoselective stepwise (8+3)-cycloaddition reaction between tropone derivatives and donor-acceptor cyclopropanes. *Org. Lett.* 15: 4928–4931.

900 Remfry, F.G.P. (1913). Perezone. *J. Chem. Soc.* 103: 1076–1088.

901 Kogl, F. and Boer, A.G. (1935). On the constitution of Perezone. *Recl. Trav. Chim. Pays-Bas* 54: 779–794.

902 Sánchez, I.H., Basurto, F., and Joseph-Nathan, P. (1984). The stereocontrol of the perezone to pipitzol transformation. *J. Nat. Prod.* 47 (2): 382–383.

903 Walls, F., Padilla, J., Joseph-Nathan, P. et al. (1965). Structures of α and β-pipitzols. *Tetrahedron Lett.* (21): 1577–1582.

904 Büchi, G. and Mak, C.P. (1977). Biomimetic syntheses of neolignans guianin, burchellin, 2-*epi*,3A-epiburchellin and futoenone. *J. Am. Chem. Soc.* 99 (24): 8073–8075.

905 Büchi, G. and Chu, P.S. (1981). A Synthesis of gymnomitrol. *Tetrahedron* 37 (25): 4509–4513.

906 Collins, J.L., Grieco, P.A., and Walker, J.K. (1997). Cationic (5+2)-cycloaddition reactions promoted by trimethylsilyl triflate in highly polar media. *Tetrahedron Lett.* 38 (8): 1321–1324.

907 Grieco, P.A. and Walker, J.K. (1997). Intramolecular cationic (5+2)-cycloaddition reactions promoted by trimethylsilyl triflate in 3.0 M lithium perchlorate-ethyl acetate: application to a formal total synthesis of (+/-)-isocomene. *Tetrahedron* 53 (26): 8975–8996.

908 Tanino, K., Kondo, F., Shimizu, T., and Miyashita, M. (2002). Stereoselective synthesis of cycloheptanone derivatives via an intermolecular (5+2)-cycloaddition reaction. *Org. Lett.* 4 (13): 2217–2219.

909 Witherell, R.D., Ylijoki, K.E.O., and Stryker, J.M. (2008). Cobalt-mediated η5-pentadienyl/alkyne (5+2)-cycloaddition. Synthesis and characterization of unbridged η^2,η^3-coordinated cycloheptadienyl complexes. *J. Am. Chem. Soc.* 130 (7): 2176–2177.

910 Garnier, E.C. and Liebeskind, L.S. (2008). Organometallic enantiomeric scaffolding: general access to 2-substituted oxa- and azabicyclo[3.2.1]octenes via a Brønsted acid catalyzed

(5+2)-cycloaddition reaction. *J. Am. Chem. Soc.* 130 (23): 7449–7458.

911 Burns, N.Z., Witten, M.R., and Jacobsen, E.N. (2011). Dual catalysis in enantioselective oxidopyrylium-based (5+2)-cycloadditions. *J. Am. Chem. Soc.* 133: 14578–14581.

912 Witten, M.R. and Jacobsen, E.N. (2014). Catalytic asymmetric synthesis of 8-oxabicyclooctanes by intermolecular (5+2)-pyrylium cycloadditions. *Angew. Chem. Int. Ed.* 53: 5912–5916.

913 Orue, A., Uria, U., Reyes, E. et al. (2015). Catalytic enantioselective (5+2)-cycloaddition between oxidopyrylium ylides and enals under dienamine activation. *Angew. Chem. Int. Ed.* 54: 3043–3046.

914 Albrecht, L., Dickmeiss, G., Acosta Cruz, F. et al. (2012). Asymmetric organocatalytic formal [2+2]-cycloadditions via bifunctional H-bond directing dienamine catalysis. *J. Am. Chem. Soc.* 134: 2543–2546.

915 Bennett, D.M., Okamoto, I., and Danheiser, R.L. (1999). Hetero (4+2)-cycloadditions of (trialkylsilyl)vinylketenes. Synthesis of α,β-unsaturated δ-valerolactones and -lactams. *Org. Lett.* 1 (4): 641–644.

916 Fang, J., Ren, J., and Wang, Z.W. (2008). $Sc(OTf)_3$-catalyzed smooth tandem (3+2)-cycloaddition/ring opening of donor-acceptor cyclopropane 1,1-diesters with enol silyl ethers. *Tetrahedron Lett.* 49 (47): 6659–6662.

917 Takasu, K., Nagao, S., and Ihara, M. (2006). Construction of highly-functionalized cyclopentanes from silyl enol ethers and activated cyclopropanes by (3+2)-cycloaddition catalyzed by triflic imide. *Adv. Synth. Catal.* 348 (16–17): 2376–2380.

918 (a) Crabtree, R.H. (2011). An organometallic future in green and energy chemistry? *Organometallics* 30: 17–19. (b) see e.g. Bauer, E.B. (2008). Recent advances in iron catalysis in organic synthesis. *Curr. Org. Chem.* 12 (16): 1341–1369. (c) Yan, T., Fweringa, B.L., and Barta, K. (2014). Iron catalyzed direct alkylation of amines with alcohols. *Nat. Commun.* 5: 5602. (d) Ludwig, J.R., Zimmerman, P., Gianino, J.B., and Schindler, C.S. (2016). Iron(III)-catalysed carbonyl-olefin metathesis. *Nature* 533 (7603): 374–379. (e) Brill, M., Lazreg, F., Cazin, C.S.J., and Nolan, S.P. (2016). Transition metal-catalyzed carboxylation of organic substrates with carbon dioxide. *Top. Organomet. Chem.* 53 (Carbon dioxide and organometallics): 225–278. (f) Gandeepan, P., Kaplaneris, N., Santoro, S. et al. (2019). Biomass-derived solvents for sustainable transition metal-catalyzed C—H activation. *ACS Sust. Chem. Eng.* 7 (9): 8023–8040.

919 Hartwig, J.F. (2009). *Organotransition Metal Chemistry: From Bonding to Catalysis*, 1e, 1–1160. Sausalito, CA: University Science Books.

920 Frenking, G. and Fröhlich, N. (2000). The nature of the bonding in transition-metal compounds. *Chem. Rev.* 100: 717–774.

921 Krapp, A., Pandey, K.K., and Frenking, G. (2007). Transition metal-carbon complexes. A theoretical study. *J. Am. Chem. Soc.* 129: 7596–7610.

922 (a) Tolman, C.A. (1972). 16 and 18 electron rule in organometallic chemistry and homogeneous catalysis. *Chem. Soc. Rev.* 1 (3): 337–353. (b) Baird, M.C. (2014). Seventeen-electron compounds in an eighteen-electron world – an historical perspective. *J. Organomet. Chem.* 751: 50–54.

923 (a) Frenking, G. (2014). Chemical bonding in transition metals. In: *The Chemical Bond: Chemical Bonding Across the Periodic Table* (ed. G. Frenking and S. Sason), 175–218. Weinheim: Wiley-VCH. (b) Landis, C.R. and Weinhold, F. (2016). 18-Electron rule and the 3c/4e hyperbonding saturation limit. *J. Comput. Chem.* 37 (2): 237–241.

924 (a) Jensen, W.B. (2005). The origin of the 18-electron rule. *J. Chem. Educ.* 82: 28. (b) Pyykkoe, P. (2006). Understanding the eighteen-electron rule. *J. Organomet. Chem.* 691 (21): 4336–4340.

925 Hill, A.F. (2002). *Organotransition Metal Chemistry*, RSC Tutorial Chemistry 7 (ed. E.W. Abel), 1–186. Cambridge: Royal Society of Chemistry.

926 Jones, C.J. (2001). *Basic Coordination Chemistry and Ligand Field Theory, in d- and f-Block Chemistry*, RSC Tutorial Chemistry 4 (ed. E.W. Abel), 1–176. Cambridge: Royal Society of Chemistry.

927 (a) Mathey, F. and Sevin, A. (1996). *Molecular Chemistry of the Transition Elements*, 1–242. Chichester: Wiley. (b) Mathey, F. (2013). Transition metal organometallic chemistry. In: *SpringerBriefs in Molecular Sciences*, 1–100. Singapore: Springer.

928 Elschenbroich, C. (2006). *Organometallics, 3rd Completely Revised and Extended Edition*, 1–817. Weinheim: Wiley-VCH.

929 Collman, J.P. and Hegedus, L.S. (1987). *Principles and Applications of Organotransition Metal*

Chemistry, 1–945. Sausalito, CA: University Science Books.

930 Collman, J.P., Hegedus, L.S., Norton, J.R., and Finke, R.G. (1987). *Principles and Applications of Organotransition Metal Chemistry*, 2e, 1–989. Oxford: Oxford University Press.

931 Lewis, G.N. (1916). The atom and the molecule. *J. Am. Chem. Soc.* 38: 762–785.

932 (a) Langmuir, I. (1919). The structure of atoms and the octet theory of valence. *Proc. Natl. Acad. Sci. U.S.A.* 5: 252–259. (b) Langmuir, I. (1919). The arrangement of electrons in atoms and molecules. *J. Am. Chem. Soc.* 41: 868–934.

933 (a) Noury, S., Silvi, B., and Gillespie, R.J. (2002). Chemical bonding in hypervalent molecules: is the octet rule relevant? *Inorg. Chem.* 41 (8): 2164–2172. (b) Gillespie, R.J. and Silvi, B. (2002). The octet rule and hypervalence: two misunderstood concepts. *Coord. Chem. Rev.* 233–234: 53–62.

934 Dewar, M.J.S. (1945). Mechanism of the benzidine and related rearrangements. *Nature* 156 (3974): 784–784.

935 Dewar, M.J.S. (1946). The mechanism of benzidine-type rearrangements, and the role of π-electrons in organic chemistry. *J. Chem. Soc.* 406–408.

936 Dewar, M.J.S. (1946). The kinetics of some benzidine rearrangements, and a note on the mechanism of aromatic substitution. *J. Chem. Soc.* 777–781.

937 Dewar, M.J.S. (1951). A review of the π-complex theory. *Bull. Soc. Chim. Fr.* 18 (3–4): C71–C79.

938 Frenking, G. (2001). Understanding the nature of the bonding in transition metal complexes: from Dewar's molecular orbital model to an energy partitioning analysis of the metal-ligand bond. *J. Organomet. Chem.* 635: 9–23.

939 Dewar, M.J.S. (1952). A molecular orbital theory of organic chemistry. 1. General principles. *J. Am. Chem. Soc.* 74: 3341–3345.

940 Saunders, M., Vogel, P., Hagen, E.L., and Rosenfeld, J. (1973). Evidence for protonated cyclopropane intermediates from studies of stable solutions of carbonium ions. *Acc. Chem. Res.* 6 (2): 53–59.

941 Vogel, P. (1985). Carbocation Chemistry. In: *Studies in Organic Chemistry*, 1–596. Amsterdam: Elsevier.

942 Walsh, A.D. (1947). Structures of ethylene oxide and cyclopropane. *Nature* 159 (4031): 165–165.

943 Chatt, J. and Duncanson, L.A. (1953). Olefin coordination compounds. 3. Infra-red spectra and structure – attempted preparation of acetylene complexes. *J. Chem. Soc.* 2939–2947.

944 Black, M., Mais, R.H.B., and Owston, P.G. (1969). Crystal and molecular structure of Zeises salt, $KPtCl_3{\cdot}C_2H_4{\cdot}H_2O$. *Acta Crystallogr., Sect. B: Struct. Crystallogr. Cryst. Chem.* B25: 1753–1759.

945 Lippard, S.J. (1969). Metal π-organic complexes: nature of bonding and their role in chemistry. *Ann. N.Y. Acad. Sci.* 153 (3): 677–688.

946 Hartley, F.R. (1972). Metal-olefin and metal-acetylene bonding in complexes. *Angew. Chem. Int. Ed.* 11 (7): 596–606.

947 (a) Hoffmann, R. (1981). Theoretical organometallic chemistry. *Science* 211 (4486): 995–1002. (b) Hoffmann, H.M.R. (1984). The cycloaddition of allyl cations to 1,3-dienes – general-method for the synthesis of 7-membered carbocycles. *Angew. Chem. Int. Ed. Engl.* 23 (1): 1–19.

948 Mingos, D.M.P. (1982). Bonding of unsaturated organic molecules to transition metals. In: *Comprehensive Organometallic Chemistry*, Chapter 19, vol. 3 (ed. G. Wilkinson, F.G.A. Stone and E.W. Abel), 1–88. Oxford: Pergamon Press.

949 (a) Gemel, C., Steinke, T., Coloka, M. et al. (2004). Transition metal chemistry of low valent group 13 organyls. *Eur. J. Inorg. Chem.* (21): 4161–4176. (b) Raubenheimer, H.G. and Schmidbaur, H. (2012). Gold chemistry guided by the isolobality concept. *Organometallics* 31 (7): 2507–2522. (c) Che, J. and Jia, G. (2013). Recent development in the chemistry of transition metal-containing metallabenzenes and metallabenzynes. *Coord. Chem. Rev.* 257 (17–18): 2491–2521. (d) Yannello, V.J. and Fredrickson, D.C. (2015). Generality of the 18-n rule: intermetallic structural chemistry explained through isolobal analogies to transition metal complexes. *Inorg. Chem.* 54 (23): 11385–11398.

950 Mond, L., Hirtz, H., and Cowap, M.D. (1910). Some new metallic carbonyls. *J. Chem. Soc* 97: 798–809.

951 Goldberg, P. and Blanchard, A.A. (1946). Dicobalt octacarbonyl, cobalt nitrosyl tricarbonyl, and cobalt tetracarbonyl hydride. *Inorg. Synth.* 2: 238–243.

952 Clark, R.J., Whiddon, S.E., and Serfass, R.E. (1968). Low pressure synthesis of cobalt carbonyls. *J. Organomet. Chem.* 11 (3): 637–640.

953 Sumner, G.G., Klug, H.P., and Alexander, L.E. (1964). Crystal structure of dicobalt octacarbonyl. *Acta Crystallogr.* 17 (6): 732–742.

954 Sweany, R.L. and Brown, T.L. (1977). Infrared spectra of matrix-isolated dicobalt octacarbonyl – evidence for 3Rd isomer. *Inorg. Chem.* 16 (2): 415–421.

955 Hurd, D.T., Sentell, G.W., and Norton, F.J. (1949). Manganese carbonyl. *J. Am. Chem. Soc.* 71 (5): 1899–1899.

956 Brimm, E.O., Lynch, M.A., and Sesny, W.J. (1954). Preparation and properties of manganese carbonyl. *J. Am. Chem. Soc.* 76 (14): 3831–3835.

957 McDowell, R.S., Yates, J.T., and Horrocks, W.D. (1961). Infrared spectrum of $Co(CO)_3NO$. *J. Chem. Phys.* 34 (2): 530–534.

958 Legzdins, P. and Malito, J.T. (1975). Organometallic nitrosyl chemistry. 1. Reactions of nitrosyl chloride with some anionic and neutral carbonyl complexes. *Inorg. Chem.* 14 (8): 1875–1878.

959 Johnson, B.F.G. and Benfield, R.E. (1978). Structures of binary carbonyls and related compounds. 1. New approach to fluxional behavior. *J. Chem. Soc., Dalton Trans.* (11): 1554–1568.

960 Adams, H.N., Fachinetti, G., and Strahle, J. (1980). Distribution of terminal and bridging CO-groups in an anionic carbonylmetal cluster as a result of ion pair formation – crystal and molecular structure of $LiCo_3(CO)_{10}{\cdot}i\text{-}Pr_2O$. *Angew. Chem. Int. Ed. Engl.* 19 (5): 404–405.

961 Kemmitt, R.D.W. and Russell, D.R. (1982). Cobalt. In: *Comprehensive Organometallic Chemistry*, Chapter 34, vol. 5 (ed. G. Wilkinson, F.G.A. Stone and E.W. Abel), 1–276. Oxford: Pergamon Press.

962 Chiang, T., Kerber, R.C., Kimball, S.D., and Lauher, J.W. (1979). (η3-Triphenylcyclopropenyl)tricarbonylcobalt. *Inorg. Chem.* 18 (6): 1687–1691.

963 Seyferth, D. (1976). Chemistry of carbon functional alkylidynetricobalt nonacarbonyl cluster complexes. *Adv. Organomet. Chem.* 14: 97–144.

964 (a) Kubas, G.J. (2001). Metal-dihydrogen and σ-bond coordination: the consummate extension of the Dewar–Chatt–Duncanson model for metal-olefin bonding. *J. Organomet. Chem.* 635: 37–68. (b) Kubas, G.J. (2007). Dihydrogen complexes as prototypes for the coordination chemistry of saturated molecules. *Proc. Natl. Acad. Sci. U.S.A.* 104: 6901–6907.

965 Kubas, G.J. (2007). Fundamentals of H_2 binding and reactivity on transition metals underlying hydrogenase function and H_2 production and storage. *Chem. Rev.* 107: 4152–4205.

966 Kubas, G.J. (2014). Activation of dihydrogen and coordination of molecular H_2 on transition metals. *J. Organomet. Chem.* 751: 33–49.

967 Belkova, N.V., Epstein, L.M., Filippov, O.A., and Shubina, E.S. (2016). Hydrogen and dihydrogen bonds in the reactions of metal hydrides. *Chem. Rev.* 116: 8545–8587.

968 (a) Tokmic, K., Markus, C.R., Zhu, L., and Fout, A.R. (2016). Well-defined cobalt(I) dihydrogen catalyst: experimental evidence for a Co(I)/Co(III) redox process in olefin hydrogenation. *J. Am. Chem. Soc.* 138 (36): 11907–11913. (b) Eguilor, B., Esteruelas, M.A., Lezaun, V. et al. (2017). Elongated dihydrogen versus compressed dihydride in osmium complexes. *Chem. Eur. J.* 23 (7): 1526–1430. (c) Goldberg, J.M., Goldberg, K.I., Heinekey, D.M. et al. (2017). Detection of an iridium-dihydrogen complex: a proposed intermediate in ionic hydrogenation. *J. Am. Chem. Soc.* 139 (36): 12638–12646.

969 Burdett, J.K., Phillips, J.R., Pourian, M.R. et al. (1987). Electronic stability of metal dihydrogen and metal-polyhydrogen complexes. *Inorg. Chem.* 26 (18): 3054–3063.

970 Sevin, A. (1981). Oxidative addition of molecular hydrogen to a d8-transition metal – a semi-empirical theoretical study. *New J. Chem.* 5 (4): 233–241.

971 Saillard, J.Y. and Hoffmann, R. (1984). C—H and H—H activation in transition metal complexes and on surfaces. *J. Am. Chem. Soc.* 106 (7): 2006–2026.

972 Hay, P.J. (1987). Ab initio theoretical studies of dihydrogen coordination vs oxidative addition of H_2 to 5-coordinate W complexes. *J. Am. Chem. Soc.* 109 (3): 705–710.

973 Nemcsok, D.S., Kovacs, A., Rayon, V.M., and Frenking, G. (2002). Nature of the transition-metal-η^2-H_2 bond in $TM(CO)_3X_2(\eta^2\text{-}H_2)$ (TM=Cr, Mo, W; X=CO, PH_3, PCl_3, PMe_3) complexes. *Organometallics* 21 (26): 5803–5809.

974 Kubas, G.J., Ryan, R.R., Swanson, B.I. et al. (1984). Characterization of the 1st examples of isolable molecular-hydrogen complexes, $Mo(CO)_3(PCy_3)_2(H_2)$, $W(CO)_3(PCy_3)_2(H_2)$, $Mo(CO)_3(P(i\text{-}Pr)_3)_2(H_2)$,$W(CO)_3(P(i\text{-}Pr)_3)_2(H_2)$ – evidence for a side-on bonded H_2 ligand. *J. Am. Chem. Soc.* 106 (2): 451–452.

975 Kubas, G.J. (1988). Molecular hydrogen complexes – coordination of a σ-bond to transition metals. *Acc. Chem. Res.* 21 (3): 120–128.

976 Crabtree, R.H. (1990). Dihydrogen complexes – some structural and chemical studies. *Acc. Chem. Res.* 23 (4): 95–101.

977 Janak, K.E., Shin, J.H., and Parkin, G. (2004). Molybdenocene trihydride complexes: influence of a [Me_2Si] ansa bridge on classical versus nonclassical nature, stability with respect to elimination of dihydrogen, and acidity. *J. Am. Chem. Soc.* 126 (40): 13054–13070.

978 Vogt, M., Pons, V., and Heinekey, D.M. (2005). Synthesis and characterization of a dicationic dihydrogen complex of iridium with a bis-carbene ligand set. *Organometallics* 24 (8): 1832–1836.

979 Crabtree, R.H. and Lavin, M. (1985). C—H and H—H bond activation – dissociative vs nondissociative binding to iridium. *J. Chem. Soc., Chem. Commun.* 12: 794–795.

980 Crabtree, R.H. and Lavin, M. (1985). $[IrH_2(H_2)_2P(C_6H_{11})_{32}]^+$ – a non-classical polyhydride complex. *J. Chem. Soc., Chem. Commun.* 23: 1661–1662.

981 Koetzle, T.F. and Schultz, A.J. (2005). Single crystal neutron diffraction: a valuable tool for probing bond activation in transition metal σ-complexes. *Top. Catal.* 32 (3–4): 251–255.

982 Heinekey, D.M., Lledos, A., and Lluch, J.M. (2004). Elongated dihydrogen complexes: what remains of the H—H bond? *Chem. Soc. Rev.* 33 (3): 175–182.

983 Brammer, L., Howard, J.A.K., Johnson, O. et al. (1991). Low-temperature neutron diffraction study of bis(tri-*para*-tolylphosphine)heptahydridorhenium(VII), $[ReH_7(P(p\text{-toluyl})_3)_2]$ – a polyhydride complex with a short H—H contact that models an intermediate stage in the oxidative addition of dihydrogen at transition metal centers. *J. Chem. Soc., Chem. Commun.* 4: 241–243.

984 Hasegawa, T., Li, Z.W., Parkin, S. et al. (1994). Determination of the structure of $[Os(\eta^2\text{-}H_2)en(2)MeCO(2)]PF_6$ by X-ray and neutron diffraction. *J. Am. Chem. Soc.* 116 (10): 4352–4356.

985 Yousufuddin, M., Wen, T.B., Mason, S.A. et al. (2005). A neutron diffraction study of $[OsClH_3(P(Ph)_3)_3]$: complex containing a highly "stretched" dihydrogen ligand. *Angew. Chem. Int. Ed.* 44 (44): 7227–7230.

986 Tal'roze, V.L. and Lyubimova, A.K. (1952). Secondary processes in the ion source of the mass spectrograph. *Dokl. Akad. Nauk* 86: 909–912.

987 Olah, G.A., Klopman, G., and Schlosberg, R.H. (1969). Chemistry in super acids. 3. Protonation of alkanes and intermediacy of alkanonium ions pentacoordinated carbon cations of CH_5^+ type. Hydrogen exchange protolytic cleavage hydrogen abstraction and polycondensation of methane ethane 2,2-dimethylpropane (neopentane) and 2,2,3,3-tetramethylbutane in $FSO_3H\text{-}SbF_5$ ("magic acid") solution. *J. Am. Chem. Soc.* 91 (12): 3261–3268.

988 Olah, G.A., Westerman, P.W., Klopman, G., and Mo, Y.K. (1972). Electrophilic reactions at single bonds. 7. Hydrogen-deuterium exchange accompanying protolysis (deuterolysis) of borohydride and aluminum-hydride anions with anhydrous strong acids – intermediacy of pentahydroboron (BH_5) and pentahydroaluminum (AlH_5) and their structural relationship with methonium ion(CH_5^+). *J. Am. Chem. Soc.* 94 (22): 7859–7862.

989 White, E.T., Tang, J., and Oka, T. (1999). CH_5^+: the infrared spectrum observed. *Science* 284 (5411): 135–137.

990 Smith, D., Adams, N.G., and Alge, E. (1982). Isotope exchange and collisional association in the reactions of CH_3^+ and its deuterated analogs with H_2, HD, and D_2. *J. Chem. Phys.* 77 (3): 1261–1268.

991 Huang, X.C., Johnson, L.M., Bowman, J.M., and McCoy, A.B. (2006). Deuteration effects on the structure and infrared spectrum of CH_5^+. *J. Am. Chem. Soc.* 128 (11): 3478–3479.

992 Huang, X.C., McCoy, A.B., Bowman, J.M. et al. (2006). Quantum deconstruction of the infrared spectrum of CH_5^+. *Science* 311 (5757): 60–63.

993 Kumar, P. and Marx, D. (2006). Understanding hydrogen scrambling and infrared spectrum of bare CH_5^+ based on ab initio simulations. *Phys. Chem. Chem. Phys.* 8: 573–586.

994 Lias, S.G., Bartmess, J.E., Liebman, J.F. et al. (1988). Gas-phase ion and neutral thermochemistry. *J. Phys. Chem. Ref. Data* 17 (Suppl. 1): 861.

995 Lias, S.G., Bartmess, J., Liebman, J. et al. (2009). Ion Energetics Data. NIST (National Institute of Standards and Technology) Chemistry WebBook: http://webbook.nist.gov/.

996 Crabtree, R.H. and Hamilton, D.G. (1986). Classical (M=Os) and nonclassical (M=Fe, Ru) polyhydride structures for the complexes $MH_4(PR_3)_3$. *J. Am. Chem. Soc.* 108 (11): 3124–3125.

997 Crabtree, R.H., Lavin, M., and Bonneviot, L. (1986). Some molecular hydrogen complexes of iridium. *J. Am. Chem. Soc.* 108 (14): 4032–4037.

998 Clark, H.C. and Smith, M.J.H. (1986). Chemistry of platinum hydrides – a Pt(II) *cis*-dihydride or a Pt(0) η^2-dihydrogen complex. *J. Am. Chem. Soc.* 108 (13): 3829–3830.

999 Duncan, J.L. and Harper, J. (1984). The structure of the diborane molecule. *Mol. Phys.* 51: 371–380.

1000 Yao, Y.S. and Hoffmann, R. (2011). BH_3 under pressure: leaving the molecular diborane motif. *J. Am. Chem. Soc.* 133: 21002–21009.

1001 Turley, J.W. and Rinn, H.W. (1969). Crystal structure of aluminum hydride. *Inorg. Chem.* 8: 18–22.

1002 Andrews, L. and Wang, X.F. (2003). The infrared spectrum of Al_2H_6 in solid hydrogen. *Science* 299: 2049–2052.

1003 Wang, X.F. and Andrews, L. (2003). Infrared spectra of gallium hydrides in solid hydrogen: $GaH_{1,2,3}$, $Ga_2H_{2,4,6}$, and the $GaH_{2,4}$- anions. *J. Phys. Chem. A* 107: 11371–11379.

1004 Jones, C. (2001). The stabilisation and reactivity of indium trihydride complexes. *J. Chem. Soc., Chem. Commun.* (22): 2293–2298.

1005 Longuet-Higgins, H.C. (1946). The structure of some electron-deficient molecules. *J. Chem. Soc.* 139–143.

1006 Shen, M.Z., Liang, C.X., and Schaefer, H.F. (1993). The tetramer of borane and its heavier valence-isoelectronic analogs – M_4H_{12} with M=B, Al, and Ga. *Chem. Phys.* 171: 325–345.

1007 Giri, R., Shi, B.F., Engle, K.M. et al. (2009). Transition metal-catalyzed C—H activation reactions: diastereoselectivity and enantioselectivity. *Chem. Soc. Rev.* 38: 3242–3272.

1008 Jazzar, R., Hitce, J., Renaudat, A. et al. (2010). Functionalization of organic molecules by transition metal-catalyzed C(sp^3)-H activation. *Chem. Eur. J.* 16: 2654–2672.

1009 Chen, D.Y.K. and Youn, S.W.C.-H. (2012). Activation: a complementary tool in the total synthesis of complex natural products. *Chem. Eur. J.* 18: 9452–9474.

1010 Kuhl, N., Hopkinson, M.N., Wencel-Delord, J., and Glorius, F. (2012). Beyond directing groups: transition metal-catalyzed C—H activation of simple arenes. *Angew. Chem. Int. Ed.* 51: 10236–10254.

1011 Gao, K. and Yoshikai, N. (2014). Low-valent cobalt catalysis: new opportunities for C—H functionalization. *Acc. Chem. Res.* 47: 1208–1219.

1012 Zhang, W., Wang, N.X., and Xing, Y.L. (2015). Advances in transition metal-catalyzed direct sp^3-C—H bond functionalization. *Synlett* 26: 2088–2098.

1013 Gensch, T., Hopkinson, M.N., Glorius, F., and Wencel-Delord, J. (2016). Mild metal-catalyzed C—H activation: examples and concepts. *Chem. Soc. Rev.* 45: 2900–2936.

1014 Beak, P. and Snieckus, V. (1982). Directed lithiation of aromatic tertiary amides – an evolving synthetic methodology for polysubstituted aromatics. *Acc. Chem. Res.* 15 (10): 306–312.

1015 Whisler, M.C., MacNeil, S., Snieckus, V., and Beak, P. (2004). Beyond thermodynamic acidity: a perspective on the complex-induced proximity effect (CIPE) in deprotonation reactions. *Angew. Chem Int. Ed.* 43 (17): 2206–2225.

1016 Gilman, H. and Morton, J.W. (1954). The metalation reaction with organolithium compounds. *Org. React.* 8: 258–304.

1017 Dimroth, O. (1898). Direct Einführung von Quecksilber in aromatische Verbindungen. *Ber. Dtsch. Chem. Ges.* 31: 2154–2156.

1018 Dimroth, O. (1902). On the mercurisation of aromatic compounds. *Ber. Dtsch. Chem. Ges.* 35: 2032–2045.

1019 Kharasch, M.S. and Isbell, H.S. (1931). The chemistry of organic gold compounds. III. Direct introduction of gold into the aromatic nucleus (preliminary communication). *J. Am. Chem. Soc.* 53 (2): 3053–3059.

1020 Chatt, J. and Davidson, J.M. (1965). Tautomerism of arene and ditertiary phosphine complexes of Ru(0) and preparation of new types of hydridocomplexes of Ru(II). *J. Chem. Soc.* 843–855.

1021 Green, M.L.H. and Knowles, P.J. (1970). Formation of a W phenyl hydride derivative from benzene. *J. Chem. Soc., Chem. Commun.* (24): 1677–1677.

1022 Foley, P. and Whitesides, G.M. (1979). Thermal generation of bis(triethylphosphine)-3,3-dimethylplatinacyclobutane from dineopentyl-bis(triethylphosphine)Pt(II). *J. Am. Chem. Soc.* 101 (10): 2732–2733.

1023 Bennett, M.A. and Milner, D.L. (1967). Chlorotris(triphenylphosphine)Ir(I) – an example of hydrogen transfer to a metal from a coordinated ligand. *J. Chem. Soc., Chem. Commun.* (12): 581–582.

1024 Gol'dshleger, N.F., Tyabin, M.B., Shilov, A.E., and Shteinman, A.A. (1969). Activation of saturated hydrocarbons. Deuterium-hydrogen exchange in solutions of transition metal complexes. *Zh. Fiz. Khim.* 43 (8): 2174–2175.

1025 Crabtree, R.H. (2015). A. E. Shilov's influence on early work in organometallic C—H activation and functionalization. *J. Organomet. Chem.* 793: 41–46.

1026 Graham, M.A., Turner, J.J., Poliakof, M., and Perutz, R.N. (1972). Intermediates in chromium carbonyl photochemistry. *J. Organomet. Chem.* 34 (2): C34–C36.

1027 Perutz, R.N. and Turner, J.J. (1975). Photochemistry of group-6 hexacarbonyls in low-temperature matrices. 3. Interaction of pentacarbonyls with noble gases and other matrices. *J. Am. Chem. Soc.* 97 (17): 4791–4800.

1028 Cooper, A.C., Clot, E., Huffman, J.C. et al. (1999). Computational and experimental test of steric influence on agostic interactions: a homologous series for Ir(III). *J. Am. Chem. Soc.* 121 (1): 97–106.

1029 Wick, D.D., Reynolds, K.A., and Jones, W.D. (1999). Evidence for methane σ-complexes in reductive elimination reactions from Tp′Rh(L)(CH_3)H. *J. Am. Chem. Soc.* 121 (16): 3974–3983.

1030 Tempel, D.J., Johnson, L.K., Huff, R.L. et al. (2000). Mechanistic studies of Pd(II)-α-diimine-catalyzed olefin polymerizations. *J. Am. Chem. Soc.* 122 (28): 6686–6700.

1031 Scherer, W. and McGrady, G.S. (2004). Agostic interactions in d(0) metal alkyl complexes. *Angew. Chem. Int. Ed.* 43 (14): 1782–1806.

1032 Dawoodi, Z., Green, M.L.H., Mtetwa, V.S.B., and Prout, K. (1982). Evidence for a direct bonding interaction between Ti and a β-C-H moiety in a Ti-ethyl compound – X-ray crystal structure of [Ti($Me_2PCH_2CH_2PMe_2$)$EtCl_3$]. *J. Chem. Soc., Chem. Commun.* (14): 802–803.

1033 (a) Brookhart, M. and Green, M.L.H. (1983). Carbon–hydrogen transition metal bonds. *J. Organomet. Chem.* 250 (1): 395–408. (b) Brookhart, M., Green, M.L.H., and Wong, L.L. (1988). Carbon–hydrogen transition metal bonds. *Prog. Inorg. Chem.* 36: 1–124.

1034 Brookhart, M., Green, M.L.H., and Parkin, G. (2007). Agostic interactions in transition metal compounds. *Proc. Natl. Acad. Sci. U.S.A.* 104: 6908–6914.

1035 Lein, M. (2009). Characterization of agostic interactions in theory and computation. *Coord. Chem. Rev.* 253: 625–634.

1036 Scherer, W., Herz, V., and Hauf, C. (2012). On the nature of β-agostic interactions: a comparison between the molecular orbital and charge density picture. In: *Electron Density and Chemical Bonding I: Experimental Charge Density Studies*, Structure and Bonding, vol. 146 (ed. D. Stalke), 159–207. Heidelberg: Springer-Verlag.

1037 (a) Mitoraj, M.P., Michalak, A., and Ziegler, T. (2007). On the nature of the agostic bond between metal centers and β-hydrogen atoms in alkyl complexes. An analysis based on the extended transition state method and the natural orbitals for chemical valence scheme (ETS-NOCV). *Organometallics* 28 (13): 3727–3733. (b) Pike, S.D., Thompson, A.L., Algarra, A.G. et al. (2012). Synthesis and characterization of a Rh (I) σ-alkane complex in the solid state. *Science* 337 (6102): 1648–1651.

1038 Grubbs, R.H. and Coates, G.W. (1996). α-Agostic interactions and olefin insertion in metallocene polymerization catalysts. *Acc. Chem. Res.* 29 (2): 85–93.

1039 Kubas, G.J. (2001). *Metal Dihydrogen and σ-Bond Complexes*, 1–472. New York: Kluwer Academic/Plenum Publishers.

1040 Labinger, J.A. and Bercaw, J.E. (2002). Understanding and exploiting C—H bond activation. *Nature* 417 (6888): 507–514.

1041 Perutz, R.N. and Procacci, B. (2016). Photochemistry of transition metal hydrides. *Chem. Rev.* 116: 8506–8544.

1042 Janowicz, A.H. and Bergman, R.G. (1982). Activation in completely saturated hydrocarbons – direct observation of M + R—H=M(R)(H). *J. Am. Chem. Soc.* 104 (1): 352–354.

1043 Hoyano, J.K. and Graham, W.A.G. (1982). Oxidative addition of the C—H bonds of neopentane and cyclohexane to a photochemically generated Ir(I) complex. *J. Am. Chem. Soc.* 104 (13): 3723–3725.

1044 Hoyano, J.K., McMaster, A.D., and Graham, W.A.G. (1983). Activation of methane by Ir complexes. *J. Am. Chem. Soc.* 105 (24): 7190–7191.

1045 Banister, J.A., Cooper, A.I., Howdle, S.M. et al. (1996). "Solvent-free" photochemical activation of CH_4, C_2H_4, and C_2H_6 by (C_5Me_5)Ir$(CO)_2$ in supercritical fluid solution. *Organometallics* 15 (7): 1804–1812.

1046 Hall, C. and Perutz, R.N. (1996). Transition metal alkane complexes. *Chem. Rev.* 96 (8): 3125–3146.

1047 Driver, T.G., Day, M.W., Labinger, J.A., and Bercaw, J.E. (2005). Mechanism of C—H bond activation of alkyl-substituted benzenes by cationic Pt(II) complexes. *Organometallics* 24 (15): 3644–3654.

1048 Sun, X.Z., George, M.W., Kazarian, S.G. et al. (1996). Can organometallic noble gas compounds be observed in solution at room temperature? A time-resolved infrared (TRIR) and UV spectroscopic study of the photochemistry of $M(CO)_6$ (M=Cr, Mo, and W) in supercritical noble gas and CO_2 solution. *J. Am. Chem. Soc.* 118 (43): 10525–10532.

1049 Bengali, A.A., Schultz, R.H., Moore, C.B., and Bergman, R.G. (1994). Activation of the C—H bonds in neopentane and neopentane-D12 by (η^5-C_5Me_5)$Rh(CO)_2$ – spectroscopic and temporal resolution of Rh-Kr and Rh-alkane complex intermediates. *J. Am. Chem. Soc.* 116 (21): 9585–9589.

1050 Arndtsen, B.A., Bergman, R.G., Mobley, T.A., and Peterson, T.H. (1995). Selective intermolecular C—H bond activation by synthetic metal complexes in homogeneous solution. *Acc. Chem. Res.* 28 (3): 154–162.

1051 McNamara, B.K., Yeston, J.S., Bergman, R.G., and Moore, C.B. (1999). The effect of alkane structure on rates of photoinduced C—H bond activation by $Cp^*Rh(CO)_2$ in liquid rare gas media: an infrared flash kinetics study. *J. Am. Chem. Soc.* 121 (27): 6437–6443.

1052 Vetter, A.J., Flaschenriem, C., and Jones, W.D. (2005). Alkane coordination selectivity in hydrocarbon activation by [Tp′Rh(CNneopentyl)]: the role of alkane complexes. *J. Am. Chem. Soc.* 127 (35): 12315–12322.

1053 Gross, C.L. and Girolami, G.S. (1998). Metal-alkane complexes. Rapid exchange of hydrogen atoms between hydride and methyl ligands in [(C_5Me_5)Os(dmpm)(CH_3)H+]. *J. Am. Chem. Soc.* 120 (26): 6605–6606.

1054 Weiller, B.H., Wasserman, E.P., Bergman, R.G. et al. (1989). Time-resolved Ir spectroscopy in liquid rare-gases – direct rate measurement of an intermolecular alkane C—H oxidative addition reaction. *J. Am. Chem. Soc.* 111 (21): 8288–8290.

1055 Sun, X.Z., Grills, D.C., Nikiforov, S.M. et al. (1997). Remarkable stability of (η^5-C_5H_5)$Re(CO)_2L$ (L = *n*-heptane, Xe, and Kr): a time-resolved infrared spectroscopic study of (η^5-C_5H_5)$Re(CO)_3$ in conventional and supercritical fluid solution. *J. Am. Chem. Soc.* 119 (32): 7521–7525.

1056 Geftakis, S. and Ball, G.E. (1998). Direct observation of a transition metal alkane complex, $CpRe(CO)_2$(cyclopentane), using NMR spectroscopy. *J. Am. Chem. Soc.* 120 (38): 9953–9954.

1057 Lin, Z.Y. (2002). Structural and bonding characteristics in transition metal-silane complexes. *Chem. Soc. Rev.* 31 (4): 239–245.

1058 Rybtchinski, B. and Milstein, D. (1999). Metal insertion into C—C bonds in solution. *Angew. Chem. Int. Ed.* 38 (7): 870–883.

1059 Van der Boom, M.E. and Milstein, D. (2003). Cyclometalated phosphine-based pincer complexes: mechanistic insight in catalysis, coordination, and bond activation. *Chem. Rev.* 103 (5): 1759–1792.

1060 Jun, C.H., Moon, C.W., Lee, H., and Lee, D.Y. (2002). Chelation-assisted carbon–carbon bond activation by Rh(I) catalysts. *J. Mol. Catal. A: Chem.* 189 (1): 145–156.

1061 Jun, C.H. (2004). Transition metal-catalyzed carbon–carbon bond activation. *Chem. Soc. Rev.* 33 (9): 610–618.

1062 Jun, C.H. and Park, J.W. (2014). Metal-organic cooperative catalysis in C—C bond activation. In: *C—C Bond Activation*, Topics in Current Chemistry, vol. 346 (ed. G. Dong), 59–83. Berlin, Heidelberg: Springer-Verlag.

1063 Jones, W.D. (2014). Mechanistic studies of transition metal-mediated C—C bond activation. In: *C—C Bond Activation*, Topics in Current Chemistry, vol. 346 (ed. G. Dong), 1–31. Berlin, Heidelberg: Springer-Verlag.

1064 Etienne, M. and Weller, A.S. (2014). Intramolecular C—C agostic complexes: C—C σ interactions by another name. *Chem. Soc. Rev.* 43: 242–259.

1065 Souillart, L. and Cramer, N. (2015). Catalytic C—C bond activations via oxidative addition to transition metals. *Chem. Rev.* 115: 9410–9464.

1066 Armentrout, P.B., Halle, L.F., and Beauchamp, J.L. (1981). Periodic trends in transition metal–hydrogen, metal–carbon, and metal–oxygen bond dissociation energies – correlation with reactivity and electronic structure. *J. Am. Chem. Soc.* 103: 6501–6502.

1067 Skinner, H.A. and Connor, J.A. (1985). Metal-ligand bond energies in organometallic compounds. *Pure Appl. Chem.* 57: 79–88.

1068 Wiedner, E.S., Chambers, M.B., Pitman, C.L. et al. (2016). Thermodynamic hydricity of transition metal hydrides. *Chem. Rev.* 115: 8655–8692.

1069 Low, J.J. and Goddard, W.A. (1984). Theoretical studies of oxidative addition and reductive elimination – $H_2 + Pt(PH_3)_2 {=} Pt(H)_2(PH_3)_2$. *J. Am. Chem. Soc.* 106 (23): 6928–6937.

1070 Low, J.J. and Goddard, W.A. (1986). Theoretical studies of oxidative addition and reductive elimination. 3. C—H and C—C reductive coupling from Pd and Pt bis(phosphine) complexes. *J. Am. Chem. Soc.* 108 (20): 6115–6128.

1071 Low, J.J. and Goddard, W.A. (1986). Theoretical studies of oxidative addition and reductive elimination. 2. Reductive coupling of H—H, H—C, and C—C bonds from Pd and Pt complexes. *Organometallics* 5 (4): 609–622.

1072 Diefenbach, A. and Bickelhaupt, F.M. (2004). Activation of H—H, C—H, C—C, and C—Cl bonds by Pd(0). Insight from the activation strain model. *J. Phys. Chem. A* 108 (40): 8460–8466.

1073 Johnson, D.F. and Weinberg, W.H. (1995). Quantification of the selective activation of C—C bonds in short-chain alkanes – reactivity of ethane, propane, isobutane, *n*-butane and neopentane on Ir(11). *J. Chem. Soc., Faraday Trans.* 91 (20): 3695–3702.

1074 Periana, R.A. and Bergman, R.G. (1984). Rapid intramolecular rearrangement of a hydridocyclopropylrhodium complex to a rhodacyclobutane – independent synthesis of the metallacycle by addition of hydride to the central carbon atom of a cationic Rh π-allyl complex. *J. Am. Chem. Soc.* 106 (23): 7272–7273.

1075 Bart, S.C. and Chirik, P.J. (2003). Selective, catalytic carbon–carbon bond activation and functionalization promoted by late transition metal catalysts. *J. Am. Chem. Soc.* 125 (4): 886–887.

1076 Rybtchinski, B., Vigalok, A., Ben-David, Y., and Milstein, D. (1996). A room temperature direct metal insertion into a nonstrained carbon–carbon bond in solution. C—C vs C—H bond activation. *J. Am. Chem. Soc.* 118 (49): 12406–12415.

1077 Gozin, M., Weisman, A., Ben-David, Y., and Milstein, D. (1993). Activation of a C—C bond in solution by transition metal insertion. *Nature* 364 (6439): 699–701.

1078 Gozin, M., Aizenberg, M., Liou, S.Y. et al. (1994). Transfer of methylene groups promoted by metal complexation. *Nature* 370 (6484): 42–44.

1079 Sundermann, A., Uzan, O., Milstein, D., and Martin, J.M.L. (2000). Selective C—C vs C—H bond activation by Rh(I) PCP pincer complexes. A computational study. *J. Am. Chem. Soc.* 122: 7095–7104.

1080 Salem, H., Ben-David, Y., Shimon, L.J.W., and Milstein, D. (2006). Exclusive C—C activation and an apparent α-H elimination with a rhodium phosphinite pincer complex. *Organometallics* 25 (9): 2292–2300.

1081 Montag, M., Efremenko, I., skin-Posner, Y. et al. (2012). Exclusive C—C oxidative addition in a Rh thiophosphoryl pincer complex and computational evidence for an η^3-C—C—H agostic intermediate. *Organometallics* 31: 505–512.

1082 Sundermann, A., Uzan, O., and Martin, J.M.L. (2001). Exclusive C—C activation in the Rh(I) PCN pincer complex. A computational study. *Organometallics* 20 (9): 1783–1791.

1083 Gandelman, M., Shimon, L.J.W., and Milstein, D. (2003). C—C versus C—H activation and versus agostic C—C interaction controlled by electron density at the metal center. *Chem. Eur. J.* 9 (18): 4295–4300.

1084 Brayshaw, S.K., Green, J.C., Kociok-Koehn, G. et al. (2006). A Rh complex with one (RhC)...C—C and one (RhH)...H—C agostic bond. *Angew. Chem. Int. Ed.* 45 (3): 452–456.

1085 Suresh, C.H. and Koga, N. (2004). Orbital interactions in the Ru olefin metathesis catalysts. *Organometallics* 23 (1): 76–80.

1086 Suresh, C.H. and Baik, M.H. (2005). α,β-(C—C—C) Agostic bonds in transition metal based olefin metathesis catalyses. *J. Chem. Soc., Dalton Trans.* (18): 2982–2984.

1087 Harvey, B.G., Mayne, C.L., Arif, A.M., and Ernst, R.D. (2005). Structural and spectroscopic demonstration of agostic C—C interactions in electron-deficient metallacyclobutanes and related cage complexes: possible implications for olefin polymerizations and metatheses. *J. Am. Chem. Soc.* 127 (47): 16426–16435.

1088 Olah, G.A., Kelly, D.P., Jeuell, C.L., and Porter, R.D. (1970). Stable carbonium ions. 98. Nonclassical cyclopropylcarbinyl cation. *J. Am. Chem. Soc.* 92 (8): 2544–2546.

1089 Olah, G.A., Porter, R.D., Kelly, D.P., and Jeuell, C.L. (1972). Stable carbocations. 114. Structure of cyclopropylcarbinyl and cyclobutyl cations. *J. Am. Chem. Soc.* 94 (1): 146–156.

1090 Staral, J.S., Yavari, I., Roberts, J.D. et al. (1978). Low-temperature ^{13}C NMR spectroscopic investigation of C_4H_7+ evidence for an equilibrium involving non-classical bicyclobutonium ion and bisected cyclopropylcarbinyl cation. *J. Am. Chem. Soc.* 100 (25): 8016–8018.

1091 Staral, J.S. and Roberts, J.D. (1978). Observation by ^{13}C NMR spectroscopy of hydride shifts in C_4H_7+ derived from cyclopropylcarbinol-1-^{13}C in SbF_5-SO_2ClF-SO_2F_2 solution. *J. Am. Chem. Soc.* 100 (25): 8018–8020.

1092 Dewar, M.J.S. and Reynolds, C.H. (1984). Ground-states of molecules. 68. The $C_4H_7^+$

potential surface. *J. Am. Chem. Soc.* 106 (21): 6388–6392.

1093 Olah, G.A., Prakash Reddy, V., and Surya Prakash, G.K. (1992). Long-lived cyclopropylcarbinyl cations. *Chem. Rev.* 92 (1): 69–95.

1094 Guan, W., Zeng, G., Kameo, H. et al. (2016). Cooperative catalysis of combined systems of transition metal complexes with Lewis acids: theoretical understanding. *Chem. Rev.* 16: 2405–2425.

1095 (a) Crabtree, R.H. (2005). *The Organometallic Chemistry of the Transition Metals*, 4e, 1–600. New York: Wiley. (b) Crabtree, R.H. (2009). *The Organometallic Chemistry of the Transition Metals*, 5e, 1–546. New York: Wiley. (c) Crabtree, R.H. (2014). *The Organometallic Chemistry of the Transition Metals*, 6e, 1–520. New York: Wiley.

1096 Gray, H.B. and Olcott, R.J. (1962). Kinetics of reactions of diethylenetriamineaquoPt(II) ion with different ligands. *Inorg. Chem.* 1 (3): 481–485.

1097 Basolo, F., Gray, H.B., and Pearson, R.G. (1960). Mechanism of substitution reactions of complex ions. 17. Rates of reaction of some Pt(II) and Pd(II) complexes with pyridine. *J. Am. Chem. Soc.* 82 (16): 4200–4203.

1098 Zingales, F., Chiesa, A., and Basolo, F. (1966). Kinetics and mechanism of substitution reactions of arene Mo tricarbonyl. *J. Am. Chem. Soc.* 88 (12): 2707–2710.

1099 Basolo, F. (1981). Mechanisms of carbon monoxide replacement in metal carbonyls. *Inorg. Chim. Acta* 50 (1): 65–70.

1100 Tilset, M., Vollhardt, K.P.C., and Boese, R. (1994). Chemistry of dinuclear fulvalene complexes – dihydrides, zwitterions, and ring-slippage complexes derived from (fulvalene)$M_2(CO)_6$ (M=Mo, W). *Organometallics* 13: 3146–3169.

1101 Bennett, M.A., Lu, Z.B., Wang, X.Q. et al. (1998). Ligand-induced ring slippage of η^6- to η^4-naphthalene. Preparation and structural characterization of Ru(η^4-C10H8)(η^4-1,5-C8H12)(L) [L=PMe3, PEt3, P(OMe)3] and of derived binuclear complexes containing bridging naphthalene, Ru2(μ-η^6:η^4-C10H8)(η^4-1,5-C8H12)2(L) [L=PEt3, P(OMe)3]. *J. Am. Chem. Soc.* 120: 10409–10415.

1102 (a) Butterick, R. III,, Ramachandran, B.M., Carroll, P.J., and Sneddon, L.G. (2006). Chemistry of Mn- and Re(tricarbadecaboranyl)(tricarbonyl) complexes: evidence for an associative mechanism of ligand substitution involving an η6–η4 cage-slippage process analogous to η5–η3-cyclopentadienyl ring-slippage. *J. Am. Chem. Soc.* 128 (28): 8626–8637. (b) Ahmed, H. and McGrady, J.E. (2008). On the role of the indenyl effect in controlling intramolecular hydride transfer in iron carbonyl complexes. *J. Organomet. Chem.* 693 (25): 3697–3702.

1103 (a) Glueck, D.S. and Bergman, R.G. (1991). Mechanism of ligand substitution in an Ir amide complex. *Organometallics* 10 (5): 1479–1486. (b) Calhorda, M.J., Romao, C.C., and Veiros, L.F. (2002). The nature of the indenyl effect. *Chem. Eur. J.* 8: 868–875. (c) Cheung, W.-M., Huang, E.K., Zhu, J. et al. (2013). Facile η5–η1 ring slippage of the cycloolefin ligands in osmocene and bis(η5-indenyl)ruthenium(II). *Inorg. Chem.* 52 (18): 10449–10455. (d) Orian, L., Swart, M., and Bickelhaupt, F.M. (2014). Indenyl effect due to metal slippage? Computational exploration of Rh-catalyzed acetylene (2+2+2)-cyclotrimerization. *ChemPhysChem* 15 (1): 219–228.

1104 Kayran, C., Kozanoglu, F., Ozkar, S. et al. (1999). Ligand substitution kinetics in M(CO)4($\eta^{2:2}$-1,5-cyclooctadiene) complexes (M=Cr, Mo, W) – substitution of 1,5-cyclooctadiene by bis(diphenylphosphino)alkanes. *Inorg. Chim. Acta* 284 (2): 229–236.

1105 Howell, J.A.S. and Burkinshaw, P.M. (1983). Ligand substitution reactions at low-valent 4-coordinate, 5-coordinate, and 6-coordinate transition metal centers. *Chem. Rev.* 83 (5): 557–599.

1106 Szymanska-Buzar, T. (1997). Photochemical reactions of group 6 metal carbonyls in catalytic transformation of alkenes and alkynes. *Coord. Chem. Rev.* 159: 205–220.

1107 Baerends, E.J. and Rosa, A. (1998). Metal-CO photodissociation in transition metal complexes: the role of ligand-field and charge-transfer excited states in the photochemical dissociation of metal-ligand bonds. *Coord. Chem. Rev.* 177: 97–125.

1108 Ellis, J.E., Faltynek, R.A., Rochfort, G.L. et al. (1980). Synthesis and properties of (triphenylphosphine)pentacarbonylvanadium(0), $V(CO)_5Ph_3P$. *Inorg. Chem.* 19 (4): 1082–1085.

1109 Shi, Q.Z., Richmond, T.G., Trogler, W.C., and Basolo, F. (1984). Mechanism of carbon monoxide substitution in metal carbonyl radicals – vanadium hexacarbonyl and its phosphine substituted derivatives. *J. Am. Chem. Soc.* 106 (1): 71–76.

1110 Shen, J.K., Freeman, J.W., Hallinan, N.C. et al. (1992). Syntheses, kinetics, and mechanism of ligand substitution reactions of 17-electron half-open chromocene carbonyl complexes. *Organometallics* 11 (10): 3215–3224.

1111 Kowaleski, R.M., Basolo, F., Trogler, W.C. et al. (1987). Syntheses, kinetics, and mechanism of ligand substitution reactions of 17-electron cyclopentadienyl and pentadienyl vanadium carbonyl complexes. *J. Am. Chem. Soc.* 109 (16): 4860–4869.

1112 Hershberger, J.W., Klingler, R.J., and Kochi, J.K. (1983). Kinetics, thermodynamics, and mechanism of the radical chain process for ligand substitution of metal carbonyls. *J. Am. Chem. Soc.* 105 (1): 61–73.

1113 Carré, F., Colomer, E., Corriu, R.J.P., and Vioux, A. (1984). Stereochemistry of the insertion of manganese into Si—H and Ge—H bonds – complexes containing a 2-electron, 3-center Mn...H...Si (Or Ge) interaction. *Organometallics* 3 (8): 1272–1278.

1114 Park, Y., Ahn, S., Kang, D., and Baik, M.H. (2016). Mechanism of Rh-catalyzed oxidative cyclizations: closed versus open shell pathways. *Acc. Chem. Res.* 49: 1263–1270.

1115 Zhou, P., Vitale, A.A., Filippo, J.S., and Saunders, W.H. (1985). Kinetic isotope effect associated with the dissociative addition of dihydrogen to *trans*-Ir(CO)Cl$(Ph_3P)_2$. *J. Am. Chem. Soc.* 107 (26): 8049–8054.

1116 Stoutland, P.O. and Bergman, R.G. (1985). Insertion of Ir into the C—H bonds of alkenes – the π-complex cannot be an intermediate. *J. Am. Chem. Soc.* 107 (15): 4581–4582.

1117 Jones, W.D. and Feher, F.J. (1984). The mechanism and thermodynamics of alkane and arene C—H bond activation in $(C_5Me_5)Rh(PMe_3)(R)(H)$. *J. Am. Chem. Soc.* 106 (6): 1650–1663.

1118 Haggin, J. (1993). Catalytic process cuts isobutene from C_4 streams. *Chem. Eng. News* 71 (36): 31–32.

1119 Braun, T. (2005). Oxidative addition of NH_3 to a transition metal complex: a key step for the metal-mediated derivatization of ammonia? *Angew. Chem. Int. Ed.* 44: 5012–5014.

1120 Teltewskoi, M., Kallaene, S.I., Braun, T., and Herrmann, R. (2013). Synthesis of Rh(I) boryl complexes: catalytic N—H activation of anilines and ammonia. *Eur. J. Inorg. Chem.* 2013: 5762–5768.

1121 Chang, Y.H., Nakajima, Y., Tanaka, H. et al. (2013). Facile N—H bond cleavage of ammonia by an Ir complex bearing a noninnocent PNP-pincer type phosphaalkene ligand. *J. Am. Chem. Soc.* 135: 11791–11794.

1122 McCarthy, S.M., Lin, Y.C., Devarajan, D. et al. (2014). Intermolecular N—H oxidative addition of ammonia, alkylamines, and arylamines to a planar σ3-phosphorus compound via an entropy-controlled electrophilic mechanism. *J. Am. Chem. Soc.* 136: 4640–4650.

1123 Velez, E., Betore, M.P., Casado, M.A., and Polo, V. (2015). N—H Activation of ammonia by [{M(μ-OMe)(cod)}$_2$] (M=Ir, Rh) complexes: a DFT study. *Organometallics* 34: 3959–3966.

1124 Wang, D.Y., Choliy, Y., Haibach, M.C. et al. (2016). Assessment of the electronic factors determining the thermodynamics of "oxidative addition" of C—H and N—H bonds to Ir(I) complexes. *J. Am. Chem. Soc.* 138: 149–163.

1125 Betoré, M.P., Casado, M.A., García-Orduña, P. et al. (2016). Oxidative addition of the N—H bond of ammonia to Ir bis(phosphane) complexes: a combined experimental and theoretical study. *Organometallics* 35: 720–731.

1126 Joslin, F.L., Johnson, M.P., Mague, J.T., and Roundhill, D.M. (1991). Synthesis and reaction chemistry of new (η^5-cyclopentadienyl)Ru(II) amine and amide complexes. *Organometallics* 10 (8): 2781–2794.

1127 Cámpora, J., Palma, P., del Río, D. et al. (2004). Synthesis and reactivity of a mononuclear parent amido Ni complex. Structures of Ni[C_6H_3-2,6-(($CH_2P(Pr)_2$)-Pr-*i*)$_2$](NH_2) and Ni[C_6H_3-2,6-(($CH_2P(Pr)_2$)-Pr-*i*)$_2$](OMe). *Organometallics* 23 (24): 5653–5655.

1128 Conner, D., Jayaprakash, K.N., Cundari, T.R., and Gunnoe, T.B. (2004). Synthesis and reactivity of a coordinatively unsaturated Ru(II) parent amido complex: studies of X—H activation (X=H or C). *Organometallics* 23 (11): 2724–2733.

1129 Zhao, J., Goldman, A.S., and Hartwig, J.F. (2005). Oxidative addition of ammonia to form a stable monomeric amido hydride complex. *Science* 307 (5712): 1080–1082.

1130 (a) Aye, K.T., Gelmini, L., Payne, N.C. et al. (1990). Stereochemistry of the oxidative addition of an epoxide to Pt(II): relevance to catalytic reactions of epoxides. *J. Am. Chem. Soc.* 112 (6): 2464–2465. (b) Jamali, S., Nabavizadeh, S.M., and Rashidi, M. (2005). Oxidative addition of methyl iodide to a new type of binuclear Pt(II) complex: a kinetic study. *Inorg. Chem.* 44 (23): 8594–8601. (c) Guo, Y.-A., Liang, T., Seung, W. et al. (2017). Nickel-catalyzed cross-coupling of vinyl dioxanones to form enantiomerically enriched cyclopropanes. *J. Am. Chem. Soc.* 139 (20): 6847–6850.

1131 Cross, R.J. (1985). Ligand substitution reactions of square-planar molecules. *Chem. Soc. Rev.* 14 (3): 197–223.

1132 Kang, J.W. and Maitlis, P.M. (1971). (Pentamethylcyclopentadienyl)Rh and -Ir complexes. 4. Oxidative addition reactions to dicarbonyl(pentamethylcyclopentadienyl)Rh and dicarbonyl(pentamethylcyclopentadienyl)Ir. *J. Organomet. Chem.* 26 (3): 393–399.

1133 Oliver, A.J. and Graham, W.A.G. (1970). Nucleophilic properties of (cyclopentadienyl)(triphenylphosphine)(carbonyl)Ir. *Inorg. Chem.* 9 (12): 2653–2657.

1134 Stille, J.K. and Lau, K.S.Y. (1977). Mechanisms of oxidative addition of organic halides to group-8 transition metal complexes. *Acc. Chem. Res.* 10 (12): 434–442.

1135 Tsuji, J. (1986). New general synthetic methods involving (π-allyl)Pd complexes as intermediates and neutral reaction conditions. *Tetrahedron* 42 (16): 4361–4401.

1136 Tsuji, J. (1995). *Palladium Reagents and Catalysts: Innovations in Organic Synthesis*, 1–574. New York: Wiley.

1137 Trost, B.M. and van Vranken, D.L. (1996). Asymmetric transition metal-catalyzed allylic alkylations. *Chem. Rev.* 96 (1): 395–422.

1138 Trost, B.M. (1996). Designing a receptor for molecular recognition in a catalytic synthetic reaction: allylic alkylation. *Acc. Chem. Res.* 29 (8): 355–364.

1139 Amatore, C., Jutand, A., Meyer, G., and Mottier, L. (1999). Evidence of the reversible formation of cationic (π-allyl)Pd(II) complexes in the oxidative addition of allylic acetates to Pd(0) complexes. *Chem. Eur. J.* 5 (2): 466–473.

1140 Amatore, C., Gamez, S., and Jutand, A. (2001). Rate and mechanism of the reversible formation of cationic (η^3-allyl)Pd complexes in the oxidative addition of allylic acetate to Pd(0) complexes ligated by diphosphanes. *Chem. Eur. J.* 7 (6): 1273–1280.

1141 Amatore, C., Bahsoun, A.A., Jutand, A. et al. (2005). Rate and mechanism of the reaction of (*E*)-PhCH=CH—CH(Ph)-OAc with Pd(0) complexes in allylic substitutions. *Organometallics* 24 (7): 1569–1577.

1142 Jutand, A. (2004). Mechanism of Pd-catalyzed reactions: role of chloride ions. *Appl. Organomet. Chem.* 18 (11): 574–582.

1143 Amatore, C., Broeker, G., Jutand, A., and Khalil, F. (1997). Identification of the effective Pd(0) catalytic species generated in situ from mixtures of $Pd(dba)_2$ and bidentate phosphine ligands. Determination of their rates and mechanism in oxidative addition. *J. Am. Chem. Soc.* 119 (22): 5176–5185.

1144 Amatore, C., Jutand, A., and Meyer, G. (1998). Unexpected bell-shaped effect of the ligand on the rate of the oxidative addition to Pd(0) complexes generated in situ from mixtures of $Pd(dba)_2$ and *para*-substituted triarylphosphines. *Inorg. Chim. Acta* 273 (1–2): 76–84.

1145 Sorensen, H.S., Larsen, J., Rasmussen, B.S. et al. (2002). Synthesis and investigation of new macrocyclic diphosphine-Pd(0) complexes based on the barbiturate binding receptor. *Organometallics* 21 (24): 5243–5253.

1146 Amatore, C., Jutand, A., Meyer, G. et al. (1998). Comparative reactivity of Pd(0) complexes generated in situ in mixtures of triphenylphosphine or tri-2-furylphosphine and $Pd(dba)_2$. *Organometallics* 17 (14): 2958–2964.

1147 Amatore, C., Bucaille, A., Fuxa, A. et al. (2001). Rate and mechanism of the oxidative addition of phenyl iodide to Pd^0 ligated by triphenylarsine: evidence for the formation of a T-shaped complex $[PhPdI(AsPh_3)]$ and for the decelerating effect of $CH_2{=}CH{-}SnBu_3$ by formation of $[Pd(0)(\eta^2{-}CH_2{=}CH{-}SnBu_3)(AsPh_3)_2]$. *Chem. Eur. J.* 7 (10): 2134–2142.

1148 Marshall, W.J. and Grushin, V.V. (2003). Palladium(II) and Pd(0) complexes of BINAP(O) (2-(diphenylphosphino)-2′-(diphenylphosphinyl)-1,1′-binaphthyl). *Organometallics* 22 (3): 555–562.

1149 Amatore, C. and Jutand, A. (2000). Anionic Pd(0) and Pd(II) intermediates in Pd-catalyzed Heck and cross-coupling reactions. *Acc. Chem. Res.* 33 (5): 314–321.

1150 Amatore, C., Carré, E., Jutand, A., and M'Barki, M.A. (1995). Rates and mechanism of the formation of zero-valent Pd complexes from mixtures of $Pd(OAc)_2$ and tertiary phosphines and their reactivity in oxidative additions. *Organometallics* 14 (4): 1818–1826.

1151 Amatore, C., Blart, E., Genêt, J.P. et al. (1995). New synthetic applications of water-soluble

acetate Pd/TPPTS catalyst generated in-situ – evidence for a true Pd(0) species intermediate. *J. Org. Chem.* 60 (21): 6829–6839.
1152 Amatore, C., Azzabi, M., and Jutand, A. (1991). Role and effects of halide ions on the rates and mechanisms of oxidative addition of iodobenzene to low-ligated zero-valent Pd complexes $Pd(0)(PPh_3)_2$. *J. Am. Chem. Soc.* 113 (22): 8375–8384.
1153 Amatore, C., Carré, E., Jutand, A. et al. (1996). Oxidative addition of aryl halides to transient anionic (σ-aryl)Pd(0) intermediates – application to Pd-catalyzed reductive coupling of aryl halides. *Chem. Eur. J.* 2 (8): 957–966.
1154 Goossen, L.J., Koley, D., Hermann, H., and Thiel, W. (2004). The mechanism of the oxidative addition of aryl halides to Pd-catalysts: a DFT investigation. *J. Chem. Soc., Chem. Commun.* (19): 2141–2143.
1155 Broering, M. and Brandt, C.D. (2003). A five coordinate Pd(II) complex stable in solution and in the solid state. *J. Chem. Soc., Chem. Commun.* (17): 2156–2157.
1156 Kulawiec, R.J. and Crabtree, R.H. (1990). Coordination chemistry of halocarbons. *Coord. Chem. Rev.* 99: 89–115.
1157 Senn, H.M. and Ziegler, T. (2004). Oxidative addition of aryl halides to Pd(0) complexes: a density functional study including solvation. *Organometallics* 23 (12): 2980–2988.
1158 Barrios-Landeros, F. and Hartwig, J.F. (2005). Distinct mechanisms for the oxidative addition of chloro-, bromo-, and lodoarenes to a bisphosphine Pd(0) complex with hindered ligands. *J. Am. Chem. Soc.* 127 (19): 6944–6945.
1159 Murphy, E.F., Murugavel, R., and Roesky, H.W. (1997). Organometallic fluorides: compounds containing carbon–metal–fluorine fragments of d-block metals. *Chem. Rev.* 97 (8): 3425–3468.
1160 Jasim, N.A., Perutz, R.N., Whitwood, A.C. et al. (2004). Contrasting reactivity of fluoropyridines at Pd and Pt: C—F oxidative addition at Pd, P—C and C—F activation at Pt. *Organometallics* 23 (26): 6140–6149.
1161 Casado, A.L., Espinet, P., and Gallego, A.M. (2000). Mechanism of the Stille reaction. 2. Couplings of aryl triflates with vinyltributyltin. Observation of intermediates. A more comprehensive scheme. *J. Am. Chem. Soc.* 122 (48): 11771–11782.
1162 Casado, A.L., Espinet, P., Gallego, A.M., and Martinez-Ilarduya, J.M. (2001). Snapshots of a Stille reaction. *J. Chem. Soc., Chem. Commun.* (4): 339–340.
1163 Jutand, A. and Negri, S. (2003). Rate and mechanism of the oxidative addition of vinyl triflates and halides to Pd(0) complexes in DMF. *Organometallics* 22 (21): 4229–4237.
1164 Roy, A.H. and Hartwig, J.F. (2003). Oxidative addition of aryl tosylates to Pd(0) and coupling of unactivated aryl tosylates at room temperature. *J. Am. Chem. Soc.* 125 (29): 8704–8705.
1165 Roy, A.H. and Hartwig, J.F. (2004). Oxidative addition of aryl sulfonates to Pd(0) complexes of mono- and bidentate phosphines. Mild addition of aryl tosylates and the effects of anions on rate and mechanism. *Organometallics* 23 (2): 194–202.
1166 Collman, J.P. and Roper, W.R. (1969). Oxidative addition of d-8 complexes. *Adv. Organomet. Chem.* 7: 53–94.
1167 Labinger, J.A. and Osborn, J.A. (1980). Mechanistic studies of oxidative addition to low-valent metal-complexes. 6. Stereochemistry at carbon in addition of alkyl halides to Ir(I). *Inorg. Chem.* 19 (11): 3230–3236.
1168 Labinger, J.A., Osborn, J.A., and Coville, N.J. (1980). Mechanistic studies of oxidative addition to low-valent metal complexes. 7. Mechanisms for addition of alkyl halides to Ir (I). *Inorg. Chem.* 19 (11): 3236–3243.
1169 Bullock, R.M., Song, J.S., and Szalda, D.J. (1996). Protonation of metal hydrides by strong acids. Formation of an equilibrium mixture of dihydride and dihydrogen complexes from protonation of Cp*Os(CO)2(H). Structural characterization of [CpW(CO)2(PMe3)(PMe3)(H)2](+)OTf(-). *Organometallics* 15 (10): 2504–2516.
1170 Cusanelli, A. and Sutton, D. (1996). *Cis* and *trans* rhenium hydride dinitrogen complexes from protonation reactions of Cp*Re(PMe3)2(N2). *Organometallics* 15 (17): 3717–3722.
1171 Brown, J.M. and Cooley, N.A. (1988). C—C Bond formation through organometallic elimination reactions. *Chem. Rev.* 88 (7): 1031–1046.
1172 Beletskaya, I.P. and Cheprakov, A.V. (2004). Metal complexes as catalysts for C—C coupling reactions. In: *Comprehensive Coordination Chemistry II* (ed. J.A. McCleverty and R.J. Meyer), 305–308. Oxford: Elsevier.
1173 Burawoy, A., Gibson, C.S., and Holt, S. (1935). The organic compounds of gold. Part V. Cyano-compounds. Preliminary experiments on the production of alkyl radicals. *J. Chem. Soc.* 1024–1030.

1174 Gent, W.L.G. and Gibson, C.S. (1949). The organic compounds of gold. 11. Diethylthiocyanatogold. *J. Chem. Soc.* 1835–1840.

1175 Foss, M.E. and Gibson, C.S. (1949). The organic compounds of gold. S12. The constitution of coordination complexes of dialkylgold compounds with diamines containing the N—C—C—N grouping. *J. Chem. Soc.* 3063–3075.

1176 Tamaki, A., Magennis, S.A., and Kochi, J.K. (1974). Catalysis by gold – alkyl isomerization, *cis-trans* rearrangement, and reductive elimination of alkylgold(III) complexes. *J. Am. Chem. Soc.* 96 (19): 6140–6148.

1177 Komiya, S., Albright, T.A., Hoffmann, R., and Kochi, J.K. (1976). Reductive elimination and isomerization of organogold complexes – theoretical studies of trialkylgold species as reactive intermediates. *J. Am. Chem. Soc.* 98 (23): 7255–7265.

1178 Tatsumi, K., Hoffmann, R., Yamamoto, A., and Stille, J.K. (1981). Reductive elimination of d^8-organotransition metal complexes. *Bull. Chem. Soc. Jpn.* 54 (6): 1857–1867.

1179 Gillie, A. and Stille, J.K. (1980). Mechanisms of 1,1-reductive elimination from Pd. *J. Am. Chem. Soc.* 102 (15): 4933–4941.

1180 Goldberg, K.I., Yan, J.Y., and Breitung, E.M. (1995). Energetics and mechanisms of C—C and C—I reductive elimination from a Pt(IV) center. *J. Am. Chem. Soc.* 117 (26): 6889–6896.

1181 Ananikov, V.P., Musaev, D.G., and Morokuma, K. (2005). Theoretical insight into the C—C coupling reactions of the vinyl, phenyl, ethynyl, and methyl complexes of Pd and Pt. *Organometallics* 24 (4): 715–723.

1182 Dierkes, P. and van Leeuwen, P.W.N.M. (1999). The bite angle makes the difference: a practical ligand parameter for diphosphine ligands. *J. Chem. Soc; Dalton Trans.* (10): 1519–1529.

1183 Brown, J.M. and Guiry, P.J. (1994). Bite angle dependence of the rate of reductive elimination from diphosphine Pd complexes. *Inorg. Chim. Acta* 220 (1–2): 249–259.

1184 Calhorda, M.J., Brown, J.M., and Cooley, N.A. (1991). The C—C bond forming step in catalytic cross-coupling – migration or elimination. *Organometallics* 10 (5): 1431–1438.

1185 Mann, G., Baranano, D., Hartwig, J.F. et al. (1998). Carbon-sulfur bond forming reductive elimination involving sp-, sp^2-, and sp^3-hybridized carbon. Mechanism, steric effects, and electronic effects on sulfide formation. *J. Am. Chem. Soc.* 120 (36): 9205–9219.

1186 Marcone, J.E. and Moloy, K.G. (1998). Kinetic study of reductive elimination from the complexes (diphosphine)Pd(R)(CN). *J. Am. Chem. Soc.* 120 (33): 8527–8528.

1187 Hayashi, T., Konishi, M., Kobori, Y. et al. (1984). Dichloro[1,1′-bis(diphenylphosphino)ferrocene] Pd-(II) – an effective catalyst for cross-coupling of secondary and primary alkyl Grignard and alkylzinc reagents with organic halides. *J. Am. Chem. Soc.* 106 (1): 158–163.

1188 Kranenburg, M., Kamer, P.C.J., and van Leeuwen, P.W.N.M. (1998). The effect of the bite angle of diphosphane ligands on activity and selectivity in Pd-catalyzed cross-coupling. *Eur. J. Inorg. Chem.* (2): 155–157.

1189 van Leeuwen, P.W.N.M., Kamer, P.C.J., Reek, J.N.H., and Dierkes, P. (2000). Ligand bite angle effects in metal-catalyzed C—C bond formation. *Chem. Rev.* 100 (8): 2741–2769.

1190 Birkholz, M.N., Freixa, Z., and van Leeuwen, P.W.N.M. (2009). Bite angle effects of diphosphines in C—C and C—X bond forming cross coupling reactions. *Chem. Soc. Rev.* 38 (4): 1099–1118.

1191 Donahue, C.M., McCollom, S.P., Forrest, C.M. et al. (2015). Impact of coordination geometry, bite angle, and *trans* influence on metal-ligand covalency in phenyl-substituted phosphine complexes of Ni and Pd. *Inorg. Chem.* 54: 5646–5659.

1192 Goliaszewski, A. and Schwartz, J. (1985). Interception of bis(η^3-allylic)Pd intermediates by maleic anhydride – a change in an overall C—C bond forming scheme using organotin reagents. *Organometallics* 4 (2): 417–419.

1193 Goliaszewski, A. and Schwartz, J. (1984). C—C Bond formation by induced elimination from unsymmetrically substituted (allyl)(allyl′)Pd complexes. *J. Am. Chem. Soc.* 106 (17): 5028–5030.

1194 Bertani, R., Berton, A., Carturan, G., and Campostrini, R. (1988). Reaction of $Pd(\eta^3—C_3H_5)_2$ with cyano-olefins – reductive coupling of allyl groups in the presence of triphenylphosphine. *J. Organomet. Chem.* 349 (1–2): 263–268.

1195 Nakamura, H., Aoyagi, K., Shim, J.G., and Yamamoto, Y. (2001). Catalytic amphiphilic allylation via bis-π-allylpalladium complexes and its application to the synthesis of medium-sized carbocycles. *J. Am. Chem. Soc.* 123 (3): 372–377.

1196 Szabo, K.J. (2000). Umpolung of the allylpalladium reactivity: mechanism and regioselectivity of the electrophilic attack on bis-allylPd complexes formed in Pd-catalyzed transformations. *Chem. Eur. J.* 6 (23): 4413–4421.

1197 Henc, B., Jolly, P.W., Salz, R. et al. (1980). Transition-metal allyls. 4. The (η^3-allyl)2M complexes of Ni, Pd and Pt – reaction with tertiary phosphines. *J. Organomet. Chem.* 191 (2): 449–475.

1198 Kuhn, O. and Mayr, H. (1999). Kinetics and mechanisms of the reactions of (π-allyl)Pd complexes with nucleophiles. *Angew. Chem. Int. Ed.* 38 (3): 343–346.

1199 Krause, J., Goddard, R., Mynott, R., and Porschke, K.R. (2001). A dinuclear Pd(I) ethynyl complex: synthesis, structure, and dynamics. *Organometallics* 20 (10): 1992–1999.

1200 Krause, J., Bonrath, W., and Porschke, K.R. (1992). ($R_2PC_2H_4PR_2$)Pd(0)(alkene) and (ethyne) complexes. *Organometallics* 11 (3): 1158–1167.

1201 Méndez, M., Cuerva, J.M., Gómez-Bengoa, E. et al. (2002). Intramolecular coupling of allyl carboxylates with allyl stannanes and allyl silanes: a new type of reductive elimination reaction? *Chem. Eur. J.* 8 (16): 3620–3628.

1202 Norton, J.R. (1979). Organometallic elimination mechanisms – studies on Os alkyls and hydrides. *Acc. Chem. Res.* 12 (4): 139–145.

1203 Bird, C.W. (1962). Synthesis of organic compounds by direct carbonylation reactions using metal carbonyls. *Chem. Rev.* 62 (4): 283–302.

1204 Cassar, L., Chiusoli, G.P., and Guerrier, F. (1973). Synthesis of carboxylic acids and esters by carbonylation reactions at atmospheric pressure using transition-metal catalysts. *Synthesis* (9): 509–523.

1205 Anderson, G.K. and Cross, R.J. (1984). Carbonyl insertion reactions of square-planar complexes. *Acc. Chem. Res.* 17 (2): 67–74.

1206 (a) Berke, H. and Hoffmann, R. (1978). Organometallic migration reactions. *J. Am. Chem. Soc.* 100 (23): 7224–7236. (b) Bernardi, F., Bottoni, A., Nicastro, M. et al. (2000). Theoretical study of the mechanism of carbonyl insertion reactions catalyzed by Ni complexes. *Organometallics* 19 (11): 2170–2178.

1207 (a) Franssen, N.M.G., Walters, A.J.C., Reek, J.N.H., and de Bruin, B. (2011). Carbene insertion into transition metal-carbon bonds: a new tool for catalytic C—C bond formation. *Catal. Sci. Technol.* 1 (2): 153–165. (b) Xia, Y., Zhang, Y., and Wang, J. (2013). Catalytic cascade reactions involving metal carbene migratory insertion. *ACS Catal.* 3 (11): 2586–2598. (c) Xia, Y., Qiu, D., and Wang, J. (2017). Transition-metal-catalyzed cross-couplings through carbene migratory insertion. *Chem. Rev.* 117 (23): 13810–13889. (d) Zhukhovitskiy, A.V., Kobylianskii, I.J., Wu, C.-Y., and Toste, F.D. (2018). Migratory insertion of carbenes into Au(III)—C bonds. *J. Am. Chem. Soc.* 140 (1): 466–474.

1208 Calderazzo, F. (1977). Synthetic and mechanistic aspects of inorganic insertion reactions – insertion of carbon monoxide. *Angew. Chem Int. Ed. Engl.* 16 (5): 299–311.

1209 Walborsky, H.M. and Allen, L.E. (1971). Stereochemistry of tris(triphenylphosphine)Rh chloride decarbonylation of aldehydes. *J. Am. Chem. Soc.* 93 (21): 5465–5468.

1210 Bly, R.S., Silverman, G.S., and Bly, R.K. (1985). Reaction of (η^5—C_5H_5)$(CO)_2$Fe(2-norbornyl) complexes and triphenylmethyl tetrafluoroborate – competitive β-elimination and redox-catalyzed migratory carbonyl insertion. *Organometallics* 4 (2): 374–383.

1211 Jernakoff, P. and Cooper, N.J. (1984). Migratory insertion within a cationic alkylidene-aryl complex of tungsten. *J. Am. Chem. Soc.* 106 (10): 3026–3027.

1212 Kletzin, H., Werner, H., Serhadli, O., and Ziegler, M.L. (1983). A model reaction of an elementary step in the Fischer-Tropsch synthesis – $[C_6Me_6RuH(C_2H_4)Pr_3]$+ from $C_6Me_6Ru(Me)_2Pr_3$. *Angew. Chem. Int. Ed. Engl.* 22 (1): 46–47.

1213 Fellmann, J.D., Schrock, R.R., and Traficante, D.D. (1982). α-Hydride vs β-hydride elimination – an example of an equilibrium between two tautomers. *Organometallics* 1 (3): 481–484.

1214 Schrock, R.R., Shih, K.Y., Dobbs, D.A., and Davis, W.M. (1995). α-Elimination can be faster than β-elimination in D2 alkyl complexes of Mo and W that contain the trimethylsilyl-substituted triamidoamine ligand. *J. Am. Chem. Soc.* 117 (24): 6609–6610.

1215 Shih, K.Y., Totland, K., Seidel, S.W., and Schrock, R.R. (1994). Spontaneous loss of molecular-hydrogen from W(IV) alkyl complexes to give alkylidyne complexes. *J. Am. Chem. Soc.* 116 (26): 12103–12104.

1216 Milstein, D. and Calabrese, J.C. (1982). Oxidative addition of unactivated epoxides to Ir (I) complexes – formation of stable *cis*-hydridoformylmethyl and hydridacylmethyl complexes. *J. Am. Chem. Soc.* 104 (13): 3773–3774.

1217 Green, M.L.H. (1978). Studies on synthesis, mechanism and reactivity of some organo-Mo and organo-W compounds. *Pure Appl. Chem.* 50 (1): 27–35.

1218 Koga, N. and Morokuma, K. (1991). Ab initio molecular orbital studies of catalytic elementary reactions and catalytic cycles of transition metal complexes. *Chem. Rev.* 91 (5): 823–842.

1219 Di, G.A., Castarlenas, R., and Oro, L.A. (2015). Mechanistic considerations on catalytic H/D exchange mediated by organometallic transition metal complexes. *C. R. Chim.* 18 (7): 713–741.

1220 Arlman, E.J. and Cossee, P. (1964). Ziegler–Natta catalysis. 3. Stereospecific polymerization of propene with the catalyst system $TiCl_3$—$AlEt_3$. *J. Catal.* 3 (1): 99–104.

1221 Watson, P.L. (1982). Ziegler–Natta polymerization – the lanthanide model. *J. Am. Chem. Soc.* 104 (1): 337–339.

1222 Soto, J., Steigerwald, M.L., and Grubbs, R.H. (1982). Concerning the mechanism of Ziegler–Natta polymerization – isotope effects on propagation rates. *J. Am. Chem. Soc.* 104 (16): 4479–4480.

1223 Clawson, L., Soto, J., Buchwald, S.L. et al. (1985). Olefin insertion in a metal alkyl in a Ziegler polymerization system. *J. Am. Chem. Soc.* 107 (11): 3377–3378.

1224 Brookhart, M. and Lincoln, D.M. (1988). Comparison of migratory aptitudes of hydride and alkyl groups in β-migratory insertion reactions of $Cp(P(OMe)_3)Rh(C_2H_4)R+$ (R=H, Et). *J. Am. Chem. Soc.* 110 (26): 8719–8720.

1225 Rix, F.C. and Brookhart, M. (1995). Energetics of migratory insertion reactions in Pd(II) acyl ethylene, alkyl ethylene, and alkyl carbonyl complexes. *J. Am. Chem. Soc.* 117 (3): 1137–1138.

1226 Eisch, J.J., Piotrowski, A.M., Brownstein, S.K. et al. (1985). Organometallic compounds of Group-III. 41. Direct observation of the initial insertion of an unsaturated hydrocarbon into the Ti—C bond of the Soluble Ziegler polymerization catalyst, Cp_2TiCl_2—$MeAlCl_2$. *J. Am. Chem. Soc.* 107 (24): 7219–7221.

1227 Kazlauskas, R.J. and Wrighton, M.S. (1982). Photochemistry of metal carbonyl alkyls – study of thermal β-hydrogen transfer in photogenerated, 16-valence-electron (alkyl)(dicarbonyl)(cyclopentadienyl)Mo and (alkyl)(dicarbonyl)(cyclopentadienyl)W complexes. *J. Am. Chem. Soc.* 104 (22): 6005–6015.

1228 Cracknell, R.B., Orpen, A.G., and Spencer, J.L. (1984). The synthesis and structure of $[Co(\eta C_5Me_5)Et(P(\textit{para}\text{-tolyl})_3)]$ cation – a model for the β-elimination transition state. *J. Chem. Soc.; Chem. Commun.* (6): 326–328.

1229 Casey, C.P., Fagan, P.J., and Miles, W.H. (1982). Synthesis and interconversions of dinuclear iron complexes with μ—CH_3, μ—CH_2, and μ—CH ligands. *J. Am. Chem. Soc.* 104 (4): 1134–1136.

1230 Schwartz, J. and Labinger, J.A. (1976). Hydrozirconation – new transition metal reagent for organic synthesis. *Angew. Chem. Int. Ed. Engl.* 15 (6): 333–340.

1231 Labinger, J.A., Hart, D.W., Seibert, W.E., and Schwartz, J. (1975). Electrophilic cleavage of C—Zr(IV) bond – comparison and contrast with other transition metal alkyl systems. *J. Am. Chem. Soc.* 97 (13): 3851–3852.

1232 Yamamoto, A. (1995). Comparison of the reactivities of neutral and cationic transition metal alkyls and hydrides. *J. Organomet. Chem.* 500 (1–2): 337–348.

1233 Crabtree, R.H., Mihelcic, J.M., and Quirk, J.M. (1979). Iridium complexes in alkane dehydrogenation. *J. Am. Chem. Soc.* 101 (26): 7738–7740.

1234 Burk, M.J., Crabtree, R.H., Parnell, C.P., and Uriarte, R.J. (1984). Selective stoichiometric and catalytic C—H bond cleavage reactions in hydrocarbons by Ir complexes. *Organometallics* 3 (5): 816–817.

1235 Arndtsen, B.A. and Bergman, R.G. (1995). Unusually mild and selective hydrocarbon C—H bond activation with positively charged Ir(III) complexes. *Science* 270 (5244): 1970–1973.

1236 Burger, P. and Bergman, R.G. (1993). Facile intermolecular activation of C—H bonds in methane and other hydrocarbons and Si—H bonds in silanes with the Ir(III) complex (Cp*) $(PMe_3)Ir(CH_3)(OTf)$. *J. Am. Chem. Soc.* 115 (22): 10462–10463.

1237 Lohrenz, J.C.W. and Jacobsen, H. (1996). Cationic Ir(III) complexes: a step towards synthetic applicability of C—H activation. *Angew. Chem. Int. Ed. Engl.* 35 (12): 1305–1307.

1238 Xu, W.W., Rosini, G.P., Gupta, M. et al. (1997). Thermochemical alkane dehydrogenation catalyzed in solution without the use of a hydrogen acceptor. *J. Chem. Soc.; Chem. Commun.* (23): 2273–2274.

1239 Hughes, R.P., Williamson, A., Sommer, R.D., and Rheingold, A.L. (2001). The first transition metal complex of tetrafluorobenzyne: $Ir(\eta^5$—$C_5Me_5)(PMe_3)(\eta^2$—$C_6F_4)$. *J. Am. Chem. Soc.* 123 (30): 7443–7444.

1240 (a) Chirik, P.J. and Bercaw, J.E. (2005). Cyclopentadienyl and olefin substituent effects on insertion and beta-hydrogen elimination with group 4 metallocenes. Kinetics, mechanism, and thermodynamics for zirconocene and hafnocene alkyl hydride derivatives. *Organometallics* 24 (22): 5407–5423. (b) Luo, X.L., Tang, D.Y., and Li, M. (2005). Quantum investigation on the mechanism of isomerization of 1-butylene catalyzed by Rh complex. *J. Mol. Struct. THEOCHEM* 731 (1–3): 139–147.

1241 (a) Xu, R., Klatt, G., Wadepohl, H., and Köppel, H. (2010). Hydrogen scrambling in $[(C_5R_5)(L)M(H)(C_2H_4)]+$ (M=Co, Rh). Relation of experimental kinetic data to the barriers of the elementary reaction steps. *Inorg. Chem.* 49 (7): 3269–3296. (b) Sauriol, F., Sonnenberg, J.F., Chadder, S.J. et al. (2010). Remarkable reactions and intermediates in titanocene(IV) chemistry: migratory insertion reactions of 2,2-disubstituted-1-alkenes, intramolecular 1,5-σ bond metathesis via ε-agostic interactions, and a rare example of a β-agostic alkyltitanocene complex. *J. Am. Chem. Soc.* 132 (38): 13357–13370. (c) Sosa Carrizo, E.D., Bickelhaupt, F.D., and Fernández, I. (2015). Factors controlling β-elimination reactions in group 10 metal complexes. *Chem. Eur. J.* 21 (41): 14362–14369.

1242 Stille, J.K., Huang, F., and Regan, M.T. (1974). Mechanism of acid chloride decarbonylation with chlorotris(triphenylphosphine)Rh(I) – stereochemistry and direction of elimination. *J. Am. Chem. Soc.* 96 (5): 1518–1522.

1243 Jolly, P.W. and Pettit, R. (1966). Evidence for a novel metal-carbene system. *J. Am. Chem. Soc.* 88 (21): 5044–5045.

1244 Green, M.L.H., Mitchard, L.C., and Swanwick, M.G. (1971). Carbene complexes of Fe, Mo, and Ru – new route to metal-carbene derivatives. *J. Chem. Soc. A* (6): 794–797.

1245 Ivin, K.J., Rooney, J.J., Stewart, C.D. et al. (1978). Mechanism for stereospecific polymerization of olefins by Ziegler–Natta catalysts. *J. Chem. Soc., Chem. Commun.* (14): 604–606.

1246 Szabo, M.J., Berke, H., Weiss, T., and Ziegler, T. (2003). Is the polymerization of linear α-olefins by transition metal-carbene complexes a viable process? A theoretical study based on density functional theory. *Organometallics* 22 (18): 3671–3677.

1247 Coates, G.W. (2002). Polymerization catalysis at the millennium: frontiers in stereoselective, metal-catalyzed polymerization. *J. Chem. Soc., Dalton Trans.* (4): 467–475.

1248 Buchmeiser, M.R. (2000). Homogeneous metathesis polymerization by well defined group VI and group VIII transition metal-alkylidenes: fundamentals and applications in the preparation of advanced materials. *Chem. Rev.* 100 (4): 1565–1604.

1249 Grubbs, R.H. and Tumas, W. (1989). Polymer synthesis and organotransition metal chemistry. *Science* 243 (4893): 907–915.

1250 O'Donnell, P.M., Brzezinska, K., Powell, D., and Wagener, K.B. (2001). "Perfect comb" ADMET graft copolymers. *Macromolecules* 34 (20): 6845–6849.

1251 Schrock, R.R., Clark, D.N., Sancho, J. et al. (1982). Multiple metal–carbon bonds. 31. Tungsten(VI)-neopentylidyne complexes. *Organometallics* 1 (12): 1645–1651.

1252 Tipper, C.F.H. (1955). Some reactions of cyclopropane, and a comparison with the lower olefins. 2. Some Pt-cyclopropane complexes. *J. Chem. Soc.* 2045–2046.

1253 Jennings, P.W. and Johnson, L.L. (1994). Metallacyclobutane complexes of the group-8 transition metals – synthesis, characterizations, and chemistry. *Chem. Rev.* 94 (8): 2241–2290.

1254 Hours, A.E. and Snyder, J.K. (2006). $PtCl_2$-promoted cyclopropane opening in (4+2+2) homo-Diels–Alder cycloadducts. *Tetrahedron Lett.* 47: 675–678.

1255 Hours, A.E. and Snyder, J.K. (2008). $PtCl_2$-mediated cyclopropane opening: a mechanistic study. *Organometallics* 27: 410–417.

1256 Wong, W., Watkins, S.F., Singer, S.J. et al. (1972). Preparation and X-ray crystallographic determination of structure of platinacyclobutenone $(Ph_3P)_2Pt[OC_3(Ph_2)]$. *J. Chem. Soc.; Chem. Commun.* (11): 672–673.

1257 Tebbe, F.N. and Harlow, R.L. (1980). Titanacyclobutenes. *J. Am. Chem. Soc.* 102 (19): 6149–6151.

1258 (a) Schlund, R., Schrock, R.R., and Crowe, W.E. (1989). Direct polymerization of acetylene to give living polyenes. *J. Am. Chem. Soc.* 111 (20): 8004–8006. (b) Fox, H.H., Wolf, M.O., Odell, R. et al. (1994). Living cyclopolymerization of 1,6-heptadiyne derivatives using well-defined alkylidene complexes – polymerization mechanism, polymer structure, and polymer properties. *J. Am. Chem. Soc.* 116 (7): 2827–2843.

1259 (a) Schrock, R.R., Luo, S.F., Lee, J.C. et al. (1996). Living polymerization of (*o*-(trimethylsilyl)phenyl)acetylene by molybdenum imido alkylidene complexes. *J. Am. Chem. Soc.* 118 (16): 3883–3895. (b) Schrock, R.R., Tonzetich, Z.J., Lichtscheidt, A.G. et al. (2008). Carboxylate-based molybdenum alkylidene catalysts: synthesis, characterization, and use as initiators for 1,6-heptadiyne cyclopolymerizations. *Organometallics* 27 (15): 3986–3995.

1260 (a) Kuzmany, H., Mehring, M., and Rogh, S. (eds.) (1988). *Springer Series in Solid State Sciences: Electronic Properties of Conjugated Polymers*, vol. 76, 1–439. New York: Springer-Verlag. (b) Masuda, T. (2017). Substituted polyacetylenes: synthesis, properties, and functions. *Polym. Rev.* 57 (1): 1–14.

1261 Schrock, R.R. (1999). Olefin metathesis by Mo imido alkylidene catalysts. *Tetrahedron* 55 (27): 8141–8153.

1262 Sheng, Y.H., Wu, Y.D., and Leszczynski, J. (2004). A theoretical study of the ring opening of metallacyclobutene derived from the addition of acetylene to Mo-alkylidenes. *Organometallics* 23 (13): 3189–3196.

1263 Diver, S.T. and Giessert, A.J. (2004). Enyne metathesis (enyne bond reorganization). *Chem. Rev.* 104 (3): 1317–1382.

1264 Park, S., Kim, M., and Lee, D. (2005). Tandem sequence of cross metathesis ring closing metathesis reaction of alkynyl silyloxy-tethered enynes. *J. Am. Chem. Soc.* 127 (26): 9410–9415.

1265 Hansen, E.C. and Lee, D. (2006). Search for solutions to the reactivity and selectivity problems in enyne metathesis. *Acc. Chem. Res.* 39 (8): 509–519.

1266 Lippstreu, J.J. and Straub, B.F. (2005). Mechanism of enyne metathesis catalyzed by Grubbs Ru-carbene complexes: a DFT study. *J. Am. Chem. Soc.* 127 (20): 7444–7457.

1267 O'Connor, J.M., Fong, B.S., Ji, H.L. et al. (1995). A new metal-mediated cyclization – conversion of metallacyclobutene and alkyne substrates to η^4-cyclopentadiene products. *J. Am. Chem. Soc.* 117 (30): 8029–8030.

1268 Hoye, T.R., Chen, K.J., and Vyvyan, J.R. (1993). Preparation of Fischer carbene complexes by alkylation of acylmetalates with alkyl iodides. *Organometallics* 12 (7): 2806–2809.

1269 Pennella, F., Banks, R.L., and Bailey, G.C. (1968). Disproportionation of alkynes. *J. Chem. Soc., Chem. Commun.* (23): 1548–1549.

1270 Bunz, U.H.F. (2005). How are alkynes scrambled? *Science* 308 (5719): 216–217.

1271 Mortreux, A. and Blanchard, M. (1974). Metathesis of alkynes by a Mo hexacarbonyl-resorcinol catalyst. *J. Chem. Soc., Chem. Commun.* (19): 786–787.

1272 Bages, S., Petit, M., Mortreux, A., and Petit, F. (1990). Study on metathesis of acetylene hydrocarbons catalyzed by the system $O_2Mo(acac)_2$-$AlEt_3$-PhOH. *J. Mol. Catal.* 59 (3): L25–L29.

1273 Wengrovius, J.H., Sancho, J., and Schrock, R.R. (1981). Multiple metal-carbon bonds. 21. Metathesis of acetylenes by W(VI)-alkylidyne complexes. *J. Am. Chem. Soc.* 103 (13): 3932–3934.

1274 Fürstner, A., Mathes, C., and Lehmann, C.W. (1999). Mo[N(*t*-Bu)(Ar)]3 complexes as catalyst precursors: in situ activation and application to metathesis reactions of alkynes and diynes. *J. Am. Chem. Soc.* 121 (40): 9453–9454.

1275 Fürstner, A., Mathes, C., and Lehmann, C.W. (2001). Alkyne metathesis: development of a novel Mo-based catalyst system and its application to the total synthesis of Epothilone A and C. *Chem. Eur. J.* 7 (24): 5299–5317.

1276 Tsai, Y.C., Diaconescu, P.L., and Cummins, C.C. (2000). Facile synthesis of trialkoxyMo(VI)-alkylidyne complexes for alkyne metathesis. *Organometallics* 19 (25): 5260–5262.

1277 Blackwell, J.M., Figueroa, J.S., Stephens, F.H., and Cummins, C.C. (2003). Enediynes via sequential acetylide reductive coupling and alkyne metathesis: easy access to well-defined Mo initiators for alkyne metathesis. *Organometallics* 22 (17): 3351–3353.

1278 (a) Zhang, W., Kraft, S., and Moore, J.S. (2003). A reductive recycle strategy for the facile synthesis of Mo(VI)-alkylidyne catalysts for alkyne metathesis. *J. Chem. Soc., Chem. Commun.* (7): 832–833. (b) Zhang, W., Kraft, S., and Moore, J.S. (2004). Highly active trialkoxyMo(VI)-alkylidyne catalysts synthesized by a reductive recycle strategy. *J. Am. Chem. Soc.* 126 (1): 329–335. (c) Zhang, W. and Moore, J.S. (2005). Reaction pathways leading to arylene ethynylene macrocycles via alkyne metathesis. *J. Am. Chem. Soc.* 127 (33): 11863–11870.

1279 Zhang, W. and Moore, J.S. (2007). Alkyne metathesis: catalysts and synthetic applications. *Adv. Synth. Catal.* 349 (1–2): 93–120.

1280 (a) Sisco, S.W., Larson, B.M., and Moore, J.S. (2014). Relaxing conformational constraints in dynamic macrocycle synthesis. *Macromolecules* 47 (12): 3829–3836. (b) Lee, S., Yang, A., Moneypenny, T.P., and Moore, J.S. (2016). Kinetically trapped tetrahedral cages via alkyne metathesis. *J. Am. Chem. Soc.* 138 (7): 2182–2185.

1281 Schrock, R.R. (1995). The alkoxide ligand in olefin and acetylene metathesis reactions. *Polyhedron* 14 (22): 3177–3195.

1282 Schrock, R.R. (2002). High oxidation state multiple metal–carbon bonds. *Chem. Rev.* 102 (1): 145–179.

1283 Strutz, H., Dewan, J.C., and Schrock, R.R. (1985). Reaction of $Mo(CCMe_3)[OCH(CF_3)_2]_3$·(dimethoxyethane) with *tert*-butylacetylene, an aborted acetylene polymerization. *J. Am. Chem. Soc.* 107 (21): 5999–6005.

1284 Zhu, J., Jia, G.C., and Lin, Z.Y. (2006). Theoretical investigation of alkyne metathesis catalyzed by W/Mo alkylidyne complexes. *Organometallics* 25 (7): 1812–1819. see also:Suresh, C.H. and Frenking, G. (2012). 1,3-Metal-carbon bonding and alkyne metathesis: DFT investigations on model complexes of Group 4, 5, and 6 transition metals. *Organometallics* 31 (20): 7171–7180.

1285 Mortreux, A., Petit, F., Petit, M., and Szymanskabuzar, T. (1995). Reactions of $W(CCMe_3)$ $(OCMe_3)_3$ with terminal alkynes – metathesis and polymerization. *J. Mol. Catal. A: Chem.* 96 (2): 95–105.

1286 Grevels, F.W., Schulz, D., and Koernerv, E. (1974). Photochemical synthesis of bis(η-methyl acrylate)tricarbonyliron. *Angew. Chem. Int. Ed. Engl.* 13 (8): 534–536.

1287 Franke, W., Schwarz, H., Thies, H. et al. (1980). An experimental and theoretical study of the degenerate carbon skeleton isomerization of the cyclopentyl cation in the gas phase. *Angew. Chem. Int. Ed. Engl.* 19 (6): 485–487.

1288 Franke, W., Schwarz, H., Thies, H. et al. (1981). Degenerate carbon skeleton isomerization of the cyclopentyl cation in the gas phase – experimental and theoretical evidence for the existence of a pyramidal $C_5H_9^+$ cation as an intermediate in the unimolecular ethylene elimination. *Chem. Ber. Recl.* 114 (8): 2808–2824.

1289 Cecchi, P., Pizzabiocca, A., Renzi, G. et al. (1993). Gas phase protonation of spiropentane – a novel entry Into the $C_5H_9^+$ potential energy surface. *J. Am. Chem. Soc.* 115 (22): 10338–10347.

1290 Szabó, K.J. and Cremer, D. (1995). Route to a kinetically stabilized protonated spirocyclopentane with a pentacoordinated carbon atom – the missing link between bicyclo[3.2.0]hept-3-yl and 7-norbornyl cation. *J. Org. Chem.* 60 (7): 2257–2259.

1291 Mango, F.D. (1975). Transition metal catalysis of pericyclic reactions. *Coord. Chem. Rev.* 15 (2–3): 109–205.

1292 Halpern, J. (2001). Organometallic chemistry at the threshold of a new millenium. Retrospect and prospect. *Pure Appl. Chem.* 73 (2): 209–220.

1293 Blum, J., Zlotogorski, C., Schwarz, H., and Hohne, G. (1978). Isolation of a metallocyclic intermediate in $[Ir(CO)_3Cl]_2$-catalyzed rearrangement of 1,3-bis-homocubane. *Tetrahedron Lett.* (37): 3501–3504.

1294 Grubbs, R.H., Miyashita, A., Liu, M., and Burk, P. (1978). Preparation and reactions of phosphine Ni(cyclopentanes). *J. Am. Chem. Soc.* 100 (8): 2418–2425.

1295 Rathke, J.W. and Muetterties, E.L. (1975). Phosphine chemistry of Fe(0) and Fe(II). *J. Am. Chem. Soc.* 97 (11): 3272–3273.

1296 (a) Fritch, J.R. and Vollhardt, K.P.C. (1979). Cyclobutadiene-metal complexes as potential intermediates in alkyne metathesis – flash vacuum pyrolysis of substituted (η^4-cyclobutadiene)(η^5-cyclopentadienyl)Co complexes. *Angew. Chem. Int. Ed. Engl.* 18 (5): 409–411. (b) Le Pahl, J., Derien, S., Demerseman, B. et al. (2005). Ruthenium-catalyzed synthesis of alkylidenecyclobutenes via head-to-head dimerization of propargylic alcohols and cyclobutadiene-ruthenium intermediates. *Chem. Eur. J.* 11 (4): 1312–1324.

1297 Negishi, E. and Takahashi, T. (1994). Patterns of stoichiometric and catalytic reactions of organozirconium and related complexes of synthetic interest. *Acc. Chem. Res.* 27 (5): 124–130.

1298 von Wilde, M.P. (1874). Vermischte Mittheilungen. *Ber. Dtsch. Chem. Ges.* 7: 352–357.

1299 Sabatier, P. and Senderens, J.B. (1897). Action of Ni on ethylene. *C.R. Hebd. Seances Acad. Sci.* 124: 616–618.

1300 Sabatier, P. and Senderens, J.B. (1903). Transformation of aldehydes and ketones in alcohols by catalytic hydrogenation. *C.R. Hebd. Seances Acad. Sci.* 137: 301–303.

1301 Lattes, A. (2000). From catalytic hydrogenation to the chemical theory of the catalysis: Paul Sabatier, genius chemist, decentralization apostle. *C.R. Acad. Sci. Sér. IIC* 3 (9): 705–709.

1302 Ipatieff, W. (1907). Katalytische Reaktionen bei hohen Temperaturen und Drucken – Reduktionskatalyse. *Ber. Dtsch. Chem. Ges.* 40 (2): 1270–1281.

1303 Ipatieff, W. (1907). Katalytische Reaktionen bei hohen Temperaturen und Drucken. Reduktionskatalyse in Gegenwart von Metalloxyden. *Ber. Dtsch. Chem. Ges.* 40 (2): 1281–1290.

1304 Knowles, W.S. (2002). Asymmetric hydrogenations (Nobel lecture). *Angew. Chem. Int. Ed.* 41 (12): 1999–2007.

1305 Knowles, W.S. (2003). Asymmetric hydrogenations (Nobel lecture 2001). *Adv. Synth. Catal.* 345: 3–13.

1306 Knowles, W.S. and Noyori, R. (2007). Pioneering perspectives on asymmetric hydrogenation. *Acc. Chem. Res.* 40: 1238–1239.

1307 Noyori, R. (2002). Asymmetric catalysis: science and opportunities (Nobel lecture). *Angew. Chem. Int. Ed.* 41 (12): 2008–2022.

1308 Noyori, R. (2003). Asymmetric catalysis: science and opportunities (Nobel lecture 2001). *Adv. Synth. Catal.* 345: 15–32.

1309 Sharpless, K.B. (2002). Searching for new reactivity (Nobel lecture). *Angew. Chem. Int. Ed.* 41 (12): 2024–2032.

1310 (a) Dang, T.P. and Kagan, H.B. (1971). Asymmetric synthesis of hydratropic acid and amino-acids by homogeneous catalytic hydrogenation. *J. Chem. Soc.; Chem. Commun.* (10): 481–481. (b) Kagan, H.B. and Dang, T.P. (1972). Asymmetric catalytic reduction with transition metal complexes. 1. Catalytic system of Rh(I) with (-)-2,3-*O*-isopropylidene-2,3-dihydroxy-1,4-bis(diphenylphosphino) butane – new chiral diphosphine. *J. Am. Chem. Soc.* 94 (18): 6429–6433. (c) Kagan, H.B., Langlois, N., and Dang, T.P. (1975). Réduction asymétrique catalysée par des complexes de métaux de transition. IV. Synthèse d'amines chirales au moyen d'un complexe de rhodium et d'isopropylidene dihydroxy-2,3-bis(diphénylphosphino)-1,4-butane (diop). *J. Organomet. Chem.* 90 (3): 353–365. (d) Golapalaiah, K. and Kagan, H.B. (2011). Use of nonfunctionalized enamides and enecarbamates in asymmetric synthesis. *Chem. Rev.* 111 (8): 4599–4657.

1311 Haleema, S., Sasi, P.V., Ibnusaud, I. et al. (2012). Enantiomerically pure compounds related to chiral hydroxy acids derived from renewable resources. *RSC Adv.* 2 (25): 9257–9285.

1312 (a) Enthaler, S., Hagemann, B., Erre, G. et al. (2006). An environmentally benign process for the hydrogenation of ketones with homogeneous iron catalysts. *Chem. Asian J.* 1: 598–604. (b) Junge, K., Schröder, K., and Beller, M. (2011). Homogeneous catalysis using Fe complexes: recent developments in selective reductions. *J. Chem. Soc., Chem. Commun.* 47: 4849–4859.

1313 (a) Kelsen, V., Wendt, B., Werkmeister, S. et al. (2013). The use of ultrasmall Fe(0) nanoparticles as catalysts for the selective hydrogenation of unsaturated C—C bonds. *J. Chem. Soc.; Chem. Commun.* 3416–3418. (b) Junge, K., Wendt, B., Zhou, S.L., and Beller, M. (2013). Iron-catalyzed reduction of carboxylic esters to alcohols. *Eur. J. Org. Chem.* 2061–2065. (c) Wienhofer, G., Westerhaus, F.A., Junge, K. et al. (2013). A molecularly defined Fe-catalyst for the selective hydrogenation of α,β-unsaturated aldehydes. *Chem. Eur. J.* 19: 7701–7707. (d) Wienhofer, G., Westerhaus, F.A., Junge, K., and Beller, M. (2013). Fast and selective Fe-catalyzed transfer hydrogenations of aldehydes. *J. Organomet. Chem.* 744: 156–159.

1314 Gieshoff, T.N., Villa, M., Welther, A. et al. (2015). Iron-catalyzed olefin hydrogenation at 1 bar H_2 with a $FeCl_3$—$LiAlH_4$ catalyst. *Green Chem.* 17: 1408–1413.

1315 Formenti, D., Topf, C., Junge, K. et al. (2016). Fe_2O_3/NGr@C- and Co—Co_3O_4/NGr@C-catalysed hydrogenation of nitroarenes under mild conditions. *Catal. Sci. Technol.* 6: 4473–4477.

1316 (a) Morris, R.H. (2015). Exploiting metal-ligand bifunctional reactions in the design of iron asymmetric hydrogenation catalysts. *Acc. Chem. Res.* 48 (5): 1494–1502. (b) Zeil, T. and Milstein, D. (2015). Hydrogenation and dehydrogenation iron pincer catalysts capable of metal-ligand cooperation by aromatization/dearomatization. *Acc. Chem. Res.* 48 (7): 1979–1994. (c) Chakraborty, S., Bhattacharya, P., Dai, H.-G., and Guan, H.-R. (2015). Nickel and iron pincer complexes as catalysts for the reduction of carbonyl compounds. *Acc. Chem. Res.* 48 (7): 1995–2003. (d) Li, Y.-Y., Yu, S.-L., Shen, W.-Y., and Gao, J.-X. (2015). Fe-, Co-, and Ni-catalyzed asymmetric transfer hydrogenation and asymmetric hydrogenation of ketones. *Acc. Chem. Res.* 48 (9): 2587–2598. (e) Bauer, G. and Hu, X. (2016). Recent developments of iron pincer complexes for catalytic applications. *Inorg. Chem. Front.* 3 (6): 741–765.

1317 Ferraccioli, R. (2017). Asymmetric hydrogenation catalyzed by bifunctional Knölker type complexes. *Lett. Org. Chem.* (7): 14, 461–471.

1318 López, R. and Royo, B. (2017). Iron N-heterocyclic carbenes in reduction reactions. *Israel J. Org. Chem.* 57 (12): 1151–1159.

1319 (a) Stephan, D.W. and Erker, G. (2010). Frustrated Lewis pairs: metal-free hydrogen activation and more. *Angew. Chem. Int. Ed.* 49: 46–76. (b) Stephan, D.W. and Erker, G. (2015). Frustrated Lewis pair chemistry: development and perspectives. *Angew. Chem. Int. Ed.* 54: 6400–6441.

1320 (a) Wei, S.M. and Du, H.F. (2014). A highly enantioselective hydrogenation of silyl enol ethers catalyzed by chiral frustrated Lewis pairs. *J. Am. Chem. Soc.* 136: 12261–12264. (b) Feng, X.Q. and Du, H.F. (2014). Metal-free asymmetric hydrogenation and hydrosilylation catalyzed by frustrated Lewis pairs. *Tetrahedron Lett.* 55: 6959–6964.

1321 Skara, G., De Vieeschouwer, F., Geerlings, P. et al. (2017). Heterolytic splitting of molecular hydrogen by frustrated and classical Lewis pairs: a unified reactivity concept. *Sci. Rep.* 7: 16024.

1322 (a) Scott, D.J., Fuchter, M.J., and Ashley, A.E. (2017). Designing effective 'frustrated Lewis pair' hydrogenation catalysts. *Chem. Soc. Rev.* 46 (19): 5689–5700. (b) Meng, W., Feng, X.-Q., and Du, H.-F. (2018). Frustrated Lewis pairs catalyzed asymmetric metal-free hydrogenations and hydrosilylations. *Acc. Chem. Res.* 51 (1): 191–201.

1323 Welch, G.C., Juan, R.R.S., Masuda, J.D., and Stephan, D.W. (2006). Reversible, metal-free hydrogen activation. *Science* 314 (5802): 1124–1126.

1324 Welch, G.C. and Stephan, D.W. (2007). Facile heterolytic cleavage of dihydrogen by phosphines and boranes. *J. Am. Chem. Soc.* 129: 1880–1881.

1325 McCahill, J.S.J., Welch, G.C., and Stephan, D.W. (2007). Reactivity of "Frustrated Lewis pairs": three-component reactions of phosphines, a borane, and olefins. *Angew. Chem. Int. Ed.* 46: 4968–4971.

1326 Chase, P.A., Welch, G.C., Jurca, T., and Stephan, D.W. (2007). Metal-free catalytic hydrogenation. *Angew. Chem. Int. Ed.* 46 (42): 8050–8053.

1327 Stephan, D.W. (2015). Frustrated Lewis pairs. *J. Am. Chem. Soc.* 137: 10018–10032.

1328 Parks, D.J. and Piers, W.E. (1996). Tris(pentafluorophenyl)boron-catalyzed hydrosilation of aromatic aldehydes, ketones, and esters. *J. Am. Chem. Soc.* 118: 9440–9441.

1329 Parks, D.J., Blackwell, J.M., and Piers, W.E. (2000). Studies on the mechanism of $B(C_6F_5)_3$-catalyzed hydrosilation of carbonyl functions. *J. Org. Chem.* 65 (10): 3090–3098.

1330 Piers, W.E., Marwitz, A.J.V., and Mercier, L.G. (2011). Mechanistic aspects of bond activation with perfluoroarylboranes. *Inorg. Chem.* 50: 12252–12262.

1331 Oestreich, M., Hermeke, J., and Mohr, J. (2015). A unified survey of Si—H and H—H bond activation catalysed by electron-deficient boranes. *Chem. Soc. Rev.* 44: 2202–2220.

1332 (a) Zhang, M. and Zhang, A.Q. (2010). Iron-catalyzed hydrosilylation reactions. *Appl. Organomet. Chem.* 24: 751–757. (b) Addis, D., Shaikh, N., Zhou, S. et al. (2010). Chemo- and stereoselective Fe-catalyzed hydrosilylation of ketones. *Chem. Asian J.* 5 (7): 1687–1691. (c) Nikonov, G.I. (2015). An iron catalyst for asymmetric alkene hydrosilylation. *ChemCatChem* 7 (13): 1918–1919.

1333 (a) Shaikh, N.S., Enthaler, S., Junge, K., and Beller, M. (2008). Iron-catalyzed enantioselective hydrosilylation of ketones. *Angew. Chem. Int. Ed.* 47: 2497–2501. (b) Bleith, T. and Gade, L.H. (2016). Mechanism of the Fe(II)-catalyzed hydrosilylation of ketones: activation of iron carboxylate precatalysts and reaction pathways of the active catalyst. *J. Am. Chem. Soc.* 138 (14): 4972–4983.

1334 Mikhailine, A., Lough, A.J., and Morris, R.H. (2009). Efficient asymmetric transfer hydrogenation of ketones catalyzed by an Fe complex containing a P—N—N—P tetradentate ligand formed by template synthesis. *J. Am. Chem. Soc.* 131: 1394–1395.

1335 (a) Morris, R.H. (2009). Asymmetric hydrogenation, transfer hydrogenation and hydrosilylation of ketones catalyzed by Fe complexes. *Chem. Soc. Rev.* 38: 2282–2291. (b) Meyer, N., Lough, A.J., and Morris, R.H. (2009). Iron(II) complexes for the efficient catalytic asymmetric transfer hydrogenation of ketones. *Chem. Eur. J.* 15: 5605–5610.

1336 Zuo, Z.-Q., Zhang, L., Leng, X.-B., and Huang, Z. (2015). Iron-catalyzed asymmetric hydrosilylation of ketones. *J. Chem. Soc.; Chem. Commun.* 51 (24): 5073–5076.

1337 Malacea, R., Poli, R., and Manoury, E. (2010). Asymmetric hydrosilylation, transfer hydrogenation and hydrogenation of ketones catalyzed by iridium complexes. *Coord. Chem. Rev.* 254: 729–752.

1338 Susse, L., Hermeke, J., and Oestreich, M. (2016). The asymmetric Piers hydrosilylation. *J. Am. Chem. Soc.* 138: 6940–6943.

1339 (a) Rylander, P.N. (1985). *Hydrogenation Methods*, 1–193. London: Academic Press Inc.

(London) Ltd. (b) Blaser, H.-U., Malan, C., Pugin, B. et al. (2003). Selective hydrogenation for fine chemicals: recent trends and new developments. *Adv. Synth. Catal.* 345 (1–2): 103–151. (c) Blaser, H.-U. (2010). The development and application of industrially viable catalysts for the selective hydrogenation of complex molecules. *Top. Catal.* 53: 997–1001. (d) Nerozzi, F. (2012). Heterogeneous catalytic hydrogenation. *Platinum Met. Rev.* 56 (4): 236–241.

1340 (a) Nishimura, S. (2001). *Handbook of Heterogeneous Catalytic Hydrogenation for Organic Synthesis*, 1–720. New York: Wiley. (b) Bond, G.S. (2005). *Metal Catalyzed Reactions of Hydrocarbons*, 1–655. New York: Springer Science & Business Media Inc.

1341 Katano, S., Kim, Y., Kawai, M., and Trenary, M. (2014). Surface hydrogenation reactions at the single-molecule level. *Chem. Rec.* 15 (5): 819–826.

1342 Bönnemann, H. and Richards, R.M. (2001). Nanoscopic metal particles – synthetic methods and potential applications. *Eur. J. Inorg. Chem.* 2455–2480.

1343 Roucoux, A., Schulz, J., and Patin, H. (2002). Reduced transition metal colloids: a novel family of reusable catalysts? *Chem. Rev.* 102 (10): 3757–3778.

1344 (a) Roucoux, A. and Philippot, K. (2007). Homogeneous hydrogenation: colloids – hydrogenation with noble-metal nanoparticles. In: *Handbook of Homogeneous Hydrogenation*, Chapter 9 (ed. J.G. de Vries and C.J. Elsevier), 217–256. Weinheim: Wiley-VCH. (b) Zang, W., Lik, G., Wang, L., and Zhang, X. (2015). Catalytic hydrogenation by noble-metal nanocrystals with well-defined facets: a review. *Catal. Sci. Technol.* 5: 2531–2553.

1345 Bulut, S., Fei, Z., Siankevich, S. et al. (2015). Aqueous-phase hydrogenation of alkenes and arenes: the growing role of nano-scale catalysts. *Catal. Today* 247: 96–103.

1346 (a) Yamada, Y.M., Yuyama, Y., Sato, T. et al. (2014). A Pd-nanoparticle and silicon-nanowire-array hybrid: a platform for catalytic heterogeneous reactions. *Angew. Chem. Int. Ed.* 53: 127–131. (b) Kainz, Q.M., Linhardt, R., Grass, R.N. et al. (2014). Palladium nanoparticles supported on magnetic carbon-coated Co nanobeads: highly active and recyclable catalysts for alkene hydrogenation. *Adv. Funct. Mater.* 24: 2020–2027.

1347 da Silva Parra, F. and Rossi Marcia, L. (2014). Palladium on magnetite: magnetically recoverable catalyst for selective hydrogenation of acetylenic to olefinic compounds. *Tetrahedron* 70: 3314–3318.

1348 (a) Rogers, S.M., Catlow, C.R.A., Chan-Thaw, C.E. et al. (2017). Tandem site- and size-controlled Pd nanoparticles for the directed hydrogenation of furfural. *ACS Catal.* 7 (4): 2266–2274. (b) Luza, L., Ranbor, C.P., Gual, A. et al. (2017). Revealing hydrogenation reaction pathways on naked gold nanoparticles. *ACS Catal.* 7 (4): 2791–2799. (c) Hu, D., Fan, W.-Q., Liu, Z., and Li, L. (2018). Three-dimensionally hierarchical Pt/C nanocomposite with ultra-high dispersion of Pt nanoparticles as a highly efficient catalyst for chemoselective cinnamaldehyde hydrogenation. *ChemCatChem* 10 (4): 779–788.

1349 (a) Astruc, D., Lu, F., and Aranzaes, J.R. (2005). Nanoparticles as recyclable catalysts: the frontier between homogeneous and heterogeneous catalysis. *Angew. Chem. Int. Ed.* 44: 7852–7872. (b) Yan, N., Xiao, C., and Kou, Y. (2010). Transition metal nanoparticle catalysis in green solvents. *Coord. Chem. Rev.* 254: 1179–1218.

1350 (a) Karini, B., Mansouri, F., and Vali, H. (2015). A highly water-dispersible/magnetically separable Pd-catalyst: selective transfer hydrogenation or direct reductive *N*-formylation of nitroarenes in water. *ChemPlusChem* 80 (12): 1750–1759. (b) Gao, Y.-X., Zhang, H.-W., Han, A. et al. (2018). Ru/ZrO_2 catalysts for transfer hydrogenation of levulinic acid with formic acid/formate mixtures: importance of support stability. *ChemistrySelect* 3 (5): 1343–1351.

1351 Gao, L., Kojima, K., and Nagashima, H. (2015). Transition metal nanoparticles stabilized by ammonium salts of hyperbranched polystyrene: effect of metals on catalysis of the biphasic hydrogenation of alkenes and arenes. *Tetrahedron* 71: 6414–6423.

1352 Augustine, R.L. (1996). *Heterogeneous Catalysis for Synthetic Chemist*, 1–446. New York: Marcel Dekker.

1353 Smith, G.V. and Notheisz, F. (1999). *Heterogeneous Catalysis in Organic Chemistry*, 1–346. San Diego, CA: Academic Press.

1354 (a) Wehrli, J.T., Baiker, A., Monti, D.M., and Blaser, H.-U. (1990). Enantioselective hydrogenation of α-ketoesters: preparation and catalytic behavior of different alumina-supported platinum catalysts modified with cinchonidine. *J. Mol. Catal.* 61 (2): 207–226. (b) Garland,

M. and Blaser, H.-U. (1990). A heterogeneous ligand-accelerated reaction – enantioselective hydrogenation of ethyl pyruvate catalyzed by cinchona-modified Pt/Al_2O_3 catalysts. *J. Am. Chem. Soc.* 112 (19): 7048–7050. (c) Blaser, H.-U. and Spindler, F. (1997). Enantioselective catalysis for agrochemicals. The case histories of (*S*)-metolachlor, (*R*)-metalaxyl and clozylacon. *Top. Catal.* 4 (3–4): 275–282.

1355 (a) Meeuwissen, J., Kuil, M., van der Burg, A.M. et al. (2009). Application of a supramolecular-ligand library for the automated search for catalysts for the asymmetric hydrogenation of industrially relevant substrates. *Chem. Eur. J.* 15: 10272–10279. (b) Pignataro, L., Bovio, C., Civera, M. et al. (2012). A library approach to the development of BenzaPhos: highly efficient chiral supramolecular ligands for asymmetric hydrogenation. *Chem. Eur. J.* 18: 10368–10381.

1356 Thomas, J.M. (1999). Design, synthesis, and in situ characterization of new solid catalysts. *Angew. Chem. Int. Ed.* 38 (24): 3589–3628.

1357 Lindlar, H. (1952). Ein Neuer Katalysator für Selektive Hydrierungen. *Helv. Chim. Acta* 35 (2): 446–450.

1358 Studer, M., Burkhardt, S., and Blaser, H.-U. (2001). Catalytic hydrogenation of chiral α-amino and α-hydroxy esters at room temperature with Nishimura catalyst without racemization. *Adv. Synth. Catal.* 343 (8): 802–808.

1359 Ando, C., Ikumoto, A., Kurokawa, H. et al. (2001). Selective hydrogenation of (*E*)-2-hexenal to (*E*)-2-hexen-1-ol over Co-based bimetallic catalysts. *Catal. Commun.* 2 (10): 323–327.

1360 Maegawa, T., Ito, N., Oono, K. et al. (2009). Bimetallic palladium-platinum-on-carbon-catalyzed H—D exchange reaction: synergistic effect on multiple deuterium incorporation. *Synthesis* (16): 2674–2678.

1361 Horiuti, I. and Polanyi, M. (1934). Exchange reactions of hydrogen on metallic catalysts. *Trans. Faraday Soc.* 30: 1164–1172.

1362 Mattson, B., Foster, W., Greimann, J. et al. (2013). Heterogeneous catalysis: the Horiuti–Polanyi mechanism and alkene hydrogenation. *J. Chem. Educ.* 90: 613–619.

1363 Zaera, F. and Somorjai, G.A. (1984). Hydrogenation of ethylene over Pt(111) single-crystal surfaces. *J. Am. Chem. Soc.* 106 (8): 2288–2293.

1364 Baro, A.M. and Ibach, H. (1981). Thermal evolution and decomposition of ethylene on Pt(111). *J. Chem. Phys.* 74 (7): 4194–4199.

1365 Oefner, H. and Zaera, F. (2002). Ethylene adsorption on Pt: kinetics, bonding, and relevance to catalysis. *J. Am. Chem. Soc.* 124 (37): 10982–10983.

1366 Bent, B.E. (1996). Mimicking aspects of heterogeneous catalysis: generating, isolating, and reacting proposed surface intermediates on single crystals in vacuum. *Chem. Rev.* 96 (4): 1361–1390.

1367 Zaera, F. and Bernstein, N. (1994). On the mechanism for the conversion of ethylene to ethylidyne on metal-surfaces – vinyl halides on Pt(111). *J. Am. Chem. Soc.* 116 (11): 4881–4887.

1368 Zaera, F. and Chrysostomou, D. (2000). Propylene on Pt(111) II. Hydrogenation, dehydrogenation, and H—D exchange. *Surf. Sci.* 457 (1–2): 89–108.

1369 Zeigarnik, A.V., Valdés-Pérez, R.E., and Temkin, O.N. (1998). Metal-catalyzed ethylene hydrogenation: the method of interactive search for multiple working hypotheses. *Langmuir* 14 (16): 4510–4516.

1370 Pery, T., Pelzer, K., Buntkowsky, G. et al. (2005). Direct NMR evidence for the presence of mobile surface hydrides on Ru nanoparticles. *ChemPhysChem* 6 (4): 605–607.

1371 García-Antón, J., Axet, M.R., Jansat, S. et al. (2008). Reactions of olefins with Ru hydride nanoparticles: NMR characterization, hydride titration, and room-temperature C—C bond activation. *Angew. Chem. Int. Ed.* 47 (11): 2074–2078.

1372 Brown, H.C. and Brown, C.A. (1963). Reaction of sodium borohydride with Ni acetate in ethanol solution – a highly selective Ni hydrogenation catalyst. *J. Am. Chem. Soc.* 85 (7): 1005–1006.

1373 Augustine, R.L., Yaghmaie, F., and van Peppen, J.F. (1984). Heterogeneous catalysis in organic chemistry. 2. A mechanistic comparison of noble-metal catalysts in olefin hydrogenation. *J. Org. Chem.* 49 (11): 1865–1870.

1374 Smith, G.V. and Swoap, J.R. (1966). Metal-catalyzed hydrogenation and exchange of cyclohexene. *J. Org. Chem.* 31 (12): 3904–3906.

1375 Mitsui, S., Senda, Y., and Saito, H. (1966). Stereoselectivity of catalysts in hydrogenation. I. Catalytic hydrogenation of 2-cyclopentylidenecyclopentanol and 2-benzylidene-1-tetralol. *Bull. Chem. Soc. Jpn.* 39 (4): 694–697.

1376 Steiner, H., Giannousis, P., Pische-Jacques, A., and Blaser, H.-U. (2000). Diastereoselective hydrogenation of

4-(phosphonomethyl)-2-pyridinecarboxylic acid. *Top. Catal.* 13 (3): 191–194.

1377 Rothwell, I.P. (1997). A new generation of homogeneous arene hydrogenation catalysts. *J. Chem. Soc., Chem. Commun.* (15): 1331–1338.

1378 Hallman, P.S., Evans, D., Osborn, J.A., and Wilkinson, G. (1967). Selective catalytic homogeneous hydrogenation of terminal olefins using tris(triphenylphosphine)hydridochlororuthenium (II) – hydrogen transfer in exchange and isomerisation reactions of olefins. *J. Chem. Soc., Chem. Commun.* (7): 305–306.

1379 Calvin, M. (1938). Homogeneous catalytic hydrogenation. *Trans. Faraday Soc.* 34 (2): 1181–1191.

1380 Iguchi, M. (1939). Studies on the catalytic oxido-reduction by some metallic complex salts. XIV. Catalytic reduction by rhodium complex salts. *Nippon Kagaku Kaishi* 60: 1787–1792.

1381 Halpern, J., Harrod, J.F., and James, B.R. (1961). Homogeneous catalytic hydrogenation of olefinic compounds. *J. Am. Chem. Soc.* 83 (3): 753–754.

1382 (a) Osborn, J.A., Jardine, F.H., Young, J.F., and Wilkinson, G. (1966). Preparation and properties of tris(triphenylphosphine)halogenorhodium(I) and some reactions thereof including catalytic homogeneous hydrogenation of olefins and acetylenes and their derivatives. *J. Chem. Soc. A* (12): 1711–1732. (b) de Vries, J.G. and Elsevier, C.J. (eds.) (2007). *The Handbook of Homogeneous Hydrogenation*, Chapter 1–45, 1–1568. Weinheim: Wiley-VCH.

1383 Jardine, F.H. and Wilkinson, G. (1967). Homogeneous catalytic hydrogenation of unsaturated aldehydes to form saturated aldehydes. *J. Chem. Soc. C* (4): 270–271.

1384 Schrock, R.R. and Osborn, J.A. (1971). Preparation and properties of some cationic complexes of Rh(I) and Rh(III). *J. Am. Chem. Soc.* 93 (10): 2397–2407.

1385 Schrock, R.R. and Osborn, J.A. (1971). Coordinatively unsaturated cationic complexes of Rh (I), Ir (I), Pd(II), and Pt(II) – generation, synthetic utility, and some catalytic studies. *J. Am. Chem. Soc.* 93 (12): 3089–3091.

1386 Halpern, J. and Wong, C.S. (1973). Hydrogenation of tris(triphenylphosphine) chlororhodium(I). *J. Chem. Soc., Chem. Commun.* (17): 629–630.

1387 Halpern, J., Okamoto, T., and Zakhariev, A. (1977). Mechanism of chlorotris(triphenylphosphine)Rh(I)-catalyzed hydrogenation of alkenes – reaction of chlorodihydridotris(triphenylphosphine)Rh (III) with cyclohexene. *J. Mol. Catal.* 2 (1): 65–68.

1388 Young, J.F., Osborn, J.A., Jardine, F.H., and Wilkinson, G. (1965). Hydride intermediates in homogeneous hydrogenation reactions of olefins and acetylenes using Rh catalysts. *J. Chem. Soc., Chem. Commun.* (7): 131–132.

1389 Halpern, J. (1980). Homogeneous catalytic hydrogenation – a retrospective account. *J. Organomet. Chem.* 200 (1): 133–144.

1390 Chan, A.S.C. and Halpern, J. (1980). Interception and characterization of a hydridoalkylrhodium intermediate in a homogeneous catalytic hydrogenation reaction. *J. Am. Chem. Soc.* 102 (2): 838–840.

1391 Halpern, J. (1982). Mechanism and stereoselectivity of asymmetric hydrogenation. *Science* 217 (4558): 401–407.

1392 Stephens, T.A. and Wilkinson, G. (1966). New complexes of Ru(II) and Ru(III) with triphenylphosphine triphenylarsine trichlorostannate pyridine and other ligands. *J. Inorg. Nucl. Chem.* 28 (4): 945–946.

1393 Rosales, M., Alvarado, Y., Gallardo, N., and Rubio, R. (1995). Kinetics and mechanisms of homogeneous catalytic reactions. 2. Hydrogenation of cyclohexane catalyzed by $[RuH(CO)(NCMe)_2(PPh_3)]BF_4$. *Transition Met. Chem.* 20 (3): 242–245.

1394 Jeske, G., Lauke, H., Mauermann, H. et al. (1985). Highly reactive organolanthanides – a mechanistic study of catalytic olefin hydrogenation by bis(pentamethylcyclopentadienyl) and related 4f complexes. *J. Am. Chem. Soc.* 107 (26): 8111–8118.

1395 Noyori, R. and Hashiguchi, S. (1997). Asymmetric transfer hydrogenation catalyzed by chiral ruthenium complexes. *Acc. Chem. Res.* 30 (2): 97–102.

1396 Brunel, J.M. (2007). Scope, limitations and mechanistic aspects in the selective homogeneous Pd-catalyzed reduction of alkenes under transfer hydrogen conditions. *Tetrahedron* 63 (18): 3899–3906.

1397 Amatore, C., Jutand, A., and M'Barki, M.A. (1992). Evidence of the formation of zero-valent Pd from $Pd(OAc)_2$ and triphenylphosphine. *Organometallics* 11 (9): 3009–3013.

1398 Amatore, C., Jutand, A., and Thuilliez, A. (2001). Formation of Pd(0) complexes from $Pd(OAc)_2$ and a bidentate phosphine ligand (dppp) and their reactivity in oxidative addition. *Organometallics* 20: 3241–3249.

1399 Jiang, Y. and Berke, H. (2007). Dehydrocoupling of dimethylamine-borane catalysed by Re complexes and its application in olefin transfer-hydrogenations. *J. Chem. Soc., Chem. Commun.* (34): 3571–3573.

1400 Yu, J.Q. and Spencer, J.B. (1997). First evidence that the mechanism of catalytic hydrogenation with homogeneous Pd and Rh catalysts is strongly influenced by substrate polarity. *J. Am. Chem. Soc.* 119 (22): 5257–5258.

1401 Gungor, M., Jardine, F.H., and Wheatley, J.D. (1988). Homogeneous deuteriation of alkenes using rhhalide[(P(aryl)$_3$)]$_3$. *J. Chem. Soc., Dalton Trans.* (6): 1653–1656.

1402 Hardick, D.J., Blagbrough, I.S., and Potter, B.V.L. (1996). Isotopic enrichment by asymmetric deuteriation. an investigation of the synthesis of deuteriated (*S*)-(-)-methylsuccinic acids from itaconic acid. *J. Am. Chem. Soc.* 118 (25): 5897–5903.

1403 Harthun, A., Selke, R., and Bargon, J. (1996). Proof of a reversible, pairwise hydrogen transfer during the homogeneously Rh(I)-catalyzed hydrogenation of α,β-unsaturated carbonic acid derivatives with in situ NMR spectroscopy and parahydrogen. *Angew. Chem. Int. Ed. Engl.* 35 (21): 2505–2507.

1404 Nelson, D.J., Li, R.B., and Brammer, C. (2005). Using correlations to compare additions to alkenes: homogeneous hydrogenation by using Wilkinson's catalyst. *J. Org. Chem.* 70 (3): 761–767.

1405 Lee, H.M., Smith, D.C., He, Z.J. et al. (2001). Catalytic hydrogenation of alkenes by the Ru-carbene complex HRu(CO)Cl(PCy3)(IMes) IMes=bis(1,3-(2,4,6-trimethylphenyl)imidazol-2-ylidene). *Organometallics* 20 (4): 794–797.

1406 Dharmasena, U.L., Foucault, H.M., dos Santos, E.N. et al. (2005). N-heterocyclic carbenes as activating ligands for hydrogenation and isomerization of unactivated olefins. *Organometallics* 24 (6): 1056–1058.

1407 Herrmann, W.A., Frey, G.D., Herdtweck, E., and Steinbeck, M. (2007). Synthesis and characterization of N-heterocyclic carbene substituted phosphine and phosphite Rh complexes and their catalytic properties in hydrogenation reactions. *Adv. Synth. Catal.* 349 (10): 1677–1691.

1408 (a) Kluwer, A.M., Koblenz, T.S., Jonischkeit, T. et al. (2005). Kinetic and spectroscopic studies of the [Pd(Ar-bian)]-catalyzed semi-hydrogenation of 4-octyne. *J. Am. Chem. Soc.* 127 (44): 15470–15480. (b) Zhao, Q.-L., Miao, X., and Wang, L.-L. (2013). Enantioselective hydrogenation of functionalized olefins catalyzed by Rh-chiral bidentatephosphite complexes. *Tetrahedron: Asymmetry* 24 (2–3): 104–107. (c) Lao, J.R., Benet-Buchholz, J., and Vidal-Ferran, A. (2014). 1,1-P—OP ligands with P-stereogenic phosphino groups in asymmetric hydrogenations and hydroformylations. *Organometallics* 33 (12): 2960–2963. (d) León, F., González-Liste, P.J., García-Garrido, S.E. et al. (2017). Broad scope synthesis of ester precursors of nonfunctionalized chiral alcohols based on the asymmetric hydrogenation of α,β-dialkyl-, α,β-diaryl-, and α-Alkyl-β-aryl-vinyl esters. *J. Org. Chem.* 82 (11): 5852–5867.

1409 Donsi, F., Cimino, S., Pirone, R. et al. (2005). Crossing the breakthrough line of ethylene production by short contact time catalytic partial oxidation. *Catal. Today* 106 (1–4): 72–76.

1410 Donsi, F., Cimino, S., Di Benedetto, A. et al. (2005). The effect of support morphology on the reaction of oxidative dehydrogenation of ethane to ethylene at short contact times. *Catal. Today* 105 (3–4): 551–559.

1411 Kung, H.H. and Kung, M.C. (1997). Oxidative dehydrogenation of alkanes over vanadium-magnesium-oxides. *Appl. Catal., A* 157 (1–2): 105–116.

1412 Knotek, P., Capek, L., Bulanek, R., and Adam, J. (2007). Vanadium supported on hexagonal mesoporous silica: active and stable catalysts in the oxidative dehydrogenation of alkanes. *Top. Catal.* 45 (1–4): 51–55.

1413 de Almeida, K.J. and Cesar, A. (2006). Methane C—H bond activation by neutral lanthanide and thorium atoms in the gas phase: a theoretical prediction. *Organometallics* 25 (14): 3407–3416.

1414 Adlhart, C. and Uggerud, E. (2007). Mechanisms for the dehydrogenation of alkanes on Pt: insights gained from the reactivity of gaseous cluster cations, Pt-n(+) n=1–21. *Chem. Eur. J.* 13 (24): 6883–6890.

1415 Bergman, R.G. (1984). Activation of alkanes with organotransition metal complexes. *Science* 223 (4639): 902–908.

1416 Baudry, D., Ephritikhine, M., Felkin, H., and Holmes-Smith, R. (1983). The selective catalytic conversion of cycloalkanes into cycloalkenes using a soluble Re polyhydride system. *J. Chem. Soc., Chem. Commun.* (14): 788–789.

1417 Gupta, M., Hagen, C., Flesher, R.J. et al. (1996). A highly active alkane dehydrogenation catalyst: stabilization of dihydrido Rh and Ir complexes by a P—C—P pincer-ligand. *J. Chem. Soc., Chem. Commun.* (17): 2083–2084.

1418 Göttker-Schnetmann, I., White, P., and Brookhart, M. (2004). Iridium bis(phosphinite) p-XPCP pincer complexes: highly active catalysts for the transfer dehydrogenation of alkanes. *J. Am. Chem. Soc.* 126 (6): 1804–1811.

1419 Yildiz, B. and Kazimi, M.S. (2006). Efficiency of hydrogen production systems using alternative nuclear energy technologies. *Int. J. Hydrogen Energy* 31 (1): 77–92.

1420 Ritter, J.A. and Ebner, A.D. (2007). State-of-the-art adsorption and membrane separation processes for hydrogen production in the chemical and petrochemical industries. *Sep. Sci. Technol.* 42 (6): 1123–1193.

1421 Düsterwald, H.G., Günnewig, J., and Radtke, P. (2007). DRIVE – the future of automotive power: fuel cells perspective. *Fuel Cells* 7 (3): 183–189.

1422 Xu, J., Yeung, C.M.Y., Ni, J. et al. (2008). Methane steam reforming for hydrogen production using low water-ratios without carbon formation over ceria coated Ni catalysts. *Appl. Catal., A* 345 (2): 119–127.

1423 (a) Chaloner, P.A., Esteruelas, M.A., Jóo, F., and Oro, L.A. (1994). *Homogeneous Hydrogenation*, Catalysis by Metal Complexes, vol. 15, 1–290. Dordrecht, Netherlands: Kluwer Academic. (b) Borodzinski, A. and Bond, G.C. (2008). Selective hydrogenation of ethyne in ethene-rich streams on Pd catalysts, part 2: steady-state kinetics and effects of Pd particle size, carbon monoxide, and promoters. *Catal. Rev. Sci. Eng.* 50 (3): 379–464.

1424 (a) Gao, L. and Rongchao, J.G. (2014). Gold nanocluster-catalyzed semihydrogenation: a unique activation pathway for terminal alkynes. *J. Am. Chem. Soc.* 136: 11347–11354. (b) Furukawa, S. and Komatsu, T. (2016). Selective hydrogenation of functionalized alkynes to (*E*)-alkenes, using ordered alloys as catalysts. *ACS Catal.* 6 (3): 2121–2125. (c) Fiorio, J.L., López, N., and Rossi, L.M. (2017). Gold/ligand-catalyzed selective hydrogenation of alkynes into *cis*-alkenes via H_2 heterolytic activation by frustrated Lewis pairs. *ACS Catal.* 7: 2973–2980. (d) Diaz de los Bernados, M., Pérez-Rodriguez, S., Guai, A. et al. (2017). Facile synthesis of NHC-stabilized Ni nanoparticles and their catalytic application in the *Z*-selective hydrogenation of alkynes. *J. Chem. Soc.; Chem. Commun.* 53 (56): 7894–7897. (e) Landge, V.G., Pitchaimani, J., Midya, S.P. et al. (2018). Phosphine-free cobalt pincer complex catalyzed *Z*-selective semi-hydrogenation of unbiased alkynes. *Catal. Sci. Technol.* 8 (2): 428–433. (f) Jagtap, S.A. and Bhanage, B.M. (2018). Ligand assisted Rh-catalyzed selective semi-hydrogenation of alkynes using syngas and molecular hydrogen. *ChemistrySelect* 3 (2): 713–718.

1425 Sánchez-Delgado, R.A., Rosales, M., Esteruelas, M.A., and Oro, L.A. (1995). Homogeneous catalysis by osmium complexes – a review. *J. Mol. Catal. A: Chem.* 96 (3): 231–243.

1426 Andriollo, A., Esteruelas, M.A., Meyer, U. et al. (1989). Kinetic and mechanistic investigation of the sequential hydrogenation of phenylacetylene catalyzed by OsHCl(CO) $[PMe(t\text{-}Bu)_2]_2$, OsHCl(CO)$[P(i\text{-}Pr)_3]_2$. *J. Am. Chem. Soc.* 111 (19): 7431–7437.

1427 Esteruelas, M.A., López, A.M., Oro, L.A. et al. (1993). Mechanism of the hydrogenation of phenylacetylene catalyzed by $[Ir(cod)(\eta^2\text{-}(i\text{-}Pr)_2PCH_2CH_2OMe)]BF_4$. *Organometallics* 12 (5): 1823–1830.

1428 Bianchini, C., Meli, A., Peruzzini, M. et al. (1992). Selective hydrogenation of 1-alkynes to alkenes catalyzed by an Fe(II) *cis*-hydride η^2-dihydrogen complex – a case of intramolecular reaction between η^2-H2 and σ-vinyl ligands. *Organometallics* 11 (1): 138–145.

1429 Sánchez-Delgado, R.A. and Rosales, M. (2000). Kinetic studies as a tool for the elucidation of the mechanisms of metal complex-catalyzed homogeneous hydrogenation reactions. *Coord. Chem. Rev.* 196: 249–280.

1430 Pelagatti, P., Venturini, A., Leporati, A. et al. (1998). Chemoselective homogeneous hydrogenation of phenylacetylene using thiosemicarbazone and thiobenzoylhydrazone Pd(II) complexes as catalysts. *J. Chem. Soc., Dalton Trans.* (16): 2715–2722.

1431 Costa, M., Pelagatti, P., Pelizzi, C., and Rogolino, D. (2002). Catalytic activity of Pd(II) complexes with tridentate nitrogen ligands in the hydrogenation of alkenes and alkynes. *J. Mol. Catal. A: Chem.* 178 (1–2): 21–26.

1432 Hauwert, P., Maestri, G., Sprengers, J.W. et al. (2008). Transfer semihydrogenation of alkynes catalyzed by a zero-valent Pd N-heterocyclic carbene complex. *Angew. Chem. Int. Ed.* 47 (17): 3223–3226.

1433 Srimani, D., skin-Posner, Y., Ben-David, Y., and Milstein, D. (2013). Iron pincer complex catalyzed, environmentally benign, *E*-selective semi-hydrogenation of alkynes. *Angew. Chem. Int. Ed.* 52: 14131–14134.

1434 López-Serrano, J., Lledós, A., and Duckett, S.B. (2008). A DFT study on the mechanism of Pd-catalyzed alkyne hydrogenation: neutral versus cationic pathways. *Organometallics* 27 (1): 43–52.

1435 (a) Durland, J.R. and Adkins, H. (1937). Hydrogenation of phenanthrene. *J. Am. Chem. Soc.* 59: 135–137. (b) Musser, D.M. and Adkins, H. (1938). The selective hydrogenation of derivatives of naphthalene and diphenyl. *J. Am. Chem. Soc.* 60: 664–669.

1436 Adkins, H., Wolff, I.A., Palvic, A., and Hutchinson, E. (1944). Preparation of derivatives of pyrrole and pyridine by hydrogenation. *J. Am. Chem. Soc.* 66: 1293–1295.

1437 Maegawa, T., Akashi, A., Yaguchi, K. et al. (2009). Efficient and practical arene hydrogenation by heterogeneous catalysts under mild conditions. *Chem. Eur. J.* 15: 6953–6963.

1438 (a) Dwivedi, A.D., Rai, R.K., Gupta, K., and Singh, S.K. (2017). Catalytic hydrogenation of arenes in water over in situ generated ruthenium nanoparticles immobilized on carbon. *ChemCatChem* 9 (11): 1930–1938. (b) Ji, P.-F., Song, Y., Drake, T. et al. (2018). Titanium(III)-oxo clusters in a metal-organic framework support single-site Co(II)-hydride catalysts for arene hydrogenation. *J. Am. Chem. Soc.* 140 (1): 433–440.

1439 Ranade, V.S., Consiglio, G., and Prins, R. (1999). Functional-group-directed diastereoselective hydrogenation of aromatic compounds. 1. *J. Org. Chem.* 64: 8862–8867.

1440 Ranade, V.S., Consiglio, G., and Prins, R. (2000). Functional-group-directed diastereoselective hydrogenation of aromatic compounds. 2. *J. Org. Chem.* 65: 1132–1138.

1441 Nishimura, S., Ohbuchi, S., Ikeno, K., and Okada, Y. (1984). Hydrogenation and hydrogenolysis. 17. The selectivities of PT Group-metals in catalytic hydrogenation of 2-naphthol and tetrahydro-2-naphthols. *Bull. Chem. Soc. Jpn.* 57: 2557–2564.

1442 Gualandi, A. and Savoia, D. (2016). Substrate induced diastereoselective hydrogenation/reduction of arenes and heteroarenes. *RSC Adv.* 6: 18419–18451.

1443 Beyerman, H.C., Eenshuistra, J., Eveleens, W., and Zweistra, A. (1959). Studies on sedum alkaloids. 5. Synthesis and optical resolution of sedamine and sedridine. *Recl. Trav. Chim. Pays-Bas* 78: 43–58.

1444 Cooke, G.A. and Fodor, G. (1968). Diastereoisomeric 2-aryl-6-methyl-3,4(1,2-piperido)-oxazines. *Can. J. Chem.* 46: 1105–1111.

1445 Kuwano, R., Sato, K., Kurokawa, T. et al. (2000). Catalytic asymmetric hydrogenation of heteroaromatic compounds, indoles. *J. Am. Chem. Soc.* 122: 7614–7615.

1446 Wang, D.S., Chen, Q.A., Lu, S.M., and Zhou, Y.G. (2012). Asymmetric hydrogenation of heteroarenes and arenes. *Chem. Rev.* 112: 2557–2590.

1447 Kuwano, R. (2015). Catalytic asymmetric hydrogenation of heteroarenes and arenes. *Phosphorus, Sulfur Silicon Relat. Elem.* 190: 715–719.

1448 Wang, D.S., Ye, Z.S., Chen, Q.A. et al. (2011). Highly enantioselective partial hydrogenation of simple pyrroles: a facile access to chiral 1-pyrrolines. *J. Am. Chem. Soc.* 133: 8866–8869.

1449 (a) Domine, M.E., Hernández-Soto, M.C., Navarro, M.T., and Pérez, Y. (2011). Pt and Pd nanoparticles supported on structured materials as catalysts for the selective reductive amination of carbonyl compounds. *Catal. Today* 172 (1): 13–20. (b) Takale, B.S., Feng, X.-J., Lu, Y. et al. (2016). Unsupported nanoporous gold catalyst for chemoselective hydrogenation reactions under low pressure: effect of residual silver on the reaction. *J. Am. Chem. Soc.* 138 (32): 10356–10364. (c) Komanoya, T., Kinemura, T., Kita, Y. et al. (2017). Electronic effect of Ru nanoparticles on efficient reductive amination of carbonyl compounds. *J. Am. Chem. Soc.* 139 (33): 11493–11499. (d) Nandi, S., Patel, P., Jakhar, A. et al. (2017). Cucurbit[6]uril-stabilized Pd nanoparticles as a highly active catalyst for chemoselective hydrogenation of various reducible groups in aqueous media. *ChemistrySelect* 2 (31): 9911–9919.

1450 (a) Zassinovich, G., Mestroni, G., and Gladiali, S. (1992). Asymmetric hydrogen transfer reactions promoted by homogeneous transition-metal catalysts. *Chem. Rev.* 92 (5): 1051–1069. (b) Klomp, D., Hanefeld, U., and Peters, J. (2007). A transfer hydrogenation including Meerwein-Ponndorf-Verley reduction. In: *Handbook for Homogeneous Hydrogenation*, Chapter 20 (ed. J.G. de Vries and C.J. Elsevier). Weinheim: Wiley-VCH. (c) Hudson, R., Hamasaka, G., Osako, T. et al. (2013). Highly efficient iron(0) nanoparticle-catalyzed hydrogenation in water in flow. *Green Chem.* 15 (8): 2141–2148.

1451 Gladiali, S. and Alberico, E. (2006). Asymmetric transfer hydrogenation: chiral ligands and applications. *Chem. Soc. Rev.* 35 (3): 226–236.

1452 Samec, J.S.M., Bäckvall, J.E., Andersson, P.G., and Brandt, P. (2006). Mechanistic aspects of transition metal-catalyzed hydrogen transfer reactions. *Chem. Soc.Rev.* 35 (3): 237–248.

1453 Hillier, A.C., Lee, H.M., Stevens, E.D., and Nolan, S.P. (2001). Cationic Ir complexes bearing imidazol-2-ylidene ligands as transfer hydrogenation catalysts. *Organometallics* 20 (20): 4246–4252.

1454 Corberán, R., Sanaú, M., and Peris, E. (2007). "Cp*Ir(III)" complexes with hemi-cleaveable ligands of the type N-alkenyl imidazolin-2-ylidene. Reactivity and catalytic properties. *Organometallics* 26 (14): 3492–3498.

1455 Hamid, M.H.S.A., Slatford, P.A., and Williams, J.M.J. (2007). Borrowing hydrogen in the activation of alcohols. *Adv. Synth. Catal.* 349 (10): 1555–1575.

1456 Gnanamgari, D., Moores, A., Rajaseelan, E., and Crabtree, R.H. (2007). Transfer hydrogenation of imines and alkenes and direct reductive amination of aldehydes catalyzed by triazole-derived Ir(I) carbene complexes. *Organometallics* 26 (5): 1226–1230.

1457 Campaña, A.G., Estévez, R.E., Fuentes, N. et al. (2007). Unprecedented hydrogen transfer from water to alkenes and alkynes mediated by Ti(III) and late transition metals. *Org. Lett.* 9 (11): 2195–2198.

1458 Coffey, R.S. (1967). A catalyst for homogeneous hydrogenation of aldehydes under mild conditions. *J. Chem. Soc., Chem. Commun.* (18): 923–923.

1459 Clapham, S.E., Hadzovic, A., and Morris, R.H. (2004). Mechanisms of the H2-hydrogenation and transfer hydrogenation of polar bonds catalyzed by Ru hydride complexes. *Coord. Chem. Rev.* 248 (21–24): 2201–2237.

1460 Ito, M. and Ikariya, T. (2007). Catalytic hydrogenation of polar organic functionalities based on Ru-mediated heterolytic dihydrogen cleavage. *J. Chem. Soc., Chem. Commun.* (48): 5134–5142.

1461 Meerwein, H. and Schmidt, R. (1925). A new process for the reduction of aldehydes and ketones. *Liebigs Ann. Chem.* 444: 221–238.

1462 Ponndorf, W. (1926). The reversible exchange of oxygen between aldehydes or ketones on the one hand and primary or secondary alcohols on the other hand. *Angew. Chem.* 39: 138–143.

1463 Verley, A. (1925). Exchange of functional groups between two molecules. Exchange of alcohol and aldehyde groups. *Bull. Soc. Chim. Fr.* 37: 537–542.

1464 Wilds, A.L. (1944). Reduction with Al alkoxides (Meerwein–Ponndorf–Verley reduction). *Org. React.* 2: 178–223.

1465 Djerassi, C. (1951). The Oppenauer oxidation. *Org. React.* 6: 207–272.

1466 Ooi, T., Ichikawa, H., and Maruoka, K. (2001). Practical approach to the Meerwein–Ponndorf–Verley reduction of carbonyl substrates with new aluminum catalysts. *Angew. Chem. Int. Ed.* 40 (19): 3610–3612.

1467 Molander, G.A. and McKie, J.A. (1993). Samarium(II) Iodide induced sequential intramolecular nucleophilic acyl substitution and stereospecific intramolecular Meerwein–Ponndorf–Verley reduction Oppenauer oxidation. *J. Am. Chem. Soc.* 115 (13): 5821–5822.

1468 Knauer, B. and Krohn, K. (1995). A reinvestigation of the Meerwein–Ponndorf–Verley reduction – a highly efficient variation using Zr catalysts. *Liebigs Ann. Chem.* (4): 677–683.

1469 Ekstroem, J., Wettergren, J., and Adolfsson, H. (2007). A simple and efficient catalytic method for the reduction of ketones. *Adv. Synth. Catal.* 349 (10): 1609–1613.

1470 Kumbhar, P.S., Sánchez-Valente, J., López, J., and Figueras, F. (1998). Meerwein–Ponndorf–Verley reduction of carbonyl compounds catalysed by Mg—Al hydrotalcite. *J. Chem. Soc., Chem. Commun.* (5): 535–536.

1471 Zhu, Y.Z., Chuah, G.K., and Jaenicke, S. (2006). Selective Meerwein–Ponndorf–Verley reduction of α,β-unsaturated aldehydes over Zr-zeolite-β. *J. Catal.* 241 (1): 25–33.

1472 Chuah, G.K., Jaenicke, S., Zhu, Y.Z., and Liu, S.H. (2006). Meerwein–Ponndorf–Verley reduction over heterogeneous catalysts. *Curr. Org. Chem.* 10 (13): 1639–1654.

1473 Linn, D.E. and Halpern, J. (1987). Roles of neutral and anionic Ru polyhydrides in the catalytic hydrogenation of ketones and arenes. *J. Am. Chem. Soc.* 109 (10): 2969–2974.

1474 Bäckvall, J.E. (2002). Transition metal hydrides as active intermediates in hydrogen transfer reactions. *J. Organomet. Chem.* 652 (1-2): 105–111.

1475 Pámies, O. and Bäckvall, J.E. (2001). Studies on the mechanism of metal-catalyzed hydrogen transfer from alcohols to ketones. *Chem. Eur. J.* 7 (23): 5052–5058.

1476 Gallezot, P. and Richard, D. (1998). Selective hydrogenation of α,β-unsaturated aldehydes. *Catal. Rev. Sci. Eng.* 40 (1–2): 81–126.

1477 Bhanage, B.M., Ikushima, Y., Shirai, M., and Arai, M. (1999). The selective formation of unsaturated alcohols by hydrogenation of α,β-unsaturated aldehydes in supercritical carbon dioxide using unpromoted Pt/Al_2O_3 catalyst. *Catal. Lett.* 62 (2-4): 175–177.

1478 Noyori, R. and Ohkuma, T. (2001). Asymmetric catalysis by architectural and functional molecular engineering: practical chemo- and stereoselective hydrogenation of ketones. *Angew. Chem. Int. Ed.* 40 (1): 40–73.

1479 Xie, J.H., Liu, X.Y., Xie, J.B. et al. (2011). An additional coordination group leads to extremely efficient chiral Ir catalysts for asymmetric hydrogenation of ketones. *Angew. Chem. Int. Ed.* 50: 7329–7332.

1480 Xie, J.B., Xie, J.H., Liu, X.Y. et al. (2011). Chiral Ir spiro aminophosphine complexes: asymmetric hydrogenation of simple ketones, structure, and plausible mechanism. *Chem. Asian J.* 6: 899–908.

1481 Yan, P.C., Zhu, G.L., Xie, J.H. et al. (2013). Industrial scale-up of enantioselective hydrogenation for the asymmetric synthesis of Rivastigmine. *Org. Process Res. Dev.* 17: 307–312.

1482 Mertens, P.G.N., Pölman, H., Ye, X. et al. (2007). Au(0) nano-colloids as recyclable quasihomogeneous metal catalysts in the chemoselective hydrogenation of α,β-unsaturated aldehydes and ketones to allylic alcohols. *Catal. Today* 122 (3–4): 352–360.

1483 Casey, C.P. and Guan, H.R. (2007). An efficient and chemoselective Fe catalyst for the hydrogenation of ketones. *J. Am. Chem. Soc.* 129 (18): 5816–5817.

1484 (a) Fleischer, S., Zhou, S.L., Junge, K., and Beller, M. (2013). General and highly efficient Fe-catalyzed hydrogenation of aldehydes, ketones, and, α,β-unsaturated aldehydes. *Angew. Chem. Int. Ed.* 52: 5120–5124. (b) Gorgas, N., Stöger, B., Veiros, L.F. et al. (2014). Efficient hydrogenation of ketones and aldehydes catalyzed by well-defined Fe(II) PNP pincer complexes: evidence for an insertion mechanism. *Organometallics* 33 (23): 6905–6914. (c) Misal Castro, L.C., Li, H.-Q., Sortais, J.-B., and Darcel, C. (2015). When iron met phosphines: a happy marriage for reduction catalysis. *Green Chem.* 17 (4): 2283–2303. (d) Gorgas, N., Stöger, B., Veiros, L.F., and Kirchner, K. (2016). Highly efficient and selective hydrogenation of aldehydes: a well-defined Fe(II)-catalyst exhibits noble-metal activity. *ACS Catal.* 6 (4): 2664–2672.

1485 Walling, C. and Bollyky, L. (1964). Homogeneous hydrogenation in absence of transition-metal catalysts. *J. Am. Chem. Soc.* 86 (18): 3750–3752.

1486 Walling, C. and Bollyky, L. (1961). Base-catalyzed homogeneous hydrogenation. *J. Am. Chem. Soc.* 83 (13): 2968–2969.

1487 Berkessel, A., Schubert, T.J.S., and Müller, T.N. (2002). Hydrogenation without a transition-metal catalyst: on the mechanism of the base-catalyzed hydrogenation of ketones. *J. Am. Chem. Soc.* 124 (29): 8693–8698.

1488 Corma, A., Iborra, S., and Velty, A. (2007). Chemical routes for the transformation of biomass into chemicals. *Chem. Rev.* 107: 2411–2502.

1489 Behr, A., Westfechtel, A., and Gómes, J.P. (2008). Catalytic processes for the technical use of natural fats and oils. *Chem. Eng. Technol.* 31: 700–714.

1490 (a) Pritchard, J., Filonenko, G.A., van Putten, R.-J. et al. (2015). Heterogeneous and homogeneous catalysis for the hydrogenation of carboxylic acid derivatives: history, advances and future directions. *Chem. Soc. Rev.* 44: 3808–3833. (b) Toyao, T., Siddiki, S.M.A.H., Morita, Y. et al. (2017). Rhenium-loaded TiO_2: a highly versatile and chemoselective catalyst for the hydrogenation of carboxylic acid derivatives and the *N*-methylation of amines using H_2 and CO_2. *Chem. Eur. J.* 23 (59): 14848–14859.

1491 (a) Bouveault, L. and Blanc, G.L. (1903). Preparation of primary alcohols through the use of corresponding acids. *C.R. Hebd. Seances Acad. Sci.* 136: 1676–1678. (b) Bouveault, L. and Blanc, G.L. (1903). Preparation of primary alcohols by means of the corresponding acids. *C.R. Hebd. Seances Acad. Sci.* 137: 60–62. (c) Bouveault, L. and Blanc, G.L. (1904). Transforming saturated monobasic acids into the corresponding primary alcohols. *Bull. Soc. Chim. Fr.* 31: 666–672.

1492 Bodnar, B.S. and Vogt, P.F. (2009). An improved Bouveault–Blanc ester reduction with stabilized alkali metals. *J. Org. Chem.* 74: 2598–2600.

1493 (a) Brown, H.C. and Ramachandran, P.V. (1996). Sixty years of hydride reductions. In: *Reductions in Organic Synthesis: Recent Advances and Practical Applications*, ACS Symposium Series, vol. 641, 1–30. American Chemical Society. (b) Yoon, N.M. (1996). Selective reduction of organic compounds with aluminum and boron hydrides. *Pure Appl. Chem.* 68: 843–848. (c) Cha, J.S. (2007). Selective reduction of carbonyl and epoxy compounds using aluminum, boron and other metal reagents. Comparison of reducing characteristics between the Meerwein–Ponndorf–Verley type reduction and metal complex hydrides reduction: a review. *Bull. Korean Chem. Soc.* 28 (12): 2162–2190. (d) Cha, J.S. (2011). Thirty six years of research on the selective reduction and hydroboration. *Bull. Korean Chem. Soc.* 32 (6): 1808–1846.

1494 Wojcik, B. and Adkins, H. (1934). Catalytic hydrogenation of amides to amines. *J. Am. Chem. Soc.* 56: 2419–2424.

1495 Smith, A.M. and Whyman, R. (2014). Review of methods for the catalytic hydrogenation of carboxamides. *Chem. Rev.* 114: 5477–5510.

1496 Adkins, H., Wojcik, B., and Covert, L.W. (1933). The catalytic hydrogenation of esters to alcohols. III. *J. Am. Chem. Soc.* 55: 1669–1676.

1497 Adkins, H. and Billica, H.R. (1948). The hydrogenation of esters to alcohols at 25-degrees-150-degrees. *J. Am. Chem. Soc.* 70: 3121–3125.

1498 Adkins, H., Burgoyne, E.E., and Schneider, H.J. (1950). The Cu—Cr oxide catalyst for hydrogenation. *J. Am. Chem. Soc.* 72: 2626–2629.

1499 Adkins, H. and Pavlic, A.A. (1947). Hydrogenation of esters to alcohols over Raney Ni. 1. *J. Am. Chem. Soc.* 69: 3039–3041.

1500 Rase, H.F. (2000). *Handbook of Commercial Catalysts: Heterogeneous Catalysts*, 1e, 1–520. Boca Raton, FL: CRC Press.

1501 Yokoyama, T. and Yamagata, N. (2001). Hydrogenation of carboxylic acids to the corresponding aldehydes. *Appl. Catal., A* 221 (1–2): 227–239.

1502 Pouilloux, Y., Autin, F., and Barrault, J. (2000). Selective hydrogenation of methyl oleate into unsaturated alcohols relationships between catalytic properties and composition of cobalt-tin catalysts. *Catal. Today* 63 (1): 87–100.

1503 Zhang, J., Leitus, G., Ben-David, Y., and Milstein, D. (2006). Efficient homogeneous catalytic hydrogenation of esters to alcohols. *Angew. Chem. Int. Ed.* 45 (7): 1113–1115.

1504 Saudan, L.A., Saudan, C.M., Debieux, C., and Wyss, P. (2007). Dihydrogen reduction of carboxylic esters to alcohols under the catalysis of homogeneous Ru complexes: high efficiency and unprecedented chemoselectivity. *Angew. Chem. Int. Ed.* 46 (39): 7473–7476.

1505 Balaraman, E., Gnanaprakasam, B., Shimon, L.J.W., and Milstein, D. (2010). Direct hydrogenation of amides to alcohols and amines under mild conditions. *J. Am. Chem. Soc.* 132: 16756–16758.

1506 Barrios-Francisco, R., Balaraman, E., Skin-Posner, Y. et al. (2013). PNN Ruthenium pincer complexes based on phosphinated 2,2′-dipyridinemethane and 2,2′-oxobispyridine. Metal-ligand cooperation in cyclometalation and catalysis. *Organometallics* 32: 2973–2982.

1507 Zhang, J., Leitus, G., Ben-David, Y., and Milstein, D. (2005). Facile conversion of alcohols into esters and dihydrogen catalyzed by new Ru complexes. *J. Am. Chem. Soc.* 127: 10840–10841.

1508 Gunanathan, C., Ben-David, Y., and Milstein, D. (2007). Direct synthesis of amides from alcohols and amines with liberation of H_2. *Science* 317: 790–792.

1509 Balaraman, E., Khaskin, E., Leitus, G., and Milstein, D. (2013). Catalytic transformation of alcohols to carboxylic acid salts and H_2 using water as the oxygen atom source. *Nat. Chem.* 5: 122–125.

1510 Poverenov, E., Gandelman, M., Shimon, L.J.W. et al. (2004). Nucleophilic de-coordination and electrophilic regeneration of "hemilabile" pincer-type complexes: formation of anionic dialkyl, diaryl, and dihydride Pt(II) complexes bearing no stabilizing π-acceptors. *Chem. Eur. J.* 10 (19): 4673–4684.

1511 Poverenov, E., Gandelman, M., Shimon, L.J.W. et al. (2005). Pincer "hemilabile" effect. PCN Pt(II) complexes with different amine "arm length". *Organometallics* 24 (6): 1082–1090.

1512 Benito-Garagorri, D. and Kirchner, K. (2008). Modularly designed transition metal PNP and PCP pincer complexes based on aminophosphines: synthesis and catalytic applications. *Acc. Chem. Res.* 41 (2): 201–213.

1513 Arakawa, H., Aresta, M., Armor, J.N. et al. (2001). Catalysis research of relevance to carbon management: progress, challenges, and opportunities. *Chem. Rev.* 101 (4): 953–996.

1514 Yang, N. and Wang, R. (2015). Sustainable technologies for the reclamation of greenhouse gas CO_2. *J. Cleaner Prod.* 103: 784–792.

1515 (a) Baiker, A. (2000). Utilization of carbon dioxide in heterogeneous catalytic synthesis. *Appl. Organomet. Chem.* 14 (12): 751–762. (b) Omae, I. (2006). Aspects of carbon dioxide utilization. *Catal. Today* 115 (1–4): 33–52.

1516 Aresta, M. and Dibenedetto, A. (2004). The contribution of the utilization option to reducing the CO_2 atmospheric loading: research needed to overcome existing barriers for a full exploitation of the potential of the CO_2 use. *Catal. Today* 98 (4): 455–462.

1517 Tada, S., Thiel, I., Lo, H.K., and Copéret, C. (2015). CO_2 hydrogenation: supported nanoparticles vs. immobilized catalysts. *Chimia* 69: 759–764.

1518 Yang, H., Zhang, C., Gao, P. et al. (2017). A review of the catalytic hydrogenation of carbon dioxide into value-added hydrocarbons. *Catal. Sci. Technol.* 7 (20): 4580–4598.

1519 Federsel, C., Jackstell, R., and Beller, M. (2010). State-of-the-art catalysts for hydrogenation of carbon dioxide. *Angew. Chem. Int. Ed.* 49 (36): 6254–6257.

1520 Federsel, C., Boddien, A., Jackstell, R. et al. (2010). A well-defined Fe catalyst for the reduction of bicarbonates and carbon dioxide to formates, alkyl formates, and formamides. *Angew. Chem. Int. Ed.* 49: 9777–9780.

1521 Moret, S., Dyson, P.J., and Laurenczy, G. (2014). Direct synthesis of formic acid from carbon dioxide by hydrogenation in acidic media. *Nat. Commun.* 5: 4017.

1522 Tlili, A., Blondiaux, E., Frogneux, X., and Cantat, T. (2015). Reductive functionalization of CO_2 with amines: an entry to formamide, formamidine and methylamine derivatives. *Green Chem.* 17: 157–168.

1523 Taheri, A. and Berben, L.A. (2016). Making C—H bonds with CO_2: production of formate by molecular electrocatalysts. *J. Chem. Soc., Chem. Commun.* 52: 1768–1777.

1524 Chheda, J.N., Huber, G.W., and Dumesic, J.A. (2007). Liquid-phase catalytic processing of biomass-derived oxygenated hydrocarbons to fuels and chemicals. *Angew. Chem. Int. Ed.* 46 (38): 7164–7183.

1525 (a) Zhou, H., Gu, J.-J., Li, P. et al. (2013). Leaf-architectured 3D hierarchical artificial photosynthetic system of perovskite titanates towards CO_2 photoreduction into hydrocarbon fuels. *Sci. Rep.* 3: 1707. (b) Haviv, E., Shimon, L.J.W., and Neumann, R. (2017). Photochemical reduction of CO_2 with visible light using a polyoxometalate as photoreductant. *Chem. Eur. J.* 23 (1): 92–95. (c) Barber, J. and Tran, P.D. (2013). From natural to artificial photosynthesis. *J. R. Soc. Interface* 10 (81): 20120984. (d) Sharma, P. and Kolhe, M.L. (2017). Review of sustainable solar hydrogen production using photon fuel on artificial leaf. *Int. J. Hydrogen Energy* 42 (36): 22704–22712. (e) Olivo, A., Ghedini, E., Paschalicchio, P. et al. (2018). Sustainable carbon dioxide photoreduction by a cooperative effect of reactor design and titania metal promotion. *Catalysts* 8 (1): 41/1–41/22.

1526 Wang, W.H., Himeda, Y., Muckerman, J.T. et al. (2015). CO_2 hydrogenation to formate and methanol as an alternative to photo- and electrochemical CO_2 reduction. *Chem. Rev.* 115: 12936–12973.

1527 Pagliaro, M., Konstandopoulos, A.G., Ciriminna, R., and Palmisano, G. (2010). Solar hydrogen: fuel of the near future. *Energy Environ. Sci.* 10: 279–287.

1528 Weatherbee, G.D. and Bartholomew, C.H. (1982). Hydrogenation of CO_2 on group-VIII metals. 2. Kinetics and mechanism of CO_2 hydrogenation on Ni. *J. Catal.* 77 (2): 460–472.

1529 Daza, Y.A. and Kuhn, J.N. (2016). CO_2 conversion by reverse water gas shift catalysis: comparison of catalysts, mechanisms and their consequences for CO_2 conversion to liquid fuels. *RSC Adv.* 6: 49675–49691.

1530 Román-Martínez, M.C., Cazorla-Amorós, D., Linares-Solano, A., and de Salinas-Martínez, Lecea, C. (1994). Carbon dioxide hydrogenation catalyzed by alkaline earth-based and Pt-based catalysts supported on carbon. *Appl. Catal., A* 116 (1–2): 187–204.

1531 Ihm, S.K., Park, Y.K., Jeon, J.K. et al. (1998). A study on methanol synthesis through CO_2 hydrogenation over Cu-based catalysts. *Stud. Surf. Sci. Catal.* 505–508.

1532 (a) Rezayee, N.M., Huff, C.A., and Sanford, M.S. (2015). Tandem amine and Ru-catalyzed hydrogenation of CO_2 to methanol. *J. Am. Chem. Soc.* 137: 1028–1031. (b) Wesselbaum, S., Moha, V., Meuresch, M. et al. (2015). Hydrogenation of carbon dioxide to methanol using a homogeneous Tu-triphos catalyst: from mechanistic investigations to multiphase catalysis. *Chem. Sci.* 6: 693–704. (c) Du, X.L., Jiang, Z., Su, D.S., and Wang, J.Q. (2016). Research progress on the indirect hydrogenation of carbon dioxide to methanol. *ChemSusChem* 9: 322–332. (d) An, B., Zhang, J.-H., Cheng, K. et al. (2017). Confinement of ultrasmall Cu/ZnO_x nanoparticles in metal-organic frameworks for selective methanol synthesis from catalytic hydrogenation of CO_2. *J. Am. Chem. Soc.* 139 (10): 3834–3840.

1533 Asif, M., Gao, X., Lv, H.-G. et al. (2018). Catalytic hydrogenation of CO_2 from 600 MW supercritical coal power plant to produce methanol: a techno-economic analysis. *Int. J. Hydrogen Energy* 43 (5): 2726–2741.

1534 (a) Yang, X.-F., Kattel, S., Senanayake, S.D. et al. (2015). Low pressure CO_2 hydrogenation to methanol over gold nanoparticles activated on a $CeOx/TiO_2$ interface. *J. Am. Chem. Soc.* 137

(32): 10104–10107. (b) see also Wu, C., Zhang, P., Zhang, Z.-F. et al. (2017). Efficient hydrogenation of CO_2 to methanol over supported subnanometer gold catalysts at tow temperature. *ChemCatChem* 9 (19): 3691–3696.

1535 Bando, K.K., Ichikuni, N., Soga, K. et al. (2000). Characterization of Rh particles and Li-promoted Rh particles in Y zeolite during CO_2 hydrogenation – a new mechanism for catalysis controlled by the dynamic structure of Rh particles and the Li additive effect. *J. Catal.* 194 (1): 91–104.

1536 Melo, C.I., Szczepanska, A., Bogel-Lukasik, E. et al. (2016). Hydrogenation of carbon dioxide to methane by Ru nanoparticles in ionic liquid. *ChemSusChem* 9: 1081–1084.

1537 Urakawa, A., Jutz, F., Laurenczy, G., and Baiker, A. (2007). Carbon dioxide hydrogenation catalyzed by a Ru dihydride: a DFT and high-pressure spectroscopic investigation. *Chem. Eur. J.* 13 (14): 3886–3899.

1538 Jessop, P.G., Ikariya, T., and Noyori, R. (1994). Homogeneous catalytic hydrogenation of supercritical carbon dioxide. *Nature* 368 (6468): 231–233.

1539 Olmos, A., Asensio, G., and Pérez, P.J. (2016). Homogeneous metal-based catalysis in supercritical carbon dioxide as reaction medium. *ACS Catal.* 6: 4265–4280.

1540 Jessop, P.G., Hsiao, Y., Ikariya, T., and Noyori, R. (1996). Homogeneous catalysis in supercritical fluids: hydrogenation of supercritical carbon dioxide to formic acid, alkyl formates, and formamides. *J. Am. Chem. Soc.* 118 (2): 344–355.

1541 Tanaka, R., Yamashita, M., and Nozaki, K. (2009). Catalytic hydrogenation of carbon dioxide using Ir(III)-pincer complexes. *J. Am. Chem. Soc.* 131: 14168–14169.

1542 Tanaka, R., Yamashita, M., Chung, L.W. et al. (2011). Mechanistic studies on the reversible hydrogenation of carbon dioxide catalyzed by an Ir-PNP complex. *Organometallics* 30: 6742–6750.

1543 Aoki, W., Wattanavinin, N., Kusumoto, S., and Nozaki, K. (2016). Development of highly active Ir-PNP catalysts for hydrogenation of carbon dioxide with organic bases. *Bull. Chem. Soc. Jpn.* 89: 113–124.

1544 Langer, R., Diskin-Posner, Y., Leitus, G. et al. (2011). Low-pressure hydrogenation of carbon dioxide catalyzed by a Fe pincer complex exhibiting noble metal activity. *Angew. Chem. Int. Ed.* 50: 9948–9952.

1545 Zell, T., Ben-David, Y., and Milstein, D. (2014). Unprecedented Fe-catalyzed ester hydrogenation. Mild, selective, and efficient hydrogenation of trifluoroacetic esters to alcohols catalyzed by an Fe pincer complex. *Angew. Chem. Int. Ed.* 53: 4685–4689.

1546 Takaya, J. and Iwasawa, N. (2008). Hydrocarboxylation of allenes with CO_2 catalyzed by silyl pincer-type Pd complex. *J. Am. Chem. Soc.* 130 (46): 15254–15255.

1547 (a) De Bellefon, C. and Fouilloux, P. (1994). Homogeneous and heterogeneous hydrogenation of nitriles in a liquid-phase – chemical, mechanistic, and catalytic aspects. *Catal. Rev. Sci. Eng.* 36 (3): 459–506. (b) Gómez, S., Peters, J.A., and Maschmeyer, T. (2002). The reductive amination of aldehydes and ketones and the hydrogenation of nitriles: mechanistic aspects and selectivity control. *Adv. Synth. Catal.* 344 (10): 1037–1057. (c) Krupka, J. (2010). Mechanism of the heterogeneously catalyzed hydrogenation of nitriles. *Chem. Listy* 104 (7): 709–723. (d) Krupka, J. and Pasek, J. (2012). Nitrile hydrogenation on solid catalysts – new insights into the reaction mechanism. *Curr. Org. Chem.* 16 (8): 988–1004. (e) Shao, Z.-H., Fu, S., Wei, M.-F. et al. (2016). Mild and selective Co-catalyzed chemodivergent transfer hydrogenation of nitriles. *Angew. Chem. Int. Ed.* 55 (47): 14653–14657. (f) Adam, R., Bheeter, C.B., Cabrero-Antonino, J.R. et al. (2017). Selective hydrogenation of nitriles to primary amines by using a cobalt phosphine catalyst. *ChemSusChem* 10 (5): 842–846. (g) Tokmic, K., Jackson, B.J., Salazar, A. et al. (2017). Cobalt-catalyzed and Lewis acid-assisted nitrile hydrogenation to primary amines: a combined effort. *J. Am. Chem. Soc.* 139 (38): 13554–13561. (h) Ferraccioli, R., Borovika, D., Surkus, A.-E. et al. (2018). Synthesis of cobalt nanoparticles by pyrolysis of vitamin B12: a non-noble-metal catalyst for efficient hydrogenation of nitriles. *Catal. Sci. Technol.* 8 (2): 499–507.

1548 Ostgard, D.J. (2008). The selective hydrogenation of nitriles to primary amines. *Special. Chem. Mag.* 28 (3): 28–31.

1549 Nieto-Marquez, A., Toledano, D., Sánchez, P. et al. (2010). Impact of nitrogen doping of carbon nanospheres on the Ni-catalyzed hydrogenation of butyronitrile. *J. Catal.* 269: 242–251.

1550 Enthaler, S., Junge, K., Addis, D. et al. (2008). A practical and benign synthesis of primary amines through Ru-catalyzed reduction of nitriles. *ChemSusChem* 1: 1006–1010.

1551 Enthaler, S., Addis, D., Junge, K. et al. (2008). A general and environmentally benign catalytic reduction of nitriles to primary amines. *Chem. Eur. J.* 14: 9491–9494.

1552 Werkmeister, S., Junge, K., and Beller, M. (2014). Catalytic hydrogenation of carboxylic acid esters, amides, and nitriles with homogeneous catalysts. *Org. Process Res. Dev.* 18: 289–302.

1553 Chakraborty, S. and Berke, H. (2014). Homogeneous hydrogenation of nitriles catalyzed by Mo and W amides. *ACS Catal.* 4: 2191–2194.

1554 Adam, R., Alberico, E., Baumann, W. et al. (2016). NNP-type pincer imidazolylphosphine Ru complexes: efficient base-free hydrogenation of aromatic and aliphatic nitriles under mild conditions. *Chem. Eur. J.* 22: 4991–5002.

1555 Mukherjee, A., Srimani, D., Chakraborty, S. et al. (2015). Selective hydrogenation of nitriles to primary amines catalyzed by a Co pincer complex. *J. Am. Chem. Soc.* 137: 8888–8891.

1556 von Braun, J., Blessing, G., and Zobel, F. (1923). Catalytic hydration under pressure in the presence of Ni salts, VI. Nitriles. *Ber. Dtsch. Chem. Ges.* 56: 1988–2001.

1557 Kindler, K. and Hesse, F. (1933). New and improved methods for the synthesis of pharmacologically important amines. VII. Synthesis of secondary and tertiary amines by the hydrogenation of nitriles. *Arch. Pharm.* 271: 439–445.

1558 Schwögler, E.J. and Adkins, H. (1939). Preparation of certain amines. *J. Am. Chem. Soc* 61: 3499–3502.

1559 Huber, W. (1944). Hydrogenation of basic nitriles with Raney Ni. *J. Am. Chem. Soc.* 66: 876–879.

1560 Allgeier, A.M. and Duch, M.W. (2001). Reactivity and surface analysis studies on the deactivation of Raney Ni during adiponitrile hydrogenation. *Chem. Ind.* 82: 229–239.

1561 Johnson, T.A. and Freyberger, D.P. (2001). Lithium hydroxide modified sponge catalysts for control of primary amine selectivity in nitrile hydrogenations. *Chem. Ind.* 82: 201–227.

1562 Fell, B. and Sojka, J. (1991). Selective hydrogenation of fats and derivatives using Ziegler-type organometallic catalysts III – selective hydrogenation of saturated and unsaturated fatty-acid nitriles to saturated and unsaturated primary and secundary amines. *Fett Wiss. Technol.-Fat Sci. Technol.* 93 (3): 79–90.

1563 Chase, P.A., Jurca, T., and Stephan, D.W. (2008). Lewis acid-catalyzed hydrogenation: $B(C_6F_5)_3$-mediated reduction of imines and nitriles with H_2. *J. Chem. Soc., Chem. Commun.* (14): 1701–1703.

1564 (a) Stephan, D.W. and Erker, G. (2013). Frustrated Lewis pair mediated hydrogenations. In: *Frustrated Lewis Pairs I. Uncovering and Understanding*, Topics in Current Chemistry, vol. 332 (ed. G. Erker and D.W. Stephan), 85–110. Stuttgart: Springer-Verlag. (b) Stephan, D.W. (2015). Frustrated Lewis pairs: from concept to catalysis. *Acc. Chem. Res.* 48: 306–316.

1565 (a) Sutherland, A.G. (2005). One or more CH bond(s) formed by substitution: reduction of C-halogen and C-chalcogen bonds. In: *Comprehensive Organic Functional Group Transformations II*, Chapter 1.01, vol. 1, 1–30. Amsterdam: Elsevier Inc. (b) Peters, D.G. (2016). Aliphatic and aromatic halides conversions. In: *Organic Electrochemistry*, 5e (ed. O. Hammerich and B. Speiser), 941–976. Boca Raton, FL: CRC Press.

1566 Cucullu, M.E., Nolan, S.P., Belderrain, T.R., and Grubbs, R.H. (1999). Catalytic dehalogenation of aryl chlorides mediated by Ru(II) phosphine complexes. *Organometallics* 18: 1299–1304.

1567 Ukisu, Y., Kameoka, S., and Miyadera, T. (2000). Catalytic dechlorination of aromatic chlorides with noble-metal catalysts under mild conditions: approach to practical use. *Appl. Catal., B* 27: 97–104.

1568 Tavoularis, G. and Keane, M.A. (1999). Gas phase catalytic dehydrochlorination and hydrodechlorination of aliphatic and aromatic systems. *J. Mol. Catal. A: Chem.* 142: 187–199.

1569 Kamiguchi, S., Arai, K., Okumura, K. et al. (2015). Solid-state Mo sulfide clusters with an octahedral metal framework as hydrogenation, dehydrogenation, and hydrogenolysis catalysts similar to the Pt group metals. *Appl. Catal., A* 505: 417–421.

1570 Ajiki, K., Hirano, M., and Tanaka, K. (2005). Rhodium-catalyzed reductive coupling of disulfides and diselenides with alkyl halides, using hydrogen as a reducing agent. *Org. Lett.* 7 (19): 4193–4195.

1571 Neumann, W.P. (1987). Tri(*n*-butyl)tin hydride as reagent in organic synthesis. *Synthesis* (8): 665–683.

1572 Pereyre, M., Quintard, J.-P., and Rahm, A. (1987). *Tin in Organic Synthesis*, 1–342. London: Butterworth.

1573 RajanBabu, T.V. (2004). *Tributylstannane, e-EROS: Encyclopedia of Reagents for Organic Synthesis*. Wiley (on line).

1574 (a) Alonso, F., Beletskaya, I.P., and Yus, M. (2002). Metal-mediated reductive hydrodehalogenation of organic halides. *Chem. Rev.* 102 (11): 4009–4091. (b) Ziffle, V.E. and Fletcher, S.P. (2014). Reduction of saturated alkyl halides to alkanes. In: *Comprehensive Organic Synthesis*, 2e, vol. 8 (ed. P. Knochel and G.A. Molander), 999–1010. Amsterdam: Elsevier, B. V.

1575 Chatgilialoglu, C. (1992). Organosilanes as radical-based reducing agents in synthesis. *Acc. Chem. Res.* 25 (4): 188–194.

1576 Chatgilialoglu, C. (1995). Structural and chemical properties of silyl radicals. *Chem. Rev.* 95 (5): 1229–1251.

1577 Boukherroub, R., Chatgilialoglu, C., and Manuel, G. (1996). $PdCl_2$-catalyzed reduction of organic halides by triethylsilane. *Organometallics* 15 (5): 1508–1510.

1578 Doyle, M.P., McOsker, C.C., and West, C.T. (1976). Silane reductions in acidic media. 7. Aluminum chloride-catalyzed hydrogen-halogen exchange between organosilanes and alkyl-halides – efficient hydrocarbon synthesis. *J. Org. Chem.* 41 (8): 1393–1396.

1579 Scott, V.J., Celenligil-Cetin, R., and Ozerov, O.V. (2005). Room-temperature catalytic hydrodefluorination of C(sp^3)-F bonds. *J. Am. Chem. Soc.* 127 (9): 2852–2853.

1580 Panisch, R., Bolte, M., and Müller, T. (2006). Hydrogen- and fluorine-bridged disilyl cations and their use in catalytic C—F activation. *J. Am. Chem. Soc.* 128 (30): 9676–9682.

1581 (a) Yang, J. and Brookhart, M. (2007). Iridium-catalyzed reduction of alkyl halides by triethylsilane. *J. Am. Chem. Soc.* 129 (42): 12656–12657. (b) Yang, J. and Brookhart, M. (2009). Reduction of alkyl halides by triethylsilane based on a cationic iridium bis(phosphinite) pincer catalyst: scope, selectivity and mechanism. *Adv. Synth. Catal.* 357 (1–2): 175–187.

1582 Gevorgyan, V., Rubin, M., Benson, S. et al. (2000). A novel $B(C_6F_5)_3$-catalyzed reduction of alcohols and cleavage of aryl and alkyl ethers with hydrosilanes. *J. Org. Chem.* 65 (19): 6179–6186.

1583 Ison, E.A., Trivedi, E.R., Corbin, R.A., and Abu-Omar, M.M. (2005). Mechanism for reduction catalysis by metal oxo: hydrosilation of organic carbonyl groups catalyzed by a Re(V) oxo complex. *J. Am. Chem. Soc.* 127 (44): 15374–15375.

1584 Du, G.D., Fanwick, P.E., and Abu-Omar, M.M. (2007). Mechanistic insight into hydrosilylation reactions catalyzed by high valent Re equivalent to X (X=O, NAr, or N) complexes: the silane (SiH) does not add across the metal-ligand multiple bond. *J. Am. Chem. Soc.* 129 (16): 5180–5187.

1585 Ollivier, C. and Renaud, P. (2001). Organoboranes as a source of radicals. *Chem. Rev.* 101 (11): 3415–3434.

1586 Renaud, P., Beauseigneur, A., Brecht-Forster, A. et al. (2007). Boron: a key element in radical reactions. *Pure Appl. Chem.* 79 (2): 223–233.

1587 Miura, K., Tomita, M., Yamada, Y., and Hosomi, A. (2007). Indium-catalyzed radical reductions of organic halides with hydrosilanes. *J. Org. Chem.* 72 (3): 787–792.

1588 Ojima, I., Nihonyan, M., Kogure, T., and Nagai, Y. (1972). Reduction of carbonyl compounds with various hydrosilane-Rh(I) complex combinations. *Bull. Chem. Soc. Jpn.* 45: 3506.

1589 Ojima, I., Nihonyan, M., and Nagai, Y. (1972). Rhodium complex-catalyzed hydrosilylation of carbonyl compounds. *J. Chem. Soc., Chem. Commun.* 938a.

1590 Bezman, S.A., Churchil, M.R., Osborn, J.A., and Wormald, J. (1971). Preparation and crystallographic characterization of a hexameric triphenylphosphinecopper hydride cluster. *J. Am. Chem. Soc.* 93: 2063–2065.

1591 Mahoney, W.S., Brestensky, D.M., and Stryker, J.M. (1988). Selective hydride-mediated conjugate reduction of α,β-unsaturated carbonyl compounds using $[(Ph_3P)CuH]_6$. *J. Am. Chem. Soc.* 110: 291–293.

1592 Lipshutz, B.H., Keith, J., Papa, P., and Vivian, R. (1998). Convenient, efficient method for conjugate reductions using catalytic quantities of Cu(I). *Tetrahedron Lett.* 39: 4627–4630.

1593 Lipshutz, B.H., Chrisman, W., Noson, K. et al. (2000). Copper hydride-catalyzed tandem 1,4-reduction/alkylation reactions. *Tetrahedron* 56: 2779–2788.

1594 Lipshutz, B.H. and Frieman, B.A. (2005). CuH in a bottle: a convenient reagent for asymmetric hydrosilylations. *Angew. Chem. Int. Ed.* 44: 6345–6348.

1595 Díez-González, S. and Nolan, S.P. (2008). Copper, silver, and gold complexes in hydrosilylation reactions. *Acc. Chem. Res.* 41: 349–358.

1596 Shaikh, N.S., Junge, K., and Beller, M. (2007). A convenient and general Fe-catalyzed hydrosilylation of aldehydes. *Org. Lett.* 9: 5429–5432.

1597 Addis, D., Shaikh, N., Zhou, S.L. et al. (2010). Chemo- and stereoselective Fe-catalyzed hydrosilylation of ketones. *Chem. Asian J.* 5: 1687–1691.

1598 Dieskau, A.P., Begouin, J.M., and Plietker, B. (2011). $Bu_4N[Fe(CO)_3(NO)]$-catalyzed hydrosilylation of aldehydes and ketones. *Eur. J. Org. Chem.* 5291–5296.

1599 Ruddy, A.J., Kelly, C.M., Crawford, S.M. et al. (2013). (*N*-Phosphinoamidinate)Fe pre-catalysts for the room temperature hydrosilylation of carbonyl compounds with broad substrate scope at low loadings. *Organometallics* 32: 5581–5588.

1600 Savela, R., Majewski, M., and Leino, R. (2014). Iron-catalyzed arylation of aromatic ketones and aldehydes mediated by organosilanes. *Eur. J. Org. Chem.* 4137–4147.

1601 Wekesa, F.S., rias-Ugarte, R., Kong, L. et al. (2015). Iron-catalyzed hydrosilylation of aldehydes and ketones under solvent-free conditions. *Organometallics* 34: 5051–5056.

1602 (a) Bleith, T. and Gade, L.H. (2016). Mechanism of the Fe(II)-catalyzed hydrosilylation of ketones: activation of Fe carboxylate precatalysts and reaction pathways of the active catalyst. *J. Am. Chem. Soc.* 138: 4972–4983. (b) Xue, B.-J., Sun, H.-J., Niu, Q.-F. et al. (2017). Catalytic hydrosilylation of carbonyl compounds by hydrido thiophenolato Fe(II) complexes. *Catal. Commun.* 94: 23–28. (c) Johnson, C. and Albrecht, M. (2017). Triazolylidene Fe(II) piano-stool complexes: synthesis and catalytic hydrosilylation of carbonyl compounds. *Organometallics* 36 (15): 2902–2913.

1603 (a) Bheeter, L.P., Henrion, M., Brelot, L. et al. (2012). Hydrosilylation of aldehydes and ketones catalyzed by an N-heterocyclic carbene-Ni hydride complex under mild conditions. *Adv. Synth. Catal.* 354: 2619–2624. (b) Wu, F.-F., Zhou, J.-N., Fang, Q. et al. (2012). Nickel(II)-dipyridylphosphine-catalyzed enantioselective hydrosilylation of ketones in air. *Chem. Asian J.* 7 (11): 2527–2530.

1604 Zheng, J.X., Darcel, C., and Sortais, J.B. (2013). A convenient Ni-catalysed hydrosilylation of carbonyl derivatives. *Catal. Sci. Technol.* 3: 81–84.

1605 Chakraborty, S., Bhattacharya, P., Dai, H.G., and Guan, H.R. (2015). Nickel and iron pincer complexes as catalysts for the reduction of carbonyl compounds. *Acc. Chem. Res.* 48: 1995–2003.

1606 Trovitch, R.J. (2014). Comparing well-defined Mn, Fe, Co, and Ni ketone hydrosilylation catalysts. *Synlett* 25: 1638–1642.

1607 Nishiyama, H. and Furuta, A. (2007). An iron-catalysed hydrosilylation of ketones. *J. Chem. Soc., Chem. Commun.* (7): 760–762.

1608 Hosokawa, S., Ito, J., and Nishiyama, H. (2010). A chiral Fe complex containing a bis(oxazolinyl)phenyl ligand: preparation and asymmetric hydrosilylation of ketones. *Organometallics* 29: 5773–5775.

1609 Flückiger, M. and Togni, A. (2011). Iron(II)-catalyzed asymmetric hydrosilylation of acetophenone. *Eur. J. Org. Chem.* 4353–4360.

1610 Zuo, Z.Q., Zhang, L., Leng, X.B., and Huang, Z. (2015). Iron-catalyzed asymmetric hydrosilylation of ketones. *J. Chem. Soc.; Chem. Commun.* 51: 5073–5076.

1611 Yu, S.L., Shen, W.Y., Li, Y.Y. et al. (2012). Iron-catalyzed highly enantioselective reduction of aromatic ketones with chiral P_2N_4-type macrocycles. *Adv. Synth. Catal.* 354: 818–822.

1612 (a) Inagaki, T., Phong, L.T., Furuta, A. et al. (2010). Iron- and cobalt-catalyzed asymmetric hydrosilylation of ketones and enones with bis(oxazolinylphenyl)amine ligands. *Chem. Eur. J.* 16: 3090–3096. (b) Yu, F., Zhang, X.-C., Wu, F.-F. et al. (2011). Cobalt(II)-catalyzed asymmetric hydrosilylation of simple ketones using dipyridylphosphine ligands in air. *Org. Biomol. Chem.* 9 (16): 5652–5654. (c) Chen, X. and Lu, Z. (2016). Iminophenyl oxazolinylphenylamine for enantioselective Co-catalyzed hydrosilylation of aryl ketones. *Org. Lett.* 18 (18): 1658–1661.

1613 Zhou, S., Junge, K., Addis, D. et al. (2009). A convenient and general Fe-catalyzed reduction of amides to amines. *Angew. Chem. Int. Ed.* 48: 9507–9510.

1614 Enthaler, S. (2010). Synthesis of secondary amines by Fe-catalyzed reductive amination. *ChemCatChem* 2: 1411–1415.

1615 (a) Castro, L.C.M., Sortais, J.B., and Darcel, C. (2012). NHC-carbene cyclopentadienyl Fe-based catalyst for a general and efficient hydrosilylation of imines. *J. Chem. Soc., Chem. Commun.* 48: 151–153. (b) Bhunia, M., Hota, P.K., Vijaykumar, C. et al. (2016). A highly efficient base-metal catalyst: chemoselective reduction of imines to amines using an abnormal-NHC—Fe(0) complex. *Organometallics* 35 (17): 2930–2937. (c) Li, B., Zhang, S.-L., Wu, W.-H. et al. (2017). Imidazolium-based ionic liquid-catalyzed hydrosilylation of imines and reductive amination of aldehydes using hydrosilane as the reductant. *RSC Adv.* 7 (51): 31795–31799.

1616 Bezlada, A., Szewczyk, M., and Mlynarski, J. (2016). Enantioselective hydrosilylation of imines catalyzed by chiral Zn acetate complexes. *J. Org. Chem.* 81: 336–342.

1617 Koller, J. and Bergman, R.G. (2012). Controlled hydrosilylation of carbonyls and imines catalyzed by a cationic Al alkyl complex. *Organometallics* 31: 2530–2533.

1618 Pérez, M., Qu, Z.W., Caputo, C.B. et al. (2015). Hydrosilylation of ketones, imines and nitriles catalysed by electrophilic phosphonium cations: functional group selectivity and mechanistic considerations. *Chem. Eur. J.* 21: 6491–6500.

1619 Liberman-Martin, A.L., Bergman, R.G., and Tilley, T.D. (2015). Lewis acidity of bis(perfluorocatecholato)silane: aldehyde hydrosilylation catalyzed by a neutral Si compound. *J. Am. Chem. Soc.* 137: 5328–5331.

1620 Ren, X.-Y. and Du, H.-F. (2016). Chiral frustrated Lewis pairs catalyzed highly enantioselective hydrosilylations of 1,2-dicarbonyl compounds. *J. Am. Chem. Soc.* 138: 810–813.

1621 Ohta, T., Kamiya, M., Kusui, K. et al. (1999). Rhodium-catalyzed reduction of esters to alcohols using diphenylsilane. *Tetrahedron Lett.* 40: 6963–6966.

1622 Ohta, T., Kamiya, M., Nobutomo, M. et al. (2005). Reduction of carboxylic acid derivatives using diphenylsilane in the presence of a Rh—PPh_3 complex. *Bull. Chem. Soc. Jpn.* 78: 1856–1861.

1623 Cheng, C. and Brookhart, M. (2012). Efficient reduction of esters to aldehydes through Ir-catalyzed hydrosilylation. *Angew. Chem. Int. Ed.* 51: 9422–9424.

1624 Fernández-Salas, J.A., Manzini, S., and Nolan, S.P. (2014). Chemoselective Ru-catalysed reduction of carboxylic acids. *Adv. Synth. Catal.* 356: 308–312.

1625 Nagashima, H. (2015). Efficient transition metal-catalyzed reactions of carboxylic acid derivatives with hydrosilanes and hydrosiloxanes, afforded by catalyst design and the proximity effect of two Si—H groups. *Synlett* 26: 866–890.

1626 Bezier, D., Venkanna, G.T., Castro, L.C.M. et al. (2012). Iron-catalyzed hydrosilylation of esters. *Adv. Synth. Catal.* 354: 1879–1884.

1627 Merel, D.S., Do, M.L.T., Gaillard, S. et al. (2015). Iron-catalyzed reduction of carboxylic and carbonic acid derivatives. *Coord. Chem. Rev.* 288: 50–68.

1628 Fernandes, A.C. and Romao, C.C. (2006). Silane/MoO_2Cl_2 as an efficient system for the reduction of esters. *J. Mol. Catal. A: Chem.* 253: 96–98.

1629 Pehlivan, L., Metay, E., Laval, S. et al. (2011). Reduction of aromatic and aliphatic esters using the reducing systems $MoO_2(acac)_2$ or $V(O)(O\text{-}i\text{-}Pr)_3$ in combination with 1,1,3,3-tetramethyldisiloxane. *Eur. J. Org. Chem.* 7400–7406.

1630 Kovalenko, O.O. and Adolfsson, H. (2015). Highly efficient and chemoselective Zn-catalyzed hydrosilylation of esters under mild conditions. *Chem. Eur. J.* 21: 2785–2788.

1631 Das, S., Li, Y.H., Lu, L.Q. et al. (2016). A general and selective Rh-catalyzed reduction of amides, *N*-acyl amino esters, and dipeptides using phenylsilane. *Chem. Eur. J.* 22: 7050–7053.

1632 Hanada, S., Tsutsumi, E., Motoyama, Y., and Nagashima, H. (2009). Practical access to amines by Pt-catalyzed reduction of carboxamides with hydrosilanes: synergy of dual Si—H groups leads to high efficiency and selectivity. *J. Am. Chem. Soc.* 131: 15032–15040.

1633 Pisiewicz, S., Junge, K., and Beller, M. (2014). Mild hydrosilylation of amides by Pt N-heterocyclic carbene catalysts. *Eur. J. Inorg. Chem.* 2014: 2345–2349.

1634 Cheng, C. and Brookhart, M. (2012). Iridium-catalyzed reduction of secondary amides to secondary amines and imines by diethylsilane. *J. Am. Chem. Soc.* 134: 11304–11307.

1635 Krackl, S., Someya, C.I., and Enthaler, S. (2012). Reductive cleavage of amides to alcohols and amines catalyzed by well-defined bimetallic Mo complexes. *Chem. Eur. J.* 18: 15267–15271.

1636 Sharma, H.K., rias-Ugarte, R., Tomlinson, D. et al. (2013). $(Me_3N)Mo(CO)_5$-catalyzed reduction of DMF by disiloxane and disilane moieties: fate of the Si-containing fragments. *Organometallics* 32: 3788–3794.

1637 Volkov, A., Tinnis, F., Slagbrand, T. et al. (2014). $Mo(CO)_6$ catalysed chemoselective hydrosilylation of α,β-unsaturated amides for the formation of allylamines. *J. Chem. Soc., Chem. Commun.* 50: 14508–14511.

1638 Hanada, S., Ishida, T., Motoyama, Y., and Nagashima, H. (2007). The Ru-catalyzed reduction and reductive *N*-alkylation of secondary amides with hydrosilanes: practical synthesis of secondary and tertiary amines by judicious choice of hydrosilanes. *J. Org. Chem.* 72: 7551–7559.

1639 Das, S., Zhou, S.L., Addis, D. et al. (2010). Selective catalytic reductions of amides and nitriles to amines. *Top. Catal.* 53: 979–984.

1640 Das, S., Addis, D., Junge, K., and Beller, M. (2011). Zinc-catalyzed chemoselective reduction of tertiary and secondary amides to amines. *Chem. Eur. J.* 17: 12186–12192.

1641 Volkov, A., Buitrago, E., and Adolfsson, H. (2013). Direct hydrosilylation of tertiary amides to amines by an in situ formed Fe/N-heterocyclic carbene catalyst. *Eur. J. Org. Chem.* 2066–2070.

1642 Dombray, T., Helleu, C., Darcel, C., and Sortais, J.B. (2013). Cobalt carbonyl-based catalyst for hydrosilylation of carboxamides. *Adv. Synth. Catal.* 355: 3358–3362.

1643 Fernandes, A.C. and Romão, C.C. (2007). Reduction of amides with silanes catalyzed by MoO_2Cl_2. *J. Mol. Catal. A: Chem.* 272: 60–63.

1644 Inamoto, Y., Kaga, Y., Nishimoto, Y. et al. (2013). Indium triiodide catalyzed reductive functionalization of amides via the single-stage treatment of hydrosilanes and organosilicon nucleophiles. *Org. Lett.* 15: 3452–3455.

1645 Kovalenko, O.O., Volkov, A., and Adolfsson, H. (2015). Mild and selective Et_2Zn-catalyzed reduction of tertiary amides under hydrosilylation conditions. *Org. Lett.* 17: 446–449.

1646 Chadwick, R.C., Kardelis, V., Lim, P., and Adronov, A. (2014). Metal-free reduction of secondary and tertiary *N*-phenyl amides by tris(pentafluorophenyl)boron-catalyzed hydrosilylation. *J. Org. Chem.* 79: 7728–7733.

1647 Blondiaux, E. and Cantat, T. (2014). Efficient metal-free hydrositylation of tertiary, secondary and primary amides to amines. *J.Chem. Soc.; Chem. Commun.* 50: 9349–9352.

1648 Huang, P.Q., Lang, Q.W., and Wang, Y.R. (2016). Mild metal-free hydrosilylation of secondary amides to amines. *J. Org. Chem.* 81: 4235–4243.

1649 Schwartz, M. and Berry, R.J. (2001). Ab initio investigation of substituent effects on bond dissociation enthalpies in siloxanes and silanols. *J. Mol. Struct. THEOCHEM* 538: 9–17.

1650 Fernández-Álvarez, F.J., Aitani, A.M., and Oro, L.A. (2014). Homogeneous catalytic reduction of CO_2 with hydrosilanes. *Catal. Sci. Technol.* 4: 611–624.

1651 Berkefeld, A., Piers, W.E., and Parvez, M. (2010). Tandem frustrated Lewis pair/tris(pentafluorophenyl)borane-catalyzed deoxygenative hydrosilylation of carbon dioxide. *J. Am. Chem. Soc.* 132: 10660–10661.

1652 Chen, J.W., Falivene, L., Caporaso, L. et al. (2016). Selective reduction of CO_2 to CH_4 by tandem hydrosilylation with mixed Al/B catalysts. *J. Am. Chem. Soc.* 138: 5321–5333.

1653 Brook, M.A. (2000). *Silicon in Organic, Organometallic and Polymer Chemistry*, 1–704. New York: Wiley.

1654 Sommer, L.H., Pietrusza, E.W., and Whitmore, F.C. (1947). Peroxide-catalyzed addition of trichlorosilane to 1-octene. *J. Am. Chem. Soc.* 69: 188.

1655 Speier, J.L., Webster, J.A., and Barnes, G.H. (1957). The addition of silicon hydrides to olefinic double bonds. 2. The use of group-VIII metal catalysts. *J. Am. Chem. Soc.* 79: 974–979.

1656 Marciniec, B. (2005). Catalysis by transition metal complexes of alkene silylation – recent progress and mechanistic implications. *Coord. Chem. Rev.* 249: 2374–2390.

1657 Islamova, R.M., Dobryin, M.V., Ivanov, D.M. et al. (2016). Bis-Nitrile and bis-dialkylcyanamide Pt(II) complexes as efficient catalysts for hydrosilylation cross-linking of siloxane polymers. *Molecules* 21, 311: 1–11.

1658 Roy, A.K. (2007). A review of recent progress in catalyzed homoneous hydrosilylation. *Adv. Organomet. Chem.* 1–59.

1659 Trögel, D. and Stohrer, J. (2011). Recent advances and actual challenges in late transition metal catalyzed hydrosilylation of olefins from an industrial point of view. *Coord. Chem. Rev.* 255: 1440–1459.

1660 Nakajima, Y. and Shimada, S. (2015). Hydrosilylation reaction of olefins: recent advances and perspectives. *RSC Adv.* 5: 20603–20616.

1661 Nesmeyanov, A.N., Freidlina, R.K., Chukovskaya, E.C. et al. (1962). Addition, substitution, and telomerization reactions of olefins in the presence of metal carbonyls or colloidal iron. *Tetrahedron* 17: 61–68.

1662 Bart, S.C., Lobkovsky, E., and Chirik, P.J. (2004). Preparation and molecular and electronic structures of Fe(0) dinitrogen and silane complexes and their application to catalytic hydrogenation and hydrosilation. *J. Am. Chem. Soc.* 126 (42): 13794–13807.

1663 Hayasaka, K., Kamata, K., and Nakazawa, H. (2016). Highly efficient olefin hydrosilylation catalyzed by Fe complexes with iminobipyridine ligand. *Bull. Chem. Soc. Jpn.* 89: 394–404.

1664 Tondreau, A.M., Atienza, C.C.H., Weller, K.J. et al. (2012). Iron catalysts for selective anti-Markovnikov alkene hydrosilylation using tertiary silanes. *Science* 335: 567–570.

1665 Tondreau, A.M., Atienza, C.C.H., Darmon, J.M. et al. (2012). Synthesis, electronic structure, and alkene hydrosilylation activity of terpyridine and bis(imino)pyridine Fe dialkyl complexes. *Organometallics* 31: 4886–4893.

1666 Peng, D.J., Zhang, Y.L., Du, X.Y. et al. (2013). Phosphinite-iminopyridine Fe catalysts for chemoselective alkene hydrosilylation. *J. Am. Chem. Soc.* 135: 19154–19166.

1667 Buslov, I., Becouse, J., Mazza, S. et al. (2015). Chemoselective alkene hydrosilylation catalyzed by Ni pincer complexes. *Angew. Chem. Int. Ed.* 54: 14523–14526.

1668 Du, X.Y., Zhang, Y.L., Peng, D.J., and Huang, Z. (2016). Base-metal-catalyzed regiodivergent alkene hydrosilylations. *Angew. Chem. Int. Ed.* 55: 6670–6674.

1669 Chen, J.H., Cheng, B., Cao, M.Y., and Lu, Z. (2015). Iron-catalyzed asymmetric hydrosilylation of 1,1-disubstituted alkenes. *Angew. Chem. Int. Ed.* 54: 4661–4664.

1670 Gibson, S.E. and Rudd, M. (2007). The role of secondary interactions in the asymmetric Pd-catalysed hydrosilylation of olefins with monophosphane ligands. *Adv. Synth. Catal.* 349: 781–795.

1671 Chen, J.H., Cao, M.Y., Cheng, B.A., and Lu, Z. (2015). Catalytic asymmetric hydrosilylation of 1,1-disubstituted alkenes. *Synlett* 26: 2332–2335.

1672 Bleith, T., Wadepohl, H., and Gade, L.H. (2015). Iron achieves noble metal reactivity and selectivity: highly reactive and enantioselective Fe complexes as catalysts in the hydrosilylation of ketones. *J. Am. Chem. Soc.* 137: 2456–2459.

1673 Chalk, A.J. and Harrod, J.F. (1965). Homogeneous catalysis. 2. Mechanism of hydrosilylation of olefins catalyzed by group 8 metal complexes. *J. Am. Chem. Soc.* 87: 16–21.

1674 Glaser, P.B. and Tilley, T.D. (2003). Catalytic hydrosilylation of alkenes by a Ru silylene complex. Evidence for new hydrosilylation mechanism. *J. Am. Chem. Soc.* 125: 13640–13641.

1675 Guczi, L., Sarkany, A., and Tetenyi, P. (1974). Hydrogenolysis of some saturated-hydrocarbons over a Pt black catalyst. *J. Chem. Soc., Faraday Trans. 1* 70: 1971–1981.

1676 Zimmer, H., Tetenyi, P., and Paal, Z. (1982). Hydrogenolysis of alkanes with quaternary carbon atoms over Pt and Ni black catalysts. *J. Chem. Soc., Faraday Trans. 1* 78: 3573–3586.

1677 Leclercq, G., Leclercq, L., and Maurel, R. (1979). Hydrogenolysis of alkanes on alumina supported Pt. Kinetic study and effect of the hydrocarbon structure. *Bull. Soc. Chim. Belg.* 88 (7–8): 599–620.

1678 Levitskii, I.I., Gyul'maliev, A.M., and Udal'tzova, E.A. (1979). Mechanism of alkane hydrogenolysis. *J. Catal.* 58 (1): 144–148.

1679 Bond, G.C. and Yide, X. (1984). Hydrogenolysis of alkanes. 1. Hydrogenolysis of ethane, propane and *n*-butane on 6-percent Pt/SiO_2 (Europt-1). *J. Chem. Soc., Faraday Trans. 1* 969–980.

1680 Lunin, V.V., Deineka, V.I., Abasov, S.I., and Plate, A.F. (1975). Hydrogenolysis of *n*-heptane and *n*-hexane in the presence of a TI-Mo-hydrogen system. *Neftekhimiya* 15 (6): 832–835.

1681 (a) Sarkany, A., Matusek, K., and Tetenyi, P. (1977). Transformation of some saturated-hydrocarbons on Ir, Rh and Ru blacks. *J. Chem. Soc., Faraday Trans. 1* 73 (11): 1699–1709. (b) Galadima, A., Wells, R.K.P., and Anderson, J.A. (2012). Hydroconversion of *n*-alkanes over carbided Rh/molybdena zirconia catalysts. *Top. Catal.* 55 (14–15): 931–939. (c) Flaherty, D.W., Hibbitts, D.D., and Iglesia, E. (2014). Metal-catalyzed C—C bond cleavage in alkanes: effects of methyl substitution on transition-state structures and stability. *J. Am. Chem. Soc.* 136 (27): 9664–9676.

1682 (a) Bond, G.C. and Hooper, A.D. (2000). Modification of Ru catalysts for alkane hydrogenolysis. *Appl. Catal., A* 191 (1-2): 69–81. (b) Bond, G.C. and García, J.J. (2017). Hydrogenolysis of alkanes: reactions of *n*-butane on Ru/zeolite catalysts. *Catal. Sci. Technol.* 7 (22): 5294–5300. (c) Nakagawa, Y., Oya, S.-i., Kanno, D. et al. (2017). Regioselectivity and reaction mechanism of Ru-catalyzed hydrogenolysis of squalane and model alkanes. *ChemSusChem* 10 (1): 189–198.

1683 Quignard, F., Lécuyer, C., Choplin, A. et al. (1992). Surface organometallic chemistry of Zr – application to the stoichiometric activation of the C—H bonds of alkanes and to the low-temperature catalytic hydrogenolysis of alkanes. *J. Mol. Catal.* 74 (1–3): 353–363.

1684 Scott, S.L. and Basset, J.M. (1994). Stoichiometric and catalytic reactivity of organometallic fragments supported on inorganic oxides. *J. Mol. Catal.* 86 (1–3): 5–22.

1685 Zakharov, V.A., Dudchenko, V.K., Kolchin, A.M., and Ermakov, Y.I. (1975). Formation of Zr hydrides and their role in polymerization of ethylene on carried organozirconium catalysts. *Kinet. Catal.* 16 (3): 808–809.

1686 Zakharov, V.A., Dudchenko, V.K., Paukshtis, E.A. et al. (1977). Formation of Zr hydrides in supported organozirconium catalysts and their role in ethylene polymerization. *J. Mol. Catal.* 2 (6): 421–435.

1687 (a) Corker, J., Lefebvre, F., Lécuyer, C. et al. (1996). Catalytic cleavage of the C—H and C—C bonds of alkanes by surface organometallic chemistry: an EXAFS and IR characterization of a Zr—H catalyst. *Science* 271 (5251): 966–969. (b) Pelletier, J.D.A. and Basset, J.-M. (2016). Catalysis by design: well-defined single-site heterogeneous catalysts. *Acc. Chem. Res.* 49 (4): 664–677.

1688 Lefebvre, F., Thivolle-Cazat, J., Dufaud, V. et al. (1999). Oxide supported surface organometallic complexes as a new generation of catalysts for carbon–carbon bond activation. *Appl. Catal., A* 182 (1): 1–8.

1689 Lefebvre, F. and Basset, J.M. (1999). Recent applications in catalysis of surface organometallic chemistry. *J. Mol. Catal. A: Chem.* 146 (1–2): 3–12.

1690 Lécuyer, C., Quignard, F., Choplin, A. et al. (1991). Surface organometallic chemistry on oxides – selective catalytic low-temperature hydrogenolysis of alkanes by a highly electrophilic Zr hydride complex supported on silica. *Angew. Chem. Int. Ed. Engl.* 30 (12): 1660–1661.

1691 Dufaud, V.R. and Basset, J.M. (1998). Catalytic hydrogenolysis at low temperature and pressure of polyethylene and polypropylene to diesels or lower alkanes by a Zr hydride supported on silica-alumina: a step toward polyolefin degradation by the microscopic reverse of Ziegler–Natta polymerization. *Angew. Chem. Int. Ed.* 37 (6): 806–810.

1692 Besedin, D.V., Ustynyuk, L.Y., Ustynyuk, Y.A., and Lunin, V.V. (2002). A theoretical DFT study of the mechanism of C—C bond hydrogenolysis in alkanes on silica-supported zirconium hydrides. *Mendeleev Commun.* (5): 173–175.

1693 Besedin, D.V., Ustynyuk, L.Y., Ustynyuk, Y.A., and Lunin, V.V. (2005). Hydrogenolysis of alkanes and olefin polymerization on the silica-supported Zr hydrides: a comparative DFT study. *Top. Catal.* 32 (1–2): 47–60.

1694 Bond, G.C. and Sheridan, J. (1952). Studies in heterogeneous catalysis. 4. The hydrogenation of cyclopropane. *Trans. Faraday Soc.* 48 (8): 713–715.

1695 Benson, J.E. and Kwan, T. (1956). Kinetics and mechanism of the hydrogenation of cyclopropane over a Ni-silica-alumina catalyst. *J. Phys. Chem.* 60 (12): 1601–1605.

1696 Sinfelt, J.H., Yates, D.J.C., and Taylor, W.F. (1965). Kinetics of reaction of cyclopropane with hydrogen over a series of silica-supported metals. *J. Phys. Chem.* 69 (6): 1877–1882.

1697 Sridhar, T.S. and Ruthven, D.M. (1970). Hydrogenation of cyclopropane over supported Ni catalysts. *J. Catal.* 16 (3): 363–369.

1698 Sridhar, T.S. and Ruthven, D.M. (1972). Kinetics and mechanism of catalytic hydrogenation of cyclopropane. *J. Catal.* 24 (1): 153–160.

1699 Capitano, A.T., Gabelnick, A.M., and Gland, J.L. (2000). Catalytic cyclopropane hydrogenation on Pt(111) using in situ soft X-ray methods. *J. Phys. Chem. B* 104 (14): 3337–3342.

1700 Gröger, C. and Musso, H. (1976). Hydrogenolysis of small ring compounds. 4. Selective hydrogenations of 1,1-disubstituted cyclopropanes. *Angew. Chem. Int. Ed.* 15 (6): 373–374.

1701 Gröger, C., Musso, H., and Rossnagel, I. (1980). Hydrogenolysis of small cycloalkanes. 7. Selective hydrogenation of 1,1-disubstituted cyclopropane derivatives. *Chem. Ber. Recl.* 113 (11): 3621–3628.

1702 Hertzsch, W. and Musso, H. (1985). Hydrogenolysis of small cycloalkanes. 11. Hydrogenation of bicyclo[1.1.0]butane-1-carboxylates and butane-2-carboxylates. *Liebigs Ann. Chem.* (7): 1485–1491.

1703 Frank, J., Konrad, G., Rossnagel, I. et al. (1987). Hydrogenolysis of small cycloalkanes. 17. Catalytic hydrogenation of dimethyl bicyclo[1.1.0]butane-2,2-dicarboxylate. *Chem. Ber. Recl.* 120 (3): 443–444.

1704 (a) Kazanskii, B.A. and Lukina, M.Y. (1950). Catalytic ring opening hydrogenation of hydrocarbons of the cyclobutane series. *Dokl. Akad. Nauk* 74 (2): 263–266. (b) Lukina, M.Y., Zotova, S.V., and Kazanskii, B.A. (1957). The hydrogenation of ethylcyclobutane and 1,1,2-trimethylcyclopropane in the presence of Pd-charcoal. *Dokl. Akad. Nauk* 116 (5): 793–796.

1705 Barbier, J., Marecot, P., and Maurel, R. (1980). Comparative study on the hydrogenolysis of cyclopropane, cyclobutane and cyclopentane on platinum/aluminum oxide. *New J. Chem.* 4 (6): 385–388.

1706 Kazanskii, B.A., Liberman, A.L., and Plate, A.F. (1947). Hydrogenation of cyclopentane with ring opening in the presence of platinized carbon. *Dokl. Akad. Nauk* 57: 571–574.

1707 Murakami, M., Amii, H., Shigeto, K., and Ito, Y. (1996). Breaking of the C—C bond of cyclobutanones by Rh(I) and its extension to catalytic synthetic reactions. *J. Am. Chem. Soc.* 118 (35): 8285–8290.

1708 Liou, S.Y., van der Boom, M.E., and Milstein, D. (1998). Catalytic selective cleavage of a strong C—C single bond by Rh in solution. *J. Chem. Soc., Chem. Commun.* (6): 687–688.

1709 Weiss, J.M. and Downs, C.R. (1920). Catalytic oxidation I – benzene. *J. Ind. Eng. Chem.* 12: 228–232.

1710 (a) Beydoun, A., Figueras, F., and De Mourgues, L.D. (1971). Catalytic oxidation of benzene on vanadium oxides. 1. Catalysis on pure vanadium oxides. *Bull. Soc. Chim. Fr.* (3): 806–810. (b) Uraz, C. and Atalay, S. (2007). Oxidation of benzene to maleic anhydride in a fluidized bed reactor. *Chem. Eng. Technol.* 30 (12): 1708–1715.

1711 Burnett, J.C., Keppel, R.A., and Robinson, W.D. (1987). Commercial production of maleic anhydride by catalytic processes using fixed bed reactors. *Catal. Today* 1 (5): 537–586.

1712 Shelmerdine, J., Popper, F., and McNeil, D. (1952). The production of phthalic anhydride by the oxidation of tar oils. *Chem. Ind.* (10): 216–216.

1713 Boger, T. and Menegola, M. (2005). Monolithic catalysts with high thermal conductivity for improved operation and economics in the production of phthalic anhydride. *Ind. Eng. Chem. Res.* 44 (1): 30–40.

1714 Suresh, A.K., Sharma, M.M., and Sridhar, T. (2000). Engineering aspects of industrial liquid-phase air oxidation of hydrocarbons. *Ind. Eng. Chem. Res.* 39 (11): 3958–3997.

1715 (a) Chavan, S.A., Srinivas, D., and Ratnasamy, P. (2002). Oxidation of cyclohexane, cyclohexanone, and cyclohexanol to adipic acid by a non-HNO_3 route over Co/Mn cluster complexes. *J. Catal.* 212: 39–45. (b) Hu, B.-Y., Yuan, Y.-J., Xiao, J. et al. (2008). Rational oxidation of cyclohexane to cyclohexanol, cyclohexanone and adipic acid with air over metalloporphyrin and cobalt salt. *J. Porphyrins Phthalocyanines* 12: 27–34. (c) Li, L., Li, H., Jin, C. et al. (2010). Surface Co silicate and CoOx cluster anchored to SBA-15: highly efficient for cyclohexane partial oxidation. *Catal. Lett.* 136: 20–27.

1716 Zhao, R., Ji, D., Lu, G.M. et al. (2004). A highly efficient oxidation of cyclohexane over Au/ZSM-5 molecular sieve catalyst with oxygen as oxidant. *J. Chem. Soc., Chem. Commun.* (7): 904–905.

1717 Skouta, R. and Li, C.J. (2008). Gold-catalyzed reactions of C—H bonds. *Tetrahedron* 64 (22): 4917–4938.

1718 Ghosh, S., Acharyya, S.S., Tiwari, R. et al. (2014). Selective oxidation of propylene to propylene oxide over Ag-supported W oxide nanostructure with molecular oxygen. *ACS Catal.* 4: 2169–2174.

1719 van Putten, R.J., van der Waal, J.C., de Jong, E. et al. (2013). Hydroxymethylfurfural, a versatile platform chemical made from renewable resources. *Chem. Rev.* 113: 1499–1597.

1720 Chheda, J.N., Roman-Leshkov, Y., and Dumesic, J.A. (2007). Production of 5-hydroxymethylfurfural and furfural by dehydration of biomass-derived mono- and poly-saccharides. *Green Chem.* 9: 342–350.

1721 Zakrzewska, M.E., Bogel-Lukasik, E., and Bogel-Lukasik, R. (2011). Ionic liquid-mediated formation of 5-hydroxymethylfurfural-A promising biomass-derived building block. *Chem. Rev.* 111: 397–417.

1722 Wu, L.Q., Song, J.L., Zhang, B.B. et al. (2014). Very efficient conversion of glucose to 5-hydroxymethylfurfural in DBU-based ionic liquids with benzenesulfonate anion. *Green Chem.* 16: 3935–3941.

1723 Teong, S.P., Yi, G.S., and Zhang, Y.G. (2014). Hydroxymethylfurfural production from bioresources: past, present and future. *Green Chem.* 16: 2015–2026.

1724 Saha, B. and bu-Omar, M.M. (2014). Advances in 5-hydroxymethylfurfural production from biomass in biphasic solvents. *Green Chem.* 16: 24–38.

1725 Yang, Y., Hu, C.W., and bu-Omar, M.M. (2012). Conversion of carbohydrates and lignocellulosic biomass into 5-hydroxymethylfurfural using $AlCl_3{\cdot}6H_2O$ catalyst in a biphasic solvent system. *Green Chem.* 14: 509–513.

1726 (a) Lan, J.H., Lin, J.C., Chen, Z.Q., and Yin, G.C. (2015). Transformation of 5-hydroxymethylfurfural (HMF) to maleic anhydride by aerobic oxidation with heteropolyacid catalysts. *ACS Catal.* 5: 2035–2041. (b) Wang, F., Yuan, Z.L., Liu, B. et al. (2016). Catalytic oxidation of biomass derived 5-hydroxymethylfurfural (HMF) over Ru(III)-incorporated Zr phosphate catalyst. *J. Ind. Eng. Chem.* 38: 181–185.

1727 (a) Gao, X.-C., Lu, X.-C., and Xu, J. (2009). Direct oxidation of benzene to phenol by dioxygen over nano-vanadium oxide. *Chin. J. Chem.* 27 (11): 2155–2158. (b) Makgwane, P.R. and Ray, S.S. (2015). Development of high-performance nanostructured V_2O_5/SnO_2 catalyst for efficient benzene hydroxylation. *Appl. Catal., A* 492: 10–22.

1728 Fang, R.Q., Luque, R., and Li, Y.W. (2016). Selective aerobic oxidation of biomass-derived HMF to 2,5-diformylfuran using a MOF-derived magnetic hollow Fe—Co nanocatalyst. *Green Chem.* 18: 3152–3157.

1729 Liu, B., Ren, Y.S., and Zhang, Z.H. (2015). Aerobic oxidation of 5-hydroxymethylfurfural into 2,5-furandicarboxylic acid in water under mild conditions. *Green Chem.* 17: 1610–1617.

1730 Zhang, Z.H., Zhen, J.D., Liu, B. et al. (2015). Selective aerobic oxidation of the biomass-derived precursor 5-hydroxymethylfurfural to 2,5-furandicarboxylic acid under mild conditions over a magnetic palladium nanocatalyst. *Green Chem.* 17: 1308–1317.

1731 Haruta, M., Kobayashi, T., Sano, H., and Yamada, N. (1987). Novel gold catalysts for the oxidation of carbon monoxide at a temperature far below 0-degrees-C. *Chem. Lett.* (2): 405–408.

1732 Hutchings, G.J. (2014). Selective oxidation using supported gold bimetallic and trimetallic nanoparticles. *Catal. Today* 238: 69–73.

1733 Kim, T.S., Gong, J., Ojifinni, R.A. et al. (2006). Water activated by atomic oxygen on Au(111) to oxidize CO at low temperatures. *J. Am. Chem. Soc.* 128 (19): 6282–6283.

1734 Chen, M., Cai, Y., Yan, Z., and Goodman, D.W. (2006). On the origin of the unique properties of supported Au nanoparticles. *J. Am. Chem. Soc.* 128 (19): 6341–6346.

1735 Li, Z.Y., Yuan, Z., Li, X.N. et al. (2014). CO oxidation catalyzed by single gold atoms supported on aluminum oxide clusters. *J. Am. Chem. Soc.* 136: 14307–14313.

1736 Widmann, D. and Behm, R.J. (2014). Activation of molecular oxygen and the nature of the active oxygen species for CO oxidation on oxide supported Au catalysts. *Acc. Chem. Res.* 47: 740–749.

1737 Hayashi, T., Tanaka, K., and Haruta, M. (1998). Selective vapor-phase epoxidation of propylene over Au/TiO_2 catalysts in the presence of oxygen and hydrogen. *J. Catal.* 178 (2): 566–575.

1738 Mul, G., Zwijnenburg, A., van der Linden, B. et al. (2001). Stability and selectivity of Au/TiO_2 and $Au/TiO_2/SiO_2$ catalysts in propene epoxidation: an in situ FT-IR study. *J. Catal.* 201 (1): 128–137.

1739 Khatib, S.J. and Oyama, S.T. (2015). Direct oxidation of propylene to propylene oxide with molecular oxygen: a review. *Catal. Rev. Sci. Eng.* 57: 306–344.

1740 Fu, Q., Saltsburg, H., and Flytzani-Stephanopoulos, M. (2003). Active nonmetallic Au and Pt species on ceria-based water-gas shift catalysts. *Science* 301 (5635): 935–938.

1741 Liu, Z.P., Jenkins, S.J., and King, D.A. (2005). Origin and activity of oxidized gold in water-gas-shift catalysis. *Phys. Rev. Lett.* 94 (19): 196102.

1742 Xu, Y.J., Landon, P., Enache, D. et al. (2005). Selective conversion of cyclohexane to cyclohexanol and cyclohexanone using a gold catalyst under mild conditions. *Catal. Lett.* 101 (3–4): 175–179.

1743 Shul'pin, G.B., Shilov, A.E., and Süss-Fink, G. (2001). Alkane oxygenation catalysed by gold complexes. *Tetrahedron Lett.* 42 (41): 7253–7256.

1744 Guan, B.T., Xing, D., Cai, G.X. et al. (2005). Highly selective aerobic oxidation of alcohol catalyzed by a Au(I) complex with an anionic ligand. *J. Am. Chem. Soc.* 127 (51): 18004–18005.

1745 Choudhary, V.R., Dumbre, D.K., Uphade, B.S., and Narkhede, V.S. (2004). Solvent-free oxidation of benzyl alcohol to benzaldehyde by *tert*-butyl hydroperoxide using transition metal containing layered double hydroxides and/or mixed hydroxides. *J. Mol. Catal. A: Chem.* 215 (1–2): 129–135.

1746 Choudhary, V.R., Dhar, A., Jana, P. et al. (2005). A green process for chlorine-free benzaldehyde from the solvent-free oxidation of benzyl alcohol with molecular oxygen over a supported nano-size gold catalyst. *Green Chem.* 7 (11): 768–770.

1747 Choudhary, V.R., Jha, R., and Jana, P. (2007). Solvent-free selective oxidation of benzyl alcohol by molecular oxygen over uranium oxide supported nano-gold catalyst for the production of chlorine-free benzaldehyde. *Green Chem.* 9 (3): 267–272.

1748 Zheng, N.F. and Stucky, G.D. (2007). Promoting gold nanocatalysts in solvent-free selective aerobic oxidation of alcohols. *J. Chem. Soc., Chem. Commun.* 3862–3864.

1749 Enache, D.I., Edwards, J.K., Landon, P. et al. (2006). Solvent-free oxidation of primary alcohols to aldehydes using $Au-Pd/TiO_2$ catalysts. *Science* 311 (5759): 362–365.

1750 Li, G., Enache, D.I., Edwards, J. et al. (2006). Solvent-free oxidation of benzyl alcohol with oxygen using zeolite-supported Au and Au-Pd catalysts. *Catal. Lett.* 110 (1–2): 7–13.

1751 Mori, K., Hara, T., Mizugaki, T. et al. (2004). Hydroxyapatite-supported Pd nanoclusters: a highly active heterogeneous catalyst for selective oxidation of alcohols by use of molecular oxygen. *J. Am. Chem. Soc.* 126 (34): 10657–10666.

1752 Kimmerle, B., Grunwaldt, J.D., and Baiker, A. (2007). Gold catalysed selective oxidation of alcohols in supercritical carbon dioxide. *Top. Catal.* 44 (1–2): 285–292.

1753 Abad, A., Concepcion, P., Corma, A., and García, H. (2005). A collaborative effect between gold and a support induces the selective oxidation of alcohols. *Angew. Chem. Int. Ed.* 44 (26): 4066–4069.

1754 Pushkarev, V.V., Kovalchuk, V.I., and D'Itri, J.L. (2004). Probing defect sites on the CeO_2 surface with dioxygen. *J. Phys. Chem. B* 108 (17): 5341–5348.

1755 Biffis, A., Cunial, S., Spontoni, P., and Prati, L. (2007). Microgel-stabilized gold nanoclusters: powerful "quasi-homogeneous" catalysts for the aerobic oxidation of alcohols in water. *J. Catal.* 251: 1–6.

1756 Biffis, A. (2001). Functionalised microgels: novel stabilisers for catalytically active metal colloids. *J. Mol. Catal. A: Chem.* 165 (1–2): 303–307.

1757 Biffis, A., Orlandi, N., and Corain, B. (2003). Microgel-stabilized metal nanoclusters: size control by microgel nanomorphology. *Adv. Mater.* 15 (18): 1551–1555.

1758 Biffis, A. and Sperotto, E. (2003). Microgel-stabilized metal nanoclusters: improved solid-state stability and catalytic activity in Suzuki couplings. *Langmuir* 19 (22): 9548–9550.

1759 Biffis, A. and Minati, L. (2005). Efficient aerobic oxidation of alcohols in water catalysed by microgel-stabilised metal nanoclusters. *J. Catal.* 236 (2): 405–409.

1760 Minati, L. and Biffis, A. (2005). A polymer support with controllable solubility in mutually immiscible solvents. *J. Chem. Soc., Chem. Commun.* (8): 1034–1036.

1761 Tsunoyama, H., Sakurai, H., Negishi, Y., and Tsukuda, T. (2005). Size-specific catalytic activity of polymer-stabilized gold nanoclusters for aerobic alcohol oxidation in water. *J. Am. Chem. Soc.* 127 (26): 9374–9375.

1762 Tsunoyama, H., Sakurai, H., and Tsukuda, T. (2006). Size effect on the catalysis of gold clusters dispersed in water for aerobic oxidation of alcohol. *Chem. Phys. Lett.* 429 (4–6): 528–532.

1763 Tsunoyama, H., Sakurai, H., Ichikuni, N. et al. (2004). Colloidal gold nanoparticles as catalyst for carbon–carbon bond formation: application to aerobic homocoupling of phenylboronic acid in water. *Langmuir* 20 (26): 11293–11296.

1764 Miyamura, H., Matsubara, R., Miyazaki, Y., and Kobayashi, S. (2007). Aerobic oxidation of alcohols at room temperature and atmospheric conditions catalyzed by reusable gold nanoclusters stabilized by the benzene rings of polystyrene derivatives. *Angew. Chem. Int. Ed.* 46 (22): 4151–4154.

1765 Gunasekaran, N. (2015). Aerobic oxidation catalysis with air or molecular oxygen and ionic liquids. *Adv. Synth. Catal.* 357: 1990–2010.

1766 (a) Sigel, R., Sigel, A., and Sigel, H. (eds.) (2007). The ubiquitous roles of cytochrome P450 proteins. In: *Metal Ions in Life Sciences*, vol. 3, 1–678. New York: Wiley. (b) Ortiz de Montellano, P.R. (2014). Role of cytochrome P450 enzymes in biotransformation. In: *Handbook of Metabolic Pathways of Xenobiotics*, vol. 1 (ed. P.W. Lee), 249–266. Chichester: Wiley.

1767 (a) Meunier, B., de Visser, S.P., and Shaik, S. (2004). Mechanism of oxidation reactions catalyzed by cytochrome P450 enzymes. *Chem. Rev.* 104 (9): 3947–3980. (b) Hrycay, E.G. and Bandiera, S.M. (2015). Monooxygenase, peroxidase and peroxygenase properties and reaction mechanisms of cytochrome P450 enzymes. *Adv. Exp. Med. Biol.* 851: 1–61.

1768 Weitzel, C. and Simonsen, H.T. (2015). Cytochrome P450-enzymes involved in the biosynthesis of mono- and sesquiterpenes. *Phytochem. Rev.* 14 (1): 7–24.

1769 Zhou, X., Chan, K., and Yeung, J.H.K. (2012). Herb-drug interactions with Danshen (Salvia miltiorrhiza): a review on the role of cytochrome P450 enzymes. *Drug Metab. Drug Interact.* 27 (1): 9–18.

1770 Güngerich, F.P. (2006). Cytochrome P450s and other enzymes in drug metabolism and toxicity. *AAPS J.* 8 (1): E101–E111.

1771 Güngerich, F.P. (2006). A malleable catalyst dominates the metabolism of drugs. *Proc. Natl. Acad. Sci. U.S.A.* 103 (37): 13565–13566.

1772 Pavek, P. and Dvorak, Z. (2008). Xenobiotic-induced transcriptional regulation of xenobiotic metabolizing enzymes of the cytochrome P450 superfamily in human extrahepatic tissues. *Curr. Drug Metab.* 9 (2): 129–143.

1773 Zanger, U.M. and Schwab, M. (2013). Cytochrome P450 enzymes in drug metabolism: regulation of gene expression, enzyme activities, and impact of genetic variation. *Pharmacol. Ther.* 138 (1): 103–141.

1774 Ortiz de Montellano, P.R. (2014). Role of cytochrome P450 enzymes in biotransformations. In: *Handbook of Metabolic Pathways of Xenobiotics*, vol. 1 (ed. P.W. Lee), 249–266. New York: Wiley.

1775 Sono, M., Roach, M.P., Coulter, E.D., and Dawson, J.H. (1996). Heme-containing oxygenases. *Chem. Rev.* 96 (7): 2841–2887.

1776 Auclair, K., Hu, Z.B., Little, D.M. et al. (2002). Revisiting the mechanism of P450 enzymes with the radical clocks norcarane and spiro[2,5]octane. *J. Am. Chem. Soc.* 124 (21): 6020–6027.

1777 Löew, G.H., Kert, C.J., Hjelmeland, L.M., and Kirchner, R.F. (1977). Active-site models of horseradish-peroxidase compound I and a cytochrome-P-450 analog – electronic-structure and electric-field Gradients. *J. Am. Chem. Soc.* 99 (10): 3534–3536.

1778 Dawson, J.H. and Sono, M. (1987). Cytochrome P-450 and chloroperoxidase – thiolate-ligated heme enzymes – spectroscopic determination of their active site structures and mechanistic implications of thiolate ligation. *Chem. Rev.* 87 (5): 1255–1276.

1779 Dawson, J.H. (1988). Probing structure/function relations in heme-containing oxygenases and peroxidases. *Science* 240 (4851): 433–439.

1780 White, R.E. (1991). The involvement of free-radicals in the mechanisms of monooxygenases. *Pharmacol. Ther.* 49 (1–2): 21–42.

1781 Hjelmeland, L.M., Aronow, L., and Trudell, J.R. (1977). Intramolecular determination of primary kinetic isotope effects in hydroxylations catalyzed by cytochrome-P-450. *Biochem. Biophys. Res. Commun.* 76 (2): 541–549.

1782 Groves, J.T., McClusky, G.A., White, R.E., and Coon, M.J. (1978). Aliphatic hydroxylation by highly purified liver microsomal cytochrome-P-450 – evidence for a carbon radical intermediate. *Biochem. Biophys. Res. Commun.* 81 (1): 154–160.

1783 Groves, J.T. (1985). Key elements of the chemistry of cytochrome-P-450 – the oxygen rebound mechanism. *J. Chem. Educ.* 62 (11): 928–931.

1784 Ortiz de Montellano, P.R. and Stearns, R.A. (1987). Timing of the radical recombination step in cytochrome-P-450 catalysis with ring-strained probes. *J. Am. Chem. Soc.* 109 (11): 3415–3420.

1785 (a) Atkinson, J.K., Hollenberg, P.F., Ingold, K.U. et al. (1994). Cytochrome P450-catalyzed hydroxylation of hydrocarbons – kinetic deuterium-isotope effects for the hydroxylation of an ultrafast radical clock. *Biochemistry* 33 (35): 10630–10637. (b) Cryle, M.J., Stuthe, J.M.U., Ortiz de Montellano, P.R., and de Voss, J.J. (2004). Cyclopropyl fatty acids implicate a radical but not a cation as an intermediate in P450BM3-catalysed hydroxylations. *J. Chem. Soc., Chem. Commun.* (5): 512–513.

1786 Hanzlik, R.P., Hogberg, K., Moon, J.B., and Judson, C.M. (1985). Intramolecular kinetic deuterium-isotope effects on microsomal hydroxylation and chemical chlorination of toluene-α-D1 and toluene-*α*,α-D2. *J. Am. Chem. Soc.* 107 (24): 7164–7167.

1787 Hanzlik, R.P. and Ling, K.H.J. (1990). Active-site dynamics of toluene hydroxylation by cytochrome-P-450. *J. Org. Chem.* 55 (13): 3992–3997.

1788 Hanzlik, R.P. and Ling, K.H.J. (1993). Active-site dynamics of xylene hydroxylation by cytochrome-P-450 as revealed by kinetic deuterium isotope effects. *J. Am. Chem. Soc.* 115 (21): 9363–9370.

1789 Karki, S.B., Dinnocenzo, J.P., Jones, J.P., and Korzekwa, K.R. (1995). Mechanism of oxidative amine dealkylation of substituted *N*,*N*-dimethylanilines by cytochrome-P-450 – application of isotope effect profiles. *J. Am. Chem. Soc.* 117 (13): 3657–3664.

1790 Manchester, J.I., Dinnocenzo, J.P., Higgins, L.A., and Jones, J.P. (1997). A new mechanistic probe for cytochrome P450: an application of isotope effect profiles. *J. Am. Chem. Soc.* 119 (21): 5069–5070.

1791 Newcomb, M., Le Tadic-Biadatti, M.-H., Putt, D.A., and Hollenberg, P.F. (1995). An incredibly fast apparent oxygen rebound rate-constant for hydrocarbon hydroxylation by cytochrome-P-450 enzymes. *J. Am. Chem. Soc.* 117 (11): 3312–3313.

1792 Newcomb, M. and Chestney, D.L. (1994). A hypersensitive mechanistic probe for distinguishing between radical and carbocation intermediates. *J. Am. Chem. Soc.* 116 (21): 9753–9754.

1793 Newcomb, M., Le Tadic-Biadatti, M.-H., Chestney, D.L. et al. (1995). A nonsynchronous concerted mechanism for cytochrome-P-450 catalyzed hydroxylation. *J. Am. Chem. Soc.* 117 (49): 12085–12091.

1794 Toy, P.H., Newcomb, M., and Hollenberg, P.F. (1998). Hypersensitive mechanistic probe studies of cytochrome P450-catalyzed hydroxylation reactions. Implications for the cationic pathway. *J. Am. Chem. Soc.* 120 (31): 7719–7729.

1795 Shaik, S.S., Filatov, M., Schröder, D., and Schwarz, H. (1998). Electronic structure makes a difference: cytochrome P-450 mediated hydroxylations of hydrocarbons as a two-state reactivity paradigm. *Chem. Eur. J.* 4 (2): 193–199.

1796 Ogliaro, F., Harris, N., Cohen, S. et al. (2000). A model "rebound" mechanism of hydroxylation by cytochrome P450: stepwise and effectively concerted pathways, and their reactivity patterns. *J. Am. Chem. Soc.* 122 (37): 8977–8989.

1797 Yoshizawa, K., Ohta, T., Eda, M., and Yamabe, T. (2000). Two-step concerted mechanism for the hydrocarbon hydroxylation by cytochrome P450. *Bull. Chem. Soc. Jpn.* 73 (2): 401–407.

1798 Karasevich, E.I., Shestakov, A.F., and Shilov, A.E. (1999). Novel mechanism of aliphatic hydroxylation in enzymatic and biomimetic systems: problems of the reaction dynamics. *Exp. Toxicol. Pathol.* 51 (4–5): 335–341.

1799 (a) Toy, P.H., Newcomb, M., Coon, M.J., and Vaz, A.D.N. (1998). Two distinct electrophilic oxidants effect hydroxylation in cytochrome P-450-catalyzed reactions. *J. Am. Chem. Soc.* 120 (37): 9718–9719. (b) Valentine, A.M., Le Tadic-Biadatti, M.-H., Toy, P.H. et al. (1999). Oxidation of ultrafast radical clock substrate probes by the soluble methane monooxygenase from Methylococcus capsulatus (Bath). *J. Biol. Chem.* 274 (16): 10771–10776. (c) Jiang, Y., He, X., and Ortiz de Montellano, P.R. (2006). Radical intermediates in the catalytic oxidation of hydrocarbons by bacterial and human cytochrome P450 enzymes. *Biochemistry* 45 (2): 533–542.

1800 (a) Newcomb, M., Shen, R., Lu, Y. et al. (2002). Evaluation of norcarane as a probe for radicals in cytochome P450- and soluble methane monooxygenase-catalyzed hydroxylation reactions. *J. Am. Chem. Soc.* 124 (24): 6879–6886. (b) Duan, L., Jogl, G., and Cane, D.E. (2016). The cytochrome P450-catalyzed oxidative rearrangement in the final step of pentalenolactone biosynthesis: substrate structure determines mechanism. *J. Am. Chem. Soc.* 138 (38): 12678–12689.

1801 (a) Meunier, B. (1992). Metalloporphyrins as versatile catalysts for oxidation reactions and oxidative DNA cleavage. *Chem. Rev.* 92: 1411–1456. (b) ten Brink, G.J., Arends, I.W.C.E., and Sheldon, R.A. (2000). Green, catalytic oxidation of alcohols in water. *Science* 287: 1636–1639. (c) Che, C.-M., Lo, V.K.-Y., Zhou, C.-Y., and Huang, J.-S. (2011). Selective functionalisation of saturated C—H bonds with metalloporphyrin catalysts. *Chem. Soc. Rev.* 40 (4): 1950–1975.

1802 Barona-Castano, J.C., Carmona-Vargas, C.C., Brocksom, T.J., and de Oliveira, K.T. (2016). Porphyrins as catalysts in scalable organic reactions. *Molecules* 21: 1–27.

1803 (a) Cai, Y., Liu, Y., Lu, Y. et al. (2008). Ionic manganese porphyrins with S-containing counter anions: mimicking cytochrome P450 activity for alkene epoxidation. *Catal. Lett.* 124 (3–4): 334–339. (b) Simões, M.M.G., Neves, C.M.B., Pires, S.M.G. et al. (2013). Mimicking P450 processes and the use of metalloporphyrins. *Pure Appl. Chem.* 89 (8): 1671–1681. (c) Liu, Q., Zhou, D.-D., Li, Z.-X. et al. (2017). Efficient and bio-inspired conversion of cellulose to formic acid catalyzed by metalloporphyrins in alkaline solution. *Chin. J. Chem.* 35 (7): 1063–1068.

1804 Groves, J.T., Nemo, T.E., and Myers, R.S. (1979). Hydroxylation and epoxidation catalyzed by Fe-porphine complexes – oxygen-transfer from iodosylbenzene. *J. Am. Chem. Soc.* 101: 1032–1033.

1805 Luo, W.P., Liu, D.W., Sun, J. et al. (2014). Effects of oxygen transfer limitation and kinetic control on biomimetic catalytic oxidation of toluene. *Chin. J. Chem. Eng.* 22: 509–515.

1806 Che, C.M., Lo, V.K.Y., Zhou, C.Y., and Huang, J.S. (2011). Selective functionalisation of saturated C—H bonds with metalloporphyrin catalysts. *Chem. Soc. Rev.* 40: 1950–1975.

1807 Swann, S. and Xanthakos, T.S. (1931). Cobaltic sulfate as an oxidizing agent. *J. Am. Chem. Soc* 53: 400–404.

1808 Onopchen, A. and Schulz, J.G.D. (1973). Electron-transfer with aliphatic substrates – oxidation of cyclohexane with Co (III) ions alone and in presence of oxygen. *J. Org. Chem.* 38 (21): 3729–3733.

1809 Zaidi, S.A.H. (1988). Liquid-phase oxidation of cyclohexanone to adipic acid catalyzed by Co and bromide ions in acetic acid. *Appl. Catal.* 42 (2): 247–254.

1810 Koshel', G.N., Koshel', S.G., Postnova, M.V. et al. (2004). Liquid-phase catalytic oxidation of methyl derivatives of biphenyl. *Kinet. Catal.* 45 (6): 821–825.

1811 Zhou, L., Xu, J., Miao, H. et al. (2005). Catalytic oxidation of cyclohexane to cyclohexanol and cyclohexanone over Co_3O_4 nanocrystals with molecular oxygen. *Appl. Catal., A* 292: 223–228.

1812 Heiba, E.I., Dessau, R.M., and Köhl, W.J. (1969). Oxidation by metal salts. V. Cobaltic acetate oxidation of alkylbenzenes. *J. Am. Chem. Soc.* 91 (24): 6830–6837.

1813 Tang, S.W. and Liang, B. (2007). Scale formation and its mechanism in the liquid-phase oxidation of toluene by air. *Ind. Eng. Chem. Res.* 46 (23): 7826–7829.

1814 Morimoto, T. and Ogata, Y. (1967). Kinetics of autoxidation of toluene catalysed by cobaltic acetate. *J. Chem. Soc. B* (1): 62–66.

1815 Druliner, J.D. and Wasserman, E. (1988). Oscillations and mechanism – catalyzed O_2 oxidation of cyclohexanone. *J. Am. Chem. Soc.* 110 (16): 5270–5275.

1816 Rahman, A., Mupa, M., and Mahamadi, C. (2016). A mini review on new emerging trends for the synthesis of adipic acid from metal-nano heterogeneous catalysts. *Catal. Lett.* 146: 788–799.

1817 Higgins, J. (1967). Theory of oscillating reactions. *Ind. Eng. Chem.* 59: 18–62.

1818 Taylor, A.F. (2002). Mechanism and phenomenology of an oscillating chemical reaction. *Prog. React. Kinet. Mech.* 27: 247–325.

1819 Schmitz, G. (2016). Historical overview of the oscillating reactions. Contribution of Professor Slobodan Anic. *React. Kinet. Mech. Catal.* 118: 5–13.

1820 McDonald, A.G. (2003). Implications of enzyme kinetics. *Biochem. Soc. Trans.* 31 (3): 719–722.

1821 Schibler, U. and Näf, F. (2005). Cellular oscillators: rhythmic gene expression and metabolism. *Curr. Opin. Chem. Biol.* 17 (2): 223–229.

1822 Zhang, X.L., Lee, C.S.M., Hayward, D.O., and Mingos, D.M.P. (2005). Oscillatory behaviour observed in the rate of oxidation of methane over metal catalysts. *Catal. Today* 105 (2): 283–294.

1823 Lersch, M. and Tilset, M. (2005). Mechanistic aspects of C—H activation by Pt complexes. *Chem. Rev.* 105 (6): 2471–2526.

1824 Zhu, H.J. and Ziegler, T. (2006). A theoretical study of the original Shilov reaction involving methane activation by Pt tetrachloride in an acidic aqueous solution. *J. Organomet. Chem.* 691 (21): 4486–4497.

1825 Yoneyama, T. and Crabtree, R.H. (1996). Palladium(II) catalyzed acetoxylation of arenes with iodosyl acetate. *J. Mol. Catal. A: Chem.* 108: 35–40.

1826 Neufeldt, S.R. and Sanford, M.S. (2012). Controlling site selectivity in Pd-catalyzed C—H bond functionalization. *Acc. Chem. Res.* 45: 936–946.

1827 Topczewski, J.J. and Sanford, M.S. (2015). Carbon–hydrogen (C—H) bond activation at Pd(IV): a frontier in C—H functionalization catalysis. *Chem. Sci.* 6: 70–76.

1828 Cook, A.K. and Sanford, M.S. (2015). Mechanism of the Pd-catalyzed arene C—H acetoxylation: a comparison of catalysts and ligand effects. *J. Am. Chem. Soc.* 137: 3109–3118.

1829 Wang, F., Xu, J., and Liao, S.J. (2002). One-step heterogeneously catalytic oxidation of *o*-cresol by oxygen to salicylaldehyde. *J. Chem.Soc.; Chem. Commun.* (6): 626–627.

1830 Goldstein, A.S., Beer, R.H., and Drago, R.S. (1994). Catalytic oxidation of hydrocarbons with O_2 or H_2O_2 using a sterically hindered Ru complex. *J. Am. Chem. Soc.* 116 (6): 2424–2429.

1831 An, Z.J., Pan, X.L., Liu, X.M. et al. (2006). Combined redox couples for catalytic oxidation of methane by dioxygen at low temperatures. *J. Am. Chem. Soc.* 128 (50): 16028–16029.

1832 Fairlamb, I.J.S. (2015). Redox-active NOx ligands in Pd-mediated processes. *Angew. Chem. Int. Ed.* 54: 10415–10427.

1833 (a) Narsimhan, K., Iyoki, K., Dinh, K., and Roman-Leshkov, Y. (2016). Catalytic oxidation of methane into methanol over Cu-exchanged zeolites with oxygen at low temperature. *ACS Cent. Sci.* 2: 424–429. (b) Ravi, M., Ranocchiari, M., and van Bokhoven, J.A. (2017). The direct catalytic oxidation of methane to methanol – a critical assessment. *Angew. Chem. Int. Ed.* 56 (52): 16464–16483.

1834 Naróg, D., Szczepanik, A., and Sobkowiak, A. (2008). Iron(II,III)-catalyzed oxidation of limonene by dioxygen. *Catal. Lett.* 120 (3–4): 320–325.

1835 Neumann, R. and Khenkin, A.M. (2006). Molecular oxygen and oxidation catalysis by phosphovanadomolybdates. *J. Chem. Soc.; Chem. Commun.* (24): 2529–2538.

1836 Yuan, Q., Deng, W.P., Zhang, Q.H., and Wang, Y. (2007). Osmium-catalyzed selective oxidations of methane and ethane with hydrogen peroxide in aqueous medium. *Adv. Synth. Catal.* 349 (7): 1199–1209.

1837 Akita, M. and Moro-Oka, Y. (1998). Reductive activation of dioxygen: a new concept for the catalytic oxidation. *Catal. Today* 44 (1–4): 183–188.

1838 Barton, D.H.R., Hay-Motherwell, R.S., and Motherwell, W.B. (1983). Further studies on the activation of the C—H bond in saturated hydrocarbons. *Tetrahedron Lett.* 24 (19): 1979–1982.

1839 Balavoine, G., Barton, D.H.R., Boivin, J. et al. (1986). An efficient electrochemical process for the oxidation of saturated hydrocarbons – the Gif–Orsay system. *J. Chem. Soc., Chem. Commun.* (23): 1727–1729.

1840 Barton, D.H.R. and Doller, D. (1992). The selective functionalization of saturated hydrocarbons – Gif chemistry. *Acc. Chem. Res.* 25 (11): 504–512.

1841 Moro-Oka, Y. (1998). Reactivities of active oxygen species and their roles in the catalytic oxidation of inactive hydrocarbon. *Catal. Today* 45 (1–4): 3–12.

1842 Tabushi, I. and Koga, N. (1979). P-450 type oxygen activation by porphyrin-manganese complex. *J. Am. Chem. Soc.* 101 (21): 6456–6458.

1843 Tabushi, I. and Yazaki, A. (1981). P-450-type dioxygen activation using H_2 colloidal Pt as an effective electron-donor. *J. Am. Chem. Soc.* 103 (24): 7371–7373.

1844 Kirillova, N.V., Kuznetsova, N.I., Kuznetsova, L.I. et al. (2002). Reductive activation of dioxygen in catalytic systems including platinum and heteropoly compounds: oxidation of cyclohexane. *Catal. Lett.* 84 (3–4): 163–168.

1845 Stahl, S.S. (2004). Palladium oxidase catalysis: selective oxidation of organic chemicals by direct dioxygen-coupled turnover. *Angew. Chem. Int. Ed.* 43 (26): 3400–3420.

1846 Piera, J. and Backvall, J.E. (2008). Catalytic oxidation of organic substrates by molecular oxygen and hydrogen peroxide by multistep electron transfer – a biomimetic approach. *Angew. Chem. Int. Ed.* 47: 3506–3523.

1847 Wendlandt, A.E., Suess, A.M., and Stahl, S.S. (2011). Copper-catalyzed aerobic oxidative C—H functionalizations: trends and mechanistic insights. *Angew. Chem. Int. Ed.* 50: 11062–11087.

1848 Shi, Z.Z., Zhang, C., Tang, C.H., and Jiao, N. (2012). Recent advances in transition-metal catalyzed reactions using molecular oxygen as the oxidant. *Chem. Soc. Rev.* 41: 3381–3430.

1849 Anson, C.W., Ghosh, S., Hammes-Schiffer, S., and Stahl, S.S. (2016). Co(salophen)-catalyzed aerobic oxidation of *p*-hydroquinone: mechanism and implications for aerobic oxidation oatalysis. *J. Am. Chem. Soc.* 138: 4186–4193.

1850 Kunai, A., Wani, T., Uehara, Y. et al. (1989). Catalytic oxygenation of benzene – catalyst design and its performance. *Bull. Chem. Soc. Jpn.* 62 (8): 2613–2617.

1851 Kunai, A., Hata, S., Ito, S., and Sasaki, K. (1986). Mechanistic study of air oxidation of benzene in sulfuric acid catalyzed by cuprous ions. *J. Org. Chem.* 51 (18): 3471–3474.

1852 Jintoku, T., Taniguchi, H., and Fujiwara, Y. (1987). Palladium-catalyzed hydroxylation of benzene with O_2 or H_2O via the C—H aromatic bond activation – preparation of phenol. *Chem. Lett.* (9): 1865–1868.

1853 Jintoku, T., Takaki, K., Fujiwara, Y. et al. (1990). Palladium-catalyzed direct oxidation of benzene with molecular oxygen to phenol. *Bull. Chem. Soc. Jpn.* 63 (2): 438–441.

1854 Jintoku, T., Nishimura, K., Takaki, K., and Fujiwara, Y. (1991). *Chem. Lett.* (1): 193.

1855 Miyake, T., Hamada, M., Sasaki, Y., and Oguri, M. (1995). Direct synthesis of phenol by hydroxylation of benzene with oxygen and hydrogen. *Appl. Catal., A* 131 (1): 33–42.

1856 Gosling, P.A., Van Esch, J.H., Hoffmann, M.A.M., and Nolte, R.J.M. (1993). Reductive activation of dioxygen by a Mn(III)-porphyrin/Rh(III)-formate catalytic system. *J. Chem. Soc.; Chem. Commun.* (5): 472–473.

1857 Mukaiyama, T. and Yamada, T. (1995). Recent advances in aerobic oxygenation. *Bull. Chem. Soc. Jpn.* 68 (1): 17–35.

1858 Bowden, K., Heilbron, I.M., Jones, E.R.H., and Weedon, B.C.L. (1946). Researches on acetylenic compounds. 1. The preparation of acetylenic ketones by oxidation of acetylenic carbinols and glycols. *J. Chem. Soc.* 39–45.

1859 Collins, J.C., Hess, W.M., and Frank, F.J. (1968). Dipyridine-Cr(VI) oxide oxidation of alcohols in dichloromethane. *Tetrahedron Lett.* 3363–3366.

1860 Harrison, I.T. (1964). Oxidation of saturated alcohols by Mn dioxide. *Proc. R. Soc. London, Ser. A* 110–110.

1861 Fétizon, M. and Golfier, M. (1968). Selective oxidation of alcohols by silver carbonate. *C.R. Hebd. Seances Acad. Sci.* 267 (15): 900–903.

1862 Corey, E.J. and Suggs, J.W. (1975). Pyridinium chlorochromate – efficient reagent for oxidation of primary and secondary alcohols to carbonyl compounds. *Tetrahedron Lett.* (31): 2647–2650.

1863 Coates, W.M. and Corrigan, J.R. (1969). Pyridine dichromate as an oxidizing agent. *Chem. Ind.* (44): 1594–1594.

1864 Corey, E.J. and Schmidt, G. (1979). Useful procedures for the oxidation of alcohols involving pyridinium dichromate in aprotic media. *Tetrahedron Lett.* (5): 399–402.

1865 Corey, E.J., Barrette, E.P., and Magriotis, P.A. (1985). A new Cr(VI) reagent for the catalytic oxidation of secondary alcohols to ketones. *Tetrahedron Lett.* 26 (48): 5855–5858.

1866 Pfitzner, K.E. and Moffatt, J.G. (1965). Sulfoxide-carbodiimide reactions. I. A facile oxidation of alcohols. *J. Am. Chem. Soc.* 87 (24): 5661–5670.

1867 Albright, J.D. and Goldman, L. (1967). Dimethyl sulfoxide-acid anhydride mixtures for oxidation of alcohols. *J. Am. Chem. Soc.* 89 (10): 2416–2423.

1868 Onodera, K., Hirano, S., and Kashimura, N. (1965). Oxidation of carbohydrates with dimethyl sulfoxide containing phosphorus pentoxide. *J. Am. Chem. Soc.* 87 (20): 4651–4652.

1869 Parikh, J.R. and von Doering, W. (1967). Sulfur trioxide in oxidation of alcohols by dimethyl sulfoxide. *J. Am. Chem. Soc.* 89 (21): 5505–5507.

1870 Mancuso, A.J. and Swern, D. (1981). Activated dimethylsulfoxide – useful reagents for synthesis. *Synthesis* (3): 165–185.

1871 Omura, K. and Swern, D. (1978). Oxidation of alcohols by activated dimethyl sulfoxide – preparative, steric and mechanistic study. *Tetrahedron* 34 (11): 1651–1660.

1872 (a) Mancuso, A.J., Huang, S.L., and Swern, D. (1978). Oxidation of long-chain and related alcohols to carbonyls by dimethyl sulfoxide activated by oxalyl chloride. *J. Org. Chem.* 43 (12): 2480–2482. (b) Tidwell, T.T. (2004). Oxidation of alcohols to carbonyl compounds via alkoxysulfonium ylides: the Moffat, Swern, and related oxidations. *Org. React.* 35: 297–555. (c) Ahmad, N.M. (2007). Swern oxidation. In: *Name Reactions for Functional Group Transformations* (ed. J.J. Li and E.J. Corey), 291–308. Hoboken, NJ: Wiley.

1873 (a) Dess, D.B. and Martin, J.C. (1983). Readily accessible 12-I-5 oxidant for the conversion of primary and secondary alcohols to aldehydes and ketones. *J. Org. Chem.* 48 (22): 4155–4156. (b) Dess, D.B. and Martin, J.C. (1991). A useful 12-I-5 triacetoxyperiodinane (the Dess–Martin periodinane) for the selective oxidation of primary or secondary alcohols and a variety of related 12-I-5 species. *J. Am. Chem. Soc.* 113 (19): 7277–7287.

1874 Holsworth, D.D. (2007). Dess-Martin periodinane oxidation. In: *Name Reactions for Functional Group Transformations* (ed. J.J. Li and E.J. Corey), 218–236. Hoboken, NJ: Wiley.

1875 Ciriminna, R. and Pagliaro, M. (2010). Industrial oxidations with organocatalyst TEMPO and its derivatives. *Org. Process Res. Dev.* 14 (1): 245–251.

1876 Lebedev, O.L. and Kazarnovskii, S.N. (1960). Kataliticheskoe Okislenie Alifaticheskikh Aminov Perekisyu Vodoroda. *Zh. Obshch. Khim.* 30 (5): 1631–1635.

1877 (a) Anelli, P.L., Biffi, C., Montanari, F., and Quici, S. (1987). Fast and selective oxidation of primary alcohols to aldehydes or to carboxylic acids and of secondary alcohols to ketones mediated by oxoammonium salts under 2-phase conditions. *J. Org. Chem.* 52 (12): 2559–2562. (b) Bogdan, A. and McQuade, D.T. (2009). A biphasic oxidation of alcohols to aldehydes and ketones using a simplified packed-bed microreactor. *Beilstein J. Org. Chem.* 5: 17. (7 pages).

1878 Zhao, M.M., Li, J., Mano, E. et al. (2005). Oxidation of primary alcohols to carboxylic acids using sodium chlorite catalyzed by TEMPO and bleach: 4-methoxyphenylacetic acid. *Org. Synth.* 81: 195–199.

1879 Corberán, V.C., González-Pérez, M.E., Martinez-González, S., and Gómez-Avilés, A. (2014). Green oxidation of fatty alcohols: challenges and opportunities. *Appl. Catal., A* 474: 211–223.

1880 Blackburn, T.F. and Schwartz, J. (1977). Homogeneous catalytic oxidation of secondary alcohols to ketones by molecular oxygen under mild conditions. *J. Chem. Soc.; Chem. Commun.* (5): 157–158.

1881 Heyns, K. and Blazejewicz, L. (1960). Über Katalytische Oxydationen. 14. Katalytische Oxydation von Primären und Sekundären Hydroxylverbindungen mit Sauerstoff am Platinkontakt in Flüssiger Phase. *Tetrahedron* 9 (1–2): 67–75.

1882 Mallat, T. and Baiker, A. (1994). Oxidation of alcohols with molecular oxygen on Pt metal catalysts in aqueous solutions. *Catal. Today* 19 (2): 247–283.

1883 Martin, J., Martin, C., Faraj, M., and Bregeault, J.M. (1984). Selective homogeneous oxidation of secondary alcohols by dioxygen catalyzed by Rh (III). *New J. Chem.* 8 (3): 141–143.

1884 Roundhill, D.M., Dickson, M.K., Dixit, N.S., and Sudha-Dixit, B.P. (1980). Catalysis by multi-metallics – catalyzed homogeneous oxidation of alcohols and ketones with molecular oxygen in the presence of hexarhodium hexadecacarbonyl and dirhenium decacarbonyl. *J. Am. Chem. Soc.* 102 (17): 5538–5542.

1885 Bressan, M., Morandini, F., Morvillo, A., and Rigo, P. (1985). Cationic complexes of Rh(I) as catalysts in the homogeneous O_2 oxidation of terminal alkenes to methyl ketones. *J. Organomet. Chem.* 280 (1): 139–146.

1886 Tang, R., Diamond, S.E., Neary, N., and Mares, F. (1978). Homogeneous catalytic oxidation of amines and secondary alcohols by molecular oxygen. *J. Chem. Soc., Chem. Commun.* (13): 562–562.

1887 Matsumoto, M. and Watanabe, N. (1984). Oxidation of allylic alcohols to unsaturated carbonyl compounds by Ru dioxide and dioxygen/Ru dixoxide. *J. Org. Chem.* 49 (18): 3435–3436.

1888 Bilgrien, C., Davis, S., and Drago, R.S. (1987). The selective oxidation of primary alcohols to aldehydes by O_2 employing a trinuclear Ru carboxylate catalyst. *J. Am. Chem. Soc.* 109 (12): 3786–3787.

1889 Genêt, J.P., Pons, D., and Juge, S. (1989). A practical catalytic procedure for oxidation of alcohols into aldehydes or ketones with $RuCl_3$-water and $Ca(OCl)_2$ or $NaIO_4$. *Synth. Commun.* 19 (9–10): 1721–1726.

1890 Wang, G.Z., Andreasson, U., and Bäckvall, J.E. (1994). Aerobic oxidation of secondary alcohols via Ru-catalyzed hydrogen transfer involving a new triple catalytic system. *J. Chem. Soc., Chem. Commun.* (9): 1037–1038.

1891 Markó, I.E., Giles, P.R., Tsukazaki, M. et al. (1997). Efficient, aerobic, Ru-catalyzed oxidation of alcohols into aldehydes and ketones. *J. Am. Chem. Soc.* 119 (51): 12661–12662.

1892 Farmer, V. and Welton, T. (2002). The oxidation of alcohols in substituted imidazolium ionic liquids using Ru catalysts. *Green Chem.* 4 (2): 97–102.

1893 Shen, H.Y., Zhou, S.C., Wei, M.H., and Zong, H.X. (2006). Selective aerobic oxidation of alcohols to aldehydes and ketones by polymer-bound Ru complexes. *React. Funct. Polym.* 66 (8): 827–831.

1894 Sigman, M.S. and Jensen, D.R. (2006). Ligand-modulated Pd-catalyzed aerobic alcohol oxidations. *Acc. Chem. Res.* 39: 221–229.

1895 Schultz, M.J. and Sigman, M.S. (2006). Recent advances in homogeneous transition metal-catalyzed aerobic alcohol oxidations. *Tetrahedron* 62: 8227–8241.

1896 Huang, Z.J., Li, F.B., Chen, B.F., and Yuan, G.Q. (2015). Sustainable catalytic oxidation of alcohols over the interface between air and water. *Green Chem.* 17: 2325–2329.

1897 Yamada, T. and Mukaiyama, T. (1989). Novel method for oxidation of secondary alcohols into ketones with molecular oxygen by using Co(II) complex catalyst. *Chem. Lett.* (3): 519–522.

1898 Wang, L., Chen, Z.X., Huang, M.Q. et al. (2016). A green route to cyclohexanone: selective oxidation of cyclohexanol promoted by non-precious catalyst of h-WO_3 nanorods. *Catal. Lett.* 146: 1283–1290.

1899 Markó, I.E., Giles, P.R., Tsukazaki, M. et al. (1996). Copper-catalyzed oxidation of alcohols to aldehydes and ketones: an efficient, aerobic alternative. *Science* 274 (5295): 2044–2046.

1900 Markó, I.E., Gautier, A., Chelle-Regnaut, I. et al. (1998). Efficient and practical catalytic oxidation of alcohols using molecular oxygen. *J. Org. Chem.* 63 (22): 7576–7577.

1901 Jiang, N. and Ragauskas, A.J. (2005). Copper(II)-catalyzed aerobic oxidation of primary alcohols to aldehydes in ionic liquid [bmpy]PF6. *Org. Lett.* 7 (17): 3689–3692.

1902 Jiang, N. and Ragauskas, A.J. (2006). Cu(II)-catalyzed selective aerobic oxidation of alcohols under mild conditions. *J. Org. Chem.* 71 (18): 7087–7090.

1903 Figiel, P.J., Leskela, M., and Repo, T. (2007). TEMPO-copper(II) diimine-catalysed oxidation of benzylic alcohols in aqueous media. *Adv. Synth. Catal.* 349 (7): 1173–1179.

1904 Mannam, S., Alamsetti, S.K., and Sekar, G. (2007). Aerobic, chemoselective oxidation of alcohols to carbonyl compounds catalyzed by a DABCO-copper complex under mild conditions. *Adv. Synth. Catal.* 349 (14–15): 2253–2258.

1905 Minisci, F., Recupero, F., Cecchetto, A. et al. (2004). Mechanisms of the aerobic oxidation of alcohols to aldehydes and ketones, catalysed under mild conditions by persistent and non-persistent nitroxyl radicals and transition metal salts – polar, enthalpic, and captodative effects. *Eur. J. Org. Chem.* (1): 109–119.

1906 Cao, Q., Dornan, L.M., Rogan, L. et al. (2014). Aerobic oxidation catalysis with stable radicals. *J. Chem. Soc., Chem. Commun.* 50 (35): 4524–4543.

1907 McCann, S.D. and Stahl, S.S. (2015). Copper-catalyzed aerobic oxidations of organic molecules: pathways for two-electron oxidation with a four-electron oxidant and a one-electron redox-active catalyst. *Acc. Chem. Res.* 48: 1756–1766.

1908 Jiang, N. and Ragauskas, A.J. (2007). Selective aerobic oxidation of activated alcohols into acids or aldehydes in ionic liquids. *J. Org. Chem.* 72 (18): 7030–7033.

1909 Mondal, A., Das, A., Adhikary, B., and Mukherjee, D.K. (2014). Palladium nanoparticles in ionic liquids: reusable catalysts for aerobic oxidation of alcohols. *J. Nanopart. Res.* 16: 1–10.

1910 Coleman, K.S., Lorber, C.Y., and Osborn, J.A. (1998). Selective catalytic oxidation of alcohols by a Ru/Cu bifunctional system using molecular oxygen. *Eur. J. Inorg. Chem.* (11): 1673–1675.

1911 Yamaguchi, K. and Mizuno, N. (2002). Supported Ru catalyst for the heterogeneous oxidation of alcohols with molecular oxygen. *Angew. Chem. Int. Ed.* 41 (23): 4538–4542.

1912 Ebitani, K., Ji, H.B., Mizugaki, T., and Kaneda, K. (2004). Highly active trimetallic Ru/CeO_2/CoO(OH) catalyst for oxidation of alcohols in the presence of molecular oxygen. *J. Mol. Catal. A: Chem.* 212 (1–2): 161–170.

1913 Marui, K., Higashiura, Y., Kodama, S. et al. (2014). Vanadium-catalyzed green oxidation of benzylic alcohols in water under air atmosphere. *Tetrahedron* 70: 2431–2438.

1914 Luo, J., Yu, H., Wang, H.J. et al. (2014). Aerobic oxidation of benzyl alcohol to benzaldehyde catalyzed by carbon nanotubes without any promoter. *Chem. Eng. J.* 240: 434–442.

1915 (a) Chen, K.X., Zhang, P.F., Wang, Y., and Li, H.R. (2014). Metal-free allylic/benzylic oxidation strategies with molecular oxygen: recent advances and future prospects. *Green Chem.* 16: 2344–2374. (b) Watanabe, H., Asano, S., Fujita, S.-i. et al. (2015). Nitrogen-doped, metal-free activated carbon catalysts for aerobic oxidation of alcohols. *ACS Catal.* 5 (5): 2886–2894. (c) Liu, C.-K., Fang, Z., Yang, Z. et al. (2015). Highly practical sodium(I)/azobenzene catalyst system for aerobic oxidation of benzylic alcohols. *RSC Adv.* 5 (97): 79699–79702. (d) Karimi, B., Vahdati, S., and Vali, H. (2016). Synergistic catalysis within TEMPO-functionalized periodic mesoporous organosilica with bridge imidazolium groups in the aerobic oxidation of alcohols. *RSC Adv.* 6 (68): 63717–63723. (e) Zhu, S.-H., Cen, Y.-L., Yang, M. et al. (2017). Probing the intrinsic active sites of modified graphene oxide for aerobic benzylic alcohol oxidation. *Appl. Catal., B* 211: 89–97. (f) Ma, J., Hong, C., Wan, Y. et al. (2017). Aerobic oxidation of secondary alcohols in water with ABNO/*tert*-butyl nitrite/KPF6 catalytic system. *Tetrahedron Lett.* 58 (7): 652–657.

1916 Smidt, J., Hafner, W., Jira, R. et al. (1959). Catalytic reactions of olefins on Pt metal compounds. *Angew. Chem. Int. Ed. Engl.* 71 (5): 176–182.

1917 Phillips, F.C. (1894). Note on a form of silver obtained in the reduction of the sulphide by hydrogen. *J. Am. Chem. Soc* 16 (10): 700–703.

1918 Gligorich, K.M. and Sigman, M.S. (2006). Mechanistic questions about the reaction of molecular oxygen with Pd in oxidase catalysis. *Angew. Chem. Int. Ed.* 45: 6612–6615.

1919 Keith, J.A., Nielsen, R.J., Oxgaard, J. et al. (2009). Comment on "Mechanism and kinetics of the Wacker process: a quantum mechanical approach". *Organometallics* 28: 1618–1619.

1920 Keith, J.A. and Henry, P.M. (2009). The mechanism of the Wacker reaction: a tale of two hydroxypalladations. *Angew. Chem. Int. Ed.* 48: 9038–9049.

1921 Stirling, A., Nair, N.N., Lledos, A., and Ujaque, G. (2014). Challenges in modelling homogeneous catalysis: new answers from ab initio molecular dynamics to the controversy over the Wacker process. *Chem. Soc. Rev.* 43: 4940–4952.

1922 Fernandes, R.A. and Chaudhari, D.A. (2014). Iron(III) Sulfate as terminal oxidant in the synthesis of methyl ketones via Wacker oxidation. *J. Org. Chem.* 79: 5787–5793.

1923 Wang, Y.F., Gao, Y.R., Mao, S. et al. (2014). Wacker-type oxidation and dehydrogenation of terminal olefins using molecular oxygen as the sole oxidant without adding ligand. *Org. Lett.* 16: 1610–1613.

1924 Henry, P.M. (1964). Kinetics of oxidation of ethylene by aqueous Pd(II) chloride. *J. Am. Chem. Soc.* 86 (16): 3246–3250.

1925 Keith, J.A., Nielsen, R.J., Oxgaard, J., and Goddard, W.A.I. (2007). Unraveling the Wacker oxidation mechanisms. *J. Am. Chem. Soc.* 129 (41): 12342–12343.

1926 Francis, J.W. and Henry, P.M. (1996). Palladium(II)-catalyzed exchange and isomerization reactions. 16. The kinetics and stereochemistry of the oxidation and isomerization of hexafluoro allylic alcohols in aqueous solution catalyzed by $PdCl_3$(pyridine) (-). *J. Mol. Catal. A: Chem.* 112 (3): 317–326.

1927 Hamed, O., Thompson, C., and Henry, P.M. (1997). Stereochemistry of the Wacker reaction: modes of addition of hydroxide, methoxide, and phenyl at high and low [Cl^-]. A study using chirality transfer. *J. Org. Chem.* 62 (21): 7082–7083.

1928 Tsuji, J., Morikawa, M., and Kiji, J. (1963). Reaction of olefin-Pd chloride complex with carbon monoxide. *Tetrahedron Lett.* (16): 1061–1064.

1929 Tsuji, J., Morikawa, M., and Kiji, J. (1963). Organic syntheses by means of noble metal compounds. 2. Syntheses of saturated carboxylic esters from olefins. *Tetrahedron Lett.* (22): 1437–1440.

1930 Tsuji, J., Imamura, S., Morikawa, M., and Kiji, J. (1964). Organic syntheses by means of noble metal compounds. 8. Catalytic carbonylation of allylic compounds with Pd chloride. *J. Am. Chem. Soc.* 86 (20): 4350–4353.

1931 Tsuji, J., Morikawa, M., and Kiji, J. (1964). Organic syntheses by means of noble metal compounds. 7. Reactions of olefin-Pd chloride complexes with carbon monoxide. *J. Am. Chem. Soc.* 86 (22): 4851–4853.

1932 Tsuji, J., Imamura, S., Morikawa, M., and Kiji, J. (1964). Organic syntheses by means of noble metal compounds. VI. Synthesis of muconic acid. *J. Am. Chem. Soc.* 86 (20): 2095–2095.

1933 Tsuji, J. and Takahashi, H. (1965). Organic syntheses by means of noble metal compounds. 12. Reaction of cyclooctadiene-Pd chloride complex with ethyl malonate. *J. Am. Chem. Soc.* 87 (14): 3275–3276.

1934 Tsuji, J. (2000). Organopalladium chemistry in the '60s and '70s. *New J. Chem.* 24 (3): 127–135.

1935 (a) Tsuji, J. (1984). Synthetic applications of the Pd-catalyzed oxidation of olefins to ketones. *Synthesis* 369–384. (b) Dong, J.J., Browne, W.R., and Feringa, B.L. (2015). Palladium-catalyzed *anti*-Markovnikov oxidation of terminal alkenes. *Angew. Chem. Int. Ed.* 54 (3): 734–744.

1936 Baiju, T.V., Gravel, E., Doris, E., and Namboothiri, I.N.N. (2016). Recent developments in Tsuji–Wacker oxidation. *Tetrahedron Lett.* 57: 3993–4000.

1937 (a) Henry, P.M. (1980). *Palladium Catalyzed Oxidation of Hydrocarbons*, Catalysis by Metal Complexes, 1–435. Dordrecht: D. Reidel Publ. Co. (b) Popp, B.V. and Stahl, S.S. (2007). Palladium-catalyzed oxidation reactions: comparison of benzoquinone and molecular oxygen as stoichiometric oxidants. In: *Organometallic Oxidation Catalysis*, Topics in Organometallic Chemistry, vol. 22 (ed. F. Meyer and C. Limberg), 149–189. Berlin, Heidelberg: Springer-Verlag. (c) Gligorich, K.M. and Sigman, M.S. (2009). Recent advancements and challenges of Pd(II)-catalyzed oxidation reactions with molecular oxygen as the sole oxidant. *J. Chem. Soc., Chem. Commun.* (26): 3854–3867.

1938 Henry, P.M. (2002). The Wacker oxidation and related intermolecular reactions involving oxygen and other group 16 atom nucleophiles: the Wacker oxidation and related asymmetric syntheses. In: *Handbook of Organopalladium Chemistry for Organic Synthesis*, Chapter V.3.1.1 (ed. E.-I. Negishi), 2119–2139. New York: Wiley.

1939 Liu, G.S. and Stahl, S.S. (2006). Highly regioselective Pd-catalyzed intermolecular aminoacetoxylation of alkenes and evidence for *cis*-aminopalladation and S_N2 C—O bond formation. *J. Am. Chem. Soc.* 128: 7179–7181.

1940 Martínez, C. and Muñiz, K. (2012). Palladium-catalyzed vicinal difunctionalization of internal alkenes: diastereoselective synthesis of diamines. *Angew. Chem. Int. Ed.* 51: 7031–7034.

1941 Mitsudome, T., Mizumoto, K., Mizugaki, T. et al. (2010). Wacker-type oxidation of internal olefins using a $PdCl_2$/*N,N*-dimethylacetamide catalyst system under Cu-free reaction conditions. *Angew. Chem. Int. Ed.* 49: 1238–1240.

1942 Morandi, B., Wickens, Z.K., and Grubbs, R.H. (2013). Regioselective Wacker oxidation of internal alkenes: rapid access to functionalized ketones facilitated by cross-metathesis. *Angew. Chem. Int. Ed.* 52: 9751–9754.

1943 Wickens, Z.K., Morandi, B., and Grubbs, R.H. (2013). Aldehyde-selective Wacker-type oxidation of unbiased alkenes enabled by a nitrite Co-catalyst. *Angew. Chem. Int. Ed.* 52: 11257–11260.

1944 Wickens, Z.K., Skakuj, K., Morandi, B., and Grubbs, R.H. (2014). Catalyst-controlled Wacker-type oxidation: facile access to functionalized aldehydes. *J. Am. Chem. Soc.* 136: 890–893.

1945 Guo, J.Y. and Teo, P.L. (2014). Anti-Markovnikov oxidation and hydration of terminal olefins. *J. Chem. Soc., Dalton Trans.* 43: 6952–6964.

1946 Stahl, S.S. (2005). Palladium-catalyzed oxidation of organic chemicals with O_2. *Science* 309: 1824–1826.

1947 Wu, W.Q. and Jiang, H.F. (2012). Palladium-catalyzed oxidation of unsaturated hydrocarbons using molecular oxygen. *Acc. Chem. Res.* 45: 1736–1748.

1948 Scheuermann, M.L. and Goldberg, K.I. (2014). Reactions of Pd and Pt complexes with molecular oxygen. *Chem. Eur. J.* 20: 14556–14568.

1949 Zhu, M.K., Zhao, J.F., and Loh, T.P. (2010). Palladium-catalyzed oxime assisted intramolecular dioxygenation of alkenes with 1 atm of air as the sole oxidant. *J. Am. Chem. Soc.* 132: 6284–6285.

1950 Chuang, G.J., Wang, W.K., Lee, E., and Ritter, T. (2011). A dinuclear Pd catalyst for α-hydroxylation of carbonyls with O_2. *J. Am. Chem. Soc.* 133: 1760–1762.

1951 Cai, X., Majumdar, S., Fortman, G.C. et al. (2011). Oxygen binding to [Pd(L)(L')] (L=NHC, L'=NHC or PR3, NHC=N-heterocyclic carbene). Synthesis and structure of a paramagnetic *trans*-[Pd(NHC)2(η^1-O2)2] complex. *J. Am. Chem. Soc.* 133: 1290–1293.

1952 Bergstrom, R.G., Wahl, G.H., and Zollinger, H. (1974). Dediazoniations of arene diazonium ions in homogenous solutions. 5. Reaction of unactivated nitrogen molecules with phenyl cations in 2,2,2-trifluoroethanol. *Tetrahedron Lett.* (34): 2975–2978.

1953 Hashida, Y., Landells, R.G.M., Lewis, G.E. et al. (1978). Dediazoniation of arenediazonium ions in homogeneous solution. 11. Evidence for 2 intermediates in reaction of 2,4,6-trimethylbenzenediazonium ion in 2,2,2-trifluoroethanol. *J. Am. Chem. Soc.* 100 (9): 2816–2823.

1954 Szele, I. and Zollinger, H. (1978). Dediazoniation of arenediazonium ions in homogeneous solution. 10. Solvent effects on reactions of benzenediazonium and 2,4,6-trimethylbenzenediazonium ion – evidence for 2 intermediates. *J. Am. Chem. Soc.* 100 (9): 2811–2815.

1955 Szele, I. and Zollinger, H. (1981). Dediazoniation of arenediazonium ions. 20. Secondary deuterium isotope effects in reactions of benzenediazonium tetrafluoroborate. *Helv. Chim. Acta* 64 (8): 2728–2732.

1956 Zollinger, H. (1983). Mechanistic investigations on reactions of molecular nitrogen with organic species. *Pure Appl. Chem.* 55 (2): 401–408.

1957 Ambroz, H.B. and Kemp, T.J. (1979). Aryl cations – new light on old intermediates. *Chem. Soc. Rev.* 8 (3): 353–365.

1958 Meisenheimer, J. (1902). The reactions of aromatic nitro bodies. *Liebigs Ann. Chem.* 323 (1/3): 205–246.

1959 Blaziak, K., Danikiewicz, W., and Makosza, M. (2016). How does nucleophilic aromatic substitution really proceed in nitroarenes? Computational prediction and experimental verification. *J. Am. Chem. Soc.* 138: 7276–7281.

1960 (a) Sanger, F. (1945). The free amino groups of insulin. *Biochem. J.* 39: 507–515. (b) Mottishaw, J.D., Erck, A.R., Kramer, J.H., and Sun, H. (2015). Electrostatic potential maps and natural bond orbital analysis: visualization and conceptualization of reactivity in Sanger's reagent. *J. Chem. Educ.* 92 (11): 1846–1852.

1961 (a) Schlosser, M. and Ruzziconi, R. (2010). Nucleophilic substitutions of nitroarenes and pyridines: new insight and new applications. *Synthesis* 2111–2123. (b) Um, I.-H., Im, L.-R., Kang, J.-S. et al. (2012). Mechanistic assessment of SNAr displacement of halides from 1-halo-2, 4-dinitrobenzenes by selected primary and secondary amines: Brønsted and Mayr analyses. *J. Org. Chem.* 72 (21): 9738–9746.

1962 Buncel, E. and Terrier, F. (2010). Assessing the superelectrophilic dimension through σ-complexation, S_NAr and Diels–Alder reactivity. *Org. Biomol. Chem.* 8: 2285–2308.

1963 Roberts, J.D., Semenow, D.A., Simmons, H.E., and Carlsmith, L.A. (1956). The mechanism of aminations of halobenzenes. *J. Am. Chem. Soc.* 78 (3): 601–611.

1964 Wittig, G. (1965). 1,2-Dehydrobenzene. *Angew. Chem. Int. Ed. Engl.* 4 (9): 731–737.

1965 Bunnett, J.F. and Zahler, R.E. (1951). Aromatic nucleophilic substitution reactions. *Chem. Rev.* 49 (2): 273–412.

1966 Bunnett, J.F. (1958). Mechanism and reactivity in aromatic nucleophilic substitution reactions. *Q. Rev.* 12 (1): 1–16.

1967 Kim, J.K. and Bunnett, J.F. (1970). Evidence for a radical mechanism of aromatic nucleophilic substitution. *J. Am. Chem. Soc.* 92 (25): 7463–7464.

1968 Kim, J.K. and Bunnett, J.F. (1970). Alkali metal promoted aromatic nucleophilic substitution. *J. Am. Chem. Soc.* 92 (25): 7464–7466.

1969 Savéant, J.M. (2000). Electron transfer, bond breaking and bond formation. *Adv. Phys. Org. Chem.* 35: 117–192.

1970 Kornblum, N., Michel, R.E., and Kerber, R.C. (1966). Radical-anions as intermediates in substitution reactions. *J. Am. Chem. Soc.* 88 (23): 5660–5662.

1971 Kornblum, N., Michel, R.E., and Kerber, R.C. (1966). Chain reactions in substitution processes which proceed via radical-anion intermediates. *J. Am. Chem. Soc.* 88 (23): 5662–5663.

1972 Kornblum, N. (1975). Substitution reactions which proceed via radical-anion intermediates. *Angew. Chem. Int. Ed. Engl.* 14 (11): 734–745.

1973 Russell, G.A. and Danen, W.C. (1966). Coupling reactions of 2-nitro-2-propyl anion. *J. Am. Chem. Soc.* 88 (23): 5663–5663.

1974 Greene, J.C., Wolfe, J.F., and Hudlicky, T. (1972). Reaction of dialkali salts of benzoylacetone with 2-chloroquinoline – evidence for an $S_{RN}1$

mechanism in heteroaromatic nucleophilic substitution. *J. Org. Chem.* 37 (20): 3199–3200.
1975 Kornblum, N., Cheng, L., Davies, T.M. et al. (1987). Substitution reactions which proceed via radical-anion intermediates. 29. Electron-transfer substitution reactions – the *para*-nitrocumyl system. *J. Org. Chem.* 52 (2): 196–204.
1976 Bowman, W.R. (1988). Reactivity of substituted aliphatic nitro-compounds with nucleophiles. *Chem. Soc. Rev.* 17 (3): 283–316.
1977 Bunnett, J.F. and Creary, X. (1974). Arylation of arenethiolate Ions by $S_{RN}1$ mechanism – convenient synthesis of diaryl sulfides. *J. Org. Chem.* 39 (21): 3173–3174.
1978 Rossi, R.A., Pierini, A.B., and Penenory, A.B. (2003). Nucleophilic substitution reactions by electron transfer. *Chem. Rev.* 103: 71–167.
1979 Bunnett, J.F. (1993). Some novel concepts in aromatic reactivity. *Tetrahedron* 49: 4477–4484.
1980 Soumillion, J.-P. (1993). Photoinduced electron transfer employing organic anions. In: *Photoinduced electron transfer employing organic anions*, Topics in Current Chemistry, vol. 168 (ed. J. Mathay), 93–141. Heidelberg: Springer-Verlag.
1981 Bunnett, J.F. and Singh, P. (1981). $S_{RN}1$ reactions of *ortho*-dibromobenzene with ketone enolate ions. *J. Org. Chem.* 46 (24): 5022–5025.
1982 Mangion, D. and Arnold, D.R. (2002). Photochemical nucleophile – olefin combination, aromatic substitution reaction. Its synthetic development and mechanistic exploration. *Acc. Chem. Res.* 35 (5): 297–304.
1983 Rossi, R.A. (2005). Photoinduced aromatic substitution reactions. In: *Synthetic Organic Photochemistry*, Molecular and Supramolecular Photochemistry, vol. 12 (ed. A.G. Griesbeck and J. Mattay), 495–527. New York: Marcel Dekker.
1984 Salmoria, G.V., Dall'Oglio, E., and Zucco, C. (1998). Aromatic nucleophilic substitutions under microwave irradiation. *Tetrahedron Lett.* 39: 2471–2474.
1985 Cherng, Y.J. (2000). Efficient nucleophilic substitution reaction of aryl halides with amino acids under focused microwave irradiation. *Tetrahedron* 56: 8287–8289.
1986 Chaouchi, M., Loupy, A., Marque, S., and Petit, A. (2002). Solvent-free microwave-assisted aromatic nucleophilic substitution – synthesis of aromatic ethers. *Eur. J. Org. Chem.* 1278–1283.
1987 Xu, Z.B., Lu, Y., and Guo, Z.R. (2003). An efficient and fast procedure for the preparation of 2-nitrophenylamines under microwave conditions. *Synlett* 564–566.
1988 Liu, C.Z., Zang, X.F., Yu, B.H. et al. (2011). Microwave-promoted TBAF-catalyzed S_NAr reaction of aryl fluorides and ArSTMS: an efficient synthesis of unsymmetrical diaryl thioethers. *Synlett* 1143–1148.
1989 Diness, F. and Fairlie, D.P. (2012). Catalyst-free *N*-arylation using unactivated fluorobenzenes. *Angew. Chem. Int. Ed.* 51: 8012–8016.
1990 Soria-Castro, S.M., Caminos, D.A., and Penenory, A.B. (2014). An expedient route to heterocycles through α-arylation of ketones and arylamides by microwave induced thermal $S_{RN}1$ reactions. *RSC Adv.* 4: 17490–17497.
1991 Savéant, J.M. (1990). Single electron transfer and nucleophilic substitution. *Adv. Phys. Org. Chem.* 26: 1–130.
1992 Savéant, J.M. (1993). Electron-transfer, bond breaking, and bond formation. *Acc. Chem. Res.* 26 (9): 455–461.
1993 Andrieux, C.P., Robert, M., Saeva, F.D., and Savéant, J.M. (1994). Passage from concerted to stepwise dissociative electron-transfer as a function of the molecular structure and of the energy of the incoming electron – electrochemical reduction of aryldialkyl sulfonium cations. *J. Am. Chem. Soc.* 116 (17): 7864–7871.
1994 Speiser, B. (1996). Electron transfer and chemical reactions – stepwise or concerted? On the competition between nucleophilic substitution and electron transfer. *Angew. Chem. Int. Ed. Engl.* 35 (21): 2471–2474.
1995 Sperotto, E., van Klink, G.P.M., van Koten, G., and de Vries, J.G. (2010). The mechanism of the modified Ullmann reaction. *J. Chem. Soc.; Dalton Trans.* 39: 10338–10351.
1996 Ullmann, F. and Bielecki, J. (1901). Synthesis in the biphenyl series. (I. Announcement). *Ber. Dtsch. Chem. Ges.* 34: 2174–2185.
1997 Ullmann, F. (1903). On a new formation-manner of diphenyl-amine derivatives. [Preliminary announcement.]. *Ber. Dtsch. Chem. Ges.* 36: 2382–2384.
1998 Ullmann, F. and Sponagel, P. (1905). Concerning the phenylisation of phenolene. *Ber. Dtsch. Chem. Ges.* 38: 2211–2212.
1999 Goldberg, I. (1906). Über Phenylirung bei Gegenwart von Kupfer als Katalysator. *Ber. Dtsch. Chem. Ges.* 39: 1691–1692.
2000 Lindley, J. (1984). Copper-assisted nucleophilic substitution of aryl halogen. *Tetrahedron* 40: 1433–1456.
2001 Kunz, K., Scholz, U., and Ganzer, D. (2003). Renaissance of Ullmann and Goldberg

reactions – progress in Cu-catalyzed C-N-, C-O- and C-S-coupling. *Synlett* 2428–2439.

2002 Sperotto, E., de Vries, J.G., van Klink, G.P.M., and van Koten, G. (2007). Ligand-free Cu(I) catalyzed N- and O-arylation of aryl halides. *Tetrahedron Lett.* 48: 7366–7370.

2003 Evano, G., Theunissen, C., and Pradal, A. (2013). Impact of Cu-catalyzed cross-coupling reactions in natural product synthesis: the emergence of new retrosynthetic paradigms. *Nat. Prod. Rep.* 30: 1467–1489.

2004 Wolfe, J.P. and Buchwald, S.L. (1997). Nickel-catalyzed amination of aryl chlorides. *J. Am. Chem. Soc.* 119: 6054–6058.

2005 Park, N.H., Teverovskiy, G., and Buchwald, S.L. (2014). Development of an air-stable Ni precatalyst for the amination of aryl chlorides, sulfamates, mesylates, and triflates. *Org. Lett.* 16: 220–223.

2006 Martin, A.R., Makida, Y., Meiries, S. et al. (2013). Enhanced activity of [Ni(NHC)CpCl] complexes in arylamination catalysis. *Organometallics* 32: 6265–6270.

2007 Lavoie, C.M., MacQueen, P.M., and Stradiotto, M. (2016). Nickel-catalyzed *N*-arylation of primary amides and lactams with activated (hetero)aryl electrophiles. *Chem. Eur. J.* 22: 18752–18755.

2008 Lavoie, C.M., MacQueen, P.M., Rotta-Loria, N.L. et al. (2016). Challenging Ni-catalysed amine arylations enabled by tailored ancillary ligand design. *Nat. Commun.* 7: 11073.

2009 Ma, D.W., Zhang, Y.D., Yao, J.C. et al. (1998). Accelerating effect induced by the structure of α-amino acid in the Cu-catalyzed coupling reaction of aryl halides with α-amino acids. Synthesis of benzolactam-V8. *J. Am. Chem. Soc.* 120: 12459–12467.

2010 Ma, D.W. and Xia, C.F. (2001). CuI-catalyzed coupling reaction of β-amino acids or esters with aryl halides at temperature lower than that employed in the normal Ullmann reaction. Facile synthesis of SB-214857. *Org. Lett.* 3: 2583–2586.

2011 Goodbrand, H.B. and Hu, N.X. (1999). Ligand-accelerated catalysis of the Ullmann condensation: application to hole conducting triarylamines. *J. Org. Chem.* 64: 670–674.

2012 Evindar, G. and Batey, R.A. (2003). Copper- and palladium-catalyzed intramolecular aryl guanidinylation: an efficient method for the synthesis of 2-aminobenzimidazoles. *Org. Lett.* 5: 133–136.

2013 Lang, F.R., Zewge, D., Houpis, I.N., and Volante, R.P. (2001). Amination of aryl halides using Cu catalysis. *Tetrahedron Lett.* 42: 3251–3254.

2014 Kwong, F.Y., Klapars, A., and Buchwald, S.L. (2002). Copper-catalyzed coupling of alkylamines and aryl iodides: an efficient system even in an air atmosphere. *Org. Lett.* 4: 581–584.

2015 Klapars, A., Antilla, J.C., Huang, X.H., and Buchwald, S.L. (2001). A general and efficient copper catalyst for the amidation of aryl halides and the *N*-arylation of nitrogen heterocycles. *J. Am. Chem. Soc.* 123: 7727–7729.

2016 Wolter, M., Klapars, A., and Buchwald, S.L. (2001). Synthesis of *N*-aryl hydrazides by Cu-catalyzed coupling of hydrazides with aryl iodides. *Org. Lett.* 3: 3803–3805.

2017 Klapars, A., Huang, X.H., and Buchwald, S.L. (2002). A general and efficient Cu catalyst for the amidation of aryl halides. *J. Am. Chem. Soc.* 124: 7421–7428.

2018 Antilla, J.C., Klapars, A., and Buchwald, S.L. (2002). The Cu-catalyzed *N*-arylation of indoles. *J. Am. Chem. Soc.* 124: 11684–11688.

2019 Antilla, J.C., Baskin, J.M., Barder, T.E., and Buchwald, S.L. (2004). Copper-diamine-catalyzed *N*-arylation of pyrroles, pyrazoles, indazoles, imidazoles, and triazoles. *J. Org. Chem.* 69: 5578–5587.

2020 Strieter, E.R., Blackmond, D.G., and Buchwald, S.L. (2005). The role of chelating diamine ligands in the Goldberg reaction: a kinetic study on the Cu-catalyzed amidation of aryl iodides. *J. Am. Chem. Soc.* 127: 4120–4121.

2021 Strieter, E.R., Bhayana, B., and Buchwald, S.L. (2009). Mechanistic studies on the Cu-catalyzed *N*-arylation of amides. *J. Am. Chem. Soc.* 131: 78–88.

2022 Kwong, F.Y. and Buchwald, S.L. (2003). Mild and efficient Cu-catalyzed amination of aryl bromides with primary alkylamines. *Org. Lett.* 5: 793–796.

2023 Shafir, A. and Buchwald, S.L. (2006). Highly selective room-temperature Cu-catalyzed C—N coupling reactions. *J. Am. Chem. Soc.* 128: 8742–8743.

2024 Altman, R.A. and Buchwald, S.L. (2006). 4,7-Dimethoxy-1,10-phenanthroline: an excellent ligand for the Cu-catalyzed *N*-arylation of imidazoles. *Org. Lett.* 8: 2779–2782.

2025 Altman, R.A., Koval, E.D., and Buchwald, S.L. (2007). Copper-catalyzed *N*-arylation of imidazoles and benzimidazoles. *J. Org. Chem.* 72: 6190–6199.

2026 Deldaele, C. and Evano, G. (2016). Room-temperature practical Cu-catalyzed amination of aryl iodides. *ChemCatChem* 8: 1319–1328.

2027 Jerphagnon, T., van Klink, G.P.M., de Vries, J.G., and Van Koten, G. (2005). Aminoarenethiolate-Cu(I)-catalyzed amination of aryl bromides. *Org. Lett.* 7: 5241–5244.

2028 Monnier, F. and Taillefer, M. (2009). Catalytic C—C, C—N, and C—O Ullmann-type coupling reactions. *Angew. Chem. Int. Ed.* 48: 6954–6971.

2029 Sperotto, E., van Klink, G.P.M., de Vries, J.G., and van Koten, G. (2010). Aminoarenethiolato-Cu(I) as (pre-)catalyst for the synthesis of diaryl ethers from aryl bromides and sequential C—O/C—S and C—N/C—S cross coupling reactions. *Tetrahedron* 66: 9009–9020.

2030 Hennessy, E.J. and Buchwald, S.L. (2002). A general and mild Cu-catalyzed arylation of diethyl malonate. *Org. Lett.* 4: 269–272.

2031 Yip, S.F., Cheung, H.Y., Zhou, Z.Y., and Kwong, F.Y. (2007). Room-temperature Cu-catalyzed α-arylation of malonates. *Org. Lett.* 9: 3469–3472.

2032 Xie, X.A., Cai, G.R., and Ma, D.W. (2005). CuI/L-proline-catalyzed coupling reactions of aryl halides with activated methylene compounds. *Org. Lett.* 7: 4693–4695.

2033 Maiti, D. and Buchwald, S.L. (2009). Orthogonal Cu- and Pd-based catalyst systems for the O- and N-arylation of aminophenols. *J. Am. Chem. Soc.* 131: 17423–17429.

2034 Beletskaya, I.P. and Cheprakov, A.V. (2012). The complementary competitors: palladium and copper in C—N cross-coupling reactions. *Organometallics* 31: 7753–7808.

2035 Cohen, T., Wood, J., and Dietz, A.G. (1974). Organocopper intermediates in exchange-reaction of aryl halides with salts of Cu(I) – possible role of Cu(III). *Tetrahedron Lett.* 3555–3558.

2036 Sung, S., Sale, D., Braddock, D.C. et al. (2016). Mechanistic studies on the Cu-catalyzed *N*-arylation of alkylamines promoted by organic soluble ionic bases. *ACS Catal.* 6: 3965–3974.

2037 Tye, J.W., Weng, Z., Johns, A.M. et al. (2008). Copper complexes of anionic nitrogen ligands in the amidation and imidation of aryl halides. *J. Am. Chem. Soc.* 130: 9971–9983.

2038 Giri, R. and Hartwig, J.F. (2010). Cu(I)-amido complexes in the Ullmann reaction: reactions of Cu(I)-amido complexes with iodoarenes with and without autocatalysis by CuI. *J. Am. Chem. Soc.* 132: 15860–15863.

2039 Zhang, S.L., Liu, L., Fu, Y., and Guo, Q.X. (2007). Theoretical study on Cu(I)-catalyzed cross-coupling between aryl halides and amides. *Organometallics* 26: 4546–4554.

2040 Zhang, S.L. and Ding, Y.Q. (2011). Theoretical study on mechanism of Cu(I)-catalyzed cross-coupling between aryl halides and alkylamines. *Organometallics* 30: 633–641.

2041 Liu, X.F., Zhang, S.L., and Ding, Y.Q. (2012). Synthesis, crystal structures and reactivity of Cu(I) amidate complexes with aryl halides: insight into Cu(I)-catalyzed Goldberg reaction. *J. Chem. Soc.; Dalton Trans.* 41: 5897–5902.

2042 Kaddouri, H., Vicente, V., Ouali, A. et al. (2009). Copper-catalyzed arylation of nucleophiles by using butadienylphosphines as ligands: mechanistic insight. *Angew. Chem. Int. Ed.* 48: 333–339.

2043 Casitas, A., King, A.E., Parella, T. et al. (2010). Direct observation of Cu(I)/Cu(III) redox steps relevant to Ullmann-type coupling reactions. *Chem. Sci.* 1: 326–330.

2044 King, A.E., Huffman, L.M., Casitas, A. et al. (2010). Copper-catalyzed aerobic oxidative functionalization of an arene C—H bond: evidence for an aryl-Cu(III) intermediate. *J. Am. Chem. Soc.* 132: 12068–12073.

2045 Hey, D.H. and Waters, W.A. (1937). Some organic reactions involving the occurrence of free radicals in solution. *Chem. Rev.* 21 (1): 169–208.

2046 Arai, S., Yamagishi, T., Ototake, S., and Hida, M. (1977). Ullmann condensation reaction of haloanthraquinone derivatives with amines in aprotic solvents. *Bull. Chem. Soc. Jpn.* 50: 547–548.

2047 Arai, S., Hida, M., Yamagishi, T., and Ototake, S. (1977). Ullmann condensation reaction of haloanthraquinone derivatives with amines in aprotic solvents. 2. Presence of an induction period in condensation with 2-aminoethanol. *Bull. Chem. Soc. Jpn.* 50: 2982–2985.

2048 Arai, S., Hida, M., and Yamagishi, T. (1978). Ullmann condensation reaction of halo-anthraquinone derivatives with amines in aprotic solvents. 3. Formation and role of Cu(II) species in condensation with 2-aminoethanol by Cu(I) catalyst. *Bull. Chem. Soc. Jpn.* 51: 277–282.

2049 Couture, C. and Paine, A.J. (1985). Mechanisms and models for homogeneous Cu-mediated ligand exchange reactions of the type – CuNu + ArX=ArNu + CuX. *Can. J. Chem.* 63: 111–120.

2050 Aalten, H.L., van Koten, G., Grove, D.M. et al. (1989). The Cu-catalyzed reaction of sodium methoxide with aryl bromides – a mechanistic study leading to a facile synthesis of anisole derivatives. *Tetrahedron* 45: 5565–5578.

2051 Bacon, R.G.R. and Hill, H.A.O. (1964). Metal ions + complexes in organic reactions. I. Substitution reactions between aryl halides + cuprous salts in organic solvents. *J. Chem. Soc.* 1097–1107.

2052 Bacon, R.G.R. and Hill, H.A.O. (1964). Metal ions + complexes in organic reactions. 2. Substitution reactions of aryl halides promoted by cuprous oxide. *J. Chem. Soc.* 1108–1112.

2053 Bacon, R.G.R. and Hill, H.A.O. (1964). Metal ions + complexes in organic reactions. 3. Reduction of 1-bromonaphthalene in some reaction systems containing Cu. *J. Chem. Soc.* 1112–1119.

2054 Litvak, V.V. and Shein, S.M. (1974). Nucleophilic substitution in aromatic series. 50. Catalysis by CuBr in reaction of substituted bromobenzenes with sodium methylate. *Zh. Org. Khim.* 10: 2360–2364.

2055 Weingarten, H. (1964). Mechanism of Ullmann condensation. *J. Org. Chem.* 29: 3624–3626.

2056 Jones, G.O., Liu, P., Houk, K.N., and Buchwald, S.L. (2010). Computational explorations of mechanisms and ligand-directed selectivities of Cu-catalyzed Ullmann-type reactions. *J. Am. Chem. Soc.* 132: 6205–6213.

2057 Zhang, S.L., Zhu, Z.Z., and Ding, Y.Q. (2012). Proposal for halogen atom transfer mechanism for Ullmann *O*-arylation of phenols with aryl halides. *J. Chem. Soc.; Dalton Trans.* 41: 13832–13840.

2058 Rout, L., Saha, P., Jammi, S., and Punniyamurthy, T. (2008). Efficient Cu(I)-catalyzed C—S cross coupling of thiols with aryl halides in water. *Eur. J. Org. Chem.* 640–643.

2059 Sperotto, E., van Klink, G.P.M., de Vries, J.G., and van Koten, G. (2008). Ligand-free Cu-catalyzed C—S coupling of aryl iodides and thiols. *J. Org. Chem.* 73: 5625–5628.

2060 Wolfe, J.P., Wagaw, S., Marcoux, J.F., and Buchwald, S.L. (1998). Rational development of practical catalysts for aromatic carbon-nitrogen bond formation. *Acc. Chem. Res.* 31: 805–818.

2061 Hartwig, J.F. (1998). Carbon-heteroatom bond forming reductive eliminations of amines, ethers, and sulfides. *Acc. Chem. Res.* 31: 852–860.

2062 Hartwig, J.F. (1998). Transition metal catalyzed synthesis of arylamines and aryl ethers from aryl halides and triflates: scope and mechanism. *Angew. Chem. Int. Ed.* 37: 2046–2067.

2063 Muci, A.R. and Buchwald, S.L. (2002). Practical Pd catalysts for C—N and C—O bond formation. In: *Cross-Coupling Reactions: A Practical Guide*, Topics in Current Chemistry, vol. 219 (ed. N. Miyaura), 131–209. Berlin, Heidelberg: Springer-Verlag.

2064 Surry, D.S. and Buchwald, S.L. (2008). Biaryl phosphane ligands in Pd-catalyzed amination. *Angew. Chem. Int. Ed.* 47 (34): 6338–6361.

2065 Hartwig, J.F. (2008). Evolution of a fourth generation catalyst for the amination and thioetherification of aryl halides. *Acc. Chem. Res.* 41: 1534–1544.

2066 Fors, B.P., Krattiger, P., Strieter, E., and Buchwald, S.L. (2008). Water-mediated catalyst preactivation: an efficient protocol for C—N cross-coupling reactions. *Org. Lett.* 10: 3505–3508.

2067 Willis, M.C., Chauhan, J., and Whittingham, W.G. (2005). A new reactivity pattern for vinyl bromides: *cine*-substitution via Pd-catalysed C—N coupling/Michael addition reactions. *Org. Biomol. Chem.* 3: 3094–3095.

2068 Newman, S.G. and Lautens, M. (2010). The role of reversible oxidative addition in selective Pd(0)-catalyzed intramolecular cross-couplings of polyhalogenated substrates: synthesis of brominated indoles. *J. Am. Chem. Soc.* 132: 11416–11417.

2069 Crawford, S.M., Lavery, C.B., and Stradiotto, M. (2013). BippyPhos: a single ligand with unprecedented scope in the Buchwald–Hartwig amination of (hetero) aryl chlorides. *Chem. Eur. J.* 19: 16760–16771.

2070 Reddy, C.V., Kingston, J.V., and Verkade, J.G. (2008). $(t\text{-Bu})_2PN{=}P(i\text{-BuNCH}_2CH_2)_3N$: new efficient ligand for Pd-catalyzed C—N couplings of aryl and heteroaryl bromides and chlorides and for vinyl bromides at room temperature. *J. Org. Chem.* 73: 3047–3062.

2071 Mann, G., Incarvito, C., Rheingold, A.L., and Hartwig, J.F. (1999). Palladium-catalyzed C—O coupling involving unactivated aryl halides. Sterically induced reductive elimination to form the C—O bond in diaryl ethers. *J. Am. Chem. Soc.* 121: 3224–3225.

2072 Willis, M.C., Taylor, D., and Gillmore, A.T. (2004). Palladium-catalyzed intramolecular O-arylation of enolates: application to benzo[b]furan synthesis. *Org. Lett.* 6: 4755–4757.

2073 Milton, E.J., Fuentes, J.A., and Clarke, M.L. (2009). Palladium-catalysed synthesis of aryl-alkyl ethers using alkoxysilanes as nucleophiles. *Org. Biomol. Chem.* 7: 2645–2648.

2074 Beydoun, K. and Doucet, H. (2011). Palladium catalytic system for inhibition of *O*-arylation type reaction and regioselective direct arylation at C_2 of phenols. *Catal. Sci. Technol.* 1: 1243–1249.

2075 Islam, M., Mondal, S., Mondal, P. et al. (2011). Open air *O*-arylation reaction of phenols with aryl halides catalyzed by polymer-anchored Cu(II) complexes. *Transition Met. Chem.* 36: 1–11.

2076 Joshi, H., Sharma, K.N., Sharma, A.K. et al. (2013). Graphene oxide grafted with $Pd_{17}Se_{15}$ nano-particles generated from a single source precursor as a recyclable and efficient catalyst for C—O coupling in *O*-arylation at room temperature. *J. Chem. Soc., Chem. Commun.* 49: 7483–7485.

2077 Torraca, K.E., Huang, X.H., Parrish, C.A., and Buchwald, S.L. (2001). An efficient intermolecular Pd-catalyzed synthesis of aryl ethers. *J. Am. Chem. Soc.* 123: 10770–10771.

2078 Burgos, C.H., Barder, T.E., Huang, X.H., and Buchwald, S.L. (2006). Significantly improved method for the Pd-catalyzed coupling of phenols with aryl halides: understanding ligand effects. *Angew. Chem. Int. Ed.* 45 (26): 4321–4326.

2079 Fernández-Rodríguez, M.A. and Hartwig, J.F. (2010). One-pot synthesis of unsymmetrical diaryl thioethers by Pd-catalyzed coupling of two aryl bromides and a thiol surrogate. *Chem. Eur. J.* 16: 2355–2359.

2080 Lee, J.Y. and Lee, P.H. (2008). Palladium-catalyzed carbon-sulfur cross-coupling reactions with In tri(organothiolate) and its application to sequential one-pot processes. *J. Org. Chem.* 73: 7413–7416.

2081 Sayah, M. and Organ, M.G. (2011). Carbon-sulfur bond formation of challenging substrates at low temperature by using Pd-PEPPSI-*i*-Pent. *Chem. Eur. J.* 17: 11719–11722.

2082 Wang, L., Zhou, W.Y., Chen, S.C. et al. (2012). A highly efficient Pd-catalyzed one-pot synthesis of unsymmetrical aryl alkyl thioethers under mild conditions in water. *Adv. Synth. Catal.* 354: 839–845.

2083 Teverovskiy, G., Surry, D.S., and Buchwald, S.L. (2011). Pd-Catalyzed synthesis of Ar-SCF_3 compounds under mild conditions. *Angew. Chem. Int. Ed.* 50: 7312–7314.

2084 Wu, J., Liu, Y.F., Lu, C.H., and Shen, Q.L. (2016). Palladium-catalyzed difluoromethylthiolation of heteroaryl bromides, iodides, triflates and aryl iodides. *Chem. Sci.* 7: 3757–3762.

2085 Potukuchi, H.K., Spork, A.P., and Donohoe, T.J. (2015). Palladium-catalyzed α-arylation of carbonyls in the de novo synthesis of aromatic heterocycles. *Org. Biomol. Chem.* 13: 4367–4373.

2086 Louie, J., Driver, M.S., Hamann, B.C., and Hartwig, J.F. (1997). Palladium-catalyzed amination of aryl triflates and importance of triflate addition rate. *J. Org. Chem.* 62: 1268–1273.

2087 Tundel, R.E., Anderson, K.W., and Buchwald, S.L. (2006). Expedited Pd-catalyzed amination of aryl nonaflates through the use of microwave-irradiation and soluble organic amine bases. *J. Org. Chem.* 71: 430–433.

2088 Fors, B.P., Watson, D.A., Biscoe, M.R., and Buchwald, S.L. (2008). A highly active catalyst for Pd-catalyzed amination reactions: cross-coupling reactions using aryl mesylates and the highly selective monoarylation of primary amines using aryl chlorides. *J. Am. Chem. Soc.* 130: 13552–13554.

2089 Navarro, O., Kaur, H., Mahjoor, P., and Nolan, S.P. (2004). Cross-coupling and dehalogenation reactions catalyzed by (N-heterocyclic carbene)Pd(allyl)Cl complexes. *J. Org. Chem.* 69: 3173–3180.

2090 Zhang, Y., Lavigne, G., and Cesar, V. (2015). Buchwald-Hartwig amination of (hetero)aryl tosylates using a well-defined N-heterocyclic carbene/Pd(II) precatalyst. *J. Org. Chem.* 80: 7666–7673.

2091 Surry, D.S. and Buchwald, S.L. (2011). Dialkylbiaryl phosphines in Pd-catalyzed amination: a user's guide. *Chem. Sci.* 2: 27–50.

2092 Lundgren, R.J., Peters, B.D., Alsabeh, P.G., and Stradiotto, M. (2010). A P,N-ligand for Pd-catalyzed ammonia arylation: coupling of deactivated aryl chlorides, chemoselective arylations, and room temperature reactions. *Angew. Chem. Int. Ed.* 49: 4071–4074.

2093 Lundgren, R.J., Sappong-Kumankumah, A., and Stradiotto, M. (2010). A highly versatile catalyst system for the cross-coupling of aryl chlorides and amines. *Chem. Eur. J.* 16: 1983–1991.

2094 Zhang, Y., Cesar, V., and Lavigne, G. (2015). Efficient and versatile Buchwald-Hartwig amination of (hetero)aryl chlorides using the

Pd-PEPPSI-i-Pr(NMe_2)$_2$ precatalyst in the presence of carbonate base. *Eur. J. Org. Chem.* 2042–2050.

2095 Maiti, D., Fors, B.P., Henderson, J.L. et al. (2011). Palladium-catalyzed coupling of functionalized primary and secondary amines with aryl and heteroaryl halides: two ligands suffice in most cases. *Chem. Sci.* 2: 57–68.

2096 Navarro, O., Marion, N., Mei, J.G., and Nolan, S.P. (2006). Rapid room temperature Buchwald–Hartwig and Suzuki–Miyaura couplings of heteroaromatic compounds employing low catalyst loadings. *Chem. Eur. J.* 12: 5142–5148.

2097 Marion, N., Navarro, O., Mei, J.G. et al. (2006). Modified (NHC)Pd(allyl)Cl (NHC=N-heterocyclic carbene) complexes for room-temperature Suzuki–Miyaura and Buchwald–Hartwig reactions. *J. Am. Chem. Soc.* 128: 4101–4111.

2098 Marion, N. and Nolan, S.P. (2008). Well-defined N-heterocyclic carbenes-Pd(II) precatalysts for cross-coupling reactions. *Acc. Chem. Res.* 41 (11): 1440–1449.

2099 Chartoire, A., Claver, C., Corpet, M. et al. (2016). Recyclable NHC catalyst for the development of a generalized approach to continuous Buchwald Hartwig reaction and workup. *Org. Process Res. Dev.* 20: 551–557.

2100 Wagner, P., Bollenbach, M., Doebelin, C. et al. (2014). *t*-BuXPhos: a highly efficient ligand for Buchwald–Hartwig coupling in water. *Green Chem.* 16: 4170–4178.

2101 Salomé, C., Wagner, P., Bollenbach, M. et al. (2014). Buchwald–Hartwig reactions in water using surfactants. *Tetrahedron* 70: 3413–3421.

2102 Chartoire, A., Boreux, A., Martin, A.R., and Nolan, S.P. (2013). Solvent-free aryl amination catalysed by [Pd(NHC)] complexes. *RSC Adv.* 3: 3840–3843.

2103 Topchiy, M.A., Asachenko, A.F., and Nechaev, M.S. (2014). Solvent-free Buchwald–Hartwig reaction of aryl and heteroaryl halides with secondary amines. *Eur. J. Org. Chem.* 3319–3322.

2104 Kosugi, M., Kameyama, M., and Migita, T. (1983). Palladium-catalyzed aromatic amination of aryl bromides with *N,N*-diethylamino-tributyltin. *Chem. Lett.* 927–928.

2105 Boger, D.L. and Panek, J.S. (1984). Palladium(0)-mediated β-carboline synthesis – preparation of the CDE ring system of lavendamycin. *Tetrahedron Lett.* 25: 3175–3178.

2106 Paul, F., Patt, J., and Hartwig, J.F. (1994). Palladium-catalyzed formation of carbon-nitrogen bonds – reaction intermediates and catalyst improvements in the hetero cross-coupling of aryl halides and tin amides. *J. Am. Chem. Soc.* 116: 5969–5970.

2107 Guram, A.S. and Buchwald, S.L. (1994). Palladium-catalyzed aromatic aminations with in-situ generated aminostannanes. *J. Am. Chem. Soc.* 116: 7901–7902.

2108 Guram, A.S., Rennels, R.A., and Buchwald, S.L. (1995). A simple catalytic method for the conversion of aryl bromides to arylamines. *Angew. Chem. Int. Ed.* 34: 1348–1350.

2109 Louie, J. and Hartwig, J.F. (1995). Palladium-catalyzed synthesis of arylamines from aryl halides – mechanistic studies lead to coupling in the absence of Sn reagents. *Tetrahedron Lett.* 36: 3609–3612.

2110 Driver, M.S. and Hartwig, J.F. (1995). A rare, low-valent alkylamido complex, a diphenylamido complex, and their reductive elimination of amines by 3-coordinate intermediates. *J. Am. Chem. Soc.* 117: 4708–4709.

2111 Hartwig, J.F., Richards, S., Baranano, D., and Paul, F. (1996). Influences on the relative rates for C—N bond-forming reductive elimination and β-hydrogen elimination of amides. A case study on the origins of competing reduction in the Pd-catalyzed amination of aryl halides. *J. Am. Chem. Soc.* 118: 3626–3633.

2112 Driver, M.S. and Hartwig, J.F. (1997). Carbon-nitrogen bond forming reductive elimination of arylamines from Pd(II)-phosphine complexes. *J. Am. Chem. Soc.* 119: 8232–8245.

2113 Widenhoefer, R.A. and Buchwald, S.L. (1996). Halide and amine influence in the equilibrium formation of Pd-tris(*o*-tolyl)phosphine mono(amine) complexes from Pd-aryl halide dimer. *Organometallics* 15: 2755–2763.

2114 Maimone, T.J. and Buchwald, S.L. (2010). Pd-catalyzed *O*-arylation of ethyl acetohydroximate: synthesis of *O*-arylhydroxylamines and substituted benzofurans. *J. Am. Chem. Soc.* 132: 9990–9991.

8

Transition-metal-catalyzed C—C bond forming reactions

8.1 Introduction

Chemical industry converts raw materials (coal, petroleum, natural gas, wood distillates, and compounds isolated from the biomass by fermentation or by a chemical process) into products of high added value (drugs, crop protection agents, dyes, fragrances, isolation and construction materials, fabrics, coatings, and many other modern utilities) either by adding functions to them (e.g. selective C—H and C=C bond oxidation) and/or by generating C—C bonds (single, double or triple). Classically, for the formation of a single C—C bond, chemists have relied upon displacement reactions by matching a nucleophile with an electrophile (e.g. Nu—H + R—X + base → Nu—R + base-H^+/X^-, or Nu—metal + R—X → Nu—R + $metal^+/X^-$; R = alkyl, alkenyl, aryl, heteroaryl, and alkynyl) [1] and on electrophilic substitutions (Ar—H + E—X ⇄ Ar—E + HX; e.g. Friedel–Crafts alkylation and acylation) [2]. Catalyzed or photocatalyzed free radical additions to C—C multiple bonds are applied more and more into fine organic synthesis (e.g. R—X (X = I, Br) + BCH=CHA → R—CH(B)—CH(X)—A), not only to generate polymers [3–10]. Aldehyde and ketone olefinations are also powerful methods to create double C=C bonds (Sections 5.3.7 and 5.3.8). These are not atom economical as they produce coproducts. The same can be said for the very much used additions of organometallic reagents to unsaturated systems (Nu—metal + A=B ⇄ Nu—A—B—metal, with A, B = carbon or heteroelement, = is a double or triple bond) [11]. Direct additions of reactants of type R—H or R—X to unsaturated compounds (R—H + A=B ⇄ R—A—B—H, R—X + A=B ⇄ R—A—B—X) are more atom economical [12]. They include the aldol (e.g. R^1CHO + $R^2CH_2COR^3$ ⇄ $R^1CH(OH)—CH_2(R^2)—COR^3$, Section 5.7.6) and related reactions such as the Claisen condensation (e.g. $R^1CH_2COOR^2 + R^3CHO$ ⇄ $R^1CH(COOR^2)—CH(OH)R^3$), the Henry reaction (e.g. $R^1CH_2NO_2 + R^2CHO$ ⇄ $R^1CH(NO_2)—CH(OH)R^2$), the Michael addition (e.g. $R^1CH_2COR^2 + CH_2$=$CHCOR^3$ → $R^1CH(COR^2)—CH_2—CH_2COR^3$), and the Mannich reaction (e.g. $R^1COCH_3 + CH_2O$ + R_2NH → $R^1COCH_2—CH_2—NR_2 + H_2O$) [13]. Under acidic conditions, alkene alkylate arenes (e.g. Ar—H + CH_2=CHR ⇄ ArCH(Me)R) and alkenes (e.g. the alkyl process, Section 8.3.1) and aldehydes hydroxyalkylate alkenes (e.g. the Prins reaction RCH=CH_2 + CH_2=O ⇄ RCH=CH—CH_2OH) [14, 15]. Very much used is the benzoin condensation (e.g. 2RCHO → RC(OH)—COR, Section 7.5.1), the Stetter reaction (e.g. $R^1CHO + R^2CH—CH=CH—COR^3$ $R^1COCH(R^2)CH—CH^2—COR^3$, Section 7.5.2), the Rauhut–Currier reaction (e.g. ArCOCH=CHCOOR + CH_2=CHCOMe → $ArCOCH_2CH(COOR)C(=CH_2)$COMe, Section 7.5.5), the Morita–Baylis–Hillman reaction (e.g. $R^1CHO + R^2CH$=CHA → $R^1CH(OH)$—C(A)=CHR^2, Section 7.5.6) [16], the Roelen hydroformylation of alkenes (RCH=CH_2 + CO + H_2 → RCH_2CH_2CHO + RCH(CHO)Me, Section 8.2.3), the related hydrogenative coupling of dienes or alkynes with carbonyl compounds (e.g. CH_2=CH(R^1)CH=CH_2 + H_2 + R^2CHO → CH_2=CH—C(Me)(R^1)—CH(OH)—R^2, Section 8.3.10), and the hydroxyalkylation of 1,3-diene, allenes, and alkynes through transfer hydrogenation (e.g. H_2C=C=CR^1(R^2) + R^3CH^2OH → CH_2=CH—C(R^1)(R^2)—CH(R^3)OH, Section 8.3.9). All these reactions are formal hydrocarbations of unsaturated compounds. Pericyclic reactions such as cycloadditions (Section 5.3), cheletropic reactions (Section 5.4), sigmatropic rearrangements (Section 5.5), especially the Claisen rearrangement and its variants (Section 5.5.9.2), and the ene-reactions (Section 5.7) are also atom economical. Aerobic didehydrogenative coupling reactions $R^1—H + R^2—H + ½O_2 → R^1—R^2 + H_2O$ are the second best atom-economical process to generate C—C bonds (Section 8.5) [17–22]. All these reactions follow either concerted or nonconcerted mechanisms. For nonconcerted reactions, reactive intermediates can be ion pairs, radicals, diradicals, carbenes, or radical ions. All these reactions can be catalyzed. This chapter

Organic Chemistry: Theory, Reactivity and Mechanisms in Modern Synthesis, First Edition. Pierre Vogel and Kendall N. Houk.

reviews some of the most useful C—C bond forming reactions catalyzed by transition metals, their salts, or complexes (first row: Ti (four valence electrons), V (5), Cr (6), Mn (7), Fe (8), Co (9), Ni (10), Cu (11), and Zn (12); second row: Zr (4), Nb (5), Mo (6), Ru (8), Rh (9), Pd (10), and Ag (11); and third row: Hf (4), Ta (5), W (6), Re (7), Os (8), Ir (9), Pt (10), and Au (11)) [23]. A few examples of reactions catalyzed by rare earth metal derivatives (Sc, Y and, the lanthanides) will also be given.

8.2 Organic compounds from carbon monoxide

Carbon monoxide is obtained by partial combustion of coal and carbon-containing materials (fuels, wood, biomass, and municipal waste), by dry reforming of methane ($CO_2 + CH_4 \rightleftarrows 2CO + 2H_2$) [24], by decarbonylation of aldehydes ($RCHO \rightarrow R{-}H + CO$) such as 5-hydroxymethylfurfural and furfural obtained from the biomass [25, 26], or by retro hydroformylation (e.g. $R{-}CH_2CH_2CHO \rightleftarrows RCH{=}CH_2 + H_2 + CO$) [27]. At 800 °C, coke and CO_2 equilibrate with CO (Boudouard reaction: $C + CO_2 \rightleftarrows 2CO$). Water reacts with hot coke in the water gas shift reaction giving the water gas ($C + H_2O \rightleftarrows H_2 + CO$). As natural gas, CH_4, is one of the most abundant, low-cost, carbon-based feed stocks available, efforts have been made to convert it directly into methanol, methyl esters, acetic acid [28], and H_2 through steam reforming ($CH_4 + 2H_2O \rightleftarrows CO_2 + 4H_2$), today the major source of this gas [29].

Carbon monoxide is a colorless, odorless, and tasteless gas that is slightly less dense than air. It is highly toxic to animals and humans (>35 ppm). It is also produced in animal metabolism in low quantities. In the atmosphere, it plays a role in the formation of ground-level ozone. Liquefaction of coal produce H_2 and CO (syngas), which can be used in a large number of industrial processes [30, 31]. Carbon monoxide can be seen as oxomethylene ($:CH_2 \Rightarrow :C{=}O:$), a nucleophilic carbene ($:C{=}O: \leftrightarrow {}^-C{\equiv}O^+$). It can thus react as a carbene (see the cheletropic reactions, Section 5.4.3, the UV irradiation ketene produces $CO + :CH_2$, an example of Norrish type I scission, Section 6.4.1) and/or as a nucleophile (see e.g. the Koch–Haaf carbonylation: [32, 33] under aqueous acidic conditions, tertiary alcohols react with CO providing the corresponding carboxylic acids *t*-alkyl-COOH; e.g. $t\text{-BuOH} + CO + H_3O^+ \rightleftarrows [t\text{-Bu}^+ + C{=}O + 2H_2O] \rightleftarrows [t\text{-Bu}{-}C{\equiv}O^+ + 2H_2O] \rightleftarrows t\text{-BuCOOH} + H_3O^+$) [34, 35]. Carbon monoxide is isolobal with carbenes, nitrenes, and 16 valence electron transition monometal complexes (Section 7.7).

8.2.1 Fischer–Tropsch reactions

The conversion of coal [30] and biomass [36–38] into liquid fuels and other chemicals is often based on the Fischer–Tropsch reactions [36]. Hydrogenation of CO into methane + water is thermodynamically favorable ($\Delta_r H° = -49.2\,\text{kcal mol}^{-1}$, $\Delta_r G° = -33.5\,\text{kcal mol}^{-1}$). The first example, methanation catalyzed by nickel ($CO + 3H_2 \rightarrow CH_4 + H_2O$), was reported by Sabatier and Senderens [39, 40]. In 1926, Fischer and Tropsch showed that linear alkenes and alkanes (as well as some oxygenates: alcohols, aldehydes, and ketones) are formed at 200–300 °C and under atmospheric pressure over Co or Fe catalysts (reactions (8.1) and (8.2)) [41, 42].

$$nCO + (2n+1)H_2 \underset{200\text{–}300\,°C}{\overset{\text{cat.}}{\rightleftarrows}} C_nH_{2n+2} + nH_2O \quad (8.1)$$

$$2nCO + (n+1)H_2 \underset{250\text{–}300\,°C}{\overset{\text{cat.}}{\rightleftarrows}} C_nH_{2n+2} + nCO_2 \quad (8.2)$$

As syngas ($CO + H_2$) is readily available, the Fischer–Tropsch process became industrially important for economies that had good supplies of cheap coal, but which lacked oil [43–46]. Iron catalysts give mainly linear alkenes and oxygenates, whereas cobalt catalysts give mostly linear alkanes. Ruthenium, one of the most active catalysts, gives high-molecular-weight hydrocarbons [47]. Rhodium catalysts make significant amounts of oxygenates in addition to hydrocarbons [48–57], whereas nickel gives mainly methane [58]. Zeolites impregnated with cobalt salts are very much studied catalysts [59–62]. Under conditions of low pressure and low temperature (≤200 °C), the major primary hydrocarbons tend to be alk-1-enes. These are hydrogenated into alkanes or isomerized into internal olefins in subsequent steps. The "primary" olefins can also be incorporated in new polymerization chains to form further secondary products. Fischer and Tropsch postulated the initial formation of surface carbides that react with hydrogen to give methylenes, which in turn polymerize. In the 1970s, X-ray studies of complexes containing organic ligands bonded to metals offered some structural models for the postulated intermediates, which, in turn, led to the first molecular hypotheses [63, 64].

A variety of labeling experiments [65–68], surface [69], and organometallic studies confirmed that, in the first steps, coordinated (chemisorbed) CO is transformed into a type of surface carbide [70–72].

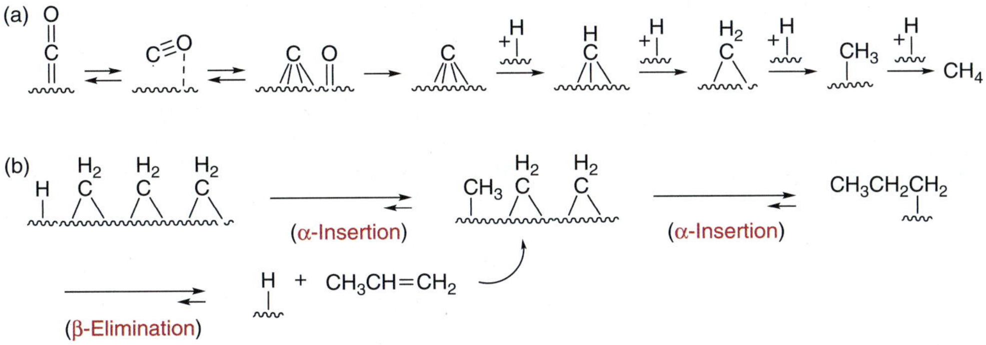

Scheme 8.1 (a) Representation of the adsorption and cleavage of CO and the stepwise hydrogenation of surface carbide giving methylene and other species. (b) Brady and Pettit alkyl mechanism for the polymerization of surface methylenes to surface alkyls.

This carbide ($C \equiv M_x$, $x = 1, 2, 3, 4$) is hydrogenated into surface methylidyne ($HC{\equiv}M_y$, $y = 1, 2, 3$), to surface methylene ($H_2C{=}M_z$, $z = 1, 2, 3$), and then to surface methyl (e.g. Me—M, M—Me—M), and finally, into methane (Scheme 8.1a). The formation of linear hydrocarbons involves a stepwise polymerization of methylenes [73–76]. A mechanism (Scheme 8.1b) by which methylenes polymerize to give alk-1-enes has been proposed by Brady and Pettit [75, 76]. It is called the alkyl mechanism in which polymerization of surface methylenes is initiated by a surface hydride (or methyl); it was proposed to proceed by coupling between formally sp^3-hybridized carbon atoms. Chain growth is terminated by β-elimination of hydride from the growing alkyl chain on the surface [63, 64, 77]. By introducing $^{13}CH_2{=}^{13}CHBr$, $(^{13}CH_2{=}^{13}CH)_4Si$, and $^{13}CH_2{=}^{13}CH_2$ to the Fischer–Tropsch reaction mixture, Maitlis and coworkers found that vinyl ($^{13}C_2$) from the three probes was incorporated into the alkanes and alkenes, but that ethyl from EtBr or Et_4Si was not incorporated [78–80]. There were no products formed incorporating only one ^{13}C atom, showing that cleavage of the $^{13}C_2$ vinyl initiator to C_1 fragments does not occur. Furthermore, while $^{13}C_2$ was incorporated into alkane and alkene products, there was no evidence for any ^{13}C incorporation into oxygenates (MeOH, CH_3CHO, and EtOH). This suggested that vinyl, not ethyl, is an integral part of the surface oligomerization mechanism (Scheme 8.2: the alkenyl mechanism). In the first step of the Fischer–Tropsch process, H_2 is activated by coordination to one, two of three metal atoms. With the help of hydride, CO is deoxygenated giving C—M species including carbide ($C \equiv M_x$), methylidyne ($HC{\equiv}M_y$), methylene ($H_2C{=}M_z$), and methyl (Me—M, M—Me—M) groups. The detailed mechanism of the conversion of $O{=}C{=}(M)_x$ into $HC{\equiv}M_x$ is not known. It can be seen as one hydride and a proton attacking the oxygen center with subsequent H_2O elimination. Then, α-insertion of a hydride leads to $HC{\equiv}M_y$. Alternatively, hydride addition and protonation to $O{=}C{=}M_y$ ($y = 1, 2, 3$) generates $HO{-}CH{=}M_y$ (C=O hydrogenation). Quantum mechanical calculations suggested that the hydride insertion into (Co surface)=C=O gives M—CHO and $M(CH_2O)$ [81]. A subsequent metal-assisted ionization of HO^- furnishes $HC{\equiv}M_y$. Hydride α-insertion into $HC{\equiv}M_y$ gives $H_2C{=}M_z$, then $Me{-}M_z$, and finally CH_4. Combination of $C \equiv M_x$ and $H_2C{=}M_z$ forms a vinylidene $CH_2{=}C{=}M_z$ species; combination of vicinal or germinal $HC{\equiv}M_y$ and $H_2C{=}M_z$ generates vinylmetal species $CH_2{=}CH{-}M_z$. Models for this reaction exist in organometallic chemistry, e.g. the formation of trimetallic alkenyl complex **2** by reaction of (μ-methylene)diruthenium **1** with dicarbonyl(cyclopentadienyl)(ethylidyne)tungsten (reaction (8.3)) [82].

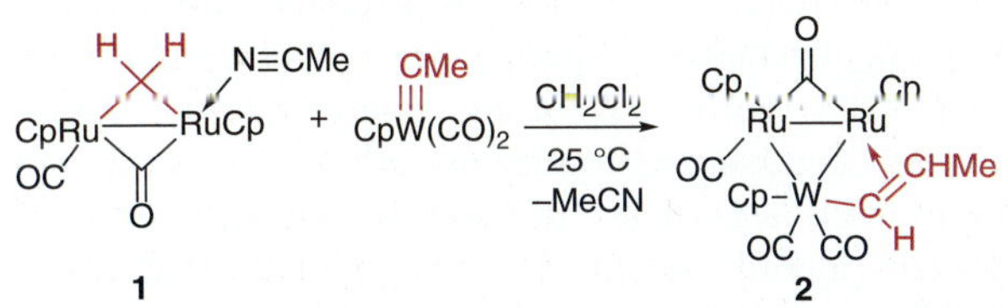

(8.3)

The second step of the Fischer–Tropsch process is the oligomerization that is carried by the surface vinyl (and alkenyl) species reacting with the surface methylenes (Scheme 8.2). The propagation is the coupling of a sp^2(alkenyl carbon: $RCH{=}M_z$) with a sp^3(methylene carbon: $M{-}CH_2{-}M$) carbon that is an easier process than $sp^3(C){-}sp^3(C)$ coupling. This coupling engenders an (allyl)metal species that undergoes an 1,3-hydrogen shift to generate a (prop-1-en-1-yl)metal moiety (MeCH=CH—M) that undergoes, on its turn, the next insertion into a methylene=M_2 species ($M{-}CH_2{-}M$), and so on.

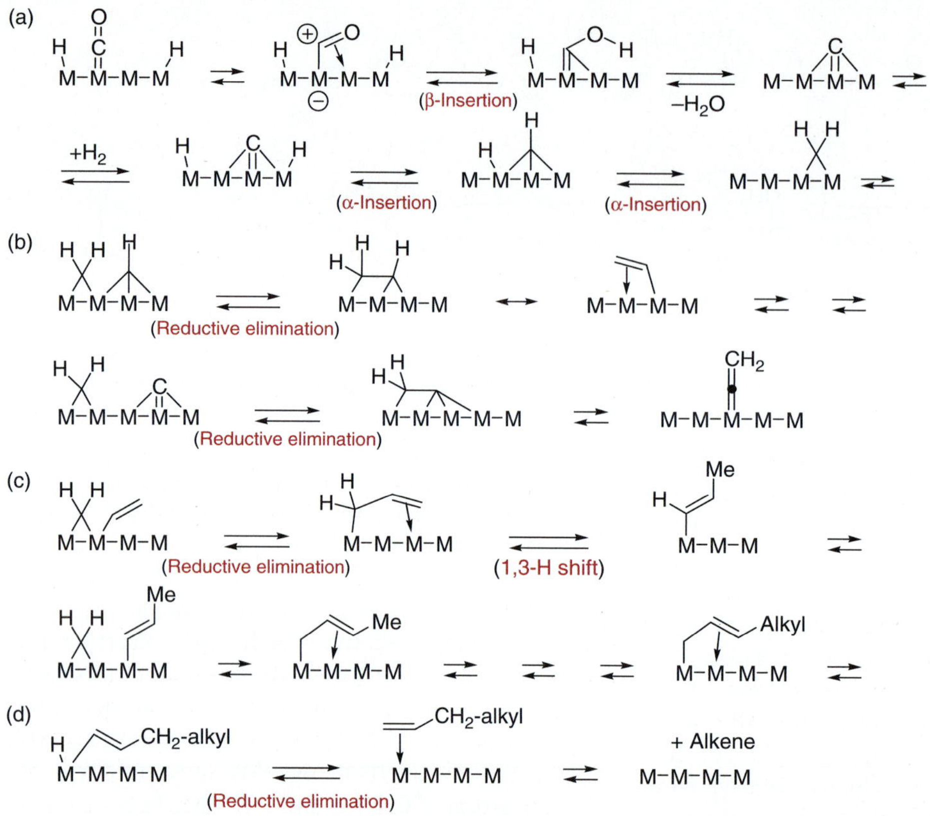

Scheme 8.2 The alkenyl mechanism for the polymerization of surface methylenes to alkenes. (a) Formation of C_1 species; (b) formation of C_2 species; (c) polymerization; and (d) chain termination.

The final step is not, as in the alkyl mechanism (Scheme 8.1b), a β-hydrogen elimination providing a terminal alkene, but a reductive elimination from R—CH=CH—M—H with the formation of the products (terminal alkene: $RCH{=}CH_2 + M$). The rate-determining step is probably the formation of surface-C_2 species (e.g. a π-complex of $RCH{=}CH_2$ with 1, 2, 3, etc., metal atoms). Under Co/Al_2O_3-catalyzed conditions, it has been found that long-chain acetylenes initiate the Fischer–Tropsch process already at 150 °C, a temperature that is too low for normal Fischer–Tropsch reactions [83]. Labeling experiments suggested that oxygenates such as EtOH arise form CO but not via methylene=M_2 [64].

8.2.2 Carbonylation of methanol

Acetic acid (ethanoic acid) is an important bulk commodity chemical with world production capacity of c. 10^7 tons/year [84–86]. Bacterial fermentation produces 25% and catalytic carbonylation of methanol (reaction (8.4)) produces 75% of it. During the 1960s, BASF developed a CoI_2-based catalyst for this reaction [87, 88], which required high pressure (680 atm) and high temperature (250 °C) [89]. It was supplanted by the Monsanto process in 1970 in which the active catalytic species is square-planar *cis*-$[Rh(CO)_2I_2]^-$ formed in situ by reaction of a Rh compound such as $RhCl_3{\cdot}(H_2O)_3$, $RhCl(CO)(PPh_3)_2$, $[Rh(CO)_2Cl_2]_2$, and 0.05 M HI, MeI, and $CaI_2{\cdot}(H_2O)_3$ (Scheme 8.3) [90]. The reaction has a selectivity of 99% in AcOH and does not require high pressure (150–200 °C, 30–60 atm) [91]. It was also observed that $IrCl_3(H_2O)_3$, $Ir_4(CO)_{12}$, and $Ir(CO)(PPh_3)_2Cl$ gave similar results as rhodium catalysts. Forster showed that $[Rh(CO)_2Cl]_2$ reacts with $[PhNMe_3]^+/I^-$ and MeI to give **4** [92]. Vacuum distillation of **5** at 200 °C liberates MeCOI and **3**. On bubbling CO through a solution of **5** at 20 °C, **6** forms, which eliminates MeCOI that then reacts with H_2O to give AcOH and HI. Methanol, on its turn, reacts with HI to generate H_2O and MeI. Complex **7** decomposes slowly at 20 °C giving **3** (Scheme 8.3). The rate-determining step of this process is the oxidative addition of MeI to $[Rh(CO)_2I_2]^-$. This is confirmed by the kinetics of the overall carbonylation process that are zero order in both reactants MeOH and CO, but first order in Rh catalyst and MeI cocatalyst. The active catalyst (**3**) has been identified by in situ high-pressure infrared spectroscopy [93].

$$\text{MeOH} + \text{CO} \xrightarrow{\text{cat.}} \text{MeCOOH} \quad (8.4)$$

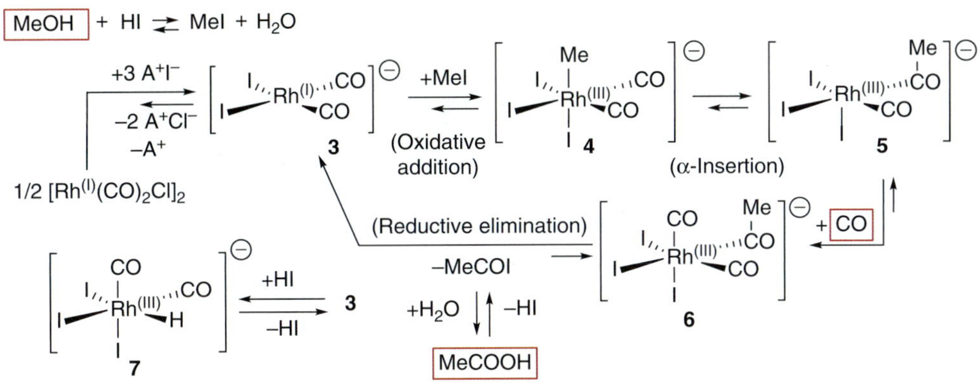

Scheme 8.3 Monsanto carbonylation of methanol: Forster's mechanism of the $[Rh(CO)_2I_2^-]A^+$-catalyzed carbonylation of MeOH (counter-cation used in this study: $A^+ = [PhNMe_3]^+$; precatalyst: $[Rh(CO)_2Cl]_2$, cocatalyst: MeI).

Haynes et al. [84, 94, 95] found that intermediate **4** ($[MeRh(CO)_2I_3]^-/A^+$) can be detected spectroscopically by using high concentration of MeI [96]. The acetyl complex **5** exists as an iodide-bridged dimer in the solid state and in noncoordinating solvents but is easily cleaved to monomeric species in MeOH. Several "immobilized" Rh catalysts have been proposed for the Monsanto acetic acid synthesis. One of them described by Drago et al. [97, 98] in 1981 fixes $[Rh(CO)_2I_2]^-$ on polymeric methylated polyvinylpyridine **8** (see the *Acetica*™ process) [99].

8:

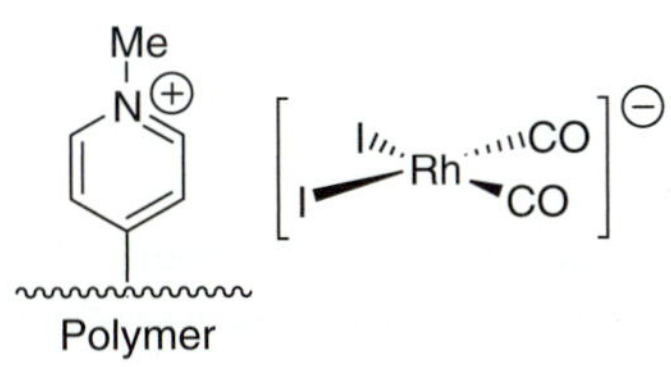

In 1996, BP Chemicals announced a new methanol carbonylation process, *Cativa*™, based on a promoted iridium/iodide catalyst that operates in several plants worldwide [100]. The catalytic cycle involves the same fundamental steps as the rhodium-catalyzed process, i.e. oxidative addition of MeI to Ir(I), followed by migratory CO insertion to form Ir(III) acetyl complex. On the basis of spectroscopic and kinetic observations, Forster proposed that two different (but linked) cycles exist, one involving neutral iridium complexes, and the other predominantly involving anionic species (Figure 8.1) [101]. At a low concentration of H_2O and iodide, the "neutral cycle" operates with $[Ir(CO)_3I]$ as the resting state and rate-determining oxidative addition of MeI to $[Ir(CO)_2I]$. At higher iodide concentrations, the "anionic cycle" predominates. Addition of MeI to $[Ir(CO)_2I_2]^-$ is c. 100 times faster than to $[Rh(CO)_2I_2]^-$ and gives a stable $[MeIr(CO)_2I_3]^-$ complex, which is the resting state of the "anionic" cycle. Under these conditions, the reaction is inhibited by iodide anions and the rate-determining step involves dissociative substitution of I^- by CO in $[MeIr(CO)_2I_3]^-$, followed by migratory CO insertion in $[MeIr(CO)_3I_2]$. The reaction is promoted by carbonyl complexes of W, Re, Ru, and Os and iodides of Zn, Cd, Hg, Ga, and In. Iodide salts LiI and Bu_4NI are catalyst poisons (reaction inhibitors). The promoters accelerate carbonylation of *cis*-$[MeIr(CO)_2I_3]^-$ by abstracting an iodide anion, allowing coordination of CO to give $[MeIr(CO)_3I_2]$. Migratory CO insertion in the latter complex is 700 times faster than in *cis*-$[MeIr(CO)_2I_3]^-$ (PhCl, 85 °C) [100].

Methanol carbonylation can also be catalyzed by Ru, Ni, Pd, and Pt complexes [102]. It has been catalyzed by metal-acid zeolites such as H—Mordenite (H—MOR) and Cu—H—Mordenite (Cu—H—MOR) [103]. Acetic anhydride is produced by the Tennessee Eastman acetic anhydride process by carbonylation of methyl acetate ($MeCOOMe + CO \rightarrow (MeCO)_2O$). It uses Rh salts as catalysts and LiI as a cocatalyst. LiI converts MeCOOMe into MeCOOLi and MeI, the latter being carbonylated into MeCOI as in Figure 8.1. Then, MeCOOLi + MeCOI generates $(MeCO)_2O$ + LiI [104]. In the presence of oxygen and over Cu-exchanged Y-zeolite, the carbonylation of methanol produces dimethyl carbonate ($CO + 2MeOH + \frac{1}{2}O_2 \rightarrow (MeO)_2CO + H_2O$) according to the mechanism proposed in Scheme 8.4. The formation of $(MeO)_2CO$ is initiated by reaction of molecularly absorbed MeOH with O_2 to form either mono-(**9**) or di-methoxide species (**10**) bound

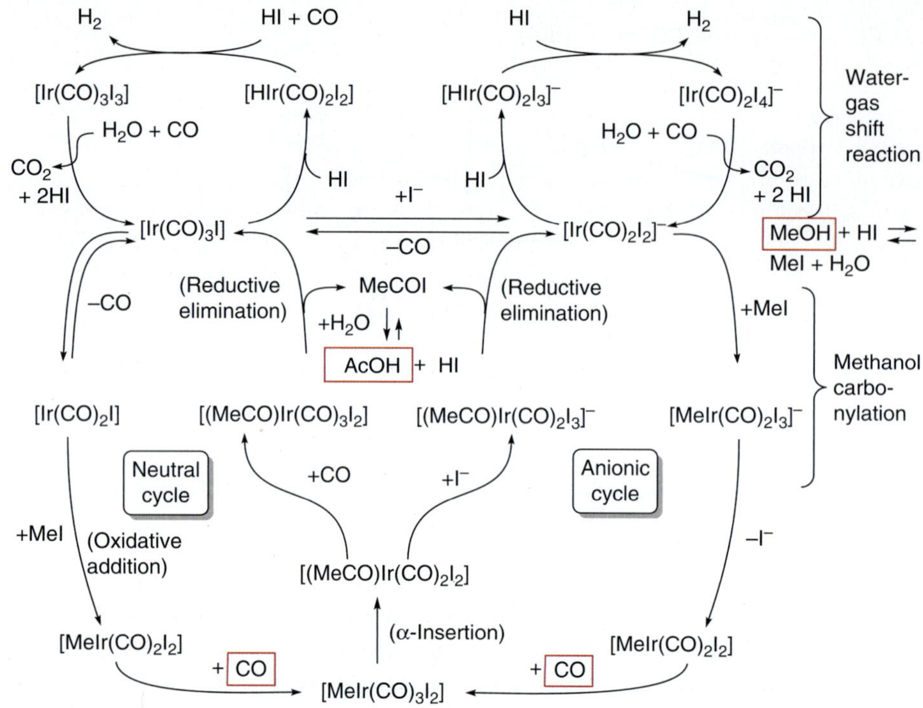

Figure 8.1 Catalytic cycles for iridium-catalyzed methanol carbonylation (MeOH + CO ⇄ AcOH). The same type of cycles explains the competing water-gas-shift reaction ($CO + H_2O \rightleftarrows H_2 + CO_2$).

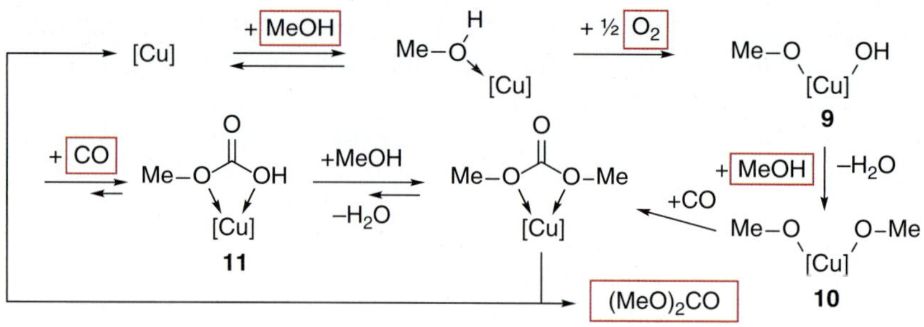

Scheme 8.4 Proposed reaction mechanism for the oxidative carbonylation of methanol producing dimethyl carbonate.

to Cu^+ cations. Reaction of the mono-methoxide species with CO produces monomethyl carbonate species **11**, which then reacts rapidly with MeOH to give $(MeO)_2CO$. A second pathway adds CO to di-methoxide species **10** to produce $(MeO)_2CO$ directly [105–108].

Problem 8.1 What products can form upon carbonylation of ethanol catalyzed by a Rh catalyst in the presence of ethyl iodide? [109]

Problem 8.2 Propose a mechanism for the hydrocarbonylation of acetic acid into propionic acid catalyzed by $[Rh(CO)_2Cl]_2$ and LiI [110].

8.2.3 Hydroformylation of alkenes

First discovered by Roelen [111] hydroformylation of alkenes (8.5) [112] has become one of the most important homogeneous catalyzed reaction in the industry [113–117]. Catalysts have been developed that permit regioselective and enantioselective reactions [118–127].

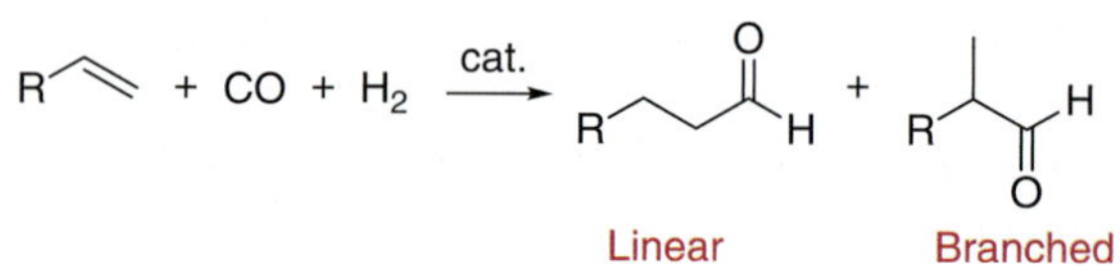

(8.5)

The basic reaction converts olefins to aldehydes (oxoproducts) in the presence of CO + H_2 (syngas) using cobalt catalyst (reaction termed "oxo process"). When the hydroformylation reaction is carried out at 150–200 °C and under 100–150 atm (c. 10–15 MPa) of syngas, almost any form of cobalt acts as a catalyst precursor for the reaction, such as $CoCO_3$ and $Co(OAc)_2$. Under these conditions, the Co(II) salt is reduced by H_2 to Co(0), followed by coordination to CO giving $Co_2(CO)_8$. Subsequently, in the presence of H_2, $Co_2(CO)_8$ is converted to $HCo(CO)_4$, which is believed to be the pivotal active catalyst [128, 129] (Heck–Breslow mechanism) [130, 131]. The room temperature stoichiometric hydroformylation (Type I reaction (8.6)) converts $HCo(CO)_4$ into $Co_2(CO)_8$. It is not analogous to the type II stoichiometric hydroformylation reaction (8.7) conducted with $Co_2(CO)_8$ and H_2 [132].

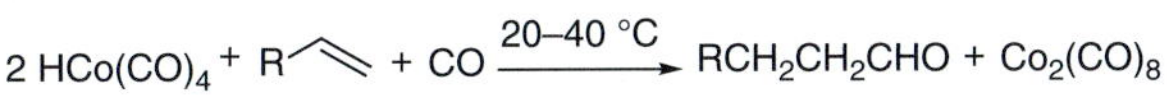

(8.6)

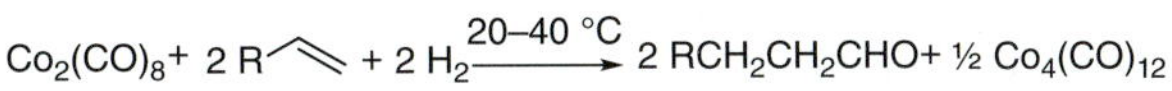

(8.7)

The type II reaction (8.7) proceeds via the formation of the hydride species (reaction (8.8)), followed by coordination of the alkene and its insertion into the M—H bond and the addition of a CO ligand, resulting in the formation of the acylcobalt complex **12** (reaction (8.9)), similar to the mechanism of type I reaction (Scheme 8.5). However, the similarity ends at this point. The presence of H_2 in type II reaction (8.7) drives the hydrogenolysis step to occur via a different pathway than that observed in the type I reaction (8.6). Although in reaction (8.6) this step is carried out by a second molecule of $HCo(CO)_4$, in reaction (8.7), this step is carried out by H_2, as evidenced by isotope (H/D) distribution in the aldehyde product as a function of $HCo(CO)_4/DCo(CO)_4$ initial ratio and H_2/D_2 ratio. Indeed, when hydroformylation of type (II) (reaction (8.7)) was carried out with a 1 : 1 mixture of H_2/D_2, a 1 : 1 mixture of $RCH_2CH_2CHO/RCH_2CH_2CDO$ was obtained whether the reaction mixture contained initially a 93 : 7 or a 8 : 92 mixture of $HCo(CO)_4/DCo(CO)_4$ (40 °C, 28 hours). Under these conditions, RCH_2CH_2CHO is not equilibrated with RCH_2CH_2CDO in the presence of D_2. Interestingly, the deuterium content in the alkyl chain of the aldehyde corresponds to the $DCo(CO)_4/HCo(CO)_4$ ratio in the original equilibrated solution, whereas the content of deuterium in the aldehyde group (CDO vs. CHO) corresponds to the initial proportion D_2/H_2 in the gas phase. These results indicate that the H atom incorporation (reaction (8.8)) into the alkyl chain and the hydrogenolysis (reaction (8.10)) of the (acyl)Co complex **12** are two independent processes involving two independent active intermediates and transition states without any coupling between them (Scheme 8.5).

The majority of industrial oxo-products are *n*-butanal and isobutyraldehyde obtained by propene hydroformylation. They are important intermediates for the production of esters and acrylates. With the discovery in 1968 that Wilkinson's complex $Rh(PPh_3)_3Cl$ catalyzes the propene hydroformylation much better than cobalt-derived complexes, a low-pressure oxo process was developed (LP OxoSM process) to produce butyraldehydes. Union Carbide built a first plant in 1976 in Puerto Rico. It operates at less than 20 atm (2 MPa) and lower reaction temperature (90–100 °C) and fewer byproducts are formed than in the classical oxo process. The active catalyst $(PPh_3)_2(CO)RhH$ (**13**) results from the reaction of $Ph_3P + CO$ with (acac)(CO)(PPh_3)Rh ("ROPAC," a stable crystalline compound, acac = $MeCOCH{=}C(O^-)Me$) (Scheme 8.6). According to Heck and Breslow, it coordinates propene and then β-hydrogen insertion

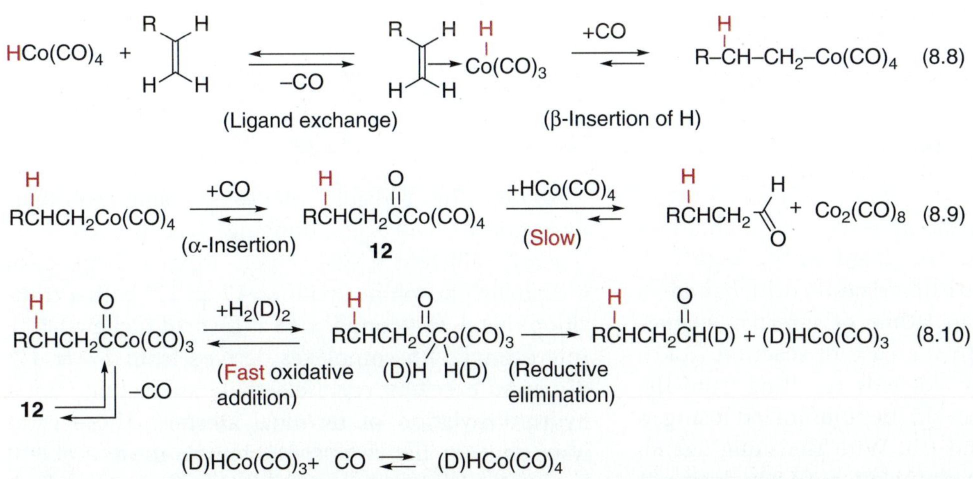

Scheme 8.5 Mechanism of the "oxo process." β-Insertion of hydrogen (8.8) and hydrogenolysis (8.9) of the acylcobalt species **12** are independent processes. Reaction (8.9) does not compete in the presence of H_2 (D_2).

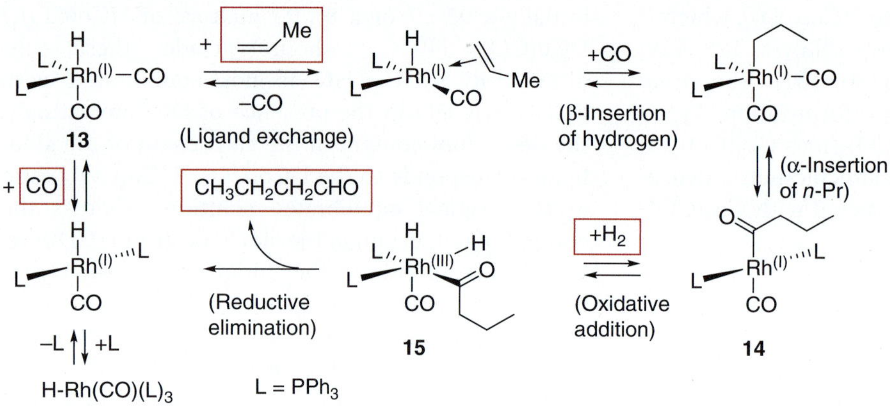

Scheme 8.6 Heck–Breslow-type mechanism of the LP OxoSM Process.

occurs, followed by α-insertion into a carbonyl group, forming the (acyl)Rh intermediate **14**. Oxidative addition of H_2 gives dihydride **15** that then undergoes irreversible reductive elimination with liberation of the oxo-product [133]. About 10% of isobutyraldehyde also forms (the β-insertion of hydrogen onto propene gives the corresponding *i*-Pr—Rh complex).

The regioselectivity *n*-butyraldehyde vs. isobutyraldehyde is c. 3–4 for the cobalt oxo process and c. 10 with the triphenylphosphine-rhodium catalyst. This regioselectivity can be improved by using other ligands than Ph_3P, for instance, with bisphosphites of type **16** [115]. Hydroformylation of acrylamide with Rh-Ph_3P catalyst (THF, 60 °C, 95 atm) exclusively gives the branched aldehyde MeCH(CHO)—$CONH_2$ [134].

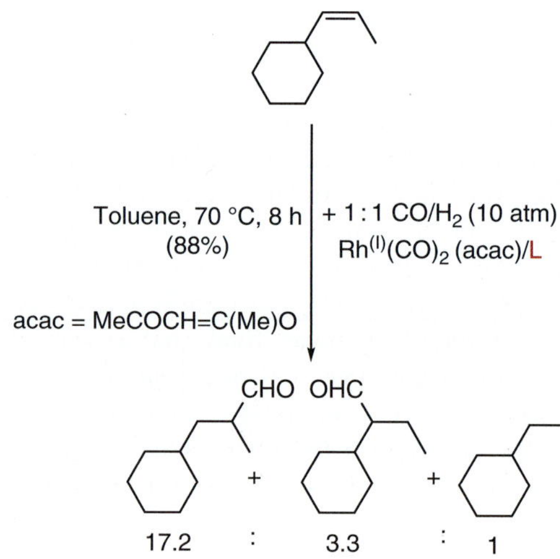

(8.11)

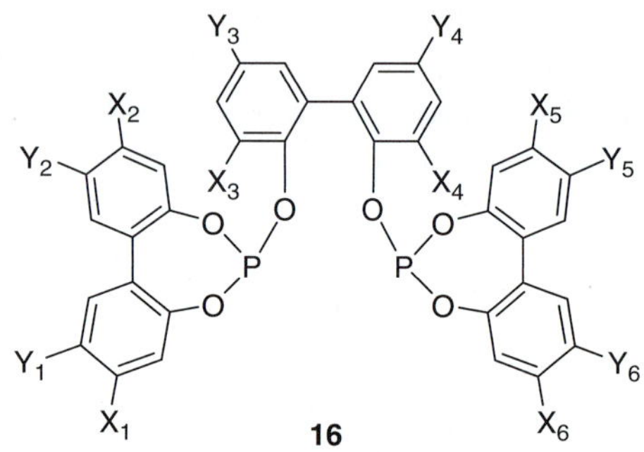

16

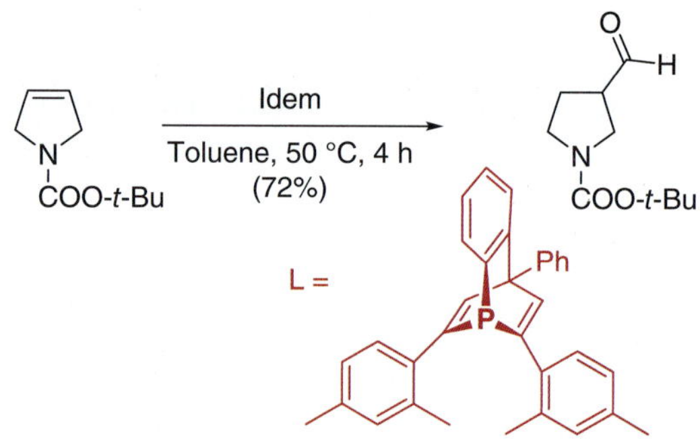

(8.12)

The high hydroformylation activity of a rhodium catalyst [135] is always associated with a high tendency to catalyze alkene isomerization, which renders the selective hydroformylation of internal alkenes impossible. However, in the case of reaction (8.11), the formation of linear aldehyde resulting from the alkene isomerization could be minimized using a phosphabarrelene ligand (L). With the same ligand, the Rh-catalyzed hydroformylation (8.12) does not lead to 2,5-dihydroazole → 2,3-dihydroazole isomerization [136–138].

Among the possible modes of supramolecular interaction, hydrogen bonding has proved particularly efficient [139]. Upon mixing 2 equiv. of 6-diphenylphosphinopyridone **17** ⇄ **17′** with a transition metal, salt complex **18** is formed (Scheme 8.7). Interestingly, Rh complexes derived from **17** ⇄ **17′** displayed excellent regioselectivity and activity upon hydroformylation of terminal alkenes. These catalysts allowed the first room temperature, ambient pressure, and regioselective hydroformylation, which is of particular use to synthetic organic chemistry [140–142].

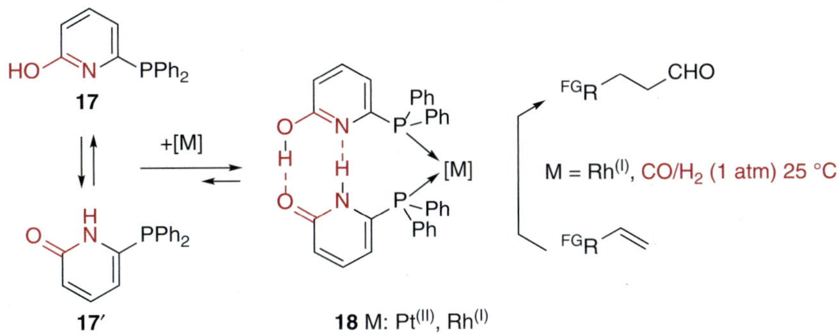

Scheme 8.7 Breit's diphosphine ligands resulting from autoassociation through hydrogen bonding.

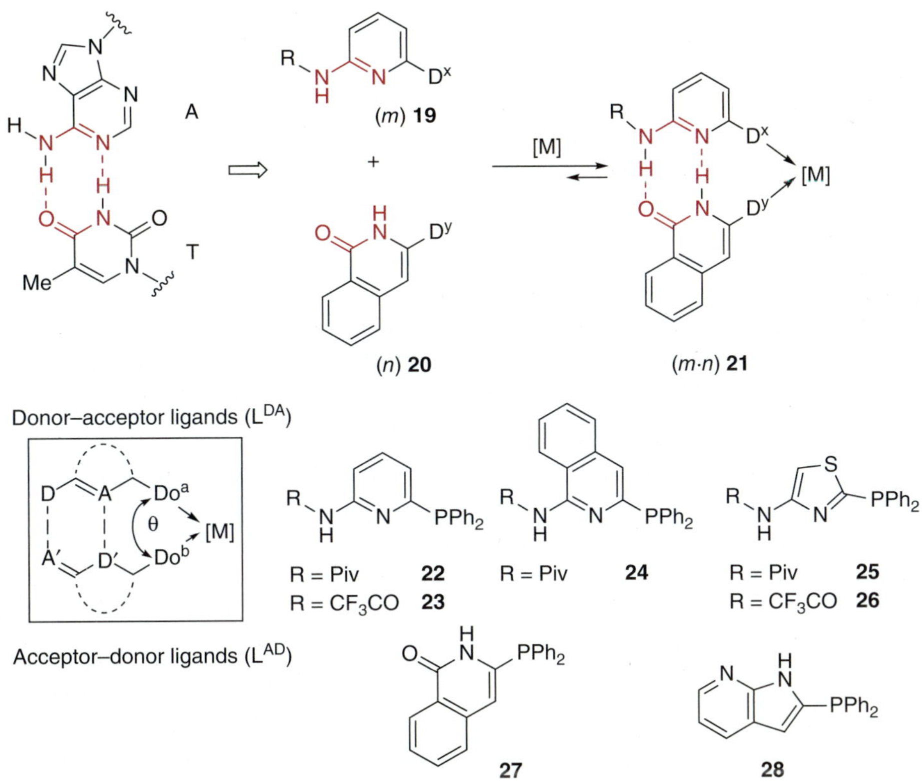

Scheme 8.8 Breit's library of ligands with complementary hydrogen bonding motifs analogous to the DNA A–T base pair.

The advantage of the self-assembly approach is its inherent possibility for combinatorial ligand library generation through mixing of two different ligands. In order to generate selectively the unsymmetrical heterodimeric ligand, a self-assembly platform based on the Watson–Crick base pairing of A and T in DNA has been developed. As an A–T base pair analog, the 2-aminopyridine **19**/isoquinolone (**20**) platform proved suitable. Thus, mixing of two monodentate ligands based on this platform in the presence of a transition metal salt led to the selective formation of the heterodimeric complexes **21** featuring a bidentate coordination mode, as proved by NMR and X-ray diffraction analysis (Scheme 8.8) [143, 144].

Screening of catalyst library made of ligands **22–28** for regioselective hydroformylation of oct-1-ene allowed elucidation of the influence of the heterocyclic self-assembly platform on catalyst properties. Thus, although all catalysts displayed higher catalytic activity compared to the industrially applied rhodium/triphenylphosphine system, regioselectivity (linear vs. branched aldehyde) varied from 89 : 11 to >99 : 1. Thus, the nature of the heterocyclic self-assembly platform influences the catalyst performance, presumably because of changes in coordination geometry and hydrogen bond strength. The best catalyst **26/27** is based on the five-membered thiazole heterocycle and gave a regioselectivity better than 99 : 1 in favor of the linear aldehyde [145, 146]. Water as cosolvent enhances both rate and regioselectivity in favor of terminal aldehydes. One-pot hydroformylation and

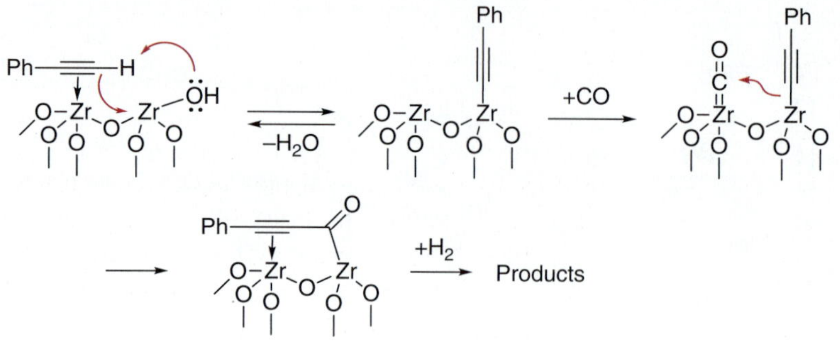

Scheme 8.9 Proposed reaction mechanism for the zirconia–silica-catalyzed hydroformylation of phenylacetylene.

hydrogenation using Rh as sole catalyst [147] or with Ru as cocatalyst yields primary alcohols in good yield and selectivity [148–150]. The use of water-soluble catalysts represent a significant advantage in homogeneous catalysis. Although catalyzed homogeneously, the technique of an "immobilized catalyst in mobile phase" offers all advantages of a heterogeneous catalysis with heterogeneous catalyst handling yields more economical approach to the oxo process, which is possible by using water-soluble catalysts such as **29** (Ruhrchemie/Rhône-Poulenc oxo process) [151]. Chiral bis-phosphine and bis-phosphite ligands (e.g. **30**) have been used for asymmetric hydroformylation reactions [152–155]. The combined application of $Rh_4(CO)_{12}$ and $HRe(CO)_5$ as precursors in the hydroformylation reaction leads to very strong synergistic rate effect. In situ spectroscopic measurements confirm the presence of both mononuclear and dinuclear intermediates such as $RCORh(CO)_4$ and $RhRe(CO)_9$ in the active catalytic system [156].

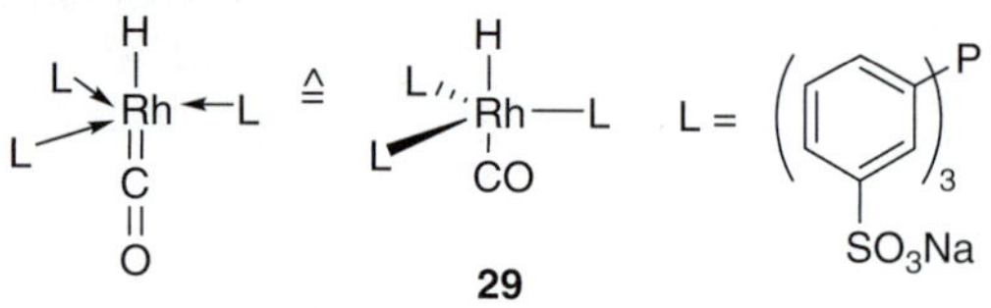

29

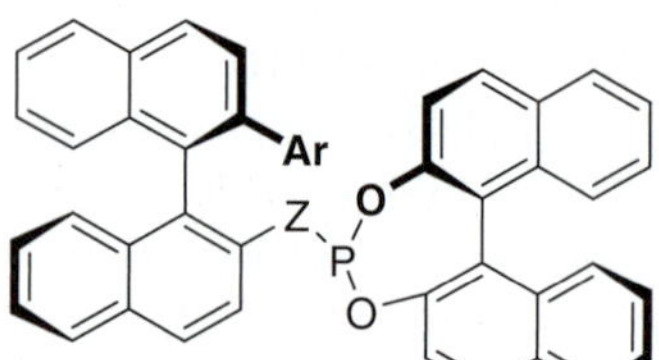

30

Z = O, NR Ar = Ph, 3-MeOC$_6$H$_4$

Hydroformylation of alkynes is a little studied reaction as acetylenes are rapidly hydrogenated into alkenes before undergoing the oxo process. Furthermore, the products of hydroformylation of alkynes, regioisomeric α,β-unsaturated aldehydes, are also hydrogenated easily under the classical conditions of the oxo process [157]. Le Floch and coworkers have reported that zirconia–silica heterogeneous catalyst can promote the hydroformylation of phenylacetylene (reaction (8.13)) according to the mechanism shown in Scheme 8.9 [158]. This mechanism is compatible with the observation that, in the presence of cyclohexene, the products of Pauson–Khand reaction are also formed (Section 8.2.7). Applying electron-poor self-assembling ligand systems with Rh catalyst hydroformylation of dialkyl- and dirarylalkynes furnished the corresponding enals [159].

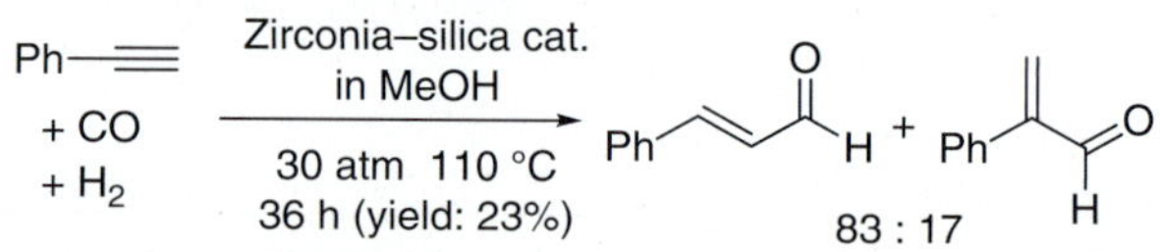

(8.13)

Alkene hydroformylations and subsequent hydrogenation of aldehydes into corresponding alcohols with CO surrogates have been realized with CO_2/H_2 utilizing $[H_4Ru_4(CO)_{12}]$ as a catalyst, with formaldehyde in the presence of $[Ni(cycoocta\text{-}1,5\text{-}diene)_2]$ + $(cyclohexyl)_3P$ [160] and with CO/H_2 generated by Ir-catalyzed dehydrogenation–decarbonylation of polyols (biomass) [161]. Retro-hydroformylation of aldehydes ($RCH_2CH_2CHO \rightarrow R{-}CH{=}CH_2 + CO + H_2$) occurs in open flasks in the presence of Rh or Ir complexes as catalysts [27].

Problem 8.3 Explain the following reaction [162].

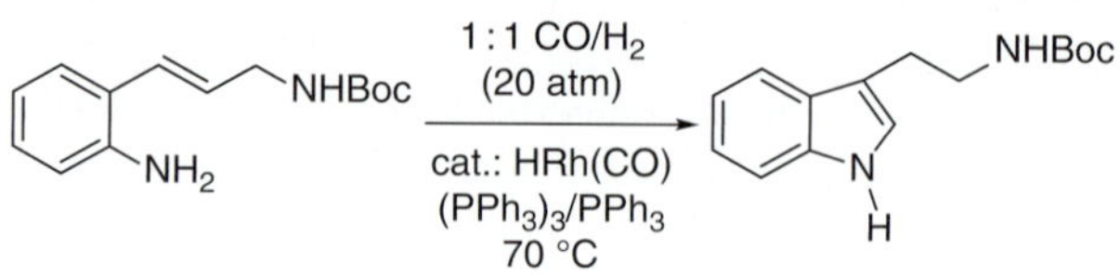

Problem 8.4 Explain the following indole synthesis [163].

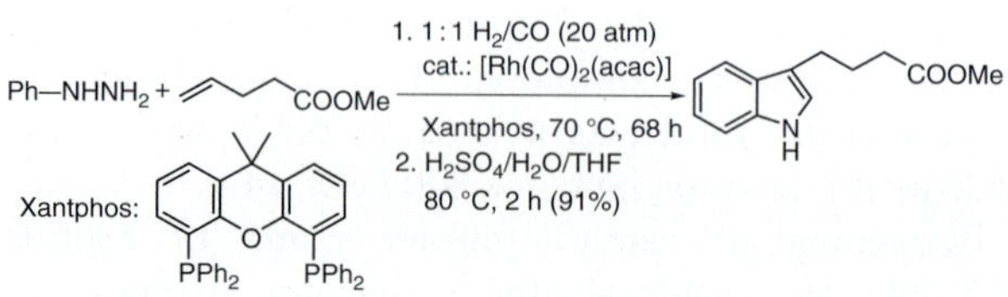

8.2.4 Silylformylation

A hydrosilane (R_3SiH) can be seen as equivalent to H_2, with the silyl group equivalent to a proton. In 1979, Murai and Sonoda reported that aldehyde (RCHO) and $HSiEt_2Me + CO$ in the presence of $Co_2(CO)_8/PPh_3$ give the corresponding α-(siloxy)carbaldehyde $RCH(OSiEt_2Me)CHO$ [164]. The silylformylation (e.g. reaction (8.14)) can also be catalyzed by $[Rh(cod)Cl]_2$ [165]. In 1991, they found that cyclic orthoesters can be combined with hydrosilanes and CO in the presence of $Co_2(CO)_8$ catalyst (0–25 °C, 1 atm) (e.g. reaction (8.15)) [166]. Silylformylation of alkenes do not give the corresponding β-silyl-carbaldehydes as the latter are isomerized into the more stable enoxysilanes (e.g. reaction (8.16)) [167–169]. Iridium siloxide complex $[Ir(cod)(PCy_3)(OSiMe_3)]$ (Cy = cyclohexyl) catalyzes the silylacarbonylation of $Me_3SiCH{=}CH_2$ with $PhMe_2SiH + CO$ giving (Z)-$Me_3SiCH_2CH{=}CHOSiMe_2Ph$ (95%) [170].

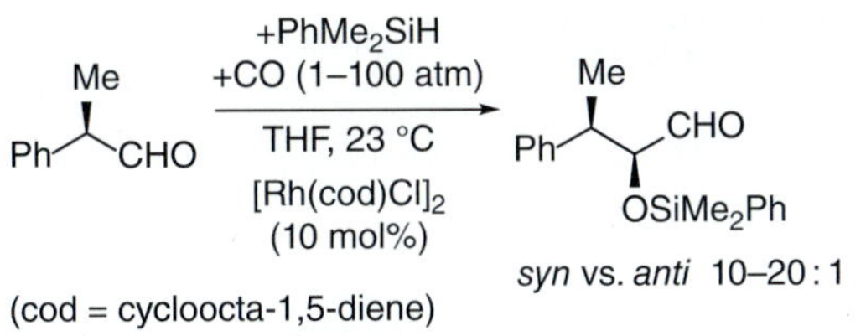

(8.14)

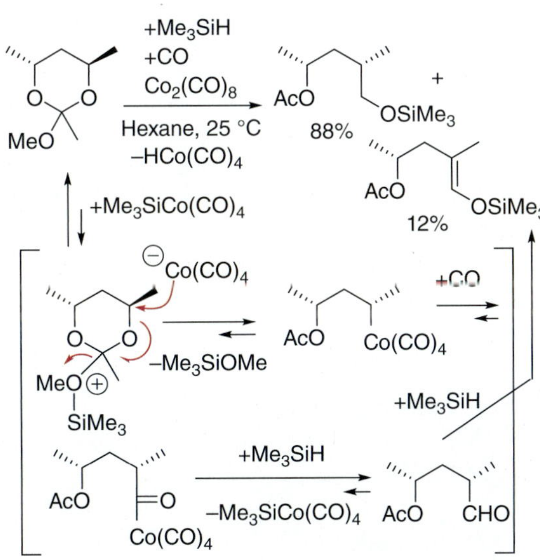

(8.15)

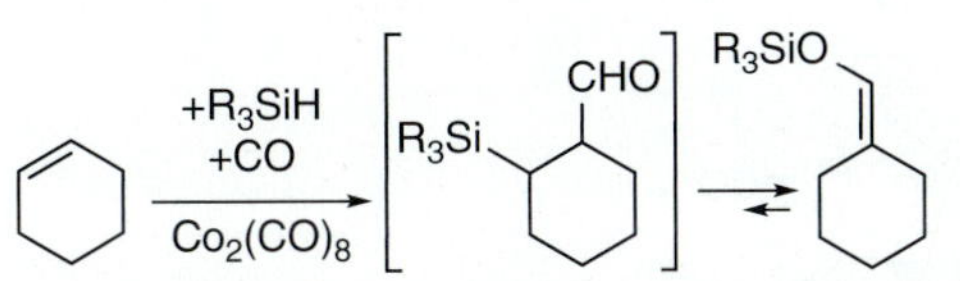

(8.16)

Leighton and coworkers have developed a Rh-catalyzed intramolecular silylformylation of alkenes (e.g. reaction (8.17)) that converts β-silyloxy-alkenes into stable β-(oxysilyl)carbaldehydes. A possible mechanism of the reaction is given in Scheme 8.10a [171, 172]. When using β-(diallyl)silyl homoallyl ethers (e.g. **31**), the Rh-catalyzed silylformylation generates a β-silylaldehyde (e.g. **32**) that undergoes an uncatalyzed intramolecular Hosomi–Sakurai allylation of the aldehyde (an example of carbonyl metalla ene-reaction, Section 5.7.5) [173–175], a reaction that is highly diastereoselective giving **33**, for instance. The method has been applied to the synthesis of polyketide fragments as illustrated with transformation **33** → **34** (Scheme 8.10b).

Matsuda et al. [176] and Ojima et al. independently discovered the silylformylation of alkynes ($R'C{\equiv}CH + CO + R_3SiH \rightarrow R'(CHO)C{=}CH{-}SiR_3$) catalyzed by rhodium ($Rh_4(CO)_{12}$) and mixed rhodium–cobalt ($Co_2Rh_2(CO)_{12}$) catalysts, respectively [177]. Quantum mechanical calculations confirmed that the $Co_2Rh_2(CO)_{12}$ affects efficient alkyne silylformylation because of a bimetallic synergism [178]. The catalytic cycle (Figure 8.2) consists of oxidative addition of hydrosilane, alkyne insertion, CO insertion, and reductive elimination. Although major bond-forming events take place at the most polarizable Rh center, cobalt also plays a role as it fixes the metal cluster on the organic substrate. It also keeps the hydride ligand temporarily attached on the metal center, thus retarding hydrosilylation that competes with silylformylation.

Ojima et al. have found that the alkyne silylformylation can also be catalyzed by $(t\text{-BuNC})_4RhCo(CO)_4$ and $Rh(acac)(CO)_3$ at 25 °C and under 1–10 atm of CO [179]. The reaction is accompanied by the formation of alkenylsilane resulting from a concurrent hydrosilylation. An efficient reusable catalyst system has been proposed that uses $[Rh(cod)][BPh_4]$ as catalyst in $[bmin][PF_6]$ ionic liquid (bmin = 1-butyl-3-methylimidazolium) [180]. Leighton and coworker utilize the alkyne silyformylation in their stereoselective synthesis of polypropionate fragments, as illustrated in Scheme 8.11 [181]. The Rh(I)-catalyzed silylformylation of but-2-yne generates enal **35**. In situ diastereoselective aldehyde crotylation with enantiomerically pure (S,S)-*trans* EZ-CrotylMix and (S,S)-*cis* EZ-CrotylMix that are commercially available provides siloxanes **36** and **37**, respectively, with good yields and enantioselectivities. Fleming–Tamao oxidation [182–184] under standard conditions [185] converts **36** and **37** into stereotriads **38** (*anti,syn*) and **39** (*anti,anti*), respectively. Under "aprotic" conditions [186, 187], stereoisomers **40** (*syn,syn*) and **41** (*syn,anti*), respectively, are obtained. Applying the enantiomers of the above crotylation agents, the enantiomers of **36**–**41** are prepared in the same way.

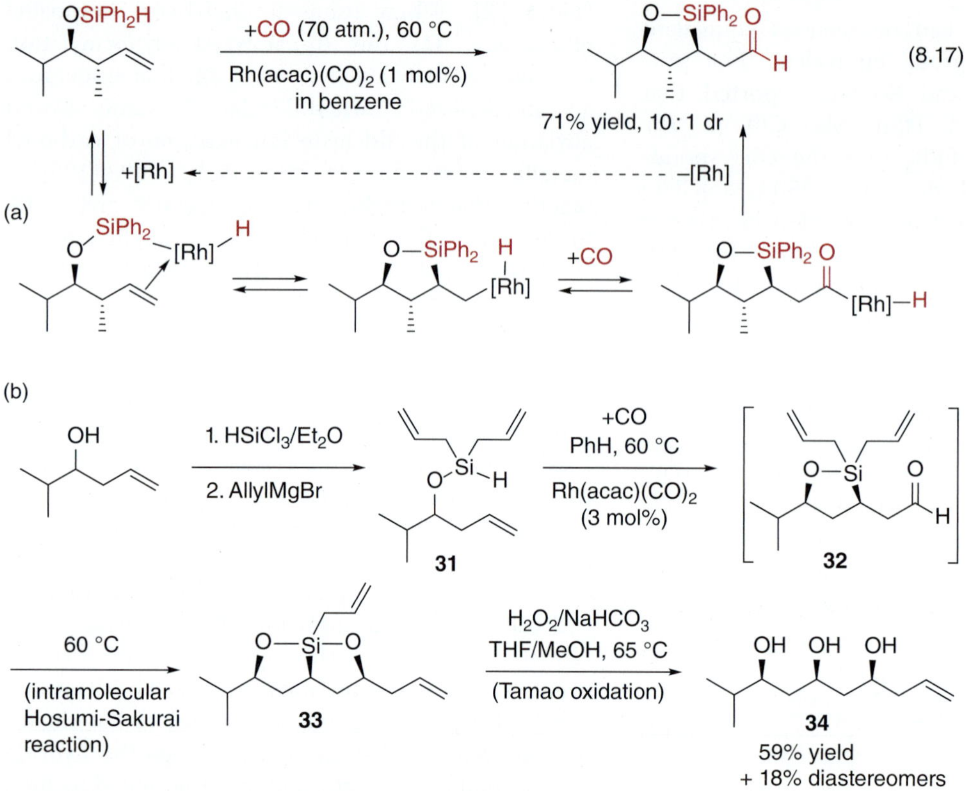

(8.17)

Scheme 8.10 Leighton's intramolecular alkene silyl formylation. (a) Proposed reaction mechanism and (b) example of synthesis of a polyketide.

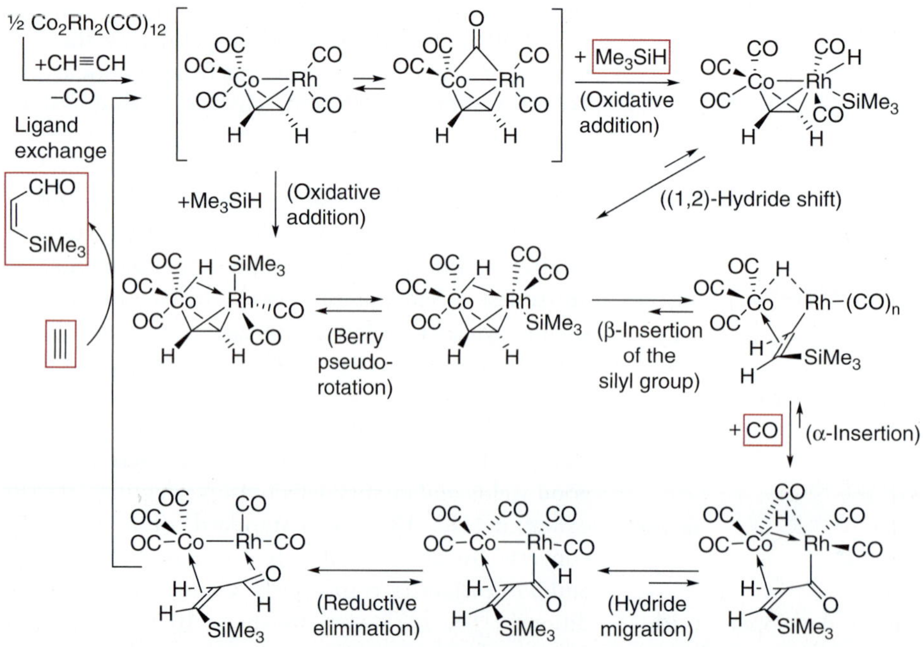

Figure 8.2 Bimetallic pathway of the alkyne silylformylation.

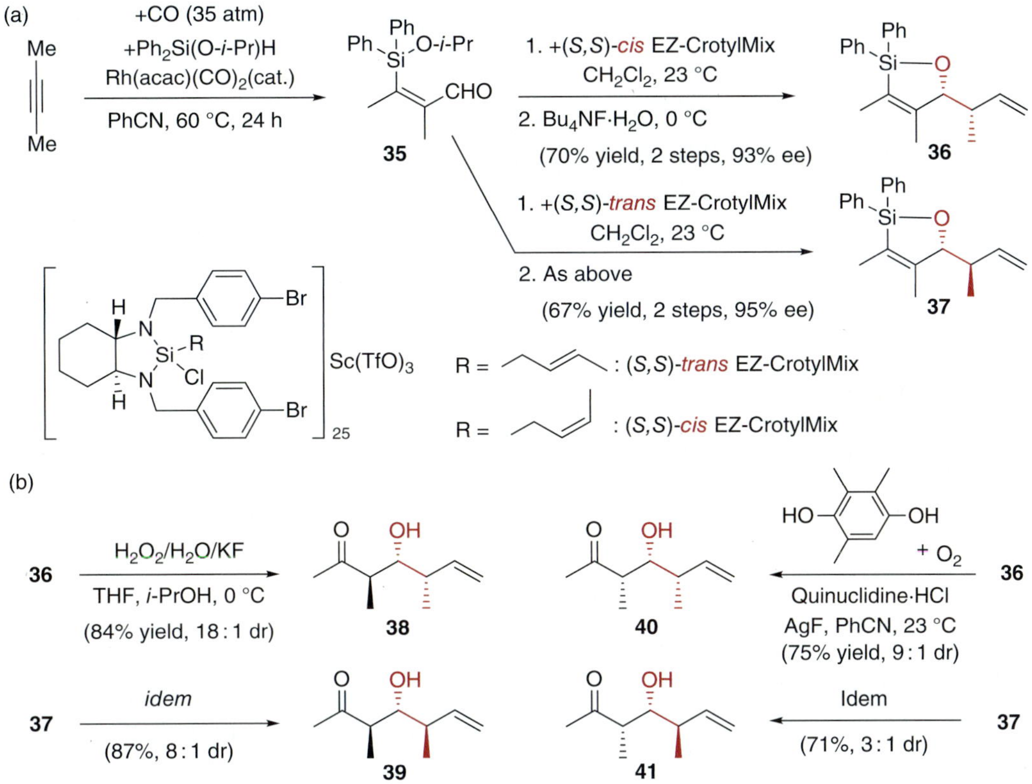

Scheme 8.11 Leighton's synthesis of polypropionate fragments: (a) preparation of stereotriads; (b) preparation of stereotriads by alkenylsilane oxidation (stereotriads; dr = ratio of diastereosiomers).

Problem 8.5 Propose a mechanism for the silylformylation reaction: $PhCH_2CHO + PhMe_2SiH + CO \rightarrow PhCH_2CH(CHO){-}OSiMe_2Ph$ catalyzed by $[Rh(cod)Cl]_2$ in THF [165].

Problem 8.6 What is a possible product of reaction $HC{\equiv}C(CH_2)_3CHO + Et_3SiH + CO$ (1 atm) catalyzed by $Rh_2Co_2(CO)_{12}$ at 25 °C? [188]

8.2.5 Reppe carbonylations

Highly branched C_5 and C_{10} carboxylic acids are made from alkenes + CO + H_2O by the Koch carboxylic acid synthesis (reaction (8.18)) [33, 189].

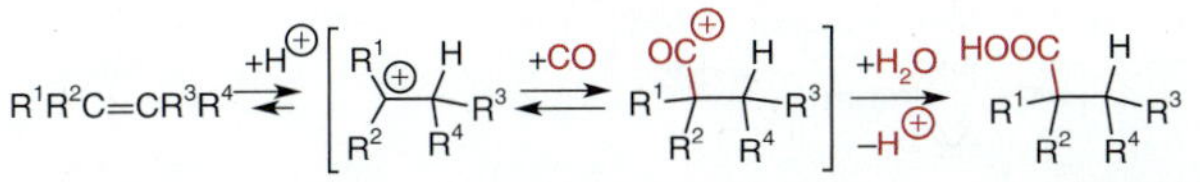

(8.18)

Most linear aliphatic carboxylic acids are made via the hydroformylation route (Section 8.2.3). The method requires two synthetic steps, the second being the oxidation of the corresponding aldehydes. One-step conversion of alkenes into linear aliphatic acids is possible by the $PdCl_2/CuCl_2$-catalyzed oxidative carbonylation (see Section 8.2.6).

Reppe's first catalytic carbonylation process converts acetylene CO and ROH to acrylic esters using $Ni(CO)_4$ and haloic acid (HX) as precatalysts (reaction (8.19)) [190–198]. As $Ni(CO)_4$ is poisonous, it is now replaced by Fe, Ru, Co, Rh, Ir, Pd, Pt, or Mo catalysts. The proposed mechanism (Scheme 8.12) involves (i) the oxidative addition of HX to $Ni(CO)_4$, (ii) coordination of acetylene to the metal and its subsequent insertion into the Ni—H bond (β-insertion of hydrogen), (iii) α-insertion of the vinyl group into a carbonyl ligand with the formation of a (acyl)metal complex that (iv) reacts with ROH (water, alcohols) to give the final acrylic product. Alternatively, (v) the unsaturated $Ni(CO)_2(vinyl)X$ intermediate coordinates a CO molecule and then undergoes α-insertion of the vinyl group into a carbonyl ligand [199]. Carbonylations of alkenes combining nucleophiles different from water and alcohols (reaction (8.20))

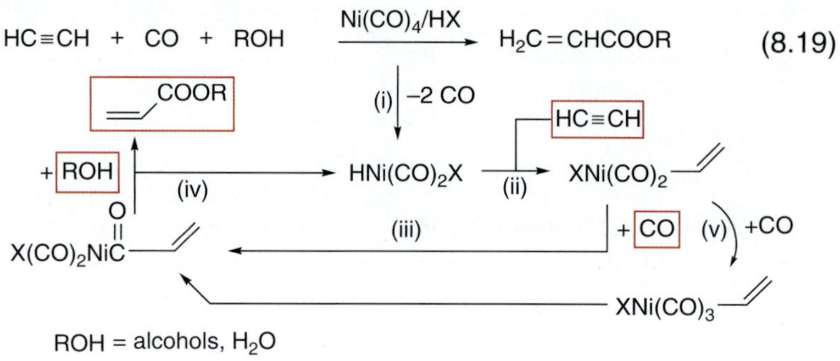

Scheme 8.12 Proposed reaction mechanisms of the original Reppe acetylene carbonylation.

and of conjugated dienes (reactions (8.21) and (8.22)) have received considerable attention also [200].

$$RCH{=}CH_2 + CO + Nu{-}H \xrightarrow{cat.} RCH_2CH_2CONu + R{-}CH(CONu){-}CH_3 \quad (8.20)$$

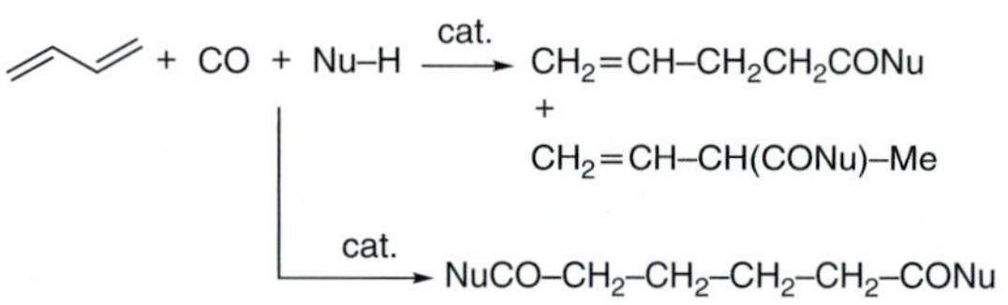

(8.21)

(8.22)

Shell developed a process converting propyne to methyl methacrylate in the presence of PdX_2/2-pyridylphosphite catalyst [201]. This reaction is inspired from earlier work of Tsuji and coworkers [202–204]. The rhodium-catalyzed HI-promoted carbonylation of MeOH to AcOH (Monsanto process, Section 8.2.2, Scheme 8.3) is not limited to MeOH, as higher alcohols (e.g. EtOH, *n*-propanol) and alkenes (e.g. $CH_2{=}CH_2$, $CH_3{-}CH{=}CH_2$) are also converted into monocarboxylic acids (e.g. CH_3CH_2COOH, $CH_3(CH_2)_2COOH$). In the case of the carbonylation of ethylene, the mechanism shown in Scheme 8.13 has been proposed. Intermediate $HRh(CO)_2I_3^-$ has been characterized by NMR spectroscopy for reactions carried out at low temperature and using $[Rh(CO)_2I_2]^-[AsPh_4]^+$ as precatalyst [205].

8.2.6 Pd(II)-mediated oxidative carbonylations

In the Wacker process (Section 7.11.5), a key step is the nucleophilic addition of water to the alkene activated by coordination to the transition metal. Complex (alkene)$PdCl_3^-$ exchanges one of its Cl^- ligand with H_2O giving (alkene)$PdCl_2(H_2O) + Cl^-$. After deprotonation, (alkene)$PdCl_2(OH)^-$ undergoes β-insertion of the OH group (*syn* nucleophilic addition) with the formation of $CH_2(OH){-}CH_2{-}PdCl_2^-$. Tsuji realized that H_2O nucleophile can be exchanged for CO, a nucleophilic reagent capable to coordinate the transition metal. This led him to invent the oxidative carbonylation of alkenes (reaction (8.23)) in 1963 [206, 207]. This reaction is made catalytic in $PdCl_2$

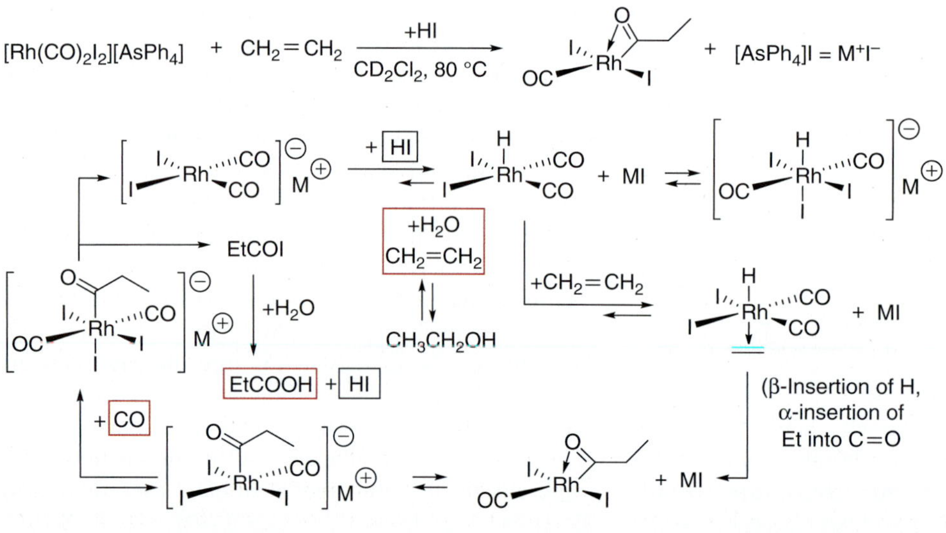

Scheme 8.13 Proposed reaction mechanism of the Rh(I)/HI-catalyzed carbonylation of ethylene (ethanol is the source of ethylene and water).

using $CuCl_2$ as a reoxidizing agent (as in the Wacker process). Using alcohol as a medium, the oxidative carbonylation of alkene (8.24) was invented. Under basic conditions, in 1969, Yukama and Tsutsumi discovered the formation of methyl cinnamate by oxidative carbonylation of styrene (reaction (8.25)). Further modifications of the reaction conditions permitted the synthesis of succinates as shown by Heck and others in 1972 (reaction (8.26)) [208, 209].

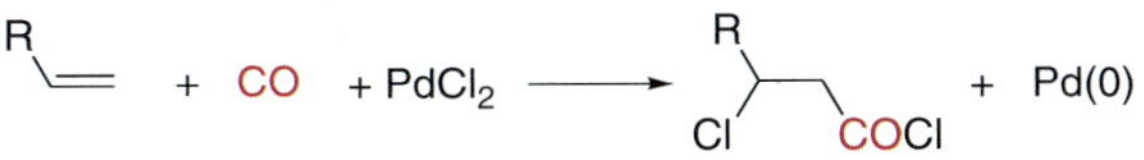

(8.23)

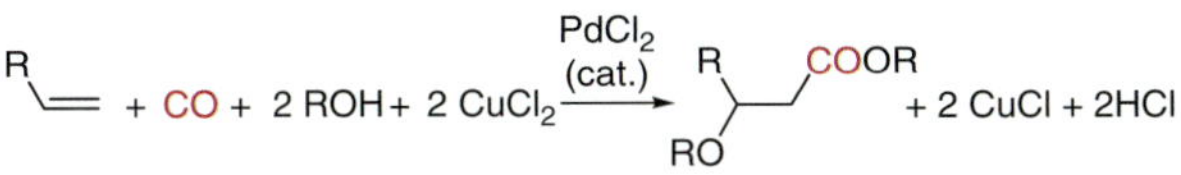

(8.24)

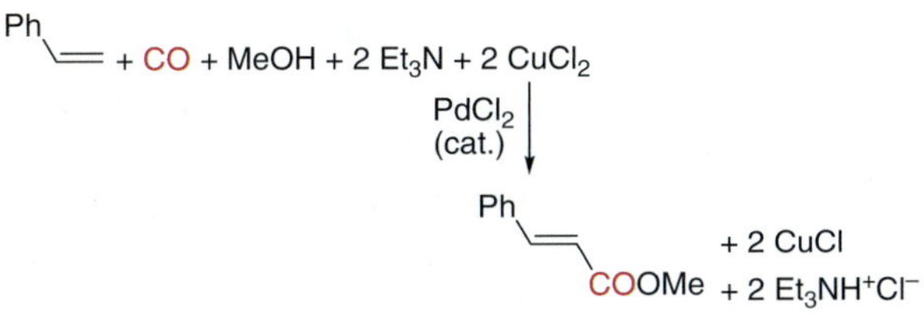

(8.25)

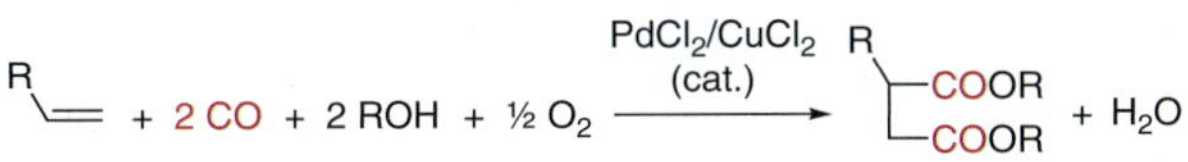

(8.26)

In all these reactions that reduce Pd(II) into metallic Pd(0), the presence of charcoal permits a better dispersion of metallic Pd and thus a more efficient reoxidation by $CuCl_2$. In fact, the initial palladium catalyst does not have to be $PdCl_2$; Pd on charcoal is more efficient and leads to much high TOF [210]. A general mechanism for these reactions is shown in Scheme 8.14. A synthesis of β-amino acid esters ($RCH{=}CH_2 + HNR'COOBn + PhI(OOCCF_3)_2 + CO$ (1 atm) $+ MeOH \rightarrow RCH(NHCOOBn)CH_2COOMe + PhI + 2CF_3COOH$) has been presented by Liu and coworkers that applies $Pd(OOCCF_3)_2$ as the catalyst [211].

Pd-catalyzed carbonylation of acetylenes gives α,β-unsaturated mono- and diesters [212], and saturated di- and triesters [213–217]. With 2 equiv. of $CuCl_2$ (oxidant), terminal alkynes are converted into acetylenecarboxylates under atmospheric pressure of CO at 25 °C, in alcohols and in the presence of 5 mol% of $PdCl_2$ ($R{-}C{\equiv}CH + CO + R'OH + 2CuCl_2 \rightarrow RC{\equiv}C{-}COOR' + 2HCl + 2CuCl$) [218]. Tsuji and Iwamoto found that carbonylation of amines give corresponding oxamides and N,N'-dialkylureas with generation of H_2 using $PdCl_2$ or Pd/C as catalyst according to reaction: $2RNH_2 + 2CO \rightarrow RNHCO{-}CONHR + RNHCONHR + H_2$ [219]. The double alkoxycarbonylation of substituted alkenes gives mixtures of regioisomers and enantiomers. Applying enantiomerically enriched Pd(II) systems containing P monodentate of bidentate, as well as P, N ligands, regioselective and enantioselective syntheses of substituted succinic diester derivatives have been realized [220–223].

Problem 8.7 (A) Tsuji and coworkers found that acetylene reacts with CO in the presence of 1 equiv. of $PdCl_2$ affording maleyl and fumaryl chloride [213]. (B) They also found that diphenylacetylene in ethanol reacts with CO in the presence of a catalytic amount of $PdCl_2$ giving α,β-diphenyl-γ-crotonolactone (66%) and ethyl α,β-diphenylmaleate (26%) in the presence of HCl [214]. Propose a mechanism for these reactions [224, 225].

8.2.7 Pauson–Khand reaction

An important method for the preparation of cyclopentenones is the Pauson–Khand reaction, a formal (2+2+1)-cycloaddition between an alkene, an alkyne, and CO [226–228]. This reaction, discovered in 1971, converts norbornene and hexacarbonyl (phenylacetylene)dicobalt complex into the corresponding cyclopentenone (reaction (8.27)) [229, 230].

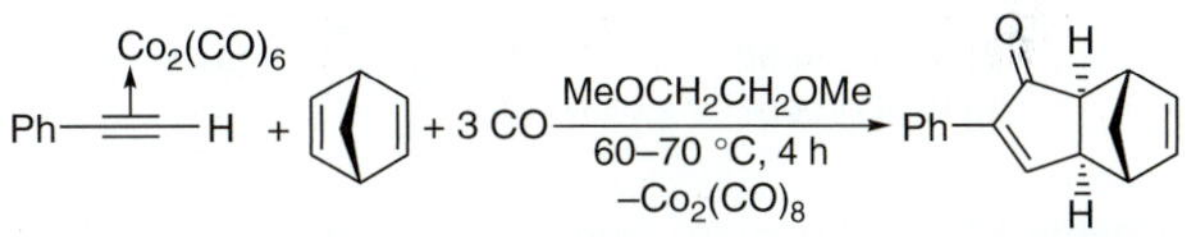

(8.27)

At the beginning, application of the Pauson–Khand reaction to other alkenes and alkynes required stoichiometric amounts of air-stable alkyne–$Co_2(CO)_6$ complexes, high temperature (50–120 °C), and CO pressure. An important improvement came with the use of N-methylmorpholine-N-oxide and trimethylamine N-oxide [231–234]. This permitted to obtain cyclopentenones at room temperature. N-oxides generate free coordination sites at cobalt by oxidative removal of CO ligands. TEMPO ((2,2,6,6-tetramethyl-1-piperidinyloxy), other name: (2,2,6,6-tetramethylpiperidin-1-yl)oxyl) also promotes stoichiometric Pauson–Khand reactions [235]. Significant rate increase of the stoichiometric Pauson–Khand reaction has been observed using primary amines as solvents or additives [236, 237]. The first intermolecular catalytic Pauson–Khand reaction

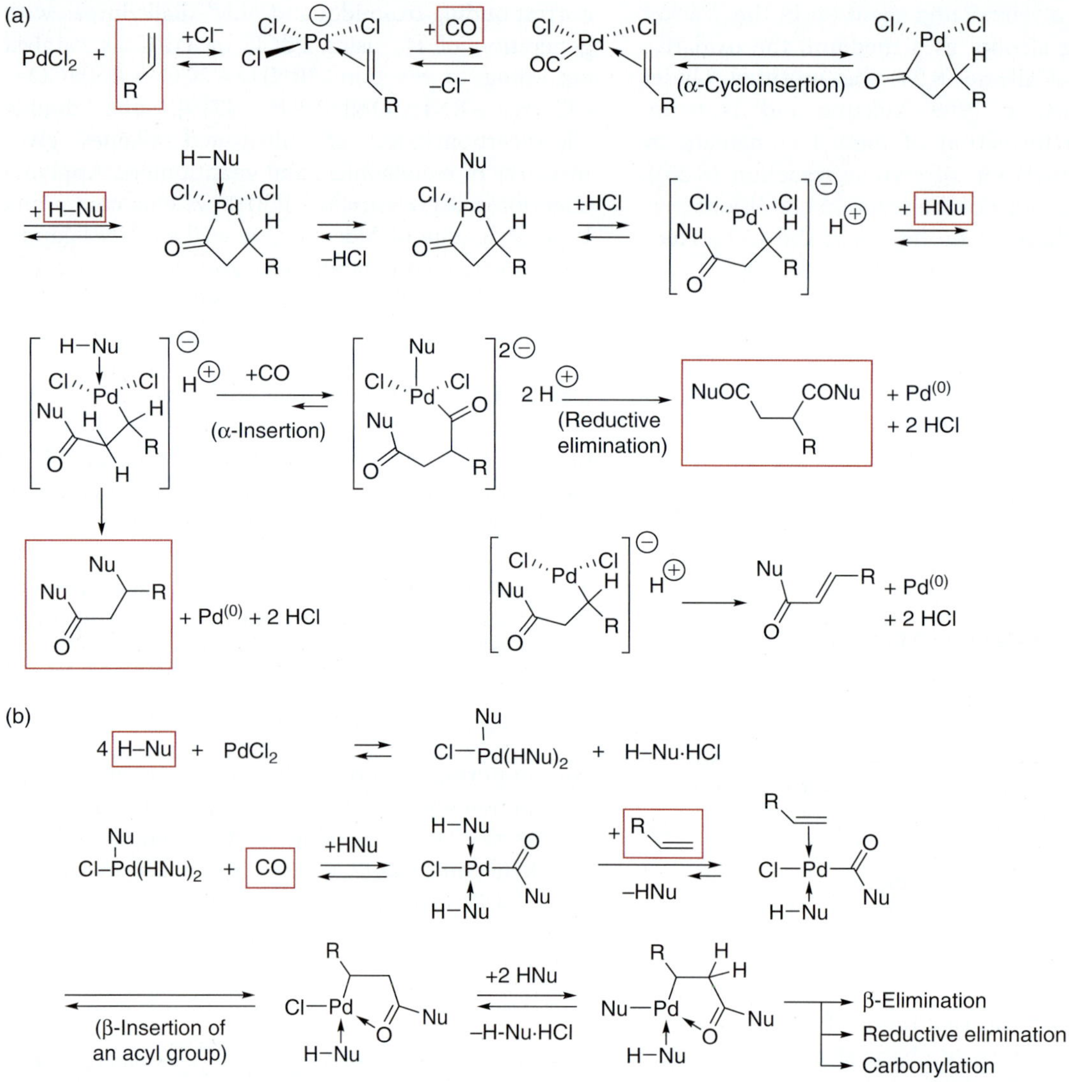

Scheme 8.14 Proposed reaction mechanism for the Pd(II)-mediated oxidative carbonylations of alkenes (see also Scheme 7.94, Section 7.11.5; HNu = H_2O, ROH, RNH_2, soft carbon nucleophiles, etc.): (a) the (alkene)$PdCl_2$ complex starts the reaction, (b) the nucleophile reacts first with $PdCl_2$.

was reported by Rautenstrauch and coworkers in 1990: hept-1-yne mixed with ethylene (40 atm) and CO (100 atm) formed 2-pentylcyclopent-2-en-1-one in 48% yield [238]. A more practical procedure was reported later by Jeong et al. who realized an intermolecular Pauson–Khand reaction at 3 atm of CO and 110 °C by stabilization of the cobalt catalyst with $P(OPh)_3$ [239]. Apart from $Co_2(CO)_8$, other sources of cobalt catalyst were proposed. For example, Sugihara and Yamaguchi used nonacarbonyl(methylidyne)tricobalt ($Co_3(CO)_9(\mu^3\text{-CH})$) as an efficient catalyst [240]. [$Co(acac)_2/NaBH_4$] under CO pressure and $Co_2(CO)_8$ in supercritical fluids have been used as catalysts [239, 241]. Pagenkopf and Livinghouse realized the catalytic Pauson–Khand reaction at 1 atm of CO with photoactivation of $Co_2(CO)_8$ (reaction (8.28)) [242].

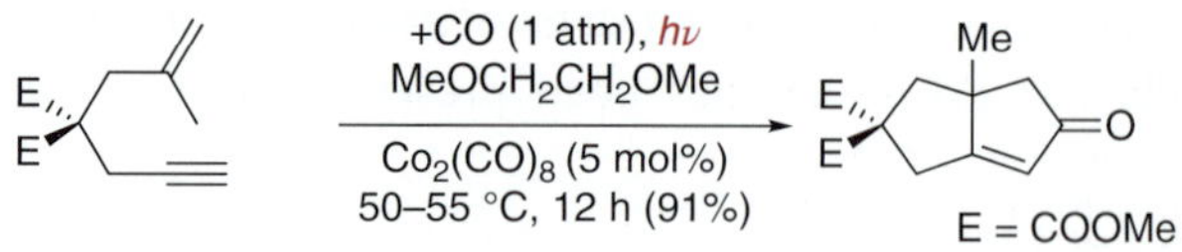

(8.28)

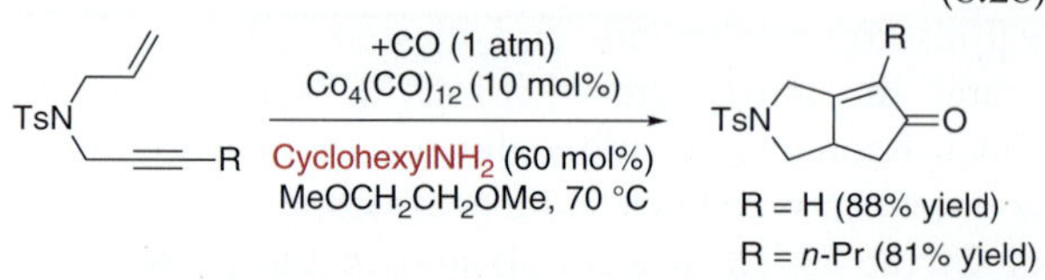

(8.29)

The first example of thermal catalytic cycloaddition under atmospheric pressure of CO was reported by

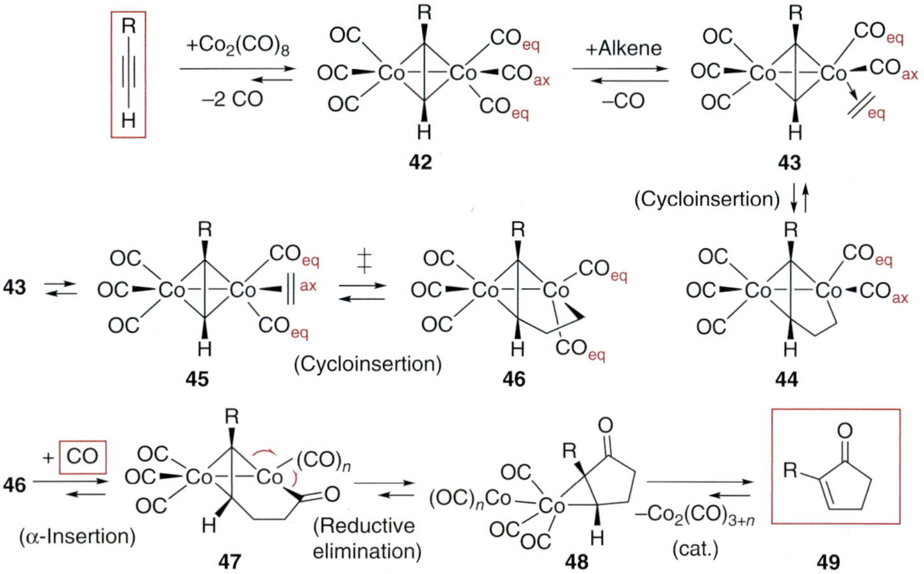

Scheme 8.15 Magnus–Príncipe mechanism of the Pauson–Khand reaction.

Rajesh and Periasamy [243]. A combination of $CoBr_3$ (0.4 equiv.) and Zn (0.43 equiv.) was used to facilitate the intermolecular reaction (110 °C). Hashimoto and coworkers showed that phosphane sulfides are able to promote the $Co_2(CO)_8$-catalyzed reaction (1 atm CO, 70 °C) [244]. Conversion of enynes to bicyclic cyclopentenones has been realized with $Co_2(CO)_8$ and tetramethylthiourea as catalysts under 1 atm of CO (benzene, 70 °C) [245]. Other Lewis bases such as phosphine oxides [246] and sulfoxides have been shown to promote the Pauson–Khand reaction [247]. In the presence of $Co_2(CO)_8$ or $Co_4(CO)_{12}$ and cyclohexylamine as catalysts, the Pauson–Khand reaction (8.29) occurs under 1 atm of CO and at 70 °C [234]. In 1985, Magnus and Príncipe proposed the mechanism shown in Scheme 8.15 [248]. This mechanism has been supported by quantum mechanical calculations on model reaction: $H{-}C{\equiv}CH + H_2C{=}CH_2 + CO \rightarrow$ 2-cyclopenten-1-one [249]. In the presence of $Co_2(CO)_8$, the tetrahedral (alkyne)$Co_2(CO)_6$ complex **42** forms. After loss of a CO ligand and coordination of the alkene intermediate, **43** is obtained in which the alkene moiety occupies an equatorial position. Subsequent coupling of the alkyne and alkene units forms the metallocycle **44** (oxidative cyclization, β-cycloinsertion, Section 7.7.14), alkyl migration (α-insertion, Section 7.7.9) into the carbonyl group as in **47**, and reductive elimination provides the cyclopentenone complex **48** that finally liberates product **49** and a (polycarbonyl)dicobalt complex that starts another catalytical cycle by coordinating the starting alkyne.

Electrospray ionization mass spectrometry has been employed to detect an alkene complex of type **43** (which undergoes cycloinsertion into **44**) or **45** (which undergoes cycloinsertion into **46**) [250, 251]. The X-ray crystal structures of **50** and **51** have been reported [252].

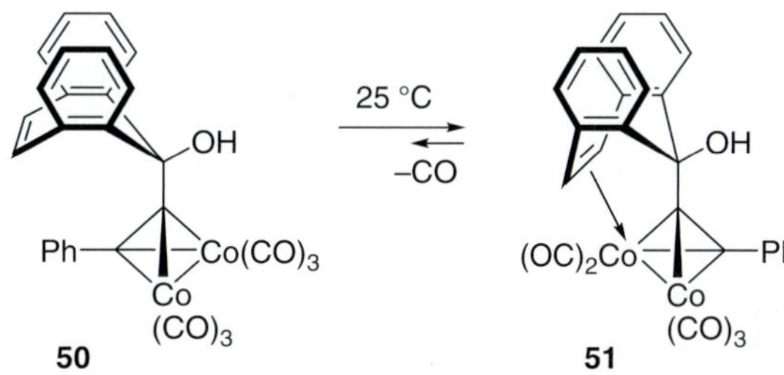

On heating **51** (toluene under reflux), no product of Pauson–Khand reaction is formed. This supports the hypothesis that alkene insertion occurs preferentially from an axial position as for **45** → **46**, and not from an equatorial position. Complex **51** represents an arrested Pauson–Khand intermediate, the consequence of its rigidity (no rotational freedom). It is important to note that all the bonding steps occur on only one cobalt atom. The other cobalt atom of the complex is presumed to act as an anchor that has additional electronic influences on the bond-forming metal atom via the Co—Co bond. Kinetics for the $Co_2(CO)_8$-catalyzed intermolecular Pauson–Khand reaction of $Me_3SiC{\equiv}CH$ + norbornadiene at a constant CO pressure gave the rate law: rate = $k[Co_2(CO)_8]^{1.3}[\text{norbornadiene}]^{0.3-1.2}[CO]^{-1.9}$. Thus, the reaction is overall zero order with respect to alkyne concentration, 1.3 order with respect to catalyst concentration, between 0.3 and 1.2 order with respect to the alkene, and −1.9 order with respect to CO concentration. The alkene insertion (e.g. **45** → **46**, Scheme 8.15) appears to be the rate-limiting step, in agreement with theoretical studies [249]. Using $Co_4(CO)_{12}$ as a catalyst, a similar rate law is observed [253]. Because of easier recycling of heterogeneous catalysts, immobilized Co-catalysts on silica gel [254, 255], molecular sieves + *t*-BuOH [256], charcoal [257], or nanoparticles have been proposed [257–264]. After simple filtration, the catalyst is reused up to 10 times without significant loss of activity. Commercially available Raney cobalt has been applied as a catalyst in intra- and intermolecular Pauson–Khand reactions [265]. The problem of regioselectivity associated with the use of nonsymmetrical alkynes and alkenes in intermolecular reactions (e.g. reaction (8.30)) is suppressed when tethering these two compounds [266]. Alternatively, playing with secondary interactions between the reacting partners through, for example, hydrogen bonding or Lewis acid–base interactions that are preserved during the C—C bond forming steps, high regioselectivity can be realized for intermolecular Pauson–Khand reactions. This concept was applied first by Krafft [267]. With alkyl or hydroxy, methoxymethyl-substituted alkenes, no regiocontrol was realized. In contrast, with alkenes containing a sulfur or nitrogen Lewis base, increased yields and good regiocontrol were obtained (reaction (8.30)) [268, 269].

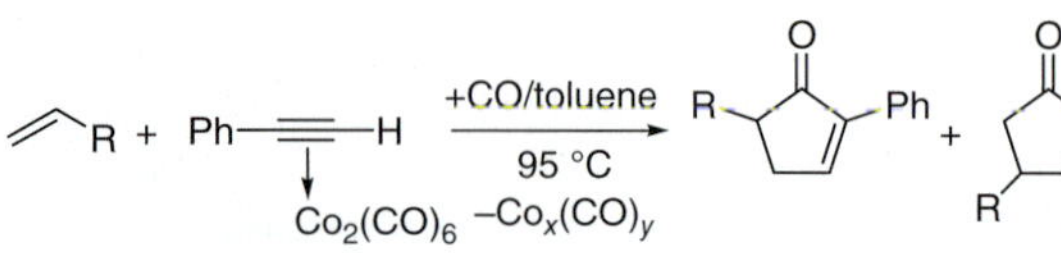

(8.30)

(a) R = n-C_6H_{13} 1 : 1 (49%)
(b) R = CH_2CH_2SMe 18 : 1 (68%)
(c) R = $CH_2CH_2NMe_2$ 5 : 1 (72%)

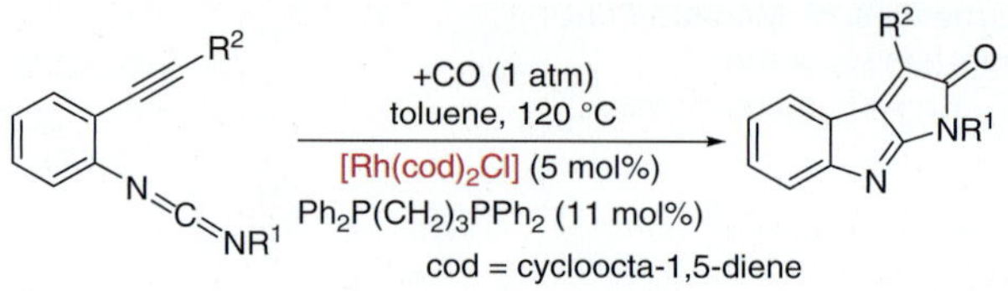

(8.31)

Apart from $Co_2(CO)_8$ and $Co_4(CO)_{12}$, $Cp_2Ti(CO)_2$ [270], $Ru_3(CO)_{12}$ [271, 272], $[RhCl(CO)_2]_2$ [273], $RhCl(PPh_3)_3$, $[RhCl(CO)(Ph_2P(CH_2)_3PPh_2)$ [274], and $Mo(CO)_3(DMF)_3$ (DMF = *N*,*N*-dimethylformamide) [275] alone, or in combination with silver salts, have been used as catalysts in the Pauson–Khand reaction. Buchwald and coworkers reported that $(t\text{-Bu})Me_2SiN{=}C$: can be used instead of CO to generate iminocyclopentenes [276–278]. In this case, $Cp_2Ti(PMe_3)_2$ and Ni(0) catalysts were used. The intramolecular Rh-catalyzed alkyne-carbodimide + CO Pauson–Khand-type reactions (8.31) have been reported [279]. Excellent diastereoselectivity has been observed for Rh(I)-catalyzed intramolecular Pauson–Khand reactions (8.32) (Scheme 8.16) [280, 281].

Yield and diastereoselectivity are higher at 1 atm than at lower CO pressure. Quantum mechanical calculations suggested two mechanisms for the origin of diastereoselectivity in which optimum selectivity can be attributed to a five- rather than a four-coordinate organorhodium complex of type **52**. The latter is favored at high pressure of CO. Oxidative cyclization **52** → **53** goes through a transition state in which gauche interactions between substituent R^1 and alkylrhodium moiety that forms are minimized (shown with **53**). This is not the case when forming the minor compound [282]. CO insertion in **53** generates **54**, which gives the major

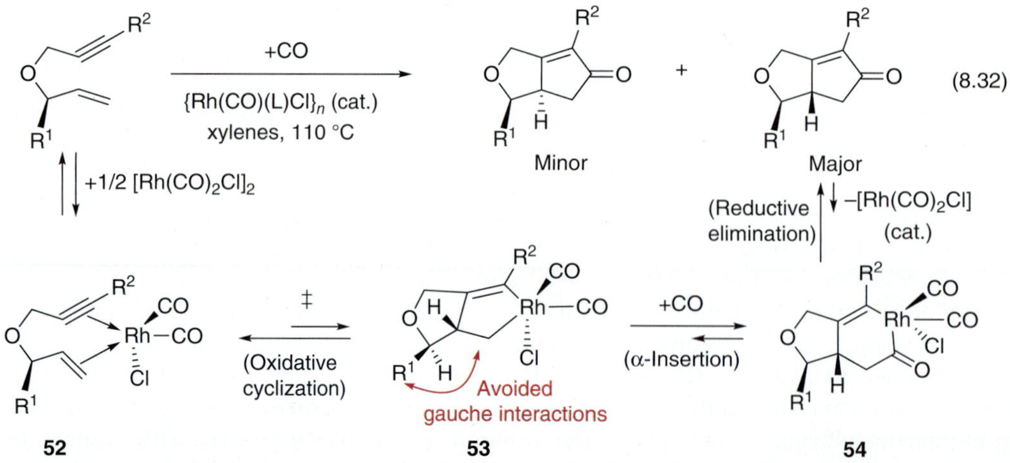

Scheme 8.16 [Rh(CO)(ligand)]Cl-catalyzed Pauson–Khand reaction.

(2+2+1)-cycloadduct after reductive elimination and recovery of the catalyst. Narasaka and Shibata reported the first intramolecular Pauson–Khand-type reaction (8.33) of allenyne using a (carbonyl)iron complex under photoirradiation [283, 284]. Independently from the length of the tether between the alkyne and allene moieties, only products implying (2+2+1)-cycloaddition of the terminal double bond of the allene are observed. Interestingly, in the case of the Ir(I)-catalyzed reactions (8.34), the internal double bond of the allene moiety is preferred for the Pauson–Khand reaction [285]. Catalyzed Pauson–Khand and allenic Pauson–Khand reactions are extremely useful synthetic tools [280, 286, 287]. The (2+2+1)-cycloaddition of allenes + CO have been catalyzed by $[Rh(CO)_2Cl]_2$ [287].

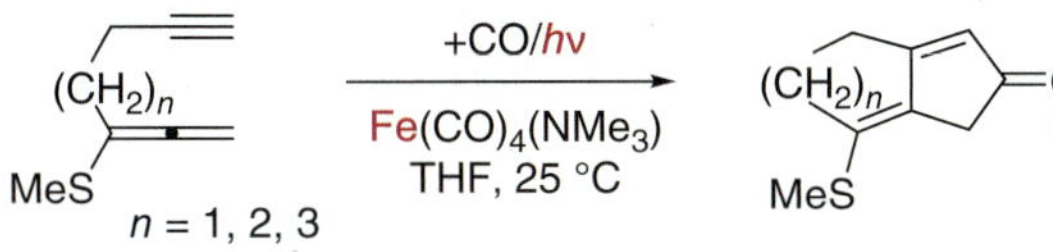

(8.33)

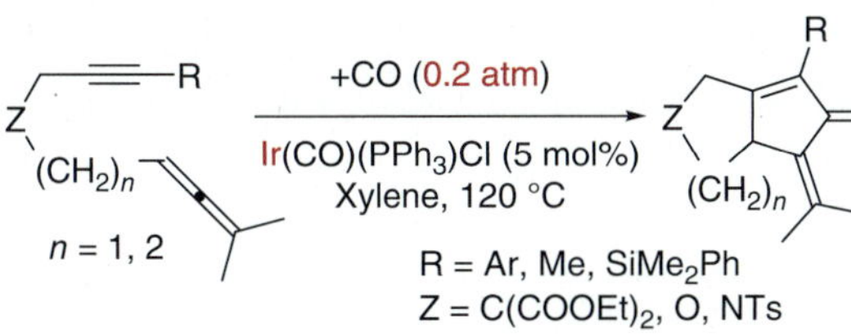

(8.34)

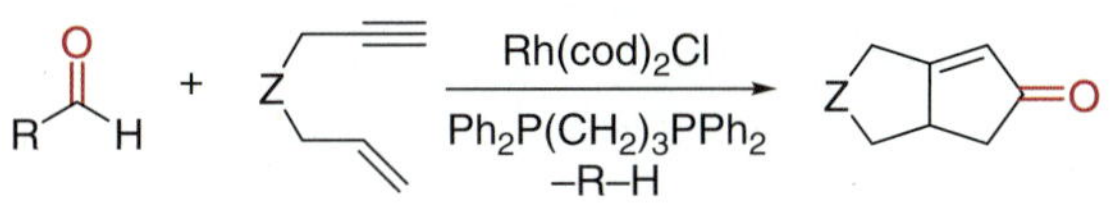

(8.35)

Formic esters have been used as CO surrogates in Rh—(I)-catalyzed Pauson–Khand reactions. The rhodium catalyst converts HCOOR into ROH and CO, the latter is then used in the (2+2+1)-cycloaddition [288]. Morimoto et al. have shown that aldehydes are a useful source of CO (RCHO + [M] ⇄ R—CO[M]—H ⇄ R—[M—H]=C=O → RH + CO + [M]) in the Rh(I)-catalyzed Pauson–Khand reaction (8.35) [289]. The best sources of CO are C_6F_5CHO, cinnamaldehyde, and formaldehyde [290, 291]. Pauson–Khand reaction of allenynes catalyzed by Co—Rh heterobimetallic nanoparticles has been reported [264]. Wender et al. have developed the Rh(I)-catalyzed (2+2+1)-cycloaddition of 1,3-dienes + alkene + CO, showing that dienes can be used instead of alkynes in the Pauson–Khand reaction [292]. Because cyclopentenone moieties are found in many natural products and compounds of biological interest, several authors have proposed catalytic enantioselective Pauson–Khand reactions [293]. An example is shown with reaction (8.36).

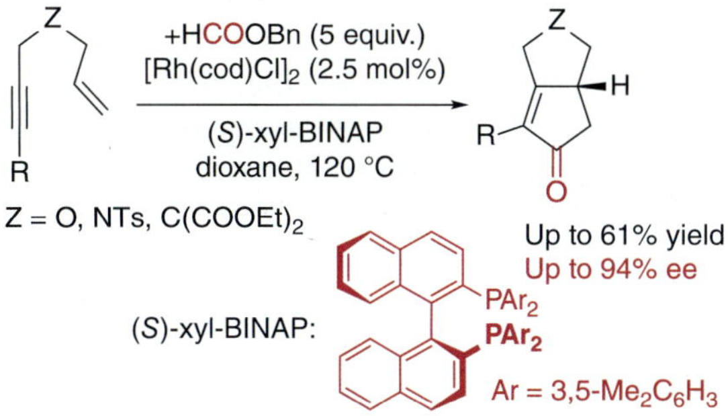

(8.36)

Problem 8.8 What is the major product **P** of the following reaction? [294]

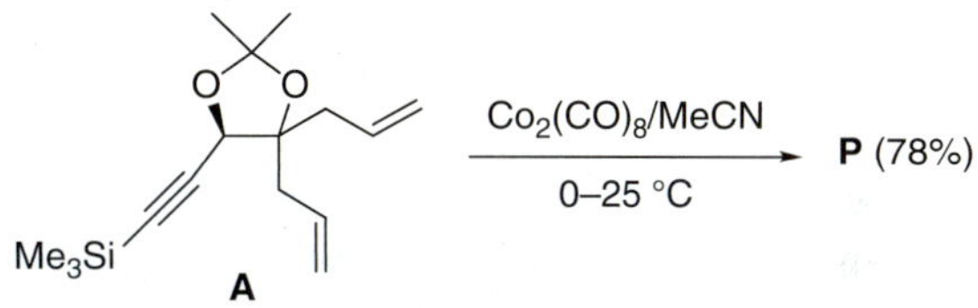

Problem 8.9 What is the product **P** of the following reaction? [295]

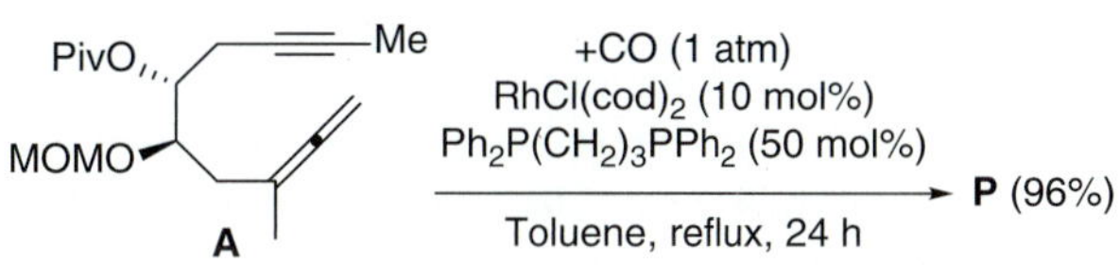

8.2.8 Carbonylation of halides: synthesis of carboxylic derivatives

In 1959, Chiusoli reported the $Ni(CO)_4$-catalyzed carbonylation of allyl chlorides, generating the corresponding β,γ-unsaturated acyl chlorides [296, 297]. Then, $PdCl_2$ and metallic Pd were found to be better catalysts for reaction (8.37), which also works with allylic esters and ethers [298–301]. With allylic carbonates [302] and allyl phosphates [303], the carbonylation reaction can be accomplished quantitatively at room temperature and under atmospheric pressure of CO. HCOOH and CH_2O can be used as nongaseous surrogates for CO [304].

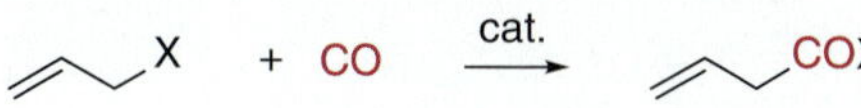

(8.37)

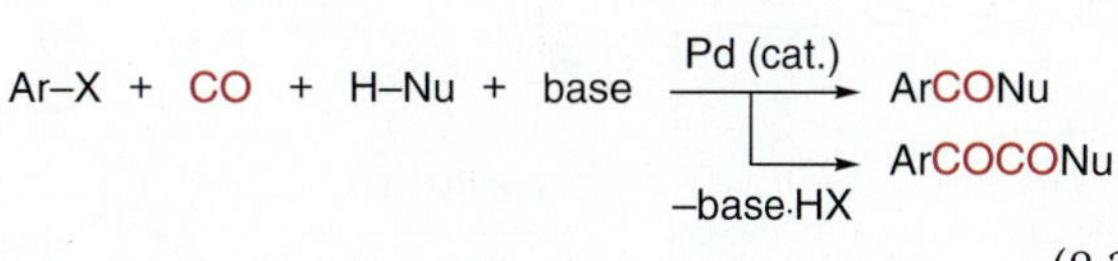

(8.38)

In 1974, Heck and coworker discovered the three-component coupling of an aryl halide, CO, and a nucleophile H—Nu (reaction (8.38)) [305]. It is now an important method for the preparation of carboxylic derivatives such as esters, α-ketoesters, amides, α-ketoamides, ketones, aldehydes, and anhydrides [306, 307]. *N*-Methoxy-*N*-methyl amides (Weinreb amides [308]) are well-established acylating agents [309, 310]. They can be obtained from aryl iodides [311] or from aryl bromides according to reaction (8.39) under atmospheric pressure of CO using $Pd(OAc)_2$ as a catalyst and a diphosphine ligand. The best results are obtained with Xantphos, which presents a large bite angle (110°) [312]. Carbonylation of alkyl halides and sulfonates ($R—CH_2—X + CO + H—Nu + base \rightarrow R—CH_2—CONu + H\text{-}base^+/X^-$) has also become routine [313].

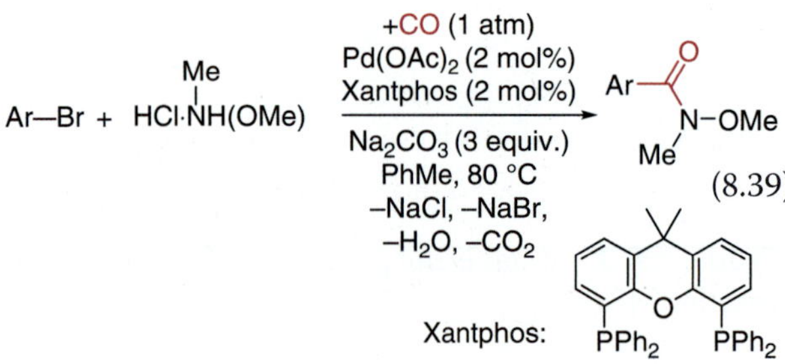

(8.39)

Among the aryl halides, aryl chlorides are the most abundant and most economical, but they are less reactive than bromides, iodides, and triflates. In 1989, Milstein and coworkers reported that the carbonylation of aryl chlorides can be carried out in the presence of $Pd[(i\text{-}Pr)_2PCH_2CH_2CH_2P(i\text{-}Pr)_2]_2$ as catalyst (150 °C, 5 atm) [314]. Later, Buchwald and coworkers developed a general practical protocol for the aminocarbonylation of aryl chlorides at atmospheric pressure of CO (reaction (8.40)), which uses electron-rich diphosphine ligands (e.g. $Cy_2P—(CH_2)_3—PCy_2$, Cy = cyclohexyl) together with sodium phenoxide as the basic additive [315]. The latter plays a dual role. Being more nucleophilic than the amine, it intercepts the acylpalladium intermediate **55** (Scheme 8.17) resulting from oxidative addition of ArCl and migratory insertion of CO; the latter leads to the formation of a phenyl ester. PhONa is also the catalyst of the aminolysis of this esters [316, 317].

Carbonylation of aryl arenesulfonates and mesylates have been catalyzed by $Pd(OAc)_2$ and diphosphine ligands [318, 319]. Efficient carbonylations (8.41) of aryl iodides, bromides, chlorides, and triflates catalyzed by $Pd(OAc)_2$/dppf under atmospheric pressure of CO has been realized in a tetraalkylphosphonium bromide (example of an ionic liquid) as a solvent [320]. In this case, after isolation of the products (esters), the ionic liquid containing the Pd(0) catalyst is dried under vacuum and recharged with a starting material for another carbonylation with minor loss in catalytical activity.

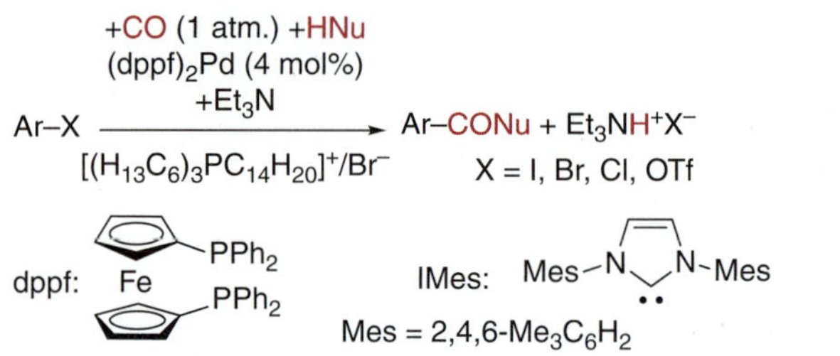

(8.41)

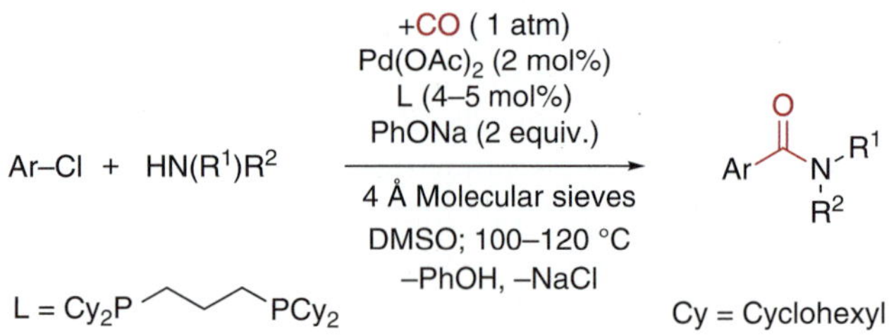

(8.40)

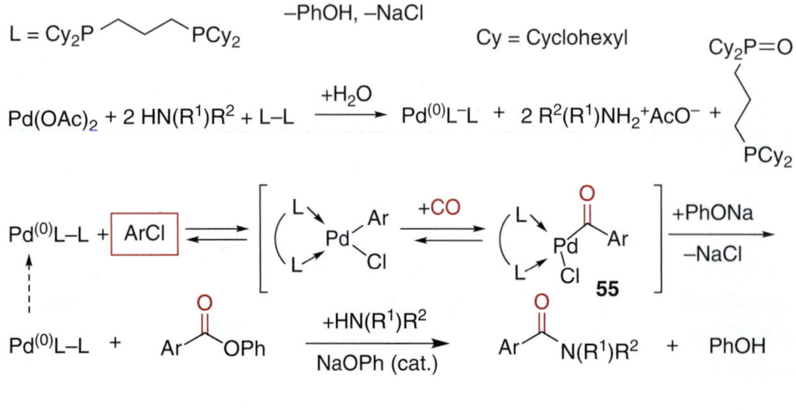

Scheme 8.17 Proposed reaction mechanism for the Pd-catalyzed aminocarbonylation of aryl chlorides.

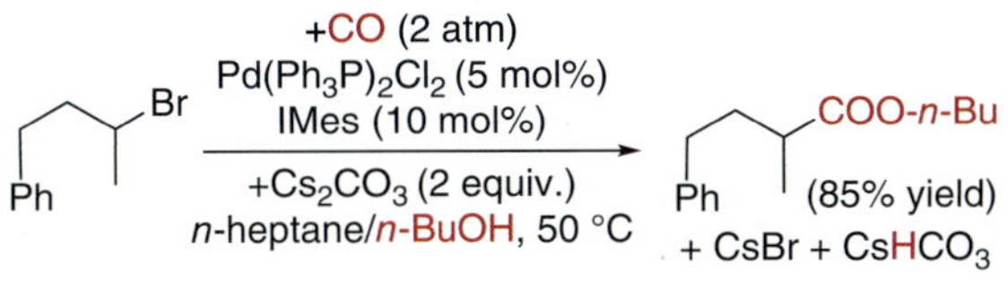

(8.42)

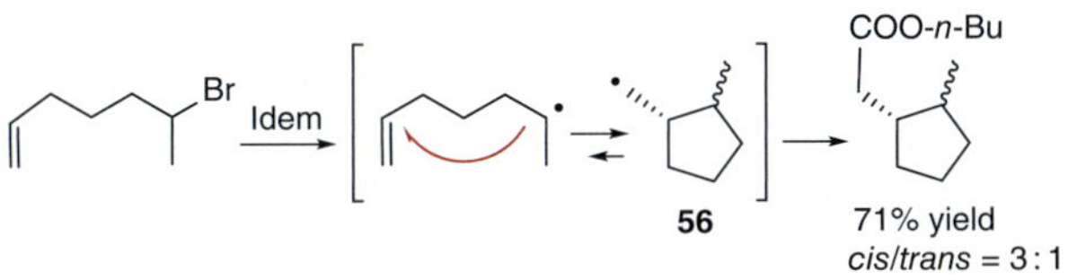

(8.43)

Unactivated alkyl bromides are carbonylated under 2 atm of CO in the presence of $Pd(Ph_3P)_2Cl_2$ and IMes (a *N*-heterocyclic carbene ligand: *N,N′*-bis(2,4,6-trimethylphenyl)imidazole-2-ylidene) as illustrated with reaction (8.42). Mechanistic studies suggested that the catalytic process involves the formation of secondary alkyl radical intermediates (Scheme 8.18) [321]. Carbonylation of hept-6-en-2-yl bromide gave a mixture of diastereomeric cyclopentyl products (reaction (8.43)) which is consistent with the formation of hept-6-en-2-yl radical intermediate that cyclizes rapidly into **56** [322]. Instead of using toxic CO, the direct carboxylation of aryl bromides can be carried out with CO_2 and Et_2Zn ($ArBr + CO_2 + Et_2Zn \rightarrow ArCOOZnEt + EtBr$) in the presence of $Pd(OAc)_2$ and a phosphine ligand as the catalyst [323].

8.2.9 Reductive carbonylation of halides: synthesis of carbaldehydes

In 1974, Schoenberg and Heck reported the first Pd-catalyzed reductive carbonylation of aryl and vinyl halides with CO utilizing H_2 as a reductant at 80–150 °C and under 80–100 atm (e.g. $ArBr + CO + H_2 + R_3N \rightarrow ArCHO + R_3NH^+/Br^-$) [324]. In 2006, Beller and coworkers presented an efficient alternative that requires only 5 atm of pressure at 100–125 °C (Scheme 8.19a) [325]. Mechanistic studies (Scheme 8.19b) showed that complexes of type $Pd(0)(CO)_mL_n$ and $Pd(Br)(H)L_2$ are resting states of the active catalysts and do not intervene directly in the catalytic cycle. The rate-determining step is the oxidative addition $PdL + ArBr \rightarrow ArPd(L)Br$. The product-forming step proceeds via base-mediated hydrogenolysis of complex of type ArCOPd(L)Br of its dimer $[ArCOPd(L)Br]_2$ [326]. The air-sensitive ligand n-$BuPAd_2$ can be substituted by the cheaper and easier to handle di(*t*-butyl)phosphinites ($(t\text{-}Bu)_2P$—OR) or by $(t\text{-}Bu)_2PCl + n\text{-}PrOH$ + TMEDA, a surrogate of the corresponding phophinite [327].

Mild reaction conditions were found using other reductants than H_2 such as Bu_3SnH [328–330], organosilanes [331–333], and formate salts [334–336]. Owing to the toxicity of CO and the requirement for high-pressure equipment in several cases, CO surrogates that can generate CO in situ under atmospheric pressure have been sought. In 2004, Cacchi et al.

Scheme 8.18 Palladium-catalyzed reactions may involve alkyl radical, Pd(I), and Pd(III) intermediates.

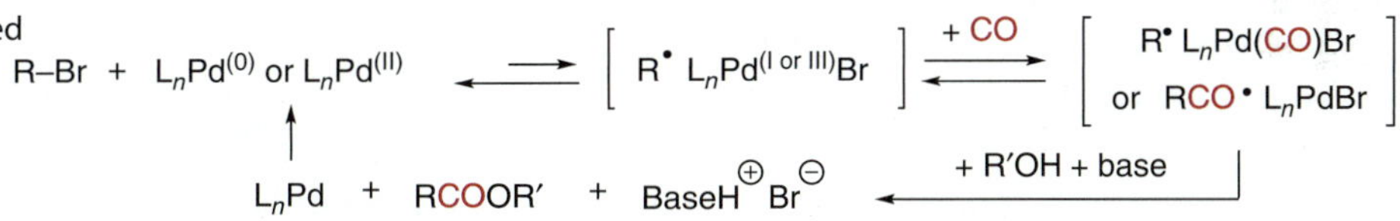

Scheme 8.19 (a) Beller's reductive carbonylation of aryl and heteroaryl bromides and (b) proposed reaction mechanism.

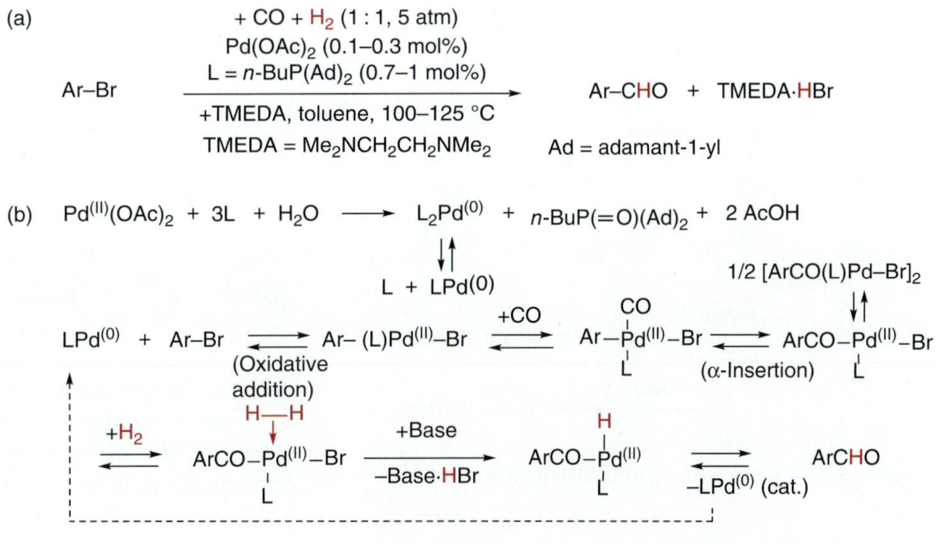

developed a Pd-catalyzed reductive carbonylation of aryl iodides using the mixed anhydride of formic and acetic acid as CO source and Et_3SiH as a reductant (reaction (8.44)) [337]. In 2013, Manabe and coworkers introduced the use of *N*-formylsaccharin as CO source (reaction (8.45)) [338], a compound used by Cossy and coworkers as an amine formylating agent [339]. The same year, Skrydstrup and coworkers reported the use of 9-methylfluorene-9-carbonyl chloride and potassium formate as the CO source [340–342].

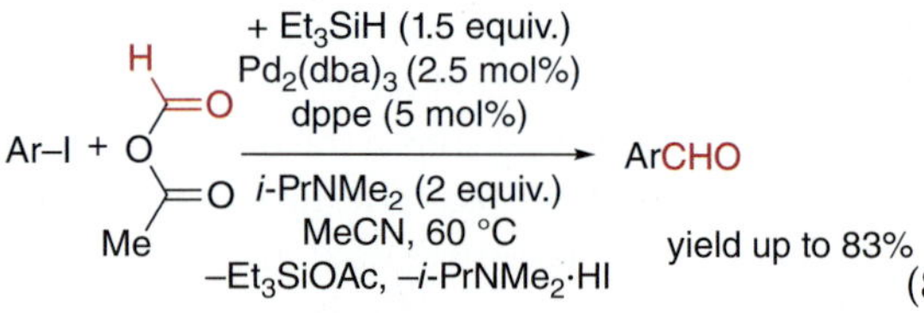
(8.44)

dppe = 1,1′-bis(diphenylphosphino)ethane

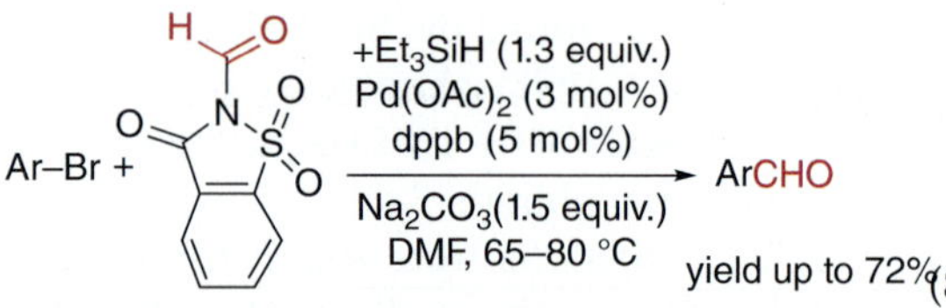
(8.45)

dppb = 1,1′-bis(diphenylphosphino)butane

In 2014, Liu and coworkers have shown that aryl iodides can be converted into the corresponding benzaldehydes (reaction (8.46)) by CO_2 (10 atm) with poly(methylhydrosiloxane) (PMHS) as a hydride source, in the presence of 1,8-diazabicyclo[5.4.0]undec-7-ene (DBU) as a base, and Pd/C as a catalyst (Scheme 8.20) [343].

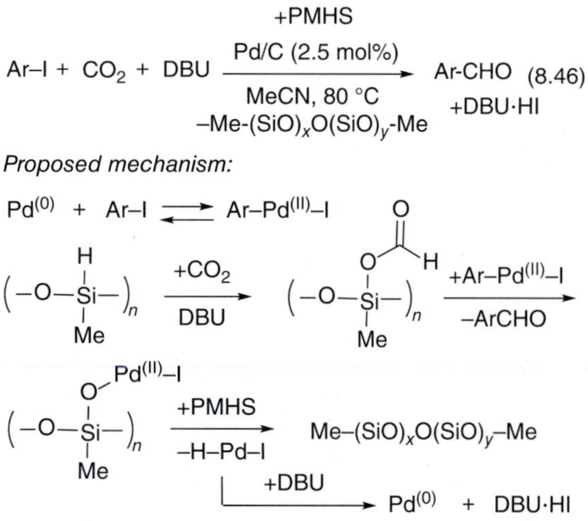

Scheme 8.20 Liu's reductive carbonylation of aryl iodide by CO_2.

Problem 8.10 Under the atmospheric pressure of CO, the following reaction gives mostly product **B**, whereas under 100 atm of CO, the ketoamide **D** becomes the dominant product [306]. Explain these results.

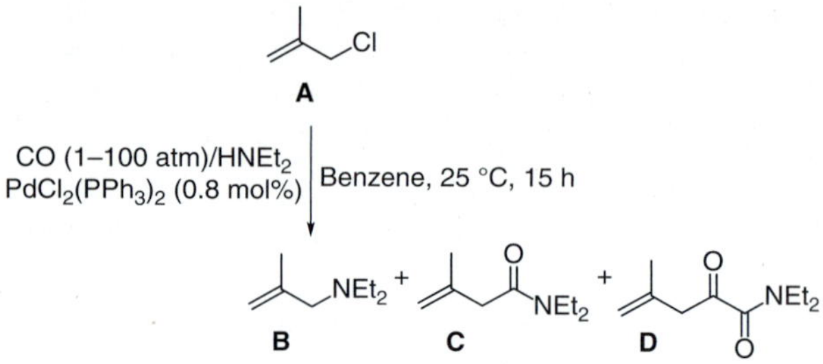

Problem 8.11 What is the major product **P** of the following reaction? [344]

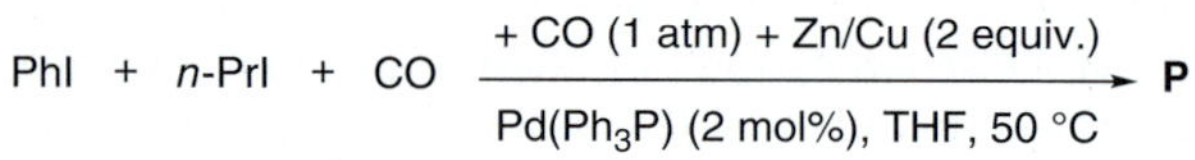

Problem 8.12 Estimate the heat of isomerization of hep-6-en-2-yl radical *trans*-(2-methylcyclopentyl) methyl radical (see reaction (8.43)) and the equilibrium constant of this isomerization under standard conditions (1 atm, 25 °C, gas phase).

8.2.10 Carbonylation of epoxides and aziridines

Carbonylation of epoxides provides direct access to β-lactones, useful synthons in organic synthesis [345–349], and precursors of poly (β-hydroxyalkanoates), important biodegradable polyesters [348, 350, 351]. Reaction (8.47) was discovered in 1957 using $Co_2(CO)_8$ as the catalyst (high pressure of CO and high temperature) [352].

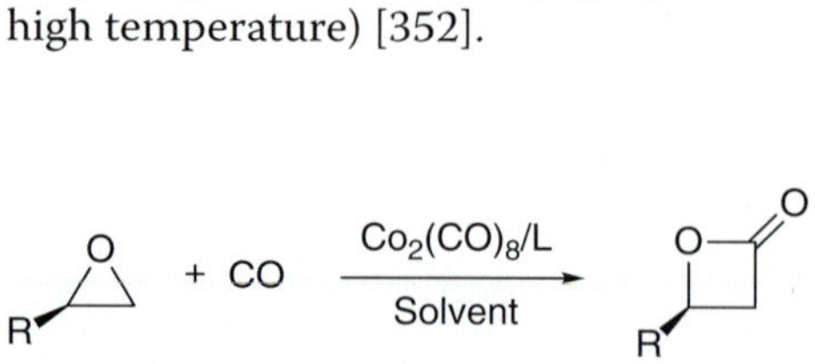
(8.47)

The key step is the nucleophilic attack of $[Co(CO)_{4-n}L_n]^-$ anion to the least sterically hindered carbon center of the epoxide (oxirane), with inversion of configuration. It was then found that $[Rh(cycloocta-1,5-diene)Cl]_2$ [353, 354] and Rh $(CO)(PPh_3)_2Cl$

catalyze the carbonylation of isoprene epoxide and styrene epoxide, respectively, under less harsh conditions (27–150 atm, 70–110 °C) [355]. Improved methods were developed by Drent and Kragtwijk [356, 357] using $Co_2(CO)_8$ and 3-hydroxypyridine and by Alper and coworkers using [PPN]$Co(CO)_4$ (PPN = bis(triphenylphosphine)iminium = $[Ph_3P{=}N{=}PPh_3]^+$) in conjunction with a Lewis acid such as $BF_3{\cdot}Et_2O$ or $SnCl_4$ in 1,2-dimethoxyethane (80 °C, 62 atm) [358]. On their side, Coates and coworkers have developed a highly active and selective porphyrin-based (OEP = octaethylporphyrin) epoxide carbonylation catalyst **57** showing excellent turnover numbers (TONs up to 10 000) and turnover frequencies (TOFs up to 1670 h^{-1}) with regioselective carbonyl insertion occurring between the oxygen and the sterically least hindered carbon center of the epoxide (steric regiocontrol) [359–361]. The same regioselectivity is observed using catalyst **58**. Interestingly, contrasteric control is obtained with catalyst **59** [362]. Enantiomerically pure catalyst **59** induces enantioselective carbonylation of *meso*-1,2-dialkyloxiranes [363].

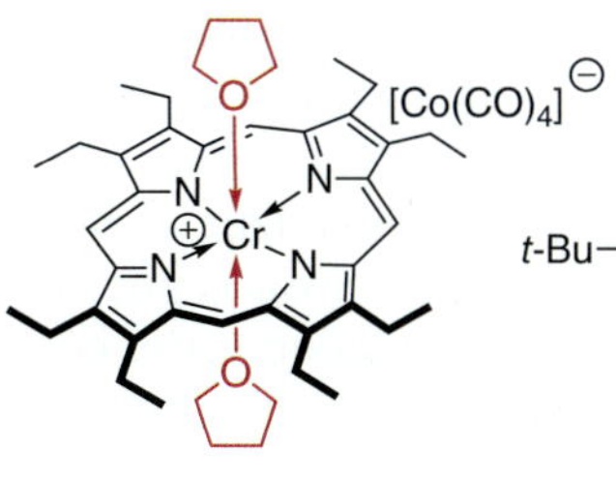

$[(OEP)Cr(THF)_2][Co(CO)_4]$: **57**

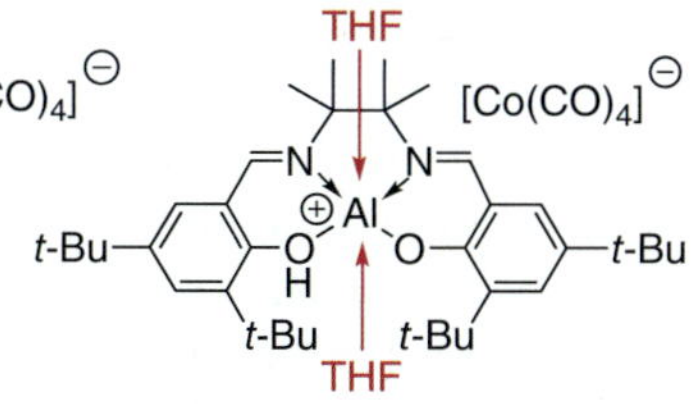

58

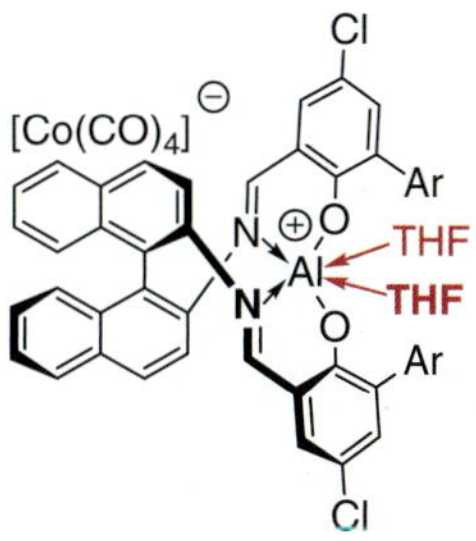

Ar = 2,4,6-$Me_3C_6H_2$ **59**

The carbonylation reaction can be run at room temperature under less than 20 atm of CO. When

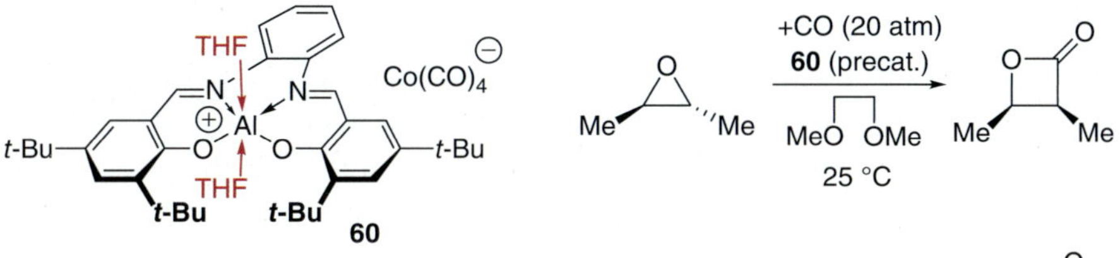

(8.48)

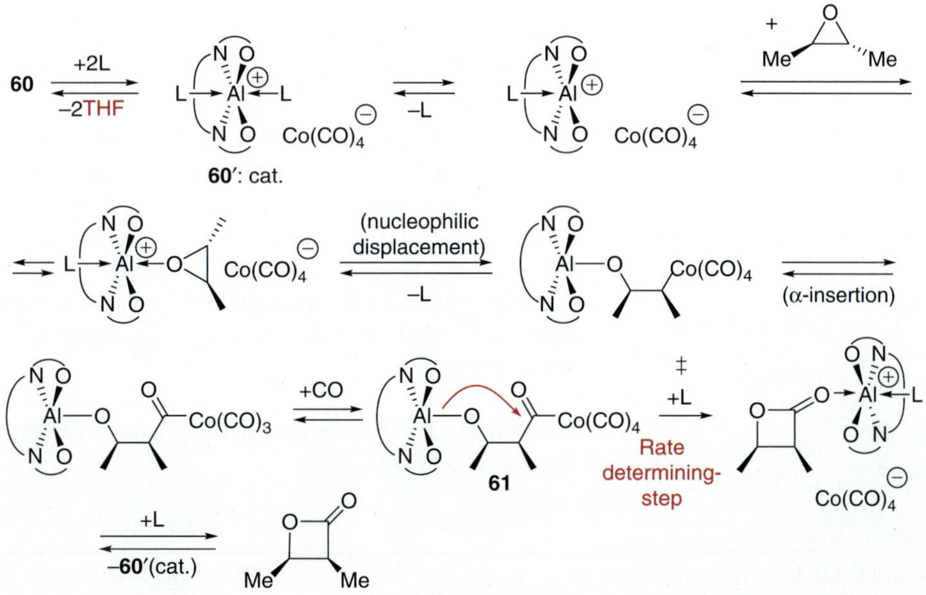

Scheme 8.21 Coates' mechanism of the epoxide carbonylation catalyzed by [Lewis acid]$^+$[$Co(CO)_4$]$^-$ catalysts such as **60**.

the carbonylation of *trans*-2,3-epoxybutane to β-valerolactone (reaction (8.48), Scheme 8.21) is performed in 1,2-dimethoxyethane solution using **60** as a catalyst, the reaction rate is independent of the epoxide concentration and CO pressure, but first order in **60**. The rate of lactone formation varies considerably in different solvents and depends primarily on the coordinating ability of the solvent. In mixtures containing THF (tetrahydrofuran), the reaction is first order in THF concentration. From spectroscopic and kinetic data, the catalyst resting state is proposed to be the neutral (β-aluminoxy)(acyl)cobalt species **61**, which can be trapped with isocyanates. As the formation of **61** from *trans*-1,2-epoxybutane, CO, and **60** is fast, lactone ring closing is the rate-determining step. The favorable influence of donating solvents such as THF is attributed to their stabilizing effect of the aluminum cation formed upon generation of the lactone [364]. For ab initio molecular dynamics simulations of reaction oxirane + CO catalyzed by BF_3 and $Co(CO)_4{}^-$, see: [365].

Two equivalents of CO can be incorporated into an epoxide. For instance, in the presence of a cobalt catalyst and a phase transfer agent, styrene oxide treated with MeI, NaOH, and CO (1 atm) gives 3-hydroxy-4-phenyl-2(5*H*)-furanone (reaction (8.49)) [366].

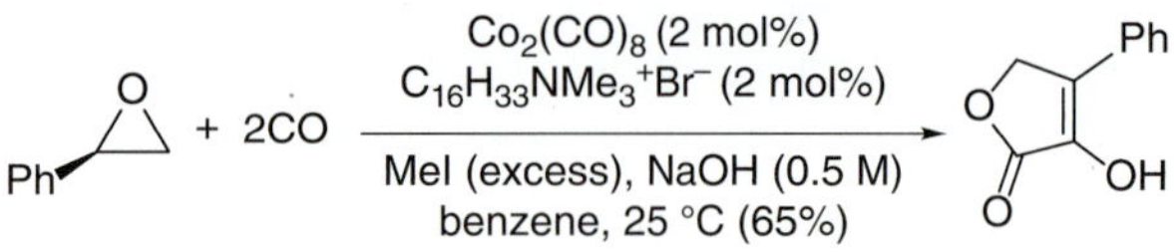

(8.49)

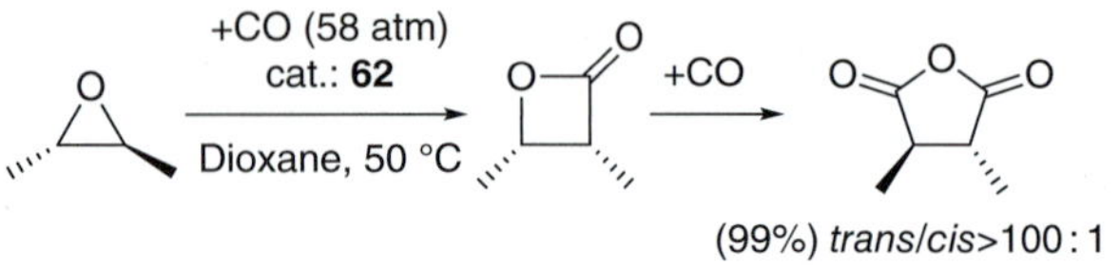

(8.50)

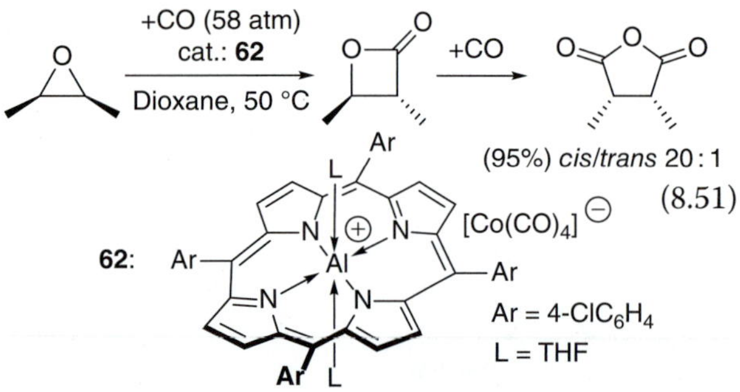

(8.51)

Using bimetallic catalyst **62**, Coates and coworkers reported the first catalytic method for the conversion of epoxides to succinic anhydrides via one-pot double carbonylation (e.g. reactions (8.50) and (8.51)) [367]. These reactions occur in two stages: first, the epoxide is carbonylated to a β-lactone, and then, the β-lactone is subsequently carbonylated to a succinic anhydride. Disubstituted and enantiomerically pure anhydrides are obtained with excellent retention of configuration. Investigation by in situ IR spectroscopy reveals that the two carbonylation stages are sequential and nonoverlapping. The rates of both epoxide and β-lactone carbonylation are independent of CO pressure and are first order in **62**. The rate of epoxide carbonylation is independent of substrate concentration and first order in donor solvent, whereas the rate of β-lactone carbonylation is first order in lactone concentration and inversely dependent on the concentration of donor solvent.

A β-lactam is the active functionality in penicillin and cephalosporin, two of the most important antimicrobial agents known [368]. Both naturally occurring and unnatural β-lactams are prepared by chemical [369–379] and enzymatic methods [380–382]. β-Lactams (azetidin-2-ones) are also important synthetic precursors of β-amino acids [383–386]. As aziridines are readily available (Section 5.4.2), their direct carbonylation represents a valuable method to prepare β-lactams [387, 388]. The reaction can be catalyzed by $[Ru(CO)_2Cl]_2$ in benzene (90 °C, 20 atm of CO) [389, 390], by $Ni(CO)_4$ using 1.5 equiv. of LiI [391] (25 °C, 1 atm CO), or by $Co_2(CO)_8$ (100 °C, 35 atm CO) [392]. In the latter case (reaction (8.52)), the active catalyst is $[Co(CO)_4]^-$. It induces nucleophilic ring opening attacking the least substituted carbon center of the aziridine moiety, with the inversion of configuration [393].

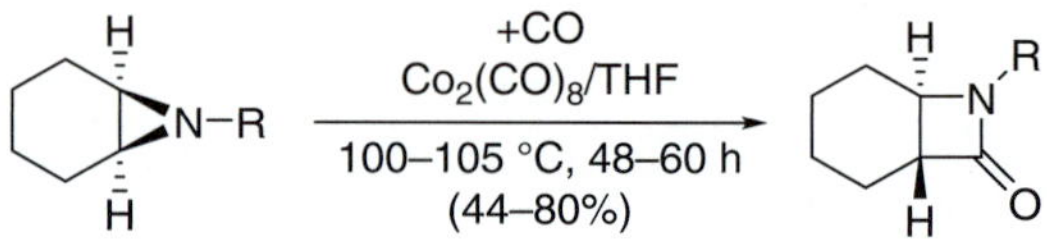

(8.52)

Catalysts such as **60** and $[Cp_2Ti(THF)_2]^+/[Co(CO)_4]^-$ are also useful for the carbonylation of aziridines into β-lactams [360]. Vinyl azirines are carbonylated under 1 atm of CO in the presence of Pd_2(dibenzylideneacetone)$_3$·$CHCl_3$ giving the corresponding *trans*-3-alkenylazetidin-2-ones as a major product (insertion of CO between the allylic C—N bond) [394].

Problem 8.13 What is product **P** of carbonylation of **A** catalyzed by **60**? [395]

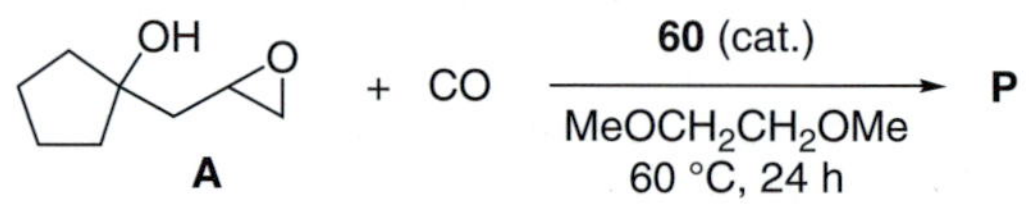

Problem 8.14 What is product **P** of the carbonylation of **A** (60 atm of CO, 2 days)? [396]

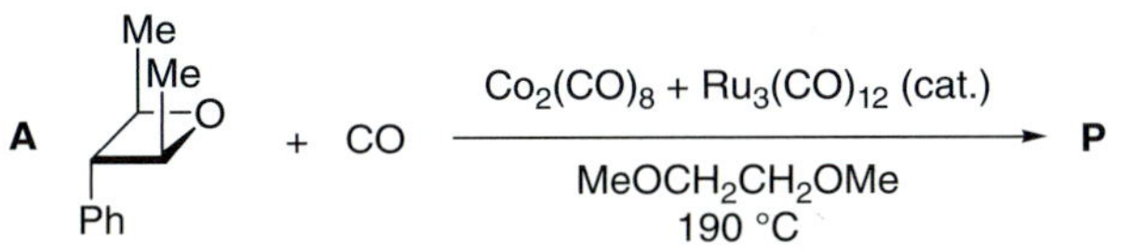

8.2.11 Hydroformylation and silylformylation of epoxides

Hydroformylation of epoxides provides an elegant and inexpensive pathway to β-hydroxyaldehydes, which are formal cross-aldol products between two aliphatic aldehydes and potentially useful intermediates in organic synthesis [397]. In the industry, hydroformylation of ethylene oxide is an attractive reaction because hydrogenation of the product, 3-hydroxypropanal gives propane-1,3-diol, which is an intermediate in the production of polyester fibers and films. To achieve this reaction, cobalt, rhodium, or iridium catalysts have been employed [348, 398]. Nozaki and coworkers [399] use $Co_2(CO)_8$/iminophosphine **63** as a catalyst for the hydroformylation of epoxides in trimethyl orthoformate (reaction (8.53)). This allows in situ protection of the carbaldehyde group as dimethylacetal, thus avoiding the formation of side products of crotonization and cross-aldol/crotonization.

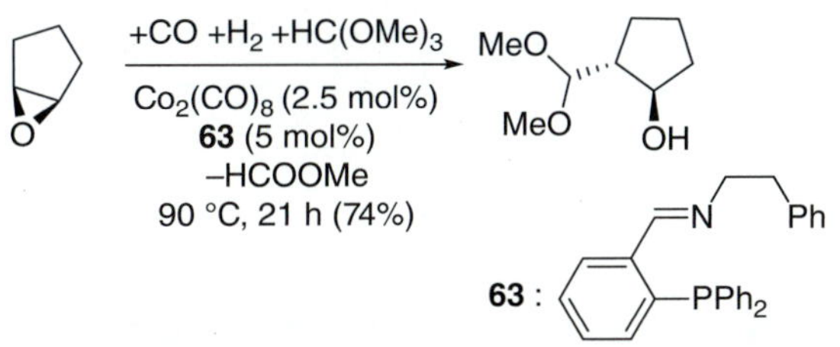

(8.53)

The possible mechanism of the $Co_2(CO)_8$-catalyzed reaction is shown in Scheme 8.22. Hydrogenation of $Co_2(CO)_8$ generates $HCo(CO)_4$. Its conjugate base attacks the least sterically hindered carbon center of the epoxide moiety with the inversion of configuration (S_N2). Subsequent carbonyl insertion gives a β-hydroxyacylcobalt species, which is hydrogenated providing a dihydride intermediate. CO coordination of the latter and subsequent reductive elimination furnishes the β-hydroxyaldehyde, which, in the presence of $HC(OMe)_3$, forms the corresponding β-hydroxyacetal.

Murai and coworkers have reported the silylformylation of epoxides to give β-silyloxyaldehydes, e.g. reaction (8.54) [400]. The regio- and stereoselectivity of the reaction is similar to the hydroformylation of epoxides presented above. It is proposed that $PhMe_2Si—Rh—H$ species generated by oxidative addition of $PhMe_2Si—H$ to $[Rh(CO)_3Cl]_2$ is the key catalyst.

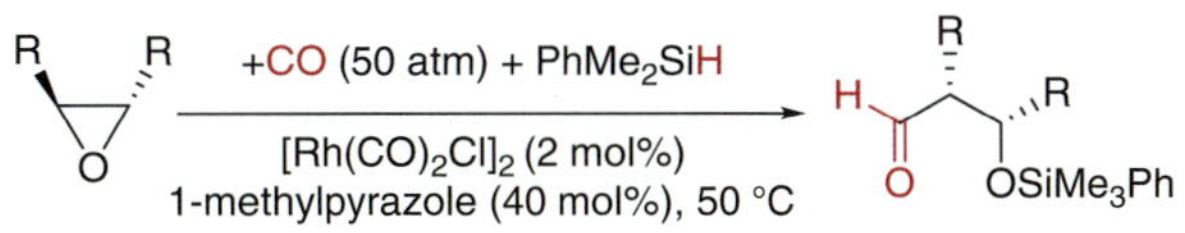

(8.54)

Problem 8.15 What is the major product of reaction of $CO + H_2$ with (2*S*)-2-benzyloxymethyloxirane catalyzed by $Co_2(CO)_8$ and ligand **63** in $HC(OMe)_3$? [399]

Problem 8.16 What is the major product of the reaction of $CO + PhMe_2SiH$ with 7-oxa-bicyclo[4.1.0] heptane catalyzed by $[Rh(CO)_2Cl]_2$ in CH_2Cl_2 containing 1-methylpyrazole? [400]

8.3 Direct hydrocarbation of unsaturated compounds

The addition of a C—H (R—H) bond across a double or triple bond is a hydrocarbation (e.g. R—H + A=E ⇄ R—A—E—H, R—H + R′C≡Z ⇄ R(R′)C=ZH). It is a classical and general method to generate C—C and C—heteroelement bonds. For the formation of C—C bonds, the unsaturated compound (A=E, R′C≡Z) is a polar carbonyl (C=O), imine (C=NR′), carbonitrile (R′C≡N), their π-conjugated derivatives

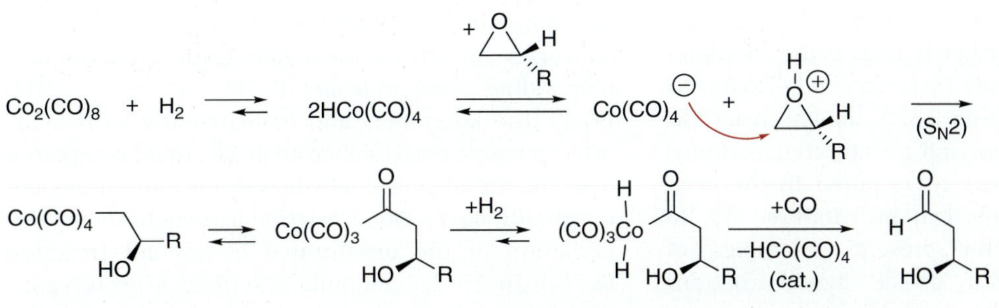

Scheme 8.22 Proposed reaction mechanism of $Co_2(CO)_8$-catalyzed hydroformylation of epoxides.

and analogs (e.g. Michael acceptors), or nonpolar alkenes, alkynes, and their conjugated derivatives. Catalyzed hydrocarbation (via C—H activation in lieu of prefunctionalization C—H → C—X → C—M) of unsaturated compounds is now an important tool of organic synthesis [401, 402]. The C—H bond activations and functionalizations of arenes, heteroarenes, alkenes, and alkanes can be divided into two general classes: those controlled by a neighboring function through steric, electronic factors, or specific coordination of the reagent or catalyst (directed C—H bond activation, the controlling function can be geminal, vicinal, or remote from the reacting C—H bond) and those occurring without such function (undirected C—H bond activation) [403]. One-step, concerted $[\pi^2s+\sigma^2s]$-cycloadditions are predicted to be difficult reactions (symmetry forbidden, small overlap between unoccupied molecular orbital/highest occupied molecular orbital (LUMO/HOMO) and large energy gaps between these reactant MOs, Section 5.3). Conformational strain required for the theoretically allowed concerted $[\pi^2a+\sigma^2s]$- or $[\pi^2s+\sigma^2a]$-cycloadditions renders these one-step reactions impossible. Usually, hydrocarbations follow multistep mechanisms involving either carbanionic intermediates of type R^- that undergo nucleophilic additions (8.55) or radical intermediates of type $R^\bullet$ that undergo radical additions (8.56) to the unsaturated compounds. Depending on the type of reactant R—H with so-called "active hydrogen," base B: in reaction (8.55) is used in stoichiometric or catalytic amount (e.g. base-induced aldol reaction, Claisen condensation, and Michael addition). The same can be said for the radical initiator $X^\bullet$ in reaction (8.56). Hydrocarbations may involve carbenium ion intermediates of type R^+ resulting from hydride transfer from R—H to a carbenium salt or another oxidant (Y^+). These reactions are thus considered as electrophilic additions (8.57) [404]. When reactant R—H is an alkene or heteroalkene, ene-reactions are possible. The allylic hydrogen atom is transferred to the enophile (A=E, R′C≡CR″, R′C≡N). They can be one-step, concerted processes or not, depending on the nature of the reactants (Section 5.7). Usually, they are catalyzed by protic or Lewis acids as the enophiles are generally Lewis bases (Section 7.6.1–7.6.5). As we shall see below (Section 8.3.2), the ene-reactions of enophiles that are not α,β-unsaturated carbonyl compounds (Lewis bases), as required in the classical Alder ene-reactions, can be catalyzed by Ru complexes. This chapter presents examples of transition-metal-catalyzed direct hydrocarbations of unsaturated systems. The methods do not produce coproducts (atom economical) and do not require tedious prefunctionalizations of the R—H reagents [12, 405–407]. Note that terminal alkynes (have relatively low pK_a, Table 1.A.23) represent a special case of reagents for hydrocarbation as their conjugate bases (acetylides) can be obtained more readily than for any other nonactivated hydrocarbons (allyl, benzyl, aryl, alkenyl, and alkyl carbanions). The hydrogen/metal exchange can be done with a catalytic amount of metal such as Cu, Rh, Pd, or Co, and this permits the direct hydroalkynylation of unsaturated compounds. In the presence of homochiral ligands, the reactions can be enantioselective and they represent one of the most efficient methods of asymmetric synthesis [408–417]. These useful reactions will not be detailed in this chapter.

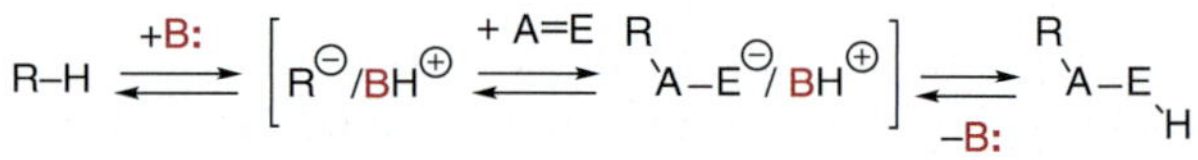

(8.55)

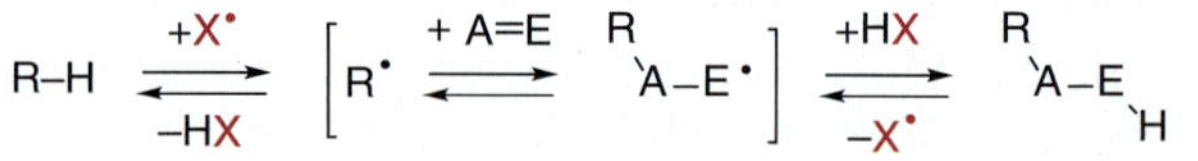

(8.56)

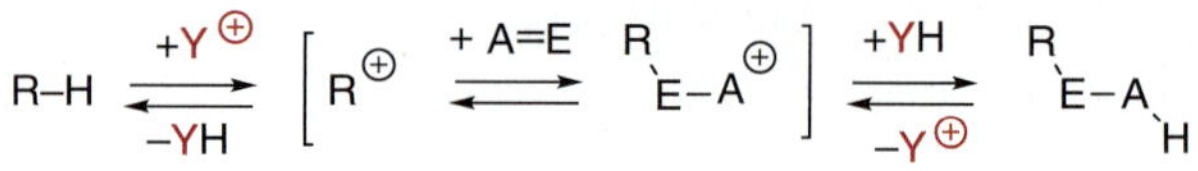

(8.57)

8.3.1 Hydroalkylation of alkenes: alkylation of alkanes

The direct functionalization of alkanes is the "Holy Grail" of chemical synthesis [418–422]. Although Nature has used monohydrogenase and other enzymes to functionalize them, alkanes have been considered nonreactive in conventional chemistry except in the presence of oxygen-centered (e.g. phthalimido-*N*-oxyl radical, PINO [423]) nitrogen-centered radicals and, fluorine, chlorine and sometimes bromine atom that can abstract a hydrogen atom from the alkane to produce the corresponding alkyl radicals ($R—H + X^\bullet \rightleftarrows R^\bullet + XH$) [424]. The latter may add (concurrently with many other possible reactions) to an unsaturated compound (e.g. alkene, alkyne, aldehyde, ketone, and imine) and eventually start a chain reaction leading to the hydrocarbation of the unsaturated compound (reaction (8.56)). In 1912, Langmuir described a prototypical heterogeneous–homogeneous reaction in which H• radical formed by interaction of H_2 with tungsten and by dissociative chemisorption and desorption into the gas phase [425]. Since then, the role of metallic surface-generated gas-phase radicals in catalytic reactions has been demonstrated in many processes [426]. For instance, when using a supersonic nozzle made of nickel or molybdenum, ethane, propane, and butane react (80 Torr, 1000 °C, 10 ms contact times) to give C_nH_m products with n extending to 12. Under the same conditions, CH_4 does not form higher hydrocarbons [427]. Aliphatic alkylation is a process whereby a branched alkane is alkylated with an alkene to produce a highly branched alkane product. Most current industrial processes are based on sulfuric acid (H_2SO_4) or hydrofluoric (HF) catalyzed alkylations of isobutane with C_3—C_5-olefins [428, 429]. For the production of high octane number motor-gasoline, 2,2,3-trimethylpentane (TMP) is obtained from but-2-ene and isobutane. Protonation of but-2-ene gives *s*-butyl cation, which isomerizes irreversibly into *t*-butyl cation ($\Delta_r H°$(*s*-Bu$^+$ ⇄ *t*-Bu$^+$, gas phase) = −16 kcal mol^{-1}, Table 1.A.14). The latter adds to but-2-ene giving 3,4,4-trimethylpent-2-yl cation that abstracts an hydride from isobutane to give TMP and *t*-butyl cation, which starts another alkene addition (Scheme 8.23) [430]. Solid phosphoric acid catalyst can be used to convert C_4-hydrocarbons obtained by the Fischer–Tropsch synthesis [431].

Methane is a particularly attractive substrate for alkane homologation as it is readily available in large amounts. Homogeneous methane conversion has focused on selective oxidation via activation with electrophilic late metal complexes in acidic media, or with reactive oxo species [432–434]. Alternative strategies involving nonoxidative mechanism such as σ-bond metathesis [435–437] have been developed for the catalytic hydromethylation of propene into isobutane (Scheme 8.24) [438]. Neopentylscandium complex **64** is prepared by addition of a pentane solution of $LiCH_2CMe_3$ to Cp^*_2ScCl (Cp* = pentamethylcyclopentadienyl) at low temperature (displacement reaction). It reacts with CH_4 to give isobutane and methylscandium complex **65**. The latter adds to propene forming **66** (an example of alkene metallacarbation). Complex **66** reacts with methane giving isobutane and catalyst **65**. Kinetic studies revealed that metallation of methane with **66** proceeds via a second-order rate law: rate = k[**66**][CH_4] ($k = 4.1 \times 10^{-4}$ M^{-1} s^{-1} at 26 °C). The primary kinetic isotopic effects k_H/k_D = 10.2 with CH_4 vs. CD_4 and k_H/k_D = 5.2 with CH_2D_2, respectively, are consistent with the linear transfer of hydrogen from CH_4 to the isobutyl group in transition state [**67**]‡ ($\Delta^‡H$ = 11.4 kcal mol^{-1}, $\Delta^‡S$ = −36 eu).

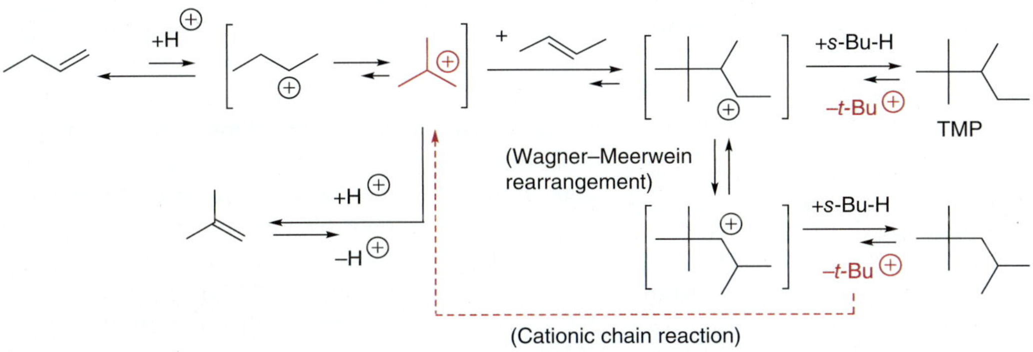

Scheme 8.23 Alkylate technology for the production of high octane number gasoline.

Scheme 8.24 Catalytic hydromethylation of propene (Cp* = η^5-Me_5C_5).

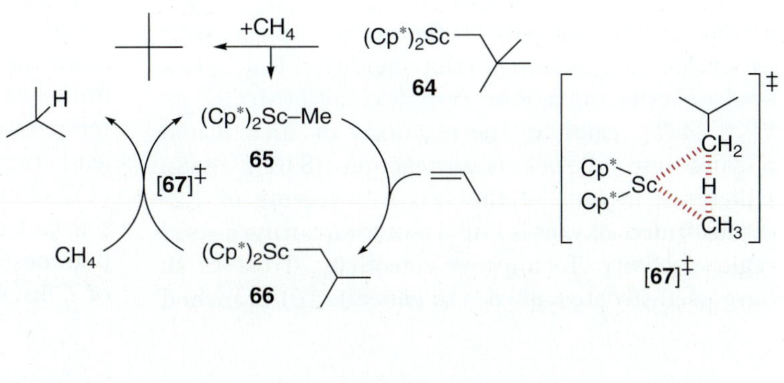

Styrenes and norbornene add to 1,3-dicarbonyl compounds giving the corresponding products of hydrocarbation in the presence of $Bi(OTf)_3$ as a catalyst [439]. Hydrocarbation of butadiene with reagents containing acidic C—H bonds are catalyzed by allyl(cyclopentadienyl)palladium and a phosphine ligand (reaction (8.58)). Using enantiomerically pure phosphine ligands such as Josiphos, asymmetric Pd-catalyzed 1,4-hydrocarbations of 1,3-dienes have been realized (e.g. reaction (8.59)) [440].

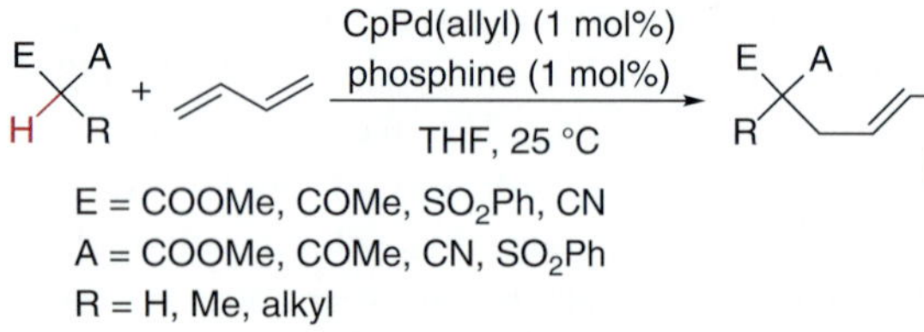

(8.58)

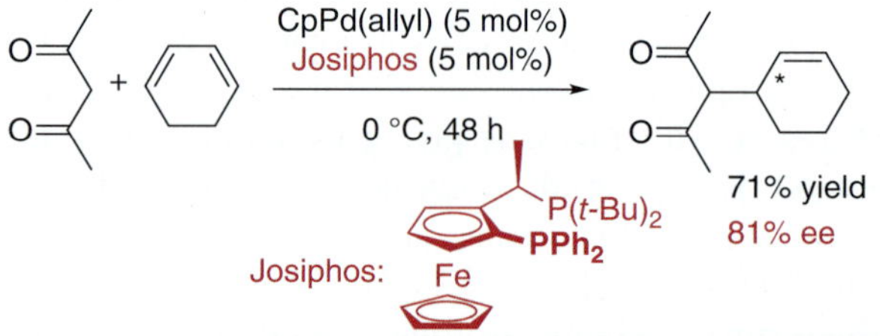

(8.59)

Intramolecular addition of 1,3-diones to nonactivated alkenes have been catalyzed by $PdCl_2(MeCN)_2$ [441]. Herzon and Hartwig have reported the hydroaminoalkylation of alkenes catalyzed by $Ta(NMe_2)_5$ in which α-(C—H) of secondary amines add across alkene double bonds [442, 443]. Tetrabenzyltitanium is a good catalyst for the activation of α-(C—H) bond of amines (e.g. reaction (8.60)) [444].

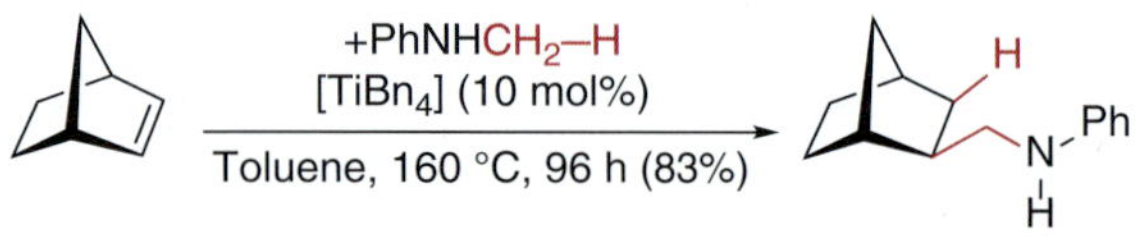

(8.60)

8.3.2 Alder ene-reaction of unactivated alkenes and alkynes

Trost et al. have found that ruthenium complexes such as (cycloocta-1,5-diene)(cyclopentadienyl)RuCl [445, 446], or better, the cationic complex $[CpRu(MeCN)_3]^+$ PF_6^- [447], catalyze ene-reactions of unactivated alkynes and alkenes (e.g. reaction (8.61)) [448]. Difference in bulk of the two substituents of 1,2-disubstituted alkynes is important for insuring a good regioselectivity. To improve selectivity, Trost et al. have used silylated alkynes to generate (*E*)-branched alkenylsilanes (reaction (8.62)) [449–451], while Lee and coworker have employed borylated alkynes (reaction (8.63)), which provide the corresponding (*Z*)-branched alkenylboronates [452].

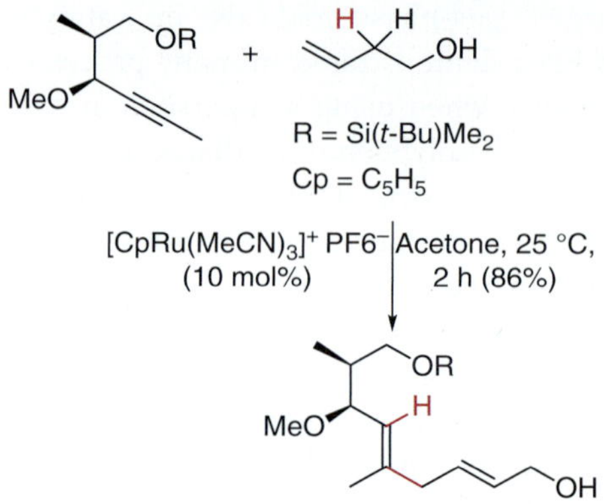

(8.61)

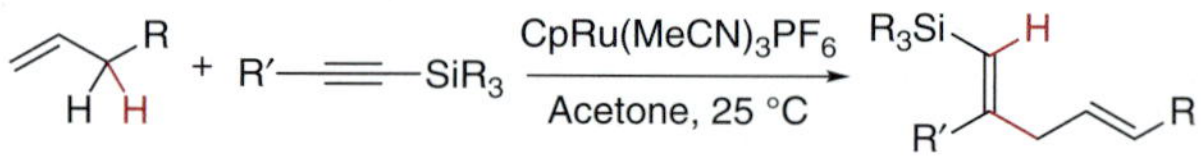

(8.62)

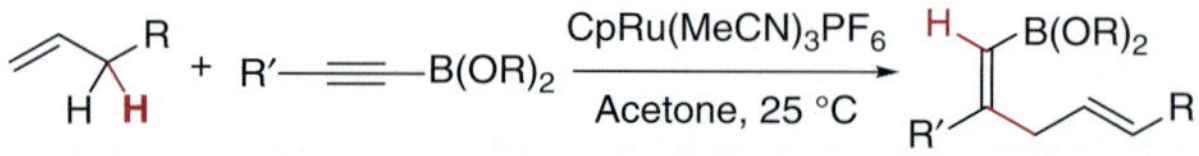

(8.63)

Quantum mechanical calculations predicted for the CpRuCl-catalyzed ene-reaction a mechanism involving the formation of a ruthenacyclopentene intermediate **68**, the rate-determining step being the oxidative coupling of the coordinated alkyne and alkene to form it (β-cycloinsertion, Section 7.7.14) [453]. When $CpRu(H_2O)^+$ or $CpRu^+$ is assumed to be the operative catalyst, a second mechanism competes with that involving ruthenacyclopentene intermediates. It involves the formation of π-allylruthenium intermediates of type **69** (Scheme 8.25) [454]. The ene-reaction of nonactivated acetylenes with alkenes has been catalyzed by $Co(dppp)Br_2/Zn/ZnI_2$ in CH_2Cl_2 (dppp = 1,3-bis(diphenylphosphino)propane) [455].

The Conia-ene reaction is the intramolecular ene reaction of unsaturated ketones and aldehydes (Section 5.7.1) [456]. The thermal reaction suffers from the requirement of high temperatures, which limits its application. It can be conducted at lower temperature, but additives such as strong base, strong acid, or photochemical activation is often needed [457–459]. Balme and coworkers have reported the CuI-catalyzed Conia-ene reactions of α-alkynic β-ketoesters under mild conditions in the presence of *t*-BuOK [460, 461]. Alternative catalytic systems

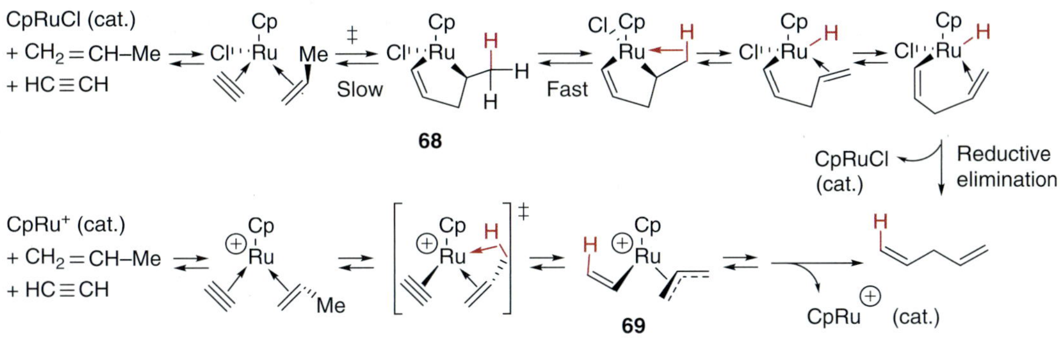

Scheme 8.25 Proposed reaction mechanisms for the Ru-catalyzed ene-reaction of unactivated alkynes and alkenes.

include Au/Ag [462–466], $Ni(acac)_2/Yb(OTf)_3$ [467], $In(NTf_2)_3$ [468, 469], and $Cu(OTf)_2{\cdot}PhH/AgBF_4$ (e.g. reactions (8.64) and (8.65)) [470, 471].

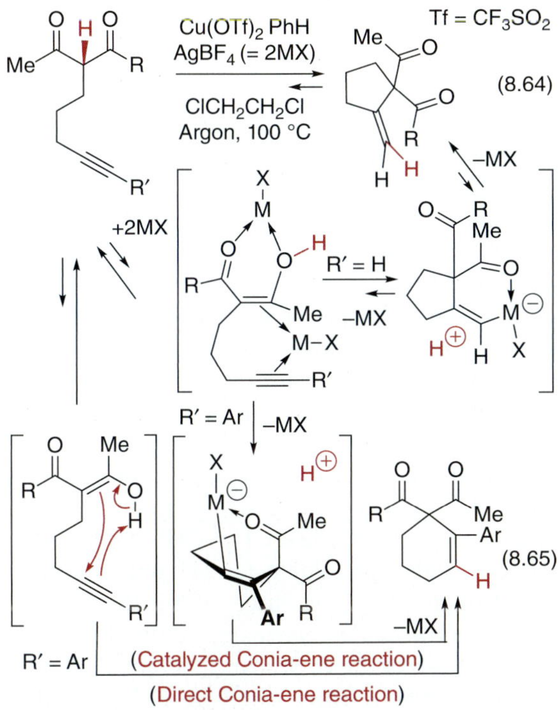

Problem 8.17 What are the major products **P**, **Q** of reactions **A** + **B** and **C** + **B** catalyzed by $CpRu(MeCN)_3PF_6$ (acetone, 25 °C) [472].

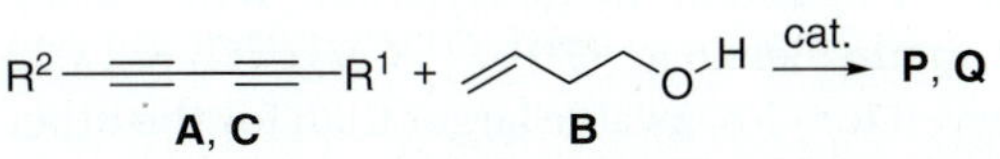

A R^1 = H, R^2 = Bu
C R^1 = $SiEt_3$, R^2 = CH_2CH_2OH

Problem 8.18 What is the main product **P** of the following reaction? [452]

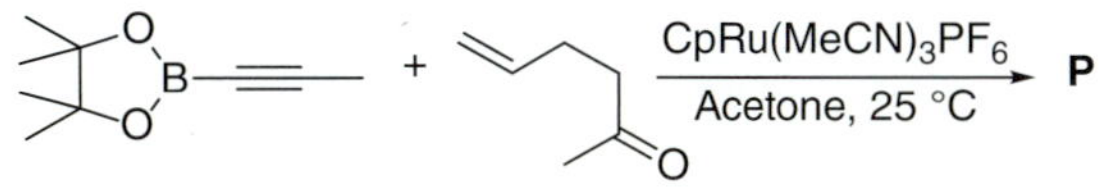

8.3.3 Hydroarylation of alkenes: alkylation of arenes and heteroarenes

The direct hydroarylation of alkenes provides alkylarenes and alkylheteroarenes (Ar—H + CH_2=CHR → Ar—CH_2—CH_2R). The reaction is catalyzed by protic acid or by transition metal complexes. Ethylbenzene (Ph—Et) is the key intermediate in the manufacture of styrene (PhCH=CH_2), one of the most important industrial monomer. In the traditional process developed in the 1930s, the alkylation combines benzene with ethylene in the presence of a Friedel–Crafts catalyst such as $AlCl_3$/HCl (160 °C). In the 1960s, supported catalysts such as BF_3/Al_2O_3 (the Alkar™ process) or Kieselguhr-supported phosphoric acid have been introduced. Nowadays, zeolites are preferred as catalysts. Cumene was a blending agent for high-octane gasoline during the Second World War. Cumene is the intermediate for the production of phenol and acetone. The first industrial process technology for cumene production used H_2SO_4 (180–240 °C) to catalyze the alkylation of benzene with propene. Nowadays, solid catalysts [473] based on zeolites permit more efficient processes (key-step: electrophilic aromatic substitution of benzene by isopropyl cation resulting from the protonation of propene) [474]. Friedel–Crafts alkylation of arenes with styrene derivatives has been catalyzed by $Bi(OTf)_3{\cdot}(H_2O)_n$ [475–477]. In 1965, Chatt and Davidson observed equilibrium (8.66) in which a π-coordinated (naphthalene)ruthenium complex undergoes oxidative C—H addition producing an arylruthenium complex, thus realizing an efficient

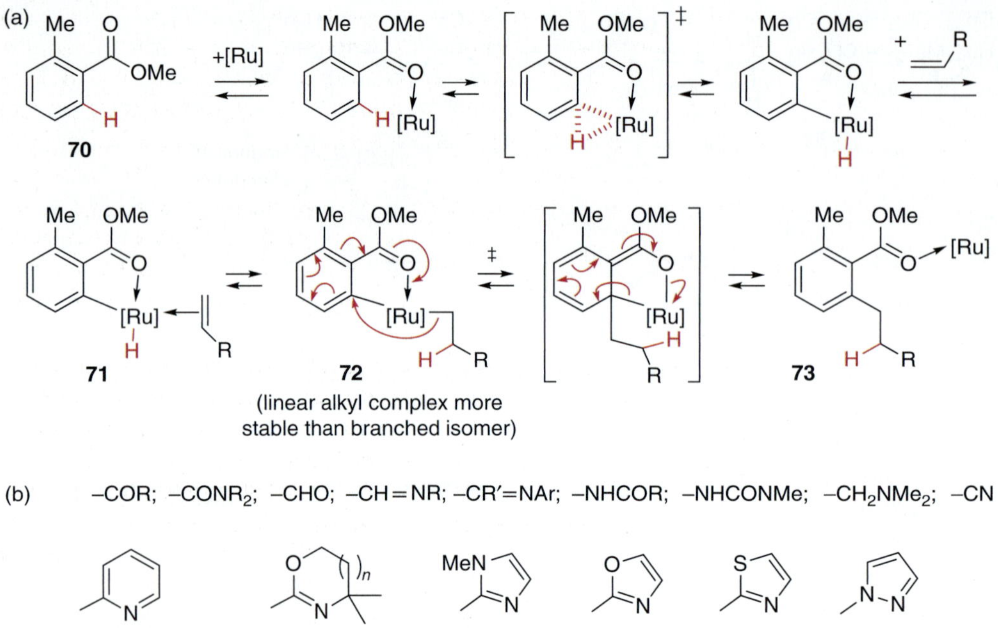

Scheme 8.26 (a) Murai's and Morokuma's mechanism of the $Ru(CO)(PPh_3)_3H_2$-catalyzed hydrocarbation of alkenes with arenes and heteroarenes and (b) examples of monodentate directing groups other than carboxylic esters.

C—H activation of an arene system [478]. In 1993, Murai and coworkers reported a first example of highly efficient intermolecular C—H/alkene coupling reaction (8.67) catalyzed by $RuH_2(CO)(PPh_3)_3$ (Scheme 8.26a) [479].

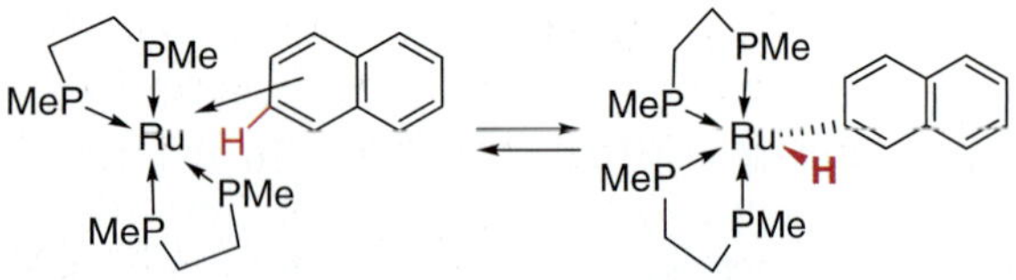

(8.66)

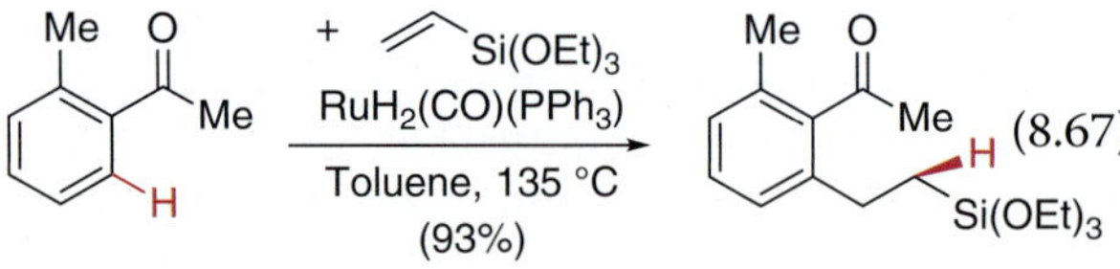

(8.67)

A variety of aromatic and heteroaromatic compounds possessing a Lewis base that can coordinate to a transition metal (Scheme 8.26b) are ortho-coupled with alkenes. The reaction conditions tolerate several functional groups such as NMe_2, OMe, F, N(Et)COMe, CF_3, COOEt, CN, acetals, and OAc. Only the C—H bond *ortho* to the Ru coordinating group is activated [479–482]. These are examples of chelation-assisted hydroarylations (lateral regiocontrol by the coordinating substituent). Pd complexes catalyze these reactions [483] as well as cheaper first-row transition metal complexes [484]. If two regioisomeric compounds can be made, the least sterically hindered *ortho* C—H bond is generally preferred for the hydrocarbation reaction. Reaction mechanism (Scheme 8.26a) is supported by deuterium and ^{13}C-kinetic studies, as well as by quantum mechanical calculations [485]. The first step is a reversible oxidative addition of the ruthenium complex to the C—H bonds of the aromatic starting material. When treating $C_6D_5COOMe + CH_2{=}CHSi(OEt)_3$ with $RuH_2(CO)(PPh_3)_3$ in toluene for 24 h under reflux, no product of coupling was observed but the two *ortho* positions of the methyl benzoate and the three vinylic C—H positions of the vinyl silane contained 0.6 H and 0.4 D, respectively. Thus, extensive H/D scrambling among these five C—H positions takes place. This implies that *ortho*-C—H (C—D) bond cleavage is not the rate-determining step and a rapid equilibrium occurs prior to the reductive elimination given the final product of alkene hydroarylation. When using methyl 2-methylbenzoate (**70**), $k(^{12}C)/k(^{13}C) = 1.033$ was observed for C-6, a value larger than for the other carbon centers of **70**. This indicates that the *ortho*-C-6 carbon center participates in the rate-determining step (**70** ⇄ **71** ⇄ **72** → **73**). Thus, the C—C bond formation (reductive elimination **72** → **73**) is the turnover-limiting step [486].

The iridium-catalyzed carbonyl-directed hydroarylation of monosubstituted alkenes reported by Bower and coworkers (e.g. reaction (8.68)) provides the corresponding branched alkylarenes (e.g. **74**) when the diphosphine ligand is dppb (1,4-bis(diphenylphosphino)butane) or $d^{F}ppb$ (1,4-bis[di(pentafluorophenyl)phosphino]butane) (with large bite angle), and the corresponding linear alkylarenes (e.g. **75**) when using dppm (1,1-bis(diphenylphosphino)methane, with small bite angle between the two P atoms) [487, 488]. Transition structures **76** and **77**, respectively, have been proposed for these reactions [489, 490].

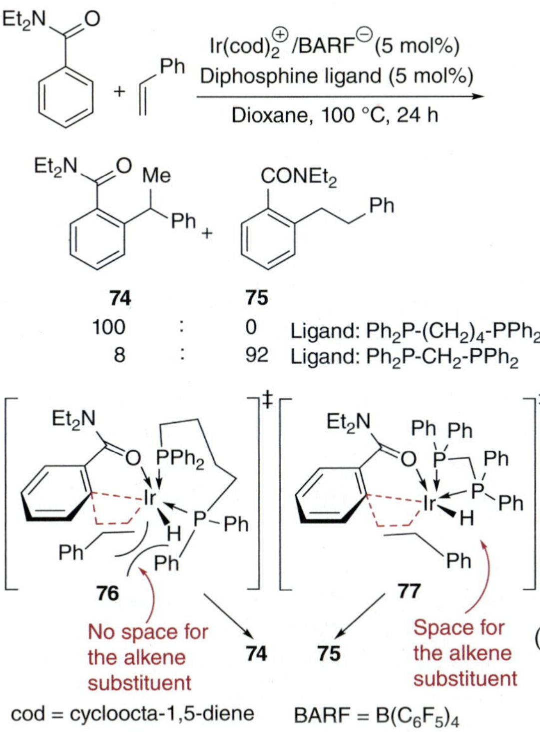

(8.68)

The Ru(II) complex TpRu(CO)(MeCN)(Me) (Tp = hydridotris(pyrazolyl)borate) reacts with furan to produce methane and TpRu(CO)(MeCN)(2-furyl). The latter catalyzes the formation of 2-ethylfuran from ethylene and furan (120 °C, 1.5–6 atm, in mesitylene) [491]. The same complex catalyzes the formation of ethylbenzene from benzene and ethylene [492–495]. Darses and coworkers reported an efficient catalytic system generated in situ from the reaction of inexpensive $[RuCl_2(p\text{-cymene})]_2$ and HCOONa, in association with a phosphine ligand [496, 497]. The system catalyzes hydroarylation of vinylsilanes, anti-Markovnikov hydroarylation of styrenes, and the formation of functionalized allylsilanes via C—H bond activation [498]. Arcadi et al. have catalyzed the addition of arenes to alkynes using $NaAuCl_4$ [499]. They also found gold-catalyzed reactions of indoles/pyrroles with 1,3-dicarbonyl compounds [500] and 7-azaindole with conjugated enones [501, 502]. Electron-rich systems such as 2-methylfuran, azulene, and 1,3,5-trimethoxybenzene add to methyl vinyl ketone in the presence of gold/silver catalyst (Michael type of additions) [503]. $RuCl_3$/AgOTf catalyzes the electrophilic intramolecular hydroarylation of distant alkene moieties [504]. Intramolecular addition of electron-rich arenes to alkenes [505] and to imines have been catalyzed with Au/Ag systems [506, 507]. Yoshikai and coworker have developed cobalt-based catalysts for the direct arylation of alkenes and alkynes [508]. Bergman and coworkers reported enantioselective cyclization (8.69) catalyzed by the rhodium complex resulting from the ligand exchange of $[RhCl(cyclooctene)_2]_2$ and an enantiomerically pure phosphoramidite ligand L* [509]. In this reaction, the Rh complex undergoes first oxidative addition into the indole C(2)—H bond, followed by hydrocarbation of the allylamine moiety. An example of enantioselective intermolecular hydropyridination of monosubstituted alkene catalyzed by rare earth metal catalyst **78** has been reported by Hou and coworkers in 2014 (reaction (8.70)). The chiral cyclopentadienyl ligand of **78** has been developed by Cramer and coworker. Applying analogous chiral cyclopentadienyl ligands, the latter group has reported several rhodium-catalyzed enantioselective intramolecular alkene arylations [510].

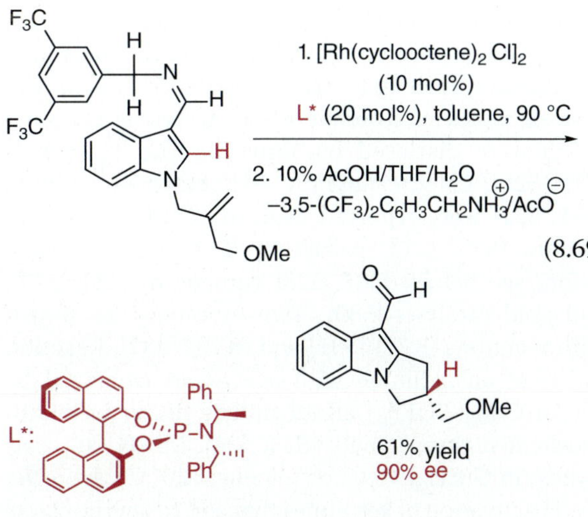

(8.69)

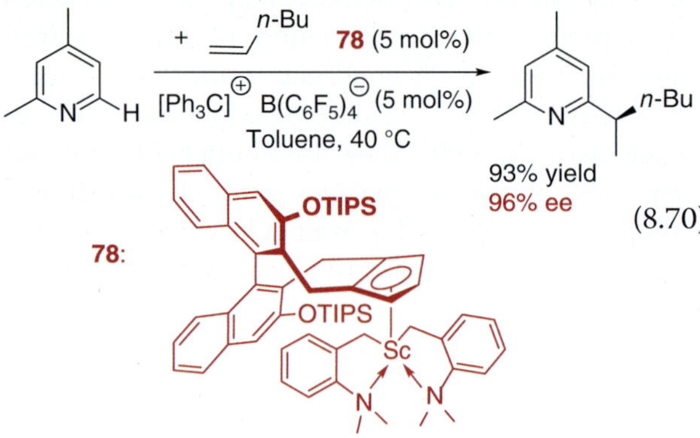

(8.70)

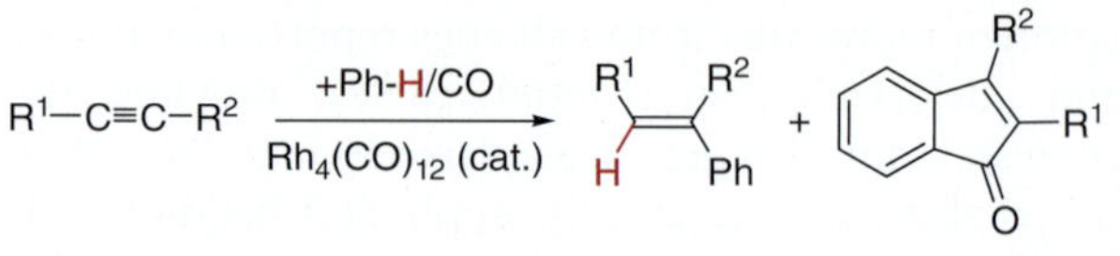

(8.71)

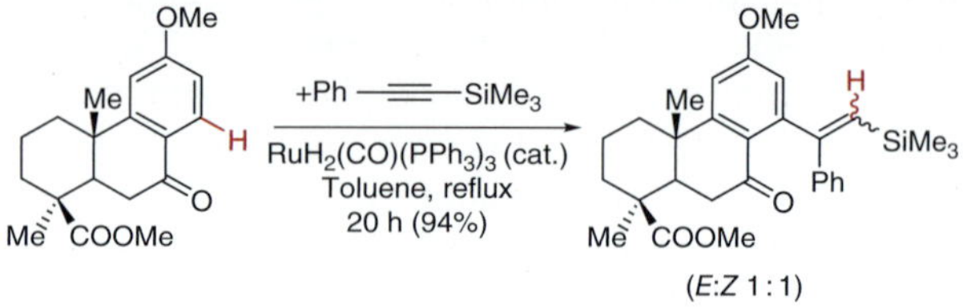

(8.72)

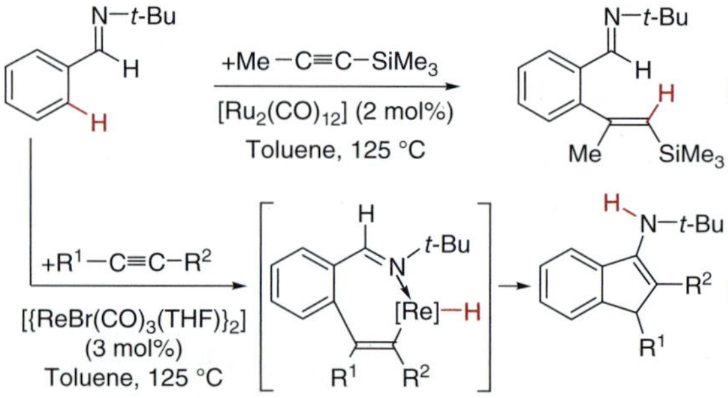

(8.73)

(8.74)

Problem 8.19 What is the main product of the reaction of benzonitrile with CH_2=CHSi(OEt)$_3$ catalyzed by $RuH_2(CO)(PPh_3)$ in toluene heated under reflux? [480]

Problem 8.20 What is the main product of the reaction of 1-tetralone with MeC≡CSiMe$_3$ catalyzed by $RuH_2(CO)(PPh_3)$ in toluene heated under reflux? [511]

Problem 8.21 What is product **P** of the following reaction? [512]

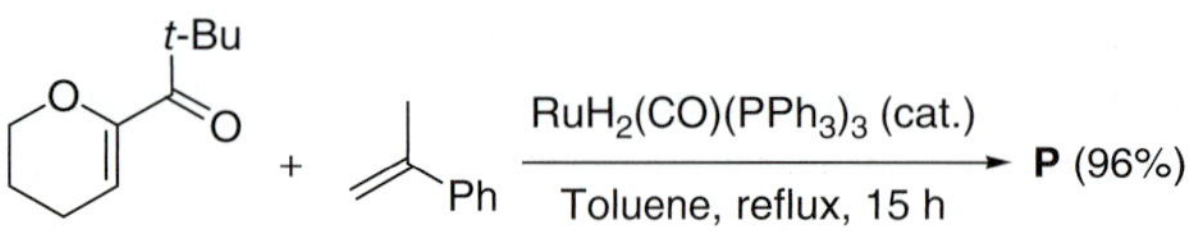

8.3.4 Hydroarylation of alkynes: alkenylation of arenes and heteroarenes

Alkenylarenes and alkenylheteroarenes are obtained by direct hydroarylations of alkynes (alkenylation of arenes and heteroarenes), which are catalyzed by protic acids or transition metal complexes [513–518]. The first examples of aromatic C—H/acetylene coupling (8.71) were disclosed by Yamazaki and coworkers [519, 520]. Later, catalytic C—H/acetylene coupling [481, 521] were reported using rhodium [522–524], iridium [525, 526], palladium (Fujiwara hydroarylation, see below) [527, 528] ruthenium [511, 529], and gold catalysts [530]. Two examples are shown with reactions (8.72) [531] and (8.73) [532]. Reaction (8.74) of an aldimine with acetylenes catalyzed by [{ReBr(CO)$_3$(THF)}$_2$] afford indene derivatives with excellent regioselectivity [533, 534]. $Mn_2(CO)_{10}$ catalyzes (in $ClCH_2CH_2Cl$ or toluene, 120 °C) the *ortho* C—H activation of ketimines that add to acrylic esters giving indane derivatives [535].

Complex $Ru_3(CO)_{12}$ with an additive such as NH_4PF_6 or $HBF_4{\cdot}OEt_2$ is an effective catalyst for the coupling reaction of arylamines and alkynes as illustrated with reaction (8.75) [536]. A possible mechanism is shown in Scheme 8.27 for aniline reacting with propyne.

Mechanistic studies on the Fujiwara alkyne hydroarylation catalyzed by $Pd(OAc)_2/CF_3COOH$ (Scheme 8.28) have suggested that the rate-determining step is not a palladium oxidative addition onto the *ortho*-C—H bond of the benzoate moiety, but the electrophilic addition of (alkyne)palladium complex intermediate. This is supported by the observation of an inverse kinetic isotopic effect $k_H/k_D = 0.64–0.82$ (25 °C) a value similar to k_H/k_D observed for the nitration of toluene and deuterated toluene ($k_H/k_D = 0.85$) and for the protonation of $PhSiMe_3/C_6D_5SiMe_3$ with $HClO_4$ ($k_H/k_D = 0.79$). Normal kinetic isotopic effects ($k_H/k_D > 1$) are observed for C—H(D) activation, for instance, $k_H/k_D = 5$ for the trifluoroacetoxylation of toluene with $Pd(OAc)_2/CF_3COOH$, $k_H/k_D = 3–5$ for the olefin arylation catalyzed by $Pd(OAc)_2/CF_3COOH$ [537].

In 1931, Kharash and Isbell reported that $AuCl_3$ reacts with benzene and derivatives to form arylgold(III) complexes [538], an example of C—H activation that occurs via electrophilic addition

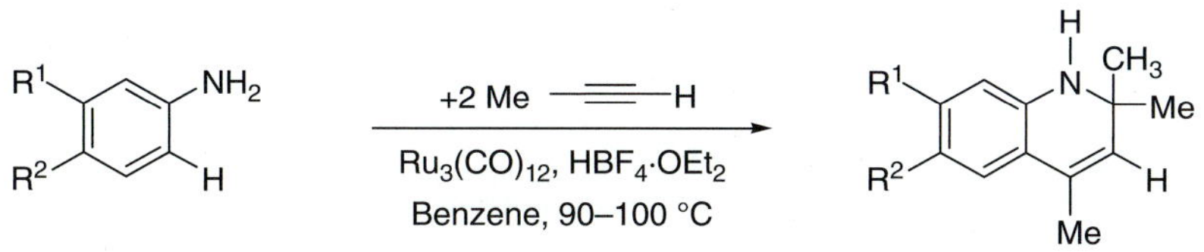

(8.75)

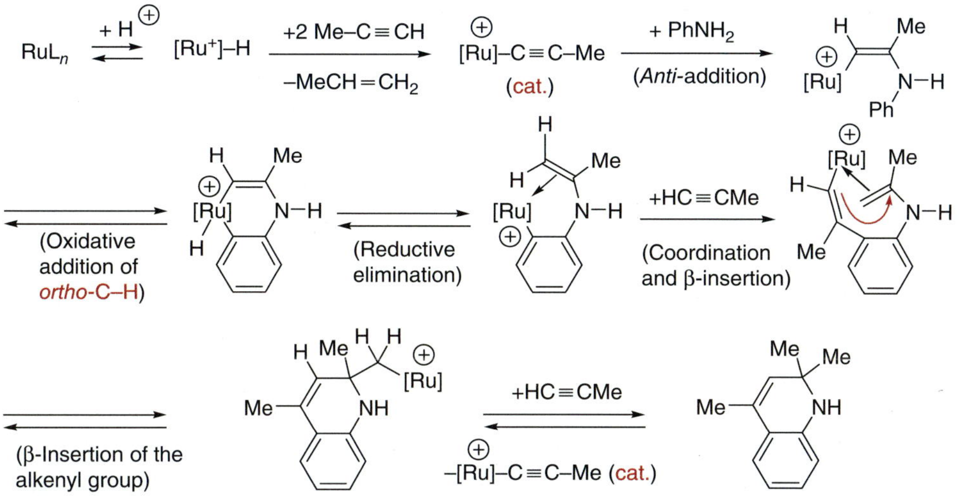

Scheme 8.27 Proposed reaction mechanism for the ruthenium-catalyzed *ortho*-C—H bond activation and cyclization reactions of arylamines with terminal alkynes.

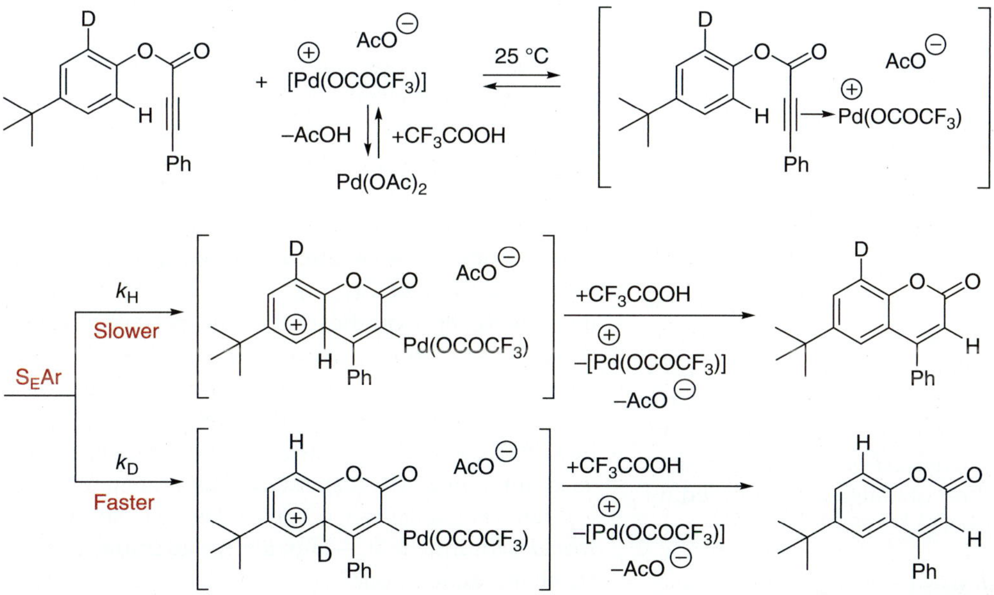

Scheme 8.28 Proposed reaction mechanism of the Fujiwara alkyne hydroarylation.

and elimination (Ar—H + $AuCl_3$ → $ArAuCl_2$ + HCl) [539]. When stabilized by a suitable ligand such as 2,6-lutidine, $PhAuCl_2$ reacts with phenylacetylene yielding diphenylacetylene [540]. In 2003, Reetz and Sommer reported the direct addition of arenes to alkynes with different gold catalysts such as $AuCl_3/AgSbF_6$ and $(Ph_3P)AuCl/BF_3{\cdot}OEt_2$ [541]. The regioselectivity is governed by electronic factors (Markovnikov-type additions are observed). The combination of $AuCl_3 + 3AgSbF_6$ and $AuCl_3 + 3AgOTf$ showed enhanced catalytic activity for the alkyne hydroarylation [542]. Echavarren and coworker reported facile annulation to seven- (e.g. reaction (8.76)) and eight-membered ring systems (e.g. reaction (8.77)) on indoles by Au-catalyzed cyclization with alkynes [543–545].

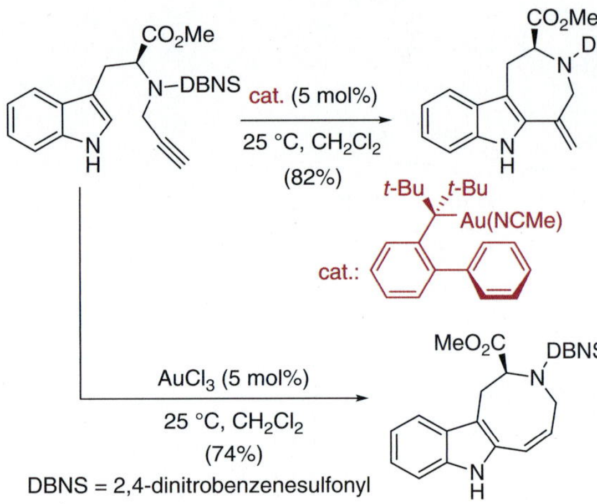

(8.76)

(8.77)

8.3.5 Hydroarylation of carbon-heteroatom multiple bonds

The direct hydroarylation of carbonyl compounds, imines, and carbonitriles has been catalyzed by transition metal complexes [407]. An early example reported by Murai and coworkers in 2002 adds *N*-methylimidazole to aldehydes in the presence of an Ir-catalyst and a silane (reaction (8.78)) [546]. In 2004, Larock and coworkers added arenes to carbonitriles producing imines that are hydrolyzed into the corresponding ketones (reaction (8.79)). In 2013, Wang and coworkers reported the Pd-catalyzed C—H addition of indole to carbonitriles producing the corresponding 3-acylindoles (reaction (8.80)). Using a cationic rhodium complex, the group of Shi catalyzed the regioselective direct hydroarylation of *N*-sulfonylaldimines (reaction (8.81)) [547]. The regioselectivity arises from a lateral control because of the pyridine moiety. In the presence of Et_3SiH, 1-methyl-2-phenyl-1*H*-imidazole undergoes *ortho*-C—H insertion of aldehydes catalyzed by $Mn(CO)_5Br$ in toluene at 115 °C [548].

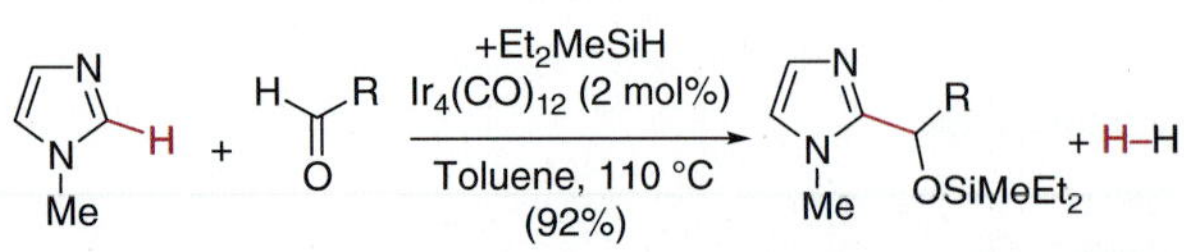

(8.78)

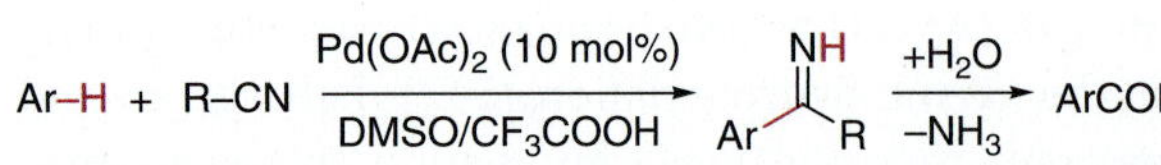

(8.79)

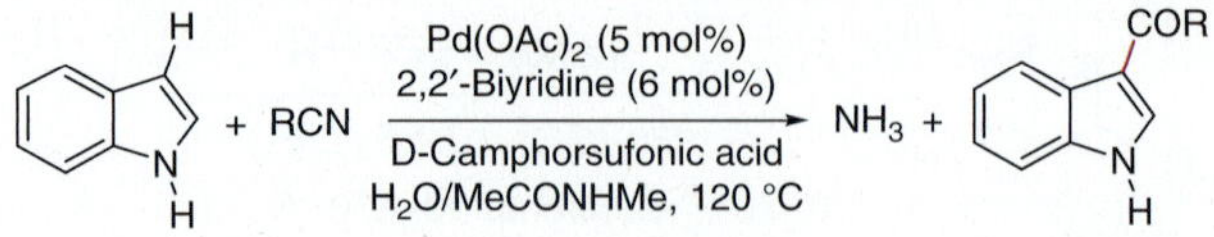

(8.80)

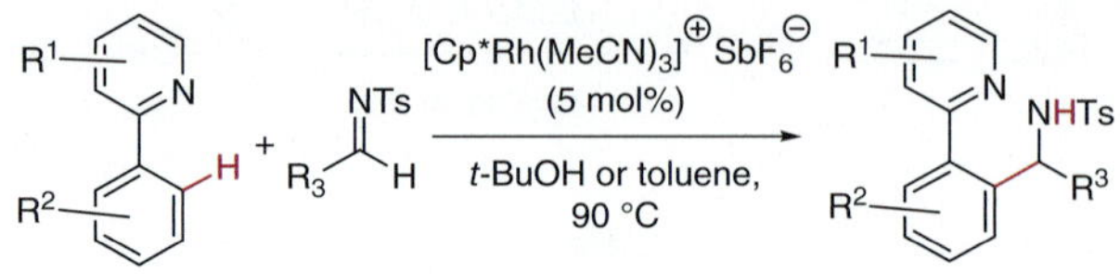

Ts = 4-MeC$_6$H$_4$SO$_2$; Cp* = pentamethylcyclopentadienyl

(8.81)

Problem 8.22 What is the main product of reaction of PhC(=NOMe)Et with Ph—N=C=O catalyzed by $[Cp^*Rh(MeCN)_3][SbF_6]_2$.in 1,2-dichloroethane at 100 °C [549].

8.3.6 Hydroalkenylation of alkynes, alkenes, and carbonyl compounds

Using dinuclear palladium complexes **82**, Tsukada et al. described the hydroalkenylations of alkenes with reagents that do not require the aid of any directing group as in the case of the transition-metal-catalyzed reactions described above (Sections 8.3.3–8.3.5) [550]. For instance, the reaction of 2-deuteroindene (**79**) and hex-3-yne (Scheme 8.29) produces a 4 : 1 mixture of **80** and **81**. The reaction of nondeuterated indene does not generate **81**, showing a substantial lower reactivity of C(2)—D than C(2)—H, in agreement with the hypothesis that the process involves direct C—H bond activation **82** + **79** → **83**. Then, hydride (deuteride) insertion onto the alkyne gives **84**. Alternatively, alkenylation **83** → **85** can compete. Both **84** and **85** undergo reductive eliminations with final product formation and regeneration of a dinuclear palladium complex **86**. The latter then reacts with indene (C—H oxidative addition) to start another reaction cycle.

In 2008, Skrydstrup and coworkers reported that Pd hydride generated in situ from $[Pd_2(dba)_3]$ + $(t\text{-}Bu)_3P$ + i-PrCOCl catalyzes the intermolecular coupling of alkynes and alkenes giving 1,3-dienes (e.g. Ar—C≡C—Ar + CH$_2$=CHR → (*Z*)—ArCH=C(Ar)—CH=CHR) [551]. Intramolecular hydroalkenylation of alkenes has been catalyzed by rhodium complexes [552]. Intermolecular asymmetric hydrovinylation of styrene has also been described (e.g. reaction (8.82)) [553].

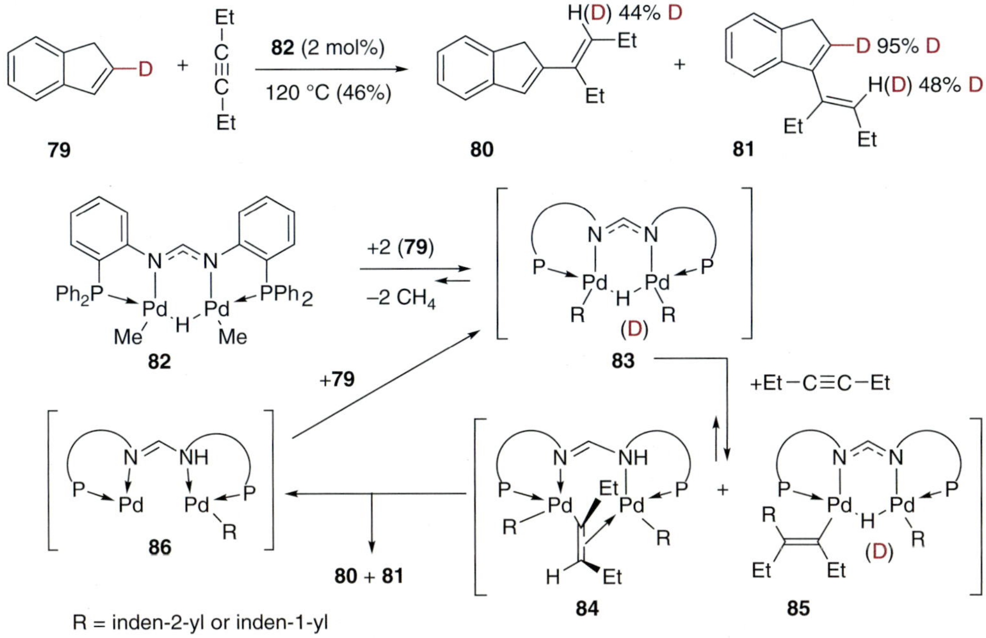

Scheme 8.29 Proposed reaction mechanism for the palladium-catalyzed hydroalkenylation of alkynes.

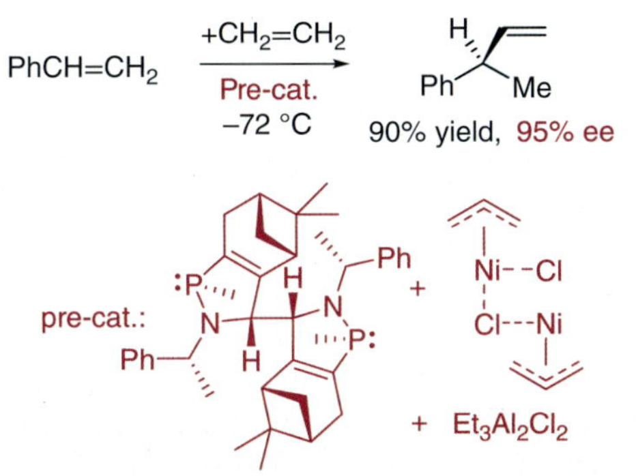

(8.82)

Ritter and coworkers have reported an intermolecular, stereoselective, and regioselective iron-catalyzed 1,4-addition of α-olefins to 1,3-dienes using 1 mol% of an iminopyridine ferrous chloride complex [554]. Highly selective cross-couplings of alkenes and aldehydes (reaction (8.83)) use Ni complexes containing electron-rich diaminocarbenes and strong π-acceptor phosphites as ligands [555].

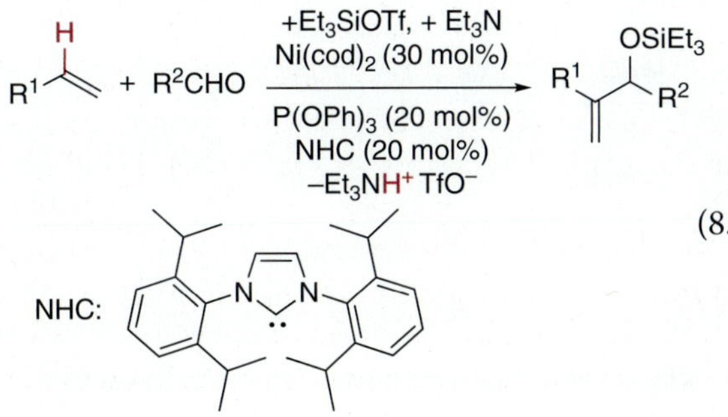

(8.83)

Problem 8.23 The following cycloisomerizations of enynes **A** (intramolecular hydroalkenylations of alkynes) have been catalyzed [556] by the Jonas' iron complex [557, 558]. Propose a mechanism for these isomerizations.

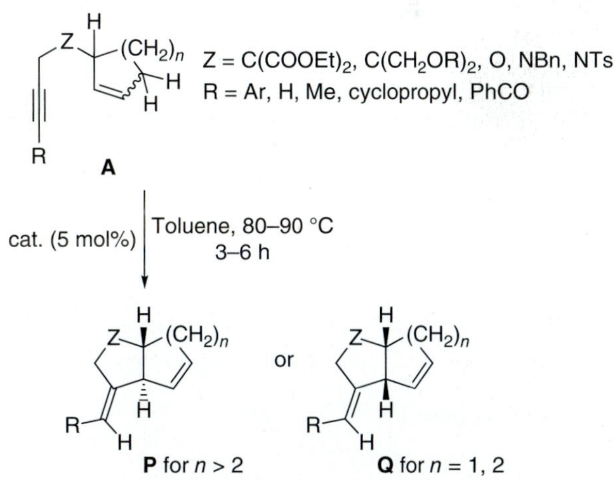

Jonas complex: $CpFe(C_2H_4)_2Li(TMEDA)$
TMEDA = $Me_2NCH_2CH_2NMe_2$

8.3.7 Hydroacylation of alkenes and alkynes

The direct addition of a C—H bond of a carbaldehyde group to a C—C multiple bond is useful to prepare various types of ketones [559–562]. In 1972, Sakai et al. reported a first intramolecular hydroacylation (isomerization of 2,3-disubstituted hex-4-enal into cyclopentanones) catalyzed either by $SnCl_4$ or by $Rh(Ph_3P)_3Cl$ [563]. In 1976, Lochow and Miller reported reaction (8.84) [564]. The same group reported in 1980 the intermolecular hydroacylation of ethylene by pent-4-enal giving heptenones that was accompanied by alkene isomerization [565]. Later, James and Young found that $Ru(PPh_3)_3Cl_2$ catalyzes the hydroacylation of ethylene with propanal to give pentan-3-one (210 °C, 18 hours, 2–4% yield) [566, 567]. Rh-catalyzed hydroacylation competes with decarbonylation (RCHO → R—H + CO) making the intermolecular hydroacylation low yielded in many instances [562]. However, tailored ligands can reduce that competition [568]. In 1978, Suggs reported that decarbonylation is suppressed for chelating aldehydes (e.g. *ortho*-(Ph_2P)—C_6H_4CHO, α,β-unsaturated aldehydes, and β-sulfido-aldehydes) [569, 570]. Watanabe and coworkers reported that $Ru_2(CO)_{12}$ catalyzes the hydroacylation of cyclohexene with PhCHO under 20 atm of CO at 200 °C and produces cyclohexyl phenyl ketone in 44% yield [571, 572]. Enantioselective syntheses of seven- and eight-membered heterocyclic compounds have been realized using [Rh(*R*,*R*)—Me—Duphos]BF_4 as a catalyst (e.g. reaction (8.85)) [573]. Enantioselective intramolecular hydroacylation of 2-(cyclohex-1-en-1-yl)benzaldehyde into benzo[h]bicyclo[4.3.0]non-8-en-7-one has been catalyzed by $[Rh(cod)Cl]_2$ and (*R*)-DTM-Segphos [574]. Enantioselective intermolecular hydroacylations have also been reported [575–578].

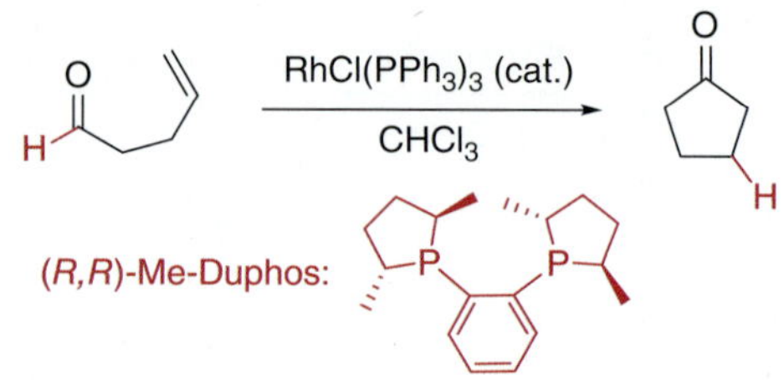

(8.84)

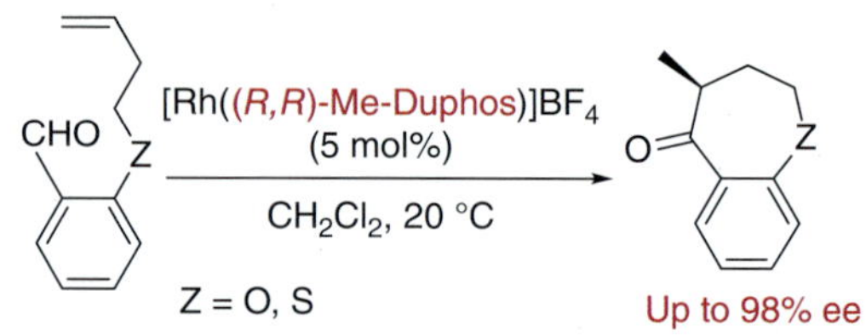

(8.85)

The rhodium-catalyzed intermolecular hydroacylation reaction is greatly accelerated using 2-amino-3-picoline (**87**) as the cocatalyst [579–581]. This is explained by the formation of intermediate aldimines **88** that undergo alkene hydroiminations, reactions that are faster than the corresponding hydroacylations (Scheme 8.30). The imines **89** so-obtained are readily hydrolyzed to liberate the corresponding ketones (the pyridine moiety of **89** catalyzes the water addition to the neighboring C=N double bond) [582].

The combination of a small bite angle diphosphine such as bis(dicyclohexylphosphino)methane and $[Rh(cod)OMe]_2$ catalyzes the hydroacylation of nonchelating aldehydes (RCHO) with 2-vinylphenols (Ar—CH=CH_2) that gives the corresponding branched arylketones (Ar—CH(Me)COR). Deuterium labeling experiments and Hammett correlation studies support the oxidative addition of Rh to

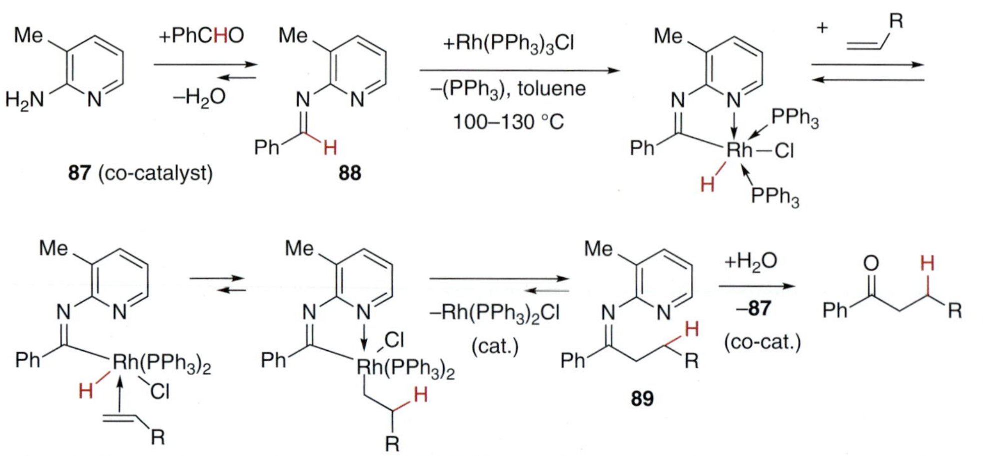

Scheme 8.30 Acceleration of rhodium-catalyzed hydroacylations of alkenes by 2-amino-3-picoline (metal–organic cooperative catalysis).

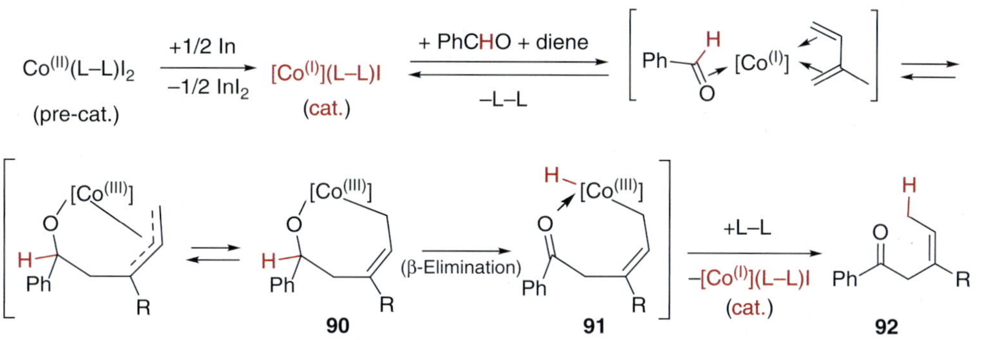

Scheme 8.31 Proposed reaction mechanism for the regioselective cobalt-catalyzed 1,4-addition of aldehyde to 1,3-dienes.

the aldehyde C—H bond as an irreversible and turnover-limiting step (RCHO + [Rh] → RCO—[Rh]—H). The small bite angle of the diphosphine ligand is crucial for lowering the barrier of this step [583, 584]. Because (acyl)rhodium intermediates tend to undergo undesired decarbonylation (RCO—[Rh]—H → R—[Rh]—H + CO), cobalt [585–589], ruthenium [571, 590, 591], nickel [592], and *N*-heterocyclic carbenes have been used as catalysts for intermolecular alkene hydroacylations [593–595]. Cobalt(II) salts and chiral diphosphine ligands catalyze the enantioselective intramolecular hydroacylation of 2-acylbenzaldehydes and 2-alkenylbenzaldehydes to afford phthalide and indanone derivatives, respectively [587].

Hydroacylation of isoprene with benzaldehyde (reaction (8.86)) occurs at 120 °C under CO atmosphere in the presence of Ru(cod)(cot) catalyst and PPh_3 as cocatalyst [596]. Other Ru catalysts have been proposed for these 1,2-additions by Krische and coworkers [597]. Aromatic aldehydes provide mostly 1,4-addition products in cobalt-catalyzed hydroacylations of 1,3-dienes (e.g. reaction (8.87)) [598].

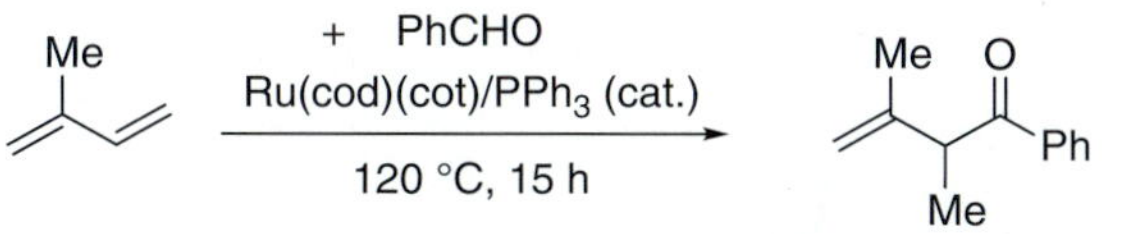

(cod = cycloocta-1,5-diene, cot = cycloocta-1,3,5-triene)

(8.86)

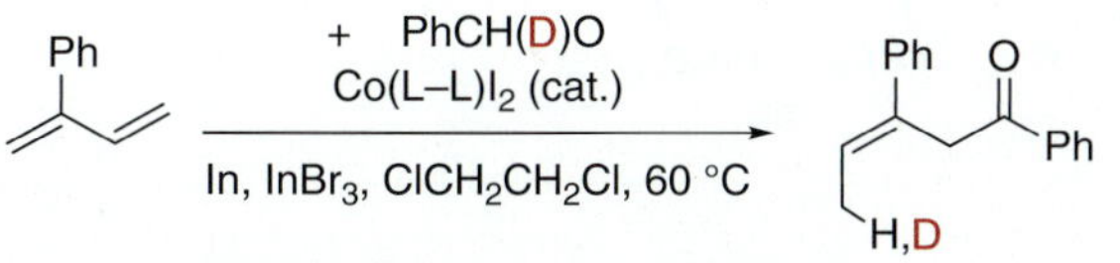

L–L = $Ar_2P(CH_2)_3PAr_2$, Ar = 2,3-$(MeO)_2C_6H_3$

(8.87)

A kinetic isotopic effect k_H/k_D = 2.7 has been observed for the competition of hydroacylation of PhCHO and PhCDO in reaction (8.87). The rate laws of these reactions are zeroth order both in aldehyde and diene concentrations. Thus, neither the oxidative addition of the cobalt catalyst to the C—H bond of the aldehyde nor the formation of a diene metal complex is the turnover-limiting step. The above kinetic deuterium isotope effect is consistent either with a β-hydride migration (**90** → **91**) or a reductive elimination (**91** → **92**) in the turnover-limiting step (Scheme 8.31) [598].

Intramolecular alkyne hydroacylation has been catalyzed by cationic rhodium complexes (e.g. reaction (8.88), dppe = 1,2-bis(diphenylphosphino)ethane) [599]. Intermolecular hydroacylation of alkynes has been catalyzed by Wilkinson's catalyst and 2-amino-3-picoline (**87**, cocatalyst) [600]. Unexpectedly, in this case, the branched α,β-enones are obtained, as illustrated with reaction (8.89).

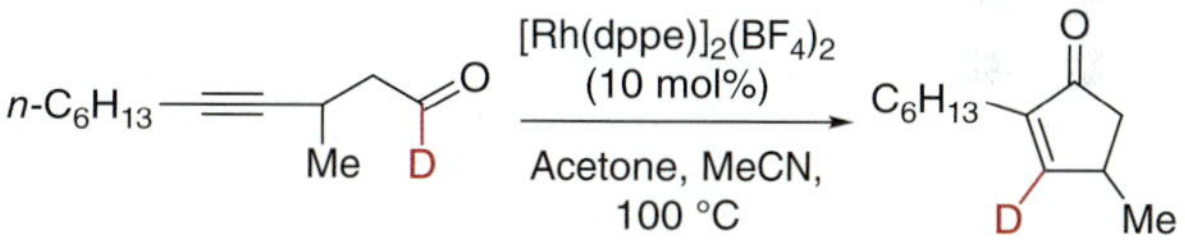

(8.88)

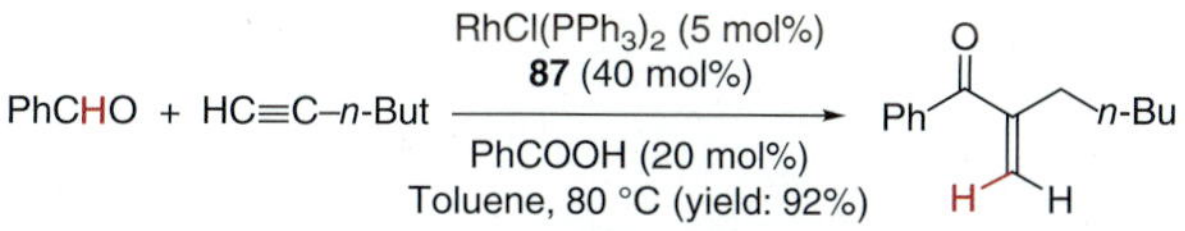

(8.89)

Problem 8.24 Propose a mechanism for the following reaction [601].

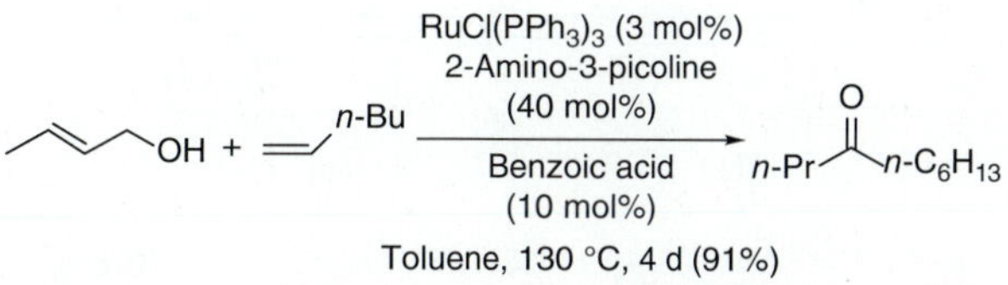

Problem 8.25 What is the main product **P** of reaction of 2-methylthiobenzaldehyde and 2-nitrophenylacetylene catalyzed by $Rh(norbornadiene)_2BF_4$ and

bis(dicyclohexylphosphino)methane in acetone at 55 °C? What is the product **Q** resulting from the treatment of **P** with Sn and aqueous HCl in EtOH? [602, 603]

8.3.8 Hydrocyanation of alkenes and alkynes

Carbonitriles are important intermediates in the preparation of materials, pharmaceuticals, cosmetics, and agrochemicals [604]. Nitriles serve as precursors for aldehydes, amines, ketones, esters, amides, and heterocyclic compounds. The simplest way to obtain them is through direct hydrocyanation of alkenes and alkynes. In 1971, Du Pont started the production of adiponitrile ($NC(CH_2)_4CN$), which combines butadiene with 2 equiv. of HCN [605–607]. Originally, it was developed using a monodentate phosphate-based Ni(0) catalysts such as $(ArO)_3P$. Hydrogenation of adiponitrile produces hexamethylenediamine ($H_2N—(CH_2)_6—NH_2$) and its reaction with adipic acid ($HOOC(CH_2)_4COOH$) gives the polyamide known as Nylon-6,6. Hydrocyanation of butadiene leads to a mixture of linear (**93**) and branched (**94**) carbonitrile, obtained in varying ratio depending on the ligand employed. In a second step, **94** is isomerized into **93**. Then, **93** is isomerized into but-3-ene-1-carbonitrile, which is hydrocyanated into adiponitrile, a reaction that requires a Lewis acid cocatalyst such as $AlCl_3$. Replacing the phosphate ligands by bidentate ligands such as bis(diphenylphosphino)ferrocene (dppf) leads to higher conversion and selectivity for **93** [608, 609]. Triptycene-based bidentate ligand **95** (first described by Hofmann and coworkers [610, 611]) has a very rigid backbone and a large bite angle. Together with $Ni(cod)_2$, it leads to selectivities up to 98% in the formation of **93**. This is due to the good ability of the (diphosphine)Ni catalyst to promote **94** → **93** isomerization (Scheme 8.32) [612]. A series of bis-phosphonite ligands with cycloalkyl spacers have been made. They yield (monochelate)Ni(cod) complexes that are highly active and selective catalysts for the hydrocyanation of styrene and butadiene [613, 614].

Nonactivated olefins undergo hydrocyanation according to reaction (8.90) in the presence of a cobalt catalyst such as **96**. The reaction conditions are very mild and tolerate functional alkenes [615]. In 1996, RajanBabu and Casalnuovo described the Ni-catalyzed enantioselective (95% ee) hydrocyanation of 2-methoxy-6-vinylnaphthalene in the presence of phenyl β-D-glucopyranoside-derived diphosphinite

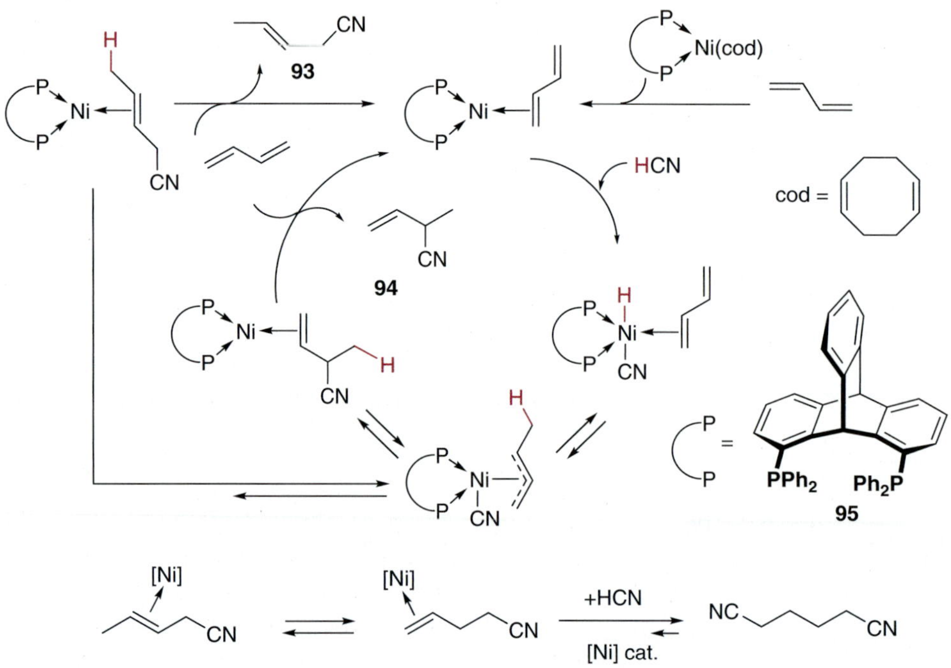

Scheme 8.32 Proposed reaction mechanism for butadiene hydrocyanation and isomerization of adducts obtained into (*E*)-pent-3-ene-1-carbonitrile with catalyst **95**. With other Ni catalysts, the latter compound is isomerized into pent-4-ene-1-carbonitrile that undergoes hydrocyanation to give adiponitrile.

ligand [616]. In the meantime, many more enantioselective hydrocyanations have been reported [617] (e.g. reaction (8.91) that uses Ni(cod)$_2$ and **97**) [618].

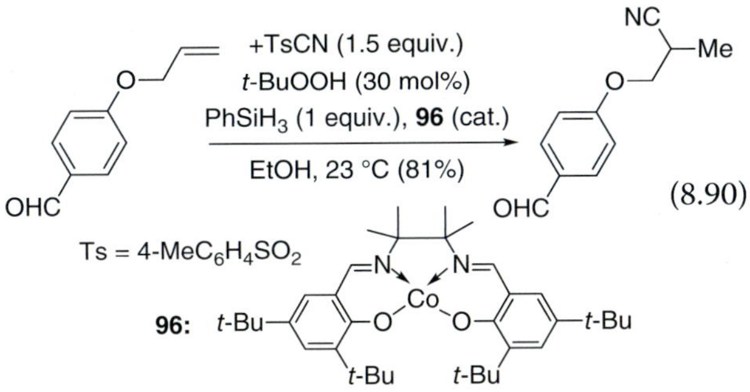

(8.90)

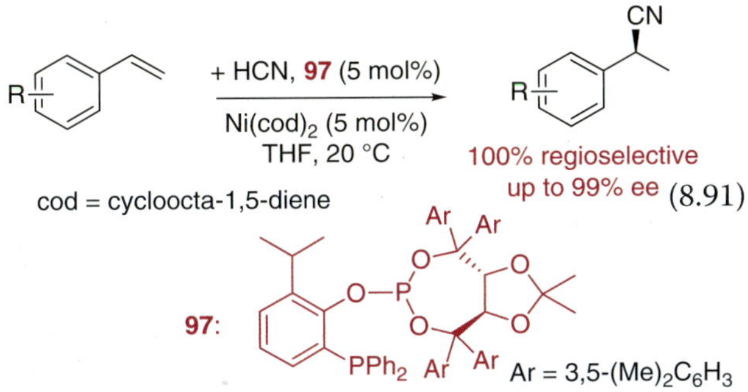

(8.91)

Because HCN is highly toxic and explosive, surrogates for it have been used such as acetone cyanohydrin, ethyl cyanoformate and Me_3SiCN [619]. Alternatively, nickel-catalyzed reversible alkene + carbonitrile reactions can be employed instead as illustrated with reaction (8.92), which gives a 91 : 9 mixture of the linear and branched carbonitrile in 97% yield [620].

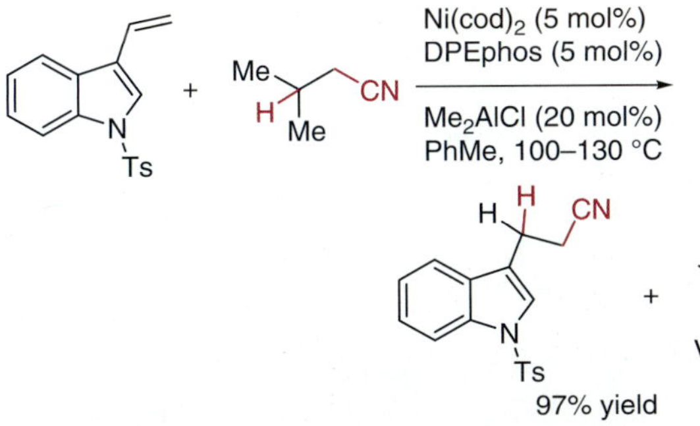

DPEphos = bis(2-diphenylphosphinophenyl)ether

(8.92)

Problem 8.26 Propose a mechanism for the nickel-catalyzed hydrocyanation of **93** into adiponitrile [621].

8.3.9 Direct reductive hydrocarbation of unsaturated compounds

Hydrogenation of carbon π-systems in the presence of carbonyl and imine compounds promotes reductive C—C coupling in a more atom-economical way than classical metallation with formation of organometallic reagents (alkene or alkyne + M—H → R^1—M; R^1—M + R^2(R^3)C=Z → R^1(R^2)(R^3)C—Z—M; R^1(R^2)(R^3)C—Z—M + H_2O → R^1(R^2)(R^3)C—ZM + MOH) [622] Catalytic systems for the hydrometallative reductive coupling of alkenes [623–625], alkynes [626–630], enones [631–640], and dienes [641–649] with carbonyl reagents has been realized using silanes, alanes, and boranes as terminal reductants. Fischer–Tropsch-type reactions (Scheme 8.2, Section 8.2.1) and Roelen hydroformylation (Scheme 8.5, Section 8.2.3) generate alkylmetal intermediates able to react with other species than hydrogen and thus allowing C—C bond formation competitively with hydrogenation. Dihydrogen (H_2) is the terminal reductant in these reactions. Suitable catalysts are those permitting heterolytic activation of H_2 (H_2 + M—X ⇄ M—H + HX) [650, 651] and generate alkylmetal hydride intermediates that do not undergo quick reductive elimination. Such catalysts are cationic Rh(I) [652] or Ir(I) complexes [653]. A first example is the catalytic hydrogen-mediated reductive intramolecular aldol condensation of enone **98** into aldol **99** (Scheme 8.33a) reported by Krische and coworkers [654]. In the presence of 1% of Rh(Ph$_3$P)Cl, aldol **99** is formed next to 6-phenyl-6-oxohexanal (**100**), the product of alkene hydrogenation of **98**. Using Rh(cod)$_2$OTf and (*p*-CF$_3$C$_6$H$_4$)$_3$P as precatalysts leads to a 1 : 1 mixture of **99** and **100**. On adding KOAc (mild base), a 10 : 1 mixture of *syn* vs. *anti*-aldol **99** is obtained in 89% yield, together with only 0.1% of **100**. The hydrogenative aldol reaction can be done intermolecularly and with high diastereo- and enantioselectivity if a suitable enantiomerically pure phosphite ligand is used (Scheme 8.33b). The first enantioselective reductive aldol reaction was reported by Taylor et al. in 2000 who coupled acrylic esters, aldehydes, and Et$_2$MeSiH (reductant) in the presence of Rh(BINAP) catalysts [636].

In the meantime, the Krische's group has developed many more metal-catalyzed hydrogenative hydrocarbations [655]. They include intermolecular coupling of enones and enals with aldehydes and ketones, 1,3-cyclohexadiene with α-keto-aldehydes, and coupling of 1,3-enynes and 1,3-diynes with α-keto-aldehydes [652, 656, 657]. Enantioselective direct reductive coupling of enynes with aromatic aldehydes [658] and α-ketoesters (reaction (8.93)) has been

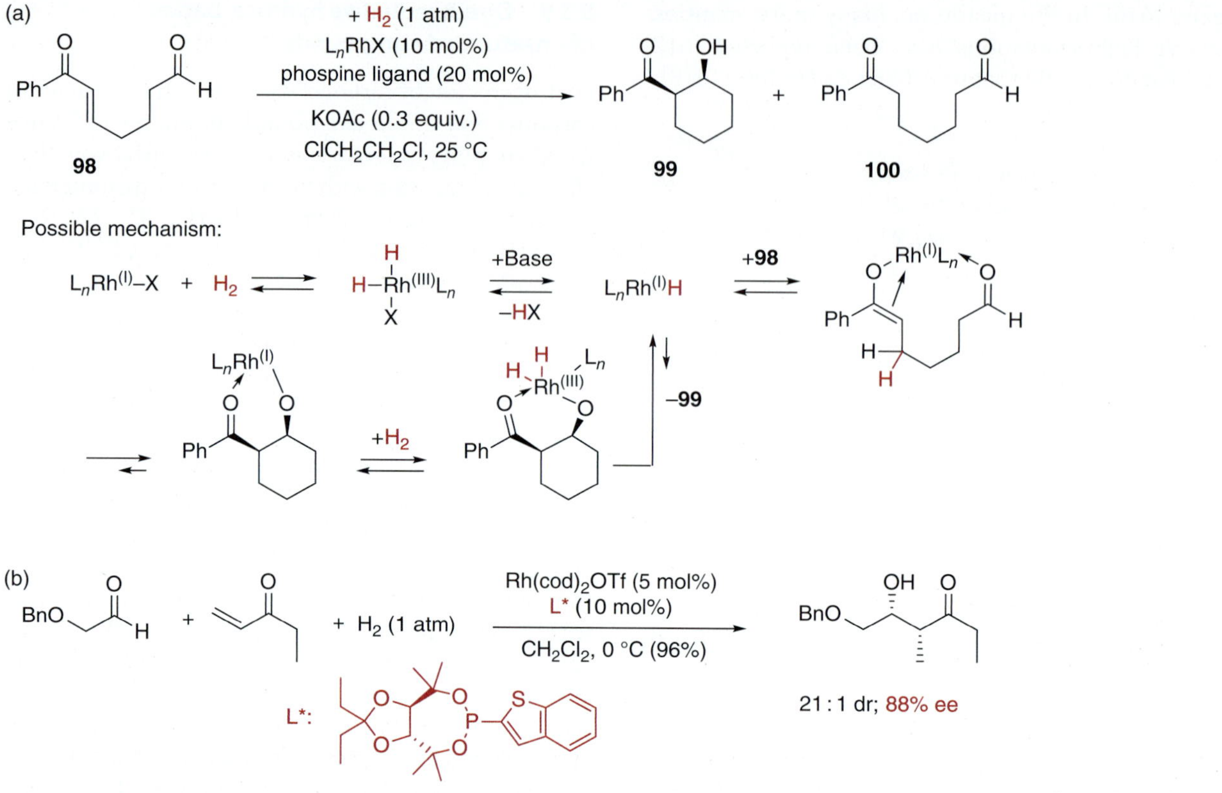

Scheme 8.33 Example of Krische's catalytic C—C bond formation via capture of a hydrogenation intermediate: (a) intramolecular hydrogenative aldol reaction and (b) example of diastereo- and enantioselective hydrogenative intermolecular aldol reaction.

reported [657, 659, 660]. Enantioselective hydrogenative cyclization of 1,6-enynes into 1-alkylidene-2-alkylcyclopentanes has also been presented [661]. Intermolecular hydroacylation of alkenes in the absence of chelation assistance has been realized using anhydrides (reaction (8.94)) [562, 662]. In 2008, the same group realized the first hydrogenative coupling of 2-vinylpyridines with *N*-sulfonylimines (e.g. reaction (8.95)) [663].

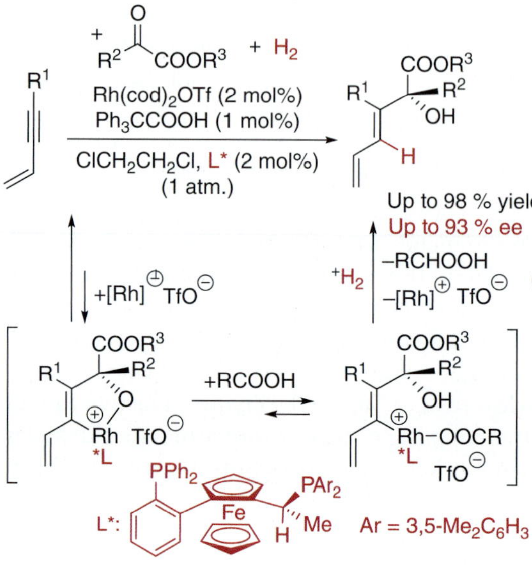

(8.93)

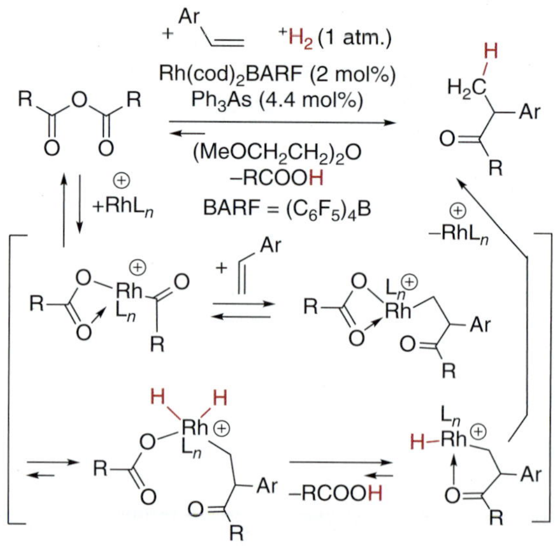

(8.94)

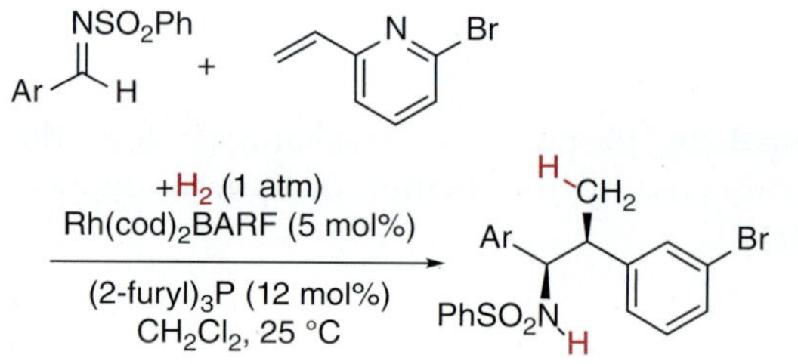

(8.95)

Dihydrogen reductant can be replaced by other reductants able to transfer an equivalent of H_2. A reductant of choice is isopropanol that produces volatile and nontoxic acetone as the coproduct in the presence of a Ru-catalyst. A first illustration is reaction (8.96) that condenses reductively allenes and aldehydes into homoallylic alcohols [664]. Alternatively, formic acid (HCOOH) can be used as the final reductant as illustrated with reaction (8.97) that realized the reductive hydrocarbation of benzaldehyde with isoprene [665]. Paraformaldehyde can be used in these reductive hydrocarbations [666]. The reductive condensation of 1,3-dienes and imines (reactions (8.98) and (8.99)) leads to the diastereoselective formation of homoallylamines [667]. Other methods of hydrocarbation of alkenes with amines are Reppe's metal-catalyzed carbonylative hydroaminomethylation $R^1CH{=}CH_2 + CO + R^2(R^3)NH + H_2 \rightarrow R^1CH_2CH_2CH_2N(R^2)R^3 + R^1CH(Me){-}CH_2N(R^2)R^3$ [198, 668, 669], metal-catalyzed alkene hydroaminomethylation $R^1CH{=}CH_2 + R^2(Me)NH \rightarrow R^1CH_2CH_2CH_2N(R^2) + R^1CH(Me)CH_2NHR^2$ [442, 443, 670, 671], and photoredox-catalyzed C—H addition of methylamines to electron-poor alkenes, e.g. $R^1OOCCH{=}CH_2 + MeN(R^2)H \rightarrow R^1OOCCH_2CH_2CH_2N(R^2)H$ [672, 673].

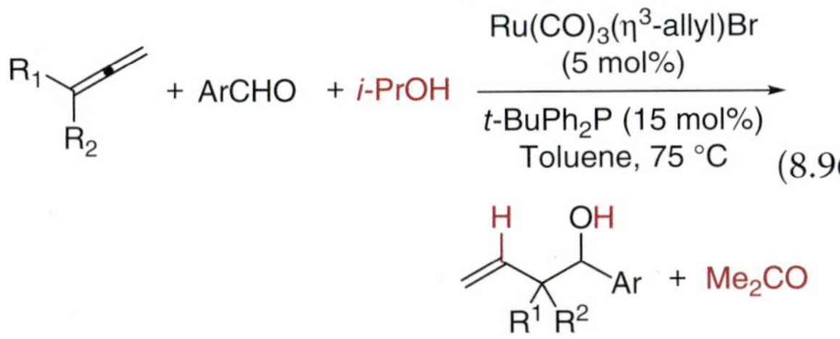

(8.96)

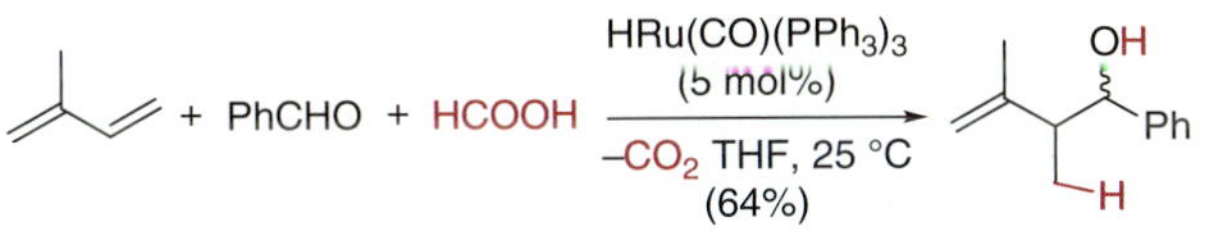

(8.97)

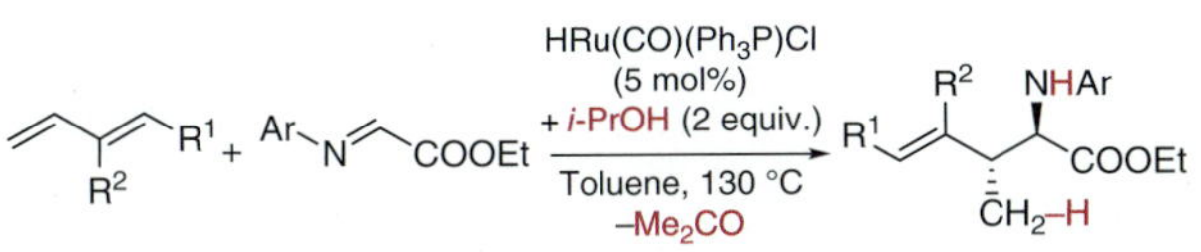

(8.98)

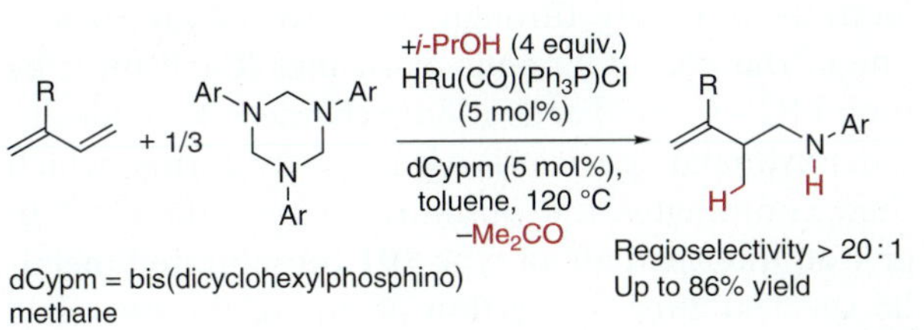

(8.99)

Problem 8.27 In the presence of an excess of isopropanol, $HRu(CO)(PPh_3)_3Cl$ and a diphosphine ligand as a catalyst, allenes **A** react with formaldimines forming homoallylamines with complete regioselectivity. Propose a mechanism for this reaction [674].

8.3.10 Direct hydrocarbation via transfer hydrogenation

In 2008, Krische and coworkers found that primary alcohols that equilibrate with their corresponding aldehydes in the presence of Ru-catalysts (transfer hydrogenation) can be reacted with (allyl)ruthenium intermediates resulting from the hydrogen transfer to 1,3-dienes (reaction (8.100)). In these reactions, the alcohol provides H_2 and the carbonyl compound for the C—C-coupling with the unsaturated compound [597, 665, 675]. This permits direct hydrocarbation reactions of type $R^1CH_2OH + R^2CH{=}CH_2 \rightarrow R^1CH(OH){-}CH(Me)R^2$ and displacement reactions of type $R^1CH_2OH + CH_2{=}CH{-}CH(R){-}OAc \rightarrow R^1CH(OH){-}CH(R){-}CH{=}CH_2 + AcOH$. Using an enantiopure diphosphine ligand, asymmetric hydrocarbation reactions have been realized as shown with reaction (8.101) [676–678]. As the reactions are regio-, diastereo-, and enantioselective, a remarkably efficient synthesis of complicated natural polyketide and polypropionate antibiotics and analogs is now available [679]. Primary alcohols react with 1,3-enynes in the presence of $HRu(CO)(PPh_3)_3Cl$ and a diphosphine giving mixtures of diastereomeric propargylic alcohols [680]. Using enantiopure diphosphine ligand and an iridium complex as a catalyst, asymmetric diastereoselective propargylations are realized as illustrated with reaction (8.102) [681, 682]. Under similar reaction conditions, methanol and 1,1-disubstituted allenes furnish higher alcohols that incorporate all-carbon quaternary centers ($CH_2{=}C{=}C(R^1)R^2 + MeOH \rightarrow CH_2{=}CH{-}C(R^1)(R^2)CH_2OH$) [683]. The direct conversion of secondary to tertiary alcohols via-Ru-catalyzed C—C coupling has been demonstrated with the prenylation and geranylation of 3-hydroxy-2-oxindoles with isoprene and myrcene, respectively [684].

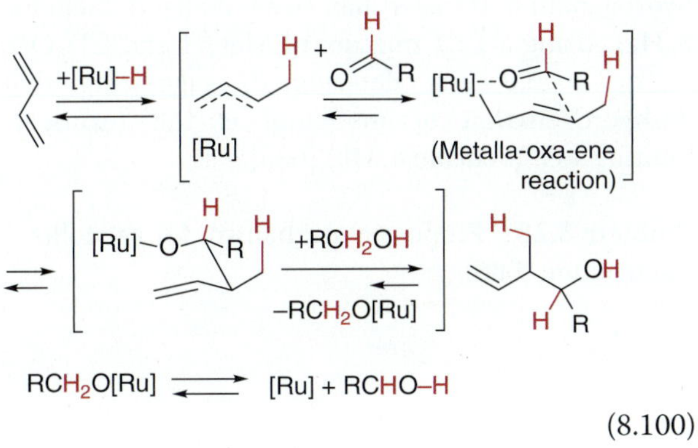

(8.100)

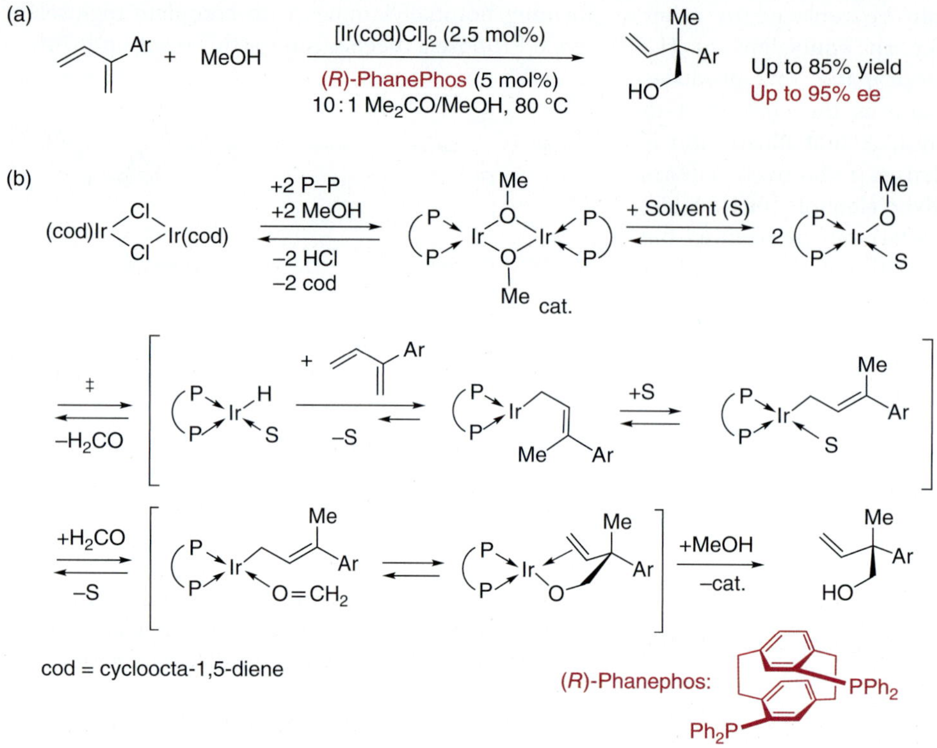

Scheme 8.34 (a) Methanol can be used in the enantioselective reductive hydrocarbation of π-systems of 1,3-dienes and (b) proposed reaction mechanism.

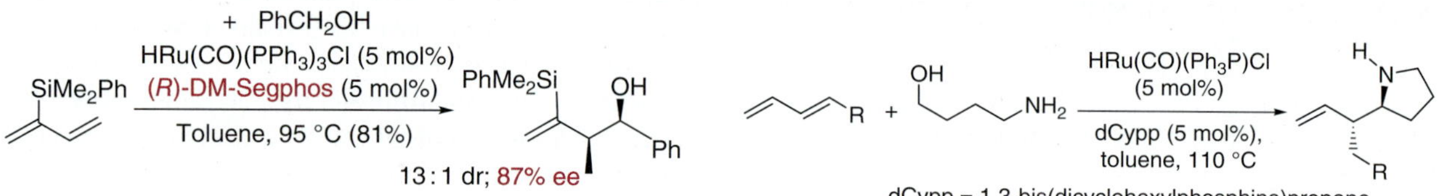

(8.101)

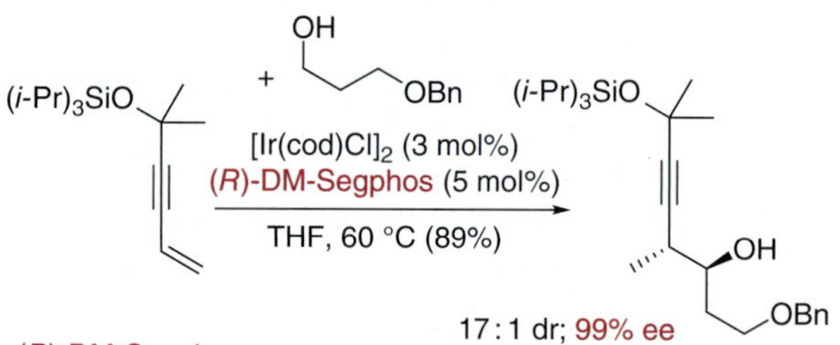

(R)-DM-Segphos = (R)-(+)-5,5′-bis[di(3,5-xylyl)phosphino]-4,4′-benzodioxole

(8.102)

Enantioselective and regioselective hydromethylation of 2-substituted butadienes via Ru-catalyzed hydrogenation transfer has been realized (Scheme 8.34a). Using a 1 : 1 mixture of MeOH and CD_3OD, $k_H/k_D = 3.0$ was determined, suggesting that deshydrogenation of methanol is the turnover-limiting step (Scheme 8.34b) [685].

Problem 8.28 Propose a mechanism for the following reaction [686].

8.4 Carbacarbation of unsaturated compounds and cycloadditions

In the near future, one shall use simple alkanes as a source of two alkyl radicals that add across unsaturated moieties such as alkenes, alkynes, carbonyl compounds, and imines (reaction (8.103)).

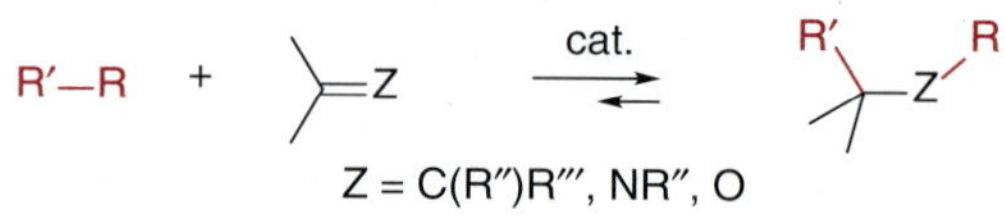

(8.103)

For success, catalysts capable to insert regioselectively (formally through oxidative addition) into one of the σ(C—C) bonds of alkanes R′—R must be found (C—C bond activation: Scheme 8.35a). Ideally, a dialkylmetal species (R—[ML$_n$]—R′) forms, which then coordinates the unsaturated reactant C=Z to give an intermediate of type **101** (unsaturated metallic species) after β-insertion of one of the two alkyl groups (e.g. R′). Intermediate **101** can be stabilized

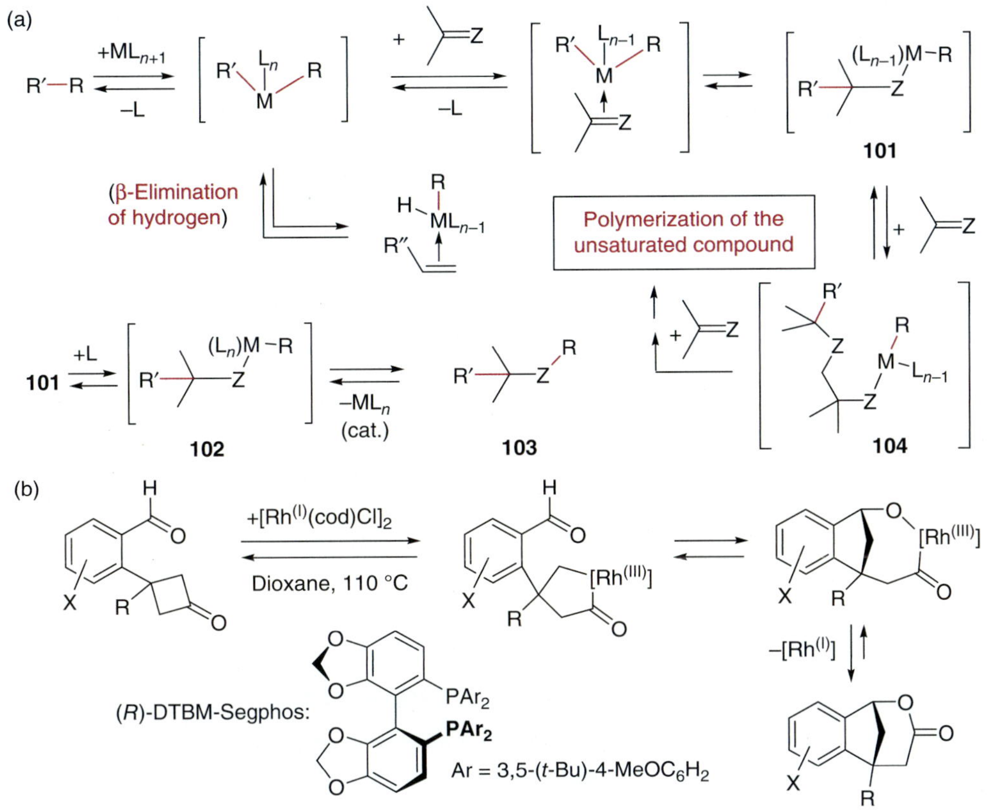

Scheme 8.35 (a) Possible routes for the carbacarbation of an unsaturated compound; competition with its polymerization and fragmentation of the alkane through deshydrocarbation, and (b) example of carbacarbation of a carbonyl moiety.

by coordination of ligand L and it equilibrates with complex **102**. The latter can then undergo a reductive elimination giving the desired product of carbacarbation **103** and the unsaturated species ML_n, the actual catalyst. If intermediate **101** coordinates to a second molecule of reactant C=Z, a polymerization process might compete. Indeed, in the case of an alkene, or alkyne, β-insertion generates another dialkylmetal species of type **104**, which might initiate the polymerization of reactant C=Z. Another condition (see e.g. the hydrogenolysis of alkanes catalyzed by $(\equiv SiO)_3ZrH$; Section 7.10) is that β-hydride elimination in the dialkylmetal intermediate R—[ML_n]—R′ should not compete. As most metal–carbon bonds of complexes involved in organometallic chemistry are much weaker (20–30 kcal mol^{-1}) than C—C bonds in alkanes (80–90 kcal mol^{-1}), usually dialkylmetal complexes undergo reductive elimination with formation of R—R′ or fragmentation through deshydrocarbation. In order to facilitate C—C bond activation by an unsaturated transition metal complex, two basic strategies can be applied. One is to start with strained hydrocarbons (R—R′), the strain energy being released during the oxidative addition of the metal complex [ML_n] in the C—C bond. An example reported in 2014 by Cramer and coworkers implies the formal intermolecular [σ^2+ π^2]-cycloaddition of the C(1)—C(2) bond of a cyclobutanone unit to a carbonyl group of an aldehyde (Scheme 8.35b) [687, 688]. Utilizing an enantiopure ligand such as DTBM-Segphos, asymmetric reactions were realized with ee up to 99.4% [688]. The second approach is to use metal complexes and reagents R—R′ providing strong M—C bonds in R—[M]—R′ such as metal cyanide bonds [689–698]. Originally, carbocyanation of alkenes and alkynes was performed by using nickel catalysts alone, but the reaction was limited to aryl and allyl cyanides. By combining the Ni-catalysts by Lewis acid cocatalysts, the carbacyanations are accelerated and the scope of nitrile reactants that can be used has been extended to alkenyl, alkynyl, and alkyl cyanides [699]. Carbacarbation of unsaturated compound via (m+n)-cycloadditions (m, n = 2, 3, 4, 5, 6, …) and cyclopropanation (cheletropic addition or (1+n)-cycloaddition) are the most powerful synthetic tools (Sections 5.3 and 5.4), especially those catalyzed by Lewis/Brønsted acids or by transition metal complexes [689–698].

8.4.1 Formal [σ^2+π^2]-cycloadditions

As mentioned above, relief of ring-strain during the oxidative addition of the ML_n + C—C → dialkylmetal complex can compensate for the thermodynamic disadvantage of the formation of two relatively weak M—C bonds. An example is given with the rhodium- and nickel-catalyzed [σ^2+π^2]-cycloaddition of biphenylene to alkenes and alkynes. Dimeric complex $[(dtbpm)RhCl]_2$ (dtbpm = bis(di-*tert*-butylphosphino) methane) reacts at 85 °C with biphenylene producing the rhodiacyclopentadiene **105**. In the presence of CO, dibenzocyclopentadienone is obtained. In the presence of an alkyne, phenanthrenes form (reaction (8.104)) together with benzenes resulting from the cyclotrimerization of the alkyne that is catalyzed by **105** [700].

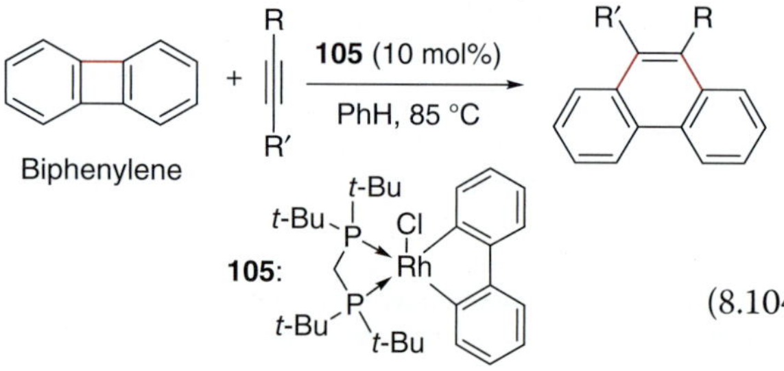

(8.104)

Other complexes are capable to activate the σ(C—C) bonds of biphenylene and to catalyze the [σ^2+π^2]-cycloadditions of biphenylene with alkynes [701, 702]. Examples are given with complexes **106** and **107**.

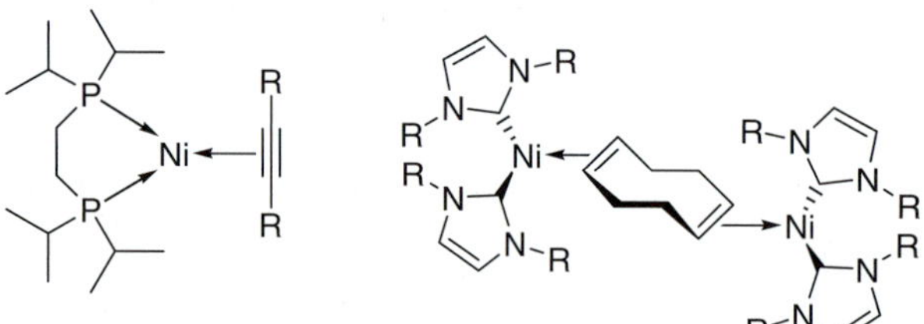

106 R = Ph, *t*-Bu, CF_3 **107** R = *n*-Pr, *i*-Pr

An earlier example of [σ^2+π^2]-cycloaddition of silacyclobutanes to acetylenes catalyzed by $Pd(PPh_3)_3Cl_2$ was reported in 1975 by Sakurai and Imai (reaction (8.105)) [703].

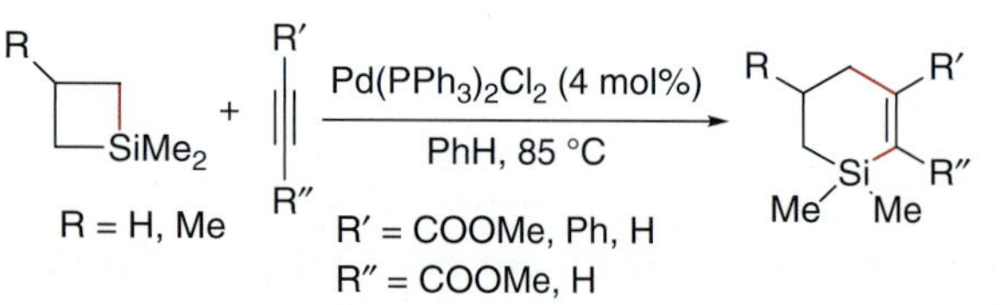

(8.105)

Cyclopropanes bearing electron-releasing and electron-withdrawing substituents on two vicinal carbon centers can equilibrate with zwitterions. The latter can be quenched with unsaturated compounds and generate five-membered ring systems (Section 8.4.6) [704–707].

Problem 8.29 What is the product **P** formed by reaction of bicyclo[2.2.1]hept-2-ene with biphenylene in the presence of 10 mol% of $[(dtbpm)RhCl]_2$ (dtbpm = $(t\text{-}Bu)_2PCH_2P(t\text{-}Bu)_2$]? [700]

Problem 8.30 What are the successive products **B** and **C** produced in the following reaction? [708, 709]

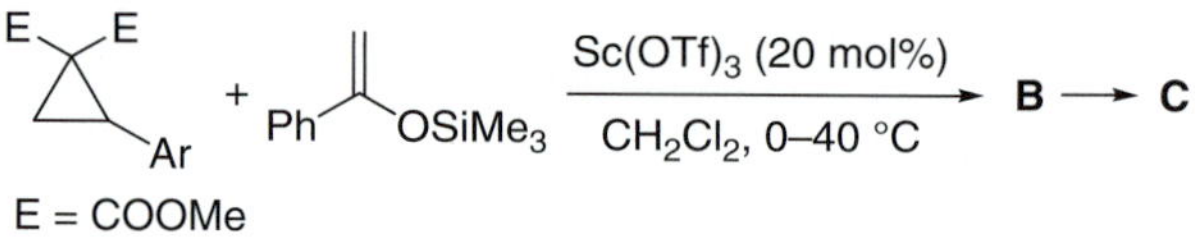

8.4.2 (2+1)-Cycloadditions

In 1885, Büchner and Curtius [710], and later Büchner and coworkers, treated several aromatic olefinic and acetylenic compounds with diazoacetic ester to form three-membered ring compounds (reaction (8.106)) under either thermal, photochemical, or catalytic conditions [711]. Early catalysts were solid metals such as bronze (precipitating Cu from $CuSO_4$ solution with Zn [712]), copper washed with HCl/H_2O [713], the activated Ullmann catalyst (prepared by washing Cu powder with ethylenediaminetetraacetic acid [714]), or metallic salts such as $CuSO_4$, $Cu[(i\text{-}PrO)_3P]Cl$ [715], $Cu[(MeO)_3P]I$ [716–718], Cu(OTf) [719], and $Pd(OAc)_2$ [720].

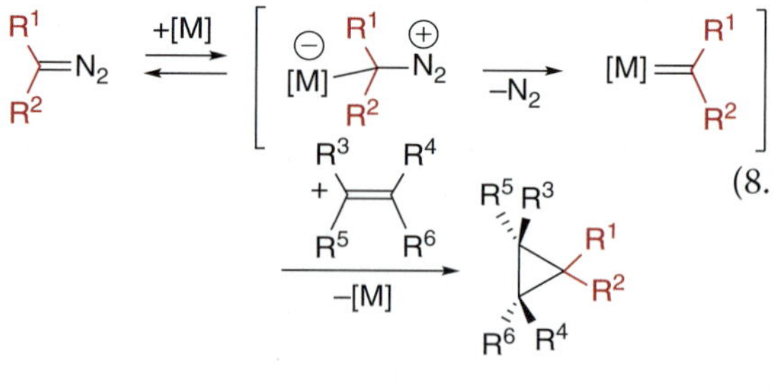

(8.106)

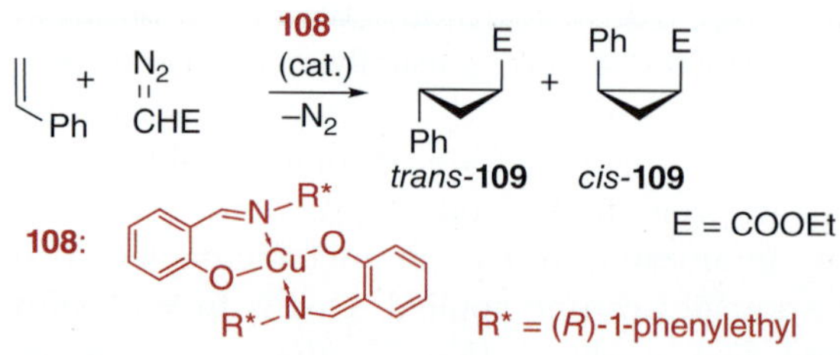

(8.107)

Highly chemoselective, diastereoselective, and enantioselective (2+1)-cycloadditions are now routine. Metal-carbenoid of type [M]=C(R^1)R^2 is formed as intermediates that undergo useful reactions such as cyclopropanation with alkenes (reaction (8.106)), cyclopropenation with alkynes, C—H and N—H insertions, and ylide formation (e.g. $R^1(R^2)C: + :X—R \rightleftarrows R^1(R^2)C^{(-)}—X^{(+)}—R$) [721–724]. Using α-diazoacetates for the metal-catalyzed cyclopropanation of terminal alkenes, *trans*-1,2-disubstituted cyclopropanes are formed selectively [725]. In 1966, Nozaki et al. presented the first example of asymmetric catalysis by means of soluble transition metal complexes (e.g. reaction (8.107)). They decomposed ethyl diazoacetate in styrene in the presence of chiral, enantiomerically pure copper complex **108** and obtained enantiomerically enriched ethyl (+)-*trans* (*trans*-**109**) and (+)-*cis*-2-phenylcyclopropanecarboxylate (*cis*-**109**) [726]. By reaction of esters of α-amino acids with Grignard reagents and combining the resulting carbinols with salicaldehyde and $Cu(OAc)_2$, complexes **110** were obtained. They have been applied to the industrial synthesis of enantiomerically enriched (>90% ee) potent insecticides such as (+)-*trans*-chrysanthemic acid, Permethrin, and the dehydropeptidase I inhibitor cilastatin [727]. An enantioselective synthesis of sesquiterpene lactone xanthatin, a cytostatic and antibiotic agent, has been realized via enantio-, regio-, and diastereoselective cyclopropanation of methyl 2-furoate [728].

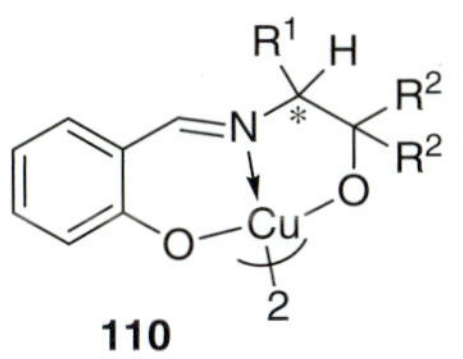

110

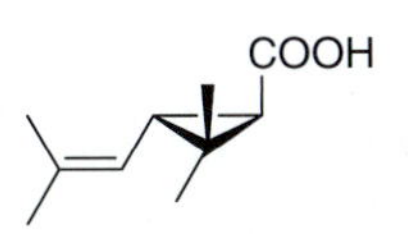

(+)-*trans*-Chrysanthemic acid

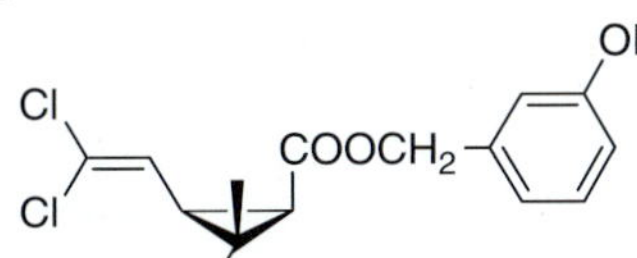

Permethrin

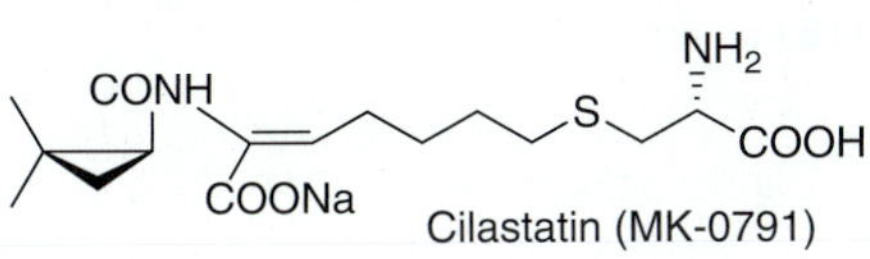

Cilastatin (MK-0791)

Kinetic studies with Cu(OTf) show that decomposition of $N_2CHCOOEt$ in the presence of hex-1-ene to produce mostly ethyl 1-butylcyclopropane-2-carboxylate and N_2 follows rate law (8.108) [719].

$$\frac{d[N_2]}{dt} = k\frac{[N_2CHCOOEt][Cu(OTf)]}{[1\text{-hexene}]} \tag{8.108}$$

The inverse dependence of rate law on alkene concentration was attributed to deactivation of Cu(OTf) by multiple coordination of the alkene. Catalysis by Cu(II) complexes involves prior reduction to Cu(I) species by diazo compounds followed by catalytic reactions of Cu(I) complexes, which imply two parts. These are the copper carbene formation and the cyclopropanation stages. Because of their high reactivity, copper carbene complexes are difficult to observe. They have been detected by ultra-low temperature NMR [729]. Other transition metal carbene species, such as those of Ru and Os (see below), have been isolated from stoichiometric reactions with diazoacetates [730, 731]. For asymmetric cyclopropanations (8.109) of substituted styrenes **111** into cyclopropanes **112** with $N_2CHCOOEt$ and **113** as catalysts (obtained by deprotonation of the corresponding bis-oxazoline with BuLi/THF and reaction with $CuCl_2$ [732]), a positive electron demand was evidenced by the Hammet correlation $\log(k_{rel}) = -0.51\,\sigma^*$ (negative ρ) suggesting substantial development of cationic character at the α-carbon center in the transition state.

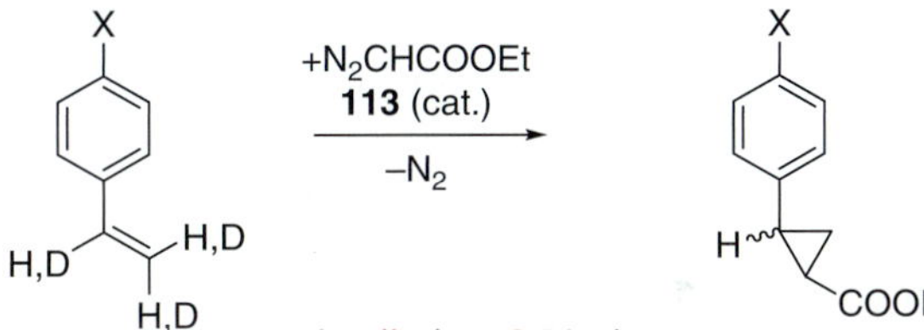

111 $\log(k_{rel}) = -0.51\,\sigma^*$ **112**

(r^2 = 0.999; X = H, MeO, Me, CF_3, NO_2)

113:

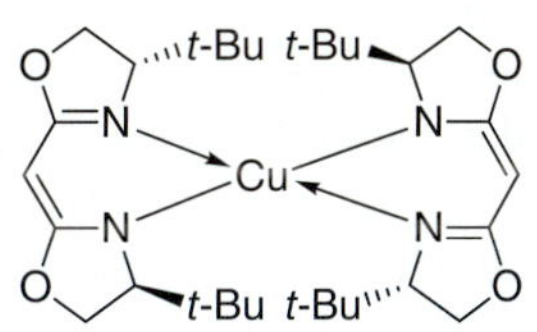

(8.109)

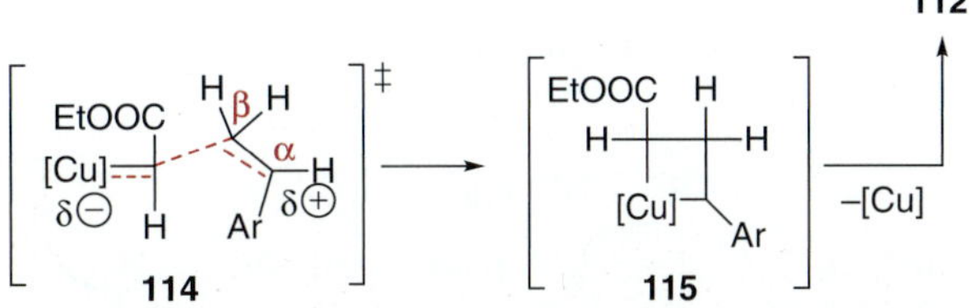

114 **115**

With α-deuterostyrene and β,β-dideuterostyrene, kinetic deuterium isotopic effects $k_H/k_D = 1.02 \pm 0.02$ and 0.95 ± 0.02, respectively, were found. They indicate that β-carbon is somewhat rehybridized in the transition state, whereas no significant effect can be seen at the α-position. The data are consistent with a concerted but asynchronous transition state **114**, where the new C—C bond to the β-position is formed well before the bond at the α-carbon [733].

Several paths can be followed from transition state **114** to products **112**. In a first one, direct formation of the cyclopropane with liberation of the copper catalyst occurs. In a second one, ring closing of **114** generates a cupracyclobutane **115**, which then undergoes fast reductive elimination to form the cyclopropane product **112** and the initial copper catalyst [734, 735]. Catalytic asymmetric cyclopropanations have been applied to the total synthesis of natural products and analogs [736–740]. In the presence of 1 mol% $Pd(OAc)_2$, CH_2N_2 reacts with terminal, strained cyclic, or acyl-substituted alkenes to generate the corresponding cyclopropanes at 0–5 °C [689, 720, 741–743]. Tomilov and coworkers proposed that $Pd(OAc)_2$ (and other Pd(II) salts) are reduced into Pd(0) species with CH_2N_2 and that the latter form π-complexes with the alkene [744]. A possible mechanism for the formation of Pd(0) species is $CH_2N_2 + PdX_2 \rightarrow CH_2{=}PdX_2 + N_2 \rightarrow X{—}CH_2{—}Pd{—}X \rightarrow Pd(0) + CH_2X_2$ [745]. Quantum mechanical calculations for the reaction of ethylene with CH_2N_2 predicted $Pd(\eta^2\text{-ethylene})_2$ to be the resting state of the catalyst and the product of $Pd(OAc)_2$ reduction with CH_2N_2. $Pd(CH_2{=}CH_2)_2$ and CH_2N_2 equilibrate with $Pd(CH_2{=}CH_2)(CH_2N_2)$ from which N_2 is eliminated in the rate-determining step. The resulting carbene complex $(\eta^2\text{-}CH_2{=}CH_2)Pd{=}CH_2$ reacts without intrinsic barrier with CH_2N_2 to give $(\eta^2\text{-}CH_2{=}CH_2)_2Pd$ and N_2 and with ethylene to give $(\eta^2\text{-}CH_2{=}CH_2)$palladacyclobutane, a facile reaction. Reductive elimination of cyclopropane from the latter complex is also facile [746]. $Pd(OAc)_2$ catalyzes the cyclopropanation of electron-deficient alkenes **117** with aryldiazocarbonyl compounds **116** giving isomeric cyclopropanes *cis-* and *trans-***118**. The reaction is highly stereoselective for acrylates, and acrolein, not at all for acrylamides and acrylonitrile (Scheme 8.36).

In the case of acrylonitrile + $PhC(N_2)COOMe$ and using $Pd(OAc)_2$ with 1,10-phenanthroline as ligand, after 10 days at 20 °C, pyrazolidine derivative **120** was isolated in 57% yield. In the absence of Pd-catalyst, pyrazolidine **119** forms slowly at 20 °C, which is then isomerized into **120** in the presence of silica gel. On heating, **119** is converted into products **121** and N_2. It is thus not excluded that instead of generating a carbenepalladium complex that cyclopropanates the alkene, the Pd-catalyst promotes the dipolar cycloaddition of the diazoacetate onto the alkene and that a subsequent thermal exclusion of N_2 from a pyrrazolidine generates the cyclopropane product [747]. Dirhodium tetraacetate ($Rh_2(OAc)_4$) is commonly used as a catalyst for the cyclopropanation with diazoacetates [748, 749]. Highly electron-deficient ligands such as trifluoroacetate lead to more electrophilic carbenoids while electron-releasing ligands such as acetamido lead to less reactive and more selective carbenoids [750]. Under standard conditions, styrene reacts with ethyl diazoacetate yielding mixtures of *trans-* and *cis-*1,2-disubstituted cyclopropanes in a diastereoselectivity < 2 : 1. Considerable enhancement in the diastereoselectivity can be achieved by using bulky ester derivatives such as $[2,6\text{-}(t\text{-}Bu)_2\text{-}4\text{-}MeC_6H_2]OOC{—}CHN_2$ in combination with dirhodium (tetraacetamide) ($Rh_2(NHCOMe)_4$) (reaction (8.110)) [750]. The cyclopropanation of a large variety of olefins with dimethyl diazomalonate is best catalyzed with $Rh_2(esp)_2$ (esp = α,α′,α′,α′-tetramethyl-1,3-benzenedipropanoate). The catalyst can also be applied to the (2+1)-cycloaddition of alkynes to generate cyclopropenes as illustrated with reaction (8.111) [751].

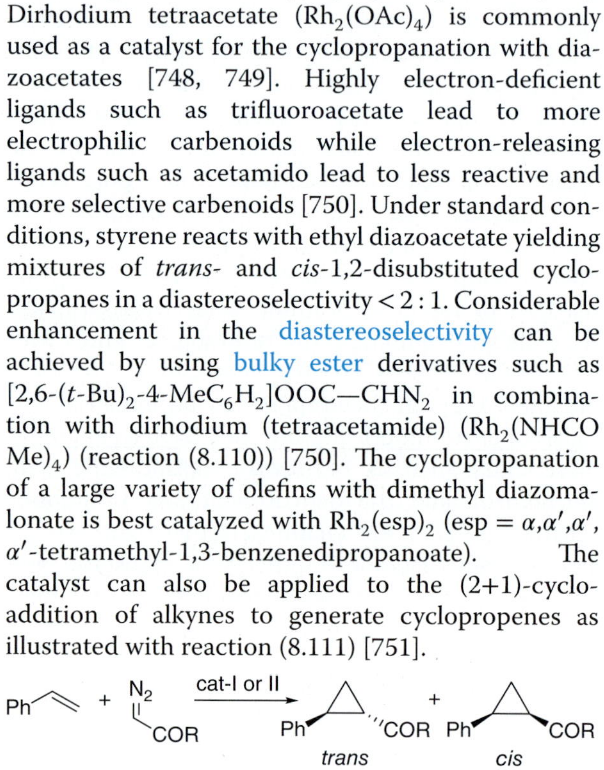

	trans		cis
R = OEt, cat-I	62	:	38
R = OEt, cat-II	60	:	40
R = 2,6-(*t*-Bu)$_2$-4-MeC$_6$H$_2$O, cat-I	84	:	16
R = 2,6-(*t*-Bu)$_2$-4-MeC$_6$H$_2$O, cat-II	98	:	2

(8.110)

cat-I: $Rh_2(OAc)_4$ cat-II: $Rh_2(NHCOMe)_4$

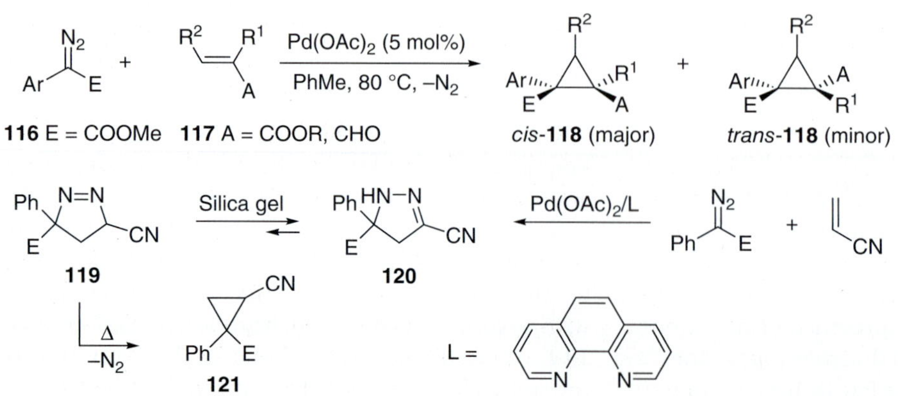

Scheme 8.36 Pd-catalyzed cyclopropanation of electron-deficient alkenes with aryldiazocarbonyl compounds.

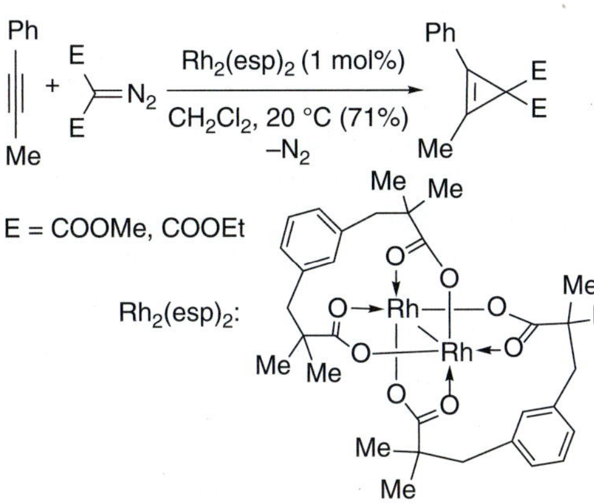

(8.111)

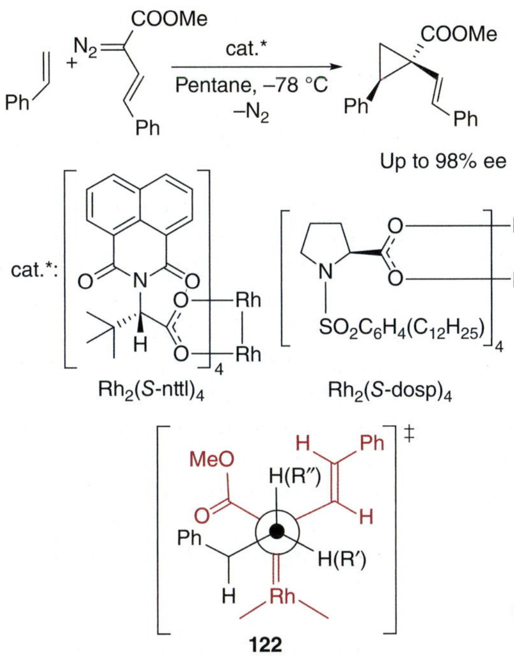

(8.112)

A very high degree of asymmetric induction is obtained when the cylopropanation is carried out in a nonpolar solvent at low temperature with $Rh_2(S\text{-nttl})_4$ [752], $Rh_2(5S\text{-MEPY})_4$ (5S-MEPY = methyl 2-pyrrolidone-5(*S*)-carboxylate), or $Rh_2(S\text{-dosp})_4$ as catalyst [753–755]. The latter is soluble in pentane at −78 °C [756]. The high diastereoselectivity of cyclopropanation (8.112) is explained by Davies with transition state **122** [756, 757]. The approach of the alkene to the (vinylcarbene)rhodium species occurs on the side of the electron-withdrawing group (COOMe) with the bulky functionality of the alkene (Ph) pointing away from the face of the metallic complex. (*E*)-Alkenes are unreactive under these conditions because they cannot avoid having a substituent pointing directly toward the rhodium atom. The fact that the diastereocontrol is enhanced by electron-donating groups on the alkene while (*E*)-alkenes are unreactive support the side-on non-synchronous model **122**. Alternative mechanism (transition state) has been proposed by Corey and coworkers who favor "face on" (2+2)-cycloaddition of alkene to the (carbene)metal [758, 759]. Based on the kinetic isotopic effects and for the reaction of terminal alkynes, Nowlan and Singleton favor "end on" (2+1)-cycloaddition of alkene to the metal carbenoid [760, 761]. Quantum mechanical calculations on a model reaction using ethylene, CH_2N_2 as reactants, $Rh_2(\text{formate})_4$ as a catalyst, and MeCl as a coordinating solvent, both (2+1) and (2+2)-pathways have been identified for the cyclopropanation step depending on the catalytic carrier involved [762].

Ruthenium(II)-based catalyst **123** reported by Nishiyama et al. [763], as well as cobalt(II)-based catalyst **124** reported by Katsuki and coworkers, has permitted highly diastereo- and enantioselective cyclopropanations using azoacetates [764]. The reactions have been carried out in water and the diazo compound can be generated in situ by reaction of ethyl glycinate with $NaNO_2$ (H_2SO_4 cat.) [765].

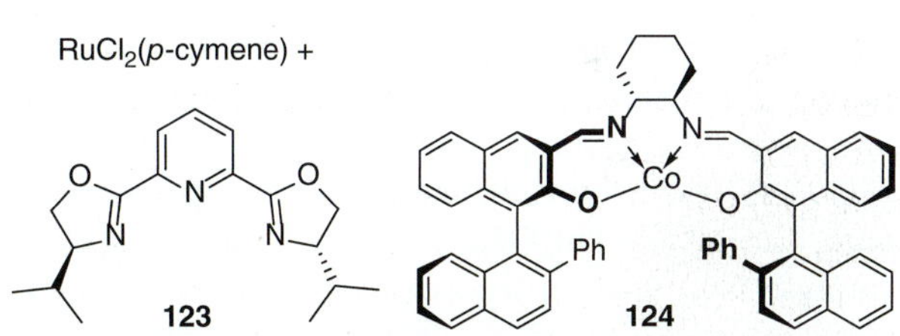

Callot et al. [766, 767] and then Kodadek and coworker [768, 769] have used rhodium porphyrins to catalyze the cyclopropanation of alkenes with diazoacetates. Iron [770], ruthenium [771–773], and cobalt porphyrins [774, 775] have also been applied as catalysts. Other catalysts have been proposed such as pentacarbonyl(η^2-(*Z*)-cyclooctene)chromium [776], (IPr)CuCl [777], (IPr)AuCl [778, 779] (IPr = 1,3-bis(2,6-diisopropylphenyl)imidazol-2-ylidene), and $AgSbF_6$ [780]. Jørgensen and coworkers reported the asymmetric cyclopropanation of *N*-tosyl iminoesters with trimethylsilyl diazomethane to produce the corresponding aziridines **125** in good diastereo- (20 : 1) and enantioselectivity (88% ee) applying $AgSbF_6$ and (*R*)-TolBINAP ((*R*)-2,2′-bis(di-*p*-tolyphosphino) binaphthyl) as catalysts (reaction (8.113)) [781].

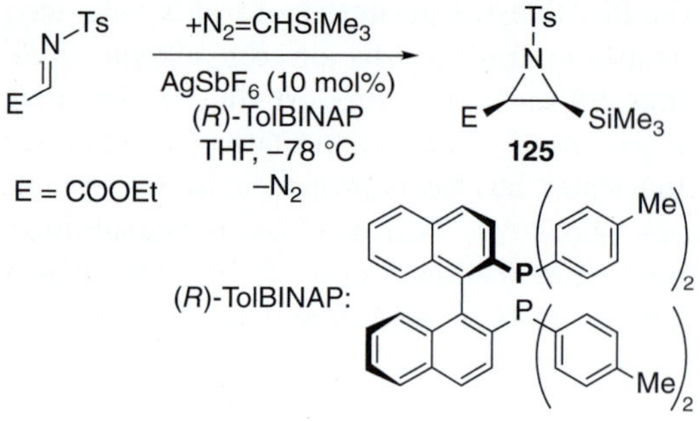

(8.113)

Triarylaminium salts such as $(4\text{-BrC}_6\text{H}_4)_3\text{N}^{\bullet +}/\text{SbCl}_6^-$ smoothly catalyze the cyclopropanation of dienes, styrenes, and tetrasubstituted alkenes by ethyl diazoacetate. The reactions are regioselective and, in the case of additions to conjugated dienes, cyclopropane periselective. A cation radical chain mechanism involving single-electron transfer (SET) from the unsaturated reactant to $(4\text{-BrC}_6\text{H}_4)_3\text{N}^{\bullet +}/\text{SbCl}_6^-$ as the initial step has been proposed [782, 783].

Because of the inherent danger of explosion of diazo compounds, other sources of metalcarbenoids have been searched. Aryliodonium ylides derived from dimethyl malonate have been used for the olefin cyclopropanations in the presence of $Rh_2(OAc)_4$ [784]. Selectivity studies revealed that the main pathway of these reactions involves metallacarbene intermediates (reaction (8.114)). The decomposition of phenyliodonium ylides with copper catalysts also seem to involve cupracarbene intermediates [785, 786]. Electron-rich alkenes are cyclopropanated with $CH_3NO_2 + PhI(OAc)_2$, a base, and $Rh_2(OAc)_4$ catalyst to provide nitrocyclopropanes [787]. Enantiomeric excesses up to 90% for 4-bromostyrene and 98% for pent-1-ene have been observed with phenyliodonium ylide derived from dimethyl malonate in the presence of chiral rhodium-based catalyst (reaction (8.115)) [788].

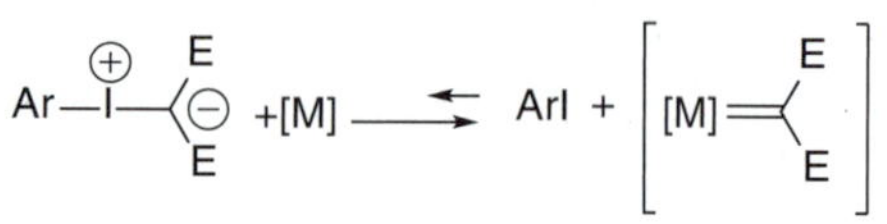

(8.114)

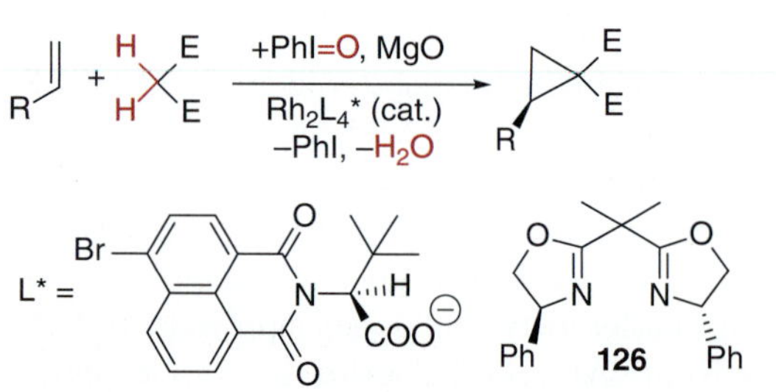

(8.115)

Using CuCl, $AgSbF_6$, and ligand **126** as catalysts, enantioselective cyclopropanation of alkenes using iodonium ylides derived from methyl nitroacetate has been realized [789]. The Simmons–Smith reaction is a powerful method for the synthesis of cyclopropanes from alkenes [790–795]. $TiCl_4$ has been found to accelerate the Simmons–Smith reaction [796]. A first catalytic, enantioselective Simmons–Smith cyclopropanation of an allylic alcohol (reaction (8.116)) has been reported by Kobayashi and coworkers [797]. They used a chiral zinc prepared in situ by reaction of the corresponding disulfonamide **127** and Et_2Zn. Transition state **128** was proposed for these reactions.

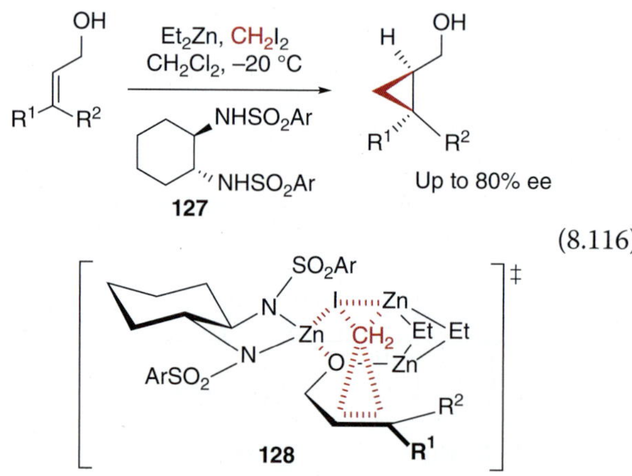

(8.116)

Carbenemetal complexes are in fact metallalkenes. They are highly polarizable and might react as electrophiles or nucleophiles depending on the nature of the metal, the ligands, and substitution of the carbene moiety. This can be expressed by the formalism (8.117).

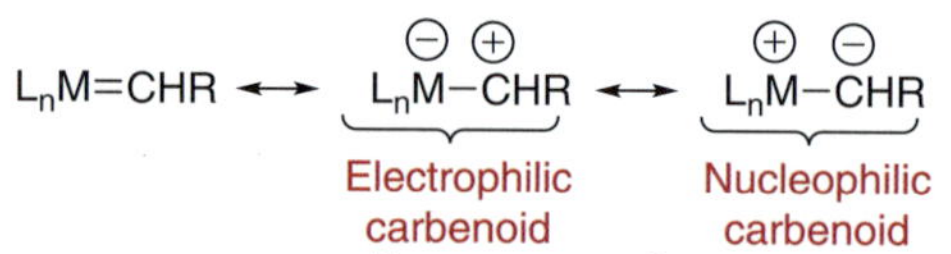

(8.117)

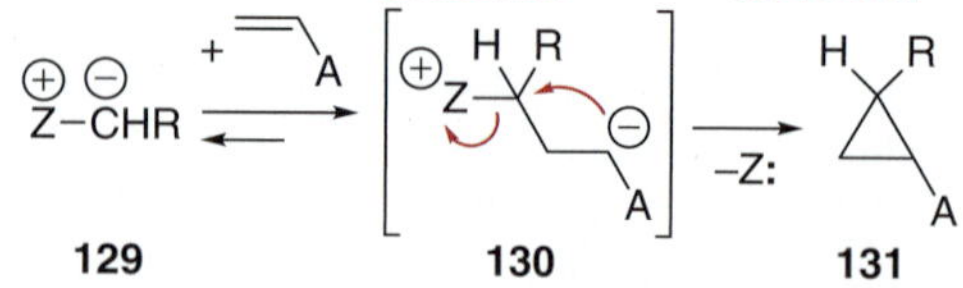

(8.118)

The dipolar limiting structures can be seen as metal–carbon ylides. With alkenes that are Michael acceptors (reaction (8.118)), carbon ylides **129** can add and generate zwitterion intermediates **130** that undergo quick γ-elimination of Z:, the latter being a good nucleofugal group. This produces the corresponding cyclopropanes **131** [798]. The most commonly employed ylides **129** are α-halocarbanions [799], sulfur ylides [800–806], selenium ylides [807–811], tellurium ylides [812–814], phosphorus ylides [815–820], arsenic ylides [821–824], and ammonium ylides [825–828]. A catalytical version of the cyclopropanation reaction via ammonium ylides has allowed the enantioselective construction of bicyclo[4.1.0]heptan-2-one and bicyclo[3.1.0]hexan-2-one derivatives [829].

Problem 8.31 What are the two products **B** (73%) and **C** (7%) formed in the following reaction? [830, 831]

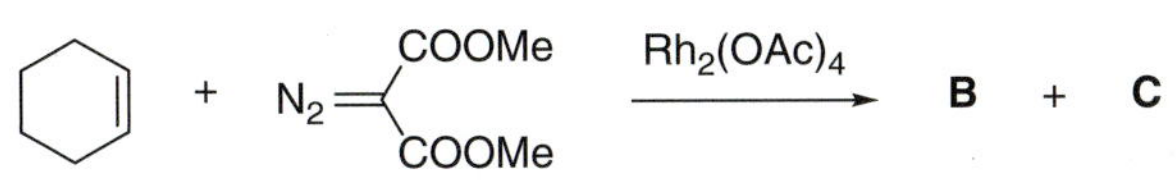

8.4.3 Ohloff–Rautenstrauch cyclopropanation

In 1976, Ohloff and coworkers at Firmenich Co, Geneva, reported the $ZnCl_2$-catalyzed isomerization of (±)-dehydrolinalyl acetate **132** into (±)-carvenone (**133**) (70%), (±)-2-acetoxycar-2-ene (**134**; 5%), and (±)-2,6,6-trimethyl-2-ethynyl-5-acetoxytetrahydropyran (**135**: 5%) [832]. Formation of **134** was explained in terms of zwitterionic intermediate **136** arising from the electrophilic activation of the alkene moiety by the Lewis acid and intramolecular alkyne addition under assistance by the acetoxy group. In 1984, Rautenstrauch, also at Firmenich, reported the $PdCl_2(MeCN)_2$-catalyzed cycloisomerization reaction (8.119) of unsaturated propargylic acetates that generates bicyclo[n.1.0]enol esters [833].

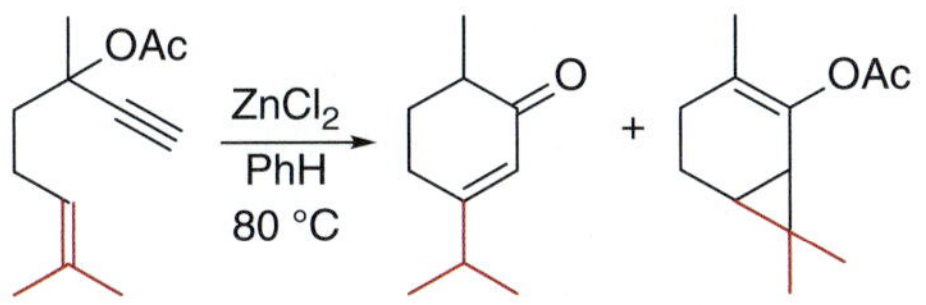

132 **133** **134**

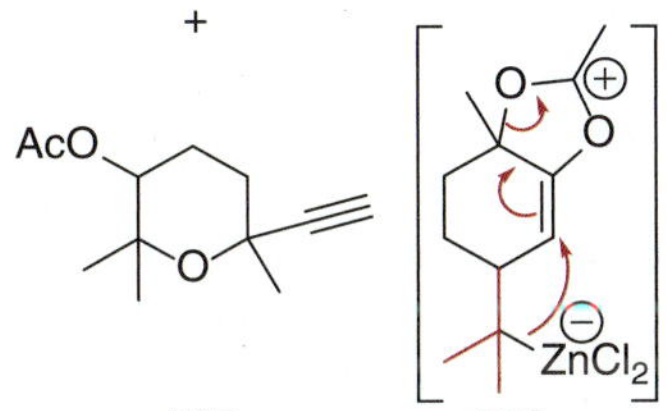

135 **136**

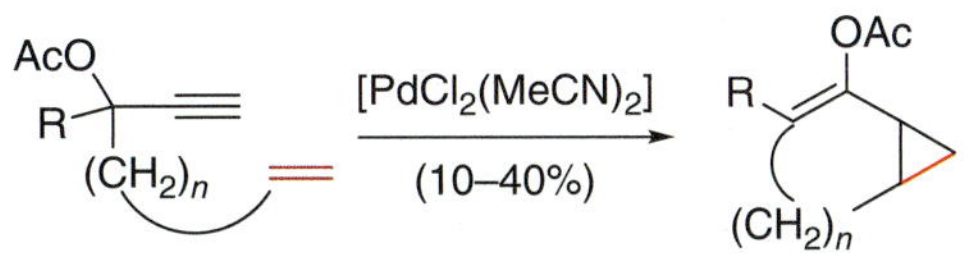

(8.119)

Since then, a large number of analogous reactions have been described [834] using either $PtCl_2$ in toluene [835], $[Au(IPr)(NCMe)]PF_6$ in CH_2Cl_2 [836], or AuCl, $AuCl_3$, or $[Au(PPh_3)Cl]/AgSbF_5$ in CH_2Cl_2 [837–839]. Cationic rhodium complex $Rh(cod)_2SbF_6$ catalyzes cyclopropanation of *N,N*-disubstituted acrylamides by propargyl esters [840]. Based on quantum mechanical calculations, the mechanism shown in Scheme 8.37 has been proposed for the $PtCl_2$-catalyzed Ohloff–Rautenstrauch intramolecular cyclopropanation [841–843].

In this mechanism, the transition metal forms a π-complex **137** with the alkyne moiety and activates it as an electrophile that can be considered to be a carbene of type **137″**. The latter undergoes intramolecular cyclopropanation of the alkene moiety leading to the bicyclic intermediate **138**. Acyloxy group migration **138** → **139** gives finally the bicyclic products and liberate the metal catalyst. Evidence in favor of the intermediacy of (alkylidene)metal complex **138** is given by intermolecular quenching with alkenes as illustrated with reaction (8.120) that uses AuCl as a catalyst in CH_2Cl_2 [842, 844, 845].

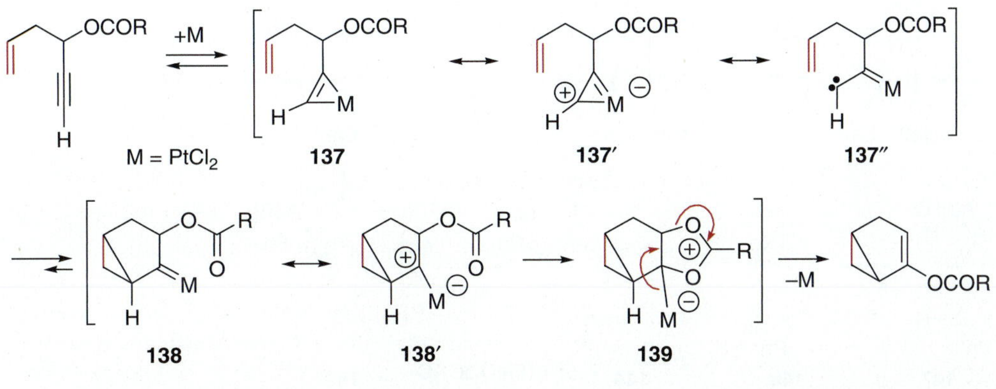

Scheme 8.37 Proposed reaction mechanism for the $PtCl_2$-catalyzed cycloisomerization of ene-propargylic esters (Ohloff–Rautenstrauch intramolecular cyclopropanation).

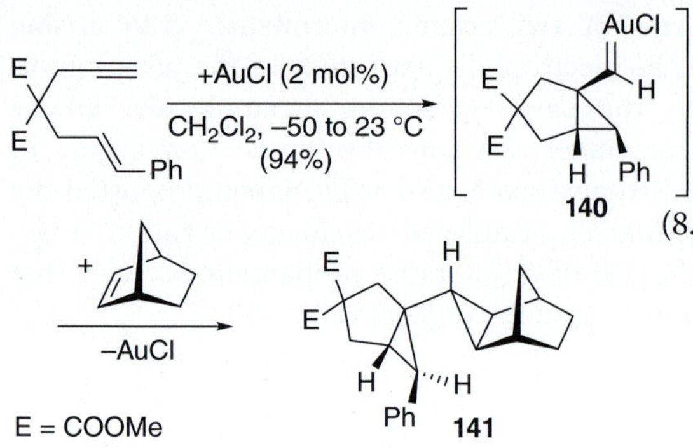

(8.120)

With other catalysts such as $[Au(PPh_3)(MeCN)]^+/SbF_6^-$ or $[2,4\text{-}(t\text{-}Bu)_2C_6H_3O]_3PAuCl$, mixtures of products **144** and **148** form together with **141**. Their formation can be interpreted in terms of the electrophilic carbenoid moiety **140** ↔ **140′**, which undergoes cyclopropylcarbinyl (**140′**, **145**) ⇄ cyclobutyl (**142**, **142′**) ⇄ homoallyl (**143**, **146** → **147**) cation rearrangements (Scheme 8.38) [846–851].

Intermediates **140**, **146** have been quenched by nucleophiles such as electron-rich arenes and heteroarenes, allylsilanes, and 1,3-dicarbonyl compounds [852]. Ohe and coworkers [853, 854] have developed an intermolecular version of the Ohloff–Rautenstrauch cyclopropanation reaction using $[Ru(CO)_3Cl_2]_2$ as the catalyst (reaction (8.121)) [855]. Toste and coworkers have reported an asymmetric version of the cyclopropanation of vinylarenes with propargyl ester, using $AgSbF_6$ as a catalyst and the homochiral ligand (*R*)-(−)-DTBM-Segphos [856].

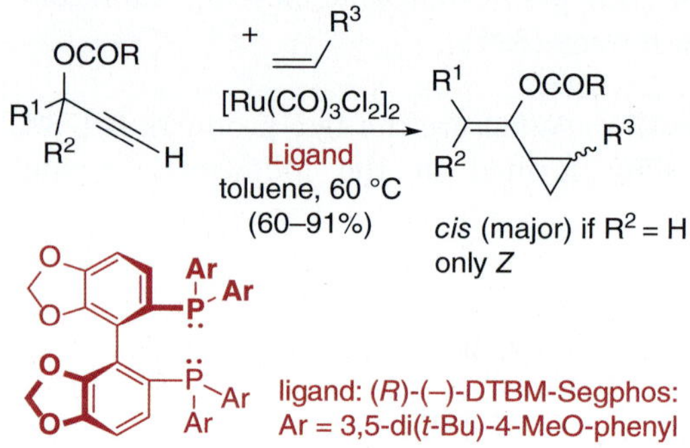

(8.121)

8.4.4 $[\pi^2+\pi^2]$-Cycloadditions

Formally "forbidden" according to the Woodward–Hoffmann rules (Section 5.3) when carried out under thermal conditions, the suprafacial/suprafacial $[\pi^2s+\pi^2s]$-cycloadditions of alkenes and/or alkynes can be induced photochemically (Sections 6.8.1 and 6.8.2). Thermal $[\pi^2s+\pi^2s]$-cycloadditions are possible for substituted cycloaddents equilibrating with relatively stable 1,4-diradical ↔ 1,4-zwitterion intermediates (Section 5.3.4). For alkenes and alkynes bearing conjugated C=O, C≡N, C=NR, NO_2, or SO_2R moieties (Michael acceptors) that can bind to Lewis and/or protic acids, their $[\pi^2s+\pi^2s]$-cycloadditions

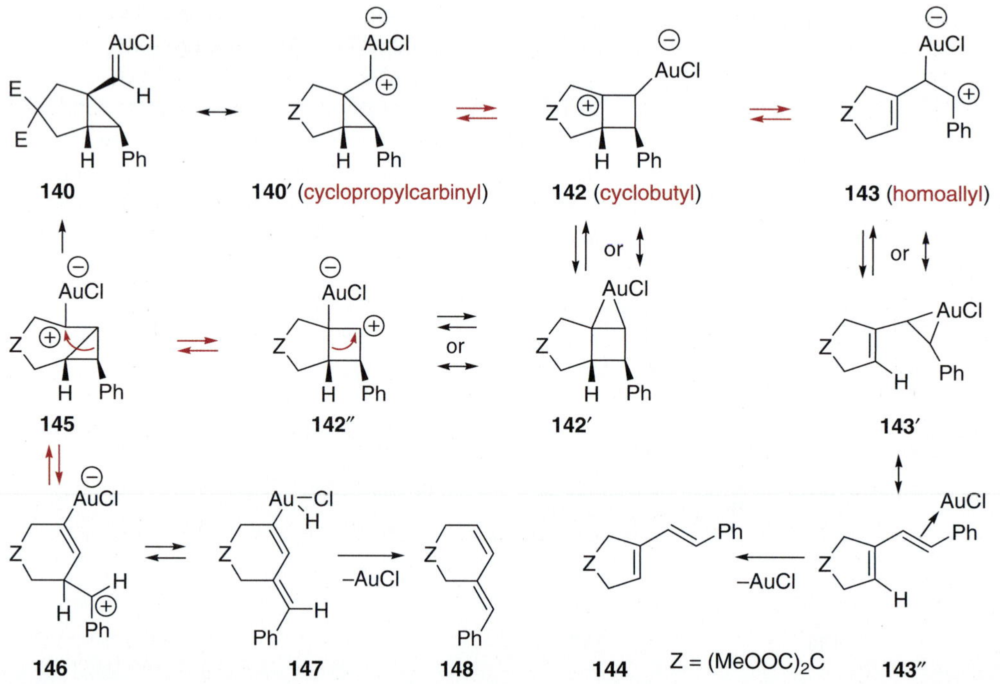

Scheme 8.38 Proposed reaction mechanisms for the gold-catalyzed enyne rearrangements into monocyclic dienes.

can occur under very mild conditions (Section 7.6.8). When one of the cycloaddents can be oxidized into its radical cation, it also undergoes smooth [π^2s+π^2s]-cycloadditions (Umpolung by single-electron transfer, see Sections 7.5.8 and 7.5.10). In this chapter, we present examples of [π^2s+π^2s]-cycloadditions of nonpolar alkenes and alkynes catalyzed by transition metal complexes.

The transition-metal-catalyzed [π^2s+π^2s]-cycloaddition of norbornadiene is known since 1957 [857]. In the presence of Ni(cycloocta-1,5-diene)$_2$ and (m-PhC$_6$H$_4$O)$_3$P, buta-1,3-diene is cyclodimerized into 1,2-*cis*-divinylcyclobutane (97%) and 4-vinylcyclohexene (2.4%) [858]. In 1972, the titanium-catalyzed cross-coupling of buta-1,3-diene with ethylene was reported for the first time by Cannell (reaction (8.122)) [859]. Strained alkenes are more reactive than acyclic alkenes and have found several applications in transition-metal-catalyzed [π^2s+π^2s]-cycloadditions. Two examples are given with reactions (8.123) [689, 860] and (8.124) that use a nickel-based catalyst [861]. The *endo*-stereoselectivity in reaction (8.123) is considered to originate from the complex intermediate **149** in which norbornadiene coordinates to the Ni metal as an *endo* bidentate ligand [862]. If a Pd(0) catalyst is used instead of Ni(0), a (3+2)-cycloaddition of norbornadiene to methylidenecyclopropane is observed (Section 8.4.6) [863]. Low-valent ruthenium complexes have been used to catalyze the suprafacial [π^2s+π^2s]-cycloadditions of norbornene and norbornadiene to various acetylenes as illustrated with reaction (8.125) [864].

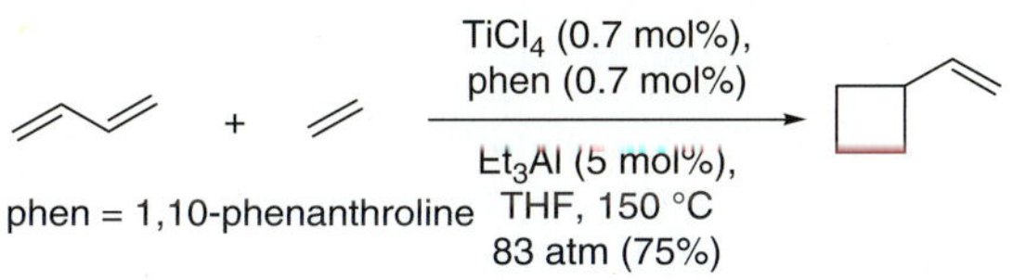

(8.122)

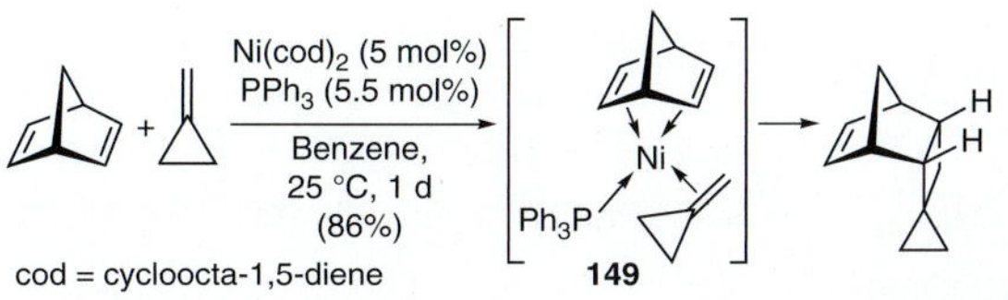

(8.123)

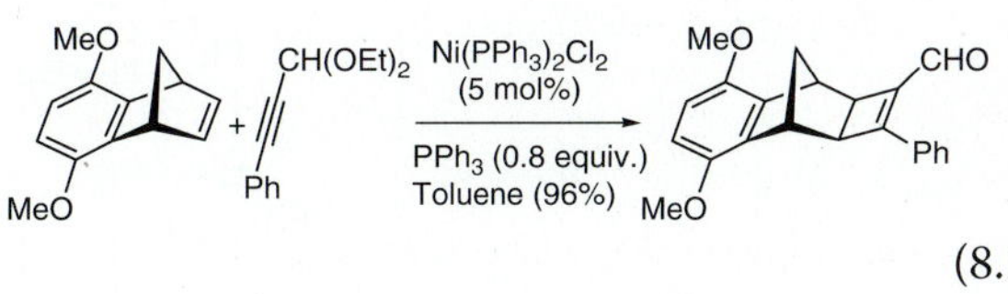

(8.124)

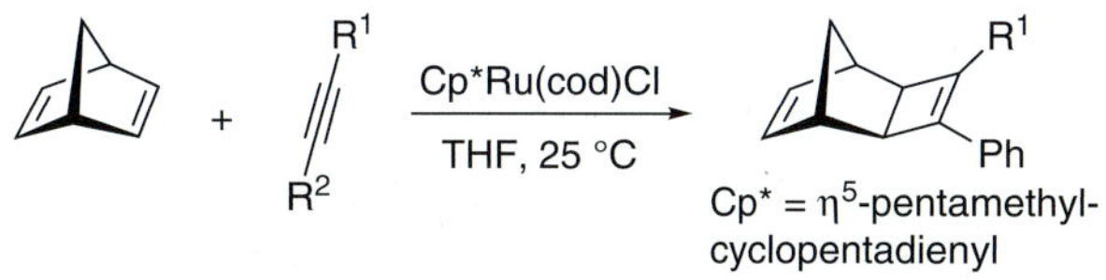

		Relative rate:
R^1 = Me	R^2 = Ph	<0.0016
R^1 = COOEt	R^2 = Ph	(1.0)
R^1 = COOEt	R^2 = *t*-Bu	No reaction
R^1 = COOEt	R^2 = *c*-hexyl	0.03
R^1 = COOEt	R^2 = CH$_2$OH	13

(8.125)

Electron-withdrawing substituents of the acetylene accelerate the reaction, whereas bulky groups retard it. The Ru-catalyzed (2+2)-cycloadditions of 7-oxabenzonorbornadienes [865], 7-azabenzonorbornadienes [866], norbornene, alkynyl sulfides and sulfones [867], and ynamides [868] have been described. Based on quantum mechanical calculations, the mechanism shown in Scheme 8.39 was predicted [869].

CoI$_2$/PPh$_3$/Zn (powder) [870, 871] and Co(dppp) Br$_2$/Zn/ZnI$_2$ (dppp = Ph$_2$P(CH$_2$)$_3$PPh$_2$) [872] catalyze the (2+2)-cycloadditions of alkynes, including derivatives bearing alkyl groups exclusively. In the presence of a (bicyclo[2.2.2]octa-2,5-diene)Rh complex and NaBPh$_4$, ynamides and nitroalkenes undergo catalytic (2+2)-cycloadditions giving cyclobutenamides [873]. Terminal alkynes and internal alkynes undergo [Re(CO)$_3$(THF)Br]$_2$-catalyzed (2+2)-cycloadditions with norbornene, norbornadiene, and benzonorbornadiene [874]. The enantioselective (2+2)-cycloaddition of alkynes possessing an ester function and norbornene derivatives has been realized using Rh(cod)[(S)-H$_8$-BINAP)]BF$_4$ as a homochiral catalyst (up to 91% ee) [875]. Trost and coworkers found that low-valent palladium catalyzes intramolecular suprafacial (2+2)-cycloadditions of enynes [876].

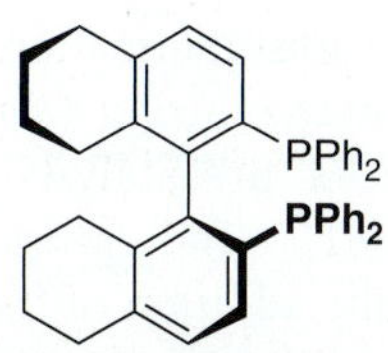

(S)-H$_8$-BINAP

Chirik and coworkers have reported the intramolecular iron-catalyzed [π^2s+π^2s]-cycloadditions of α,ω-dienes into bicyclo[3.2.0]heptane derivatives using (i-PrPDI)Fe(N$_2$)$_2$ as the catalyst (reaction (8.126)) [877]. The reaction is stereospecific (suprafacial/suprafacial) as shown for a (Z,Z)-dideuterated dialkene. The catalyst (i-PrDPI)Fe(N$_2$)$_2$) exchanges

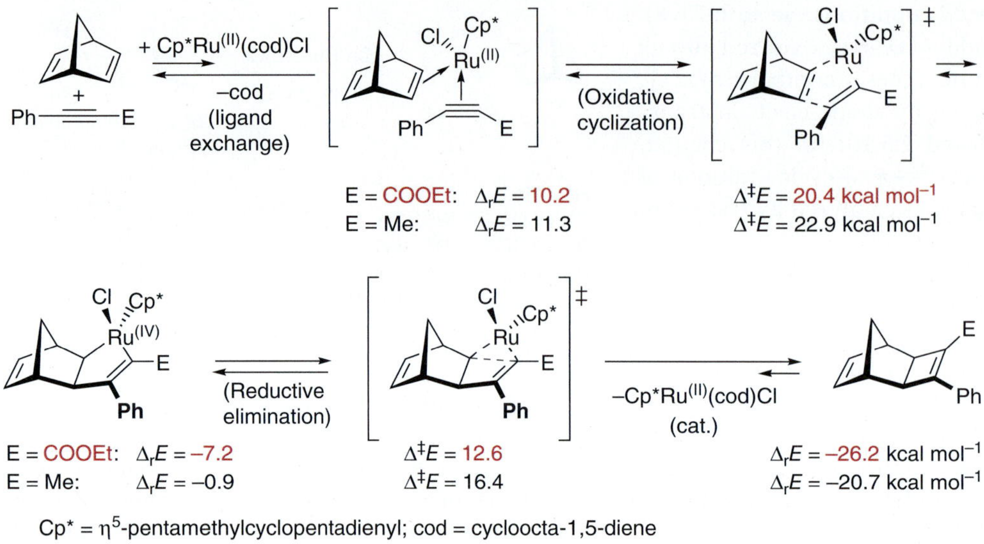

Scheme 8.39 Mechanism predicted by quantum mechanical calculations for the Ru-catalyzed (2+2)-cycloadditions of alkynes to norbornadiene.

N_2 for CO, alkenes, and H_2. It also catalyzes the hydrogenation of alkenes, alkynes [878, 879], and azides [880].

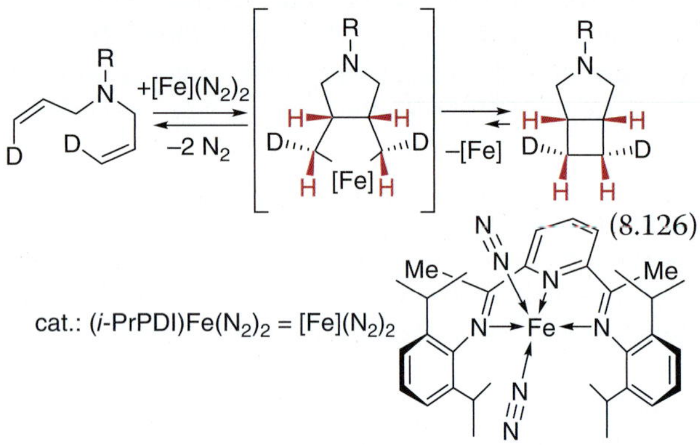

(8.126)

Toste and coworkers reported a gold-catalyzed and enantioselective cycloisomerization of allenenes into (alkylidene)cyclobutanes using (*R*)-DTBM-Segphos/AuCl and $AgBF_4$ as precatalysts (e.g. **150** → **151**; Scheme 8.40) [881]. The proposed mechanism involves nucleophilic addition of the allene moiety of **150** to the electrophilic LAu^+ catalyst generating a cationic intermediate **152** that undergoes rearrangement with the formation of stereoisomeric benzyl cation intermediates **153** and **154**. The latter can be quenched with methanol (e.g. **153** + MeOH → **155**). In the absence of methanol, one of them (**154**) undergoes intramolecular carbation of the vinylgold intermediate via the tertiary alkyl cation **156** (stabilized by metal hyperconjugation). This cation undergoes β-elimination of the metallic moiety LAu^+, providing the (alkylidene)cyclobutane product **151**.

Problem 8.32 What are the major products **P** and **Q** of the reactions of *N*-phenylmaleimide with dimethyl norbornadiene-2,3-dicarboxylate and with 2-methoxynorbornadiene catalyzed by $Ni(cod)_2$ and PPh_3 in toluene at 80–110 °C? [689]

Problem 8.33 Estimate the heat of the $[\pi^2+\pi^2]$-cycloaddition of norbornadiene + PhC≡CCOOMe.

8.4.5 (3+1)-Cycloadditions

Noncatalyzed (3+1)-cycloadditions are not common. The very few examples reported involve an isonitrile and an ylide such as dipolar azomethine imine [882] and azomethine ylide (e.g. reaction (8.127)) [883].

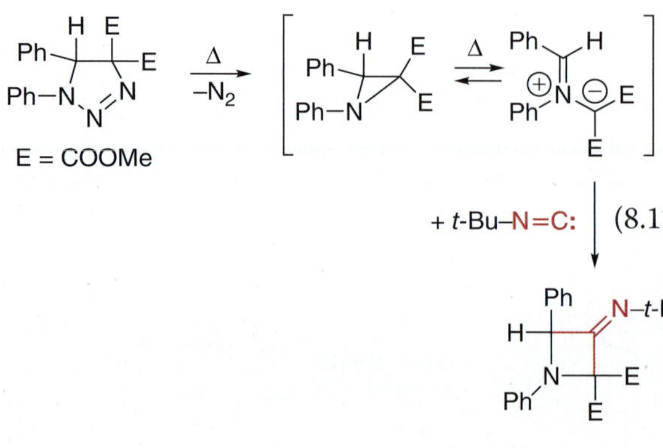

(8.127)

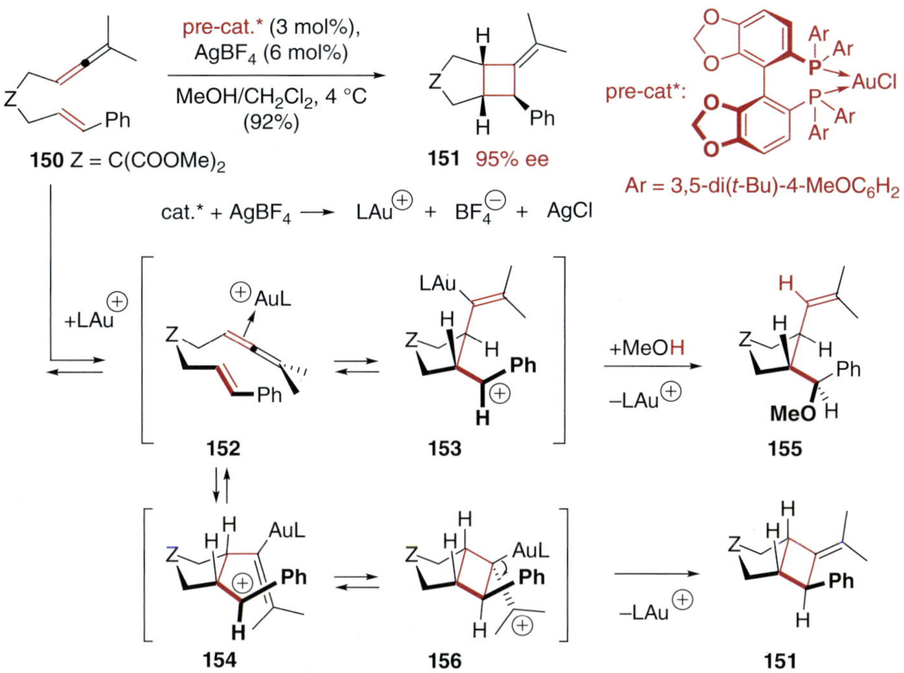

Scheme 8.40 Proposed reaction mechanism of a catalyzed asymmetric intramolecular [π^2+π^2]-cycloaddition of an alkene and an allene induced by an enantiopure homochiral cationic gold catalyst.

A rare case of catalyzed (3+1)-cycloaddition has been reported by Kurahashi and de Meijere who inserted CO into the C(1)—C(2) bond of (alkylidene)cyclopropanes using $Co_2(CO)_8$ as a catalyst [884]. The cobalt catalyst was found to be much more efficient than $Cr(CO)_6$, $Mo(CO)_6$, $W(CO)_6$, and $Fe(CO)_5$. An example is given with reaction (8.128) and possible mechanisms are proposed in Scheme 8.41. These reactions are analogous to the carbonylation of epoxides and aziridines that produce β-lactones and β-lactams, respectively (Section 8.2.10).

8.4.6 (3+2)-Cycloadditions

Five-membered ring compounds can be formed by cycloadditions of two-centered unsaturated moieties to three-centered reagents contributing with 2σ electrons such as cyclopropanes, with 2π electrons such as allyl cationic systems or vinylcarbenes, or with 4π electrons as for ylides resulting from the formal combination sextet species Y with a Lewis base of type X=Z: (Scheme 8.42). Alternatively, five-membered rings can be obtained by cheletropic addition of 1,3-dienes (Sections 5.4.4 and 5.4.5), by (2+2+1)-cycloadditions such as the Pauson–Khand reaction (Section 8.2.7), or by annulation reactions based on Umpolung by an amine, a phosphine, or a diaminocarbene (Section 7.5.3–7.5.7).

Lewis acids catalyze (3+2)-cycloadditions of push–pull 1,2-disubstituted cyclopropanes (Section 7.6.9). For other cyclopropanes, a transition metal must be used as a catalyst [885, 886]. The latter must be able to add across one of its σ(C—C) bonds to generate a metallacyclobutane, which can then be combined with an alkene or an alkyne to permit a cycloinsertion, producing a metallacyclohexane or metallacyclohexene. A final reductive elimination produces the five-membered ring and releases the metallic catalyst. This mechanism (Scheme 8.43) has been inferred originally from the observations of Cassar and Halpern that the reaction of $[Rh(CO)_2Cl]_2$ with quadricyclane generates stable rhodacyclopentanone derivative **157** [887].

With Ni(N≡C—Me)$_2$ as a catalyst, quadricyclane adds at 20 °C to acrylonitrile, acrylic esters, maleic esters, and fumaric esters giving stereoisomeric adducts that are not those obtained under thermal conditions as illustrated with reaction (8.129), a formal [σ^2s+π^2s]-cycloaddition. The uncatalyzed thermal reaction (8.130) is a formal [σ^2s+σ^2s+π^2s]-cycloaddition [888].

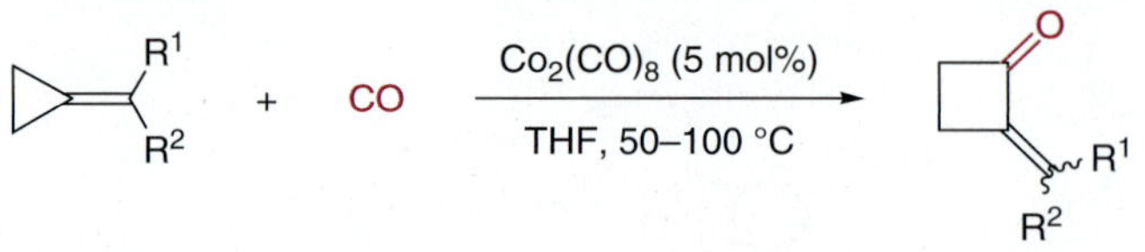

(8.128)

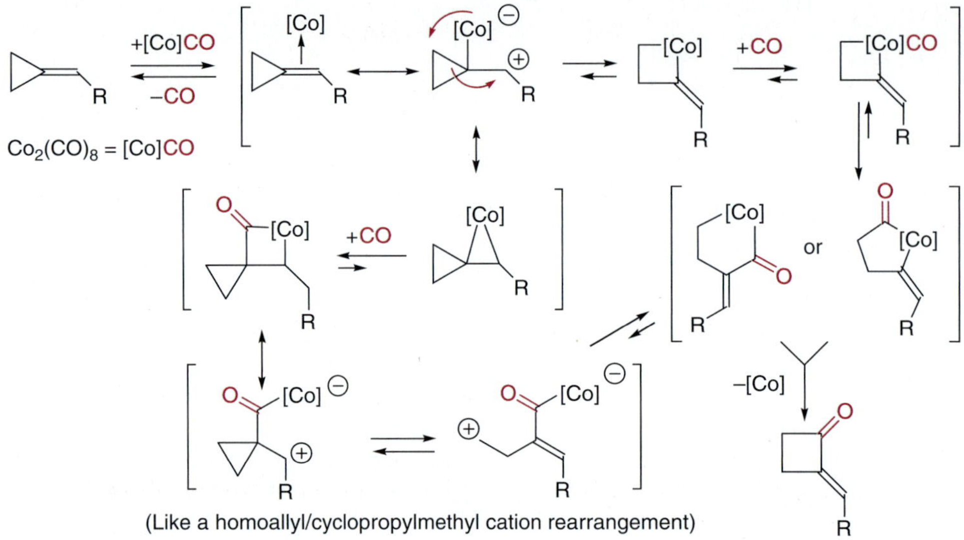

Scheme 8.41 Proposed reaction mechanisms for the $Co(CO)_8$-catalyzed (3+1)-cycloaddition of CO and allylidenecyclopropanes.

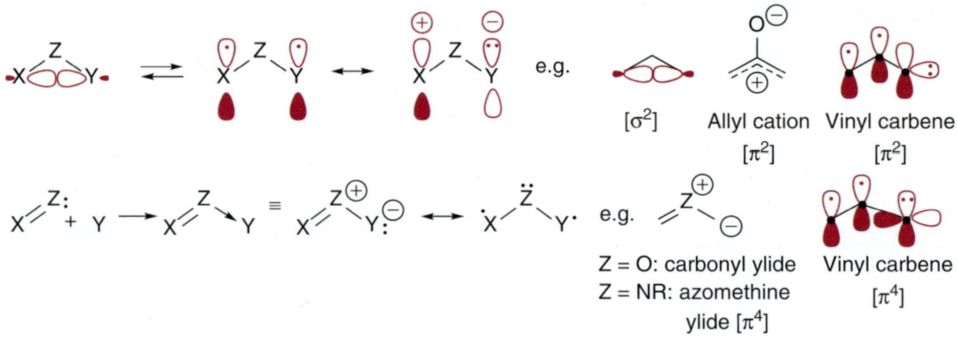

Scheme 8.42 Possible three-centered reagents that can undergo (3+2)-cycloadditions with alkenes, alkynes, carbonyl, and imine moieties.

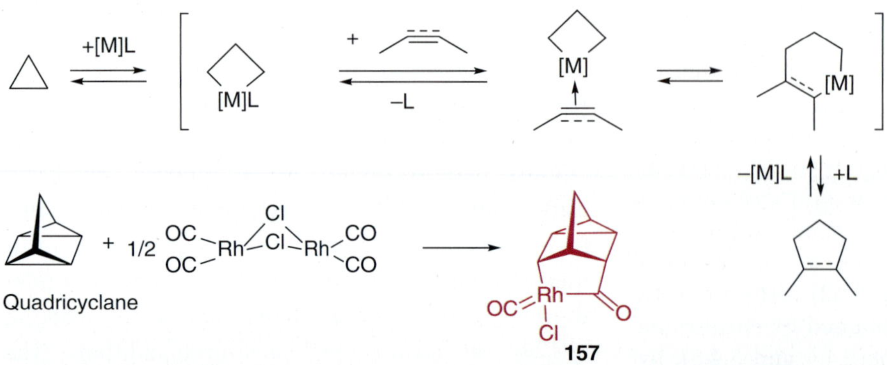

Scheme 8.43 Probable mechanism of the transition-metal-catalyzed (3+2)-cycloadditions of cyclopropane to π-systems.

Scheme 8.44 (a) Tsuji's Pd-catalyzed intermolecular (3+2)-cycloaddition of vinylcyclopropanes bearing electron-withdrawing substituents to Michael acceptors; (b) example of Plietker's iron-catalyzed (3+2)-cycloaddition of a vinylcyclopropane to an electron-poor imines; and (c) examples of enantiopure chiral phosphine ligands used in Pd-catalyzed (3+2)-cycloadditions of alkenylcyclopropanes.

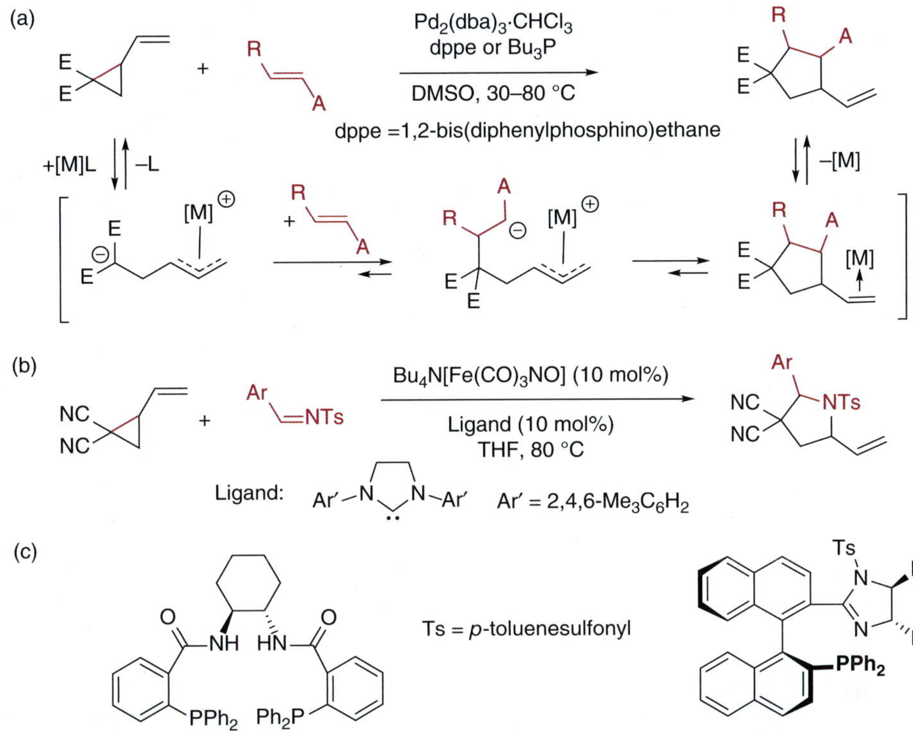

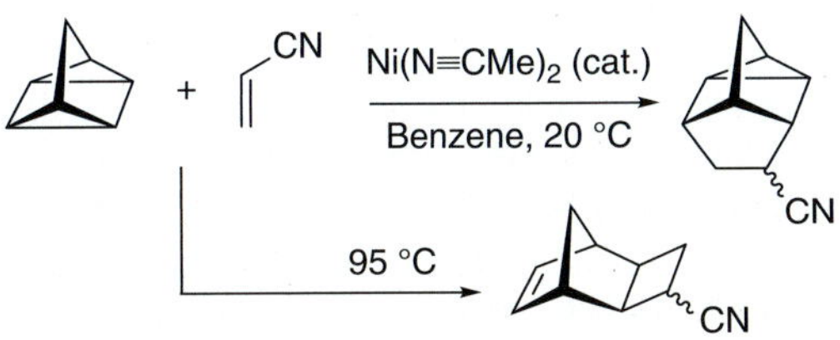

(8.129)

(8.130)

In 1985, Tsuji and coworkers found that vinylcyclopropanes undergo Pd-catalyzed intermolecular formal (3+2)-cycloadditions with α,β-unsaturated carboxylic esters and ketones giving vinylcyclopentanes (Scheme 8.44a) [889]. In 2012, Plietker and coworkers reported that these reactions can be catalyzed by nucleophilic iron complexes derived from $Bu_4N^+/[Fe(CO)_3(NO)]^-$ and a diaminocarbene. The method can be applied to the synthesis of pyrrolidines using *N*-tosyl arylimines (Scheme 8.44b). Trost et al. realized enantioselective dynamic kinetic Pd-catalyzed (3+2)-cycloadditions of vinylcyclopropanes to azalactones and to Meldrum's acid alkylidenes using enantiopure chiral ligands such as **158** [890, 891]. Shi and coworkers also reported enantioselective Pd-catalyzed (3+2)-cycloadditions of vinylcyclopropanes with β,γ-unsaturated α-keto carboxylic esters employing ligand **159** (Scheme 8.44c) [892, 893]. Intramolecular (3+2)-cycloadditions of nonactivated vinylcyclopropanes, alkenes, and alkynes (substituted with alkyl and/or aryl groups) have been catalyzed by Rh complexes by Yu and coworkers [894, 895], including an asymmetric version using (*R*)-H_8-BINAP as a ligand [896].

Transition-metal-catalyzed formal (3+2)-cycloadditions of methylidenecyclopropanes to alkenes, alkynes, aldehydes, and imines give five-membered carbocycles and heterocycles [886, 897, 898]. An early example (reaction (8.131), E = COOR) was reported by Noyori et al. [899, 900]. The cycloaddition of methylidenecyclopropane to methyl maleate and methyl fumarate catalyzed by Ni(N≡C—Me)$_2$ in benzene at 60 °C gives a mixture of products demonstrating that both σ(C(1)—C(2)) and σ(C(2)—C(3)) bond insert the metal catalyst [901].

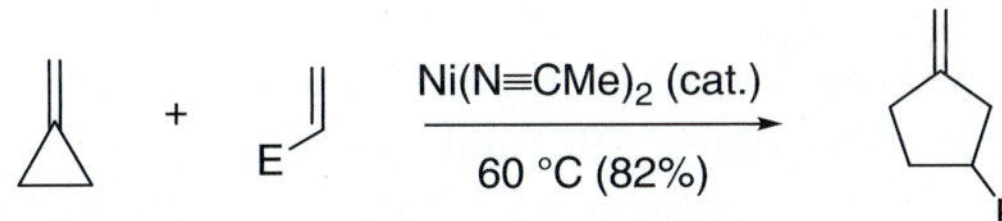

(8.131)

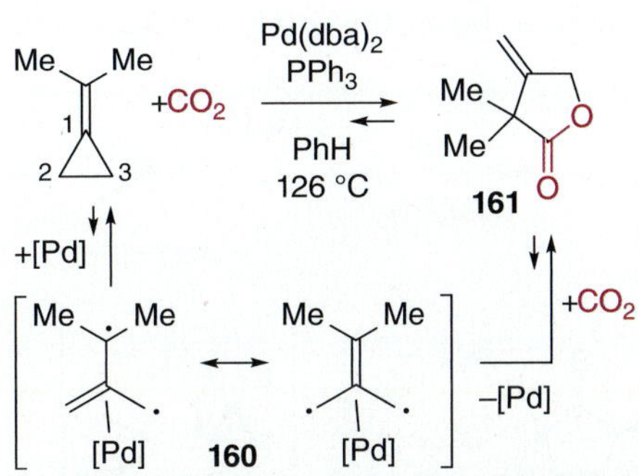

(8.132)

In 1979, Inoue et al. presented examples of CO_2 fixation (e.g. reaction (8.132)) involving the (3+2)-cycloadditions of alkylidenecyclopropanes to CO_2 catalyzed by a Pd catalyst (dba = dibenzylideneacetone) [902]. The selective formation of **161** is best explained by invoking the formation of a η^4-trimethylenemethane palladium complex of type **160** as a reactive intermediate. Stable trimethylidenemethane (TMM) complexes of Fe [903, 904] and Mo are known [905]. Trost and coworkers showed that (TMM) palladium complexes **163** are generated by reacting 2-trimethylsilylmethylallyl acetate (**162**) with Pd(0) complexes [906–908]. The TMM complexes generated in situ add to electrodeficient alkenes (Michael acceptors) [909, 910], aldehydes [911–913], ynones [914], reactive ketones [915, 916], and *N*-tosylimines [917]. As the (3+2)-cycloaddition with dimethyl maleate produces *cis/trans* mixtures of methylidenecyclopentanes **165**, a stepwise, nonconcerted mechanism was suggested. Intermediate **163** reacts as a nucleophile generating a zwitterionic species **164** that has the time for rotation about the σ(EC—CE) bond before its ring closure to produce cycloadducts **165** (Scheme 8.45) [918, 919].

TMM-Pd complexes have been generated by using other precursors than **162**. For instance, mesylate **166** was used to generate methylidenepyrrolidine **167** (reaction (8.133)) [920]. Tsuji and coworkers showed that the tosyl-substituted allyl carbonate **168** reacts with ethyl acrylate to give methylidene–cyclopentane **169** (reaction (8.134)) without losing the tosyl group (driving force being the formation of HOCOOEt that decomposes into CO_2 + EtOH) [921, 922].

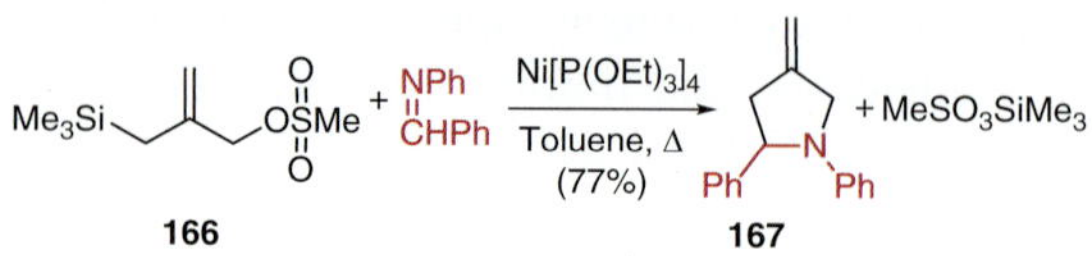

(8.133)

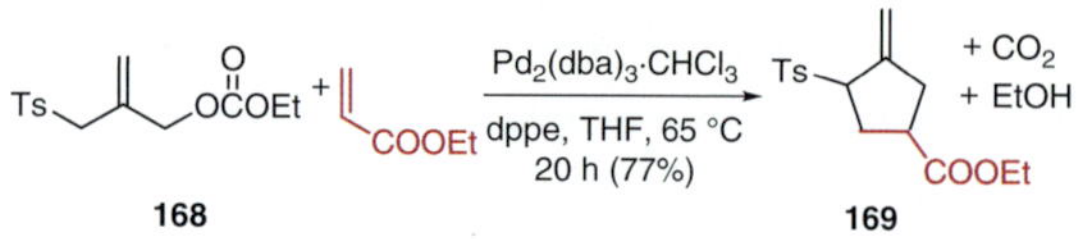

(8.134)

As seen with reaction (8.131), methylidenecyclopropane undergoes Ni-catalyzed (3+2)-cycloadditions. They proceed via oxidative addition of the metal across both the σ(C(1)—C(2)) and σ(C(2)—C(3)) bonds generating nickellacyclobutane intermediates. The competition between these two modes of metal insertion depends on the nature of the ligands of the nickel catalyst [923, 924]. The earliest examples of intramolecular methylidenecyclopropane cycloaddition were reported independently by Motherwell and coworkers (e.g. reaction (8.135)) [925] and Nakamura and coworker (e.g. reaction (8.136)) [926].

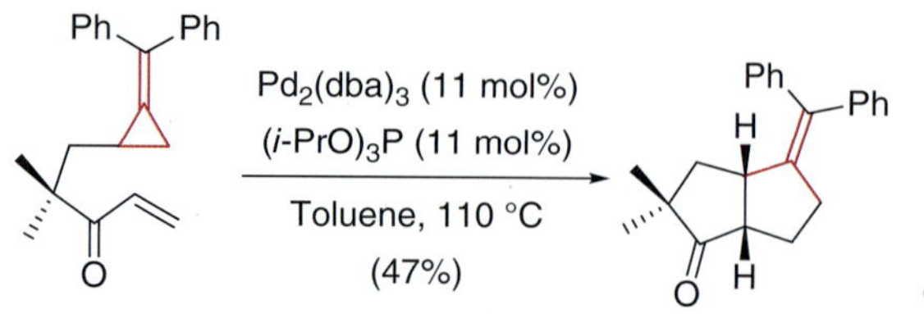

(8.135)

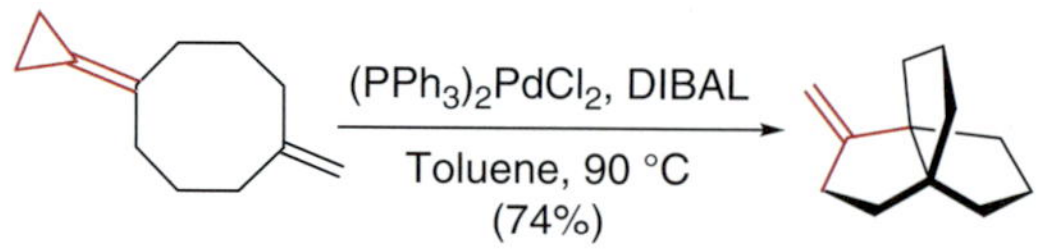

(8.136)

Lautens et al. studied the stereocontrol of metal-catalyzed intramolecular (3+2)-cycloadditions of methylidenecyclopropane to alkynes [927]. With dideuterated compound **170** and a Pd-catalyst, they found a highly stereoselective reaction forming the *cis*-adduct **173**. The results were interpreted in terms of intermediates **171** → **172** ↔ **172′** (Scheme 8.46). This mechanism is supported by quantum mechanical calculations [928]. Pd-catalyzed intramolecular (3+2)-cycloadditions of (alk-5-enylidene)cyclopropanes have also been reported [929].

Ethyl diazopyruvate **174** reacts in the presence of a Cu(II)-catalyst to form a copper carbenoid, which adds to alkenes to form cyclopropanes (Section 8.4.2). Wenkert and coworkers showed that the intermediate metal carbenoid can also serve as a 1,3-dipole **175** ↔ **175′** that adds stereoselectively to dihydrofuran **176** generating cycloadduct **177** (reaction (8.137)) [930, 931].

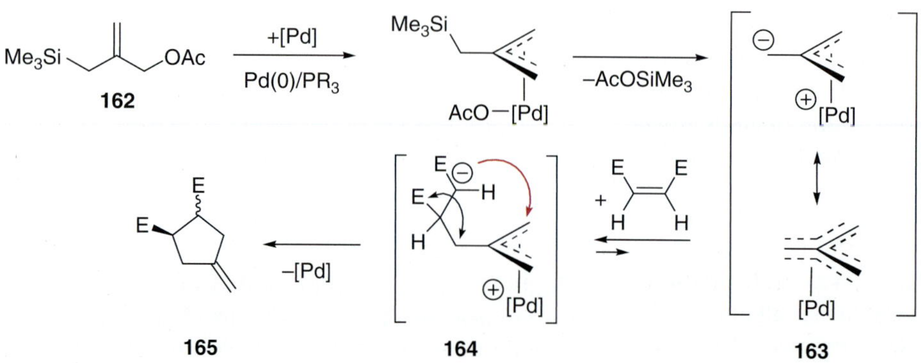

Scheme 8.45 Proposed reaction mechanism of Trost's Pd-catalyzed (3+2)-cycloaddition.

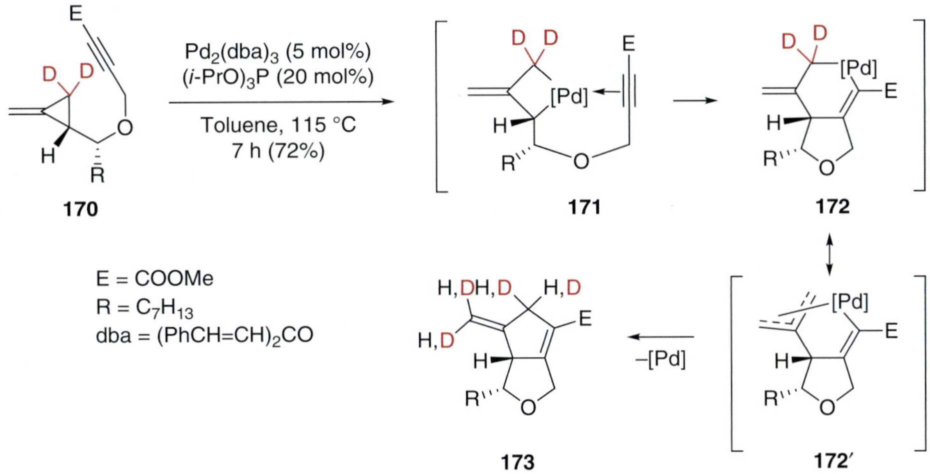

Scheme 8.46 Proposed reaction mechanism of the Pd-catalyzed intramolecular (3+2)-cycloaddition of a methylidenecyclopropane to an alkyne.

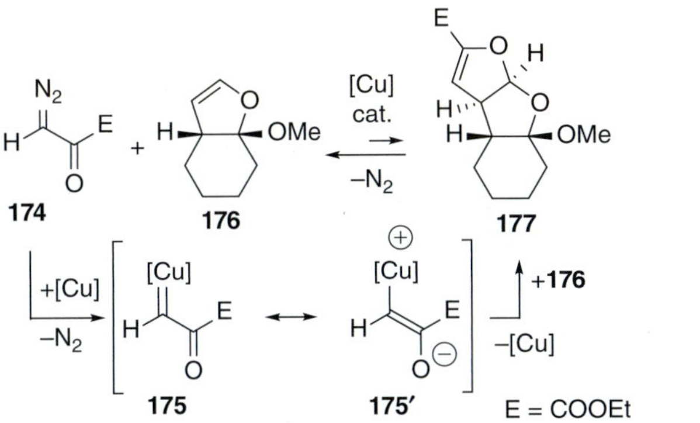

(8.137)

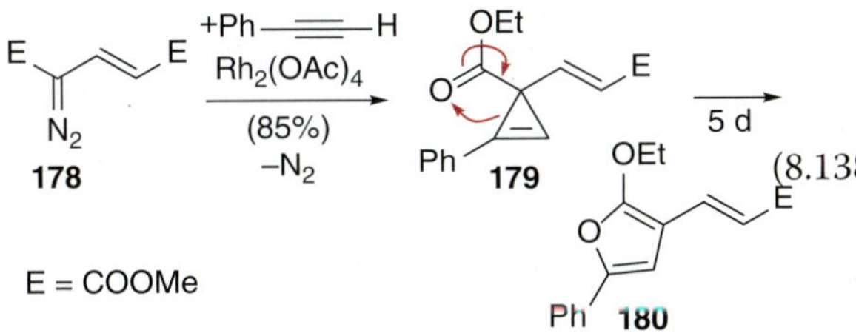

(8.138)

Other diazo compounds undergo similar (3+2)-cycloadditions to electron-rich alkenes in the presence of Rh-catalysts [932–935]. When **178** is treated with $Rh_2(OAc)_4$, the product of cyclopropanation **179** of phenylacetylene is isolated. Rearrangement of **179** into furan **180** (reaction (8.138)) occurs in the absence of a catalyst, but the rate of this isomerization increases in the presence of $Rh_2(OAc)_4$. This suggested that the vinylcarbenoid intermediate generates five-membered ring products in two-step processes, rather than through a direct (3+2)-cycloaddition. In 1987, Aumann and Uphoff found that pentacarbonyl[(methoxy)benzylidene]metal (Cr, Mo, W) and phenylpropadiene react to give the corresponding dibenzylidenecyclopentane cycloadducts **182** via (trimethylidenemethane)metal complex intermediates **181** (reaction (8.139)) [936–938]. On their side, Barluenga et al. reacted chromium alkenylcarbene complexes **183** with allenes in the presence of Wender's catalyst ([Rh(naphthalene)(cod)][SbF_6]) or $[Rh(cod)Cl]_2$ and obtained products **184** (reaction (8.140)) [939]. A broad range of functionalized 5-alkylidenecyclopentene derivatives have been obtained using all kinds of substituted allenes.

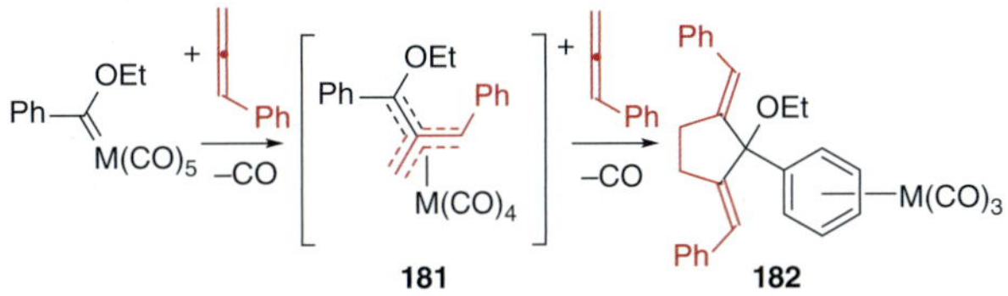

(8.139)

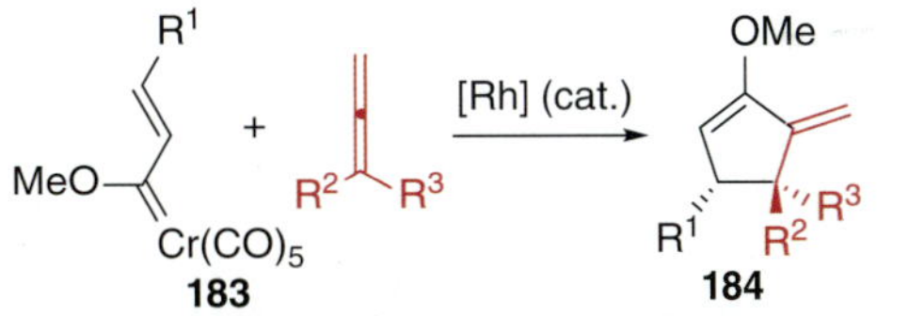

(8.140)

A formal intramolecular (3+2)-cycloaddition of arene-yne **185** has been catalyzed by $(Ph_3P)AuCl + AgSbF_6$ (Scheme 8.47). The proposed mechanism involves first a Au^+-induced alkenylation of one of the arene rings followed by protonation of the second alkyne moiety that generates a dienyl cationic intermediate **186** that undergoes electrocyclization into **187** [940].

Diastereoselective reductive (3+2)-cycloaddition of allenes to conjugated enones has been catalyzed by $CoI_2(dppe)/ZnI_2$ using 2.7 equiv. of Zn (Scheme 8.48) [941].

Cationic Rh(I) complex such as that resulting from the combination of $[Rh(CO)_2Cl]_2$ with $AgSbF_6$ under CO atmosphere (1 atm) catalyzes the regioselective tandem heterocyclization/(3+2)-cycloaddition of 2-(1-alkynyl)-2-alken-1-ones with alkynes bearing at least one electron-withdrawing group A (Scheme 8.49a). A possible mechanism is given in Scheme 8.49b [942].

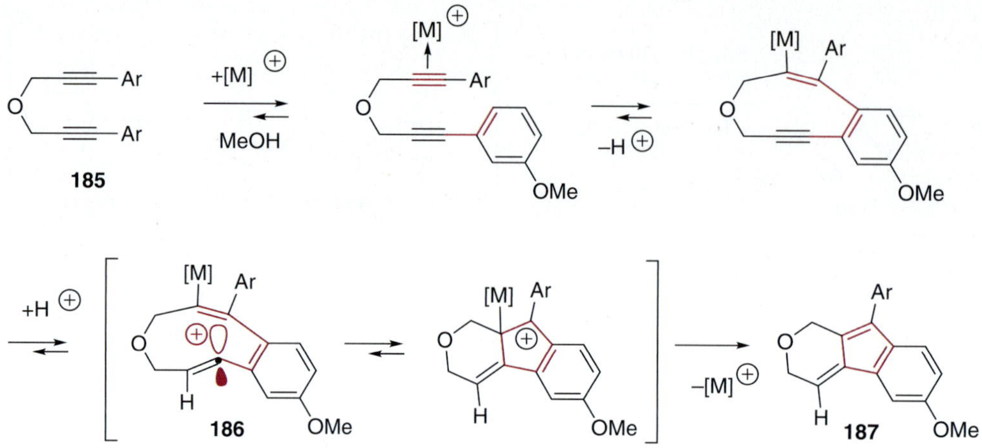

Scheme 8.47 Gold-catalyzed intramolecular electrocyclization.

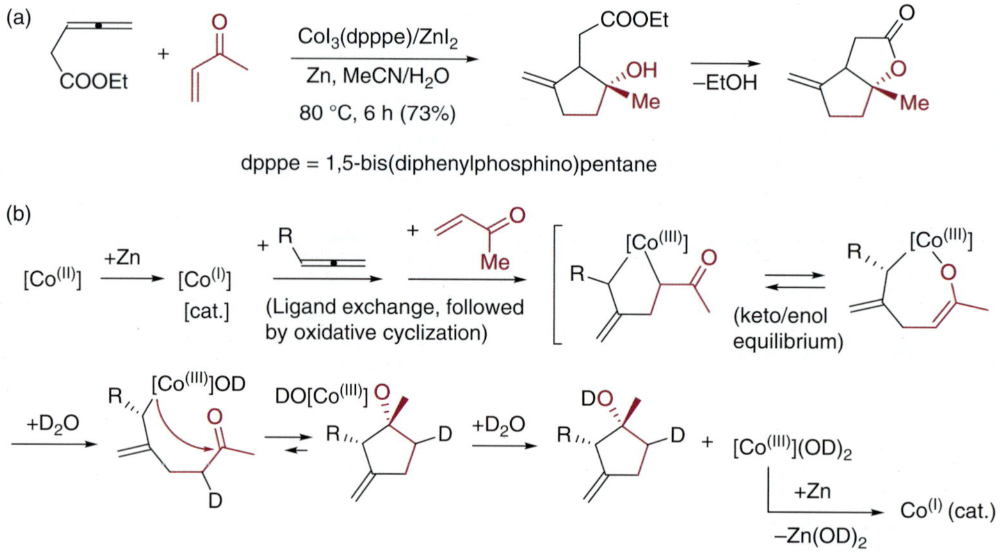

Scheme 8.48 (a) Reductive Co-catalyzed (3+2)-cycloaddition of allenes to enones and (b) proposed reaction mechanism.

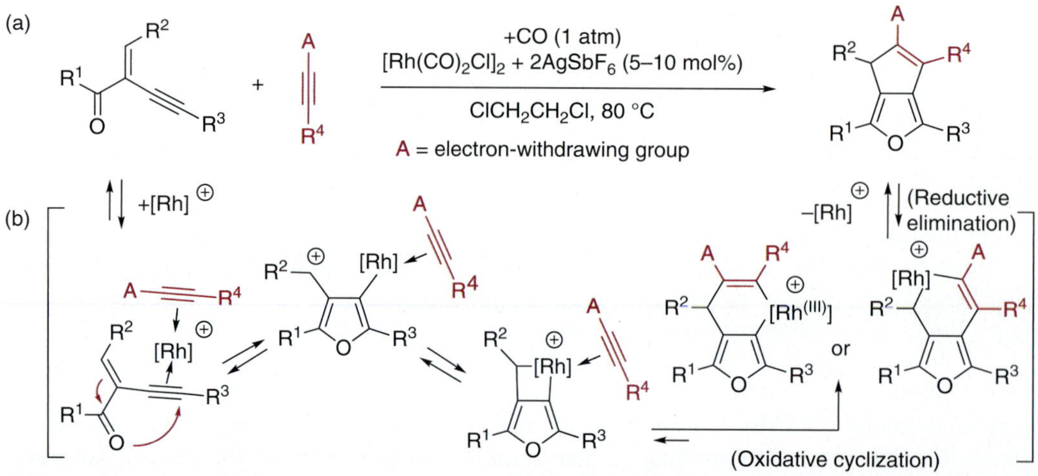

Scheme 8.49 (a) Tandem heterocyclization/(3+2)-cycloaddition catalyzed by a cationic Rh(I) complex and (b) proposed reaction mechanism.

Problem 8.34 What is the main product of the following reaction? [943, 944]

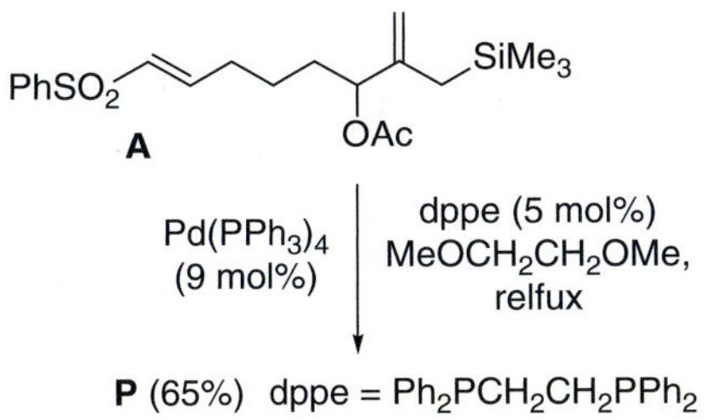

Problem 8.35 Propose a mechanism for reaction (8.140) using alkyl CH=C=CHCOOMe.

8.4.7 (4+1)-Cycloadditions

Any one-center reagent that can accept two further electrons in its valence shell can undergo (4+1)-cheletropic additions with 1,3-dienes, generating the corresponding five-membered ring compounds (Section 5.4). One-center reagents are carbenes (:C(R_1)R_2) and any species analogous to them such as carbon monoxide (:C=O), isonitriles (:C=NR), and for noncarbon systems, nitrenes, N=O, S, SO, SO_2, singlet O_2, germylenes, and s-ylidenes. Several syntheses of five-membered ring systems start with a cyclopropanation into a vinylcyclopropane, followed by a (1,3)-sigmatropic shift as developed by the Danheiser et al. [945, 946] and Hudlicky groups [947, 948] in the early 1980s [949]. A first example of catalyzed direct (4+1)-cycloaddition or (4+1)-cheletropic addition of carbon monoxide has been reported by Sigman and Eaton [950, 951]. They reacted conjugated bis-allenes with CO in the presence of 10 mol% $Fe(CO)_5$, which produced the corresponding cyclopentanones (reaction (8.141)). Murai and coworkers reported the first catalytic carbonylation of α,β-unsaturated imines that produced unsaturated γ-lactams (reaction (8.142)) [952]. Murakami et al. [953, 954] first reported the Rh(I)-catalyzed carbonylation of vinylallenes. Using an enantiopure Rh(I)-catalyst [955], enantiomerically enriched cyclopentanones were obtained as illustrated with reaction (8.143) [956]. $[Pt(cod)_2]$/(*R*,*R*-Me or Et-Duphos) catalysts were also found to induce asymmetric (4+1)-carbonylations [957]. Diastereoselective (4+1)-cycloaddition of CO with (alkenyl)propargyl acetates are catalyzed by $[Rh(CO)_2Cl]_2$ [958].

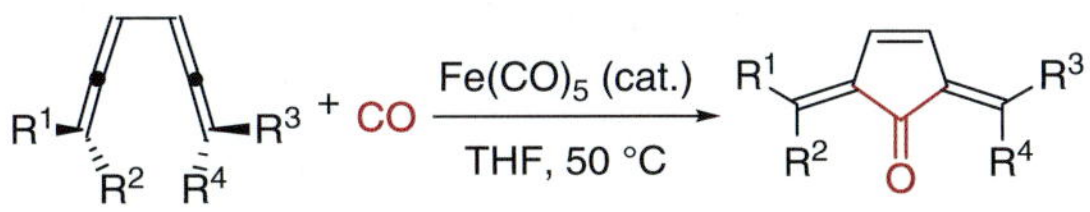

(8.141)

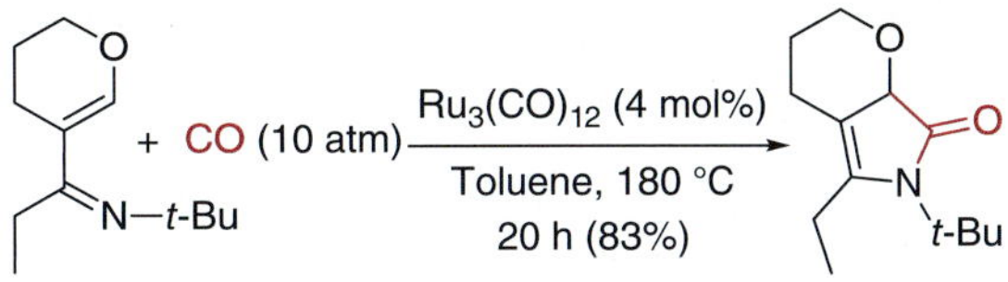

(8.142)

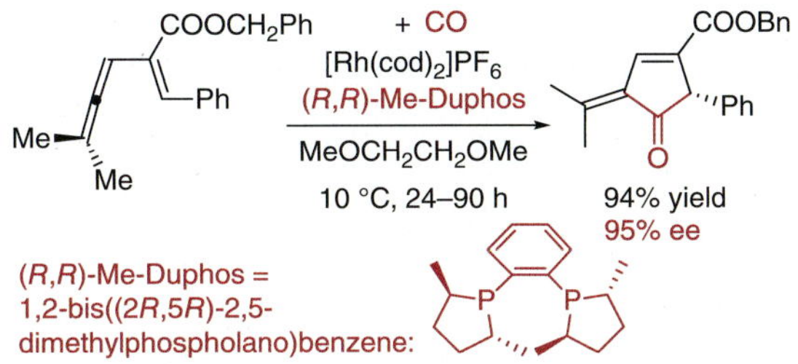

(8.143)

Rh(I)-catalyzed carbonylation of 3-acyl-1,4-enynes (isomers of vinylallenes) also produces cyclopentenones [959]. Fischer complexes ([(alkoxyalkenyl) carbene]$M(CO)_5$; M = Cr, Mo, W) react with 1,3-dienes without catalyst giving mixtures of product of (3+2) and (4+1)-cycloadditions [960]. Cu-catalyzed (4+1)-cycloaddition of α,β-acetylenic ketones with α-diazo esters realizes a direct route to highly substituted furans as illustrated with reaction (8.144) [961]. Dihydrofurans are obtained enantioselectively by Cu-catalyzed reactions of α-diazoesters with α-benzylidene-β-ketoesters [962]. Ito et al. reported the Et_2AlCl-mediated (4+1)-cycloaddition of α,β-unsaturated carbonyl compounds with methylisocyanide giving unsaturated *N*-substituted iminolactones, which are converted into γ-butyrolactones [963]. Chatani and coworkers have described the $GaCl_3$-catalyzed (4+1)-cycloaddition of α,β-unsaturated ketones with isocyanides leading to furan-2(3*H*)-imines (reaction (8.145)) [964, 965].

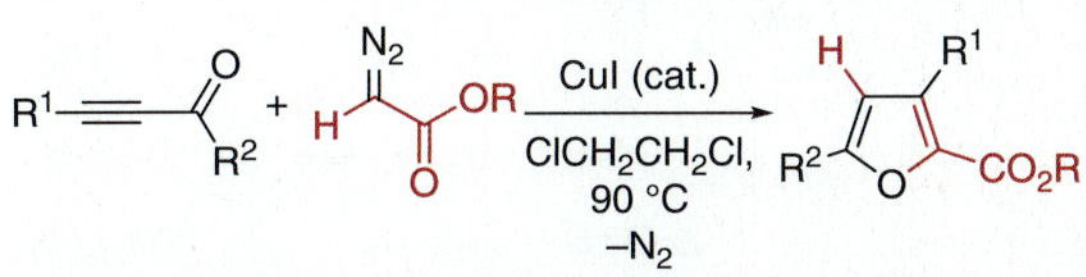

(8.144)

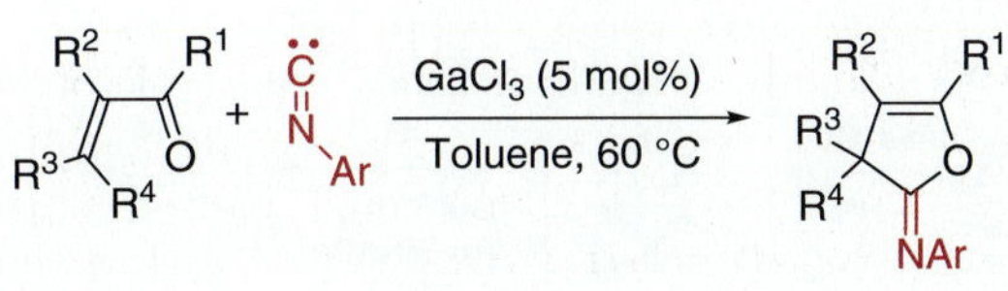

(8.145)

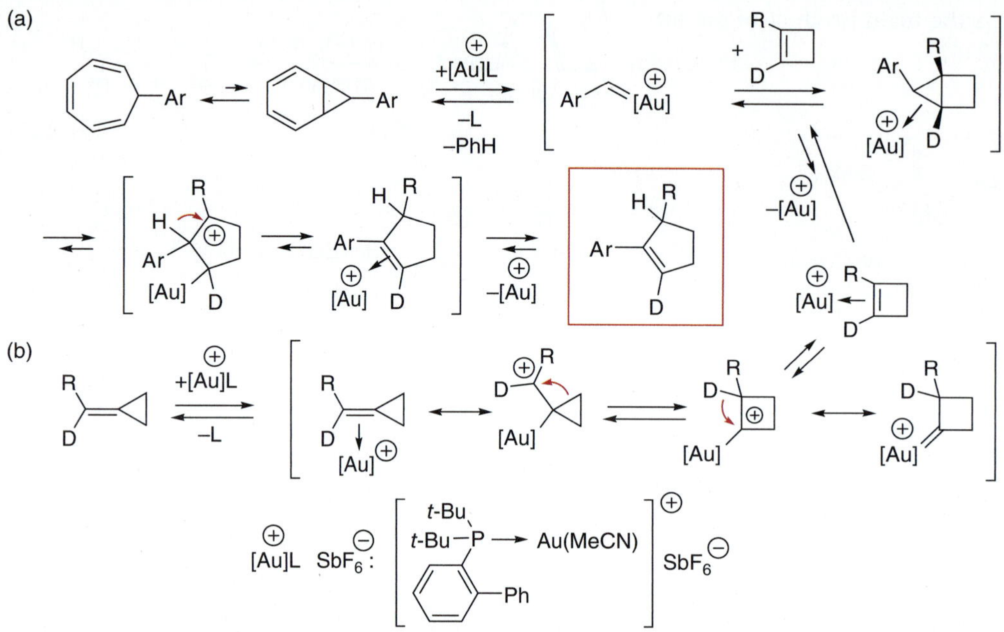

Scheme 8.50 Au(I)-catalyzed formal (4+1)-cycloaddition of cyclobutenes and alkylidenecyclopropanes with 7-arylcyclohepta-1,3,5-trienes. (a) Cyclopropanation of cyclobutene and (b) Au(I)-catalyzed isomerization of alkylidenecyclopropanes into cyclobutenes.

Rh(II) [966], Pd(II) [967], and Au(I)-catalyzed formal (4+1)-cycloadditions of α-diazoesters to propargyl alcohols provide the corresponding 2,5-dihydrofurans (reaction (8.146)) probably via the formation of oxonium ylide intermediates that are rearranged into allenyl alcohols [968]. In situ generated azoalkenes react with sulfur ylides in the presence of $Cu(OTf)_2$ as a catalyst. Formal (4+1)-cycloadditions of azoalkenes with acylcarbenes have been realized. In the presence of an enantiopure chiral diphosphine ligand, the reactions can be enantioselective (e.g. reaction (8.147)) [969]. Iron-catalyzed decarboxylative (4+1)-cycloadditions of benzoxazinones and sulfur ylides give functionalized indoline products [970].

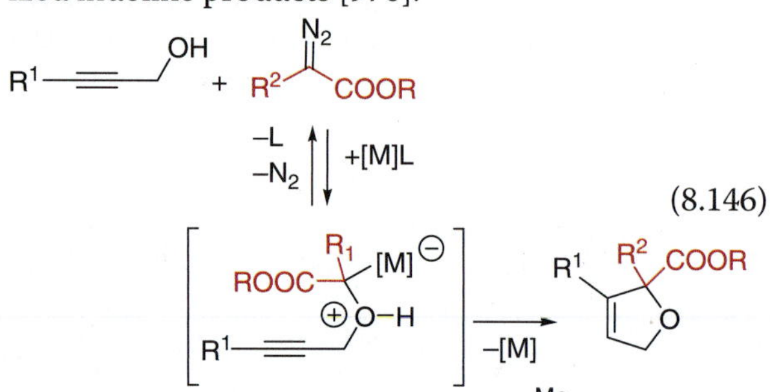

(8.146)

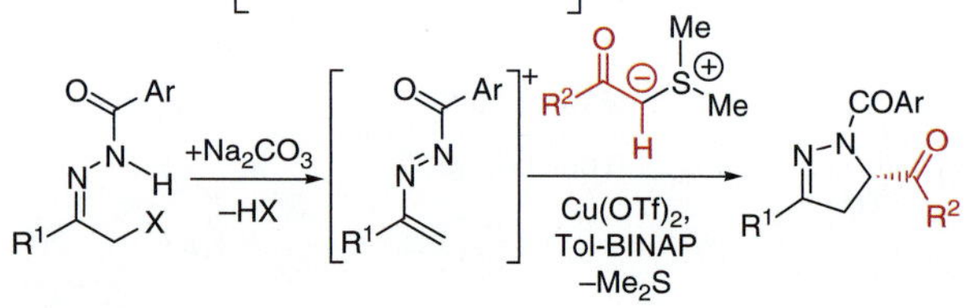

(8.147)

In 2014, Echavarren and coworkers reported the gold(I)-catalyzed reaction of cyclobutenes with 7-arylcyclohepta-1,3,5-trienes that provides benzene and 1-arylcyclopentene derivatives (Scheme 8.50a) [971]. 7-Substituted cycloheptatrienes undergo Au(I)-catalyzed retro-Büchner reactions to form benzene and (carbene)Au(I) complexes [972]. The latter cyclopropanate the cyclobutenes giving the corresponding bicyclo[2.1.0]heptanes that are, on their turn, isomerized into cyclopentenes under the action of the Au(I)-catalyst. The same catalyst isomerizes alkylidenecyclopropanes into cyclobutenes. Thus, alkylidenecyclopropanes can be used in these formal (4+1)-cycloadditions (Scheme 8.50b).

In dichloroethylene and at 80 °C, cationic gold catalyst $Au(Ph_3P)NTf_2$ catalyzes the cycloisomerization of *β*,β-diaryl-substituted *ortho*-(alkynyl)styrenes (Scheme 8.51) [973].

Problem 8.36 Propose a mechanism for reaction (8.143) [955].

Problem 8.37 Propose a mechanism for reaction (8.144).

Problem 8.38 Propose a mechanism for reaction (8.145).

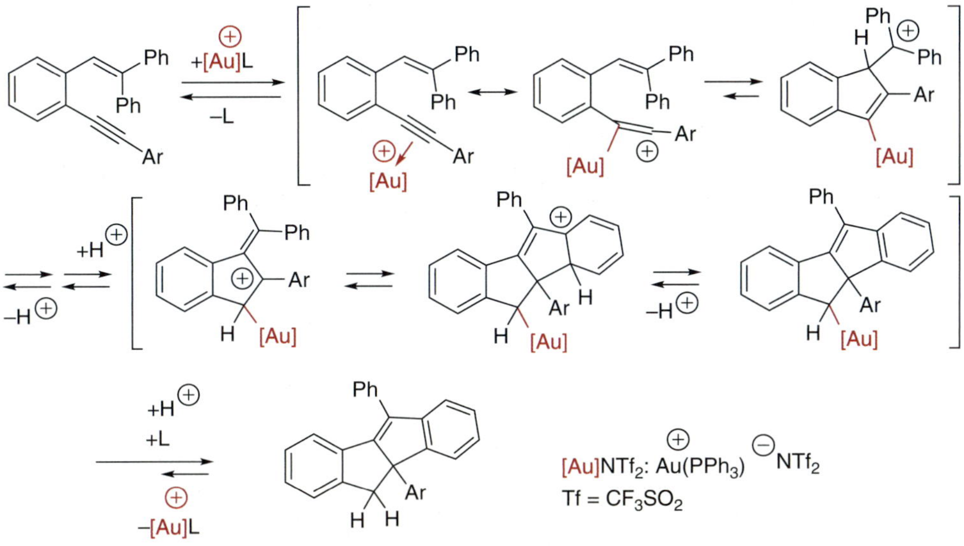

Scheme 8.51 Formal intramolecular (4+1)-cycloaddition of β,β-diaryl *o*-(alkynyl)styrenes catalyzed by cationic Au(I).

8.4.8 (2+2+1)-Cycloadditions

The most studied (2+2+1)-cycloaddition is the Pauson–Khand reaction (Section 8.2.7). Classically, the reaction requires an alkyne as one of the two π-reactants giving cyclopent-2-enones. In 1994, Negishi et al. reported the (2+2+1)-cyclocarbonylation of 1-phenylhepta-1,6-diene into 2-phenylbicyclo[3.3.0]octan-3-one in a stepwise process involving a zirconacyclopentane intermediate [974]. More recently, Wender and coworkers developed $[Rh(CO)_2Cl]_2$-catalyzed (2+2+1)-cycloaddition of 1,3-dienes, alkynes, and CO, in which the dienes served as alkene π-moieties [292, 975, 976]. On their side, the groups of Itoh and coworker [977] and Mukai and coworker [287] have reported $[Rh(CO)_2Cl]_2$-catalyzed (2+2+1)-cyclocarbonylations (8.148) and (8.149), respectively, in which one of the π-systems is an allene moiety. The (2+2+1)-cyclocarbonylation of bis(sulfonylallene)s has also been reported (reaction (8.150)) [978].

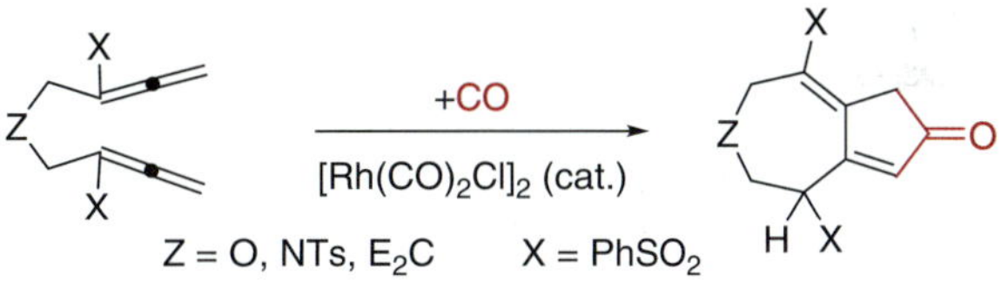

(8.150)

Similar to the Pauson–Khand reaction, the (2+2+1)-cocyclization of isocyanates, alkynes, and CO that generates polysubstituted maleimides (reaction (8.151)) is catalyzed by $Ru_3(CO)_{12}$ [979] or by $Fe(CO)_5$ [980]. Intermolecular (2+2+1)-cycloaddition of alkyne, alkene, and isocyanate is catalyzed by Ni-catalysts (e.g. reactions (8.152), which probably involves an intermediate of type HNi(L)—N(R^3)—CO—C(R^1)=C(R^2)—CH=CHE) [981]. Employing phenyl formate as the CO surrogate, Ni(0) complexes catalyze (2+2+1)-cycloadditions of alkynes and imines leading to α,β-unsaturated γ-lactams (reaction (8.153)). Instead of an alkyne, norbornene can be used as a cycloaddent [982].

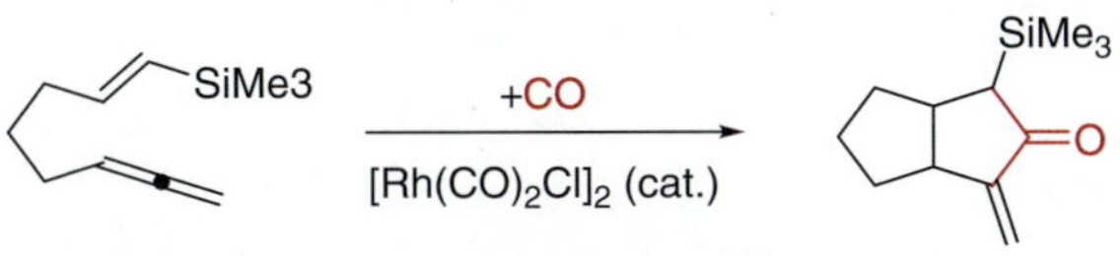

(8.148)

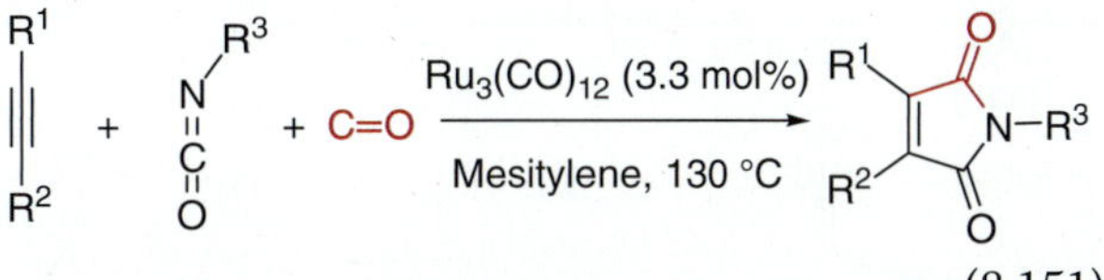

(8.151)

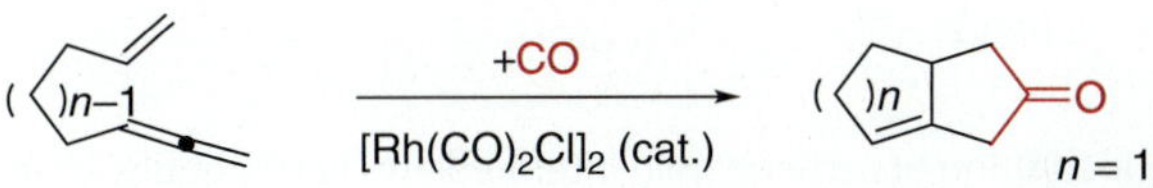

n = 1,2

(8.149)

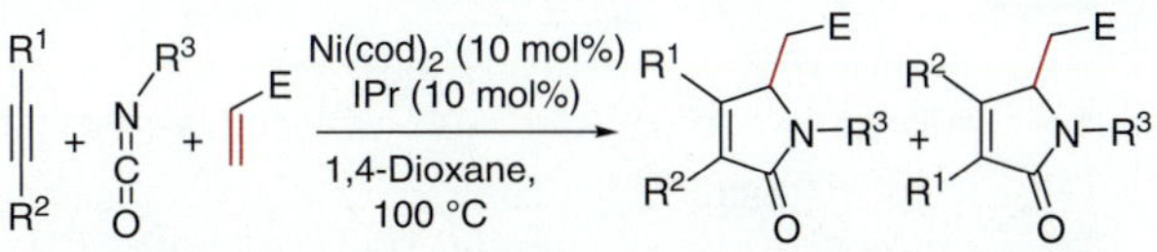

IPr = 1,3-bis(2,6-diisopropylphenyl)imidazol-2-ylidene, E = COOR

(8.152)

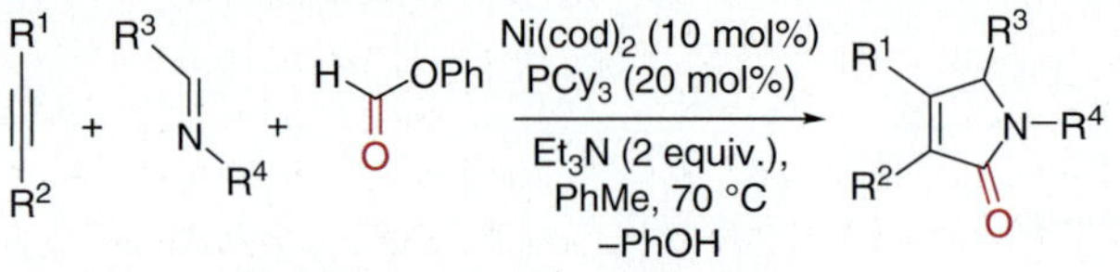

cod = cycloocta-1,5-diene, Cy = cyclohexyl

(8.153)

Gold-catalyzed cycloisomerization of 1,n-enynes involves gold carbene intermediates (e.g. **140**, reaction (8.120), Section 8.4.3) [845–851]. Gold carbene intermediates are also involved in the Au-catalyzed 1,n-diyne cycloisomerizations [983–986]. In the case of reaction (8.154), carbenoid intermediates **188** have been quenched by 2 equiv. of aldehydes providing 4-(cyclohexa-1,3-dienyl)-1.3-dioxolanes in a formal (2+2+1)-cycloaddition as described by Rao and Chan (Scheme 8.52) [987].

Problem 8.39 $Fe(CO)_5$-mediated carbonylative (5+1)-cycloaddition of vinylcyclopropane generates cyclohexanone [988, 989]. Ir-catalyzed carbonylative (5+1)-cycloaddition of allenylcyclopropanes gives 2-alkyline-3-cyclohexen-1-ones [990]. Propose a reaction mechanism for the $Ru_3(CO)_{12}$-catalyzed carbonylation of cyclopropyl—C(=N—*t*-Bu)-Ph into 3,4-dihydro-1-(*t*-butyl)-6-phenyl-2(1*H*)-pyridinone [991].

8.4.9 [$\pi^4+\pi^2$]-Cycloadditions of unactivated cycloaddents

Diels–Alder reaction or [π^4s+π^2s]-cycloaddition (Sections 5.3.9–5.3.16 and 7.6.6) is the most widely used method to prepare six-membered rings. Other common classes of reactions to generate six-membered rings are the [π^2s+π^2s+π^2s]-cycloadditions (Section 8.4.10), the (3+3)-cycloadditions (Section 8.4.11), Dötz-type (3+2+1)-cycloadditions (that are not catalytical reactions), and catalyzed (3+2+1)-cycloadditions (Section 8.4.12) (see also Problem 8.39 for carbonylative (5+1)-cycloadditions).

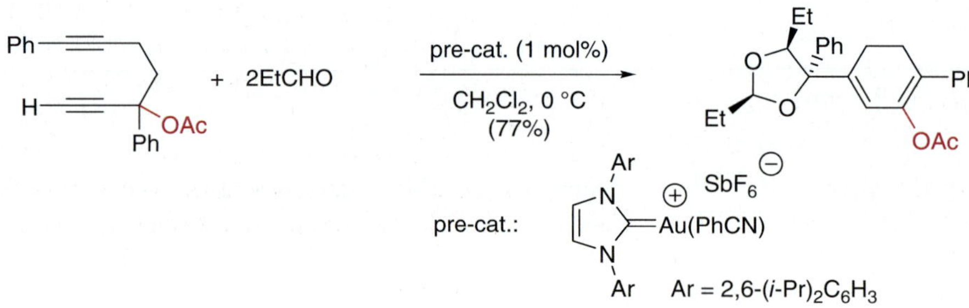

(8.154)

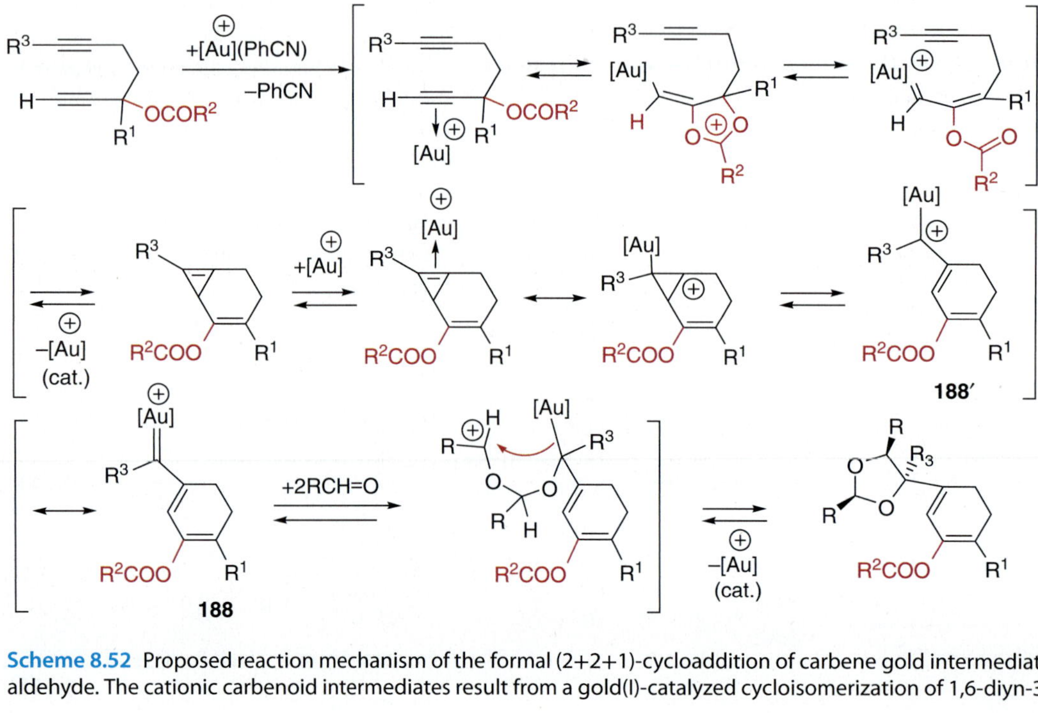

Scheme 8.52 Proposed reaction mechanism of the formal (2+2+1)-cycloaddition of carbene gold intermediates with 2 equiv. of an aldehyde. The cationic carbenoid intermediates result from a gold(I)-catalyzed cycloisomerization of 1,6-diyn-3-yl esters.

Conjugated 1,3-dienes and dienophiles bearing substituents that can stabilize 1,4-diradical intermediates and/or 1,4-zwitterionic intermediates undergo facile thermal Diels–Alder additions. Lewis and Brønsted acids are good catalysts for Diels–Alder reactions involving one cycloaddent bearing one or several substituents that can act as Lewis base (Section 7.6.6). This is valid for stepwise cycloadditions undergoing through zwitterionic intermediates, as well as for concerted cycloadditions. Transition-metal-based Lewis acids provide good catalysts; some of them can function in the presence of water, which is not possible with harder Lewis acids that generate complexes that are protic acids. The latter can destroy the cycloaddents and/or induce their polymerization. Generally, an electron-rich transition metal can be converted into a Lewis acid by incorporation of electron-withdrawing ligands and ligands with hard donor atoms. Thus, $[CpFe(CO)(P(OMe)_3)(THF)]BF_4$ [992], $[Ru(salen)(NO)(H_2O)]^+$, $[Ti(Cp^*)_2(H_2O)_2]^{++}$, $TiCp_2(CF_3SO_3)_2$, and $ZrCp_2(CF_3SO_3)_2$ are efficient catalysts for the Diels–Alder addition of conjugated α,β-unsaturated ketones, aldehydes, esters, and carbonitriles [993]. Nonactivated 1,3-dienes and dienophiles such as unsubstituted or alkyl substituted cycloaddents react with difficulty under thermal conditions and cannot be accelerated with Lewis or Brønsted acids. One option is to make one of the two cycloaddents more electron poor by single-electron transfer. Therefore, one speaks of hole-catalyzed cycloadditions (Section 7.5.8) [994–996]. In principle, electrophilic transition metal complexes are able to abstract one electron from a dienophile (which can be a 1,3-diene, an alkene or an alkyne). They can catalyze $[\pi^4+\pi^2]$-cycloadditions in a similar way as $Ar_3N^{\bullet+}/SbCl_6^-$ [997]. Most of the transition-metal-catalyzed $[\pi^4+\pi^2]$-cycloadditions are usually explained in terms of stepwise mechanisms in which the transition metal plays its template effect in bringing the diene and dienophile in its coordination sphere [689]. In fact, this is common to all transition-metal-catalyzed cycloadditions. In the case of $[\pi^4+\pi^2]$-cycloadditions, this leads to the formation of metallacycloheptadiene (Scheme 8.53a), or metallacycloheptene intermediates (Scheme 8.53b), that undergo reductive eliminations with liberation of the cycloadducts.

In 1968, Carbonaro et al. reported probably the first example of transition-metal-catalyzed Diels–Alder addition of butadiene to but-2-yne in the presence of low-valent iron species such as Fe(cyclooctatetraene)$_2$, $FeCl_3 + i\text{-}PrMgCl$, or $Fe(acac)_3 + Et_3Al$ [998]. The reaction produces 1,2-dimethylcyclohexa-1,4-diene. In 1974, Garratt and Wyatt showed that low-valent nickel complexes generated by reduction of $Ni(acac)_2$ with Et_3Al reverse the reactivity of butadiene + methyl sorbate, such as the "diene" compound (4π-electron cycloaddent) was the butadiene and the "dienophile" the π(C(4)—C(5)) double bond of the sorbate (reaction

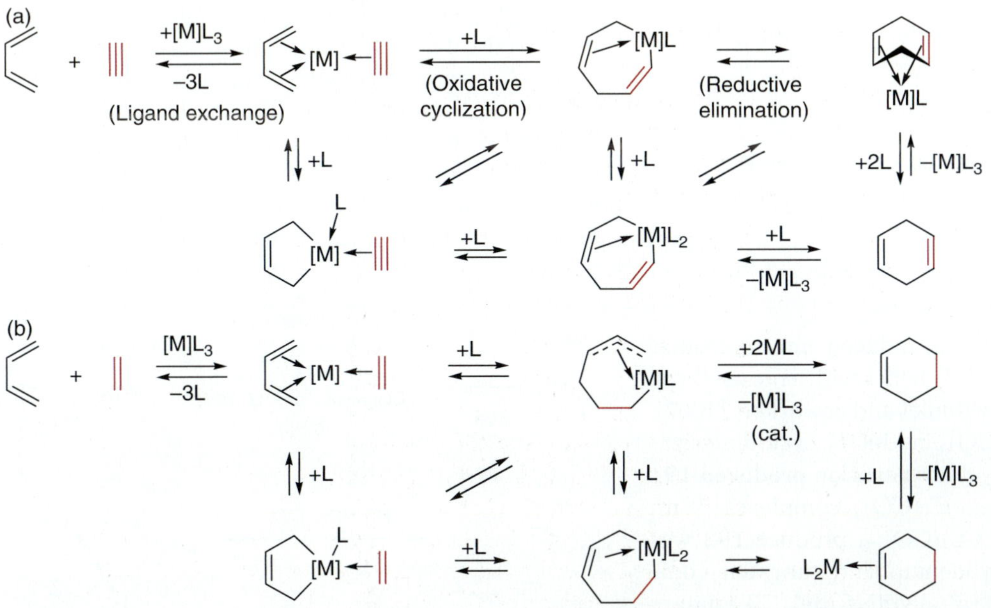

Scheme 8.53 The possible mechanisms of transition-metal-catalyzed cycloadditions of 1,3-dienes (a) to acetylenes and (b) to alkenes.

(8.155)) in contrast with the thermal reaction in which methyl sorbate acts as the 4π-electron (diene) partner (reaction (8.156)) [999].

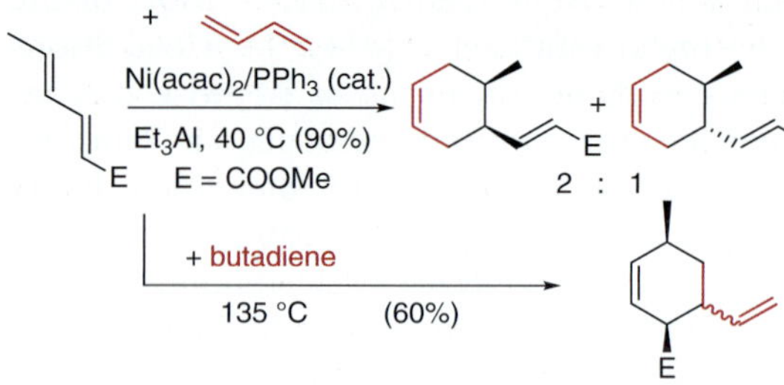

(8.155)

(8.156)

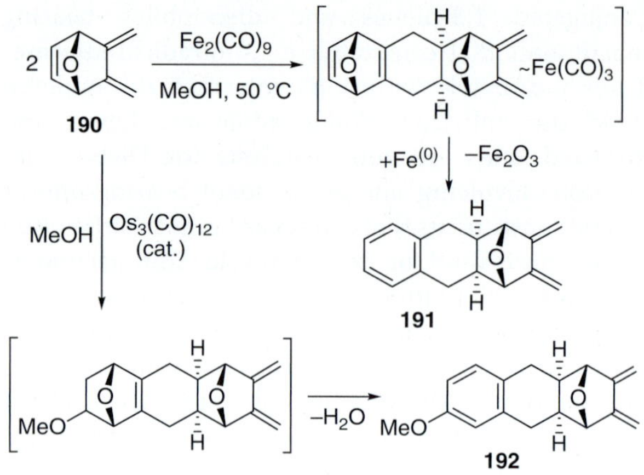

In 1979, Genêt and Ficini reported the $[\pi^4+\pi^2]$-cycloaddition of butadiene to ynamines catalyzed by $FeCl_3 + i$-PrMgCl (in Et_2O, −20 to +20 °C) [1000]. The same year, Wender and Jenkins described Ni(0)-catalyzed intramolecular $[\pi^4+\pi^2]$-cycloaddition of 1,3-dienes and alkynes (see below) [1001]. Rhodium(I) complexes [1002, 1003] and $[Ir(cod)Cl]_2$ also catalyze these intramolecular cycloadditions [1004]. In 1990, Dolor and Vogel reported the $Rh(Ph_3P)Cl$-catalyzed intermolecular $[\pi^4+\pi^2]$-cycloaddition of 1,3-dienes with propargylic alcohols [1005]. In 1983 already, tom Dieck and Diercks had found that the (diazadiene) iron complex **189** catalyzes the $[\pi^4+\pi^2]$-cycloaddition of isoprene with various alkynes (reaction (8.157)) [1006].

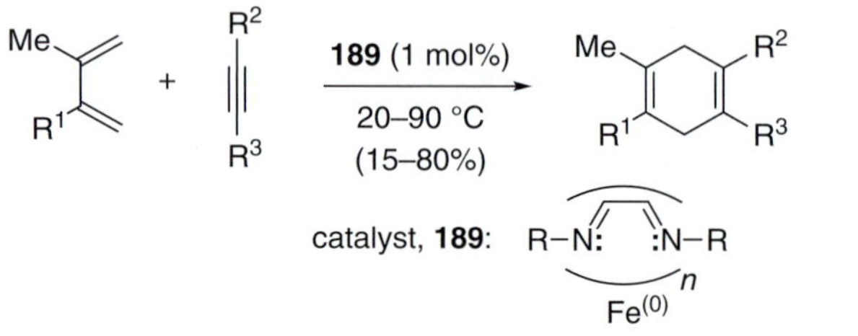

(8.157)

An early example of iron-induced intermolecular Diels–Alder addition of 1,3-dienes with a monoalkene was reported in 1982 by Roulet and coworkers [1007]. In the presence of $Fe_2(CO)_9$ in MeOH, cyclodimerization of **190** followed by deoxygenation produced **191** and its *endo* + (*exo*-diene)$Fe(CO)_3$ complexes. With a catalytic amount of $Os_3(CO)_{12}$, product **192** was obtained [1008]. Molybdenum and tungsten complexes of type $[M(CO)_2(\eta^3\text{-allyl})(NCMe)_2Cl]$ catalyze the intramolecular Diels–Alder addition of terminal alkene with a furan moiety [1009].

Heating cyclooctatetraene (cot) in heptane with $[Ru_4H_4(CO)_{12}]$ yields $Ru(CO)_3(C_{16}H_{16})$ and $Ru_2(CO)_5(C_{16}H_{16})$ complexes as well as the formal $[\pi^2+\pi^4]$-cycloadduct of cot [1010]. Wender and coworkers [1001, 1011, 1012] studied the intramolecular dienyne Diels–Alder reactions catalyzed by $Ni(cod)_2$ and $(2\text{-PhC}_6H_4O)_3P$ or other nickel catalysts [1013]. Although dienyne **193** is decomposed on heating to 140 °C without undergoing intramolecular Diels–Alder addition, in the presence of a nickel(0) catalyst derived from $Ni(acac)_2/Et_2AlOEt/(2\text{-FC}_6H_4O)_3P$, smooth cycloaddition giving **194** occurs at 72 °C [1013]. The Ni(0)-catalyzed cycloaddition is relatively insensitive to substitution of the dienyne, therefore accommodating a greater range of substrates than would be expected for the thermal Diels–Alder addition, or reactions catalyzed by protic or Lewis acids. Although proceeding though a multistep pathway, the metal-catalyzed cycloaddition occurs with the "normal" stereoselectivity (suprafacial/suprafacial: $[\pi^2s+\pi^4s]$, Section 5.3.2). The cycloadditions can be applied to construct cyclohexenes with angular methyl groups. Enantioselective Rh-catalyzed intramolecular diene-ene Diels–Alder reactions have been reported [1014] (e.g. reaction (8.158)) [1002].

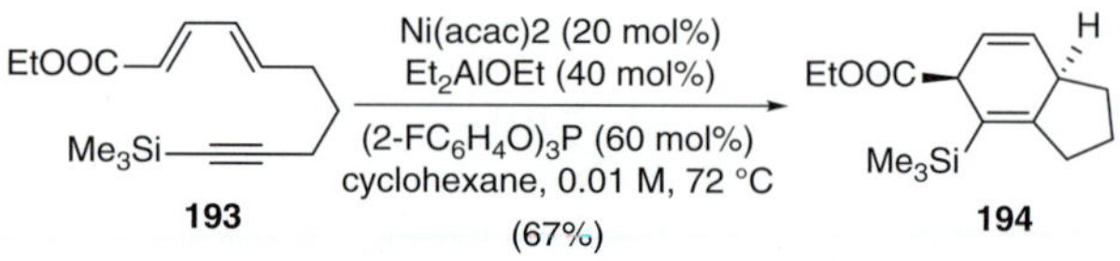

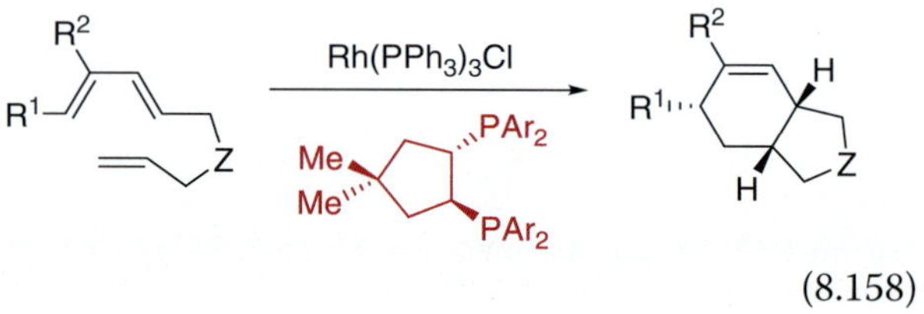

(8.158)

In the presence of cationic Co(L—L)(solvent)$_n^+$ as a catalyst, phenylacetylene adds to isoprene (and other 2-substituted butadienes) to give the corresponding cyclohexa-1,4-dienes, the *para*-adduct **195** being preferred with L—L = dppe($Ph_2PCH_2CH_2PPh_2$) and the *meta*-adduct **196** being the major adduct with L—L = imin [1015]. Quantum mechanical calculations suggested that the reaction takes place in a stepwise manner, starting with the formation of complex [Co(L—L)(isoprene)(PhC≡CH)]$^+$ (**197**). The rate-determining step is the oxidative cyclization yielding the corresponding metallacycloheptadiene intermediate **198**. A second, faster step converts **196** into complex **197**. Direct conversion of **195** into **197** is calculated to have higher activation barriers than the two-step **195** → **196** → **197** processes. Solvent molecules then displace the diene in **199** giving the cycloadduct **196** and the active catalyst that exchanges two solvent molecules for the starting diene and dienophile (Scheme 8.54). The different regioselectivity (*meta* vs. *para*) arises form steric interactions of the bidentate ligand L—L with isoprene and phenylacetylene moieties in complexes **197** that determine the distance between the carbon atoms forming the C—C bond and the relative free enthalpies of the starting complexes **197**. The first C—C bond formed in the rate-determining step of the [Co(dppe)]$^+$-catalyzed reaction yielding the *para*-adduct **195** is the C(4)—C(1′) bond, and for the *meta*-adduct **196**, it is the C(1)—C(1′) bond. The opposite order is calculated for the [Co(imin)]$^+$-catalyzed reactions, where the C(1)—C(2′) bond formation is the initial step toward the *para*-product, whereas the C(4)—C(2′) bond is formed first in the reaction giving the *meta*-product [1016].

Fürstner and Stimson reported the gold-catalyzed intramolecular Diels–Alder addition of dienynes **200** into cycloadducts **201** [1017]. In this case, the mechanism (Scheme 8.55a) differs significantly from that shown in Scheme 8.54. Coordination of the alkyne moiety to the cationic catalyst [(Ph_3P)Au]SbF_6 (made from (Ph_3P)AuCl + $AgSbF_6$) is proposed to promote an intramolecular cyclopropanation. This reaction has been described to involve carbenoid intermediate of type **202″**, then follows a fast metalla-Cope rearrangement producing metallacycloheptadiene intermediate of type **204**. A final reductive elimination liberates the product **201** and the catalytically

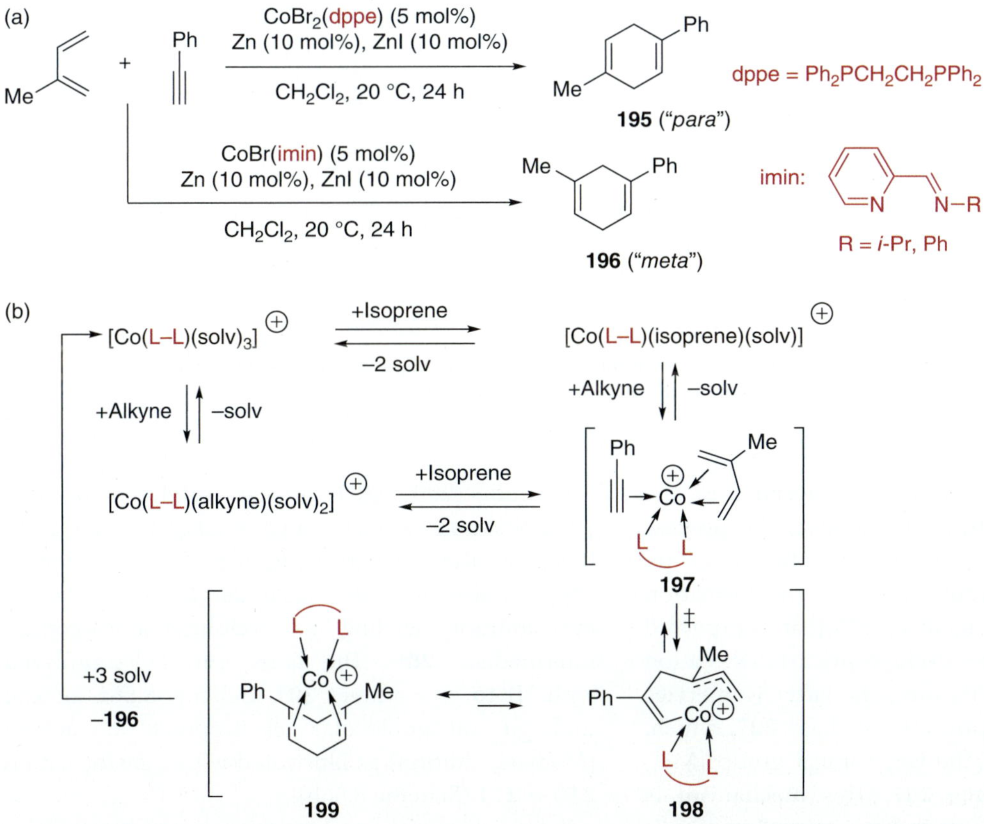

Scheme 8.54 (a) Regioselectivity of the catalyzed [$\pi^4+\pi^2$]-cycloaddition of isoprene to phenylacetylene depends on the nature of the [Co(L—L)(solv)$_3$]$^+$-catalyst and (b) suggested mechanism (solv = solvent molecule: CH_2Cl_2).

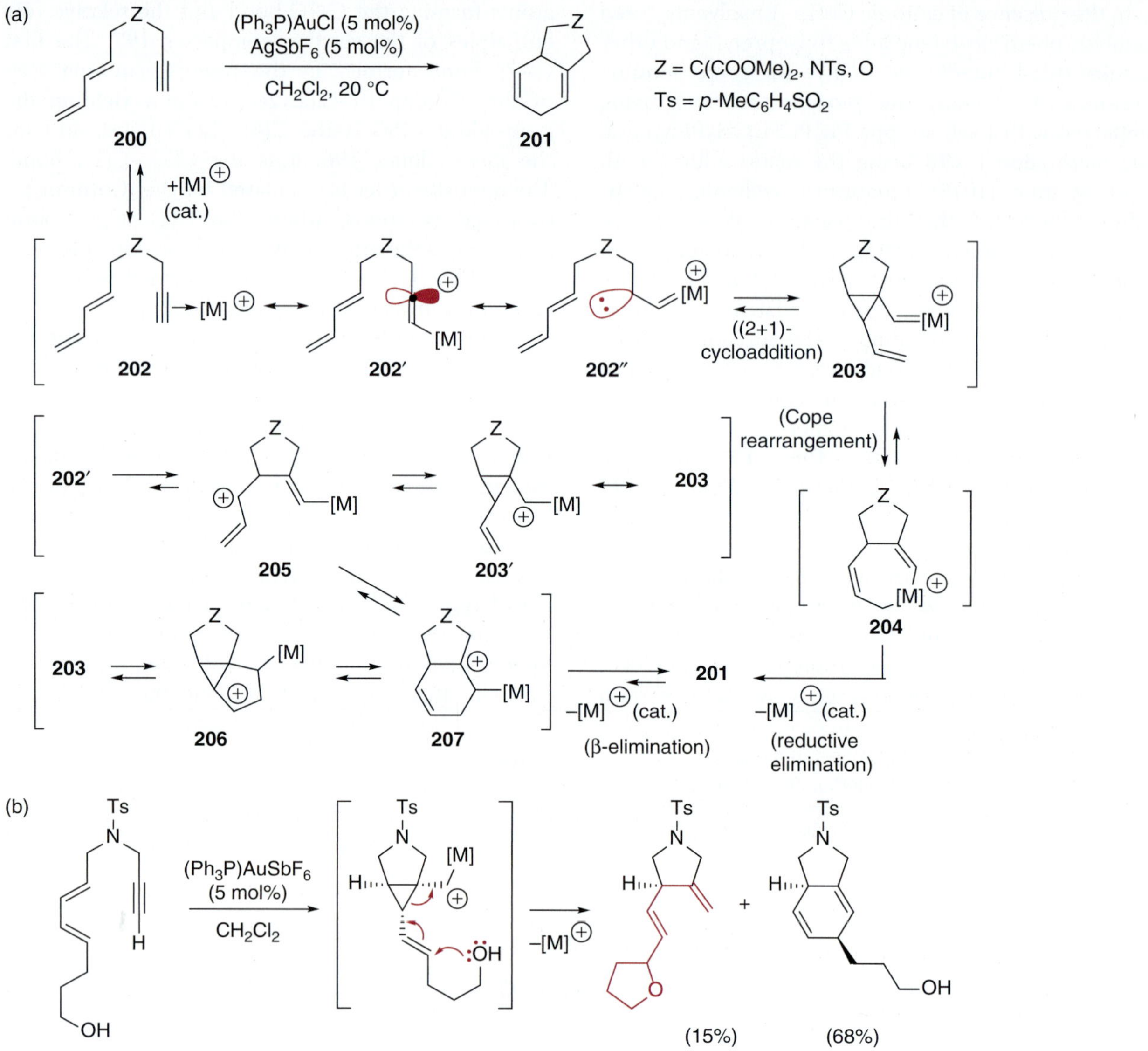

Scheme 8.55 (a) Proposed mechanism for the [Au+]-catalyzed intramolecular [$\pi^4+\pi^2$]-cycloaddition of dienynes and (b) experimental support for the intermediacy of (alkenyl)cyclopropylcarbinyl cation intermediates.

active species $[M]^+$. Alternatively, because of the cationic character of the gold catalyst, the process might generate cyclopropylcarbinyl cation intermediate **203** through a route involving the formation first of the cation intermediate **205** that is expected to isomerize into the cyclopropylcarbinyl cation intermediate **206**. On its turn, the latter isomerizes into the homoallyl cation intermediate **207**, which, upon β-elimination of the electrofugal group $[M]^+$, produces cyclohexadiene **201**. This mechanism is supported by the observation that cationic intermediates can be trapped intramolecularly by an alcoholic function (Scheme 8.55b).

Reaction **200** → **201** is also catalyzed by CuI (10 mol%) and Et_3N (1 equiv.) in CH_2Cl_2 at 20 °C. In this case, the mechanism implies the formation of copper(I)acetylide **208** that undergoes fast [$\pi^4+\pi^2$]-cycloaddition leading to cyclohexadienyl–copper intermediate **209**. The latter can be protolyzed with HI to give adduct **201** (Scheme 8.56a) or can undergo intramolecular electrophilic quenching (Michael addition) as illustrated with transformation **210** → **211** (Scheme 8.56b).

Regio-, diastereo-, and enantioselective intermolecular (4+2)-carbocyclizations of vinylarylaldehydes with alkenes and alkynes leading to substituted

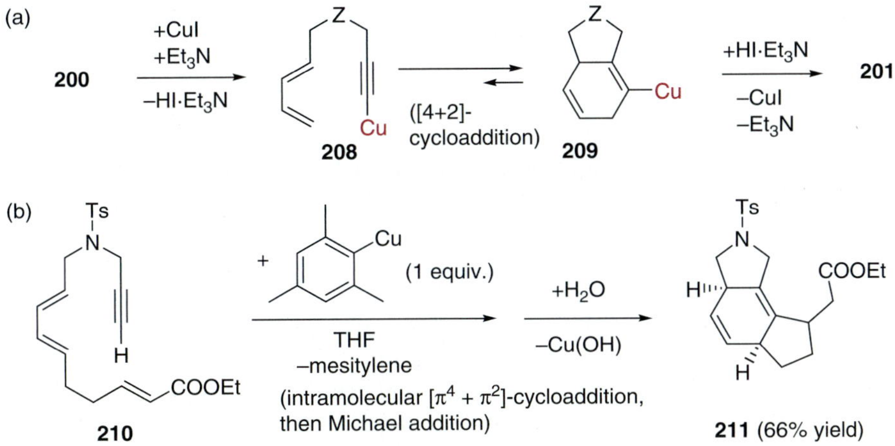

Scheme 8.56 (a) CuI-catalyzed intramolecular [π^4+π^2]-cycloaddition of dienyne and (b) it can be coupled with an electrophilic quenching of an alkenylcopper intermediate (Diels–Alder/alkylation cascade).

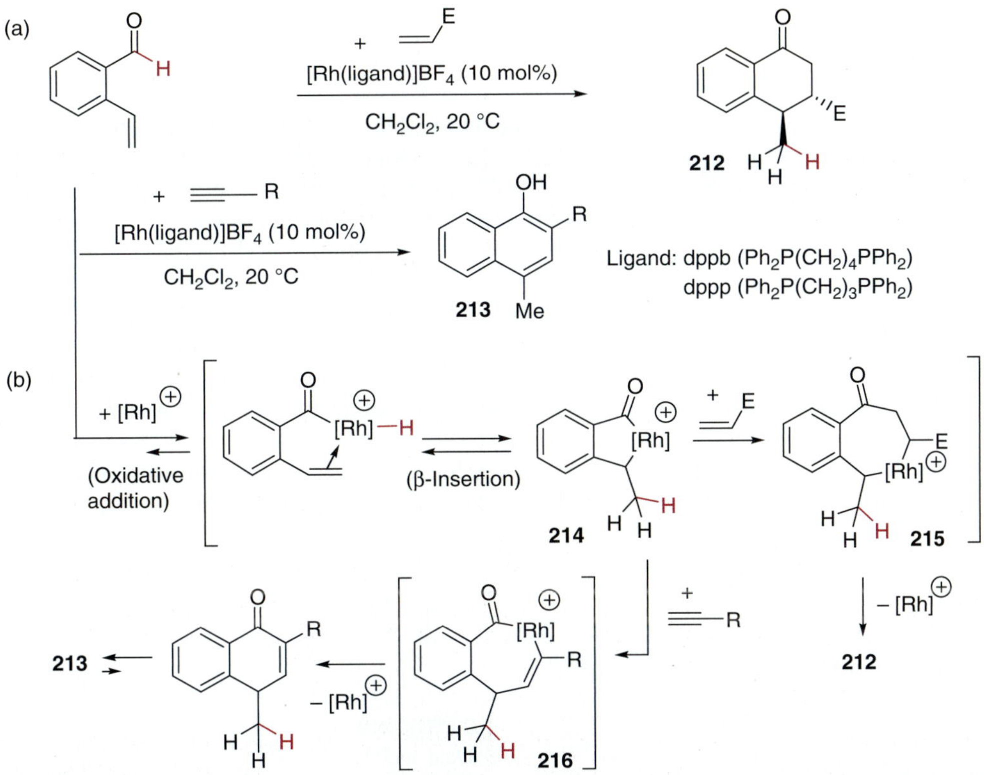

Scheme 8.57 (a) Examples of Rh(I)-catalyzed (4+2)-carbocyclizations of *o*-vinylbenzaldehyde and (b) proposed reaction mechanism.

tetralones **212** and 1-naphthols **213** have been developed by using a cationic rhodium(I)/dppb or dppp complex as a catalyst (Scheme 8.57a) [1018]. A possible mechanism involves an oxidative addition of the cationic Rh(I) complex onto the carbaldehyde C—H bond, followed by an hydrogen β-insertion onto the *ortho* alkenyl moiety giving **214**. The latter coordinates the alkene or alkyne and undergoes β-insertion generating seven-membered intermediate **215** or **216**, respectively. These intermediates liberate the products of (4+2)-carbacyclization through reductive eliminations (Scheme 8.57b).

Problem 8.40 Hilt et al. have developed an enantioselective version of the Co-catalyzed intermolecular Diels–Alder addition, for instance, **A** → **P** [1019]. Give a mechanism for the two successive reactions.

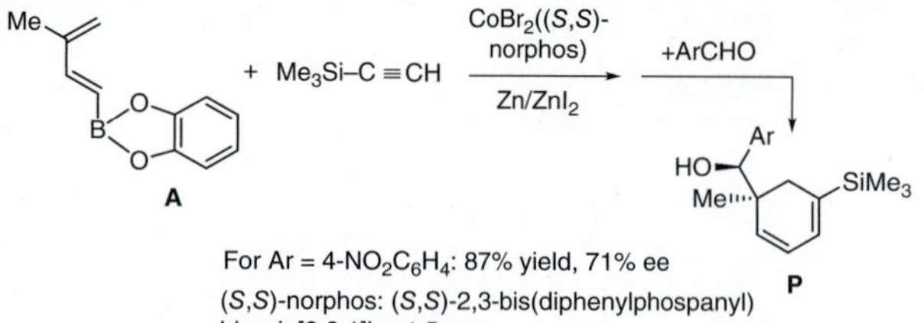

For Ar = 4-$NO_2C_6H_4$: 87% yield, 71% ee
(*S*,*S*)-norphos: (*S*,*S*)-2,3-bis(diphenylphospanyl)bicyclo[2.2.1]hept-5-ene

Problem 8.41 Propose a total synthesis of (±)-alloyohimban (**P**) and (±)-yohiman (**Q**) starting from tryptamine (**A**), hexa-3,5-dienoyl chloride (**B**), and propargyl bromide **C** [1020].

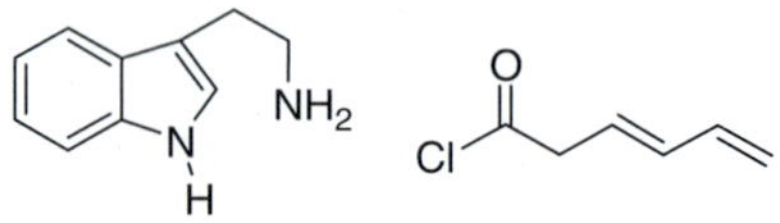

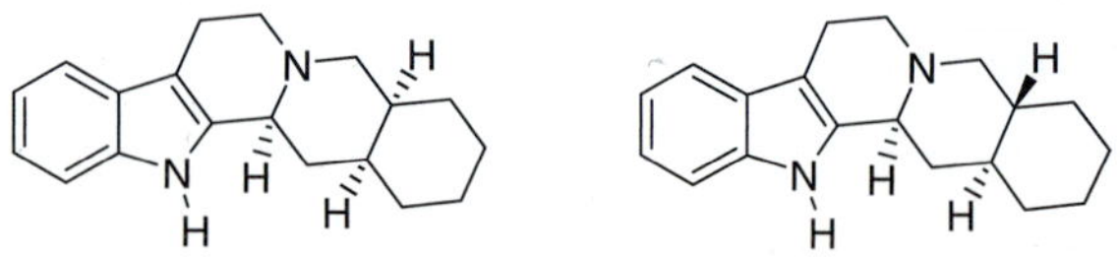

8.4.10 (2+2+2)-Cycloadditions

Transition-metal-catalyzed (2+2+2)-cycloadditions produce important six-membered carbo- and heterocycles. The cycloaddents can be alkynes alone, or mixtures of alkynes, alkenes, allenes, nitriles, imines, aldehydes, ketones, and heterocumulenes. The reactions can be intramolecular, intermolecular, or crossed [1021–1029]. Enantioselective versions of the (2+2+2)-cyclizations are numerous [1030–1043]. Although the thermal [π^2s+π^2s+π^2s]-cycloadditions are symmetry allowed [1044] (Section 5.3.2) and realize "aromatic" Hückel 6 electron arrays in their transition states (Section 5.3.3), entropy cost to bring three cycloaddents together (Section 2.9), as well as high enthalpic activation energy associated with the condensation of nonactivated π-cycloaddents makes the thermal carbocyclizations very difficult [1045]. The reduced pressure pyrolysis of cyclododeca-1,5,9-triyne (**217**) gives hexaradialene (**219**). The reaction does not undergo through a intramolecular [π^2s+π^2s+π^2s]-cycloaddition via 1,2:3,4:5,6-tricyclobutabenzene (**218**). This was demonstrated by the pyrolysis of the 1,10-(^{13}C)$_2$-labeled **217** that produces the 1,4-(^{13}C)-labeled [6]-radialene and none of the 1,2-(^{13}C)-labeled isotomer [1045]. The rearrangement is a sequence of three (3,3)-sigmatropic rearrangements (Scheme 8.58). Even though the thermal (2+2+2)-cycloaddition **217** → **218** is exothermic by c. −37 kcal mol^{-1}, and has a reaction entropy variation near zero, it does not occur.

The use of a transition metal catalyst enables the entropy constraints of intermolecular cyclotrimerizations to be circumvented by coordination of the reactants to the metal complex in a stepwise process (template effect). Thus, metal-catalyzed cyclotrimerization of acetylenes into benzene derivatives ($\Delta_r H°$(3HC≡CH → PhH) = −143.8 kcal mol^{-1}) can be achieved with many transition metals including Ni, Co, Pd, Cr, Rh, Fe, Ru, Ti, Ta, and Nb

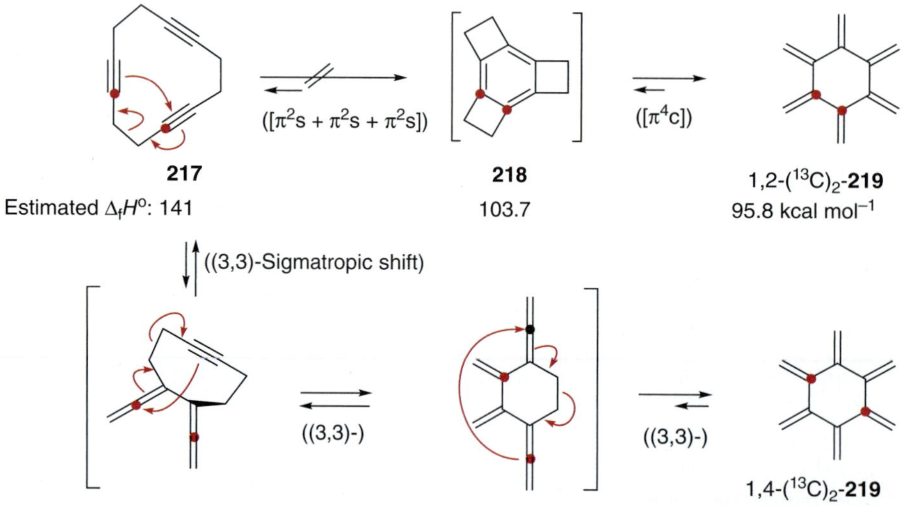

Scheme 8.58 Thermolysis of cyclododeca-1,5,9-triyne produces hexaradialene (1,2,3,4,5,6-hexakis(methylidene)cyclohexane), not via [π^2s+π^2s+π^2s]-cycloaddition followed by a triple electrocyclic ring opening, but via three successive (3,3)-sigmatropic rearrangements.

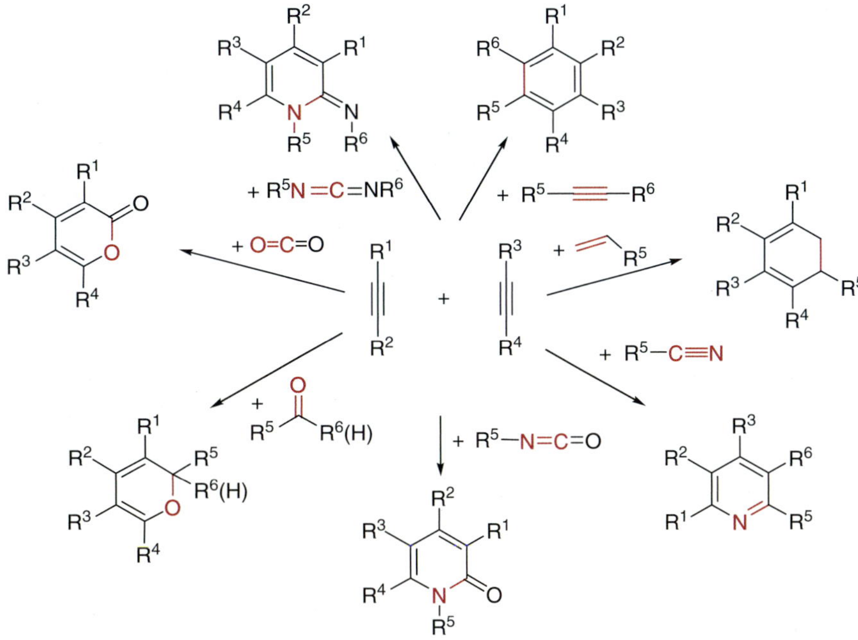

Scheme 8.59 Transition-metal-catalyzed formal (2+2+2)-cycloadditions.

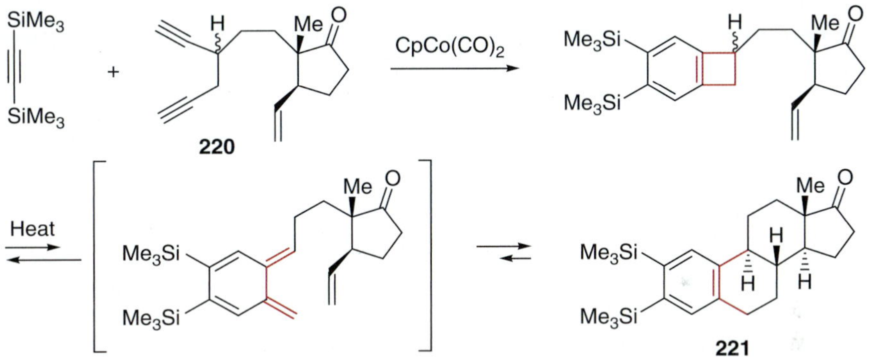

Scheme 8.60 Vollhardt's synthesis of steroids.

[689, 690, 1046–1048]. In 1866, Berthelot prepared benzene on heating acetylene in a copper tube (reaction (8.159)) [1049]. In 1948, Reppe and Vetter catalyzed this cyclotrimerization with $(Ph_3P)_2Ni(CO)_2$ (60–70 °C). Since then, this cyclotrimerization has been modified to incorporate alkenes [1050], nitriles [1051, 1052], isocyanates [1053], aldehydes, ketones [1054], CO_2, imines, and diimides (Scheme 8.59).

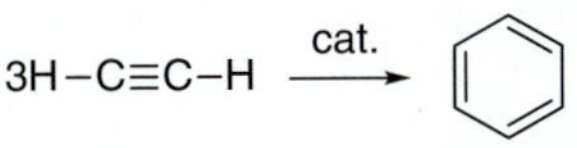

(8.159)

Vollhardt and coworker have proposed a five-step synthesis of estratrienone via **221** that employs the cycloaddition of a hexa-1,5-diyne **220** with $Me_3Si—C≡C—SiMe_3$ catalyzed by $CpCo(CO)_2$ (Scheme 8.60) [1055].

Although $CpCo(CO)_2$ is the most commonly used cobalt catalyst, the use of CO-free catalysts have been reported, such as $FeCl_3$/zinc "diaminocarbene" [1056] or $FeCl_2$ ligated with 2-(benzimidazolyl)-6-(1-arylimino)pyridine in the presence of Zn metal [1057]. Salts of Co, Pd, and Ni have also been shown to catalyze intramolecular cyclotrimerization to afford substituted benzenes. Zinc metal reduces the salts into low-valent metal species that are stabilized under the form of diaminocarbene complexes. An example is given with reaction (8.160) [1056].

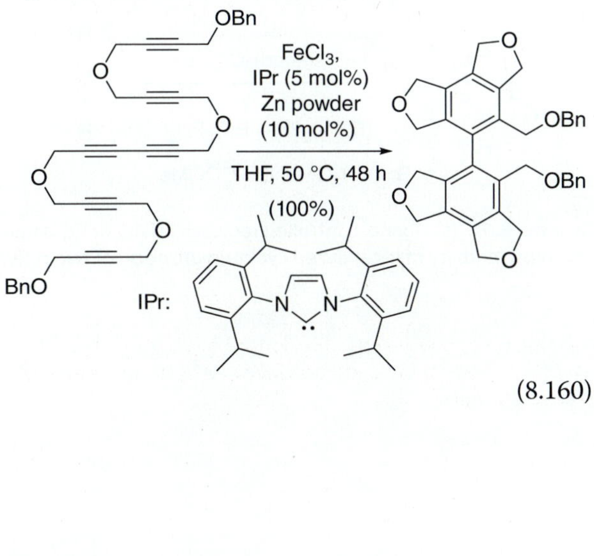

(8.160)

Okamoto and coworkers have shown that iminomethylpyridine ligands in the presence of $CoCl_2 \cdot (H_2O)_6$/Zn catalyst system cyclotrimerize terminal alkynes to yield 1,2,4-trisubstituted benzene rings selectively (Scheme 8.61a) [1058, 1059]. Similar results have been obtained by Hilt et al. when dicyclohexylimines are used as ligands with a $CoBr_2/Zn/ZnI_2$ catalyst system [1060–1062]. When catalyzed by $NbCl_3$(dme) and $TaCl_3$(dme) (dme = 1,2-dimethoxyethane), the cyclotrimerization of terminal alkynes give mixture of 1,3,5- and 1,2,4-trisubstituted benzenes. The cyclotrimerization favors head-to-tail cycloadducts when catalyzed with diniobium complexes **222** [1063].

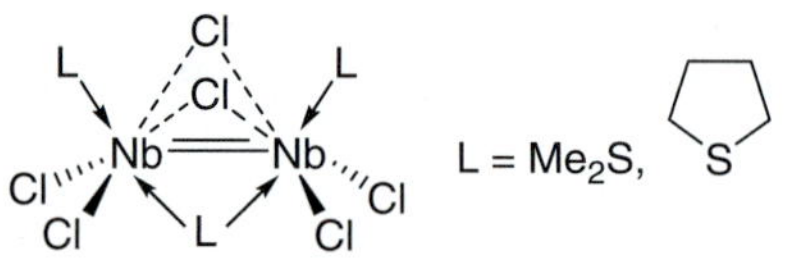

222

Interestingly, when 1,2-bis(4-methoxyphenyl)thioethane is used as a ligand, the choice of solvent has a profound effect on the regioselectivity. It varies from 1 : 6.2 in CH_2Cl_2 to 22.9 : 1 in MeCN for the 1,2,5- vs. 1,3,5-regioisomers. Polysubstituted benzenes are produced in high yields when a variety of terminal and internal alkynes are used with the $[Ir(cod)_2Cl]_2$ + diphosphine ligand system as a catalyst (Scheme 8.61b) [1064, 1065]. Takai and coworkers have found that the cycloaddition of terminal alkynes to β-keto-esters **223** is catalyzed by $Mn(CO)_5Br$ in the presence of molecular sieves (reaction (8.161)) and give benzene derivatives **224** [1066].

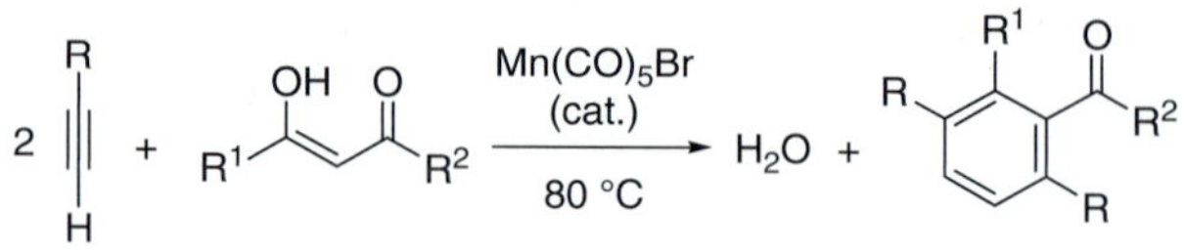

223 **224**

(8.161)

One possible mechanism (formation of metallacyclopentadienes of type **225** undergoing Diels–Alder reaction to generate metallanorbornadiene intermediates of type **226**) for the transition-metal-catalyzed cyclotrimerization is presented in Scheme 8.62a. An alternative mechanism (formation of metallacycloheptatriene intermediates of type **229**) is shown in Scheme 8.62b for the Cp*RuCl-catalyzed synthesis of pyridines [1067]. Quantum mechanical

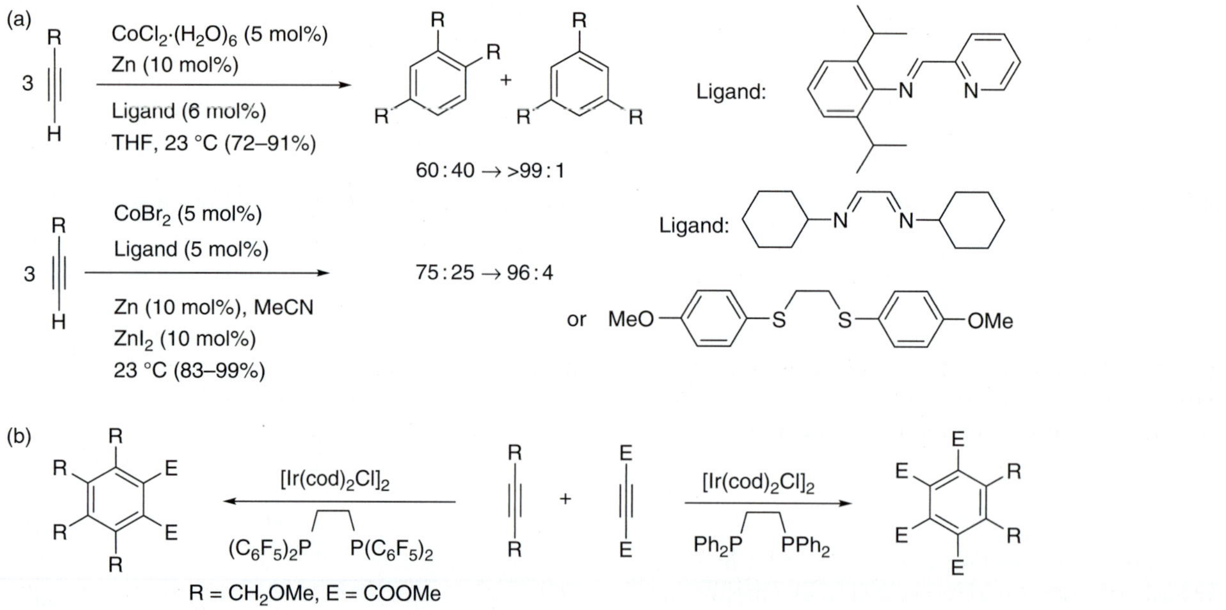

Scheme 8.61 (a) Ligand-controlled regioselectivity of Co-catalyzed terminal alkyne cyclotrimetrizations and (b) ligand-controlled chemioselectivity of Ir-catalyzed cyclotrimerization of two differently substituted alkynes.

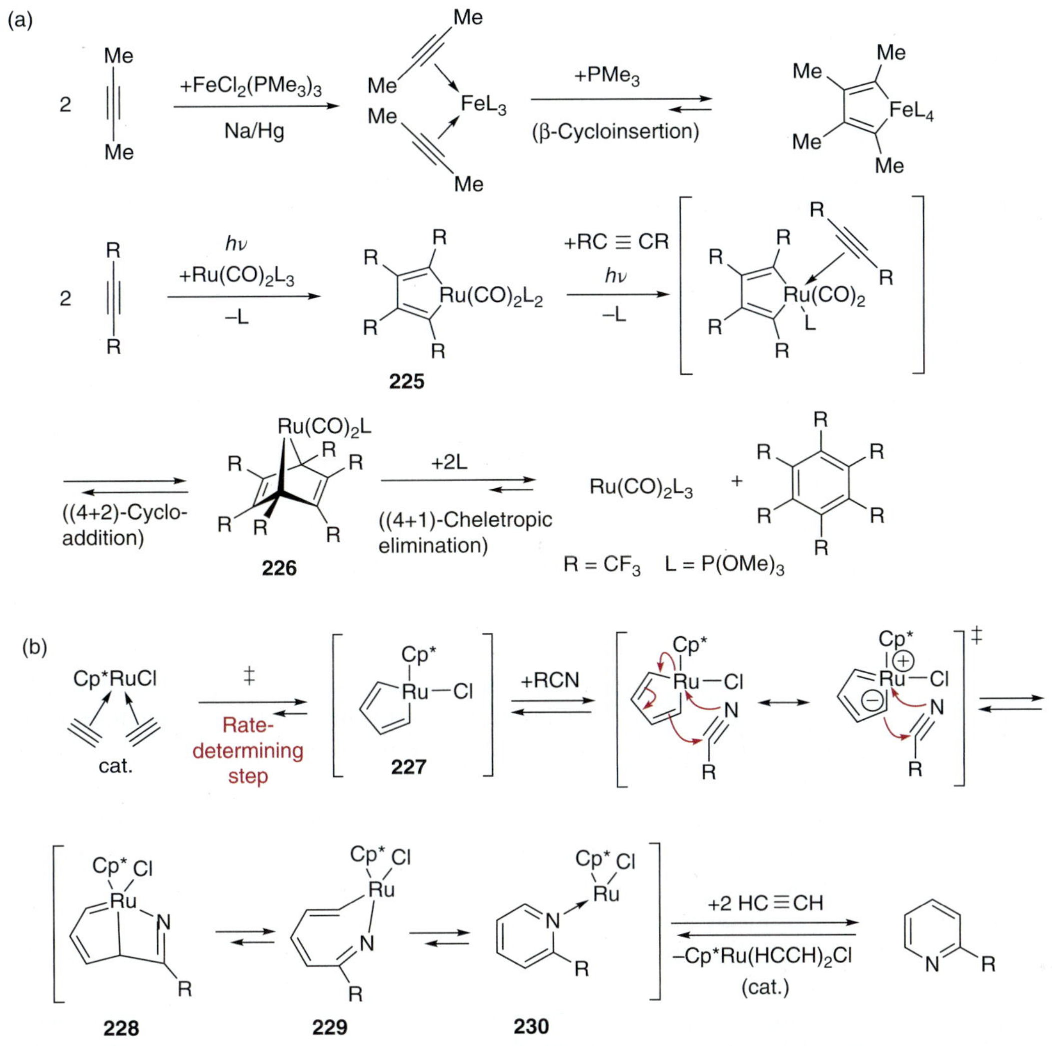

Scheme 8.62 (a) Model reactions for a possible mechanism (7-metallanorbornadiene hypothesis) of transition-metal-catalyzed alkyne (2+2+2)-cyclotrimerization and (b) Yamamoto's mechanism (7-metallacycloheptatriene hypothesis) for the Cp*RuCl-catalyzed (2+2+2)-cycloaddition of acetylene with a carbonitrile (pyridine synthesis).

calculations for the CpRuCl-catalyzed cyclodimerization of acetylene with CF_3CN (Scheme 8.62b) suggested that the reaction starts with the oxidative cyclization of two acetylene molecules producing a ruthenacyclopentadiene intermediate **227**. Formal (2+2)-cycloaddition of **227** with the carbonitrile gives rise to the azaruthenabicyclo[3.2.0]heptatriene **228** and its transformation into the final η^1-pyridine complex **230** takes place via the azaruthenacycloheptatriene **229**. The rate-limiting step is the initial oxidative cyclization. In the case of CpCo- and CpRh-catalyzed reactions, the oxidative coupling of one alkyne moiety with a nitrile moiety can also be the rate-determining step [1068].

Gutnov et al. have obtained enantiomerically enriched 2-arylpyridine (reaction (8.162)) utilizing a menthyl-derived cobalt complex as the catalyst [1069]. The irradiation of the reaction mixture with visible light (λ_{irr} = 350–500 nm) or sunlight enables to perform the reaction under mild conditions. Intermediate **231** is proposed to be formed preferentially, which leads to the enantioselective step.

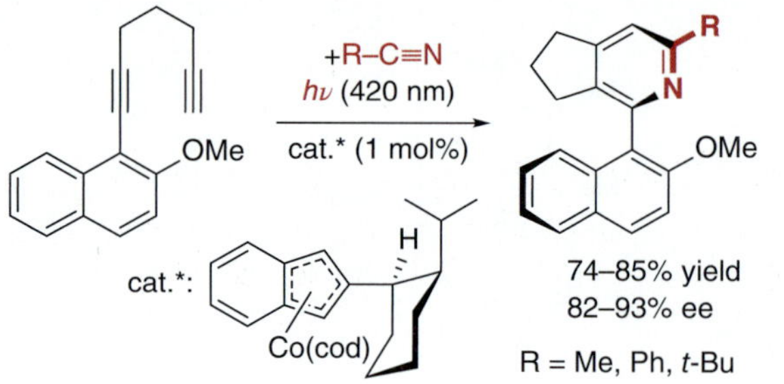

(8.162)

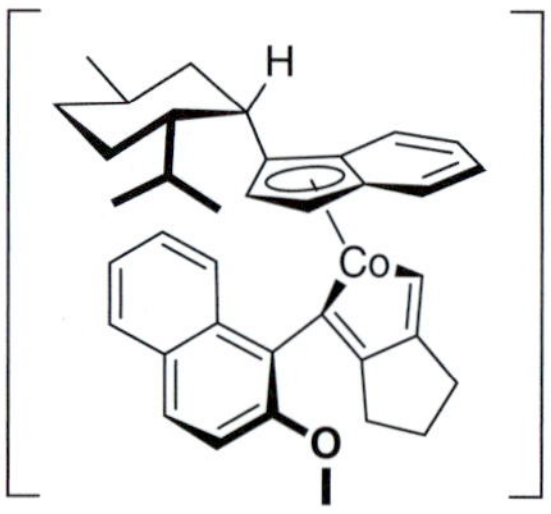

231

Shibata et al. have obtained enantiomerically enriched axially chiral bis-aryl compounds (up to 99% ee) for the (2+2+2)-cycloaddition of an α,ω-diyne (possessing *ortho*-substituted naphthalene moieties on their terminus) and a disubstituted alkyne catalyzed by $[IrCl(cod)]_2$ (0.5 mol%) + (*S*,*S*)-Meduphos (1,2-bis(2,5-dimethylphospholano)benzene (1 mol%)) [1070]. Axially chiral phthalides have been obtained with ee up to 87% for the $[Rh\{(S)\text{-}H_8\text{-BINAP}\}]BF_4$-catalyzed (2+2+2)-cycloaddition of α,ω-diynes and terminal alkynes [1031, 1032, 1071]. The (2+2+2)-cycloaddition of two alkyne units (or one diyne) with one alkene moiety results in the formation of cyclohexa-1,3-dienes [1072–1080]. Shibata et al. have used a chiral rhodium complex to prepare optically active cyclohexa-1,3-dienes (reaction (8.163)) [1081]. Asymmetric induction has been reported for the (2+2+2)-cycloadditions of 1,6-diynes and isocyanates (giving polysubstituted α-pyridones) catalyzed by chiral rhodium complexes [1082]. There are few reports of (2+2+2)-carbocyclization of enynes and alkenes or intramolecular reaction of dienynes [445, 1034, 1083]. Cp*Ru(cod)Cl catalyzes the conversion of 1,11-diene-6-ynes into tricyclo[7.3.0.0^{2,6}]dodec-1-enes (reaction (8.164)) [1050].

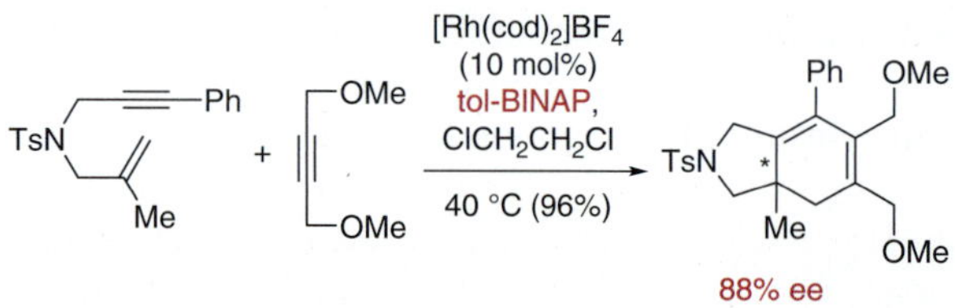

(8.163)

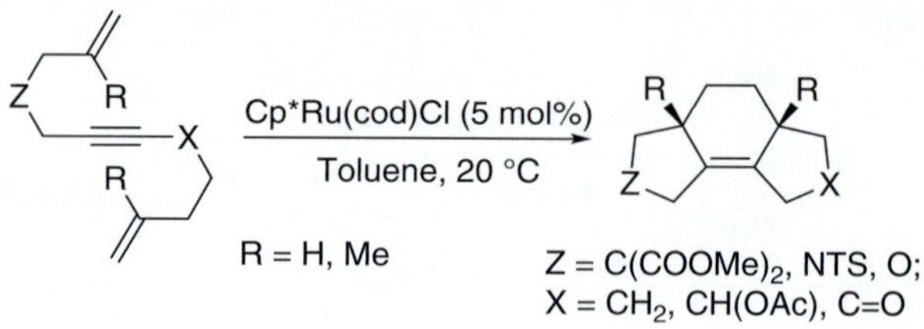

(8.164)

In 1962, Schrauzer and Eichler reported the cyclodimerization of norbornadiene catalyzed by Ni(0) complexes such as $Ni(CH_2{=}C(CN)_2)_2$ that gives several cyclodimers, including products of formal $[\pi^2s+\pi^2s]$- and $[\pi^2s+\pi^2s+\pi^2s]$-cycloaddition [1084]. Although uncatalyzed cycloadditions of electron-deficient alkynes to norbornadiene usually produce homo-Diels–Alder adducts (reaction (8.165)), acetylene without strong electron-withdrawing substituents is unreactive [1085]. Several transition metal complexes such as $Ni(cod)_2$ [1086], $Co(acac)_2$/dppe (2 mol%) + Et_2AlCl (8 mol%) [1087, 1088], (diazadiene)$FeCl_2$ [1089], $CoI_2(PPh_3)$/Zn [1090], and (norbornadiene)$RuCl_2(PPh_3)$ catalyze these (2+2+2)-cycloadditions [1091].

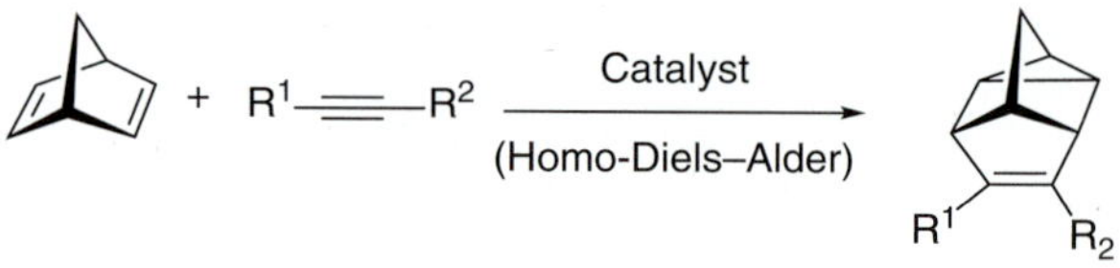

(8.165)

The first examples of enantioselective homo-Diels–Alder reactions (8.166) were reported independently by Lautens and Brunner in 1990 [1088, 1092–1094]. For instance, with PhC≡CH, norbornadiene gives the corresponding (2+2+2)-cycloadduct in 99% yield and 99% ee with $Co(acac)_2$ + Et_2AlCl + (+)-norphos ((2*S*,3*S*)-2,3-dimethyl-2,3-bis(diphenylphosphino) bicyclo[2.2.1]hept-5-ene). In 1993, Trost et al. reported the Ru-catalyzed "bishomo-Diels–Alder addition" of cyclooct-1,5-diene (cod) with various alkynes giving (2+2+2)-cycloadducts (reaction (8.166)) [1095]. Quantum mechanical calculations suggested that the cycloaddition occurs via an intermediate formed by the active species $[CpRu(cod)]^+$ and an alkyne substrate molecule. The first C—C coupling step is the rate-determining step [1096]. In the presence of Ni(0) precatalysts and a *N*-heterocyclic carbene (NHC) enynes, **232** react with aldehydes affording the corresponding ring-opened tautomers **233** (reaction (8.167)), rather than pyrans. With ketones, enynes **232** react to give the expected dihydropyrans **234** (reaction (8.168)).

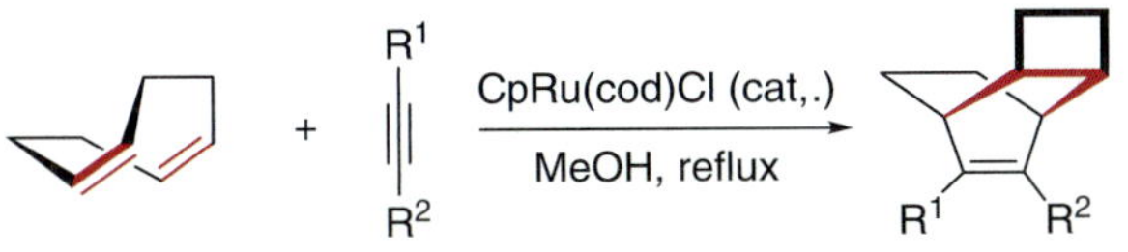

R^1 = H, Me, Et, CH_2OR;
R^2 = CH_2OR, COOMe, CH_2CH_2OR, C_3H_6COOMe

(8.166)

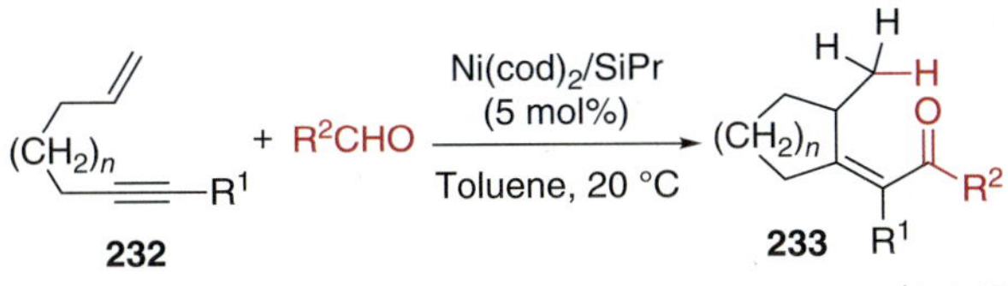

(8.167)

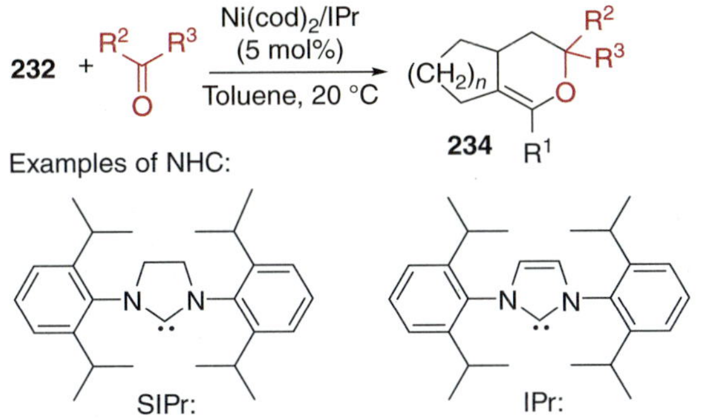

(8.168)

Problem 8.42 Kesenheimer and Groth have utilized the $CpCo(C_2H_4)_2$-catalyzed (2+2+2)-cycloaddition in the synthesis of the core of several antibiotics, especially the angucyclines [1097]. Propose a reaction sequence converting **A** into (−)-8-O-methyl-tetrangomycin (**P**).

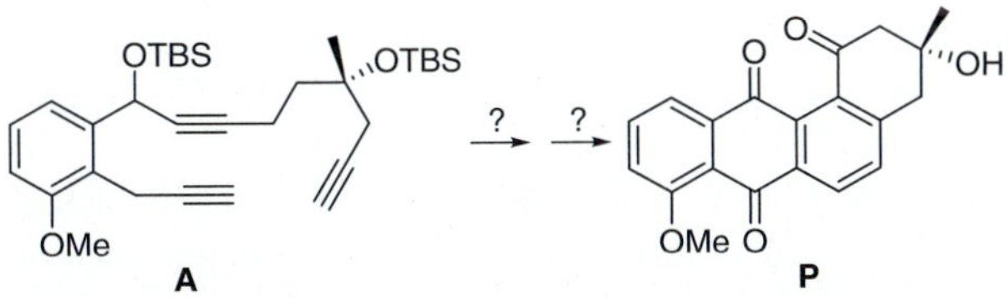

Problem 8.43 Propose a mechanism for the Ru-catalyzed (2+2+2)-cycloaddition (8.166) [1096].

8.4.11 (3+3)-Cycloadditions

Stepwise (3+3)-cycloaddition reactions have been rare before 1990 but are now becoming a useful approach to benzene and six-membered heterocyclic compounds. The reactions can rely on organocatalysis (e.g. phosphine [1098] and *N*-heterocyclic carbenes [1099, 1100]), Lewis acid catalysis (Section 7.6.9), or transition metal catalysis [1101–1104]. A first example of metal-assisted (3+3)-cycloaddition was reported by Wojcicki and coworkers in 1983 (reaction (8.169)) [1105]. It involves an allyliron complex **235** reacting with $S[NS(O)_2Me]_2$, a reaction that proceeds via initial formation of a (3+2)-cycloadduct **236** followed by a rearrangement into **237**. In the presence of Pd(0) catalyst, activated aziridine **238** undergoes formal (3+3)-cycloadditions with **162** to give the corresponding piperidine **239** (reaction (8.170)) [1106–1109]. The reaction is proposed to take place by addition of zwitterionic palladium complex **163** to **238**.

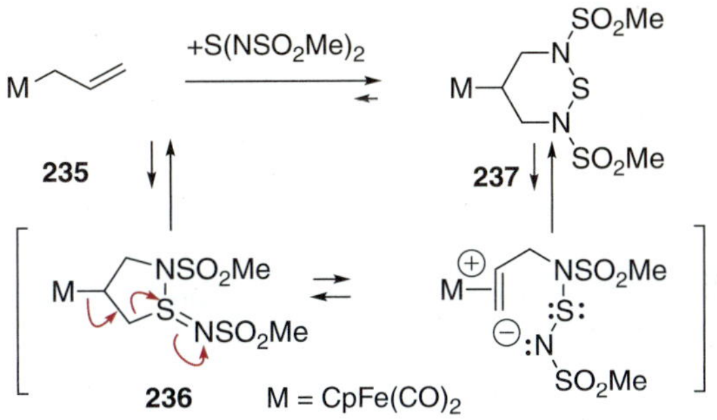

(8.169)

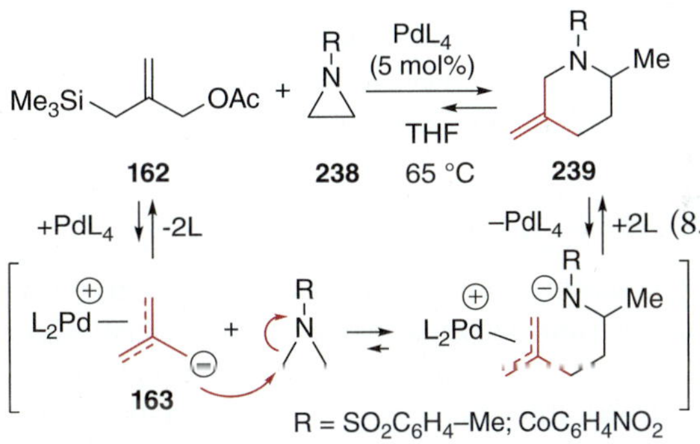

(8.170)

Using aryl-substituted derivatives of **162**, nitrones **240** and an enantiomerically pure Pd-catalyst Hayashi and coworkers reported in 2007 asymmetric (3+3)-cycloadditions producing oxazines **241** (Scheme 8.63a; *trans*/*cis* ratio 83 : 11, up to 93% ee for the *trans*-products) [1110]. Rhodium enolcarbenoid are obtained readily from 2-diazo-3-silyloxy-3-butene-carboxylic esters or from donor–acceptor cyclopropenes. They react as dipoles onto other dipoles such as nitrones, hydrazones, ylide derivatives, nitrile oxides, oximes, and imines [1103]. Applying chiral Hashimoto's catalyst $Rh_2(S\text{-}PTA)_4$ [1111, 1112], Doyle and coworkers reported in 2011 enantioselective formal (3+3)-cycloadditions shown in

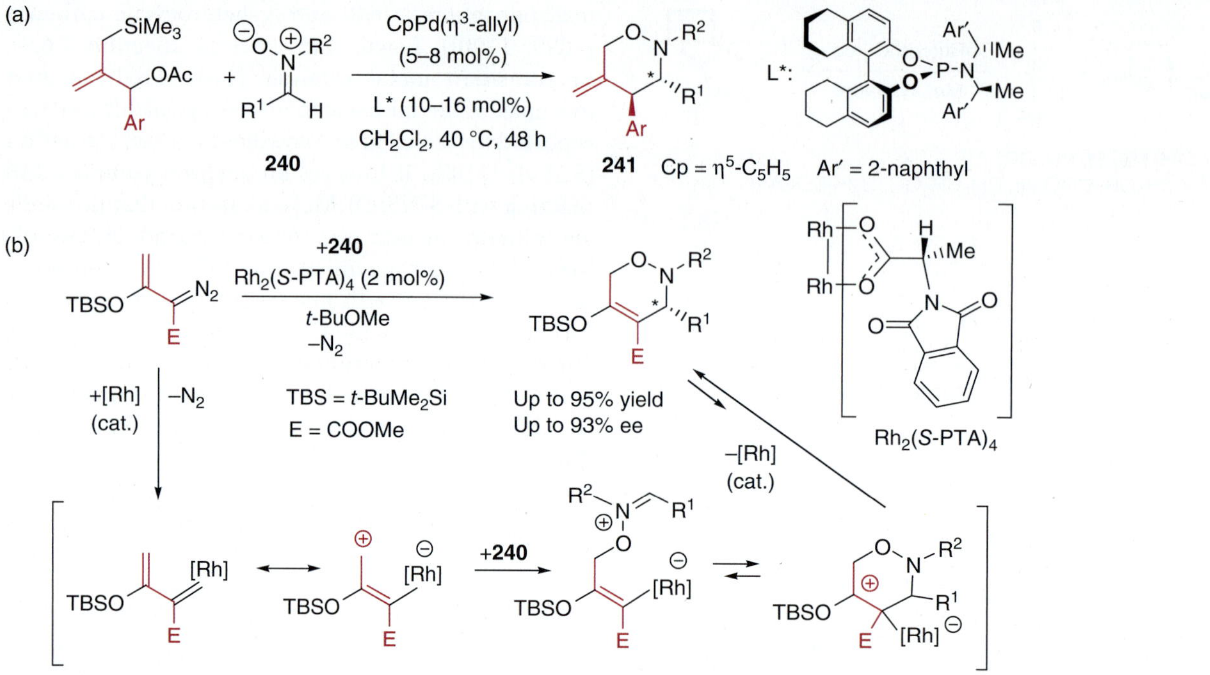

Scheme 8.63 Asymmetric syntheses of 1,2-oxazines: (a) Hayashi's palladium-catalyzed and (b) Doyle's rhodium-catalyzed formal (3+3)-cycloadditions.

Scheme 8.63b [1113]. Vinyl/enol gold carbenoids (e.g. PhCH=C(OR)—CH=[Au$^+$]/SbF$_6^-$) are obtained by Au(I)-induced rearrangement of propargyl esters and acetals. These species undergo all kinds of cyclizations including (3+3)-cycloadditions with azomethine imines [1102] and nitrones [1104].

8.4.12 (3+2+1)-Cycloadditions

In 2004, Chung and coworkers reported the Rh-catalyzed carbonylative formal (3+2+1)-cycloaddition (8.171) [1114]. α-Pyrones have been prepared by Ru-catalyzed intermolecular (3+2+1)-cycloaddition of α,β-unsaturated ketones with silylacetylenes and CO (reaction (8.172)) [1115]. Related to this reaction is the Pd-catalyzed carbonylative (3+2+1)-cycloaddition of 1,3-cyclohexadiones and alkynes to give chromene-2,5-diones [1116].

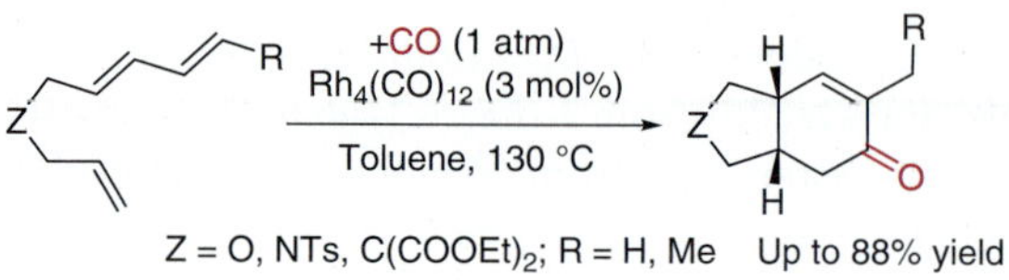

(8.171)

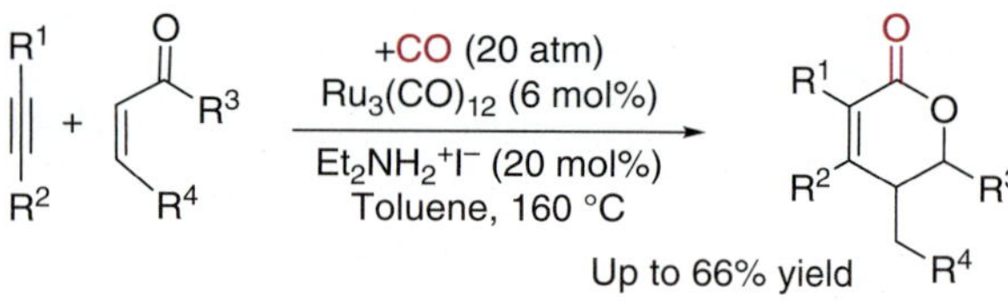

(8.172)

Vinylcyclopropanes are two-, three-, or five-carbon cycloaddents depending on other cycloaddents and reaction conditions [707]. They can undergo catalyzed (2+n)-, (3+n)-, and (5+n)-cycloadditions [886] as well as (3+2+1)-cycloadditions with CO (reaction (8.173): homologous Pauson–Khand reaction) developed by Yu and coworkers [1117–1119]. Rh(I) catalyzes the carbonylative (3+2+1)-cycloadditions of alkenylidenecyclopropanes and alkenes as illustrated with reaction (8.174). The reaction involves probably trimethylidenemethane metallacycle intermediate [1120]. When an alkyne moiety is present instead of an alkene as a cocycloaddent, phenols are obtained [1121, 1122]. With cyclopropyl(homoallyl) amines, the Rh(I)-catalyzed carbonylative cyclization is stereoselective as exemplified with reaction (8.175).

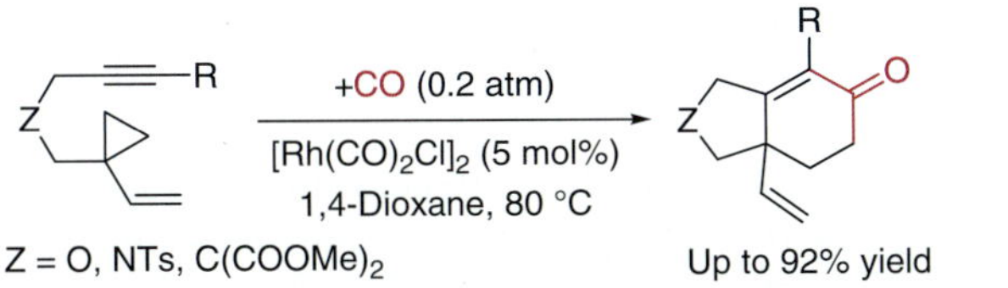

(8.173)

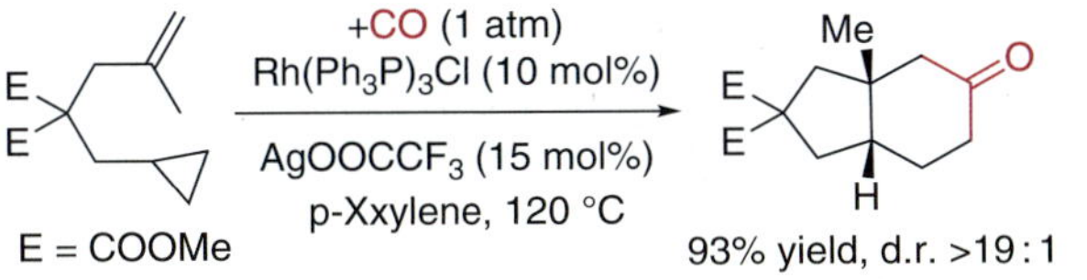

(8.174)

(8.175)

Problem 8.44 Propose a mechanism for the Rh-catalyzed carbonylative (3+2+1)-cycloaddition (8.171).

Problem 8.45 Propose a mechanism for the following [$HCo(CO)_4$]-catalyzed three-component synthesis of 1,3-oxazinan-4-ones [1123].

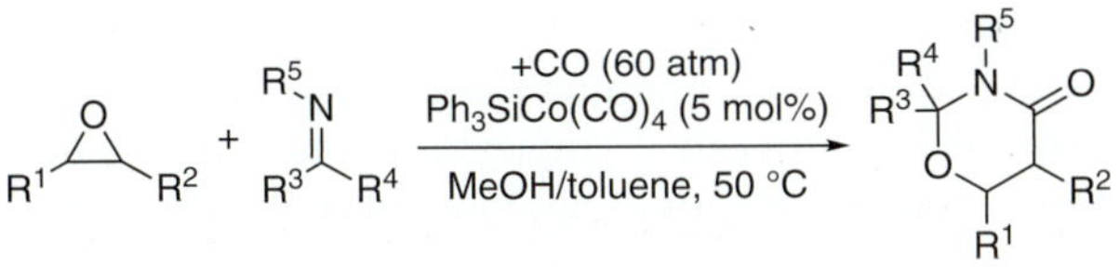

8.4.13 (4+3)-Cycloadditions

In principle, seven-membered rings can be constructed either through (n+m)-cycloadditions (with m + n = 7), through (x+y+z)-cycloadditions (with x + y + z = 7), or through (2+2+2+1)-cycloadditions [1124]. The (4+3)-cycloadditions can use allylic cations, or equivalent three-center species, and 1,3-dienes as cycloaddents [1125]. These reactions can follow multistep mechanisms or not [1126–1129]. Indeed, one-step, concerted [π^4s+π^2s]-cycloaddition of allyl cation and butadiene is permitted according to the Woodward–Hoffmann rules (Section 5.3.2) and is predicted to realize an "aromatic" six-electron array by the PMO theory in its transition state (Section 5.3.3) [1130–1133]. Binger et al. reported the Pd(0)-catalyzed (4+3)-cycloaddition of methylidenecyclopropane to dimethyl muconate **242** (reaction (8.176)) [923, 1134, 1135]. With Pd(0), methylidenecyclopropane generates a (trimethylidenemethane) palladium complex intermediate. With Pd[(*i*-PrO)$_3$P]$_4$ as catalysts, Trost et al. observed a mixture of five- and seven-membered carbocycles in nearly equal amounts (reaction (8.177)) [1136]. Heating **243** in the presence of a Pd-catalyst leads to mixtures of (4+3)- and (3+2)-intramolecular cycloadducts **244** and **245** (reaction (8.178)). Their proportion depends on the nature of the phosphite or phosphine ligand [1137] (see also Ref. [1138]).

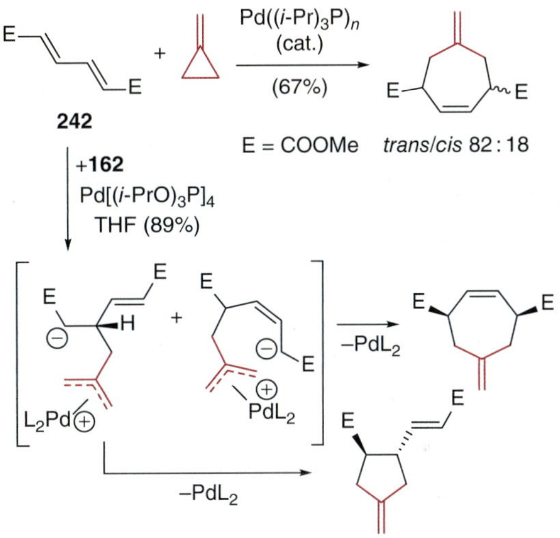

(8.176)

(8.177)

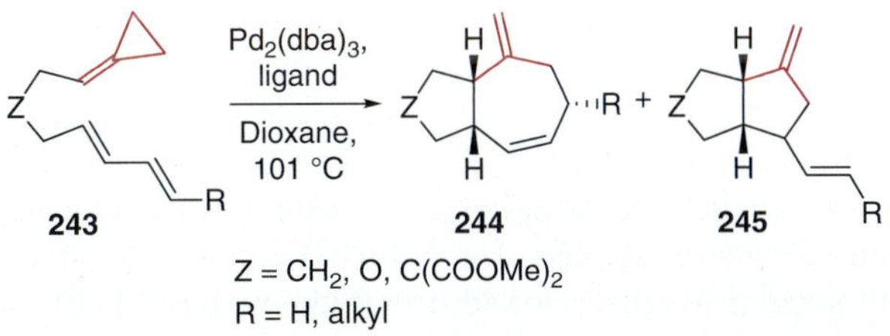

(8.178)

Hayashi and coworkers have used γ-methylidene-δ-valerolactones **246** as an equivalent of four-carbon unit in the Pd(0)-catalyzed (4+3)-cycloaddition with nitrones **247** to produce 1,2-oxazepine **248** [1139, 1140]. When applying enantiomerically pure phosphoramidite ligand, (*S*,*R*,*R*)-**249** diastereoselective

(c. 85 : 15) and enantioselective (4+3)-cycloadditions were observed (reaction (8.179); up to 96% ee). Mascareñas and coworkers reported on the AuCl- and $PtCl_2$-catalyzed intramolecular (4+3)-cycloaddition between 1,3-dienes and allenes. Upon coordination of the electrophilic transition metal, the allene moiety forms an allylic cation species **250** that undergoes the (4+3)-cycloaddition with the diene unit (reaction (8.180)) [1141].

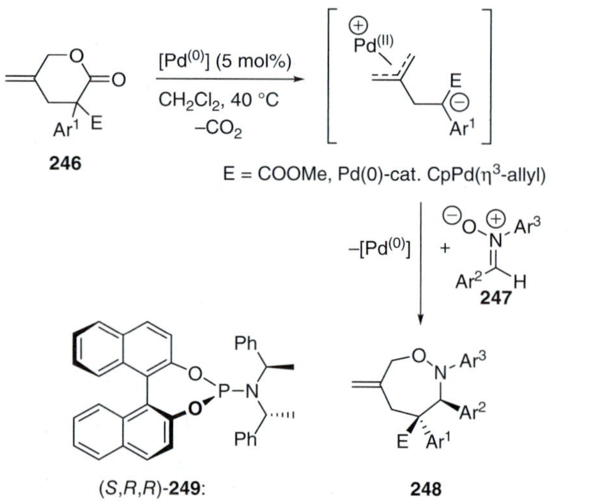

(8.179)

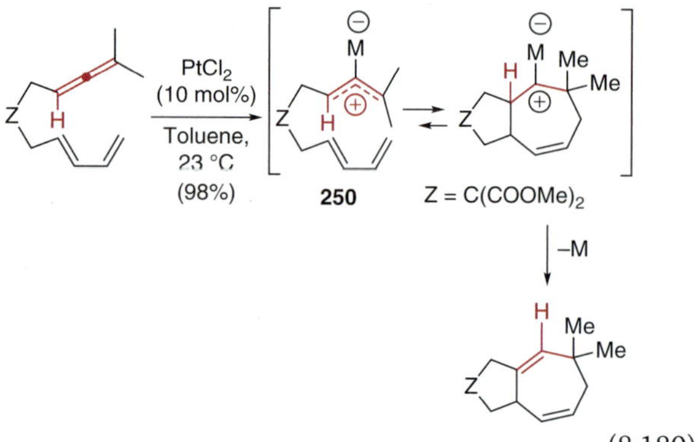

(8.180)

On their side, Shapiro and Toste have obtained dihydroazepines **252** by gold(III)-catalyzed intermolecular (4+3)-cycloaddition (Scheme 8.64) [1142]. The electrophilic gold complex generates an allyl cationic intermediate of type **253**, resulting from the 1,2-rearrangement of the propargyl ester **251** [1143, 1144] that shows a reactivity analogous to that reported for electrophilic stabilized vinylcarbenoids (see **253′**) [853, 856, 1145–1148].

The intramolecular (4+3)-cycloadditions of allenes with furan or diene moieties have been catalyzed by Au(I) and Ag(I) salts [1149, 1150]. Rh-catalyzed intermolecular (4+3)-cycloaddition of furan with a cyclopentenyl diazocarboxylate have been reported [1151]. Asymmetric synthesis of tropanes by Rh-catalyzed has been realized (e.g. reaction (8.181)) through (4+3)-cycloadditions of pyrroles with a methyl 2-(siloxyvinyl)diazoacetate [1152]. The reactions proceed via tandem cyclopropanations (Sections 5.4.1 and 8.4.1) followed by Cope rearrangements (Section 5.5.9.6).

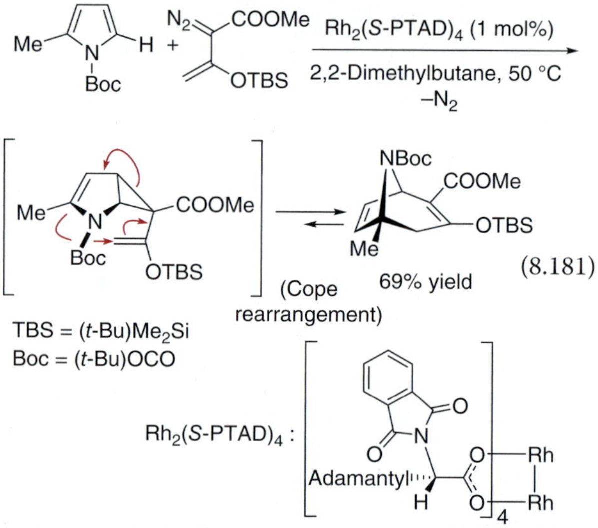

Lewis acid-catalyzed intramolecular (4+3)-cycloadditions of methylidene aziridines (equilibrating with 2-aminoallyl cation moieties) with 1,3-dienes or furan units have been reported [1153]. Using an enantiomerically pure dimine ligand, CuOTf- and $AgSbF_6$-catalyzed enantioselective (4+3)-cycloaddition of nitrogen-stabilized oxyallyl cations derived from allenamides to furans has been developed [1154]. Regio-, diastereo-, and enantioselective (4+3)-cycloadditions between vinylcarbenoids (generated from the corresponding vinyldiazoacetates) and 1,3-dienes giving cyclohepta-1,4-dienes have been catalyzed by enantiopure dirhodium catalysts such as Rh_2(*S*-PTDA)$_4$ and analogs [1155]. In 1993, Huffman and Liebeskind reported the Rh(I)-catalyzed cycloisomerization of 4-cyclopropylcyclobut-2-en-1-ones into cyclohepta-2,4-diene-1-ones and cyclohepta-2,5-diene-1-ones. The reaction can be seen as an intramolecular version of a (4+3)-cycloaddition of cyclopropanes (three-carbon cycloaddent) to cyclobutenones that formally equilibrate with alkenylketene moieties (four-carbon cycloaddent) [1156].

Problem 8.46 Saito et al. have reported on the $Ni(cod)_2/PPh_3$-catalyzed (3+2+2)-cycloaddition of ethyl cyclopropylideneacetate to terminal alkynes [1157] and on the $Ni(cod)_2$/topp (tri-O-biphenylyl phosphite)-catalyzed (4+3)-cycloaddition of ethyl

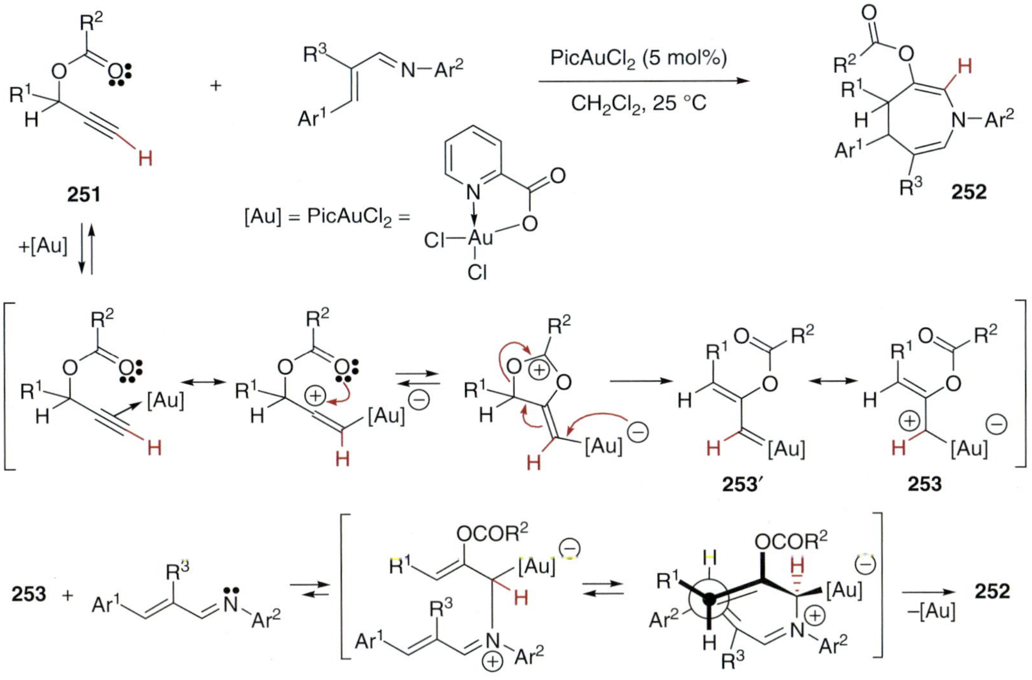

Scheme 8.64 Proposed mechanism for the Au(III)-catalyzed intermolecular (4+3)-cycloaddition of α,β-unsaturated imines with propargyl esters.

cyclopropylideneacetate to 1,3-diene [1158]. Propose a mechanism for the latter reaction.

Problem 8.47 What is the major product **P** of reaction **A** + **B**? [1137]; see also: [1138]

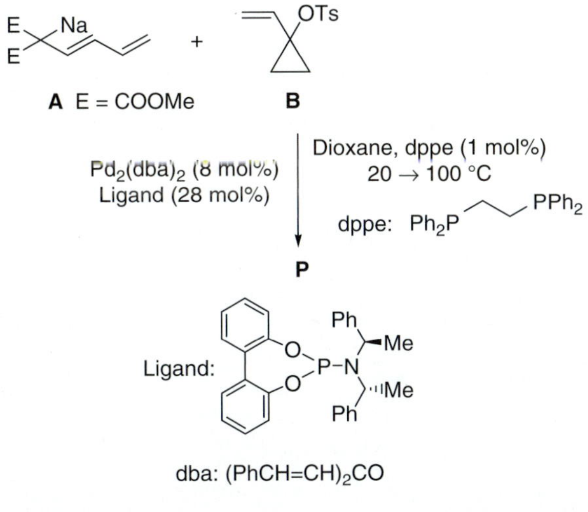

Problem 8.48 The reaction of **A** with $Me_3SiCH{=}N_2$ in the presence of $Ni(cod)_2$ as a catalyst at 60 °C produces **B** (69%) with high stereoselectivity. On heating **B** in toluene at 105 °C, product **P** is obtained, which realizes a formal (4+2+1)-cycloaddition of **A** with $Me_3SiCH{=}N_2$. What is product **P**? Propose a mechanism for the formation of **B** [1159].

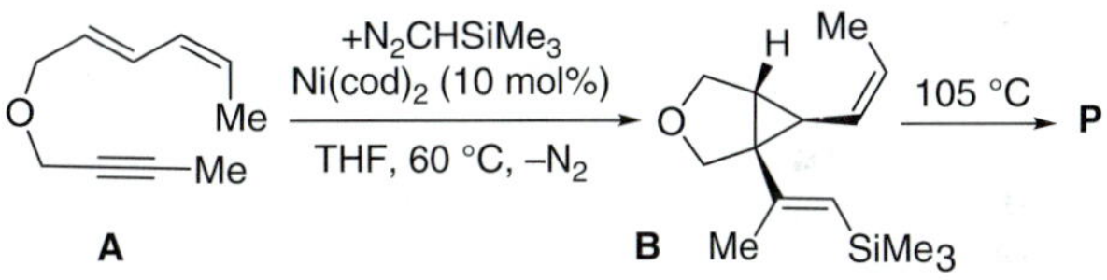

8.4.14 (5+2)-Cycloadditions

As seen in the preceding chapters, three-center cycloaddents can be allyl moieties (and heteroanalogs such as enols, enamines), cyclopropanes (and heteroanalogs such as oxiranes and aziridines), alkenylcarbenoids, allenes, or propargyl esters. When conjugated with a π-system (e.g. C=C, C≡C, C=O, C=NR), these three-center cycloaddents can react as five-center cycloaddents [1160]. In 1959, a homolog of the Diels–Alder addition was reported involving the thermal (5+2)-cycloaddition of a vinylcyclopropane and maleic anhydride to form a seven-membered ring product [1161]. In 1967, Stogryn and Brois reacted 2-vinylaziridine with hexafluoro-2-butyne that gave at 25 °C 2,3-bis(trifluoro)-3,4-dihydro-(7*H*)-azepine [1162]. In 1995, Wender et al. reported the first examples of transition-metal-catalyzed intramolecular (5+2)-cycloadditions (reaction (8.182)) of simple vinylcyclopropanes and alkynes using Wilkinson's catalyst ($(Rh(PPh_3)_3Cl)$) [1163].

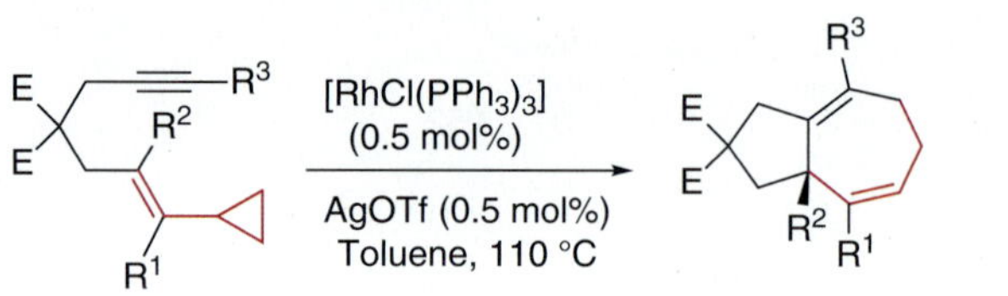

(8.182)

The intermolecular (5+2)-cycloadditions were possible with silyloxy-substituted vinylcyclopropanes **254** and terminal alkynes applying $[Rh(CO)_2Cl]_2$ (5 mol%) as a catalyst. Possible mechanisms for these reactions are given in Scheme 8.65 [1164]. The regioselectivity of these reactions is controlled first by steric factors and then by electronic factors [1165].

Electron-rich, electron-poor, conjugate internal and terminal alkynes, even acetylene itself, were found to provide cycloadducts in good to excellent yields. $[Rh(CO)_2Cl]_2$ also catalyzes intramolecular (5+2)-cycloadditions of substituted vinylcyclopropanes and alkynes [1166]. The method has been applied to the asymmetric total synthesis of (+)-Aphanamol I for which an intramolecular (5+2)-cycloaddition of vinylcyclopropane and allene moieties was engaged (e.g. reaction (8.183)) [1167, 1168].

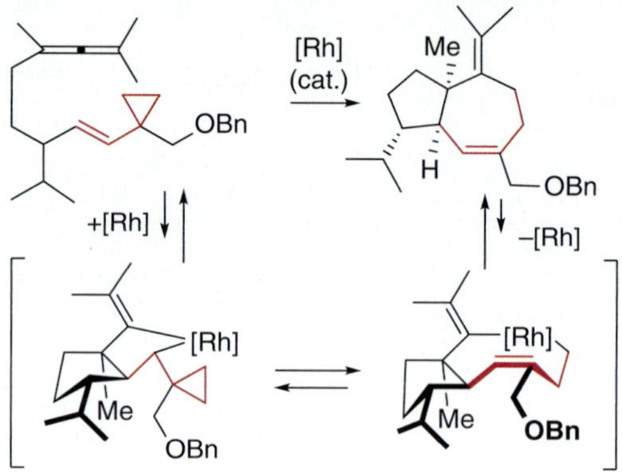

(8.183)

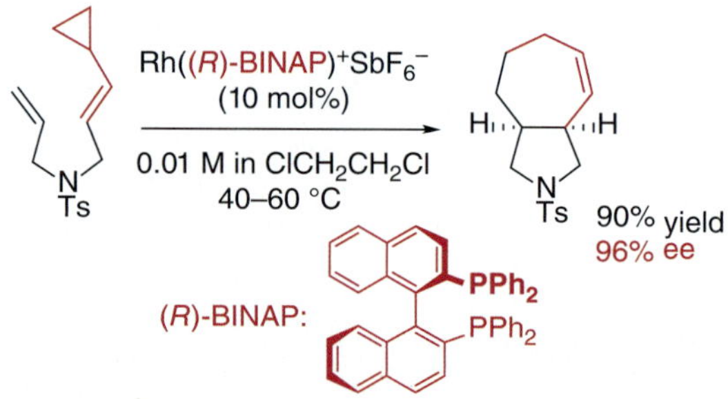

(8.184)

$[(Naphthalene)Rh(cod)]^+/SbF_6^-$ was found to be an exceptionally good catalyst [1169]. In 2002, the Wender group first reported a transition-metal-catalyzed hetero-(5+2)-cycloaddition of cyclopropylimines with dimethyl acetylenedicarboxylate giving the

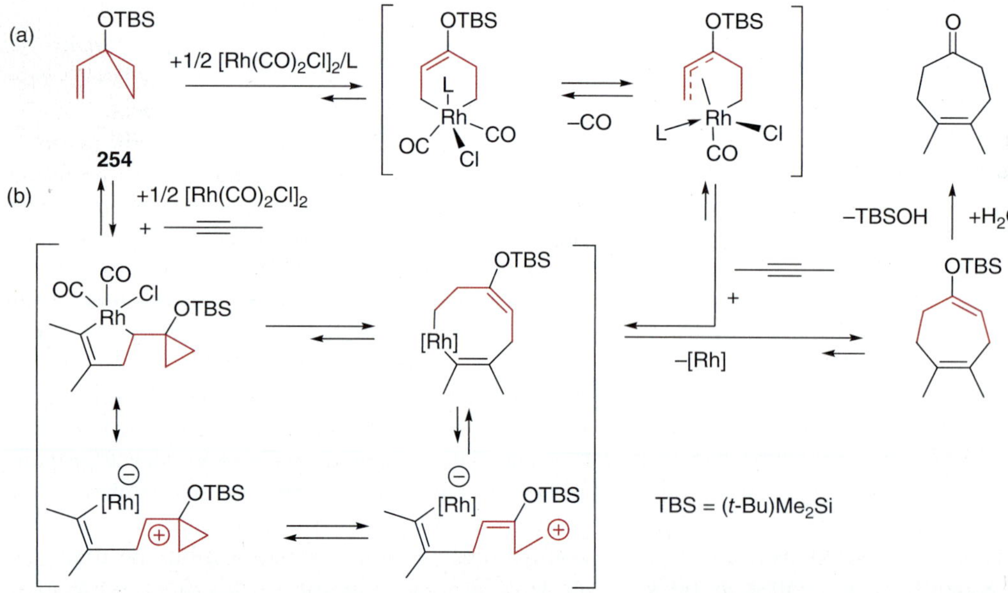

Scheme 8.65 Wender's proposals for the mechanism of the intermolecular Rh(I)-catalyzed (5+2)-cycloaddition of vinylcyclopropanes to alkynes: (a) a metallacyclopentene is formed first, followed by alkyne insertion, and (b) a metallacyclopentene is formed first, followed by a ring expansion (a kind of cyclopropylmethyl ⇄ homoallyl rearrangement).

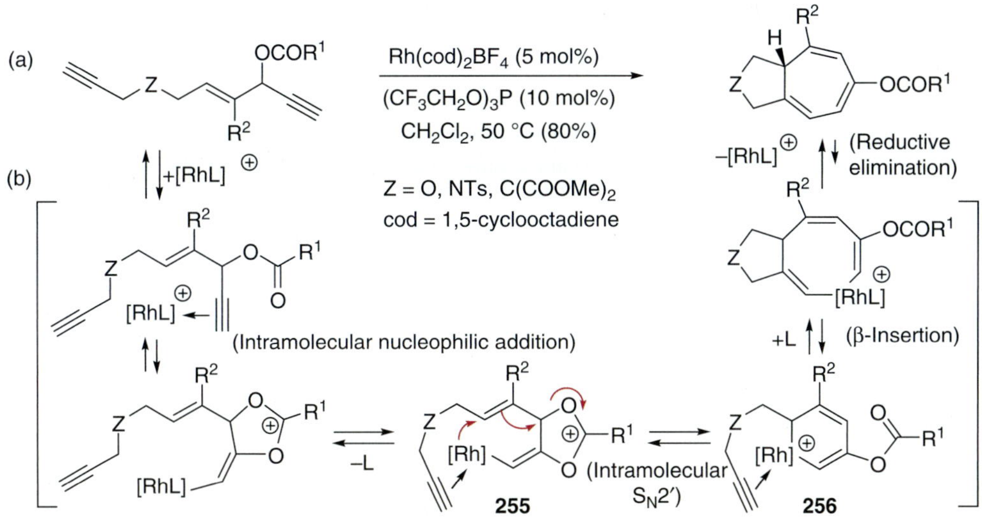

Scheme 8.66 (a) Rh(I)-catalyzed intramolecular (5+2)-cycloaddition of 3-acyloxydec-4-ene-1,9-diynes to 4-acyloxybicyclo[5.3.0]deca-1,3,5-trienes and (b) proposed reaction mechanism.

corresponding dihydroazepines [1170]. Possible mechanisms have been discussed [1171–1174]. The intra- and intermolecular (5+2)-cycloadditions have been catalyzed by rhodium complex of *N*-heterocyclic carbenes (NHC) of type [Rh(NHC)Cl(cod)]/$AgSbF_6$ [1002] and by $[Rh(cod)Cl]_2/AgSbF_6$ in fluorinated alcohols such as CF_3CH_2OH and $(CF_3)_2CHOH$ [1175]. An asymmetric intramolecular (5+2)-cycloaddition (8.184) has also been reported by the Wender group [1176]. On their side, Trost and coworkers have found that $CpRu(N{\equiv}CMe)_3{}^+/PF_6{}^-$ [445, 1177] is an effective catalyst for the intramolecular (5+2)-cycloaddition of vinylcyclopropane to alkyne units at room temperature [1178, 1179]. The reaction shows excellent diastereoselectivity as illustrated with reaction (8.185) [1180, 1181]. Quantum mechanical calculations favor a mechanism that involves an initial ene-yne oxidative cyclization to form a ruthenacyclopentene intermediate followed by cleavage of the cyclopropane (ring expansion) and reductive elimination (see Scheme 8.65b) [453a]. For the Rh(I)-catalyzed (5+2)-cycloadditions, the calculations favor initial formation of a metallacyclohexene intermediate resulting from the oxidative addition of Rh(I) to vinylcyclopropane, followed by alkyne insertion and reductive elimination (see Scheme 8.65a) [1165]. These (5+2)-cycloadditions have also been catalyzed by Ir(I) complexes [1182]. The groups of Louie [1183] and Fürstner [1184] have used Ni and Fe complexes, respectively, to catalyze these reactions. Mukai and coworkers have reported the Rh(I)-catalyzed intramolecular (5+2)-cycloadditions of allenylcyclopropanes and alkynes [1185]. As seen earlier (Section 7.6.10), Lewis acids can also promote (5+2)-cycloadditions. They have found several applications in the total synthesis of natural products [1186].

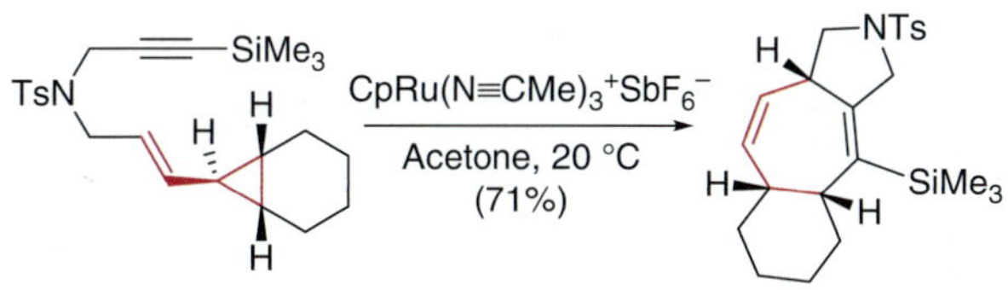

(8.185)

In 2011, Tang and coworkers reported the Rh(I)-catalyzed intramolecular (5+2)-cycloaddition of 3-acyloxydec-4-ene-1,9-diynes to 4-acyloxybicyclo[5.3.0]deca-1,3,5-trienes (Scheme 8.66a) [1187]. The proposed mechanism (Scheme 8.66b) involves the formation of cationic intermediate of type **255** (see the Ohloff–Rautenstrauch cyclopropanation, Section 8.4.3), which undergoes cyclization into **256**. Cycloinsertion of the alkyne moiety and reductive elimination provides the cycloheptatriene product [1188]. Under similar reaction conditions, the intramolecular (5+2)-cycloadditions of 3-acyloxy-1,4-enyne to alkenes provide the corresponding bicyclo[5.2.0]decadienes with high stereoselectivity [1189]. Similarly, 6-oxodec-1-ene-4,9-diyn-3-yl esters are cycloisomerized into 10-oxobicyclo[5.3.0.]deca-1,3,6-trien-2-yl esters [1189].

Azomethine ylides are prepared by Rh(II)-catalyzed reaction between pyridines and 1-sulfonyltriazoles. These five-center cycloaddents add to acetylenedicarboxylic esters forming 3,10a-dihydropyrido[1,2-d][1,4]diazepines (reaction (8.186)) [1190].

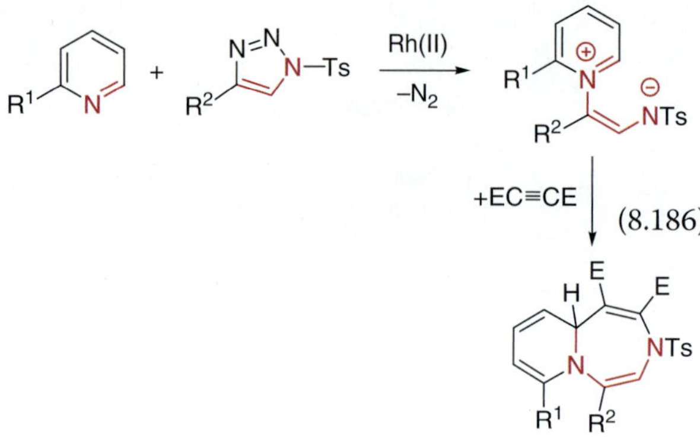

(8.186)

Problem 8.49 What is the main product of the following reaction? [1191]

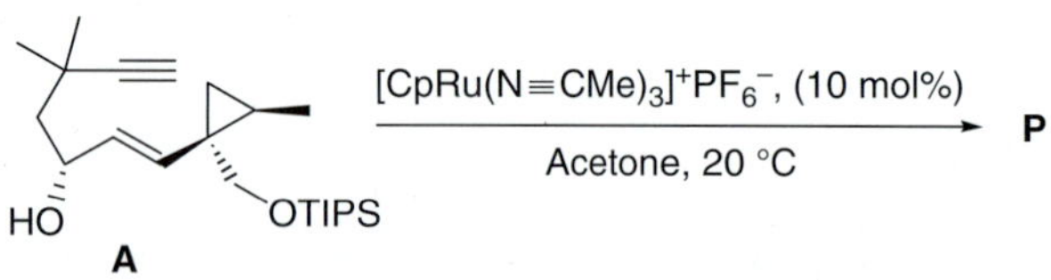

Problem 8.50 Using three transition-metal-catalyzed reactions convert **A** into **P** [1192, 1193].

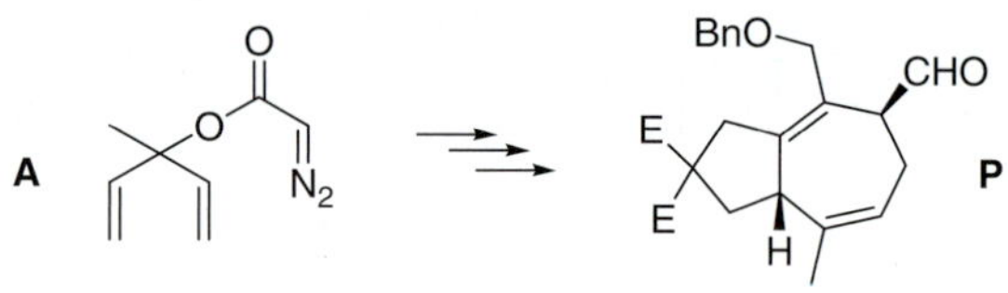

Problem 8.51 Propose a mechanism for the $[Rh(CO)_2Cl]_2$-catalyzed (5+2)-cycloaddition of vinylcyclopropane + ethyne [1194].

Problem 8.52 What is the product **P** of the following reaction?

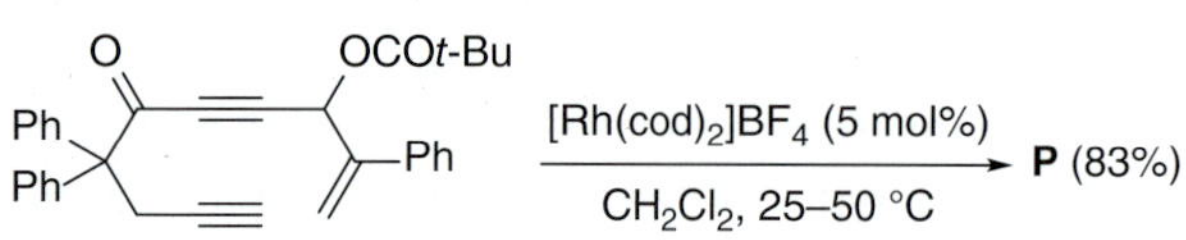

Problem 8.53 Propose a mechanism for the following (5+2)-cycloannulation [1195].

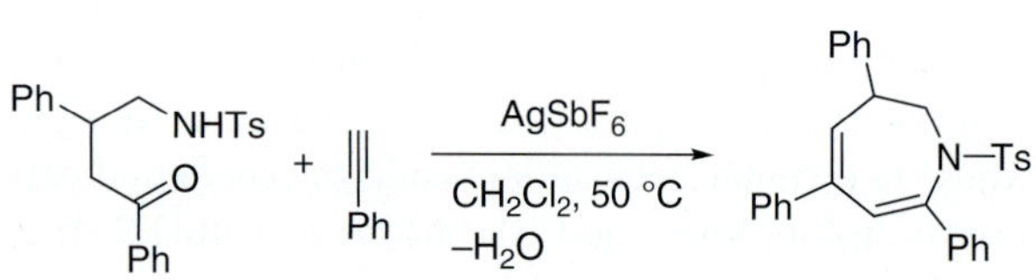

Problem 8.54 In the presence of CO, a transition-metal-catalyzed (2+2+2)-cycloadditions can become (2+2+2+1)-cycloadditions that permit the three-component construction of seven-membered ring systems. What is the main product **P** of the following reaction and propose a reaction mechanism? [1196]

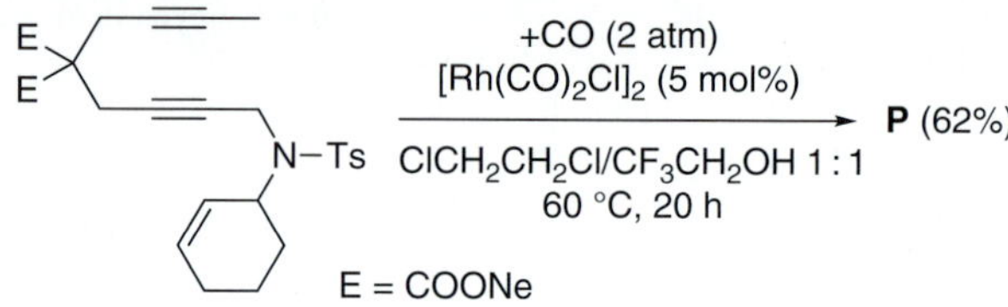

8.4.15 (4+4)-Cycloadditions

An early example of $[\pi^4+\pi^4]$-cycloaddition is the photodimerization of anthracene [1197, 1198], initially describe by Fritzsche in 1867 [1199]. Since then, a large number of photoinduced and thermal $[\pi^4+\pi^4]$-cycloadditions have been reported [1200]. The Ni(0)-catalyzed cyclodimerization of butadiene is known since 1954 [1201]. Wilke and coworkers showed that cycloocta-1,5-diene (cod) can be isolated in 96% yield using nickel in the presence of a phosphine ligand (reaction (8.187)) [1202].

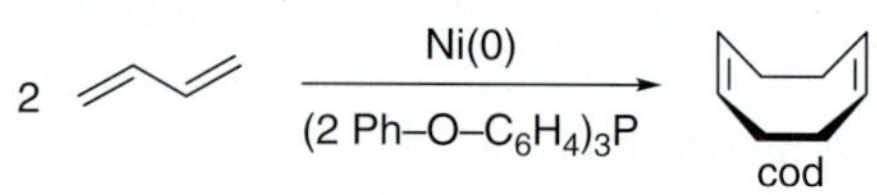

(8.187)

Selective Ni-catalyzed $[\pi^4+\pi^4]$-cyclodimerization of 1-substituted butadienes into *trans*-3,4-disubstituted cycloocta-1,5-dienes have been demonstrated [1203]. Wender et al. reported several examples of Ni(0)-catalyzed intramolecular (4+4)-cycloaddtions [1204–1207]. tom Diek and Dietrich showed that iron 1,4-diazadiene complexes also induce intermolecular $[\pi^4+\pi^4]$-cycloadditions [1208, 1209]. Intramolecular Ni-catalyzed $[\pi^4+\pi^4]$-cycloadditions using $Ni(cod)_2/Ph_2P$ have been described by Wender and coworkers (reaction (8.188)) [1210, 1211]. An intramolecular $[\pi^4+\pi^4]$-cycloaddition strategy has been proposed for the efficient synthesis of dicyclopenta[a,d]cyclooctenes [1207]. With the enantiomerically pure iron catalyst, **257** isoprene adds to (*E*)-piperilene to give (4*S*)-2,4-dimethylcycloocta-1,5-diene (reaction (8.189)) [1212].

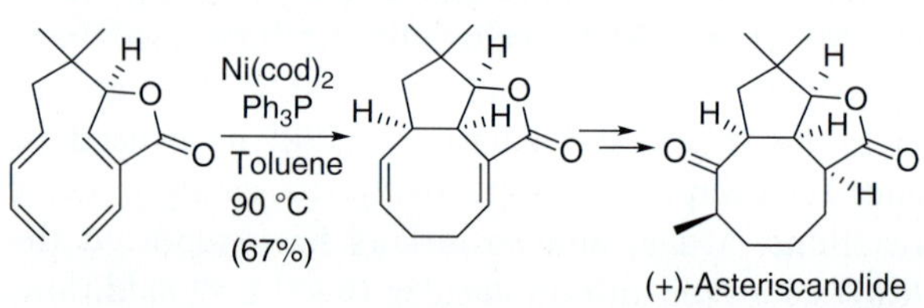

(8.188)

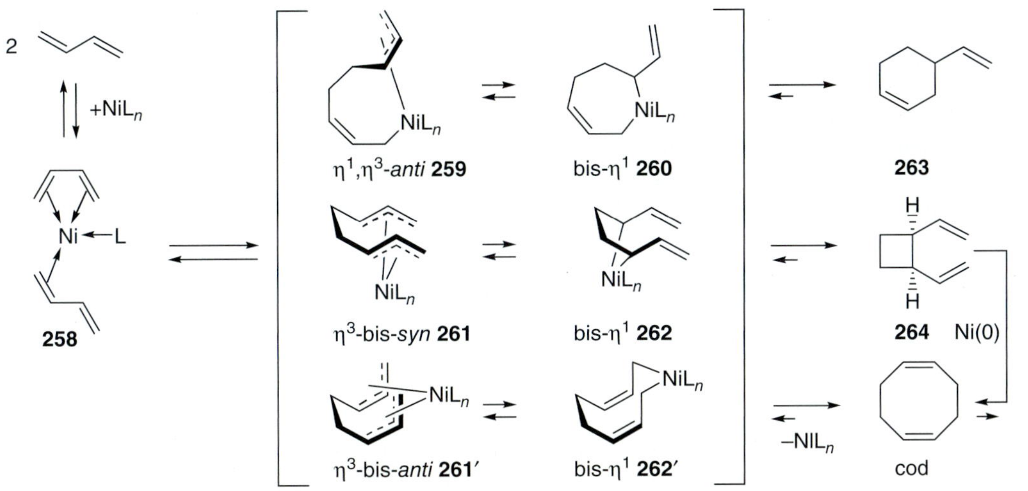

Scheme 8.67 Proposed reaction mechanism for the Ni(0)-catalyzed $[\pi^4+\pi^4]$-cyclodimerization of butadiene.

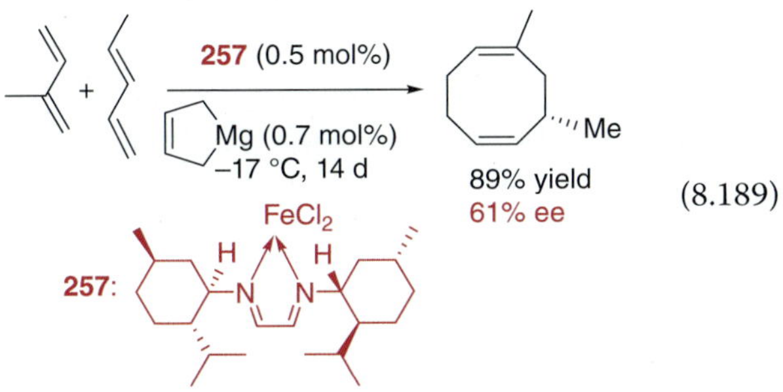

(8.189)

A possible mechanism of the $[\pi^4+\pi^4]$-cyclodimerization of butadiene is shown in Scheme 8.67. Complexation of the two dienes to nickel gives **258**. Subsequent oxidative cyclization is proposed to generate the key intermediates allyl-σ (**259**), bis-σ (**260**, **262**, and **262′**), and bis-allyl complexes (**261**, **261′**) that appear to be in dynamic equilibrium. They lead to three different cycloadducts that are 4-vinylcyclohexene (**263**), *cis*-1,2-divinylcyclobutane (**264**), and cod (cycloocta-1,5-diene). At room temperature, **264** undergoes Ni(0)-catalyzed Cope rearrangement to cod. It is possible to isolate up to 36% yield of **264** when using tris(2-biphenyl)phosphite as a ligand [858, 1213–1215].

In 1993, Huffman and Liebeskind reported the Rh(I)-catalyzed cycloisomerization of 4-cyclobutyl-2 cyclobuten-1-ones into 2,4- and 3,5-cyclooctadien-1-ones, a reaction that can be seen as an intramolecular (4+4)-cycloaddition of cyclobutane moieties (four-carbon cycloaddents) to cyclobutenones (which equilibrate formally with alkenylketenes, four-carbon cycloaddents) [1156].

Problem 8.55 In the presence of 2 equiv. of $TiCl_4$ in CH_2Cl_2, **A** is converted into "special" compound **P**. Propose a structure for **P** [1216].

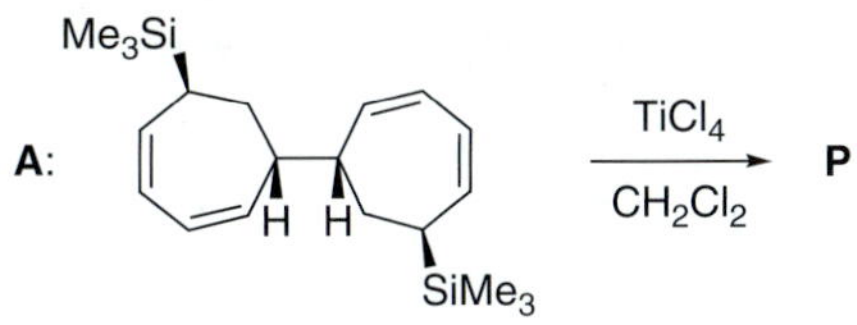

Problem 8.56 In the presence of $[P(t\text{-Bu})_2(ortho\text{-biphenyl})]AuCl/AgSbF_6$ (2.5 mol%) in 1,2-dichloroethane at 35 °C, 2-(4-chlorophenyl)oxetane reacts with *t*-butyl 3-phenylprop-2-yne-1-carboxylic ester giving a product **P** of (4+4)-annulation. What is **P** and propose a reaction mechanism [1217].

8.4.16 (4+2+2)-Cycloadditions

In 1970, Carbonara and coworkers reported the iron-catalyzed reaction of butadiene with norbornadiene (reaction (8.190)) that gives a mixture containing the $[\pi^4+\pi^2s+\pi^2s]$-cycloadduct **265** next to the $[\pi^2+\pi^2s+\pi^2s]$-cycloadduct **266** [1218]. A catalytic system based on $CoCl_2$/dppe/Et_2AlCl in toluene at 75 °C gave mostly **265** in 68% yield [1219]. The same system induces the $[\pi^4+\pi^2+\pi^2]$-cycloaddition of norbornadiene to substituted butadienes [743]. Lautens et al. presented enantioselective $[\pi^4+\pi^2+\pi^2]$-cycloadditions catalyzed by $Co(acac)_2$ or $Co(acac)_3$ in the presence of an enantiopure phosphine and a reducing agent (Et_2AlCl) [1220, 1221]. Hilt et al. studied the chemoselectivity ((2+2+2)-cyclotrimerization vs. (4+2)-cycloaddition, vs. (4+2+2)-cycloaddition) of phenyl acetylene + isoprene in the presence of

$CoBr_2$(dppe) and Mg metal. A preference for the cyclotrimerization of phenylacetylene was found for diimine cobalt complex in MeCN, whereas the Diels–Alder reaction was favored with Co(dppe) complex in CH_2Cl_2. In the presence of precatalyst **267**, Fe, Zn, and ZnI_2 in CH_2Cl_2 cycloadduct **268** were obtained in 65% yield (reaction (8.191)) [1061, 1222].

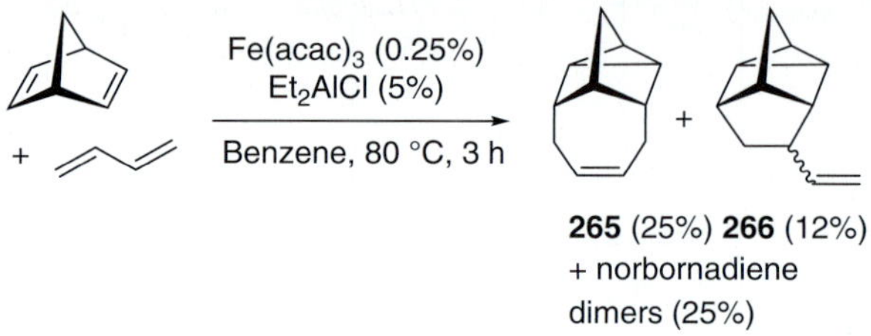

(8.190)

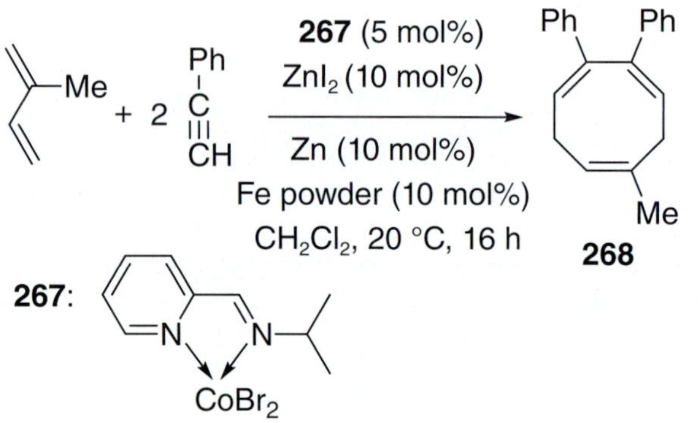

(8.191)

Several other dienes and acetylenes undergo $[\pi^4+\pi^2+\pi^2]$-cycloadditions [1222] catalyzed, for instance, by [Rh(norbornadiene)Cl]$_2$/$AgSbF_5$/phosphine [1223], and by similar complexes [1224–1226]. An enantioselective synthesis of bicyclic azocines has been realized by Rh(I)-catalyzed (4+2+2)-cycloaddition of terminal alkynes and dienyl isocyanates [1227]. In the presence of a Ni(0) catalyst, cyclobutanones react with 1,6-diynes to produce bicyclo[6.3.0]undeca-1,7-dien-3-ones. With 1,6-enynes, bicyclo[6.3.0]undeca-1-en-3-ones are obtained (e.g. reaction (8.192)) [1228]. Enantioselective synthesis of *cis*-fused bicyclo[6.3.0]undecanes bearing exocyclic *s-cis*-butadiene moieties has been realized (e.g. reaction (8.193)) through Rh(I)-catalyzed (4+2+2)-cycloadditions of deca-1,2,7,9-tetraenes and allenecarboxylic esters [1229].

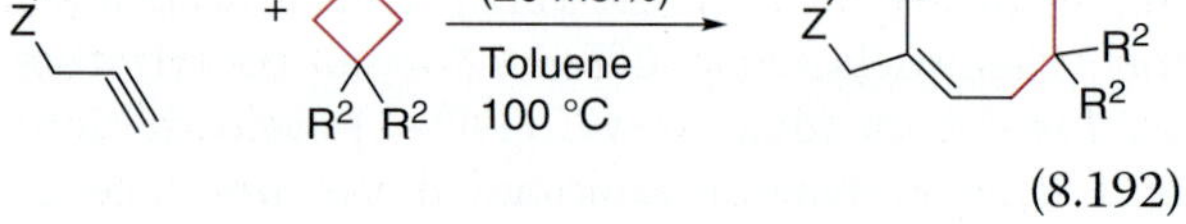

(8.192)

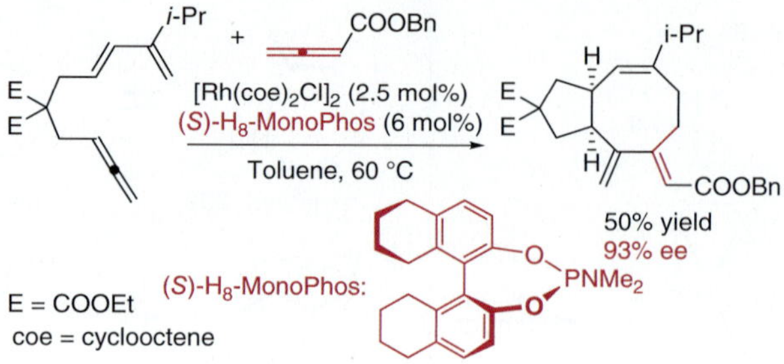

(8.193)

Problem 8.57 In the presence of $Co(acac)_2$, Et_2AlCl and an enantiomerically pure diphosphine ((*R*)-1,2-bis(diphenylphosphino)propane), norbornadiene, and isoprene undergo a (4+2+2)-cycloaddition. What is the main product of this reaction? [1221]

Problem 8.58 Isoprene and 2,2-bis(propargyl) malonic diesters undergo [Cp*Ru(MeCN)$_3$]PF_6-catalyzed (2+2+2)- and (4+2+2)-cycloadditions. What are the possible products and the mechanism of their formation? [1230]

Problem 8.59 In the presence of [Rh(cod)Cl]$_2$ (2.5 mol%) and AgOTf (5 mol%) in toluene at 20 °C, 2 equiv. of **A** add to 2,3-dimethylbutadiene giving product **P** in 98% yield. What is **P**? [1225]

A: 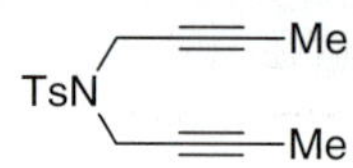

8.4.17 (6+2)-Cycloadditions

Tropone derivatives undergo intramolecular $[\pi^6s+\pi^2s]$-cycloadditions onto unactivated alkene when photolyzed [1231–1233]. In 1974, Pettit and coworkers reported the photoinduced (6+2)-cycloaddition of (tricarbonyl)cycloheptatrieneiron to dimethyl acetylenedicarboxylate producing a (tricarbonyl)diene iron complex of dimethyl bicyclo[4.2.1]nona-2,5,7-triene-7,8-dicarboxylate [1234]. In 1983, Mach et al. found the $TiCl_4$-catalyzed (6+2)-cycloadditons (8.194) [1235, 1236]. In 2005, Buono and coworkers reported the first Co(I)-catalyzed $[\pi^6s+\pi^2s]$-cycloadditions of cycloheptatriene with terminal alkynes. Enantioselective cycloadditions were observed with BINOL phosphoramidite ligand **269** (reaction (8.195)) [1237]. $PtCl_2$ catalyzes the (6+2)-cycloadditions of alkynes tethered to cycloheptatriene [1238]. The (6+2)-cycloadditions of cycloheptatriene with (internal) 1,2-diarylacetylenes have been catalyzed by [Rh(CO)$_2$Cl]$_2$/CuI/Ph_3P (xylene, 120 °C) and those of cycloocta-1,3,5,7-tetraene with allenes by CoI_2/1,2-bis(diphenylphosphino)ethane/Zn/ZnI_2 [1239].

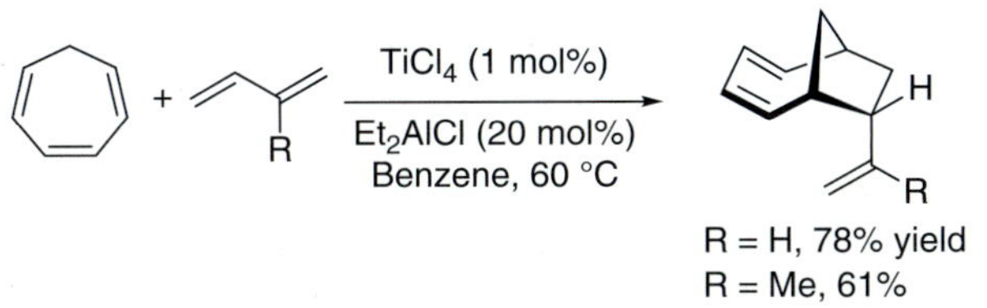

(8.194)

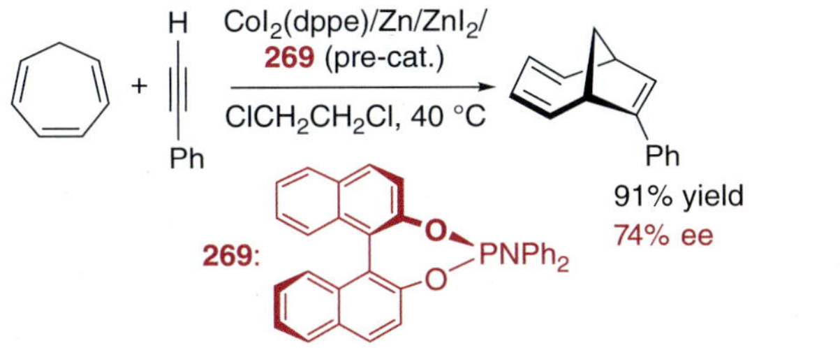

(8.195)

Cyclobutanes have a similar ring strain as cyclopropanes (Section 2.6.1) and can be used as four-carbon cycloaddents in transition-metal-catalyzed (4+2+2)-cycloadditions. This strategy has been employed by Wender et al. for the cycloisomerizations (8.196) [1240]. The proposed reaction mechanism involves first an oxidative cyclization (β-cycloinsertion) of the two alkene units into a metallacyclopentane intermediate, followed by β-elimination of the acyl group (acyl group migration to the metal) and reductive elimination [1241]. Rh(I)-catalyzed intramolecular (6+2)-cycloadditions of allenylcyclobutanes + alkynes (e.g. reaction (8.197)) have been reported by Mukai and coworkers [1242].

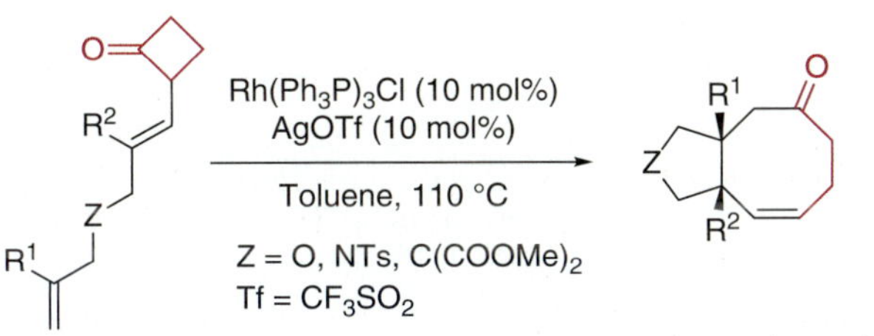

(8.196)

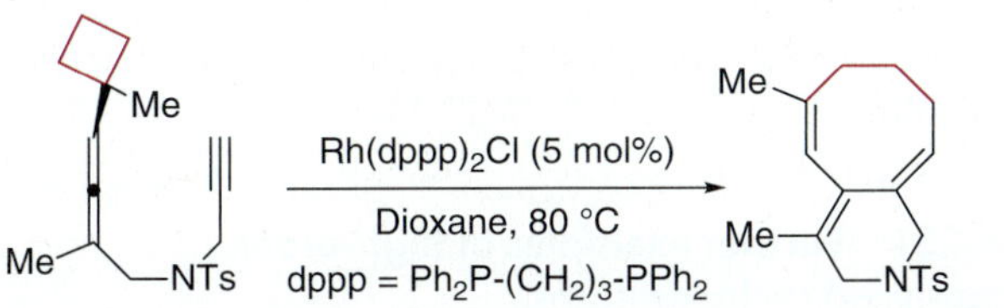

(8.197)

Problem 8.60 The Co(I)-catalyzed (6+2)-cycloaddition of cyclooctatetraene to alkynes affords bicyclo[4.2.2]deca-2,4,7,9-tetraenes **P**. Another type of cycloadducts **Q** forms concurrently. What is it? Propose a mechanism for these reactions [1243, 1244].

Problem 8.61 The Rh(I)-catalyzed cyclization **A** → **P** represents a formal intramolecular (6+2)-cycloaddition of a 4-allenal with an alkyne moiety (IMes = 1,3-bis(2,4,6-trimethylphenyl)imidazole-2-ylidene, cod = cycloocta-1,5-diene). Propose a mechanism for this reaction [1245, 1246].

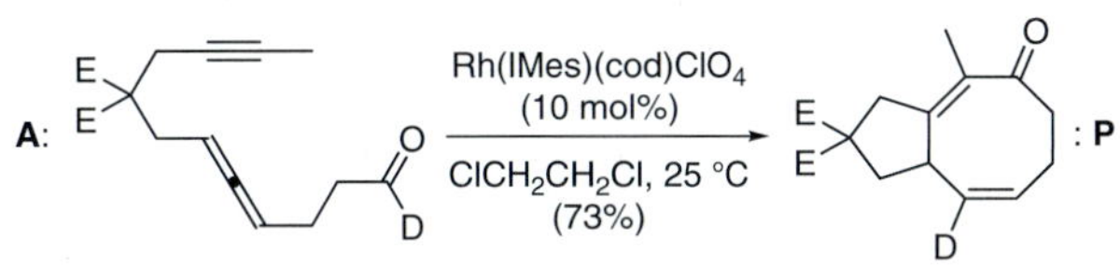

8.4.18 (2+2+2+2)-Cycloadditions

In 1911, Willstätter and Waser reported the first synthesis (in 10 steps) of cyclooctatetraene (cot) [1247]. The first one-step $[\pi^2s+\pi^2s+\pi^2s+\pi^2s]$-cyclotetramerization of acetylene into cot was discovered by Reppe and coworkers [1248, 1249]. This reaction is catalyzed by $NiBr_2/CaC_2$, $Ni(acac)_2$, or $Ni(cot)_2$. When the cycloaddition employs substituted alkynes, mixtures of various cycloadducts are obtained. tom Diek and coworkers developed a regioselective method using sterically hindered (1,2-diazadiene)nickel complexes such as **270** [1250–1255]. An example is given with reaction (8.198) for which a TON of 500 was reported.

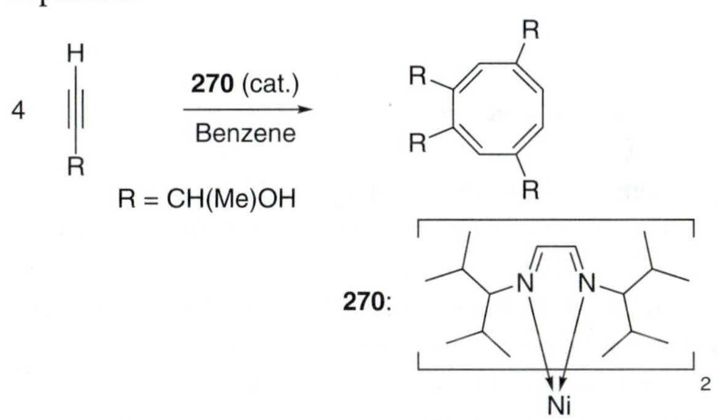

(8.198)

Based on the analysis of the ^{13}C-labeling in cyclooctatetraene obtained by Ni-catalyzed cyclotetramerization of singly ^{13}C-labeled ethyne, Colborn and Vollhardt ruled out mechanisms involving the formation of cyclobutadienes and benzenes as intermediates, as well as metathesis (Section 8.6.2) mechanisms [1256]. The ^{13}C-label distribution was consistent with a concerted or stepwise coupling mechanism of the four alkyne units. 1,3,5,7-Tetra-*tert*-butylcyclooctatetraene (**272**) has been prepared by heating under reflux in DMF the dimeric cycloadduct **271** obtained by cyclodimerization of 1,3-bis-*tert*-butyl cyclobutadiene generated by $AlBr_3$-promoted $[\pi^2s+\pi^2s]$-cycloaddition of 3,3-dimethylbut-1-yne [1257].

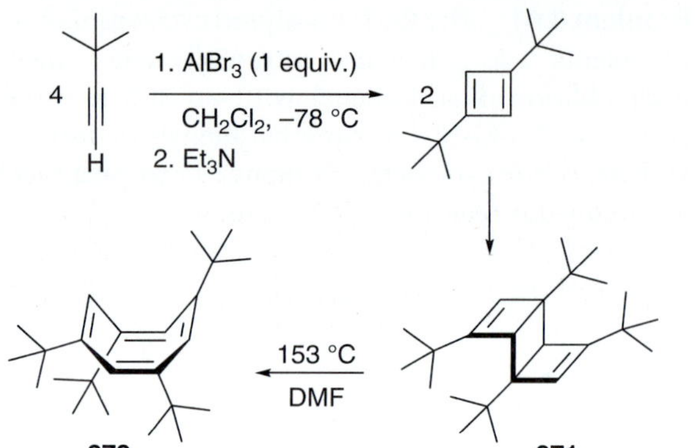

272 271

Acetylene trimerizes to benzene on the (111) face of metallic copper, as it does on the (100) and (110) planes. However, Cu(111) also yields butadiene and cyclooctatetraene (cot). Gaseous benzene is formed by a surface reaction rate-limited process, whereas butadiene and cyclooctatetraene are formed by desorption rate-limited process. A pericyclic mechanism involving dimerization of metallacyclopentadienes is proposed to account for the formation of cot [1258]. Nickel(0) [1259] and rhodium(I) complexes catalyze the cyclodimerization of hepta-1,6-diynes and octa-1,7-diynes giving symmetrical 1,2,5,6-tetrasubstituted cyclooctatetraenes [1260].

8.4.19 (5+2+1)-Cycloadditions

In 2002, Wender et al. reported the first example of transition-metal-catalyzed (5+2+1)-cycloaddition generating cyclooctenedione intermediates (e.g. reaction (8.199)) that undergo facile intramolecular aldol reactions giving the corresponding 1-hydroxybicyclo[3.3.0]oct-7-en-4-ones [1261]. An intramolecular version (e.g. reaction (8.200)) that condenses 1-cyclopropyl-1,7-heptadienes with CO has been reported by Yu and coworkers [1262]. The reaction has been applied to the synthesis of hirsutenes [1263] and (+)-asteriscanolide [1264, 1265].

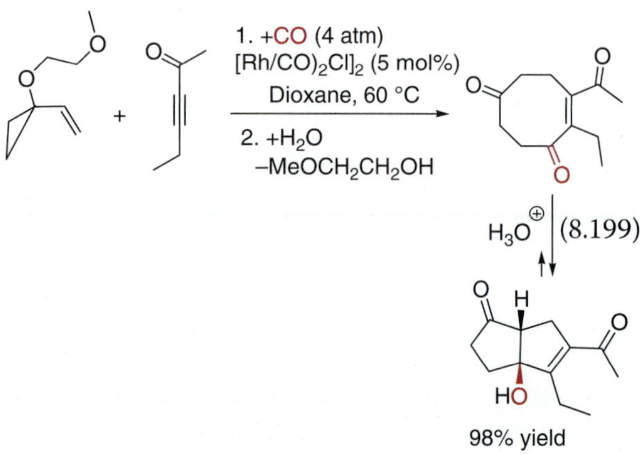

(8.199)

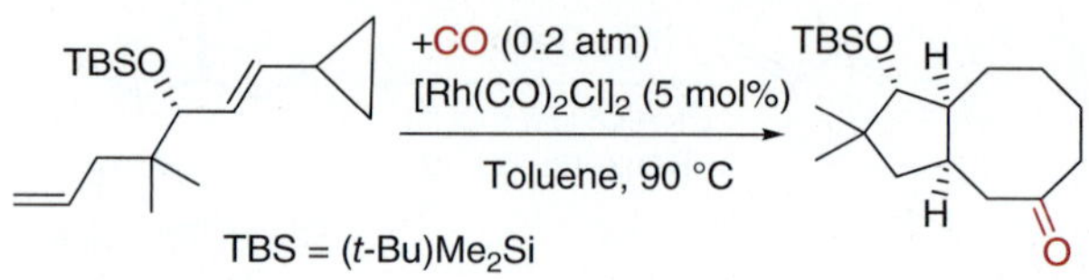

(8.200)

8.4.20 (7+1)-Cycloadditions

1-Cyclopropyl-1,3-dienes can react as seven-carbon cycloaddents. In 2011, Yu and coworkers described the cyclocarbonylation (8.201) that realizes an example of transition-metal-catalyzed (7+1)-cycloaddition [1266]. Benzocyclobutenes equilibrate with the corresponding *o*-xylylenes (Section 1.12.5). Thus, alkenylbenzocyclobutenes also realize formally seven-carbon cycloaddents and can react with CO in formal (7+1)-cycloadditions as shown by the Yu group (e.g. reaction (8.202)) [1267]. In 2015, Bower and coworkers reported an efficient synthesis of functionalized azocanes that features a Rh(I)-catalyzed (7+1)-heterocycloaddition (e.g. reaction (8.203)).

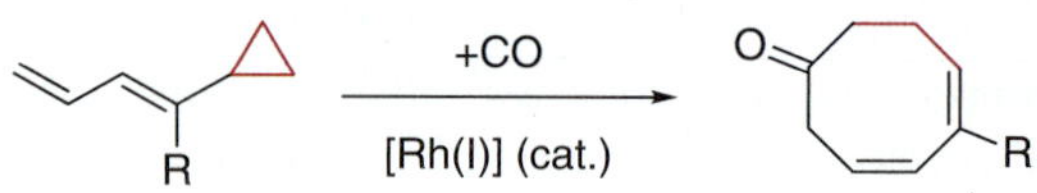

(8.201)

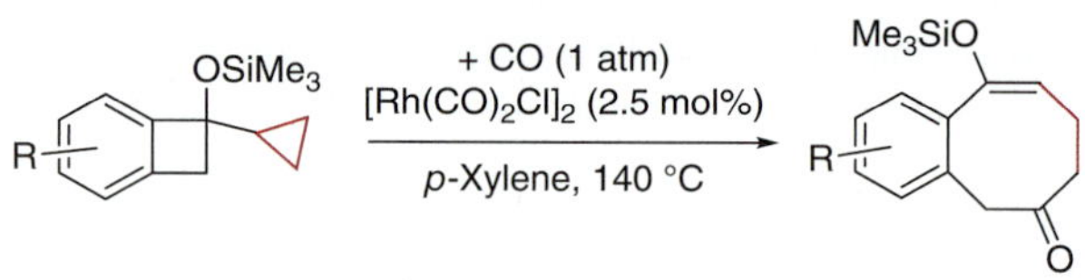

(8.202)

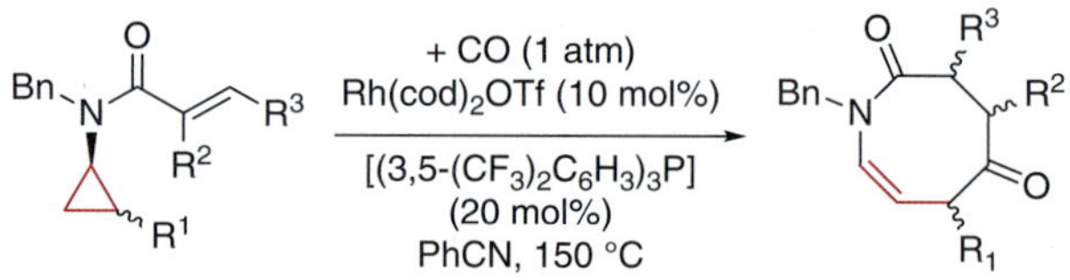

(8.203)

Problem 8.62 Propose a mechanism for reaction (8.203).

8.4.21 Further examples of high-order catalyzed cycloadditions

In 2005, Wender et al. reported a four-component synthesis of hydroxyindanones through Rh(I)-catalyzed (5+1+2+1)-cycloaddition that generates 1,6-cyclononadien-1,4-dione intermediates (reaction (8.204)) [1268]. In 2010, Saito and coworkers presented an efficient synthesis of nine-membered carbocycles using the Ni(0)-catalyzed (4+3+2)-

cycloaddition of ethyl cyclopropylideneacetate and dienynes (reaction (8.205)) [1269–1271]. In 2012, Mukai et al. reported the intramolecular Rh(I)-catalyzed isomerization of 1-methyl-1-(oct-1,2-dien-7-yn-1-yl)cyclopentane into 7-methyl-1,7,9-bicyclo [7.4.0]tricatrienes (reaction (8.206)) [1272]. The reaction involves a metallacyclopentene intermediate that undergoes an intramolecular C(sp^3)—C(sp^3)-bond oxidative addition (in fact a β-C-elimination), thus realizing a formal (7+2)-cycloaddition [1273]. The Rh(II)-catalyzed denitrogenative coupling of *N*-alkylsulfonyl 1,2,3-triazoles with 1,3,5-trioxane generates trioxazonines (reaction (8.207)) [1274].

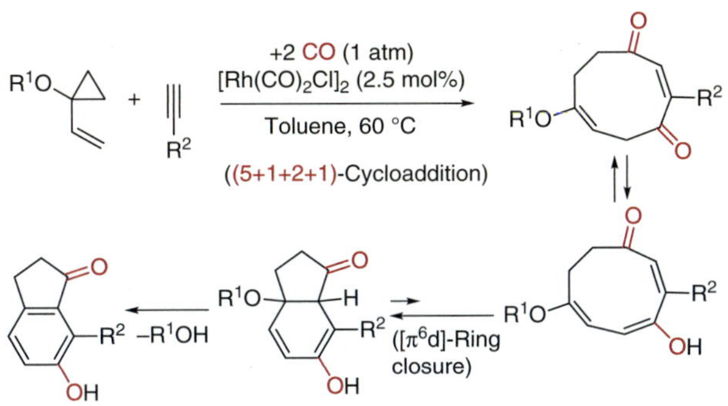

(8.204)

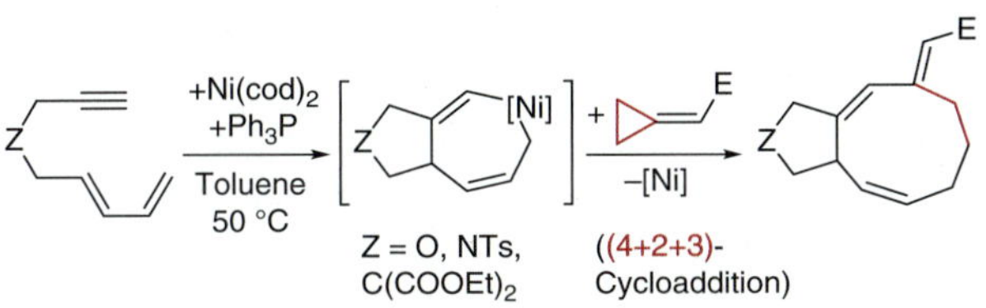

(8.205)

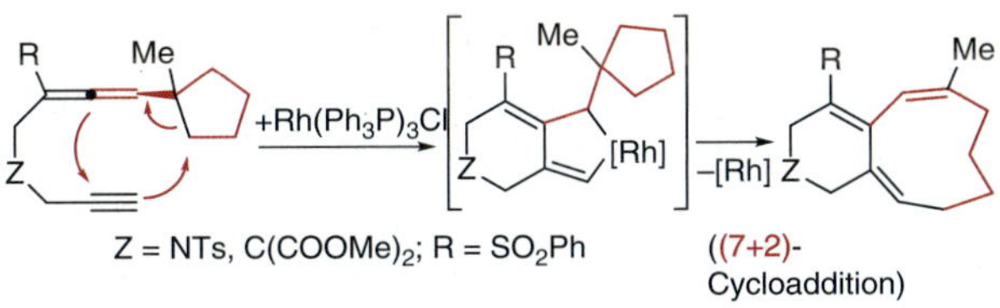

(8.206)

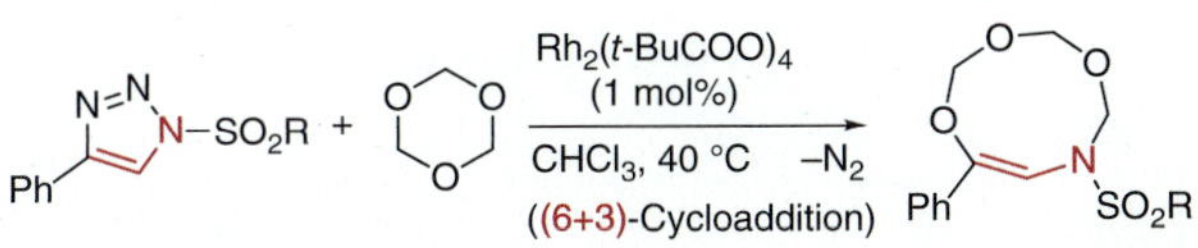

(8.207)

Thermal [π^6s+π^4s]-cycloadditions are allowed processes according to the Woodward–Hoffman rules. Tropone and cyclopentadiene adds at 80 °C giving a single (6+4)-cycloadduct in quantitative yield [1275]. In the case of reaction (8.208), the cycloaddition has been catalyzed by $ZnCl_2$ [1276]. Mach et al. observed that a (6+4)-cycloadduct predominated when 2,3-dimethylbutadiene was reacted with cycloheptatriene in the presence of $TiCl_4$ and a reducing agent (Et_2AlCl) [1235, 1236]. $FeCl_3$ catalyzes the intramolecular [π^6s+π^4s]-cycloaddition (8.209). With Ti(O-*i*-Pr)$_2$Cl$_2$ and (*S*)-BINOL as catalysts, the cycloadduct was obtained in 80% yield and 40% ee [1277].

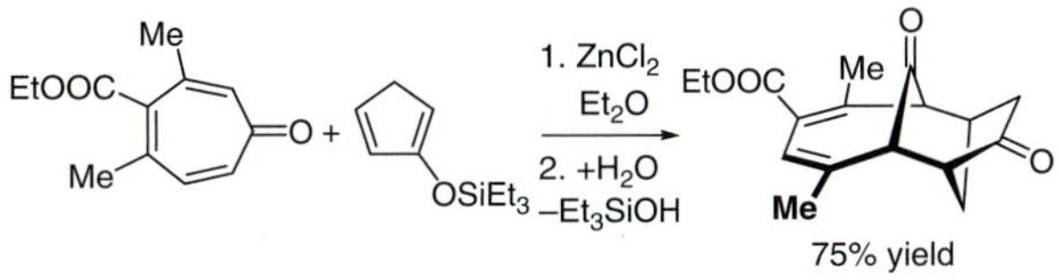

(8.208)

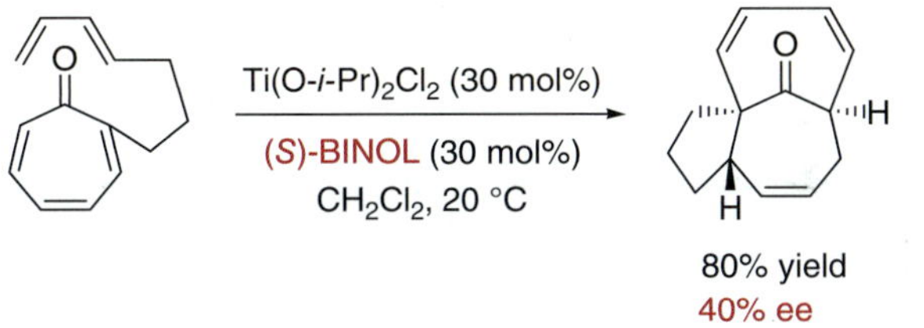

(8.209)

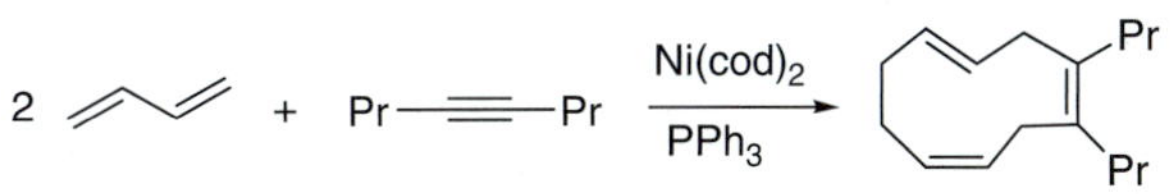

(8.210)

Baran and coworkers proposed a Ni(0)-catalyzed butadiene + butadiene + alkyne cooligomerization that would mimic Nature's cyclase in producing terpenes. With Ni(cod)$_2$/PPh$_3$ catalyst, they found that butadiene and 4-octyne gives (1*Z*,4*E*,8*E*)-1,3-dipropylcyclodecatriene with high stereoselectivity (reaction (8.210)) [1278]. According to quantum mechanical calculations, and for Ni(Me$_3$P) catalyst, the reaction is found to combine first one *s-cis*-butadiene and one *s-trans*-butadiene unit into complex **273** equilibrating with a nickellanonadiene **274**. Then, the latter loses the phosphine ligand and inserts the alkyne in the rate-determining step, affording a nickellaundecatriene **275** that undergoes reductive elimination producing (1*Z*,4*E*,8*E*)-1,3-dipropylcyclodecatriene (Scheme 8.68). Competition of the (2+2+2)-cyclotrimerization of the alkyne is the highest for electron-poor alkynes and those bearing hydroxyalkyl subtituents [1279].

In 1956, Wilke reported the cyclotrimerization (8.211) of butadiene giving (*Z,E,E*)-, (*E,E,E*)-, and (*Z,Z,E*)-cyclododeca-1,5,9-triene (**276**) [1280]. It

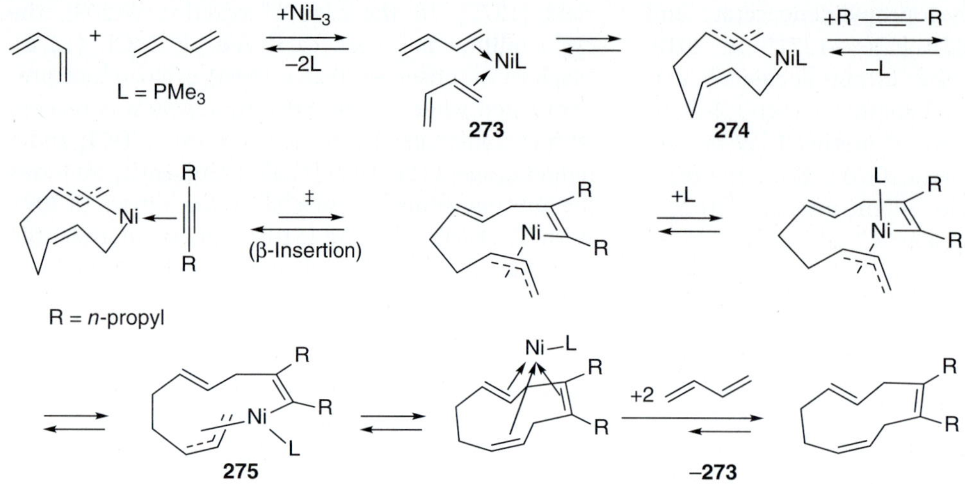

Scheme 8.68 Possible mechanism of Baran's Ni(0)-catalyzed (4+4+2)-cycloaddition of butadiene and oct-4-yne.

is catalyzed by Ziegler–Natta catalysts such as $Et_3Al_2Cl_3/TiCl_4$ [1281, 1282].

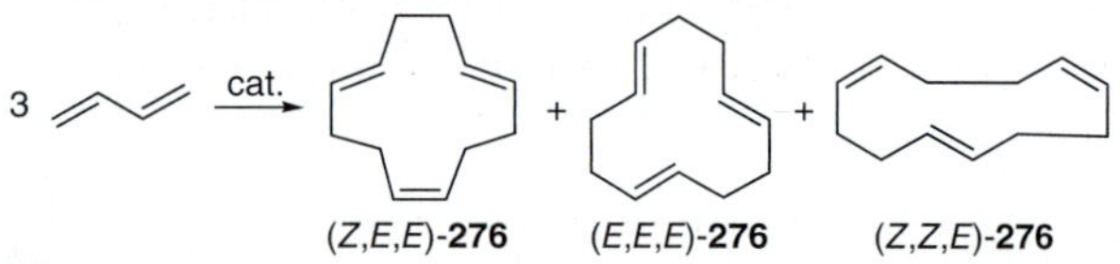

(8.211)

Products **276** are the industrial precursors of nylon-12 and 1,12-dodecanediesters [1283–1285]. Ratio of the three isomeric cyclotrimers **276** depends on the transition metal, type of ligands, and reaction parameters [1286]. Highly selective formation of (*Z,E,E*)-**276** is induced by various titanium catalysts such as (benzene)$Ti(AlCl_4)_2$ [1287–1289]. Proposed reaction mechanism of the $Et_2AlCl/TiCl_4$-catalyzed cyclotrimerization of butadiene is shown in Scheme 8.69.

In 1963, Wilke and coworkers reported that the reduction of Ni(II) salts in the presence of butadiene produces a catalyst that induces the cyclotrimerization of butadiene [1290, 1291]. Ni(0) complexes anchored on divinylphosphine-substituted styrene-divinylbenzene polymers catalyze predominantly the cyclodimerization of butadiene into 4-vinylcyclohexene and cycloocta-1,5-diene (cod). Homogeneous Ni(0) compounds coordinated to α,α'-dipyridyl [1292] or Schiff bases [1293], as well as Ni(0) anchored on linear polyvinylpyridines, also induce the cyclotrimerization of butadiene [1294]. The condensation reaction of Ni vapor and 6,6-dimethylfulvene produces a mixture of 4,4,8,8-tetramethyl-1,4,5,8- and 1,4,7,8-tetrahydro-*syn*-indacenes [1295].

Problem 8.63 What is the main product of reaction of tropone and cyclohexa-1,3-diene in the presence of $FeCl_3$ (25 mol%)? [1277]

Problem 8.64 Werner and coworkers have found that cationic $[Rh(I)(\eta^4\text{-butadiene})_2(PR_3)]^+X^-$ ($X = CF_3SO_3, (C_6F_5)_4B$; R = *i*-Pr. cyclohexyl) catalyzes the cyclotetramerization of butadiene giving predominantly (1*Z*,5*Z*,9*E*,13*E*)-cyclohexadecatetraene [1296]. Propose a mechanism for this reaction [1297].

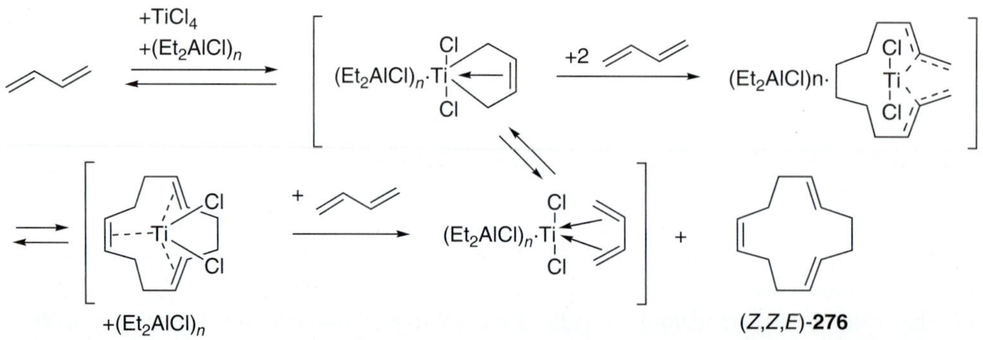

Scheme 8.69 Proposed reaction mechanism of Wilke's cyclotrimerization ($[\pi^4+\pi^4+\pi^4]$-cycloaddition) of butadiene.

8.4.22 Annulations through catalytic intramolecular hydrometallation

Hydroalkylation (Section 8.3.1), hydroalkenylation (Section 8.3.6), and hydroarylation (Section 8.3.3) of alkenes, alkynes (Section 8.3.4), or carbon-heteroatom multiple bonds (Section 8.3.5), as well as hydroacylation of alkenes and alkynes (Section 8.3.6), can be diverted into annulations if the hydrocarbation reagent (R—H) contains a multiple bond (Scheme 8.70).

An example of annulation reaction outlined in Scheme 8.70d is given with the enantioselective (4+2)-carbocyclization (8.212) reported by Tanaka et al. [1298]. An example of an intramolecular version of the process schematized in Scheme 8.70e is given with reaction (8.213) developed by Sato and coworkers. The reaction realizes a formal (6+2)-cycloaddition [1245]. An intermolecular version of this reaction using terminal alkynes has also been reported [1246].

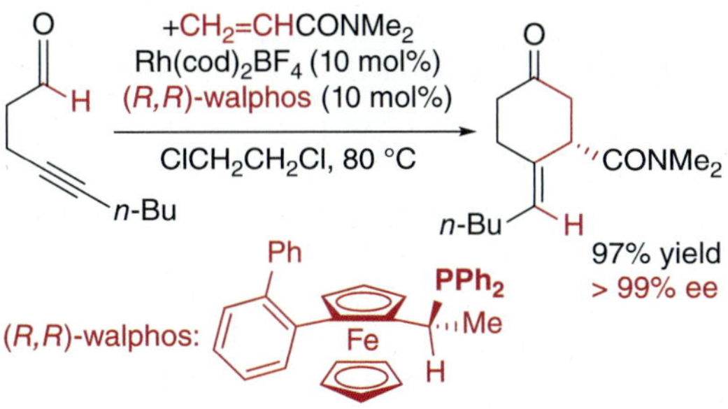

(8.212)

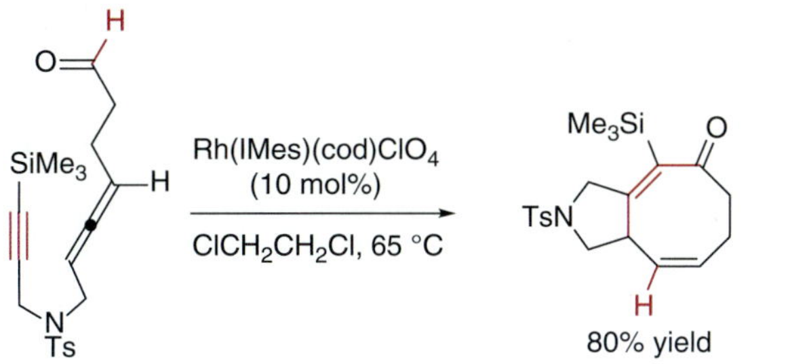

(8.213)

Problem 8.65 Propose a reaction mechanism for the Rh-catalyzed isomerization (8.213)

8.4.23 Oxidative annulations

The oxidative annulations (8.214)–(8.216) are highly exothermic. Reactions (8.214) and (8.215) represent (2+1)- and (4+2)-carbo-annulations, respectively, and reaction (8.216) represents a (3+2)-heterocarbo-annulations.

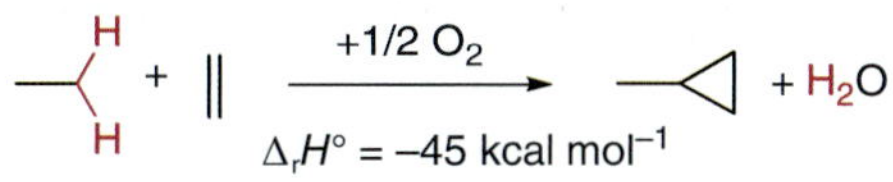

(8.214)

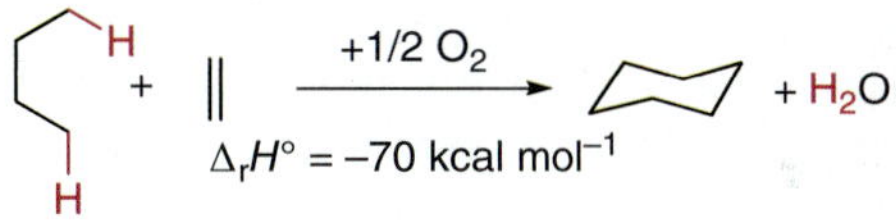

(8.215)

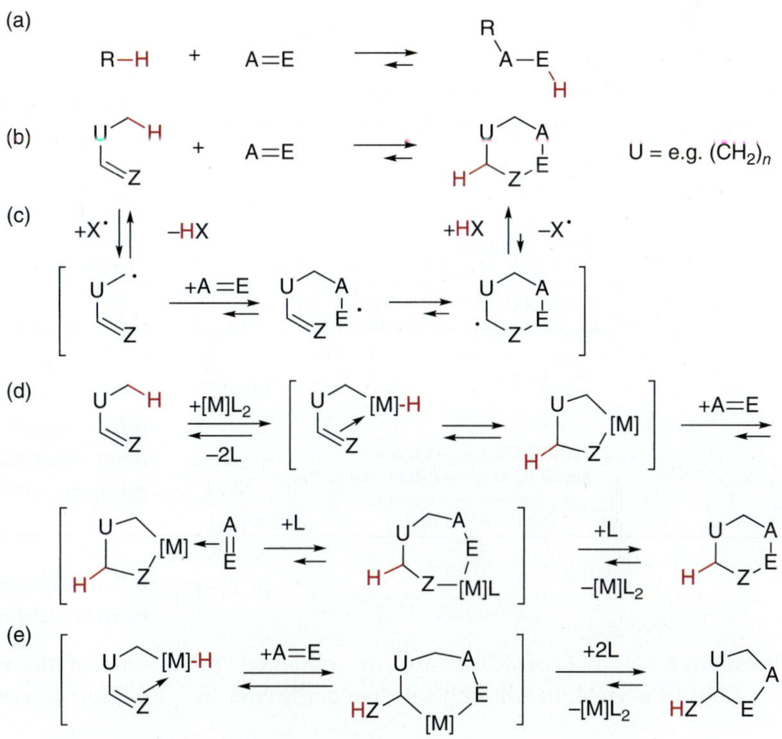

Scheme 8.70 (a) General hydrocarbation reaction; (b) hydrocarbation diverted into an annulation reaction; (c) possible radical-catalyzed annulation; (d) transition-metal-catalyzed annulation through intramolecular hydrometallation; and (e) alternative mode of annulation through intramolecular hydrometallation (A=E can a double or triply bonded cycloaddent, A, and Z are carbon or heteroatomic moieties).

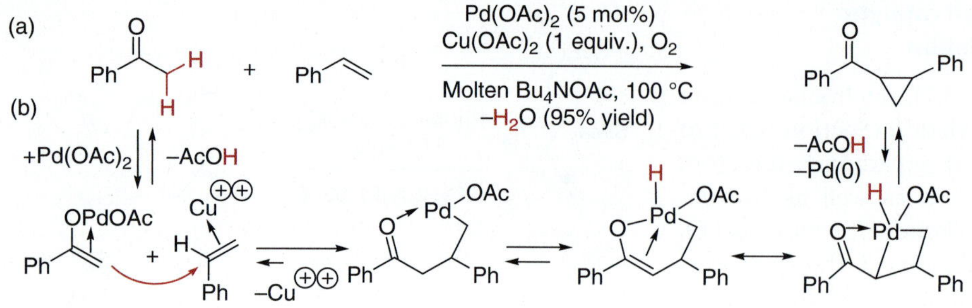

$Pd(0) + 2Cu(OAc)_2 \longrightarrow Pd(OAc)_2 + 2Cu(OAc)$

$2Cu(OAc) + 1/2O_2 + 2AcOH \longrightarrow 2Cu(OAc)_2 + H_2O$

Scheme 8.71 (a) Cotugno's oxidative (2+1)-carbo-annulation and (b) hypothetical reaction mechanism.

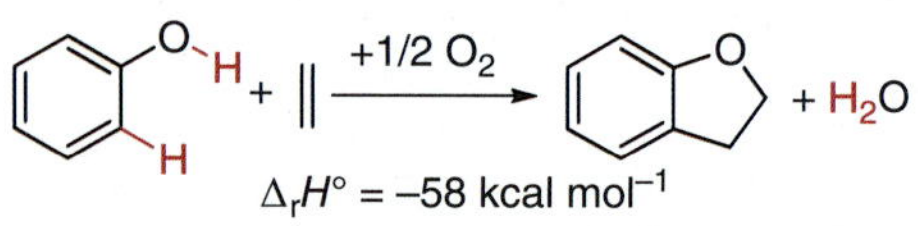

(8.216)

In the presence of a suitable oxidant (e.g. O_2 + Cu(II) salts as catalysts, $(BzO)_2$), transition metals able to catalyze hydrocarbation may find a way to do it twice and help to "burn" the two hydrogen atoms that must be eliminated in an oxidative annulation process. An example of oxidative (2+1)-carbo-annulation reported by Cotugno et al. is shown in Scheme 8.71 [1299].

The indole synthesis (reaction (8.217)) pioneered in 2008 by Fagnou and coworkers is an example of oxidative (3+2)-heterocarbo-annulation that condenses anilides with alkynes [1300]. The method can be applied to the synthesis of pyrroles [1301, 1302]. In 2007, Uera et al. presented a (4+2)-heterocarbo-annulation in which Rh(III) catalyzes the oxidative coupling of benzoic acid with internal alkynes into the corresponding isocoumarins (reaction (8.218)) [1303, 1304]. In the meantime, all kinds of (m+n)-carbo- and heterocarbo-annulations have been developed and they represent useful tools for organic synthesis [1305].

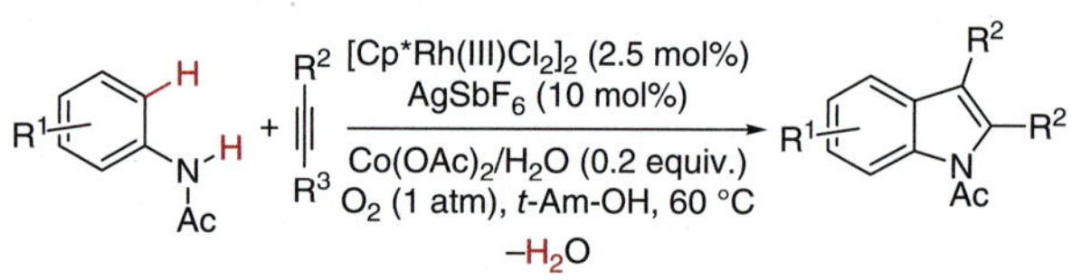

(8.217)

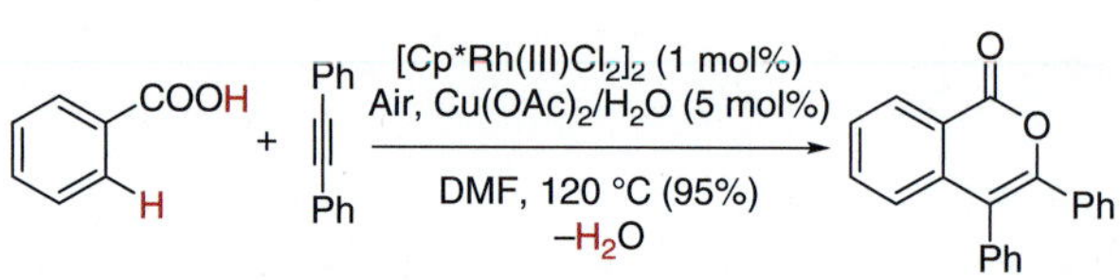

(8.218)

Problem 8.66 Benzo[b]oxepines are obtained by reaction of *o*-vinylphenol with internal acetylenes in the presence of $[Cp^*RhCl_2]_2$ (2.5 mol%), $Cu(OAc)_2/H_2O$ (50 mol%), and air in MeCN at 55 °C [1306]. Propose a mechanism for this oxidative (5+2)-heterocarba-annulation.

8.5 Didehydrogenative C—C-coupling reactions

Ideally, one wishes to eliminate 1 equiv. of H_2 from two different organic compounds with high chemoselectivity, regioselectivity, and stereoselectivity and high tolerance for functional groups. As these reactions are endothermic in most cases (e.g. $\Delta_r H°$ ($2\ CH_4 \rightleftarrows Me{-}Me + H_2$) = +15.4 kcal mol^{-1}; $\Delta_r H°$($PhH + CH_3{-}CH_3 \rightleftarrows Ph{-}Et + H_2$) = +7.7 kcal mol^{-1}), an oxidant is required to convert H_2 into H_2O, H_2O_2, or another product of oxidation that brings the necessary driving force. This route is being developed a lot as it is environment friendly, especially if O_2 or air is the oxidant [20, 1307–1319], and an iron salt as the catatyst [1320, 1321] (e.g. $Ar^1(Ar^2)CH_2 + Ar^3{-}H$ + 2,3-dichloro-5,6-dicyanobenzoquinone → $Ar^1(Ar^2)CH{-}Ar^3$ + 2,3-dichloro-5,6-dicyanohydroquinone, catalyzed by $FeCl_2$, in $ClCH_2CH_2Cl$, 100 °C [1322]). Furthermore, aerobic didehydrogenative coupling reaction $R^1{-}H + R^2{-}H + ½O_2 \rightarrow R^1{-}R^2 + H_2O$ is the second best atom-economical process to generate C—C bonds [17–22]. In the presence of suitable metal oxide catalysts and O_2, methane has been converted into ethane at 850 °C [1323–1326]. CO_2 has also been used as an oxidant in the direct conversion of methane into ethane and ethylene [1327].

8.5.1 Glaser–Hay reaction: oxidative alkyne homocoupling

One of the oldest known oxidative coupling is the Glaser reaction (8.219) in which a terminal alkyne reacts with basic $(CuCl)_2$ to give a 1,3-diyne upon air oxidation [1328, 1329].

$$2\,R\text{–}C{\equiv}CH + \tfrac{1}{2}\,O_2 \xrightarrow[NH_4OH]{(CuCl)_2} R\text{–}C{\equiv}C\text{–}C{\equiv}C\text{–}R + H_2O \tag{8.219}$$

1,3-Diyne derivatives are important materials in biological, polymer, and material sciences. Cu(II) salts in stoichiometric amounts (Glaser's conditions [1328, 1329],) or as catalysts in the presence of O_2 and an appropriate nitrogen base (Hay's conditions [1330, 1331]) have been used to promote oxidative alkyne homocouplings [1332–1338]. Base-free aerobic homocouplings of alkynes have been catalyzed by oxidized gold nanoparticles [1339]. Conditions for "green" oxidative alkyne homocoupling using CuCl as the catalyst in DMSO, without Pd, ligand, and base, have been reported [1340, 1341]. Oxidative alkyne homocoupling often competes with the Sonagashira reaction (R—C≡CH + Ar—I + Et_3NH → R—C≡C—Ar + Et_3NH^+/I^-, Section 8.8.8) catalyzed by Pd complexes, with or without Cu(I) salt as the cocatalyst [1342]. In 1964, Bohlmann et al. proposed that the reaction proceeds via the formation of alkynyldicopper(II) intermediates of type $[Cu_2(\mu\text{-}C{\equiv}CR)_2]$, which react further to give the 1,3-diyne products directly [1343]. Mizuno and coworkers have found that the dicopper silicotungstate **277** is an effective homogeneous catalyst for the oxidative homocoupling of various kinds of alkynes [1344]. Catalyst **277** (0.2 mol%) can easily be recovered after the reaction and reused with conservation of its catalytic performance (TON = 468, 2PhC≡CH + O_2 → PhC≡C—C≡CPh + H_2O, in PhCN, 100 °C, 1 atm O_2, yield 92%). A possible reaction mechanism is shown in Scheme 8.72.

8.5.2 Oxidative C—C cross-coupling reactions

In 1967, Moritani and Fujiwara reported that $PdCl_2$ promotes the formation of stilbene from styrene and benzene [1345] (oxidative alkenylation of aryl C—H) [1346]. Since then, several related Pd(II)-mediated C—H functionalizations have been developed, including reactions using air or O_2 as an oxidant [1347–1357], as well as Ru- [1358, 1359] and Rh(III)-catalyzed alkenylations of aryl(heteroaryl) C—H moieties [1360]. For instance, PhOMe + (*E*)-PhCH=CHCOOEt + ½O_2 → $MeOC_6H_4$—C(Ph)=CHCOOEt + H_2O is catalyzed by $Pd(OAc)_2$ and PhCOOH at 90 °C. Similarly, the oxidative coupling of arenes with α,β-unsaturated aldehydes has been catalyzed with $Pd(OAc)_2/H_4PMo_{11}V_1O_{40}$ (molybdovanadophosphoric acid) using O_2 (1 atm) as the oxidant, giving the corresponding (*E*)-β-arylenals [1361, 1362]. On oxidative arylation of alkenes with pyridines, *N*-oxides occur exclusively at C(2) of the heterocycle. The reaction is catalyzed by $Pd(OAc)_2/K_2CO_3$ or pyridine and uses Ag_2CO_3 as the oxidant [1363]. $Pd(OAc)_2$ catalyzes the regioselective β-arylation of enamides with simple arenes ($Cu(OAc)_2$ as oxidant) [1364]. Intramolecular Fujiwara–Moritani arylations (or oxidative Mizoroki–Heck reactions, see Section 8.8.7) of nonactivated alkenes have been reported [1365] as well as the enyne construction from alkynes via oxidative Pd(II)-catalyzed coupling [1366]. A possible mechanism implies reactions of Pd(0) with O_2. This reaction has been described first by Wilke et al. in 1967 with the preparation of $(Ph_3P)_2Pd(\eta^2\text{-}O_2)$ from $(Ph_3P)_4Pd$. Solutions of η^2-peroxo complexes form $Ph_3P{=}O$ over time [1367]. Aerobic (O_2) oxidative C—C cross-coupling reactions between benzofuran (C(2)—H), indole (C(2)—H), and benzene derivatives have been catalyzed by

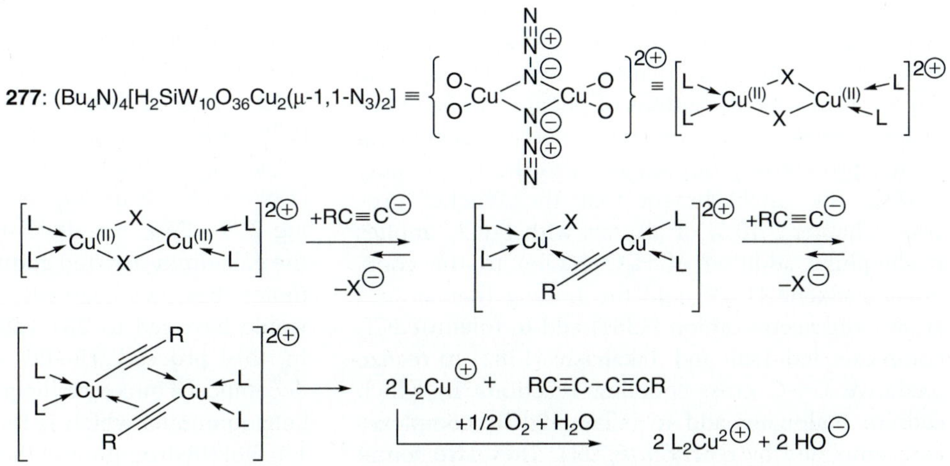

Scheme 8.72 Bohlmann's reaction mechanism for the Cu^{2+}-catalyzed oxidative homocoupling of alkynes (L = amine, WO_6; X = N_3).

$Pd(OAc)_2 + H_3PMo_{11}VO_{40}$ (10 mol%) and $Pd(OAc)_2$ (20 mol%) + CuOAc (1 equiv.) [1368]. Pd-catalyzed oxidative carbocyclizations are useful didehydrogenative intramolecular C—C cross-coupling reactions. Many of them use stoichiometric oxidants such as *t*-BuOOH, H_2O_2, benzoquinone, MnO_2, $Cu(OAc)_2$, $CuBr_2$, $Cu(OTf)_2$, $AgNO_3$, AgF, $AgOOCCF_3$, or $PhI(OAc)_2$. Others use air or pure O_2 with a Pd complex as the catalyst, with or without copper salts as cocatalysts [1369, 1370]. An example reported in 1998 by Toyota et al. is shown with reaction (8.220) [1371]. It is based on an earlier Pd-catalyzed aerobic carbocyclization reaction discovered by Ito et al. [1372]. These reactions involve (alkene)Pd(II)X_2 complex intermediates that are electrophilic and add a nucleophile (Nu-E) such as an arene (Fujiwara–Moritani), an enoxysilane (Ito-Toyota), or water (Wacker process, Section 7.11.5). Then, the corresponding adducts, (alkyl)Pd(II) intermediates, undergo β-elimination with the formation of Pd(0) + HX + EX. Pd(0) can be reoxidized into Pd(II) salts by an oxidant, for instance, by CuX_2 + ½O_2 or by benzoquinone and $AgNO_3$.

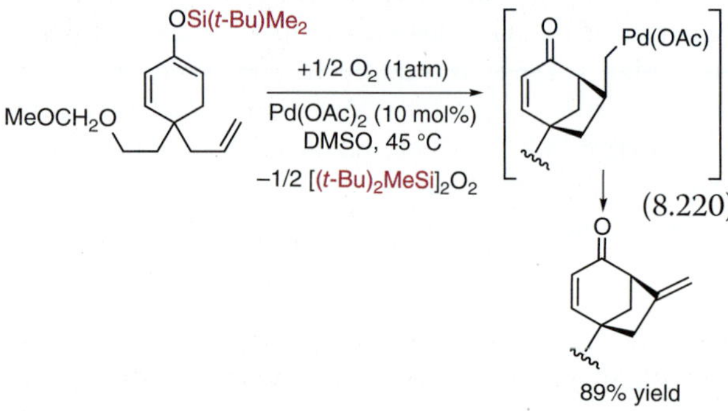

(8.220)

The reaction can be intermolecular. In the presence of a cationic Pd(II) complex and 2% of PTS (polyoxyethanyl α-tocopherol sebacate, a nonionic surfactant introduced by Lipschutz et al. and that helps to solubilize organic compounds in water, micelle formation), the oxidative couplings have been realized in water [1373, 1374]. A lateral control [1375] by the *ortho*-acetamido group is responsible of the observed regioselectivity [1376]. The Fujiwara–Moritani reactions of phenol can be either *ortho*-, *meta*- [1377], or *para*-selective depending on the protective group attached to the OH moiety [1378]. The establishment that the Wacker process (Chapter 7.10.5) of alkenes with $PdCl_2$ implies nucleophilic addition of H_2O (^-OH) to the coordinated alkene [1379] and the finding that amines [1380] and acetate anion [1381] add to (olefin)$PdCl_2$ complexes led Tsuji and Takahashi [1382] to realize oxidative C—C cross-coupling reactions in which sodium malonates add to (alkene)$PdCl_2$ complexes as exemplified by reaction (8.221). They also found that $(allyl)_2Pd_2Cl_2$ adds diethyl sodiomalonate (reaction (8.222)) [1383]. In 1973, Trost and Dietsch reported the first asymmetric version of the oxidative allylation of sodiomalonates (reaction (8.223)) in which the asymmetric induction is realized in the presence of enantiomerically pure ligands such as (+)-*o*-anisylcyclohexylmethylphosphine and (−)-sparteine (up to 24% ee) [1384].

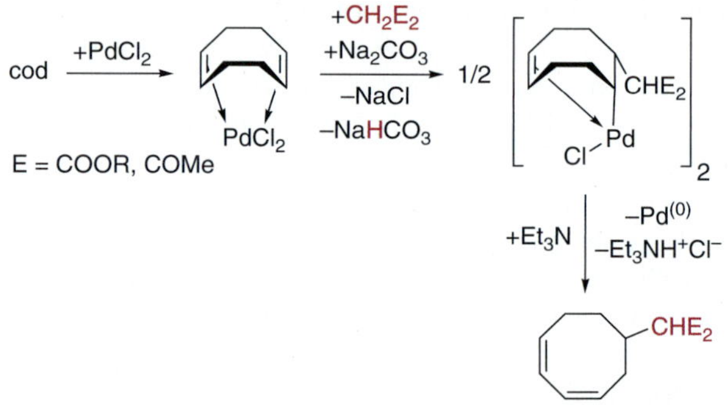

(8.221)

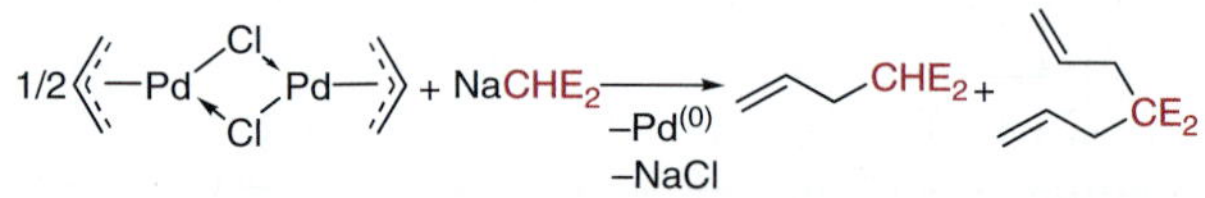

(8.222)

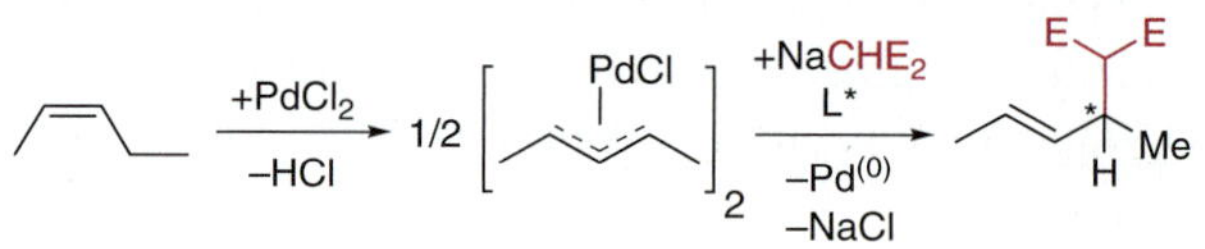

(8.223)

All reactions (8.221)–(8.223) are stoichiometric in $PdCl_2$. They can be made catalytical in PdX_2 if a stoichiometric oxidant is used such as benzoquinone (see Scheme 7.90, Section 7.11.2) [1385]. The quinone-based palladium-catalyzed allylic acetoxylation stands out as a practical process for the preparation of allylic alcohol derivatives [1386–1389]. Bäckvall and coworkers found that the oxidative acetylation of 1,2-dideuterocyclohexene with $Pd(OAc)_2$ (cat.) and benzoquinone gives a 1 : 1 mixture of 1,2-dideutero- (**280**) and 2,3-dideutero-1-acetoxycyclohex-2-ene (**279**), which can be explained by the formation of (η^3-allyl)palladium complex **278** as an intermediate [1390, 1391]. It undergoes nucleophilic addition giving **279** + **280**. An alternative mechanism involving the palladium-assisted addition of AcOH to the olefin moiety has thus been ruled out (Scheme 8.73a) as it would have led to **281** + **280** and not to **279** + **280**. In this process, H—Pd—OAc is formed, and it decomposes into metallic palladium or it coordinates benzoquinone, which is then reduced into benzene-1,4-diol (hydroquinone) (Scheme 8.73b) [1392].

Scheme 8.73 (a) Bäckvall's mechanism of Pd(II)-catalyzed allylic acetoxylation of cyclohexene and (b) proposed reaction mechanism of the oxidation HPdOAc + AcOH into $Pd(OAc)_2$ by benzoquinone.

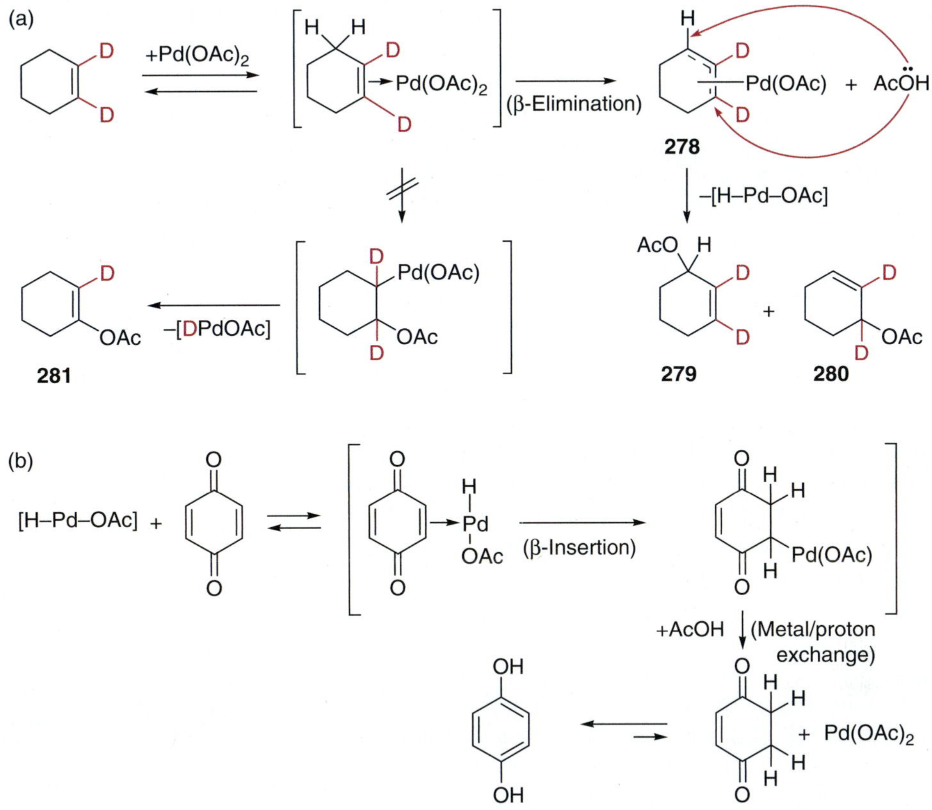

In 2008, Shi and coworkers found that benzoquinone can be employed as an oxidant with [1,2-bis(benzylsulfinyl)ethane]palladium diacetate (**282**) as a catalyst for direct intra- and intermolecular allylic alkylations [1393]. In the case of reaction (8.224), a kinetic deuterium isotopic effect $k_H/k_D = 2.7$ was observed, which supports a mechanism in which the rate-determining step is the oxidative addition of $Pd(OAc)_2$ into the allylic C—H bond, generating allylic palladium intermediate similar to **278** (Scheme 8.73a).

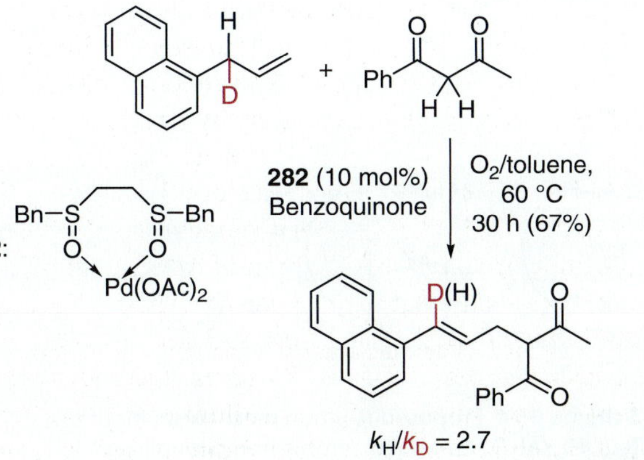

(8.224)

8.5.3 Oxidative aryl/aryl homocoupling reactions

Biaryl, heteroaryl/aryl, and heteroaryl/heteroaryl compounds are of importance because they are used as building blocks of many natural compounds and analogs, they are found in many drugs, and intervene as chiral ligands for a wide range of metallic catalysts [1394–1406]. Biaryl products have been prepared by oxidative C—C homocoupling between two arenes [1407] and through catalyzed direct arylation of arenes (Section 8.8.9). The coupling of two aromatic compounds by treatment with a Lewis acid and a Brønsted acid is called the Scholl reaction, which was disclosed in 1910 [1408]. It has been utilized for intramolecular oxidative cyclodidehydrogenation to generate polyaromatic hydrocarbons (e.g. hexabenzocoronenes, reaction (8.225)). The Scholl reaction uses oxidants such as $FeCl_3$ [1409], $CuCl_2$, $Cu(OTf)_2 + AlCl_3$, $Tl(O_2CCF_3)_3$ in CF_3COOH, $BF_3{\cdot}OEt_2$, $Pb(OAc)_4/BF_3{\cdot}OEt_2$ in MeCN, $Et_3O^+SbCl_6^-$, $SbCl_5$, $MoCl_5$, and DDQ + CH_3SO_3H [1410]. The involvement of arene radical cation intermediates (by SET to form the arene to the oxidant) has been probed [1411].

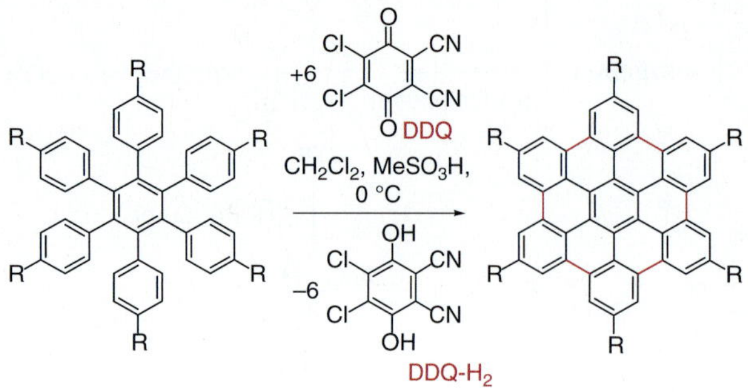

(8.225)

Pd(II) salts can also be used as oxidants [1412]. In 1966, Davidson and Triggs [1413, 1414] and later Yoshimoto and Itatani [1415–1417] found conditions employing catalytic quantities of $Pd(OAc)_2$ under high pressures (9–50 atm) of O_2 [1418]. The use of 1,10-phenanthroline (phen) as a ligand was shown to have a beneficial effect on the oxidative coupling of dimethyl phthalate to tetramethyl biphenyl-3,3′,4,4′-tetracarboxylate, an important intermediate in the production of polyimide films [1419, 1420]. Oxidative homocoupling of thiophene [1421] and of arylpyridines has also been described [1422]. The oxidative homocoupling of naphth-2-ols is one of the most useful methods for the synthesis of BINOL derivatives. It can be carried out with various kinds of stoichiometric oxidants such as Fe, Cu, Mn, and Ti salts [1423, 1424], or better, using air in the presence of homogeneous [1425–1429] or heterogeneous catalysts [1430–1433]. A combination of $FeCl_3{\cdot}(H_2O)_6$ and alumina, montmorillonite K-10, or silica gel gives solid Lewis acids that have been applied by Li and coworkers as catalysts: air was used as oxidant to prepare (±)-1,1′-bi-2-naphthol (IUPAC: (±)-1,1′-binaphthalene-2,2′-diol) and derivatives (reaction (8.226)) [1434].

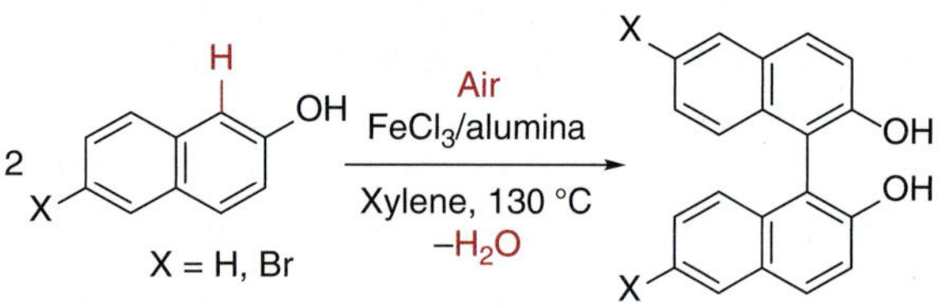

(8.226)

Ding et al. showed that oxidative cross-coupling of β-naphthol (IUPAC: naphthalen-2-ol, 2-naphthol) and β-naphthylamine (IUPAC: naphthalen-2-amine) can be carried out in a two-phase system using $FeCl_3$ and air [1424, 1435]. Ruthenium hydroxide supported by alumina ($Ru(OH)_3/Al_2O_3$) is an effective heterogeneous catalyst for the aerobic biaryl homocoupling of β-naphthols and substituted phenols in water (reactions (8.227) and (8.228)) [1436].

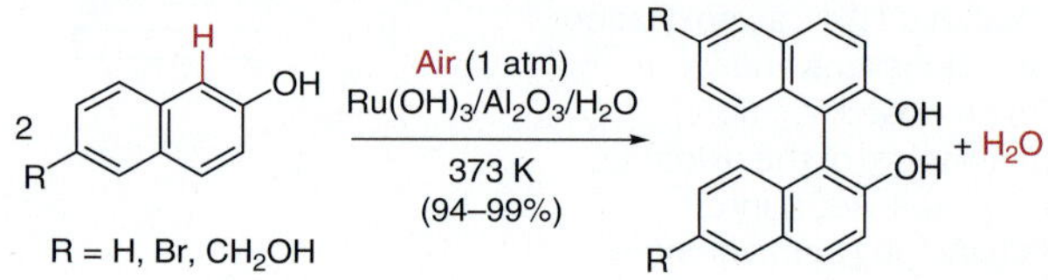

(8.227)

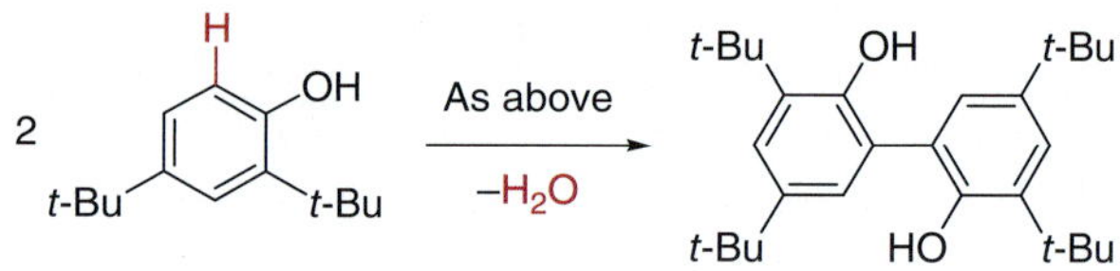

(8.228)

The result of competitive coupling reactions (e.g. β-naphthol + methyl 3-hydroxynaphthalene-2-carboxylate reacting with $O_2/Ru(OH)_3/Al_2O_3$ giving essentially BINOL) and esr studies suggested that the coupling proceeds via the homocoupling of two aryl radicals and the $Ru(OH)_3/Al_2O_3$ catalyst acts as a one-electron oxidant (Scheme 8.74).

A Ru-catalyzed oxidative homocoupling of 2-arylpyridines via C—H activation has been reported. The reaction has also been carried out with stoichiometric $FeCl_3$ [1437]. In 2001, an enantioselective oxidative homocoupling of β-naphthols catalyzed by CuI in the presence of a chiral diamines has been presented by Kozlowski and coworkers [1438, 1439]. Alternative methods have used Ru(II)-salen catalysts [1440] or vanadium(IV)-based catalysts [1441, 1442]. The liquid-phase oxidative homocoupling of β-naphthol into (±)-BINOL has been carried out in CH_2Cl_2 solvent over vanadium catalysts supported on inorganic oxides using O_2 as an oxidant [1443]. Simple benzene and substituted derivatives undergo direct oxidative coupling reactions to give biphenyl and substituted derivatives using $PhI(OAc)_2$ as an oxidant and a gold catalyst such as $HAuCl_4$ or $AuCl(Ph_3P)$ in AcOH [1444, 1445].

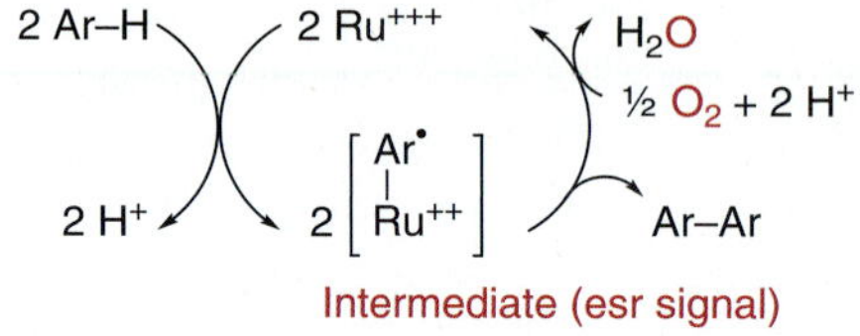

Scheme 8.74 Proposed reaction mechanism for the $Ru(OH)_3/Al_2O_3$-catalyzed aerobic homocoupling of 2-naphthols and phenols.

8.5.4 Oxidative aryl/aryl cross-coupling reactions

In 1976, Kozhevnikov found that cross-coupling of furan and thiophene to yield 2-(2-furyl)thiophene occurred in the presence of stoichiometric $Pd(OAc)_2/Cu(OAc)_2/CaCl_2$ (an example of oxidative heteroaryl/heteroaryl cross-coupling) [1446]. In 1981, Itahara reported the $Pd(OAc)_2$-mediated direct arylation of pyrroles and indoles with arenes (an example of oxidative aryl/heteroaryl cross-coupling) [1447]. In 1982, Nakamura et al. found the $Pd(OAc)_2$-mediated direct arylation of isoxazole with arenes [1448]. In 2006, Lu and coworkers reported the synthesis of unsymmetrical biaryls from simple arenes using $K_2S_2O_8$ as an oxidant and $Pd(OAc)_2/CF_3COOH$ as a catalyst (examples of oxidative aryl/aryl cross-couplings) [1449]. They showed that tuning the concentration of CF_3COOH was crucial for selective C—C cross-coupling reactions. It was proposed that a first electrophilic attack of $Pd(OAc)_2$ onto the arene in excess (Ar^1—H) gives Ar^1—Pd(AcOH)—OAc. This is followed by a second electrophilic attack of Ar^1—Pd (AcOH)—OAc onto the electron-rich arene Ar^2—H giving an intermediate undergoing reductive elimination producing Ar^1—Ar^2, Pd(0), and 2 AcOH. A last step oxidizes Pd(0) into $Pd(OAc)_2$. Fagnou and coworkers found that the direct arylation of indoles with simple arenes occurs either at C(2) or C(3) center of the indole depending on reaction conditions ($Pd(OOCCF_3)_2$ cat., $Cu(OAc)_2$ oxidant, CsOPiv base, microwave heating at 110 °C) [1450, 1451]. On their side, DeBoef et al. used $Pd(OAc)_2 + H_4PMo_{11}VO_{40}$ as the catalyst and O_2 as the oxidant (3 atm) for the oxidative C—C cross-coupling of benzofuran and benzene to give 2-phenylbenzofuran in 84% yield [1368]. After prolonged reaction time, 2,3-diphenylbenzofuran is formed. Using Ag_2CO_3 as the oxidant and K_2CO_3 as the base, $Pd(OAc)_2$ + 1,10-phenanthroline catalyzes (toluene, 130 °C) the regioselective oxidative coupling of benzene and pyridines forming the corresponding 3-phenylpyridines regioselectively [1452]. Oxidative coupling of arenesulfinates with arenes ($Ar^1—SO_2Na + Ar^2—H + 2Cu(OAc)_2 \rightarrow Ar^1—Ar^2 + 2CuOAc + AcONa + AcOH + SO_2$) has been catalyzed by $Pd(PhCN)_2Cl_2$ [1453]. Efficient and highly regioselective oxidative cross-coupling reactions can be realized relying upon directing groups. The latter coordinate the metal catalyst to affect the first C—H activation regioselectively. Then, the resulting $Ar^1M(X)L_n$ species react with another arene Ar^2—H, generally present in large excess, to lead to the second C—H activation giving intermediates of type $Ar^1—M(L)_x—Ar^2$ that lead to Ar^1—Ar^2 (e.g. reaction (8.229)) [1454].

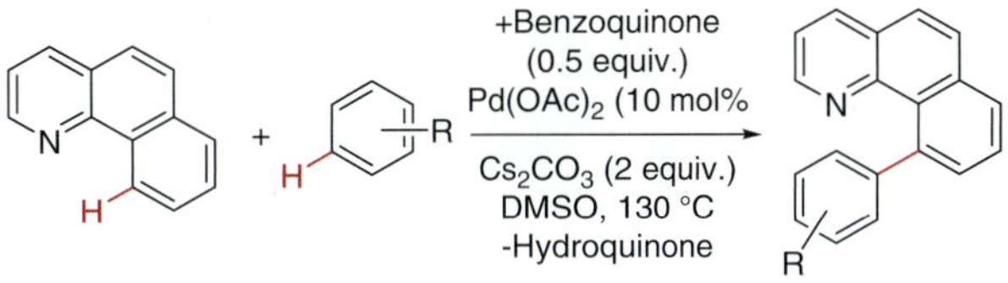

(8.229)

Regioselective oxidative cross-coupling of 2-arylpyridines, 8-methyl and 8-benzylquinoline, 2-phenylquinoline, and *N*-aryl amides has been realized using $(Ph_2I)^+BF_4^-$ as oxidant in the presence of $Pd(OAc)_2$ catalyst in AcOH/toluene at 100 °C [1455]. For the benzoquinone-promoted palladium-catalyzed oxidative cross-coupling of benzo[h]quinoline (Ar^1—H) with 1,2-dimethoxybenzene (Ar^2—H), Hull and Sanford have established a mechanism that first involves the regioselective formation (lateral control by the pyridine moiety) of a dimeric arylpalladium intermediate of type $(Ar^1—Pd—OAc)_2$ [1456]. The latter undergoes a reversible oxidative addition of Ar^2—H followed by coordination of benzoquinone and then reductive elimination to release the biaryl product Ar^1—Ar^2 + hydroquinone. The reaction follows a first-order rate law with respect to the concentration of Ar^2—H and exhibits saturation kinetics with respect to benzoquinone (BQ). The order is 0.5 with respect to complex $(Ar^1—Pd—OAc)_2$. This indicates cleavage of the native dimer to a monomeric species before the rate-determining step. Saturation kinetics with respect to BQ is due to the formation of a steady-state (SS) intermediate that undergoes subsequent reaction with BQ. On adding 0.1–0.5 equiv. of AcOH in the presence of 1 equiv. of BQ, the initial reaction rate shows inverse-first-order dependence on the concentration of AcOH. This suggests that AcOH is lost during formation of the SS intermediate, consistently with the mechanism proposed. When using CD_3CO_2D in the absence of BQ, ^{2}H-NMR spectroscopic analysis showed significant D incorporation into 1,2-dimethoxybenzene (Ar^2—H), presumably via reversible C—H activation/deuteration. Under saturation conditions ([BQ] > 0.15 M), the kinetic deuterium isotopic effect $k_H(Ar^2—H)/k_D(Ar^2—D)$ was found to be 3.39 ± 0.10, which indicates that the rate-determining step involves significant C—H bond breaking in Ar^2—H. When $Pd(OCOAr)_2$ was used instead of $Pd(OAc)_2$, the Hammett plot obtained from variously substituted palladium benzoates showed a negative ρ value (-0.94 ± 0.16) consistently with some positive charge buildup in the transition state for the C—H activation. When replacing

1,2-dimethoxybenzene by a 1 : 1 mixture of 2,6-dimethyl-1-methoxybenzene and 2,6-dimethyl-1-nitrobenzene, it was found that the electron-rich arene (Ar^2—H) reacted slightly faster than the electron-poor arene. This proves that Ar^2—H deprotonation cannot be a mechanism for its activation. Furthermore, as this competition experiment shows a very modest difference in reactivity (1.3 : 1), S_EAr mechanism can also be rejected. Thus, C—H activation in this reaction proceeds by an intermediate mechanism that resembles σ-bond metathesis (see Section 7.7.4) [1457]. Pyridine, pyrimidine, pyrazole, oxazolidine [1458], and amide functions [1459, 1460] have also been used as directing groups to control the regioselectivity of the first C—H activation. Most commonly used catalysts are Pd(II) salts or Rh(III) complexes. Alone, Cu(II) salts in stoichiometric amount, or Cu(II) salts as catalysts under aerobic conditions, can induce the oxidative C—H/C—H coupling reactions. Ni-based catalysts are also used [1461].

In 2012, Russell and coworkers reported a useful oxidative cross-coupling of simple electron-rich arenes with aryltrimethylsilanes catalyzed by Ph_3P/Au(I) salts using $PhI(OAc)_2$ as the stoichiometric oxidant ($Ar^1—H + Ar^2SiMe_3 + PhI(OAc)_2 \rightarrow Ar^1—Ar^2 + PhI + AcOH + Me_3SiOAc$) [1462]. The method permits the regioselective intermolecular arylation of heterocyclics at room temperature [1463]. Tethering the arene and arylsilane reagents permits to generate five- to nine-membered rings regioselectively. Mechanistic studies suggested a first step that oxidizes the Au(I) catalyst: $LAu(I)—X + PhIX_2 \rightarrow LAu(III)X_3 + PhI$. Then, Ar^2SiMe_3 undergoes C—Si auration: $LAu(III)X_3 + Ar^2—SiMe_3 \rightarrow Ar^2(L)Au(III)X_2 + Me_3SiX$. The latter Au(III) complex does the C—H auration of Ar^1—H giving $Ar^2(L)Au(III)(Ar^1)X + HX$. A final reductive elimination liberates the product of coupling $Ar^1—Ar^2$ and LAu(I)X, the catalyst [1464, 1465].

8.5.5 TEMPO-cocatalyzed oxidative C—C coupling reactions

In 2008, Studer and coworkers reported that TEMPO (2,2,4,4-(tetramethylpyridin-1-yl)oxyl) catalyzes the aerobic homocoupling reaction of aryl, alkenyl, and alkynylmagnesium halides ($2RMgX + ½O_2 \rightarrow R—R + XMg—O—MgX$) [1466]. They also found that TEMPO catalyzes the aerobic C—C cross-coupling reaction between nitrones and alkynyl-Grignard reagents (reaction (8.230)) [1467]. In 2012, Jiao and coworkers reported the TEMPO-catalyzed aerobic C—C-coupling between two $C(sp^3)$—H bonds [1468]. Sonobe et al. reported the oxidative cross-coupling of 9,10-dihydroacridine with protected glycine derivatives (reaction (8.231)) [1469]. An example of aerobic di. dehydrogenative cross-coupling involving $C(sp^2)$—H and $C(sp^3)$—H bonds is given with reaction (8.232). Cu-catalyzed aerobic oxidations are promoted by Cu salts in the presence of TEMPO and a base [1470, 1471]. Enantioselective homocoupling (8.233) [1472, 1473] has been realized using (*R*)-BINAM/CuCl + TEMPO as the catalyst (BINAM = 1,1′-binaphthalene-2,2′-dimamine) [1474]. Instead of nitroxyl radicals such as TEMPO and ketoABNO, stable triarylamminium salts (e.g. $(p\text{-}BrC_6H_4)_3N^{\bullet+}/SbCl_6^-$) can be used as catalysts for aerobic oxidative C—C coupling reactions [1475].

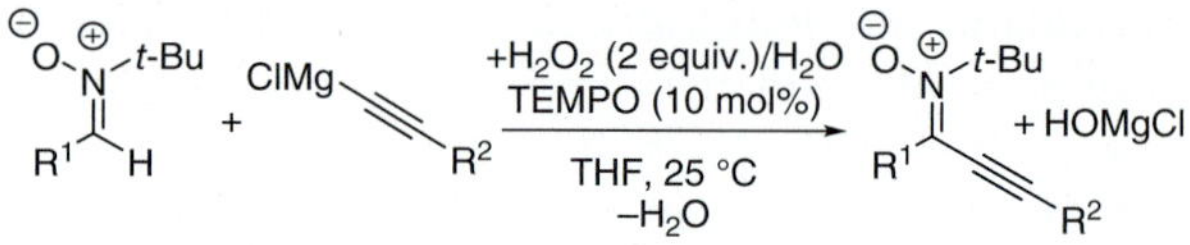

(8.230)

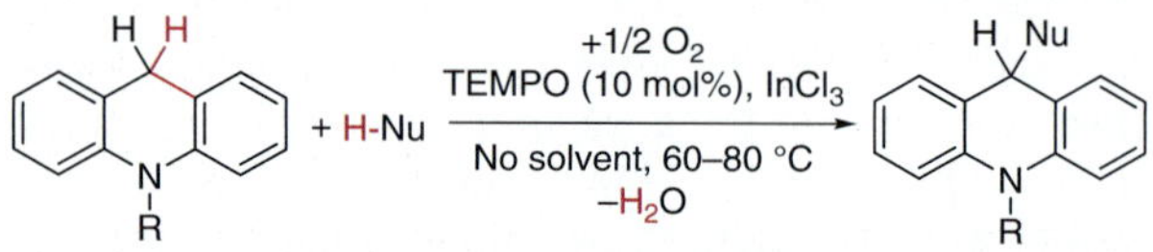

(8.231)

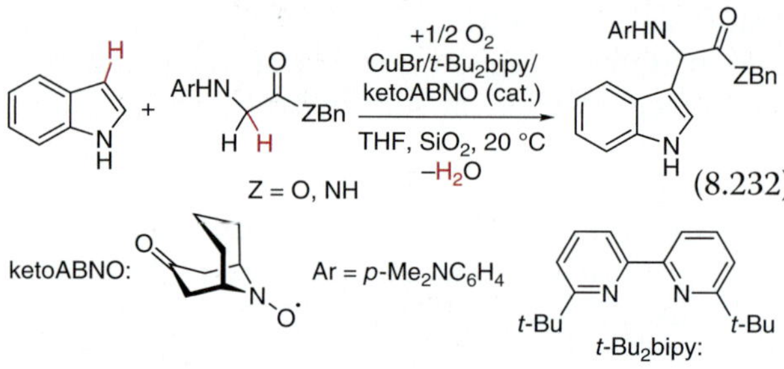

(8.232)

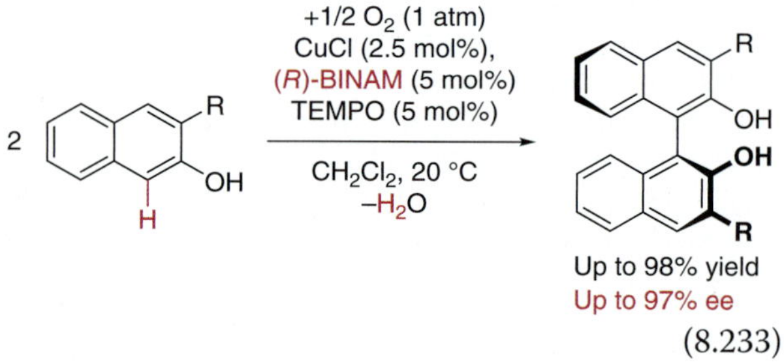

(8.233)

Oxidative α-arylations of *N*,*N*-dimethylanilines with indoles, indolizine, imidazole, and *N*,*N*-dimethylaniline ($Ar^1—H + H—CH_2N(Me)Ar^2 + ½O_2 \rightarrow Ar^1—CH_2N(Me)Ar^2 + H_2O$) have been catalyzed by CuBr (5 mol%) in MeCN at 50 °C [1310]. CuI (5 mol%, in AcOH, 80 °C) catalyzes the cross-didehydrogenative Mannich-type reaction of tertiary amines with methyl ketones ($R^1(R^2)N—CH_2R^3 + H—CH_2COR^4 + ½O_2 \rightarrow R^1(R^2)N—CH(R^3)\text{-}CH_2—COR^4 + H_2O$) [1476–1478].

Problem 8.67 In the presence of 5 mol% of CuBr/3,3′-bis(t-bulyl)-2,2′-bipyridine/ketoABNO and 30 mol% of p-(dimethylamino)pyridine, molecular oxygen (1 atm) promotes the oxidative Strecker reaction between HCN and N-(p-methoxyphenyl) cyclohexylmethyl amine at 50 °C [1469]. What is the main product of this reaction and propose a mechanism for it.

8.5.6 Oxidative aminoalkylation of alkynes and active C—H moieties

In 1932, de Paolini and Ribet reported the dealkylation of tertiary amines with benzoyl peroxide [1479]. In 1946, Horner and Schwenk gave an explanation for the acceleration effect of N,N-dimethylaniline on the benzoyl peroxide-promoted polymerization of styrene [1480]. They then proposed the mechanism shown in Scheme 8.75 (X = Bz) for the amine dealkylation [1481, 1482]. Hydrogen transfer from the oxidant can occur in one step (direct hydrogen transfer, **283** → **285**) or in two steps via the radical cation intermediate **284**. The latter induces polymerization of alkenes; they can be quenched by $^3O_2$ [1483]. Collapse of **285** with BzO$^\bullet$ generates the aminal derivatives **287** (X = Bz) or, by electron transfer, iminium salts of type **286**. The latter react with H_2O to produce the secondary amines **288** and aldehydes R′CHO.

Oxidative dealkylations of tertiary amines with H_2O_2 catalyzed by horseradish peroxidase [1484, 1485], lignin peroxidase [1486–1488], and other cytochrome P450 enzymes follow a similar mechanism in which M=O species oxidize the amines as X—O$^\bullet$ [1489–1492]. In 1988, Murahashi et al. found that ruthenium salts and complexes catalyze the oxidation of tertiary amines with t-BuOOH to give the corresponding α-aminoalkyl t-butyl ethers (**287**, X = O—t-Bu) that are hydrolyzed into **288** + R′CHO [1493]. Nucleophiles other than H_2O can be reacted with the iminium intermediates **286**, including carbon nucleophiles such as allylsilanes and HCN [1494–1499]. With $RuCl_3$ catalyst, the oxidant can be H_2O_2 or O_2 [1494]. In 1989, Miura and coworkers reported the O_2 oxidation of N,N-dimethylaniline (60 °C, MeCN) catalyzed by iron salts to give N-methylformanilide and several other products arising from the reactions of aminoalkyl radical intermediates of type of **285** [1500, 1501]. Using $CuCl_2/O_2$ and terminal alkynes, they obtained a 27–43% yield of the corresponding products of oxidative C—C cross-coupling, N-methyl-N-propargylanilines [1502]. Li and coworkers, using t-BuOOH as the oxidant and CuBr as the catalyst, have realized better yielding oxidative coupling of tertiary amines with arylacetylenes [20, 1503], nitromethane [1504, 1505], indoles [1506], malonates, naphthols, and with other carbon nucleophiles [1507]. An alternative method using NBS (N-bromosuccimide) as an oxidant and CuBr as a catalyst has been presented by Fu and coworkers [1508]. A better yielded method has been proposed by Volla and Vogel; it uses $FeCl_2$ as a catalyst and $(t\text{-BuO})_2$ as an oxidant [1509]. The reaction is highly chemoselective for tertiary amines. The methylamino group reacts faster than alkyl and benzylamino moieties (e.g. reaction (8.234)).

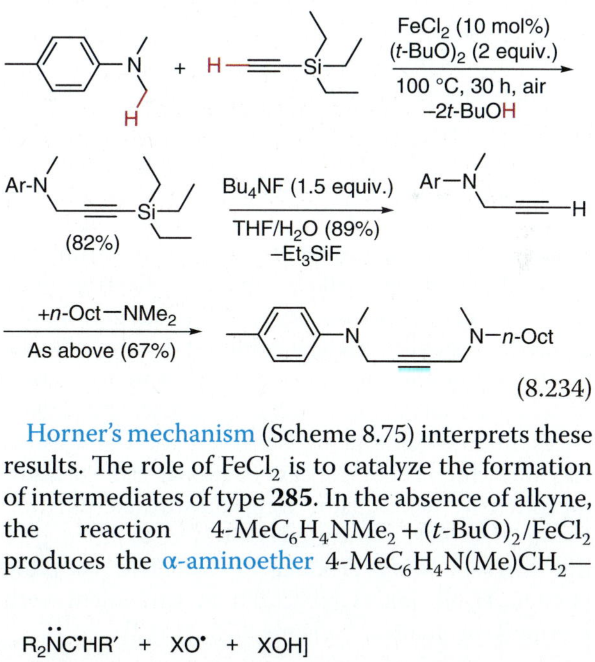

(8.234)

Horner's mechanism (Scheme 8.75) interprets these results. The role of $FeCl_2$ is to catalyze the formation of intermediates of type **285**. In the absence of alkyne, the reaction 4-MeC$_6$H$_4$NMe$_2$ + $(t\text{-BuO})_2$/FeCl$_2$ produces the α-aminoether 4-MeC$_6$H$_4$N(Me)CH$_2$—

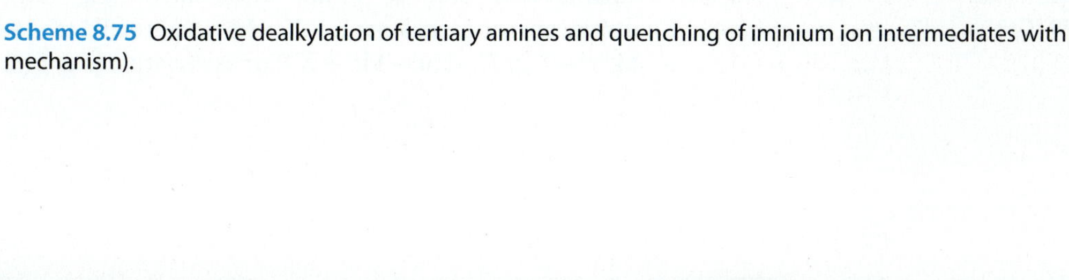

Scheme 8.75 Oxidative dealkylation of tertiary amines and quenching of iminium ion intermediates with nucleophiles (Horner's mechanism).

O—t-Bu. The latter does not react with phenylacetylene in the absence of $FeCl_2$, showing that $FeCl_2$, which acts as a Lewis acid promoter, induces the S_N1 cleavage of the α-aminoethers of type **287** into iminium ion intermediates of type **286** [1510]. In the absence of $FeCl_2$, $(t\text{-BuO})_2 + 4\text{-MeC}_6H_4NMe_2$ did not react to produce the α-aminoether of type **287**. This demonstrates that $FeCl_2$ catalyzes its formation, even though it was reported that dialkyl peroxides are unreactive with $FeCl_2$ [1511]. Reaction $ArNMe_2 + (t\text{-BuO})_2/FeCl_2$ might not involve homolysis, $(t\text{-BuO})_2 \rightarrow 2t\text{-BuO}^\bullet$. The C—C bond forming step involves reaction of iminium ions of type **286** with acetylide anions. When a 1 : 1 mixture of Ph—C≡C—D and n-C_6H_{13}C≡C—H was heated to 100 °C for 12 hours in the presence of 2 equiv. of Et_3N, a 0.5 : 0.5 : 0.5 : 0.5 mixture of Ph—C≡C—H, Ph—C≡CD, n-hex-C≡C—D, and n-hex-C≡C—H was formed. The same experiment run with 10% $FeCl_2$ and in the absence of Et_3N also led to H/D exchange between the two terminal acetylenes, but the reaction was not complete after 12 hours at 100 °C. When using $FeCl_2$ and Et_3N, the exchange was complete at 100 °C after 12 hours. These experiments show not only that tertiary amines are able to deprotonate the alkynes but that $FeCl_2$ itself can generate iron acetylide intermediates. The latter reaction is slower than deprotonation by Et_3N. Any mechanism involving addition of iminium ion intermediates to the alkynes in the rate-determining step (with formation of alkenyl cation intermediates) can be ruled out because differences in the initial reaction rates as a function of the nature of the alkyne (electron-rich (R = 4-$MeOC_6H_4$, alkyl, silyl) or electron-poor (R = COOEt)) are not observed. If the reaction should involve aminoalkyl radical addition to the alkynes in the rate-determining step (with the formation of alkenyl radical intermediates), significant differences in rates should be observed for the various alkynes, which is not the case (a similar rate constant for phenyl-, hexyl-, and ester-substituted ethyne). Using $PhCONH_2$, 4-$MeC_6H_4NMe_2$ reacted to give the corresponding product of oxidative coupling (4-$MeC_6H_4N(Me)CH_2NHCOPh$ in agreement with Horner's mechanism (Scheme 8.75) [1512].

Problem 8.68 Give a mechanism for reaction:

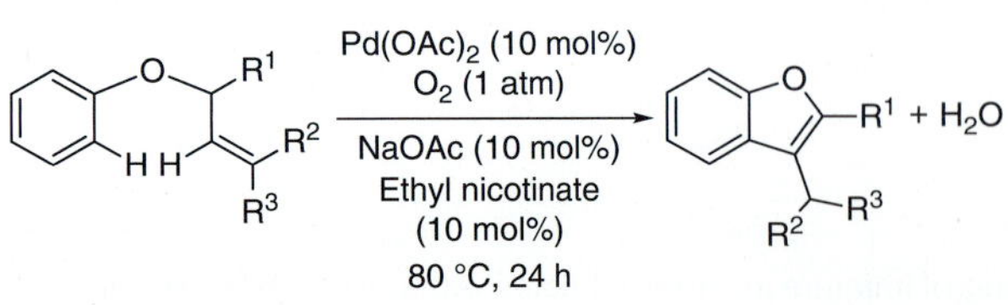

Problem 8.69 What is the main product **P** for the oxidative cross-coupling of **A** and **B**?

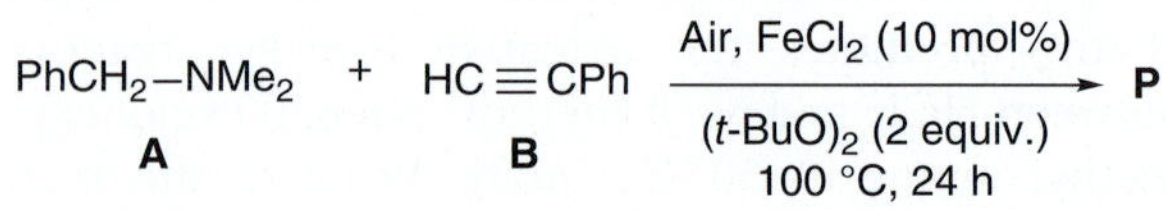

Problem 8.70 Propose a mechanism for the oxidative cross-coupling **A** + **B** → **P** [1514].

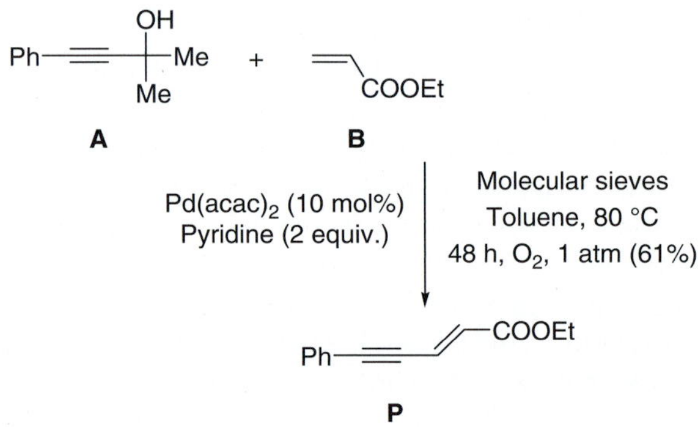

Problem 8.71 What are the possible products of the following reactions? [1504]

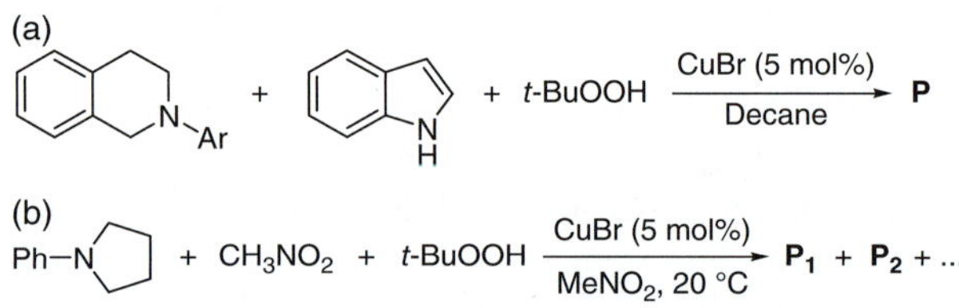

8.6 Alkane, alkene, and alkyne metathesis

A metathetical reaction is a reaction in which two or more reagents exchange parts.

They are quite common such as acid/base neutralizations (e.g. HCl(aq) + NaOH(aq) → NaCl(aq) + H_2O (liquid)), silver chloride precipitation (e.g. $AgNO_3$(aq) + NaCl(aq) → AgCl(solid) + $NaNO_3$(aq)), nucleophilic displacement reactions (Section 3.9.3, e.g. i-PrOH + HCl/$MgCl_2$/H_2O ⇄ i-PrCl + $MgCl_2$/H_2O), and light or radical-induced chlorination of hydrocarbons (R—H + Cl_2 ⇄ R—Cl + HCl). If uncatalyzed, these reactions can be classified according to their mechanisms that can be either dissociative (e.g. S_N1 displacement reactions: R—X + M^+/Nu^- ⇄ R^+/X^- + M^+/Nu^- ⇄ R—Nu + M^+ + X^-) or associative (e.g. S_N2

displacement reactions: R—X + M^+/Nu^- ⇄ [Nu···R···X]$^-$/M^+ ⇄ R—Nu + M^+ + X^-, Wittig–Gilman metal/halogen exchange reaction, e.g. BuLi + R—Br ⇄ [R—Br—Bu]$^-$/Li^+ ⇄ BuBr + R—Li) [1515, 1516] and may or may not involve single-electron transfers (SET; e.g. Ar—X + M^+/Nu^- ⇄ $[Ar—X]^{•-}$ + M^+/$Nu^•$ ⇄ $Ar^•$ + MX + $Nu^•$ ⇄ Ar—Nu + MX, or $Ar^•$ + M^+/Nu^- ⇄ $[ArNu]^{•-}$ + M^+; $[ArNu]^{•-}$ + Ar—X ⇄ ArNu + $[Ar—X]^{•-}$, aromatic radical-nucleophilc displacement, chain mechanism [1517–1520]). If catalyzed, these reactions follow multistep mechanisms composed of a succession of associative and/or dissociative processes. The Wittig olefination (Section 5.3.7) is also a metathetical reaction. It exchanges doubly bonded molecular fragments (e.g. Ph_3P=CHR^1 + R^2CH=O → Ph_3PO + R^1CH=CHR^2), whereas the reactions mentioned above exchange singly bonded molecular moieties. Alkane (R^1—R^2 + R^3—R^4 ⇄ R^1—R^3 + R^2—R^4), alkene (R^1CH=CHR^2 + R^3CH=CHR^4 ⇄ R^1CH=CHR^3 + R^2CH=CHR^4), and alkyne (R^1C≡CR^2 + R^3C≡CR^4 ⇄ R^1C≡CR^3 + R^2C≡CR^4) metathesis are presented in this chapter together with alkene/alkyne cross-metathesis (R^1CH_2CH=CHR^2 + H—C≡C—R^3 ⇄ R^1CH_2—CH=CH—C(R^3)=CHR^2). In general, these reactions are catalyzed by heterogeneous and/or homogeneous catalysts containing transition metals.

8.6.1 Alkane metathesis

Alkane metathesis permits a better valorization of oil and natural gas (methane). This has an important impact on energy consumption and the environment. In the Chevron process that combines heterogeneous dehydrogenation/hydrogenation catalysts (Pt/Al_2O_3) and olefin metathesis catalysts (WO_3/SiO_2, Section 8.6.2), high temperature and high pressure (400 °C, 60 bar) are necessary to convert alkanes into their lower and higher analogs (reaction (8.235)).

$$2C_nH_{2n+2} \rightleftarrows C_{n-i}H_{2(n-i)+2} + C_{n+i}H_{2(n+i)+2} \quad n \geq 2, i = 1, 2, \ldots, (n-1) \quad (8.235)$$

In 1997, Basset and coworkers reported the conversion of alkanes at low temperature and pressure, typically 150 °C and 1 bar. The catalyst is a silica-supported tantalum hydride [1521]. The reaction between methane and propane generates ethane [1522]. Further development has led to W hydride supported on alumina that showed TON = 120 after 120 hours under 0.8 bar of propane [1523–1528]. The carbon–carbon bond breaking and bond making steps correspond to those of alkene metathesis, which involve formal (2+2)-cycloreversion and (2+2)-cycloaddition. Silica-supported complexes such as [(≡SiO)metal(≡NAr)(=CH-*t*-Bu)(CH_2-*t*-Bu)] also catalyze the conversion of linear alkanes into their lower and higher homologs [1529]. In the presence of $W(H)_3$/Al_2O_3 catalyst, 2-methylpropane and 2-methylpropene are transformed into 2,3-dimethylbutane and 2,3-dimethylbutenes with high selectivity [1530]. It was proposed that an alkyl species is formed by C—H activation of alkanes, which, by a subsequent β-H or α-H transfer step, yields the alkene or a (carbene)metal hydride intermediate, the necessary propagating species for alkene metathesis (Scheme 8.76). The corresponding reverse step yields the alkane metathesis products [1531, 1532].

In 2006, Goldman and coworkers reported a homogeneous catalytic system combining Kaska Ir complexes **289** (pincer complexes) with Schrock olefin metathesis complexes **290** (Section 8.6.2) such as **[Mo]-1** and **W-Si2** (Scheme 8.77) [1533]. Several other catalysts have been evaluated and

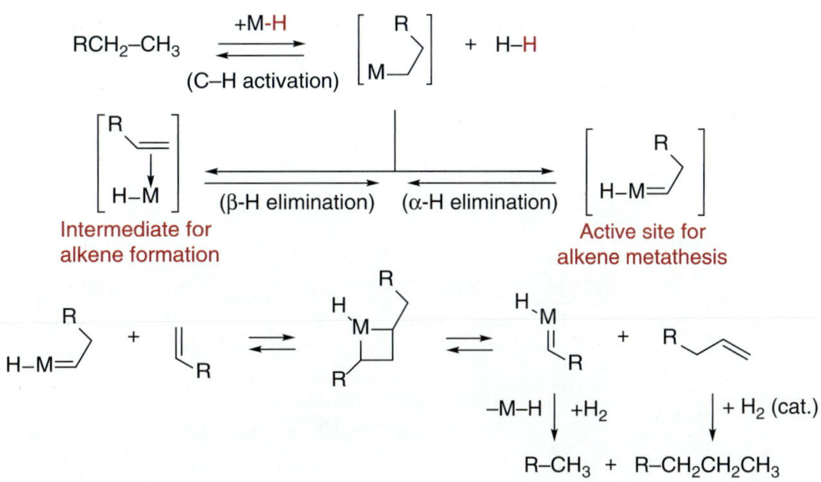

Scheme 8.76 Proposed reaction mechanism for the alkane metathesis using Group 5 and Group 6 hydride catalysts.

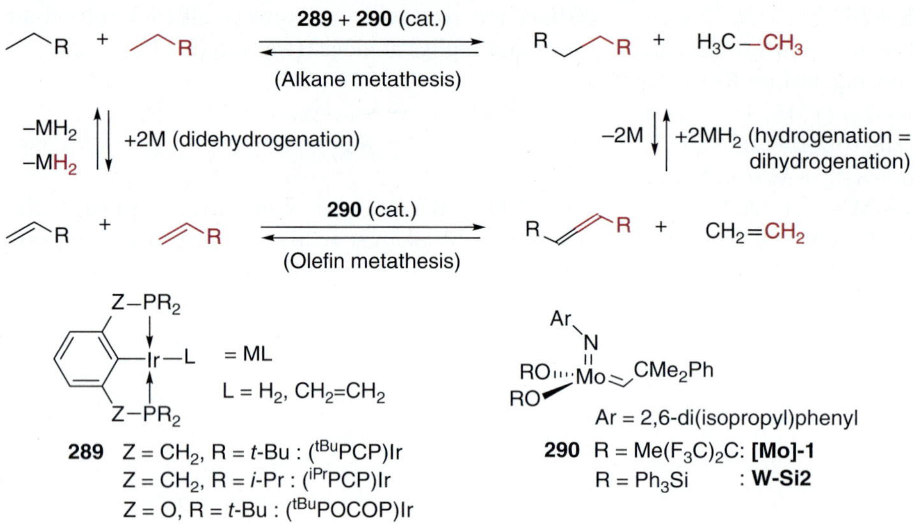

Scheme 8.77 Metathesis of alkanes by tandem alkane didehydrogenation and olefin metathesis

$W(NAr)(CHR)(OSiPh_3)_2$ (Ar = 2,6-diisopropylphenyl) seems to be quite promising [1534, 1535].

8.6.2 Alkene metathesis

Alkene metathesis is widely used in organic synthesis [1536–1543], in the synthesis of natural and biologically active compounds [1544–1546], and in polymer chemistry for the formation of carbon–carbon double bonds (Scheme 8.78) [1547, 1548]. Alkene metathesis permits the transformation of biomass-derived unsaturated compounds into valuable products [1549, 1550]. Basically, it is an alkylidene exchange between two reacting olefins, mediated by transition metal alkylidene complexes [1551, 1552]. Most exploited processes are ring-closing metathesis (RCM), acyclic diene metathesis (ADMET), ring-opening metathesis polymerization (ROMP), cross-metathesis (CM), ring-opening cross-metathesis (ROCM), ring-rearrangement metathesis (RRM), ring-closing enenyne metathesis (RCEM) [1539, 1553–1564], and intermolecular (cross) alkene/alkyne metathesis (IEM) [1565–1567]. Most catalysts are homogeneous, but success has been met by having them on solid supports, which make their recovery and product purification more economical [1568–1575].

Alkene–carbonyl cross-metathesis (not developed here) is under exploration and has disclosed interesting new synthetic possibilities [1576].

In 2005, Chauvin, Grubbs and Schrock were awarded the Nobel Prize in chemistry for the development of the metathesis method in organic synthesis. Hundreds of homogeneous alkene metathesis catalysts have been developed of which more than 60 are commercially available [1577]. Catalysts shown in Figure 8.3 have been most successful. The first catalysts were originally designed for Ziegler polymerization. In 1956, it was found at Du Pont that

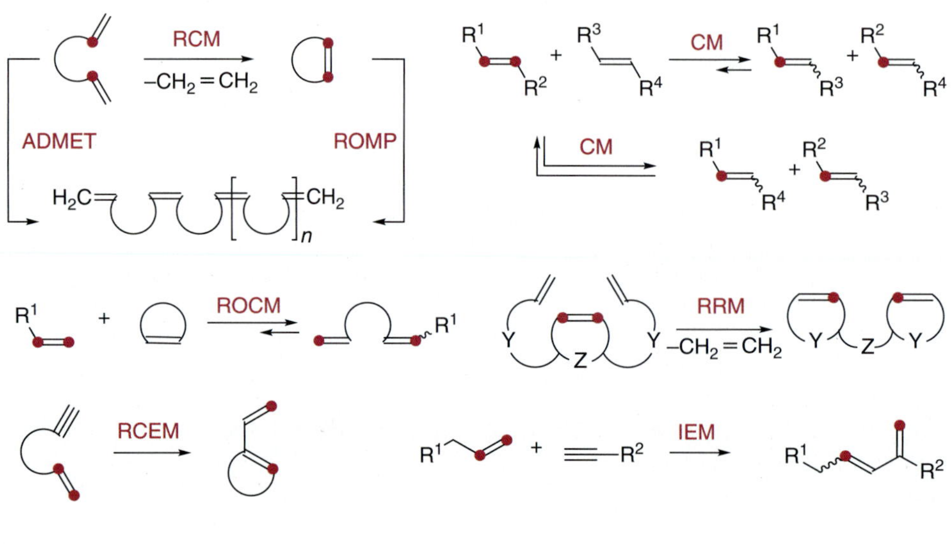

Scheme 8.78 Widely applied alkene metathesis reactions.

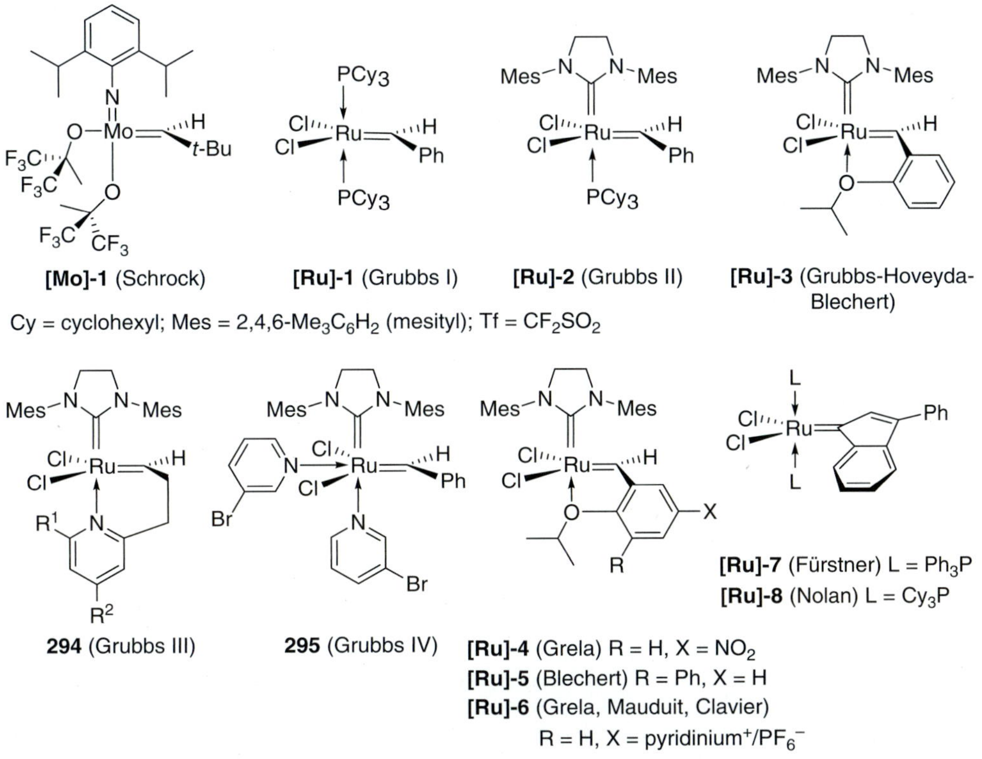

Figure 8.3 Some of the most used catalysts for alkene metathesis.

propene produces a copolymer of propene and ethylene when heated in the presence of a solid Mo/Al catalyst. The gas phase of the reaction mixture was made of ethylene, propene, and but-1-ene [1578]. The discovery in 1967 by Calderon et al. that ring-opening polymerization and the disproportionation of acyclic alkenes was essentially the same process led to coining of the term "olefin metathesis" [1579]. Early catalysts such as $WCl_6/EtAlCl_2$ were used for the polymerization of cyclopentadiene. Other systems included WCl_6/Bu_4Sn, $WOCl_4/EtAlCl_2$, MoO/SiO_2, and Re_2O_7/Al_2O_3. By the late 1960s, the Phillips group had developed the triolefin process (reaction (8.236)) [1580, 1581].

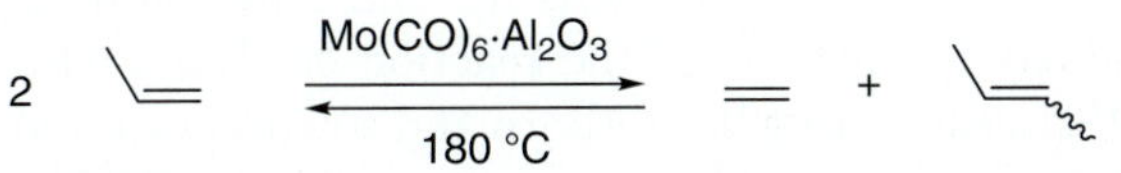

(8.236)

This led to the development of the Lummus ABB process, which converts ethylene into propylene through cross-metathesis with (*E*)- and (*Z*)-but-2-enes on W-based catalysts [1582]. This discovery was then turned into an alkane conversion process first by Phillips [1583] and then by Chevron by combining heterogeneous didehydrogenation/dihydrogenation and olefin metathesis catalysts [1584], which allows a given alkane to be converted into its lower and higher homologs (Section 8.6.1). The initially proposed mechanism for alkene metathesis was that of a pair-wise exchange of alkylidene through a cyclobutane complex in which two alkenes coordinated to the metal exchange their alkylidene groups (reaction (8.237): Calderon's proposal) [1579].

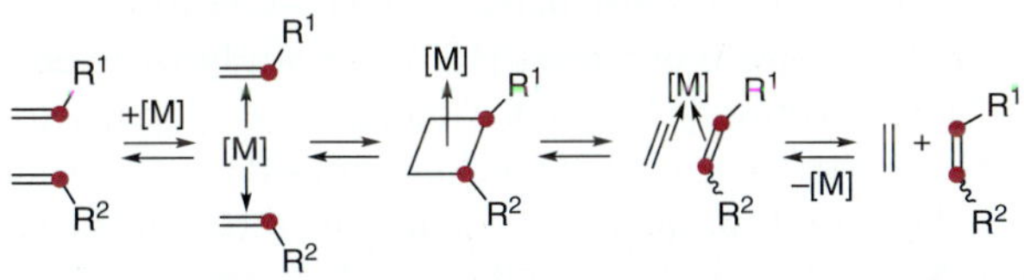

(8.237)

In 1971, Hérisson and Chauvin observed that in some cases, not only the alkenes resulting from the pair-wise exchange of the two alkylidenes of the starting olefins were formed but also alkenes resulting from the homocoupling of all possible alkylidene fragments were formed at the beginning of the reaction. They proposed their "carbene" mechanism (Scheme 8.79) [1585]. In 1975, Schrock characterized the first "tantalum ylide" $(Me_3CCH_2)_3Ta{=}CHCMe_3$ [1586]. It was then found that Ta complexes bearing alkoxy ligands led to alkene metathesis.

In 2005, Piers and coworker reported the first direct observation of a ruthenium metallacycle **292** obtained

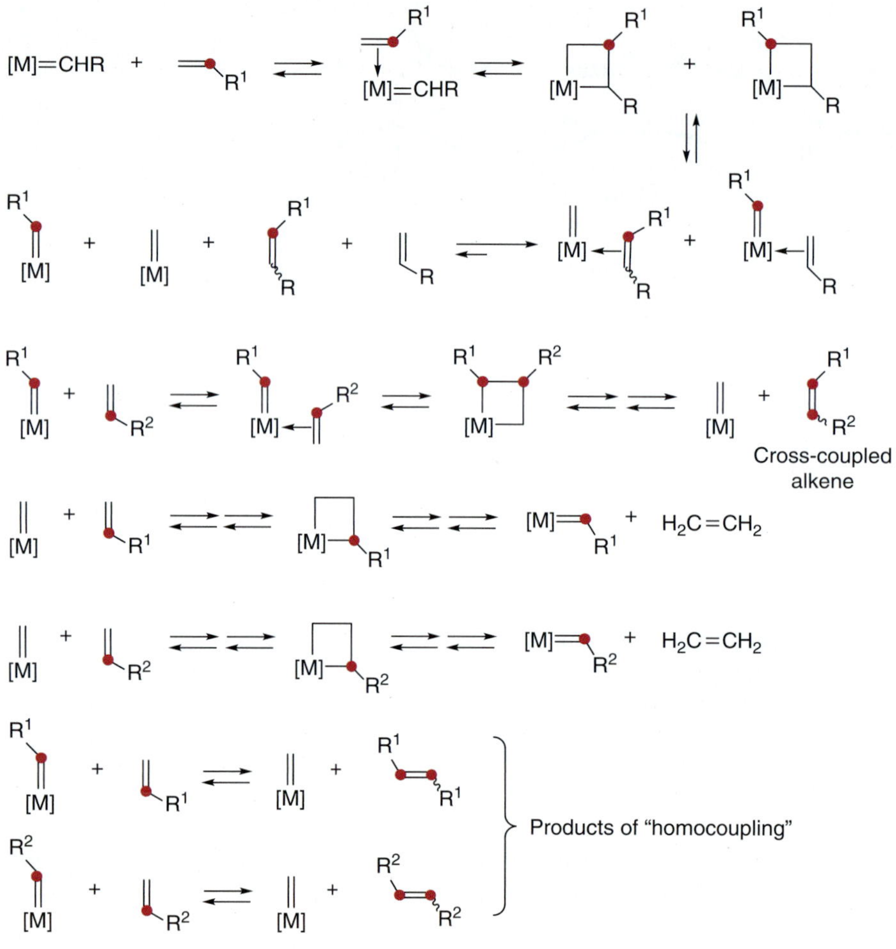

Scheme 8.79 Hérisson's and Chauvin's "carbene" mechanism (see also Scheme 7.62, Section 7.7.11, for possible competitive reactions).

by reaction of **291** with ethylene. When using carbon doubly labeled ethylene, $^{13}CH_2$=$^{13}CH_2$, triply labeled **292**-$^{13}C_3$ and mono-labeled $[^{13}CH_2$=CH—$PCy_3]^+$/$[B(C_6F_5)_4]^-$ were obtained [1587]. A study by Wenzel and Grubbs [1588] on a complex analogous to **292** showed that ruthenacyclobutanes derived from ethylene and propene are dynamic structures on the NMR time scale (equilibria **292** ⇄ **293** ⇄ **293′**) [1589, 1590].

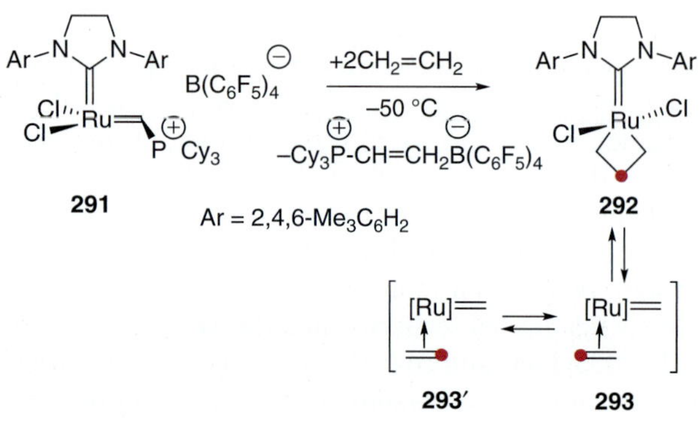

In 1973, Casey and Burkhard demonstrated a metathesis like exchange between a Fischer carbene and an electron-rich alkene [1591]. Katz and McGinnis [1592], and Grubbs et al. [1593] showed that a mixture of octa-1,7-diene and 1,1,8,8-tetradeuterioocta-1,7-diene gives products of cross-coupling together with cyclohexene + ethylene already at the beginning of the reaction. As required for the carbene mechanism, a statistical distribution of ethylene-d_4, ethylene-d_2, and ethylene-d_0 is obtained at any time of the reaction catalyzed with $PhWCl_3/AlCl_3$. In 1982, Osborn and coworkers found that tungsten alkylidene complex *t*-BuCH=W(Cl_2)(OCH_2—*t*-Bu)$_2$ + $AlCl_3$ produces a cationic complex able to catalyze alkene metathesis [1594]. In the late 1980s, Schrock et al. developed the molybdenum catalyst **[Mo]-1** (Figure 8.3) which, for the first time, allowed metathesis on substrates bearing functional groups and with olefins bearing electron-releasing or electron-withdrawing substituents [1595]. The Schrock catalyst **[Mo]-1** and analogs are 14-valence electron complexes that do not require ligand de-coordination to react with alkenes and to form the corresponding molybdenacyclobutanes. The Mo complexes are air and moisture sensitive but can be stabilized in different ways [1596], for instance, by coordinating them with 2,2′-bipyridine as shown by Fürstner and coworkers. The active catalysts are regenerated in situ by reaction with $ZnCl_2$ [1597]. In 1988, Novak and Grubbs catalyzed the ring-opening metathesis polymerization of 7-oxanorbornene derivatives with $RuCl_2$·(hydrate) in water, the latter acting as a cocatalyst [1598, 1599]. This permitted Kiessling and coworkers [1600] to prepare neoglycopolymer in water [1601]. The first well-defined ruthenium metathesis precatalyst $Ph_2C{=}CH{=}Ru(PPh_3)_2Cl_2$ was reported by Grubbs and coworkers [1602]. On substituting Ph_3P by more nucleophilic phosphines such as $(i\text{-Pr})_3P$ or Cy_3P (Cy = cyclohexyl), more active catalysts were obtained. This was also true for simple alkylideneruthenium complexes that led to the development of **[Ru]-1** (Figure 8.3; Grubbs I catalyst), which is tolerated by a large number of functional groups [1603, 1604]. Herrmann and coworkers were the first to use diaminocarbene (NHC) ligands in alkene metathesis [1605]. Independently, the groups of Nolan [1606–1608], Grubbs [1609, 1610], and Herrmann [1611, 1612] combined labile phosphine ligands for rapid metathesis initiation with nonlabile stabilizing NHC ligand, resulting in heteroleptic complexes such as **[Ru]-2** (Figure 8.3; Grubbs II: "second-generation" catalyst). The role of steric and electronic effects of NHC ligands toward the enhanced activity of catalysts on replacing phosphines by NHC ligands has been evidenced theoretically and experimentally [1609, 1613–1619]. It has also been observed that the substituents on nitrogen centers of the NHC ligands affect the catalyst activity [1620–1622]. In 2000, the groups of Hoveyda [1623] and Blechert [1624] simultaneously reported the phosphine-free catalyst **[Ru]-3**. This catalyst shows improved activity toward electron-deficient alkenes, such as acrylonitriles, fluorinated alkenes, and activity for trisubstituted alkene synthesis. Grubbs and coworkers also developed the tethered pyridine complex **294** and the bis(3-bromopyridyl)ruthenium complex **295** [1620]. On their side, Grela and coworkers prepared analogs of **[Ru]-3** with substituted benzylidene moiety, for instance **[Ru]-4**, which shows enhanced catalytic activity compared with **[Ru]-3**, probably by weakening the isopropoxy → ruthenium bond strength [1625]. This catalyst was used in the synthesis of (−)-securinine and of a precursor of BILN2061, a cyclic peptide that shows antihepatitis C virus activity [1626]. The latter synthesis requires a macrocyclization involving a terminal alkene onto vinylcyclopropane moiety. It was done in a biphasic catalytic system composed of ionic liquid $[bmim]PF_6$ (1-butyl-3-methylimidazolium hexafluorophosphase), in supercritical CO_2 at 70 °C and under 400 bar [1627, 1628].

Compared with earlier solid catalysts, well-defined homogeneous catalysts (Figure 8.3) present the advantage of working at rather low temperatures. However, they suffer from their instability that make them decompose and contaminate the final products with metallic impurities. Furthermore, they can induce other reactions than alkene metathesis, such as alkene isomerization [1629, 1630]. Thus, further development is expected to resolve these problems. One possible route is to immobilize the homogeneous catalysts by incorporating them in polymers or on surfaces [1631]. Examples are shown with **296** [1632] and **297** [1532].

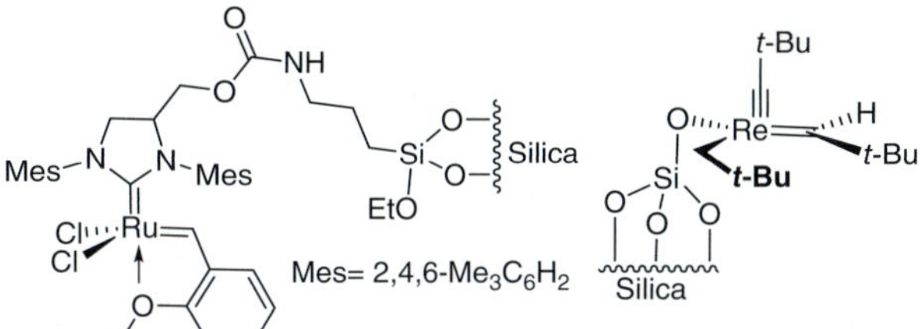

296 (Monge–Marat) **297** (Basset)

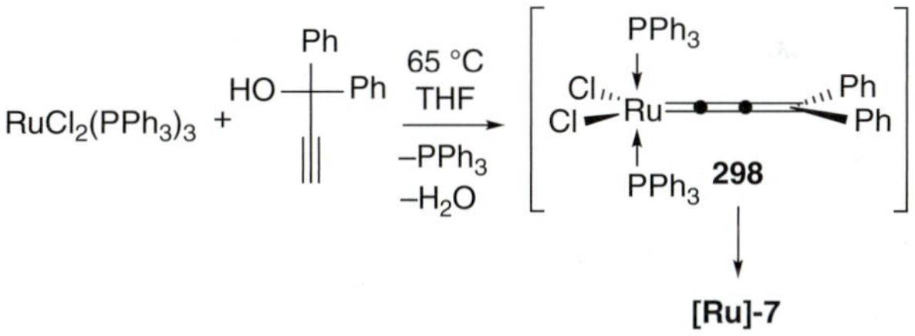

(8.238)

Heating $RuCl_2(PPh_3)_2$ with 1,1-diphenylpropargyl alcohol in THF gives a diphenylallenylidene complex **298** that rearranges into indenylidene complex **[Ru]-7** (reaction (8.238)) [1633–1637]. The triphenylphosphine ligands in **[Ru]-7** can be exchanged for more nucleophilic ligands such as Cy_3P, which gives **[Ru]-8**. Cationic allenylidene arene-ruthenium complex **300** obtained from **299** can be isolated. It rearranges into **301**, a complex superior to second- generation Grubbs catalyst (**[Ru]-2**) in the ROMP of cyclopentene (at −40 °C, TON ~ 10 000) and of cyclooctene (20 °C, 5 minutes, TOF 1 056 000 h^{-1}) [1638]. It is also superior to **[Ru]-2** in the RCM in terms of yield and (*E*) vs. (*Z*) stereoselectivity [1639, 1640].

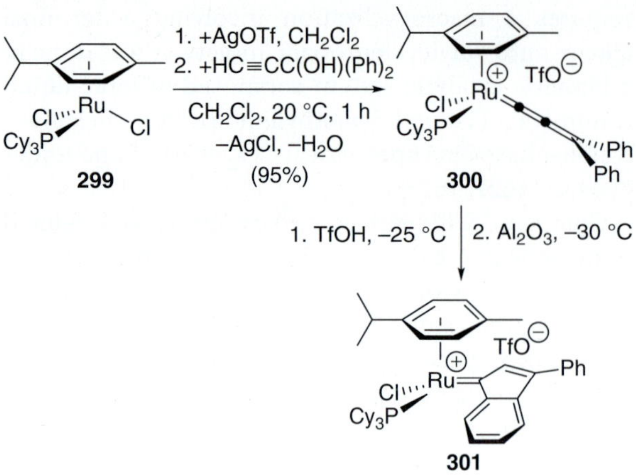

In 2003, Grubbs and coworkers proposed general rules for reactivity in alkene cross-metathesis [1641]. They designed four classes of olefins based on their ability (relative rate constants) to homocouple, for each Grubbs I (**[Ru]-1**) and Grubbs II (**[Ru]-2**) and Schrock catalyst (**[Mo]-1**). Type I alkenes give fast homocoupling, whose product of homocoupling participates in secondary metathesis (fast reversible metathesis). Type II alkenes are slowly homocoupled and their products of homocoupling are unreactive. Type III alkenes give no product of homocoupling (e.g. for steric reasons). Type IV alkenes are noninhibitory spectators to alkene metathesis. Three general trends are observed: 1) metathesis between Type I alkenes gives a statistical outcome, 2) metathesis between two alkenes of the same type (not Type I) gives nonselective outcome, and most importantly, 3) that alkenes of two different types would, for a given catalyst, give a selective cross-metathesis (heterocoupling). For instance, three-component synthesis of dienone **302** (reaction (8.239)) has been realized using Grubbs II catalyst **[Ru]-2**, isobutene (Type III alkene), hexa-1,5-diene (Type I alkene), and vinyl methyl ketone (Type II alkene). The Blechert group [1642] found that variation of the structure of the diaminocarbene ligand of **[Ru]-2** and **[Ru]-3** affects the (*E*)/(*Z*) stereoselectivity.

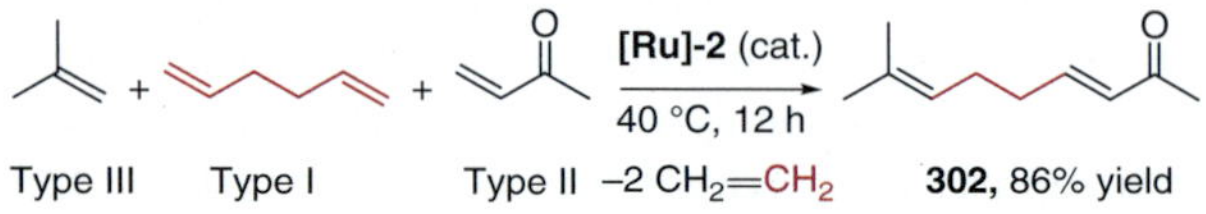

(8.239)

Cross-metathesis (CM, Scheme 8.71) is usually (*E*)-stereoselective owing to the reversibility of the reaction. The selective formation of the less stable (*Z*)-alkenes requires the formation of a cyclic compound through ring-closing metathesis (RCM). However, catalysts have been developed that promote the formation of (*Z*)-alkenes through CM under conditions of kinetic control [1643]. This implies that those catalysts do not promote the (*Z*)/(*E*) isomerization and the reverse CM [1644]. Insect pheromone **304** has been prepared by Grubbs and coworker by CM of hex-1-ene and non-8-en-1-yl acetate (**303**; reaction (8.240)) applying catalyst **305** [1645].

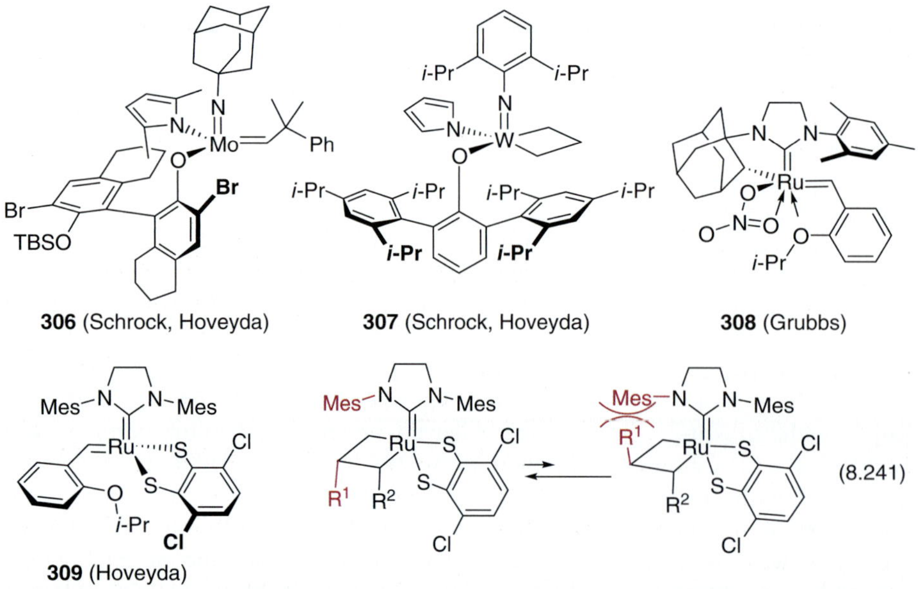

(8.241)

Figure 8.4 Examples of catalysts of (*Z*)-stereoselective alkene metatheses.

Other catalysts for (*Z*)-stereoselective CM have been proposed [1646–1655]. Examples are given in Figure 8.4. Possible origin of the (*Z*)-stereoselectivity for catalysts **306–309** are gauche interactions in the substituted metallacyclobutane intermediates and/or in the transition state leading to them (see e.g. equilibrium (8.241)) [1656]. However, stereoelectronic factors also play a role on the stereoselectivity when substituents are small [1657].

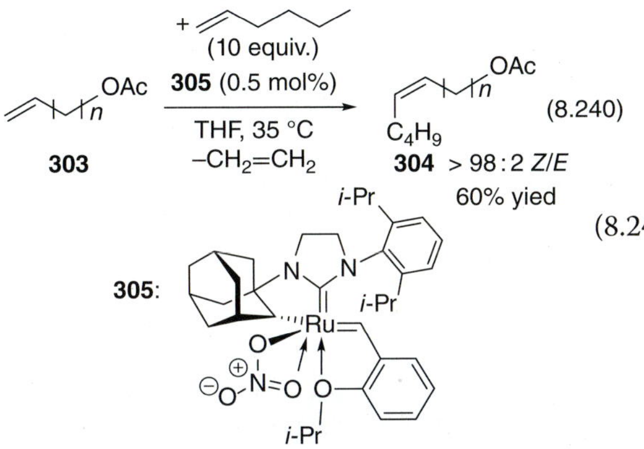

(8.240)

Highly enantioselective cross-coupling metatheses [1658, 1659] have been observed applying enantiomerically pure Grubbs catalysts of type **310** [1660, 1661], Hoveyda's catalysts of type **311** [1662] or Schrock's catalysts of type **312** [1663].

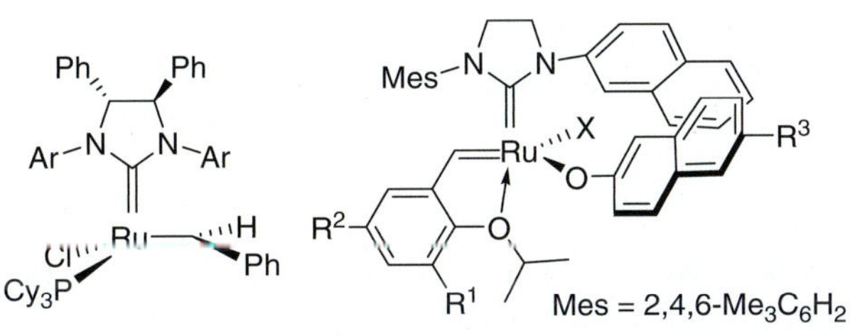

310 (Grubbs)
Ar = 2-(*i*-Pr)C_6H_4
Ar = 2-(*i*-Pr)-4-MeO-5-(*t*-Bu)C_6H_2
Cy = cyclohexyl

311 (Hoveyda) X = Cl,I
R^1 = H, OMe, NO_2
R^2 = H, Ph
R^3 = H, CF_3

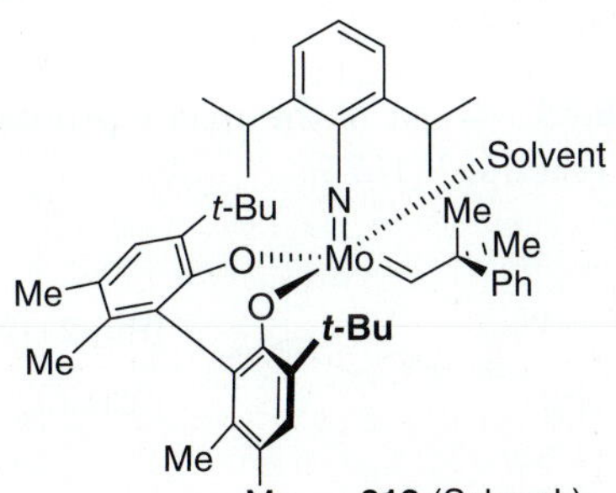

312 (Schrock)

An example of enantioselective ring-closing olefin metathesis is given with reaction (8.242) [1664].

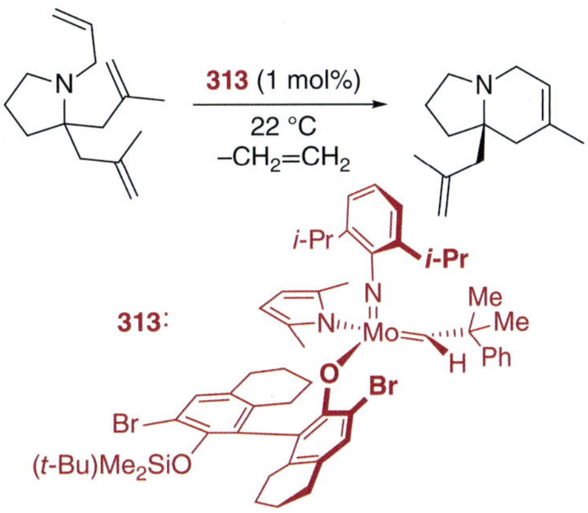

(8.242)

Problem 8.72 Styrene (in excess) and 8-oxabicyclo [3.2.1]oct-6-ene-3-one equilibrate with a compound **P** in the presence of 1 mol% of Grubbs II catalyst (**[Ru]-2**). What is **P** and give a possible mechanism [1665].

Problem 8.73 What is product **P** of the two following successive reactions? [1666]

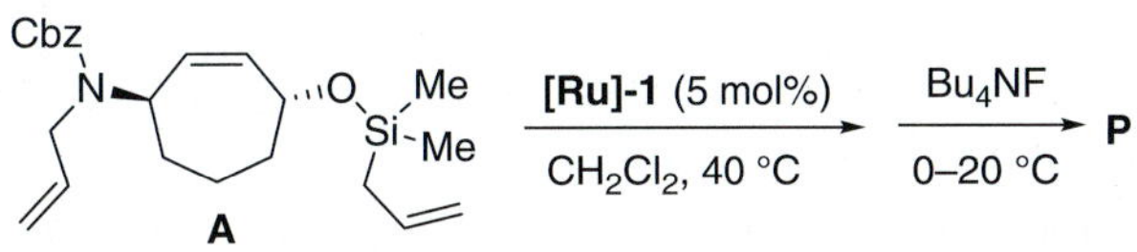

8.6.3 Enyne metathesis: alkene/alkyne cross-metathesis

Complexes **[Ru]-1**, **[Ru]-2**, **[Ru]-3**, and **295** (Figure 8.3) have been commonly applied in the alkene/alkyne cross-metathesis that generates 1,3-dienes (Scheme 8.80a) [1566, 1667], a reaction reported first by Katz and Sivavec [1668]. In fact, it corresponds to the formal splitting of the alkene into two carbene fragments that add across the alkyne. The intermolecular process is called cross-enyne metathesis, whereas the intramolecular process is called ring-closing enyne metathesis. A possible mechanism of the enyne metathesis is shown in Scheme 8.80b [1565, 1567]. It features the rapid reaction of the ruthenium catalyst with the alkene that generates (via a ruthenacyclobutane intermediate **313**) styrene and another alkylideneruthenium complex. The latter then coordinates the alkyne and forms

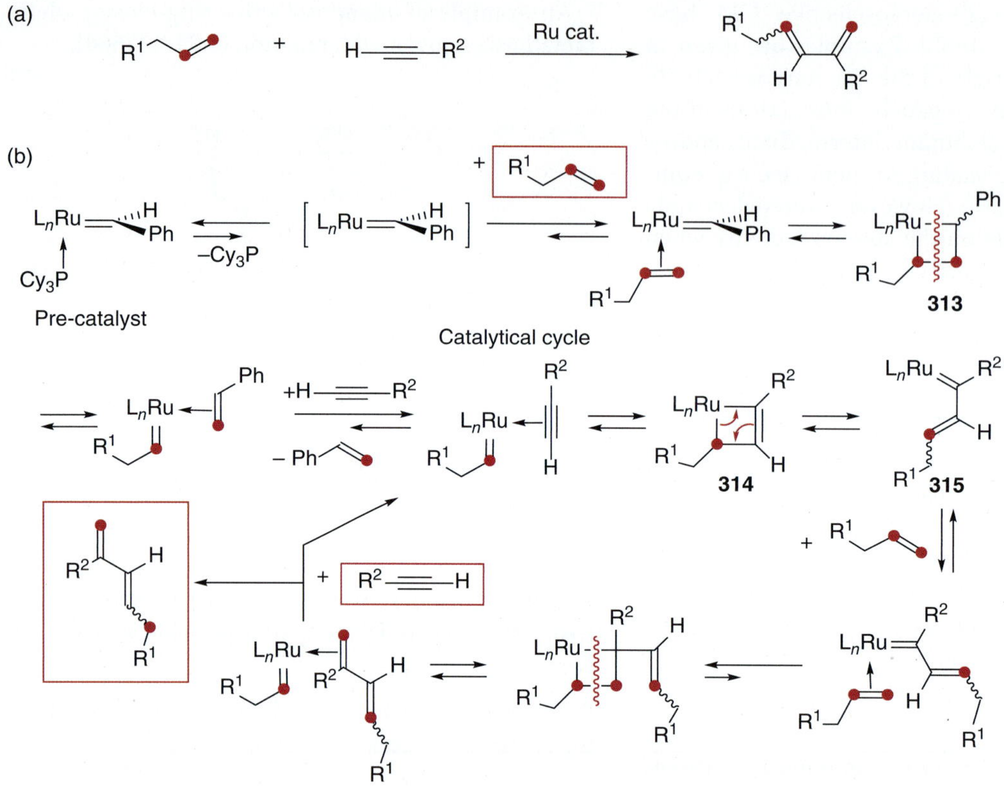

Scheme 8.80 (a) Intermolecular alkene/alkyne cross-metathesis and (b) proposed reaction mechanism.

a metallacyclobutene intermediate of type **314** (see however Scheme 7.65, Section 7.7.12) that isomerizes into vinylcarbeneruthenium intermediate **315** [1669]. The latter then reacts with the alkene to give the final 1,3-diene.

An example of application of alkyne/alkene cross-metathesis is given with reaction (8.243) that produces cyclohepta-1,3-dienes by ring expansion of cyclopentene [1670].

Reaction (8.244) gives an example of ring-closing enyne metathesis. In the presence of ethylene, better yields are obtained as it helps the regeneration of the active catalyst [1671, 1672]. It also suppresses alkyne polymerization [1673]. In case of $R^1C{\equiv}CR^2 + CH_2{=}CH_2$, the corresponding dienes $R^1C({=}CH_2){-}C({=}CH_2){-}R^2$ are produced [1672].

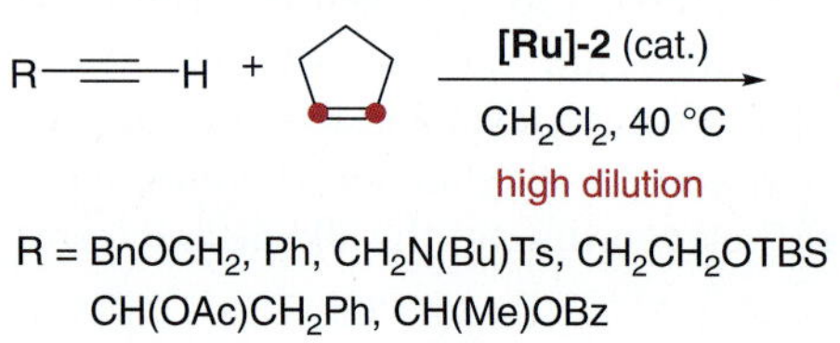

(8.243)

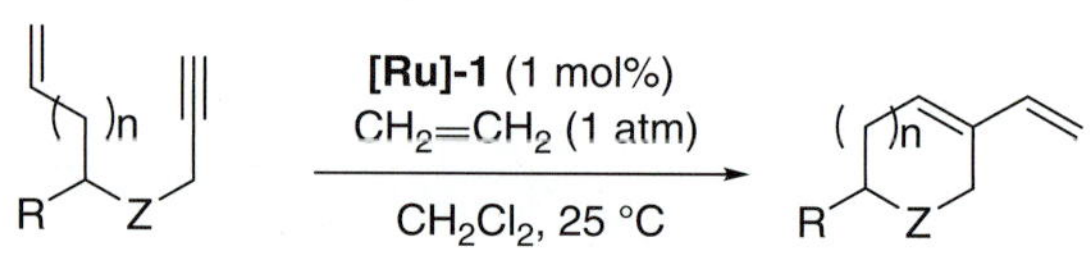

(8.244)

Problem 8.74 What is the major product **P** of the following reaction? [1674]

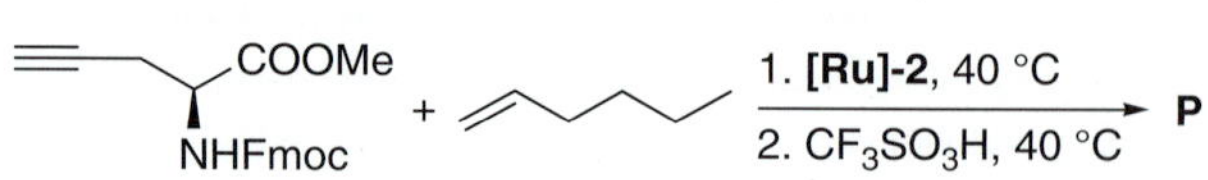

Problem 8.75 What is the major product **P** of the following reaction? [1675]

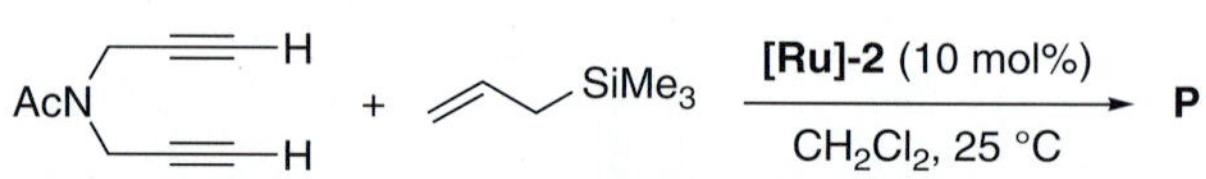

Problem 8.76 Give a mechanism for the following reaction [1676].

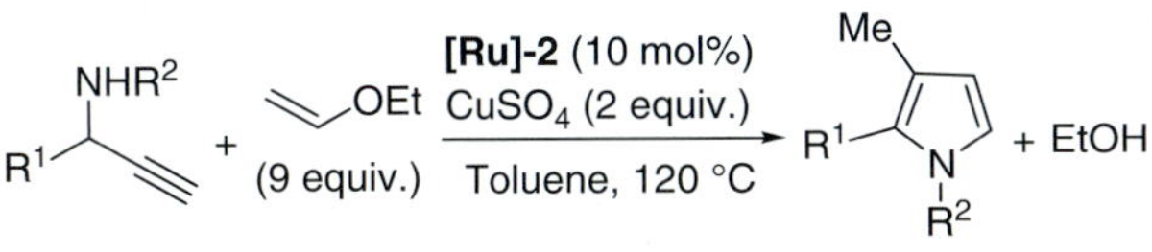

8.6.4 Alkyne metathesis

The discovery of alkyne metathesis and possible mechanisms of reaction have been presented in Section 7.7.13. For the metathesis of simple alkynes (without Lewis base functional groups), a mixture of Mo(CO)$_9$ and *p*-chlorophenol works as a simple catalyst in 1,2-dichlorethane and leads to high yield of symmetrical alkynes (reaction (8.245)) [1677].

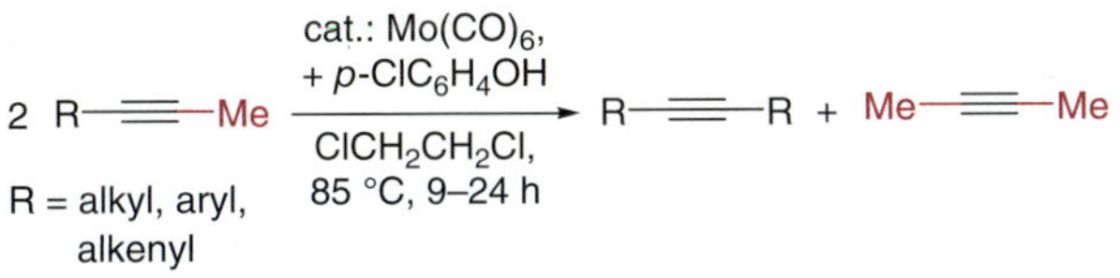

(8.245)

Alkyne metathesis is now applied to the total synthesis of complex natural products and materials [1538, 1544, 1555, 1558, 1678–1680]. Among the various Schrock alkylidynes that qualify as catalysts, tungsten neopentylidyne **316** (or **[W]-1**) has found the broadest use (Figure 8.5). Complex **316** and its variants **317** develop by Tamm and coworkers are air- and moisture-sensitive but affect alkyne metathesis reactions already at room temperature [1681–1685]. The combination of an electron-releasing imidazolin-2-imido ligand with two electron-withdrawing alkoxide ligands seems to be important for an efficient catalyst. Because of their Lewis acidity, these catalysts are not compatible with all substrates containing donor sites [1686], although ring-closing metathesis of peptide-derived alkynes has been successful [1687]. Superior tolerance is observed for molybdenum alkylidynes of type **318** and their alkoxide analogs [1680, 1688]. Nevertheless, catalysts **318** require inert conditions as they can activate small molecules including N$_2$ [1689]. More precatalysts such as **319–322** have been developed by the Fürstner's group (Figure 8.5) [1690, 1691]. Complex **322** is air-stable [1692, 1693].

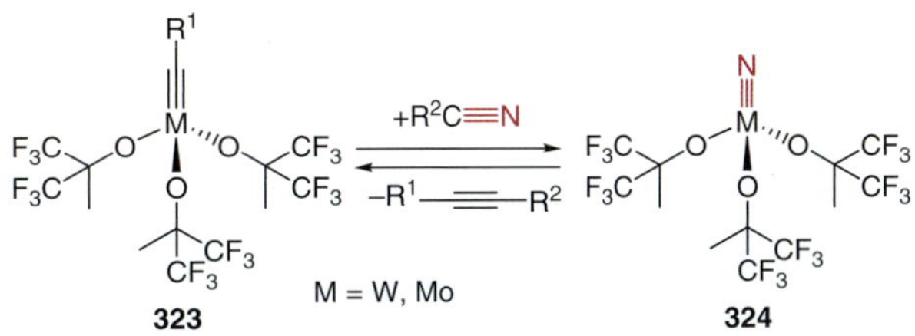

(8.246)

Complexes **323** equilibrate with nitride complex **324** when mixed with a carbonitrile (equilibrium (8.246)). Fürstner's precatalysts [1694] are very practical and tolerate multifunctional compounds for alkyne metathesis as illustrated with the synthesis of gallicynoic acid (reaction (8.247), an example of intermolecular cross-metathesis) and of the epothilone

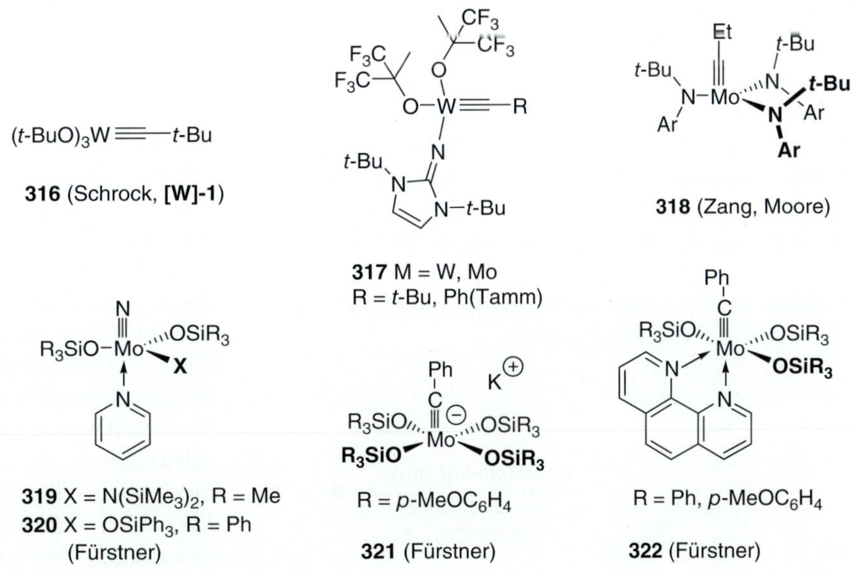

Figure 8.5 Established precatalysts for alkyne metathesis.

A precursor **325** (reaction (8.248)), an example of macrocyclization.

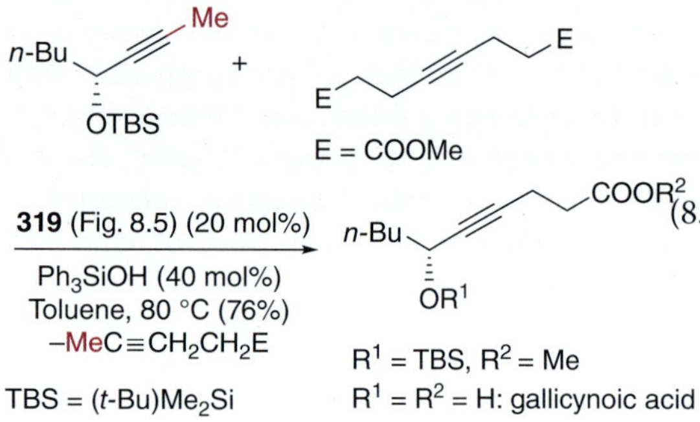

(8.247)

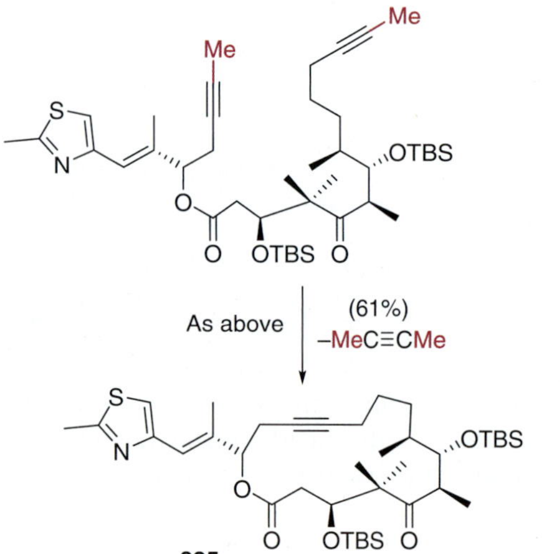

(8.248)

325

Complex **319** reacts with aromatic aldehydes to give the corresponding nitriles (nitrogen transfer reagent). The tungsten nitride complexes (N≡W(OC$(CF_3)_2Me)_3$(dme), $[Li(dme)_2][N{\equiv}W(OC(CF_3)_2Me)_4]$, and $[N{\equiv}W(OCMe_2CF_3)_3]_3$ prepared by salt elimination from $[N{\equiv}WCl_3]_4$ (dme = $MeOCH_2CH_2OMe$)) catalyze the cross-metathesis of 3-hexyne with nitriles to form propionitrile and the corresponding alkynes [1695]. Quantum mechanical calculations support a (2+2)-cycloaddition/cycloreversion mechanism analogous to that of alkyne metathesis [1696].

Problem 8.77 What are the main products **P** + **Q** of the following reactions? [1697]

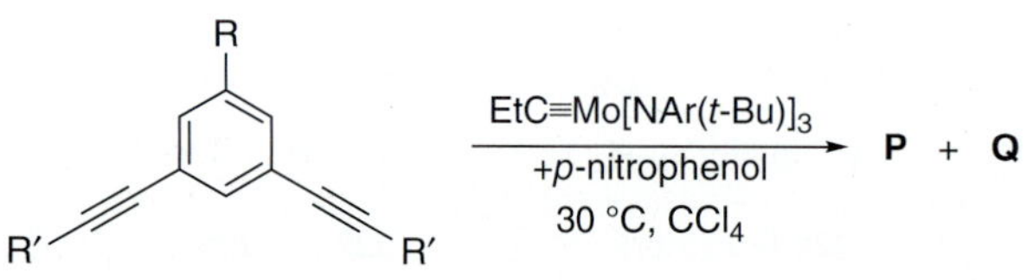

Problem 8.78 What products are formed in the following reaction? [1696]

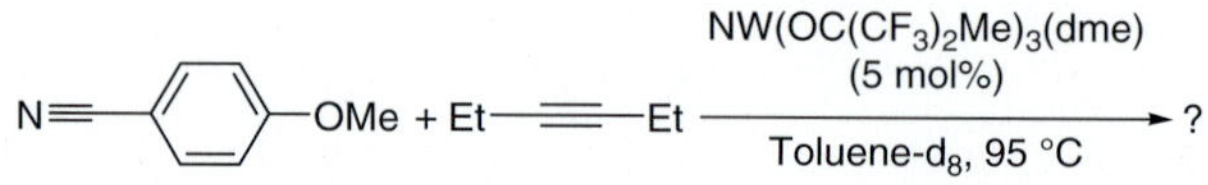

8.7 Additions of organometallic reagents

Classically, C—C bond formation is based on six paradigms: (i) a carbon nucleophile Nu—H (forms a carbanion intermediate Nu^- with an appropriate base) or Nu—M (organometallic reagent) adds to a double or triple bond containing at least one carbon atom (e.g. reaction (8.249), (ii) a carbon nucleophile Nu—H(M) displaces a nucleofugal moiety X from a carbon electrophile R—X (reaction (8.250)), (iii) electrophilic additions and substitutions, (iv) cycloadditions (Section 5.3), (v) ene-reactions (Section 5.7), and (vi) aldehyde and ketone olefinations that can be seen as cycloadditions (Sections 5.3.7 and 5.3.8). Noncatalyzed nucleophilic additions are often more facile than displacement reactions because doubly or triply bonded electrophiles have higher electron affinities (Table 1.A.22) than corresponding singly bonded electrophiles such as alkyl halides or esters (e.g. sulfonates and phosphonates). The latter can undergo substitutions following S_N1, S_N2, or elimination/addition mechanisms. Alternatively, a mechanism that implies single-electron transfer (SET) from the nucleophile to the electrophilic reagents can be followed. In the latter case, free radicals might be generated that may produce unwanted coproducts and polymers. This can also occur with addition reactions. The nucleophilic reagent (Nu—H + base or Nu—M) can induce a concurrent elimination of HX from the electrophilic partner R—X with the formation of a nonreactive alkenes, alkynes, or cycloalkanes. Substrates of the type vinyl-X and aryl-X (X = nucleofugal group) do not undergo direct S_N1 or S_N2 reactions readily. However, the nucleophilic substitutions can occur via an addition/elimination mechanism (e.g. XCH=CHCOOMe + Nu—M ⇄ Nu—C(X)H—CH=C(OM)OMe ⇄ NuCH=CHCOOMe + MX; Ar—X + Nu—M ⇄ cyclohexadienyl anion/M^+ (Meisenheimer complex) ⇄ Ar—Nu + MX), or, for aromatic nucleophilic substitutions (S_NAr), via an elimination/addition mechanism (e.g. Ar—X + NuM ⇄ aryne + NuH + MX ⇄ Ar—Nu + MX). Alternatively, and as already mentioned, the substitution can occur via a SET mechanism (e.g. Ar—X + Nu—M ⇄ $[Ar{-}X]^{\bullet -}$ +

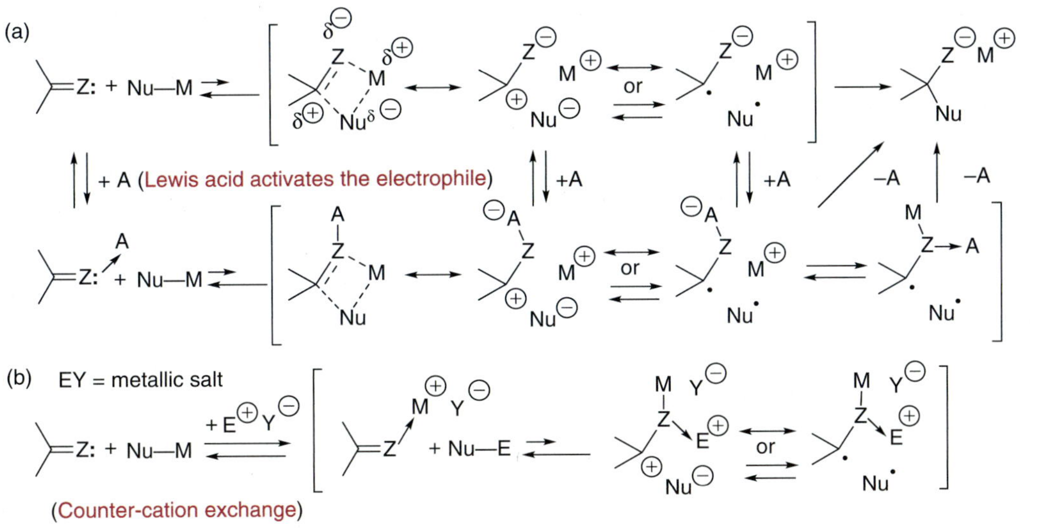

Figure 8.6 Representation of the transition states of the addition of an organometallic nucleophile Nu—M to an electrophile (C=Z, C=O, C=NR, C=S, C≡N, C=C, C≡C, etc.) and possible mode of catalysis (a) by a Lewis acid A and (b) by a salt E^+/Y^- that exchanges M^+ for E^+ in the organometallic nucleophile Nu—M. The diradical structures can be either limiting structures of the addition transition states or be reactive intermediates (radical mechanisms).

$Nu^\bullet + M^+ \rightleftarrows Ar^\bullet + MX + Nu^\bullet$; $Ar^\bullet + Nu{—}M \rightleftarrows [Ar{—}Nu]^{\bullet-} + M^+$; $[Ar{—}Nu]^{\bullet-} + Ar{—}X \rightleftarrows [Ar{—}X]^{\bullet-} + Ar{—}Nu$). Today, catalytic routes exist for almost any types of addition (8.249) and displacement reaction (8.250).

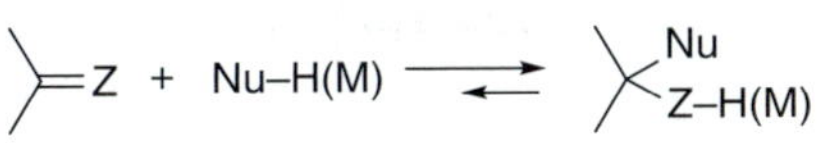
(8.249)

$$R{-}X + Nu{-}H(M) \rightleftarrows R{-}Nu + H{-}X(M) \qquad (8.250)$$

Organometallic compounds are powerful nucleophiles because of their polarity ($R{—}M \leftrightarrow R^-/M^+$) and low ionization energies (IE, Section 1.8). The metal M might also bind to Z of the doubly or triply bonded electrophile, or to X of the singly bonded electrophile (R—X). This leads to (i) a template effect that brings the nucleophile closer to the electrophilic carbon center; in the same time, the coordination of M (M^+, or M—Nu) to the electrophile increases (ii) its polarity and (iii) its electron affinity. Transition states of nucleophilic additions represented in Figure 8.6 suggest a number of ways to catalyze the nucleophilic additions: (i) add a Lewis acid A that binds to the electrophile better than to Nu—M, (ii) add a species (e.g. a complex or a salt with at least two free coordination sites) that reinforces the template effect as its binds to both the nucleophile and the electrophile, (iii) exchange the metal M of the nucleophile Nu—M by another metal, a metallic complex, or counterion (e.g. bulky ammonium and phosphonium ion) that reinforces the template effect, and/or (iv) that renders the single-electron transfer from the nucleophile to the electrophile easier. The diradical structures in Figure 8.6 can be limiting structures of the transition states or true intermediates of addition reactions. In the latter cases, free, neutral radical $Nu^\bullet$ may or may not diffuse out the solvent cage. Today, methods are being developed that use nonacidic C—H bonds in arenes, alkenes, and alkanes as nucleophiles in catalyzed direct addition (Section 8.7) and displacement reactions (Section 8.8). These reactions are often classified as direct C—H functionalization.

The first compound containing a carbon–metal bond was discovered by Cadet in 1760. Heating a mixture of As_2O_3 with AcOK gave a volatile, stinking product called cacodyl oxide, $(Me_2As)_2O$ ($As_2O_3 + 4MeCOOK \rightarrow (Me_2As)_2O + 2K_2CO_3 + 2CO_2$).

In 1849, Frankland obtained diethylzinc ($3EtI + 3Zn \rightarrow Et_2Zn + EtZnI + ZnI_2$), a pyrophoric compound [1698]. In 1852, he described the preparation of dimethylmercury ($2MeBr + 2Na/Hg \rightarrow Me_2Hg + 2NaBr$). On their side, König and Schweizer obtained tetraethyllead (the antiknocking agent added to gasoline since 1922) ($4EtI + 4Na/Pb \rightarrow Et_4Pb + 4NaI$). The same method allowed them to prepare Et_3Sb and Et_3Bi [1699]. The first organotin compound, Et_4Sn, was obtained by Löwig [1700]. These alkyl metals are difficult to handle and are relatively unreactive with organic compounds [1701]. In 1859, Hallwachs and Schafarik prepared the first organoaluminum compound $Et_3Al_2I_3$ from metallic Al and EtI. It took almost 100 years before Ziegler et al. discovered the synthetic and catalytical potential of organoaluminum compounds. He was awarded the Nobel Prize in

1963 together with Natta for their discovery of the so-called "Mülheim atmospheric polyethylene process" [1703] and stereospecific polymerizations of propene and other α-olefins [1704–1708]. In 1863, Friedel and Crafts reacted $SiCl_4$ with Et_2Zn and obtained Et_4Si [1709]. The work of Frankland inspired Reformatsky who showed in 1887 that zinc enolates derived from the treatment of α-bromoesters with zinc ($BrCH_2COOEt + Zn \rightarrow CH_2{=}C(OEt)OZnBr$) are useful nucleophilic reagents adding to aldehydes. In 1899, Barbier replaced zinc by magnesium in reactions that convert ketones and alkyl iodides into carbinols [1710]. Grignard studied this reaction in detail and demonstrated in 1900 that the Barbier reactions involve the formation of RMgX (Grignard reagents). They are also strong bases able to deprotonate acids with $pK_a < 40$ [1711]. The Grignard reagents are almost universally accessible and at the same time are more reactive than zinc, boron, aluminum, tin, silicon, or lead organometallic reagents; consequently, they have been one of the favorite tools in preparative chemistry for one century [1712–1714]. In 1917, Schlenk and Holtz discovered ethylithium [1715]. In 1930, Ziegler and Colonius developed the preparation of alkylithium by direct metallation of alkyl halides [1716]. The reaction works much better with lithium containing 0.02% sodium than with pure lithium because the sodium helps the electron transfer implied in the process ($R{-}X + (Li)_n \rightleftarrows R^\bullet + LiX + (Li)_{n-1} \rightarrow R{-}Li + Li_{(n-1)}X$) [1717].

8.7.1 Additions of Grignard reagents

A classical method of formation of Grignard reagents ($RX + Mg \rightarrow RMgX$, in Et_2O, THF) [1713, 1718–1723]) takes place at the magnesium metal surface [1724]. Reaction commences with a single-electron transfer (e.g. $RBr + (Mg)_n \rightarrow RBr^{\bullet-} + (Mg)_n^{\bullet+}$) and decomposition of the radical anion ($RBr^{\bullet-} \rightarrow R^\bullet + Br^-$), followed by rapid combination of the organic radical with a magnesium ion ($R^\bullet + (Mg)_n^\bullet + Br^- \rightarrow RMgBr + (Mg)_{n-1}$) [1725–1728]. When hex-5-en-1-yl bromide reacts with Mg, small amounts of cyclized products are formed together with products of radical coupling [1729–1732]. Often Grignard reagents are written as RMgX. However, their actual structures in solution are far more complicated (see below) [1733]. The order of reactivity of metallic Mg with organic halides is RI > RBr > RCl. Unreactive halides toward magnesium shavings or chips may react by using activated forms of Mg such as the distilled metal [1734], Mg solubilized with anthracene [1735] or by reducing Mg salts with Na or K metal (Rieke's activated Mg) [1736, 1737]. Even alkyl fluorides, which are unreactive under normal conditions, form Grignard reagents with Rieke's activated magnesium. Sonication or mechanical pretreatment can also be used to activate magnesium [1738, 1739]. Anthracenemagnesium selectively reduces C_6F_{12} or C_6F_6 and $CF_3C_6F_{11}$ or $CF_3C_6F_5$ to C_6F_5MgF and $CF_3C_6F_4MgF$, respectively [1740]. Knochel and coworkers have found that addition of LiCl to RMgCl produces highly active reagents of the type $RMgCl_2^-/Li^+$ and R_2MgCl^-/Li^+ that undergo fast Br/Mg exchange reactions [1741–1743]. A particularly efficient way to achieve this conversion is the treatment of bromoarenes with *i*-PrMgCl·LiCl (e.g. ArBr + *i*-PrMgCl·LiCl + 2THF → $[Ar{-}Br{-}i\text{-}Pr]^-/Li^+ + MgCl_2(THF)_2 \rightarrow$ ArMgCl·LiCl + *i*-PrBr + 2THF) [1744–1749]. Mg/I exchange can also be realized in the same way [1750]. Br/Mg exchange is faster than Cl/Mg exchange reaction and, very interestingly, it is faster than nucleophilic addition to ester, carbonitrile moieties, thus adding a new dimension to Grignard reagent chemistry [1714]. In 2012, Hayashi and coworkers reported that $FeCl_3/CuBr/Bu_3P$ catalyzes the alkene/Grignard exchange cyclopentyl-Br + $RCH{=}CH_2 \rightleftarrows$ cyclopentene + RCH_2CH_2MgBr (via β-elimination of H and subsequent alkene hydromagnesiation, an exchange reaction described in 1971 by Tamura and Kochi [1751]) and that (alk-1-yl)MgBr adds to alkynes to generate (alkenyl)MgBr (e.g. $R^1MgBr + R^2C{\equiv}CR^3 \rightarrow R^1(R^2)C{=}C(R^3)MgBr$) in the presence of $Fe(acac)_3/CuBr/Bu_3P/Me_2NCH_2CH_2NMe_2$ as a catalyst [1752]. On their side, Nakamura and coworkers found the $FeCl_2$-catalyzed hydromagnesiation of diarylacetylenes and diynes with $EtMgBr/Et_2O$ as an hydride donor (e.g. Ph—C≡C—Ph + EtMgBr → (*E*)-PhCH=C(Ph)MgBr + $CH_2{=}CH_2$) [1753].

As the C—Mg bonds in Grignard reagents have a polar character, they may dissociate into (alkyl anion)$^-$/MgX$^+$ ion pairs and undergo racemization [1754, 1755]. Contrary to the highly reactive alkali organometallics, the Grignard reagents do not require low temperature and special techniques for their preparation and handling. Without catalyst, they add readily to aldehydes, ketones, esters, acyl halides, nitriles, epoxides, and their vinylogs (Michael acceptors), as well as to CO_2, pyridines, and strained olefins (reaction (8.249)) such as cyclopropenes [1756–1758] and 7-*syn*-norbornenol [1759, 1760]. In solution, Grignard reagents exist as Schlenk equilibria (8.251) [1761]. Mixtures of various species including dimeric and trimeric aggregates are formed [1762, 1763]. The degree of aggregation depends on the halogen (X), concentration, nature of the alkyl or aryl group R,

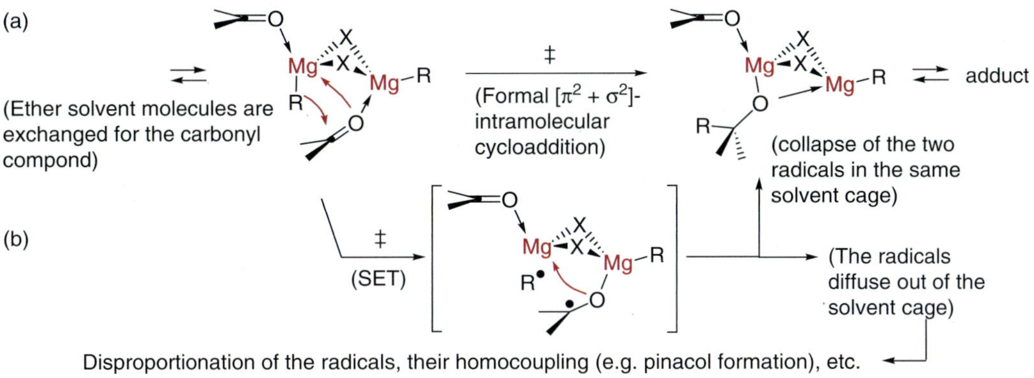

Scheme 8.81 Two possible limiting mechanisms for the addition of Grignard reagents to carbonyl compounds: (a) the polar mechanism and (b) the SET mechanism.

temperature, solvent, and additives [1764, 1765]. For simple alkyl or aryl magnesium chlorides in Et_2O, the predominant species is a solvated halogen-bridged dimer [1766, 1767]. For low concentration (<0.1 M) in Et_2O, the alkylmagnesium bromides are monomeric and dimeric at higher concentration (>0.5 M) [1768].

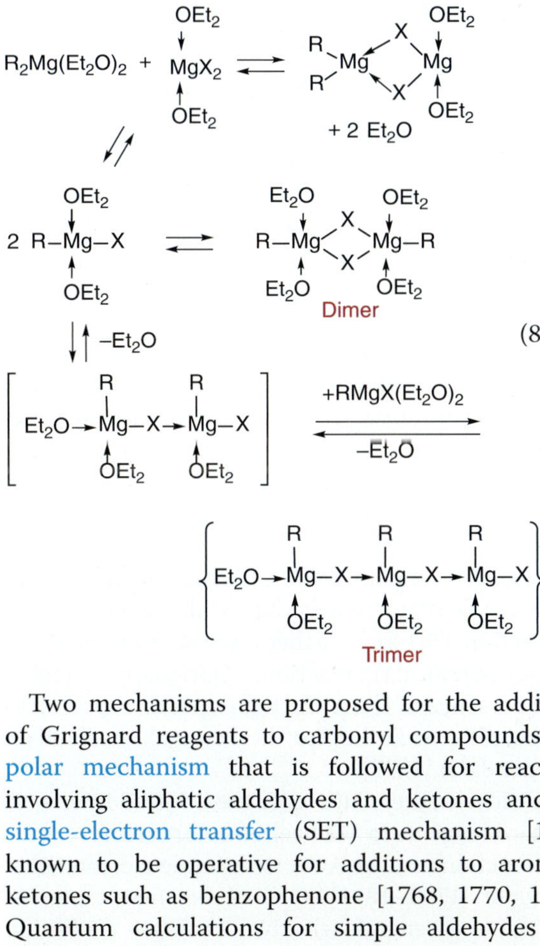

(8.251)

Two mechanisms are proposed for the additions of Grignard reagents to carbonyl compounds: the polar mechanism that is followed for reactions involving aliphatic aldehydes and ketones and the single-electron transfer (SET) mechanism [1769] known to be operative for additions to aromatic ketones such as benzophenone [1768, 1770, 1771]. Quantum calculations for simple aldehydes and MeMgCl in Me_2O confirm these hypotheses [1772]. In the polar mechanism (Scheme 8.81a), the Grignard reagent reacts as a dihalide-bridged dimeric species. The carbonyl compound replaces ether molecules that solvate the magnesium centers leading to a relatively strong Mg· · ·O interaction (Lewis acid/base complexation). This brings the alkyl (or aryl) group of the Grignard reagent in proximity to the carbon of the carbonyl function. Small alkyl groups are transferred to the carbonyl group simultaneously with the formation of a magnesium alcoholate. If the alkyl (or aryl) group of the Grignard reagent is large or if the ketone is an aromatic or a polarizable substituent (can accept an electron readily), then a single-electron transfer occurs before the migration of the alkyl (or aryl) group (SET mechanism, Scheme 8.81b). This generates a diradical intermediate that collapses into the product of addition concurrently with the diffusion of the radicals out of the solvent cage. In such a case, unwanted coproducts are formed.

Kinetic secondary deuterium isotope effects have been used as criteria for distinguishing between a polar or SET mechanism for the addition of alkylmagnesium bromide to benzaldehyde. In the case of reaction of allylboronate that adds to benzaldehyde and α-deuterated benzaldehyde according to a concerted, cyclic transition state ("carbonyl-ene reaction" with transfer of the boron moiety from carbon to oxygen center, Section 5.7.5), $k_H/k_D = 0.85$ in toluene at 24 °C and 0.74 in THF at 78 °C [1773]. The reactions of alkylmagnesium bromide and alkyllithium with PhCHO and PhCDO in ether at −78 °C showed normal isotope effects k_H/k_D = 1.04 and 1.095, respectively. This is ascribed to a SET mechanism generating ion radical pairs in the rate-determining step [1774–1776]. In the presence of enantiopure Cu or Ti catalyst, 1,2-addition of Grignard reagents to aldehydes and ketones can produce enantio-enriched alcohols [1777].

In 1940, Kharash and Tawney observed that catalytic amounts of copper halides favor 1,4-addition over the 1,2-addition of Grignard reagents to α,β-unsaturated ketones [1778]. Later, House et al. demonstrated that Grignard reagents undergo metal/metal exchange with cuprous salts generating organocuprous compounds responsible for the high 1,4-regioselectivity [1779, 1780]. Gilman and Kirby found in 1941 that other metals than copper catalyze the conjugate additions of Grignard reagents (e.g. R_2Cd, R_2Zn, and R_3Al) [1781]. Combining enantiomerically pure ligands with copper or zinc catalysts, enantioselective 1,4-additions of Grignard reagents to enones have been realized [1782–1788]. Ma and coworkers have reported that $Fe(acac)_3$ catalyzes the selective 1,4-addition of Grignard reagent to allenoates (reaction (8.252)) [1789]. The mechanism of this reaction is under debate. It is not excluded that the actual catalyst is a reduced form of iron resulting from the reaction of the Grignard reagent with $Fe(acac)_3$, such as $R_xFe(MgX)_y$ [1790, 1791].

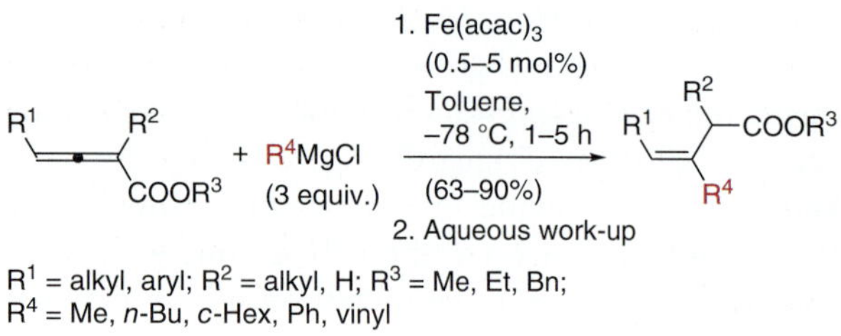

(8.252)

A change in 1,2- vs. 1,4-regiochemistry occurs also in the presence of complexed zinc salts where the reactive species could be a zincate R_3ZnMgX [1792, 1793]. For reactions under thermodynamic control (products are equilibrated with reactants), the 1,2- vs. 1,4-regioselectivity is determined by the relative stability of the 1,2- and 1,4-adduct (magnesium alcoholate or enolate, respectively). The 1,2-addition gives a magnesium alk-2-enylalkoxide (reaction ((8.253)), whereas the 1,4-addition produces a more stable magnesium enolate (reaction (8.254)). Therefore, 1,4-addition is favored over 1,2-addition under conditions of thermodynamic control. The free energy difference $\Delta_rG(8.254) - \Delta_rG(8.253)$ is estimated to $-11\,\text{kcal mol}^{-1}$ for the addition of an hydride to crotonaldehyde in H_2O (compare the gas-phase heats of formation of butanal and of (*E*)-but-2-enol and consider that the latter compounds have the same pK_a in water (Table 1.A.23, reactions (8.255) and (8.256)). With other nucleophiles than RMgX, the difference in heats of 1,4- vs. 1,2-addition may diminish, for instance, when the alkoxide resulting from the 1,2-addition enjoys significant enthalpic anomeric effect because of polar geminal disubstitution (Section 2.7.9).

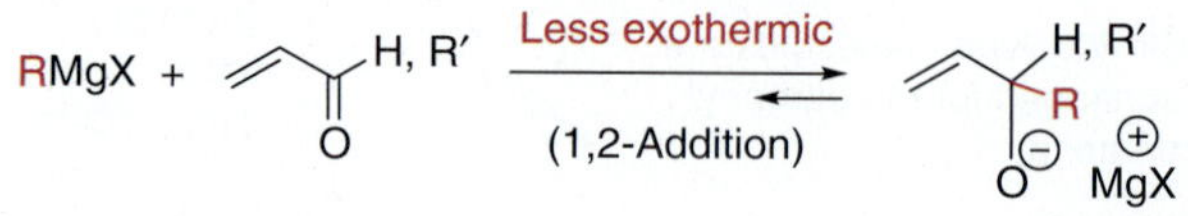

(8.253)

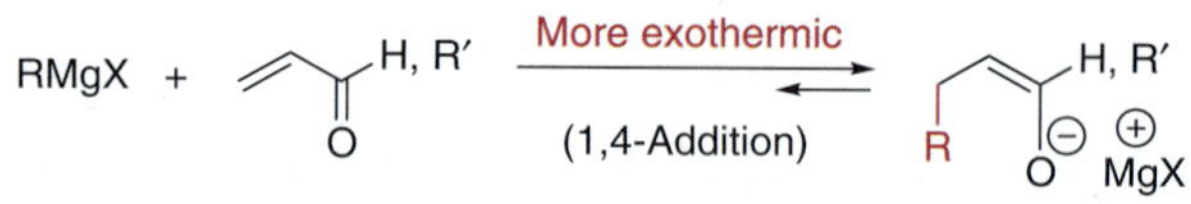

(8.254)

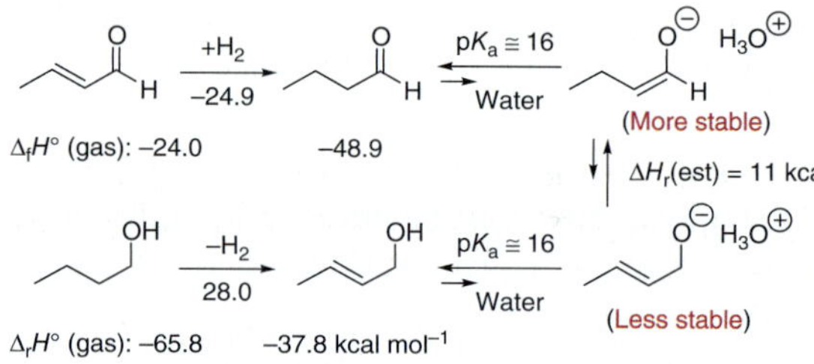

(8.255)

(8.256)

Considering the above thermochemical considerations, it is a surprise that Grignard reagents usually prefer 1,2-additions (unless steric hindrance retards the 1,2-addition [1794–1796]) to 1,4-additions in the absence of catalyst. For alkylmagnesium halides adding by the polar mechanism, the 1,2-addition is favored kinetically because there is a strong electrostatic interaction between the oxygen center of the carbonyl function and the magnesium atom of the Grignard reagent that makes this addition quick and irreversible. This brings the alkyl groups into the proximity of the carbonyl group of the α,β-unsaturated carbonyl compound as shown in Scheme 8.81a and it favors 1,2-addition. In the case of a SET mechanism (Scheme 8.81b), the combination of radical $R^\bullet$ with center C(1) of the allyl-OMgX radical is faster than migration of $R^\bullet$ to center C(3) of this allylic radical (see Section 8.8.1 for a further discussion on the intermediacy of radicals in reactions of Grignard reagents). In the presence of Cu(I), the R group is transported to position C(3) according to the Feringa's mechanism (Scheme 8.82) [1797].

Enantioselective Cu(I)-catalyzed 1,4-additions of Grignard reagents to Michael acceptors are now available (reactions called ACA = asymmetric conjugate additions) [1798–1811]. One of the first examples of catalyzed enantioselective conjugate addition [1812] was reported by Lippard and coworkers [1813]. It was found that HMPA ($(Me_2N)_3P{=}O$)

Scheme 8.82 Feringa's mechanism of the Cu(I)-catalyzed conjugate addition of Grignard reagents (homochiral diphosphine ligand: e.g. (*R,S*)-Josiphos, (*R,S*)-Taniaphos, up to 99% ee).

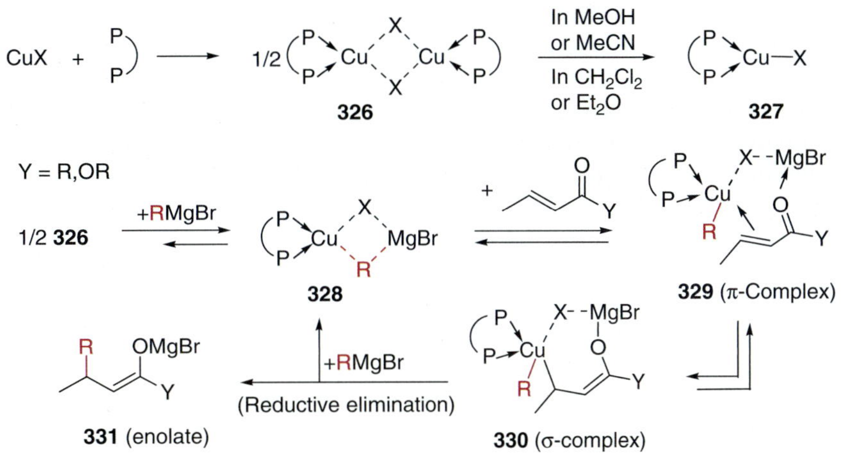

and silyl reagents accelerate the 1,4-addition of the intermediate cuprates and provide improvement for enantioselectivity [1814]. In 1991, van Koten and coworkers introduced arenethiolate complexes as homochiral catalysts [1815]. Zhou and Pfaltz obtained good enantioselectivities with catalysts containing a L-proline-derived phosphine ligand (reaction (8.257)) [1816]. Taddol-derived phosphine–phosphite ligands in racemic 2-methyltetrahydrofuran as solvent also led to good enantioselectivities (reaction (8.258)) [1800]. Asymmetric conjugate additions of Cu(I) salts and homochiral diaminocarbenes [1817, 1818] permit enantioselective 1,4-additions of Grignard reagents. Asymmetric Rh(I)-catalyzed 1,4-additions of organometallics are also much used [1819].

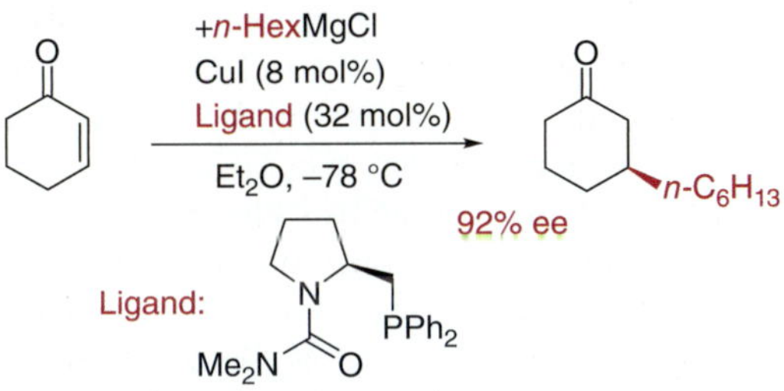

(8.257)

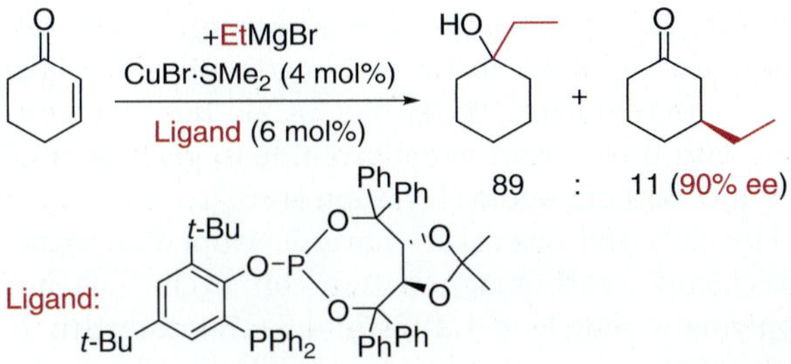

(8.258)

With 3-alkylcyclohex-2-enone, the conjugate addition of Grignard reagents generates 3,3-dialkylcyclohexanone. This reaction can also be carried out with homochiral Cu-based catalysts leading to good enantioselectivities [1820, 1821]. The enantioselectivity depends very much on solvent (bad in THF, good in toluene and *t*-BuOMe). Enantio- and regioselective additions of EtMgBr to alkenyl-*N*-heteroaromatics (Ar—CH=CHR) to give the corresponding product of alkylation (Ar—CH_2—CH(R)Et) has been catalyzed by CuBr·SMe_2 and a homochiral diphosphine [1822]. On the basis of kinetic, spectroscopic, and electrochemical analysis, Feringa and coworkers have proposed the mechanism shown in Scheme 8.82 for the Cu(I)-catalyzed enantioselective 1,4-additions of Grignard reagents to *α*,β-unsaturated carbonyl compounds. Dimeric complex of diphosphine ligand and Cu(I) salt **326** is equilibrated with the monomeric complex **327**. Reaction with the Grignard reagent forms the doubly bridged complex **328** that coordinates the double C=C bond of the *α*,β-unsaturated carbonyl compound giving π-complex **329**. The latter undergoes a β-insertion of the alkyl group R producing the σ-complex **330**. The rate-determining step is the reductive elimination **330** → **331** (adduct) + **328** (catalyst). Yield and enantioselectivity depend on the Schlenk equilibrium [1797].

Reactions of Grignard reagents with *α*,*β*,*γ*,δ-unsaturated carbonyl compounds (ketones, carboxylic esters, and carboxamides) usually give mixtures of products unless a suitable catalyst is applied. Copper-catalyzed enantioselective conjugate additions of organometallics to extended Michael acceptors are now routine [1810]. With copper catalysts 1,4-additions have been reported [1823, 1824] as well as selective 1,6-additions to 2,4-dienoates [1825–1828], 2,4-dienamides [1824], 2,4-dienones [1796, 1829–1832], and enynoates [1833, 1834]. Interestingly, 1,6-selective addition of aryl Grignard reagents to *α*,*β*,*γ*,δ-unsaturated carboxylic esters and carboxamides can be achieved with iron catalysts (reaction (8.259)) [1835, 1836]. The *anti* stereoselectivity of the 1,6-addition with respect to the carboxamide group is attributed to the formation of an intermediate dieneiron complex **332**. Enantioselective regiodivergent 1,4 vs. 1,6 copper-catalyzed conjugate additions have applied homochiral diaminocarbene ligands with $Cu(OTf)_2$ as a catalyst [1837]. Catalytic enantioselective 1,6-conjugate addition of Grignard reagents to linear dienoates have applied $CuBr{\cdot}SMe_2$ and homochiral ferrocenyl-based phosphines [1834, 1838]. Cu(I)-catalyzed enantioselective 1,2-additions of Grignard reagents to carbonyl compounds [1839] and imines have been reported [1840, 1841].

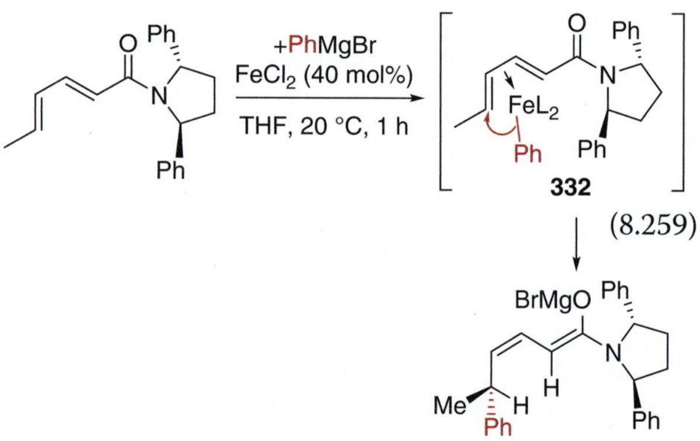

(8.259)

In general, alkenes and alkynes that are not Michael acceptors do not add Grignard reagents without a catalyst. In principle, the carbomagnesiation of alkenes and alkynes realizes a useful way to produce new Grignard reagents. In 1924, Job and Reich demonstrated that $NiCl_2$ catalyzes the addition of PhMgBr to ethylene giving $PhCH_2CH_2MgBr$ that can be quenched with CO_2 to produce 3-phenylpropionic acid [1842]. In the meantime, several examples of catalyzed alkene and alkyne carbomagnesiation have been reported [1843–1848]. In 1978, Duboudin et al. reported $(Ph_3P)_2NiCl_2$-catalyzed carbomagnesiations with MeMgBr and PhMgBr of unactivated alkynes such as PhC≡CEt and EtC≡C—Et [1849]. Acetylene reacts with Grignard reagents by hydrogen/metal exchange: RMgX + HC≡CH → HC≡CMgX + R—H. However, in the presence of 5% $CuBr{\cdot}OEt_2$, *syn*-carbometallation (8.260) is observed [1850]. In 2005, using $Fe(acac)_3$/CuBr/Bu_3P as a catalyst, Hayashi and coworkers realized the *syn*-arylmagnesiation of dialkylacetylenes [1851]. In 2007, Oshima and coworkers catalyzed the *syn*-arylmagnesiation of unactivated acetylene with $CrCl_2$ and pivalic acid (reaction (8.261)) [1852]. In 2001, de Armas and Hoveyda catalyzed alkylmagnesiation of styrene with Cp_2ZrCl_2 (titanocenedichloride) and using primary and secondary alkyl tosylates as the alkyl source. Reaction (8.262) proceeds through a π-complex of type (alkene)$ZrCp_2$ [1853]. Fox and coworkers have described the Cu-catalyzed carbomagnesiation of alkyl, alkenyl, alkynyl, and aryl Grignard reagents to 3-hydroxymethylcyclopropenes with high regio- and diastereoselectivity. The cyclopropylmetals so obtained can be trapped with a variety of electrophiles to generate highly substituted cyclopropanes [1854]. Polysubstituted conjugated enynes have been prepared by addition of Grignard reagents to 1,3-diynes in the presence of $FeCl_2$ catalyst [1855]. Enantioselective carbomagnesiation of nonactivated alkenes has also been developed [1856].

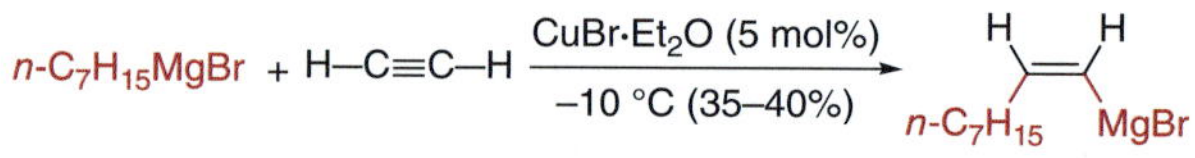

(8.260)

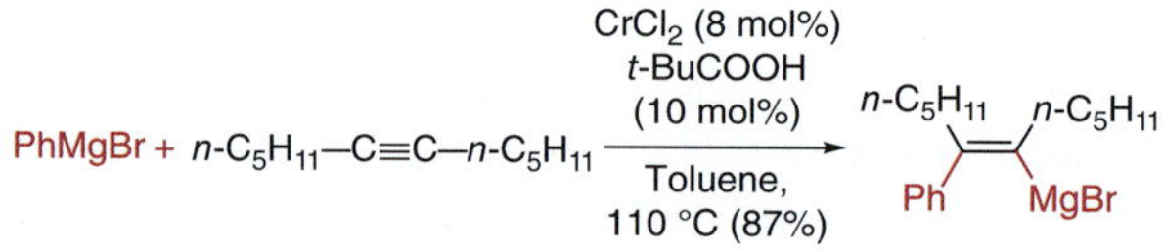

(8.261)

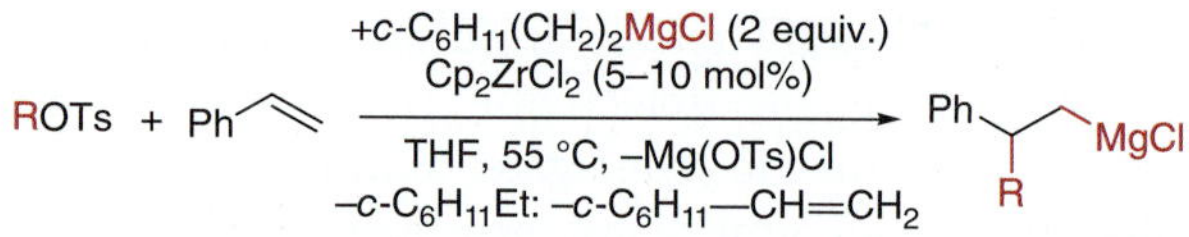

(8.262)

Regioselective double alkylation of styrenes with alkyl Grignard reagents and alkyl bromides having a β-heteroatom functional group has been realized by the use of 5 mol% of Cp_2TiCl_2 in THF (Scheme 8.83). Cp_2TiCl_2 reacts with 3 equiv. of n-BuMgCl to generate first Cp_2TiCl, Cp_2TiBu, and finally $[Cp_2TiBu_2]^-/MgCl^+$. Electron transfer from the latter to the β-bromoethyl ether in the presence of BuMgCl leads to butyl radical, which adds to styrene giving a benzyl radical that is trapped with Cp_2Ti (arising from reaction $Cp_2TiBu_2 \rightarrow Cp_2Ti$ + but-1-ene + butane). Subsequent Ti/Mg metal/metal exchange with n-BuMgCl gives intermediate PhCH(MgCl)CH_2-Bu that couples with the β-bromoethyl bromide to give the final product **333** of double alkylation [1857].

Hoveyda and coworkers have developed an enantioselective carbomagnesiation of cyclic alkenes applying ethylene-1,2-bis(η^5-4,5,6,7-tetrahydro-1-indenyl)zirconium dichloride ([EBTHI]$ZrCl_2$) as the homochiral catalyst (Scheme 8.84) [1858]. EtMgCl reacts with the latter to give an ethylenezirconium complex (**334**) that coordinates the alkene giving **335**. Then, oxidative cyclization occurs giving a zirconacyclopentane intermediate **336** that adds

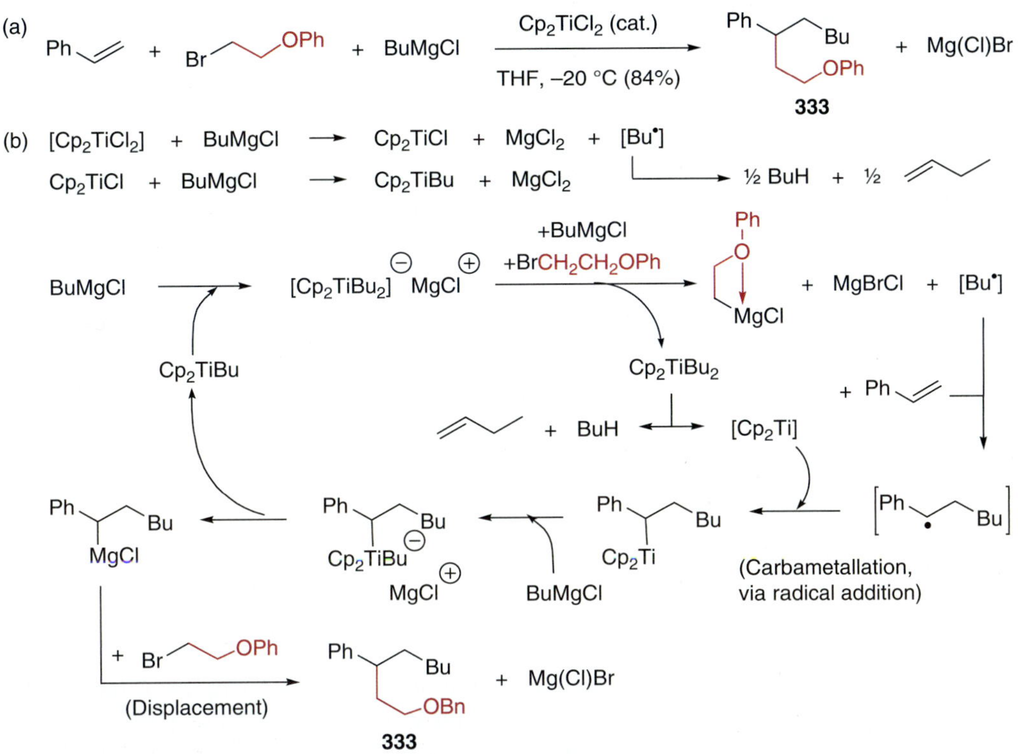

Scheme 8.83 (a) Regioselective double alkylation of styrene with alkyl Grignard and alkyl bromides and (b) proposed reaction mechanism that involves alkyl radical formation and its addition to the alkene.

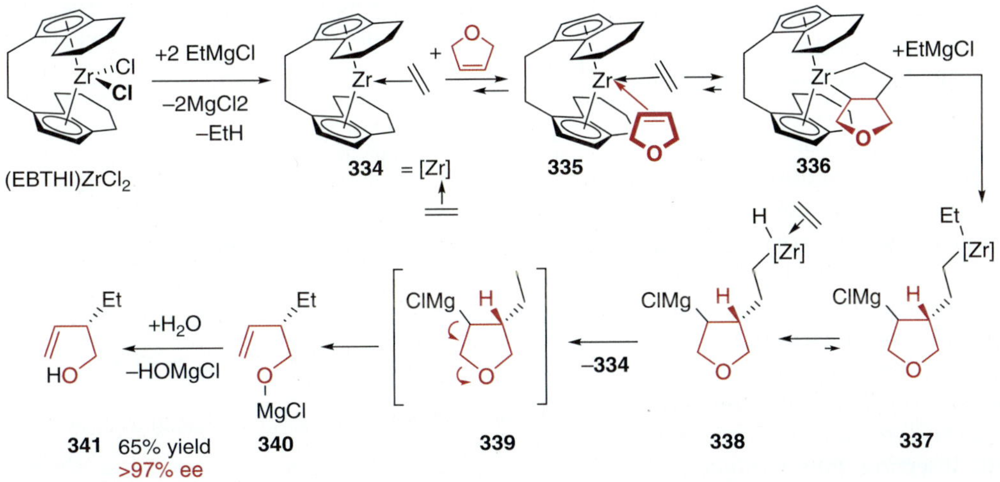

Scheme 8.84 Hoveyda's asymmetric carbomagnesiation of cyclic allylic ethers.

EtMgCl to give **337**. Subsequent β-elimination engenders **338**, which then undergoes reductive elimination generating the ethylenezirconium complex **334** and the product of carbomagnesiation **339**. A β-elimination gives a magnesium alkene-alcoholate **340** that, under aqueous work-up, liberates the product of alkylation **341**. The overall process can be seen as the result of an indirect, enantioselective S_N2' displacement of the ethereal moiety of 2,5-dihydrofuran by EtMgCl.

Problem 8.79 Propose a synthesis of compound **P** [1801].

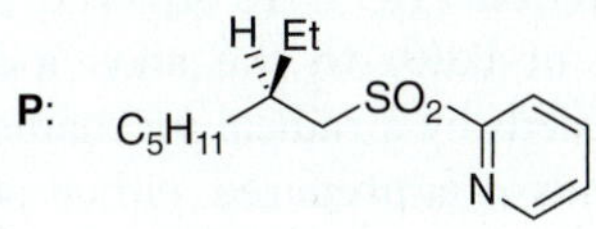

Problem 8.80 Propose a synthesis of 2-hexyl-1-phenyloct-1-ene starting with hex-1-ene and 1-phenyloct-1-yne [1752].

8.7.2 Additions of alkylzinc reagents

Organozinc derivatives have much covalent character in the carbon–metal bond and are thus less reactive than organomagnesium and organoalkali reagents. This low reactivity allows for the presence of many functional groups (e.g. ester, keto, cyano, and nitro) in organozinc compounds [1859–1866]. Although dialkyl-, dialkenyl-, and diarylzinc reagents (R_2Zn) are highly pyrophoric, corresponding RZnHal reagents can be handled easily applying Schlenck techniques. Recently, Knochel and coworkers have developed the bench-stable pivalates RZn(OPiv) [1867, 1868]. These reagents contain Mg and Li salts and are better described by $RZnX{\cdot}Mg(OPiv)_2{\cdot}LiCl$ [1869]. Reactions of organozinc compounds can be promoted by ultrasounds [1870–1874] or under microwave irradiation [1875–1880]. Combined with lithium amides such as lithium 2,2,6,6-tetramethylpiperidide, they become powerful metallation agents for hydrogen/metal exchange [1881–1884]. A broad range of zinc organometallics have been prepared by the direct insertion of zinc (added in the form of Zn dust) to a polyfunctional alkyl iodide [1885], by using highly activated "Rieke-zinc" with an alkyl bromide [1886–1888]. Interestingly, activated organic halides such as benzylic, allylic, or propargylic halides undergo the insertion reaction far more readily, allowing the use of the corresponding bromide, or in several cases, the corresponding chloride or phosphate [1859–1863, 1889, 1890]. Although organozinc reagents are relatively unreactive, they undergo fast transmetallations with most transition metal salts (except those of manganese and lanthanides). Knochel and coworkers found that addition of LiCl to zinc powder considerably facilitates its insertion into various aryl and heteroaryl iodides and bromides [1891]. Using directing groups, functionalized aryl and heteroaryl zinc reagents have been obtained regioselectively from poly iodo- and bromo-aryls and heteroaryls [1892]. Organozinc halides of type RZnHal are also obtained using a Pd- or Ni-catalyzed halide–zinc exchange reaction [1893, 1894]. The initial oxidative addition of Pd(0) to the alkyl iodide or alkyl bromide proceeds by a radical mechanism. Dialkylzinc compounds are prepared either starting from alkyl iodides ($2R{-}I + Et_2Zn \rightarrow R_2Zn + 2EtI$) or more attractively from olefins via hydroboration ($RCH{=}CH_2 + R'_2BH \rightarrow RCH_2{-}CH_2BR'_2$; $2RCH_2{-}CH_2BR'_2 + Et_2Zn \rightarrow (RCH_2CH_2)_2Zn + 2EtBR'_2$) [1895, 1896]. Alkenylzinc derivatives can be prepared via dialkylhydroboration of acetylenes and subsequent transmetallation with Et_2Zn [1897–1899]. In general, the reaction conditions for the boron–zinc exchange are so mild that a number of sensitive functional groups such as acrylic ester, nitro, diethyl malonate, and enol ether are tolerated [1900]. The functionalized organozinc compounds react smoothly with various electrophiles including aldehydes, ketones, and their α,β-unsaturated analogs in the presence of a transition metal catalyst. In 1996, Noyori and coworkers reported a first example of selective 1,4-addition of Et_2Zn to cycloalk-2-enones catalyzed by $CuCN + BnNHSO_2Ph$. The intermediate ethylzinc enolate was quenched with an electrophile such as PhCHO (aldol reaction) or allyl acetate (ketone α-allylation) [1901]. In 2006, Marshall et al. reported that selective 1,4-addition of diakylzinc to α,β-unsaturated aldehydes can be catalyzed by $Pd(OAc)_2 + Ph_3P$ or $Me_2CuCNLi_2$ [1902]. Asymmetric versions of the reaction have been developed [1859b, 1903–1906]. An example is given with tandem addition/fluorination of acyclic β-ketoesters with dialkylzinc reagents catalyzed by $Cu(OTf)_2$ and homochiral phosphoramidate ligand **342** (reaction (8.263)) reported in 2011 by Ma and coworkers [1907].

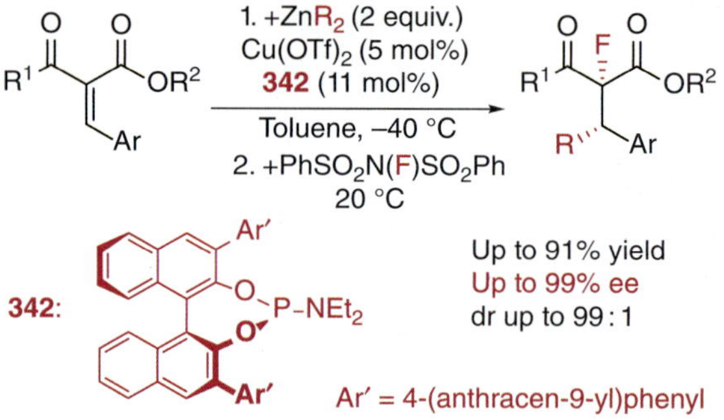

(8.263)

Tanaka and coworkers reported the Cu-catalyzed carbozincation of alkynylsulfoxides [1908], Marek and coworkers the CuI-catalyzed additions (8.264) [1909], and Xie et al. the ethyl- and methylzincation of alkynylsulfones [1910]. Ethylzincation of yamides has been catalyzed by a Rh(I) complex (reaction (8.265)) [1911, 1912]. In 1997, Knochel and coworkers reported the *syn*-phenylzincation (*Z*/*E* ratio > 99 : 1) of 1-phenylbut-1-yne with Ph_2Zn catalyzed by $Ni(acac)_2$ (reaction (8.266)) [1913]. After iodination of the alkenylzinc compound so obtained and subsequent Negishi coupling (Section 8.8.2) with *para*-BrZn—C_6H_4—O—$CH_2CH_2NMe_2$, tamoxifen was prepared ((*Z*)-Ph(Et)C=C(Ph)-*p*-$C_6H_4OCH_2CH_2NMe_2$·HCl, selective estrogen receptor modulator) [1848, 1914]. The alkenylzincations of internal alkynes with (1-phenylvinyl)$_2$Zn catalyzed by CoF_2 and a diphosphine ligand are stereoselective and lead to the formation of 1,3-dienes after aqueous work-up [1915].

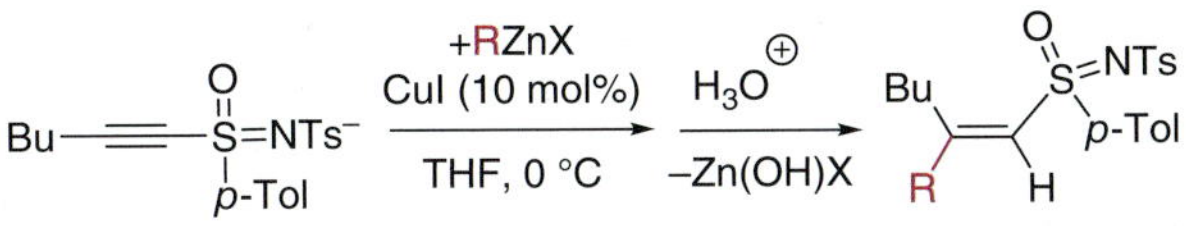

(8.264)

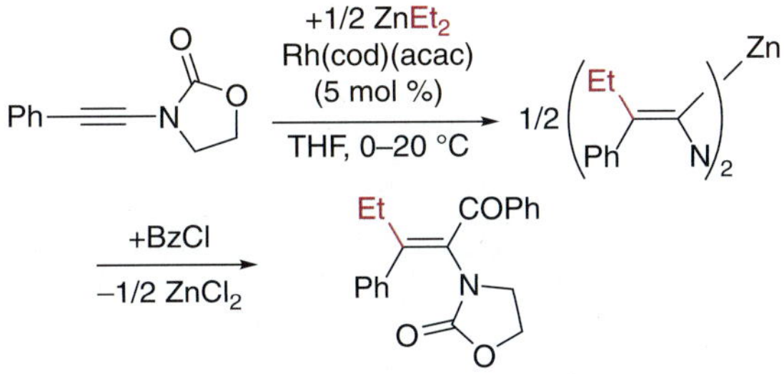

(8.265)

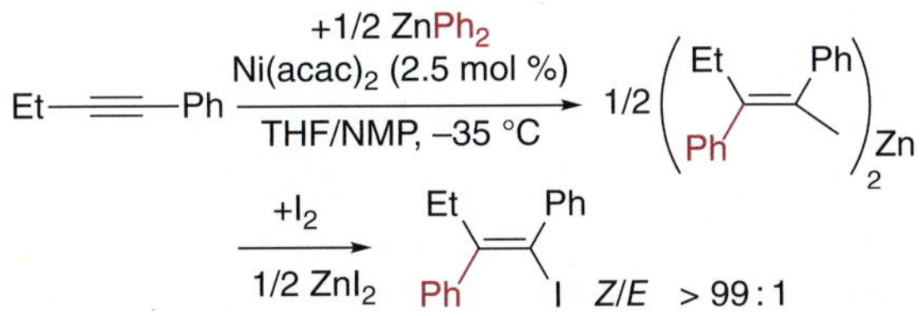

(8.266)

Problem 8.81 What are the products of reaction of benzoxanorbornadiene with Ph_2Zn in the presence of $FeCl_3$ (1 mol%) and 1,2-[(4-FC_6H_4)$_2$P]$_2C_6H_4$ (2 mol%) in THF/toluene at 0 °C and subsequent acidic aqueous work-up? [1916]

8.7.3 Additions of organoaluminum compounds

The properties of aluminum reagents derive mainly from their high Lewis acidity. Trialkylaluminum (R_3Al) monomers equilibrate with doubly bridged dimeric species, as borane equilibrates with diborane, which is related to the tendency of the Al atom to complete electron octets. Bonds of Al with oxygen or the halogens are very strong: the Al—O bond energy amounts to c. 140 kcal mol^{-1}. Because of this, nearly all organoaluminum compounds are particularly reactive with O_2 and often ignite spontaneously in the air. The high oxyphilicity of organoaluminum compounds is of great value in the design of selective synthetic reactions [1917–1919]. Not only synthetic organic chemistry and catalysis have profited from organoaluminum compounds but also inorganic hightech materials such as ceramics, semiconductors, zeolites, and others [1920]. $EtAlCl_2$ and Et_2AlCl can be made from Al metal and EtCl. Et_3Al is formed in a process wherein metallic Al is first hydrogenated, and the resulting products are added to ethylene to form Et_3Al. When isobutene is reacted with Al + H_2, triisobutylaluminum is formed: essentially no (*i*-Bu)$_3$Al addition to isobutene occurs (steric hindrance). Dialkylaluminum hydride (e.g. diisobutylaluminum hydride: DIBAL-H) can be prepared from the corresponding trialkylaluminum compounds on heating with hydrogen and metallic aluminum. Grignard and organoalkali reagents convert $AlCl_3$ into triorganylaluminum species [1921, 1922]. Dialkylaluminum hydrides add to alkenes (hydroalumination), a method well suited for preparing a wide variety of unsymmetrical alkylaluminums of type R_2AlR' [1921]. Similarly, hydroalumination of alkynes produces alkenylaluminums (*syn*-addition under kinetic control). Some trialkylaluminums add to alkenes and alkynes (carboalumination). The latter reaction can be catalyzed by Cp_2ZrCl_2 as discovered by Negishi and coworkers [1923, 1924]. It is applicable to functionalized terminal acetylenes (Scheme 8.85a) and is highly regio- (anti-Markovnikov) and stereoselective (*syn*-addition) [1925, 1926]. As neither preformed $MeZrCp_2Cl$ without Me_3Al nor Me_3Al alone would undergo the carbometallation, a bimetallic mechanism (Scheme 8.85b) was proposed. With Et_3Al, the Zr-catalyzed carboalumination of alkynes produced aluminacyclopentenes **344** (Scheme 8.85c). The bimetallic catalyst **343** has been isolated [1927].

With Cp_2ZrCl_2 as a catalyst, the methylalumination of alkenes is problematic because of concurrent alkene polymerization and formation of other undesired products. However, with the bulkier Erker's (−)-dichlorobis(neomenthylindenyl)zirconium (**345**) as a catalyst, alk-1-ene carboalumination in a polar solvent such as CH_2Cl_2 or $MeCHCl_2$ gives the corresponding product of anti-Markovnikov addition in good yield and high enantioselectivity (reaction (8.267)) [1928]. This reaction is sometimes called ZACA = zirconium-catalyzed asymmetric carboalumination) [1929, 1930]. In an apolar solvent

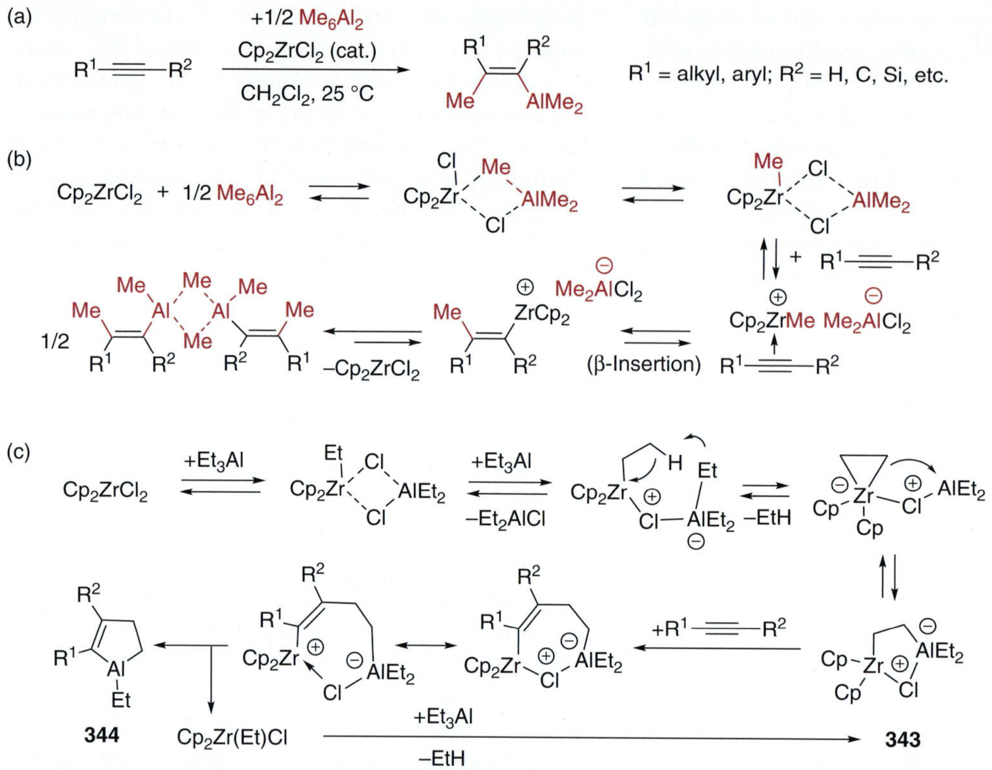

Scheme 8.85 (a) Zr-catalyzed methylalumination of alkynes (called sometimes ZMA reaction); (b) proposed reaction mechanism; and (c) Zr-catalyzed ethylalumination of alkynes.

such as hexane, alkenes react with Et_3Al producing aluminacyclopentane derivatives (reaction (8.268)).

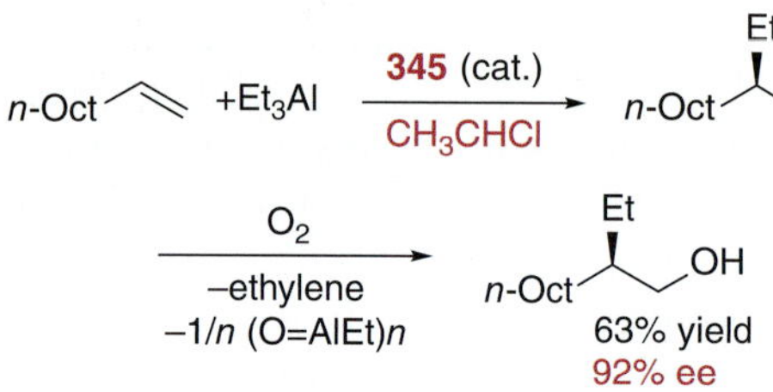

(8.267)

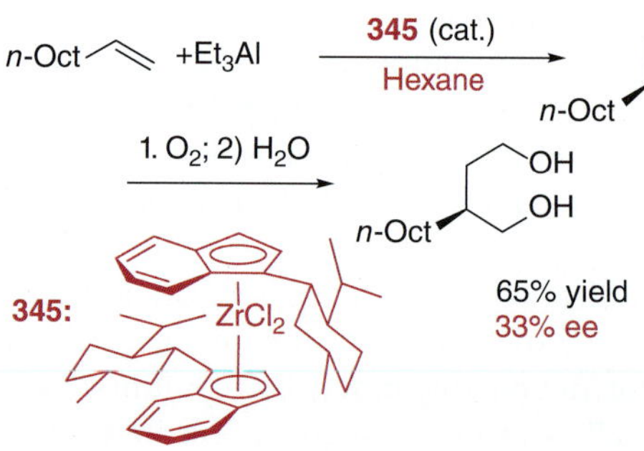

(8.268)

First examples of Cu(I)-catalyzed asymmetric conjugate additions of $AlMe_3$ using enantiopure 2-aryloxazoline ligand and $(t\text{-Bu})Me_2SiOTf$ as a cocatalyst were reported by Iwata and coworkers in 1966. Three years later, Woodward and coworkers avoided the use of silyl trilate applying a binaphthol heterodonor S,O ligand **346** that contains both a hard (OH) and a soft (SH) Lewis base able to coordinate bimetallic Al/Cu species. With Me_3Al and other organometallic reagents, modest enantioselectivities were obtained (reaction (8.269)) [1931]. In the meantime, a large number of alternative chiral ligands have been proposed [1932–1935]. An efficient method was developed by Alexakis and coworkers applying phosphoramidite ligands (e.g. **347**) and copper thiophene-2-carboxylate as catalyst (reaction (8.270)) [1906, 1936]. In 2011, Hoveyda and coworkers showed that a wide range of silyl substituted alkenylaluminum reagents add to 3-substituted cycloalk-2-enones in the presence of $CuCl_2 \cdot H_2O$ and diaminocarbene **348** (reaction (8.271)) [1937]. Rh(I)(binap) complexes catalyze enantioselective additions of arylaluminum reagents to cyclic α,β-unsaturated *N*-tosyl ketimines. Depending on the solvent and substituents at the ring, the reaction occurs either in a 1,2- or in a 1,4-manner [1938].

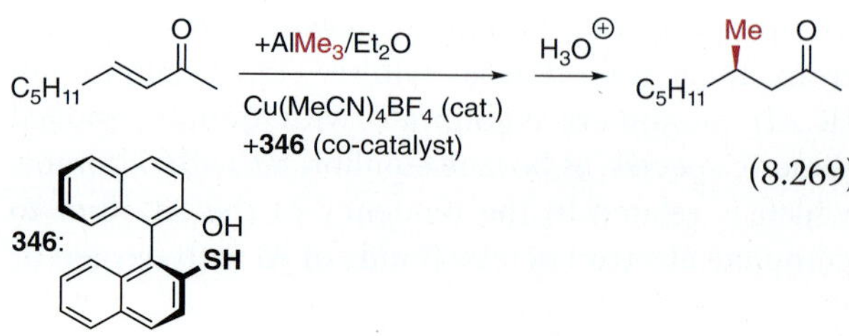

(8.269)

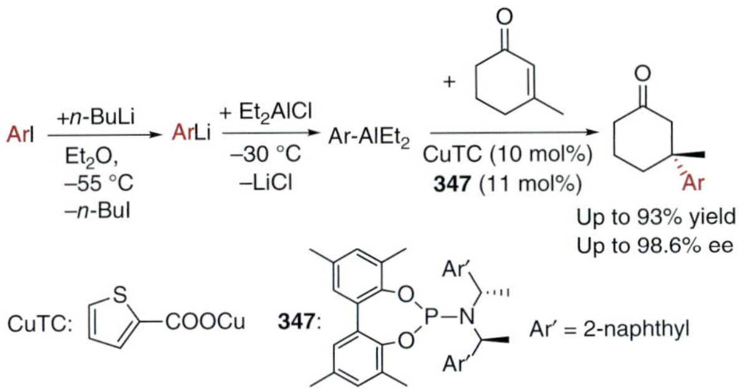

(8.270)

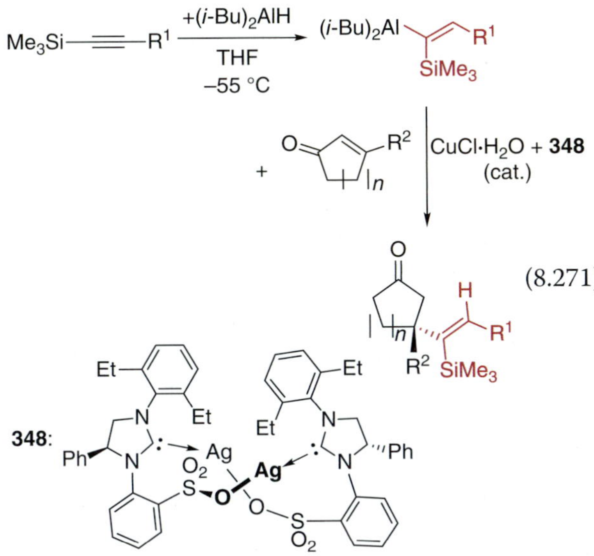

(8.271)

Problem 8.82 What is the product of the following two-step reaction [1939].

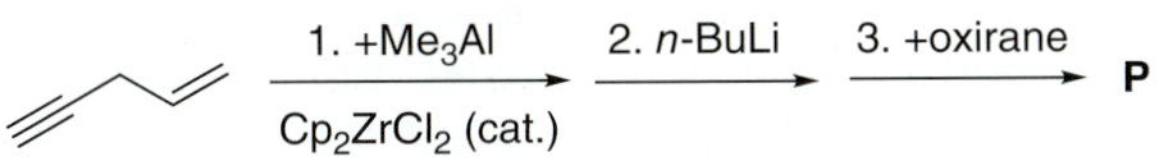

Problem 8.83 Propose a method to convert bis(allyl)dimethylsilane into **P** [1929, 1940].

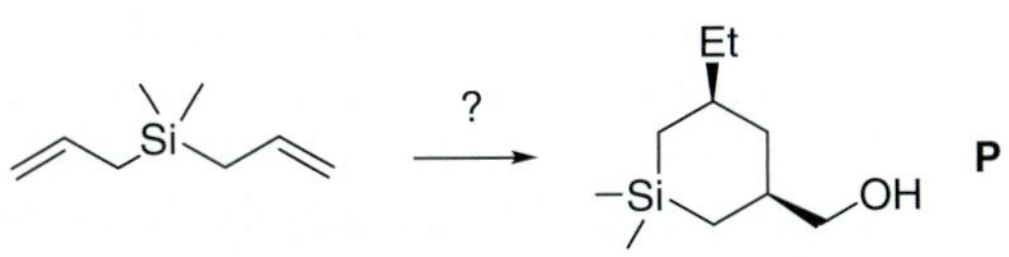

8.7.4 Additions of organoboron, silicium, and zirconium compounds

Like organoaluminum reagents, organoborons (boranes) are Lewis acids [1941, 1942]. They have much lower reactivity than Grignard and organolithium reagents but are more reactive than silanes (organosilicons). Nevertheless, they are extremely useful in C—C forming reactions such as the Miyaura–Suzuki coupling reaction (Section 8.8.4) and as allylating agents in carbonyl-bora-ene reactions (Section 5.7.5). In 1956, Brown (was awarded the Nobel Prize in 1979 together with G. Wittig) and Subba Rao reported that diborane, B_2H_6, reacts with alkenes in ether solvents to form trialkylboranes ((alkyl)$_3$B)) [1943–1945]. Alkynes also react with diborane giving the corresponding tris(alkenyl)boranes. These additions are catalyzed by transition metals; they are highly *syn*-selective giving anti-Markovnikov products with the boron occupying the least substituted carbon center [1946]. They follow concerted mechanisms with four-centered transition states. Cycloocta-1,5-diene reacts with borane to give 9-borabicyclo[3.3.1]nonane = 9-BBN-H, whereas 2,3-dimethylbut-2-ene reacts with borane to give *tert*-hexylborane = ThexBH$_2$. The latter and other substituted boranes can be used for the hydroboration of alkenes and alkynes [1947, 1948]. Boronic acid derivatives (RB(OH)$_2$, RB(OR′)$_2$) are pivotal in organic synthesis as they tolerate great functional tolerance [1949–1951]. They can be prepared by transition-metal-catalyzed hydroboration [1952, 1953], for instance with pinacolborane (HBpin: HB(O—CMe$_2$—CMe$_2$—O); prepared by: pinacol + $BH_3 \cdot Me_2S$ → HBpin + H_2 + Me_2S, 0–25 °C) [1954]. The Cu-catalyzed addition of bis(pivalato)diboron (B_2(pin)$_2$) to propynoic esters gives the corresponding 3-borylated prop-2-enoic esters [1955, 1956]. Terminal alkynes with a substituent at C(3) able to bind temporarily with the boron reagent (e.g. alcohol, amine [1957], sulfone [1958], and directing group) are borylated at C(2) with B_2(pin)$_2$) in the presence of a Cu-catalyst [1959–1964]. Transmetallation can also be applied to generate organoboranes. Chloro-bis(diisopropylamino)borane is a reagent of choice for the borylation of organomagnesium and organolithium reagents, giving boronic derivatives (e.g. RMgX or RLi + ClBN(*i*-Pr)$_2$ → RBN(*i*-Pr)$_2$ + MgCl or LiCl, R = alkyl, aryl,alkynyl, N_2=CH) [1965, 1966]. Arylboronic acids (ArB(OH)$_2$) can also be prepared from BH_3 and arylmagnesium halides [1967, 1968] or by borylation of aryllithium derivatives with trimethoxyborane [1969, 1970]. Arylboronates (ArB(OR)$_2$) are obtained in a one-step procedure from the corresponding aryl halides and haloboronic ester via an intermediate arylzinc species [1971]. Dibromo and dichlorovinylboranes have been prepared by reaction of vinylsilanes with BBr_3 and BCl_3, respectively [1972]. Organodiborons can be prepared by Pd(0)-catalyzed addition of bis(pinacolato)diboron (B_2(pin)$_2$) to alkynes to give (*Z*)-bis(boryl)alkenes [1973]. In the presence of Pt(0)-catalyst, 1,3-dienes react with $B_2(pin)_2$ to give 1,4-bis(boryl)but-2-ene and bis(boryl)alkane derivatives [1974]. Monohydroboration of 1,3-dienes with catecholborane can be catalyzed by $NiCl_2$ or $CoCl_2$ + diphosphine ligands [1975].

In the presence of CO, organoboranes undergo Pd-catalyzed carbonylative reactions (Suzuki–Myaura carbonylation, e.g. $Ar^1—B(OH)_2 + R^2—I + CO \rightarrow Ar^1COAr^2 + B(OH)_2I$) [1976–1978]. In 1997, Miyaura and coworkers opened a new route to β-substituted ketones with the Rh(I)/diphosphine 1,4-addition of arylboron reagents to α,β-unsaturated ketones [1979]. An asymmetric version of the reaction was realized later (e.g. reaction (8.272)) by Hayashi and coworkers [1980, 1981]. The method has been extended to the preparation of 2-arylphosphonates (reaction (8.273)) [1982]. Asymmetric synthesis of α-aminoesters of type $ArCH_2—CH(NHR)COOMe$ has been realized by enantioselective 1,4-addition of arylboronics [1983] or by addition of $ArBF_3K$ to CH_2=C(NHR)COOMe and subsequent enantioselective protonation [1984]. Enantioselective 1,4-addition of alkyboranes (obtained by hydroboration of terminal alkenes with 9-BBN-H) has been catalyzed by a Cu + N-heterocyclic carbene complex [1985, 1986]. Regioselective 1,4-additions of organoboronic acids have been catalyzed by $Ni(cod)_2 + Ph_3P$ in dioxane/water at 100 °C [1987] and by CuOTf + homochiral phosphoramidite ligands [1988]. Enantioselective Rh(I)-catalyzed 1,4-addition of tetrarylborates (obtained readily by $NaBF_4 + 4ArMgBr \rightarrow NaBAr_4 + 2MgBr_2 + 2MgF_2$) to β,β-disubstituted α,β-unsaturated esters has been reported by Shintani and Hayashi in 2011 (reaction (8.274)) [1989].

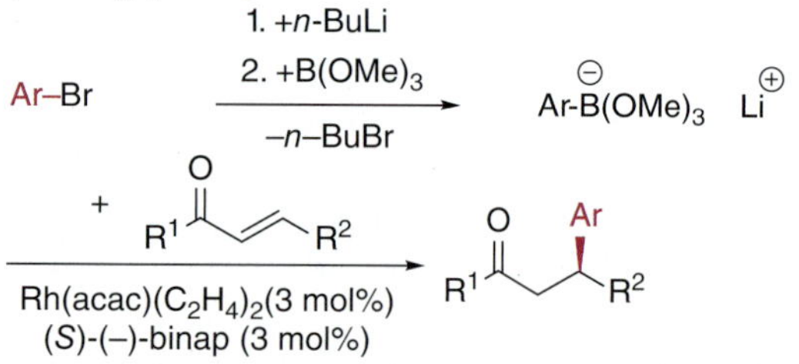

(8.272)

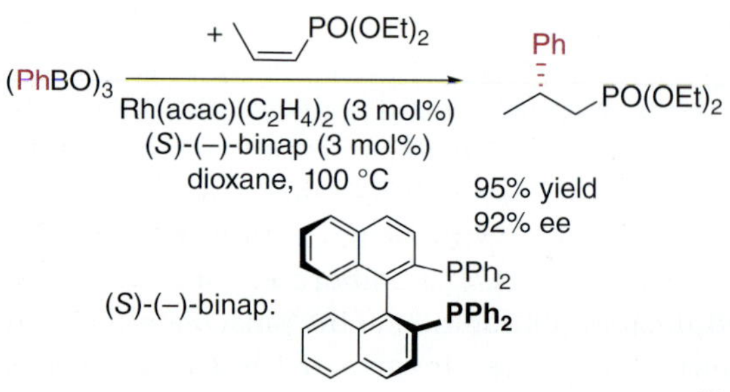

(8.273)

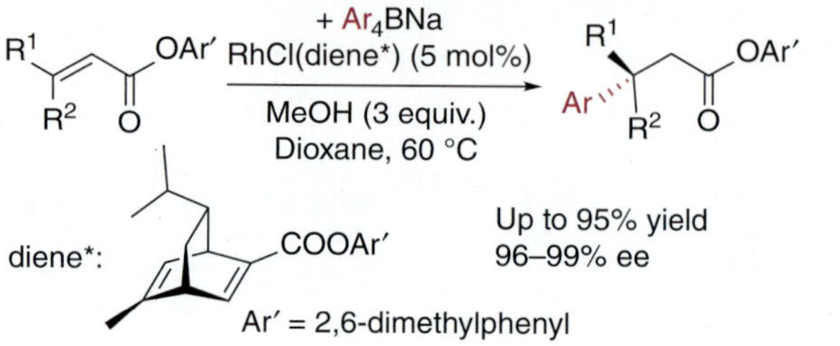

(8.274)

In the presence of Ni(0) catalysts, trialkylboranes and boronic acids add to alkynes, and the adducts so obtained add to imines. This realizes an efficient three-component synthesis of allyl amines (reaction (8.275)). In this case, the catalyst is (cycloocta-1,5-diene)$_2$nickel and a trialkylphosphine ligand (e.g. $(c\text{-}C_6H_{11})_3$ P and $(c\text{-}C_5H_9)_3P$). Using an enantiomerically pure ligand such as (S)-(+)-(neomenthyl) diphenylphosphane, optically active alkyl amines are obtained (up to 42% ee) [1990].

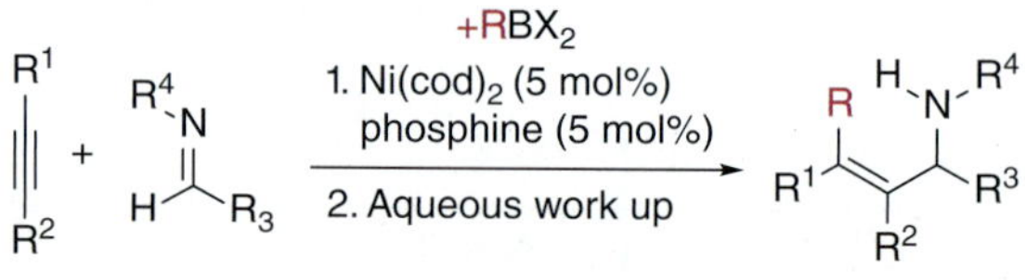

(X = OH, R)
R^1 = aryl, alkyl, R^2 = alkyl, H; R^3 = aryl, alkyl
Exclusive *syn*-addition across alkyne (>97 : 3)
Compatible with ketones, esters, and hydroxylic solvents

(8.275)

Silicon–carbon bonds are almost apolar. There are similarities in the behavior of boron and silicon, especially for the high stability of bonds these elements make with oxygen and fluorine. Both carbon and silicon derivatives have a tendency to form anionic derivatives with electropositive metals, and these undergo bimolecular (associative) nucleophilic displacement reactions. Substitution at silicon is much easier than at carbon [1941]. As silicon is more electropositive than carbon, the β(Si—C) bond stabilizes a carbenium ion by hyperconjugation much better than a β(C—C) or β(C—H) bond [1991–1996]. This property, together with the high affinity of silicon for oxygen and fluorine renders organosilicon reagents very useful in organic synthesis [1997–2008], as illustrated with the Hosomi–Sakurai reaction discovered in 1976 (carbonyl sila-ene reaction; Section 5.7.5) in which allylsilanes react with a wide variety of aldehydes and ketones activated with $TiCl_4$, or another Lewis acid, giving the corresponding homoallylic alcohols after aqueous work-up [173–175]. One general method of synthesis of organosilanes is the displacement of a silyl chloride by a Grignard of organolithium

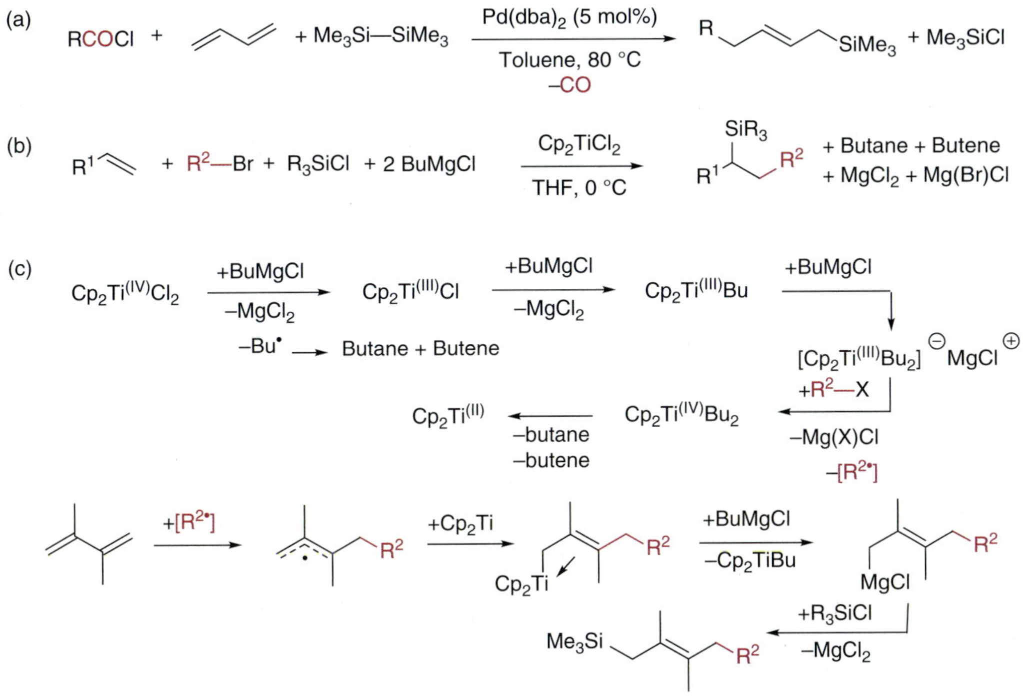

Scheme 8.86 Examples of three-component carbosilylations. (a) Pd(0)-catalyzed 1,3-diene decarbonylative carbosilylation; (b) alkenes and dienes combine with alkyl halides and silyl chlorides in the presence of a Grignard reagent and Cp_2TiCl_2 as a catalyst; and (c) proposed mechanism for the latter reaction.

reagent (e.g. furan + n-BuLi → α-furyl-Li + n-BuH; α-furylLi + Me_3SiCl → α-furyl-$SiMe_3$ + LiCl). A second is the transition-metal-catalyzed hydrosilylation of alkenes, alkynes, and allenes that, in general, provide terminal silanes (anti-Markovnikov adducts; e.g. R—CH=CH_2 + $MeSiCl_2H$ → $RCH_2CH_2SiMeCl_2$, catalyzed by H_2PtCl_6; CH_2=C=CH_2 + Et_3SiH → CH_2=$CHCH_2SiEt_3$ catalyzed by $Pd(Ph_3P)_4$/maleic anhydride)) [2009–2011]. In some cases, Markovnikov adducts are formed selectively (lateral control, e.g. HC≡C$(CH_2)_2$OAc + Et_3SiH → CH_2=C$(SiEt_3)$—$(CH_2)_3$OAc when catalyzed with (cyclopentadienyl) $Ru(MeCN)_3{}^+/PF_6{}^-$ in CH_2Cl_2) [2012]. Ni-catalysts have been applied in the hydrosilylation of alkenes [2013, 2014]. Ni nanoparticles generated from Ni(O—t-Bu)$_2$·(KCl)$_x$ catalyze the anti-Markovnikov hydrosilylation of unactivated alkenes [2015]. Combinations of transition metal and chiral amine catalyze enantioselective conjugate silyl additions to α,β-unsaturated aldehydes [2016]. Lewis acids such $AlCl_3$, $EtAlCl_2$, or $HfCl_4$ and Brønsted acids such as proton-exchanged Montmorillonite catalyze the ene-reactions of allylsilanes with alkenes and alkynes [2017, 2018]. Silanes that cannot enter in sila-ene reactions undergo carbosilylation less readily but have been observed for the intramolecular *trans*-vinylation and *trans*-arylation of unactivated alkynes using $EtAlCl_2$ as a catalyst [2019]. In 1995, Tsuji and coworkers reported the three-component 1,4-carbosilylation of 1,3-dienes using disilanes and acyl chlorides catalyzed by Pd(dibenzylidenacetone)$_2$ (Scheme 8.86a)) [2020]. The same catalyst induces the carbosilylation of 1,1-dimethylallene with halides RX and $Bu_3SnSiMe_3$ giving Me_2C=C(R)CH_2SiMe_3 + Bu_3SnX [2021]. In 2000, Terao and coworkers found that Cp_2TiCl_2 and a Grignard reagent catalyze the carbosilylation of alkenes and dienes using alkyl halides and chlorosilanes (Scheme 8.67b). A radical mechanism (Scheme 8.86c) has been proposed based on the observation of the esr spectrum of $[Cp_2Ti(III)Bu_2]^-/MgCl^+$ formed as an intermediate [2022]. Allene carbosilylation with chlorosilanes and Grignard reagents has been catalyzed by $Pd(dba)_2$ (THF, 25 °C) [2023].

Hydrozirconation of terminal alkenes with Schwartz reagent (Cp_2ZrHCl) [2024] provides corresponding (alk-1-yl)Zr$(Cp)_2$Cl compounds [2025, 2026]. In the presence of CuCl and an enantiopure ligand (e.g. phosphoramidite (R)-**349**), the alkylzirconium reagents generated in situ undergo asymmetric conjugate additions [2027–2029] (e.g. reaction (8.276)) [2030].

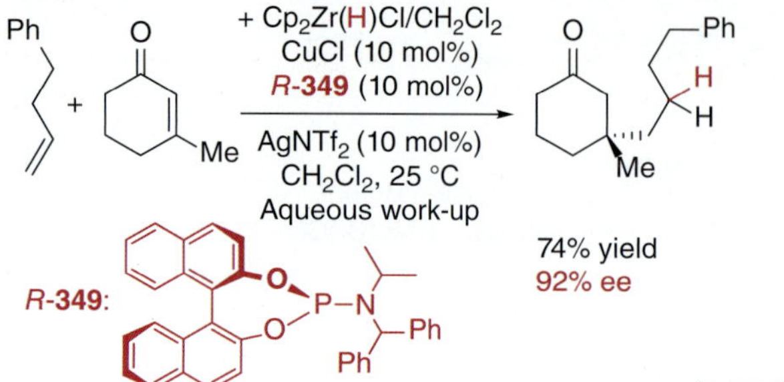

(8.276)

Problem 8.84 Propose a mechanism for the Pd(0)-catalyzed carbosilylation shown in Scheme 8.86a.[2021]

8.8 Displacement reactions

The transition-metal-catalyzed formal displacement reactions described in this chapter are summarized here-below. They are among the most applied tools to construct C—C bonds (Ar = aryl, het-aryl). In many cases, enantioselective versions of these cross-coupling reactions are available [2031]. Palladium-based catalysts are among the most active. Because of Pd toxicity and cost of its removal from final products, procedures are being developed that use minute amounts of this metal [2032].

Metallic carbon nucleophiles

Reaction		Name
R^1—M + R^2—X → R^1—R^2 + MX	e.g. PhMgBr + vinylBr	Kharash (1943)
R^1—M + $R^2CH{=}CHCH_2$—X → $R^2CH{=}CHCH_2$—R^1 + MX		Tsuji–Trost (1965)
	M = MgHal	Kumada–Tamao–Corriu (1972)
	M = ZnHal	Negishi (1977)
	M = SnR_3	Stille (1978)
	M = BY_2, BY_3^-/K^+	Suzuki–Miyaura (1979)
	M = SiR_3	Hiyama (1988)
$R^1ZnI + R^2COSEt \rightarrow R^1COR^2 + EtSZnI$		Fukuyama (1998)
$R^1B(OH)_2 + R^2COSR^3 \rightarrow R^1COR^3 + R^2SB(OH)_2$		Liebeskind–Srogl (2000)
$ArB(OR^1)_2 + R^2COH(R^3)$—X → $R^2COH(R^3)$—Ar + $B(OR^1)_2X$		Gossen (2001)

C—H nucleophiles (activated C(sp^3)—H, C(sp^2)—H, C(sp)—H)

Reaction	Name
$(ROOC)_2CH_2$ + Ar—X → $(ROOC)_2CH$—Ar + HX	Ullmann–Hurtley (1929)
R^1—H (pK_a < 20) + $R^2CH{=}CHCH_2$—X → $R^2CH{=}CHCH_2$—R^1 + HX	Tsuji–Trost (1965)
$R^1CH{=}C(R^2)$—H + Ar—X → $R^1CH{=}C(R^2)Ar + Ar(R^1)C{=}C(R^2)H$ + HX	Mizoroki–Heck (1971)
RC≡CH + Ar—X → RC≡CAr + HX	Sonogashira–Hagihara (1975)
Ar^1—H + Ar^2—X → Ar^1—Ar^2 + HX	Catellani–Chiusoli (1985)
$R^1CH_2COR^2$ + Ar—X → $R^1CH(Ar)COR^2$ + HX	Buchwald–Hartwig–Miura (1997)

Nonactivated C(sp^3)—H nucleophiles

Reaction	Name
$R^1(R^2)(R^3)C$—H + Ar—X → $R^1(R^2)(R^3)C$—Ar + HX	Dyker (1994)
R^1CH_2—CH_2R^2 + Br—C≡C—R^3 R → $R^1CH_2CH(R^2)$—C≡C—R^3 + HBr	Chatani (2011)

The precatalysts can be soluble (homogeneous) or not (heterogeneous) and be all kinds of transition metal salts, dispersed solids, or nanoparticles combined or not with all kinds of ligands and/or cocatalysts. The "actual catalyst" or the catalytically active species might be the result of a reaction between the metallic precatalyst with one of the reactant, with the ligand, with the cocatalyst (e.g. a base, an acid), and/or with the medium including traces of water (e.g. $Pd(II)X_2 + 3R_3P + H_2O \rightarrow Pd(0)(R_3P)_2 + R_3P{=}O + 2HX$). Detailed mechanisms of transition-metal-catalyzed reactions have been established in a few cases only, mostly for Pd-catalyzed couplings involving homogeneous precatalysts and phosphine ligands [2033]. For some reactions, it has been possible to prove the formation of radical intermediates. The latter may result in side reactions or participate fully to the catalytic cycle of the main reaction [2034–2036]. The Grignard reagents have been the most powerful nucleophilic reagents since 1900, we therefore start with them.

8.8.1 Kharash cross-coupling and Kumada–Tamao–Corriu reaction

In 1941, Kharasch and coworkers showed that transition metal salts affect the course and mechanism of the reactions of Grignard reagents [1778, 2037, 2038]. For instance, in the presence of CuCl or $MnCl_2$ (2 mol%) reaction, $MeMgBr + Ph_2CO$ produces 93–95% of the expected carbinol ($Ph_2CH(Me)OH$), whereas in the presence of $FeCl_3$ or $CoCl_2$, the yield of carbinol falls to 21% and 2%, respectively, and benzopinacol ($Ph_2C(OH)C(OH)Ph_2$) forms competitively in 65% and 93% yield, respectively [2039]. In 1943, they reported that $CoCl_2$ catalyzes the formation of alkanes and alkenes, suggesting the formation of alkyl radical intermediates from the alkylmagnesium halides [2040]. This was also found with alkyllithiums reacting with alkyl halides [2041]. The same year, Kharasch and Fuchs reported that $CoCl_2$, $CrCl_2$, and CuCl catalyze the C—C-cross-coupling of arylMgBr with alkenyl bromides (e.g. reaction (8.277)), which is often called the Kharash cross-coupling [2042] (not to be confounded with the Kharasch allylic oxidation using peresters, or Kharasch–Sosnovsky reaction discovered in 1958 [2043–2047], and the Kharasch addition discovered in 1945 ($RCH{=}CH_2 + CXCl_3 \rightarrow RCH(X){-}CH_2CCl_3$)) [2048–2051].

$$PhMgBr + CH_2{=}CHBr \xrightarrow[-MgBr_2]{CuCl_2\ (5\ mol\%)} Ph{-}CH{=}CH_2 \quad \text{56\% yield} \tag{8.277}$$

In many cases, the Kharasch cross-coupling is accompanied by the formation of products of homocoupling (Wurtz reaction) in proportion that varies with the catalyst ($CoCl_2$, $FeCl_3$, $NiCl_2$, CuCl, and $MnCl_2$) [2052]. Disproportionation also competes [2053, 2054]. For instance, (cyclopentyl)MgBr in Et_2O reacts with 5% $MnCl_2$ to give cyclopentane and cyclopentene. Silver salts also catalyze the Kharasch coupling, the oxidative homocoupling, and disproportionation, as shown by Tamura and Kochi [1751, 2055]. Cu(I) salts catalyze the cross-coupling with less disproportionation [2054]. Nevertheless, the reaction is not general and fails with secondary and tertiary alkyl halides [2056, 2057]. Early reports by Kharasch and Reinmuth [2058] and then by Tamura and Kochi [2053, 2057, 2059] indicated that these cross-coupling reactions can be catalyzed by iron salts. This discovery remained dormant because in 1972, very efficient and versatile Ni-catalyzed cross-couplings of aryl or alkenyl halides with Grignard reagents were reported simultaneously by the groups of Kumada in Japan [2060] and Corriu in France [2061]. This became the Kumada–Tamao–Corriu (KTC) reaction, which can also be catalyzed by Pd-based catalysts [2062–2065]. The reaction works well with the primary and secondary alkyl, aryl, alkenyl, and allylic Grignard reagents (R^1MgHal). A wide variety of electrophilic partners R—X can be used that include alkyl, alkenyl, and aryl halides (R^2-Hal), as well as alkenyl and aryl sulfonates, sulfates, phosphates, carbamates, carbonates, and aryl ethers [2066–2068]. In 1979, Wenkert et al. reported the $Ni(Ph_3P)_2Cl_2$-catalyzed cross-coupling reaction: 2-methoxynaphtalene + PhMgBr → 2-phenylnaphthalene + Mg(MeO)Br [930]. Since then, it was found that nickel catalysts with $(cyclohexyl)_3P$, $(i\text{-}Pr)_3P$ [2069] or *N*-heterocyclic carbene ligands improve the scope of the reaction [2070, 2071]. Catalysts such as $NiCl_2 + Ph_2P(CH_2)_2PPh_2$ (dppe) are very effective with simple aryl and alkyl Grignard reagents. For the cross-coupling of alkenyl halides and allylic Grignard reagents, $NiCl_2 + Me_2P(CH_2)_2PMe_2$ is often preferred and for alkyl/alkyl cross-coupling, $Ni(acac)_2$ + diphosphine and/or TMEDA gives good results [2072]. When using sterically hindered Grignard reagents and electrophilic substrates, Ph_3P might be a better ligand. In 2009, Hu and coworkers applied the Ni(II)—Cl pincer complex $[(^{Me}NN_2)NiCl]$ to catalyze the sp^3(C)-sp^3(C) coupling of nonactivated and functionalized alkyl halides with Grignard reagents, under very mild conditions (−35 °C) (reaction (8.278)) [2073, 2074]. The same complex catalyzes the cross-coupling of alkynyl Grignard and alkyl halides [2075]. The pincer Ni complex do not promote β-elimination from alkylmetal intermediates readily [2076]. Radical intermediates have been evidenced in the Ni-catalyzed reactions [2077].

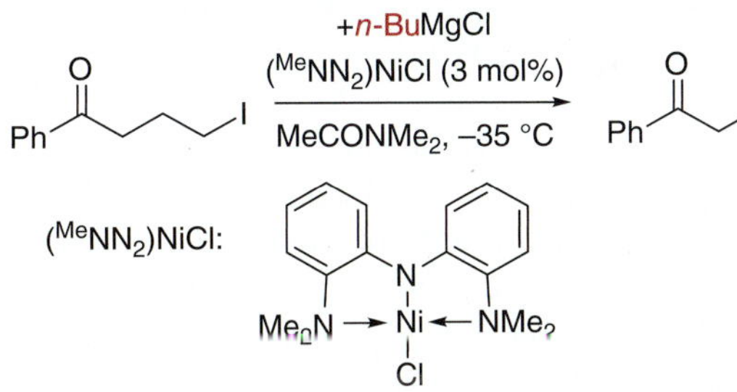

(8.278)

One possible mechanism for the KTC reaction is shown in Scheme 8.87. The catalytically active species $Ni(0)L_2$ or $Pd(0)L_2$ (L = phosphine ligand) is generated by reduction of $NiCl_2$ ($PdCl_2$) with 2 equiv. of Grignard reagents: $NiCl_2$ ($PdCl_2$) + 2RMgX + 2L → L_2NiR_2 (L_2PdR_2) + 2Mg(Cl)X. Reductive elimination of R—R provides $Ni(0)L_2$ ($Pd(0)L_2$). Oxidative addition (Section 7.7.8) ArCl + $Ni(0)L_2$ ($Pd(0)L_2$) gives $ArNi(II)L_2Cl$ ($ArPd(II)L_2Cl$). Transmetallation $ArNi(II)L_2Cl$ ($ArPd(II)L_2Cl$) + RMgCl generates Mg(Cl)X + $L_2Ni(II)(Ar)(R)Cl$ (ArPd(II)(R)Cl). Reductive elimination (Section 7.7.8) from the latter furnishes Ar—R and liberates $L_2Ni(0)$ ($L_2Pd(0)$), which starts another catalytic cycle.

The $^{12}C/^{13}C$ kinetic isotope effects for Ni(0)(dppp)-catalyzed reaction of *ortho*-chlorotoluene with

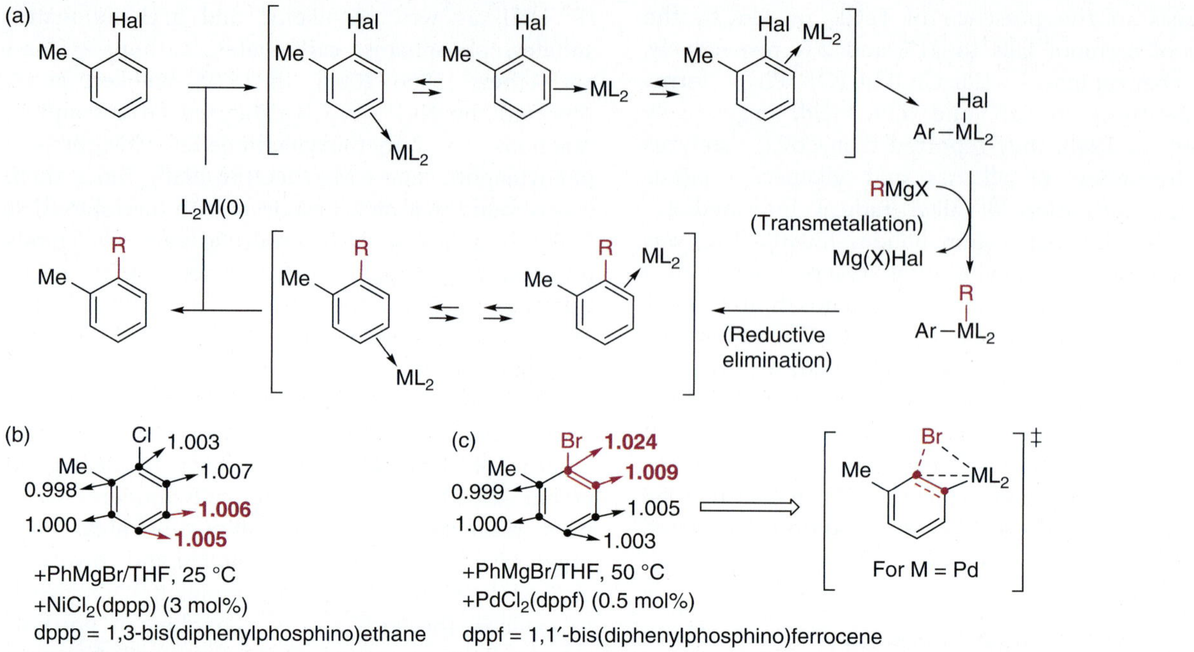

Scheme 8.87 (a) Possible mechanism of the Kumada–Tamao–Corriu (KTC) reaction. (b) $^{12}C/^{13}C$ kinetic isotopic effects for Ni-catalyzed reaction with *o*-chlorotoluene. (c) $^{12}C/^{13}C$ kinetic isotopic effects for the Pd-catalyzed reaction of *o*-bromotoluene.

PhMgBr (Scheme 8.87b)) are relatively small (1.003–1.009) and distributed only on the less hindered side of the benzene ring [2078]. This is consistent with a first irreversible step, which is the π-complexation of $L_2Ni(0)$ catalyst by the aromatic ring on its less hindered C=C bonds. Once the (haloarene)Ni(0) π-complex is formed, it does not dissociate and proceeds quickly to intramolecular oxidative addition. This has been ascribed to the d-electron donative property of the $L_2Ni(0)$ catalytic species [2079]. Pd(0) is a weaker d-electron donor and the (haloarene)$Pd(0)L_2$ complexes become less stable than the corresponding (haloarene)$Ni(0)L_2$ complexes. It is thus the oxidative addition step that follows the ligand exchange that becomes the first irreversible step as demonstrated by the $^{12}C/^{13}C$ kinetic isotope effects (Scheme 8.87c) for the reaction of *o*-bromotoluene with PhMgBr catalyzed by $PdCl_2$(dppf). In this case, a relatively large kinetic isotope effect (1.024) is found for C(1), and relatively small effects (1.003–1.005) are found for C(4) and C(5). For dihalogenated heteroarenes, many examples of site-selective cross-coupling have been reported [2080]. Highly *ortho*-selective cross-coupling of dichlorobenzene derivatives with Grignard reagents have been realized using a combination of $Pd_2(dba)_3$ and PCy_3 (dba = dibenzylideneacetone, Cy = cyclohexyl). The use of hydroxylated terphenylphosphines (reaction (8.279)) further improved yields of the KTC reactions of dichlorophenol and dichloroaniline [2081]. High preference of alkylation at the *ortho* position of the hydroxyl group has been reported for Ni-catalyzed cross-coupling reactions of dihalophenols with alkyl Grignard reagents. Reactions of 2,4-dihalophenols, with various combinations of F, Cl, and Br afford *ortho*-cross-coupled products selectively (reaction (8.280)) [2082]. KTC cross-couplings of sterically hindered aryl chlorides have been catalyzed by $[Pd(\mu\text{-}Cl)(Cl)(NHC)]_2$ where NHC stands for N-heterocyclic carbene [2083]. Cross-coupling of (2-methyl-1-naphthyl)magnesium bromide with 2-methyl-1-naphthyl bromide at −15 °C in the presence of Ni-catalyst prepared in situ from $NiBr_2$ and (*S*)-1-[(*R*)-2-(diphenylphosphino)ferrocenyl]ethyl methyl ether gave (*R*)-(−)-2,2-dimethyl-1,1′-binaphthyl in 69% yield and 95% ee (reaction (8.281)) [2084]. Reaction (8.282) is another example of asymmetric KTC reaction [2085].

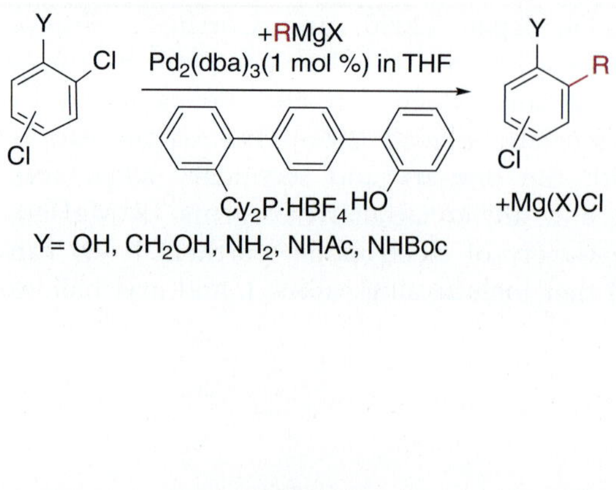

(8.279)

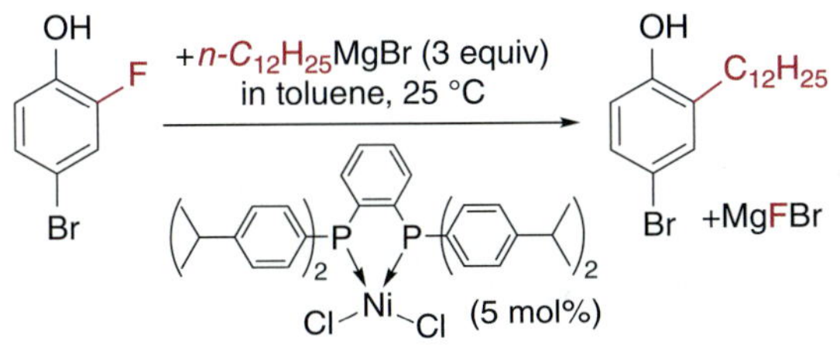

(8.280)

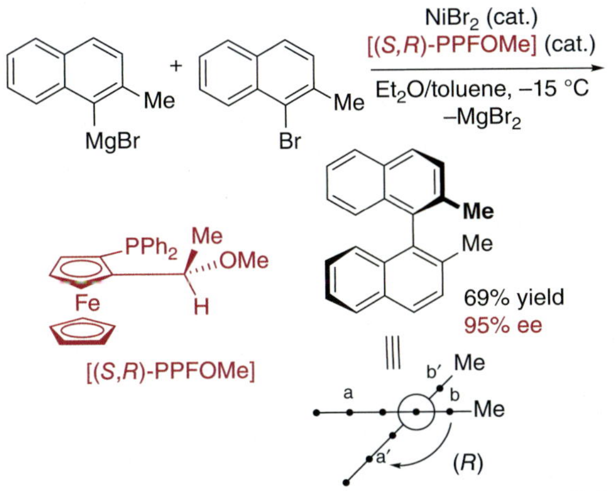

(8.281)

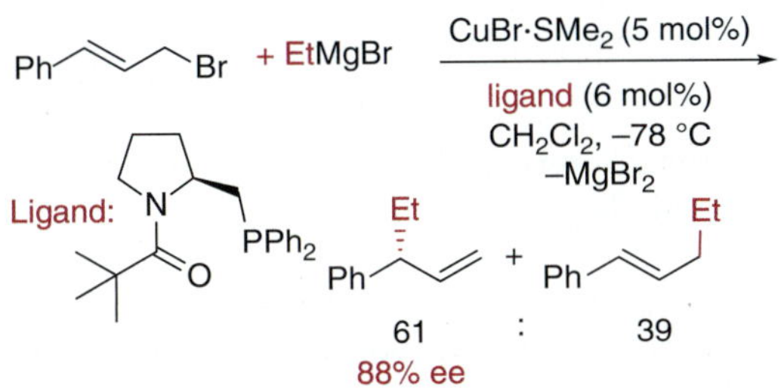

(8.282)

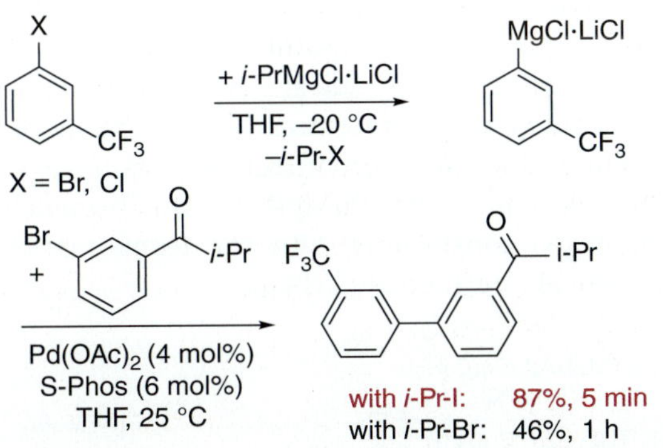

(8.283)

In 2009, Manolikakes and Knochel reported that KTC reactions are accelerated by *i*-Pr—I. This permits the rapid cross-couplings of a wide range of functionalized aryl- and heteroarylmagnesium reagents with aryl bromides using $Pd(OAc)_2$ + S-Phos (2-dicyclohexylphosphanyl-2′,6′-dimethoxyphenyl) as catalyst (e.g. reaction (8.283)) [2086]. Sensitive substrates such as carboxylic esters, carbonitriles, and ketones are tolerated in these reactions (transmetallation Mg/Pd being faster than Grignard addition to these functions at 25 °C). As various alkyl iodides such as MeI or neopentyl iodide lead to a similar rate acceleration, whereas no rate acceleration was observed with aryl triflates, a radical mechanism has been invoked to interpret the results [2087–2098]. Indeed, when 2-bromo-1-(but-3-enyl)benzene was submitted to the KTC reaction catalyzed by PEPPSI-IPr [(1,3-bis(2,6-diisopropylphenyl)imidazol-2-ylidene](3-chloropyridin)palladium(II)dichloride], the major product was the expected product of C—C cross-coupling, the bis-aryl compound **350** in the absence of any additive (Scheme 8.88a). In the presence of 1.1 equiv. of *i*-Pr—I, the reaction was much faster and the major product resulted from intramolecular radical addition consistently with the formation of radical intermediates (Scheme 8.88b) that lead to the cyclized product **351**. It was verified that [2-(but-3-enyl)phenyl]MgCl does not give any product of cyclization. Radical intermediates have also been evidenced in alkyl/alkyl cross-coupling catalyzed by $Ni(acac)_2$ + TMEDA in THF at 20 °C [2072]. In 2014, Tilley and Lipschutz reported that the cross-couplings PhMgBr + ArHal → Ph—Ar + Mg(Hal)Br are catalyzed by Me_3Si(dipp)N—Ni(II)—N(dipp)$SiMe_3$ (dipp = 2,6-di(isopropyl)phenyl). The latter complex reacts with MeMgBr giving $[Me_3Si(dipp)N—Ni(II)(Me)—N(dipp)SiMe_3]^-/MgBr^+$. With PhMgBr, $[Me_3Si(dipp)N—Ni(II)(Ph)—N(dipp)SiMe_3]^-/MgBr^+$ is generated. The latter reduces the aryl halide (ArHal) with the formation of aryl radical ($Ar^\bullet$) and $[Me_3Si(dipp)N—Ni(III)(Ph)—N(dipp)SiMe_3]$ + Mg(Hal)Br. The aryl radical adds to $[Me_3Si(dipp)N—Ni(II)(Ph)—N(dipp)SiMe_3]^-/MgBr^+$ giving $[Me_3Si(dipp)N—Ni(III)(Ar)(Ph)—N(dipp)SiMe_3]^-/MgBr^+$. Reductive elimination provides product Ph—Ar and $[Me_3Si(dipp)N—Ni(I)—N(dipp)SiMe_3]^-MgBr^+$. The latter Ni(I) complex reacts with $[Me_3Si(dipp)N—Ni(III)(Ph)—N(dipp)SiMe_3]$ formed earlier equilibrating with $Me_3Si(dipp)N—Ni(II)—N(dipp)SiMe_3$ + $[Me_3Si(dipp)N—Ni(II)(Ph)—N(dipp)SiMe_3]^-/MgBr^+$. Under well-chosen conditions, the above Ni(I), Ni(II), and Ni(III) complexes are observable [2099].

Because iron precatalysts are inexpensive and benign to the environment, they are now preferred to Ni and Pd [2100]. In 1944, Vavon and Mottez observed the $FeCl_3$-catalyzed cross-coupling of ArMgCl with organic halides [2101]. In 1953, Cook and coworkers reported the $FeCl_3$-catalyzed

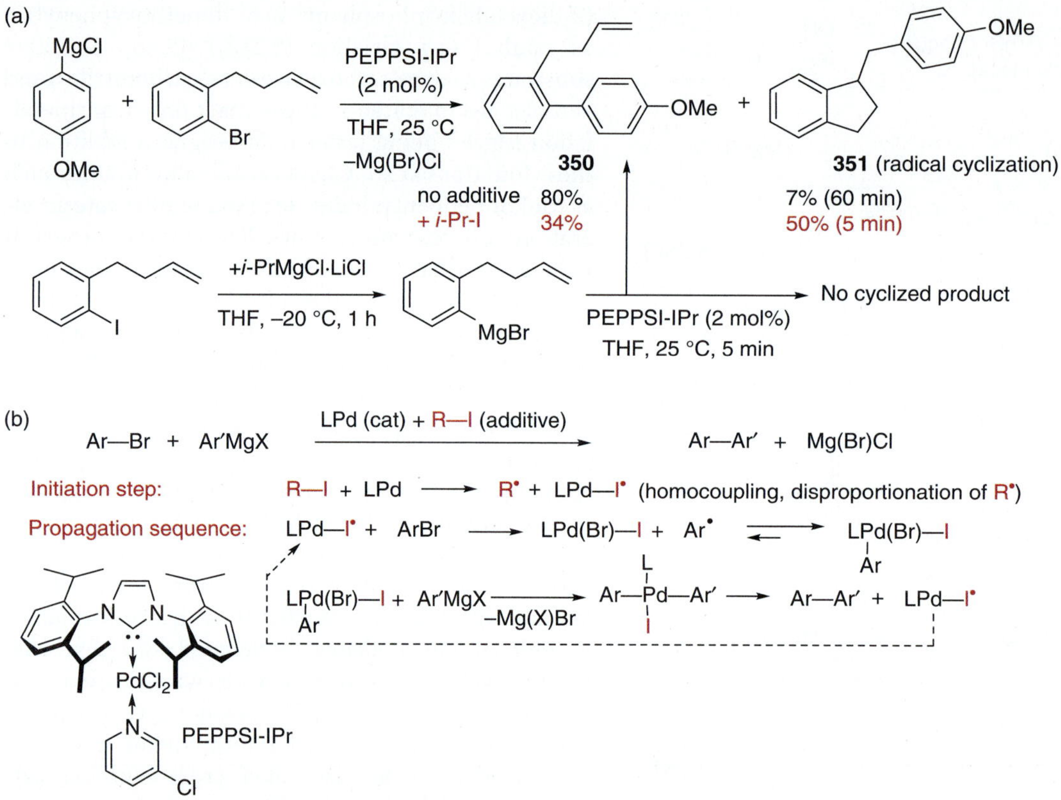

Scheme 8.88 (a) Additive *i*-Pr—I accelerates the Pd-catalyzed KTC reaction and (b) proposed radical mechanism for the reaction of an aryl bromide with an arylmagnesium reagent.

cross-coupling of Grignard reagents with acyl chloride to give the corresponding ketones [2102]. This reaction has been rediscovered by Marchese and coworkers in 1984 [2103]. During the early seventies, Tamura and Kochi carried out mechanistic studies on the Fe-catalyzed Kharash reaction [2053, 2104]. In 1982, Julia and coworkers reported the Fe-catalyzed C—C-cross-coupling of Grignard reagents with vinyl sulfones [2105, 2106]. In 1983, Molander and coworkers presented the Fe-catalyzed cross-coupling of aryl Grignard reagents with alkenyl halides [2107]. In 1986, Naso and coworkers found the Fe-catalyzed cross-coupling of alkenyl sulfides with Grignard reagents [2108] and, in 2005, Itami and coworkers reported the cross-coupling $ArMgBr + CH_2{=}CHSPh \rightarrow Ar{-}CH{=}CH_2 + Mg(PhS)Br$ catalyzed by $Fe(acac)_3$ [2109]. Several other groups have reported Fe-catalyzed KTC reactions [1791, 2100, 2110–2139].

The Jonas' complex $[Li(TMEDA)]_2[Fe(C_2H_4)_4]$[557] (TMEDA=$N,N',N'',N'''$-tetramethylethylenediamine) is an excellent catalyst for these cross-couplings. Under these conditions, halides react faster than ketones, esters, and carbonitriles with the organometallic species (e.g. reaction (8.284)). Mechanisms implying radical intermediates cannot be ruled out, at least not in the case of reaction (8.285) [2115]. The reaction of iron salts with Grignard reagents is known to generate reduced forms of iron complexes [2140, 2141]. Hölzer and Hofmann observed almost complete retention of configuration in reaction (8.286) when catalyzed by $NiCl_2$ or $PdCl_2$, whereas partial racemization occurred with $Fe(acac)_3$ and $Co(acac)_3$. They proposed that reversible (alkyl)[Fe](vinyl) ⇄ alkyl radical + (vinyl)[Fe]• equilibrium accounts for the racemization. Furthermore, it was found that (*R*)-2-bromooctane was racemized upon reaction with PhMgBr in the presence of Fe-catalysts, consistently with the formation of oct-2-yl radical intermediate [2142].

$$PhMgBr \xrightarrow[\text{THF, }-20\,°C\text{ (94\%)}]{+BrCH_2CH{=}CHCOOEt,\ [Li(TMEDA)]_2[Fe(C_2H_4)_4]\text{ (5 mol\%)}} Ph{-}CH_2CH{=}CHCOOEt + MgBr_2 \tag{8.284}$$

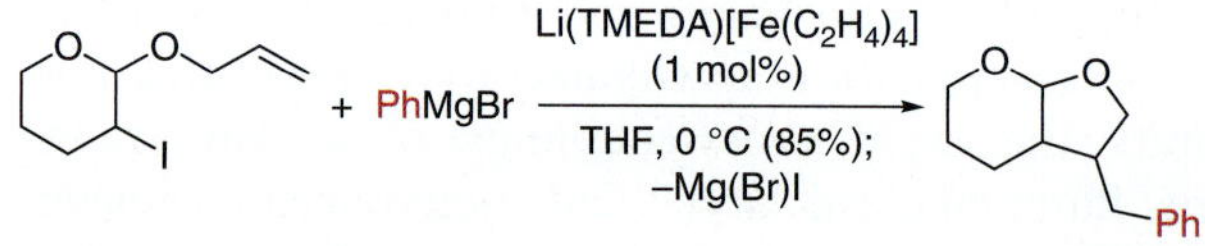

(8.285)

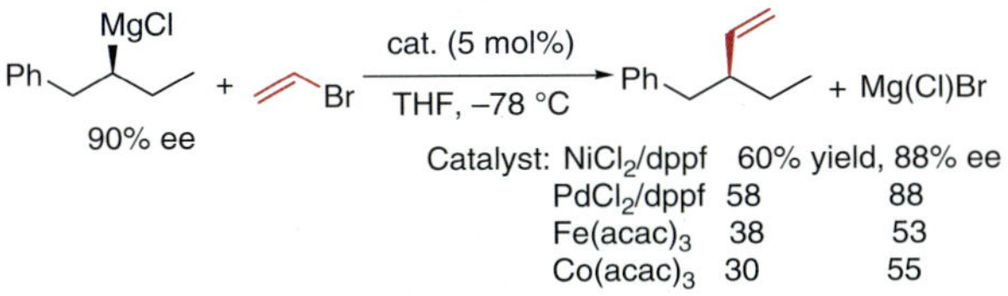

(8.286)

Knochel and coworkers found that difficult $Fe(acac)_3$-catalyzed KTC cross-couplings of ArMgCl with aryl chlorides are better yielded on metallating ArMgCl with CuCN·LiCl into Ar—Cu(CN)MgCl + LiCl in THF at −20 °C. Under these conditions, ketones do not add the Grignard reagents [2143, 2144]. On their side, Nakamura and coworkers found that $FeF_3 \cdot H_2O$ + the chlorohydrate salt of a diaminocarbene catalyzes the cross-couplings $Ar^1Cl + Ar^2MgBr \rightarrow Ar^1—Ar^2 + Mg(Br)Cl$ [2123]. For the Fe-catalyzed KTC reaction, Bedford and coworkers proposed a radical mechanism similar to that retained for the Kharasch coupling in which a reduced form of iron [Fe(n)] transfers an electron to the halides (R—Hal + [Fe(n)] → $R^\bullet$ + Hal—[Fe(n+1)]; Hal—[Fe(n+1)] + ArMgX → Mg(X)Hal + Ar—[Fe(n+1)]; and Ar—[Fe(n+1)] + $R^\bullet$ → Ar—R + [Fe(n)]) [2124–2127]. Radical intermediates of type $RCH^\bullet$—(COOR′) have also been invoked by Nakamura and coworkers in the case of the asymmetric coupling RCH(Cl)COOR′ + ArMgBr → RCH(Ar)COOR′ + MgBrCl catalyzed by $FeCl_3$ and a homochiral diphosphine ligand [2145]. The nature of the actual iron catalyst is still under debate [2137, 2146]. Mononuclear complexes of type $(diphosphine)_2Fe(I)X$ [2147] and -ate complexes of type $R_3Fe^-/MgX(TMEDA)^+$ have been characterized when Grignard reagents are added to $FeCl_3$/diphosphine and $FeCl_3$/TMEDA mixtures, respectively [2148]. In the 1970s, Kochi observed (esr) the formation of $S = ½$ iron species during reaction $FeCl_3$ + MeMgBr. In 2016, Neidig and coworkers isolated and characterized it to be $[MgCl(THF)_5][Fe_8Me_{12}]$, a octanuclear complex of iron. Alone it reacts poorly with PhCH=CHBr giving PhCH=CHMe in low yield. However, in the presence of MeMgBr, the product of cross-coupling is obtained in 99% yield. This suggests that polynuclear iron complex intermediates and/or nanoparticles of iron can catalyze the KTC reaction, being the actual catalysts or other precatalysts [2149, 2150].

Not only organohalides can be used as electrophiles in the KTC reaction. Electrophilic partners such as triflates (e.g. reaction (8.287)) [1791, 2117, 2151–2153], acetates and carbonates [2154] (e.g. reaction (8.288)) [2155], phosphates (e.g. reaction (8.289)) [2156], and sulfonates (e.g. cyclopentyl-OTs + $Ph(CH_2)_3MgBr$ → cyclopentyl-$(CH_2)_3Ph$ + Mg(OTs)Br, Cu catalyst) [2157] have been used successfully. In the case of 4-bromocyclopent-1-en-1-yl triflate, selective alkynylation of the secondary alkyl bromide occurs in the presence of $FeCl_2$ + 1,2-di[bis(3,5-di(*t*-butyl)phenyl) phosphino]benzene (reaction (8.290)), whereas displacement of the cycloalkenyl triflates is favored in the presence of $FeCl_3$ and LiBr (reaction (8.291)) [2158].

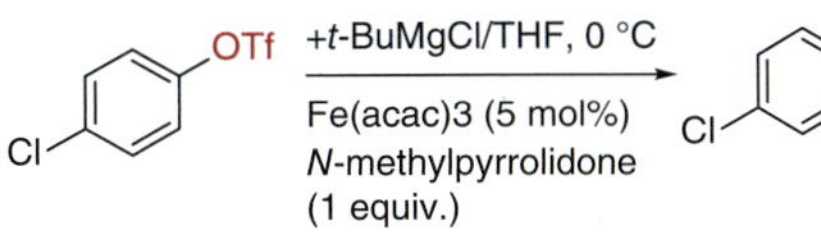

(8.287)

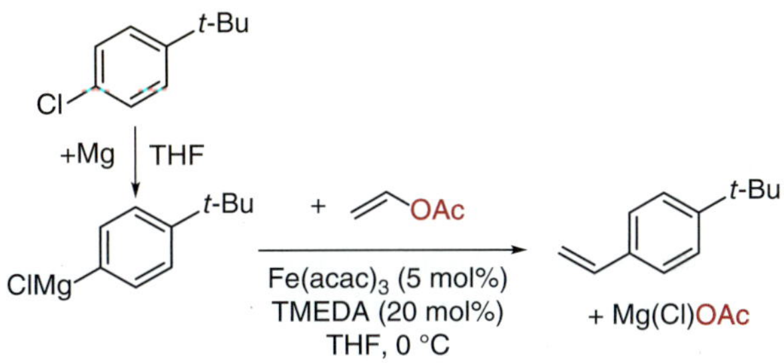

(8.288)

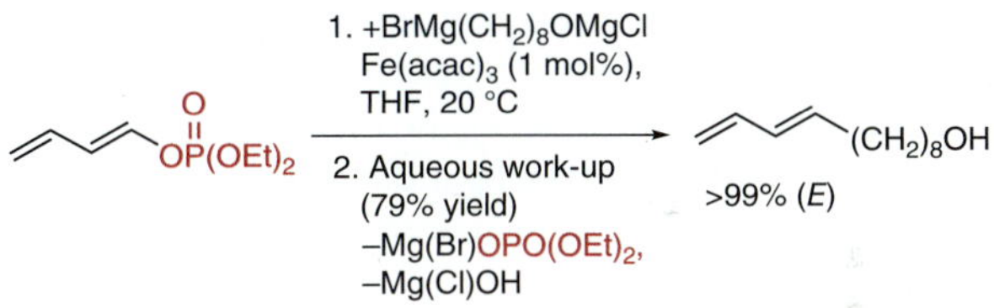

(8.289)

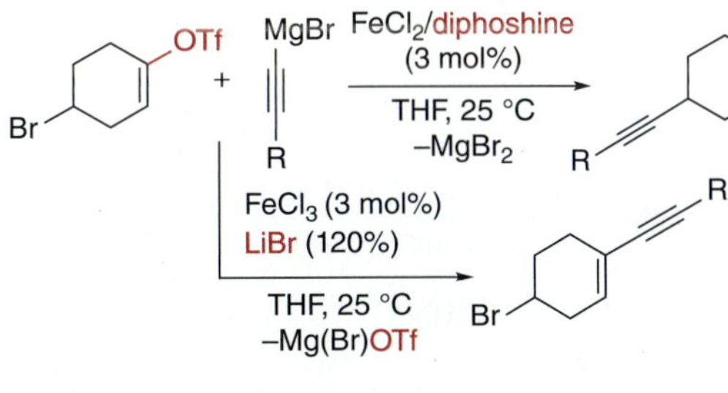

(8.290)

(8.291)

Arene- and alkanesulfonyl chlorides are inexpensive starting materials. When they are reacted with Grignard reagents, they form the corresponding sulfones (Scheme 8.89a) [2159]. However, in the presence of $Pd(P(t\text{-}Bu)_3)_2$ catalyst and under desulfinylative conditions (heating to 70 °C), only products of C—C-homocoupling of the Grignard reagents are observed (Scheme 8.89a). However, upon addition of an equivalent of $ZnCl_2$, the corresponding

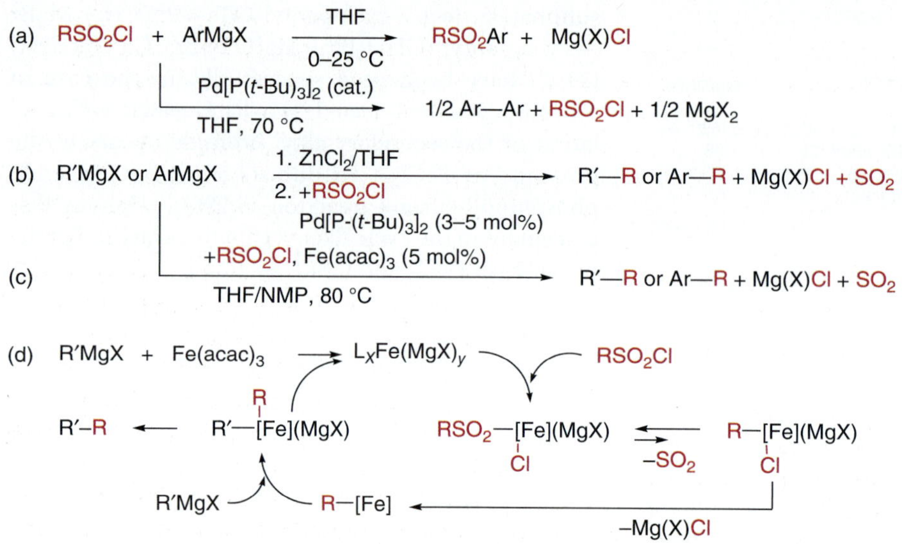

Scheme 8.89 Arene-, alkene-, and alkanesulfonyl chlorides are suitable electrophilic substrates for desulfinylative KTC reactions. (a) Formation of sulfones and of products of homocoupling in the presence of Pd-catalyst; (b) if the Grignard reagent is converted into the corresponding organozinc reagent and Pd-catalyst is applied, desulfinylative cross-coupling occurs on heating; (c) if only $Fe(acac)_3$ is used as a catalyst, desulfinylative cross-coupling occurs on heating; and (d) possible mechanism of the Fe-catalyzed process.

organozinc reagents are generated that undergo the Pd-catalyzed desulfinylative Negishi cross-coupling reaction (Scheme 8.89b) [2160] (Section 8.8.2). Arene, phenylmethane, and alkenesulfonyl chlorides are suitable electrophilic substrates for aryl and benzylzinc halides. Volla and Vogel have shown in 2008 that alkane- and arenesulfonyl chlorides undergo smooth desulfinylative C—C-cross-coupling reactions with Grignard reagents in the presence of catalytic amounts of $Fe(acac)_3$ (acetylacetonate) without the need of any extra ligands (Scheme 8.89c). A possible reaction mechanism is given in Scheme 8.89d [2161].

When the Grignard reagent was added dropwise to the heated (80 °C) mixture of RSO_2Cl, $Fe(acac)_3$/THF/NMP (*N*-methylpyrrolidone), the cross-coupling reaction is faster than addition of the Grignard carbonyl moieties (e.g. reaction (8.292)).

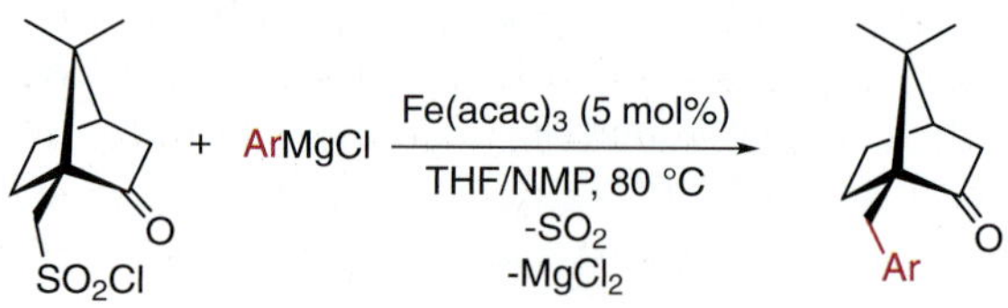

(8.292)

$Co(acac)_3$ catalyzes the KTC reaction of alkenyl triflates [2162], $CoCl_2\cdot(LiI)_2(TMEDA)_4$ (TMEDA = $Me_2NCH_2CH_2NMe_2$) catalyzes the coupling of primary and secondary inactivated alkyl bromides with alkyl Grignard reagents [2163], and $Co(acac)_3/PPh_3$ catalyzes the sp^2(C)-sp^2(C) cross-coupling of alkenyl triflates with aryl and alkenyl Grignard reagents [2164]. Aryl bromides react with alkyl Grignard reagents in the presence of *N*,*N*,*N*′,*N*′-tetramethyl-1,3-propanediamine and $CoCl_2$ + N-heterocyclic carbene as a catalyst [2165]. The $CoCl_2$ + TMEDA system catalyzes the alkylation of aryl Grignard reagents with primary and secondary alkyl bromides; the reaction tolerates ester, amide, and keto groups [2166]. Carbon–carbon bond formation between Grignard reagents and alkyl sulfonates requires magnesium/metal exchange. For instance, equal amounts of $CuBr\cdot SMe_2$, LiBr, and LiSPh in THF at 0 °C realizes a soluble copper catalyst able to promote the C—C cross-coupling between primary, secondary, tertiary alkyl, aryl, vinyl, and allylic Grignard reagents and tosylates [2157].

Problem 8.85 What are the four hydrocarbons formed in the reaction of *n*-BuMgBr + *n*-nonyl-Br in the presence of $Fe(OAc)_2$ (1 mol%) and $[o\text{-}(Ph_2P)C_6H_4]_2O$ (1 mol%)? [2167]

Problem 8.86 What is the major product of the following reaction? [2072]

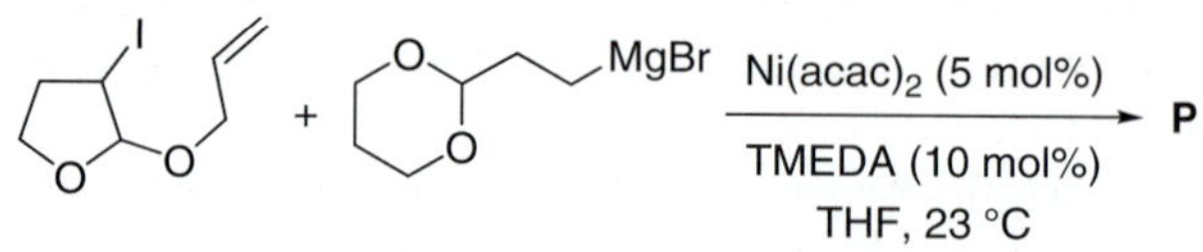

8.8.2 Negishi cross-coupling

The Negishi C—C cross-coupling (8.293) discovered in 1977 [2168] is a widely used reaction. It combines organozinc nucleophiles with organic electrophiles using catalysts that are Ni or Pd complexes containing phosphine, diphosphine, tertiary diamine, or a *N*-heterocyclic carbene ligand [2168–2174]. In 2010, together with Akira Suzuki (Section 8.8.4) and Richard F. Heck (Section 8.8.7), Ei-ichi Negishi received the Nobel Prize in Chemistry for his work on Pd-catalyzed C—C cross-coupling reactions [2175].

$$R^1\text{–ZnY} + R^2\text{–X} \rightarrow R^1\text{–}R^2 + \text{Zn(Y)X} \quad (8.293)$$

$$\begin{aligned} R^1 &= \text{alkenyl, aryl, allyl, alkylbenzyl,} \\ &\quad \text{homoallyl, homopropargyl} \\ R^2 &= \text{alkenyl, aryl, allyl, alkynyl, propargyl} \\ Y &= \text{Cl, Br, I, OCO–}t\text{-Bu} \\ X &= \text{Cl, Br, I, SO}_2\text{Cl, OR} \end{aligned}$$

Recent studies have demonstrated that Cu [2176–2178], Co [2179–2182], or Fe-catalysts [2183–2186] can be used instead of Ni or Pd. Furthermore, aryl fluorides (Scheme 8.90) [2187, 2188], alkenyl and aryl sulfonates, carboxylic esters [2066, 2067, 2189], and aryl ethers [2190] can be electrophilic partners. In 2006, Dubbaka and Vogel showed that inexpensive sulfonyl chlorides undergo desulfinylative cross-couplings [2191, 2192] with organozinc reagents in the presence of $Pd[P(t\text{-Bu})_3]_2$ catalyst in boiling THF ($R^1ZnCl + R^2SO_2Cl \rightarrow R^1—R^2 + ZnCl_2 + SO_2$) [2160]. With acyl chlorides, ketones are obtained in good yield ($R^1ZnX + R^2COCl \rightarrow R^1COR^2 + Zn(X)Cl$) [2193–2195], a reaction that can apply Fe-catalysts [2196]. In 2015, Hu and coworkers presented a stereoselective synthesis of conjugated enynes through anti-Markovnikov $FeBr_2$-catalyzed reductive carbozincation of terminal alkynes (e.g. $Ar^1—C{\equiv}CH + R—I + Zn \rightarrow (E)—Ar(ZnI)C{=}CHR$) and subsequent CuI-catalyzed Negishi coupling ($Ar^1(ZnI)C{=}CHR + Br—C{\equiv}C—Ar^2 \rightarrow (E)—Ar^2C{\equiv}C(Ar^1){=}CHR + Zn(Br)I$. In a similar way, arylalkenes ($Ar^1(ZnI)C{=}CHR + Br—Ar^3 \rightarrow Ar^3(Ar^1)C{=}CHR + Zn(Br)I$) were obtained, the Negishi reaction being catalyzed by $Ni(cod)_2$ + 1,1′-bipyridine [2184].

In the Fukuyama coupling, thioesters are reacted with a large variety of organozinc reagents in the presence of a Pd- or Ni-catalyst [2197], also producing ketones ($R^1ZnX + R^2COSR^3 \rightarrow R^1COR^2 + Zn(SR^3)X$) [2195, 2198, 2199]. Diaryl ketones have been prepared by Ni-catalyzed C—N cleavage with arylzinc reagents ($Ar^1ZnCl + Ar^2N(COO—t\text{-Bu})_2 \rightarrow Ar^1COAr^2 + Zn[N(Boc)_2]Cl$) [2175c]. In the presence of alkali halides, zincate species of types $[R^1ZnX_2]^-$ and $[R^1ZnX_3]^{2-}$ form that are transmetallated more rapidly than R^1ZnX [2200]. As mentioned earlier (Section 8.7.2), organozinc reagents of type RZnY and R_2Zn are air and moisture sensitive and thus require special handling techniques, except for pivalates of type RZnOPiv (R = aryl, heteroaryl, alkyl) that are bench-stable compounds. Their cross-couplings with aryl and heteroaryl iodides, bromides, and chlorides have been realized in the presence of high-performance catalysts such as PEPPSI-IPr (see Scheme 8.88) and analogs [1868, 2201]. Polyfunctional organozinc compounds have been used extensively in cross-coupling reactions as illustrated in Scheme 8.90 [1885, 2168, 2170–2173, 2202].

The general mechanism (Scheme 8.91a) of Pd- and Ni-catalyzed Negishi coupling involves three key steps. After the formation of the catalytically active [M(0)] complex by reduction of the transition metal salt by the organozinc reagent, an oxidative addition equilibrates $[M(0)] + R^2—X$ with adduct $R^2—[M(II)]—X$ [2203]. The latter undergoes transmetallation (metal/metal exchange) with R^1ZnHal giving $Zn(Hal)X + R^2—[M(II)]—R^1$. A final reductive elimination provides the product of cross-coupling $R^1—R^2$ and liberates the catalyst [M(0)]. This mechanism is similar to that proposed for some Pd- and Ni-catalyzed KTC reactions $Ar—Hal + RMgX \rightarrow Ar—R + Mg(Hal)X$ (Scheme 8.87). It is supported by quantum mechanical calculations for the cross-coupling $Ar^1ZnCl + Ar^2Cl \rightarrow Ar^1—Ar^2 + ZnCl_2$ catalyzed by $Ni(Cy_2P(CH_2)_3PCy_2)$(toluene) [2204]. Depending on the nature of the electrophilic partner, catalyst, and reaction conditions, other mechanisms have been proposed, including radical mechanisms similar or not to that presented in Scheme 8.88b. For the Ni-catalyzed cross-coupling of alkylzinc halides with aryl iodides, quantum mechanical calculations by Cárdenas and coworkers supported the occurrence of a catalytic cycle involving transmetallation of alkylzinc halide to Ni(I) species (e.g. **352** + RZnX → **353**, Scheme 8.91b) followed by oxidative addition of the aryl iodide and formation of Ni(III) intermediates **354** and subsequent reductive elimination that gives the final product of cross-coupling and liberates the catalytically active Ni(I) species. Alternatively, oxidative addition of the Ni(I) intermediate to Ar—X produces another Ni(III) complex (e.g. **352** + ArX → **355**) that undergoes transmetallation. The Ni(I) intermediate might be formed by dismutation $[Ni(0)] + [Ni(II)]X_2 \rightarrow 2[Ni(I)]X$ or by SET $Ar—[Ni(II)]—R + ArX \rightarrow Ar—[Ni(III)]—R^{\bullet+} + [ArX]^{\bullet-} \rightarrow Ar—R + [Ni(I)]^+ + [ArX]^{\bullet-}$ [2205]. For the stereoconvergent phenylation of a racemic, secondary propargyl bromide with Ph_2Zn in the presence of $NiCl_2$ + (−)-*i*-Pr-pybox, Fu and Schley proposed the intermediacy of the corresponding propargylic radical (Scheme 8.91c) [2206].

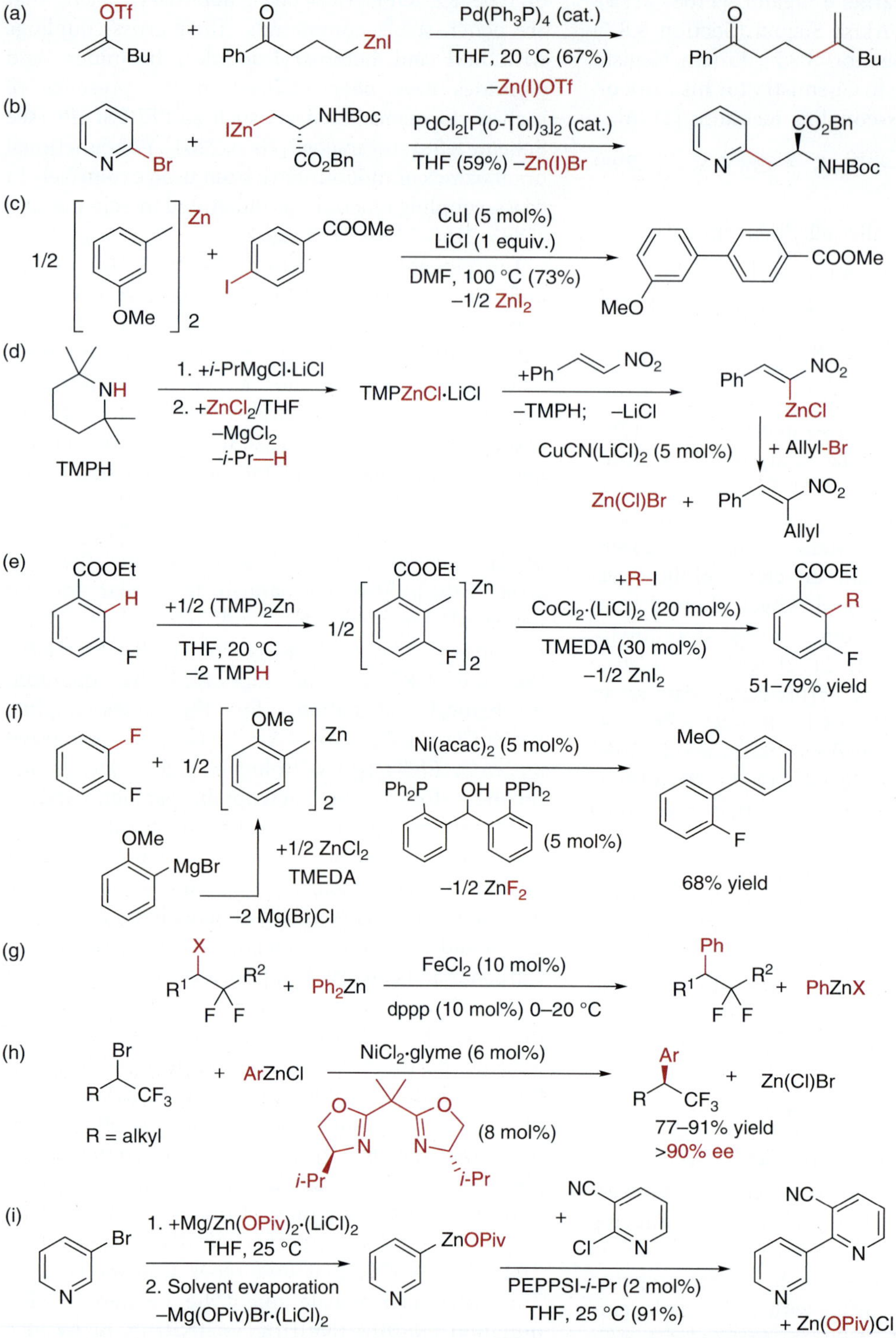

Scheme 8.90 Selected examples of Negishi cross-coupling reactions. (a) Alkenyl triflate (enol trific ester) as electrophilic partner, (b) heteroaryl bromide as electrophilic partner, (c) aryl iodide as electrophilic partner, (d) allyllation of an alkenylzinc reagent, (e) alkyl iodide as electrophilic partner, (f) aryl fluoride as electrophilic partner, (g) iron-catalyzed arylation of alkyl halide, (h) asymmetric arylation of alkyl bromides, and (i) heteroaryl chloride as electrophile partner reacting with an arylzinc pivalate.

Scheme 8.91 (a) Classical reaction mechanism proposed for the Ni-catalyzed Negishi coupling, formation of [Ni(0)] and [Ni(II)] complexes; (b) formation of [Ni(I)] and [Ni(III)] intermediates in the cross-coupling of alkylzinc compounds and aryl iodides; and (c) proposed radical mechanism for the enantioselective Ni-catalyzed phenylation of a racemic propargylic bromide.

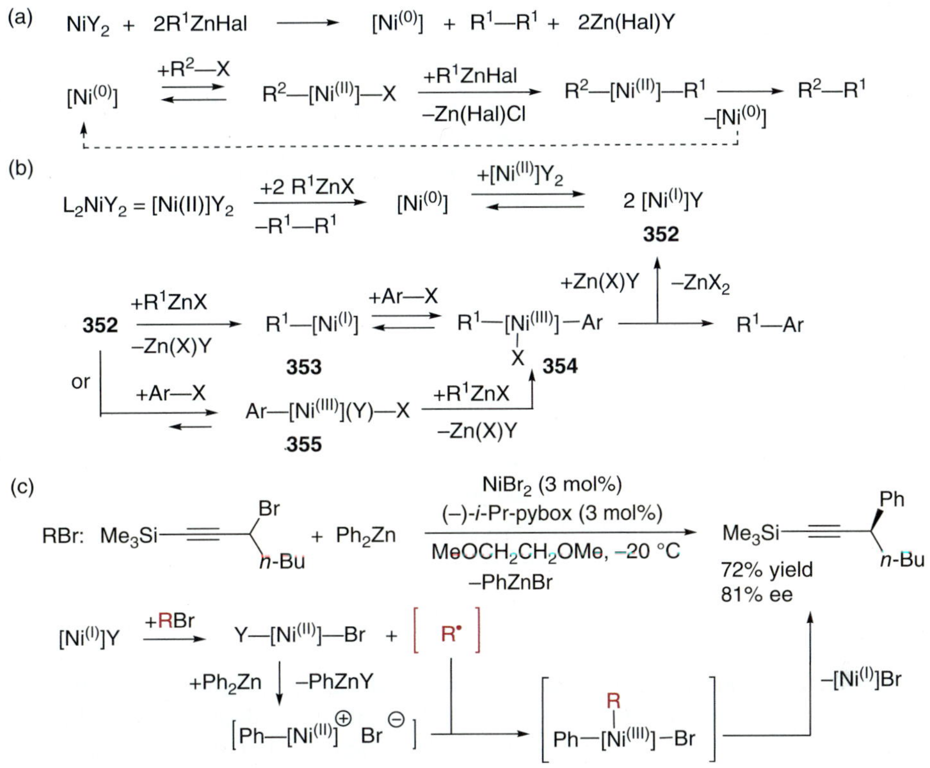

Problem 8.87 Propose syntheses for the protected α-amino acids **A**, **B**, and **C** (Bn = $PhCH_2$; Boc = *t*-BuOCO) [2207–2210].

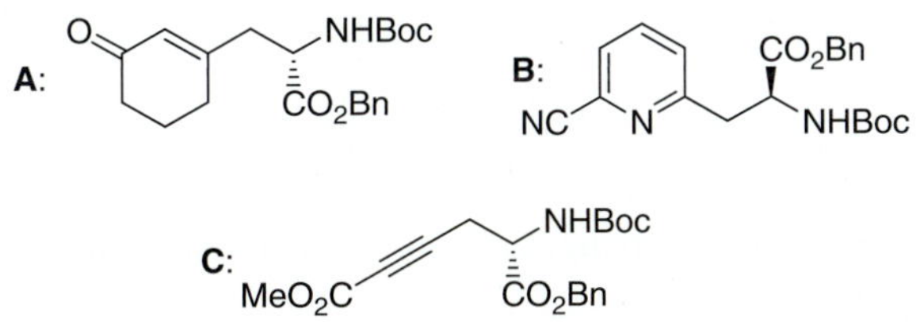

8.8.3 Stille cross-coupling and carbonylative Stille reaction

Tin is one of the earliest metals known to man and has influenced the course of his way of life (tin plates, solder, antifouling agents, paint biocides, fungicides, antifeedants, acaricides, wood preservatives, stone preservation agents, etc.) [2211]. The most common form of tin in organotin compounds is tetravalent and tetrahedral. The covalent radius of tin is large (2.14 Å) and bonds to the tin atom are long (Sn—C: 2.2 Å; Sn—H: 1.7 Å) and mostly covalent, but with high polarizability. It is not surprising that organotin compounds exhibit reactivities in both homolytic and heterolytic senses ($DH°(R_3SnMe \rightleftarrows R_3Sn + Me^\bullet) \approx 50\ kcal\ mol^{-1}$). Stabilization of an electron-deficient center at the β-position, known as the "β-stannyl effect," [2212] is more important than the "β-silyl effect" (Section 4.8.6). Most organotin compounds are stable in air, insensitive to water, and soluble in ordinary organic solvents [2213–2217]. Unfortunately, they are toxic, much more so than silicon, boron, zinc, magnesium, or alkali organometallics [2218–2220]. However, special methods exist to reduce their undesired impact on health and environment [2221]. The alkyl groups in R_4Sn are usually introduced onto tin by tetraalkylation of $SnCl_4$ with an organometallic reagent. The various alkyltin chlorides, R_nSnCl_{4-n}, are prepared by the Kocheshkov reaction in which R_4Sn and $SnCl_4$ in the appropriate proportions are heated together (c. 200 °C). Other functional groups can be introduced by Cl substitution. Metal hydrides reduce the trialkyltin chlorides into the trialkyltin hydrides from which the hexaalkylditins R_3SnSnR_3 and the transient trialkylstannyl radicals $R_3Sn^\bullet$ can be obtained. Similarly, the dialkyltin dichlorides are reduced to the dialkyltin dihydrides that are a source of the oligomeric dialkyltins $(R_2Sn)_n$ and the transient dialkylstannylenes R_2Sn:. Reaction of alkyltin chlorides with the appropriate nucleophiles gives alkyltin alkoxides, amides, thiolates, carboxylates, etc. The presence of these electronegative groups on tin renders the metal susceptible to coordination by Lewis bases. Simple tetrahedral four-coordination is the exception rather than the rule. Hydrolysis of organotin halides gives the organotin oxides, and in the case of dihalides and trihalides, the reaction proceeds through a series of well-characterized intermediate hydrolysis products. Reactions of Me_3SnNa with (*R*)-2-octyl tosylate, chloride, and bromide proceed with predominant inversion of configuration. Displacement of bromide with Me_3SnNa affords the product with the poorest enantiomeric excess. Competing S_N2 and free radical processes are believed to be responsible. However, the corresponding displacements by Ph_3SnLi or Ph_3SnNa are stereospecific with complete inversion of configuration. Enones such as α,β-unsaturated ketones yield 1,4-adducts with trialkylstannyllithium compounds [2222]. Alkenyl triflates (trifluoromethanesulfonates) are displaced with retention of configuration with $(Bu_3Sn)_2$ $Cu(CN)Li_2$ [2223]. Similarly, the complex $(Ph_3Sn)_2Zn$ $(Me_2NCH_2CH_2NMe_2)$ reacts with alkenyl halides and triflates to yield the corresponding alkenylstannanes in the presence of $Pd(Ph_3P)_4$ with retention. Allylic compounds can be obtained through metal/metal exchange of allyl Grignard reagents with Ph_3SnCl in THF [2224]. Direct coupling of allyl halides with Bu_3SnCl can be effected with magnesium metal in the presence of a catalytic amount of lead bromide [2225]. An alternative coupling procedure uses zinc in aqueous THF containing NH_4Cl [2226]. Enantiomerically enriched allenic stannanes can be derived from enantiomerically pure propargylic alcohols (S_N2' reaction) [2227]. α-Oxygenated alkenylstannanes are very useful synthetic intermediates. Asymmetric stannylation can be carried out by chelation-directed deprotonation of primary carbamates in the presence of (−)-spartein followed by addition of Me_3SnCl (e.g. reaction (8.294)) [2228] Nonracemic α-aminoorganostannanes have also been developed [2229].

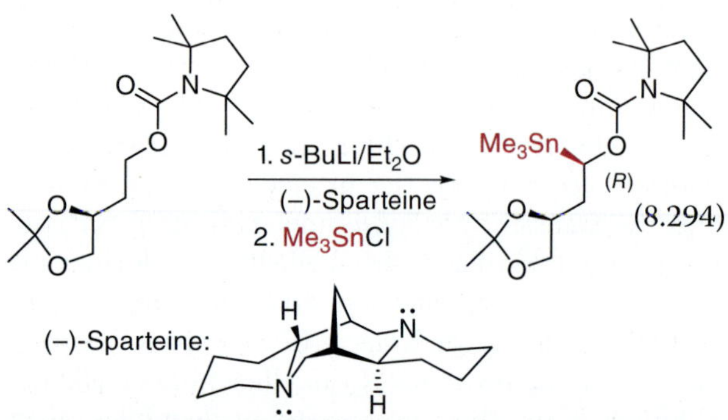

(8.294)

The cross-coupling of organostannanes (R^1—SnR_3) with organic electrophiles (R^2—X) is known as the Stille cross-coupling or Stille reaction (8.295). Both the nucleophilic and electrophilic partners can be of type C(sp^3), C(sp^2), or C(sp) [2230–2233]. The leaving group can be halides, sulfonates, phosphates [2234], and diazonium salts [2235]. In the presence of carbon monoxide (reaction (8.296)), ketones are obtained according to the carbonylative Stille reaction [2230, 2236–2241]. In 2003, Dubbaka and Vogel reported that sulfonyl chlorides can be used in the desulfinylative Stille (reaction (8.297) and carbonylative Stille coupling (reaction (8.298)) [2240]. Generally, the Stille reaction applies a Pd precatalyst with a bulky phosphine, chelating, pincer, or carbene ligand (e.g. Pd-PEPPSI-Ipent [2242], see Scheme 8.88), although examples of Stille reactions using only a Cu-catalyst are known [2243–2245]. In many cases, Pd-catalyzed Stille reactions are more efficient (faster, better yielded) in the presence of additives such as CuI [2246, 2247], chloride [2248] or fluoride anion [2249–2253], CuF [2254, 2255], or gold cocatalyst [2256–2259]. The Stille reaction is a mild process that tolerates a wide variety of functional groups and thus has been applied to the construction of complicated compounds, including natural products [2260–2264].

$$R^1{-}SnR_3 + R^2{-}X \rightarrow R^1{-}R^2 + R_3SnX \tag{8.295}$$

$$(n\text{-Bu})_3SnR^1 + R^2{-}X + CO \rightarrow R^1{-}CO{-}R^2 + Bu_3SnX \tag{8.296}$$

R^1 = alkyl, allyl, benzyl, alkenyl, alkynyl, aryl, heteroaryl

R^2 = alkyl, allyl, benzyl, alkenyl, aryl, heroaryl, acyl

R = alkyl

X = Cl, Br, I, OSO_2R'(R' = CF_3, Me, C_6F_5, pyrazolyl), $OPO(OR')_2$, N_2^+/Y^-

$$(n\text{-Bu})_3SnR^1 + R^2{-}SO_2Cl \rightarrow R^1{-}R^2 + Bu_3SnCl + SO_2 \tag{8.297}$$

$$(n\text{-Bu})_3SnR^1 + R^2{-}SO_2Cl + CO \rightarrow R^1{-}CO{-}R^2 + Bu_3SnCl + SO_2 \tag{8.298}$$

In 1976, Eaborn and coworkers reported the first Pd-catalyzed reductive homocoupling $2Ar{-}Br + Bu_3SnSnBu_3 \rightarrow Ar{-}Ar + 2Bu_3SnBr$ (7–53% yield) [2265]. A year later, Migita and coworkers realized the first cross-coupling $R^1COCl + R_4Sn \rightarrow R^1COR + R_3SnCl$ catalyzed by $Rh(Ph_3P)_3Cl$ [2266] or by $Pd(Ph_3P)_4$ [2267]. They also disclosed arylation

ArX + allylSnR$_3$ → Ar-allyl + R$_3$SnX catalyzed by Pd(Ph$_3$P)$_4$ [2268]. In 1978, Milstein and Stille presented a high yielding synthesis of ketones by cross-coupling (8.299) of further acyl chlorides with organostannanes (Me$_4$Sn, Bn$_4$Sn, (n-Bu)$_4$Sn, and Ph$_4$Sn) catalyzed by BnPd(Ph$_3$P)$_2$Cl/HMPA (hexamethylphosphoramide) [2269, 2270]. A similar ketone synthesis was developed in 2003 by Liebeskind and coworkers based on the Pd-catalyzed cross-coupling of thioesters with organostannanes (R^{1}COS—p-tolyl + Bu$_3$Sn—R^2 → R^1—CO—R^2 + Bu$_3$Sn—S—p-tolyl) [2271]. In 1979, Milstein and Stille reported a cross-coupling reaction (8.300) that is also catalyzed by BnPd(Ph$_3$P)$_2$Cl/HMPA. They proposed a first mechanism for it that is analogous to that presented above for the Negishi coupling (Scheme 8.91a): (i) oxidative addition [Pd] + R^2—X ⇄ R^2[Pd]X, (ii) transmetallation R^2[Pd]X + R^{1}SnR$_3$ ⇄ R^2[Pd]R^1 + R$_3$SnX, and (iii) reductive elimination R^2[Pd]R^1 → R^1—R^2 + [Pd] ([Pd] = catalytically active complex) [2272]. Using trialkylvinyl, trialkylbenzyl, trialkylvinyl, and trialkylphenylstannane, the alkyl group is not transferred: the benzyl, vinyl (e.g. reaction (8.301)), and phenyl groups are transferred more quickly. With optically active α-deuteriobenzyl bromide reacting with Me$_4$Sn, α-deuterioethylbenzene is formed with inversion of configuration in this case (see below Scheme 8.93) [2273].

$$R_4Sn + R^1COCl \rightarrow R^1COR + R_3SnCl \quad (8.299)$$

$$R_4Sn + R^2\text{–}Br \rightarrow R^2\text{–}R + R_3SnBr \quad (8.300)$$

$$(n\text{-Bu})_3Sn\text{–}CH = CH_2 + R^2\text{–}Br \rightarrow R^2\text{–}CH = CH_2 + (n\text{-Bu})_3SnBr \quad (8.301)$$

Depending on the nature of the electrophile, the ligands of the Pd-catalyst, other additives and solvent, the oxidative addition of R^2—X might occur with full or partial inversion or retention of configuration. For instance, in the presence of [Pd(η^3-allyl)Cl] as a catalyst and maleic anhydride as an additive, allyl chloride **356** reacts with (n-Bu)$_3$SnPh and (n-Bu)$_3$Sn(vinyl) mostly with retention in less coordinating solvent such as CH$_2$Cl$_2$,THF, or benzene, and with complete inversion in coordinating solvents such as DMSO or MeCN (Scheme 8.92a). Oxidative addition with inversion can be described as a S$_N$2 displacement (Scheme 8.92b), whereas oxidative addition with retention implies a cyclic three-center transition state preceded or not by the formation of a σ-complex of type **357** (Scheme 8.92c) [2274]. The S$_N$2 mechanism is favored for electron-rich Pd complexes [2275]. This is the case in the coordinating solvent and/or in the presence of anion Y$^-$ that combines with the catalyst forming an anionic species of type (L$_n$PdY)$^-$ [2276, 2277]. Oxidative addition of Pd(PPh$_3$)$_4$ to PhI first gives cis-[ArPd(PPh$_3$)$_2$I], which is quickly isomerized into $trans$-[ArPd(PPh$_3$)$_2$I] under the conditions of the Stille reaction. Buchwald and coworkers have developed a highly active system based on Pd(OAc)$_2$ precatalyst and a biaryl monophosphine ligand (e.g. XPhos) [2278]. Stille coupling of aryl chlorides and sterically hindered stannanes become possible with this type of catalyst (e.g. Scheme 8.92d,e) [2279].

Usually, the catalytically active complexes are considered to be bisligated, PdL$_2$. Quantum mechanical calculations estimate monoligated PdL complexes to be more reactive in oxidative additions than corresponding PdL$_2$ complexes [2280–2284]. According to Schoenebeck and Houk, the higher reactivity of monoligated PdL complexes arises from the fact that they do not have to distort as much as PdL$_2$ to reach the transition state of the oxidative addition [2285]. Kinetic studies reported by Hartwig and coworkers revealed that the rate constants of the irreversible oxidative additions Ph—X (X = Cl, Br, I) + PdL$_2$ (L = trialkylphosphine) → L$_2$Pd(Ph)X depend on the nature of the phenyl halide and not so much on the bulk of the phosphine ligand L. For large phosphine (e.g. (t-Bu)$_3$P, Cy$_3$P), the products are L + T-shaped complexes LPd(Ph)X. The latter can equilibrate with their dimers. The observed rate constants of reaction PhI + PdL$_2$ does not depend on [L] and that of reaction PhI + Pd(Cy$_3$P) (Cy = cyclohexyl) has an inverse dependence on [Cy$_3$P], which demonstrates that bisligated PdL$_2$ reacts with PhI. In the case of the less reactive chlorobenzene, the monoligated complexes PdL add to PhCl as indicated by the inverse dependence on [L] [2286]. Quantum mechanical calculations on reactions Pd[(t-Bu)$_3$P]$_2$ + PhX (X = Br, I)) suggested that the rate-determining steps correspond to associative displacements L$_2$Pd + PhX → [L$_2$Pd· · ·PhX]‡ → LPd(PhX) + L, and for Pd[(t-Bu)$_3$P]$_2$ + PhCl, the rate-determining step corresponds to the oxidative addition LPd(PhX) + L → LPd(Ph)X + L [2284].

Depending on the nature of the organotin reagents and reaction conditions, alkyl group with chiral α-carbon center is transferred either with inversion (e.g. Scheme 8.93a) [2287] or with retention of configuration (e.g. Scheme 8.93b) [2288]. For the transmetallation step, "open mechanism" (Scheme 8.93c) explains inversion of configuration and an alternative "cyclic mechanism" (Scheme 8.93e) is invoked to interpret retention of configuration. In the presence of an additive or coordinating solvent, an "ionic open mechanism" (Scheme 8.93d) may also intervene

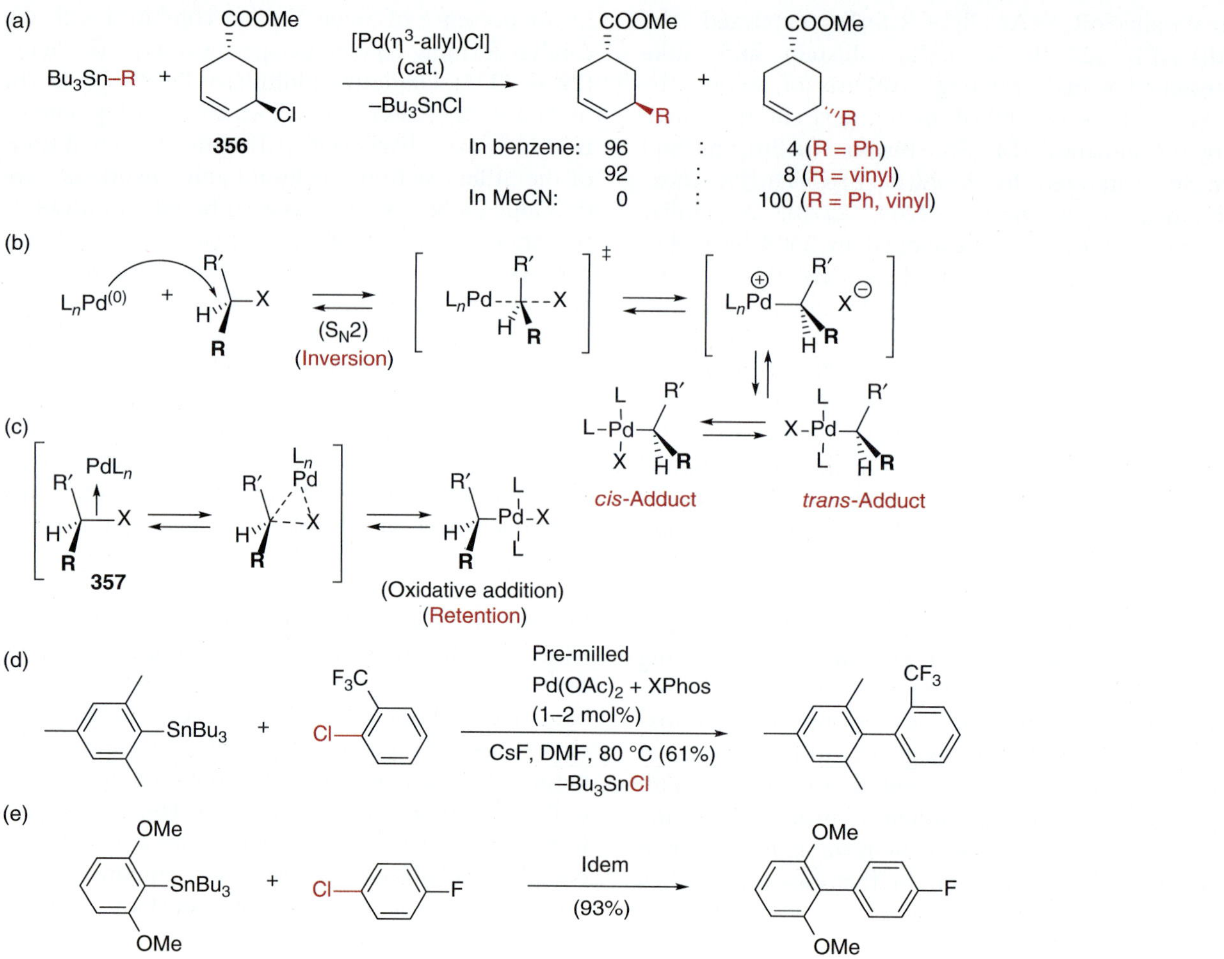

Scheme 8.92 (a) Examples of Pd-catalyzed Stille cross-coupling reactions; (b) the oxidative addition occurs with inversion and (c) with retention of configuration depending on the reaction conditions; and (d, e) examples of reactions engaging bulky organostannanes and aryl chlorides (XPhos = 1-(dicyclohexylphosphino)-2-[2,4,6-tris(isopropyl)phenyl]benzene, see Scheme 8.95).

[2233, 2274]. The open mechanism (Scheme 8.93c) is favored for badly coordinating anionic ligands (e.g. X = TfO = triflate), which create highly electrophilic Pd species, leading to fast transmetallation. The counter-anion X^- is often totally displaced by a coordinating solvent S (e.g. HMPA (hexamethylphosphoramide), NMP (*N*-methyl-2-pyrrolidone), DMF, and THF) producing a cationic complex of type $L_2Pd(R^2)(S)^+/X^-$. The cyclic mechanism is favored when halide X can make a bridge with tin and when ligand L can dissociate readily (e.g. L = Ph_3As, $(\alpha\text{-furyl})_3P$). Free phosphine is detrimental, but it can be sequestered with Cu(I) salts (e.g. CuI, CuBr, Cu(I) thiophen-2-carboxylate = CuTC) [2289]. If the cyclic mechanism is followed, it often implies that transmetallation is the rate-determining step of the overall Stille cross-coupling reaction. The rate at which organostannanes are transmetallated follows the order alkynyl > alkenyl > aryl > heteroaryl > allyl > benzyl > α-alkoxyalkyl > alkyl.

Reductive elimination terminates the Pd-catalyzed Stille reaction. This step involves a cyclic three-center transition state that requires *cis*-$L_nPd(R^1)R^2$ intermediates (Section 7.7.8). In general, the rate constant of the reductive elimination follows the order $C(sp^2)—C(sp^2) > C(sp^2)—C(sp^3) > C(sp^3)—C(sp^3)$. Calculated activation free energies ($\Delta^{\ddagger}G$ in kcal mol^{-1}) for $LPd(Me_3P)Me_2 \rightarrow Me—Me + LPd(Me_3P)$ are 28.6 (L = Me_3P), 27.0 (L = MeCN), 21.7 (L = ethylene), 10.0 (L = (*E*)-NC—CH=CH—CN), 8.6 (L = maleic anhydride), and 5.9 (L = benzoquinone). They show that additives such as electron-poor alkenes (ancillary ligands) can accelerate the reaction [2290].

In 1990, Liebeskind and Fengl pointed out the sluggish Pd-catalyzed Stille cross-coupling reaction work much better in the presence of Cu(I) salts [2246]. In 1996, Allred and Liebeskind reported

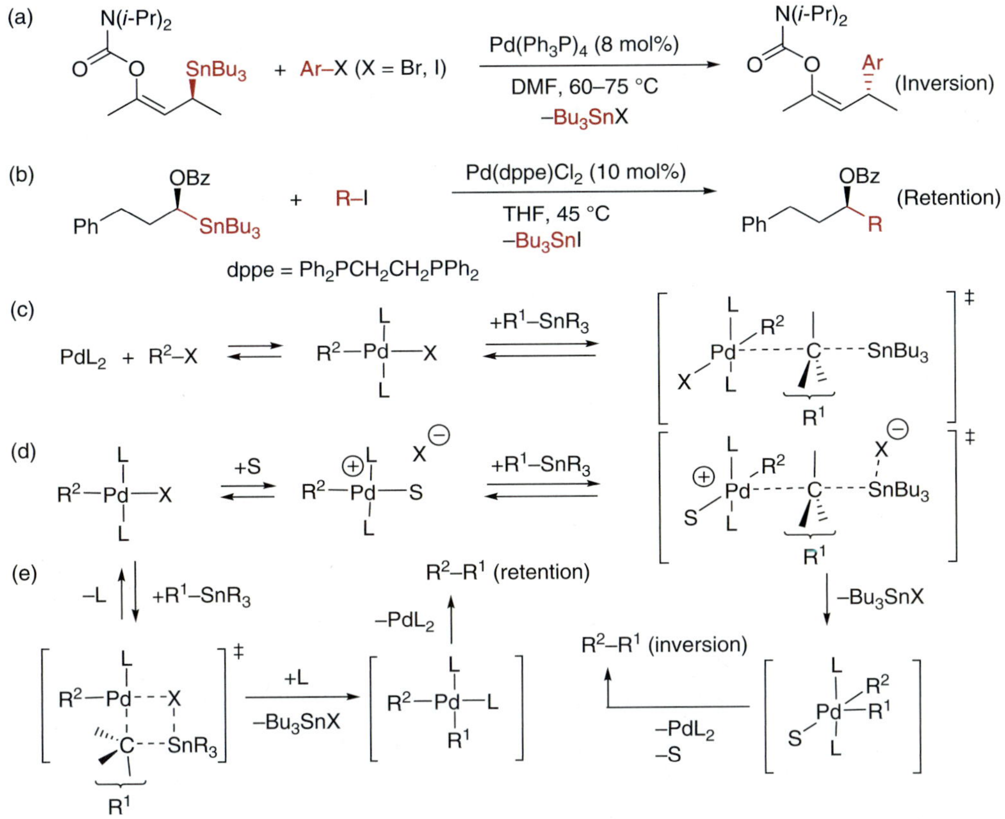

Scheme 8.93 Examples of Pd-catalyzed Stille cross-coupling reactions with enantiomerically enriched organostannanes leading to (a) inversion and (b) retention of configuration; (c) proposed "open," (d) "ionic open mechanism," and (e) "cyclic reaction mechanism" of the transmetallation.

the 2-thiophenylCOOCu-mediated (1.5 equiv.) Stille cross-coupling [2291]. Falck and coworkers reported that Cu alone catalyzes the cross-coupling of α-stannylated benzoate with *S*-ethyl chloroformates, which occurs with retention of configuration (Scheme 8.94a). Sulfur substituents on either coupling partners lead to quicker reactions and better yields. This suggests that chelated species are formed during the catalytic cycle [2292]. Alkynylation of α-stannylated thionocarbamates also leads to retention of configuration (Scheme 8.94b). A possible mechanism is given in Scheme 8.94c. The groups of Liebeskind and Falck disclosed the Cu-catalyzed cross-coupling of α-stannylated thionocarbamates **358** with α-amino acid-derived thioesters **359** that produces corresponding ketones **360** with high stereoselectivity and retention of configuration at the thionocarbamate-bearing stereocenter (Scheme 8.94c). A possible mechanism of the latter reaction is shown in Scheme 8.94d.

Problem 8.88 Propose a mechanism for the carbonylative Stille reaction (8.296).

Problem 8.89 What is the product formed on treating 2-iodophenyl (trimethylstannyl)methyl ether with $Pd(Ph_3P)_4$ in toluene at 40 °C [2293].

8.8.4 Suzuki–Miyaura cross-coupling

Suzuki–Miyaura cross-coupling reaction (8.302) is the most widely applied catalyzed C—C bond forming reaction. Its success originates from a combination of mild reaction conditions and large functional group variation tolerance [2294–2300]. It is applied in the total synthesis of natural products [2301–2305] and to the synthesis of pharmaceuticals, agrochemicals [2306, 2307], and polymers [2308–2311]. After amide bond formation, Suzuki–Miyaura coupling is the most applied reaction in medicinal chemistry [1405, 1406]. The reaction can be run in aqueous media [2312, 2313] or without a solvent [2314]. Contrary to organotin, organoboron reagents are environmentally benign. It was discovered in 1979 by Suzuki and coworkers with the cross-coupling reactions (8.302) of organoboranes (R^1-$B(R')_2$) such as (alken-1-yl)boranes with alken-1-yl, alkyn-1-yl

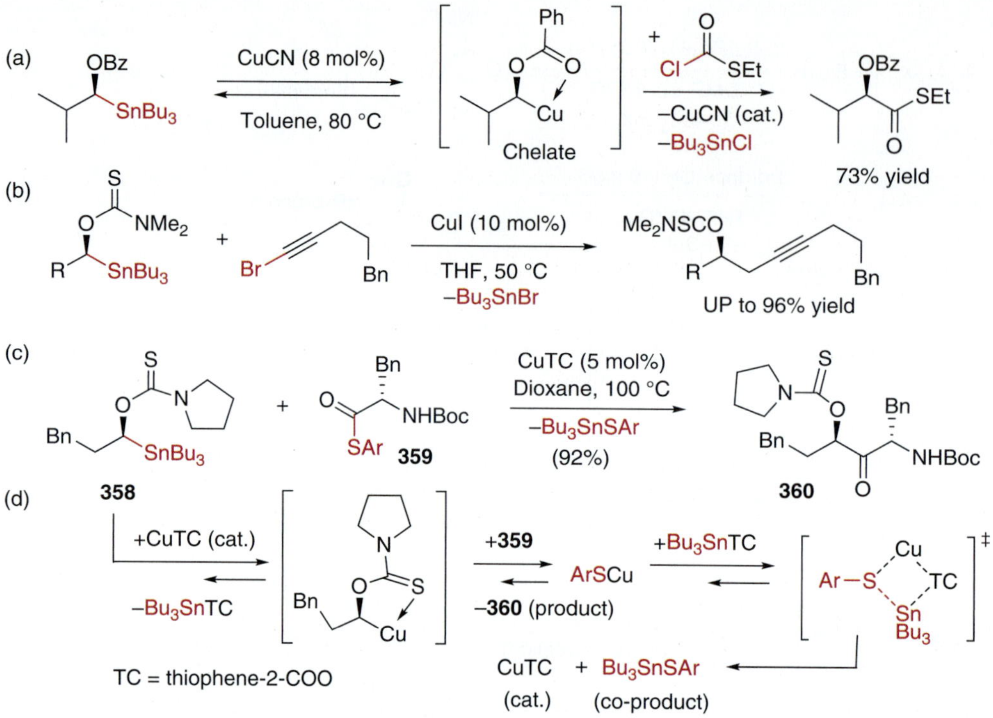

Scheme 8.94 (a–c) Examples of Cu-catalyzed Stille cross-coupling reactions and (d) proposed reaction mechanism for reaction **358** + **359** → **360** + Bu_3SnSAr.

[2315], and aryl iodides and aryl bromides applying a Pd precatalyst and a base [2316]. It was then extended successfully to C—C cross-coupling between inexpensive boronic acids (R^1—$B(OH)_2$), esters ($R^1B(OR)_2$, e.g. catechol and pinacol) (reaction (8.303)), organotrifluoroborate salts [2317–2319] (e.g. R^1—BF_3^-/K^+ that can be prepared by reaction R^1—$B(OH)_2 + 2KHF_2 \rightleftarrows R^1$—$BF_3^-/K^+ + KF + 2H_2O$), *N*-coordinated boronates (e.g. R^1—$B[(OCH_2CH_2)_2NMe]$ and R^1—$B[OCOCH_2)_2NMe]$), boronate salts (e.g. R^1—$B(OR)_3^-Li^+$) [2320], and electrophilic partners (R^2—X) such as chlorides [2321], sulfonates [2322], sulfonyl chlorides (reaction (8.304)) [2323], sulfones [2324], and diazonium salts [2235]. Thioesters react with boronic acids forming ketones. This is the Liebeskind–Srogl reaction (8.305) [2325, 2326]. Frequently used bases are K_2CO_3, *t*-BuOK, Cs_2CO_3, K_3PO_4, and NaOH. The reactivity order for aryl electrophiles is ArI > $ArSO_2Cl$ > ArOTf > ArBr ≫ ArCl ≫ ArF. In the presence of CO, organoboranes undergo Pd-catalyzed carbonylative reactions (Suzuki–Miyaura carbonylation (8.306)) [1976–1978].

$$R^1\text{–}BR_2 + R^2\text{–}X \rightarrow R^1\text{–}R^2 + X\text{–}BR_2 \quad (8.302)$$

$$R^1\text{–}B(OR)_2 + R^2\text{–}X \rightarrow R^1\text{–}R^2 + X\text{–}B(OR)_2 \quad (8.303)$$

$$X = F, Cl, Br, I, OTf, OSO_2R, OR', N_2^+/BF_4^-, \times NMe_3^+/TfO^-, SR'$$

$$R^1B(OH)_2 + R^2SO_2Cl \rightarrow R^1\text{–}R^2 + ClB(OH)_2 + SO_2 \quad (8.304)$$

$$R^1B(OH)_2 + R^2COSR^3 \rightarrow R^1\text{–}CO\text{–}R^2 + R^3SB(OH)_2 \quad (8.305)$$

$$R^1B(OH)_2 + R^2Hal + CO \rightarrow R^1\text{–}CO\text{–}R^2 + HalB(OH)_2 \quad (8.306)$$

$$R^1B(OH)_2 + R^2COOR^3 \rightarrow R^1\text{–}R^2 + R^3OB(OH)_2 + CO \quad (8.307)$$

Until 1998, triarylphosphine ligands (Ar_3P) were used with Pd precatalysts. Buchwald and coworkers developed electron-rich and bulky monophosphine ligands of the type dialkyl(diaryl)phosphines such as SPhos and XPhos [2327]. An application is given by the synthesis of the amphotericin B skeleton developed by Burke and coworkers (Scheme 8.95) [2328]. These ligands enhance the rate of both the oxidative addition (R^2—X + LPd(0) → R^2—LPd(II)—X) and reductive elimination step R^2—LPd(II)—R^1 → R^1—R^2 + LPd(0) (see Scheme 8.96 for a possible reaction mechanism).

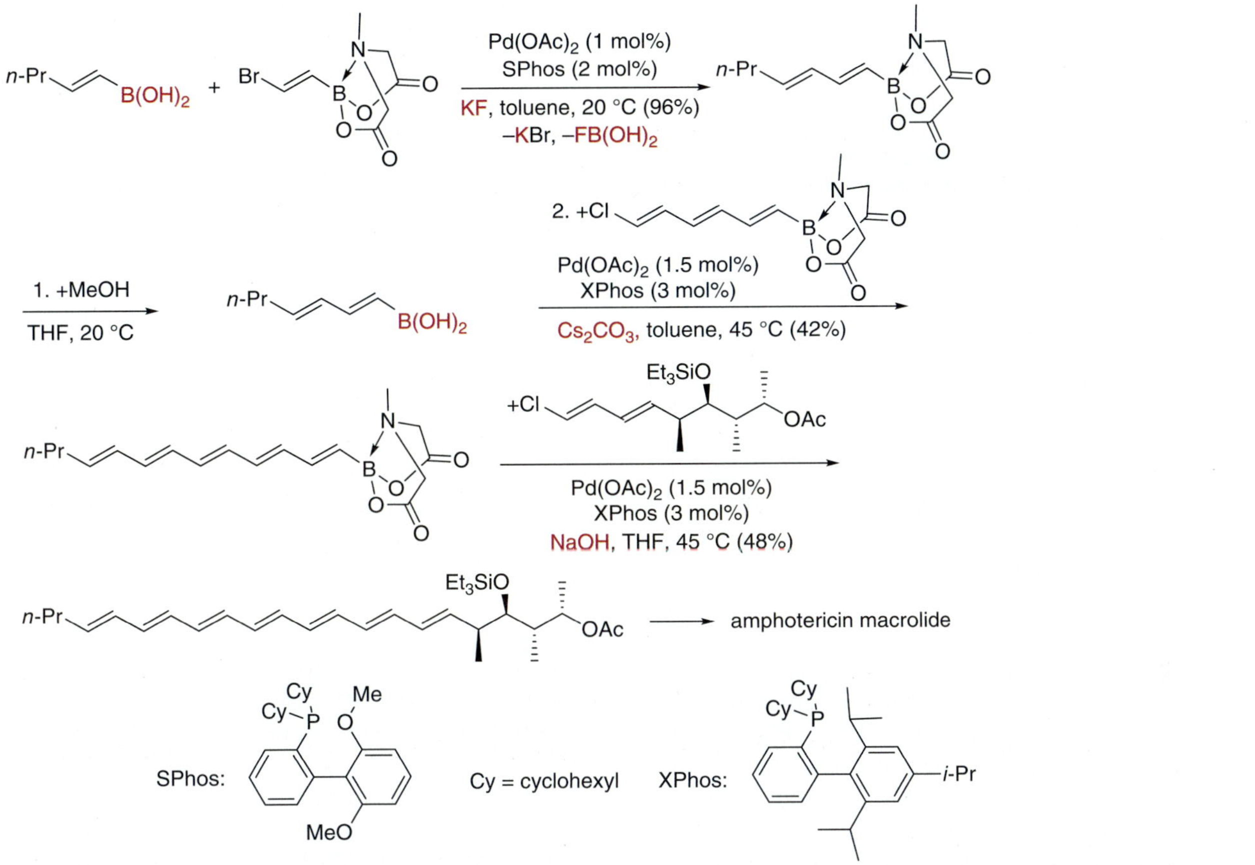

Scheme 8.95 Iterative Pd-catalyzed Suzuki–Miyaura reactions applying Buchwald's dialkyl(diaryl) monophosphine ligands.

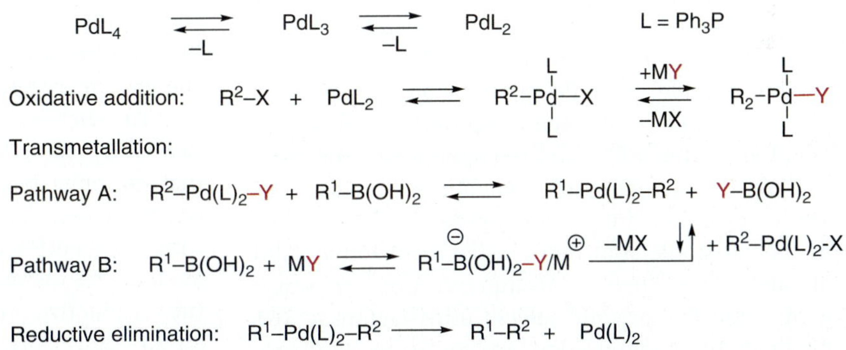

Scheme 8.96 General reaction mechanism proposed for the Pd-catalyzed Suzuki–Miyaura cross-coupling of boronic acids ($R^1B(OH)_2$) with halides (R^2—X).

Because of the cost and toxicity of Pd, other catalysts have been sought [2329, 2330]. The first Ni-catalyzed Suzuki–Miyaura coupling has been reported in 1995 by Percec and coworkers using boronic acids and aryl mesylates [2331]. Electrophilic partners such aryl chlorides [2332], fluorides [2333–2335], aryl ethers, esters, carbamates, carbonates, sulfamates [2336], and phosphates [2337] that do not react or react sluggishly under Pd-catalyzed reaction conditions may react in the presence of Ni-catalysts [2066, 2332, 2338, 2339]. As shown by Hu and coworkers, pincer–Ni complexes catalyze the cross-coupling of alkyl–alkyl [2340] and alkyl–aryl systems [2341]. Decarbonylative coupling of areneboronic acids with phenyl carboxylates (reaction (8.307)) has been catalyzed by $Ni(OAc)_2$ and $(n\text{-Bu})_3P$ (toluene, Na_2CO_3, 150 °C) [2342]. Ni catalysis permits the use of carboxamides as electrophilic partners (acyl-X equivalents) [2343, 2344] (e.g. reaction (8.308)) [2345].

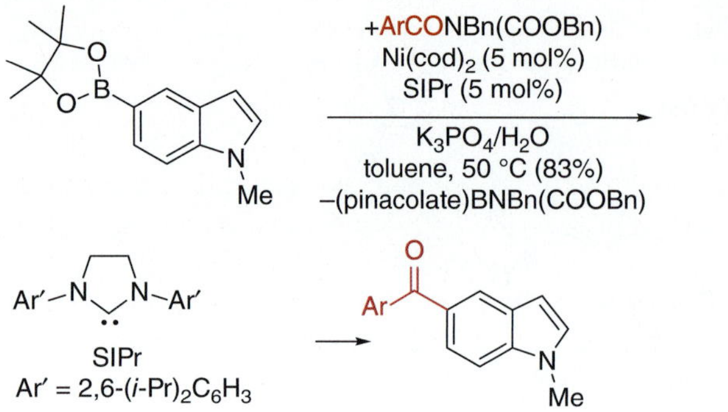

(8.308)

Suzuki–Miyaura reactions catalyzed by Ru [2346], as well as by Cu [1964, 2132, 2347–2352] and Fe, have been reported [2137, 2353–2356]. Bedford and coworkers found that iron–phosphine complexes (e.g. $FeCl_2$ + 1-(diisopropyl)phosphino-2-diphenylphosphinobenzene) prove to be good precatalysts for the cross-coupling of alkyl, benzyl, and allyl halides not only with aryl triorganoborate salts but also with related aluminum, gallium, indium, and thallium-based nucleophiles [2357]. ESR studies showed that $(diphosphine)_2Fe(I)$—X species are formed during the reaction. With very bulky diphosphine ligands, (diphosphine)Fe(I)—X species are generated instead. These species may or may not be responsible for the catalysis. As Pd can be much more active than Cu or Fe-based catalysts, one needs to verify that traces of Pd in the latter metals, at the end, are not responsible of the catalytic effect [2358]. Stabilized nanoparticles of Pd [2359–2361] and recoverable solid Pd-catalysts on polymers or other supports have been proposed [2362–2372]. Poly(2-aminophenol)-stabilized nanoparticles of gold catalyze the coupling of areneboronic acids with aryl halides, including aryl chlorides [2373]. A simple, bench-stable, and inexpensive recoverable solid Pd-catalyst for Suzuki–Miyaura reactions in water or in renewable organic solvent ethyl lactate is **361** obtained by condensation of melanine·$Pd(OAc)_2$ with formaldehyde (reaction (8.309)). Felbinac, an anti-inflammatory drug, is obtained via reaction (8.310) [2374].

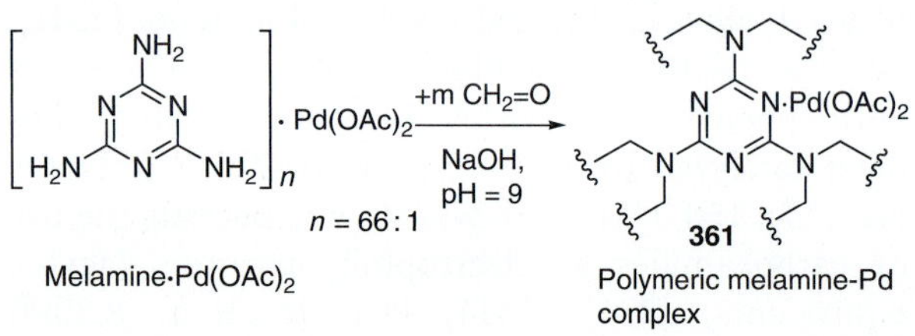

(8.309)

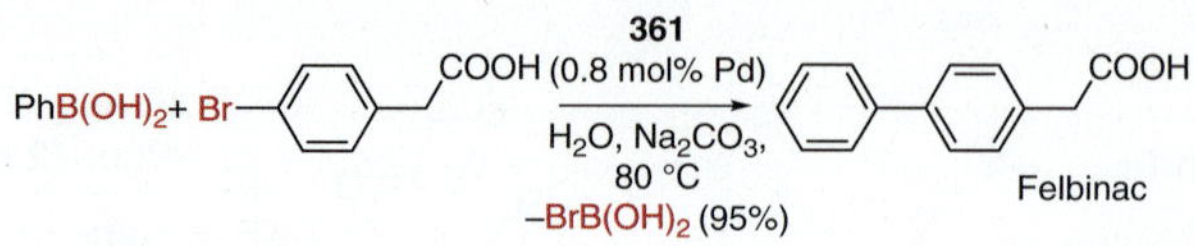

(8.310)

As for Pd-catalyzed Negishi (Section 8.8.2, Scheme 8.91a) and Stille cross-coupling reaction (Section 8.8.3, Scheme 8.93), the Suzuki–Miyaura reaction follows the general mechanism that starts with oxidative addition of a coordinatively unsaturated [Pd(0)] complex onto the electrophilic partner forming adduct R^2—[Pd(II)]—X and transmetallation with the organoboron reagent and final reductive elimination (Scheme 8.96). The leaving group X in adduct R^2—[Pd(II)]—X can be exchanged for anion Y^- brought by the added base M^+Y^- (Y = OH, F; M = Na, K, Cs): R^2—[Pd(II)]—X+MY⇄R^2—[Pd(II)]—Y + MX. A priori transmetallation of R^2—[Pd(II)]—X, or of R^2—[Pd(II)]—Y, can occur either with organoboron or boronic derivative (B with a sextet of electrons: Lewis acid) or with boronate (ate) complex R^1—$B(OH)_2Y^-/M^+$ (B with an octet of electrons: Lewis base) [2375]. For reaction p-$MeC_6H_4B(OH)_2$ + Ph—I → p-Me—C_6H_4—Ph + I—$B(OH)_2$ in THF/H_2O + K_2CO_3 (80 °C) catalyzed by $Pd(Ph_3P)_4$ (1 mol%), Carrow and Hartwig have established that the reaction of phenylpalladium hydroxide complex $[Ph(Ph_3P)Pd(OH)]_2$ with p-toluyl$B(OH)_2$ (pathway A, Scheme 8.96) is much faster than the reaction of phenylpalladium iodide $Ph(Ph_3P)_2Pd$—I with p-toluyl$B(OH)_3^-/K^+$ (pathway B, Scheme 8.96) [2376]. On their side, Amatore and coworkers have shown for reaction Ph-$B(OH)_2$ + p-N≡C—C_6H_4Br (= Ar—Br) → p-N≡C—C_6H_4—Ph (= Ar—Ph) + $BrB(OH)_2$ in DMF/H_2O catalyzed by $Pd(Ph_3P)_4$ that anionic bases M^+/Y^- play multiple antagonistic roles. Two are positive: 1) formation of *trans*-Ar—$Pd(Ph_3P)_2$—Y that reacts with $PhB(OH)_2$ in the rate-determining step and 2) catalysis of the reductive elimination from intermediates *cis*-Ar—$Pd(Ph_3P)_2$—Ph → Ar—Ph + $Pd(Ph_3P)_2$. Two roles are negative as they lead to the formation of less reactive boronate complexes $PhB(OH)_2Y^-/M^+$ and to complexation of the Y group in Ar-$Pd(Ph_3P)_2$ Y by cation M^+ [2377–2380]. The mechanistic studies available do not exclude pathway B (Scheme 8.96) in other Suzuki–Miyaura reactions [2376, 2381]. The general reaction mechanism implies that the catalytically active [Pd(0)] complex bears two phosphine ligands (e.g. L = Ph_3P); [Pd(0)] = $Pd(0)L_2$ is a 14-valence electron palladium complex. It must be noted that a solvent molecule might coordinate this species. Furthermore, when a bulky monophosphine ligand is used, it might form a monoligated complex Pd(0)L as catalytically active species. The latter might be coordinated with one or two solvent molecules. This indicates that the boronic derivatives, ligand, base, and solvent must be well chosen for a successful Suzuki–Miyaura cross-coupling reaction.

In 2000, Yin and Buchwald reported an efficient asymmetric synthesis of atropisomeric biaryls with up to 92% ee using $Pd(OAc)_2$ precatalyst with chiral ligand Kenphos (reaction (8.311)). Quantum mechanical calculations suggested that the high atroposelectivity arises from a combination of weak H-bonding effects and steric interactions in the transition state of the final reductive elimination of the catalytic process, the step that determines the stereoselectivity [2382]. Several other chiral ligands have been proposed since then for the enantioselective Suzuki–Miyaura coupling of arylboronic acids with aryl halides [2383–2386].

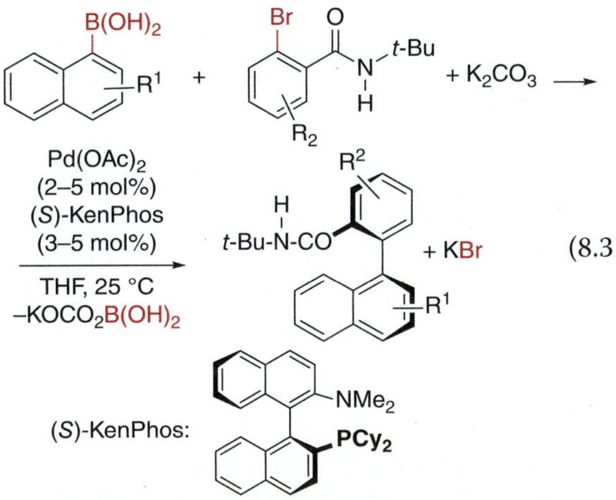

(8.311)

Cross-coupling reactions of alkyl reagents have been slow to develop because alkyl halides undergo oxidative additions with transition metal complexes much less readily than alkenyl and aryl halides. Moreover, alkylmetal intermediates undergo all kinds of side reactions such as β-elimination of hydrogen, protodehalogenation, and protodeboronation. On their side, alkenyl and arylmetal complexes enjoys extra-stabilization because of electron transfer between the π-system and d-orbitals of the transition metal. Despite these problems, the first coupling of an alkylborane with aryl or alkenyl halides was reported by Suzuki and Miyaura in 1986 [2387]. In 2002, Fu and coworkers found that tosylate **362** reacts with $Pd(0)/(t\text{-Bu})_2PMe$ to give alkylpalladium **363** in the absence of organoborane. In most cases, the oxidative addition is a S_N2 displacement reaction with inversion of configuration. The subsequent β-elimination gives a 3 : 1 ratio of (*Z*)-*t*-BuCD=CDH and (*Z*)-*t*-BuCH=CDH (k_H/k_D = 3 : 1). In the presence of PhB-9-BBN, the cross-coupled product **364** is obtained (Scheme 8.97a) with retention of configuration. This implies a cyclic transition state for the transmetallation (see Scheme 8.93e) [2375]. Alkylborane **365** undergoes Pd-catalyzed cross-coupling with 2-iodocyclohex-2-enone giving **366** also with retention of configuration (Scheme 8.97b) [2388]. Pd-catalyzed arylation of enantiopure (diarylmethyl)boronates to generate enantio-enriched triarylmethanes ($Ar^1(Ar^2)(Ar^3)CH$) also occur with retention [2389]. In 2011, Jarvo and Taylor studied Ni-catalyzed alkyl/alkyl Suzuki–Miyaura cross-coupling of deuterium-labeled alkylboranes with alkyl bromides. They found that transmetallation from boron to nickel also occurs with retention of configuration. Furthermore, they observed that alkylnickel intermediates are configurationally stable under the cross-coupling conditions [2390]. The same year, Sugimone and coworkers reported that it was possible to direct the stereochemical course of the Pd-catalyzed arylation of (amidobenzyl)borons **367** by the acidic additive. With phenol, the reaction occurs with inversion, whereas with the Lewis acid A = $Zr(O\text{—}i\text{-Pr})_4/i\text{-PrOH}$, retention of configuration is observed. Transition structures **368** and **369** have been proposed to explain the results (Scheme 8.97c) [2391]. In 2010, Molander and coworkers reported the stereospecific arylation of secondary alkyl trifluoroborates **370** (Scheme 8.97d). As for the arylation of **367**, boron in transition structure **371** benefits from intramolecular coordination with the amide group, which leads to transmetallation with inversion of configuration and also restricts the formation of *syn*-periplanar arrangement of Pd and β-hydrogen in the corresponding transmetallated intermediate, thus avoiding concurrent β-elimination [2392]. This group also reported the stereospecific cross-coupling of aryl chlorides with organotrifluoroborates containing an α-benzyloxy group [2393].

Problem 8.90 What is the main product **P** of the following reaction in 1 : 1 methoxycyclopentane/H_2O at 105 °C? [2393]

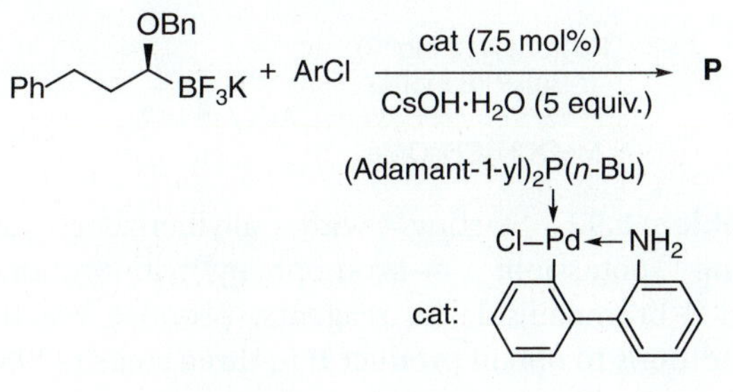

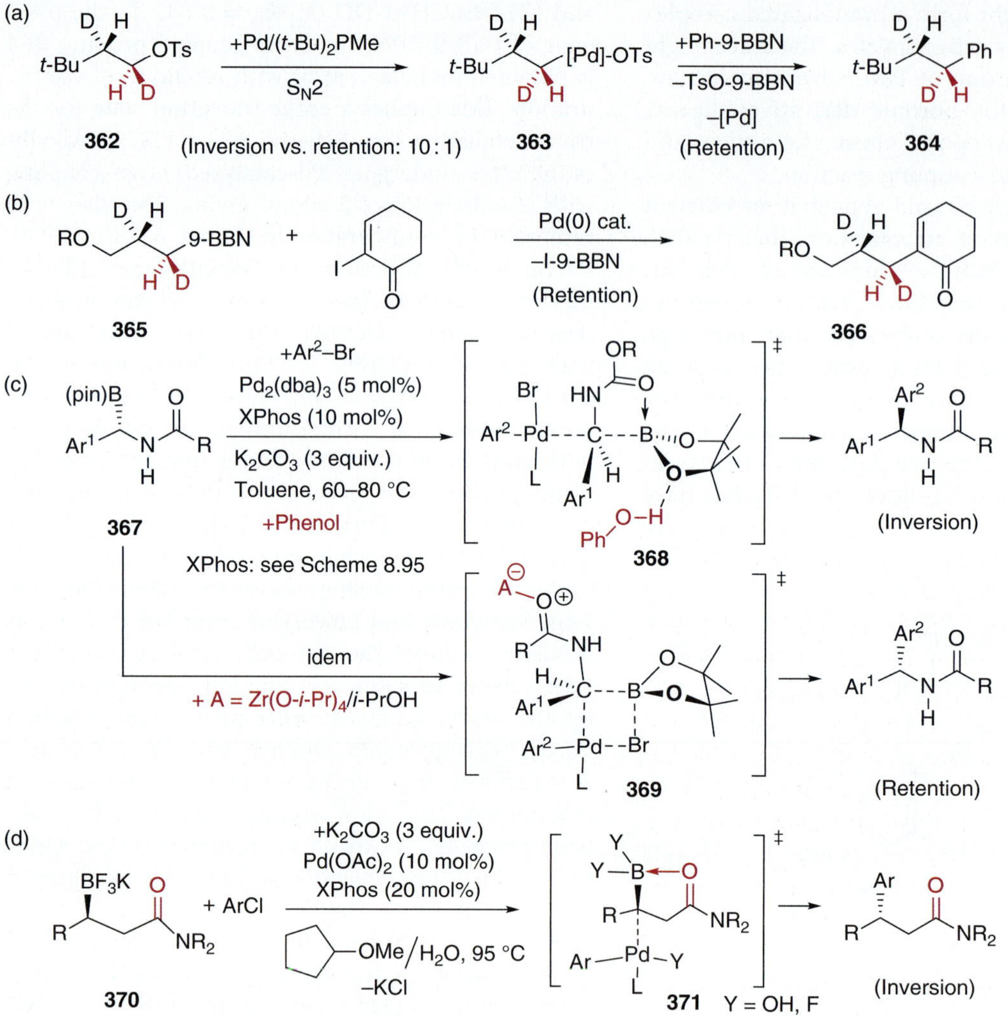

Scheme 8.97 Stereoselective Pd-catalyzed Suzuki–Miyaura reactions: (a) dialkyl(phenyl)boron with an alkyl tosylate, (b) alkylboronate with an alkenyl iodide, (c) (acetamidobenzyl)boronates with aryl bromides involving inversion or retention of configuration depending on the acidic additive, and (d) invertive arylation of β-trifluoroboratoamides.

Problem 8.91 Aryltriazenes in combination with a Lewis acid undergo Pd-catalyzed cross-coupling reactions with arylboronic acids. Propose a mechanism for the following reaction and give the coproducts of reaction [2394].

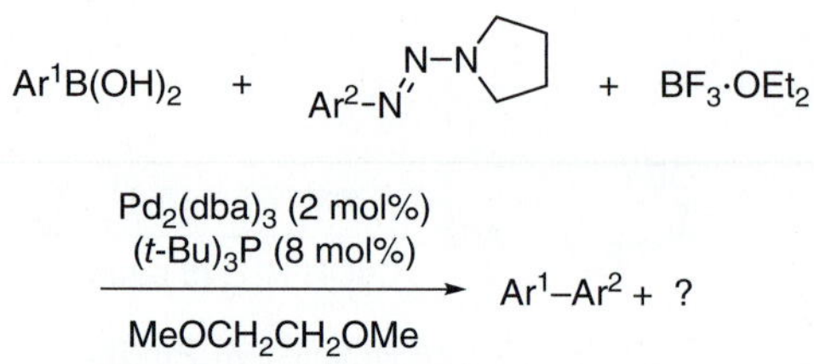

Problem 8.92 Starting with allylbenzene, and using potassium (4-bromophenyl)trifluoroborate and 4-bromoanisole as reagents, propose reaction conditions to obtain product **P** in three steps [2395].

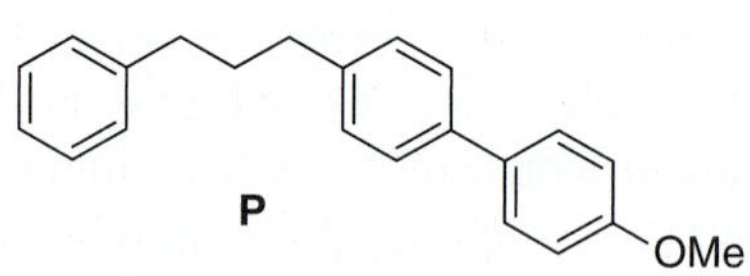

P

8.8.5 Hiyama cross-coupling

Silica and silicates form the major component of the Earth's crust. The element was isolated in 1811 by Gay-Lussac and Thénard ($SiF_4 + 4K \rightarrow 4KF + Si$). It forms an extremely strong bond with oxygen and fluorine ($DH°(F_3Si^•/F^•) = 160$ kcal mol^{-1}); $DH°(Me_3Si^•/X^•)$ for X = OH (128), Cl (113), NHMe (100), Br (96), H (90.3), Me (89.4), Me_3Si (80.5), and I (77)) [2396, 2397]. Generally, organosilanes are obtained via displacement of organohalides with Grignard or lithium reagents (e.g. furan + n-BuLi → α-furyl-Li + Bu—H, α-furyl-Li + Me_3SiCl → α-furyl-$SiMe_3$ + LiCl) or by transition-metal-catalyzed hydrosilylation of alkenes, allenes, and alkynes (Section 8.7.4), but alternative methods are being developed, such as deshydrogenative silylation catalyzed by pincer transition metal complexes (e.g. alkane + H—$Si(OSiMe_3)_2Me$ + t-BuCH=CH_2 → alk-1-yl-$Si(OSiMe_3)_2Me$ + t-$BuCH_2$—CH_3) [2398]. Organosilicons are stable, have low toxicity, and their handling is easy. Silicon–carbon bonds are almost apolar. The nucleophilicity of organosilicons is strongly enhanced in the presence of a silicophilic agents such as HO^- or F^- anion that equilibrate with the corresponding silicate (ate) salts. In 1978, Kumada and coworkers found that alkenylfluorosilicates (obtained by additions: R—C≡CH + $HSiCl_3$ → (*E*)-RCH=$CHSiCl_3$ catalyzed by H_2PtCl_6, followed by treatment with an excess of KF) undergo cross-coupling reactions with allyl halides (displacement reactions) in the presence of a Pd-catalyst (reaction (8.312)) [2399]. Similar cross-couplings with aryl and alkenyl halides were reported in 1982; however, yields remained modest. In 1988, Hiyama and coworkers demonstrated that simple tetracoordinate organosilicon reagents like alkenyl- and alkynyl(trimethyl)silanes undergo Pd-catalyzed cross-couplings with aryl iodides in the presence of tris(dimethylamino)sulfonium difluoro(trimethyl)silicate (TASF=$(Me_2N)_3S^+/Si(Me)_3F_2^-$, e.g. reaction (8.313)) [2400]. TASF is the fluoride anion source that adds to the silane to generate a pentacoordinate silicate. The latter then undergoes transmetallation with Pd(0) (R^1—SiY_3F^- + R_2-[Pd(II)]—X → R^1—[Pd(II)]—R^2 + SiY_3FX^-). Other source of F^- can be applied, such as tetrabutyl ammonium fluoride (TBAF), $Bu_4N^+/Ph_3SiF_2^-$, CsF, and KF. Alkoxides can also be employed to generate pentacoordinate silicates. Thereafter, a large number of C—C bond forming reactions have been applied to the cross-coupling of organosilicons with all kinds of organic electrophilic partners (R^2—X). When using silanols, the F^- source is not required anymore. The silanol is converted into the corresponding potassium silanolate that undergoes direct Pd-catalyzed cross-coupling with aryl halides. This is called the Hiyama–Denmark coupling [2401, 2402]. Under these conditions, (*t*-butyl)dimethylsilyl ether are not cleaved (e.g. reaction (8.314)) [2403–2405]. When using alkenyl-, alkynyl-, or aryltrialkylsilanes, the alkyl groups do not undergo cross-coupling and are lost in the coproducts. In 2005, Hiyama proposed the use of organo[2-hydroxymethyl)phenyl]dimethylsilanes of type **372** as universal cross-coupling reagents [2406]. The *ortho*-hydroxymethylphenyl group activates the Si center intramolecularly with a base such as K_2CO_3 (reaction (8.315)). The coproduct **373** can be recycled into reagent **372** (for a better atom economy). Trifluoromethylated compounds are very much sought for their interesting properties in medicine and agriculture. In 2010. Buchwald and coworkers reported the direct trifluoromethylation of aryl chlorides with trifluoro(triethyl)silane using a Pd/BrettPhos catalyst and KF (reaction (8.316)) [2407]. As for other catalyzed displacement reactions, sulfonyl chlorides can be used (DMF/dioxane, 100 °C) as electrophilic partners in Pd-catalyzed desulfinylative Hiyama couplings ($Ar^1Si(OEt)_3$ + Bu_4NF + Ar^2—SO_2Cl → Ar^1—Ar^2 + SO_2 + $(EtO)_3SiF$ + Bu_4NCl) [2408].

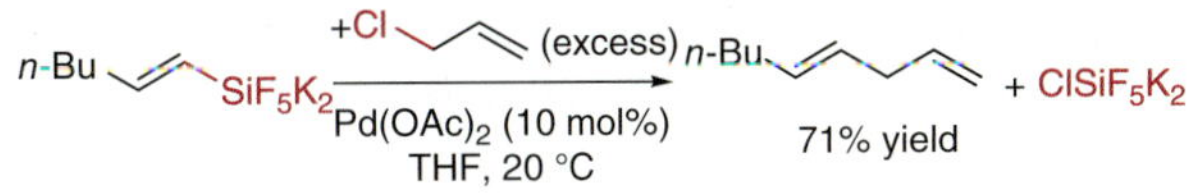

(8.312)

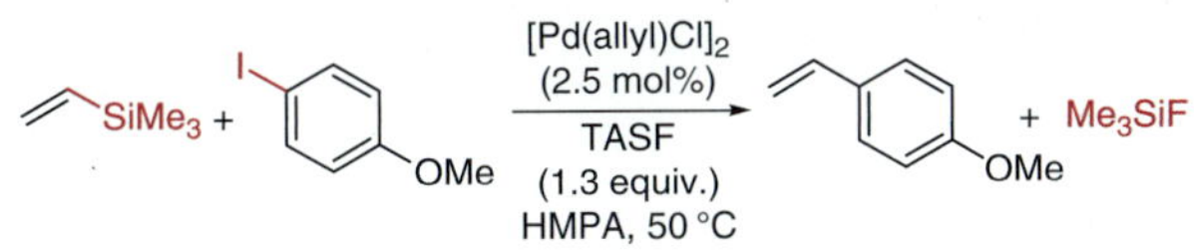

(8.313)

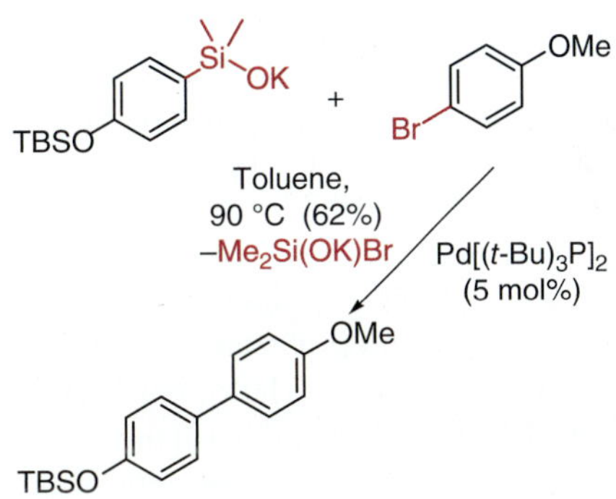

(8.314)

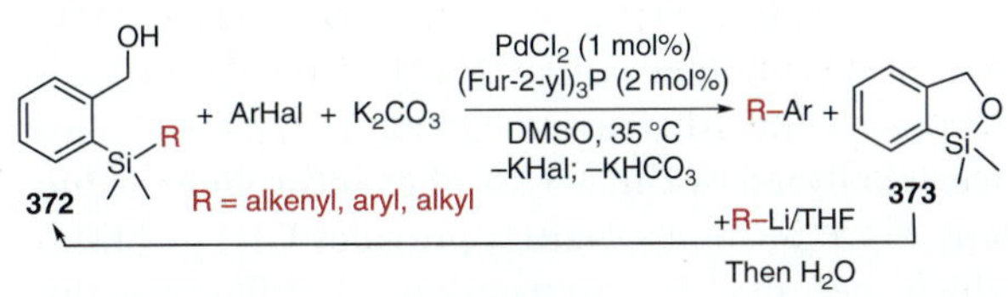

(8.315)

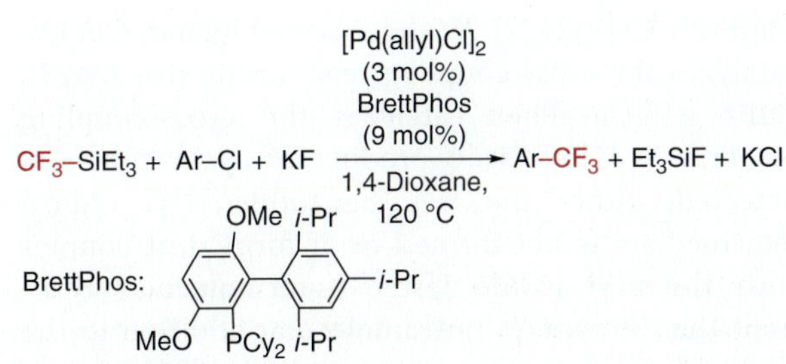

(8.316)

Although Pd catalysts are the most versatile, Ni- and Cu-based catalysts have been employed. In 2004, Fu and Powell found that $NiBr_2$ + 4,7-diphenyl-1,10-phenanthroline catalyzes the cross-coupling of $PhSiF_3$ (+ CsF) and *cis*-4-bromo-1-chlorocyclohexane giving a 55/45 mixture of *cis*- and *trans*-4-phenyl-1-chlorocyclohexane. The reaction is assumed to involve a selective homolytic cleavage of the C—Br bond and the 4-chlorocyclohexyl radical intermediate which results in quenching by a Ph-[Ni(III)] species without sterereoselectivity [2409]. In the presence of enantiopure diamine **374**, enantioselective Ni-catalyzed cross-coupling (8.317) has been observed. Asymmetric induction is realized during the C—C bond forming step, which probably involves a α-ketoalkyl radical intermediate [2410].

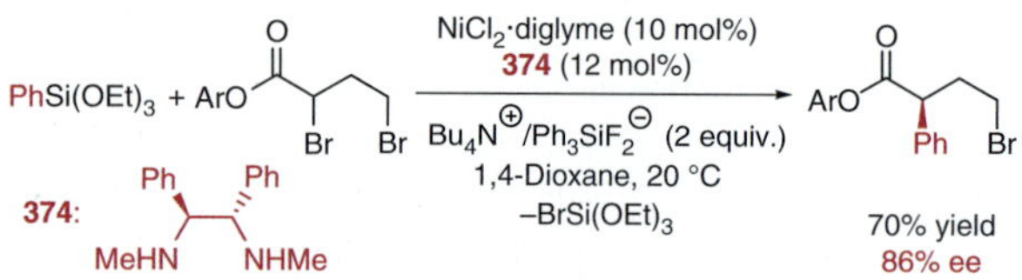

(8.317)

CuI·$Ph_2P(o\text{-}Me_2NC_6H_4)$ catalyzes the cross-coupling of aryltriethoxysilanes ($Ar^1Si(OEt)_3$ + KF or CsF) with aryl iodides (Ar^2—I). A possible mechanism implies transmetallation of $Ar^1Si(OEt)_3F^-/K^+$ by CuI (or CuF) and subsequent oxidative addition of Ar^2—I giving Ar^1—Cu(III)(iodide)—Ar^2 + $(EtO)_3Si$—F + KI. A final reductive elimination provides Ar^1—Ar^2 and the catalyst CuI [1471, 2411]. CuI or $Cu(MeCN)_4BF_4$ catalyzes the cross-couplings of alkenyl(triethoxy)silanes [2412] and alkynyl(trimethyl)silanes with aryl halides [2413]. Cu(I)-catalyzed vinylation of benzyl halides [2414] and allyl halides with vinylsilanes have been presented [2415]. The trifluoromethylation of aryl iodides [2416], aryl bromides [2417, 2418], and aryldiazonium salts [2419] with CF_3—$SiMe_3$ (Ruppert–Prakash reagent [2420, 2421], + KF) have been catalyzed by CuI + 1,10-phenanthroline. Reduction of CF_3SiMe_3 by $NaBH_4$ provides CHF_2—$SiMe_3$, which permits the preparation of difluoromethylarenes (CHF_2SiMe_3 + CsF + ArI → CHF_2—Ar + Me_3SiF + CsI) [2422]. Without added ligand, CuOAc catalyzes the cross-coupling reaction (8.318) [2423]. $CuBr_2$ + PhDavePhos catalyzes the cross-coupling reaction (8.319): As no product of cyclization was detected, either the free benzothien-2-yl radical intermediate is not formed or, if formed, it couples with the aryl iodide faster (intermolecular reaction) than it cyclizes (intramolecular addition to the alkene moiety). One cannot exclude the formation of *p*-methoxyphenyl radical that would add at C(2) of the benzothiophene system. In the presence of a radical scavenging reagent, lower yield of the cross-coupling reaction was observed [2424].

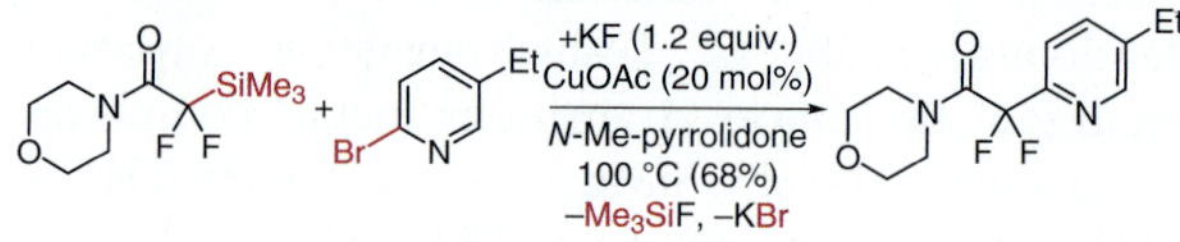

(8.318)

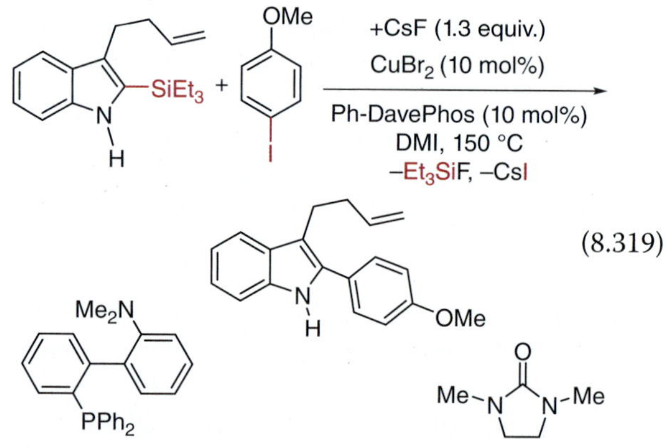

(8.319)

Ph-DavePhos DMI

Problem 8.93 Give the product and coproducts of the following cross-coupling reaction and propose a reaction mechanism.

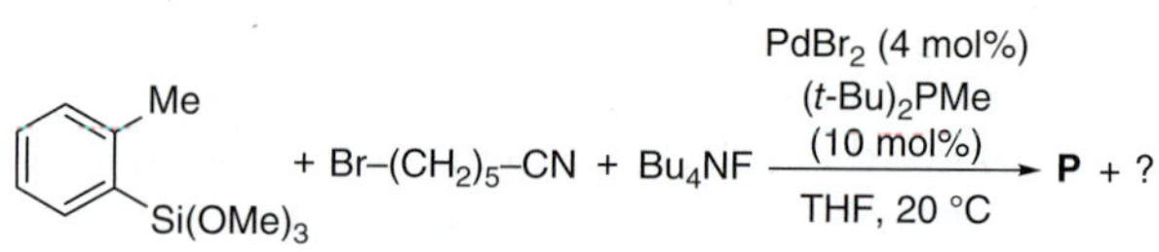

8.8.6 Tsuji–Trost reaction: allylic alkylation

For the Wacker process (Section 7.11.5) that oxidizes ethylene into acetaldehyde, Smidt and coworkers had shown in 1962 that hydroxide anion adds to a (ethylene)$PdCl_2$ complex [2425]. In 1965, Tsuji and Takahashi found that carbon nucleophiles such as the conjugate base of diethyl malonate add to (cycloocta-1,5-diene)$PdCl_2$ complex [1382]. The same year, they reported the reaction of $[(allyl)PdCl]_2$ with the same nucleophile that gives a mixture of mono and doubly allylated diethyl malonate (allyl-$CH(COOEt)_2$ + $(allyl)_2C(COOEt)_2$) [1383]. In 1973, Trost and Fullerton introduced the use of phosphine ligands [2426], which significantly increased the scope of the Pd-catalyzed allylic alkylation reaction [2427–2436]. A wide range of allylic electrophilic partners (with decreasing reactivity: Cl > NR_3^+/X^- > NO_2 > SO_2Ph > $OPO(OR)_2$ > OAc > OCO_2R > OPh > OH, allyl alcohols [2437]) and nucleophilic partners (e.g. malonates, malodinitrile, enolates, enamines [2438], metallated diarylmethanes [2439], and also N- and O-nucleophiles, see below) may be efficiently combined to form corresponding products **377** or **378** in a regioselective and stereoselective manner. For example, Pd-catalyzed processes typically favor nucleophilic substitution at the least substituted allylic terminus (**377**) regardless whether **375** or **376** is the starting material (reaction (8.320)). Phosphine ligands can be exchanged for N- and S-based ligands, and cocatalysts can be used. For instance, direct dehydrative cross-coupling of allylic alcohols with terminal alkynes giving 1,4-enynes is catalyzed by $Pd(Ph_3P)_4 + Ti(O\text{-}i\text{-}Pr)_4$ and 2-dimethylamino-2′-di(*t*-butyl)phosphinobiphenyl [2440]. Although Pd permits the highest versatility, other metals have been applied as precatalysts [2429, 2441–2450].

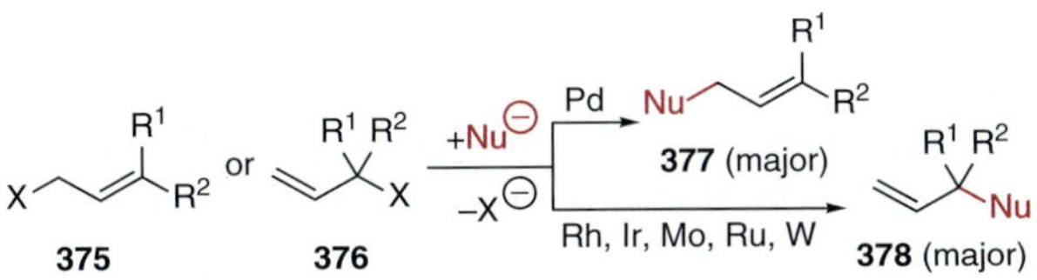

(8.320)

Transition-metal-catalyzed nucleophilic substitutions of allylic substrates proceed via allylmetal intermediates. The structure of such intermediates can vary along the continuum between the η^1-allyl complex **379**, the enyl ($\sigma+\pi$)-complex **380**, and the η^3-allyl complex **381** depending on the metal, its ligands, and the nature of the substituents (equilibria (8.321)) [2434, 2441].

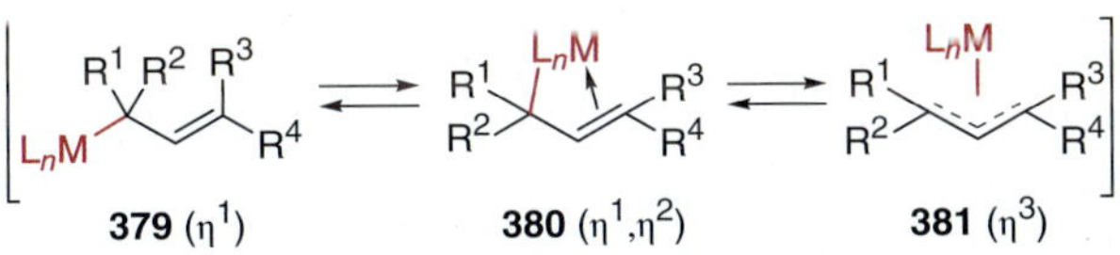

(8.321)

The regioselectivity of the ensuing nucleophilic attack is therefore dictated by a combination of steric and electronic factors that vary with the intermediate complex and the nucleophile [2451]. However, it is generally possible to selectively prepare either **377** or **378** through a judicious selection of catalyst and allylic substrate and ligand. Steric factors favor **377** irrespective of the starting allyl derivatives. With Ru [2452–2456], Mo [2457–2459], Rh [2441, 2442, 2460–2463], Ir [2464–2467], and W [2468, 2469], products **378** are favored.

The oxidative addition of Pd(0) complexes to allylic acetates is reversible and is carried out in the presence of a base, whereas for allyl carbonates, carbamates, aryl ethers, or vinyl epoxides, the reaction proceeds under neutral conditions [2470]. The stereochemistry of the reaction depends on the nature of the nucleophile. With "soft" nucleophiles, the external attack at carbon center on the face opposite to Pd of the allylpalladium complex intermediate (e.g. **382**) is favored (pathway A) (Scheme 8.98). As the oxidative addition proceeds with inversion of configuration (S_N2), the product (e.g. **383**) with overall retention of configuration is obtained. Carbanions of type R—C⁻(X)(Y), where R = H, allyl or aryl, and X and Y = COOR′, COR′, SR′, SOR′, SO_2R', NO_2, CN, belong to that category of nucleophiles (pK_a of conjugate acids: 10–20 range). Enolates [2471], oxygen, sulfur, and nitrogen nucleophiles also add to the allyl moiety. In the case of "hard" nucleophiles such as organometallic derivatives of Mg, Al, Zr, Sn, and B, the initial attack proceeds directly at the palladium atom (giving intermediate **384**) followed by an intramolecular delivery resulting in overall inversion of configuration giving corresponding product **385** (pathway B) (Scheme 8.98).

Both enantiomerically enriched allyl acetates **386** and **387** react with sodium dimethylmalonate in the presence of palladium catalyst in THF giving products **388** and **389** in a 92 : 8 ratio (reaction (8.322)) [2472, 2473]. The regioselectivity does not

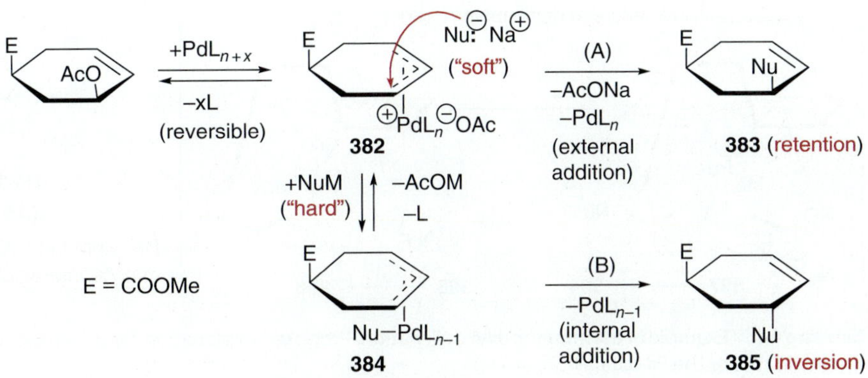

Scheme 8.98 Stereoselectivity of the Pd(0)-catalyzed allylic alkylation depends on the nature of nucleophile (Nu^-/Na^+: "soft"; Nu—M: "hard").

depend on the location of the leaving group. It can be affected by the nature of the ligand, and for diphosphine ligand by its bite-angle β_n [2474, 2475]. The steric hindrance at the palladium center increases with wider bite angles and thus induces an increased selectivity at larger bite angle (dppe (1,2-bis(diphenylphosphino)ethane): $\beta_n = 78.1°$, dppf (1,1′-bis(diphenylphosphino)ferrocene): β_n: 99°, Xantphos (4,5-bis(diphenylphosphino)-9,9-dimethylxanthene): $\beta_n = 110°$) [2476]. Control of the regioselectivity is observed in the intramolecular versions of the allylic alkylation leading to medium-sized rings. As shown with reaction (8.323), complete regioselectivity (no 10-membered ring) and stereoselectivity (no (Z)-cycloalkene) is found [2477].

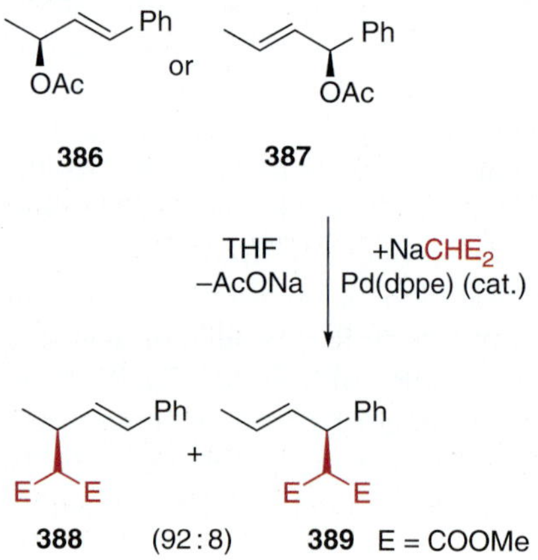

(8.322)

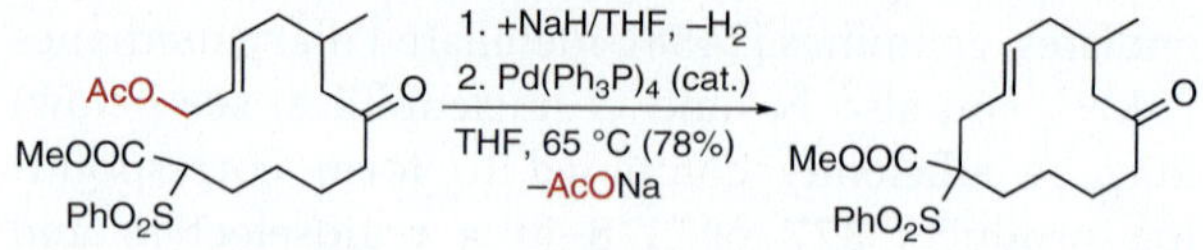

(8.323)

Enantiomerically pure chiral ligands such as (*S*,*S*)-**390** [2478], (*S*,*S*)-**391** [2479, 2480], (*R*)-MeO-MOP ((*R*)-2-(diphenylphosphino)-2-methoxy-1,1′-binaphthyl) [2481], (*S*,*S*)-**392** [2482], (*S*,*S*)-**393** [2483], (*S*)-*t*-Bu-Phos [2484, 2485], and diphosphoramidite derived from D-xylose [2486] have permitted high level of enantioselectivity for allylic alkylations with stabilized carbanions ("soft" nucleophiles) and various N-, O-, and S-nucleophiles [2436, 2487–2490].

Racemic isoprene monoepoxide reacts with ethyl acetylacetate in the presence of $Pd_2(dba)_3{\cdot}CHCl_3$ and ligand (*S*,*S*)-**390** giving products **394**, **395**, and **396** (Scheme 8.99). The two diastereomeric (π-allyl)Pd complex intermediates **397** and **398** are proposed to be formed and to react with the nucleophile. In **398**, the hydroxymethyl moiety repels the chiral ligand. This favors diastereomeric intermediate **397** [2487]. Interestingly, the addition of $Bu_4N^+/Si(Ph)_3F_2^-$ leads to rate enhancement and better regioselectivity. Asymmetric allylic alkylation permits the easy construction of all-carbon quaternary stereocenters [2491].

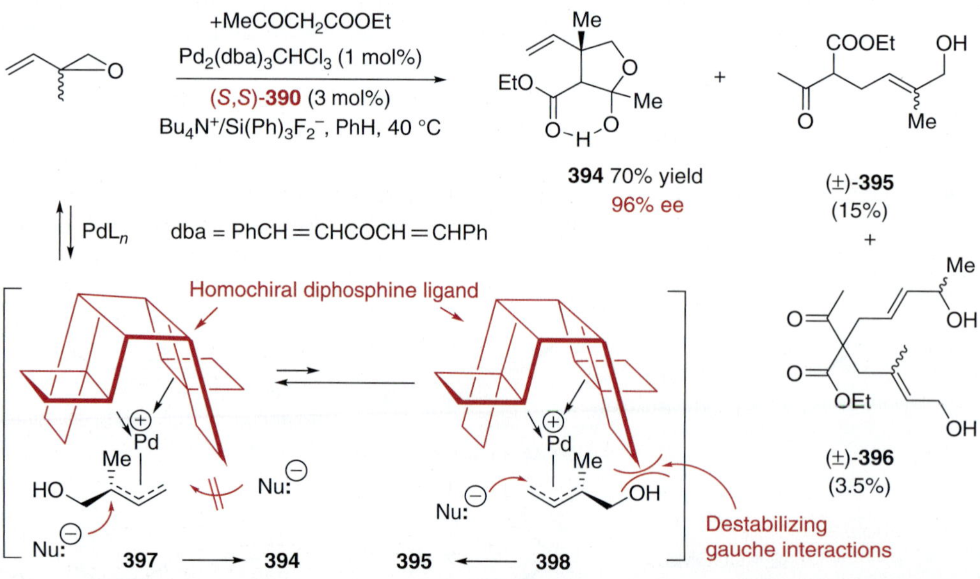

Scheme 8.99 Example of asymmetric allylic alkylation. Proposed explanation for the enantioselectivity induced by the enantiopure ligand (in red) of the Pd-catalyst.

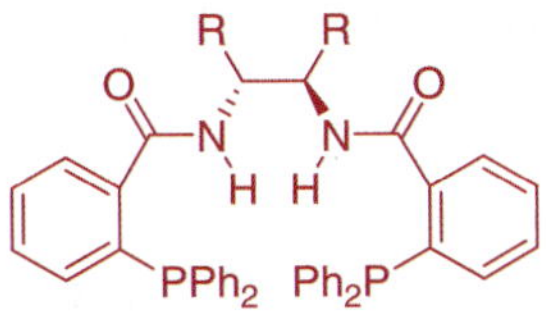

R, R = $(CH_2)_4$: (*S*,*S*)-**390**
R = Ph: (*S*,*S*)-**391**

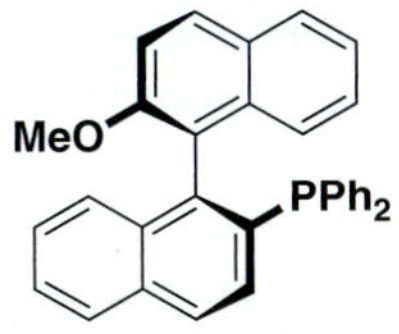

(*R*)-MeOMOP

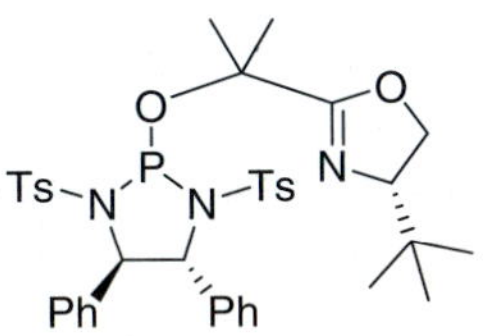

(*S*,*S*)-**393**

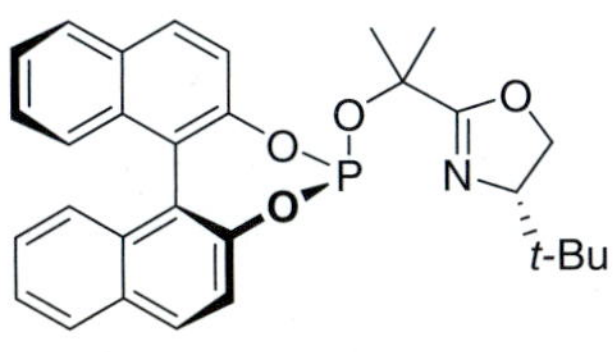

(*S*,*S*)-**392**

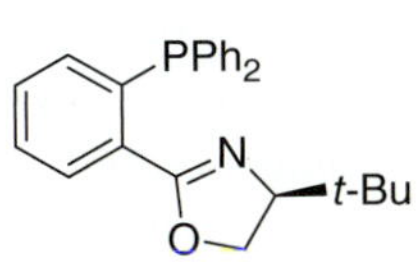

(*S*)-*t*-Bu-Phos

Although nucleophiles usually attack the terminal carbon center of the allyl Pd moiety, addition onto the central carbon giving palladacyclobutane intermediates and, after reductive elimination of palladium, the corresponding cyclopropane derivatives has been observed, first by Hegedus and coworkers in 1980 [2492]. Several other examples have been reported since then [2493–2499], and the existence of palladacyclobutane intermediates was confirmed by their isolation [2500]. This mode of attack can occur with strong σ-donor ligands such as TMEDA (*N*,*N*,*N*′,*N*′-tetramethylethylene diamines) and with less stabilized carbanions such as enolates of amides, esters, ketones, and sulfonamides. An example is given with reaction (8.324) [2501–2503].

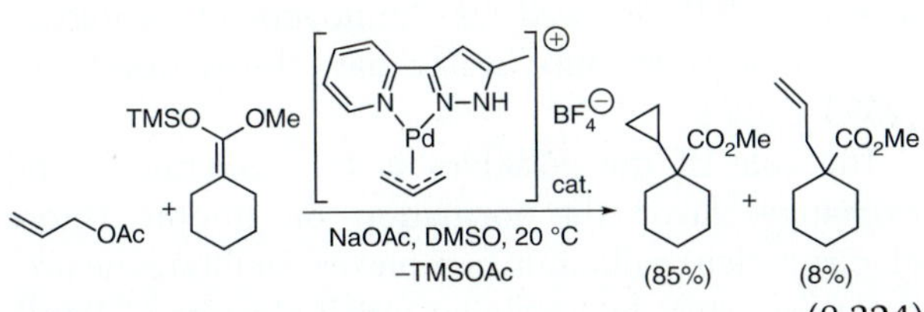

(8.324)

Problem 8.94 What is the major product **P** of the following reaction? [2504]

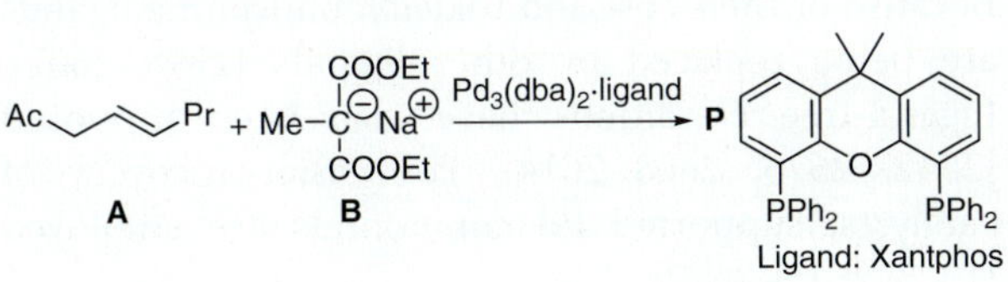

Ligand: Xantphos

8.8.7 Mizoroki–Heck coupling

The Mizoroki–Heck coupling (8.325) is the second most used C—C bond-forming reaction after the Suzuki–Miyaura coupling (Section 8.8.4) [23, 2505]. It is one of the modern tools of C—H functionalization [2506, 2507]. Formally, the arylation of alkenes is a displacement reaction in which an alkene (R—H) displaces an aryl electrophile (Ar—X) to form R—Ar and HX. As we shall see, the reaction mechanism does not involve the formation of alkenyl carbanions as intermediates. All sorts of Pd compounds have been utilized as precatalysts. Less expensive and more environment-friendly metals such as Ni, Co, Cu, and Fe can be used as catalysts [2508]. Transition-metal-free coupling reactions of aryl halides promoted by alkali alcoholates alone or in combination with amines are being developed [2509, 2510]. Related to the Mizoroki–Heck coupling is the Pd-catalyzed acylation of aryl bromides [2511] and chlorides [2512] with aldehydes (Ar—Hal + $R^1(R^2)$CHCHO+pyrrolidine ⇄ Ar—Hal + pyrrolidine enamine+H_2O⇄ArCOCH(R^2)R^1+H-pyrrolidinium$^+$ /Hal$^-$) developed by Xiao and coworkers [2513]. In 2002, Willis and coworkers reported on the intramolecular "inverse Heck cross-coupling" of alkenyl triflates with arenes catalyzed with $Pd(OAc)_2$ + phosphine ligand (THF/H_2O, 25 °C) [2514, 2515]. Using a taddol-based phosphoramidite ligand, Albicker and Cramer realized enantioselective alkenylations of 6,6-bis(CH_2-aryl)cychohex-1-en-1-yl triflates [2516].

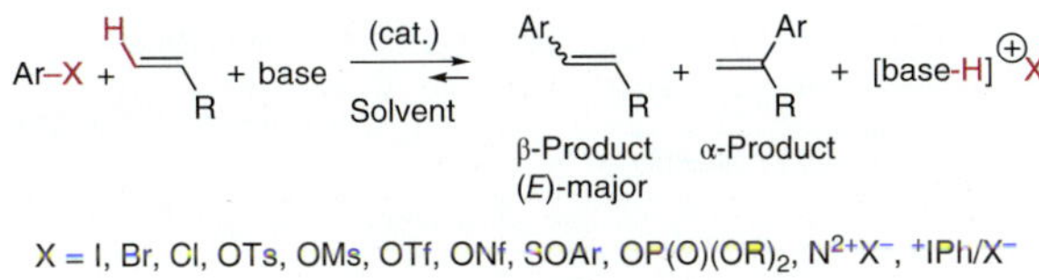

X = I, Br, Cl, OTs, OMs, OTf, ONf, SOAr, OP(O)(OR)$_2$, $N^{2+}X^-$, $^+IPh/X^-$

(8.325)

In the late 1960s, Heck reported that arylated alkenes are formed by reaction of alkenes with stoichiometric amounts of [ArPdCl] or [ArPdOAc], generated in situ by combining ArHgCl with $PdCl_2$, or ArHgOAc with $Pd(OAc)_2$, respectively [2517–2519]. Independently, Mizoroki and coworkers in 1971 [2520] and Heck and Nolly in 1972 [2521] discovered that aryl iodides can be used as a substitute for aryl-mercury compounds in the presence of catalytic amount of palladium. The Pd-catalyzed reaction (8.325) was then developed to allow coupling of less reactive halides, such as bromides [2522], chlorides [2251, 2321, 2523–2525], and phenol esters such as triflates [2526–2528], tosylates [2529, 2530], presylates [2531], nonaflates

($C_4F_9SO_2O$—) [2532], phosphates [2533], arylsulfoxides [2534], and aryldiazonium salts [2535]. On their side, aroyl chlorides (reaction (8.326)), aromatic acid anhydrides (reaction (8.327)) [2536] and esters (reaction (8.328)) [2537], and free aromatic carboxylic acids [2538, 2539] undergo decarbonylative couplings [2540, 2541]. Similarly, arenesulfonyl chlorides undergo desulfinylative Pd-catalyzed Mizoroki–Heck couplings (reaction (8.326)) [2192, 2542, 2543]. Using alkenyl electrophilic partners, alkenylation of olefins are realized readily (reactions (8.329) and (8.330)) [2531, 2533, 2535, 2544, 2545]. With chiral ligands on palladium, asymmetric induction is possible [2546–2556].

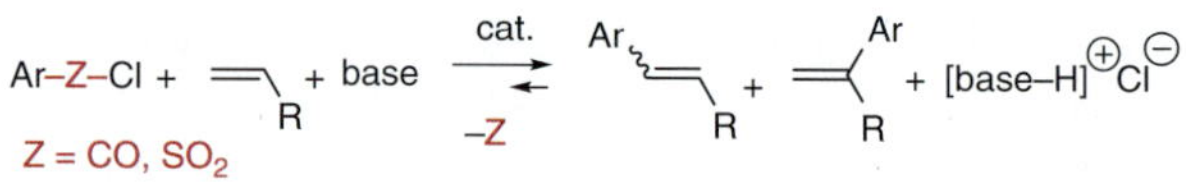

(8.326)

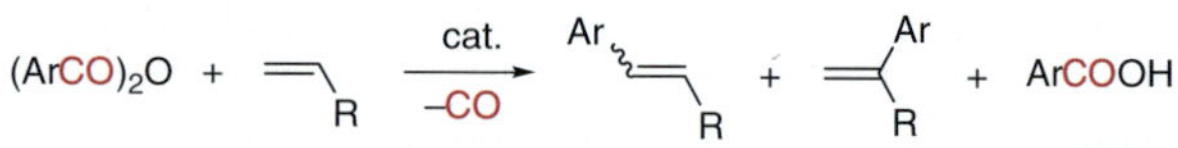

(8.327)

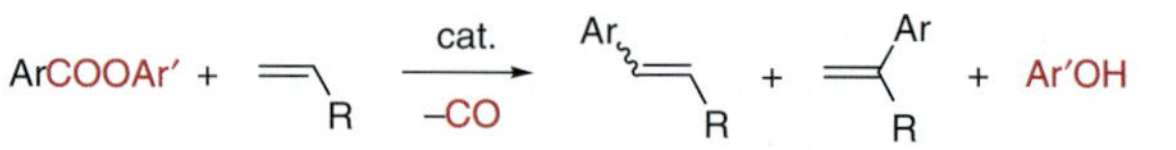

(8.328)

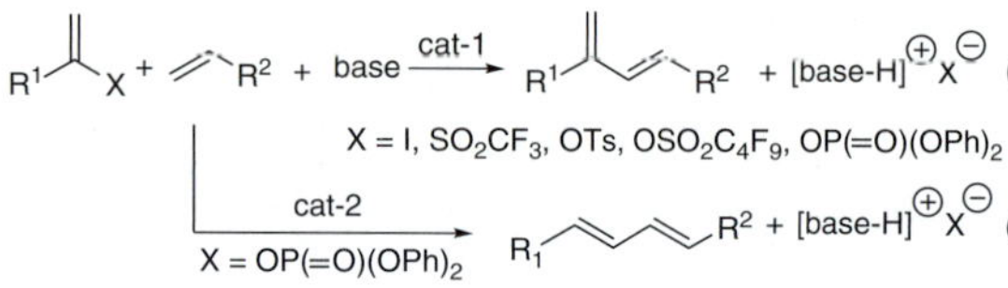

(8.329)

(8.330)

The first patented application of the Mizoroki–Heck reaction is the synthesis of stilbene by reaction of PhCl or $PhNO_2$ with styrene [2557, 2558]. In 1996, Hoechst patented the synthesis of functionalized β-arylacrylic esters (reaction (8.331)) in which the palladacycle **399** is the catalyst obtained by reaction of $Pd(OAc)_2$ with $(t\text{-Bu})_3P$ [2559–2562].

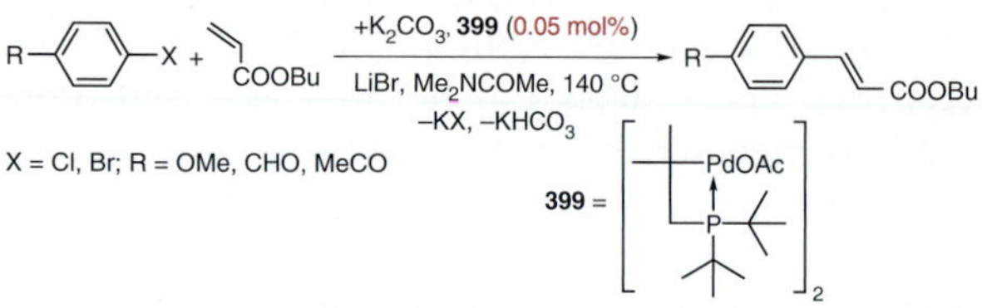

(8.331)

Usually, the Pd-catalyzed Mizoroki–Heck reaction is performed in polar solvent (DMF, Me_2NCOMe, MeCN) containing some water [1373, 2563] and using Pd concentration of 0.1–10 mol% along with stoichiometric amount of an organic (Et_3N, Cy_2NMe, DABCO (1,4-diazabicyclo[2.2.2]octane), isoquinoline, and tetramethylguanidine) or inorganic base (NaOAc, KOAc, K_2CO_3, Na_2CO_3, Cs_2CO_3, Ag_2CO_3, and $NaHCO_3$) [2564]. Ionic liquids, fluorous phase, and supercritical fluids such as sc-CO_2 are also used [2565]. A large number of heterogeneous catalysts [2033] have been applied such as Pd nanoparticles [2566–2575], Pd on charcoal [2576], Pd on SiO_2 [2577], Pd on molecular sieves [2578], Pd supported on polymers [2579], and two-dimensional Pd nanoparticles encapsulated in graphite, very useful in the arylation of electron-poor alkenes (tend to polymerize in the presence of base through anionic polymerization) [2580]. Most homogeneous catalysts are made of $Pd(OAc)_2$, $PdCl_2$, $Pd(OCOCF_3)_2$, $Pd_2(dba)_3{\cdot}CHCl_3$ and a phosphine (e.g. Ph_3P, $(t\text{-Bu})_3P$, $(t\text{-Bu})_2PH$, (adamant-1-yl)$_2$PH, (adamant-1-yl)$_3$P, diphosphine (e.g. **400–404**), Hartwig's electron-rich ligands [2581–2583]), tetraphosphine (e.g. Doucet–Santelli's ligands Tedicyp, **405** [2584],), diaminocarbenes (e.g. **406–408**) [2585, 2586], or other nonphosphine ligands such as **409–412** and phosphites. These ligands stabilize Pd(0) species and retard the formation of metallic particles (loss of TON). Catalytic systems often contain additives (e.g. LiCl, LiBr, NaBr, KCl, $AgBF_4$, Bu_4NCl, Bu_4NBr, and $(n\text{-oct})_3MeN^+/Cl^-$) that amplify performance. The Jeffrey-Larock family of protocols uses simple Pd(II) salts as precatalysts in aqueous solvents in the presence of Bu_4NBr or Bu_4NCl and LiCl [2535, 2587–2590]. Nonanionic phase transfer agents such as polyethyleneglycols have also been utilized [2591]. Heterogeneous catalysts [2592] as well as homogeneous catalysts for reactions in pure water have been developed [2593–2595].

The role of the additives is to accelerate ligand exchange, favor the formation of anionic (more electron-rich) palladium complexes, facilitate the preactivation step, and stabilize small clusters of Pd(0), thus preventing the formation of metallic precipitate. Silver(I) and thalium(I) salts are used to quench halide anion and thus favor ligand exchange (e.g. alkene binding to the palladium) [2548, 2596, 2597]. Because of their cost and toxicity, phosphine ligands are being replaced by other ligands [2598–2607]. Ligand-free conditions have also been proposed [2542, 2576, 2608–2614]. For easier recovery of catalysts, supported Pd compounds are employed [1373, 2615–2622].

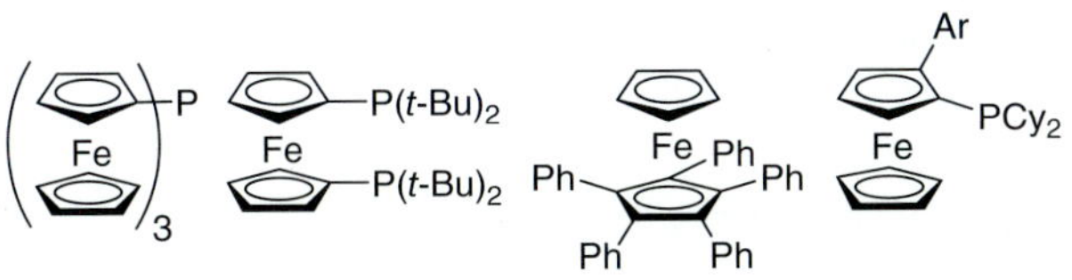

400 **401** **402** **403**

Cy = cyclohexyl

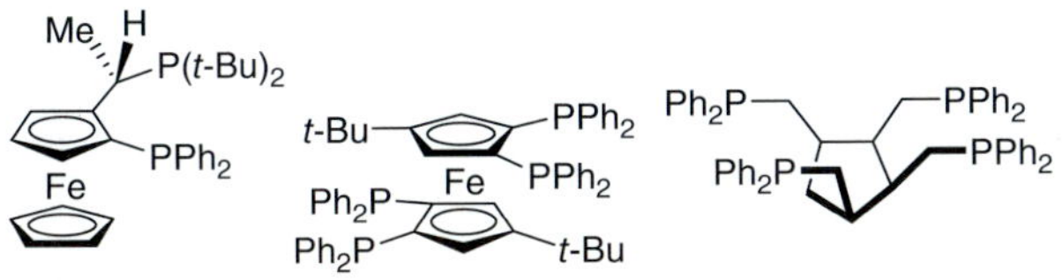

404 **405** Tedicyp

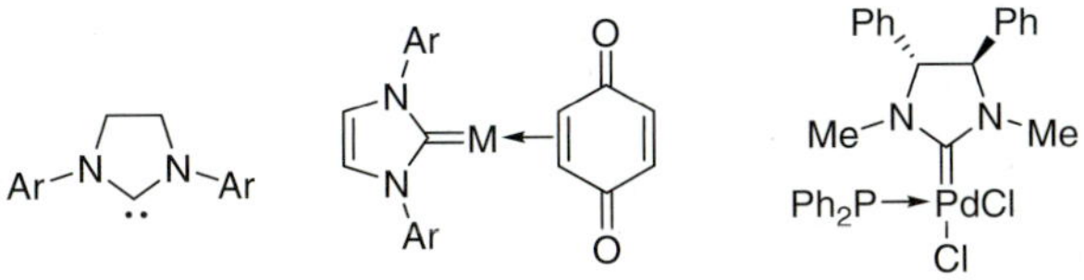

406 **407** M = Pd, Pt, Ni **408**

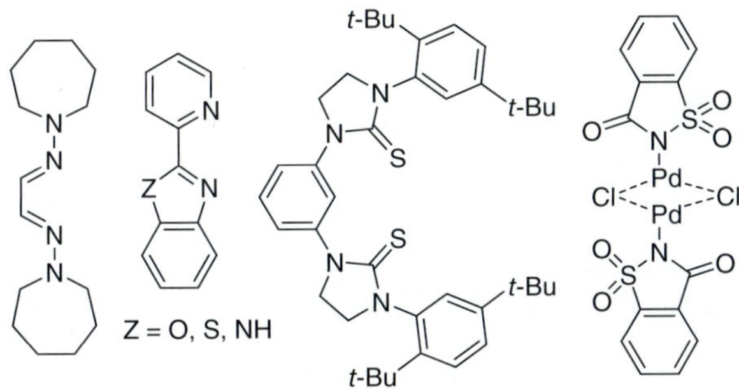

409 **410** **411** **412**

Several mechanisms have been proposed for the Pd-catalyzed Mizoroki–Heck reaction [2623]. As for other Pd-catalyzed C—C cross-coupling reactions using electrophilic reagents (Sections 8.8.1–8.8.6), the first step is their oxidative additions to the active catalytical species. In 1998, Shaw proposed that the latter are Pd(II) complexes adding onto Ar—X giving corresponding Ar—[Pd(IV)]—X intermediates [2624–2626]. In most cases, however, the first step is proposed to be an oxidative addition of Pd(0) complexes onto Ar—X giving Ar—[Pd(II)]—X intermediates [2013, 2577, 2627–2630]. The mechanism depends on the nature of the electrophilic partner, on the nature of the ligands and additives, and on solvent and temperature [2623, 2631]. *A priori*, radical intermediates cannot be excluded in Pd-catalyzed reactions. As already seen, they are common for Ni-, Co-, and Cu-catalyzed reactions. The mechanism shown in Scheme 8.100 has been proposed by Amatore and Jutand for the coupling reaction using $Pd(OAc)_2/PPh_3$ catalyst and Et_3N as stoichiometric base.

For the reaction PhI + styrene in comparable concentrations, the overall carbopalladation (complexation of the alkene and subsequent aryl β-insertion) from $PhPd(OAc)(PPh_3)$ is the rate-determining step of the catalytic cycle. This mechanism shows the retarding role of AcO^- because it stabilizes $Pd(0)(PPh_3)_2$ by forming $Pd(0)(PPh_3)_2(OAc)^-$. The base (Et_3N) indirectly also stabilizes $Pd(0)(PPh_3)_2(OAc)^-$ by preventing its decomposition by protons to the more reactive bent $Pd(0)(PPh_3)_2$ (oxidative addition of PhI). AcO^- retards the reaction by formation of $(alkene)Pd(0)(PPh_3)_2(OAc)^-$, which is nonreactive. The slow carbopalladation step is accelerated by the base and by AcO^- that generates $ArPd(II)(OAc)(PPh_3)_2$, which in turn is more reactive than $ArPd(II)(PPh_3)_2I$. The base, the alkene, and AcO^- play the same dual role: they retard the oxidative addition and accelerate the carbopalladation. For reactive aryl precursors (Ar—I), this dual effect favors the efficiency of the catalytic process by making the rate of the oxidative addition closer to the rate of the carbopalladation [2632–2634]. The mechanism shown in Scheme 8.100 is also applicable to reactions using aryl triflates as electrophilic partners [2635] and other palladium precursors such as $Pd(dba)_2/PPh_3$, $PdCl_2(PPh_3)_2$, and $Pd(PPh_3)_4$ and NaOAc or KOAc as a base. Acetate anion combines with $Pd(0)(PPh_3)_4$ and forms $(Pd(0)(PPh_3)_2(OAc)^-$ or reacts with $ArPd(II)(PPh_3)_2I$ to generate the more reactive $ArPd(II)(PPh_3)_2OAc$ [2636]. When using aryl chlorides (Ar—Cl) as electrophilic partners, the rate-determining step is the oxidative addition $ArCl + Pd(PPh_3)_2(OAc)^-$. This reaction competes with the complex formation $(alkene)Pd(PPh_3)_2(OAc)^-$, which is nonreactive. Thus, the alkene inhibits the overall process. To overwhelm this problem, the alkene has to be added slowly to the reaction mixture in order to maintain a low concentration of it, or use ligands L rendering $Pd(PPh_3)_2(OAc)^-$ faster in its oxidative addition with the electrophilic partner. This is realized using electron-rich ligands such as **400–406**.

Under conditions of kinetic control, the β-insertion of the aryl group is *syn* selective, as in the β-elimination forming the final arylalkene (Schemes 8.100 and 8.101). This is demonstrated with reaction (8.332) [2524].

At 120 °C, a 1 : 1 mixture of (*E*)- and (*Z*)-**413** is formed (thermodynamic control), showing that the β-hydrogen elimination is reversible. This explains why (*E*)-arylalkenes are major products (reactions (8.325)–(8.328)). The (*E*)-stereoselectivity of the arylation can be changed into (*Z*)-stereoselectivity by matrix imprinting [2637]. For the process

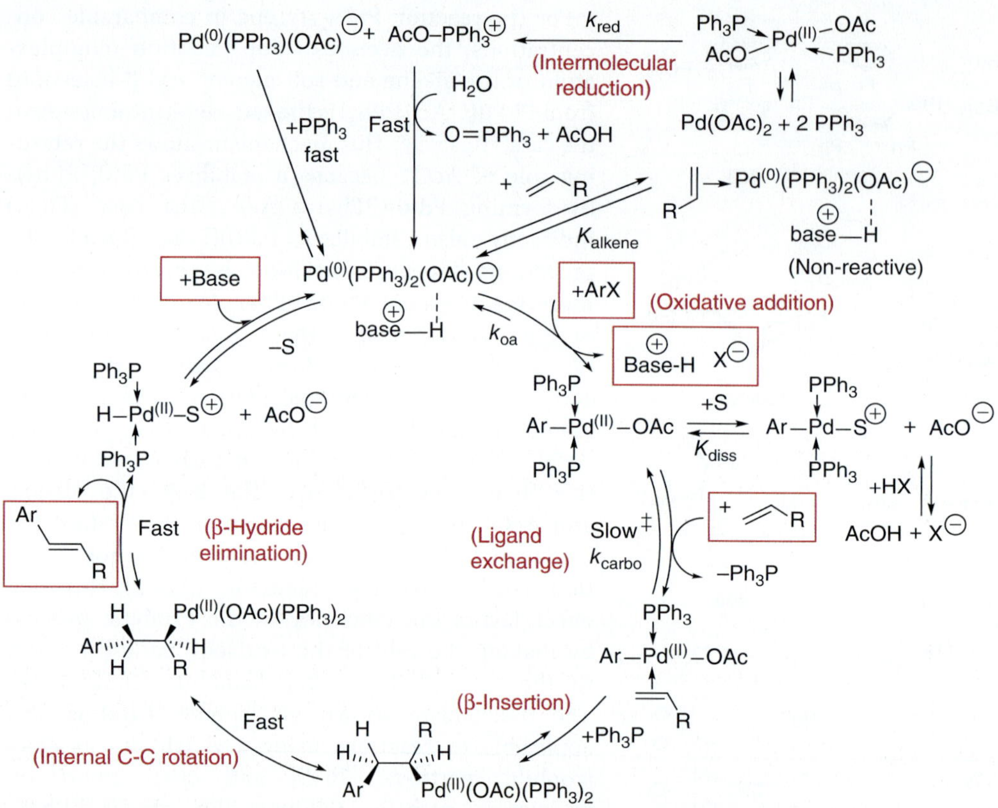

Scheme 8.100 Reaction mechanism of the $Pd(OAc)_2/PPh_3$-catalyzed Mizoroki–Heck coupling as proposed by Amatore and Jutand ("anionic cycle": involves anionic intermediates). Rate and equilibrium constants in DMF at 25 °C when ArX = PhI, R = Ph, and base = NEt_3: k_{red} 4×10^{-4} s^{-1}, $K_{alkene} = 61$ M^{-1}, $k_{oa} = 140$ M^{-1} s^{-1} (65 M^{-1} s^{-1} in the presence of 3 equiv. of NEt_3), $K_{diss} = 1.4\times10^{-3}$ M, $k_{carbo} = 5\times10^{-5}$ s^{-1} (10^{-4} M^{-1} s^{-1} in the presence of 3 equiv. of NEt_3), k_{carbo} is the apparent rate constant of the overall carbopalladation process, which involves complexation and insertion of the alkene.

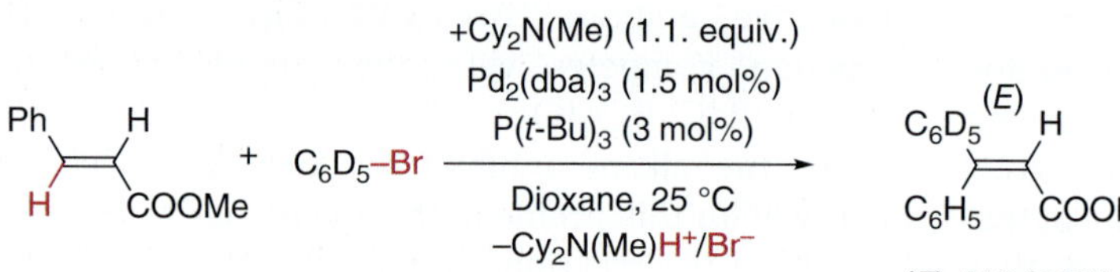

(E)-**413** (94%)

(8.332)

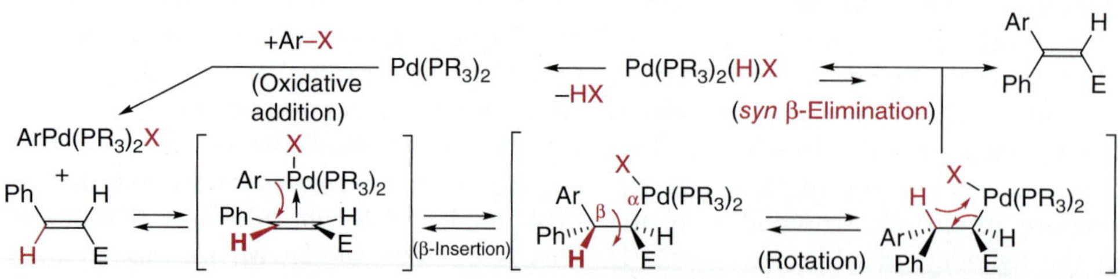

414

Scheme 8.101 *Syn* selectivity for the β-insertion (arylation) and the β-elimination of hydride in Pd-catalyzed Mizoroki–Heck coupling (assuming a "neutral cycle," not like the "anionic cycle" of Scheme 8.100).

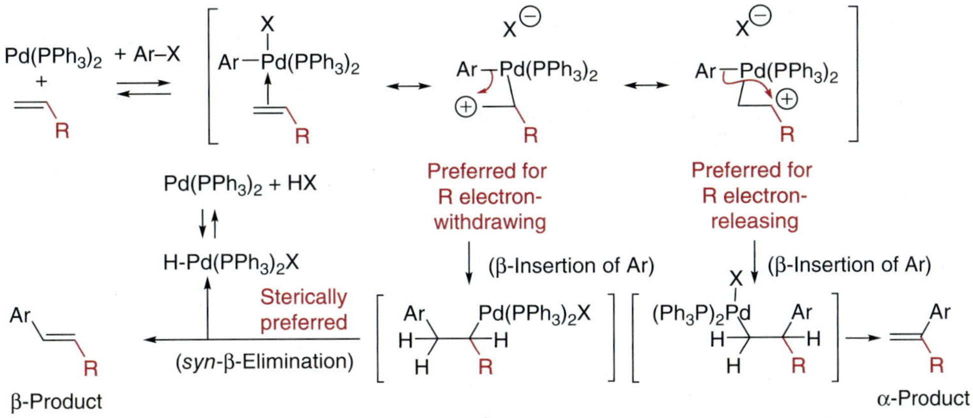

Scheme 8.102 Simple interpretation for the regioselectivity of Pd-catalyzed Mizoroki–Heck reactions as a function of the nature of the alkene and σ-ligand X. If X is slightly electronegative, one speaks about neutral route or "neutral cycle" for the arylation and if X is highly electronegative (e.g. BF_4^-), one speaks about cationic route or "cationic cycle" for the arylation (compare with the "anionic cycle," Scheme 8.100).

to occur, the alkylpalladium complex resulting from the β-insertion of the aryl group must be able to rotate about the $\sigma(C_\alpha—C_\beta)$ bond of **414** in order to bring a β-(C—H) bond synperiplanar with the α-C—Pd bond. When this rotation is blocked, as for cyclic alkenes, the β'-(C—H) bond will align better with the α-C—Pd bond and undergo β-hydride elimination, thus leading to arylated products in which the C=C double bond has migrated as exemplified with reaction (8.333) [2638]. Yield and regioisomeric product ratio depend on the nature of the catalyst.

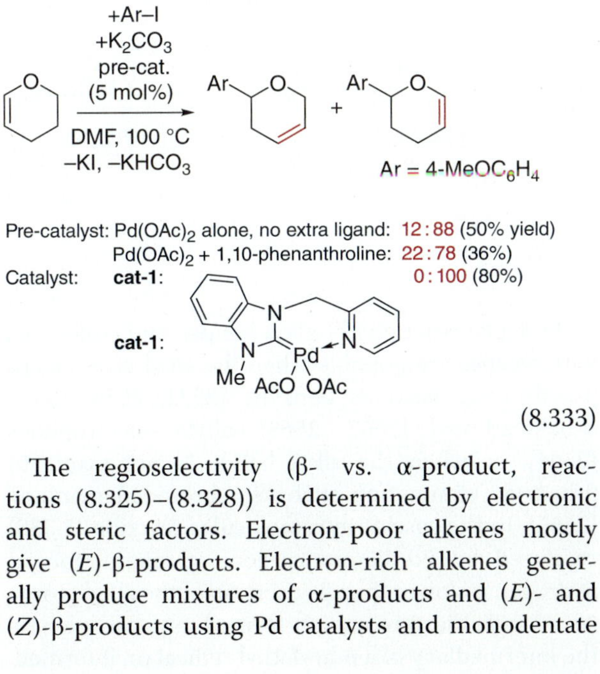

(8.333)

The regioselectivity (β- vs. α-product, reactions (8.325)–(8.328)) is determined by electronic and steric factors. Electron-poor alkenes mostly give (*E*)-β-products. Electron-rich alkenes generally produce mixtures of α-products and (*E*)- and (*Z*)-β-products using Pd catalysts and monodentate ligands (or without ligands) [2639–2641]. The regioselectivity can be improved in favor of the α-products by increasing the electron demand of the palladium in the π-complex it makes with the alkene. The latter depends on the nature of the Pd—X moiety. For an electronegative X group, heterolysis Pd—X ⇄ Pd^+/X^- is possible (cationic route, or "cationic cycle") and thus it increases the electron demand. The consequence of this electronic effect can be represented by viewing the (alkene)$Pd(Ph_3P)_2X$ complex as a palladacyclopropane in which the positive charge is partially delocalized on the carbon centers of the alkene moiety. For alkenes with electron-withdrawing substituents (e.g. acrylic esters), the positive charge is unstable at the α-carbon center, thus favoring the β-insertion of the aryl group at the β-carbon center of the alkene, leading to β-product, whatever the nature of X in Pd—X ("neutral cycle" or "cationic cycle"). On the contrary, for alkenes with electron-releasing substituents (e.g. enol ethers and enamines), the positive charge is delocalized at the α-carbon center rather than at the β-carbon center, making the aryl group able to overcome the greater steric hindrance it encounters during the β-insertion. As a consequence, this favors the formation of α-products with regioselectivity that increases with the electronegativity of X (Scheme 8.102). Examples of regioselective Pd-catalyzed Mizoroki–Heck reactions are given with reaction (8.334).

In order to realize a reaction that proceeds with high electron demand with enol ethers or enamides, strongly coordinating bidentate ligands, such as dppp

(1,3-bis(diphenyl(phosphino)propane) (see e.g. reaction (8.334) [2639]) or dmphen (2,9-dimethyl-1,10-phenanthroline), must be used [2623, 2642]. With weakly coordinating σ-ligand X such as triflate or tosylate, the reaction proceeds with high electron demand ("cationic route," Scheme 8.102) [2530]. In contrast, with aryl halides, the electron demand is lower ("neutral route"). To achieve a high α-selectivity, silver or thallium tetrafluoroborate, or hexachloroantimonate, should be used as an additive to substitute X (Br, I) of the (alkene)Pd$(PPh_3)_2$X complex by the BF_4^- or $SbCl_6^-$ anion. The degree of heterolytic dissociation of Pd—X in (alkene)Pd$(PPh_3)_2$X complexes also depends on the ionizing power of the solvent. Thus, α-selectivity is generally improved in ionic liquids [2643, 2644] or water [2645, 2646]. With less reactive aryl chlorides, Ar—Cl, high α-selectivity has been obtained for the arylation of electron-rich alkenes using dppp as a ligand and Et_3NH^+/BF_4^- as an additive [2647]. Regioselective vinylation of electron-rich alkenes with vinyl halides has been reported for Pd-catalyzed reaction using 1,3-bis(diphenylphosphino)propane monoxide (dpppO) as a ligand, without the need of halide scavengers (e.g. $AgBF_4$, $TlBF_4$), ionic liquids, or ionic additives [2648]. In 2008, Xiao and coworkers have found that electron-rich alkenes are α-arylated with aryl bromides in good yield and high regioselectivity using ethyleneglycol as a solvent (and no other additive) and a base (Et_3N, 140 °C) and Pd$(OAc)_2$/dppp (1,3-bis(diphenylphosphino)propane as a catalyst [2649]. Aryl chlorides can also be used and require a more electron-rich ligand such as 4-MeO-dppp (1,3-bis[di(4-methoxyphenyl)phosphino]propane) (e.g. reaction (8.335)) [2513, 2650].

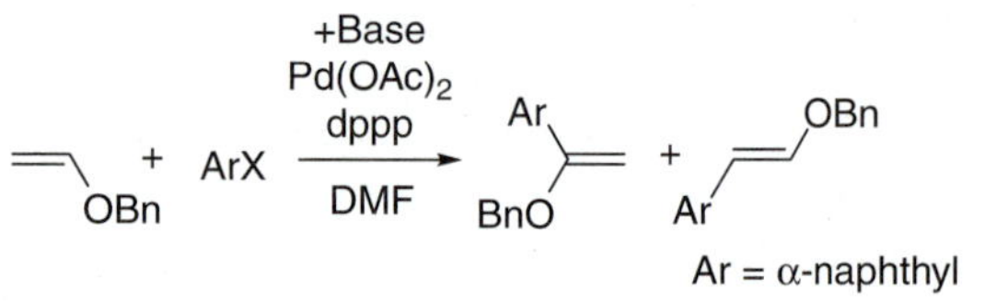

X	Base	T	t	Ratio (yield)
X = I	Base: Et_3N	100 °C	18 h	67 : 37 (63%)
X = I	Base: TlOAc	80 °C	1 h	>99 : 1 (92%)
X = OTf	Base: Et_3N	60 °C	0.5 h	>99 : 1 (93%)

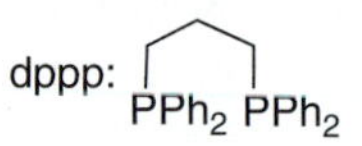

(8.334)

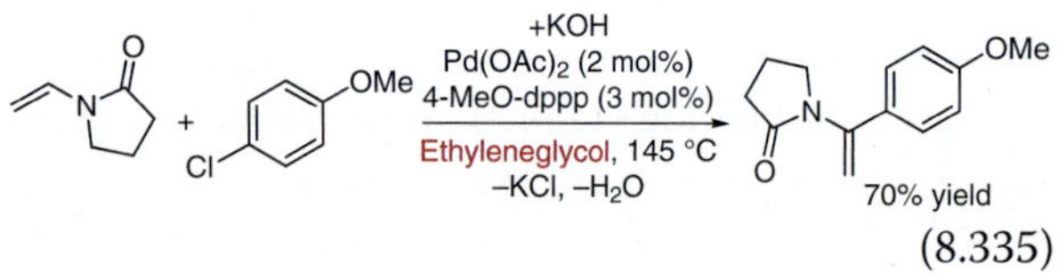

(8.335)

As seen with reaction (8.333), Mizoroki–Heck arylation of cycloalkenes generates products with an asymmetric carbon center. In the presence of enantiomerically pure diphosphine ligand, high enantiomeric excesses have been reported for both intermolecular [2651] and intramolecular arylations and alkenylations of cyclic alkenes [2550–2552]. The first example of asymmetric intermolecular Mizoroki–Heck reaction (8.336) was reported by Hayashi and coworkers in 1991 [2652]. An example of asymmetric intramolecular vinylation is shown with reaction (8.337) [2653, 2654].

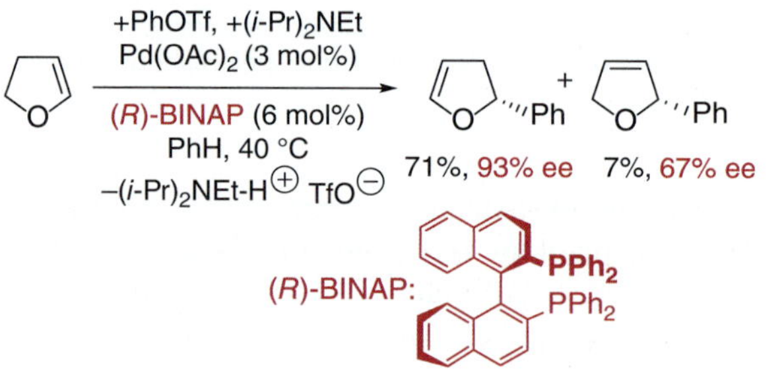

(8.336)

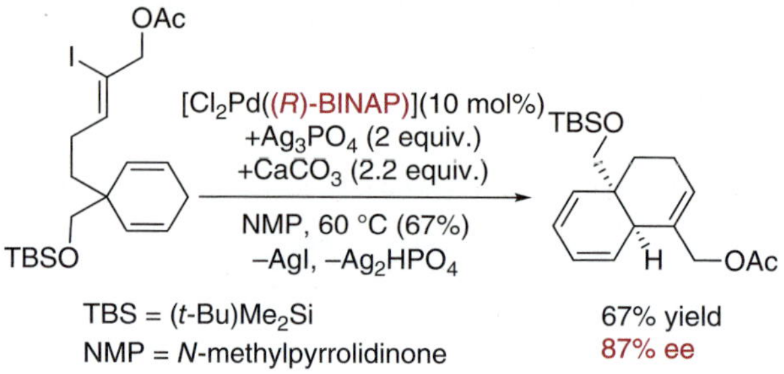

(8.337)

Heck-like reactions of alkyl halides and sulfonates with alkenes are possible when the alkyl group lacks β-hydrogens, such as benzylic [2521, 2655–2661], α-halocarbonyl [2662, 2663], allylic electrophiles [2664], or when the alkyl halide is not prone to β-hydride elimination such as 1-bromoadamantane [2665]. In the case of intramolecular reaction (8.338) catalyzed by Pd/diaminocarbene, β-elimination of hydrogen is slower than β-insertion of the alkyl group. As a single diastereomer is formed, one can exclude the intermediacy of a β-arylethyl radical or, if formed, one must admit that rotation about its single C—C bond is slower than addition to the alkene [2666].

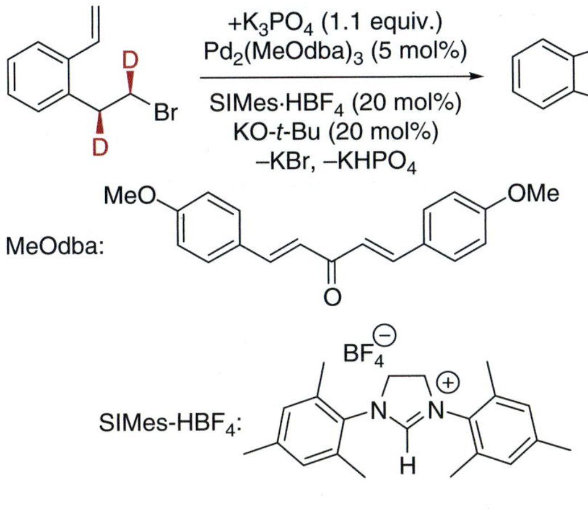

(8.338)

Replacement of Pd for more economical and environment-friendly metals is an active field of research [2508]. In 1986, Umani-Ronchi and coworkers found that $NiCl_2(Ph_3P)_2$ catalyzes the cross-coupling of aryl halides and styrene (yield < 55%) in the presence of an excess of metallic Zn [2667]. Since then, several studies on Ni-catalyzed Mizoroki–Heck couplings have been reported [2630, 2668]. A 98% yield of ethyl 4-(4′-hydroxyphenyl)cinnamate was obtained in the $NiCl_2(Ph_3P)(H_2O)$-catalyzed coupling of 4-bromo-4′-hydroxybiphenyl in the presence of an inorganic base [2669]. An efficient, two-step synthesis of aryl methyl ketones (reaction (8.339)) has been developed by Skrydstrup and coworkers. The high activity of $Ni(cod)_2$/dppf/Cy_2NMe combination is explained by the formation of cationic Ni(II) intermediates that favor the α-regioselective insertion of the aryl group (Ar-$[Ni(II)]^+(CH_2{=}CHOR)/TfO^- \rightarrow$ Ar(OR)CH—CH_2—$[Ni(II)]^+/TfO^-$) and the subsequent β-hydride elimination (Ar(RO)CH—CH_2—$[Ni(II)]^+/TfO^- \rightarrow$ Ar(RO)C=CH_2 + H—Ni(II)—OTf) [2670]. In 2012, Watson and coworkers found that $Ni(cod)_2$/dppf activates the C—O bond of aryl and alkenyl pivalates (reaction (8.340)) [2671]. Jamison and coworkers have used the system $Ni(cod)_2/Cy_2PPh/Et_3SiOTf/Et_3N$ to catalyze α-regioselective benzylations (8.341) of ethylene and terminal alkenes [2672].

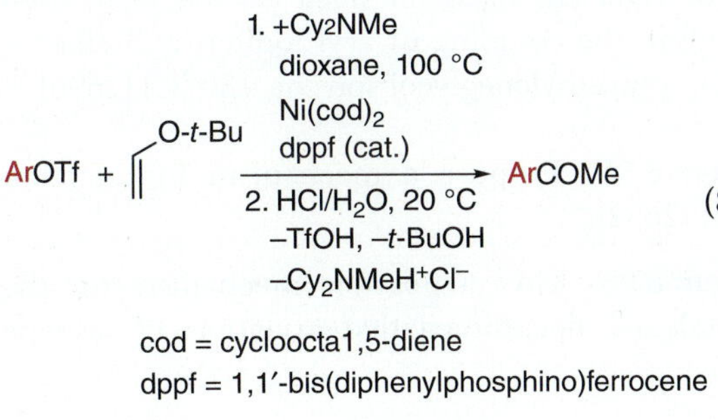

(8.339)

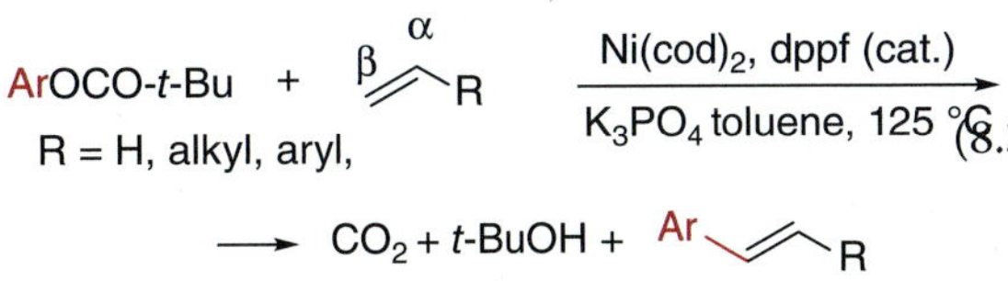

(8.340)

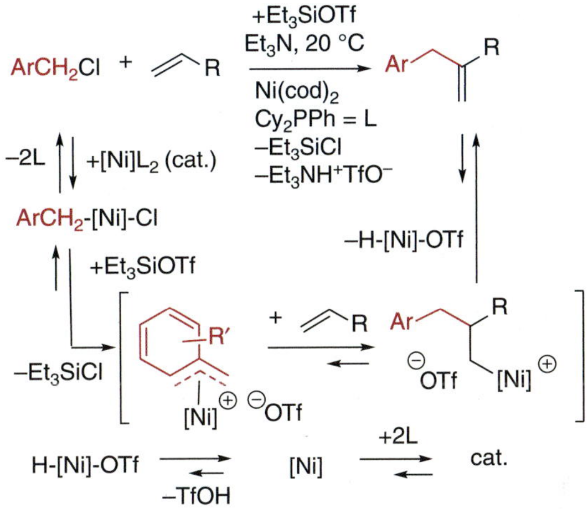

(8.341)

In 1991, Branchaud and coworkers used cobalamine $[Co(dmgH)_2Py]Cl$ to catalyze the alkylation of styrene (e.g. *s*-Bu—Br+PhCH=CH_2→(*E*)-*s*-Bu—CH=CHPh). The reaction is activated by light or a reductant such as Zn. It involves the addition of alkyl radical to the alkene [2673]. In 2002, Oshima and coworkers used $CoCl_2$(dpph) (dpph = 1,6-bis(diphenylphosphino) hexane) and Me_3SiCH_2MgCl to catalyze the cross-coupling of bromocyclohexane with styrene giving (*E*)-β-cyclohexylstyrene (e.g. reaction (8.342)) [2674]. The reaction has been extended to a variety of alkyl bromides and chlorides [2675], epoxides [2676], and aziridines [2677]. The reaction of cyclopropylmetyl bromide with styrene gives 1-phenylhexa-1,5-diene in 50% yield (c-C_3H_5—CH_2Br + PhCH=CH_2 → CH_2=CH—CH_2—CH_2—CH=CHPh + HBr), suggesting the intermediacy of alkyl radicals that add to the alkene. This hypothesis is confirmed by the cyclization/coupling reaction (8.343). The reaction of 1,2-dibromoethane with styrene produces (*E,E*)-Ph—CH=CH—CH_2CH_2—CH=CHPh, which indicates that 2-bromoethylnickel species are not formed as they are expected to undergo fast bromide elimination. A possible mechanism (Scheme 8.103) implies SET from $Co(II)Cl_2$(dpph) to the alkyl bromide (R—Br) giving $Co(III)Cl_2$(dpph)—Br and a radical anion $[R{-}Br]^{\bullet-}$. The latter is fragmentated into Br^- and the alkyl radical $R^\bullet$. Note that the Grignard reagent might also reduce R—Br. The alkyl radical adds to styrene regioselectively forming an alkylbenzyl radical. The latter combines

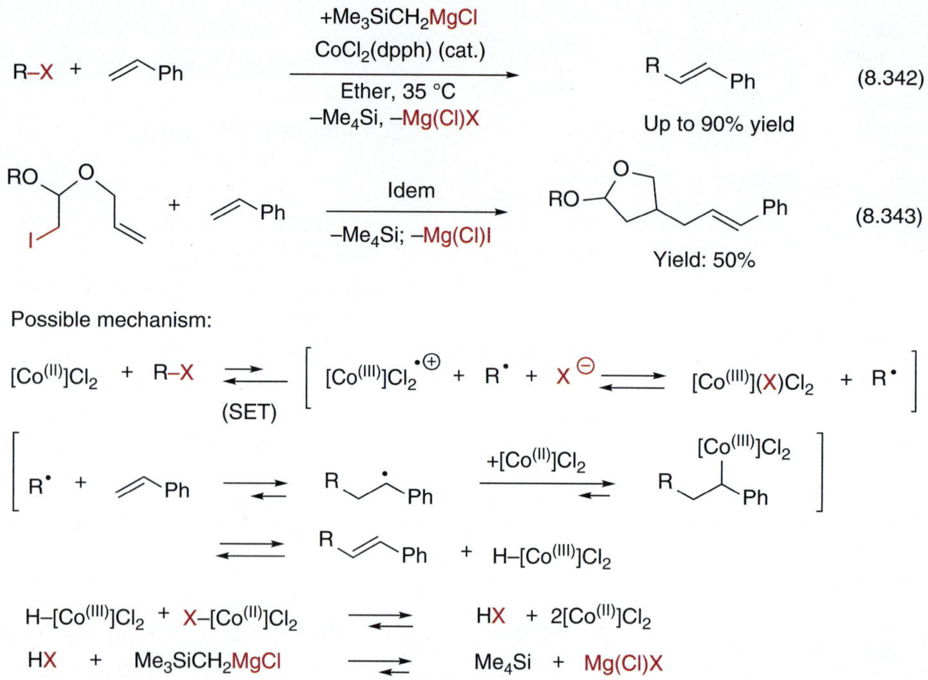

(8.342)

(8.343)

Scheme 8.103 Proposed reaction mechanism of the Co(II)-catalyzed Mizoroki–Heck coupling in the presence of stoichiometric amount of Me_3SiCH_2MgCl (base).

with $Co(II)Cl_2$(dppph). Subsequent elimination provides the final product of Mizoroki–Heck coupling and complex H—Co(III)Cl_2(dpph) that equilibrates with HCl + Co(I)Cl(dpph). Combination HCl + Co(I)Cl(dpph) + Co(III)Cl_2(dpph)—Br produces HBr + 2Co(II)Cl_2(dpph). Irreversible protolysis of the Grignard reagent gives Me_4Si + Mg(Br)Cl. With many Pd- or Ni-based catalysts, cross-coupling reactions of simple alkyl electrophiles fail because the β-hydrogen elimination from the corresponding alkylmetal intermediates is more facile [2678]. This is not the case when these intermediates can be avoided and radical intermediates are involved instead in the process [2679].

Copper-based catalysts that promote cross-couplings often favor the formation of radical intermediates [2680]. In 1997, Iyer and coworkers found the first Cu-catalyzed Mizoroki–Heck coupling using CuI or CuBr as a catalyst, with K_2CO_3 as a base in hot *N*-methylpyrrolidinone (150 °C). Milder conditions were then developed by Li and coworkers who used CuI/DABCO as a catalyst and K_2CO_3 in boiling ethanol [2681]. Copper bronze in ionic liquids was found active for Mizoroki–Heck couplings [2682]. In 2013, Nishikata and coworkers reported the coupling (8.344) of substituted styrenes with tertiary bromides containing an electron-withdrawing group A. The reaction is inhibited by TEMPO confirming the intermediacy of relatively stable radicals of type $R^1(R^2)(A)C^{\bullet}$ [2683].

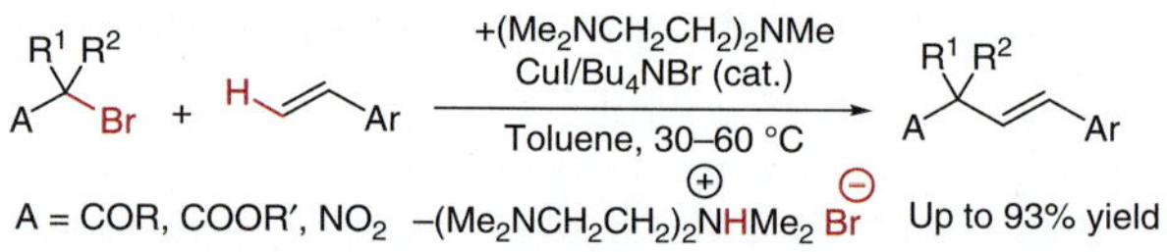

(8.344)

Mizoroki–Heck couplings have been catalyzed by titanocene [2684] and by Ru complexes [2346, 2685–2688]. In 2008, Vogel and coworkers reported that aryl and heteroaryl iodides and bromides undergo stereoselective C—C cross-coupling with styrenes and electron-rich alkenes in DMSO (solvent), *t*-BuOK (base), and $FeCl_2$ (precatalyst) without ligands. Yields are higher when proline or picolinic acid are used as ligands [2689]. A heterogeneous and recoverable form of $Fe(acac)_3$ fixed on silica gel has been used to catalyze the coupling of aryl iodides and alkenes (K_2CO_3, polyethyleneglycol solvent, 130 °C) [2690].

Problem 8.95 Propose a mechanism for reaction (8.335) [2652].

Problem 8.96 Give a possible mechanism for the Rh-catalyzed decarbonylative arylation of styrene [2691].

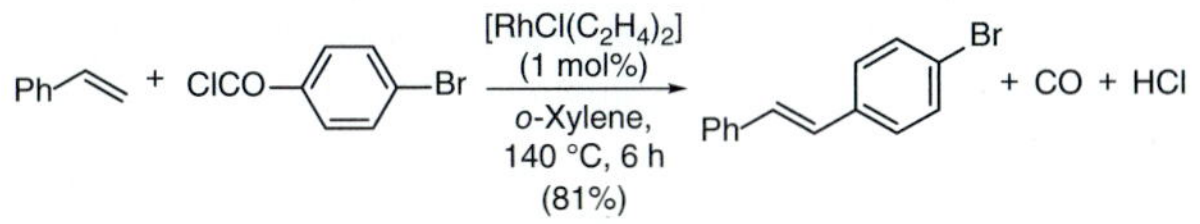

Problem 8.97 What are the possible products of the following reaction **A** → **P**? [2655]

A:

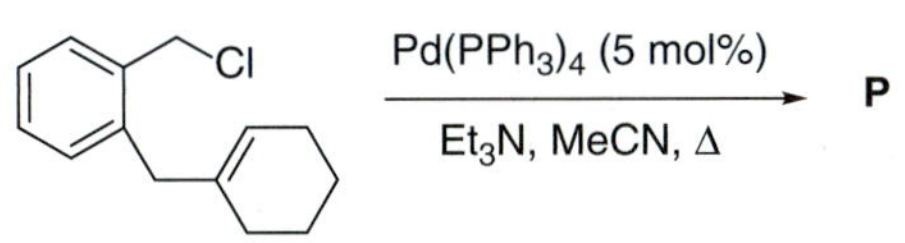

Problem 8.98 What is the major product **P** of the Ni-catalyzed cross-couplings of **A** + **B**? [2692]

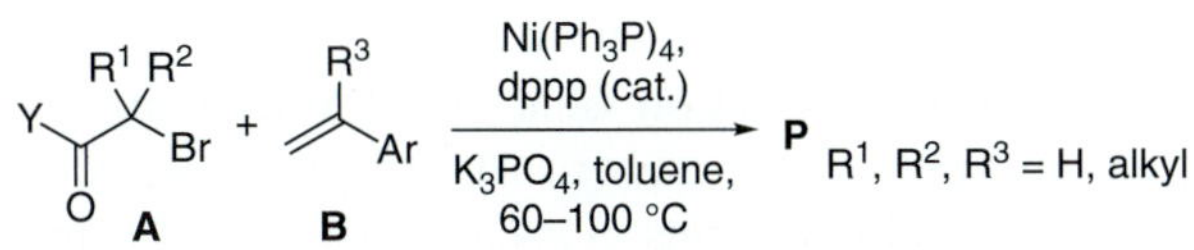

Problem 8.99 Propose a reaction mechanism for the Cu-catalyzed couplings (8.344) [2683].

8.8.8 Sonogashira–Hagihara cross-coupling

Classically, terminal alkynes are arylated by Ar—X (synthesis of arylalkynes)) and are alkenylated (synthesis of conjugated enynes) by alkenyl-X in the presence of a Pd precatalyst (e.g. $Pd(Ph_3P)_4$, $Pd(Ph_3P)_2Cl_2$, $Pd(dppe)Cl_2$, $Pd(dppp)Cl_2$, $Pd(dppf)Cl_2$), $Pd(PhCN)_2Cl_2$, $Pd[(MeCON(2\text{-pyrimidyl})_2]$, Pd(*N*-heterocyclic $carbene)_2Br_2$), and a Cu(I) salt (e.g. CuI) as a cocatalyst (reaction (8.345)). A stoichiometric amount of a base (e.g. Et_3N, Et_2NH, $(i\text{-}Pr)_2NH$, pyrrolidine, piperidine, morpholine, K_2CO_3, and Cs_2CO_3) is necessary to neutralize the coproduct HX [2693–2696]. As shown by Amatore and coworkers, the $Pd(II)X_2$ precatalysts are reduced into catalytically active Pd(0) species by the amine and/or phosphine ligand [2697]. The amines may intervene in the catalytical cycle (Scheme 8.104a)) [2698]. Primary [2699] and secondary alkyl halides have also been alkynylated in the presence of Pd-based precatalysts [1702]. Decarbonylative cross-couplings with acyl chlorides (reaction (8.346)) [2701–2703] and desulfinylative arylations using inexpensive arenesulfonyl chlorides (reaction (8.347), catalyzed by $Pd(dba)_2/(t\text{-}Bu)_3P + CuI$ in boiling THF) are also known [2704]. Although much less efficient than Pd-based precatalysts, Fe(II), Fe(III), Ru(I), Co(II), Ni(II), Cu(I), Ag(I), Au(I), and In(III) salts combined

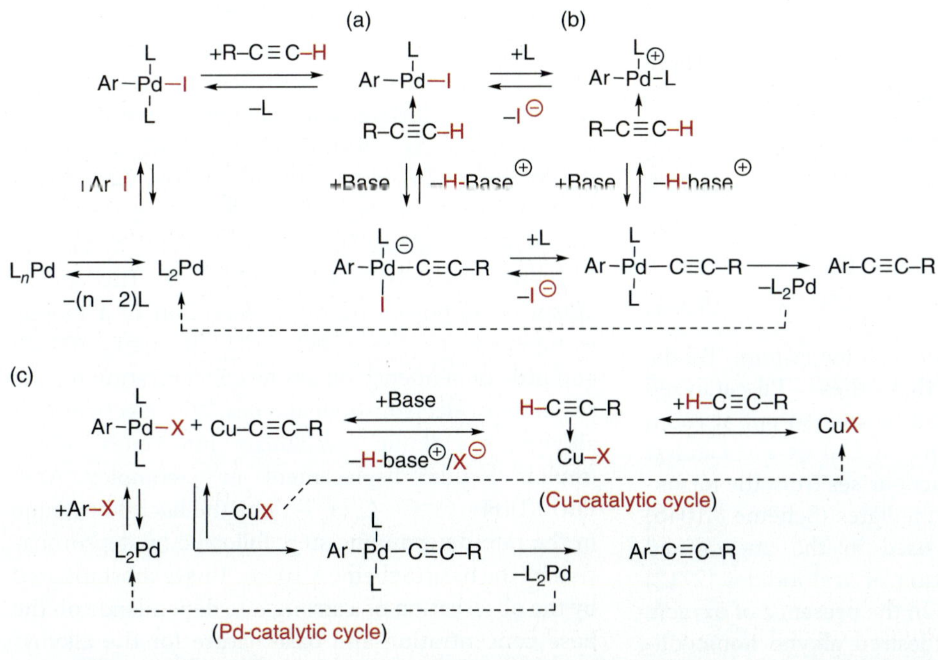

Scheme 8.104 (a) Proposed anionic mechanism, (b) cationic mechanism of the Pd-catalyzed Cassar–Dieck–Heck displacement of aryl iodides by terminal alkynes, and (c) proposed reaction mechanism of the Pd- and Cu-catalyzed Sonogashira–Hagihara cross-coupling of aryl halides and pseudo-halides with terminal alkynes.

with appropriate ligands have been found to catalyze the alkynylation of C(sp^2) electrophilic partners [2705]. Copper nanoparticles stabilized by DMF catalyze Sonogashira–Hagihara cross-coupling reactions [2706]. Alkynylation of electron-poor aryl fluorides does not require any catalyst as the displacement reaction (S_NAr) can follow the addition/elimination mechanism (via Meisenheimer complex intermediates and cyclohexadienyl anions) [2707].

$$R^1\text{–C} \equiv \text{C–H} + R^2\text{–X} + \text{base} \rightarrow R^1\text{–C} \equiv \text{C–Ar}^2 + \text{H–base}^+/X^- \quad (8.345)$$

R^1 =aryl, heteroaryl, alkenyl, alkyl, SiR_3

R^2 =aryl, heteroaryl, alkenyl, 1° and 2° -alkyl

X =I, Br, Cl, OTf, SPh, N_2^+/BF_4^-, PhI^+/X^-

$$R^1\text{–C} \equiv \text{C–H} + R^2\text{COCl} + \text{base} \rightarrow R^1\text{–C} \equiv \text{C–}R^2 + \text{CO} + \text{H–Base}^+/\text{Cl}^- \quad (8.346)$$

R^2 = aryl, heteroaryl, RCO

$$\text{R–C} \equiv \text{C–H} + \text{ArSO}_2\text{Cl} + \text{K}_2\text{CO}_3 \rightarrow \text{R–C} \equiv \text{C–Ar} + \text{SO}_2 + \text{KCl} + \text{KHCO}_3 \quad (8.347)$$

In 1975, Cassar reported the $Pd(PPh_3)_4$-catalyzed arylation (8.348) [2708]. On their side, Dieck and Heck found the $Pd(PPh_3)_2(OAc)_2$-catalyzed cross-coupling of terminal alkynes with aryl, heterocyclic, and vinylic bromides and iodides (reaction (8.349)) [2709]. These reactions required heating to 100 °C.

$$\text{R–C} \equiv \text{C–H} + \text{ArX} + \text{NaOMe} \rightarrow \text{R–C} \equiv \text{C–Ar} + \text{NaX} + \text{MeOH} \quad (8.348)$$

$$\text{R–C} \equiv \text{C–H} + \text{ArX(alkenyl–X)} + \text{Et}_3\text{N} \rightarrow \text{R–C} \equiv \text{C–Ar(alkenyl)} + \text{Et}_3\text{NH}^+/\text{X}^- \quad (8.349)$$

A few months later in1975, Sonogashira, Tohda, and Hagihara showed that these Pd-catalyzed cross-coupling reactions can be carried out at room temperature in the presence of CuI as a cocatalyst [2710]. This rate enhancement arises from the formation of R—C≡C—Cu intermediates (Scheme 8.104b) [2711], the nucleophiles used in the uncatalyzed Castro–Stephens alkynylation of aryl iodides [2712] and alkenyl iodides [2713]. In the presence of oxygen, the latter leads to the undesired alkyne homocouplings (Hay–Glaser reaction, Section 8.5.1). Recent developments use noncopper cocatalysts such as AgI and AuCl that are less prone to oxidation [2696].

In 1975, Heck proposed a first mechanism for the Cu-free Pd-catalyzed alkynylation of aryl iodides (Cassar–Dieck–Heck alkynylation) that first implies the oxidative addition of PdL_2 onto Ar—I (as for other Pd-catalyzed displacement reactions) and then carbopalladation of the alkyne by adduct Ar—Pd(L_2)—I to give intermediates of type R—Pd(L_2)C=C(Ar)H [2709]. In 2008, Mårtensson and coworkers prepared such complexes and found that they do not produce the expected products of coupling R—C≡C—Ar under the Cassar–Dieck–Heck reaction conditions [2714]. In an alternative mechanism proposed by Soheili and coworkers in 2003, the acetylene π-complex Ar—Pd(L_2)(RC≡CH)—I eliminates HI under the action of the base producing R—C≡C—Ar + H-base+/I^- and liberates the catalytically active species PdL_2 (Scheme 8.104) [2715]. Jutand and coworkers found that the alkyne inhibits the oxidative addition step and that the amine, depending on its coordinating ability, may accelerate it due to the formation of a PdL(amine) complex (L = phosphine) that adds to Ar—I faster than PdL_2 [2698]. For alkynylations of p-$CF_3C_6H_4$—I and using $Pd(dba)_3{\cdot}CHCl_3/AsPh_3$ as the precatalyst in MeOH, Mårtensson and coworkers found two different linear Hammet plots $\log(k_Y/k_H) = \rho\sigma_I$ for the couplings with p-Y—C_6H_4—C≡CH. With electron-rich alkynes (Y = H, Me, MeO, Me_2N), a negative ρ value was obtained, whereas for electron-poor alkynes (Y = H, Cl, MeCO, NO_2), a positive ρ value was observed. This was interpreted in terms of an anionic mechanism (Scheme 8.104a) for the reactions of electron-poor alkynes: deprotonation of π-complex Ar—Pd(L)(I)(H—C≡C—C_6H_4—Y) by the amine generates an anionic intermediate [Ar—Pd(L)(I)—C≡C—C_6H_4—Y]$^-$ in the rate-determining step, which is followed by substitution of I^- by a ligand L giving Ar—Pd(L_2)—C≡C—C_6H_4—Y. This mechanism is supported by the observation of a kinetic isotopic effect k_H(R—C≡C—H)/k_D(R—C≡C—D) ≈ 2 and little dependence on amine concentration for the reactions of electron-poor alkynes. With electron-rich alkynes, a cationic mechanism intervenes, which implies iodide displacement in π-complex Ar—Pd(L)(I)(H—C≡C—C_6H_4—Y) by the ligand or amine in the rate-determining step, followed by deprotonation by the base (Scheme 8.104b). This is substantiated by the observation of a strong rate dependence on the base concentration and base nature for the alkynylations with electron-rich alkynes [2714]. Quantum mechanical calculations on model reactions support these hypotheses [2716].

In the presence of a Cu(I) salt as a cocatalyst, the accepted mechanism is shown in Scheme 8.104c. Displacement reactions of type Ar—I + H—C≡C—Ph + base → Ar—C≡C—Ph + H-base$^+$/I$^-$ can be catalyzed by $Pd(PPh_3)_4$ and CuI in $(i\text{-}Pr)_2NH$ (base and solvent, 0–20 °C). The transmetallation Ar—[Pd]—I + Cu—C≡C—Ph ⇄ Ar—[Pd]—C≡C—Ph + CuI is the rate-determining step. In 2013, Lei and coworkers reported zero-order kinetic dependences on the concentration of aryl iodide and alkyne and first-order kinetic dependences on the concentration of both the Pd precatalyst and the Cu cocatalyst, indicating synergistic catalysis. Hammett correlation plots recorded for these reactions gave a positive ρ value (1.29) for Ar—I, indicating that intermediates Ar—[Pd]—I act as electrophiles, and a negative ρ value (−1.65) for Cu—C≡C—Ar′ showing that copper acetylides act as nucleophiles. The more Ar is electron poor and the more Ar′ is electron rich, the faster the transmetallation. If reductive elimination Ar—[Pd(II)]—C≡C—Ar′-I → Ar—C≡C—Ar′+[Pd(0)] were the rate-limiting step instead of the transmetallation, the rate of the coupling would not depend on the concentration of the Cu salt, which is not the case. ^{31}P-NMR confirmed that adducts of type Ar—[Pd]—I are the resting states. For reaction p-EtOOC—C_6H_4—I + H—C≡C—Ph + $(i\text{-}Pr)_2NH$ → p-EtOOC—C_6H_4—C≡C—Ph + $(i\text{-}Pr)_2NH_2{}^+$/I$^-$ using $Pd(PPh_3)_4$ (1 mol%) and CuI (2 mol%), $\Delta^{\ddagger}H$ = 8.8 kcal mol^{-1} was determined between −25 and 25 °C [2717]. For other Sonogashira–Hagihara couplings, the turnover-limiting step has been proposed to be the oxidative addition $Pd(0)L_2$ + Ar—X ⇄ Ar—Pd(II)(L_2)—X or PdL ($PdLX^-$) + Ar—X ⇄ Ar—Pd(L)—X (⇄ Ar—Pd(L)$X_2{}^-$) for bulky ligands L [2280] as activation enthalpies ($\Delta^{\ddagger}H$) correlate with the HOMO energies of the aryl halides [2705, 2718].

Classically, alkylation of a terminal alkyne is carried out by displacement of an alkyl electrophile R—X by alkali metal acetylides R′C≡CM, a method that is intolerant to functional groups and needs low temperature. It competes with base-induced elimination from R—X. With R′—C≡C—MgX, the coupling reaction can be activated by Co- [2719], Fe- [2138, 2158], or Ni-based catalysts [2075]. The catalyzed coupling of alkylmetals with alkynyl halides is an alternative method that requires preactivation of the alkyne [2720, 2721]. The groups of Fu in 2003 [2699], and of Glorius in 2006, reported the direct coupling of terminal alkynes and alkyl electrophiles with Pd precatalysts combined with N-heterocyclic carbene ligands [1702]. Later, Ni-based catalysts have been found to catalyze these alkyne alkylations with good functional tolerance [2722–2724]. In 2015, Hu and coworkers presented the pincer complex **415** that catalyzes the alkylation of terminal alkynes with primary alkyl iodides and bromides at room temperature (Scheme 8.105a) [2725]. The rate law of coupling oct-1-yne + 1-iodooctane + t-BuOLi → n-hex-C≡C-n-oct + t-BuOH+ LiI is zero order with respect to the concentrations of oct-1-yne, 1-iodooctane, the base (t-BuOLi), and cocatalyst (CuI) and first order with respect to the concentration of catalyst **415**. This may suggest that the reductive elimination of type R′—[Ni(II)]—C≡CR → R′—C≡C—R + [Ni(0)] is the turnover-determining step. However, it was found that the rate of reaction is not increased for couplings using Ar—C≡C—H with electron-withdrawing substituents on the phenyl ring. A rate decrease is observed instead, thus suggesting that the reductive elimination is not the rate-determining step. Reaction of **415** + Ph—C≡C—H + t-BuOLi in the presence of CuI produces acetylide **416**, which has been fully characterized. In the absence of Ph—C≡C—H and with or without CuI, **416** reacts with 1-iodooctane at room temperature in DMF with much less efficiency than **415** + CuI + Ph—C≡C—H + 1-iodooctane. Interestingly, on adding 1 equiv. of Ph—C≡C—H to the above mixture, the efficiency increased (rate and yield) and reached that of the coupling catalyzed by **415**. Without t-BuOLi, the reaction does not occur. In the absence of CuI, the reaction is slower, indicating that Ph—C≡C—Cu reacts faster than Ph—C≡C—Li. The mechanism (Scheme 8.105b) proposed by Hu and coworkers involves equilibrium **416** + Ph—C≡C—H + CuI + t-BuOLi ⇄ **417**+ LiI + t-BuOH. The rate-determining step is the decoordination of the MeN—$(CH_2)_2$ side arm giving complex **418** that undergoes oxidative addition to the alkyl iodide. The coupling is inhibited by TEMPO, consistently with a radical mechanism for the oxidative addition of the Ni(II) intermediate. Final reductive elimination gives the coupling product Ph—C≡C-n-octyl and liberates the catalytically active species **416**. CuI or another Cu(I)X species coordinates one of the Ni acetylide moieties in **417** and **418** to account for the zero-order rate law with respect to the concentration of CuI.

$Fe(acac)_2$ + CuI + inorganic base (K_3PO_4, $CsCO_3$) induces alkynylations of aryl iodides (140 °C, DMSO; N-methylpyrrolidone). No extra ligand is required [2726–2728]. In the absence of any Cu(I) salt, $Fe(acac)_2$ + 2,2′-bipyridine [2729], $FeCl_3$ + MeNHCH$_2$CH$_2$NHMe [2730], or $FeCl_3$ + Ph_3P catalyzes cross-couplings of terminal alkynes and aryl iodides (Cs_2CO_3, toluene, 135 °C) [2731].

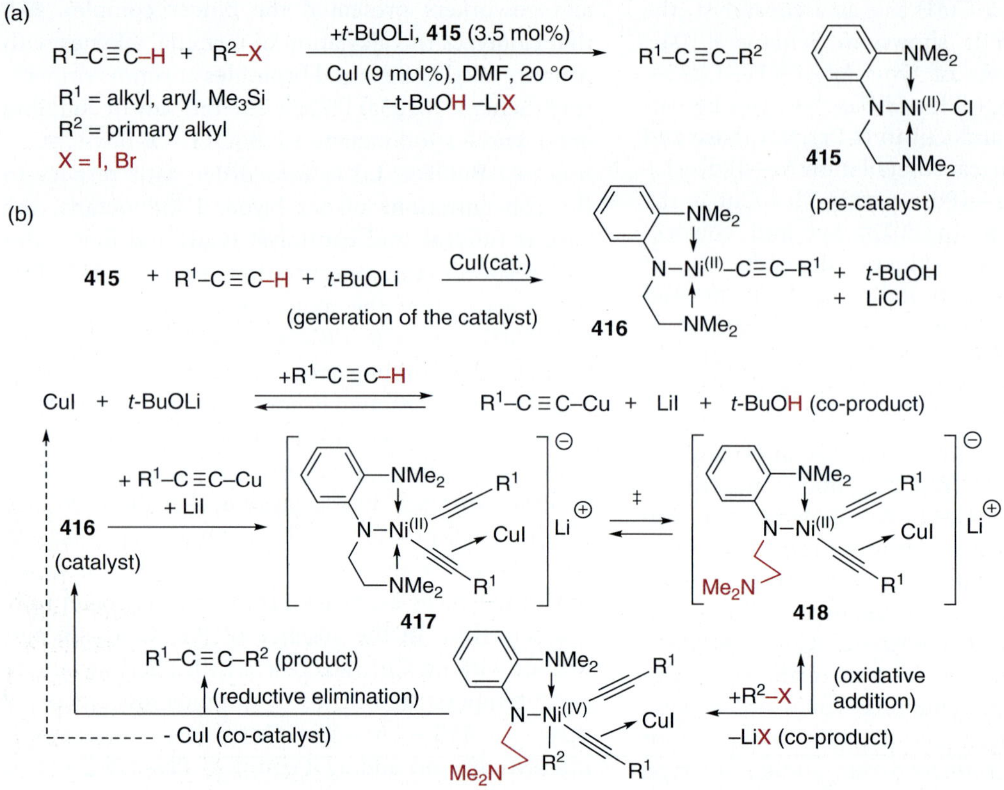

Scheme 8.105 (a) Hu's Ni-catalyzed alkylation of terminal alkynes and (b) proposed reaction mechanism.

Pd-, Cu-, and amine-free cross-couplings have been catalyzed by Sm powder in hot polyethyleneglycol [2732]. Co nanoparticles with $Ph_3P + CuI$ catalyze Sonogashira–Hagihara couplings (*N*-methylpyrrolidone, K_2CO_3, 120 °C) [2733]. Supported gold and gold nanoparticles [2734–2739] as well as rhodium nanoparticles have been proposed as catalysts [2740]. Pd impurities might be responsible of the catalytic effects in all cross-couplings applying non-Pd precatalysts [2741–2743]. The Castro–Stephens and Sonogashira–Hagihara C—C cross-coupling reactions are examples of nucleophilic alkynylations. Other examples are direct displacements of halides and epoxides by ethynylmetals. In 2002, Yamaguchi and coworkers reported the $GaCl_3$-*ortho*-alkynylation of phenols with chloroacetylenes [2744]. Since then, several electrophilic alkynylations of arenes and heteroarenes ("inverse Sonogashira–Hagihara cross-couplings") have been reported [2745–2747]. Waser and coworkers have used tri(*i*-propyl)silylethynyl-1,2-benziodoxol-3(1*H*)-one (($(i\text{-Pr})_3Si—C≡C—X$ equivalent) [2748] as electrophilic acetylene in the AuCl-catalyzed alkynylation of electron-rich heterocyclics such as indoles, pyrroles [2749], thiophenes [2750], and tryptophan in peptides [2751]. They also reported *para*-regioselective alkynylation of anilines [2752].

Problem 8.100 Propose a reaction mechanism for the following decarboxylative Sonogashira–Hagihara coupling [2753].

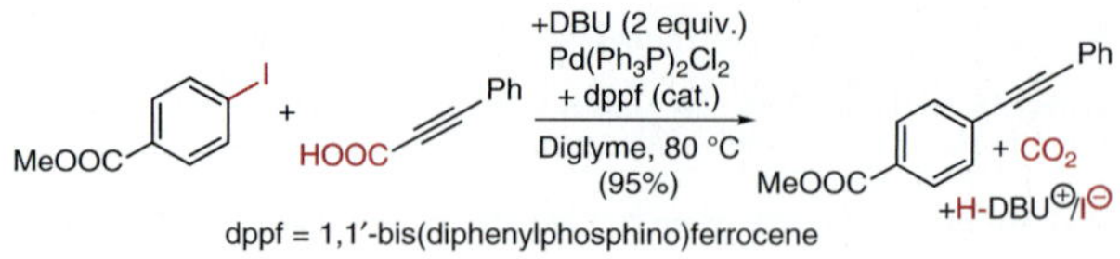

8.8.9 Arylation of arenes(heteroarenes) with aryl(heteroaryl) derivatives

Related to the Mizoroki–Heck arylation of alkenes is the direct arylation (heteroarylation) of arenes (or heteroaranes: $Ar^2—H$) with aryl halides, phenol sulfonic esters, and diaryliodonium salts ($Ar^1—X$) to generate biaryl, aryl-heteroaryl, and bi(hetero)aryl systems ($Ar^1—Ar^2$) that are predominant features in many pharmaceutically relevant and biologically active compounds [2754–2759], and in material sciences (reaction (8.350)) [2756, 2760–2768]. Most used catalysts are Pd based [2759, 2769, 2770], but other less expensive transition metals such as Ru [2771–2773], Au [2774, 2775], Cu [2776], Ni [2777], Co [2778], and Fe [2777, 2779], together with suitable ligands and additives, are capable to catalyze the direct arylations of arenes [2780]. Desulfinylative arylation of heteroaromatics ($Ar^1—H + Ar^2SO_2Cl \rightarrow Ar^1—Ar^2 + SO_2$) have been reported [2781–2783]. Aryl chlorides can arylate arenes using Ru(II) catalysts [2784]. In some cases, no transition metal catalyst is necessary [2785], as bases such as *t*-BuOK [2786] or EtOK + 1,10-phenanthroline [2787] can induce the displacement reactions upon heating [2510, 2777, 2788, 2789]. Copper-catalyzed arylation with diaryliodonium salts has been described [2790]. The intermolecular version (8.350) of the reaction is difficult to control, because the arene presents several C—H bond for the coupling reaction. This difficulty may not exist for intramolecular reactions. Using a taddol-derived phosphoramidite ligand and pivalic acid as an additive, Saget and Cramer obtained dibenzazepinones with quaternary stereocenters enantioselectively (reaction (8.351)) [2791].

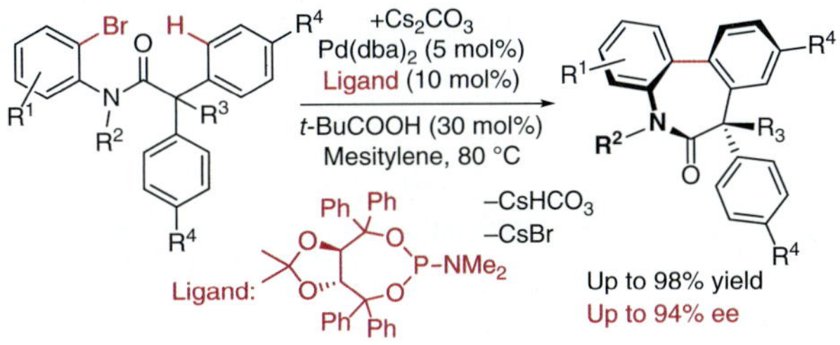

(8.351)

In 1985, Catellani and Chiusoli reacted PhBr with norbornene in the presence of $Pd(Ph_3P)_4$ catalyst and *t*-BuOK as the base [2792]. The product of Mizoroki–Heck arylation was not observed (disfavored synperiplanar C—H/Pd—C bonds in the intermediate **419** resulting from the β-insertion of the phenyl group into the bicyclic alkene moiety (Scheme 8.106)) but **420** was formed instead in 65% yield.

With indene, de Meijere obtained product **421** resulting from a triple "domino" arylation (reaction (8.352)) [2793, 2794]. The Catellani reaction demonstrated for the first time the possibility of Pd-catalyzed direct arylation of arenes [2795–2798]. It is now a powerful tool for the construction of multisubstituted arenes that are found in natural products, drugs, and drug candidates [2799–2807].

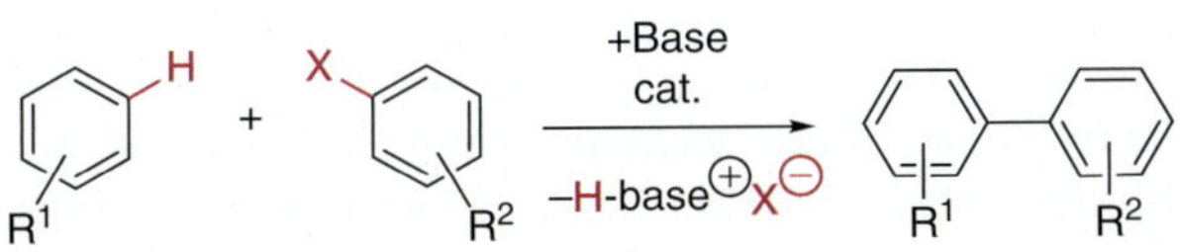

(8.350)

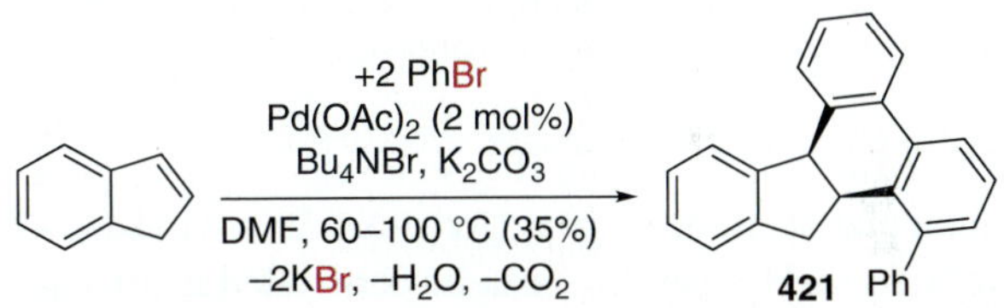

(8.352)

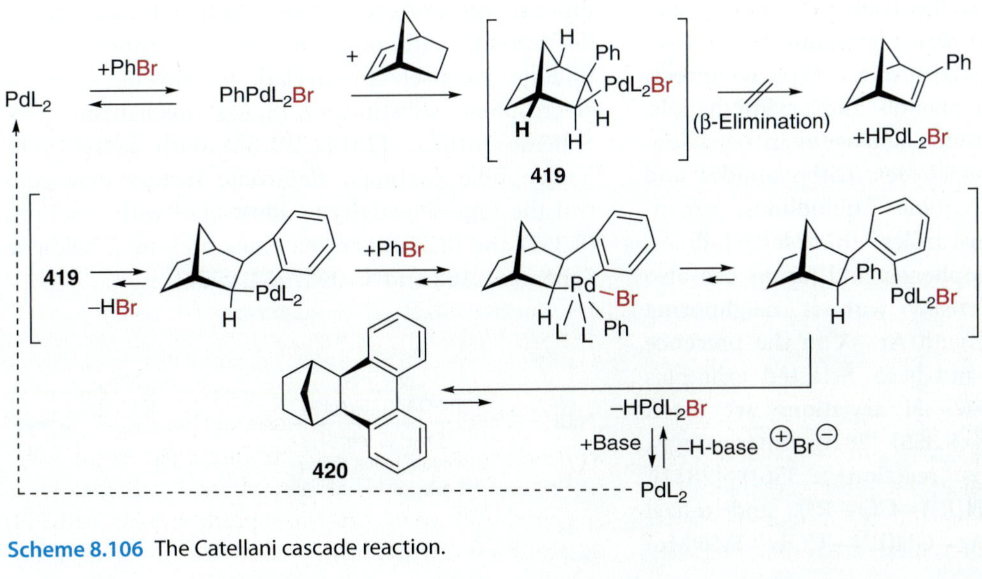

Scheme 8.106 The Catellani cascade reaction.

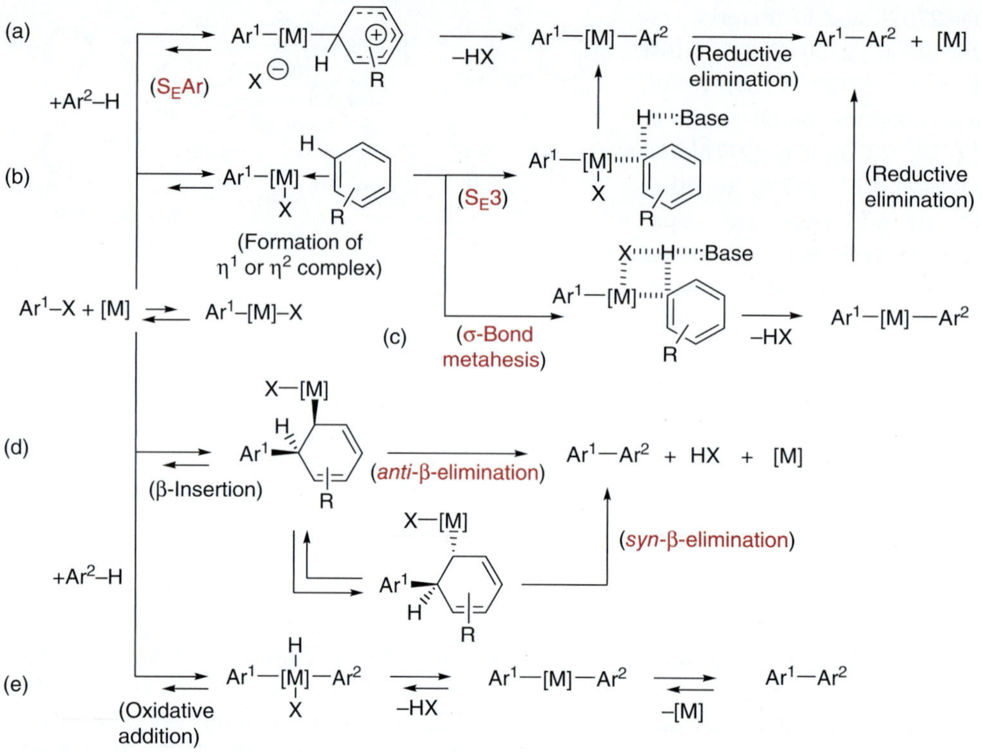

Scheme 8.107 Hypothetical reaction mechanisms of the transition-metal-catalyzed direct arylation of arenes with aryl halides and pseudo-halides. (a) Aromatic electrophilic substitution initiated by the heterolysis of the arylmetalhalide into $Ar^1[M]^+$/X-ion pair followed by HY elimination and a reductive elimination, (b) formation of an arenemetal complex followed by a base induced concerted proton elimination, (c) σ-bond metathesis, (d) β-insertion of the arylmetal complex onto one double bond of the arene followed by a β-eliminationof H-[M]-X, (e) oxidative addition of the arylmetal halide onto the C—H bond of the substrate followed by two successive reductive eliminations.

One method by which the regioselectivity of intermolecular arene arylations (Ar^1—X+Ar^2—H+base → Ar^1—Ar^2 + H-base$^+$/X$^-$) can be controlled is through the use of directing groups [2808, 2809]. Common directing groups bear a lone pair of electrons that can coordinate to the transition metal catalyst (form chelates) to direct arylation via a five- or six-membered metallacycle. Typical directing groups are OH in phenols and arylmethanols, carbonyl groups of ketones, aldehydes in benzaldehydes and phenylacetaldehydes, carboxamides and amides, or imines. Pyridines, quinolines, oxazolines, imidazolines, pyrazoles thiazoles, indoles, azoles, isooxazoles, thiophenes, and furans can also be arylated regioselectively without neighboring group assistance [403] with Ar—X in the presence of a suitable catalyst and base. Selected examples of Pd-catalyzed *ortho*-C—H arylations are given with reactions (8.353)–(8.355) [2810]. Interestingly, *N*-aryl pivanilides (e.g. reaction (8.356)) [2811], β-arylketones (Ar—CH(R^1)—CO—R^2), and α-aryl Weinreb amides (Ar—CH(R^1)—CON(OMe)Me) are arylated with diaryliodonium salts regioselectively at their *meta*-positions when catalyzed by $Cu(OTf)_2$ [2812]. These reactions have been proposed to follow a Heck type mechanism (see Scheme 8.107d) [2813, 2814]. Under the same conditions, electron-rich arenes (phenol and aniline derivatives) without coordinating carbonyl group undergo *para*-selective arylations following aromatic electrophilic substitution (S_EAr) mechanisms (see Scheme 8.107a) [2814, 2815]. With heterocyclic nucleophilic partners, electronic factors may control the regioselectivity as illustrated with reactions (8.357) and (8.358) and reactions presented below in Schemes 8.108 and 8.109 [2754, 2776].

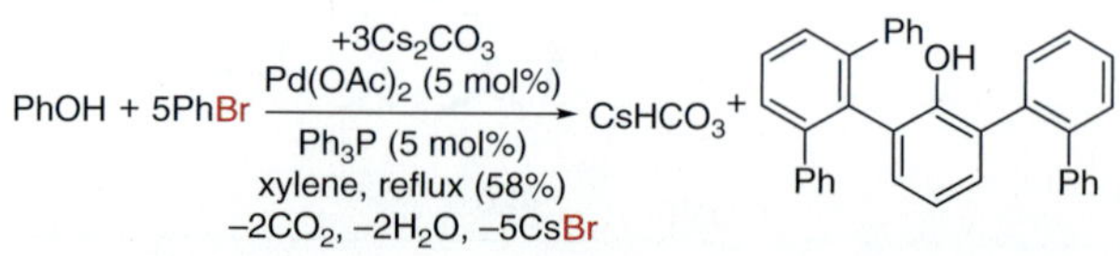

(8.353)

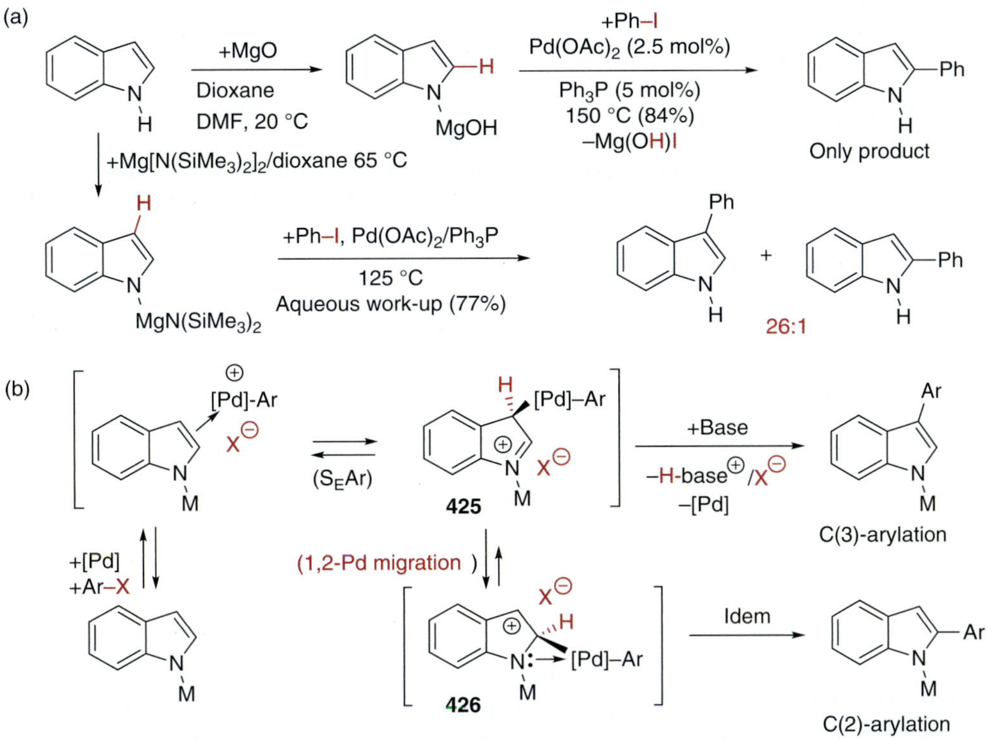

Scheme 8.108 Ames' Pd-catalyzed arylation of indoles. (a) Regioselectivity depends on the nature of *N*-metallated indole and (b) aromatic electrophilic substitution mechanism involving 1,2-migration of Pd.

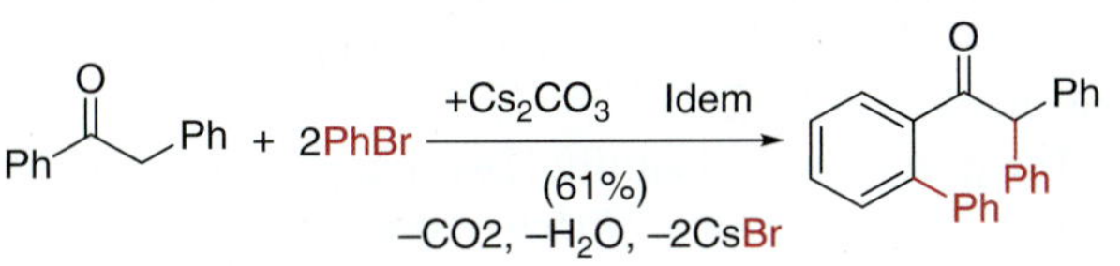

(8.354)

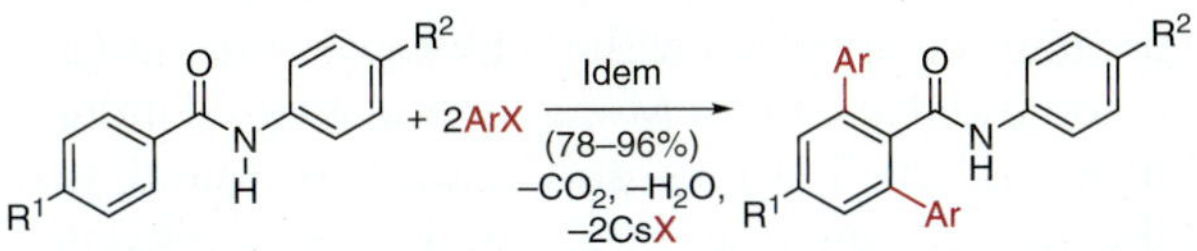

(8.355)

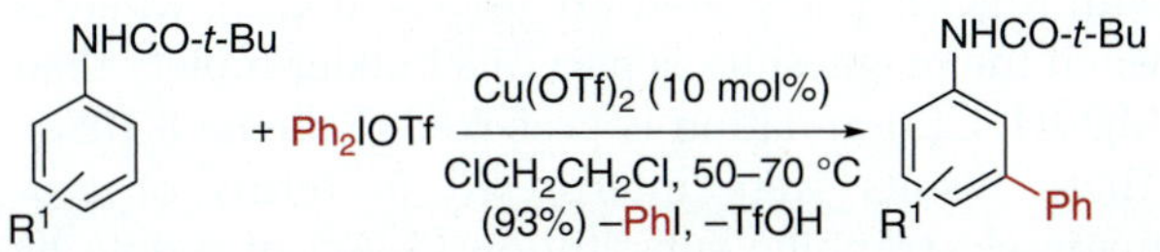

(8.356)

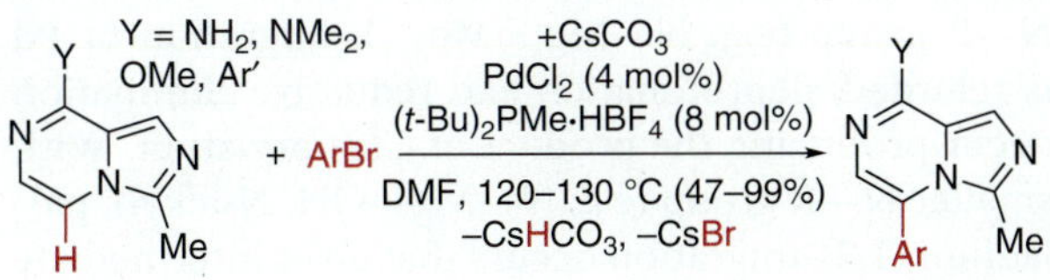

(8.357)

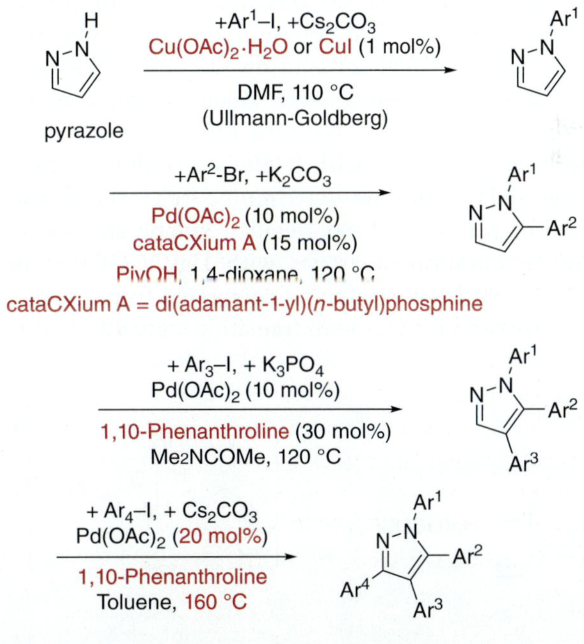

Scheme 8.109 Four successive regioselective arylations of pyrazole that do not require any protection/deprotection step.

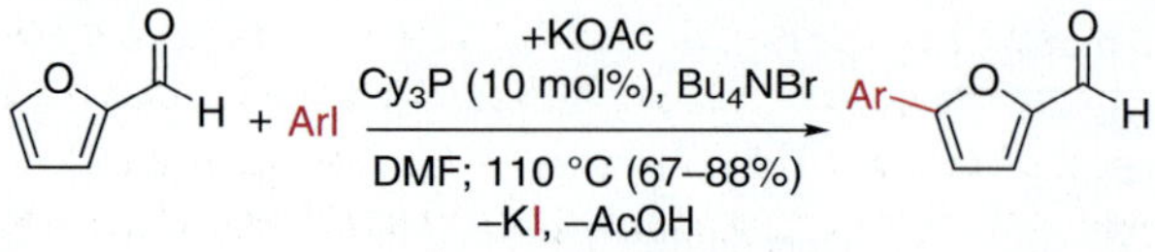

(8.358)

The reaction mechanism of direct arylation of arenes (Ar^2—H) implies oxidative addition of the low-valent transition metal into the aryl halide (e.g. Ar^1—X + [Pd(0)] ⇄ Ar^1—[Pd(II)]—X), followed by one of the possible carbon–carbon bond forming processes (Scheme 8.107) that include (a) electrophilic aromatic substitution (S_EAr), (b) a concerted S_E3 process assisted by a base, (c) a σ-bond metathesis also assisted by a base, (d) a Heck-type carbometallation (complexation of the Ar^2—H with Ar^1—[Pd(II)]—X followed by β-insertion of Ar^1 and subsequent *anti*-β-hydride elimination, or subsequent isomerization and *syn*-β-hydride elimination), or (e) a C(sp^2)—H bond oxidative addition [521, 2816–2818]. The actual reaction mechanism depends on the nature of both reactants Ar^1—X and Ar^2—H, the transition metal catalyst, solvent, base, and ligand used [2759, 2819].

For long, the S_EAr mechanism, which implies the formation of a cyclohexadienyl cation intermediate in the rate-determining step, was the preferred one [2820–2822]. However, in 1967, Kresge and Brennan found a large kinetic isotopic effect $k_H(C_6H_6)/k_D(C_6D_6) = 6.0 \pm 0.1$ (25 °C) for the mercuration of benzene with $[Hg(OAc)]^+/ClO_4^-$, which suggested that the rate-determining step is the proton transfer [2823]. In 1980, detailed kinetic studies on the mercuration of arenes with $Hg(OCOCF_3)_2$ in CF_3COOH led Roberts and coworkers to propose a base-assisted process with transition state **423** that is preceded by a fast reversible step forming an arene π-complex **422** [2824]. The mechanism is similar to that proposed by Winstein and Traylor in 1955 for the acetolyses of dialkymercury compounds [2825].

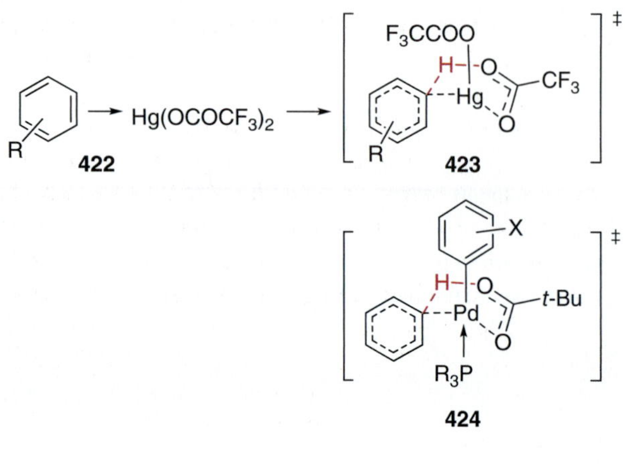

In 2003, Hennessy and Buchwald found a kinetic isotopic effect $k_H/k_D = 4$ for the intramolecular Pd-catalyzed alkylation of $C_6H_5(C_6D_5)$—N(R)—$COCH_2$—Cl + Et_3N → oxindole + Et_3H^+/Cl^- (toluene, 80 °C) [2826]. A similar isotopic effect of 3.5 was reported by Fagnou and coworkers in 2004 for intramolecular Pd-catalyzed arylation [2827], which is compatible with a rate-determining step in which the C—H bond is broken (Scheme 8.107b: S_E3 or σ-bond metathesis: concerted metallation/deprotonation mechanisms.) [2828] Consistently with such mechanisms (and not with the S_EAr mechanism), the same group [2829] and that of Echavarren [2817, 2830] found that arenes substituted by electron-withdrawing substituents react the fastest. They proposed the proton abstraction mechanism for these Pd-catalyzed arylations in which either X (of the electrophilic arylating agent) or the external base assists the C—H cleavage in the transition state (Scheme 8.107b) [2275]. In 2006, Lafrance and Fagnou developed an efficient intermolecular arylation of benzene, which incorporates pivalic acid as an additive. The latter promotes the C—H bond cleavage in transition state **424** [2831]. Gold(I)-mediated C—H activation of arenes also follows the concerted metallation/deprotonation mechanism [2775].

The groups of Nakamura [1448] and Ames in 1982 [2832] and of Ohta in 1990 have conducted pioneering work on the inter- and intramolecular Pd-catalyzed direct C—H arylation of heteroaromatics [2833]. In 1995, the latter group reported the regioselective Pd-catalyzed arylation with 2-chloropyrazines at C(2) of indole and at C(3) of *N*-methyl and *N*-tosyl indoles [2834]. In 2005, Ames and coworkers found that Pd-catalyzed phenylation of indole and *N*-alkylated indoles with Ph—I occurs regioselectively at C(2). The conjugate base of indole obtained by treatment with MgO is phenylated exclusively at C(2), whereas when the magnesium is part of a bulkier moiety than MgOH, C(3)-arylation is preferred (Scheme 8.108a). These results were interpreted in terms of aromatic electrophilic substitution (S_EAr) of indole by Ar—[Pd(II)]—X giving the cationic intermediate **425** (Scheme 8.108b, see also Scheme 8.107a). With bulky N—R group (e.g. N—Mg($SiMe_3$)$_2$), migration of Pd is retarded, deprotonation and reductive elimination occur producing the product of C(3)-arylation. With smaller N—R group (e.g. N—Mg—OH, N-alkyl), palladium (1,2)-migration occurs that gives intermediate **426**, the driving force being the stabilization of Pd through coordination by the amino group. Then, deprotonation and reductive elimination lead to the product of C(2)-arylation. These hypotheses are supported by kinetic studies. For the C(2)-phenylation of N-methylindole with Ph—I, the rate law is zero order with respect to the concentration of Ph—I, and first order with respect to the concentrations of N-methylindole and catalyst. Hammet plot for reactions of 6-substituted N-methylindoles gave negative ρ value (−0.71) consistently with the formation of benzylic cation intermediates of type **426** in the rate-determining step. A surprisingly large deuterium kinetic isotopic effect k_H/k_D-C(3) = 1.6 for 3-deuterio-N-methylindole was found even though H—C(3) is not cleaved during the reaction. A smaller kinetic isotopic effect k_H/k_D-C(2) = 1.2 was measured for 2-deuterio-N-methylindole, indicating that deprotonation of C(2)—H is not the rate determining step but follows it [2822]. With indole N-metallated by EtMgBr and using Pd(dba)$_2$ as the precatalyst in toluene at 100 °C, Beletskaya and coworkers reported that arylation with Ph—Br prefers C(3) using $(t\text{-Bu})_3$P ligand and C(2) using XantPhos ligand [2835]. Phenylation of indole with Ph—Br in the presence of Pd(OAc)$_2$/(Ph$_2$P)$_2$CH$_2$ as a precatalyst and LiOH as a base ("on water" reaction, 110 °C) gives preferentially 3-phenylindole, whereas using Ph—I and KOAc under the same reaction conditions leads preferentially to 2-phenyllindole [2836].

Regioselective direct C(sp^2)—H arylation and heteroarylation of heterocyclics is an active topic of research [2837]. Selective sequential arylation of heterocyclic compounds represents a powerful tool for drug discovery especially when protection/deprotection steps are not necessary [2756, 2838]. An example is given with the four successive arylations of pyrazole (Scheme 8.109) presented by Fuse and coworkers in 2015 [2838].

Problem 8.101 What is the main product **P** of the following reaction? [2839]

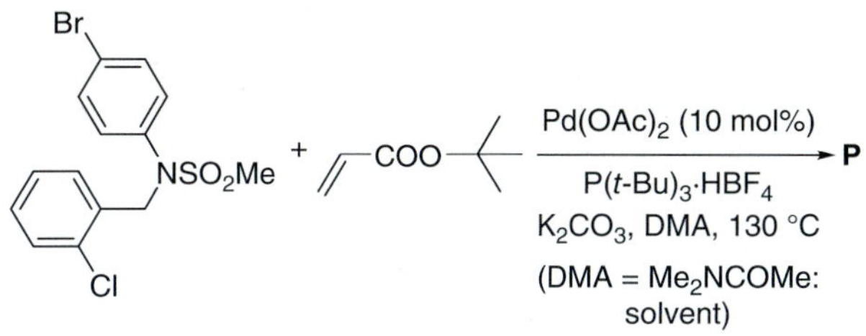

8.8.10 α-Arylation of carbonyl compounds and nitriles

The α-aryl and α-heteroaryl carbonyl units are found in many pharmaceuticals and natural compounds with biological activities [2755, 2840–2843]. α-Aryl esters and amides are useful intermediates in organic synthesis that can be converted into α-aryl-carboxylic acids and β-aryl-amines [2844–2847]. One of the most convenient methods to generate these compounds is the one-step Pd-catalyzed intermolecular α-arylation of enolates of Buchwald [2848], Hartwig [2849], and Miura and coworkers [2850] disclosed in 1997 (Scheme 8.110a). The method initially developed for ketones has been extended to carboxylic esters, carboxamides, and carbonitriles [2845, 2851–2853] and uses aryl chlorides, bromides, iodides, or pseudohalides as electrophilic partners; a Pd precatalyst; and a phosphine ligand. Ni(0) precatalysts with suitable phosphine ligands have also been employed [2846]. For instance, in the presence of the catalyst generated in situ from Ni(cod)$_2$, 3,4-bis(dicyclohexylphosphino)thiophene, and K$_3$PO$_4$, various carboxylic esters and amides undergo α-arylation with aryl pivalates (ArOCO—t-Bu) in toluene at 150 °C [2854]. Room temperature α-arylation of malonic esters with aryl iodides or

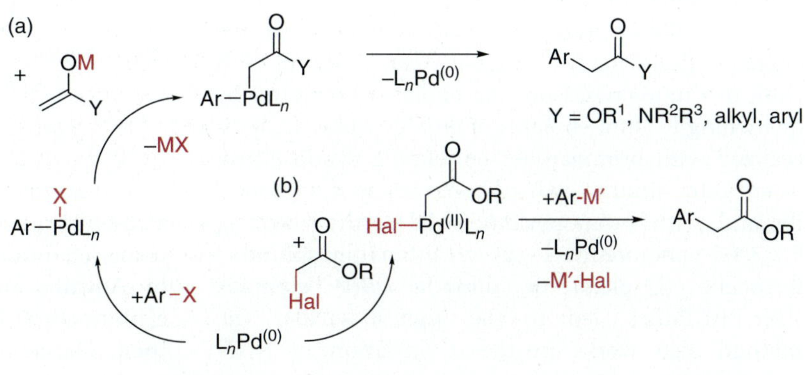

Scheme 8.110 (a) Hartwig and Buchwald routes to α-arylated carbonyl compounds via enolates and (b) Goosen's route involving arylative transmetallation (M′ = B(OR)$_2$) of a Pd-enolate (Hal = Br).

bromides is catalyzed by CuI + 2-picolinic acid [2855]. The arylation of activated methylene compounds in the presence of copper metal or copper salts has been disclosed in 1929 by Hurtley [2856]. The process is now called the Ullmann–Hurtley C—C cross-coupling reaction [2857–2861]. In 2001, Goosen condensed α-halo esters to Pd(0) species; a subsequent transmetallation with areneboronic esters and reductive elimination generates α-aryl esters (Scheme 8.110b) [2862]. An early intramolecular displacement reaction of an enolate with an aryl iodide (catalyzed by $Ni(cod)_2$) had been presented already in 1975 by Semmelhack [2863]. In 1997, Natsume and coworkers found examples of intramolecular enolate arylations by aryl triflates catalyzed by $PdCl_2(Ph_3P)_2$) [2864].

An example of ester α-arylation is shown below with Buchwald's [2853] synthesis of the *tert*-butyl ester **427** of (±)-naproxen (anti-inflammatory agent discovered by Syntex in 1968) [2865].

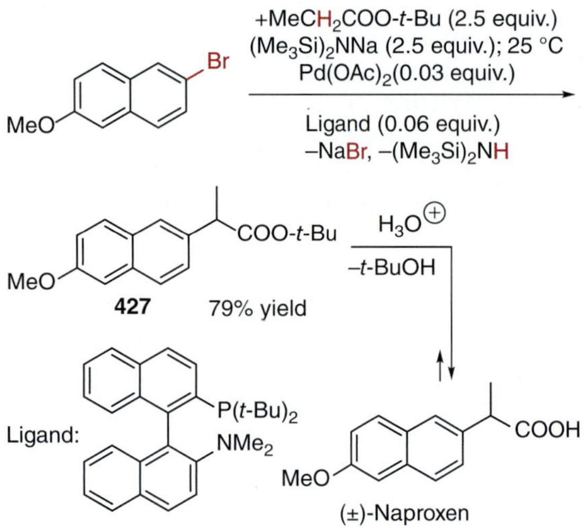

In a similar approach, Hartwig (83% yield) uses $Pd_2(dba)_3$ as a catalyst (1%, dba = (*E,E*)-dibenzylideneacetone), 1.1 equiv. of EtCOO—*t*-Bu, and 2.3 equiv. of $(Me_3Si)_2NNa$ [2851]. The intermolecular Pd-catalyzed α-arylation of zinc amide enolates (Reformatsky reagents generated in situ from α-bromoamides or zinc enolates generated by quenching lithium enolates of amides with $ZnCl_2$) is realized with bromoarenes containing cyano, nitro, ester, keto, fluoro, hydroxyl, or amino functionality and with bromopyridines. Pd-catalyst bearing 1,2,3,4,5-pentamethyl-1′-(di-*tertio*-butylphosphino) ferrocene (Q-phos) or dimeric Pd(I) complex $\{[P(t\text{-Bu})_3PdBr\}_2$ led to the highest yields. The method also works in the α-vinylation of *N,N*-diethylpropionamide with 2-bromopropene (reaction (8.359)) [2852].

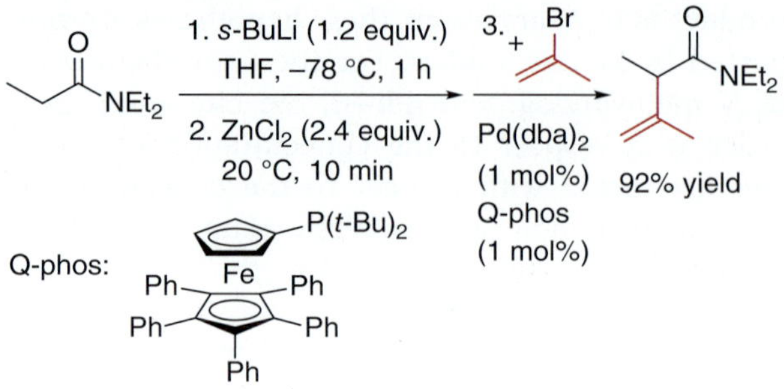

(8.359)

The first example of intermolecular catalytic asymmetric α-arylation of ketone was reported in 1998 by Buchwald and coworkers using (*S*)-BINAP and $Pd_2(dba)_3$ as a catalyst (reaction (8.360)) [2866].

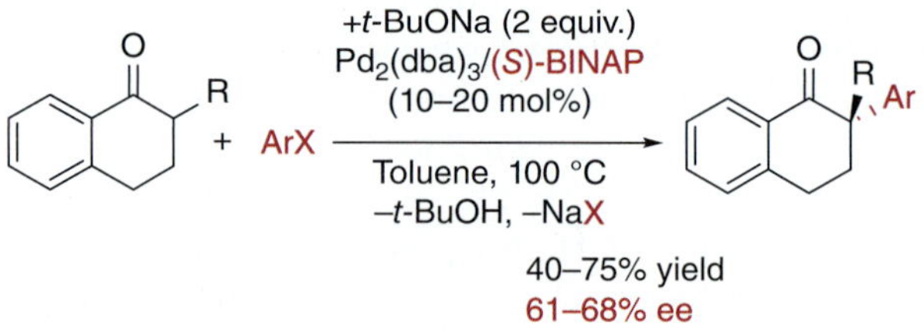

(8.360)

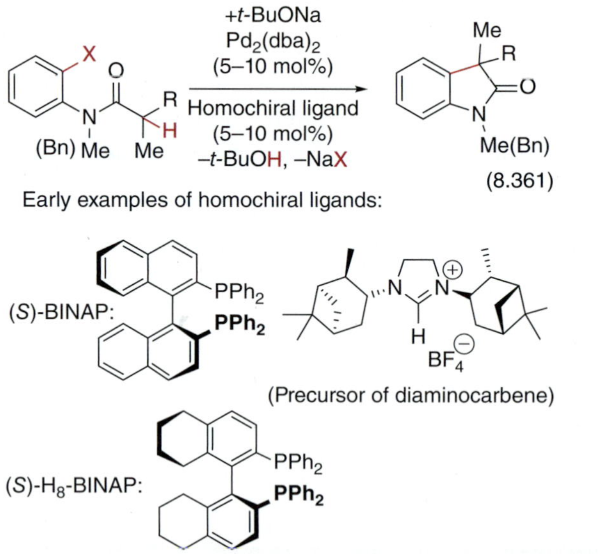

(8.361)

Asymmetric intramolecular enolate arylations have also been reported by Hartwig and coworkers [2867] using enantiomerically pure diaminocarbenes and by Aoyama and coworkers using (*S*)-H_8-BINAP as a homochiral ligand (reaction (8.361)) [2868]. Since then, a large number of enantioselective α-arylations and α-heteroarylations of carbonyl compounds have been reported [2755, 2846, 2869]. Enoxysilanes derived from lactones or acyl oxazolidones (reaction (8.362)) are arylated enantioselectively with diaryliodonium salts in the presence of homochiral copper catalyst **428** [2870].

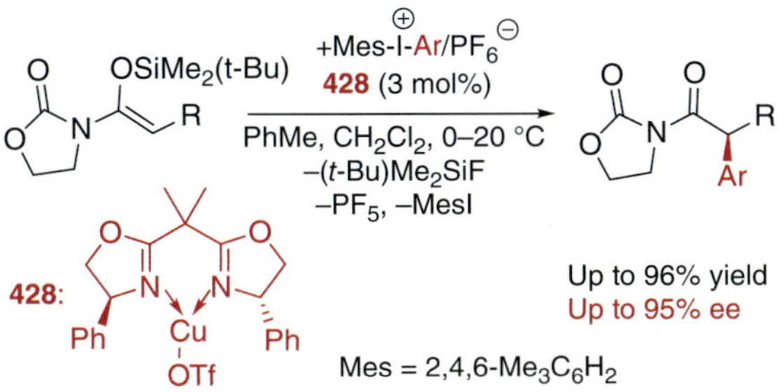

(8.362)

The C—C cross-coupling reactions presented in this chapter are formal displacements of type Ar—X + R—H (with an acidic C—H bond) + base → Ar—R + H-base$^+$/X$^-$. Reactions of type Ar—H + R—X → Ar—R + HX correspond to classical electrophilic aromatic substitutions. They can also be catalyzed by a transition metal complex if β-elimination from R—[M]—X intermediates can be retarded or avoided by an adequate choice of R and catalyst. An example is the synthesis of oxindoles from α-chloroacetanilides (Ar—N(R^1)—COCH$_2$—Cl) catalyzed by Pd(OAc)$_2$ + *o*-Ph—C$_6$H$_4$—P(Bu)$_2$ (Et$_3$N, toluene, 80 °C) developed by Hennessy and Buchwald [2826]. In principle, if the reaction involves a radical rather than R—[M]—X intermediates, β-elimination can be avoided.

Problem 8.102 Devise a one-pot synthesis of benzofurans from 2-bromophenols and ketones [2871].

8.8.11 Direct arylation and alkynylation of nonactivated C—H bonds in alkyl groups

Direct remote C(sp^3)—H functionalization of alkyl groups emerges now as a powerful tool to construct C—C bonds [2872, 2873]. In 1994, Dyker reported the Pd-catalyzed reaction (8.363) showing the possibility of annulation reactions via direct intramolecular arylation of primary C(sp^3)—H [2874]. In 2003, Baudoin and coworkers extended the concept to a practical level using 2-BrC$_6$H$_4$—CH(R^1)(R^2) that produce the corresponding benzocyclobutenes [2875–2877]. In 2005, Daugulis and coworkers reported the Pd-catalyzed intermolecular β-arylation of the ethyl group of 2-ethylpyridine by an aryl iodide (reaction (8.364)) [2878] and the γ-arylation of picolinamides (reaction (8.365)). The pyridine ring is proposed to be the directing group in these direct regioselective C(sp^3)—H arylations. The same group then found examples of Pd-catalyzed β-arylation of *N*-(quinolin-8-yl) propionamide [2879] and of *N*-(2-methylthio)phenyl propionamide [2880, 2881]. An application to the synthesis of non-natural amino acids is shown with reaction (8.366) [2881]. An example of Fe-catalyzed C(sp^3)—H β-arylation (reaction (8.367)) has been reported by Nakamura and coworkers in 2013 [2882].

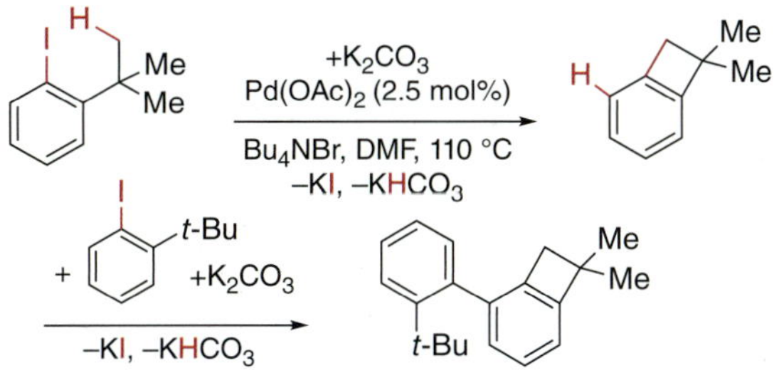

(8.363)

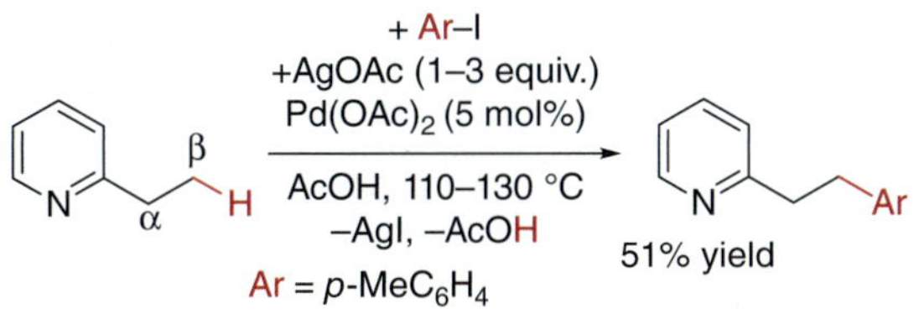

(8.364)

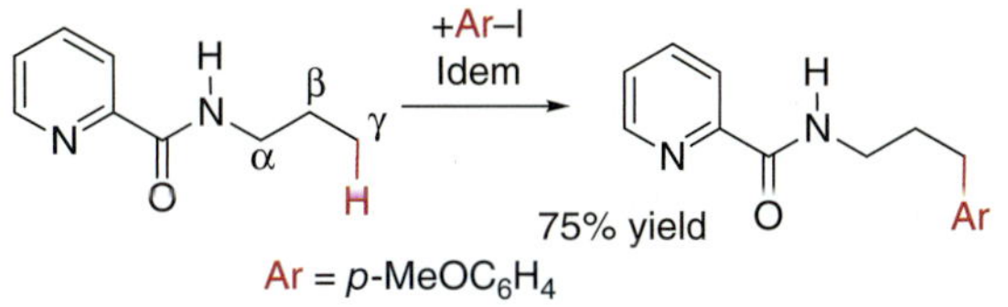

(8.365)

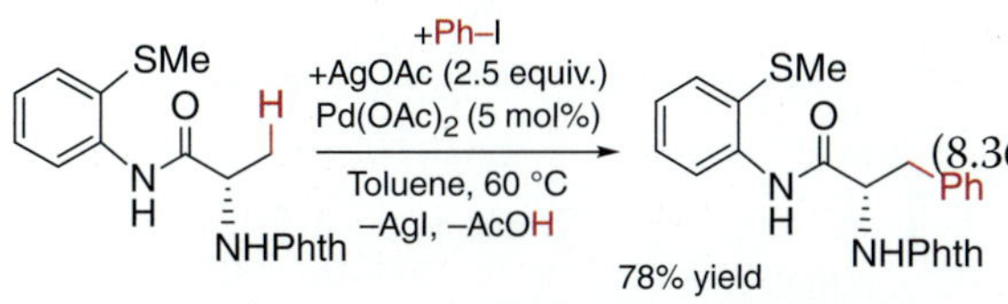

(8.366)

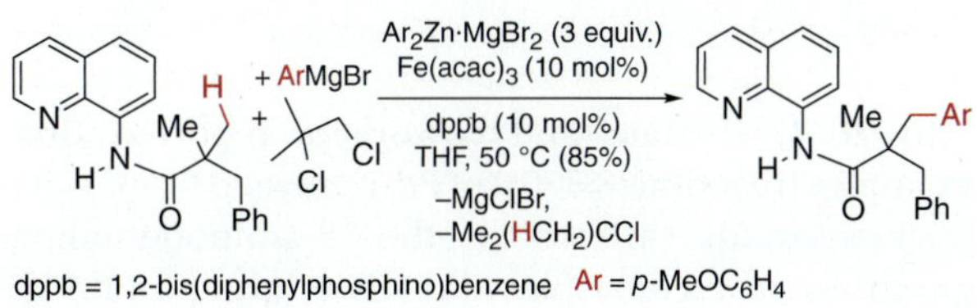

(8.367)

In 2007, Fagnou and coworkers reported efficient intramolecular Pd-catalyzed C(sp^3)—H arylation (reaction (8.368)) and demonstrated the importance of an additive such as pivalic acid in the C—H activation, which is the rate-determining step (transition state **429**) as indicated by a kinetic isotopic effect $k_H/k_D = 5.4 \pm 0.3$ at 115 °C [2883]. The method has been applied to the synthesis of cyclobutarenes, indanes, indolines, dihydrobenzofurans, and indanones using aryl chlorides [2884–2888]. In the presence of enantiopure ligands, enantioselective intramolecular C(sp^3)—H arylations have been observed [2888–2891]. An example is given with reaction (8.369) reported by Cramer and coworkers in 2012 and for which they showed that additive **430** (a bulky carboxylic acid like pivalic acid) together with (*R,R*)-SagePhos monophosphine ligand leads to the highest enantioselectivities [2892]. This is consistent with the Fagnou's hypothesis (transition state **429**) about the cooperative role of the added carboxylic acid (cocatalyst).

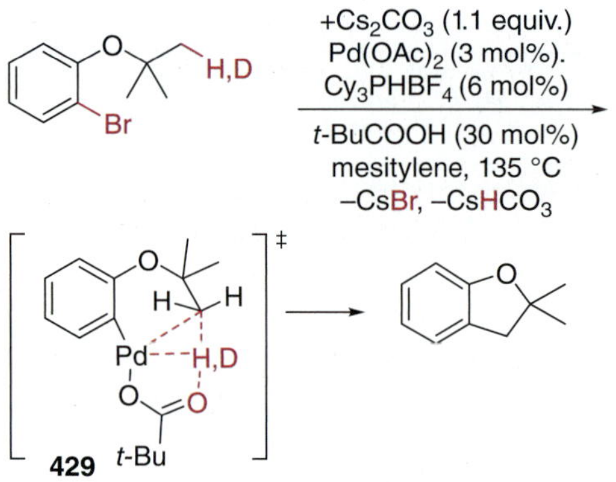

(8.368)

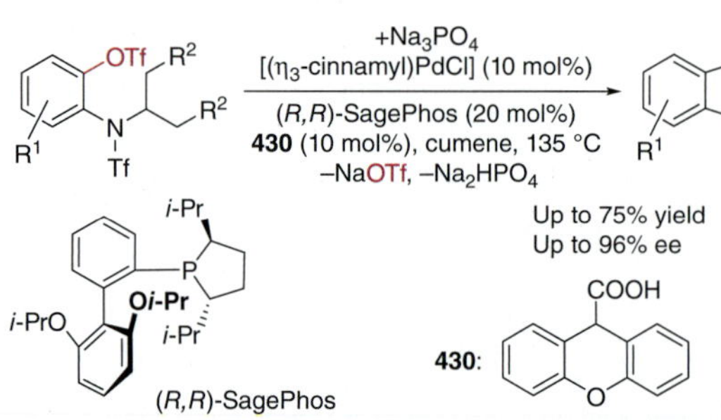

(8.369)

In 2011, Chatani and coworkers reported first examples (reaction (8.370)) of Pd-catalyzed C(sp^3)—H β-alkynylations in which the 8-aminoquinoline group is a removable directing group [2893]. On their side, Yu and coworkers used amides of type p-$CF_3C_6F_4$—NHCO—CH(R)—Me for their C(sp^3)—H β-alkynylations [2894].

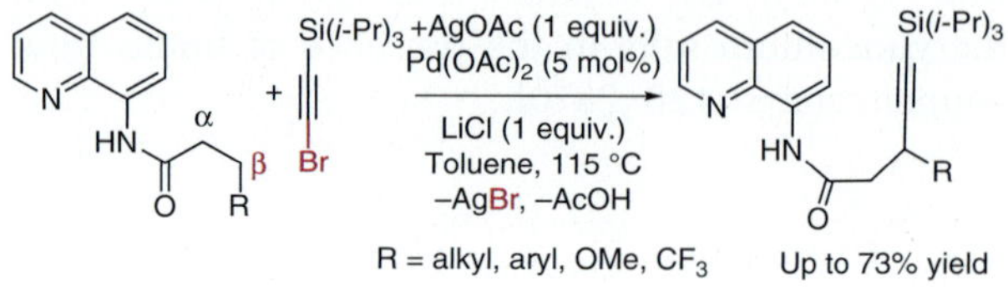

(8.370)

Problem 8.103 What is the main product **P** of the following reaction? [2875]

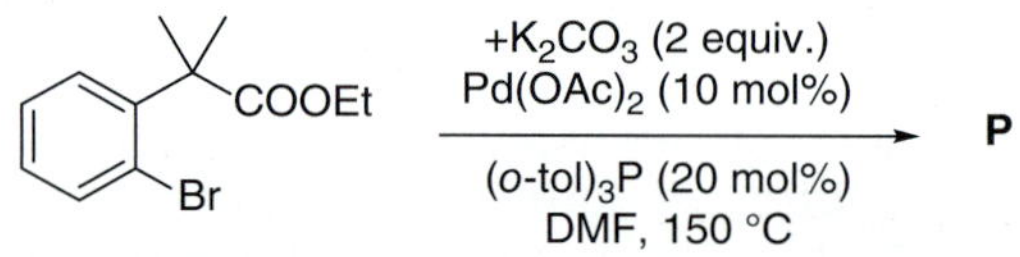

8.8.12 Direct alkylation of nonactivated C—H bonds in alkyl groups

Direct displacements of type R^1—H + R^2—X + base → R^1—R^2 + H-base$^+$/X^- are nucleophilic substitutions that require an activated (acidic) C—H bond in R^1—H (nucleophile). In principle, the electrophilic partner, R^2—X, can undergo an oxidative addition with a suitable transition metal complex forming adduct intermediate R^2—[M]—X, which can then activate C—H bonds in R^1—H. The regioselectivity and chemoselectivity of the reaction (which of the nonactivated C—H bond of R^1—H will be substituted?) can be controlled by a suitable directing group that chelates R^2—[M]—X. This concept has been illustrated in the preceding chapters for R^2 = alkenyl, aryl, heteroaryl, and alkynyl. When R^2 is an alkyl with a β-C—H bond, β-elimination (converts R^2—X into an alkene + HX) competes and impedes the catalyzed displacement reaction. In 2013, Daugulis and coworkers reported the intermolecular Pd-catalyzed β-benzylation (8.371) [2895]. On their side, Ge and coworkers discovered the Ni-catalyzed intermolecular β-butylation (8.372) [2896]. These are 8-aminoquinoline-assisted direct β-alkylations of nonacidic C(sp^3)—H of alkyl groups. In 2014, Cramer and coworkers developed a new asymmetric synthesis of highly valuable β-lactams based on a Pd-catalyzed intramolecular α-alkylation with chloroacetamides (reaction (8.373)). The process requires bulky taddol phosphoramidite ligand **431** and 1-adamantyl carboxylic acid as a cocatalyst [2897].

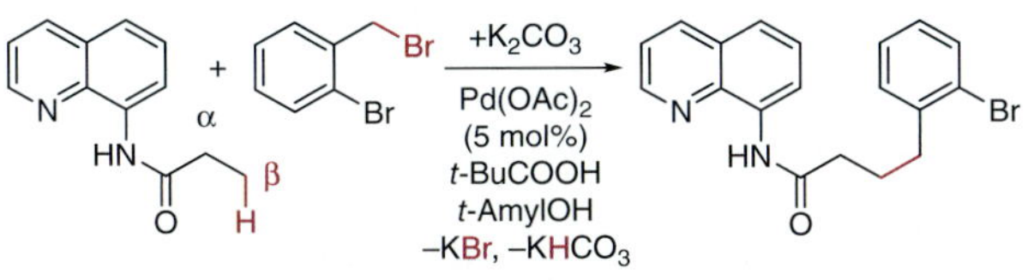

(8.371)

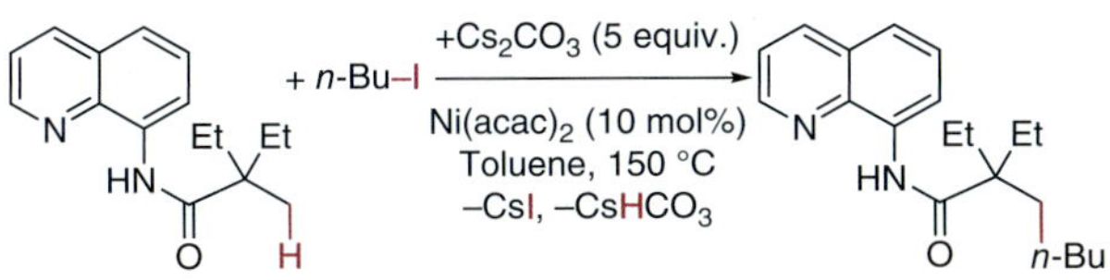

(8.372)

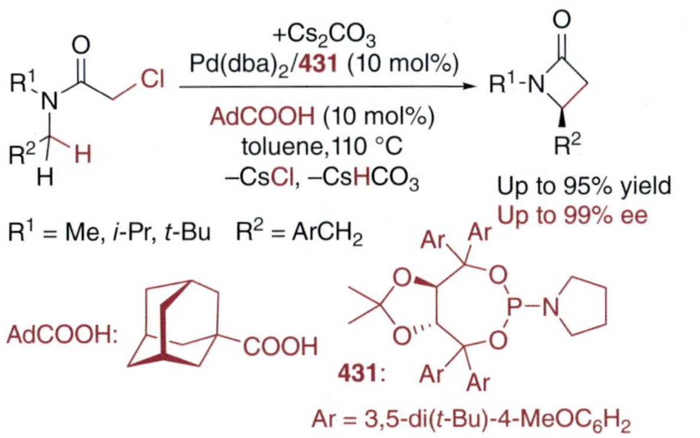

(8.373)

References

1 Smith, M.B. (2013). Aliphatic substitution, nucleophilic and organometallic. In: *March's Advanced Organic Chemistry, Reactions, Mechanisms, and Structure*, 373–568. Hoboken, NJ: Wiley.

2 Smith, M.B. (2013). Aromatic substitution, electrophilic. In: *March's Advanced Organic Chemistry, Reactions, Mechanisms, and Structure*, 569–648. Hoboken, NJ: Wiley.

3 Renaud, P. and Sibi, M.P. (2008). *Radicals in Organic Synthesis*, Wiley Online Library Ed. Weinheim: Wiley-VCH.

4 Rowlands, G.J. (2009). Radicals in organic synthesis. Part 1. *Tetrahedron* 65 (42): 8603–8655.

5 Rowlands, G.J. (2010). Radicals in organic synthesis: Part 2. *Tetrahedron* 66 (9): 1593–1636.

6 Tebben, L. and Studer, A. (2011). Nitroxides: applications in synthesis and in polymer chemistry. *Angew. Chem. Int. Ed.* 50 (22): 5034–5068

7 Quiclet-Sire, B. and Zard, S.Z. (2011). Fun with radicals: some new perspectives for organic synthesis. *Pure Appl. Chem.* 83 (3): 519–551.

8 Lapointe, G., Kapat, A., Weidner, K., and Renaud, P. (2012). Radical azidation reactions and their application in the synthesis of alkaloids. *Pure Appl. Chem.* 84 (7): 1633–1641.

9 (a) Sebren, L.J., Devery, J.J., and Stephenson, C.R. (2014). Catalytic radical domino reactions in organic synthesis. *ACS Catalysis* 4 (2): 703–716. (b) Liautard, V. and Landais, Y. (2015). *Multicomponent Reactions in Organic Synthesis* (ed. J. Zhu, Q. Wang and M.-X. Wang), 401–438. Weinheim: Wiley-VCH.

10 Denes, F., Pichowicz, M., Povie, G., and Renaud, P. (2014). Thiyl radicals in organic synthesis. *Chem. Rev.* 114 (5): 2587–2693.

11 Smith, M.B. (2013). Addition to carbon–carbon multiple bonds. In: *Addition to Carbon–Carbon Multiple Bonds*, 859–902. Hoboken, NJ: Wiley.

12 Trost, B.M. (1991). The atom economy - a search for synthetic efficiency. *Science* 254 (5037): 1471–1477.

13 Smith, M.B. (2013). Carbon attack by active hydrogen compounds. In: *March's Advanced Organic Chemistry*, 1141–1188. Hoboken, NJ: Wiley.

14 Arundale, E. and Mikeska, L.A. (1952). The olefin aldehyde condensation - the Prins reaction. *Chem. Rev.* 51: 505–555.

15 Overman, L.E. and Velthuisen, E.J. (2006). Scope and facial selectivity of the Prins-pinacol synthesis of attached rings. *J. Org. Chem.* 71: 1581–1587.

16 Li, C.J. (2010). The development of catalytic nucleophilic additions of terminal alkynes in water. *Acc. Chem. Res.* 43: 581–590.

17 Hendrickson, J.B. (1975). Systematic synthesis design. 4. Numerical codification of construction reactions. *J. Am. Chem. Soc.* 97: 5784–5800.

18 Newhouse, T., Baran, P.S., and Hoffmann, R.W. (2009). The economies of synthesis. *Chem. Soc. Rev.* 38: 3010–3021.

19 Gaich, T. and Baran, P.S. (2010). Aiming for the ideal synthesis. *J. Org. Chem.* 75: 4657–4673.

20 (a) Li, C.J. (2009). Cross-dehydrogenative coupling (CDC): exploring C—C bond formations beyond functional group transformations. *Acc. Chem. Res.* 42 (2): 335–344. (b) He, C. and Lei, A. (2013). Application of nontoxic Fe salts in oxidative C—C coupling reactions. In: *Comprehensive Inorganic Chemistry II*, vol. 6 (ed. J. Reedjik and K. Poeppelmeier), 521–547. Amsterdam, The Netherlands: Elsevier. B. V.

21 Girard, S.A., Knauber, T., and Li, C.J. (2014). The cross-dehydrogenative coupling of C(sp^3)-H bonds: a versatile strategy for C—C bond formations. *Angew. Chem. Int. Ed.* 53 (1): 74–100.

22 Narayan, R., Matcha, K., and Antonchick, A.P. (2015). Metal-free oxidative C—C bond formation through C—H bond functionalization. *Chem. Eur. J.* 21: 14678–14693.

23 Corbet, J.-P. and Mignani, G. (2006). Selected patented cross-coupling reaction technologies. *Chem. Rev.* 106 (7): 2651–2710.

24 Pakhare, D. and Spivey, J. (2014). A review of dry CO_2 reforming of methane over noble metal catalysts. *Chem. Soc. Rev.* 43: 7813–7837.

25 Geilen, F.M.A., vom Stein, T., Engendahl, B. et al. (2011). Highly selective decarbonylation of 5-(hydroxymethyl)furfural in the presence of compressed carbon dioxide. *Angew. Chem. Int. Ed.* 50: 6831–6834.

26 Meng, Q.W., Qiu, C.W., Ding, G.Q. et al. (2016). Role of alkali earth metals over Pd/Al_2O_3 for decarbonylation of 5-hydroxymethylfurfural. *Catal. Sci. Technol.* 6: 4377–4388.

27 Kusumoto, S., Tatsuki, T., and Nozaki, K. (2015). The retro-hydroformylation reaction. *Angew. Chem. Int. Ed.* 54: 8458–8461.

28 Periana, R.A., Mironov, O., Taube, D. et al. (2003). Catalytic, oxidative condensation of CH_4 to CH_3COOH in one step via C—H activation. *Science* 301 (5634): 814–818.

29 Nachhammer, O., Bode, A., and Hormuth, W. (2016). Financial and ecological evaluation of hydrogen production on large scale. *Chem. Eng. Technol.* 39: 1185–1193.

30 Jin, E.L., Zhang, Y.L., He, L.L. et al. (2014). Indirect coal to liquid technologies. *Appl. Catal., A* 476: 158–174.

31 Wu, X.F., Fang, X.J., Wu, L.P. et al. (2014). Transition-metal-catalyzed carbonylation reactions of olefins and alkynes: a personal account. *Acc. Chem. Res.* 47: 1041–1053.

32 Koch, H. and Haaf, W. (1958). Synthese von Carbonsäuren aus Alkoholen vom Typ des Neopentylalkohols. *Angew. Chem. Ind. Ed.* 70: 311.

33 Koch, H. and Haaf, W. (1958). Über die Synthese verzweigter Carbonsauren nach der Ameisensaure-Methode. *Liebigs Ann. Chem.* 618: 251–266.

34 Li, J.J. (2009). Koch–Haaf carbonylation. In: *Name Reactions*, 319. Berlin, Heidelberg: Springer-Verlag.

35 Fukuyama, T., Mukai, Y., and Ryu, I. (2011). Koch–Haaf reaction of adamantanols in an acid-tolerant hastelloy-made microreactor. *Beilstein J. Org. Chem.* 7: 1288–1293.

36 Zhang, Q.H., Cheng, K., Kang, J.C. et al. (2014). Fischer–Tropsch catalysts for the production of hydrocarbon fuels with high selectivity. *ChemSusChem* 7: 1251–1264.

37 (a) Galadima, A. and Muraza, O. (2015). Hydroisomerization of sustainable feedstock in biomass-to-fuel conversion: a critical review. *Int. J. Energy Res.* 39: 741–759. (b) Galadima, A. and Muraza, O. (2015). Waste to liquid fuels: potency, progress and challenges. *Int. J. Energy Res.* 39: 1451–1478.

38 Choudhury, H.A., Chakma, S., and Moholkar, V.S. (2015). Biomass gasification integrated Fischer–Tropsch synthesis: perspectives, opportunities and challenges. In: *Recent Advances in Thermochemical Conversion of Biomass*, (ed. A. Pandy, T. Bhaskar, M Stöcker, and R. Sukumaran), 383–435. Amsterdam, The Netherlands: Elsevier, B. V.

39 Sabatier, P. and Senderens, J.B. (1902). New methane synthesis. *C.R. Hebd. Séances Acad. Sci.* 134: 514–516.

40 Sabatier, P. and Senderens, J.B. (1902). Direct hydrogenation of carbon oxides in the presence of various split metals. *C.R. Hebd. Séances Acad. Sci.* 134: 689–691.

41 Fischer, F. and Tropsch, H. (1926). The synthesis of petroleum at atmospheric pressures from gasification products of coal. *Brennst.-Chem.* 7: 97–104.

42 Fischer, F. and Tropsch, H. (1926). The direct synthesis of petroleum hydrocarbons with standard pressure. (First report.). *Ber. Dtsch. Chem. Ges.* 59: 830–831.

43 Espinoza, R.L., Steynberg, A.P., Jager, B., and Vosloo, A.C. (1999). Low temperature Fischer–Tropsch synthesis from a Sasol perspective. *Appl. Catal., A* 186 (1–2): 13–26.

44 Pichler, H. (1952). 25 years of synthesis of gasoline by catalytic conversion of carbon monoxide and hydrogen. *Adv. Catal.* 4: 271–341.

45 Dry, M.E. (1990). The Fischer–Tropsch process - commercial aspects. *Catal. Today* 6 (3): 183–206.

46 Jager, B. and Espinoza, R. (1995). Advances in low-temperature Fischer–Tropsch synthesis. *Catal. Today* 23 (1): 17–28.

47 Vannice, M.A. (1975). Catalytic synthesis of hydrocarbons from H_2/CO mixtures over Group-8 metals. 1. Specific activities and product distributions of supported metals. *J. Catal.* 37 (3): 449–461.

48 Bhasin, M.M., Bartley, W.J., Ellgen, P.C., and Wilson, T.P. (1978). Synthesis gas conversion over supported Rh- and Rh/Fe catalysts. *J. Catal.* 54 (2): 120–128.

49 Ichikawa, M. (1978). Catalysis by supported metal crystallites from carbonyl clusters.1. Catalytic methanol synthesis under mild conditions over supported Rh, Pt, and Ir crystallites prepared from Rh, Pt, and Ir carbonyl cluster compounds deposited on ZnO and MgO. *Bull. Chem. Soc. Jpn.* 51 (8): 2268–2272.

50 Ichikawa, M. (1979). Catalysis by supported metal crystallites prepared from metal-carbonyl clusters. 3. Catalytic vapor phase hydroformylation of olefins over Rh, bimetallic Rh-Co, and Co crystallites prepared from the carbonyl cluster compounds dispersed on zinc-oxide. *J. Catal.* 56 (1): 127–129.

51 Watson, P.R. and Somorjai, G.A. (1981). The hydrogenation of carbon monoxide over Rh oxide surfaces. *J. Catal.* 72 (2): 347–363.

52 Watson, P.R. and Somorjai, G.A. (1982). The formation of oxygen-containing organic-molecules by the hydrogenation of carbon monoxide using a lanthanum rhodate catalyst. *J. Catal.* 74 (2): 282–295.

53 Kip, B.J., Hermans, E.G.F., van Wolput, J.H.M.C. et al. (1987). Hydrogenation of carbon monoxide over Rh-silica catalysts promoted with Mo-oxide and Th-oxide. *Appl. Catal.* 35 (1): 109–139.

54 Lavalley, J.C., Saussey, J., Lamotte, J. et al. (1990). Infrared study of carbon monoxide hydrogenation over Rh-ceria and Rh-silica catalysts. *J. Phys. Chem.* 94 (15): 5941–5947.

55 Bowker, M. (1992). On the mechanism of ethanol synthesis on Rh. *Catal. Today* 15 (1): 77–100.

56 Quyoum, R., Berdini, V., Turner, M.L. et al. (1998). Mechanistic studies of methylene chain propagation in the Fischer–Tropsch synthesis. *J. Catal.* 173: 355–365.

57 Prieto, G., Concepcion, P., Martinez, A., and Mendoza, E. (2011). New insights into the role of the electronic properties of oxide promoters in Rh-catalyzed selective synthesis of oxygenates from synthesis gas. *J. Catal.* 280: 274–288.

58 Wang, J.B., Kawazoe, Y., Sun, Q. et al. (2016). The selectivity and activity of catalyst for CO hydrogenation to methanol and hydrocarbon: a comparative study on Cu, Co and Ni surfaces. *Surf. Sci.* 645: 30–40.

59 Xing, C., Yang, G.H., Wu, M.B. et al. (2015). Hierarchical zeolite Y supported co- bifunctional catalyst for facile tuning of the product distribution of Fischer–Tropsch synthesis. *Fuel* 148: 48–57.

60 Peng, X.B., Cheng, K., Kang, J.C. et al. (2015). Impact of hydrogenolysis on the selectivity of the Fischer–Tropsch synthesis: diesel fuel production over mesoporous zeolite-Y-supported Co nanoparticles. *Angew. Chem. Int. Ed.* 54: 4553–4556.

61 Sineva, L.V., Asalieva, E.Y., and Mordkovich, V.Z. (2015). Role of zeolite in the synthesis of liquid hydrocarbons from CO and H_2 on a composite Co-catalyst. *Catal. Ind.* 7: 245–252.

62 Asalieva, E.Y., Kul'chakovskaya, E.V., Sineva, L.V. et al. (2016). Fischer–Tropsch synthesis with Co-catalyst and zeolite multibed arrangement. *Pet. Chem.* 56: 275–280.

63 (a) Deluzarche, A., Hindermann, J.P., Kieffer, R. et al. (1982). Mechanism of Fischer–Tropsch reactions in heterogeneous catalysis. *Bull. Soc. Chim. Fr.* (9–10): 329–336. (b) Maitlis, P.M., Quyoum, R., Long, H.C., and Turner, M.L. (1999). Towards a chemical understanding of the Fischer–Tropsch reaction: alkene formation. *Appl. Catal., A* 186 (1–2): 363–374. (c) Maitlis, P.M. (2004). Fischer–Tropsch, organometallics, and other friends. *J. Organomet. Chem.* 689 (24): 4366–4374.

64 (a) Peña, D., Cognigni, A., Nemayer, T. et al. (2018). Identification of carbon species on iron-based catalysts during Fischer–Tropsch synthesis. *Appl. Catal., A* 554: 10–23. (b) Chen, W., Zijlstra, B., Filot, I.A.W. et al. (2018). Mechanism of carbon monoxide dissociation on a cobalt Fischer–Tropsch catalyst. *ChemCatChem* 10 (1): 136–140.

65 Wentrcek, P.R., Wood, B.J., and Wise, H. (1976). Role of surface carbon in catalytic methanation. *J. Catal.* 43 (1–3): 363–366.

66 Araki, M. and Ponec, V. (1976). Methanation of carbon monoxide on Ni and Ni-Co alloys. *J. Catal.* 44 (3): 439–448.

67 (a) Rabo, J.A., Risch, A.P., and Poutsma, M.L. (1978). Reactions of carbon monoxide and hydrogen on Co, Ni, Ru, and Pd metals. *J. Catal.* 53 (3): 295–311. (b) McNab, A.L., McCue, A.J., Dionisi, D., and Anderson, J.A. (2017). Quantification and qualification by in-situ FTIR of species formed on supported-cobalt catalysts during the Fischer–Tropsch reaction. *J. Catal.* 353: 286–294. (c) McNab, A.L., McCue, A.J., Dionisi, D., and Anderson, J.A. (2017). Combined quantitative FTIR and online GC study

of Fischer–Tropsch catalysts. *J. Catal.* 353: 295–304.

68 Biloen, P., Helle, J.N., and Sachtler, W.M.H. (1979). Incorporation of surface carbon into hydrocarbons during Fischer–Tropsch synthesis - mechanistic implications. *J. Catal.* 58 (1): 95–107.

69 Kaminsky, M.P., Winograd, N., Geoffroy, G.L., and Vannice, M.A. (1986). Direct Sims observation of methylidyne, methylene, and methyl intermediates on a Ni(111) methanation catalyst. *J. Am. Chem. Soc.* 108 (6): 1315–1316.

70 Shriver, D.F. and Sailor, M.J. (1988). Transformations of carbon monoxide and related ligands on metal ensembles. *Acc. Chem. Res.* 21 (10): 374–379.

71 Bradley, J.S. (1983). The chemistry of carbidocarbonyl clusters. *Adv. Organomet. Chem.* 22: 1–58.

72 Chisholm, M.H., Hammond, C.E., Johnston, V.J. et al. (1992). Metal alkoxides - models for metal oxides. 17. Reductive cleavage of carbon monoxide to carbide and oxide by ditungsten and tetratungsten alkoxides - crystal and molecular-structures of $W_4(\mu\text{-}_4\text{-C})(OCH_2\text{-C-Pen})_{14}$, $W_4(\mu\text{-}_4\text{-C})(O)(OCH_2\text{-}t\text{-Bu})_{12}$, and $W_4(\mu\text{-}_4\text{-C})(O)(O\text{-}i\text{-Pr})_{12}$. *J. Am. Chem. Soc.* 114 (18): 7056–7065.

73 Biloen, P. (1980). Mechanism of the Fischer–Tropsch synthesis. *Recl. Trav. Chim. Pays-Bas* 99 (2): 33–38.

74 Biloen, P. and Sachtler, W.M.H. (1981). Mechanism of hydrocarbon synthesis over Fischer–Tropsch catalysts. *Adv. Catal.* 30: 165–216.

75 Brady, R.C. and Pettit, R. (1980). Reactions of diazomethane on transition-metal surfaces and their relationship to the mechanism of the Fischer–Tropsch reaction. *J. Am. Chem. Soc.* 102 (19): 6181–6182.

76 Brady, R.C. and Pettit, R. (1981). On the mechanism of the Fischer–Tropsch reaction - the chain propagation step. *J. Am. Chem. Soc.* 103 (5): 1287–1289.

77 Maitlis, P.M. and Zanotti, V. (2009). The role of electrophilic species in the Fischer–Tropsch reaction. *J. Chem. Soc., Chem. Commun.* 1619–1634.

78 Turner, M.L., Long, H.C., Shenton, A. et al. (1995). The alkenyl mechanism for Fischer–Tropsch surface methylene polymerization - the reactions of vinylic probes with CO/H_2 over Rh-catalysts. *Chem. Eur. J.* 1 (8): 549–556.

79 Ma, F., Sunley, G.J., Saez, I.M., and Maitlis, P.M. (1990). Evidence for vinylic intermediates in the Fischer–Tropsch reaction to give alkenes and alkanes. *J. Chem. Soc., Chem. Commun.* (18): 1279–1280.

80 Turner, M.L., Byers, P.K., Long, H.C., and Maitlis, P.M. (1993). Vinyl initiation of Fischer–Tropsch polymerization over Rh. *J. Am. Chem. Soc.* 115 (10): 4417–4418.

81 Inderwildi, O.R., Jenkins, S.J., and King, D.A. (2008). Fischer–Tropsch mechanism revisited: alternative pathways for the production of higher hydrocarbons from synthesis gas. *J. Phys. Chem. C* 112 (5): 1305–1307.

82 (a) Davies, D.L., Parrott, M.J., Sherwood, P., and Stone, F.G.A. (1987). Chemistry of polynuclear metal-complexes with bridging carbene or carbyne ligands. 55. Coupling of methylene and alkylidyne groups at a trimetal center - crystal structures of $[Ru_2W(\mu\text{-}\omega\text{-}\eta\text{-}trans\text{-CH=C(H)}C_6H_4Me\text{-}_4)(\mu\text{-CO})(CO)_3(\eta\text{-}C_5H_5)_2(\eta\text{-}C_5Me_5)]\cdot CH_2Cl_2$ and $[Ru_2W(\mu_3\text{-}CCH_2C_6H_4Me\text{-}_4)(\mu\text{-CO})_3(CO)(\eta\text{-}C_5H_5)_3]$. *J. Chem. Soc., Dalton Trans.* (5): 1201–1208. (b) see also: Saez, I.M., Meanwell, N.J., Nutton, A. et al. (1986). Decomposition reactions of dimethyldi-μ-methylenebis(η-pentamethylcyclopentadienyl) dirhodium and their relation to the mechanism of the Fischer–Tropsch reactions; the formation of propylene from three C_1 ligands. *J. Chem. Soc., Dalton Trans.* (8): 1565–1575.

83 Zhang, Y.L., Hou, L., Tierney, J.W., and Wender, I. (2005). Acetylenes as probes in the Fischer–Tropsch reaction. *Top. Catal.* 32 (3–4): 125–133.

84 Haynes, A. (2006). Acetic acid synthesis by catalytic carbonylation of methanol. *Top. Organomet. Chem.* 18: 179–205.

85 Thomas, C.M. and Süss-Fink, G. (2003). Ligand effects in the Rh-catalyzed carbonylation of methanol. *Coord. Chem. Rev.* 243 (1–2): 125–142.

86 Haynes, A. (2010). Catalytic methanol carbonylation. *Adv. Catal.* 53: 1–45.

87 Hohenschutz, H., Von Kutepow, N., and Himmele, W. (1966). Acetic acid process. *Hydrocarbon Process.* 45 (11): 141–144.

88 Von Kutepow, N., Himmele, W., and Hohenschutz, H. (1965). Die Synthese von Essigsäure aus Methanol und Kohlenoxyd. *Chem. Ing. Tech.* 37 (4): 383–388.

89 Mizoroki, T. and Nakayama, M. (1965). Absorption spectrum of Co(II) ion catalyst in reaction of methanol with carbon monoxide and hydrogen under a high pressure. *Bull. Chem. Soc. Jpn.* 38 (11): 1876–1880.

90 Maitlis, P.M., Haynes, A., Sunley, G.J., and Howard, M.J. (1996). Methanol carbonylation revisited: thirty years on. *J. Chem. Soc., Dalton Trans.* (11): 2187–2196.

91 Paulik, F.E. and Roth, J.F. (1968). Novel catalysts for low-pressure carbonylation of methanol to acetic acid. *J. Chem. Soc., Chem. Commun.* (24): 1578.

92 Forster, D. (1976). Mechanism of a Rh-complex-catalyzed carbonylation of methanol to acetic acid. *J. Am. Chem. Soc.* 98 (3): 846–848.

93 Forster, A. (1977). Mechanism of Rh-complex-catalyzed carbonylation of methanol to acetic acid. *Ann. N.Y. Acad. Sci.* 295: 79–82.

94 Haynes, A., Mann, B.E., Gulliver, D.J. et al. (1991). Direct observation of $MeRh(CO)_2I_3^-$ - the key intermediate in Rh-catalyzed methanol carbonylation. *J. Am. Chem. Soc.* 113 (22): 8567–8569.

95 Haynes, A., Mann, B.E., Morris, G.E., and Maitlis, P.M. (1993). Mechanistic studies on Rh-catalyzed carbonylation reactions - spectroscopic detection and reactivity of a key intermediate, $[MeRh(CO)_2I_3^-]$. *J. Am. Chem. Soc.* 115 (10): 4093–4100.

96 Kinnunen, T. and Laasonen, K. (2001). The active catalytic species and its isomerisation in the catalytic carbonylation of methanol - a density functional study. *J. Mol. Struct. THEOCHEM* 540: 91–100.

97 Drago, R.S., Nyberg, E.D., and El A'mma, A.G. (1981). Investigation of unique metal-complexes produced with polystyrene-2,2′-bipyridine and several hydrogenation catalysts derived from this Co-polymer. *Inorg. Chem.* 20: 2461–2466.

98 Drago, R.S., Nyberg, E.D., El A'mma, A.G., and Zombeck, A. (1981). Ionic attachment as a feasible approach to heterogenizing anionic solution catalysts - carbonylation of methanol. *Inorg. Chem.* 20 (3): 641–644.

99 Yoneda, N., Kusano, S., Yasui, M. et al. (2001). Recent advances in processes and catalysts for the production of acetic acid. *Appl. Catal., A* 221 (1–2): 253–265.

100 Haynes, A., Maitlis, P.M., Morris, G.E. et al. (2004). Promotion of Ir-catalyzed methanol carbonylation: mechanistic studies of the Cativa process. *J. Am. Chem. Soc.* 126 (9): 2847–2861.

101 Forster, D. (1979). Kinetic and spectroscopic studies of the carbonylation of methanol with an iodide-promoted Ir-catalyst. *J. Chem. Soc., Dalton Trans.* (11): 1639–1645.

102 Tonde, S.S., Kelkar, A.A., Bhadbhade, M.M., and Chaudhari, R.V. (2005). Isolation and characterization of an iodide bridged dimeric Pd-complex in carbonylation of methanol. *J. Organomet. Chem.* 690 (6): 1677–1681.

103 Blasco, T., Boronat, M., Concepción, P. et al. (2007). Carbonylation of methanol on metal-acid zeolites: evidence for a mechanism involving a multisite active center. *Angew. Chem. Int. Ed.* 46 (21): 3938–3941.

104 Zoeller, J.R., Agreda, V.H., Cook, S.L. et al. (1992). Eastman-chemical-company acetic anhydride process. *Catal. Today* 13: 73–91.

105 King, S.T. (1997). Oxidative carbonylation of methanol to dimethyl carbonate by solid-state ion-exchanged CuY catalysts. *Catal. Today* 33: 173–182.

106 Zhang, Y. and Bell, A.T. (2008). The mechanism of dimethyl carbonate synthesis on Cu-exchanged zeolite Y. *J. Catal.* 255 (2): 153–161.

107 Nam, J.K., Choi, M.J., Cho, D.H. et al. (2013). The influence of support in the synthesis of dimethyl carbonate by Cu-based catalysts. *J. Mol. Catal. A: Chem.* 370: 7–13.

108 Ding, X.S., Dong, X.M., Kuang, D.T. et al. (2014). Highly efficient catalyst $PdCl_2$-$CuCl_2$-KOAc/AC@Al_2O_3 for gas-phase oxidative carbonylation of methanol to dimethyl carbonate: preparation and reaction mechanism. *Chem. Eng. J.* 240: 221–227.

109 Yacob, S., Kilos, B.A., Barton, D.G., and Notestein, J.M. (2016). Vapor phase ethanol carbonylation over Rh supported on zeolite 13X. *Appl. Catal., A* 520: 122–131.

110 Serp, P., Chateau, L., Feurer, R. et al. (1998). Rhodium-catalyzed hydrocarbonylation of acetic acid into higher acids. *J. Mol. Catal. A: Chem.* 136: 269–278.

111 Martin, F.; Roelen, O.; Feisst, W. 1938. Hydrocarbons from hydrogenation of a carbon oxide. US Patent 2127127.

112 Roelen, O. (1948). Eine neue Synthese von Aldehyden und deren Derivaten, ausgehend von Olefinen, Kohlenoxyd und Wasserstoff. *Angew. Chem.* 60 (3): 62–62.

113 Cornils, B., Hermann, W.A., and Rasch, M. (1994). Otto Roelen, Pioneer in industrial homogeneous catalysis. *Angew. Chem. Int. Ed. Engl.* 33: 2144–2163.

114 Wiese, K.-D. and Obst, D. (2008). Hydroformylation. *Top. Organomet. Chem.* 18: 1–33.

115 Breit, B. (2007). Recent advances in alkene hydroformylation. *Top. Curr. Chem.* 279: 139–172.

116 Breit, B. (2007). Synthesis by hydroformylation of alkenes. In: *Science of Synthesis* (ed. R. Brückner), 277–317. Stuttgart: Thieme Verlag.

117 Tudor, R. and Ashley, M. (2007). Enhancement of industrial hydroformylation processes by the adoption of Rh-based catalyst. Part I. Development of the LP Oxo process to the commercial stage. *Platinum Met. Rev.* 51 (3): 116–126.

118 Godard, C., Perandones, B.F., Gual, A., and Claver, C. (2013). Asymmetric carbonylation. In: *Comprehensive Inorganic Chemistry II* (ed. J.P.K. Reedijk), 383–411. Amsterdam, The Netherlands: Elsevier.

119 Abrams, M.L., Foarta, F., and Landis, C.R. (2014). Asymmetric hydroformylation of (*Z*)-enamides and enol esters with Rh-bisdiazaphos catalysts. *J. Am. Chem. Soc.* 136: 14583–14588.

120 Risi, R.M., Maza, A.M., and Burke, S.D. (2015). Asymmetric hydroformylation-initiated tandem sequences for syntheses of (+)-patulolide C, (−)-pyrenophorol, (+)-decarestrictine L, and (+)-Prelog–Djerassi lactone. *J. Org. Chem.* 80: 204–216.

121 Garcia-Simon, C., Gramage-Doria, R., Raoufmoghaddam, S. et al. (2015). Enantioselective hydroformylation by a Rh-catalyst entrapped in a supramolecular metallocage. *J. Am. Chem. Soc.* 137: 2680–2687.

122 Vidal-Ferran, A., Mon, I., Bauza, A. et al. (2015). Supramolecularly regulated ligands for asymmetric hydroformylations and hydrogenations. *Chem. Eur. J.* 21: 11417–11426.

123 Yu, Z.Y., Eno, M.S., Annis, A.H., and Morken, J.P. (2015). Enantioselective hydroformylation of 1-alkenes with commercial Ph-BPE ligand. *Org. Lett.* 17: 3264–3267.

124 Deng, Y.C., Wang, H., Sun, Y.H., and Wang, X. (2015). Principles and applications of enantioselective hydroformylation of terminal disubstituted alkenes. *ACS Catal.* 5: 6828–6837.

125 Pongracz, P., Kollar, L., and Mika, L.T. (2016). A step towards hydroformylation under sustainable conditions: platinum-catalysed enantioselective hydroformylation of styrene in γ-valerolactone. *Green Chem.* 18: 842–847.

126 You, C., Wei, B., Li, X.X. et al. (2016). Rhodium-catalyzed desymmetrization by hydroformylation of cyclopentenes: synthesis of chiral carbocyclic nucleosides. *Angew. Chem. Int. Ed.* 55: 6511–6514.

127 Pongracz, P., Kollar, L., and Mika, L.T. (2016). Enantioselective hydroformylation of 2-and 4-substituted styrenes with $PtCl_2[(R)\text{-BINAP}] + SnCl_2$ 'in situ' catalyst. *J. Organomet. Chem.* 824: 118–123.

128 Orchin, M., Kirch, L., and Goldfarb, I. (1956). Evidence for the presence of co-hydrocarbonyl under conditions of the Oxo reaction. *J. Am. Chem. Soc.* 78 (20): 5450–5451.

129 Kirch, L. and Orchin, M. (1958). The intermediate Co-hydrocarbonylolefin complex in the Oxo reaction. *J. Am. Chem. Soc.* 80 (16): 4428–4429.

130 Breslow, D.S. and Heck, R.F. (1960). Mechanism of the hydroformylation of olefins. *Chem. Ind.* (17): 467–467.

131 Heck, R.F. and Breslow, D.S. (1961). Reaction of Co-hydrotetracarbonyl with olefins. *J. Am. Chem. Soc.* 83 (19): 4023–4027.

132 Tannenbaum, R. and Bor, G. (2004). Isotope effects in the hydroformylation of olefins with Co-carbonyls as catalysts. *J. Phys. Chem. A* 108 (34): 7105–7111.

133 Marko, L. (1974). Hydroformylation of olefins with carbonyl derivatives of the noble metals as catalysts. In: *Aspects of Homogeneous Catalysis*, Chapter 1, vol. 2 (ed. R. Ugo), 3–55. Heidelberg: Springer-Verlag.

134 García, L., Claver, C., Diéguez, M., and Masdeu-Bultó, A.M. (2006). A highly selective synthesis of 3-hydroxy-2-methylpropionamide involving a one-pot tandem hydroformylation-hydrogenation sequence. *J. Chem. Soc., Chem. Commun.* (2): 191–193.

135 Sparta, M., Borve, K.J., and Jensen, V.R. (2007). Activity of Rh-catalyzed hydroformylation: added insight and predictions from theory. *J. Am. Chem. Soc.* 129 (27): 8487–8499.

136 Breit, B. and Fuchs, E. (2004). Phosphabarrelene-rhodium complexes as highly active catalysts for isomerization free hydroformylation of internal alkenes. *J. Chem. Soc., Chem. Commun.* (6): 694–695.

137 Piechaczyk, O., Doux, M., Ricard, L., and Le Floch, P. (2005). Synthesis of 1-phosphabarrelene phosphine sulfide substituted Pd(II) complexes: application in the catalyzed Suzuki cross-coupling process and in the allylation of secondary amines. *Organometallics* 24 (6): 1204–1213.

138 Fuchs, E., Keller, M., and Breit, B. (2006). Phosphabarrelenes as ligands in Rh-catalyzed hydroformylation of internal alkenes essentially free of alkene isomerization. *Chem. Eur. J.* 12 (26): 6930–6939.

139 Breit, B. and Seiche, W. (2006). Self-assembly of bidentate ligands for combinatorial homogeneous catalysis based on an A–T base pair model. *Pure Appl. Chem.* 78 (2): 249–256.

140 Breit, B. and Seiche, W. (2003). Hydrogen bonding as a construction element for bidentate donor ligands in homogeneous catalysis: regioselective hydroformylation of terminal alkenes. *J. Am. Chem. Soc.* 125 (22): 6608–6609.

141 Breit, B. and Seiche, W. (2005). Self-assembly of bidentate ligands for combinatorial homogeneous catalysis based on an A–T base-pair model. *Angew. Chem. Int. Ed.* 44 (11): 1640–1643.

142 Straub, A.T., Otto, M., Usui, I., and Breit, B. (2013). Room temperature ambient pressure (RTAP)-hydroformylation in water using a self-assembling ligand. *Adv. Synth. Catal.* 355: 2071–2075.

143 Gellrich, U., Seiche, W., Keller, M., and Breit, B. (2012). Mechanistic insights into a supramolecular self-assembling catalyst system: evidence for hydrogen bonding during Rh-catalyzed hydroformylation. *Angew. Chem. Int. Ed.* 51: 11033–11038.

144 Fuchs, D., Rousseau, G., Diab, L. et al. (2012). Tandem Rh-catalyzed hydroformylation-hydrogenation of alkenes by employing a cooperative ligand system. *Angew. Chem. Int. Ed.* 51: 2178–2182.

145 Waloch, C., Wieland, J., Keller, M., and Breit, B. (2007). Self-assembly of bidentate ligands for combinatorial homogeneous catalysis: methanol stable platforms analogous to the adenine-thymine base pair. *Angew. Chem. Int. Ed.* 46 (17): 3037–3039.

146 Smejkal, T. and Breit, B. (2008). A supramolecular catalyst for regioselective hydroformylation of unsaturated carboxylic acids. *Angew. Chem. Int. Ed.* 47 (2): 311–315.

147 Diebolt, O., Müller, C., and Vogt, D. (2012). "On-water" Rh-catalyzed hydroformylation for the production of linear alcohols. *Catal. Sci. Technol.* 2: 773–777.

148 Takahashi, K., Yamashita, M., Ichihara, T. et al. (2010). High-yielding tandem hydroformylation/hydrogenation of a terminal olefin to produce a linear alcohol using a Rh/Ru dual catalyst system. *Angew. Chem. Int. Ed.* 49: 4488–4490.

149 Takahashi, K., Yamashita, M., and Nozaki, K. (2012). Tandem hydroformylation/hydrogenation of alkenes to normal alcohols using Rh/Ru dual catalyst or Ru single component catalyst. *J. Am. Chem. Soc.* 134: 18746–18757.

150 Yuki, Y., Takahashi, K., Tanaka, Y., and Nozaki, K. (2013). Tandem isomerization/hydroformylation/hydrogenation of internal alkenes to *n*-alcohols using Rh/Ru dual- or ternary-catalyst systems. *J. Am. Chem. Soc.* 135: 17393–17400.

151 Kohlpaintner, C.W., Fischer, R.W., and Cornils, B. (2001). Aqueous biphasic catalysis: Ruhrchemie/Rhone-Poulenc Oxo process. *Appl. Catal., A* 221 (1–2): 219–225.

152 Tanaka, R., Nakano, K., and Nozaki, K. (2007). Synthesis of α-heteroarylpropanoic acid via asymmetric hydroformylation catalyzed by Rh(I)-(*R*,*S*)-BINAPHOS and the subsequent oxidation. *J. Org. Chem.* 72: 8671–8676.

153 Klosin, J. and Landis, C.R. (2007). Ligands for practical Rh-catalyzed asymmetric hydroformylation. *Acc. Chem. Res.* 40 (12): 1251–1259.

154 Perandones, B.F., Godard, C., and Claver, C. (2013). Asymmetric hydroformylation. In: *Hydroformylation for Organic Synthesis*, Topics in Current Chemistry (ed. M. Taddei and A. Mann), 79–115. Berlin, Heidelberg: Springer-Verlag.

155 Allmendinger, S., Kinuta, H., and Breit, B. (2015). Easily accessible TADDOL-derived bisphosphonite ligands: synthesis and application in the asymmetric hydroformylation of vinylarenes. *Adv. Synth. Catal.* 357: 41–45.

156 Li, C., Chen, L., and Garland, M. (2007). Synchronicity of mononuclear and dinuclear events in homogeneous catalysis. Hydroformylation of cyclopentene using $Rh_4(CO)_{12}$ and $HRe(CO)_5$ as precursors. *J. Am. Chem. Soc.* 129 (43): 13327–13334.

157 Botteghi, C. and Salomon, C. (1974). Conclusive evidence for direct hydroformylation of triple bond catalyzed by chiral Rh-complexes. *Tetrahedron Lett.* 15 (49–50): 4285–4288.

158 Göttmann, F., Le Floch, P., and Sanchez, C. (2006). Hybrid bidentate ligand for functional recognition: an application to regioselective C=C double bond hydrogenation. *J. Chem. Soc., Chem. Commun.* (19): 2036–2038.

159 Agabekov, V., Seiche, W., and Breit, B. (2013). Rhodium-catalyzed hydroformylation of alkynes employing a self-assembling ligand system. *Chem. Sci.* 4: 2418–2422.

160 Wu, L.P., Liu, Q., Jackstell, R., and Beller, M. (2014). Carbonylations of alkenes with CO surrogates. *Angew. Chem. Int. Ed.* 53: 6310–6320.

161 Verendel, J.J., Nordlund, M., and Andersson, P.G. (2013). Selective metal-catalyzed transfer of H_2 and CO from polyols to alkenes. *ChemSusChem* 6: 426–429.

162 Dong, Y. and Busacca, C.A. (1997). Indoles from *o*-haloanilines: syntheses of tryptamines and tryptophols via regioselective hydroformylation of functionalized anilines. *J. Org. Chem.* 62 (19): 6464–6465.

163 Schmidt, A.M. and Eilbracht, P. (2005). Synthesis of pharmacologically relevant indoles with amine side chains via tandem hydroformylation/Fischer indole synthesis. *J. Org. Chem.* 70 (14): 5528–5535.

164 Murai, S. and Sonoda, N. (1979). Catalytic reactions with hydrosilane and carbon monoxide. *Angew. Chem. Int. Ed. Engl.* 18 (11): 837–846.

165 Wright, M.E. and Cochran, B.B. (1996). Silylformylation of carbonyl compounds: a study of substrate, catalyst, and reaction conditions. *Organometallics* 15 (1): 317–324.

166 Chatani, N., Kajikawa, Y., Nishimura, H., and Murai, S. (1991). $Co_2(CO)_8$-catalyzed reaction of cyclic orthoesters with a hydrosilane and carbon monoxide - novel method for homologation of 1,2-diols, 1,3-diols, and 1,4-diols. *Organometallics* 10 (1): 21–23.

167 (a) Colleuille, Y. (1966). ω-aminonitriles. French Patent FR 1458642. (b) (1966). Colleuille, ω-aminonitriles Y. *Chem. Abstr* 65: 8959d.

168 Chalk, A.J. and Harrod, J.F. (1967). Homogeneous catalysis. 4. Some reactions of silicon hydrides in presence of co-carbonyls. *J. Am. Chem. Soc.* 89 (7): 1640–1647.

169 Seki, Y., Kawamoto, K., Chatani, N. et al. (1991). Reaction of terminal olefins with a hydrosilane and with carbon monoxide catalyzed by complexes of Co, Ru, and Rh. *J. Organomet. Chem.* 403 (1–2): 73–84.

170 Kownacki, I., Marciniec, B., Szubert, K., and Kubicki, M. (2005). Silylcarbonylation of vinylsilanes catalyzed by Ir(I) siloxide complexes. *Organometallics* 24 (25): 6179–6183.

171 Leighton, J.L. and Chapman, E. (1997). Rhodium-catalyzed intramolecular silylformylation of alkenes. *J. Am. Chem. Soc.* 119 (50): 12416–12417.

172 Zacuto, M.J., O'Malley, S.J., and Leighton, J.L. (2003). Tandem silylformylation-allyl(crotyl)silylation: a new approach to polyketide synthesis. *Tetrahedron* 59 (45): 8889–8900.

173 Hosomi, A. and Sakurai, H. (1976). Syntheses of γ,δ-unsaturated alcohols from allylsilanes and carbonyl-compounds in presence of $TiCl_4$. *Tetrahedron Lett.* 17 (16): 1295–1298.

174 Hosomi, A. (1988). Characteristics in the reactions of allylsilanes and their applications to versatile synthetic equivalents. *Acc. Chem. Res.* 21: 200–206.

175 (a) Biamonte, M.A. (2009). Sakurai allylation. In: *Name Reactions for Homologation, Part I* (ed. J.J. Li), 539–575. New York: Wiley. (b) Lade, J.J., Pardeshi, S.D., Vadagaonkar, K.S. et al. (2017). The remarkable journey of catalysts from stoichiometric to catalytic quantity for allyltrimethylsilane inspired allylation of acetals, ketals, aldehydes and ketones. *RSC Adv.* 7 (13): 8011–8033.

176 Matsuda, I., Ogiso, A., Sato, S., and Izumi, Y. (1989). An efficient silylformylation of alkynes catalyzed by $Rh_4(CO)_{12}$. *J. Am. Chem. Soc.* 111 (6): 2332–2333.

177 Ojima, I., Ingallina, P., Donovan, R.J., and Clos, N. (1991). Silylformylation of 1-hexyne catalyzed by Rh/Co mixed-metal carbonyl clusters. *Organometallics* 10 (1): 38–41.

178 Yoshikai, N., Yamanaka, M., Ojima, I. et al. (2006). Bimetallic synergism in alkyne silylformylation catalyzed by a Co/Rh mixed-metal cluster. *Organometallics* 25 (16): 3867–3875.

179 Ojima, I., Donovan, R.J., Eguchi, M. et al. (1993). Silylformylation catalyzed by Rh and Rh/Co mixed metal complexes and its application to the synthesis of pyrrolizidine alkaloids. *Tetrahedron* 49 (25): 5431–5444.

180 Okazaki, H., Kawanami, Y., and Yamamoto, K. (2001). The silylformylation of simple 1-alkynes catalyzed by $[Rh(cod)][BPh_4]$ in an ionic liquid, $[Bmim][PF_6]$, under biphasic conditions: an efficiently reusable catalyst system. *Chem. Lett.* (7): 650–651.

181 Foley, C.N. and Leighton, J.L. (2014). Beyond the Roche Ester: a new approach to polypropionate stereotriad synthesis. *Org. Lett.* 16: 1180–1183.

182 Fleming, I., Henning, R., and Plaut, H. (1984). The phenyldimethylsilyl group as a masked form of the hydroxy group. *J. Chem. Soc., Chem. Commun.* 29–31.

183 Fleming, I., Henning, R., Parker, D.C. et al. (1995). The phenyldimethylsilyl group as a masked hydroxy group. *J. Chem. Soc., Perkin Trans. 1* 317–337.

184 Tamao, K., Kumada, M., and Maeda, K. (1984). Silafunctional compounds in organic synthesis. 21. Hydrogen peroxide oxidation of

alkenyl(alkoxy)silanes. *Tetrahedron Lett.* 25: 321–324.

185 Jones, G.R. and Landais, Y. (1996). The oxidation of the carbon-silicon bond. *Tetrahedron* 52: 7599–7662.

186 Harrison, T.J., Rabbat, P.M.A., and Leighton, J.L. (2012). An "aprotic" Tamao oxidation/*syn*-selective tautomerization reaction for the efficient synthesis of the C(1)–C(9) fragment of fludelone. *Org. Lett.* 14: 4890–4893.

187 Foley, C.N. and Leighton, J.L. (2015). A highly stereoselective, efficient, and scalable synthesis of the C(1)–C(9) fragment of the epothilones. *Org. Lett.* 17: 5858–5861.

188 Ojima, I., Tzamarioudaki, M., and Tsai, C.Y. (1994). Extremely chemoselective silylformylation and silylcarbocyclization of alkynals. *J. Am. Chem. Soc.* 116 (8): 3643–3644.

189 (a) Möller, K.E. (1964). The Koch synthesis of branched carboxylic acids. I. Reaction at high pressures. *Brennst.-Chem.* 45 (5): 129–132. (b) Haaf, W. (1964). The Koch synthesis of branched carboxylic acids. II. Reaction at normal pressure (formic acid method). *Brennst.-Chem.* 45 (7): 209–212. (c) Mayor, Y. (1966). Koch synthesis. Actual and possible industrial applications. *Ind. Chim. (Paris)* 53 (589): 213–217. (d) Hasselberg, J. and Behr, A. (2016). Saturated branched fatty compounds: proven industrial processes and new alternatives. *Eur. J. Lipid Sci. Technol.* 118 (1): 36–46.

190 Kiss, G. (2001). Palladium-catalyzed Reppe carbonylation. *Chem. Rev.* 101 (11): 3435–3456.

191 Brennführer, A., Neumann, H., and Beller, M. (2009). Palladium-catalyzed carbonylation reactions of alkenes and alkynes. *ChemCatChem* 1: 28–41.

192 Gehrtz, P.H., Hirschbeck, V., Ciszek, B., and Fleischer, I. (2016). Carbonylations of alkenes in the total synthesis of natural compounds. *Synthesis* 48: 1573–1596.

193 Reppe, W., Magin, A., Schuster, C. et al. (1953). Carbonylierung. 1. Über die Umsetzung von Acetylen mit Kohlenoxyd und Verbindungen mit Reaktionsfahigen Wasserstoffatomen Synthesen α,β-ungesättigter Carbonsäuren und Ihrer Derivate. *Liebigs Ann. Chem.* 582 (1–2): 1–37.

194 Reppe, W., Kröper, H., Pistor, H.J., and Schlenk, H. (1953). Carbonylierung. 2. Carbonsäuren und Ihre Derivate aus Olefinischen Verbindungen und Kohlenoxyd. *Liebigs Ann. Chem.* 582 (1–2): 38–71.

195 Reppe, W., Kröper, H., von Kutepow, N. et al. (1953). Carbonylierung. 3. Umsetzung von Alkoholen und Offenen Äthern mit Kohlenoxyd zu Carbonsauren. *Liebigs Ann. Chem.* 582 (1–2): 72–86.

196 Reppe, W., Kröper, H., Pistor, J., and Weissbarth, O. (1953). Carbonylierung. 4. Einwirkung von Kohlenoxyd und Wasser auf cyclische Äther. *Liebigs Ann. Chem.* 582 (1–2): 87–116.

197 Reppe, W., Schuster, C., Keller, H. et al. (1953). Carbonylierung. 5. Zur Kenntnis der Metallcarbonyle und Metallcarbonylwasserstoffe. *Liebigs Ann. Chem.* 582 (1–2): 116–132.

198 Reppe, W. and Vetter, H. (1953). Carbonylierung. 6. Synthesen mit Metallcarbonylwasserstoffen. *Liebigs Ann. Chem.* 582 (1–2): 133–161.

199 De Angelis, F., Sgamellotti, A., and Re, N. (2000). Density functional study of the Reppe carbonylation of acetylene. *Organometallics* 19 (20): 4104–4116.

200 Hosaka, S. and Tsuji, J. (1971). Organic syntheses by means of noble metal compounds. 46. Catalytic carbonylation of conjugated dienes. *Tetrahedron* 27 (16): 3821–3829.

201 Drent, E., Arnoldy, P., and Budzelaar, P.H.M. (1994). Homogeneous catalysis by cationic Pd-complexes - precision catalysis in the carbonylation of alkynes. *J. Organomet. Chem.* 475 (1–2): 57–63.

202 Tsuji, J., Morikawa, M., and Kiji, J. (1963). Organic syntheses by means of noble metal compounds. 2. Syntheses of saturated carboxylic esters from olefins. *Tetrahedron Lett.* (22): 1437–1440.

203 Tsuji, J., Hosaka, S., Kiji, J., and Susuki, T. (1966). Organic syntheses by means of noble metal compounds. 14. Carbonylation of cyclooctadienes. *Bull. Chem. Soc. Jpn.* 39 (1): 141–145.

204 Nogi, T. and Tsuji, J. (1969). Organic syntheses by means of noble metal compounds. 39. Palladium-catalysed carbonylation of propargyl compounds. *Tetrahedron* 25 (17): 4099–4108.

205 Roe, D.C., Sheridan, R.E., and Buñel, E.E. (1994). The characterization of $HRh(CO)_2I_3^-$ and $EtCORh(CO)_2I_3^-$. Key intermediates in Rh-catalyzed carbonylation of ethylene. *J. Am. Chem. Soc.* 116 (3): 1163–1164.

206 Tsuji, J., Morikawa, M., and Kiji, J. (1963). Reaction of (olefin)Pd chloride complex with carbon monoxide. *Tetrahedron Lett.* 4 (16): 1061–1064.

207 Tsuji, J., Morikawa, M., and Kiji, J. (1964). Organic syntheses by means of noble metal

compounds. 7. Reactions of (olefin)Pd chloride complexes with carbon monoxide. *J. Am. Chem. Soc.* 86 (22): 4851–4853.

208 Fenton, D.M. and Steinwand, P.J. (1972). Noble-metal catalysis. 1. Synthesis of succinates from olefins. *J. Org. Chem.* 37 (12): 2034–2035.

209 Heck, R.F. (1972). Dicarboalkoxylation of olefins and acetylenes. *J. Am. Chem. Soc.* 94 (8): 2712–2716.

210 Mahaim, C., Carrupt, P.A., Hagenbuch, J.P. et al. (1980). An efficient synthesis of 2,3,5,6-tetramethylidene-7-oxanorbornane. *Helv. Chim. Acta* 63 (5): 1149–1157.

211 Cheng, J.S., Qi, X.X., Li, M. et al. (2015). Palladium-catalyzed intermolecular aminocarbonylation of alkenes: efficient access of β-amino acid derivatives. *J. Am. Chem. Soc.* 137: 2480–2483.

212 Cao, Q., Hughes, N.L., and Muldoon, M.J. (2016). Synthesis of 2-alkynoates by Pd(II)-catalyzed oxidative carbonylation of terminal alkynes and alcohols. *Chem. Eur. J.* 22: 11982–11985.

213 Tsuji, J., Imamura, S., Morikawa, M., and Kiji, J. (1964). Organic syntheses by means of noble metal compounds. VI. Synthesis of muconic acid. *J. Am. Chem. Soc.* 86 (20): 2095–2095.

214 Tsuji, J. and Nogi, T. (1966). Organic syntheses by means of noble metal compounds. 22. Palladium-catalyzed carbonylation of diphenylacetylene. *J. Am. Chem. Soc.* 88 (6): 1289–1292.

215 Tsuji, J. and Nogi, T. (1966). Organic syntheses by means of noble metal compounds. 25. Palladium-catalyzed carbonylation of ethyl acetylenemono- and -dicarboxylates. *J. Org. Chem.* 31 (8): 2641–2643.

216 Mori, K., Mizoroki, T., and Ozaki, A. (1975). Carbonylation of acetylenic compounds catalyzed by Pd-black. *Chem. Lett.* (1): 39–42.

217 Murray, T.F. and Norton, J.R. (1979). Design and mechanism of Pd catalysts for synthesis of methylene lactones by cyclocarbonylation of acetylenic alcohols. *J. Am. Chem. Soc.* 101 (15): 4107–4119.

218 Tsuji, J., Takahashi, M., and Takahashi, T. (1980). Facile synthesis of acetylenecarboxylates by the oxidative carbonylation of terminal acetylenes catalyzed by $PdCl_2$ under mild conditions. *Tetrahedron Lett.* 21 (9): 849–850.

219 Tsuji, J. and Iwamoto, N. (1966). Organic synthesis by means of noble metal compounds. Palladium-catalyzed carbonylation of amines. *J. Chem. Soc., Chem. Commun.* (12): 380–380.

220 Nefkens, S.C.A., Sperrle, M., and Consiglio, G. (1993). Palladium-catalyzed enantioselective bis-alkoxycarbonylation of olefins. *Angew. Chem. Int. Ed. Engl.* 32: 1719–1720.

221 Sperrle, M. and Consiglio, G. (1999). Palladium-catalysed enantioselective bis-alkoxycarbonylation of 1-olefins. Synthesis of optically active 2-substituted-butanedioates. *J. Mol. Catal. A: Chem.* 143: 263–277.

222 Wang, L.L., Kwok, W.H., Wu, J. et al. (2003). Enantioselective bis-alkoxycarbonylation of styrene catalyzed by novel chiral dipyridylphosphine cationic Pd(II) complexes. *J. Mol. Catal. A: Chem.* 196: 171–178.

223 Liang, B., Liu, J., Gao, Y.X. et al. (2007). Synthesis of thiourea-oxazolines, a new class of chiral S,N-heterobidentate ligands: application in Pd-catalyzed asymmetric bis(methoxycarbonylation) of terminal olefins. *Organometallics* 26: 4756–4762.

224 Gabriele, B., Salerno, G., Costa, M., and Chiusoli, G.P. (1995). Combined oxidative and reductive carbonylation of terminal alkynes with Pd iodide/thiourea catalysts. *J. Organomet. Chem.* 503 (1): 21–28.

225 Gabriele, B., Salerno, G., Costa, M., and Chiusoli, G.P. (2003). Recent developments in the synthesis of heterocyclic derivatives by PdI_2-catalyzed oxidative carbonylation reactions. *J. Organomet. Chem.* 687 (2): 219–228.

226 Geis, O. and Schmalz, H.G. (1998). New developments in the Pauson–Khand reaction. *Angew. Chem. Int. Ed. Engl.* 37 (7): 911–914.

227 Gibson, S.E. and Mainolfi, N. (2005). The intermolecular Pauson–Khand reaction. *Angew. Chem. Int. Ed.* 44 (20): 3022–3037.

228 (a) Strübing, D. and Beller, M. (2006). Pauson–Khand reaction. In: *Catalytic Carbonylation Reactions*, Topics in Organometallic Chemistry, vol. 18 (ed. M. Beller), 165–178. Heidelberg: Springer-Verlag. (b) Ricker, J.D. and Geary, L.M. (2017). Recent advances in the Pauson–Khand reaction. *Top. Catal.* 60 (8): 609–619.

229 Khand, I.U., Knox, G.R., Pauson, P.L., and Watts, W.E. (1971). Cobalt-induced cleavage reaction and a new series of arenecobalt carbonyl complexes. *J. Chem. Soc., Chem. Commun.* (1): 36–36.

230 Khand, I.U., Knox, G.R., Pauson, P.L. et al. (1973). Organocobalt complexes. 2. Reaction of acetylenehexacarbonyldicobalt complexes, $(R1CCR_2)Co_2(CO)_6$, with norbornene and its derivatives. *J. Chem. Soc., Perkin Trans. 1* (9): 977–981.

231 Shambayati, S., Crowe, W.E., and Schreiber, S.L. (1990). *N*-oxide promoted Pauson–Khand cyclizations at room-temperature. *Tetrahedron Lett.* 31 (37): 5289–5292.

232 Jeong, N., Chung, Y.K., Lee, B.Y. et al. (1991). A dramatic acceleration of the Pauson–Khand reaction by trimethylamine *N*-oxide. *Synlett* (3): 204–206.

233 Gordon, A.R., Johnstone, C., and Kerr, W.J. (1995). Mild and efficient *N*-oxide promoted Pauson–Khand reactions with ethylene. *Synlett* (10): 1083–1084.

234 Krafft, M.E., Boñaga, L.V.R., and Hirosawa, C. (2001). Practical Co(carbonyl) catalysis in the thermal Pauson–Khand reaction: efficiency enhancement using Lewis bases. *J. Org. Chem.* 66 (9): 3004–3020.

235 Lagunas, A., Payeras, A., Jimeno, C., and Pericas, M.A. (2005). TEMPO-promoted Pauson–Khand reaction. Single-electron activation of Co-carbonyl bonds? *Org. Lett.* 7 (14): 3033–3036.

236 Sugihara, T., Yamada, M., Ban, H. et al. (1997). Rate enhancement of the Pauson–Khand reaction by primary amines. *Angew. Chem. Int. Ed. Engl.* 36 (24): 2801–2804.

237 Sugihara, T., Okada, Y., Yamaguchi, M., and Nishizawa, M. (1999). A novel carbonylative decomplexation of alkyne-dicobalt hexacarbonyls: hydrocarbamoylation of alkynes. *Synlett* (6): 768–770.

238 Rautenstrauch, V., Megard, P., Conesa, J., and Kuster, W. (1990). 2-Pentylcyclopent-2-en-1-one by catalytic Pauson–Khand reaction. *Angew. Chem. Int. Ed. Engl.* 29 (12): 1413–1416.

239 Jeong, N., Hwang, S.H., Lee, Y.S., and Chung, Y.K. (1994). Catalytic version of the intramolecular Pauson–Khand reaction. *J. Am. Chem. Soc.* 116 (7): 3159–3160.

240 Sugihara, T. and Yamaguchi, M. (1998). The Pauson–Khand reaction catalyzed by the methylidynetricobalt nonacarbonyl cluster. *J. Am. Chem. Soc.* 120 (41): 10782–10783.

241 Jeong, N., Hwang, S.H., Lee, Y.W., and Lim, J.S. (1997). Catalytic Pauson–Khand reaction in super critical fluids. *J. Am. Chem. Soc.* 119 (43): 10549–10550.

242 Pagenkopf, B.L. and Livinghouse, T. (1996). Photochemical promotion of the intramolecular Pauson–Khand reaction. A new experimental protocol for cobalt-catalyzed (2+2+1)-cycloadditions. *J. Am. Chem. Soc.* 118 (9): 2285–2286.

243 Rajesh, T. and Periasamy, M. (1999). A sub-stoichiometric version of the Pauson–Khand reaction. *Tetrahedron Lett.* 40 (4): 817–818.

244 Hayashi, M., Hashimoto, Y., Yamamoto, Y. et al. (2000). Phosphane sulfide/octacarbonyldicobalt-catalyzed Pauson–Khand reaction under an atmospheric pressure of carbon monoxide. *Angew. Chem. Int. Ed.* 39 (3): 631–633.

245 Tang, Y.F., Deng, L.J., Zhang, Y.D. et al. (2005). Tetramethylthiourea/$Co_2(CO)_8$-catalyzed Pauson–Khand reaction under balloon pressure of CO. *Org. Lett.* 7 (4): 593–595.

246 Billington, D.C., Helps, I.M., Pauson, P.L. et al. (1988). The effect of ultrasound and of phosphine and phosphine-oxides on the Khand reaction. *J. Organomet. Chem.* 354 (2): 233–242.

247 Chung, Y.K., Lee, B.Y., Jeong, N. et al. (1993). Promoters for the (alkyne)hexacarbonyldicobalt-based cyclopentenone synthesis. *Organometallics* 12 (1): 220–223.

248 Magnus, P. and Príncipe, L.M. (1985). Origins of 1,2-stereoselectivity and 1,3-stereoselectivity in dicobaltoctacarbonyl alkene alkyne cyclizations for the synthesis of substituted bicyclo[3.3.0]octenones. *Tetrahedron Lett.* 26 (40): 4851–4854.

249 Yamanaka, M. and Nakamura, E. (2001). Density functional studies on the Pauson–Khand reaction. *J. Am. Chem. Soc.* 123 (8): 1703–1708.

250 Gimbert, Y., Lesage, D., Milet, A. et al. (2003). On early events in the Pauson–Khand reaction. *Org. Lett.* 5 (22): 4073–4075.

251 Lesage, D., Milet, A., Memboeuf, A. et al. (2014). The Pauson–Khand mechanism revisited: origin of CO in the final product. *Angew. Chem. Int. Ed.* 53: 1939–1942.

252 Banide, E.V., Müller-Bunz, H., Manning, A.R. et al. (2007). X-ray crystal structure of an alkene-pentacarbonyldicobalt-alkyne complex: isolation of a stable Magnus-type Pauson–Khand reaction intermediate. *Angew. Chem. Int. Ed.* 46 (16): 2907–2910.

253 Cabot, R., Lledó, A., Revés, M. et al. (2007). Kinetic studies on the Co-catalyzed norbornadiene intermolecular Pauson–Khand reaction. *Organometallics* 26 (5): 1134–1142.

254 Kim, S.W., Son, S.U., Lee, S.I. et al. (2000). Cobalt on mesoporous silica: the first heterogeneous Pauson–Khand catalyst. *J. Am. Chem. Soc.* 122 (7): 1550–1551.

255 Park, K.H., Son, S.U., and Chung, Y.K. (2002). Sequential actions of Pd and Co nanoparticles immobilized on silica: one-pot synthesis of bicyclic enones by catalytic allylic alkylation and Pauson–Khand reaction. *Org. Lett.* 4 (24): 4361–4363.

256 Blanco-Urgoiti, J., Abdi, D., Dominguez, G., and Peréz-Castells, J. (2008). Efficient catalytic protocol for the Pauson–Khand reaction. *Tetrahedron* 64 (1): 67–74.

257 Son, S.U., Lee, S.I., and Chung, Y.K. (2000). Cobalt on charcoal: a convenient and inexpensive heterogeneous Pauson–Khand catalyst. *Angew. Chem. Int. Ed.* 39 (22): 4158–4160.

258 Park, K.H. and Chung, Y.K. (2005). Immobilized Co/Rh heterobimetallic nanoparticle-catalyzed Pauson–Khand-type reaction. *Adv. Synth. Catal.* 347 (6): 854–866.

259 Park, K.H. and Chung, Y.K. (2005). The Pauson–Khand-type reaction catalyzed by transition metal nanoparticles. *Synlett* (4): 545–559.

260 Kim, S.W., Son, S.U., Lee, S.S. et al. (2001). Colloidal Co nanoparticles: a highly active and reusable Pauson–Khand catalyst. *J. Chem. Soc., Chem. Commun.* (21): 2212–2213.

261 Son, S.U., Lee, S.I., Chung, Y.K. et al. (2002). The first intramolecular Pauson–Khand reaction in water using aqueous colloidal Co nanoparticles as catalysts. *Org. Lett.* 4 (2): 277–279.

262 Park, K.H., Son, S.U., and Chung, Y.K. (2003). Immobilized heterobimetallic Ru/Co nanoparticle-catalyzed Pauson–Khand-type reactions in the presence of pyridylmethyl formate. *J. Chem. Soc., Chem. Commun.* (15): 1898–1899.

263 Müller, J.L., Klankermayer, J., and Leitner, W. (2007). Poly(ethylene glycol) stabilized Co nanoparticles as highly active and selective catalysts for the Pauson–Khand reaction. *J. Chem. Soc.; Chem. Commun.* (19): 1939–1941.

264 Park, J.H., Kim, S.Y., Kim, S.M. et al. (2007). Allenic Pauson–Khand reaction catalyzed by Co/Rh heterobimetallic nanoparticles. *Synlett* (3): 453–459.

265 Müller, J.L., Rickers, A., and Leitner, W. (2007). Raney cobalt: an effective and recyclable catalyst for the Pauson–Khand reaction. *Adv. Synth. Catal.* 349 (3): 287–291.

266 Schore, N.E. and Croudace, M.C. (1981). Preparation of bicyclo[3.3.0]oct-1-en-3-one and bicyclo[4.3.0]non-1(9)-en-8-one via intramolecular cyclization of α,ω-enynes. *J. Org. Chem.* 46 (26): 5436–5438.

267 Krafft, M.E. (1988). Regiocontrol in the intermolecular Co-catalyzed olefin acetylene cycloaddition. *J. Am. Chem. Soc.* 110 (3): 968–970.

268 Krafft, M.E., Juliano, C.A., Scott, I.L. et al. (1991). The directed Pauson–Khand reaction. *J. Am. Chem. Soc.* 113 (5): 1693–1703.

269 Krafft, M.E. and Juliano, C.A. (1992). Studies on the use of bidentate ligands in the directed Pauson–Khand reaction. *J. Org. Chem.* 57 (19): 5106–5115.

270 Hicks, F.A., Kablaoui, N.M., and Buchwald, S.L. (1996). Titanocene-catalyzed cyclocarbonylation of enynes to cyclopentenones. *J. Am. Chem. Soc.* 118 (39): 9450–9451.

271 Kondo, T., Suzuki, N., Okada, T., and Mitsudo, T. (1997). First Ru-catalyzed intramolecular Pauson–Khand reaction. *J. Am. Chem. Soc.* 119 (26): 6187–6188.

272 Morimoto, T., Chatani, N., Fukumoto, Y., and Murai, S. (1997). $Ru_3(CO)_{12}$-catalyzed cyclocarbonylation of 1,6-enynes to bicyclo[3.3.0]octenones. *J. Org. Chem.* 62 (11): 3762–3765.

273 Koga, Y., Kobayashi, T., and Narasaka, K. (1998). Rhodium-catalyzed intramolecular Pauson–Khand reaction. *Chem. Lett.* (3): 249–250.

274 Jeong, N. (1998). Rhodium(I)-catalyzed intramolecular Pauson–Khand reaction. *Organometallics* 17 (17): 3642–3644.

275 Adrio, J., Rivero, M.R., and Carretero, J.C. (2005). Mild and efficient Mo-mediated Pauson–Khand-type reaction. *Org. Lett.* 7 (3): 431–434.

276 Berk, S.C., Grossman, R.B., and Buchwald, S.L. (1993). Titanocene-catalyzed conversion of enynes to bicyclic cyclopentenones. *J. Am. Chem. Soc.* 115 (11): 4912–4913.

277 Zhang, M.H. and Buchwald, S.L. (1996). A Ni(0)-catalyzed process for the transformation of enynes to bicyclic cyclopentenones. *J. Org. Chem.* 61 (14): 4498–4499.

278 Hicks, F.A., Kablaoui, N.M., and Buchwald, S.L. (1999). Scope of the intramolecular titanocene-catalyzed Pauson–Khand type reaction. *J. Am. Chem. Soc.* 121 (25): 5881–5898.

279 Saito, T., Sugizaki, K., Otani, T., and Suyama, T. (2007). Rhodium-catalyzed intramolecular alkyne-carbodiimide Pauson–Khand-type reaction. *Org. Lett.* 9 (7): 1239–1241.

280 Shibata, T. (2006). Recent advances in the catalytic Pauson–Khand-type reaction. *Adv. Synth. Catal.* 348 (16–17): 2328–2336.

281 Peréz-Castells, J. (2006). Cascade reactions involving Pauson–Khand and related processes. In: *Metal Catalyzed Cascade Reactions*, Topics in Organometallic Chemistry, vol. 19, 207–257. Heidelberg: Springer-Verlag.

282 Wang, H., Sawyer, J.R., Evans, P.A., and Baik, M.H. (2008). Mechanistic insight into the diastereoselective Rh-catalyzed Pauson–Khand reaction: role of coordination number in stereocontrol. *Angew. Chem. Int. Ed.* 47 (2): 342–345.

283 Narasaka, K. and Shibata, T. (1994). Conversion of 1-(ω-alkynyl)-1,2-propadienyl sulfides to bicyclic dienones by the use of Fe-carbonyl complex. *Chem. Lett.* (2): 315–318.

284 Shibata, T., Koga, Y., and Narasaka, K. (1995). Intramolecular and intermolecular allene-alkyne coupling reactions by the use of $Fe(CO)_4(NMe_3)$. *Bull. Chem. Soc. Jpn.* 68 (3): 911–919.

285 Shibata, T., Kadowaki, S., Hirase, M., and Takagi, K. (2003). Iridium complex-catalyzed Pauson–Khand-type reaction of allenynes. *Synlett* (4): 573–575.

286 Alcaide, B. and Almendros, P. (2004). The allenic Pauson–Khand reaction in synthesis. *Eur. J. Org. Chem.* (16): 3377–3383.

287 Inagaki, F. and Mukai, C. (2006). Rhodium(I)-catalyzed intramolecular Pauson–Khand-type (2+2+1)-cycloaddition of allenenes. *Org. Lett.* 8 (6): 1217–1220.

288 Lee, H.W., Chan, A.S.C., and Kwong, F.Y. (2007). Formate as a CO surrogate for cascade processes: Rh-catalyzed cooperative decarbonylation and asymmetric Pauson–Khand-type cyclization reactions. *J. Chem. Soc., Chem. Commun.* (25): 2633–2635.

289 Morimoto, T., Fuji, K., Tsutsumi, K., and Kakiuchi, K. (2002). CO-transfer carbonylation reactions. A catalytic Pauson–Khand-type reaction of enynes with aldehydes as a source of carbon monoxide. *J. Am. Chem. Soc.* 124 (15): 3806–3807.

290 Shibata, T., Toshida, N., and Takagi, K. (2002). Rhodium-complex catalyzed Pauson–Khand-type reaction with aldehydes as a CO source. *J. Org. Chem.* 67 (21): 7446–7450.

291 Fuji, K., Morimoto, T., Tsutsumi, K., and Kakiuchi, K. (2003). Aqueous catalytic Pauson–Khand-type reactions of enynes with formaldehyde: transfer carbonylation involving an aqueous decarbonylation and a micellar carbonylation. *Angew. Chem. Int. Ed.* 42 (21): 2409–2411.

292 Wender, P.A., Croatt, M.P., and Deschamps, N.M. (2004). Rhodium(I)-catalyzed (2+2+1)-cycloadditions of 1,3-dienes, alkenes, and CO. *J. Am. Chem. Soc.* 126 (19): 5948–5949.

293 Simeonov, S.P., Nunes, J.P.M., Guerra, K. et al. (2016). Synthesis of chiral cyclopentenones. *Chem. Rev.* 116: 5744–5893.

294 Ishikawa, T., Ishii, H., Shimizu, K. et al. (2004). Diastereoselective total synthesis of isocarbacyclin from L-ascorbic acid. *J. Org. Chem.* 69 (23): 8133–8135.

295 Hirose, T., Miyakoshi, N., and Mukai, C. (2008). Total synthesis of (+)-achalensolide based on the Rh(I)-catalyzed allenic Pauson–Khand-type reaction. *J. Org. Chem.* 73 (3): 1061–1066.

296 Chiusoli, G.P. (1959). Reactions of allylic chlorinated derivatives with CO and acetylene. I. Synthesis of unsaturated carboxyl compounds with CO. *Chem. Ind.* 41: 503–506.

297 Chiusoli, G.P. (1960). Neuere Synthesen mit Kohlenmonoxyd. *Angew. Chem.* 72 (2): 74–76.

298 Dent, W.T., Whitfield, G.H., and Long, R. (1964). Carbonylation of allylic chlorides. *J. Chem. Soc.* 1588–1594.

299 Long, R. and Whitfield, G.H. (1964). Reaction of some π-allylic Pd chloride complexes with carbon monoxide. *J. Chem. Soc.* 1852–1853.

300 Tsuji, J., Imamura, S., Morikawa, M., and Kiji, J. (1964). Organic syntheses by means of noble metal compounds. 8. Catalytic carbonylation of allylic compounds with Pd chloride. *J. Am. Chem. Soc.* 86 (20): 4350–4353.

301 Tsuji, J. (1969). Carbon–carbon bond formation via Pd complexes. *Acc. Chem. Res.* 2 (5): 144–152.

302 Tsuji, J., Sato, K., and Okumoto, H. (1984). Palladium-catalyzed decarboxylation/carbonylation of allylic carbonates to form β,γ-unsaturated esters. *J. Org. Chem.* 49 (8): 1341–1344.

303 (a) Murahashi, S.I., Imada, Y., Taniguchi, Y., and Higashiura, S. (1988). Palladium(0)-catalyzed carbonylation of allyl phosphates and allyl acetates - selective synthesis of β,γ-unsaturated esters. *Tetrahedron Lett.* 29 (39): 4945–4948. (b) Carvajal, M.A., Miscione, G.P., Novoa, J.J., and Bottoni, A. (2005). DFT computational study of the mechanism of allyl chloride carbonylation catalyzed by Pd complexes. *Organometallics* 24 (9): 2086–2096.

304 Cao, J., Zheng, Z.-J., Xu, Z., and Xu, L.-W. (2017). Transition-metal-catalyzed transfer carbonylation with HCOOH or HCHO as non-gaseous C1 source. *Coord. Chem. Rev.* 336: 43–53.

305 Schönberg, A. and Heck, R.F. (1974). Palladium-catalyzed amidation of aryl, heterocyclic, and vinylic halides. *J. Org. Chem.* 39 (23): 3327–3331.

306 (a) Yamamoto, A. (1995). Palladium-catalyzed double and single carbonylation of aryl halides and allylic compounds. *Bull. Chem. Soc. Jpn.* 68 (2): 433–446. (b) Neumann, H. and Schranck, J. (2014). Carbonylation of aryl and vinyl halides. In: *Science of Synthesis, C-1 Building Blocks in Organic Synthesis* (ed. P.W.N.M. Van Leeuwen), 67–93. Stuttgart: Georg Thieme Verlag. (c) Fang, W.-W., Zhu, H.-B., Deng, Q.-Y. et al. (2014). Design and development of ligands for palladium-catalyzed carbonylation reactions. *Synthesis* 46 (13): 1689–1708. (d) Li, W.-F. and Wu, X.-F. (2017). Palladium-catalyzed carbonylative synthesis of six-membered heterocycles from aryl halides. In: *Transition Metal Catalyzed Carbonylative Synthesis of Heterocycles*, Topics in Heterocyclic Chemistry, vol. 42, 55–87. Heidelberg: Springer-Verlag. (e) Nielsen, D.U., Lian, Z., Lindhardt, A.T., and Skrydstrup, T. (2017). *Transition Metal Catalyzed Carbonylative Synthesis of Heterocycles*, Topics in Heterocyclic Chemistry, vol. 42, 89–99. Heidelberg: Springer-Verlag.

307 Hu, Y.H., Liu, J., Lu, Z.X. et al. (2010). Base-induced mechanistic variation in Pd-catalyzed carbonylation of aryl iodides. *J. Am. Chem. Soc.* 132: 3153–3158.

308 Nahm, S. and Weinreb, S.M. (1981). *N*-methoxy-*N*-methylamides as effective acylating agents. *Tetrahedron Lett.* 22 (39): 3815–3818.

309 Sibi, M.P. (1993). Chemistry of *N*-methoxy-*N*-methylamides - applications in synthesis - a review. *Org. Prep. Proced. Int.* 25 (1): 15–40.

310 Khlestkin, V.K. and Mazhukin, D.G. (2003). Recent advances in the application of *N,O*-dialkylhydroxylamines in organic chemistry. *Curr. Org. Chem.* 7 (10): 967–993.

311 Takacs, A., Petz, A., and Kollar, L. (2010). High-yielding synthesis of Weinreb amides via homogeneous catalytic carbonylation of iodoalkenes and iodoarenes. *Tetrahedron* 66: 4479–4483.

312 (a) Zhuang, L.H., Wai, J.S., Embrey, M.W. et al. (2003). Design and synthesis of 8-hydroxy-[1,6]naphthyridines as novel inhibitors of HIV-1 integrase in vitro and in infected cells. *J. Med. Chem.* 46 (4): 453–456. (b) Martinelli, J.R., Freckmann, D.M.M., and Buchwald, S.L. (2006). Convenient method for the preparation of Weinreb amides via Pd-catalyzed aminocarbonylation of aryl bromides at atmospheric pressure. *Org. Lett.* 8 (21): 4843–4846.

313 Wu, L.-P., Fang, X.-J., Liu, Q. et al. (2014). Palladium-catalyzed carbonylative transformation of C(sp^3)-X bonds. *ACS Catal.* 4 (9): 2977–2989.

314 Ben-David, Y., Portnoy, M., and Milstein, D. (1989). Chelate assisted, Pd-catalyzed efficient carbonylation of aryl chlorides. *J. Am. Chem. Soc.* 111: 8742–8744.

315 Martinelli, J.R., Clark, T.P., Watson, D.A. et al. (2007). Palladium-catalyzed aminocarbonylation of aryl chlorides at atmospheric pressure: the dual role of sodium phenoxide. *Angew. Chem. Int. Ed.* 46 (44): 8460–8463.

316 Garrou, P.E. and Heck, R.F. (1976). Mechanism of carbonylation of halo(bisligand)organoPt(II), Pd(II), and Nid(II) complexes. *J. Am. Chem. Soc.* 98 (14): 4115–4127.

317 Milstein, D. (1988). Aspects of intermediacy of carbalkoxymetal complexes in CO reactions. *Acc. Chem. Res.* 21 (11): 428–434.

318 Cai, C.X., Rivera, N.R., Balsells, J. et al. (2006). An efficient catalyst for Pd-catalyzed carbonylation of aryl arenesulfonates. *Org. Lett.* 8 (22): 5161–5164.

319 Munday, R.H., Martinelli, J.R., and Buchwald, S.L. (2008). Palladium-catalyzed carbonylation of aryl tosylates and mesylates. *J. Am. Chem. Soc.* 130 (9): 2754–2755.

320 Lapidus, A.L. and Eliseev, O.L. (2010). Catalytic carbonylation in ionic liquids. *Solid Fuel Chem.* 44: 197–202.

321 Sargent, B.T. and Alexanian, E.J. (2016). Palladium-catalyzed alkoxycarbonylation of unactivated secondary alkyl bromides at low pressure. *J. Am. Chem. Soc.* 138: 7520–7523.

322 Liu, Q., Dong, X., Li, J. et al. (2015). Recent advances on Pd radical involved reactions. *ACS Catal.* 5: 6111–6137.

323 Correa, A.M.R. (2009). Palladium-catalyzed direct carboxylation of aryl bromides with carbon dioxide. *J. Am. Chem. Soc* 131: 15974–15975.

324 Schönberg, A. and Heck, R.F. (1974). Palladium-catalyzed formylation of aryl, heterocyclic, and vinylic halides. *J. Am. Chem. Soc.* 96: 7761–7764.

325 Klaus, S., Neumann, H., Zapf, A. et al. (2006). A general and efficient method for the formylation of aryl and heteroaryl bromides. *Angew. Chem. Int. Ed.* 45: 154–158.

326 Sergeev, A.G., Spannenberg, A., and Beller, M. (2008). Palladium-catalyzed formylation of aryl bromides: elucidation of the catalytic cycle of an industrially applied coupling reaction. *J. Am. Chem. Soc.* 130: 15549–15563.

327 Neumann, H., Kadyrov, R., Wu, X.F., and Beller, M. (2012). Palladium-catalyzed reductive carbonylation of aryl bromides with phosphinite ligands. *Chem. Asian J.* 7: 2213–2216.

328 Baillargeon, V.P. and Stille, J.K. (1983). Direct conversion of organic halides to aldehydes with carbon-monoxide and Sn hydride catalyzed by Pd. *J. Am. Chem. Soc.* 105: 7175–7176.

329 Baillargeon, V.P. and Stille, J.K. (1986). Palladium-catalyzed formylation of organic halides with carbon-monoxide and Sn hydride. *J. Am. Chem. Soc.* 108: 452–461.

330 Smith, A.B., Kurti, L., Davulcu, A.H., and Cho, Y.S. (2007). Development of a scalable synthesis of a common eastern tricyclic lactone for construction of the nodulisporic acids. *Org. Process Res. Dev.* 11: 19–24.

331 Kotsuki, H., Datta, P.K., and Suenaga, H. (1996). An efficient procedure for Pd-catalyzed hydroformylation of aryl/enol triflates. *Synthesis* 470–472.

332 Behenna, D.C.S., Stockdill, J.L., and Stoltz, B.M. (2007). Synthesis of the carbocyclic core of zoanthenol: implementation of an unusual acid-catalyzed cyclization. *Angew. Chem. Int. Ed.* 46: 4077–4080.

333 Ashfield, L. and Barnard, C.F.J. (2007). Reductive carbonylation - an efficient and practical catalytic route for the conversion of aryl halides to aldehydes. *Org. Process Res. Dev.* 11: 39–43.

334 Pri-Bar, I. and Buchman, O. (1984). Reductive formylation of aromatic halides under low-carbon monoxide pressure catalyzed by transition-metal compounds. *J. Org. Chem.* 49: 4009–4011.

335 Ben-David, Y., Portnoy, M., and Milstein, D. (1989). Formylation of aryl chlorides catalyzed by a Pd-complex. *J. Chem. Soc., Chem. Commun.* 1816–1817.

336 Cai, M.Z., Zhao, H., Zhou, J., and Song, C.S. (2002). Formylation of aryl halides catalyzed by a silica-supported phosphine Pd-complex. *Synth. Commun.* 32: 923–926.

337 Cacchi, S., Fabrizi, G., and Goggiamani, A. (2004). Palladium-catalyzed synthesis of aldehydes from aryl iodides and acetic formic anhydride. *J. Comb. Chem.* 6: 692–694.

338 Ueda, T., Konishi, H., and Manabe, K. (2013). Palladium-catalyzed reductive carbonylation of aryl halides with *N*-formylsaccharin as a CO source. *Angew. Chem. Ind. Ed.* 52: 8611–8615.

339 Cochet, T., Bellosta, V., Greiner, A. et al. (2011). *N*-formylsaccharin: a new formylating agent. *Synlett* 1920–1922.

340 Korsager, S., Nielsen, D.U., Taaning, R.H., and Skrydstrup, T. (2013). Access to β-keto esters by Pd-catalyzed carbonylative coupling of aryl halides with monoester potassium malonates. *Angew. Chem. Ind. Ed.* 52: 9763–9766.

341 Korsager, S., Taaning, R.H., Lindhardt, A.T., and Skrydstrup, T. (2013). Reductive carbonylation of aryl halides employing a two-chamber reactor: a protocol for the synthesis of aryl aldehydes including ^{13}C- and D-isotope labeling. *J. Org. Chem.* 78: 6112–6120.

342 Korsager, S., Taaning, R.H., and Skrydstrup, T. (2013). Effective Pd-catalyzed hydroxycarbonylation of aryl halides with substoichiometric carbon monoxide. *J. Am. Chem. Soc.* 135: 2891–2894.

343 Yu, B., Zhao, Y.F., Zhang, H.Y. et al. (2014). Pd/C-catalyzed direct formylation of aromatic iodides to aryl aldehydes using carbon dioxide as a C1 resource. *J. Chem. Soc., Chem. Commun.* 50: 2330–2333.

344 Tamaru, Y., Ochiai, H., Yamada, Y., and Yoshida, Z. (1983). Unsymmetrical ketone synthesis via Pd-catalyzed carbonylation of organic halides. *Tetrahedron Lett.* 24 (36): 3869–3872.

345 Pommier, A. and Pons, J.M. (1995). Synthesis of natural 2-oxetanones. *Synthesis* 729–744.

346 Lowe, C. and Vederas, J.C. (1995). Naturally occurring β-lactones - occurrence, syntheses and properties - a review. *Org. Prep. Proced. Int.* 27: 305–346.

347 Wang, Y.C., Tennyson, R.L., and Romo, D. (2004). β-Lactones: intermediates for natural product total synthesis and new transformations. *Heterocycles* 64: 605–658.

348 Nakano, K. and Nozaki, K. (2006). Carbonylation of epoxides. In: *Catalytic Carbonylation Reactions*, Topics in Organometallic Chemistry, vol. 18 (ed. M. Beller), 223–238. Heidelberg: Springer-Verlag.

349 Kramer, J.W., Rowley, J.M., and Coates, G.W. (2015). *Ring Expanding Carbonylation of Epoxides. Organic Reactions*, Wiley Online Library, vol. 86, 1–843. New York.

350 Rieth, L.R., Moore, D.R., Lobkovsky, E.B., and Coates, G.W. (2002). Single-site (β-diiminate)Zn catalysts for the ring-opening polymerization of β-butyrolactone and β-valerolactone to

poly(3-hydroxyalkanoates). *J. Am. Chem. Soc.* 124: 15239–15248.

351 Coulembier, O., Degee, P., Hedrick, J.L., and Dubois, P. (2006). From controlled ring-opening polymerization to biodegradable aliphatic polyester: especially poly(β-malic acid) derivatives. *Prog. Polym. Sci.* 31: 723–747.

352 Seon, M. and Leleu, J. (1957). Hydracrylic acid esters. US 278226 19570219.

353 Aumann, R. and Ring, H. (1977). Organic syntheses with transition-metal complexes. 5. δ-Lactones by carbonylation of vinyloxiranes. *Angew. Chem. Int. Ed. Engl.* 16 (1): 50–50.

354 Aumann, R., Ring, H., Krüger, C., and Goddard, R. (1979). Organic syntheses by means of transition-metal complexes. 8. Synthesis of unsaturated δ-lactones from vinyloxiranes and carbon-monoxide via transition-metal complexes. *Chem. Ber.* 112 (11): 3644–3671.

355 Kamiya, Y., Kawato, K., and Ohta, H. (1980). The reaction of small ring compound with carbon monoxide the carbonylation of oxirane. *Chem. Lett.* (12): 1549–1552.

356 Drent, E. and Kragtwijk, E. (1994). Process and catalysts for the carbonylation of epoxides into lactones or β-hydroxycarboxylic acid derivatives. Eur. Pat. Appl. EP 577206 A2.

357 Drent, E.; Kragtwijk, E. Process and catalysts for the carbonylation of epoxides into lactones or β-hydroxycarboxylic acid derivatives *Chem. Abstr.* 1994, 120, 191517c.

358 Lee, J.T., Thomas, P.J., and Alper, H. (2001). Synthesis of β-lactones by the regioselective, cobalt and Lewis acid catalyzed carbonylation of simple and functionalized epoxides. *J. Org. Chem.* 66 (16): 5424–5426.

359 Getzler, Y.D.Y.L., Mahadevan, V., Lobkovsky, E.B., and Coates, G.W. (2002). Synthesis of β-lactones: a highly active and selective catalyst for epoxide carbonylation. *J. Am. Chem. Soc.* 124 (7): 1174–1175.

360 Mahadevan, V., Getzler, Y.D.Y.L., and Coates, G.W. (2002). [Lewis acid]$^+$[$Co(CO)_4$]$^-$ complexes: a versatile class of catalysts for carbonylative ring expansion of epoxides and aziridines. *Angew. Chem. Int. Ed.* 41 (15): 2781–2784.

361 Schmidt, J.A.R., Lobkovsky, E.B., and Coates, G.W. (2005). Chromium(III)(octaethylporphyrinato) tetracarbonylcobaltate: a highly active, selective, and versatile catalyst for epoxide carbonylation. *J. Am. Chem. Soc.* 127 (32): 11426–11435.

362 Mulzer, M. and Coates, G.W. (2014). Carbonylation of *cis*-disubstituted epoxides to *trans*-β-lactones: catalysts displaying steric and contrasteric regioselectivity. *J. Org. Chem.* 79: 11851–11862.

363 Mulzer, M., Lamb, J.R., Nelson, Z., and Coates, G.W. (2014). Carbonylative enantioselective meso-desymmetrization of *cis*-epoxides to *trans*-beta-lactones: effect of salen-ligand electronic variation on enantioselectivity. *J. Chem. Soc., Chem. Commun.* 50: 9842–9845.

364 Church, T.L., Getzler, Y.D.Y.L., and Coates, G.W. (2006). The mechanism of epoxide carbonylation by [Lewis acid]$^+$/[$Co(CO)_4$]$^-$ catalysts. *J. Am. Chem. Soc.* 128 (31): 10125–10133.

365 Stirling, A., Iannuzzi, M., Parrinello, M. et al. (2005). β-Lactone synthesis from epoxide and CO: reaction mechanism revisited. *Organometallics* 24 (10): 2533–2537.

366 Alper, H., Arzoumanian, H., Petrignani, J.F., and Saldana-Maldonado, M. (1985). Phase-transfer catalyzed double carbonylation of styrene oxides. *J. Chem. Soc., Chem. Commun.* (6): 340–341.

367 Rowley, J.M., Lobkovsky, E.B., and Coates, G.W. (2007). Catalytic double carbonylation of epoxides to succinic anhydrides: catalyst discovery, reaction scope, and mechanism. *J. Am. Chem. Soc.* 129 (16): 4948–4960.

368 Page, M.G.P. (2011). *Beta-Lactam Antibiotics in Antibiotic Discovery and Development*, Chapter 3 (ed. T.J. Dougherty and M.J. Pucci), 79–117. New York, Heidelberg, Dordrecht, London: Springer-Verlag.

369 Mukerjee, A.K. and Srivastava, R.C. (1973). Syntheses of β-lactams. *Synthesis* (6): 327–346.

370 Sharma, S.D. and Mehra, U. (1988). Some recent developments in the synthesis of β-lactams via cycloadditions. *J. Sci. Ind. Res.* 47 (8): 451–459.

371 Parrick, J. and Mehta, L.K. (1999). Four-membered ring systems. In: *Progress in Heterocyclic Chemistry*, vol. 11 (ed. G.W. Gribble and T.L. Gilchrist), 87–101.

372 Magriotis, P.A. (2001). Recent progress in the enantioselective synthesis of β-lactams: development of the first catalytic approaches. *Angew. Chem. Int. Ed.* 40 (23): 4377–4379.

373 France, S., Weatherwax, A., Taggi, A.E., and Lectka, T. (2004). Advances in the catalytic, asymmetric synthesis of β-lactams. *Acc. Chem. Res.* 37 (8): 592–600.

374 Pal, R., Ghosh, S.C., Chandra, K., and Basak, A. (2007). Synthesis of β-lactams using the Kinugasa reaction. *Synlett* (15): 2321–2330.

375 Wurz, R.P. (2007). Chiral dialkylaminopyridine catalysts in asymmetric synthesis. *Chem. Rev.* 107 (12): 5570–5595.

376 Furman, B., Kaluza, Z., Stencel, A. et al. (2007). β-Lactams from carbohydrates. *Top. Organomet. Chem.* 7: 101–132.

377 Singh, G.S., D'hooghe, M., and De Kimpe, N. (2011). Synthesis and reactivity of spiro-fused β-lactams. *Tetrahedron* 67: 1989–2012.

378 Singh, G.S. and Sudheesh, S. (2014). Advances in synthesis of monocyclic β-lactams. *ARKIVOC* 337–385.

379 Dong, K.Y., Qiu, L.H., and Xu, X.F. (2015). Selective synthesis of β-lactams via catalytic metal carbene C—H insertion reactions. *Curr. Org. Chem.* 20: 29–40.

380 Volpato, G., Rodrigues, R.C., and Fernandez-Lafuente, R. (2010). Use of enzymes in the production of semi-synthetic penicillins and cephalosporins: drawbacks and perspectives. *Curr. Med. Chem.* 17: 3855–3873.

381 Ao, Y.F., Leng, D.H., Wang, D.X. et al. (2014). Efficient synthesis of highly enantiopure β-lactam derivatives from biocatalytic transformations of amides and nitriles. *Tetrahedron* 70: 4309–4316.

382 Sklyarenko, A.V., El'darov, M.A., Kurochkina, V.B., and Yarotsky, S.V. (2015). Enzymatic synthesis of β-lactam acids (review). *Appl. Biochem. Microbiol.* 51: 627–640.

383 Palomo, C.; Olarbide, M. β-Lactam ring opening: a useful entry to amino-acids and relevant nitrogen containing compounds. Banik, B. K., Ed.; Topics in Heterocyclic Chemistry, *Heterocyclic Scaffolds I* Springer-Verlag, Heidelberg, 2010, Vol. 22, pp. 211–259.

384 Mikami, K., Fustero, S., Sanchez-Rosello, M. et al. (2011). Synthesis of fluorinated β-amino acids. *Synthesis* 3045–3079.

385 Forro, E. and Fülöp, F. (2012). Recent lipase-catalyzed hydrolytic approaches to pharmacologically important β- and γ-amino acids. *Curr. Med. Chem.* 19 (36): 6178–6187.

386 Ojima, I., Zuniga, E.S., and Seitz, J.D. (2013). Advances in the use of enantiopure β-lactams for the synthesis of biologically active compounds of medicinal interest. In: *β-Lactams: Unique Structures of Distinction for Novel Molecules*, Topics in Heterocyclic Chemistry (Vol. ed. B.K. Banik and Series ed. B.U.W. Maes), vol. 30, 1–64. Heidelberg: Springer-Verlag.

387 McNally, A., Haffemayer, B., Collins, B.S.L., and Gaunt, M.J. (2014). Palladium-catalyzed C—H activation of aliphatic amines to give strained nitrogen heterocycles. *Nature* 510: 129–133.

388 Ibrahim, H. (2014). Carbon monoxide insertion into aziridines. In: *Science of Synthesis, C-1 Building Blocks in Organic Synthesis*, vol. 1 (ed. P.W.N.M.E. van Leeuwen), 193–206. Stuttgart: Georg Thieme Verlag.

389 Alper, H., Urso, F., and Smith, D.J.H. (1983). Regiospecific metal-catalyzed ring expansion of aziridines to β-lactams. *J. Am. Chem. Soc.* 105 (22): 6737–6738.

390 Calet, S., Urso, F., and Alper, H. (1989). Enantiospecific and stereospecific Rh(I)-catalyzed carbonylation and ring expansion of aziridines - asymmetric synthesis of β-lactams and the kinetic resolution of aziridines. *J. Am. Chem. Soc.* 111 (3): 931–934.

391 Chamchaang, W. and Pinhas, A.R. (1988). A one-pot conversion of an aziridine to a β-lactam using Nickel tetracarbonyl. *J. Chem. Soc., Chem. Commun.* (11): 710–711.

392 Piotti, M.E. and Alper, H. (1996). Inversion of stereochemistry in the $Co_2(CO)_8$-catalyzed carbonylation of aziridines to β-lactams. The first synthesis of highly strained *trans*-bicyclic β-lactams. *J. Am. Chem. Soc.* 118 (1): 111–116.

393 Davoli, P., Forni, A., Moretti, I. et al. (2001). On the effect of ring substituents in the carbonylation of aziridines. *Tetrahedron* 57 (9): 1801–1812.

394 Fontana, F.T., Tron, G.C., Barbero, N. et al. (2010). Stereoselective synthesis of *trans*-β-lactams by Pd-catalysed carbonylation of vinyl aziridines. *J. Chem. Soc., Chem. Commun.* 46: 267–269.

395 Kramer, J.W., Joh, D.Y., and Coates, G.W. (2007). Carbonylation of epoxides to substituted 3-hydroxy-δ-lactones. *Org. Lett.* 9 (26): 5581–5583.

396 Wang, M.D., Calet, S., and Alper, H. (1989). Regiospecific carbonylation and ring expansion of thietanes and oxetanes catalyzed by Co- and or Ru-carbonyls. *J. Org. Chem.* 54 (1): 20–21.

397 Takahashi, K. and Nozaki, K. (2014). Hydroformylation of epoxides. In: *Science of Synthesis, C-1 Building Blocks in Organic Synthesis*, vol. 2 (ed. P.W.N.M.E. van Leeuwen), 207–217. Stuttgart: Georg Thieme Verlag.

398 Yokokawa, C., Watanabe, Y., and Takegami, Y. (1964). Studies of the organic reactions of metal carbonyl. 3. The hydroformylation of olefin oxides. *Bull. Chem. Soc. Jpn.* 37 (5): 677–679.

399 Nakano, K., Katayama, M., Ishihara, S. et al. (2004). An alternative route to protected aldols:

cobalt-catalyzed hydroformylation of epoxides and in situ protection of beta-hydroxyaldehydes by $HC(OMe)_3$. *Synlett* (8): 1367–1370.

400 Fukumoto, Y., Chatani, N., and Murai, S. (1993). Rhodium-catalyzed ring-opening silylformylation of epoxides leading to β-siloxy aldehydes. *J. Org. Chem.* 58 (16): 4187–4188.

401 Engle, K.M., Mei, T.S., Wasa, M., and Yu, J.Q. (2012). Weak coordination as a powerful means for developing broadly useful C—H functionalization reactions. *Acc. Chem. Res.* 45: 788–802.

402 (a) Colby, D.A., Tsai, A.S., Bergman, R.G., and Ellman, J.A. (2012). Rhodium-catalyzed chelation-assisted C—H bond functionalization reactions. *Acc. Chem. Res.* 45: 814–825. (b) Wang, P-S., Shen, M-L., and Gong, L-Z (2018). Transition-metal-catalyzed asymmetric allylation of carbonyl compounds with unsaturated hydrocarbons. *Synthesis* 50 (5): 956–967.

403 Hartwig, J.F. and Larsen, M.A. (2016). Undirected, homogeneous C—H bond functionalization: challenges and opportunities. *ACS Cent. Sci.* 2: 281–292.

404 Laub, H.A., Evano, G., and Mayr, H. (2014). Hydrocarbation of C—C bonds: quantification of the nucleophilic reactivity of ynamides. *Angew. Chem. Ind. Ed.* 53: 4968–4971.

405 Trost, B.M. (1995). Atom economy - a challenge for organic synthesis - homogeneous catalysis leads the way. *Angew. Chem. Int. Ed. Engl.* 34 (3): 259–281.

406 Trost, B.M. and Krische, M.J. (1998). Transition-metal-catalyzed cycloisomerizations. *Synlett* 1–16.

407 (a) Zhang, X.S., Chen, K., and Shi, Z.J. (2014). Transition-metal-catalyzed direct nucleophilic addition of C—H bonds to carbon-heteroatom double bonds. *Chem. Sci.* 5: 2146–2159. (b) Hummel, J.R., Boerth, J.A., and Ellman, J.A. (2017). Transition-metal-catalyzed C—H bond addition to carbonyls, imines, and related polarized π bonds. *Chem. Rev.* 117 (13): 9163–9227.

408 Knöpfel, T.F., Zarotti, P., Ichikawa, T., and Carreira, E.M. (2005). Catalytic, enantioselective, conjugate alkyne addition. *J. Am. Chem. Soc.* 127: 9682–9683.

409 Trost, B.M. and Weiss, A.H. (2009). The enantioselective addition of alkyne nucleophiles to carbonyl groups. *Adv. Synth. Catal.* 351: 963–983.

410 Yazaki, R., Kumagai, N., and Shibasaki, M. (2010). Direct catalytic asymmetric conjugate addition of terminal alkynes to α,β-unsaturated thioamides. *J. Am. Chem. Soc.* 132: 10275–10277.

411 Yazaki, R., Kumagai, N., and Shibasaki, M. (2011). Cooperative activation of alkyne and thioamide functionalities; direct catalytic asymmetric conjugate addition of terminal alkynes to α,β-unsaturated thioamides. *Chem. Asian J.* 6: 1778–1790.

412 Nishimura, T., Sawano, T., and Hayashi, T. (2009). Asymmetric synthesis of β-alkynyl aldehydes by Rh-catalyzed conjugate alkynylation. *Angew. Chem. Ind. Ed.* 48: 8057–8059.

413 Nishimura, T., Sawano, T., Tokuji, S., and Hayashi, T. (2010). Rhodium-catalyzed asymmetric conjugate alkynylation of nitroalkenes. *J. Chem. Soc., Chem. Commun.* 46: 6837–6839.

414 Fillion, E. and Zorzitto, A.K. (2009). Enantioselective Rh-catalyzed conjugate alkynylation of 5-benzylidene Meldrum's acids with TMS-acetylene. *J. Am. Chem. Soc.* 131: 14608–14609.

415 Nishimura, T., Sawano, T., Ou, K.Y., and Hayashi, T. (2011). Cobalt-catalyzed conjugate addition of silylacetylenes to α,β-unsaturated ketones. *J. Chem. Soc., Chem. Commun.* 47: 10142–10144.

416 Sawano, T., Ashouri, A., Nishimura, T., and Hayashi, T. (2012). Cobalt-catalyzed asymmetric 1,6-addition of (triisopropylsilyl)-acetylene to α,β,γ,δ-unsaturated carbonyl compounds. *J. Am. Chem. Soc.* 134: 18936–18939.

417 Villarino, L., García-Fandino, R., López, F., and Mascarenas, J.L. (2012). Palladium-catalyzed conjugate addition of terminal alkynes to enones. *Org. Lett.* 14: 2996–2999.

418 Arndtsen, B.A., Bergman, R.G., Mobley, T.A., and Peterson, T.H. (1995). Selective intermolecular carbon-hydrogen bond activation by synthetic metal complexes in homogeneous solution. *Acc. Chem. Res.* 28 (3): 154–162.

419 Shilov, A.E. and Shul'pin, G.B. (1997). Activation of C—H bonds by metal complexes. *Chem. Rev.* 97 (8): 2879–2932.

420 Suzuki, H., Inagaki, A., Matsubara, K., and Takemori, T. (2001). Alkane activation on a multimetallic site. *Pure Appl. Chem.* 73 (2): 315–318.

421 Labinger, J.A. and Bercaw, J.E. (2002). Understanding and exploiting C—H bond activation. *Nature* 417 (6888): 507–514.

422 Perutz, R.N. and Sabo-Etienne, S. (2007). The sigma-CAM mechanism: σ-Complexes as the basis of σ-bond metathesis at late-transition-metal centers. *Angew. Chem. Int. Ed.* 46 (15): 2578–2592.

423 Recupero, F. and Punta, C. (2007). Free radical functionalization of organic compounds catalyzed by *N*-hydroxyphthalimide. *Chem. Rev.* 107 (9): 3800–3842.

424 Fokin, A.A. and Schreiner, P.R. (2003). Metal-free, selective alkane functionalizations. *Adv. Synth. Catal.* 345 (9–10): 1035–1052.

425 Langmuir, I. (1912). A chemically active modification of hydrogen. *J. Am. Chem. Soc.* 34: 1310–1325.

426 Lunsford, J.H. (1989). The role of surface-generated gas phase radicals in catalysis. *Langmuir* 5 (1): 12–16.

427 Shebaro, L., Bhalotra, S.R., and Herschbach, D. (1997). Molecular beam chemistry: formation of benzene and other higher hydrocarbons from small alkanes and alkenes in a catalytic supersonic nozzle. *J. Phys. Chem. A* 101 (36): 6775–6780.

428 Hommeltoft, S.I. (2001). Isobutane alkylation - recent developments, and future perspectives. *Appl. Catal., A* 221 (1–2): 421–428.

429 (a) Jones, D.S.J. and Pujadó, P.E. (eds.) (2006). *Handbook of Petroleum Processing*, 1–1353. Dodrecht, The Netherlands: Springer-Verlag. (b) Wang, H., Meng, X.-Z., Zhao, G., and Zhang, S.-J. (2017). Isobutane/butene alkylation catalyzed by ionic liquids: a more sustainable process for clean oil production. *Green Chem.* 19 (6): 1462–1489.

430 Wang, P., Wang, D., Xu, C.M. et al. (2007). Ab initio calculations on the mechanism of isobutane and 2-butene alkylation reaction catalyzed by hydrofluoric acid. *Catal. Today* 125 (3–4): 263–269.

431 de Klerk, A. and de Vaal, P.L. (2008). Alkylate technology selection for Fischer–Tropsch syncrude refining. *Ind. Eng. Chem. Res.* 47 (18): 6870–6877.

432 Crabtree, R.H. (2001). Alkane C—H activation and functionalization with homogeneous transition metal catalysts: a century of progress - a new millennium in prospect (vol. 19, p. 2437, 2001). *J. Chem. Soc., Dalton Trans.* (19): 2951.

433 Periana, R.A., Taube, D.J., Gamble, S. et al. (1998). Platinum catalysts for the high yield oxidation of methane to a methanol derivative. *Science* 280 (5363): 560–564.

434 Merkx, M., Kopp, D.A., Sazinsky, M.H. et al. (2001). Dioxygen activation and methane hydroxylation by soluble methane monooxygenase: a tale of two irons and three proteins. *Angew. Chem. Int. Ed.* 40 (15): 2782–2807.

435 Watson, P.L. and Parshall, G.W. (1985). Organolanthanides in catalysis. *Acc. Chem. Res.* 18 (2): 51–56.

436 Thompson, M.E., Baxter, S.M., Bulls, A.R. et al. (1987). α-Bond metathesis for C—H bonds of hydrocarbons and Sc-H, Sc-alkyl, Sc-aryl bonds of permethylscandocene derivatives - evidence for noninvolvement of the π-system in electrophilic activation of aromatic and vinylic C—H bonds. *J. Am. Chem. Soc.* 109 (1): 203–219.

437 Jordan, R.F. and Taylor, D.F. (1989). Zirconium-catalyzed coupling of propene and α-picoline. *J. Am. Chem. Soc.* 111 (2): 778–779.

438 Sadow, A.D. and Tilley, T.D. (2003). Homogeneous catalysis with methane. A strategy for the hydromethylation of olefins based on the nondegenerate exchange of alkyl groups and σ-bond metathesis at scandium. *J. Am. Chem. Soc.* 125 (26): 7971–7977.

439 Rueping, M., Nachtsheim, B.J., and Kuenkel, A. (2007). An efficient metal-catalyzed hydroalkylation. *Synlett* (9): 1391–1394.

440 Leitner, A., Larsen, J., Steffens, C., and Hartwig, J.F. (2004). Palladium-catalyzed addition of mono- and dicarbonyl compounds to conjugated dienes. *J. Org. Chem.* 69 (22): 7552–7557.

441 Pei, T. and Widenhoefer, R.A. (2001). Palladium-catalyzed intramolecular addition of 1,3-diones to unactivated olefins. *J. Am. Chem. Soc.* 123 (45): 11290–11291.

442 Herzon, S.B. and Hartwig, J.F. (2007). Direct, catalytic hydroaminoalkylation of unactivated olefins with *N*-alkyl arylamines. *J. Am. Chem. Soc.* 129 (21): 6690–6691.

443 Herzon, S.B. and Hartwig, J.F. (2008). Hydroaminoalkylation of unactivated olefins with dialkylamines. *J. Am. Chem. Soc.* 130 (45): 14940–14941.

444 Prochnow, I., Kubiak, R., Frey, O.N. et al. (2009). Tetrabenzyltitanium: an improved catalyst for the activation of C(sp^3)-H bonds adjacent to nitrogen atoms. *ChemCatChem* 1 (1): 162–172.

445 Trost, B.M. and Indolese, A. (1993). Ruthenium-catalyzed addition of alkenes to acetylenes. *J. Am. Chem. Soc.* 115 (10): 4361–4362.

446 Trost, B.M., Indolese, A.F., Müller, T.J.J., and Treptow, B. (1995). A Ru-catalyzed addition of alkenes to alkynes. *J. Am. Chem. Soc.* 117 (2): 615–623.

447 Trost, B.M. and Toste, F.D. (1999). A new Ru-catalyst for alkene/alkyne coupling. *Tetrahedron Lett.* 40 (44): 7739–7743.

448 Trost, B.M., Gunzner, J.L., Dirat, O., and Rhee, Y.H. (2002). Callipeltoside A: total synthesis, assignment of the absolute and relative configuration, and evaluation of synthetic analogues. *J. Am. Chem. Soc.* 124 (35): 10396–10415.

449 Trost, B.M., Machacek, M., and Schnaderbeck, M.J. (2000). A regioselective Ru-catalyzed alkene-alkyne coupling. *Org. Lett.* 2 (12): 1761–1764.

450 Trost, B.M. and Machacek, M.R. (2002). An efficient one-pot enantio- and diastereoselective synthesis of heterocycles. *Angew. Chem. Int. Ed.* 41 (24): 4693–4697.

451 Trost, B.M., Machacek, M.R., and Ball, Z.T. (2003). Ruthenium-catalyzed vinylsilane synthesis and cross-coupling as a selective awroach to alkenes: benzyldimethylsilyl as a robust vinylmetal functionality. *Org. Lett.* 5 (11): 1895–1898.

452 Hansen, E.C. and Lee, D.S. (2005). Synthesis of β,β-disubstituted vinyl boronates via the Ru-catalyzed Alder ene reaction of borylated alkynes and alkenes. *J. Am. Chem. Soc.* 127 (10): 3252–3253.

453 (a) Hong, X., Trost, B.M., and Houk, K.N. (2013). Mechanism and origins of selectivity in Ru(II)-catalyzed intramolecular (5+2)-cycloadditions and ene reactions of vinylcyclopropanes and alkynes from density functional theory. *J. Am. Chem. Soc.* 135 (17): 6588–6600. (b) Liu, R., Ni, Z.-J., Giordano, L., and Tenaglia, A. (2016). Ruthenium-catalyzed hydroalkynylative cyclization of 1,6-enynes induced by substituent effects. *Org. Lett.* 18 (16): 4040–4043.

454 Chen, H. and Li, S.H. (2005). Mechanism of Ru-catalyzed Alder ene-type reaction: a theoretical study. *Organometallics* 24 (5): 872–884.

455 Hilt, G. and Treutwein, J. (2007). Cobalt-catalyzed Alder-ene reaction. *Angew. Chem. Int. Ed.* 46 (44): 8500–8502.

456 Conia, J.M. and Le Perchec, P. (1975). Thermal cyclization of unsaturated carbonyl-compounds. *Synthesis* (1): 1–19.

457 Stammler, R. and Malacria, M. (1994). New Co-catalyzed cycloisomerization of β-ketoester ε-acetylenic. *Synlett* (1): 92–92.

458 Cruciani, P., Stammler, R., Aubert, C., and Malacria, M. (1996). New Co-catalyzed cycloisomerization of ε-acetylenic β-keto esters. Application to a powerful cyclization reactions cascade. *J. Org. Chem.* 61 (8): 2699–2708.

459 Renaud, J.L., Aubert, C., and Malacria, M. (1999). Cobalt-mediated cycloisomerization of δ-substituted ε-acetylenic β-ketoesters construction of angular triquinane by a sequence ene/Pauson–Khand reactions. *Tetrahedron* 55 (16): 5113–5128.

460 Fournet, G., Balme, G., Van Hemelryck, B., and Goré, J. (1990). Stereospecific synthesis of arylidene and allylidene cyclopentanes by a Pd-catalyzed cyclization. *Tetrahedron Lett.* 31 (36): 5147–5150.

461 Fournet, G., Balme, G., and Goré, J. (1991). Synthesis of (*E*)-arylidene and allylidene cyclopentanes by annelation catalyzed by a Pd(0) complex. *Tetrahedron* 47 (32): 6293–6304.

462 Kennedy-Smith, J.J., Staben, S.T., and Toste, F.D. (2004). Gold(I)-catalyzed Conia-ene reaction of β-ketoesters with alkynes. *J. Am. Chem. Soc.* 126 (14): 4526–4527.

463 Staben, S.T., Kennedy-Smith, J.J., and Toste, F.D. (2004). Gold(I)-catalyzed 5-endo-dig carbocyclization of acetylenic dicarbonyl compounds. *Angew. Chem. Int. Ed.* 43 (40): 5350–5352.

464 Mezailles, N., Ricard, L., and Gagosz, F. (2005). Phosphine Au(I) bis-(trifluoromethanesulfonyl)imidate complexes as new highly efficient and air-stable catalysts for the cycloisomerization of enynes. *Org. Lett.* 7 (19): 4133–4136.

465 Ochida, A., Ito, H., and Sawamura, M. (2006). Using triethynylphosphine ligands bearing bulky end caps to create a holey catalytic environment: application to Au(I)-catalyzed alkyne cyclizations. *J. Am. Chem. Soc.* 128 (51): 16486–16487.

466 Pan, J.H., Yang, M., Gao, Q. et al. (2007). Bulky thioureas as new ligands for Au(I)-catalyzed cyclization of acetylenic 1,3-dicarbonyl compounds. *Synthesis* (16): 2539–2544.

467 Gao, Q., Zheng, B.F., Li, J.H., and Yang, D. (2005). Nickel(II)-catalyzed Conia-ene reaction of 1,3-dicarbonyl compounds with alkynes. *Org. Lett.* 7 (11): 2185–2188.

468 Tsuji, H., Yamagata, K., Itoh, Y. et al. (2007). Indium-catalyzed cycloisomerization of ω-alkynyl-β-ketoesters into six- to fifteen-membered rings. *Angew. Chem. Int. Ed.* 46 (42): 8060–8062.

469 Itoh, Y., Tsuji, H., Yamagata, K. et al. (2008). Efficient formation of ring structures utilizing multisite activation by In-catalysis. *J. Am. Chem. Soc.* 130 (50): 17161–17167.

470 Deng, C.L., Song, R.J., Guo, S.M. et al. (2007). Copper/silver-co-catalyzed Conia-ene reaction of

linear β-alkynic beta-ketoesters. *Org. Lett.* 9 (24): 5111–5114.

471 Deng, C.L., Zou, T., Wang, Z.Q. et al. (2009). Copper/silver-co-catalyzed Conia-ene reactions of 2-alkynic 1,3-dicarbonyl compounds. *J. Org. Chem.* 74 (1): 412–414.

472 Cho, E.J. and Lee, D. (2007). Selectivity in the Ru-catalyzed Alder-ene reactions of di- and triynes. *J. Am. Chem. Soc.* 129 (21): 6692–6693.

473 Tanabe, K. and Hölderich, W.F. (1999). Industrial application of solid acid-base catalysts. *Appl. Catal., A* 181 (2): 399–434.

474 Perego, C. and Ingallina, P. (2002). Recent advances in the industrial alkylation of aromatics: new catalysts and new processes. *Catal. Today* 73 (1–2): 3–22.

475 Le Rouzo, G., Morel-Grepet, M., and Simonato, J.P. (2006). Synthesis of 4-*tert*-octylphenol and 4-cumylphenol by metal triflate and metal triflimidate catalysts. *J. Chem. Res.* (8): 521–522.

476 Rueping, M., Nachtsheim, B.J., and Scheidt, T. (2006). Efficient metal-catalyzed hydroarylation of styrenes. *Org. Lett.* 8 (17): 3717–3719.

477 Sun, H.B., Li, B., Hua, R.M., and Yin, Y.W. (2006). An efficient and selective hydroarylation of styrenes with electron-rich arenes, catalyzed by Bi(III) chloride and affording Markovnikov adducts. *Eur. J. Org. Chem.* (18): 4231–4236.

478 Chatt, J. and Davidson, J.M. (1965). Tautomerism of arene and ditertiary phosphine complexes of Ru(0) and preparation of new types of hydridocomplexes of Ru(II). *J. Chem. Soc.* 843–855.

479 Murai, S., Kakiuchi, F., Sekine, S. et al. (1993). Efficient catalytic addition of aromatic carbon-hydrogen bonds to olefins. *Nature* 366 (6455): 529–531.

480 Kakiuchi, F., Sonoda, M., Tsujimoto, T. et al. (1999). The Ru-catalyzed addition of C—H bonds in aromatic nitriles to olefins. *Chem. Lett.* (10): 1083–1084.

481 Kakiuchi, F. and Murai, S. (2002). Catalytic C—H/olefin coupling. *Acc. Chem. Res.* 35 (10): 826–834.

482 Kakiuchi, F. and Chatani, N. (2003). Catalytic methods for C—H bond functionalization: application in organic synthesis. *Adv. Synth. Catal.* 345: 1077–1101.

483 Lyons, T.W. and Sandford, M.S. (2010). Palladium-catalyzed ligand-directed C—H functionalization reactions. *Chem. Rev.* 110: 1147–1169.

484 Yoshikai, N. (2011). Cobalt-catalyzed, chelation-assisted C—H bond functionalization. *Synlett* 1047–1051.

485 Matsubara, T., Koga, N., Musaev, D.G., and Morokuma, K. (2000). Density functional study on highly *ortho*-selective addition of an aromatic C—H bond to olefins catalyzed by a $Ru(H)_2(CO)(PR_3)_3$ complex. *Organometallics* 19 (12): 2318–2329.

486 Kakiuchi, F. and Chatani, N. (2004). Activation of inert C—H bonds. *Top. Organomet. Chem.* 11: 45–79.

487 Crisenza, G.E.M., McCreanor, N.G., and Bower, J.F. (2014). Branch-selective, Ir-catalyzed hydroarylation of monosubstituted alkenes via a cooperative destabilization strategy. *J. Am. Chem. Soc.* 136: 10258–10261.

488 Crisenza, G.E.M., Sokolova, O.O., and Bower, J.F. (2015). Branch-selective alkene hydroarylation by cooperative destabilization: iridium-catalyzed *ortho*-alkylation of acetanilides. *Angew. Chem. Int. Ed.* 54: 14866–14870.

489 Bhalla, G., Bischof, S.M., Ganesh, S.K. et al. (2011). Mechanism of efficient anti-Markovnikov olefin hydroarylation catalyzed by homogeneous Ir(III) complexes. *Green Chem.* 13: 69–81.

490 Huang, G.P. and Liu, P. (2016). Mechanism and origins of ligand-controlled linear versus branched selectivity of iridium-catalyzed hydroarylation of alkenes. *ACS Catal.* 6: 809–820.

491 Pittard, K.A., Lee, J.P., Cundari, T.R. et al. (2004). Reactions of TpRu(CO)(NCMe)(Me) (Tp = hydridotris(pyrazolyl)borate) with heteroaromatic substrates: stoichiometric and catalytic C—H activation. *Organometallics* 23 (23): 5514–5523.

492 Lail, M., Arrowood, B.N., and Gunnoe, T.B. (2003). Addition of arenes to ethylene and propene catalyzed by Ru. *J. Am. Chem. Soc.* 125: 7506–7507.

493 Foley, N.A., Lail, M., Lee, J.P. et al. (2007). Comparative reactivity of TpRu(L)(NCMe)Ph (L = CO or PMe_3): impact of ancillary ligand L on activation of carbon-hydrogen bonds including catalytic hydroarylation and hydrovinylation/oligomerization of ethylene. *J. Am. Chem. Soc.* 129: 6765–6781.

494 Foley, N.A., Lee, J.P., Ke, Z.F. et al. (2009). Ru(II) catalysts supported by hydridotris(pyrazolyl)borate for the hydroarylation of olefins: reaction scope, mechanistic studies,

and guides for the development of improved catalysts. *Acc. Chem. Res.* 42: 585–597.

495 Andreatta, J.R., McKeown, B.A., and Gunnoe, T.B. (2011). Transition-metal-catalyzed hydroarylation of olefins using unactivated substrates: recent developments and challenges. *J. Organomet. Chem.* 696: 305–315.

496 Martinez, R., Chevalier, R., Darses, S., and Genêt, J.P. (2006). A versatile Ru-catalyst for C—C bond formation by C—H bond activation. *Angew. Chem Int. Ed.* 45: 8232–8235.

497 Martinez, R., Simon, M.O., Chevalier, R. et al. (2009). C—C bond formation via C—H bond activation using an in situ generated Ru catalyst. *J. Am. Chem. Soc.* 131 (22): 7887–7895.

498 Simon, M.O., Martinez, R., Genêt, J.P., and Darses, S. (2009). Ruthenium-catalyzed C—H bond activation of Michael acceptors: an unusual reactivity leading to allylsilanes. *Adv. Synth. Catal.* 351 (1–2): 153–157.

499 Arcadi, A., Bianchi, G., Chiarini, M. et al. (2004). Gold-catalyzed conjugate addition type reaction of indoles with α,β-enones. *Synlett* (6): 944–950.

500 Arcadi, A., Alfonsi, M., Bianchi, G. et al. (2006). Gold-catalysed direct couplings of indoles and pyrroles with 1,3-dicarbonyl compounds. *Adv. Synth. Catal.* 348 (3): 331–338.

501 Alfonsi, M., Arcadi, A., Bianchi, G. et al. (2006). Gold-catalyzed C-3-alkylation of 7-azaindoles through Michael-type addition to α,β-enones. *Eur. J. Org. Chem.* (10): 2393–2402.

502 Li, Z.G., Shi, Z.J., and He, C. (2005). Addition of heterocycles to electron deficient olefins and alkynes catalyzed by Au(III). *J. Organomet. Chem.* 690 (23): 5049–5054.

503 Aguilar, D., Contel, M., Navarro, R., and Urriolabeitia, E.P. (2007). Organogold(III) iminophosphorane complexes as efficient catalysts in the addition of 2-methylfuran and electron-rich arenes to methyl vinyl ketone. *Organometallics* 26: 4604–4611.

504 Youn, S.W., Pastine, S.J., and Sames, D. (2004). Ru(III)-catalyzed cyclization of arene-alkene substrates via intramolecular electrophilic hydroarylation. *Org. Lett.* 6: 581–584.

505 Watanabe, T., Oishi, S., Fujii, N., and Ohno, H. (2007). Gold-catalyzed hydroarylation of allenes: a highly regioselective carbon–carbon bond formation producing six-membered rings. *Org. Lett.* 9: 4821–4824.

506 Luo, Y.M. and Li, C.J. (2004). A highly efficient Au/Ag-catalyzed addition of arenes to imines. *J. Chem. Soc., Chem. Commun.* (17): 1930–1931.

507 Youn, S.W. (2006). Development of the Pictet–Spengler reaction catalyzed by $AuCl_3$/AgOTf. *J. Org. Chem.* 71 (6): 2521–2523.

508 Gao, K. and Yoshikai, N. (2014). Low-valent Co-catalysis: new opportunities for C—H functionalization. *Acc. Chem. Res.* 47: 1208–1219.

509 Wilson, R.M., Thalji, R.K., Bergman, R.G., and Ellman, J.A. (2006). Enantioselective synthesis of a PKC inhibitor via catalytic C—H bond activation. *Org. Lett.* 8 (8): 1745–1747.

510 Ye, B.H. and Cramer, N. (2015). Chiral cyclopentadienyls: enabling ligands for asymmetric Rh(III)-catalyzed C—H functionalizations. *Acc. Chem. Res.* 48: 1308–1318.

511 Kakiuchi, F., Yamamoto, Y., Chatani, N., and Murai, S. (1995). Catalytic addition of aromatic C—H bonds to acetylenes. *Chem. Lett.* (8): 681–682.

512 Kakiuchi, F., Tanaka, Y., Sato, T. et al. (1995). Catalytic addition of olefinic C—H bonds to olefins. *Chem. Lett.* (8): 679–680.

513 Vasil'ev, A.V. (2009). Alkenylation of aromatic compounds. *Russ. J. Org. Chem.* 45: 1–17.

514 Ackermann, L.; Vicente, R. Ruthenium-catalyzed direct arylations through C—H bond cleavages. In *C—H Activation.* Yu, J.-Q.; Shi, Z., Eds.; Topics in Current Chemistry, Springer-Verlag, Heidelberg, 2010, pp. 211–229.

515 Wang, X.L., Zhou, L.H., and Lu, W.J. (2010). Hydroarylation of alkynes via aryl C—H bond cleavage. *Curr. Org. Chem.* 14: 289–307.

516 Yamamoto, Y. (2014). Synthesis of heterocycles via transition-metal-catalyzed hydroarylation of alkynes. *Chem. Soc. Rev.* 43: 1575–1600.

517 Boyarskiy, V.P., Ryabukhin, D.S., Bokach, N.A., and Vasilyev, A.V. (2016). Alkenylation of arenes and heteroarenes with alkynes. *Chem. Rev.* 116: 5894–5986.

518 Manikandan, R. and Jeganmohan, M. (2015). Recent advances in the Ru-catalyzed hydroarylation of alkynes with aromatics: synthesis of trisubstituted alkenes. *Org. Biomol. Chem.* 13: 10420–10436.

519 Hong, P., Cho, B.R., and Yamazaki, H. (1979). Reactions of benzenes with acetylenes catalyzed by Rh-carbonyl under carbon monoxide. *Chem. Lett.* (4): 339–342.

520 Hong, P., Cho, B.R., and Yamazaki, H. (1980). Addition of 5-membered aromatic heterocyclic-compounds to acetylenes catalyzed by tetrarhodium dodecacarbonyl - syntheses of vinyl substituted aromatic heterocyclic compounds. *Chem. Lett.* (5): 507–510.

521 Ritleng, V., Sirlin, C., and Pfeffer, M. (2002). Ru-, Rh-, and Pd-catalyzed C—C bond formation involving C—H activation and addition on unsaturated substrates: reactions and mechanistic aspects. *Chem. Rev.* 102 (5): 1731–1769.

522 Halbritter, G., Knoch, F., Wolski, A., and Kisch, H. (1994). Transition-metal complexes of diazenes. 32. Functionalization of aromatic azo compounds by the Co-catalyzed, regioselective double addition of tolane - 2,6-Distilbenylazobenzenes and 2,3-dihydrocinnolines. *Angew. Chem. Int. Ed. Engl.* 33 (15–16): 1603–1605.

523 Aulwurm, U.R., Melchinger, J.U., and Kisch, H. (1995). Transition-metal complexes of diazenes. 35. Synthesis of 1-(arylamino)indoles by Rh-catalyzed addition of alkynes to 1,2-diaryldiazenes. *Organometallics* 14 (7): 3385–3395.

524 Hayashi, T., Inoue, K., Taniguchi, N., and Ogasawara, M. (2001). Rhodium-catalyzed hydroarylation of alkynes with arylboronic acids: 1,4-shift of Rh from (2-aryl-1-alkenyl)Rh to (2-alkenylaryl)Rh intermediate. *J. Am. Chem. Soc.* 123 (40): 9918–9919.

525 Sakaguchi, S., Kubo, T., and Ishii, Y. (2001). A three-component coupling reaction of aldehydes, amines, and alkynes. *Angew. Chem. Int. Ed.* 40 (13): 2534–2536.

526 Satoh, T., Nishinaka, Y., Miura, M., and Nomura, M. (1999). Iridium-catalyzed regioselective reaction of 1-naphthols with alkynes at the *peri*-position. *Chem. Lett.* (7): 615–616.

527 Jia, C.G., Piao, D.G., Oyamada, J.Z. et al. (2000). Efficient activation of aromatic C—H bonds for addition to C—C multiple bonds. *Science* 287 (5460): 1992–1995.

528 Jia, C.G., Kitamura, T., and Fujiwara, Y. (2001). Catalytic functionalization of arenes and alkanes via C—H bond activation. *Acc. Chem. Res.* 34 (8): 633–639.

529 Hashimoto, Y., Hirano, K., Satoh, T. et al. (2012). Ruthenium(II)-catalyzed regio- and stereoselective hydroarylation of alkynes via directed C—H functionalization. *Org. Lett.* 14: 2058–2061.

530 Skouta, R. and Li, C.J. (2008). Gold-catalyzed reactions of C—H bonds. *Tetrahedron* 64 (22): 4917–4938.

531 Harris, P.W.R., Rickard, C.E.F., and Woodgate, P.D. (1999). Ruthenium-catalysed *ortho* vinylation of aromatic ketones with alkynes; unexpected cyclopentaannulation. *J. Organomet. Chem.* 589 (2): 168–179.

532 Kakiuchi, F., Sato, T., Tsujimoto, T. et al. (1998). New protocol for the site-selective alkylation and vinylation of aromatic compounds. Catalyst-specific reactions. *Chem. Lett.* (10): 1053–1054.

533 Kuninobu, Y., Kawata, A., and Takai, K. (2005). Rhenium-catalyzed formation of indene frameworks via C—H bond activation: (3+2)-annulation of aromatic aldimines and acetylenes. *J. Am. Chem. Soc.* 127 (39): 13498–13499.

534 Kuninobu, Y., Tokunaga, Y., Kawata, A., and Takai, K. (2006). Insertion of polar and nonpolar unsaturated molecules into carbon-rhenium bonds generated by C—H bond activation: synthesis of phthalimidine and indene derivatives. *J. Am. Chem. Soc.* 128 (1): 202–209.

535 Liu, W.P., Zell, D., John, M., and Ackermann, L. (2015). Manganese-catalyzed synthesis of *cis*-β-amino acid esters through organometallic C—H activation of ketimines. *Angew. Chem. Int. Ed.* 54: 4092–4096.

536 Yi, C.S. and Yun, S.Y. (2005). Scope and mechanistic study of the Ru-catalyzed *ortho*-C—H bond activation and cyclization reactions of arylamines with terminal alkynes. *J. Am. Chem. Soc.* 127 (48): 17000–17006.

537 Tunge, J.A. and Foresee, L.N. (2005). Mechanistic studies of Fujiwara hydroarylation. C—H activation versus electrophilic aromatic substitution. *Organometallics* 24 (26): 6440–6444.

538 Kharasch, M.S. and Isbell, H.S. (1931). The chemistry of organic gold compounds. III. Direct introduction of gold into the aromatic nucleus (Preliminary communication). *J. Am. Chem. Soc.* 53 (2): 3053–3059.

539 de Graaf, P.W.J., Boersma, J., and van der Kerk, G.J.M. (1976). Preparation and properties of arylgold compounds - scope and limitations of auration reaction. *J. Organomet. Chem.* 105 (3): 399–406.

540 Fuchita, Y., Utsunomiya, Y., and Yasutake, M. (2001). Synthesis and reactivity of arylgold(III) complexes from aromatic hydrocarbons via C—H bond activation. *J. Chem. Soc., Dalton Trans.* (16): 2330–2334.

541 Reetz, M.T. and Sommer, K. (2003). Gold-catalyzed hydroarylation of alkynes. *Eur. J. Org. Chem.* (18): 3485–3496.

542 Shi, Z.J. and He, C. (2004). Efficient functionalization of aromatic C—H bonds catalyzed by Au(III) under mild and solvent-free conditions. *J. Org. Chem.* 69 (11): 3669–3671.

543 Ferrer, C. and Echavarren, A.M. (2006). Gold-catalyzed intramolecular reaction of indoles with alkynes: facile formation of eight-membered rings and an unexpected allenylation. *Angew. Chem. Int. Ed.* 45 (7): 1105–1109.

544 Ferrer, C., Amijs, C.H.M., and Echavarren, A.M. (2007). Intra- and intermolecular reactions of indoles with alkynes catalyzed by Au. *Chem. Eur. J.* 13 (5): 1358–1373.

545 Dorel, R.E. and M, A. (2015). Gold(I)-catalyzed activation of alkynes for the construction of molecular complexity. *Chem. Rev.* 115: 9028–9072.

546 Fukumoto, Y., Sawada, K., Hagihara, M. et al. (2002). $[Ir_4(CO)_{12}]$-catalyzed coupling reaction of imidazoles with aldehydes in the presence of a hydrosilane to give 2-substituted imidazoles. *Angew. Chem. Int. Ed.* 41: 2779–2781.

547 Li, Y., Li, B.J., Wang, W.H. et al. (2011). Rhodium-catalyzed direct addition of aryl C—H bonds to *N*-sulfonyl aldimines. *Angew. Chem. Int. Ed.* 50: 2115–2119.

548 Kuninobu, Y., Nishina, Y., Takeuchi, T., and Takai, K. (2007). Manganese-catalyzed insertion of aldehydes into a C—H bond. *Angew. Chem. Int. Ed.* 46: 6518–6520.

549 Zhou, B., Hou, W., Yang, Y.X., and Li, Y.C. (2013). Rhodium(III)-catalyzed amidation of aryl ketone *O*-methyl oximes with isocyanates by C—H activation: convergent synthesis of 3-methyleneisoindolin-1-ones. *Chem. Eur. J.* 19: 4701–4706.

550 Tsukada, N., Setoguchi, H., Mitsuboshi, T., and Inoue, Y. (2006). Hydroalkenylation of alkynes catalyzed by dinuclear Pd complexes via C—H bond activation. *Chem. Lett.* 35 (10): 1164–1165.

551 Lindhardt, A.T., Mantel, M.L.H., and Skrydstrup, T. (2008). Palladium-catalyzed intermolecular ene-yne coupling: development of an atom efficient Mizoroki–Heck-type reaction. *Angew. Chem. Int. Ed.* 47: 2668–2672.

552 Lewis, J.C., Bergman, R.G., and Ellman, J.A. (2008). Direct functionalization of nitrogen heterocycles via Rh-catalyzed C—H bond activation. *Acc. Chem. Res.* 41 (8): 1013–1025.

553 Goossen, L.J. (2002). Asymmetric hydrovinylation: new perspectives through use of modular ligand systems. *Angew. Chem. Int. Ed.* 41 (20): 3775–3778.

554 Moreau, B., Wu, J.Y., and Ritter, T. (2009). Iron-catalyzed 1,4-addition of α-olefins to dienes. *Org. Lett.* 11 (2): 337–339.

555 Ho, C.Y. and Jamison, T.F. (2007). Highly selective coupling of alkenes and aldehydes catalyzed by $[Ni(NHC)\{P(OPh)_3\}]$: synergy between a strong σ-donor and a strong π-acceptor. *Angew. Chem. Int. Ed.* 46 (5): 782–785.

556 Fürstner, A., Martin, R., and Majima, K. (2005). Cycloisomerization of enynes catalyzed by Fe(0)-ate complexes. *J. Am. Chem. Soc.* 127 (35): 12236–12237.

557 Jonas, K., Schieferstein, L., Krüger, C., and Tsay, Y.H. (1979). Tetrakis(ethylene)irondilithium and bis(η4-1,5-cyclooctadiene)irondilithium. *Angew. Chem. Int. Ed. Engl.* 18 (7): 550–551.

558 Jonas, K. and Krüger, C. (1980). Alkali metal/transition-metal π-complexes. *Angew. Chem. Int. Ed. Engl.* 19 (7): 520–537.

559 Jun, C.H., Jo, E.A., and Park, J.W. (2007). Intermolecular hydroacylation by transition-metal complexes. *Eur. J. Org. Chem.* 1869–1881.

560 Park, Y.J., Park, J.W., and Jun, C.H. (2008). Metal-organic cooperative catalysis in C—H and C—C bond activation and its concurrent recovery. *Acc. Chem. Res.* 41 (2): 222–234.

561 Willis, M.C. (2010). Transition-metal-catalyzed alkene and alkyne hydroacylation. *Chem. Rev.* 110: 725–748.

562 Leung, J.C. and Krische, M.J. (2012). Catalytic intermolecular hydroacylation of C—C π-bonds in the absence of chelation assistance. *Chem. Sci.* 3: 2202–2209.

563 Sakai, K., Oda, O., Nakamura, N., and Ide, J. (1972). Synthetic studies on prostanoids 1: Synthesis of methyl 9-oxoprostanoate. *Tetrahedron Lett.* 1287–1290.

564 Lochow, C.F. and Miller, R.G. (1976). Transition-metal-promoted aldehyde-alkene addition reactions. *J. Am. Chem. Soc.* 98 (5): 1281–1283.

565 Vora, K.P., Lochow, C.F., and Miller, R.G. (1980). Rhodium-catalyzed hydroacylation of ethylene with 4-pentenals - reactions of 4-hexenal-1-D. *J. Organomet. Chem.* 192 (2): 257–264.

566 James, B.R. and Young, C.G. (1983). The asymmetric cyclization of substituted pent-4-enals by a chiral Rh(phosphine) catalyst. *J. Chem. Soc., Chem. Commun.* (21): 1215–1216.

567 James, B.R. and Young, C.G. (1985). Catalytic decarbonylation, hydroacylation, and resolution of racemic pent-4-enals using chiral bis(di-*tertiary*-phosphine) complexes of Rh (I). *J. Organomet. Chem.* 285 (1–3): 321–332.

568 Roy, A.H., Lenges, C.P., and Brookhart, M. (2007). Scope and mechanism of the intermolecular addition of aromatic aldehydes to olefins

catalyzed by Rh(I)-olefin complexes. *J. Am. Chem. Soc.* 129: 2082–2093.

569 Suggs, J.W. (1978). Isolation of a stable acyl-rhodium(III) hydride intermediate formed during aldehyde decarbonylation - hydroacylation. *J. Am. Chem. Soc.* 100: 640–641.

570 Suggs, J.W. (1979). Activation of aldehyde C—H bonds to oxidative addition via formation of 3-methyl-2-aminopyridyl aldimines and related compounds - rhodium-based catalytic hydroacylation. *J. Am. Chem. Soc.* 101: 489.

571 Kondo, T., Tsuji, Y., and Watanabe, Y. (1987). Ruthenium complex catalyzed intermolecular hydroacylation of olefins. *Tetrahedron Lett.* 28 (49): 6229–6230.

572 Kondo, T., Akazome, M., Tsuji, Y., and Watanabe, Y. (1990). Ruthenium complex catalyzed intermolecular hydroacylation and transhydroformylation of olefins with aldehydes. *J. Org. Chem.* 55 (4): 1286–1291.

573 Coulter, M.M., Dornan, P.K., and Dong, V.M. (2009). Rhodium-catalyzed intramolecular olefin hydroacylation: enantioselective synthesis of seven- and eight-membered heterocycles. *J. Am. Chem. Soc.* 131: 6932–6933.

574 Johnson, K.F., Schneider, E.A., Schumacher, B.P. et al. (2016). Rhodium-catalyzed enantioselective intramolecular hydroacylation of trisubstituted alkenes. *Chem. Eur. J.* 22: 15619–15623.

575 Shibata, Y. and Tanaka, K. (2009). Rhodium-catalyzed highly enantioselective direct intermolecular hydroacylation of 1,1-disubstituted alkenes with unfunctionalized aldehydes. *J. Am. Chem. Soc.* 131: 12552–12553.

576 von Delius, M., Le, C.M., and Dong, V.M. (2012). Rhodium(phosphoramidite)-catalyzed alkene hydroacylation: mechanism and octaketide natural product synthesis. *J. Am. Chem. Soc.* 134: 15022–15032.

577 Bouisseau, A., Glancy, J., and Willis, M.C. (2016). Two-component assembly of thiochroman-4-ones and tetrahydrothiopyran-4-ones using a Rh-catalyzed alkyne hydroacylation/thio-conjugate-addition sequence. *Org. Lett.* 18: 5676–5679.

578 Bouisseau, A., Gao, M., and Willis, M.C. (2016). Traceless Rh-catalyzed hydroacylation using alkyl aldehydes: the enantioselective synthesis of aryl ketones. *Chem. Eur. J.* 22: 15624–15628.

579 Jun, C.H., Lee, D.Y., and Hong, J.B. (1997). Hydroacylation of 1-alkene with heteroaromatic aldehyde by Rh(I) and additives. *Tetrahedron Lett.* 38 (38): 6673–6676.

580 Jun, C.H., Lee, H., and Hong, J.B. (1997). Chelation-assisted intermolecular hydroacylation: direct synthesis of ketone from aldehyde and 1-alkene. *J. Org. Chem.* 62 (5): 1200–1201.

581 Jun, C.H., Lee, D.Y., Lee, H., and Hong, J.B. (2000). A highly active catalyst system for intermolecular hydroacylation. *Angew. Chem. Int. Ed.* 39 (17): 3070–3072.

582 Jo, E.A. and Jun, C.H. (2006). One-pot synthesis of oxo acid derivatives by Rh(I)-catalyzed chelation-assisted hydroacylation. *Eur. J. Org. Chem.* 2504–2507.

583 Murphy, S.K., Bruch, A., and Dong, V.M. (2014). Substrate-directed hydroacylation: rhodium-catalyzed coupling of vinylphenols and nonchelating aldehydes. *Angew. Chem. Int. Ed.* 53: 2455–2459.

584 Murphy, S.K., Bruch, A., and Dong, V.M. (2015). Mechanistic insights into hydroacylation with non-chelating aldehydes. *Chem. Sci.* 6: 174–180.

585 Lenges, C.P. and Brookhart, M. (1997). Co(I)-catalyzed inter- and intramolecular hydroacylation of olefins with aromatic aldehydes. *J. Am. Chem. Soc.* 119: 3165–3166.

586 Lenges, C.P., White, P.S., and Brookhart, M. (1998). Mechanistic and synthetic studies of the addition of alkyl aldehydes to vinylsilanes catalyzed by Co(I) complexes. *J. Am. Chem. Soc.* 120: 6965–6979.

587 Yang, J.F. and Yoshikai, N. (2014). Cobalt-catalyzed enantioselective intramolecular hydroacylation of ketones and olefins. *J. Am. Chem. Soc.* 136: 16748–16751.

588 Meng, Q.X. and Wang, F. (2016). Mechanisms and reactivity differences for the Co-catalyzed enantioselective intramolecular hydroacylation of ketones and alkenes: insights from density functional calculations. *J. Mol. Model.* 22 (3): 60.

589 Yang, J.F. and Yoshikai, N. (2016). Cobalt-catalyzed annulation of salicylaldehydes and alkynes to form chromones and 4-chromanones. *Angew. Chem. Int. Ed.* 55: 2870–2784.

590 Fukuyama, T., Doi, T., Minamino, S. et al. (2007). Ruthenium hydride catalyzed regioselective addition of aldehydes to enones to give 1,3-diketones. *Angew. Chem. Int. Ed.* 46: 5559–5561.

591 Meng, Q.X., Su, P.Y., Wang, F., and Zhu, S.H. (2016). Substituent effect and ligand exchange control the reactivity in Ru(II)-catalyzed hydroacylation of isoprenes and aldehydes: a DFT study. *J. Theor. Comput. Chem.* 15: 1650019.

592 Xiao, L.J., Fu, X.N., Zhou, M.J. et al. (2016). Nickel-catalyzed hydroacylation of styrenes with simple aldehydes: reaction development and mechanistic insights. *J. Am. Chem. Soc.* 138: 2957–2960.

593 Hirano, K., Biju, A.T., Piel, I., and Glorius, F. (2009). *N*-Heterocyclic carbene-catalyzed hydroacylation of unactivated double bonds. *J. Am. Chem. Soc.* 131: 14190–14191.

594 DiRocco, D.A. and Rovis, T. (2011). Organocatalytic hydroacylation of unactivated alkenes. *Angew. Chem. Int. Ed.* 50: 7982–7983.

595 Schedler, M., Wang, D.S., and Glorius, F. (2013). NHC-catalyzed hydroacylation of styrenes. *Angew. Chem. Int. Ed.* 52: 2585–2589.

596 Kondo, T., Hiraishi, N., Morisaki, Y. et al. (1998). First intermolecular hydroacylation of 1,3-dienes with aldehydes catalyzed by Ru. *Organometallics* 17 (10): 2131–2134.

597 Shibahara, F., Bower, J.F., and Krische, M.J. (2008). Diene hydroacylation from the alcohol or aldehyde oxidation level via Ru-catalyzed C—C bond-forming transfer hydrogenation: synthesis of β,γ-unsaturated ketones. *J. Am. Chem. Soc.* 130: 14120–14122.

598 Chen, Q.A., Kim, D.K., and Dong, V.M. (2014). Regioselective hydroacylation of 1,3-dienes by Co-catalysis. *J. Am. Chem. Soc.* 136: 3772–3775.

599 Tanaka, K. and Fu, G.C. (2001). A versatile new method for the synthesis of cyclopentenones via an unusual Rh-catalyzed intramolecular trans hydroacylation of an alkyne. *J. Am. Chem. Soc.* 123 (46): 11492–11493.

600 Jun, C.H., Lee, H., Hong, J.B., and Kwon, B.I. (2002). Efficient and selective hydroacylation of 1-alkynes with aldehydes by a chelation-assisted catalytic system. *Angew. Chem. Int. Ed.* 41 (12): 2146–2147.

601 Lee, D.Y., Moon, C.W., and Jun, C.H. (2002). Synthesis of aliphatic ketones from allylic alcohols through consecutive isomerization and chelation-assisted hydroacylation by a Rh-catalyst. *J. Org. Chem.* 67 (11): 3945–3948.

602 Neuhaus, J.D., Morrow, S.M., Brunavs, M., and Willis, M.C. (2016). Diversely substituted quinolines via Rh-catalyzed alkyne hydroacylation. *Org. Lett.* 18: 1562–1565.

603 Neuhaus, J.D. and Willis, M.C. (2016). Homogeneous Rh(I) catalysis in de novo heterocycle syntheses. *Org. Biomol. Chem.* 14: 4986–5000.

604 Pollack, P., Romeder, G., Hageborn, F., and Gelboin, H.P. (2012). Nitriles. In: *Ullmann's Encyclopedia of Industrial Chemistry*, 251–265. Weinheim: Wiley-VCH.

605 Tolman, C.A. (1977). Steric effects of phosphorus ligands in organometallic chemistry and homogeneous catalysis. *Chem. Rev.* 77 (3): 313–348.

606 Tolman, C.A., McKinney, R.J., Seidel, W.C. et al. (1985). Homogeneous Ni-catalyzed olefin hydrocyanation. *Adv. Catal.* 33: 1–46.

607 Tolman, C.A. (1986). Steric and electronic effects in olefin hydrocyanation at Du Pont - a scientific and industrial success story. *J. Chem. Educ.* 63 (3): 199–201.

608 Fischer, J. and Siegel, W. (1999). Preparation of monoolefinic C5 mononitrile mixtures by hydrocyanation in the presence of a catalyst containing at least one metallocene-phosphorus(III)-nickel(0) complex. Ger. Offen. Patent, DE 19733682 A1 19996211.

609 Fischer, J. and Siegel, W. (1999). Preparation of monoolefinic C5 mononitrile mixtures by hydrocyanation in the presence of a catalyst containing at least one metallocene-phosphorus(III)-nickel(0) complex. *Chem. Abstr.* 130: 169818.

610 Ahlers, W., Röper, M.L., Hofmann, P. et al. (2001). Phosphorus, arsenic and antimony compounds based upon diaryl-anellated bicyclo[2.2.n] parent substances, their preparation, catalysts containing them and use of the catalysts. PCT Int. Appl. Patent, WO 01/058589 A1.

611 Ahlers, W., Hofmann, P., Warth, D.C.M., and Paciello, R. (2001). Phosphorus, arsenic and antimony compounds based upon diaryl-anellated bicyclo[2.2.n] parent substances, their preparation, catalysts containing them and use of the catalysts. *Chem. Abstr.* 135: 168206.

612 Bini, L., Müller, C., Wilting, J. et al. (2007). Highly selective hydrocyanation of butadiene toward 3-pentenenitrile. *J. Am. Chem. Soc.* 129 (42): 12622–12623.

613 Göthlich, A.P.V., Tensfeldt, M., Rothfuss, H. et al. (2008). Novel chelating phosphonite ligands: syntheses, structures, and Ni-catalyzed hydrocyanation of olefins. *Organometallics* 27 (10): 2189–2200.

614 Krill, S. (2002). *Applied Homogeneous Catalysis with Organometallic Compounds*, 2e, vol. 1. Weinheim: Wiley-VCH.

615 Gaspar, B. and Carreira, E.M. (2007). Mild Co-catalyzed hydrocyanation of olefins with tosyl cyanide. *Angew. Chem. Int. Ed.* 46 (24): 4519–4522.

616 RajanBabu, T.V. and Casalnuovo, A.L. (1996). Role of electronic asymmetry in the design of

new ligands: the asymmetric hydrocyanation reaction. *J. Am. Chem. Soc.* 118: 6325–6326.

617 Kurono, N. and Ohkuma, T. (2016). Catalytic asymmetric cyanation reactions. *ACS Catalysis* 6: 989–1023.

618 Falk, A., Cavalieri, A., Nichol, G.S. et al. (2015). Enantioselective Ni-catalyzed hydrocyanation using chiral phosphine-phosphite ligands: recent improvements and insights. *Adv. Synth. Catal.* 357: 3317–3320.

619 Nemoto, K., Nagafuchi, T., Tominaga, K., and Sato, K. (2016). Efficient Ni-catalyzed hydrocyanation of alkenes using acetone cyanohydrin as a safer cyano source. *Tetrahedron Lett.* 57: 3199–3203.

620 Fang, X.J., Yu, P., and Morandi, B. (2016). Catalytic reversible alkene-nitrile interconversion through controllable transfer hydrocyanation. *Science* 351: 832–836.

621 Bäckvall, J.E. and Andell, O.S. (1986). Stereochemistry and mechanism of Ni-catalyzed hydrocyanation of olefins and conjugated dienes. *Organometallics* 5 (11): 2350–2355.

622 Hassan, A. and Krische, M.J. (2011). Unlocking hydrogenation for C—C bond formation: a brief overview of enantioselective methods. *Org. Process Res. Dev.* 15: 1236–1242.

623 Kablaoui, N.M. and Buchwald, S.L. (1995). Reductive cyclization of enones by a Ti catalyst. *J. Am. Chem. Soc.* 117: 6785–6786.

624 Kablaoui, N.M. and Buchwald, S.L. (1996). Development of a method for the reductive cyclization of enones by a Ti catalyst. *J. Am. Chem. Soc.* 118: 3182–3191.

625 Crowe, W.E. and Rachita, M.J. (1995). Titanium-catalyzed reductive cyclization of δ,ε-unsaturated ketones and aldehydes. *J. Am. Chem. Soc.* 117: 6787–6788.

626 Tang, X.Q. and Montgomery, J. (1999). Nickel catalysis in the stereoselective preparation of quinolizidine, pyrrolizidine, and indolizidine alkaloids: total synthesis of (+)-allopumiliotoxin 267A. *J. Am. Chem. Soc.* 121: 6098–6099.

627 Huang, W.S., Chan, J., and Jamison, T.F. (2000). Highly selective catalytic intermolecular reductive coupling of alkynes and aldehydes. *Org. Lett.* 2: 4221–4223.

628 Colby, E.A. and Jamison, T.F. (2003). P-chiral, monodentate ferrocenyl phosphines, novel ligands for asymmetric catalysis. *J. Org. Chem.* 68: 156–166.

629 Miller, K.M., Huang, W.S., and Jamison, T.F. (2003). Catalytic asymmetric reductive coupling of alkynes and aldehydes: enantioselective synthesis of allylic alcohols and alpha-hydroxy ketones. *J. Am. Chem. Soc.* 125: 3442–3443.

630 Takai, K., Sakamoto, S., and Isshiki, T. (2003). Regioselective reductive coupling of alkynes and aldehydes leading to allylic alcohols. *Org. Lett.* 5: 653–655.

631 Revis, A. and Hilty, T.K. (1987). Novel synthesis of β-siloxy esters by condensation of carbonyls and trimethylsilane with α,β-unsaturated esters catalyzed by $RhCl_3$. *Tetrahedron Lett.* 28: 4809–4812.

632 Isayama, S. and Mukaiyama, T. (1989). Cobalt(II)-catalyzed coupling reaction of α,β-unsaturated compounds with aldehydes by the use of phenylsilane - new method for preparation of β-hydroxy nitriles, amides, and esters. *Chem. Lett.* 2005–2008.

633 Matsuda, I., Takahashi, K., and Sato, S. (1990). Rhodium-catalyzed direct coupling of α,β-unsaturated ketone, aldehyde, and trialkylsilane - an easy access to regio-defined aldol derivatives. *Tetrahedron Lett.* 31: 5331–5334.

634 Kiyooka, S., Shimizu, A., and Torii, S. (1998). A mild aldol reaction of aryl aldehydes through Pd-catalyzed hydrosilation of α,β-unsaturated carbonyl compounds with trichlorosilane. *Tetrahedron Lett.* 39: 5237–5238.

635 Ooi, T., Doda, K., Sakai, D., and Maruoka, K. (1999). Unique property of Co(I) chloride as a radical initiator as well as a Lewis acid: application to CuCl-catalyzed aldol reaction of α,β-unsaturated ketones with Bu_3SnH. *Tetrahedron Lett.* 40: 2133–2136.

636 Taylor, S.J., Duffey, M.O., and Morken, J.P. (2000). Rhodium-catalyzed enantioselective reductive aldol reaction. *J. Am. Chem. Soc.* 122: 4528–4529.

637 Zhao, C.X., Duffey, M.O., Taylor, S.J., and Morken, J.P. (2001). Enantio- and diastereoselective reductive aldol reactions with Ir-pybox catalysts. *Org. Lett.* 3: 1829–1831.

638 Baik, T.G., Luis, A.L., Wang, L.C., and Krische, M.J. (2001). Diastereoselective Co-catalyzed aldol and Michael cycloreductions. *J. Am. Chem. Soc.* 123: 5112–5113.

639 Emiabata-Smith, D., McKillop, A., Mills, C. et al. (2001). Some preliminary studies on a novel Rh(I)-catalysed tandem hydrosilylation-intramolecular aldol reaction. *Synlett* 1302–1304.

640 Huddleston, R.R., Cauble, D.F., and Krische, M.J. (2003). Borane-mediated aldol cycloreduction of monoenone monoketones: diastereoselective

formation of quaternary centers. *J. Org. Chem.* 68: 11–14.

641 Sato, Y., Takimoto, M., Hayashi, K. et al. (1994). Novel stereoselective cyclization via (π-allyl)Ni complex generated from 1,3-diene and hydride Ni-complex. *J. Am. Chem. Soc.* 116: 9771–9772.

642 Sato, Y., Takimoto, M., and Mori, M. (1996). Remarkable regio-controlled effect of 1,3-diene as a ligand on Ni-promoted cyclization. *Tetrahedron Lett.* 37: 887–890.

643 Takimoto, M., Hiraga, Y., Sato, Y., and Mori, M. (1998). Nickel-catalyzed regio- and stereo-selective synthesis of homoallylic alcohol derivatives from dienes and aldehydes. *Tetrahedron Lett.* 39: 4543–4546.

644 Kimura, M., Ezoe, A., Shibata, K., and Tamaru, Y. (1998). Novel and highly regio- and stereoselective Ni-catalyzed homoallylation of benzaldehyde with 1,3-dienes. *J. Am. Chem. Soc.* 120: 4033–4034.

645 Kimura, M., Fujimatsu, H., Ezoe, A. et al. (1999). Nickel-catalyzed homoallylation of aldehydes and ketones with 1,3-dienes and complementary promotion by diethylzinc or triethylborane. *Angew. Chem. Int. Ed.* 38: 397–400.

646 Sato, Y., Saito, N., and Mori, M. (2000). A novel asymmetric cyclization of ω-formyl-1,3-dienes catalyzed by a zerovalent Ni-complex in the presence of silanes. *J. Am. Chem. Soc.* 122: 2371–2372.

647 Kimura, M., Shibata, K., Koudahashi, Y., and Tamaru, Y. (2000). Nickel(0)-catalyzed regio- and stereoselective coupling reaction of Me_3B, isoprene, and aldehydes. *Tetrahedron Lett.* 41: 6789–6793.

648 Kimura, M., Ezoe, A., Tanaka, S., and Tamaru, Y. (2001). Nickel-catalyzed homoallylation of aldehydes in the presence of water and alcohols. *Angew. Chem. Int. Ed.* 40: 3600–3602.

649 Loh, T.P., Song, H.Y., and Zhou, Y. (2002). Nickel-catalyzed homoallylation reaction of aldehydes with 1,3-dienes: stereochemical and mechanistic studies. *Org. Lett.* 4: 2715–2717.

650 Brothers, P.J. (1981). Heterolytic activation of hydrogen by transition metal complexes. In: *Progress in Inorganic Chemistry*, vol. 28 (ed. S.J. Lippard), 1–61. Hoboken, NJ: Wiley.

651 Jeske, G., Lauke, H., Mauermann, H. et al. (1985). Highly reactive organolanthanides - a mechanistic study of catalytic olefin hydrogenation by bis(pentamethylcyclopentadienyl) and related 4f complexes. *J. Am. Chem. Soc.* 107 (26): 8111–8118.

652 Jang, H.Y. and Krische, M.J. (2004). Catalytic C—C bond formation via capture of hydrogenation intermediates. *Acc. Chem. Res.* 37: 653–661.

653 Bower, J.F. and Krische, M.J. (2011). Formation of C—C bonds via Ir-catalyzed hydrogenation and transfer hydrogenation. In: *Iridium Catalysis* (ed. P.G. Andersson), 107–138. Berlin, Heidelberg: Springer-Verlag.

654 Jang, H.Y., Huddleston, R.R., and Krische, M.J. (2002). Reductive generation of enolates from enones using elemental hydrogen: catalytic C—C bond formation under hydrogenative conditions. *J. Am. Chem. Soc.* 124: 15156–15157.

655 Nguyen, K.D., Park, B.Y., Luong, T. et al. (2016). Metal-catalyzed reductive coupling of olefin-derived nucleophiles: reinventing carbonyl addition. *Science* 354: 300–U88.

656 Jang, H.Y. and Krische, M.J. (2004). Catalytic hydrogen-mediated cross-coupling of enones and carbonyl compounds: aldol condensation by hydrogenation. *Eur. J. Org. Chem.* 3953–3958.

657 Ngai, M.Y., Kong, J.R., and Krische, M.J. (2007). Hydrogen-mediated C—C bond formation: a broad new concept in catalytic C—C coupling. *J. Org. Chem.* 72: 1063–1072.

658 Komanduri, V. and Krische, M.J. (2006). Enantioselective reductive coupling of 1,3-enynes to heterocyclic aromatic aldehydes and ketones via Rh-catalyzed asymmetric hydrogenation: mechanistic insight into the role of Brønsted acid additives. *J. Am. Chem. Soc.* 128: 16448–16449.

659 Kong, J.R., Ngai, M.Y., and Krische, M.J. (2006). Highly enantioselective direct reductive coupling of conjugated alkynes and α-ketoesters via Rh-catalyzed asymmetric hydrogenation. *J. Am. Chem. Soc.* 128: 718–719.

660 Cho, C.W. and Krische, M.J. (2006). Enantioselective reductive coupling of alkynes and α-keto aldehydes via Rh-catalyzed hydrogenation: an approach to Bryostatin substructures. *Org. Lett.* 8: 891–894.

661 Jang, H.Y., Hughes, F.W., Gong, H.G. et al. (2005). Enantioselective reductive cyclization of 1,6-enynes via Rh-catalyzed asymmetric hydrogenation: C—C bond formation precedes hydrogen activation. *J. Am. Chem. Soc.* 127: 6174–6175.

662 Hong, Y.T., Barchuk, A., and Krische, M.J. (2006). Branch-selective intermolecular hydroacylation: hydrogen-mediated coupling of anhydrides to styrenes and activated olefins. *Angew. Chem. Int. Ed.* 45: 6885–6888.

663 Komanduri, V., Grant, C.D., and Krische, M.J. (2008). Branch-selective reductive coupling of 2-vinyl pyridines and imines via Rh-catalyzed C—C bond forming hydrogenation. *J. Am. Chem. Soc.* 130: 12592.

664 Ngai, M.Y., Rucas, E., and Krische, M.J. (2008). Ruthenium-catalyzed C—C bond formation via transfer hydrogenation: branch-selective reductive coupling of allenes to paraformaldehyde and higher aldehydes. *Org. Lett.* 10: 2705–2708.

665 Shibahara, F., Bower, J.F., and Krische, M.J. (2008). Ruthenium-catalyzed C—C bond forming transfer hydrogenation: carbonyl allylation from the alcohol or aldehyde oxidation level employing acyclic 1,3-dienes as surrogates to preformed allyl metal reagents. *J. Am. Chem. Soc.* 130: 6338–6339.

666 Sam, B., Breit, B., and Krische, M.J. (2015). Paraformaldehyde and methanol as C_1 feedstocks in metal-catalyzed C—C couplings of π-unsaturated reactants: beyond hydroformylation. *Angew. Chem. Int. Ed.* 54: 3267–3274.

667 Oda, S., Franke, J., and Krische, M.J. (2016). Diene hydroaminomethylation via Ru-catalyzed C—C bond forming transfer hydrogenation: beyond carbonylation. *Chem. Sci.* 7: 136–141.

668 Eilbracht, P., Barfacker, L., Buss, C. et al. (1999). Tandem reaction sequences under hydroformylation conditions: new synthetic applications of transition metal catalysis. *Chem. Rev.* 99: 3329–3365.

669 Eilbracht, P. and Schmidt, A. (2006). Synthetic applications of tandem reaction sequences involving hydroformylation. In: *Catalytic Carbonylation Reactions*, Topics in Organometallic Chemistry, vol. 18 (ed. M. Beller), 65–95. Berlin, Heidelberg: Springer-Verlag.

670 Chong, E., García, P., and Schafer, L.L. (2014). Hydroaminoalkylation: early-transition-metal-catalyzed α-alkylation of amines. *Synlett* 46: 2884–2896.

671 Dörfler, J., Preuss, T., Brahms, C. et al. (2015). Intermolecular hydroaminoalkylation of alkenes and dienes using a Ti-mono(formamidinate) catalyst. *J. Chem. Soc., Dalton Trans.* 44: 12149–12168.

672 Miyake, Y., Ashida, Y., Nakajima, K., and Nishibayashi, Y. (2012). Visible-light-mediated addition of α-aminoalkyl radicals generated from α-silylamines to α,β-unsaturated carbonyl compounds. *J. Chem. Soc., Chem. Commun.* 48 (55): 6966–6968.

673 Miyake, Y., Nakajima, K., and Nishibayashi, Y. (2012). Visible-light-mediated utilization of α-aminoalkyl radicals: addition to electron-deficient alkenes using photoredox catalysts. *J. Am. Chem. Soc.* 134 (7): 3338–3341.

674 Oda, S., Sam, B., and Krische, M.J. (2015). Hydroaminomethylation beyond carbonylation: allene-imine reductive coupling by Ru-catalyzed transfer hydrogenation. *Angew. Chem. Int. Ed.* 54: 8525–8528.

675 (a) Bower, J.F., Kim, I.S., Patman, R.L., and Krische, M.J. (2009). Catalytic carbonyl addition through transfer hydrogenation: a departure from preformed organometallic reagents. *Angew. Chem. Int. Ed.* 48: 34–46. (b) Kim, S.W., Zhang, W.-D., and Krische, M.J. (2017). Catalytic enantioselective carbonyl allylation and propargylation via alcohol-mediated hydrogen transfer: merging the chemistry of Grignard and Sabatier. *Acc. Chem. Res.* 50 (9): 2371–2380.

676 Zbieg, J.R., Moran, J., and Krische, M.J. (2011). Diastereo- and enantioselective Ru-catalyzed hydrohydroxyalkylation of 2-silyl-butadienes: carbonyl *syn*-crotylation from the alcohol oxidation level. *J. Am. Chem. Soc.* 133: 10582–10586.

677 (a) Zbieg, J.R., Yamaguchi, E., McInturff, E.L., and Krische, M.J. (2012). Enantioselective C—H crotylation of primary alcohols via hydrohydroxyalkylation of butadiene. *Science* 336: 324–327. (b) McInturff, E.L., Yamaguchi, E., and Krische, M.J. (2012). Chiral-anion-dependent inversion of diastereo- and enantioselectivity in carbonyl crotylation via Ru-catalyzed butadiene hydrohydroxyalkylation. *J. Am. Chem. Soc.* 134: 20628–20631.

678 Ketcham, J.M., Shin, I., Montgomery, T.P., and Krische, M.J. (2014). Catalytic enantioselective C—H functionalization of alcohols by redox-triggered carbonyl addition: borrowing hydrogen, returning carbon. *Angew. Chem. Int. Ed.* 53: 9142–9150.

679 Feng, J.-J., Kasun, Z.A., and Krische, M.J. (2016). Enantioselective alcohol C—H functionalization for polyketide construction: unlocking redox-economy and site-selectivity for ideal chemical synthesis. *J. Am. Chem. Soc.* 138 (17): 5467–5478.

680 Patman, R.L., Williams, V.M., Bower, J.F., and Krische, M.J. (2008). Carbonyl propargylation from the alcohol or aldehyde oxidation level employing 1,3-enynes as surrogates to preformed allenylmetal reagents: a Ru-catalyzed C—C bond-forming transfer hydrogenation. *Angew. Chem. Int. Ed.* 47: 5220–5223.

681 Geary, L.M., Woo, S.K., Leung, J.C., and Krische, M.J. (2012). Diastereo- and enantioselective Ir-catalyzed carbonyl propargylation from the alcohol or aldehyde oxidation level: 1,3-Enynes as allenylmetal equivalents. *Angew. Chem. Int. Ed.* 51: 2972–2976.

682 Nguyen, K.D., Herkommer, D., and Krische, M.J. (2016). Ruthenium-BINAP catalyzed alcohol C—H *tert*-prenylation via 1,3-enyne transfer hydrogenation: beyond stoichiometric carbanions in enantioselective carbonyl propargylation. *J. Am. Chem. Soc.* 138: 5238–5241.

683 Moran, J., Preetz, A., Mesch, R.A., and Krische, M.J. (2011). Iridium-catalyzed direct C—C coupling of methanol and allenes. *Nat. Chem.* 3: 287–290.

684 Chen, T.Y. and Krische, M.J. (2013). Regioselective Ru-catalyzed hydrohydroxyalkylation of dienes with 3-hydroxy-2-oxindoles: prenylation, geranylation, and beyond. *Org. Lett.* 15: 2994–2997.

685 Nguyen, K.D., Herkommer, D., and Krische, M.J. (2016). Enantioselective formation of all-carbon quaternary centers via C—H functionalization of methanol: iridium-catalyzed diene hydrohydroxymethylation. *J. Am. Chem. Soc.* 138: 14210–14213.

686 Chen, T.Y., Tsutsumi, R., Montgomery, T.P. et al. (2015). Ruthenium-catalyzed C—C coupling of amino alcohols with dienes via transfer hydrogenation: redox-triggered imine addition and related hydroaminoalkylations. *J. Am. Chem. Soc.* 137: 1798–1801.

687 Souillart, L., Parker, E., and Cramer, N. (2014). Highly enantioselective Rh(I)-catalyzed activation of enantiotopic cyclobutanone C—C bonds. *Angew. Chem. Int. Ed.* 53: 3001–3005.

688 Souillart, L. and Cramer, N. (2014). Highly enantioselective Rh(I)-catalyzed carbonyl carboacylations initiated by C—C bond activation. *Angew. Chem. Int. Ed.* 53: 9640–9644.

689 Lautens, M., Klute, W., and Tam, W. (1996). Transition metal-mediated cycloaddition reactions. *Chem. Rev.* 96 (1): 49–92.

690 Ojima, I., Tzamarioudaki, M., Li, Z.Y., and Donovan, R.J. (1996). Transition metal-catalyzed carbocyclizations in organic synthesis. *Chem. Rev.* 96 (2): 635–662.

691 Hegedus, L.S. (1997). Transition metals in organic synthesis. Highlights for the year 1995. *Coord. Chem. Rev.* 161: 129–255.

692 Dell, C.P. (1997). Cycloadditions in synthesis. *Contemp. Org. Synth.* 4 (2): 87–117.

693 Fruhauf, H.W. (1997). Metal-assisted cycloaddition reactions in organotransition metal chemistry. *Chem. Rev.* 97 (3): 523–596.

694 Filippini, M.H. and Rodriguez, J. (1999). Synthesis of functionalized bicyclo[3.2.1]octanes and their multiple uses in organic chemistry. *Chem. Rev.* 99 (1): 27–76.

695 Yet, L. (2000). Metal-mediated synthesis of medium-sized rings. *Chem. Rev.* 100 (8): 2963–3008.

696 (a) Nakao, Y. (2014). Catalytic C-CN bond activation. In: *C-C-Bond Activation*, Top. Curr. Chem. (Vol. ed. G. Dong), vol. 346, 33–58. Heildelberg: Springer Verlag; (b) Wu, A., Chen, Q., Liu, W. et al. (2018). Transition-metal-free arylation of benzoxazoles with aryl nitriles. *Org. Chem. Front.* 5 (11): 1811–1814.

697 Hartley, R.C. and Caldwell, S.T. (2000). Novel methods for the synthesis of three-, four-, five-, six- and seven-membered, saturated and partially unsaturated carbocycles. *J. Chem. Soc., Perkin Trans. 1* (4): 477–501.

698 Wender, P.A., Bi, F.C., Gamber, G.G. et al. (2002). Toward the ideal synthesis. New transition metal-catalyzed reactions inspired by novel medicinal leads. *Pure Appl. Chem.* 74 (1): 25–31.

699 Nakao, Y. (2012). Nickel/Lewis acid-catalyzed carbocyanation of unsaturated compounds. *Bull. Chem. Soc. Jpn.* 85: 731–745.

700 Iverson, C.N. and Jones, W.D. (2001). Rhodium-catalyzed activation and functionalization of the C—C bond of biphenylene. *Organometallics* 20 (26): 5745–5750.

701 Müller, C., Lachicotte, R.J., and Jones, W.D. (2002). Catalytic C—C bond activation in biphenylene and cyclotrimerization of alkynes: increased reactivity of P,N- versus P,P-substituted nickel complexes. *Organometallics* 21 (9): 1975–1981.

702 Schaub, T., Backes, M., and Radius, U. (2006). Nickel(0) complexes of *N*-alkyl-substituted *N*-heterocyclic carbenes and their use in the catalytic carbon–carbon bond activation of biphenylene. *Organometallics* 25 (17): 4196–4206.

703 Sakurai, H. and Imai, T. (1975). Chemistry of organosilicon compounds. 83. Novel [σ+π]-cycloaddition of silacyclobutanes with acetylenes catalyzed by Pd-complexes. *Chem. Lett.* (8): 891–894.

704 Horiguchi, Y., Suehiro, I., Sasaki, A., and Kuwajima, I. (1993). Diastereoselective (3+2)-cycloaddition of methyl 2-phenylthiocyclopropyl ketone with enol silyl

ethers - synthesis of functionalized cyclopentanes. *Tetrahedron Lett.* 34 (38): 6077–6080.

705 Monti, H., Rizzotto, D., Léandri, G., and Monti, J.P. (1994). A novel cyclopentane annulation by (3+2)-cycloaddition of substituted methylenecyclopropyl ketones with allyltrimethylsilane. *Tetrahedron Lett.* 35 (18): 2885–2888.

706 Yadav, V.K. and Sriramurthy, V. (2004). Formal (3+2)-addition of acceptor-substituted cyclopropylmethylsilanes with aryl acetylenes. *Angew. Chem. Int. Ed.* 43 (20): 2669–2671.

707 Souillart, L. and Cramer, N. (2015). Catalytic C—C bond activations via oxidative addition to transition metals. *Chem. Rev.* 115: 9410–9464.

708 Fang, J., Ren, J., and Wang, Z.W. (2008). $Sc(OTf)_3$-catalyzed smooth tandem (3+2)-cycloaddition/ring opening of donor-acceptor cyclopropane 1,1-diesters with enol silyl ethers. *Tetrahedron Lett.* 49 (47): 6659–6662.

709 Takasu, K., Nagao, S., and Ihara, M. (2006). Construction of highly-functionalized cyclopentanes from silyl enol ethers and activated cyclopropanes by (3+2)-cycloaddition catalyzed by triflic imide. *Adv. Synth. Catal.* 348 (16–17): 2376–2380.

710 Büchner, E. and Curtius, T. (1885). Synthesis of ketonic acids from aldehydes and ethyl diazoacetate. *Ber. Dtsch. Chem. Ges.* 18: 2371–2377.

711 Dave, V. and Warnhoff, E.W. (1970). Reactions of diazoacetic esters with alkenes, alkynes, heterocyclic and aromatic compounds. *Org. React.* 18: 217–401.

712 Hodgkins, J.E. and Flores, R.J. (1963). Deamination of 7-aminobicyclo[4.1.0]heptane. *J. Org. Chem.* 28 (12): 3356–3359.

713 LaLonde, R.T. and Tobias, M.A. (1964). Carbon–carbon bond fission in cyclopropanes. 4. Acid-promoted opening of three-membered ring in *exo*-7-methylbicyclo[4.1.0]heptane + *exo*-6-methylbicyclo[3.1.0]hexane. *J. Am. Chem. Soc.* 86 (19): 4068–4073.

714 Lewin, A.H., Zovko, M.J., Rosewater, W.H., and Cohen, T. (1967). Use of complexing agents in activation of Cu for Ullmann biaryl synthesis. *J. Chem. Soc., Chem. Commun.* (2): 80–81.

715 Nishizawa, Y. (1961). Studies on organophosphorus compounds. 2. On phosphite Cu halide complex compounds. *Bull. Chem. Soc. Jpn.* 34 (8): 1170–1178.

716 Arbusoff, A. Phosphoric acid ester. *Ber. Dtsch. Chem. Ges.* 1905, 38, 1171–1173.

717 Moser, W.R. (1969). Mechanism of Cu-catalyzed addition of diazoalkanes to olefins. I. Steric effects. *J. Am. Chem. Soc.* 91 (5): 1135–1140.

718 Moser, W.R. (1969). Mechanism of Cu-catalyzed addition of diazoalkanes to olefins. 2. Electronic effects. *J. Am. Chem. Soc.* 91 (5): 1141–1146.

719 Salomon, R.G. and Kochi, J.K. (1973). Copper(I) catalysis in cyclopropanations with diazo compounds - role of olefin coordination. *J. Am. Chem. Soc.* 95 (10): 3300–3310.

720 Paulissen, R., Teyssié, P., and Hubert, A.J. (1972). Transition-metal-catalyzed cyclopropanation of olefins. *Tetrahedron Lett.* 13 (15): 1465–1466.

721 Adams, J. and Spero, D.M. (1991). Rhodium(II)-catalyzed reactions of diazo-carbonyl compounds. *Tetrahedron* 47 (10–11): 1765–1808.

722 Calter, M.A. (1997). Transition-metal-catalyzed, asymmetric reactions of diazo compounds. *Curr. Org. Chem.* 1 (1): 37–70.

723 Doyle, M.P. and Forbes, D.C. (1998). Recent advances in asymmetric catalytic metal(carbene) transformations. *Chem. Rev.* 98 (2): 911–935.

724 Davies, H.M.L. and Antoulinakis, E.G. (2001). Recent progress in asymmetric intermolecular C—H activation by Rh(carbenoid) intermediates. *J. Organomet. Chem.* 617 (1): 47–55.

725 Lebel, H., Marcoux, J.F., Molinaro, C., and Charette, A.B. (2003). Stereoselective cyclopropanation reactions. *Chem. Rev.* 103 (4): 977–1050.

726 Nozaki, H., Moriuti, S., Takaya, H., and Noyori, R. (1966). Asymmetric induction in carbenoid reaction by means of a dissymmetric Cu-chelate. *Tetrahedron Lett.* 7 (43): 5239–5244.

727 Aratani, T. (1985). Catalytic asymmetric synthesis of cyclopropanecarboxylic acids - an application of chiral Cu(carbenoid) reaction. *Pure Appl. Chem.* 57 (12): 1839–1844.

728 Bergmann, A. and Reiser, O. (2014). Enantioselective synthesis of xanthatin. *Chem. Eur. J.* 20: 7613–7615.

729 Straub, B.F. and Hofmann, P. (2001). Copper(I) carbenes: the synthesis of active intermediates in Cu-catalyzed CS cyclopropanation. *Angew. Chem. Int. Ed.* 40 (7): 1288–1290.

730 Smith, D.A., Reynolds, D.N., and Woo, L.K. (1993). Cyclopropanation catalyzed by Os(porphyrin) complexes. *J. Am. Chem. Soc.* 115 (6): 2511–2513.

731 Park, S.B., Sakata, N., and Nishiyama, H. (1996). Aryloxycarbonylcarbene complexes of [bis(oxazolinyl)pyridine]Ru as active intermediates in asymmetric catalytic cyclopropanations. *Chem. Eur. J.* 2 (3): 303–306.

732 Lowenthal, R.E., Abiko, A., and Masamune, S. (1990). Asymmetric catalytic cyclopropanation of olefins - [bis(oxazoline)]Cu complexes. *Tetrahedron Lett.* 31 (42): 6005–6008.

733 Rasmussen, T., Jensen, J.F., Østergaard, N. et al. (2002). On the mechanism of the Cu-catalyzed cyclopropanation reaction. *Chem. Eur. J.* 8 (1): 177–184.

734 Özen, C. and Tüzün, N.S. (2008). A DFT study on the mechanism of cyclopropanation via $Cu(acac)_2$-catalyzed diazo ester decomposition. *Organometallics* 27 (18): 4600–4610.

735 Grubbs, R.H. (1979). Mechanism of olefin metathesis and cyclopropanation. *Inorg. Chem.* 18 (9): 2623–2626.

736 Donaldson, W.A. (2001). Synthesis of cyclopropane containing natural products. *Tetrahedron* 57 (41): 8589–8627.

737 Honma, M., Takeda, H., Takano, M., and Nakada, M. (2009). Development of catalytic asymmetric intramolecular cyclopropanation of α-diazo-ß-keto sulfones and applications to natural product synthesis. *Synlett* 1695–1712.

738 Ruppel, J.V., Gauthier, T.J., Snyder, N.L. et al. (2009). Asymmetric Co(II)-catalyzed cyclopropanation with succinimidyl diazoacetate: general synthesis of chiral cyclopropyl carboxamides. *Org. Lett.* 11: 2273–2276.

739 Nakada, M. (2014). Enantioselective total syntheses of cyathane diterpenoids. *Chem. Rec.* 14: 641–662.

740 Cao, Z.Y. and Zhou, J. (2015). Catalytic asymmetric synthesis of polysubstituted spirocyclopropyl oxindoles: organocatalysis versus transition metal catalysis. *Org. Chem. Front.* 2: 849–858.

741 Kottwitz, J. and Vorbrüggen, H. (1975). Formation of cyclopropanes from strained alkenes with diazomethane/Pd(II) acetate. *Synthesis* (10): 636–637.

742 Suda, M. (1981). Cyclopropanation of terminal olefins using diazomethane/Pd(II) acetate. *Synthesis* (9): 714–714.

743 Doyle, M.P. (1986). Catalytic methods for metal carbene transformations. *Chem. Rev.* 86 (5): 919–940.

744 Tomilov, Y.V., Bordakov, V.G., Dolgii, I.E., and Nefedov, O.M. (1984). Reaction of diazoalkanes with unsaturated compounds. 2. Cyclopropanation of olefins by diazomethane in the presence of Pd-compounds. *Bull. Acad. Sci. USSR Div. Chem. Sci.* 33 (3): 533–538.

745 McCrindle, R., Arsenault, G.J., Farwaha, R. et al. (1989). Preparation of (chloromethyl)Pd(II) derivatives from complexes of $PdCl_2$ by reaction with diazomethane or bis(chloromethyl)Hg. *J. Chem. Soc., Dalton Trans.* (5): 761–766.

746 Straub, B.F. (2002). Pd(0) mechanism of Pd-catalyzed cyclopropanation of alkenes by CH_2N_2: a DFT study. *J. Am. Chem. Soc.* 124 (47): 14195–14201.

747 Chen, S.F., Ma, J., and Wang, J.B. (2008). Palladium-catalyzed cyclopropanation of electron-deficient olefins with aryldiazocarbonyl compounds. *Tetrahedron Lett.* 49 (48): 6781–6783.

748 Paulissen, R., Reimlinger, H., Hayez, E. et al. (1973). Transition-metal catalyzed reactions of diazocompounds. 2. Insertion in hydroxylic bond. *Tetrahedron Lett.* 14 (24): 2233–2236.

749 Hubert, A.J., Noels, A.F., Anciaux, A.J., and Teyssié, P. (1976). Rhodium(II) carboxylates - novel highly efficient catalysts for cyclopropanation of alkenes with alkyl diazoacetates. *Synthesis* (9): 600–602.

750 Doyle, M.P., Bagheri, V., Wandless, T.J. et al. (1990). Exceptionally high *trans* (*anti*) stereoselectivity in catalytic cyclopropanation reactions. *J. Am. Chem. Soc.* 112 (5): 1906–1912.

751 González-Bobes, F., Fenster, M.D.B., Kiau, S. et al. (2008). Rhodium-catalyzed cyclopropanation of alkenes with dimethyl diazomalonate. *Adv. Synth. Catal.* 350 (6): 813–816.

752 Müller, P., Bernardinelli, G., Allenbach, Y.F. et al. (2004). Selectivity enhancement in the Rh(II)-catalyzed cyclopropanation of styrene with (silanyloxyvinyl)diazoacetates. *Org. Lett.* 6 (11): 1725–1728.

753 Doyle, M.P., Pieters, R.J., Martin, S.F. et al. (1991). High enantioselectivity in the intramolecular cyclopropanation of allyl diazoacetates using a novel Rh(II)-catalyst. *J. Am. Chem. Soc.* 113 (4): 1423–1424.

754 Lin, W. and Charette, A.B. (2005). Rhodium-catalyzed asymmetric intramolecular cyclopropanation of substituted allylic cyanodiazoacetates. *Adv. Synth. Catal.* 347 (11–13): 1547–1552.

755 Negretti, S., Cohen, C.M., Chang, J.J. et al. (2015). Enantioselective dirhodium(II)-catalyzed cyclopropanations with trimethylsilylethyl and trichloroethyl aryldiazoacetates. *Tetrahedron* 71: 7415–7420.

756 Davies, H.M.L., Bruzinski, P.R., Hutcheson, D. et al. (1996). Asymmetric cyclopropanations by Rh(II) *N*-(arylsulfonyl)prolinate catalyzed decomposition of vinyldiazomethanes in the presence of alkenes. Practical enantioselective synthesis of the four stereoisomers of 2-phenylcyclopropan-1-amino acid. *J. Am. Chem. Soc.* 118 (29): 6897–6907.

757 Davies, H.M.L. (1998). Rhodium-stabilized vinylcarbenoid intermediates in organic synthesis. *Curr. Org. Chem.* 2 (5): 463–488.

758 Lou, Y., Horikawa, M., Kloster, R.A. et al. (2004). A new chiral Rh(II)-catalyst for enantioselective (2+1)-cycloaddition. Mechanistic implications and applications. *J. Am. Chem. Soc.* 126 (29): 8916–8918.

759 Lou, Y., Remarchuk, T.P., and Corey, E.J. (2005). Catalysis of enantioselective [2+1]-cycloaddition reactions of ethyl diazoacetate and terminal acetylenes using mixed-ligand complexes of the series Rh2(RCO2)n (L4-n). Stereochemical heuristics for ligand exchange and catalyst synthesis. *J. Am. Chem. Soc.* 127 (41): 14223–14230.

760 Nowlan, D.T.I. and Singleton, D.A. (2005). Mechanism and origin of enantioselectivity in the $Rh_2(OAc)(DPTI)_3$-catalyzed cyclopropenation of alkynes. *J. Am. Chem. Soc.* 127 (17): 6190–6191.

761 Nowlan, D.T.I., Gregg, T.M., Davies, H.M.L., and Singleton, D.A. (2003). Isotope effects and the nature of selectivity in Rh-catalyzed cyclopropanations. *J. Am. Chem. Soc.* 125 (51): 15902–15911.

762 Howell, J.A.S. (2007). On the mechanism of the Rh(II)-catalysed cyclopropanation of alkenes. *J. Chem. Soc., Dalton Trans.* (11): 1104–1114.

763 Nishiyama, H., Itoh, Y., Matsumoto, H. et al. (1994). New chiral Ru[bis(oxazolinyl)pyridine] catalyst - efficient asymmetric cyclopropanation of olefins with diazoacetates. *J. Am. Chem. Soc.* 116 (5): 2223–2224.

764 Niimi, T., Uchida, T., Irie, R., and Katsuki, T. (2000). Cu(II)(salen)-catalyzed highly *cis*- and enantioselective cyclopropanation. *Tetrahedron Lett.* 41 (19): 3647–3651.

765 Wurz, R.P. and Charette, A.B. (2002). Transition metal-catalyzed cyclopropanation of alkenes in water: catalyst efficiency and in situ generation of the diazo reagent. *Org. Lett.* 4 (25): 4531–4533.

766 Callot, H.J. and Piechocki, C. (1980). Cyclopropanation using Rh(III)(porphyrins) - large *cis*- vs. *trans*-selectivity. *Tetrahedron Lett.* 21 (36): 3489–3492.

767 Callot, H.J., Metz, F., and Piechocki, C. (1982). Sterically crowded cyclopropanation catalysts - *syn*-selectivity using Rh(III)(porphyrins). *Tetrahedron* 38 (15): 2365–2369.

768 Bartley, D.W. and Kodadek, T. (1993). Identification of the active catalyst in the Rh(porphyrin)-mediated cyclopropanation of alkenes. *J. Am. Chem. Soc.* 115 (5): 1656–1660.

769 Brown, K.C. and Kodadek, T. (1992). A transition state model for the Rh(porphyrin)-catalyzed cyclopropanation of alkenes by diazo esters. *J. Am. Chem. Soc.* 114 (21): 8336–8338.

770 Wolf, J.R., Hamaker, C.G., Djukic, J.P. et al. (1995). Shape and stereoselective cyclopropanation of alkenes catalyzed by Fe(porphyrins). *J. Am. Chem. Soc.* 117 (36): 9194–9199.

771 Lo, W.C., Che, C.M., Cheng, K.F., and Mak, T.C.W. (1997). Catalytic and asymmetric cyclopropanation of styrenes catalyzed by Ru(porphyrin) and porphycene complexes. *J. Chem. Soc., Chem. Commun.* (13): 1205–1206.

772 Teng, P.F., Lai, T.S., Kwong, H.L., and Che, C.M. (2003). Asymmetric inter- and intramolecular cyclopropanations of alkenes catalyzed by Rh(D-4-porphyrin): a comparison of Rh- and Ru-centered catalysts. *Tetrahedron: Asymmetry* 14 (7): 837–844.

773 Ferrand, Y., Le Maux, P., and Simonneaux, G. (2004). Highly enantioselective synthesis of cyclopropylphosphonates catalyzed by chiral Ru(porphyrins). *Org. Lett.* 6 (18): 3211–3214.

774 Chen, Y., Huang, L., Ranade, M.A., and Zhang, X.P. (2003). Fe(III) and Ru(II)(porphyrin) complex catalyzed selective olefination of aldehydes with ethyl diazoacetate. *J. Org. Chem.* 68 (9): 3714–3717.

775 Huang, L.Y., Chen, Y., Gao, G.Y., and Zhang, X.P. (2003). Diastereoselective and enantioselective cyclopropanation of alkenes catalyzed by Co(porphyrins). *J. Org. Chem.* 68 (21): 8179–8184.

776 Hahn, N.D., Nieger, M., and Dötz, K.H. (2004). Reactions of complex ligands. Part 97. Highly regio- and diastereoselective Cr(0)-catalyzed cyclopropanation of 1-alkoxy-1,3-dienes with diazo compounds. *Eur. J. Org. Chem.* (5): 1049–1056.

777 Fructos, M.R., Belderrain, T.R., Nicasio, M.C. et al. (2004). Complete control of the chemoselectivity in catalytic carbene transfer reactions from ethyl diazoacetate: an (N-heterocyclic carbene)Cu system that suppresses diazo coupling. *J. Am. Chem. Soc.* 126 (35): 10846–10847.

778 Fructos, M.R., Belderrain, T.R., de Frémont, P. et al. (2005). Au-catalyst for carbene-transfer reactions from ethyl diazoacetate. *Angew. Chem. Int. Ed.* 44 (33): 5284–5288.

779 Fedorov, A., Moret, M.E., and Chen, P. (2008). Gas phase synthesis and reactivity of a Au(carbene) complex. *J. Am. Chem. Soc.* 130 (28): 8880–8881.

780 Thompson, J.L. and Davies, H.M.L. (2007). Enhancement of cyclopropanation chemistry in the Ag-catalyzed reactions of aryldiazoacetates. *J. Am. Chem. Soc.* 129 (19): 6090–6091.

781 Juhl, K., Hazell, R.G., and Jørgensen, K.A. (1999). Catalytic enantioselective formation of aziridines from α-imino esters. *J. Chem. Soc., Perkin Trans. 1* (16): 2293–2297.

782 Bauld, N.L., Stufflebeme, G.W., and Lorenz, K.T. (1989). Cation-radical pericyclic reactions - cyclopropanation. *J. Phys. Org. Chem.* 2 (8): 585–601.

783 Bauld, N.L. and Yueh, W. (1994). Mechanistic criteria for the cation-radical vs electrophilic mechanistic distinction. *J. Am. Chem. Soc.* 116 (19): 8845–8846.

784 Müller, P. and Fernández, D. (1995). Carbenoid reactions in Rh(II)-catalyzed decomposition of iodonium ylides. *Helv. Chim. Acta* 78 (4): 947–958.

785 Müller, P. and Boléa, C. (2002). The enantioselectivity and the stereochemical course of Cu-catalyzed intramolecular C—H insertions of phenyliodonium ylides. *Helv. Chim. Acta* 85 (2): 483–494.

786 Müller, P. (2004). Asymmetric transfer of carbenes with phenyliodonium ylides. *Acc. Chem. Res.* 37 (4): 243–251.

787 Bonge, H.T. and Hansen, T. (2007). Cyclopropanation of olefins with nitromethane. *Synlett* (1): 55–58.

788 Müller, P. and Ghanem, A. (2004). Rh(II)-catalyzed enantioselective cyclopropanation of olefins with dimethyl malonate via in situ generated phenyliodonium ylide. *Org. Lett.* 6 (23): 4347–4350.

789 Moreau, B. and Charette, A.B. (2005). Expedient synthesis of cyclopropane α-amino acids by the catalytic asymmetric cyclopropanation of alkenes using iodonium ylides derived from methyl nitroacetate. *J. Am. Chem. Soc.* 127 (51): 18014–18015.

790 Emschwiller, G. (1929). Effect of the Zn/Cu couple on methylene iodide. *C.R. Hebd. Séances Acad. Sci.* 188: 1555–1557.

791 Simmons, H.E. and Smith, R.D. (1958). Bismethylenedioxy Steroids. 2. Synthesis of 9-α-methylhydrocortisone and 9-α-methylprednisolone. *J. Am. Chem. Soc.* 80 (19): 5322–5324.

792 Simmons, H.E., Caims, T.L., Vladuchick, S.A., and Hoiness, C.M. (1973). Cyclopropanes from unsaturated compounds, methylene iodide, and Zn/Cu couple. *Org. React.* 20: 1–131.

793 Motherwell, W.B. and Nutley, C.J. (1994). The role of Zn-carbenoids in organic synthesis. *Contemp. Org. Synth.* 1 (4): 219–241.

794 Charette, A.B. and Beauchemin, A. (2001). Simmons-smith cyclopropanation reaction. *Org. React.* 58: 1–415.

795 Conia, J.M. (1975). Cyclopropanation of silyl enol ethers - powerful synthetic tool. *Pure Appl. Chem.* 43 (3–4): 317–326.

796 Bax, A. and Summers, M.F. (1986). H-1 and C-13 assignments from sensitivity-enhanced detection of heteronuclear multiple-bond connectivity by 2D multiple quantum NMR. *J. Am. Chem. Soc.* 108 (8): 2093–2094.

797 Takahashi, H., Yoshioka, M., Ohno, M., and Kobayashi, S. (1992). A catalytic enantioselective reaction using a C_2-symmetrical disulfonamide as a chiral ligand - cyclopropanation of allylic alcohols by the Et_2Zn/CH_2I_2-disulfonamide system. *Tetrahedron Lett.* 33 (18): 2575–2578.

798 Sun, X.L. and Tang, Y. (2008). Ylide-initiated Michael addition/cyclization reactions beyond cyclopropanes. *Acc. Chem. Res.* 41 (8): 937–948.

799 Hakam, K., Thielmann, M., Thielmann, T., and Winterfeldt, E. (1987). Reactions with indole derivatives. 55. An enantiodivergent route to both vincamine enantiomers. *Tetrahedron* 43 (9): 2035–2044.

800 Krollpfeiffer, F. and Hartmann, H. (1950). Über Spaltungen und Umlagerungen von Phenacyl-Sulfoniumsalzen. *Chem. Ber.* 83 (1): 90–98.

801 Corey, E.J. and Chaykovsky, M. (1962). Dimethylsulfoxonium methylide. *J. Am. Chem. Soc.* 84 (5): 867–868.

802 Corey, E.J. and Chaykovsky, M. (1965). Dimethyloxosulfonium methylide ($(CH_3)_2SOCH_2$) and dimethylsulfonium methylide ($(CH_3)_2SCH_2$) formation and application to organic synthesis. *J. Am. Chem. Soc.* 87 (6): 1353–1364.

803 Koenig, H., Metzger, H., and Seelert, C. (1965). Über Schwefel-Ylide. 8. Umsetzungen des

Dimethyl-Oxo-Sulfoniummethylids mit Olefinen Aromaten und Alkylierungsmitteln. *Chem. Ber.* 98 (11): 3712–3732.

804 Kaiser, C., Trost, B.M., Beeson, J., and Weinstock, J. (1965). Preparation of some cyclopropanes and stable sulfoxonium ylides from dimethylsulfoxonium methylide. *J. Org. Chem.* 30 (11): 3972–3975.

805 Payne, G.B. (1967). Cyclopropanes from reactions of ethyl (dimethylsulfuranylidene)acetate with α,β-unsaturated compounds. *J. Org. Chem.* 32 (11): 3351–3355.

806 Aggarwal, V.K. and Winn, C.L. (2004). Catalytic, asymmetric sulfur ylide-mediated epoxidation of carbonyl compounds: scope, selectivity, and applications in synthesis. *Acc. Chem. Res.* 37 (8): 611–620.

807 Lotz, W.W. and Gosselck, J. (1973). Synthesis and reactions of dimethylphenacylselenonium ylide. *Tetrahedron* 29 (6): 917–919.

808 Tamagaki, S., Akatsuka, R., and Kozuka, S. (1977). Additions of dibenzyl selenoxides and selenonium ylides to an electron-deficient ethylene and acetylene. *Bull. Chem. Soc. Jpn.* 50 (6): 1641–1642.

809 Magdesieva, N.N., Le Nguyen, N., and Koloskova, N.M. (1977). Selenium ylides in heteroaroylcyclopropane synthesis. *Zh. Org. Khim.* 13 (5): 1010–1012.

810 Halazy, S. and Krief, A. (1981). Synthetic routes to cyclopropylidenecarbinols and to cyclopropylidienes. *Tetrahedron Lett.* 22 (22): 2135–2138.

811 Hori, M., Kataoka, T., Shimizu, H. et al. (1990). Synthesis and reactions of a stable 2-selenanaphthalene, 1-cyano-2-methyl-2-selenanaphthalene. *J. Org. Chem.* 55 (8): 2458–2463.

812 Zhong, Q., Shao, J.G., and Liu, C.Q. (1992). Study on the reactions of telluronium ylide. 8. A Facile synthesis of substituted cyclopropanes. *Synth. Commun.* 22 (4): 489–495.

813 Han, L.B., Kambe, N., Ryu, I., and Sonoda, N. (1993). Synthesis of enones and cyclopropanes by the reaction of telluronium ylides generated from bis(2-oxoalkyl)Te dichlorides. *Chem. Lett.* (3): 561–564.

814 Tang, Y., Chi, Z.F., Huang, Y.Z. et al. (1996). Highly stereoselective synthesis of vinyl and ethynylcyclopropane 1,1,-dicarboxylic esters via semi-stabilized Te ylides. *Tetrahedron* 52 (26): 8747–8754.

815 Bestmann, H.J. and Seng, F. (1962). Reaktion von Triphenylphosphin-Alkylenen mit aktivierten Doppelbindungen. *Angew. Chem.* 74 (4): 154–155.

816 Bestmann, H.J., Schmid, G., and Kisielowski, L. (1982). Synthesis of 3-membered rings by means of phosphorus ylids. *Isr. J. Chem.* 22 (1): 45–52.

817 Avery, T.D., Haselgrove, T.D., Rathbone, T.J. et al. (1998). New synthetic methodology utilising 1,2-dioxines and stabilised phosphorus ylides: a highly diastereoselective cyclopropanation reaction. *J. Chem. Soc., Chem. Commun.* (3): 333–334.

818 Palmer, F.N. and Taylor, D.K. (2000). The first practical approach to optically pure cyclopropanes derived from *trans* γ-hydroxy enones. *J. Chem. Soc., Perkin Trans. 1* 9: 1323–1325.

819 Krysiak, J., Lyon, C., Baceiredo, A. et al. (2004). Stable optically pure phosphino(silyl)carbenes: reagents for highly enantioselective cyclopropanation reactions. *Chem. Eur. J.* 10 (8): 1982–1986.

820 Redon, S., Leleu, S., Pannecoucke, X. et al. (2008). New synthesis and cyclopropanation of α-phenylselanyl α,β-unsaturated ketones with non-stabilized phosphorus ylides. *Tetrahedron* 64 (39): 9293–9304.

821 Trippett, S. and Walker, M.A. (1971). Some reactions of arsenic ylides. *J. Chem. Soc. C* (6): 1114–1117.

822 Huang, Y.Z., Xu, Y.Y., and Li, Z. (1982). Reactivity of acetylmethylene triphenylarsorane. *Org. Prep. Proced. Int.* 14 (6): 373–380.

823 Huang, Y.Z. and Shen, Y.C. (1982). Arsonium ylides. *Adv. Organomet. Chem.* 20: 115–157.

824 Shi, Z.J., Liu, W.X., Cao, W.G. et al. (2007). Synthesis and structure elucidation of a series of spiro cyclopropane derivative. *Fenxi Huaxue* 35 (12): 1779–1782.

825 Bhattacharjee, S.S., Ila, H., and Junjappa, H. (1982). Cyanotrimethylammonium methylid, formation and reaction with α,β-unsaturated carbonyl compounds - novel cyclopropanation with an ammonium ylid. *Synthesis* (4): 301–303.

826 Jonczyk, A. and Konarska, A. (1999). Generation and reactions of ammonium ylides in basic two-phase systems: convenient synthesis of cyclopropanes, oxiranes and alkenes substituted with electron-withdrawing groups. *Synlett* (7): 1085–1087.

827 Kowalkowska, A., Sucholbiak, D., and Jonczyk, A. (2005). Generation and reaction of ammonium ylides in basic two-phase systems. *Eur. J. Org. Chem.* (5): 925–933.

828 Couty, F., David, O., Larmanjat, B., and Marrot, J. (2007). Strained azetidinium ylides: new reagents for cyclopropanation. *J. Org. Chem.* 72 (3): 1058–1061.

829 Bremeyer, N., Smith, S.C., Ley, S.V., and Gaunt, M.J. (2004). An intramolecular organocatalytic cyclopropanation reaction. *Angew. Chem. Int. Ed.* 43 (20): 2681–2684.

830 Peace, B.W., Carman, F., and Wulfman, D.S. (1971). Soluble Cu(I)-catalyzed decomposition of dimethyl diazomalonate and di-*tert*-butyl diazomalonate in presence of some cycloalkenes. *Synthesis* (12): 658–661.

831 Peace, B.W. and Wulfman, D.S. (1973). Preparation and reactions of diazomalonic esters. *Synthesis* (3): 137–145.

832 Strickler, H., Davis, J.B., and Ohloff, G. (1976). $ZnCl_2$-catalyzed cyclization of dehydrolinalyl acetate. *Helv. Chim. Acta* 59 (4): 1328–1332.

833 Rautenstrauch, V. (1984). 2-Cyclopentenones from 1-ethynyl-2-propenyl acetates. *J. Org. Chem.* 49 (5): 950–952.

834 Marco-Contelles, J. and Soriano, E. (2007). Recent developments in the metal-catalyzed reactions of metallocarbenoids from propargylic esters. *Chem. Eur. J.* 13 (5): 1350–1357.

835 Mainetti, E., Mouries, V., Fensterbank, L. et al. (2002). The effect of a hydroxy protecting group on the $PtCl_2$-catalyzed cyclization of dienynes - a novel, efficient, and selective synthesis of carbocycles. *Angew. Chem. Int. Ed.* 41 (12): 2132–2135.

836 Marion, N., de Frémont, P., Lemière, G. et al. (2006). Au(I)-catalyzed cycloisomerization of 1,5-enynes bearing a propargylic acetate: formation of unexpected bicyclo[3.1.0]hexene. *J. Chem. Soc., Chem. Commun.* (19): 2048–2050.

837 Mamane, V., Gress, T., Krause, H., and Fürstner, A. (2004). Pt- and Au-catalyzed cycloisomerization reactions of hydroxylated enynes. *J. Am. Chem. Soc.* 126 (28): 8654–8655.

838 Sperger, C.A., Tungen, J.E., and Fiksdahl, A. (2011). Gold(I)-catalyzed reactions of propargyl esters with vinyl derivatives. *Chem. Eur. J.* 17: 3719–3722.

839 Qian, D.Y. and Zhang, J.L. (2015). Gold-catalyzed cyclopropanation reactions using a carbenoid precursor toolbox. *Chem. Soc. Rev.* 44: 677–698.

840 Shibata, Y., Noguchi, K., and Tanaka, K. (2010). Cationic Rh(I)-complex-catalyzed (3+2)- and (2+1)-cycloadditions of propargyl esters with electron-deficient alkynes and alkenes. *J. Am. Chem. Soc.* 132: 7896–7898.

841 Soriano, E. and Marco-Contelles, J. (2007). On accounting for the stereoselective control of the metal-catalyzed Rautenstrauch cyclopropanation by computational methods. *J. Org. Chem.* 72 (7): 2651–2654.

842 Soriano, E. and Marco-Contelles, J. (2009). Mechanistic insights on the cycloisomerization of polyunsaturated precursors catalyzed by Pt and Au complexes. *Acc. Chem. Res.* 42: 1026–1036.

843 Huang, G.P., Xie, K.L., Lee, D.S., and Xia, Y.Z. (2012). Reactivity of alkynyl metal carbenoids: DFT study on the Pt-catalyzed cyclopropanation of propargyl ester containing 1,3-diynes. *Org. Lett.* 14: 3850–3853.

844 López, S., Herrero-Gómez, E., Pérez-Galán, P. et al. (2006). Gold(I)-catalyzed intermolecular cyclopropanation of enynes with alkenes: trapping of two different gold carbenes. *Angew. Chem. Int. Ed.* 45 (36): 6029–6032.

845 Echavarren, A.M., Hashmi, A.S.K., and Toste, F.D. (2016). Gold catalysis - steadily increasing in importance. *Adv. Synth. Catal.* 358: 1347.

846 Jiménez-Núñez, E. and Echavarren, A.M. (2008). Gold-catalyzed cycloisomerizations of enynes: a mechanistic perspective. *Chem. Rev.* 108: 3326–3350.

847 Fürstner, A. (2009). Gold and Pt catalysis - a convenient tool for generating molecular complexity. *Chem. Soc. Rev.* 38: 3208–3221.

848 Hashmi, A.S.K. (2010). Homogeneous Au catalysis beyond assumptions and proposals-characterized intermediates. *Angew. Chem. Int. Ed.* 49: 5232–5241.

849 Rudolph, M. and Hashmi, A.S.K. (2012). Gold catalysis in total synthesis - an update. *Chem. Soc. Rev.* 41: 2448–2462.

850 Garayalde, D. and Nevado, C. (2012). Gold containing and gold generated 1,n-dipoles as useful platforms toward cycloadditions and cyclizations. *ACS Catal.* 2: 1462–1479.

851 Pflasterer, D. and Hashmi, A.S.K. (2016). Gold catalysis in total synthesis - recent achievements. *Chem. Soc. Rev.* 45: 1331–1367.

852 Amijs, C.H.M., López-Carrilo, V., Raducan, M. et al. (2008). Gold(I)-catalyzed intermolecular addition of carbon nucleophiles to 1,5-and 1,6-enynes. *J. Org. Chem.* 73 (19): 7721–7730.

853 Miki, K., Ohe, K., and Uemura, S. (2003). Ruthenium-catalyzed cyclopropanation of alkenes using propargylic carboxylates as precursors of vinylcarbenoids. *J. Org. Chem.* 68 (22): 8505–8513.

854 Miki, K., Uemura, S., and Ohe, K. (2005). Transition metal-catalyzed reactions using alkynes as

precursors of carbene and vinylidene complexes. *Chem. Lett.* 34 (8): 1068–1073.

855 Tenaglia, A. and Marc, S. (2006). $CpRuCl(PPh_3)_2$-catalyzed cyclopropanation of bicyclic alkenes with tertiary propargylic acetates. *J. Org. Chem.* 71 (9): 3569–3575.

856 Johansson, M.J., Gorin, D.J., Staben, S.T., and Toste, F.D. (2005). Gold(I)-catalyzed stereoselective olefin cyclopropanation. *J. Am. Chem. Soc.* 127 (51): 18002–18003.

857 Vallarino, L. (1957). Carbonyl complexes of Rh. 1. Complexes with triarylphosphines, triarylarsines, and triarylstibines. *J. Chem. Soc.* 2287–2292.

858 Heimbach, P. and Brenner, W. (1967). Catalysed Cope rearrangement of *cis*-1,2-divinylcyclobutane. *Angew. Chem. Int. Ed. Engl.* 6 (9): 800–801.

859 Cannell, L.G. (1972). Cyclodimerization of ethylene and 1,3-butadiene to vinylcyclobutane - homogeneous Ti catalysts. *J. Am. Chem. Soc.* 94 (19): 6867–6869.

860 Huang, D.J., Rayabarapu, D.K., Li, L.P. et al. (2000). Nickel-catalyzed (2+2)-cycloaddition of alkynes with activated cyclic alkenes: synthesis and novel ring expansion studies of cyclobutene products. *Chem. Eur. J.* 6 (20): 3706–3713.

861 Bernardi, F., Bottoni, A., and Rossi, I. (1998). A DFT investigation of ethylene dimerization catalyzed by Ni(0) complexes. *J. Am. Chem. Soc.* 120 (31): 7770–7775.

862 Noyori, R., Ishigami, T., Hayashi, N., and Takaya, H. (1973). Transition-metal-catalyzed [2+2] cross-addition of olefins - nickel(0)-catalyzed cycloaddition of norbornadiene and methylenecyclopropane. *J. Am. Chem. Soc.* 95 (5): 1674–1676.

863 Binger, P. and Schuchardt, U. (1980). Reactions of methylenecyclopropane. 3. Cooligomerization of methylenecyclopropane with allenes on Pd(0) catalysts. *Chem. Ber.* 113 (3): 1063–1071.

864 Jordan, R.W., Villeneuve, K., and Tam, W. (2006). Study on the reactivity of the alkyne component in Ru-catalyzed [2+2]-cycloadditions between an alkene and an alkyne. *J. Org. Chem.* 71 (15): 5830–5833.

865 Burton, R.R. and Tam, W. (2006). Ruthenium-catalyzed [2+2]-cycloadditions between C1-substituted 7-oxanorbornadienes and alkynes. *Tetrahedron Lett.* 47 (40): 7185–7189.

866 Burton, R.R. and Tam, W. (2007). Ruthenium(II)-catalyzed cyclization of azabenzonorbornadienes with alkynes. *Org. Lett.* 9 (17): 3287–3290.

867 Riddell, N. and Tam, W. (2006). Ruthenium-catalyzed (2+2)-cycloadditions of alkynyl sulfides and alkynyl sulfones. *J. Org. Chem.* 71 (5): 1934–1937.

868 Villeneuve, K., Riddell, N., and Tam, W. (2006). Ruthenium-catalyzed (2+2)-cycloadditions of bicyclic alkenes and ynamides. *Tetrahedron* 62 (16): 3823–3836.

869 Liu, P., Tam, W., and Goddard, J.D. (2007). Ruthenium-catalyzed (2+2)-cycloadditions between substituted alkynes and norbornadiene: a theoretical study. *Tetrahedron* 63 (32): 7659–7666.

870 Chao, K.C., Rayabarapu, D.K., Wang, C.C., and Cheng, C.H. (2001). Cross (2+2)-cycloaddition of bicyclic alkenes with alkynes mediated by Co complexes: a facile synthesis of cyclobutene derivatives. *J. Org. Chem.* 66 (26): 8804–8810.

871 Rayabarapu, D.K. and Cheng, C.H. (2007). New catalytic reactions of oxa- and azabicyclic alkenes. *Acc. Chem. Res.* 40 (10): 971–983.

872 Treutwein, J. and Hilt, G. (2008). Cobalt-catalyzed (2+2)-cycloaddition. *Angew. Chem. Int. Ed.* 47 (36): 6811–6813.

873 Smith, D.L., Chidipudi, S.R., Goundry, W.R., and Lam, H.W. (2012). Rhodium-catalyzed (2+2)-cycloaddition of ynamides with nitroalkenes. *Org. Lett.* 14: 4934–4937.

874 Kuninobu, Y., Yu, P., and Takai, K. (2007). Rhenium-catalyzed (2+2)-cycloadditions of norbornenes with internal and terminal acetylenes. *Chem. Lett.* 36 (9): 1162–1163.

875 Shibata, T., Takami, K., and Kawachi, A. (2006). Rh-catalyzed enantioselective (2+2)-cycloaddition of alkynyl esters and norbornene derivatives. *Org. Lett.* 8 (7): 1343–1345.

876 Trost, B.M., Yanai, M., and Hoogsteen, K. (1993). A Pd-catalyzed (2+2)-cycloaddition - mechanism of a Pd-catalyzed enyne metathesis. *J. Am. Chem. Soc.* 115 (12): 5294–5295.

877 Bouwkamp, M.W., Bowman, A.C., Lobkovsky, E., and Chirik, P.J. (2006). Iron-catalyzed $[\pi^2+\pi^2]$-cycloaddition of α,ω-dienes: the importance of redox-active supporting ligands. *J. Am. Chem. Soc.* 128 (41): 13340–13341.

878 Bart, S.C., Lobkovsky, E., and Chirik, P.J. (2004). Preparation and molecular and electronic structures of Fe(0)-dinitrogen and silane complexes and their application to catalytic hydrogenation

and hydrosilation. *J. Am. Chem. Soc.* 126 (42): 13794–13807.

879 Archer, A.M., Bouwkamp, M.W., Cortez, M.P. et al. (2006). Arene coordination in [bis(imino)pyridine]Fe complexes: identification of catalyst deactivation pathways in Fe-catalyzed hydrogenation and hydrosilation. *Organometallics* 25 (18): 4269–4278.

880 Bart, S.C., Lobkovsky, E., Bill, E., and Chirik, P.J. (2006). Synthesis and hydrogenation of [bis(imino)pyridine]Fe(imides). *J. Am. Chem. Soc.* 128 (16): 5302–5303.

881 Luzung, M.R., Mauleón, P., and Toste, F.D. (2007). Gold(I)-catalyzed (2+2)-cycloaddition of allenenes. *J. Am. Chem. Soc.* 129 (41): 12402–12403.

882 Deyrup, J.A. (1971). (3+1)-Cycloaddition. *Tetrahedron Lett.* 12 (24): 2191–2192.

883 Burger, K., Marschke, G., and Manz, F. (1982). 4-Membered ring syntheses through (3+1)-cycloaddition. 4. Synthesis of 3-imino-azetidines from azomethine ylides and isonitriles. *J. Heterocycl. Chem.* 19 (6): 1315–1317.

884 Kurahashi, T. and de Meijere, A. (2005). Cyclopropyl building blocks for organic synthesis. Part 121. C—C Bond activation by octacarbonyldicobalt: (3+1)-Co-cyclizations of methylenecyclopropanes with carbon monoxide. *Angew. Chem. Int. Ed.* 44 (48): 7881–7884.

885 Rubin, M., Rubina, M., and Gevorgyan, V. (2007). Transition-metal chemistry of cyclopropenes and cyclopropanes. *Chem. Rev.* 107: 3117–3179.

886 Gao, Y., Fu, X.F., and Yu, Z.X. (2014). Transition-metal-catalyzed cycloadditions of cyclopropanes for the synthesis of carbocycles; C—C activation in cyclopropanes. In: *C—C Bond Activation*, Topics in Current Chemistry, vol. 346 (ed. G.-B. Dong), 195–232. Berlin, Heidelberg: Springer-Verlag.

887 Cassar, L. and Halpern, J. (1970). Oxidative addition of quadricyclene to di-μ-chlorotetracarbonyldirhodium(I) and the mechanism of the Rh(I)-catalysed isomerization of quadricyclene to norbornadiene. *J. Chem. Soc., Chem. Commun.* (17): 1082–1083.

888 Noyori, R., Umeda, I., Kawauchi, H., and Takaya, H. (1975). Nickel-catalyzed reactions involving strained bonds. 12. Nickel(0)-catalyzed reaction of quadricyclane with electron-deficient olefins. *J. Am. Chem. Soc.* 97 (4): 812–820.

889 Shimizu, I., Ohashi, Y., and Tsuji, J. (1985). Palladium-catalyzed (3+2)-cycloaddition reaction of vinylcyclopropanes with α,β-unsaturated esters or ketones. *Tetrahedron Lett.* 26: 3825–3828.

890 Trost, B.M., Morris, P.J., and Sprague, S.J. (2012). Palladium-catalyzed diastereo- and enantioselective formal (3+2)-cycloadditions of substituted vinylcyclopropanes. *J. Am. Chem. Soc.* 134: 17823–17831.

891 Trost, B.M. and Morris, P.J. (2011). Palladium-catalyzed diastereo- and enantioselective synthesis of substituted cyclopentanes through a dynamic kinetic asymmetric formal (3+2)-cycloaddition of vinylcyclopropanes and alkylidene azlactones. *Angew. Chem. Int. Ed.* 50: 6167–6170.

892 Mei, L.Y., Wei, Y., Xu, Q., and Shi, M. (2013). Diastereo- and enantioselective construction of oxindole-fused spirotetrahydrofuran scaffolds through Pd-catalyzed asymmetric (3+2)-cycloaddition of vinylcyclopropanes and isatins. *Organometallics* 32: 3544–3556.

893 Mei, L.Y., Wei, Y., Xu, Q., and Shi, M. (2012). Palladium-catalyzed asymmetric formal (3+2)-cycloaddition of vinylcyclopropanes and β,γ-unsaturated α-keto esters: an effective route to highly functionalized cyclopentanes. *Organometallics* 31: 7591–7599.

894 Jiao, L., Ye, S.Y., and Yu, Z.X. (2008). Rh(I)-catalyzed Intramolecular (3+2)-cycloaddition of *trans*-vinylcyclopropane-enes. *J. Am. Chem. Soc.* 130 (23): 7178–7179.

895 Li, Q., Jiang, G.J., Jiao, L., and Yu, Z.X. (2010). Reaction of α-ene-vinylcyclopropanes: Type II. Intramolecular (5+2)-cycloaddition or (3+2)-cycloaddition? *Org. Lett.* 12: 1332–1335.

896 Lin, M., Kang, G.Y., Guo, Y.A., and Yu, Z.X. (2012). Asymmetric Rh(I)-catalyzed intramolecular (3+2)-cycloaddition of 1-yne-vinylcyclopropanes for bicyclo[3.3.0] compounds with a chiral quaternary carbon stereocenter and density functional theory study of the origins of enantioselectivity. *J. Am. Chem. Soc.* 134: 398–405.

897 Suzuki, T. and Fujimoto, H. (2000). Theoretical study of the mechanisms of palladation of methylenecyclopropane and (3+2)-cycloadditions. *Inorg. Chem.* 39: 1113–1119.

898 Nakamura, I. and Yamamoto, Y. (2002). Transition metal-catalyzed reactions of methylenecyclopropanes. *Adv. Synth. Catal.* 344: 111–129.

899 Noyori, R., Odagi, T., and Takaya, H. (1970). Nickel(0)-catalyzed reaction of methylenecyclopropanes with olefins - a novel [$\sigma^2+\pi^2$]-cycloaddition. *J. Am. Chem. Soc.* 92 (19): 5780–5781.

900 Noyori, R., Kumagai, Y., and Takaya, H. (1974). Nickel(0)-catalyzed cycloaddition of bicyclo[2.1.0]pentane and olefins - contrasting stereochemistry of thermal and transition-metal catalyzed reactions. *J. Am. Chem. Soc.* 96 (2): 634–636.

901 Noyori, R., Kumagai, Y., Umeda, I., and Takaya, H. (1972). Nickel-catalyzed reactions involving strained σ-bonds. 4. Nickel(0)-catalyzed reaction of methylenecyclopropane with olefins orientation and stereochemistry. *J. Am. Chem. Soc.* 94 (11): 4018–4020.

902 Inouc, Y., Hibi, T., Satake, M., and Hashimoto, H. (1979). Reaction of methylenecyclopropanes with carbon dioxide catalyzed by Pd(0)-complexes - synthesis of 5-membered lactones. *J. Chem. Soc., Chem. Commun.* (22): 982–982.

903 Becker, Y., Eisenstadt, A., and Shvo, Y. (1976). Reaction of transition-metal carbonyls with heterocyclic systems. 3. S_N2'-Type reaction of complexed π-allyl systems. *Tetrahedron* 32 (17): 2123–2126.

904 Donaldson, W.A., Hossain, M.A., and Cushnie, C.D. (1995). Iron-mediated C—C bond formation - preparation of (trimethylenemethane)Fe complexes via reaction of weak carbon nucleophiles with in situ generated cross-conjugated pentadienyl cations - nucleophilic attack on (trimethylenemethane)IFe complexes with carbanions. *J. Org. Chem.* 60 (6): 1611–1618.

905 Barnes, S.G. and Green, M. (1980). Synthesis and mechanism of formation of dicarbonyl(η^4-trimethylenemethane)(η^5-pentamethylcyclopentadienyl)Mo tetrafluoroborate. *J. Chem. Soc., Chem. Commun.* (6): 267–268.

906 Trost, B.M. and Chan, D.M.T. (1979). 2-Acetoxymethyl-3-allyltrimethylsilane and Pd(0) - source of trimethylenemethane-Pd complex. *J. Am. Chem. Soc.* 101 (21): 6432–6433.

907 Trost, B.M. and Chan, D.M.T. (1983). Palladium-mediated cycloaddition approach to cyclopentanoids - mechanistic studies. *J. Am. Chem. Soc.* 105 (8): 2326–2335.

908 Trost, B.M. and Miller, M.L. (1988). Criteria for concertedness in cycloadditions. *J. Am. Chem. Soc.* 110 (11): 3687–3689.

909 Trost, B.M. (1986). (3+2)-Cycloaddition approaches to 5-membered rings via trimethylenemethane and its equivalents. *Angew. Chem. Int. Ed. Engl.* 25 (1): 1–20.

910 Trost, B.M. and Parquette, J.R. (1994). A TMM cycloaddition strategy to the bicyclo[6.3.0]undecyl ring system - a total synthesis of 11-hydroxyjasionone. *J. Org. Chem.* 59 (25): 7568–7569.

911 Trost, B.M. and Bonk, P.J. (1985). On the nucleophilic nature of a TMM-PdL2 intermediate - a direct Pd-catalyzed addition of trimethylenemethane to heteroatom unsaturation. *J. Am. Chem. Soc.* 107 (26): 8277–8279.

912 Trost, B.M., King, S.A., and Schmidt, T. (1989). Palladium-catalyzed trimethylenemethane reaction to form methylenetetrahydrofurans - aldehyde and ketone substrates and the Sn effect. *J. Am. Chem. Soc.* 111 (15): 5902–5915.

913 Nakamura, I., Oh, B.H., Saito, S., and Yamamoto, Y. (2001). Novel (3+2)-cycloaddition of alkylidenecyclopropanes with aldehydes catalyzed by Pd. *Angew. Chem. Int. Ed.* 40 (7): 1298–1300.

914 Trost, B.M., Sharma, S., and Schmidt, T. (1993). In(+3) as a chemoselectivity switch for TMM-PdL_2 cycloadditions to ynones. *Tetrahedron Lett.* 34 (45): 7183–7186.

915 Trost, B.M., King, S.A., and Nanninga, T.N. (1987). Heterocyclic acceptors in diastereoselective Pd-mediated (3+2)-cycloadditions. *Chem. Lett.* (1): 15–18.

916 Trost, B.M. and Schneider, S. (1989). Pyrones as substrates for Pd-catalyzed (4+3)-cycloadditions. *Angew. Chem. Int. Ed. Engl.* 28 (2): 213–215.

917 Trost, B.M. and Marrs, C.M.A. (1993). (3+2) and (4+3)-cycloaddition approach to N-heterocycles via Pd-catalyzed TMM reactions with imines. *J. Am. Chem. Soc.* 115 (15): 6636–6645.

918 Trost, B.M. and Chan, D.M.T. (1979). New conjunctive reagents - 2-Acetoxymethyl-3-allyltrimethylsilane for methylenecyclopentane annulations catalyzed by Pd(0). *J. Am. Chem. Soc.* 101 (21): 6429–6432.

919 Trost, B.M. and Chan, D.M.T. (1983). Palladium-mediated cycloaddition approach to cyclopentanoids - introduction and initial studies. *J. Am. Chem. Soc.* 105 (8): 2315–2325.

920 Jones, M.D. and Kemmitt, R.D.W. (1986). Reactions between trimethylenemethane metal-complexes and the carbon nitrogen double-bond - Ni- and Pd-catalyzed synthesis of pyrrolidines. *J. Chem. Soc., Chem. Commun.* (15): 1201–1203.

921 Shimizu, I., Ohashi, Y., and Tsuji, J. (1984). Palladium-catalyzed (3+2)-cycloaddition reaction using 2-(sulfonylmethyl) allyl or 2-(cyanomethyl)allyl carbonate. *Tetrahedron Lett.* 25 (45): 5183–5186.

922 Yamamoto, A., Ito, Y., and Hayashi, T. (1989). Asymmetric (3+2)-cycloaddition of 2-(sulfonylmethyl)-2-propenyl carbonate catalyzed by chiral (ferrocenylphosphine)Pd complexes. *Tetrahedron Lett.* 30 (3): 375–378.

923 (a) Binger, P. and Buech, H.M. (1987). Cyclopropenes and methylenecylopropanes as multifunctional reagents in transition-metal-catalyzed reactions. In: *Small Ring Compounds*, Org. Synth. 2. Top. Curr. Chem., vol. 135, 77–151. Heildelberg: Springer-Verlag; (b) Rubin, M., Rubina, M., and Gevorgyan, V. (2007). Transition Metal Chemistry of Cyclopropenes and Cyclopropanes. *Chem. Rev.* 107 (7): 3117–3179.

924 Binger, P. and Schuchardt, U. (1981). Reactions of methylenecyclopropanes. 5. Palladium(0)-catalyzed codimerization of methylenecyclopropane with methyl acrylates and 1,2-ethylenedicarbonic esters. *Chem. Ber.* 114 (10): 3313–3324.

925 Lewis, R.T., Motherwell, W.B., and Shipman, M. (1988). Observations on the intramolecular Pd(0)-catalyzed (3+2)-cycloaddition of diphenylmethylenecyclopropanes. *J. Chem. Soc., Chem. Commun.* (14): 948–950.

926 Yamago, S. and Nakamura, E. (1988). Synthesis of [3.3.3]propellanes by exocyclic trans-annular cycloaddition of olefinic methylenecyclopropanes. *J. Chem. Soc., Chem. Commun.* (16): 1112–1113.

927 Lautens, M., Ren, Y., and Delanghe, P.H.M. (1994). Stereochemical control in metal-catalyzed (3+2)-cycloadditions of methylenecyclopropanes. *J. Am. Chem. Soc.* 116 (19): 8821–8822.

928 García-Fandiño, R., Gulías, M., Castedo, L. et al. (2008). Palladium-catalyzed (3+2)-cycloaddition of alk-5-ynylidenecyclopropanes to alkynes: a mechanistic DFT study. *Chem. Eur. J.* 14 (1): 272–281.

929 Gulías, M., García, R., Delgado, A. et al. (2006). Palladium-catalyzed (3+2) intramolecular cycloaddition of alk-5-enylidenecyclopropanes. *J. Am. Chem. Soc.* 128 (2): 384–385.

930 Wenkert, E., Alonso, M.E., Buckwalter, B.L., and Sanchez, E.L. (1983). Short syntheses of furan and catechol derivatives - a synthesis of hydrourushiol. *J. Am. Chem. Soc.* 105 (7): 2021–2029.

931 Alonso, M.E., Jano, P., Hernández, M.I. et al. (1983). A dihydro-α-furoic ester synthesis by the catalyzed reactions of ethyl diazopyruvate with enol ethers. *J. Org. Chem.* 48 (18): 3047–3050.

932 Pirrung, M.C. and Zhang, J.C. (1992). Asymmetric dipolar cycloaddition reactions of diazo-compounds mediated by a (binaphtholphosphate)Rh catalyst. *Tetrahedron Lett.* 33 (40): 5987–5990.

933 Alonso, M.E., Morales, A., and Chitty, A.W. (1982). Studies on the origin of dihydrofurans from α-diazocarbonyl compounds - concerted 1,3-dipolar cycloaddition vs nonsynchronous coupling in the Co(chelate) catalyzed reactions of α-diazodicarbonyl compounds with electron-rich olefins. *J. Org. Chem.* 47 (19): 3747–3754.

934 Davies, H.M.L. and Romines, K.R. (1988). Direct synthesis of furans by (3+2)-cycloadditions between Rh(II) acetate stabilized carbenoids and acetylenes. *Tetrahedron* 44 (11): 3343–3348.

935 Pirrung, M.C., Zhang, J.C., and Morehead, A.T. Jr., (1994). Dipolar cycloaddition of cyclic Rh-carbenoids to digonal carbon - synthesis of isoeuparin. *Tetrahedron Lett.* 35 (34): 6229–6230.

936 Aumann, R. and Uphoff, J. (1987). Organic syntheses with transition-metal complexes. 22. Trimethylenemethane complexes of Cr, Mo, and W by novel methylenation of allenes with carbene complexes - cyclopentanes by (3+2)-cycloaddition of trimethylenemethane complexes to allenes. *Angew. Chem. Int. Ed. Engl.* 26 (4): 357–359.

937 Aumann, R. and Melchers, H.D. (1988). Organic syntheses with transition-metal complexes. 30. (Trimethylenemethane)Fe complexes via methylenation of allenes with (carbene)Fe complexes. *J. Organomet. Chem.* 355 (1–3): 351–358.

938 Aumann, R. and Trentmann, B. (1989). Organic syntheses via transition-metal complexes. 36. Monofunctional [tri(methylene)methane]Cr and -Fe complexes by addition of allenes to M=C bonds of carbene complexes. *Chem. Ber.* 122 (10): 1977–1982.

939 Barluenga, J., Vicente, R., Barrio, P. et al. (2004). Metal-controlled selective (3+2)-cyclization reactions of alkenyl Fischer carbene complexes and allenes. *J. Am. Chem. Soc.* 126 (19): 5974–5975.

940 Lian, J.J., Chen, P.C., Lin, Y.P. et al. (2006). Gold-catalyzed intramolecular (3+2)-cycloaddition of arenyne-yne functionalities. *J. Am. Chem. Soc.* 128 (35): 11372–11373.

941 Chang, H.T., Jayanth, T.T., and Cheng, C.H. (2007). Cobalt-catalyzed diastereoselective reductive (3+2)-cycloaddition of allenes and enones. *J. Am. Chem. Soc.* 129 (14): 4166–4167.

942 Gao, H.Y. and Zhang, J.L. (2012). Cationic Rh(I)-catalyzed regioselective tandem heterocyclization/(3+2)-cycloaddition of 2-(1-alkynyl)-2-alken-1-ones with alkynes. *Chem. Eur. J.* 18: 2777–2782.

943 Trost, B.M. and Chan, D.M.T. (1982). Intramolecular carbocyclic (3+2)-cycloaddition via organopalladium intermediates. *J. Am. Chem. Soc.* 104 (13): 3733–3735.

944 Trost, B.M., Grese, T.A., and Chan, D.M.T. (1991). Intramolecular Pd-catalyzed trimethylenemethane cycloadditions - initial studies. *J. Am. Chem. Soc.* 113 (19): 7350–7362.

945 Danheiser, R.L., Martinezdavilla, C., Auchus, R.J., and Kadonaga, J.T.A. (1981). Stereoselective synthesis of cyclopentene derivatives from 1,3-dienes. *J. Am. Chem. Soc.* 103: 2443–2446.

946 Danheiser, R.L., Bronson, J.J., and Okano, K. (1985). Carbanion-accelerated vinylcyclopropane rearrangement - application in a general, stereocontrolled annulation approach to cyclopentene derivatives. *J. Am. Chem. Soc.* 107: 4579–4581.

947 Hudlicky, T. and Koszyk, F.J. (1980). Selectivity in retro-ene vs cyclopentene rearrangements of a *cis*-methylvinylcyclopropane. *Tetrahedron Lett.* 21: 2487–2490.

948 Hudlicky, T., Koszyk, F.J., Dochwat, D.M., and Cantrell, G.L. (1981). Intramolecular cyclopentene annulation. 2. Synthesis and C-13 nuclear magnetic resonance spectroscopy of bicyclo[4.3.0]non-7-en-2-ones. *J. Org. Chem.* 46: 2911–2915.

949 Widenhoefer, R.A.A. (2009). Pd-catalyzed formal (4+1)-annulation: a new approach to cyclopentene construction. *Angew. Chem. Int. Ed.* 48: 6950–6952.

950 Sigman, M.S. and Eaton, B.E. (1994). The first Fe-mediated catalytic carbon-nitrogen bond formation - (4+1)-Cycloaddition of allenyl imines and carbon monoxide. *J. Org. Chem.* 59 (24): 7488–7491.

951 Sigman, M.S. and Eaton, B.E. (1996). Catalytic Fe-mediated (4+1)-cycloaddition of diallenes with carbon monoxide. *J. Am. Chem. Soc.* 118 (47): 11783–11788.

952 Morimoto, T., Chatani, N., and Murai, S. (1999). The first catalytic carbonylative [4+1]-cycloaddition using a 1,3-conjugated system. A new transformation of α,β-unsaturated imines to unsaturated γ-lactams catalyzed by $Ru_3(CO)_{12}$. *J. Am. Chem. Soc.* 121 (8): 1758–1759.

953 Murakami, M., Itami, K., and Ito, Y. (1996). A study on Rh-vinylallene complexes leading to a new reaction, Rh-catalyzed carbonylative (4+1)-cycloaddition. *Angew. Chem. Int. Ed. Engl.* 34 (23–24): 2691–2694.

954 Murakami, M., Itami, K., and Ito, Y. (1996). Synthesis of (vinylallene)Rh(III) complex of planar structure: perfect π->σ conversion of 1,3-diene system. *J. Am. Chem. Soc.* 118 (46): 11672–11673.

955 Meng, Q.X., Li, M., and Zhang, J.S. (2005). The computational study on the mechanism of Rh(I)-catalyzed asymmetric carbonylative (4+1)-cycloaddition with (*R*,*R*)-Me-DuPHOS-type ligand. A DFT study. *J. Mol. Struct. THEOCHEM* 726 (1–3): 47–54.

956 Murakami, M., Itami, K., and Ito, Y. (1997). Rhodium-catalyzed asymmetric (4+1)-cycloaddition. *J. Am. Chem. Soc.* 119 (12): 2950–2951.

957 Murakami, M., Itami, K., and Ito, Y. (1999). Catalytic asymmetric (4+1)-cycloaddition of vinylallenes with carbon monoxide: reversal of the induced chirality by the choice of metal. *J. Am. Chem. Soc.* 121 (17): 4130–4135.

958 Chen, W., Tay, J.H., Yu, X.Q., and Pu, L. (2012). Diastereoselective (4+1)-cycloaddition of alkenyl propargyl acetates with CO catalyzed by $[RhCl(CO)_2]_2$. *J. Org. Chem.* 77: 6215–6222.

959 Li, X.X., Huang, S.Y., Schienebeck, C.M. et al. (2012). Rhodium-catalyzed carbonylation of 3-acyloxy-1,4-enynes for the synthesis of cyclopentenones. *Org. Lett.* 14: 1584–1587.

960 Barluenga, J., López, S., and Flórez, J. (2003). (3+2)- and (4+1)-cycloaddition reactions of Fischer alkoxy(alkenyl)carbene complexes with electronically neutral 1,3-dienes. *Angew. Chem. Int. Ed.* 42 (2): 231–233.

961 Zhao, L.B., Guan, Z.H., Han, Y. et al. (2007). Copper-catalyzed (4+1)-cycloadditions of α,β-acetylenic ketones with diazoacetates to form trisubstituted furans. *J. Org. Chem.* 72 (26): 10276–10278.

962 Zhou, J.L., Wang, L.J., Xu, H. et al. (2013). Highly enantioselective synthesis of multifunctionalized dihydrofurans by Cu-catalyzed asymmetric (4+1)-cycloadditions of α-benzylidene-β-ketoester with diazo compound. *ACS Catal.* 3: 685–688.

963 Ito, Y., Kato, H., and Saegusa, T. (1982). A new approach for stereoselective synthesis of γ-butyrolactones. *J. Org. Chem.* 47 (4): 741–743.

964 Chatani, N., Oshita, M., Tobisu, M. et al. (2003). A $GaCl_3$-catalyzed (4+1)-cycloaddition of α,β-unsaturated carbonyl compounds and isocyanides leading to unsaturated γ-lactone derivatives. *J. Am. Chem. Soc.* 125 (26): 7812–7813.

965 Oshita, M., Yamashita, K., Tobisu, M., and Chatani, N. (2005). Catalytic (4+1)-cycloaddition of α,β-unsaturated carbonyl compounds with isocyanides. *J. Am. Chem. Soc.* 127 (2): 761–766.

966 Li, Z.J., Boyarskikh, V., Hansen, J.H. et al. (2012). Scope and mechanistic analysis of the enantioselective synthesis of allenes by Rh-catalyzed tandem ylide formation/(2,3)-sigmatropic rearrangement between donor/acceptor carbenoids and propargylic alcohols. *J. Am. Chem. Soc.* 134: 15497–15504.

967 Shi, T.D., Guo, X., Teng, S.H., and Hu, W.H. (2015). Pd(II)-catalyzed formal (4+1)-cycloaddition reactions of diazoacetates and aryl propargyl alcohols to form 2,5-dihydrofurans. *J. Chem. Soc., Chem. Commun.* 51: 15204–15207.

968 Wang, J., Yao, X.B., Wang, T. et al. (2015). Synthesis of 2,5-dihydrofurans via a Au(I)-catalyzed formal (4+1)-cycloaddition of α-diazoesters and propargyl alcohols. *Org. Lett.* 17: 5124–5127.

969 Chen, J.R., Dong, W.R., Candy, M. et al. (2012). Enantioselective synthesis of dihydropyrazoles by formal (4+1)-cycloaddition of in situ-derived azoalkenes and sulfur ylides. *J. Am. Chem. Soc.* 134: 6924–6927.

970 Wang, Q., Qi, X., Lu, L.Q. et al. (2016). Iron-catalyzed decarboxylative (4+1)-cycloadditions: exploiting the reactivity of ambident Fe-stabilized intermediates. *Angew. Chem. Int. Ed.* 55: 2840–2844.

971 Wang, Y.H., Muratore, M.E., Rong, Z.T., and Echavarren, A.M. (2014). Formal (4+1)-cycloaddition of methylenecyclopropanes with 7-aryl-1,3,5-cycloheptatrienes by triple Au(I)-catalysis. *Angew. Chem. Int. Ed.* 53: 14022–14025.

972 Wang, Y.H., McGonigal, P.R., Herle, B. et al. (2014). Gold(I)-carbenes by retro-Büchner reaction: generation and fate. *J. Am. Chem. Soc.* 136: 801–809.

973 Sanjuan, A.M., Virumbrales, C., García-García, P. et al. (2016). Formal (4+1)-cycloadditions of β,β-diaryl-substituted *ortho*-(alkynyl)styrenes through Au(I)-catalyzed cycloisomerization reactions. *Org. Lett.* 18: 1072–1075.

974 Negishi, E., Choueiry, D., Nguyen, T.B. et al. (1994). Nonconcerted paths for reactions of (alkene)zirconocene complexes. *J. Am. Chem. Soc.* 116: 9751–9752.

975 Croatt, M.P. and Wender, P.A. (2010). The diene effect: the design, development, and mechanistic investigation of metal-catalyzed diene-yne, diene-ene, and diene-allene (2+2+1)-cycloaddition reactions. *Eur. J. Org. Chem.* 19–32.

976 Wender, P.A., Croatt, M.P., and Deschamps, N.M. (2006). Metal-catalyzed (2+2+1)-cycloadditions of 1,3-dienes, allenes, and CO. *Angew. Chem. Int. Ed.* 45: 2459–2462.

977 Makino, T. and Itoh, K. (2004). Rhodium complex-catalyzed cycloisomerization of allenenes: exo and endo cyclization depending on the auxiliary ligands. *J. Org. Chem.* 69: 395–405.

978 Kawamura, T., Inagaki, F., Narita, S. et al. (2010). Rhodium(I)-catalyzed intramolecular carbonylative (2+2+1)-cycloadditions and cycloisomerizations of bis(sulfonylallene)s. *Chem. Eur. J.* 16: 5173–5183.

979 Kondo, T., Nomura, M., Ura, Y. et al. (2006). Ruthenium-catalyzed (2+2+1)-cocyclization of isocyanates, alkynes, and CO enables the rapid synthesis of polysubstituted maleimides. *J. Am. Chem. Soc.* 128 (46): 14816–14817.

980 Mathur, P., Joshi, R.K., Rai, D.K. et al. (2012). One pot synthesis of maleimide and hydantoin by $Fe(CO)_5$ catalyzed (2+2+1)-co-cyclization of acetylene, isocyanate and CO. *J. Chem. Soc., Dalton Trans.* 41: 5045–5054.

981 Ozawa, T., Horie, H., Kurahashi, T., and Matsubara, S. (2010). Nickel-catalyzed (2+2+1)-cycloaddition of alkynes, acrylates and isocyanates. *J. Chem. Soc., Chem. Commun.* 46: 8055–8057.

982 Hoshimoto, Y., Ohata, T., Sasaoka, Y. et al. (2014). Nickel(0)-catalyzed (2+2+1)-carbonylative cycloaddition of imines and alkynes or norbornene leading to γ-lactams. *J. Am. Chem. Soc.* 136: 15877–15880.

983 Rao, W.D., Koh, M.J., Kothandaraman, P., and Chan, P.W.H. (2012). Gold-catalyzed cycloisomerization of 1,7-diyne benzoates to indeno[1,2-c]azepines and azabicyclo[4.2.0]octa-1(8),5-dienes. *J. Am. Chem. Soc.* 134: 10811–10814.

984 Rao, W.D., Koh, M.J., Li, D. et al. (2013). Gold-catalyzed cycloisomerization of 1,6-diyne carbonates and esters to

2,4a-dihydro-1*H*-fluorenes. *J. Am. Chem. Soc.* 135: 7926–7932.

985 Day, D.P. and Chan, P.W.H. (2016). Gold-catalyzed cycloisomerizations of 1,n-diyne carbonates and esters. *Adv. Synth. Catal.* 358: 1368–1384.

986 Asiri, A.M. and Hashmi, A.S.K. (2016). Gold-catalyzed reactions of diynes. *Chem. Soc. Rev.* 45: 4471–4503.

987 Rao, W.D. and Chan, P.W.H. (2014). Gold-catalyzed (2+2+1)-cycloaddition of 1,6-diyne carbonates and esters with aldehydes to 4-(cyclohexa-1,3-dienyl)-1,3-dioxolanes. *Chem. Eur. J.* 20: 713–718.

988 Sarel, S., Felzenst, A., Victor, R., and Yovell, J. (1974). Photoreactions of a methylene-spirane and dispirane with $Fe(CO)_5$ - σ,π-complex formation from a double cyclopropane rearrangement. *J. Chem. Soc., Chem. Commun.* 1025–1026.

989 Liu, C.H., Zhuang, Z., Bose, S., and Yu, Z.X. (2016). $Fe_2(CO)_9$-mediated (5+1)-cycloaddition of vinylcyclopropanes and CO for the synthesis of α,β-cyclohexenones. *Tetrahedron* 72: 2752–2755.

990 Murakami, M., Itami, K., Ubukata, M. et al. (1998). Iridium-catalyzed (5+1)-cycloaddition: allenylcyclopropane as a five-carbon assembling unit. *J. Org. Chem.* 63: 4–5.

991 Kamitani, A., Chatani, N., Morimoto, T., and Murai, S. (2000). Carbonylative (5+1)-cycloaddition of cyclopropyl imines catalyzed by Ru(carbonyl) complex. *J. Org. Chem.* 65: 9230–9233.

992 Olson, A.S., Seitz, W.J., and Hossain, M.M. (1991). Transition-metal catalysis of the Diels–Alder reaction. *Tetrahedron Lett.* 32 (39): 5299–5302.

993 Hollis, T.K., Odenkirk, W., Robinson, N.P. et al. (1993). Homogeneous catalysis - transition-metal based Lewis acid catalysts. *Tetrahedron* 49 (25): 5415–5430.

994 Bellville, D.J., Wirth, D.D., and Bauld, N.L. (1981). The cation-radical catalyzed Diels–Alder reaction. *J. Am. Chem. Soc.* 103 (3): 718–720.

995 Pabon, R.A., Bellville, D.J., and Bauld, N.L. (1983). Cation-radical Diels–Alder reactions of electron-rich dienophiles. *J. Am. Chem. Soc.* 105 (15): 5158–5159.

996 Yueh, W. and Bauld, N.L. (1995). Criteria for the cation-radical vs electrophilic mechanistic distinction - the aminium salt-catalyzed Diels–Alder reaction. *J. Chem. Soc., Perkin Trans. 2* (5): 871–873.

997 Eberson, L., Olofsson, B., and Svensson, J.O. (1992). Electron-transfer reactions in organic-chemistry. 21. New catalysts for the Diels–Alder dimerization of 1,3-cyclohexadiene - Brønsted, Lewis and redox catalysts. *Acta Chem. Scand.* 46 (10): 1005–1015.

998 Carbonaro, A., Greco, A., and Dall'Asta, G. (1968). Oligomerization catalysts. 3. Cyclocodimerization of conjugated dienes with acetylenic hydrocarbons catalyzed by Fe(0)-complexes. Synthesis of 1,2-diphenyl-1,4-cyclohexadiene. *J. Org. Chem.* 33 (10): 3948–3050.

999 Garratt, P.J. and Wyatt, M. (1974). Regiospecific control in formation of cyclohexenes from dienes - catalyzed and noncatalyzed reactions of butadiene and 2,3-dimethylbutadiene with methyl sorbate. *J. Chem. Soc., Chem. Commun.* (7): 251–251.

1000 Genêt, J.P. and Ficini, J. (1979). Cycloaddition of ynamines with Fe(0)-catalyzed butadiene - synthesis of β,γ-unsaturated and α,β-unsaturated 1,4-cyclohexadienamines and cyclohexenones. *Tetrahedron Lett.* 17: 1499–1502.

1001 Wender, P.A. and Jenkins, T.E. (1989). Nickel-catalyzed intramolecular [4+2] dienyne cycloadditions - an efficient new method for the synthesis of polycycles containing cyclohexa-1,4-dienes. *J. Am. Chem. Soc.* 111 (16): 6432–6434.

1002 Lee, S.I., Park, S.Y., Park, J.H. et al. (2006). Rhodium N-heterocyclic carbene-catalyzed $[\pi^4+\pi^2]$- and (5+2)-cycloaddition reactions. *J. Org. Chem.* 71 (1): 91–96.

1003 Huang, S.Y., Li, X.X., Lin, C.L. et al. (2012). Rhodium-catalyzed 1,3-acyloxy migration and subsequent intramolecular $[\pi^4+\pi^2]$-cycloaddition of vinylallene and unactivated alkyne. *J. Chem. Soc., Chem. Commun.* 48: 2204–2206.

1004 Tigchelaar, A. and Tam, W. (2012). Iridium-catalyzed intramolecular $[\pi^4+\pi^2]$-cycloadditions of alkynyl halides. *Beilstein J. Org. Chem.* 8: 1765–1770.

1005 Dolor, R.Z. and Vogel, P. (1990). Diels–Alder additions of 1,3-dienes to propargylic alcohols catalyzed by the Wilkinsons catalyst. *J. Mol. Catal.* 60: 59–63.

1006 tom Dieck, H. and Diercks, R. (1983). Diels–Alder reactions of dienes with alkynes catalyzed by diazadieneiron. *Angew. Chem. Int. Ed. Engl.* 22 (10): 778–779.

1007 Vioget, P., Vogel, P., and Roulet, R. (1982). Cycloaddition of a 1,3-diene to a carbon–carbon double bond assisted by Fe-carbonyls. Synthesis of an anthracyclinone precursor. *Angew. Chem.* 94 (6): 454–455.

1008 Vioget, P., Bonivento, M., Roulet, R., and Vogel, P. (1984). Monoolefin and diene cycloaddition induced by transition-metal carbonyls - cyclodimerization of 5,6-dimethylidene-7-oxabicyclo[2.2.1]hept-2-ene catalyzed by dodecacarbonyltriosmium. *Helv. Chim. Acta* 67 (6): 1638–1646.

1009 Bailey, M.S., Brisdon, B.J., Brown, D.W., and Stark, K.M. (1983). Transition-metal complexes as catalysts in Diels–Alder reactions. *Tetrahedron Lett.* 24 (29): 3037–3040.

1010 Humphries, A.P. and Knox, S.A.R. (1978). Formal Diels–Alder dimerization of cyclooctatetraene induced by (carbonyl)Ru complexes. *J. Chem. Soc., Dalton Trans.* (11): 1514–1523.

1011 Wender, P.A., Jenkins, T.E., and Suzuki, S. (1995). Transition-metal-catalyzed intramolecular $[\pi^4+\pi^2]$-diene allene cycloadditions - a convenient synthesis of angularly substituted ring systems with provision for catalyst-controlled chemocomplementarity and stereocomplementarity. *J. Am. Chem. Soc.* 117 (6): 1843–1844.

1012 Wender, P.A. and Smith, T.E. (1995). Transition-metal-catalyzed intramolecular $[\pi^4+\pi^2]$-cycloadditions - initial studies on stereochemistry and on Steroid and Vitamin-D analog syntheses. *J. Org. Chem.* 60 (10): 2962–2963.

1013 Wender, P.A. and Smith, T.E. (1998). Transition metal-catalyzed intramolecular $[\pi^4+\pi^2]$-cycloadditions: mechanistic and synthetic investigations. *Tetrahedron* 54 (7): 1255–1275.

1014 McKinstry, L. and Livinghouse, T. (1994). On the asymmetric Rh(I)-catalyzed $[\pi^4+\pi^2]$-cycloisomerization reaction - electronic and torsional ligand control of absolute stereoselection. *Tetrahedron* 50 (21): 6145–6154.

1015 Hilt, G., Janikowski, J., and Hess, W. (2006). *meta*-Directing Co-catalyzed Diels–Alder reactions. *Angew. Chem. Int. Ed.* 45 (31): 5204–5206.

1016 Mörschel, P., Janikowski, J., Hilt, G., and Frenking, G. (2008). Ligand-tuned regioselectivity of a Co-catalyzed Diels–Alder reaction. A theoretical study. *J. Am. Chem. Soc.* 130 (28): 8952–8966.

1017 Fürstner, A. and Stimson, C.C. (2007). Two manifolds for metal-catalyzed intramolecular Diels–Alder reactions of unactivated alkynes. *Angew. Chem. Int. Ed.* 46 (46): 8845–8849.

1018 Tanaka, K., Hojo, D., Shoji, T. et al. (2007). Rh-catalyzed $[\pi^4+\pi^2]$-carbocyclization of vinylarylaldehydes with alkenes and alkynes leading to substituted tetralones and 1-naphthols. *Org. Lett.* 9 (11): 2059–2062.

1019 Hilt, G., Hess, W., and Harms, K. (2006). Asymmetric Co-catalyzed Diels–Alder reactions of a boron-functionalized 1,3-diene with alkynes and allylboration with aldehydes. *Org. Lett.* 8 (15): 3287–3290.

1020 Wender, P.A. and Smith, T.E. (1996). Transition metal-catalyzed intramolecular $[\pi^4+\pi^2]$-cycloadditions: a novel method for the assembly of nitrogen heterocycles and its application to Yohimban alkaloid synthesis. *J. Org. Chem.* 61 (3): 824–825.

1021 Saito, S. and Yamamoto, Y. (2000). Recent advances in the transition-metal-catalyzed regioselective approaches to polysubstituted benzene derivatives. *Chem. Rev.* 100 (8): 2901–2915.

1022 Varela, J.A. and Saa, C. (2003). Construction of pyridine rings by metal-mediated (2+2+2)-cycloaddition. *Chem. Rev.* 103: 3787–3801.

1023 Kotha, S., Brahmachary, E., and Lahiri, K. (2005). Transition metal catalyzed (2+2+2)-cycloaddition and application in organic synthesis. *Eur. J. Org. Chem.* 4741–4767.

1024 Chopade, P.R. and Louie, J. (2006). (2+2+2)-Cycloaddition reactions catalyzed by transition metal complexes. *Adv. Synth. Catal.* 348 (16–17): 2307–2327.

1025 Inglesby, P.A. and Evans, P.A. (2010). Stereoselective transition metal-catalysed higher-order carbocyclisation reactions. *Chem. Soc. Rev.* 39: 2791–2805.

1026 Weding, N. and Hapke, M. (2011). Preparation and synthetic applications of alkene complexes of group 9 transition metals in (2+2+2)-cycloaddition reactions. *Chem. Soc. Rev.* 40: 4525–4538.

1027 Dominguez, G. and Perez-Castells, J. (2016). Alkenes in (2+2+2)-cycloadditions. *Chem. Eur. J.* 22: 6720–6739.

1028 Aubert, C., Fensterbank, L., García, P. et al. (2011). Transition metal catalyzed cycloisomerizations of 1,n-allenynes and -allenenes. *Chem. Rev.* 111: 1954–1993.

1029 Oonishi, Y., Hato, Y., and Sato, Y. (2015). Rhodium(I)-catalyzed (2+2+2)-cycloaddition between allene, alkyne, and imine via a strained azarhodacycle intermediate. *Adv. Synth. Catal.* 357: 3033–3039.

1030 Ikeda, S., Kondo, H., Arii, T., and Odashima, K. (2002). Catalytic enantioselective intermolecular (2+2+2)-cycloaddition of an alkene and two alkynes. *J. Chem. Soc., Chem. Commun.* 2422–2423.

1031 Nishida, G., Suzuki, N., Noguchi, K., and Tanaka, K. (2006). Enantioselective synthesis of tetra-*ortho*-substituted axially chiral biaryls through Rh-catalyzed double (2+2+2)-cycloaddition. *Org. Lett.* 8 (16): 3489–3492.

1032 Tanaka, K., Nishida, G., Wada, A., and Noguchi, K. (2004). Enantioselective synthesis of axially chiral phthalides through cationic [$Rh^1(H_8$-binap)]-catalyzed cross alkyne cyclotrimerization. *Angew. Chem. Int. Ed.* 43 (47): 6510–6512.

1033 Shibata, T. and Tsuchikama, K. (2005). Ir-catalyzed almost perfect enantioselective synthesis of helical polyaryls based on an axially-chiral sequence. *J. Chem. Soc., Chem. Commun.* 6017–6019.

1034 Shibata, T. and Tahara, Y.K. (2006). Enantioselective intramolecular (2+2+2)-cycloaddition of 1,4-diene-ynes: a new approach to the construction of quaternary carbon stereocenters. *J. Am. Chem. Soc.* 128 (36): 11766–11767.

1035 Tanaka, K., Takeishi, K., and Noguchi, K. (2006). Enantioselective synthesis of axially chiral anilides through Rh-catalyzed (2+2+2)-cycloaddition of 1,6-diynes with trimethylsilylynamides. *J. Am. Chem. Soc.* 128: 4586–4587.

1036 Yu, R.T. and Rovis, T. (2006). Enantioselective Rh-catalyzed (2+2+2)-cycloaddition of alkenyl isocyanates and terminal alkynes: application to the total synthesis of (+)-lasubine II. *J. Am. Chem. Soc.* 128: 12370–12371.

1037 Tanaka, K. (2007). Cationic Rh(I)/BINAP-type bisphosphine complexes: versatile new catalysts for highly chemo-, regio-, and enantioselective (2+2+2)-cycloadditions. *Synlett* 1977–1993.

1038 Tanaka, K., Otake, Y., Sagae, H. et al. (2008). Highly regio-, diastereo-, and enantioselective (2+2+2)-cycloaddition of 1,6-enynes with electron-deficient ketones catalyzed by a cationic Rh(I)/H(8)-binap complex. *Angew. Chem. Int. Ed.* 47: 1312–1316.

1039 Shibata, T. and Tsuchikama, K. (2008). Recent advances in enantioselective (2+2+2)-cycloaddition. *Org. Biomol. Chem.* 6: 1317–1323.

1040 Tanaka, K. (2009). Transition-metal-catalyzed enantioselective (2+2+2)-cycloadditions for the synthesis of axially chiral biaryls. *Chem. Asian J.* 4: 508–518.

1041 Amatore, M. and Aubert, C. (2015). Recent advances in stereoselective (2+2+2)-cycloadditions. *Eur. J. Org. Chem.* 265–286.

1042 Yoshizaki, S., Nakamura, Y., Masutomi, K. et al. (2016). Rhodium-catalyzed asymmetric (2+2+2)-cycloaddition of 1,6-enynes with cyclopropylideneacetamides. *Org. Lett.* 18: 388–391.

1043 Aida, Y., Sugiyama, H., Uekusa, H. et al. (2016). Rhodium-catalyzed asymmetric (2+2+2)-cycloaddition of α,ω-diynes with unsymmetrical 1,2-disubstituted alkenes. *Org. Lett.* 18: 2672–2675.

1044 Woodward, R.B. and Hoffmann, R. (1970). *The Conservation of Orbital Symmetry*, 1–178. New York: Academic Press.

1045 Dower, W.V. and Vollhardt, K.P.C. (1982). Mechanism of the isomerization of 1,5,9-cyclododecatriyne to hexaradialene - 1,2-3,4-5,6-Tricyclobutabenzene is not an intermediate. *J. Am. Chem. Soc.* 104 (24): 6878–6879.

1046 Hartung, J.B. Jr., and Pedersen, S.F. (1990). Synthesis and characterization of trihaloniobium alkyne complexes. *Organometallics* 9 (5): 1414–1417.

1047 Oshiki, T. and Takai, K. (2002). Cyclotrimerization of alkynes using tantalum complexes and the complex catalysts. *Jpn. Kokai Tokkyo Koho* 2002060354.

1048 Galan, B.R. and Rovis, T. (2009). Beyond Reppe: building substituted arenes by (2+2+2)-cycloadditions of alkynes. *Angew. Chem. Int. Ed.* 48 (16): 2830–2834.

1049 Berthelot, M. (1866). Action de la chaleur sur quelques carbures d'hydrogène. *C.R. Hebd. Séances Acad. Sci.* 62: 905–909.

1050 Tanaka, D., Sato, Y., and Mori, M. (2007). Ruthenium-catalyzed (2+2+2)-cocyclization of diene-yne. *J. Am. Chem. Soc.* 129 (25): 7730–7731.

1051 Heller, B. and Hapke, M. (2007). The fascinating construction of pyridine ring systems by transition metal-catalysed (2+2+2)-cycloaddition reactions. *Chem. Soc. Rev.* 36 (7): 1085–1094.

1052 Varela, J.A. and Saa, C. (2008). Recent advances in the synthesis of pyridines by transition-metal-catalyzed (2+2+2)-cycloaddition. *Synlett* (17): 2571–2578.

1053 Oberg, K.M., Lee, E.E., and Rovis, T. (2009). Regioselective Rh-catalyzed intermolecular (2+2+2)-cycloaddition of alkynes and isocyanates to form pyridones. *Tetrahedron* 65 (26): 5056–5061.

1054 Rodríguez-Otero, J., Montero-Campillo, M.M., and Cabaleiro-Lago, E.M. (2008). Density functional theory study of Ru(II)-catalyzed (2+2+2)-cycloaddition of 1,6-diynes with tricarbonyl compounds. *J. Phys. Chem. A* 112 (35): 8116–8120.

1055 Funk, R.L. and Vollhardt, K.P.C. (1977). Cobalt-catalyzed Steroid synthesis. *J. Am. Chem. Soc.* 99 (16): 5483–5484.

1056 Saino, N., Kogure, D., and Okamoto, S. (2005). Intramolecular cyclotrimerization of triynes catalyzed by N-heterocyclic carbene-$CoCl_2$/Zn or -$FeCl_3$/Zn. *Org. Lett.* 7 (14): 3065–3067.

1057 Liu, Y.B., Yan, X.Y., Yang, N.F., and Xi, C.J. (2011). Highly regioselective cyclotrimerization of terminal alkynes catalyzed by Fe(II)-complexes bearing 2-(benzimidazolyl)-6-(1-(arylimino)ethyl)pyridines. *Catal. Commun.* 12: 489–492.

1058 Saino, N., Amemiya, F., Tanabe, E. et al. (2006). A highly practical instant catalyst for cyclotrimerization of alkynes to substituted benzenes. *Org. Lett.* 8 (7): 1439–1442.

1059 Goswami, A., Ito, T., and Okamoto, S. (2007). Efficient activation of 2-Iminomethylpyridine/Co-based alkyne (2+2+2)-cycloaddition catalyst by addition of a silver salt. *Adv. Synth. Catal.* 349 (14–15): 2368–2374.

1060 Hilt, G., Vogler, T., Hess, W., and Galbiati, F. (2005). A simple Co-catalyst system for the efficient and regioselective cyclotrimerisation of alkynes. *J. Chem. Soc., Chem. Commun.* (11): 1474–1475.

1061 Hilt, G., Hess, W., Vogler, T., and Hengst, C. (2005). Ligand and solvent effects on Co(I)-catalysed reactions: alkyne dimerisation versus (2+2+2)-cyclotrimerisation versus Diels–Alder reaction versus (4+2+2)-cycloaddition. *J. Organomet. Chem.* 690 (23): 5170–5181.

1062 Hilt, G., Hengst, C., and Hess, W. (2008). Solvent-dependent regiochemical cyclotrimerisation of phenylacetylene with Co-catalysts containing disulfide ligands: a case study. *Eur. J. Org. Chem.* (13): 2293–2297.

1063 Kakeya, M., Fujihara, T., Kasaya, T., and Nagasawa, A. (2006). Dinuclear Nb(III)-complexes $[\{NbCl_2(L)\}_2(\mu\text{-}Cl)_2(\mu\text{-}L)]$ (L = tetrahydrothiophene, dimethyl sulfide): preparation, molecular structures, and the catalytic activity for the regioselective cyclotrimerization of alkynes. *Organometallics* 25 (17): 4131–4137.

1064 Tanaka, K. and Shirasaka, K. (2003). Highly chemo- and regioselective intermolecular cyclotrimerization of alkynes catalyzed by cationic Rh(I)/modified BINAP. *Org. Lett.* 5 (24): 4697–4699.

1065 Tanaka, K., Toyoda, K., Wada, A. et al. (2005). Chemo- and regioselective intermolecular cyclotrimerization of terminal alkynes catalyzed by cationic rhodium(I)/modified BINAP complexes: application to one-step synthesis of paracyclophanes. *Chem. Eur. J.* 11 (4): 1145–1156.

1066 Kuninobu, Y., Takata, H., Kawata, A., and Takai, K. (2008). Rhenium-catalyzed synthesis of multisubstituted aromatic compounds via C—C single-bond cleavage. *Org. Lett.* 10 (14): 3133–3135.

1067 Yamamoto, Y., Kinpara, K., Saigoku, F. et al. (2005). Cp*RuCl-catalyzed (2+2+2)-cycloadditions of α,ω-diynes with electron-deficient carbon–heteroatom multiple bonds leading to heterocycles. *J. Am. Chem. Soc.* 127 (2): 605–613.

1068 Dazinger, G., Torres-Rodrigues, M., Kirchner, K. et al. (2006). Formation of pyridine from acetylenes and nitriles catalyzed by RuCpCl, CoCp, and RhCp derivatives - a computational mechanistic study. *J. Organomet. Chem.* 691 (21): 4434–4445.

1069 Gutnov, A., Heller, B., Fischer, C. et al. (2004). Cobalt(I)-catalyzed asymmetric (2+2+2)-cycloaddition of alkynes and nitriles: synthesis of enantiomerically enriched atropoisomers of 2-arylpyridines. *Angew. Chem. Int. Ed.* 43 (29): 3795–3797.

1070 Shibata, T., Fujimoto, T., Yokota, K., and Takagi, K. (2004). Iridium complex-catalyzed highly enantio- and diastereoselective (2+2+2)-cycloaddition for the synthesis of axially chiral teraryl compounds. *J. Am. Chem. Soc.* 126 (27): 8382–8383.

1071 Tanaka, K., Nishida, G., Ogino, M. et al. (2005). Enantioselective synthesis of axially chiral biaryls through Rh-catalyzed complete intermolecular cross-cyclotrimerization of internal alkynes. *Org. Lett.* 7 (14): 3119–3121.

1072 Singleton, D.M. (1973). Catalyzed cycloaddition of acetylenes to cyclohexene, formation of 2-1 adducts. *Tetrahedron Lett.* (15): 1245–1248.

1073 Grigg, R., Scott, R., and Stevenson, P. (1988). Rhodium(I)-catalyzed reactions of enynes - linear coupling vs cycloaddition. *J. Chem. Soc., Perkin Trans. 1* (6): 1365–1369.

1074 Chang, C.A., King, J.A. Jr., and Vollhardt, K.P.C. (1981). Intermolecular Co-mediated (2+2+2)-cycloadditions - regiospecific one-step construction of bicyclic dienes from α,ω-enynes and alkynes. *J. Chem. Soc., Chem. Commun.* (2): 53–55.

1075 Trost, B.M. and Tanoury, G.J. (1987). Intramolecular carbametallations - a (2+2+2)-cycloaddition as evidence for a palladacyclopentene intermediate. *J. Am. Chem. Soc.* 109 (15): 4753–4755.

1076 Rhyoo, H.Y., Lee, B.Y., Yu, H.K.B., and Chung, Y.K. (1994). Study of the reactivity of $ClCo(PPh_3)_3$. *J. Mol. Catal.* 92 (1): 41–49.

1077 Oh, C.H., Sung, H.R., Jung, S.H., and Lim, Y.M. (2001). Rhodium catalyzed (2+2+2)-cyclizations of 1,6-enynes. *Tetrahedron Lett.* 42 (32): 5493–5495.

1078 Yamamoto, Y., Kuwabara, S., Ando, Y. et al. (2004). Pd(0)-catalyzed cyclization of electron-deficient enynes and enediynes. *J. Org. Chem.* 69 (20): 6697–6705.

1079 Chiusoli, G.P., Costa, M., and Zhou, Z.M. (1992). Selectivity control in Co(0)-catalyzed cycloaddition reactions. *Gazz. Chim. Ital.* 122 (11): 441–449.

1080 Kezuka, S., Okado, T., Niou, E., and Takeuchi, R. (2005). Iridium complex-catalyzed reaction of 1,6-enynes: cycloaddition and cycloisomerization. *Org. Lett.* 7 (9): 1711–1714.

1081 Shibata, T., Arai, Y., and Tahara, Y. (2005). Enantioselective construction of quaternary carbon centers by catalytic (2+2+2)-cycloaddition of 1,6-enynes and alkynes. *Org. Lett.* 7 (22): 4955–4957.

1082 Tanaka, K., Wada, A., and Noguchi, K. (2005). Rhodium-catalyzed chemo-, regio-, and enantioselective (2+2+2)-cycloaddition of alkynes with isocyanates. *Org. Lett.* 7 (21): 4737–4739.

1083 Seo, J., Chui, H.M.P., Heeg, M.J., and Montgomery, J. (1999). Novel chemoselectivity and stereochemical aspects of Ni-catalyzed (2+2+2)-cycloadditions. *J. Am. Chem. Soc.* 121 (2): 476–477.

1084 Schrauzer, G.N. and Eichler, S. (1962). Zur Kenntnis von Bis-Acrylnitril-Nickel(0). 9. Katalytische Reaktionen mit Norbornadien. *Chem. Ber.* 95 (11): 2764–2763.

1085 Sasaki, T., Eguchi, S., Sugimoto, M., and Hibi, F. (1972). Studies on reactions of isoprenoids. 17. Cycloaddition reactions of norbornadiene with some unsymmetrically substituted dienophiles - competitive ionic additions with homo-Diels–Alder reactions. *J. Org. Chem.* 37 (14): 2317–2320.

1086 Lautens, M., Edwards, L.G., Tam, W., and Lough, A.J. (1995). Nickel-catalyzed $[\pi^2+\pi^2+\pi^2]$ (Homo-Diels–Alder) and $[\pi^2+\pi^2]$-cycloadditions of bicyclo[2.2.1]hepta-2,5-dienes. *J. Am. Chem. Soc.* 117 (41): 10276–10291.

1087 Lyons, J.E., Myers, H.K., and Schneider, A. (1978). Selective homo-Diels–Alder addition of acetylenic hydrocarbons to norbornadiene catalyzed by a Co-complex. *J. Chem. Soc., Chem. Commun.* (15): 636–638.

1088 Lautens, M., Lautens, J.C., and Smith, A.C. (1990). Catalytic asymmetric induction in the homo-Diels–Alder reaction. *J. Am. Chem. Soc.* 112 (14): 5627–5628.

1089 tom Dieck, H., Mallien, M., and Diercks, R. (1989). Diazadienes as control ligands in homogeneous catalysis. 15. Diazadiene-Fe(0)-catalyzed synthesis of tetracyclononenes from norbornadiene and internal alkynes. *J. Mol. Catal.* 51 (1): 53–58.

1090 Duan, I.F., Cheng, C.H., Shaw, J.S. et al. (1991). Homo-Diels–Alder cycloadditions catalyzed by Co-triphenylphosphine/Zn systems. *J. Chem. Soc., Chem. Commun.* (19): 1347–1348.

1091 Tenaglia, A. and Giordano, L. (2004). Ruthenium(II)-catalyzed homo-Diels–Alder reactions of disubstituted alkynes and norbornadiene. *Tetrahedron Lett.* 45 (1): 171–174.

1092 Brunner, H., Muschiol, M., and Prester, F. (1990). Asymmetric catalysts. 55. Optically active deltacyclanes and deltacyclenes by enantioselective catalysis. *Angew. Chem. Int. Ed. Engl.* 29 (6): 652–653.

1093 Brunner, H. and Prester, F. (1991). Asymmetrical catalysis. 70. Highly enantioselective Co-catalyzed homo-Diels–Alder reaction of norbornadiene with acetylenes. *J. Organomet. Chem.* 414 (3): 401–409.

1094 Pardigon, O. and Buono, G. (1993). Enantioselective synthesis of deltacyclenes using a CoI_2/Zn catalytic system. *Tetrahedron: Asymmetry* 4 (9): 1977–1980.

1095 Trost, B.M., Imi, K., and Indolese, A.F. (1993). 1,5-Cyclooctadiene as a bis-homodiene partner in a metal-catalyzed (4+2)-cycloaddition. *J. Am. Chem. Soc.* 115 (19): 8831–8832.

1096 Huang, X. and Lin, Z.Y. (2003). Density functional theory studies of Ru-catalyzed bis-Diels–Alder cycloaddition of 1,5-cyclooctadiene with alkynes. *Organometallics* 22 (26): 5478–5484.

1097 Kesenheimer, C. and Groth, U. (2006). Total synthesis of (−)-8-*O*-methyltetrangomycin (MM 47755). *Org. Lett.* 8 (12): 2507–2510.

1098 Na, R.S., Jing, C.F., Xu, Q.H. et al. (2011). Phosphine-catalyzed annulations of azomethine imines: allene-dependent (3+2), (3+3), (4+3), and (3+2+3)-pathways. *J. Am. Chem. Soc.* 133: 13337–13348.

1099 Chan, A. and Scheidt, K.A. (2007). Highly stereoselective formal (3+3)-cycloaddition of enals and azomethine imines catalyzed by N-heterocyclic carbenes. *J. Am. Chem. Soc.* 129 (17): 5334–5335.

1100 Zhu, T.S., Zheng, P.C., Mou, C.L. et al. (2014). Benzene construction via organocatalytic formal (3+3)-cycloaddition reaction. *Nat. Commun.* 5: 5027, 1–6.

1101 Shintani, R. and Hayashi, T. (2006). Palladium-catalyzed (3+3)-cycloaddition of trimethylenemethane with azomethine imines. *J. Am. Chem. Soc.* 128: 6330–6331.

1102 Shapiro, N.D., Shi, Y., and Toste, F.D. (2009). Gold-catalyzed (3+3)-annulation of azomethine imines with propargyl esters. *J. Am. Chem. Soc.* 131: 11654–11655.

1103 Xu, X.F. and Doyle, M.P. (2014). The (3+3)-cycloaddition alternative for heterocycle syntheses: catalytically generated metalloenolcarbenes as dipolar adducts. *Acc. Chem. Res.* 47: 1396–1405.

1104 Evjen, S. and Fiksdahl, A. (2016). Gold(I)-catalysed (3+3)-cycloaddition of propargyl acetals and nitrones. *Tetrahedron* 72: 3270–3276.

1105 Leung, T.W., Christoph, G.G., and Wojcicki, A. (1983). A novel metal-assisted (3+3)-cycloaddition reaction of ligated η1-allyl. *Inorg. Chim. Acta* 76 (5–6): L281–L282.

1106 Bambal, R.B. and Kemmitt, R.D.W. (1989). (3+3)-Cycloaddition of trimethylenemethane to activated aziridines - palladium-catalyzed synthesis of piperidines. *J. Organomet. Chem.* 362 (1–2): C18–C20.

1107 Hedley, S.J., Moran, W.J., Prenzel, A.H.G.P. et al. (2001). Synthesis of functionalised piperidines through a (3+3)-cycloaddition strategy. *Synlett* (10): 1596–1598.

1108 Hedley, S.J., Moran, W.J., Price, D.A., and Harrity, J.P.A. (2003). Development of a (3+3)-cycloaddition strategy toward functionalized piperidines. *J. Org. Chem.* 68 (11): 4286–4292.

1109 Goodenough, K.M., Moran, W.J., Raubo, P., and Harrity, J.P.A. (2005). Development of a flexible approach to Nuphar alkaloids via two enantiospecific piperidine-forming reactions. *J. Org. Chem.* 70 (1): 207–213.

1110 Shintani, R., Park, S., Duan, W.L., and Hayashi, T. (2007). Palladium-catalyzed asymmetric (3+3)-cycloaddition of trimethylenemethane derivatives with nitrones. *Angew. Chem. Int. Ed.* 46 (31): 5901–5903.

1111 Anada, M., Watanabe, N., and Hashimoto, S. (1998). Highly enantioselective construction of the key azetidin-2-ones for the synthesis of carbapenem antibiotics via intramolecular C—H insertion reactions of α-methoxycarbonyl-α-diazoacetamides catalysed by chiral dirhodium(II) carboxylates. *J. Chem. Soc., Chem. Commun.* 1517–1518.

1112 Anada, M. and Hashimoto, S. (1998). Enantioselective synthesis of 4-substituted 2-pyrrolidinones by site-selective C—H insertion of α-methoxycarbonyl-α-diazoacetanilides catalyzed by dirhodium(II) tetrakis[N-phthaloyl-(*S*)-*tert*-leucinate]. *Tetrahedron lett.* 39: 79–82.

1113 Wang, X.C., Xu, X.F., Zavalij, P.Y., and Doyle, M.P. (2011). Asymmeric formal (3+3)-cycloaddition reactions of nitrones with electrophilic vinylcarbene intermediates. *J. Am. Chem. Soc.* 133: 16402–16405.

1114 Lee, S.I., Park, J.H., Chung, Y.K., and Lee, S.G. (2004). Rhodium-catalyzed cabonylative (3+2+1)-cycloaddition reaction: catalytic formation of bicyclic byclohexenones from trienes and carbon monoxide. *J. Am. Chem. Soc.* 126: 2714–2715.

1115 Fukuyama, T., Higashibeppu, Y., Yamaura, R., and Ryu, I. (2007). Ru-catalyzed intermolecular (3+2+1)-cycloaddition of α,β-unsaturated ketones with silylacetylenes and carbon monoxide leading to α-pyrones. *Org. Lett.* 9: 587–589.

1116 Wu, B. and Hua, R.M. (2010). Palladium-catalyzed (3+2+1)-cyclocarbonylative coupling of 1,3-cyclohexanediones, alkynes, and carbon monoxide: an atom-economic route to chromene-2,5-dione derivatives. *Tetrahedron Lett.* 51: 6433–6435.

1117 Jiao, L., Lin, M., Zhuo, L.G., and Yu, Z.X. (2010). Rh(I)-Catalyzed ((3+2)+1)-cycloaddition of 1-yne/ene-vinylcyclopropanes and CO: homologous Pauson–Khand reaction and total synthesis of (+/-)-α-agarofuran. *Org. Lett.* 12: 2528–2531.

1118 Feng, Y. and Yu, Z.X. (2015). Formal synthesis of (+/-)-galanthamine and (+/-)-lycoramine using Rh(I)-catalyzed ((3+2)+1)-cycloaddition of

1-ene-vinylcyclopropane and CO. *J. Org. Chem.* 80: 1952–1956.

1119 Bose, S., Yang, J., and Yu, Z.X. (2016). Formal synthesis of Gracilamine using Rh(I)-catalyzed (3+2+1)-cycloaddition of 1-yne-vinylcyclopropanes and CO. *J. Org. Chem.* 81: 6757–6765.

1120 Mazumder, S., Shang, D.J., Negru, D.E. et al. (2012). Stereoselective Rh-catalyzed (3+2+1)-carbocyclization of alkenylidenecyclopropanes with carbon monoxide: theoretical evidence for a trimethylenemethane metallacycle intermediate. *J. Am. Chem. Soc.* 134: 20569–20572.

1121 Evans, P.A., Burnie, A.J., and Negru, D.E. (2014). Rhodium-catalyzed ((3+2)+1)-carbocyclization reactions of alkynylidenecyclopropanes with carbon monoxide: regiospecific construction of polysubstituted phenols. *Org. Lett.* 16: 4356–4359.

1122 Kim, S. and Chung, Y.K. (2014). Rhodium-catalyzed carbonylative ((3+2)+1)-cycloaddition of alkyne-tethered alkylidenecyclopropanes to phenols in the presence of carbon monoxide. *Org. Lett.* 16: 4352–4355.

1123 Liu, L.X. and Sun, H.L. (2014). $[HCo(CO)_4]$-Catalyzed three-component cycloaddition of epoxides, imines, and carbon monoxide: facile construction of 1,3-oxazinan-4-ones. *Angew. Chem. Int. Ed.* 53: 9865–9869.

1124 Bennacer, B., Fujiwara, M., Lee, S.Y., and Ojima, I. (2005). Silicon-initiated carbonylative carbotricyclization and (2+2+2+1)-cycloaddition of enediynes catalyzed by Rh-complexes. *J. Am. Chem. Soc.* 127: 17756–17767.

1125 Nguyen, T.V., Hartmann, J.M., and Enders, D. (2013). Recent synthetic strategies to access seven-membered carbocycles in natural product synthesis. *Synthesis* 45: 845–873.

1126 Cramer, C.J. and Barrows, S.E. (2000). Quantum chemical characterization of cycloaddition reactions between 1,3-butadiene and oxyallyl cations of varying electrophilicity. *J. Phys. Org. Chem.* 13 (3): 176–186.

1127 Cramer, C.J. and Barrows, S.E. (1998). Quantum chemical characterization of cycloaddition reactions between the hydroxyallyl cation and dienes of varying nucleophilicity. *J. Org. Chem.* 63 (16): 5523–5532.

1128 Harmata, M. and Schreiner, P.R. (2001). Intramolecular (4+3)-cycloadditions. A theoretical analysis of simple diastereoselectivity in reactions of alkoxyallylic cations and furans. *Org. Lett.* 3 (23): 3663–3665.

1129 Cramer, C.J., Harmata, M., and Rashatasakhon, P. (2001). Intramolecular (4+3)-cycloadditions. Theoretical and experimental evaluation of *endo/exo* preferences of a cyclopentenyl cation. *J. Org. Chem.* 66 (16): 5641–5644.

1130 Rigby, J.H. and Pigge, F.C. (1997). (4+3)-Cycloaddition reactions. *Org. React.* 51: 351–478.

1131 Harmata, M. (1997). Intramolecular (4+3)-cycloaddition reactions. *Adv. Cycloaddit.* 4: 41–86.

1132 Harmata, M. (1997). Intramolecular cycloaddition reactions of allylic cations. *Tetrahedron* 53 (18): 6235–6280.

1133 Harmata, M. (2006). Asymmetric catalytic (4+3)-cycloaddition reactions. *Adv. Synth. Catal.* 348 (16–17): 2297–2306.

1134 Binger, P., Doyle, M.J., and Benn, R. (1983). Metallacycloalkanes. 6. 5-Membered to 7-membered (2,2′-bipyridine)nickelacycloalkanes from methylenecyclopropane and Ni(0) compounds. *Chem. Ber.* 116 (1): 1–10.

1135 Binger, P., Brinkmann, A., and Wedemann, P. (1983). Reactions of methylenecyclopropanes. 7. Nickel(0)-catalyzed cycloadditions of methylenecyclopropanes with alkyl acrylates. *Chem. Ber.* 116 (8): 2920–2930.

1136 Trost, B.M., Nanninga, T.N., and Chan, D.M.T. (1982). Regiochemistry of cycloaddition of (trimethylenemethane)Pd intermediates to muconic esters. *Organometallics* 1 (11): 1543–1545.

1137 Gulías, M., Duran, J., López, F. et al. (2007). Palladium-catalyzed (4+3) intramolecular cycloaddition of alkylidenecyclopropanes and dienes. *J. Am. Chem. Soc.* 129 (36): 11026–11027.

1138 Craft, D.T. and Gung, B.W. (2008). The first transannular (4+3)-cycloaddition reaction: synthesis of the ABCD ring structure of Cortistatins. *Tetrahedron Lett.* 49 (41): 5931–5934.

1139 Shintani, R., Murakami, M., and Hayashi, T. (2007). γ-Methylidene-δ-valerolactones as a coupling partner for cycloaddition: palladium-catalyzed (4+3)-cycloaddition with nitrones. *J. Am. Chem. Soc.* 129 (41): 12356–12357.

1140 Shintani, R., Murakami, M., and Hayashi, T. (2008). γ-Methylidene-δ-valerolactones as a coupling partner for cycloaddition: palladium-catalyzed (4+3)-cycloaddition with nitrones. *Pure Appl. Chem.* 80 (5): 1135–1140.

1141 Trillo, B., López, F., Gulías, M. et al. (2008). Platinum-catalyzed intramolecular (4C+3C)-cycloaddition between dienes and allenes. *Angew. Chem. Int. Ed.* 47 (5): 951–954.
1142 Shapiro, N.D. and Toste, F.D. (2008). Synthesis of azepines by a Au-catalyzed intermolecular (4+3)-annulation. *J. Am. Chem. Soc.* 130 (29): 9244–9245.
1143 Marion, N. and Nolan, S.P. (2007). Propargylic esters in Au catalysis: access to diversity. *Angew. Chem. Int. Ed.* 46 (16): 2750–2752.
1144 Correa, A., Marion, N., Fensterbank, L. et al. (2008). Golden carousel in catalysis: the cationic Au/propargylic ester cycle. *Angew. Chem. Int. Ed.* 47 (4): 718–721.
1145 Gorin, D.J., Dube, P., and Toste, F.D. (2006). Synthesis of benzonorcaradienes by Au(I)-catalyzed (4+3)-annulation. *J. Am. Chem. Soc.* 128 (45): 14480–14481.
1146 Gorin, D.J., Watson, I.D.G., and Toste, F.D. (2008). Fluorenes and styrenes by Au(I)-catalyzed annulation of enynes and alkynes. *J. Am. Chem. Soc.* 130 (12): 3736–3737.
1147 Amijs, C.H.M., López-Carrillo, V., and Echavarren, A.M. (2007). Gold-catalyzed addition of carbon nucleophiles to propargyl carboxylates. *Org. Lett.* 9 (20): 4021–4024.
1148 Witham, C.A., Mauleón, P., Shapiro, N.D. et al. (2007). Gold(I)-catalyzed oxidative rearrangements. *J. Am. Chem. Soc.* 129 (18): 5838–5839.
1149 Gung, B.W. and Craft, D.T. (2009). Gold-catalyzed transannular (4+3)-cycloaddition reactions. *Tetrahedron Lett.* 50 (22): 2685–2687.
1150 Mauleón, P., Zeldin, R.M., González, A.Z., and Toste, F.D. (2009). Ligand-controlled access to [π^4+π^2] and (4+3)-cycloadditions in Au-catalyzed reactions of allene-dienes. *J. Am. Chem. Soc.* 131 (18): 6348–6349.
1151 Krainz, T., Chow, S., Korica, N. et al. (2015). Rhodium-catalyzed (4+3)-cycloaddition to furans: direct access to functionalized bicyclo[5.3.0]decane derivatives. *Eur. J. Org. Chem.* 41–44.
1152 Reddy, R.P. and Davies, H.M.L. (2007). Asymmetric synthesis of tropanes by Rh-catalyzed (4+3)-cycloaddition. *J. Am. Chem. Soc.* 129 (34): 10312–10313.
1153 Prié, G., Prévost, N., Twin, H. et al. (2004). A Lewis acid catalyzed intramolecular (4+3)-cycloaddition route to polycyclic systems that contain a seven-membered ring. *Angew. Chem. Int. Ed.* 43 (47): 6517–6519.
1154 Huang, J. and Hsung, R.P. (2005). Chiral Lewis acid-catalyzed highly enantioselective (4+3)-cycloaddition reactions of nitrogen-stabilized oxyallyl cations derived from allenamides. *J. Am. Chem. Soc.* 127 (1): 50–51.
1155 Guzman, P.E., Lian, Y.J., and Davies, H.M.L. (2014). Reversal of the regiochemistry in the Rh-catalyzed (4+3)-cycloaddition between vinyldiazoacetates and dienes. *Angew. Chem. Int. Ed.* 53: 13083–13087.
1156 Huffman, M.A. and Liebeskind, L.S. (1993). Rhodium(I)-catalyzed intramolecular carbocyclic ring fusion - a new approach to medium-sized-ring ketones. *J. Am. Chem. Soc.* 115: 4895–4896.
1157 Saito, S., Masuda, M., and Komagawa, S. (2004). Nickel-catalyzed intermolecular (3+2+2)-cocyclization of ethyl cyclopropylideneacetate and alkynes. *J. Am. Chem. Soc.* 126 (34): 10540–10541.
1158 Saito, S. and Takeuchi, K. (2007). Nickel-catalyzed (4+3)-cycloaddition of ethyl cyclopropylideneacetate and 1,3-dienes. *Tetrahedron Lett.* 48 (4): 595–598.
1159 Ni, Y. and Montgomery, J. (2006). Synthetic studies and mechanistic insight in Ni-catalyzed (4+2+1)-cycloadditions. *J. Am. Chem. Soc.* 128: 2609–2614.
1160 Jiao, L. and Yu, Z.X. (2013). Vinylcyclopropane derivatives in transition-metal-catalyzed cycloadditions for the synthesis of carbocyclic compounds. *J. Org. Chem.* 78: 6842–6848.
1161 Sarel, S. and Breuer, E. (1959). A Novel conjugative 1,5-addition reaction involving the vinylcyclopropane system. *J. Am. Chem. Soc.* 81 (24): 6522–6523.
1162 Stogryn, E.L. and Brois, S.J. (1967). Valence isomerization of 1.2-divinylaziridines. Synthetic and kinetic studies. *J. Am. Chem. Soc.* 89: 605–609.
1163 Wender, P.A., Takahashi, H., and Witulski, B. (1995). Transition-metal-catalyzed (5+2)-cycloadditions of vinylcyclopropanes and alkynes - a homolog of the Diels–Alder reaction for the synthesis of 7-membered rings. *J. Am. Chem. Soc.* 117 (16): 4720–4721.
1164 Wender, P.A., Rieck, H., and Fuji, M. (1998). The transition metal-catalyzed intermolecular (5+2)-cycloaddition: the homologous Diels–Alder reaction. *J. Am. Chem. Soc.* 120 (42): 10976–10977.
1165 Liu, P., Sirois, L.E., Cheong, P.H.Y. et al. (2010). Electronic and steric control of regioselectivities in Rh(I)-catalyzed (5+2)-cycloadditions:

experiment and theory. *J. Am. Chem. Soc.* 132: 10127–10135.

1166 Wender, P.A. and Sperandio, D. (1998). A new and selective catalyst for the (5+2)-cycloaddition of vinylcyclopropanes and alkynes. *J. Org. Chem.* 63 (13): 4164–4165.

1167 Wender, P.A. and Zhang, L. (2000). Asymmetric total synthesis of (+)-Aphanamol I based on the transition metal catalyzed (5+2)-cycloaddition of allenes and vinylcyclopropanes. *Org. Lett.* 2 (15): 2323–2326.

1168 Wegner, H.A., de Meijere, A., and Wender, P.A. (2005). Transition metal-catalyzed intermolecular (5+2) and (5+2+1)-cycloadditions of allenes and vinylcyclopropanes. *J. Am. Chem. Soc.* 127 (18): 6530–6531.

1169 Wender, P.A. and Williams, T.J. (2002). [(arene)Rh(cod)](+) complexes as catalysts for (5+2)-cycloaddition reactions. *Angew. Chem. Int. Ed.* 41 (23): 4550–4553.

1170 Wender, P.A., Pedersen, T.M., and Scanio, M.J.C. (2002). Transition metal-catalyzed hetero-(5+2)-cycloadditions of cyclopropyl imines and alkynes: dihydroazepines from simple, readily available starting materials. *J. Am. Chem. Soc.* 124 (51): 15154–15155.

1171 Yu, Z.X., Wender, P.A., and Houk, K.N. (2004). On the mechanism of $[Rh(CO)_2Cl]_2$-catalyzed intermolecular (5+2) reactions between vinylcyclopropanes and alkynes. *J. Am. Chem. Soc.* 126 (30): 9154–9155.

1172 Yu, Z.X., Cheong, P.H.Y., Liu, P. et al. (2008). Origins of differences in reactivities of alkenes, alkynes, and allenes in $[Rh(CO)_2Cl]_2$-catalyzed (5+2)-cycloaddition reactions with vinylcyclopropanes. *J. Am. Chem. Soc.* 130 (8): 2378–2379.

1173 Montero-Campillo, M.M., Cabaleiro-Lago, E.M., and Rodríguez-Otero, J. (2008). Density functional theory study of rhodium-catalyzed hetero-(5+2)-cycloaddition of cyclopropyl imine derivatives and alkynes. *J. Phys. Chem. A* 112 (38): 9068–9074.

1174 Zhu, L., Qi, X.T., and Lan, Y. (2016). Rhodium-catalyzed hetero-(5+2)-cycloaddition of vinylaziridines and alkynes: a theoretical view of the mechanism and chirality transfer. *Organometallics* 35: 771–777.

1175 Saito, A., Ono, T., and Hanzawa, Y. (2006). Cationic Rh(I) catalyst in fluorinated alcohol: mild intramolecular cycloaddition reactions of ester-tethered unsaturated compounds. *J. Org. Chem.* 71 (17): 6437–6443.

1176 Wender, P.A., Haustedt, L.O., Lim, J. et al. (2006). Asymmetric catalysis of the (5+2)-cycloaddition reaction of vinylcyclopropanes and pi-systems. *J. Am. Chem. Soc.* 128 (19): 6302–6303.

1177 Trost, B.M. and Müller, T.J.J. (1994). Butenolide synthesis based upon a contra-electronic addition in a Ru-catalyzed Alder ene reaction - synthesis and absolute configuration of (+)-ancepsenolide. *J. Am. Chem. Soc.* 116 (11): 4985–4986.

1178 Trost, B.M., Toste, F.D., and Shen, B.H. (2000). Ruthenium-catalyzed intramolecular (5+2)-cycloadditions. *J. Am. Chem. Soc.* 122 (10): 2379–2380.

1179 Trost, B.M. and Shen, H.C. (2000). On the regioselectivity of the Ru-catalyzed intramolecular (5+2)-cycloaddition. *Org. Lett.* 2 (16): 2523–2525.

1180 Trost, B.M. and Shen, H.C. (2001). Constructing tricyclic compounds containing a seven-membered ring by Ru-catalyzed intramolecular (5+2)-cycloaddition. *Angew. Chem. Int. Ed.* 40 (12): 2313–2316.

1181 Trost, B.M., Shen, H.C., Horne, D.B. et al. (2005). Syntheses of seven-membered rings: ruthenium-catalyzed intramolecular (5+2)-cycloadditions. *Chem. Eur. J.* 11 (8): 2577–2590.

1182 Melcher, M.C., von Wachenfeldt, H., Sundin, A., and Strand, D. (2015). Iridium-catalyzed carbocyclizations: efficient (5+2)-cycloadditions of vinylcyclopropanes and alkynes. *Chem. Eur. J.* 21: 531–535.

1183 Zuo, G. and Louie, J. (2005). Selectivity in Ni-catalyzed rearrangements of cyclopropylenynes. *J. Am. Chem. Soc.* 127: 5798–5799.

1184 Fürstner, A., Majima, K., Martin, R. et al. (2008). A cheap metal for a "noble" task: preparative and mechanistic aspects of cycloisomerization and cycloaddition reactions catalyzed by low-valent Fe-complexes. *J. Am. Chem. Soc.* 130: 1992–2004.

1185 Inagaki, F., Sugikubo, K., Miyashita, Y., and Mukai, C. (2010). Rhodium(I)-catalyzed intramolecular (5+2)-cycloaddition reactions of alkynes and allenylcyclopropanes: construction of bicyclo[5.4.0]undecatrienes and bicyclo[5.5.0]dodecatrienes. *Angew. Chem. Int. Ed.* 49: 2206–2210.

1186 Ylijoki, K.E.O. and Stryker, J.M. (2013). (5+2)-Cycloaddition reactions in organic and natural product synthesis. *Chem. Rev.* 113: 2244–2266.

1187 Shu, X.Z., Huang, S.Y., Shu, D.X. et al. (2011). Interception of a rautenstrauch intermediate by alkynes for (5+2)-cycloaddition: rhodium-catalyzed cycloisomerization of 3-acyloxy-4-ene-1,9-diynes to bicyclo[5.3.0]decatrienes. *Angew. Chem. Int. Ed.* 50: 8153–8156.

1188 Shu, X.Z., Li, X.X., Shu, D.X. et al. (2012). Rhodium-catalyzed intra- and intermolecular (5+2)-cycloaddition of 3-acyloxy-1,4-enyne and alkyne with concomitant 1,2-acyloxy migration. *J. Am. Chem. Soc.* 134: 5211–5221.

1189 Shu, X.Z., Schienebeck, C.M., Li, X.X. et al. (2015). Rhodium-catalyzed stereoselective intramolecular (5+2)-cycloaddition of 3-acyloxy 1,4-enyne and alkene. *Org. Lett.* 17: 5128–5131.

1190 Lee, D.J., Han, H.S., Shin, J., and Yoo, E.J. (2014). Multicomponent (5+2)-cycloaddition reaction for the synthesis of 1,4-diazepines: isolation and reactivity of azomethine ylides. *J. Am. Chem. Soc.* 136: 11606–11609.

1191 Trost, B.M., Hu, Y., and Horne, D.B. (2007). Total synthesis of (+)-frondosin A. Application of the Ru-catalyzed (5+2)-cycloaddition. *J. Am. Chem. Soc.* 129 (38): 11781–11790.

1192 Ashfeld, B.L. and Martin, S.F. (2005). Enantioselective syntheses of Tremulenediol A and Tremulenolide A. *Org. Lett.* 7 (20): 4535–4537.

1193 Ashfeld, B.L., Miller, K.A., Smith, A.J. et al. (2007). Features and applications of $[Rh(CO)_2Cl]_2$-catalyzed alkylations of unsymmetrical allylic substrates. *J. Org. Chem.* 72 (24): 9018–9031.

1194 Hong, X., Stevens, M.C., Liu, P. et al. (2014). Reactivity and chemoselectivity of allenes in Rh(I)-catalyzed intermolecular (5+2)-cycloadditions with vinylcyclopropanes: allene-mediated rhodacycle formation can poison Rh(I)-catalyzed cycloadditions. *J. Am. Chem. Soc.* 136: 17273–17283.

1195 Zhou, M.B., Song, R.J., Wang, C.Y., and Li, J.H. (2013). Synthesis of azepine derivatives by silver-catalyzed (5+2)-cycloaddition of γ-amino ketones with alkynes. *Angew. Chem. Int. Ed.* 52: 10805–10808.

1196 Kaloko, J.J., Gary, Y.H., and Ojima, T.I. (2009). One-step formation of fused tetracyclic skeletons from cyclohexene-diynes and carbon monoxide through Rh(I)-catalyzed (2+2+2+1)-cycloaddition reaction. *J. Chem. Soc., Chem. Commun.* 4569–4571.

1197 Becker, H.D. (1990). Excited state reactivity and molecular topology relationships in chromophorically substituted anthracenes. In: *Advances in Photochemistry*, vol. 15 (ed. D.H. Volman, G.S. Hammond and K. Gollnick), 139–227. New York: Wiley.

1198 Becker, H.D. (1993). Unimolecular photochemistry of anthracenes. *Chem. Rev.* 93 (1): 145–172.

1199 Fritzsche, J. (1867). Über die festen Kohlenwasserstoffe des Steinkohlentheers. *J. Prakt. Chem.* 101: 333–343.

1200 Sieburth, S.M. and Cunard, N.T. (1996). The (4+4)-cycloaddition and its strategic application in natural product synthesis. *Tetrahedron* 52 (18): 6251–6282.

1201 Reed, H.W.B. (1954). The catalytic cyclic polymerisation of butadiene. *J. Chem. Soc.* 1931–1941.

1202 Brenner, W., Heimbach, P., Hey, H. et al. (1969). Catalytic conversion of olefins. 3. Synthesis of *cis,cis*-1,5-cyclooctadiene and *cis*-1,2-divinylcyclobutane. *Liebigs Ann. Chem.* 727: 161–182.

1203 Tenaglia, A., Brun, P., and Waegell, B. (1985). Nickel-catalyzed oligomerization of functionalized conjugated dienes. *J. Organomet. Chem.* 285 (1–3): 343–357.

1204 Wender, P.A. and Ihle, N.C. (1986). Nickel-catalyzed intramolecular (4+4)-cycloadditions - a new method for the synthesis of polycycles containing 8-membered rings. *J. Am. Chem. Soc.* 108: 4678–4679.

1205 Wender, P.A. and Ihle, N.C. (1987). Nickel-catalyzed intramolecular (4+4)-cycloadditions. 2. Allylic stereoinduction and modeling studies in the preparation of bicyclo[6.4.0]dodecadienes. *Tetrahedron Lett.* 28: 2451–2454.

1206 Wender, P.A. and Tebbe, M.J. (1991). Nickel(0)-catalyzed intramolecular (4+4)-cycloadditions. 5. The type-II reaction in the synthesis of bicyclo[5.3.1]undecadienes. *Synthesis* 1089–1094.

1207 Wender, P.A., Nuss, J.M., Smith, D.B. et al. (1997). Transition metal catalyzed cycloadditions: an intramolecular (4+4)-cycloaddition strategy for the efficient synthesis of dicyclopenta[a,d]cyclooctene 5-8-5 ring systems. *J. Org. Chem.* 62 (15): 4908–4909.

1208 tom Dieck, H. and Dietrich, J. (1984). Diazadienes as controlling ligands in catalysis. 5. Synthesis of chiral diazadienes R—N=Cr′—Cr′=N—R. *Chem. Ber.* 117 (2): 694–701.

1209 tom Dieck, H. and Dietrich, J. (1985). Diazadienes as control ligands in homogeneous catalysis. 11. Selectivity and mechanism of diene cyclodimerization on Fe(0)-complexes. *Angew. Chem. Int. Ed. Engl.* 24 (9): 781–783.

1210 Wender, P.A., Ihle, N.C., and Correia, C.R.D. (1988). Nickel-catalyzed intramolecular (4+4)-cycloadditions. 4. Enantioselective total synthesis of (+)-asteriscanolide. *J. Am. Chem. Soc.* 110 (17): 5904–5906.

1211 Wender, P.A., Croatt, M.P., and Witulski, B. (2006). New reactions and step economy: the total synthesis of (+/-)-salsolene oxide based on the type II transition metal-catalyzed intramolecular (4+4)-cycloaddition. *Tetrahedron* 62 (32): 7505–7511.

1212 Baldenius, K.U., tom Dieck, H., König, W.A. et al. (1992). Enantioselective syntheses of cyclopentanoid compounds from isoprene and *trans*-1,3-pentadiene. *Angew. Chem. Int. Ed. Engl.* 31 (3): 305–307.

1213 Wilke, G. (1988). Contributions to organo-Ni chemistry. *Angew. Chem. Int. Ed. Engl.* 27 (1): 185–206.

1214 Gugelchuk, M.M. and Houk, K.N. (1994). Stereoselective organometallic reactions - a force-field study of π-allyl intermediates in Ni(0)-catalyzed cycloadditions. *J. Am. Chem. Soc.* 116 (1): 330–339.

1215 Norrby, P.O., Aakermark, B., Haeffner, F. et al. (1993). Molecular mechanics (MM2) parameters for the (η3-allyl)Pd moiety. *J. Am. Chem. Soc.* 115 (11): 4859–4867.

1216 Aouf, C., El Abed, D., Giorgi, M., and Santelli, M. (2007). Unexpected Ti-catalyzed formal intramolecular (4+4)-cycloaddition of 1,1′-bi[(*S**,*S**)-6-(trimethylsilyl)cyclohepta-2,4-dien-1-yl]. *Tetrahedron Lett.* 48 (29): 4969–4972.

1217 Sahani, R.L. and Liu, R.S. (2016). Gold-catalyzed (4+3) and (4+4)-annulation reactions of *t*-butyl propiolate derivatives with epoxides and oxetanes for the construction of 1,4-dioxepane and 1,5-dioxocane cores. *J. Chem. Soc., Chem. Commun.* 52: 7482–7485.

1218 Greco, A., Carbonaro, A., and Dall'Asta, G. (1970). Catalytic norbornadiene-butadiene and norbornadiene-1,1-dimethylallene codimerization. *J. Org. Chem.* 35 (1): 271–274.

1219 Carbonaro, A., Cambisi, F., and Dall'Asta, G. (1971). Catalytic behavior of some Ziegler–Natta catalysts in norbornadiene-butadiene codimerization. *J. Org. Chem.* 36 (10): 1443–1445.

1220 Lautens, M., Tam, W., and Sood, C. (1993). Enantioselective Co-catalyzed [π^4+π^2+π^2]-cycloadditions. *J. Org. Chem.* 58 (17): 4513–4515.

1221 Lautens, M., Tam, W., Lautens, J.C. et al. (1995). Co-catalyzed [π^2+π^2+π^2] (homo-Diels–Alder) and [π^4+π^2+π^2]-cycloadditions of bicyclo[2.2.1]hepta-2,5-dienes. *J. Am. Chem. Soc.* 117 (26): 6863–6879.

1222 Hilt, G. and Janikowski, J. (2008). Cobalt-catalyzed [π^4+π^2+π^2]-cycloaddition for the synthesis of 1,3,6-cyclooctatrienes. *Angew. Chem. Int. Ed.* 47 (28): 5243–5245.

1223 Gilbertson, S.R. and DeBoef, B. (2002). Rhodium-catalyzed [π^4+π^2+π^2]-cycloaddition and alkyne insertion: a new route to eight-membered rings. *J. Am. Chem. Soc.* 124 (30): 8784–8785.

1224 Evans, P.A., Robinson, J.E., Baum, E.W., and Fazal, A.N. (2002). Intermolecular transition metal-catalyzed [π^4+π^2+π^2]-cycloaddition reactions: a new approach to the construction of eight-membered rings. *J. Am. Chem. Soc.* 124 (30): 8782–8783.

1225 Lee, S.I., Park, S.Y., and Chung, Y.K. (2006). Rhodium-catalyzed [π^4+π^2+π^2]-cycloaddition reaction of two enynes or diynes with one diene to give eight-membered ring compounds. *Adv. Synth. Catal.* 348 (16–17): 2531–2539.

1226 Canlas, G.M.R. and Gilbertson, S.R. (2014). [π^4+π^2+π^2]-Cycloaddition catalyzed by a new cationic Rh-bisphosphine monooxide complex. *J. Chem. Soc., Chem. Commun.* 50: 5007–5010.

1227 Yu, R.T., Friedman, R.K., and Rovis, T. (2009). Enantioselective Rh-catalyzed [π^4+π^2+π^2]-cycloaddition of dienyl isocyanates for the synthesis of bicyclic azocine rings. *J. Am. Chem. Soc.* 131: 13250–13251.

1228 Ashida, S. and Murakami, M. (2008). Nickel-catalyzed [π^4+π^2+π^2]-type annulation reaction of cyclobutanones with diynes and enynes. *Bull. Chem. Soc. Jpn.* 81: 885–893.

1229 Lainhart, B.C. and Alexanian, E.J. (2015). Enantioselective synthesis of *cis*-fused cyclooctanoids via Rh(I)-catalyzed [π^4+π^2+π^2]-cycloadditions. *Org. Lett.* 17: 1284–1287.

1230 Varela, J.A., Castedo, L., and Saa, C. (2003). "Formal" Ru-catalyzed [π^4+π^2+π^2]-cycloaddition of 1,6-diynes to 1,3-dienes: formation of cyclooctatrienes vs vinylcyclohexadienes. *Org. Lett.* 5 (16): 2841–2844.

1231 Feldman, K.S., Come, J.H., Freyer, A.J. et al. (1986). Synthesis of 8-membered carbocycles via intramolecular [π^6+π^2]-photocycloaddition of alkenyltropones. *J. Am. Chem. Soc.* 108 (6): 1327–1328.

1232 Feldman, K.S., Come, J.H., Kosmider, B.J. et al. (1989). Mechanistic and preparative studies of the intramolecular photocyclization of methylated 2-(4-pentenyl)tropones. *J. Org. Chem.* 54 (3): 592–601.

1233 Feldman, K.S., Wu, M.J., and Rotella, D.P. (1990). Total synthesis of (+/-)-dactylol and related studies. *J. Am. Chem. Soc.* 112 (23): 8490–8496.

1234 Davis, R.E., Dodds, T.A., Hseu, T.H. et al. (1974). [π^2+π^6]-Cycloaddition reactions between ligands coordinated to an Fe atom. *J. Am. Chem. Soc.* 96 (24): 7562–7564.

1235 Mach, K., Antropiusova, H., Sedmera, P. et al. (1983). [π^6s+π^2s]-Cycloadditions catalyzed by the $TiCl_4$-Et_2AlCl system. *J. Chem. Soc., Chem. Commun.* (15): 805–806.

1236 Mach, K., Antropiusova, H., Petrusová, L. et al. (1984). [π^6s+π^2s]-Cycloadditions catalyzed by Ti-complexes. *Tetrahedron* 40 (17): 3295–3302.

1237 Achard, M., Tenaglia, A., and Buono, G. (2005). First Co(I)-catalyzed [π^6+π^2]-cycloadditions of cycloheptatriene with alkynes. *Org. Lett.* 7 (12): 2353–2356.

1238 Tenaglia, A. and Gaillard, S. (2008). $PtCl_2$-Catalyzed [π^6+π^2]-cycloaddition of alkynes tethered to cycloheptatriene. *Angew. Chem. Int. Ed.* 47 (13): 2454–2457.

1239 D'yakonov, V.A., Kadikova, G.N., Gazizullina, G.F. et al. (2015). Co(I)-catalyzed [π^6+π^2]-cycloadditions of 1,2-dienes to 1,3,5,7-cyclooctatetraene. *Tetrahedron Lett.* 56: 2005–2007.

1240 Wender, P.A., Correa, A.G., Sato, Y., and Sun, R. (2000). Transition metal-catalyzed [π^6+π^2]-cycloadditions of 2-vinylcyclobutanones and alkenes: a new reaction for the synthesis of eight-membered rings. *J. Am. Chem. Soc.* 122: 7815–7816.

1241 Montero-Campillo, M.M., Rodríguez-Otero, J., and Cabaleiro-Lago, E.M. (2008). On the mechanism of Rh-catalyzed [π^6+π^2]-cycloaddition of 2-vinylcyclobutanones and alkenes. *Tetrahedron* 64: 6215–6220.

1242 Inagaki, F., Sugikubo, K., Oura, Y., and Mukai, C. (2011). Rh(I)-Catalyzed [π^6+π^2]-cycloaddition of alkyne-allenylcyclobutanes: a new entry for the synthesis of bicyclo[6.m.0] skeletons. *Chem. Eur. J.* 17: 9062–9065.

1243 Achard, M., Mosrin, M., Tenaglia, A., and Buono, G. (2006). Cobalt(I)-catalyzed [π^6+π^2]-cycloadditions of cyclooctatetra(tri)ene with alkynes. *J. Org. Chem.* 71 (7): 2907–2910.

1244 Ura, Y., Utsumi, T., Tsujita, H. et al. (2006). Formation of divalent ruthenacycles via oxidative cyclization of 1,3,5-cyclooctatriene with maleic anhydride or maleimides: an intermediate for the transition metal-mediated [π^6+π^2]-cycloaddition of 1,3,5-trienes and alkenes. *Organometallics* 25 (12): 2934–2942.

1245 Oonishi, Y., Hosotani, A., and Sato, Y. (2011). Rh(I)-Catalyzed formal [π^6+π^2]-cycloaddition of 4-allenals with alkynes or alkenes in a tether. *J. Am. Chem. Soc.* 133: 10386–10389.

1246 Oonishi, Y., Hosotani, A., and Sato, Y. (2012). Construction of monocyclic eight-membered rings: intermolecular Rh(I)-catalyzed [π^6+π^2]-cycloaddition of 4-allenals with alkynes. *Angew. Chem. Int. Ed.* 51: 11548–11551.

1247 Willstätter, R. and Waser, E. (1911). On cyclooctatetanes. (Fifth report on the cyclooctane sequence). *Ber. Dtsch. Chem. Ges.* 44: 3423–3445.

1248 (a) Reppe, W., Schlichting, O., Klager, K., and Töpel, T. (1948). Cyclisierende polymerisation von Acetylen. 1. Über Cyclooctatetraen. *Liebigs Ann. Chem.* 560 (1–2): 1–92. (b) Hummel, H.G. (1984). Walter Reppe (1892–1969). *Chem. Ber.* 117 (3): I–XXI.

1249 (a) Colborn, R.E. and Vollhardt, K.P.C. (1986). The mechanism of the cyclooctatetraene synthesis from ethyne employing nickel catalysts. *J. Am. Chem. Soc.* 108 (18): 5470–5477. (b) Straub, B.F. and Gollub, C. (2004). Mechanism of Reppe's nickel-catalyzed ethyne tetramerization to cyclooctatetraene: a DFT study. *Chem. Eur. J.* 10 (12): 3081–3090.

1250 tom Dieck, H., Svoboda, M., and Kopf, J. (1978). Diazadiene Ni-complexes. 1. Bis(diazadiene) Ni(0) and bromo-diazadiene Ni(I). *Z. Naturforsch., B* 33B (12): 1381–1385.

1251 Diercks, R., Stamp, L., and tom Dieck, H. (1984). Diazadienes as controlling ligands in homogeneous catalysis. 7. Catalytic synthesis and structure of a cyclooctatetraene with twofold vicinal substitution - 1,4,5,8-Tetrakis [(*para*-tolyloxy)methyl]-1,3,5,7-cyclooctatetraene. *Chem. Ber.* 117 (5): 1913–1919.

1252 Diercks, R. and tom Dieck, H. (1985). Diazadienes as controlling ligands in homogeneous catalysis. 9. Catalytic cyclotetramerization of propynoic esters. *Chem. Ber.* 118 (2): 428–435.

1253 tom Dieck, H., Munz, C., and Müller, C. (1987). Diazadienes and ligand control in homogeneous catalysis. 13. Cocyclotetramerization of 2 different alkynes to cyclooctatetraene in the presence of palladacyclopentadiene. *J. Organomet. Chem.* 326 (1): C1–C5.

1254 Diercks, R., Stamp, L., Kopf, J., and tom Dieck, H. (1984). Elementary steps in catalytic 1-alkyne coupling to substituted cyclooctatetraenes. *Angew. Chem.* 96 (96): 891–895.

1255 Walther, D., Braun, D., Schulz, W., and Rosenthal, U. (1989). Selective cyclotetramerization of propargylic alcohol - catalytic reactions with high selectivity and reaction rate by Ni-complexes. *Z. Anorg. Allg. Chem.* 577 (10): 270–282.

1256 Colborn, R.E. and Vollhardt, K.P.C. (1981). Mechanism of the Reppe cyclooctatetraene synthesis from ethyne - a labeling experiment. *J. Am. Chem. Soc.* 103 (20): 6259–6261.

1257 Boussie, T.R. and Streitwieser, A. Jr., (1993). Two complementary syntheses of symmetrically-tetrasubstituted cyclooctatetraenes. *J. Org. Chem.* 58 (9): 2377–2380.

1258 Kyriakou, G., Kim, J., Tikhov, M.S. et al. (2005). Acetylene coupling on Cu(111): Formation of butadiene, benzene, and cyclooctatetraene. *J. Phys. Chem. B* 109 (21): 10952–10956.

1259 Wender, P.A. and Christy, J.P. (2007). Nickel(0)-catalyzed (2+2+2+2)-cycloadditions of terminal diynes for the synthesis of substituted cyclooctatetraenes. *J. Am. Chem. Soc.* 129: 13402–13403.

1260 Nasrallah, D.J. and Croatt, M.P. (2014). Rhodium(I)-catalyzed (2+2+2+2)-cycloaddition of diynes to form cyclooctatetraenes. *Eur. J. Org. Chem.* 2014: 3767–3772.

1261 Wender, P.A., Gamber, G.G., Hubbard, R.D., and Zhang, L. (2002). Three-component cycloadditions: the first transition metal-catalyzed (5+2+1)-cycloaddition reactions. *J. Am. Chem. Soc.* 124: 2876–2877.

1262 Huang, F., Yao, Z.K., Wang, Y. et al. (2010). Rh(I)-Catalyzed two-component ((5+2)+1)-cycloaddition approach toward [5-8-5] ring systems. *Chem. Asian J.* 5: 1555–1559.

1263 Fan, X.H., Tang, M.X., Zhuo, L.G. et al. (2009). An expeditious and high-yield formal synthesis of hirsutene using Rh(I)-catalyzed ((5+2)+1)-cycloaddition. *Tetrahedron Lett.* 50: 155–157.

1264 Liang, Y., Jiang, X., and Yu, Z.X. (2011). Enantioselective total synthesis of (+)-asteriscanolide via Rh(I)-catalyzed ((5+2)+1) reaction. *J. Chem. Soc., Chem. Commun.* 47: 6659–6661.

1265 Wang, Y. and Yu, Z.X. (2015). Rhodium-catalyzed (5+2+1)-cycloaddition of ene-vinylcyclopropanes and CO: reaction design, development, application in natural product synthesis, and inspiration for developing new reactions for synthesis of eight-membered carbocycles. *Acc. Chem. Res.* 48: 2288–2296.

1266 Yao, Z.K., Li, J.J., and Yu, Z.X. (2011). Rh-Catalyzed (7+1)-cycloaddition of buta-1,3-dienylcyclopropanes and CO for the synthesis of cyclooctadienones. *Org. Lett.* 13: 134–137.

1267 Fu, X.F., Xiang, Y., and Yu, Z.X. (2015). Rh(I)-Catalyzed benzo/(7+1)-cycloaddition of cyclopropyl-benzocyclobutenes and CO by merging thermal and metal-catalyzed C—C bond cleavages. *Chem. Eur. J.* 21: 4242–4246.

1268 Wender, P.A., Gamber, G.G., Hubbard, R.D. et al. (2005). Multicomponent cycloadditions: the four-component (5+1+2+1)-cycloaddition of vinylcyclopropanes, alkynes, and CO. *J. Am. Chem. Soc.* 127: 2836–2837.

1269 Saito, S., Maeda, K., Yamasaki, R. et al. (2010). Synthesis of nine-membered carbocycles by the (4+3+2)-cycloaddition reaction of ethyl cyclopropylideneacetate and dienynes. *Angew. Chem. Int. Ed.* 49: 1830–1833.

1270 Yamasaki, R., Ohashi, M., Maeda, K. et al. (2013). Ni-Catalyzed (4+3+2)-cycloaddition of ethyl cyclopropylideneacetate and dienynes: scope and mechanistic insights. *Chem. Eur. J.* 19: 3415–3425.

1271 Yamasaki, R., Kato, K., Hanitani, D. et al. (2013). Synthesis of monocyclic nine-membered compounds by the (4+3+2)-cycloaddition-bond cleavage strategy. *Tetrahedron Lett.* 54: 3507–3509.

1272 Mukai, C., Ohta, Y., Oura, Y. et al. (2012). $C(sp^3)$-$C(sp^3)$ and $C(sp^3)$-H bond activation of 1,1-disubstituted cyclopentane. *J. Am. Chem. Soc.* 134: 19580–19583.

1273 Huang, G.P. (2015). Mechanism and selectivity in Rh-catalyzed (7+2)-cycloaddition and cyclopropanation/cyclization of allenylcyclopentane-alkynes: metallacycle-directed C(sp)-$C(sp^3)$ vs $C(sp^3)$-H activation. *J. Org. Chem.* 80: 7564–7571.

1274 Pospech, J., Ferraccioli, R., Neumann, H., and Beller, M. (2015). Rhodium(II)-catalyzed annulation of azavinyl carbenes through ring-expansion of 1,3,5-trioxane: rapid access to nine-membered 1,3,5,7-trioxazonines. *Chem. Asian J.* 10: 2623–2629.

1275 Rigby, J.H. (1997). (6+4)-Cycloaddition reactions. *Org. React.* 49: 331–425.

1276 Isakovic, L., Ashenhurst, J.A., and Gleason, J.L. (2001). Application of Lewis acid catalyzed tropone (6+4)-cycloadditions to the synthesis of the core of CP-225,917. *Org. Lett.* 3 (26): 4189–4192.

1277 Rigby, J.H. and Fleming, M. (2002). Construction of the ingenane core using an Fe(III) or Ti(IV) Lewis acid-catalyzed intramolecular (6+4)-cycloaddition. *Tetrahedron Lett.* 43 (48): 8643–8646.

1278 Holte, D., Goetz, D.C., and Baran, P.S. (2012). An approach to mimicking the sesquiterpene cyclase phase by Ni-promoted diene/alkyne cooligomerization. *J. Org. Chem.* 77: 825–842.

1279 Hong, X., Holte, D., Goetz, D.C. et al. (2014). Mechanism, reactivity, and selectivity of Ni-catalyzed (4+4+2)-cycloadditions of dienes and alkynes. *J. Org. Chem.* 79: 12177–12184.

1280 Wilke, G. (1957). Synthesen in der Cyclododekanreihe. *Angew. Chem.* 69 (11): 397–398.

1281 Wilke, G. (1962). Reactions of new π-complex catalysts. *Adv. Chem. Ser.* 34: 137–144.

1282 Wilke, G., Bogdanovic, B., Borner, P. et al. (2009). Cyclooligomerization of butadiene and transition metal π-complexes. *Angew. Chem. Int. Ed. Engl.* 2 (3): 105–115.

1283 (a) Weissermel, K. and Arpe, H.-J. (1978). Industrielle Organische Chemie. In: *Bedeutende Vor- und Zwischenprodukte*, 2e. Weinheim: Verlag Chemie. (b) Arpe, H.-J. (2007). Industrielle Organische Chemie. In: *Bedeutende Vor- und Zwischenprodukte*, 3e, 1–493. Weinheim: Wiley-VCH.

1284 Cabiac, A., Delahay, G., Durand, R. et al. (2007). The influence of textural and structural properties of Pd/carbon on the hydrogenation of *cis,trans,trans*-1,5,9-cyclododecatriene. *Appl. Catal., A* 318: 17–21.

1285 Mahajan, S.S., Sharma, M.M., and Sridhar, T. (2007). Uncatalyzed oxidation of 1,5,9-cyclododecatriene with molecular oxygen. *Ind. Eng. Chem. Res.* 46 (10): 3057–3062.

1286 Lewandowski, G., Bartkowiak, M., and Milchert, E. (2009). Optimization of the 1,3-butadiene cyclotrimerization process by the catalytic system $Et_3Al_2Cl_3$/TiCl(4): the influence of the parameters on the selectivity of 1,5,9-cyclododecatriene formation. *Catal. Lett.* 127 (1–2): 213–221.

1287 Baker, R. (1973). π-Allylmetal derivatives in organic synthesis. *Chem. Rev.* 73 (5): 487–530.

1288 Yasuda, H. and Nakamura, A. (1987). Diene, alkyne, alkene, and alkyl complexes of early transition-metals - structures and synthetic applications in organic and polymer chemistry. *Angew. Chem. Int. Ed. Engl.* 26 (8): 723–742.

1289 Polácek, J., Antropiusova, H., Petrusová, L., and Mach, K. (1991). Titanium-catalyzed cyclotrimerization of butadiene. 3. Formation of catalytic bis(di-μ-chloroaluminate)-Ti(II) complexes in $TiCl_4$-$EtXAlCl_3$-X (X=1-2) systems. *J. Mol. Catal.* 70 (1): 9–28.

1290 Breil, H., Heimbach, P., Kröner, M. et al. (1963). Synthese von Cyclododecatrienen-(1,5,9). 1. Mitt. Über die katalytische Umwandlung von Olefinen. *Makromol. Chem.* 69: 18–40.

1291 Bogdanovic, B., Heimbach, P., Kröner, M. et al. (1969). Catalytic conversion of olefins. 2. On course of reaction in cyclotrimerization of 1,3-butadiene. *Liebgs Ann. Chem.* 727: 143–160.

1292 Saito, T., Uchida, Y., Misono, A. et al. (1966). Diethyldipyridylnickel. Preparation characterization and reactions. *J. Am. Chem. Soc.* 88 (22): 5198–5201.

1293 Wu, C.Y. and Swift, H.E. (1972). Diolefin reactions catalyzed by transition metal-Schiff base complexes. *J. Catal.* 24 (3): 510–520.

1294 Schuchardt, U. and Santos Dias, F. (1985). Highly selective cyclotrimerization of butadiene to cyclododecatriene catalyzed by polymer-anchored Ni(0). *J. Mol. Catal.* 29 (1): 145–149.

1295 Nguyen, H., Sawyer, J.F., Sayer, B.G., and McGlinchey, M.J. (1979). Nickel atom promoted dimerization of dimethylfulvene - a formal (6+6)-cycloaddition. *J. Am. Chem. Soc.* 101 (8): 2203–2204.

1296 Bosch, M., Brookhart, M.S., Ilg, K., and Werner, H. (2000). The Rh-catalyzed cyclotetramerization of butadiene: isolation and structural characterization of an intermediate. *Angew. Chem. Int. Ed.* 39 (13): 2304–2307.

1297 Tobisch, S. and Werner, H. (2004). [(RhL)-L-I]-catalyzed cyclotetramerization of 1,3-butadiene: a theoretical investigation of alternative mechanistic paths for the generation of the [Rh(III)(octadienediyl)(PR3)]+ complex. *J. Chem. Soc., Dalton Trans.* (18): 2963–2968.

1298 Tanaka, K., Hagiwara, Y., and Noguchi, K. (2005). Rhodium-catalyzed regio- and enantioselective intermolecular (4+2)-carbocyclization of 4-alkynals with *N,N*-dialkyl acrylamides. *Angew. Chem. Int. Ed.* 44: 7260–7263.

1299 Cotugno, P., Monopoli, A., Ciminale, F. et al. (2014). Palladium-catalyzed cross-coupling of styrenes with aryl methyl ketones in ionic liquids: direct access to cyclopropanes. *Angew. Chem. Int. Ed.* 53: 13563–13567.

1300 Stuart, D.R., Bertrand-Laperle, M., Burgess, K.M.N., and Fagnou, K. (2008). Indole synthesis via Rh-catalyzed oxidative coupling of acetanilides and internal alkynes. *J. Am. Chem. Soc.* 130: 16474–16475.

1301 Stuart, D.R., Alsabeh, P., Kuhn, M., and Fagnou, K. (2010). Rhodium(III)-catalyzed arene and alkene C=H Bond functionalization leading to indoles and pyrroles. *J. Am. Chem. Soc.* 132: 18326–18339.

1302 Huestis, M.P., Chan, L.N., Stuart, D.R., and Fagnou, K. (2011). The vinyl moiety as a handle for regiocontrol in the preparation of unsymmetrical 2,3-aliphatic-substituted indoles and pyrroles. *Angew. Chem. Int. Ed.* 50: 1338–1341.

1303 Ueura, K., Satoh, T., and Miura, M. (2007). An efficient waste-free oxidative coupling via regioselective C=H bond cleavage: Rh/Cu-catalyzed reaction of benzoic acids with alkynes and acrylates under air. *Org. Lett.* 9 (7): 1407–1409.

1304 Ueura, K., Satoh, T., and Miura, M. (2007). Rhodium- and Ir-catalyzed oxidative coupling of benzoic acids with alkynes via regioselective C—H bond cleavage. *J. Org. Chem.* 72 (14): 5362–5367.

1305 Gulías, M. and Mascareñas, J.L. (2016). Metal-catalyzed annulations through activation and cleavage of C—H bonds. *Angew. Chem. Int. Ed.* 55: 11000–11019.

1306 Seoane, A., Casanova, N., Quinones, N. et al. (2014). Straightforward assembly of benzoxepines by means of a Rh(III)-catalyzed C—H functionalization of *o*-vinylphenols. *J. Am. Chem. Soc.* 136: 834–837.

1307 Morimoto, K., Hirano, K., Satoh, T., and Miura, M. (2010). Rhodium-catalyzed oxidative coupling/cyclization of 2-phenylindoles with alkynes via C—H and N—H bond cleavages with air as the oxidant. *Org. Lett.* 12: 2068–2071.

1308 Patureau, F.W. and Glorius, F. (2010). Rh-Catalyzed olefination and vinylation of unactivated acetanilides. *J. Am. Chem. Soc.* 132: 9982–9983.

1309 Kim, S.H., Yoon, J., and Chang, S. (2011). Palladium-catalyzed oxidative alkynylation of heterocycles with terminal alkynes under air conditions. *Org. Lett.* 13: 1474–1477.

1310 Huang, L.H., Niu, T.M., Wu, J., and Zhang, Y.H. (2011). Copper-catalyzed oxidative cross-coupling of *N*,*N*-dimethylanilines with heteroarenes under molecular oxygen. *J. Org. Chem.* 75: 1759–1766.

1311 Stowers, K.J., Fortner, K.C., and Sanford, M.S. (2011). Aerobic Pd-catalyzed C(sp^3)-H olefination: a route to both N-heterocyclic scaffolds and alkenes. *J. Am. Chem. Soc.* 133: 6541–6544.

1312 Xie, J., Li, H.M., Zhou, J.C. et al. (2012). A highly efficient Au-catalyzed oxidative C—C coupling from C—H bonds using air as oxidant. *Angew. Chem. Int. Ed.* 51: 1252–1255.

1313 Zheng, L. and Wang, J.H. (2012). Direct oxidative coupling of arenes with olefins by Rh-catalyzed C—H activation in air: observation of a strong cooperation of the acid. *Chem. Eur. J.* 18: 9699–9704.

1314 Gulzar, N., Schweitzer-Chaput, B., and Klussmann, M. (2014). Oxidative coupling reactions for the functionalisation of C—H bonds using oxygen. *Catal. Sci. Technol.* 4: 2778–2796.

1315 Baars, H., Unoh, Y., Okada, T. et al. (2014). Rhodium-catalyzed intramolecular dehydrogenatiye aryl-aryl coupling using air as terminal oxidant. *Chem. Lett.* 43: 1782–1784.

1316 Yang, X.G., Liu, Z.Q., Sun, C.L. et al. (2015). Palladium-catalyzed oxidative cross-coupling of α-cyanoketene dithioacetals with olefins. *Chem. Eur. J.* 21: 14085–14094.

1317 Gao, D.W., Gu, Q., and You, S.L. (2016). An enantioselective oxidative C—H/C—H cross-coupling reaction: highly efficient method to prepare planar chiral ferrocenes. *J. Am. Chem. Soc.* 138: 2544–2547.

1318 Zhang, Y., Wei, B.W., Wang, W.X. et al. (2017). Direct vinylogous oxidative cross-dehydrogenative coupling of 4-nitroisoxazoles with *N*-aryl tetrahydroisoquinolines in water under air conditions. *RSC Adv.* 7: 1229–1232.

1319 Varun, B.V., Dhineshkumar, J., Bettadapur, K.R. et al. (2017). Recent advancements in dehydrogenative cross-coupling reactions for C—C bond formation. *Tetrahedron Lett.* 58: 803–824.

1320 Jia, F. and Li, Z.P. (2014). Iron-catalyzed/mediated oxidative transformation of C—H bonds. *Org. Chem. Front.* 1: 194–214.

1321 Su, B., Cao, Z.C., and Shi, Z.J. (2015). Exploration of earth-abundant transition metals (Fe, Co, and Ni) as catalysts in unreactive chemical bond activations. *Acc. Chem. Res.* 48: 886–896.

1322 Li, Y.Z., Li, B.J., Lu, X.Y. et al. (2009). Cross dehydrogenative arylation (CDA) of a benzylic C—H bond with arenes by Fe catalysis. *Angew. Chem. Int. Ed.* 48: 3817–3820.

1323 Buyevskaya, O., Wolf, D., and Baerns, M. (1994). Surface processes in the catalytic oxidative coupling of methane to ethane. *Recl. Trav. Chim. Pays-Bas* 113 (10): 459–464.

1324 Martin, G.A. and Mirodatos, C. (1995). Surface-chemistry in the oxidative coupling of methane. *Fuel Process. Technol.* 42 (2–3): 179–215.

1325 Stoukides, M. (2006). Methane conversion to C_2 hydrocarbons in solid electrolyte membrane reactors. *Res. Chem. Intermed.* 32 (3–4): 187–204.

1326 Kondratenko, N.V. (2008). Fundamentals of the oxidative conversion of methane to higher hydrocarbons and oxygenates. *DGMK Tagungsbericht* 45–58.

1327 Ohtsuka, Y., Asamil, K., and Wang, Y. (1999). Novel utilization of carbon dioxide for direct conversion of methane to ethane and ethylene. *J. Chin. Inst. Chem. Eng.* 30 (6): 439–447.

1328 Glaser, C. (1869). Beiträge zur Kenntniss des Acetenylbenzols. *Ber. Dtsch. Chem. Ges.* 2: 422–424.

1329 Glaser, C.A. (1870). Untersuchungen über einige Derivate der Zimmtsäure. *Ann. Chem. Pharm.* 154: 137–171.

1330 Hay, A.S. (1960). Oxidative coupling of acetylenes. *J. Org. Chem.* 25 (7): 1275–1276.

1331 Hay, A.S. (1962). Oxidative coupling of acetylenes. 2. *J. Org. Chem.* 27 (9): 3320–3321.

1332 Siemsen, P., Livingston, R.C., and Diederich, F. (2000). Acetylenic coupling: a powerful tool in molecular construction. *Angew. Chem. Int. Ed.* 39 (15): 2633–2657.

1333 Fomina, L., Vazquez, B., Tkatchouk, E., and Fomine, S. (2002). The Glaser reaction mechanism. A DFT study. *Tetrahedron* 58 (33): 6741–6747.

1334 Gil-Mólto, J. and Nájera, C. (2005). Palladium(II) chloride and a (dipyridin-2-ylmethyl)amine-derived Pd(II) chloride complex as highly efficient catalysts for the synthesis of alkynes in water or in NMP and of diynes in the absence of reoxidant. *Eur. J. Org. Chem.* (19): 4073–4081.

1335 Li, J.H., Liang, Y., and Xie, Y.X. (2005). Efficient Pd-catalyzed homocoupling reaction and Sonogashira cross-coupling reaction of terminal alkynes under aerobic conditions. *J. Org. Chem.* 70 (11): 4393–4396.

1336 Zheng, Q.W., Hua, R.M., and Wan, Y.Z. (2010). An alternative CuCl-piperidine-catalyzed oxidative homocoupling of terminal alkynes affording 1,3-diynes in air. *Appl. Organomet. Chem.* 24: 314–316.

1337 Niu, X.J., Li, C.J., Li, J., and Jia, X.S. (2012). Importance of bases on the Cu-catalyzed oxidative homocoupling of terminal alkynes to 1,4-disubstituted 1,3-diynes. *Tetrahedron Lett.* 53: 5559–5561.

1338 van Gelderen, L., Rothenberg, G., Calderone, V.R. et al. (2013). Efficient alkyne homocoupling catalysed by Cu immobilized on functionalized silica. *Appl. Organomet. Chem.* 27: 23–27.

1339 Boronat, M., Laursen, S., Leyva-Pérez, A. et al. (2014). Partially oxidized gold nanoparticles: a catalytic base-free system for the aerobic homocoupling of alkynes. *J. Catal.* 315: 6–14.

1340 Jia, X.S., Yin, K., Li, C.J. et al. (2011). Copper-catalyzed oxidative alkyne homocoupling without palladium, ligands and bases. *Green Chem.* 13: 2175–2178.

1341 Yin, K., Li, C.J., Li, J.A., and Jia, X.S. (2011). CuCl-catalyzed green oxidative alkyne homocoupling without Pd, ligands and bases. *Green Chem.* 13: 591–593.

1342 Bandini, M., Luque, R., Budarin, V., and Macquarrie, D.J. (2005). Aryl alkynylation versus alkyne homocoupling: unprecedented selectivity switch in Cu, phosphine and solvent-free heterogeneous Pd-catalysed couplings. *Tetrahedron* 61: 9860–9868.

1343 Bohlmann, F., Schönowsky, H., Grau, G., and Inhoffen, E. (1964). Polyacetylenverbindungen. 52. Über den Mechanismus der oxydativen Dimerisierung von Acetylenverbindungen. *Chem. Ber.* 97 (3): 794–800.

1344 Kamata, K., Yamaguchi, S., Kotani, M. et al. (2008). Efficient oxidative alkyne homocoupling catalyzed by a monomeric dicopper-substituted silicotungstate. *Angew. Chem. Int. Ed.* 47 (13): 2407–2410.

1345 Moritani, I. and Fujiwara, Y. (1967). Aromatic substitution of styrene-$PdCl_2$ complex. *Tetrahedron Lett.* (12): 1119–1122.

1346 Zhou, L. and Lu, W. (2014). Towards ideal synthesis: alkenylation of aryl C—H bonds by a Fujiwara–Moritani reaction. *Chem. Eur. J.* 20: 634–642.

1347 Shue, R.S. (1971). Catalytic coupling of aromatics and olefins by homogeneous Pd(II)-compounds under oxygen. *J. Chem. Soc., Chem. Commun.* (23): 1510–1511.

1348 Dams, M., De Vos, D.E., Celen, S., and Jacobs, P.A. (2003). Toward waste-free production of Heck products with a catalytic Pd system under oxygen. *Angew. Chem. Int. Ed.* 42 (30): 3512–3515.

1349 Matsumoto, T. and Yoshida, H. (2000). Oxidative arylation of ethylene with benzene to produce styrene. *Chem. Lett.* (9): 1064–1065.

1350 Yang, F.L., Ma, X.T., and Tian, S.K. (2012). Oxidative Mizoroki–Heck-type reaction of arylsulfonyl hydrazides for a highly regio- and stereoselective synthesis of polysubstituted alkenes. *Chem. Eur. J.* 18: 1582–1585.

1351 Weng, J.Q., Yu, Z.Q., and Zhang, G.F. (2012). Direct oxidative coupling between unfunctionalized arene and olefin. *Progr. Chem.* 24: 523–544.

1352 Gigant, N. and Bäckvall, J.E. (2014). Aerobic direct C—H arylation of non-biased olefins. *Org. Lett.* 16: 4432–4435.

1353 Deb, A., Bag, S., Kancherla, R., and Maiti, D. (2014). Palladium-catalyzed aryl C—H olefination with unactivated, aliphatic alkenes. *J. Am. Chem. Soc.* 136: 13602–13605.

1354 Liu, C., Yuan, J.W., Gao, M. et al. (2015). Oxidative coupling between two hydrocarbons: an update of recent C–H functionalizations. *Chem. Rev.* 115: 12138–12204.

1355 Yuen, O.Y., So, C.M., and Kwong, F.Y. (2016). Open-air oxidative Mizoroki–Heck reaction of arylsulfonyl hydrazides with alkenes. *RSC Adv.* 6: 27584–27589.

1356 Kannaboina, P., Kumar, K.A., and Das, P. (2016). Site-selective intermolecular oxidative C-3 alkenylation of 7-azaindoles at room temperature. *Org. Lett.* 18: 900–903.

1357 Han, S.J., Kim, H.T., and Joo, J.M. (2016). Direct C—H alkenylation of functionalized pyrazoles. *J. Org. Chem.* 81: 689–698.

1358 Kozhushkov, S.I. and Ackermann, L. (2013). Ruthenium-catalyzed direct oxidative alkenylation of arenes through twofold C—H bond functionalization. *Chem. Sci.* 4: 886–896.

1359 Bruneau, C. and Dixneuf, P.H. (2016). Ruthenium(II)-catalysed functionalisation of C—H bonds with alkenes: alkenylation versus alkylation. In: *C—H Bond Activtion and Catalytic Functionalization I*, Topics in Organometallic Chemistry, vol. 55 (ed. P.H. Dixneuf and H. Doucet), 137–188. Switzerland: Springer International Publishing.

1360 Xue, X., Xu, J.B., Zhang, L.J. et al. (2016). Rhodium(III)-catalyzed direct C—H olefination of arenes with aliphatic olefins. *Adv. Synth. Catal.* 358: 573–583.

1361 Yamada, T., Sakaguchi, S., and Ishii, Y. (2005). Oxidative coupling of benzenes with α,β-unsaturated aldehydes by the $Pd(OAc)_2$[molybdovanadophosphoric acid]O_2 system. *J. Org. Chem.* 70 (14): 5471–5474.

1362 Chen, X., Engle, K.M., Wang, D.H., and Yu, J.Q. (2009). Palladium(II)-catalyzed C—H activation/C—C cross-coupling reactions: versatility and practicality. *Angew. Chem. Int. Ed.* 48 (28): 5094–5115.

1363 Cho, S.H., Hwang, S.J., and Chang, S. (2008). Palladium-catalyzed C—H functionalization of pyridine *N*-oxides: highly selective alkenylation and direct arylation with unactivated arenes. *J. Am. Chem. Soc.* 130 (29): 9254–9256.

1364 Pankajakshan, S., Xu, Y.H., Cheng, J.K. et al. (2012). Palladium-catalyzed direct C—H arylation of enamides with simple arenes. *Angew. Chem. Int. Ed.* 51: 5701–5705.

1365 Zhang, H.M., Ferreira, E.M., and Stoltz, B.M. (2004). Direct oxidative Heck cyclizations: intramolecular Fujiwara–Moritani arylations for the synthesis of functionalized benzofurans and dihydrobenzofurans. *Angew. Chem. Int. Ed.* 43 (45): 6144–6148.

1366 Hadi, V., Yoo, K.S., Jeong, M., and Jung, K.W. (2009). Expeditious enyne construction from alkynes via oxidative Pd(II)-catalyzed Heck-type coupling. *Tetrahedron Lett.* 50 (20): 2370–2373.

1367 Wilke, G., Schott, H., and Heimbach, P. (1967). Oxygen complexes of zerovalent Ni, Pd and Pt. *Angew. Chem. Int. Ed. Engl.* 6 (1): 92–93.

1368 Dwight, T.A., Rue, N.R., Charyk, D. et al. (2007). C—C bond formation via double C—H functionalization: aerobic oxidative coupling as a method for synthesizing heterocoupled biaryls. *Org. Lett.* 9: 3137–3139.

1369 Deng, Y.Q., Persson, A.K.A., and Bäckvall, J.E. (2012). Palladium-catalyzed oxidative carbocyclizations. *Chem. Eur. J.* 18: 11498–11523.

1370 Deng, Y.Q., Bartholomeyzik, T., Persson, A.K.A. et al. (2012). Palladium-catalyzed oxidative arylating carbocyclization of allenynes. *Angew. Chem. Int. Ed.* 51: 2703–2707.

1371 Toyota, M., Wada, T., Fukumoto, K., and Ihara, M. (1998). Total synthesis of (+/-)-methyl atis-16-en-19-oate via homoallyl-homoallyl radical rearrangement. *J. Am. Chem. Soc.* 120: 4916–4925.

1372 Ito, Y., Aoyama, H., Hirao, T. et al. (1979). Cyclization reactions via oxo-(π-allyl)palladium(II) intermediates. *J. Am. Chem. Soc.* 101: 494–496.

1373 Lipshutz, B.H. and Taft, B.R. (2008). Heck couplings at room temperature in nanometer aqueous micelles. *Org. Lett.* 10 (7): 1329–1332.

1374 Lipshutz, B.H., Chung, D.W., and Rich, B. (2008). Sonogashira couplings of aryl bromides: room temperature, water only, no Cu. *Org. Lett.* 10: 3793–3796.

1375 Rouquet, G. and Chatani, N. (2013). Catalytic functionalization of C(sp^2)-H and C(sp^3)-H bonds by using bidentate directing groups. *Angew. Chem. Int. Ed.* 52: 11726–11743.

1376 Nishikata, T. and Lipschutz, B.H. (2010). Cationic Pd(II)-catalyzed Fujiwara–Moritani reactions at room temperature in water. *Org. Lett.* 12: 1972–1975.

1377 Bag, S. and Maiti, D. (2016). Palladium-catalyzed olefination of aryl C—H bonds by using directing scaffolds. *Synthesis* 48: 804–815.

1378 Patra, T., Bag, S., Kancherla, R. et al. (2016). Palladium-catalyzed directed *para* C—H functionalization of phenols. *Angew. Chem. Int. Ed.* 55: 7751–7755.

1379 Smidt, J., Sedlmeier, J., Hafner, W. et al. (1962). Olefinoxydation mit Palladiumchlorid-Katalysatoren. *Angew. Chem.* 74 (3): 93–102.

1380 Stern, E.W. and Spector, M.L. (1961). Reaction of olefin-Pd(II) chloride complexes with nucleophiles - a new vinylation. *Proc. Chem. Soc. London* 370–370.

1381 Moiseev, I.I., Vargaftik, M.N., and Syrkin, J.K. (1960). The mechanism of the reaction of Pd salts with olefins in hydroxyl-containing solutions. *Dokl. Akad. Nauk SSSR* 133 (2): 377–380.

1382 Tsuji, J. and Takahashi, H. (1965). Organic syntheses by means of noble netal compounds. 12. Reaction of (cyclooctadiene)Pd chloride complex with ethyl malonate. *J. Am. Chem. Soc.* 87 (14): 3275–3276.

1383 Tsuji, J., Takahashi, H., and Morikawa, M. (1965). Organic syntheses by means of noble metal compounds. 17. Reaction of (π-allyl)Pd chloride with nucleophiles. *Tetrahedron Lett.* (49): 4387–4388.

1384 Trost, B.M. and Dietsch, T.J. (1973). New synthetic reactions - asymmetric induction in allylic alkylations. *J. Am. Chem. Soc.* 95 (24): 8200–8201.

1385 Stahl, S.S. (2004). Palladium oxidase catalysis: selective oxidation of organic chemicals by direct dioxygen-coupled turnover. *Angew. Chem. Int. Ed.* 43 (26): 3400–3420.

1386 Heumann, A., Reglier, M., and Waegell, B. (1982). Oxidation with Pd salts - stereospecific and regiospecific acetoxylation of 4-vinylcyclohexene derivatives. *Angew. Chem. Int. Ed. Engl.* 21 (5): 366–367.

1387 Principato, B., Maffei, M., Siv, C. et al. (1996). Palladium-catalysed allylic acetoxylation of dialkyl allyl phosphonates. *Tetrahedron* 52 (6): 2087–2096.

1388 Aakermark, B., Larsson, E.M., and Oslob, J.D. (1994). Allylic carboxylations and lactonization using benzoquinone and hydrogen-peroxide or *tert*-butyl hydroperoxide as oxidants. *J. Org. Chem.* 59 (19): 5729–5733.

1389 Bates, R. (2012). *Organic Synthesis Using Transition Metals*, 2e, 1–446. New York: Wiley.

1390 Grennberg, H., Simon, V., and Bäckvall, J.E. (1994). Evidence for a (π-allyl)Pd intermediate in the quinone-based Pd-catalyzed allylic acetoxylation. *J. Chem. Soc., Chem. Commun.* (3): 265–266.

1391 Grennberg, H. and Bäckvall, J.E. (1998). Mechanism of Pd-catalyzed allylic acetoxylation of cyclohexene. *Chem. Eur. J.* 4 (6): 1083–1089.

1392 Khabibulin, V.R., Kulik, A.V., Oshanina, I.V. et al. (2007). Mechanism of the oxidative carbonylation of terminal alkynes at the equivalent to C—H bond in solutions of Pd-complexes. *Kinet. Catal.* 48 (2): 228–244.

1393 Lin, S., Song, C.X., Cai, G.X. et al. (2008). Intra/intermolecular direct allylic alkylation via Pd(II)-catalyzed allylic C—H activation. *J. Am. Chem. Soc.* 130 (39): 12901–12903.

1394 Bringmann, G., Mortimer, A.J.P., Keller, P.A. et al. (2005). Atroposelective synthesis of axially chiral biaryl compounds. *Angew. Chem. Int. Ed.* 44 (34): 5384–5427.

1395 Solomon, E.I., Sundaram, U.M., and Machonkin, T.E. (1996). Multicopper oxidases and oxygenases. *Chem. Rev.* 96 (7): 2563–2605.

1396 Noyori, R. and Takaya, H. (1990). BINAP - an efficient chiral element for asymmetric catalysis. *Acc. Chem. Res.* 23 (10): 345–350.

1397 Pu, L. (1998). 1,1′-Binaphthyl dimers, oligomers, and polymers: molecular recognition, asymmetric catalysis, and new materials. *Chem. Rev.* 98 (7): 2405–2494.

1398 Kocovsky, P., Vyskocil, S., and Smrcina, M. (2003). Non-symmetrically substituted 1,1′-binaphthyls in enantioselective catalysis. *Chem. Rev.* 103 (8): 3213–3245.

1399 Chen, Y., Yekta, S., and Yudin, A.K. (2003). Modified BINOL ligands in asymmetric catalysis. *Chem. Rev.* 103 (8): 3155–3211.

1400 Quirk, J., Thornton, M., and Kirkpatrick, P. (2003). Fresh from the pipeline - rosuvastatin calcium. *Nat. Rev. Drug Discov.* 2 (10): 769–770.

1401 Carey, J.S., Laffan, D., Thomson, C., and Williams, M.T. (2006). Analysis of the reactions used for the preparation of drug candidate molecules. *Org. Biomol. Chem.* 4 (12): 2337–2347.

1402 Horton, D.A., Bourne, G.T., and Smythe, M.L. (2003). The combinatorial synthesis of bicyclic privileged structures or privileged substructures. *Chem. Rev.* 103 (3): 893–930.

1403 Ackermann, L., Vicente, R., Kapdi, A. R. Transition-metal-catalyzed direct arylation of (hetero)arenes by C—H bond cleavage. *Angew. Chem. Int. Ed.* 2009, 48, 9792–9826.

1404 Liu, C., Zhang, H., Shi, W., and Lei, A.W. (2011). Bond formations between two nucleophiles: transition metal catalyzed oxidative cross-coupling reactions. *Chem. Rev.* 111: 1780–1824.

1405 Boström, J. and Brown, D.G. (2016). Stuck in a rut with old chemistry. *Drug Discov. Today* 21: 701–703.

1406 Brown, D.G. and Bostrom, J. (2016). Analysis of past and present synthetic methodologies on medicinal chemistry: where have all the new reactions gone? *J. Med. Chem.* 59: 4443–4458.

1407 You, S.L. and Xia, J.B. (2010). Palladium-catalyzed aryl-aryl bond formation through double C—H activation. *Top. Curr. Chem.* 292: 165–194.

1408 Scholl, R. and Mansfeld, J. (1910). *Meso*-Benzdianthron(helianthron), *meso*-Naphthodianthron, and a new way of receiving flavanthren. *Ber. Dtsch. Chem. Ges.* 43: 1734–1746.

1409 Sarhan, A.A.O. and Bolm, C. (2009). Iron(III) chloride in oxidative C—C coupling reactions. *Chem. Soc. Rev.* 38 (9): 2730–2744.

1410 Zhai, L.Y., Shukla, R., and Rathore, R. (2009). Oxidative C—C bond formation (Scholl Reaction) with DDQ as an efficient and easily recyclable oxidant. *Org. Lett.* 11 (15): 3474–3477.

1411 Hammerich, O. and Parker, V.D. (1984). Kinetics and mechanisms of reactions of organic cation-radicals in solution. *Adv. Phys. Org. Chem.* 20: 55–189.

1412 Van Helden, R. and Verberg, G. (1965). Oxidative coupling of aromatic compounds with Pd Salts. *Recl. Trav. Chim. Pays-Bas* 84 (9–10): 1263–1273.

1413 Davidson, J.M. and Triggs, C. (1966). Reactions of Pd-complexes with benzene and toluene. *Chem. Ind.* (11): 457–457.

1414 Davidson, J.M. and Triggs, C. (1968). Reaction of metal ion complexes with hydrocarbons. I. Palladation and some other new electrophilic substitution reactions. Preparation of Pd(I). *J. Chem. Soc. A* (6): 1324–1328.

1415 Iataaki, H. and Yoshimoto, H. (1973). Palladium-catalyzed syntheses of aromatic coupling compounds. *J. Org. Chem.* 38 (1): 76–79.

1416 Yoshimoto, H. and Itatani, H. (1973). Palladium-catalyzed coupling reaction of aromatic-compounds. *Bull. Chem. Soc. Jpn.* 46 (8): 2490–2492.

1417 Yoshimoto, H. and Itatani, H. (1973). Palladium-catalyzed competitive reaction of aromatic-compounds. *J. Catal.* 31 (1): 8–12.

1418 Wang, D., Izawa, Y., and Stahl, S.S. (2014). Pd-Catalyzed aerobic oxidative coupling of arenes: evidence for transmetalation between two Pd(II)-aryl intermediates. *J. Am. Chem. Soc.* 136: 9914–9917.

1419 Itatani, H., Shiotani, A., and Yokota, A. (1981). Method of producing biphenyltetracarboxylic esters. US Patent 4,292,435.

1420 Shiotani, A., Itatani, H., and Inagaki, T. (1986). Selective coupling of dimethyl phthalate with Pd catalysts at atmospheric pressure. *J. Mol. Catal.* 34 (1): 57–66.

1421 Takahashi, M., Masui, K., Sekiguchi, H. et al. (2006). Palladium-catalyzed C—H homocoupling of bromothiophene derivatives and synthetic application to well-defined oligothiophenes. *J. Am. Chem. Soc.* 128: 10930–10933.

1422 Hull, K.L., Lanni, E.L., and Sanford, M.S. (2006). Highly regioselective catalytic oxidative coupling reactions: synthetic and mechanistic investigations. *J. Am. Chem. Soc.* 128 (43): 14047–14049.

1423 Toda, F., Tanaka, K., and Iwata, S. (1989). Oxidative coupling reactions of phenols with $FeCl_3$ in the solid state. *J. Org. Chem.* 54 (13): 3007–3009.

1424 Ding, K.L., Wang, Y., Zhang, L.J. et al. (1996). A novel two-phase oxidative coupling of 2-naphthols suspended in aqueous Fe^{3+} solutions. *Tetrahedron* 52 (3): 1005–1010.

1425 Noji, M., Nakajima, M., and Koga, K. (1994). A new catalytic system for aerobic oxidative coupling of 2-naphthol derivatives by the use of CuCl-amine complex - a practical synthesis of binaphthol derivatives. *Tetrahedron Lett.* 35 (43): 7983–7984.

1426 Hwang, D.R., Chen, C.P., and Uang, B.J. (1999). Aerobic catalytic oxidative coupling of 2-naphthols and phenols by $VO(acac)_2$. *J. Chem. Soc., Chem. Commun.* (13): 1207–1208.

1427 Sharma, V.B., Jain, S.L., and Sain, B. (2003). Methyltrioxorhenium-catalyzed aerobic oxidative coupling of 2-naphthols to binaphthols. *Tetrahedron Lett.* 44 (13): 2655–2656.

1428 Yadav, J.S., Reddy, B.V.S., Gayathri, K.U., and Prasad, A.R. (2003). [Bmim]PF_6/$RuCl_3 \cdot H_2O$: a novel and recyclable catalytic system for the oxidative coupling of β-naphthols. *New J. Chem.* 27 (12): 1684–1686.

1429 Umare, P.S. and Tembe, G.L. (2004). Catalytic oxidative coupling of 2-naphthol using metal β-diketonates. *React. Kinet. Catal. Lett.* 82 (1): 173–178.

1430 Sakamoto, T., Yonehara, H., and Pac, C. (1997). Catalytic activities of $CuSO_4/Al_2O_3$ in dehydrogenation of arenes by dioxygen. *J. Org. Chem.* 62 (10): 3194–3199.

1431 Kantam, M.L., Kavita, B., and Figueras, F. (1998). Oxidative coupling of 2-naphthols catalysed by clays-supported iron catalysts. *Catal. Lett.* 51 (1–2): 113–115.

1432 Fujiyama, H., Kohara, I., Iwai, K. et al. (1999). Liquid-phase oxidation of 2,6-di-*tert*-butylphenol with Cu-impregnated MCM-41 catalysts in the presence of alkali metals. *J. Catal.* 188 (2): 417–425.

1433 Iwai, K., Yamauchi, T., Hashimoto, K. et al. (2003). Clean synthesis of 3,3′,5,5′-*tetra*-tert-butyl-4,4′-diphenoquinone from the oxidative coupling of 2,6-di-*tert*-butylphenol catalyzed by alkali-promoted Cu—Mg—Al hydrotalcites in the presence of molecular oxygen. *Chem. Lett.* 32 (1): 58–59.

1434 Li, T.S., Duan, H.Y., Li, B.Z. et al. (1999). Novel oxidative coupling of 2-naphthols to 1,1′-bi-2-naphthols catalysed by solid Lewis acids using atmospheric oxygen as oxidant. *J. Chem. Soc., Perkin Trans. 1* (3): 291–294.

1435 Ding, K.L., Xu, Q.G., Wang, Y. et al. (1997). Novel two-phase oxidative cross-coupling of the two-component molecular crystal of 2-naphthol and 2-naphthylamine. *J. Chem. Soc., Chem. Commun.* (7): 693–694.

1436 Matsushita, M., Kamata, K., Yamaguchi, K., and Mizuno, N. (2005). Heterogeneously catalyzed aerobic oxidative biaryl coupling of 2-naphthols and substituted phenols in water. *J. Am. Chem. Soc.* 127 (18): 6632–6640.

1437 Guo, X.Y., Deng, G.J., and Li, C.J. (2009). Ruthenium-catalyzed oxidative homo-coupling of 2-arylpyridines. *Adv. Synth. Catal.* 351 (13): 2071–2074.

1438 Li, X.L., Yang, J., and Kozlowski, M.C. (2001). Enantioselective oxidative biaryl coupling reactions catalyzed by 1,5-diazadecalin metal complexes. *Org. Lett.* 3 (8): 1137–1140.

1439 Li, X.L., Hewgley, J.B., Mulrooney, C.A. et al. (2003). Enantioselective oxidative biaryl coupling reactions catalyzed by 1,5-diazadecalin metal complexes: efficient formation of chiral functionalized BINOL derivatives. *J. Org. Chem.* 68 (14): 5500–5511.

1440 Irie, R., Masutani, K., and Katsuki, T. (2000). Asymmetric aerobic oxidative coupling of 2-naphthol derivatives catalyzed by photo-activated chiral (NO)Ru(II)-salen complex. *Synlett* (10): 1433–1436.

1441 Luo, Z.B., Liu, Q.Z., Gong, L.Z. et al. (2002). Novel achiral biphenol-derived diastereomeric oxovanadium(IV) complexes for highly enantioselective oxidative coupling of 2-naphthols. *Angew. Chem. Int. Ed.* 41 (23): 4532–4535.

1442 Chu, C.Y. and Uang, B.J. (2003). Catalytic enantioselective coupling of 2-naphthols by new chiral oxovanadium complexes bearing a self accelerating functional group. *Tetrahedron: Asymmetry* 14 (1): 53–55.

1443 Ikeda, T., Misawa, N., Ichihashi, Y. et al. (2005). Liquid-phase oxidative coupling of 2-naphthol by vanadium catalysts supported on MCM-41. *J. Mol. Catal. A: Chem.* 231 (1–2): 235–240.

1444 Kar, A., Mangu, N., Kaiser, H.M., and Tse, M.K. (2009). Gold-catalyzed direct oxidative coupling reactions of non-activated arenes. *J. Organomet. Chem.* 694 (4): 524–537.

1445 Kar, A., Mangu, N., Kaiser, H.M. et al. (2008). A general Au-catalyzed direct oxidative coupling of non-activated arenes. *J. Chem. Soc., Chem. Commun.* 386–388.

1446 Kozhevnikov, I.V. (1976). Oxidative coupling of furan derivatives to bifurans catalyzed by Pd(II). *React. Kinet. Catal. Lett.* 5 (4): 415–419.

1447 Itahara, T. (1981). Arylation of *N*-acyl-pyrroles and *N*-acyl-indoles with arenes and Pd acetate. *J. Chem. Soc., Chem. Commun.* (5): 254–255.

1448 Nakamura, N., Tajima, Y., and Sakai, K. (1982). Direct phenylation of isoxazoles using Pd-catalysts synthesis of 4-phenylmuscimol. *Heterocycles* 17: 235–245.

1449 Li, R.S., Jiang, L., and Lu, W.J. (2006). Intermolecular cross-coupling of simple arenes via C—H activation by tuning concentrations of arenes and TFA. *Organometallics* 25 (26): 5973–5975.

1450 Stuart, D.R. and Fagnou, K. (2007). The catalytic cross-coupling of unactivated arenes. *Science* 316 (5828): 1172–1175.

1451 Stuart, D.R., Villemure, E., and Fagnou, K. (2007). Elements of regiocontrol in Pd-catalyzed oxidative arene cross-coupling. *J. Am. Chem. Soc.* 129: 12072–12073.

1452 Gao, G.L., Xia, W., Jain, P., and Yu, J.Q. (2016). Pd(II)-Catalyzed C3-selective arylation of pyridine with (hetero)arenes. *Org. Lett.* 18: 744–747.

1453 Liu, B., Guo, Q., Cheng, Y. et al. (2011). Palladium-catalyzed desulfitative C—H arylation of heteroarenes with sodium sulfinates. *Chem. Eur. J.* 17: 13415–13419.

1454 Hull, K.L. and Sanford, M.S. (2007). Catalytic and highly regioselective cross-coupling of aromatic C—H substrates. *J. Am. Chem. Soc.* 129: 11904–11905.

1455 Kalyani, D., Deprez, N.R., Desai, L.V., and Sanford, M.S. (2005). Oxidative C—H activation/C—C bond forming reactions: synthetic scope and mechanistic insights. *J. Am. Chem. Soc.* 127 (20): 7330–7331.

1456 Hull, K.L. and Sanford, M.S. (2009). Mechanism of benzoquinone-promoted Pd-catalyzed oxidative cross-coupling reactions. *J. Am. Chem. Soc.* 131 (28): 9651–9653.

1457 Davies, D.L., Donald, S.M.A., and Macgregor, S.A. (2005). Computational study of the mechanism of cyclometalation by palladium acetate. *J. Am. Chem. Soc.* 127 (40): 13754–13755.

1458 Xia, J.B. and You, S.L. (2007). Carbon–carbon bond formation through double sp^2 C—H activations: synthesis of ferrocenyl oxazoline derivatives. *Organometallics* 26: 4869–4871.

1459 Li, B.H., Tian, S.L., Fang, Z., and Shi, Z.H. (2008). Multiple C—H activations to construct biologically active molecules in a process completely free of organohalogen and organometallic components. *Angew. Chem. Int. Ed.* 47 (6): 1115–1118.

1460 Brasche, G., García-Fortanet, J., and Buchwald, S.L. (2008). Twofold C—H functionalization: Pd-catalyzed *ortho* arylation of anilides. *Org. Lett.* 10 (11): 2207–2210.

1461 Cheng, Y.Y., Wu, Y.M., Tan, G.Y., and You, J.S. (2016). Nickel catalysis enables oxidative $C(sp^2)$-H/$C(sp^2)$-H cross-coupling reactions between two heteroarenes. *Angew. Chem. Int. Ed.* 55: 12275–12279.

1462 Ball, L.T., Llyod-Jones, G.C., and Russell, C.A. (2012). Gold-catalyzed direct arylation. *Science* 337: 1644–1648.

1463 Cresswell, A.J. and Llyod-Jones, G.C. (2016). Room-temperature Au-catalysed arylation of heteroarenes: complementarity to Pd catalysis. *Chem. Eur. J.* 22: 12641–12645.

1464 Corrie, T.J.A., Ball, L.T., Russell, C.A., and Lloyd-Jones, G.C. (2017). Au-catalyzed biaryl coupling to generate 5-to 9-membered rings: turnover-limiting reductive elimination versus π-complexation. *J. Am. Chem. Soc.* 139: 245–254.

1465 Kramer, S.R. (2016). Recent advances in Au-catalyzed intermolecular aryl C—H functionalization. *Chem. Eur. J.* 22: 15584–15598.

1466 Maji, M.S., Pfeifer, T., and Studer, A. (2008). Oxidative homocoupling of aryl, alkenyl, and alkynyl Grignard reagents with TEMPO and dioxygen. *Angew. Chem. Int. Ed.* 47 (49): 9547–9550.

1467 Murarka, S. and Studer, A. (2011). Transition metal-free TEMPO-catalyzed oxidative cross-coupling of nitrones with alkynyl-Grignard reagents. *Adv. Synth. Catal.* 353: 2708–2714.

1468 Zhang, B., Cui, Y.X., and Jiao, N. (2012). Metal-free TEMPO-catalyzed oxidative C—C bond formation from $C(sp^3)$-H bonds using molecular oxygen as the oxidant. *J. Chem. Soc., Chem. Commun.* 48: 4498–4500.

1469 Sonobe, T., Oisaki, K., and Kanai, M. (2012). Catalytic aerobic production of imines en route to mild, green, and concise derivatizations of amines. *Chem. Sci.* 3: 3249–3255.

1470 Uber, J.S., Vogels, Y., van den Helder, D. et al. (2007). Pyrazole-based ligands for the [copper-TEMPO]-mediated oxidation of benzyl alcohol to benzaldehyde and structures of the Cu coordination compounds. *Eur. J. Inorg. Chem.* 2007 (26): 4197–4206.

1471 Gurung, S.K., Thapa, S., Shrestha, B., and Giri, R. (2014). Copper-catalyzed cross-coupling of aryl- and heteroaryltriethoxysilanes with aryl and heteroaryl iodides and bromides. *Synthesis* 46: 1933–1937.

1472 Wang, H. (2010). Recent advances in asymmetric oxidative coupling of 2-naphthol and its derivatives. *Chirality* 22 (9): 827–837.

1473 Hamashima, Y. and Sodecka, M. (2015). Asymmetric cross dehydrogenative-coupling reactions. *RSC Green Chem.* 26: 133–152.

1474 Alamsetti, S.K., Poonguzhali, E., Ganapathy, D., and Sekar, G. (2013). Enantioselective oxidative coupling of 2-naphthol derivatives by Cu-(*R*)-1,1′-binaphthyl-2,2′-diamine-TEMPO catalyst. *Adv. Synth. Catal.* 355: 2803–2808.

1475 Huo, C.D., Wang, C., Sun, C.G. et al. (2013). Triarylaminium salt-initiated aerobic double

Friedel–Crafts reaction of glycine derivatives with indoles. *Adv. Synth. Catal.* 355: 1911–1916.

1476 Shen, Y.M., Li, M., Wang, S.Z. et al. (2009). An efficient Cu-catalyzed oxidative Mannich reaction between tertiary amines and methyl ketones. *J. Chem. Soc., Chem. Commun.* 953–955.

1477 Ishii, Y., Sakaguchi, S., and Iwahama, T. (2001). Innovation of hydrocarbon oxidation with molecular oxygen and related reactions. *Adv. Synth. Catal.* 343 (5): 393–427.

1478 Galli, C., Gentili, P., and Lanzalunga, O. (2008). Hydrogen abstraction and electron transfer with aminoxyl radicals: synthetic and mechanistic issues. *Angew. Chem. Int. Ed.* 47: 4790–4796.

1479 de Paolini, I. and Ribet, G. (1932). The behavior of benzoyl peroxide toward amino compounds. *Gazz. Chim. Ital.* 62: 1041–1048.

1480 Horner, L. and Schwenk, E. (1950). Über die Polymerisation von Vinylverbindungen durch das System - Diacylperoxyd - Tertiares Amin. *Liebigs Ann. Chem.* 566 (1): 69–84.

1481 Horner, L. and Junkermann, H. (1955). Studien zum Ablauf der Substitution. 7. Beitrag zum kinetischen Nachweis von Durchgangsradikalen im System - Tertiares Amin-Dibenzoylperoxyd. *Liebigs Ann. Chem.* 591 (1–2): 53–68.

1482 Horner, L. and Kirmse, W. (1955). Studien zum Ablauf der Substitution. 9. Zur oxydativen Entalkylierung Gemischt Substituierter tertiarer Amine durch Dibenzoylperoxyd. *Liebigs Ann. Chem.* 597 (1): 48–65.

1483 Lalevee, J., Allonas, X., Fouassier, J.P., and Ingold, K.U. (2008). Absolute rate constants for some intermolecular reactions of α-aminoalkylperoxyl radicals. Comparison with alkylperoxyls. *J. Org. Chem.* 73 (17): 6489–6496.

1484 Galliani, G., Rindone, B., and Marchesini, A. (1978). Horseradish peroxidase-catalyzed oxidation of aromatic tertiary-amines with hydrogen peroxide. *J. Chem. Soc., Perkin Trans. 1* (5): 456–460.

1485 Galliani, G. and Rindone, B. (1980). Formation of superoxide radical-anion in the horseradish peroxidase-catalyzed oxidation of 3 aromatic tertiary-amines with hydrogen peroxide. *J. Chem. Soc., Perkin Trans. 2* (1): 1–3.

1486 Guengerich, F.P. and MacDonald, T.L. (1984). Chemical mechanisms of catalysis by Cytochromes-P-450 - a unified view. *Acc. Chem. Res.* 17 (1): 9–16.

1487 Baciocchi, E., Gerini, M.F., Lanzalunga, O. et al. (2001). Isotope-effect profiles in the oxidative *N*-demethylation of *N,N*-dimethylanilines catalysed by lignin peroxidase and a chemical model. *Eur. J. Org. Chem.* (12): 2305–2310.

1488 Baciocchi, E., Gerini, M.F., and Lapi, A. (2002). O-18 incorporation in the oxidation of *N*-methylcarbazole by lignin peroxidase and a model compound: a mechanistic insight into the oxidative N-demethylation of aromatic tertiary amines. *J. Chem. Soc., Chem. Commun.* (9): 946–947.

1489 de Ortiz, Montellano, P.R. (ed.) (1995). *Cytochrome P-450: Structure, Mechanism and Biochemistry*, 2e, 1–652. New York and London: Plenum.

1490 Cuppoletti, A., Galli, C., Gentili, P., and Petride, H. (2002). A radical route in the biomimetic oxidation of a tertiary amine. *J. Phys. Org. Chem.* 15 (9): 672–675.

1491 Urlacher, V.B., Bell, S.G., and Wong, L.L. (2007). The bacterial Cytochrome P450 monooxygenases: P450cam and P450BM-3. In: *Modern Biooxidation* (ed. R.D. Schmid and V.B. Urlacher), 99–122. New York: Wiley.

1492 Chiavarino, B., Cipollini, R., Crestoni, M.E. et al. (2008). Probing the compound I-like reactivity of a bare high-valent oxo Fe(porphyrin) complex: the oxidation of tertiary amines. *J. Am. Chem. Soc.* 130 (10): 3208–3217.

1493 Murahashi, S., Naota, T., and Yonemura, K. (1988). Ruthenium-catalyzed Cytochrome-P-450 type oxidation of tertiary amines with alkyl hydroperoxides. *J. Am. Chem. Soc.* 110 (24): 8256–8258.

1494 Murahashi, S.I., Nakae, T., Terai, H., and Komiya, N. (2008). Ruthenium-catalyzed oxidative cyanation of tertiary amines with molecular oxygen or hydrogen peroxide and sodium cyanide: C(sp^3)-H bond activation and carbon–carbon bond formation. *J. Am. Chem. Soc.* 130 (33): 11005–11012.

1495 Wilson, R.B. Jr., and Laine, R.M. (1985). Transalkylation reaction - homogeneous catalytic formation of C—N bonds. *J. Am. Chem. Soc.* 107 (2): 361–369.

1496 Chatani, N., Asaumi, T., Yorimitsu, S. et al. (2001). $Ru_3(CO)_{12}$-catalyzed coupling reaction of C(sp^3)-H bonds adjacent to a nitrogen atom in alkylamines with alkenes. *J. Am. Chem. Soc.* 123 (44): 10935–10941.

1497 DeBoef, B., Pastine, S.J., and Sames, D. (2004). Cross-coupling of C(sp^3)-H bonds and alkenes: catalytic cyclization of alkene-amide substrates. *J. Am. Chem. Soc.* 126 (21): 6556–6557.

1498 Pastine, S.J., Gribkov, D.V., and Sames, D. (2006). C(sp^3)-H Bond arylation directed by amidine protecting group: α-arylation of pyrrolidines and piperidines. *J. Am. Chem. Soc.* 128 (44): 14220–14221.

1499 Catino, A.J., Nichols, J.M., Nettles, B.J., and Doyle, M.P. (2006). The oxidative Mannich reaction catalyzed by dirhodium caprolactamate. *J. Am. Chem. Soc.* 128 (17): 5648–5649.

1500 Murata, S., Miura, M., and Nomura, M. (1989). Oxidation of 3-substituted or 4-substituted N,N-dimethylanilines with molecular oxygen in the presence of either $FeCl_3$ or [Fe(salen)]OAc. *J. Org. Chem.* 54 (19): 4700–4702.

1501 Murata, S., Miura, M., and Nomura, M. (1989). Iron-catalyzed oxidation of *N,N*-dimethylaniline with molecular oxygen. *J. Chem. Soc., Chem. Commun.* (2): 116–118.

1502 Murata, S., Teramoto, K., Miura, M., and Nomura, M. (1993). Copper-catalyzed oxidative coupling of 4-substituted *N,N*-dimethylanilines with terminal alkynes under molecular-oxygen. *J. Chem. Res. S* (10): 434–434.

1503 Li, Z.P. and Li, C.J. (2004). CuBr-Catalyzed efficient alkynylation of C(sp^3)-H bonds adjacent to a nitrogen atom. *J. Am. Chem. Soc.* 126 (38): 11810–11811.

1504 Li, Z.P. and Li, C.J. (2005). Highly efficient Cu-catalyzed nitro-Mannich type reaction: cross-dehydrogenative-coupling between C(sp^3)-H bond and C(sp^3)-H bond. *J. Am. Chem. Soc.* 127 (11): 3672–3673.

1505 Baslé, O. and Li, C.J. (2007). Copper-catalyzed oxidative alkylation of C(sp^3)-H bond adjacent to a nitrogen atom using molecular oxygen in water. *Green Chem.* 9 (10): 1047–1050.

1506 Li, Z.P. and Li, C.J. (2005). CuBr-catalyzed direct indolation of tetrahydroisoquinolines via cross-dehydrogenative coupling between C(sp^3)-H and C(sp^2)-H bonds. *J. Am. Chem. Soc.* 127 (19): 6968–6969.

1507 Li, Z.P., Bohle, D.S., and Li, C.J. (2006). Cu-catalyzed cross-dehydrogenative coupling: a versatile strategy for C—C bond formations via the oxidative activation of C(sp^3)-H bonds. *Proc. Natl. Acad. Sci. U.S.A.* 103 (24): 8928–8933.

1508 Niu, M., Yin, Z., Fu, H. et al. (2008). Copper-catalyzed coupling of tertiary aliphatic amines with terminal alkynes to propargylamines via C—H activation. *J. Org. Chem.* 73 (10): 3961–3963.

1509 Volla, C.M. and Vogel, P. (2009). Chemoselective C—H bond activation: ligand and solvent-free Fe-catalyzed oxidative C—C cross-coupling of tertiary amines with terminal, alkynes. Reaction scope and mechanism. *Org. Lett.* 11 (8): 1701–1704.

1510 Deng, G.J. and Li, C.J. (2008). Peroxide-mediated efficient addition of cycloalkanes to imines. *Tetrahedron Lett.* 49 (39): 5601–5604.

1511 Sawyer, D.T., Sobkowiak, A., and Matsushita, T. (1996). Metal [ML(x); M=Fe, Cu, Co, Mn]/hydroperoxide-induced activation of dioxygen for the oxygenation of hydrocarbons: oxygenated Fenton chemistry. *Acc. Chem. Res.* 29 (9): 409–416.

1512 Li, P.H., Zhang, Y.C., and Wang, L. (2009). Iron-catalyzed ligand-free three-component coupling reactions of aldehydes, terminal alkynes, and amines. *Chem. Eur. J.* 15 (9): 2045–2049.

1513 Rong, Y., Li, R., and Lu, W. (2007). Palladium(II)-catalyzed coupling of *p*-xylene via regioselective C—H activation in TFA. *Organometallics* 26 (18): 4376–4378.

1514 Nishimura, T., Araki, H., Maeda, Y., and Uemura, S. (2003). Palladium-catalyzed oxidative alkynylation of alkenes via C—C bond cleavage under oxygen atmosphere. *Org. Lett.* 5 (17): 2997–2999.

1515 Gilman, H., Langham, W., and Jacoby, A.L. (1939). Metalation as a side reaction in the preparation of organolithium compounds. *J. Am. Chem. Soc.* 61: 106–109.

1516 Wittig, G. and Fuhrmann, G. (1940). On the performance of halogenised anisole towards phenyllithium (V Announcement on the reaction measure of phenyl-lithium). *Ber. Dtsch. Chem. Ges.* 73: 1197–1218.

1517 Kornblum, N., Michel, R.E., and Kerber, R.C. (1966). Chain reactions in substitution processes which proceed via radical-anion intermediates. *J. Am. Chem. Soc.* 88 (23): 5662–5663.

1518 Kornblum, N., Michel, R.E., and Kerber, R.C. (1966). Radical-anions as intermediates in substitution reactions. *J. Am. Chem. Soc.* 88 (23): 5660–5662.

1519 Kornblum, N. (1975). Substitution reactions which proceed via radical-anion intermediates. *Angew. Chem. Int. Ed. Engl.* 14 (11): 734–745.

1520 Russell, G.A. and Danen, W.C. (1966). Coupling reactions of 2-nitro-2-propyl anion. *J. Am. Chem. Soc.* 88 (23): 5663–5663.

1521 Vidal, V., Theolier, A., Thivolle-Cazat, J., and Basset, J.M. (1997). Metathesis of alkanes catalyzed by silica-supported transition metal hydrides. *Science* 276 (5309): 99–102.

1522 Soulivong, D., Copéret, C., Thivolle-Cazat, J. et al. (2004). Cross-metathesis of propane and methane: a catalytic reaction of C—C bond cleavage of a higher alkane by methane. *Angew. Chem. Int. Ed.* 43: 5366–5369.

1523 Copéret, C., Maury, O., Thivolle-Cazat, J., and Basset, J.M. (2001). σ-Bond metathesis of alkanes on a silica-supported Ta(V)(alkyl)(alkylidene) complex: first evidence for alkane cross-metathesis. *Angew. Chem. Int. Ed.* 40 (12): 2331–2334.

1524 Le Roux, E., Chabanas, M., Baudouin, A. et al. (2004). Detailed structural investigation of the grafting of [Ta(=CHtBu)(CH_2-*t*-Bu)$_3$] and [Cp*$TaMe_4$] on silica partially dehydroxylated at 700 °C and the activity of the grafted complexes toward alkane metathesis. *J. Am. Chem. Soc.* 126 (41): 13391–13399.

1525 Le Roux, E., Taoufik, M., Copéret, C. et al. (2005). Development of W-based heterogeneous alkane metathesis catalysts through a structure-activity relationship. *Angew. Chem. Int. Ed.* 44 (41): 6755–6758.

1526 Basset, J.M., Copéret, C., Lefort, L. et al. (2005). Primary products and mechanistic considerations in alkane metathesis. *J. Am. Chem. Soc.* 127 (24): 8604–8605.

1527 Solans-Monfort, X., Clot, E., Copéret, C., and Eisenstein, O. (2005). d(0) Re-based olefin metathesis catalysts, Re(=CR)(=CHR)(X)(Y): the key role of X and Y ligands for efficient active sites. *J. Am. Chem. Soc.* 127 (40): 14015–14025.

1528 Schinzel, S., Chermette, H., Copéret, C., and Basset, J.M. (2008). Evaluation of the carbene hydride mechanism in the carbon–carbon bond formation process of alkane metathesis through a DFT study. *J. Am. Chem. Soc.* 130 (25): 7984–7987.

1529 Blanc, F., Thivolle-Cazat, J., Basset, J.M., and Copéret, C. (2008). Structure-reactivity relationship in alkane metathesis using well-defined silica-supported alkene metathesis catalyst precursors. *Chem. Eur. J.* 14 (29): 9030–9037.

1530 Merle, N., Stoffelbach, F., Taoufik, M. et al. (2009). Selective and unexpected transformations of 2-methylpropane to 2,3-dimethylbutane and 2-methylpropene to 2,3-dimethylbutene catalyzed by an alumina-supported tungsten hydride. *J. Chem. Soc., Chem. Commun.* (18): 2523–2525.

1531 (a) Basset, J.M., Copéret, C., Soulivong, D. et al. (2006). From olefin to alkane metathesis: a historical point of view. *Angew. Chem. Int. Ed.* 45 (37): 6082–6085. (b) Basset, J.M., Copéret, C., Soulivong, D. et al. (2010). Metathesis of alkanes and related reactions. *Acc. Chem. Res.* 43: 323–334. (c) Basset, J.-M., Callens, E., and Riache, N. (2015) Alkane metathesis. In *Hanbook of Metathesis, 2nd Ed.* (ed. R. H. Grubbs) 33–70. Weinheim: Wiley-VCH. (d) Diaz, E., Restrepo, A., and Nunez-Zarur, F. (2018). Reactivity of a silica-supported Mo alkylidene catalyst toward alkanes: A DFT study on the metathesis of propane. *Organometallics* 37 (13): 2023–2036.

1532 (a) Samantaray, M.K., Dey, R., Kavitake, S., and Basset, J.-M. (2016). New concept of C—H and C—C bond activation via surface organometallic chemistry. In: *C–H Bond Activation and Catalytic Functionalization II*, Topics in Organometallic Chemistry, vol. 56, 155–187. Springer International Publishing. (b) Nuñez-Zarur, F., Solans-Montfort, X., and Restrepo, A. (2017). Mechanistic insights into alkane metathesis catalyzed by silica-supported tantalum hydrides: a DFT study. *Inorg. Chem.* 56 (17): 10458–10473. (c) Gordon, C.P., Yamamoto, K., Searles, K. et al. (2018). Metal alkyls programmed to generate metal alkylidenes by α-H abstraction: prognosis from NMR chemical shift. *Chem. Sci.* 9 (7): 1912–1918.

1533 Goldman, A.S., Roy, A.H., Huang, Z. et al. (2006). Catalytic alkane metathesis by tandem alkane dehydrogenation olefin metathesis. *Science* 312 (5771): 257–261.

1534 Bailey, B.C., Schrock, R.R., Kundu, S. et al. (2009). Evaluation of Mo- and W-metathesis catalysts for homogeneous tandem alkane metathesis. *Organometallics* 28 (1): 355–360.

1535 Haibach, M.C., Kundu, S., Brookhart, M., and Goldman, A.S. (2012). Alkane metathesis by tandem alkane-dehydrogenation-olefin-metathesis catalysis and related chemistry. *Acc. Chem. Res.* 45: 947–958.

1536 Fürstner, A. (2000). Olefin metathesis and beyond. *Angew. Chem. Int. Ed.* 39 (17): 3012–3043.

1537 Trnka, T.M. and Grubbs, R.H. (2001). The development of L_2X_2Ru=CHR olefin metathesis catalysts: an organometallic success story. *Acc. Chem. Res.* 34 (1): 18–29.

1538 Schrock, R.R. (2002). High oxidation state multiple metal–carbon bonds. *Chem. Rev.* 102 (1): 145–179.

1539 Schrock, R.R. and Hoveyda, A.H. (2003). Molybdenum and Tungsten imido alkylidene complexes as efficient olefin-metathesis catalysts. *Angew. Chem. Int. Ed.* 42 (38): 4592–4633.

1540 (a) Vougioukalakis, G.C. and Grubbs, R.H. (2010). Ruthenium-based heterocyclic carbene-coordinated olefin metathesis catalysts. *Chem. Rev.* 110: 1746–1787. (b) Mauduit, M., Schmid, T., Dumas, A. et al. (2017). From environmentally friendly reusable ionic-tagged Ru-based complexes to industrially relevant homogeneous catalysts: toward a sustainable olefin metathesis. *Synlett* 28 (7): 773–798.

1541 Lozano-Vila, A.M., Monsaert, S., Bajek, A., and Verpoort, F. (2010). Ruthenium-based olefin metathesis catalysts derived from alkynes. *Chem. Rev.* 110: 4865–4909.

1542 (a) Deraedt, C., d'Halluin, M., and Astruc, D. (2013). Metathesis reactions: recent trends and challenges. *Eur. J. Inorg. Chem.* 2013: 4881–4908. (b) Montgomery, T.P., Ahmed, T.S., and Grubbs, R.H. (2017). Stereoretentive olefin metathesis: an avenue to kinetic selectivity. *Angew. Chem. Int. Ed.* 56 (37): 11024–11036.

1543 Higman, C.S., Lummiss, J.A., and Fogg, D.E. (2016). Olefin metathesis at the dawn of implementation in pharmaceutical and specialty chemicals manufacturing. *Angew. Chem. Int. Ed.* 55: 3552–3565.

1544 (a) Fürstner, A. (2011). Metathesis in total synthesis. *J. Chem. Soc., Chem. Commun.* 47: 6505–6511. (b) Fürstner, A. (2015). Alkyne metathesis in organic synthesis. In: *Modern Alkyne Chemistry*, Chapter 4 (ed. B.M. Trost and C.-L. Li), 69–111. Weinheim: Wiley-VCH. (c) Meng, Z. and Fuerstner, A. (2019). Total Synthesis of (-)-Sinulariadiolide. A transannular approach. *J. Am. Chem. Soc.*, 141 (2): 805–809. (d) Ehrhorn, H. and Tamm, M. Well-defined alkyne metathesis catalysts: Developments and recent applications. *Chem. Eur. J.*, 25 (13): 3190–3208.

1545 (a) Cossy, J. (2014). Application in the synthesis of natural and biologically active molecules. In: *Olefin Metathesis: Theory and Practice* (ed. K.E. Grela), 287–310. Hoboken, NJ: Wiley. (b) Lecourt, C., Dhambri, S., Allievi, L. et al. (2018). Natural products and ring-closing metathesis: synthesis of sterically congested olefins. *Nat. Prod. Rep.* 35 (1): 105–124.

1546 Fürstner, A. (2014). From understanding to prediction: gold- and Pt-based π-acid catalysis for target oriented synthesis. *Acc. Chem. Res.* 47: 925–938.

1547 (a) Knall, A.C. and Slugovc, C. (2014). Olefin metathesis polymerization. In: *Olefin Metathesis; Theory and Practice* (ed. K.E. Grela), 269–286. Hoboken, NJ: Wiley. (b) Slugovc, C. (2014). Industrial applications of olefin metathesis polymerization. In: *Olefin Metathesis: Theory and Practice* (ed. K.E. Grela), 329–334. Hoboken, NJ: Wiley.

1548 (a) Buchmeister, M. P. Tandem ring-opening metathesis - vinyl insertion polymerization: fundamentals and application to functional polyolefins. *Macromolecular Rapid Communications*, Wiley-VCH, Weinheim, 2017, 38, no. 6), 221199; (b) Tezuka, Y. (2017). Topological polymer chemistry designing complex macromolecular graph constructions. *Acc. Chem. Res.* 50 (11): 2661–2672. (c) Chen, Y., Abdellatif, M.M., and Nomura, K. (2018). Olefin metathesis polymerization: some recent developments in the precise polymerizations for synthesis of advanced materials (by ROMP, ADMET). *Tetrahedron* 74 (6): 619–643.

1549 Nickel, A. and Pederson, R.L. (2014). Commercial potential of olefin metathesis of renewal feedstocks. In: *Olefin Metathesis: Theory and Practice* (ed. K.E. Grela), 335–348. Hoboken, NJ: Wiley.

1550 Dixneuf, P.H., Bruneau, C., and Fischmeister, C. (2016). Alkene metathesis catalysis: a key for transformations of unsaturated plant oils and renewable derivatives. *Oil Gas Sci. Technol. – Rev. D IFP Energies Nouvelles* 71 (2): 19, 1–21.

1551 Bronner, S.M., Herbert, M.B., Patel, P.R. et al. (2014). Ru-based *Z*-selective metathesis catalysts with modified cyclometalated carbene ligands. *Chem. Sci.* 5: 4091–4098.

1552 Nelson, D.J., Manzini, S., Urbina-Blanco, C.A., and Nolan, S.P. (2014). Key processes in Ru-catalysed olefin metathesis. *J. Chem. Soc., Chem. Commun.* 50: 10355–10375.

1553 Nicolaou, K.C., Bulger, P.G., and Sarlah, D. (2005). Metathesis reactions in total synthesis. *Angew. Chem. Int. Ed.* 44 (29): 4490–4527.

1554 Grubbs, R.H. (2004). Olefin metathesis. *Tetrahedron* 60 (34): 7117–7140.

1555 Schrock, R.R. (2006). Multiple metal-carbon bonds for catalytic metathesis reactions (Nobel lecture). *Angew. Chem. Int. Ed.* 45 (23): 3748–3759.

1556 Connon, S.J. and Blechert, S. (2003). Recent developments in olefin cross-metathesis. *Angew. Chem. Int. Ed.* 42 (17): 1900–1923.

1557 Deshmukh, P.H. and Blechert, S. (2007). Alkene metathesis: the search for better catalysts. *J. Chem. Soc., Dalton Trans.* (24): 2479–2491.

1558 Schrock, R.R. and Czekelius, C. (2007). Recent advances in the syntheses and applications of Mo- and W-alkylidene and alkylidyne catalysts for the metathesis of alkenes and alkynes. *Adv. Synth. Catal.* 349 (1–2): 55–77.

1559 (a) Grubbs, R.H. (ed.) (2003). *Handbook of Metathesis*, vol. 1, 2 and 3. Weinheim: Wiley-VCH. (b) Grubbs, R.H., Wenzel, A.G., O'Leary, D.J., and Khosravi, E. (eds.) (2015). *Handbook of Metathesis*, 2e, 3 volume set; Vol.1: Catalyst Development and Mechanism; Vol. 2: Applications in Organic Synthesis; Vol. 3: Polymer Synthesis. Weinheim: Wiley-VCH. (c) Vanderwal, C. D. and Atwood, B. R. (2017). Recent advances in alkene metathesis for natural product synthesis. Striking achievements resulting from increased sophistication in catalyst design and synthesis strategy. *Aldrichimica Acta* 50 (1): 17–31.

1560 Astruc, D. (2005). The metathesis reactions: from a historical perspective to recent developments. *New J. Chem.* 29 (1): 42–56.

1561 Thayer, A.M. (2007). Optimism prevails for fine chemicals. *Chem. Eng. News* 85: 34–38.

1562 Donohoe, T.J., Fishlock, L.P., and Procopiou, P.A. (2008). Ring-closing metathesis: novel routes to aromatic heterocycles. *Chem. Eur. J.* 14 (19): 5716–5726.

1563 Zukowska, K. and Grela, K. (2014). Cross metathesis. In: *Olefin Metathesis: Theory and Practice* (ed. K.E. Grela), 39–84. Hoboken, NJ: Wiley.

1564 Van Lierop, B.J., Lummiss, J.A.M., and Fogg, D.E. (2014). Ring-closing methathesis. In: *Olefin Metathesis: Theory and Practice* (ed. K.E. Grela), 85–152. Hoboken, NJ: Wiley.

1565 Diver, S.T. and Giessert, A.J. (2004). Enyne metathesis (enyne bond reorganization). *Chem. Rev.* 104 (3): 1317–1382.

1566 Diver, S.T. (2006). Metal carbenes in enyne metathesis: synthetic and mechanistic studies. *J. Mol. Catal. A: Chem.* 254 (1–2): 29–42.

1567 Diver, S.T. and Griffiths, J.R. (2014). Ene-yne metathesis. In: *Olefin Metathesis:Theory and Practice* (ed. K.E. Grela), 153–186. Hoboken, NJ: Wiley.

1568 Hamad, F.B., Kai, C., Cai, Y. et al. (2013). Solid supported Ru-complexes for olefin metathesis. *Curr. Org. Chem.* 17: 2592–2608.

1569 Wang, D. and Astruc, D. (2014). Magnetically recoverable Ru-catalysts in organic synthesis. *Molecules* 19: 4635–4653.

1570 Chen, S.W., Zhang, Z.C., Ma, M. et al. (2014). Supported Ru(carbene) catalyst on ionic magnetic nanoparticles for olefin metathesis. *Org. Lett.* 16: 4969–4971.

1571 Mougel, V., Pucino, M., and Copéret, C. (2015). Strongly σ-donating thiophenoxide in silica-supported W-oxo catalysts for improved 1-alkene metathesis efficiency. *Organometallics* 34: 551–554.

1572 Gawande, M.B., Monga, Y., Zboril, R., and Sharma, R. (2015). Silica-decorated magnetic nanocomposites for catalytic applications. *Coord. Chem. Rev.* 288: 118–143.

1573 Riache, N., Callens, E., Talbi, K., and Basset, J.-M. (2016). Unsaturated fatty acid esters metathesis catalyzed by silica supported WMe5. *J. Mol. Catal. A: Chem.* 411: 344–348.

1574 Suriboot, J., Bazzi, H.S., and Bergbreiter, D.E. (2016). Supported catalysts useful in ring-closing metathesis, cross metathesis, and ring-opening metathesis polymerization. *Polymers* 8 (4): 140.

1575 (a) Yuan, J., Fracaroli, A.M., and Klemperer, W.G. (2016). Convergent synthesis of a metal-organic framework supported olefin metathesis catalyst. *Organometallics* 35: 2149–2155. (b) Franzen, R.G. (2016). Metathesis reactions on solid phase: towards new synthesis challenges. In: *Topics in Catalysis*, vol. 59, 1143–1150. Springer International Publishing.

1576 (a) Ludwig, J.R., Zimmerman, P.M., Gianino, J.B., and Schindler, C.S. (2016). Iron(III)-catalysed carbonyl-olefin metathesis. *Nature* 533 (7603): 374–379. (b) Ma, L., Li, W.-J., Xi, H. et al. (2016). $FeCl_3$-Catalyzed ring-closing carbonyl-olefin metathesis. *Angew. Chem. Int. Ed.* 55 (35): 10410–10413. (c) Hennessy, E.T. and Jacobsen, E.N. (2016). Organometallic chemistry. A new metathesis. *Nat. Chem.* 8 (8): 741–742. (d) Ludwig, J.R., Phan, S., McAtee, C.C. et al. (2017). Mechanistic investigations of the Fe(III) -catalyzed carbonyl-olefin metathesis reaction. *J. Am. Chem. Soc.* 139 (31): 10832–10842. (e) Ludwig, J.R. and Schindler, C.S. (2017). Lewis acid catalyzed carbonyl-olefin metathesis. *Synlett* 28 (13): 1501–1509.

1577 Grela, K. (2014). *Olefin Metathesis: Theory and Practice*, 1–608. Hoboken, NJ: Wiley.

1578 (a) Anderson, A.W. and Merckling, N.G. (1955). Polymeric bicyclo[2.2.1]hept-2-ene. US Patent 2721189. (b) Anderson, A.W. and Merckling, N.G. (1956). Polymeric bicyclo[2.2.1]hept-2-ene. *Chem. Abstr* 50: 3008.

1579 Calderon, N., Chen, H.Y., and Scott, K.W. (1967). Olefin metathesis - a noval reaction for skeletal transformations of unsaturated hydrocarbons. *Tetrahedron Lett.* (34): 3327–3329.

1580 Banks, R.L. and Bailey, G.C. (1964). Olefin disproportionation - new catalytic process. *Ind. Eng. Chem. Prod. Res. Dev.* 3 (3): 170–173.

1581 Heckelsberg, L.F., Banks, R.L., and Bailey, G.C. (1969). Olefin disproportionation catalysts. *Ind. Eng. Chem. Prod. Res. Dev.* 8 (3): 259–261.

1582 Mol, J.C. (2004). Industrial applications of olefin metathesis. *J. Mol. Catal. A: Chem.* 213 (1): 39–45.

1583 Heckelsberg, L.F. and Banks, R.L. (Philips Petroleum Co.) (1969). Dehydrogenation and disproportionation of hydrocarbons. US Patent 3445541 A.

1584 Hughes, T.R. (Chevron Research Co.) (1971). Comparison of saturated hydrocarbons. Ger. Offen. Patent, DE 2049164 A.

1585 Hérisson, J.L. and Chauvin, Y. (1971). Transformation catalysis of olefins by W-complexes. 2. Telomerization of cyclic olefins in presence of acyclic olefins. *Makromol. Chem.* 141: 161–176.

1586 Schrock, R.R. (1975). 1st Isolable transition-metal methylene complex and analogs - characterization, mode of decomposition, and some simple reactions. *J. Am. Chem. Soc.* 97 (22): 6577–6578.

1587 Romero, P.E. and Piers, W.E. (2005). Direct observation of a 14-electron ruthenacyclobutane relevant to olefin metathesis. *J. Am. Chem. Soc.* 127 (14): 5032–5033.

1588 Wenzel, A.G. and Grubbs, R.H. (2006). Ruthenium metallacycles derived from 14-electron complexes. New insights into olefin metathesis intermediates. *J. Am. Chem. Soc.* 128 (50): 16048–16049.

1589 Anderson, D.R., Hickstein, D.D., O'Leary, D.J., and Grubbs, R.H. (2006). Model compounds of Ru(alkene) intermediates in olefin metathesis reactions. *J. Am. Chem. Soc.* 128 (26): 8386–8387.

1590 Rocklage, S.M., Fellmann, J.D., Rupprecht, G.A. et al. (1981). Multiple metal-carbon bonds. 19. How Nb- and Ta-complexes of the type-M($CHCMe_3$)(Pr_3)$_2$$Cl_3$ can be modified to give olefin metathesis catalysts. *J. Am. Chem. Soc.* 103 (6): 1440–1447.

1591 (a) Casey, C.P. and Burkhardt, T.J. (1973). (Diphenylcarbene)(pentacarbonyl)tungsten(0). *J. Am. Chem. Soc.* 95 (17): 5833–5834. (b) Casey, C.P. and Burkhardt, T.J. (1974). Reactions of (diphenylcarbene)(pentacarbonyl)tungsten(0) with alkenes - role of metal-carbene complexes in cyclopropanation and olefin metathesis reactions. *J. Am. Chem. Soc.* 96 (25): 7808–7809.

1592 Katz, T.J. and McGinnis, J. (1975). Mechanism of olefin metathesis reaction. *J. Am. Chem. Soc.* 97 (6): 1592–1594.

1593 Grubbs, R.H., Burk, P.L., and Carr, D.D. (1975). Consideration of mechanism of olefin metathesis reaction. *J. Am. Chem. Soc.* 97 (11): 3265–3267.

1594 Kress, J., Wesolek, M., and Osborn, J.A. (1982). Tungsten(IV)(carbenes) for the metathesis of olefins - direct observation and identification of the chain carrying carbene complexes in a highly active catalyst system. *J. Chem. Soc., Chem. Commun.* (9): 514–516.

1595 Schrock, R.R., Murdzek, J.S., Bazan, G.C. et al. (1990). Synthesis of Mo(imido)(alkylidene) complexes and some reactions involving acyclic olefins. *J. Am. Chem. Soc.* 112 (10): 3875–3886.

1596 Ondi, L., Nagy, G.M., Czirok, J.B. et al. (2016). From box to bench: air-stable Mo catalyst tablets for everyday use in olefin metathesis. *Org. Process Res. Dev.* 20 (10): 1709–1716.

1597 Heppekausen, J.F.A. (2011). Rendering Schrock-type Mo(alkylidene) complexes air stable: user-friendly precatalysts for alkene metathesis. *Angew. Chem. Int. Ed.* 50: 7829–7832.

1598 Novak, B.M. and Grubbs, R.H. (1988). Catalytic organometallic chemistry in water - the aqueous ring-opening metathesis polymerization of 7-oxanorbornene derivatives. *J. Am. Chem. Soc.* 110 (22): 7542–7543.

1599 Novak, B.M. and Grubbs, R.H. (1988). The ring-opening metathesis polymerization of 7-oxabicyclo[2.2.1]hept-5-ene derivatives - a new acyclic polymeric ionophore. *J. Am. Chem. Soc.* 110 (3): 960–961.

1600 Mortell, K.H., Weatherman, R.V., and Kiessling, L.L. (1996). Recognition specificity of neoglycopolymers prepared by ring-opening metathesis polymerization. *J. Am. Chem. Soc.* 118 (9): 2297–2298.

1601 Burtscher, D. and Grela, K. (2009). Aqueous olefin metathesis. *Angew. Chem. Int. Ed.* 48 (3): 442–454.

1602 Nguyen, S.T., Johnson, L.K., Grubbs, R.H., and Ziller, J.W. (1992). Ring-opening metathesis polymerization (Romp) of norbornene by a group-VIII carbene complex in protic media. *J. Am. Chem. Soc.* 114 (10): 3974–3975.

1603 Schwab, P., Grubbs, R.H., and Ziller, J.W. (1996). Synthesis and applications of $RuCl_2(=CHR')(PR_3)_2$: the influence of the alkylidene moiety on metathesis activity. *J. Am. Chem. Soc.* 118 (1): 100–110.

1604 Schwab, P., France, M.B., Ziller, J.W., and Grubbs, R.H. (1995). A series of well-defined metathesis catalysts - synthesis of [$RuCl_2$(=CHR′)(Pr_3)$_2$] and its reactions. *Angew. Chem. Int. Ed. Engl.* 34 (18): 2039–2041.

1605 Weskamp, T., Schattenmann, W.C., Spiegler, M., and Herrmann, W.A. (1998). A novel class of Ru-catalysts for olefin metathesis. *Angew. Chem. Int. Ed.* 37 (18): 2490–2493.

1606 Huang, J.K., Stevens, E.D., Nolan, S.P., and Petersen, J.L. (1999). Olefin metathesis-active Ru-complexes bearing a nucleophilic carbene ligand. *J. Am. Chem. Soc.* 121 (12): 2674–2678.

1607 Huang, J.K., Schanz, H.J., Stevens, E.D., and Nolan, S.P. (1999). Influence of sterically demanding carbene ligation on catalytic behavior and thermal stability of Ru olefin metathesis catalysts. *Organometallics* 18 (25): 5375–5380.

1608 Clavier, H. and Nolan, S.P. (2007). N-Heterocyclic carbene and phosphine Ru(indenylidene) precatalysts: a comparative study in olefin metathesis. *Chem. Eur. J.* 13 (28): 8029–8036.

1609 Scholl, M., Trnka, T.M., Morgan, J.P., and Grubbs, R.H. (1999). Increased ring closing metathesis activity of Ru-based olefin metathesis catalysts coordinated with imidazolin-2-ylidene ligands. *Tetrahedron Lett.* 40 (12): 2247–2250.

1610 Chatterjee, A.K., Morgan, J.P., Scholl, M., and Grubbs, R.H. (2000). Synthesis of functionalized olefins by cross and ring-closing metatheses. *J. Am. Chem. Soc.* 122 (15): 3783–3784.

1611 Ackermann, L., Fürstner, A., Weskamp, T. et al. (1999). Ruthenium(carbene) complexes with imidazolin-2-ylidene ligands allow the formation of tetrasubstituted cycloalkenes by RCM. *Tetrahedron Lett.* 40 (26): 4787–4790.

1612 Weskamp, T., Kohl, F.J., Hieringer, W. et al. (1999). Highly active Ru catalysts for olefin metathesis: the synergy of N-heterocyclic carbenes and coordinatively labile ligands. *Angew. Chem. Int. Ed.* 38 (16): 2416–2419.

1613 Bielawski, C.W. and Grubbs, R.H. (2000). Highly efficient ring-opening metathesis polymerization (ROMP) using new Ru catalysts containing N-heterocyclic carbene ligands. *Angew. Chem. Int. Ed.* 39 (16): 2903–2906.

1614 Scholl, M., Ding, S., Lee, C.W., and Grubbs, R.H. (1999). Synthesis and activity of a new generation of Ru-based olefin metathesis catalysts coordinated with 1,3-dimesityl-4,5-dihydroimidazol-2-ylidene ligands. *Org. Lett.* 1 (6): 953–956.

1615 Occhipinti, G., Bjorsvik, H.R., and Jensen, V.R. (2006). Quantitative structure-activity relationships of Ru catalysts for olefin metathesis. *J. Am. Chem. Soc.* 128 (21): 6952–6964.

1616 Straub, B.F. (2005). Origin of the high activity of second-generation Grubbs catalysts. *Angew. Chem. Int. Ed.* 44 (37): 5974–5978.

1617 Straub, B.F. (2007). Ligand influence on metathesis activity of Ru(carbene) catalysts: a DFT study. *Adv. Synth. Catal.* 349 (1–2): 204–214.

1618 Correa, A. and Cavallo, L. (2006). The elusive mechanism of olefin metathesis promoted by (NHC) Ru-based catalysts: a trade between steric, electronic, and solvent effects. *J. Am. Chem. Soc.* 128 (41): 13352–13353.

1619 Tonner, R., Heydenrych, G., and Frenking, G. (2007). Bonding analysis of N-heterocyclic carbene tautomers and phosphine ligands in transition-metal complexes: a theoretical study. *Chem. Asian J.* 2 (12): 1555–1567.

1620 Love, J.A., Morgan, J.P., Trnka, T.M., and Grubbs, R.H. (2002). A practical and highly active Ru-based catalyst that effects the cross-metathesis of acrylonitrile. *Angew. Chem. Int. Ed.* 41 (21): 4035–4037.

1621 Chatterjee, A.K. and Grubbs, R.H. (1999). Synthesis of trisubstituted alkenes via olefin cross-metathesis. *Org. Lett.* 1 (11): 1751–1753.

1622 Mathew, J., Koga, N., and Suresh, C.H. (2008). C—H bond activation through σ-bond metathesis and agostic interactions: deactivation pathway of a Grubbs second-generation catalyst. *Organometallics* 27 (18): 4666–4670.

1623 Garber, S.B., Kingsbury, J.S., Gray, B.L., and Hoveyda, A.H. (2000). Efficient and recyclable monomeric and dendritic Ru-based metathesis catalysts. *J. Am. Chem. Soc.* 122 (34): 8168–8179.

1624 Gessler, S., Randl, S., and Blechert, S. (2000). Synthesis and metathesis reactions of a phosphine-free (dihydroimidazole)(carbene)Ru complex. *Tetrahedron Lett.* 41 (51): 9973–9976.

1625 Michrowska, A., Bujok, R., Harutyunyan, S. et al. (2004). Nitro-substituted Hoveyda–Grubbs Ru(carbenes): enhancement of catalyst activity through electronic activation. *J. Am. Chem. Soc.* 126 (30): 9318–9325.

1626 Honda, T., Namiki, H., Kaneda, K., and Mizutani, H. (2004). First diastereoselective chiral synthesis of (−)-securinine. *Org. Lett.* 6 (1): 87–89.

1627 Sledz, P., Mauduit, M., and Grela, K. (2008). Olefin metathesis in ionic liquids. *Chem. Soc. Rev.* 37 (11): 2433–2442.

1628 Farina, V., Zeng, X.Z., Wei, X.D. et al. (2009). The chemistry of Ru(cyclopropylmethylidene) complexes: mechanistic studies and synthetic implications for the ring-closing metathesis reaction. *Catal. Today* 140 (1–2): 74–83.

1629 Courchay, F.C., Sworen, J.C., Ghiviriga, I. et al. (2006). Understanding structural isomerization during Ru-catalyzed olefin metathesis: a deuterium labeling study. *Organometallics* 25 (26): 6074–6086.

1630 Donohoe, T.J., O'Riordan, T.J.C., and Rosa, C.P. (2009). Ruthenium-catalyzed isomerization of terminal olefins: applications to synthesis. *Angew. Chem. Int. Ed.* 48 (6): 1014–1017.

1631 Copéret, C. and Basset, J.M. (2007). Strategies to immobilize well-defined olefin metathesis catalysts: supported homogeneous catalysis vs. surface organometallic chemistry. *Adv. Synth. Catal.* 349 (1–2): 78–92.

1632 Monge-Marcet, A., Pleixats, R., Cattoen, X., and Man, M.W.C. (2012). Sol–gel immobilized Hoveyda–Grubbs complex through the NHC ligand: a recyclable metathesis catalyst. *J. Mol. Catal. A: Chem.* 357: 59–66.

1633 Selegue, J.P. (1982). Synthesis and structure of $[Ru(C_3Ph_2)(PMe_3)_2(Cp)][PFf_6]$, a cationic diphenylallenylidene complex. *Organometallics* 1 (1): 217–218.

1634 Harlow, K.J., Hill, A.F., and Wilton-Ely, J.D.E.T. (1999). The first co-ordinatively unsaturated Group 8 allenylidene complexes: insights into Grubbs' vs. Dixneuf–Fürstner olefin metathesis catalysts. *J. Chem. Soc., Dalton Trans.* (3): 285–291.

1635 Fürstner, A., Hill, A.F., Liebl, M., and Wilton-Ely, J.D.E.T. (1999). Coordinatively unsaturated Ru(allenylidene) complexes: highly effective, well defined catalysts for the ring-closure metathesis of α,ω-dienes and dienynes. *J. Chem. Soc., Chem. Commun.* (7): 601–602.

1636 Péron, D., Romero, A., and Dixneuf, P.H. (1995). Functional Ru(II)(allenylidene) and diynyl (arene) derivatives formed by activation of a diyne via a Ru=C=C=C=C=Cr(2) intermediate. *Organometallics* 14 (7): 3319–3326.

1637 Boeda, F., Clavier, H., and Nolan, S.P. (2008). Ruthenium(indenylidene) complexes: powerful tools for metathesis transformations. *J. Chem. Soc., Chem. Commun.* (24): 2726–2740.

1638 Castarlenas, R., Vovard, C., Fischmeister, C., and Dixneuf, P.H. (2006). Allenylidene-to-indenylidene rearrangement in (arene)Ru complexes: a key step to highly active catalysts for olefin metathesis reactions. *J. Am. Chem. Soc.* 128 (12): 4079–4089.

1639 Fürstner, A. and Thiel, O.R. (2000). Formal total synthesis of (−)-balanol: concise approach to the hexahydroazepine segment based on RCM. *J. Org. Chem.* 65 (6): 1738–1742.

1640 Fürstner, A., Stelzer, F., Rumbo, A., and Krause, H. (2002). Total synthesis of the turrianes and evaluation of their DNA-cleaving properties. *Chem. Eur. J.* 8 (8): 1856–1871.

1641 Chatterjee, A.K., Choi, T.L., Sanders, D.P., and Grubbs, R.H. (2003). A general model for selectivity in olefin cross metathesis. *J. Am. Chem. Soc.* 125 (37): 11360–11370.

1642 Vehlow, K., Maechling, S., and Blechert, S. (2006). Ruthenium metathesis catalysts with saturated unsymmetrical N-heterocyclic carbene ligands. *Organometallics* 25 (1): 25–28.

1643 Fürstner, A. (2013). Teaching metathesis "simple" stereochemistry. *Science* 341: 1357–1366.

1644 Werrel, S., Walker, J.C.L., and Donohoe, T.J. (2015). Application of catalytic Z-selective olefin metathesis in natural product synthesis. *Tetrahedron Lett.* 56: 5261–5268.

1645 Herbert, M.B. and Grubbs, R.H. (2015). Z-selective cross metathesis with Ru-catalysts: synthetic applications and mechanistic implications. *Angew. Chem. Int. Ed.* 54: 5018–5024.

1646 Meek, S.J., O'Brien, R.V., Llaveria, J. et al. (2011). Catalytic Z-selective olefin cross-metathesis for natural product synthesis. *Nature* 471: 461–466.

1647 Keitz, B.K., Endo, K., Patel, P.R. et al. (2012). Improved Ru-catalysts for *Z*-Selective olefin metathesis. *J. Am. Chem. Soc.* 134: 693–699.

1648 Liu, P., Xu, X.F., Dong, X.F. et al. (2012). *Z*-Selectivity in olefin metathesis with chelated Ru-catalysts: computational studies of mechanism and selectivity. *J. Am. Chem. Soc.* 134: 1464–1467.

1649 Khan, R.K.M., Torker, S., and Hoveyda, A.H. (2013). Readily accessible and easily modifiable Ru-based catalysts for efficient and *Z*-selective ring-opening metathesis polymerization and ring-opening/cross-metathesis. *J. Am. Chem. Soc.* 135: 10258–10261.

1650 Occhipinti, G., Hansen, F.R., Tornroos, K.W., and Jensen, V.R. (2013). Simple and highly *Z*-selective Ru-based olefin metathesis catalyst. *J. Am. Chem. Soc.* 135: 3331–3334.

1651 Occhipinti, G., Koudriavtsev, V., Tornroos, K.W., and Jensen, V.R. (2014). Theory assisted development of a robust and *Z*-selective olefin metathesis catalyst. *J. Chem. Soc., Dalton Trans.* 43: 11106–11117.

1652 Hoveyda, A.H. (2014). Evolution of catalytic stereoselective olefin metathesis: from ancillary transformation to purveyor of stereochemical identity. *J. Org. Chem.* 79: 4763–4792.

1653 Koh, M.J., Khan, R.K.M., Torker, S. et al. (2015). High-value alcohols and higher oxidation state compounds by catalytic *Z*-selective cross-metathesis. *Nature* 517: 181–186.

1654 Quigley, B.L. and Grubbs, R.H. (2014). Ruthenium-catalysed *Z*-selective cross-metathesis of allylic substituted olefins. *Chem. Sci.* 5: 501–506.

1655 Koh, M.J., Nguyen, T.T., Zhang, H.M. et al. (2016). Direct synthesis of *Z*-alkenyl halides through catalytic cross-metathesis. *Nature* 531: 459–465.

1656 Ibrahem, I., Yu, M., Schrock, R.R., and Hoveyda, A.H. (2009). Highly *Z*- and enantioselective ring-opening/cross-metathesis reactions catalyzed by stereogenic-at-Mo(adamantylimido) complexes. *J. Am. Chem. Soc.* 131: 3844–3845.

1657 Torker, S., Koh, M.J., Khan, R.K.M., and Hoveyda, A.H. (2016). Regarding a persisting puzzle in olefin metathesis with Ru-complexes: why are transformations of alkenes with a small substituent *Z*-selective? *Organometallics* 35: 543–562.

1658 Kress, S. and Blechert, S. (2012). Asymmetric catalysts for stereocontrolled olefin metathesis reactions. *Chem. Soc. Rev.* 41: 4389–4408.

1659 Stenne, B. and Collins, S.K. (2014). Enantioselective olefin metathesis. In: *Olefin Metathesis: Theory and Practice* (ed. K.E. Grela), 233–268. Hoboken, NJ: Wiley.

1660 Seiders, T.J., Ward, D.W., and Grubbs, R.H. (2001). Enantioselective Ru-catalyzed ring-closing metathesis. *Org. Lett.* 3 (20): 3225–3228.

1661 Berlin, J.M., Goldberg, S.D., and Grubbs, R.H. (2006). Highly active chiral Ru catalysts for asymmetric cross- and ring-opening cross metathesis. *Angew. Chem. Int. Ed.* 45 (45): 7591–7595.

1662 van Veldhuizen, J.J., Garber, S.B., Kingsbury, J.S., and Hoveyda, A.H. (2002). A recyclable chiral Ru catalyst for enantioselective olefin metathesis. Efficient catalytic asymmetric ring-opening/cross-metathesis in air. *J. Am. Chem. Soc.* 124 (18): 4954–4955.

1663 Allaert, B., Dieltiens, N., Ledoux, N. et al. (2006). Synthesis and activity for ROMP of bidentate Schiff base substituted second generation Grubbs catalysts. *J. Mol. Catal. A: Chem.* 260 (1–2): 221–226.

1664 Malcolmson, S.J., Meek, S.J., Sattely, E.S. et al. (2008). Highly efficient Mo-based catalysts for enantioselective alkene metathesis. *Nature* 456 (7224): 933–937.

1665 Wright, D.L., Usher, L.C., and Estrella-Jiménez, M. (2001). Unusual influence of substituents on ring-opening metathesis reactions. *Org. Lett.* 3 (26): 4275–4277.

1666 Zaminer, J., Stapper, C., and Blechert, S. (2002). A new and efficient synthesis of (−)-indolizidine 167B by tandem metathesis. *Tetrahedron Lett.* 43: 6739–6741.

1667 (a) Diver, S.T. and Giessert, A.J. (2004). Enyne metathesis: a catalytic, cross-selective diene synthesis. *Synthesis* 466–471. (b) Le Ravalec, V., Fischmeister, C., and Bruneau, C. (2009). First transformation of unsaturated fatty esters involving enyne cross-metathesis. *Adv. Synth. Catal.* 351 (7–8): 1115–1122. (c) Dragutan, I., Dragutan, V., Demonceau, A., and Delaude, L. (2013). Enabling access to diverse bioactive molecules through enyne metathesis concepts. *Curr. Org. Chem.* 17 (22): 2678–2720. (d) Castagno, D. (2016). Enyne metathesis reactions in the synthesis of small ring heterocycles. *Targets in Heterocyclic Systems* 20: 222–246. (e) Jung, K., Kang, E.-H., Sohn, J.-H., and Choi, T.-L. (2016). Highly β-selective cyclopolymerization of 1,6-heptadiynes and ring-closing enyne metathesis reaction using Grubbs *Z*-selective catalyst: unprecedented regioselectivity for Ru-based catalysts. *J. Am. Chem. Soc.* 138 (35): 11227–11233. (f) Sole, D., Bennasar, M.-L., Roca, T., and Valldosera, M. (2016). Exploration of ring-closing enyne metathesis for the synthesis of azonino[5,4-b]indoles. *Eur. J. Org. Chem.* (7): 1355–1366.

1668 Katz, T.J. and Sivavec, T.M. (1985). Metal-catalyzed rearrangement of alkene alkynes and the stereochemistry of metallacyclobutene ring-opening. *J. Am. Chem. Soc.* 107 (3): 737–738.

1669 Torres, O. and Pla-Quintana, A. (2016). The rich reactivity of transition metal carbenes with alkynes. *Tetrahedron Lett.* 57 (35): 3881–3891.

1670 Kulkarni, A.A. and Diver, S.T. (2003). Cycloheptadiene ring synthesis by tandem intermolecular enyne metathesis. *Org. Lett.* 5 (19): 3463–3466.

1671 Mori, M., Sakakibara, N., and Kinoshita, A. (1998). Remarkable effect of ethylene gas in the intramolecular enyne metathesis of terminal alkynes. *J. Org. Chem.* 63: 6082–6083.

1672 Gregg, T.M., Keister, J.B., and Diver, S.T. (2013). Inhibitory effect of ethylene in ene-yne metathesis: the case for ruthenacyclobutane resting states. *J. Am. Chem. Soc.* 135: 16777–16780.

1673 Grotevendt, A.G.D., Lummiss, J.A.M., Mastronardi, M.L., and Fogg, D.E. (2011). Ethylene-promoted versus ethylene-free enyne metathesis. *J. Am. Chem. Soc.* 133: 15918–15921.

1674 Kalbarczyk, K.P. and Diver, S.T. (2009). Enyne metathesis/Brønsted acid-promoted heterocyclization. *J. Org. Chem.* 74 (5): 2193–2196.

1675 Stragies, R., Schuster, M., and Blechert, S. (1999). A novel Ru-catalysed tandem diyne cycloisomerisation-cross metathesis process. *J. Chem. Soc., Chem. Commun.* (3): 237–238.

1676 Chachignon, H., Scalacci, N., Petricci, E., and Castagnolo, D. (2015). Synthesis of 1,2,3-substituted pyrroles from propargylamines via a one-pot tandem enyne cross metathesis/cyclization reaction. *J. Org. Chem.* 80: 5287–5295.

1677 Maraval, V., Lepetit, C., Caminade, A.M. et al. (2006). Alkyne metathesis: toward simplicity and efficiency. *Tetrahedron Lett.* 47 (13): 2155–2159.

1678 Mortreux, A. and Coutelier, O. (2006). Alkyne metathesis catalysts: scope and future. *J. Mol. Catal. A: Chem.* 254 (1–2): 96–104.

1679 (a) Fürstner, A. and Davies, P.W. (2005). Alkyne metathesis. *J. Chem. Soc., Chem. Commun.* (18): 2307–2320. (b) Zhang, W. and Moore, J.S. (2007). Alkyne metathesis: catalysts and synthetic applications. *Adv. Synth. Catal.* 349 (1–2): 93–120.

1680 Ortiz, N., Yu, C., Jin, Y.-H., and Zhang, W. (2017). Poly(aryleneethynylene)s: properties, applications and synthesis through alkyne metathesis. *Top. Curr. Chem.* 375 (4): 1–24.

1681 Beer, S., Brandhorst, K., Hrib, C.G. et al. (2009). Experimental and theoretical investigations of catalytic alkyne cross-metathesis with (imidazolin-2-iminato)W(alkylidyne) complexes. *Organometallics* 28 (5): 1534–1545.

1682 Haberlag, B., Wu, X.A., Brandhorst, K. et al. (2010). Preparation of (imidazolin-2-iminato)Mo and W(benzylidyne) complexes: a new pathway to highly active alkyne metathesis catalysts. *Chem. Eur. J.* 16: 8868–8877.

1683 Wu, X., Daniliuc, C.G., Hrib, C.G., and Tamm, M. (2011). (Phosphoraneiminato)W(alkylidyne) complexes as highly efficient alkyne metathesis catalysts. *J. Organomet. Chem.* 696: 4147–4151.

1684 Wu, X.A. and Tamm, M. (2011). Recent advances in the development of alkyne metathesis catalysts. *Beilstein J. Org. Chem.* 7: 82–93.

1685 Haberlag, B., Freytag, M., Jones, P.G., and Tamm, M. (2014). Tungsten and Molybdenum 2,4,6-trimethylbenzylidyne complexes as robust pre-catalysts for alkyne metathesis. *Adv. Synth. Catal.* 356: 1255–1265.

1686 Fürstner, A., Guth, O., Rumbo, A., and Seidel, G. (1999). Ring closing alkyne metathesis. Comparative investigation of two different catalyst systems and application to the stereoselective synthesis of olfactory lactones, azamacrolides, and the macrocyclic perimeter of the marine alkaloid nakadomarin A. *J. Am. Chem. Soc.* 121 (48): 11108–11113.

1687 Ghalit, N., Poot, A.J., Fürstner, A. et al. (2005). Ring-closing alkyne metathesis approach toward the synthesis of alkyne mimics of thioether A-, B-, C-, and DE-ring systems of the antibiotic nisin Z. *Org. Lett.* 7 (14): 2961–2964.

1688 Zhang, W., Kraft, S., and Moore, J.S. (2004). Highly active trialkoxyMo(VI)(alkylidyne) catalysts synthesized by a reductive recycle strategy. *J. Am. Chem. Soc.* 126 (1): 329–335.

1689 Cummins, C.C. (1998). Reductive cleavage and related reactions leading to Mo-element multiple bonds: new pathways offered by three-coordinate Mo(III). *J. Chem. Soc., Chem. Commun.* (17): 1777–1786.

1690 Heppekausen, J., Stade, R., Goddard, R., and Fürstner, A. (2010). Practical new silyloxy-based alkyne metathesis catalysts with optimized activity and selectivity profiles. *J. Am. Chem. Soc.* 132: 11045–11057.

1691 Fürstner, A. (2013). Alkyne metathesis on the rise. *Angew. Chem. Int. Ed.* 52: 2794–2819.

1692 Lhermet, R. and Fürstner, A. (2014). Cross-metathesis of terminal alkynes. *Chem. Eur. J.* 20: 13188–13193.

1693 (a) Fürstner, A. (2014). Catalysis for total synthesis: a personal account. *Angew. Chem. Int. Ed.* 53: 8587–8598. (b) Schaubach, S., Gebauer, K., Ungeheuer, F. et al. (2016). A two-component alkyne metathesis catalyst system with an improved substrate scope and functional group tolerance: development and applications to natural product synthesis. *Chem. Eur. J.* 22: 8494–8507.

1694 Bindl, M., Stade, R., Heilmann, E.K. et al. (2009). Molybdenum nitride complexes with Ph_3SiO ligands are exceedingly practical and tolerant pre-catalysts for alkyne metathesis and efficient nitrogen transfer agents. *J. Am. Chem. Soc.* 131 (27): 9468–9470.

1695 (a) Geyer, A.M., Gdula, R.L., Wiedner, E.S., and Johnson, M.J.A. (2007). Catalytic nitrile-alkyne cross-metathesis. *J. Am. Chem. Soc.* 129 (13): 3800–3801. (b) Boyd, J.P., Schlangen, M., Grohmann, A., and Schwarz, H. (2008). Room-temperature alkyne - nitrile metathesis and unambiguous proof for the existence of a high-valent iron-nitrido dication in the gas phase. *Helv. Chim. Acta* 91 (8): 1430–1434. (c) Wiedner, E.S., Gallagher, K.J., Johnson, M.J.A., and Kampf, J.W. (2011). Synthesis of molybdenum nitrido complexes for triple-bond metathesis of alkynes and nitriles. *Inorg. Chem.* 50 (13): 5936–5945.

1696 Geyer, A.M., Wiedner, E.S., Gary, J.B. et al. (2008). Synthetic, mechanistic, and computational investigations of nitrile/alkyne cross-metathesis. *J. Am. Chem. Soc.* 130 (28): 8984–8999.

1697 Zhang, W. and Moore, J.S. (2004). Arylene ethynylene macrocycles prepared by precipitation-driven alkyne metathesis. *J. Am. Chem. Soc.* 126 (40): 12796–12796.

1698 Seyferth, D. (2001). Zinc alkyls, Edward Frankland, and the beginnings of main-group organometallic chemistry. *Organometallics* 20: 2940–2955.

1699 Thayer, J.S. (1975). Organometallic chemistry. Historical perspective. *Adv. Organomet. Chem.* 13: 1–45.

1700 Löwig, C. (1852). Über Zinnäthyle. *Justus Liebigs Ann. Chem.* 84: 308–333.

1701 Caseri, W. (2014). Initial organotin chemistry. *J. Organomet. Chem.* 751: 20–24.

1702 Hallwachs, W. and Schafarik, A. (1859). Ueber die Verbindungen der Erdmetalle mit organischen Radicalen. *Liebigs Ann. Chem.* 109 (2): 206–209.

1703 Ziegler, K., Holzkamp, E., Breil, H., and Martin, H. (1955). The Mulheim normal pressure polyethylene process. *Angew. Chem.* 67 (19-2): 541–547.

1704 Natta, G. (1956). Stereospecific catalyses and isotactic polymers. *Angew. Chem.* 68 (12): 393–403.

1705 Ziegler, K. (1960). *Organometallic Chemistry.* New York: Rheinhold.

1706 (a) (1975). Molecukar Catalysis A: Chem. New York. (b) Wilke, G. (2003). Fifty years of Ziegler catalyst: consequences and development of an invention. *Angew. Chem. Int. Ed.* 42 (41): 5000–5008.

1707 (a) Boor, J. (1979). *Ziegler-Natta Catalysts and Polymerizations.* New York: Academic Press. (b) Hoff, R. and Mathers, R.T. (eds.) (2010). *Handbook of Transition Metal Polymerization Catalysts (one-line ed.).* Wiley.

1708 Takasu, A., Ishii, M., Inai, Y., and Hirabayashi, T. (2001). Highly *threo* diastereoselective anionic polymerization of (*E,E*)-methyl sorbate catalyzed by a bulky organoaluminum Lewis acid. *Macromolecules* 34 (19): 6548–6550.

1709 Friedel, C. and Crafts, J.M. (1863). Über einige neue organische Verbindungen des Siliciums und das Atomgewicht dieses Elementes. *Liebigs Ann. Chem.* 127: 28–32.

1710 Barbier, P. (1899). Synthesis of dimethylheptenol. *C.R. Hebd. Séances Acad. Sci.* 128: 110–111.

1711 Grignard, V. (1900). Sur quelques nouvelles combinaisons organométalliques du magnésium et leur application à des synthèses d'alcools et d'hydrocarbures. *C.R. Hebd. Séances Acad. Sci.* 130: 1322–1324.

1712 Lindsell, W.E. (1982). Magnesium, calcium, strontium and barium. In: *Comprehensive Organometallic Chemistry* (ed. G. Wilkinson, F.G.A. Stone and E.W. Abel), 155–252. Amsterdam, The Netherlands: Elsevier Science Ltd.

1713 (a) Richey, H.G. Jr., (ed.) (2000). *Grignard Reagents New Developments*, vol. 4, 1–418. Chichester: Wiley. (b) Bartolo, N.D., Read, J.A., Valentin, E.M., and Woerpel, K.A. (2017). Reactions of allylmagnesium halides with carbonyl compounds: reactivity, structure, and mechanism. *Synthesis* 49 (15): 3237–3246. (c) Westerhausen, M., Koch, A., Goerls, H., and Krieck, S. (2017). Heavy Grignard reagents: synthesis, physical and structural properties, chemical behavior, and reactivity. *Chem. Eur. J.* 23 (7): 1456–1483.

1714 Seyferth, D. (2009). The Grignard reagents. *Organometallics* 28 (6): 1598–1605.

1715 Schlenk, W. and Holtz, J. (1917). The simplest metal-organic alkali compounds. *Ber. Dtsch. Chem. Ges.* 50: 262–274.

1716 Ziegler, K. and Colonius, H. (1930). Experiments on the alkali-organic compounds. V. A convenient synthesis of a simple lithium alkyle. *Liebigs Ann. Chem.* 479: 135–149.

1717 Mallan, J.M. and Bebb, R.L. (1969). Metalations by organolithium compounds. *Chem. Rev.* 69 (5): 693–755.

1718 Bickelhaupt, F. (1994). Organomagnesium chemistry - nearly 100 years but still fascinating. *J. Organomet. Chem.* 475 (1–2): 1–14.

1719 Wakefield, B.J. (1995). *Organomagnesium Methods in Organic Synthesis*, 1–249. London: Academic Press.

1720 Hofmann, R.W. (2003). The quest for chiral Grignard reagents. *Chem. Soc. Rev.* 32 (4): 225–230.

1721 Knochel, P., Dohle, W., Gommermann, N. et al. (2003). Highly functionalized organomagnesium reagents prepared through halogen-metal exchange. *Angew. Chem. Int. Ed.* 42 (36): 4302–4320.

1722 Garst, J.E. and Soriaga, M.R. (2004). Grignard reagent formation. *Coord. Chem. Rev.* 248 (7–8): 623–652.

1723 Sapountzis, L., Dube, H., Lewis, R. et al. (2005). Synthesis of functionalized nitroarylmagnesium halides via an iodine-magnesium exchange. *J. Org. Chem.* 70 (7): 2445–2454.

1724 Teerlinck, C.E. and Bowyer, W.J. (1996). Reactivity of magnesium surfaces during the formation of Grignard reagents. *J. Org. Chem.* 61 (3): 1059–1064.

1725 Rogers, H.R., Hill, C.L., Fujiwara, Y. et al. (1980). Mechanism of formation of Grignard-reagents - kinetics of reaction of alkyl halides in diethyl ether with magnesium. *J. Am. Chem. Soc.* 102 (1): 217–226.

1726 Garst, J.F., Deutch, J.E., and Whitesides, G.M. (1986). Radical isomerization during Grignard-reagent formation - a quantitative treatment. *J. Am. Chem. Soc.* 108 (9): 2490–2491.

1727 Ashby, E.C. and Oswald, J. (1988). Concerning the mechanism of Grignard-reagent formation - evidence for radical escape and return to the surface of Mg. *J. Org. Chem.* 53 (26): 6068–6076.

1728 Silverman, G.S. and Rakita, P.E. (eds.) (1996). *Handbook of Grignard Reagents*, 1–708. Monticello, NY: Marcel Dekker, Inc.

1729 Walling, C. and Cioffari, A. (1970). Radical cyclization during autoxidation of 5-hexenylmagnesium bromide. *J. Am. Chem. Soc.* 92 (22): 6609–6611.

1730 Garst, J.F. and Swift, B.L. (1989). Mechanism of Grignard reagent formation - comparisons of D-model calculations with experimental product yields. *J. Am. Chem. Soc.* 111 (1): 241–250.

1731 Walling, C. (1991). The nature of radicals involved in Grignard reagent formation. *Acc. Chem. Res.* 24 (9): 255–256.

1732 Garst, J.F., Ungvary, F., Batlaw, R., and Lawrence, K.E. (1991). Solvent attack in Grignard reagent formation from bromocyclopropane and 1-bromohexane in diethyl ether. *J. Am. Chem. Soc.* 113 (14): 5392–5397.

1733 Mori, T. and Kato, S. (2009). Grignard reagents in solution: theoretical study of the equilibria and the reaction with a carbonyl compound in diethyl ether solvent. *J. Phys. Chem. A* 113 (21): 6158–6165.

1734 Fürstner, A. (1993). Chemistry of and with highly reactive metals. *Angew. Chemi. Int. Ed. Engl.* 32 (2): 164–189.

1735 Bogdanovic, B., Bons, P., Konstantinovic, S. et al. (1993). Diorganomagnesium compounds from Mg, hydrogen, and 1-alkenes and their application in synthesis. *Chem. Ber.* 126 (6): 1371–1383.

1736 Rieke, R.D. and Bales, S.E. (1974). Activated metals. 4. Preparation and reactions of highly reactive Mg metal. *J. Am. Chem. Soc.* 96 (6): 1775–1781.

1737 Rieke, R.D. (1977). Preparation of highly reactive metal powders and their use in organic and organometallic synthesis. *Acc. Chem. Res.* 10 (8): 301–306.

1738 Luche, J.L. and Damiano, J.C. (1980). Ultrasounds in organic syntheses. 1. Effect on the formation of lithium organometallic reagents. *J. Am. Chem. Soc.* 102 (27): 7926–7927.

1739 Baker, K.V., Brown, J.M., Hughes, N. et al. (1991). Mechanical activation of Mg turnings for the preparation of reactive Grignard reagents. *J. Org. Chem.* 56 (2): 698–703.

1740 Beck, C.M., Park, Y.J., and Crabtree, R.H. (1998). Direct conversion of perfluoroalkanes and perfluoroarenes to perfluoro Grignard reagents. *J. Chem. Soc., Chem. Commun.* (6): 693–694.

1741 Krasovskiy, A. and Knochel, P. (2004). A LiCl-mediated Br/Mg exchange reaction for the preparation of functionalized aryl- and heteroarylmagnesium compounds from organic bromides. *Angew. Chem. Int. Ed.* 43 (25): 3333–3336.

1742 Ren, H.J., Krasovskiy, A., and Knochel, P. (2005). Preparation of cyclic alkenylmagnesium reagents via an iodine/magnesium exchange. *J. Chem. Soc., Chem. Commun.* (4): 543–545.

1743 Krasovskiy, A., Krasovskaya, V., and Knochel, P. (2006). Mixed Mg/Li amides of the type $R_2NMgCl{\cdot}LiCl$ as highly efficient bases for the regioselective generation of functionalized aryl and (heteroaryl)Mg compounds. *Angew. Chem. Int. Ed.* 45 (18): 2958–2961.

1744 Krasovskiy, A., Straub, B.F., and Knochel, P. (2006). Highly efficient reagents for Br/Mg exchange. *Angew. Chem. Int. Ed.* 45 (1): 159–162.

1745 Hauk, D., Lang, S., and Murso, A. (2006). Minimization of side reactions in bromine magnesium exchanges with *i*-PrMgCl/LiCl and *s*-BuMgCl/LiCl mixtures. *Org. Process Res. Dev.* 10 (4): 733–738.

1746 Hirner, S., Paenknin, O., Edefuhr, M., and Somfai, P. (2008). Synthesis of aryl glycines by the α-arylation of Weinreb amides. *Angew. Chem. Int. Ed.* 47 (10): 1907–1909.

1747 Rauhut, C.B., Melzig, L., and Knochel, P. (2008). *Meta-* and *para*-difunctionalization of arenes via a sulfoxide/Mg exchange reaction. *Org. Lett.* 10 (17): 3891–3894.

1748 Duan, X.F., Ma, Z.Q., Zhang, F., and Zhang, Z.B. (2009). Magnesiation of pyridine *N*-oxides via iodine or bromine/Mg exchange: a useful tool for functionalizing pyridine *N*-oxides. *J. Org. Chem.* 74 (2): 939–942.

1749 Krasovskiy, A., Tishkov, A., del Amo, V. et al. (2006). Transition-metal-free homocoupling of organomagnesium compounds. *Angew. Chem. Int. Ed.* 45 (30): 5010–5014.

1750 Martin, R. and Buchwald, S.L. (2007). Pd-catalyzed Kumada-Corriu cross-coupling reactions at low temperatures allow the use of Knochel-type Grignard reagents. *J. Am. Chem. Soc.* 129 (13): 3844–3845.

1751 Tamura, M. and Kochi, J.K. (1971). Reactions of Grignard reagents with transition metal halides - coupling, disproportionation, and exchange with olefins. *Bull. Chem. Soc. Jpn.* 44: 3063–3073.

1752 Shirakawa, E., Ikeda, D., Masui, S. et al. (2012). Iron-copper cooperative catalysis in the reactions of alkyl Grignard reagents: exchange reaction with alkenes and carbometalation of alkynes. *J. Am. Chem. Soc.* 134: 272–279.

1753 Ilies, L., Yoshida, T., and Nakamura, E. (2012). Iron-catalyzed chemo- and stereoselective hydromagnesiation of diarylalkynes and diynes. *J. Am. Chem. Soc.* 134: 16951–16954.

1754 Fraenkel, G., Dix, D.T., and Adams, D.G. (1964). Inversion rate in a primary Grignard reagent. *Tetrahedron Lett.* (43–44): 3155–3158.

1755 Maercker, A. and Geuss, R. (1971). Inversion of secondary cyclic Grignard compounds. *Angew. Chem. Int. Ed.* 10 (4): 270–271.

1756 Richey, H.G. Jr., and Bension, R.M. (1980). Stereochemistry of addition of allylic Grignard reagents to 3-(hydroxymethyl)cyclopropenes. *J. Org. Chem.* 45 (25): 5036–5042.

1757 Nakamura, M., Isobe, H., and Nakamura, E. (2003). Cyclopropenone acetals - synthesis and reactions. *Chem. Rev.* 103 (4): 1295–1326.

1758 Rubin, M., Rubina, M., and Gevorgyan, V. (2006). Recent advances in cyclopropene chemistry. *Synthesis* (8): 1221–1245.

1759 Richey, H.G. and Wilkins, C.W. (1980). Stereochemistry of addition of the allyl Grignard reagent to hydroxybicyclo[2.2.1]hept-2-enes. *J. Org. Chem.* 45: 5027–5036.

1760 Richey, H.G., Wilkins, C.W., and Bension, R.M. (1980). Addition of *n*-butyllithium to hydroxybicyclo[2.2.1]hept-2-enes. *J. Org. Chem.* 45: 5042–5047.

1761 Schlenk, W. and Schlenk, W. Jr., (1929). The constitution of Grignard Mg compounds. *Ber. Dtsch. Chem. Ges.* 62: 920–924.

1762 Evans, D.F. and Fazakerl, V. (1968). Proton resonance studies of nature of aryl Grignard reagents in solution. *J. Chem. Soc., Chem. Commun.* (16): 974–975.

1763 Parris, G.E. and Ashby, E.C. (1971). Composition of Grignard compounds. 7. Composition of methylmagnesium and *tert*-butylmagnesium halides and their dialkylmagnesium analogs in diethyl ether and tetrahydrofuran as inferred from nuclear magnetic resonance spectroscopy. *J. Am. Chem. Soc.* 93 (5): 1206–1213.

1764 Liu, Y.D., S, C., Yu, S.L. et al. (2010). Catalytic highly enantioselective alkylation of aldehydes with deactivated Grignard reagents and synthesis of bioactive intermediate secondary arylpropanols. *J. Org. Chem.* 75: 6869–6878.

1765 Zong, H., Huang, H.Y., Liu, J.F. et al. (2012). Added-metal-free catalytic nucleophilic addition of Grignard reagents to ketones. *J. Org. Chem.* 77: 4645–4652.

1766 Ashby, E.C. (1980). A detailed description of the mechanism of reaction of Grignard reagents with ketones. *Pure Appl. Chem.* 52 (3): 545–569.

1767 Axten, J., Troy, J., Jiang, P. et al. (1994). An ab initio molecular-orbital study of the Grignard reagents CH_3MgCl and $[CH_3MgCl]_2$ - the Schlenk equilibrium. *Struct. Chem.* 5 (2): 99–108.

1768 Ashby, E.C., Laemmle, J., and Neumann, H.M. (1974). Mechanisms of Grignard reagent addition to ketones. *Acc. Chem. Res.* 7 (8): 272–280.

1769 Nugent, W.A., Bertini, F., and Kochi, J.K. (1974). Electron transfer from organolithium and Grignard reagents - reactions with peroxides. *J. Am. Chem. Soc.* 96: 4945–4954.

1770 Maruyama, K. and Katagiri, T. (1986). Mechanism of the Grignard reaction - reaction of benzil. *J. Am. Chem. Soc.* 108 (20): 6263–6270.

1771 Yamataka, H., Matsuyama, T., and Hanafusa, T. (1989). Rate determining step of the reactions of benzophenone with various Grignard reagents. *J. Am. Chem. Soc.* 111 (13): 4912–4918.

1772 Yamazaki, S. and Yamabe, S. (2002). A computational study on addition of Grignard reagents to carbonyl compounds. *J. Org. Chem.* 67 (26): 9346–9353.

1773 Gajewski, J.J., Bocian, W., Brichford, N.L., and Henderson, J.L. (2002). Secondary deuterium kinetic isotope effects in irreversible additions of allyl reagents to benzaldehyde. *J. Org. Chem.* 67 (12): 4236–4240.

1774 Gajewski, J.J., Bocian, W., Harris, N.J. et al. (1999). Secondary deuterium kinetic isotope effects in irreversible additions of hydride and carbon nucleophiles to aldehydes: a spectrum of transition states from complete bond formation to single electron transfer. *J. Am. Chem. Soc.* 121 (2): 326–334.

1775 Inoue, A., Shinokubo, H., and Oshima, K. (2000). Formation of organomagnesium compounds via EtMgBr-mediated radical cyclization of allyl β-iodoacetals. *Org. Lett.* 2 (5): 651–653.

1776 (a) Bodineau, N., Mattalia, J.M., Thimokhin, V. et al. (2000). Formation of Grignard reagents from aryl halides: effective radical probes hint at a nonparticipation of dianions in the mechanism. *Org. Lett.* 2 (15): 2303–2306. (b) Garst, J.F. and Ungvary, F. (2001). In Grignard reagent formation from cyclopropyl bromide in diethyl ether, trapping by DCPH is consistent with diffusing cyclopropyl radical intermediates. *Org. Lett.* 3 (4): 605–606.

1777 Collados, J.F., Sola, R., Harutyunyan, S.R., and Macia, B. (2016). Catalytic synthesis of enantiopure chiral alcohols via addition of Grignard reagents to carbonyl compounds. *ACS Catal.* 6 (3): 1952–1970.

1778 Kharasch, M.S. and Tawney, P.O. (1941). Factors determining the course and mechanisms of Grignard reactions. II. The effect of metallic compounds on the reaction between isophorone and methylmagnesium bromide. *J. Am. Chem. Soc.* 63: 2308–2315.

1779 House, H.O., Respess, W.L., and Whitesides, G.M. (1966). Chemistry of carbanions. 12. Role of Cu in conjugate addition of organometallic reagents. *J. Org. Chem.* 31 (10): 3128–3141.

1780 Reetz, M.T. and Kindler, A. (1995). The Kharasch reaction revisited - CuX_3Li_2-catalyzed conjugate addition reactions of Grignard-reagents. *J. Organomet. Chem.* 502 (1–2): C5–C7.

1781 Gilman, H. and Kirby, R.H. (1941). Addition reactions of organometallic compounds with conjugated systems. *J. Am. Chem. Soc.* 63: 2046–2048.

1782 Jansen, J.F.G.A. and Feringa, B.L. (1990). Enantioselective conjugate addition of Grignard reagents to enones catalyzed by chiral Zn(II) complexes. *J. Org. Chem.* 55 (13): 4168–4175.

1783 Zhou, Q.L. and Pfaltz, A. (1994). Chiral mercaptoaryl-oxazolines as ligands in enantioselective Cu-catalyzed 1,4-additions of Grignard reagents to enones. *Tetrahedron* 50 (15): 4467–4478.

1784 Stangeland, E.L. and Sammakia, T. (1997). New chiral ligands for the asymmetric Cu-catalyzed conjugate addition of Grignard reagents to enones. *Tetrahedron* 53 (48): 16503–16510.

1785 Seebach, D., Jaeschke, G., Pichota, A., and Audergon, L. (1997). Enantioselective 1,4-addition of aliphatic Grignard reagents to enones catalyzed by readily available Cu(I) thiolates derived from TADDOL - preliminary communication. *Helv. Chim. Acta* 80 (8): 2515–2519.

1786 Feringa, B.L. (2000). Phosphoramidites: marvellous ligands in catalytic asymmetric conjugate addition. *Acc. Chem. Res.* 33 (6): 346–353.

1787 Feringa, B.L., Badorrey, R., Pena, D. et al. (2004). Copper-catalyzed asymmetric conjugate addition of Grignard reagents to cyclic enones. *Proc. Natl. Acad. Sci. U.S.A.* 101 (16): 5834–5838.

1788 López, F., Harutyunyan, S.R., Minnaard, A.J., and Feringa, B.L. (2004). Copper catalyzed enantioselective conjugate addition of Grignard reagents to acyclic enones. *J. Am. Chem. Soc.* 126 (40): 12784–12785.

1789 Lu, Z., Chai, G.B., and Ma, S.M. (2007). Iron-catalyzed highly regio- and stereoselective conjugate addition of 2,3-allenoates with Grignard reagents. *J. Am. Chem. Soc.* 129 (47): 14546–14547.

1790 Bogdanovic, B. and Schwickardi, M. (2000). Transition metal catalyzed preparation of Grignard compounds. *Angew. Chem. Int. Ed.* 39 (24): 4610–4612.

1791 Fürstner, A., Leitner, A., Méndez, M., and Krause, H. (2002). Iron-catalyzed cross-coupling reactions. *J. Am. Chem. Soc.* 124 (46): 13856–13863.

1792 Jansen, J.F.G.A. and Feringa, B.L. (1989). Zinc(II)-catalyzed conjugate addition of Grignard reagents to α,β-unsaturated ketones. *J. Chem. Soc., Chem. Commun.* (11): 741–742.

1793 Kjonaas, R.A. and Vawter, E.J. (1986). Regioselective 1,4-addition to α,β-unsaturated ketones with Grignard reagents mediated by (*N*,*N*,*N*′,*N*′-tetramethylethylenediamine)Zn(II) chloride. *J. Org. Chem.* 51 (21): 3993–3996.

1794 Marker, R.E., Crooks, H.M., and Sterols, C.X.L.I.V. (1942). Some 16-alkyl-pregnenolones and progesterones. *J. Am. Chem. Soc.* 64: 1280–1281.

1795 Solo, A.J. and Gardner, J.O. (1971). Agents for alkylating steroid hormone receptors. 2. 16α-substituted progesterone derivatives. *J. Med. Chem.* 14 (3): 222–225.

1796 Modi, S.P., Gardner, J.O., Milowsky, A. et al. (1989). Conjugate addition of Grignard reagents to enones and dienones. *J. Org. Chem.* 54 (10): 2317–2321.

1797 Harutyunyan, S.R., López, F., Browne, W.R. et al. (2006). On the mechanism of the Cu-catalyzed enantioselective 1,4-addition of Grignard reagents to α,β-unsaturated carbonyl compounds. *J. Am. Chem. Soc.* 128: 9103–9118.

1798 López, F., Minnaard, A.J., and Feringa, B.L. (2007). Catalytic enantioselective conjugate addition with Grignard reagents. *Acc. Chem. Res.* 40: 179–168.

1799 Ruiz, B.M., Geurts, K., Fernández-Ibáñez, M.A. et al. (2007). Highly versatile enantioselective conjugate addition of Grignard reagents to α,β-unsaturated thioesters. *Org. Lett.* 9: 5123–5126.

1800 Robert, T., Velder, J., and Schmalz, H.G. (2008). Enantioselective Cu-catalyzed 1,4-addition of Grignard reagents to cyclohexenone using Taddol-derived phosphine-phosphite ligands and 2-methyl-THF as a solvent. *Angew. Chem. Int. Ed.* 47 (40): 7718–7721.

1801 Bos, P.H., Minnaard, A.J., and Feringa, B.L. (2008). Catalytic asymmetric conjugate addition of Grignard reagents to α,β-unsaturated sulfones. *Org. Lett.* 10 (19): 4219–4222.

1802 Wang, S.Y. and Loh, T.P. (2010). Highly enantioselective Cu(I)-Tol-BINAP-catalyzed asymmetric conjugate addition of Grignard reagents to α,β-unsaturated esters. *J. Chem. Soc., Chem. Commun.* 46: 8694–8703.

1803 Teichert, J.F. and Feringa, B.L. (2011). Catalytic asymmetric conjugate addition of Grignard reagents to coumarins: synthesis of versatile chiral building blocks. *J. Chem. Soc., Chem. Commun.* 47: 2679–2681.

1804 Naeemi, Q., Robert, T., Kranz, D.P. et al. (2011). Chiral phosphine-phosphite ligands in the enantioselective 1,4-addition of Grignard reagents to α,β-unsaturated carbonyl compounds. *Tetrahedron: Asymmetry* 22: 887–892.

1805 Bilcik, F., Drusan, M., Marak, J., and Sebesta, R. (2012). Enantioselective one-pot conjugate addition of Grignard reagents to cyclic enones followed by amidomethylation. *J. Org. Chem.* 77: 760–765.

1806 Mao, B., Fañanás-Mastral, M., and Feringa, B.L. (2013). Asymmetric conjugate addition of Grignard reagents to pyranones. *Org. Lett.* 15: 286–289.

1807 Reddy, P. and Bandichhor, R. (2013). Enantioselective Grignard addition to nitroolefin. *Tetrahedron Lett.* 54: 3911–3915.

1808 Schoonen, A.K., Fernández-Ibáñez, M.A., Fañanás-Mastral, M. et al. (2014). Chiral amides via Cu-catalysed enantioselective conjugate addition. *Org. Biomol. Chem.* 12: 36–41.

1809 Germain, N. and Alexakis, A. (2015). Formation of contiguous quaternary and tertiary stereocenters by sequential asymmetric conjugate addition of Grignard reagents to 2-substituted enones and Mg-enolate trapping. *Chem. Eur. J.* 21: 8597–8606.

1810 Schmid, T.E., Drissi-Amraoui, S., Crevisy, C. et al. (2015). Copper-catalyzed asymmetric conjugate addition of organometallic reagents to extended Michael acceptors. *Beilstein J. Org. Chem.* 11: 2418–2434.

1811 Tang, F.X. and Ye, J.Y. (2015). Recent Advances in Cu-catalyzed asymmetric conjugate addition of Grignard reagents. *Chin. J. Org. Chem.* 35: 1414–1427.

1812 Harutyunyan, S.R., den Hartog, T., Geurts, K. et al. (2008). Catalytic asymmetric conjugate addition and allylic alkylation with Grignard reagents. *Chem. Rev.* 108 (8): 2824–2852.

1813 Villacorta, G.M., Rao, C.P., and Lippard, S.J. (1988). Synthesis and reactivity of binuclear tropocoronand and related organo-Cu(I) complexes - catalytic enantioselective conjugate addition of Grignard reagents to 2-cyclohexen-1-one. *J. Am. Chem. Soc.* 110 (10): 3175–3182.

1814 Ahn, K.H., Klassen, R.B., and Lippard, S.J. (1990). Enantioselective conjugate addition of Grignard reagents to enones catalyzed by chiral cuprate complexes. *Organometallics* 9 (12): 3178–3181.

1815 Lambert, F., Knotter, D.M., Janssen, M.D. et al. (1991). On the way to chiral Cu(I)(arenethiolate) catalysts for the enantioselective conjugate addition of MeLi and MeMgI to benzylideneacetone. *Tetrahedron: Asymmetry* 2 (11): 1097–1100.

1816 Zhou, Q.L. and Pfaltz, A. (1993). Chiral mercaptoaryl-oxazolines as ligands in asymmetric catalysis - enantioselective Cu-catalyzed 1,4-addition of Grignard reagents to α,β-unsaturated ketones. *Tetrahedron Lett.* 34 (48): 7725–7728.

1817 Germain, N., Magrez, M., Kehrli, S. et al. (2012). Chiral NHC ligands for the Cu-catalyzed asymmetric conjugate addition of Grignard reagents. *Eur. J. Org. Chem.* 5301–5306.

1818 Li, J. and Xu, L. (2015). Novel C_2-symmetric six-membered NHCs for asymmetric Cu-catalysed conjugate addition of Grignard reagents to 3-methylhexenone. *Tetrahedron* 71: 2858–2862.

1819 Heravi, M.M., Dehghani, M., and Zadsirjan, V. (2016). Rh-catalyzed asymmetric 1,4-addition reactions to α,β-unsaturated carbonyl and related compounds: an update. *Tetrahedron: Asymmetry* 27: 513–588.

1820 Matsumoto, Y., Yamada, K.I., and Tomioka, K. (2008). C_2 symmetric chiral NHC ligand for asymmetric quaternary carbon constructing Cu-catalyzed conjugate addition of Grignard reagents to 3-substituted cyclohexenones. *J. Org. Chem.* 73 (12): 4578–4581.

1821 Goncalves-Contal, S., Gremaud, L., Palais, L. et al. (2016). Cu-catalyzed enantioselective conjugate addition to α,β-unsaturated aldehydes with various organometallic reagents. *Synthesis* 48: 3301–3308.

1822 Jumde, R.P., Lanza, F., Veenstra, M.J., and Harutyunyan, S.R. (2016). Catalytic asymmetric addition of Grignard reagents to alkenyl substituted aromatic N-heterocycles. *Science* 352: 433–437.

1823 Yamamoto, Y., Yamamoto, S., Yatagai, H. et al. (1982). Lewis acid mediated reactions of organocopper reagents - entrainment in the conjugate addition to α,β-unsaturated ketones, esters, and acids via the $RCu{\cdot}BF_3$ system. *J. Org. Chem.* 47 (1): 119–126.

1824 Barbot, F., Kadib-Elban, A., and Miginiac, P. (1983). 1,4-Addition of Grignard reagents and 1,6-addition of Li organocuprates to *N,N*-diethylsorbamide. *Tetrahedron Lett.* 24 (46): 5089–5090.

1825 Corey, E.J., Kim, C.U., Chen, R.H.K., and Takeda, M. (1972). Method for stereospecific synthesis of 1,3-dienes and 1,4-dienes via organocopper reagents. *J. Am. Chem. Soc.* 94 (12): 4395–4396.

1826 Corey, E.J. and Chen, R.H.K. (1973). Addition of organocopper reagents to δ-2,4-dienoic esters - highly stereoselective route to trisubstituted and tetrasubstituted olefins. *Tetrahedron Lett.* 14 (18): 1611–1614.

1827 Ganem, B. (1974). Conjugate addition of (1-trialkylsilylpropynyl)Cu derivatives - new synthesis of 1,5-allenic and acetylenic olefins. *Tetrahedron Lett.* 15 (51/52): 4467–4470.

1828 Wipf, P. and Grenon, M. (2006). Toward the total synthesis of lophotoxin - new methodologies and synthetic strategies. *Can. J. Chem.* 84 (10): 1226–1241.

1829 Barbot, F., Kadib-Elban, A., and Miginiac, P. (1983). Study of the action of organometallics on the ketone $CH_3—CO—CH{=}CHCH{=}CHCH_3$ - application to the preparation of β-ethylenic ketones from Li organocuprates. *J. Organomet. Chem.* 255 (1): 1–9.

1830 Corey, E.J. and Boaz, N.W. (1985). The reactions of combined organocuprate chlorotrimethylsilane reagents with conjugated carbonyl compounds. *Tetrahedron Lett.* 26 (49): 6019–6022.

1831 Liu, H., Gayo, L.M., Sullivan, R.W. et al. (1994). Synthetic utility of alkenylcyclobutenedione monoketals - Michael additions and the synthesis of a natural benzofuranosesquiterpenequinone. *J. Org. Chem.* 59 (12): 3284–3288.

1832 Ürdingen, M. and Krause, N. (2000). Preparative and mechanistic studies on Cu-mediated 1,6-addition reactions to a 2,4-dienone. *Tetrahedron* 56 (18): 2799–2804.

1833 Krause, N. and Gerold, A. (1997). Regio- and stereoselective syntheses with organocopper reagents. *Angew. Chem. Int. Ed. Engl.* 36 (3): 187–204.

1834 Meng, F.K., Li, X.B., Torker, S. et al. (2016). Catalytic enantioselective 1,6-conjugate additions of propargyl and allyl groups. *Nature* 537: 387–393.

1835 Fukuhara, K. and Urabe, H. (2005). Iron-catalyzed 1,6-addition of aryl Grignard reagents to 2,4-dienoates and -dienamides. *Tetrahedron Lett.* 46 (4): 603–606.

1836 Okada, S., Arayama, K., Murayama, R. et al. (2008). Iron-catalyst-switched selective conjugate addition of Grignard reagents: $\alpha,\beta,\gamma,\delta$-unsaturated amides as versatile templates for asymmetric three-component coupling processes. *Angew. Chem. Int. Ed.* 47 (36): 6860–6864.

1837 Hénon, H., Mauduit, M., and Alexakis, A. (2008). Regiodivergent 1,4 versus 1,6 asymmetric Cu-catalyzed conjugate addition. *Angew. Chem. Int. Ed.* 47 (47): 9122–9124.

1838 den Hartog, T., Harutyunyan, S.R., Font, D. et al. (2008). Catalytic enantioselective 1,6-conjugate addition of Grignard reagents to linear dienoates. *Angew. Chem. Int. Ed.* 47 (2): 398–401.

1839 Rong, J., Pellegrini, T., and Harutyunyan, S.R. (2016). Synthesis of chiral tertiary alcohols by CuI-catalyzed enantioselective addition of organomagnesium reagents to ketones. *Chem. Eur. J.* 22: 3558–3570.

1840 Tomioka, K., Yamada, K.-I., and Yamamoto, Y. (2014). Copper-catalyzed asymmetric addition reaction of imines. In: *Copper-Catalyzed Asymmetric Synthesis* (ed. A. Alexakis, N. Krause and S. Woodward), 239–266. Weinheim: Wiley-VCH.

1841 Rong, J.W., Collados, J.F., Ortiz, P. et al. (2016). Catalytic enantioselective addition of Grignard reagents to aromatic silyl ketimines. *Nat. Commun.* 7: 13780.

1842 Job, A. and Reich, R. (1924). Catalytic activation of ethylene by organometallic Ni. *C.R. Hebd. Séances Acad. Sci.* 179: 330–332.

1843 Normant, J.F. and Alexakis, A. (1981). Carbometallation (C-Metallation) of alkynes - stereospecific synthesis of alkenyl derivatives. *Synthesis* 841–870.

1844 Yamamoto, Y. and Asao, N. (1993). Selective reactions using allylic metals. *Chem. Rev.* 93 (6): 2207–2293.

1845 Fallis, A.G. and Forgione, P. (2001). Metal mediated carbometallation of alkynes and alkenes containing adjacent heteroatoms. *Tetrahedron* 57: 5899–5913.

1846 Flynn, A.B. and Ogilvie, W.W. (2007). Stereocontrolled synthesis of tetrasubstituted olefins. *Chem. Rev.* 107: 4698–4745.

1847 Itami, K. and Yoshida, J.I. (2008). Carbomagnesiation reactions. In: *The Chemistry of Organomagnesium Compounds* (ed. Z. Rappoport and I. Marek), 631–679. Chichester: Wiley.

1848 Murakami, K. and Yorimitsu, H. (2013). Recent advances in transition metal catalyzed intermolecular carbomagnesiation and carbozincation. *Beilstein J. Org. Chem.* 9: 278–302.

1849 Duboudin, J.G., Jousseaume, B., and Saux, A. (1978). Reactivity of organomagnesium with acetylenic compounds in presence of Ni-complex $(PPh_3)_2NiCl_2$. *J. Organomet. Chem.* 162: 209–222.

1850 Alexakis, A., Cahiez, G., and Normant, J.F. (1979). Vinyl organocopper compounds. 9. Addition of organic Cu-derivatives to acetylene. *J. Organomet. Chem.* 177: 293–298.

1851 Shirakawa, E., Yamagami, T., Kimura, T. et al. (2005). Arylmagnesiation of alkynes catalyzed cooperatively by Fe- and Cu-complexes. *J. Am. Chem. Soc.* 127: 17164–17165.

1852 Murakami, K., Ohmiya, H., Yorimitsu, H., and Oshima, K. (2007). Chromium-catalyzed arylmagnesiation of alkynes. *Org. Lett.* 9: 1569–1571.

1853 de Armas, J. and Hoveyda, A.H. (2001). Zr-catalyzed electrophilic carbomagnesation of aryl olefins. Mechanism-based control of Zr/Mg ligand exchange. *Org. Lett.* 3: 2097–2100.

1854 Yan, N., Liu, X.Z., and Fox, J.M. (2008). Facially selective and regioselective carbometalation of cyclopropenes by aryl Grignard reagents. *J. Org. Chem.* 73 (2): 563–568.

1855 Ilies, L., Yoshida, T., and Nakamura, E. (2014). Synthesis of polysubstituted enynes through Fe-catalyzed carbomagnesiation of conjugated diynes. *Synlett* 25: 527–530.

1856 Marek, I. (1999). Enantioselective carbometallation of unactivated olefins. *J. Chem. Soc., Perkin Trans. 1* 535–544.

1857 Terao, J., Kato, Y., and Kambe, N. (2008). Titanocene-catalyzed regioselective alkylation of styrenes with Grignard reagents using β-bromoethyl ethers, thioethers, or amines. *Chem. Asian J.* 3 (8–9): 1472–1478.

1858 Morken, J.P., Didiuk, M.T., and Hoveyda, A.H. (1993). Zirconium-catalyzed asymmetric carbomagnesation. *J. Am. Chem. Soc.* 115 (15): 6997–6998.

1859 (a) Hirose, T. and Kodama, K. (2014). Recent advances in organozinc reagents. In: *Comprehensive Organic Synthesis*, 2e, vol. 1 (ed. P. Knochel and G.A. Molander), 204–266. Amsterdam, The Netherlands: Elsevier, B. V. (b) Kim, J.H., Ko, Y.O., Bouffard, J., and Lee, S.-G. (2015). Advances in tandem reactions with organozinc reagents. *Chem. Soc. Rev.* 44 (8): 2489–2507. (c) Dilman, A.D. and Levin, V.V. (2016). Advances in the chemistry of organozinc reagents. *Tetrahedron Lett.* 57 (36): 3986–3992.

1860 (a) Knochel, P. and Singer, R.D. (1993). Preparation and reactions of polyfunctional organozinc reagents in organic synthesis. *Chem. Rev.* 93 (6): 2117–2188. (b) Knochel, P., Millot, N., Rodriguez, A.L., and Tucker, C.E. (2001). Preparation and applications of functionalized organozinc compounds. *Org. React.* 58: 417–731.

1861 Knochel, P. (1995). Stereoselective reactions mediated by functionalized diorganozincs. *Synlett* (5): 393–403.

1862 Knochel, P., Almena-Perea, J.J., and Jones, P. (1998). Organozinc mediated reactions. *Tetrahedron* 54 (29): 8275–8319.

1863 Knochel, P., Calaza, M.I., and Hupe, E. (2004). Carbon–carbon bond-forming reactions mediated by organozinc reagents. In:

Metal-Catalyzed Cross-Coupling, 2e (ed. A. de Meijere and F. Dietrich), 619–670. Weinheim: Wiley-VCH.

1864 Knochel, P. and Jones, P. (eds.) (1999). *Organozinc Reagents, A practical Approach*, 1–370. Oxford: University Press Oxford.

1865 Ramon, D.J. and Yus, M. (2004). Chiral tertiary alcohols made by catalytic enantioselective addition of unreactive Zn reagents to poorly electrophilic ketones? *Angew. Chem. Int. Ed.* 43 (3): 284–287.

1866 Rimkus, A. and Sewald, N. (2004). Conjugate addition of organozinc compounds to nitroolefins. *Synthesis* (1): 135–146.

1867 Bernhardt, S., Manolikakes, G., Kunz, T., and Knochel, P. (2011). Preparation of solid salt-stabilized functionalized organozinc compounds and their application to cross-coupling and carbonyl addition reactions. *Angew. Chem. Int. Ed.* 50: 9205–9209.

1868 Greshock, T.J., Moore, K.P., McClain, R.T. et al. (2016). Synthesis of complex drug-like molecules by the use of highly functionalized bench-stable organozinc reagents. *Angew. Chem. Int. Ed.* 55: 13714–13718.

1869 Hernán-Gómez, A., Herd, E., Hevia, E. et al. (2014). Organozinc pivalate reagents: segregation, solubility, stabilization, and structural insights. *Angew. Chem. Int. Ed.* 53: 2706–2710.

1870 Petrier, C., De Souza Barbosa, J., Dupuy, C., and Luche, J.L. (1985). Ultrasound in organic-synthesis. 7. Preparation of organozinc reagents and their Ni-catalyzed reactions with α,β-unsaturated carbonyl compounds. *J. Org. Chem.* 50 (26): 5761–5765.

1871 Luche, J.L. (1987). Organometallic sonochemistry - successes, problems and by-products. *Ultrasonics* 25 (1): 40–44.

1872 Aukrust, I.R., Nilsen, N.O., Romming, C., and Skattebøl, L. (1998). Reactions of myrtenylZn bromide with carbonyl compounds. Regio- and diastereo-selectivity. *Acta Chem. Scand.* 52 (4): 385–390.

1873 Durant, A., Delplancke, J.L., Libert, V., and Reisse, J. (1999). Sonoelectroreduction of metallic salts: a new method for the production of reactive metallic powders for organometallic reactions and its application in organozinc chemistry. *Eur. J. Org. Chem.* (11): 2845–2851.

1874 Miller, A.O., Krasnov, V.I., Peters, D. et al. (2000). Perfluorozinc aromatics by direct insertion of Zn into C—F or C—Cl bonds. *Tetrahedron Lett.* 41 (20): 3817–3819.

1875 Abramovitch, R.A. (1991). Applications of microwave-energy in organic chemistry - a review. *Org. Prep. Proced. Int.* 23 (6): 683–711.

1876 Hu, Y.L., Yu, J.H., Yang, S.Y. et al. (1999). Substitution reaction of nitro group on α-nitrostyrene by organozinc halides under microwave irradiation. *Synth. Commun.* 29 (7): 1157–1165.

1877 Perreux, L. and Loupy, A. (2001). A tentative rationalization of microwave effects in organic synthesis according to the reaction medium, and mechanistic considerations. *Tetrahedron* 57 (45): 9199–9223.

1878 Wilson, N.S. and Roth, G.P. (2002). Recent trends in microwave-assisted synthesis. *Curr. Opin. Drug Discov. Dev.* 5 (4): 620–629.

1879 Nüchter, M., Ondruschka, B., Bonrath, W., and Gum, A. (2004). Microwave assisted synthesis - a critical technology overview. *Green Chem.* 6 (3): 128–141.

1880 De la Hoz, A., Díaz-Ortiz, A., and Moreno, A. (2004). Selectivity in organic synthesis under microwave irradiation. *Curr. Org. Chem.* 8 (10): 903–918.

1881 Clegg, W., Dale, S.H., Hevia, E. et al. (2006). Pre-metalation structural insights into the use of alkali-metal-mediated zincation for directed *ortho*-metalation of a tertiary aromatic amide. *Angew. Chem. Int. Ed.* 45 (15): 2370–2374.

1882 Clegg, W., Dale, S.H., Harrington, R.W. et al. (2006). Post-metalation structural insights into the use of alkali-metal-mediated zincation for directed ortho-metalation of a tertiary aromatic amide. *Angew. Chem. Int. Ed.* 45 (15): 2374–2377.

1883 Wittig, G., Meyer, F.J., and Lange, G. (1951). Über das Verhalten von Diphenylmetallen als Komplexbildner. *Liebigs Ann. Chem.* 571 (3): 167–201.

1884 Mulvey, R.E., Mongin, F., Uchiyama, M., and Kondo, Y. (2007). Deprotonative metalation using ate compounds: synergy, synthesis, and structure building. *Angew. Chem. Int. Ed.* 46 (21): 3802–3824.

1885 Boudier, A., Bromm, L.O., Lotz, M., and Knochel, P. (2000). New applications of polyfunctional organometallic compounds in organic synthesis. *Angew. Chem. Int. Ed.* 39 (24): 4414–4435.

1886 Guijarro, A., Rosenberg, D.M., and Rieke, R.D. (1999). The reaction of active Zn with organic bromides. *J. Am. Chem. Soc.* 121 (17): 4155–4167.

1887 Zhu, L., Wehmeyer, R.M., and Rieke, R.D. (1991). The direct formation of functionalized alkyl(aryl)Zn halides by oxidative addition of highly reactive Zn with organic halides and their reactions with acid-chlorides, α,β-unsaturated ketones, and allylic, aryl, and vinyl halides. *J. Org. Chem.* 56 (4): 1445–1453.

1888 Rieke, R.D., Hanson, M.V., Brown, J.D., and Niu, Q.J. (1996). Direct formation of secondary and tertiary alkylzinc bromides and subsequent Cu(I)-mediated couplings. *J. Org. Chem.* 61 (8): 2726–2730.

1889 Berk, S.C., Yeh, M.C.P., Jeong, N., and Knochel, P. (1990). Preparation and reactions of functionalized benzylic organometallics of Zn and Cu. *Organometallics* 9 (12): 3053–3064.

1890 Jubert, C. and Knochel, P. (1992). Preparation of polyfunctional nitro olefins and nitroalkanes using the Cu-Zn reagents RCu(CN)ZnI. *J. Org. Chem.* 57 (20): 5431–5438.

1891 Krasovskiy, A., Malakhov, V., Gavryushin, A., and Knochel, P. (2006). Efficient synthesis of functionalized organozinc compounds by the direct insertion of Zn into organic iodides and bromides. *Angew. Chem. Int. Ed.* 45 (36): 6040–6044.

1892 Boudet, N., Sase, S., Sinha, P. et al. (2007). Directed *ortho* insertion (DoI): a new approach to functionalized aryl and heteroaryl Zn reagents. *J. Am. Chem. Soc.* 129 (41): 12358–12359.

1893 Stadtmüller, H., Vaupel, A., Tucker, C.E. et al. (1996). Stereoselective preparation of polyfunctional cyclopentane derivatives by radical Ni- or Pd-catalyzed carbozincations. *Chem. Eur. J.* 2 (10): 1204–1220.

1894 Stadtmüller, H., Lentz, R., Tucker, C.E. et al. (1993). Pd-catalyzed iodine/Zn exchange reactions - a new Pd-mediated intramolecular carbozincation of alkenes. *J. Am. Chem. Soc.* 115 (15): 7027–7028.

1895 Rozema, M.J., Eisenberg, C., Lütjens, H. et al. (1993). Enantioselective preparation of polyfunctional secondary allylic alcohols using functionalized dialkylzincs prepared by a Cu(I)-catalyzed iodine/Zn exchange reaction. *Tetrahedron Lett.* 34 (19): 3115–3118.

1896 Rozema, M.J., Sidduri, A.R., and Knochel, P. (1992). Preparation of functionalized dialkylzinc reagents via an Iodine/Zn exchange reaction - highly enantioselective synthesis of functionalized secondary alcohols. *J. Org. Chem.* 57 (7): 1956–1958.

1897 Srebnik, M. (1991). Stereospecific preparation of trisubstituted allylic alcohols by alkene transfer from B to Zn. *Tetrahedron Lett.* 32 (22): 2449–2452.

1898 Oppolzer, W. and Radinov, R.N. (1992). Catalytic asymmetric synthesis of secondary (*E*)-allyl alcohols from acetylenes and aldehydes via (1-alkenyl)Zn intermediates - preliminary communication. *Helv. Chim. Acta* 75 (1): 170–173.

1899 Oppolzer, W. and Radinov, R.N. (1993). Synthesis of (*R*)-(−)-muscone by an asymmetrically catalyzed macrocyclization of an ω-alkynal. *J. Am. Chem. Soc.* 115 (4): 1593–1594.

1900 Langer, F., Schwink, L., Devasagayaraj, A. et al. (1996). Preparation of functionalized dialkylzincs via a B/Zn exchange. Reactivity and catalytic asymmetric addition to aldehydes. *J. Org. Chem.* 61 (23): 8229–8243.

1901 Kitamura, M., Miki, T., Nakano, K., and Noyori, R. (1996). Conjugate addition of diorganozincs to α,β-unsaturated ketones catalyzed by a Cu(I)-sulfonamide combined system. *Tetrahedron Lett.* 37: 5141–5144.

1902 Marshall, J.A., Herold, M., Eidam, H.S., and Eidam, P. (2006). Pd- and Cu-catalyzed 1,4-additions of organozinc compounds to conjugated aldehydes. *Org. Lett.* 8: 5505–5508.

1903 Thaler, T. and Knochel, P. (2009). Copper-catalyzed asymmetric Michael addition of Mg, Zn, and Al organometallic reagents: efficient synthesis of chiral molecules. *Angew. Chem. Int. Ed.* 48: 645–648.

1904 Jerphagnon, T., Pizzuti, M.G., Minnaard, A.J., and Feringa, B.L. (2009). Recent advances in enantioselective Cu-catalyzed 1,4-addition. *Chem. Soc. Rev.* 38: 1039–1075.

1905 Palais, L., Babel, L., Quintard, A. et al. (2010). Copper-catalyzed enantioselective 1,4-addition to α,β-unsaturated aldehydes. *Org. Lett.* 12: 1988–1991.

1906 Hawner, C. and Alexakis, A. (2010). Metal-catalyzed asymmetric conjugate addition reaction: formation of quaternary stereocenters. *J. Chem. Soc., Chem. Commun.* 46: 7295–7306.

1907 Wang, L., Meng, W., Zhu, C.L. et al. (2011). The long-arm effect: influence of axially chiral phosphoramidite ligands on the diastereo- and enantioselectivity of the tandem 1,4-addition/fluorination. *Angew. Chem. Int. Ed.* 50: 9442–9446.

1908 Maezaki, N., Sawamoto, H., Yoshigami, R. et al. (2003). Geometrically selective synthesis of functionalized β,β-disubstituted vinylic sulfoxides by

Cu-catalyzed conjugate addition of organozinc reagents to 1-alkynyl sulfoxides. *Org. Lett.* 5: 1345–1347.

1909 Sklute, G., Bolm, C., and Marek, I. (2007). Regio- and stereoselective Cu-catalyzed carbozincation reactions of alkynyl sulfoximines and sulfones. *Org. Lett.* 9: 1259–1261.

1910 Xie, M.H., Lin, G.F., Zhang, J.H. et al. (2010). Regio- and stereoselective synthesis of tetrasubstituted allylic alcohols by three-component reaction of acetylenic sulfone, dialkylzinc, and aldehyde. *J. Organomet. Chem.* 695: 882–886.

1911 Gourdet, B., Rudkin, M.E., Watts, C.A., and Lam, H.W. (2009). Preparation of multisubstituted enamides via Rh-catalyzed carbozincation and hydrozincation of ynamides. *J. Org. Chem.* 74: 7849–7858.

1912 Gourdet, B. and Lam, H.W. (2009). Stereoselective synthesis of multisubstituted enamides via Rh-catalyzed carbozincation of ynamides. *J. Am. Chem. Soc.* 131: 3802–3803.

1913 Studemann, T. and Knochel, P. (1997). New Ni-catalyzed carbozincation of alkynes: a short synthesis of (*Z*)-tamoxifen. *Angew. Chem. Int. Ed.* 36: 93–95.

1914 Cann, R.O., Waltermire, R.E., Chung, J.C. et al. (2010). Process development for a large scale stereoselective synthesis of (*Z*)-(1-bromobut-1-ene-1,2-diyl)dibenzene, a key intermediate of a selective estrogen receptor modulator. *Org. Process Res. Dev.* 14: 1147–1152.

1915 Wu, J.L. and Yoshikai, N. (2016). Cobalt-catalyzed alkenylzincation of unfunctionalized alkynes. *Angew. Chem. Int. Ed.* 55: 336–340.

1916 Ito, S., Itoh, T., and Nakamura, M. (2011). Diastereoselective carbometalation of oxa- and azabicyclic alkenes under Fe-catalysis. *Angew. Chem. Int. Ed.* 50: 454–457.

1917 Maruoka, K. and Yamamoto, H. (1988). Organoaluminums in organic synthesis. *Tetrahedron* 44 (16): 5001–5032.

1918 (a) Oishi, M. (2004). Product subclass 9: triorganoaluminum compounds. In: *Science of Synthesis, Thieme Chemistry*, vol. 7, (ed. D. Bellus, E. N. Jacobsen, S.V. Ley, R. Noyori, M. Regitz, P. J. Reider, E. Schaumann, I. Shinkai, E. J. Thomas, B. M. Trost) 261–385. Stuttgart: Georg Thieme Verlag. (b) Wehmschulte, R.J. (2013). Low valent organoaluminum (+I, +II) species. In: *Modern Organoaluminum Reagents*, Topics in Organometallic Chemistry, vol. 41, (ed. S. Woodward, S. Dagorne) 91–124. Berlin, Heidelberg: Springer-Verlag. (c) Knochel, P., Blümke, T., Groll, K., and Chen, Y.-H. (2013). Preparation of organoalanes for organic synthesis. In: *Topics in Organometallic Chemistry. Modern Organoaluminum Reagents*, vol. 41, (ed. S. Woodward, S. Dagorne) 173–186. Berlin, Heidelberg: Springer-Verlag.

1919 Equey, O. and Alexakis, A. (2004). Enantioselective opening of cyclic vinyl epoxides with organoaluminum reagents catalysed by Cu salts. *Tetrahedron: Asymmetry* 15 (10): 1531–1536.

1920 Witt, M. and Roesky, H.W. (2000). Organoaluminum chemistry at the forefront of research and development. *Curr. Sci.* (Indian Academy of Sciences, Satheesh, S. K. (ed.)) 78 (4): 410–430.

1921 (a) Sato, S. (2007). Aluminum. In: *Comprehensive Organometallic Chemistry III*, vol. vol. 9 (ed. D.M.P. Mingos and R.H. Crabtree), 245–296; (b)Eisch, J.J. (2015). Emergence of electrophilic alumination as the counterpart of established nucleophilic lithiation: an academic sojourn in organometallics with William Kaska as fellow traveler. *J. Chem. Soc. Dalton Trans.* 44 (15): 6671–6679.

1922 Lewinski, J. and Wheatley, A.E.H. (2013). Simple trivalent organoaluminum species: perspectives on structure, bonding, and reactivity. In: *Modern Organoaluminum Reagents*, Topics in Organometallic Chemistry, vol. 41 (ed. S. Woodward and S. Dagorne), 1–58. Heidelberg: Springer-Verlag.

1923 Okukado, N. and Negishi, E. (1978). One-step conversion of terminal acetylenes into terminally functionalized (*E*)-3-methyl-2-alkenes via Zr-catalyzed carboalumination - simple and selective route to terpenoids. *Tetrahedron Lett.* 2357–2360.

1924 van Horn, D.E. and Negishi, E. (1978). Selective carbon–carbon bond formation via transition-metal catalysis. 8. Controlled carbometalation - reaction of acetylenes with organoalane-tirconocene dichloride complexes as a route to stereo-defined and regio-defined trisubstituted olefins. *J. Am. Chem. Soc.* 100: 2252–2254.

1925 Rand, C.L., van Horn, D.E., Moore, M.W., and Negishi, E. (1981). Selective carbon–carbon bond formation via transition-metal catalysis. 19. A versatile and selective route to difunctional trisubstituted (*E*)-alkene synthons via Zr-catalyzed carboalumination of alkynes. *J. Org. Chem.* 46: 4093–4096.

1926 Yoshida, T. and Negishi, E. (1981). Selective carbon–carbon bond formation via transition-metal catalysis. 21. Mechanism of the Zr-catalyzed carboalumination of alkynes - evidence for direct carboalumination. *J. Am. Chem. Soc.* 103: 4985–4987.

1927 Negishi, E., Kondakov, D.Y., Choueiry, D. et al. (1996). Multiple mechanistic pathways for Zr-catalyzed carboalumination of alkynes. Requirements for cyclic carbometalation processes involving C—H activation. *J. Am. Chem. Soc.* 118: 9577–9588.

1928 Erker, G., Aulbach, M., Knickmeier, M. et al. (1993). The role of torsional isomers of planarly chiral nonbridged bis(indenyl)metal type complexes in stereoselective propene polymerization. *J. Am. Chem. Soc.* 115: 4590–4601.

1929 Kondakov, D.Y. and Negishi, E. (1995). Zirconium-catalyzed enantioselective methylalumination of monosubstituted alkenes. *J. Am. Chem. Soc.* 117: 10771–10772.

1930 Xu, S.Q. and Negishi, E. (2016). Zirconium-catalyzed asymmetric carboalumination of unactivated terminal alkenes. *Acc. Chem. Res.* 49: 2158–2168.

1931 Bennett, S.M.W., Brown, S.M., Muxworthy, J.P., and Woodward, S. (1999). Copper-catalysed asymmetric conjugate addition of organometallic reagents to linear enones using thiourethane ligands. *Tetrahedron Lett.* 40: 1767–1770.

1932 Pamies, O.; Dieges, M. Conjugate addition of organoaluminum species to Michael acceptors and related processes. Woodward, S.; Dagorne, S., Eds. Topics in Organometallic Chemistry, Springer-Verlag, Berlin, Heidelberg, 2013, Vol. 41 (*Modern Organoaluminum Reagents*), pp. 277–306.

1933 Endo, K., Hamada, D., Yakeishi, S., and Shibata, T. (2013). Effect of multinuclear Cu/Al complexes in highly asymmetric conjugate addition of Me_3Al to acyclic enones. *Angew. Chem. Int. Ed.* 52: 606–610.

1934 Dabrowski, J.A., Villaume, M.T., and Hoveyda, A.H. (2013). Enantioselective synthesis of quaternary carbon stereogenic centers through Cu-catalyzed conjugate additions of aryl- and alkylaluminum reagents to acyclic trisubstituted enones. *Angew. Chem. Int. Ed.* 52: 8156–8159.

1935 Cottet, P., Bleschke, C., Capdevila, M.G. et al. (2016). Catalytic enantioselective total synthesis of riccardiphenol B. *Adv. Synth. Catal.* 358: 417–425.

1936 Hawner, C., Li, K.Y., Cirriez, V., and Alexakis, A. (2008). Copper-catalyzed asymmetric conjugate addition of (aryl)Al reagents to trisubstituted enones: construction of aryl-substituted quaternary centers. *Angew. Chem. Int. Ed.* 47: 8211–8214.

1937 May, T.L., Dabrowski, J.A., and Hoveyda, A.H. (2011). Formation of vinyl-, vinylhalide- or acyl-substituted quaternary carbon stereogenic centers through NHC—Cu-catalyzed enantioselective conjugate additions of Si-containing vinylaluminums to β-substituted cyclic enones. *J. Am. Chem. Soc.* 133: 736–739.

1938 Hirner, S., Kolb, A., Westmeier, J. et al. (2014). Rhodium-catalyzed enantioselective addition of organoaluminum reagents to *N*-tosyl ketimines. *Org. Lett.* 16: 3162–3165.

1939 Zhu, G.G. and Negishi, E. (2008). 1,4-Pentenyne as a five-carbon synthon for efficient and selective syntheses of natural products containing 2,4-dimethyl-1-penten-1,5-ylidene and related moieties by means of Zr-catalyzed carboalumination of alkynes and alkenes. *Chem. Eur. J.* 14: 311–318.

1940 Kondakov, D.Y. and Negishi, E. (1996). Zirconium-catalyzed enantioselective alkylalumination of monosubstituted alkenes proceeding via noncyclic mechanism. *J. Am. Chem. Soc.* 118: 1577–1578.

1941 Eischenbroich, C. (2006). *Organometallics*, 3e, 1–817. Weinheim: Wiley-VCH.

1942 (a) Fernández, E. and Whiting, A. (eds.) (2015). *Synthesis and Application of Organoboron Compounds*, vol. 49, 1–331. Switzerland: Springer International Publishing. (b) Xu, L., Zhang, S., and Li, P.-F. (2015). Boron-selective reactions as powerful tools for modular synthesis of diverse complex molecule. *Chem. Soc. Rev.* 44 (24): 8848–8858. (c) Fyfe, J.W.B. and Watson, A.J.B. (2017). Recent developments in organoboron chemistry: old dogs, new tricks. *Chem* 3 (1): 31–55.

1943 Brown, H.C. and Subba Rao, B.C. (1956). A new technique for the conversion of olefins into organoboranes and related alcohols. *J. Am. Chem. Soc.* 78 (21): 5694–5695.

1944 Brown, H.C. and Subba Rao, B.C. (1957). Hydroboration of olefins - a remarkably fast room-temperature addition of diborane to olefins. *J. Org. Chem.* 22 (9): 1136–1137.

1945 Brown, H.C. (1962). *Hydroboration*, 1–290. New York: W. A. Benjamin Inc.

1946 Beletskaya, I. and Pelter, A. (1997). Hydroborations catalysed by transition metal complexes. *Tetrahedron* 53: 4957–5026.

1947 (a) Housecroft, C.E. (1995). Compounds with three- or four-coordinate boron, emphasing cyclic sytems. In: *Comprehensive Organometallic Chemistry II* (ed. E.W. Abel, R.G.A. Stone and G. Wilkinson), 129–195. Oxford: Pergamon. (b) Herberich, G.E. (1995). Boron rings LIgated to metals. In: *Comprehensive Organometallic Chemistry II* (ed. E.W. Abel, R.G.A. Stone and G. Wilkinson), 197–216. Oxford: Pergamon.

1948 Yamamoto, Y. and Miyaura, N. (2008). Metal-catalyzed reactions of organoboron compounds in organic syntheses. *J. Synth. Org. Chem. Jpn.* 66: 194–204.

1949 Hall, D.G. (2005). *Structure, Properties, and Preparation of Boronic Acid Derivatives. Overview of Their Reactions and Applications in Boronic Acids: Preparation ad Applications in Organic Synthesis and Medicine* (ed. D.G. Hall), 1–99. Weinheim: Wiley-VCH.

1950 Lou, S. and Schaus, S.E. (2008). Asymmetric Petasis reactions catalyzed by chiral biphenols. *J. Am. Chem. Soc.* 130: 6922–6923.

1951 Gao, F., Carr, J.L., and Hoveyda, A.H. (2012). Copper-catalyzed enantioselective allylic substitution with readily accessible carbonyl- and acetal-containing vinylboron reagents. *Angew. Chem. Int. Ed.* 51: 6613–6617.

1952 Yoshida, K. and Hayashi, T. (2005). *Rhodium-Catalyzed Additions of Boronic Acids to Alkenes and Carbonyl Compounds in Boronic Acids: Preparation and Applications in Organic Synthesis and Medicine* (ed. D.G. Hall), 171–203. Weinheim: Wiley-VCH.

1953 Trost, B.M. and Ball, Z.T. (2005). Addition of metalloid hydrides to alkynes: hydrometallation with boron, silicon, and tin. *Synthesis* 853–887.

1954 Tucker, C.E., Davidson, J., and Knochel, P. (1992). Mild and stereoselective hydroborations of functionalized alkynes and alkenes using pinacolborane. *J. Org. Chem.* 57: 3482–3485.

1955 Takahashi, K., Ishiyama, T., and Miyaura, N. (2001). A borylcopper species generated from bis(pinacolato)diboron and its additions to α,β-unsaturated carbonyl compounds and terminal alkynes. *J. Organomet. Chem.* 625: 47–53.

1956 Lee, J.E., Kwon, J., and Yun, J. (2008). Copper-catalyzed addition of diboron reagents to α,β-acetylenic esters: efficient synthesis of β-boryl-α,β-ethylenic esters. *J. Chem. Soc., Chem. Commun.* 733–734.

1957 Jang, H., Zhugralin, A.R., Lee, Y., and Hoveyda, A.H. (2011). Highly selective methods for synthesis of internal α-vinylboronates through efficient NHC—Cu-catalyzed hydroboration of terminal alkynes. Utility in chemical synthesis and mechanistic basis for selectivity. *J. Am. Chem. Soc.* 133: 7859–7871.

1958 Moure, A.L., Mauleon, P., Arrayas, R.G., and Carretero, J.C. (2013). Formal regiocontrolled hydroboration of un-biased internal alkynes via borylation/allylic alkylation of terminal alkynes. *Org. Lett.* 15: 2054–2057.

1959 Semba, K., Fujihara, T., Terao, J., and Tsuji, Y. (2012). Copper-catalyzed highly regio- and stereoselective directed hydroboration of unsymmetrical internal alkynes: controlling regioselectivity by choice of catalytic species. *Chem. Eur. J.* 18: 4179–4184.

1960 Park, J.K., Ondrusek, B.A., and McQuade, D.T. (2012). Regioselective catalytic hydroboration of propargylic species using Cu(I)—NHC complexes. *Org. Lett.* 14: 4790–4793.

1961 Moure, A.L., Arrayas, R.G., Cardeñas, D.J. et al. (2012). Regiocontrolled CuI-catalyzed borylation of propargylic-functionalized internal alkynes. *J. Am. Chem. Soc.* 134: 7219–7222.

1962 Alfaro, R., Parra, A., Alemán, J. et al. (2012). Copper(I)-catalyzed formal carboboration of alkynes: synthesis of tri- and tetrasubstituted vinylboronates. *J. Am. Chem. Soc.* 134: 15165–15168.

1963 Alfaro, R., Parra, A., Alemán, J., and Tortosa, M. (2013). New methods in organic synthesis through Cu-catalyzed borylation reactions: stereoselective synthesis of 1,4-diols and vinylboronates. *Synlett* 24: 804–812.

1964 Zhou, Y.Q., You, W., Smith, K.B., and Brown, M.K. (2014). Copper-catalyzed cross-coupling of boronic esters with aryl iodides and application to the carboboration of alkynes and allenes. *Angew. Chem. Int. Ed.* 53: 3475–3479.

1965 Chavant, P.Y. and Vaultier, M. (1993). Preparation of some organo-bis(diisopropylamino)boranes and their application to the synthesis of oxazaborolidines. *J. Organomet. Chem.* 455 (1–2): 37–46.

1966 Arthur, M.P., Goodwin, H.P., Baceiredo, A. et al. (1991). Synthesis of stable boryl-substituted diazomethane and nitrilimines. *Organometallics* 10 (9): 3205–3210.

1967 Kabalka, G.W., Sastry, U., Sastry, K.A.R. et al. (1983). Synthesis of arylboronic acids via the reaction of borane with arylmagnesium halides. *J. Organomet. Chem.* 259 (3): 269–274.

1968 Kabalka, G.W., Varma, R.S., Gai, Y.Z., and Baldwin, R.M. (1986). A new route to iodine-labeled *N*-isopropyl iodoamphetamine via organoboranes. *Tetrahedron Lett.* 27 (33): 3843–3844.

1969 Sharp, M.J. and Snieckus, V. (1985). Synthetic connections to the aromatic directed metalation reaction - unsymmetrical biaryls by Pd-catalyzed cross coupling of directed metalation-derived arylboronic acids with aryl halides. *Tetrahedron Lett.* 26 (49): 5997–6000.

1970 Altman, J., Boehnke, H., Steigel, A., and Wulff, G. (1986). *N*-Substituted-2-carboxamidophenylboronic acid anhydrides. *J. Organomet. Chem.* 309 (3): 241–246.

1971 Claudel, S., Gosmini, C., Paris, J.M., and Périchon, J. (2007). A novel transmetallation of arylzinc species into arylboronates from aryl halides in a Barbier procedure. *J. Chem. Soc., Chem. Commun.* (35): 3667–3669.

1972 Mikhail, I. and Kaufmann, D. (1990). Optically-active vinylboronates. *J. Organomet. Chem.* 398 (1–2): 53–57.

1973 Pubill-Ulldemolins, C., Bonet, A., Bo, C. et al. (2010). A new context for Pd-mediated B-addition reaction: an open door to consecutive functionalization. *Org. Biomol. Chem.* 8: 2667–2682.

1974 Ishiyama, T. and Miyaura, H. (1999). Synthesis of organoboron compounds via the transition metal-catalyzed addition and coupling reaction of (alkoxo)diborons. *J. Synth. Org. Chem. Jpn.* 57: 503–511.

1975 Zaidlewicz, M. and Meller, J. (1997). Syntheses with organoboranes. 7. Monohydroboration of conjugated dienes with catecholborane catalyzed by complexes of Ni(II) chloride and Co(II) chloride with diphosphines. *Tetrahedron Lett.* 38: 7279–7282.

1976 Ishiyama, T., Miyaura, N., and Suzuki, A. (1991). Palladium-catalyzed carbonylative cross-coupling reaction of iodoalkanes with 9-alkyl-9-BBN derivatives - a direct and selective synthesis of ketones. *Tetrahedron Lett.* 32 (47): 6923–6926.

1977 Ishiyama, T., Kizaki, H., Miyaura, N., and Suzuki, A. (1993). Synthesis of unsymmetrical biaryl ketones via Pd-catalyzed carbonylative cross-coupling reaction of arylboronic acids with iodoarenes. *Tetrahedron Lett.* 34 (47): 7595–7598.

1978 Zhu, Z.L., Zhang, W.Q., and Gao, Z.W. (2016). Suzuki-Miyaura carbonylative reaction in the synthesis of biaryl ketones. *Prog. Chem.* 28: 1626–1633.

1979 Sakai, M., Hayashi, H., and Myaura, N. (1997). Rhodium-catalyzed conjugate addition of aryl- or 1-alkenylboronic acids to enones. *Organometallics* 16 (20): 4229–4231, see also: Hargrave, J.D., Allen, J.C., and Frost, C.G. (2010). Alternatives to organoboron reagents in Rh-catalyzed conjugate additions. *Chem. Asian J.* 2010 (5): 386–396.

1980 Takaya, Y., Senda, T., Kurushima, H. et al. (1999). Rhodium-catalyzed asymmetric 1,4-addition of arylboron reagents to α,β-unsaturated esters. *Tetrahedron: Asymmetry* 10: 4047–4056.

1981 Takaya, Y., Ogasawara, M., and Hayashi, T. (1999). Rhodium-catalyzed asymmetric 1,4-addition of arylboron compounds generated in situ from aryl bromides. *Tetrahedron Lett.* 40: 6957–6961.

1982 Hayashi, T., Senda, T., Takaya, Y., and Ogasawara, M. (1999). Rhodium-catalyzed asymmetric 1,4-addition to 1-alkenylphosphonates. *J. Am. Chem. Soc.* 121: 11591–11592.

1983 Reetz, M.T., Moulin, D., and Gosberg, A. (2001). BINOL-based diphosphonites as ligands in the asymmetric Rh-catalyzed conjugate addition of arylboronic acids. *Org. Lett.* 3: 4083–4085.

1984 Navarre, L., Martinez, R., Genêt, J.P., and Darses, S. (2008). Access to enantioenriched α-amino esters via Rh-catalyzed 1,4-addition/enantioselective protonation. *J. Am. Chem. Soc.* 130: 6159–6169.

1985 Yoshida, M., Ohmiya, H., and Sawamura, M. (2012). Enantioselective conjugate addition of alkylboranes catalyzed by a Cu—N-heterocyclic carbene complex. *J. Am. Chem. Soc.* 134: 11896–11899.

1986 Sawamura, M., Ohmiya, H., and Yoshida, M. (2012). Enantioselective conjugate addition of alkylboranes catalyzed by a Cu—N-heterocyclic carbene complex. *Synthesis* 44: A115–A117.

1987 Hong, Y.C., Gandeepan, P., Mannathan, S. et al. (2014). Alkene-assisted Ni-catalyzed regioselective 1,4-addition of organoboronic acid to dienones: a direct route to all-carbon quaternary centers. *Org. Lett.* 16: 2806–2809.

1988 Wu, C.L., Yue, G.Z., Nielsen, C.D.T. et al. (2016). Asymmetric conjugate addition of organoboron reagents to common enones using Cu catalysts. *J. Am. Chem. Soc.* 138: 742–745.

1989 Shintani, R. and Hayashi, T. (2011). Rhodium-catalyzed asymmetric 1,4-addition of Na tetraarylborates to *β*,β-disubstituted *α*,β-unsaturated esters. *Org. Lett.* 13: 350–352.

1990 Patel, S.J. and Jamison, T.F. (2003). Catalytic three-component coupling of alkynes, imines, and organoboron reagents. *Angew. Chem. Int. Ed.* 42 (12): 1364–1367.

1991 Sommer, L.H. and Whitmore, F.C. (1946). Organo-silicon compounds. 3. α-Chloroalkyl and β-chloroalkyl silanes and the unusual reactivity of the latter. *J. Am. Chem. Soc.* 68 (3): 485–487.

1992 Sommer, L.H., Tyler, L.J., and Whitmore, F.C. (1948). Organosilicon compounds. 17. Reactions of allyltrimethylsilane. *J. Am. Chem. Soc.* 70 (9): 2872–2874.

1993 Sommer, L.H., Bailey, D.L., and Whitmore, F.C. (1948). Organosilicon compounds. 16. Further studies of β-eliminations involving Si. *J. Am. Chem. Soc.* 70 (9): 2869–2872.

1994 Kirby, A.J. (1996). *Stereoelectronic Effects*, vol. 36, 1–96. Oxford: Oxford University Press.

1995 Lambert, J.B. (1990). The Interaction of Si with positively charged carbon. *Tetrahedron* 46 (8): 2677–2689.

1996 Lambert, J.B. and Chelius, E.C. (1990). The β-effect of Si in the synperiplanar geometry. *J. Am. Chem. Soc.* 112 (22): 8120–8126.

1997 Fleming, I., Barbero, A., and Walter, D. (1997). Stereochemical control in organic synthesis using Si-containing compounds. *Chem. Rev.* 97 (6): 2063–2192.

1998 Stevens, B.D. and Nelson, S.G. (2005). Tandem Sakurai-aldol addition reactions as a route to structurally complex carbocycles. *J. Org. Chem.* 70 (11): 4375–4379.

1999 Yamamoto, H. and Boxer, M.B. (2007). Super Brønsted acid catalysis in organic synthesis. *Chimia* 61 (5): 279–281.

2000 Lee, P.H. (2007). Indium and gallium-mediated addition reactions. *Bull. Korean Chem. Soc.* 28 (1): 17–28.

2001 Ollevier, T. and Li, Z.Y. (2006). The first catalytic Sakurai reaction of *N*-alkoxycarbonylamino sulfones with allyltrimethylsilane. *Org. Biomol. Chem.* 4 (24): 4440–4443.

2002 Kampen, D. and List, B. (2006). Efficient Brønsted acid catalyzed Hosomi–Sakurai reaction of acetals. *Synlett* (16): 2589–2592.

2003 Williams, D.R., Morales-Ramos, A.I., and Williams, C.M. (2006). Reactivity studies of 3,3-bis(trimethylsilyl)-2-methyl-1-propene in Lewis acid-catalyzed allylation reactions. *Org. Lett.* 8 (20): 4393–4396.

2004 Pospisil, J., Kumamoto, T., and Marko, I.E. (2006). Highly diastereoselective silyl-modified Sakurai multicomponent reaction. *Angew. Chem. Int. Ed.* 45 (20): 3357–3360.

2005 Sellars, J.D., Steel, P.G., and Turner, M.J. (2006). Hosomi–Sakurai reactions of silacyclic allyl silanes. *J. Chem. Soc., Chem. Commun.* (22): 2385–2387.

2006 Wadamoto, M. and Yamamoto, H. (2005). Silver-catalyzed asymmetric Sakurai–Hosomi allylation of ketones. *J. Am. Chem. Soc.* 127 (42): 14556–14557.

2007 Howarth, J., James, P., and Dai, J.F. (2004). An exploration of the catalytic Sakurai reaction in the moisture stable ionic liquids [bmim]PF_6 and [bmim]BF_4. *J. Mol. Catal. A: Chem.* 214 (1): 143–146.

2008 Zhang, H.J., Priebbenow, D.L., and Bolm, C. (2013). Acylsilanes: valuable organosilicon reagents in organic synthesis. *Chem. Soc. Rev.* 42: 8540–8571.

2009 Armitage, D.A. (1982). Organosilanes. In: *Comprehensive Organometallic Chemistry*, Chapter 6.1, vol. 2 (ed. G. Wilkinson, F.G.A. Stone and E.W. Abel), 1–203. Oxford: Pergamon Press.

2010 (a) Colvin, E.W. (1981). Silicon in organic synthesis. In: *Butterworths Monographs in Chemistry and Chemical Engineering*, 1–348; (b) Jutzi, P. (2005). The many faces of organosilicon chemistry. *Nachtrichten Chem.* 53 (11): 1116–1120; (c) Sakurai, H. (2011). Silicon organosilicon chemistry. In: *Encyclopedia of Inorganic and Boorganic Chemistry*, 1–30. Wiley; (d) Baehr, S., Xue, W., and Oestreich, M. C(sp3)-Si cross coupling. *ACS Catal.* 2019 (1): 9, 16–24.

2011 Nakajima, Y. and Shimada, S. (2015). Hydrosilylation reaction of olefins: recent advances and perspectives. *RSC Adv.* 5: 20603–20616.

2012 Trost, B.M. and Ball, Z.T. (2001). Markovnikov alkyne hydrosilylation catalyzed by Ru complexes. *J. Am. Chem. Soc.* 123 (50): 12726–12727.

2013 Buslov, I., Becouse, J., Mazza, S. et al. (2015). Chemoselective alkene hydrosilylation catalyzed by Ni-pincer complexes. *Angew. Chem. Int. Ed.* 54: 14523–14526.

2014 Busbov, I., Keller, S.C., and Hu, X.L. (2016). Alkoxy hydrosilanes as surrogates of gaseous silanes for hydrosilylation of alkenes. *Org. Lett.* 18: 1928–1931.

2015 Buslov, I., Song, F., and Hu, X.L. (2016). An easily accessed Ni nanoparticle catalyst for alkene hydrosilylation with tertiary silanes. *Angew. Chem. Int. Ed.* 55: 12295–12299.

2016 Ibrahem, I., Santoro, S., Himo, F., and Córdova, A. (2011). Enantioselective conjugate silyl additions to *α*,β-unsaturated aldehydes catalyzed by combination of transition metal and chiral amine catalysts. *Adv. Synth. Catal.* 353: 245–252.

2017 Motokura, K. and Baba, T. (2012). An atom-efficient synthetic method: carbosilylations of alkenes, alkynes, and cyclic acetals using Lewis and Brønsted acid catalysts. *Green Chem.* 14: 565–579.

2018 Motokura, K., Matsunaga, S., Noda, H. et al. (2012). Water-accelerated allylsilylation of alkenes using a proton-exchanged montmorillonite catalyst. *ACS Catal.* 2: 1942–1946.

2019 Asao, N., Shimada, T., Shimada, T., and Yamamoto, Y. (2001). Lewis acid-catalyzed stereoselective carbosilylation. Intramolecular *trans*-vinylsilylation and *trans*-arylsilylation of unactivated alkynes. *J. Am. Chem. Soc.* 123: 10899–10902.

2020 Obora, Y., Tsuji, Y., and Kawamura, T. (1995). 1,4-Carbosilylation of 1,3-dienes via Pd-catalyzed 3-component coupling reaction. *J. Am. Chem. Soc.* 117: 9814–9821.

2021 Wu, M.Y., Yang, F.Y., and Cheng, C.H. (1999). Carbosilylation of allenes catalyzed by Pd complexes: a new efficient route to substituted allylic silanes. *J. Org. Chem.* 64: 2471–2474.

2022 Nii, S., Terao, J., and Kambe, N. (2000). Titanocene-catalyzed carbosilylation of alkenes and dienes using alkyl halides and chlorosilanes. *J. Org. Chem.* 65 (17): 5291–5297.

2023 Fujii, Y., Terao, J., Kuniyasu, H., and Kambe, N. (2007). Silylation and alkylation of allenes using chlorosilanes and alkyl halides in the presence of Pd catalyst and Grignard reagents. *J. Organomet. Chem.* 692: 375–381.

2024 Schwartz, J. and Labinger, J.A. (1976). Hydrozirconation - new transition-metal reagent for organic synthesis. *Angew. Chem. Int. Ed. Engl.* 15 (6): 333–340.

2025 Buchwald, S.L., La Maire, S.J., Nielsen, R.B. et al. (1987). A Modified procedure for the preparation of $Cp_2Zr(H)Cl$ (Schwartz's reagent). *Tetrahedron Lett.* 28 (34): 3895–3898.

2026 Buchwald, S.L., La Maire, S.J., Nielsen, R.B. et al. (1993). Schwartz's reagent. *Org. Synth.* 71: 77–82.

2027 Sidera, M., Roth, P.M.C., Maksymowicz, R.M., and Fletcher, S.P. (2013). Formation of quaternary centers by Cu-catalyzed asymmetric conjugate addition of alkylzirconium reagents. *Angew. Chem. Int. Ed.* 52: 7995–7999.

2028 (a) Maksymowicz, R.M., Sidera, M., Roth, P.M.C., and Fletcher, S.P. (2013). A Convenient catalytic asymmetric conjugate addition reaction to enones using alkylzirconium reagents. *Synthesis* 45: 2662–2668. (b) Maciver, E.E., Maksymowicz, R.M., Wilkinson, N. et al. (2014). Asymmetric conjugate addition of alkylzirconium reagents to α,β-unsaturated lactones. *Org. Lett.* 16: 3288–3291.

2029 Arnold, D.M., Krainz, T., and Wipf, P. (2015). One-pot hydrozirconation/copper-catalyzed conjugate addition of alkylzirconocenes to enones. *Org. Synth.* 92: 277–295.

2030 Roth, P.M.C., Sidera, M., Maksymowicz, R.M., and Fletcher, S.P. (2014). Copper-catalyzed asymmetric conjugate addition of alkylzirconium reagents to cyclic enones to form quaternary centers. *Nat. Protoc.* 9: 104–111.

2031 (a) Cherney, A.H., Kanduce, N.T., and Reisman, S.E. (2015). Enantioselective and enantiospecific transition-metal-catalyzed cross-coupling reactions of organometallic reagents to construct C—C bonds. *Chem. Rev.* 115: 9587–9652. (b) Yang, H., Yang, X., and Tang, W. (2016). Transition-metal catalyzed asymmetric carbon–carbon cross-coupling with chiral ligands. *Tetrahedron* 72: 6143–6174.

2032 Roy, D. and Uezumi, Y. (2018). Recent advances in Pd-catalyzed cross-coupling reactions at ppm to ppb molar catalyst loadings. *Adv. Synth. Catal.* 360 (4): 602–625.

2033 Phan, N.T.S., van der Sluys, M., and Jones, C.W. (2006). On the nature of the active species in Pd-catalyzed Mizoroki–Heck and Suzuki–Miyaura couplings - homogeneous or heterogeneous catalysis, a critical review. *Adv. Synth. Catal.* 348: 609–679.

2034 Jahn, U. (2012). Radicals in transition metal catalyzed reactions? Transition metal catalyzed radical reactions? A fruitful interplay anyway. Part 1. Radical catalysis by Group 4 to Group 7 elements. In: *Radicals in Synthesis III*, Topics in Current Chemistry, vol. 320 (ed. M. Heinrich and A. Gansäuer), 121–189. Berlin, Heidelberg: Springer-Verlag.

2035 Jahn, U. (2012). Radicals in transition metal catalyzed reactions? Transition metal catalyzed radical reactions? A fruitful interplay anyway. Part 2. Radical catalysis by Group 8 and 9 elements. In: *Radicals in Synthesis III*, Topics in Current Chemistry, vol. 320 (ed. M. Heinrich and A. Gansäuer), 191–322. Berlin, Heidelberg: Springer-Verlag.

2036 Jahn, U. (2012). Radicals in transition metal catalyzed reactions? Transition metal catalyzed radical reactions? A fruitful interplay anyway.

Part 3. Catalysis by Group 10 and 11 elements and bimetallic catalysis. In: *Radicals in Synthesis III*, Topics in Current Chemistry, vol. 320 (ed. M. Heinrich and A. Gansäuer), 323–451. Berlin, Heidelberg: Springer-Verlag.

2037 Kharasch, M.S., Kleiger, S.C., Martin, J.A., and Mayo, F.R. (1941). Factors determining the course and mechanisms of Grignard reactions. I. Preliminary study: the effects of metallic compounds on some Grignard-carbonyl interactions. *J. Am. Chem. Soc.* 63: 2305–2307.

2038 Kharasch, M.S. and Fields, E.K. (1941). Factors determining the course and mechanisms of Grignard reactions. IV. The effect of metallic halides on the reaction of aryl Grignard reagents and organic halides. *J. Am. Chem. Soc.* 63: 2316–2320.

2039 Kharasch, M.S. and Lambert, F.L. (1941). Factors determining the course and mechanisms of Grignard reactions. III. The effect of metallic halides on the reaction between benzophenone and methylmagnesium bromide. *J. Am. Chem. Soc.* 63: 2315–2316.

2040 Kharasch, M.S., Lewis, D.W., and Reynolds, W.B. (1943). Factors determining the course and mechanism of Grignard reactions. VII. Analysis of gases formed during the reaction of phenylmagnesium bromide and organic halides in the presence of cobaltous halides. *J. Am. Chem. Soc.* 65: 493–495.

2041 Kharasch, M.S., Lewis, D.W., and Reynolds, W.B. (1943). Factors determining the course and mechanisms of Grignard reactions. IX. The effect of metallic halides on the reaction of organolithium compounds with organic halides. *J. Am. Chem. Soc.* 65: 498–500.

2042 Kharasch, M.S. and Fuchs, C.F. (1943). Factors influencing the course and mechanisms of Grignard reactions. XI. The effect of metallic halides on the reaction of Grignard reagents with vinyl halides and substituted vinyl halides. *J. Am. Chem. Soc.* 65: 504–507.

2043 Kharasch, M.S. and Sosnovsky, G. (1958). The reactions of *tert*-butyl perbenzoate and olefins - a stereospecific reaction. *J. Am. Chem. Soc.* 80: 756.

2044 Eames, J. and Watkinson, M. (2001). Catalytic allylic oxidation of alkenes using an asymmetric Kharasch–Sosnovsky reaction. *Angew. Chem. Int. Ed.* 40: 3567–3571.

2045 Le Bras, J. and Muzart, J. (2003). Amino acid/Cu-catalyzed enantioselective allylic benzoyloxylation of olefins in water promoted by diethylene glycol. *Tetrahedron: Asymmetry* 14: 1911–1915.

2046 Wei, W., Zhang, C., Xu, Y., and Wan, X.B. (2011). Synthesis of *tert*-butyl peresters from aldehydes by Bu_4NI-catalyzed metal-free oxidation and its combination with the Kharasch–Sosnovsky reaction. *J. Chem. Soc., Chem. Commun.* 47: 10827–10829.

2047 García-Cabeza, A.L., Moreno-Dorada, F.J., Ortega, M.J., and Guerra, F.M. (2016). Copper-catalyzed oxidation of alkenes and heterocycles. *Synthesis* 48: 2323–2342.

2048 Kharasch, M.S., Jensen, E.V., and Urry, W.H. (1945). Addition of CCl_4 and $CHCl_3$ to olefins. *Science* 102 (2640): 128–128.

2049 Minisci, F. (1975). Free-radical additions to olefins in presence of redox systems. *Acc. Chem. Res.* 8 (5): 165–171.

2050 Tallarico, J.A., Malnick, L.M., and Snapper, M.L. (1999). New reactivity from $(PCy_3)_2Cl_2Ru{=}CHPh$: a mild catalyst for Kharasch additions. *J. Org. Chem.* 64: 344–345.

2051 Fernández-Zumel, M.A., Thommes, K., Kiefer, G. et al. (2009). Atom-transfer radical addition reactions catalyzed by RuCp* complexes: a mechanistic study. *Chem. Eur. J.* 15: 11601–11607.

2052 Sayles, D.C. and Kharasch, M.S. (1961). Synthesis of symmetrically substituted ethane derivatives. *J. Org. Chem.* 26: 4210–4214.

2053 Tamura, M. and Kochi, J. (1971). Iron catalysis in reaction of Grignard reagents with alkyl halides. *J. Organomet. Chem.* 31 (3): 289–309.

2054 Tamura, M. and Kochi, J.K. (1972). Copper-catalyzed coupling of Grignard reagents and alkyl halides in tetrahydrofuran solutions. *J. Organomet. Chem.* 42: 205–228.

2055 Tamura, M. and Kochi, J.K. (1971). Mechanism of Ag-catalyzed reaction of Grignard Reagents with alkyl halides. *J. Am. Chem. Soc.* 93: 1483–1485.

2056 Tamura, M. and Kochi, J.K. (1971). Coupling of Grignard reagents with organic halides. *Synth. Int. J. Methods Synth. Org. Chem.* 303–305.

2057 Kochi, J.K. (1974). Electron-transfer mechanisms for organometallic intermediates in catalytic reactions. *Acc. Chem. Res.* 7 (10): 351–360.

2058 Kharasch, M.S. and Reinmuth, O. (1954). *Grignard Reactions of Nonmetallic Substances*. New York: Prentice-Hall.

2059 Tamura, M. and Kochi, J. (1971). Vinylation of Grignard reagents - catalysis by iron. *J. Am. Chem. Soc.* 93 (6): 1487–1489.

2060 Tamao, K., Sumitani, K., and Kumada, M. (1972). Selective carbon–carbon bond formation by cross-coupling of Grignard reagents with organic halides - catalysis by Ni-phosphine complexes. *J. Am. Chem. Soc.* 94 (12): 4374–4376.

2061 Corriu, J.R. and Masse, J.P. (1972). Activation of Grignard reagents by transition-metal complexes - new and simple synthesis of *trans*-stilbenes and polyphenyls. *J. Chem. Soc., Chem. Commun.* (3): 144–144.

2062 Yoshikai, N., Mashima, H., and Nakamura, E. (2005). Nickel-catalyzed cross-coupling reaction of aryl fluorides and chlorides with Grignard reagents under Ni/Mg bimetallic cooperation. *J. Am. Chem. Soc.* 127 (51): 17978–17979.

2063 Limmert, M.E., Roy, A.H., and Hartwig, J.F. (2005). Kumada coupling of aryl and vinyl tosylates under mild conditions. *J. Org. Chem.* 70 (23): 9364–9370.

2064 Ackermann, L., Born, R., Spatz, J.H., and Meyer, D. (2005). Efficient aryl-(hetero)aryl coupling by activation of C—Cl and C—F bonds using Ni complexes of air-stable phosphine oxides. *Angew. Chem. Int. Ed.* 44 (44): 7216–7219.

2065 Heravi, M.M. and Hajiabbasi, P. (2012). Recent advances in Kumada–Tamao–Corriu cross-coupling reaction catalyzed by different ligands. *Monatsh. Chem.* 143: 1575–1592.

2066 Yu, D.G., Li, B.J., and Shi, Z.J. (2010). Exploration of new C—O electrophiles in cross-coupling reactions. *Acc. Chem. Res.* 43 (12): 1486–1495.

2067 Rosen, B.M., Quasdorf, K.W., Wilson, D.A. et al. (2011). Nickel-catalyzed cross-couplings involving carbon-oxygen bonds. *Chem. Rev.* 111: 1346–1416.

2068 Hu, X.L. (2011). Nickel-catalyzed cross coupling of non-activated alkyl halides: a mechanistic perspective. *Chem. Sci.* 2: 1867–1886.

2069 Dankwardt, J.W. (2004). Nickel-catalyzed cross-coupling of aryl Grignard reagents with aromatic alkyl ethers: an efficient synthesis of unsymmetrical biaryls. *Angew. Chem. Int. Ed.* 43: 2428–2432.

2070 Iglesias, M.J., Prieto, A., and Nicasio, M.C. (2012). Kumada–Tamao–Corriu coupling of heteroaromatic chlorides and aryl ethers catalyzed by (IPr)Ni(allyl)Cl. *Org. Lett.* 14: 4318–4321.

2071 Tobisu, M. and Chatani, N. (2015). Cross-couplings using aryl ethers via C—O bond activation enabled by Ni catalysts. *Acc. Chem. Res.* 48: 1717–1726.

2072 Guisán-Ceinos, M., Soler-Yanes, R., Collado-Sanz, D. et al. (2013). Ni-Catalyzed cascade cyclization Kumada alkyl-alkyl cross-coupling. *Chem. Eur. J.* 19: 8405–8410.

2073 Vechorkin, O. and Hu, X.L. (2009). Nickel-catalyzed cross-coupling of non-activated and functionalized alkyl halides with alkyl Grignard reagents. *Angew. Chem. Int. Ed.* 48 (16): 2937–2940.

2074 Vechorkin, O., Proust, V., and Hu, X.L. (2009). Functional group tolerant Kumada–Corriu–Tamao coupling of non-activated alkyl halides with aryl and heteroaryl nucleophiles: catalysis by a Ni-pincer complex permits the coupling of functionalized Grignard reagents. *J. Am. Chem. Soc.* 131: 9756–9766.

2075 Vechorkin, O., Godinat, A., Scopelliti, R., and Hu, X.L. (2011). Cross-coupling of non-activated alkyl halides with alkynyl Grignard reagents: a Ni-pincer complex as the catalyst. *Angew. Chem. Int. Ed.* 50: 11777–11781.

2076 Breitenfeld, J., Vechorkin, O., Corminboeuf, C. et al. (2010). Why are (NN_2)Ni pincer complexes active for alkyl-alkyl coupling: β-H Elimination is kinetically accessible but thermodynamically uphill. *Organometallics* 29: 3686–3689.

2077 Breitenfeld, J., Wodrich, M.D., and Hu, X.L. (2014). Bimetallic oxidative addition in Ni-catalyzed alkyl-aryl Kumada coupling reactions. *Organometallics* 33: 5708–5715.

2078 Yoshikai, N., Matsuda, H., and Nakamura, E. (2008). Ligand exchange as the first irreversible step in the Ni-catalyzed cross-coupling reaction of Grignard reagents. *J. Am. Chem. Soc.* 130 (46): 15258–15259.

2079 Massera, C. and Frenking, G. (2003). Energy partitioning analysis of the bonding in $L_2TM—C_2H_2$ and $L_2TM—C_2H_4$ (TM=Ni, Pd, Pt; $L_2=(PH_3)_2$, $(PMe_3)_2$, $H_2PCH_2PH_2$, $H_2P(CH_2)_2PH_2$). *Organometallics* 22 (13): 2758–2765.

2080 Schroeter, S., Stock, C., and Bach, T. (2005). Regioselective cross-coupling reactions of multiple halogenated nitrogen-, oxygen-, and sulfur-containing heterocycles. *Tetrahedron* 61 (9): 2245–2267.

2081 Ishikawa, S. and Manabe, K. (2007). Highly ortho-selective cross-coupling of dichlorobenzene derivatives with Grignard reagents. *Org. Lett.* 9 (26): 5593–5595.

2082 Wang, J.R. and Manabe, K. (2009). High *ortho* preference in Ni-catalyzed cross-coupling of halophenols with alkyl Grignard reagents. *Org. Lett.* 11 (3): 741–744.

2083 Hartmann, C.E., Nolan, S.P., and Cazin, C.S.J. (2009). Highly active $[Pd(\mu\text{-}Cl)(Cl)(NHC)]_2$ (NHC=N-Heterocyclic Carbene) in the cross-coupling of Grignard reagents with aryl chlorides. *Organometallics* 28 (9): 2915–2919.

2084 Hayashi, T., Hayashizaki, K., Kiyoi, T., and Ito, Y. (1988). Asymmetric synthesis catalyzed by chiral ferrocenylphosphine transition-metal complexes. 6. Practical asymmetric synthesis of 1,1′-binaphthyls via asymmetric cross-coupling with a chiral [(alkoxyalkyl)ferrocenyl]monophosphine Ni catalyst. *J. Am. Chem. Soc.* 110 (24): 8153–8156.

2085 Selim, K.B., Yamada, K., and Tomioka, K. (2008). Copper-catalyzed asymmetric allylic substitution with aryl and ethyl Grignard reagents. *J. Chem. Soc., Chem. Commun.* (41): 5140–5142.

2086 Manolikakes, G. and Knochel, P. (2009). Radical catalysis of Kumada cross-coupling reactions using functionalized Grignard reagents. *Angew. Chem. Int. Ed.* 48 (1): 205–209.

2087 Corey, E.J., Semmelhack, M.F., and Hegedus, L.S. (1968). Course of allylic coupling reactions involving (allyl)Ni complexes. *J. Am. Chem. Soc.* 90 (9): 2416–2417.

2088 Corey, E.J., Hegedus, L.S., and Semmelhack, M.F. (1968). Interconversion of lower valence states of (π-allyl)Ni complexes. A new synthesis of (π-allyl)Ni derivatives. *J. Am. Chem. Soc.* 90 (9): 2417–2418.

2089 Hegedus, L.S. and Waterman, E.L. (1974). Reactions of (π-2-methallyl)Ni bromide with methylbenzoquinones - evidence for electron-transfer. *J. Am. Chem. Soc.* 96 (21): 6789–6791.

2090 Hegedus, L.S. and Miller, L.L. (1975). Reaction of (π-allyl)Ni bromide complexes with organic halides - stereochemistry and mechanism. *J. Am. Chem. Soc.* 97 (2): 459–460.

2091 Hegedus, L.S. and Thompson, D.H. (1985). Reactions of organic halides with (π-allyl)Ni halide complexes - a mechanistic study. *J. Am. Chem. Soc.* 107 (20): 5663–5669.

2092 Elson, I.H., Morrell, D.G., and Kochi, J.K. (1975). Electron-transfer from d10 species - possible initiation for free-radical pathways in oxidative additions. *J. Organomet. Chem.* 84 (1): C7–C10.

2093 Tsou, T.T. and Kochi, J.K. (1979). Mechanism of biaryl synthesis with Ni-complexes. *J. Am. Chem. Soc.* 101 (25): 7547–7560.

2094 Labinger, J.A., Kramer, A.V., and Osborn, J.A. (1973). Mechanistic studies of oxidative addition to low-valent metal-complexes. 2. Reactions of α-bromo esters. *J. Am. Chem. Soc.* 95 (23): 7908–7909.

2095 Kramer, A.V., Labinger, J.A., Bradley, J.S., and Osborn, J.A. (1974). Mechanistic studies of oxidative addition to low-valent metal-complexes. 3. Mechanism of formation of Pt to carbon bonds. *J. Am. Chem. Soc.* 96 (22): 7145–7147.

2096 Kramer, A.V. and Osborn, J.A. (1974). Mechanistic studies of oxidative addition to low-valent metal-complexes. 4. Observation of CIDNP effects in Pt(0) and Pd(0) reactions. *J. Am. Chem. Soc.* 96 (25): 7832–7833.

2097 Kochi, J.K. (1978). *Organometallics Mechanisms and Catalyses*, 1–623. New York: Academic Press.

2098 Astruc, D. (1995). *Electron Transfer and Radical Processes in Transition-Metal*, 1–652. New York: Wiley-VCH.

2099 Lipschutz, M.I. and Tilley, T.D. (2014). Carbon–carbon cross-coupling reactions catalyzed by a two-coordinate Ni(II)(bisamido) complex via observable Ni(I), Ni(II), and Ni(III) intermediates. *Angew. Chem. Int. Ed.* 53: 7290–7294.

2100 (a) Guérinot, A. and Cossy, J. (2016). Iron-catalyzed C—C cross-couplings using organometallics. *Top. Curr. Chem.* 374 (4): 1–74. (b) Piontek, A., Bisz, E., and Szostak, M. (2018). Iron-catalyzed cross-couplings in the synthesis of pharmaceuticals: In pursuit of sustainability. *Angew. Chem. Int. Ed.* 57 (35): 11116–11128; (c) Sears, J.D.; Neate, P.G.N.; Neidig, M.L. Intermediates and mechanism in iron-catalyzed cross-coupling *J. Am. Chem. Soc.* 2018, 140 (38), 11872–11883 (d) Neidig, M.L., Carpenter, S.H., Curran, D.J. et al. (2019). Development and evolution of mechanistic understanding in iron-catalyzed cross-coupling. *Acc. Chem. Res.* 52 (1): 140–150.

2101 Vavon, G. and Mottez, P. (1944). Action des dérivés halogénés sur les magnésiens aromatiques en présence de chlorure ferrique. *C.R. Hebd. Séances Acad. Sci.* 218: 557–559.

2102 Percival, W.C., Wagner, R.B., and Cook, N.C. (1953). Grignard reactions. 21. The synthesis of aliphatic ketones. *J. Am. Chem. Soc.* 75: 3731–3734.

2103 Fiandanese, V., Marchese, G., Martina, V., and Ronzini, L. (1984). Iron-catalyzed cross-coupling reactions of acyl chlorides with Grignard reagents - a mild, general, and convenient synthesis of aliphatic and aromatic ketones. *Tetrahedron Lett.* 25 (42): 4805–4808.

2104 Smith, R.S. and Kochi, J.K. (1976). Mechanistic studies of Fe catalysis in cross-coupling of alkenyl halides and Grignard reagents. *J. Org. Chem.* 41 (3): 502–509.

2105 Fabre, J.L., Julia, M., and Verpeaux, J.N. (1982). Cross-coupling reaction of vinylic sulfones with Grignard reagents in the presence of transition metal salts. Stereoselective preparation of trisubstitued olefins. *Tetrahedron Lett.* 23 (24): 2469–2472.

2106 Alvarez, E., Cuvigny, T., Du Penhoat, C.H., and Julia, M. (1988). Syntheses with sulfones. 48. Stereoselective synthesis of 2-isopropyl-1,4-dienes through the Fe-catalyzed cross-coupling reaction of 2-benzenesulfonyl-1,4-dienes and isopropylmagnesium chloride. *Tetrahedron* 44 (1): 111–118.

2107 Molander, G.A., Rahn, B.J., Shubert, D.C., and Bonde, S.E. (1983). Iron-catalyzed cross-coupling reactions - synthesis of arylethenes. *Tetrahedron Lett.* 24 (49): 5449–5452.

2108 Fiandanese, V., Miccoli, G., Naso, F., and Ronzini, L. (1986). Stereospecific synthesis of alkenyl sulfides by cross-coupling reactions of secondary alkyl Grignard reagents with (*Z*)-1-bromo-2-phenylthioethene or (*E*)-1-bromo-2-phenylthioethene in the presence of transition-metal catalysts. *J. Organomet. Chem.* 312: 343–348.

2109 Itami, K., Higashi, S., Mineno, M., and Yoshida, J. (2005). Iron-catalyzed cross-coupling of alkenyl sulfides with Grignard reagents. *Org. Lett.* 7 (7): 1219–1222.

2110 Cahiez, G., Chavant, P.Y., and Metais, E. (1992). Organomanganese(II) reagents. 24. A new simple route to furanic ketones - preparation of elsholtzione, naginata ketone and perilla ketone. *Tetrahedron Lett.* 33 (36): 5245–5248.

2111 Cahiez, G. and Marquais, S. (1996). Highly chemo- and stereoselective Fe-catalyzed alkenylation of organomanganese reagents. *Tetrahedron Lett.* 37 (11): 1773–1776.

2112 Cahiez, G. and Avedissian, H. (1998). Highly stereo- and chemoselective Fe-catalyzed alkenylation of organomagnesium compounds. *Synthesis* (8): 1199–1205.

2113 Cahiez, G., Habiak, V., Duplais, C., and Moyeux, A. (2007). Iron-catalyzed alkylations of aromatic Grignard reagents. *Angew. Chem. Int. Ed.* 46 (23): 4364–4366.

2114 Cahiez, G., Duplais, C., and Moyeux, A. (2007). Iron-catalyzed alkylation of alkenyl Grignard reagents. *Org. Lett.* 9 (17): 3253–3254.

2115 Martin, R. and Fürstner, A. (2004). Cross-coupling of alkyl halides with aryl Grignard reagents catalyzed by a low-valent Fe complex. *Angew. Chem. Int. Ed.* 43 (30): 3955–3957.

2116 Fürstner, A., Krause, H., and Lehmann, C.W. (2006). Unusual structure and reactivity of a homoleptic "super-ate" complex of Fe: implications for Grignard additions, cross-coupling reactions, and the Kharasch deconjugation. *Angew. Chem. Int. Ed.* 45 (3): 440–444.

2117 Scheiper, B., Bonnekessel, M., Krause, H., and Fürstner, A. (2004). Selective Fe-catalyzed cross-coupling reactions of Grignard reagents with enol triflates, acid chlorides, and dichloroarenes. *J. Org. Chem.* 69 (11): 3943–3949.

2118 Dohle, W., Kopp, F., Cahiez, G., and Knochel, P. (2001). Fe(III)-Catalyzed cross-coupling between functionalized (aryl)Mg compounds and alkenyl halides. *Synlett* (12): 1901–1904.

2119 Duplais, C., Bures, F., Sapountzis, I. et al. (2004). An efficient synthesis of diaryl ketones by iron-catalyzed arylation of aroyl cyanides. *Angew. Chem. Int. Ed.* 43 (22): 2968–2970.

2120 Nakamura, M., Hirai, A., and Nakamura, E. (2000). Iron-catalyzed olefin carbometalation. *J. Am. Chem. Soc.* 122 (5): 978–979.

2121 Nakamura, M., Matsuo, K., Ito, S., and Nakamura, B. (2004). Iron-catalyzed cross-coupling of primary and secondary alkyl halides with aryl Grignard reagents. *J. Am. Chem. Soc.* 126 (12): 3686–3687.

2122 Nakamura, M., Ito, S., Matsuo, K., and Nakamura, E. (2005). Iron-catalyzed chemoselective cross-coupling of primary and secondary alkyl halides with arylzinc reagents. *Synlett* (11): 1794–1798.

2123 Hatakeyama, T. and Nakamura, M. (2007). Iron-catalyzed selective biaryl coupling: remarkable suppression of homocoupling by the fluoride anion. *J. Am. Chem. Soc.* 129 (32): 9844–9845.

2124 Bedford, R.B., Bruce, D.W., Goodby, J.W., and Hird, M. (2004). Iron(III) salen-type catalysts for the cross-coupling of aryl Grignards with alkyl halides bearing β-hydrogens. *J. Chem. Soc., Chem. Commun.* 24: 2822–2823.

2125 Bedford, R.B., Bruce, D.W., Frost, R.M., and Hird, M. (2005). Simple Fe-amine catalysts for the cross-coupling of aryl Grignards with alkyl halides bearing β-hydrogens. *J. Chem. Soc., Chem. Commun.* (33): 4161–4163.

2126 Bedford, R.B., Betham, M., Bruce, D.W. et al. (2006). Iron nanoparticles in the coupling of alkyl halides with aryl Grignard reagents. *J. Chem. Soc., Chem. Commun.* (13): 1398–1400.

2127 Bedford, R.B., Betham, M., Bruce, D.W. et al. (2006). Iron-phosphine, -phosphite, -arsine, and -carbene catalysts for the coupling of primary and secondary 2-alkyl halides with aryl Grignard reagents. *J. Org. Chem.* 71 (3): 1104–1110.

2128 Guérinot, A., Reymond, S., and Cossy, J. (2007). Iron-catalyzed cross-coupling of alkyl halides with alkenyl Grignard reagents. *Angew. Chem. Int. Ed.* 46 (34): 6521–6524.

2129 Shinokubo, H. and Oshima, K. (2004). Transition metal-catalyzed carbon–carbon bond formation with Grignard reagents - novel reactions with a classic reagent. *Eur. J. Org. Chem.* (10): 2081–2091.

2130 Bolm, C., Legros, J., Le Paih, J., and Zani, L. (2004). Iron-catalyzed reactions in organic synthesis. *Chem. Rev.* 104 (12): 6217–6254.

2131 Necas, D. and Kotora, M. (2006). Application of with Fe-compound catalysis in organic synthesis. *Chem. Listy* 100 (11): 967–973.

2132 Gøgsig, T.M., Lindhardt, A.T., and Skrydstrup, T. (2009). Heteroaromatic sulfonates and phosphates as electrophiles in Fe-catalyzed cross-couplings. *Org. Lett.* 11: 4886–4888.

2133 Agrawal, T. and Cook, S.P. (2013). Iron-catalyzed cross-coupling reactions of alkyl Grignards with aryl sulfamates and tosylates. *Org. Lett.* 15: 96–99.

2134 Agrawal, T. and Cook, S.P. (2014). Iron-catalyzed coupling of aryl sulfannates and aryl/vinyl tosylates with aryl Grignards. *Org. Lett.* 16: 5080–5083.

2135 Chen, X., Quan, Z.J., and Wang, X.C. (2015). Iron-catalyzed cross-coupling of heteroaromatic tosylates with alkyl and aryl Grignard reagents. *Appl. Organomet. Chem.* 29: 296–300.

2136 Xing, T., Zhang, Z., Da, Y.X. et al. (2015). Iron-catalyzed Kumada cross-coupling reactions of pyrimidin-2-yl phosphates: an efficient approach to C_2-functionalized pyrimidines. *Asian J. Org. Chem.* 4: 538–544.

2137 Bedford, R.B. (2015). How low does Fe go? Chasing the active species in Fe-catalyzed cross-coupling reactions. *Acc. Chem. Res.* 48: 1485–1493.

2138 Cheung, C.W., Ren, P., and Hu, X.L. (2014). Mild and phosphine-free Fe-catalyzed cross-coupling of non-activated secondary alkyl halides with alkynyl Grignard reagents. *Org. Lett.* 16: 2566–2569.

2139 Bauer, G., Wodrich, M.D., Scopelliti, R., and Hu, X.L. (2015). Iron pincer complexes as catalysts and intermediates in alkyl aryl Kumada coupling reactions. *Organometallics* 34: 289–298.

2140 Shilov, A.E., Shilova, A.K., Kvashina, E.F., and Vorontsova, T.A. (1971). Isolation of a binuclear complex intermediate in reaction of molecular nitrogen with Ti compounds. *J. Chem. Soc., Chem. Commun.* (24): 1590–1591.

2141 Jezowska-Trzebiatowska, B. and Sobota, P. (1972). Reaction mechanism and properties of compounds formed in system $FeCl_3$+Mg+THF+N_2. *J. Organomet. Chem.* 46 (2): 339–343.

2142 Hölzer, B. and Hofmann, R.W. (2003). Kumada–Corriu coupling of Grignard reagents, probed with a chiral Grignard reagent. *J. Chem. Soc., Chem. Commun.* (6): 732–733.

2143 Sapountzis, I., Lin, W.W., Kofink, C.C. et al. (2005). Iron-catalyzed aryl-aryl cross-couplings with Mg-derived Cu reagents. *Angew. Chem. Int. Ed.* 44 (11): 1654–1657.

2144 Ila, H., Baron, O., Wagner, A.J., and Knochel, P. (2006). Functionalized Mg organometallics as versatile intermediates for the synthesis of polyfunctional heterocycles. *J. Chem. Soc., Chem. Commun.* 583–593.

2145 Jin, M., Adak, L., and Nakamura, M. (2015). Iron-catalyzed enantioselective cross-coupling reactions of α-chloroesters with aryl Grignard reagents. *J. Am. Chem. Soc.* 137: 7128–7134.

2146 Cassani, C., Bergonzini, G., and Wallentin, C.J. (2016). Active species and mechanistic pathways in Fe-catalyzed C—C-bond forming cross-coupling reactions. *ACS Catal.* 6: 1640–1648.

2147 Bedford, R.B., Carter, E., Cogswell, P.M. et al. (2013). Simplifying Fe(phosphine) catalysts for cross-coupling reactions. *Angew. Chem. Int. Ed.* 52: 1285–1288.

2148 Bedford, R.B., Brenner, P.B., Carter, E. et al. (2014). TMEDA in Fe-catalyzed Kumada coupling: amine adduct versus homoleptic "ate" complex formation. *Angew. Chem. Int. Ed.* 53: 1804–1808.

2149 Muñoz, S.B. III, Daifuku, S.L., Brennessel, W.W., and Neidig, M.L. (2016). Isolation, characterization, and reactivity of $Fe_8Me_{12}^-$: Kochi's S=1/2 species in Fe-catalyzed cross-couplings with MeMgBr and Fe(III) salts. *J. Am. Chem. Soc.* 138: 7492–7495.

2150 Bedford, R.B. (2016). Homogeneous catalysis. Eight irons hit the right spin. *Nat. Chem.* 8: 904–905.

2151 Fürstner, A. and Leitner, A. (2002). Iron-catalyzed cross-coupling reactions of alkyl-Grignard reagents with aryl chlorides, tosylates, and triflates. *Angew. Chem. Int. Ed.* 41 (4): 609–612.

2152 Le Marquand, P., Tsui, G.C., Whitney, J.C.C., and Tam, W. (2008). Iron-catalyzed cross-coupling reactions between a bicyclic alkenyl triflate and Grignard reagents. *J. Org. Chem.* 73 (19): 7829–7832.

2153 Dunet, G. and Knochel, P. (2006). Iron-catalyzed cross-coupling between alkenyl and dienyl sulfonates and functionalized (aryl)Cu reagents. *Synlett* 407–410.

2154 Li, B.J., Xu, L., Wu, Z.H. et al. (2009). Cross-coupling of alkenyl/aryl carboxylates with Grignard reagent via Fe-catalyzed C—O bond activation. *J. Am. Chem. Soc.* 131: 14656–14657.

2155 Greenhalgh, M.D., Kolodziej, A., Sinclair, F., and Thomas, S.P. (2014). Iron-catalyzed hydromagnesiation: synthesis and characterization of benzylic Grignard reagent intermediate and application in the synthesis of ibuprofen. *Organometallics* 33: 5811–5819.

2156 Cahiez, G., Habiak, V., and Gager, O. (2008). Efficient preparation of terminal conjugated dienes by coupling of dienol phosphates with Grignard reagents under Fe catalysis. *Org. Lett.* 10 (12): 2389–2392.

2157 Burns, D.H., Miller, J.D., Chan, H.K., and Delaney, M.O. (1997). Scope and utility of a new soluble Cu catalyst [CuBr/LiSPh-/LiBr/THF]: a comparison with other Cu catalysts in their ability to couple one equivalent of a Grignard reagent with an alkyl sulfonate. *J. Am. Chem. Soc.* 119 (9): 2125–2133.

2158 Hatakeyama, T., Okada, Y., Yoshimoto, Y., and Nakamura, M. (2011). Tuning chemoselectivity in Fe-catalyzed Sonogashira-type reactions using a bisphosphine ligand with peripheral steric bulk: selective alkynylation of nonactivated alkyl halides. *Angew. Chem. Int. Ed.* 50: 10973–10976.

2159 Gilman, H. and Fothergill, R.E. (1929). The reaction between sulfochlorides and organomagnesium halides. *J. Am. Chem. Soc.* 51: 3501–3508.

2160 Dubbaka, S.R. and Vogel, P. (2006). Palladium-catalyzed desulfinylative Negishi C—C bond forming cross-couplings of sulfonyl and organozinc chlorides. *Tetrahedron Lett.* 47 (20): 3345–3348.

2161 Volla, C.M.R. and Vogel, P. (2008). Iron-catalyzed desulfinylative C—C cross-coupling reactions of sulfonyl chlorides with Grignard reagents. *Angew. Chem. Int. Ed.* 47: 1305–1307.

2162 Shirakawa, E., Sato, T., Imazaki, Y. et al. (2007). Cobalt-catalyzed cross-coupling of alkynyl Grignard reagents with alkenyl triflates. *J. Chem. Soc., Chem. Commun.* (43): 4513–4515.

2163 Cahiez, G., Chaboche, C., Duplais, C. et al. (2008). Cobalt-catalyzed cross-coupling reaction between functionalized primary and secondary alkyl halides and aliphatic Grignard reagents. *Adv. Synth. Catal.* 350 (10): 1484–1488.

2164 Shirakawa, E., Imazaki, Y., and Hayashi, T. (2008). Cobalt-catalyzed coupling of alkenyl triflates with aryl and alkenyl Grignard reagents. *Chem. Lett.* 37 (6): 654–655.

2165 Hamaguchi, H., Uemura, M., Yasui, H. et al. (2008). Cobalt-catalyzed cross-coupling reactions of aryl bromides with alkyl Grignard reagents. *Chem. Lett.* 37 (11): 1178–1179.

2166 Cahiez, G., Chaboche, C., Duplais, C., and Moyeux, A. (2009). A new efficient catalytic system for the chemoselective Co-catalyzed cross-coupling of aryl Grignard reagents with primary and secondary alkyl bromides. *Org. Lett.* 11 (2): 277–280.

2167 Dongol, K.G., Koh, H., Sau, M., and Chai, C.L.L. (2007). Iron-catalysed C(sp^3)-C(sp^3) cross-coupling reactions of unactivated alkyl halides with alkyl Grignard reagents. *Adv. Synth. Catal.* 349 (7): 1015–1018.

2168 Negishi, E., King, A.O., and Okukado, N. (1977). Selective carbon–carbon bond formation via transition-metal catalysis. 3. Highly selective synthesis of unsymmetrical biaryls and diarylmethanes by Ni-catalyzed or Pd-catalyzed reaction of aryl derivatives and benzylzinc derivatives with aryl halides. *J. Org. Chem.* 42 (10): 1821–1823.

2169 Negishi, E. (1982). Palladium-catalyzed or Ni-catalyzed cross-coupling - a new selective method for carbon–carbon bond formation. *Acc. Chem. Res.* 15 (11): 340–348.

2170 Kobayashi, M. and Negishi, E. (1980). A palladium-catalyzed stereospecific substitution-reaction of homoallylzincs with β-bromo-substituted α,β-unsaturated carbonyl derivatives - a highly selective synthesis of mokupalide. *J. Org. Chem.* 45 (25): 5223–5225.

2171 Negishi, E., Bagheri, V., Chatterjee, S. et al. (1983). Selective carbon–carbon bond formation via transition-metal catalysis. 36. Palladium-catalyzed acylation of organozincs and other organometallics as a convenient route to ketones. *Tetrahedron Lett.* 24 (47): 5181–5184.

2172 Grey, R.A. (1984). A palladium-catalyzed synthesis of ketones from acid-chlorides and organozinc compounds. *J. Org. Chem.* 49 (12): 2288–2289.

2173 Rossi, R., Bellina, F., Carpita, A., and Gori, R. (1995). New catalyst precursors constituted of $AsPh_3$ and Pd on carbon or Pd(II) acetate as efficient promoters of selective cross-coupling reactions between functionalized alkenyl halides and arylzinc or 1-alkynylzinc chlorides. *Synlett* (4): 344–346.

2174 (a) Shi, S.C. and Szostak, M. (2016). Efficient synthesis of diaryl ketones by Ni-catalyzed Negishi cross-coupling of amides by carbon–nitrogen bond cleavage at room temperature accelerated by a solvent effect. *Chem. Eur. J.* 22 (30): 10420–10424. (b) Roesner, S. and Buchwald, S.L. (2016). Continuous-flow synthesis of biaryls by Negishi cross-coupling of fluoro- and trifluoromethyl-substituted (hetero)arenes. *Angew. Chem. Int. Ed.* 55 (35): 10463–10467. (c) Shi, S.C. and Szostak, M. (2016). Nickel-catalyzed diaryl ketone synthesis by N—C cleavage: direct Negishi cross-coupling of primary amides by site-selective *N,N*-di-Boc activation. *Org. Lett.* 18 (22): 5872–5875. (d) Xia, T.-T., He, L., Liu, Y.A. et al. (2017). Palladium-catalyzed cross-coupling of ethyl bromodifluoroacetate with aryl bromides or triflates and cross-coupling of ethyl bromofluoroacetate with aryl iodides. *Org. Lett.* 19 (10): 2610–2613. (e) Shi, S.-C. and Szostak, M. (2017). Nickel-catalyzed Negishi cross-coupling of *N*-acylsuccinimides: stable, amide-based, twist-controlled acyl-transfer reagents via N—C activation. *Synthesis* 49 (16): 3602–3608. (f) Brittain, W.D.G. and Cobb, S.:.L. (2018). Negishi cross-couplings in the synthesis of amino acids. *Org. Biomol. Chem.* 16 (1): 10–20. (g) Borges-González, J. and Martín, T. (2018). Efficient synthesis of benzocyclotrimer analogues by Negishi cross-coupling and intramolecular nucleophilic substitution. *J. Chem. Soc., Chem. Commun.* 54 (4): 362–365.

2175 Negishi, E.I. (2011). Magical power of transition-metals: past, present, and future (Nobel Lecture). *Angew. Chem. Int. Ed.* 50: 6738–6764.

2176 Malosh, C.F. and Ready, J.M. (2004). Catalytic cross-coupling of alkylzinc halides with α-chloroketones. *J. Am. Chem. Soc.* 126: 10240–10241.

2177 Thapa, S., Kafle, A., Gurung, S.K. et al. (2015). Ligand-free Cu-catalyzed Negishi coupling of alkyl-, aryl-, and alkynylzinc reagents with heteroaryl iodides. *Angew. Chem. Int. Ed.* 54: 8236–8240.

2178 Thapa, S., Vangala, A.S., and Giri, R. (2016). Copper-catalyzed Negishi coupling of diarylzinc reagents with aryl iodides. *Synthesis* 48: 504–511.

2179 Ohmiya, H., Tsuji, T., Yorimitsu, H., and Oshima, K. (2004). Cobalt-catalyzed cross-coupling reactions of alkyl halides with allylic and benzylic Grignard reagents and their application to tandem radical cyclization/cross-coupling reactions. *Chem. Eur. J.* 10: 5640–5648.

2180 Struzinski, T.H., von Gohren, L.R., and MacArthur, A.H.R. (2009). Modified Co(II)(acetylacetonate) complexes as catalysts for Negishi-type coupling reactions: influence of ligand electronic properties on catalyst activity. *Transition Met. Chem.* 34: 637–640.

2181 Hammann, J.M., Haas, D., and Knochel, P. (2015). Cobalt-catalyzed Negishi cross-coupling reactions of (hetero)arylzinc reagents with primary and secondary alkyl bromides and iodides. *Angew. Chem. Int. Ed.* 54: 4478–4481.

2182 Haas, D., Hammann, J.M., Lutter, F.H., and Knochel, P. (2016). Mild Co-catalyzed Negishi cross-couplings of (hetero)arylzinc reagents with (hetero)aryl halides. *Angew. Chem. Int. Ed.* 55: 3809–3812.

2183 Hatakeyama, T., Nakagawa, N., and Nakamura, M. (2009). Iron-catalyzed Negishi coupling toward an effective olefin synthesis. *Org. Lett.* 11: 4496–4499.

2184 Cheung, C.W. and Hu, X.L. (2015). Stereoselective synthesis of trisubstituted alkenes through sequential Fe-catalyzed reductive *anti*-carbozincation of terminal alkynes and base-metal-catalyzed Negishi cross-coupling. *Chem. Eur. J.* 21: 18439–18444.

2185 Jia, W., Zhao, L.Z., Wei, H.X. et al. (2016). Iron-catalyzed cross-coupling of secondary alkyl chloride and the alkynes Grignard reagent. *Chin. J. Org. Chem.* 36: 1060–1064.

2186 Adams, C.J., Bedford, R.B., Carter, E. et al. (2012). Iron(I) in Negishi cross-coupling reactions. *J. Am. Chem. Soc.* 134: 10333–10336.

2187 Wu, D. and Wang, Z.X. (2014). P,N,N-Pincer Ni-catalyzed cross-coupling of aryl fluorides and chlorides. *Org. Biomol. Chem.* 12: 6414–6424.

2188 Sun, A.D., Leung, K., Restivo, A.D. et al. (2014). Nickel-catalyzed $C(sp^2)$-$C(sp^3)$ bond formation by carbon-fluorine activation. *Chem. Eur. J.* 20: 3162–3168.

2189 Li, B.J., Yu, D.G., Sun, C.L., and Shi, Z.J. (2011). Activation of "inert" alkenyl/aryl C—O bond and its application in cross-coupling reactions. *Chem. Eur. J.* 17: 1728–1759.

2190 Wang, C., Ozaki, T., Takita, R., and Uchiyama, M. (2012). Aryl ether as a Negishi coupling partner: an approach for constructing C—C bonds under mild conditions. *Chem. Eur. J.* 18: 3482–3485.

2191 Dubbaka, S.R. and Vogel, P. (2005). Organosulfur compounds: electrophilic reagents in transition-metal-catalyzed carbon–carbon bond forming reactions. *Angew. Chem. Int. Ed.* 44 (47): 7674–7684.

2192 Ortgies, D.H., Hassanpour, A., Chen, F. et al. (2016). Desulfination as an emerging strategy in Pd-catalyzed C—C coupling reactions. *Eur. J. Org. Chem.* 408–425.

2193 Aissa, C., Riveiros, R., Ragot, J., and Fürstner, A. (2003). Total syntheses of amphidinolide T1, T3, T4, and T5. *J. Am. Chem. Soc.* 125: 15512–15520.

2194 Iwai, T., Nakai, T., Mihara, M. et al. (2009). Pd-Catalyzed cross-coupling reactions of pyridine carboxylic acid chlorides with alkylzinc reagents. *Synlett* 1091–1094.

2195 Kim, S.H. and Rieke, R.D. (2011). Preparation of aryl ketones via Ni-catalyzed Negishi-coupling reactions with acid chlorides. *Tetrahedron Lett.* 52: 1523–1526.

2196 Benischke, A.D., Leroux, M., Knoll, I., and Knochel, P. (2016). Iron-catalyzed acylation of polyfunctionalized aryl- and benzylzinc halides with acid chlorides. *Org. Lett.* 18: 3626–3629.

2197 Tokuyama, H., Yokoshima, S., Yamashita, T., and Fukuyama, T. (1998). A novel ketone synthesis by a Pd-catalyzed reaction of thiol esters and organozinc reagents. *Tetrahedron Lett.* 39: 3189–3192.

2198 Mori, Y. and Seki, M. (2007). A practical synthesis of multifunctional ketones through the Fukuyama coupling reaction. *Adv. Synth. Catal.* 349: 2027–2038.

2199 Cherney, A.H. and Reisman, S.E. (2014). Pd-Catalyzed Fukuyama cross-coupling of secondary organozinc reagents for the direct synthesis of unsymmetrical ketones. *Tetrahedron* 70: 3259–3265.

2200 Mccann, L.C., Hunter, H.N., Clyburne, J.A.C., and Organ, M.G. (2012). Higher-order zincates as transmetalators in alkyl-alkyl Negishi cross-coupling. *Angew. Chem. Int. Ed.* 51: 7024–7027.

2201 Pompeo, M., Froese, R.D.J., Hadei, N., and Organ, M.G. (2012). Pd-PEPPSI-IPent(Cl): a highly effective catalyst for the selective cross-coupling of secondary organozinc reagents. *Angew. Chem. Int. Ed.* 51: 11354–11357.

2202 Erdik, E. (1992). Transition-metal-catalyzed reactions of organozinc reagents. *Tetrahedron* 48 (44): 9577–9648.

2203 Xin, J., Zhang, G.H., Deng, Y. et al. (2015). Which one is faster? A kinetic investigation of Pd and Ni-catalyzed Negishi-type oxidative coupling reactions. *J. Chem. Soc., Dalton Trans.* 44: 19777–19781.

2204 Nicolas, E., Ohleier, A., D'Accriscio, F. et al. (2015). "(Diphosphine)Ni"-catalyzed Negishi cross-coupling: an experimental and theoretical study. *Chem. Eur. J.* 21: 7690–7694.

2205 Phapale, V.B., Gusan-Ceinos, M., Buñuel, E., and Cárdenas, D.J. (2009). Nickel-catalyzed cross-coupling of (alkyl)Zn halides for the formation of C(sp^2)-C(sp^3) bonds: scope and mechanism. *Chem. Eur. J.* 15: 12681–12688.

2206 Schley, N.D. and Fu, G.C. (2014). Nickel-catalyzed Negishi arylations of propargylic bromides: a mechanistic investigation. *J. Am. Chem. Soc.* 136: 16588–16593.

2207 Jackson, R.F.W., Wishart, N., Wood, A. et al. (1992). Preparation of enantiomerically pure protected 4-oxo-α-amino acids and 3-aryl-α-amino acids from serine. *J. Org. Chem.* 57 (12): 3397–3404.

2208 Jackson, R.F.W., Wishart, N., and Wythes, M.J. (1993). A new α-amino-acid γ-anion equivalent - direct synthesis of enantiomerically pure protected amino-acids. *Synlett* (3): 219–220.

2209 Dunn, M.J., Jackson, R.F.W., Pietruszka, J. et al. (1993). Preparation of a serine-derived organozinc reagent in tetrahydrofuran - synthesis of novel, enantiomerically pure allenic, acetylenic and heteroaryl amino-acids. *Synlett* (7): 499–500.

2210 Dunn, M.J., Jackson, R.F.W., Pietruszka, J., and Turner, D. (1995). Synthesis of enantiomerically pure unsaturated α-amino-acids using serine-derived Zn/Cu reagents. *J. Org. Chem.* 60 (7): 2210–2215.

2211 (a) Blunden, S.J., Cusack, P.A., and Hill, R. (1985). *The Industrial uses of Tin Chemicals*, 1–337. Burlington House, London: The Royal Society of Chemistry. (b) Evans, C.J. (1998). Industrial uses of tin chemicals. In: *Chemistry of Tin* (ed. D.J. Smith), 442–479. Dordredcht, The Netherlands: Springer.

2212 Lambert, J.B., Wang, G.T., and Teramura, D.H. (1988). Interaction of the C—Ge or C—Sn bond with positive charge on a β-carbon. *J. Org. Chem.* 53 (23): 5422–5428.

2213 (a) Davies, A.G. and Smith, P.J. (1982). Tin organometallics. In: *Comprehensive Organometallic Chemistry*, Chapter 11, 1e, vol. 2 (ed. G. Wilkinson, F.G.A. Stone and E.W. Abel), 519–627. Oxford: Pergamon Press. (b) Davies, A.G. (2007). Tin organometallics. In: *Comprehensive Organometallic Chemistry*, Fundamentals to Applications, Chapter 3.14, 3e, vol. 3 (ed. R.H. Crabtree and D.M.P. Mingos), 809–883. Amsterdam, The Netherlands: Elsevier Ltd.

2214 Sato, T. (1995). Tin, in. In: *Comprehensive Organometallic Chemistry*, Chapter 8, 2e, vol. 11 (ed. E.W. Abel, F.G.A. Stone and G. Wilkinson), 355–387. Oxford: Pergamon.

2215 Marshall, J.A. (2002). *Organometallics in Synthesis, A Manual*, Chapter 2, 2e (ed. M. Schlosser), 353–464. New York: Wiley.

2216 Davies, A.G. (2004). *Organotin Chemistry*, 2e, 1–426. Weinheim: Wiley-VCH.

2217 Hiyama, T. (2013). Organosilicium and organotin chemistry. In: *Organometallics in Synthesis, A Manual*, Chapter 3, 3e (ed. M. Schlosser), 373–544. Weinheim: Wiley-VCH.

2218 (a) Baguley, P.A. and Walton, J.C. (1998). Flight from the tyranny of tin: the quest for practical radical sources free from metal encumbrances. *Angew. Chem. Int. Ed.* 37 (22): 3073–3082. (b) Parsons, A. (2002). A race against tin. *Chem. Br.* 38 (2): 42–44.

2219 Cruz, A., Anselmo, A.M., Suzuki, S., and Mendo, S. (2015). Tributyltin (TBT): a review on microbial resistance and degradation. *Crit. Rev. Environ. Sci. Technol.* 45: 970–1006.

2220 Niu, L., Li, Y.L., and Li, Q.S. (2014). Medicinal properties of organotin compounds and their limitations caused by toxicity. *Inorg. Chim. Acta* 423: 2–13.

2221 Le Grognec, E., Chrétien, J.M., Zammattio, F., and Quintard, J.P. (2015). Methodologies limiting or avoiding contamination by organotin residues in organic synthesis. *Chem. Rev.* 115: 10207–10260.

2222 Still, W.C. (1977). Stannylation-destannylation - new syntheses of carbonyl compounds via organotin intermediates. *J. Am. Chem. Soc.* 99 (14): 4836–4838.

2223 Gilbertson, S.R., Challener, C.A., Bos, M.E., and Wulff, W.D. (1988). An examination of the coupling of vinyl and aryl triflates with stannyl cuprates for the purpose of providing regioselective access to vinyl lithiums. *Tetrahedron Lett.* 29 (38): 4795–4798.

2224 Seyferth, D. and Weiner, M.A. (1961). Preparation of organolithium compounds by transmetalation reaction. 3. Allyllithium and methallyllithium. *J. Org. Chem.* 26 (12): 4797–4800.

2225 Tanaka, H., Hai, A.K.M.A., Ogawa, H., and Torii, S. (1993). Mg/$PbBr_2$ bimetal redox-promoted stannation of propargyl, allyl, vinyl, and aryl halides. *Synlett* (11): 835–836.

2226 Carofiglio, T., Marton, D., and Tagliavini, G. (1992). New simple route to allylstannanes by Zn-mediated coupling of allyl bromides with Bu_3SnCl or Bu_2SnCl_2 in an H_2O (NH_4Cl)/THF medium. *Organometallics* 11 (9): 2961–2963.

2227 Ruitenberg, K., Westmijze, H., Kleijn, H., and Vermeer, P. (1984). Convenient synthesis of (triphenyl)germyl and (triphenyl)stannyl substituted allenes. *J. Organomet. Chem.* 277 (2): 227–234.

2228 Helmke, H. and Hoppe, D. (1995). Chelation-directed asymmetric lithiation and C-substitution of 1,2,4-butanetriol acetonide. *Synlett* (9): 978–980.

2229 Coeffard, V., Beaudet, I., Quintard, J.P., and Le Grognec, E. (2010). Preparation of enantiomerically enriched α-aminoorganostannanes and their applications in stereoselective synthesis. *Chirality* 22: 864–869.

2230 Stille, J.K. (1986). The Pd-catalyzed cross-coupling reactions of organotin reagents with organic electrophiles. *Angew. Chem. Int. Ed. Engl.* 25 (6): 508–523.

2231 Farina, V., Krishnamurthy, V., and Scott, W.J. (1997). The Stille reaction. *Org. React.* 50: 1–652.

2232 Hartwig, J.F. (2009). *Organotransition Metal Chemistry: From Bonding to Catalysis*, 1e, 1–1160. Sausalito, CA: University Science Books.

2233 Cordovilla, C., Bartolomé, C., Martínez-Ilarduya, J.M., and Espinet, P. (2015). The Stille reaction, 38 years later. *ACS Catal.* 5: 3040–3053.

2234 So, C.M. and Kwong, F.Y. (2011). Palladium-catalyzed cross-coupling reactions of aryl mesylates. *Chem. Soc. Rev.* 40: 4963–4972.

2235 Roglans, A., Pla-Quintana, A., and Moreno-Mañas, M. (2006). Diazonium salts as substrates in Pd-catalyzed cross-coupling reactions. *Chem. Rev.* 106: 4622–4643.

2236 Jeanneret, V., Meerpoel, L., and Vogel, P. (1997). C-Glycosides and C-disaccharide precursors through carbonylative Stille coupling reactions. *Tetrahedron Lett.* 38 (4): 543–546.

2237 Laio, B.Q. and Negishi, E. (2000). Development of a method for the synthesis of α-substituted α,β-unsaturated lactones based on Stille-type Pd-catalyzed carbonylation of (Z)-ω-iodoalkenols. An efficient and selective synthesis of (+)-hamabiwalactone B. *Heterocycles* 52 (3): 1241–1249.

2238 Mazzola, R.D. Jr., Giese, S., Benson, C.L., and West, F.G. (2004). Improved yields with added Cu(I) salts in carbonylative Stille couplings of sterically hindered vinylstannanes. *J. Org. Chem.* 69 (1): 220–223.

2239 Steunenberg, P., Jeanneret, V., Zhu, Y.H., and Vogel, P. (2005). C(1 -> 4)-Linked disaccharides through carbonylative Stille cross-coupling. *Tetrahedron: Asymmetry* 16 (2): 337–346.

2240 Dubbaka, S.R. and Vogel, P. (2003). Palladium-catalyzed Stille cross-couplings of sulfonyl chlorides and organostannanes. *J. Am. Chem. Soc.* 125 (50): 15292–15293.

2241 Dubbaka, S.R., Steunenberg, P., and Vogel, P. (2004). Aryl and arylmethyl C-glycosides through desulfitative Stille and carbonylative stille cross-coupling of tinglycals and sulfonyl chlorides. *Synlett* (7): 1235–1238.

2242 Dowlut, M., Mallik, D., and Organ, M.G. (2010). An efficient low-temperature Stille–Migita cross-coupling reaction for heteroaromatic compounds by Pd-PEPPSI-IPent. *Chem. Eur. J.* 16: 4279–4283.

2243 Kang, S.K., Kim, J.S., Yoon, S.K. et al. (1998). Copper-catalyzed coupling of polymer bound iodide with organostannanes. *Tetrahedron Lett.* 39: 3011–3012.

2244 Li, J.H., Tang, B.X., Tao, L.M. et al. (2006). Reusable Cu-catalyzed cross-coupling reactions of aryl halides with organotins in inexpensive ionic liquids. *J. Org. Chem.* 71: 7488–7490.

2245 Wang, M.Y. and Lin, Z.Y. (2010). Stille cross-coupling reactions of alkenylstannanes with alkenyl iodides mediated by Cu(I) thiophene-2-carboxylate: a density functional study. *Organometallics* 29: 3077–3084.

2246 Liebeskind, L.S. and Fengl, R.W. (1990). 3-Stannylcyclobutenediones as nucleophilic cyclobutenedione equivalents - synthesis of substituted cyclobutenediones and cyclobutenedione monoacetals and the beneficial effect of catalytic CuI on the Stille reaction. *J. Org. Chem.* 55: 5359–5364.

2247 Farina, V., Kapadia, S., Krishnan, B. et al. (1994). On the nature of the Cu effect in the Stille cross-coupling. *J. Org. Chem.* 59 (20): 5905–5911.

2248 Verbeeck, S., Meyers, C., Franck, P. et al. (2010). Dual effect of halides in the Stille reaction: in situ halide metathesis and catalyst stabilization. *Chem. Eur. J.* 15: 12831–12837.

2249 Littke, A.F. and Fu, G.C. (1999). The first general method for Stille cross-couplings of aryl chlorides. *Angew. Chem. Int. Ed.* 38: 2411–2413.

2250 Grasa, G.A. and Nolan, S.P. (2001). Palladium/imidazolium salt catalyzed coupling of aryl halides with hypervalent organostannates. *Org. Lett.* 3: 119–122.

2251 Littke, A.F., Schwarz, L., and Fu, G.C. (2002). Pd/P(t-Bu)$_3$: a mild and general catalyst for Stille reactions of aryl chlorides and aryl bromides. *J. Am. Chem. Soc.* 124 (22): 6343–6348.

2252 Ariafard, A. and Yates, B.F. (2009). Subtle balance of ligand steric effects in Stille transmetalation. *J. Am. Chem. Soc.* 131: 13981–13991.

2253 (a) Hervé, M., Lefèvre, G., Mitchell, E.A. et al. (2015). On the triple role of fluoride ions in Pd-catalyzed Stille reactions. *Chem. Eur. J.* 21: 18401–18406. (b) Grimaud, L. and Jutand, A. (2017). Role of fluoride ions in Pd-catalyzed cross-coupling reactions. *Synthesis* 49 (6): 1182–1189.

2254 Mee, S.P.H., Lee, V., and Baldwin, J.E. (2004). Stille coupling made easier - the synergic effect of Cu(I) salts and the fluoride ion. *Angew. Chem. Int. Ed.* 43 (9): 1132–1136.

2255 Mee, S.P.H., Lee, V., and Baldwin, J.E. (2005). Significant enhancement of the Stille reaction with a new combination of reagents CuI with CsF. *Chem. Eur. J.* 11: 3294–3308.

2256 Pérez-Temprano, M.H., Casares, J.A., de Lera, A.R. et al. (2012). Strong metallophilic interactions in the Pd-arylation by Au(aryls). *Angew. Chem. Int. Ed.* 51: 4917–4920.

2257 Delpozo, J., Casares, J.A., and Espinet, P. (2013). The decisive role of ligand metathesis in Au/Pd bimetallic catalysis. *J. Chem. Soc., Chem. Commun.* 49: 7246–7248.

2258 Delpozo, J., Carrasco, D., Pérez-Temprano, M.H. et al. (2013). Stille coupling involving bulky groups feasible with Au cocatalyst. *Angew. Chem. Int. Ed.* 52: 2189–2193.

2259 Carrasco, D., García-Melchor, M., Casares, J.A., and Espinet, P. (2016). Dramatic mechanistic switch in Sn/Au-I group exchanges: transmetalation vs. oxidative addition. *J. Chem. Soc., Chem. Commun.* 52: 4305–4308.

2260 Smith, A.B. and Ott, G.R. (1998). Total syntheses of (−)-macrolactin A, (+)-macrolactin E, and (−)-macrolactinic acid: an exercise in Stille cross-coupling chemistry. *J. Am. Chem. Soc.* 120: 3935–3948.

2261 De Souza, M.V.N. (2006). Stille reaction: an important tool in the synthesis of complex natural products. *Curr. Org. Synth.* 3: 313–326.

2262 Heravi, M.M., Hashemi, E., and Azimian, F. (2014). Recent developments of the Stille reaction as a revolutionized method in total synthesis. *Tetrahedron* 70: 7–21.

2263 Rummelt, S.M., Preindl, J., Sommer, H., and Fürstner, A. (2015). Selective formation of a trisubstituted alkene motif by *trans*-hydrostannation/Stille coupling: application to the total synthesis and late-stage modification of 5,6-dihydrocineromycin-B. *Angew. Chem. Int. Ed.* 54: 6241–6245.

2264 Asaba, T., Katoh, Y., Urabe, D., and Inoue, M. (2015). Total synthesis of crotophorbolone. *Angew. Chem. Int. Ed.* 54: 14457–14461.

2265 Azarian, D., Dua, S.S., Eaborn, C., and Walton, D.R.M. (1976). Reactions of organic halides with R_3M-MR_3 compounds (M=Si,Ge,Sn) in presence of [tetrakis(triarylphosphine)]Pd. *J. Organomet. Chem.* 117: C55–C57.

2266 Kosugi, M., Shimizu, Y., and Migita, T. (1977). Reaction of allyltin compounds. 2. Facile preparation of allyl ketones via allyltins. *J. Organomet. Chem.* 129: C36–C38.

2267 Kosugi, M., Shimizu, Y., and Migita, T. (1977). Alkylation, arylation, and vinylation of acyl chlorides by means of organotin compounds in presence of catalytic amounts of [tetrakis(triphenylphosphine)]Pd(0). *Chem. Lett.* 1423–1424.

2268 Kosugi, M., Sasazawa, K., Shimizu, Y., and Migita, T. (1977). Reactions of allyltin compounds. 3. Allylation of aromatic halides with allyltributyltin in presence of [tetrakis(triphenylphosphine)]Pd(0). *Chem. Lett.* 301–302.

2269 Milstein, D. and Stille, J.K. (1978). General, selective, and facile method for ketone synthesis from acid-chlorides and organotin compounds catalyzed by Pd. *J. Am. Chem. Soc.* 100: 3636–3638.

2270 Milstein, D. and Stille, J.K. (1979). Mild, selective, general method of ketone synthesis from acid-chlorides and organotin compounds catalyzed by Pd. *J. Am. Chem. Soc.* 44: 1613–1618.

2271 Wittenberg, R., Srogl, J., Egi, M., and Liebeskind, L.S. (2003). Ketone synthesis under neutral conditions. Cu(I) diphenylphosphinate-mediated, Pd-catalyzed coupling of thiol esters and organostannanes. *Org. Lett.* 5: 3033–3035.

2272 Milstein, D. and Stille, J.K. (1979). Palladium-catalyzed coupling of tetraorganotin compounds with aryl and benzyl halides - synthetic utility and mechanism. *J. Am. Chem. Soc.* 101: 4992–4998.

2273 Stille, J.K. and Lau, K.S.Y. (1977). Mechanisms of oxidative addition of organic halides to Group-8 transition-metal complexes. *Acc. Chem. Res.* 10 (12): 434–442.

2274 Espinet, P. and Echavarren, A.M. (2004). The mechanisms of the Stille reaction. *Angew. Chem. Int. Ed.* 43 (36): 4704–4734.

2275 García-Melchor, M., Braga, A.A.C., Lledós, A. et al. (2013). Computational perspective on Pd-catalyzed C—C cross-coupling reaction mechanisms. *Acc. Chem. Res.* 46: 2626–2634.

2276 Kozuch, S., Amatore, C., Jutand, A., and Shaik, S. (2005). What makes for a good catalytic cycle? A theoretical study of the role of an anionic Pd(0) complex in the cross-coupling of an aryl halide with an anionic nucleophile. *Organometallics* 24: 2319–2330.

2277 Goossen, L.J., Koley, D., Hermann, H.L., and Thiel, W. (2005). Mechanistic pathways for oxidative addition of aryl halides to Pd(0) complexes: a DFT study. *Organometallics* 24: 2398–2410.

2278 Martin, R. and Buchwald, S.L. (2008). Palladium-catalyzed Suzuki–Miyaura cross-coupling reactions smploying dialkylbiaryl phosphine ligands. *Acc. Chem. Res.* 41 (11): 1461–1473.

2279 Naber, J.R. and Buchwald, S.L. (2008). Palladium-catalyzed Stille cross-coupling reaction of aryl chlorides using a pre-milled Pd acetate and XPhos catalyst system. *Adv. Synth. Catal.* 350: 957–961.

2280 Ahlquist, M. and Norrby, P.O. (2007). Oxidative addition of aryl chlorides to monoligated Pd(0): a DFT-SCRF study. *Organometallics* 26: 550–553.

2281 Li, Z., Fu, Y., Guo, Q.X., and Liu, L. (2008). Theoretical study on monoligated Pd-catalyzed cross-coupling reactions of aryl chlorides and bromides. *Organometallics* 27: 4043–4049.

2282 Gourlaouen, C., Ujaque, G., Lledós, A. et al. (2009). Why is the Suzuki–Miyaura cross-coupling of sp^3-carbons in α-bromo sulfoxide systems fast and stereoselective? A DFT study on the mechanism. *J. Org. Chem.* 74: 4049–4054.

2283 Besora, M., Gourlaouen, C., Yates, B., and Maseras, F. (2011). Phosphine and solvent effects on oxidative addition of CH_3Br to $Pd(PR_3)$ and $Pd(PR_3)_2$ complexes. *J. Chem. Soc., Dalton Trans.* 40: 11089–11094.

2284 McMullin, C.L., Jover, J., Harvey, J.N., and Fey, N. (2010). Accurate modelling of Pd(0) + PhX oxidative addition kinetics. *J. Chem. Soc., Dalton Trans.* 39: 10833–10836.

2285 Schoenebeck, F. and Houk, K.N. (2010). Ligand-controlled regioselectivity in Pd-catalyzed cross-coupling reactions. *J. Am. Chem. Soc.* 132: 2496–2497.

2286 Barrios-Landeros, F., Carrow, B.P., and Hartwig, J.F. (2009). Effect of ligand steric properties and halide identity on the mechanism for oxidative addition of haloarenes to trialkylphosphine Pd(0) complexes. *J. Am. Chem. Soc.* 131: 8141–8154.

2287 Kalkofen, R. and Hoppe, D. (2006). First example of an enantiospecific sp^3-sp^2 Stille coupling of a chiral allylstannane with aryl halides. *Synlett* 1959–1961.

2288 Goli, M., He, A.Y., and Felck, J.R. (2011). Pd-catalyzed cross-coupling of α-(acyloxy)-tri-*n*-butylstannanes with alkenyl, aryl, and heteroaryl electrophiles. *Org. Lett.* 13: 344–346.

2289 Roth, G.P., Farina, V., Liebeskind, L.S., and Penacabrera, E. (1995). Optimized Stille coupling reactions catalyzed by Pd on Carbon with CuI as co-catalyst. *Tetrahedron Lett.* 36: 2191–2194.

2290 Pérez-Rodríguez, M., Braga, A.A.C., Garcia-Melchor, M. et al. (2009). C—C Reductive elimination in Pd Complexes, and the role of coupling additives. A DFT study supported by experiment. *J. Am. Chem. Soc.* 131: 3650–3657.

2291 Allred, G.D. and Liebeskind, L.S. (1996). Copper-mediated cross-coupling of organostannanes with organic iodides at or below room temperature. *J. Am. Chem. Soc.* 118: 2748–2749.

2292 Falck, J.R., Patel, P.K., and Bandyopadhyay, A. (2007). Stereospecific cross-coupling of α-(thiocarbamoyl)organostannanes with alkenyl, aryl, and heteroaryl iodides. *J. Am. Chem. Soc.* 129: 790–793.

2293 Cardenas, D.J., Mateo, C., and Echavarren, A.M. (1994). Synthesis of oxapalladacycles and azapalladacycles from organostannanes. *Angew. Chem. Int. Ed.* 33: 2445–2447.

2294 Suzuki, A. (1982). Organoborates in new synthetic reactions. *Acc. Chem. Res.* 15 (6): 178–184.

2295 Suzuki, A. (1983). Some aspects of organic synthesis using organoborates. *Top. Curr. Chem.* 112: 67–115.

2296 Miyaura, N. and Suzuki, A. (1995). Palladium-catalyzed cross-coupling reactions of organoboron compounds. *Chem. Rev.* 95 (7): 2457–2483.

2297 Suzuki, A. (1999). Recent advances in the cross-coupling reactions of organoboron derivatives with organic electrophiles, 1995–1998. *J. Organomet. Chem.* 576 (1–2): 147–168.

2298 Hassan, J., Sévignon, M., Gozzi, C. et al. (2002). Aryl-aryl bond formation one century after the discovery of the Ullmann reaction. *Chem. Rev.* 102 (5): 1359–1469.

2299 Agrofoglio, L.A., Gillaizeau, I., and Saito, Y. (2003). Palladium-assisted routes to nucleosides. *Chem. Rev.* 103: 1875–1916.

2300 (a) Sun, H.Y. and Hall, D.G. (2015). At the forefront of the Suzuki–Miyaura reaction: advances in stereoselective cross-couplings. In: *Synthesis and Applications of Organoboron Compounds* (ed. E. Fernández and A. Whiting), 221–242. Berlin, Heidelberg: Springer-Verlag. (b) Rygus, J.P.G. and Crudden, C.M. (2017). Enantiospecific and iterative Suzuki–Miyaura cross-couplings. *J. Am. Chem. Soc.* 139 (50): 18124–18137. (c) Del Zotto, A. and Zuccaccia, D. (2017). Metallic palladium, PdO, and palladium supported on metal oxides for the Suzuki–Miyaura cross-coupling reaction: a unified view of the process of formation of the catalytically active species in solution. *Catal. Sci. Technol.* 7 (18): 3934–3951. (d) Len, C., Bruniaux, S., Delbecq, F., and Parmar, V.S. (2017). Palladium-catalyzed Suzuki–Miyaura cross-coupling in continuous flow. *Catalysts* 7 (5): 146/1–146/23.

2301 Chemler, S.R., Trauner, D., and Danishefsky, S.J. (2001). The *B*-alkyl Suzuki–Miyaura cross-coupling reaction: development, mechanistic study, and applications in natural product synthesis. *Angew. Chem. Int. Ed.* 40: 4544–4568.

2302 Sasaki, M. and Fuwa, H. (2004). Total synthesis of polycyclic ether natural products based on Suzuki–Miyaura cross-coupling. *Synlett* 1851–1874.

2303 Nicolaou, K.C., Bulger, P.G., and Sarlah, D. (2005). Palladium-catalyzed cross-coupling reactions in total synthesis. *Angew. Chem. Int. Ed.* 44 (29): 4442–4489.

2304 Selepe, M.A. and Van Heerden, F.R. (2013). Application of the Suzuki–Miyaura reaction in the synthesis of flavonoids. *Molecules* 18: 4739–4765.

2305 Rossi, R., Bellina, F., Lessi, M. et al. (2015). Recent applications of phosphane-based Pd catalysts in Suzuki–Miyaura reactions involved in total syntheses of natural products. *Curr. Org. Chem.* 19: 1302–1409.

2306 Torborg, C. and Beller, M. (2009). Recent applications of Pd-catalyzed coupling reactions in the pharmaceutical, agrochemical, and fine chemical industries. *Adv. Synth. Catal.* 351 (18): 3027–3043.

2307 Verdelet, T., Mercey, G., Correa, N. et al. (2011). Straightforward and efficient synthesis of 3-benzyloxy-4-bromopicolinate ester and 3-benzyloxy-5-bromopicolinate ester, common building blocks for pharmaceuticals and agrochemicals. *Tetrahedron* 67: 8757–8762.

2308 Schluter, A.D. (2001). The tenth anniversary of Suzuki polycondensation (SPC). *J. Polym. Sci., Part A: Polym. Chem.* 67: 1533–1556.

2309 Beinhoff, M., Karakaya, B., and Schluter, A.D. (2003). Synthesis of low generation phenylenealkylene dendrons as nonpolar building blocks for a dendrimer construction set. *Synthesis* 79–90.

2310 Yokoyama, A., Suzuki, H., Kubota, Y. et al. (2007). Chain-growth polymerization for the synthesis of polyfluorene via Suzuki–Miyaura coupling reaction from an externally added initiator unit. *J. Am. Chem. Soc.* 129: 7236–7237.

2311 Yokozawa, T., Suzuki, R., Nojima, M. et al. (2011). Precision synthesis of poly(3-hexylthiophene) from catalyst-transfer Suzuki–Miyaura coupling polymerization. *Macromol. Rapid Commun.* 32: 801–806.

2312 Casalnuovo, A.L. and Calabrese, J.C. (1990). Palladium-catalyzed alkylations in aqueous-media. *J. Am. Chem. Soc.* 112: 4324–4330.

2313 Shaughnessy, K.H. and DeVasher, R.B. (2005). Palladium-catalyzed cross-coupling in aqueous media: recent progress and current applications. *Curr. Org. Chem.* 9: 585–604.

2314 Asachenko, A.F., Sorochkina, K.R., Dzhevakov, P.B. et al. (2013). Suzuki–Miyaura cross-coupling under solvent-free conditions. *Adv. Synth. Catal.* 355: 3553–3557.

2315 Miyaura, N., Yamada, K., and Suzuki, A. (1979). New stereospecific cross-coupling by the Pd-catalyzed reaction of 1-alkenylboranes with 1-alkenyl or 1-alkynyl halides. *Tetrahedron Lett.* 20: 3437–3440.

2316 Miyaura, N. and Suzuki, A. (1979). Stereoselective synthesis of arylated (*E*)-alkenes by the reaction of alk-1-enylboranes with aryl halides in the presence of Pd catalyst. *J. Chem. Soc., Chem. Commun.* 866–867.

2317 Molander, G.A. and Biolatto, B. (2003). Palladium-catalyzed Suzuki–Miyaura cross-coupling reactions of potassium aryl- and heteroaryltrifluoroborates. *J. Org. Chem.* 68 (11): 4302–4314.

2318 Molander, G.A., Yun, C.S., Ribagorda, M., and Biolatto, B. (2003). B-alkyl Suzuki–Miyaura cross-coupling reactions with air-stable potassium alkyltrifluoroborates. *J. Org. Chem.* 68: 5534–5539.

2319 Molander, G.A. and Sandrock, D.L. (2009). Potassium trifluoroborate salts as convenient, stable reagents for difficult alkyl transfers. *Curr. Opin. Drug Discov. Dev.* 12: 811–823.

2320 Lennox, A.J.J. and Lloyd-Jones, G.C. (2014). Selection of boron reagents for Suzuki–Miyaura coupling. *Chem. Soc. Rev.* 43: 412–443.

2321 Littke, A.F. and Fu, G.C. (2002). Palladium-catalyzed coupling reactions of aryl chlorides. *Angew. Chem. Int. Ed.* 41 (22): 4176–4211.

2322 Jacks, T.E., Belmont, D.T., Briggs, C.A. et al. (2004). Development of a scalable process for CI-1034, an endothelin antagonist. *Org. Process Res. Dev.* 8: 201–212.

2323 Dubbaka, S.R. and Vogel, P. (2004). Palladium-catalyzed Suzuki–Miyaura cross-couplings of sulfonyl chlorides and boronic acids. *Org. Lett.* 6 (1): 95–98.

2324 Rouhi, A.M. (2004). Fine chemicals. *Chem. Eng. News* 82: 49–51.

2325 Liebeskind, L.S. and Srogl, J. (2000). Thiol ester-boronic acid coupling. A mechanistically unprecedented and general ketone synthesis. *J. Am. Chem. Soc.* 122: 11260–11261.

2326 Prokopcova, H. and Kappe, C.O. (2009). The Liebeskind–Srogl C—C cross-coupling reaction. *Angew. Chem. Int. Ed.* 48: 2276–2286.

2327 Buchwald, S.L. and Martin, R. (2008). Palladium-catalyzed Suzuki–Miyaura cross-coupling reactions employing dialkylbiaryl phosphine ligands. *Acc. Chem. Res.* 41 (11): 1461–1473.

2328 Lee, S.J., Gray, K.C., Paek, J.S., and Burke, M.D. (2008). Simple, efficient, and modular syntheses of polyene natural products via iterative cross-coupling. *J. Am. Chem. Soc.* 130: 466–468.

2329 Urawa, Y., Miyazawa, M., Ozeki, N., and Ogura, K. (2003). A novel methodology for efficient removal of residual Pd from a product of the Suzuki–Miyaura coupling with polymer-supported ethylenediamine derivatives. *Org. Process Res. Dev.* 7: 191–195.

2330 Egorova, K.S. and Ananikov, V. (2016). Which metals are green for catalysis? Comparison of the toxicities of Ni, Cu, Fe, Pd, Pt, Rh, and Au salts. *Angew. Chem. Int. Ed.* 55: 12150–12162.

2331 Percec, V., Bae, J.Y., and Hill, D.H. (1995). Aryl mesylates in metal-catalyzed homocoupling and cross-coupling reactions. 2. Suzuki-type Ni-catalyzed cross-coupling of aryl arenesulfonates and aryl mesylates with arylboronic acids. *J. Org. Chem.* 60: 1060–1065.

2332 Inada, K. and Miyaura, N. (2000). Synthesis of biaryls via cross-coupling reaction of arylboronic acids with aryl chlorides catalyzed by $NiCl_2$/triphenylphosphine complexes. *Tetrahedron* 56: 8657–8660.

2333 Tobisu, M., Xu, T., Shimasaki, T., and Chatani, N. (2011). Nickel-catalyzed Suzuki–Miyaura reaction of aryl fluorides. *J. Am. Chem. Soc.* 133: 19505–19511.

2334 Ohashi, M., Saijo, H., Shibata, M., and Ogoshi, S. (2013). Palladium-catalyzed base-free Suzuki–Miyaura coupling reactions of fluorinated alkenes and arenes via a Pd fluoride key intermediate. *Eur. J. Org. Chem.* 443–447.

2335 Malineni, J., Jerozek, R.L., Zhang, N., and Percec, V. (2016). An indefinitely air-stable σ-Ni(II) precatalyst for quantitative cross-coupling of unreactive aryl halides and mesylates with aryl neopentylglycolboronates. *Synthesis* 48: 2795–2807.

2336 Quasdorf, K.W., Riener, M., Petrova, K.V., and Garg, N.K. (2009). Suzuki–Miyaura coupling of aryl carbamates, carbonates, and sulfamates. *J. Am. Chem. Soc.* 131: 17748–17749.

2337 Chen, H., Huang, Z.B., Hu, X.M. et al. (2011). Nickel-catalyzed cross-coupling of aryl phosphates with arylboronic acids. *J. Org. Chem.* 76: 2338–2344.

2338 Han, F.S. (2013). Transition-metal-catalyzed Suzuki–Miyaura cross-coupling reactions: a remarkable advance from Pd to Ni catalysts. *Chem. Soc. Rev.* 42: 5270–5298.

2339 Mesganaw, T. and Garg, N.K. (2013). Ni- and Fe-Catalyzed cross-coupling reactions of phenol derivatives. *Org. Process Res. Dev.* 17: 29–39.

2340 García, P.M.P., Di Franco, T., Orsino, A. et al. (2012). Nickel-catalyzed diastereoselective alkyl-alkyl Kumada coupling reactions. *Org. Lett.* 14: 4286–4289.

2341 Di Franco, T., Boutin, N., and Hu, X.L. (2013). Suzuki–Miyaura cross-coupling reactions of unactivated alkyl halides catalyzed by a Ni-pincer complex. *Synthesis* 45: 2949–2958.

2342 Muto, K., Yamaguchi, J., Musaev, D.G., and Itami, K. (2015). Decarbonylative organoboron cross-coupling of esters by Ni catalysis. *Nat. Commun.* 6: 7508.

2343 Shi, S.C., Meng, G.R., and Szostak, M. (2016). Synthesis of biaryls through Ni-catalyzed Suzuki–Miyaura coupling of amides by carbon-nitrogen bond cleavage. *Angew. Chem. Int. Ed.* 55: 6959–6963.

2344 Liu, L., Chen, P., Sung, Y. et al. (2016). Mechanism of Ni-catalyzed selective C—N bond activation in Suzuki–Miyaura cross-coupling of amides: a theoretical investigation. *J. Org. Chem.* 81: 11686–11696.

2345 Weires, N.A., Baker, E.L., and Garg, N.K. (2016). Nickel-catalysed Suzuki–Miyaura coupling of amides. *Nat. Chem.* 8: 75–79.

2346 Na, Y., Park, S., Han, S.B. et al. (2004). Ruthenium-catalyzed Heck-type olefination and Suzuki coupling reactions: studies on the nature of catalytic species. *J. Am. Chem. Soc.* 126 (1): 250–258.

2347 Li, J.H., Li, J.L., and Xie, Y.X. (2007). TBAB-promoted ligand-free Cu-catalyzed cross-coupling reactions of aryl halides with arylboronic acids. *Synthesis* 984–988.

2348 Yang, C.T., Zhang, Z.Q., Liu, Y.C., and Liu, L. (2011). Copper-catalyzed cross-coupling reaction of organoboron compounds with primary alkyl halides and pseudohalides. *Angew. Chem. Int. Ed.* 50: 3904–3907.

2349 Sun, Y.Y., Yi, J., Lu, X. et al. (2014). Cu-Catalyzed Suzuki–Miyaura reactions of primary and secondary benzyl halides with arylboronates. *J. Chem. Soc., Chem. Commun.* 50: 11060–11062.

2350 Gurung, S.K., Thapa, S., Kafle, A. et al. (2014). Copper-catalyzed Suzuki–Miyaura coupling of arylboronate esters: transmetalation with (PN)CuF and identification of intermediates. *Org. Lett.* 16: 1264–1267.

2351 Handa, S., Smith, J.D., Hageman, M.S. et al. (2016). Synergistic and selective Cu/ppm Pd-catalyzed Suzuki–Miyaura couplings: in water, mild conditions, with recycling. *ACS Catal.* 6: 8179–8183.

2352 Ren, P., Stern, L.A., and Hu, X.L. (2012). Copper-catalyzed cross-coupling of functionalized alkyl halides and tosylates with secondary and tertiary alkyl Grignard reagents. *Angew. Chem. Int. Ed.* 51: 9110–9113.

2353 Hatakeyama, T., Hashimoto, T., Kondo, Y. et al. (2010). Iron-catalyzed Suzuki–Miyaura coupling of alkyl halides. *J. Am. Chem. Soc.* 132: 10674–10676.

2354 Nakagawa, N., Hatakeyama, T., and Nakamura, M. (2015). Iron-catalyzed Suzuki–Miyaura coupling reaction of unactivated alkyl halides with Li alkynylborates. *Chem. Lett.* 44: 486–488.

2355 Bedford, R.B., Brenner, P.B., Carter, E. et al. (2014). Expedient Fe-catalyzed coupling of alkyl, benzyl and allyl halides with arylboronic esters. *Chem. Eur. J.* 20: 7935–7938.

2356 Bedford, R.B., Gallagher, T., Pye, D.R., and Savage, W. (2015). Towards Fe-catalysed Suzuki biaryl cross-coupling: unusual reactivity of 2-halobenzyl halides. *Synthesis* 47: 1761–1765.

2357 Bedford, R.B., Brenner, P.B., Carter, E. et al. (2014). Iron(phosphine)-catalyzed cross-coupling of tetraorganoborates and related Group 13 nucleophiles with alkyl halides. *Organometallics* 33: 5767–5780.

2358 Bedford, R.B., Nakamura, M., Gower, N.J. et al. (2009). Iron-catalysed Suzuki coupling? A cautionary tale. *Tetrahedron Lett.* 50: 6110–6111.

2359 Pérez-Lorenzo, M. (2012). Palladium nanoparticles as efficient catalysts for Suzuki Cross-coupling reactions. *J. Phys. Chem. Lett.* 3: 167–174.

2360 Balanta, A., Godard, C., and Claver, C. (2011). Pd nanoparticles for C—C coupling reactions. *Chem. Soc. Rev.* 40: 4973–4985.

2361 Dong, W.H., Zhang, L., Wang, C.H. et al. (2016). Palladium nanoparticles embedded in metal-organic framework derived porous carbon: synthesis and application for efficient Suzuki–Miyaura coupling reactions. *RSC Adv.* 6: 37118–37123.

2362 Uozumi, Y. (2004). Recent progress in polymeric Pd catalysts for organic synthesis. *Immobiliz. Catal.* 242: 77–112.

2363 He, H.S., Yan, J.J., Shen, R.X. et al. (2006). Non-cross-linked polystyrene-supported triphenylphosphine-microencapsulated palladium: an efficient and recyclable catalyst for Suzuki–Miyaura reactions. *Synlett* 563–566.

2364 Solodenko, W., Mennecke, K., Vogt, C. et al. (2006). Polyvinylpyridine, a versatile solid phase for coordinative immobilisation of Pd precatalysts - applications in Suzuki–Miyaura reactions. *Synthesis* 1873–1881.

2365 Glegola, K., Framery, E., Pietrusiewicz, K.M., and Sinou, D. (2006). Efficient recycling in Suzuki–Miyaura reactions of aryl chlorides with arylboronic acids using polymer-supported aryldicyclohexylphosphine. *Adv. Synth. Catal.* 348: 1728–1733.

2366 Li, J.H., Hu, X.C., and Me, Y.X. (2006). Polymer-supported DABCO-Pd complex as a stable and reusable catalyst for room temperature Suzuki–Miyaura cross-couplings of aryl bromides. *Tetrahedron Lett.* 47: 9239–9243.

2367 Liao, Y.T., He, L.S., Huang, J. et al. (2010). Magnetite nanoparticle-supported coordination polymer nanofibers: synthesis and catalytic application in Suzuki–Miyaura coupling. *ACS Appl. Mater. Interfaces* 2: 2333–2338.

2368 Das, P., Sharma, D., Shil, A.K., and Kumari, A. (2011). Solid-supported Pd nano and microparticles: an efficient heterogeneous catalyst for ligand-free Suzuki–Miyaura cross coupling reaction. *Tetrahedron Lett.* 52: 1176–1178.

2369 Ding, S.Y., Gao, J., Wang, Q. et al. (2011). Construction of covalent organic framework for catalysis: Pd/COF-LZU1 in Suzuki–Miyaura coupling reaction. *J. Am. Chem. Soc.* 133: 19816–19822.

2370 Yamada, Y.M.A., Sarkar, S.M., and Uozumi, Y. (2012). Self-assembled poly(imidazole-Pd): highly active, reusable catalyst at parts per million to parts per billion levels. *J. Am. Chem. Soc.* 134: 3190–3198.

2371 Heravi, M.M., Hashemi, E., Beheshtiha, Y.S. et al. (2014). $PdCl_2$ on modified poly(styrene-co-maleic anhydride): a highly active and recyclable catalyst for the Suzuki–Miyaura and Sonogashira reactions. *J. Mol. Catal. A: Chem.* 394: 74–82.

2372 Sharma, R.K., Dutta, S., Sharma, S. et al. (2016). Fe_3O_4 (iron oxide)-supported nanocatalysts: synthesis, characterization and applications in coupling reactions. *Green Chem.* 18: 3184–3209.

2373 Han, J., Liu, Y., and Guo, R. (2009). Facile synthesis of highly stable Au nanoparticles and their unexpected excellent catalytic activity for Suzuki–Miyaura cross-coupling reaction in water. *J. Am. Chem. Soc.* 131: 2060–2061.

2374 Edwards, G.A., Trafford, M.A., Hamilton, A.E. et al. (2014). Melamine and melamine-formaldehyde polymers as ligands for Pd and application to Suzuki–Miyaura cross-coupling reactions in sustainable solvents. *J. Org. Chem.* 79: 2094–2104.

2375 Matos, K. and Soderquist, J.A. (1998). Alkylboranes in the Suzuki–Miyaura coupling: stereochemical and mechanistic studies. *J. Org. Chem.* 63: 461–470.

2376 Carrow, B.P. and Hartwig, J.F. (2011). Distinguishing between pathways for transmetalation in Suzuki–Miyaura reactions. *J. Am. Chem. Soc.* 133: 2116–2119.

2377 Amatore, C., Jutand, A., and Le Duc, G. (2011). Kinetic data for the transmetalation/reductive elimination in Pd-catalyzed Suzuki–Miyaura reactions: unexpected triple role of hydroxide Ions used as base. *Chem. Eur. J.* 17: 2492–2503.

2378 Amatore, C., Jutand, A., and Le Duc, G. (2012). The triple role of fluoride ions in Pd-catalyzed Suzuki–Miyaura reactions: unprecedented transmetalation from [$ArPdFL_2$] complexes. *Angew. Chem. Int. Ed.* 51: 1379–1382.

2379 Amatore, C., Jutand, A., and Le Duc, G. (2012). Mechanistic origin of antagonist effects of d-catalyzed Suzuki–Miyaura reactions. *Chem. Eur. J.* 18: 6616–6625.

2380 Amatore, C., Le Duc, G., and Jutand, A. (2013). Mechanism of Pd-catalyzed Suzuki–Miyaura reactions: multiple and antagonistic roles of anionic "bases" and their counter-cations. *Chem. Eur. J.* 19: 10082–10093.

2381 Lennox, A.J.J. and Llyod-Jones, G.C. (2013). Transmetalation in the Suzuki–Miyaura coupling: the fork in the frail. *Angew. Chem. Int. Ed.* 52: 7362–7370.

2382 Shen, X.Q., Jones, G.O., Watson, D.A. et al. (2010). Enantioselective synthesis of axially chiral biaryls by the Pd-catalyzed Suzuki–Miyaura reaction: substrate scope and quantum mechanical investigations. *J. Am. Chem. Soc.* 132: 11278–11287.

2383 Bermejo, A., Ros, A., Fernández, R., and Lassaletta, J.M. (2008). C_2-Symmetric bis-hydrazones as ligands in the asymmetric Suzuki–Miyaura cross-coupling. *J. Am. Chem. Soc.* 130: 15798–15799.

2384 Uozumi, Y., Matsuura, Y., Arakawa, T., and Yamada, Y.M.A. (2009). Asymmetric Suzuki–Miyaura coupling in water with a chiral Pd catalyst supported on an amphiphilic resin. *Angew. Chem. Int. Ed.* 48: 2708–2710.

2385 Yamamoto, T., Akai, Y., Nagata, Y., and Suginome, M. (2011). Highly enantioselective synthesis of axially chiral biarylphosphonates: asymmetric Suzuki–Miyaura coupling using high-molecular-weight, helically chiral polyquinoxaline-based phosphines. *Angew. Chem. Int. Ed.* 50: 8844–8847.

2386 Suginome, M., Yamamoto, T., Nagata, Y. et al. (2012). Catalytic asymmetric synthesis using chirality-switchable helical polymer as a chiral ligand. *Pure Appl. Chem.* 84: 1759–1789.

2387 Miyaura, N., Ishiyama, T., Ishikawa, M., and Suzuki, A. (1986). Palladium-catalyzed cross-coupling reactions of B-alkyl-9-BBN or trialkylboranes with aryl and 1-alkenyl halides. *Tetrahedron Lett.* 27: 6369–6372.

2388 Ridgway, B.H. and Woerpel, K.A. (1998). Transmetalation of alkylboranes to Pd in the Suzuki coupling reaction proceeds with retention of stereochemistry. *J. Org. Chem.* 63: 458–460.

2389 Matthew, S.C., Glasspole, B.W., Eisenberger, P., and Crudden, C.M. (2014). Synthesis of enantiomerically enriched triarylmethanes by enantiospecific Suzuki–Miyaura cross-coupling reactions. *J. Am. Chem. Soc.* 136: 5828–5831.

2390 Taylor, B.L.H. and Jarvo, E.R. (2011). Stereochemistry of transmetalation of alkylboranes in Ni-catalyzed alkyl-alkyl cross-coupling reactions. *J. Org. Chem.* 76: 7573–7576.

2391 Awano, T., Ohmura, T., and Suginome, M. (2011). Inversion or retention? Effects of acidic additives on the stereochemical course in enantiospecific Suzuki–Miyaura coupling of α-(acetylamino)benzylboronic esters. *J. Am. Chem. Soc.* 133: 20738–20741.

2392 Sandrock, D.L., Jena-Gerard, L., Chen, C.Y. et al. (2010). Stereospecific cross-coupling of secondary alkyl β-trifluoroboratoamides. *J. Am. Chem. Soc.* 132: 17108–17110.

2393 Molander, G.A. and Wisniewski, S.R. (2012). Stereospecific cross-coupling of secondary organotrifluoroborates: potassium 1-(benzyloxy)alkyltrifluoroborates. *J. Am. Chem. Soc.* 134: 16856–16868.

2394 Saeki, T., Son, E.C., and Tamao, K. (2004). Boron trifluoride induced Pd-catalyzed cross-coupling reaction of 1-aryltriazenes with areneboronic acids. *Org. Lett.* 6: 617–619.

2395 Molander, G.A. and Sandrock, D.L. (2008). Orthogonal reactivity in boryl-substituted organotrifluoroborates. *J. Am. Chem. Soc.* 130: 15792–15793.

2396 Walsh, R. (1981). Bond-dissociation energy values in Si-containing compounds and some of their implications. *Acc. Chem. Res.* 14: 246–252.

2397 Grant, D.J. and Dixon, D.A. (2009). Heats of formation and bond dissociation energies of the halosilanes, methylhalosilanes, and halomethylsilanes. *J. Phys. Chem. A* 113: 3656–3661.

2398 Komiyama, T., Minami, Y., and Hiyama, T. (2017). Recent advances in transition-metal-catalyzed synthetic transformations of organosilicon reagents. *ACS Catal.* 7: 631–651.

2399 Yoshida, J.I., Tamao, K., Takahashi, M., and Kumada, M. (1978). Organofluorosilicates in organic synthesis. 4. Stereoselective preparation of 1,4-dienes by Pd-catalyzed allylation of (*E*)-alkenylpentafluorosilicates - application to total synthesis of (+/−)-recifeiolide. *Tetrahedron Lett.* 2161–2164.

2400 Hatanaka, Y. and Hiyama, T. (1988). Cross-coupling of organosilanes with organic halides mediated by Pd catalysts and tris(diethylamino)sulfonium difluorotrimethylsilicate. *J. Org. Chem.* 53 (4): 918–920.

2401 (a) Denmark, S.E., Baird, J.D., and Regens, C.S. (2008). Palladium-catalyzed cross-coupling of five-membered heterocyclic silanolates. *J. Org. Chem.* 73: 1440–1455. (b) Denmark, S.E. and Regens, C.S. (2008). Palladium-catalyzed cross-coupling reactions of organosilanols and their salts: practical alternatives to B- and Sn-based methods. *Acc. Chem. Res.* 41 (11): 1486–1499.

2402 Denmark, S.E. and Ambrosi, A. (2015). Why you really should consider using Pd-catalyzed cross-coupling of silanols and silanolates. *Org. Process Res. Dev.* 19: 982–994.

2403 Denmark, S.E., Smith, R.C., Chang, W.T.T., and Muhuhi, J.M. (2009). Cross-coupling reactions of aromatic and heteroaromatic silanolates with aromatic and heteroaromatic halides. *J. Am. Chem. Soc.* 131: 3104–3118.

2404 Tymonko, S.A., Smith, R.C., Ambrosi, A., and Denmark, S.E. (2015). Mechanistic significance of the Si—O—Pd bond in the Pd-catalyzed cross-coupling reactions of alkenylsilanolates. *J. Am. Chem. Soc.* 137: 6192–6199.

2405 Tymonko, S.A., Smith, R.C., Ambrosi, A. et al. (2015). Mechanistic significance of the Si—O—Pd bond in the Pd-catalyzed cross-coupling reactions of arylsilanolates. *J. Am. Chem. Soc.* 137: 6200–6218.

2406 Nakao, Y., Imanaka, H., Sahoo, A.K. et al. (2005). Alkenyl- and aryl[2-(hydroxymethyl)phenyl]dimethylsilanes: an entry to tetraorganosilicon reagents for the Si-based cross-coupling reaction. *J. Am. Chem. Soc.* 127: 6952–6953.

2407 Cho, E.J., Senecal, T.D., Kinzel, T. et al. (2010). The Pd-catalyzed trifluoromethylation of aryl chlorides. *Science* 328: 1679–1681.

2408 Zhang, W., Liu, F., Li, K., and Zhao, B.L. (2014). Pd-catalyzed desulfitative Hiyama coupling with sulfonyl chlorides. *Appl. Organomet. Chem.* 28: 379–381.

2409 Powell, D.A. and Fu, G.C. (2004). Nickel-catalyzed cross-couplings of organosillicon reagents with unactivated secondary alkyl bromides. *J. Am. Chem. Soc.* 126: 7788–7789.

2410 Dai, X., Strotman, N.A., and Fu, G.C. (2008). Catalytic asymmetric Hiyama cross-couplings of racemic α-bromo esters. *J. Am. Chem. Soc.* 130: 3302–3303.

2411 Gurung, S.K., Thapa, S., Vangala, A.S., and Giri, R. (2013). Copper-catalyzed Hiyama coupling of (hetero)aryltriethoxysilanes with (hetero)aryl iodides. *Org. Lett.* 15: 5378–5381.

2412 Cornelissen, L., Lefrancq, M., and Riant, O. (2014). Copper-catalyzed cross-coupling of vinylsiloxanes with bromoalkynes: synthesis of enynes. *Org. Lett.* 16: 3024–3027.

2413 Nishihara, Y., Noyori, S., Okamoto, T. et al. (2011). Copper-catalyzed Sila–Sonogashira–Hagihara cross-coupling reactions of alkynylsilanes with aryl iodides under Pd-free conditions. *Chem. Lett.* 40: 972–974.

2414 Cornelissen, L., Cirriez, V., Vercruysse, S., and Riant, O. (2014). Copper-catalyzed Hiyama cross-coupling using vinylsilanes and benzylic electrophiles. *J. Chem. Soc., Chem. Commun.* 50: 8018–8020.

2415 Cornelissen, L., Vercruysse, S., Sanhadji, A., and Riant, O. (2014). Copper-catalyzed vinylsilane allylation. *Eur. J. Org. Chem.* 2014: 35–38.

2416 Oishi, M., Kondo, H., and Amii, H. (2009). Aromatic trifluoromethylation catalytic in Cu. *J. Chem. Soc., Chem. Commun.* 1909–1911.

2417 Morimoto, H., Tsubogo, T., Litvinas, N.D., and Hartwig, J.F. (2011). A broadly applicable Cu reagent for trifluoromethylations and perfluoroalkylations of aryl iodides and bromides. *Angew. Chem. Int. Ed.* 50: 3793–3798.

2418 Morstein, J., Hou, H.Y., Cheng, C., and Hartwig, J.F. (2016). Trifluoromethylation of arylsilanes with [(phen)CuCF3]. *Angew. Chem. Int. Ed.* 55: 8054–8057.

2419 Danoun, G., Bayarmagnai, B., Grunberg, M.F., and Goossen, L.J. (2013). Sandmeyer trifluoromethylation of arenediazonium tetrafluoroborates. *Angew. Chem. Int. Ed.* 52: 7972–7975.

2420 Prakash, G.K.S., Krishnamurti, R., and Olah, G.A. (1989). Synthetic methods and reactions. 141. Fluoride-induced trifluoromethylation of carbonyl compounds with trifluoromethyltrimethylsilane ($TMSCF_3$) - a trifluoromethide equivalent. *J. Am. Chem. Soc.* 111: 393–395.

2421 Prakash, G.K.S., Panja, C., Vaghoo, H. et al. (2006). Facile synthesis of TMS-protected trifluoromethylated alcohols using trifluoromethyltrimethylsilane ($TMSCF_3$) and various nucleophilic catalysts in DMF. *J. Org. Chem.* 71: 6806–6813.

2422 Fier, P.S. and Hartwig, J.F. (2012). Copper-mediated difluoromethylation of aryl and vinyl iodides. *J. Am. Chem. Soc.* 134: 5524–5527.

2423 Arlow, S.I. and Hartwig, J.F. (2016). Synthesis of aryldifluoroamides by Cu-catalyzed cross-coupling. *Angew. Chem. Int. Ed.* 55: 4567–4572.

2424 Komiyama, T., Minami, Y., and Hiyama, T. (2016). Aryl(triethyl)silanes for biaryl and teraryl synthesis by Cu(II)-catalyzed cross-coupling reaction. *Angew. Chem. Int. Ed.* 55: 15787–15791.

2425 Smidt, J., Hafner, W., Jira, R. et al. (1962). The oxidation of olefins with $PdCl_2$ catalysts. *Angew. Chem. Int. Ed.* 1: 80–88.

2426 Trost, B.M. and Fullerton, T.J. (1973). New synthetic reactions - allylic alkylation. *J. Am. Chem. Soc.* 95: 292–294.

2427 Trost, B.M. (1977). Organopalladium intermediates in organic synthesis. *Tetrahedron* 33 (20): 2615–2649.

2428 Trost, B.M. (1980). New rules of selectivity - allylic alkylations catalyzed by Pd. *Acc. Chem. Res.* 13 (11): 385–393.

2429 Tsuji, J. (1990). *Organic Synthesis with Palladium Compounds*, Reactivity and Structure Concepts in Organic Chemistry, vol. 10, 1–210. Berlin, Heidelberg: Springer-Verlag.

2430 Trost, B.M. and Verhoeven, T.R. (1982). Organopalladium compounds in organic synthesis and in catalysis. In: *Comprehensive Organometallic Chemistry*, Chapter 57, 1e, vol. 8 (ed. G. Wilkinson, F.G.A. Stone and E.W. Abel), 799–938. Oxford: Pergamon.

2431 Tsuji, J. and Minami, I. (1987). New synthetic reactions of allyl alkyl carbonates, allyl β-keto carboxylates, and allyl vinylic carbonates catalyzed by Pd complexes. *Acc. Chem. Res.* 20 (4): 140–145.

2432 Yamaguchi, Y., Suzuki, Y., Matsumoto, S. et al. (2016). Palladium-catalyzed asymmetric allylic alkylation using C_2-symmetric chiral bidentate bis(*N*-heterocyclic carbene) ligands with the *o*-xylylene framework. *Chem. Lett.* 45 (7): 798–800.

2433 Chen, X.-S., Hou, C.-J., and Hu, J.-P. (2016). Chiral phosphine-phosphoramidite ligands in asymmetric catalysis. *Synth. Commun.* 46 (11): 917–941.

2434 Trost, B.M. and van Vranken, D.L. (1996). Asymmetric transition metal-catalyzed allylic alkylations. *Chem. Rev.* 96 (1): 395–422.

2435 (a) Trost, B.M. and Crawley, M.L. (2003). Asymmetric transition-metal-catalyzed allylic alkylations: applications in total synthesis. *Chem. Rev.* 103 (8): 2921–2943. (b) Engelin, C.J. and Fristrp, P. (2011). Palladium catalyzed allylic C—H alkylation: a mechanistic perspective. *Molecules* 16: 951–969.

2436 Kammerer, C., Prestat, G., Madec, D., and Poli, G. (2014). Synthesis of γ-lactams and γ-lactones via intramolecular Pd-catalyzed allylic alkylations. *Acc. Chem. Res.* 47: 3439–3447.

2437 Qiang, J.-L. and Jiang, G. (2017). Recent advances in Pd-catalyzed Tsuji–Trost-type allylic alkylation with allylic alcohols. *Curr. Catal.* 6 (1): 25–30.

2438 Ibrahem, I. and Córdova, A. (2006). Direct catalytic intermolecular α-allylic alkylation of aldehydes by combination of transition-metal and organocatalysis. *Angew. Chem. Int. Ed.* 45: 1952–1956.

2439 Sha, S.C., Zhang, J.D., Carroll, P.J., and Walsh, P.J. (2013). Raising the pKa limit of "soft" nucleophiles in Pd-catalyzed allylic substitutions: application of diarylmethane pronucleophiles. *J. Am. Chem. Soc.* 135: 17602–17609.

2440 Li, Y.X., Xuan, Q.Q., Liu, L. et al. (2013). A Pd(0)-Catalyzed direct dehydrative coupling of terminal alkynes with allylic alcohols to access 1,4-enynes. *J. Am. Chem. Soc.* 135: 12536–12539.

2441 Evans, P.A. and Nelson, J.D. (1998). Conservation of absolute configuration in the acyclic Rh-catalyzed allylic alkylation reaction: evidence for an enyl ($\sigma+\pi$) organorhodium intermediate. *J. Am. Chem. Soc.* 120 (22): 5581–5582.

2442 Evans, P.A. and Leahy, D.K. (2003). Regioselective and enantiospecific Rh-catalyzed allylic alkylation reactions using Cu(I) enolates: synthesis of (−)-sugiresinol dimethyl ether. *J. Am. Chem. Soc.* 125 (30): 8974–8975.

2443 Osborn, J.A., Jardine, F.H., Young, J.F., and Wilkinson, G. (1966). Preparation and properties of tris(triphenylphosphine)halogenorhodium(1) and some reactions thereof including catalytic homogeneous hydrogenation of olefins and acetylenes and their derivatives. *J. Chem. Soc. A* (12): 1711–1732.

2444 Takeuchi, R. and Kashio, M. (1998). Iridium complex-catalyzed allylic alkylation of allylic esters and allylic alcohols: unique regio- and stereoselectivity. *J. Am. Chem. Soc.* 120 (34): 8647–8655.

2445 Blacker, A.J., Clarke, M.L., Loft, M.S. et al. (2000). Platinum-catalysed allylic alkylation: reactivity, enantioselectivity, and regioselectivity. *Chem. Eur. J.* 6 (2): 353–360.

2446 Trost, B.M. and Andersen, N.G. (2002). Utilization of Mo- and Pd-catalyzed dynamic kinetic asymmetric transformations for the preparation of tertiary and quaternary stereogenic centers: a concise synthesis of Tipranavir. *J. Am. Chem. Soc.* 124 (48): 14320–14321.

2447 Trost, B.M. and Jiang, C.H. (2003). Pd-catalyzed asymmetric allylic alkylation. A short route to the cyclopentyl core of viridenomycin. *Org. Lett.* 5 (9): 1563–1565.

2448 Bartels, B., García-Yebra, C., and Helmchen, G. (2003). Asymmetric Ir(I)-catalysed allylic alkylation of monosubstituted allylic acetates with phosphorus amidites as ligands. *Eur. J. Org. Chem.* (6): 1097–1103.

2449 Murphy, K.E. and Hoveyda, A.H. (2003). Enantioselective synthesis of α-alkyl-*β*,γ-unsaturated esters through efficient Cu-catalyzed allylic alkylations. *J. Am. Chem. Soc.* 125 (16): 4690–4691.

2450 Trost, B.M. (2015). Metal catalyzed allylic alkylation: its development in the Trost laboratories. *Tetrahedron* 71: 5708–5733.

2451 Trost, B.M. and Hung, M.H. (1984). On the regiochemistry of metal-catalyzed allylic alkylation - a model. *J. Am. Chem. Soc.* 106 (22): 6837–6839.

2452 Kondo, T., Ono, H., Satake, N. et al. (1995). Nucleophilic and electrophilic allylation reactions - synthesis, structure, and ambiphilic reactivity of (*η*3-allyl)Ru (II) complexes. *Organometallics* 14 (4): 1945–1953.

2453 Morisaki, Y., Kondo, T., and Mitsudo, T.A. (1999). Ruthenium-catalyzed allylic substitution of cyclic allyl carbonates with nucleophiles. Stereoselectivity and scope of the reaction. *Organometallics* 18 (23): 4742–4746.

2454 Trost, B.M., Fraisse, P.L., and Ball, Z.T. (2002). A stereospecific Ru-catalyzed allylic alkylation. *Angew. Chem. Int. Ed.* 41 (6): 1059–1061.

2455 Kawatsura, M., Ata, F., Wada, S. et al. (2007). Ruthenium-catalysed linear-selective allylic alkylation of allyl acetates. *J. Chem. Soc., Chem. Commun.* 298–300.

2456 Kitamura, M., Miyata, K., Seki, T. et al. (2013). CpRu-catalyzed asymmetric dehydrative allylation. *Pure Appl. Chem.* 85: 1121–1132.

2457 Trost, B.M. and Lautens, M. (1982). Molybdenum catalysts for allylic alkylation. *J. Am. Chem. Soc.* 104 (20): 5543–5545.

2458 Trost, B.M., Dogra, K., Hachiya, I. et al. (2002). Designed ligands as probes for the catalytic binding mode in Mo-catalyzed asymmetric allylic alkylation. *Angew. Chem. Int. Ed.* 41 (11): 1929–1932.

2459 Trost, B.M. (2012). Pd- and Mo-Catalyzed asymmetric allylic alkylation. *Org. Process Res. Dev.* 16: 185–194.

2460 Tsuji, J., Minami, I., and Shimizu, I. (1984). Allyation of carbonucleophiles with allylic carbonates under neutral conditions catalyzed by Rh complexes. *Tetrahedron Lett.* 25 (45): 5157–5160.

2461 Evans, P.A. and Nelson, J.D. (1998). Regioselective Rh-catalyzed allylic alkylation with a modified Wilkinson's catalyst. *Tetrahedron Lett.* 39 (13): 1725–1728.

2462 Takeuchi, R. and Kitamura, N. (1998). Rhodium complex-catalysed allylic alkylation of allylic acetates. *New J. Chem.* 22 (7): 659–660.

2463 Hayashi, T., Okada, A., Suzuka, T., and Kawatsura, M. (2003). High enantioselectivity in Rh-catalyzed allylic alkylation of 1-substituted 2-propenyl acetates. *Org. Lett.* 5 (10): 1713–1715.

2464 Helmchen, G., Dahnz, A., Dübon, P. et al. (2007). Iridium-catalysed asymmetric allylic substitutions. *J. Chem. Soc., Chem. Commun.* (7): 675–691.

2465 Takeuchi, R. (2002). Iridium complex-catalyzed highly selective organic synthesis. *Synlett* (12): 1954–1965.

2466 Takeuchi, R. and Kezuka, S. (2006). Iridium-catalyzed formation of carbon–carbon and carbon-heteroatom bonds. *Synthesis* (20): 3349–3366.

2467 Hethcox, J.C., Shockley, S.E., and Stoltz, B.M. (2016). Iridium-catalyzed diastereo-, enantio-, and regioselective allylic alkylation with prochiral enolates. *ACS Catal.* 6: 6207–6213.

2468 Trost, B.M. and Hung, M.H. (1983). Tungsten-catalyzed allylic alkylations - new avenues for selectivity. *J. Am. Chem. Soc.* 105 (26): 7757–7759.

2469 Lehmann, J. and Lloyd-Jones, G.C. (1995). Regiocontrol and stereoselectivity in W(bipyridine)-catalyzed allylic alkylation. *Tetrahedron* 51 (32): 8863–8874.

2470 Tsuji, J. (1986). New general synthetic methods involving (π-allyl)Pd complexes as intermediates and neutral reaction conditions. *Tetrahedron* 42 (16): 4361–4401.

2471 Braun, M. and Meier, T. (2006). Tsuji–Trost allylic alkylation with ketone enolates. *Angew. Chem. Int. Ed.* 45 (42): 6952–6955.

2472 Hayashi, T., Yamamoto, A., and Hagihara, T. (1986). Stereochemistry and regiochemistry in Pd-catalyzed nucleophilic substitution of optically active (*E*)-allyl and (*Z*)-allyl acetates. *J. Org. Chem.* 51 (5): 723–727.

2473 Togni, A. (1991). Asymmetric allylic alkylation catalyzed by Pd(sparteine) complexes. *Tetrahedron: Asymmetry* 2 (7): 683–690.

2474 (a) Birkholz, M.N., Freixa, Z., and van Leeuwen, P.W.N.M. (2009). Bite angle effects of diphosphines in C—C and C—X bond forming cross coupling reactions. *Chem. Soc. Rev.* 38 (4): 1099–1118. (b) Jover, J., Fey, N., Harvey, J.N. et al. (2012). Expansion of the ligand knowledge base for chelating P,P-donor ligands (LKB-PP). *Organometallics* 31 (15): 5302–5306.

2475 van Leeuwen, P.W.N.M., Kamer, P.C.J., Reek, J.N.H., and Dierkes, P. (2000). Ligand bite angle effects in metal-catalyzed C—C bond formation. *Chem. Rev.* 100 (8): 2741–2769.

2476 Norsikian, S. and Chang, C.W. (2009). Control of the regioselectivity in Pd(0)-catalyzed allylic alkylation. *Curr. Org. Synth.* 6 (3): 264–289.

2477 Trost, B.M. and Verhoeven, T.R. (1978). Stereocontrolled synthesis of (+/-)-recifeiolide via organopalladium chemistry. *Tetrahedron Lett.* (26): 2275–2278.

2478 Trost, B.M., Krische, M.J., Radinov, R., and Zanoni, G. (1996). On asymmetric induction in allylic alkylation via enantiotopic facial discrimination. *J. Am. Chem. Soc.* 118 (26): 6297–6298.

2479 Trost, B.M. and Jiang, C.H. (2001). Atom economic asymmetric creation of quaternary carbon: regio- and enantioselective reactions of a vinylepoxide with a carbon nucleophile. *J. Am. Chem. Soc.* 123 (51): 12907–12908.

2480 You, S.L. and Dai, L.X. (2006). Enantioselective Pd-catalyzed decarboxylative allylic alkylations. *Angew. Chem. Int. Ed.* 45 (32): 5246–5248.

2481 Hayashi, T., Kawatsura, M., and Uozumi, Y. (1997). Regio- and enantioselective allylic alkylation catalysed by a chiral (monophosphine)Pd complex. *J. Chem. Soc., Chem. Commun.* (6): 561–562.

2482 Prétôt, R. and Pfaltz, A. (1998). New ligands for regio- and enantiocontrol in Pd-catalyzed allylic alkylations. *Angew. Chem. Int. Ed.* 37 (3): 323–325.

2483 Helmchen, G. and Pfaltz, A. (2000). Phosphinooxazolines - a new class of versatile, modular P,N-ligands for asymmetric catalysis. *Acc. Chem. Res.* 33 (6): 336–345.

2484 Mohr, J.T., Behenna, D.C., Harned, A.M., and Stoltz, B.M. (2005). Deracemization of quaternary stereocenters by Pd-catalyzed enantioconvergent decarboxylative allylation of racemic β-ketoesters. *Angew. Chem. Int. Ed.* 44: 6924–6927.

2485 Reeves, C.M., Behenna, D.C., and Stoltz, B.M. (2014). Development of (trimethylsilyl)ethyl ester protected enolates and applications in Pd-catalyzed enantioselective allylic alkylation: intermolecular cross-coupling of functionalized electrophiles. *Org. Lett.* 16: 2314–2317.

2486 Raluy, E., Diéguez, M., and Pàmies, O. (2007). Sugar-based diphosphoroamidite as a promising new class of ligands in Pd-catalyzed asymmetric allylic alkylation reactions. *J. Org. Chem.* 72 (8): 2842–2850.

2487 Trost, B.M., Machacek, M.R., and Aponick, A. (2006). Predicting the stereochemistry of diphenylphosphino benzoic acid (DPPBA)-based Pd-catalyzed asymmetric allylic alkylation reactions: a working model. *Acc. Chem. Res.* 39 (10): 747–760.

2488 Lu, Z. and Ma, S.M. (2008). Metal-catalyzed enantioselective allylation in asymmetric synthesis. *Angew. Chem. Int. Ed.* 47 (2): 258–297.

2489 Leitner, A., Shu, C.T., and Hartwig, J.F. (2005). Effects of catalyst activation and ligand steric properties on the enantioselective allylation of amines and phenoxides. *Org. Lett.* 7: 1093–1096.

2490 García-Mancheño, O., Priego, J., Cabrera, S. et al. (2003). 1-Phosphino-2-sulfenylferrocenes as planar chiral ligands in enantioselective Pd-catalyzed allylic substitutions. *J. Org. Chem.* 68: 3679–3686.

2491 Hong, A.Y. and Stoltz, B.M. (2013). The construction of all-carbon quaternary stereocenters by use of Pd-catalyzed asymmetric allylic alkylation reactions in total synthesis. *Eur. J. Org. Chem.* 2745–2759.

2492 Hegedus, L.S., Darlington, W.H., and Russell, C.E. (1980). Cyclopropanation of ester enolates by π-allylpalladium chloride complexes. *J. Org. Chem.* 45 (25): 5193–5196.

2493 Hoffmann, H.M.R., Otte, A.R., and Wilde, A. (1992). Nucleophilic attack at the central carbon atom of (π-allyl)Pd complexes - formation of α-cyclopropyl esters. *Angew. Chem. Int. Ed. Engl.* 31 (2): 234–236.

2494 Wilde, A., Otte, A.R., and Hoffmann, H.M.R. (1993). Cyclopropanes via nucleophilic attack at the central carbon of (π-allyl)Pd complexes. *J. Chem. Soc., Chem. Commun.* (7): 615–616.

2495 Carfagna, C., Galarini, R., Linn, K. et al. (1993). Nucleophilic attack at the central allyl carbon atom in $[(\eta_3\text{-allyl})ML_2]^+$ complexes (M=Pd, Pt) - experimental facts and new theoretical insights. *Organometallics* 12 (8): 3019–3028.

2496 Otte, A.R., Wilde, A., and Hoffmann, H.M.R. (1994). Cyclopropanes by nucleophilic attack of mono-substituted and diaryl-substituted (η_3-allyl)Pd complexes - aryl effect and stereochemistry. *Angew. Chem. Int. Ed. Engl.* 33 (12): 1280–1282.

2497 Formica, M., Musco, A., Pontellini, R. et al. (1993). Ambivalence of nucleophilic attack on central and terminal allyl carbon atoms of $[(\eta_3\text{-allyl})ML_2]^+$ (M=Pd Or Pt) Complexes. *J. Organomet. Chem.* 448 (1–2): C6–C9.

2498 Carfagna, C., Mariani, L., Musco, A. et al. (1991). The regioselectivities and stereoselectivities of the reaction of allyl acetates and silyl ketene acetals catalyzed by Pd(0) complexes - a new route to cyclopropane derivatives. *J. Org. Chem.* 56 (12): 3924–3927.

2499 Rudler, H., Harris, P., Parlier, A. et al. (2001). Transition-metal catalyzed synthesis of δ-hydroxy-γ-lactones from bis(trimethylsilyl) ketene acetals and allylic acetates via γ-unsaturated carboxylic acids. Comments on the formation of α-cyclopropyl carboxylic acids. *J. Organomet. Chem.* 624 (1–2): 186–202.

2500 Hoffmann, H.M.R., Otte, A.R., Wilde, A. et al. (1995). Isolation and X-ray crystal structure of a palladacyclobutane - insight into the mechanism of cyclopropanation. *Angew. Chem. Int. Ed. Engl.* 34 (1): 100–102.

2501 Satake, A. and Nakata, T. (1998). Novel (η3-allyl)Pd(pyridinylpyrazole) complex: synthesis, reactivity, and catalytic activity for cyclopropanation of ketene silyl acetal with allylic acetates. *J. Am. Chem. Soc.* 120 (40): 10391–10396.

2502 Satake, A., Koshino, H., and Nakata, T. (1999). Cyclopropanation of ketene silyl acetals with allylic acetates using (η3-allyl)Pd(pyridinylpyrazole) catalysts. *Chem. Lett.* (1): 49–50.

2503 Satake, A., Kadohama, H., Koshino, H., and Nakata, T. (1999). Asymmetric cyclopropanation of ketene silyl acetal with allylic acetate catalyzed by a Pd complex. *Tetrahedron Lett.* 40 (18): 3597–3600.

2504 van Haaren, R.J., Oevering, H., Coussens, B.B. et al. (1999). On the influence of the bite angle of bidentate phosphane ligands on the regioselectivity in allylic alkylation. *Eur. J. Inorg. Chem.* (8): 1237–1241.

2505 (a) Oestreich, M. (ed.) (2009). *The Mizoroki-Heck Reaction*, 1–587. Chichester: Wiley. (b) de Vries, J.C. (2012). Palladium-catalysed coupling reactions. In: *Organometallics as Catalysts in the Fine Chemical Industry*, Topics in Organometallic Chemistry, vol. 42, 1–34. Berlin, Heidelberg: Springer-Verlag.

2506 McMurray, L., O'Hara, O.F., and Gaunt, M.J. (2011). Recent developments in natural product synthesis using metal-catalysed C—H bond functionalisation. *Chem. Soc. Rev.* 40: 1885–1898.

2507 (a) Yamaguchi, J., Yamaguchi, A.D., and Itami, K.C.-H. (2012). Bond functionalization: emerging synthetic tools for natural products and pharmaceuticals. *Angew. Chem. Int. Ed.* 51: 8960–9009. (b) Santos, C.I.M., Barata, J.F.B., Faustino, M.A.F. et al. (2013). Revisiting Heck–Mizoroki reactions in ionic liquids. *RSC Adv.* 3 (42): 19219–19238. (c) Yang, J., Zhao, H.-W., He, J., and Zhang, C.-P. (2018). Pd-catalyzed Mizoroki–Heck reactions using fluorine-containing agents as the cross-coupling partners. *Catalysts* 8 (1): 23/1–23/35.

2508 Wang, S.S. and Yang, G.Y. (2016). Recent developments in low-cost TM-catalyzed Heck-type reactions (TM=transition metal, Ni, Co, Cu, and Fe). *Catal. Sci. Technol.* 6: 2862–2876.

2509 Shirakawa, E. and Hayashi, T. (2012). Transition-metal-free coupling reactions of aryl halides. *Chem. Lett.* 41: 130–134.

2510 Sun, C.L. and Shi, Z.J. (2014). Transition-metal-free coupling reactions. *Chem. Rev.* 114: 9219–9280.

2511 Ruan, J., Saidi, O., Iggo, J.A., and Xiao, J. (2008). Direct acylation of aryl bromides with aldehydes by Pd catalysis. *J. Am. Chem. Soc.* 130: 10510–10511.

2512 Colbon, P., Ruan, J., Purdie, M., and Xiao, J. (2010). Direct acylation of aryl chlorides with aldehydes by Pd(pyrrolidine) co-catalysis. *Org. Lett.* 12: 3670–3673.

2513 Ruan, J.W. and Xiao, J.L. (2011). From α-arylation of olefins to acylation with aldehydes: a journey in regiocontrol of the Heck reaction. *Acc. Chem. Res.* 44: 614–626.

2514 Willis, M.C., Claverie, C.K., and Mahon, M.F. (2002). Tandem intermolecular Suzuki coupling/intramolecular vinyl triflate/arene coupling. *J. Chem. Soc., Chem. Commun.* 832–833.

2515 Cruz, A.C.F., Miller, N.D., and Willis, M.C. (2007). Intramolecular Pd-catalyzed direct arylation of alkenyl triflates. *Org. Lett.* 9: 4391–4393.

2516 Albicker, M.R. and Cramer, N. (2009). Enantioselective Pd-catalyzed direct arylations at ambient temperature: access to indanes with quaternary stereocenters. *Angew. Chem. Int. Ed.* 48: 9139–9142.

2517 Heck, R.F. (1968). Arylation methylation and carboxyalkylation of olefins by Group 8 metal derivatives. *J. Am. Chem. Soc.* 90 (20): 5518–5526.

2518 Heck, R.F. (1969). Mechanism of arylation and carbomethoxylation of olefins with organopalladium compounds. *J. Am. Chem. Soc.* 91 (24): 6707–6714.

2519 Heck, R.F. (1971). Electronic and steric effects in olefin arylation and carboalkoxylation reactions with organopalladium compounds. *J. Am. Chem. Soc.* 93 (25): 6896–6901.

2520 Mizoroki, T., Mori, K., and Ozaki, A. (1971). Arylation of olefin with aryl iodide catalyzed by Pd. *Bull. Chem. Soc. Jpn.* 44 (2): 581–581.

2521 Heck, R. and Nolley, J.P. Jr., (1972). Palladium-catalyzed vinylic hydrogen substitution reactions with aryl, benzyl, and styryl halides. *J. Org. Chem.* 37 (14): 2320–2322.

2522 Dieck, H.A. and Heck, R.F. (1974). (Organophosphine)Pd complexes as catalysts for vinylic hydrogen substitution reactions. *J. Am. Chem. Soc.* 96 (4): 1133–1136.

2523 Ben-David, Y., Portnoy, M., Gozin, M., and Milstein, D. (1992). Palladium-catalyzed vinylation of aryl chlorides - chelate effect in catalysis. *Organometallics* 11 (6): 1995–1996.

2524 Littke, A.F. and Fu, G.C. (2001). A versatile catalyst for Heck reactions of aryl chlorides and aryl bromides under mild conditions. *J. Am. Chem. Soc.* 123 (29): 6989–7000.

2525 Bedford, R.B., Cazin, C.S.J., and Holder, D. (2004). The development of Pd catalysts for C—C and C-heteroatom bond forming reactions of aryl chloride substrates. *Coord. Chem. Rev.* 248: 2283–2321.

2526 Jutand, A. and Mosleh, A. (1995). Rate and mechanism of oxidative addition of aryl triflates to zero valent Pd complexes - evidence for the formation of cationic (σ-aryl)Pd complexes. *Organometallics* 14 (4): 1810–1817.

2527 Chen, Q.Y. and Yang, Z.Y. (1986). Palladium-catalyzed reaction of phenyl fluoroalkanesulfonates with alkynes and alkenes. *Tetrahedron Lett.* 27 (10): 1171–1174.

2528 Andersson, C.M. and Hallberg, A. (1988). Regioselective Pd-catalyzed arylation of vinyl ethers with 4-nitrophenyl triflate - control by addition of halide ions. *J. Org. Chem.* 53 (9): 2112–2114.

2529 Fu, X.Y., Zhang, S.Y., Yin, J.G. et al. (2002). First examples of a tosylate in the Pd-catalyzed Heck cross-coupling reaction. *Tetrahedron Lett.* 43 (4): 573–576.

2530 Gøgsig, T.M., Lindhardt, A.T., Dekhane, M. et al. (2009). Heteroaromatic tosylates as electrophiles in regioselective Mizoroki–Heck coupling reactions with electron-rich olefins. *Chem. Eur. J.* 15 (24): 5950–5955.

2531 Hansen, A.L. and Skrydstrup, T. (2005). Regioselective Heck couplings of α,β-unsaturated tosylates and mesylates with electron-rich olefins. *Org. Lett.* 7 (25): 5585–5587.

2532 Webel, M. and Reissig, H.U. (1997). Heck reactions starting from silyl enol ethers - a simple one-pot nonaflation-coupling procedure for the synthesis of 1,3-dienes. *Synlett* (10): 1141–1142.

2533 Hansen, A.L., Ebran, J.P., Ahlquist, M. et al. (2006). Heck coupling with nonactivated alkenyl tosylates and phosphates: examples of effective 1,2-migrations of the alkenyl Pd(II) intermediates. *Angew. Chem. Int. Ed.* 45 (20): 3349–3353.

2534 García-Ruano, J.L., Alemán, J., and Paredes, C.G. (2006). Oxidative addition of Pd(0) to Ar-SO_2R bonds: Heck-type reactions of sulfones. *Org. Lett.* 8 (13): 2683–2686.

2535 Jeffery, T. (1984). Palladium-catalyzed vinylation of organic halides under solid liquid-phase transfer conditions. *J. Chem. Soc., Chem. Commun.* (19): 1287–1289.

2536 Stephan, M.S., Teunissen, A.J.J.M., Verzijl, G.K.M., and de Vries, J.G. (1998). Heck reactions without salt formation: aromatic carboxylic anhydrides as arylating agents. *Angew. Chem. Int. Ed. Engl.* 37 (5): 662–664.

2537 Goossen, L.J. and Paetzold, J. (2002). Pd-catalyzed decarbonylative olefination of aryl esters: towards a waste-free Heck reaction. *Angew. Chem. Int. Ed.* 41 (7): 1237–1241.

2538 Myers, A.G., Tanaka, D., and Mannion, M.R. (2002). Development of a decarboxylative palladation reaction and its use in a Heck-type olefination of arene carboxylates. *J. Am. Chem. Soc.* 124 (38): 11250–11251.

2539 Forgione, P., Brochu, M.C., St-Onge, M. et al. (2006). Unexpected intermolecular Pd-catalyzed cross-coupling reaction employing heteroaromatic carboxylic acids as coupling partners. *J. Am. Chem. Soc.* 128 (35): 11350–11351.

2540 Blaser, H.U. and Spencer, A. (1982). The Pd-catalyzed arylation of activated alkenes with aroyl chlorides. *J. Organomet. Chem.* 233 (2): 267–274.

2541 Hu, P., Kan, J., Su, W.P., and Hong, M.C. (2009). $Pd(O_2CCF_3)_2$/Benzoquinone: a versatile catalyst system for the decarboxylative olefination of arenecarboxylic acids. *Org. Lett.* 11 (11): 2341–2344.

2542 Dubbaka, S.R. and Vogel, P. (2005). Palladium-catalyzed desulfitative Mizoroki–Heck couplings of sulfonyl chlorides with mono- and disubstituted olefins: Rh-catalyzed desulfitative Heck-type reactions under phosphine- and base-free conditions. *Chem. Eur. J.* 11 (9): 2633–2641.

2543 Dubbaka, S.R., Zhao, D.B., Fei, Z.F. et al. (2006). Palladium-catalyzed desulfitative Mizoroki–Heck coupling reactions of sulfonyl chlorides with olefins in a nitrile-functionalized ionic liquid. *Synlett* (18): 3155–3157.

2544 Madden, K.S., David, S., Knowles, J.P., and Whiting, A. (2015). Heck–Mizoroki coupling of vinyliodide and applications in the synthesis of dienes and trienes. *J. Chem. Soc., Chem. Commun.* 51: 11409–11412.

2545 Ebran, J.P., Hansen, A.L., Gøgsig, T.M., and Skrydstrup, T. (2007). Studies on the Heck reaction with alkenyl phosphates: can the 1,2-migration be controlled? Scope and limitations. *J. Am. Chem. Soc.* 129: 6931–6942.

2546 Sato, Y., Sodeoka, M., and Shibasaki, M. (1989). Catalytic asymmetric C—C bond formation - asymmetric synthesis of *cis*-decalin derivatives by Pd-catalyzed cyclization of prochiral alkenyl iodides. *J. Org. Chem.* 54 (20): 4738–4739.

2547 Carpenter, N.E., Kucera, D.J., and Overman, L.E. (1989). Palladium-catalyzed polyene cyclizations of trienyl triflates. *J. Org. Chem.* 54 (25): 5846–5848.

2548 Sato, Y., Sodeoka, M., and Shibasaki, M. (1990). On the role of Ag salts in asymmetric Heck-type reaction - a greatly improved catalytic asymmetric synthesis of *cis*-decalin derivatives. *Chem. Lett.* (10): 1953–1954.

2549 Morel, A., Silarska, E., Trzeciak, A.M., and Pernak, J. (2013). Palladium-catalyzed asymmetric Heck arylation of 2,3-dihydrofuran - effect of prolinate salts. *J. Chem. Soc., Dalton Trans.* 42: 1215–1222.

2550 Shibasaki, M. and Ohshima, T. (2009). Desymmetrizing Heck reactions. In: *The Mizoroki–Heck Reaction* (ed. M. Oestreich), 463–483. Chichester: Wiley.

2551 Link, J.T. and Wada, C.K. (2009). Intramolecular enantioselective Mizoroki–Heck reactions. In: *The Mizoroki–Heck Reaction* (ed. M. Oestreich), 433–462. Chichester: Wiley.

2552 Dounay, A.B. and Overman, L.E. (2009). The asymmetric intramolecular Mizoroki–Heck reaction in natural product total synthesis. In: *The Mizoroki–Heck Reaction* (ed. M. Oestreich), 533–568. Chichester: Wiley.

2553 Lucciola, D. and Keay, B.A. (2011). Further developments of an enantioselective Pd-catalyzed polyene cyclization: surprising solvent and ligand effects. *Synlett* 1618–1622.

2554 Yang, Z.G. and Zhou, J.R. (2012). Palladium-catalyzed, asymmetric Mizoroki–Heck reaction of benzylic electrophiles using phosphoramidites as chiral ligands. *J. Am. Chem. Soc.* 134: 11833–11835.

2555 Oestreich, M. (2014). Breaking news on the enantioselective intermolecular Heck reaction. *Angew. Chem. Int. Ed.* 53: 2282–2285.

2556 Zhang, Q.S., Wan, S.L., Chen, D. et al. (2015). Palladium-catalyzed asymmetric intermolecular Mizoroki–Heck reaction for construction of a chiral quaternary carbon center. *J. Chem. Soc., Chem. Commun.* 51: 12235–12238.

2557 Shue, R.S. (1973). Olefin arylation and olefin coupling in the presence of palladium carboxylates. US Patent 3775511A 19731127.

2558 Shue, R. S. (1974). Olefin coupling in the presence of palladium carboxylates. US Patent 3855329A 19741217.

2559 Geissler, H., Gross, P., and Guckes, B. (1998). Preparation of new palladaphosphacyclobutanes as catalysts. German Patent DE 19647584 A1 19980520.

2560 Geissler, H., Gross, P., and Guckes, B. (1998). Preparation of new palladaphosphacyclobutanes as catalysts. *Chem. Abstr.* 129: 28078.

2561 Rampf, F. and Eckert, M. (2003). Arylation of olefins. WO2003048107.

2562 Rampf, F. and Eckert, M. (2003). Arylation of olefins. *Chem. Abstr.* 139: 36330.

2563 Li, C.J. (2005). Organic reactions in aqueous media with a focus on carbon-carbon bond formations: a decade update. *Chem. Rev.* 105 (8): 3095–3165.

2564 Alonso, F., Beletskaya, I.P., and Yus, M. (2005). Non-conventional methodologies for transition-metal catalysed carbon-carbon coupling: a critical overview. Part 1. The Heck reaction. *Tetrahedron* 61 (50): 11771–11835.

2565 Bonrath, W., Létinois, U., Netscher, T., and Schütz, J. (2009). Mizoroki–Heck reactions:

modern solvent systems and reaction techniques. In: *The Mizoroki–Heck Reaction* (ed. M. Oestreich), 495–531. Chichester: Wiley.

2566 Trzeciak, A.M. and Ziolkowski, J.J. (2007). Monomolecular, nanosized and heterogenized Pd catalysts for the Heck reaction. *Coord. Chem. Rev.* 251 (9–10): 1281–1293.

2567 Astruc, D. (2007). Palladium nanoparticles as efficient green homogeneous and heterogeneous carbon–carbon coupling precatalysts: a unifying view. *Inorg. Chem.* 46 (6): 1884–1894.

2568 Biffis, A., Zecca, M., and Basato, M. (2001). Palladium metal catalysts in Heck C—C coupling reactions. *J. Mol. Catal. A: Chem.* 173 (1–2): 249–274.

2569 Jiang, J.Z. and Cai, C. (2006). In situ formation of dispersed Pd nanoparticles in microemulsion: efficient reaction system for ligand-free Heck reaction. *J. Colloid Interface Sci.* 299 (2): 938–943.

2570 Nikbin, N., Ladlow, M., and Ley, S.V. (2007). Continuous flow ligand-free Heck reactions using monolithic Pd(0) nanoparticles. *Org. Process Res. Dev.* 11 (3): 458–462.

2571 Djakovitch, J., Koehler, K., and de Vries, J.G. (2007). The role of Pd nanoparticles as catalysts for carbon–carbon coupling reactions. In: *Nanoparticles and Catalysis* (ed. D. Astruc), 303–348. Weinheim: Wiley-VCH.

2572 Niembro, S., Shafir, A., Vallribera, A., and Alibes, R. (2008). Palladium nanoparticles supported on an organic–inorganic fluorinated hybrid material. Application to microwave-based Heck reaction. *Org. Lett.* 10 (15): 3215–3218.

2573 Tao, R.T., Miao, S.D., Liu, Z.M. et al. (2009). Pd nanoparticles immobilized on sepiolite by ionic liquids: efficient catalysts for hydrogenation of alkenes and Heck reactions. *Green Chem.* 11 (1): 96–101.

2574 Liu, S.Y., Jiang, X.Z., and Zhuo, G.L. (2008). Heck reaction catalyzed by colloids of delaminated Pd-containing layered double hydroxide. *J. Mol. Catal. A: Chem.* 290 (1–2): 72–78.

2575 Sydnes, M.O. (2014). Use of nanoparticles as catalysts in organic synthesis - cross-coupling reactions. *Curr. Org. Chem.* 18: 312–326.

2576 Palmisano, G., Bonrath, W., Boffa, L. et al. (2007). Heck reactions with very low ligandless catalyst loads accelerated by microwaves or simultaneous microwaves/ultrasound irradiation. *Adv. Synth. Catal.* 349 (14+15): 2338–2344.

2577 Huang, L. and Wong, P.K. (2010). Recent progress on the studies of the true catalyst in the Heck reaction with supported Pd particles. *Curr. Org. Synth.* 7: 599–613.

2578 Martin-Aranda, R.M. and Cejka, J. (2010). Recent advances in catalysis over mesoporous molecular sieves. *Top. Catal.* 53: 141–153.

2579 Kann, N. (2010). Recent applications of polymer supported organometallic catalysts in organic synthesis. *Molecules* 15: 6306–6331.

2580 Walter, J., Heiermann, J., Dyker, G. et al. (2000). Hexagonal or quasi two-dimensional Pd nanoparticles - tested at the Heck reaction. *J. Catal.* 189 (2): 449–455.

2581 Driver, M.S. and Hartwig, J.F. (1996). A second-generation catalyst for aryl halide amination: mixed secondary amines from aryl halides and primary amines catalyzed by $(DPPF)PdCl_2$. *J. Am. Chem. Soc.* 118 (30): 7217–7218.

2582 Hartwig, J.F., Shelby, Q., and Kataoka, N. (2002). Catalyst for aromatic or vinylic C—O, C—N, and C—C bond formation. WO 011883.

2583 Hartwig, J.F., Shelby, Q., and Kataoka, N. (2002). Catalyst for aromatic or vinylic C—O, C—N, and C—C bond formation. *Chem. Abstr.* 136: 185757.

2584 Doucet, H. and Santelli, M. (2006). *cis,cis,cis*-1,2,3,4-Tetrakis(diphenylphosphinomethyl) cyclopentane: tedicyp, an efficient ligand in Pd-catalysed reactions. *Synlett* (13): 2001–2015.

2585 Beletskaya, I.P. and Cheprakov, A.V. (2008). Focus on catalyst development and ligand design. In: *The Mizoroki–Heck Reaction* (ed. M. Oestreich), 51–132. Chichester: Wiley.

2586 Fortman, G.C. and Nolan, S.P. (2011). N-heterocyclic carbene (NHC) ligands and Pd in homogeneous cross-coupling catalysis: a perfect union. *Chem. Soc. Rev.* 40: 5151–5169.

2587 Jeffery, T. (1985). Highly stereospecific Pd-catalyzed vinylation of vinylic halides under solid–liquid phase-transfer conditions. *Tetrahedron Lett.* 26 (22): 2667–2670.

2588 Jeffery, T. and Galland, J.C. (1994). Tetraalkylammonium salts in Heck-type reactions using an alkali-metal hydrogen carbonate or an alkali-metal acetate as the base. *Tetrahedron Lett.* 35 (24): 4103–4106.

2589 Jeffery, T. (1994). Heck-type reactions in water. *Tetrahedron Lett.* 35 (19): 3051–3054.

2590 Larock, R.C. and Babu, S. (1987). Synthesis of nitrogen-heterocycles via Pd-catalyzed intramolecular cyclization. *Tetrahedron Lett.* 28 (44): 5291–5294.

2591 Sauvagnat, B., Lamaty, F., Lazaro, R., and Martinez, J. (2001). Highly selective intermolecular Heck reaction for the soluble polymer supported synthesis of glutamic acid analogues. *Tetrahedron* 57 (48): 9711–9718.

2592 Lamblin, M., Nassar-Hardy, L., Hierso, J.C. et al. (2010). Recyclable heterogeneous Pd catalysts in pure water: sustainable developments in Suzuki, Heck, Sonogashira and Tsuji–Trost reactions. *Adv. Synth. Catal.* 352: 33–79.

2593 Slamani, A., Demir, S., and Ozdemir, I. (2012). Use of benzimidazolium salts for in situ generation of Pd catalysts in Heck reactions in water. *Catal. Commun.* 29: 141–144.

2594 Potier, J., Menuel, S., Rousseau, J. et al. (2014). Multifunctional cyclodextrin-based N,N-bidentate ligands for aqueous Heck arylation. *J. Appl. Catal. A: Chem.* 479: 1–8.

2595 Roberts, G.M., Zhang, S.Y., Zhao, Y., and Woo, L.K. (2015). Improving reactivity and selectivity of aqueous-based Heck reactions by the local hydrophobicity of phosphine ligands. *Tetrahedron* 71: 8263–8270.

2596 Burns, B., Grigg, R., Santhakumar, V. et al. (1992). Pd-catalyzed tandem cyclization - anion capture processes. 1. Background and hydride ion capture by (alkyl)Pd and (π-allyl)Pd species. *Tetrahedron* 48 (35): 7297–7320.

2597 Batsanov, A.S., Knowles, J.P., and Whiting, A. (2007). Mechanistic studies on the Heck–Mizoroki cross-coupling reaction of a hindered vinylboronate ester as a key approach to developing a highly stereoselective synthesis of a C1—C7 (*Z,Z,E*)-triene synthon for viridenomycin. *J. Org. Chem.* 72 (7): 2525–2532.

2598 Park, S.B. and Alper, H. (2003). Highly efficient, recyclable Pd(II) catalysts with bisimidazole ligands for the Heck reaction in ionic liquids. *Org. Lett.* 5 (18): 3209–3212.

2599 Yang, D., Chen, Y.C., and Zhu, N.Y. (2004). Sterically bulky thioureas as air- and moisture-stable ligands for Pd-catalyzed Heck reactions of aryl halides. *Org. Lett.* 6 (10): 1577–1580.

2600 Cui, X., Li, J.A., Zhang, Z.P. et al. (2007). Pd(quinoline-8-carboxylate)2 as a low-priced, phosphine-free catalyst for Heck and Suzuki reactions. *J. Org. Chem.* 72 (24): 9342–9345.

2601 Zhang, X.M., Xi, Z.X., Liu, A.L., and Chen, W.Z. (2008). Facile and efficient olefination of aryl halides catalyzed by a Pd complex containing a heteroarene functionalized N-heterocyclic carbene. *Organometallics* 27 (17): 4401–4406.

2602 Karimi, B. and Enders, D. (2006). New N-heterocyclic carbene Pd complex/ionic liquid matrix immobilized on silica: application as recoverable catalyst for the Heck reaction. *Org. Lett.* 8 (6): 1237–1240.

2603 Li, H.J. and Wang, L. (2006). Triethanolamine as an efficient and reusable base, ligand and reaction medium for phosphane-free Pd-catalyzed Heck reactions. *Eur. J. Org. Chem.* 5099–5102.

2604 Chandrasekhar, S., Narsihmulu, C., Sultana, S.S., and Reddy, N.R. (2002). Poly(ethylene glycol) (PEG) as a reusable solvent medium for organic synthesis. Application in the Heck reaction. *Org. Lett.* 4 (25): 4399–4401.

2605 Zhou, L. and Wang, L. (2006). Functionalized ionic liquid as an efficient and recyclable reaction medium for phosphine-free Pd-catalyzed Heck reaction. *Synthesis* (16): 2653–2658.

2606 Montoya, V., Pons, J., Branchadell, V. et al. (2008). Highly efficient pyridylpyrazole ligands for the Heck reaction. A combined experimental and computational study. *Organometallics* 27 (6): 1084–1091.

2607 Cai, Y.Q., Lu, Y., Liu, Y. et al. (2008). Efficient Heck reactions catalyzed by a Pd/diol-imidazolium salt in aerial atmosphere. *Catal. Commun.* 9 (6): 1209–1213.

2608 de Vries, A.H.M., Mulders, J.M.C.A., Mommers, J.H.M. et al. (2003). Homeopathic ligand-free Pd as a catalyst in the Heck reaction. A comparison with a palladacycle. *Org. Lett.* 5 (18): 3285–3288.

2609 Yao, Q.W., Kinney, E.P., and Yang, Z. (2003). Ligand-free Heck reaction: $Pd(OAc)_2$ as an active catalyst revisited. *J. Org. Chem.* 68 (19): 7528–7531.

2610 Chen, F.X., Lin, I., Li, H. et al. (2007). Alkaline earth metal oxides as solid bases for ligand-free Heck reaction. *Catal. Commun.* 8 (12): 2053–2058.

2611 Gordillo, A., de Jesús, E., and López-Mardomingo, C. (2007). Consecutive Pd-catalyzed Hiyama–Heck reactions in aqueous media under ligand-free conditions. *J. Chem. Soc., Chem. Commun.* 4056–4058.

2612 Vogel, M.A.K., Stark, C.B.W., and Lyapkalo, I.M. (2007). An expedient variant of Heck reaction of alkenyl nonaflates: homogeneous ligand-free Pd catalysis at room temperature. *Adv. Synth. Catal.* 349 (7): 1019–1024.

2613 Du, L.H. and Wang, Y.G. (2007). Microwave-promoted Heck reaction using $Pd(OAc)_2$ as catalyst under ligand-free and solvent-free conditions. *Synth. Commun.* 37 (1–3): 217–222.

2614 Ambrogio, I., Fabrizi, G., Cacchi, S. et al. (2008). Unusual selectivity-determining factors in the phosphine-free Heck arylation of allyl ethers. *Organometallics* 27 (13): 3187–3195.

2615 Papp, A., Galbács, G., and Molnár, A. (2005). Recyclable ligand-free mesoporous heterogeneous Pd catalysts for Heck coupling. *Tetrahedron Lett.* 46 (45): 7725–7728.

2616 Huang, L., Wang, Z., Ang, T.P. et al. (2006). A novel SiO_2 supported Pd metal catalyst for the Heck reaction. *Catal. Lett.* 112 (3–4): 219–225.

2617 Kantam, M.L., Roy, S., Roy, M. et al. (2006). Nanocrystalline Mg oxide stabilized Pd(0): an efficient heterogeneous catalyst for Heck and Sonogashira coupling of aryl halides. *Synlett* (17): 2747–2750.

2618 Zhou, H., Zhuo, G.L., and Jiang, X.Z. (2006). Heck reaction catalyzed by Pd supported on LDH-F hydrotalcite. *J. Mol. Catal. A: Chem.* 248 (1–2): 26–31.

2619 Budarin, V.L., Clark, J.H., Luque, R. et al. (2008). Palladium nanoparticles on polysaccharide-derived mesoporous materials and their catalytic performance in C—C coupling reactions. *Green Chem.* 10 (4): 382–387.

2620 Demel, J., Park, S.E., Cejka, J., and Stepnicka, P. (2008). The use of Pd nanoparticles supported with MCM-41 and basic (Al)MCM-41 mesoporous sieves in microwave-assisted Heck reaction. *Catal. Today* 132 (1–4): 63–67.

2621 Baruwati, B., Guin, D., and Manorama, S.V. (2007). Pd on surface-modified $NiFe_2O_4$ nanoparticles: a magnetically recoverable catalyst for Suzuki and Heck reactions. *Org. Lett.* 9 (26): 5377–5380.

2622 Liu, H.Q., Wang, L., and Li, P.H. (2008). Highly efficient and recyclable Pd catalyst anchored on organic–inorganic hybrid material: application in the Heck reaction. *Synthesis* (15): 2405–2411.

2623 Jutand, A. (2009). Mechanisms of the Mizoroki–Heck reaction. In: *The Mizoroki–Heck Reaction* (ed. M. Oestreich), 1–50. Chichester: Wiley.

2624 Shaw, B.L., Perera, S.D., and Staley, E.A. (1998). Highly active, stable, catalysts for the Heck reaction; further suggestions on the mechanism. *J. Chem. Soc., Chem. Commun.* 1361–1362.

2625 Shaw, B.L. (1998). Speculations on new mechanisms for Heck reactions. *New J. Chem.* 22: 77–79.

2626 Julía-Hernández, F., Arcas, A., and Vicente, J. (2012). Providing support in favor of the existence of a Pd(II)/Pd(IV) catalytic cycle in a Heck-type reaction. *Chem. Eur. J.* 17: 7780–7786.

2627 Bohm, V.P.W. and Herrmann, W.A. (2001). Mechanism of the Heck reaction using a phosphapalladacycle as the catalyst: classical versus Pd(IV) intermediates. *Chem. Eur. J.* 7 (19): 4191–4197.

2628 Allolio, C. and Strassner, T. (2014). Palladium complexes with chelating bis-NHC ligands in the Mizoroki–Heck reaction mechanism and electronic effects, a DFT study. *J. Org. Chem.* 79: 12096–12105.

2629 Guest, D., da Silva, V.H.M., Batista, A.P.D. et al. (2015). (N-heterocyclic carbene)-palladate complexes in anionic Mizoroki–Heck coupling cycles: a combined experimental and computational study. *Organometallics* 34: 2463–2470.

2630 da Siva, V.H.M., Braga, A.A.C., and Cundari, T.R. (2016). N-Heterocyclic carbene based Ni and Pd complexes: a DFT comparison of the Mizoroki–Heck catalytic cycles. *Organometallics* 35: 3170–3181.

2631 Knowles, J.P. and Whiting, A. (2007). The Heck–Mizoroki cross-coupling reaction: a mechanistic perspective. *Org. Biomol. Chem.* 5 (1): 31–44.

2632 Amatore, C. and Jutand, A. (2000). Anionic Pd(0) and Pd(II) intermediates in Pd-catalyzed Heck and cross-coupling reactions. *Acc. Chem. Res.* 33 (5): 314–321.

2633 Amatore, C., Carre, E., Jutand, A., and Medjour, Y. (2002). Decelerating effect of alkenes in the oxidative addition of phenyl iodide to Pd(0) complexes in Heck reactions. *Organometallics* 21 (21): 4540–4545.

2634 Jutand, A. (2004). Dual role of nucleophiles in Pd-catalyzed Heck, Stille, and Sonogashira reactions. *Pure Appl. Chem.* 76 (3): 565–576.

2635 Jutand, A. and Mosleh, A. (1997). Ni- and Pd-catalyzed homocoupling of aryl triflates. Scope, limitation, and mechanistic aspects. *J. Org. Chem.* 62 (2): 261–274.

2636 Amatore, C., Jutand, A., and M'Barki, M.A. (1992). Evidence of the formation of zero valent Pd from $Pd(OAc)_2$ and triphenylphosphine. *Organometallics* 11 (9): 3009–3013.

2637 Tsvelikhovsky, D., Pessing, D., Avnir, D., and Blum, J. (2008). Forcing a *cis*-product by matrix imprinting: heck reaction catalyzed by Pd acetate entrapped within *cis*-imprinted sol–gel derived silicates. *Adv. Synth. Catal.* 350 (18): 2856–2858.

2638 Jarusiewicz, J., Yoo, K.S., and Jung, K.W. (2009). Highly regioselective Heck coupling reactions of aryl halides and dihydropyran in the presence of an NHC-pyridine ligand. *Synlett* (3): 482–486.

2639 Cabri, W. and Candiani, I. (1995). Recent developments and new perspectives in the Heck reaction. *Acc. Chem. Res.* 28 (1): 2–7.

2640 von Schenck, H., Åkermark, B., and Svensson, M. (2003). Electronic control of the regiochemistry in the Heck reaction. *J. Am. Chem. Soc.* 125 (12): 3503–3508.

2641 Zhang, L.J., Jiang, Z., Dong, C.N. et al. (2014). Palladium-catalyzed highly regioselective Mizoroki–Heck arylation of allylamines with aryl chlorides. *ChemCatChem* 6: 311–318.

2642 Nilsson, P., Olofsson, K., and Larhed, M. (2009). Focus on regioselectivity and product outcome in organic synthesis. In: *The Mizoroki–Heck Reaction* (ed. M. Oestreich), 133–162. Chichester: Wiley.

2643 Mo, J., Xu, L.J., and Xiao, J.L. (2005). Ionic liquid-promoted, highly regioselective Heck arylation of electron-rich olefins by aryl halides. *J. Am. Chem. Soc.* 127 (2): 751–760.

2644 Liu, S.F., Thomson, N., Pettman, A. et al. (2008). Ionic liquids as solvent for regioselective arylation of α-substituted allylic alcohols by aryl bromides. *J. Mol. Catal. A: Chem.* 279 (2): 210–217.

2645 Vallin, K.S.A., Larhed, M., and Hallberg, A. (2001). Aqueous DMF-K_2CO_3 as a substitute for Tl and Ag additives in the Pd-catalyzed conversion of aryl bromides to acetyl arenes. *J. Org. Chem.* 66 (12): 4340–4343.

2646 Arvela, R.K., Pasquini, S., and Larhed, M. (2007). Highly regioselective internal Heck arylation of hydroxyalkyl vinyl ethers by aryl halides in water. *J. Org. Chem.* 72: 6390–6396.

2647 Mo, J. and Xiao, J.L. (2006). The Heck reaction of electron-rich olefins with regiocontrol by hydrogen-bond donors. *Angew. Chem. Int. Ed.* 45 (25): 4152–4157.

2648 McConville, M., Saidi, O., Blacker, J., and Xiao, J.L. (2009). Regioselective Heck vinylation of electron-rich olefins with vinyl halides: is the neutral pathway in operation? *J. Org. Chem.* 74 (7): 2692–2698.

2649 Hyder, Z.R., W, J., and Xiao, J.L. (2008). Hydrogen-bond-directed catalysis: faster, regioselective and cleaner Heck arylation of electron-rich olefins in alcohols. *Chem. Eur. J.* 14: 5555–5566.

2650 Ruan, J.W., Iggo, J.A., Berry, N.G., and Xiao, J.L. (2010). Hydrogen-bonding-promoted oxidative addition and regioselective arylation of olefins with aryl chlorides. *J. Am. Chem. Soc.* 132: 16689–16699.

2651 Coyne, A.G., Fitzpatrick, M.O., and Guiry, P.J. (2009). Ligand design for intermolecular asymmetric Mizoroki–Heck reactions. In: *The Mizoroki–Heck Reaction* (ed. M. Oestreich), 405–431. Chichester: Wiley.

2652 Ozawa, F., Kubo, A., and Hayashi, T. (1992). Palladium-catalyzed asymmetric arylation of 2,3-dihydrofuran - 1,8-Bis(dimethylamino)naphthalene as an efficient base. *Tetrahedron Lett.* 33 (11): 1485–1488.

2653 Sato, Y., Watanabe, S., and Shibasaki, M. (1992). Further-studies on a catalytic asymmetric-synthesis of decalin derivatives. *Tetrahedron Lett.* 33 (18): 2589–2592.

2654 Hii, K.K., Claridge, T.D.W., Brown, J.M. et al. (2001). The intermolecular asymmetric Heck reaction: mechanistic and computational studies. *Helv. Chim. Acta* 84: 3043–3056.

2655 Wu, G.Z., Lamaty, F., and Negishi, E. (1989). Metal-promoted cyclization. 26. Palladium-catalyzed cyclization of benzyl halides and related electrophiles containing alkenes and alkynes as a novel route to carbocycles. *J. Org. Chem.* 54 (11): 2507–2508.

2656 Pan, Y., Zhang, Z.G., and Hu, H.W. (1992). Vinylation of alkyl halides catalyzed by Pd catalyst. *Synth. Commun.* 22 (14): 2019–2029.

2657 Pan, Y., Zhang, Z.Y., and Hu, H.W. (1995). Vinylation of benzylic quaternary ammonium-salts catalyzed by Pd. *Synthesis* (3): 245–247.

2658 Kumar, P. (1997). Synthesis of substituted 1-tetralones. *Org. Prep. Proced. Int.* 29 (4): 477–480.

2659 Wang, L.S., Pan, Y., Jiang, X., and Hu, H.W. (2000). Palladium-catalyzed reaction of α-chloromethylnaphthalene with olefins. *Tetrahedron Lett.* 41 (5): 725–727.

2660 Higuchi, K., Sawada, K., Nambu, H. et al. (2003). A convenient synthesis of the beraprost intermediate: a useful method for introducing a C3 unit at the benzyl position. *Org. Lett.* 5 (20): 3703–3704.

2661 Narahashi, H., Yamamoto, A., and Shimizu, I. (2004). Heck-type benzylation of olefins with benzyl trifluoroacetates. *Chem. Lett.* 33 (3): 348–349.

2662 Mori, M., Oda, I., and Ban, Y. (1982). Cyclization of α-haloamide with internal double-bond by use of the low-valent metal complex. *Tetrahedron Lett.* 23 (50): 5315–5318.

2663 Glorius, F. (2003). Palladium-catalyzed Heck-type reaction of 2-chloro acetamides with olefins. *Tetrahedron Lett.* 44 (30): 5751–5754.

2664 Oppolzer, W. (1989). Intramolecular, stoichiometric (Li, Mg, Zn) and catalytic (Ni, Pd, Pt) metallo-ene reactions in organic synthesis. *Angew. Chem. Int. Ed. Engl.* 28 (1): 38–52.

2665 Braese, S., Waegell, B., and de Meijere, A. (1998). Palladium-catalyzed coupling reactions of 1-bromoadamantane with styrenes and arenes. *Synthesis* (2): 148–152.

2666 Firmansjah, L. and Fu, G.C. (2007). Intramolecular Heck reactions of unactivated alkyl halides. *J. Am. Chem. Soc.* 129: 11340–11341.

2667 Boldrini, G.P., Savoia, D., Tagliavini, E. et al. (1986). Nickel-catalyzed coupling of activated alkenes with organic halides. *J. Organomet. Chem.* 301: C62–C64.

2668 Lebedev, S.A., Lopatina, V.S., Petrov, E.S., and Beletskaya, I.P. (1988). Condensation of organic bromides with vinyl compounds catalyzed by Ni complexes in the presence of Zn. *J. Organomet. Chem.* 344 (2): 253–259.

2669 Kelkar, A.A., Hanaoka, T., Kubota, Y., and Sugi, Y. (1994). The vinylation of bromobiphenyls using homogeneous Ni catalysts. *Catal. Lett.* 29: 69–75.

2670 Gøgsig, T.M., Kleimark, J., Lill, S.O.N. et al. (2012). Mild and efficient Ni-catalyzed Heck reactions with electron-rich olefins. *J. Am. Chem. Soc.* 134: 443–452.

2671 Ehle, A.R., Zhou, Q., and Watson, M.P. (2012). Nickel(0)-catalyzed Heck cross-coupling via activation of aryl C—OPiv bonds. *Org. Lett.* 14: 1202–1205.

2672 Matsubara, R., Gutierrez, A.C., and Jamison, T.F. (2011). Nickel-catalyzed Heck-type reactions of benzyl chlorides and simple olefins. *J. Am. Chem. Soc.* 133: 19020–19023.

2673 Branchaud, B.P. and Detlefsen, W.D. (1991). Cobaloxime-catalyzed radical alkyl-styryl couplings. *Tetrahedron Lett.* 32 (44): 6273–6276.

2674 Ikeda, Y., Nakamura, T., Yorimitsu, H., and Oshima, K. (2002). Cobalt-catalyzed Heck-type reaction of alkyl halides with styrenes. *J. Am. Chem. Soc.* 124 (23): 6514–6515.

2675 Affo, W., Ohmiya, H., Fujioka, T. et al. (2006). Cobalt-catalyzed trimethylsilylmethylmagnesium-promoted radical alkenylation of alkyl halides: a complement to the Heck reaction. *J. Am. Chem. Soc.* 128 (24): 8068–8077.

2676 Ikeda, Y., Yorimitsu, H., Shinokubo, H., and Oshima, K. (2004). Cobalt-mediated Mizoroki–Heck-type reaction of epoxide with styrene. *Adv. Synth. Catal.* 346 (13–15): 1631–1634.

2677 Cerai, G.P. and Morandi, B. (2016). Atom economical Co-catalysed regioselective coupling of epoxides and aziridines with alkenes. *J. Chem. Soc., Chem. Commun.* 52: 9769–9772.

2678 Rudolph, A. and Lautens, M. (2009). Secondary alkyl halides in transition-metal-catalyzed cross-coupling reactions. *Angew. Chem. Int. Ed.* 48: 2656–2670.

2679 Yorimitsu, H. and Oshima, K. (2006). New synthetic reactions catalyzed by Co complexes. *Pure Appl. Chem.* 78 (2): 441–449.

2680 Evano, G., Blanchard, N., and Toumi, M. (2008). Copper-mediated coupling reactions and their applications in natural products and designed biomolecules synthesis. *Chem. Rev.* 108: 3054–3131.

2681 Li, J.H., Wang, D.P., and Xie, Y.X. (2005). CuI/DABCO as a highly active catalytic system for the Heck-type reaction. *Tetrahedron Lett.* 46 (30): 4941–4944.

2682 Calo, V., Nacci, A., Monopoli, A. et al. (2005). Copper bronze catalyzed Heck reaction in ionic liquids. *Org. Lett.* 7 (4): 617–620.

2683 Nishikata, T., Noda, Y., Fujimoto, R., and Sakashita, T. (2013). An efficient generation of a functionalized tertiary-alkyl radical for Cu-catalyzed tertiary-alkylative Mizoroki–Heck type reaction. *J. Am. Chem. Soc.* 135: 16372–16375.

2684 Terao, J. and Kambe, N. (2006). Transition metal-catalyzed C—C bond formation reactions using alkyl halides. *Bull. Chem. Soc. Jpn.* 79 (5): 663–672.

2685 Farrington, E.J., Brown, J.M., Barnard, C.F.J., and Rowsell, E. (2002). Ruthenium-catalyzed oxidative Heck reactions. *Angew. Chem. Int. Ed.* 41 (1): 169–171.

2686 Horniakova, J., Nakamura, H., Kawase, R. et al. (2005). Pyridine-derived Ru and Pt complexes immobilized on ordered mesoporous silica as catalysts for Heck vinylation. *J. Mol. Catal. A: Chem.* 233 (1–2): 49–54.

2687 Yorke, J., Wan, L., Xia, A.B., and Zheng, W.J. (2007). Catalytic applications of 1,2,4-diazaphospholide-based Ru complexes in the Heck reaction. *Tetrahedron Lett.* 48 (50): 8843–8845.

2688 Oi, S., Sakai, K., and Inoue, Y. (2005). Ruthenium-catalyzed arylation of 2-alkenylpyridines with aryl bromides: alternative *E,Z*-selectivity to Mizoroki–Heck reaction. *Org. Lett.* 7: 4009–4011.

2689 Loska, R., Volla, C.M., and Vogel, P. (2008). Iron-catalyzed Mizoroki–Heck cross-coupling reaction with styrenes. *Adv. Synth. Catal.* 350 (18): 2859–2864.

2690 Hajipour, A.R. and Azizi, G. (2013). Iron-catalyzed cross-coupling reaction: recyclable heterogeneous iron catalyst for selective olefination of aryl iodides in poly(ethylene glycol) medium. *Green Chem.* 15: 1030–1034.

2691 Sugihara, T., Satoh, T., Miura, M., and Nomura, M. (2003). Rhodium-catalyzed Mizoroki–Heck-type arylation of alkenes with aroyl chlorides under phosphane- and base-free conditions. *Angew. Chem. Int. Ed.* 42 (38): 4672–4674.

2692 Liu, C., Tang, S., Liu, D. et al. (2012). Nickel-catalyzed Heck-type alkenylation of secondary and tertiary α-carbonyl alkyl bromides. *Angew. Chem. Int. Ed.* 51: 3638–3641.

2693 Alami, M., Ferri, F., and Linstrumelle, G. (1993). An efficient Pd-catalyzed reaction of vinyl and aryl halides or triflates with terminal alkynes. *Tetrahedron Lett.* 34: 6403–6406.

2694 Sonogashira, K. (2002). Development of Pd/Cu catalyzed cross-coupling of terminal acetylenes with C(sp^2)-halides. *J. Organomet. Chem.* 653: 46–49.

2695 Chinchilla, R. and Nájera, C. (2007). The Sonogashira reaction: a booming methodology in synthetic organic chemistry. *Chem. Rev.* 107 (3): 874–922.

2696 Chinchilla, R. and Nájera, C. (2011). Recent advances in Sonogashira reactions. *Chem. Soc. Rev.* 40: 5084–5121.

2697 Amatore, C., Blart, E., Genêt, J.P. et al. (1995). New synthetic applications of water-soluble acetate Pd/Tppts catalyst generated in situ - evidence for a true Pd(0) species intermediate. *J. Org. Chem.* 60 (21): 6829–6839.

2698 Tougerti, A., Negri, S., and Jutand, A. (2007). Mechanism of the Cu-free Pd-catalyzed Sonagashira reactions: multiple role of amines. *Chem. Eur. J.* 13: 666–676.

2699 Eckhardt, M. and Fu, G.C. (2003). The first applications of carbene ligands in cross-couplings of alkyl electrophiles: Sonogashira reactions of unactivated alkyl bromides and iodides. *J. Am. Chem. Soc.* 125: 13642–13643.

2700 Altenhoff, G., Wurtz, S., and Glorius, F. (2006). The first Pd-catalyzed Sonogashira coupling of unactivated secondary alkyl bromides. *Tetrahedron Lett.* 47: 2925–2928.

2701 Willy, B. and Müller, T.J.J. (2008). Consecutive multi-component syntheses of heterocycles via Pd/Cu-catalyzed generation of alkynones. *ARKIVOC* 195–208.

2702 Merkul, E., Boersch, C., Frank, W., and Müller, T.J.J. (2009). Three-component synthesis of *N*-Boc-4-iodopyrroles and sequential one-pot alkynylation. *Org. Lett.* 11: 2269–2272.

2703 Merkul, E., Oeser, T., and Müller, T.J.J. (2009). Consecutive three-component synthesis of ynones by decarbonylative Sonogashira coupling. *Chem. Eur. J.* 15: 5006–5011.

2704 Dubbaka, S.R. and Vogel, P. (2004). Palladium-catalyzed desulfitative Sonogashira–Hagihara cross-couplings of arenesulfonyl chlorides and terminal alkynes. *Adv. Synth. Catal.* 346 (13–15): 1793–1797.

2705 Plenio, H. (2008). Catalysts for the Sonogashira coupling - the crownless again shall be king. *Angew. Chem. Int. Ed.* 47: 6954–6956.

2706 Oha, H., Kitai, K., Suzuki, T., and Obora, Y. (2017). *N*,*N*-Dimethylformamide-stabilized Cu nanoparticles as a catalyst precursor for Sonogashira–Hagihara cross-coupling. *RSC Adv.* 7 (37): 22869–22874.

2707 (a) Prüger, B., Holmeister, G.E., Jacobsen, C.B. et al. (2010). Transition-metal-free formal Sonogashira coupling and α-carbonyl arylation reactions. *Chem. Eur. J.* 16: 3783–3790. (b) Jin, G.Y., Zhang, X.X., and Cao, S. (2013). Transition-metal-free Sonogashira-type cross-coupling of alkynes with fluoroarenes. *Org. Lett.* 15: 3114–3117.

2708 Cassar, L. (1975). Synthesis of aryl-substituted and vinyl-substituted acetylene derivatives by use of Ni and Pd complexes. *J. Organomet. Chem.* 93: 253–257.

2709 Dieck, H.A. and Heck, F.R. (1975). Palladium-catalyzed synthesis of aryl, heterocyclic and vinylic acetylene derivatives. *J. Organomet. Chem.* 93: 259–263.

2710 Sonogashira, K., Tohda, Y., and Hagihara, N. (1975). A convenient synthesis of acetylenes - catalytic substitutions of acetylenic hydrogen with bromoalkenes, iodoarenes, and bromopyridines. *Tetrahedron Lett.* 16 (50): 4467–4470.

2711 Wang, X.B., Song, Y.M., Qu, J.P., and Luo, Y. (2017). Mechanistic insights into the Cu-cocatalyzed Sonogashira cross-coupling reaction: key role of an anion. *Organometallics* 36: 1042–1048.

2712 Stephens, R.D. and Castro, C.E. (1963). Substitution of aryl iodides with Cu(I) acetylides - a synthesis of tolanes and heterocyclics. *J. Org. Chem.* 28: 3313–3315.

2713 Burdon, J., Coe, P.L., Marsh, C.R., and Tatlow, J.C. (1967). Reactions of organocopper compounds with halogeno-olefins. *J. Chem. Soc., Chem. Commun.* 1259–1260.

2714 Ljungdahl, T., Bennur, T., Dallas, A. et al. (2008). Two competing mechanisms for the Cu-free Sonogashira cross-coupling reaction. *Organometallics* 27: 2490–2498.

2715 Soheili, A., Baneze-Walker, J., Murry, J.A. et al. (2003). Efficient and general protocol for the Cu-free Sonogashira coupling of aryl bromides at room temperature. *Org. Lett.* 5: 4191–4194.

2716 García-Melchor, M., Pacheco, M.C., Nájera, C. et al. (2012). Mechanistic exploration of the Pd-catalyzed Cu-free Sonogashira reaction. *ACS Catal.* 2: 135–144.

2717 He, C., Ke, J., Xu, H., and Lei, A.W. (2013). Synergistic catalysis in the Sonogashira coupling reaction: quantitative kinetic investigation of transmetalation. *Angew. Chem. Int. Ed.* 52: 1527–1530.

2718 an der Heiden, M.R., Plenio, H., Immel, S. et al. (2008). Insights into Sonogashira cross-coupling by high-throughput kinetics and descriptor modeling. *Chem. Eur. J.* 14: 2857–2866.

2719 Ohmiya, H., Yorimitsu, H., and Oshima, K. (2006). Cobalt-mediated cross-coupling reactions of primary and secondary alkyl halides with 1-(trimethylsilyl)ethenyl- and 2-trimethylsilylethynylmagnesium reagents. *Org. Lett.* 8: 3093–3096.

2720 Cahiez, G., Gager, O., and Buendia, J. (2010). Copper-catalyzed cross-coupling of alkyl and aryl Grignard reagents with alkynyl halides. *Angew. Chem. Int. Ed.* 49: 1278–1281.

2721 Thaler, T., Guo, L.N., Mayer, P., and Knochel, P. (2011). Highly diastereoselective C(sp^3)-C(sp) cross-coupling reactions between 1,3-and 1,4-substituted cyclohexylzinc reagents and bromoalkynes through remote stereocontrol. *Angew. Chem. Int. Ed.* 50: 2174–2177.

2722 Vechorkin, O., Barmaz, D., Proust, V., and Hu, X.L. (2009). Ni-catalyzed Sonogashira coupling of nonactivated alkyl halides: orthogonal functionalization of alkyl iodides, bromides, and chlorides. *J. Am. Chem. Soc.* 131: 12078–12079.

2723 Yi, J., Lu, X., Sun, Y.Y. et al. (2013). Nickel-catalyzed Sonogashira reactions of non-activated secondary alkyl bromides and iodides. *Angew. Chem. Int. Ed.* 52: 12409–12413.

2724 Gallego, D., Bruck, A., Irran, E. et al. (2013). From bis(silylene) and bis(germylene) pincer-type Ni(II) complexes to isolable intermediates of the Ni-catalyzed Sonogashira cross-coupling reaction. *J. Am. Chem. Soc.* 135: 15617–15626.

2725 García, P.M.P., Ren, P., Scopelliti, R., and Hu, X.L. (2015). Nickel-catalyzed direct alkylation of terminal alkynes at room temperature: a hemilabile pincer ligand enhances catalytic activity. *ACS Catal.* 5: 1164–1171.

2726 Mao, J.C., Xie, G.L., Wu, M.Y. et al. (2008). Ligand-free Fe/Cu-cocatalyzed alkynylation coupling reactions. *Adv. Synth. Catal.* 350: 2477–2482.

2727 Volla, C.M. and Vogel, P. (2008). Iron/copper-catalyzed C—C cross-coupling of aryl iodides with terminal alkynes. *Tetrahedron Lett.* 49 (41): 5961–5964.

2728 Huang, H., Jiang, H.L., Chen, K.X., and Liu, H. (2008). Efficient Fe/Cu-cocatalyzed alkynylation of aryl iodides with terminal alkynes. *J. Org. Chem.* 73: 9061–9064.

2729 Pan, C.D., Luo, F., Wang, W.H. et al. (2009). Iron-catalysed Sonogashira reactions. *J. Chem. Res. (S)* 478–481.

2730 Carril, M., Correa, A., and Bolm, C. (2008). Iron-catalyzed Sonogashira reactions. *Angew. Chem. Int. Ed.* 47 (26): 4862–4865.

2731 Sawant, D.N., Tambade, P.J., Wagh, Y.S., and Bhanage, B.M. (2010). $FeCl_3/PPh_3$-catalyzed Sonogashira coupling reaction of aryl iodides with terminal alkynes. *Tetrahedron Lett.* 51: 2758–2761.

2732 Mao, J.C., Wu, M.Y., Xie, G.L., and Ji, S.J. (2009). Samarium powder-catalyzed Pd-free and ligand-free Sonogashira coupling reactions. *Adv. Synth. Catal.* 351: 2101–2106.

2733 Feng, L., Liu, F., Sun, P., and Bao, J. (2008). Sonogashira reaction of aryl halides with terminal alkynes catalyzed by cobalt hollow nanospheres. *Synlett* 1415–1417.

2734 González-Arellano, C., Abad, A., Corma, A. et al. (2007). Catalysis by Au(I) and Au(III): a parallelism between homo- and heterogeneous catalysts for Cu-free Sonogashira cross-coupling reactions. *Angew. Chem. Int. Ed.* 46: 1536–1538.

2735 González-Arellano, C., Corma, A., Iglesias, M., and Sánchez, F. (2008). Soluble Au and Pd complexes heterogenized on MCM-41 are effective and versatile catalysts. *Eur. J. Inorg. Chem.* 1107–1115.

2736 Li, P., Wang, L., Wang, M., and You, F. (2008). Gold(I) iodide catalyzed Sonogashira reactions. *Eur. J. Org. Chem.* 5946–5951.

2737 Corma, A., Juárez, R., Boronat, M. et al. (2011). Gold catalyzes the Sonogashira coupling reaction without the requirement of Pd impurities. *J. Chem. Soc., Chem. Commun.* 47: 1446–1448.

2738 Kyriakou, G., Beaumont, S.K., Humphrey, S.M. et al. (2010). Sonogashira coupling catalyzed by Au nanoparticles: does homogeneous or heterogeneous catalysis dominate? *ChemCatChem* 2: 1444–1449.

2739 Beaumont, S.K., Kyriakou, G., and Lambert, R.M. (2010). Identity of the active site in Au nanoparticle-catalyzed Sonogashira coupling of phenylacetylene and iodobenzene. *J. Am. Chem. Soc.* 132: 12246–12248.

2740 Kanuru, V.K., Humphrey, S.M., Kyffin, J.M. et al. (2009). Evidence for heterogeneous Sonogashira coupling of phenylacetylene and iodobenzene catalyzed by well defined Rh nanoparticles. *J. Chem. Soc., Dalton Trans.* 7602–7605.

2741 Lauterbach, T., Livendahl, M., Rosellón, A. et al. (2010). Unlikeliness of Pd-free Au(I)-catalyzed Sonogashira coupling reactions. *Org. Lett.* 12: 3006–3009.

2742 Buchwald, S.L. and Bolm, C. (2009). On the role of metal contaminants in catalyses with $FeCl_3$. *Angew. Chem. Int. Ed.* 48: 5586–5587.

2743 Thome, I., Nijs, A., and Bolm, C. (2012). Trace metal impurities in catalysis. *Chem. Soc. Rev.* 41: 979–987.

2744 Kobayashi, K., Arisawa, M., and Yamaguchi, M. (2002). $GaCl_3$-catalyzed ortho-ethynylation of phenols. *J. Am. Chem. Soc.* 124: 8528–8529.

2745 Dudnik, A.S. and Gevorgyan, V. (2010). Formal Inverse Sonogashira reaction: direct alkynylation of arenes and heterocycles with alkynyl halides. *Angew. Chem. Int. Ed.* 49: 2096–2098.

2746 Messaoudi, S., Brion, J.D., and Alami, M. (2010). Transition-metal-catalyzed direct C—H alkenylation, alkynylation, benzylation, and alkylation of (hetero)arenes. *Eur. J. Org. Chem.* 6495–6516.

2747 Brand, J.P. and Waser, J. (2012). Electrophilic alkynylation: the dark side of acetylene chemistry. *Chem. Soc. Rev.* 41: 4165–4179.

2748 Waser, J. (2016). Benziodoxol(on)e reagents as tools in organic synthesis: the background behind the discovery at the laboratory of catalysis and organic synthesis. *Synlett* 27: 2761–2773.

2749 Brand, J.P., Charpentier, J., and Waser, J. (2009). Direct alkynylation of indole and pyrrole heterocycles. *Angew. Chem. Int. Ed.* 48: 9346–9349.

2750 Brand, J.P. and Waser, J. (2010). Direct alkynylation of thiophenes: cooperative activation of TIPS-EBX with Au and Brønsted acids. *Angew. Chem. Int. Ed.* 49: 7304–7307.

2751 Tolnai, G.L., Brand, J.P., and Waser, J. (2016). Gold-catalyzed direct alkynylation of tryptophan in peptides using TIPS-EBX. *Beilstein J. Org. Chem.* 12: 745–749.

2752 Brand, J.P. and Waser, J. (2012). *Para*-selective Au-catalyzed direct alkynylation of anilines. *Org. Lett.* 14: 744–747.

2753 Pyo, A., Kim, Y.H., Park, K. et al. (2012). Mechanistic study of Pd-catalyzed decarboxylative coupling of phenylpropiolic acid and aryl iodide. *Appl. Organomet. Chem.* 26: 650–654.

2754 Zhao, D.B., You, J.S., and Hu, C.W. (2011). Recent progress in coupling of two heteroarenes. *Chem. Eur. J.* 17: 5466–5492.

2755 Rossi, R., Lessi, M., Manzini, C. et al. (2016). Direct (hetero) arylation reactions of (hetero) arenes as tools for the step-and atom-economical synthesis of biologically active unnatural compounds including pharmaceutical targets. *Synthesis* 28: 3821–3862.

2756 Pitts, A.K., O'Hara, F., Snell, R.H., and Gaunt, M.J. (2015). A concise and scalable strategy for the total synthesis of dictyodendrin B based on sequential C—H functionalization. *Angew. Chem. Int. Ed.* 54: 5451–5455.

2757 Liang, Y. and Wnuk, S.F. (2015). Modification of purine and pyrimidine nucleosides by direct C—H bond activation. *Molecules* 20: 4874–4901.

2758 Gayakhe, V., Sanghvi, Y.S., Fairlamb, I.J.S., and Kapdi, A.R. (2015). Catalytic C—H bond functionalisation of purine and pyrimidine nucleosides: a synthetic and mechanistic perspective. *J. Chem. Soc., Chem. Commun.* 51: 11944–11960.

2759 Alberico, D., Scott, M.E., and Lautens, M. (2007). Aryl-aryl bond formation by transition-metal-catalyzed direct arylation. *Chem. Rev.* 107 (1): 174–238.

2760 Mercier, L.G. and Leclerc, M. (2013). Direct (hetero)arylation: a new tool for polymer chemists. *Acc. Chem. Res.* 46: 1597–1605.

2761 Zhang, J.X., Kang, D.Y., Barlow, S., and Marder, S.R. (2012). Transition metal-catalyzed C—H activation as a route to structurally diverse di(arylthiophenyl)-diketopyrrolopyrroles. *J. Mater. Chem.* 22: 21392–21394.

2762 Abdo, N.I., El-Shehawy, A.A., El-Barbary, A.A., and Lee, J.S. (2012). Palladium-catalyzed direct C—H arylation of thieno[3,4-b]pyrazines: synthesis of advanced oligomeric and polymeric materials. *Eur. J. Org. Chem.* 5540–5551.

2763 Okamoto, K., Zhang, J.X., Housekeeper, J.B. et al. (2013). Arylation reaction: atom efficient and greener syntheses of π-conjugated small molecules and macromolecules for organic electronic materials. *Macromolecules* 46: 8059–8078.

2764 Segawa, Y., Maekawa, T., and Itami, K. (2015). Synthesis of extended π-systems through C—H activation. *Angew. Chem. Int. Ed.* 54: 66–81.

2765 Bura, T., Morin, P.O., and Leclerc, M. (2015). En route to defect-free polythiophene derivatives by direct heteroarylation polymerization. *Macromolecules* 48: 5614–5620.

2766 Marzano, G., Kotowski, D., Babudri, F. et al. (2015). Tin-free synthesis of a ternary random copolymer for BHJ solar cells: direct (hetero)arylation versus Stille polymerization. *Macromolecules* 48: 7039–7048.

2767 Bura, T., Blaskovits, J.T., and Leclerc, M. (2016). Direct (hetero)arylation polymerization: trends and perspectives. *J. Am. Chem. Soc.* 138: 10056–10071.

2768 Morin, P.O., Bura, T., and Leclerc, M. (2016). Realizing the full potential of conjugated polymers: innovation in polymer synthesis. *Mater. Horiz.* 3: 11–20.

2769 McGlacken, G.P. and Bateman, L.M. (2009). Recent advances in aryl-aryl bond formation by direct arylation. *Chem. Soc. Rev.* 38: 2447–2464.

2770 Shibahara, F. and Murai, T. (2013). Direct C—H arylation of heteroarenes catalyzed by Pd/N-based ligand complexes. *Asian J. Org. Chem.* 2: 624–636.

2771 Arockiam, P.B., Bruneau, C., and Dixneuf, P.H. (2012). Ruthenium(II)-catalyzed C—H bond activation and functionalization. *Chem. Rev.* 112: 5879–5918.

2772 Ackermann, L. (2014). Robust Ru (II)-catalyzed C—H arylations: carboxylate assistance for the efficient synthesis of angiotensin-II-receptor blockers. *Org. Process Res. Dev.* 19: 260–269.

2773 Seki, M. (2016). A new catalytic system for the Ru-catalyzed C—H arylation reactions and its application in the practical syntheses of pharmaceutical agents. *Org. Process Res. Dev.* 20: 867–877.

2774 Li, M. and Hua, R.M. (2009). Gold(I)-catalyzed direct C—H arylation of pyrazine and pyridine with aryl bromides. *Tetrahedron Lett.* 50: 1478–1481.

2775 Lu, P.F., Booman, T.C., Slawin, A.M.Z., and Larrosa, I. (2010). Gold(I)-mediated C—H activation of arenes. *J. Am. Chem. Soc.* 132: 5580–5581.

2776 Daugulis, O., Do, H.Q., and Shabashov, D. (2009). Palladium- and Cu-catalyzed arylation of C—H bonds. *Acc. Chem. Res.* 42: 1074–1086.

2777 Mousseau, J.J. and Charette, A.B. (2013). Direct functionalization processes: a journey from Pd to Cu to Fe to Ni to metal-free coupling reactions. *Acc. Chem. Res.* 46: 412–424.

2778 Ackermann, L. (2014). Cobalt-catalyzed C—H arylations, benzylations, and alkylations with organic electrophiles and beyond. *J. Org. Chem.* 79: 8948–8954.

2779 Liu, W., Cao, H., and Lei, A.W. (2010). Iron-catalyzed direct arylation of unactivated arenes with aryl halides. *Angew. Chem. Int. Ed.* 49: 2004–2008.

2780 Kapdi, A.R. (2014). Organometallic aspects of transition-metal catalysed regioselective C—H bond functionalisation of arenes and heteroarenes. *J. Chem. Soc., Dalton Trans.* 43 (8): 3021–3034.

2781 Skhiri, A., Beladhria, A., Yuan, K. et al. (2015). Pd-Catalysed direct arylation of heteroaromatics using (poly)halobenzenesulfonyl chlorides as coupling partners: one step access to (poly)halo-substituted bi(hetero)aryls. *Eur. J. Org. Chem.* 4428–4436.

2782 Hfaiedh, A., Yuan, K., Ben Ammar, H. et al. (2015). Eco-friendly solvents for Pd-catalyzed desulfitative C—H bond arylation of heteroarenes. *ChemSusChem* 8: 1794–1804.

2783 Yuan, K., Soule, J.-F., and Doucet, H. (2015). Functionalization of C—H bonds via metal-catalyzed desulfitative coupling: an alternative tool for access to aryl- or alkyl-substituted (hetero)arenes. *ACS Catal.* 5: 978–991.

2784 Zha, G.F., Qin, H.L., and Kantchev, E.A.B. (2016). Ruthenium-catalyzed direct arylations with aryl chlorides. *RSC Adv.* 6: 30875–30885.

2785 Lei, A.W., Liu, W., Liu, C., and Chen, M. (2010). Arylation of unactivated arenes. *J. Chem. Soc., Dalton Trans.* 39: 10352–10361.

2786 Yanagisawa, S., Ueda, K., Taniguchi, T., and Itami, K. (2008). Potassium *t*-butoxide alone can promote the biaryl coupling of electron-deficient nitrogen heterocycles and haloarenes. *Org. Lett.* 10: 4673–4676.

2787 Liu, W., Liu, R.L., and Bi, Y.L. (2015). Potassium ethoxide/phenanthroline promoted chemoselective direct C—H arylation of unactivated arenes with aryl iodides. *Tetrahedron* 71: 2622–2628.

2788 Yanagisawa, S. and Itami, K. (2011). *tert*-Butoxide-mediated C—H bond arylation of aromatic compounds with haloarenes. *ChemCatChem* 3: 827–829.

2789 (a) Chan, T.L., Wu, Y.N., Choy, P.Y., and Kwong, F.Y. (2013). A radical process towards the development of transition-metal-free aromatic C—C bond-forming reactions. *Chem. Eur. J.* 19: 15802–15814. (b) Mehta, V.P. and Punji, B. (2013). Recent advances in transition-metal-free direct C—C and C-heteroatom bond forming reactions. *RSC Adv.* 3: 11957–11986.

2790 Fañanás-Mastral, M. (2017). Copper-catalyzed arylation with diaryliodonium salts. *Synthesis* 49 (9): 1905–1930.

2791 Saget, T. and Cramer, N. (2013). Enantioselective C—H arylation strategy for functionalized dibenzazepinones with quaternary stereocenters. *Angew. Chem. Int. Ed.* 52: 7865–7868.

2792 Catellani, M. and Chiusoli, G.P. (1985). Palladium-catalyzed synthesis of 1,2,3,4,4*a*,12*b*-hexahydro-1,4-methanotriphenylenes. *J. Organomet. Chem.* 286 (1): C13–C16.

2793 Reiser, O., Weber, M., and de Meijere, A. (1989). Selective 1-3 coupling of norbornene and iodobenzene - a facile synthesis of benzo[e]pyrenes with annelated cycloaliphatic moieties. *Angew. Chem. Int. Ed. Engl.* 28 (8): 1037–1038.

2794 de Meijere, A. and Bräse, S. (1999). Palladium in action: domino coupling and allylic substitution reactions for the efficient construction of complex organic molecules. *J. Organomet. Chem.* 576 (1–2): 88–110.

2795 Catellani, M. and Ferioli, L. (1996). An improved synthesis of 1,4-*cis,exo*-hexa- or tetrahydromethano- or -ethanobiphenylene derivatives catalyzed by Pd complexes. *Synthesis* 769–772.

2796 Catellani, M., Frignani, F., and Rangoni, A. (1997). A complex catalytic cycle leading to a regioselective synthesis of *o,o'*-disubstituted vinylarenes. *Angew. Chem. Int. Ed. Engl.* 36: 119–122.

2797 Catellani, M., Moli, E., and la Ca, N. (2008). Catalytic sequential reactions involving palladacycle-directed aryl coupling steps. *Acc. Chem. Res.* 41: 1512–1522.

2798 Martins, A., Mariampillai, B., and Lautens, M. (2010). Synthesis in the key of catellani: norbornene-mediated *ortho* C—H functionalization. In: *C–H Activation*, Topics in Current Chemistry, vol. 292 (ed. J.-Q. Yu and Z. Shi), 1–33. Heidelberg, Berlin: Springer-Verlag.

2799 Ferraccioli, R., Carenzi, D., Motti, E., and Catellani, M. (2006). A simple catalytic synthesis of condensed pyridones from *o*-bromoarylcarboxamides involving *ipso* substitution via palladacycles. *J. Am. Chem. Soc.* 128: 722–723.

2800 Sui, X.W., Zhu, R., and Gu, Z.H. (2013). Palladium-catalyzed chemoselective Catellani *ortho*-arylation reactions and their applications in natural products synthesis. *Synlett* 24: 2023–2031.

2801 Sui, X.W., Zhu, R., Li, G.C. et al. (2013). Pd-catalyzed chemoselective Catellani *ortho*-arylation of iodopyrroles: rapid total synthesis of rhazinal. *J. Am. Chem. Soc.* 135: 9318–9321.

2802 Weinstabl, H., Suhartono, M., Qureshi, Z., and Lautens, M. (2013). Total synthesis of (+)-linoxepin by utilizing the Catellani reaction. *Angew. Chem. Int. Ed.* 52: 5305–5308.

2803 Jiao, L. and Bach, T. (2013). Palladium-catalyzed direct C—H alkylation of electron-deficient pyrrole derivatives. *Angew. Chem. Int. Ed.* 52: 6080–6083.

2804 Narbonne, V., Retailleau, P., Maestri, G., and Malacria, M. (2014). Diastereoselective synthesis of dibenzoazepines through chelation on Pd(IV) intermediates. *Org. Lett.* 16: 628–631.

2805 Sickert, M., Weinstabl, H., Peters, B. et al. (2014). Intermolecular domino reaction of two aryl iodides involving two C—H functionalizations. *Angew. Chem. Int. Ed.* 53: 5147–5151.

2806 Qureshi, Z., Schlundt, W., and Lautens, M. (2015). Introduction of hindered electrophiles via C—H functionalization in a Pd-catalyzed multicomponent domino reaction. *Synthesis* 47: 2446–2456.

2807 Huang, Y.Z., Zhu, R., Zhao, K., and Gu, Z.H. (2015). Palladium-catalyzed Catellani *ortho*-acylation reaction: an efficient and regiospecific synthesis of diaryl ketones. *Angew. Chem. Int. Ed.* 54: 12669–12672.

2808 Zhang, F.Z. and Spring, D.R. (2014). Arene C—H functionalisation using a removable/modifiable or a traceless directing group strategy. *Chem. Soc. Rev.* 43: 6906–6919.

2809 Schranck, J., Tlilli, A., and Beller, M. (2014). Functionalization of remote C—H bonds: expanding the frontier. *Angew. Chem. Int. Ed.* 53: 9426–9428.

2810 Wang, J.X., McCubbin, J.A., Jin, M.Z. et al. (2008). Palladium-catalyzed direct Heck arylation of dual π-deficient/π-excessive heteroaromatics. Synthesis of C-5 arylated imidazo[1,5-a]pyrazines. *Org. Lett.* 10 (14): 2923–2926.

2811 Phipps, R.J. and Gaunt, M.J. (2009). A *meta*-selective Cu-catalyzed C—H bond arylation. *Science* 323: 1593–1597.

2812 Duong, H.A., Gilligan, R.E., Cooke, M.L. et al. (2011). Cu(II)-catalyzed *meta*-selective direct arylation of α-aryl carbonyl compounds. *Angew. Chem. Int. Ed.* 50: 463–466.

2813 Chen, B., Hou, X.L., Li, Y.X., and Wu, Y.D. (2011). Mechanistic understanding of the unexpected *meta* selectivity in Cu-catalyzed anilide C—H bond arylation. *J. Am. Chem. Soc.* 133: 7668–7671.

2814 Wang, X.C., Gong, W., Fang, L.Z. et al. (2015). ligand-enabled *meta*-C—H activation using a transient mediator. *Nature* 519: 334–338.

2815 Ciana, C.L., Phipps, R.J., Brandt, J.R. et al. (2011). A highly *para*-selective Cu(II)-catalyzed direct arylation of aniline and phenol derivatives. *Angew. Chem. Int. Ed.* 50: 458–462.

2816 Lersch, M. and Tilset, M. (2005). Mechanistic aspects of C—H activation by Pt complexes. *Chem. Rev.* 105 (6): 2471–2526.

2817 García-Cuadrado, D., de Mendoza, P., Braga, A.A.C. et al. (2007). Proton-abstraction mechanism in the Pd-catalyzed intramolecular arylation: substituent effects. *J. Am. Chem. Soc.* 129: 6880–6886.

2818 Balcells, D., Clot, E., and Eisenstein, O. (2010). C—H Bond activation in transition metal species from a computational perspective. *Chem. Rev.* 110: 749–823.

2819 Ackermann, L. (2011). Carboxylate-assisted transition-metal-catalyzed C—H bond functionalizations: mechanism and scope. *Chem. Rev.* 111: 1315–1345.

2820 Catellani, M. and Chiusoli, G.P. (1992). Palladacycle formation by electrophilic aromatic-substitution, as monitored by ^{1}H-NMR spectroscopy. *J. Organomet. Chem.* 425: 151–154.

2821 Martín-Matute, B., Mateo, C., Cárdenas, D.J., and Echavarren, A.M. (2001). Intramolecular C—H activation by (alkyl)Pd(II) complexes: insights into the mechanism of the Pd-catalyzed arylation reaction. *Chem. Eur. J.* 7: 2341–2348.

2822 Lane, B.S., Brown, M.A., and Sames, D. (2005). Direct Pd-catalyzed C-2 and C-3 arylation of indoles: a mechanistic rationale for regioselectivity. *J. Am. Chem. Soc.* 127: 8050–8057.

2823 Kresge, A.J. and Brennan, J.F. (1967). Perchloric acid-catalyzed aromatic mercuration in acetic acid solution. 2. Substitution process. *J. Org. Chem.* 32: 752–755.

2824 Fung, C.W., Khorramdelvahed, M., Ranson, R.J., and Roberts, R.M.G. (1980). Kinetics and mechanism of mercuriation of aromatic compounds by Hg trifluoroacetate in CF_3COOH. *J. Chem. Soc., Perkin Trans. 2* 267–272.

2825 Winstein, S. and Traylor, T.G. (1955). Mechanisms of reaction of organomercurials. 2. Electrophilic substitution on saturated carbon - acetolysis of dialkylmercury compounds. *J. Am. Chem. Soc.* 77: 3747–3752.

2826 Hennessy, E.J. and Buchwald, S.L. (2003). Synthesis of substituted oxindoles from α-chloroacetanilides via Pd-catalyzed C—H functionalization. *J. Am. Chem. Soc.* 125: 12084–12085.

2827 Campeau, L.C., Parisien, M., Leblanc, M., and Fagnou, K. (2004). Biaryl synthesis via direct arylation: establishment of an efficient catalyst for intramolecular processes. *J. Am. Chem. Soc.* 126: 9186–9187.

2828 Lapointe, D. and Fagnou, K. (2010). Overview of the mechanistic work on the concerted metallation-deprotonation pathway. *Chem. Lett.* 39: 1119–1126.

2829 Lafrance, M., Rowley, C.N., Woo, T.K., and Fagnou, K. (2006). Catalytic intermolecular direct arylation of perfluorobenzenes. *J. Am. Chem. Soc.* 128: 8754–8756.

2830 García-Cuadrado, D., Braga, A.A.C., Maseras, F., and Echavarren, A.M. (2006). Proton abstraction mechanism for the Pd-catalyzed intramolecular arylation. *J. Am. Chem. Soc.* 128: 1066–1067.

2831 Lafrance, M. and Fagnou, K. (2006). Palladium-catalyzed benzene arylation: incorporation of catalytic pivalic acid as a proton shuttle and a key element in catalyst design. *J. Am. Chem. Soc.* 128: 16496–16497.

2832 Ames, D.E. and Bull, D. (1982). Some reactions of 3-halogenocinnolines catalyzed by Pd compounds. *Tetrahedron* 38: 383–387.

2833 Ohta, A., Akita, Y., Ohkuwa, T. et al. (1990). Palladium-catalyzed arylation of furan, thiophene, benzo[b]furan and benzo[b]thiophene. *Heterocycles* 31: 1951–1958.

2834 Akita, Y., Inoue, A., Yamamoto, K. et al. (1985). Palladium-catalyzed coupling reaction of chloropyrazines with indole. *Heterocycles* 23: 2327–2333.

2835 Nikulin, M.V., Lebedev, A.Y., Voskoboinikov, A.Z., and Beletskaya, I.P. (2008). Role of the ligand in the Pd-catalyzed C-2 and C-3 arylation of the indole Mg salt. *Dokl. Chem.* 423: 326–329.

2836 Joucla, L., Batail, N., and Djakovitch, L. (2010). "On Water" direct and site-selective Pd-catalyzed C—H arylation of (NH)-indoles. *Adv. Synth. Catal.* 352: 2929–2936.

2837 Théveau, L., Schneider, C., Fruit, C., and Hoarau, C. (2016). Orthogonal Pd-catalyzed direct C—H bond arylation of heteroaromatics with aryl halides. *ChemCatChem* 8: 3183–3194.

2838 Fuse, S., Morita, T., Johmoto, K. et al. (2015). Directing/protecting-group-free synthesis of tetraaryl substituted pyrazoles through four direct arylations on an unsubstituted pyrazole scaffold. *Chem. Eur. J.* 21: 14370–14375.

2839 Leclerc, J.P., André, M., and Fagnou, K. (2006). Heck, direct arylation, and hydrogenation: two or three sequential reactions from a single catalyst. *J. Org. Chem.* 71 (4): 1711–1714.

2840 Shen, T.Y. (1972). Perspectives in nonsteroidal antiinflammatory agents. *Angew. Chem. Int. Ed. Engl.* 11 (6): 460–472.

2841 Wright, W.B., Press, J.B., Chan, P.S. et al. (1986). Thromboxane synthetase inhibitors and antihypertensive agents. 1. *N*-[(1*H*-Imidazol-1-yl)alkyl]aryl amides and *N*-[(1*H*-1,2,4-triazol-1-yl)alkyl]aryl amides. *J. Med. Chem.* 29 (4): 523–530.

2842 Edmondson, S., Danishefsky, S.J., Sepp-Lorenzino, L., and Rosen, N. (1999). Total synthesis of spirotryprostatin A, leading to the discovery of some biologically promising analogues. *J. Am. Chem. Soc.* 121 (10): 2147–2155.

2843 Venkatesan, H., Davis, M.C., Altas, Y. et al. (2001). Total synthesis of SR 121463 A, a highly potent and selective vasopressin V_2: receptor antagonist. *J. Org. Chem.* 66 (11): 3653–3661.

2844 Hama, T., Liu, X.X., Culkin, D.A., and Hartwig, J.F. (2003). Palladium-catalyzed α-arylation of esters and amides under more neutral conditions. *J. Am. Chem. Soc.* 125: 11176–11177.

2845 Culkin, D.A. and Hartwig, J.F. (2003). Palladium-catalyzed α-arylation of carbonyl compounds and nitriles. *Acc. Chem. Res.* 36 (4): 234–245.

2846 Bellina, F. and Rossi, R. (2010). Transition metal-catalyzed direct arylation of substrates with activated sp^3-hybridized C—H bonds and some of their synthetic equivalents with aryl halides and pseudohalides. *Chem. Rev.* 110: 1082–1146.

2847 Johansson, C.C.C. and Colacot, T.J. (2010). Metal-catalyzed α-arylation of carbonyl and related molecules: novel trends in C—C bond formation by C—H bond functionalization. *Angew. Chem. Int. Ed.* 49: 676–707.

2848 Palucki, M. and Buchwald, S.L. (1997). Palladium-catalyzed α-arylation of ketones. *J. Am. Chem. Soc.* 119: 11108–11109.

2849 Hamann, B.C. and Hartwig, J.F. (1997). Palladium-catalyzed direct α-arylation of ketones. Rate acceleration by sterically hindered chelating ligands and reductive elimination from a transition metal enolate complex. *J. Am. Chem. Soc.* 119: 12382–12383.

2850 Satoh, T., Kawamura, Y., Miura, M., and Nomura, M. (1997). Palladium-catalyzed regioselective mono- and diarylation reactions of 2-phenylphenols and naphthols with aryl halides. *Angew. Chem. Int. Ed. Engl.* 36: 1740–1742.

2851 Lee, S., Beare, N.A., and Hartwig, J.F. (2001). Palladium-catalyzed α-arylation of esters and protected amino acids. *J. Am. Chem. Soc.* 123 (34): 8410–8411.

2852 Hama, T., Culkin, D.A., and Hartwig, J.F. (2006). Palladium-catalyzed intermolecular α-arylation of Zn amide enolates under mild conditions. *J. Am. Chem. Soc.* 128 (15): 4976–4985.

2853 Moradi, W.A. and Buchwald, S.L. (2001). Palladium-catalyzed α-arylation of esters. *J. Am. Chem. Soc.* 123 (33): 7996–8002.

2854 Koch, E., Takise, R., Studer, A. et al. (2015). Ni-catalyzed α-arylation of esters and amides with phenol derivatives. *J. Chem. Soc., Chem. Commun.* 51: 855–857.

2855 Yip, S.F., Cheung, H.Y., Zhou, Z.Y., and Kwong, F.Y. (2007). Room-temperature Cu-catalyzed α-arylation of malonates. *Org. Lett.* 9: 3469–3472.

2856 Hurtley, W.R.H. (1929). Replacement of halogen in *ortho*-bromobenzoic acid. *J. Chem. Soc.* 1870–1873.

2857 Setsune, J., Matsukawa, K., and Kitao, T. (1982). Synthesis of benzofuran-2-one derivatives by Cu(I)-promoted coupling reactions of *ortho*-bromophenol with active methylene compounds. *Tetrahedron Lett.* 23: 663–666.

2858 Suzuki, H., Kobayashi, T., Yoshida, Y., and Osuka, A. (1983). A simple one-pot conversion of aryl halides into arylacetonitriles. *Chem. Lett.* 193–194.

2859 Suzuki, H., Qui, Y., Inoue, J. et al. (1987). Synthesis of α-arylated phenylsulfonylacetonitriles - a useful precursor for substituted α-arylalkanoic acid and their derivatives. *Chem. Lett.* 887–890.

2860 Hennessy, E.J. and Buchwald, S.L. (2002). A general and mild Cu-catalyzed arylation of diethyl malonate. *Org. Lett.* 4: 269–272.

2861 Xie, X.A., Cai, G.R., and Ma, D.W. (2005). CuI/L-proline-catalyzed coupling reactions of aryl halides with activated methylene compounds. *Org. Lett.* 7: 4693–4695.

2862 Goossen, L.J. (2001). Pd-Catalyzed synthesis of arylacetic acid derivatives from boronic acids. *J. Chem. Soc., Chem. Commun.* (7): 669–670.

2863 Semmelhack, M.F., Chong, B.P., Stauffer, R.D. et al. (1975). Total synthesis of cephalotaxus alkaloids - problem in nucleophilic aromatic-substitution. *J. Am. Chem. Soc.* 97: 2507–2516.

2864 Muratake, H. and Natsume, M. (1997). Palladium-catalyzed intramolecular α-arylation of aliphatic ketones. *Tetrahedron Lett.* 38: 7581–7582.

2865 Harrison, I.T., Lewis, B., Nelson, P. et al. (1970). Nonsteroidal antiinflammatory agents. 1. 6-Substituted 2-naphthylacetic acids. *J. Med. Chem.* 13 (2): 203–205.

2866 Åhman, J., Wolfe, J.P., Troutman, M.V. et al. (1998). Asymmetric arylation of ketone enolates. *J. Am. Chem. Soc.* 120 (8): 1918–1919.

2867 Lee, S. and Hartwig, J.F. (2001). Improved catalysts for the Pd-catalyzed synthesis of oxindoles by amide α-arylation. Rate acceleration, use of aryl chloride substrates, and a new carbene ligand for asymmetric transformations. *J. Org. Chem.* 66 (10): 3402–3415.

2868 Arao, T., Kondo, K., and Aoyama, T. (2006). Asymmetric construction of quaternary carbon stereocenter by Pd-catalyzed intramolecular α-arylation. *Chem. Pharm. Bull.* 54: 1743–1744.

2869 Burtoloso, A.C.B. (2009). Catalytic enantioselective α-arylation of carbonyl compounds. *Synlett* (2): 320–327.

2870 Harvey, J.S., Simonovich, S.P., Jamison, C.R., and MacMillan, D.W.C. (2011). Enantioselective α-arylation of carbonyls via Cu(I)-bisoxazoline catalysis. *J. Am. Chem. Soc.* 133: 13782–13785.

2871 Potukuchi, H.K., Spork, A.P., and Donohoe, T.J. (2015). Palladium-catalyzed α-arylation of carbonyls in the de novo synthesis of aromatic heterocycles. *Org. Biomol. Chem.* 13: 4367–4373.

2872 (a) Mei, T.S., Kou, L., Ma, S. et al. (2012). Heterocycle formation via Pd-catalyzed C—H tunctionalization. *Synthesis* 44: 1778–1791. (b) Qiu, G.Y.S. and Wu, J. (2015). Transition metal-catalyzed direct remote C—H functionalization of alkyl groups via C(sp^3)-H bond activation. *Org. Chem. Front.* 2: 169–178. (c) Wang, X., Zhu, L.-Z., Chen, S. et al. (2015). Nickel-catalyzed direct C(sp^3)-H arylation of aliphatic amides with thiophenes. *Org. Lett.* 17 (21): 5228–5231. (d) Luo, F.-H., Yang, J., Li, Z.-K. et al. (2016). Design and synthesis of 2-methyl-7-aminobenzoxazole as auxiliary in the Pd(II)-catalyzed arylation of a β-positioned C(sp^3)-H bond. *Adv. Synth. Catal.* 358 (6): 887–893.

2873 Liu, Y.-B. and Ge, H.-B. (2017). Site-selective C—H arylation of primary aliphatic amines enabled by a catalytic transient directing group. *Nat. Chem.* 9 (1): 26–32.

2874 Dyker, G. (1994). Transition-metal-catalyzed annelation reactions. 5. Pd-Catalyzed C—H activation of *tert*-butyl groups - a simple synthesis of 1,2-dihydrocyclobutabenzene derivatives. *Angew. Chem. Int. Ed. Engl.* 33: 103–105.

2875 Baudoin, O., Herrbach, A., and Gueritte, F. (2003). The Pd-catalyzed C—H activation of benzylic *gem*-dialkyl groups. *Angew. Chem. Int. Ed.* 42: 5736–5740.

2876 Hitce, J. and Baudoin, O. (2007). Substituted benzocarbocycles by Pd-catalyzed cascade reactions featuring a C(sp^3)-H activation step. *Adv. Synth. Catal.* 349: 2054–2060.

2877 Chaumontet, M., Piccardi, R., Audic, N. et al. (2008). Synthesis of benzocyclobutenes by Pd-catalyzed C—H activation of methyl groups: method and mechanistic study. *J. Am. Chem. Soc.* 130: 15157–15166.

2878 Shabashov, D. and Daugulis, O. (2005). Catalytic coupling of C—H and C—I bonds using pyridine as a directing group. *Org. Lett.* 7: 3657–3659.

2879 Zaitsev, V.G., Shabashov, D., and Daugulis, O. (2005). Highly regioselective arylation of C(sp^3)-H bonds catalyzed by Pd acetate. *J. Am. Chem. Soc.* 127: 13154–13155.

2880 Shabashov, D. and Daugulis, O. (2010). Auxiliary-assisted Pd-catalyzed arylation and alkylation of sp^2 and sp^3 C—H bonds. *J. Am. Chem. Soc.* 132: 3965–3972.

2881 Tran, L.D. and Daugulis, O. (2012). Nonnatural amino acid synthesis by using C—H bond functionalization methodology. *Angew. Chem. Int. Ed.* 51: 5188–5191.

2882 Shang, R., Ilies, L., Matsumoto, A., and Nakamura, E. (2013). β-Arylation of carboxamides via Fe-catalyzed C(sp^3)-H bond activation. *J. Am. Chem. Soc.* 135: 6030–6032.

2883 Lafrance, M., Gorelsky, S.I., and Fagnou, K. (2007). High-yielding Pd-catalyzed intramolecular alkane arylation: reaction development and mechanistic studies. *J. Am. Chem. Soc.* 129: 14570–14571.

2884 Watanabe, T., Oishi, S., Fujii, N., and Ohno, H. (2008). Palladium-catalyzed C(sp^3)-H activation of simple alkyl groups: direct preparation of indoline derivatives from *N*-alkyl-2-bromoanilines. *Org. Lett.* 10: 1759–1762.

2885 Rousseaux, S., Gorelsky, S.I., Chung, B.K.W., and Fagnou, K. (2010). Investigation of the mechanism of C(sp^3)-H bond cleavage in Pd(0)-catalyzed intramolecular alkane arylation adjacent to amides and sulfonamides. *J. Am. Chem. Soc.* 132: 10692–10705.

2886 Rousseaux, S., Davi, M., Sofack-Kreutzer, J. et al. (2010). Intramolecular Pd-catalyzed alkane C—H arylation from aryl chlorides. *J. Am. Chem. Soc.* 132: 10706–10716.

2887 Rousseaux, S., Liegault, B., and Fagnou, K. (2012). Palladium(0)-catalyzed cyclopropane C—H bond functionalization: synthesis of quinoline and tetrahydroquinoline derivatives. *Chem. Sci.* 3: 244–248.

2888 Nakanishi, M., Katayev, D., Besnard, C., and Kündig, E.P. (2011). Fused indolines by Pd-catalyzed asymmetric C—C coupling involving an unactivated methylene group. *Angew. Chem. Int. Ed.* 50: 7438–7441.

2889 Anas, S., Cordi, A., and Kagan, H.B. (2011). Enantioselective synthesis of 2-methylindolines by Pd-catalysed asymmetric C(sp^3)-H activation/cyclisation. *J. Chem. Soc., Chem. Commun.* 47: 11483–11485.

2890 Saget, T. and Cramer, N. (2012). Palladium(0)-catalyzed enantioselective C—H arylation of cyclopropanes: efficient access to functionalized tetrahydroquinolines. *Angew. Chem. Int. Ed.* 51: 12842–12845.

2891 Saget, T. and Cramer, N. (2014). Enantioselective Pd(0)-catalyzed C—H arylation strategy for chiral heterocycles. *Pure Appl. Chem.* 86: 265–272.

2892 Saget, T., Lemouzy, S.J., and Cramer, N. (2012). Chiral monodentate phosphines and bulky carboxylic acids: cooperative effects in Pd-catalyzed enantioselective C(sp^3)-H functionalization. *Angew. Chem. Int. Ed.* 51: 2238–2242.

2893 Ano, Y., Tobisu, M., and Chatani, N. (2011). Palladium-catalyzed direct ethynylation of C(sp^3)-H bonds in aliphatic carboxylic acid derivatives. *J. Am. Chem. Soc.* 133: 12984–12986.

2894 He, J., Wasa, M., Chan, K.S.L., and Yu, J.Q. (2013). Palladium(0)-catalyzed alkynylation of C(sp^3)-H bonds. *J. Am. Chem. Soc.* 135: 3387–3390.

2895 Nadres, E.T., Santos, G.I.F., Shabashov, D., and Daugulis, O. (2013). Scope and limitations of auxiliary-assisted, Pd-catalyzed arylation and alkylation of sp^2 and sp^3 C—H bonds. *J. Org. Chem.* 78: 9689–9714.

2896 Wu, X.S., Zhao, Y., and Ge, H.B. (2014). Nickel-catalyzed site-selective alkylation of unactivated C(sp^3)-H bonds. *J. Am. Chem. Soc.* 136: 1789–1792.

2897 Pedroni, J., Boghi, M., Saget, T., and Cramer, N. (2014). Access to β-lactams by enantioselective Pd(0)-catalyzed C(sp^3)-H alkylation. *Angew. Chem. Int. Ed.* 53: 9064–9067.

Index

Organic Chemistry: Theory, Reactivity and Mechanisms in Modern Synthesis, First Edition. Pierre Vogel and Kendall N. Houk.

D

E

L

M

N

O

Q

R

S

T

U

V